HOYER - KREUTER
TECHNOLOGICAL
DICTIONARY

of the terms used in

Crafts and Industries, Engineering and Engineering Science, Mining and Metallurgy; Raw and Finished Materials, Material Testing, Semi-Finished and Finished Products; Electrical and Communication Engineering; Metrology; Cinematograph Engineering, Ordnance, Optical, Medical, and Sanitary Engineering; Safety Engineering; Civil Engineering and Chemical Technology; Agriculture and Forestry; Foodstuffs, Textile and Clothing Industries; Trade, Banking, and Fair Practice; Traffic, Motoring and Motor Engineering, Shipbuilding and Navigation; Patents, Law, and Customs, and in many other fields of Engineering and Commercial Activity

Sixth
Completely Revised Edition

Edited by

Dr.-Ing. E. h. Alfred Schlomann

with the support of the Deutscher Verband Technisch-Wissenschaftlicher Vereine, the Verein deutscher Ingenieure, and numerous Industrial Firms in Germany and abroad

Volume II

English — German — French

1932
Published by Julius Springer · Berlin W 9

Librairie Polytechnique	The Machinery Publishing Co.	The Industrial Press
Ch. Béranger	London W. C. 1	New York
Paris · 15 Rue des Saints-Pères	52—54 High Holborn	140—148 Lafayette Street

ISBN-13:978-3-642-98543-0   e-ISBN-13:978-3-642-99357-2
DOI: 10.1007/978-3-642-99357-2

Softcover reprint of the hardcover 1st edition 1932

To Dr.-Ing. E. h. **Waldemar Hellmich**
Director of the Verein deutscher Ingenieure

I dedicate this book in grateful appreciation of the energetic and helpful support he has given to my work on the Technical Dictionaries

PREFACE.

The signal success of the Hoyer-Kreuter Dictionary in all its previous editions, and the appreciation with which it has been received throughout the world, are eloquent proof of its all-round utility.

Since the fifth edition appeared, the number of dictionaries, both general and special, has grown very rapidly. It was felt, then, that the time had come to supplement the numerous Technical and Special Dictionaries, as well as the large General Dictionaries, by a new and comprehensive dictionary covering broadly the whole field of Technology and its allied subjects. Only a work of this kind can make up for the natural shortcomings of the Special Dictionaries, shortcomings which arise from the necessity imposed upon the dictionary maker of confining himself to a certain closely defined field and discarding all terms which, though liable to be required at any time in connection with a special subject, appertain to the wide field of General Engineering, Commerce, Law, or other allied subjects. It is only in an "all-purpose" Technical Dictionary, in conjunction with Special Dictionaries, that a complete vocabulary is found covering the Crafts, Civil and Mechanical Engineering, Applied Sciences, Industry and Agriculture, Trade and Commerce, Transport and Communications, Political Economy, Customs Law, etc., etc.

The sixth edition of the Hoyer-Kreuter Dictionary is intended for all those who find that an ordinary general dictionary does not meet their requirements, but who need for their work a General Technical Dictionary. It will prove equally invaluable to all users of Special Dictionaries who, in their professional work, are not satisfactorily served by the vocabularies contained in these dictionaries, but require a General Technical Dictionary for their correspondence, for study, or for general reading.

The new edition is a complete revision and very wide extension of the fifth edition, which appeared 28 years ago. Special consideration has been given to the inclusion of the terms and phrases used in: Technology in general; Industrial Manufacture, Technical Chemistry, Chemical Technology, and Agriculture; Engineering and Materials of Construction; Raw Materials, Semi-finished and Finished Products, including the necessary machinery, apparatus, and contrivances; Trade, Transport and Communications, Customs Practice, and National and International Law, as far as these terms are considered requisite for the international interchange of thought and international industrial relations.

In agreement with the *Deutscher Verband Technisch-Wissenschaftlicher Vereine* and the *Verein deutscher Ingenieure*, I carried out this work *pari passu* with my current work on the Illustrated Technical Dictionaries. In compiling and verifying the technical terms, I followed the same method which had proved so satisfactory in connection with the Illustrated Technical Dictionaries, and enlisted the services of a great number of foreign experts, and the co-operation of large industrial firms in Germany, Great Britain, U. S. A., and France.

In preparing the manuscript, I felt that the elimination of terms from the extremely valuable contents of the earlier edition, and the addition of new terms, should not be left to chance, but should follow a definite plan. The first thing to do was to select the key-words, and then supplement them, on a definite system, by adding the modifications, combinations, phrases, etc., deemed necessary for practical purposes. I trust that the result — roughly 100 000 terms in each of the three languages — is a dictionary that will be in every way useful and satisfactory to its users.

The search for the terms presented no difficulties, seeing that there is scarcely any article to-day that is not industrially manufactured, hardly any substance that does not serve, in some form or other, as an engineering material, hardly any activity that is not in some way related to Engineering or Industry. The bulk of the material embodied in the dictionary was taken from such sources as the contents of scientific and economic journals and treatises, glossaries compiled by industrial firms, individual contributions of some of my associates, prospectuses, Customs Schedules, catalogues of exhibitions, Standard Specifications, lists of goods, and statistical publications.

The translations of the terms and phrases suggested in the first place were carefully checked, and revised where necessary, by reference to exact definitions, and the same applies to the terms taken from printed documents such as prospectuses issued by industrial firms, or official documents published in several languages. Despite every care, incongruities or errors may, of course, have crept in. Certain terms and phrases can be translated in different ways, depending on the peculiarities of style of the translator. This dictionary cannot, of course, exhaust all such possibilities, so I have tried in every case to give one translation which is good and clearly conveys the meaning of the original.

In cases where, in one language, a particular term is applicable to several professions, industries, etc. while the foreign equivalent varies according to the industry, etc. in which it is used, the particular branch of industry is noted in parenthesis in accordance with the "Explanations of Abbreviations and Signs" given at the beginning of the book. This reference has, however, been omitted where there can be no doubt as to the industry, etc. covered by the term.

Expressions which are preferred, or are used almost exclusively, in the United States, are marked (A).

I would like to sincerely thank all the individuals, institutions, firms, etc. for their generous assistance in placing their excellent and valuable material at my disposal. Their ready co-operation has enabled me to include in this dictionary a large number of technical terms which have not, to my knowledge, as yet appeared in any dictionary devoted to Engineering, Agriculture, Commerce, or Communications. The number of those who, by making valuable original contributions available to me, have added to the contents of this new edition, is too large to permit of all their names being recorded here. Mr. *Benno Kirschbaum*, President of the Kirschbaum Bureau of Translation, Inc., of Philadelphia, Pa., enlisted the co-operation of a large section of the North American industrials by calling attention to the object and scope of this dictionary, and induced them to send me valuable contributions. Herr *Martiny*, Chief Engineer with the *Fried. Krupp* Company of Essen, and Herr *Adolf Glückmann*, Chief Engineer with the "Bamag" Co. of Berlin, as also Herr *August Gerritzen* of Essen, were particularly helpful in furthering our work. To all these gentlemen I am greatly indebted and would like here to record my deep gratitude.

Several standard dictionaries were consulted for the purpose of checking and completing the contents, chief among them being the following: J. O. Kettridge, *Dictionary of Technical Terms and Phrases*, Routledge, London; Webster, *New International Dictionary of the English Language*, Bell & Sons, London; *Dictionnaire Encyclopédique de la Langue Française*, Larousse, Paris; *Volkswirtschaftliches Wörterbuch*, Julius Springer, Berlin; *Der grosse Duden*, Bibliographisches Institut, Leipzig; Muret-Sanders, *Enzyklopädisches Wörterbuch*, Langenscheidt, Berlin; Sachs-Villatte, *Enzyklopädisches Wörterbuch*, Langenscheidt, Berlin; Webel, *Technical and Scientific Dictionary*, Routledge, London; Eitzen, *Wörterbuch der Handelssprache*, Haessel, Leipzig; *Handwörterbuch des elektrischen Fernmeldewesens*, Julius Springer, Berlin; *Répertoire Technologique des Noms d'Industries et de Professions* (Ministère du Travail et de la Prévoyance Sociale), Berger-Levrault & Cie., Paris-Nancy; *Sammlungen Technischer Ausdrücke der Fried. Krupp A.-G.*, Essen; *Sammlung technischer Fachwörter im Gebrauch der Meguin A.-G.*, Butzbach, Hessen (Berlin-Anhaltische Maschinenbau A.-G., Berlin); *Vocabulaire Téléphonique International en Sept Langues, Texte officiel établi par le Comité Consultatif International des Communications téléphoniques à grande distance*, Librairie de l'Enseignement Technique, Léon Eyrolles, Paris.

For proof-reading, I have had the able assistance of Herr *J. Busch*, Hamburg, Engineering Chemist, Lecturer and Licenced Interpreter for the French language; Herr *Friedrich Göbber*, Hamburg; Fräulein *Alice Steinberg*, Hamburg; Mr. *Douglas Carpenter*, London; Mr. *Lewis L. Sell*, Ph. D., New York, and several other gentlemen. The corrections made by them far exceeded the scope of the usual corrections, for, in many cases, they suggested changes which, being based on a thorough knowledge of the subject and of the languages, must be considered as essential improvements.

The *Deutscher Verband Technisch-Wissenschaftlicher Vereine* and the *Verein deutscher Ingenieure* readily agreed to our making use of the great stock of material collected by the editors of the Illustrated Technical Dictionaries, thereby acknowledging the great significance of lexicographical work to engineering. The gratitude I owe them for their courtesy will, I am sure, be shared by very many engineers at home and abroad. Engineering, Industry, and Commerce must also be considerably indebted to the *Publishing House of Julius Springer*, Berlin, for making possible the issue of this work and so rendering a valuable service to the furtherance of international economic and intellectual relations.

In placing this new and revised edition of the Hoyer-Kreuter Technological Dictionary in the hands of the public, I trust it will be of service in international scientific, technical and industrial relations and help to make easier the work of numerous colleagues.

Berlin, February 1932   **A. Schlomann.**

Note. — The numerals appearing in Vol. I (German-English-French) at the end of each group of terms and the reference letters printed in front of each term, have been added for the purpose of enabling supplementary volumes in additional languages, now in preparation and to be issued at a later date, to be used in conjunction with Vol. I of this Dictionary.

Explanation of the Abreviations and Signs – Erklärung der Abkürzungen und Zeichen – Explication des Abréviations et des Signes.

Zeichen Signs Signes	Fachgebiet Speciality Spécialité	Deutsch	English	Français
A	—	amerikanisch	American	américain
Acc	accumulators	Sammler	—	accumulateurs
Acc	accumulateurs	Sammler	accumulators	—
Acoust	acoustics	Akustik	—	acoustique
Acoust	acoustique	Akustik	acoustics	—
Aero	aeronautics	Luftfahrwesen	—	aéronautique
Aéro	aéronautique	Luftfahrwesen	aeronautics	—
Agr	agriculture	Ackerbau; Landwirtschaft	agriculture	agriculture
Aig	fabrication des aiguilles	Nadelfertigung	needle manufacture	—
Airpl	construction of airplanes	Flugzeugbau	—	construction d'avions
Akust	Akustik	—	acoustics	acoustique
Amid	fabrication d'amidon	Stärkeherstellung	starch manufacture	—
Arch	architecture	Baukunst	architecture	architecture
Arch hydr	architecture hydraulique	Wasserbauwesen	hydraulic architecture	—
Arch nav	architecture navale	Schiffbau	shipbuilding	—
Arm	service de l'armement	Waffenwesen	service of arms	—
Arm	service of arms	Waffenwesen	—	service de l'armement
Arp	arpentage	Feldmeßkunde	surveying	—
Astro	Astronomie	—	astronomy	astronomie
Astro	astronomie	Astronomie	astronomy	—
Astro	astronomy	Astronomie	—	astronomie
Assur	assurance	Versicherungswesen	insurance	—
Aufber	Aufbereitung	—	ore dressing; coal dressing	préparation des minerais ou des houilles
Auto	Automobilbau; Kraftfahrwesen	—	construction of automobiles; automobilism	construction des automobiles; automobilisme
Auto	construction des automobiles; automobilisme	Automobilbau; Kraftfahrwesen	construction of automobiles; automobilism	—
Auto	construction of automobiles; automobilism	Automobilbau; Kraftfahrwesen	—	construction des automobiles; automobilisme
Aut tel	automatic telephony	Selbstanschlußbetrieb; automatische Telefonie	—	téléphonie automatique
Avion	construction d'avions	Flugzeugbau	construction of airplanes	—
Bäck	Bäckerei	—	bakery	boulangerie
Bak	bakery	Bäckerei	—	boulangerie
Bank	banking	Bankwesen	—	banque
Bank	Bankwesen	—	banking	banque
Banqu	banque	Bankwesen	banking	—
Basket	basketry	Korbmacherei	—	vannerie
Bât	bâtiments	Bauwesen	building	—
Batt d'or	battage d'or	Goldschlägerei	gold beating	—
Bauk	Baukunst	—	architecture	architecture
Bauw	Bauwesen	—	building	bâtiments
Bekleid	Bekleidung	—	clothing	vêtements
Beleucht	Beleuchtungswesen	—	lighting	éclairage
Bergb	Bergbau	—	mining	exploitation de mines
Bildh	Bildhauerei	—	sculpture	sculpture
Blanch	blanchiment	Bleicherei	bleaching	—
Bleach	bleaching	Bleicherei	—	blanchiment
Bleich	Bleicherei	—	bleaching	blanchiment
Boil	steam boilers	Dampfkessel	—	chaudières à vapeur
Bookb	bookbinding	Buchbinderei	—	reliure
Bookkeep	bookkeeping	Buchhaltung	—	comptabilité
Booksell	bookseller	Buchhändler	—	libraire
Börse	Börsenwesen	—	exchange	bourse
Bot	Botanik	—	botany	botanique
Bot	botanique	Botanik	botany	—
Bot	botany	Botanik	—	botanique
Böttch	Böttcherei; Küferei	—	cooperage	tonnellerie
Bouch	boucherie	Fleischerei	butchery	—
Boul	boulangerie	Bäckerei	bakery	—
Bourse	bourse	Börsenwesen	exchange	—
Bout	fabrication de boutons	Knopfmacherei	button manufacture	—
Brass	brasserie	Brauerei	brewery	—
Brau	Brauerei	—	brewery	brasserie
Brenn	Brennerei	—	distillery	distillerie
Brev	brevets d'invention	Patentwesen	patenting	—
Brew	brewery	Brauerei	—	brasserie
Brick	brick works	Ziegelei	—	briqueterie
Bridge	bridge building	Brückenbau	—	construction de ponts
Brique	briqueterie	Ziegelei	brick works	—
Bross	brosserie	Bürstenfertigung	brush manufacture	—
Brük-kenb	Brückenbau	—	bridge building	construction de ponts
Brunn	Brunnenbau	—	well boring	construction de puits
Brush	brush manufacture	Bürstenfertigung	—	brosserie
Buchb	Buchbinderei	—	bookbinding	reliure
Buchdr	Buchdruckerei; Druckerei; Grafik	—	printing	imprimerie
Buchhalt	Buchhaltung	—	bookkeeping	comptabilité
Buchhändl	Buchhändler	—	bookseller	libraire
Build	building	Bauwesen	—	bâtiments
Bureau	fourniture de bureau; machines de bureau	Bürobedarf; Büromaschinen	office furnishing; office machines	—
Büro	Bürobedarf; Büromaschinen	—	office furnishing; office machines	fourniture de bureau; machines de bureau
Bürst	Bürstenfertigung	—	brush manufacture	brosserie
Butch	butchery	Fleischerei	—	boucherie
Button	button manufacture	Knopfmacherei	—	fabrication de boutons
Cand	candle manufacture	Lichtfertigung	—	fabrication de chandelles
Carp	carpentry	Zimmerei	—	charpenterie
Carross	carrosserie	Wagenbauerei	coachmaking	—
Chand	fabrication de chandelles	Lichtfertigung	candle manufacture	—
Chap	chapellerie	Hutmacherei	hatting	—
Charb	charbonnière	Köhlerei	charcoal burning	—
Charc	charcoal burning	Köhlerei	—	charbonnière
Charp	charpenterie	Zimmerei	carpentry	—
Chaud	chaudières à vapeur	Dampfkessel	steam boilers	—
Chaudr	chaudronnerie de cuivre	Kupferschmiede	coppersmith's trade	—
Chauf vent	chauffage et ventilation	Heizung und Lüftung	heating and ventilating	—
Chem	Chemie	—	chemistry	chimie
Chem	chemistry	Chemie	—	chimie
Chem d f	chemin de fer	Eisenbahnwesen	railway	—
Chim	chimie	Chemie	chemistry	—
Clockm	clockmaking	Uhrmacherei	—	horlogerie
Cloth	clothing	Bekleidung	—	vêtements
Clothm	cloth manufacture	Tuchmacherei	—	manufacture de draps
Clout	clouterie	Nagelschmiede	nailsmith's trade	—
Coachm	coachmaking	Wagenbauerei	—	carrosserie
Coal dress	coal dressing	Kohleaufbereitung	—	préparation des houilles
Coin	coinage	Münzwesen	—	monnayage
Commerce	commerce	Handel	trade	—
Comptabl	comptabilité	Buchhaltung	bookkeeping	—
Constr pont	construction de ponts	Brückenbau	bridge building	—

Zeichen Signs Signes	Fachgebiet Speciality Spécialité	Deutsch	English	Français
Coop	cooperage	Böttcherei	—	tonnellerie
Cop- persm	coppersmith's trade	Kupfer- schmiede	—	chaudronnerie de cuivre
Corderie	corderie	Seilerei	rope making	—
Cordo	cordonnerie	Schuhmacherei	shoe making	—
Corr	corroyeur	Gerber	currier	—
Coutel	coutellerie	Messer- schmiede	cutler's trade	—
Cou- vreur	couvreur	Dachdecker	tiler	—
Curr	currier	Gerber	—	corroyeur
Cutl	cutler's trade	Messer- schmiede	—	coutellerie
Dachd	Dachdecker	—	tiler	couvreur
Dampfk	Dampfkessel	—	steam boilers	chaudières à vapeur
Dampfm	Dampf- maschinen	—	steam engines	machines à vapeur
Dessin	dessin	Zeichnen	drawing	—
Distill	distillerie	Brennerei	distillery	—
Distill	distillery	Brennerei	—	distillerie
Doreur	doreur	Vergolder	gilder	—
Douane	douane	Zollwesen	duty	—
Drahtz	Drahtzieherei	—	wire drawing	tréfilerie
Drap	manufacture de draps	Tuchmacherei	cloth manu- facture	—
Drawing	drawing	Zeichnen	—	dessin
Drechsl	Drechslerei	—	turnery	tournerie
Dreh	Dreherei	—	turnery	tournerie
Duty	duty	Zollwesen	—	douane
Dyer	dyer	Färber	—	teinturier
Earthw	earthwork	Erdarbeiten	—	terrassement
Eclai- rage	éclairage	Beleuchtungs- wesen	lighting	—
Ec pol	économie poli- tique	Volkswirt- schaft	political econo- my	—
Eisenb	Eisenbahn- wesen	—	railway	chemin de fer
Electr	électricité	Elektrizität	electricity	—
Electr	electricity	Elektrizität	—	électricité
Elektr	Elektrizität	—	electricity	électricité
El Leit	Elektrische Leitungen	—	electric lines	lignes électri- ques
El line	electric lines	Elektrische Leitungen	—	lignes électri- ques
Engr	engraving	Gravierkunst	—	gravure
Erdarb	Erdarbeiten	—	earthwork	terrassement
Erdk	Erdkunde	—	geography	géographie
Ex- change	exchange	Börsenwesen	—	bourse
Expl min	exploitation de mines	Bergbau	mining	—
f	—	weiblich	feminine	féminin
Färb	Färber	—	dyer	teinturier
Farr	farriery	Hufschmiede	—	maréchalerie
Feil	Feilenhauerei	—	file cutting	taillage de limes
Feldm	Feldmeßkunde	—	surveying	arpentage
Ferbl	ferblanterie	Klempnerei	tinman's trade	—
Fernm	Fernmelde- technik	—	telephone; telegraph	téléphone; télégraphe
Feuerl	Feuerlösch- wesen	—	fire extinguish- ing	service d'incen- die
Feuerw	Feuerwerkerei	—	pyrotechnics	pyrotechnique
Fil	filature	Spinnerei	spinning	—
File	file cutting	Feilenhauerei	—	taillage de limes
Film	fabrication de films; pré- sentation de films	Filmfertigung; Filmvorfüh- rung	film manu- facture; film demonstra- tion	—
Film	Filmfertigung; Filmvorfüh- rung	—	film manu- facture; film demonstra- tion	fabrication de films; pré- sentation de films
Film	film manu- facture; film demonstra- tion	Filmfertigung; Filmvorfüh- rung	—	fabrication de films; pré- sentation de films
Fire ext	fire extinguish- ing	Feuerlösch- wesen	—	service d'in- cendie
Fisch	Fischerei	—	fishery	pêche
Fish	fishery	Fischerei	—	pêche
Fleisch	Fleischerei	—	butchery	boucherie
Flugz	Flugzeugbau	—	construction of airplanes	construction d'avions
Fond	fonderie	Gießerei	foundry	—
Fond caract	fonderie de ca- ractères	Schriftgießerei	letter foundry	—
Fcrest	forestier	Forstwesen	forestry	—
Fcrest	forestry	Forstwesen	—	forestier
Forg	forge	Schmiede	forge	—
Form	Formerei	—	moulding	moulage
Forstw	Forstwesen	—	forestry	forestier
Found	foundry	Gießerei	—	fonderie
Funkw	Funkwesen	—	radiocasting	radiodiffusion

Zeichen Signs Signes	Fachgebiet Speciality Spécialité	Deutsch	English	Français
Galv	galvanotech- nics	Galvanotech- nik	—	galvano-tech- nique
Galv	Galvanotech- nik	—	galvanotech- nics	galvano-tech- nique
Galv	galvano-tech- nique	Galvanotech- nik	galvanotech- nics	galvano-tech- nique
Gard	gardening	Gärtnerei	—	jardinage
Gärtn	Gärtnerei	—	gardening	jardinage
Geol	Geologie	—	geology	géologie
Géol	géologie	Geologie	geology	géologie
Geom	Geometrie	—	geometry	géométrie
Geom	geometry	Geometrie	—	géométrie
Géom	géométrie	Geometrie	geometry	—
Geogr	geography	Erdkunde	—	géographie
Géogr	géographie	Erdkunde	geography	—
Gerb	Gerber	—	currier	corroyeur
Gieß	Gießerei	—	foundry	fonderie
Gild	gilder	Vergolder	—	doreur
Glasf	Glasfertigung; Glasschlei- ferei	—	glass manu- facture; glass grinding	verrerie; polis- sage de verre
Glassm	glass manufac- ture; glass grinding	Glasfertigung; Glasschlei- ferei	—	verrerie; polis- sage de verre
Gold-b	gold beating	Goldschlägerei	—	battage d'or
Goldschl	Goldschlägerei	—	gold beating	battage d'or
Gold- schm	Goldschmiede- kunst	—	goldsmith's art	orfèvrerie
Goldsm	goldsmith's art	Goldschmiede- kunst	—	orfèvrerie
Grav	Gravierkunst	—	engraving	—
Grav	gravure	Gravierkunst	engraving	gravure
Grav bois	gravure sur bois	Holzschneide- kunst	wood engrav- ing	—
Handel	Handel	—	trade	commerce
Hatt	hatting	Hutmacherei	—	chapellerie
Haush	Haushalt	—	household	ménage
Heat vent	heating and ventilating	Heizung und Lüftung	—	chauffage et ventilation
Heilk	Heilkunde; Medizin	—	medical science	science médicale
Heiz Lüft	Heizung und Lüftung	—	heating and ventilating	chauffage et ventilation
Holz- bearb	Holzbearbei- tung	—	wood working	travail du bois
Holz- schn	Holzschneide- kunst	—	wood engrav- ing	gravure sur bois
Horl	horlogerie	Uhrmacherei	watchmaking; clockmaking	—
Househ	household	Haushalt	—	ménage
Huf- schm	Hufschmiede	—	farriery	maréchalerie
Hutm	Hutmacherei	—	hatting	chapellerie
Hydr	hydraulics	Hydraulik	—	hydraulique
Hydr	Hydraulik	—	hydraulics	hydraulique
Hydr	hydraulique	Hydraulik	hydraulics	—
Hydr arch	hydraulic architecture	Wasserbau- wesen	—	architecture hydraulique
Impr	imprimerie	Buchdruckerei	printing	—
Impr tiss	imprimerie de tissus	Zeugdruckerei	textile printing	—
Insur	insurance	Versicherungs- wesen	—	assurance
Jard	jardinage	Gärtnerei	gardening	—
Jewel	jeweller	Juwelier	—	joaillier
Joail	joaillier	Juwelier	jeweller	—
Join	joiner's trade	Tischlerei	—	menuiserie
Juw	Juwelier; Steinschnei- der	—	jeweller	joaillier
Klempn	Klempnerei	—	tinman's trade	ferblanterie
Knopfm	Knopf- macherei	—	button manu- facture	fabrication de boutons
Köhl	Köhlerei	—	charcoal burning	charbonnière
Korbm	Korbmacherei	—	basketry	vannerie
Kriegsg	Kriegsgerät	—	war material	matériel de guerre
Kupfer- schm	Kupfer- schmiede	—	coppersmith's trade	chaudronnerie de cuivre
Lace-m	lace manu- facture	Posamenterie	—	passementerie
Lamin	laminoir	Walzwerk	rolling mill	—
Landw	Landwirt- schaft; Ackerbau	—	agriculture	agriculture
Lav	laverie	Wäscherei	washery	—
Letter-f	letter foundry	Schriftgießerei	—	fonderie de caractères
Libr	libraire	Buchhändler	bookseller	—
Lichtb	Lichtbildnerei; Fotografie	—	photography	photographie
Lichtf	Lichtfertigung	—	candle manu- facture	fabrication de chandelles

Zeichen / Signs / Signes	Fachgebiet / Specialty / Spécialité	Deutsch	English	Français
Lighting	lighting	Beleuchtungswesen	—	éclairage
Ligne él	lignes électriques	Elektrische Leitungen	electric lines	—
Lim	taillage de limes	Feilenhauerei	file cutting	—
Loc	locomotive	Lokomotive	locomotive	locomotive
Locksm	locksmith's trade	Schlosserei	—	serrurerie
Lok	Lokomotive	—	locomotive	locomotive
Luftf	Luftfahrwesen	—	aeronautics	aéronautique
m		männlich	masculine	masculin
Maç	maçonnage	Maurerei	masonry	—
Mach	construction de machines	Maschinenbau	construction of machines	—
Mach	construction of machines	Maschinenbau	—	construction de machines
Mach écr	machines à écrire	Schreibmaschinen	typewriters	—
Mach out	machines-outils	Werkzeugmaschinen	machine-tools	—
Mach tool	machine-tools	Werkzeugmaschinen	—	machines-outils
Mach vap	machines à vapeur	Dampfmaschinen	steam engines	—
Maréch	maréchalerie	Hufschmiede	farriery	—
Mail	mail	Post	—	poste
Mal	Malerei		painting	peinture
Mar	marine	Seewesen	marine	marine
Mas	masonry	Maurerei	—	maçonnage
Masch	Maschinenbau	—	construction of machines	construction de machines
Mat guerre	matériel de guerre	Kriegsgerät	war material	—
Math	mathematics	Mathematik	—	mathématiques
Math	Mathematik	—	mathematics	mathématiques
Math	mathématiques	Mathematik	mathematics	—
Maur	Maurerei	—	masonry	maçonnage
Méc	mécanique	Mechanik	mechanics	—
Mech	mechanics	Mechanik	—	mécanique
Mech	Mechanik	—	mechanics	mécanique
Med	medical science	Medizin; Heilkunde	—	science médicale
Méd	science médicale	Medizin; Heilkunde	medical science	—
Men	menuiserie	Tischlerei	joiner's trade	—
Ménage	ménage	Haushalt	household	—
Messerschm	Messerschmiede	—	cutler's trade	coutellerie
Met	Metallkunde; Hüttenkunde	—	metallurgy	métallurgie
Métal	métallurgie	Metallkunde; Hüttenkunde	metallurgy	—
Metal	metallurgy	Metallkunde; Hüttenkunde	—	métallurgie
Meteor	meteorology	Meteorologie; Wetterkunde	—	météorologie
Météor	météorologie	Meteorologie; Wetterkunde	meteorology	—
Meun	meunerie	Müllerei	millering	—
Mill	millering	Müllerei	—	meunerie
Min	Mineralogie	—	mineralogy	minéralogie
Mine	mining	Bergbau	—	exploitation de mines
Miner	mineralogy	Mineralogie	—	minéralogie
Minér	minéralogie	Mineralogie	mineralogy	—
Monn	monnayage	Münzwesen	coinage	—
Mot	Verbrennungsmotoren	—	combustion motors	moteurs à combustion
Mot	combustion motors	Verbrennungsmotoren	—	moteurs à combustion
Mot	moteurs à combustion	Verbrennungsmotoren	combustion motors	—
Moul	moulage	Formerei	moulding	—
Mould	moulding	Formerei	—	moulage
Müller	Müllerei	—	millering	meunerie
Münzw	Münzwesen	—	coinage	monnayage
Mus	Musik; Musikinstrumente	—	music; musical instruments	musique; instruments de musique
Mus	music; musical instruments	Musik; Musikinstrumente	—	musique; instruments de musique
Mus	musique; instruments de musique	Musik; Musikinstrumente	music; musical instruments	—
n		sächlich	neuter	neutre
Nadl	Nadelfertigung	—	needle manufacture	fabrication des aiguilles
Nagelschm	Nagelschmiede	—	nailsmith's trade	clouterie
Nailsm	nailsmith's trade	Nagelschmiede	—	clouterie
Nav	navigation	Schiffahrt	navigation	navigation
Needl	needle manufacture	Nadelfertigung	—	fabrication des aiguilles
Office	office furnishing; office machines	Bürobedarf; Büromaschinen	—	fourniture de bureau; machines de bureau
Opt	Optik	—	optics	optique
Opt	optics	Optik	—	optique
Opt	optique	Optik	optics	—
Ore dress	ore dressing	Erzaufbereitung	—	préparation des minerais
Orf	orfèvrerie	Goldschmiedekunst	goldsmith's art	—
Outil	outils	Werkzeuge	tools	—
Paint	painting	Malerei	—	peinture
Pap	Papierfertigung	—	paper manufacture	papeterie
Pap	paper manufacture	Papierfertigung	—	papeterie
Pap	papeterie	Papierfertigung	paper manufacture	—
Passem	passementerie	Posamenterie	lace manufacture	—
Pat	Patentwesen	—	patenting	brevets d'invention
Pat	patenting	Patentwesen	—	brevets d'invention
Pav	pavage	Pflasterei	paving	—
Pav	paving	Pflasterei	—	pavage
Pêch	pêche	Fischerei	fishery	—
Peint	peinture	Malerei	painting	—
Pflast	Pflasterei	—	paving	pavage
Phot	photographie	Fotografie	photography	—
Phot	photography	Fotografie	—	photographie
Phys	Physik	—	physics	physique
Phys	physics	Physik	—	physique
Phys	physique	Physik	physics	—
pl		Mehrzahl	plural	pluriel
Pol ec	political economy	Volkswirtschaft	—	économie politique
Pompe	construction des pompes	Pumpenbau	construction of pumps	—
Porcel	fabrication de porcelaine	Porzellanherstellung	porcelain manufacture	—
Porcel	porcelain manufacture	Porzellanherstellung	—	fabrication de porcelaine
Porzel	Porzellanherstellung	—	porcelain manufacture	fabrication de porcelaine
Posam	Posamenterie	—	lace manufacture	passementerie
Post	Post	—	mail	poste
Poste	poste	Post	mail	—
Pot	poterie	Töpferei	pottery	—
Pott	pottery	Töpferei	—	poterie
Poudr	poudrerie	Pulverherstellung	powder manufacture	—
Powd	powder manufacture	Pulverherstellung	—	poudrerie
Prép	préparation des houilles ou des minerais	Aufbereitung	coal *or* ore dressing	—
Print	printing	Buchdruckerei	—	imprimerie
Puits	construction de puits	Brunnenbau	well boring	—
Pulv	Pulverherstellung	—	powder manufacture	poudrerie
Pump	Pumpenbau	—	construction of pumps	construction des pompes
Pump	construction of pumps	Pumpenbau	—	construction des pompes
Pyrot	pyrotechnics	Feuerwerkerei	—	pyrotechnique
Pyrot	pyrotechnique	Feuerwerkerei	pyrotechnics	—
Radio	radiocasting	Funkwesen	—	radiodiffusion
Radio	radiodiffusion	Funkwesen	radiocasting	—
Railw	railway	Eisenbahnwesen	—	chemin de fer
Reliure	reliure	Buchbinderei	bookbinding	—
Road	road building	Straßenbau	—	construction de routes
Roll mill	rolling mill	Walzwerk	—	laminoir
Ropem	rope making	Seilerei	—	corderie
Rout	construction de routes	Straßenbau	road building	—
s	—	siehe	see	voyez
Saddl	saddler's trade	Sattlerei	—	sellerie
Sägew	Sägewerk	—	sawmill	scierie
Sailm	sail making	Segelmacherei	—	voilerie
Salt	salt producing	Salzgewinnung	—	production de sel
Salzgew	Salzgewinnung	—	salt producing	production de sel

Zeichen Signs Signes	Fachgebiet Speciality Spécialité	Deutsch	English	Français	Zeichen Signs Signes	Fachgebiet Speciality Spécialité	Deutsch	English	Français
Samm-ler	Akkumula-toren	—	accumulators	accumulateurs	Terr	terrassement	Erdarbeiten	earthwork	—
Sattl	Sattlerei	—	saddler's trade	sellerie	Textil	Textilindustrie	—	textile industry	industrie textile
Savonn	savonnerie	Seifensiederei	soap works	—	Textile	industrie textile	Textilindustrie	textile industry	—
Sawm	sawmill	Sägewerk	—	scierie	Textile	textile industry	Textilindustrie	—	industrie textile
Schiff	Schiffahrt	—	navigation	navigation	Text print	textile printing	Zeugdruckerei	—	imprimerie de tissus
Schiffb	Schiffbau	—	shipbuilding	architecture navale	Tiler	tiler	Dachdecker	—	couvreur
Schloss	Schlosserei	—	locksmith's trade	serrurerie	Tinm	tinman's trade	Klempnerei	—	ferblanterie
Schmied	Schmiede; Hammer-werk	—	forge	forge	Tischl	Tischlerei	—	joiner's trade	menuiserie
Schneid	Schneiderei; Näherei	—	tailoring	métier du tailleur	Tiss	tisseranderie	Weberei	weaving	—
Schreib	Schreib-maschinen	—	typewriters	machines à écrire	Tobacco	tobacco manu-facture	Tabakverarbei-tung	—	manufacture de tabac
Schriftg	Schriftgießerei	—	letter foundry	fonderie de caractères	Tonn	tonnellerie	Böttcherei	cooperage	—
Schuhm	Schuhmacherei	—	shoe making	cordonnerie	Tool	tools	Werkzeuge	—	outils
Scier	scierie	Sägewerk	sawmill	—	Töpf	Töpferei	—	pottery	poterie
Sculpt	sculpture	Bildhauerei	sculpture	sculpture	Tourn	tournerie	Drechslerei; Dreherei	turnery	—
Seew	Seewesen	—	marine	marine	Trade	trade	Handel	—	commerce
Segelm	Segelmacherei	—	sail making	voilerie	Traffic	traffic	Verkehrswesen	—	trafic
Seide	Seidenferti-gung	—	silk manu-facture	fabrication de soie	Trafic	trafic	Verkehrswesen	traffic	—
Seif	Seifensiederei	—	soap works	savonnerie	Trav bois	travail du bois	Holzbearbei-tung	wood working	—
Seil	Seilerei	—	rope making	corderie	Tréfil	tréfilerie	Drahtzieherei	wire drawing	—
Sel	production de sel	Salzgewinnung	salt producing	—	Tuchm	Tuchmacherei	—	cloth manu-facture	manufacture de draps
Selbst-anschl	Selbstan-schlußbetrieb; automati-tische Telefo-nie	—	automatic telephony	téléphonie automatique	Tunn	Tunnelbau	—	tunneling	construction de tunnels
Sell	sellerie	Sattlerei	saddler's trade	—	Tunn	tunneling	Tunnelbau	—	construction de tunnels
Serr	serrurerie	Schlosserei	locksmith's trade	—	Tunn	construction de tunnels	Tunnelbau	tunneling	—
Serv inc	service d'incen-die	Feuerlösch-wesen	fire extinguish-ing	—	Turn	turnery	Drechslerei; Dreherei	—	tournerie
Shipb	shipbuilding	Schiffbau	—	architecture navale	Typewr	typewriters	Schreib-maschinen	—	machines à écrire
Shoem	shoe making	Schuhmacherei	—	cordonnerie	Uhrm	Uhrmacherei	—	watchmaking; clockmaking	horlogerie
Silk	silk manu-facture	Seidenferti-gung	—	fabrication de soie	v	—	siehe	see	voyez
Soap	soap works	Seifensiederei	—	savonnerie	Vann	vannerie	Korbmacherei	basketry	—
Soie	fabrication de soie	Seidenferti-gung	silk manu-facture	—	Vergold	Vergolder	—	gilder	doreur
Spinn	Spinnerei	—	spinning	filature	Verkehr	Verkehrswesen	—	traffic	trafic
Spinn	spinning	Spinnerei	spinning	filature	Verr	verrerie; polissage de verre	Glasfertigung; Glasschlei-ferei	glass manu-facture; glass grind-ing	—
Sport	Sport	Sport	sport	sport	Versich	Versicherungs-wesen	—	insurance	assurance
Starchm	starch manu-facture	Stärkeher-stellung	—	fabrication d'amidon	Vêtem	vêtements	Bekleidung	clothing	—
Stärkef	Stärkeher-stellung	—	starch manu-facture	fabrication d'amidon	Voilerie	voilerie	Segelmacherei	sail making	—
Steam eng	steam engines	Dampfma-schinen	—	machines à vapeur	Volksw	Volkswirt-schaft	—	political economy	économie politique
Steinm	Steinmetz	—	stone cutter	taille-pierre	Wafenw	Waffenwesen	—	service of arms	service de l'armement
Stone cutter	stone cutter	Steinmetz	—	taille-pierre	Wagenb	Wagenbauerei	—	coachmaking	carrosserie
Straßenb	Straßenbau	—	road building	construction de routes	Walzw	Walzwerk	—	rolling mill	laminoir
Sucr	fabrication de sucre	Zuckerher-stellung	sugar manufac-ture	—	War mat	war material	Kriegsgerät	—	matériel de guerre
Sugar	sugar manufac-ture	Zuckerher-stellung	—	fabrication de sucre	Wäsch	Wäscherei	—	washery	laverie
Surv	surveying	Feldmeß-kunde	—	arpentage	Wash	washery	Wäscherei	—	laverie
Tabac	manufacture de tabac	Tabakverarbei-tung	tobacco manu-facture	—	Wasserb	Wasserbau-wesen	—	hydraulic architecture	architecture hydraulique
Tabak	Tabakverarbei-tung	—	tobacco manu-facture	manufacture de tabac	Watchm	watchmaking	Uhrmacherei	—	horlogerie
Tail	tailoring	Schneiderei	—	métier du tailleur	Weav	weaving	Weberei	—	tisseranderie
Taill	métier du tailleur	Schneiderei	tailoring	—	Web	Weberei	—	weaving	tisseranderie
Taille-pierre	taille-pierre	Steinmetz	stone cutter	—	Well	well boring	Brunnenbau	—	construction de puits
Techn	Technik; Technologie	—	technics; technology	technique; technologie	Werkz	Werkzeuge	—	tools	outils
Techn	technics; technology	Technik; Technologie	—	technique technologie	Werk-zeug-masch	Werkzeug-maschinen	—	machine-tools	machines-outils
Techn	technique; technologie	Technik; Technologie	technics; technology	—	Wetter	Wetterkunde; Meteorologie	—	meteorology	météorologie
Teint	teinturier	Färber	dyer	—	Wiredr	wire drawing	Drahtzieherei	—	tréfilerie
Tel	telephone; telegraph	Fernmelde-technik	—	téléphone; télégraphe	Wood engr	wood engrav-ing	Holzschneide-kunst	—	gravure sur bois
Tél	téléphone; télégraphe	Fernmelde-technik	telephone; telegraph	—	Wood work-ing	wood working	Holzbearbei-tung	—	travail du bois
Tél aut	téléphonie automatique	Automatische Telefonie; Selbstan-schlußbetrieb	automatic telephony	—	Zeichn	Zeichnen	—	drawing	dessin
					Zeugdr	Zeugdruckerei	—	textile printing	imprimerie de tissus
					Ziegel	Ziegelei	—	brick works	briqueterie
					Zimm	Zimmerei	—	carpentry	charpente-rie
					Zollw	Zollwesen	—	duty	douane
					Zuck	Zuckerherstel-lung	—	sugar manu-facture	fabrication de sucre

A

A pole (El line) ‖ Spitzbock m. ‖ poteau m. couple.

abaca ‖ Abaka m.; Manilahanf m. ‖ abaca m.; chanvre m. de Manille.

abacus ‖ Rechenbrett n. ‖ abaque m. / ~ (Arch) ‖ Abakus m.; Kapitäldeckplatte f. ‖ abaque m.; abaco m.; tailloir m.

abaft (Aero; Mar) ‖ achter ‖ derrière; en arrière de.

abandon, to ‖ aufgeben ‖ abandonner; renoncer. / ~ a mine ‖ ein Bergwerk n. auflassen oder verlassen ‖ abandonner une mine.

abandonment ‖ Verzicht m.; Verzichtleistung f. ‖ abandonnement m.; renonciation f.; désistement m.

abate, to (Meteor) ‖ sich beruhigen; nachlassen ‖ se calmer; mollir.

abated ‖ herabgesetzt ‖ réduit; diminué; rabattu.

abatement ‖ Preisermäßigung f.; Nachlaß m.; Abschlag m. ‖ réduction f. de prix; rabais m.

abattoir, public ‖ städtischer Schlachthof m. ‖ abattoir m. municipal.

abbey ‖ Abtei f. ‖ abbaye f.

abbreviate, to ‖ abkürzen ‖ abréger.

abbreviation ‖ Abkürzung f.; Kurzzeichen n. ‖ abréviation f. / composition of ~s (Print) ‖ Abbreviaturensatz m. ‖ composition f. avec beaucoup d'abréviations.

abbreviation key (Typewr) ‖ Abkürzungstaste f. ‖ touche f. d'abréviation.

ABC telegraph ‖ ABC-Telegraf m. ‖ cadran m.

abeam of (Aero; Mar) ‖ dwars ‖ par le travers.

abelmosk seed ‖ Abelmoschkorn n. ‖ semence f. d'abelmosch.

aberration (Opt) ‖ Abweichung f.; Aberration f. ‖ aberration f.; écart m. / amplitude of ~ ‖ Größe f. der Abweichung ‖ amplitude f. d'aberration. / angle of ~ ‖ Aberrationswinkel m. ‖ angle m. d'aberration. / chromatic ~ of the object glass ‖ Farbenabweichung f. oder chromatische Abweichung f. des Objektivs ‖ aberration f. chromatique de l'objectif. / extraaxial ~ ‖ Aberration f. außerhalb der Achse ‖ aberration f. extra-axiale ou en dehors de l'axe. / ~ of light ‖ Abirrung f. des Lichts ‖ aberration f. de la lumière. / spherical ~ of the object ‖ sphärische Abweichung f. des Objektivs ‖ aberration f. de sphéricité de l'objectif.

ability to pay ‖ Solvenz f. ‖ solvabilité f.

ablation ‖ Abtragung f. ‖ ablation f.

able to put up security ‖ kautionsfähig ‖ capable de fournir un cautionnement. / ~ to work ‖ arbeitsfähig ‖ apte au travail; capable de travailler.

able-bodied seaman ‖ befahrener Seemann m. ‖ homme m. de mer.

abnormal ‖ anormal; abnorm ‖ anormal.

abolish, to ‖ außer Gebrauch m. setzen ‖ mettre hors d'emploi m.

aboard (Mar) ‖ an Bord m. ‖ à bord m.

abortion forceps pl. ‖ Abortuszange f. ‖ pince f. d'avortement ou à faux germes.

about ‖ zirka; ungefähr ‖ environ; à peu près. / to go ~ (Aero; Mar) ‖ den Kurs m. ändern ‖ virer de bord. / ~ sledge ‖ Poßhammer m. ‖ marteau m. à devant; marteau m. à frapper devant.

above ‖ über; oben ‖ au-dessus. / ~ (Trade) ‖ mehr als ‖ plus de; plus que. / rollers pl. arranged one ~ the other ‖ übereinander liegende Walzen fpl. ‖ cylindres mpl. placés l'un au-dessus de l'autre. / ~ average ‖ über dem Durchschnitt m. ‖ au-dessus de la moyenne. / ~ ground ‖ oberirdisch ‖ au-dessus du sol. / ~ the horizon ‖ über dem Horizont m. ‖ au-dessus de l'horizon. / ~ the sluice stream ‖ stromaufwärts von der Schleuse f. ‖ en amont m. de l'écluse. / ~ sea level ‖ über dem Meeresspiegel m. ‖ au-dessus du niveau de la mer. / ~ water ‖ über Wasser n. ‖ au-dessus de l'eau f.

abradant ‖ Schleifmittel n. ‖ abrasif m. / bedding the journal in the bearing by means of an ~ ‖ Einschleifen n. des Zapfens in das Lager ‖ rodage m. à l'émeri du tourillon dans le coussinet.

abrade, to ‖ schleifen ‖ roder; meuler; rectifier à la meule.

abrasion ‖ Abschaben n. ‖ abrasion f.; meulage m.; raclage m. / ~ of the insulation (Electr) ‖ Abscheuerung f. der Isolation ‖ arrachement m. de la couche isolante. / ~ testing machine ‖ Abnutzungsprüfmaschine f. ‖ machine f. à vérifier l'usure.

abrasive ‖ Schleifmittel n. ‖ abrasif m. / ~ hardness ‖ Schneidhärte f. ‖ trempe f. active ou d'outil coupant. / ~ machine (Textile) ‖ Schmirgelmaschine f. ‖ machine f. à meuler à l'émeri.

abraum salts pl. ‖ Abraumsalze npl. ‖ sels mpl. de déblai.

abroad ‖ im Ausland n. ‖ à l'étranger m.

abroad ‖ Ausland n. ‖ étranger m. / passport for ~ ‖ Auslandspaß m. ‖ passeport m. pour l'étranger.

abrotanum herb ‖ Eberraute f. ‖ herbe f. d'abrotane.

abrupt ‖ schroff; abschüssig; jäh ‖ abrupt; brusque; escarpé. / ~ sharp angle ‖ scharf abgesetzte Stelle f.; scharfer Ansatz m. ‖ entaillé à angle vif.

abscissa ‖ Abszisse f. ‖ abscisse f. / axis of ~s ‖ Abszissenachse f. ‖ axe m. des abscisses.

absence ‖ Mangel m. ‖ absence f.; manque m. / ~ of current (Electr) ‖ Stromlosigkeit f. ‖ absence f. ou manque m. de courant. / ~ of haze in the object ‖ Schleierfreiheit f. des Bildes ‖ absence f. de voiles dans l'image. / in the ~ of ‖ mangels; in Ermangelung f. von ‖ à défaut m. de; faute de.

absinth ‖ Absinth m.; Wermut m. ‖ absinthe m. / dry extract of ~ ‖ trockener Absinthauszug m. ‖ extrait m. d'absinthe sec.

absolute ‖ absolut; unabhängig ‖ absolu. / to become ~ ‖ Rechtskraft f. erlangen ‖ avoir force f. de loi. / ~ alcohol ‖ reiner Alkohol m. ‖ alcool

m. absolu. / ~ movement ‖ absolute Bewegung f. ‖ mouvement m. absolu. / ~ pressure ‖ absoluter Druck m. ‖ pression f. absolue. / ~ speed ‖ absolute Geschwindigkeit f. ‖ vitesse f. absolue. / ~ system of measures ‖ absolutes Maßsystem n. ‖ système m. absolu de mesures. / ~ temperature ‖ absolute Temperatur f. ‖ température f. absolue. / ~ tenacity ‖ Zugfestigkeit f. ‖ résistance f. à la traction. / ~ unit ‖ absolute Einheit f. ‖ unité f. absolue. / ~ vacuum ‖ absolute Luftleere f. ‖ vide m. parfait or absolu. / ~ velocity see absolute speed. / ~ zero ‖ absoluter Nullpunkt m. ‖ zéro m. absolu.

absolutely accurate ‖ haargenau; haarscharf ‖ rigoureusement exact.

absorb, to ‖ absorbieren; aufsaugen; aufnehmen ‖ absorber. / ~ a gas ‖ ein Gas n. verschlucken ‖ absorber un gaz. / ~ heat ‖ Wärme f. aufnehmen ‖ absorber la chaleur. / ~ a shock ‖ einen Stoß m. auffangen ‖ amortir un choc.

absorbability ‖ Absorbierbarkeit f. ‖ absorbabilité f.

absorbable ‖ absorbierbar ‖ absorbable.

absorbed radiation ‖ absorbierte Strahlung f. ‖ radiation f. absorbée.

absorbent liquid ‖ Absorptionsflüssigkeit f. ‖ liquide m. d'absorption. / ~ material ‖ Absorptionsmittel n. ‖ absorbant m.

absorber ‖ Absorptionsgefäß n. ‖ cuve f. à colonne liquide ou à liquide; récipient m. ou vase m. d'absorption; absorbeur m.

absorbing apparatus ‖ Absorptionsapparat m. ‖ appareil m. d'absorption. / ~ capacity of a lens ‖ Absorptionswert m. einer Linse ‖ capacité f. d'absorption d'une lentille. / ~ effect ‖ absorbierende Wirkung f. ‖ action f. absorbante. / ~ filter of neutral glass ‖ neutrales Blendglas n. ‖ verre m. sombre à teinte neutre. / ~ fluid ‖ absorbierende Schicht f. ‖ couche f. absorbante. / ~ gas mask ‖ Absorptionsgasmaske f. ‖ masque m. à gaz absorbant. / dark ~ glass ‖ dunkles Absorptionsglas n. ‖ verre m. absorbant foncé. / ~ liquid see absorbent liquid / ~ power ‖ Absorptionsvermögen n.; Absorptionsfähigkeit f. ‖ pouvoir m. absorbant. / ~ wedge (Opt) ‖ Absorptionskeil m. ‖ coin m. absorbant.

absorptiometer ‖ Absorptiometer n. ‖ absorptiomètre m.

absorption ‖ Absorption f. ‖ absorption f. / atmospheric ~ ‖ atmosphärische Absorption f. ‖ absorption f. atmosphérique. / chemical ~ ‖ chemische Absorption f. ‖ absorption f. chimique. / ~ of energy ‖ Kraftverbrauch m. ‖ absorption f. de force. / ~ of gases ‖ Absorption f. der Gase ‖ absorption f. des gaz. / ~ of heat ‖ Wärmeaufnahme f. ‖ absorption f. de chaleur. / ~ of hydrogen by metals ‖ Absorption f. des Wasserstoffes durch Metall ‖ absorption f. de l'hydrogène par les métaux. / ~ of light ‖ Lichtabsorption f. ‖ absorption f. de lumière. / luminous ~ ‖ Lichtabsorption f. ‖ absorp-

tion f. lumineuse. / ~ of shocks || Stoß-dämpfung f. || amortissement m. des chocs / thermal ~ || Wärmeabsorption f. || absorption f. thermale. / ~ of water || Wasseraufnahme f. || absorption f. d'eau.

absorption apparatus || Absorptionsapparat m. || appareil m. d'absorption. / ~ band || Absorptionsstreifen m.; Absorptionsbande f. || bande f. d'absorption. / ~ bands pl. of the colouring bodies of blood || Absorptionsbanden fpl. von Blutfarbstoffen || bandes fpl. d'absorption des colorants du sang. / ~ bottle || Absorptionsflasche f. || récipient m. d'absorption. / ~ capacity || Absorptionsvermögen n.; Schluckfähigkeit f. || pouvoir m. ou puissance f. d'absorption. / ~ cell see absorption vessel. / ~ circuit || Absorptionskreis m. || circuit m. absorbant. / coefficient of ~ || Absorptionskoeffizient m. || coefficient m. d'absorption. / ~ colouring || Färbung f. durch Einsaugung; Einsaugungsfärbung f. || coloration f. par absorption. / ~ cooling machine || Absorptionskühlmaschine f. || machine f. à froid à absorption. / ~ current || Absorptionsstrom m. || courant m. d'absorption. / degree of ~ || Absorptionsfähigkeit f. || degré m. d'absorption. / ~ line || Absorptionslinie f. || raie f. d'absorption. / ~ liquid || Absorptionsflüssigkeit f. || liquide m. d'absorption. / ~ machine || Absorptionsmaschine f. || machine f. à absorption. / ~ method || Absorptionsverfahren n. || méthode f. absorptiométrique. / ~ plant || Absorptionsanlage f. || installation f. d'absorption. / ~ spectrum || Absorptionsspektrum n. || spectre m. d'absorption. / ~ tube || Absorptionsröhre f.; Eudiometer n. || tube m. d'absorption; eudiomètre m.

absorption vessel || Absorptionsgefäß n. || cuve f. à colonne liquide ou à liquide; récipient m. ou vase m. d'absorption; absorbeur m. / ~ of variable fluid depth || Absorptionsgefäß n. mit veränderlicher Schichtdicke || récipient m. d'absorption à colonne liquide d'épaisseur variable.

absorptive || absorbierend || absorptif.

abstract, to || destillieren; entziehen || distiller; enlever.

abstract mathematics pl. || reine Mathematik f. || mathématiques fpl. pures.

abstract || Rechnungsauszug m. || relevé m. de compte.

abstraction of water || Wasserentziehung f. || soustraction f. de l'eau.

abundance || Überfluß m. || abondance f. / ~ of short wave rays in the red free light || Reichhaltigkeit f. des Rotfreilichtes an kurzwelligen Strahlen || richesse f. de la lumière exempte de rouge en rayons de petite longueur d'onde. / ~ of seams || Flözreichtum m. || richesse f. en couches.

abundant || reichhaltig; ausgiebig || abondant; riche; bien assorti.

aburton || querschiffs || par le travers.

abuse || Mißbrauch m. || abus m.

abut, to || mit den Enden npl. stumpf zusammenfügen || abouter.

abutment (Bridge) || Widerlager n. || culée f. / ~ pier || Widerlagspfeiler m. || pile f. de culée.

acacia || Akazie f. || acacia m. / common ~ || unechte Akazie f.; Schotendorn m. || faux-acacia m. || Akazienblatt n. || feuille f. d'acacia. / ~ oil || Akazienöl n. || essence f. ou huile f. d'acacia. / ~ pod ||

Akazienschote f. || gousse f. d'acacia. / ~ wood || Akazienholz n. || bois m. d'acacia.

academy || Akademie f. || académie f. / ~ of architecture || Bauschule f. || école f. d'architecture. / veterinary ~ || tierärztliche Hochschule f. || école f. vétérinaire supérieure.

acaroid gum || Akaroidharz n. || gomme f. acaroïde.

accelerate, to || beschleunigen || accélérer. / ~ combustion by a blower || die Verbrennung f. durch ein Gebläse beschleunigen || activer la combustion par une machine soufflante.

accelerated || beschleunigt || accéléré. / uniformly ~ || gleichmäßig beschleunigt || accéléré uniformément. / ~ motion || beschleunigte Bewegung f. || mouvement m. accéléré. / ~ velocity || beschleunigte Geschwindigkeit f. || vitesse f. accélérée.

accelerating device for double inclines || Beschleunigungsvorrichtung f. für Ablaufberge || dispositif m. d'accélération pour plans inclinés.

acceleration || Beschleunigung f. || accélération f. / angular ~ || Winkelbeschleunigung f. || accélération f. angulaire. / ~ of gravity || Erdbeschleunigung f. || accélération f. ou intensité f. de la pesanteur. / negative ~ || Verzögerung f. || accélération f. négative; retardation f. / normal ~ || normale Beschleunigung f. || accélération f. normale. / tangential ~ || tangentiale Beschleunigung f. || accélération f. tangentielle. / uniform ~ || gleichmäßige Beschleunigung f. || accélération f. uniforme.

accelerator || Beschleunigungsvorrichtung f. || accélérateur m. / ~ pedal || Beschleunigungsfußhebel m.; Akzeleratorpedal n. || pédale f. d'accélérateur.

accelerometer || Beschleunigungsmesser m. || accéléromètre m.

accentuate, to ~ the contrasts pl. || die Kontrastwirkung f. erhöhen || obtenir des contrastes mpl. plus vifs.

accept, to || honorieren; annehmen; akzeptieren || honorer; accepter; payer; accueillir. / ~ a bill || einen Wechsel m. annehmen || accepter un effet; honorer ou accueillir une traite. / ~ on original conditions || zu ursprünglichen Bedingungen fpl. annehmen || accepter aux conditions fpl. originales. / ~ on terms of letter || zu brieflichen Bedingungen fpl. annehmen || accepter aux conditions fpl. de la lettre.

acceptance || Annahme f. || acceptation f. / ~ (Bank) || Akzept n. || acceptation f.; traite f. acceptée. / ~ (Trade) || Zusage f. || assentiment m.; promesse f. / to present a bill for ~ || einen Wechsel m. zum Akzept vorzeigen || présenter une traite à l'acceptation. / refusal of ~ || Annahmeverweigerung f. || refus m. d'acceptation. / ~ in blank || Blankoakzept n. || acceptation f. à découvert. / ~ of material || Werkstoffabnahme f. || réception f. de matériaux. / partial ~ || Teilakzept n. || acceptation f. partielle. / specification for ~ || Abnahmevorschrift f. || instruction f. de réception. / ~ test || Übernahmeprobe f.; Abnahmeprüfung f. || essai m. de réception. / official ~ test || amtliche Abnahmeprüfung f. || essai n. de réception officiel. / trial run for ~ (Nav) || Abnahmeprobefahrt f. || essai m. de recette.

accepted bill || Akzept n. || acceptation f.; traite f. acceptée.

acceptor || Akzeptant m. || acceptant m.; accepteur m.

access || Zutritt m.; Zugang m.; Zugänglichkeit f. || accès m.; entrée f.; abord m. / easy of ~ || leicht zugänglich || d'un accès facile.

accessibility || Zugänglichkeit f. || accessibilité f. / ~ of all parts || Zugänglichkeit f. aller Teile || accessibilité f. de tous les organes. / easy ~ || leichte Zugänglichkeit f. || accessibilité f. facile. / ~ of the types || Zugänglichkeit f. der Typen || accessibilité f. des caractères.

accessible || zugänglich || accessible; abordable.

accessories pl. || Zubehör n.; Zubehörteile mpl. || accessoires mpl. / automobile ~ || Automobilzubehörteile mpl. || accessoires mpl. d'automobiles. / boiler ~ || Kesselzubehörteile mpl. || accessoires mpl. de chaudières. / ~ of centrifugal machines || Zentrifugalzubehör n. || accessoires mpl. pour centrifuges. / ~ for coach building || Wagenbauteile mpl. || accessoires mpl. de carrosserie. / ~ for wireless apparatusses || Radiozubehör n. || accessoires mpl. pour appareils de téléphonie et télégraphie sans fil.

accessory see also accessories || Zubehörteil m. || accessoire m.; garniture f.; équipement m.; fourniture f. / ~ apparatus || Nebenapparat m. || appareil m. auxiliaire. / ~ box || Zubehörkasten m. || caisse f. à accessoires. / ~ parts || Zubehörteile mpl. || pièces fpl. accessoires.

accident || Unglücksfall m.; Unfall m. || accident m. / first aid treatment in case of ~s || erste Hilfe f. bei Unglücksfällen || premier secours m. en cas d'accidents. / ~ to machinery || Maschinenunfall m.; Maschinenschaden m. || accident m. à la machine. / ~ in a mine || Grubenunfall m. || accident m. à la mine. / ~ to plant || Betriebsunfall m. || accident m. de service à l'usine. / to prevent ~s || Unfälle mpl. verhüten oder vermeiden || prévenir des accidents mpl. / ~ at sea || Gefahren fpl. zur See; Seeschaden m. || fortune f. de mer.

accidental error || Zufälligkeitsfehler m. || erreur f. accidentelle.

accident insurance || Unfallversicherung f. || assurance f. contre les accidents.

accidently || durch Zufall m. || par hasard m.

accident preventives pl. || Unfallverhütungsmaßregeln fpl. || mesures fpl. préventives des accidents. / ~ signalling system || Unfallmeldeanlage f. || installation f. pour l'annonce des accidents.

acclimatization || Einheimischwerden n.; Akklimatisierung f. || acclimatation f.

acclimatize, to || anpassen; einheimisch machen; akklimatisieren || acclimater; accoutumer.

acclimatized || akklimatisiert || acclimaté.

accommodate, to (Bank) || unterbringen || loger; placer; caser. / ~ (Opt) || akkommodieren; in Übereinstimmung f. bringen || accommoder. / ~ the workmen || die Belegschaft f. unterbringen || loger la population ouvrière.

accommodating connection for extension stations || Anpassungsschaltung f. für Nebenstellen || circuit m. d'accommodation pour installations supplémentaires.

accommodating power of the eyes || Anpassungsfähigkeit f. der Augen || amplitude f. de l'accommodation des yeux.

accommodation ‖ Akkommodation f. ‖ accommodation f. / power of ~ ‖ Akkommodationsfähigkeit f. ‖ pouvoir m. d'accommodation. / to provide persons of a village with fresh ~ elsewhere ‖ Einwohner mpl. eines Dorfes umsiedeln ‖ déménager des habitants d'un village. / relaxed state of ~ ‖ Akkommodationsruhe f. ‖ relâchement m. de l'accommodation. / state of ~ ‖ Akkommodationszustand m. ‖ état m. d'accommodation.

accommodation bill ‖ Gefälligkeitswechsel m.; Gefälligkeitsakzept n. ‖ effet m. de complaisance.

accompanying gun ‖ Begleitgeschütz n. ‖ canon m. d'accompagnement.

accomplish, to ‖ zuwegebringen ‖ venir à bout m. de; accomplir; terminer.

accord ‖ Übereinstimmung f. ‖ concordance f.; accord m.; conformité f.

accord (Mus) ‖ Akkord m. ‖ accord m.

accordance, in ~ with ... ‖ vorschriftsgemäß; in Übereinstimmung f. mit ... ‖ conformément à; en conformité f. avec ...

according to custom ‖ gewohnheitsmäßig; usancemäßig ‖ selon l'usage m.; d'usage m. ‖ ~ to directions ‖ vorschriftsmäßig ‖ conforme aux prescriptions fpl. / ~ to the usual practice (Trade) ‖ usancegemäß ‖ selon l'habitude f.

accordion ‖ Ziehharmonika f.; Handharmonika f.; Akkordion n. ‖ accordéon m.

account, to ‖ Rechenschaft f. ablegen ‖ rendre compte m.

account ‖ Kostenanschlag m. ‖ devis m. estimatif; estimation f. / ~ (Trade) ‖ Rechnung f. ‖ calcul m.; compte m.; opération f.; facture f.; note f.; mémoire m. ‖ balance of ~ ‖ Rechnungssaldo m.; solde m. (de compte) / bank(ing) ~ ‖ Bankkonto n. ‖ compte m. en banque. / capital ~ ‖ Kapitalkonto n. ‖ compte m. capital. / to charge to the ~ ‖ anrechnen ‖ compter; mettre ou passer en compte m. / ~ of charges ‖ Spesenrechnung f. ‖ compte m. de frais. / to check items in an ~ ‖ die Posten mpl. eines Kontos prüfen ‖ pointer ou vérifier les articles mpl. d'un compte. / to draw up an ~ ‖ eine Rechnung f. ausschreiben ‖ établir un compte. / extract of ~ ‖ Kontoauszug m. ‖ extrait m. de compte. / final ~ ‖ Schlußrechnung f. ‖ compte m. définitif. / ~ for ‖ Rechnung f. über ‖ compte m. pour. / to give an ~ ‖ Rechenschaft f. ablegen ‖ rendre compte m. / to keep an ~ ‖ ein Konto n. unterhalten ‖ tenir un compte. / to make ~s pl. agree ‖ die Rechnungen fpl. in Übereinstimmung bringen ‖ mettre d'accord les comptes mpl. / to make out an ~ ‖ eine Rechnung f. aufstellen oder ausschreiben ‖ établir ou dresser un compte; faire la facture. / money received on ~ ‖ Rechnungsabzahlung f. ‖ versement m. / on ~ ‖ abschläglich ‖ à compte m. / opening of ~ ‖ Kontoeröffnung f. ‖ ouverture f. d'un compte. / to pay on ~ ‖ auf Abschlag m. zahlen; anzahlen ‖ payer à compte m.; donner en acompte m. / payment on ~ ‖ Abschlagszahlung f. ‖ acompte m.; payement m. partiel ou par acompte. / to place to ~ ‖ in Ansatz m. bringen ‖ mettre en ligne f. de compte; passer en compte m. / to render an ~ ‖ eine Rechnung f. senden ‖ remettre un compte. / ~ rendered ‖ Rechenschaftsbericht m. ‖ compte m. rendu. / statement of ~ ‖ Rechnungsauszug m. ‖ extrait

m. ou relevé m. de compte; bordereau m. / to take into ~ ‖ berücksichtigen ‖ considérer; prendre en considération f.

accountable ‖ haftbar ‖ responsable.

accountant (Bank) ‖ Rendant m. ‖ trésorier m.; caissier m. / ~ (Bookkeep) ‖ Buchhalter m.; Bücherrevisor m. ‖ comptable m.; teneur m. de livres. / mining ~ ‖ Rechnungsführer m. der Gruben ‖ comptable m. des mines.

accountancy ‖ Rechnungswesen n. ‖ comptabilité f. / ~ machine ‖ Buchhaltungsmaschine f. ‖ machine f. à faire la comptabilité.

account book ‖ Kontobuch n.; Geschäftsbuch n. ‖ livre m. commercial ou de comptes. / ~ day ‖ Lohntag m. ‖ jour m. de paye ou du payement.

accounted ‖ in Rechnung f. gestellt ‖ facturé.

accounting ‖ Rechnungswesen n. ‖ comptabilité f.

accounts pl. ‖ Rechnungswesen n. ‖ comptabilité f.

account sales ‖ Verkaufsrechnung f. ‖ compte m. de vente. / ~ sheet ‖ Kontoblatt n. ‖ relevé m. de compte.

accredited agent ‖ bevollmächtigter Vertreter m. ‖ agent m. accrédité.

accumulate, to ‖ aufspeichern; ansammeln ‖ accumuler. / ~ (Money) ‖ auflaufen ‖ s'enfler; s'accumuler. / ~ energy ‖ Energie f. aufspeichern ‖ accumuler de l'énergie f.

accumulated interest ‖ aufgelaufene Zinsen mpl. ‖ intérêts m. pl. accumulés.

accumulation ‖ Stauung f. ‖ accumulation f. / ~ of snow-drift in cuttings ‖ Schneeverwehung f. der Einschnitte ‖ enneigement m. des tranchées.

accumulator ‖ Sammler m.; Akkumulator m. ‖ accumulateur m. / alcaline ~ ‖ alkalischer Sammler m. ‖ accumulateur m. alcalin. / to charge an ~ ‖ einen Sammler m. aufladen ‖ charger un accumulateur. / charging an ~ ‖ Aufladung f. eines Sammlers ‖ chargement m. d'un accumulateur. / compressed air type ~ ‖ Druckluftspeicher m. ‖ accumulateur m. à charge d'air comprimé. / dry ~ ‖ Trockensammler m. ‖ accumulateur m. sec. / electric ~ ‖ elektrischer Sammler m. oder Akkumulator m. ‖ accumulateur m. électrique. / fluid ~ ‖ Akkumulator m. mit Säurefüllung ‖ accumulateur m. à acide. / grid-type ~ ‖ Gittersammler m. ‖ accumulateur m. à grillage. / hydraulic ~ ‖ hydraulischer Sammler m.; Druckwasserspeicher m. ‖ château m. d'eau de pression; accumulateur m. hydraulique. / iron-nickel ~ ‖ Eisennickelsammler m. ‖ accumulateur m. au fer et au nickel. / liquid power ~ ‖ Flüssigkeitskraftspeicher m. ‖ accumulateur m. hydraulique d'énergie. / to overcharge the ~ ‖ den Sammler m. überladen ‖ surcharger l'accumulateur m. / pocket ~ ‖ Taschensammler m. ‖ accumulateur m. de poche. / portable ~ ‖ transportabler oder beweglicher Sammler m. ‖ accumulateur m. transportable. / ~ for quick charging ‖ Sammler m. für schnelle Ladung ‖ accumulateur m. à formation rapide ou à charge rapide. / ~ for slow charging ‖ Sammler m. für langsame Ladung ‖ accumulateur m. à formation lente ou à charge lente. / stationary ~ ‖ stationärer Sammler m. ‖ accumulateur m. stationnaire. / steam ~ ‖ Dampfspeicher m. ‖ ac-

cumulateur m. de vapeur. / transportable ~ ‖ transportabler Sammler m. ‖ accumulateur m. transportable. / weight ~ ‖ Gewichtskraftspeicher m. ‖ accumulateur m. à poids.

accumulator acid ‖ Sammlersäure f.; Füllsäure f. ‖ acide m. de remplissage ou pour accumulateurs. / ~ battery ‖ Sammlerbatterie f. ‖ batterie f. d'accumulateurs. / ~ box ‖ Sammlergefäß n. ‖ bac m. ou récipient m. ou cuve f. d'accumulateur. / ~ box with guide channels ‖ Rippenglasgefäß n. für Sammler ‖ bac m. d'accumulateur en verre à surfaces cannelées. / ~ car ‖ Sammlerwagen m. ‖ automotrice f. ou voiture f. à accumulateurs. / ~ charging ‖ Sammlerladung f. ‖ charge f. d'accumulateur. / ~ charging apparatus ‖ Sammlerladeapparat m. ‖ installation f. de recharge d'accumulateurs. / ~ erector ‖ Sammlermontör m. ‖ monteur m. d'accumulateurs. / ~ factory ‖ Sammlerfabrik f. ‖ fabrique f. d'accumulateurs. / ~ grid plate ‖ Sammlergitterplatte f. ‖ plaque f. à grillage d'accumulateur. / ~ jar ‖ Sammlergefäß n. ‖ bac m. ou récipient m. ou cuve f. ou vase m. d'accumulateur. / ~ plate ‖ Sammlerplatte f. ‖ plaque f. d'accumulateur. / ~ room ‖ Sammlerraum m. ‖ salle f. des accumulateurs. / ~ tank ‖ Sammlerkasten m. ‖ caisse f. d'accumulateur. / ~ testing instrument ‖ Sammlerprüfer m. ‖ appareil m. à essayer les accumulateurs.

accuracy ‖ Genauigkeit f. ‖ précision f.; exactitude f.; justesse f. / ~ of a balance ‖ Genauigkeit f. einer Wage ‖ précision f. d'une balance. / ~ of calculation ‖ Rechnungsgenauigkeit f. ‖ précision f. de calcul. / degree of ~ ‖ Genauigkeitsgrad m. ‖ degré m. de précision. / with a considerable degree of ~ ‖ von großer Genauigkeit f. ‖ d'une exactitude considérable. / ~ of construction ‖ Herstellungsgenauigkeit f. ‖ exactitude f. d'exécution. / to increase the ~ ‖ die Genauigkeit f. erhöhen ‖ augmenter la précision. / ~ of measurement ‖ Genauigkeit f. der Messung ‖ précision f. de la mesure. / high degree of ~ of measurement ‖ große Meßgenauigkeit f. ‖ haut degré m. de précision de mesure. / want of ~ ‖ Mangel m. an Genauigkeit ‖ manque m. de précision.

accurate ‖ genau ‖ précis; exacte. / absolutely ~ ‖ haargenau; haarscharf ‖ rigoureusement exact. / ~ map ‖ genaue Landkarte f. ‖ carte f. exacte. / ~ wire drawing works pl. ‖ Genaudrahtzieherei f.; Präzisionsdrahtzieherei f. ‖ tréfilerie f. de précision. / ~ yarn quadrant ‖ Genauigkeitsgarnwage f. ‖ romaine f. de précison.

accusation ‖ Anklage f. ‖ accusation f.; prévention f.; incrimination f.

accuse, to ‖ anklagen ‖ accuser; dénoncer.

accuser ‖ Ankläger m. ‖ accusateur m.

accustomed to the sea (Persons) ‖ seefest ‖ exempt du mal de mer; pratiqué à la mer.

acescent (Chem) ‖ sauer werdend ‖ acescent.

acetaldehyde ‖ Azetaldehyd n. ‖ acétaldéhyde m.

acetate ‖ Azetat n.; essigsaures Salz n. ‖ acétate m. / ~ of alumina ‖ essigsaure Tonerde f. ‖ acétate m. d'alumine. / ~ of ammonium ‖ Ammoniumazetat n. ‖ acétate m. d'ammoniaque. / ~ of barium ‖ Bariumazetat n. ‖ acétate m. de barium.

/ ~ of lead ‖ Bleiazetat n.; Bleizucker m. ‖ acétate m. de plomb. / plumbic ~ ‖ Bleizucker m. ‖ sucre m. de Saturne. / ~ of potassium ‖ essigsaures Kalium n. ‖ acétate m. de potasse. / ~ lacquer ‖ Azetatlack m. ‖ laque m. d'acétate.

acetic ‖ essigsauer ‖ acéteux; acétique.

acetic acid ‖ Essigsäure f. ‖ acide m. acétique. / glacial ~ ‖ Eisessig m. ‖ acide m. acétique cristallisable.

acetic acid apparatus ‖ Essigsäureapparat m. ‖ appareil m. à fabriquer l'acide acétique. / ~ bacterium ‖ Essigsäurebakterium n. ‖ bactérie f. d'acide acétique. / ~ determination ‖ Essigsäurebestimmung f. ‖ dosage m. de l'acide acétique. / ~ fermentation ‖ Essigsäuregärung f. ‖ fermentation f. acétique. / ~ plant ‖ Essigsäureanlage f. ‖ installation f. pour l'acide acétique.

acetic aldehyde ‖ Azetaldehyd n. ‖ aldéhyde m. acétique. / ~ anhydride ‖ Essigsäureanhydrid n. ‖ anhydride m. acétique. / ~ ester see ~ ether. / ~ ether ‖ Essigäther m.; Essigester m.; Essigsäureäthylester m. ‖ éther m. acétique.

acetification ‖ Essigbildung f. ‖ acétification f.

acetimeter ‖ Säuremesser m. ‖ acidimètre m.

acetone ‖ Azeton n. ‖ acétone f. / ~ bisulfite ‖ Azetonbisulfit n. ‖ bisulfite m. d'acétone. / ~ oil ‖ Azetonöl n. ‖ huile f. d'acétone.

acetous ‖ essigsauer ‖ acéteux; acétique.

acetylable ‖ azetylierbar ‖ acétylable.

acetylate, to ‖ azetylieren ‖ acétyler.

acetyl cellulose lacquer ‖ Azetylzelluloselack m. ‖ laque m. de cellulose acétylée. / ~ chloride ‖ Chlorazetyl n.; Azetylchlorid n. ‖ chlorure m. d'acétyle.

acetylene ‖ Azetylen n. ‖ acétylène m. / compressed ~ ‖ komprimiertes Azetylen n. ‖ acétylène m. comprimé. / dissolved ~ ‖ gelöstes Azetylen n. ‖ acétylène m. dissous. / liquid ~ ‖ verflüssigtes Azetylen n. ‖ acétylène m. liquide ou en solution.

acetylene apparatus ‖ Azetylenapparat m. ‖ appareil m. à acétylène. / ~ black ‖ Azetylenschwarz n. ‖ noir m. d'acétylène. / ~ Bunsen burner ‖ Azetylenbunsenbrenner m. ‖ bec m. ou brûleur m. de Bunsen à l'acétylène. / ~ burner ‖ Azetylenbrenner m. ‖ bec m. ou brûleur m. à acétylène. / ~ cutting ‖ Azetylenschneidverfahren n. ‖ découpage m. à l'acétylène. / ~ cutting plant ‖ Azetylenschneidanlage f. ‖ installation f. de découpage à l'acétylène.

acetylene gas ‖ Azetylengas n. ‖ gaz m. (d')acétylène. / ~ bottle ‖ Azetylengasflasche f. ‖ bouteille f. à gaz acétylène. / ~ burner ‖ Azetylengasbrenner m. ‖ bec m. à gaz acétylène. / ~ headlight see ~ searchlight. / ~ lighting ‖ Azetylengasbeleuchtung f. ‖ éclairage m. par acétylène. / ~ plant ‖ Azetylengasanlage f. ‖ installation f. pour fabriquer le gaz acétylène. / ~ producing plant ‖ Azetylengasentwicklungsanlage f. ‖ installation f. pour la production de gaz acétylène. / ~ searchlight ‖ Azetylenscheinwerfer m. ‖ phare m. à acétylène. / ~ works pl. ‖ Azetylengasanstalt f. ‖ usine f. de gaz acétylène.

acetylene generator ‖ Azetylenentwickler m.; Azetylenerzeuger m. ‖ générateur m. d'acétylène. / ~ lamp ‖ Azetylenlampe f.

‖ lampe f. à acétylène. / ~ lantern ‖ Azetylenlaterne f. ‖ lanterne f. à acétylène. / ~ lighting ‖ Azetylenbeleuchtung f. ‖ éclairage m. à l'acétylène. / ~ oxyhydrogen fusing burner ‖ Azetylensauerstoffschneidbrenner m. ‖ brûleur m. acétylène-oxyhydrique à découper. / ~ plant ‖ Azetylenanlage f. ‖ installation f. d'acétylène. / ~ producing plant ‖ Azetylenerzeugungsanlage f. ‖ installation f. pour la production d'acétylène. / ~ purifying agent ‖ Azetylenreinigungsmasse f. ‖ matière f. d'épuration pour le gaz acétylène. / ~ purifying and drying mass ‖ Azetylenreinigungs- und -trocknungsmasse f. ‖ matière f. d'épuration et de séchage pour acétylène. / ~ searchlight see acetylene gas searchlight. / ~ tetrachloride ‖ Azetylentetrachlorid n. ‖ tétrachlorure m. d'acétylène. / ~ welding ‖ Azetylenschweißverfahren n. ‖ soudage m. à l'acétylène. / ~ welding plant ‖ Azetylenschweißanlage f. ‖ installation f. de soudage à acétylène.

acetylsalicylic acid ‖ Azetylsalizylsäure f. ‖ acide m. acétylsalicylique.

ache ‖ Schmerz m. ‖ douleur f.

achievable ‖ ausführbar ‖ praticable; exécutable.

achromatic ‖ farblos; achromatisch ‖ achromatique. / non ~ lens ‖ nicht achromatische Linse f. ‖ lentille f. non achromatique. / ~ glass lens ‖ Glaschromat m. ‖ objectif m. achromatique en verre. / ~ lens ‖ achromatische Linse f. ‖ lentille f. achromatique. / ~ objective ‖ Achromat m. ‖ objectif m. achromatique. / ~ quartz and fluorite lens ‖ Quarzfluoritachromatlinse f. ‖ objectif m. quartz-fluorine.

achromatism ‖ Farblosigkeit f.; Achromatismus m. ‖ achromatisme m.

achromatization (Opt) ‖ Achromatisierung f. ‖ achromatisation f.

achromatize, to (Opt) ‖ achromatisieren ‖ achromatiser.

achromatopsy ‖ Farbenblindheit f. ‖ achromatopsie f.; daltonisme m.

achroodextrin ‖ Achroodextrin n. ‖ achroodextrine f.

acicular ‖ nadelförmig ‖ aciculaire.

acid ‖ sauer ‖ acide. / ~ to litmus ‖ sauer auf Lackmus m. reagierend ‖ acide au tournesol m. / ~ openhearth furnace ‖ saurer Martinofen m. ‖ four m. Martin acide. / with an ~ reaction ‖ sauer reagierend ‖ à réaction f. acide.

acid ‖ Säure f. ‖ acide m. / acetic ~ ‖ Essigsäure f. ‖ acide m. acétique. / antimonious ~ ‖ antimonige Säure f. ‖ acide m. antimonieux. / arsenic ~ ‖ Arseniksäure f. ‖ acide m. arsénique. / arsenious ~ ‖ arsenige Säure f. ‖ acide m. arsénieux. / battery ~ ‖ Sammlersäure f. ‖ acide m. pour accumulateurs. / benzoic ~ ‖ Benzoesäure f. ‖ acide m. benzoïque. / carbonic ~ ‖ Kohlensäure f. ‖ acide m. carbonique. / chlorhydric ~ ‖ Chlorwasserstoffsäure f.; Salzsäure f. ‖ acide m. chlorhydrique ou hydrochlorique. / concentrated ~ ‖ konzentrierte Säure f. ‖ acide m. concentré. / density of ~ (Acc) ‖ Säuredichte f. ‖ densité f. de l'acide. / diluted ~ ‖ verdünnte Säure f. ‖ acide m. dilué. / excess of ~ ‖ Säureüberschuß m. ‖ excédent m. d'acide. / fatty ~ ‖ Fettsäure f. ‖ acide m. gras. / formic ~ ‖ Ameisensäure f. ‖ acide m. formique. / to free from ~ ‖ entsäuern ‖ neutraliser. / fuming sulphuric ~ ‖ rauchende

Schwefelsäure f. ‖ acide m. sulfurique fumant. / glycerinated sulphuric ~ ‖ Glyzerinschwefelsäure f. ‖ acide m. sulfurique glycériné. / highly concentrated ~ ‖ stark konzentrierte Säure f. ‖ acide m. fortement concentré. / hydriodic ~ ‖ Jodwasserstoffsäure f. ‖ acide m. hydriodique ou iodhydrique. / hydrochloric ~ see chlorhydric. / hydrocyanic ~ ‖ Blausäure f. ‖ acide m. cyanhydrique. / hypochlorous ~ ‖ unterchlorige Säure f. ‖ acide m. hypochloreux. / inorganic ~ ‖ anorganische Säure f. ‖ acide m. inorganique. / lactic ~ ‖ Milchsäure f. ‖ acide m. lactique. / liquid carbonic ~ ‖ flüssige Kohlensäure f. ‖ acide m. carbonique liquide. / manganic ~ ‖ Mangansäure f. ‖ acide m. manganique. / mellitic ~ ‖ Honigsäure f. ‖ acide m. mellitique. / mineral ~ ‖ mineralische Säure f. ‖ acide m. minéral. / nitro-muriatic ~ ‖ Königswasser n.; Salpetersalzsäure f. ‖ acide m. nitro-muriatique; eau f. régale. / oleic ~ ‖ Ölsäure f. ‖ acide m. oléique. / organic ~ ‖ organische Säure f. ‖ acide m. organique. / oxalic ~ ‖ Oxalsäure f. ‖ acide m. oxalique. / palmitic ~ ‖ Palmitinsäure f. ‖ acide m. palmitique. / persulphuric ~ ‖ Überschwefelsäure f. ‖ acide m. persulfurique. / picric ~ ‖ Pikrinsäure f. ‖ acide m. picrique. / pyrogallic ~ ‖ Pyrogallussäure f. ‖ acide m. pyrogallique. / pyrolignic ~ ‖ Holzessig m. ‖ acide m. pyroxylique; vinaigre m. de bois. / pyromucic ~ ‖ brenzlige Schleimsäure f. ‖ acide m. pyromucique. / residuary ~ ‖ Abfallsäure f. ‖ acide m. résiduaire. / salicylic ~ ‖ Salizylsäure f. ‖ acide m. salicylique. / stearic ~ ‖ Stearinsäure f. ‖ acide m. stéarique. / the steel will not corrode under the action of ~s ‖ der Stahl m. ist widerstandsfähig gegen (den Angriff von) Säure ‖ l'acier m. résiste à l'influence d'acides. / strongly concentrated ~ see highly concentrated ~. / sulphuric ~ ‖ Schwefelsäure f. ‖ acide m. sulfurique. / sulphurous ~ ‖ schwefelige Säure f. ‖ acide m. sulfureux. / tannic ~ ‖ Gerbsäure f.; Gerbstoff m.; Tannin n. ‖ tannin m.; acide m. tannique. / tartaric ~ ‖ Weinsäure f. ‖ acide m. tartrique. / uric ~ ‖ Harnsäure f. ‖ acide m. urique. / weak ~ ‖ schwache Säure f. ‖ acide m. faible.

acid anhydride ‖ Säureanhydrid n. ‖ anhydride m. d'acide. / ~ bath ‖ Säurebad n. ‖ bain m. d'acide. / ~ bottle ‖ Säureflasche f. ‖ bouteille f. pour acides. / ~ car ‖ Säurewagen m. ‖ wagon m. pour acides. / ~ chloride ‖ Säurechlorid n. ‖ chlorure m. d'acide. / ~ density ‖ Säuredichte f. ‖ densité f. de l'acide. / ~ determination ‖ Säurebestimmung f. ‖ dosage m. de l'acidité. / ~ fume ‖ Säuredampf m. ‖ vapeur f. acide; buée f. corrosive. / ~ hose ‖ Säureschlauch m. ‖ tuyau m. flexible pour acides. / ~ hydroextractor for cloth carbonizing ‖ Säurezentrifuge f. für Tuchkarbonisation ‖ essoreuse f. à succion d'acide pour la carbonisation des draps.

acidiferous ‖ säurehaltig ‖ acidifère.

acidifiable base ‖ säurefähige Base f. ‖ base f. acidifiable.

acidification ‖ Säuerung f.; Säuern n.; Ansäuerung f. ‖ acidification f.

acidify, to ‖ säuern; ansäuern; sauer machen ‖ aciduler; acidifier.

acidifying plant, rope (Weav) ‖ Strangsäureeinrichtung f. ‖ installation f. à aciduler en boyaux.

acidimeter ‖ Säuremesser m. ‖ acidimètre m.; pèse-acide m.

acidity ‖ saure Beschaffenheit f.; Säure f. ‖ constitution f. acide; acidité f.

acid liquor man (Bleach) ‖ Beizer m. ‖ décapeur m. / ~ mixture (Acc) ‖ Säuregemisch n. ‖ mélange m. d'acides.

acidness ‖ Säure f. ‖ acidité f.

acid number ‖ Säurezahl f. ‖ indice m. d'acide.

acidol chrome colour ‖ Azidolchromfarbstoff m. ‖ couleur f. chrome d'acidole. / ~ dye ‖ Azidolfarbe f. ‖ couleur f. à l'acidole.

acid potassium oxalate ‖ Kleesalz n. ‖ sel m. d'oseille; bioxalate m. de potasse. / ~ process (Met) ‖ saures Verfahren n. ‖ procédé m. acide.

acid-proof ‖ säurefest; säurebeständig ‖ anti-acide; résistant *ou* inattaquable aux acides. / ~ brick ‖ säurefester Stein m. ‖ brique f. résistant aux acides. / ~ castings pl. ‖ säurefester Guß m. ‖ moulage m. à l'épreuve des acides. / special fire-proof and ~ castings pl. ‖ feuer- und säurebeständiger Sonderguß m. ‖ fonte f. moulée spéciale résistante à l'influence du feu et aux effets des acides. / ~ clothes pl. for workmen ‖ Säurekleidung f. für Arbeiter ‖ habillement m. d'ouvrier inattaquable aux acides. / ~ clothing ‖ Säureanzug m.; säurefester Anzug m. ‖ habit m. *ou* costume m. inattaquable aux acides. / ~ ferro-silicon castings pl. ‖ säurebeständiger Siliziumeisenguß m. ‖ moulages mpl. en ferro-silicium résistants aux (effets des) acides. / ~ fittings pl. ‖ säurebeständige Armatur f. ‖ garniture f. à l'épreuve d'acides. / ~ floor ‖ säurefester Fußboden m. ‖ plancher m. résistant aux acides. / ~ jar ‖ säurefestes Gefäß n. ‖ vase m. résistant aux acides. / ~ lining ‖ säurefeste Auskleidung f. ‖ revêtement m. résistant aux acides. / ~ paint ‖ säurebeständiger Anstrich m. ‖ peinture f. inattaquable aux acides *ou* anti-acide. / ~ pan *see* ~ jar. / ~ steel ‖ säurebeständiger *oder* säurefester Stahl m. ‖ acier m. résistant aux (effets des) acides. / ~ tube ‖ säurebeständiges Rohr n. ‖ tuyau m. résistant aux acides. / ~ wire ‖ säurebeständiger Draht m. ‖ fil m. inattaquable aux acides.

acid protection room ‖ Säureschutzraum m. ‖ espace m. protecteur de l'acide. / ~ pump ‖ Säurepumpe f. ‖ pompe f. à acide.

acid-resisting *see* acid-proof.

acid separator ‖ Säureabscheider m. ‖ séparateur m. d'acide. / ~ siphon ‖ Säureheber m. ‖ siphon m. à l'acide. / ~ slag ‖ saure Schlacke f. ‖ laitier m. acide. / ~ splitting ‖ Säurespaltung f. ‖ coupure f. acide. / ~ steeping bowl ‖ Säureeinweichbottich m. ‖ cuve f. d'acide. / ~ tar ‖ säurereicher Teer m. ‖ goudron m. acide. / ~ test ‖ Säureprüfung f.; Säureprobe f. ‖ essai m. de l'acide. / ~ tower ‖ Säureturm m. ‖ colonne f. à acides.

acidulate, to ‖ säuern ‖ acidifier; aciduler.

acidulous spring ‖ Sauerbrunnen m.; Säuerling m. ‖ eaux fpl. acidules *ou* gazeuses.

acid value ‖ Säurezahl f. ‖ indice m. d'acide. / ~ valve ‖ Säureventil n. ‖ soupape f. pour acides. / ~ vat ‖ Säurebottich m. ‖ cuve f. à acide.

acierate, to ‖ verstählen ‖ aciérer.

acknowledge, to ~ the receipt ‖ den Empfang m. anzeigen ‖ accuser (la) réception.

acknowledgment of receipt ‖ Empfangsbestätigung f. ‖ accusé m. de réception.

aclinic ‖ aklinisch ‖ aclinique.

aconite root ‖ Eisenhutwurzel f. ‖ tubercules fpl. d'aconit.

aconitine ‖ Akonitin n. ‖ aconitine f.

acorn ‖ Eichel f. ‖ gland m. / ~ coffee ‖ Eichelkaffee m. ‖ café m. de glands. / ~ oil ‖ Eichelkernöl n. ‖ huile f. de glands.

acoustic ‖ akustisch ‖ acoustique. / ~ chamber ‖ Schallkammer f. ‖ chambre f. de résonance. / ~ frequency ‖ Hörfrequenz f. ‖ fréquence f. acoustique. / ~ funnel ‖ Schallbecher m. ‖ pavillon m. / ~ instrument ‖ akustisches Instrument n. ‖ instrument m. acoustique. / ~ method of measuring altitude ‖ Schallhöhenmessung f. ‖ mesure f. acoustique d'altitude; sondage m. aérien.

acoustics pl. ‖ Lehre f. vom Schall; Akustik f. ‖ acoustique f.

acoustic signalling device ‖ Schallsignalapparat m. ‖ signaux-avertisseur m. à voix. / ~ telegraphy ‖ Gehörtelegrafie f. ‖ télégraphie f. acoustique. / ~ vault (build) ‖ Schallgewölbe n. ‖ voûte f. acoustique.

acquire, to ~ an undertaking ‖ eine Unternehmung f. erwerben ‖ acquérir une entreprise.

acquisition ‖ Erwerbung f.; Erwerb m. ‖ acquisition f.

acquittal ‖ Freispruch m. ‖ acquittement m.

acquittance ‖ Quittung f. ‖ acquit m.; quittance f.; reçu m.

acquitted ‖ freigesprochen ‖ acquitté.

acre ‖ Acker m.; Morgen m. Land ‖ acre f.

acreage ‖ Flächenraum m. in Morgen ‖ extension f. en acres.

acridine colour ‖ Akridinfarbstoff m. ‖ couleur f. d'acridine.

across (Mar) ‖ dwars ‖ par le travers. / ~ the direction of rolling ‖ quer zur Walzrichtung ‖ transversalement au laminage.

act, to ~ upon (Chem) ‖ reagieren auf ‖ réagir sur.

act ‖ Handlung f.; Tat f.; Werk n. ‖ action f.; acte m. / ~ (Document) ‖ Akte f.; Aktenstück n.; Urkunde f. ‖ document m.; pièce f. / ~ of God ‖ höhere Gewalt f. ‖ force f. majeure.

acted, to be ~ upon by (Chem) ‖ angegriffen sein von ‖ être attaqué par.

acting ‖ handelnd ‖ agissant. / double ~ ‖ doppeltwirkend ‖ à double action f. *ou* effet m. / oppositely ~ ‖ entgegengesetzt wirkend ‖ agissant en sens m. opposé *ou* contraire. / single ~ ‖ einfach wirkend ‖ à simple action f. *ou* effet m.

actinic ‖ aktinisch ‖ actinique. / ~ ray ‖ aktinischer Strahl m. ‖ rayon m. actinique.

actinism ‖ Lichtstrahlenwirkung f. ‖ actinisme m.

actinium ‖ Aktinium n. ‖ actinium m.

actinometer ‖ Aktinometer n. ‖ actinomètre m.

action ‖ Wirkung f.; Einwirkung f. ‖ action f. / ~ (Law) ‖ Rechtsstreit m. ‖ affaire f. judiciaire. / ~ (Mach) ‖ Gang m. ‖ fonction f.; marche f. / back ~ ‖ Rücklauf m. ‖ marche f. arrière. / to bring an ~ ‖ einen Prozeß m. anstrengen ‖

intenter un procès. / chemical ~ ‖ chemische Umsetzung f. ‖ transformation f. chimique. / to come into ~ ‖ in Betrieb m. kommen ‖ se mettre en marche f. / ~ contrary to morality ‖ gegen die guten Sitten verstoßende Handlung f. ‖ action f. contraire aux bonnes mœurs. / cooling ~ ‖ abkühlende Wirkung f. ‖ effet m. de refroidissement. / ~ for damages ‖ Schadenersatzklage f. ‖ action f. en réparation du préjudice *ou* du dommage causé. / ~ at a distance ‖ Fernwirkung f. ‖ action f. à distance. / ~ of the earth on the magnetic needle ‖ kosmische Einwirkung f. auf die Magnetnadel ‖ influence f. cosmique sur l'aiguille aimantée. / for high-speed ~ (Mach) ‖ schnell laufend ‖ à grande vitesse f. / for medium-speed ~ (Mach) ‖ mäßig schnell laufend ‖ à vitesse f. modérée. / ~ of points (Electr) ‖ Spitzenwirkung f.; Spitzenentladung f. ‖ pouvoir m. des pointes. / positive ~ ‖ tatsächliche Wirkung f. ‖ action f. matérielle *ou* effective. / to put out of ~ ‖ außer Betrieb m. setzen ‖ arrêter. / secondary ~ ‖ Nebenwirkung f. ‖ action f. secondaire. / for slow-speed ~ (Mach) ‖ langsam laufend ‖ à petite vitesse f. / to be without ~ (Chem) ‖ ohne Einwirkung f. sein ‖ être sans action f.

actionable ‖ klagbar; einklagbar ‖ exigible; actionnable.

action cycle ‖ Arbeitsperiode f. ‖ période f. de travail. / manner of ~ ‖ Wirkungsweise f. ‖ manière f. de l'action. / principle of ~ and reaction ‖ Prinzip n. der Wirkung und Gegenwirkung ‖ principe m. de l'action et de la réaction. / ~ radius ‖ Aktionsradius m. ‖ rayon m. d'action. / sphere of ~ ‖ Arbeitsfeld n. ‖ champ m. d'activité. / ~ turbine ‖ Gleichdruckturbine f.; Aktionsturbine f. ‖ turbine f. à action. / ~ wheel ‖ Aktionsrad n. ‖ couronne f. à action.

active ‖ rege; rührig; tätig; aktiv ‖ actif; vif; animé. / ~ force (Mech) ‖ Wucht f. ‖ force f. vive. / ~ mass (Acc) ‖ wirksame Masse f. ‖ matière f. active. / ~ paste (Acc) ‖ aktive Masse f. ‖ masse f. active. / ~ surface of the carbon ‖ aktive Oberfläche f. der Kohle ‖ surface f. active du charbon.

activity ‖ Tätigkeit f.; Wirksamkeit f. ‖ activité f. / to extend the field of activities of a company ‖ das Tätigkeitsfeld n. einer Gesellschaft erweitern ‖ agrandir la sphère d'activité *ou* le champ de travail d'une société. / ~ of vaporisation ‖ Verdampfungsschnelligkeit f. ‖ activité f. de la vaporisation.

actual ‖ wirklich ‖ actuel; réel. / ~ size ‖ natürliche Größe f. ‖ grandeur f. naturelle.

actuate, to ‖ in Gang m. bringen *oder* setzen ‖ mettre en marche f.

actuating rod ‖ Regelstange f.; Antriebstange f. ‖ tirant m. de commande.

actuation ‖ Betätigung f. ‖ commande f.

acute (Geom) ‖ spitz ‖ aigu.

acute-angled ‖ spitzwinklig ‖ à angles aigus; oxygone. / ~ triangle ‖ spitzwinkliges Dreieck n. ‖ triangle m. acutangle.

acyclic ‖ azyklisch ‖ acyclique.

adamant ‖ Diamant m. ‖ diamant m.

adapt, to ‖ anpassen ‖ adapter.

adaptability ‖ Anpassungsfähigkeit f. ‖ faculté f. d'adaption. / ~ to the smelting

process ‖ Schmelzwürdigkeit f. ‖ exploitabilité f. par la fusion.

adaptation ‖ Anpassung f. ‖ adaptation f. / dark ~ ‖ Dunkelanpassung f. ‖ adaptation f. à l'obscurité. / ~ of impedance ‖ Scheinwiderstandsanpassung f. ‖ adaptation f. d'impédance.

adapted, dark ~ eye ‖ dunkel adaptiertes Auge n. ‖ œil m. adapté à l'obscurité.

adapter ‖ Verlängerungsstück n.; Einsatzstück n.; Paßstück n. ‖ allonge f.; pièce f. de rapport. / ~ to stand ‖ Stativaufsatz m. ‖ montant m. d'un pied. / ~ to tripod see ~ to stand. / ~ device ‖ Adaptierungseinrichtung f. ‖ dispositif m. d'adaption.

adaption see adaptation.

add, to ‖ anfügen; anschließen ‖ annexer; joindre; ajouter. / ~ (Bookkeep) ‖ summieren ‖ additionner. / ~ (Chem) ‖ hinzugeben; zusetzen ‖ ajouter. / ~ to (Chem) ‖ sich anlagern; anreihen ‖ se fixer sur. / ~ ores ‖ Erz n. nachsetzen ‖ ajouter du mineral. / ~ up ‖ addieren; zusammenzählen ‖ additionner.

added, not ~ up ‖ nicht addiert ‖ non additionné.

adding device ‖ Addierwerk n. ‖ totalisateur m. / each of these ~s is combined with a drawer ‖ jedes dieser Addierwerke npl. ist mit einer Schublade verbunden ‖ chacun de ces totalisateurs mpl. est combiné avec un tiroir. / reading the ~s by simple pressure of key ‖ Ablesen n. der Addierwerke durch einfachen Druck auf die Taste ‖ lecture f. des totalisateurs par simple pression sur la touche. / ~ visible or non-visible as desired ‖ Addierwerk n. nach Wahl ablesbar oder verdeckt ‖ totalisateur m. à choix visible ou recouvert. / ~ under cover ‖ Addierwerk n. unter Verschluß ‖ totalisateur m. sous clé.

adding machine ‖ Addiermaschine f. ‖ machine f. à additionner; addiateur m.; additionneuse f. / duplex ~ ‖ Addiermaschine f. mit zwei Zählwerken ‖ machine f. à additionner à deux compteurs. / single counter ~ ‖ Addiermaschine f. mit einem Zählwerk ‖ machine f. à additionner à un seul compteur.

adding and writing machine, combined ‖ vereinigte Schreib- und Rechenmaschine f. ‖ machine f. à écrire et machine f. à calculer combinées.

adding mechanism ‖ Additionsmechanismus m. ‖ mécanisme m. additionneur; dispositif m. additif. / cash register with ~ ‖ Registrierkasse f. mit Addierwerk ‖ caisse f. enregistreuse avec totalisateur.

addition ‖ Zusatz m. ‖ addition f.; adjonction f. / ~ (Bank) ‖ Zuschuß m. ‖ supplément m.; secours m. en argent; allocation f. / ~ (Build) ‖ Anbau m. ‖ annexe f. / after ~ of ‖ nach Zusatz m. von ‖ après addition f. de. / ~ of colour ‖ Farbzusatz m. ‖ addition f. de matière colorante. / with the ~ ‖ zuzüglich ‖ en plus.

additional air ‖ Zusatzluft f. ‖ air m. complémentaire ou supplémentaire. / ~ article ‖ Zusatzartikel m. ‖ article m. supplémentaire ou additionnel. / ~ breaker ‖ Nachbrecher m. ‖ reconcasseur m. / ~ building ‖ Nebengebäude n. ‖ annexe f.; bâtiment m. secondaire. / ~ burden ‖ Überdruck m. ‖ surcharge m. / ~ charge (Acc) ‖ Nachladung f. ‖ rechargement m.; chargement m. supplémentaire. /

~ expenditure ‖ Mehraufwand m.; Mehrausgabe f. ‖ surcroît m. de dépense. / ~ fee ‖ Zuschlaggebühr f. ‖ surtaxe f. / ~ income ‖ Nebeneinkünfte fpl.; Nebeneinnahme f. ‖ revenus mpl. accessoires; casuel m. / ~ instrument (Tel) ‖ Zusatzeinrichtung f. ‖ organe m. accessoire. / ~ multiple (Tel) ‖ Ansatzfeld n. ‖ section f. additionnelle. / ~ order ‖ Nachbestellung f. ‖ commande f. ou ordre m. supplémentaire. / ~ pipe ‖ Ansatzrohr n. ‖ tuyau m. additionnel. / ~ plant ‖ Nebenanlage f. ‖ hall m. accessoire; annexe f. / ~ postage ‖ Strafporto n. ‖ surtaxe f. postale. / ~ rack (Tel) ‖ Leitungszusatzgestell n. ‖ bâti m. secondaire. / ~ set (Electr) ‖ Zusatzaggregat n. ‖ dynamo f. surélévatrice. / ~ stress ‖ Zusatzspannung f. ‖ tension f. additionnel. / ~ voltage ‖ Zusatzspannung f. ‖ tension f. additionnelle. / ~ water ‖ Zusatzwasser n. ‖ eau f. additionnelle. / ~ supplying of ~ water ‖ Zusatzwasserlieferung f. ‖ fourniture f. de l'eau supplémentaire.

addition wheel ‖ Summenrad n. ‖ roue f. de sommation.

additively ‖ addierend ‖ additivement. / to combine ~ ‖ addieren ‖ s'ajouter.

address, to ‖ anreden ‖ adresser la parole.

address ‖ Anrede f. ‖ titre m.; vedette f. / ~ (Letter) ‖ Anschrift f. ‖ adresse f. / ~ in case of need ‖ Notanschrift f. ‖ adresse f. provisoire ou au besoin. / ~ full ~ of the consignee ‖ genaue Versandanschrift f. ‖ adresse f. exacte pour l'expédition. / telegraphic ~ ‖ Drahtanschrift f. ‖ adresse f. télégraphique.

address card ‖ Anschriftkarte f. ‖ carte f. d'adresse.

addressing chest ‖ Anschriftenlade f. ‖ tiroir m. à adresses.

addressing machine ‖ Anschriftenmaschine f.; Adressiermaschine f. ‖ machine f. à adresses ou à adresser. / type for ~ ‖ Anschriftenmaschinentype f. ‖ type m. pour machine à imprimer des adresses. / ~ stencil ‖ Anschriftenmaschinenschablone f. ‖ stencil m. de machine à adresser.

addressing plate ‖ Anschriftenplatte f. ‖ châssis m. ou planche f. à adresses.

address office ‖ Anschriftenbüro n. ‖ bureau m. d'adresses. / ~ panel ‖ Feld n. für die Anschrift ‖ champ m. d'adresse. / ~ plate see addressing plate.

adept ‖ sachkundig; eingeweiht; erfahren ‖ expert; compétent.

adhere, to ‖ innehalten ‖ observer. / ~ to the tongue (Mine) ‖ an der Zunge f. hängen ‖ adhérer à la langue.

adherence ‖ Adhäsion f.; Haftenbleiben n. ‖ adhérence f.; adhésion f.

adherent ‖ haftend ‖ adhérent. / firmly ~ ‖ fest anhängend ‖ adhérent solidement.

adhering dirt ‖ anhaftender Schmutz m. ‖ crasse f. adhérente.

adhesion ‖ Haftfestigkeit f.; Adhäsion f. ‖ adhérence f.; adhésion f. / area of ~ ‖ Haftfläche f. ‖ surface f. d'adhérence. / ~ between brick and mortar ‖ Haften n. zwischen Stein und Mörtel ‖ adhérence f. de la brique au mortier. / limit of ~ ‖ Adhäsionsgrenze f. ‖ limite f. d'adhérence.

adhesion culture ‖ Adhäsionskultur f. ‖ culture f. adhérente. / ~ force ‖ Haftvermögen n.; Adhäsionskraft f. ‖ force f. d'adhérence. / ~ railway ‖ Reibungsbahn f.; Adhäsionseisenbahn f. ‖ chemin m. de fer à adhérence. / ~ weight of a

locomotive ‖ Adhäsionsgewicht n. einer Lokomotive / poids m. adhérent d'une locomotive.

adhesive, very (Chem) ‖ fest haftend ‖ très adhérent.

adhesive ‖ Bindemittel n.; Klebstoff m. ‖ adhésif m.; matière f. collante. / ~ grease ‖ Adhäsionsfett n. ‖ graisse f. adhésive ou adhérente. / ~ patch see ~ plaster. / ~ plaster ‖ Heftpflaster n.; Kleb(e)pflaster n.; Klebetaft m. / taffetas m. collant ou anglais ou d'Angleterre; emplâtre m. (collant); sparadrap m. / ~ power ‖ Adhäsionskraft f. ‖ pouvoir m. adhérent; adhérence f. / layer of ~ rubber ‖ Klebgummischicht f. ‖ couche f. de caoutchouc collante. / ~ substance ‖ Klebstoff m. ‖ matière f. collante; liant m.; colle f. / ~ weight ‖ Adhäsionsgewicht n. ‖ poids m. adhérent.

adiabatic ‖ adiabatisch ‖ adiabatique. / ~ compression ‖ adiabatische Verdichtung f. ‖ compression f. adiabatique. / ~ curve ‖ Adiabate f. ‖ adiabate f.; courbe f. adiabatique. / ~ expansion ‖ adiabatische Expansion f. ‖ détente f. adiabatique.

adipocere ‖ Fettwachs n.; Leichenfett n. ‖ adipocire m.; gras m. des cadavres.

adipose ‖ fettig; ölig ‖ onctueux; graisseux; adipeux.

adipous see adipose.

adit (Mine) ‖ Erbstollen m.; Förderstollen m. ‖ galerie f. à ciel ouvert ou de roulage. / to clear an ~ (Mine) ‖ einen Stollen m. aufräumen ‖ saigner une areine. / deep ~ (Mine) ‖ Wasserlösungsstollen m.; Grundstollen m. ‖ galerie f. d'écoulement. / end of an ~ (Mine) ‖ Stollenort m. ‖ extrémité f. d'une galerie. / to run an ~ (Mine) ‖ einen Stollen m. treiben ‖ percer une galerie.

adit drainage (Mine) ‖ Wasserstollen m. ‖ galerie f. d'écoulement. / ~ end (Mine) ‖ Abbaustoß m. ‖ front m. ou fond m. de taille. / ~ level (Mine) ‖ Stollenzugang m. ‖ galerie f. à flanc de coteau; areine f.; arcine f.

adjacent ‖ anliegend ‖ adjacent; collant. / to be ~ to ... ‖ grenzen an ... ‖ confiner à ...; approcher de ...; toucher à ... / ~ angle ‖ Nebenwinkel m. ‖ angle m. adjacent ou contigu.

adjoining ‖ anliegend; anstoßend ‖ adjacent; collant; contigu. / ~ angle ‖ Nebenwinkel m. ‖ angle m. adjacent ou contigu. / ~ concession (Mine) ‖ Nachbarfeld m. ‖ champ m. voisin. / ~ post ‖ Hilfspfosten m. ‖ poteau m. de soutien ou de réserve. / ~ rail ‖ Anschlagschiene f. ‖ rail m. fixe.

adjournment ‖ Vertagung f. ‖ ajournement m.

adjudicate, to ‖ verdingen ‖ mettre en adjudication f.

adjunct ‖ Zusatz m.; Zubehör n. ‖ adjonction f.; adjoint m.; addition f.; supplément m.

adjust, to ‖ einpassen; anpassen; einstellen. ‖ ajuster; adapter. / ~ (Mach) ‖ aufstellen; montieren; adjustieren ‖ dresser; ajuster. / ~ (Opt) ‖ einstellen ‖ mettre au point; ajuster. / ~ (Print) ‖ gerade richten; unterlegen ‖ façonner. / ~ the columns (Print) ‖ die Spalten fpl. brechen ‖ ajuster les colonnes fpl. / ~ to the frequency ‖ auf die Frequenz f. einstellen ‖ ajuster à la fréquence. / ~ the

ignition || die Zündung einstellen || régler l'allumage m. / ~ an instrument || ein Instrument f. justieren *oder* rektifizieren || régler un instrument. / ~ levels pl. || auf gleiches Niveau n. einstellen || égaliser les niveaux mpl. / ~ a piece to || ein Stück n. anpassen || ajuster une pièce sur. / ~ the rail || die Schiene richten || redresser le rail.

adjustable (Mach) || nachstellbar; verstellbar; stellbar || ajustable; réglable. / ~ cam || verstellbarer Nocken m. || came f. réglable. / ~ coil instrument (Electr) || Drehspulinstrument n. || instrument m. de mesure à cadre tournant. / ~ condenser (Radio) || Drehkondensator m. || condensateur m. variable. / ~ cutter || einsetzbare Schneidzunge f. || languette f. rapportable. / ~ in all directions || nach allen Richtungen einstellbar || orientable en tous sens. / ~ firing || regelbare Feuerung f. || foyer m. réglable. / ~ grate || verstellbarer Rost m. || grille f. déplaçable. / ~ head rest || verstellbarer Kopfbügel m. || arc m. de tête réglable. / ~ pin || Stellzapfen m. || tourillon m. réglable. / ~ pole support || Deichselausgleichvorrichtung f. || équilibreur m. de timon. / ~ ring || verstellbarer Ring m. || anneau m. ajustable. / ~ starter (Electr) || Regelanlasser m. || appareil m. de mise en marche réglable.

adjusted || geordnet || ajusté.

adjuster || Ordner m. || ajusteur m.

adjusting (Mach; Mech) || Adjustierung f. || ajustage m. / ~ the dies pl. || Gesenkjustierung f. || réglage m. des matrices. / electrical ~ || elektrisches Einstellen n. || ajustage m. *ou* réglage m. électrique. / ~ of the forms (Print) || Rapport m.; Zusammenstimmung f. der Formen || rapport m. / ~ of the springs || Richten n. der Federn || dressage m. des ressorts.

adjusting apparatus || Stellvorrichtung f. || appareil m. de mise au point. / ~ balance || Justierwage f. || ajustoir m. / ~ block || Justierklotz m.; Schrötlingseisen n. || bilboquet m. / ~ device || Nachstellvorrichtung f. || dispositif m. de réglage *ou* de rattrapage de jeu. / ~ file || Genauigkeitsfeile f. || lime f. à ajuster *ou* pour le travail de précision. / ~ gauge || Einstellehre f. || calibre m. de réglage. / ~ gauge of a screw clamp shape || schraubzwingenähnliche Einstellehre f. || calibre m. à mâchoires filetées. / ~ ledge || Verstelleiste f. || listeau m. déplaçable. / ~ lever || Verstellhebel m. || levier m. orientable *ou* déplaçable. / ~ machine || Justiermaschine f. || machine f. d'ajustage. / ~ machinery || Justierwerk n. || atelier m. d'ajustage. / ~ nut || Stellmutter f.; Nachstellmutter f. || écrou m. tendeur *ou* à trous *ou* de fixage. / ~ pivot || Einstellzapfen m. || pivot m. de réglage. / ~ ring || Stellring m. || anneau m. de serrage; bague f. de butée *ou* d'arrêt. / ~ ring in one piece || ungeteilter Stellring m. || bague f. d'arrêt en une pièce. / ~ ring bolt || Stellringbolzen m. || boulon m. de bagues d'arrêt. / ~ rod || Verstellstange f. || tige f. de réglage. / ~ roller || Einstellwalze f. || tambour m. de réglage. / ~ screw || Stellschraube f. || vis f. de réglage *ou* de rappel. / finely ~ screw || Feinstellschraube f. || vis f. pour mise au point précise. / ~ screw of axle box wedges || Achslagerstellkeilschraube f. ||

vis f. de serrage des boîtes d'essieu. / ~ shop || Zurichterei f. || atelier m. d'ajustage *ou* pour le dressage. / ~ slide (Tel) || Schwächungsanker m. || fer m. doux mobile. / ~ strip || Beigabe f. || cale f. à rattrapage du jeu; lardon m. d'ajustage. / ~ tools pl. || Einstellen der Werkzeuge || mise f. au point *ou* ajustage m. des outils. / ~ wedge for axle boxes || Achslagerstellkeil m. || coin m. de serrage *ou* de réglage des boîtes d'essieu. / ~ wedge of crosshead || Kreuzkopfkeil m. || clavette f. de réglage de crosse; coin m. d'ajustage de crosse *ou* de serrage de crosse.

adjustment (Mach) || Einstellung f. || ajustage m. / ~ (Opt) || Einstellen n. || mise f. au point. / ~ of the brushes (Electr) || Bürsteneinstellung f. || réglage m. *ou* décalage des balais. / chute ~ || Rutscheneinstellung f. || réglage m. du plan incliné. / coarse ~ || Grobeinstellung f. || mise f. au point rapide. / ~ of cutters || Messerverstellung f. || réglage m. des lames. / ~ of the distance between cylinder and types || Einstellung f. des Abstandes zwischen Walze und Typen || réglage m. d'écartement entre le cylindre et les caractères. / ~ in elevation || Höhenverstellung f. || déplacement m. en hauteur. / fine ~ || Feineinstellung f. || mise f. au point lente *ou* micrométrique *ou* précise. / ~ to a fixed point of the scale || Einstellung f. auf einen bestimmten Strich der Skale || mise f. au point sur un trait déterminé du cadran. / hand-operated ~ || Einstellung f. von Hand || ajustage m. à main. / ~ for illumination || Beleuchtungseinstellung f. || réglage m. pour l'éclairage. / inclinable ~ || Schrägeinstellung f. || ajustage m. oblique. / initial ~ || Nulleinstellung f. || mise f. au point zéro. / mechanical ~ || maschinelle Verstellung f. || réglage m. mécanique. / ~ for faster or slower movement || Umstellung f. auf rascheren oder langsameren Gang || réglage m. pour obtenir une marche plus *ou* moins rapide. / no further ~ || keine weitere Nachstellung f. || sans plus de réglage m. / ~ for observation || Beobachtungseinstellung f. || réglage m. pour l'observation. / ~ of slide || Stößelverstellung f. || réglage m. du coulisseau. / ~ of stroke || Hubverstellung f. || réglage m. de la course. / ~ of top cutter || Obermesserverstellung f. || réglage m. de la lame supérieure. / precise ~ of the type || genaue Einstellung f. der Type || amenée f. du caractère au point précis. / ~ to zero of the adding machine || Nullstellung f. der Additionsmaschine || mise f. de la machine à additionner au point zéro.

adjustment device || Nachstellvorrichtung f. || dispositif m. de réglage *ou* de rattrapage de jeu. / ~ scale || Einstellskale f. || échelle f. de calage *ou* d'ajustage.

administer, to || verwalten || administrer.

administered || verwaltet || administré.

administrate, to || verwalten || administrer; gérer.

administration || Verwaltung f. || administration f. / ~ central of a concern || Hauptverwaltung f. eines Konzerns || administration f. centrale d'un groupe. / ~ of customs || Zollverwaltung f. || administration f. douanière. / ~ of a mine || Bergwerksverwaltung f. || administration f. d'une mine.

administrator || Verwalter m. || administrateur m. / ~ of the estate || Liquidator m. || liquidateur m.

admissible || zulässig || admissible. / ~ wear || zulässige Abnützung f. || usure f. admissible.

admission || Zutritt m. || accès m.; entrée f. / ~ (Steam) || Aufnahme f.; Einströmung f. || admission f.; entrée f.; introduction f. / axial ~ || axiale Beaufschlagung f. || admission f. *ou* introduction f. axiale. / ~ of manufacturing || Fabrikationsfreigabe f. || permis m. général de production; admission f. de la fabrication. / tangential ~ || tangentiale Zuführung f. || amenée f. tangentielle.

admission cam || Einlaßnocken m. || came f. d'admission. / ~ period || Einströmungsperiode f. || période f. d'admission. / ~ pipe || Dampfeinströmungsrohr n. || tuyau m. d'admission de vapeur. / ~ pressure || Anfangsdruck m. || pression f. d'admission.

admission valve || Einlaßventil n. || soupape f. d'admission. / ~ box || Einlaßventilgehäuse n. || chapelle f. de soupape d'admission. / ~ cone || Einlaßventilkegel m. || cône m. de soupape d'admission. / ~ lever || Einlaßventilhebel m. || levier m. de soupape d'admission. / ~ rod || Einlaßventilstange f. || tige f. de commande de la soupape d'admission. / ~ roller || Einlaßventilrolle f. || galet m. de la soupape d'admission.

admission velocity || Eintrittsgeschwindigkeit f. || vitesse f. d'admission.

admit, to ~ compressed air || Druckluft f. zuführen || introduire de l'air comprimé.

admittance || Zutritt m. || accès m.; entrée f. / ~ (Electr) || Leitwert m. || admittance f. / indicial ~ || Kennleitwert m. || admittance f. indicatrice.

admixed substance || zugemischter Stoff m. || substance f. ajoutée en mélangeant.

admixing material || Zusatzstoff m. || matière f. d'addition *ou* à ajouter.

admixtion || Beimischung f. || admixtion f.

admixture of clay || Tonbeimengung f. || addition f. d'argile. / disturbing ~ || störende Beimengung f. || constituant m. perturbateur.

adobe || Luftziegel m.; Lehmstein m.; ungebrannter Ziegel m. || brique f. crue *ou* séchée à l'air.

adopted || angenommen || adopté.

adoption of the valveless type of two-stroke engine || ventillose Zweitaktbauart f. || adoption f. d'un moteur à deux temps du système sans soupapes.

Adrianople red || Merinorot n.; Türkischrot n. || rouge m. turc; rouge m. d'Adrianople; rouge m. des Indes.

adrift || treibend im Wasser; triftig || en dérive f.

adularia || edler Feldspat m.; Adular m.; Mondstein m. || adulaire m.; spath m. adulaire.

adulterant || Fälschungsstoff m. || adultérant m.

adulterate, to || verfälschen || falsifier; altérer; adultérer.

adulterated spirit || denaturierter Alkohol m. || alcool m. dénaturé.

adulteration || Verfälschung f. || adultération f.; falsification f.

adunc || hakenförmig || à crochet m.

advance, to || aufschlagen || renchérir. / ~ money on security || lombardieren ||

prêter sur gages. / ~ one step ‖ um einen Schritt m. vorrücken ‖ avancer un pas.
advance (Mach) ‖ Voreilung f. ‖ avance f. / ~ (Tool) ‖ Vorschub m. ‖ avancement m. / ~ (Trade) ‖ Aufschlag m. ‖ hausse f.; augmentation f.; renchérissement m. / ~ of the ignition ‖ Vorzündung f. ‖ avance f. à l'allumage. / in ~ ‖ im voraus ‖ d'avance f. / ~ against merchandise ‖ Vorschuß m. auf Waren ‖ avance f. sur marchandises. / ~ in prices (Trade) ‖ Preissteigerung f.; Hausse f. ‖ renchérissement m.; hausse f.
advance mechanism ‖ Vorschubmechanismus m. ‖ dispositif m. d'avancement. / ~ money (Mar) ‖ Handgeld n. beim Heuern ‖ avance f. prime d'engagement. / payment in ~ ‖ Vorauszahlung f. ‖ payement m. d'avance ou par anticipation. / ~ rod ‖ Vorschubstange f. ‖ tige f. d'avancement.
advancing, to be (Price) ‖ anziehen ‖ hausser; augmenter.
advantage ‖ Gewinn m.; Vorteil m.; Nutzen m.; Vorsprung m. ‖ avantage m.
advantageous ‖ vorteilbringend; vorteilhaft ‖ avantageux; profitable; lucratif. / to be ~ ‖ Vorteil m. bringen ‖ tourner à l'avantage.
adventure ‖ gewagtes Unternehmen n. ‖ affaire f. risquée; aventure f. / ~ (Mine) ‖ Gewerkschaft f. ‖ société f. des exploitants. / ~ in a mine ‖ Kux m. ‖ action f. de mine.
adventurer (Mar) ‖ Schmugglerschiff n. ‖ interlope m.; aventurier m. / ~ (Mine) ‖ Kuxeninhaber m. ‖ actionnaire m. de mines.
adventurine feldspar ‖ Avanturinfeldspat m.; Sonnenstein m. ‖ feldspath m. aventurine.
adverse ‖ widrig ‖ adverse. / ~ balance ‖ Unterbilanz f. ‖ déficit m.
advertise, to ‖ anzeigen; bekanntmachen; Propaganda f. machen ‖ publier; annoncer; faire de la propagande.
advertisement ‖ Annonce f.; Anzeige f.; Inserat n. ‖ annonce f. ‖ Bekanntmachung f. ‖ avis m.; publication f. ‖ Reklame f. ‖ réclame f.; propagande f. / ~ for show windows ‖ Reklameschaufensterklopfer m. ‖ figurine f. automatique de réclame pour vitrines.
advertisement board ‖ Anzeigetafel f. ‖ tableau-annonce m. / ~ broker ‖ Anzeigenakquisitör m. ‖ courtier m. d'annonces. / canvasser for ~s ‖ Annoncensammler m. ‖ courtier m. d'annonces. / ~ illumination ‖ Reklamebeleuchtung f. ‖ éclairage m. de réclame. / ~ maker-up (Print) ‖ Anzeigenmetteur m. ‖ annoncier m. / ~ office ‖ Reklameanstalt f. ‖ bureau m. de réclame. / ~ pressure ‖ Anzeigenabzug m. ‖ épreuve f. d'annonce. / ~ setter ‖ Anzeigensetzer m. ‖ compositeur m. d'annonces.
advertiser ‖ Inserent m. ‖ personne f. qui fait une annonce; publicateur m.
advertising ‖ Reklame f. ‖ réclame f. / electric ~ ‖ Lichtreklame f. ‖ réclame f. lumineuse. / luminous ~ ‖ Lichtreklame f. ‖ réclame f. lumineuse.
advertising aim ‖ Werbezweck m. ‖ fin f. de propagande. / ~ art ‖ Werbekunst f.; Reklamekunst f. ‖ art m. de propagande. / ~ article ‖ Werbeartikel m.; Reklameartikel m. ‖ objet m. de propagande; article m. de réclame.

advertising board ‖ Reklameschild n. ‖ enseigne m. / luminous ~ ‖ Leuchtschild n. ‖ enseigne m. lumineuse.
advertising calendar ‖ Reklamekalender m. ‖ calendrier-réclame m. / ~ clock ‖ Reklameuhr f. ‖ horloge-réclame f. / cost of ~ ‖ Insertionskosten pl. ‖ frais mpl. d'insertion. / ~ expenses pl. ‖ Werbekosten pl. ‖ frais mpl. d'acquisition. / ~ and gratuitous articles pl. ‖ Reklame- und Zugabeartikel mpl. ‖ articles mpl. de réclame et à donner par-dessus le marché. / ~ journal ‖ Anzeigenblatt n. ‖ feuille f. d'avis ou d'annonces. / ~ lettering ‖ Reklameschrift f.; Reklamebeschriftung f.; Reklamebuchstaben mpl.; Werbeschrift f. ‖ caractères mpl. pour réclame. / ~ magazine ‖ Reklamezeitschrift f.; Werbezeitschrift f.; Werbeschrift f. ‖ journal m. ou revue f. de réclame. / ~ office ‖ Annoncenbüro n. ‖ office f. de publicité. / ~ pencil ‖ Reklamebleistift m. ‖ crayon-réclame m. / ~ pillar ‖ Plakatsäule f.; Litfaßsäule f.; Anschlagsäule f.; Reklamesäule f. ‖ colonne-affiche f. / ~ pocket knife ‖ Reklametaschenmesser n. ‖ canif m. de réclame. / ~ poster ‖ Reklameplakat n. ‖ affiche f. réclame. / ~ printed matter ‖ Reklamedrucksache f. ‖ imprimé m. commerciel ou de réclame. / ~ ribbon ‖ Reklameband n. ‖ ruban m. de réclame. / ~ ruler ‖ Reklamelineal n. ‖ règle f. de réclame. / ~ scale ‖ Reklamemaßstab m. ‖ règle f. divisée de réclame. / ~ sign-(board) ‖ Reklameschild n. ‖ enseigne f. pour la réclame. / ~ success ‖ Werbeerfolg m.; Reklameerfolg m. ‖ succès m. de propagande. / ~ tariff ‖ Anzeigentarif m. ‖ tarif m. d'annonces. / ~ van ‖ Reklamewagen m. ‖ voiture f. réclame.
advertize ... see advertise ...
advice ‖ Ratschlag m. ‖ conseil m. / ~ (Exchange) ‖ Meldung f. ‖ annonce f.; rapport m. / ~ (Trade) ‖ Avis m. ‖ avis m. / letter of ~ ‖ Avisbrief m. ‖ lettre f. d'avis.
advise, to ‖ anraten; beraten ‖ conseiller. / ~ (Trade) ‖ avisieren; ankündigen ‖ aviser; donner avis m.; notifier.
adviser ‖ Ratgeber m. ‖ conseilleur m. / legal ~ ‖ Sachwalter m.; Rechtsbeistand m. ‖ homme m. d'affaires; avoué m.
advisory board ‖ Verwaltungsrat m. ‖ conseil m. d'administration; administrateur m. / ~ member ‖ Beisitzer m. ‖ assesseur m.
adze, to ~ the sleepers pl. (Railw) ‖ die Schwellen fpl. einblatten ‖ entailler les traverses fpl. / ~ the timber (Carp) ‖ das Holz dechseln ‖ dresser le bois à l'herminette.
adze (Carp) ‖ Dachsbeil n.; Querbeil n.; Dechsel f. ‖ assette f.; asseau m.; hachette f.; herminette f. / ice ~ ‖ Eispickel m. ‖ herminette f. à glace.
aequator ‖ Äquator m. ‖ équateur m.
aerate, to (Chem) ‖ lüften; mit Kohlensäure f. sättigen ‖ aérer; carbonater; saturer d'acide carbonique. / ~ in the vat ‖ im Gärbottich m. aufziehen ‖ aérer en cuve f.
aerated (Chem) ‖ lufthaltig ‖ aéré. / ~ water ‖ kohlensäurehaltiges Wasser n. ‖ eau f. gazeuse. / ~ water machinery ‖ Mineralwassermaschine f. ‖ machine f. à eau gazeuse.
aeration ‖ Sättigung f. mit Luft ‖ aération f.
aerial ‖ luftig; oberirdisch ‖ aérien.

aerial (Radio) ‖ Antenne f.; Luftleiter m.; Luftdraht m. ‖ antenne f.; aérien m. / bow-net ~ ‖ Reusenantenne f. ‖ antenne f. en forme de nasse. / cross-coil ~ ‖ Kreuzrahmenantenne f. ‖ antenne f. à cadre double. / directional ~ ‖ gerichtete Antenne f. ‖ antenne f. dirigée; aérien m. à ondes dirigées. / dumb ~ ‖ verstimmte Antenne f. ‖ antenne f. muette. / fan ~ ‖ Fächerantenne f. ‖ antenne f. en éventail. / frame ~ ‖ Rahmenluftleiter m.; Rahmenantenne f. ‖ antenne f. en éventail. / guide tube for ~ ‖ Antennenführungsrohr n. ‖ tube m. guide d'antenne. / horizontal ~ ‖ wagrechter Luftleiter m. ‖ antenne f. horizontale. / instruction for the construction of ~s ‖ Antennenbauvorschrift f. ‖ ordonnance f. relative aux antennes. / L-(shaped) ~ ‖ L-Antenne f. ‖ antenne f. (en forme) L. / limbs pl. of the ~ ‖ Glieder npl. des Luftleiters ‖ tronçons m. de l'aérien. / natural vibration of ~s ‖ Eigenschwingung f. von Antennen ‖ oscillation f. propre d'antennes. / plain ~ ‖ einfacher Sender m.; einfacher Luftleiter m. ‖ dispositif m. d'émission directe; antenne f. simple. / plane ~ ‖ Flächenantenne f. ‖ antenne f. en nappe. / receiving ~ ‖ Empfangsdraht m.; Empfangsantenne f. ‖ antenne f. de réception. / star ~ ‖ sternförmiger Luftleiter m.; Sternantenne f. ‖ antenne f. en étoile. / trailing ~ ‖ Hängeantenne f. ‖ antenne f. suspendue. / transmitting ~ ‖ Sendeantenne f.; Sendeluftleiter m. ‖ antenne f. émetteuse ou d'émission. / umbrella ~ ‖ Schirm-(netz)antenne f. ‖ antenne f. en parapluie. / vertical wire ~ ‖ lineare Antenne f. ‖ antenne f. linéaire. / wire ~ ‖ Drahtantenne f. ‖ aérien m. filiforme.
aerial attack ‖ Fliegerangriff m. ‖ attaque f. d'avions. / ~ bomb ‖ Fliegerbombe f. ‖ bombe f. aérienne ou d'avion. / ~ bomb aiming device ‖ Luftbombenzielvorrichtung f. ‖ appareil m. de pointage pour lancement de bombes aériennes. / ~ bombardment ‖ Luftbombardement n. ‖ bombardement m. aérien. / ~ cable ‖ oberirdisches Kabel n.; Luftkabel n. ‖ câble m. aérien. / ~ cable pole ‖ Luftkabelmast m. ‖ appui m. ou mât m. de câble aérien. / ~ cableway ‖ Drahtseilbahn f.; Luftseilbahn f. ‖ transport m. aérien ou voie f. aérienne par câble; funiculaire m.; chemin m. de fer funiculaire. / ~ camera ‖ Luftbildkamera f. ‖ appareil m. photographique aérien. / ~ capacity ‖ Antennenkapazität f.; Antennenreichweite f. ‖ capacité f. ou portée f. de l'antenne. / ~ change-over switch ‖ Luftdrahtumschalter m. ‖ commutateur m. d'antenne. / ~ circuit ‖ Luftleiterkreis m.; Antennenkreis m. ‖ circuit m. d'antenne. / ~ compass ‖ Luftfahrtkompaß m. ‖ compas m. de navigation aérienne. / ~ conduit ‖ Oberleitung f. ‖ conduite f. aérienne. / ~ crossing (El line) ‖ Luftkreuzung f. ‖ croisement m. aérien; traversée f. aérienne. / ~ cut-out ‖ Freileitungssicherung f. ‖ coupe-circuit m. aérien. / ~ drum ‖ Antennentrommel f. ‖ rouet m. d'antenne. / ~ frog ‖ Luftweiche f. ‖ changement m. de voie aérien; aiguillage m. aérien. / ~ heating switch ‖ Heizumschalter m. der Antenne ‖ commutateur m. d'échauffement d'antenne. / effective ~ height ‖ wirksame Antennenhöhe f.

‖ hauteur f. effective d'antenne. / ~ inductance ‖ Antenneninduktanz f. ‖ inductance f. d'antenne. / ~ lead-in ‖ Freileitungseinführung f. ‖ entrée f. de lignes aériennes. / ~ line ‖ oberirdische Leitung f.; Freileitung f. ‖ ligne f. aérienne (fil) m. aérien. / ~ loading coil ‖ Antennenverlängerungsspule f. ‖ self m. de syntonisation d'antenne; bobine f. de rallonge d'antenne. / ~ map ‖ Luftbildplan m.; Luftbildkarte f. ‖ carte f. ou plan m. d'images aériennes. / ~ mapping camera ‖ Luftmeßbildkamera f. ‖ chambre f. aéro-photogrammétrique. / ~ navigation ‖ Luftschiffahrt f. ‖ navigation f. aérienne / ~ network ‖ Freileitungsanlage f. ‖ installation f. de ligne aérienne. / ~ nitrogen ‖ Luftstickstoff m. ‖ azote m. atmosphérique. / ~ observation ‖ Lufterkundung f.; Luftbeobachtung f. ‖ observation f. aérienne. / ~ observer ‖ Flugzeugbeobachter m.; Franz m. ‖ observateur m. aérien. / ~ perspective ‖ Luftperspektive f. ‖ perspective f. aérienne. / ~ photography ‖ Fliegerfotografie f.;‖ photographie f. aérienne. / ~ propeller ‖ Luftschraube f. ‖ aéro-propulseur m.; hélice f. d'avion. / ~ radiation ‖ Antennenstrahlung f. ‖ rayonnement m. de l'antenne. / ~ railway ‖ Hängebahn f. ‖ chemin de fer m. aérien. / ~ region above sites ‖ Luftraum m. über Grundstücken ‖ espace m. d'air au-dessus de terrains. / ~ resistance ‖ Antennenwiderstand m. ‖ résistance f. d'antenne. / ~ rope man (Mine) ‖ Seilbahnlader m. ‖ chargeur m. du chemin de fer aérien. / ~ ropeway ‖ Drahtseilbahn f. ‖ funiculaire m.; chemin m. de fer funiculaire; transporteur m. aérien à câble. / ~ short wave condenser ‖ Antennenverkürzungskondensator m. ‖ condensateur m. de raccourcissement d'antenne. / ~ syntonizing mean ‖ Antennenabgleichmittel n. ‖ moyen m. de syntonisation d'antenne. / ~ torpedo ‖ Lufttorpedo m. ‖ torpille f. aérienne. / ~ tramway ‖ Drahtseilbahn f. ‖ chemin de fer m. funiculaire. / ~ tuning condenser ‖ Kondensator m. zur Luftleiterabstimmung ‖ condensateur m. de syntonisation d'antenne. / ~ tuning inductance ‖ Induktanz f. zur Luftleiterabstimmung ‖ inductance f. à syntoniser le circuit d'antenne. / ~ view ‖ Flugzeugaufnahme f. ‖ vue f. prise de l'avion. / ~ warfare ‖ Luftkrieg m. ‖ guerre f. aérienne. / ~ wire (Electr) ‖ Freileitung f. ‖ fil m. aérien. / ~ wire (Radio) ‖ Antennendraht m. ‖ fil m. d'antenne.

aeriform ‖ luftförmig; gasförmig ‖ aériforme; gazeux.

aero ‖ Luftfahrzeug n. ‖ aéronef m.

aerobic bacterium ‖ aerobes Bakterium n. ‖ bactérie f. aérobique.

aerodrome ‖ Flugplatz m.; Flughafen m. ‖ aéro-port m.; port m. aérien; aérodrome m.

aerodynamical ‖ aerodynamisch ‖ aérodynamique. / ~ balance ‖ aerodynamisches Gleichgewicht n. ‖ balance f. aérodynamique. / ~ volume displacement ‖ Luftverdrängung f. ‖ déplacement m. d'air.

aerodynamics pl. ‖ Aerodynamik f.; Luftdrucklehre f.; Dynamik f. luftförmiger Körper ‖ aérodynamique f.

aerodyne ‖ Luftfahrzeug n. schwerer als Luft ‖ aéronef m. plus lourd que l'air.

aero engine ‖ Flugmotor m. ‖ aéro-moteur m.; moteur m. pour avions.

aerofoil ‖ Tragfläche f. ‖ aile f.; plan m.

aerography ‖ Luftbeschreibung f. ‖ aérographie f.

aerolite ‖ Meteorstein m. ‖ aérolithe m.; météorolithe m.; pierre f. météorique; météorite f.

aerological measuring, instrument for ‖ aerologisches Meßgerät n. ‖ instrument m. pour mesures aérologiques.

aerology ‖ aeronautische Wetterkunde f. ‖ aérologie f.

aeromechanics pl. ‖ Aeromechanik f. ‖ aéromécanique f.

aerometer ‖ Aeromesser m. ‖ aéromètre m.

aerometric instrument ‖ Luftmesser m. ‖ compteur d'air.

aeronaut ‖ Luftschiffer m. ‖ aéronaute m.; aérostier m.

aeronautical compass ‖ Luftfahrtkompaß m. ‖ compas m. de navigation aérienne. / ~ establishment ‖ Luftschiffstation f. ‖ établissement m. d'aérostation. / ~ map ‖ Flugkarte f. ‖ carte f. aéronautique. / ~ station (Radio) ‖ Flughafenfunkstelle f. ‖ station f. aéronautique.

aeronautics pl. ‖ Aeronautik f.; Luftschiffahrt f.; Luftfahrwesen n. ‖ aéronautique f.; navigation f. aérienne.

aerophysical measurement, method of ‖ aerophysikalisches Meßverfahren n. ‖ méthode f. de mesures aérophysiques.

aeroplane ‖ Flugzeug n.; Flugmaschine f.; Aeroplan m. ‖ aéroplane m.; avion m. / accessories pl. for ~s ‖ Flugzeugzubehörteile mpl. ‖ accessoires mpl. d'avions. / air service ~ ‖ Arbeitsflugzeug n. ‖ avion m. de travail aérien. / allmetal ~ ‖ Ganzmetallflugzeug n. ‖ avion m. entièrement métallique. / bombing ~ ‖ Bombenflugzeug n. ‖ avion m. à bombes ou de bombardement. / civil ~ ‖ Zivilflugzeug n. ‖ avion m. civile. / commercial ~ ‖ Verkehrsflugzeug n. ‖ avion m. de transport commercial. / ~ for dusting crops ‖ Streuflugzeug n. ‖ avion m. de saupoudrage ou d'épandage. / double-motor ~ ‖ Zweimotorenflugzeug n. ‖ appareil m. bimoteur. / double-plane ~ ‖ Zweidecker m.; Doppeldecker m. ‖ appareil m. biplan; biplan m. / duck type ~ ‖ Entenflugzeug n. ‖ avion-canard m. / feeder line ~ ‖ Zubringerflugzeug n. ‖ avion m. de ligne auxiliaire. / fighting ~ ‖ Schlachtflugzeug n. ‖ avion m. de combat. / freight ~ ‖ Frachtflugzeug n. ‖ avion m. de fret. / goods-carrying ~ see aeroplane, freight. / large ~ ‖ Großflugzeug n. ‖ gros avion m. / light ~ ‖ Leichtflugzeug n. ‖ avion m. léger. / mail ~ ‖ Postflugzeug n. ‖ avion m. postal. / ~ with mid-set wing ‖ Mitteldecker m. ‖ avion m. à ailes demie-surélevées. / night ~ ‖ Nachtflugzeug n. ‖ avion m. nocturne. / photographic ~ ‖ Luftbildflugzeug n. ‖ avion m. de photographie. / private ~ ‖ Privatflugzeug n. ‖ avion m. privé. / reconnaissance ~ see aeroplane, scout. / sanitary ~ ‖ Sanitätsflugzeug n. ‖ avion m. sanitaire. / school ~ ‖ Schulflugzeug n. ‖ avion m. école. / scout ~ ‖ Aufklärungsflugzeug n. ‖ avion m. éclaireur ou de reconnaissance. / ski ~ ‖ Kufenflugzeug n.; Schneeflugzeug n. ‖ avion m. skieur ou à skis. / sporting ~ ‖ Sportflugzeug n. ‖ avion m. de sport. / subdivision of ~ types ‖ Einteilung f.

der Flugzeuge ‖ classification f. des avions. / tent for ~ ‖ Flugzeugzelt n. ‖ tente f. pour aéroplane. / torpedo ~ ‖ Torpedoflugzeug n. ‖ avion m. torpilleur. / training ~ ‖ Übungsflugzeug n. ‖ avion m. d'entraînement. / transport ~ ‖ Transportflugzeug n. ‖ avion m. de transport. / wire for ~s ‖ Flugzeugdraht m. ‖ fil m. pour aéroplanes. / ~ of wood construction ‖ Holzflugzeug n. ‖ avion m. en bois.

aeroplane cable ‖ Flugzeugseil n. ‖ câble m. pour aéroplanes. / ~ carrier ‖ Flugzeugmutterschiff n. ‖ navire m. porte-avion. / ~ dimensions pl. ‖ Flugzeugmasse npl. ‖ dimensions fpl. de l'avion. / ~ engine ‖ Flugmotor m. ‖ moteur m. d'aéroplane. / ~ formation ‖ Flugzeugverband m. ‖ formation f. des avions. / ~ glider ‖ Gleitflieger m. ‖ planeur m. / ~ hangar ‖ Flugzeugschuppen m. ‖ hangar m. pour aéroplanes. / ~ industry ‖ Flugzeugindustrie f. ‖ industrie f. aéronautique. / ~ meteorograph ‖ Flugzeugmeteorograf m. ‖ météorographe m. pour avions. / ~ motor work ‖ Flugmotorenwerk n. ‖ usine f. à moteurs d'avions. / ~ radio set ‖ funkentelegrafische Bordanlage f. ‖ télégraphie f. sans fil de l'avion. / ~ roofing material ‖ Flugzeugbespannungsstoff m. ‖ matériel m. ou tissu m. à couvrir les aéroplanes. / ~ science ‖ Flugzeugkunde f. ‖ science f. de l'avion. / ~ station (Radio) ‖ Flugzeugstation f. ‖ poste f. d'aéroplane. / ~ tyre ‖ Flugzeugluftreifen m. ‖ pneu m. d'avion. / ~ varnish ‖ Flugzeuglack m. ‖ vernis m. pour aéroplanes. / wire for ~s ‖ Flugzeugdraht m. ‖ fil m. pour aéroplanes.

aeroscope ‖ Aeroskop n. ‖ aérosope m.

aerostat ‖ Luftballon m. ‖ aérostat m.

aerostatic ‖ aerostatisch ‖ aérostatique.

aerostatics pl. ‖ Aerostatik f. ‖ aérostatique f.

aerostation ‖ Ballonschiffahrt f. ‖ aérostation f.

aero survey photograph ‖ Luftmeßbild n. ‖ photographie f. de mesure aérienne; aérophotogramme m.

aeruginous ‖ rostig ‖ rouillé; enrouillé.

affect, to ‖ beeinflussen ‖ influencer; influer sur. / ~ (Chem) ‖ verändern ‖ altérer.

affected by the air ‖ luftempfindlich ‖ altérable à l'air m. / the pipes are ~ by the water used ‖ die Röhren fpl. werden vom Abwasser angegriffen ‖ les tubes mpl. sont attaqués par l'eau employée.

affection, inflammatory ‖ entzündliche Erkrankung f. ‖ maladie f. inflammatoire.

affiche ‖ Affiche f. ‖ affiche f.

affidavit ‖ eidesstattliche Versicherung f. ‖ déclaration f. écrite et affirmée par serment.

affiliated company ‖ Zweiggesellschaft f. ‖ compagnie f. affiliée. / ~ concern ‖ Zweiggeschäftsbetrieb m. ‖ entreprise f. affiliée.

affiliation ‖ Angliederung f. ‖ affiliation f.

affinage (Met) ‖ Affinierung f.; Läutern n. ‖ affinage m.

affinity, chemical ‖ chemische Verwandtschaft f. ‖ affinité f. chimique. / elective ~ (Chem) ‖ Wahlverwandtschaft f. ‖ affinité f. élective.

affirm, to ‖ eidlich versichern ‖ jurer; attester sous serment.

affliction of the eye ‖ Augenkrankheit f. ‖ maladie f. des yeux.

afloat ‖ flott ‖ à flot. / to get a stranded ship ~ ‖ ein gestrandetes Schiff abbringen *oder* abarbeiten ‖ déséchouer *ou* déchouer *ou* relever *ou* remettre à flot un vaisseau échoué.

aft (Aero; Mar) ‖ achtern ‖ derrière; en arrière de.

afterbody (Shipb) ‖ Achterschiff n.; Hinterschiff n. ‖ arrière m.; poupe m.

afterdamp ‖ Nachschwaden m.; Schwaden m. ‖ mofette f.

after-fermentation ‖ Nachgärung f. ‖ fermentation f. secondaire. / ~-hatchway (Shipb) ‖ Hinterluke f. ‖ écoutille f. de poupe.

afterhold (Shipb) ‖ Achterraum m. ‖ cale f. arrière.

aftermath ‖ Grumt n. ‖ regain m.

after-part of a vessel ‖ Gatt n. eines Schiffes ‖ cul m. d'un vaisseau.

after-piece of the rudder ‖ Ruderhacke f. ‖ safran m. de gouvernail.

afterwine ‖ Tresterwein m. ‖ piquette f.

aft-gate (hydr arch) ‖ unteres Schleusentor n.; Untertor n. ‖ porte f. d'aval *ou* de mouille.

agalloch ‖ Adlerholz n. ‖ bois m. d'aigle.

agalmatolite ‖ Bildstein m. ‖ agalmatolite f.

agar-agar ‖ Agar-Agar n. ‖ agar-agar m.

agate ‖ Achat m. ‖ agate f. / ~ ball ‖ Achatkugel f. ‖ bille f. d'agate. / ~ cap ‖ Achathütchen n. ‖ chape f. d'agate. / ~ cutter ‖ Achatschneider m. ‖ tailleur m. d'agates. / ~ drill ‖ Achatbohrer m. ‖ foret m. en agate. / ~ driller ‖ Achatbohrer m. ‖ perceur m. d'agates. / ~ goods pl. ‖ Achatwaren fpl. ‖ articles mpl. en agate. / ~ grinding factory ‖ Achatschleiferei f. ‖ atelier m. de polissage des agates. / ~ mortar ‖ Achatmörser m. ‖ mortier m. d'agate. / ~ tip ‖ Achatfuß m. ‖ pied m. en agate.

agave, fiber of ‖ Agavefaser f. ‖ fibre f. d'agave.

age, to ‖ altern ‖ vieillir.

age (Mach) ‖ Lebensdauer f. ‖ durée f. / ~ class ‖ Altersklasse f. ‖ classe f. d'âge.

aged, uneven ‖ von ungleichem Alter n. ‖ d'âges mpl. différents.

ageing (Met) ‖ Alterung f. ‖ vieillissement m.

agency ‖ Agentur f. ‖ agence f. succursale. / ~ (Goods) ‖ Niederlage f.; Warenlager n. ‖ dépôt m.; entrepôt m.; magasin m. / ~ and commission ‖ Agentur f. und Kommission f. ‖ agents mpl. et commissaires mpl.

agenda ‖ Tagebuch n. ‖ agenda m.

agent (Chem) ‖ wirkendes Mittel n. ‖ agent m. / oxidizing ~ ‖ Oxydationsmittel n. ‖ agent m. d'oxydation. / purifying ~ ‖ Reinigungsmittel n. ‖ agent m. de purification. / reducing ~ ‖ reduzierendes Mittel n. ‖ agent m. de réduction.

agent (Diplomatic) ‖ Geschäftsträger m. ‖ chargé m. d'affaires *ou* d'un état. / ~ (Trade) ‖ Vertreter m. ‖ agent m. / accredited ~ ‖ bevollmächtigter Agent m. ‖ agent m. accrédité. / ~ with power of attorney ‖ Bevollmächtigter m. ‖ mandataire m.; plénipotentiaire m.

agglomerated coal ‖ Preßkohle f.; zusammengepreßte Kohle f. ‖ charbon m. aggloméré.

agglomerating plant (Min; Met) ‖ Agglomerieranlage f. ‖ installation f. d'agglomération.

aggregate ‖ Aggregat n. ‖ agrégat m.

aggregation, state of ‖ Aggregatszustand m. ‖ état m. d'aggrégat.

aging see ageing.

agio ‖ Agio n.; Aufgeld n. ‖ agio m.

agitate, to (Chem) ‖ schütteln ‖ agiter.

agitating of the bath solution ‖ Bewegung f. der Badflüssigkeit ‖ agitation f. du bain. / ~ arm ‖ Rührarm m. ‖ bras m. agitateur. / ~ machine see agitator.

agitator ‖ Rührwerk n. ‖ agitateur m.; remueur m. / ~ shoe ‖ Schlämmschuh m. ‖ sabot m. de lavage.

agree, to ‖ übereinstimmen ‖ être concordant *ou* d'accord; convenir. / ~ on see to agree upon. / ~ to ‖ billigen; genehmigen; zustimmen; gutheißen; einwilligen in ‖ approuver; consentir à; adhérer à; agréer. / ~ upon ‖ vereinbaren; übereinkommen ‖ tomber d'accord; s'accorder; convenir de; stipuler. / ~ with ‖ zusagen; gefallen ‖ être du goût de; convenir à. / ~ with an opinion ‖ sich einer Ansicht f. anschließen; mit einer Meinung f. übereinstimmen ‖ se rallier à une opinion.

agreement ‖ Zustimmung f. ‖ consentement m.; approbation f.; assentiment m. / ~ (Trade) ‖ Übereinstimmung f.; Vertrag m. ‖ concordance f.; accord m.; conformité f.; contrat m. / to come to an ~ with a person ‖ sich mit jemandem verständigen ‖ s'entendre avec quelqu'un. / to conclude an ~ ‖ einen Vertrag m. abschließen ‖ conclure un marché *ou* un traité; contracter. / to enter into an ~ ‖ ein Übereinkommen n. treffen ‖ convenir un arrangement. / ~ already in force ‖ laufender Vertrag m. ‖ contrat m. en cours. / to make an ~ see agreement, to conclude an. / ~ by meter ‖ Akkord m. nach Maß ‖ marché m. au mètre. / ~ by piece ‖ Akkord m. nach Stückzahl ‖ marché m. à la pièce. / ~ provisional ~ ‖ Vorvertrag m. ‖ accord m. provisoire. / ~ for service ‖ Arbeitsverhältnis n. ‖ contrat m. de travail. / to request the termination of an ~ of service ‖ die Aufhebung f. des Arbeitsverhältnisses verlangen ‖ exiger la résiliation f. du contrat de travail. / ~ of two results ‖ Übereinstimmung f. zweier Ergebnisse ‖ concordance f. de deux résultats. / verbal ~ ‖ mündliche Vereinbarung f. ‖ accord m. verbal. / working ~ ‖ Interessengemeinschaft f. ‖ convention f.

agreement clause ‖ Vertragsklausel f. ‖ clause f. de contrat. / stipulations pl. of the ~ ‖ Bestimmungen fpl. des Vertrages ‖ stipulations fpl. du contrat.

agricultural chemistry ‖ Agrikulturchemie f. ‖ chimie f. appliquée à l'agriculture. / ~ college ‖ Ackerbauschule f. ‖ école f. d'agriculture. / ~ day-labourer ‖ landwirtschaftlicher Tagelöhner m. ‖ journalier m. agricole; manœuvrier m. agricole. / ~ distillery ‖ landwirtschaftliche Brennerei f. ‖ distillerie f. agricole. / ~ district ‖ landwirtschaftliches Gebiet n. ‖ région f. agricole. / ~ expert ‖ Agraringenieur m. ‖ ingénieur m. agricole. / ~ implement ‖ landwirtschaftliches Gerät n. ‖ instrument m. agricole; outil m. de labourage. / ~ implement drawn by motor ‖ Motoranhängegerät n. für die Landwirtschaft ‖ machine f. d'agriculture remorquée par moteur. / small ~ implement ‖ landwirtschaftliches Kleingerät n. ‖ instrument m. aratoire de petit modèle.

agricultural machine ‖ landwirtschaftliche Maschine f. ‖ machine f. agricole *ou* pour l'agriculture. / one horse ~ ‖ Einspännerlandmaschine f. ‖ machine f. agricole pour simple attelage. / two-horse ~ ‖ Zweispännerlandmaschine f. ‖ machine f. agricole pour double attelage. / ~ factory ‖ Fabrik f. für landwirtschaftliche Maschinen ‖ fabrique f. de machines agricoles.

agricultural product ‖ Landeserzeugnis n.; landwirtschaftliches Erzeugnis n. ‖ produit m. agricole. / ~ roller ‖ Ackerwalze f. ‖ rouleau m. à labourer. / ~ service ‖ landwirtschaftlicher Betrieb m. ‖ service m. d'exploitation agricole. / ~ work ‖ Landarbeit f. ‖ travail m. des champs; travail m. aux champs.

agriculture ‖ Ackerbau m.; Landwirtschaft f. ‖ agriculture f.

agriculturist ‖ Ackerbaukundiger m.; Gutsbesitzer m.; Landwirt m. ‖ agronome m.; agriculteur m.

aground, to be (Mar) ‖ auf Grund m. sitzen; festsitzen; festgekommen sein ‖ être échoué; être mouillé par la quille; être au sec. / to run ~ ‖ auf Grund m. geraten ‖ échouer.

ahead (Mar) ‖ vorwärts ‖ en avant. / easy ~ ‖ langsam vorwärts ‖ en avant doucement. / full speed ~ ‖ Volldampf m. vorwärts ‖ en avant toute vitesse. / half speed ~ ‖ mit halber Kraft f. vorwärts ‖ à mi-vitesse f. en avant. / slow ~ see ahead, easy.

ahead drum ‖ Vorwärtstrommel f. ‖ tambour m. à marche avant. / ~ position ‖ Vorwärtsstellung f. ‖ position f. d'avance. / ~ running ‖ Vorwärtsgang m. ‖ marche f. en avant. / ~ and astern running ‖ Vor- und Rückwärtsgang m. ‖ marche f. avant et marche f. arrière. / ~ turbine ‖ Vorwärtsturbine f. ‖ turbine f. à marche avant.

aid, first ~ station in case of accidents ‖ Verbandstelle f. für erste Hilfe in Unglücksfällen ‖ poste m. de premier secours en cas d'accidents. / ~ to vision ‖ Sehhilfsmittel n. ‖ appareil m. pour l'amélioration de la vue.

aid fund ‖ Unterstützungsfonds m. ‖ fond m. de secours.

aileron (Aero) ‖ Querruder n.; Quersteuer n. ‖ aileron m. / washed out ~ ‖ Querruder n. mit verjüngten Enden ‖ aileron m. en fuite. / ~ cable ‖ Querruderkabel n. ‖ câble m. de commande d'aileron. / ~ crank ‖ Querruderhebel m. ‖ guignole f. de l'aileron.

aim, to ‖ zielen ‖ viser.

aiming device, aerial bomb ‖ Luftbombenzielvorrichtung f. ‖ appareil m. de pointage pour lancement de bombes aériennes.

aiming telescope ‖ Visierfernrohr n. ‖ lunette f. de visée.

air, to ‖ lüften ‖ aérer.

air ‖ Luft f. ‖ air m. / to allow ~ to enter ‖ Luft f. eintreten lassen ‖ faire rentrer l'air m. / ambient ~ ‖ umgebende Luft f. ‖ air m. ambiant. / away from the ~ ‖ unter Luftabschluß m. ‖ à l'abri m. de l'air. / bad ~ see air, foul. / carburetted ~ ‖ Gasgemisch n. ‖ air m. carburé. / cold ~ ‖ Kaltluft f. ‖ air m. froid. / ~ for combustion ‖ Verbrennungsluft f. ‖ air m. comburant. / compressed ~ ‖ Druckluft f.; Preßluft f. ‖ air m. comprimé. / cooled

~ ‖ gekühlte Luft f. ‖ air m. refroidi. / to displace the ~ (Chem) ‖ entlüften ‖ chasser l'air m. / drawing-off ~ ‖ Abluft f. ‖ air m. d'échappement. / ~ being excluded ‖ unter Luftausschluß m. ‖ a l'abri m. de l'air. / to expose to the ~ ‖ der Luft f. aussetzen; auslüften ‖ éventer; mettre à l'air. / foul ~ ‖ gebrauchte Wetter npl. ‖ air m. vicié; gaz mpl. délétères. / free from ~ (Chem) ‖ luftfrei ‖ exempt ou purgé d'air m. / ~ of great oxygen contents pl. ‖ sauerstoffreiche Luft f. ‖ air m. enrichi d'oxygène. / half in the ~ ‖ halb in der Luft f. ‖ mi-aérien. / hot ~ ‖ Heißluft f. ‖ air m. chaud. / to keep from contact with the ~ ‖ unter Luftausschluß m. aufbewahren ‖ conserver à l'abri m. de l'air. / liquid ~ ‖ flüssige Luft f. ‖ air m. liquide. / ~ in mines ‖ Grubenluft f. ‖ air m. (des galeries) de mine. / moved ~ ‖ bewegte Luft f. ‖ air m. agité ou en mouvement. / in the open ~ ‖ frei ohne Dach n.; im Freien n. ‖ à l'air m. libre; en plein air. / oppressive ~ ‖ matte Wetter npl. ‖ air m. lourd ou méphitique. / to prevent the ~ from entering ‖ das Eindringen n. von Luft verhindern ‖ empêcher l'air m. de pénétrer. / to renew the ~ ‖ lüften ‖ aérer. / saturated ~ ‖ gesättigte Luft f. ‖ air m. saturé. / ~ sucked ‖ angesaugte Luft f. ‖ air m. aspiré. / surrounding ~ ‖ Außenluft f. ‖ air m. extérieur; atmosphère f. / transparent ~ ‖ sichtiges Wetter n. ‖ temps m. clair. / upcurrent of ~ ‖ Aufwind m. ‖ courant m. d'air ascendant. / ventilating ~ ‖ Bewetterungsluft f. ‖ air m. de ventilation. / vital ~ ‖ Lebensluft f. ‖ air m. vital. / volume of ~ drawn in ‖ angesaugte Luftmenge f. ‖ volume m. d'air aspiré. / weight of ~ drawn in ‖ angesaugtes Luftgewicht n. ‖ poids m. d'air aspiré.

air admission port ‖ Lufteinlaßschlitz m. ‖ lumière f. d'admission d'air. / ~ **balloon** ‖ Luftballon m. ‖ ballon m. aérostatique. / ~ **bath** ‖ Luftbad n. ‖ bain m. d'air. / ~ **bed** ‖ Luftmatratze f. ‖ matelas m. à air. / ~ **bladder of fish** ‖ Schwimmblase f. des Fisches ‖ vessie f. natatoire de poisson. / ~ **blast** ‖ Gebläse n. ‖ soufflerie f. / ~ **bottle** ‖ Luftflasche f. ‖ bouteille f. à air. / ~ **bound** (Pipe line) ‖ Lufttasche f. ‖ poche f. d'air.

air brake ‖ Luftdruckbremse f. ‖ frein m. à air (comprimé). / ~ **equipment** ‖ Luftbremsausrüstung f. ‖ équipement m. du frein à air. / ~ **valve** ‖ Luftdruckbremsventil n. ‖ clapet m. ou soupape f. de frein à air comprimé.

air braking ‖ Luftdämpfung f. ‖ amortissement m. à air. / ~ **break switch** ‖ Schalter m. mit Luftunterbrechung ‖ commutateur m. à air. / ~ **bubble** ‖ Luftblase f. ‖ bulle f. d'air. / ~ **buffer** ‖ Luftpuffer m. ‖ tampon m. atmosphérique ou pneumatique. / ~ **cable** ‖ Luftkabel n. ‖ câble m. aérien.

air case of the chimney of a steam ship ‖ Schornsteinmantel m. eines Dampfschiffes ‖ chemise f. de la cheminée d'un bateau à vapeur. / ~ **of the funnel** ‖ Mantel m. eines Schornsteins; Rauchfangmantel m. ‖ chemise f. de la cheminée. / ~ **of a life boat** ‖ Luftkasten m. eines Rettungsbootes ‖ caisse f. à air d'un canot de sauvetage.

air chamber (Mill) ‖ Windkammer f. ‖ chambre f. de soufflures. / ~ (Pump)

‖ Windkessel m. ‖ réservoir m. à air.

air changing valve ‖ Luftwechselklappe f. ‖ clapet m. d'aération. / ~ **channel** ‖ Luftkanal m. ‖ canal m. d'air. / ~ **chimney** ‖ Ansaugschlot m. ‖ cheminée f. d'aspiration. / ~ **chuck** ‖ Druckluftfutter n. ‖ mandrin m. à serrage pneumatique. / ~ **circulation** ‖ Luftumlauf m. ‖ circulation f. d'air. / ~ **cock** ‖ Entlüftungshahn m. ‖ robinet m. d'air. / swinging ~ **column** ‖ pendelnde Luftsäule f. ‖ colonne f. d'air alternative.

air compressor ‖ Luftverdichter m. ‖ compresseur m. d'air. / single stage ~ ‖ einstufige Luftpumpe f. ‖ pompe f. à air à une phase. / ~ **with a delivery per minute of x cubic meters of compressed air at y atmospheres pressure** ‖ Luftverdichter m. mit minütlicher Leistung von x Kubikmeter auf y Atmosphären gepreßter Luft ‖ compresseur m. d'air refoulant par minute x mètres cubes d'air comprimé à y atmosphères.

air condenser ‖ Luftverdichter m.; Luftkondensator m. ‖ aéro-condenseur m.; condensateur m. à air.

air conduit (Mine) ‖ Wetterlutte f. ‖ buse f. d'airage ou d'aérage. / ~ **adjusting apparatus** (Mine) ‖ Wetterluttenrichtapparat m. ‖ appareil m. à dresser les tuyaux d'aération. / ~ **ventilator** (Mine) ‖ Luttenventilator m. ‖ ventilateur m. de galerie.

air-cooled ‖ luftgekühlt ‖ à refroidissement m. d'air; refroidi par air ou à l'air. / ~ **engine** see ~ **motor**. / ~ **machine gun** ‖ luftgekühltes Maschinengewehr n. ‖ mitrailleuse f. à refroidissement d'air. / ~ **motor** ‖ luftgekühlter Motor m. ‖ moteur m. à refroidissement d'air; moteur m refroidi par air. / ~ **transformer** ‖ luft. gekühlter Transformator m. ‖ transformateur m. à air.

air cooler ‖ Luftkühler m. ‖ aéro-refroidisseur m.; rafraîchisseur m. ou réfrigérant m. à air.

air-cooling ‖ Luftkühlung f. ‖ refroidissement m. par air. / forced ~ ‖ Preßluftkühlung f. ‖ refroidissement m. à air comprimé. / ~ **plant** ‖ Luftkühlanlage f. ‖ installation f. de rafraîchissement d'air.

air-core inductance ‖ Induktanz f. mit Luftkern ‖ inductance f. sans fer. / ~ **protecting choke** ‖ Impedanzspule m. mit Luftkern ‖ bobine f. de réactance sans noyau de fer. / ~ **transformer** ‖ Lufttransformator m. ‖ transformateur m. à air.

air corps pl. ‖ Luftstreitkräfte fpl. ‖ forces fpl. (militaires) aériennes.

aircraft ‖ Luftfahrzeug n. ‖ aéronef m. / ~ **heavier than air** ‖ Luftfahrzeug n. schwerer als Luft ‖ aéronef m. plus lourd que l'air. / ~ **lighter than air** ‖ Luftfahrzeug n. leichter als Luft ‖ aéronef m. plus léger que l'air.

aircraft armament ‖ Flugzeugbestückung f. ‖ armement m. d'aéroplane. / ~ **bomb** ‖ Fliegerbombe f. ‖ bombe f. aérienne ou d'avion. / ~ **carrier** ‖ Flugzeugmutterschiff n.; Flugzeugträger m. ‖ navire m. porte-aéronefs; bâtiment m. porte-avions. / **constructional element of the** ~ ‖ Flugzeugbauelement n. ‖ élément m. de la construction des avions. / ~ **radio** ‖ Fliegerfunkerei f. ‖ radiotélégraphie f. d'aéroplane. / ~ **shed** ‖ Flugzeughalle f. ‖ han-

gar m. pour avions. / ~ **station** (Radio) ‖ Flugzeugfunkstelle f. ‖ station f. T.S.F. d'aéronef.

air cruiser ‖ schweres Kampfflugzeug n.; Luftkreuzer m. ‖ croiseur m. aérien.

air current ‖ Zug m.; Zugluft f.; Luftströmung f. ‖ courant m. atmosphérique ou d'air; écoulement m. d'air. / ~ (Mine) ‖ einfallender Wetterstrom m. ‖ courant m. d'air. / reversible ~ ‖ umkehrbarer Luftstrom m. ‖ courant m. d'air reversible. / warm ~ ‖ warmer Luftstrom m. ‖ courant m. d'air chaud.

air cushion ‖ Luftkissen n.; Luftpolster n.; Luftpuffer m. ‖ coussin m. pneumatique; matelas m. d'air; amortisseur m. à air. / ~ **damper** ‖ Lufteinlaßklappe f. ‖ clapet m. d'entrée d'air. / ~ **damping** ‖ Luftdämpfung f. ‖ amortissement m. par l'air. / ~ **damping plant** ‖ Luftbefeuchtungsanlage f. ‖ installation f. d'humectation de l'air. / ~ **density** ‖ Luftdichte f. ‖ densité f. d'air. / ~ **discharge pipe** ‖ Entlüftungsrohr n. ‖ tuyau m. de dégagement de l'air. / ~ **door** (Mine) see **air gate**. / ~ **drain** ‖ Luftkanal m. ‖ évent m. / ~ **draught** ‖ Luftzug m. ‖ appel m. d'air.

air-dried ‖ lufttrocken; an der Luft f. getrocknet ‖ séché à l'air. / ~ (Brew) ‖ geschwelkt ‖ fané. / ~ **malt** ‖ Luftmalz n. ‖ malt m. séché à l'air.

air-dry ‖ lufttrocken ‖ séché à l'air.

air-drying ‖ Lufttrocknung f. ‖ séchage m. à l'air. / ~ **by means of absorbents** ‖ Lufttrocknung f. durch Absorptionsmittel ‖ séchage m. de l'air à l'aide de corps absorbants.

air duct ‖ Luftkanal m. ‖ conduite f. ou conduit m. ou gaine f. à air. / ~ **eddy** ‖ Luftwirbel m. ‖ remous m. ou tourbillon m. d'air. / ~ **escape** ‖ Wetterlutte f. ‖ buse f. d'aérage. / ~ **exhaust(er)** ‖ Luftabzug m. ‖ évent m. / ~ **feeder** ‖ Luftzuführungsrohr n. ‖ manche f. à air. / ~ **fighting** ‖ Luftkampf m. ‖ combat m. aérien. / ~ **filter** ‖ Luftfilter n.; Luftreiniger m. ‖ filtre m. à air. / ~ **flow** see ~ **current**. / ~ **flow pattern** ‖ Luftströmungsbild n. ‖ forme f. d'écoulement de filet d'air. / ~ **force** (Phys) ‖ Luftkraft f. ‖ force f. aérodynamique; réaction f. de l'air. / ~ **forces** pl. (War impl) ‖ Luftstreitkräfte fpl. ‖ forces fpl. militaires aériennes. / ~ **friction** ‖ Luftreibung f. ‖ frottement m. de l'air. / ~ **furnace** ‖ Windofen m. ‖ fourneau m. à vent. / ~ **gallery** (Mine) ‖ Wetterstrecke f. ‖ galerie f. d'aérage.

air gap ‖ Luftspalt m.; Luftzwischenraum m. ‖ espace m. d'air. / ~ (Electr; Magnet) ‖ Interferrikum n. ‖ entrefer m.

air gas ‖ Luftgas n. ‖ gaz m. à l'air. / ~ **generating plant** ‖ Luftgaserzeugungsanlage f. ‖ installation f. de formation du gaz d'air.

air gate (Mine) ‖ Wettertür f. ‖ **porte f.** d'aérage. / main ~ ‖ Hauptwetterstrecke f. ‖ voie f. d'aérage.

air gun ‖ Luftgewehr n. ‖ fusil m. pneumatique ou à vent ou à air comprimé.

air-hardening ‖ Lufthärtung f.; Erhärten n. an der Luft ‖ durcissement m. ou trempe f. à l'air. / ~ **lime** ‖ Luftkalk m. ‖ chaux f. durcissant à l'air.

air head see **air gate**.

air heater ‖ Winderhitzer m.; Lufterhitzer m. ‖ appareil m. à air chaud; aéro-chauf-

feur m.; récupérateur m. d'air. / ~-man ‖ Winderhitzerarbeiter m. ‖ ouvrier m. de récupérateur.

air heating ‖ Luftheizung f. ‖ chauffage m. à l'air (chaud). / ~ apparatus ‖ Luftheizungsgerät n. ‖ appareil m. pour le chauffage à l'air; calorifère m. à air chaud. / ~ plant ‖ Luftheizungsanlage f. ‖ installation f. de chauffage à air.

airhole (Build) ‖ Luftloch n.; Luftschacht m. ‖ évent m.; éventoir m.; éventouse f.; ventouse f.; soupirail m. / ~ (Found) ‖ Gußblase ‖ soufflure f. / ~(Glass oven) ‖ Luftloch n.; Zugloch ‖ bonichon m. / ~ (Met) ‖ Windfang m. ‖ éventouse f. / ~ (Mine) ‖ Wetterlutte f. ‖ buse f. d'aérage ou d'airage. / ~ of a charcoal pile ‖ Luftkanal m. oder Zug m. oder Abzucht f. eines Kohlenmeilers ‖ évent m. en maçonnerie d'un fourneau de carbonisation. / ~ of a mould ‖ Pfeife f. einer Gußform ‖ évent m. d'un moule.

air humidifier ‖ Luftanfeuchter m.; Luftbefeuchter m. ‖ humidificateur m. d'air. / ~ humidifying plant ‖ Luftbefeuchtungsanlage f. ‖ installation f. pour humidifier l'air.

airing ‖ Lüftung f.; Lüften n. ‖ ventilation f. / ~ of mines ‖ Wetterwirtschaft f. ‖ service m. d'aérage d'une mine. / ~ plant ‖ Belüftungsanlage f.; Entlüftungsanlage f. ‖ installation f. d'aération.

air injection compressor ‖ Einblaseluftpumpe f. ‖ compresseur m. d'injection d'air. / ~ Diesel engine ‖ Dieselmotor m. mit Drucklufteinspritzung ‖ moteur m. Diesel à injection pneumatique. / engine for injection of the fuel by air ‖ Einblasemaschine f. ‖ moteur m. du type à injection pneumatique.

air inlet ‖ Lufteintritt m. ‖ entrée f. d'air. / ~ pipe ‖ Lufteinlaßrohr n. ‖ tube m. de prise d'air.

air isolation ‖ Luftisolation f. ‖ isolement m. à air. / ~ kiln ‖ Luftdarre f. ‖ touraille f. à air.

air layer of high humidity ‖ wasserreiche Luftschicht f. ‖ couche f. d'air humide ou riche en eau. / ~ of low humidity ‖ wasserarme Luftschicht f. ‖ couche f. d'air sèche ou pauvre en eau.

airless injection ‖ luftlose Einspritzung f. ‖ injection f. mécanique. / ~ Diesel engine ‖ kompressorloser Dieselmotor m. ‖ moteur m. Diesel sans compresseur.

air level ‖ Libelle f.; Setzwage f. ‖ niveau m. à bulle d'air. / ~ liquefying plant ‖ Anlage f. zur Erzeugung von flüssiger Luft ‖ installation f. pour la liquéfaction de l'air. / ~ lock ‖ Luftdrossel f. ‖ soupape f. de retenue d'air. / ~ mail ‖ Luftpost f.; Flugpost f. ‖ poste f. aérienne. / ~ mattress ‖ Luftmatratze f. ‖ matelas m. hermétique ou à air.

air moistening apparatus ‖ Luftbefeuchtungsgerät n. ‖ appareil m. pour la humidification d'air. / ~ chamber ‖ Luftbefeuchtungskammer f. ‖ chambre f. d'humidification d'air.

air mortar ‖ Luftmörtel m. ‖ mortier m. durcissant à l'air. / ~ navigation ‖ Luftschiffahrt f. ‖ navigation f. aérienne. / ~ nose see air nozzle. / ~ nozzle ‖ Luftdüse f. ‖ gicleur m. ou ajutage m. à air; buse f. d'air.

air-operated chuck ‖ Druckluftfutter n. ‖ mandrin m. pneumatique.

air outlet conduit ‖ Abluftkanal m. ‖ canal m. d'évacuation d'air. / ~ passage ‖ Luftdurchgang m. ‖ passage m. d'air.

air pipe see airhole.

air piping ‖ Luftleitungsrohr n. ‖ conduite f. d'air. / high pressure ~' ‖ Hochdruckluftleitung f. ‖ tuyauterie f. d'air à haute pression.

air pistol ‖ Luftpistole f.; Abblasepistole f. ‖ pistolet m. souffleur ou à vent ou pneumatique.

airplane see aeroplane.

air pocket ‖ Luftsack m. ‖ poche f. d'air. / ~ in piping ‖ Luftsack m. in der Leitung ‖ bulle f. d'air dans la conduite.

airport ‖ Flughafen m.; Flugzeughafen m.; Lufthafen m.; Flugplatz m. ‖ aéroport m.; aérodrome m.; port m. aérien; gare f. aérienne. / commercial ~ ‖ Verkehrsflughafen m. ‖ aéroport m. civil. / custom's ~ ‖ Zollflughafen m. ‖ aéroport m. douanier. / municipal ~ ‖ Stadtflughafen m. ‖ aéroport m. de ville.

air post ‖ Flugpost f. ‖ poste f. aérienne. / ~ preheater ‖ Luftvorwärmer m.; Winderhitzer m. ‖ réchauffeur m. d'air; appareil m. à air chaud.

air pressure ‖ Luftdruck m.; Atmosphärendruck m. ‖ pression f. atmosphérique ou de l'atmosphère ou barométrique ou d'air. / ~ at ground level ‖ Bodenluftdruck m. ‖ pression f. de l'air au sol ou à la surface de la terre. / to record the ~ ‖ den Barometerstand m. aufzeichnen ‖ enregistrer la pression d'air.

air pressure boiler ‖ Luftdruckkessel m. ‖ chaudière f. à pression atmosphérique. / ~ compensation ‖ Luftdruckkompensation f. ‖ compensation f. de la compression d'air. / ~ gauge ‖ Luftdruckmesser m. ‖ manomètre m. d'air. / ~ recorder ‖ Winddruckmesser m. ‖ enregistreur m. de la pression d'air. / ~ valve ‖ Preßluftventil n. ‖ soupape f. à air comprimé.

air-proof see also air-tight ‖ hermetisch; luftdicht ‖ hermétique.

air propeller (Aero) see airscrew. / ~ (Ventilation) ‖ Ventilator m. ‖ ventilateur m.; aérateur m.

air pump (Auto) ‖ Luftpumpe f. ‖ gonflepneus m. / ~ (Mach) ‖ Luftpumpe f. ‖ pompe f. pneumatique ou à air. / doublebarrelled ~ ‖ zweistiefelige Luftpumpe f. ‖ machine f. pneumatique à deux corps. / scavenging ~ ‖ Spülluftpumpe f. ‖ pompe f. de balayage. / starting ~ ‖ Anlaßluftpumpe f. ‖ pompe f. à air de mise en marche.

air pump governor ‖ Luftpumpenregler m. ‖ régulateur m. de la pompe à air. / ~ valve ‖ Luftpumpenventil n. ‖ clapet m. de pompe f. à air.

air purifier ‖ Luftreiniger m. ‖ filtre m. à air; épurateur m. d'air. / ~ (Mill) ‖ Windfege f. ‖ nettoyeur m. à air.

air purifying apparatus ‖ Luftverbesserungsgerät n. ‖ appareil m. d'épuration de l'air.

air quantity required for combustion ‖ zur Verbrennung notwendige Luftmenge f. ‖ quantité f. d'air nécessaire à la combustion. / ~ supplied ‖ zugeführte Luftmenge f. ‖ quantité f. d'air admise.

air raft ‖ Schlauchboot m. ‖ radeau m. pneumatique. / ~ rail traffic ‖ Flugeisenbahnverkehr m. ‖ trafic m. aéroferroviaire. / ~ regulation ‖ Luftregelung f. ‖ réglage m. d'air. / ~ relief cock ‖ Ent-

lüftungshahn m. ‖ robinet m. de purge d'air. / ~ relief screw (Mot) ‖ Entlüftungsschraube f. ‖ vis f. de purge. / ~ reservoir ‖ Luftkasten m. ‖ réservoir m. à air. / ~ resistance ‖ Luftwiderstand m. resistance f. de l'air.

airscape ‖ Luftmeßbild n. ‖ photographie f. de mesure aérienne; aérophotogramme m.

airscrew (Aero) ‖ Luftschraube f.; Luftpropeller m. ‖ hélice f. aérienne; propulseur m. aérien. / pusher ~ ‖ Druckschraube f.; Druckpropeller m. ‖ hélice f. propulsive. / tractor ~ ‖ Zugschraube f.; Zugpropeller m. ‖ hélice f. tractive.

airscrew blade ‖ Luftschraubenblatt n. ‖ pâle f. de l'hélice aérienne. / ~ blade face ‖ Vorderseite f. der Luftschraube ‖ face f. de l'hélice aérienne. / ~ blade leading edge ‖ Leitkante f. der Luftschraube ‖ arête f. directrice de l'hélice aérienne. / ~ hub ‖ Luftschraubennabe f. ‖ moyeu m. de l'hélice aérienne.

air-seasoned wood ‖ lufttrockenes Holz n. ‖ bois m. séché à l'air.

air separator ‖ Windsichter m. ‖ séparateur m. à vent ou à air. / ~ for water plants ‖ Wasserentlüftungsanlage f. ‖ séparateur m. d'air pour installations d'eau.

air service aeroplane ‖ Arbeitsflugzeug n. ‖ avion m. de travail aérien. / ~ shaft (Mine) ‖ Luftschacht m.; Wetterschacht m.; Windschacht m. ‖ bure f. ou buse f. ou puits m. d'airage; puits m. d'aérage; avaleresse f.

airship ‖ Luftschiff n. ‖ dirigeable m. / the ~ descends ‖ das Luftschiff n. fällt oder sinkt ‖ le dirigeable m. descend. / engine driven ~ ‖ Motorluftschiff n. ‖ dirigeable m. à moteur. / non-rigid ~ ‖ Prallluftschiff n.; unstarres Luftschiff n. ‖ dirigeable m. souple; souple m. / rigid ~ ‖ Starrluftschiff n. ‖ dirigeable m. rigide; rigide m. / the ~ sinks see the ~ descends. / small ~ ‖ Kleinluftschiff n. ‖ dirigeable m. de petites dimensions.

airship body ‖ Luftschiffkörper m. ‖ carène f. du dirigeable. / ~ hull ‖ Luftschiffhülle f. ‖ enveloppe f. du dirigeable. / ~ shed ‖ Luftschiffhalle f. ‖ hangar m. pour dirigeables. / ~ station ‖ Luftschiffstation f. ‖ station f. de ballon dirigeable. / ~ type ‖ Luftschifftyp m. ‖ type m. du dirigeable.

air sluice (Mine) ‖ Luftschleuse f. ‖ sas m. à air. / ~ sounding (A) ‖ Schallhöhenmessung f. ‖ mesure f. acoustique d'altitude; sondage m. aérien.

air space between bars ‖ Rostspalt m. ‖ entrefer m. de grille. / paper cored cable with ~ ‖ Papierluftraumkabel n. ‖ câble m. sous papier et couche d'air. / ~ paper insulation ‖ Papierluftraumisolierung f. ‖ isolement m. à air et au papier.

air speed ‖ Luftgeschwindigkeit f. ‖ vitesse f. aérodynamique. / ~ indicator ‖ Luftgeschwindigkeitsmesser m. ‖ indicateur m. de vitesse d'air; anémomètre m.

air and steam mixture ‖ Luft- und Dampfmischung f. ‖ mélange m. d'air et de vapeur. / ~ stove ‖ Luftwärmeofen m. ‖ calorifère f. à air chaud. / ~ strainer ‖ Luftfilter n.; Luftreinigungssieb n.; Luftsieb n. ‖ filtre m. à air. / ~ strangler ‖ Lufteinlaßklappe f. ‖ clapet m. d'entrée d'air.

air stratum of high humidity ‖ wasserreiche Luftschicht f. ‖ couche f. d'air

riche en eau *ou* humide. / isothermal ~ ||
Luftschicht f. gleicher Wärme || couche
f. d'air isotherme. / ~ of low humidity ||
wasserarme Luftschicht f. || couche f.
d'air pauvre en eau *ou* sèche.

air suction box || Saugtopf m. || pot m.
d'aspiration. / ~ device || Luftabsaug-
vorrichtung f. || dispositif m. d'aspiration
d'air. / ~ valve || Luftsaugeventil n. ||
soupape f. d'aspiration d'air.

air supply || Luftzufuhr f. || admission f.
d'air. / ~ (Mine) || Wetterführung f. ||
aérage m.; airage m. / ~ tube || Luft-
ansaugrohr n. || tube m. d'admission
d'air

air survey camera || Luftbildmeßkamera f.
|| chambre f. aéro-photogrammétrique. /
~ target || Luftziel n. || objectif m. *ou* but
m. de tir aérien. / ~ temperature || Luft-
wärme f. || température f. de l'air. / ~
thermometer || Luftthermometer n. || ther-
momètre m. à air.

air-tight || luftdicht; hermetisch || imper-
méable à l'air; fermé hermétiquement;
hermétique; étanche à air. / ~ cover ||
luftdichter Deckel m. || couvercle m.
hermétique.

air-traffic rules pl. || Luftverkehrsvor-
schriften fpl. || règlements mpl. de cir-
culation aérienne.

air tube (auto) || Luftschlauch m. || chambre
f. à air. / ~ (Mine) || Luftröhre f.; Wetter-
lutte f. || buse f. d'airage *ou*| d'aérage.

air turbidity || Lufttrübung f. || obscurcisse-
ment m. *ou* assombrissement m. de l'air.
/ ~ tuyere (Mine) || Wetterlutte f. || buse
f. d'aérage. / ~ units pl. || Luftstreit-
kräfte fpl. || forces fpl. militaires aérien-
nes.

air tyre || Luftreifen m. || pneu m.; pneu-
matique m.; bandage m. pneumatique.
/ giant ~ || Riesenluftreifen m. || pneu
m. gros *ou* ballon.

air valve (Build) (Found) || Luftkanal m.;
Windfang m.; Zugloch n. || évent m.;
éventouse f.; éventoir m. / ~ (Mach) ||
Luftventil n. || soupape f. à air.

air vane || Windflügel m. || aube f. *ou*
palette f. à vent; ailette f. aérienne. /
~ velocity || Luftgeschwindigkeit f. ||
vitesse f. de l'air.

air vent || Dunstloch n. || soupirail m.;
ventouse f.; évent m. / ~ for tank ||
Tankentlüftung f. || évent m. du réservoir.
/ ~ cock || Entlüftungshahn m. || robinet
m. d'échappement *ou* de purge d'air.

air vessel || Windkessel m. || réservoir m.
à air.

airvoid interstellar space || luftleerer Wel-
tenraum m. || espace m. de l'univers vide
d'air.

air volume displacement || Luftverdrän-
gung f. || déplacement m. d'air. / ~
vortex || Luftwirbel m. || tourbillon m.
ou remous m. d'air. / ~ war || Luftkrieg
m. || guerre f. aérienne. / ~ warning
service || Luftwarnungsdienst m. ||
service m. d'avertissement météorologique pour la navigation aérienne. / ~ wave
|| Luftwelle f. || vague f. d'air; onde f.
aérienne.

airway (Aero) || Luftweg m.; Flugweg m. ||
voie f. *ou* route f. aérienne. / ~ (Mine) ||
Wetterstrecke f. || galerie f. d'airage *ou*
d'aérage.

airway beacon || Flugstreckenfeuer n. ||
phare m. de ligne. / intermediate ~ ||
Zwischenfeuer n. || phare m. intermé-

diaire. / principal ~ || Hauptfeuer n. ||
phare m. principal.

airway lighting || Flugstreckenbefeuerung
f. || balisage m. de ligne. / ~ man ||
Wetterstreckenhauer m. || coupeur m.
de troussage. / ~ map || Luftwegkarte f. ||
carte f. de route *ou* de ligne aérienne. /
~ weather report || Streckenwettermel-
dung f. || avis m. *ou* rapport m. météoro-
logique de la ligne aérienne.

air wetting plant see ~ humidifying plant.
/ ~ whirl || Luftwirbel m. || tourbillon m.
ou remous m. d'air. / ~ whistle || Luft-
pfeife f. || sifflet m. à air.

airworthiness || Lufttüchtigkeit f. || tenue f.
en l'air.

aisle (Build) || Flügel || aile f.

à-jour fabric || à-jour-Ware f. || tissu m.
à jour.

ajowan seed || Ajowansamen m. || graine f.
d'ajowan.

alabandine || Manganblende f.; Mangan-
glanz m. || alabandine f.; manganèse m.
sulfuré.

alabandite see alabandine.

alabaster || Alabaster m. || albâtre m. /
raw ~ || roher Alabaster m. || albâtre m.
brut. / rough hewed ~ || grob behauener
Alabaster m. || albâtre m. équarri.

alabaster articles pl. || Alabasterwaren fpl.
|| objets mpl. en albâtre. / ~ glass ||
Alabasterglas n.; Milchglas n. || verre m.
alabastrin *ou* opale. / ~ goods pl. || Ala-
basterwaren fpl. || articles mpl. en al-
bâtre. / ~ lamp || Alabasterbeleuchtungs-
gegenstand m.; Alabasterlampe f. || ar-
ticle m. d'éclairage en albâtre. / ~ mill ||
Alabastermühle f. || moulin m. d'albâtre.
/ ~ quarry || Alabasterbruch m. || carrière
f. d'albâtre.

alarm || Alarm m.; Alarmsignal n.; Warn-
zeichen n. || alarme f. / ~ (Apparatus)
|| Warngerät n.; Alarmapparat m. || aver-
tisseur m.; appareil m. d'alarme. / ~
(Watchm) see also alarm clock || Wecker
m. || sonnerie f. d'alarme; reveil m. / ~
with drop indicator disc || Wecker m. mit
Fallscheibe || sonnerie f. d'alarme à volet.
/ electric ~ || elektrischer Alarmapparat
m. || avertisseur m. électrique; appareil
m. d'alarme électrique. / exhaust ~ ||
Auspuffpfeife f. || sifflet m. sur l'échappe-
ment. / maximum ~ (Tel) || Maximal-
melder m. || avertisseur m. à maximum.
/ polarized ~ || polarisierter Wecker m.
(Watchm) sonnerie f. || d'appel polarisée.
/ tramway ~ || Signalhorn n. für Straßen-
bahnen || avertisseur m. pour tramways.

alarm apparatus see also alarm || Alarm-
gerät n.; Warngerät n.; Signalapparat
m. || appareil m. d'alarme; avertisseur
m. / ~ for watchmen's control systems ||
Alarmapparat m. in Wächterkontrollan-
lagen || appareil m. d'alarme pour in-
stallations de contrôleurs de rondes.

alarm bell || Alarmglocke f. || tocsin m.;
sonnerie f. d'alarme; trembleur m. /
~ of counting machine || Zählglocke f. ||
sonnette-compteur f.

alarm and signal bell (Railw) || Läutewerks-
glocke f. || cloche f. de sonnerie d'aver-
tissement.

alarm clock || Weckeruhr f.; Wecker m. ||
réveil m.; réveille-matin m. / ~ with mu-
sic || Weckeruhr f. mit Musik || réveil m.
avec musique. / ~ with stroke || Wecker-
uhr f. mit Schlag || réveil m. avec son-
nerie.

alarm cork || Knallkorken m. || bouchon m.
détonant.

alarm device || Alarmvorrichtung f. || dis-
positif m. d'alarme / automatic ~ || selbst-
tätiger Alarmschalter m. || avertisseur
m. automatique. / ~ for coin-boxes (Tel) ||
Alarmeinrichtung f. für Kassiervorrich-
tungen || installation f. d'alarme pour
les postes à prépaiement. / ~ for fire bri-
gades (Tel) || Alarmeinrichtung f. für die
Feuerwehr || installation f. d'alarme pour
le corps de pompiers.

alarm equipment || Alarmeinrichtung f. ||
équipement m. d'alarme. / ~ fuse ||
Alarmsicherung f. || fusible m. avertis-
seur.

alarming apparatus see alarm apparatus.

alarm mechanism || Weckerwerk n. || mou-
vement m. de réveil *ou* à sonnerie. / ~
plant || Alarmanlage f. || installation f.
d'alarme.

alarm relay || Alarmschütz n.; Alarmrelais
n. || relais m. de contrôle. / ~ of filament
current || Heizstromalarmrelais n. || re-
lais m. de contrôle du courant de chauf-
fage. / ~ of grid potential || Gitteralarm-
relais n. || relais m. de contrôle du poten-
tiel de grille. / ~ for plate potential ||
Anodenalarmrelais n. || relais m. de con-
trôle du potentiel de plaque.

alarm signal || Alarmzeichen n.; Alarm-
signal n.; Warnungszeichen n. || alarme
f. / bell ~ system || Alarmanlage f. ||
sonnerie f. d'alarme. / ~ thermometer ||
Alarmthermometer n. || thermomètre m.
avertisseur. / ~ watch || Taschenuhr f. mit
Weckvorrichtung; Taschenwecker m. ||
montre-réveil f. / ~ whistle || Alarm-
pfeife f. || sifflet m. d'alarme.

albite || Natronfeldspat m.; Albit m. || al-
bite f.

albolit || Kunstelfenbein n.; Albolith m. ||
albolite f.; ivoire m. artificiel.

album || Album n. || album m. / ~ for por-
traits || Album n. für Personenaufnah-
men || album m. pour portraits. / ~ for
postcards || Postkartenalbum n. || album
m. pour cartes postales. / ~ for talking
machine records || Schallplattenalbum
n. || album m. pour disques de machines
parlantes.

albumen || Albumin n.; Eiweißstoff m.;
Eiweiß n. || albumine f. / coagulable ~ ||
gerinnbares *oder* koagulierbares Eiweiß
n. || albumine f.-coagulable.

albumen glue || Eiweißleim m. || colle f. al-
buminoïde. / ~ paper || Albuminpapier n.
|| papier m. albumineux. / ~ tannate ||
gerbsaures Albumin n. || tannate m. d'al-
bumine. / ~-turbid || eiweißtrüb || trouble
d'albumine. / ~ turbidity || Eiweißtrü-
bung f. || trouble m. d'albumine *ou* des
matières azotées.

album factory || Albumfabrik f. || fabrique
f. d'albums. / ~ fittings pl. || Album-
beschläge mpl. || ferrures fpl. *ou* garni-
tures fpl. pour albums.

albumin see albumen.

albuminate || Albuminat n.; Eiweißverbin-
dung f. || albuminate m.

albuminoid substance || Eiweißstoff m. ||
matière f. albuminoïde.

albuminous matter see albuminoid substance.

albumose || Albumose f. || albumose f.

alburnum (Wood) || Splint m. || aubier m.

alchemist || Alchimist m.; Goldmacher m. ||
alchimiste m.

alcohol || trinkbarer Sprit m.; Alkohol m. || alcool m. / absolute ~ || reiner Alkohol m. || alcool m. absolu. / allyl ~ || Allylalkohol m. || alcool m. allylique. / amylic ~ || Amylalkohol m.; Fuselöl n. || alcool m. amylique. / aromatic ~ || aromatisierter Alkohol m. || alcool m. aromatisé. / butyl ~ || Butylalkohol m. || alcool m. butylique. / crude ~ || unreiner Alkohol m. || alcool m. brut. / denatured ~ || denaturierter Spiritus m. || alcool m. dénaturé. / ethyl ~ || Äthylalkohol m. || alcool m. éthylique. / methylic ~ || Methylalkohol m. || alcool m. méthylique. / monohydric ~ || einwertiger Alkohol m. || monoalcool m. / polyhydric ~ || mehrwertiger Alkohol m. || polyalcool m. / propyl ~ || Propylalkohol m. || alcool m. propylique.

alcohol amount see alcohol contents. / ~ apparatus || Alkoholapparat m. || appareil m. à alcool.

alcoholate || Alkoholauszug m.; Alkoholat n. || alcoolate m.

alcohol burner || Spirituskocher m. || fourneau m. à alcool. / ~ carburettor || Spiritusvergaser m. || carburateur m. à alcool. / ~ contents pl. || Alkoholgehalt m. || degré m. alcoolique; taux m. d'alcool; teneur f. en alcool. / ~ derivative || Alkoholderivat n. || produit m. dérivé de l'alcool. / ~ determination || Alkoholbestimmung f. || dosage m. de l'alcool. / ~ distillery || Spiritusbrennerei f. || distillerie f. d'alcool. / ~ factor || Alkoholfaktor m. || facteur m. pour le calcul de l'alcool.

alcoholic || alkoholisch || alcoolique. / ~ ferment || Alkoholgärungspilz m. || ferment m. alcoolique. / ~ fermentation || Alkoholgärung f. || fermentation f. alcoolique. / ~ liquid || weingeisthaltige Flüssigkeit f. || liquide m. ou liqueur f. alcoolique. / ~ preparation || Alkoholpräparat n. || alcoolat m. / ~ solution || Alkohollösung f.; alkoholische Lösung f. || solution f. alcoolique; alcoolat m. / ~ strength see alcohol contents. / ~ yeast || Alkoholhefe f. || levûre f. alcoolique.

alcohol lamp || Alkohollampe f. || lampe f. à alcool ou à esprit-de-vin.

alcoholmeter || Alkoholmesser m.; Alkoholometer n.; Alkoholwage f. || alcoolomètre m.

alcoholmetry || Alkoholometrie f. || alcoolométrie f.

alcoholometer see alcoholmeter.

alcohol preparation || Alkoholpräparat n. || alcoolat m. / ~ plant || Spirituspräparateherstellungsanlage f. || installation f. à fabriquer les préparations à l'alcool.

alcohol-proof cement || alkoholfester Kitt m. || mastic m. résistant à (l'action de) l'alcool.

alcohol soldering lamp || Spirituslötlampe f. || lampe f. à alcool pour le soudage. / ~ table || Alkoholtabelle f. || table f. pour le calcul de l'alcool. / ~ thermometer || Alkoholthermometer n.; Weingeistthermometer n. || thermomètre m. à alcool.

alcoholysis || Umesterung f.; Alkoholyse f. || alcoolyse f.

alcove || Alkoven m. || alcove f.

aldehyde || Aldehyd n. || aldéhyde m. / acetic ~ || Azetaldehyd n. || aldéhyde m. acétique.

aldehydic || aldehydisch || aldéhydrique.

alder || Erle f. || aune m.; aulne m. / common ~ || Schwarzerle f.; Roterle f. || aune m. commun.

alder bark || Erlenrinde f. || écorce f. d'aune. / ~ wood || Erlenholz n.; Elsenholz n. || bois m. d'aune; aune m.

ale || Ale n. || ale m.

alecost oil || Balsamkrautöl n. || essence f. de balsamite.

alee || leewärts || sous le vent.

ale-house || Schenke f. || cabaret m.

alembic (Chem) || Helm m. und Kolben m.; Destillierkolben m. || alambic m.

alert || rührig || actif.

alertness || Rührigkeit f. || activité f.

alfa (Pap) || Spartogras n.; Esparto n.; Alfa f. || sparte m.

alfenide || Alfenid n. || alfénide m. / ~ goods pl. || Alfenidwaren fpl. || articles mpl. en alfénide.

alga || Tang m.; Seegras n.; Alge f. || algue f.; fucus. / medicinal ~ || Alge f. zum Heilgebrauch || algue f. médicinale.

algaroba resin || Algarobaharz n. || résine f. algarrabo.

algebra || Algebra f.; Buchstabenrechnung f. || algèbre f.

algebraical || algebraisch || algébrique. / ~ sign || algebraisches Zeichen n. || signe m. algébrique.

algol colour || Algolfarbstoff m. || colorant m. algol.

alidade (Surv) || Diopterlineal n. || alidade f. / ~ transporter || Dioptertransportör m. || transporteur m. à alidade.

alien || ausländisch || étranger.

alienation || Enteignung f. || expropriation f. / partial ~ || Teilenteignung f. || expropriation f. partielle.

alight, to || herabfallen; sich niederlassen || tomber; descendre. / ~ on earth (Aero) || landen; auf dem Boden landen || atterrir. / ~ on sea (Aero) || auf dem Meere n. landen oder niedergehen || amerrir.

alighting (Aero) || Landen n.; Landung f. || atterrissage m.; descente f. à terre. / ~ on earth || Bodenlandung f. || atterrissage m. sur le sol. / ~ on sea || Anwassern n.; Wasserlandung f.; Wasserung f. || amérissage m.

align, to || ausrichten || aligner; redresser.

alignment || Einfluchtung f.; Flucht f. || alignement m. / to check ~ || die Ausrichtung f. prüfen || repérer l'alignement. / out of ~ || schlecht ausgerichtet || désaligné. / regular ~ of the letters || Zeilengeradheit f. || rectitude f. des lignes. / ~ of telegraph-poles || Richtung f. der Telegrafenstangen || alignement m. des poteaux télégraphiques.

aliment || Nahrungsmittel n.; Speise f. || produit m. alimentaire.

alimentary industry || Nahrungsmittelindustrie f. || industrie f. alimentaire. / ~ paste mould || Form f. für Teigwaren || moule m. à pâtes alimentaires.

aline, to see to align.

aliphatic series || Fettreihe f. || série f. des corps gras.

alisonite (Miner) || Kupferbleiglanz m. || alisonite f.

alive || lebhaft; lebend; lebendig || vif. / ~ (Electr) || stromführend || sous courant.

alizarin || Alizarin n. || alizarine f. / ~ colouring || Alizarinfarbstoff m. || colorant m. d'alizarine. / ~ dye see ~ colouring. / ~ ink || Alizarintinte f. || encre f. d'alizarine.

/ ~ lake || Alizarinlack m. || laque f. d'alizarine.

alizarine see alizarin.

alkalescence || Alkaleszenz f.; Neigung f. zum Alkalischwerden || alcalescence f.

alkalescency see alkalescence.

alkalescent || alkalisierend || alcalescent.

alkali || Alkali n. || alcali m. / volatile ~ || Salmiakgeist m. || alcali m. volatil.

alkali-chloride electrolysis || Alkalichloridelektrolyse f. || électrolyse f. de chlorures alcalins.

alkali metal || Alkalimetall n. || métal m. alcalin.

alkaline || alkalisch; laugenartig || alcalin. / to make ~ || alkalisieren; alkalisch machen || alcaliser; alcaliniser.

alkaline bath || alkalisches Bad || bain m. alcalin. / ~ cell (Elektr) || Alkalizelle f. || pile f. alcaline. / ~ metal see alkali metal. / ~ salt || Alkalisalz n.; Abraumsalz n. || sel m. alcalin ou de déblai.

alkaline solution || alkalische Lösung f. || solution f. alcaline. / ~ caustic ~ (Chem) || ätzalkalische Lösung f. || solution f. alcaline caustique.

alkalinity || Alkaleszenz f. || alcalinité f.

alkalinization || Alkalischmachen n. || alcalinisation f.

alkali-proof || alkalibeständig || inattaquable aux alcalis.

alkalize, to || alkalisieren || alcaliser.

alkaloid || Alkaloid n. || alcaloïde m.

alkalous see alkaline.

alkanet root see alkanna root.

alkanna root || Alkannawurzel f. || racine f. d'orcanette.

alkekengi berry || Judenkirschenbeere f. || baie f. d'alkékenge.

allanite (Miner) || Allanit m. || allanite f.

allemontite (Miner) || Arsen(ik)antimon n.; Arsenikspießglanz m. || allemontite f.

alley (Avenue of trees) || Allee f.; Baumreihe f. || allée f.; avenue f. / ~ (Small street) || Durchgang m.; kleine Gasse f. || passage m. / blind ~ || Sackgasse f. || impasse f.; cul-de-sac m.

alliance || Bündnis n.; Verbindung f. || union f.; alliance f.

alligation || Legierung f. || alliage m.

allmetal aeroplane || Ganzmetallflugzeug n. || avion m. entièrement métallique. / ~ construction || Ganzmetallbau m. || construction f. entièrement métallique.

allocate, to (Tel) || vormerken || prendre rang.

allocation of frequencies || Wellenverteilung f. || distribution f. des fréquences.

allotment || Auslosung f. || tirage m. au sort. / ~ (Distribution) || Zuteilung f. || partage m.; répartition f. / ~ (Mine) || Grubenfeld n. || concession f. de mine; lot m.; terrain m. alloué. / ~ for iron ore || Eisensteinfeld n. || concession f. pour minerai de fer. / ~ of ore || Erzfeld n. || champ m. d'exploitation de minerai; gîte m. métallifère. / underground worked ~ || Tiefbaubetrieb m. || exploitation f. souterraine; concession f. avec excavation souterraine.

allotment letter || Zuteilungsanzeige f. || lettre f. de répartition.

allotropic || allotropisch || allotropique.

allotropy || Allotropie f. || allotropie f.

allotted || zugeteilt || réparti.

allotter (Aut tel) || Wählersucher m. || distributeur m. des appels.

allow, to ‖ bewilligen ‖ accorder; concéder. / ~ (Trade) ‖ anrechnen; in Abzug bringen; vergüten ‖ mettre en compte; rembourser. / ~ rebate ‖ Rabatt m. bewilligen ‖ accorder un rabais. / ~ to set (Cement) ‖ abbinden ‖ faire la prise.

allowance ‖ Ration f. ‖ ration f. / ~ (Concession) ‖ Bewilligung f. ‖ concession f.; vote m. / ~ (Deficiency in weight of coin) ‖ Toleranz f. ‖ remède m.; tolérance f. / ~ (Discount) ‖ Nachlaß m.; Rabatt m.; Preisermäßigung f. ‖ remise f.; rabais m.; diminution f. / ~ (Free play) ‖ Spielraum m. ‖ jeu m.; liberté f.; chasse f.; tolérance f.; supplément m. pour l'ajustage. / ~ (Coin) ‖ Remedium n. ‖ tolérance f.; remède m. / ~ for loss in weight ‖ Frachtrückvergütung f.; Refaktie f. ‖ réfaction f. / ~ for postage ‖ Portovergütung f. ‖ indemnité f. pour frais de correspondance. / short ~ ‖ kleine Ration f. ‖ ration f. diminuée ou réduite. / ~ for special duty ‖ Dienstzulage f. ‖ indemnité f. de service. / ~ for tare ‖ Abzug m. für Tara ‖ bonification f. sur la tare. / ~ for waste (Mar) ‖ Überration f.; Zuschlagration f. ‖ dixième m. / ~ in weight (Coin) ‖ Remedium n. am Schrot ‖ remède m. ou tolérance f. de poids.

alloy, to ‖ legieren ‖ allier. / ~ gold with an other metal ‖ das Gold karatieren ‖ faire l'alliage de l'or avec un autre métal.

alloy (Chem) ‖ Legierung f. ‖ alliage m. / ~ (Coin; Met) ‖ Feingehalt m.; Korn n.; Mischungsverhältnis n. ‖ titre m. / ~ that does not deteriorate with age ‖ alterungsfreie Legierung f. ‖ alliage m. n'accusant pas de vieillissement. / ~ and standard (Coin) ‖ Schrot n. und Korn n. der Münzen ‖ alliage m. et titre m. / antimony ~ ‖ antimonhaltige Legierung f. ‖ alliage m. antimonié. / fusible ~ ‖ schmelzbare Legierung f. ‖ alliage m. fusible. / ~ of gold with copper ‖ rote Karatierung f. ‖ alliage m. de l'or avec le cuivre. / ~ of gold with copper and silver ‖ gemischte Karatierung f. ‖ alliage m. de l'or avec le cuivre et l'argent. / ~ of gold with silver ‖ weiße Karatierung f. ‖ alliage m. de l'or avec l'argent. / heat-conducting ~ ‖ wärmeleitende Legierung f. ‖ alliage m. transcalorique. / high-grade ~ see alloy of high percentage. / ~ of high percentage ‖ hochprozentige Legierung f. ‖ alliage m. à haute teneur ou à haut pourcentage. / ~ of lead ‖ Bleilegierung f. ‖ alliage m. au plomb. / lead bismuth ~ ‖ Bleiwismutlegierung f. ‖ alliage m. plombbismuth. / lead tin ~ ‖ Bleizinnlegierung f. ‖ alliage m. plomb-étain. / ~ of low percentage ‖ niedrigprozentige Legierung f. ‖ alliage m. à faible teneur. / ~ of metals ‖ Metallegierung f.; Komposition f. ‖ alliage m. de métaux; composition f. / no ~ ‖ ohne Zusatz m. ‖ sans alliage m. / ~ of palladium ‖ Palladiumlegierung f. ‖ palladure f. / steel ~ ‖ Stahllegierung f. ‖ alliage m. d'acier.

alloyed ‖ legiert ‖ allié. / ~ superrefined steel ‖ legierter Edelstahl m. ‖ acier m. spécial allié.

alloying of gold with other metals ‖ Karatierung f. des Goldes ‖ alliage m. de l'or avec un autre métal.

alloying element ‖ Legierungskörper m. ‖ constituant m. d'un alliage.

alloy steel ‖ Legierungsstahl m.; legierter Stahl m. ‖ acier m. spécial; alliage m. d'acier.

all-round box tip wagon ‖ Rundkipper m.; Kastenrundkipper m. ‖ wagonnet m. à caisse basculante dans toutes les directions. / ~ machine ‖ universell brauchbare Maschine f. ‖ machine f. universelle.

alluvial ‖ alluvial; angespült ‖ alluvionnaire; d'alluvion f. / ~ basin ‖ Schwemmmulde f. ‖ creux m. rempli de matières alluviales. / ~ cone ‖ Schuttkegel m.; Schutthalde f. ‖ cône m. de déjection. / ~ deposits pl. ‖ angeschwemmter Boden m.; Anschwemmung f. ‖ terre f. alluviale; alluvions fpl. / ~ gold ‖ Seifengold n.; Alluvialgold n. ‖ or m. retiré des alluvions. / ~ ore deposit ‖ Erzseife f. ‖ depôt m. metallifère alluvionnaire. / ~ sand ‖ Flußsand m.; Schwemmsand m. ‖ sable m. de ravine ou de rivière. / ~ soil see alluvium.

alluvion see alluvium.

alluvium (Geol) ‖ Alluvium n. ‖ alluvion f. / ~ (Alluvial soil) ‖ angeschwemmtes Land n.; angespülter Boden m.; Anschwemmung f.; Angeschwemmtes n.; alluviale Ablagerung f.; Alluvium n.; Anspülung f.; Versandung f. ‖ atterrissement m.; dépot m.; alluvion f.

all-weather body (Auto) ‖ Allwetterkarosserie f. ‖ voiture f. décapotable.

allyl ‖ Allyl n. ‖ allyle m. / ~ alcohol ‖ Allylalkohol m. ‖ alcool m. allylique.

allylic alcohol see allyl alcohol. / ~ mustard oil ‖ Allylsenföl n. ‖ huile f. allylique de moutarde.

almanac ‖ Kalender m. ‖ almanach m. / astronomical ~ see nautical ~. / nautical ~ ‖ Schiffskalender m.; astronomischer Kalender m. ‖ connaissance f. des temps; almanach m. nautique.

almandine ‖ Karfunkel m.; roter Granat m. ‖ escarboucle m.; grenat m. pyrope.

alman forge ‖ Luppenschmiede f. ‖ allemanderie f. / ~ furnace ‖ Gekrätzofen m. ‖ fourneau m. à fondre l'arcot.

almond ‖ Mandel f. ‖ amande f. / bitter ~ ‖ bittere Mandel f. ‖ amande f. amère. / burnt ~s pl. ‖ gebrannte Mandeln fpl. ‖ amandes fpl. pralinées. / hard-shelled ~ ‖ hartschalige Mandel f. ‖ amande f. en coque dure. / shelled ~ ‖ abgeschälte Mandel f. ‖ amande f. pelée. / soft-shelled ~ ‖ Krachmandel f. ‖ amande f. en coque tendre.

almond blancher ‖ Mandelbleicher m. ‖ blanchisseur m. d'amandes. / ~ furnace see alman furnace. / ~ grinding machine ‖ Mandelreibmaschine f. ‖ machine f. à râper les amandes. / ~ milk ‖ Mandelmilch f. ‖ lait m. d'amande. / ~ oil ‖ Mandelöl n. ‖ huile f. d'amandes. / ~ paste ‖ Mandelteig m. ‖ pâte f. d'amande. / ~ peeling machine ‖ Mandelschälmaschine f. ‖ machine f. à peler les amandes; râpeuse f. mécanique pour amandes. / ~ soap ‖ Mandelseife f. ‖ savon m. de pâte d'amandes; amandine f.

aloe ‖ Aloe f. ‖ aloès m. / ~ extract ‖ Aloeauszug m.; Aloeextrakt m. ‖ extrait m. d'aloès. / ~ fiber ‖ Aloefaser f. ‖ fibre f. d'aloès. / ~ hemp ‖ Aloehanf m. ‖ chanvre m. d'aloès. / ~ wood ‖ Paradiesholz n. ‖ bois m. d'aloès.

aloft ‖ oben ‖ en haut. / ~! ‖ hoch! ‖ haut!; en haut!

along ‖ längs; entlang ‖ le long. / ~ the rolling direction ‖ längs der Walzrichtung f. ‖ dans le sens du laminage. / ~ shore ‖ längs der Küste f. ‖ le long de la côte.

alongside ‖ nebenstehend ‖ en marge f. / ~ (Mar) ‖ längsseits ‖ bord m. à bord; le long du bord.

alow ‖ unten ‖ bas; dessous.

alpaca (Met) ‖ Alpaka n.; Alpakasilber n. ‖ alpaca m. / ~ (Bot) ‖ Alpako n.; Pako n. ‖ alpaga m.; alpaca m. / ~ (Spinn) ‖ Alpaka m. oder n.; Alpakastoff m. ‖ alpaga m.; alpaca m. / ~ hair ‖ Alpakahaar n. ‖ poil m. d'alpaga. / ~ silver goods pl. ‖ Alpakasilberwaren fpl. ‖ articles mpl. en alpaca argenté. / ~ spinner ‖ Alpakaspinner m. ‖ fileur m. d'alpaga. / ~ weaver ‖ Alpakaweber m. ‖ tisseur m. d'alpaga. / ~ wool ‖ Alpakawolle f.; Alpakowolle f. ‖ laine f. d'alpaga.

alphabet for the blind ‖ Blindenschrift f. ‖ alphabet m. des aveugles. / ~ Morse ‖ Morsealphabet n. ‖ alphabet m. Morse. / Russian ~ ‖ russisches Alphabet n. ‖ alphabet m. russe.

alphabetical order ‖ abeceliche oder alphabetische Reihenfolge f. ‖ ordre m. alphabétique. / in strict ~ ‖ in streng abecclicher Reihenfolge ‖ dans l'ordre strictement alphabétique.

alphabetize, to ‖ abecelich ordnen; alphabetisieren ‖ ranger par ordre alphabétique.

alphabet perforating machine for sweat band ‖ Alphabetperforiermaschine f. für Hutleder ‖ machine f. pour perforer les initiales dans le cuir de chapeau.

alpine railway ‖ Gebirgsbahn f. ‖ chemin m. de fer alpin.

alquifou ‖ Glaserzerz n.; Bleiglanz m. ‖ alquifoux m.

altar ‖ Altar m. ‖ autel m. / ~ ornaments pl. ‖ Altarschmuck m. ‖ parement m. d'autel.

altazimuth comet finder ‖ azimutaler Kometensucher m. ‖ chercheur m. de comètes azimutales. / ~ telescope ‖ azimutales Fernrohr n. ‖ lunette f. azimutale. / telescope in ~ mount ‖ azimutal montiertes Fernrohr n. ‖ lunette f. à monture azimutale.

alter, to ‖ abändern; verändern ‖ changer; modifier; corriger.

alteration (Build) ‖ Umänderung f.; Renovierung f.; Erneuerung f. ‖ restauration f.; renovation f. / a complete ~, modernization and enlargement of the plant was begun ‖ ein vollständiger Umbau m. der Anlage wurde begonnen ‖ il fut procédé à une reconstruction et à la modernisation de l'usine.

alternate, to ‖ abwechseln ‖ alterner. / ~ the spokes ‖ die Speichen fpl. versetzt anordnen ‖ alterner les rais mpl.

alternate-working kiln ‖ Ofen m. mit unterbrochenem Betriebe ‖ four m. discontinu.

alternating bevel teeth pl. (Saw) ‖ wechselseitiger Schrägschliff m. ‖ affûtage m. à biseaux.

alternating current ‖ Wechselstrom m. ‖ courant m. alternatif. / forced ~ ‖ erzwungener Wechselstrom m. ‖ courant m. alternatif contraint. / free ~ ‖ freier Wechselstrom m. ‖ courant m. alternatif libre. / three-phase ~ ‖ Drehstrom m. ‖ courant m. triphasé.

alternating current ammeter ‖ Strommesser m. für Wechselstrom ‖ ampèremeter m. pour courant alternatif. / ~ arc lamp ‖ Wechselstrombogenlampe f. ‖ lampe f. à arc à courant alternatif. / ~ bell ‖ Wechselstromwecker m. ‖ sonnerie f. à courant alternatif. / ~ bridge ‖ Wechselstrommeßbrücke f. ‖ pont m. de mesure à courant alternatif. / ~ calling relay ‖ Rufwechselstromrelais n. ‖ relais m. récepteur d'appel.

alternating - current - continuous - current converter ‖ Wechselstromgleichstromeinankerumformer m. ‖ convertisseur m. ou commutatrice f. à courant alternatif en courant continu. / ~ dynamotor ‖ Wechselstromgleichstromumformer m. ‖ dynamoteur m. à courant alternatif en courant continu.

alternating current field ‖ Wechselstromfeld n. ‖ champ m. alternatif. / ~ generator ‖ Wechselstromgenerator m.; Wechselstromdynamomaschine f. ‖ alternateur m.; dynamo f. ou génératrice f. à courant alternatif. / ~ heating ‖ Wechselstromheizung f. der Elektronenröhren ‖ chauffage m. par courant alternatif. / ~ machine ‖ Wechselstrommaschine f. ‖ machine f. à courant alternatif. / ~ measuring set ‖ Wechselstrommeßgerät n. ‖ instrument m. de mesure en courant alternatif. / ~ motor ‖ Wechselstrommotor m. ‖ moteur m. à courant alternatif; alternomoteur m. / ~ relay ‖ Wechselstromrelais n. ‖ relais m. à courant alternatif. / ~ shunt motor ‖ Wechselstromnebenschlußmotor m. ‖ moteur m. à courant alternatif ou alternomoteur m. en dérivation. / ~ starter ‖ Wechselstromanlasser m. ‖ démarreur m. à courant alternatif. / ~ traction ‖ Wechselstromfahrbetrieb m. ‖ traction f. à courant alternatif. / ~ transformer ‖ Wechselstromtransformator m. ‖ transformateur m. à courant alternatif. / ~ voltmeter ‖ Spannungsmesser m. für Wechselstrom ‖ voltmètre m. pour courant alternatif.

alternating perforation ‖ Zickzacklochung f. ‖ perforation f. en quinconce. / ~ voltage (Electr) ‖ Wechselspannung f. ‖ tension f. alternative.

alternation (Electr) ‖ Polwechsel m.; Stromwechsel m. ‖ alternance f.

alternator ‖ Wechselstromerzeuger m.; Wechselstromdynamomaschine f. ‖ alternateur m. / direct coupled ~ ‖ unmittelbar gekuppelter Generator m. ‖ alternateur m. directement accouplé. / disc type ~ ‖ Generator m. mit Scheibenanker ‖ alternateur m. à rotor en disque. / drum type ~ ‖ Generator m. mit Trommelanker ‖ alternateur m. à rotor en tambour. / external pole ~ ‖ Außenpolgenerator m. ‖ alternateur m. à pôles extérieurs. / internal pole ~ ‖ Innenpolgenerator m. ‖ alternateur m. à pôles intérieurs. / over-excited ~ ‖ übererregter Generator m. ‖ alternateur m. surexcité. / polyphase ~ ‖ Mehrphasenwechselstromgenerator m. ‖ alternateur m. polyphasé. / rope-driven ~ ‖ Generator m. für Seilantrieb ‖ alternateur m. à commande par câbles. / separately excited ~ ‖ Generator m. mit Fremderregung ‖ alternateur m. à excitation séparée. / single-phase ~ ‖ Einphasengenerator m. ‖ alternateur m. monophasé. / three-phase ~ ‖ Drehstromgenerator m. ‖ alternateur m.

triphasé. / turbine ~ ‖ Turbogenerator m ‖ turboalternateur m. / two-phase ~ ‖ Zweiphasenwechselstromgenerator m. ‖ alternateur m. biphasé.

alterno-motor, monophase ‖ Einphasenwechselstrommotor m. ‖ moteur m. à courant alternatif monophasé. / three-phase ~ ‖ Dreiphasenwechselstrommotor m. ‖ moteur m. à courant alternatif triphasé.

altimeter ‖ Höhenmesser m. ‖ altimètre m. / recording ~ ‖ Höhenschreiber m. ‖ altimètre m. enregistreur.

altimetry ‖ Höhenmessung f.; Höhenmeßkunde f. ‖ hypsométrie f.

altitude (Astron) ‖ Höhe f. (eines Sternes) ‖ hauteur f.; élévation f. / ~ (Geogr) ‖ Höhe f. ‖ altitude f.; élévation f. / acoustic method of measuring ~ ‖ Schallhöhenmessung f. ‖ mesure f. acoustique d'altitude; sondage m. aérien. / cruising ~ (Aviation) ‖ Fahrhöhe f. ‖ hauteur f. de voyage. / ~ of a place ‖ Höhenlage f. eines Ortes ‖ altitude f. ou hauteur f. d'un lieu. / ~ of the pole (Astron) ‖ Polhöhe f. ‖ élévation f. ou hauteur f. du pôle; latitude f. / ~ above sea-level ‖ Höhe f. über dem Meeresspiegel; Seehöhe f. ‖ altitude f.; hauteur f. au-dessus du niveau de la mer. / ~ of the sun ‖ Sonnenstand m. ‖ position f. du soleil.

altitude control (Airship) ‖ Höhensteuerung f. ‖ dispositif m. de dosage réglable des gaz. / ~ correction (Astron) ‖ Polhöhenkorrektion f. ‖ rectification f. de la latitude. / ~ flight ‖ Höhenflug m. ‖ vol m. d'altitude.

alum, to (Dyer) ‖ mit Alaunwasser n. beizen ‖ aluner.

alum ‖ Alaun m. ‖ alun m. / magnesia ~ ‖ Talkerdealaun m. ‖ alun m. de magnésie. / pure ~ ‖ gediegener Alaun m. ‖ alun m. vierge. / talc ~ ‖ Magnesiaalaun m. ‖ alun m. de magnésie.

alum earth ‖ Alaunerde f.; Tonerde f. ‖ alumine f.; oxyde m. d'aluminium; terre f. alumineuse. / ~ quarry ‖ Alaunerdegrube f. ‖ carrière f. de terre alumineuse.

alumen see alum.

alumina ‖ Tonerde f. ‖ alumine f. / ~ acetate ‖ essigsaure Tonerde f. ‖ acétate m. d'alumine. / ~ chlorate ‖ chlorsaure Tonerde f. ‖ chlorate m. d'alumine. / ~ formiate ‖ ameisensaure Tonerde f. ‖ formiate m. d'alumine. / ~ hydrate ‖ Tonerdehydrat n. ‖ alumine f. hydratée. / ~ nitrate ‖ salpetersaure Tonerde f. ‖ nitrate m. d'alumine. / ~ oleate ‖ ölsaure Tonerde f. ‖ oléate m. d'alumine. / ~ oxide of ~ ‖ Aluminoxyd n. ‖ oxyde m. d'alumine. / ~ preparation plant ‖ Tonerdeaufbereitungsanlage f. ‖ installation f. de préparation d'alumine. / ~ recovery of ~ ‖ Gewinnung f. reiner Tonerde ‖ production f. d'alumine. / ~ resinate ‖ harzsaure Tonerde f. ‖ résinate m. d'alumine. / ~ sulphate ‖ schwefelsaure Tonerde f. ‖ sulfate m. d'alumine.

aluminate of soda ‖ Natriumaluminat n. ‖ aluminate m. de soude.

aluminium ‖ Aluminium n. ‖ aluminium m. / equipment for producing ~ ‖ Aluminiumgewinnungseinrichtung f. ‖ installation f. de production de l'aluminium. / secondary ~ ‖ Aluminium n. zweiter Schmelzung ‖ aluminium m. de deuxième fusion.

aluminium acetate ‖ Aluminiumazetat n. ‖ acétate m. d'alumine. / ~ alloy ‖ Aluminiumlegierung f. ‖ alliage m. d'aluminium. / ~ apparatus ‖ Aluminiumapparat m. ‖ appareil m. d'aluminium. / ~ article for tourists ‖ Aluminiumreisegerät n. ‖ article m. de tourisme en aluminium. / ~ box ‖ Aluminiumdose f. ‖ boîte f. en aluminium. / ~ bronze ‖ Aluminiumbronze f. ‖ bronze m. d'aluminium. / ~ cable ‖ Aluminiumkabel n. ‖ câble m. en aluminium. / ~ capsule ‖ Aluminiumkapsel f. ‖ capsule f. en aluminium. / ~ casting ‖ Aluminiumguß m. ‖ fonte f. d'aluminium. / ~ cell ‖ Aluminiumzelle f. ‖ élément m. en aluminium. / ~ cell rectifier ‖ Aluminiumzellengleichrichter m. ‖ redresseur m. avec anode en aluminium. / ~ chloride ‖ Chloraluminium n.; Aluminiumchlorid n. ‖ chlorure m. d'aluminium. / ~ clip ‖ Aluminiumklemme f. ‖ pince f. en aluminium. / ~ colour ‖ Aluminiumfarbe f. ‖ couleur f. d'aluminium. / ~ cooking sets pl. ‖ Aluminiumkochgeschirr n. ‖ batterie f. de cuisine ou bouillottes fpl. de cuisine en aluminium. / ~ cover ‖ Aluminiumbekleidung f. ‖ revêtement m. en aluminium. / ~ fluoride ‖ Fluoraluminium n. ‖ fluorure m. d'aluminium. / ~ foil ‖ Blattaluminium n.; Aluminiumfolie f. ‖ aluminium m. en feuilles. / ~ frame ‖ Aluminiumgestell n. ‖ bâti m. en aluminium. / ~ knife, fork and spoon ‖ Aluminiumbesteck n. ‖ couvert m. en aluminium. / ~ mounting (Spectacles) ‖ Aluminiumfassung f. ‖ monture f. en aluminium. / ~ oxide ‖ Aluminiumoxyd n.; Tonerde f. ‖ alumine f.; terre f. d'alumine; oxyde m. d'aluminium. / ~ plate (Household) ‖ Aluminiumgeschirr n. ‖ batterie f. de cuisine ou vaisselle f. en aluminium. / ~ plate (Met) ‖ Aluminiumblech n. ‖ planche f. ou plaque f. ou feuille f. en aluminium. / ~ polish ‖ Aluminiumputzmittel n. ‖ produit m. à nettoyer l'aluminium. / ~ powder ‖ Aluminiumpulver n. ‖ poudre f. d'aluminium. / ~ printing ‖ Aluminiumdruck m. ‖ impression f. sur aluminium; algraphie f. / ~ rolling mill ‖ Aluminiumwalzwerk n. ‖ laminoir m. à aluminium. / ~ sheet ‖ Aluminiumblech n. ‖ feuille f. d'aluminium. / ~ soldering ‖ Aluminiumlötung f. ‖ soudure f. d'aluminium. / ~ sport's article ‖ Aluminiumsportartikel m. ‖ article m. de sport en aluminium. / ~ stopper ‖ Aluminiumstöpsel m. ‖ bouchon m. en aluminium. / ~ tray ‖ Aluminiumschale f. ‖ cuvette f. en aluminium. / ~ ware ‖ Aluminiumwaren fpl. ‖ objets mpl. en aluminium. / ~ welding ‖ Aluminiumschweißung f. ‖ soudage m. d'aluminium. / ~ wire ‖ Aluminiumdraht m. ‖ fil m. d'aluminium. / ~ works pl. ‖ Aluminiumhütte f. ‖ fonderie f. d'aluminium.

aluminothermic ‖ aluminothermisch ‖ aluminothermique.

aluminothermics pl. ‖ Aluminothermie f. ‖ aluminothermie f.

aluminous ‖ alaunhaltig ‖ alumineux.

aluminum see aluminium.

alum mine ‖ Alaunbergwerk n. ‖ alunière f.; carrière f. ou mine f. d'alun. / ~ ore ‖ Alaunerz n. ‖ minerai m. d'alun. / ~ pit see alum mine. / ~ slate ‖ Alaunschiefer m. ‖ schiste m. alumineux ou alunifère; ampélite f. alumineuse. / ~ slate over-

burnt in torrefying ‖ überröstetes Alaunschiefererz n. ‖ schiste m. alumineux torréfié. / ~ steep (Curr) ‖ Alaunbrühe f. ‖ étoffe f. d'alun et de sel. / ~ stone ‖ Alaunstein m. ‖ pierre f. alumineuse. / ~ works pl. ‖ Alaunhütte f.; Alaunsiederei f. ‖ alunière f.; alunerie f.

alunite (Miner) ‖ Alaunstein m. ‖ alunite f.

amadou ‖ Feuerschwamm m.; Zündschwamm m. ‖ amadou m.

amalgam, to see to amalgamate.

amalgam ‖ Amalgam n. ‖ amalgame m. / cupriferous ~ ‖ kupferhaltiges Amalgam n. ‖ amalgame m. cuivrique. / ~ of gold ‖ Goldamalgam n. ‖ amalgame m. aurique ou d'or. / ~ of palladium ‖ Palladiumamalgam n. ‖ palladure f. de mercure.

amalgamate, to (Chem) ‖ amalgamieren; verquicken ‖ amalgamer. / ~ (Pol Econ) ‖ zusammenlegen ‖ fusionner.

amalgamated mine ‖ konsolidiertes Bergwerk n. ‖ mine f. fusionnée ou consolidée.

amalgamating cask ‖ Quickfaß n. ‖ tonneau m. d'amalgamation. / ~ device for gold ores ‖ Amalgamiereinrichtung f. für Golderze ‖ installation f. pour amalgamer les minerais d'or. / ~ mill ‖ Quickmühle f. ‖ moulin m. à amalgamer. / ~ pan ‖ Amalgamierpfanne f. ‖ cuve f. d'amalgamation. / ~ skin ‖ Quickbeutel m.; Quicksack m. ‖ chamois m. / ~ table ‖ Amalgamiertisch m. ‖ table f. d'amalgamation.

amalgamation ‖ Amalgamierung f. ‖ amalgamation f. / ~ of enterprises in the form of a joint-stock company ‖ Zusammenfassung f. von Unternehmungen in Form einer Aktiengesellschaft ‖ groupement m. des installations par la fondation d'une société anonyme. / ~ of interests ‖ Interessengemeinschaft f. ‖ convention f.; syndicat m.; communauté f. d'intérêts. / ~ of mines ‖ Konsolidierung f. von Bergwerken ‖ réunion f. ou fusion f. ou consolidation f. de mines. / rational ~ of the existing works ‖ rationelle Zusammenfassung f. der bestehenden Anlagen ‖ fusion f. rationnelle d'installations déjà existantes. / ~ of similar enterprises ‖ Zusammenlegen n. gleichartiger Betriebe ‖ fusion f. d'exploitations de même nature.

amalgamation apparatus ‖ Amalgamier(ungs)vorrichtung f. ‖ appareil m. d'amalgamation.

amalgamator ‖ Amalgamierer m. ‖ amalgamateur m.

amalgam catcher ‖ Amalgamfänger m. ‖ collecteur m. de mercure. / ~ cleaner ‖ Reinigungspfanne f. ‖ cuve f. de purification. / ~ distilling furnace ‖ Amalgamdestillationsofen m. ‖ four m. pour la distillation des amalgames. / ~ drum ‖ Amalgamtrommel f. ‖ tambour m. à amalgame. / ~ filter ‖ Amalgamfilter n. ‖ filtre m. pour amalgame. / ~ gilding ‖ Quecksilbervergoldung f.; Feuervergoldung f. ‖ dorure f. au mercure ou au feu.

amalgamize, to see to amalgamate.

amalgam press ‖ Amalgampresse f. ‖ presse f. pour amalgame. / ~ solution ‖ Amalgambad n. ‖ bain m. d'amalgamation.

amateur fisherman ‖ Gelegenheitsfischer m. ‖ pêcheur m. amateur. / ~ photographer ‖ Amatörfotograf m. ‖ photographe m. amateur.

amaurosis ‖ schwarzer Star m. ‖ cataracte f. noire.

amber ‖ Bernstein m. ‖ ambre m. jaune; succin m. / artificial ~ ‖ Kunstbernstein m. ‖ ambre m. jaune artificiel. / regenerated ~ ‖ regenerierter Bernstein m. ‖ ambre m. recomposé. / yellow ~ ‖ gelber Amber m. ‖ ambre m. jaune.

amber fisherman ‖ Bernsteinfischer m. ‖ pêcheur m. d'ambre. / ~ goods pl. ‖ Bernsteinwaren fpl. ‖ objets mpl. en ambre.

ambergris ‖ graue Ambra f.; Ambra f. ‖ ambre m. gris.

amber jewellery ‖ Bernsteinschmuck m. ‖ parure f. en ambre jaune. / ~ mine ‖ Bernsteinfundort m. ‖ extraction f. ou mine f. d'ambre. / ~ miner ‖ Bernsteingräber m. ‖ extracteur m. d'ambre. / ~ oil ‖ Bernsteinöl n.; Ambraöl n. ‖ huile f. de succin ou d'ambre. / ~ turner ‖ Bernsteindrechsler m. ‖ tourneur m. sur ambre. / ~ varnish ‖ Bernsteinlack m.; Bernsteinfirnis m. ‖ vernis m. au succin; laque f. d'ambre. / ~ ware ‖ Bernsteinwaren fpl. ‖ articles mpl. en ambre.

ambient air ‖ umgebende Luft f. ‖ air m. ambiant. / ~ temperature ‖ Umgebungstemperatur f. ‖ température f. ambiante.

amblyopic ‖ schwachsichtig ‖ amblyope.

ambreada ‖ falsche Bernsteinkoralle f. ‖ ambréade f.

ambroid ‖ Ambroid n. ‖ ambroïde m.

ambroin ‖ Ambroin n. ‖ ambroïne f.

ambulance see ambulance car. / ~ attendant ‖ Heilgehilfe m. ‖ aide-chirurgien m. / ~ box ‖ Verbandkasten m. ‖ boîtier m.; coffret m. à pansement.

ambulance car ‖ Krankenwagen m. ‖ voiture f. ou caisson m. d'ambulance; ambulance f. / ~ for persons affected with infectious diseases ‖ Krankenwagen m. für ansteckend Erkrankte ‖ voiture f. d'ambulance pour les personnes atteintes de maladies contagieuses. / ~ motor ~ ‖ Kraftwagen m. für Krankenbeförderung; Krankentransportauto n.; Krankenautomobil n. ‖ ambulance f. automobile; automobile m. d'ambulance. / ~ for slightly and seriously injured ‖ Wagen m. für Leicht- und Schwerverletzte ‖ voiture f. d'ambulance pour des personnes légèrement et grièvement blessées.

ambulance carriage see ambulance car. / ~ cart see ambulance car. / ~ service ‖ Krankenbeförderungswesen n. ‖ transport m. des malades. / ~ wagon see ambulance car.

amelioration ‖ Verbesserung f. ‖ amélioration f.; correction f.; réforme f. / ~ (Trade) ‖ Preissteigerung f. ‖ haussement m. de prix.

amends pl. ‖ Entschädigung f. ‖ dédommagement m.; indemnité f.; rémunération f. / to make ~ ‖ entschädigen ‖ dédommager; indemniser.

American chair ‖ Klappstuhl m. ‖ siège m. pliant. / ~ heel ‖ Keilferse f. ‖ talon m. américain.

amethyst ‖ Amethyst m. ‖ améthyste m.

ametropic ‖ fehlsichtig ‖ amétrope.

amianthus ‖ Amiant m.; Asbest m. ‖ amiante m. / ~ in fibres ‖ Amiant m. in Fasern ‖ amiante m. en fibres.

amide (Chem) ‖ Amid n. ‖ amide m. / determination of the ~s ‖ Amidenbestimmung f. ‖ dosage m. des amides.

amidobenzoic acid ‖ Amidobenzoesäure f. ‖ acide m. amidobenzoïque.

amidonaphthol sulphonic acid ‖ Amidonaphtholsulfosäure f. ‖ acide m. sulfonique d'amido-naphtol.

amidships ‖ mittschiffs ‖ au milieu m. du navire.

amino acid ‖ Aminosäure f. ‖ acide m. aminoïque.

ammeter ‖ Amperemesser m.; Amperemeter n.; Strommesser m. ‖ ampèremètre m.; ammètre m. / aperiodic ~ ‖ aperiodischer Strommesser m. ‖ ampèremètre m. apériodique. / continuous current ~ ‖ Gleichstromamperemesser m. ‖ ampèremètre m. pour courant continu. / direct current ~ see continuous current ~. / edgewise ~ ‖ hochkant angeordneter Amperemesser m. ‖ ampèremètre m. monté de champ. / high-tension ~ ‖ Hochspannungsstrommesser m. ‖ ampèremètre m. à haute tension. / hot-wire ~ ‖ Hitzdrahtamperemesser m.; Hitzdrahtstrommesser m. ‖ ampèremètre m. thermique ou à fil chaud. / iron-clad ~ ‖ gekapselter Strommesser m. ‖ ampèremètre m. blindé. / moving coil ~ ‖ Drehspulstrommesser m. ‖ ammètre m. à bobine mobile. / moving iron ~ ‖ Dreheisenstrommesser m. ‖ ampèremètre m. à fer doux. / pocket ~ ‖ Taschenstrommesser m. ‖ ampèremètre m. de poche. / portable ~ ‖ tragbarer Strommesser m.; Montagestrommesser m. ‖ ammètre m. portatif. / precision ~ ‖ Genauigkeitsstrommesser m. ‖ ampèremètre m. de précision. / recording ~ ‖ Registrierstrommesser m.; schreibender Strommesser m. ‖ ampèremètre m. enregistreur. / ~ without zero ‖ Strommesser m. mit unterdrücktem Nullpunkt ‖ ampèremètre m. d'échelle à zéro buté.

ammonia see also ammonium ‖ Ammoniak n. ‖ ammoniaque f. / anhydrous ~ ‖ wasserfreies Ammoniak n. ‖ ammoniaque f. anhydre. / aqua ~ see liquid ~. / liquid ~ ‖ flüssiges Ammoniak n. ‖ ammoniaque f. liquide. / synthetic ~ ‖ synthetisches Ammoniak n. ‖ ammoniaque f. synthétique.

ammonia absorption cooling machine ‖ Ammoniakabsorptionsmaschine f. ‖ machine f. à ammoniaque à absorption. / ~ acetate ‖ essigsaures Ammoniak n. ‖ acétate m. d'ammoniaque. / ~ alum ‖ Ammoniakalaun m. ‖ alun m. ammoniacal ou d'ammoniaque. / ~ bicarbonate ‖ doppeltkohlensaures Ammoniak n. ‖ bicarbonate m. d'ammoniaque. / ~ bifluoride ‖ Ammonbifluorid n. ‖ bifluorure m. d'ammoniaque.

ammoniacal ‖ ammoniakalisch ‖ ammoniacal. / ~ gas ‖ Ammoniakgas n. ‖ gaz m. ammoniac. / ~ gum ‖ Ammoniakgummi m. ‖ gomme f. ammoniaque. / ~ liquor ‖ Gaswasser n. ‖ eau f. gazeuse ou ammoniacale. / ~ salt ‖ Salmiak m.; Ammoniaksalz m. ‖ sel m. ammoniac.

ammonia carbonate ‖ kohlensaures Ammoniak n. ‖ carbonate m. d'ammoniaque. / ~ chloride ‖ Ammoniumchlorid n. ‖ chlorure m. d'ammonium. / ~ citrate ‖ zitronensaures Ammoniak n. ‖ citrate m. d'ammoniaque. / ~ compression machine ‖ Ammoniakverdichtungsmaschine f.; Ammoniakkompressionsmaschine f. ‖ machine f. à compression d'ammoniaque.

ammonia compressor ‖ Ammoniakverdichter m. ‖ compresseur m. à ammoniaque.

/ compound system ~ || zweistufiger Ammoniakverdichter m. || compresseur m. à ammoniaque à deux étages. / single stage ~ || einstufiger Ammoniakverdichter m. || compresseur m. à ammoniaque à un seul étage.

ammonia cooling machine || Ammoniakkältemaschine f. || machine f. frigorifique à ammoniaque. / ~ distilling || Ammoniakdestillation f. || distillation f. d'ammoniaque. / radiating ~ distributor || rotierender Ammoniakverteiler m. || distributeur m. automatique d'ammoniaque. / ~ drum || Ammoniakflasche f. || bonbonne f. d'ammoniaque. /~ formiate || ameisensaures Ammoniak n. || formiate m. d'ammoniaque. / ~ gas || Ammoniakgas n. || gaz m. ammoniacal; gaz m. ammoniac. / ~ hydrate || Ammoniakflüssigkeit f. || solution f. aqueuse d'ammoniaque; ammoniaque f. liquide. / ~ hydrochlorate || Ammoniumchlorid n. || chlorhydrate m. d'ammoniaque. / ~ ice machine || Ammoniakeismaschine f. || machine f. à glace à ammoniaque. / ~ inlet valve || Ammoniakeinlaßventil n. || vanne f. d'entrée d'ammoniaque. / ~ lactate || milchsaures Ammoniak n. || lactate m. d'ammoniaque. / ~ liquor || Ammoniakwasser n. || eau f. ammoniacale. / ~ magnesium phosphate || phosphorsaure Ammoniakmagnesia f. || phosphate m. ammoniacomagnésique. / ~ molybdate || molybdänsaures Ammoniak n. || molybdate m. d'ammoniaque. / ~ muriate || Salmiak m. || sel m. ammoniac. / ~ nitrate || salpetersaures Ammoniak n. || nitrate m. ou azotate m. d'ammoniaque. / ~ nitrogen || Ammoniakstickstoff m. || azote m. ammoniacal. / ~ outlet valve || Ammoniakauslaßventil n. || vanne f. de sortie d'ammoniaque. / ~ persulphate || überschwefelsaures Ammoniak n.; Ammonpersulfat n. || persulfate m. ou persulfure m. d'ammoniaque. / ~ phosphate || phosphorsaures Ammoniak n. || phosphate m. d'ammoniaque. / ~ producing plant || Ammoniakgewinnungsanlage f. || installation f. de production ou de récupération de l'ammoniaque. / ~ receiver || Ammoniakkessel m.; Ammoniakbehälter m. || récipient m. ou réservoir m. à ammoniaque. / ~ recovery plant see ~ producing plant. / ~ sesquicarbonate || Ammoniumkarbonat n. || carbonate m. d'ammoniaque. / ~ solution || wässeriges Ammoniak n.; Salmiakgeist m.; Salmiakspiritus m. || esprit m. de sel ammoniac; solution f. ammoniacale. / ~ sulphate || schwefelsaures Ammon n.; Ammoniumsulfat n. || sulfate m. d'ammoniaque. / ~ sulphate plant || Ammoniumsulfatanlage f. || installation f. de sulfate d'ammoniaque. / ~ tartrate || weinsteinsaures Ammoniak n. || tartrate m. d'ammoniaque. / ~ vanadate || vanadinsaures Ammoniak n. || vanadate m. d'ammoniaque. / ~ water || Ammoniakwasser n.; wässeriges Ammoniak n. || eau f. ammoniacale ou d'ammoniaque. / ~ works pl. || Ammoniakanlage f. || installation f. pour la production d'ammoniaque.

ammonite || Ammonit n. || ammonit m.

ammonium see also ammonia || Ammonium n.; Ammoniak n. || ammonium m.; ammoniaque f. / ~ acetate || Ammoniumazetat n.; essigsaures Ammo-

niak n. || acétate m. d'ammoniaque. / ~ alum || Ammoniakalaun m. || alun m. ammoniacal. / ~ bifluoride || Ammonbifluorid n. || bifluorure m. d'ammoniaque. / ~ carbonate || kohlensaures Ammoniak n.; Hirschhornsalz n. || carbonate m. d'ammoniaque. / ~ chloride || Salmiak n.; Ammoniumchlorid n. || chlorure m. d'ammonium. / ~ citrate || zitronensaures Ammoniak n. || citrate m. d'ammoniaque. / ~ fluoride || Fluorammonium n. || fluorure m. d'ammonium. / ~ formiate || ameisensaures Ammoniak n. || formiate m. d'ammoniaque. / ~ hydroxide || Ammoniumhydroxyd n.; Ätzammoniak n. || ammoniaque f.; hydroxyde m. d'ammonium. / ~ lactate || milchsaures Ammoniak n. || lactate m. d'ammoniaque. / ~ magnesium phosphate || phosphorsaure Ammoniakmagnesia f. || phosphate m. ammoniaco-magnésique. / ~ molybdate || Ammoniummolybdat n.; molybdänsaures Ammoniak n. || molybdate m. d'ammoniaque. / ~ nitrate || Ammonsalpeter m.; salpetersaures Ammoniak n. || nitrate m. ou azotate m. d'ammoniaque. / ~ nitrogen || Ammoniakstickstoff m. || azote m. ammoniacal. / ~ oxalate || Ammoniumoxalat n. || oxalate m. d'ammoniaque. / ~ persulphate || überschwefelsaures Ammoniak n. || persulfate m. d'ammoniaque. / ~ phosphate || Ammoniumphosphat n.; phosphorsaures Ammoniak n. || phosphate m. d'ammoniaque. / ~ salt || Ammoniumsalz n. || sel m. d'ammonium. / ~ sodium phosphate || Natriumammoniumphosphat n. || phosphate m. de soude ammoniacal. / ~ sulphate || Ammoniumsulfat n. || sulfate m. d'ammoniaque. / ~ sulphate plant || Ammoniumsulfatanlage f. || installation f. d'usines de sulfate d'ammonium. / ~ sulphate saltpetre || Ammonsulfatsalpeter m. || sulfonitrate m. d'ammoniaque. / ~ sulphide || Schwefelammonium n. || sulfure m. d'ammonium. / ~ superphosphate || Ammoniaksuperphosphat n. || superphosphate m. d'ammoniaque. / ~ tartrate || weinsteinsaures Ammoniak n. || tartrate m. d'ammoniaque. / ~ vanadate || vanadinsaures Ammoniak n. || vanadate m. d'ammoniaque.

ammunition || Munition f. || munition f. / charged ~ || gefüllte Munition f. || munition f. chargée. / complete round of ~ || vollständiges Geschoß n. || projectile m. complet. / to prevent the chemical disintegration of the ~ || chemische Zersetzungsprozesse mpl. in der Munition verhindern || prévenir la décomposition chimique de la munition.

ammunition box || Patronenkasten m. || boîte f. aux cartouches. / ~ chamber || Munitionskammer f. || chambre f. de munition. / ~ drum || Patronentrommel f. || tambour m. à cartouches. / ~ machinery || Munitionsmaschine f. || machine f. pour la fabrication des munitions.

amorphous || amorph; unkristallinisch; gestaltlos || amorphe; à l'état amorphe. / ~ sugar || amorpher Zucker m. || sucre m. d'orge.

amortisation || Tilgung f. || amortissement m.

amount || Menge f. || quantité f. / ~ (Money) || Betrag m. || montant m.; encaisse f. / ~ (Ware) || Bestand m. || état m.; effectif m.; inventaire m. / ~ of

alcohol || Alkoholgehalt m. || degré m. alcoolique; taux m. d'alcool; teneur f. en alcool. / ~ of ascending current of air || Aufwindwert m. || coefficient m. d'ascendance du vent. / ~ of balance || Saldobetrag m. || montant m. du solde. / ~ of copper used || Kupferaufwand m. || cuivre m. employé. / ~ of freight || Frachtbetrag m. || montant m. du fret. / gross ~ || Bruttobetrag m. || montant m. brut. / ~ of indemnity || Entschädigungsbetrag m. || montant m. de l'indemnité. / ~ of metal in ores || Metallgehalt m. der Erze || aloi m. / net ~ || Nettobetrag m. || montant m. net. / ~ of ozone in the air || Ozongehalt m. der Luft || teneur m. ou richesse f. de l'air en ozone; quantité f. d'ozone dans l'air. / ~ of rainfall || Niederschlagsmenge f. || quantité f. de pluie tombée ou de pluies. / ~ of resistance || Widerstandswert m. || valeur f. de la résistance. / ~ of revenue || Nutzungswert m. || valeur f. de rapport. / ~ of the settling (Build) || Sackmaß n.; Senkungsmaß n.; Senkungszuschuß m. || tassement m. / total ~ || Gesamtbetrag m. || somme f. totale. / total ~ of delivery of a pump || Gesamtliefermenge f. einer Pumpe || débit m. total d'une pompe. / ~ of upward current of air || Aufwindwert m. || coefficient m. d'ascendance du vent.

amounting to... || sich belaufend auf... || se montant à...

amount key (Cash register) || Betragtaste f. || touche f. de montant.

amperage || Stromstärke f. || ampérage m.

ampere || Ampere n. || ampère m.

ampere-hour || Amperestunde f. || ampère-heure f. / ~ meter || Amperestundenmesser m.; Amperestundenzähler m. || ampère-heure-mètre m.

amperemeter see also ammeter || Amperemesser m. || ampèremètre m.

Ampere's rule || Amperesche Regel f. || règle f. d'Ampère.

ampere second || Amperesekunde f. || ampère-seconde f. / ~ turn || Amperewindung f. || ampère-tour m.

amphibian (Aero) || Amphibienflugzeug n. || avion m. amphibie; amphibie f.

amphibole (Miner) || Hornblende f.; Amphibol m. || hornblende f.; amphibole m. / ~ quarry || Hornblendegrube f. || carrière f. d'hornblende.

ample means pl. || reichliche Mittel npl. || moyens mpl. abondants.

amplification (Tel) || Verstärkung f. || amplification f. / ~ of the electron tube || Verstärkerwirkung f. der Elektronenröhre || amplification f. du tube électrique. / frequency dependency of ~ (Tel) || Verstärkungsverlauf m. || dépendance f. de l'amplification des fréquences. / ~ of gauge in the curves (Railw) || Spurerweiterung f. in den Kurven || élargissement m. de la voie dans les courbes. / total ~ (Tel) || Gesamtverstärkung f. || amplification f. totale.

amplification factor || Verstärkungsfaktor m. || coefficient m. d'amplification. / ~ figure (Tel) || Verstärkungsziffer f. || chiffre m. d'amplification. / ~ testing set (Tel) || Prüfkasten m. für Verstärkerämter || appareil m. pour mesurer l'amplification.

amplifier (Tel) || Verstärker m. || amplificateur m. / equalizing ~ || entzerrender

Verstärker m. ‖ amplificateur m. de correction. / primary ~ ‖ Grundverstärker m. ‖ amplificateur m. d'entrée. / rectifying ~ ‖ Richtverstärker m. ‖ amplificateur-redresseur m. / resistance capacity coupled ~ ‖ Verstärker m. mit Widerstandskapazitätskopplung ‖ amplificateur m. à accouplement par capacité de résistance. / super-heterodyne ~ ‖ Zwischenfrequenzverstärker m. ‖ amplificateur m. à fréquence intermédiaire. / three stage ~ ‖ Dreifachverstärker m. ‖ amplificateur m. à trois étages. / transformer coupled ~ ‖ Verstärker m. mit Übertragerkopplung ‖ amplificateur m. à transformateur. / two-stage ~ ‖ zweistufiger Verstärker m. ‖ amplificateur m. à deux étages. / two-valve intermediate ~ ‖ Doppelrohrzwischenverstärker m. ‖ amplificateur m. intermédiaire à deux tubes.

amplifier equipment ‖ Verstärkeranlage f. ‖ installation f. d'amplification. / ~ noises pl. ‖ Pfeifen n. der Verstärkerröhren ‖ sifflement m. des amplificateurs. / ~ triode see ~ valve. / ~ valve ‖ Verstärkerröhre f. ‖ triode f. ou lampe f. amplificatrice. / ~ valve socket (Tel) ‖ Verstärkerröhrenfassung f. ‖ douille f. pour triodes amplificatrices.

amplify, to (Electr) ‖ den Strom m. verstärken ‖ augmenter le courant m.

amplifying without distortion (Radio) ‖ verzerrungsfreie Verstärkung f. ‖ amplification f. sans distorsion.

amplifying tube see amplifier valve. / ~ voltmeter ‖ Röhrenvoltmeter n. ‖ voltmètre m. à triode.

amplitude ‖ Amplitude f.; Schwingungsweite f. ‖ amplitude f. / ~ of accommodation ‖ Akkommodationsbreite f. ‖ amplitude f. de l'accommodation. / ~ of oscillation ‖ Schwingungsweite f. ‖ amplitude f. d'oscillation. / ~ of the pendulum swing ‖ Pendelausschlag m.; Ausschlag m. des Pendels ‖ amplitude f. de l'oscillation du pendule. / southern ~ (Astron) ‖ Mittagsweite f. ‖ amplitude f. méridionale. / ~ of swell (Hydr arch) ‖ Stauweite f. ‖ amplitude f. du remous. / ~ of swing ‖ Schwingungsweite f.; Schwingungsamplitude f. ‖ amplitude f. d'oscillation.

amplitude curve ‖ Amplitudenkurve f. ‖ courbe f. des amplitudes.

ampulla ‖ Ampulle f.; Phiole f.; Salbengefäß n. ‖ ampoule f.; fiole f. / ~ for hypodermic injection ‖ Ampulle f. für Hauteinspritzung ‖ ampoule f. pour injection hypodermique.

amygdaline ‖ mandelartig ‖ amygdalin.

amygdaloidal ‖ mandelförmig ‖ amygdaliforme.

amyl ‖ Amyl n. ‖ amyle m.

amylaceous matters pl. ‖ mehlige Stoffe mpl. ‖ matières fpl. amylacées.

amyl acetate ‖ Amylazetat n.; essigsaures Amyl n. ‖ acétate m. d'amyle; amyle m. acétique. / ~ acetate collodion ‖ Amylazetatkollodium n. ‖ collodion m. d'acétate d'amyle. / ~ alcohol see amylic alcohol. / ~ benzoate ‖ benzoesaures Amyl n. ‖ benzoate m. d'amyle. / ~ butyrate ‖ butylsaures Amyl n. ‖ butyrate m. d'amyle.

amylene, hydrated ‖ Amylenhydrat n. ‖ amylène m. hydraté.

amyl formiate ‖ ameisensaures Amyl n. ‖ formiate m. d'amyle.

amylic alcohol ‖ Fuselöl n.; Amylalkohol m. ‖ alcool m. amylique ou d'amyle ou de queue; huile f. de queue; fusol m.

amylocellulose ‖ Amylozellulose f. ‖ amylocellulose f.

amylodextrine ‖ Amylodextrin n. ‖ amylodextrine f.

amyl propionate ‖ propionsaures Amyl n. ‖ propionate m. d'amyle. / ~ salicylate ‖ salizylsaures Amyl n. ‖ salicylate m. d'amyle.

amylum ‖ Weizenstärke f. ‖ amidon m. de froment.

amyl valerianate ‖ baldriansaures Amyl n. ‖ valérianate m. d'amyle.

anaerobic ‖ anaerob ‖ anaérobique. / ~ bacterium ‖ anaerobes Bakterium n. ‖ bactérie f. anaérobique.

analyse, to see to analize.

analysis (Chem) ‖ Analyse f.; Zerlegung f.; Untersuchung f. ‖ analyse f. / ~ (Math) ‖ Analysis f. ‖ analytical ~ ‖ analytische Analyse f. oder Untersuchung f. ‖ analyse f. analytique. / average ~ ‖ Durchschnittsbestimmung f. ‖ analyse f. moyenne. / bacteriological ~ ‖ bakteriologische Analyse f. ‖ analyse f. bactériologique. / ~ by means of a blow-pipe ‖ Lötrohranlage f. ‖ analyse f. pyrognostique. / chemical ~ ‖ chemische Analyse f. ‖ analyse f. chimique. / ~ in the dry way ‖ Analyse f. auf trockenem Wege ‖ analyse f. par voie sèche. / electrolytical ~ ‖ elektrolytische Analyse f. ‖ analyse f. électrolytique. / elementary ~ ‖ Elementaranalyse f. ‖ analyse f. élémentaire ou ultime. / ~ of flue gases ‖ Rauchgasanalyse f. ‖ analyse f. des gaz de fumée. / gravimetrical ~ see analysis by weight. / ~ in the humid way ‖ Analyse f. auf nassem Wege ‖ analyse f. par voie humide. / microchemical ~ ‖ mikrochemische Analyse f. ‖ analyse f. microchimique. / optical ~ of a gas mixture ‖ optische Analyse f. eines Gasgemisches ‖ analyse f. optique d'un mélange de gaz. / physiological ~ ‖ physiologische Analyse f. ‖ analyse f. physiologique. / ponderal ~ see analysis by weight. / pyrognostical ~ ‖ Lötrohranalyse f.; pyrognostische Analyse f. ‖ analyse f. pyrognostique. / qualitative ~ ‖ qualitative Analyse f. ‖ analyse f. qualitative. / quantitative ~ ‖ quantitative Analyse f. ‖ analyse f. quantitative; dosage m. / rapid ~ ‖ Schnellanalyse f. ‖ analyse f. rapide. / spectral ~ ‖ Spektralanalyse f. ‖ analyse f. spectrale. / spectral ~ of dye stuffs ‖ Spektralanalyse f. von Farbstoffen ‖ analyse f. spectrale des colorants. / spectroscopic ~ see analysis, spectral. / spectrum ~ see analysis, spectral. / thermal ~ ‖ thermische Analyse f. ‖ analyse f. thermique. / volumetrical ~ ‖ volumetrische Analyse f.; Maßanalyse f. ‖ analyse f. volumétrique. / ~ by weight ‖ Gewichtsanalyse f. ‖ analyse f. en poids; analyse f. pondérale. / zymotechnical ~ ‖ gärungsphysiologische oder zymotechnische Analyse f. ‖ analyse f. zymotechnique.

analyst ‖ Analytiker m. ‖ analyste m.

analytic see analytical.

analytical ‖ analytisch ‖ analytique. / ~ balance ‖ chemische Wage f. ‖ balance f. chimique. / ~ geometry ‖ analytische Geometrie f. ‖ géométrie f. analytique.

analytically pure ‖ analysenrein ‖ analytiquement pur.

analyze, to ‖ analysieren; chemisch zerlegen ‖ analyser.

analyzer ‖ Analysator m. ‖ analyseur m. / flue gas ~ ‖ Rauchgasprüfer m. ‖ analyseur m. des gaz de fumée.

anastatic printing ‖ anastatischer Druck m. ‖ réimpression f. anastatique; palingraphie f.

anastigmatic ‖ anastigmatisch ‖ anastigmatique. / ~ folding magnifier ‖ anastigmatische Einschlaglupe f. ‖ loupe f. fermante anastigmatique. / ~ lens ‖ anastigmatische Linse f. ‖ lentille f. anastigmatique. / ~ magnifier ‖ anastigmatische Lupe f. ‖ loupe f. anastigmatique. / ~ object glass ‖ anastigmatisches Objektiv n. ‖ objectif m. anastigmatique.

anatomical ‖ anatomisch ‖ anatomique. / ~ institute ‖ anatomisches Institut n. ‖ salle f. d'anatomie. / ~ preparation ‖ anatomisches Präparat n. ‖ préparation f. anatomique. / ~ structure of the eye ‖ anatomischer Bau m. des Auges ‖ anatomie f. de l'œil.

anatomy ‖ Anatomie f. ‖ anatomie f.

anatta (Dyer) ‖ Orlean m. ‖ rocou m.; anotto m. / ~ seed ‖ Orleankörner npl. ‖ graines fpl. de rocou. / ~ works pl. ‖ Orleanfärberei f. ‖ rocouerie f.

anatto see anatta.

anchor, to ‖ verankern ‖ ancrer. / ~ a ship (Mar) ‖ ein Schiff m. vor Anker legen ‖ mettre un navire à l'ancre.

anchor (Build) ‖ Anker m.; Zugband n. ‖ ancre f.; moufle f.; tirant m. / ~ (Nav.; Watchm) ‖ Anker m. ‖ ancre f. / backing ~ ‖ Kattanker m. ‖ empennelle f. / to clear for coming to ~ ‖ zum Ankern n. klarmachen ‖ faire péneau m. / ~ down the stream (Pont) ‖ Windanker m. ‖ ancre f. d'aval. / to drag the ~ ‖ vor Anker m. treiben ‖ chasser sur l'ancre m. / rail ~ (Railw) ‖ Klemme f. ‖ dispositif m. d'ancrage. / to ride at ~ ‖ vor Anker liegen ‖ être au mouillage. / spare ~ (Mar) ‖ Notanker m. ‖ ancre f. de la cale. / to weigh the ~ ‖ den Anker m. lichten ‖ lever l'ancre m.

anchorage ‖ Ankerplatz m. ‖ mouillage m.

anchor bearing ‖ Ankerpeilung f. ‖ relèvement m. à l'ancre. / ~ bolt (Build) ‖ Fundamentanker m. ‖ boulon m. de fondation. / ~ buoy ‖ Ankerboje f. ‖ bouée f. d'ancre. / ~ chain ‖ Ankerkette f. ‖ câblechaîne m.; chaîne f. d'ancre. / ~ clip (Electr) ‖ Bandagehalter m. am Anker ‖ fermeture f. de frette; joint m. de frette. / ~ escapement (Watch) ‖ Ankerunruhe f.; Ankerhemmung f. ‖ échappement m. à ancre.

anchoring buttress (Bridge) ‖ Ankerpfeiler m. ‖ culée f. d'ancrage. / ~ fascine ‖ Ankerfaschine f. ‖ fascine f. d'ancrage ou de retraite ou à ancre. / ~ ground ‖ Ankergrund m.; Ankerplatz m.; Anlegeplatz m. ‖ mouillage m.; fond m. à mouillage. / ~ masonry (Bridge) ‖ Ankermauerwerk n. ‖ maçonnerie f. d'ancrage. / ~ pile (Build) ‖ Ankerpfahl m. ‖ pilot m. d'ancrage. / ~ place see ~ ground. / ~ pole ‖ Abspannmast m. / poteau m. de rappel.

anchor log (For line poles) ‖ Ankerklotz m. ‖ billot m. d'ancre. / ~ plate ‖ Ankerplatte f. ‖ contreplaque f.; plaque f. d'ancrage. / ~ screw ‖ Ankerschraube f. ‖ boulon m. d'ancrage.

2*

anchor-shaped piece (Watch) || Anker m. || ancre m.

anchor stake (Nav) || Ankerpfahl m. || piquet m. de retenue.

anchor testing machine || Ankerprüfmaschine f. || machine f. à essayer les ancres. / ~ with sliding weight balance || Ankerprüfmaschine f. mit Laufgewichtswage || machine f. à éprouver les ancres avec balance à poids curseur.

anchor tie beam (Carp) || Zugbalken m. || tirant m. / ~ winch || Ankerwinde f. || cabestan m. d'ancre.

anchovy || Anschovis f.; Anschove f.; Sardelle f. || anchois m.; sardine f. mise en sel. / ~ oil || Anschovisöl n. || anchois m.

ancon (Build) || Konsole f.; Kragstück n. || corbeau m.; console f.; ancone f.

anemogram || Windmesserschaubild n. || anémogramme f.

anemograph || Anemograf m.; schreibender Windstärkemesser m. || anémographe m. / fan wheel ~ see rotating wheel vane ~. / rotating wheel vane ~ || Windradanemograf m. || anémographe m. à moulinet.

anemometer || Luftgeschwindigkeitsmesser m.; Windstärkemesser m.; Windmesser m.; Anemometer n. || anémomètre m.; indicateur m. de vitesse du vent. / ~ of cup type || Schalenkreuzwindmesser m. || anémomètre m. à ailes hémisphériques. / registering ~ || Windgeschwindigkeitsschreiber m. || anémotachographe m. / ~ of windmill type || Flügelradwindmesser m. || anémomètre m. à ailettes.

anemometry || Windmessung f. || anémométrie f.

anemoscope || Windzeiger m.; Windrichtungsmesser m. || anémoscope m.

aneroid barometer || Aneroidbarometer n. || baromètre m. anéroïde.

anethol || Anethol n.; Fenchelöl n. || anéthol m.; huile f. de fenouil.

angelica || Angelika f. || angélique f. / ~ oil || Angelikaöl n. || essence f. d'archangélique. / ~ root || Angelikawurzel f. || racine f. d'archangélique.

angel's hair || Engelshaar n. || cheveux mpl. d'ange.

angle, to || angeln || pêcher à la ligne.

angle (Geom; Build; etc.) || Winkel m.; Ecke f. || angle m. / ~ (Tube) || Knie n.; Kniestück n. || coude m. / an ~ of about x degrees || ein Winkel m. von ca. x Grad || un angle m. voisin de x degrés. / acute ~ || spitzer Winkel m. || angle m. aigu. / adjacent ~ || Nebenwinkel m. || angle m. adjacent. / adjoining ~ see adjacent ~. / ~ of advance || Voreilungswinkel m. || angle m. d'avance. / alternate ~ || Wechselwinkel m. || angle m. alterne. / ~ of altitude || Höhenwinkel m. || angle m. d'ascension. / ~ of bosh (Met) || Rastwinkel m. || angle m. des étalages. / ~ of brush displacement || Bürstenverstellungswinkel m. || angle m. de décalage des balais. / clearance ~ (Turning steel) || Anstellwinkel m. || angle m. de coupe. / ~ of climb (Aero) || Steigwinkel m. || angle m. de montée. / ~ of convergence || Konvergenzwinkel m. || angle m. de convergence. / ~ between cranks || Kurbelversetzung f. || angle m. de calage des manivelles. / critical ~ || Grenzwinkel m. || angle m. limite. / ~ of crossing (Railw) || Herzstückneigung f.; Kreuzungsverhältnis n.; Kreuzungswinkel m. || déviation f. du cœur; angle m.

du croisement. / ~ in a curve || Knick m. in einer Kurve || angle m. dans une courbe. / cut-off ~ (Build) || abgestumpfter Winkel m.; abgestumpfte Ecke f. || pan m. coupé. / cutting ~ || Schneidwinkel m. || angle m. de coupe. / dead ~ || toter Winkel m. || angle m. mort. / ~ of declination || Abweichungswinkel m. || angle m. de déclinaison. / ~ of deflection || Ausschlagwinkel m. || angle m. d'écartement. / ~ of elevation (Surv) || Erhöhungswinkel m.; Erhebungswinkel m. || angle m. d'élévation ou de hauteur. / ~ of emergence || Austrittswinkel m. || angle m. d'émergence. / to express the ~ in seconds of angle || den Winkel m. in Winkelsekunden ausdrücken || exprimer l'angle en secondes d'angle. / external ~ (Build) || ausspringende Ecke f. des Gesimses || angle m. saillant. / ~ of the flank || Flankenwinkel m. || angle m. de flanc. / ~ of the flanks of thread || Flankenwinkel m. des Gewindes || angle m. des flancs du filet. / ~ of friction || Reibungswinkel m.; Ruhewinkel m. || angle m. de frottement. / ~ of the frog (Railw) || Herzwinkel m. || angle m. du croisement. / gliding ~ (Aero) || Gleitwinkel m. || angle m. de vol plané. / horary ~ (Nav) || Stundenwinkel m. || angle m. horaire. / ~ of incidence || Einfallwinkel m. || angle m. d'incidence. / ~ of inclination || Neigungswinkel m. || angle m. d'inclinaison. / ~ of jib (Shovel) || Auslegerneigung f. || inclinaison f. de la flèche. / ~ of lag || Verzögerungswinkel m. || angle m. de retard. / ~ of lag of phase || Phasennacheilungswinkel m. || angle m. de retard de phase. / ~ of lead || Voreilungswinkel m. || angle m. d'avance. / ~ of lead of phase || Phasenvoreilungswinkel m. || angle m. d'avance de phase. / ~ of lock || Ausschlagwinkel m. der Räder || angle m. de braquage. / to measure the ~ in radians || den Winkel m. im Bogenmaß messen || mesurer l'angle en mesure d'arc. / measured ~ || gemessener Winkel m. || angle m. mesuré. / oblique ~ || schiefer Winkel m. || angle m. oblique. / obtuse ~ || stumpfer Winkel m. || angle m. obtus. / opposite ~ || Gegenwinkel m. || angle m. opposé; angle m. correspondant. / optical ~ || Gesichtswinkel m.; Sehwinkel m.; optischer Winkel m. || angle m. optique ou visuel. / ~ of phase difference || Phasenverschiebungswinkel m. || angle m. de déphasage. / ~ of pitch (Aero) || Steigungswinkel m. || angle m. de tangage. / ~ of position || Kurswinkel m. || angle m. de route. / to read off the ~ || den Winkelwert m. ablesen || lire la valeur de l'angle. / re-entering ~ (Build) || einspringender Winkel m. || angle m. encoignure f. / ~ of reflection || Reflexionswinkel m.; Rückstrahlungswinkel m. || angle m. de réflexion. / refracting ~ || brechender Winkel m. || angle m. réfringent. / ~ of refraction || Brechungswinkel m. || angle m. de réfraction. / ~ of repose || Reibungswinkel m.; Ruhewinkel m. || angle m. de frottement. / right ~ || rechter Winkel m. || angle m. droit. / ~ of rotation of the crank || Kurbeldrehwinkel m. || angle m. de rotation de la manivelle. / sharp ~ (Mach) || scharf abgesetzte Stelle f. || épaulement m. à angle aigu. / ~ of sight || Gesichtswinkel m. || angle m. de visée. /

solid ~ (Crystal) || Kristallecke f.; räumlicher Winkel m. || angle m. solide; sommet m.; stéradian m. / ~ of squint for one eye || Schielablenkung f. für ein Auge || strabisme m. monoculaire. / ~ of strabism || Schielwinkel m. || angle m. strabique. / ~ of strike (Miner) || Streichwinkel m. || angle m. de direction. / ~ of the talus || Abdachungswinkel m. || angle m. de la pente. / ~ of taper || Konizität f. des Kegels || conicité f. ou angle m. du cône. / tipping ~ || Kippwinkel m. || angle m. de bascule. / vertical ~ (Geom) || Scheitelwinkel m. || angle m. opposé au sommet; angle m. vertical. / ~ of view || Blickwinkel m. || angle m. de visée. / visual ~ see optical ~. / ~ of yaw (Aero) || Gierungswinkel m. || angle m. de lacet.

angle bar || Winkel m. (aus Eisen oder Stahl); Winkeleisen n. || cornière f. / ~ bevel (Join) || Schmiege f.; Schrägmaß n.; Gehrmaß n. || équerre f. à mitre. / ~ bracket || Winkelkonsole f. || équerre f.; support m. cornier. / ~ crane || Winkelkran m. || grue f. à support triangulaire.

angled || winkelig || à angle; angulaire.

angle-drive || Winkelantrieb m. || commande f. par engrenage d'angle. / ~ fishplate || Winkellasche f. || éclisse f. cornière. / ~ forming machine || Eckenbiegemaschine f. || machine f. pour pliage d'angles. / ~ frame with sharp and round corners || Winkelrahmen m. mit scharfen und runden Ecken || châssis m. cornier à coins vifs et arrondis. / ~ gauge || Winkelmesser m. || goniomètre m.

angle-iron (Build) || Eckband n.; Winkelband n. || cornière f.; équerre f. d'angle. / ~ (Met) || Winkeleisen n. || fer m. cornière ou d'angle; cornière f. / unequal ~ || ungleichschenkliges Winkeleisen n. || cornière f. à ailes inégales.

angle-iron bending machine || Winkeleisenbiegemaschine f. || machine f. à cintrer les cornières. / ~ bracket (Railw) || Eckwinkel m. || cornière f. de renforcement. / ~ guide || Winkeleisenführung f. || guidage m. par cornières. / ~ hoop || Winkeleisenring m. || anneau m. en cornière. / ~ shear || Winkeleisenschere f. || cisaille f. à cornières.

angle lever (Mech) || Winkelhebel m. || coude m. pliant ou d'angle; levier m. courbé ou coudé. / ~ measuring instrument || Winkelmesser m.; Winkelmeßgerät n. || instrument m. à mesurer les angles. / ~ meter see ~ measuring instrument. / ~ piece (Build) || Befestigungswinkel m. || équerre f. de fixation. / ~ plate || Winkelblech n.; Eckblech n. || support m. d'équerre; tôle f. en angle; équerre f. d'angle ou en tôle; plaque f. cornière. / ~ pole (El line) || Eckmast m. || poteau m. ou support m. d'angle. / ~ port (Shipb) || Eckpforte f. || sabord m. d'angle. / ~ pulley (Ropeway) || Ablenkungsrolle f. || poulie f. de déviation.

angler || Angler m. || pêcheur m. à la ligne.

angle reading || Ablesung f. des Winkels || lecture f. de l'angle. / ~ ring || Winkelring m. || anneau m. cornier ou de cornière. / ~ ring flange || Winkelflansch m. || bride f. à cornières. / ~ sheet iron see angle plate.

anglesite || Bleiglas n. || anglésite f.; cristal m.

angle steel bar || Winkelstahlschiene f. || traverse f. en cornière d'acier. / ~ stop

(Preventing creeping of rails) ‖ Stütz-winkel m.; Stemmwinkel m. ‖ équerre f. d'arrêt. / ~ table (Mach Tool) ‖ Winkel-tisch m. ‖ console f. de table. / ~ valve ‖ Eckventil n. ‖ soupape f. d'équerre ou d'angle. / ~ wheel ‖ Winkelrad n. ‖ roue f. d'angle.

angling accessories pl. ‖ Angelgerät n. ‖ articles mpl. de pêche. / ~ line ‖ Angel-schnur f. ‖ échampeau m.; palancre f.; palangre f.

angular ‖ winkelig; winkelförmig; eckig ‖ angulaire; angulé. / ~ acceleration ‖ Winkelbeschleunigung f. ‖ accélération f. angulaire. / ~ adjustment of tilting table ‖ Schrägeinstellung f. des Werkzeug-tisches ‖ ajustage m. incliné de la table d'ouvrage. / ~ advance ‖ Winkelvoreilung f. ‖ avancement m. angulaire. / ~ dis-persion ‖ Winkeldispersion f. ‖ disper-sion f. à angle. / ~ displacement ‖ Win-kelverschiebung f. ‖ déplacement m. an-gulaire. / ~ fish plate (Railw) ‖ Winkel-lasche f. ‖ éclisse f. cornière. / ~ iron see angle iron. / ~ motion ‖ Winkelbewe-gung f. ‖ mouvement m. angulaire. / ~ point (Geom) ‖ Scheitel m. eines Win-kels ‖ sommet m. d'un angle. / ~ speed variation ‖ Veränderung f. der Winkel-geschwindigkeit ‖ variation f. de vitesse angulaire. / ~ thread ‖ Spitzgewinde n. ‖ filet m. triangulaire. / ~ variation ‖ Winkelabweichung f. ‖ variation f. an-gulaire. / ~ velocity (Elektr) ‖ Winkel-geschwindigkeit f.; Kreisfrequenz f. ‖ vitesse f. ou vélocité f. angulaire.

angulated ‖ eckig; winkelig ‖ angulaire; angulé.

angulometer ‖ Winkelmesser m. ‖ grapho-mètre m.

anhydride ‖ Anhydrid n. ‖ anhydride m. / ~ acid ~ ‖ Säureanhydrid n. ‖ anhydride m. d'acide.

anhydrite ‖ Anhydrit m.; wasserfreier Gips m.; Muriazit m. ‖ anhydrite f.; chaux f. hydro-sulfatée.

anhydrous (Chem) ‖ wasserfrei ‖ anhydre; anhydride. / ~ acid ‖ wasserfreie Säure f. ‖ acide m. anhydre. / ~ ammonia ‖ wasserfreies Ammoniak n. ‖ ammonia-que f. anhydre. / ~ carbonic acid ‖ Kohlensäureanhydrid n. ‖ anhydride m. carbonique. / ~ lime ‖ Ätzkalk m.; Branntkalk m.; gebrannter Kalk m. ‖ oxyde m. de calcium; chaux f. vive ou caustique ou calcinée.

aniline ‖ Anilin n. ‖ aniline f. / ~ blue ‖ Anil-linblau n. ‖ bleu m. d'aniline. / ~ colour ‖ Anilinfarbe f. ‖ couleur f. ou colorant m. d'aniline. / ~ colouring see ~ dye. / ~ dye ‖ Anilinfarbstoff m.; Anilinfarbe f. ‖ cou-leur f. d'aniline; teinture f. à base d'ani-line. / ~ dye solution ‖ Anilinfarbstoff-lösung f. ‖ solution f. de couleur d'ani-line. / ~ lac ‖ Anilinlack m. ‖ laque f. d'aniline. / ~ oil ‖ Anilinöl n. ‖ huile f. d'aniline. / ~ red ‖ Anilinrot n. ‖ rouge m. d'aniline. / ~ salt ‖ Anilinsalz n. ‖ sel m. d'aniline.

animal ‖ tierisch; animalisch ‖ animal. / ~ colouring matter ‖ tierischer Farbstoff m. ‖ matière f. colorante d'origine animale. / ~ charcoal ‖ Beinschwarz n.; Knochen-kohle f. ‖ charbon m. d'os; noir m. ani-mal. / ~ essential oil ‖ ätherisches Öl n. tierischen Ursprungs ‖ huile f. essentielle animale. / ~ fatty substance ‖ tierischer Fettstoff m. ‖ corps m. gras d'origine ani-

male. / ~ grease ‖ tierisches Fett n. ‖ graisse f. animale.

animal hair ‖ Tierhaar n. ‖ poil m. / ~ curl-ing machine ‖ Tierhaarkräuselmaschine f. ‖ ratineuse f. pour poils d'animaux.

animal kingdom ‖ Tierreich n. ‖ règne m. animal. / ~ oil ‖ tierisches Öl n. ‖ huile f. animale ou d'origine animale. / ~ of ori-gin ‖ tierischen Ursprunges m. ‖ d'origine f. animale. / ~ produce ‖ tierisches Er-zeugnis n. ‖ produit m. (du règne) animal. / ~ refuse ‖ tierischer Abfall m. ‖ déchet m. d'animaux. / ~ wax ‖ tierisches Wachs n. ‖ cire f. animale.

animal ‖ Tier n. ‖ animal m. / dead ~ ‖ totes Tier n. ‖ animal m. mort. / draught ~ ‖ Zugtier n. ‖ animal m. ou bête f. de trait. / service by draught ~s ‖ Zugtier-betrieb m. ‖ service m. par animaux de trait. / slaughtered ~ ‖ geschlachtetes Tier n. ‖ animal m. abattu.

animal-trap ‖ Tierfalle f. ‖ trappe f. pour animaux.

animated ‖ lebhaft ‖ vif; animé.

anion ‖ Anion n. ‖ anion m.

anise ‖ Anis m. ‖ anis m. / ~ camphor ‖ Aniskampfer m. ‖ camphre m. ani-sique.

aniseed oil ‖ Anisöl n.; Anissamenöl n. ‖ essence f. d'anis; huile f. de graines d'anis.

anise oil see aniseed oil. / ~ seed ‖ Anis-samen m. ‖ semence f. d'anis. / ~ seed oil see aniseed oil.

anisette ‖ Anislikör m. ‖ anisette f.

anisic acid ‖ Anissäure f. ‖ acide m. anisi-que. / ~ aldehyde ‖ Anisaldehyd n. ‖ aldé-hyde m. anisique.

anisidine ‖ Anisidin n. ‖ anisidine f.

anisotropic ‖ anisotropisch ‖ anisotrope.

anisotropy ‖ Anisotropie f. ‖ anisotropie f.

ankle ‖ Fußgelenk n.; Fußknöchel m. ‖ cheville f. du pied.

annabergite ‖ Nickelblüte f.; arseniksaurer Nickel m. ‖ nickel m. arséniaté.

Annat barley ‖ Annatgerste f. ‖ orge f. d'Annat.

anneal, to (Glassm) ‖ kühlen ‖ recuire. / ~ (Met) ‖ ausglühen; anlassen ‖ recuire; faire revenir. / ~ the iron sheets pl. ‖ die Blechtafeln fpl. ausglühen ‖ recuire les tôles fpl. de fer. / ~ the steel ‖ den Stahl m. ausglühen ‖ recuire l'acier m. / ~ the wire ‖ den Draht m. ausglühen ‖ recuire le fil.

annealed ‖ geglüht ‖ recuit. / ~ sheet iron ‖ ausgeglühtes Eisenblech n. ‖ tôle f. de fer recuite. / ~ wire ‖ ausgeglühter Draht m. ‖ fil m. recuit.

annealing (Glassm) ‖ Kühlen n. ‖ recuit m. / ~ (Met) ‖ Ausglühen n.; Atätmen n. ‖ re-cuit m.; recuite f.; recuison m. / by (means of) hardening and subsequent ~ higher resisting qualities may be obtained ‖ durch Härten n. und nachheriges An-lassen n. werden höhere Festigkeiten fpl. erzielt ‖ on obtient des résistances fpl. plus élevées quand on fait revenir l'acier m. après une trempe.

annealing box ‖ Glühgefäß n.; Glühkasten m.; Glühtopf m. ‖ caisse f. à recuire ou à cémentation; pot m. de cémentation ou de recuite / cast ~ ‖ gegossener Glühtopf m. ‖ pot m. de cémentation en fonte. / forged ~ ‖ geschmiedeter Glühtopf m. ‖ pot m. de cémentation en fer forgé.

annealing colour ‖ Anlauffarbe f. ‖ couleur f. de recuite.

annealing furnace (Met) ‖ Glühofen m.; Temperofen m.; Ausglühflammofen m. ‖ four m. à recuire; four m. de cémenta-tion. / ~ (Glass Painting) ‖ Kühlofen m. ‖ fourneau m. de reuite. 1 ~ installation ‖ Temperofenanlage f. ‖ installation f. de four de cémentation.

annealing and hardening furnace ‖ Glüh-und Härteofen m. ‖ four m. à recuire et à tremper. / ~ installation ‖ Glüh-anlage f. ‖ installation f. à recuire. / ~ oven see ~ furnace. / ~ pot for file ma-kers ‖ Härtetopf m. für Feilenfertigung ‖ auge f. à tremper les limes. / ~ shop ‖ Glühhaus n. ‖ hall m. de recuit.

annex, to ‖ anfügen; anschließen ‖ an-nexer; joindre; ajouter.

annex ‖ Zusatz m.; Anhang m.; Nachtrag m.; Beiblatt n. ‖ annexe f.

annexe (Build) ‖ Nebenanlage f.; Neben-gebäude n. ‖ hall m. accessoire; annexe f.; bâtiment m. secondaire.

annexed (Build) ‖ angebaut ‖ annexé. / ~ (in letters) ‖ anliegend ‖ ci-joint. / ~ building ‖ Nebenbau m. ‖ annexe m. / as is shown in the ~ sketch ‖ wie neben-stehende Abbildung zeigt ‖ le croquis ci-contre démontre bien que ...

annotta see anatta.

annotto see anatta.

announce, to ‖ ankündigen; anzeigen ‖ an-noncer; publier; faire savoir.

announcement ‖ Bekanntmachung f. ‖ avis m.; publication f.

annoyance ‖ Schererei f.; Störung f. ‖ tracasserie f.; ennuis mpl.

annual balance of accounts ‖ Jahres-abschluß m. ‖ bilan m.; balance f. / ~ bill ‖ Jahresrechnung f. ‖ compte m. annuel. / ~ flow of a barrage ‖ Jahres-wassermenge f. einer Talsperre ‖ quan-tité f. d'eau annuelle. / ~ output ‖ Jahres-leistung f. ‖ production f. annuelle. / ~ precipitation ‖ jährliche Niederschlags-menge f. ‖ quantité f. annuelle de con-densation ou de pluies. / ~ report ‖ Jahresbericht m.; jährlicher Geschäfts-bericht m. ‖ rapport m. annuel. / ~ ring (Wood) ‖ Jahresring m. ‖ cerne m. / ~ wood with small ~ rings ‖ engringiges Holz n.; Holz n. mit engen Jahresringen ‖ bois m. aux cernes étroits. / ~ statement see ~ balance. / ~ subscription ‖ Jahres-beitrag m. ‖ contribution f. annuelle. / ~ writing-off of depreciations ‖ jährliche Abschreibung f. ‖ amortissement m. annuel.

annuity ‖ Annuität f.; Jahresrente f. ‖ annuité f.; rente f. annuelle. / „~ due" (Patent) ‖ „Taxe fällig" ‖ „annuité due". / ~ bond ‖ Rentenbrief m. ‖ titre m. de rente.

annul, to ‖ austilgen; tilgen; annullieren ‖ casser; annuler; destituer.

annular ‖ ringförmig ‖ annulaire. / ~ cog wheel ‖ Zahnrad n. mit innerer Ver-zahnung ‖ roue f. dentée intérieure. / ~ furnace see ~ kiln. / ~ kiln (Brick) ‖ Ring-ofen m. ‖ four m. annulaire. / ~ saw ‖ Kronsäge f.; Ringsäge f. ‖ scie f. cylindri-que.

annulate; annulated see annular.

annulled ‖ getilgt; annulliert ‖ annulé.

annulment ‖ Nichtigkeitserklärung f. ‖ an-nulation f.; cassation f.

annunciator (Tel) ‖ Signaltafel f. ‖ annon-ciateur m.; indicateur m. téléphonique. / automatic fire-alarm ~ ‖ selbsttätiger

Feuermelder m. || annonciateur m. automatique de feu.

annunciator board || Fallscheibenapparat m. || tableau-indicateur m. / ~ disc || Fallscheibe f. || clapet m.

anode || Anode f.; positiver Pol m. || anode f.; plaque f.; pôle m. positif. / ~ bag || Anodenbeutel m. || sac m. à anodes. / ~ battery || Anodenbatterie f. || batterie f. d'anode; pile f. pour radiophonie. / ~ cell || Anodenzelle f. || vase m. anodique. / ~ circuit || Anodenkreis m. || circuit m. plaque. / ~ continuous current || Anodengleichstrom m. || courant m. continu anodique. / ~ current || Anodenstrom m. || courant m. anodique ou de plaque. / ~ discharge || Anodenentladung f. || décharge f. de l'anode. / ~ rays pl. || Anodenstrahlen mpl. || lumière f. anodique. / ~ reaction || Anodenrückwirkung f. || réaction f. de l'anode. / ~ rod || Anodenstange f. || barre f. d'anodes. / ~ screening grid || Anodenschutznetz n.; Anodenschutzgitter n. || écran m. de plaque; seconde grille f.; grille f. protectrice d'anode. / ~ voltage || Anodenspannung f. || tension f. anodique ou de plaque.

anodic current density || spezifische Anodenstromstärke f. || densité f. anodique du courant. / ~ mud || Anodenschlamm m. || boue f. anodique.

anomal || anomal; unregelmäßig; regelwidrig || anomal.

anomaly || Unregelmäßigkeit f.; Abweichung f. von der Regel; Anomalie f.; Regelwidrigkeit f. || anomalie f. / ~ during the fermentation || Gärungserscheinung f. || anomalie f. en cours de fermentation.

anorthite || Anorthit m. || anorthite f.

answer, to || antworten; erwidern; beantworten || répondre; répliquer. / ~ (Tel) || abfragen || demander.

answer || Antwort f. || réponse f.; réplique f.

answerable for ... || verantwortlich für ... || responsable de ... / to be ~ for ... || gutsagen für ... || répondre de ...

answer back mechanism (Tel) || Rückfragevorrichtung f. || poste m. mixte ou à double appel.

answering (Tel) || Abfragen n. || demande f. / ~ a bell signal (Railw) || Rückmeldung f. || réplique f. / ~ board || Abfrageamt n. || groupe f. de départ. / ~ cord || Abfrageschnur f. || cordon m. de réponse ou de demande. / ~ jack (Tel) || Abfrageklinke f. || jack m. de demande ou de réponse. / ~ jack panel || Abfragefeld n. || champ m. de jacks de demande. / ~ plug || Abfragestöpsel m.; Antwortstöpsel m.; Meldestöpsel m. || fiche f. de demande ou de réponse; fiche f. annotatrice. / ~ position || Abfrageplatz m.; Abfragestellung f. || position f. de demande ou de réponse.

antagonist || Gegenpartei f. || parti m. opposé; adversaire m.; opposition f.

antagonistic || entgegengesetzt wirkend || antagoniste. / ~ spring (Tel) || Abreißfeder f. || ressort m. antagoniste.

antechamber (Bridge) || Luftschleuse f. || sas m. à air. / ~ system of injection || Vorkammereinspritzung f. || injection f. par précombustion.

antedate, to || vordatieren || antidater.

ant egg || Ameisenei n. || œuf m. de fourmi.

antenna (Radio) see also aerial || Antenne f.; Luftleiter m.; Luftdraht m. || antenne f. / artificial ~ || künstliche Antenne f.;

künstlicher Luftleiter m. || antenne f. artificielle. / earth ~ || Erdantenne f. || antenne f. basse. / effective height of ~ || wirksame Antennenhöhe f. || hauteur f. effective de l'antenne. / elevated ~ || Hochantenne f. || antenne f. extérieure. / extended ⊤-shaped ~ || verlängerter ⊤-Luftleiter m. || antenne f. en ⊤ à branches horizontales prolongées. / fan-shaped ~ || Fächerluftleiter m. || antenne f. en éventail. / horizontal extension of ~ || horizontale Ausbreitung f. des Luftleiters || branche f. horizontale de l'antenne. / liability for damages caused by ~s || Antennenhaftung f. || responsabilité f. des dommages causés par des antennes. / Marconi ~ || Marconi-Antenne f.; Marconi-Luftleiter m. || antenne f. Marconi. / simple ~ || einfacher Luftleiter m. || antenne f. simple. / trailing ~ || Hängeantenne f. || antenne f. suspendue.

antenna coil || Antennenspule f. || bobine f. d'antenne. / ~ cording || Antennenlitze f. || câble m. ou cordon m. ou brin m. d'antenne. / ~ duct || Antennendurchführung f. || traversée f. d'antennes. / ~ input || der Antenne zugeführte Leistung f. || énergie f. absorbée par l'antenne. / ~ insulator || Antennenisolator m. || isolateur m. pour les antennes. / ~ material || Antennenwerkstoff m.; Werkstoff m. für Luftleiter || matériel m. d'antenne. / ~ proximity zone || Nahbereich m. der Antenne || zone f. d'action immédiate. / ~ resistance (Radio) || Luftleiterwiderstand m. || résistance f. d'antenne. / ~ switch with fuse || Luftleiterschalter m. mit Sicherung || interrupteur m. d'antenne avec coupe-circuit fusible. / ~ terminal || Luftleiterklemme f. || serre-fil m. d'antenne; borne f. d'antenne. / ~ wire || Antennendraht m.; Antennenlitze f. || fil m. ou câble m. d'antenne.

anthracene || Anthrazen n. || anthracène m. / ~ dye || Anthrazenfarbe f. || couleur f. d'anthracène. / ~ oil || Anthrazenöl n. || huile f. d'anthracène; huile f. anthracénique. / ~ pitch || Anthrazenpech n. || poix f. d'anthracène.

anthracite || Anthrazit m.; Kohlenblende f. || anthracite m. / columnar ~ || stengeliger Anthrazit m. || houille f. bacillaire.

anthracite coal || Anthrazitkohle f.; Glanzkohle f. || anthracite m.; houille f. luisante ou sèche. / ~ mine || Anthrazitbergwerk n. || mine f. d'anthracite. / ~ trebles pl. || Anthrazitnußkohle f. || noisettes fpl. d'anthracite.

anthraquinone || Anthrachinon n. || anthraquinone f.

anti-aircraft artillery || Flugabwehrartillerie f. || artillerie f. antiaérienne.

anti-aircraft-defence || Flugzeugabwehr f.; Flugabwehr f. || défense f. antiaérienne. / terrestrial ~ || Flugabwehr f. von der Erde aus || défense f. terrestre antiaérienne.

anti-aircraft fire control instrument || Flugabwehrfeuerkommandogerät n. || instrument m. de direction du tir antiaérien. / ~ firing || Flugabwehrschießen n.; ~ firing table || Flugabwehrschußtafel f. || table f. de tir antiaérien. / ~ gun || Flugabwehrkanone f.; Flak f.; Luftabwehrgeschütz n. || canon m. antiaérien ou contre-avion. / ~ machine gun || Flug-

abwehrmaschinengewehr n. || mitrailleuse f. de défense contre avions.

antiarthritic oil || Gichtöl n. || huile f. antiarthritique. / ~ wadding || Gichtwatte f. || ouate f. antiarthritique.

antiasthmatic cigarette || Asthmazigarette f. || cigarette f. antiasthmatique.

anticathode || Antikathode f. || anticathode f.

antichlor || Antichlor n. || antichlore m.

anticipate, to || zuvorkommen || devancer; prévenir.

anticipation || Vorschuß m. || avance f. / payment by or in ~ || Vorauszahlung f.; Abschlagszahlung f.; Akontozahlung f. || payement m. d'avance ou par anticipation.

anticlinal (Geol) || antiklinal; mit entgegengesetzter Neigung f. || anticlinal. / ~ flexure || Schichtensattel m. || anticlinal m.; selle f. des roches. / ~ formation || Sattelbildung f. || formation f. anticlinale.

anti-clockwise motion || Linksdrehung f. || tournant m. à gauche. / ~ rotating system || Linkssystem n. || trièdre m. à gauche.

anticoherer (Tel) || Antikohärer m. || anticohéreur m.

anti-corrosive composition || Rostschutzmittel n. || enduit m. anti-corrosif.

anti-creeper (Railw) || Schienenklemme f. gegen das Wandern der Schienen || dispositif m. d'ancrage des rails.

anti-dazzling screen (Auto) || Blendschutzscheibe f. || écran m. antiéblouissant.

anti-distortion device (Tel) || Entzerrer m. compensateur m. de distorsion.

anti-flatulent powder for cattle || Viehpulver n. (gegen Trommelsucht) || poudre f. météorifuge.

anti-fouling composition (Shipb) || fäulnisverhütende Farbe f.; Rostschutzfarbe f.; Unterwasserfarbe f. || peinture f. sous-marine. / ~ paint see ~ composition. / ~ painting (Shipb) || fäulnisverhütender Anstrich m.; Schutzanstrich m.; Unterwasseranstrich m. || enduit m. préservatif (pour carènes de navires).

anti-freezing mixture || kältebeständiges Gemisch n. || mélange m. anti-réfrigérant. / ~ solution || Gefrierschutzlösung f. || solution f. anti-réfrigérante.

anti-friction || Mittel n. gegen Reibung; Antifriktion f. || anti-friction f. / ~ apparatus || Antifriktionsapparat m. || appareil-antifriction m. / ~ grease || Heißachsenschmiere f. || graisse f. pour essieux chauds. / ~ metal || Antifriktionsmetall n.; Lagermetall n. || métal m. antifriction ou antifrictionnaire ou pour coussinets.

anti-incrustant composition || Kesselsteinverhütungsmittel n. || anti-tartre m.; composition f. anti-tartreuse ou détartreuse.

anti-induction design (Tel) || Induktionsschutzplan m. || projet m. d'anti-induction. / ~ rule (Tel) || Induktionsschutzvorschrift f. || règle f. d'anti-induction.

anti-inductive circuit || Induktionsschutzschaltung f. || ligne f. de contre-tension. / ~ protection for telephone circuits || Induktionsschutz m. für Fernsprechleitungen || anti-induction f. des circuits téléphoniques.

antilogarithm || Antilogarithmus m. || antilogarithme m.

antimonial ethiops ‖ Spießglanzmohr m. | éthiops m. antimonial. / ~ lead ‖ Antimonblei n. ‖ plomb m. à l'antimoine; plomb m. antimonieux. / ~ nickel ‖ Antimonnickel m.; Nickelantimonglanz m.; Nickelspießglanz m. ‖ nickel m. antimonieux; ullmannite f. / ~ silver ‖ Antimonsilber n.; Spießglanzsilber n. ‖ dicrasite f.; argent m. antimonial. / ~ sulphuretted lead-ore ‖ Spießglanzblei n.; Bournonit m. ‖ bournonite f.; mine f. d'antimoine noir.

antimoniate ‖ Antimonsalz n.; antimonsaures Salz n. ‖ antimoniate m. / ~ of potassium ‖ antimonsaures Kalium n. ‖ antimoniate m. de potasse.

antimonic acid ‖ Antimonsäure f. ‖ acide m. antimonique.

antimonious acid ‖ antimonige Säure ‖ acide m. antimonieux. / ~ lead see antimonial lead.

antimonite (Chem) ‖ antimonigsaures Salz n. ‖ antimonite m. / ~ (Miner) ‖ Antimonglanz m.; Grauspießglanzerz n. ‖ stibine f.; stibite f.; antimonite f.

antimony ‖ Antimon n. ‖ antimoine m. / native ~ ‖ gediegenes Antimon n.; Spießglanz m. ‖ antimoine m. natif. / red ~ ‖ Rotspießglanzerz n. ; Antimonblende f. ‖ kermésite f.; antimoine m. rouge.

antimony alloy ‖ Antimonlegierung f. ‖ alliage m. antimonié. / ~ bath ‖ Antimonbad n. ‖ bain m. d'antimoniage ou d'antimoine. / ~ bloom ‖ Antimonblüte f. ‖ antimoine m. oxydé ou blanc. / ~ chloride ‖ Chlorantimon n.; Antimontrichlorid n.; Antimonbutter f. ‖ trichlorure m. ou beurre m. d'antimoine. / ~ fluoride ‖ Fluorantimon n. ‖ fluorure m. d'antimoine. / ~ glance ‖ Antimonglanz m.; Grauspießglanzerz n. ‖ stibine f.; stibite f.; antimonite f. / ~ glass ‖ Antimonglas n. ‖ verre m. ou sulfure m. d'antimoine. / ~ lactate ‖ milchsaures Antimon n. ‖ lactate m. d'antimoine. / ~ lead ‖ Antimonblei n. ‖ plomb m. antimonial. / ~ mine ‖ Antimonbergwerk n. ‖ mine f. d'antimoine. / ~ ochre ‖ Antimonocker m.; Spießglanzocker m. ‖ cervantite f.; antimoine m. oxydé terreux.

antimony ore ‖ Antimonerz n. ‖ minerai m. d'antimoine. / grey ~ (Miner) ‖ Antimonglanz m.; Grauspießglanzerz n. ‖ stibine f.; stibite f.; antimonite f. / red ~ ‖ Rotspießglanz m.; Spießglanzblende f.; Antimonblende f.; Mineralkermes m.; Kermes m.; Pyrantimonit m. ‖ antimoine m. rouge.

antimony oxide ‖ Antimonoxyd n. ‖ oxyde m. d'antimoine. / ~ pentachloride ‖ Antimonchlorid n. ‖ pentachlorure m. d'antimoine.

antimony sulphide ‖ Schwefelantimon n. ‖ sulfure m. d'antimoine. / black ~ ‖ schwarzes Schwefelantimon n. ‖ sulfure m. d'antimoine noir. / gold-coloured ~ see red ~. / red ~ ‖ rotes Schwefelantimon n. ‖ pentasulfure m. d'antimoine; sulfure m. doré d'antimoine.

antimony trichloride ‖ Antimonchlorür n. ‖ trichlorure m. d'antimoine. / ~ works pl. ‖ Antimonhütte f. ‖ fonderie f. d'antimoine.

antinode of potential ‖ Spannungsgegenknoten m. ‖ antinœud m. de tension.

antipyrine ‖ Antipyrin n. ‖ antipyrine f.

antique oil ‖ antikes Öl n. ‖ huile f. antique.

antiquities pl. ‖ Antiquitäten fpl.; Altertümer npl. ‖ antiquités fpl. / ~ repairer ‖ Antiquitätenausbesserer m. ‖ réparateur m. d'antiquités. / ~ trade ‖ Antiquitätenhandel m. ‖ commerce m. d'antiquités.

anti-rust ‖ Rostschutz m. ‖ antirouille f.

anti-rusting composition ‖ Rostschutzmittel n. ‖ enduit m. antirouille. / ~ paint ‖ Eisenschutzfarbe f.; Rostschutzfarbe f. ‖ peinture f. antirouille.

anti-scaling composition ‖ Kesselsteinlösemittel n. ‖ désincrustant m.

antiseptic ‖ fäulniswidrig; antiseptisch ‖ antiseptique. / ~ composition ‖ antiseptische Masse f. ‖ matière f. antiseptique. / ~ dressing ‖ antiseptischer Verbandstoff m. ‖ pansement m. antiseptique. / ~ wadding ‖ antiseptische Watte f. ‖ ouate f. antiseptique.

antiseptic ‖ Antiseptikum n.; fäulniswidriges Mittel n. ‖ antiseptique m.

anti-spraying insulator ‖ Sprühschutzisolator m. ‖ isolateur m. protecteur contre effluve.

anti-tank artillery ‖ Kampfwagenabwehrartillerie f. ‖ artillerie f. de défense contre les chars d'assaut. / ~ gun ‖ Kampfwagenabwehrgeschütz n. ‖ canon m. antichar.

antler goods pl. ‖ Gegenstände mpl. aus Geweihen ‖ objets mpl. en bois de cerf. / ~ trade ‖ Geweihindustrie f. ‖ industrie f. de bois de cerf.

anvil (Forg) ‖ Amboß m. ‖ enclume f. / ~ (Met) ‖ Prellklotz m.; Prallstock m.; Reitel m.; Stoßreitel m. ‖ billot m.; rabat m.; battoir m. / ~ (Surv) ‖ Meßtisch m. ‖ planchette f.; support m. / lengthwise adjustable ~ (Surv) ‖ nach der Länge verstellbarer Meßtisch m. ‖ planchette f. ou support m. réglable en longueur. / vertically adjustable ~ (Surv) ‖ senkrecht verstellbarer Meßtisch m. ‖ support m. réglable en hauteur. / ~ file-cutting ‖ Feilenhauamboß m. ‖ enclume f. à tailler les limes; tas m. / grooved ~ ‖ Gesenkamboß m.; Senkstock m. ‖ enclume f. cannelée ou sillonnée. / hand ~ ‖ Handamboß m. ‖ enclumette f. / mower's ~ ‖ Dengelamboß m. ‖ enclumette f. / chaploir m. / ~ with one arm ‖ Galgenamboß m.; Hornamboß m.; Sperrhorn n. ‖ enclume f. à potence; bigorne f. / rising ~ ‖ doppelter Hornamboß m.; Sperrhorn n.; Spitzamboß m. ‖ bigorne f.; enclume f. à deux cornes. / small ~ ‖ Handamboß m.; kleiner Amboß m. ‖ tasseau m.; bigorneau m.; enclumette f. / tinman's ~ ‖ Klempneramboß m. ‖ Spengleramboß m. ‖ enclumette f. de ferblantier.

anvil beak ‖ Amboßhorn n. ‖ corne f. ou bigorne f. d'enclume. / ~ bed ‖ Schabotte f.; Amboßuntersatz m.; Amboßfutter n.; Amboßstock m. ‖ chabotte f.; billot m. ou socle m. ou semelle f. de l'enclume. / ~ block ‖ Schmiedestock m.; Schabotte f.; Schmiedeblock m. ‖ tronchet m.; chabotte f.; billot m. de forge. / ~ chisel ‖ Schrotmeißel m.; Setzeisen n. ‖ tranche f. / ~ face ‖ Amboßbahn f. ‖ face f. ou table f. de l'enclume. / ~ horn see anvil beak. / ~ insertion piece ‖ Amboßeinsatz m. ‖ enclumette f. de faulx. / ~ inset stake ‖ Stöckel m. ‖ tas m. à queue. / ~ plate see ~ stand see ~ bed. / ~ stock see ~ bed.

apart, to take ‖ zerlegen; auseinandernehmen ‖ démonter. / ~ from … ‖ abgesehen von … ‖ à part de …

apatite ‖ Apatit m.; Phosphorit m. ‖ apatite f.; phosphorite f.

aperient tablet ‖ Purgiertablette f.; Abführtablette f. ‖ tablette f. purgative.

aperiodic ‖ aperiodisch ‖ apériodique. / ~ circuit (Radio) ‖ aperiodischer Kreis m. ‖ circuit m. apériodique. / ~ system ‖ aperiodisches System n. ‖ système m. apériodique. / ~ voltmeter ‖ aperiodisches Voltmeter n. ‖ voltmètre m. apériodique.

aperiodically coupled ‖ aperiodisch gekoppelt ‖ accouplé apériodique.

apertometer ‖ Apertometer n. ‖ apertomètre m.

aperture ‖ Öffnung f.; Schlitz m. ‖ aperture f.; fente f. / ~ of an angle ‖ Winkelöffnung f. ‖ ouverture f. d'un angle. / free ~ (Opt) ‖ freie Öffnung f. ‖ ouverture f. libre. / ratio of the ~ of the lens and the focal length ‖ Verhältnis n. der Objektivöffnung zur Brennweite ‖ rapport m. entre l'ouverture de l'objectif et la distance focale. / numerical ~ (Opt) ‖ numerische Apertur f. ‖ ouverture f. numérique. / ~ of the object glass ‖ Objektivöffnung f. ‖ diamètre m. de l'objectif. / ~ of a wall ‖ Maueröffnung f. ‖ jour m. ou ouverture f. d'un mur. / ~ of a window ‖ Fensteröffnung f. ‖ jour m. ou baie f. de fenêtre.

aperture angle (Opt) ‖ Öffnungswinkel m. ‖ angle m. d'ouverture. / ~ diaphragm ‖ Aperturblende f. ‖ diaphragme m. d'ouverture.

apertured ‖ geschlitzt ‖ fendu.

aperture ratio (Opt) ‖ Öffnungsverhältnis n. ‖ ouverture f. relative. / ~ stop see ~ diaphragm.

aphelion ‖ Sonnenferne f. ‖ aphélie f.

aphrite ‖ Schaumkalk m.; Aphrit m. ‖ aragonite f. nacrée.

aphrizite ‖ Aphrizit m. ‖ aphrizite m.

aphronitre ‖ Mauersalpeter m.; Salpeterschaum m. ‖ aphronitre f.

apiarist ‖ Bienenzüchter m.; Imker m. ‖ apiculteur m.

apicultural implements pl. ‖ Imkereigerät n. ‖ ustensiles mpl. d'apiculture.

apiculture ‖ Imkerei f. ‖ apiculture f.

aplanatic ‖ aplanatisch ‖ aplanétique. / ~ folding magnifier ‖ aplanatische Einschlaglupe f. ‖ loupe f. fermante aplanétique. / ~ magnifier ‖ aplanatische Lupe f. ‖ loupe f. aplanétique.

apochromatic objective ‖ Apochromat n. ‖ objectif m. apochromatique. / ~ telescope objective ‖ apochromatisches Fernrohrobjektiv n. ‖ objective m. de lunette apochromatique.

apoglucic acid ‖ Apogluzinsäure f. ‖ acide m. apoglucique.

apomecometer ‖ Fernmesser m. ‖ apomécomètre m.

apophyllite ‖ Fischaugenstein m. ‖ apophyllite f.

apothecary ‖ Pharmazeut m.; Apotheker m. ‖ pharmacien m.; apothicaire m. / ~ glasses pl. for apothecaries ‖ Apothekerglaswaren fpl. ‖ verreries fpl. de pharmacie. / ~'s shop ‖ Apotheke f. ‖ pharmacie f. / furnitures pl. for apothecaries shops ‖ Apothekeneinrichtung f. ‖ installation f. de pharmacies.

apparatus ‖ Apparat m.; Gerät n.; Vorrichtung f. ‖ appareil m.; mécanisme m.; dispositif m. / automatic ~ ‖ selbsttätiger Apparat m. ‖ appareil m. automatique.

/ cast-iron ~ ‖ gußeiserner Apparat m. ‖ appareil m. en fonte. / ~ for controlling the generation of power ‖ Apparat m. zur Überwachung der Energieerzeugung ‖ appareil m. pour le contrôle de la production d'énergie. / ~ for counter steam ‖ Dampfbremsapparat m. ‖ appareil m. de freinage à contre-vapeur. / ~ for demonstration ‖ Vorführungsapparat m. ‖ appareil m. de démonstration. / ~ for dividing ‖ Apparat m. zum Einteilen ‖ appareil m. à diviser. / ~ for domestic use ‖ Haus(signal)apparat m. ‖ appareil m. pour les appartements. / ~ for drying with evaporation ‖ Eindampfapparat m. ‖ appareil m. d'évaporation. / electric(al) ~ ‖ elektrisches Gerät n. ‖ appareil m. électrique. / electromedical ~ ‖ elektromedizinischer Apparat m. ‖ appareil m. d'électricité médicale. / for heating ‖ Apparat m. zum Erhitzen ‖ appareil m. de chauffage ou à chauffer. / ~ for high-frequency ‖ Hochfrequenzapparat m. ‖ appareil m. pour le traitement à haute fréquence. / ~ for malting ‖ Malzkeimapparat m. ‖ germoir m. mécanique. / ~ for peeling wood ‖ Holzschälapparat m. ‖ appareil m. à écorcer les troncs d'arbres; appareil m. à enlever l'écorce des arbres. / ~ for registering rapidly passing events or occurrences ‖ Registriergerät n. für schnell verlaufende Vorgänge ‖ appareil m. enregistreur pour événements rapides. / scientific ~ ‖ wissenschaftlicher Apparat m. ‖ appareil m. scientifique. / ~ for taking-off the gases ‖ Gasentziehungsapparat m. ‖ appareil m. de prise des gaz. / ~ for taking-off the top gases of blast furnaces ‖ Gichtgasfang m. ‖ appareil m. pour prendre les gaz des hauts-fourneaux. / transmitting and receiving ~ (Tel) ‖ Sende- und Empfangsapparat m. ‖ appareil m. d'émission et de réception.

apparatus single plug ‖ Geräteeinzelstecker m. ‖ contact m. à une fiche pour appareils. / ~ wagon ‖ Apparatekarren m. ‖ voiture f. portant les appareils. / ~ works pl. ‖ Armaturenfabrik f. ‖ usine f. de garnitures.

apparel (Mar) ‖ Gien n.; Schwertakel n. ‖ caliorne f.; palan m. à caliorne. / ladies' ~ ‖ Damenkonfektion f. ‖ confection f. pour dames.

apparent ‖ anscheinend ‖ en apparence f. / ~(Seeming)‖scheinbar ‖ apparent; virtuel. / ~ component (Electr) ‖ Scheinwert m. ‖ valeur f. apparente. / ~ conductivity ‖ scheinbares Leitvermögen n. ‖ conductance f. apparente. / ~ extract ‖ scheinbarer Extraktgehalt m. ‖ teneur f. apparente en extrait. / ~ gravity see ~ extract. / ~ image ‖ scheinbares Bild n. ‖ image f. virtuelle. / ~ motion ‖ scheinbare Bewegung f. ‖ mouvement m. apparent. / ~ power ‖ scheinbare Leistung f.; Scheinleistung f. ‖ débit m. apparent; puissance f. apparente. / ~ reduction of weight ‖ scheinbare Gewichtsabnahme f. ‖ diminution f. de poids apparente.

appartment, suit of ~s pl. ‖ Zimmerflucht f. ‖ enfilade f. (de pièces).

appeal (Law) ‖ Berufung f. ‖ appellation f. / ~ (Trade) ‖ Reklamation f. ‖ réclamation f. / to give notice of ~ ‖ Berufung f. einlegen ‖ interjeter appel m. / to lodge an ~ ‖ Beschwerde f. erheben ‖ porter plainte. / right of ~ ‖ Einspruchsrecht n. ‖ droit m. d'opposition.

appearance ‖ Augenschein m. ‖ vue f.; apparence f. / ~ (Condition) ‖ äußere Beschaffenheit f. ‖ aspect m. extérieur. / ~ of cracks ‖ Bildung f. von Rissen ‖ formation f. de fissures. / glazed ~ ‖ glänzendes Aussehen n. ‖ apparence f. luisante. / neat ~ ‖ hübsches Aussehen n. ‖ forme f. plaisante et agréable.

appellant ‖ Berufungskläger m. ‖ appelant m.

appending label ‖ Anhänger m. ‖ étiquette f. attache.

appendix ‖ Füllansatz m. ‖ manche f. de gonflement m. ou de remplissage ou d'appendice.

appertaining, music ~ thereto ‖ Begleitmusik f. ‖ musique f. d'accompagnement.

apple ‖ Apfel m. ‖ pomme f. / core of ~ ‖ Kerngehäuse n. vom Apfel ‖ trognon m. de pomme. / ~ peel ‖ Apfelschale f. ‖ pelure f. de pomme. / ~ pulp ‖ gepreßter Apfel m. ‖ pâte f. de pommes. / ~ sugar ‖ Apfelzucker m. ‖ sucre m. de pommes.

apple tree ‖ Apfelbaum m. ‖ pommier m. / wild ~ ‖ Holzapfelbaum m. ‖ pommier m. sauvage.

appliance ‖ Gerät n.; Vorrichtung f. ‖ appareil m.; dispositif m.; engin m. / automatic control ~ ‖ selbsttätige Steuervorrichtung f. ‖ appareil m. automatique de contrôle. / ~ for boilers ‖ Kesselbedarfsartikel m. ‖ accessoire m. de chaudières.

applicable ‖ verwendbar; anwendbar ‖ applicable; employable; utilisable.

application ‖ Anwendung f. ‖ emploi m.; utilité f.; application f. / ~ (Demand) ‖ Antrag m.; Gesuch n.; Anliegen n. ‖ demande f. / ~ of complex quantities to alternating current problems ‖ komplexes Rechnen n. mit Wechselströmen ‖ calcul m. de phénomènes alternatifs au moyen de quantités complexes. / ~ for a post ‖ Stellengesuch n. ‖ demande f. de place. / ~ of the tariff ‖ Anwendung f. des Zolltarifs ‖ application f. du tarif de douane. / form of ~ ‖ Antragsformular n. ‖ demande f.; proposition f. / letter of ~ ‖ Bewerbungsschreiben n. ‖ demande f. de place. / time of ~ ‖ Anmeldefrist f. ‖ terme m. de notification.

applied-art museum ‖ Kunstgewerbemuseum n. ‖ musée f. d'art industriel.

applied chemistry ‖ angewandte Chemie f. ‖ chimie f. appliquée. / ~ E. M. F. (Electr) ‖ angeschlossene E. M. K. f. ‖ F. E. M. f. appliquée.

apply, to ‖ anwenden ‖ appliquer. / ~ (To put on) ‖ auflegen ‖ poser ou mettre sur. / ~ colours pl. ‖ die Farben fpl. auftragen ‖ appliquer les couleurs fpl. / ~ for ‖ sich bewerben um‖rechercher; essayer d'obtenir; demander. / ~ gold ‖ die Goldblätter npl. auflegen ‖ dorer dans la dorure en détrempe. / ~ the glue to (Bookb) ‖ den Leim m. auftragen ‖ enduire de colle f. / ~ the handles (Pott) ‖ das Geschirr garnieren ‖ garnir la vaisselle; habiller les pots. / ~ to ... ‖ ansetzen ‖ mettre; appliquer; ajuster. / ~ to ... (To appeal) ‖ sich wenden an ... ‖ s'adresser à .... / ~ the vermeil (Gild; Paint) ‖ hellen ‖ vermeillonner.

applying colours pl. (Paint) ‖ Auftragen n. der Farben ‖ colorisation f.

appoint, to ‖ anstellen ‖ assigner une poste; engager.

appointment (Arrangement; Prescription) ‖ Anordnung f.; Verordnung f.; Vorschrift f.; Bestimmung f. ‖ arrangement m.; ordonnance f.; prescription f.; règlement m. / ~ (Calling; Nomination) ‖ Anstellung f.; Ernennung f.; Berufung f. ‖ placement m.; nomination f.; appel m. / ~ (Decision) ‖ Bestimmung f.; Festsetzung f. ‖ détermination f.; destination f.; fixation f.; définition f. / ~ (Post; Situation) ‖ Amt n.; Stelle f.; Stellung f.; Posten m.; Anstellung f. ‖ emploi m.; poste m.; place f.; charge f.; position f.

appointments pl. ‖ Einrichtung f.; Ausrüstung f. ‖ équipement m.; installation f. / ~ (Equipment; Dressing) ‖ Kleidung f.; Ausrüstung f. ‖ habillement m.; équipement m. / ~ horse ~ ‖ Pferdegeschirr n.; Geschirr n.; Beschirrung f. ‖ harnais m. (pour chevaux); harnachement m. ou équipement m. d'un cheval.

appraisal of damage ‖ Schadens(ab)schätzung f. ‖ taxation f. du dommage.

appraise, to ‖ taxieren; abschätzen ‖ taxer; estimer; evaluer.

appraisement ‖ Abschätzung f. ‖ estimation f.; taxation f.

appraiser ‖ Abschätzer m.; Taxator m. ‖ estimateur m.; taxateur m. / licensed ~ ‖ konzessionierter Taxator m. ‖ commissaire-priseur m. avec licence. / sworn ~ ‖ gerichtlich vereidigter Gutachter m. ‖ expert m. juré.

appreciation ‖ Wertschätzung f. ‖ appréciation f.

apprentice ‖ Lehrling m. ‖ apprenti m. / ~ (Mar) ‖ Jungmatrose m.; Halbmann m. ‖ novice m.

apprentice's premium ‖ Lehrgeld n. ‖ salaire m. d'apprentissage.

apprenticeship ‖ Lehre f. ‖ apprentissage m. / ~ (Time) ‖ Lehrzeit f. ‖ temps m. d'apprentissage.

apprentice training room ‖ Lehrlingswerkstatt f. ‖ atelier m. d'apprentissage.

apricot ‖ Aprikose f. ‖ abricot m. / ~ gum ‖ Aprikosenbaumgummi m. ‖ gomme f. d'abricotier. / ~ kernel ‖ Aprikosenkern m. ‖ noyau m. d'abricot. / ~ kernel oil ‖ Aprikosenkernöl n. ‖ huile f. de noyaux d'abricots ou d'abricotier. / ~ stone see ~ kernel. / ~ tree ‖ Aprikosenbaum m. ‖ abricotier m.

approach ‖ Zugangsstraße f. ‖ chemin m. d'accès. / elevated ~ ‖ Rampe f. ‖ rampe f. d'accès. / ~ of a train ‖ Herannahen n. eines Zuges ‖ approche f. d'un train.

approaching buoy ‖ Ansegelungsboje f. ‖ bouée f. d'atterrage.

approach ladder ‖ Zugangsleiter f. ‖ échelle f. d'accès. / ~ locking (Railw) ‖ Anrücksperre f. ‖ blocage m. par approche. / ~ ramp ‖ Zufahrtrampe f. ‖ rampe f. d'accès. / ~ section (Railw) ‖ Anrückabschnitt m.; Annäherungsabschnitt m. ‖ canton m. bloqueur.

approbation of a project ‖ Genehmigung f. eines Projektes ‖ approbation f. d'un projet.

appropriate ‖ zweckmäßig; sachdienlich; sachgemäß ‖ convenable; pratique. / ~ treatment ‖ sachgemäße Behandlung f. ‖ traitement m. pratique.

appropriateness, to make with ‖ sachgemäß herstellen ‖ préparer judicieusement.

appropriation ‖ Bewilligung f. ‖ concession f.; vote m. / ~ of money ‖ Geldbewilligung f. ‖ allocation f. d'argent.

approval ‖ Zustimmung f. ‖ consentement m.; approbation f.; assentiment m. / to meet with ~ ‖ Anklang m. finden ‖ être bien accueilli; être reçu avec faveur f. / on ~ ‖ probeweise ‖ à titre m. d'essai. / to send on ~ ‖ zur Ansicht f. schicken ‖ envoyer au choix m. *ou* à condition f.

approved ‖ genehmigt; gut befunden; zugelassen ‖ approuvé; accordé; agréé.

approximate, to ‖ aufrunden ‖ arrondir un nombre.

approximate ‖ angenähert ‖ approché. / ~ amount ‖ ungefährer Betrag m. ‖ montant m. approximatif. / ~ calculation ‖ ungefähre Berechnung f. ‖ calculation f. approximative. / ~ estimate ‖ ungefährer Anschlag m. ‖ évaluation f. approximative. / ~ figures pl. ‖ annähernde Ziffern fpl. ‖ chiffres mpl. approximatifs. / ~ price ‖ ungefährer Preis m. ‖ prix m. approximatif. / ~ range of measurement ‖ angenäherter Meßbereich m. ‖ étendue f. approximative des mesures. / ~ regulation ‖ Grobregelung f. ‖ réglage m. approximatif. / ~ value ‖ Näherungswert m. ‖ valeur f. approchée.

approximation ‖ Annäherung f. ‖ approximation f. / ~ method ‖ Näherungsmethode f.; Näherungsverfahren n. ‖ méthode f. d'approximation.

approximative ‖ annähernd ‖ approximatif.

apron ‖ Schürze f.; Knieleder n. ‖ tablier m. / leather ~ ‖ Lederschürze f. ‖ tablier m. en cuir. / ~ clasp ‖ Schürzenhaken m. ‖ agrafe f. pour tabliers. / ~ material ‖ Schürzenstoff m. ‖ étoffe f. pour tabliers. / stuff for ~s ‖ Schürzenstoff m. ‖ étoffe f. pour tabliers.

apse (Arch) ‖ Chornische f.; Altarnische f. ‖ apse f.; tribunal m. d'église.

apyrite ‖ Lithiumturmalin m. ‖ apyrite f.

apyrous ‖ feuerbeständig ‖ réfractaire; apyre.

aqua ammonia ‖ flüssiges Ammoniak n. ‖ ammoniaque f. liquide.

aqua fortis ‖ Gelbbrennsäure f. ‖ eau f. forte mordante. / ~ ‖ Scheidewasser n.; Salpetersäure f. ‖ eau-forte f.; acide m. azotique *ou* nitrique. / gilder's ~ ‖ Königswasser n. ‖ acide m. nitro-muriatique; eau f. régale.

aqua regalis ‖ Königswasser n. ‖ eau f. régale. / ~ regia *see* ~ regalis. / ~ regis *see* ~ regalis.

aquarium glass ‖ Aquarienglas n. ‖ verre m. pour aquariums.

aquatics pl. ‖ Wassersport m. ‖ sport m. nautique; yachting m.

aquatic sport *see* aquatics.

aquatinta-engraving ‖ Kupferstich m. in Tuschmanier ‖ gravure f. à l'aqua-tinta; gravure f. imitant le dessin au lavis.

aqueduct ‖ Aquadukt n.; Wasserleitung f. ‖ aqueduc m.

aqueous ‖ wasserhaltig; wässerig ‖ aqueux.

aqueous-alcoholic ‖ wässerigalkoholisch ‖ hydroalcoolique.

aqueous ammonia ‖ wässeriges Ammoniak n.; Salmiakgeist m.; Salmiakspiritus m. ‖ esprit m. de sel ammoniac. / ~ solution ‖ wässerige Lösung f. ‖ solution f. aqueuse. / ~ vapour ‖ Wasserdampf m. ‖ vapeur f. d'eau.

Arabic figure ‖ arabische Ziffer f. ‖ chiffre m. arabe.

arabinose ‖ Gummizucker m. ‖ arabinose m.

arable land ‖ Ackerboden m. ‖ sol m.; terroir m.; terre f. arable *ou* labourable *ou* végétable.

arachis oil ‖ Erdnußöl n. ‖ huile f. d'arachide.

arœometer ‖ Aräometer n.; hydrostatische Senkwage f. ‖ aréomètre m.

aragonite ‖ Schalenkalk m. ‖ aragonite f.

arbitrate, to ‖ schlichten; schiedsrichterlich entscheiden ‖ arranger; aplanir; arbitrer.

arbitrated ‖ schiedsrichterlich entschieden ‖ arbitré.

arbitration f. ‖ schiedsrichterliche Entscheidung f. ‖ arbitrage m.; parère m. / ~ (Exchange) ‖ Arbitrage f. ‖ arbitrage m. / court of ~ ‖ Schiedsgericht n. ‖ tribunal m. arbitral. / ~ of exchange ‖ Wechselarbitrage f. ‖ arbitrage m. des changes. / to refer to ~ ‖ einen Schiedsvertrag m. eingehen ‖ conclure un traité d'arbitrage. / to submit to ~ ‖ sich dem Schiedsgericht m. unterwerfen ‖ se soumettre à l'arbitrage m.

arbitrator ‖ Schiedsrichter m. ‖ arbitre m.; dispacheur m. / ~'s award ‖ Schiedsspruch m. ‖ sentence f. arbitrale; arbitrage m.

arbor (Bot) ‖ Baum m. ‖ arbre. / ~ (Mach) ‖ Spindel f.; Welle f.; Drehstift m. ‖ arbre m.; essieu m.; fuseau m.; axe m. / ~ (Tool) ‖ Dorn m.; Bolzen m. ‖ mandrin m. / ~ of balance wheel ‖ Unruhewelle f. ‖ axe m. du balancier. / cutter ~ ‖ Fräsbolzen m. ‖ mandrin m. porte-fraise *ou* porte-pièce. / left handed ~ (Watchm) ‖ Linkser m. ‖ arbre m. à rebours. / ~ for polishing balances ‖ Unruhepolierdrehstift m. ‖ arbre m. à polir les balanciers. / work ~ ‖ Aufspanndorn m.; Aufspannbolzen m. ‖ mandrin m. porte-pièce.

arboriculture ‖ Baumzucht f. ‖ arboriculture f.

arboriculturist ‖ Baumgärtner m. ‖ arboriculteur m.

arbor press ‖ Dorneintreibepresse f. ‖ presse f. à mandriner.

arbour, iron ‖ eiserne Gartenlaube f. ‖ berceau m. *ou* tonnelle f. en fer.¹

arc ‖ Bogen m.; Kreisbogen m. ‖ arc m. / ~ (Arch) ‖ Rundung f.; Bogen m. ‖ cintre m. / ~ (Electr) ‖ (elektrischer) Lichtbogen m. ‖ arc m. (électrique *ou* voltaïque). / ~ (Lamp) *see also* arc lamp ‖ Bogenlampe f. ‖ lampe f. à arc; arc m. / breaking of an ~ ‖ Unterbrechung f. eines Lichtbogens ‖ rupture f. d'un arc. / ~ with choking coil (Lamp) ‖ Lichtbogenlampe f. mit Drosselspule ‖ lampe f. à arc avec bobine de réactance. / ~ on closing circuit (Electr) ‖ Schließungsbogen m. ‖ arc m. de fermeture. / ~ with coloured effect (Lamp) ‖ Effektlichtbogenlampe f. ‖ lampe f. à arc coloré. / damping of the ~ ‖ Dämpfung f. des Lichtbogens ‖ amortissement m. de l'arc. / direct current ~ (Lamp) ‖ Gleichstrombogenlampe f. ‖ arc m. à courant continu. / double carbon ~ (Lamp) ‖ Doppelbogenlampe f. ‖ arc m. à double jeu de charbons. / ~ *see* arc (Electr). / extinction of the ~ (Electr) ‖ Auslöschen n. des Lichtbogens ‖ extinction f. de l'arc. / flaming electric ~ ‖ elektrischer Flammenbogen m. ‖ arc m. électrique à flamme. / graduated ~ ‖ Gradbogen m.; Limbus m. ‖ arc m. gradué; limbe m. / heat of the ~ ‖ Lichtbogenwärme f. ‖

température f. de l'arc. / ~ with inclined carbons (Lamp) ‖ Bogenlampe f. mit schrägstehenden Kohlen ‖ lampe f. à arc à charbons inclinés. / length of ~ ‖ Lichtbogenlänge f. ‖ longueur f. de l'arc. / open ~ ‖ nackter Lichtbogen m. ‖ arc m. en air libre. / ~ of oscillation ‖ Schwingungsbogen m. ‖ arc m. des oscillations. / raising and lowering gear for ~s (Arc lamps) ‖ Bogenlampenaufzug m. ‖ moulinet m. de hissage pour lampes à arcs. / semi-enclosed ~ (Lamps) ‖ halboffene Bogenlampe f. ‖ lampe f. à arc demi-clos. / singing ~ ‖ singender *oder* musikalischer Lichtbogen m. ‖ arc m. chantant *ou* musical. / spluttering of ~ ‖ Sprühen n. des Lichtbogens ‖ crachement m. de l'arc. / ~ of time (Astro) ‖ Zeitbogen m. ‖ arc m. du temps. / toothed ~ (Mach) ‖ Zahnbogen m. ‖ arc m. *ou* secteur m. denté. / unsteady ~ ‖ flackernder Lichtbogen m. ‖ arc m. vacillant. / wandering of the ~ ‖ Wandern n. des Lichtbogens ‖ migration f. de l'arc.

arc arrester ‖ Bogenblitzableiter m. ‖ coupe-arc m. / ~ breaker ‖ Funkenlöscher m. ‖ souffleur m. d'arc.

arch (Arch) ‖ Rundung f.; Bogen m. ‖ cintre m.; arc m.; arche f. / ~ (Build) ‖ Gewölbe n. ‖ voûte f. / ~ (Glassm) ‖ Ofen m. ‖ arche f. / ~ (Mine) ‖ Ortung f.; Querschlag m. ‖ taillement m.; galerie f. traverse. / Arabian ~ ‖ arabischer Hufeisenrundbogen m. ‖ arc-cintre m. arabe. / ~ of a bridge ‖ Brückenbogen m. ‖ arc m. *ou* arche f. *ou* voûte f. de pont. / centre ~ (Bridge) ‖ Mittelbogen m. ‖ arche f. maîtresse. / chief ~ (Bridge) ‖ Hauptbogen m. ‖ arche f. maîtresse. / diagonal ~ ‖ Gratbogen m.; Kreuzbogen m. ‖ arc m. arêtier *ou* diagonal; croisée f. d'ogive. / ~ for drainage pipes ‖ Ablaufrohrbogen m. ‖ tube m. coudé de sortie. / flood ~ ‖ Flutbrücke f. ‖ avant-pont m. / foiled ~ ‖ Nasenbogen m.; Kleebogen m. ‖ arc m. lobé. / ~ of a furnace ‖ Ofengewölbe n. ‖ arche f. d'un fourneau. / Gothic ~ *see* arch, pointed. / groined ~ *see* arch, diagonal. / head ~ ‖ Obergurt m. ‖ arc-doubleau m. supérieur. / horseshoe ~ ‖ maurischer Bogen m. ‖ arc m. en fer à cheval. / lanceolated ~ ‖ Hufeisenspitzbogen m. ‖ ogive f. lancéolée. / longitudinal ~ (Build) ‖ Längengurt m. ‖ arc-formeret m. du long d'une voûte. / Moorish ~ ‖ maurischer Bogen m. ‖ arc m. en fer à cheval. / over ~ ‖ Obergurt m. ‖ arc-doubleau m. supérieur. / pier ~ ‖ freistehender Längengurt m.; Scheidebogen m. ‖ arc-bornant m. / pointed ~ ‖ Spitzbogen m. ‖ arc m. pointu *ou* aigu *ou* gothique *ou* à l'ogive; ogive f. / rammed concrete ~ ‖ Stampfbetongewölbe n. ‖ voûte f. en béton damé. / relieving ~ ‖ Türbogen m. ‖ arc m. de porte *ou* de décharge. / splayed ~ ‖ ausgeschrägter Bogen m. ‖ arc m. ébrasé. / triumphal ~ ‖ Triumphbogen m. ‖ arc m. de triomphe. / trussed ~ ‖ Fachwerkbogen m. ‖ poutre f. en arc. / ~ of a vault ‖ Gewölbbogen m. ‖ arc m. *ou* arceau m. d'une voûte. / wall ~ ‖ Schildbogen m. ‖ arc-formeret m. du long d'un mur. / ~ in a wall ‖ Mauerbogen m. ‖ arc m. *ou* arceau m. dans un mur. / window ~ ‖ Fensterbogen m. ‖ arc m. de fenêtre.

archæological || archäologisch || archéologique.

arch bridge *see* arched bridge. / ~ centering || Bogenschalung f. || coffrage m. de voûtes.

arched || gekrümmt || courbe; courbé. / ~ (Arch) || bogenförmig; gewölbt || arqué; cintré; en arc; en cintre; envoûté. / slightly ~ || leicht gekrümmt || bombé.

arched bridge || Bogenbrücke f. || pont m. en arc. / ~ buttress || Gewölbpfeiler m. || appui m.; contrefort m. / ~ plate || Tonnenblech n. || tôle f. cintrée *ou* à voûte. / ~ roof || Tonnendach n.; tonnenförmiges Dach n. || toit m. en berceau *ou* en tonnelle.

archil || Orseille f. || orseille f.

Archimedean screw || archimedische Schraube f.; Schnecke f. || vis f. d'Archimède. / ~ screw drill || Drillbohrer m. || porteforet m.; foret m. à vis d'Archimède.

arching (Geol) || Aufwölbung f. || formation f. en voûte. / ~ (Joiner) || Schweifung f. || cambrure f.; bouge m.

architect || Architekt m.; Baumeister m. || architecte m. / assistant ~ || Bauführer m. || aide architecte m. / government's ~ || Bauinspektor m. || inspecteur m. des travaux publics; ingénieur m. *ou* architecte m. du gouvernement. / naval ~ || Marineingeniör m. || ingénieur m. des constructions navales *ou* de la marine.

architectonic(s) || Bauwissenschaft f.; Lehre f. von der Baukunst || architectonique f.

architectural school || Baugewerkschule f. || école f. d'architecture.

architecture || Architektur f.; Baukunst f. || architecture f.; art [m. de bâtir. / hydraulic ~ || Wasserbaukunst f. || architecture f. hydraulique. / naval ~ || Schiffbaukunst f. || architecture f. navale. / academy of ~ || Bauschule f. || école f. d'architecture.

architrave (Arch) || Architrav[m.; Hauptbalken m. || architrave f. épistyle m. / ~ (Door; Window) || Einfassung f. || châssis m.

archives pl. || Archiv n. || archives fpl.

arch jamb || Hauptpfeiler m. || jambage m.; maître m. pilier. / ~ like || bogenförmig || arqué; cintré; en arc m.; en cintre m. / ~ pier (Arch) || Kämpferpfeiler m. eines Bogens || pied-droit m. d'un arc. / ~ pillar *see* ~ jamb. / ~ stone || Keilstein m.; Gewölbstein m. || voussoir m. / ~ supporter of flat foot || Plattfußeinlage f. || support m. pour pieds plats. / ~ wall || Grundmauer f.; Stützmauer f. || jambage m. de pierres. / ~ work || Gewölbbogen m. || voûte f.

arc ignition || Lichtbogenzündung f. || allumage m. par arc.

arcing over (Electr) || Feuern n.; Funken n. || coups mpl. de feu. / ~ step || Lichtbogenschritt m. || pas m. de l'arc. / ~ tip || Lichtbogenspitze f. || pointe f. de l'arc.

arc lamp *see also* arc || Bogenlampe f. || lampe f. à arc; arc m. / ~ with adjustable focus || Bogenlampe f. mit einstellbarem Brennpunkt || lampe f. à arc à foyer réglable. / ~ copying || Kopierbogenlampe f. || lampe f. à arc pour travaux de copie. / differential ~ || Differentialbogenlampe f. || lampe f. à arc différentielle. / enclosed ~ || Dauerbrandbogenlampe f.; geschlossene Bogenlampe f. || lampe f. à arc en vase clos *ou* à arc enfermé. / ~ with fixed focus *see*

focu(s)sing ~. / focus(s)ing ~ || Bogenlampe f. mit festem *oder* unveränderlichem Brennpunkt; Fixpunktbogenlampe f. || lampe f. à arc à foyer fixe *ou* à arc à point lumineux fixe. / machine for several ~s connected in series || Maschine f. für hintereinander geschaltete Bogenlampen || machine f. alimentant plusieurs lampes montées en série. / motor ~ || Motorbogenlampe f. || lampe f. à arc à moteur. / self-regulating ~ || selbstregelnde Bogenlampe f. || lampe f. à arc auto-régulatrice *ou* à arc à réglage automatique. / shunt type ~ || Nebenschlußbogenlampe f. || lampe f. à arc en dérivation. / ~ with variable focus || Bogenlampe f. mit veränderlichem Brennpunkt || lampe f. à arc à foyer variable.

arc lamp accessories pl. || Bogenlampenzubehör n. || accessoire m. pour lampes à arc. / ~ carbon || Bogenlampenkohle f.; Lichtbogenkohle f. || charbon m. *ou* crayon m. d'une lampe à arc. / ~ case || Bogenlampenlaterne f. || boîte f. de lampe à arc. / ~ globe || Bogenlampenglocke f. || globe m. de lampe à arc. / wire-netting for ~ globes || Drahtgeflecht n. einer Bogenlampenglocke || filet m. de sûreté d'un globe de lampe à arc. / ~ radiator || Bogenlichtbestrahlungsapparat m. || appareil m. de radiation à lumière à arc. / ~ starter || Lampenanlasser m. || résistance f. d'allumage.

arc light || Bogen(licht)lampe f.; Bogenlicht n. || lampe f. à arc; lumière f. à arc. / ~ bath || Bogenlichtbad n. || bain m. de lumière à arc.

arc lighting dynamo || Bogenlichtdynamo f. || dynamo f. pour lumière à arc.

arc oscillation || Lichtbogenschwingung f. || oscillation f. d'arc.

arc resistance || Lichtbogenwiderstand m. || résistance f. de l'arc. / ~ furnace || Lichtbogenwiderstandsofen m. || four m. à résistance de l'arc.

arc spectrum || Bogenspektrum n. || spectre m. de l'arc.

arc-striking mechanism || Lichtbogenbildner m. || amorceur m. d'arc.

arc transmitter || Lichtbogensender m. || émetteur m. à arc.

arc welding || Lichtbogenschweißung f. || soudure f. à l'arc. / ~ apparatus || Lichtbogenschweißapparat m. || appareil m. à souder à l'arc électrique.

area (Build) || lichter Raum m.; Raum m. im Lichten || aire f. / ~ (Geom) || Inhalt m. || aire f.; capacité f. / ~ (Surv) || Flächeninhalt m.; Flächenraum m.; Fläche f. || aire f.; surface f.; superficie f.; étendue f. / calculation of ~ || Flächenberechnung f. || calcul m. des aires. / ~ of contact || Kontaktfläche f. || surface f. de contact. / cooled ~ || gekühlte Bodenfläche f. || superficie f. refroidie. / ~ of the cross section || Querschnittfläche f. || aire f. de la coupe transversale. / ~ of low pressure || Tiefdruckgebiet n. || régime m. de dépression *ou* de basses pressions. / ~ of protection || Schutzkreis m. || cercle m. de protection. / ~ served by a crane || Reichweite f. eines Kranes || portée f. d'une grue. / ~ supplied *see* ~ of supply. / ~ of supply (Electr) || Verbrauchsgebiet n. || zone f. desservie *ou* de consommation. / ~ of a surface || Inhalt m. einer Fläche || aire f. d'une surface. / ~ to be surveyed || Vermes-

sungsgebiet n. || terrain m. soumis à l'arpentage. / total ~ || Gesamtoberfläche f. || surface f. totale.

areca nut || Arekanuß f. || noix f. palmiste *ou* d'arec.

argental mercury || Silberamalgam n. || mercure m. argental; amalgame m. d'argent.

argentan || Neusilber n.; Argentan n. || argentan m.; argent m. d'Allemagne; packfong m.; maillechort m.

argentate || Silberoxydverbindung f. || argentate m.

argentic || silberhaltig || argentique. / ~ bromide || Bromsilber n. || bromure m. d'argent. / ~ nitrate || salpetersaures Silber n.; Höllenstein m. || nitrate m. d'argent; pierre f. infernale.

argentiferous || silberhaltig; silberführend || argentifère.

argentine || silberartig || argentin. / ~ water || Versilberungsflüssigkeit f. || eau f. argentine.

argentine || Schaumkalk m.; Aphrit m. || aragonite f. nacrée.

argentpyrite || Silberkies m.; Argentopyrit m. || argentopyrite f.

argillaceous || lehmig || argileux. / ~ earth || Tonerde f. || terre f. argileuse. / fireproof goods pl. formed from ~ mass || aus toniger Masse geformte feuerfeste Waren fpl. || ouvrages mpl. réfractaires moulés en masse argileuse. / ~ schist || Tonschiefer m. || schiste m. argileux. / ~ shale *see* ~ schist.

argillite || Argillit m.; Tonschiefer m. || argillite f.

argillous || tonartig || argileux.

argument || Beweis m. || preuve f.; argument m.

argyrose || Argentit m. || argyrose f.

arithmetical || arithmetisch || arithmétique. ~ sign || arithmetisches Zeichen n. || signe m. arithmétique.

arithmometer || Rechenmaschine f. || arithmomètre m.; machine f. à calculer.

arkose || Kaolinsandstein m. || grès m. feldspathique; arcose f.

arm, to ~ a furnace || einen Ofen m. beschlagen || armer un fourneau.

arm || Arm m. || bras m. / ~ (Weapon) || Waffe f. || arme f. / ~ of a balance || Wagebalken m. || fléau m. / ball-jointed ~ || Kugelgelenkarm m. || fléau m. à articulation à rotule; bras m. à joints sphériques. / ~ of a horseshoe-magnet || Schenkel m. eines Hufeisenmagneten || branche f. d'un aimant en fer à cheval. / ~ of lever || Hebelarm m. || bras m. de levier. / ~ of precision || Präzisionswaffe f. || arme f. de précision. / ~ of a river || Flußarm m. || bras m. d'une rivière. / secondary ~ of a canal || Zweigkanal m. || embranchement m. d'un canal. / small ~ || Handwaffe f. || arme f. portative. / steering gear ~ || Lenkstockhebel m. || levier m. de direction. / steering knuckle ~ || Lenkschenkel m. || levier m. de commande de la fusée. / ~ of a wheel || Radarm m.; Radspeiche f. || rais m. *ou* rayon m. d'une roue. / ~ of a wheel with section in form of a cross || Radarm m. mit kreuzförmigem Querschnitt || rais m. d'une roue en forme de croix.

armament || Kriegsausrüstung f.; Bewaffnung f. || armement m. / to reduce ~s || abrüsten || démobiliser; désarmer. / ~ of the tank || Bestückung f. des Kampf-

wagens ‖ armement m. de char de combat. / office of ~ ‖ Heereswaffenamt n. | office m. d'armement. / ~ restriction | Rüstungsbeschränkung f. ‖ limitation f. des armements.

armature ‖ Beschlag m.; Armatur f. | armature f.; ferrure f. / ~ (Cable) ‖ Bewehrung f.; Armierung f.; Armatur f. | armature f. / ~ (Electr) ‖ Anker m.; Rotor m.; Läufer m. ‖ induit m.; rotor m. / alternator ~ (Electr) ‖ Wechselstromanker m. ‖ induit m. d'alternateur. / balanced ~ (Electr) ‖ ausbalanzierter Anker m. ‖ induit m. équilibré. / barwound ~ ‖ Stabanker m. ‖ induit m. à barres. / bipolar ~ (Electr) ‖ zweipoliger Anker m. ‖ induit m. bipolaire. / closed slot ~ (Electr) ‖ Anker m. mit geschlossenen Nuten ‖ induit m. à canaux fermés. / cylindrical ring ~ (Electr) ‖ Zylinderringanker m. ‖ induit m. en anneau cylindrique. / the ~ drops (Tel) ‖ das Relais n. fällt ab ‖ l'armature f. retombe. / drum-wound ~ (Electr) ‖ Trommelanker m. ‖ induit m. en tambour. / external pole ~ (Electr) ‖ Außenpolanker m. ‖ induit m. à pôles extérieurs. / inner pole ~ (Electr) ‖ Innenpolanker m. ‖ induit m. à pôles intérieurs. / intensity ~ (Electr) ‖ Intensitätsanker m. ‖ induit m. enroulé à haute tension. / ~ of a magnet ‖ Anker m. eines Magnets ‖ armature f. d'un aimant. / magnetic ~ ‖ Weicheisenanker m. ‖ induit m. de fer magnétique; armature f. en fer doux. / multipolar ~ (Electr) ‖ mehrpoliger Anker m. ‖ induit m. multipolaire. / open slot ~ (Electr) ‖ Anker m. mit offenen Nuten ‖ induit m. à rainures ouvertes. / ~ with pilot-windings (Electr) ‖ Stufenanker m. ‖ armature f. à étages. / radial ~ (Electr) ‖ Polanker m.; Sternanker m. ‖ armature f. à pôles. / revolving ~ (Electr) ‖ rotierender Anker m. ‖ induit m. tournant. / ring-wound ~ (Electr) ‖ Ringanker m. ‖ induit m. en anneau. / short-circuited ~ (Electr) ‖ Kurzschlußanker m. ‖ induit m. en court-circuit. / shuttle ~ (Electr) ‖ Doppel-T-Anker m. ‖ armature f. en double T. / slotted ~ (Electr) ‖ Nutenanker m. ‖ induit m. à rainures. / squirrel cage ~ (Electr) ‖ Käfiganker m.; Kurzschlußanker m. ‖ induit m. à cage d'écureuil.

armature ampere-turns pl. ‖ Ankeramperewindungen fpl. ‖ ampères-tours mpl. d'induit. / ~ bar (Electr) ‖ Ankerstab m. ‖ barre f. d'induit. / ~ binding wire (Electr) ‖ Ankerbindedraht m. ‖ frette f. d'induit. / ~ body (Electr) ‖ Ankerkörper m. ‖ corps m. de l'induit. / ~ coil (Electr) ‖ Ankerspule f. ‖ bobine f. d'induit. / ~ core (Electr) ‖ Ankerkern m. ‖ noyau m. d'induit. / ~ and transformer-core building press ‖ Presse f. für den Anker- und Transformatorenbau ‖ presse f. à serrer les tôles pour induits et transformateurs. / ~ core disc (Electr) ‖ Ankerblech n. ‖ disque m. en tôle pour l'induit. / ~ cross (Electr) ‖ Ankerstern m. ‖ étoile f. d'induit. / ~ current (Electr) ‖ Läuferstrom m. ‖ courant m. de l'induit. / ~ drum (Electr) ‖ Trommelanker m. ‖ induit m. en cylindre. / ~ field (Electr) ‖ Ankerfeld n. ‖ champ m. de l'induit. / ~ flux (Electr) ‖ Ankerkraftfluß m. ‖ flux m. d'induit. / ~ gap (Electr) ‖ Ankerbohrung f. ‖ ouverture f.

d'induit. / ~ head flange (Electr) ‖ vorderer Ankerflansch m. ‖ flasque m. fixe de l'induit. / ~ hub (Electr) ‖ Ankerbuchse f. ‖ manchon m. d'induit. / ~ induction (Electr) ‖ Ankerinduktion m. ‖ induction f. dans l'induit. / ~ iron (Electr) ‖ Ankereisen n. ‖ fer m. d'induit. / ~ leakage (Electr) ‖ Ankerstreuung f. ‖ dispersion f. d'induit. / ~ leakage flux (Electr) ‖ Ankerstreufluß m. ‖ flux m. d'induit de dispersion. / ~ line of force (Electr) ‖ Ankerkraftlinie f. ‖ ligne f. de force de l'induit. / ~ pinion (Electr) ‖ Ankerzahnrad n. ‖ pignon m. d'induit. / heavy ~ punching (Electr) ‖ großes oder starkes Ankerblech n. ‖ grande tôle f. d'induit. / ~ reaction (Electr) ‖ Ankergegenwirkung f.; Ankerrückwirkung f. ‖ réaction f. d'induit. / ~ shaft (Electr) ‖ Ankerwelle f. ‖ arbre m. d'induit. / ~ slip (Electr) ‖ Ankerschlüpfung f.; Ankerschlupf m. ‖ glissement m. de l'induit. / ~ spider (Electr) ‖ Ankerstern m. ‖ crossillon m. d'induit. / ~ stray flux (Electr) ‖ Ankerstreufluß m. ‖ flux m. de dispersion dans l'induit. / ~ tooth (Electr) ‖ Ankerzahn m. ‖ dent f. d'induit. / ~ winding (Electr) ‖ Ankerwickelung f. ‖ enroulement ou bobinage m. d'induit.

armchair ‖ Armstuhl m.; Sessel m. ‖ fauteuil m. / swivel ~ ‖ Armstuhl m. mit verstellbarer Lehne ‖ fauteuil m. basculant.

arm clip (El line) ‖ Ziehband n. für Querträger ‖ bride f. / ~ file ‖ Armfeile f. ‖ carreau m.; lime f. à bras.

arming press (Bookb) ‖ Deckelpresse f. ‖ presse f. de carton.

armistice ‖ Waffenstillstand m. ‖ armistice m.

armlet ‖ Armband n. ‖ bracelet m.

armored see armoured.

armour ‖ Panzer m.; Armierung f. ‖ cuirasse f.; blindage m.; armature f. / ~ of a cable ‖ Schutzhülle f. eines Kabels ‖ envelope f. protectrice ou armature f. d'un câble. / hard steel ~ of safes ‖ Hartstahlpanzerung f. der Geldschränke ‖ blindage m. en acier au manganèse des coffres-forts. / ~ of the tank ‖ Panzer m. des Kraftwagens ‖ cuirasse f. de char de combat.

armour bending press ‖ Panzerplattenbiegepresse f. ‖ presse f. à cintrer les blindages.

armour-cased (Shipb) ‖ gepanzert ‖ cuirassé.

armoured ‖ armiert ‖ armé. / ~ cable ‖ Panzerdraht m. ‖ câble m. blindé. / ~ car body ‖ gepanzerter Wagenaufbau m. ‖ caisse f. cuirassée du char. / ~ concrete ‖ Eisenbeton m.; armierter Beton m. ‖ béton m. armé. / ~ cruiser ‖ Panzerkreuzer m. ‖ croiseur m. cuirassé. / ~ deck ‖ Panzerdeck m. ‖ pont m. cuirassé. / ~ fighting vehicle (Arm) ‖ gepanzertes Kampffahrzeug n. ‖ véhicule m. de combat blindé. / ~ forces pl. ‖ Panzerstreitkräfte fpl. ‖ forces fpl. militaires cuirassées. / ~ motor (Electr) ‖ Panzermotor m. ‖ moteur m. cuirassé. / ~ motor car ‖ gepanzertes Kampffahrzeug n. ‖ véhicule m. de combat blindé. / ~ street car ‖ Straßenpanzerwagen m. ‖ char m. cuirassé de rue. / ~ tractor ‖ gepanzerter Schlepper m. ‖ tracteur m. ou remorqueur m. blindé. / ~ train ‖ Panzerzug m. ‖ train m. blindé.

armourer ‖ Büchsenmacher m.; Waffenschmied m. ‖ armurier m. / ~ of cables ‖ Kabelpanzermaschinenarbeiter m. ‖ conducteur m. de machine à armer les câbles.

armouring see also armature ‖ Armatur f.; Armierung f. ‖ armature f. / ~ of cables ‖ Panzerung f. der Kabel ‖ armature f. des câbles.

armouring machine for cables ‖ Kabelarmierungsmaschine f.; Kabelbewehrungsmaschine f. ‖ machine f. à armer les câbles.

armour plate ‖ Panzerplatte f. ‖ plaque f. de blindage ou de cuirasse. / compound ~ ‖ Verbundstahlpanzerblech n. ‖ tôle f. de blindage en acier compound.

armour-plated ‖ gepanzert ‖ blindé; cuirassé.

armour plate rolling mill ‖ Panzerplattenwalzwerk n. ‖ laminoir m. à blindages; laminage m. de plaques de blindage. / ~ plating (Shipb) ‖ Panzerung f. ‖ cuirassement m. / ~ steel ‖ Panzerstahl m. ‖ acier m. de blindage.

arm rest (Auto) ‖ Armlehne f. ‖ accoudoir m.; bras m. de fauteuil. / ~ (Med) ‖ Armstütze f. ‖ appui-main m.; soutien-bras m.

arms pl. ‖ Waffen fpl. ‖ armes fpl. / accessories pl. for ~ ‖ Waffenzubehörteile mpl. ‖ accessoires mpl. d'armes. / steel for ~ ‖ Waffenstahl m. ‖ acier m. pour armes.

arm saw ‖ Handsäge f. ‖ scie f. à main. / ~ sling ‖ Armgurt m. ‖ brassière f. / ~ support ‖ Armstütze f. ‖ appui-bras m. / ~ support for templets ‖ Schablonenhalter m. ‖ support m. de la trousse.

army, mechanized ‖ verkraftetes Heer n. ‖ armée f. mécanique.

army baker ‖ Militärbäcker m. ‖ boulanger m. de garnison. / ~ cable (Electr) ‖ Armeekabel n. ‖ câble m. de l'armée. / ~ cloth ‖ Militärtuch n. ‖ drap m. de troupe. / ~ industrial college ‖ Heereswirtschaftshochschule f. ‖ école f. industrielle de l'armée. / ~ mechanization and motorization ‖ Heeresverkraftung f. ‖ mécanisation f. et motorisation f. d'armée. / ~ telephone ‖ Armeefernsprecher m. ‖ téléphone m. de l'armée. / ~ telescope ‖ Gewehrzielfernrohr n. ‖ hausse f. à lorgnette pour fusils.

arnica flower ‖ Arnikablüte f. ‖ fleur f. d'arnique. / ~ root ‖ Arnikawurzel f. ‖ racine f. d'arnique.

arnotto ‖ Orlean m. ‖ rocou m.; roucou m.; anotto m.

aroma (Wine) ‖ Blume f. ‖ bouquet m.

aromatic ‖ gewürzig; aromatisch ‖ aromatique. / ~ alcohol ‖ aromatisierter Alkohol m. ‖ alcool m. aromatisé. / ~ distillate ‖ aromatisches Destillat n. ‖ eau f. distillée aromatique. / ~ distiller ‖ Gewürzdestillateur m. ‖ distillateur m. d'essences ou d'aromates. / ~ oil ‖ aromatisches Öl n. ‖ huile f. aromatique. / ~ pastille ‖ Serailkügelchen n. ‖ pastille f. du sérail. / ~ powder ‖ wohlriechender Puder m. ‖ poudre f. de senteur. / ~ vinegar ‖ Toilettenessig m. ‖ vinaigre m. parfumé ou de toilette. / ~ water ‖ aromatisches Wasser n. ‖ eau m. aromatique.

aromatize, to ‖ aromatisieren ‖ aromatiser.

aromatized meal ‖ aromatisiertes Mehl n. ‖ farine f. aromatisée.

arrack ‖ Arrak m. ‖ arack m.; rack m.

arrange, to ‖ akkordieren; übereinkommen ‖ accorder; tomber d'accord m.

/ ~ (To mount) ‖ aufstellen ‖ mettre ; placer. / ~ (To regulate) ‖ ordnen ‖ assortir; mettre en ordre m.; régler. / ~ (Build) ‖ einfluchten ‖ dresser à la ligne; dresser d'alignement; enligner; aligner. / ~ clearly ‖ übersichtlich gruppieren ‖ grouper de façon synoptique. / ~ after the fashion of a pendulum ‖ pendelnd lagern ‖ articuler comme un pendule. / ~ to oscillate ‖ pendelnd lagern ‖ articuler comme un pendule.

arranged, the flues pl. are ~ in the boiler end ‖ die Rauchrohre npl. sind in den Kesselboden eingebaut ‖ les tubes mpl. de fumée sont montés dans le fond de la chaudière. / ~ side by side ‖ nebeneinander angeordnet ‖ installé l'un à côté de l'autre. / staggered ~ ‖ versetzt angeordnet ‖ installé en quinconce. / ~ in tandem ‖ hintereinander angeordnet ‖ installés pl. en tandem. / well ~ ‖ übersichtlich ‖ sommaire; clair; net.

arrangement (Appointment) ‖ Anordnung f. ‖ disposition f. / ~ (Agreement) ‖ Übereinkommen n.; Übereinkunft f.; Vereinbarung f. ‖ arrangement m.; convention f. / ~ (Accommodation) ‖ Vergleich m. ‖ comparaison f.; accord m. / ~ (Combination) ‖ Zusammenstellung f. ‖ rapprochement m.; combinaison f.; association f. / ~ (Trade) ‖ Ausgleich m. ‖ arrangement m. / amicable ~ ‖ gütlicher Vergleich m. ‖ arrangement m. à l'amiable. / brake lever ~s pl. ‖ Bremsgestänge n. ‖ timonerie f. du frein. / ~ for central code recording ‖ Sammelmorseeinrichtung f. ‖ dispositif m. d'enregistrement central par signaux Morse. / to come to an ~ ‖ in's Reine n. kommen ‖ s'arranger. / ~ with creditors ‖ Abfindung f. mit Gläubigern ‖ arrangement m. à l'amiable. / internal ~ ‖ innere Einrichtung f. ‖ aménagement m. intérieur. / to make ~s ‖ Anstalten fpl. machen ‖ se préparer à. / to make a private ~ with one's creditors ‖ sich mit seinen Gläubigern mpl. abfinden ‖ s'arranger; passer un arrangement avec ses créanciers. / ~ of rivets ‖ Nieteinteilung f.; Nietanordnung f. ‖ disposition f. ou distribution f. des rivets. / ~ in straight rows ‖ reihenförmige Anordnung f. ‖ disposition f. par files. / temporary ~s ‖ zeitweilige Anordnungen fpl. ‖ arrangements mpl. temporaires. / whole ~ ‖ Gesamtanordnung f. ‖ disposition f. d'ensemble.

arranging, bar ~ device ‖ Stabordner m. ‖ classeur m. de barres. / ~ station (Railw) ‖ Rangierbahnhof m.; Verschiebebahnhof m. ‖ gare f. de triage; station f. de composition et de décomposition; station f. de formation.

arrears pl. ‖ Außenstände mpl.; Rückstände mpl. ‖ arrérages mpl.; actifs créances fpl. / to be in ~ with ‖ ins Stocken n. geraten ‖ être interrompu. / ~ of interest ‖ rückständige Zinsen mpl. ‖ intérêts mpl. moratoires.

arrest, to ‖ anhalten; arretieren ‖ arrêter; stopper.

arrested fermentation ‖ Rastgärung f. ‖ fermentation f. d'arrêt ou d'interruption.

arrester ‖ Blitzableiter m. ‖ éclateur m.; parafoudre m.; paratonnerre m. / disc type ~ ‖ Scheibenblitzableiter m. ‖ parafoudre m. à disques. / electrolytic ~ ‖ elektrolyti-

scher Blitzableiter m. ‖ parafoudre m. électrolytique. / horn type ~ ‖ Hörnerblitzableiter m. ‖ parafoudre m. à cornes. / lightning ~ see arrester. / plate type ~ ‖ Plattenblitzableiter m. ‖ parafoudre m. à lames. / point type ~ ‖ Spitzenblitzableiter m. ‖ parafoudre m. à pointes. / spark gap type of ~ ‖ Funkenstreckenblitzableiter m. ‖ parafoudre m. à distance d'explosion.

arrester board ‖ Blitzschutztafel f. ‖ panneau m. parafoudre.

arresting cam ‖ Auflaufnocken m. ‖ came f. d'arrêt. / ~ device ‖ Anschlagvorrichtung f.; Hubbegrenzer m. ‖ dispositif m. d'arrêt ou de butée.

arris (Build) ‖ ausspringende Ecke f.; Kante f. ‖ angle m. saillant. / ~ (Roof) ‖ Grat m.; Gratlinie f. ‖ arête f.; ligne f. d'arête. / ~ beam (Carp) ‖ Gratbalken m.; Gratstichbalken m. ‖ coyer m.

arrival ‖ Ankunft ‖ arrivée f. / ~ curve (Tel) ‖ Empfangskurve f. ‖ courbe f. à l'arrivée. / port of ~ ‖ Ankunftshafen m. ‖ port m. d'arrivée. / time of ~ ‖ Ankunftszeit f. ‖ arrivée f.; heure f. d'arrivée.

arrivals pl. of corn ‖ Getreidezufuhr f. ‖ arrivage m. de blés.

arrive, to (Railw) ‖ ankommen ‖ arriver.

arrowroot ‖ Arrowroot n. ‖ arrow-root m.

arsenal ‖ Zeughaus n. ‖ arsenal m.

arseniate ‖ arseniksaures Salz n.; Arsensalz n. ‖ arséniate m. / ~ of cobalt ‖ Kobaltblüte f.; arseniksaurer Kobalt m.; roter Erdkobalt m.; Erythrin m. ‖ cobalt m. arsénié; fleur f. erythrine f.; mine f. de cobalt en efflorescence de cobalt. / ~ of copper ‖ Holzkupfererz n. ‖ cuivre m. arséniaté fibreux. / ~ of lead ‖ Arsenbleierz n. ‖ plomb m. arséniaté. / ~ of lime ‖ Pharmakolith m. ‖ arséniate m. de chaux; chaux f. arséniatée; pharmocolithe f.

arsenic ‖ Arsen n.; Arsenik n. ‖ arsenic m. / disulphide of ~ ‖ rotes Schwefelarsen n.; Realgar n. ‖ arsenic m. sulfuré rouge; réalgar m. / flaky ~ ‖ Fliegenstein m.; Näpfchenkobalt m. ‖ arsenic m. noir ou écailleux. / metallic ~ ‖ Scherbenkobalt m. ‖ arsenic m. noir. / native ~ ‖ Scherbenkobalt m.; gediegenes Arsen n.; Fliegenstein m. ‖ arsenic m. natif. / red ~ see arsenic, disulphide of. / trisulphide of ~ ‖ gelbes Schwefelarsen n.; Auripigment n.; Orpiment n. ‖ arsenic m. sulfuré jaune. / white ~ ‖ Giftmehl n. ‖ arsenic m. blanc; poudre f. d'arsenic. / yellow sulphide of ~ see arsenic, trisulphide of.

arsenical ‖ arsenhaltig ‖ arsenical. / ~ acid ‖ Arsensäure f. ‖ acide m. arsénique. / ~ cadmia ‖ Giftstein m. ‖ cadmie f. arsenicale. / ~ copper ‖ Arsenkupfer n. ‖ cuivre m. arsénical. / ~ iron see ~ pyrites. / ~ nickel-ore ‖ Arsennickelglanz m. ‖ diosmose f.; nickel m. gris ou arsénio-sulfure. / ~ pyrites ‖ Arsenikkies m. ‖ arséniosulfure m. de fer; fer m. arsenical; mine f. arsenicale blanche.

arsenic bath ‖ Arsenbad n. ‖ bain m. d'arséniage. / ~ bloom ‖ Arsenikblüte f.; weißer Arsenik m. ‖ acide m. arsénieux natif; arsenic m. blanc ou oxydé. / ~ copper ‖ Arsenkupfer n. ‖ cuivre m. arsenical. / red ~ glass ‖ rotes Schwefelarsen n.; Realgar n. ‖ arsenic m. sulfuré rouge;

réalgar m. / ~ mine ‖ Arsenikerzbergwerk n. ‖ mine f. d'arsenic / ~ ore ‖ Arsenikerz n. ‖ minerai m. arsénifère. / ~ pentasulphide ‖ Arsensulfid n. ‖ pentasulfure m. d'arsenic. / ~ powder ‖ Giftmehl n. ‖ arsenic m. blanc; poudre f. d'arsenic. / ~ sulphide ‖ Schwefelarsen n. ‖ trisulfure m. d'arsenic. / ~ trisulphide ‖ Arsensulfür n. ‖ trisulphide m. d'arsenic. / ~ works pl. ‖ Arsenikhütte f. ‖ fabrique f. d'arsenic.

arseniferous ‖ arsenführend ‖ arsénifère.

arsenious acid ‖ arsenige Säure f. ‖ acide m. arsénieux. / ~ anhydride ‖ Giftmehl n. ‖ anhydride m. arsénieux. / ~ oxide ‖ Arsenik n.; Arsenikblume f. ‖ arsenic m.; anhydride m. arsénieux. / ~ trisulphide ‖ gelbes Schwefelarsen n.; Auripigment n.; Orpiment n. ‖ arsenic m. sulfuré jaune.

arsenite ‖ arsenigsaures Salz n. ‖ arsénite m. / ~ of potassium ‖ arsensaures Kalium n. ‖ arsénite m. de potasse.

arseniurated see arseniuretted.

arseniuretted hydrogen ‖ Arsenwasserstoff m. ‖ hydrogène m. arsénié.

arsenolamprite ‖ Arsenglanz m. ‖ arsenolamprite f.

arsenolite ‖ Arsenikkalk; Arsenblüte f. ‖ arsénolite f.

arsenopyrite ‖ Arsenkies m. ‖ mispickel m.; pyrite f. arsenicale; fer m. arsenical; arsénopyrite f.

art ‖ Kunst f. ‖ art m. / applied ~ ‖ Kunstgewerbe n. ‖ art m. industriel. / ~ of building ‖ Architektur f.; Baukunst f. ‖ architecture f.; art m. de bâtir. / castings pl. for objects of ~s ‖ Kunstguß m. ‖ objets mpl. d'art en fonte (moulée). / ~ ceramic ‖ Kunstkeramik f. ‖ art m. céramique. / ecclesiastical ~ ‖ kirchliche Kunst f. ‖ art m. ecclésiastique. / ~ of engraving on copper ‖ Kupferstechkunst f. ‖ chalcographie f. / ~ of glass making ‖ Glasmacherkunst f. ‖ verrerie f. / object of ~ ‖ Kunstgegenstand m. ‖ objet m. d'art. / small ~ furniture ‖ Kleinkunstmöbel npl. ‖ meubles mpl. mignons. / ~ of surveying mines ‖ Markscheidekunst f. ‖ géometrie f. souterraine. / works pl. of applied ~ ‖ kunstgewerbliche Arbeiten fpl. ‖ travaux mpl. d'art industriel.

art bronze ware ‖ Gegenstand m. aus Kunstbronze ‖ objet m. d'art en bronze. / ~ ceramics pl. ‖ Kunstkeramik f. ‖ céramique f. d'art. / ~ embroiderer ‖ Kunststicker m. ‖ brodeur m. d'art.

artery ‖ Arterie f.; Pulsader f. ‖ artère f. / ~ forceps ‖ Arterienklemme f. ‖ pince f. hémostatique.

artesian well ‖ artesischer Brunnen m. ‖ puits m. artésien.

art foundry ‖ Kunstgießerei f. ‖ fonderie f. d'art.

arthrosporic bacterium ‖ arthrospores Bakterium n. ‖ bactérie f. arthrospore.

artichoke ‖ Artischocke f. ‖ artichaut m.

article, sewed ‖ genähter Gegenstand m.; Konfektion f. ‖ article m. confectionné. / technical ~ ‖ technischer Gegenstand m. oder Bedarfsartikel m. ‖ article m. technique. / ~ of value ‖ Wertgegenstand m. ‖ objet m. précieux. / very popular ~ ‖ Zugartikel m. ‖ articleréclame f.

articles pl. of association ‖ Gesellschaftsstatuten npl. ‖ statuts mpl. sociaux. / ~ for barbers ‖ Barbierbedarfsartikel mpl. ‖

fournitures fpl. pour coiffeurs. / ~ for carnival || Karnevalsartikel mpl. || articles mpl. de carneval. / chemical-pharmaceutical ~ || chemisch-pharmazeutische Artikel mpl. || articles mpl. chimiques-pharmaceutiques. / ~ for consumption || Genußmittel n. || comestible m. / ~ for daily use || Konsumartikel mpl. || articles mpl. de consommation. / devotional ~ || Devotionalien pl. || articles mpl. de piété. / ~ of export || Exportartikel mpl.; Ausfuhrgegenstände mpl. || articles mpl. d'exportation. / ~ for hairdressers || Frisörbedarfsartikel mpl. || fournitures fpl. pour coiffeurs. / ~ made on lathes || Drehereiwaren fpl. || articles mpl. tournés. / ~ for making a noise || Radauartikel mpl. || articles mpl. pour faire du bruit. / ~ totally or partly composed of metal-filaments || ganz oder teilweise aus Metallfäden bestehende Waren fpl. || articles mpl. composés entièrement ou partiellement de fils métalliques. / ~ of mineral materials || Waren fpl. aus mineralischen Stoffen || produits mpl. en matières minérales. / ~ for social clubs || Vereinsartikel mpl. || articles mpl. de tombola pour sociétés. / stationery ~ made of cardboard || Schreibwarenartikel mpl. aus Pappmaché || articles mpl. de bureau en carton pâte.

articulated coupling || Gelenkkupplung f. || joint m. articulé. / ~ locomotive || kurvenbewegliche Lokomotive f. || locomotive f. articulée. / ~ pipe || Gelenkschlauch m. || tuyau m. articulé. / ~ rack || Gelenkzahnstange f. || crémaillère f. articulée. / ~ rod || Gelenkstange f. || tige f. articulée. / ~ tank locomotive || Drehgestelltenderlokomotive f. || locomotivetender f. à essieux couplés articulés.

articulation (Anatomy) || Gelenk n. || jointure f. / ~ (Geol) || Gefüge n.; Struktur f. || structure f.; grainure f.; texture f. / ~ (Tel) || Verständlichkeit f. || netteté f. / ~ by ball and socket || Kugelgelenk n. || articulation f. à rotule; joint m. sphérique ou à boulet. / ~ of letters (Tel) || Lautverständlichkeit f. || netteté f. pour les lettres. / ~ of sentences (Tel) || Satzverständlichkeit f. || netteté f. pour les phrases. / sufficient ~ (Tel) || hinreichende Sprachgüte f. || bonne qualité f. de transmission. / ~ of syllables (Tel) || Silbenverständlichkeit f. || netteté f. pour les syllabes. / ~ of words (Tel) || Wortverständlichkeit f. || netteté f. pour les mots. / ~ piece || Bremsgelenkstück n.; Gelenkstück n. || pièce f. d'articulation.

artificer || Handwerker m. || homme m. de métier; ouvrier m.; artisan m.

artificial || künstlich; nachgemacht || artificiel; factive. / ~ balancing line (Tel) || Leitungsnachbildung f. || ligne f. artificielle. / plant for ~ chimney draught || Schornsteinanlage f. mit künstlichem Zug || installation f. de tirage artificiel pour les cheminées. / ~ cooling || künstliche Kühlung f. || rafraîchissement m. artificiel. / ~ cotton || Kunstbaumwolle f. || coton m. artificiel. / ~ cotton making machine || Kunstbaumwollefertigungsmaschine f. || machine f. de fabrication du coton artificiel. / ~ denture || Kunstgebiß n. || denture f. artificielle; prothèse f. dentaire. / ~ drying || künstliches Trocknen n. || séchage m. artificiel.

artificial eye || künstliches Auge n.; Glasauge n. || œil artificiel. / ~ maker || Glasaugenmacher m. || oculariste m.

artificial flower || Kunstblume f.; künstliche Blume f. || fleur f. artificielle. / ~ maker || Blumenarbeiter m. || fleuriste m. en fleurs artificielles. / ~ mounter || Kunstblumenarbeiterin f. || monteuse f. en fleurs artificielles. / ~ painter || Blumenblättermaler m. || peintre m. en feuilles de fleurs artificielles.

artificial fog || künstlicher Nebel m. || brouillard m. artificiel. / ~ gem || Glasstein m.; künstlicher oder unechter Edelstein m.; Similistein m. || pierre f. précieuse artificielle; pierre f. fausse. / ~ grinding mill || Kunstschleifstein m. || meule f. artificielle. / ~ horn producing machine || Kunsthornherstellungsmaschine f. || machine f. à fabriquer la corne artificielle.

artificial ice || künstliches Eis n. || glace f. artificielle. / ~ skating rink || Kunsteisbahn f. || piste f. de patinage (sur glace) artificielle.

artificial insulating material || künstlicher Isolationsstoff m. || isolants mpl. synthétiques.

artificial leather cardboard || Kunstlederpappe f. || carton m. de cuir artificiel. / ~ factory || Kunstlederfabrik f. || fabrique f. de cuir artificiel.

artificial limbs pl. || künstliche Glieder npl. || membres mpl. artificiels. / ~ magnet || künstlicher Magnet m. || aimant m. artificiel. / ~ marble || Gipsmarmor m. || marbre m. factice. / machine for making ~ materials || Kunst(werk)stoffeherstellungsmaschine f. || machine f. à fabriquer des matières artificielles. / ~ mountain sun || künstliche Höhensonne f. || soleil m. d'altitude artificiel.

artificial silk || Kunstseide f. || soie f. artificielle. / ~ mixed with natural silk || mit natürlicher Seide gemischte Kunstseide f. || soie f. artificielle à l'état de mélange avec de la soie naturelle. / ~ mixed with other spun materials || mit anderen Spinnstoffen gemischte Kunstseide f. || soie f. artificielle à l'état de mélange avec d'autres fibres textiles.

artificial silk factory || Kunstseidenfabrik f. || fabrique f. de soie artificielle.

artificial silk fibre || Kunstseidenfaser f. || fibre f. de soie artificielle. / ~ carded || gekrempelte Kunstseidenfaser f. || fibre f. de soie artificielle cardée. / combed ~ || gekämmte Kunstseidenfaser f. || fibre f. de soie artificielle peignée. / ~ in ribbons || Kunstseidenfaser f. in Bändern || fibre f. de soie artificielle en rubans. / ~ in rovings see ~ in slubbings. / ~ in slubbings || Kunstseidenfaser f. in Vorgespinsten || fibre f. de soie artificielle en mèches.

artificial silk, machine for the manufacture of ~ || Kunstseideherstellungsmaschine f. || machine f. de fabrication de la soie artificielle. / ~ waste || Kunstseidenabfall m. || déchet m. de soie artificielle. / ~ yarn || Kunstseidengarn n. || fil m. de soie artificielle.

artificial skating rink see artificial ice scating rink. / ~ slate || künstlicher Schiefer m. || ardoise f. artificielle.

artificial stone || Kunststein m. || pierre f. artificielle ou factice.

artificial stone articles pl. || Kunststeinwaren fpl. || objets mpl. en pierre artificielle. / ~ bruising mill || Kunststeinschrotmühle f. || moulin m. concasseur à meules artificielles. / ~ factory || Kunststeinfabrik f. || fabrique f. de pierres artificielles. / ~ industry || Kunststeinindustrie f. || industrie f. de produits en pierre artificielle. / ~ making machine || Kunststeinherstellungsmaschine f. || machine f. à fabriquer des pierres artificielles.

artificial tooth || Kunstzahn m.; künstlicher Zahn m. || dent f. artificielle. / ~ violet oil || synthetisches Veilchenöl n. || essence f. artificielle de violette. / ~ wood articles pl. || Xylolithwaren fpl.; Steinholzwaren fpl. || objets mpl. en bois artificiel ou en xylolithe.

artificial wool making machine || Kunstwolleherstellungsmaschine f. || machine f. de fabrication de la laine artificielle. / ~ manufactory || Kunstwollefabrik f. || fabrique f. de laine artificielle.

artillery, antitank || Kampfwagenabwehrartillerie f. || artillerie f. de défense contre les chars d'assaut. / caterpillar ~ || Raupenartillerie f. || artillerie f. sur chenilles. / close support ~ || direkte Unterstützungsartillerie f. || artillerie f. d'appui direct. / mountain ~ || Gebirgsartillerie f. || artillerie f. de montagne. / tractor drawn ~ || kraftgeschleppte Artillerie f. || artillerie f. à tracteur.

artillery cable (Tel) || Artilleriekabel n. || câble m. d'artillerie ou de campagne. / ~ caterpillar tractor || Artillerieraupenschlepper m. || tracteur m. d'artillerie à chenilles. / ~ committee || Artilleriekomitee n. || comité m. d'artillerie.

artillery fire control device || Artilleriefeuerleitgerät n. || appareil m. de conduite du tir de l'artillerie. / direction of ~ || Feuerleitung f. || direction f. ou conduite f. du tir.

artisan || Handwerker m. || homme m. de métier; ouvrier m.; artisan m.

artistic || künstlerisch || artistique. / ~ engraving || künstlerischer Stich m. || gravure f. artistique. / ~ establishment for ceramic printing in colours || keramische Buntdruckerei f. || atelier m. artistique à imprimer des produits céramiques en couleurs. / ~ finish || künstlerische Ausführung f. || modèle m. artistique. / ~ forged piece || Kunstschmiedestück n. || pièce f. d'art en fer forgé. / ~ foundry || Kunstgießerei f. || fonderie f. artistique. / ~ furniture || Kunstmöbel n. || meuble m. artistique. / ~ glass || Kunstglas n. || verre m. d'art. / ~ glass ware || Kunstglasware f. || verrerie f. artistique. / ~ lamp || kunstvolle Lampe f. || lampe f. artistique. / ~ tin article || Zinnfigur f. || objet m. d'art en étain. / ~ turned work || Kunstdrechslerarbeit f. || travail m. d'art tourné.

artist's paint || Künstlerfarbe f. || couleur f. fine pour artistes.

art paper || Kunstdruckpapier n. || papier m. pour impression artistique. / ~ pottery || Kunsttöpferei f. || poterie f. artistique. / ~ printing office / Kunstdruckerei f. || imprimerie f. artistique. / ~ publishers pl. || Kunstverlag m. || éditeurs mpl. d'art. / ~ publishing establishment || Kunstverlaganstalt f. || maison f. de publications et librairies artistiques.

/ ~ square weaver ‖ Kunstweber m. ‖ tisseur m. d'art. / ~ stained glass ‖ gemaltes Fenster m. ‖ vitrail m. d'art. / ~ work ‖ Kunstbau f. ‖ ouvrage m. d'art.

as soon as possible ‖ tunlichst bald ‖ le plus tôt possible. / ~ to . . . ‖ hinsichtlich . . .; . . . betreffend; was . . . anbetrifft ‖ à l'égard de . . .; au sujet de . . .; concernant . . .

asa fœtida ‖ stinkender Asant m.; Asafötida f. ‖ asa-fœtida f.

asbestos ‖ Asbest m. ‖ amiante m.; asbeste m.; filasse f. de montagne. / flaked ~ ‖ Flockenasbest m. ‖ amiante m. floconneux. / lignous ~ ‖ Holzasbest m. ‖ asbeste m. ligniforme.

asbestos base ‖ Asbestunterlage f. ‖ couche f. d'asbeste. / ~ board ‖ Asbestpappe f. ‖ carton m. en amiante. / ~ cement ‖ Asbestzement m. ‖ ciment m. d'asbeste. / ~ cloth ‖ Asbestgewebe n.; Asbesttuch n. ‖ tissu m. ou toile f. d'amiante. / ~ concrete ‖ Asbestbeton m. ‖ béton m. d'amiante. / ~ cord ‖ Asbestschnur f. ‖ corde f. ou cordon m. en amiante; tresse f. d'amiante. / ~ covering ‖ Asbestbekleidung f. ‖ revêtement m. en asbeste. / ~ covering of floors ‖ Asbestbodenbelag m. ‖ couverture f. de plancher en asbeste. / ~ diaphragm ‖ Asbestdiaphragma n.; Asbestscheidewand f. ‖ diaphragme m. en amiante. / ~ dish ‖ Asbestschale f. ‖ capsule f. d'amiante. / ~ fabric see ~ cloth. / ~ fiber ‖ Asbestfaser f. ‖ fibre f. d'amiante. / ~ glove ‖ Asbesthandschuh m. ‖ gant m. en amiante. / ~ goods pl. ‖ Asbestwaren fpl. ‖ articles mpl. en amiante. / ~ iron wire gauze ‖ Asbesteisendrahtnetz n. ‖ toile f. métallique amiantée. / ~ layer ‖ Asbesteinlage f. ‖ couche f. d'amiante ou d'asbeste. / ~ manufacture ‖ Asbestfabrikat n.; Asbesterzeugnis n. ‖ article m. d'amiante. / ~ manufacturer ‖ Asbestfabrikant m. ‖ fabricant m. d'amiante. / ~ milk (Chem) ‖ Asbestaufschlemmung f. ‖ lait m. d'amiante. / ~ paper ‖ Asbestpapier n. ‖ papier m. en amiante. / ~ plate ‖ Asbestscheibe f.; Asbestplatte f. ‖ disque m. d'asbeste; plaque f. d'amiante. / ~ powder ‖ Asbestpulver n. ‖ poudre f. d'amiante. / ~ putty ‖ Asbestkitt m. ‖ mastic m. d'amiante. / ~ quarry ‖ Asbestgrube f. ‖ carrière f. d'amiante. / ~ ring ‖ Asbestring m. ‖ anneau m. en amiante. / ~ rope ‖ Asbestschnur f.; Asbestseil n. ‖ corde f. ou cordon m. en amiante. / ~ sheet ‖ Asbestplatte f. ‖ plaque f. d'amiante. / ~ slate ‖ Asbestschiefer m. ‖ ardoise f. d'amiante ou d'asbeste. / ~ slate manufacturing plant ‖ Asbestschieferherstellungsanlage f. ‖ installation f. pour la fabrication des ardoises d'asbeste. / ~ spinning machine ‖ Asbestspinnereimaschine f. ‖ machine f. à filer l'asbeste. / ~ stopper ‖ Asbeststöpsel m. ‖ bouchon m. en amiante. / ~ stuff ‖ Asbestleinwand f. ‖ tissu m. d'amiante. / ~ tray see ~ plate. / ~ weaver ‖ Asbestweber m. ‖ tisseur m. d'amiante. / ~ wire ‖ Asbestdraht m. ‖ fil m. isolé à l'amiante. / ~ wool ‖ Asbestfaser f.; Asbestflocken fpl. ‖ floche f. d'amiante. / ~ yarn ‖ Asbestfaden m. ‖ fil m. en amiante.

ascend, to ‖ in die Höhe steigen; aufsteigen ‖ monter; ascendre; s'élever en altitude.

ascending current (of air) ‖ Aufwind m. ‖ vent m. ascendant. / ~ part of a letter

(Print) ‖ Oberlänge f. einer Letter ‖ queue f. de dessus d'un caractère. / ~ and descending pipe lines for heating water ‖ Steig- und Falleitungen fpl. für Heizwasser ‖ tuyauteries fpl. pour faire monter et descendre l'eau de chauffage. / ~ step (Mining) ‖ Firststoß m. ‖ gradin m. renversé.

ascension ‖ Aufsteigen n. ‖ élévation f.; ascension f. / right ~ (Astro) ‖ Rektaszension f. ‖ ascension f. droite.

ascensor, hydraulic ‖ hydraulischer Aufzug m. ‖ ascenseur m. hydraulique.

ascent (Loading ramp) ‖ Rampe f.; Auffahrt f. ‖ rampe f.; chemin m. en talus. / ~ (Step) ‖ Aufgang m.; Stiege f.; Treppe f. ‖ montée f.; escalier m. / ~ (Topogr) ‖ Steigung f.; Gefälle f.; Neigung f. ‖ montée f.; pente f.; declivité f. / movable ~ (Railw) ‖ tragbare oder transportable Rampe f. ‖ rampe f. roulante.

ascertain, to ‖ vergewissern ‖ confirmer.

ascertaining the visibility ‖ Sehschärfenbestimmung f. ‖ determination f. de l'acuité visuelle.

asclepiad, Syrian ‖ syrische Seidenpflanze f. ‖ asclépiade f. de Syrie.

aseismic district ‖ erdbebenfreies Gebiet n. / domaine m. aséismique.

aseptic ‖ aseptisch; keimfrei ‖ aseptique.

ash ‖ Esche f. ‖ frêne m.

ash see also ashes ‖ Asche f. ‖ cendre f. / bone ~ ‖ Knochenasche f. ‖ cendre f. d'os. / charcoal ~ ‖ Holzkohlenasche f. ‖ fraisil m. de charbon de bois. / coal ~ ‖ Kohlenasche f. ‖ cendre f. de charbon ou de houille. / fly ~ see ashes, fly. / gold ~ ‖ Goldasche f. ‖ cendre f. d'orfèvre. / lead ~ ‖ Bleiasche f. ‖ cendre f. de plomb. / pearl ~ ‖ Perlasche f. ‖ perlasse f. / pit coal ~ ‖ Steinkohlenasche f. ‖ cendre f. de charbon; fraisil m. / quick ~ see ashes, fly. / soda ~ ‖ Sodaasche f. ‖ cendre f. de soude. / wood ~ ‖ Holzasche f. ‖ cendre f. de bois.

ash bath ‖ Aschenbad n. ‖ bain m. de cendres. / ~ bin see ash box. / ~ boiler ‖ Aschensieder m. ‖ lessiveur m. de cendres.

ash box see also ash pan ‖ Asch(en)kasten m.; Aschfall m. ‖ cendrier m. / ~ damper ‖ Aschkastenklappe f. ‖ clapet m. du cendrier. / ~ door ‖ Aschkastentür f. ‖ porte f. du cendrier.

ash bunker ‖ Aschenbunker m. ‖ trémie f. à cendres. / ~ concrete ‖ Löschebeton m. ‖ béton m. de cendres. / ~ contents pl. see ashes, contents of. / ~ conveying plant ‖ Aschenförderanlage f. ‖ installation f. pour le transport des cendres. / ~ conveyor ‖ Aschenförderer m.; Aschentransporteinrichtung f. ‖ dispositif m. ou appareil m. à transporter les cendres; transporteur m. à cendres. / ~ ejector ‖ Aschenausbläser m. ‖ escarbilleur m.

ashes pl. see also ash ‖ Asche f. ‖ cendres fpl.; escarbilles fpl. / ~ pl. (Met) ‖ Gekrätz n.; Krätze f.; Metallabfälle mpl. ‖ déchets mpl. métalliques; terre f. de monnaie; crasse f.; cendres fpl. / blue ~ ‖ Blauasche f.; Kupferblau n. ‖ cendre f. bleue native; cendres fpl. bleues; bleu m. de montagne. / coal ~ see ash, coal. / flaky ~ see fly ~. / fly ~ ‖ Flugasche f. ‖ cendre f. mouvante ou volante ou volatile; cendres fpl. folles. / flying ~ see fly ~. / flue ~ see fly ~. / free from ~ ‖ aschenfrei ‖ exempt de cendres fpl.

/ light ~ see fly ~. / melted ~ (Chem) ‖ Pottaschenfluß m.; rohe Pottasche f.; Ochras m. ‖ salin m.; cassoudes fpl.; casottes fpl. / quick ~ see ashes, fly. / volatile ~ see fly ~. / volcanic ~ ‖ vulkanische Asche f.; vulkanischer Sand m. ‖ cendres fpl. volcaniques; sable m. volcanique.

ashes, contents pl. of ‖ Aschengehalt m. ‖ teneur f. en cendres. / determination of ~ ‖ Aschenbestimmung f. ‖ détermination f. des cendres. / percentage of ~ ‖ Prozentsatz m. an Asche; Aschengehalt m. ‖ pourcentage m. ou teneur f. en cendres.

ash filler ‖ Aschenauflader m. ‖ chargeur m. de cendres. / ~ handling plant see ash conveying plant.

ash hoist ‖ Aschheißmaschine f. ‖ monte-escarbilles m. / automatic ~ ‖ Aschenejektor m. ‖ éjecteur m. d'escarbilles.

ash hole ‖ Aschenkasten m.; Aschenloch n.; Aschenkeller m. ‖ cendrier m.; trou m. ou fosse f. à cendres. / ~ keel ‖ Eschenholzkiel m. ‖ quille f. en bois de frêne.

ashlar ‖ Haustein m.; Quader(stein) m. ‖ moellon m. d'appareil; pierre f. de taille. / small ~ for mosaics ‖ Bruchstein m. für Mosaik ‖ moellon m. pour mosaïque.

ashlar breaker ‖ Steinbrecher m. ‖ casseur m. de moellons.

ashlaring ‖ Dachverschalung f. ‖ revêtement m. du toit.

ashlar quarry ‖ Mauersteinbruch m. ‖ carrière f. de moellons. / ~ (stone) work ‖ Quadermauerwerk n.; Hausteingemäuer n. ‖ maçonnerie f. vive ou en pierres de taille.

ash-leaved maple ‖ eschenähnlicher Ahorn m. ‖ érable m. négundo.

ash leaves pl. ‖ Eschenblätter npl. ‖ feuilles fpl. de frêne.

ashler see ashlar.

ashore, to be ~ (Mar) ‖ festsitzen; festgekommen sein ‖ être échoué; être mouillé par la quille.

ash pan ‖ Asch(en)kasten m.; Aschfall m. ‖ caisse f. à cendres; cendrier m. / ~ damper ‖ Aschfallklappe f. ‖ clapet m. du cendrier; porte f. à rabattement du cendrier. / ~ door ‖ Aschfalltür f. ‖ porte f. du cendrier. / ~ dump see ~ damper. / ~ gear ‖ Aschfallzug m. ‖ tringle f. de commande des portes du cendrier. / ~ slide ‖ Aschkastenschieber m. ‖ porte f. glissante du cendrier.

ash pit (Loc) ‖ Asch(en)grube f.; Asch(en)kasten m. ‖ fosse f. à cendres ou d'extinction; cendrier m. / ~ (Railw) ‖ Reinigungsgrube f.; Löschgrube f.; Feuergrube f. ‖ fosse f. à piquer le feu.

ash plate of a fining-forge ‖ Hinterzacken m.; Aschenzacken m. des Frischherdes ‖ taque f. de fond; herre f.; haire f.; rustine f.; plaque f. ou taque f. de rustine. / ~ removal plant ‖ Entaschungsanlage f. ‖ installation f. pour l'enlèvement des cendres. / ~ removing installation see ~ removal plant. / ~ scuttle (Shipb) ‖ Aschenpforte f. ‖ sabord m. à escarbilles. / ~ stop ‖ Aschenschieber m. ‖ registre m. de cendrier. / ~ transporter ‖ Aschenförderer m. ‖ transporteur m. de cendre. / ~ tray ‖ Aschenbecher m. ‖ cendrier m.

ashtree ‖ Esche f. ‖ frêne m. / ~ oil ‖ Eschenholzöl n. ‖ huile f. de frêne. / ~ wood see ash wood.

ash washer ‖ Aschenwascher m. ‖ lavoir m. à scories. / ~ washing plant ‖ Aschenwäsche f. ‖ lavage m. des cendres. / ~ wheeler (Min) ‖ Aschenschlepper m. ‖ brouetteur m. de cendres.

ash wo̜od ‖ Eschenholz n. ‖ bois m. de frêne. / ~ handle ‖ Werkzeugheft n. aus Eschenholz ‖ manche m. en frêne.

aslope ‖ abdachig ‖ incliné; en pente; en talus.

asp ‖ Espe f.; Zitterpappel f. ‖ tremble m.

asparagine ‖ Asparagin n. ‖ asparagine f.

asparagus ‖ Spargel m. ‖ asperge f. / ~ cultivation ‖ Spargelzüchterei f. ‖ aspergerie f. / ~ grower ‖ Spargelplantagenbesitzer m. ‖ asparagiculteur m. / ~ knife ‖ Spargelstecher m. ‖ coupe-asperges m.

aspartic acid ‖ Asparaginsäure f. ‖ acide m. aspartique.

asphalt, to ‖ asphaltieren ‖ bitumer; asphalter.

asphalt ‖ Erdpech n.; Asphalt m.; Asiphaltstein m.; Bergpech n. ‖ bitume m. solide; asphalte m.; goudron m. minéral. / artificial ~ ‖ künstlicher Asphalt m. ‖ asphalte m. artificiel. / cast ~ ‖ Schmelzasphalt m. ‖ asphalte m. ou fondu coulé. / compressed ~ ‖ Stampfasphalt m. ‖ asphalte m. comprimé. / crude ~ ‖ Asphaltgestein n. ‖ roche f. asphaltique./ melted ~ ‖ Gußasphalt m. ‖ asphalte m. coulé. / warping pipes in ~ ‖ Asphaltieren n. von Röhren ‖ enveloppement m. en asphalte de tubes. / proper ~ ‖ eigentlicher Asphalt m. ‖ asphalte m. proprement dit.

asphalt boiler ‖ Asphaltkochkessel m. ‖ chaudière f. d'asphalte. / ~ box for cable armouring machines ‖ Asphaltkasten m. für Kabelarmierungsmaschinen ‖ récipient m. d'asphalte pour machines à armer les câbles. / ~ cement ‖ Asphaltmastix m. ‖ mastic m. d'asphalte. / ~ concrete ‖ Asphaltbeton m. ‖ béton m. d'asphalte. / ~ conduit ‖ Asphaltrohr n. ‖ tube m. ou conduit m. en asphalte. / ~ fuel having a high ~ content ‖ asphaltreicher Brennstoff m. ‖ combustible m. riche en bitume. / ~ covering ‖ Asphaltdecke f. ‖ revêtement m. bitumé.

asphalted cable ‖ asphaltiertes Kabel n. ‖ câble m. asphalté.

asphalt extraction ‖ Asphaltgewinnung f. ‖ extraction f. d'asphalte. / ~ floor ‖ Asphaltestrich m.; Asphaltboden m. ‖ aire f. en asphalte. / ~ furnace ‖ Asphaltofen m. ‖ four m. à asphalte.

asphaltic felt ‖ Asphaltfilz m. ‖ feutre m. asphalté. / ~ lime stone ‖ Asphaltkalkstein m. ‖ calcaire m. asphaltique. / ~ roofing-board ‖ Dachpappe f. ‖ carton m. bitumé pour toitures.

asphalt macadam ‖ Asphaltmakadam m. ‖ route f. asphaltée. / ~ mill ‖ Asphaltmühle f. ‖ moulin m. à asphalte. / ~ mortar ‖ Asphaltmörtel m. ‖ mortier m. d'asphalte.

asphaltos see asphaltum.

asphalt paste for filling ‖ Asphaltkompoundmasse f. ‖ mastic m. d'asphalte. / ~ pavement ‖ Asphaltpflaster n. ‖ pavé m. d'asphalte. / ~ paving see pavement. / ~ plate ‖ Asphaltplatte f. ‖ carreau m. en asphalte. / ~ powder ‖ pulverisierter Asphalt m. ‖ asphalte m. en poudre. / ~ product ‖ Asphaltprodukt n. ‖ produit m. d'asphalte. / ~ slab

manufacturing press ‖ Presse f. zur Herstellung von Asphaltplatten ‖ presse f. pour la fabrication de plaques en asphalte. / ~ trade, plant for the ‖ Asphaltindustrieanlage f.; Asphaltverarbeitungsanlage f. ‖ installation f. industrielle de l'asphalte; installation f. à travailler l'asphalte.

asphaltum ‖ Asphalt m.; Erdpech n.; Judenpech n. ‖ asphalte m.; bitume m. solide ou de Judée; goudron m. minéral. / ~ paper ‖ Asphaltpapier n. ‖ papier m. asphalté. / ~ product ‖ Asphalterzeugnis n. ‖ produit m. d'asphalte.

asphalt varnish ‖ Asphaltlack m. ‖ vernis m. à l'asphalte. / ~ work ‖ Asphaltarbeit f. ‖ ouvrage m. ou pavage m. en asphalte.

aspirate, to ‖ aufsaugen; ansaugen; absaugen ‖ aspirer.

aspirating cylinder ‖ Ansaugzylinder m. ‖ cylindre m. aspirateur. / ~ dredger ‖ Saugbagger m. ‖ drague f. suceuse ou aspirante. / ~ hole (Ventilator) ‖ Saugöffnung f. ‖ ouïe f. / ~ mouth see ~ hole.

aspiration duct ‖ Absaugkanal m. ‖ canal m. d'aspiration.

aspirator ‖ Aspirator m. ‖ aspirateur m. ‖ water-jet ~ ‖ Wasserluftpumpe f. ‖ trompe f. à eau.

asp wood ‖ Espenholz n. ‖ bois m. de tremble.

ass ‖ Esel m. ‖ âne m. / ~ of a block (Mar) ‖ Herd m. eines Blocks ‖ cul m. ou queue f. d'une poulie.

assay, to ‖ probieren; prüfen ‖ essayer; éprouver. / ~ an ore ‖ eine Erzprobe f. ausführen ‖ essayer un minerai.

assay ‖ Probe f.; Prüfung f.; Proben n. ‖ essai m.; épreuve f.; prise f. d'essai. / ~ (Sample) ‖ Probe f.; das zu Prüfende ‖ échantillon m.; spécimen m. / the ~ averages ... ‖ die Probe f. ergibt durchschnittlich ... ‖ l'essai m. donne une moyenne ... / dry ~ ‖ Trockenprobe f.; Proben n. auf trockenem Wege ‖ essai m. par la voie sèche. / fire ~ see dry ~. / ~ of metal ‖ Metallprobe f. ‖ essai m. ou vérification f. du métal. / ~ by the moist way see wet ~. / ore ~ ‖ Erzprobe f. ‖ essai m. du minerai. / wet ~ ‖ Proben n. auf nassem Wege ‖ essai m. par la voie humide.

assay balance (Coin) ‖ Goldwage f.; Probierwage f. ‖ balance f. d'essayeur; biquet m.; trébouchet m. / ~ certificate ‖ Metallprobenattest n. ‖ certificat m. d'essai.

assay crucible ‖ Probiertiegel m. ‖ creuset m. d'essai. / pear-shaped ~ ‖ Probiertute f.; Tute f. ‖ têt m. à la fusion.

assayer ‖ Münzwardein m. ‖ Probierer m.; Prober m. ‖ essayeur m. / ~'s forceps pl. ‖ Probierzange f. ‖ pince f. d'essayeur. / ~'s room ‖ Probiergaden m.; Probierlaboratorium n.; Probierraum m. ‖ chambre f. d'essayeur; essaierie f. / ~'s tongs pl. see ~'s forceps.

assay furnace (Chem) ‖ Muffel- oder Probierofen m.; Kapellenofen m. ‖ fourneau m. à coupelle ou d'essayeur ou docimastique. / ~ grain (Met) ‖ Probekorn n. ‖ culot m. ou grain m. d'essai.

assaying ‖ Probemachen n.; Probieren n. ‖ essai m.; essayage m. / ~ (Art) ‖ Probierkunst f. ‖ docimasie f.; docimastique f.; art m. de l'essayeur ou d'essayeur. / ~ by the cupel ‖ Probe f. durch Ab-

treiben; Kapellenprobe f. ‖ coupellation f. en petit essai m. de coupelle. ‖ ~ of ore ‖ Erzprobe f. ‖ essai m. du minerai. / precious metal ~ ‖ Edelmetallprobe f. ‖ essai m. de métaux précieux. / ~ for water ‖ Näßprobe f. ‖ essai m. pour l'humidité.

assaying balance ‖ Probierwage f. ‖ balance f. d'essai. / ~ drop (Glass) ‖ Probetropfen m. ‖ larme f. d'essai. / ~ furnace ‖ Probierofen m. ‖ fourneau m. d'essai. / method of ~ ‖ Probiermethode f. ‖ méthode f. d'essai. / ~ vessel ‖ Probiergefäß n. ‖ vase m. d'essai.

assay lead ‖ Probierblei n.; Kornblei n. ‖ plomb m. d'essai ou en grains. / ~ master ‖ Oberwardein m. ‖ essayeur m. en chef. / ~ office ‖ Probiergaden m.; Probierlaboratorium n.; Probierraum m. ‖ chambre f. d'essayeur; essaierie f. / ~ porringer ‖ Schlackenscherben m. ‖ écuelle f. à scorifier.

assay-sample ‖ Probe f.; Probestück n.; Versuchsprobe f. ‖ échantillon m.; spécimen m. / ~ standard ‖ Normalversuchsprobe f. ‖ échantillon m. normal.

assay spoon (Chem) ‖ Probierlöffel m. ‖ éprouvette f.; cuiller f. d'essai. / ~ test ‖ Probiergefäß n. ‖ têt m.; test m. / ~ ton ‖ Probiertonne f. ‖ ton m. d'essai. / ~ trowel ‖ Probekelle f. ‖ truelle f. d'essai. / ~ weight ‖ Probiergewicht n. ‖ poids m. pour l'essai ou d'essai.

ass breeder ‖ Eselzüchter m. ‖ éleveur m. d'ânes. / ~ breeding ‖ Eselzucht f. ‖ élevage m. d'ânes.

assemblage ‖ Montage f.; Anschluß m. ‖ assemblage m.; montage m.; jonction f. / ~ (Carp) ‖ Verbindung f. ‖ jonction f.; assemblage m. / ~ with key piece (Carp) ‖ Schurzwerk n. ‖ assemblage m. à clef.

assemble, to (Carp) ‖ abbinden ‖ assembler; joindre. / ~ (Mach) ‖ montieren; aufstellen ‖ monter; assembler. / ~ (Miner) ‖ sich scharen ‖ se réunir. / ~ butt-on-butt (Carp) ‖ anpfropfen ‖ joindre en about; abouter. / ~ a piping ‖ eine Leitung f. verlegen ‖ poser une conduite; placer une tuyauterie.

assembled ball bearing ‖ fertig montiertes Kugellager n. ‖ roulement m. à billes monté.

assembling ‖ Zusammenbau m.; Zusammensetzarbeit f. ‖ assemblage m. / ~ (Carp) ‖ Abbinden n.; Verbinden n. ‖ assemblage m. / ~ work ‖ Zusammenstellarbeit f. ‖ assemblage m.

assembly ‖ Zusammenstellung f. ‖ assemblage m. / ~ (Design) ‖ Zusammenstellungszeichnung f. ‖ dessin m. d'ensemble. / general ~ ‖ Generalversammlung f. ‖ assemblée f. générale.

assembly hall ‖ Versammlungssaal m. ‖ salle f. d'assemblée ou de réunion ou de séance. / ~ room ‖ Versammlungsraum m. ‖ salle f. d'assemblée.

assent ‖ Zustimmung f. ‖ consentement m.; approbation f.; assentiment m. / subject to your immediate ~ ‖ bei umgehender Zusage f. ‖ sous condition (de la réception) de votre consentement m. par retour du courrier.

assertion ‖ Angabe f. ‖ déclaration f.; indication f.

asses hiring-out ‖ Eselvermietung f. ‖ location f. d'ânes.

assess, to ‖ abschätzen ‖ estimer. / ~ ‖ besteuern ‖ imposer; taxer.

assessable || steuerpflichtig || imposable.
assessed at || abgeschätzt auf || imposé à; taxé à.
assessment || Einschätzung f.; Abschätzung f. | taxation f. / ~ by the tax payer || Selbsteinschätzung f. || déclaration f. de ses revenues par l'imposable.
assessor || Beisitzer m. || assesseur m. / (Appraiser) || Taxator m. || taxateur m.; commissaire-priseur m.
asset || Aktivwert m. || valeur f. effective. / ~ item || Aktivposten m. || position f. active.
assets pl. || Aktiva pl. || actif m. / ~ (Money's worth) || Geldeswert m. || valeur f. marchande ou en argent.
assign, to || zedieren || céder.
assign || Rechtsnachfolger m. || ayant-cause m.; ayant-droit m.
assignee || Mandatar m. || mandataire m.
assignment || Übertragung f. || transfert m. / deed of ~ || Abtretungsurkunde f. || acte m. de transfert; transfert m.; assignation f.; cession f.
assimilate, to || assimilieren || assimiler.
assimilation || Assimilation f.; Assimilieren n. || assimilation f.
assist, to || Vorschub m. leisten || favoriser; aider; assister. / ~ (to advise) || raten || conseiller; porter remède; deviner. / ~ (to aid) || aushelfen; helfen || aider. / ~ the voltage || die Spannung erhöhen || élever ou augmenter la tension.
assistance || Aushilfe f. || coup m. de main.
assistant || Assistent m.; Gehilfe m.; Helfer m. || assistant m.; aide m.; auxiliaire m. / ~ (Managing) || Betriebsassistent m. || aide m. / ~ (Workman) || Hilfsarbeiter m. || aide-ouvrier m. / surveyor's ~ || Vermessungshelfer m. || aide m. de l'arpenteur.
assistant engine || Hilfsmaschine f. || machine f. auxiliare ou de renfort ou pilote. / ~ engine-driver || Hilfsmaschinist m. || aide-machiniste m. / ~ engineer || Hilfsmaschinist m. || aide-mécanicien m. / ~ mine conveyor || Markscheidersteiger m. || aide-arpenteur m. / ~ stoker || Hilfsheizer m. || aide-machiniste m.
assisted by || mit Hilfe f. von || aidé par; assisté de.
assizer || Eichinspektor m. || inspecteur m. des poids et mesures.
association || Genossenschaft f. || association f. ou société f. coopérative; syndicat m. / ~ (Syndicate) || Konsortium n. || consortium m.; syndicat m.; groupe m. / ~ for the promotion of mental and social culture || Bildungsverein m. || société f. pour la propagation de connaissance et de culture générale. / ~ for the study of ... || Studiengesellschaft f. für ... || société f. d'études de ... / articles pl. of ~ || Gesellschaftsstatuten npl. || statuts mpl. sociaux.
assort, to || sondern || assortir; trier.
assorted || assortiert || assorti. / ~ ores pl. || assortierte Erze npl. || minerais mpl. assortis.
assorting sieve || Sortiersieb n. || crible m. de triage.
assortment || Zusammenstellung f. || rapprochement m.; combinaison f.; association f. / ~ (Booksell) || Sortiment n. || assortiment m. / mixed ~ || gemischtes Sortiment n. || assortiment m. varié.
ass skin || Schreibtafelpergament n.; Pergament n. || parchemin m. (animal); vélin m.

assurance see also insurance || Versicherung f. || assurance f.
assured || zuversichtlich || confiant. / ~ (insured) || versichert || assuré.
astatic || astatisch || astatique. / ~ couple || astatisches Nadelpaar n. || couple m. d'aiguilles astatique.
astaticity see astatization.
astatization (Tel) || Astasierung f. || astatisation f.; astacité f.
astern (Ship) || achteraus; im oder am Hinterteil n. des Schiffes || en arrière du navire. / ~ (Nav) || rückwärts || arrière. / to go ~ || achteraus fahren; mit der Maschine f. rückwärts gehen || faire machine f. arrière. / going ~ of a ship || Rücklauf m. eines Schiffes || marche f. arrière ou culée f. d'un vaisseau.
astern drum || Rückwärtstrommel f. || tambour m. à marche arrière. / ~ running || Rückwärtsgang m. || marche f. en arrière. / ~ running of the propeller || Rückwärtslauf m. der Schraube || marche f. en arrière de l'hélice. / ~ turbine || Rückwärtsturbine f. || turbine f. de marche arrière.
asthma preparation || Asthmapräparat n. || médicament m. contre l'asthma.
astigmatic || astigmatisch || astigmatique. / ~ corrected objective || astigmatisch korrigiertes Objektiv n. || objectif m. astigmate corrigé. / ~ eye || astigmatisches Auge n. || œil m. astigmate. / axis test disk for ~ eyes || Achsensehprüfscheibe f. für astigmatische Augen || goniomètre m. pour l'astigmatisme ou pour des yeux astigmates.
astigmatism || Astigmatismus m. || astigmatisme m. / total ~ || Totalastigmatismus m. || astigmatisme m. total.
astrakanite (Miner) || Astrakanit m. || astrakanite f.
astrakhan (Fur) || Astrachanfell n.; Astrachan m. || astrakan m. / imitation ~ || unechter Astrachan m. || imitation f. d'astrakan.
astral lamp || Astrallampe f. || lampe f. astrale.
astro camera for photographing stars and nebulae || Astrokamera f. zur Fotografie von Sternen und Nebelflecken || chambre f. noire astro-photographique pour photographier les étoiles et les nébuleuses.
astrograph || Astrograf m. || astrographe m. / ~ with reflector system || Astrograf m. mit Spiegelsystem || astrographe m. à miroir.
astronautics pl. || Fahrt f. in den Weltenraum; Raumflug m.; Kosmonautik f. || super-aviation f.; navigation f. extra-atmosphérique.
astronomical clock || astronomische Uhr f. || horloge f. astronomique. / ~ instrument || astronomisches Instrument n. || instrument m. astronomique ou d'astronomie. / ~ observation || astronomische Beobachtung f.; Sternbeobachtung f. || observation f. céleste ou astronomique. / ~ spectrograph || astronomischer Spektrograf m. || spectrographe m. astronomique. / ~ telescope || astronomisches Fernrohr n. || lunette f. astronomique. / ~ telescope objective || astronomisches Fernrohrobjektiv n. || objectif m. de lunette astronomique.
astronomy || Astronomie f.; Sternkunde f. || astronomie f.

astro-optical equipment || astronomische Optik f. || optique f. astronomique. / ~ glass polishing shop || Poliersaal m. für astronomische Optik || atelier m. pour le polissage des objectifs astronomiques.
astrophotographic || astrofotografisch || astrophotographique. / ~ camera || astrofotografische Kamera f. || chambre f. astrophotographique. / ~ double camera || astrofotografische Doppelkamera f. || double-chambre f. astrophotographique. / ~ objective || astrofotografisches Objektiv n.; Objektiv n. für Astrofotografie || objectif m. astrophotographique ou pour l'astrophotographie.
astrophotography || Astrofotografie f. || astrophotographie f. / measuring apparatus for ~ || Meßapparat m. für astrofotografische Aufnahmen || appareil m. pour la mensuration des clichés astrophotographiques.
astrophysical || astrophysikalisch || astrophysique.
astro-spectroscope || Astrospektroskop n.; Sternspektroskop n. || spectroscope m. astronomique.
asunder || auseinander; entzwei; getrennt || en pièces; en morceaux; en deux. / machine taken ~ || zerlegte oder auseinandergenommene Maschine f. || machine f. démontée.
asymmetrical || asymmetrisch; nicht symmetrisch; ungleichförmig || asymétrique. / ~ structure (Miner) || asymmetrischer Bau m. || structure f. asymétrique.
asymmetry || Asymmetrie f.; Unebenmäßigkeit f. || asymétrie f.
asymptote || Asymptote f. || asymptote f.
asymptotical || asymptotisch || asymptotique.
asynchronous || asynchron || asynchrone. / ~ alternator (Electr) || Asynchrongenerator m. || alternateur m. asynchrone. / ~ disc discharger (Radio) || asynchrone Scheibenfunkenstrecke f. || éclateur m. à disque asynchrone. / ~ motor || Asynchronmotor m. || moteur m. asynchrone. / ~ remote controlled power station || ferngesteuetes Asynchronspeicherwerk n. || usine f. d'énergie à machines asynchrones commandée à distance.
at works pl. || ab Werk n. || prise à la fabrique.
athwart (Mar) || dwars; quer || par le travers. / ~ the stream (Nav) || querstroms || à travers le ou au travers du courant.
athwart-hawse || quer vor dem Bug || par le travers des écubiers. / ~ sea || Dwarssee f. || mer f. de travers. / ~ -ships || querschiffs || par le travers.
atlas || Atlas m.; Kartenwerk n. || atlas m.
atmometer || Verdunstungsmesser m. || évaporimètre m.; atmomètre m.
atmosphere (Meteor) || Lufthülle f.; Atmosphäre f. || atmosphère f. / standard ~ || Normalatmosphäre f. || atmosphère f. normale ou standard. / surrounding ~ || Außenluft w. || air m. extérieur; atmosphère f. / to take from surrounding ~ || Luft f. ansaugen || prendre l'air m. de l'ambiant. / ~ of vapour || Dampfatmosphäre f. || atmosphère f. de vapeur.
atmospheric || atmosphärisch || atmosphérique. / ~ absorption (Radio) || atmosphärische Absorption f. || absorption f. atmosphérique. / ~ brake || Luftbremse

f. || frein m. atmosphérique. / ~ condition || Wetterlage f. || état m. *ou* situation f. atmosphérique. / ~ conditions pl. || Wetterverhältnisse npl. || conditions fpl. atmosphériques. / ~ electricity || Luftelektrizität f. || électricité f. atmosphérique. / ~ equilibrium || atmosphärisches Gleichgewicht n. || équilibre m. atmosphérique. / ~ phenomenon || atmosphärische Erscheinung f. || phénomène m. atmosphérique.

atmospheric pressure || Luftdruck m.; Atmosphärendruck m. || pression f. atmosphérique *ou* de l'atmosphère *ou* barométrique. / the ~ decreases || der Luftdruck nimmt ab || la pression atmosphérique baisse *ou* diminue. / the ~ increases || der Luftdruck nimmt zu || la pression atmosphérique hausse.

atmospheric water || Niederschlagswasser n. || eau f. de pluie; eau f. météorique. / ~ water vapour || Luftwasserdampf m. || vapeur f. d'eau contenue dans l'air.

atmospheric(s) pl. (Disturbance; Radio) || Luftstörung f.; atmosphärische Störung f. || perturbation f. atmosphérique.

atoll || (ringförmige) Koralleninsel f.; Atoll n. || attoll m.; île f. de coraux.

atom || Atom n. || atome m. / weight of an ~ *see* atomic weight.

atom gram || Atomgramm n. || atome-gramme m.

atomic || atomisch || atomique. / ~ dispersion || Atomdispersion f. || dispersion f. atomique. / ~ heat || Atomwärme f. || chaleur f. atomique.

atomic weight || Atomgewicht n.; Mischungsgewicht n. || poids m. atomique. / small ~ || niedriges Atomgewicht n. || poids m. atomique faible. / table of ~s || Atomgewichtstabelle f. || tableau m. des poids atomiques des éléments.

atomicity || Atomwertigkeit f.; Atomigkeit f.; Wertigkeit f. || atomicité f.

atomistic || atomistisch || atomistique.

atomization of water || Wasserzerstäubung f. || pulvérisation f. de l'eau.

atomize, to || zerstäuben || atomiser; pulvériser. / ~ by compressed air || zerstäuben n. mittels Druckluft f. || atomiser par air m. comprimé. / ~ as a fan-shaped spray || fächerförmig zerstäuben || pulvériser sous forme f. d'éventail. / ~ the fuel thoroughly || den Brennstoff m. zu einem feinen Ölnebel zerstäuben || pulvériser le combustible en un fin nuage d'huile.

atomizer || Zerstäuber m. || atomiseur m.; pulvérisateur m.; vaporisateur m. / compressed-air ~ || Druckluftzerstäuber m. || appareil m. aéro-pulvérisateur. / oil ~ || Ölzerstäuber m. || pulvérisateur m. d'huile. / two nozzle ~ || Zweidüsenzerstäuber m. || pulvérisateur m. à deux gicleurs.

atomizer cone || Zerstäuberkegel m. || champignon m. du pulvérisateur.

atomizing by compressed air || Zerstäubung f. durch Druckluft || pulvérisation f. par air comprimé. / ~ plant || Zerstäubungsanlage f. || installation f. de pulvérisation.

atom number || Atomzahl f. || nombre m. atomique. / weight of an ~ *see* atomic weight.

at once || sofort; auf der Stelle f. || immédiatement; sur-le-champ.

atrophy || Atrophie f. || atrophie f.

atropin || Atropin n. || atropine f.

attach, to || anfügen; befestigen; anheften || annexer; joindre; ajouter; attacher; assujettir. / ~ (Oneself) || sich anschließen || s'attacher; se joindre.

attachable yellow glass (Phot) || aufsteckbares Gelbglas n. || verre m. jaune adaptable.

attached to ... (Chem) || gebunden an ... || fixé à ...

attaching machine for stitched books || Broschüreneinhängemaschine f. || machine f. à assujettir les brochures. / ~ plug (Tel) || Anschlußstöpsel m. || fiche f. de prise de poste.

attachment (Fastening) || Befestigung f. || attache f.; fixation f. / ~ (Implement) || Zusatzgerät n.; Hilfsvorrichtung f. || appareil m. complémentaire; organe m. auxiliaire. / ~ collar || Befestigungslasche f. || collier m. de fixation. / ~ flange || Befestigungsflansch m. || collerette f. de fixation. / ~ rail || Befestigungsschiene f. || rail m. *ou* barre f. *ou* ailette f. de fixation. / ~ right || Beschlagsrecht n. || droit m. de saisie.

attack, to (Chem) || anfressen; angreifen; zerfressen || attaquer; corroder.

attack || Angriff m.; Angreifen n. || attaque f. / means pl. of ~ (Mine) || Aufschlußmittel n. || moyen m. de désagrégation.

attackability (Chem) || Angreifbarkeit f. || attaquabilité f.

attackable (Chem) || angreifbar || attaquable.

attacked || angegriffen || attaqué.

attacking (Chem) || Anfressung f. || attaque f.; corrosion f.

attar of roses || Rosenöl n. || essence f. de rose.

attemperator coil for fermenting tuns || Gärbottichkühler m. || réfrigérant m. pour cuves de fermentation.

attend, to ~ to the machine || die Maschine bedienen || conduire la machine. / ~ from two sides || von zwei Seiten fpl. bedienen || servir de deux côtés mpl.

attendance || Bedienung f. || réglage m.; service m. / easy ~ || leichte Bedienung f. || réglage m. *ou* service m. facile. / with little need of ~ || anspruchslos in der Wartung f. || d'un entretien m. réduit au minimum. / no ~ || bedienungslos || sans service.

attendance button (Radio) || Bedienungsknopf m. || bouton m. de réglage. / ~ crew || Bedienungsmannschaft f. || hommes mpl. de service. / ~ regulations pl. || Bedienungsvorschrift f. || instructions fpl. de service; notice f. de réglage.

attendant seat || Krankenwärtersitz m. || siège m. de l'infirmier. / ~ path (Railw) || Fußpfad m. || chemin m. de service; sentier m. / ~ station (Tel) || Vermittlungsstelle f. || bureau m. téléphonique.

attention (Attendance) || Wartung f. || soin m. / ~ (Attentiveness) || Aufmerksamkeit f. || attention f. / early ~ || Vorsorge f. || prévoyance f.

attenuation || Schwächung f.; Abnahme f.; Verringerung f. || atténuation f. / ~ (Brew) || Vergärung f. || fermentation f.; tion f. / ~ (Tel) || räumliche Dämpfung f.; Dämpfung f. || affaiblissement m.; atténuation f. / effective ~ (Tel) || wirksame Dämpfung f. || affaiblissement m. effectif. / final ~ (Brew) || Endvergärung f. || atténuation f. finale. / to-

tal ~ (Tel) || Gesamtdämpfung f.; Dämpfungsmaß n. || affaiblissement m. total; équivalent m. de transmission.

attenuation constant (Tel) || Dämpfungskonstante f. || constante f. d'affaiblissement. / ~ degree (Brew) || Vergärungsgrad m. || degré m. de fermentation *ou* d'atténuation. / ~ determination (Brew) || Vergärungsgradbestimmung f. || détermination f. de l'atténuation. / ~ equalizer (Tel) || Dämpfungsausgleicher m.; Entzerrer m. || égaliseur m. d'atténuation. / ~ equilibration (Tel) || Dämpfungsausgleichung f. || compensation f. de la distorsion. / ~ equivalent *see* ~ measure. / ~ measure (Tel) || Dämpfungsmaß n. || équivalent m. de transmission. / ~ test (Tel) || Dämpfungsprüfung f. || essai m. de l'affaiblissement.

attestation || Beglaubigung f. || attestation f.; témoignage m.

attested || beglaubigt || attesté. / ~ copy || beglaubigte Abschrift f. || copie f. certifiée.

attitude of flight || Fluglage f. || position f. *ou* régime m. en vol.

attle (Mine) || taubes Gestein n.; Gangmasse f. || gangue f.; matière f. *ou* roche f. stérile.

attorney || Bevollmächtigter m.; Sachwalter m. || mandataire m.; avoué m.; plénipotentiaire m.; agréé m.; homme m. d'affaires. / to confer powers pl. of ~ *see* to give power of ~. / full power of ~ || Generalvollmacht f. || plein pouvoir m.; procuration f. générale. / to give power of ~ || bevollmächtigen; Vollmacht erteilen || autoriser; donner (plein) pouvoir m.

attorney's fee || Rechtsanwaltshonorar n. || honoraires mpl. de l'avocat.

attract, to (Electr) || anziehen || attirer; solliciter.

attraction (Electr) || Anziehung f. || attraction f. / ~ power of a magnet || Anziehungskraft f. eines Magnets || force f. d'attraction d'un aimant. / ~ strength *see* attraction power.

attractive || ansprechend || agréable. / ~ force *see* attraction power.

attractiveness || Reiz m. || attrait m.

attrition || Reibung f. || attrition f.; frottement m. / ~ product || Ausmahlerzeugnis n. || produit m. de convertissage.

auburn || kastanienbraun || brun châtain.

auction, to ~ under compulsion || zwangsweise versteigern || faire une enchère forcée.

auction || Versteigerung f.; Auktion f. || vente f. publique *ou* à l'encan *ou* aux enchères; encan m.; enchères fpl. / public ~ || öffentliche Versteigerung f. || encan m. public; enchère f. publique. / to sell by ~ || verauktionieren || mettre à l'encan *ou* à l'enchère. / to be sold by ~ || zur Auktion f. kommen || passer aux enchères fpl. / timber ~ || Holzversteigerung f. || vente f. du bois aux enchères.

auctioneer || Auktionator m.; Versteigerer m. || commissaire-priseur m.

auction sale *see* auction.

audibility (Tel) || Verständigung f. || qualité f. de transmission. / ~ factor || Hörbarkeitsfaktor m. || facteur m. d'audibilité.

audible || hörbar || audible. / ~ call || akustisches Signal n. || appel m. phonique. / incoming ~ current || ankommender

Hörstrom m. ‖ courant m. de transmission arrivant.

audience ‖ Auditorium n.; Zuhörerschaft f. ‖ auditoire m.; auditeurs mpl.

audion ‖ Audion n. ‖ audion m.

audion receiver ‖ Audionempfänger m. ‖ récepteur m. à l'audion. / ~ with back-coupling ‖ Audionempfänger mit Rückkopplung ‖ récepteur m. à lampe amplificatrice.

audion tube ‖ Audionröhre f. ‖ lampe f. audion.

audit, to ‖ revidieren ‖ faire la révision.

auditor ‖ Rendant m. ‖ trésorier m.; caissier m. / ~ (Controller of accounts) ‖ Rechnungsrevisor m. ‖ commissaire m. des comptes.

auditorium ‖ Auditorium n.; Zuschauerraum m.; Hörsaal m.; Vortragssaal m. ‖ salle f. (d'audition); auditoire m.

auger ‖ Hohlbohrer m. ‖ tarière f. ouverte. / ~ (Carp) ‖ Holzbohrer m.; Zimmermannsbohrer m. ‖ foret m. à bois; mèche f. pour bois. / ~ (Earthw) ‖ Sandbohrer m.; Sandkelle f.; Schöpfbohrer m. ‖ cuillère f. ou tarière f. à mâchoire. / ~ for boring barrels ‖ Gewehrlaufbohrer m. ‖ foret m. ou alésoir m. ou mouche f. de l'arquebusier. / earth ~ see auger, ground. / funnel-shaped ~ ‖ Trichterbohrer m. ‖ tarière f. en entonnoir. / great ~ ‖ Stangenbohrer m. ‖ tarière f. / ground ~ ‖ Erdbohrer m. ‖ tarière f. pour le sol; sonde f. / long eye ~ ‖ Stangenbohrer m. ‖ tarière f. torse. / mechanic ~ (Min) ‖ Tiefbohrapparat m. ‖ équarrissoir m. de puits de mine. / screw ~ ‖ Schneckenbohrer m. ‖ foret m. hélicoïdal; tarière f. à spirale ou à vis ou en hélice; tarière f. hélicoïdale ou tordue ou torse. / ~ for sleepers ‖ Schwellenbohrer m. ‖ tarière f. pour traverses. / twisted ~ see auger, screw.

auger bit (Carp; Join) ‖ Löffelbohrer m.; Hohlbohrer m. mit Zahn ‖ mèche-cuiller f.; tarière f. à ou en cuiller. / ~ bore ‖ Zapfenloch n. ‖ enlaçure f. / ~ gouge ‖ Löffelbohrer m. ‖ tarière f. à cuiller. / ~ hole see ~ bore.

augite ‖ Pyroxen m.; Augit m. ‖ pyroxène m.; augite f. / granular ~ ‖ Kokkolith m. ‖ coccolithe f.; pyroxène m. coccolithe ou granuliforme; pierre f. à noyaux. / ~ porphyry ‖ Melaphyr m.; schwarzer Porphyr m.; Augitporphyr m. ‖ mélaphyre m.

aural forceps (Med) ‖ Ohrpinzette f. ‖ pince f. à oreilles.

aurate ‖ goldsaures Salz n. ‖ aurate m. / ~ of ammonia ‖ Knallgold n.; Goldoxydammoniak n. ‖ or m. fulminant; aurate m. d'ammoniaque.

aureole ‖ Lichthülle f. ‖ auréole f.

auric ‖ goldhaltig ‖ aurifère.

auric acid ‖ Goldsäure f. ‖ acide m. aurique. / ~ chloride ‖ Goldchlorid n.; Chlorgold n. ‖ or m. potable; sesqui-chlorure m. d'or.

aurichalcite ‖ Messingblüte f. ‖ aurichalcite f.

auric oxide ‖ Goldoxyd n. ‖ oxyde m. d'or.

auricular fibrillation ‖ Vorhofflimmern n. ‖ fibrillation f. auriculaire.

auriferous ‖ goldhaltig; goldführend ‖ aurifère.

auripigment ‖ Rauschgelb n. ‖ deutosulfure m. d'arsenic.

auronal dye ‖ Auronalfarbe f. ‖ colorant m. auronal.

aurora borealis ‖ Polarlicht n. ‖ aurore f. boréale; lumière f. polaire.

aurous cyanide ‖ Goldcyanür n. ‖ protocyanure m. d'or.

austenite ‖ Austenit n. ‖ austénite f.; hartilite f.

australite ‖ Australit n. ‖ australite m.

authentic ‖ authentisch ‖ authentique. / ~ zuverlässig; glaubhaft ‖ éprouvé; certain.

authentication ‖ Beglaubigung f. ‖ attestation f.

authenticity ‖ Zuverlässigkeit f. ‖ certitude f.; authenticité f.; sûreté f.; solidité f.

author ‖ Autor m.; Urheber m.; Verfasser m.; Schriftsteller m. ‖ auteur m.; écrivain m.

authority ‖ Sachverständiger m. ‖ expert m. / ~ within a district ‖ Territorialgewalt f. ‖ pouvoir m. territorial. / government ~ ‖ Regierungsbehörde f. ‖ autorité f. du gouvernement. / local ~ ‖ Ortsbehörde f. ‖ autorité f. de localité. / ~ to prospect ‖ Schürfbefugnis f. ‖ autorisation f. de faire des fouilles.

authorization ‖ Bevollmächtigung f.; Mandat n. ‖ autorisation f.; plein pouvoir m.; procuration f.; mandat m.

authorize, to ‖ bevollmächtigen ‖ autoriser; donner plein pouvoir m.

authorized agent ‖ Bevollmächtigter m. ‖ mandataire m.; plénipotentiaire m. / ~ person ‖ Mandatar m. ‖ mandataire m.

author's proof ‖ Verfasserverbesserung f. ‖ correction f. d'auteur.

auto see automobile.

autobus ‖ Kraftomnibus m. ‖ omnibus m. automobile.

auto car see automobile or motor car.

autochrom plate ‖ Autochromplatte f. ‖ plaque f. autochrome. / ~ printing ‖ Autochromdruck m. ‖ simili-gravure f.; impression f. en photochromie.

autoclave ‖ Autoklav m. ‖ autoclave m.

autocollimating grating spectroscope ‖ Gitterspektroskop n. mit Autokollimation ‖ spectroscope m. à réseau autocollimateur. / ~ spectroscope ‖ Autokollimationsspektroskop n. ‖ spectroscope m. autocollimateur.

autogenous ‖ autogen ‖ autogène. / to weld ~ly ‖ autogen schweißen ‖ souder par le procédé autogène.

autogenous cutting ‖ autogenes Schneiden n. ‖ coupage m. ou découpage m. ou coupure f. ou coupe f. autogène. / ~ and welding ‖ autogenes Schneiden n. und Schweißen n. ‖ coupure f. et soudure f. autogènes. / ~ machine ‖ autogene Schneidmaschine f. ‖ machine f. pour le coupage autogène. / ~ plant ‖ autogene Schneidanlage f. ‖ installation f. de coupage autogène. / ~ shop ‖ Werk n. für autogenes Schneiden ‖ atelier m. de coupure-autogène.

autogenous fusing burner ‖ Autogenschneidbrenner m. ‖ brûleur m. à coupure autogène. / ~ installation ‖ autogene Schneidanlage f. ‖ installation f. de coupage autogène.

autogenous soldering ‖ autogenes Löten n. ‖ soudure f. ou brasure f. autogène.

autogenous welding ‖ autogene Schweißung f. ‖ soudure f. ou soudage m. autogène. / ~ installation ‖ autogene Schweißanlage f. ‖ installation f. de soudure autogène. / ~ machine ‖ autogene Schweißmaschine

f. ‖ machine f. pour la soudure autogène. / ~ plant ‖ autogene Schweißanlage f. ‖ installation f. de soudure autogène. / ~ and cutting plant ‖ Autogenschweiß- und -schneidanlage f. ‖ installation f. de soudure et de coupage autogène. / ~ shop ‖ Werkstatt f. für autogene Schweißung ‖ atelier m. de soudure autogène.

autogiro ‖ Windmühlenflugzeug n. ‖ autogyre m.

autograph artist ‖ Autograph m. ‖ autographiste m.

autographic colour ‖ Autografiefarbe f. ‖ couleur f. autographique. / ~ ink ‖ Autografentinte f. ‖ encre f. autographique. / ~ printing ‖ Autografie f. ‖ autographie f. / ~ printing paper ‖ Autografiepapier n. ‖ papier m. autographique. / ~ writing-ink ‖ lithografische Tusche f. ‖ encre f. autographique.

autographical see autographic.

autography see autographic printing.

auto-jigger (Electr) ‖ Autojigger m. ‖ autojigger.

automatic ‖ selbsttätig; automatisch ‖ automatique. / fully ~ ‖ vollselbsttätig; vollautomatisch ‖ complètement automatique.

automatic (Machine) see automatic lathe.

automatic branch exchange (Aut tel) ‖ Selbstanschlußnebenamt n.; Automatennebenamt n. ‖ bureau m. automatique annexe. / ~ center punch ‖ Federdruckkörner m. ‖ pointeau m. automatique à ressort. / ~ central office see ~ exchange. / ~ charging (Accumulators) ‖ selbsttätiges Aufladen n. ‖ charge f. automatique. / continuous furnace with ~ charging and cooling of the work for use in the production of large quantities ‖ Ofen m. mit selbsttätiger Beschickung und Abkühlung des Massengutes ‖ four m. à alimentation et refroidissement automatiques pour la production en grandes séries. / ~ charging device (Met) ‖ selbsttätige Aufgabevorrichtung f. ‖ mécanisme m. de chargement automatique. / ~ circuit breaker ‖ Selbstunterbrecher m.; Selbstausschalter m. ‖ interrupteur m. ou coupe-circuit m. ou déclencheur m. automatique. / ~ coupling ‖ selbsttätige Kupplung f. ‖ attelage m. automatique. / ~ cut-out see ~ circuit breaker. / ~ cutting-out ‖ selbsttätige Abschaltung f. ‖ déconnexion f. automatique. / ~ cutter-grinder ‖ Fräserschärfautomat m. ‖ machine f. automatique à affûter les fraises. / ~ discharger ‖ Selbstentleerer m. ‖ déchargeur m. automatique. / ~ disconnecting ‖ Selbstauslösung f. ‖ débrayage m. automatique. / ~ door closer ‖ selbsttätiger Türschließer m. ‖ ferme-porte m. automatique. / ~ exchange (Aut tel) ‖ Selbstanschlußamt m.; Automatenamt n. ‖ bureau m. téléphonique automatique. / ~ field break switch ‖ selbsttätiger Magnetausschalter m. ‖ interrupteur m. automatique d'excitation. / ~ fire alarm ‖ selbsttätiger Feuermelder m. ‖ avertisseur m. d'incendie automatique. / ~ furnace ‖ selbsttätige Feuerung f. ‖ foyer m. automatique. / ~ goods lift ‖ selbsttätiger Lastenaufzug; Lastenaufzug m. ohne Führerbegleitung ‖ monte-charge m. ne nécessitant pas de conducteur en permanence. / ~ hot water apparatus ‖ Heißwasserautomat m. ‖ ap-

pareil m. automatique à eau chaude. /
~ keyway milling machine ‖ selbsttätige
Keilnutenfräsmaschine f. ‖ machine f.
automatique à fraiser les rainures de
clavettes.

automatic lathe ‖ Drehautomat m.; selbsttätige Drehbank f. ‖ tour m. automatique. / single purpose ~ ‖ Sonderzweckautomat m. ‖ tour m. automatique pour
usages spéciaux.

automatic machine *see also* automaton ‖
Automat m.; automatische Maschine f. ‖
machine f. automatique; automate m. /
~ (Penny-in-the-slot) ‖ Selbstverkäufer
m. ‖ distributeur m. automatique.

automatic oiler ‖ automatischer Öler m.;
Selbstöler m. ‖ graisseur m. automatique. / ~ padlock ‖ Vorhängeschloß n.
mit Rückspringbügel ‖ cadenas m. automatique. / ~ receiver (Tel) ‖ selbsttätiger
Empfänger m. ‖ récepteur m. automatique. / ~ refrigerator ‖ Kälteautomat m. ‖
réfrigérateur m. automatique; automate m. frigorifique. / ~ regulating
of clockwork ‖ selbsttätige Regelung f.
des Uhrwerkes ‖ réglage m. automatique
du mouvement d'horlogerie. / ~ rifle ‖
Selbstladegewehr n. ‖ fusil m. automatique. / ~ screw cutting lathe ‖ Schrauben-
automat m. ‖ tour m. automatique pour
boulonnerie *ou* à vis. / ~ sharpening machine for cold and hot saws ‖ Kalt- und
Warmsägenschärfautomat m. ‖ machine
f. automatique à affûter les scies à
chaud et à froid. / ~ sharpening machine for side milling cutters ‖ Scheibenfräserselbstschärfer m. ‖ machine f. automatique à affûter les fraises-disques. / ~
starter (Electr) ‖ Selbstanlasser m. ‖
démarreur m. automatique. / ~ stopping ‖
Selbstabstellung f. ‖ arrêt m. automatique. / ~ subscriber's station (Aut tel) ‖
automatische Teilnehmerstation f. ‖ poste
m. automatique. / ~ suction pump for
the removal of fecals ‖ Selbstsaugpumpe
f. für die Fäkalienabfuhr ‖ pompe f.
aspirante automatique pour l'enlèvement
des matières fécales. / ~ switch ‖ selbsttätiger Schalter m.; Selbstschalter m. ‖
interrupteur m. automatique. / ~ switch
with zero-voltage release ‖ Selbstschalter
m. mit Nullspannungsauslösung ‖ interrupteur m. automatique à déclenchement
à voltage nul. / ~ telephone exchange ‖
Selbstanschlußamt n. ‖ bureau m. central automatique. / ~ telephone station ‖
Fernsprechautomat m. ‖ poste m. téléphonique automatique. / ~ telephone
system ‖ Selbstanschlußsystem n.; Wäh-
lersystem n. ‖ système m. de téléphonie
automatique. / ~ tipper (wagon) ‖ Selbstentlader m.; selbsttätiger Kippwagen m.
‖ wagonnet m. basculeur à déchargement
automatique. / ~ tipping device ‖ selbsttätige Kippvorrichtung f. ‖ dispositif m.
basculant de déchargement automatique. / ~ valve ‖ Selbstschlußventil n. ‖
soupape f. à fermeture automatique. /
~ wagon tipping device ‖ selbsttätiger
Wagenkipper m. ‖ basculeur m. automatique pour wagons.

automatization ‖ Automatisierung f. ‖ au-
tomatisation f.

automaton *see also* automatic machine ‖
Automat m. ‖ automate m. / the ~ is self-
winding ‖ die Automateneinrichtung hat
selbsttätigen Aufzug ‖ le déclenchement
se remonte automatiquement. / ~ of

telescope ‖ Automateneinrichtung f. des
Fernrohres ‖ déclenchement m. automatique de la lunette d'approche. /
factory of automatons or of automata ‖
Automatenfabrik f. ‖ fabrique f. d'automates.

automobile *see also* motor car ‖ Kraftwagen m.; Automobil n. ‖ automobile
m.; auto m.; voiture f. automobile. ‖
electric ~ ‖ Elektromobil n.; elektrischer
Kraftwagen m. ‖ électromobile m. /
petrol-electric ~ ‖ benzin-elektrischer
Kraftwagen m. ‖ voiture f. pétroléo-
électrique.

automobile accessory ‖ Kraftwagenzube-
hörteil m.; Kraftwagenzubehör n. ‖ accessoire m. pour automobile. / ~ body ‖
Karosserie f.; Kraftwagenaufbau m. ‖
carosserie f. / ~ body sheet ‖ Karosserie-
blech n. ‖ tôle f. pour carosseries. / ~
breaker ‖ Motorsteinbrecher m. ‖ concasseur m. à moteur. / ~ castings pl. ‖
Autoguß m. ‖ fonte f. pour autos. / ~
clock ‖ Kraftwagenuhr f. ‖ horloge f.
pour automobiles. / ~ clothes pl. ‖ Kraft-
fahrerbekleidung f. ‖ vêtement m. pour
automobilistes. / ~ construction ‖ Kraftwagenbau m. ‖ construction f. des automobiles. / special machine for construction
of ~s ‖ Automobilbausondermaschine f. ‖
machine f. spéciale pour la construction
des automobiles. / ~ delivery van ‖ Liefer-
kraftwagen m.; Lieferauto n. ‖ automobile
f. de livraison. / ~ exhibition ‖ Automobilausstellung f. ‖ exposition f. automobile
ou d'automobiles. / ~ fittings pl. ‖ Kraft-
wagenbeschläge mpl. ‖ ferrures fpl. d'automobiles. / ~ frame ‖ Kraftwagenrahmen m.; Kraftwagengestell n. ‖ châssis
m. d'automobile. / ~ gearing ‖ Kraft-
wagengetriebe n. ‖ engrenage f. d'automobile. / ~ goggles pl. ‖ Kraftfahrerbrille f.; Autobrille f. ‖ lunettes fpl.
d'auto. / ~ head ‖ Kraftwagenverdeck n.
‖ dais m. pour automobiles. / ~ heating ‖
Kraftwagenheizung f. ‖ chauffage m.
d'automobile. / ~ horn ‖ Autohupe f. ‖
corne f. *ou* trompe f. pour automobiles. /
~ ignition ‖ Kraftwagenzündung f.; Automobilzündung f. ‖ allumage m. pour
automobiles. / ~ leather ‖ Automobilleder n. ‖ cuir m. pour automobiles. / ~
lighting ‖ Kraftwagenbeleuchtung f. ‖
éclairage m. pour automobiles. / ~ mirror ‖ Kraftwagenspiegel m. ‖ avertisseur
m. *ou* espion m. pour automobiles. / ~ oil
‖ Kraftwagenöl n.; Autoöl n. ‖ huile f.
pour automobiles. / ~ parts pl. ‖ Kraft-
wagenteile mpl.; Automobilteile mpl. ‖
accessoires mpl. pour automobiles. / ~
railway car ‖ Triebwagen m. ‖ automotrice f. de chemin de fer. / ~ road ‖
Autostraße f.; Kraftwagenstraße f. ‖
route f. pour automobiles. / ~ road making ‖ Automobilstraßenbau m. ‖ construction f. de routes pour automobiles. /
~ top *see* ~ body. / ~ watch *see* ~ clock.

automobilism ‖ Kraftfahrwesen n.; Auto-
mobilismus m. ‖ automobilisme m.

automolite (Miner) ‖ Zinkspinell m.; Automolit m. ‖ spinelle m. zincifère; gahnite f.

automotive ordnance ‖ Kraftartillerie f. ‖
artillerie f. mécanique.

auto-room (Aut tel) ‖ Wählerraum m. ‖
salle f. d'installations automatiques.

auto-switch with air release ‖ Selbstschalter m. mit Luftauslösung ‖ disjoncteur
m. à rupture par l'air.

autotype *see* autotypy.

autotypy ‖ Autotypie f. ‖ autotypie f.;
similigravure f.

autumnal equinox ‖ Herbsttagundnacht-
gleiche f.; Herbstäquinoktium n.; Herbst-
punkt m. ‖ équinoxe m. d'automne;
point m. équinoxial de l'automne. / ~
point *see* ~ equinox.

autumn crop ‖ Herbsternte f. ‖ moisson f.
ou récolte f. d'automne. / ~ timber ‖
Herbstholz n. ‖ bois m. d'automne.

auxiliary apparatus ‖ Hilfsvorrichtung f. ‖
appareil m. accessoire. / ~ boiler ‖ Hilfskessel m. ‖ chaudière f. auxiliaire. / ~ chain
(Railw) ‖ Sicherheitskette f. ‖ chaîne f.
de sûreté. / ~ compensation set (Electr) ‖
Zusatzausgleichsaggregat n. ‖ survolteur m. de compensation. / ~ cruiser ‖
Hilfskreuzer m. ‖ croiseur m. auxiliaire. /
~ engine (Electr) ‖ Hilfsmaschine f. ‖
machine f. auxiliaire *ou* pilote. / ~ grid
voltage (Radio) ‖ Hilfsgitterspannung f.
‖ tension f. de grille auxiliaire. / ~ hoisting gear ‖ Hilfshubwerk n. ‖ treuil m. *ou*
chariot m. auxiliaire. / ~ hoisting tackle
see ~ hoisting gear. / ~ ignition device ‖
Hilfszündvorrichtung f. ‖ dispositif m.
d'allumage auxiliaire. / ~ keel (Shipb) ‖
Schlingerkiel m.; Kimmkiel m.; Seitenkiel m. ‖ quille f. latérale *ou* auxiliaire
ou de dérive. / ~ light ‖ Zusatzfeuer n. ‖
feu m. auxiliaire.

auxiliary machine ‖ Hilfsmaschine f. ‖ machine f. auxiliaire. / ~ for metallurgical
works ‖ Hüttenwerkhilfsmaschine f. ‖
machine f. auxiliaire pour chantiers métallurgiques. / ~ for rolling mills ‖ Walz-
werkhilfsmaschine f. ‖ machine f. auxiliaire pour laminoirs. / ~ for ships ‖
Schiffshilfsmaschine f. ‖ machine f. auxiliaire pour bateaux.

auxiliary main office (Tel) ‖ Hilfsknoten-
amt n. ‖ bureau m. auxiliaire de transit. /
~ mirror ‖ Hilfsspiegel m. ‖ miroir m.
supplémentaire. / ~ phase ‖ Hilfsphase f. ‖
phase f. auxiliaire. / ~ phase winding ‖
Hilfsphasenwicklung f. ‖ enroulement m.
de phase auxiliaire. / ~ railway ‖ Hilfsbahn f. ‖ chemin m. de fer provisoire. / ~
set on board ‖ Hilfsaggregat n. für
Schiffe ‖ groupe m. auxiliaire pour navires. / ~ shops pl. and installations pl. ‖
Hilfsbetriebe mpl. ‖ services mpl. auxiliaires. / ~ switch (Tel) ‖ Nebenschalter
m. ‖ commutateur m. supplémentaire. /
~ tank ‖ Hilfsbehälter m. ‖ réservoir m.
auxiliaire. / ~ transmitting system ‖
Hilfsübertragersystem n. ‖ système m.
de transformateurs auxiliaires. / ~ unit
see ~ set. / ~ vessel ‖ Hilfsschiff n. ‖ bâtiment m. auxiliaire. / ~ wing (Aero) ‖
Hilfsflügel m. ‖ aile f. auxiliaire.

avail, to ‖ in Anspruch m. nehmen ‖ occuper.

availability ‖ Gültigkeitsdauer f. ‖ durée f.
de validité.

available ‖ verwendbar ‖ employable; utilisable; applicable. / ~ assets pl. ‖ verfügbare Aktiva npl. ‖ actif m. disponible.
/ ~ energy ‖ verfügbare Energie f. ‖ énergie f. disponible.

avalanche ‖ Lawine f.; Schneesturz m. ‖
avalanche f.; chute f. d'avalanche. / ~ of
sand and stones ‖ Mure f.; Schuttstrom
m.; Rüfe f. ‖ avalanche f. de boue et de
pierres. / ~ of snow ‖ Schneelawine f. ‖
avalanche f. de neige. / ~ baffle **works**
pl. (Railw) ‖ Lawinenschutzmauer f. ‖
mur m. pare-avalanche.

avenue (Drive) || Anfahrt f. || approche f.; rampe f.

average || Durchschnitt m. || moyenne f. / ~ (Mar) || Seebeschädigung f.; Havarie f. || avarie f. / above the ~ || über dem Durchschnitt || au-dessus de la moyenne. / at an ~ see average, on an. / ~s pl. of the current production || Durchschnittswerte mpl. aus der laufenden Fertigung || chiffres mpl. moyens de la fabrication courante. / gross ~ (Mar) || allgemeine Havarie f. || avarie f. commune. / to make good ~ || Entschädigung f. für Havarie leisten || indemniser des avaries fpl. / on an ~ || durchschnittlich || en moyenne. / petty ~ (Mar) || kleine Havarie f. || menue avarie. / the real ~s pl. are only half these quantities || die wirklichen Durchschnittsziffern fpl. sind nur halb so groß || en réalité on atteint à peine la moitié de ces chiffres moyens. / to recover ~ || Ersatz m. für Havarie erhalten || recouvrer des avaries fpl. / simple ~ (Mar) || einfache Havarie f. || avarie f. simple. / small ~ (Mar) || kleine Havarie f. || menue avarie f.

average age || Durchschnittsalter n. || âge m. moyen. / ~ amount || Durchschnittsbetrag m. || moyenne f.; avérage m. / ~ analysis || Durchschnittsbestimmung f. || analyse f. moyenne. / ~ assay || Durchschnittsprobe f. || essai m. moyen. / ~ cost per ... || Durchschnittspreis m. je ... || prix m. moyen par ... / ~ deflection || mittlerer Ausschlag m. || déviation f. moyenne. / ~ figures pl. of the current production || Durchschnittswert m. aus der laufenden Fertigung || chiffres mpl. moyens de la fabrication courante. / ~ flow of a barrage || Mittelwasser n. einer Talsperre || eau f. moyenne d'un barrage. / ~ load || durchschnittliche Belastung f. || charge f. moyenne. / ~ number of men || durchschnittliche Stärke f. der Belegschaft || effectif m. moyen du personnel. / ~ pitch (Electr) || mittlerer Wicklungsschritt m. || pas m. moyen. / ~ price || Durchschnittspreis m. || prix m. moyen. / ~ speed || Durchschnittsgeschwindigkeit f. || vitesse f. moyenne ou commerciale. / the ~ values pl. have been inserted in the sketch || die Mittelwerte mpl. sind in der Abbildung eingetragen || les valeurs fpl. moyennes se trouvent indiquées dans la figure.

averaging || durchschnittlich betragend || approximativement; donnant une moyenne de.

averuncator || Baumschere f. || sécateur m.

aviation || Flugwesen n. || aviation f. / commercial ~ || Verkehrsflugwesen n. || aviation f. commerciale.

aviation course illumination || Flugplatzbeleuchtung f. || éclairage m. pour le champ d'aviation. / ~ field || Flugfeld n. || champ m. d'aviation. / ~ industry || Flugzeugindustrie f. || industrie f. aéronautique. / ~ weather service || Luftwarnungsdienst m. || service m. d'avertissement météorologique pour la navigation aérienne.

aviator || Luftschiffer m. || aviateur m. / ~'s outfit || Fliegerausrüstung f. || équipement m. d'aviateur.

avion || Kampfflugzeug n. || avion m. de combat.

award, to ~ a medal to || prämiieren || primer.

awash || Überstauung f. || submersion f.

awl || Ahle f.; Pfriem m. || alêne f.; poinçon m.; perçoir m. / ~ (Mar) || Marleisen n.; Marlspieker m. || épissoir m.; marprime f.; poinçon m. à épisser. / broad ~ || flache Ahle f. || alêne f. plate. / common ~ (Saddl) || Sattlerahle f. || alêne f. à coudre. / ~ for correcting types (Print) || Korrigierahle f. || alêne f. à correction. / ~ for embroidering || Stecher m. zum Sticken || poinçon m. à broder. / ~ works pl. || Ahlenfabrik f. || fabrique f. d'alênes.

awning (Build) || Markise f. || marquise f. / ~ (Ship) || Sonnensegel n. || tente f. / ~ (Vehicle) || Wagenplane f. || bâche f.; prélart m. / ~ maker || Zeltenmacher m. || bâcheur m. / ~ stanchion (Mar) || Sonnensegelstütze f. || montant m. des tentes.

axe, to ~ a quarry stone (Build) || einen Bruchstein m. bossieren || piquer un moellon.

axe || Axt f.; Beil n. || cognée f.; hache f. / ~ (Agr) || Hacke f. || hache f. / broad ~ (Carp) || Breitbeil n.; Breitaxt f. || épaule f. de mouton; doloire f. des charpentiers. / ~ for cutting wood || Holzaxt f. || marlin m. / joint hook shaped ~ (Carp) || Bundaxt f. || tire-boucher m. / miner's ~ || Keilhaue f. || pic m. / paring ~ || schwere Schälaxt f. || cognée f. à blanchir. / wood cleaver's ~ || Schrotaxt f. || cognée f. de bûcheron.

axial || achsrecht; axial || axial. / to move in ~ direction || axial verschieben || décaler axialement. / ~ flow turbine || Achsialturbine f. || turbine f. axiale.

axially symmetrical || achsensymmetrisch || symétrique par rapport à l'axe. / ~ eye || achsensymmetrisches Auge n. || œil m. symétrique par rapport à l'axe.

axinite || Glasstein m. || axinite f.

axis || Achse f.; Mittellinie f. || axe m. / ~ of abscissæ || Abszissenaxe f.; x-Achse f. || axe m. des abscisses ou de x. / ~ of the bore (Arm) || Seelenachse f. einer Feuerwaffe || axe m. de l'âme d'un canon. / ~ of the cone || Kegelachse f. || axe m. du cône. / conjugate ~ (Hyperbola; Ellipse) || Nebenachse f. || axe m. non transverse; axe m. imaginaire; second axe m. / ~ of a crystal || Kristallachse f. || axe m. de cristal. / ~ of the earth || Erdachse f. || axe m. de la terre. / ~ of eddy || Wirbelachse f. || axe m. du tourbillon. / ~ of the eye || Augenachse f. || axe m. de l'œil. / ~ of flotation (Mech) || Schwimmachse f. || axe m. de flottaison. / ~ of the heel-post of a lock-gate || Wendeachse f. eines Schleusentores || pivot m. d'une porte d'écluse. / ~ of a hinge || Türangelzapfen m. || pivot m. / ~ of incidence || Einfallslot n. || axe m. d'incidence; normale f. / ~ that bears throughout its length || der ganzen Länge nach geführte Achse f. || axe m. guidé sur toute sa longueur. / ~ of the lens || Linsenachse f. || axe m. de la lentille. / longitudinal ~ (Math) || Längsachse f. || axe m. longitudinal. / ~ lying inclined in space || im Raume geneigt liegende Achse f. || axe m. incliné dans l'espace. / major ~ (Ellipse) || Hauptachse f.; große Symmetrieachse f. || grand axe m. / minor ~ (Ellipse) || Nebenachse f. || petit axe m. / optic ~ || Sehachse f. || axe m. optique

ou visuel. / ~ of ordinates || Ordinatenachse f. || axe m. des ordonnées. ~/ of oscillation || Schwingungsachse f. || axe m. d'oscillation. / principal ~ || Hauptachse f. || axe m. principal. / ~ of revolution || Rotationsachse f.; Umdrehungsachse f. || axe m. de révolution ou de rotation. / to rotate round its own ~ || um die eigene Achse drehen || tourner ou tourillonner autour de son axe propre. / ~ of rotation see ~ of revolution. / secondary ~ || Nebenachse f. || axe m. imaginaire ou non principal ou non traverse; second axe m. / ~ that bears at separate points || an einzelnen Stellen geführte Achse f. || axe m. guidé en certains points. / to set the ~ of the telescope parallel to the plane of the instrument || die Fernrohrachse f. zur Instrumentenebene parallel einstellen || rendre l'axe m. de la lunette parallèle au plan du limbe. / ~ of sight || Visierachse f. || axe m. de visée. / transversal ~ || Querachse f. || axe m. transversal. / vertical ~ || Vertikalachse f. || axe m. vertical. / ~ of vision || Gesichtsachse f.; Sehachse f. || axe m. optique ou visual. / visual ~ see axis of vision. / ~ of vortex || Wirbelachse f. || axe m. du tourbillon. / to take off the weight from the ~ || die Achse f. entlasten || décharger l'axe m.

axis test disk for astigmatic eyes || Achsensehprüfscheibe f. für astigmatische Augen || goniomètre m. pour l'astigmatisme ou pour des yeux astigmates.

axle || Radachse f.; Achse f. || essieu m.; arbre m. / ~ (Coachm) || Wagenachse f. || essieu m. (de voiture). / adjustable ~ || Lenkachse f. || essieu m. orientable. / braking ~ || Bremsachse f. || essieu m. freiné. / the ~ is broken || die Achse f. ist gebrochen || l'essieu m. est brisé. / cambered ~ || gestürzte Achse f. || essieu m. arqué. / ~ of car || Wagenachse f. || essieu m. de voiture. / cardan ~ || Kardanwelle f. || arbre m. à cardans. / carrying ~ || Laufachse f.; Tragachse f. || essieu m. porteur. / channel section ~ || Achse f. mit U-Querschnitt || essieu m. à profil en U. / coupled ~ || Kuppelachse f. || essieu m. (ac)couplé. / ~ coupled with a dynamo shaft || mit einer Dynamowelle gekuppelte Achse f. || essieu m. accouplé avec un arbre de dynamo. / crank ~ || Kropfachswelle f.; Kurbelachse f.; gekröpfte Achse f. || essieu m. coudé. / cranked ~ see axle, crank. / the ~ was distorted without showing any incipient fracture || die Achse wurde verdreht, ohne einen Anbruch zu zeigen || l'arbre m. fut tordu sans présenter une amorce de fissure. / divided ~ || geteilte Achse f. || essieu m. sectionné. / driving ~ || Triebachse f.; Motorachse f. || essieu m. moteur. / dropped ~ || gekröpfte Achse f. || essieu m. coudé. / fixed ~ || Festachse f. || essieu m. fixe. / flexible ~ approved by the German railway association || Vereinslenkachse f. || essieu m. de l'union des chemins de fer allemands. / floating ~ || fliegende Achse f. || essieu m. flottant. / forked ~ || Gabelachse f. || essieu m. chapé. / free flexible ~ || freie Lenkachse f. || essieu m. orientable libre. / front ~ || Vorderachse f. || essieu m. d'avant. / full floating ~ || vollfliegende Achse f. || essieu m. tout flottant. / half of ~ || Halbachse f. || demi-essieu m. / hollow ~ ||

Hohlachse f. || essieu m. creux. / manufacture of hollow ~s by pressing || Herstellung f. von Hohlachsen durch Pressung || fabrication f. des essieux creux étirés par pression hydraulique. / iron ~ || eiserne Radachse f. *oder* Wagenachse f. || essieu m. en fer. / jointed cross shaft ~ (Auto) || Achse f. mit Querkardanwellen; Schwingachse f. || essieu m. à cardans transversaux. / loose ~ || Blindachse f. || faux-essieu m. / the ~s pl. are manufactured in any desired diameter || die Achsen fpl. werden in jeder gewünschten Stärke hergestellt || les essieux mpl. se fabriquent de tout diamètre. / mounted ~ || montierte Achse f. || essieu m. monté. / oblique crank ~ || schrägschenklige Kropfachswelle f. || essieu m. coudé à corps oblique. / passing ~ || durchgehende Achse f. || arbre m. traversant. / ~ of the pulley || Blocknagel m. || essieu m. de la poulie. / rear ~ (Auto) || Hinterachse f. || essieu m. d'arrière. / rear ~ (Coachm) || Hinterachsbrücke f. || pont m. arrière. / rear ~ tube || Hinterachsrohr n. || tube m. du pont arrière. / flared tube of rear ~ (Auto) || Hinterachstrichter m. || cône m. tubulaire du pont arrière. / semi-floating ~ || halbfliegende Achse f. || essieu m. demi-flottant. / sliding ~ || Schiebestange f.; Gleitachse f. || essieu m. à glissement. / solid ~ || Vollachse f. || essieu m. massif. / solid drawn ~ || gezogene Achse f. || essieu m. étiré. / steel-cast crank ~ || Stahlgußkurbelwelle f. || essieu m. coudé d'acier moulé. / steering ~ (Auto) || Lenkachse f. || essieu m. directeur. / straight ~ || gerade Achse f. || essieu m. droit. / T-beam ~ || Achse f. mit Doppel-T-Querschnitt || essieu m. à profil en double-T. / tender ~ (Railw) || Tenderachse f. || essieu m. de tender. / three quarter floating ~ || dreiviertelfliegende Achse f. || essieu m. flottant trois quarts. / the ~ is (through) bored from end to end || die Achse f. ist mit durchgehender Bohrung versehen || l'essieu m. est percé sur toute sa longueur. / tubular ~ || Rohrachse f. || essieu m. tubulaire.

axle arm || Halslager n. einer Welle || fusée f. / ~ base || Achsabstand m.; Achsstand m. || écartement m. des essieux; empattement m.

axle bearing || Achslager n. || boîte f. d'essieu. / ~ lining *see* ~ step. / ~ step || Achslagerschale f. || coussinet m. de boîte d'essieu.

axle body || Achskörper m. || corps m. de l'essieu. / ~ bore || Achsbohrung f.; Achsloch n. || alésage m. de l'essieu.

axle box || Achsbuchse f.; Achslager n. || boîte f. d'essieu *ou* à graisse *ou* à huile. / ~ body || Achslagergehäuse n. || corps m. de boîte à huile *ou* de boîte d'essieu. / ~ cover || Achsbuchsendeckel m. || couvercle m. de boîte à graisse. / ~ guide ||

Achsbuchsenführung f. || guides mpl. de la boîte à graisse. / ~ keep || Achslagerunterkasten m. || partie f. inférieure de boîte d'essieu. / ~ liner || Achslagergleitplatte f. || fourrure de boîte d'essieu.

axle box slide || Achslagerführung f. || guide m. de la boîte d'essieu. / front ~ || vordere Achslagerführung f. || guide m. de la boîte d'essieu avant. / oblique ~ || schräge Achslagerführung f. || guide m. de la boîte d'essieu oblique. / rear ~ || hintere Achslagerführung f. || guide m. de la boîte d'essieu arrière.

axle box wedge || Achslagerstellkeil m. || coin m. de serrage *ou* de réglage des boîtes d'essieu.

axle bracket on motor cars || Achsstütze f. an Kraftwagen || support m. d'essieu d'automobiles. / ~ cap || Achskappe f.; Radkapsel f. || chapeau m. d'essieu *ou* de moyeu. / ~ changing device || Achswechselvorrichtung f. || dispositif m. de rechange d'essieux.

axle collar || Achsbund m. || collet m. de l'essieu. / ~ forged from the solid || aus dem Vollen geschmiedeter Bundring m. || collet m. d'essieu venu de forge. / shoulder of the ~ || Anlauf m. am Achsbund || congé m. du collet de l'essieu.

axle compressor || Achskompressor m. || compresseur m. d'essieux. / ~ cutting-off and centring machine || Achsenabstechund Zentrierbank f. || tour m. à tronçonner et centrer les essieux. / ~ drive bevel wheel || Tellerrad n. || couronne f. d'angle du différentiel. / ~ end || Wellenzapfen m. || tourillon m. / ~ fairing || Achsverkleidung f. || carénage m. de l'essieu. / ~ fracture || Achsenbruch m. || rupture f. d'essieu. / ~ friction || Achsenreibung f. || frottement m. de l'essieu. / ~ grease || Achsenschmiere f. || graisse f. pour les essieux. / ~ guard || Achshalter m.; Achsgabel f. || plaque f. de garde. / half of ~ || Halbachse f. || demi-essieu m. / ~ hole || Achsloch n.; Achsbohrung f. || alésage m. de l'essieu. / ~ housing || Achsgehäuse n. || carter m. de l'essieu.

axle journal || Achsschenkel m.; Achszapfen m.; Achshals m.; Achsstummel m. || fusée f. d'essieu; tourillon m.; tronçon m. coudé. / fracture of the ~ || Achsschenkelbruch m. || rupture f. de la fusée d'un essieu. / ~ grinding machine || Achsschenkelschleifmaschine f. || machine f. à rectifier les fusées d'essieu.

axleless undercarriage || achsenloses Fahrgestell n. || châssis m. sans essieu.

axle light || Achsenbeleuchtung f. || éclairage m. par l'axe. / ~ load || Achsbelastung f. || charge f. de l'essieu *ou* par essieu. / ~ middle lathe || Achsmittendrehbank f. || tour m. à charioter les corps d'essieux. / ~ neck || Achsschenkel m. || fusée f. de l'essieu. / ~ pin || Splint m. || cheville f. d'essieu. / pressure of ~ || Achsdruck

m. || charge f. *ou* pression f. de l'essieu. / ~ ring || Achsring m. || anneau m. de bout d'essieu. / ~ roughing lathe || Achsschrupppdrehbank f. || tour m. à dégrossir les essieux. / ~ shaft || Differentialseitenwelle f. || arbre m. de roue motrice. / ~ slide *see* axle box slide. / ~ spanner *see* axle wrench.

axletree || Radachse f.; Achse f. || essieu m. / ~ (Wind mill) || Flügelwelle f.; Königswelle f.; Kronenwelle f. || arbre m. / fore ~ || Vorderachse f. || essieu m. de devant. / ~ of a plough || Pflughaupt n. || essieu m. de la charrue.

axletree arm || Achsschenkelzapfen m. || fusée f. d'essieu. / ~ bed || Achsfutter n. || corps m. d'essieu. / ~ box || Achsbüchse f. || boîte f. d'essieu *ou* à graisse *ou* de roue.

axle tube || Achsrohr n. || tube m. *ou* enveloppe f. de l'essieu.; tube m. fourreau. / ~ weight on the ~ *see* axle load. / ~ wrench || Achsenschlüssel m. || clef f. pour essieux.

Axminster carpet || Axminsterteppich m. || velvet-carpet m.; tapis-chenille m.

azimuth || Azimut m.; Scheitelkreis m. || azimut m.; cercle m. vertical. / slow motion of the telescope in ~ || Feinbewegung f. des Fernrohres in Azimut || mouvement m. lent horizontal de la lunette dans l'azimut.

azimuthal || azimutal || azimutal.

azimuth circle || Azimutkreis m. || cercle m. de relèvement. / ~ correction || Azimutalkorrektion f. || rectification f. de l'azimut. / ~ diaphragm || Azimutblende f. || diaphragme m. azimutal. / ~ finder || Azimutzeiger m. || indicateur m. d'azimut.

azo dye || Azofarbstoff m. || colorant m. azoïque.

azotate || salpetersaures Salz n.; Nitrat n. || azotate m.; nitrate m.

azote || Stickstoff m. || azote m.; nitrogène m.

azotic acid || Scheidewasser n.; Salpetersäure f. || eau-forte f.; acide m. azotique *ou* nitrique.

azotite || Nitrit n.; salpetrigsaures Salz n. || azotite m.; nitrite.

azotometer || Azotometer n. || azotomètre m.

azure || himmelblau || azuré.

azure || Lasurblau n. || azur m.; outremer m. / ~ blue || Azurblau n. || bleu m. d'azur. / ~ copper ore || Kupferlasur f.; Azurit m.; Chessylit m. || azurite f.; cuivre m. carbonaté bleu; azur m. de cuivre. / ~ dyer || Himmelblaufärber m. || azureur m. / ~ stone || Lapislazuli m. || lapis-lazuli m.; pierre f. d'azur; lazulite m.

azurite || Kupferlasur f.; Azurit m.; Chessylit m. || azurite f.; cuivre m. carbonaté bleu; azur m. de cuivre. / earthy ~ || Bergblau n. || azurite f. terreuse.

azyme bread || Matze f. || pain m. azyme.

B

B-battery ‖ Anodenbatterie f. ‖ batterie f. de plaque.

B-position (Tel) ‖ Verbindungsplatz m. ‖ position f. B.

babbit, to ‖ mit Weißmetall n. ausgießen ‖ garnir de métal m. blanc.

babbit ‖ Lagerfutter n. ‖ fourrure f. *ou* douille f. de coussinet.

babbitted bearing ‖ Weißmetallager n. ‖ coussinet m. en métal antifriction.

babbit metal ‖ Weißmetall n.; Babbitmetall n. ‖ métal m. blanc *ou* antifriction *ou* à coussinets; régule m.

baby articles pl. ‖ Babyartikel mpl. ‖ articles mpl. pour bébés.

baby Bessemer steel ‖ Kleinbessemerstahl m. ‖ acier m. Bessemer du petit convertisseur. / ~ plant ‖ Kleinbessemerei f. ‖ petite installation f. Bessemer.

baby bottle ‖ Milchflasche f. ‖ bouteille f. à lait. / ~ carriage ‖ Kinderwagen m. ‖ voiture f. d'enfant. / ~ linen ‖ Kinderwäsche f.; Erstlingsausstattung f. ‖ lingerie f. *ou* linge m. pour nourissons.

baby tank (Arm) ‖ Kleinkampfwagen m. ‖ tankette f.

bac (Brew) ‖ Kühlschiff n.; Kühlstock m.; Kühle f. ‖ bac m. refroidissoir *ou* rafraîchissoir. / ~ (Ship) ‖ Fähre f.; Fährkahn m.; Prahm m. ‖ bac m.; passe-cheval m.; pont m. flottant.

Bachelor of Commerce ‖ Diplomkaufmann m. ‖ diplômé m. de l'école de commerce.

bacillus ‖ Bazillus m. ‖ bacille m.

back, to (Horse) ‖ hufen; rückwärts gehen ‖ reculer; faire reculer. / ~ (Bank) ‖ indossieren ‖ endosser. / ~ a book (Bookb) ‖ einen Buchrücken abpressen ‖ endosser un livre. / ~ the shell (Galvano) ‖ hintergießen ‖ doubler. / ~ up *see* ~ the shell. / ~ up the water ‖ das Wasser stauen ‖ retenir *ou* élever les eaux fpl.

back ‖ rückwärts ‖ arrière; à rebours. / from front to ~ ‖ von vorn nach hinten ‖ d'avant en arrière.

back (Anatomy) ‖ Rücken m. ‖ dos m. / ~ (Brew) *see* bac. / ~ (Coachm) ‖ Hinterteil m. am Stoße der Nabe ‖ gros bout m. du moyeu. / ~ (Join) ‖ Rückenlehne f.; Rücklehne f. ‖ dossier m. / ~ (Min) ‖ Gangspalte f.; Spalte f. ‖ fente f.; fissure f. / ~ (Phot) ‖ Kassette f. ‖ châssis m. / ~ (Weav) ‖ Futter n. ‖ doublure f. / ~ of the blast furnace ‖ Hinterknobben m. eines Hochofens ‖ rustine f. d'un haut-fourneau. / ~ of a book ‖ Buchrücken m. ‖ dos m. d'un livre. / ~ for brushes ‖ Bürstenholz n. ‖ bois m. de brosses. / ~ of a chimney ‖ Hinterwand f. oder Rückenplatte f. eines Kamins ‖ contre-cœur m. de cheminée; plaque f. de feu. / fixed ~ ‖ feste Rücklehne f. ‖ dossier m. fixe. / ~ of a gallery ‖ Firste f. eines Stollens ‖ faîte m. d'une galerie. / glued ~ of the booklet ‖ beleimter Rücken m. der Broschüre ‖ dos m. de la brochure enduit de colle. / ~ of a hand saw ‖ Rücken m. einer Handsäge ‖ dos m. *ou* dosseret m. *ou* dossier f. d'une scie à main. / hinged ~

back arch (Build) ‖ Laibungsbogen m. einer Tür ‖ arrière-voussure f. / ~ band ‖ Tragriemen m.; Gabeltragriemen m. ‖ dossière f. / ~ board (Mar) ‖ Gegenschlagbug m.; Schlingerschlagbug m. ‖ mauvaisbord m. / ~ board (Shipb) ‖ Lehnbrett n. ‖ dossier m. d'un canot. / ~ boxes pl. (Print) ‖ oberer Schriftkasten m. ‖ haut m. de casses. / ~ callipers pl. ‖ Taster m. mit Zahnbogen ‖ compas m. d'épaisseur à crémaillère. / ~ contact ‖ Ruhekontakt m. ‖ contact m. de repos. / ~ coupling (Radio) ‖ Rückkopplung f. ‖ réaction f. / ~ cover (Cylinder) ‖ Hinterdeckel m. ‖ couvercle m. arrière. / ~ cushion ‖ Rückenkissen n. ‖ coussin m. de dossier. / ~ dike ‖ Hinterdamm m.; Hinterdeich m.; Achterdeich m. ‖ arrière-digue f. *ou* -bord m. / ~ discharge (Radio) ‖ Rückentladung f. ‖ retour m.; décharge f. en retour. / ~ door ‖ Hintertür f. ‖ fausse-porte f.; poterne f.; arrière-porte f. / ~ drain ‖ Hintergraben m. ‖ fossé m. en arrière d'une digue. / ~ end plate (Boil) ‖ Hinterboden m.; Hinterwand f. ‖ fond m. arrière.

backed note ‖ Ladeschein m. ‖ permis m. d'embarquement.

backfall (Pap) ‖ Sattel m. oder Kropf m. oder Berg m. des Papierzeughollanders ‖ gorge f. d'une pile à cylindre. / breasting ~ *see* backfall.

back-firing of the flame ‖ Rückschlagen der Flamme ‖ retour m. de flamme.

back-flow ‖ Rückströmung f.; courant m. *ou* écoulement m. de retour. / ~ focal plane ‖ hintere Brennebene f. ‖ plan m. focal postérieur. / ~-fork ‖ Hintergabel f. ‖ fourche f. arrière.

backgammon ‖ Puffspiel n. ‖ jaquet m.

back-gear, to throw into ‖ das Vorgelege einrücken ‖ renverser la marche.

background ‖ Hintergrund m. ‖ fond m.

backhouse ‖ Hintergebäude n. ‖ bâtiment m. de derrière.

backing (Build) ‖ Nachmauerung f.; Hintermauerung ‖ remplissage m. à l'arrière d'une voûte. / ~ (Hydr arch) ‖ Hinterfüllung f. einer Ufermauer ‖ terre-plein m. / ~ (Print) ‖ Bedrucken n. der Rückseite ‖ retiration f. / ~ (Trade) ‖ Rückendeckung f. ‖ couverture f.; garantie f. / ~ higher than the wall (Hydr arch) ‖ Hinterfüllung f. mit Auflast ‖ terre-plein m. surchargé. / ~ to the left ‖ Linksablenkung f. ‖ déviation f. vers la gauche. / ~ of the wind ‖ Zurückspringen n. des Windes ‖ rebroussement m. du vent. / ~ wood ‖ (Shipb) ‖ Panzerunterlage f. ‖ matelas m. en bois.

backing metal (Letter-f) ‖ Hintergießmetall n. ‖ métal m. de doublage.

backing-off lathe ‖ Hinterdrehbank f. ‖ tour m. à dépouiller.

backing oven ‖ Brennofen m. ‖ fourneau m. de calcinage. / ~ plate (Acc) ‖ Stützplatte f. ‖ plaque f. d'appui; contre-plateau m. / ~ press (Bookb) ‖ Rückenpresse f. ‖ presse f. à endosser.

back-joiner (Bookb) ‖ Einhänger m. ‖ emboîteur m.

back-kick ‖ Rückschlag m. ‖ choc m. en arrière; contre-coup m. / ~ (Electr) ‖ Rückentladung f. ‖ retour m.; décharge f. en retour.

back-knife gauge lathe ‖ Schablonendrehbank f. mit Hintermesser ‖ tour m. à gabarit à outil à l'arrière.

back lash (Mach) ‖ toter Gang m.; Spielraum m. ‖ jeu m. (mort). / ~ of fascine work ‖ Rücklage f. des Packwerks ‖ couche f. ordinaire. / ~ lens ‖ Hinterlinse f. ‖ lentille f. arrière. / ~light ‖ Rücklicht n. ‖ feu m. d'arrière. / ~ liner (Bookb) ‖ Buchrückenpresser m. ‖ endosseur m. / ~ magnetization ‖ Gegenmagnetisierung f. ‖ aimantation f. antagoniste. / ~ part (Met) ‖ Hintergestell n. eines Hochofens ‖ creuset m. de derrière. / ~ pedalling brake ‖ Rücktrittbremse f. ‖ frein m. à contre-pédale. / ~ piece of a centre (Carp) ‖ Kranzstück n. eines Lehrbogens ‖ courbe f. de cintre; veau m. de cintre. / ~ pier (Arch) ‖ zurücktretender Kämpfer m. ‖ fausse-alette f. / ~ plan ‖ Hinteransicht f. ‖ élévation f. de derrière.

back-plate (Boil) ‖ Hinterboden m.; Hinterwand f. ‖ fond m. arrière. / ~ of a fining-forge ‖ Hinterzacken m. *oder* Aschenzacken m. des Frischherdes ‖ taque f. de fond; herre f.; haire f.; rustine f.; plaque f. *ou* taque f. de rustine. / ~ of a locomotive firebox ‖ Rückwand f. der Lokomotivfeuerbuchse; Stehkesselrückwand f. ‖ face f. *ou* plaque f. arrière de la boîte à feu de locomotive.

back-pressure ‖ Gegendruck m. ‖ contre-pression f. / ~ bottle filler ‖ Gegendruckflaschenfüllapparat m. ‖ soutireuse f. de bouteilles à contre-pression. / ~ turbine ‖ Gegendruckturbine f. ‖ turbine f. à contrepression. / ~ valve ‖ Rückschlagventil n. ‖ soupape f. de retenue.

back-print, to ‖ wiederdrucken ‖ imprimer au verso.

back puppet (Turn) ‖ Hinterdocke f. ‖ poupée f. de derrière. / ~ quilt ‖ Rückenpolster n. eines Sitzes ‖ dossier m. / ~ release (Aut tel) ‖ Rückauslösung f. ‖ débrayage m. inverse. / ~ rest ‖ Rückenlehne f. ‖ dossier m. / ~ saw ‖ Rückensäge f.; Fuchsschwanz m. mit Rücken ‖ scie f. à dos *ou* à derrière. / ~ seat ‖ Rücksitz m. ‖ siège m. d'arrière. / ~ shield ‖ Windschutzscheibe f. für Hintersitze ‖ pare-brise m. pour places arrières. / ~ side ‖

Rückseite f. ‖ dos m. / ~-sided (Wood) ‖
windschief ‖ gauche. / ~ space key ‖
Rücktaste f.; Rücklauftaste f. ‖ touche
f. de marche arrière. / ~-spring lock ‖
Schnappschloß n.; Bastardschloß n.;
Schnippschnapp n. ‖ serrure f. bâtarde. /
~ square ‖ Anschlagwinkel m. ‖ équerre
f. à branche épaisse; équerre f. épaulée. /
~ stand ‖ Hintergestell n. ‖ support m.
d'arrière. / ~ starling (Bridge) ‖ Pfeilerhinterhaupt n.; Pfeilersterz m.; Talpfeilerkopf m. ‖ arrière-bec m.; bec m.
d'aval.

backstay (Shipb) ‖ Pardun n. ‖ galhauban
m. / ~ chain (Bridge) ‖ Rückhaltskabel n. ‖
câble m. d'amarre *ou* de retenue. / ~ insulator ‖ Pardunenisolator m. ‖ isolateur m. d'arrêt.

back stop (Mach Tool) ‖ Gegenhalter m. ‖
butée f. / ~ stroke ‖ Rückschlag m.;
Rückwärtshub m. ‖ choc m. de retour;
coup m. d'arrière du piston.

back-tenter (Pap) ‖ Papiermaschinenwär-
ter m. ‖ aide-conducteur m. de machine
à papier. / ~ (Spinn) ‖ Spulensetzer m. ‖
poseur m. de bobines.

back tool (Bookb) ‖ Filet n. ‖ filet m.; fer
m. à fileter. / ~ type parachute ‖ Rückenkissenfallschirm m. ‖ parachute f. sac-
dorsal. / ~ tyre ‖ Hinterradreifen m. ‖
pneu m. arrière. / ~ view ‖ Rückansicht
f. ‖ vue f. de derrière. / ~ wall ‖ Rückwand f. ‖ paroi f. arrière.

backward, to topple ‖ rückwärts umfallen ‖
tomber à la renverse. / ~ motion ‖ Rückwärtsbewegung f. ‖ mouvement m. de
recul. / ~ movement ‖ Rückgang m. ‖
marche f. en arrière. / ~ pitch (Armature
winding) ‖ Rückwärtsschritt m. ‖ pas m.
en arrière.

backwards ‖ rückwärts ‖ en arrière.

back-washer (Wool) ‖ Plätter m.; Plätt-
maschinenarbeiter m. ‖ lisseur m.

back-water (Mar) ‖ Stauwasser n. ‖ eaux
fpl. dormantes.

back-wheel brake ‖ Hinterradbremse f. ‖
frein m. de la roue arrière. / ~ rim brake ‖
Hinterradfelgenbremse f. ‖ frein m. sur
la jante de roue arrière.

back wing (Arch) ‖ Hinterflügel m. ‖ arrière-corps m. d'un bâtiment. / ~ wool ‖
Rückenwolle f. ‖ laine f. mère *ou* prime;
mère f. laine.

bacon ‖ Speck m. ‖ lard m. / fresh ~ ‖ frischer Speck m. ‖ lard m. frais. / ~ curer ‖
Speckräucherer m. ‖ saleur de lard m.

bacteria pl. ‖ Bakterien fpl. ‖ bactéries fpl.
/ dead ~ ‖ Bakterienleichen fpl. ‖ cadavres
mpl. de bactéries.

bacteria colony ‖ Bakterienkolonie f. ‖ colonie f. de bactéries. / ~ contamination ‖
Bakterieninfektion f. ‖ infection f. bactérienne.

bacterial cloudiness ‖ Bakterientrübung f. ‖
trouble m. de bactéries. / ~ culture ‖ Bakterienkultur f. ‖ culture f. de bactéries. /
~ form ‖ Bakterienform f. ‖ forme f. de
bactérie. / ~ turbidity ‖ Bakterientrübung f. ‖ trouble m. de bactéries.

bacteria spore ‖ Bakterienspore f. ‖ spore
f. de bactérie.

bacteriological ‖ bakteriologisch ‖ bactériologique. / ~ analysis ‖ bakteriologische
Analyse f. ‖ analyse f. bactériologique. /
~ apparatus ‖ bakteriologischer Apparat
m. ‖ appareil m. de bactériologie. / ~ culture ‖ Bakterienzüchtung f. ‖ culture f.
bactérienne. / ~ research ‖ bakteriolo-

gische Untersuchung f. ‖ recherche f.
bactériologique. / ~ set ‖ bakteriologisches Besteck n. ‖ trousse f. bactériologique.

bacteriologist ‖ Bakteriologe m. ‖ bactério-
logue m.

bacteriology ‖ Bakteriologie f. ‖ bactériologie f.

bacterium ‖ Bakterium ‖ bactérie f. / aerobic ~ ‖ aerobes Bakterium n. ‖ bactérie f.
aérobique. / anaerobic ~ ‖ anaerobes
Bakterium n. ‖ bactérie f. anaérobique. /
arthrosporic ~ ‖ arthrospores Bakterium
n. ‖ bactérie f. arthrospore. / exosporic ~
‖ exospores Bakterium n. ‖ bactérie f.
exospore. / putrefactive ~ ‖ Fäulnisbak-
terium n. ‖ bactérie f. putride. / slimy ~ ‖
schleimbildendes Bakterium n. ‖ bactérie f. visqueuse. / thread ~ ‖ Fadenbakterium n. ‖ bactérie f. filamenteuse.

bad air (Min) ‖ gebrauchte Wetter npl. ‖
air m. vicié. / ~ condition ‖ Schadhaftigkeit f. ‖ mauvais état m. / ~ luck ‖ Un-
glück n. ‖ malheur m. / ~ mortar ‖ Halbmörtel m. ‖ mortier m. bâtard. / ~ weather
‖ Unwetter n. ‖ gros temps m. / ~ wool ‖
Ausschußwolle f. ‖ abat-chauvée f.

badge ‖ Abzeichen n. ‖ marque f.; insigne
m. / ~ (Auto) ‖ Wagenschild n. ‖ panon-
ceau m. / enamelled ~ ‖ Kennzeichen n.
in Email ‖ insigne m. en émail.

badge factory ‖ Abzeichenfabrik f. ‖ fabrique f. d'insignes *ou* de marques. / ~ pin ‖
Abzeichennadel f. ‖ broche f. d'insignes.

badger's fat ‖ Dachsfett n. ‖ graisse f. de
blaireau. / ~ skin ‖ Dachsfell n. ‖ peau
f. de blaireau.

badigeon ‖ Gipsmörtel m.; Stuckmörtel m.
‖ badigeon m.; mortier-stuc m. / yellow
~ ‖ Mauergelb n. ‖ badigeon m.; badigeon m. jaune.

bad-part, to take in ‖ übelnehmen ‖ prendre
en mauvaise part.

baffle, to ~ the gas ‖ das Gas drosseln ‖
chicaner le gaz.

baffle plate ‖ Stauscheibe f. ‖ chicane f.;
culot m.

bag, to ~ away (Nav) ‖ nach Lee sacken ‖
aller en dérive.

bag ‖ Beutel m.; Täschchen n. ‖ pochette f.;
bourse f.; petit sac m. / cycle ~ ‖ Fahrradtasche f. ‖ sacoche f. de bicyclette. /
leather ~ ‖ Ledertasche f. ‖ sac m. en
cuir. / to pack in ~s pl. ‖ sacken ‖ mettre
en sac m.; ensacher. / paper ~ ‖ Papierbeutel m.; Tüte f. ‖ sac m. en papier. /
sailor's ~ ‖ Kleidersack m. ‖ sac m. de
matelot. / school ~ ‖ Schultasche f. ‖ sac
m. d'école; porte-cahiers m. / straw ~ ‖
Strohtasche f. ‖ sac m. en paille.

bag conveyor ‖ Sacktransportör m. ‖ transporteur m. de sacs. / ~ cutter ‖ Sack-
schneider m. ‖ coupeur m. de sacs. / ~
dusting machine ‖ Sackausstäuber m. ‖
batteuse f. de sacs. / ~ filter (Sug) ‖
Beutelfilter m. ‖ filtre m. à poche.

baggage ‖ Gepäck n. ‖ bagage m. / ~ car
‖ Gepäckwagen m.; Packwagen m. ‖
fourgon m.; wagon m. à bagages. / ~
check ‖ Gepäckschein m. ‖ bulletin m.
de bagages. / ~ train ‖ Güterzug m. ‖
train m. de marchandises. / ~ wagon ‖
(bedeckter) Güterwagen m. ‖ wagon m.
à marchandises (couvert *ou* fermé).

bagging hopper ‖ Einsacktrichter m. ‖ ensachoir m.

bag-maker ‖ Sackmacher m. ‖ confectionneur m. de sacs. / ~ (Milit) ‖ Täschner m.

‖ musettier m. / ~'s goods pl. ‖ Täschner-
waren fpl. ‖ articles mpl. de maroquinerie.

bag making machine ‖ Taschenherstel-
lungsmaschine f. ‖ machine f. pour fabriquer des pochettes. / ~ net ‖ Sacknetz n.;
Senke f.; Hamen m. ‖ rets m. à sac; filet
m. d'achat; ableret m.; ablerette f. /
~ scoop ‖ Löffelbagger m. ‖ cure-môle m.;
drague f. à cuiller. / ~ sewer ‖ Sacknäher
m. ‖ couseur m. de sacs. / ~ spoon *see* ~
scoop. / ~ weaver ‖ Sackweber m. ‖
tisseur m. de sacs.

bail, to ~ out water with buckets ‖ Wasser
mit Schöpfeimern schöpfen ‖ baqueter les
eaux.

bail ‖ Henkel m. ‖ anse f.

bailiff ‖ Gerichtsvollzieher m. ‖ huissier m.

bailing ladle ‖ Schöpfer m.; Schöpfgefäß
n. ‖ godet-puisoir m.

bails pl. (Trade) ‖ Aval m. ‖ avals mpl.

bail scoop ‖ Schaufel f.; Schippe f.; Wurfschaufel f. ‖ pelle f.

bait, to ‖ ködern ‖ amorcer.

bait ‖ Köder m. ‖ appât m.; amorce f.

baize ‖ Fries m.; Flaus m. ‖ frise f.

bake, to ‖ backen ‖ cuire; frire. / ~ the clay
‖ Ton m. brennen ‖ cuire l'argile. / ~ a
wall ‖ mit Füllsteinen mpl. mauern; hin-
termauern ‖ bloquer.

baked cheese ‖ Backsteinkäse m. ‖ fromage
m. cuit.

bakehouse ‖ Brotbäckerei f. ‖ boulangerie
f.; manutention f. / machine for ~s ‖
Maschine f. für Bäckereien ‖ machine f.
pour boulangers.

bakelite ‖ Bakelit n. ‖ bakélite f.

bakelized paper ‖ Hartpapier n. ‖ carton
m. bakélisé.

bake oven *see* baking oven.

baker ‖ Bäcker m.; Brotbäcker m. ‖ boulanger m. / foreman ~ ‖ Bäckereiwerk-
meister m. ‖ brigadier m. d'une boulangerie.

baker's machine ‖ Bäckereimaschine f. ‖
machine f. pour boulangerie. / ~ stove ‖
Backofen m. ‖ four m. pour pâtissiers. /
~ tool made of wood ‖ Bäckergerät n. aus
Holz ‖ outil m. en bois pour boulangers. /
~ ware ‖ Backware f.; Backwerk n. ‖ articles mpl. de boulangerie.

bakery ‖ Bäckerei f. ‖ boulangerie f. /
army ~ ‖ Militärbäckerei f. ‖ boulangerie
f. militaire. / mechanical ~ ‖ mechanische
Bäckerei f. ‖ boulangerie f. mécanique.

bakery machine ‖ Bäckereimaschine f. ‖
machine f. de boulangerie.

baking (Pott) ‖ Brennen n. ‖ cuisson f. /
~ of the coke ‖ Backen n. des Kokses ‖
agglutination f. *ou* cuisson f. du coke.

baking coal ‖ backende Kohle f. ‖ charbon
m. collant. / non-~ coal ‖ Magerkohle f. ‖
houille f. maigre. / ~ device ‖ Bäckerei-
einrichtung f. ‖ équipement m. de boulangerie. / ~ dough ‖ Backmasse f. ‖ pâte
f. de boulangerie. / ~ engine *see* bakery
machine. / ~ experiment ‖ Backversuch
m. ‖ essai m. de cuisson. / ~ hood ‖ Back-
haube f. ‖ petit four m. à pâtisserie. /
~ mould ‖ Backform f.; Kuchenform f. ‖
forme f. pour boulangeries *ou* à pâtisserie. / ~ oil ‖ Backöl n. ‖ huile f. à frire.

baking oven ‖ Backofen m. ‖ four m. (de
boulangerie *ou* à cuire). / combination
steam ~ ‖ Kombinationsdampfbackofen
m. ‖ four m. combiné à vapeur. / electrical ~ ‖ Elektrobackofen m. ‖ four m.
électrique. / peel steam ~ heated by

wood ‖ Einschießdampfbackofen m. für Holzheizung ‖ four m. à vapeur pour enfourner à chauffage au bois. / steam ~ ‖ Dampfbackofen m. ‖ four m. à vapeur. / ~ heated by wood ‖ Dampfbackofen m. für Holzheizung ‖ four m. à vapeur pour chauffage au bois. / transportable ~ ‖ transportabler Backofen m. ‖ four m. de boulangerie transportable. / tunnel ~ for bread factory ‖ Tunnelbackofen m. für Brotfertigung ‖ four m. à tunnel pour la panification. / ~ with confectionery sole mounted above ‖ Backofen m. mit aufgebautem Konditorherd ‖ four m. surmonté d'un four pâtissier.

baking plate ‖ Backblech n. ‖ tôle f. à pâtisserie. / ~ powder ‖ Backpulver n. ‖ poudre f. à cuire; levain m. en poudre. / ~ tin see ~ plate.

balalaika ‖ Balalaika f. ‖ balalaika f.

balance, to (Techn) ‖ ausbalanzieren; ins Gleichgewicht n. bringen; abgleichen ‖ balancer; équilibrer; contre-peser. / ~ (Accounts) ‖ saldieren; ausgleichen ‖ balancer; solder; arrêter un compte. / ~ (Books) ‖ abschließen ‖ régler. / ~ (Scale) ‖ einspielen ‖ balancer. / ~ the instrument ‖ das Instrument auswuchten ‖ équilibrer l'instrument. / ~ by static and dynamic methods ‖ statisch und dynamisch auswuchten ‖ équilibrer par les méthodes fpl. statiques et dynamiques.

balance ‖ Wage f.; Balkenwage f. ‖ balance f. / ~ (Mech) ‖ Gleichgewicht n. ‖ équilibre m. / ~ (Mill) ‖ Balanz f.; Balanzhaue f. ‖ anille f. / ~ (Surplus) ‖ Überschuß m. ‖ surplus m.; bénéfice m. / ~ (Trade) ‖ Saldo m.; Restbestand m. ‖ solde m.; reliquat m. / ~ (Watchm) ‖ Unruhe f. ‖ balancier m. d'une montre. / ~ of account ‖ Rechnungssaldo m. ‖ reliquat m. de solde de compte. / adverse ~ ‖ Unterbilanz f. ‖ déficit. / all kind of ~s ‖ Wagen fpl. aller Art ‖ balances fpl. de toutes sortes. / analytical ~ ‖ Analysenwage f.; Präzisionswage f. ‖ balance f. chimique ou de précision. / annual ~ ‖ Jahresabschluß m. ‖ bilan m. / to approve the ~ ‖ die Bilanz f. genehmigen ‖ approuver la balance. / balancing ‖ Abgleich m. ‖ équilibre m.; balance f. / bent-lever ‖ Zeigerwage f. ‖ balance f. à cadran. / ~ carried forward ‖ Saldovortrag m. ‖ solde m. porté à nouveau. / ~ carried to the credit of new account ‖ Saldovortrag m. auf neue Rechnung ‖ solde m. reporté au crédit de nouveau compte. / chemical ~ ‖ chemische Wage f. ‖ balance f. chimique. / common ~ ‖ Balkenwage f.; Krämerwage f. ‖ balance f. ordinaire. / to correct disturbances pl. of the ~ ‖ Gleichgewichtsstörungen fpl. beseitigen ‖ corriger des défauts mpl. d'équilibre. / counter ~ ‖ Entlastungsvorrichtung f. ‖ dispositif m. de décharge ou de mise au repos. / ~ by counterweights ‖ Auswuchtung f. durch Gegengewichte ‖ équilibrage m. par contrepoids. / Danish ~ ‖ Schnellwage f. mit festem Gewicht ‖ balance f. romaine à contrepoids fixe. / to disturb the ~ ‖ das Gleichgewicht stören ‖ troubler ou rompre l'équilibre m. / dynamic ~ ‖ dynamisches Gleichgewicht n. ‖ équilibre m. dynamique. / electrodynamic ~ ‖ elektrodynamische Wage f. ‖ balance f. électrodynamique.

/ grain ~ ‖ Getreidewage f. ‖ balance f. pèse-grains ou à grains. / hydrostatical ~ ‖ hydrostatische Wage f. ‖ balance f. hydrostatique. / indicator ~ ‖ Zeigerwage f. ‖ balance f. à cadran. / ~ of interest ‖ Zinsensaldo m. ‖ solde m. des intérêts. / kitchen ~ ‖ Haushaltwage f.; Küchenwage f.; Wirtschaftswage f. ‖ balance f. de cuisine. / metallometric ~ ‖ metallometrische Wage f. ‖ balance f. métallométrique. / parcel ~ ‖ Paketwage f. ‖ balance f. à paquets. / pharmaceutical ~ ‖ Tarierwage f. ‖ trébuchet m. / precision ~ ‖ chemische Wage f.; Präzisionswage f. ‖ balance f. chimique ou de précision. / Roman ~ ‖ Schnellwage f. mit Laufgewicht ‖ balance f. romaine ou à poids curseur; crochet m.; romaine f. / ~ with sack filler and lifter ‖ Schütt- und Absackwage f. ‖ balance f. à remplissage et à pesage. / sensible ~ ‖ empfindliche Wage f. ‖ balance f. sensible. / ~ of slide ‖ Stößelausbalanzierung f. ‖ équilibre m. du coulisseau. / sliding weight ~ ‖ Laufgewichtswage f. ‖ balance f. à poids curseur. / spring ~ ‖ Federwage f. ‖ bascule f. (à ressort). / ~ by springs ‖ Auswuchtung f. durch Pufferfedern ‖ équilibrage m. par ressorts. / static ~ ‖ statisches Gleichgewicht n. ‖ équilibre m. statique. / statistical ~ ‖ statistische Bilanz f. ‖ balance f. statistique. / steelyard ~ ‖ Brückenwage f. ‖ romaine f. / to strike a ~ ‖ eine Bilanz f. aufstellen oder ziehen ‖ dresser un bilan. / Swedish ~ see Danish ~. / test ~ ‖ Prüfwage f. ‖ trébuchet m. / torsion ~ ‖ Drehwage f. ‖ balance f. de torsion. / trial ~ ‖ Bruttobilanz f. ‖ bilan m. brut. / voltametric ~ ‖ voltametrische Wage f. ‖ balance f. voltamétrique. / wagon ~ ‖ Waggonwage f. ‖ bascule f. à wagon.

balance arm ‖ Wagebalken m. ‖ fléau m. de balance. / ~ beam see ~ arm. / ~ boss (Mill) ‖ Hauenbuchse f. ‖ manchon m. d'anille. / ~ case ‖ Wagekasten m. ‖ cage m. de la balance. / ~ crane ‖ Kran m. mit Gegengewicht ‖ grue f. à contrepoids.

balanced by a counterweight ‖ durch ein Gegengewicht n. ausbalanziert ‖ équilibré par un contrepoids. / ~ differential transformer ‖ Ausgleichsübertrager m. ‖ transformateur m. d'équilibre. / ~ motor ‖ ausgewuchteter Motor m. ‖ moteur m. équilibré. / ~ slide ‖ Entlastungsschieber m. ‖ tiroir m. équilibré. / ~ three-winding transformer see ~ differential transformer.

balance error ‖ Abgleichfehler m. ‖ défaut m. d'équilibrage; déséquilibre m. duplex. / ~ frame of a draw bridge ‖ Schwengel m. oder Wippe f. einer Aufzugbrücke ‖ bascule f. à fléau; flèche f. / ~ lever ‖ Hebel m. mit Gegengewicht ‖ levier m. à contrepoids. / ~ lever (Hand press) ‖ Schwengel m. ‖ balancier m. / ~ pan ‖ Wagschale m. ‖ plateau m. de balance.

balancer ‖ Ausgleichsmaschine f. ‖ machine f. à équilibrer.

balance sheet ‖ Jahresabschluß m.; Bilanz f. ‖ fin f. de l'année; balance f.; bilan m. / to make up a ~ ‖ eine Bilanz f. aufstellen ‖ dresser un bilan.

balance spring (Watchm) ‖ Unruhefeder f. ‖ ressort m. de balancier. / ~ test ‖ Gleichgewichtsprüfung f. ‖ mesure f. de l'équilibre. / ~ tester (Tel) ‖ Nachbil-

dungsprüfer m. ‖ appareil m. pour vérifier les équilibreurs. / ~ testing set (Tel) ‖ Nachbildungssatz m. ‖ appareillage m. d'équilibrage. / ~ vice (Watchm) ‖ Spindelklöbchen n. ‖ étau m. à queue d'horloger.

balance wheel (Watchm) ‖ Unruhe f.; Hemmungsrad n. ‖ balancier m.; roue f. de rencontre. / compensating ~ ‖ kompensierende Unruhe f. ‖ balancier m. compensé.

balance wheel manufacturing ‖ Unruheräderherstellung f. ‖ fabrication f. de balanciers. / ~ pliers ‖ Räderklemme f. ‖ pince f. aux roues de rencontre. / ~ spring ‖ Unruhefeder f. ‖ ressort m. du balancier.

balance weight ‖ Gegengewicht n. ‖ contrepoids m. / ~ of the reversing ‖ Gegengewicht n. der Steuerung ‖ contrepoids m. de changement de marche.

balancing (Trade) ‖ Aufrechnung f. ‖ compensation f. / ~ (Techn) ‖ Abgleichen n.; Abgleich m. ‖ balancement m.; balance f.; équilibre m. / ~ of accounts (Closing) ‖ Rechnungsabschluß m. ‖ arrêté m. ou règlement m. de compte. / ~ of accounts (Verifying) ‖ Gegenrechnung f. ‖ décompte m.; vérification f.; compte-courant m. / ~ of cash accounts ‖ Kassenabschluß m. ‖ clôture f. de la caisse. / ~ of circuits ‖ Leitungsabgleichverfahren n.; Symmetrierung f. ‖ équilibrage m. des circuits. / ~ by condensers ‖ Kondensatorabgleich m. ‖ équilibrage m. condensateur. / ~ by counterweight ‖ Ausgleichung f. durch Gegengewicht ‖ équilibrage m. par contre-poids. / pneumatic ~ for weighing machines ‖ Luftdruckentlastung f. für Wagen ‖ mise au repos pneumatique pour bascules. / ~ two-way repeaters ‖ Abgleichung f. für Verstärker ‖ équilibrage m. des amplificateurs à deux fils.

balancing aerial ‖ Abstimmungsantenne f. ‖ antenne f. de compensation. / ~ apparatus ‖ Nullinstrument n. ‖ appareil m. de zéro. / ~ battery ‖ Ausgleichsbatterie f. ‖ batterie f. d'équilibrage. / ~ condenser ‖ Ausgleichkondensator m. ‖ condensateur m. d'équilibrage. / ~ device ‖ Balanziervorrichtung f. ‖ dispositif m. à équilibrer. / ~ generator ‖ Ausgleichgenerator m. ‖ égalisatrice f.; compensatrice f.

balancing machine (Adjusting) ‖ Ausrichtmaschine f. ‖ appareil m. à équilibrer. / ~ (Counterbalancing) ‖ Auswuchtmaschine f.; Wuchtmaschine f. ‖ machine f. à équilibrage dynamique ou à vérifier l'équilibrage.

balancing, method of ~ (Cable) ‖ Abgleichverfahren n. ‖ mode m. d'équilibrage. / method of ~ (Tel) ‖ Nachbildungsverfahren n. ‖ méthode f. d'équilibrage.

balancing network (Tel) ‖ Leitungsnachbildung f. ‖ circuit m. d'équilibrage; reproduction f. de la ligne; équilibreur m. / ~ rack (Tel) ‖ Nachbildungsgestell n. ‖ bâti m. des équilibreurs.

balancing tester (Tel) ‖ Abgleichprüfer m. ‖ appareil m. à essayer l'équilibrage. / ~ transformer ‖ Ausgleichtransformator m. ‖ autotransformateur-égalisateur m. / ~ and extension tube of a telescope ‖ Balanzierrohr n. und Verlängerungsrohr n. eines Fernrohres ‖ pièce f. intermédiaire pour équilibrer et allonger le tube

d'une lunette. / ~ wheel || Schwungrad n. || volant m.

balas rubis || Rubinspinell m. || rubis spinelle m. *ou* balais m.

balata || Balata f. || balata m. / ~ mixed with other materials || mit anderen Stoffen gemischte Balata f. || balata m. mélangé d'autres matières. / raw ~ || rohe Balata f. || balata m. brut.

balata belt || Balataband n.; Baumwollriemen m. || courroie f. en coton *ou* en Balata. / ~ driving belt || Balatatreibriemen m. || courroie f. de commande en balata. / ~ trade machine || Balataindustriemaschine f. || machine f. pour l'industrie du balata.

balcony || Balkon m. || balcon m. / ~ (Shipb) || Hintergalerie f.; Achtergalerie f. || Galerie f.

balcony door || Balkontür f. || porte f. de balcon. / ~ support || Balkonträger m. || colonne f. méniane. / ~ window || Balkonfenster n. || fenêtre f. du balcon.

bale, to || in Ballen mpl. verpacken; ballen || mettre en balles fpl.; emballer.

bale || Kollo n.; Stückgut n. || colli(s) m. / ~ (Packing) || Verpackung f. || emballage m. / hop ~ || Hopfenballen m. || balle f. de houblon. / ~ of paper || Papierballen m. || balle f. de papier.

bale breaker (Spinn) || Ballenbrecher m. || déchireur m. de balle; casse-ballot m.

baleen || Fischbein n. || fanon m. de baleine.

bale-goods pl. || Güter npl. in Ballen || marchandises fpl. en balles. / ~ opener || Ballenöffner m. || ouvreuse f. de balles. / ~ packing press || Ballenpresse f. || presse f. à balles.

baling of the hop || Ballen n. des Hopfens || mise f. en balle du houblon.

baling press || Ballenpresse f.; Packpresse f.; Paketierpresse f. || presse f. à emballer *ou* à paqueter. / scrap ~ || Schrottpaketierpresse f. || presse f. à paqueter la mitraille. / vertical ~ || stehende Ballenpresse f. || presse f. verticale à emballer.

balk || Balken m.; Bohle f. || poutre f. de bois.

ball, to (Sea) || sich beruhigen || calmer; se calmer; mollir. / ~ together || sich zusammenballen || se condenser.

ball || Kugel f. || bille f.; boule f. / ~ (Glassm) || Külbchen n.; Ballen m.; Posten m. || paraison f. / ~ (Met) || Klumpen m.; Deul m.; Luppe f.; Frischluppe f.; Schrei m. || massé m.; masset m.; loupe f.; balle f.; lopin m.; boule f. / ~ (Toy) || Ball m. || balle f. / charcoal-steel ~ || Luppe f. von gefrischtem Stahl || massé m. d'acier affiné. / ~ of fined steel *see* ball (Met). / to form into a ~ || kugeln || rouler; arrondir. / ~ of the governor || Schwungkugel f. || boule f. du régulateur. / hollow ~ || Hohlkugel f. || boule f. creuse. / india rubber ~ || Gummiball m. || balle f. en caoutchouc. / india rubber ~ machine || Gummiballmaschine f. || machine f. pour faire des balles en caoutchouc. / ~ of rags (Copperplate print) || Filzbällchen n. || tampon m. *ou* bouchon m. / shingled ~ (Met) || gezängte Luppe f. || lopin m. cinglé; massiau m.; masseau m. / small shingled ~ (Met) || kleine gezängte Luppe f. || masselet m. / ~ of steel *see* ball (Met). / ~ of yarn || Garnknäuel m. || pelote f. *ou* peloton m. de fil.

ball and socket joint || Kugelgelenk n. || joint m. à boulets *ou* à billes; articulation f. à rotule. / ~ bearing || Kugelgelenklager m. || palier f. articulé à billes.

ballast, to (Nav) || mit Ballast m. beladen || lester. / ~ (Road) || (be)schottern || ballaster. / ~ a road || eine Straße beschottern || ballaster une voie. / ~ a ship || ein Schiff beladen || lester un vaisseau.

ballast (Nav) || Ballast m. || lest m.; ballast m.; ballastage m. / ~ (Railw) || Bettungsstoff m.; Schüttungsmaterial n. || matière f. de ballastage; ballast m. / ~ for cruising || Fahrtballast m. || lest m. de voyage *ou* pour le voyage *ou* de navigation. / gravel ~ (Railw) || Kiesbettung f. || ballast m. en gravier. / railroad ~ || Eisenbahnschotter m. || balast m. pour voies ferrées. / shifting ~ (Mar) || fliegender Ballast m. || lest m. volant.

ballast digging || Schotterbruch m. || extraction f. de cailloux.

ballasted up || ausgewogen || équilibré.

ballast elevator || Schotterbecherwerk n. || noria f. à ballast.

ballasting || Kiesschüttung f.; Beschotterung f. || couche f. de gravier; ballastage m. / ~ of a road || Steinbettung f. einer Chaussee || empierrement m. d'une voie.

ballasting material (Railw) || Bettungsmaterial n. || matière f. de ballastage. / ~ material *see* ballast.

ballast lighter || Baggerprahm m.; Modderprahm m. || cure-môle m.; prame f. *ou* chaland m. à transporter la vase. / ~ pit || Schottergrube f. || ballastière f. / ~ pump || Ballastpumpe f. || gueuse f.; petit cheval m. de lest d'eau. / ~ sand || Ballastsand m. || sable m. *ou* gravier m. de ballast. / ~ screener || Schottertrommel f. || tambour m. à ballast. / ~ train || Schotterzug m. || train m. de ballast *ou* de ballastage. / ~ wagon || Kieswagen m.; Schotterwagen m. || wagon m. à gravier.

ball bearing || Kugellager n. || palier m. à billes; roulement m. à palier m. *ou* boîte f. *ou* coussinet m. à billes. / ~ assembled || fertig montiertes Kugellager n. || roulement m. à billes monté. / ~ for cycles || Fahrradkugellager n. || palier m. à billes pour bicyclettes. / double thrust ~ with spherical seating rings || Doppeldruckkugellager m. mit Einstellscheiben || butée f. double à billes avec sièges sphériques et contre-plaques. / single thrust ~ with flat seating || flaches Druckkugellager n. || butée f. à billes avec siège plan. / thrust ~ || Druckkugellager n.; Kugeldrucklager n. || roulement m. de butée.

ball bearing cup || Kugellagergehäuse n.; Kugellagerschale f. || bague f. de roulement à billes; coussinet m. sphérique. / ~ spindle || Kugellagerspindel f. || broche f. sur roulement à billes. / ~ steel || Kugellagerstahl m. || acier m. pour coussinets à billes. / ~ turntable || Kugeldrehscheibe f. || plaque f. tournante à billes.

ball cartridge || Kugelpatrone f. || cartouche f. à balle. / ~ cock || Schwimmerventil n. || robinet-flotteur m. / ~ connecting branch || Kugelstutzen m. || tubulure f. à boulets. / ~ cup || Kugelschale f. || cuvette f. / ~ cutting machine || Kugelschneidmaschine f. || coupeuse f. à boules; tour m. à découper les billes.

baller (Spinn) || Wicklerin f.; Knaulwicklerin f. || peloteuse f.

ball grinding machine || Kugelschleifmaschine f. || machine f. à réctifier les billes. / ~ gudgeon || Kugelzapfen m. || tourillon m. à boulet. / ~ handle || Kugelgriff m. || poignée f. à bille.

balling (Spinn) || Wickeln n. || pelotonnage m. / ~ furnace (Met) || Schweißofen m. || four m. *ou* fourneau m. à réchauffer *ou* à souder le fer. / ~ machine (Spinn) || Knäuelwickelmaschine f.; Wickelmaschine f. || machine f. à peloter; peloteuse f.; pelotonneuse f.

ballistic || ballistisch || balistique. / ~ form of projectile || ballistische Form des Geschosses || forme f. balistique du projectile. / ~ instrument || ballistisches Instrument n. || instrument m. balistique.

ballistics pl. || Ballistik f.; Wurflehre f. || balistique f.

ball joint || Kugelgelenk n. || joint m. sphérique *ou* à boulet *ou* à rotule.

ball-jointed arm || Kugelgelenkarm m. || bras m. à joints sphériques.

ball measuring stand || Kugelmeßhalter m. || support m. pour le calibrage des billes.

ball mill || Kugelmühle f. || moulin m. à boulets *ou* à billes; broyeur m. à boules. / grinding plate for ~s || Mahlplatte f. für Kugelmühlen || plaque f. triturante pour moulins à boulets. / ~ with screens || Siebkugelmühle f. || moulin m. tamiseur à boulets. / ~ for wet grinding || Naßkugelmühle f. || moulin m. à boulets pour le broyage par voie humide.

ball net || Ballnetz n. || filet m. à balles. / ~ nut || Kugelmutter f. || écrou m. sphérique.

ballonet || Ballonet n. || ballonet m. / flabby ~ || unpralles Ballonet n. || ballonnet m. flasque.

balloon || Luftballon m. || aérostat m.; ballon m. / ~ without aeronaut || unbemannter Ballon m. || ballon m. nonmonté *ou* sans pilote. / captive ~ || Fesselballon m. || ballon m. captif. / free ~ || Luftballon m. || ballon m. libre. / kite ~ || Drachenballon m. || ballon m. cerf-volant; drachenballon m. / sounding ~ || Registrierballon m.; Versuchsballon m. || ballon m. enregistreur *ou* explorateur; ballon-sonde m.

balloon anchor || Ballonanker m. || ancre f. du ballon. / ~ barrage || Ballonsperre f. || barrage m. de ballons. / basket of the ~ || Ballonkorb m. || nacelle f. de ballon. / ~ conductor (Tel) || Ballonader f. || conducteur m. à ballon. / element with a ~ (Electr) || Ballonelement n. || pile f. à ballon. / ~ envelope || Ballonhülle f. || enveloppe f. du ballon. / ~ envelope fabric || Ballonstoff m. || étoffe f. à ballon. / nacelle of the ~ *see* basket of the ~. / ~ net || Ballonnetz n. || filet m. de ballon.

ballooning || Ballontechnik f. || technique f. aérostatique.

balloon silk || Ballonseide f. || soie f. pour ballons. / ~ tyre || Ballonreifen m. || pneu m. ballon. / ~ wire *see* ~ conductor.

ball packing press || Ballenpackpresse f. || presse f. à emballer. / ~ pin || Kugelbolzen m. || boulon m. à rotule. / ~ pivot || Kugelzapfen m. || tourillon m. à boulet; pivot m. à rotule. / ~ polishing drum || Kugelpoliertrommel f. || tambour m. lisseur de billes. / ~ press || Kugelpresse f. || presse f. à billes.

ball race || Laufring m.; Kugelkorb m.; Kugelring m.; Kugelspur f.; Kugelkranz

m.; Kugelkäfig m. ‖ bague f. *ou* rondelle f. à billes; anneau m. de roulement; cage f. à billes. / ~ bearing ‖ Ringkugellager n. ‖ palier m. à roulements annulaires. / ~ steel ‖ Kugelstahl m. ‖ acier m. à billes.

ball retaining valve ‖ Kugelrückschlagventil n. ‖ soupape f. de retenue à boulet.

ball-shaped end ‖ kugelig ausgebildetes Ende n. ‖ extrémité f. sphérique. / ~ vessel ‖ kugelförmiges Gefäß n. ‖ vase m. sphérique.

ball socket ‖ Kugelpfanne f. ‖ coussinet m. sphérique. / ~ soda ‖ rohe Soda f. ‖ soude f. brute. / ~ spark gap ‖ Kugelfunkstrecke f. ‖ éclateur m. à électrodes sphériques. / ~ stage (Opt) ‖ Kugeltisch m. ‖ surplatine f. hémisphérique. / ~ stage microscope ‖ Kugelmikroskop n. ‖ microscope m. à articulations. / ~ support ‖ Kugelhalter m. ‖ support m. à billes. / ~ test for hardening ‖ Kugeldruckprobe f. ‖ épreuve f. de pression par empreinte. / ~ tester ‖ Fallhärteprüfer m. ‖ scléroscope m.; mouton m. à bille. / ~ testing machine ‖ Kugelprüfmaschine f. ‖ machine f. à essayer les billes.

ball thrust apparatus for cylinders and tubes ‖ Kugeldruckprüfapparat m. für Zylinder und Rohre ‖ appareil m. d'essai à bille pour cylindres et tuyaux. / ~ for rails ‖ Kugeldruckprüfapparat m. für Schienen ‖ appareil m. d'essai à bille pour rails. / ~ for rolls ‖ Kugeldruckprüfapparat m. für Walzen ‖ appareil m. d'essai à bille pour rouleaux. / ~ for wheel rims ‖ Kugeldruckprüfapparat m. für Radkränze ‖ appareil m. d'essai à bille pour bandages de roues.

ball thrust bearing ‖ Kugeldrucklager n.; Druckkugellager n. ‖ palier m. de butée à billes; crapaudine f. à billes. / ~ rapid press ‖ Kugeldruckschnellpresse f. ‖ presse f. rapide à billes. / ~ testing machine ‖ Kugeldruckprüfmaschine f. ‖ machine f. à essayer à la bille.

ball top-attachment (Brew) ‖ Destillieraufsatz m. ‖ condenseur m. à boules. / ~ valve ‖ Kugelventil n. ‖ soupape f. à boulet. / ~ washer ‖ Kugelwascher m. ‖ laveur m. à boulets. / ~ winder ‖ Knaulwickler m. ‖ pelotonneur m. / ~ winding machine ‖ Knäuelwickelmaschine f. ‖ peloteuse f.

balm *see* balsam.

balm-mint ‖ Gartenmelisse f. ‖ mélisse f. / ~ leaves pl. ‖ Melissenblätter npl. ‖ feuilles fpl. de mélisse.

balsam ‖ Balsam m. ‖ baume m. / ~ of copaiba ‖ Kopaibabalsam m. ‖ baume m. de copahu. / ~ of Gilead ‖ Gileadbalsam m. ‖ baume m. de Gilead. / ~ of Mecca ‖ Mekkabalsam m.; balsam judaicum ‖ baume m. de Judée. / natural ~ ‖ natürlicher Balsam m. ‖ baume m. naturel. / ~ of Peru ‖ Perubalsam m. ‖ baume m. du Pérou. / ~ of Tolu ‖ Tolubalsam m. ‖ baume m. de tolu.

Baltic ‖ Ostsee f. ‖ mer f. baltique.

Baltimore yellow ‖ Baltimoregelb n. ‖ jaune m. Baltimore.

baluster ‖ Geländerdocke f.; Geländersäule f.; Geländerstab m.; Gangpfosten m.; Baluster m. ‖ balustre m.; potelet m. / ~ maker ‖ Geländermacher m. ‖ rampiste m. / ~ railing ‖ Geländer n. mit Docken ‖ balustrade f.

balusters pl. ‖ Geländer n.; Treppengeländer n. ‖ balustrade f.; rampe f.

balustrade ‖ Brüstung f.; Brustlehne f.; Geländer n. ‖ balustrade f.; rampe f.; clôture f.

bamboo ‖ Bambus m.; Bambusrohr n. ‖ bambou m. / ~ divided in slips ‖ gespaltener *oder* in Streifen zerlegter Bambus m. ‖ bambou m. débité en lames *ou* refendu. / simply splitted ~ ‖ einfach gespaltener Bambus m. ‖ bambou m. simplement refendu.

bamboo articles pl. ‖ Bambuswaren fpl. ‖ articles mpl. en bambou. / ~ filament ‖ Bambusfaser f. ‖ fibre f. de bambou. / ~ furniture ‖ Bambusmöbel n. ‖ meuble m. en bambou. / ~ ware *see* ~ articles. / ~ worker ‖ Bambusarbeiter m. ‖ bamboutier m.

banana ‖ Banane f. ‖ banane f. / meal of ~s ‖ Bananenmehl n. ‖ farine f. de bananes. / ~ plug (Radio) ‖ Bananenstecker m. ‖ fiche f. banane.

band ‖ Band n. ‖ ruban m.; bande f.; ceinture f. / ~ (Bookb) ‖ Heftschnur f. ‖ ficelle f.; nerf m. / ~ (Build) ‖ Leiste f.; Fase f. ‖ bande f.; face f. / ~ (Driving belt) ‖ Riemen m. ohne Ende; Transmissionsriemen m.; Treibriemen m. ‖ courroie f. (de commande *ou* sans fin). / ~ (Electr) ‖ Bandage f. ‖ bandage m. / ~ (Glass) ‖ Haftblei n. ‖ attache f.; lien m. / ~ (Ropem) ‖ Schnur f. ‖ corde f. / ~ of armature (Electr) ‖ Ankerbandage f. ‖ bandage m. de l'induit. / ~ for driving belts ‖ Bahn f. für Treibriemen ‖ bande f. pour courroies de transmission. / endless ~ ‖ Transportband n. ohne Ende; laufendes Band n. ‖ tapis m. roulant; transporteur m. continu. / small ~ (Arch) ‖ Bändchen n.; Leistchen n.; Plättchen n.; Riemchen n.; Saum m.; Steg m. ‖ bandelette f.; filet m.; listel m. carré; réglet m. / ~ spring ~ ‖ Federbund m. ‖ bride f. de ressort.

bandage ‖ Binde f.; Bandage f.; Verband m. ‖ bande f.; bandage m.; pansement m. / box for ~s ‖ Verbandkasten m. ‖ boîte f. à pansement; boîtier m. / ~ sewer ‖ Bandagennäherin f. ‖ couseuse f. de bandages.

bandaging material ‖ Verbandstoff m. ‖ matière f. à pansement.

bandagist ‖ Bandagist m. ‖ bandagiste m.

band brake ‖ Bandbremse f. ‖ frein m. à ruban et vis *ou* à collier.

band conveyer ‖ Bandtransportör m. ‖ transporteur m. à courroie. / travelling ~ ‖ fahrbarer Bandförderer m. ‖ transporteûr m. à bande sur roues.

band coupling ‖ Bandkupplung f. ‖ accouplement m. à ruban. / ~ elevator ‖ Gurtbecherwerk n. ‖ élévateur m. à bande.

bander (Tail) ‖ Besatznäher m. ‖ couseur m. de bandes. / ~ (Weav) ‖ Wrapper m. ‖ placeur m. de bandes.

banderole (Surv) ‖ Meßfähnchen n.; Meßflagge f. ‖ banderole f.

banderoling machine ‖ Banderoliermaschine f. ‖ machine f. à banderoler *ou* à coller les banderoles.

banding steel ‖ Stahlband n. ‖ feuillard m. *ou* bande f. en acier.

band iron *see also* hoop iron ‖ Bandeisen n. ‖ fer m. feuillard *ou* en rubans. / to draw ~ ‖ Bandeisen n. ziehen ‖ étirer le

fer feuillard. / ~ hoop ‖ Bandeisenreif m. ‖ cercle m. en feuillard.

band joint ‖ Bandgelenk n. ‖ articulation f. à ruban. / ~ making machine ‖ Bandmaschine f. ‖ machine f. à faire les rubans. / ~ pulley ‖ Schnurrad n. ‖ roue f. à corde.

bands pl. (Min) ‖ Bergemittel n. ‖ laie f.; banc m. de schiste (entre deux veines). / ~ for doors and windows ‖ Tür- und Fensterbänder npl. ‖ ferrures fpl. *ou* charnières fpl. de portes et de fenêtres.

band saw ‖ Bandsäge f. ‖ scie f. à ruban. / ~ for logs ‖ Blockbandsäge f. ‖ scie f. à ruban pour débiter les grumes. / ~ for cutting metal ‖ Bandsäge f. zum Schneiden von Metallen ‖ scie f. à ruban pour métaux. / ~ for rolling mills ‖ Walzwerkbandsäge f. ‖ scie f. à ruban pour laminoirs.

band saw guide ‖ Bandsägeführung f. ‖ guidage m. pour scies à ruban. / ~ pulley ‖ Bandsägerolle f. ‖ rouleau m. *ou* poulie f. de scies à ruban. / electric ~ soldering device ‖ elektrische Bandsägenlötvorrichtung f. ‖ dispositif m. électrique de brasure pour scies à ruban.

band sawyer ‖ Bandsäger m. ‖ scieur m. à la scie à ruban.

band steel ‖ Bandstahl m. ‖ acier m. feuillard. / ~ for pens ‖ Schreibfederbandstahl m. ‖ acier m. feuillard pour plumes.

band string (Bookb) ‖ Heftschnur f. ‖ ficelle f.; nerf m. / ~ width ‖ Bandbreite f. ‖ largeur f. de ruban.

band width (Bookb) *see* band. / ~ width ‖ Bandbreite f. ‖ largeur f. de bande.

banister ‖ Geländerdocke f.; Geländersäule f. ‖ balustre m.; potelet m.

banisters pl. ‖ Treppengeländer n. ‖ rampe f. *ou* balustrade f. d'escalier.

banjo ‖ Banjo n. ‖ banjo m.

bank ‖ Bank f. ‖ banc m. / ~ (Aero) ‖ Querneigung f. ‖ virage m. / ~ (Hydr arch) ‖ Flußdamm m.; Kai m.; Pier m. ‖ quai m. / ~ (Mar) ‖ Untiefe f.; Flach n. ‖ basse f.; haut-fond m. / ~ (Min) ‖ Bank f. couche f. / ~ (Print) ‖ Laufbrett n. der Presse ‖ berceau m. de la presse. / ~ (River) ‖ Ufer n.; Flußufer n. ‖ bord m.; rive f. / ~ (Road) ‖ Damm m. ‖ chaussée f. / ~ (Trade) ‖ Bank f. ‖ banque f. / ~ (Weav) ‖ Schweifgestell n. ‖ cannelier m.; candré m. de l'ourdisseur. / ~ of boilers ‖ Kesselbatterie f. ‖ batterie f. de chaudières. / cooperative ~ ‖ Genossenschaftsbank f. ‖ banque f. coopérative. / ~ of a drawing shaft (Min) ‖ Hängebank f. ‖ margelle f. *ou* recette f. d'un puits d'extraction; palier m. de déchargement; pas m. de bure. / savings ~ ‖ Sparkasse f. ‖ caisse f. d'épargne. / ~ of shells ‖ Muschelbank f. ‖ banc m. de coquilles. / ~ of thunder clouds ‖ Gewitterbank f. ‖ banc m. d'orage. / workmen's savings ~ ‖ Werkssparkasse f. ‖ caisse f. d'épargne d'usine.

bank account ‖ Bankkonto n. ‖ compte m. de banque. / ~ balance ‖ Banksaldo m. ‖ balance f. de banque. / ~ bill ‖ Kassenanweisung f.; Kassenschein m. ‖ bon m. de caisse; billet m. de banque. / ~ commission ‖ Bankprovision f. ‖ commission f. de banque. / ~ credit ‖ Bankkredit m. ‖ crédit m. de banque. / ~ defences pl. ‖ Uferbefestigung f. ‖ défenses fpl. de rive. / director of a ~

see bank manager. / ~ draft || Bankanweisung f. || traite f. de banque.

banked-up, power basin with ~ water level for a power plant || Betriebsbecken n. mit Stau für eine Kraftanlage || réservoir m. de service avec barrage d'eau pour une usine.

banker || Wechsler m.; Bankier m. || changeur m.; banquier m. / ~'s guarantee | Bankgarantie f. || garantie f. de banque.

bank, failure of a ~ || Zusammenbruch m. einer Bank || krach m. de banque. / ~ furniture || Bankeinrichtung f. || ameublement m. de banque.

banking || Bankwesen n. || les banques fpl. / ~ (Bullion trade) || Geldhandel m. | trafic m. de l'argent; change m. / ~ account || Bankguthaben n. || crédit m. en banque. / ~ business || Geldgeschäft n. Wechselgeschäft n. || affaire f. d'argent *ou* de banque. / ~ establishment || Bank f. || banque f. / ~ house || Bankhaus n. | maison f. de banque; banque f.

banking-up a river || Aufstau m. eines Flusses || haussement m. de niveau d'un fleuve; barrage m. d'une rivière.

bank manager || Bankdirektor m. || directeur m. *ou* gouverneur m. d'une banque. / ~ messenger || Kassenbote m. || garçon m. de recette.

banknote || Banknote f. || billet m. de banque. / ~ (Cheque) || Kassenanweisung f.; Kassenschein m.; Bankanweisung f. || bon m. de caisse. / ~ case || Geldscheintasche f. || porte-feuille m. / ~ circulation || Banknotenumlauf m. || circulation f. des billets de banque.

bank paper || Bankpapier n. || valeurs fpl. de banque. / ~ rate (of discount) || Bankzinsfuß m.; Bankdiskont m. || cours m. de banque; escompte m. officiel. / report of a ~ || Bankbericht m. || rapport m. d'une banque. / return of a ~ *see* report of a ~. / ~ rider (Min) || Bremsbergführer m. || conducteur m. de plan incliné; cayateur m.

bankrupt || bank(e)rott; insolvent; zahlungsunfähig || insolvable; failli. / to become ~ || in Konkurs m. geraten || tomber en faillite f.; faillir.

bankruptcy || Konkurs m.; Bank(e)rott m.; Insolvenz f. || faillite f.; banqueroute f.; insolvabilité f. / to file petition of ~ || Konkurs m. anmelden || demander la liquidation judiciaire; se déclarer en faillite. / assignee in ~ || Konkursverwalter m. || syndic m. de faillite. / ~ court || Konkursgericht n. || cour f. de faillites. / declaration of ~ || Konkursanmeldung f. || déclaration f. de faillite. / ~ proceedings pl. || Konkursverfahren n. || procédure f. en matière de faillite.

bankrupt's estate || Konkursmasse f. || masse f. *ou* actif m. de la faillite.

bank share || Bankaktie f. || action f. de banque.

banner || Banner n.; Fahne f. || bannière f. / ~ (Railw) || Scheibensignal n. || disque m. / ~ fittings pl. || Fahnenbeschläge mpl. || armatures fpl. de drapeaux.

baobab oil || Affenbrotbaumöl n. || huile f. de baobab.

baptistery || Taufkapelle f. || baptistère m.

bar || Stange f.; Stab m. || barre f.; barreau m.; bâton m.; tige f.; perche f. / ~ (Build) || Brechstange f. || anspect m.; levier m. (de fer). / ~ (Chain) || Kettensteg m. ||

étai m. de maillon. / ~ (Furnace) || Roststab m. || barreau m.; barre f. / ~ (Inn) || Schenktisch m. || buffet m. / ~ (Locksm) || Riegel m.; Schubriegel m.; Schloßriegelschaft m. || verrou m.; barre f.; queue f. / ~ (Mar) || Spake f. || manivelle f.; barre f.; anspect m. / ~ (Met) || Barren m.; Barre f.; Schiene f. || barre f.; ingot m.; bloom m. / ~ (Railw) || Schranke f. || barrière f. / ~ of bit (Saddl) || Schaumstange f. || barre f. d'un mors de bride. / ~ at the bottom of a cask || Riegel m. *oder* Querholz n. eines Faßbodens || barre f. de fond d'un tonneau. / brake truss ~ || Bremsdreieck n. || levier m. de frein en triangle. / ~ of the coining press || Schwengel m. der Prägepresse || barre f. du balancier. / cross ~ || Querbalken m. || barre f. transversale. / cross ~s pl. (Build) || Gatter n.; Gitter n.; Gitterwerk n. || grille f.; treille f.; treillis m. / ~ crosswise from the ends of the axle journals || Querprobe f. aus den Enden der Achsschenkel || éprouvette f. en travers des bouts des fusées d'essieu. / ~ cut lengthwise from the axle || in der Längsrichtung der Achse entnommener Probestab m. || éprouvette f. en long des essieux. / driving ~ (Loom) || Treibstange f. || barre f. du métier. / flat ~ || Platine f. || larget m. / ~ of a fly press || Schwunghebel m. einer Schraubenpresse || balancier m. *ou* verge f. du balancier à vis. / gear shift ~ (Auto) || Schaltstange f. || tige f. de commande des fourchettes. / hammered ~ || geschmiedeter Stab m. || barre f. forgée. / handle ~ || Lenkstange f. || guidon m. / handle ~ bender || Lenkstangenbieger m. || courbeur m. de guidons. / knocking-over ~ (Weav) || Abschlagschiene f. || barre f. d'abattage. / merchant iron ~ || Handelseisen n. || fer m. marchand. / moulded ~ (Arch) || Ziereisen n. || barre f. moulurée; fer m. à dessins. / muck ~ *see* puddle(d) ~. / puddle(d) ~ || Rohbarren m.; Rohschiene f. || fer m. ébauché; ébauché m.; barre f. de fer brut. / rolling mill for sharpening ~s || Anspitzwalzwerk n. für Stangen || laminoir m. à appointer des barres. / round ~ || runder Barren m. *oder* Stab m. || barreau m. rond; barre f. ronde; rond m. / ~ of any section || Stange f. beliebigen Querschnitts || barre f. à profil quelconque. / spliced ~ to prevent creeping of the rails || Stemmlasche f. gegen Schienenwandern || éclisse f. épaulée pour empêcher le cheminement des rails. / square ~ || Vierkantbarren m. || barre f. carrée. / twisted ~ || gedrehter Stab m. || barre f. tordue. / welding ~ || Schweißstab m. || baguette f. à souder.

bar arranging device || Stabordner m. || classeur m. de barres.

barbed needle || Bartnadel f. || aiguille f. à barbe.

barbed wire || Stacheldraht m. || ronce f. artificielle; fil m. (de fer) barbelé. / ~ entanglement || Stacheldrahtverhau m. || abatis m. de fil barbelé. / ~ making machine || Stacheldrahtmaschine f. || machine f. à fabriquer les fils de fer barbelé.

bar bending device || Stabbiegevorrichtung f. || appareil m. à cintrer les barres.

barberry bud || Berberitzenknospe f. || bouton m. d'épine-vinette. / ~ juice || Berberitzensaft m. || jus m. d'épine-vinette. / ~ wood || Berberitzenholz n.;

Sauerdornholz n. || bois m. d'épinevinette.

barber surgeon || Heilgehilfe m. || aide-chirurgien m.

bar collet chuck || Stangenspannpatrone f. || pince-barre m. / ~ copper || Stangenkupfer n. || cuivre m. en barres. / ~ cutting machine || Blockschere f. || machine f. *ou* cisaille f. à couper les lingots. / ~ drawing bench || Stangenziehbank f. || banc m. à étirer les barres.

bare blank || nu; découvert. / ~ cable || blankes Kabel n. || câble m. nu. / ~ electrode || nackte Elektrode f. || électrode f. nue. / ~ rope || blankes Seil n. || corde f. nue; câble m. nu. / ~ spots fermentation || Gärung f. mit kahlen Stellen || fermentation f. à places chauves. / ~ welding electrode *see* ~ welding rod. / ~ welding rod || nackter Schweißstab m. || barre f. *ou* électrode f. de soudure nue. / ~ wire || blanker *oder* nackter Draht m. || fil m. (métallique) nu.

bar equipment || Schankhalleneinrichtung f. || équipement m. (de halle) de débit de boissons.

bar feed || Stangenvorschub m. || avance f. des barres.

bar frame (Loc) || Rostrahmen m. || cadre m. *ou* support m. de grille. / ~ forged in one piece || in einem Stück geschmiedeter Rostrahmen m. || châssis m. à barreaux forgé d'une seule pièce.

bar furniture (Inn) || Ausschankeinrichtung f. || installation f. de débit.

bargain (Mine) || Gedinge n. || forfait m.

barge (Ship) || Schute f.; Prahm m.; Schaluppe f. || prame f.; barque f.; barge f.; chaloupe f.; chaland m. / ~ course (Build) || Traufschicht f.; Traufziegelreihe f. || battellement m.

bargeman || Prahmknecht m. || batelier m.

bar handle || Stangengriff m. || poignée f. de tige. / ~ head (Shipb) || Spakenkopf m. || tête f. de barre.

baric chloride || Bariumchlorid n. || chlorure m. de baryum.

bar iron || Stabeisen n.; Stangeneisen n. || fer m. en barres. / fashioned ~ *see* section ~. / flat ~ || Flacheisen n. || fer m. méplat. / notched ~ || Zaineisen n. || verge f. crénelée; fer-carillon m.; barre f. de fer crêpée. / round ~ || Rundeisen n. || fer m. en barres rondes; fer m. rond. / section ~ || Formeisen n.; Fassoneisen n. || fer m. en barres façonné; fer m. profilé. / square ~ || Quadrateisen n.; Vierkanteisen n. || fer m. carré *ou* plat rectangulaire.

bar iron bundling machine || Stabeisenbündelmaschine f. || machine f. à mettre les fers en barres en faisceaux. / ~ cutter || Stabeisenschere f. || cisailles fpl. pour fers en barres. / ~ twisting machine || Stabeisenverwindemaschine f. || machine f. à torsader les fers en barres.

barite || Barytin m.; Schwerspat m. || barytine f.

barium || Barium n. || baryum m. / ~ acetate || Bariumazetat n.; essigsaurer Baryt m. || acétate m. de barium *ou* de baryte. / ~ arseniate || arsensaures Barium n. || arséniate m. de baryte. / ~ borate || borsaures Barium n. || borate m. de baryte. / ~ carbonate || kohlensaures Barium n.; Bariumkarbonat n. || carbonate m. de baryte. / ~ chlorate || chlorsaures Barium n. || chlorate m. de baryte.

barium chloride ‖ Bariumchlorid n.; Chlorbarium n. ‖ chlorure m. de barium. / solution of ~ ‖ Bariumchloridlösung f. ‖ solution f. de chlorure de barium.

barium chromate ‖ chromsaures Barium n. ‖ chromate m. de baryte. / ~ hydrate ‖ Ätzbaryt m.; Bariumhydrat n. ‖ baryte f. hydratée. / ~ hyperoxide ‖ Bariumsuperoxyd n. ‖ hyperoxyde m. barytique. / ~ nitrate ‖ salpetersaures Barium n. ‖ nitrate m. de baryte. / ~ oxide ‖ Bariumoxyd n. ‖ oxyde m. de barium. / ~ peroxide ‖ Bariumsuperoxyd n. ‖ bioxyde m. de barium. / ~ phosphate ‖ phosphorsaures Barium n. ‖ phosphate m. de baryte. / ~ platinocyanide ‖ Bariumplatincyanür n. ‖ platinocyanrue m. de barium. / ~ sulphate ‖ schwefelsaures Barium n. ‖ sulfate m. de barium ou de baryte. / ~ sulphide ‖ Bariumsulfid n.; Schwefelbarium n. ‖ sulfure m. de barium. / ~ yellow ‖ Barytgelb n. ‖ jaune m. de baryte.

bark, to (Curr) ‖ mit Lohe gerben; ablohen; lohgerben ‖ tanner. / ~ (trees) ‖ abschälen; entrinden; abrinden ‖ écorcer.

bark (Botany) ‖ Rinde f.; Borke f. ‖ écorce f. / ~ (Curr) ‖ Lohe f.; Gerberlohe f.; tan m.; écorce f. / ~ (Shipb) ‖ Barke f. ‖ barque f. / fivemasted ~ ‖ Fünfmastbark f. ‖ cinq-mâts barque f. / to strip the ~ from the wood ‖ abborken ‖ écorcer. / three-masted ~ ‖ Dreimastbark f. ‖ trois-mâts barque f.

bark beetle ‖ Borkenkäfer m. ‖ scarabée m.; bostriche m. / ~ cutter ‖ Rindenschneider m. ‖ coupe-écorce m. / ~ drying press / Lohetrockenpresse f. ‖ presse f. à sécher le tan.

barked wood ‖ Schälholz n.; geschältes Holz n. ‖ bois m. écorcé.

bark grinding mill ‖ Lohmühle f. ‖ moulin m. à (piler le) tan.

barking (Forest) ‖ Abborken n. ‖ écorçage m. / wood ~ ‖ Holzabrinden n. ‖ écorçage m. du bois.

barking axe ‖ Schälbeil n. ‖ hache-écorce m.

bark mill see bark grinding mill. / ~ peeler ‖ Abrinder m. ‖ écorceur m.; déchireur m. d'écorce. / ~ peeling machine ‖ Rindenschälmaschine f. ‖ machine f. à décortiquer. / ~ plait ‖ Rindengeflecht n. ‖ tresse f. d'écorce. / ~ tanning ‖ Lohgerben n. ‖ tannage m. aux écorces.

bar lathe ‖ Prismadrehbank n. ‖ tour m. à banc prismatique.

barley ‖ Gerste f. ‖ orge f. / battledore ~ ‖ Fächergerste f. ‖ orge f. en éventail. / bruised ~ ‖ Gerstenschrot n. ‖ mouture f. d'orge. / four-rowed ~ ‖ vierzeilige Gerste f. ‖ orge f. à quatre rangs. / seed ~ ‖ Aussaatgerste f. ‖ orge f. de semence. / six-rowed ~ ‖ sechsreihige oder sechszeilige Gerste f. ‖ orge f. à six rangs. / sowing ~ see seed ~. / steeped ~ ‖ geweichte Gerste f. ‖ orge f. trempée. / summer ~ ‖ Sommergerste f. ‖ orge f. d'été. / two-rowed ~ ‖ zweizeilige Gerste f. ‖ orge f. à deux rangs. / winter ~ ‖ Wintergerste f. ‖ escourgeon m.

barley chaff ‖ Gerstenspelze f. ‖ balle f. d'orge. / ~ cleaner ‖ Gerstenputzmaschine f. ‖ machine f. à nettoyer l'orge. / ~ cleaner and separator ‖ Gerstenputz- und Sortiermaschine f. ‖ machine f. à nettoyer et à trier l'orge; trieur m. nettoyeur d'orge. / ~ cleaner and sorter

with broken grains separator ‖ Gerstenreinigungs-, Sortier- und Halbkörnerauslesemaschine f. ‖ machine f. à nettoyer et à trier l'orge et à séparer les demi-grains. / ~ cleaning machine see ~ cleaner. / ~ cleaning plant ‖ Gerstenreinigungsanlage f.; Gerstenputzerei f. ‖ installation f. pour le nettoyage de l'orge. / ~ corn ‖ Gerstenkorn n. ‖ grain m. d'orge. / ~ and malt dressing plant ‖ Gerste- und Malzputzerei f. ‖ installation f, de nettoyage de l'orge et du malt. / ~ duty on ‖ Gerstensteuer f. ‖ impôt m. sur l'orge.

barley examination ‖ Gerstenuntersuchung f. ‖ analyse f. d'orge. / section cutter for ~ ‖ Gerstenprüfer m. ‖ farinatome m.; coupe-grains pour l'orge.

barley fat ‖ Gerstenfett n. ‖ graisse f. d'orge. / ~ husk see ~ chaff. / ~ loft ‖ Gerstenboden m. ‖ grenier m. à orge. / ~ malt ‖ Gerstenmalz n. ‖ malt m. d'orge. / ~ meal ‖ Gerstenmehl n. ‖ farine f. d'orge. / ~ seed oil ‖ Gerstensamenöl n. ‖ huile f. d'orge. / ~ separator ‖ Gerstensortiermaschine f. ‖ trieur m. d'orge. / ~ shell see ~ chaff. / ~ soaking trough ‖ Gerstenweiche f. ‖ auge f. de trempage pour orge; cuve f. mouilloire pour l'orge. / ~ starch ‖ Gerstenstärke f. ‖ amidon m. d'orge. / ~ steep ‖ Gerstenquellstock m.; Gerstenweichstock m. ‖ cuve f. mouilloire ou à tremper pour l'orge. / ~ steeping tank see ~ steep. / ~ straw ‖ Gerstenstroh n. ‖ paille f. d'orge. / ~ tester ‖ Gerstenprüfer m. ‖ farinatome m.; coupe-grains m. pour l'orge. / ~ washing plant ‖ Gerstenwäscherei f. ‖ installation f. de lavage de l'orge.

bar link (Chain) ‖ Kettenglied n.; Schake f. mit Steg ‖ maillon m. à étai. / ~ (Mach tool) ‖ Mitnehmerzapfen m. ‖ pivot m.

bar loom ‖ Bandwebstuhl m. ‖ métier m. à la barre.

barm ‖ Kunsthefe f.; Bärme f. ‖ levure f.; levain m. / beer ~ ‖ Bierhefe f. ‖ levure f. de bière. / German ~ ‖ Preßhefe f. ‖ levure f. sèche ou pressée. / pressed ~ see German ~.

bar magnet ‖ Stabmagnet m. ‖ barreau m. aimanté.

barnacle (Mar) ‖ Schülpe f. ‖ coquillage m.; cravans mpl.

barogram ‖ Barogramm n. ‖ barogramme m.

barograph ‖ Barograf m. ‖ baromètre m. enregistreur.

barometer ‖ Barometer n.; Wetterglas n. ‖ baromètre m. / aneroid ~ ‖ Aneroidbarometer n. ‖ baromètre m. anéroïde. / bulb ~ ‖ Gefäßbarometer n. ‖ baromètre m. à cuvette. / portable ~ ‖ Reisebarometer n. ‖ baromètre m. portatif. / siphon ~ ‖ Heberbarometer n. ‖ baromètre m. à siphon. / standard ~ ‖ Prüfungsluftdruckmesser m. ‖ baromètre m. d'essai ou de contrôle. / test ~ ‖ Prüfungsluftdruckmesser m. ‖ baromètre m. d'essai ou de contrôle. / wheel ~ ‖ Zeigerbarometer n. ‖ baromètre m. à cadran.

barometer, height of the ‖ Barometerstand m.; Luftdruckstand m. ‖ hauteur f. barométrique.

barometric(al) ‖ barometrisch ‖ barométrique. / ~ column ‖ Barometersäule f. ‖ colonne f. barométrique. / ~ height see barometer height. / ~ pressure ‖ (barometrischer) Luftdruck m.; Atmosphären-

druck m. ‖ pression f. atmosphérique ou de l'atmosphère ou barométrique.

barometry ‖ Luftdruckmessung f. ‖ barométrie f.

baro-thermometer ‖ Barothermometer n. ‖ baro-thermomètre m.

barque (Shipb) see bark.

barrack ‖ Baracke f. ‖ baraque f. / decomposable ~ of corrugated iron ‖ zerlegbare Wellblechbaracke f. ‖ baraque f. de tôle ondulée démontable. / wooden ~ ‖ Holzbaracke f. ‖ baraque f. en bois.

barrack building ‖ Barackenbau m. ‖ construction f. de baraquements. / ~ factory ‖ Barackenfabrik f. ‖ fabrique f. de baraques. / ~ hospital ‖ Krankenbaracke f. ‖ hôpital-baraquement m.

barracks pl. ‖ Kaserne f. ‖ caserne f.

barrage ‖ Talsperre f. ‖ barrage-réservoir m.; barrage m. de la vallée. / balloon ~ ‖ Ballonsperre ‖ barrage m. de ballons. / catchment area of a ~ ‖ Einzugsgebiet n. einer Talsperre ‖ superficie f. des terrains contribuants d'un barrage; domaine m. d'affluence d'un barrage. / creeping (firing) ~ ‖ Feuerwalze f. ‖ barrage m. de tir roulant.

barrage fire ‖ Sperrfeuer n. ‖ tirs mpl. de barrage. / ~ power station ‖ Talsperrenkraftwerk n. ‖ usine f. de barrage d'eau.

barred debt ‖ verjährte Schuld f. ‖ dette f. caduque.

barrel, to ‖ eintonnen; in Fässer npl. füllen ‖ entonner; embariller; enfutailler; mettre en tonneaux ou en barils ou en fûts.

barrel ‖ Faß n.; Tonne f. ‖ baril m.; tonneau m.; fût m.; futaille f. / ~ (Coop) ‖ Gebinde n. ‖ barrique f.; futaille f. / ~ (Gun) ‖ Lauf m.; Rohr n. ‖ canon m. / ~ (Watchm) ‖ Federgehäuse n. ‖ Federhaus n. ‖ barillet m. / ~ of a bell (Found) ‖ Glockenkörper m. ‖ corps m. d'une cloche. / boiler ~ (Loc) ‖ Langkessel m. ‖ corps m. cylindrique. / cast steel ~ (Gun) ‖ Gußstahllauf m. ‖ canon m. en acier fondu. / gun ~ ‖ Gewehrlauf m.; Flintenlauf m. ‖ canon m. de fusil. / gun ~ (Cannon) ‖ Geschützrohr n.; Kanonenlauf m.; Kanonenrohr n. ‖ canon m. / iron ~ ‖ eisernes Faß n. ‖ tonneau m. en fer. / ~ of a musket see gun ~. / packing ~ ‖ Packfaß n. ‖ tonneau m. d'emballage. / paper ~ ‖ Faß n. aus Papier ‖ fût m. en carton. / ~ of a pump ‖ Pumpenzylinder m.; Pumpenstiefel m. ‖ corps m. ou cylindre m. de pompe. / rifled ~ (Gun) ‖ gezogener Flintenlauf m. oder Büchsenlauf m. ‖ canon m. rayé ou rainé ou carabiné. / smooth ~ (Gun) ‖ glatter Lauf m. ‖ canon m. lisse. / standard gauge ~ ‖ Normalfaß n. ‖ baril m. normal. / transport ~ ‖ Transportfaß n. ‖ fût m. d'expédition; tonneau m. ou baril m. de transport. / wooden ~ ‖ hölzernes Faß n. ‖ tonneau m. en bois.

barrel brazer (Arm) ‖ Lauflöter m. ‖ braseur m. de canons. / ~ brush ‖ Faßbürste f. ‖ brosse f. pour fûts. / ~ car ‖ Faßwagen m. ‖ voiture f. à tonneaux. / ~ cleaning machine ‖ Faßreinigungsmaschine f. ‖ machine f. à nettoyer les tonneaux. / ~ cock ‖ Faßhahn m. ‖ cannette f. de tonneau. / ~ compass ‖ Trommelkompaß m. ‖ compas-tambour m. / ~ fender ‖ Faßaufschlagfender m. ‖ défense f. de barils. / ~ filling machine ‖ Faßfüllmaschine f. ‖ machine f. à remplir

les tonneaux. / ~ fine borer (Arm) ‖ Laufbohrer m. ‖ aléseur m. de canons. / ~ forger (Arm) ‖ Laufschmied m. ‖ forgeur m. de canons. / ~ hoist ‖ Faßaufzug m. ‖ élévateur m. pour fûts. / ~ howel (Coop) ‖ Dechsel f. ‖ herminette f. / ~ maker (Arm) ‖ Gewehrlaufmacher m. ‖ canonnier-armourier m. / ~ maker (Coop) ‖ Faßbinder m. ‖ tonnelier m. / ~ organ ‖ Drehorgel f.; Leierkasten m. ‖ orgue f. de Barbarie ou à manivelle. / ~ packing machine ‖ Faßpackmaschine f. ‖ machine f. à empaqueter en tonneaux. / ~ plug ‖ Kaliberzylinder m. ‖ cylindre m. vérificateur ou à calibrer. / ~ purifying plant ‖ Faßreinigungsanlage f. ‖ installation f. à nettoyer les tonneaux. / ~ raising jack ‖ Faßwinde f. ‖ cric m. relève fût. / ~ rinser ‖ Faßspülgerät n. ‖ rinceuse f. à fûts. / ~ roof ‖ Tonnendach n.; tonnenförmiges Dach n. ‖ toit m. en berceau ou en tonnelle. / ~ scrubber ‖ Faßbürstmaschine f.; Faßwaschmaschine f. ‖ brosseuse f. pour fûts; machine f. à brosser les fûts; appareil m. à laver les fûts. / ~ setter (Arm) ‖ Laufrichter m. ‖ dresseur m. de canons. / ~-shaped ‖ tonnenförmig ‖ bombé; en forme de baril. / ~-shaped roller bearing ‖ Tonnenlager n. ‖ roulement m. à rouleaux bombés. / ~ stand ‖ Faßbock m. ‖ porte-fût m. / ~ testing machine ‖ Faßprüfmaschine f. ‖ machine f. à essayer les tonneaux. / ~ vault ‖ Tonnengewölbe n. ‖ voûte f. en tonnelle. / (cask and) ~ wagon ‖ Faßtransportwagen m. ‖ wagonnet m. ou haquet m. pour le transport des barils. / ~ washer ‖ Faßwaschapparat m. ‖ appareil m. à laver les fûts. / ~ washing room (Brew) ‖ Schwankhalle f. ‖ salle f. de rinçage. / ~ wheel (Watchm) ‖ Walzenrad n. ‖ roue f. de tambour.

barren, getting (Mine) ‖ Vertaubung f. ‖ appauvrissement m. / ~ of seams ‖ flözleer ‖ sans couches fpl.

barren rock (Mine) ‖ Abraum m. ‖ découverte f.; mort-terrain m.; déblai m.; lit m. de décombres. / ~ track ‖ Taubfeld n. ‖ terrain m. stérile. / ~ zone ‖ taube Zone f. ‖ zône f. stérile.

barrenness ‖ Taubheit f. ‖ stérilité f.

barret ‖ Barett n. ‖ béret m.; toque f.

barrier (Build) ‖ Stangengeländer n.; Brüstung f. ‖ barrière f.; balustrade f.; rampe f.; garde-fou m.; parapet m. / ~ (Fortification) ‖ Aufschüttung f.; Verschanzung f. ‖ barrière f. / ~ (Railw) ‖ Schranke f. ‖ clôture f.; barrière f. / ~ without railings ‖ gitterloses Geländer n. ‖ garde-fou m. sans barreaux. / ~ of railway ‖ Eisenbahnwegschranke f. ‖ barrière f. de chemin de fer. / ~ with a rod ‖ Stangenbarriere f. ‖ barrière f. à lisse suspendu.

barrier gate ‖ Gattertor n. ‖ barrière f. de sortie. / ~ guard (Electr) ‖ Schutznetzgitter n. ‖ grille f. protectrice.

barring engine (Steam eng) ‖ Drehmaschine f.; Schaltmaschine f.; Schwungmaschine f.; Anlaßmaschine f. ‖ servomoteur m. de lancement ou de démarrage ou de mise en marche. / ~ gear ‖ Drehvorrichtung f.; Stellwerk n.; Schaltwerk n. ‖ appareil m. de lancement ou de mise en marche ou de démarrage.

barrister ‖ Anwalt m. ‖ avoué m.; avocat m.

bar room ‖ Schenkstube f. ‖ estaminet m.; taverne f.

barrow (Mine) ‖ Halde f. ‖ halde f. / wheel ~ ‖ Karren m.; Karre f.; Schubkarren m. ‖ brouette f.

barrow man (Mine) ‖ Fördermann m.; Schlepper m.; Wagenstößer m. ‖ esclaunneur m.; rouleur m. / ~ tongs pl. for rails ‖ Schienentragzange f. ‖ portetenaille m. pour rails. / ~ way (Mine) ‖ Laufbrett n.; Laufbohle f. ‖ planche f. de chemin; marche-pied m.

bars pl. ‖ Stabeisen n. ‖ fer m. en barres. / ~ for windows ‖ Fensterstäbe mpl. ‖ barreaux mpl. pour croisées.

bar shape ‖ Stangenform f.; Stabform f. ‖ forme de barre. / ~-shaped ‖ stabförmig ‖ en forme de barre. / ~ sharpening machine ‖ Stangenanspitzmaschine f. ‖ machine f. à appointer à froid les barres. / ~ sharpening rolling mill ‖ Anspitzwalzwerk n. für Stangen ‖ laminoir m. à appointer les barres. / ~ shearing machine ‖ Stabeisenschere f. ‖ cisaille f. à fers en barres. / ~ silver ‖ Silber n. in Barren ‖ argent m. en barres ou en lingots. / ~ stop ‖ Stangenanschlag m. ‖ butée f. pour la barre.

barter ‖ Tausch m. ‖ échange m.; troc m.

bartering ‖ Umtausch m. ‖ échange m.

bar tin ‖ Stangenzinn n. ‖ étain m. en verges. / ~ wimble ‖ Riegelbohrer m. ‖ barroir m.

barwood extract ‖ Rotholzextrakt m. ‖ extract m. de bois rouge.

bar working machine, full-automatic ‖ Stangenvollautomat m. ‖ machine f. automatique à travailler les barres. / semi-automatic ~ ‖ Stangenhalbautomat m. ‖ machine f. demi-automatique à travailler les barres.

bar-wound armature (Electr) ‖ Anker m. mit Stabwicklung ‖ induit m. à barres.

baryta ‖ Baryt m.; Schwerspat m.; Schwererde f.; Baryterde f.; Bariumoxyd n. ‖ baryte f. (sulfatée). / ~ lye ‖ Barytlauge f. ‖ lessive f. de baryte. / ~ mill ‖ Barytmühle f. ‖ moulin m. à baryte. / ~ mine ‖ Barytbruch m. ‖ carrière f. de baryte. / ~ white ‖ Barytweiß n. ‖ blanc m. de baryte.

barytes see baryta.

barytiferous oxide of manganese ‖ Psilomelan m.; Hartmanganerz n.; schwarzer Glaskopf m. ‖ psilomélane m.; hydroxyde m. de manganèse barytifère ou de manganèse oxydé; manganèse m. barytique hydraté.

basal ‖ basisch; fundamental ‖ basique. / ~ solid angle (Min) ‖ Seitenecke f. oder Randecke f. eines Kristalls ‖ angle m. ou sommet m. latéral d'un cristal.

basal bristle ‖ Basalborste f. ‖ brosse f. ou soie f. dorsale.

basalt ‖ Basalt m. ‖ basalte m. / columnar~ ‖ Säulenbasalt m. ‖ basalte m. en colonnes. / ~ in tubular masses ‖ Basaltschiefer m. ‖ basalte m. schisteux.

basaltic ‖ basalthaltig ‖ basaltique. / ~ lava ‖ Basaltlava f. ‖ lave f. basaltique. / ~ tuff ‖ Basalttuff m. ‖ tuf m. basaltique. / ~ wacke ‖ Basaltwacke f. ‖ wackite f. basaltique.

basaltiform ‖ basaltartig ‖ basaltoïde.

basalt quarry ‖ Basaltbruch m. ‖ carrière f. de basalte. / ~ slate ‖ Basaltschiefer m. ‖ basalte m. schisteux. / ~ works pl. ‖ Basaltwerk n. ‖ carrières fpl. de basalte.

bascule bridge ‖ Klappbrücke f. ‖ pont m. à bascule; pont-levis m. à trappe.

base (Met) ‖ unecht; falsch; unedel ‖ commun. / ~ metal ‖ unedles Metall n. ‖ métal m. commun.

base (Arch) ‖ Base f.; Basis f.; Fuß m.; Fußgestell n.; Grundfläche f. ‖ base f.; ligne f. de base; pied m. / ~ (Chem) ‖ Base f.; Lauge f. ‖ base f. / ~ (Dentistry) see baseplate. / ~ (Geom) ‖ Grundlinie f.; Grundfläche f. ‖ base f. / ~ (Glow lamp) ‖ Sockel m. ‖ culot m. / ~ (Logarithm) ‖ Grundzahl f.; Basis f. ‖ base f. / ~ (Mech) ‖ Böckchen n. ‖ support m. / ~ (Min) ‖ Kristallendfläche f. ‖ base f. / ~ (Surv) ‖ Netzpunkt m.; Richtpunkt m.; Fixpunkt m. ‖ point m. de repère. / acidifiable ~ ‖ säurefähige Base f. ‖ base f. acidifiable. / ~ of a column ‖ Fuß m. oder Basis f. einer Säule; Piedestal n.; Säulenständer m. ‖ piédestal m. ou base f. d'une colonne. / ~ of the dam ‖ Sohle f. der Sperrmauer ‖ base f. du mur de barrage. / welded and flanged ~ of the dome ‖ geschweißter und gebördelter Domunterteil m. ‖ embase f. du dôme soudée et bridée. / ~ extended round a building ‖ Sockel m. eines Gebäudes ‖ socle m. ou base f. ou embasement m. d'un bâtiment. / ~ of the inkstand ‖ Tintenfaßhalter m. ‖ socle m. de l'encrier. / insulating ~ ‖ Isoliersockel m. ‖ base f. isolante. / metal ~ ‖ Metallfuß m. ‖ pied m. en métal. / ~ of a pole ‖ Mastsockel m. ‖ socle m. de poteau. / rail ~ (Railw) ‖ Schienenfuß m. ‖ patin m. du rail. / width of ~ of rail ‖ Fußbreite f. der Schiene ‖ largeur f. du patin de rail. / ~ for railways ‖ Unterlage f. für Eisenbahnschienen ‖ traverse f. pour voies ferrées. / ~ of a rock ‖ Grundmasse f. eines Gesteins ‖ pâte f. (première) d'un minéral composé. / salifiable ~ ‖ salzfähige Base f. ‖ base f. salifiable. / ~ of a slope ‖ Grundfläche f. einer Böschung ‖ empattement m. d'un talus. / ~ of stand ‖ Standfläche f. des Stativs ‖ base f. du pied. / ~ of verification ‖ Hilfsstandlinie f. ‖ base f. de vérification. / ~ of wall ‖ Mauerfuß m. ‖ pied m. de mur. / ~ of a wedge (Mech) ‖ Kopf m. oder Rücken m. eines Keiles ‖ tête f. d'un coin.

base circle ‖ Grundkreis m. ‖ cercle m. primitif. / ~ court ‖ Hühnerhof m.; Geflügelhof m.; Wirtschaftshof m. ‖ basse-cour f.

Basel blue ‖ Baslerblau n. ‖ bleu m. de Bâle.

baseless ‖ grundlos ‖ dénué de fondement.

base line (Surv) ‖ Standlinie f.; Grundlinie f.; Basis f. ‖ ligne f. de base; base f. ~ lines pl. ‖ Polygonzug m.; Standlinienzug m. ‖ cheminement m. périmétrique. / ~ load of production ‖ Grundlast f. der Förderleistung ‖ charge f. de base de la production. / ~ load power station ‖ Grundkraftwerk n. ‖ usine f. de base.

basement (Build) ‖ Fundament n.; Grundbau m. ‖ fondation f. / ~ (Dwelling) ‖ Erdgeschoß n. ‖ rez-de-chaussée m.

baseplate ‖ Grundplatte f. ‖ plaque f. d'assise ou de fondation. / ~ (Dentistry) ‖ Gebißplatte f.; Gaumenplatte f. ‖ plaque f. base ou de dentier. / cast ~ (Dentistry) ‖ gegossene Gaumenplatte f. ‖ plaque f. base moulée. / ~ for dental prosthesis ‖ Platte f. für künstliche Gebisse ‖ plaque f. base pour dentiers. / ~ for frog (Railw) ‖ Herzstückunterlagsplatte f. ‖ plaque f. de sur-

haussement pour cœur. / swaged ~ (Dentistry) || gepreßte Gaumenplatte f. || plaque f. base emboutie. / swaging of ~s (Dentistry) || Pressen n. von Gebißplatten || emboutissage m. de plaques base pour dentiers. / ~ for the turntable || Grundplatte f. der Drehscheibe || plaque f. d'assise d'une plaque tournante.

base-table (Arch) || Sockelplatte f. || dalle f. d'embasement.

base tin || Halbzinn n. || étain m. bas.

basic || basisch || basique. / ~ lining (Met) || basisches Futter n. || revêtement m. basique. / resistant ~ lining (Met) || haltbares basisches Futter n. || revêtement m. basique durable. / ~ openhearth furnace || basischer Martinofen m. || four m. Martin basique. / ~ patent || Hauptpatent n. || brevet m. principal. / ~ pig || Phosphorroheisen n. || fonte f. phosphoreuse. / ~ process (Met) || basisches Verfahren n. || procédé m. basique. / ~ slag || basische Schlacke f.; Thomasschlacke f. || laitier m. basique; scorie f. Thomas. / ~ steel || Thomasstahl m. || acier m. Thomas.

basicity || Basizität f. || basicité f.

basil (Curr) || Schafleder n. || basane f. / ~ (Scissors) || Schleiffläche f.; schiefe Schneide f.; Zuschärfungsfläche f. || biseau m. / ~ of planing tool || Gehrung f. oder Schräge des Hobeleisens || basile m. du fer de rabot.

basilene dye || Basilenfarbstoff m. || couleur f. de basilène.

basil leaves pl. || Basilikumkraut n. || feuilles fpl. de basilicon. / ~ oil || Basilikumöl n. || essence f. de basilicon.

basin || Becken n.; Schale f.; Schüssel f. || bassin m.; cuvette f. / ~ (Hydr arch) || Wasserbecken n.; Wasserbehälter m.; Bassin n. || bassin m.; réservoir m. / ~ (Glassm) || Schleifschale f. || boule f.; débordoir m.; sphère f. / ~ of a balance || Wagschale f. || bassin m. ou plateau m. d'une balance. / coal ~ (Geol) || Kohlenbecken n. || bassin m. houiller. / evaporating ~ || Abdampfschale f.; Abdampfgefäß n. || capsule f. évaporatoire ou à évaporation. / ~ of a harbour || Hafenbecken n. || bassin m. d'un port. / ~ for a pump storage station || Becken n. für ein Pumpspeicherwerk || réservoir m. d'une usine d'accumulation par pompage. / turning ~ || Wendebecken n. || bassin m. de virement.

basis see base.

basket || Korb m. || panier m. / ~ (Airship) || Gondel f. || nacelle f. / ~ (Dredger) || Baggereimer m. || benne f.; godet m.; cuiller f. / ~ of the balloon || Ballonkorb m. || nacelle f. de ballon. / cable ~ || Kabelkorb m. || benne f. à câble. / ~ of a capital (Arch) || Kern m. oder Glocke f. eines Kapitäls || vase m. ou corbeille f. ou tambour m. de chapiteau. / ~ for carboys || Säureballonkorb m. || panier m. à bonbonnes. / ~ filled with gravel (Hydr arch) || Senkkorb m. || panier m. rempli de gravier. / ~ of a hydro-extractor || Schleuderkorb m. || panier m. d'essoreuse. / ~ for industrial purposes || Industriekorb m. || vannerie f. industrielle. / observation ~ || Fesselballonkorb m. || nacelle f. d'observation. / ~ for office || Briefkorb m. || corbeille f. de bureau. / shaving ~ || Spankorb m. || corbeille f. en copeaux de bois. / wood ~ || Korb m. aus Holz || corbeille f. en bois refendu.

basket embroiderer || Korbstickerin f. || brodeuse f. sur paniers. / ~ funnel || Abzugsschrank m. || chapelle f.; hotte f. / ~ furniture || Korbmöbel pl. || vannerie f. d'ameublement; meubles mpl. en osier ou jonc.

basket maker || Korbmacher m. || vannier m. / ~'s rush || Korbmacherbinse f.; Spartgras n. || épart m.; espart m. / ~'s supply || Korbmacherbedarfsartikel mpl. || articles mpl. pour vanneries.

basket mender || Korbflicker m. || raccommodeur m. de paniers. / ~ platter || Korbflechter m. || tresseur m. de paniers. / fine ~ platter || Kunstkorbflechter m. || tresseur de paniers artistiques.

basketry || Korbmacherei f. || vannerie f.

basket varnish || Korblack m. || vernis m. pour vannerie.

basketware || Korbwaren fpl. || vannerie f.; paniers mpl. / metal ~ || Blechkorbwaren fpl. || vannerie f. métallique.

basketwork || Flechtwerk n.; Strauchwerk n. || clayonnage m.

basket works pl. || Korbwarenfabrik f. || fabrique f. de vannerie.

basonic axis of a sharp-folded seam (Mine) || Muldenlinie f. || ennoyage m.; crochon m.

basquill bolt || Fensterwirbel m.; Wirbel m.; Zugriegel m. || passe-quille f.; crémone f.; tourniquet m.

bas-relief || Bas-Relief n. || bas-relief m.

bass see bast.

basset (Mine) || das Ausgehende || affleurement m.

bassoon || Fagott n.; Baßflöte f. || basson m.

bassorin || Bassorin n. || bassorine f.

bass-viol bow || Baßbogen m. || archet m. de basses.

basswood || amerikanische Linde f. || tilleul m. d'Amérique.

bast || Bast m.; Pflanzenbast m. || liber m.; bastin m.; sparte f.; écorce f. végétale.

bastard cut (File) || Bastardhieb m. || taille f. bâtarde; moyenne taille f. / ~ file || Bastardfeile f.; Vorfeile f. || lime f. bâtarde. / ~ title of a book || Schmutztitel m. || faux-titre m.; avant-titre m.

baster (Tail) || Hefter m. || bâtisseur m.

bast fibre || Bastfaser f. || fibre f. d'écorce.

bastite (Miner) || Bastit m.; Schillerspat m.; Schillerstein m.; Diallag (m.) || bastite f.; diallage m. (métalloïde).

bast mat || Bastmatte f. || natte f. d'écorce. / ~ rope || Bastseil n. || corde f. de liber.

basting thread (Tail) || Heftfaden m. || bâti m.; faufil m.

bat (Hatt) || Fach n. || plateau m.

batch (Met) || Satz m.; Gicht f.; Schicht f. || charge f.; fournée f. / ~ (Glassm) || Glassatz m.; Fritte f.; Glasmasse f. || composition f. du verre; fritte f. / ~ of bricks || Ziegelbrand m. (ein Brand Ziegel) || cuite f. de briques.

batcher (Spinn) || Batscher m. || batcheur m.

batching machine || Aufwickelmaschine f. || machine f. enrouleuse de tissus. / ~ for the decatizer || Dekatierwickelmaschine f. || enrouleuse f. de décatissage. / ~ for the decatizer for the batching of goods with wrappers || Dekatierwickelmaschine f. zum Aufwickeln der Stoffe mit Zwischenläufertuch || enrouleuse f. de décatissage pour l'enroulement des étoffes avec doublier.

bath || Bad n. || bain m. / ~ (Tub) see also bath tub || Badewanne f.; Wanne f. ||

baignoire f.; cuve f. / acid ~ (Met) || saures Bad n. || bain m. acide. / ~ of autogenous welded iron || autogen geschweißte Eisenwanne f. || cuve f. en fer soudée à l'autogène. / ~ of earthenware || Steingutwanne f. || cuve f. en grès. / electrolytic ~ || elektrolytisches Bad n. || bain m. électrolytique. / electroplating ~ || galvanoplastisches Bad n. || bain m. galvanoplastique. / fixing ~ || Fixierbad n. || bain m. fixateur. / ~ of hard glass || Wanne f. aus Hartglas || cuve f. en verre fondu. / establishment for medicinal ~s || Anstalt f. für Heilbäder; medizinische Badeanstalt f. || établissement m. pour bains médicaux. / oil ~ || Ölbad n.; Ölkasten m. || bain m. ou chambre f. d'huile. / ordinary ~ || Vollbad n.; Wannenbad n. || bain m. de baignoire. / ~ of pitchpine || Holzwanne f. aus Pitchpine || cuve f. en bois de pitchpin. / to prepare a ~ || ein Bad n. ansetzen || préparer un bain. / wooden ~ || Holzwanne f. || cuve f. en bois.

bath clock (Galv) || Baduhr f. || horloge m. (de bain) électrolytique. / ~ current regulator (Galv) || Badstromregler m. || régulateur m. de courant pour bain. / ~ gas stove || Gasbadeofen m. || four m. à bain à gaz. / ~ heater || Badwärmer m. || chauffe-bain m. / ~ house || Badeanstalt f.; Badehaus n.; Bad n. || établissement m. de bains. / public ~ house || Volksbad n.; Stadtbad n. || bains mpl. publics.

bathing accommodation || Badeeinrichtung f. || installation f. de bain; appareil m. balnéatoire. / ~ beach || Strandbad n. || plage f. / ~ cap || Badekappe f. || bonnet m. de bain. / rubber ~ cap || Gummibademütze f. || bonnet m. de bain en caoutchouc. / ~ drawers pl. || Badehose f. || caleçon m. de bain. / ~ dress || Badeanzug m. || costume m. de bain. / ~ establishment || Badeanstalt f. || établissement m. de bains. / ~ extract || Badeextrakt m. || extrait m. pour bains. / ~ gown || Bademantel m. || peignoir m.; baigneuse f. / ~ place see ~ establishment. / ~ preparation || Badepräparat n.; Bademittel n. || préparation f. pour bains. / ~ room see bath room. / ~ salt || Badesalz n. || sel m. pour bains. / ~ shoe || Badeschuh m. || espadrille f. / ~ suit || Badeanzug m. || costume m. de bain. / ~ tricot || Badetrikot n. || tricot m. de bain. / ~ tub see bath tub.

bathometer || Tiefenmesser m. || bathomètre m.

bath oven see bath stove. / ~ plant || Badeeinrichtung f. || installation f. de bain.

bath room || Badezimmer n.; Badestube f.; Badekammer f. || salle f. ou cabinet m. de bain. / ~ installation || Badezimmereinrichtung f. || installation f. de salle de bain. / ~ stove || Badezimmerofen m. || chauffe-bain m.; poêle f. pour bains ou pour cabinets de bain.

bath salt || Badezusatz m. || produit m. additionnel pour le bain. / ~ soap || Badeseife f. || savon m. pour bain. / ~ stone quarry || Muschelkalksteinbruch m. || carrière f. de calcaire coquillier. / ~ stove || Badeofen m. || chauffe-bain m.; poêle f. chauffe-bains; poêle f. ou four m. ou foyer m. à bain. / gas ~ stove || Gasbadeofen m. || four m. à bain à gaz. / ~ towel || Badetuch n. || peignoir m.

bath tub ‖ Badewanne f.; Wanne f. ‖ baignoire f.; cuve f. / enamelled cast iron ~ ‖ gußeiserne emaillierte Badewanne f. ‖ baignoire f. en fonte émaillée. / enamelled sheet-iron ~ ‖ Badewanne f. aus emailliertem Blech ‖ baignoire f. en tôle émaillée. / ~ of glazed stone-plates ‖ Badewanne f. aus glasierten Steinplatten ‖ baignoire f. en plaques de grès vernissées. / ~ of pure nickel ‖ Badewanne f. aus Reinnickel ‖ baignoire f. en nickel pur. / zinc ~ ‖ Zinkbadewanne f.; Zinkwanne f. ‖ baignoire f. ou cuve f. en zinc.

bath vapour (Galv) ‖ Badnebel m. ‖ vapeur f. du bain. / ~ warmer ‖ Badwärmer m. ‖ chauffe-bain m. / ~ water plant ‖ Badewasseranlage f. ‖ installation f. pour eau de bain.

batik ‖ Batik n. ‖ batik m. / ~ colour ‖ Batikfarbe f. ‖ couleur f. batik.

bat's-wing burner ‖ Fledermausbrenner m.; Schlitzbrenner m. ‖ bec m. fendu.

batt bearing with taper clamping sleeve ‖ Spannhülsenlager n. ‖ roulement m. à douille de serrage.

batten, to ‖ mit Latten befestigen ‖ latter. / ~ the cotton ‖ die Baumwolle klopfen ‖ battre le coton. / ~ the hatches (Mar) ‖ die Luken fpl. schalken; die Persenningsleisten fpl. aufnageln ‖ clouer les lattes des prélarts d'écoutille.

batten ‖ Latte f.; Leiste f.; Haftlatte f. ‖ latte f.; volige f.; tringle f. / ceiling ~ (Shipb) ‖ Wegerungsplanke f. ‖ bordage m. du revêtement intérieur; lambris m.; vaigre f. / ~ of a loom ‖ Lade f. eines Webstuhles ‖ battant m. d'un métier.

batter, to ‖ aushämmern ‖ battre; marteler.

batter ‖ Böschung f. ‖ fruit m. latéral; talus m. / ~ for freeing castings ‖ Putztrommel f. ‖ tambour m. à nettoyer la fonte. / ~ of a wall ‖ Böschung f. oder Schmiege f. oder Schräge f. einer Mauer ‖ adossement m. ou pente f. ou talus m. d'un mur.

battered face of a wall (Hydr arch) ‖ Böschungsfläche f. einer Ufermauer ‖ parement m. d'un mur de quai.

battering of a wall ‖ Ausloten n. einer Wand ‖ gauchissement m. d'une muraille.

batter level ‖ Neigungsmesser m.; Klinometer m. ‖ clinomètre m.

battery ‖ Batterie f. ‖ batterie f. / ~ of accumulators ‖ Akkumulatorenbatterie f. ‖ batterie f. d'accumulateurs. / balancing ~ ‖ Ausgleichbatterie f. ‖ batterie f. d'équilibrage. / ~ of boilers ‖ Kesselgruppe f. ‖ batterie f. de chaudières; batterie f. génératrice de vapeur. / to boost the ~ ‖ die Batterie verstärken ‖ renforcer la charge. / ~ boosting ~ ‖ Zusatzbatterie f. ‖ batterie f. survoltrice. / ~ of bottle cells (Electr) ‖ Flaschenbatterie f. ‖ batterie f. de piles-bouteilles. / to break up a ~ ‖ eine Batterie auseinandernehmen ‖ démonter une pile. / buffer ~ ‖ Pufferbatterie f. ‖ batterie f. tampon. / to change the ~ ‖ die Batterie auswechseln ‖ changer la batterie. / to charge the ~ ‖ die Batterie aufladen ‖ charger la batterie. / ~ of coke ovens ‖ Koksofenbatterie f. ‖ batterie f. de fours à coke. / to discharge the ~ ‖ die Batterie entladen ‖ décharger la batterie. / dry ~ ‖ Trockenbatterie f. ‖ pile f. sèche. / ~ for field telegraph ‖ Feldtelegrafenbatterie f. ‖ batterie f. pour le

service télégraphique de campagne. / ~ of floodlight ‖ Landebahnleuchte f. ‖ projecteur m. ou batterie f. d'atterrissage. / ~ of hardening boilers ‖ Härtekesselbatterie f. ‖ batterie f. d'autoclaves. / high-voltage ~ ‖ Starkstromelement n. ‖ pile f. galvanique; élément m. à courant de haute intensité. / ignition ~ ‖ Zündbatterie f. ‖ batterie f. d'allumage. / interchangeable ~ ‖ auswechselbare Batterie f. ‖ batterie f. interchangeable. / to load a ~ ‖ eine Batterie speisen ‖ charger une batterie. / to maintain a ~ ‖ eine Batterie f. speisen ‖ entretenir une pile. / to milk the ~ ‖ die Batterie zum Kochen bringen ‖ survolter la batterie. / motorized ~ (Arm) ‖ motorisierte Batterie f. ‖ batterie f. motorisée. / portable ~ ‖ tragbare Batterie f. ‖ batterie f. portative. / ~ for power-cars ‖ Triebwagenbatterie f. ‖ batterie f. pour chariots à force-motrice. / run-down ~ ‖ erschöpfte Batterie f. ‖ batterie f. déchargée. / to run the ~ down ‖ die Batterie zu stark entladen ‖ épuiser la batterie trop fortement. / running down of ~ ‖ Erschöpfung f. der Batterie ‖ épuisement m. de la batterie. / secondary ~ ‖ Sekundärbatterie f. ‖ batterie f. secondaire. / ~ of slaking drums ‖ Löschtrommelbatterie f. ‖ batterie f. de tambours extincteurs. / stationary ~ ‖ stationäre Batterie f. ‖ batterie f. fixe. / storage ~ ‖ Akkumulatorenbatterie f. ‖ batterie f. d'accumulateur. / storage ~ for lighting ‖ Beleuchtungsbatterie f. ‖ batterie f. d'éclairage. / subsidiary ~ ‖ Verstärkungsbatterie f. ‖ pile f. auxiliaire. / thermoelectric ~ ‖ thermo-elektrische Säule f.; Thermosäule f. ‖ pile f. thermoélectrique. / train lighting ~ ‖ Zugbeleuchtungsbatterie f. ‖ batterie f. pour éclairage des trains. / transportable ~ ‖ transportable Batterie f. ‖ batterie f. mobile.

battery acid ‖ Akkumulatorensäure f. ‖ acide m. pour accumulateurs. / ~ bar ‖ Batterieschiene f. ‖ barre f. d'amenée de voltage. / ~ boiler ‖ Batteriekessel m. ‖ générateur m. multibouilleur. / ~ box ‖ Batteriekasten m. ‖ caisse f. de batterie. / ~ cable ‖ Batteriekabel n. ‖ câble m. de la prise de courant de batterie. / ~ calling machine ‖ Batterierufmaschine f. ‖ machine f. d'appel à batterie centrale. / ~ charger ‖ Ladesatz m. ‖ groupe m. de charge. / ~ commutator ‖ Kommutator m.; Polwechsler m. ‖ commutateur m. des pôles d'une pile. / ~ cupboard ‖ Batterieschrank m. ‖ armoire f. à batterie. / ~ cut-off ‖ Abschaltbatterie f. ‖ batterie f. de déconnexion. / ~ dies pl. ‖ Pochstempel m. ‖ matrice f. de pilon. / ~ floor stand ‖ Bodengestell n. für Bleisammler ‖ plateforme f. de batterie. / ~ ignition ‖ Batteriezündung f. ‖ allumage m. par accumulateur. / ~ jar ‖ Batterieglas n. ‖ récipient m. en verre pour batteries. / ~ knife ‖ Batterieschaber m. ‖ racloir m. de pile. / ~ lead ‖ Batterieleitung f. ‖ conducteur m. d'amenée de voltage. / ~ line ‖ Batterieleitung f. ‖ fil m. conducteur de batterie. / ~ maintenance ‖ Instandhaltung f. der Batterie ‖ entretien m. de la batterie. / ~ man ‖ Batteriearbeiter m. ‖ surveillant m. de batterie. / ~ motor car ‖ Akkumulatorenwagen m. ‖ voiture f. à accumulateurs. / ~ mud ‖ Elementschlamm m. ‖ dépôt m. de l'élément. / ~ plug ‖ Batteriestecker m. ‖

fiche f. pour batteries. / ~ resistance (electr) ‖ Batteriewiderstand m. ‖ résistance f. intérieure de la pile. / ~ room ‖ Akkumulatorenraum m. ‖ salle f. d'accumulateurs. / ~ shoe ‖ Pochschuh m. ‖ sabot m. de pilon. / ~ stand ‖ Batteriegestell n. ‖ étagère f. de batterie. / ~ stop ‖ Batteriekontakt m. ‖ contact m. d'une pile. / ~ switch ‖ Zellenschalter m. ‖ réducteur-adjoncteur m. / ~ switchboard ‖ Batterieschalttafel f. ‖ tableau m. de distribution des piles. / ~ switch conductor ‖ Zellenschalterleitung m. ‖ conducteur m. de réducteur-adjoncteur. / ~ terminal ‖ Batterieklemme f. ‖ borne f. de batterie. / ~ tester ‖ Batteriemeßinstrument n.; Batterieprüfer m. ‖ vérificateur m. ou voltmètre m. de batterie. / ~ testing instrument ‖ Batterieprüfinstrument n. ‖ instrument m. à essayer la batterie. / ~ wire ‖ Batteriedraht m. ‖ fil m. pour batterie.

battik see batik.

batting (Cotton) ‖ Klopfen n. ‖ battage m.

battle cruiser ‖ Panzerkreuzer m. ‖ grand croiseur m. de bataille; cuirassé m. rapide.

battledore barley ‖ Fächergerste f. ‖ orge f. en éventail.

battle ship ‖ Linienschiff n. ‖ cuirassé m. d'escadre.

baulk, to ~ timber ‖ Holz n. aus dem Groben behauen; berappen ‖ dégauchir le bois.

bauxite ‖ Bauxit m. ‖ bauxite f. / ~ brick ‖ Bauxitstein m. ‖ brique f. en bauxite. / ~ kiln ‖ Bauxitofen m. ‖ four m. pour bauxite. / ~ quarry ‖ Bauxitgrube f. ‖ carrière f. de bauxite.

bavin ‖ Reisigbündel n.; Welle f. ‖ bourrée f.; cotret m.; fagot m.; faisceau m.; fascine f.

bay of a bridge ‖ Brückenbogen m. ‖ travée f. / ~ of a door ‖ Türnische f. ‖ baie f. de porte. / ~ of joists ‖ Fach n. oder Feld n. oder Raum m. zwischen zwei Balken ‖ claire-voie f.; travée f. / ~ of a lock ‖ Vorschleuse f. ‖ tête f. d'écluse. / main ~ (Build) ‖ Hauptschiff n. ‖ travée f. principale. / ~ of a shaft ‖ Schachtfeld n.; Schachtverzug m. ‖ intervalle m. de puits. / ~ of the shop (Build) ‖ Werkstattschiff n. ‖ travée f. d'un hall. / ~ of a sluice (Hydr arch) ‖ Haupt n. einer Schleuse ‖ tête f. d'écluse. / ~ of a window ‖ Fensterausschnitt m.; Fensternische f. ‖ baie f. ou échancrure f. de fenêtre.

bayed, x-~ building ‖ mehrschiffiges Gebäude n. ‖ bâtiment m. à x-travées.

bay laurel oil ‖ Lorbeeröl n. ‖ essence f. de laurier.

bayonet ‖ Bajonett n.; Seitengewehr n. ‖ baïonnette f. / ~ cap ‖ Bajonettsockel m. ‖ culot m. à baïonnette. / ~ catch ‖ Bajonettverschluß m. ‖ verrouillage m. ou monture f. à baïonnette. / ~ clutch ‖ Bajonettgriff m. ‖ manette f. à douille de baïonnette. / ~ fixing see ~ catch. / ~ girder ‖ Bajonetträger m. ‖ traverse f. à baïonnette. / ~ joint ‖ Bajonettverschluß m.; Bajonettverbindung. f. ‖ fermeture f. ou joint m. à baïonnette. / ~ lampholder see ~ socket. / ~ socket ‖ Bajonettfassung f. ‖ douille f. à la baïonnette.

bay salt ‖ Seesalz n.; Siedesalz n. ‖ sel m. marin ou de mer; salmare f. / ~ stall ‖ Fensterbank f. ‖ siège m. ou banc m. à

l'embrassement; banc m. de fenêtre *ou* de croisée *ou* dans la baie; carolle f. / ~ work (Carp) ‖ Bundwand f.; Fachwand f.; Riegelwand f. ‖ cloison f. en charpente; pans mpl. de bois; paroi f. en clayonnage *ou* en colombage.

baza(a)r ‖ Bazar m. ‖ bazar m.

beach ‖ Gestade n.; Strand m. ‖ plage f.; rivage m. / raised ~ gehobene Küste f.; Steilküste f. ‖ plage f. soulevée.

beach deposit ‖ Uferwall m. ‖ rempart m. côtier.

beaching trolley for seaplanes ‖ Wasserflugzeugschleppwagen m. ‖ remorque f. pour hydroplanes.

beacon (Mar) ‖ Feuerbake f.; Leuchtfeuer n.; Leuchtturm m. ‖ fanal m.; phare m.; tour f. à feu. / airway ~ ‖ Flugstreckenfeuer n. ‖ phare m. de ligne. / intermediate airway ~ ‖ Zwischenfeuer n. für Flugstrecken ‖ phare m. intermédiaire de ligne. / principal airway ~ ‖ Hauptfeuer n. für Flugstrecken ‖ phare m. principal de ligne. / location ~ ‖ Ansteuerungsfeuer n. ‖ phare m. de terrain. / triple beam rotary ~ ‖ dreistrahliges Drehfeuer n. ‖ feu m. tournant à trois faisceaux.

beacon buoy ‖ Bakentonne f.; Leuchtboje f. ‖ bouée f. à la balise.

beaconing ‖ Bebakung f. ‖ balisage m.

bead, to ~ the edge of a tube ‖ das Rohr umbördeln ‖ dudgeonner le tube.

bead ‖ Perle f. ‖ perle f. / ~ (Locksm) ‖ Sieke f. ‖ rebord m. / ~ (Roll) ‖ Wulst m. ‖ bourrelet m.; renflement m. / artificial ~ ‖ künstliche Perle f. ‖ perle f. artificielle. / ~ of artificial horn ‖ Perle f. aus Kunsthorn ‖ perle f. en corne artificielle. / ~ for chaplets ‖ Perle f. *oder* Kugel f. für Rosenkränze ‖ perle f. à chapelets. / clay ~ ‖ Tonperle f. ‖ perle f. d'argile. / ~ of glass ‖ Glasperle f.; Stickperle f.; Venezianer *oder* venezianische Perle f. ‖ perle f. de Venise. / mourning ~ ‖ Trauerperle f. ‖ perle f. de deuil. / to press a ~ on ‖ auf der Presse bördeln ‖ border à la presse. / ~ of rim ‖ Felgenrand m. ‖ rebord m. de jante. / steel ~ ‖ Stahlperle f. ‖ perle f. d'acier. / welding ~ ‖ Schweißperle f. ‖ perle f. de soudure.

bead bag ‖ Perltasche f. ‖ sac m. en perles. / ~ crown ‖ Perlenkranz m. ‖ couronne f. en perles. / ~ curtain ‖ Perlvorhang m. ‖ rideau m. en perles.

beaded bag *see* bead bag.

beaded-edge tyre ‖ Wulstreifen m. ‖ pneu m. à talon.

bead flower ‖ Perlenblume f. ‖ fleur f. en perles. / ~ fringe ‖ Perlfranse f. ‖ frange f. de perles.

beading hand tool ‖ Rohrstemmeisen n.; Rohrstemmeißel m. ‖ matoir m. à tubes. / ~ machine ‖ Siekenmaschine f. ‖ machine f. à border. / ~ press ‖ Bördelpresse f.; Kümpelpresse f. ‖ presse f. à emboutir. / ~ work ‖ Perlenstickerei f. ‖ broderie f. en perles.

bead moulding (Arch) ‖ Perlstab m. ‖ batonnet m. perlé. / ~ platter ‖ Perlenflechter m. ‖ tisseur m. de perles.

beads pl. ‖ Rosenkranz m. ‖ rosaire m.; chapelet m. / ~ (Arch) *see* bead moulding.

bead underplate ‖ Perluntersatz m. ‖ dessous m. en perles. / ~ work (Working) ‖ Perlenstickerei f.; Perlennäherei f. ‖

broderie f. de perles. / ~ wreath ‖ Perlkranz m. ‖ couronne f. de perles.

beak (Zool) ‖ Schnabel m. ‖ bec m. / ~ (Of a receptacle) ‖ Schnabel m.; Schneppe f.; Tülle f. ‖ bec m.; gueule f. / ~ of anvil ‖ Amboßhorn n. ‖ bigorne f. / ~ of the sock ‖ Pflugnase f. ‖ nez m. d'une charrue.

beaker ‖ Becherglas n.; Kochbecher m. ‖ gobelet m. de verre.

beakerglass *see* beaker.

beakhead (Ship) ‖ Galion n.; Schiffsschnabel m. ‖ poulaine f.; éperon m.

beak iron ‖ Sperrhorn n.; doppelter Hornamboß m. ‖ bigorne f.; enclume f. à deux cornes.

beam, to ~ the warp ‖ die Kette bäumen ‖ ensoupler la chaîne.

beam (Balance) ‖ Wagebalken m. ‖ fléau m. / ~ (Build; Shipb) ‖ Balken m.; Träger m. ‖ poutre f.; poutrelle f.; solive f.; support m. en fer; fer m. à plancher. / ~ (Crane) ‖ Ausleger m. ‖ flèche f.; bras m.; fauconneau m. / ~ (Of light) ‖ Lichtstrahl m. ‖ faisceau m. de lumière. / ~ (Plough) ‖ Pflugbaum m. ‖ flèche f. *ou* arbre m. *ou* haie f. d'une charrue. / ~ (Steam eng) ‖ Balanzier m.; Schwinghebel m. ‖ balancier m. / ~ (Weav) ‖ Baum m.; Weberbaum m. ‖ ensouple f. / ~ (Windmill) ‖ Rute f. ‖ bras m. / arched ~ (Carp) ‖ Krümmer m.; gekrümmter Balken m. ‖ poutre f. courbée; poutre cintrée. / armed ~ ‖ armierter Balken m. ‖ poutre f. armée. / bed plate ~ ‖ Grundlagerbalken m. ‖ poutre f. de fondation. / bent ~ *see* arched ~. / ~ bound with iron bars *see* armed ~. / built ~ with keys ‖ verdübelter Träger m. ‖ paire f. de solives armée. / built-up ~ ‖ zusammengesetzter Balken m. ‖ poutre f. rapportée. / cross ~ ‖ Querbalken m. ‖ poutre f. transversale; traverse f.; barrot m. / curved ~ *see* arched ~. / deck ~ (Shipb) ‖ Deckbalken m. ‖ bau m.; barrot m. de pont. / divided ~ (Weav) ‖ geteilter Baum m. ‖ ensouple f. divisée. / grooved ~ (Carp) ‖ ausgefalzter Balken m. ‖ poutre f. rainurée. / half-~ (Shipb) ‖ Halbbalken m. ‖ barrotin m. / head ~ ‖ Holm m. ‖ sommier m. / hollow ~ ‖ Hohlbalken m. ‖ poutre f. creuse / horizontal ~ of a crane ‖ Kranbalken m.; Kranarm m.; Kranschnabel m.; Rollenholm m. ‖ volée f. *ou* fauconneau m. *ou* flèche f. d'une grue. / indented built ~ ‖ Zahnbalken m.; verzahnter Balken m. ‖ poutre f. à dents de scie *ou* à crémaillère. / ~ of light ‖ Lichtstrahlenbüschel n.; Lichtstrahl m. ‖ faisceau m. de rayons *ou* de lumière. / ladder ~ ‖ Leiterbalken m.; Leiterbaum m. ‖ échelier m.; arbre m. de l'échelle. / longitudinal ~ (Bridge) ‖ Längsträger m. ‖ longrine f.; longeron m. / main ~ (Carp) ‖ Längsbalken m. ‖ drome f. / movable ~ (Railw) ‖ Wiegeträger m.; Wiegebalken m. ‖ traverse f. mobile. / pressed movable ~ ‖ gepreßter Wiegebalken m. ‖ traverse f. mobile emboutie. / to put-in a ~ ‖ einen Balken m. einziehen ‖ encastrer une poutre. / rolled iron ~ ‖ Walzeisenträger m. ‖ poutre f. de fer laminé. / roof and floor ~ ‖ Querträger m. ‖ entretoise f. / sloping ~ ‖ Sparren m. ‖ abalétrier m. / squared ~ with shots ‖ vollkantiger Balken m. ‖ poutre f. à vive arête. / T-~ ‖ T-Eisen n.

‖ fer m. T. / tip ~ ‖ Kippwelle f. ‖ arbre m. de versement *ou* de pivotement. / upper-deck ~ (Shipb) ‖ Oberdecksbalken m. ‖ barrot m. du pont supérieur; bau m. des gaillards. / wooden ~ (Curr) ‖ Streichbaum m.; Schäbbaum m.; Gerbebaum m. ‖ chevalet m. des tanneurs.

beam and sail gear (Windmill) ‖ Rutenzeug n. ‖ armature f. *ou* garniture f. des bras.

beam-bending machine ‖ Balkenbiegemaschine f. ‖ machine f. à cintrer les baux *ou* les poutrelles. / ~ and straightening machine ‖ Balkenbiege- und Richtmaschine f. ‖ machine f. à cintrer et à dresser les poutres. / ~ press ‖ Balkenbiegepresse f. ‖ presse f. à cintrer les poutrelles.

beam, camber of ~ (Shipb) ‖ Decksbalkenbucht f. ‖ bouge m. de barrot. / ~ carrier (Weav) ‖ Kettenbaumträger m. ‖ porteur m. d'ensouples. / ~ compasses pl. ‖ Stangenzirkel m. ‖ trusquin m.; compas m. à verge. / ~ end ‖ Balkenkopf m. ‖ tête f. de barrot. / ~ engine *see* ~ steam engine.

beamer (Silk) ‖ Anrüster m. ‖ entaqueur m. / ~ (Weav) ‖ Kettenaufleger m. ‖ monteur m. de chaînes. / ~'s assistant (Weav) ‖ Hilfscherer m. ‖ aide-warpeur m.

beaming of warps ‖ Kettenaufziehen n. ‖ montage m. de chaînes.

beaming machine (Weav) ‖ Bäummaschine f. ‖ ensouple f. / ~ minder ‖ Maschinenbäumer m. ‖ ensoupleur m. à la machine.

beam radio ‖ Richtungsfunk m. ‖ radiotélégraphie f. dirigée.

beam, round of ~ *see* beam, camber of.

beams pl. ‖ Gebälk n. ‖ charpente f.

beam scale ‖ Balkenwage f. ‖ balance f. (à fléaux). / equal-armed ~ ‖ gleicharmige Balkenwage f. ‖ balance f. à fléaux égaux.

beam steam engine ‖ Balanzierdampfmaschine f.; Schwinghebeldampfmaschine f. ‖ machine f. à vapeur à balancier.

bean ‖ Bohne f. ‖ haricot m.; fève f. / to shell ~s pl. ‖ Bohnen fpl. aushülsen ‖ écosser des haricots mpl.

bean cutting machine ‖ Bohnenschneidmaschine f. ‖ machine f. à couper les haricots. / ~ form ‖ Nierenform f.; Bohnenform f. ‖ forme f. haricot. / ~ meal ‖ Bohnenmehl n. ‖ farine f. de fèves. / ~ ore ‖ Bohnerz n. ‖ minerai m. de fer en grains; limonite f. / ~ shot (Met) ‖ Kupfergranalien fpl. ‖ cuivre m. en graines *ou* en dragées.

bear, to ‖ tragen ‖ porter. / ~ (Nav) ‖ peilen ‖ sonder; mesurer. / ~ in shore (Nav) ‖ nach Land zu halten ‖ porter sur *ou* vers la terre.

bear (Met) ‖ Sau f.; Eisenklumpen m.; Eisensau f.; Ofensau f. ‖ loup m.; bloc m.; renard m.

bearberry leaves pl. ‖ Bärentraubenblätter npl. ‖ feuilles fpl. de busserole.

beard (Bot) ‖ Spelze f. ‖ barbe f. / ~ (Print) ‖ Fleisch n. (am Buchstaben) ‖ blanc m. / feather ~ ‖ Federfahne f. ‖ barbe f. de plume.

beard soap ‖ Rasierseife f. ‖ savon m. pour la barbe.

bearer ‖ Überbringer m. ‖ porteur m. / ~ (Bridge) ‖ Unterzug m. ‖ solive f. transversale. / ~ (Mach) ‖ Stütze f.; Träger m. ‖ appui m.; support m. / ~ (Print) ‖

Kolumnenträger m. ‖ porte-page m. / ~ (Windmill) ‖ Auflageknagge f. ‖ tasseau n. de support. / ~ of a cheque ‖ Scheckinhaber m. ‖ porteur m. d'un chèque. / swan-neck ~ (Railw) ‖ Schwanenhalsträger m. ‖ traverse f. en col de cygne.

bearer cable ‖ Tragseil n. ‖ câble m. porteur.

bearers pl. (Roll mill) ‖ Ständer mpl.; Ständergerüst n. ‖ cages fpl.; fermes fpl.; montants mpl.; piliers mpl.

bear fat ‖ Bärenfett n. ‖ graisse f. d'ours.

bearing (Aero; Nav) ‖ Peilung f. ‖ relèvement m. / ~ (Build) ‖ Auflager n.; Lager n. ‖ appui m. / ~ (Geol; Mining) ‖ Streichen n. eines Ganges *oder* Flözes ‖ direction f. d'une veine. / ~ (Lock gate) ‖ Lagerpfanne f. ‖ crapaudine f. / ~ (Mach) ‖ Lager n.; Lagerschale f. ‖ palier m.; coussinet m.; roulement m. / ~ (Railw) ‖ Rahmenstuhl m. ‖ support m. / ~ (Range) ‖ Tragweite f. ‖ portée f. / ~ (Spinn) ‖ Spindellagerung f. ‖ support m. / adjustable ~ ‖ nachstellbares Lager n. ‖ coussinet m. réglable *ou* ajustable. / babbitted ~ ‖ Weißmetallager n. ‖ coussinet m. en métal antifriction. / ball ~ ‖ Kugellager n. ‖ roulement m. *ou* palier m. à billes; coussinet m. sphérique *ou* à billes. / ball race ~ ‖ Ringkugellager n. ‖ roulement m. à billes. / ball and socket joint ~ ‖ Kugelgelenklager n. ‖ palier m. articulé à billes. / ball thrust ~ ‖ Kugeldrucklager n. ‖ butée f. à billes. / barrel shaped roller ~ ‖ Tonnenlager n. ‖ roulement m. à rouleaux bombés. / ~ of a beam ‖ freitragende Länge n. *oder* Tragweite f. eines Balkens ‖ portée f. d'une poutre. / ~ of bedplate ‖ Grundplattenlager n. ‖ support m. de la plaque de fondation. / big end ~ *see* connecting rod ~. / ~ for bogie pin (Railw) ‖ Drehzapfenlager n. ‖ crapaudine f. du bogie. / camshaft ~ ‖ Nockenwellenlager n. ‖ palier m. de l'arbre à cames. / closed ~ ‖ Stecklager n. ‖ palier m. fermé. / ~ of the coast ‖ Verlauf m. der Küste ‖ gisement m. de la côte. / collar ~ ‖ Halslager n. ‖ coussinet m. à collets; palier m. à collets. / collar step ~ ‖ Ringspurlager n. ‖ crapaudine f. annulaire. / collar thrust ~ ‖ Kammlager n. ‖ palier m. de butée à cannelures. / compass ~ (Navig) ‖ Kompaßpeilung f. ‖ relèvement m. au compas. / cone ~ ‖ Kegellager n. ‖ coussinet m. conique. / conical roller ~ ‖ konisches Rollenlager n. ‖ roulement m. à rouleaux coniques. / connecting-rod ~ ‖ Pleuelstangenlager n. ‖ coussinet m. de la bielle *ou* de tête de bielle. / ~ cooled by water ‖ wassergekühltes Lager n. ‖ coussinet m. refroidi à circulation d'eau. / crank ~ ‖ Kurbellager n. ‖ palier m. de manivelle. / crankpin ~ ‖ Kurbelzapfenlager n. ‖ coussinet m. de tête de bielle. / crankshaft ~ ‖ Kurbelwellenlager n. ‖ palier m. de vilebrequin; palier m. de l'arbre de manivelle. / cup and cone ~ ‖ Kegelkugellager n. ‖ roulement m. à billes par cône et cuvettes. / disc ~ ‖ Scheibenlager n. ‖ palier m. à disque. / external ~ ‖ Außenlager n. ‖ palier m. extérieur. / footstep ~ ‖ Fußlager n. ‖ crapaudine f. inférieure. / guide roller ~ ‖ Führungsrollenlager n. ‖ palier m. à rouleaux de guidage. / half of the ~ ‖ Lagerhälfte f. ‖ moitié f. de palier. / hanger ~ ‖ Hängelager n. ‖ palier m. suspendu. / heated ~ ‖ warm-

gelaufenes Lager n. ‖ coussinet m. chauffé. / intermediate ~ ‖ Zwischenlager n. ‖ palier m. intermédiaire. / to line a ~ with . . . ‖ ein Lager n. ausfüttern mit . . . ‖ garnir un palier de . . . / lining of the ~ ‖ Lagerausguß m. ‖ garniture f. de palier. / lubricating ~ ‖ Schmierlager n. ‖ palier-graisseur m. / magnetic ~ (Nav) ‖ magnetische Peilung f. ‖ relèvement m. magnétique. / movable ~ (Build) ‖ bewegliche Auflagerung f. ‖ appui m. mobile. / plain ~ *see* slide ~. / radial clutch ~ ‖ Kupplungsringlager n. ‖ roulement m. annulaire d'embrayage. / ~ by a radiocompass station ‖ Fremdpeilung f. ‖ relèvement m. par une station radiogoniométrique. / ring oiler ~ ‖ Ringschmierlager n. ‖ palier m. à graissage par bagues. / roller ~ ‖ Rollenlager n. ‖ palier m. *ou* roulement m. à rouleaux. / self-aligning roller ~ *see* barrel-shaped roller ~. / self-lubricating ~ ‖ Schmierlager n.; selbstschmierendes Lager n. ‖ palier-graisseur m.; coussinet m. autograisseur. / self-lubricating phosphorbronze ring ~ ‖ Phosphorbronzeringschmierlager n. ‖ coussinet m. en bronze phosphoreux garni de graisseurs automatiques à bagues. / self-oiling ~ *see* self-lubricating ~. / side ~ ‖ Längslager n. ‖ palier m. longitudinal. / single-row rigid ball journal ~ ‖ einreihiges Querkugellager n. ‖ roulement m. à simple rangée de billes. / single-row self-aligning ball journal ~ ‖ einreihiges Querkugellager n. mit Einstellring ‖ roulement m. à simple rangée de billes et à rotule. / slide ~ ‖ Gleitlager n. ‖ palier m. lisse. / small end ~ ‖ Kolbenbolzenlager n. ‖ coussinet m. de pied de bielle. / spherical ~ *see* ball ~. / spindle ~ ‖ Zapfenlager n. ‖ palier m. d'un tourillon. / spring ~ (Auto) ‖ Federlager n. ‖ bride f. de ressort formant palier. / ~ by stars ‖ Gestirnpeilung f. ‖ relèvement m. d'un astre. / steering knuckle thrust ~ ‖ Druckkugellager n. des Vorderradlenkzapfens ‖ butée f. de pivot. / step ~ ‖ Spurlager n. ‖ crapaudine f. / stuffing box ~ ‖ Stopfbuchsenlager n. ‖ palier m. à presseétoupes. / ~ by the sun ‖ Sonnenpeilung f. ‖ relèvement m. du soleil. / to take a ~ ‖ orten ‖ s'orienter. / to take the ~s in the space ‖ sich im Raume m. orientieren ‖ s'orienter dans l'espace. / to take up ~s ‖ die Lager npl. nachstellen ‖ rattraper le jeu. / taper roller ~ ‖ Kegelrollenlager n. ‖ roulement m. à rouleaux coniques. / three-point ~ ‖ Dreipunktlagerung f. ‖ montage m. sur trois points. / thrust ~ ‖ Drucklager n. ‖ palier m. de butée. / thrust ball ~ ‖ Druckkugellager n. ‖ roulement m. de butée (à billes); butée f. à billes. / thrust clutch ~ ‖ Drucklager n. der Kupplung ‖ butée f. d'embrayage. / tilting ~ ‖ Kipplager n.; Klapplager n. ‖ palier m. renversable *ou* culbutable. / true ~ (Nav) ‖ rechtweisende Peilung f. ‖ relèvement m. vrai. / water-cooled ~ *see* bearing cooled by water.

bearing axle ‖ Tragachse f. ‖ essieu m. porteur. / ~ bar ‖ Rostträger m.; Rostbalken m. eines Feuerrostes ‖ support m.; sommier m.; chevalet m.; traverse f. d'une grille. / ~ block (Build) ‖ Kämpferstein m. ‖ imposte f.; sommier m. / ~ bolt ‖ Tragschraube f. ‖ boulon-porteur m.

bearing bracket ‖ Lagerbock m. ‖ chaise f. *ou* chevalet m. de palier; porte-coussinet m. / ~ (Electr mach) ‖ Lagerschild m. ‖ bouclier m. *ou* plateau m. de palier. / ~ type (Electr mach) ‖ Lagerschildtype f. ‖ type f. de bouclier à palier.

bearing bush ‖ Lagerbuchse f.; Lagerschale f. ‖ coussinet m. / adjustable ~ ‖ einstellbare Lagerschale f. ‖ coussinet m. réglable. / half of ~ ‖ Lagerschalenhälfte f. ‖ moitié f. du coussinet.

bearing cap ‖ Lagerdeckel m. ‖ couvercle m. *ou* chapeau m. de palier. / to lift the ~ ‖ den Lagerdeckel abheben ‖ enlever le chapeau *ou* le couvercle du palier.

bearing columns pl. (Blast furnace) ‖ Rahmensäulen fpl. ‖ cadres-colonnes fpl. / ~ felt ‖ Lagerfilz m. ‖ feutre m. de graissage. / ~ field ‖ Peilfeld n. ‖ champ m. de relèvement. / ~ friction loss ‖ Lagerreibungsverlust m. ‖ perte f. par le frottement des coussinets. / half of the ~ ‖ Lager(schalen)hälfte f. ‖ moitié f. du palier. / ~ lining *see* bearing bush. / ~ metal ‖ Lagermetall n. ‖ Weißmetall n. ‖ métal m. antifriction *ou* blanc *ou* à coussinets. / ~ pedestal *see* bearing bracket. / ~ pile (Build) ‖ Rostpfahl m. ‖ pilot m. de support; pilotis m. de grillage.

bearing plate ‖ Unterlagplatte f. ‖ plaque f. de fondation *ou* de support. / ~ (Electr) *see* bearing bracket.

bearing power ‖ Tragfähigkeit f. ‖ puissance f.; force f. portative. / ~ rail ‖ Tragschiene f. ‖ rail m. d'appui.

bearings . . . *see* bearing . . .

bearing shaft ‖ einlagerige Welle f. ‖ arbre m. à un palier. / ~ shell *see* ~ bush.

bearing spring ‖ Tragfeder f. ‖ ressort m. de suspension. / ~ for locomotives ‖ Tragfeder f. für Lokomotiven ‖ ressort m. de suspension pour locomotives. / ~ for railway vehicles ‖ Eisenbahntragfeder f. ‖ ressort m. de suspension pour voitures de chemins de fer.

bearing stand *see* bearing bracket. / ~ strap (Saddl) ‖ Schweberiemen m. (eines Geschirrs) ‖ branche f. d'avaloire. / ~ strength of a rail ‖ Tragfähigkeit f. einer Schiene ‖ puissance f. d'un rail. / ~ stud ‖ Anschlagnocke f.; Stützknagge f. ‖ butée f. d'aiguille.

bearing surface (Aero) ‖ Lauffläche f. ‖ surface f. de roulement. / ~ (Build) ‖ Auflagerfläche f. ‖ surface f. portante *ou* d'appui. / ~ (Mach) ‖ Lagerfläche f.; Auflagefläche f. des Lagers ‖ surface f. portante de palier.

bears pl. (Met) *see* bear.

bear's-breech oil ‖ Bärenklauöl n. ‖ essence f. de berce.

beat, to ‖ schlagen; abklopfen; klopfen; pochen ‖ battre; frapper. / ~ close (Print) ‖ Schwärze f. mit dem Ballen gleichmäßig auftragen ‖ encrer uniformément. / ~ the cocoons pl. ‖ die Kokons mpl. stauchen *oder* schlagen ‖ battre les cocons mpl. / ~ the cotton ‖ die Baumwolle klopfen *oder* schlagen ‖ battre le coton. / ~ down (Earthw) ‖ feststampfen; (mit einer Ramme) stampfen; rammen ‖ fouler; damer. / ~ down the pavement ‖ das Pflaster besetzen *oder* rammen *oder* stoßen ‖ damer *ou* battre le pavé. / ~ the floor ‖ den Estrich rammen ‖ battre l'aire. / ~ the form (Print) ‖

Schwärze f. auftragen ‖ toucher *ou* encrer la forme. / ~ the ground away (Mining) ‖ herein gewinnen; in Angriff nehmen; verhauen ‖ abattre; percer le terrain. / ~ the hemp ‖ den Hanf m. klopfen ‖ battre *ou* piler le chanvre. / ~ in (A wedge etc.) ‖ eintreiben ‖ chasser. / ~ in a nail ‖ einen Nagel m. einschlagen ‖ enfoncer un clou. / ~ the ink *see* ~ the form. / ~ the lathe (Weav) ‖ die Lade f. anschlagen ‖ frapper le battant. / ~ out ‖ ausschweifen; ausbauchen; erweitern ‖ écolleter. / ~ out the bosses ‖ ausbeulen ‖ débosseler. / ~ out the bosses of copper sheathing ‖ Kupferblech n. austreiben ‖ dresser une planche de cuivre. / ~ out iron ‖ Eisen n. ausschmieden *oder* plätten *oder* flachschmieden ‖ battre *ou* forger *ou* aplatir le fer. / ~ out the scythe ‖ die Sense f. dengeln ‖ chapler la faux. / ~ out a vessel ‖ ein Gefäß aushämmern *oder* schweifen ‖ bosseler un vase. / ~ a proof ‖ einen Bürstenabzug m. machen; einen Korrekturbogen m. abziehen ‖ tirer une épreuve avec la brosse.

beat ‖ Schlag m. ‖ coup m.; battement m. / ~ (Phys) ‖ Schwebung f. ‖ battement m. / ~ (Watch) ‖ Ticken n.; Schlag m. ‖ tic-tac m.

beaten cobwork (Build) ‖ Piseebau m. ‖ construction f. en pisé *ou* en terre battue. / ~ gold ‖ Blattgold n. ‖ or m. en feuilles. / ~ piece (Met) ‖ gekumpeltes Stück n. ‖ pièce f. emboutie. / ~ silver ‖ Blattsilber n. ‖ feuilles fpl. d'argent.

beater (Bleach) ‖ Prätschmaschine f.; Pantschmaschine f. ‖ battoir m.; plateau m. à battoir. / ~ (Clothm) ‖ Walkhammer m.; Klopfstock m. ‖ fouloir m.; gaulette f. / ~ (Hatt) ‖ Klopfer m.; Schlägel m.; Klöpfel m. ‖ battoir m.; maillet m.; enfonçoir m. / ~ (Mason) ‖ Krücke f.; Kalkhacke f. ‖ rabot m.; bouloir m. / ~ (Pap) ‖ Stampfe f. ‖ pilon m. / ~ (Spinn) *see also* beating machine ‖ Schlagmaschine f.; Flackmaschine f. ‖ batteur m. / ~ of the agitator (Brew) ‖ Flügel m. des Rührwerks ‖ aile f. d'agitateur. / ~ for upholstered furniture ‖ Möbelklopfer m. ‖ bat-meuble m.

beater cross *see* beating cross. / ~ detacher ‖ Schlägerdetaschör m. ‖ détacheur m. à batteurs.

beaterman (Pap) ‖ Holländermüller m. ‖ gouverneur m.

beater pick ‖ Stopfhacke f.; Spitzhacke f. ‖ pioche f. à bourrer; pic m. / ~ row (Breaker) ‖ Stabreihe f. ‖ rangée f. de barres.

beaters pl. (Mill) ‖ Schlägerwerk n. ‖ batteurs mpl.

beating of the cotton ‖ Klopfen n. der Baumwolle ‖ battage m. du coton.

beating arm ‖ Schlagleiste f. ‖ battoir m.

beating brush (Print) ‖ Abklopfbürste f.; Abziehbürste f. ‖ brosse f. à mouler *ou* à épreuves. / to strike off a proof with the ~ ‖ einen Bürstenabzug m. machen ‖ tirer à la brosse (une épreuve).

beating cross ‖ Schlagkreuz n. ‖ croisillon m. portant des marteaux. / ~ mill ‖ Schlagkreuzmühle f. ‖ moulin m. à pales rotatives; broyeur m. de percussion à crosse.

beating-down of piles ‖ Rammen n. oder Einrammen n. der Pfähle ‖ pilotage m.

beating machine (Spinn) ‖ Schlagmaschine f. ‖ batteuse f. / ~ with x beating rods ‖ Klopfmaschine f. mit x Klopfstäben ‖ machine f. à battre avec x bâtons batteurs.

beating mill (Clothm) ‖ Stampfkalander m.; Stoßkalander m.; Schlagmühle f. ‖ calandre f. à pilons. / ~ opener (Spinn) ‖ Klopfwolf m. ‖ loup-batteur m. / ~ plant for sacks ‖ Sackausklopfanlage f. ‖ installation f. de battage des sacs. / ~ rod ‖ Klopfstab m. ‖ bâton m. batteur. / ~ shoe (Mining) ‖ Pochschuh m. ‖ sabot m. de bocards. / ~ vat ‖ untere Küpe f.; Schlageküpe f. ‖ batterie f.

beat receiving *see* ~ reception. / ~ reception ‖ Schwebungsempfang m. ‖ réception f. par battements; réception f. hétérodyne.

beaver fur ‖ Biberpelz m. ‖ fourrure f. de castor. / ~ hair ‖ Biberhaar n. ‖ poil m. de castor.

becket (Mar) ‖ Handpferd n. ‖ sauvegarde f.

become, to ~ clear (Meteor) ‖ aufheitern; aufklaren ‖ s'éclaircir. / ~ darker (Dyer) ‖ dunkeln; nachdunkeln ‖ se foncer. / ~ hot by friction ‖ heißwerden (durch Reibung) ‖ s'échauffer (par friction). / ~ out of center ‖ unrundwerden ‖ s'ovaliser.

bed, to ‖ festmachen ‖ fixer. / ~ the sleepers on the ballast ‖ die Schwellen unterstopfen ‖ bourrer le ballast sous la traverse.

bed ‖ Bett n. ‖ lit m. / ~ (Geol) *see also* layer *or* stratum *or* seam. / ~ (Layer) ‖ Auftrag m.; Lage f.; Schicht f. ‖ couche f. / ~ (Tool) ‖ Matrize f.; Unterlage f.; Lochscheibe f. ‖ matrice f.; perçoir m. / axletree ~ (Coachm) ‖ Achsfutter n. ‖ support m. d'essieu en bois. / box-shaped ~ ‖ Kastengestell n. bâti m. formant caisson. / ~ for the breech (Gun) ‖ Schwanzschraubenlager n. ‖ logement m. de la culasse. / ~ of broken stones (Road) ‖ Schotterbett n. ‖ empierrement m. / camp ~ ‖ Feldbett n. ‖ lit m. de camp. / casting ~ ‖ Gießbett n. ‖ lit m. de coulée. / the ~s pl. crop out (Mining) ‖ die Schichten fpl. gehen zu Tage aus ‖ les couches fpl. affleurent. / forked ~ (Mot) ‖ Gabelrahmen m. ‖ bâti m. en fourche. / in ~s pl. (Geol) ‖ lagenweise; lagerförmig ‖ par couches fpl.; en couches. / iron ~ ‖ eisernes Bett n.; Eisenbett n. ‖ lit m. en fer. / lawns pl. and flower ~s pl. ‖ gärtnerische Anlagen fpl. ‖ parterres fpl. de fleurs et de gazon. / ~ of the machine ‖ Maschinenbett n. ‖ banc m. de la machine. / ~ of the main-shaft ‖ Hauptwellenlager n. ‖ coussinet m. de l'arbre principal de commande. / ~ of masonry ‖ Mauerbettung f. ‖ lit m. de maçonnerie. / natural ~ *see* bed (Geol). / ~ of a pavement ‖ Planum n. einer zu pflasternden Straße ‖ aire f. du pavé; plateforme f. / pig ~ ‖ Gießbett n. ‖ lit m. de coulée. / ~ of rails ‖ Schienenbett n. ‖ lit m. de rails. / ~ of the reverberatory furnace ‖ Herd m. des Flammofens ‖ foyer m. du four à réverbère. / ~ of a river ‖ Flußbett n. ‖ lit m. de fleuve *ou* de rivière; fond m. d'une rivière. / ~ of road metal *see* ~ of broken stones. / roller cooling ~ ‖ Rollenkühlbett n. ‖ rafraîchisseur m. *ou* refroidissoir m. à rouleaux. / ~ of sand ‖ Sandbett n. ‖ lit m.

de sable. / ~ of a sluice ‖ Schleusenbett n.; Schleusenboden m. ‖ plancher m. *ou* radier m. d'une écluse. / ~ with straps (Furniture) ‖ Gurtbett n. ‖ lit m. de sangle. / ~ of straw-sheaves (Build) ‖ Schaubenlage f. ‖ rangée f. de javelles. / to trace the ~s pl. along their course (Mining) ‖ die Schichten fpl. in ihrem Verlauf verfolgen ‖ suivre les couches fpl. dans leurs parcours. / ~ of the twyer / Lager n. der Windform ‖ logement m. de la tuyère.

bed-built (Build) ‖ Lagerfuge f.; Bettungsfuge f.; Ruhefuge f. ‖ joint m. de lit *ou* d'assise.

bed clothes pl. ‖ Bettzeug n.; Bettwäsche f. ‖ literie f.; lingerie f. de lit. / material for ~ ‖ Bettbezugstoff m. ‖ étoffe f. pour lingerie de lit.

bed damask ‖ Bettdamast m. ‖ damassé m. de lit.

bedded in . . . ‖ gelagert auf . . .; gelagert in . . . ‖ supporté par . . .; appuyé sur . . .; logé dans . . .; monté sur . . . / ~ rock ‖ geschichtetes Gestein n. ‖ roche f. stratifiée.

bedder (Mill) ‖ Bodenstein m. ‖ gisante f.; meule f. dormante.

bed die *see* bed (Tool).

bedding ‖ Betten npl.; Bettzeug n. ‖ literie f. / ~ (Geol) *see* bed (Geol). / ~ of pipes / Rohrverlegung f. ‖ montage m. des tuyaux. / ~ the track ‖ Verfüllen n. der Gleise ‖ soulèvement m. des rails.

bedding disinfection ‖ Desinfektion f. von Bettzeug ‖ désinfection f. de literie. / ~ material ‖ Bettstoff m. ‖ étoffe f. de literie.

bed down ‖ Bettdaune f. ‖ duvet m. de lit.

bedew, to ‖ betauen ‖ (se) couvrir de rosée.

bed feather ‖ Bettfeder f. ‖ plume f. de lit. / ~ cleaning ‖ Bettfedernreinigung f. ‖ nettoyage m. des plumes de lit. / ~ cleaning machine ‖ Bettfedernreinigungsmaschine f. ‖ machine f. à nettoyer les plumes de lit. / ~ manufacturing machine ‖ Bettfedernverarbeitungsmaschine f. ‖ machine f. pour apprêter des duvets de lit.

bed fittings pl. ‖ Bettbeschläge mpl. ‖ ferrures fpl. de lits. / ~ flannel ‖ Bettflanell n. ‖ flanelle f. de lit. / ~ hinge ‖ Betthaken m. ‖ crampon m. de lit. / ~ linen ‖ Bettwäsche f.; Bettleinen n. ‖ linge m. *ou* lingerie f. de lit. / embroidered ~ linen ‖ gestickte Bettwäsche f. ‖ lingerie f. brodée de lit. / ~ pan ‖ Bettpfanne f. ‖ bassinoire f.

bed-pillow ‖ Pfühl n.; Federbett n. ‖ lit m. de plume; édredons mpl. et oreillers mpl. / ~ factory ‖ Federbettenfabrik f. ‖ fabrique f. de lits de plumes.

bedplate (Bridge; Railw) ‖ Unterlagplatte f.; Stoßplatte f.; Auflagerplatte f.; Stuhlplatte f. ‖ platine f. *ou* selle f. pour rails; plaque f. d'assise *ou* d'appui. / ~ (Bruiser) ‖ Bodenplatte f. ‖ plaque f. de pipe. / ~ (Build) ‖ Grundplatte f.; Fundamentrahmen m.; Gründungsrahmen m. ‖ embase f.; châssis m. de base; cadre m. *ou* plaque f. de fondation. / ~ (Mach) ‖ Sohlplatte f. ‖ plaque-semelle f. / common ~ ‖ gemeinsame Grundplatte f. ‖ plaque f. de fondation *ou* base f. commune.

bedplate beam ‖ Grundlagerbalken m. ‖ poutre f. de fondation. / ~ bearing ‖ Grundplattenlager n. ‖ support m. de la plaque de fondation.

bedrock ‖ Muttergestein n. ‖ roche f. de fond.

bedroom suite ‖ Schlafzimmereinrichtung f. ‖ installation f. de chambres à coucher.

bed rug ‖ Bettvorleger m. ‖ descente f. de lit. / ~ slide ‖ Bettschiene f. ‖ coulisse f. de lits.

bedspread ‖ Bettdecke f. ‖ couverture f. de lit.

bedstead ‖ Bettstelle f. ‖ couchette f.; bois m. de lit. / folding ~ ‖ Klappbettstelle f. ‖ lit m. cage ou de sangle. / ship ~ ‖ Schiffsbettstelle f. ‖ lit m. ou couchette f. pour bateaux.

bedstone (Mill) ‖ Bodenstein m. ‖ gisante f.; meule f. dormante.

bed succession (Geol) ‖ Schichtenfolge f. ‖ suite f. des couches. / ~ system (Geol) ‖ Formation f. ‖ formation f. terrain m. / ~ thickness (Geol) ‖ Mächtigkeit f. der Schicht ‖ épaisseur f. de la couche.

bee ‖ Biene f. ‖ abeille f.

beech ‖ Rotbuche f.; Buche f. ‖ hêtre m. / copper ~ ‖ Blutbuche f. ‖ hêtre m. rouge. / red ~ see beech.

beech bark ‖ Buchenrinde f. ‖ écorce f. de hêtre. / ~ coal ‖ Buchenholzkohle f. ‖ charbon m. de hêtre.

beechnut ‖ Buchecker f. ‖ faîne f. / ~ oil ‖ Bucheckernöl n. ‖ huile f. de faîne.

beech-tar ‖ Buchenholzteer m. ‖ goudron m. de hêtre. / ~ oil ‖ Buchenholzteeröl n. ‖ huile f. de goudron de hêtre. / ~ pitch ‖ Buchenholzpech n. ‖ poix f. de goudron de hêtre.

beech-wood ‖ Buchenholz n. ‖ bois m. de hêtre. / ~ dowel ‖ Buchenholzdübel m. ‖ goujon m. de hêtre.

beef ‖ Ochsenfleisch n. ‖ bœuf m. / ~ marrow fat ‖ Rindermarkfett n. ‖ graisse f. de moelle de bœuf. / ~ room (Shipb) ‖ Fleischladeraum m. ‖ cale f. à salaison. / ~ tallow ‖ Rindertalg m. ‖ suif f. de bœuf.

beehive ‖ Bienenstock m. ‖ ruche f. (d'abeilles). / wheeled ~ ‖ Bienenwanderwagen m. ‖ ruche f. sur roues ou sur voiture.

beehive oven (Met) ‖ Bienenkorbofen m. ‖ four m. à ruche.

beekeeper ‖ Imker m.; Bienenzüchter m. ‖ apiculteur m.; éleveur m. d'abeilles.

beekeeping ‖ Bienenzucht f.; Imkerei f. ‖ apiculture f.

beer ‖ Bier n. ‖ bière f. / Bavarian ~ ‖ Bayrisches Bier n.; Bayrischbier n. ‖ bière f. bavaroise. / bitter ~ ‖ Bitterbier n. ‖ bière f. amère. / bottled ~ ‖ Flaschenbier n. ‖ bière f. en bouteilles. / brown ~ ‖ Braunbier n. ‖ bière f. brune. / carbonated ~ ‖ karbonisiertes Bier n. ‖ bière f. artificiellement saturée d'acide carbonique. / double ~ ‖ Doppelbier n. ‖ bière f. double. / ~ on draught ‖ Faßbier n. ‖ bière f. en fût. / export ~ ‖ Ausfuhrbier n.; Exportbier n. ‖ bière f. d'exportation. / ~ for keeping see lager ~. / lager ~ ‖ Lagerbier n. ‖ bière f. de garde (reposée). / malt ~ ‖ Malzbier n. ‖ bière f. de malt. / residue ~ in chip cask ‖ Abseihbier n. ‖ fonds mpl. de cuves ou de foudres. / small ~ ‖ Dünnbier n. ‖ bière f. de table; petite bière f. / strong ~ ‖ Lagerbier n.; Doppelbier n. ‖ bière f. forte; bière f. de garde; bière f. double. / table ~ ‖ Einfachbier n. ‖ bière f. de table; petite bière f. / wheaten ~ ‖

Weizenbier n. ‖ bière f. de blé. / white ~ ‖ Weißbier n. ‖ bière f. blanche.

beer barrel ‖ Bierfaß n. ‖ fût m. à bière. / ~ barrel stave ‖ Bierfaßdaube f. ‖ douve f. pour tonneau à bière. / ~ conduit cleaning apparatus ‖ Reinigungsapparat m. für Bierleitungen ‖ appareil m. à nettoyer les tubes de soutirage de bière.

bee-rearing ‖ Bienenzucht f. ‖ apiculture f. / ~ implement ‖ Bienenzuchtgerät n. ‖ instrument m. d'apiculture.

beer exporter ‖ Bierausführer m. ‖ exportateur m. de bière. / ~ filling apparatus ‖ Bieranzapfvorrichtung f. ‖ appareil m. à débiter la bière. / ~ filling pressure apparatus ‖ Bierdruckvorrichtung f.; Bierdruckapparat m. ‖ pression f. ou pompe f. à bière; appareil m. à soutirer la bière sous pression. / ~ filter ‖ Bierfilter n. ‖ filtre m. à bière. / ~ glass ‖ Bierglas n. ‖ verre m. à bière; bock m. / ~ glass undercover ‖ Bierglasuntersatz m. ‖ dessous m. pour verre à bière. / ~ hall ‖ Bierhalle f. ‖ buvette f. / ~ house ‖ Schenke f. ‖ cabaret m. / ~ impregnating apparatus ‖ Bierimprägniervorrichtung f. ‖ appareil m. d'imprégnation de la bière. / ~ jug ‖ Bierkrug m. ‖ cruchon m. à bière. / ~ pitcher see ~ glass. / ~ pressure and distributing apparatus see ~ filling pressure apparatus. / ~ pressure regulator ‖ Bierdruckregler m. ‖ régulateur m. de la pression de la bière. / ~ siphon ‖ Biersiphon m. ‖ siphon m. à bière. / ~ store ‖ Bierniederlage f. ‖ dépôt m. de bière. / ~ warmer ‖ Bierwärmer m. ‖ chauffe-bière m. / ~ wort ‖ Bierwürze f. ‖ moût m. de bière. / ~ yeast ‖ Bierhefe f.; Bierzeug m.; Samenhefe f. ‖ levure f. ou levain m. de bière.

beeswax ‖ Bienenwachs n. ‖ cire f. d'abeilles. / ~ bleacher ‖ Wachsbleicher m. ‖ blanchisseur m. de cire d'abeilles. / ~ bleaching works pl. ‖ Wachsbleicherei f. ‖ blanchisserie f. de cire d'abeilles. / ~ boiler ‖ Wachsschmelzer m. ‖ fondeur m. de cire d'abeilles.

beet see also beetroot ‖ Runkelrübe f.; Bete f.; Rübe f.; Rübe f. ‖ betterave f.; rave f. / red ~ ‖ rote Rübe f. ‖ betterave f.

beet juice ‖ Rübensaft m. ‖ jus m. de betterave. / ~ mill ‖ Rübensaftfabrik f. ‖ râperie f. de betteraves.

beetle (Pav) ‖ Ramme f.; Hoye f.; Pflasterramme f.; Jungfer f. ‖ hie f.; demoiselle f.; dame f. / ~ (Shipb) ‖ Ramme f. für Kielklötze ‖ blin m. / paving ~ see beetle.

beetling of cloth ‖ Schlagen n. von Geweben ‖ beetlage m. de tissus.

beetling engine ‖ Stampfkalander m.; Stoßkalander m.; Schlagmühle f. ‖ calandre f. à pilons. / rotary ~ ‖ Quetschmangel f. ‖ calandre f. rotative.

beetling mill see beetling engine.

beet potash ‖ Rübenpottasche f. ‖ salins mpl. de betteraves. / ~ pulp ‖ Rübenpulpe f. ‖ pulpe f. de betteraves. / ~ rasping station ‖ Rübensaftfabrik f. ‖ râperie f. de betteraves.

beetroot see also beet ‖ Runkelrübe f.; betterave f. / ~ coffee ‖ Runkelrübenkaffee m. ‖ café m. de betteraves. / ~ cutter ‖ Rübenschneider m. ‖ couperacines m. / ~ distillery ‖ Rübenbrennerei f. ‖ distillerie f. de betteraves. / ~ drying ‖ Rübentrockenanstalt f. ‖ sécherie f. de betteraves. / ~ molasses

distillery ‖ Rübensirupbrennerei f. ‖ distillerie f. de mélasse. / ~ rasp ‖ Runkelnreibe f. ‖ râpe f. pour betteraves. / ~ spirit ‖ Melassesprit m. ‖ alcool m. de betteraves. / ~ storing ‖ Rübeneinmieten n. ‖ ensilage des betteraves f. / ~ sugar see beet sugar.

beet sugar ‖ Rübenzucker m. ‖ sucre m. de betteraves. / ~ manufactory ‖ Rübenzuckerfabrik f.; Zuckerfabrik f. ‖ fabrique f. de sucre de betteraves. / ~ manufacturer ‖ Rübenzuckerfabrikant m. ‖ fabricant m. betteraviste.

beforehand payment ‖ Vorauszahlung f.; Vorschußzahlung f. ‖ payement m. d'avance ou par anticipation.

begin, to ~ digging ‖ einschlagen; schürfen ‖ commencer à faire une fouille. / ~ the shift (Mine) ‖ die Schicht antreten ‖ commencer le travail.

beginning of fermentation of the wort ‖ Ankommen n. der Würze ‖ entrée f. en fermentation du moût.

Behrens electrometer (Electr) ‖ Säulenelektroskop n. ‖ électroscope m. de Behrens à pile sèche.

belay, to ~ a vessel ‖ ein Schiff n. belegen oder festmachen ‖ amarrer un navire.

belaying cleat ‖ Kreuzklampe f.; Belegklampe f. ‖ taquet m. d'amarrage ou à branches. / ~ pin (Shipb) ‖ Karvelnagel m.; Belegnagel m.; Kavielnagel m. ‖ cabillot m.; cheville f. de tournage.

belfry ‖ Glockenstuhl m. ‖ beffroi m.; cage f. de clocher; clocheton m. / ~ (Steeple) ‖ Glockenturm m. ‖ clocher m. / ~ arch ‖ Schalloch n. ‖ baie f. de clocher; ouie f.

bell ‖ Klingel f. ‖ Schelle f.; (kleine) Glocke f. ‖ sonnette f.; (petite) cloche f.; timbre m. / ~ (Electr) ‖ Wecker m.; Alarmglocke f. ‖ sonnerie f. d'appel ou d'alarme. / ~ (Typewr; Cash) ‖ Läutewerk n. ‖ sonnerie f. (d'annonce). / ~ (Windinstrument) ‖ Schalltrichter m.; Stürze f. ‖ pavillon m. / alarm and signal ~ (Railw) ‖ Läutewerksglocke f. ‖ cloche f. de sonnerie. / cast steel ~ ‖ Glocke f. aus Stahlguß ‖ cloche f. en acier coulé. / clock ~ ‖ Uhrglocke f. ‖ cloche f. d'horloge. / closed-circuit ~ ‖ Ruhestromglocke f. ‖ sonnerie f. à courant fermé. / constantly active ~ see continuously sounding ~. / continuously sounding ~ ‖ Rasselglocke f.; Rasselwecker m.; Fortschellklingel f. ‖ sonnerie f. d'appel continu; sonnerie f. continue. / diving ~ ‖ Taucherglocke f. ‖ cloche f. de plongeur. / double-stroke ~ ‖ Doppelschlagwecker m. ‖ sonnerie f. à deux battants. / electric ~ ‖ elektrische Klingel f.; elektrisches Läutewerk n. ‖ sonnette f. ou sonnerie f. électrique. / float-ing ~ (Of the regulator of blast) ‖ schwimmende Glocke f. ‖ cloche f. flottante. / fog ~ ‖ Nebelglocke f. ‖ cloche f. de brouillard. / little ~ ‖ Schelle f. ‖ sonnette f.; grelot m. / ~s pl. for locomotives ‖ Lokomotivläutewerk n. ‖ cloches fpl. pour locomotives. / magneto-electric ~ ‖ Induktionswecker m. ‖ sonnerie f. magnétoélectrique. / polarized ~ ‖ polarisierter Wecker m. ‖ sonnerie f. d'appel polarisée. / reply ~ (Railw) ‖ Rückmeldeläutewerk n. ‖ sonnerie f. de réplique ou de répétition. / ship's ~ ‖ Schiffsglocke f. ‖ cloche f. du bord. / single-stroke ~ ‖ Einschlagglocke f. ‖ sonnerie f. à un coup. / small ~ see

little ~. / tramway warning ~ || Straßen-
bahnglocke f. || timbre m. avertisseur
pour tramways.
bell accumulator || Klingelakkumulator m.
|| accumulateur m. de sonnerie.
belladonna || Belladonna f. || belladone f. /
~ extract || Belladonnaauszug m. || ex-
trait m. de belladone.
bell apparatus || Glockenapparat m. || ap-
pareil m. à cloche. / ~ arch see belfry
arch. / ~ buoy || Glockenboje f.; Glocken-
tonne f. || bouée f. à cloche ou à sonnerie. /
~ button || Glockenknopf m.; Klingel-
knopf m. || bouton m. de sonnerie. / ~
clapper || Glockenklöppel m. || battant m.
d'une cloche. / ~ clapper ring || Klöppel-
ring m. einer Glocke; Glockenring m. ||
bélière f. de cloche. / ~ compass ||
Glockenkompaß m. || compas-cloche m. /
~ cord || Klingelschnur f. || cordon m. de
sonneries.
bell crank || Schwengel m. oder Arm m. an
der Glocke || manivelle f. de cloche. / ~
of link motion (Loc) || Winkelhebel m.
der Lokomotivsteuerung || levier m.
coudé de relevage. / ~ drive (Loc) || Win-
kelantrieb m. || commande f. à équerre. /
~ lever (Loc) || Bremswinkel m. || équerre
f. de renvoi.
bell crusher || Glockenmühle f. || moulin m.
à cône; broyeur m. à cloche.
Belleville boiler || Bellevillekessel m. ||chau-
dière f. Belleville. / ~ spring || Belleville-
feder f. || rondelle f. Belleville.
bell founder || Glockengießer m. || fondeur
m. de cloches. / ~ foundry || Glocken-
gießerei f. || fonderie f. de cloches. / ~
framing see belfry.
bellite || Bellit n. || bellit m.
bell item counter (Office mach) || Glocken-
zählwerk n. || sonnette-compteur f.
bell jar || Glasglocke f.; Glasballon m. ||
cloche f. ou ballon m. de verre. / ~ (Phys)
|| Vakuumglocke f. || cloche f. à vide.
bell metal || Glockenmetall n.; Glocken-
speise f. || métal m. de cloches; bronze m.
à cloches.
bell-mouthed || schalltrichterförmig || à
trompe.
bellows pl. || Blasebalg m. || soufflet m. / ~
(Blast-eng) see also blast or blower || Ge-
bläse n. || soufflerie f. / blowpipe ~ || Löt-
rohrgebläse n. || chalumeau m. à soufflet.
/ organ ~ || Blasebalg m. einer Orgel ||
soufflet m. d'orgue. / single ~ || Hand-
blasebalg m. || soufflet m. simple. /
wooden ~ (Mach) || hölzernes Balgen-
gebläse n. || soufflet m. de bois.
bellows blow-pipe || Lötbrenner m. || chalu-
meau m. / ~ frame (Forg) || Blasebalg-
gerüst n.; Balgengerüst n. || chevalet m.
du soufflet. / ~ lever (Forg) || Blasebalg-
schwengel m. || branloire f. d'un soufflet
de forge. / ~ maker || Blasebalgmacher
m.; Bälgemacher m. || souffletier m.
bell process (Chem) || Glockenverfahren n. ||
procédé m. à cloches. / ~ pull see ~
wire.
bell-push (Tel) || Klingeltaste f.; Ruftaste
f. || bouton m. de sonnerie ou d'appel. /
constant current ~ || Ruhestromtaste f. ||
bouton m. de sonnerie à courant con-
tinu. / repeating ~ || Läutetaste f. oder
Ruftaste f. mit Rücksignal || bouton m.
de sonnerie à répétition.
bell-ringing machine || Glockenläut(e)ma-
schine f. || machine f. à sonner les
cloches. / electric ~ || elektrische Glocken-

läutmaschine f. || machine f. électrique
à sonner les cloches.
bell rope see also bell wire || Klingelschnur
f.; Klingelzug m.; Glockenstrang m. ||
cordon m. de sonnette ou de cloche.
bell-shaped glass jar see bell jar. / ~ in-
sulator || Glockenisolator m. || isolateur
m. à cloche. / ~ magnet || Glocken-
magnet m. || aimant m. en forme de
cloche; aimant m. campanulé.
bell signal (Typewr etc.) || Glockensignal
n.; Glockenzeichen n.; Klingelzeichen n. ||
signal m. d'avertissement du timbre;
son m. du timbre. / ~ stop (Electr) ||
Weckerausschalter m. || interrupteur m.
de sonnerie. / ~ system || Glockenanlage
f. || sonnerie f. / ~ swipe || Schwengel m.
oder Arm m. an der Glocke || manivelle f.
ou battant m. d'une cloche. / ~ transfor-
mer || Klingeltransformator m. || trans-
formateur m. pour sonnerie. / ~ vessel
for clocks || Glockenschale f. für Wecker ||
cloche f. de reveils. / ~ wire || Klingel-
schnur f.; Klingelzug m. || cordon m.
de sonnette ou de cloche. / ~ wire lever
|| Klingelzughebel m. || levier m. ou
renvoi m. d'une sonnette.
belly, to || bauchig sein; sich ausbauchen ||
(se) forjeter; avoir du renflement; se
bomber.
belly || Unterleib m.; Bauch m. || bas-ventre
m.; ventre m. / ~ of the crucible ||
Tiegelbauch m. || ventre m. du creuset. /
~ of a shaft furnace || Kohlensack m.
eines Schachtofens || ventre m. d'un
four à cuve.
belly board (Mus) || Klangholz n.; Reso-
nanzholz n.; Instrumentenholz n. || bois
m. de résonance. / ~ hammer (Forg) ||
Brusthammer m. || marteau m. frontal.
bellying || Ausbauchung f. || bombement
m.; renflement m.
below the dam / unterhalb der Sperrmauer / en aval du barrage m. / ~
ground (Mining) || unter Tag(e) || au
fond.
belt (Driving band) || Riemen m.; Treib-
riemen m. || courroie f. (de commande).
/ ~ (Geogr) || geografischer Gürtel m.;
Zone f. || zône f. / ~ (Girdle) || Gürtel m.;
Gurt m. || ceinture f. / ~ (Motor car) ||
Absetzstreifen m. || ralingue f. / ~
(Transporting band) see conveyor ~. /
~ of best leather || Kernledertreibriemen
m. ||courroie f. cuir corroyé. / camel hair ~
|| Kamelhaarriemen m. || courroie f. en
poil de chameau. / chain ~ || Ketten-
riemen m. || courroie f. à chaîne; chaîne
f. de transmission. / cloth ~ || Zeuggürtel
m. || ceinture f. en tissu. / cone ~ || Keil-
riemen m. || courroie f. en forme de coin.
/ conveying ~ see conveyor ~. / conveyor
~ || Förderband n.; Transportband m. ||
Fördergurt m. || bande f. transporteuse;
courroie f. de transport; ruban m. trans-
porteur. / cotton ~ || Baumwollriemen m.
|| courroie f. de coton. / crossed ~ || ge-
kreuzter oder geschränkter Riemen m. ||
courroie f. croisée. / endless ~ || endloser
Riemen m. || courroie f. sans fin. / exces-
sively stretched ~ || zu straff gespannter
Riemen || courroie f. trop tendue. / the ~
expands || der Riemen dehnt sich || la
courroie s'allonge. / fan ~ || Windflügel-
riemen m. || courroie f. de ventilateur. /
furniture ~ || Möbelgurt m. || sangle f. de
meuble. / inclined ~ (For transporting)
Steigband n. || bande f. de transport

inclinée. / laminated ~ || Lamellenriemen
m. || courroie f. à plis multiples. / leather ~
|| Ledertreibriemen m.; Lederriemen m. /
courroie f. en cuir. / ~ for motor cycles ||
Motorradriemen m. || courroie f. pour
motocyclettes. / open ~ (Mach) || offener
Riemen m. || courroie f. ouverte. / ~ for
plaids || Plaidriemen m. || courroie f. pour
plaids. / to regulate the tension of the ~
|| den Riemen nachspannen || régler la
tension de la courroie. / rubber ~ (Mach)
|| Gummitreibriemen m. || courroie f.
en caoutchouc. / rubber ~ (Girdle) ||
Gummigürtel m. || ceinture f. élastique. /
safety ~ (Aero) || Anschnallgurt m. ||
ceinture f. en sangle. / to shorten the ~ ||
den Riemen kürzen || raccourcir la cour-
roie. / sorting ~ (Dress ore) || Leseband
n. || bande f. de triage. / two-ply ~ ||
Doppelriemen m. || courroie f. double.
belt brake || Bandbremse f. || frein m. à
courroie. / ~ cement || Riemenkitt m. ||
colle f. de courroie. / ~ cloud || Band-
wolke f. || nuage m. en bande. / ~ com-
position || Riemenfett n. || apprêt m. de
courroie. / ~ conveyance || Bandförde-
rung f.; Gurtförderung f. || transport m.
par bande.
belt conveyor || Gurtförderer m.; Band-
förderer m.; Bandtransportör m. || trans-
porteur m. à courroie ou à bande ou
à ruban. / portable ~ for transporting
sacks || fahrbarer Bandförderer m. für
Säckestapelung || transporteur m. roulant
à bande pour l'empilage de sacs.
belt coupling || Riemenkupplung f. || ac-
couplement m. par courroie.
belt drive || Riemenbetrieb m.; Riemen-
antrieb m. || commande f. par courroie. /
~ with automatic and instantaneous
stop motion || Riemenantrieb m. mit
selbsttätiger Momentausrückung || com-
mande f. par courroie avec dispositif de
débrayage instantané automatique. /
double-sided ~ || zweiseitiger Riemen-
antrieb m. || commande f. double par
courroies; commande f. par courroies
des deux côtés. / single-pulley ~ || Ein-
scheibenriemenantrieb m. || commande f.
par monopoulie.
belt-driven || durch Riemen angetrieben;
mit Riemenantrieb (versehen) || actionné
par courroie; pourvu de commande par
courroie. / ~ dynamo || Dynamo f. für
Riemenantrieb || génératrice f. à com-
mande par courroie. / ~ machine || mit
Riemen angetriebene Maschine f. || ma-
chine f. commandée par courroie.
belt driving see belt drive. / ~ drum ||
Riementrommel f. || tambour m. à cour-
roie. / ~ dryer || Bandtrockner m. || sé-
cheur m. à bande.
belted see belt-driven.
belt element || Bandelement n. || élément
m. de bande. / ~ fastener || Riemenver-
binder m. || agrafe f. de courroie. / ~ fork
|| Riemengabel f. || fourchette f. d'em-
brayage ou de débrayage ou de courroie.
/ ~ gearing || Riemenvorgelege n. || renvoi
m. à courroie. / ~ grease || Treibriemen-
fett n.; Riemenfett n. || graisse f. pour
courroies. / ~ guide || Ein- und Ausrück-
vorrichtung f. für Treibriemen; Riemen-
führer m.; Riemenleiter m. || guide m.
(de) courroie; fourche f. de guidage de
courroies. / ~ guider see ~ fork.
belting see also belt || Treibriemen m. ||
courroie f. de commande. / camel hair ~

|| Kamelhaarriemen m. || courroie f. de poil de chameau. / rubber ~ || Gummitreibriemen m. || courroie f. en caoutchouc. / ~ pendulum hydro-extractor || Riemenpendelzentrifuge f. || essoreuse f. oscillante à courroie.

belt joint || Riemenverbindung f. || attache f. de courroie. / ~ leather || Treibriemenleder n. || cuir m. à transmission *ou* à courroies. / ~ lubricant || Riemenschmiere f. || lubrifiant m. pour courroies. / ~ maker || Riemenmacher m. || ceinturonnier m. / ~ making machine || Riemenherstellungsmaschine f. || machine f. pour fabriquer des courroies. / ~ mounter || Riemenaufleger m. || monte-courroie m. / ~ pull || Riemenzug m. || tension f. de la courroie. / ~ pulley || Riemenscheibe f.; Gurtscheibe f. || poulie f. (de courroie *ou* de transmission). / ~ pulley fly wheel || Riemenschwungrad n. || poulie-volant f. / ~ punch || Riemenlocher m. || emporte-pièce m. pour courroies. / ~ reverse || Riemenumsteuerung f. || renversement m. de marche par courroies. / ~ rivet || Riemenniete f. || rivet m. de courroie. / ~ rocker || Riemenwippe f. || galet m. tendeur à contrepoids. / ~ sanding machine || Riemensandschleifmaschine f. || machine f. à polir à courroie. / ~ saw || Bandsäge f. || scie f. à ruban; scie f sans fin; scie à lame continue. / ~ shifter || Riemenausrücker m.; Ausrücker m.; Riemenabsteller m. || débrayeur m. de la courroie; passe-courroie m. / ~ slip || Riemenschlupf m. || glissement m. de la courroie. / ~ stretcher || Riemenspanner m. || tendeur m. de courroie. / ~ stretching slide || Riemenspannschlitten m. || chariot m. tendeur. / ~ tension || Riemenspannung f. || tension f. de la courroie. / ~ tension adjusting device *see* ~ stretcher. / ~ thickness || Riemenstärke f. || épaisseur f. de la courroie. / ~ tightener *see* ~ stretcher. / ~ tightening device || Riemenspannvorrichtung f. || dispositif m. tendeur de courroie. / ~ wear || Riemenverschleiß m. || usure f. de la courroie.

bench || Arbeitstisch m.; Werkbank f. || banc m.; établi m. / ~ (Glassm) || Glasmacherstuhl m. || banc m. du verrier. / ~ (Join) || Hobelbank f. || établi m. de menuisier. / ~ (Law) || Gericht n. || tribunal m.; cour f. de justice; juridiction f. / ~ (Mine) || Bank f. || strate f. / ~ (Seat) || Bank f. || banc m. / cooper's ~ || Schneidbank f. || Schnitzbank f. || chevalet m. des tonneliers. / inclined ~ || Schrägrampe f. || rampe f. inclinée. / longitudinal ~ || Längsbank f. || banquette f. longitudinale. / optical ~ || optische Bank f. || banc m. d'optique. / school ~ || Schulbank f. || banc m. d'école.

bench anvil || Bankamboß m. || enclumot m. (de banc); enclumeau m. / ~ axe (Carp) || Handaxt f.; Bankaxt f.; Zimmeraxt f. || hache f. à main / ~ hammer || Bankhammer m. || marteau m. d'établi. / ~ hand || Werkbankarbeiter m. || ouvrier m. d'établi. / ~ mark || Niveaufixpunkt m.; Fixpunkt m. || repère m.; point m. fixé *ou* de repère. / ~ moulder || Bankformer m. || mouleur m. d'établi. / ~ plane (Coop) || Bankhobel m. || colombe f. à joindre. / ~ rammer (Mould) || Bankstampfer m. || fouloir m. d'établi. / ~ reel (Sailm) || Garnwinde f. || dévidoir

m.; tournette f. / ~ screw (Join) || Bankzange f. || presse f. d'établi.

bench vice || Bankschraubstock m. || étau m. à agrafe *ou* d'établi. / ~ with block || Bankschraubstock m. mit Amboß || étau m. à agrafe avec tas. / ~ with screw clamp || Schraubstock m. mit Schraubzwinge || étau m. à griffe *ou* avec presse.

bend, to || biegen; umbiegen; aufbiegen; krümmen || cintrer; plier; couder; courber. / ~ (To sag) || durchbiegen || fléchir. / ~ (To warp) || sich biegen; rundbiegen || ployer; plier. / ~ at angles || kröpfen || couder. / ~ while cold || kalt biegen || courber à froid. / ~ and turn the funnel || einen Schornstein schleifen *oder* schräg führen || dévoyer une cheminée. / ~ by hand || aus freier Hand f. biegen || courber à la main. / ~ a horseshoe || ein Hufeisen n. krümmen *oder* verengern || voûter un fer à cheval. / ~ horseshoes pl. upwards || ein Hufeisen aufbiegen || genéter un fer à cheval. / ~ while hot || warm biegen || courber à chaud. / ~ round || rundbiegen || rouler; cintrer. / ~ a spring || eine Feder spannen || bander un ressort. / ~ in tempering (steel) || sich beim Härten verziehen || courber *ou* se courber *ou* se voiler à la trempe. / ~ a type bar || einen Typenhebel verbiegen || fausser une tige à caractères.

bend || Krümmung f.; Biegung f. || courbe f.; courbure f.; cintre m. / ~ (Rope) || Schleife f.; Schlinge f. || anneau m.; ganse f.; œil m. / ~ (Tube) || Muffenbogen m.; Bogenrohr n. || raccord m. courbé; coude m. / ~ for exhaust || Auspuffkrümmer m. || coude m. d'échappement. / normal ~ (Press) || Normalbug m. || pli m. normal. / outward ~ || Ufervorsprung m. || partie f. convexe de la rive. / places pl. marked by sharp ~s || Stellen fpl., an denen die Form scharfe Krümmungen aufweist || parties fpl. de la forme accusant des courbes très accentuées. / ~ of a river || Flußkrümmung f. || coude m. *ou* sinuosité f. d'une rivière. / outward ~ of a river || ausgehende Flußkrümmung f. || convexité f. *ou* partie f. convexe du cours d'une rivière. / round ~ (Press) || Rundbug m. || pli m. arrondi. / ~ of a street || Straßenbiegung f. || tournant m. d'une rue.

bendable || biegbar; verbiegbar || pliable; faussable; susceptible de se fausser.

bender || Bieger m. || plieur m.; cintreur m. / rail ~ || Schienenbiegemaschine f. || machine f. à cintrer les rails. / tube ~ || Rohrbiegemaschine f. || machine f. à cintrer les tubes.

bend face plate (Railw) || Schienenrichtplatte f. || table f. en fonte pour le dressage des rails.

bending || Biegung f. || flexion f.; courbure f. / conical ~ of boiler shells || konisches Einrollen n. von Kesselschüssen || roulage m. conique des tôles de chaudières. / ~ of conical shells || Runden n. konischer Blechschüsse || roulage m. des tôles coniques. / cylindrical ~ of boiler shells / zylindrisches Einrollen n. von Kesselschüssen || roulage m. cylindrique des tôles de chaudières. / piece sharply notched before ~ || vor dem Biegen scharf eingekerbtes Stück n. || pièce f. nettement entaillée avant le pliage.

bending and forming machine || Biege- und Formmaschine f. || machine f. à cintrer

et à façonner. / horizontal ~ press || horizontale Biege- und Formpresse f. || bulldozer m. à cintrer et façonner.

bending and straightening machine || Biege- und Richtmaschine f. || machine f. à cintrer et à dresser. / horizontal ~ press || horizontale Biege- und Richtpresse f. || presse f. horizontale à cintrer et à dresser.

bending apparatus *see* bending machine. / ~ arm *see* ~ beam. / ~-back point (Math) || Rückkehrpunkt m. || point m. de rebroussement. / ~ beam || Biegebalken m.; Biegearm m.; Biegestange f. || bras m. à courber ou de cintrage. / ~ beam support || Biegestangenunterstützung f. || support m. du bras de cintrage. / ~ form || Biegungsform f. || forme f. de cintrage. / ~ jaw || Biegebacke f.; Biegewange f. || mâchoire f. de pliage. / ~ load || Biegebelastung f. || charge f. de flexion.

bending machine || Biegemaschine f. || machine f. à cintrer. / ~ for corrugated iron || Wellblechbiegemaschine f. || machine f. à plier les tôles ondulées. / ~ for keel plates || Kielplattenbiegemaschine f. || machine f. à cintrer les plaques de quille. / laminated spring ~ || Blattfederbiegemaschine f. || machine f. à cintrer les ressorts à lames. / ~ for mast plates || Mastplattenbiegemaschine f. || machine f. à cintrer les plaques pour mâts de navires. / ~ for pasteboard and vulcanized fibre || Biegemaschine f. für Pappe und Vulkanfiber || machine f. à plier le carton et la fibre vulcanisée. / pipe elbow ~ || Knierohrbiegemaschine f. || machine f. à faire les coudes des tuyaux de poêle. / plate ~ || Blechbiegemaschine f. || machine f. à cintrer les tôles. / four-roller plate ~ || Vierwalzenblechbiegemaschine f. || machine f. à cintrer les tôles à quatre cylindres. / three-roller plate ~ || Dreiwalzenblechbiegemaschine f. || machine f. à rouler *ou* à cintrer les tôles à trois cylindres. / ~ for reinforced concrete || Betonbiegemaschine f. || machine f. à cintrer le béton. / round iron ~ || Rundeisenbieger m. || machine f. à plier le fer rond. / sectional iron ~ || Biegemaschine f. für Formeisen || machine f. à cintrer les fers profilés. / stave ~ || Daubenbiegemaschine f. || plieuse f. pour douves. / ~ for tyres || Reifenbiegemaschine f. || machine f. à cintrer des cercles. / wire ~ || Drahtbiegemaschine f. || machine f. à cintrer les fils métalliques plieuse f. pour fils métalliques. / ~ hand || Biegemaschinenarbeiter m. || courbeur m. *ou* cintreur m. à la machine.

bending mandrel || Biegedorn m. || mandrin m. de pliage. / ~ moment || Biegungsmoment n. || moment m. fléchissant *ou* de flexion. / ~ pliers pl. || Biegezange f. || pince f. à cintrer. / ~ point || Biegestelle f. || point m. de pliage.

bending press || Biegepresse f. || presse f. à cintrer *ou* à courber; cintreuse f. / hydraulic ~ of x tons pressing power || hydraulische Biegepresse f. mit x ts Druckkraft || presse f. hydraulique à cintrer d'une force de x tonnes. / pipe ~ || Rohrbiegepresse f. || presse f. à dresser et à cintrer les tubes.

bending radius || Biegungsradius m. || rayon m. de cintrage. / ~ roll || Biegewalze f. || cylindre m. de cintrage.

bending slabs pl. (Shipb) ‖ Richtplatte f.; Richtplattenplan m. ‖ plate-forme f. à dresser; marbre m. / ~ pl. and scrive boards pl. ‖ Richtplatten- und Spantenplan m. ‖ plaques fpl. pour courber les couples.

bending speed ‖ Biegegeschwindigkeit f. ‖ vitesse f. de cintrage.

bending strain ‖ Biegeverformung f. ‖ déformation f. due à la flexion.

bending strength ‖ Biegungsfestigkeit f.; Biegefestigkeit f.; Biegungswiderstand m. ‖ résistance f. à la flexion.

bending stress ‖ Biegungsbeanspruchung f.; Biegebeanspruchung f. ‖ effort m. fléchissant ou transversal ou de flexion; travail m. à la flexion. / ~ by wind pressure ‖ Beanspruchung f. auf Biegung durch den Winddruck ‖ effort m. de flexion par la poussée du vent. / ~ durability ‖ Dauerbiegefestigkeit f. ‖ résistance f. continue à la flexion.

bending test ‖ Biegeprobe f. ‖ essai m. de pliage ou de flexion ou de ployage. / plate ~ ‖ Blechbiegeprobe f. ‖ essai m. de ployage de la tôle. / ~ by shock ‖ Schlagbiegeversuch m. ‖ essai m. de flexion par choc. / ~ by shock of a notched bar ‖ Schlag- und Biegeprobe f. mit scharf eingekerbtem Stab ‖ essai m. de choc et de ployage sur barreau entaillé.

bending tongs pl. see ~ pliers. / ~ vibration testing machine ‖ Biegeschwingungsprüfmaschine f. ‖ machine f. à mesurer les vibrations dues à la flexion. / sample of ~ work ‖ Biegeprobe f.; Biegemuster n. ‖ échantillon m. de pliage.

bend leather ‖ Kernleder n. ‖ cuir m. de choix ou de la meilleure qualité. / ~ test see bending test.

beneaped, to become (Mar) ‖ bei der Ebbe trocken werden ‖ assécher.

beneficial fodder ‖ bekömmliches Futter n. ‖ nourriture f. digestible ou profitable.

benefit of the law ‖ Rechtswohltat f. ‖ bénéfice m. de la loi.

Bengal inflammables pl. ‖ bengalische Zündwaren fpl. ‖ inflammables mpl. de Bengale. / ~ matches pl. ‖ bengalische Zündhölzer npl. ‖ allumettes fpl. du Bengale. / ~ stripes pl. (Weav) ‖ Gingham m.; englische oder schottische Leinwand f. ‖ guingamp m.; guingan m.

bent ‖ gebogen; krumm ‖ courbe; courbé; tordu. / ~ at right angles (Mach) ‖ gekröpft ‖ coudé. / ~ while cold ‖ im kalten Zustande m. verbogen ‖ ployé à froid m. / ~ while cold under the hammer ‖ unter dem Hammer in kaltem Zustande gebogen ‖ courbé à froid sous le marteau. / the shaft is ~ ‖ die Welle f. ist verbogen ‖ l'arbre m. est faussé. / ~ without showing any defects ‖ verbogen, ohne daß sich schadhafte Stellen zeigten ‖ déformé sans qu'il y eût des parties endommagées. / ~ under tup ‖ unter dem Fallwerk n. geschlagen ‖ essayé au mouton.

bent crank with handle ‖ gekröpfte Handkurbel f. ‖ manivelle f. à main coudée.

bent lever ‖ Kniehebel m. ‖ levier m. coudé ou brisé. / ~ balance ‖ Zeigerwage f. ‖ balance f. à cadran.

bent link ‖ gekröpftes Glied n. ‖ membre m. coudé; articulation f. coudée. / ~ rib ‖ gebogener Balken m. ‖ solive f. courbée. / ~ stave for repairs ‖ Flickdaube f. ‖

douve f. brute; merrain m. / ~ timber (Shipb) ‖ Bugholz n. ‖ bois m. cintré.

bent tube ‖ Bogenrohr n.; Knierohr n. ‖ tube m. courbé; coude m. / ~ type of water-tube boiler ‖ krummrohriger Wasserrohrkessel m. ‖ chaudière f. aquatubulaire à tubes cintrés.

bent-up rail foot (Railw) ‖ aufgebogener Schienenfuß m. ‖ patin m. recourbé.

bent wood furniture ‖ Möbel npl. aus gebogenem Holz; Wiener Möbel npl. ‖ meubles mpl. en bois ployé ou courbé ou cintré.

benzaldehyde ‖ Benzaldehyd n.; Bittermandelöl n. ‖ benzaldéhyde m.; aldéhyde m. benzoïque.

benzene see also benzine or benzol(e) ‖ Benzol n; Benzin n. ‖ benzène m.; benzine f.; benzol m. / crude ~ ‖ Rohbenzol n. ‖ benzol m. brut.

benzene distiller ‖ Benzindestillierer m. ‖ raffineur m. de benzine. / ~ distilling apparatus ‖ Benzindestillierapparat m. ‖ appareil m. de distillation de benzine. / ~ gas generator ‖ Benzingaserzeuger m. ‖ générateur m. de gaz de benzine.

benzine see also benzene ‖ Benzin n. ‖ essence f.; benzine f. / perfumed ~ ‖ wohlriechendes Benzin n. ‖ benzine f. parfumée.

benzine blow lamp ‖ Benzinlötlampe f. ‖ lampe f. à essence pour souder. / ~ burner ‖ Benzinbrenner m. ‖ brûleur m. à essence. / ~ chloride ‖ Chlorbenzin n. ‖ chlorure m. de benzine. / ~ engine ‖ Benzinmotor m. ‖ machine f. à essence. / ~ plant ‖ Benzinanlage f. ‖ installation f. d'usines de benzine. / ~ soap ‖ Benzinseife f. ‖ savon m. à la benzine. / ~ still ‖ Benzindestillierapparat m. ‖ appareil m. de distillation de benzine. / ~ tank ‖ Benzinbehälter m. ‖ réservoir m. d'essence.

benzoate ‖ benzoesaures Salz n.; Benzoesalz n. ‖ benzoate m.

benzoic acid ‖ Benzoesäure f. ‖ acide m. benzoïque.

benzoin ‖ Benzoe f. ‖ benjoin m. / ~ gum ‖ Benzoeharz n. ‖ gomme f. de benjoin.

benzol(e) ‖ Benzol n. ‖ benzol m.; benzène m. / crude ~ ‖ Rohbenzol n. ‖ benzol m. brut. / ~ for degreasing ‖ Benzol n. für Fettextraktion ‖ benzol m. de dégraissage.

benzol(e) chloride ‖ Chlorbenzol n. ‖ chlorure m. de benzol. / ~ crane ‖ Benzolkran m. ‖ grue f. à benzol. / ~ distilling plant ‖ Benzoldestillierungsanlage f. ‖ installation f. de distillation de benzol. / ~ extraction device ‖ Benzolgewinnungsanlage f. ‖ installation f. d'extraction de benzol. / ~ plant ‖ Benzolfabrikanlage f. ‖ installation f. de fabrique de benzol. / ~ recovery plant ‖ Benzolwiedergewinnungsanlage f. ‖ appareil m. de recupération des benzols. / ~ recuperation ‖ Benzolgewinnung f. ‖ récupération f. du benzol. / ~ solution ‖ Benzollösung f. ‖ solution f. benzénique ou de benzol. / ~ varnish ‖ Benzollack m. ‖ vernis m. au benzol.

benzoyl chloride ‖ Chlorbenzoyl n. ‖ chlorure m. de benzoyle.

benzyl acetate ‖ essigsaures Benzyl n. ‖ acétate m. de benzyle. / ~ benzoate ‖ benzoesaures Benzyl n. ‖ benzoate m. de benzyle. / ~ chloride ‖ Chlorbenzyl n. ‖ chlorure m. de benzyle. / ~ formiate ‖

ameisensaures Benzyl n. ‖ formiate m. de benzyle.

beret ‖ Barett n. ‖ barrette f.; béret m. / knitted ~ ‖ Trikotmütze f. ‖ béret m. tricoté.

bergamot oil ‖ Bergamottöl n. ‖ essence f. de bergamotte.

Berlin blue ‖ Berlinerblau n. ‖ bleu m. de Prusse. / ~ ware ‖ Gußeisenschmuck m. ‖ bijouterie f. en fonte de fer.

berlin(e) (Motor car) ‖ Limusine f. mit Innenlenkung und abgeteiltem Führersitz ‖ berline f. / ~ (Coach) ‖ Berline f. berline f.; limousine f.

berry ‖ Beere f. ‖ baie f. / berries pl. for dyeing ‖ Färbebeeren fpl. ‖ baies fpl. pour teinture. / edible ~ ‖ eßbare Beere f. ‖ baie f. comestible. / ~ of viburnum ‖ Wasserholderbeere f. ‖ baie f. de viorne.

berry wax ‖ Pflanzenwachs n. ‖ cire f. végétale. / ~ wine ‖ Beerenwein m. ‖ vin m. de baies.

berth, to ~ in the dock ‖ docken ‖ amener dans un dock; entrer dans un bassin.

berth (Mar) ‖ Koje f.; Schlafkoje f. ‖ couchette f. / ~ (Nav) ‖ Ankergrund m.; Liegeplatz m. ‖ fond m.; mouillage m. / building ~ (Shipb) ‖ Helling f. ‖ cale f. / the ~s are covered throughout their lengths with glazed iron-frame constructions ‖ die Hellinge fpl. sind ihrer ganzen Ausdehnung nach mit verglasten, in Eisenfachwerk ausgeführten Hallen überdacht ‖ sur les cales fpl. s'élèvent des halls en charpente métallique vitrée, qui abritent toute la longueur des bassins. / ~ of the crew (Mar) ‖ Mannschaftsraum m. ‖ poste m. de l'équipage. / on the ~ it is possible to lay down vessels of up to x meters in length and y meters in width ‖ auf der Helling können Schiffe bis zu x m Länge und y m Breite gebaut werden ‖ sur la cale on peut mettre en chantier des navires de x mètres de longueur et ayant jusqu'à y mètres de largeur. / ~ roofed over ‖ überdachte Helling f. ‖ cale f. couverte. / the shed is built with its axis at right angles to the ~s ‖ man hat die Halle quer zu den Hellingen angeordnet ‖ le hall m. s'étend transversalement à l'axe des cales. / the ~s take the form of trough-shaped basins ‖ die Hellinge fpl. sind als trogförmige Becken ausgeführt ‖ les cales fpl. sont des bassins en forme d'auge.

berthierite ‖ Eisenantimonglanz m. ‖ berthiérite f.

berthing manœuvre (Airship) ‖ Einbringemanöver n. ‖ manœuvre m. de campement ou de rentrée.

berth mole ‖ Hellingmole f. ‖ môle m. de cale.

beryl ‖ Smaragd m.; edler Beryll m. ‖ émeraude f.

beryllium ‖ Beryllium n.; Berylliummetall n. ‖ glucinium m.; béryllium m.

besom ‖ Besen m. ‖ balai m.

bespoke tailor ‖ Maßschneider m. ‖ tailleur m. sur mesure.

Bessemer converter ‖ Bessemerbirne f. ‖ convertisseur m. Bessemer. / ~ converter refining process ‖ Bessemerbirnenverfahren n. ‖ affinage m. au convertisseur Bessemer. / ~ iron ‖ Bessemereisen n. ‖ fer m. Bessemer.

bessemerization see Bessemer-process.

Bessemer-process || Bessemerverfahren n. || procédé m. Bessemer. / to convert pig-iron into steel by ~ || bessemern || traiter la fonte crue par le procédé Bessemer. / production of steel by the ~ || Herstellung f. von Stahl im Bessemerverfahren || fabrication f. de l'acier par le procédé Bessemer.

Bessemer steel || Bessemerstahl m. || acier m. Bessemer. / baby ~ || Kleinbessemerstahl m. || acier m. Bessemer du petit convertisseur. / baby ~ plant || Kleinbessemerei f. || petite installation f. Bessemer.

Bessemer steel works pl. || Bessemerwerk n. || aciérie f. Bessemer.

best work (Dress ore) || Scheiderz n. || minerai m. riche; minerai m. de scheidage.

beswing bran (Mill) || Flugkleie f. || soufflure f. de son.

betel nut || Betelnuß f. || noix f. de bétel.

Bethlehem crib || Weihnachtskrippe f. || crèche f. de Noël.

beton see also concrete || Beton m. || béton m. / ~ funnel || Betonrutsche f. || trémie f. à beton.

between-deck (Shipb) || Zwischendeck n. || entre-pont m.; faux-pont m.

bevel, to (Carp; Join; Stone-c) || abkanten; facettieren; ausschärfen || délarder; écorner; émousser; tailler en chanfrein; dresser en biais; chanfreiner; facetter. / ~ (Shipb; Carp) || mit der Schmiege oder dem Schmiegstock messen; schmiegen || mesurer avec la fausse équerre; équerrer.

bevel || schiefwinkelig; schräg || biais; conique.

bevel || Schmiege f.; Gehrmaß n.; Schrägmaß n.; Schrägmodel m.; Schrägwinkel m.; Stellwinkel m. || angle m. oblique; béveau m.; biveau m.; fausse équerre f.; sauterelle f.; équerre f. pliante. / ~ (Arch) || Fase f. || facette f. / ~ (Join) || Gehrung f.; Gierung f.; Gehre f. || onglet m.; anglet m.; biais m.; biaisement m. / ⊤-~ || doppelte Schmiege f. || sauterelle f. en ⊤.

bevel cant || abgeschrägte Kante f. || chanfrein m.; biseau m.

bevel cutting || Anschneiden n. von Stemmkanten || chanfreinage m. de bords matés. / ~ (Carp; Join) || Schmiegschnitt m. || fausse-coupe f.

bevel edge cutting machine for welding || Schweißkantenschere f. || cisaille f. circulaire à chanfreiner les bords des tôles pour soudure.

bevel gear || konisches Getriebe n.; Kegelradgetriebe n.; Kegelräderwerk n. || engrenage m. conique ou à roues coniques ou à roues d'angles; roues fpl. d'angle; pignons mpl. coniques. / ~ cutting machine || Kegelräderfräsmaschine f. || machine f. à tailler les roues coniques. / ~ grinding machine || Kegelradschleifmaschine f. || machine f. à rectifier les engrenages coniques.

bevel gearing see bevel gear.

bevel gear planing machine || Kegelradhobelmaschine f. || machine f. à raboter les engrenages coniques. / automatic ~ roughing-out machine with horizontal work arbor || selbsttätige Kegelradvorfräsmaschine mit horizontalem Aufspannbolzen || machine f. automatique à ébaucher les engrenages coniques avec mandrin porte-pièce horizontal. / ~

wheel with wooden teeth || Kegelrad n. mit Holzzähnen || roue f. conique à denture de bois.

bevelled || abgeschrägt || biseauté; chanfreiné. / ~ (Bevel-angled) || schiefwinklig; schräg || biais; conique. / ~ track section (Railw) || Trapezjoch n. || châssis m. trapéziforme. / ~ wheel || Kegelrad n. || roue f. conique.

bevelling (Shipb) || Schmiege f. || angle m. d'équerrage.

bevelling board (Join) || Gehrungsstoßlade f. || machine f. à onglet. / ~ (Shipb) || Schmiegbrett n. || planchette f. d'équerrage.

bevelling machine || Zuschärfmaschine f. || machine f. à chanfreiner. / nut ~ || Mutternabkantmaschine f. || machine f. à chanfreiner les écrous. / ~ for pasteboard || Pappenabschrägmaschine f. || machine f. à biseauter le carton. / ~ for ship beams || Schmiegemaschine f. für Schiffsspanten || machine f. à équerrer les couples.

bevelling pin (Shipb) || Schmiegkolben m. || taquet m. d'équerrage.

bevel rule see bevel. / ~ square see bevel.

bevel wheel || Kegelrad n.; konisches Rad n. || roue f. conique ou d'angle. / pair of ~s || Kegelräderpaar n. || pair f. de roues coniques.

bevel wheel planing machine || Kegelradhobelmaschine f. || raboteuse f. à tailler des pignons coniques. / ~ toothing || Kegelradverzahnung f. || denture f. de roues coniques.

beverage || Getränk n. || boisson f. / distilled ~ || destilliertes Getränk n. || boisson f. distillée. / fermented ~ || gegorenes Getränk n. || boisson f. fermentée. / non-alcoholic ~ || alkoholfreies Getränk n. || boisson f. non-alcoolique ou antialcoolique. / trade of ~s || Getränkeindustrie f. || industrie f. des boissons.

bevil see bevel.

beware! || Achtung! Obacht! || gare! attention! / ~! (Packing) || Vorsicht! zerbrechlich! || fragile!

bezel (Jewel) || Fassung f. || chaton m.

bhang || Haschisch n. || hachi(s)ch m.

biatomic || zweiatomig || biatomique.

biaxial || zweiachsig || biaxe; à deux axes mpl. / ~ crystal || zweiachsiger Kristall m. || cristal m. biaxe. / ~ telescope with reversible spirit levels || biaxiales Fernrohr n. mit Wendelibelle || lunette f. biaxiale munie d'une libelle réversible.

biaxiality || Zweiachsigkeit f. || biaxie f.

bib || Latz m. || corset m.; bavette f.

bibasic || zweibasisch || bibasique.

biborate of soda || Borax m. || borax m.; borate m. de soude.

bicarbonate || zweifach kohlensaures Salz n.; Bikarbonat n. || bicarbonate m. / ~ of potassium || doppelkohlensaures Kali n.; Kaliumbikarbonat n. || bicarbonate m. de potasse. / ~ of soda || doppeltkohlensaures Natron n. || bicarbonate m. de soude.

bichromate || Bichromat n. || bichromate m. / ~ of potassium || doppeltchromsaures Kali n.; Kaliumbichromat n. || bichromate m. de potasse.

bichromate battery || Chromsäureelement n. || élément m. au bichromate de potasse.

bichromate gelatine || Chromgelatine f. || gélatine f. chromatée.

Bickford fuse || Zündschnur f. || mèche f. de sûreté.

bick iron || Sperrhorn n.; Bankhorn n. || enclume f. bigorne; bigorne f.

biconcave || bikonkav || biconcave.

biconvex || bikonvex || biconvexe.

bicycle || Zweirad n.; Fahrrad n. || bicyclette f.; cycle m.; vélocipède m. / ~ for children || Kinderfahrrad n. || bicyclette f. pour enfants. / folding ~ || zusammenlegbares Fahrrad n. || bicyclette f. pliante. / lady's ~ || Damenrad n. || bicyclette f. de dame. / man's ~ || Herrenrad n. || bicyclette f. d'homme.

bicycle accessories pl. || Fahrradzubehörteile mpl. || accessoires mpl. pour bicyclettes. / ~ chain || Fahrradkette f. || chaîne f. à bicyclette. / ~ crane || Velozipedkran m. || grue-vélocipède f. / ~ frame || Fahrradrahmen m. || cadre m. de bicyclette. / ~ handle || Fahrradgriff m. || poignée f. à bicyclette. / ~ lamp || Fahrradlaterne f. || lanterne f. de bicyclette. / ~ lubricator || Fahrradöler m. || graisseur m. pour bicyclettes. / ~ motor || Fahrradmotor m. || moteur m. de bicyclette. / ~ pump || Fahrradpumpe f. || pompe f. de bicyclette. / ~ saddle || Fahrradsattel m. || selle f. de bicyclette. / ~ saddle making machine || Fahrradsättelherstellungsmaschine f. || machine f. à fabriquer les selles de bicyclettes. / ~ spoke || Fahrradspeiche f. || rayon m. de bicyclette. / ~ tyre || Fahrrad(luft)-schlauch m.; Fahrradpneumatik m. || pneumatique f. de bicyclette.

bid, to || auffordern || inviter; engager. / ~ over || überbieten || renchérir; surenchérir.

bid || Gebot n.; Angebot n. || offre f.; première offre f.; enchère f. / higher ~ || Übergebot n. || surenchère f. / highest ~ || Meistgebot n.; Höchstgebot n. || dernière enchère f.; offre f. la plus élevée.

bier || Bahre f.; Totenbahre f. || bière f.

bifilar || bifilar || bifilaire.

bifilar magnetometer || Bifilarmagnetometer n. || magnétomètre m. bifilaire. / ~ suspension || Doppelfadenaufhängung f. || suspension f. bifilaire. / ~ winding || Bifilarwicklung f. || enroulement m. bifilaire.

bifilary (Phys) || Bifilare f. || bifilaire m.

bifocal glass || Zweistärkenbrille f.; Bifokalbrille f. || lunette f. à double foyer. / break in the vision of a ~ || Bildsprung m. in einer Zweistärkenbrille || saut m. de l'image à la lunette à double foyer.

bifocal lens || Bifokalglas n. || verre m. ou lentille f. à double foyer.

bifurcate, to || gabeln || bifurquer.

bifurcation || Gabelteilung f.; Gabelung f. || fourchure f.; bifurcation f. / ~ station (Tel) || Gabelamt n. || station f. de bifurcation.

bight of a river || Flußkrümmung f.; Rack n. || coude m. ou sinuosité f. d'une rivière. / ~ of a rope (Shipb) || doppelter Part m. eines Taues || double m. d'une manœuvre.

bilateral drive || doppelseitiger Antrieb m. || commande f. des deux côtés.

bilberry || Heidelbeere f.; Blaubeere f. || myrtille f.; airelle f. / evaporated bilberries pl. || getrocknete Heidelbeeren fpl. || airelles fpl. sèches.

bilberry wine || Heidelbeerwein m. || syrop m. de myrtilles.

bilge (Shipb) || Bilge f.; Flach n.; Kimm m. || sentine f.; bouchain m. / ~ block (Shipb) || Stapelklotz m. || savate f. / ~ coads pl. of the cradle || Schlitten-

balken mpl. *oder* Schlittenkufen fpl. des Ablaufgerüstes ‖ coites fpl. courantes; couettes fpl.; anguilles fpl. du ber. / ~ keel ‖ Schlingerkiel m.; Kimmkiel m.; Seitenkiel m. ‖ quille f. latérale *ou* auxiliaire *ou* de dérive. / ~ keelson ‖ Seitenkielschwein n. ‖ carlingue f. des bouchains. / ~ pump ‖ Lenzpumpe f.; Sodpumpe f.; Bilgepumpe f. ‖ pompe f. de cale. / ~ strake ‖ Kimmgang m.; Kimmungsplanke f. ‖ virure f. de bouchain. / ~ water ‖ Schlagwasser n. ‖ eau f. de fond de cale. / ~ ways pl. *see* ~ coads.

bill, to ‖ registrieren; einordnen ‖ enregistrer.

bill ‖ Schein m.; Bescheinigung f.; Zettel m.; Quittung f. ‖ billet m.; certificat m.; attestation f.; quittance f. / ~ (Account) ‖ Rechnung f.; Nota f.; Faktura f.; Quittung f. ‖ compte m.; calcul m.; note f.; mémoire m.; facture f. / ~ (Earthw; Agr) ‖ Erdhacke f.; Erdhaue f.; Karst m.; Hacke f.; Spitzhacke f. ‖ houe f.; hoyau m.; pic m.; pioche f. / ~ (Garden) *see* gardener's ~. / ~ (Law) ‖ Schriftstück n.; Rechtsschrift f. ‖ écrit m.; pièce f.; document m. / ~ (Parliament) ‖ Gesetzentwurf m.; Gesetzvorlage f.; Bill f. ‖ projet m. de loi; bill m. / ~ (Placard) ‖ Anschlagzettel m.; Anschlag m.; Aushängezettel m. ‖ affiche f.; placard m.; écriteau m. / ~ (Programme) ‖ Programm n.; Theaterzettel m. ‖ programme m.; prospectus m. / ~ (Trade) *see also* bill of exchange ‖ Wechsel m. ‖ lettre f. de change. / ~ (Zoology) ‖ Schnabel m. ‖ bec m. / ~ of the anchor ‖ Ankerspitze f.; Ankerklaue f. ‖ pointe f. *ou* bec m. de l'ancre. / annual ~ ‖ Jahresrechnung f. ‖ compte m. annuel. / ~ of carriage *see* ~ of freight. / commercial ~ *see* trade ~. / ~ of cost(s) ‖ Kostenrechnung f. ‖ compte m. des frais. / counter ~ (Trade) ‖ Gegenwechsel m. ‖ rechange m.; retraite f.; retour m. / ~ of the course of exchange ‖ Kurszettel m. ‖ bulletin m. *ou* cote f. de la bourse. / customer's ~ (Trade) ‖ Kundenwechsel m. ‖ billet m. donné en payement par un client. / ~ for debt ‖ Schuldschein m.; Schuldverschreibung f. ‖ billet m.; obligation f.; reconnaissance f. / ~ of delivery ‖ Liefer(ungs)schein m. ‖ reçu m. / ~ of deposit ‖ Depositenschein m. ‖ récépissé m. d'un montant déposé. / ~ of entry ‖ Einfuhrschein m.; Zolleinfuhrschein m. ‖ congé m. *ou* permis m. d'entrée.

bill of exchange ‖ Wechsel m.; Tratte f. ‖ lettre f. de change; traite f.; papier m.; effet m.; disposition f. / to accept a ~ ‖ einen Wechsel m. annehmen *oder* akzeptieren ‖ accepter un effet; honorer *ou* accueillir une traite. / to cash a ~ ‖ einen Wechsel m. einlösen *oder* auslösen *oder* honorieren ‖ acquitter *ou* racheter *ou* honorer une lettre de change. / commercial ~ *see* bill, trade. / counter ~ *see* bill, counter. / customer's ~ *see* bill, customer's. / discounted ~ ‖ diskontierter Wechsel m. ‖ effet m. escompté. / to draw a ~ ‖ einen Wechsel m. ausstellen *oder* ausschreiben *oder* ziehen ‖ tirer une lettre de change. / drawn ~ ‖ gezogener Wechsel m.; Tratte f. ‖ traite f. / ~ drawn in favour of … ‖ Wechsel m. ausgestellt für … ‖ effet m. tiré au nom de … / ~ drawn on … ‖ Wechsel m.

gezogen auf … ‖ effet m. tiré sur … / ~ falling due on … ‖ Wechsel m., fällig am …'‖ effet m. payable le … / first ~ ‖ Primawechsel m. ‖ première f. de change; première lettre f. de change. / foreign ~ *see* bill, foreign. / guarantee ~ *see* bill, guarantee. / to honour a ~ *see* to cash a ~. / ~ payable at … *see* ~falling due on … / to present a ~ for acceptance ‖ einen Wechsel m. zum Akzept vorzeigen *oder* vorlegen ‖ présenter une traite à l'acception. / to renew a ~ ‖ einen Wechsel m. verlängern *oder* prolongieren *oder* erneuern ‖ prolonger *ou* atermoyer une lettre de change. / renewal of a ~ ‖ Prolongation f. *oder* Verlängerung f. eines Wechsels ‖ renouvellement m. *ou* prolongation f. d'une lettre de change. / ~ at sight ‖ Sichtwechsel m. ‖ effet m. *ou* billet m. à vue. / ~ at x days sight ‖ Wechsel m. auf x Tage Sicht ‖ billet m. *ou* traite f. à x jours de vue. / to sign a ~ ‖ einen Wechsel m. unterschreiben *oder* unterzeichnen ‖ souscrire une lettre de change. / sole ~ *see* bill, sole. / ~ to square ‖ Ausgleichswechsel m. ‖ appoint m. / trade ~ *see* bill, trade.

bill of expenses incurred ‖ Spesenrechnung f. ‖ compte m. des frais. / ~ of fare ‖ Speisekarte f. ‖ carte f.; menu m. / foreign ~ (Trade) ‖ Ausland(s)wechsel m. ‖ lettre f. de change sur l'étranger. / ~ of a fount ‖ Schriftzettel m. ‖ police f. pour une fonte de caractères. / ~ of freight ‖ Frachtbrief m.; Frachtschein m. ‖ lettre f. de voiture; connaissement m. / gardener's ~ ‖ Hippe f.; Gärtnerhippe f.; (gekrümmtes) Gartenmesser n. ‖ serpe f. de jardinier. / guarantee ~ ‖ Kautionswechsel m. ‖ cautionnement m. / ~ of health ‖ Gesundheitspaß m. ‖ brevet m. de santé. / ~ of indictment ‖ Anklageschrift f. ‖ acte f. d'accusation. / ~ of invoice ‖ Einkaufsrechnung f. ‖ compte m. d'achat. / ~ of lading ‖ Konnossement n.; Frachtbrief m.; Verladeschein m. ‖ connaissement m.; police f. *ou* certificat m. de chargement. / ~ of material ‖ Stückliste f.; Stückverzeichnis n. ‖ spécification f. / sickle ~ ‖ Hippe f.; serpillon f. / sole ~ (Trade) ‖ Solawechsel m. ‖ seule f. de change; seule lettre f. de change. / ~ of tonnage (Shipb) ‖ Meßbrief m. ‖ certificat m. de jaugeage *ou* du tonnage. / trade ~ ‖ Kundenwechsel m.; Warenwechsel m. ‖ papier m. commercial; effet m. de commerce. / vine ~ ‖ Winzerhippe f. ‖ serpe f. de vigneron. / way ~ *see* ~ of freight. / ~ of weight ‖ Gewichtsnota f. ‖ note f. du poids. / wood-cutter's ~ ‖ Holzhauerhippe f. ‖ serpe f. de bûcheron.

bill bearer ‖ Zettelverteiler m. ‖ porteur m. d'affiches. / ~ broker ‖ Wechselagent m.; Wechselmakler m. ‖ agent m. de change. / ~ discount ‖ Wechseldiskont m. ‖ escompte m. des effets. / ~ discounter ‖ Wechseldiskontierer m. ‖ banquier m. d'escompte.

billet ‖ Holzscheit n. ‖ bûche f. / ~ (Met) ‖ Rohschiene f.; Knüppel m. ‖ barre f. brute; billette f. / the steel enters the market in the shape of ~s ‖ der Stahl m. wird in Form von Knüppeln in den Handel gebracht ‖ l'acier m. se vend en billettes. / ~ of wood ‖ Holzscheit n. ‖ bûche f.

billet boring machine ‖ Knüppelbohrmaschine f. ‖ foreuse f. pour billettes. / ~ mill ‖ Knüppelwalzwerk n. ‖ laminoir m. à billettes. / ~ press to reduce scrap bundles ‖ Blockpresse f. zum Nachpressen von Schrottpaketen ‖ presse f. à blocs à réduire les balles de mitrailles. / ~ roll ‖ Vorwalze f. ‖ cylindre m. ébaucheur. / ~ rounds pl. ‖ Knüppelholz n. ‖ rondins mpl. / ~ saw ‖ Scheitersäge f. ‖ scie f. à bûches. / ~ shearing machine ‖ Knüppelschere f. ‖ cisaille f. à billettes. / ~ shears pl. *see* ~ shearing machine. / ~ web ‖ Schüttersäge f. ‖ scie f. à bûches. / ~ wood *see* ~ rounds.

billhook ‖ Hippe f. ‖ serpe f.; serpette f. / Vosges pattern ~ ‖ Hippe f. in Wasgenwaldform ‖ serpe f. façon vosges.

billiard ‖ Billard n. ‖ billard m. / ~ ball ‖ Billardball m.; Billardkugel f. ‖ bille f. de billard. / ~ chalk ‖ Billardkreide f. ‖ craie f. de billard. / ~ cloth ‖ Billardtuch n. ‖ drap m. de billard. / ~ cue ‖ Billardqueue n. ‖ queue f. de billard. / ~ cushion ‖ Billardbande f. ‖ bande f. de billard. / ~ joiner ‖ Billardschreiner m. ‖ menuisier m. en billards. / ~ requisites pl. ‖ Billardzubehör n. ‖ accessoires mpl. pour billards. / ~ table ‖ Billardtisch m.; Billard n. ‖ billard m. / ~ table maker ‖ Billardtafelmacher m. ‖ billardier m. / ~ table varnisher ‖ Billardtafellackierer m. ‖ vernisseur m. de billards. / ~ trimmer ‖ Billardgarnierer m. ‖ garnisseur m. de billards. / ~ works pl. ‖ Billardfabrik f. ‖ fabrique f. de billards.

bill jobber ‖ Börsenmakler m. ‖ agent m. de change.

billon ‖ Scheidemünze f. ‖ billon m.; monnaie f. de billon; petite monnaie f. / ~ of silver ‖ Pagament n. ‖ mélange m.; alliage m.

billon silver ‖ Scheidemünzsilber n. ‖ argent m. de billon.

bill printer ‖ Plakatdrucker m. ‖ imprimeur m. d'affiches. / ~ stamp ‖ Wechselstempel m. ‖ timbre m. de change *ou* de traite.

billy (Spinn) ‖ Vorspinnmaschine f. ‖ métier m. *ou* fileuse f. en gros; banc m. à broches.

bimetal ‖ Bimetall n. ‖ bi-métal m.

bimetallic thermometer ‖ Doppelmetallwärmegradmesser m. ‖ thermomètre m. bimétallique.

bimetallism ‖ Doppelwährung f. ‖ bimétallisme m.

bin ‖ Behälter m.; Kasten m.; Kiste f. ‖ réservoir m.; boîte f.; boîtier m.; tonneau m. / ~ (Mill) ‖ Vorratsraum m.; Behälter m. ‖ trémie f.; boisseau m.; casier m.; raw material ~ ‖ Erztasche f. ‖ poche f. à minerais. / storage ~ *see* raw material ~.

binaries pl. ‖ Doppelstern m. ‖ étoile f. double. / resolving very close ~ ‖ Auflösung f. engster Doppelsterne ‖ séparation f. des étoiles doubles très approchées.

binary ‖ binär ‖ binaire. / ~ lens ‖ Zwillingslinse f. ‖ lentille f. double.

bind, to (Bookb) ‖ binden; einbinden ‖ relier. / ~ (Met) ‖ fressen ‖ gripper. / ~ in calf (Bookb) ‖ in Franzband m. binden ‖ relier en veau m. / ~ with iron hoops ‖ mit eisernen Ringen beschlagen ‖ fretter. / ~ with iron work (Build) ‖ mit Eisen n. beschlagen ‖ ferrer; armer. / ~ off (Textile) ‖ ketteln ‖ remmailler; entrelacer.

binder ‖ Bindemittel n. ‖ liant m. / ~ (Agr) ‖ Garbenbinder m. ‖ moissonneuse-lieuse f. / ~ (Wire) ‖ Drahtheftklammer f. ‖ attache-lettre m.; crochet m. / grain ~ ‖ Bindemäher m. ‖ moissonneuse-lieuse f.

binder canvas (Grain binder) ‖ Binder-tuch n. ‖ toile f. de moissonneuse-lieuse. / ~'s press ‖ Heftlade f. ‖ cousoir m.

bindery ‖ Buchbinderei f.; Buchbinder-werkstatt f. ‖ atelier m. de relieur.

bindheimite ‖ Antimonbleispat m. ‖ blei-niérite f.

binding (Band) ‖ Einfassungsborte f. ‖ ruban m. à border. / ~ (Bookb) ‖ Einband m. ‖ reliure f. / ~ by means of cardan shafts ‖ Gelenkwellenverbindung f. ‖ assemblage m. par arbres à cardan. / parchment ~ ‖ Pergamenteinband m. ‖ reliure f. en par-chemin.

binding apparatus ‖ Bandeinfaßapparat m. ‖ appareil m. à border. / ~ attach-ment (Grain binder) ‖ Bindevorrichtung f. ‖ lieur m. / ~ bond ‖ Verbindungsband n. ‖ bande f. d'attache. / ~ clamp ‖ Klemme f.; borne f.; serre-fils m. / ~ edge of the sheet ‖ Einheftkante f. des Blattes ‖ bord m. de la feuille à relier ou où se trouvent les perforations.

binding-in of line wires ‖ Binden n. des Leitungsdrahtes ‖ ligature f. du fil de ligne.

binding joist (Carp) ‖ Hauptbalken m.; Binderbalken m. ‖ maîtresse f. poutre. / ~ leaves pl. (Weav) ‖ Liagekämme mpl. ‖ lisses fpl. de liage. / ~ machine (Agr) ‖ Garbenbinder m. ‖ moissonneuse-lieuse f. / ~ material ‖ Bindemittel n. ‖ matière f. agglutinante.

binding-off machine ‖ Kettelmaschine f. ‖ remmailleuse f.

binding post ‖ Verbindungsklemme f.; Klemmschraube f. ‖ borne f. de con-nexion; serre-fil m. / ~ press for books ‖ Buchbinderpresse f. ‖ presse f. à relier. / ~ recess (Electr) ‖ Bandagenute f. ‖ creux m. de frette ou de bandage. / ~ ring (Office) ‖ Heftring m. ‖ anneau m. de fixation pour les feuillets. / ~ screw ‖ Verbindungsschraube f.; Klemme f.; Klemmschraube f. ‖ boulon m. clavette; borne f.; vis f. de serrage. / ~ table (Moving machine) ‖ Bindetisch m. ‖ table f. de liage. / ~ thread ‖ Bindfaden m.; Spagat m.; Bindegarn n. ‖ ficelle f. d'emballage; ficelle f. d'attache. / ~ threads pl. (Weav) ‖ Bindefäden mpl. ‖ Liage f. ‖ liage m. / ~ wire ‖ Bindedraht m. ‖ fil m. d'archal ou de ligature ou à lier.

binnacle ‖ Kompaßhäuschen n.; Kompaß-nachthaus n. ‖ habitacle m.; gésole f. / ~ light ‖ Kompaßlaterne f. ‖ fanal m. d'habitacle.

binocular ‖ beidäugig; binokular ‖ bin-oculaire. / ~ corneal microscope ‖ bin-okulares Hornhautmikroskop n. ‖ micro-scope m. cornéen binoculaire.

binocular eyepiece ‖ binokulares Okular n. ‖ oculaire m. binoculaire. / astronomical ~ without erecting the image ‖ binokulares Okular n. für astronomische Beobach-tungen ohne Bildumkehrung ‖ oculaire m. binoculaire pour les observations astronomiques sans redressement. / ter-restrial ~ which erects the image into the natural position ‖ binokulares Okular n. für terrestrische Beobachtungen mit Bildumkehrung ‖ oculaire m. binocu-

laire pour les observations terrestres re-dressant l'image.

binocular glass ‖ Fernglas n. ‖ jumelles fpl. / ~ look-out telescope ‖ binokulares Aussichtsfernrohr n. ‖ lunette f. d'ap-proche binoculaire. / ~ observation ‖ zweiäugige Beobachtung f. ‖ observation f. binoculaire. / ~ prism telescope ‖ Dop-pelprismenfernrohr n. ‖ jumelles fpl. à prismes. / ~ telescopic magnifier ‖ bin-okulare Fernrohrlupe f. ‖ téléloupe f. binoculaire. / ~ telescopic magnifiers pl. ‖ Doppelfernrohrlupe f. ‖ téléloupes fpl. binoculaires. / ~ telescopic ophthalmo-scope magnifier ‖ binokulare Ophthal-moskopfernrohrlupe f. ‖ téléloupe f. bin-oculaire de l'ophtalmoscope. / ~ tube ‖ binokularer Tubus m.; Doppeltubus m. ‖ tube m. binoculaire; double-tube m.

binocular ‖ Fernglas n.; Feldstecher m. ‖ binocle m.; lunette f.; jumelles fpl. (de campagne). / prism ~ ‖ Prismenfeld-stecher m. ‖ jumelles fpl. à prismes.

bin wagon ‖ Kübelwagen m. ‖ wagon m. à benne.

biochemical preparation ‖ biochemisches Präparat n. ‖ préparation f. biochimique.

biochemistry ‖ Biochemie f. ‖ biochimie f.

biological cleaning plant for waste water ‖ biologische Abwasserkläranlage f. ‖ in-stallation f. de nettoyage biologique des eaux résiduaires. / ~ phenomena ‖ Le-benserscheinungen fpl. ‖ phénomènes mpl. biologiques.

biotite ‖ schwarzer Glimmer m.; Magnesia-glimmer m. ‖ mica m. magnésien ou noir.

bioxide ‖ Dioxyd n. ‖ bioxyde m. / ~ of man-ganese ‖ Manganhyperoxyd n.; Weich-mangan n.; Graumangan n. ‖ manganèse m. oxydé ou oxydé gris ou oxydé mé-talloïde; peroxyde m. de manganèse.

biplane ‖ Doppeldecker m. ‖ biplan m.

bipolar ‖ zweipolig ‖ bipolaire. / ~ machine ‖ zweipolige Maschine f. ‖ machine f. bi-polaire.

birch ‖ Birkenholz n.; Birkenreisig n. ‖ bouleau m. / ~ bark ‖ Birkenrinde f. ‖ écorce f. de bouleau. / ~ bark oil ‖ Bir-kenrindenöl n. ‖ essence f. d'écorce de bouleau. / ~ broom ‖ Reisigbesen m. ‖ balai m. de bouleau. / ~ bud oil ‖ Bir-kenknospenöl n. ‖ huile f. de boutons de bouleau. / ~ leave ‖ Birkenblatt n. ‖ feuille f. de bouleau. / ~ oil ‖ Birkenöl n. ‖ es-sence f. de bouleau. / ~ tar ‖ Birkenteer m. ‖ goudron m. de bouleau. / ~ tar oil ‖ Birkenteeröl n.; Juchtenöl n. ‖ huile f. de goudron de bouleau. / ~ tree ‖ Birke f. ‖ bouleau m. / ~ wood ‖ Birkenholz n. ‖ bois m. de bouleau.

bird, naturalized ~ ‖ naturalisierter Vogel m. ‖ oiseau m. naturalisé. / preserved ~ ‖ konservierter Vogel m. ‖ oiseau m. con-servé. / stuffed ~ ‖ ausgestopfter Vogel m. ‖ oiseau m. empaillé.

bird-cage ‖ Vogelkäfig m. ‖ cage m. d'oiseau; volière f. / ~ support ‖ Vogel-käfigständer m. ‖ support m. pour cages d'oiseaux.

bird's beak ‖ Vogelschnabel m. ‖ bec m. d'oiseau. / ~ eye view ‖ Ansicht f. aus der Vogelschau ‖ vue f. à vol d'oiseau. / ~ feather ‖ Vogelfeder f. ‖ plume f. d'oiseau. / ~ flight ‖ Vogelflug m. ‖ vol m. des oiseaux. / ~ head ‖ Vogelkopf m. ‖ tête f. d'un oiseau. / ~ nest (Mar) ‖ Krähennest n. ‖ cage f.; échauguette f. / ~ seed ‖ Geflügelfutter n. ‖ nourriture

f. de volaille. / ~ skin ‖ Vogelbalg m. ‖ peau f. d'oiseau; oiseau m. en peau.

biretta ‖ Barett n. ‖ béret m.

biscuit ‖ Biskuit n.; Keks m. ‖ biscuit m. / ~ (Mar) ‖ Schiffszwieback m.; Hartbrot n. ‖ biscuit m. (de mer). / ~ (Porcel) ‖ unglasiertes Porzellan n. ‖ biscuit m. / ship ~ see biscuit (Mar).

biscuit-baking, to give the ~ to porcelain ‖ das Porzellan verglühen ‖ cuire en dé-gourdi la porcelaine.

biscuit factory ‖ Biskuitfabrik f. ‖ fabrique f. de biscuits. / ~ industry ‖ Biskuitindu-strie f. ‖ industrie f. de la biscuiterie.

biscuit-ware (Porcel) ‖ unglasiertes Por-zellan n. ‖ biscuit m. / ~ dipper (Pott) ‖ Porzellanglasierer m. ‖ trempeur m. de biscuit.

bisect, to ‖ halbieren ‖ diviser ou partager en deux (parties égales).

bisector ‖ Halbierende f. ‖ bissectrice f.

bishop, to ~ the balls pl. or rollers pl. (Print) ‖ die Ballen mpl. oder Walzen fpl. schwärzen ‖ encrer les balles fpl., les tampons mpl. ou les rouleaux mpl.

bisilicate ‖ Bisilikat n. ‖ bisilicate m.

bismuth ‖ Wismut n.; Aschblei n. ‖ bis-muth m.; étain m. de glace. / acicular ~ ‖ Belonit m.; Nadelerz n. ‖ bismuth m. sulfuré plumbo-cuprifère.

bismuth mine ‖ Wismuterzbergwerk n. ‖ mine f. de bismuth. / ~ salt ‖ Wismut-salz n. ‖ sel m. de bismuth. / ~ silicate ‖ Kieselwismut m.; Wismutblende f.; Eulytin m. ‖ eulytine f.; bismuth m. silicaté. / ~ subnitrate ‖ Wismutweiß n. ‖ blanc m. de fard ou de perle.

bissextile year ‖ Schaltjahr n. ‖ année f. intercalaire ou bissextile.

bisulphate ‖ Bisulfat n. ‖ bisulfate m. / ~ of potash ‖ doppeltschwefelsaures Ka-lium n. ‖ bisulfate m. de potasse. / ~ of soda ‖ saures schwefelsaures Natron n. ‖ bisulfate m. de soude.

bisulphide of carbon ‖ Schwefelkohlenstoff m. ‖ sulfure m. de carbone. / ~ of potas-sium ‖ zweifachschwefelsaures Kalium n. ‖ bisulfure m. de potassium.

bisulphite ‖ Bisulfit n. ‖ bisulfite m.

bit ‖ Stück n.; Stückchen n. ‖ morceau m.; fragment m.; pièce f.; tranche f. / ~ (Cutting edge of tools) ‖ Schneide f. (von Werkzeugen) ‖ tranchant m.; tail-lant m.; fil m.; coupant m. / ~ (Drill) ‖ Bohreisen n.; Bohrspitze f.; Bohreinsatz m.; Bohrer m. ‖ foret m.; mèche f. / ~ (Key) ‖ Schlüsselbart m. ‖ panneton m. / ~ (Mine) ‖ Steinbohrer m.; Bergbohrer m. ‖ trépan m.; bit m.; fleuret m. / ~ (Plane) ‖ Hobeleisen n. ‖ fer m. de rabot. / ~ (Rein; Bridle) ‖ Gebiß n. ‖ Pferde-gebiß n. ‖ Kandare f. ‖ mors m. / ~ (Soldering copper) ‖ Kolben m.; Löt-kolbenkopf m. ‖ cuivre m.; porte-goutte m. / ~ (Vice; Tongs) ‖ Backe f.; Maul n. ‖ mors m.; bouche f.; mâchoire f.

bit of an axe ‖ Schneide f. einer Axt ‖ tranchant m. d'une hache. / bore ~ see ~ of a drill. / ~ of the bridle ‖ Mund-stück n. oder Gebiß n. des Zaumes ‖ embouchure f. de la bride. / center ~ ‖ Zentrum(s)bohrer m. ‖ foret m. ou mèche f. à centre; foret m. à trois pointes. / ~ of a chisel ‖ Schneide f. eines Meißels ‖ tranchant m. d'un ciseau. / common ~ ‖ Spitzbohrer m. ‖ foret m. à langue d'aspic. / double-cutting ~ ‖ zweischnei-diger Bohrer m. ‖ mèche f. à deux tran-

chants. / ~ of a drill ‖ Schneide f. eines Bohrers; Bohrschneide f. ‖ mèche f. de foret. / duck-nose ~ ‖ halbelliptischer Löffelbohrer m. ‖ mèche f. à cuiller creusée en gouge. / earth-boring ~ ‖ Erdbohrer m. ‖ sonde f.; tarière f. pour le sol ou à trépan. / finishing ~ ‖ Schlichtbohrer m. ‖ alésoir m. / ~ with four wings (Stone-pick) ‖ Steinhaue f.; Steinbohrer m. ‖ bonnet m. carré ou de prêtre; percemeule m.; casse-pierres m. / ground ~ see earth-boring ~. / half-turn twist ~ see half-twist ~. / half-twist ~ ‖ Schneckenbohrer m. ‖ tarière f. hélicoïdale ou mitorse. / ~ of a knife ‖ Schneide f. eines Messers ‖ coupant m. d'un couteau. / pianomaker's ~ ‖ Klavierbauerbohrer m. ‖ mèche f. pour facteur de pianos. / plough ~ ‖ starkes Nuthobeleisen n. ‖ fer m. du bouvet à approfondir renforcé. / screw ~ with helical cutters ‖ Schneckenbohrer m. ‖ mèche f. torse à hélice. / sharp-pointed ~ (Pyrot) ‖ Spitzbohrer m. ‖ égravoir m. / square ~ ‖ vierschneidiger Kronenbohrer m. ‖ fleuret m.; perçoir m. à couronne. / ~ of tongs ‖ Zangenmaul n. ‖ bouche f. d'une tenaille. / tube ~ ‖ Kanonenbohrer m. ‖ mèche f. à canons. / twist ~ ‖ Spiralbohrer m. ‖ foret m. hélicoïdal; mèche f. torse. / twisted-eye ~ ‖ Öhrbohrer m. ‖ tarière f. à douille.

bitartrate ‖ zweifach oder doppeltweinsaures Salz n.; Bitartrat n. ‖ bitartrate m.; tartrate m. acide. / ~ of potash ‖ doppeltweinsaures Kali n.; Weinstein m. ‖ bitartrate m. de potasse; crème m. de tartre; tartre m.

bit-brace see bit-stock.

bit-bridle ‖ Stangenzaun m.; Stangengebiß n.; Kandare f. ‖ mors m. de bride.

bite, to (Mach) ‖ fassen; greifen; eingreifen; einschneiden ‖ mordre; prendre. / ~ (Met) ‖ beizen; ätzen; zerfressen; angreifen ‖ corroder; attaquer; mordre. / ~ (Print) ‖ einschneiden ‖ inciser.

bite ‖ Greifen n.; Fassen n. ‖ mordant m. / ~ (Fishing) ‖ Köder m. ‖ amorce f.; appât m.

bit-holder ‖ Bohrhalter m. ‖ porte-mèche m.; porte-foret m.

biting ‖ scharf; beißend; schneidend ‖ mordant; pénétrant.

bit-mouth see bit (Rein; Bridle).

bit-stock ‖ Brustbohrer m.; Brustleier f.; Faustleier f. ‖ vilebrequin m.

bitten see biting.

bitter ‖ bitter ‖ amer.

bitter almond oil ‖ Bittermandelöl n. ‖ essence f. d'amandes amères. / ~ soap ‖ Bittermandelseife f. ‖ savon m. d'amandes amères. / ~ water ‖ Bittermandelwasser n. ‖ eau f. d'amandes amères.

bitter cordial see bitters. / ~ earth ‖ Bittererde f.; Magnesia f. ‖ magnesie f. / ~ orange peel ‖ Pomeranzenschale f. ‖ écorce f. d'oranges amères.

bitter principle ‖ Bitterstoff m. ‖ principe m. amer; amer m. / to eliminate the ~s pl. ‖ entbittern ‖ enlever les principes mpl. amers.

bitters pl. ‖ Magenbitter m. ‖ bitter m.; liqueur f. stomachique. / ~ manufacturer ‖ Magenbitterfabrikant m. ‖ fabricant m. de bitter.

bitter salt ‖ Bittersalz n. ‖ sel m. amer; sulfate m. de magnésie. / ~ water see bitter water.

bitter spar ‖ Bitterspat m.; Rautenspat m. ‖ chaux f. carbonatée magnésifère; dolomie f. / ~ water ‖ Bitterwasser n. ‖ eau f. amère ou magnésifère. / ~ wood ‖ Quassiaholz n.; Bitterholz n. ‖ bois m. de quassia amara.

bitts pl. (Shipb) ‖ Beting f.; Mastknecht m. ‖ bitte f.; biton m.

bitumen ‖ Bitumen n.; Asphalt m.; Erdpech n. ‖ bitume m.; asphalte m. / compact ~ ‖ Asphalt m.; Erdpech n.; Judenpech n. ‖ asphalte m.; bitume m. solide; goudron m. minéral; bitume m. de Judée. / contents pl. of ~ in pit coal ‖ Bitumengehalt m. der Steinkohle ‖ teneur f. en produits bitumineux de la houille. / soft ~ ‖ weiches Bitumen n. ‖ bitume m. mou. / solid ~ ‖ festes Erdpech n. ‖ bitume m. solide ou dur.

bitumen pavement ‖ Bitumendecke f. ‖ revêtement m. bitumineux. / ~ road ‖ Bitumenstraße f. ‖ rue f. bitumée. / ~ work ‖ Asphaltarbeit f. ‖ pavage m. en asphalte.

bituminate, to ‖ mit Asphalt m. bestreichen ‖ bituminer.

bituminous (Coal) ‖ fett; bituminös ‖ bitumineux; gras. / ~ coal ‖ Fettkohle f. ‖ houille f. bitumineuse ou grasse. / non-~ coal ‖ Magerkohle f. ‖ charbon m. maigre. / ~ lignite ‖ ölreiche Braunkohle f. ‖ lignite m. riche en huile. / ~ material ‖ bituminöser Rohstoff m. ‖ matière f. bitumineuse. / ~ pitch ‖ Braunkohlenteerpech n.; Asphaltpech n. ‖ poix f. d'asphalte; poix f. de goudron de lignite. / ~ rock ‖ bituminöses Gestein n. ‖ roche f. bitumineuse. / ~ slate ‖ Brandschiefer m. ‖ schiste m. bitumineux. / ~ slate extraction ‖ Brandschiefergewinnung f. ‖ extraction f. de schiste bitumineux. / ~ tar ‖ Braunkohlenteer m. ‖ goudron m. de lignite.

bitumized road ‖ asphaltierte Straße f. ‖ route f. bitumée.

bivalent ‖ zweiwertig ‖ bivalent.

black, to ‖ schwärzen ‖ noircir. / ~ the mould (Found) ‖ die Form stäuben; die Form mit Kohlenstaub bestreuen ‖ saupoudrer le moule. / ~ the balls (Print) ‖ die Ballen mpl. schwärzen ‖ encrer les balles fpl. ou les tampons mpl. / ~ the loam-mould ‖ die Lehmform f. schlichten oder schwärzen ‖ noircir le moule de terre. / ~ the rollers (Print) ‖ die Walzen fpl. schwärzen ‖ encrer les rouleaux mpl. / ~ shoes pl. ‖ Stiefel mpl. wichsen oder putzen ‖ cirer ou brosser des chaussures fpl. / ~ the inside of the telescope ‖ das Innere des Fernrohres matt schwärzen ‖ noircir en mat l'intérieur de la lunette.

black ‖ schwarz ‖ noir. / ~ (Crude iron) ‖ übergar ‖ lamailleux.

black ‖ Schwarz n.; schwarze Farbe f. ‖ noir m. / ~ (Blackening) ‖ Schwärze f. ‖ noir m. / animal ~ ‖ tierisches Schwarz n. ‖ noir m. animal. / burnt ~ ‖ gebrannter Kienruß m. ‖ noir m. de fumée calciné. / deep ~ ‖ Tiefschwarz n.; schweres Schwarz n. ‖ noir m. foncé ou chargé. / Frankfort ~ see German ~. / German ~ ‖ Frankfurter Schwarz n. ‖ noir m. d'Allemagne. / heavy ~ see deep ~. / mineral ~ ‖ Mineralschwarz n. ‖ noir m. minéral. / Parisian ~ ‖ Pariser Schwarz n. ‖ noir m. de Paris. / vegetable ~ ‖ pflanzliches Schwarz n. ‖ noir m. végétal. / vine ~ ‖ Rebenschwarz n. ‖ noir m. de vigne.

black-alder bark ‖ Faulbaumrinde f. ‖ écorce f. de bourdaine. / ~ berry ‖ Faulbaumbeere f. ‖ baie f. de bourdaine. / ~ tree ‖ Faulbaum m. ‖ bourdaine f.

black amber ‖ Fuchsambra m. ‖ ambre m. noir. / ~ ball (Dyer) ‖ Schwarzkugel f. ‖ boule f. de noir. / ~ balls pl. (Chem) ‖ rohe Soda f.; Rohsoda f. ‖ soude f. brute.

black-band (Miner) ‖ Kohleneisenstein m. ‖ black-band m.; clavai m.

blackbat (Miner) ‖ Brandschiefer m. ‖ pyroschiste m.; schiste m. bitumineux.

blackberry ‖ Brombeere f. ‖ mûre f. de ronce; mûre f. sauvage. / ~ leaves pl. ‖ Brombeerblätter npl. ‖ feuilles fpl. de ronce.

blackboard ‖ Schulwandtafel f. ‖ tableau m. noir ou pour écoles.

black bulb thermometer ‖ Schwarzkugelthermometer n. ‖ thermomètre m. à boule noire ou noircie. / ~ cinder (Blast furnace) ‖ Rohschlacke f.; Hochofenschlacke f. ‖ laitier m. de la fonte terne. / ~ colours pl. ‖ Schwarzfarben fpl. ‖ couleurs fpl. noires.

black-currant ‖ schwarze Johannisbeere f. ‖ cassis m.

black-dye ‖ Schwärze f. ‖ couleur f. noire; noir m. colorant. / ~ for leather ‖ Lederschwärze f. ‖ noir m. pour cuir.

black-edge ‖ Trauerrand m. ‖ cadre m. noir.

blacken, to see to black.

black-enamelled ‖ schwarz emailliert ‖ émaillé noir.

blackening see blacking.

blacker (Shoem) ‖ Schwarzfärber m. ‖ metteur m. au noir.

black-fish ‖ Sepia f.; Tintenfisch m. ‖ sèche f.; seiche f.; cornet m.; calmar m.

blacking ‖ Schwärzen n.; Schwärzung f. ‖ noircissement m.; noircissage m. / ~ (Matter) ‖ Schwärze f.; Schlichte f. ‖ noir m.; enduit m. / ~ (For shoes) ‖ Schwärze f.; Wichse f.; Schuhwichse f. ‖ cirage m. / ~ of iron pieces ‖ Schwärzen n. von Eisenteilen ‖ noircissage m. des pièces de fer. / ~ of the moulds ‖ Schwärzen n. der Lehmformen ‖ noircissement m. des moules. / ~ for shoes see blacking.

blacking brush ‖ Wichsbürste f.; Schuhbürste f. ‖ brosse f. à cirer. / ~ machine for iron pieces ‖ Schwärzmaschine f. für Eisenteile ‖ machine f. à noircir les pièces de fer. / ~ swab sprayer (Found) ‖ Schwärzeverteiler m. ‖ pinceau m. à noircir.

black iron mica ‖ Schwarzstein m. ‖ pierre f. noire; roche-noire f. / ~ japan see ~ varnish.

black-lead ‖ Graphit m.; Pottlot n.; Reißblei n. ‖ graphite m.; plombagine f.; crayon m. de mine. / ~ crucible ‖ Reißbleitiegel m. ‖ creuset m. de plombagine; creuset en graphite. / ~ melting pot see ~ crucible.

black-leading machine ‖ Graphitiermaschine f. ‖ machine f. à graphiter.

blackleg ‖ Streikbrecher m. ‖ briseur m. de grève. / ~s pl. ‖ Klauenseuche f. ‖ piétin m.

black letter ‖ gotische Schrift f.; Gotisch f. ‖ caractères mpl. gothiques ou allemandes. / ~ letter-press ink ‖ Schwärze f.; Druckerschwärze f. ‖ encre f. noire d'imprimerie.

black malt ‖ Farbmalz n. ‖ farbmalz m.; malt m. grillé ou torréfié. / ~ cooling sieve ‖ Farbmalzkühlsieb n. ‖ tamis m. à

refroidir le malt torréfié. / ~ extract ||
Farbmalzauszug m. || extrait m. de malt
torréfié. / ~ roasting drum || Farbmalz-
röstmaschine f. || torréfacteur m. à malt.
/ ~ succedaneum || Farbmalzsurrogat n.
|| succédané m. du malt torréfié.

black mould || Humus m.; Gartenerde f. ||
humus m.; terre f. végétale. / ~ nickel
bath || Schwarznickelbad n. || bain m.
de nickelage noir. / ~ plate see ~ sheet.

black-pot attendant (Petroleum) || Öl-
kesselwärter m. || charbonnier m. cui-
seur.

black salt || rohe oder schwarze Pottasche
f.; Pottaschenfluß m.; Ochras m. || salin
m. / ~ sheet || Schwarzblech n. || tôle f.
noire. / ~ sheet-iron see ~ sheet. / ~ silver
|| Melanglanz m.; Stephanit m.; Schwarz-
gültigerz m. || argent m. noir; argent m.
sulfuré fragile; argent m. antimonié sulfuré noir. / ~ silver-glance see ~ silver.

blacksmith || Hammerschmied m.; Grob-
schmied m.; Hufschmied m. || maréchal
m. ferrant; forgeron m. / carriage ~ ||
Wagenschmied m. || forgeron m. en voi-
tures.

blacksmith's coal || Schmiedekohle f. || char-
bon m. de forge. / ~ work || Grob-
schmiedearbeit f. || taillanderie f.; ou-
vrage m. du forgeron.

black squall || Gewitterbö f. || grain m. noir.
/ ~ varnish || Asphaltlack m.; Asphalt-
firnis m. || vernis m. noir ou d'asphalte;
laque f. à l'asphalte.

black-wash, to ~ see to black.

black wash (Found) || Schlichte f.; Schwärze
f. || enduit m. noir. / ~ work see black-
smith's work.

bladder, animal || tierische Blase f. || vessie
f. animale.

bladder green || Saftgrün n.; Blasengrün n.
|| vert m. de vessie ou de sève ou d'iris.

bladdery fermentation || Blasengärung f. ||
fermentation f. bulleuse.

blade || Blatt n. || feuille f. / ~ (Clothm) ||
Schermesser n.; Tuchscherblatt n. ||
couteau m.; feuille f. des forces; lame f.
/ ~ (Knife) || Klinge f. || lame f. / ~ (Railw)
|| Weichenzunge f. || aiguille f. de change-
ment de voie. / ~ (Saw) || Blatt n. || lame f.
/ ~ (Stirrer) || Flosse f. || lamelle f. / ~
(Turbine) || Turbinenschaufel f.; Schaufel
f. || aube f. / ~ of a chaff cutter || Häcksel-
messer n. || lame f. de coupe-paille. /
guide ~ || Leitschaufel f. || pale f. direc-
trice. / ~ with handle || Klinge f. mit
Griff || lame f. à poignée. / hollow ground
~ || hohlgeschliffene Klinge f. || lame f.
evidée ou à gouttières. / immovable ~
(Cloth shearing machine) || Lieger m.;
Kontermesser n. || contre-couteau m.;
femelle f. d'une tondeuse. / the ~s pl.
are interchangeable (Turbine) || die Tur-
binenschaufeln fpl. lassen sich auswech-
seln (sind auswechselbar) || les aubes fpl.
sont interchangeables. / knife ~ || Messer-
klinge f. || lame f. de couteau. / lower ~
(Cloth) see immovable ~. / narrow ~ ||
schmale Klinge f. || lame f. étroite. / oar ~
|| Ruderblatt n.; Ruderschaufel f. || pale
f. ou pelle f. d'aviron. / ~ for pencil shar-
peners || Klinge f. für Bleistiftspitzer ||
lame f. de taille-crayons. / ~ of a plough ||
Pflugschar f.; Schar f. || soc m. de char-
rue. / ~ of a propeller || Propellerflügel m.
|| pale f. d'hélice. / razor ~ || Rasiermesser-
klinge f. || lame f. de rasoir. / reserve ~ ||
Ersatzklinge f. || lame f. de rechange. /

reversing ~ (Turbine) || Umkehrschaufel
f. || aube f. d'inversion (de flux). / sabre ~
|| Säbelklinge f. || lame f. de sabre. / saw ~
|| Sägeblatt n.; Sägeklinge f. || lame f.
ou feuille f. de scie. / scissors ~ || Scheren-
klinge f. || lame f. de ciseaux. / screw ~
(Shipb) || Schraubenflügel m. || pale f.
ou aile f. d'hélice. / set of ~s (Turbine) ||
Schaufelsatz m. || jeu m. d'aubes. / ~ of
shears || Scherblatt n. || lame f.; tran-
chant m.; mâchoire f. / ~ of a shovel ||
Schaufelblatt n. || pale f. de pelle. / steam
turbine ~ || Dampfturbinenschaufel f. ||
aube f. de turbines à vapeur. / switch ~
(Electr) || Schaltermesser n. || lame f.
d'interrupteur. / sword ~ || Fechtklinge
f.; Schwertklinge f. || lame f. (d'épée).

blade file || Spaltfeile f. || lime f. à clef;
lime f. à refendre. / ~ form || Blattform f.
|| forme f. de lame. / ~ pitch (Turbine) ||
Schaufelteilung f. || pas m. de l'aubage. /
~ section (Turbine) || Schaufelquerschnitt
m. || profil m. de l'aube. / set of ~s (Tur-
bine) || Schaufelsatz m. || jeu m. d'aubes.
/ ~ switch (Electr) || Messerschalter m. ||
interrupteur m. à couteau. / ~ wheel
(Turbine) || Schaufelrad n.; Laufrad n. ||
roue f. à aubes; roue f. mobile; rotor
m.

blading (Turbine) || Schaufelung f.; Be-
schaufelung f. || aubage m.; ailetage m. /
~ of a steam turbine rotor || Beschaufe-
lung f. eines Dampfturbinenläufers ||
mise f. en place de l'aubage d'une roue
motrice de turbine à vapeur.

blame, to || rügen; tadeln || blâmer.

blame || Rüge f.; Tadel m. || blâme m.

blameless || untadelhaft; tadellos || irré-
prochable.

blanchards pl. || Leinwand f. aus gebleich-
tem Garn || blanchard m.

blank || blanko; unausgefüllt; unausgefer-
tigt || en blanc.

blank (Coin) || Münzplatte f.; Scheibe f. für
Münzen || flan m. / ~ (Formulary) || For-
mular n. || formule f.; formulaire m.
/ ~ (Met) || Rohling m. || ébauche f. /
~ (Print) || Durchschuß m. || interligne f.
/ ~ already machined ~ (Mach; Met) ||
vorgearbeitetes Werkstück n. || pièce f.
ébauchée.

blank acceptance || Blankoakzept n. || ac-
ceptation f. à découvert. / ~ credit ||
Blankokredit m. || crédit m. à découvert.
/ ~ endorsement || Blankoindossament n.
|| endossement m. en blanc.

blanket (Print) || Druckfilz m.; Filzunter-
lage f.; Drucktuch n. || blanchet m.
(de drap). / ~ (Met) || Schutzdecke f.
über flüssigem Metall || couverture f. /
~ (Woollen coverlet) || Wolldecke f.;
wollene Bettdecke f. || couverture f. de
laine. / cotton ~ || Baumwoll(molton)-
decke f. || couverture f. en (molleton)
coton. / half-woollen ~ || Halbwolldecke
f. || couverture f. mi-laine. / quilted ~ ||
Steppdecke f. || couverture f. ouatée ou
piquée; courte-pointe f. / woollen ~ see
blanket.

blanket maker || Deckenmacher m. || cou-
verturier m. / ~ sewer || Deckennäherin
f. || couseuse f. de couvertures. / ~ sluice
(Dress. ore) || Planherd m.; Planenherd
m. || table f. à toile. / ~ weaver || Decken-
weber m. || tisseur m. de couvertures.

blank flange || Deckelflansch m. || Blind-
flansch m. || bride f. aveugle ou d'ob-
turation.

blanking machine || Aushaumaschine f. ||
machine f. à découper.

blank key || Blanktaste f. || touche f. du
blanc. / ~ letter of attorney || Blanko-
vollmacht f. || blanc-seing m. / ~ line
(Print) || weiße Zeile f. || ligne f. de blanc.
/ ~ material (Print) || Blindmaterial n.;
Ausschluß m.; Durchschuß m. || blancs
mpl.

blankness || Leere f.; Weiße f. || vide m.

blank page || Blankseite f.; erste Seite f.
|| fausse-page f.; page f. blanche.

blanks pl. (Print) see blank material.

blank-trial || Leerversuch m. || essai m.
témoin.

blast, to || (mit Pulver) sprengen || faire
sauter. / ~ a bore-hole || schießen; spren-
gen || faire sauter les rocs. / ~ the cinders
pl. after tapping || die Schlacke f. nach
dem Abstich ausblasen || flamber le
creuset d'un haut-fourneau après la
coulée. / ~ a mine || einen Sprengschuß
abfeuern; eine Mine sprengen || faire
sauter une mine; faire jouer une mine.

blast (Explosion) || Schuß m.; Spreng-
schuß m.; Sprengung f. || explosion f.;
action f. de faire sauter. / ~ (Explosive
charge) || Mine f.; Sprengmine f.; Spreng-
ladung f. || mine f.; fourneau m.; charge
f. d'explosif. / ~ (Met) || Gebläse n.;
Gebläseluft f.; Gebläsewind m.; Wind
m. || soufflerie f.; vent m. de soufflerie;
air m. / cold ~ (Met) || kalte Gebläseluft
f.; kalter Wind m. || air m. froid. / hot ~
(Met) || heißer Wind m.; erhitzte oder
Gebläseluft f. || air m. chaud. / the ~ is on
|| das Gebläse arbeitet || le vent va; la
soufflerie marche. / ~ of the steam-
whistle || Pfiff m. oder Signal n. mit der
Dampfpfeife || coup m. de sifflet. / the
~ stops || das Gebläse ist abgestellt ||
le vent est arrêté. / ~ of wind || Windstoß
m. || bouffée f. de vent.

blast air || Gebläseluft f. || air m. d'une
soufflerie / ~ plant || Sprengluftanlage f.
|| installation f. de sautage au gaz com-
primé.

blast box || Windkessel m. || boîte f. à vent.
/ ~ connection || Düsenstock m. || porte-
vent m. / ~-cooling furnace for bottle
industry || Zugkühlofen m. für die Fla-
schenindustrie || four m. de refroidisse-
ment à tirage pour l'industrie des bou-
teilles. / ~ cylinder || Gebläsezylinder m.
|| cylindre m. soufflant ou de la soufflerie
ou du compresseur.

blast engine || Gebläsemaschine f.; Gebläse
n. || machine f. soufflante; soufflerie f. /
~ for a furnace || Gebläseanlage f. für
Hochöfen || soufflante f. de haut-four-
neau. / screw ~ || Spiralgebläse n.;
Schraubengebläse n. || machine f. souf-
flante à vis d'Archimède; cagniardelle f.

blast-firing || Sprengen n. (mit Pulver) ||
sautage m. / electric ~ || elektrische
Minenzündung f. || mise f. à feu élec-
trique des mines; inflammation f. ou
sautage m. électrique des amorces.

blast-furnace || Hochofen m.; Gebläse-
(schacht)ofen m. || haut-fourneau m.;
fourneau m. à soufflet. / ~ without ap-
pendices || Hochofenrumpf m. || masse f.
ou cuve f. du haut-fourneau. / ~ with
closed front || Hochofen m. mit ge-
schlossener Brust || haut-fourneau m. à
poitrine fermée. / ~ for mixed fuel ||
Hochofen m. für gemischten Brennstoff
|| haut-fourneau m. au mélange ou à

combustible mélangé. / ~ with open front ‖ Hochofen m. mit offener Brust ‖ haut-fourneau m. à poitrine ouverte. / ~ for raw coal ‖ Hochofen m. für rohe Steinkohle ‖ haut-fourneau m. marchant à la houille brute. / to set to work the ~ ‖ den Hochofen m. in Betrieb setzen ‖ mettre le fourneau en marche.

blast-furnace armour ‖ Hochofenpanzer m. ‖ blindage m. de haut-fourneau. / ~ blowing plant ‖ Hochofengebläseanlage f. ‖ installation f. des machines-soufflantes à moteurs à gaz de hauts-fourneaux. / ~ cement ‖ Hochofenzement m. ‖ ciment m. du haut-fourneau. / ~ charging device ‖ Hochofenbeschickungsvorrichtung f. ‖ appareil m. de chargement pour hauts-fourneaux. / ~ charging hoist ‖ Hochofenbegichtungsanlage f. ‖ dispositif m. de chargement du gueulard pour hauts-fourneaux. / ~ cinder ‖ Hochofenschlacke f. ‖ scorie f. du haut-fourneau; laitier m. / ~ coke ‖ Hochofenkoks m. ‖ coke m. de haut-fourneau. / ~ cone ‖ Gichtglocke f. ‖ cloche f. de haut-fourneau. / ~ elevator ‖ Gichtaufzug m. ‖ élévateur m. de haut-fourneau. / ~ equipment ‖ Hochofenanlage f. ‖ installation f. de hauts-fourneaux. / ~ fittings pl. ‖ Hochofenarmatur f. ‖ armature f. pour hauts-fourneaux. / ~ framework ‖ Hochofengerüst n. ‖ charpente f. de haut-fourneau. / ~ gas ‖ Hochofengas n.; Gichtgas n. ‖ gaz m. du gueulard *ou* de hauts-fourneaux. / ~ gas blowing engine ‖ Hochofengasgebläsemaschine f. ‖ machine-soufflante f. à moteurs à gaz de hauts-fourneaux. / ~ gas-engine ‖ Gichtgasmaschine f.; Gichtgasmotor m. ‖ moteur m. à gaz de haut-fourneau. / ~ gas purifying plant ‖ Hochofengasreinigungsanlage f. ‖ installation f. d'épuration de gaz de hauts-fourneaux. / ~ hoist ‖ Gichtaufzug m. ‖ monte-charge m. pour hauts-fourneaux. / ~ jacket ‖ Hochofenmantel m.; Hochofenpanzer m. ‖ blindage m. pour hauts-fourneaux.

blast-furnace plant ‖ Hochofenanlage f. ‖ installation f. de hauts-fourneaux. / ~ composed of x furnaces ‖ Hochofenanlage f. von x Hochöfen ‖ installation f. de x hauts-fourneaux.

blast-furnace slag ‖ Hochofenschlacke f. ‖ laitier m. de haut-fourneau. / ~ in pieces ‖ Hochofenstückschlacke f. ‖ laitier m. du haut-fourneau en morceaux.

blast-furnace slag cement ‖ Hochofenschlackenzement m. ‖ ciment m. de laitier du haut-fourneau. / ~ granulating plant ‖ Granulierungsanlage f. für Hochofenschlacke ‖ installation f. à granuler le laitier de haut-fourneau.

blast gate (Found) ‖ Windschieber m. ‖ régistre m. *ou* vanne f. de réglage du vent. / ~ heating apparatus ‖ Winderhitzungsvorrichtung f. ‖ appareil m. pour le chauffage du vent; appareil m. à air chaud. / ~ heating stove *see* ~ heating apparatus.

blasting ‖ Sprengen n. (mit Pulver) ‖ sautage m. / ~ (Technics) ‖ Sprengtechnik f. ‖ technique f. de faire sauter les explosifs.

blasting cap ‖ Sprengkapsel f.; Zündkapsel f. ‖ détonateur m.; capsule f. fulminante. / ~ charge ‖ Sprengladung f. ‖ charge f. explosive. / ~ oil ‖ Nitroglyzerin n. ‖ nitroglycérine f. / ~ powder ‖ Sprengpulver n. ‖ poudre f. de mine.

blast-lamp (Glassm) ‖ Gebläse n. ‖ chalumeau m. / ~ (Glazing) ‖ Einbrennlampe f. ‖ lampe f. à souder *ou* à chalumeau. / oxyhydrogen ~ ‖ Knallgasgebläse n. ‖ chalumeau m. oxyhydrique.

blast machine *see* ~ engine. / ~ main ‖ Hauptwindleitung f. ‖ conduite f. de vent.

blast-pipe (Blower) ‖ Düsenrohr n.; Windleitung f. ‖ tuyère f.; porte-vent m. / ~ (Cupola furnace) ‖ Windleitung f. ‖ conduite f. de vent. / ~ (Loc) ‖ Auslaßrohr n.; Dampfauslaßrohr n.; Ausblasrohr n. ‖ échappement m.; tuyau m. d'échappement de vapeur. / steam ~ *see* blast pipe (Loc).

blast pressure ‖ Gebläsedruck m.; Windpressung f. ‖ pression f. du vent. / ~ pump ‖ Wasserstrahlluftpumpe f. ‖ aspirateur m. d'air marchant à l'eau; trompe f. à eau. / ~ stone (Blower) ‖ Windstein m. ‖ contre-vent m. / ~ tank ‖ Windkessel m. ‖ boîte f. à vent. / ~ valve ‖ Windschieber m. ‖ vanne f. à air.

blaze, to ‖ flammen; aufflammen; lodern ‖ flamber. / ~ off the steel ‖ den Stahl m. abbrennen ‖ recuire l'acier m. par le flambage; décaper l'acier m.

blea (Join) ‖ Mondring m.; falscher Splint m.; Kernschäle f. ‖ faux aubier m.

bleach, to ‖ bleichen; entfärben ‖ blanchir; décolorer.

bleached sand ‖ Bleichsand m. ‖ sable m. blanchi.

bleacher ‖ Bleicher m. ‖ blanchisseur m. / grass ~ ‖ Naturbleicher m. ‖ blanchisseur m. sur pré.

bleach field ‖ Rasenbleiche f.; Rasenbleichplatz m. ‖ blanchisserie f. sur pré. / ~ green *see* ~ field.

bleaching ‖ Bleichen n.; Bleiche f. ‖ blanchiment m. / chemical ~ ‖ chemisches Bleichen n.; Schnellbleiche f. ‖ blanchiment m. chimique. / ~ of cotton goods ‖ Bleichen n. von Baumwollwaren ‖ blanchiment m. de tissus de coton. / flax ~ ‖ Flachsbleiche f. ‖ blanchiment m. de lin. / fur ~ ‖ Bleichen n. von Pelzen ‖ décoloration f. de fourrures. / grass ~ ‖ Rasenbleiche f. ‖ blanchiment m. sur pré. / ~ of jute ‖ Bleichen n. der Jute ‖ blanchiment m. de jute. / ~ of linen ‖ Bleichen n. von Leinenwaren ‖ blanchiment m. de toiles de lin. / ~ of silk ‖ Seidenbleiche f.; Bleichen n. der Seide ‖ blanchiment m. de la soie. / ~ of textiles ‖ Bleichen n. von Textilwaren ‖ blanchiment m. de fibres textiles. / ~ of wool ‖ Bleichen n. der Wolle ‖ blanchiment m. de la laine.

bleaching agent (Chem) ‖ Entfärbungsmittel n. ‖ décolorant m. / ~ (Washing) ‖ Bleichmittel n. ‖ agent m. de blanchiment; produit m. à blanchir.

bleaching apparatus ‖ Bleichvorrichtung f. ‖ appareil m. à blanchir. / ozone ~ ‖ Ozonbleichvorrichtung f. ‖ appareil m. de blanchiment à l'ozone.

bleaching boiler ‖ Bleichkessel m. ‖ chaudière f. à blanchiment. / ~ clay ‖ Bleicherde f. ‖ terre f. à blanchir. / ~ clay extraction plant ‖ Bleicherdeextraktionsanlage f. ‖ installation f. pour l'extraction des huiles récupérables des terres décolorantes. / ~ earth *see* ~ clay. / ~ electrolyzer ‖ Bleichelektrolysör m. ‖ électrolyseur m. de blanchiment. / ~ field *see* bleach field. / ~ ground *see*

bleach field. / ~ machine ‖ Bleichereimaschine f. ‖ machine f. pour le blanchiment. / ~ machine for textile goods ‖ Bleichereimaschine f. für Textilien ‖ machine f. de blanchissage pour textiles. / ~ machine operator ‖ Maschinenbleicher m. ‖ blanchisseur m. à la machine.

bleaching-out process ‖ Ausbleichverfahren n. ‖ procédé m. de blanchiment.

bleaching plant ‖ Bleichanlage f.; Entfärbungsanlage f. ‖ installation f. de blanchiment. / ~ powder ‖ Bleichkalk m.; Chlorkalk m.; Bleichpulver n. ‖ chlorure m. de chaux; poudre f. à blanchir. / ~ soda ‖ Bleichsoda f. ‖ soude f. à blanchir. / ~ water (Chem) ‖ Bleichwasser n. ‖ eau f. de Javelle. / ~ works pl. *see* bleach works pl.

bleach works pl. ‖ Bleicherei f. ‖ blanch(iss)erie f.

bleeder ‖ Ölstandhahn m. ‖ robinet m. de niveau d'huile.

blend, to ‖ ~ the colours ‖ die Farben fpl. einmischen ‖ détremper les couleurs.

blende (Miner) ‖ Zinkblende f.; Blende f. ‖ zinc m. sulfuré; blende f. / botryoidal ~ ‖ Schalenblende f. ‖ zinc m. sulfuré concrétionné. / fibrous ~ ‖ Strahlenblende f. ‖ zinc m. sulfuré; wurtzite f.

blended (Distill) ‖ verschnitten ‖ coupé; mêlé; mélangé.

blende mine ‖ Zinkblendegrube f. ‖ mine f. de blende. / ~ roasting furnace ‖ Blenderöstofen m. ‖ four m. à calciner la blende.

blight (Bot) ‖ Meltau m.; Rost m.; Brand m. ‖ rouille f.

blimp ‖ unstarres Kleinluftschiff n. ‖ dirigeable m. souple de petites dimensions.

blind, to (Build) ‖ verkleiden; verblenden ‖ revêtir.

blind (Opt) ‖ blind ‖ aveugle. / ~ through cataract ‖ starblind ‖ cataracté. / ~ colour ~ ‖ farbenblind ‖ daltonien.

blind ‖ Markise f.; Sonnendach n. ‖ marquise f. / metal ~ ‖ Metallrolladen m. ‖ fermeture f. métallique. / roller ~ ‖ Rolladen m. ‖ jalousie f. / Venetian ~ ‖ Sommerladen m.; Sonnenblende f. ‖ persienne f. (en bois). / Venetian ~ (Phot) ‖ Jalousieverschluß m. ‖ fermeture f. à jalousie. / wood window ~ ‖ Holzrolladen m. ‖ jalousie f. en bois.

blind alley ‖ Sackgasse f. ‖ impasse f.; cul-de-sac m. / ~ anchor ‖ einarmiger Hafenanker m. ‖ ancre f. borgne. / ~ calculating machine ‖ unsichtbar schreibende Rechenmaschine f. ‖ machine f. à calculer à point d'impression invisible. / ~ -coal ‖ Anthrazit m. ‖ anthracite m. / ~ -flying ‖ Blindfliegen n. ‖ vol m. aveugle.

blinding (Opt) ‖ Blendung f.; Blenden n. ‖ éblouissement m. / ~ (Road) ‖ Sanddecke f. ‖ couche f. de sable; ensablement m.

blind institute ‖ Blindenanstalt f. ‖ hospice m. des aveugles.

blindness ‖ Blindheit f. ‖ cécité f. / ~ caused by a cataract ‖ Starblindheit f. ‖ cécité f. qui résulte d'une cataracte.

blind strap ‖ Rolladengurt m. ‖ sangle f. de persiennes. / ~ track ‖ totes Gleis n. ‖ voie f. en cul-de-sac *ou* en impasse. / ~ wall ‖ blinde Mauer f. *oder* Fassade f.; Blendfassade f. ‖ mur m. orbre *ou* aveugle; façade f. feinte. / ~ wireless traffic ‖ Blindfunkverkehr m. ‖ trafic m. unilatéral.

blinker *see* blinkers / ~ beacon ‖ Blickfeuer n.; Blinkfeuer n. ‖ phare m. à éclipse.

blinkers pl. (For horses) ‖ Scheuklappen fpl.; Scheuleder n. ‖ œillères fpl.

blister (Found) ‖ Gußblase f.; Galle f.; Lucke f. ‖ paille f.; soufflure f. de fonte; défaut m. *ou* défectuosité f. dans la fonte. / ~ (Glass) ‖ Glasblase f. ‖ bosse f. / the ingots are full of ~s ‖ der Guß ist voller Blasen ‖ les blocs mpl. d'acier fondu sont pleins de soufflures.

blistered (Found; Glass) ‖ blasig; luckig ‖ vésiculé; venteux; bulleux; lacuneux. / ~ casting ‖ poröser Guß m. ‖ fonte f. spongieuse. / ~ steel *see* blister-steel. / not ~ steel ‖ blasenloser Stahl m. ‖ acier m. sans ampoules.

blister-steel ‖ Blasenstahl m.; Brennstahl m.; Zementstahl m. ‖ acier m. poule *ou* boursouflé *ou* de cémentation.

blizzard ‖ Schneegestöber n. ‖ tourbillon m. *ou* tourmente f. de neige.

bloater ‖ Bückling m.; geräucherter Hering m. ‖ hareng m. saur *ou* fumé.

block, to (To stop up) ‖ verstopfen ‖ obstruer. / ~ (Print) ‖ aufklotzen; klotzen ‖ clouer. / ~ out timber ‖ Holz n. schneiden *oder* zuschneiden ‖ couper *ou* débiter *ou* découper le bois. / ~ up the freestones ‖ die Werksteine mpl. versetzen ‖ poser les pierres fpl. de taille. / ~ up a road ‖ eine Straße sperren f. ‖ barrer une route.

block ‖ Block m. ‖ bloc m.; massif m. / ~ (Build) ‖ Häuserblock m. ‖ île f.; îlot m.; bloc m. de maisons. / ~ (Forg) ‖ Hammerstock m. ‖ billot m.; tronchet m.; chabotte f. de l'enclume. / ~ (Hatt) ‖ Hutform f.; Hutstock m. ‖ forme f. de chapeau. / ~ (Mach) ‖ Richtklotz m. ‖ taquet m. d'équerrage. / ~ (Print) ‖ Druckstock m.; Bildstock m. ‖ Klischee n. ‖ cliché m. / ~ (Pulley) ‖ Flaschenzug m.; Flasche f.; Rollenkloben m.; Kloben m. ‖ moufle f.; poulie f. / ~ (Railw) *see* block station. / ~ (Shipb) ‖ Kielklotz m.; Stapelklotz m.; Stapelblock m. ‖ tain m.; tin m. / ~ (Shoem) ‖ Lochholz n. ‖ billot m. / ~ (Support) ‖ Stativ n.; Gestell n. ‖ support m. / bottom ~ (Pulley) ‖ Unterflasche f. ‖ moufle m. inférieur. / cargo ~ (Mar) ‖ Ladeblock m.; Ladekloben m. ‖ poulie f. de charge. / date ~ ‖ Notizblock m. ‖ bloc-notes m. / erratic ~ ‖ Findling m. ‖ bloc m. erratique. / flat ~ (Pulley) ‖ Plattblock m. ‖ poulie f. plate. / guide ~s pl. (Mach) ‖ Geradführungsbacken fpl. ‖ coulisseaux mpl.; patins mpl. / incoming ~ (Tel) ‖ Anfangssperre f. ‖ bloc m. d'entrée. / metal ~ for hats ‖ Hutform f. aus Metall ‖ forme f. en métal pour la chapellerie. / ~ of an organ-pipe ‖ Pfeifenblock m.; Pfeifenboden m.; Pfeifenkern m. ‖ noyau m. *ou* pied m. d'un tuyau d'orgue. / pillow ~ (Mach) ‖ Lager n. einer horizontalen Welle ‖ coussinet m. *ou* palier m. *ou* grain m. d'un tourillon. / pulley ~ ‖ Hülse f. *oder* Gehäuse n. eines Flaschenzugs ‖ chape f. *ou* corps m. d'une moufle *ou* d'une poulie. / rough-pressing of discs from ~s ‖ Vorpressen n. der Blöcke zu Scheiben ‖ ébauchage m. des blocs en disques à la presse. / rounded ~ of the cope (Build) ‖ runder Deckstein m. in einer Mauerabdeckung; Sattelstein m. ‖ tablette f. en bahut. / section ~ (Railw)

‖ Durchgangsblockwerk n.; Streckenblock m. ‖ block m. de section. / ~ of shares ‖ Aktienpaket n. ‖ tranché f. d'actions. / sliding ~ ‖ Gleitbacke f. ‖ bloc m. coulissant. / stretching ~ ‖ Spannbock m. ‖ support m. de tendeur. / ~ for tackles ‖ Flaschenzugkloben m. ‖ poulie f. à moufle. / thrust ~ ‖ Drucklager n. ‖ palier m. de butée. / ~ of wood ‖ Holzblock m.; Holzklotz m. ‖ bloc m. de bois; pièce f. de bois. / wooden ~ for hatters ‖ Hutform f. aus Holz ‖ forme f. en bois pour chapeaux.

blockade, to ‖ blockieren; absperren; sperren ‖ bloquer; faire le blocus. / ~ a track (Railw) ‖ ein Gleis sperren ‖ bloquer une voie.

blockade ‖ Blockade f.; Absperrung f. ‖ blocus m.

block almanac ‖ Abreißkalender m. ‖ calendrier m. en feuilles. / ~ and lock (Tel) ‖ Blockfeld n. ‖ champ m. de bloc. / ~ apparatus (Tel) ‖ Blockwerk n. ‖ appareil m. de bloc. / ~ barrier ‖ Blockwall m. ‖ rempart m. de blocs. / ~ battery (Electr) ‖ Blockbatterie f. ‖ batterie f. block. / ~ bond (Mason) ‖ Blockverband m. ‖ liaison f. anglaise. / ~ brake ‖ Klotzbremse f. ‖ frein m. à sabot. / ~ chain ‖ Blockkette f. ‖ chaîne f. plate *ou* à blocs. / ~ circuit (Radio) ‖ Sperrkreis m. ‖ circuit m. piège d'onde. / clearing a ~ (Railw) ‖ Blockfreigabe f. ‖ déblocage m. / ~ condenser ‖ Blockkondensator m.; Abschlußkondensator m.; Sperrkondensator m. ‖ condensateurs mpl. en série; condensateur m. terminal *ou* fixe. / ~ condenser for signalling purposes ‖ Rufsperrkondensator m. ‖ condensateur m. bloquant les courants d'appel. / ~ cutter (Engrav) ‖ Druckformengravör m. ‖ graveur m. sur bois pour impression.

blocked (Lock) ‖ verriegelt ‖ verrouillé. / ~ (Railw) ‖ geblockt ‖ bloqué.

block hammer (Mach) ‖ Stampfe f.; Stempel m.; Ramme f. ‖ marteau m. pilon *ou* vertical. / ~-house ‖ Blockhaus n. ‖ maison f. en bois blindé *ou* en troncs d'arbres; blockhaus m.

blocking (Railw) ‖ Verblocken n.; Blockung f.; Blockierung f. ‖ blocage m.; serrage m.; verrouillage m. / ~ (Trade) ‖ Sperre f. ‖ suspension f.; blocus m. / automatic ~ (Railw) ‖ selbsttätige Blockung f. ‖ blocage m. automatique. / electric ~ ‖ elektrische Blockung f. ‖ blocage m. électrique. / mechanical ~ ‖ mechanische Blockung f. ‖ blocage m. mécanique. / section ~ ‖ Streckenblockung f. ‖ blocage m. de section.

blocking axe (Mar) ‖ Schiffszimmermannsaxt f. ‖ hache f. de charpentier de navire. / ~ condenser *see* block condenser. / ~ current (Tel) ‖ Blockstrom m. ‖ courant m. de blocage. / ~ inductor (Tel; Railw) ‖ Blockinduktor m. ‖ inducteur m. de bloquage. / ~ machine (Woodw) ‖ Blockschneidemaschine f. ‖ machine f. à découper les blocs. / ~ means pl. (Tel) ‖ Sperrorgan n. ‖ membre m. de blocage. / ~ roller (Railw) ‖ Blockrolle f. ‖ poulie f. de block.

blocking system ‖ Blocksystem n.; Blocksignalsystem n. ‖ block-système m. / ~ for manual operating ‖ Handblocksystem n. ‖ block-système m. à main. /

~ plant ‖ Blockanlage f. ‖ installation f. du bloc-système.

block instrument case (Tel) ‖ Blockschrank m. ‖ armoire f. de poste automatique.

blockmaker (Hatt) ‖ Hutformmacher m. ‖ formier m. en bois. / ~ (Print) ‖ Abklatscher m.; Galvanoplastiker m. ‖ clicheur m.

blockmaking (Print) ‖ Klischieren n.; Abklatschen n. ‖ clichage m.

blockmotor ‖ Blockmotor m. ‖ moteur m. monobloc.

block pavement ‖ Klotzpflaster n.; Holzpflaster n. ‖ pavement m. en bois; pavé m. de bois. / ~ press ‖ Blockpresse f. ‖ presse f. à mouler. / ~ printing machine (Text print) ‖ Modelldruckmaschine f. ‖ machine f. à planche.

block-signal, to (Railw) ‖ das Haltezeichen geben ‖ couvrir *ou* bloquer la voie.

block signal (Railw) ‖ Blocksignal n. ‖ signal m. de bloc; signal-bloqueur m. / ~ signalling apparatus (Railw) ‖ Blocksignalvorrichtung f. ‖ appareil m. de bloc. / ~ station ‖ Blockstation f.; Blockstelle f. ‖ blockstation f.; station f. *ou* poste m. de block. / ~ switch ‖ Blockierschalter m. ‖ enclencheur m. de bloquage. / ~ system *see* blocking system.

block tackle ‖ Flaschenzug m. ‖ palan m. / hand-operated ~ ‖ Handflaschenzug m. ‖ palan m. à main.

block telephone ‖ Blockfernsprecher m. ‖ téléphone m. de block. / ~ tin ‖ Blockzinn n. ‖ étain m. en saumons *ou* en blocs / ~ tool ‖ Kastenwerkzeug n. ‖ semelle f. à combinaisons. / ~ transformer (Railw) ‖ Blocktransformator m. ‖ transformateur m. de ligne. / ~ type (Print) ‖ Blockschrift f. ‖ lettres fpl. égyptiennes. / ~ wall *see* block barrier. / ~ wire stitching machine ‖ Blockdrahtheftmaschine f. ‖ machine f. à coudre les blocs au fil de fer.

blockwood pavement *see* block pavement.

blood ‖ Blut n. ‖ sang m. / dried ~ ‖ getrocknetes Blut n.; Blutmehl n. ‖ sang m. (des)séché *ou* coagulé.

blood black ‖ Blutkohle f. ‖ charbon m. de sang. / ~ cell *see* ~ corpuscle. / ~ charcoal *see* blood black. / ~ corpuscle ‖ Blutkörperchen n.; Blutzelle f. ‖ globule m. du sang. / ~ corpuscle counting apparatus ‖ Blutkörperzählapparat m. ‖ appareil m. à compter les globules du sang. / ~ drying device ‖ Bluttrockner m. ‖ appareil m. à sécher le sang. / ~ drying installation ‖ Bluttrocknungsanstalt f. ‖ fabrique f. de sang coagulé *ou* de sang desséché. / ~ examination ‖ Blutuntersuchung f. ‖ examen m. du sang. / ~ heat ‖ Körperwärme f.; Blutwärme f. ‖ température f. du sang. / ~ meal ‖ Blutmehl n.; getrocknetes Mehl n. ‖ sang m. desséché *ou* coagulé. / ~ orange ‖ Blutapfelsine f. ‖ orange f. rouge.

blood-poisoning ‖ Blutvergiftung f. ‖ empoisonnement m. du sang. / to protect against ~ ‖ vor Blutvergiftung f. schützen ‖ protéger contre les infections fpl. du sang.

blood-red ‖ Blutrot n. ‖ rouge-sang m. / ~ (Met) ‖ dunkle Rotglut f. ‖ rouge m. sombre. / ~ heat ‖ Rotglühhitze f. ‖ chaude f. rouge sang.

blood solution ‖ Blutlösung f. ‖ solution f. du sang. / ~ spectrum ‖ Blutspektrum n. ‖ spectre m. du sang. / ~ -stone (Miner) ‖

Blutstein m.; Roteisenstein m. ‖ hématite f.; pierre f. sanguine. / ~ vessel ‖ Blutgefäß n. ‖ vaisseau m. sanguin.

bloom, to ‖ vorhämmern ‖ battre; marteler; forger préalablement.

bloom (Met) ‖ Luppe f.; Klumpen m.; Deul m.; Bloom m. ‖ bloom m.; loupe f.; pain m.; masset m. / slab ~ ‖ Bramme f. ‖ brame f.

bloomer (Curr) ‖ Narbenabstoßer m. ‖ effleureur m. de cuir.

bloomers pl. ‖ Schlupfhose f.; Schlüpfer m. ‖ culotte f. pour dames.

bloomery ‖ Luppenfrischhütte f. ‖ bloomerie f. / ~ fire ‖ Rennfeuer n. ‖ basfoyer m.; forge f. catalane.

blooming machine ‖ Luppenmühle f. ‖ squeezer m. rotatif. / ~ mill ‖ Blockstraße f.; Vorstraße f.; Blockwalzwerk n. ‖ blooming m.; train m. dégrossisseur; laminoir m. à blooms *ou* de serrage. / ~ mill for welding ‖ Schweißwalzwerk n. ‖ laminoir-soudeur m.; trainsoudeur m. / ~ roll ‖ Vorwalze f.; Luppenwalze f.; Puddelwalze f. ‖ cylindre m. ébaucheur *ou* du blooming. / ~ rolling-mill ‖ Puddelwalzwerk n.; Rohschienenwalzwerk n. ‖ train m. ébaucheur; train m. de puddlage; laminoir m. ébaucheur.

bloom shears pl. ‖ Blockschere f. ‖ cisailles fpl. à lingots *ou* à blooms. ~ tongs pl. *see* bloom shears pl. / ~ wagon ‖ Luppenwagen m. ‖ wagon m. à loupes.

blot, to (Print) ‖ unsauber abziehen ‖ mâchurer.

blot ‖ Klecks m.; Tintenfleck m. ‖ tâche f.; paté m.

blotter ‖ Tintenlöscher m. ‖ buvard m. / ~ holder ‖ Löschblatthalter m. ‖ portebuvard m.

blotting book *see* ~ pad. / ~ pad ‖ Löschpapierblock m.; Schreibunterlage f. *oder* Schreibmappe f. mit Löschpapier ‖ sousmain m. buvard.

blotting-paper ‖ Löschpapier n.; Fließpapier n. ‖ papier m. buvard; papier m. brouillard. / ~ highly absorptive ~ ‖ schnelltrocknendes Löschpapier n. ‖ papier m. buvard séchant rapidement. / sheet of ~ ‖ Löschblatt n. ‖ feuille f. de papier brouillard.

blouse ‖ Bluse f. ‖ blouse f. / ~ flannel ‖ Blusenflanell m. ‖ flanelle f. pour blouses. / ~ material ‖ Blusenstoff m. ‖ étoffe f. pour blouses. / ~ protector ‖ Blusenschoner m. ‖ protecteur m. de blouse.

blow, to ‖ blasen; wehen ‖ souffler. / ~ down the furnace ‖ den Hochofen m. ausgehen lassen *oder* ausblasen ‖ mettre le fourneau hors de marche *ou* hors de feu; refroidir le fourneau. / ~ a fuse ‖ eine Sicherung durchbrennen ‖ fondre *ou* sauter un fusible. / ~ glass ‖ Glas n. blasen ‖ souffler le verre. / ~ in *or* into ‖ einblasen ‖ souffler; insouffler. / ~ in air under the grate by means of steam blast ‖ Luft f. mittels Dampfstrahls unter den Rost einblasen ‖ insouffler de l'air m. au-dessous de la grille au moyen d'un jet de vapeur. / ~ in the blast ‖ das Gebläse n. anlassen ‖ donner le vent. / ~ in the blast furnace ‖ den Hochofen m. anblasen ‖ mettre à feu le hautfourneau; allumer le haut-fourneau. / ~ off with gun powder ‖ mit Pulver n. absprengen ‖ faire sauter à la poudre. / ~ off the steam ‖ den Dampf m. abblasen ‖ dégager *ou* vider *ou* larguer la vapeur. / ~ out ‖ ausblasen ‖ purger. / ~ out (To cause to swell) ‖ aufblasen ‖ gonfler; enfler. / ~ out the dust from engines ‖ den Staub aus Maschinen ausblasen ‖ nettoyer les machines au jet d'air. / ~ out the furnace *see* to blow down the furnace / ~ through ‖ durchblasen ‖ souffler à travers. / ~ up a bridge ‖ eine Brücke sprengen ‖ faire sauter un pont. / ~ up a ship ‖ ein Schiff in die Luft sprengen ‖ faire sauter un navire.

blow (Emission of air etc) ‖ Blasen n.; Wehen n. ‖ soufflage m. / ~ (Stroke) ‖ Schlag m. ‖ choc m. / total duration of ~ ‖ gesamte Schlagdauer f. ‖ durée f. totale du choc. / ~ of the hammer (Typewr) ‖ Hammerschlag m. ‖ coup m. de marteau. / the ~s pl. are struck on the hub ‖ der Schlag erfolgt auf die Nabe der Räder ‖ le coup est porté sur le moyeu des roues. / number of ~s per second (Typewr) ‖ Anschläge mpl. je Sekunde ‖ nombre m. des frappes par seconde.

blower *see also* blowing engine ‖ Gebläse n. ‖ souffleur m.; soufflerie f.; soufflet m.; machine f. soufflante. / ~ (Man) ‖ Blasemeister m. ‖ régleur m. de vent. / blowpipe ~ ‖ Lötrohrgebläse n. ‖ chalumeau m. à soufflet / centrifugal ~ ‖ Schleudergebläse n. ‖ soufflerie f. centrifuge. / ~ for the drier ‖ Trocknergebläse n. ‖ soufflante f. du sécheur. / efficiency of delivery of the ~ *see* efficiency. / electric motor ~ ‖ Gebläse n. mit elektrischem Antrieb ‖ soufflerie f. à moteur électrique. / helical ~ ‖ Propellergebläse n. ‖ soufflerie f. à hélice. / high-pressure ~ ‖ Hochdruckgebläse n. ‖ soufflerie f. à haute pression. / piston ~ ‖ Kolbengebläse n. ‖ soufflerie f. à piston. / Root's ~ ‖ Kapselgebläse n. ‖ ventilateur m. de Root. / rotary ~ ‖ rotierendes *oder* umlaufendes Gebläse n. ‖ soufflerie f. rotative. / soot ~ ‖ Rußbläser m. ‖ soufflet m. à suie. / ~ for steel works ‖ Stahlwerkgebläse n. ‖ soufflerie f. d'aciéries. / steam-jet ~ ‖ Dampfstrahlgebläse n. ‖ souffleur m. à vapeur. / turbo ~ ‖ Turbogebläse n. ‖ soufflerie f. centrifuge. / ~ actuated by water-power ‖ Wasserstrahlgebläse n. ‖ soufflerie f. alimentée par une trompe à eau.

blower efficiency ‖ Förderleistung f. des Gebläses ‖ débit m. du ventilateur. / ~ engineer ‖ Gebläsemaschinist m. ‖ mécanicien m. des machines soufflantes. / ~ output *see* efficiency. / ~ regulating man ‖ Gebläsewärter m. ‖ régleur m. de vent.

blow gun ‖ Abblas(e)pistole f. ‖ pistolet m. souffleur. / ~ hole ‖ Gußblase f. ‖ soufflure f. de fonte; bouillon m.

blowing current (Electr) ‖ Abschmelzstromstärke f. ‖ intensité f. (du courant) de fusion. / ~ cylinder ‖ Gebläsezylinder m. ‖ cylindre m. soufflant. / ~ device for charging shaft furnaces with fuel and charges ‖ Einblasevorrichtung f. für Brennstoff und Schmelzstoff in Schachtöfen ‖ dispositif m. d'injection du combustible et du fondant pour fourneaux à cuve.

blowing-down of a furnace ‖ Ausblasen n. *oder* Niederblasen eines Ofens ‖ mise f. hors feu d'un fourneau.

blowing-down device (Blast-furnace) ‖ Ausblasevorrichtung f. ‖ dispositif m. pour la mise hors feu.

blowing-engine *see also* blower ‖ Gebläse n.; Gebläsemaschine f. ‖ soufflerie f.; souffleur m.; machine f. soufflante. / ~ for blast-furnaces ‖ Hochofengebläse n. ‖ soufflante f. de hauts-fourneaux. / ~ worked by blast-furnace gas ‖ Hochofengasgebläsemaschine f. ‖ machine f. soufflante à moteurs à gaz de hauts-fourneaux. / compound steam ~ ‖ Verbunddampfgebläsemaschine f. ‖ machine f. soufflante à vapeur compound. / cylinder ~ ‖ Zylindergebläsemaschine f. ‖ machine f. soufflante à piston. / ~ with slide-valves ‖ Schiebergebläse n. ‖ machine f. soufflante à tiroir.

blowing fan ‖ Blasventilator m. ‖ ventilateur m. soufflant. / central ~ installation ‖ zentrale Winderzeugungsanlage f. ‖ installation f. centrale de production de vent. / ~ iron (Glassm) ‖ Glasmacherpfeife f.; Pfeife f. ‖ canne f.; felle f. / ~ machine for glass-blowing ‖ Glasblasmaschine f. ‖ machine f. à souffler le verre.

blowing-out (Blast-furnace) *see* blowingdown. / ~ (Electr) *see* blow-out. / ~ (Steam) *see* blow-off.

blowing plant for blast-furnaces ‖ Hochofengebläseanlage f. ‖ installation f. des machines-soufflantes à moteurs à gaz de hauts-fourneaux. / ~ wedge (Quarry) ‖ Sprengkeil m. ‖ moellonier m.

blow-lamp (Soldering) ‖ Lötlampe f. ‖ lampe f. à souder.

blown ‖ aufgeblasen; aufgebläht ‖ boursofflé. / ~ (Found) ‖ blasig; porig; porös ‖ venteux; vésiculé; loupé; poreux. / ~ glass ‖ geblasenes Glas n. ‖ verre m. soufflé. ~ glass article ‖ Glasbläsereigegenstand m. ‖ article m. en verre soufflé.

blown (Join) *see* blea.

blow-off (Tyre) ‖ Platzen n. ‖ éclatement m. / ~ cock (Steam) ‖ Abblasehahn m. ‖ robinet m. de vidange *ou* de purge. / ~ pipe ‖ Ausblaserohr n. ‖ tuyau-purgeur m.; tubulure f. d'évacuation. / ~ valve ‖ Ausblaseventil n. ‖ soupape f. de purge.

blow-out (Electr) ‖ Funkenlöschung f. ‖ soufflage m. *ou* étouffement d'étincelles. / magnetic ~ ‖ magnetische Funkenlöschung f. ‖ soufflage m. magnétique d'étincelles.

blow-out coil ‖ Löschspule f.; Funkenlöschspule f. ‖ bobine f. souffleuse d'étincelles; bobine f. à étouffer les étincelles; bobine f. pare-étincelles.

blowpipe ‖ Lötrohr n. ‖ chalumeau m. / ~ with bellows ‖ Lötrohrgebläse n. ‖ chalumeau m. à soufflet / ~ with a gas holder *see* ~ with bellows. / hand ~ ‖ Handgebläse n.; Handlötrohr n. ‖ chalumeau m. à main. / oxyhydrogen ~ ‖ Knallgasgebläse n. ‖ chalumeau m. oxyhydrique.

blowpipe analysis ‖ Lötrohranalyse f. ‖ analyse f. au chalumeau. / ~ assay ‖ Lötrohrprobe f.; Lötrohrversuch m. ‖ essai m. au chalumeau. / ~ flame ‖ Lötflamme f.; Stichflamme f. ‖ flamme f. du chalumeau *ou* à souder; feu m. du chalumeau / ~ lamp ‖ Lötrohrlampe f. ‖ lampe f. du chalumeau. / ~ lead solderer ‖ Bleilöter ﹑m. ‖ soudeur m. de plomb au chalumeau. / ~ nipple *see* ~ nozzle. / ~ nozzle ‖ Lötrohrspitze f. ‖ buse f. du chalumeau. / ~ proof *see* ~ assay. / ~ reagents pl. ‖ Lötrohrreagenzien npl. ‖ réactifs mpl. au

chalumeau. / ~ solderer ‖ Lötrohrlöter m. ‖ soudeur m. au chalumeau. / ~ test see ~ assay.

blowpiping see blowpipe assay.

blow struck on a tapering mandril ‖ Schlag m. auf einen Dorn ‖ coup m. sur un mandrin.

blow-through valve ‖ Durchblaseventil n. ‖ soupape f. de purge.

blubber (Whale) ‖ Walfischspeck m. ‖ lard m. de baleine. / ~ (Fish oil) ‖ Fischtran m. ‖ huile f. de poisson.

blue, to ‖ bläuen; blau färben ‖ bleuir.

blue ‖ blau ‖ bleu.

blue ‖ Blau n.; blaue Farbe f. ‖ bleu m. / Chinese ~ ‖ Pariserblau n. ‖ bleu m. de Paris. / laundry ~ ‖ Waschblau n. ‖ bleu m. pour azurage du linge. / night ~ ‖ Nachtblau n. ‖ bleu m. de nuit. / Prussian ~ ‖ Berlinerblau n.; Preußischblau n. ‖ bleu m. de Prusse. / Saxon ~ ‖ Sächsischblau n. ‖ bleu m. de Saxe.

blue ash ‖ Blauasche f.; Bergblau n.; Kalkblau n. ‖ cendre f. bleue. / ~ black ‖ Blauschwarz n. ‖ noir m. bleu. / ~ black (Drawing) ‖ Reißkohle f.; Zeichenkohle f. ‖ crayon m. de charbon; charbon m. de saule. / ~ copper ‖ Kupferindig(o) m. ‖ cuivre m. sulfureux; covellite f.

blue cross shell ‖ Blaukreuzgeschoß n. ‖ obus m. à la croix bleue.

blue glass ‖ Blauglas n. ‖ verre m. bleu. / ~ green see bluish-green. / ~ lead ore ‖ Blaubleierz n. ‖ mine f. de plomb bleu; pyromorphite f. / ~ metal ‖ blauer Kupferstein m. ‖ matte f. bleue; métal m. bleu. / ~ pencil ‖ Blaustift m. ‖ crayon m. bleu.

blue-print ‖ Blaupause f. ‖ calque m. bleu; bleu m. / ~ lamp ‖ Lichtpauslampe f.; Kopierlampe f. ‖ lampe f. pour héliogravure ou à photocalquer. / ~ paper ‖ Blaupapier n.; Pauspapier n.; Lichtpauspapier n. ‖ papier m. calque; papier n. à imprimer en bleu.

blue sap (Of wood) ‖ Blaufäule f. ‖ pourriture f. bleue. / ~ stone ‖ Blaukugel f. ‖ boule f. de bleu. / ~ verditer see blue ash. / ~ violet ‖ Blauviolett n. ‖ bleu violet. / ~ vitriol ‖ Kupfervitriol n. ‖ couperose f. bleue; vitriol m. bleu; sulfate m. de cuivre.

bluing ‖ Bläuen n.; Blauwerden n. ‖ bleuissage m. / ~ (Of wood) see blue sap. / ~ machine for tempering iron pieces ‖ Bläumaschine f. zum Anlassen von Eisenteilen ‖ machine f. à bleuir pour recuire les pièces de fer. / ~ salt ‖ Bläusalz n. ‖ sel m. à bleuir.

bluish ‖ bläulich ‖ bleuâtre.

bluish-green ‖ blaugrün ‖ vert bleu ou bleuâtre. / ~ glass ‖ blaugrün gefärbtes Glas n. ‖ verre m. d'un bleu-vert pâle.

bluish-grey ‖ blaugrau ‖ gris bleuâtre.

bluish tinge see bluing.

blunder, to ‖ pfuschen ‖ bousiller; tricher.

blunder ‖ Mißgriff m. ‖ méprise f.

blunderer ‖ Stümper m. ‖ bousilleur m.

blunt, to (Carp) ‖ abkanten ‖ écorner. / ~ (Tools) ‖ abstumpfen ‖ émousser. / ~ glass ‖ Glas n. blind machen ‖ émousser le verre.

blunt ‖ abgestumpft; stumpf ‖ émoussé; obtus. / ~ angle (Build) ‖ abgestumpfte Ecke f. ‖ entrecoupe f. / ~ sound ‖ Knopfsonde f. ‖ sonde f. à bouts olivaires.

blur, to (Opt) ‖ verschwimmen ‖ se brouiller.

---

blurred (Opt) ‖ verschwommen ‖ flou. / ~ image ‖ unscharfes Bild n. ‖ image f. floue ou manquant de netteté.

boa ‖ Boa f. ‖ boa m.

boar, wild ~ breeder ‖ Eberhalter m. ‖ éleveur m. de verrats.

board, to (Build) ‖ bedielen; verschalen ‖ planchéier; lambrisser. / ~ (Curr) ‖ krispeln ‖ rebrousser; crêpir.

board (Bank) ‖ Kollegium n. ‖ collège m.; corps m. enseignant; conseil m. / ~ (Bookb) ‖ Pappe f.; Pappdeckel m. ‖ carton m. / ~ (Carp) ‖ Diele f.; Bohle f.; Brett n. ‖ ais m.; planche f. (épaisse). / ~ (Dwelling) ‖ Einzelwohnung f. ‖ logement m. / ~ (Joiner) ‖ Tischblatt n.; Tischplatte f. ‖ table f.; tablette f. / ~ (Mar) ‖ Bord m.; Schiffsseite f. ‖ bord m. / ~ (Mine) ‖ Abbaustrecke f. ‖ voie f.; taille f.; coistresse f.; paroi f. / ~ (Trade) ‖ Ausschuß m. ‖ comité m. / back ~ (Mould) ‖ Formbrett n.; Mantelbrett n. ‖ échantillon m.; gabarit m. / the company is worked by a ~ of managing directors ‖ den Vorstand m. der Gesellschaft bildet ein Direktorium ‖ la gérance des affaires de la société est confiée à un comité de direction. / to be represented on the ~ of a company ‖ im Präsidium n. einer Gesellschaft vertreten sein ‖ faire partie f. de la présidence d'une société. / ~ for composed types (Print) ‖ Satzbrett n. ‖ planche f. pour caractères composés. / cutting-out ~ ‖ Zuschneidetisch m. ‖ écofrai m. / drawing ~ ‖ Reißbrett n. ‖ planche f. à dessiner. / ~ of direction ‖ Direktion f. ‖ direction f.; directoire m. / ~ of directors ‖ Aufsichtsrat m. ‖ conseil m. d'administration. / dripping ~ ‖ Abtropfbrett n. ‖ égouttoir m. / ~ driven from a board gate (Mine) ‖ aus einer Diagonalen angesetzte Förderstrecke f. ‖ coistresse f. de montées. / ~ driven from an inclined gallery (Mine) ‖ aus einer einfallenden Strecke angesetzte Förderstrecke f. ‖ coistresse f. de vallées. / ~ for electricity meters ‖ Zählerplatte f. ‖ tableau m. de compteurs d'électricité. / enamelled ~ ‖ Kreidepapier n. ‖ papier m. porcelaine. / fireproof ~ ‖ feuersichere Schalttafel f. ‖ tableau m. incombustible. / floor ~ (Auto) ‖ wagerechtes Fußbrett n. ‖ plancher m. horizontal. / free on ~ (fob) ‖ frei an Bord m. ‖ franc à bord m. / glazed ~ ‖ Glanzpappe f. ‖ carton m. glacé. / ~ of health ‖ Gesundheitsamt n. ‖ conseil m. d'hygiène; office f. de santé. / ~ for instruments ‖ Instrumentenbrett n. ‖ planchette f. à instruments. / leather ~ ‖ Lederpappe f. ‖ carton m. cuir. / ~ under the lifting-wires (Weav) ‖ Platinenbrett n. ‖ planche f. à collet. / ~ of management ‖ Vorstand m. ‖ direction f. / ~ marble ‖ Marmortafel f. ‖ tableau m. en marbre. / meter ~ ‖ Zählertafel f. ‖ tableau m. aux compteurs. / moulding-out ~ see back ~. / oiled ~ ‖ geölte Deckpappe f. ‖ couverture f. en carton huilé. / ~ perpendicular to the heading (Mine) ‖ zur Diagonalen senkrechte Abbaustrecke f. ‖ coistresse f. ou costresse f. de gralles. / planed ~ ‖ gehobeltes Brett n. ‖ planche f. rabotée. / ~ of the plane-table (Surv) ‖ Meßtischplatte f. ‖ tablette f.; tablette f. / pressing ~ ‖ Preßspan m. ‖ carton m. pour apprêt de drap. / to put in ~s ‖ kartonieren ‖ cartonner. / ~ for regular customer's tables ‖

Stammtischtafel f. ‖ tableau m. pour tables d'habitués. / rough ~ ‖ rauhes Brett n. ‖ planche f. brute. / running ~ ‖ Laufbrett n. ‖ marchepied m. / small ~ ‖ Brettchen n. ‖ planchette f. / sound ~ (Arch) ‖ Kanzeldach n. ‖ abat-voix m. / terminal ~ (Tel) ‖ Knopfleiste f. ‖ planchette f. à borne. / toe ~ (Auto) ‖ geneigtes Fußbrett n. ‖ plancher m. incliné. / ~ of trade ‖ Handelsministerium n. ‖ ministère m. du commerce. / ~ of workings ‖ Bauamt n. ‖ intendance f. des bâtiments.

board bell (Tel) ‖ Schrankwecker m. ‖ sonnerie f. de bureau. / ~ chronometer ‖ Borduhr f. ‖ horloge f. ou montre f. de bord. / ~ cutter (Bookb) ‖ Kartonschneider m. ‖ coupeur m. de carton. / ~ cutter (Carp) ‖ Brettsäger m.; Brettschneider m. ‖ scieur m. de long. / ~ drop hammer ‖ Brettfallhammer m. ‖ mouton m. à planche.

boarded ceiling ‖ getäfelte Decke f. ‖ plafond m. lambrissé. / ~ floor ‖ Dielung f.; Bretterfußboden m. ‖ plancher m. / ~ floor of a turntable ‖ Abdeckung f. einer Drehscheibe ‖ plancher m. d'une plaque tournante.

boarding ‖ Bretterverkleidung f.; Dielen n.; Täfelung f. ‖ planchéiage m. / ~ of a roof ‖ Bretterschalung f. eines Daches ‖ couverture f. d'un toit.

boarding floor ‖ gedielter Fußboden m.: Bretterfußboden m. ‖ plancher m.; sol m. planchéié ou en planches. / ~ house (where the unmarried workmen are on the board lodging terms) ‖ Arbeiterheim n. (in dem unverheiratete Arbeiter Wohnung und Kost erhalten) ‖ maison-pension f. (où les célibataires sont logés et nourris). / ~ joist (Carp) ‖ Dielenbalken m. ‖ poutrelle f. de plancher. / ~ machine (Curr) ‖ Krispelmaschine f. ‖ machine f. à rebrousser; marguerite f. mécanique.

boardings pl. (Carp) ‖ Schalbretter npl.; Schallatten fpl. ‖ couchis m.; madriers mpl.

board mechanic ‖ Bordmontör m. ‖ monteur m. de bord. / ~ meeting ‖ Verwaltungsratsversammlung f. ‖ assemblée f. du conseil d'administration. / ~ partition ‖ Bretterwand f. ‖ cloison f. de planches. / ~ receiving set ‖ Bordempfangsgerät n. ‖ poste-récepteur m. de bord. / ~ roof ‖ Bretterdach n. ‖ toit m. de planches. / ~ saw ‖ Fourniersäge f.; Brettersäge f. ‖ scie f. à refendre ou de placage. / ~ sawyer ‖ Dielenschneider m. ‖ scieur m. de planches. / ~ sole (Shoem) ‖ Pappsohle f. ‖ semelle f. en carton. / ~ watch ‖ Borduhr f. ‖ horloge f. ou montre f. de bord.

boat ‖ Boot n.; Kahn m.; Schiffchen n. ‖ bateau m.; canot m.; embarcation f.; nacelle f.; barque f.; esquif m. / clinker built ~ ‖ Klinkerboot n. ‖ canot m. bordé à clin. / collapsible ~ see collapsing ~. / collapsing ~ ‖ Faltboot n.; zusammenklappbares Boot n. ‖ canot m. pliable ou pliant. / decked ~ ‖ gedecktes Boot n. ‖ bateau m. ponté / decked cabin ~ ‖ gedecktes Kajütboot n. ‖ canot m. ponté à cabine. / dinghy ~ ‖ Ruderboot n. ‖ youyou m.; bateau m. à rames. / dismountable ~ ‖ zerlegbarer Kahn m. ‖ canot m. démontable. / electrically driven ~ ‖ elektrisch angetriebenes Boot n. ‖

bateau m. à commande électrique. / ferro-concrete ~ || Schiff n. aus Eisenbeton || bateau m. en béton armé. / ferry ~ || Fähre f. || canot m. de passage. / ferry ~ (Railw) Eisenbahnfähre f. || bac m. porte-train; ferry-boat m. / fire ~ || Spritzendampfer m. || bateau-pompe m. / flying ~ || Wasserflugzeug n.; Flugboot n. || hydro-aéroplane m.; hydravion m. / folding ~ see collapsing ~. / gliding ~ || Wassergleitboot n. || hydroglisseur m. / gun ~ || Kanonenboot n. || canonnière f. / house ~ || Hausboot n. || bateau-maison m. / jolly ~ || Jolle f.; kleines Boot n. || barque f. / keeled bottom of a ~ || gekielter Bootsboden m. || fond m. de bateau en quille. / life ~ || Rettungsboot n. || canot m. ou embarcation f. de sauvetage. / motor ~ || Motorboot n. || bateau m. ou canot m. automobile. / motor ~ of sport type || Sportmotorboot n. || motobateau m. de sport. / open ~ || offenes Boot n. || bateau m. non-ponté. / open ~ with cabin || offenes Kajütboot n. || canot m. non-ponté à cabine. / pleasure ~ || Vergnügungsboot n. || bateau m. de plaisance. / pulling ~ see rowing ~. / river ~ || Flußschiff n. || bateau m. de fleuve. / rowing ~ || Ruderboot n. || bateau m. à rames. / ~ for rowing sport || Rudersportboot n. || canot m. de sport à avirons. / sailing ~ || Segelboot n. || bateau m. à voiles. / ~ of a ship || Beiboot m. eines Schiffes || canot m. d'un navire. / small ~ || Kahn m. || bateau m.; barque f.; nacelle f.; esquif m. / steam ferry ~ || Fährdampfer m. || vapeur m. de passage. / steam launch ~ || Dampfbarkasse f. || chaloupe f. à vapeur. / tow ~ || Schleppkahn m. || remorqueur m.; bateau m. de remorque. / tug ~ see tow ~. / undecked ~ see open ~.

boat ... see also boat's ... / ~ bow of ~ || Bootsbug m. || proue f. de canot. / ~ bridge || Pontonbrücke f. || pont m. de bateaux. / ~ builder || Bootsbauer m. || constructeur m. de bateaux ou d'embarcations. / ~ building || Bootsbauerei f.; Bootsbau m. || construction de canots. / ~ carriage || Bootswagen m. || haquet m. à nacelle. / ~ compass || kleiner Kompaß m. || volet m. / ~ furnishing || Bootsausrüstung f. || équipement m. de canots. / ~ hoist || Bootsheißmaschine f. || monte-embarcations m. / ~ hook || Bootshaken m. || croc m. de batelier; gaffe f. / to put off a launch with a ~ hook || eine Barkasse f. mit einem Bootshaken abstoßen || déborder une chaloupe avec une gaffe. / ~ house || Bootsschuppen m.; Bootshaus n. || hangar m. de bateau. / ~ house with club rooms || Bootshaus n. || club-house m. des canotiers.

boating || Floßbrücke f. || pont m. de radeaux.

boat, keel of ~ || Bootskiel m. || quille f. de canot. / ~ letting out || Bootsverleihung f. || location f. de bateaux. / ~-man || Bootsgast m. || canotier m. / ~ motor || Bootsmotor m. || moteur m. pour bateaux. / ~ nail || Bootsnagel m. || clou m. pour bordage de canot.

boat's ... see also boat ... / ~ awning || Bootssonnensegel n. || tente f. d'embarcation. / ~ cooking stove || Bootskombüse f. || fourneau m. d'embarcation. / ~ crew || Bootsbesatzung f. || équipage m. d'un canot ou d'une embarcation.

boat seaplane see boat, flying.
boat's flag || Bootsflagge f. || pavillon m. d'embarcation. / ~ locker || Bootskasten m. || coffre m. d'embarcation. / ~ mast || Bootsmast m. || mât m. d'embarcation. / ~ rig || Bootstakelage f. || gréement m. d'embarcation. / ~ sail || Bootssegel n. || voile f. d'embarcation. / ~ shed || Bootswerft f. || chantier m. de construction des embarcations. / ~ sling || Heißstropp m. || patte f. d'embarcation. / ~ tackle || Bootstakel n. || palan m. d'embarcation.
boatswain || Bootsmann m. || maître m. de manœuvre. / chief ~ || Oberbootsmann m. || premier maître m. de manœuvre.
boatswain's mate || Bootsmannsmaat m. || aide-bosseman m.; second-maître m.
boat-tail bullet || Geschoß n. mit verjüngtem Bodenteil; Stromliniengeschoß n. || balle f. fuselée; obus m. à culot rétreint.
boat towing || Schiffstreideln n. || halage m. de bateaux. / ~ varnish || Bootslack m. || vernis m. pour bateaux. / ~ wagon || Bootswagen m.; Pontonwagen m. || haquet m. à nacelle ou à ponton. / ~ wright || Kahnbauer m. || constructeur m. de canots.
bob || Büschel n.; Bündel n. || poignée f.; touffe f. / ~ (Mar) || Senkblei n. || sonde f.; plomb m. / ~ of a fishing line || Schwimmer m. einer Angelschnur || flotteur m. d'une ligne. / ~ of a pendulum || Gewicht n. oder Linse f. eines Pendels || lentille f. d'un pendule. / ~ plumb ~ see bob (Mar).
bobbin || Spule f.; Rolle f. || bobine f.; rouleau m. / cross wound ~ || Spule f. mit Kreuzwicklung f. || bobine f. à fil croisé. / flanged ~ || Scheibenspule f. || bobine f. à rebords. / ~ of hard paper || Spule f. aus Hartpapier || bobine f. en papier durci. / paper ~ || Papierspule f. || bobine f. en carton. / warper's ~ || Zettelspule f. || bobine f. d'ourdissoir. / ~ for winding the sewing-cotton || Rolle f. zum Aufwickeln von Nähgarn || bobine f. pour enrouler les fils à coudre. / wooden ~ || Holzspule f. || bobine f. ou bobinot m. en bois.
bobbin board || Spulenbrett n. || râtelier m. / ~ boy (Weav) || Spulenträger m. || porteur m. de bobines. / ~ carrier (Spinn) || Spulensammler m. || ramasseur m. de bobines. / ~ doffer (Spinn) || Abzieherin f. || leveuse f. de bobines. / ~ frame (Spinn) || Spulmaschine f. || bobinoir m.
bobbing (Barometer) || plötzliche Schwankung f. || oscillation f.
bobbin machine || Klöppelmaschine f. || machine f. à fuseaux. / machine for making ~s || Maschine f. zur Spulenherstellung || machine f. pour la fabrication des bobines. / ~ net || Tüll m. || tulle m. / double-ended ~ pressing-on machine || doppelendige Aufdrückmaschine f. für Spulenflanschen || presse f. double à serrer les disques de bobines. / ~ procedure || Spulenverfahren n. || méthode f. de filature avec bobines. / ~ soaker (Weav) || Garnanfeuchter m. || mouilleur m. d'époules. / ~ stand || Spulengestell n. || porte-bobines m. / ~ thread || Spulenfaden m. || fil m. de bobine. / winding machine for ~s || Bobinenspulmaschine f. || machine f. à bobiner. / ~ work || Klöppelarbeit f. || travail m. au fuseau.
bobbling of the axis || Schlottern n. der Achse || vacillation f. de l'axe.
bobsleigh || Rodelschlitten m. || luge f.

bodice || Korsett n.; Mieder n. || corsage m.; corset m. / ~ maker || Taillenarbeiterin f. || corsagière f.
bodkin (Print) || Punkturspitze f.; Ahle f. || pointe f.; languette f. / ~ (Saddl; Shoem) || Pfriem m. || alêne f.; poinçon m. / sailmaker's ~ || Segelmacherpfriem m. || marprime f. ou poinçon m. du voilier.
body (Coachm) || Karosserie f.; Wagenkasten m. || carrosserie f. / ~ (Colour) || Deckkraft f. || opacité f. / ~ (High furnace) || Schachtraum m.; Kernschacht m. || vide m.; cuve f. / ~ (Insur) || Kasko n. || corps m. et quille. / ~ (Phys) || Körper m.; Rumpf m. || corps m.; solide m. / ~ (Pott) || Scherben m. || pâte f. / ~ (Print) || Schriftkegel m. || corps m. / all-steel ~ || Ganzstahlkarosserie f. || carrosserie f. tout acier. / all-weather ~ || Allwetterkarosserie f. || voiture f. décapotable. / ~ for automobiles || Automobilkarosserie f. || carrosserie f. pour automobiles. / ~ of axle || Achskörper m. || corps m. de l'essieu. / axle-box ~ || Achslagergehäuse n. || corps m. de boîte d'essieu. / ~ of a bell || Glockenkörper m. || corps m. d'une cloche. / ~ of a boiler || Körper m. eines Kessels; Kesselkörper m. || corps m. d'une chaudière. / ~ of car || Wagenkasten m.; Karosserie f.; Wagenaufbau m. || carrosserie f. / ~ of a colour || Deckkraft f. einer Farbe || opacité f. d'une couleur. / ~ of a column || Säulenschaft m. || fût m. ou vif m. ou tronc m. d'une colonne. / compound ~ || zusammengesetzter Körper m. || corps m. composé. / ~ with concealed hood || Karosserie f. mit versenktem Dach || carrosserie f. à capote invisible. / dark ~ || dunkler Körper m. || corps m. de couleur sombre ou foncée. / elementary ~ || Grundstoff m. || corps m. élémentaire. / floating ~ || schwimmender Körper m. || corps m. flottant. / fluid ~ || flüssiger Körper m. || corps m. fluide. / framework of ~ || Gerippe n. der Karosserie || squelette m. de caisse. / gaseous ~ || gasförmiger Körper m. || corps m. gazeux. / ~ heavier than air || Körper m. schwerer als Luft || corps m. plus lourd que l'air. / ~ heavier than water || Körper m. schwerer als Wasser; sanker Körper m. || corps m. plus pesant que l'eau. / immersed ~ || eingetauchter Körper m. || corps m. immergé. / ~ of the instrument || Instrumentenkörper m. || corps m. de l'instrument. / interchangeable ~ || auswechselbare Karosserie f. || carrosserie f. interchangeable. / ~ of least resistance || Körper m. des kleinsten Widerstandes || corps m. de la moindre résistance. / ~ of a letter || Schriftkegel m. (Letterndicke in Richtung der Buchstabenhöhe gemessen) || corps m. de lettre; force f. de corps d'un caractère. / liquid ~ || tropfbar flüssiger Körper m. || corps m. liquide. / ~ of the main rod || Schaft m. der Treibstange || corps m. de bielle. / ~ of miners || Knappschaft f. || corps m. des mineurs. / ~ of a pump || Pumpenzylinder m.; Pumpenstiefel m. || corps m. ou cylindre m. de pompe. / ~ of railway || Bahnkörper m. || corps m. de la voie. / rigid ~ || starrer Körper m. || corps m. rigide. / rotor ~ (Electr) || Ankerkörper m. || corps m. d'induit. / simple ~ (Chem) || einfacher Körper m.; Grundstoff m. || élément m.; substance f. élé-

mentaire. / sliding roof ~ || Karosserie f.
mit verschiebbarem Dach || carrosserie f.
à toit mobile. / solid ~ || fester Körper m.
|| corps m. solide. / streamline ~ || Strom-
linienkarosserie f. || carrosserie f.à lignes
fuyantes. / ~ of the strongest form ||
Körper m. von gleichem Widerstande ||
corps m. d'égale résistance. / sunshine ~
‖ Limusine f. mit vorn zurückschieb-
barem Dach || carrosserie f. à avant
transformable. / testing ~ || Probefahrt-
karosserie f. || carrosserie f. d'essai. /
vitreous ~ || Glaskörper m. || corps m.
vitré. / wooden ~ || Holzkörper m. ||
pièce f. en bois.
body colour || Deckfarbe f. || couleur f. opa-
que ou à la gouache. / ~ flanging device ||
Zargenbördeleinrichtung f. || dispositif
m. pour le tombage des bords. / ~ form-
ing machine || Zargenbiegemaschine f. ||
machine f. à former les corps de boîtes.
/ framework of ~ || Gerippe n. der Karos-
serie || squelette m. de caisse. / ~ ironer
(Coachm) || Kastenbeschläger m. || fer-
reur m. en voitures. / ~ linen || Leib-
wäsche f. || linge m. de corps. / ~ part
(Coachm) || Kastenteil m. || partie f. de
caisse. / ~ plan (Shipb) || Spantenriß m.
|| plan m. des couples ou des membrures.
/ ~ resistance (Aero) || Rumpfwiderstand
m. || résistance f. du corps. / ~ spring ||
Achsenfeder f. || ressort m. d'essieu. /
~ spring (Coachm) || Tragfeder f.; Wa-
genfeder f. || ressort m. de voiture.
bog || Morast m. || terrain f. vaseux; vase
f. / ~ (Build) || Senkgrube f.; Senkloch n.
|| puisard m.; puits m. / ~ (Geol) || Moor
n. || marais m.
bog bean leaves pl. || Bitterkleeblätter npl.
|| feuilles fpl. de trèfle d'eau.
boggy ground || Torfboden m.; sumpfiger
Boden m. || sol m. tourbeux ou maré-
cageux.
Boghead coal || Bogheadkohle f. || boghead
m.
bogie || Drehgestell n. || bogie f.; truck m.
pivotant. / pivoted ~ || Drehschemel m. ||
train m. de roues pivotant. / pressed
steel ~ || Drehgestell n. aus gepreßten
Blechen || bogie f. emboutie en tôles.
bogie brake || Drehgestellbremse f. || frein
m. de bogie. / ~ car || Drehgestellwagen
m. || voiture f. à bogies. / ~ crane || Roll-
kran m. || grue f. à chariot. / ~ engine ||
kurvenbewegliche Lokomotive f.; Loko-
motive f. mit beweglichem Radgestell ||
locomotive f. à jeu transversal d'essieu
ou à inscription en courbe. / ~ frame ||
bewegliches Radgestell n. || train m. de
quatre roues mobile autour d'une cheville.
bogie-land || Sumpfland n.; Marschland n.
|| pays m. marécageux. / to drain a piece
of ~ || ein versumpftes Grundstück n.
entwässern || dessécher ou drainer un
terrain marécageux.
bogie locomotive see bogie engine.
bogie spring, elliptical ~ || Doppelfeder f.
für Drehgestelle || ressort m. à pincette
pour bogies. / ~ for locomotives || Dreh-
gestellfeder f. für Lokomotiven || ressort
m. pour bogies de locomotive.
bogie wagon (Crane) || Fahrgestell n.;
Laufwagen m. || chariot m. / ~ (Railw) ||
Drehgestellwagen m. || wagon m. à bo-
gies.
bogie wheel (Loc) || Laufrad n. || roue f.
porteuse.

bog iron ore || Limonit m.; Raseneisenstein
m.; Sumpferz n. || limonite f.; mine f. de
fer à fleur de terre; mine f. de marais. / ~
ore see ~ iron ore. / ~ peat || Rasentorf m.
|| tourbe f. des marais.
boil, to || kochen; abkochen; aufkochen;
sieden || cuire; faire bouillir; bouillir. /
~ briskly || durchkochen; lebhaft kochen
|| faire bouillir fortement; bouillir vive-
ment. / ~ the clear liquor (Sugar) || das
Klärsel kochen || cuire le sirop. / ~ down
(Chem) || einkochen || concentrer. / ~ the
juice || den Saft m. einkochen || évaporer
ou cuire le jus. / ~ linseed oil upon
litharge || Firnis m. kochen || lithargyrer
l'huile de lin. / ~ off || auskochen '||
chasser par ébullition. / ~ out (Chem)
|| auskochen || bouillir complètement. /
~ over || übersieden || déborder. / ~
the silk || die Seide kochen oder ent-
schälen || cuire ou décreuser la soie. /
~ together || verkochen || faire bouillir
ensemble. / ~ water || das Wasser ab-
kochen || bouillir l'eau.
boiled || abgekocht || bouilli. / ~ (Silk) ||
gekocht; geschält || décreusé; cuit.
boiler || Kessel m. || chaudière f. / ~
(Build) || Heizkessel m. || chaudière
f. de chauffage. / ~ (Dyer) || Küpe f. ||
cuve f. de teinture. / ~ (Househ) || Koch-
topf m. || marmite f. / ~ (Industry) ||
Kochkessel m.; Siedekessel m. || bouil-
loire f.; chaudron m.; chaudière f. à
cuire. / ~ (Mach) || Kessel m.; Dampf-
kessel m. || chaudière f. à vapeur ou
générateur. / ~ (Sugar) || Siedepfanne f.;
Läuterungskessel m. || chaudière f. de
raffinage. / armature of a ~ || Kesselgarni-
tur f. || garniture f. d'une chaudière à
vapeur. / automatically controlled ~ ||
selbsttätig arbeitender Kessel m. || chau-
dière f. à réglage automatique. / battery
of ~s || Kesselbatterie f. || batterie f. de
chaudières à vapeur. / bed of the ~ ||
Kessellager n. || fondation f. d'une
chaudière. / bleaching ~ || Bleichkessel
m. || chaudière f. à blanchiment. /
body of a ~ || Kesselkörper m. || corps
m. de la chaudière à vapeur. / cast-
iron ~ || gußeiserner Kessel m. || chau-
dière f. en fonte. / ~ for central heating ||
Zentralheizungskessel m. || chaudière f.
pour le chauffage central. / ~ for coal
dust combustion || Kohlenstaubkessel m.
|| chaudière f. chauffée au poussier de
charbon. / combination of flue and
smoke-tube ~ || kombinierter Flammrohr-
rauchrohrkessel m. || chaudière f. com-
binée de tube-foyer et de tubes de fu-
mée. / copper ~ || Kupferkessel m. || chau-
dière f. en cuivre. / ~ with copper fire-
box || Kessel m. mit kupferner Feuer-
buchse || chaudière f. avec boîte à feu
en cuivre. / Cornish ~ || Cornwallkessel
m.; Einflammrohrkessel m. || chaudière
f. à un seul tube-foyer. / countercurrent
~ || Gegenstromkessel m. || chaudière f.
système contre-courant. / cross-tube ~ ||
Quersieder m. || générateur m. transver-
sal; chaudière f. à bouilleurs transver-
saux. / cylindrical ~ || Zylinderkessel m. ||
chaudière f. cylindrique. / cylindrical
part of a ~ (Loc) || Langkessel m. || corps
m. cylindrique de la chaudière. / double
~ || Doppelkessel m. || chaudière f. double.
/ double-ended ~ || Doppelender m.;
Doppelkessel m. || chaudière f. à foyers
aux deux extrémités. / double-flue ~ ||

Zweiflammrohrkessel m. || chaudière f.
à deux foyers intérieurs. / double-shell ~ ||
doppelwandiger Kessel m. || chaudière f.
à double paroi. / double-storied ~ ||
Etagenkessel m. || chaudière f. à étages.
evaporating ~ || Abdampfkessel m. || chau-
dière f. ou bassine f. d'évaporation. / ex-
haust heat ~ || Abwärmekessel m. || chau-
dière f. utilisant la chaleur d'échappe-
ment. / experimental ~ || Versuchskessel
m. || chaudière f. d'essai. / fire-tube ~ ||
Feuerrohrkessel m. || chaudière f. multi-
tubulaire à tubes de fumée. / fixed ~ ||
ortsfester Kessel m. || chaudière f. fixe. /
flue ~ || Flammrohrkessel m. || chaudière
f. à tube-foyer. / garniture of a ~ || Kes-
selgarnitur f. || garniture f. d'une chau-
dière à vapeur. / hardening ~ || Härte-
kessel m. || bassin m. de trempe. / high-
pressure ~ || Hochdruckkessel m. || chau-
dière f. à haute pression. / hot water ~ ||
Heißwasserkessel m. || chaudière f. pour
installations d'eau chaude. / internal
flue ~ || Feuerbuchsenkessel m. || chau-
dière f. à foyer intérieur. / lagging of a ~
|| Kesselbekleidung f. || enveloppe f. en
tôle pour chaudières. / land type of ~ ||
Kessel m. für ortsfesten Betrieb || chau-
dière f. fixe. / ~ with large water space ||
Großwasserraumkessel m. || chaudière f.
à grand volume d'eau. / linen ~ || Wasch-
kessel m. || chaudière f. pour le lessivage. /
locomotive ~ || Lokomotivkessel m. ||
chaudière f. de locomotive. / loco-type
~ see locomotive ~. / low-pressure ~ ||
Niederdruckkessel m. || chaudière f. à
basse pression. / main ~ || Hauptkessel m.
|| chaudière f. principale. / marine ~ ||
Schiffskessel m. || chaudière f. marine. /
mean-pressure ~ || Mitteldruckkessel m.
|| chaudière f. à moyenne pression. /
multiple-stage ~ || Etagenkessel m. ||
chaudière f. à étages. / multitubular
~ || Röhrenkessel m. || chaudière f.
multitubulaire. / multitubular ~ with
removable fire box and smoke tu-
bes || ausziehbarer Flammrohrsiederöh-
renkessel m. || chaudière f. avec boîte à
feu et faisceau tubulaire amovible. / one-
flue ~ || Einflammrohrkessel m. || chau-
dière f. à un seul tube-foyer ou à un seul
foyer. / one-seam rivetted ~ || Kessel m.
mit einer Nietnaht || chaudière f. à une
rangée de rivets. / open ~ || offener
Kessel m. || chaudière f. découverte. /
portable ~ || beweglicher oder trans-
portabler Kessel m. || chaudière f. trans-
portable ou locomobile. / principal ~ ||
Hauptkessel m. || chaudière f. principale.
/ rag ~ || Hadernkocher m. || chaudière
f. à haillons. / regulation of ~s || Kessel-
regelung f. || réglage m. de chaudières. /
scouring ~ (Bleach) || Bäuchkessel m. ||
chaudière f. de coulage. / semi-portable ~
|| Lokomobilkessel m. || chaudière f. mi-
fixe. / single ended ~ || Einendkessel m. ||
chaudière f. à un seul foyer. / smoke-
tube ~ || Rauchröhrenkessel m. || chaudière
f. à tubes de fumée. / standard ~ || Normal-
kessel m. || chaudière-type m. / stationary
~ || Landkessel m.; Kessel m. für orts-
festen Betrieb || chaudière f. terrestre ou
fixe. / steam ~ || Dampfkessel m. || chau-
dière f. à vapeur. / ~ with steep tubes ||
Steilrohrkessel m. || chaudière f. tubu-
laire verticale. / stirring ~ || Rührkessel
m. || chaudière f. à agitateur. / super-
heater ~ || Kessel m. mit eingebautem

Überhitzer ‖ chaudière f. avec surchauffeur intérieur. / training ~ ‖ Lehrkessel m. ‖ chaudière f. d'instruction. / tubular ~ ‖ Röhrenkessel m. ‖ chaudière f. tubulaire *ou* à tubes. / upper ~ ‖ Oberkessel m. ‖ chaudière f. supérieure. / vertical ~ ‖ stehender Kessel m. ‖ chaudière f. verticale. / ~ with large vertical tubes ‖ weitrohriger Steilrohrkessel m. ‖ chaudière f. à gros tubes verticaux. / vertical tube ~ ‖ stehender Röhrenkessel m. ‖ chaudière f. tubulaire verticale. / vulcanizing ~ ‖ Vulkanisierkessel m. ‖ chaudière f. de vulcanisation. / water capacity of a ~ ‖ Wasserinhalt m. *oder* Wasserfassungsvermögen n. eines Kessels. / cubage m. d'eau d'une chaudière. / water-tube ~ ‖ Wasserrohrkessel m. ‖ chaudière f. aquatubulaire *ou* à tubes d'eau.

boiler acceptance ‖ Kesselabnahme f. ‖ réception f. d'une chaudière. / ~ acceptance test ‖ Kesselabnahmeprüfung f. ‖ épreuve f. de réception d'une chaudière. / ~ anti-scaling composition *see* composition. / armature of a ~ ‖ Kesselgarnitur f.; Kesselarmatur f. ‖ garniture f. d'une chaudière à vapeur. / ~ attendance ‖ Kesselbedienung f. ‖ service m. de la chaudière. / ~ attendant ‖ Kesselwärter m. ‖ alimenteur m. (de chaudières); chauffeur m. / bank of ~s *see* battery of ~s. / barrel of a ~ *see* body of a ~. / ~ barrel (Loc) ‖ Langkessel m. ‖ chaudière f. cylindrique. / ~ base *see* boiler end. / battery of ~s ‖ Kesselbatterie f. ‖ batterie f. de chaudières à vapeur. / ~ bearer ‖ Kesselträger m. ‖ support m. de la chaudière. / bed of a ~ ‖ Kessellager n.; Kesselfundament n. ‖ fondation f. d'une chaudière. / bedding of a ~ *see* bed of a ~. / body of a ~ ‖ Kesselkörper m. ‖ corps m. d'une chaudière à vapeur. / ~ bottom *see* boiler end. / ~ braces pl. ‖ Verstrebungen fpl. eines Dampfkessels ‖ renforts mpl. d'une chaudière. / ~ bunker ‖ Kesselbunker m. ‖ soute f. de la chaudière. / capacity of a ~ ‖ Fassungsvermögen n. *oder* Rauminhalt m. eines Kessels; Kesselinhalt m. ‖ volume m. *ou* capacité f. *ou* contenance f. d'une chaudière. / ~ car ‖ Kesselwagen m. ‖ wagon m. citerne. / ~ casing ‖ Kesselbekleidung f. ‖ chemise f. de chaudière. / ~ cleading ‖ Kesselverkleidung f. ‖ garniture f. *ou* enveloppe f. d'une chaudière. / ~ cleaner ‖ Kesselreiniger m. ‖ nettoyeur m. de chaudière. / ~ cleaning ‖ Kesselreinigung f. ‖ nettoyage m. de chaudière. / ~ clothes pl. *see* boiler worker, clothes for. / ~ coal ‖ Kesselkohle f. ‖ charbon m. de chaudière. / ~ coaling plant ‖ Kesselbekohlungsanlage f. ‖ installation f. pour alimenter les foyers des chaudières. / ~ composition ‖ Kesselsteinverhütungsmittel n. ‖ enduit m. anti-incrustant pour chaudières; tartrifuge m.; anti-incrustant m.; sélénifuge m. / ~ construction ‖ Bau m. des Kessels ‖ construction f. des chaudières; travail m. de chaudronnerie. / ~ construction material ‖ Kesselbaumaterial n. ‖ matériaux mpl. pour la construction des chaudières. / ~ covering ‖ Kesselisolierung f.; Kesselverkleidung f. ‖ revêtement m. pour chaudières; garniture f. *ou* enveloppe f. d'une chaudière. / ~ cradle ‖ Kesselbock

m. ‖ chevalet m. / ~ dome ‖ Kesseldom m. ‖ dôme m. de chaudière. ‖ ~ dress *see* boiler worker, clothes for. / efficiency of a ~ ‖ Leistungsfähigkeit f. eines Kessels ‖ puissance f. d'une chaudière.

boiler end ‖ Kesselboden m. ‖ fond m. de chaudière. / ~ less than x mm diameter ‖ Kesselboden m. unter x mm Durchmesser ‖ fond m. de chaudière d'un diamètre inférieur à x mm. / dished ~ ‖ gewölbter Kesselboden m. ‖ fond m. bombé. / flanged ~ ‖ umgezogener Kesselboden m. ‖ fond m. de chaudière à collet. / inward-flanged ~ ‖ eingehalster Kesselboden m. ‖ fond m. à collet intérieur. / machine-flanged ~ ‖ maschinell umgezogener Kesselboden m. ‖ fond m. de chaudière bridé à la machine. / flat ~ ‖ flacher Kesselboden m. ‖ fond m. plat.

boiler end, cylindrical seam on ~ ‖ Rundnaht f. am Kesselboden ‖ joint m. circulaire du fond de chaudière. / ~ flange ‖ Krempe f. des Kesselbodens ‖ bord m. du fond de chaudière. / flanging the ~ ‖ Bördeln n. des Kesselbodens ‖ bridage m. *ou* emboutissage m. du fond de chaudière. / ~ plate ‖ Kesselbodenblech n. ‖ tôle f. pour fonds de chaudières.

boiler explosion ‖ Kesselexplosion f. ‖ explosion f. de chaudière. / ~ feed ‖ Kesselbekohlung f.; Kesselspeisung f. ‖ alimentation f. des chaudières. / ~ feed device ‖ Kesselspeisevorrichtung f. ‖ appareil m. alimentaire des chaudières. / ~ feeder ‖ Kesselwärter m. ‖ alimenteur m. (de chaudières). / ~ feeding plant ‖ Kesselbekohlungsanlage f.; Kesselspeiseanlage f. ‖ installation f. d'alimentation des chaudières. / ~ feeding pump ‖ Kesselspeisepumpe f. ‖ pompe f. d'alimentation pour chaudières. / ~ fittings pl. ‖ Kesselarmaturen fpl. ‖ accessoires mpl. pour chaudières. / ~ flue ‖ Fuchs m. ‖ carneau m. / ~ fluid *see* ~ composition. / ~ forge ‖ Kesselschmiede f. ‖ chaudronnerie f.; atelier m. de chaudronnerie. / ~ forging job *see* boiler work. / ~ fur *see* boiler scale. / ~ furnace ‖ Kesselfeuerung f. ‖ foyer m. de chaudière. / ~ furniture ‖ Kesselgarnitur f. ‖ garniture f. d'une chaudière. / garniture of a ~ *see* ~ furniture. / ~ grate ‖ Kesselrost m. ‖ grille f. pour chaudières. ‖ ~ holder ‖ Kesselträger m. ‖ support m. d'une chaudière.

boiler house ‖ Kesselhaus n. ‖ chaufferie f.; bâtiment m. *ou* salle f. *ou* chambre f. de chaudières. / equipment for ~es ‖ Kesselhauseinrichtung f. ‖ installation f. de chaufferies. / ~ meter ‖ Kesselhausinstrument n. ‖ instrument m. de mesure pour salle de chaudières.

boiler incrustation *see* boiler scale. / ~ inspection ‖ Kesselrevision f.; Dampfkesselprüfung f.; Kesseluntersuchung f. ‖ visite f. *ou* revision f. des chaudières. / ~ inspection association ‖ Dampfkesselüberwachungsverein m.; Kesselrevisionsverein m. ‖ association f. pour la surveillance des générateurs de vapeur. / ~ inspector ‖ Kesselprüfer m.; Kesselaufseher m.; Kesselrevisor m. ‖ inspecteur m. de chaudières. / ~ insurance ‖ Kesselversicherung f. ‖ assurance f. pour chaudières. / lagging of ~ ‖ Kesselbekleidung f. ‖ enveloppe f. (en tôle) d'une chaudière. / ~ maker ‖ Kesselschmied m. ‖ chaudronnier m.; constructeur m.

de chaudières. / ~ making ‖ Kesselfertigung f. ‖ grosse chaudronnerie f. / ~ masonry ‖ Kesseleinmauerung f. ‖ emmurage m. *ou* maçonnerie f. de chaudières. / ~ mountings pl. ‖ Kesselgarnitur f. ‖ garniture f. d'une chaudière. / ~ output ‖ Kesselleistung f. ‖ débit m. (de vapeur) d'une chaudière. / ~ plant ‖ Kesselanlage f. ‖ installation f. *ou* batterie f. de chaudières. / ~ plate ‖ Kesselplatte f.; Kesselblech n. ‖ tôle f. de chaudière. / ~ pressure ‖ Kesseldruck m. ‖ pression f. dans la chaudière. / regulation of ~s ‖ Kesselregelung f. ‖ réglage m. de chaudières à vapeur. / ~ repairer ‖ Kesselflicker m. ‖ chaudronnier réparateur m. / ~ repairing ‖ Kesselausbesserung f.; Kesselreparatur f. ‖ réparation f. *ou* raccommodage m. des chaudières. / ~ rinsing pump ‖ Kesselspülpumpe f. ‖ pompe f. à laver les chaudières. / ~ room ‖ Heizraum m. ‖ chaufferie f.

boiler scale ‖ Kesselstein m.; Pfannenstein m. ‖ incrustations fpl.; tartre m.; couche f. de tartre; calcin m.; sédiments mpl. / incrustation of ~ ‖ Kesselsteinansatz m. ‖ dépôt m. de calcin, précipitation f. de tartre. / to knock-off the ~ ‖ den Kesselstein m. abklopfen ‖ détartrer *ou* piquer *ou* écailler une chaudière.

boiler scaling ‖ Kesselsteinabmeißeln n. ‖ piquage m. des chaudières. / ~ scaling hammer ‖ Kesselhammer m. ‖ marteau m. de piquage. / ~ sections working machine ‖ Kesselgliederbearbeitungsmaschine f. ‖ machine f. à travailler les pièces de chaudières. / ~ sediment *see* boiler scale. / ~ setting ‖ Kessellagerung f. ‖ disposition f. *ou* logement m. de la chaudière.

boiler shell ‖ Kesselmantel m. ‖ corps m. cylindrique *ou* chemise f. *ou* enveloppe f. de la chaudière. / ~ drilling machine ‖ Kesselmantelbohrmaschine f. ‖ machine f. à percer pour tôles de chaudières. / ~ plate ‖ Kesselwand f. ‖ paroi f. de la chaudière.

boiler smith ‖ Kesselschmied m. ‖ chaudronnier m. / ~ solvent ‖ Kesselsteinlösemittel n.; Kesselsteinbeseitigungsmittel n. ‖ induit m. désincrustant pour chaudières; désincrustant m. / ~ switchboard ‖ Kesselwarte f. ‖ poste m. de surveillance pour chaudières. / ~ testing *see also* ~ inspection ‖ Kesselrevision f.; Kesselprüfung f. ‖ révision f. *ou* surveillance f. des chaudières.

boiler thermometer with fixed immersion tube ‖ Kesselthermometer n. mit festem Tauchrohr ‖ thermomètre m. pour chaudières avec tube d'introduction fixe. / ~ with two branches ‖ Kesselthermometer n. mit Doppelstutzen ‖ thermomètre m. pour chaudières avec tube double.

boiler tube ‖ Kesselrohr n.; Siederohr n. ‖ tube m. de chaudière; bouilleur m. / ~ cleaner ‖ Kesselrohrreiniger m. ‖ appareil m. à décrasser les tubes de chaudière. / cleaning apparatus for the packing space of ~ closures ‖ Dichtungsflächenreiniger m. für Siederohrverschlüsse ‖ nettoyeur m. de surfaces d'étoupement des tubes bouilleurs. / ~ cleaning drum ‖ Kesselrohrreinigungstrommel f. ‖ tambour-laveur m. pour tuyaux de chaudières. / ~ and smoke-tube working machine ‖ Siederohr- und

Rauchrohrbearbeitungsmaschine f. || machine f. pour le travail des tubes bouilleurs et des tubes à fumée.

boiler, volume of a ~ *see* capacity of a ~. / water capacity of a ~ || Wasserinhalt m. *oder* Wasserfassungsvermögen n. eines Kessels || cubage m. d'eau d'une chaudière. / ~ work || Kesselschmiedearbeit f. || travail m. de chaudronnerie *ou* du chaudronnier; chaudronnerie f. / ~ worker || Kesselarbeiter m. || chaudronnier m. / clothes pl. for ~ workers || Kesselanzug m. || habillement m. pour chaudronniers. / ~ works pl. || Kesselfabrik f.; Kesselwerke npl.; Kesselschmiede f. || chaudronnerie f.; atelier m. de chaudronnerie.

boiling || siedend || bouillant. / high ~ || hochsiedend || à point d'ébullition élevé. / low ~ || tiefsiedend; leichtsiedend || à bas point d'ébullition. / stable when ~ || kochbeständig || stable à l'ébullition.

boiling || Kochen n.; Sieden n. || cuisson f.; ébullition f. / ~ the rags (Pap) || Kochen n. der Lumpen || cuisson f. des haillons. / ~ of saltpetre (Chem) || Salpetersud m. || cuite f. *ou* évaporation f. du salpêtre. / ~ of the silk || Entschälen n. *oder* Kochen n. der Seide || cuisson f. *ou* décreusage m. de la soie. / slag ~ (Met) || Schlackenfrischen n.; Schlackenpuddeln n. || puddlage m. gras *ou* en scories. / ~ of sugar juice || Verkochen n. des Zuckersaftes; Kochen n. des Klärsels || concentration f. des jus sucrés; cuite f. du sirop. / ~ of the wort || Würzkochen n. || cuisson f. de la bière.

boiling apparatus || Kochapparat m. || appareil m. de cuisson *ou* à bouillir. / ~ burner || Rundbrenner m.; Ringbrenner m. || fourneau m. à gaz; couronne f. de gaz. / ~ down || Einkochen n. || évaporation f. / ~ down apparatus made from pure nickel with double wall || doppelwandiger Einkochapparat m. aus Reinnickel || appareil m. d'évaporation en nickel pur à double paroi. / ~ fermentation || kochende Gärung f. || fermentation f. tumultueuse. / ~ conical flask || Becherkolben m. || fiole f. conique à ébullition. / ~ heat || Siedehitze f. || température f. d'ébullition. / ~ kettle || Kochkessel m. || marmite f. / ~ liquor || Siedelauge f.; kochende Lauge f. || lessive f. bouillante; eaux fpl. de cuite. / ~ and crabbing machine || Koch- und Fixiermaschine f.; Brennbock m. || machine f. à cuire et fixer. / ~ out || Abbrühen n.; Abdämpfen n. || échaudage m. / ~ period || Siedezeit f. || temps m. de mise en ébullition. / ~ point || Siedepunkt m. || point m. d'ébullition. / ~ room (Sugar) || Siederei f. || laboratoire m. à sucre. / ~ test || Kochprobe f. || essai m. de cuisson. / ~ tube || Reagenzrohr n. || tube m. à essai. / ~ vessel || Kochkessel m.; Aufkochgefäß n. || bouilleur m.; chaudière f. à bouillir.

bolero || Damenjäckchen n. || boléro m.

bolide (Astron) || Feuerball m.; Feuerkugel f. || météore m.; aérolithe m.; bolide m.

bollard (Hydr arch) || Poller m.; Schiffshalter m. auf Kaimauern || borne f. d'amarrage. / ~ (Shipb) || Poller m.; Bäting f. || bitte f.; bitton m. / ~s pl. (Harbour) || Duckdalbe f. || estacade f.

bolometer || Bolometer n. || bolomètre m.

bolometric || bolometrisch || bolométrique.

bolster || Polster n.; Unterlage f.; Kissen n. || rembourrement m.; coussin m. / ~ (Build) || Schalbrett n.; Schallatte f.; Schalholz n. || planche f. de boisage; bois m. de couchis. / ~ (Carp) || Sattelholz n.; Schirrbalken m. || corbeau m.; racinal m.; sous-poutre f.; sous-longueron m. / ~ (Locksm) || Lochscheibe f.; Lochring m. || perçoir m. / hind ~ (Coachm) || Hinterachsschale f. || hausse f.; sellette f. de derrière. / rigid ~ (Railw) || fester Schemel m. || traverse f. fixe. / ~ of a spindle || Halslager n. einer Spindel || collet m. d'une broche. / swing ~ || Pendelwiege f. || balancier m. transversal. / swivelling ~ (Railw) || Drehschemel m. || traverse f. mobile.

bolster lip (Coachm) || Kotlöffel m. || cacheboue m.; couvre-moyeu m.

bolt, to (Mach) || anbolzen; verbolzen; mit Schrauben befestigen || cheviller; boulonner. / ~ (Mill) || sieben || cribler; tamiser. / ~ together || durch Schraubenbolzen mpl. verbinden || assembler par boulons mpl. / ~ up || verriegeln || verrouiller; fermer au verrou.

bolt (Locksm) || Riegel m.; Schubriegel m. || verrou m.; barre f.; pêne m. / ~ (Mach) || Bolzen m.; Schraube f. || boulon m.; cheville f.; vis f. / ~ (Railw) || Laschenbolzen m. || boulon m. d'éclisse. / basquill ~ || (Window) || Wirbel m.; Fensterwirbel m. || tourniquet m.; crémone f. / clinched ~ || vernieteter Bolzen m. || boulon m. rivé. / cold press for forming and upsetting ~s || Kaltpresse f. zum Formen und Stauchen von Bolzen || presse f. à froid pour former et fouler des boulons. / to cotter a ~ || einen Bolzen m. versplinten || goupiller un boulon m. / countersunk ~ || versenkter Bolzen m. || boulon m. noyé. / countersunk-headed ~ || Bolzen m. mit versenktem Kopf || boulon m. à tête noyée. / cup-headed ~ || Bolzen m. mit halbrundem Kopf || boulon m. à tête bombée. / dormant ~ (Locksm) || hebende Falle f. || pêne m. dormant. / ~ driven through the tyre || durchgehender Bolzen m. des Radreifens || boulon m. traversant le bandage de part en part. / fish ~ *see* track ~. / with flanged nut (Railw) || Schraube f. mit Bundmutter || vis f. avec écrou à collet. / flat ~ (Locksm) || glatter Riegel m. || verrou m. plat. / flat-headed ~ || Scheibenbolzen m. || boulon m. à tête plate. / flat sliding ~ || Vierkantschubriegel m. || pêne m. plat. / forelock ~ || Schließbolzen m.; Splintbolzen m.; Keilbolzen m. || boulon m. à clavette *ou* à goupille. / form ~ *see* foundation ~. / foundation ~ (Mach) || Grundbolzen m. || boulon m. de fondation. / half-turning ~ || halbtouriger Riegel m. || pêne m. à demi-tour. / ~ with a handle || Riegel m. mit Griff? Schwanzriegel m. || verrou m. à queue. / ~ with head and nut || Mutterschraube f.; Bolzen m. mit Kopf und Mutter || boulon m. à tête et écrou. / ~ of a hinge || Dorn m. *oder* Stift m. eines Scharnierbandes || cheville f. d'une charnière. / king ~ || Königszapfen m. || pivot m. / ~ with large head || Bolzen m. mit großem Kopf || boulon m. à forte *ou* grosse tête. / ~ leaded-in || eingebleiter Bolzen m. || boulon m. scellé au plomb. / lock ~ || Schloßriegel m. || pêne m. / to loosen a ~ || einen Bolzen m. lockern || dégager *ou* déserrer un boulon m. / ~

with machined head || Bolzen m. mit bearbeitetem Kopf || boulon m. avec tête usinée. / ~ with nut || Bolzen m. mit Mutter || boulon m. à écrou. / panheaded ~ || Bolzen m. mit konischem Kopf || boulon m. à tête conique. / passing ~ || durchgehender Bolzen m. || boulon m. traversant. / pin ~ || Federbolzen m. || boulon m. à ressort. / plain ~ || glatter Bolzen m. || boulon m. uni *ou* lisse. / pointed ~ || Scharfbolzen m. || boulon m. *ou* cheville f. à pointe aiguë. / rivetted ~ || vernieteter Bolzen m. || boulon m. rivé. / round-headed ~ || Bolzen m. mit rundem Kopf || boulon m. à tête ronde. / safety sliding ~ || Sicherheitsschubriegel m. || pêne m. de sûreté. / screw ~ || Schraubenbolzen m.; Schraube f. || boulon m. à vis *ou* fileté; vis f. / sliding ~ || Schubriegel m. || pêne m.; targette f. / slotted ~ || geschlitzter Bolzen m. || boulon m. fendu. / snug ~ || Nasenschraube f. || boulon m. à ergot. / spring shackle ~ || Federbolzen m. || boulon m. à ressort. / square-headed ~ || Bolzen m. mit Vierkantkopf || boulon m. à tête carrée. / T-headed ~ || Hammer(kopf)bolzen m.; Hammerschraube f. || boulon m. à tête à T. / ~ with both ends threaded || an beiden Enden mit Schraubengewinde versehener Bolzen m. || boulon m. fileté aux deux extrémités. / to tighten well a ~ || eine Schraube f. gut anziehen || serrer bien un boulon m. / track ~ (Railw) || Laschenschraube f.; Laschenbolzen m. || boulon m. d'éclisse. / ~ of a turntable || Sperrklaue f. der Drehscheibe || crapaud m. de la plaque tournante. / wedge ~ || Nachstellkeil m. || vis f. à coin de serrage. / wing-headed ~ || Flügelschraube f. || boulon m. à ailettes. / ~ for windows || Fensterriegel m. || verrou m. pour fenêtres.

bolt and nut || Schraubenbolzen m. mit Mutter || boulon m. à écrou. / ~ making machine || Maschine f. zur Herstellung von Schrauben und Muttern || machine f. pour la fabrication des boulons et écrous.

bolt chain || Bolzenkette f. || chaîne f. à boulons. / ~ chisel || Kreuzmeißel m. / bec m. d'âne; bédane m. / ~ cutter || Bolzenschneider m. || coupe-boulon m.

bolted || verriegelt || boulonné. / the castings are ~ directly to the foundation || die Gußstücke npl. sind mit der Unterlage verbolzt || les pièces fpl. de fonte sont boulonnées sur le socle même.

boltel || Rundstab m. || boudin m.; bosel m.; baguette f. (ronde).

bolter (Mill) || Sichtmaschine f.; Beutlerei f.; Beutel m. || bluterie f.; bluteau m.; blutoir m. / size of the ~ || Siebgröße f. || numéro m. du tamis.

bolter-up (Shipb) || Verbolzer m. || boulonneur m.

bolt forging machine || Bolzenschmiedemaschine f. || machine f. à forger les boulons. / ~ head || Bolzenkopf m. || tête f. de boulon. / ~ header || Nageleisen n. || cloutière f. / ~ head forging machine || Bolzenkopfschmiedemaschine f. || machine f. à forger les têtes de boulons. / pitch circle of ~ hole || Lochkreis m. || cercle m. de forage des trous de boulons.

bolting (Mill) || Beuteln n. || blutage m. / ~ (Shipb) || Verbolzung f. || chevillage m.

bolting cloth || Beuteltuch n. || tissu m. pour bluteries. / woollen ~ || Wollenbeuteltuch n. || toile f. en laine.

5 *

bolting hutch (Mill) ‖ Mehlkasten m. ‖ houche f. à mouture; récipient m. à boulange. / ~ machine (Mill) ‖ Sichtmaschine f.; Beutlerei f.; Beutel m. ‖ bluterie f.; bluteau m.; blutoir m. / ~ mill ‖ Beutelmaschine f. ‖ bluterie f.; machine f. à cribler ou à tamiser. / ~ work (Mill) ‖ Siebzeug n. ‖ tamis m.

bolt keeper ‖ Zuhaltung f. ‖ arrêt m. de pêne. / ~ lock ‖ Riegelschloß n. oder Schubriegelschloß n. ‖ serrure f. à pêne. / ~ maker ‖ Bolzenmacher m. ‖ boulonnier m.; fabricant m. de boulons. / ~ milling machine ‖ Bolzenschaftfräsmaschine f. ‖ machine f. à fraiser les tiges de boulons. / ~ nab ‖ Schließblech n. ‖ moraillon m.; nappe f.; fermoir m.

boltrope (Mar) ‖ Liek n. ‖ ralingue f. / to sew the ~ to a sail ‖ lieken ‖ ralinguer. / ~ twine see ~ yarn. / ~ yarn ‖ Liekgarn n. ‖ fil m. à ralingue.

bolt screw-cutting machine ‖ Bolzenschneidmaschine f. ‖ machine f. à tarauder les boulons. / ~ screwing ‖ Bolzenverschraubung f. ‖ joint m. par boulonnage. / ~ shearing machine ‖ Bolzenschere f. ‖ cisaille f. à boulons. / ~ staple ‖ Riegelhaken m.; Riegelhaspe f.; Riegelkrampe f. ‖ verterelle f. de verrou; auberon m. / ~ thread-cutting machine ‖ Bolzengewindeschneidmaschine f. ‖ machine f. à fileter les tiges de boulons.

bomb ‖ Bombe f. ‖ bombe f. / aerial ~ ‖ Fliegerbombe f. ‖ bombe f. aérienne ou d'avion. / aircraft ~ see aerial ~. / calorimetric ~ ‖ kalorimetrische Bombe f. ‖ bombe f. calorimétrique. / chemical ~ ‖ Gasbombe f. ‖ bombe f. à gaz. / demolition ~ ‖ Zerstörungsbombe f. ‖ bombe f. de démolisation. / fragmentation ~ ‖ Splitterbombe f. ‖ bombe f. à fragmentation. / smoke ~ ‖ Rauchbombe f. ‖ bombe f. fumigène.

bombardment (Electr) ‖ Auftreffen n. (der Elektronen auf die Antikathode) ‖ bombardement m. / ~ by airplane ‖ Luftbombardement n. ‖ bombardement m. aérien.

Bombay hemp ‖ Bombayhanf m. ‖ bombax m.

bombazet, tweeled ‖ Merino m. ‖ mérinos m.

bombazine ‖ Bombasin m. ‖ alepine f.

bomber ‖ Bombenflugzeug n. ‖ avion m. à bombes ou de bombardement. / day ~ ‖ Tagbombenflugzeug n. ‖ avion m. de bombardement de jour. / large ~ ‖ Großbombenflugzeug n. ‖ gros avion m. de bombardement. / night ~ ‖ Nachtbombenflugzeug n. ‖ avion m. de bombardement de nuit.

bomb furnace ‖ Kanonenofen m. ‖ poêle m. cylindrique en fonte; bloc m.

bombing aeroplane see bomber.

bombproof, to cover ‖ bombensicher eindecken ‖ mettre à l'abri ou à l'épreuve des bombes.

bomb rack ‖ Bombenabwurfvorrichtung f. ‖ lance-bombe f. / ~ release lever ‖ Bombenabwurfhebel m. ‖ levier m. de lancement des bombes. / ~ sight ‖ Bombenabwurfzielgerät n. ‖ viseur m. de bombardement.

bonbon ‖ Bonbon m. ‖ bonbon m. / ~ machine ‖ Bonbonmaschine f. ‖ machine f. à la fabrication de bonbons.

bond (Build) ‖ Verband m. ‖ appareil m. de maçonnerie; liaison f. / ~ (Bank) ‖ Wertpapier n. ‖ valeur f.; effet m.; titre m.; papier m. de valeur ou de crédit. /

~ (Bank; Obligation) ‖ Schuldverschreibung f.; Obligation f. ‖ obligation f. / ~ (Exchange) ‖ Revers m. ‖ acte m. de reverse. / ~ (Trade) ‖ Garantieschein m. ‖ cautionnement m.; promesse f. de garantie. / ~ cross ~ (Build) ‖ Kreuzverband m. ‖ liaison f. à croisettes. / elastic ~ ‖ vegetabilische Bindung f. ‖ agglomérant m. végétal. / form of ~ ‖ Bindungsform f. ‖ mode m. de liaison. / government ~ ‖ Staatsschuldverschreibung f. ‖ obligation f. d'État. / in ~ ‖ unter Verschluß m. ‖ sous clef; entreposé. / issue of ~s ‖ Ausgabe f. oder Emission f. von Schuldverschreibungen ‖ émission f. d'obligations. / longitudinal ~ (Build) ‖ Längsverband m. ‖ appareil m. de pierres posées en longueur. / ~ in masonry ‖ Mauerverband m. ‖ liaison f. ou assemblage m. de maçonnerie. / old English ~ (Build) ‖ Blockverband m. ‖ liaison f. anglaise. / ~ surety ~ (Trade) ‖ Verpflichtungsschein m. ‖ cautionnement m. / vitrified ~ ‖ keramische Binde f. ‖ agglomérant m. céramique. / welded ~ (Railw) ‖ geschweißter Schienenstoß m. ‖ joint m. brasé.

bond circular ‖ Kurszettel m. ‖ bulletin m. de la bourse.

bonded ‖ unter Zollverschluß ‖ entreposé. / ~ warehouse ‖ Transitlager n. ‖ dépôt m. de marchandises transitaires.

bonding strength ‖ Bindefähigkeit f. ‖ pouvoir m. agglutinant.

bondsman ‖ Kohlenwagenfüller m. ‖ chargeur m. de wagonnets ou de berlines.

bond stone (Mas) ‖ Binder m. ‖ boutisse f.

bone ‖ Knochen m.; Bein m. ‖ os m. / calcined ~s ‖ gebrannte Knochen mpl. ‖ Knochenasche f. ‖ os mpl. calcinés; cendre f. d'os. / cattle ~ ‖ Tierknochen m. ‖ os m. de bestiaux. / crushed ~s ‖ Knochenschrot m. ‖ os mpl. concassés ou granulés. / dissolved ~s ‖ gelöste Knochen mpl. ‖ os mpl. dissous. / ground ~s ‖ gemahlene Knochen mpl.; Knochenmehl n. ‖ os mpl. moulus. / made of ~ ‖ knöchern ‖ d'os m.; en os m.

bone ashes pl. ‖ Knochenasche f. ‖ cendre f. d'os. / ~-backed brushes pl. ‖ Bürstenbinderwaren fpl. in Knochen ‖ brosserie f. montée sur os. / ~ bed (Geol) ‖ Knochenbrekzie f. ‖ brèche f. osseuse. / ~ black ‖ Knochenkohle f.; Tierkohle; Beinschwarz n. ‖ charbon m. ou noir m. d'os; charbon ou noir animal. / ~ boiler ‖ Knochensieder m. ‖ bouilleur m. d'os. / ~ breaker ‖ Knochenbrecher m. ‖ concasseur m. d'os. / ~ button ‖ Knochenknopf m.; Beinknopf m. ‖ bouton m. d'os. / ~ calciner ‖ Knochenbrenner m. ‖ brûleur m. d'os. / ~ carving production ‖ Beinschnitzerei f. ‖ sculpté Beinware f. ‖ objet m. sculpté en os. / ~ charcoal ‖ Knochenkohle f.; Tierkohle f. ‖ noir m. d'os; charbon m. animal. / ~ charring kiln ‖ Knochenverkohlungsofen m. ‖ four m. de carbonisation d'os.

bone coal ‖ Knochenkohle f. ‖ charbon m. d'os. / ~ cooking drum ‖ Knochenkohlekochbottich m. ‖ tonneau m. de cuisson à charbon d'os. / ~ glowing furnace ‖ Knochenkohleglühofen m. ‖ four m. à charbon d'os. / ~ steaming apparatus ‖ Knochenkohledämpfer m. ‖ évaporateur m. à charbon d'os. / ~ washer ‖ Knochenkohlewäsche f. ‖ laveur m. de charbon d'os.

bone comb ‖ Knochenkamm m. ‖ peigne m. en os. / ~ crusher ‖ Knochenbrecher m. ‖ broyeur m. ou concasseur m. pour os. / ~ cutter ‖ Knochenschneider m.; Beinschnitzer m. ‖ débiteur m. ou découpeur m. ou scieur m. ou tabletier m. d'os. / ~ cutting ‖ Beinschneiden n. ‖ débitage m. d'os. / ~ dust see also ~ meal ‖ Knochenmehl n. ‖ poudre f. noire; poudre f. ou farine f. ou poussière f. d'os. / ~ dust factory ‖ Knochenmehlfabrik f. ‖ fabrique f. de farine d'os. / ~ earth ‖ Kapellenasche f.; Kläre f. ‖ cendre f. de coupelle; claire f. / ~ engraver ‖ Beingravör m. ‖ graveur m. sur os. / ~ fat ‖ Knochenfett n. ‖ suif m. d'os. / ~ flour see ~ dust. / ~ glass ‖ Milchglas n. ‖ verre m. opale. / ~ glue ‖ Knochenleim m. ‖ colle f. ou gélatine f. d'os; ostéocolle f.; osséine f. / ~ goods pl. ‖ Knochenwaren fpl.; Beinwaren fpl. ‖ tabletterie f. d'os; articles mpl. en os. / ~ grease ‖ Knochenfett n. ‖ graisse f. d'os; suif m. d'os. / ~ grease extracting apparatus ‖ Knochenentfettungsapparat m. ‖ appareil m. de dégraissage des os. / ~ handle ‖ Horngriff m. ‖ manche m. d'os. / ~ handle filer ‖ Knochenheftfeiler m. ‖ limeur m. de manches en os. / ~ label ‖ Knochenschild n. ‖ étiquette f. en os. / ~ label maker ‖ Knochenschilderarbeiter m. ‖ découpeur m. d'étiquettes en os. / ~ lace ‖ Klöppelspitze f. ‖ dentelle f. au fuseau. / ~ manure ‖ Knochendünger m. ‖ engrais m. d'os. / ~ meal see also ~ dust ‖ Knochenmehl n. ‖ poudre f. d'os. / ~ meal plant ‖ Knochenmehlanlage f. ‖ installation f. d'usine de poudre d'os. / ~ mill ‖ Knochenmühle f. ‖ moulin m. à os. / ~ nail (Med) ‖ Knochennagel m. ‖ clou m. à os. / ~ oil ‖ Knochenöl n. ‖ huile f. animale ou d'os. / ~ oil boiler ‖ Knochenölsieder m. ‖ fabricant m. d'huile d'os. / ~ plate (Med) ‖ Knochenverbindungsschiene f. ‖ attelle f. à os. / ~ powder see ~ dust. / ~ sawing ‖ Beinsägen n. ‖ sciage m. d'os. / ~ sawyer ‖ Knochensäger m. ‖ scieur m. d'os. / ~ superphosphate ‖ Knochensuperphosphat n. ‖ superphosphate m. d'os. / ~ turner ‖ Knochendrechsler m. ‖ tourneur m. sur os. / ~ working plant ‖ Knochenverarbeitungsanlage f. ‖ installation f. de préparation d'os.

boning rod ‖ Nivellierkreuz n.; Fluchtstab m. ‖ nivelette f.; voyant m.; porte-lanterne f. / ~ stick see boning rod.

bonnet (Cloth) ‖ Haube f. ‖ bonnet m.; chapeau m. / ~ (Mar) ‖ Haube f. eines Segels ‖ bonnette f. maillée. / ~ (Mot) ‖ Motorhaube f. ‖ capot f. / ~ catch ‖ Haubenhalter m. ‖ attache-capot m.

bonus (Bank) ‖ Superdividende f. ‖ boni m. / ~ (Office) ‖ Gehaltserhöhung f.; Gehaltszulage f. ‖ augmentation f. de traitement. / ~ (Trade) ‖ Zugabe f. ‖ supplément m. / premium ~ system ‖ Prämienlohnsystem n. ‖ système m. de salaire à prime.

bony ‖ knochig ‖ osseux.

book, to ‖ buchen; anschreiben ‖ passer en écriture f.; noter; prendre note f. / ~ baggage ‖ Gepäck n. abfertigen ‖ enregistrer des bagages. / ~ the luggage (To make book) ‖ das Gepäck aufgeben ‖ faire enregistrer les bagages mpl.

book ‖ Buch n. ‖ livre m. / ~ (Weav) ‖ Organdy m.; Organdin m.; Mull m. ‖

organdie f.; mulle f. / ~ of arrivals ‖ Eingangsbuch n. ‖ livre m. des entrées. / bound ~ ‖ gebundenes Buch n. ‖ livre m. relié. / ~ in boards ‖ Pappband m. ‖ livre m. relié en carton; reliure f. en carton. / ~ bound in calf ‖ in Leder gebundenes Buch n ‖ livre m. relié en veau ou en basane. / ~ in cloth ‖ in Leinwand f. gebundenes Buch n. ‖ livre m. relié en toile. / ~ of commission ‖ Auftragsbuch n. ‖ livre m. de commande. / to enter in the ~s ‖ buchen ‖ passer en écriture f.; noter; prendre note f. / ~ for entering receipts ‖ Einnahmebuch n. ‖ livre m. des recettes. / old ~s ‖ alte Bücher npl. ‖ vieux livres mpl. / ~ of patterns ‖ Musterbuch n. ‖ livre m. ou collection f. d'échantillons. / ~ of samples see ~ of patterns. / ~ in sheets ‖ Buch n. in losen Bogen ‖ livre m. a feuilles détachées ou en feuilles. / stitched ~ ‖ broschiertes Buch n. ‖ livre m. broché.

book attaching machine ‖ Broschüreneinhängemaschine f. ‖ machine f. à assujettir les brochures. / back of the ~ ‖ Buchrücken m. ‖ dos m. du livre. / balancing of ~s ‖ Kassenabschluß m. ‖ clôture f. de la caisse.

bookbinder ‖ Buchbinder m. ‖ relieur m. / journeyman ~ ‖ Buchbindergesell m. ‖ compagnon m. relieur.

bookbinder's calico ‖ Buchbinderleinen n.; Buchbinderkaliko n. ‖ toile f. de relieur; calicot m. pour reliures. / ~ colour printing press ‖ Buchbinderfarbendruckpresse f. ‖ presse f. pour impression en couleurs pour ateliers de reliure. / ~ colours pl. ‖ Farben fpl. für Buchbinder ‖ couleurs fpl. pour relieurs. / ~ goods pl. ‖ Buchbinderwaren fpl. ‖ ouvrages mpl. de relieur. / ~ sewing table ‖ Heftlade f. ‖ cousoir m. / ~ shop ‖ Buchbinderei f.; Buchbinderwerkstatt f. ‖ atelier m. de relieur. / ~ wire ‖ Heftdraht m. ‖ fil m. à relier.

bookbindery see bookbinder's shop.

bookbinding ‖ Buchbinderei f. ‖ reliure f. / ~ (Art) ‖ Buchbinderkunst f. ‖ art m. de relieur. / ~ for publishing houses ‖ Verlagsbuchbinderei f. ‖ reliure f. pour librairies d'éditeurs.

bookbinding cord ‖ Buchbinderschnur f. ‖ tranchefile f. pour reliure. / ~ glue ‖ Buchbinderleim m. ‖ colle f. de relieur. / ~ machine ‖ Buchbindereimaschine f. ‖ machine f. de reliure.

book carrier ‖ Buchhülle f. ‖ enveloppe f. du livre. / ~ case ‖ Bücherschrank m.; Büchergestell n. ‖ bibliothèque f.; étagère f. pour livres. / ~ craft ‖ Buchgewerbe n. ‖ industries fpl. du livre.

bookdebt ‖ Buchschuld f. ‖ dette f. simple.

book-edge marbler (Bookb) ‖ Schnittmarmorierer m. ‖ marbreur m. sur tranches.

booker ‖ Buchhalter m. ‖ comptable m.; teneur m. de livres. / ~ of tobacco-leaves ‖ Tabakblattbucher m. ‖ capseur m. des feuilles de tabac.

book gilding ‖ Buchvergolderei f. ‖ dorure f. en reliure. / ~ glueing machine ‖ Broschürenleimmaschine f. ‖ machine f. à coller les brochures.

booking ‖ Buchen n.; Eintragen n.; Einschreiben n. ‖ enregistrement m. ‖ ~ of luggage ‖ Gepäckannahme f. ‖ enregistrement m. des bagages.

booking machine ‖ Buchungsmaschine f. ‖ machine f. de comptabilité. / ~ office ‖ Fahrkartenschalter m. ‖ guichet m.

book ink ‖ Buchdruckfarbe f. ‖ encre f. pour l'impression du gros œuvre. / ~ keeper ‖ Buchhalter m. ‖ teneur m. des livres; comptable m.

bookkeeping ‖ Buchhaltung f.; Buchführung f. ‖ tenue f. des livres; comptabilité f. / ~ by double entry ‖ doppelte Buchhaltung f. oder Buchführung f. ‖ tenue f. des livres en partie double. / loose leaf ~ ‖ Buchführung f. auf losen Blättern ‖ comptabilité f. sur feuilles détachées. / ~ by single entry ‖ einfache Buchhaltung f. oder Buchführung f. ‖ tenue f. des livres en partie simple.

bookkeeping machine ‖ Buchhaltungsmaschine f. ‖ machine f. à tenir les livres. / ~ work ‖ Buchhaltungsarbeiten fpl. ‖ travaux mpl. comptables.

booklet ‖ Broschüre f. ‖ catalogue m.; brochure f. / ~ prepared from loose sheets ‖ Broschüre f. aus losen Bogen ‖ brochure f. formée de feuilles détachées. / ~ prepared from stitched packs ‖ Broschüre f. aus gehefteten Blocks ‖ brochure f. formée de blocs piqués ou cousus. / ~ stitched sideways ‖ seitlich geheftete Broschüre f. ‖ brochure f. agrafée latéralement. / ~ covering machine ‖ Broschüreneinhängemaschine f. ‖ machine f. à couvrir la brochure.

book linen ‖ Buch(binder)leinen n. ‖ toile f. de relieur; librets mpl. / ~ manufacture ‖ Buchgewerbe n. ‖ industries fpl. du livre. / ~ mark ‖ Lesezeichen n. ‖ signet m.; liseuse f. / ~ muslin (Weav) ‖ Organdy m.; Organdin m.; Mull m. ‖ organdie f.

book-post, by ‖ unter Kreuzband n. ‖ sous bande f.

book printing establishment ‖ Werkdruckerei f. ‖ établissement m. d'impression des livres. / ~ machine ‖ Buchdruckereimaschine f. ‖ machine f. à imprimeries de livres. / ~ scriptures pl. ‖ Buchdruckschriften fpl. ‖ lettres fpl. pour imprimeries.

bookseller ‖ Sortimentsbuchhändler m. ‖ libraire m.; marchand m. de livres. / ~'s goods pl. ‖ Buchhändlerwaren fpl. ‖ articles mpl. de librairie. / ~'s shop ‖ Buchhandlung f. ‖ librairie f. / second-hand ~'s shop ‖ Antiquariatsbuchhandlung f. ‖ librairie f. d'occasion. / technical ~'s shop ‖ technische Buchhandlung f. ‖ librairie f. technique. / ~'s trade ‖ Buchhandel m. ‖ commerce m. des livres.

book selling ‖ Buchhandel m. ‖ commerce m. des livres. / ~ setter ‖ Buchsetzer m. ‖ labeurier m. / ~ shelf see book stand.

bookshop see bookseller's shop.

book stand ‖ Büchergestell n.; Bücherbrett n. ‖ tablettes fpl.; étagère f. (à livres). / ~ stitching ‖ Buchheften n. ‖ brochage m. / ~ support ‖ Buchstütze f. ‖ support m. pour livres. / ~ value ‖ Buchwert m. ‖ valeur f. dont les livres font foi.

boom (Exchange) ‖ Hausse f. ‖ hausse f. / ~ (Mar) ‖ Baum m.; Spiere f. ‖ espar m.; mâtereau m. / ~ of a harbour ‖ Hafenbaum m. ‖ Hafenschlengel m. ‖ estacade f. ou barre f. d'un port. / horizontal ~ of a girder ‖ Längsband n. ‖ semelle f. / main ~ (Mar) ‖ Giekbaum m. ‖ baume f.; bôme f.; gui m. / mizzen ~ see main ~.

/ spencer ~ see main ~. / studdingsail ~ ‖ Leesegelsspier f. ‖ bout-dehors m.; boute-hors m. de la bonnette. / ~ of a truss (Bridge) ‖ Gurt m.; die Gurtung eines Brückenträgers ‖ semelle f. d'une poutre. / trysail ~ see main ~. / upper ~ (Bridge) ‖ Obergurt m. ‖ semelle f. supérieure.

boomsail ‖ Baumsegel n. ‖ brigantine f.; voile f. goëlette bordant sur gui; voile f. à bôme.

boon (Flax; Hemp) ‖ holziger Kern m. ‖ tige f. ligneuse.

boost, to ~ the battery ‖ die Batterie verstärken ‖ renforcer la charge.

boosted circuit ‖ Zusatzstromkreis m. ‖ circuit m. survolté.

booster (Electr) ‖ Zusatzmaschine f.; Zusatzdynamo f. ‖ dynamo f. supplémentaire ou auxiliaire; survolteur-dévolteur m. / ~ aggregate ‖ Zusatzaggregat n. ‖ survolteur m. / ~ battery ‖ Zusatzbatterie f. ‖ batterie f. survoltrice. / ~ brake ‖ Vakuumbremse f. ‖ servo-frein m. à dépression. / ~ radiator ‖ Zusatzkühler m.; Hilfskühler m. ‖ radiateur m. additionnel ou auxiliaire. / ~ transformer ‖ Zusatztransformator m.; Spannungserhöher m. ‖ (transformateur-) survolteur m.; autotransformateur-élévateur m.

boot ‖ Stiefel m.; Schuh m. ‖ botte f.; soulier m. / ~ (Envelope) ‖ Schutzhülle f. ‖ enveloppe f. / double-soled ~ ‖ Stiefel m. mit Doppelsohlen ‖ botte f. à doubles semelles. / electrically heated ~ ‖ Heizstiefel m. ‖ bottine f. chauffée électriquement. / laced ~s pl. ‖ Stiefeletten fpl. ‖ bottines fpl. / part of ~s ‖ Schuhbestandteil m. ‖ partie f. de chaussures. / patent leather ~ ‖ lackierter Stiefel m. ‖ botte f. de cuir verni. / single-soled ~ ‖ Stiefel m. mit einfacher Sohle ‖ botte f. à simple semelle. / sporting ~ ‖ Sportstiefel m. ‖ chaussure f. de sport. / waterproof ~ ‖ Wasserstiefel m. ‖ botte f. imperméable.

boot and shoe machine ‖ Maschine f. für Schuhfabrikation ‖ machine f. pour la fabrication des chaussures. / ~ maker ‖ Schuhmacher m.; Schuster m. ‖ bottier m.; cordonnier m. / ~ protector ‖ Sohlenschoner m. ‖ protège-semelles m.

boot button ‖ Schuhknopf m. ‖ bouton m. pour souliers. / ~ cardboard ‖ Brandpappe f. ‖ carton m. pour la seconde semelle. / ~ cleaning machine ‖ Stiefelputzmaschine f. ‖ cireuse-polisseuse f.

bootee ‖ Halbschuh m. ‖ savate f.; bottine f. / ~ leg ‖ Stiefelettenschaft m. ‖ tige f. de bottine.

boot factory ‖ Schuhfabrik f. ‖ fabrique f. de chaussures. / ~ grease ‖ Stiefelschmiere f. ‖ graisse f. pour chaussures.

booth in a fair ‖ Marktbude f. ‖ boutique f.; loge f. de (la) foire.

boot-heel ‖ Stiefelabsatz m.; Absatz m. ‖ talon m. / ~ works pl. ‖ Absatzfabrik f. ‖ fabrique f. de talons.

boot-hose ‖ (lederne) Gamasche f.; Stiefelstrumpf m. ‖ guêtre f.

boot irons pl., machine for making ‖ Stiefeleisenherstellungsmaschine f. ‖ machine f. à fabriquer les fers à botte.

bootjack ‖ Stiefelknecht m. ‖ tire-bottes m.

bootlace ‖ Schuhriemen m. ‖ cordon m. ou lacet m. de chaussures. / leather ~ ‖ Lederschnürsenkel m.; Schnürriemen m. aus Leder ‖ lacet m. en cuir.

bootlast ‖ Schuhleisten m. ‖ forme f. pour souliers *ou* chaussures. / **~ hollowing machine** ‖ Schuhleistenaushöhlmaschine f. ‖ machine f. à évider les embauchoirs à chaussures.

bootleg ‖ Stiefelschaft m. ‖ tige f. de botte.

bootmaker ‖ Stiefelmacher m. ‖ bottier m. / **bespoke ~** ‖ Maßschuster m. ‖ cordonnier m. sur mesure. / **orthopædic ~** ‖ orthopädischer Schuhmacher m. ‖ bottier m. orthopédiste.

bootmaking, machine for ‖ Maschine f. für Stiefelfertigung ‖ machine f. cordonnière.

boot polish ‖ Schuhkrem m. ‖ crème f. pour chaussures.

boots pl. and shoes pl. ‖ Schuhwaren fpl.; Schuhwerk n. ‖ chaussures fpl.; chaussure f. / **~ of caoutchouc** ‖ Schuhwaren fpl. aus Kautschuk ‖ chaussures fpl. en caoutchouc; galoches fpl. / **coarse ~** ‖ grobes Schuhwerk n. ‖ chaussure f. grossière. / **fur ~** ‖ Pelzschuhwerk n. ‖ chaussure f. fourrée. / **~ of leather with leather soles** ‖ Schuhwaren fpl. aus Leder mit Ledersohlen ‖ chaussures fpl. en cuir avec semelles en cuir. / **~ of leather with wooden soles** ‖ Schuhwaren fpl. aus Leder mit Holzsohlen ‖ chaussures fpl. en cuir avec semelles en bois.

boot sole ‖ Stiefelsohle f.; Schuhsohle f. ‖ semelle f. / **~ top** ‖ Stiefelschaft m.; Stiefelstulpe f. ‖ retroussis m. *ou* tige f. de botte. / **~ tree** ‖ Stiefelholz n.; Leisten m. ‖ embauchoir m. / **~ twine** ‖ Schusterzwirn m. ‖ fil m. poissé.

bopping (Barometer) ‖ plötzliche Schwankung f. ‖ oscillation f.

boracic acid ‖ Borsäure f. ‖ acide m. borique.

boracite ‖ Borazit m. ‖ magnésie f. boratée; borate m. de magnésie; boracite f.

borage flower ‖ Boretschblüte f. ‖ fleur f. de bourrache.

borate ‖ boraxsaures oder borsaures Salz n. ‖ borate m. / **~ of lime** ‖ Hayesin m.; Hydroborocalcit m. ‖ hayesénite f. / **~ of magnesia** ‖ Borazit m. ‖ magnésie f. boratée; borate m. de magnésie; boracite f.

borax ‖ Borax m. ‖ borax m.; borate m. *ou* biborate m. de soude. / **crystallized ~** ‖ kristallisierter Borax m. ‖ borax m. cristallisé. / **ground ~** ‖ gemahlener Borax m. ‖ borax m. pulvérisé. / **native ~** ‖ borsaures Natrium n.; Borax m.; Tinkal m. ‖ soude f. boratée; borate m. de soude; tincal m. / **to remove the ~ by dilute sulphuric acid after soldering** ‖ den Borax m. nach dem Löten abbeizen ‖ dérocher le borax après le soudage.

borax box ‖ Lötbüchse f.; Boraxbüchse f. ‖ boraxoir m.; borasseau m.; rochoir m. / **~ varnish** ‖ Boraxfirnis m. ‖ vernis m. de borax.

bord-and-pillar system (Mine) ‖ Pfeilerbau m. ‖ méthode f. des massifs courts; méthode f. des piliers et galeries.

border, to ‖ umbördeln ‖ border. / **~ (Net)** ‖ säumen ‖ enlarmer. / **~ (Ornament)** ‖ einfassen ‖ entourer. / **~ (Textile)** ‖ einfassen ‖ border; ourler. / **~ on . . .** ‖ angrenzen; grenzen an . . . ‖ confiner à . . .; approcher de . . .; toucher à . . .

border ‖ Rand m.; Kante f. ‖ bord m.; tranche f. / **~ (Bordering)** ‖ Einfassung f. ‖ bordure f.; encadrement m. / **~ (Gard)** ‖ Rabatte f. ‖ plate-bande f.; bordure f. / **~ (Join)** ‖ Zarge f. ‖ bord m. / **~ (Min)** ‖ Markscheide f. ‖ borne f. / **~ (Print)** ‖ Randverzierung f.; Zierleiste f. ‖ bordure f.; vignette f. / **~ (Road)** ‖ Gradsteine mpl.; Randsteine mpl. ‖ bordure f. (d'une chaussée). / **~ of a door panel** ‖ Füllungsglieder npl.; Friesglieder npl.; bordure f. *ou* moulure f. autour d'un panneau. / **~ of the finger nail** ‖ Fingernagellimbus m. ‖ lit m. de l'ongle digital. / **~ for flower beds** ‖ Blumenbeeteinfassung f. ‖ châssis m. de parterres de fleurs. / **~ of a salt pan** ‖ Pfannbord m.; Bord m. einer Salzpfanne ‖ bord m. d'une chaudière à sel; versat m. d'une chaudière à sel.

border bed see **border** (Gard). / **~ drawings** pl. ‖ Randzeichnungen fpl. ‖ dessins mpl. du *ou* au bord.

bordering ‖ anstoßend ‖ contigu. / **~ machine** ‖ Bördelmaschine f. ‖ machine f. à border. / **~ press** ‖ Bördelpresse f.; Kümpelpresse f. ‖ presse f. à emboutir *ou* à border.

border mesh of a net ‖ Randmasche f. eines Netzes ‖ enlarme f. d'un filet. / **~ piling** ‖ Spundwand f. ‖ cours m. de planches.

borders pl. ‖ Besatzartikel mpl. ‖ bordures fpl.

border stone ‖ Grenzstein m. ‖ borne f.

bore, to ‖ bohren; durchbohren ‖ forer; percer. / **~ a mineral** ‖ ein Mineral n. erbohren ‖ découvrir un minéral par sondage. / **~ out the stay bolts** pl. ‖ die Stehbolzen mpl. abbohren ‖ enlever les entretoises fpl. / **~ through** ‖ durchbohren ‖ forer; percer. / **~ up a cylinder** ‖ einen Zylinder ausbohren ‖ aléser un cylindre. / **the axle is hole bored from end to end** ‖ die Achse f. ist mit durchgehender Bohrung versehen ‖ l'essieu m. est percé sur toute sa longueur. / **the tubes** pl. **were to be furnished bored in the rough** ‖ die Rohre npl. waren vorgebohrt zu liefern ‖ les tubes devaient être livrés avec un forage d'ébauchage.

bore ‖ Bohrung f.; Bohren n. ‖ alésage m.; forage m. / **~ (Diameter)** ‖ Bohrung f.; Bohrlochdurchmesser m. ‖ diamètre m. du trou; forure f. / **~ (Arm)** ‖ Seele f. ‖ âme f. / **~ (Nailsm)** ‖ Nageldocke f.; Nageleisen n.; Nagelform f. ‖ clouère f.; étampe f. de cloutier. / **cylinder ~** ‖ Zylinderbohrung f. ‖ diamètre m. du cylindre. / **~ of dividing spindle** ‖ Teilspindelbohrung f. ‖ alésage m. de broche de la poupée à diviser. / **the ~ had not increased** ‖ die Bohrung hatte sich nicht mehr vergrößert ‖ l'alésage m. ne s'était plus élargi.

bore bit ‖ Bohreisen n.; Bohreinsatz m. ‖ fer m. à forer; foret m.; mèche f. / **~ chips** pl. ‖ Bohrspäne mpl. ‖ bûchilles fpl.; alésures fpl.; copeaux mpl. de foret. / **~ crown** ‖ Bohrkrone f. ‖ tête f. de mèche. / **~ dust** ‖ Bohrmehl n. ‖ poussière f. de foret; farine f. de foret. / **~ frame** ‖ Bohrgerüst n. ‖ chevalet m. de sondage *ou* de forage.

bore-hole (Min) ‖ Bohrloch n. ‖ trou m. de sondage. / **check ~** ‖ Kontrollbohrloch n. ‖ sondage m. de contrôle. / **drilling of the ~ out of the hard steel blocks** ‖ Ausbohren n. der Seele aus harten Stahlblöcken ‖ perforage m. des blocs d'acier d'une grande dureté. / **to put down a ~** (Min) ‖ eine Bohrung f. ausführen ‖ exécuter un sondage. / **to start a ~** (Min) ‖

ein Bohrloch n. ansetzen ‖ commencer un sondage. / **~ bottom** (Min) ‖ Bohrgrund m.; Bohrort m. ‖ fond m. de sondage *ou* d'un trou de mine. / **~ pump** ‖ Bohrlochpumpe f. ‖ pompe f. de sondage.

borer ‖ Bohrarbeiter m. ‖ sondeur m.; foreur m. / **~ (Tool)** ‖ Bohrer m. ‖ foret m.; mèche f. / **~ with circular bit** (Join) ‖ Kreisbohrer m.; Kreisausheber m. ‖ coupe-cercle m. / **cross-mouthed ~** (Min) ‖ Kreuzbohrer m.; Kronenbohrer m. ‖ fleuret m. à tête carrée; pistolet m. à pointe carrée. / **long ~** ‖ Lang(loch)bohrer m. esseret m. / **long ~** (Min) ‖ Abbohrer m. ‖ aiguille f.; barre f. *ou* fer m. à mine; fleuret m. / **master ~** (Min) ‖ Bohrmeister m. ‖ chef foreur m. / **square ~** ‖ vierschneidiger Kronenbohrer m. ‖ perçoir m. à couronne.

bore rod (Min) ‖ Bohrgestänge n. ‖ tige f. de sonde *ou* de sondage. / **~ specimen** ‖ Bohrprobe f. ‖ carotte f. de sondage.

boric acid ‖ Borsäure f. ‖ acide m. borique. / **flaky ~** ‖ Borsäure f. in Schuppen ‖ acide m. borique en paillettes.

boric ointment ‖ Borsalbe f. ‖ onguent m. borique; vaseline f. boriquée.

boring ‖ Bohrarbeit f.; Lochung f.; Bohrung f. ‖ perçage m.; forage m.; alésage m.; sondage m. / **~ (Chips)** ‖ Drehspan m. ‖ tournure f. / **to find a layer by ~** ‖ eine Schicht f. anbohren ‖ découvrir une couche par la sonde. / **~ of the plate spring** ‖ Bohrung f. der Scheibenfeder ‖ trou m. de la rondelle Belleville. / **~ by means of rods** ‖ Gestängebohren n. ‖ sondage m. à tige rigide. / **~ by means of a rope** ‖ Seilbohren n. ‖ sondage m. chinois *ou* au câble. / **attachment for ~ taper holes** ‖ Konischbohrapparat m. ‖ appareil m. à aléser cône. / **deep ~** ‖ Tiefbohrung f. ‖ sondage m. *ou* forage m. *ou* fonçage m. à grande profondeur. / **deep ~ tools** pl. ‖ Tiefbohrgerät n.; Tiefbohrwerkzeuge npl. ‖ outils mpl. à forer *ou* de sondage en grande profondeur. / **experimental ~** ‖ Bohrversuch m. ‖ essai m. de sondage.

boring apparatus (Min) ‖ Bohrzeug n. ‖ appareil m. de sondage. / **cylinder ~** (Mach) ‖ Zylinderbohrapparat m. ‖ appareil m. à aléser les cylindres.

boring bar (Mach) ‖ Bohrspindel f. ‖ arbre m. porte-mèche. / **~ (Tool)** ‖ Bohrstange f. ‖ barre f. d'alésage. / **~ of a boring machine** ‖ Bohrspindel f. einer Bohrmaschine ‖ arbre m. d'une machine à aléser.

boring bit ‖ Bohrschneide f. ‖ tranchant m. de mèche. / **~ chisel** ‖ Bohrmeißel m. ‖ trépan m. de sondage; burin m.; cassepierre m. / **~ company** ‖ Bohrgesellschaft f. ‖ société f. de sondages. / **~ cutter** ‖ Bohrmesser n. ‖ lame f. d'alésage. / **~ device** ‖ Bohrvorrichtung f. ‖ dispositif m. à percer. / **~ frame** ‖ Bohrgerüst n. ‖ chevalet m. de sondage. / **~ hammer** ‖ Bohrhammer m. ‖ marteau m. perforateur. / **~ head** ‖ Bohrreitstock m. ‖ contrepoupée f. de perçage. / **~ lathe** see also **boring machine** ‖ Bohr(dreh)bank f.; Bohrmaschine f. ‖ tour m. à aléser; tour-alésoir m.

boring machine ‖ Bohrmaschine f.; Bohrbank f. ‖ machine f. à percer *ou* à forer *ou* à aléser; tour-alésoir m.; perceuse f.; aléseuse f.; forerie f. / **cylinder ~** ‖ Zylinderbohrmaschine f. ‖ machine f. à

aléser les cylindres. / double-table ~ || Bohrmaschine f. mit zwei Tischen || perceuse f. à deux tables. / horizontal ~ || Wagerechtbohrmaschine f. || machine f. horizontale à aléser; forerie f. horizontale. / ingot ~ || Blockausbohrmaschine f. || foreuse f. pour bloom et blocs. / pegholes ~ || Dübellochbohrmaschine f. || machine f. à forer les trous de goujons. / ~ with self-centering jaws || Bohrmaschine f. mit selbstzentrierenden Klemmbacken || machine f. à aléser avec mâchoires à centrage automatique. / shaft ~ || Wellenbohrbank f. || tour m. à forer les arbres. / vertical ~ || Bohrmaschine f. mit stehender Bohrwelle || aléseuse f. verticale.

boring and milling machine || Bohr- und Fräsmaschine f. || machine f. à aléser et à fraiser. / ~ and reaming machine || Bohrdrehbank f. || tour m. à percer et à aléser. / ~ and turning machine || kombinierte Bohr- und Drehbank f. || tour m. alésoir. / ~ and turning machine with horizontal face plate || Bohrwerk n. und Drehwerk n. mit wagerechter Planscheibe || aléseuse f. et tour m. à plateau horizontal. / vertical ~ and turning machine || Karusselldrehbank f. || tour m. alésoir à plateau horizontal.

boring machine hand || Ausbohrmaschinenarbeiter m. || aléseur m. à la machine.

boring oil || Bohröl n. || huile f. pour perçage *ou* taraudage. / ~ pipe box || Anbohrapparat m.; Anbohrschelle f. || boîte f. de tuyau à percer. / ~ pressure || Bohrdruck m. || pression f. de perçage.

boring rod || Bohrstange f. || barre f. d'alésage. / ~ (Mach) || Bohrspindel f. || arbre m. porte-mèche. / ~ (Min) || Bohrgestänge n.; Bohrstock m. || tige f. *ou* canne f. *ou* barre f. de sondage; tige f. de sonde. / ~ snout || Bohrstangenhalter m. || porte-barre d'alésage m. / system of ~s (Min) || Bohrgestänge n.; Gestänge n. für Tiefbohrungen || tiges fpl. pour les sondages.

borings pl. (Min) || Bohrmehl n. || poussière f. de perçage; farine f. de forage. / ~ (Met) || Bohrspäne mpl. || bûchilles fpl.; alésures fpl.; copeaux mpl. de forage.

boring tools pl. || Bohrwerkzeug n. || outils mpl. de forage. / ~ (Min) || Gestänge n. für Bohrwerkzeuge; Bohrgestänge n. || barre f. de sondage. / ~ for wells || Brunnenbohrgerät n. || outils mpl. de creusage pour puits.

boring tower || Bohrturm m. || tour f. à forer *ou* de sondage. / ~ tube || Bohrrohr n. || tuyau m. de forage.

borne by || gelagert auf *oder* in || supporté par; appuyé sur; logé dans; monté sur.

borofluoric acid || Borflußsäure f. || acide m. borofluorique.

boron || Bor n. || bore m. / ~ steel || Borstahl m. || acier m. au bore.

boshes pl. (Blast furnace) || Rast f. || étalage m. / ~ and hearth || Rast f. und Gestell n.; Unterschacht m. || grand foyer m.

boss (Forg) || Gesenk n. || estampe f.; étampe f.; matrice f. / ~ (Forg; Swage block) || Gesenkplatte f.; Lochplatte f. || tas-étampe f. / ~ (Mas) || Mörteltrog m. || auge f. à mortier. / balance ~ (Mill) || Hauenbuchse f. || manchon m. d'anille. / propeller ~ (Shipb) || Schraubennabe f. || moyeu m. d'hélice.

botanic (al) collection || botanische Sammlung f. || collection f. botanique. / ~

investigation || botanische Untersuchung f. || recherche f. botanique. / ~ loupe || botanische Lupe f. || loupe f. pour botanistes.

botanist || Botaniker m. || botaniste m.

botany || Botanik f.; Pflanzenkunde f. || botanique f.

botryoidal blende || Schalenblende f. || zinc m. sulfuré concrétionné.

bottle, to || auf Flaschen fpl. ziehen; in Flaschen fpl. abfüllen || embouteiller; soutirer; mettre en bouteilles fpl.

bottle || Flasche f. || bouteille f.; flacon m. / air ~ || Luftflasche f. || bouteille f. à air. / broken ~ || zerbrochene Flasche f. || bouteille f. cassée. / ~ for compressed air || Druckluftbehälter m. || cylindre m. à air comprimé. / corked ~ || verkorkte Flasche f. || bouteille f. bouchée. / earthenware ~ see stoneware ~. / ~ with narrow mouth || enghalsige Flasche f. || bouteille f. à goulot étroit. / narrownecked ~ see bottle with narrow mouth. / ~ of paper || Papierflasche f. || bouteille f. de papier. / ~ with screw cap || Flasche f. mit Schraubdeckel || bouteille f. à bouchon fileté. / steel ~ || Stahlflasche f. || bouteille f. en acier. / stone ~ || Kruke f. || cruchon m. / stoneware ~ || Steingutflasche f.; Steinzeugflasche f.; Steinkrug m. || bouteille f. en grès; bonbonne f. en poterie; buchon m. de grès. / varnished stoneware ~ || glasierte Stein(gut)flasche f. || bonbonne f. en poterie vernissée. / washing ~ || Waschflasche f. || flacon m. laveur. / water motor for cleaning ~s || Wassermotor m. zur Flaschenreinigung || moteur m. hydraulique pour le rinçage des bouteilles. / widenecked ~ || Flasche mit weitem Hals; weithalsige Flasche f. || bouteille f. à large goulot.

bottle basket || Flaschenkorb m. || panier m. à bouteilles. / ~ battery (Electr) || Flaschenelement n. || pile f. bouteille. / ~ beer see bottled beer. / ~ blower || Flaschenbläser m. || souffleur m. de bouteilles. / ~ blowing machine || Flaschenblasmaschine f. || machine f. à souffler les bouteilles. / ~ bottom moulder || Flaschenbodenformer m. || mouleur m. de fonds de bouteilles. / ~ box || Flaschenkasten m. || panier m. *ou* casier m. à bouteilles; porte-bouteilles m. / ~ breakage || Flaschenbruch m. || bris m. *ou* casse f. de bouteilles. / ~ brush || Flaschenbürste f. || goupillon m. / ~ brush head || Flaschenbürstenkopf m. || tête f. de brosse à bouteilles. / ~ brushing machine || Flaschenbürstmaschine f. || machine f. à brosser les bouteilles. / ~ cap || Flaschenkapsel f. || capsule f. pour bouteilles. / ~ capping machine || Flaschenkapselmaschine f. || machine f. à capsuler les bouteilles. / ~ capsule || Flaschenkapsel f. || capsule f. à boucher *ou* de bouteille. / ~ capsuling machine || Flaschenverkapselmaschine f. || capsuleur m. de bouteilles. / ~ car || Flaschenwagen m. || voiture f. à bouteilles. / ~ case || Flaschenkasten m. || caisse f. à bouteilles; porte-bouteilles m. / ~ casting machine || Flaschengießmaschine f. || machine f. à couler les bouteilles. / ~ cell || Flaschenelement n. || pile f. bouteille. / ~ cellar || Flaschenkeller m. || cave f. aux bouteilles. / ~ cellarage machine || Flaschenkellereimaschine f. || machine

f. pour caves à bouteilles. / ~ charger || Flaschenfüllmaschine f. || machine f. à remplir les bouteilles. / ~ cleaner || Flaschenspülmaschine f. || rince-bouteilles m.; machine f. à rincer les bouteilles. / ~ cleaning machine || Flaschenreinigungsmaschine f. || machine f. laveuse pour bouteilles. / ~ cleaning plant || Flaschenreinigungsanlage f. || installation f. pour le rinçage des bouteilles. / ~ cork || Flaschenkork m. || bouchon m. de bouteille. / ~ corker || Flaschenpfropfmaschine f. || bouche-bouteilles m. / ~ corking machine || Flaschenverkorkmaschine f. || machine f. à boucher les bouteilles.

bottled beer || Flaschenbier n. || bière f. en bouteilles.

bottle draining box || Flaschenauslaufkasten m. || panier-égouttoir m. pour bouteilles. / ~ draining truck || Flaschenabtropfgestell n. || égouttoir m. pour bouteilles. / ~ emptier || Flaschenausgießer m. || bouchon verseur m. pour bouteilles. / ~ fastening machine || Flaschenverschließmaschine f. || machine f. à fermer les bouteilles. / ~ filler || Flaschenfüllmaschine f. || machine f. à remplir les bouteilles. / ~ filling machine see ~ filler. / ~ foil || Flaschenkapsel f. || capsule f. pour bouteilles. / ~ glass || Flaschenglas n. || verre m. à bouteilles. / ~-gourd || Flaschenkürbis m. || calebasse f. / ~-gourd pip || Flaschenkürbiskern m. || pépin m. de calebasse. / ~ hamper || Flaschenkorb m. || panier m. à bouteilles. / ~ industry || Flaschenindustrie f. || industrie f. des bouteilles. / ~ jack || Flaschenwinde f. || vérin m. à bouteille. / ~ label || Flaschenetikette f. || étiquette f. pour bouteille. / ~ labelling machine || Flaschenetikettiermaschine f. || machine f. à étiqueter les bouteilles. / ~ labelling and dating machine || Flaschenetikettier- und Datiermaschine f. || machine f. à étiqueter et à dater les bouteilles. / insulated ~ manufacturing machine || Isolierflaschenherstellungsmaschine f. || machine f. pour fabriquer des bouteilles isolantes. / ~ mould || Form f. für Flaschen || moule m. pour flaconneries. / ~ mouth see bottle neck.

bottle neck || Flaschenhals m. || col m. de bouteille. / narrow ~ || enger Flaschenhals m. || goulot m. (étroit). / ~ drawing end || Rohrangel f. || queue f. d'étirage sur tubes. / ~ moulder || Flaschenhalsformer m. || mouleur m. de cols de bouteilles.

bottle package || Flaschenkiste f. || caisse f. à bouteilles. / ~ pasteurizing apparatus || Flaschenpasteurisierapparat m. || appareil m. à pasteuriser les bouteilles; pasteurisateur m. à bouteilles.

bottler || Flaschenfüller m. || embouteilleur m.

bottle rack (Brew) || Flaschenkasten m. || caisse f. à bouteilles. / ~ ring maker || Flaschenmundstückformer m. || poseur m. de bagues à bouteilles. / ~ rinsing plant || Flaschenspülerei f. || atelier m. de rinçage des bouteilles. / ~ shop || Flaschen(verkaufs)keller m. || cave f. où l'on vend des bouteilles. / ~ shutter || Flaschenverschluß n. || fermeture f. de bouteille. / ~ sorter || Flaschensortierer m. || trieur m. de bouteilles. / ~ stand || Flaschenständer m. || porte-bouteilles m.

/ ~ stand (Cupboard) ‖ Flaschenschrank m. ‖ armoire f. à bouteilles. / ~ sterilizer ‖ Flaschensterilisierapparat m. ‖ appareil m. à stériliser les bouteilles; stérilisateur m. de bouteilles.

bottle stopper ‖ Flaschenkork m. ‖ bouchon m. à bouteilles. / ~ with lever ‖ Flaschenhebelverschluß m. ‖ fermeture f. à levier pour bouteilles. / mechanical ~ ‖ mechanischer Flaschenverschluß m. ‖ bouchon m. mécanique. / porcelain ~ ‖ Flaschenverschluß m. aus Porzellan ‖ bouton m. en porcelaine pour bouchons.

bottle stopping machine ‖ Flaschenverschließmaschine f. ‖ machine f. à boucher les bouteilles. / ~ testing apparatus ‖ Flaschenprobapparat m. ‖ appareil m. à essayer la résistance des bouteilles. / ~ transport cart ‖ Flaschentransportwagen m. ‖ haquet m. ou chariot m. pour le transport de bouteilles. / ~ tray ‖ Flaschenbrett n. ‖ porte-bouteilles m. / ~ uncorking device ‖ Flaschenentkorker m. ‖ débouchoir m. de bouteilles.

bottle washer (Rubber washer) ‖ Flaschengummischeibe f. ‖ rondelle f. de caoutchouc pour bouteilles. / ~ (Machine) ‖ Flaschenspülmaschine f. ‖ machine f. à rincer les bouteilles. / ~ (Man) ‖ Flaschenreiniger m.; Flaschenspüler m. ‖ rinceur m. de bouteilles.

bottle washing ‖ Flaschenreinigung f. ‖ nettoyage m. des bouteilles. / ~ washing machine ‖ Flaschenwaschmaschine f.; Flaschenspülmaschine f.‖rince-bouteilles m.; machine f. à laver ou à rincer les bouteilles. / ~ wax ‖ Flaschenlack m. ‖ cire f. à cacheter les bouteilles. / ~ wheel-barrow ‖ Flaschenkarren m. ‖ chariot m. à bouteilles. / ~ wire ‖ Flaschendraht m. ‖ fil m. à fermer les bouteilles. / ~ wiring machine ‖ Flaschenverdrahtungsmaschine f. ‖ machine f. à agrafer les bouteilles. / ~ wrapper ‖ Flaschenhülse f. ‖ paillon m. à bouteilles.

bottling ‖ Flaschenfüllung f.; Abziehen n. auf Flaschen ‖ soutirage m.; embouteillage m.; mise f. en bouteilles. / ~ apparatus ‖ Flaschenfüllapparat m. ‖ appareil m. à remplir les bouteilles ou de soutirage. / ~ department ‖ Flaschenbierabteilung f.; Flaschenfüllerei f.; Flaschenkeller m. ‖ canetterie f. / ~fit ‖ flaschenreif ‖ bon à être mise en bouteilles.

bottling machine ‖ Flaschenfüllmaschine f. ‖ soutireuse f.; machine f. à soutirer ou à remplir les bouteilles ou de mise en bouteilles. / ~ for mineral water ‖ Mineralwasserfüllmaschine f. ‖ doseuse f. pour eaux minérales. / revolving ~ ‖ rotierende Flaschenfüllmaschine f. ‖ soutireuse f. à bouteilles rotative. / bottling and corking machine ‖ Flaschenfüll- und Korkmaschine f. ‖ tireuse-boucheuse f. à bouteilles.

bottling machinery see bottling machine. / ~ plant ‖ Flaschenfüllanlage f. ‖ installation f. de soutirage. / ~ room ‖ Flaschenbierabzieraum m. ‖ atelier m. de mise en bouteilles.

bottom, to ~ casks ‖ Fässer npl. ausböden oder verbodmen ‖ mettre des fonds mpl. dans des tonneaux.

bottom ‖ Boden m. ‖ fond m. / ~ (Geol) ‖ Flußbett n. ‖ lit m.; fond m. / ~ (Geom) ‖ Grundfläche f.; Basis f. ‖ base f.; plan m. inférieur. / ~ (Mar) ‖ Grund m.; Boden m. ‖ fond m. / ~ (Min) ‖ Sohle

f. ‖ sol m.; fond m.; base f. / ~ of a bottle ‖ Flaschenboden m. ‖ cul m. de bouteille. / ~ of a cask ‖ Faßboden m. ‖ fond m. de tonneau ou de fût. / ~ of a channel ‖ Grundbett n. eines Flusses; Sohle f. eines Flußbettes ‖ fond m. d'un lit. / conical-shaped ~ (Dress ore) ‖ Spitzkasten m. ‖ bac m. à fond conique. / conical-shaped ~ (Tank) ‖ Spitze f. eines Behälters ‖ fond m. conique d'un réservoir. / double ~ ‖ Doppelboden m. ‖ double fond m. / fathomable ~ (Mar) ‖ lotbarer Grund m. ‖ fond m. sondable. / flat ~ ‖ Flachboden m. ‖ fond m. plat. / foreign ~ (Mar) ‖ fremdes Schiff n. ‖ bâtiment m. étranger ou de propriété étrangère. / ~ of a furnace ‖ Sohle f. eines Ofens ‖ sole f. d'un fourneau. / gable ~ ‖ Sattelboden m. ‖ fond m. en dos d'âne. / gravel ~ see gravelly ~. / gravelly ~ ‖ Kiesgrund m.; Kieselgrund m. ‖ fond m. de gravier. / hinged ~ ‖ Bodenklappe f.; Klappboden m. ‖ trappe f. de fond; fond m. mobile ou à rabattement. / wagon with hopper ~ ‖ Wagen m. mit Bodentrichter ‖ wagon m. à fond en trémie. / ~ of a lock ‖ Schleusenbett n.; Schleusenboden m. ‖ plancher m. ou radier m. d'une écluse. / ~ in the moulding room ‖ Herd m. in der Formerei ‖ sole f. au moulage. / muddy ~ ‖ Moddergrund m.; Schlammgrund m.; Schlickgrund m. ‖ fond m. de vase ou mou; vase f. molle. / oozy ~ see muddy ~. / ~ of the packing case ‖ Kistenboden m. ‖ fond m. de la caisse. / ~ of a perpendicular ‖ Fußpunkt m. eines Lotes ‖ pied m. d'une perpendiculaire. / perforated ~ ‖ Siebboden m. ‖ fond m. en tôle perforée. / ~ of a river ‖ Grund m. eines Flusses ‖ fond m. d'une rivière. / the ~s pl. are made up ready for rivetting (Boil) ‖ die Böden mpl. werden fertig zum Nieten geliefert ‖ les fonds mpl. sont fournis prêts au rivetage. / rounded ~ ‖ runder Boden m. ‖ fond m. arrondi. / sandy ~ ‖ Sandgrund m. ‖ fond m. de sable. / ~ of the sea ‖ Meeresboden m. ‖ fond m. de la mer. / self-discharging wagon for discharging from the ~ ‖ Bodenselbstentlader m. ‖ wagon m. à déchargement automatique par le fond. / ~ of the shaft ‖ Schachtsohle f.; Schachttiefstes n. ‖ fond m. du puits. / ~ of a shell ‖ Geschoßboden m. ‖ culot m. du projectile. / shingly ~ see gravelly ~. / ~ of the ship ‖ Schiffsboden m. ‖ fond m. du navire. / ~ of a shot see ~ of a shell. / from the ~ towards the top ‖ von unten nach oben ‖ de bas en haut. / ~ of trough ‖ Muldentiefstes n. ‖ fond m. du synclinal. / ~ of wagon ‖ Wagenboden m. ‖ fond m. de wagon; plancher m.

bottom ballasting ‖ untere Schüttung f. ‖ balastage m. première couche. / ~ bar (Saddl) ‖ Schaumstange f. ‖ barre f. d'un mors de bride. / ~ block ‖ Unterflasche f. ‖ moufle m. inférieur. / ~ boards pl. (Shipb) ‖ Remis m.; Rennlatten fpl. ‖ paracloses fpl. / ~ box (Bucket elevator) ‖ Schöpftrog m. ‖ caisse f. à la partie inférieure de la noria. / ~ bush (Mot) ‖ Grundbüchse f. ‖ boîte f. à clapets. / ~ clack (Pump) ‖ Saugventil n. ‖ soupape f. d'aspiration. / ~ clack (Steam eng) ‖ Einlaßventil n. ‖ soupape f. d'aspiration. / ~ cover of the cylinder ‖ Zylinderboden m. ‖ fond m. du cylindre.

/ ~ covering ‖ Bodenbelag m. ‖ pavé m. / ~ die ‖ Matrize f.; Unterlage f. ‖ matrice f. / ~ discharging wagon ‖ Bodenentleerer m. ‖ wagon m. déchargeant par le fond. / ~ fermentation ‖ Untergärung f. ‖ fermentation f. avec dépôt. / ~ flange ‖ Fußflansch m. ‖ bride f. inférieure. / ~ flange of a fire box ‖ Feuerbuchsbodenring m. ‖ cadre m. inférieur d'un foyer. / ~ flap (Railw) ‖ Bodenklappe f. ‖ trappe f. de fond; fond m. à rabattement. / ~ gear ‖ erste Geschwindigkeit f. ‖ première vitesse f. / ~ heating ‖ Bodenbeheizung f. ‖ chauffage m. par le fond.

bottoming (Road) ‖ Packlage f.; Grundbau m. ‖ empierrement m. de base; blocage m. / ~ (Sug) ‖ Decken n. ‖ terrage m.

bottom iron ‖ Schaleneisen n. ‖ carcas m. / ~ lift (Min) ‖ Saugsatz m. eines Pumpenschachtes ‖ pompe f. inférieure élévatoire installée au fond des puits. / ~ maker (Met) ‖ Bodenarbeiter m.; Tiegelofenmann m. ‖ dameur m. de fonds; ouvrier m. au four à creuset. / ~ millstone ‖ Bodenstein m. ‖ gisante f.; meule f. dormante. / ~ outlet pipe of a dam ‖ Grundablaßrohr n. einer Sperrmauer ‖ conduite f. de fond ou de vidange d'un mur du barrage. / ~ outlet sluice ‖ Grundablaßschütz n. ‖ vanne f. de vidange de fond. / ~ part (Found) ‖ Kastenhälfte f. ‖ demi-châssis m. / ~ plank (Shipb) ‖ Bodenplanke f. ‖ bordage m. de fond.

bottom plate (Locksm) ‖ Schloßboden m. ‖ fond m. de serrure. / ~ (Mach) ‖ Bodenblech n.; Sohlplatte f. ‖ tôle f. de plancher; plaque f. de fondation. / ~ (Typewr) ‖ Bodenplatte f. ‖ plaque f. de fond.

bottom prism (Acc) ‖ Bodenprisma n. ‖ prisme support m. de fond.

bottom roller ‖ Unterwalze f. ‖ cylindre m. inférieur. / ~ fixed ~ ‖ ortsfeste Unterwalze f. ‖ cylindre m. inférieur fixe. / supported ~ ‖ unterstützte Unterwalze f. ‖ cylindre m. inférieur supporté.

bottoms pl. ‖ Faßgeläger n.; Faßhefe f.; Niederschlag m. ‖ dépôt m. ou fonds mpl. ou lie f. de foudre.

bottom shaft ‖ Untermesserwelle f. ‖ arbre m. inférieure à lames. / ~ sleeper ‖ Bodenschwelle f. ‖ traverse f. de plancher. / ~ slide drawing press ‖ Geschirrziehpresse f. mit beweglichem Tisch ‖ presse f. à emboutir à table mobile. / ~ swage (Locksm) ‖ Unterteil m. des Gesenks ‖ dessous m. de l'estampe. / ~ tumbler (Dredger) ‖ untere Eimertrommel f. ‖ tambour m. du bas. / ~ view ‖ Grundriß m. ‖ plan m. / ~ wooden frame ‖ Bodenholzgestell n. ‖ plateforme f. en bois.

bouillon capsule ‖ Bouillonwürfel m. ‖ capsule f. pour bouillon.

boulangerite ‖ Antimonbleiblende f.; Schwefelantimonblei n. ‖ boulangérite f.

boulder ‖ Geröll n. ‖ caillou m. / erratic ~ ‖ Findling m. ‖ bloc m. erratique.

boulder flint ‖ Flintstein m. ‖ silex m.

boulders pl. ‖ Geschiebeblöcke mpl. ‖ moraine f. / movement of ~ ‖ Geschiebeführung f. ‖ transport m. de débris minéraux.

boulder-stones pl. ‖ Gerölle n.; Geschiebe n. ‖ cailloux mpl. roulés; galet m.

boule (Join) ‖ Möbeleinlage f.; Einlegeholz n.; Einlegemetall n. ‖ bois m. ou métal m. de marqueterie.

bound, to ‖ begrenzen ‖ borner; limiter.

bound ‖ gebunden ‖ lié. / ~ (Bookb) ‖ gebunden ‖ relié. / ~ book ‖ gebundenes Buch n. ‖ livre m. relié. / ~ masonry ‖ in Verband aufgeführtes Mauerwerk n. ‖ maçonnerie f. en liaison. / ~-off (Cement) ‖ abgebunden ‖ pris.

bound (Border stone) ‖ Grenzstein m.; Markstein m. ‖ borne f. / ~ (Limit) ‖ Grenze f.; Schranke f. ‖ limite f.; borne f. / ~ (Shock) ‖ Anprall m.; Aufprall m. ‖ choc m.; collision f.

boundaries pl. of improved river channels ‖ Normallinien fpl. oder Baulinien fpl. oder Streichlinien fpl. der Flußufer ‖ tracé m. des rives d'une rivière.

boundary ‖ Grenze f. ‖ limite f.; borne f.; frontière f. / ~ (Min) ‖ Markscheide f. ‖ borne f. d'une mine. / ~ layer ‖ Grenzschicht f. ‖ couche f. limite. / ~ light ‖ Umrandungsfeuer n. ‖ feu m. de délimitation de terrain. / ~ line ‖ Grenzlinie f.; Grenze f. ‖ ligne f. de séparation ou de démarcation. / ~ post (Surv) ‖ Markpfahl m. ‖ croix f.; poteau m. servant de borne. / ~ stone (Surv) ‖ Feldstein m.; Grenzstein m.; Malstein m.; Markstein m. ‖ borne f. / ~ vortex ‖ Randwirbel m. ‖ tourbillon m. de bord.

bound-off (Cement) ‖ abgebunden ‖ pris.

bounty (Mar) ‖ Handgeld n. beim Heuern ‖ avance f. prime d'engagement. / ~ on exportation ‖ Ausfuhrprämie f. ‖ prime f. d'exportation.

bouquet ‖ Strauß m.; Blumenstrauß m. ‖ bouquet m.

bourette yarn ‖ Bourrettegarn n. ‖ fil m. de bourrette de soie.

bournonite ‖ Bournonit m.; Spießglanzbleierz n.; Antimonkupferglanz m.; Schwarzspießglanzerz n. ‖ antimoine m. sulfuré plombo-cuprifère; bournonite f.

bourrette see bourette.

bout (Weav) ‖ Gang m. beim Scheren der Kette ‖ portée f.

bovril ‖ Fleischextrakt m. ‖ extrait m. de viande.

bow ‖ Bogen m.; Rundung f. ‖ courbe f.; courbure f.; cintre m. / ~ (Arm) ‖ Bogen m. ‖ arc m. / ~ (Drawing) ‖ Kurvenlineal n. ‖ pistolet m. / ~ (Shipb) ‖ Bug m. ‖ avant m. / ~ of boat ‖ Bootsbug m. ‖ proue f. de la coque. / ~ of the case (Watchm) ‖ Gehäusebügel m. ‖ anneau m. de la boîte. / ~ of a key ‖ Schlüsselraute f.; Schlüsselring m. ‖ anneau m. d'une clef. / ~ for lady's bag ‖ Damentaschenbügel m. ‖ monture f. de sac à main. / lean ~ ‖ scharfer Schiffsbug m. ‖ proue f. fine ou maigre. / ~ for leather ware ‖ Bügel m. für Lederwaren ‖ monture f. pour articles de cuir. / violin ~ ‖ Geigenbogen m. ‖ archet m. / violoncello ~ ‖ Cellobogen m. ‖ archet m. de violoncelles.

bow compasses pl. ‖ Null(en)zirkel m.; Bogenzirkel m. ‖ compas m. à pompe; petit balustre m. / ~ divider ‖ Teilzirkel m. ‖ compas m. de précision.

bowel ‖ Darm m. ‖ boyau m. / ~s pl. ‖ Eingeweide n. ‖ entrailles fpl. / ~ washer ‖ Darmputzer m. ‖ laveur m. de boyaux.

bower of furs ‖ Haarfacher m. ‖ arçonneur m. de poils.

bow file ‖ Raumfeile f.; Riffelfeile f.; Bogenfeile f. ‖ lime f. à archet; riflard m.; rifloir m. / ~-heavy ‖ buglastig ‖ chargé à la proue; lourd du nez; à nez lourd.

bowl ‖ Schale f.; Schüssel f.; Napf m.; Terrine f.; Bowle f. ‖ coupe f.; bol m.; écuelle f.; jatte f.; tasse f.; cuvette f.; terrine f. / ~ (Pott) ‖ Pfeifenkopf m. ‖ fourneau m. ou tête de la pipe. / calender ~ ‖ Kalanderwalze f. ‖ cylindre m. de calandres. / delivering ~ (Weav) ‖ Abzugswalze f. ‖ rouleau m. de tirage ou d'appel. / wooden ~ ‖ Holzschüssel f. ‖ sébille f.

bowline ‖ Bogenlinie f. ‖ arceau m.; courbure f. du cintre. / ~ (Mar) ‖ Buline f.; Boleine f.; Bulin ‖ bouline f. / on a ~ (Nav) ‖ am Winde ‖ près du vent.

bowline knot (Mar) ‖ Pfahlstich m. ‖ nœud m. d'agui à élingue. / ~ (Sailm) ‖ Augstich m. ‖ nœud m. de chaise.

bowling-green ‖ Rasenplatz m.; Ballspielplatz m. ‖ boulingrin m.

bow net ‖ Fischreuse f.; Reuse f. ‖ bire f.; nasse f.; panier m. / crayfish ~ ‖ Krebskorb m.; Krebsreuse f. ‖ nasse f. à écrevisses. / fish ~ ‖ Fischreuse f. ‖ nasse f. à poissons. / osier ~ ‖ Reuse f. aus Weidengeflecht ‖ nasse f. en osier. / wire ~ ‖ Drahtreuse f. ‖ nasse f. en fil de fer.

bow net aerial ‖ Reusenantenne f. ‖ antenne f. en forme de nasse. / ~ plaiter ‖ Reusenmacher m. ‖ tresseur m. de nasses.

bow oar (Mar) ‖ Bugriemen m. ‖ aviron m. de brigadier. / ~ pen ‖ Zirkelfeder f. ‖ tire-ligne m. de compas. / ~ pencil ‖ Zirkelstift m. ‖ branche f. du compas portant le crayon. / ~ saw ‖ Bügelsäge f.; Bogensäge f. ‖ scie f. à étrier ou à archet.

bowse, to (Shipb) ‖ auftaljen ‖ palanguer. / ~ taut (Mar) ‖ steifholen ‖ haler raide.

bow spacer ‖ Meßzirkel m. ‖ compas m. de mesure.

bowsprit ‖ Bugspriet n. ‖ beaupré m.

bowtell (Arch) ‖ Rundstab m.; Stab m. ‖ boudin m.; bosel m.; baguette f.

bow wave ‖ Bugwelle f. ‖ vague f. de proue.

box, to ~ a wheel ‖ ein Rad n. ausbuchsen ‖ emboîter une roue.

box ‖ Kiste f.; Kasten m.; Schachtel f.; Dose f. ‖ caisse f.; boîte f. / ~ (Build) ‖ Verschlag m. ‖ retranchement m. / ~ (Cable) ‖ Muffe f. ‖ manchon m. / ~ (Envelope) ‖ Kapsel f.; Gehäuse n. ‖ enveloppe f. / ~ (Found) ‖ Formkasten; Gießkasten m.; Gießlade f. ‖ châssis m.; châssis m. de moulage. / ~ (Jewels) ‖ Etui n. ‖ étui m. / ~ (Luggage) ‖ Koffer m. ‖ malle f.; coffre m. / ~ (Theatre) ‖ Theaterloge f. ‖ loge f. / ~ (Trade) ‖ Verpackung f.; Emballage f. ‖ emballage m. / ~ (Trunk; Tress) ‖ Lade f. ‖ caisse f.; coffre m.; tiroir m. / ~ (Weav) ‖ Schützenkasten m. ‖ boîte f. pour la navette. / air-tight closed ~ ‖ luftdicht verschlossener Behälter m. ‖ bidon m. hermétiquement fermé. / annealing ~ ‖ Glühtopf m. ‖ pot m. à recuire ou de cémentation. / axle ~ ‖ Achslager n. ‖ boîte f. d'essieu. / axle ~ body ‖ Achslagergehäuse n. ‖ corps m. de boîte d'essieu. / ~ for bandages ‖ Verbandkasten m. ‖ boîte f. à pansement; boîtier m. / ~ for bottles ‖ Flaschenkasten m. ‖ caisse f. à bouteilles. / brass ~ ‖ Messinghülse f. ‖ boîte f. ou douille f. en laiton. / ~ of bricks ‖ Steinbaukasten m. ‖ jeu m. de construction; boîte f. de construction en pierre. / cast annealing ~ ‖ gegossener Glühtopf m. ‖ pot m. de cémentation ou

à recuire fondu. / charging ~ ‖ Lademulde f. ‖ récipient m. de chargement. / clock ~ ‖ Uhrgehäuse n. für Großuhren ‖ caisse f. d'horloges. / ~ of coils (Electr) ‖ Widerstandskasten m. ‖ boîte f. de résistance. / ~ of compasses ‖ Reißzeug n. ‖ étui m. de mathématiques. / ~ of distribution (Electr) ‖ Abzweigkasten m. ‖ boîte f. de distribution. / elastic axle ~ ‖ federndes Achslager n. ‖ boîte f. d'essieu à ressorts. / ~ of fire door ‖ Feuerzarge f. ‖ encadrement m. de porte de fourneau. / forged annealing ~ ‖ geschmiedeter Glühtopf m. ‖ pot m. de cémentation à recuire forgé. / ~ of freight car ‖ Wagenkasten m. ‖ caisse f. de voiture. / ~ of games ‖ Baukasten m. ‖ boîte f. de construction. / glass ~ (Acc) ‖ Glasgefäß n. ‖ récipient m. en verre. / hard paper ~ ‖ Hartpapierdose f. ‖ boîte f. en papier durci. / ~ for horse shoe smithing ‖ Hufbeschlagstand m. ‖ cabine f. pour maréchalerie. / in ~es pl. ‖ in Büchsen fpl. ‖ en boîtes fpl. / inside ~ ‖ Innenlager n. ‖ coussinet m. intérieur. / ~ used for the joining of the lower and the upper spring (Coachm) ‖ Kapsel f. zur Verbindung der Ober- und Unterfeder ‖ boîte f. d'assemblage du ressort supérieur et du ressort inférieur. / journal ~ (Railw) ‖ Achslager n. ‖ boîte f. d'essieu. / junction ~ (Electr) ‖ Abzweigdose f. ‖ boîte f. de raccordement ou de dérivation. / latticed ~ ‖ Gitterloge f. ‖ loge f. grillée. / ~ of leads ‖ Bleiminenbüchse f. ‖ étui m. à mines. / letter ~ ‖ Briefkasten m. ‖ boîte f. à lettres. / ~ lined with plate ‖ innen mit Blech beschlagener Wagenkasten m. ‖ caisse f. doublée intérieurement de tôle. / lubricating ~ ‖ Schmierlager n. ‖ palier m. graisseur. / musical ~ ‖ Spieldose f. ‖ boîte f. à musique. / nave ~ (Coachm) ‖ Nabenbuchse f. ‖ boîte f. de roue. / outside ~ ‖ Außenlager n. ‖ coussinet m. extérieur. / ~ for postage stamps ‖ Briefmarkenschachtel f. ‖ boîte f. pour timbresposte. / ~ for provisions (Mar) ‖ Proviantkiste f. ‖ caisson m.; caisse f. de provision. / removable ~ (Wagon) ‖ Wagenkasten m. zum Abheben ‖ caisse f. (de voiture) enlevable ou amovible. / rigid axle ~ ‖ festes Achslager n. ‖ boîte f. d'essieu fixe. / screw ~ ‖ Schraubenmutter f. ‖ écrou m. / ~ with side flaps folding down (Wagon) ‖ Wagenkasten m. mit Seitenklappen ‖ caisse f. (de voiture) avec portes de côté rabattables. / small ~ ‖ Kästchen n. ‖ coffret m. / ~ for spares ‖ Ersatzteilkasten m. ‖ boîte f. pour pièces de rechange. / ~ for sponge of papier-maché ‖ Schwammdose f. aus Pappmaché ‖ boîte f. à éponge en carton pâte. / steering gear ~ ‖ Lenkgehäuse n. ‖ carter m. de direction. / swill ~ ‖ Müllkasten m.; Mülleimer m. ‖ boîte f. à ordures. / tempering ~ ‖ Glühtopf m. ‖ pot m. à recuire. / tin ~ ‖ Blechemballage f.; Blechbüchse f.; Blechdose f. ‖ boîte f. métallique. / ~ of vice ‖ Schraubstockhülse f. ‖ boîte f. d'étau. / wagon ~ ‖ Wagenkasten m. ‖ caisse f.; caisse f. ou corps m. de voiture. / the ~ of the wagon is arranged for being tipped ‖ der Wagenkasten ist kippbar ‖ la caisse est à rabattement. / wheel ~ (Coachm) ‖ Nabenbuchse f. ‖ boîte f. de roue. / wooden ~ (Smaller) ‖ Holzschachtel f.

box 74 bracket seat

|| boîte f. en bois refendu. / wooden ~
(Greater) || Holzkiste f.; Holzkasten m. ||
caisse f. *ou* coffret m. en bois.

box board || Kistenbrett n. || planche f. de
caisse. /~branding machine || Kistenbret-
terbrennmaschine f. || machine f. à mar-
quer au feu les planches de caisses. / ~
making machine || Kistenbrettermaschine
f. || machine f. à faire les planches de
caisses. / ~ printing machine || Kisten-
bretterbedruckmaschine f. || machine f.
à marquer les planches des caisses.

box car || Kastenwagen m. || chariot m. à
caisse. / ~ cart || Kastenkarren m. ||
charrette f. à caisse. / ~ casting || Kasten-
guß m. || coulage m. *ou* coulée f. en
châssis. / closing machine for ~es ||
Kistenverschlußmaschine f.; Schachtel-
schließmaschine f. || machine f. à fermer
les boîtes. / ~ compound || Muffen-
ausgußmasse f. || masse f. *ou* mastic
m. de remplissage pour manchons. /
~ desk || Kastenpult n. || coffre-pupitre
m. / ~ end of the main rod || Treib-
stangenbügel m. || tête f. de bielle en
étrier.

boxer of sleepers || Schwellenstopfer m.;
Stopfer m. || bourreur m. de traverses.

box filling machine || Dosenfüllmaschine f.
|| machine f. à remplir les boîtes. / ~
fittings pl. || Kistenbeschläge mpl. / ~
garnitures fpl. *ou* ferrures fpl. de caisse.
/ folding machine for ~es || Schachtelfalt-
maschine f. || plieuse f. pour boîtes; ma-
chine f. à plier les boîtes. / ~ form ||
Kastenform f. || forme f. caisse. / ~form
bed || Kastenbett n. || bâti m. en forme
de caisson. / universal folding, round-
ing and ~ forming machine || Universal-
abkant-, Rund- und Kastenbiegemaschi-
ne f. || machine f. universelle à plier,
rouler et former des boîtes. / ~ freight
car (Railw) || Kastenwagen m.; bedeck-
ter Güterwagen m. || wagon m. à caisse
ou à marchandises couvert *ou* fermé. / ~
girder || Kastenträger m. || poutre f. à
caisson. / gumming machine for ~es ||
Schachtelklebemaschine f. || colleuse f.
pour boîtes; machine f. à coller les boîtes.
/ ~ head (Electr) || Dosenendverschluß
m. || tête f. de boîte multiple. / ~ hold
|| Kistengriff m. || manche f. de caisse.

boxing of the sleepers (Railw) || Stopfen
n. der Schwellen || bourrage m. des
traverses. / ~ material || Schüttungs-
material n.; Bettungsmaterial n. || balast
m. / ~ pick || Stopfhacke f. || pioche f. à
bourrer. / ~ pole (Railw) || Stopfstange
f. || batte f. à bourrer.

box key (Locksm) || Aufsteckschlüssel m. ||
clef f. à douille. / ~ kite || Kastendrachen
m. || cerf-volant m. cellulaire. / ~ lock
|| Kastenschloß n.; Kofferschloß n.;
Schrankschloß n. || serrure f. d'armoire
ou de coffre. / locking device for ~es ||
Kistenverschluß m. || fermeture f. de
caisses. / ~ maker || Kistenmacher m. ||
caissier m. / ~ making machine || Kisten-
fabrikationsmaschine f. || machine f.
pour la fabrication de caisses. / ~ nailer
|| Kistennagler m. || cloueur m. de caisses.
/ ~ nailing machine || Kistennagel-
maschine f. || machine f. à clouer les
caisses. / ~ opener || Büchsenöffner m.;
Konservenbüchsenöffner m. || clef f. à
ouvrir les boîtes de conserves. / ~ plate
(Acc) || Kastenplatte f. || plaque f. de
caisse. / ~ prizer (Cigar) || Wickelpresser
m. || presseur m. de cigares. / ~ relay ||
Dosenrelais n. || relais m. à boîte. / ~
respirator || Gasmaske f. || masque m.
contre les gaz. / ~ rib || Kastenrippe f.
nervure f. en caisson. / ~ root || Bux-
baumwurzel f. || racine f. de buis. / ~
scales pl. || Kastenwage f. || balance
f. à caisse. / ~ screw || Schraubenbüchse
f. || douille f. taraudée.

box-shaped || kastenförmig || en forme de
boîte. / ~ bed || Kastengestell n. || bâti
m. formant caisson. / ~ steel plates pl.
pressed by hydraulic power || hydrau-
lisch gepreßte kastenförmige Stahlbleche
npl. || tôles fpl. d'acier embouties sous
la presse hydraulique à la forme de
boîtes.

box spanner || Schienenschraubenschlüssel
m. || clé f. à tire-fonds. / ~ spar || Kasten-
holm m.; Kastenspiere f. || longeron m.
en caisson. / ~ staple (Locksm) || Schließ-
kappe f.; überbauter Schließhaken m. ||
gâche f. / ~ switch || Dosenschalter m.;
Drehschalter m. || interrupteur m. rota-
tif. / ~ table || Kastentisch m. || table f.
cubique. / ~ terminal (Electr) || Dosen-
endverschluß m. || boîte f. terminale. / ~
tool || Stichelhaus n. || porte-outil m. à
charioter. / ~ tool for tool block || Block-
stichelhaus n. || porte-outil m. pour bloc
central. / ~ top || Kartonnagendeckel m. ||
couvercle m. de cartonnages. / ~ trail
carriage (Arm) || Kastenlafette f. || affût
m. rigide *ou* monoflèche. / ~tree || Buchs-
baum m. || buis m. / ~-type window ||
Kastenfenster n. || fenêtre f. à caisson. /
~ upper beam || Kastenoberwange f. ||
sommier m. supérieur. / ~ wagon ||
Gepäck- oder Packwagen m. || wagon
m. à bagages. / ~ wood || Buchsbaum-
holz n. || (bois m. de) buis m.

boy (Mar) || Schiffsjunge m. || mousse m.

boy's clothing || Kinderbekleidung f. ||
vêtements mpl. pour enfants. / ~ tai-
lor || Knabenkleidermacher m.; Knaben-
schneider m. || tailleur m. pour enfants.

brace, to (To cramp) || klammern || cram-
ponner. / ~ (To strut) || verspannen;
ausstreben || croisillonner; entretoiser. /
~ aback (Mar) || gegenbrassen || contre-
brasser. / ~ full (Mar) || abbrassen || bras-
ser à porter; décharger les voiles. / ~
together (Print) || mit Klammern fpl.
verbinden || accolader.

brace (Bit stock) || Brustleier f.; Brust-
bohrer m. || vilebrequin m. / ~ (Build) ||
Tragband n.; Versteifung f.; Strebe f.;
Verstrebung f. || entretoise f.; croisillon
m.; étrésillon m. / ~ (Print) || Akkolade
f.; (zusammenfassende) Klammer f.;
geschweifte Klammer f. || accolade f. /
diagonal ~ || Diagonalstrebe f. || entre-
toise f. diagonale; bras m. de croix
St. André. / diagonal ~s pl. || diagonale
Verstrebung f. || contreventement m. en
diagonale. / double ~ (El line) || Zangen-
verbindung f. || barre f. double. / ~ of
a gutter || Haken m. einer Dachrinne ||
ferrement m. d'une gouttière. / St. Andrew's cross ~ *see* diagonal ~. / stiffening
~ || Verstärkungsstrebe f. || entretoise f.
de renforcement. / vertical ~ || Vertikal-
verband m. || entretoisement m. vertical.

brace buckle || Hosenträgerschnalle f. ||
boucle f. de bretelles.

braced || versteift || entretoisé. / ~ wing
(Aero) || verstrebter Flügel m. || aile f.
entretoisé *ou* à mât *ou* à montant.

brace head || Krückelstück n.; Bohrkrückel
m. || manche m. de manœuvre; manivelle
f.; tourne-à-gauche m.

braceless (Aero) || verspannungslos; un-
verspannt || sans haubanage m.

bracelet || Armband n. || bracelet m.

braces pl. || Hosenträger m.; Tragband n. ||
bretelles fpl. / elastic ~ || Gummihosen-
träger mpl. || bretelles fpl. en caout-
chouc.

bracing (Build) || Verspreizung f. || entre-
toisement m. des cloisons. / cross ~
(Bridge) || Querverbindung f.; Quer-
versteifung f.; Windverkreuzung f. ||
contreventement m.; cours m. d'entretoises; pièces fpl. de pont. / diagonal ~
(Aero) || Kreuzverspannung f. || croisillons
mpl. d'incidence.

bracing cable || Fangkabel n.; Spannkabel
n. || hauban m. inférieur de sustentation;
câble m. tendeur. / ~ rope (Pont) ||
Spanntau n. || traversière f.; amarre f.;
écharpe f.

bracket (Carp) || Sattelholz n.; Schirr-
balken m. || corbeau m.; racinal m.;
sous-poutre f. || sous-longeron m. / ~
(Lighting) || Armleuchter m. || girandole
f.; chandelier m. à bras. / ~ (Print) ||
eckige Klammer f. || crochet m. / ~
(Support) || Arm m.; Ausleger m.; Wand-
arm m.; Winkelstütze f. || applique f.;
console f.; potence f. / angle ~ || Winkel-
konsole f. || console f. à équerre. /
bearing ~ || Lagerbock m. || chevalet m.;
chaisse f. de palier. / ~ of the boiler ||
Kesselstütze f. || patte f. de sustentation
ou support m. de la chaudière. / double
~ || Doppelausleger m. || potence f. *ou*
support m. double. / electric light ~ ||
Wandarm m. *oder* Arm m. für elektri-
sche Beleuchtung || branche f. de lampe
électrique. / front-spring ~ (Auto) || Vor-
derfederbock m. || support m. de ressort
avant. / fulcrum ~ || Hebelträger m. ||
support m. de point fixe. / guide ~ ||
Führungsbuchse f. || boîte f. de guidage. /
hook-shaped ~ (Insulator) || Hakenstütze
f. || support m. à crochet. / insulator ~ ||
Isolatorstütze f. || support m. d'isola-
teur. / lamp ~ || Lampenarm m.; Lam-
penstütze f.; Laternenhalter m. || sup-
port m. de lampe. / lantern ~ (Railw)
|| Laternenstütze f. Laternenhalter m. ||
porte-lanterne. m. / ~ of a lever-draw-
bridge || Schwungbaum m. an einer
Zugbrücke || flèche f. de pont-levis à
bascule. / link ~ (Loc) || Kulissenlager
n. || attache m. de coulisse. / motor ~ ||
Motorgrundplatte f. || plaque f. de fonda-
tion du moteur. / rear-spring ~ (Auto)
|| Hinterfederbock m. || support m. de
ressort arrière. / spring ~ (Auto) || Feder-
bock m. || support m. de ressort. / ~
stamped-out || gepreßte Konsole f. ||
console f. emboutie. / street-lantern ~ ||
Laternenstütze f. || poteau m. *ou* sup-
port m. de lanterne / tipping ~ (Railw)
|| Abrollbock m. || chevalet m. de bas-
cule. / wall ~ || Wand(lager)bock m.;
Wandkonsole f.; Wandlagerstuhl m. ||
console f.

bracket clock || Standuhr f. || pendule f. /
~ crane || Konsolkran m. || grue f. à
console. / ~ fan || Konsolfächer m. ||
ventilateur m. mural. / ~ joint (Railw) ||
Winkellasche f. || éclisse-cornière f. /
~ projection || Auskragung f. || saillie f. /
~ seat || Klappsitz m. || strapontin m.

/ ~ stud ‖ Hebelträgerschraube f. ‖ prisonnier m. de support de point fixe. / ~ support ‖ Kragstütze f. ‖ console f. d'encorbellement. / ~ travelling crane ‖ Wandlaufkran m. ‖ grue-console f. roulante.

brackish water ‖ brackiges Wasser n.; Brackwasser n. ‖ eau f. saumâtre.

bractea (Bot) ‖ Deckblatt n.; Deckschuppe f.; Doldenblatt n. ‖ bractée f.

brad ‖ Stift m. ‖ pointe f.

bradawl ‖ flache Ahle f.; Bindeahle f. ‖ alêne f.

braid, to ‖ klöppeln; umklöppeln ‖ tresser.

braid ‖ Flechtschnur f.; Tresse f.; Litze f.; Zopf m.; Umklöppelung f. ‖ tresse f.; passe-poil m.; galon m.; natte f.

braided ‖ umklöppelt tressé. / ~ cord ‖ geflochtene *oder* geschlagene Leine f. ‖ corde f. tressée; drisse f.; corde f. en fils tordus. / ~ incandescent mantle ‖ geklöppelter Glühstrumpf m. ‖ manchon m. tricoté à incandescence. / ~ line *see* ~ cord. / ~ wire ‖ umklöppelter Draht m. ‖ fil m. tressé. / ~ wire (Radio) ‖ Litzendraht m. ‖ litzendraht m.

braider ‖ Schnurmacher m. ‖ ganseur m.

braiding (Of wire) ‖ Umklöppelung f. ‖ tressage m. / cotton ~ ‖ Baumwollumklöppelung f. ‖ tressage m. de coton. / impregnated ~ ‖ getränkte *oder* imprägnierte Umklöppelung f. ‖ tressage m. imprégné. / ~ with spun yarn ‖ Gespinstumflechtung f. ‖ tressage m. de filet.

braiding machine (Lacem) ‖ Flechtmaschine f. ‖ machine f. à lacets *ou* à tresser. / ~ (Wire) ‖ Litzenmaschine f.; Umflechtmaschine f.; Klöppelmaschine f. ‖ machine f. à tresser; tresseuse f. mécanique. / high-speed ~ ‖ Schnellflechtmaschine f. ‖ machine f. à tresser rapide. / wire ~ ‖ Drahtumflechtmaschine f. ‖ machine f. à tresser le fil.

brail (Mar) ‖ Geitau n. ‖ cargue f.; cargue-point f.

brailed-in (Mar) ‖ in der Gei f. hängend; aufgegeit ‖ cargué.

Braille alphabet ‖ Braillealphabet n. ‖ alphabet m. Braille.

brain ‖ Gehirn n. ‖ cerveau m.

brake, to ‖ abbremsen; bremsen ‖ retenir au frein; freiner. / ~ (Flax; Hemp) *see also* to break ‖ brechen; braken ‖ macquer; broyer; briser; teiller. / capable of being braked ‖ bremsbar ‖ freinable. / ~ the clockwork ‖ das Uhrwerk n. bremsen ‖ freiner le mouvement m. d'horlogerie. / ~ the flax *or* hemp by hammers *or* stamps ‖ Flachs m. *oder* Hanf m. mit Schlägeln *oder* Stampen brechen ‖ piler le lin *ou* le chanvre.

brake ‖ Bremse f. ‖ frein m. / ~ (Flax; Hemp) *see also* break ‖ Breche f.; Flachsbreche f.; broie f.; macque f.; brisoir m.; tillotte f. / air ~ *see* air-pressure ~. / air-pressure ~ ‖ Druckluftbremse f.; Luftdruckbremse f. ‖ frein m. à air comprimé. / to apply the ~s pl. ‖ die Bremsen fpl. anziehen ‖ serrer les freins mpl. / atmospheric ~ *see* vacuum ~. / automatic ~ ‖ selbsttätige Bremse f. ‖ frein m. automatique. / back-pedalling ~ ‖ Rücktrittbremse f. ‖ frein m. à contre-pédale. / back-wheel ~ *see* rear-wheel ~. / band ~ ‖ Bandbremse f. ‖ frein m. à collier *ou* à ruban. / belt ~ ‖ Riemenbremse f.; Bandbremse f. ‖ frein m. à courroie. / booster ~ (Aut) ‖ Vakuumbremse f. ‖ servo-frein m. à vide. / car ~ ‖ Wagenbremse f. ‖ frein m. pour voitures. / ~ of the carriage (Typewr) ‖ Wagenbremsvorrichtung f. ‖ frein m. sur le chariot. / centrifugal ~ ‖ Zentrifugalbremse f. ‖ frein m. centrifuge. / clutch ~ ‖ Kupplungsbremse f. ‖ frein m. d'embrayage. / compressed-air ~ *see* airpressure ~. / continuous ~ ‖ Dauerbremse f. ‖ frein m. permanent. / countersteam ~ ‖ Gegendampfbremse f. ‖ frein m. à contre-vapeur. / counterweight ~ ‖ Wurfhebelbremse f. ‖ frein m. à levier à contrepoids. / differential ~ ‖ Differentialbremse f. ‖ frein m. de différentiel *ou* sur le différentiel. / differential shaft ~ *see* differential ~. / double-acting ~ ‖ doppeltwirkende Bremse f. ‖ frein m. à double effet. / dynamometrical ~ ‖ Bremsdynamometer n. ‖ frein m. dynamométrique; dynamomètre m. à frein. / electric ~ ‖ elektrische Bremse f. ‖ frein m. électrique. / emergency ~ ‖ Notbremse f. ‖ frein m. de secours *ou* de détresse. / expansion band ~ ‖ Expansionsbandbremse f. ‖ frein m. à ruban extensible. / fan ~ ‖ Flügelbremse f. ‖ moulinet m. / flywheel ~ ‖ Schwungradbremse f. ‖ frein m. sur le volant. / foot ~ ‖ Tritthebelbremse f.; Fußbremse f. ‖ frein m. à pédale *ou* à pied; frein m. à levier à pédale. / foot-operated ~ ‖ Fußtrittbremse f. ‖ frein m. commandé par pédale. / four-shoe ~ ‖ Vierklotzbremse f. ‖ frein m. à quatre sabots. / four-wheel ~ ‖ Vierradbremse f. ‖ frein m. sur les quatre roues. / front-wheel ~ ‖ Vorderradbremse f. ‖ frein m. de roue avant. / hand ~ (Auto) ‖ Handbremse f. ‖ frein m. à main *ou* de secours. / hand lever ~ ‖ Handhebelbremse f. ‖ frein m. à levier à main. / hub ~ ‖ Nabenbremse f. ‖ frein m. sur le moyeu. / hydraulic ~ ‖ Flüssigkeitsbremse f.; frein m. hydraulique. / inner ~ ‖ Innenbremse f. ‖ frein m. intérieur. / load pressure ~ ‖ Lastdruckbremse f. ‖ frein m. actionné par le poids de la charge. / magnetic ~ ‖ magnetische Bremse f. ‖ frein m. magnétique. / mechanical ~ ‖ mechanische Bremse f. ‖ frein m. mécanique. / metal-to-metal ~ ‖ Metallbackenbremse f. ‖ frein m. à mâchoires métalliques. / muzzle ~ (Gun) ‖ Mündungsbremse f. ‖ frein m. de bouche. / outer ~ ‖ Außenbremse f. ‖ frein m. extérieur. / passenger emergency ~ *see* emergency ~. / pneumatic ~ *see* air-pressure ~. / prony ~ ‖ Bremszaum m. ‖ frein m. prony. / to put on the ~s pl. ‖ die Bremsen fpl. anziehen; bremsen ‖ enrayer *ou* caler *ou* serrer le frein; freiner. / railway-car ~ ‖ Wagenbremse f. ‖ frein m. pour wagons. / rear-wheel ~ ‖ Hinterradbremse f. ‖ frein m. de roue arrière. / to release the ~ ‖ die Bremse lösen *oder* lüften ‖ débloquer *ou* desserrer le frein; défreiner. / rim ~ ‖ Felgenbremse f. ‖ frein m. sur jante. / safety ~ ‖ Sicherheitsbremse f. ‖ frein m. de sûreté. / screw ~ ‖ Spindelbremse f. ‖ frein m. à vis. / second ~ of flax ‖ Schleppracke f.; Schrubb-Breche f. ‖ broie f. seconde. / self-acting ~ ‖ selbsttätige Bremse f. ‖ frein m. automatique *ou* automoteur. / short-circuit ~ ‖ Kurzschlußstrombremse f. ‖ frein m. par court-circuit. / six-wheel ~ ‖ Sechsradbremse f. ‖ frein m. aux six roues. / slide-armature ~ ‖ Verschiebeankerbremse f. ‖ frein m. à induit balladeur. / sprag ~ ‖ Knüppelbremse f. ‖ freinage m. à gourdin. / steam ~ ‖ Dampfbremse f. ‖ frein m. à vapeur. / to throw off the ~s pl. *see* to release the ~. / transmission ~ ‖ Getriebebremse f. ‖ frein m. de mécanisme *ou* à engrenages. / tyre ~ ‖ Reifenbremse f. ‖ frein m. sur pneu. / vacuum ~ ‖ Vakuumbremse f.; Luftbremse f. ‖ frein m. à vide. / water-cooled ~ ‖ wassergekühlte Bremse f. ‖ frein m. à refroidissement d'eau. / wheel ~ ‖ Radbremse f. ‖ frein m. à roue.

brake apparatus ‖ Bremsvorrichtung f. ‖ appareil m. de freinage. / ~ arm ‖ Bremshebel m. ‖ levier m. de frein.

brake band ‖ Bremsband n. ‖ ruban m. *ou* bande f. de frein; collier m. de frein. / ~ of cotton straps ‖ Bremsband n. aus Baumwollgurten ‖ ruban m. à frein en sangles de coton. / ~ coupling ‖ Bremsbandkupplung f. ‖ accouplement m. à friction par bande. / ~ lining ‖ Bremsbandbelag m. ‖ garniture f. de la bande de frein.

brake beam (Windmill) ‖ Preßbalken m. ‖ sommier m. du frein. / ~ block ‖ Bremsschuh m.; Bremsblock m.; Bremsklotz m. ‖ patin m. *ou* sabot m. de frein. / ~ box ‖ Bremskasten m. ‖ boîte f. à frein. / ~ cable ‖ Bremskabel n.; Bremsseil n. ‖ câble m. de frein. / ~ cam ‖ Bremsnocken m.; Bremsdaumen m. ‖ came f. de frein. / ~ chain ‖ Bremskette f. ‖ chaîne f. de frein. / ~ cheek ‖ Bremsbacke f. ‖ mâchoire f. de frein. / ~ circuit ‖ Bremsstromkreis m. ‖ circuit m. de freinage. / ~ compensating lever (Aut) ‖ Bremsausgleichhebel m. ‖ levier m. compensateur du frein. / ~ compensating shaft (Aut) ‖ Bremsausgleichwelle f. ‖ arbre m. compensateur de frein. / ~ control ‖ Bremskontrolle f. ‖ manœuvre f. du freinage. / ~ cord (Typewr) ‖ Bremsschnur f. ‖ câble m. de frein. / ~ coupling ‖ Bremskupplung f. ‖ accouplement m. à frein. / ~ crank ‖ Bremskurbel f. ‖ manivelle f. de serrage *ou* de frein. / ~ cylinder ‖ Bremswalze f. ‖ cylindre m. de freinage. / ~ disc ‖ Bremsscheibe f. ‖ disque m. *ou* poulie f. de frein. / ~ drum ‖ Bremstrommel f.; Bremsscheibe f. ‖ tambour m. *ou* poulie f. de frein. / ~ dynamo ‖ Bremsdynamomaschine f. ‖ dynamo-frein m. / ~ equalizer ‖ Bremsausgleicher m. ‖ compensateur m. de frein. / ~ equipment ‖ Bremsausrüstung f. ‖ garniture f. d'appareils de frein. / quick-acting ~ equipment ‖ Schnellbremsausrüstung f. ‖ garniture f. d'appareils de frein rapide. / ~ gear ‖ Bremsschaltung f. ‖ timonerie f. de frein. / ~ handle ‖ Bremshebelgriff m. ‖ manette f. de frein. / ~ hand lever ‖ Handbremshebel m. ‖ levier m. de frein à main. / ~ hand wheel ‖ Bremsrad n. ‖ volant m. à main du freinage. / ~ hangers support ‖ Bremsgehängeträger m. ‖ support m. de la suspension de frein. / ~ horse power ‖ Bremsleistung f. in PS; Bremspferdestärke f.; gebremste Pferdestärke f. ‖ cheval m. *ou* C. V. (mesuré) au frein; puissance f. du frein. / ~ intermediate shaft ‖ Bremszwischenwelle f. ‖ arbre m. intermédiaire de frein.

brake lever ‖ Bremshebel m. ‖ levier m. de frein. / ~ arrangement ‖ Bremsgestänge n. ‖ timonerie f. du frein.

/ ~ guide ‖ Bremshebelführung f. ‖ guide du levier de frein.

brake lining ‖ Bremsbelag m.; Bremsfutter n. ‖ fourrure f. ou garniture f. de frein. / ~ magnet ‖ Bremslüftungsmagnet m.; Bremsmagnet m. ‖ électro m. de démarrage du frein; électro m. de frein; électro-aimant m. de freinage. / ~ maker (Coachm) ‖ Bremsenmacher m. ‖ monteur m. de freins.

brakeman (Railw) ‖ Bremser m. ‖ garde-frein m. / ~ (Mining) ‖ Fördermaschinist m. ‖ machiniste m. d'extraction. / ~'s cabin ‖ Bremserhäuschen n. ‖ guérite f. / ~'s platform ‖ Bremserstand m. ‖ poste m. ou cabine f. du garde-frein.

brake operating lever ‖ Bremsbetätigungshebel m. ‖ levier m. de commande du frein. / ~ pedal ‖ Bremsfußhebel m.; Bremspedal n. ‖ pédale m. de frein. / ~ pendant of a carriage ‖ Bremsgehängeteil m. eines Wagens ‖ dispositif m. de suspension des freins d'une voiture. / ~ post (Windmill) ‖ Preßstiel m. ‖ montant m. du frein. / ~ power ‖ Bremskraft f. ‖ puissance f. du frein ou au frein ou de freinage. / ~ pressure ‖ Bremsdruck m. ‖ pression f. de freinage. / ~ pulley ‖ Bremsscheibe f. ‖ poulie f. de frein. / ~ pull rod (Motor car) ‖ Bremszugstange f. ‖ tige f. de serrage de frein. / ~ pull rod (Railw) ‖ Bremslasche f. ‖ joue f. de bielle du frein. / ~ pump (Mar) ‖ Pumpe f. mit einem Geckstock ‖ pompe f. à bringuebale. / ~ short range of travel ‖ kurzer Bremsweg m. ‖ faible course f. de freinage. / ~ ratched wheel (Railw) ‖ Sperrad n. ‖ roue f. à rochet du frein. / ~ regulator ‖ Bremsregler m. ‖ régulateur m. du frein. / ~ resistance ‖ Bremswiderstand m. ‖ résistance f. de freinage. / additional ~ resistance ‖ Zusatzbremswiderstand m. ‖ résistance f. additionnelle ou supplémentaire de freinage. / ~ ring ‖ Bremsring m. ‖ bague f. de freinage. / ~ rod ‖ Bremsstange f. ‖ tige f. de frein. / ~ rods pl. ‖ Bremsgestänge n. ‖ triangle f. du frein. / ~ screw ‖ Bremsspindel f. ‖ vis f. du frein. ~ screw with nut ‖ Bremsspindel f. mit Mutter ‖ vis f. de frein avec écrou. / ~ shaft ‖ Bremswelle f. ‖ arbre m. de frein; barre f. d'accouplement des freins; tige f. de raccordement des freins.

brake shoe ‖ Bremsschuh m.; Hemmschuh m.; Bremsklotz m. ‖ patin m. ou sabot m. ou porte-sabot m. de frein. / inner ~ ‖ Innenbacke f. einer Bremse ‖ mâchoire f. de frein intérieure. / wooden ~ ‖ hölzerner Bremsklotz m. ‖ sabot m. en bois.

brake shoe holder ‖ Bremsbackenhalter m. ‖ porte-sabot m. / ~ lining ‖ Bremsbackenbelag m. ‖ garniture f. de mâchoire du frein.

brake sieve (Mining) ‖ Setzsieb n. ‖ crible m. / ~ slipper ‖ Bremsschlitten m. ‖ patin m. de frein.

brakesman see brakeman.

brake spindle ‖ Bremsspindel f. ‖ vis f. de frein. / ~ spring ‖ Bremsfeder f. ‖ ressort m. de frein. / ~ test ‖ Bremsversuch m.; Bremsprobe f. ‖ épreuve f. sous l'action des freins; essai m. au frein. / ~ toggle ‖ Bremsnocken m. ‖ came f. de frein. / ~ toggle shaft ‖ Bremsnockenwelle f. ‖ axe f. de came de frein. / ~ toothed segment ‖ Bremszahnsegment n. ‖ sec-

teur m. denté de frein. / ~ truss bar ‖ Bremsdreieck n. ‖ levier m. en triangle de frein. / ~ tube ‖ Bremswelle f. ‖ tube m. de frein. / equalizing ~ valve ‖ Bremsventil n. mit Ausgleichvorrichtung ‖ robinet m. de décharge égalisatrice. / ~ wire ‖ Bremsseil n. ‖ câble m. palonnier.

braking ‖ Bremsung f. ‖ freinage m. / eddy-current ~ ‖ Wirbelstrombremsung f. ‖ freinage m. à courants parasites. / gradual ~ ‖ langsames Bremsen n. ‖ freinage m. progressif. / short-circuit ~ ‖ Kurzschlußbremsung f. ‖ freinage m. par court-circuit. / ~ by short-circuiting armature ‖ Ankerkurzschlußbremsung f. ‖ freinage m. par court-circuit d'induit.

braking action ‖ Bremswirkung f.; Bremsvorgang m. ‖ action f. du frein; effet m. de freinage. / ~ club ‖ Bremsknüppel m. ‖ barre f. d'enrayement. / ~ contact ‖ Bremskontakt m. ‖ contact m. de freinage. / ~ cylinder ‖ Bremszylinder m. ‖ cylindre m. de frein. / ~ device ‖ Hemmvorrichtung f. ‖ dispositif m. d'enrayage. / ~ distance ‖ Bremsweg m. ‖ distance f. parcourue pendant le freinage; chemin m. parcouru après l'application des freins. / ~ effect ‖ Bremswirkung f. ‖ effet m. des freins. / ~ incline (Mining) ‖ Bremsberg m. ‖ plan m. incliné de traînage. / ~ magnet ‖ Bremsmagnet m. ‖ électro m. de freinage. / ~ moment ‖ Bremsmoment n. ‖ moment m. de freinage. / ~ period ‖ Bremsdauer f. ‖ période f. de freinage. / ~ position ‖ Bremsstellung f. ‖ position f. de freinage. / ~ power ‖ Bremsleistung f. ‖ puissance f. du frein ou de freinage; effort m. de freinage. / ~ surface ‖ Bremsfläche f. ‖ surface f. de frein(age). / ~ work ‖ Bremsarbeit f. ‖ travail m. de freinage.

Bramah lock ‖ Bramahschloß n. ‖ serrure f. à pompe.

bran ‖ Kleie f. ‖ son m. / broad ~ ‖ grobe Kleie f. ‖ gros son m. / ~ of grit ‖ Grützenkleie f. ‖ remoulage m. / small ~ ‖ feine Kleie f. ‖ petit son m.

branch, to ‖ abzweigen ‖ brancher sur. / ~ a current ‖ einen Strom m. abzweigen ‖ dériver un courant. ‖ ~ off ‖ sich abzweigen ‖ se détacher; s'embrancher.

branch ‖ Ast m.; Zweig m. ‖ branche f. ; rameau m. / ~ (Build) ‖ Flügel m. ‖ aile f. / ~ (Electr) ‖ Abzweigleitung f. ‖ conducteur m. de branchement. / ~ (Office) ‖ Filiale f.; Zweiggeschäft n. ‖ succursale f.; agence f.; filiale f. / ~ (Profession) ‖ Fach n. ‖ branche f. / ~es pl. for bands ‖ Zweige mpl. für Bindezwecke ‖ branches fpl. pour liens. / ~ of a bank see branch-bank. / ~ of business ‖ Geschäftszweig m. ‖ branche f. de commerce; spécialité f. / ~ of a concern ‖ Niederlassung f. eines Konzerns ‖ succursale f. d'un groupe. / connecting ~ ‖ Einsatzstutzen m.; Verbindungsrohr n. ‖ tubulure f. de connexion ou de communication. / ~ of a course (Min) ‖ Trumm n. eines Ganges ‖ veine f. d'un filon ramifié. / foreign ~ ‖ Auslandsfiliale f. ‖ succursale f. étrangère. / to form ~es pl. ‖ sich verzweigen ‖ se ramifier. / to free from ~es pl. ‖ entästen ‖ émonder. / ~ of industry ‖ Erwerbszweig m. ‖ branche f. d'industrie. / ~ of river ‖ Flußarm m. ‖ branche f. d'une rivière. / ~ of trade see ~ of industry.

branch-bank ‖ Filialbank f.; Zweigbank f.; Bankfiliale f. ‖ succursale f. de banque; banque f. succursale.

branch box (Electr) ‖ Abzweigkasten m.; Abzweigmuffe f. ‖ boîte f. ou manchon m. de branchement. / ~ catch (Wood pulp) ‖ Astfänger m. ‖ arrête-nœud m. ‖ ~ catcher (Wood pulp) ‖ Astfänger ‖ arrête-nœud m. / ~ circuit (Electr) ‖ Verzweigungsschaltung f. ‖ circuit m. de bifurcation.

branched currents pl. ‖ Zweigströme mpl.; verzweigte Ströme mpl. ‖ courants mpl. bifurqués. / ~ wire ‖ abgezweigter Draht m. ‖ fil m. de déviation.

bran-chest ‖ Kleienkasten m.; Schrotkasten m. ‖ dodinage m.

branch establishment ‖ Zweiganstalt f. ‖ succursale f. / ~ exchange (Tel) ‖ Nebenamt n. ‖ bureau m. secondaire.

branching (Railw) ‖ Verzweigung f.; Abzweigung f. ‖ bifurcation f.; embranchement m.; ramification f. / ~ of the river ‖ Stromabzweigung f.; Stromzerteilung f. ‖ branchement m. de la rivière.

branching-off (Railw) ‖ Abzweigung f. ‖ embranchement m.; bifurcation f.

branching pipe ‖ Zweigrohr n. ‖ tuyau m. d'embranchement.

branch knot ‖ Astknoten m.; Astknorren m. ‖ nœud m. d'une branche. / ~ line ‖ Zweigbahn f.; Gleisanschluß m.; Gleisabzweigung f.; Nebenlinie f. ‖ embranchement m.; raccordement m. à des voies; ligne f. latérale ou secondaire. / ~ office ‖ Filialgeschäft n.; Zweiggeschäft n.; Filiale f. ‖ succursale f. (d'une maison); filiale f. / ~ office (Tel) ‖ Unterzentrale f. ‖ sous-station f. / ~ piece ‖ Abzweigstück m. ‖ pièce f. d'embranchement ou de jonction. / ~ rot ‖ Astfäule f. ‖ pourriture f. des branches. / ~-T (Electr) ‖ Abzweigmuffe f.; T-Dose f. ‖ boîte f. de connexion en T. / ~-T (Tube) ‖ Rohrabzweigstück n. ‖ raccord m. en T. / ~ terminal ‖ Abzweigklemme f. ‖ borne f. de dérivation ou de branchement. / ~ timber ‖ Astholz n. ‖ bois m. des branches. / ~ track of the works ‖ Gleisabzweigung f. zum Werk ‖ branchement m. de voie d'usine. / ~ tube ‖ Zweigleitung f. ‖ tube m. de dérivation. / ~ wire ‖ Drahtabzweigung f. ‖ fil m. de déviation. / ~ wood see ~ timber. / ~ works pl. ‖ Zweigfabrik f. ‖ fabrique f. succursale; succursale f.

brand, to ‖ markieren; signieren ‖ marquer. / ~ (with a hot iron) ‖ einbrennen ‖ marquer au fer chaud.

brand ‖ Abzeichen n.; Marke f.; Signatur f. ‖ marque f.; insigne m.; étiquette f. / ~ (Factory) ‖ Fabrikmarke f.; Fabrikzeichen n. ‖ marque f. de fabrique. / every ~ ‖ alle Sorten fpl. ‖ toutes les sortes fpl. / ~ of steel ‖ Stahlmarke f. ‖ marque f. d'acier.

branding machine, box board ‖ Kistenbretterbrennmaschine f. ‖ machine f. à marquer au feu les planches de caisses. / packing box ~ ‖ Packkistenbrennmaschine f. ‖ machine f. à marquer au fer rouge les caisses d'emballage.

branding oven ‖ Brennofen m. ‖ four m. à chauffer les fers à marquer. / ~ stamp ‖ Brenneisen n. ‖ fer m. à marquer.

brand-new see bran-new.

bran duster ‖ Kleienbürste f.; Kleiebürstmaschine f. ‖ brosse f. à son.

brandy ‖ Branntwein m.; Trinkbranntwein m. ‖ eau-de-vie f. / ~ distiller ‖ Schnapsbrenner m. ‖ distillateur m. de vin; brandevinier m. / ~ distillery ‖ Branntweinbrennerei f. ‖ distillerie f.; brûlerie f.; distillerie f. d'eau de vie.

brank ‖ Buchweizen m. ‖ sarrasin m.

bran-new ‖ nagelneu ‖ flambant neuf.

branning ‖ Schebeckenbleiche f.; Buntbleiche f. ‖ passage m.; sonage m.

branny fibrous stock ‖ Übergangsgrieß m. ‖ refus m. de sasseur.

bran roll mill ‖ Kleiepreßstuhl m. ‖ aplatisseur m. de son. / ~ tub (Needl) ‖ Rollfaß n.; Scheuerfaß n. ‖ frottoir m. tambour m. de nettoyage ou de polissage.

bran vinegar, crude (Met) ‖ Sauerwasser n. ‖ eau f. sure.

brasque ‖ Kohlengestübbe n. ‖ brasque f.

brass (Arm) ‖ Geschützbronze f.; Kanonenmetall n. ‖ bronze m. ou métal m. à canons. / ~ (Mach) ‖ Metallfutter n.; Lagerschale m. ‖ coussinet m.; douille f. / ~ (Met) ‖ Gelbguß m.; Messing n.; Latun n. ‖ laiton m.; cuivre m. jaune; archal m. / to cover with ~ ‖ mit Messing n. überziehen ‖ laitonner. / to heighten the yellow colour of ~ in nitric acid ‖ Messing n. abbrennen ‖ décaper le laiton. / impure ~ ‖ Rohmessing n.; Stückmessing n. ‖ laiton m. brut. / lacquered ~ ‖ lackiertes Messing n. ‖ laiton m. laqué. / to line ~es pl. with white metal ‖ Lagerschalen fpl. mit Weißmetall ausgießen ‖ antifrictionner les coussinets mpl. / nickel-plated ~ ‖ vernickeltes Messing n. ‖ laiton m. nickelé. / polished ~ ‖ blankes oder poliertes Messing n. ‖ laiton m. poli. / red ~ ‖ Rotguß m.; Tombak m.; Rotmessing n. ‖ tombac m.; laiton m. rouge; bronze m. / solid ~ ‖ Massivmessing n. ‖ laiton m. massif. / spherical ~ ‖ Kugellagergehäuse n.; Kugellagerschale f. ‖ cage m. de roulement à billes; coussinet m. sphérique. / straightened ~ ‖ geglättetes oder gerichtetes Messing(blech) n. ‖ laiton m. plané. / yellow ~ ‖ Gelbguß m.; Messing n. ‖ laiton m.; cuivre m. jaune.

brass articles pl. ‖ Messingwaren fpl. ‖ objets mpl. en cuivre jaune. / ~ band ‖ Messingband n. ‖ bande f. en laiton. / ~ bar ‖ Messingstange f. ‖ barre f. en laiton. / ~ bath ‖ Messingbad n. ‖ bain m. de laitonisage. / ~ binding wire ‖ Messingbindedraht m. ‖ fil m. laiton. / ~ box ‖ Messinghülse f. ‖ douille f. en laiton. / ~ brace ‖ Messingklammer f. ‖ agrafe f. en laiton. / ~ bushes pl. ‖ Pfannenlager n. ‖ coussinet m. / ~ -cased ‖ mit Messing n. verkleidet ‖ plaqué ou garni de laiton m. / ~ casing ‖ Messingfassung f. ‖ enveloppe f. ou gaine f. de laiton. / ~ chain ‖ Messingkette f. ‖ chaîne f. en laiton. / ~ clip (Office) ‖ Messingbriefklammer f. ‖ attache f. en laiton. / ~ conduit tube ‖ Messingisolierrohr n. ‖ tube m. isolant de laiton. / ~ corner ‖ Messingecke f. ‖ coin m. de laiton. / ~ engraved cylinder (Pap) ‖ Musterwalze f.; Gaufrierwalze f. ‖ gaufroir m. ~ dipping plant ‖ Gelbbrennanlage f. ‖ installation f. de dérochage. / ~ disc ‖ Messingscheibe f. ‖ disque f. en laiton. / ~ double salt ‖ Messingdoppelsalz n. ‖ sel m. double de laiton. / ~ finishing machine ‖ Messingschlichtmaschine f. ‖ machine f. à tourner le laiton. / ~ foil ‖ Messingfolie f.; Rauschgold n. ‖ laiton m. en feuilles; clinquant m. / ~ forge ‖ Mes-

singhütte f.; Messingwerk n. ‖ usine f. à laiton. / ~ founder ‖ Gelbgießer m.; Erzgießer m. ‖ fondeur m. de laiton ou en bronze. / ~ foundry ‖ Gelbgießerei f.; Messinggießerei f.; Bronzegießerei f. ‖ fonderie f. de laiton ou de cuivre jaune ou de bronze.

brassing ‖ Vermessingen n. ‖ laitonnage m.

brass ingots pl. ‖ Messing n. in Barren ‖ barres fpl. de laiton. / ~ latten ‖ Messingblech n. ‖ planches fpl. de laiton; laiton m. en feuilles ou en lames. / ~ mounting ‖ Messingfassung f. ‖ monture f. ou douille f. en laiton. / gilt ~ mounting ‖ fein vergoldete Messingfassung f. ‖ monture f. en cuivre doré. / oxidized ~ mounting ‖ oxydierte Messingfassung f. ‖ monture f. en cuivre oxydé. / ~ nail ‖ Messingstift m. ‖ clou m. en laiton. / ~ pan ‖ Messingpfanne f. ‖ chaudière f. de laiton. / ~ pattern maker ‖ Bronzemodellör m. ‖ modeleur m. pour bronze. / ~ plate (Met.) see ~ latten. / ~ plate (Build; Mach) ‖ Messingplatte f.; Messingschild n. ‖ plaque f. métallique ou de laiton. / ~ plating plant ‖ Vermessingungsanlage f. ‖ installation f. de laitonisage. / ~ powder ‖ Messingpulver n. ‖ laiton m. en poudre. / ~ profile ‖ Messingprofil n. ‖ profilé m. de laiton. / ~ reglet (Print) ‖ Messinglinie f. ‖ réglette f. en laiton. / ~ rod ‖ Messingstange f. ‖ tige f. en laiton. / ~ rods pl. (Met) ‖ Messing n. in Stangen ‖ barres fpl. de laiton; laiton m. en verges. / ~ rolling mill ‖ Messingwalzwerk n. ‖ laminoir m. à laiton. / ~ rule for printers ‖ Messinglinie f. für den Buchdruck ‖ filet m. en laiton pour imprimeries. / ~ -sheathed insulating conduit ‖ Isolierrohr n. mit Messingüberzug ‖ tube m. isolateur doublé de laiton. / ~ sheet ‖ Messingblech n. ‖ feuille f. en laiton. / ~ sheet hammered spring hard ‖ federhart gehämmertes Messingblech n. ‖ tôle f. de laiton écrouie. / ~ shell ‖ Messinghülse f. ‖ douille f. ou capsule f. en laiton. / ~ solder ‖ Messingschlaglot n. ‖ soudure f. de laiton; brasure f. / ~ soldering ‖ Messinglötung f. ‖ soudure f. au laiton. / ~ thumb screw ‖ Messingflügelschraube f. ‖ vis f. à ailettes en laiton. / ~ tube ‖ Messingröhre f. ‖ tuyau m. de laiton; tube m. en laiton. / ~ tubing ‖ Messingrohr n. ‖ tube f. de laiton. / ~ turner ‖ Gelbgußdreher m.; Messingdreher m. ‖ tourneur m. en cuivre jaune. / ~ types pl. ‖ Messingschrift f. ‖ caractères mpl. typographiques en laiton. / ~ ware ‖ Messingware f. ‖ dinanderie f.; articles mpl. en laiton. / ~ weight ‖ Messinggewicht n. ‖ poids m. en cuivre ou en laiton. / ~ wind instrument ‖ Blechblasinstrument n. ‖ instrument m. à vent en métal; instrument m. de cuivre.

brass wire ‖ Messingdraht m.; Tombakdraht m. ‖ fil m. de laiton ou d'archal. / black ~ ‖ schwarzer Messingdraht m. ‖ fil m. de laiton noir. / clear ~ ‖ lichter oder blanker Messingdraht m. ‖ fil m. de laiton clair. / ~ covered with silk ‖ seidenübersponnener Messingdraht m. ‖ fil m. de laiton recouvert en soie.

brass work see brass ware.

brattice (Mine) ‖ Wetterscheider m.; Schachtscheider m. ‖ cloison f. d'aérage ou de séparation.

bratticeman (Mine) ‖ Wetterarbeiter m. ‖ constructeur m. de barrages.

brattice way (Mine) ‖ Firstenstrecke f. ‖ galerie f. cuvelée.

brayer (Print) ‖ Farbläufer m.; Farbreiber m. ‖ broyeur m.

braze, to ‖ hartlöten; löten ‖ braser; souder.

brazed ‖ hartgelötet ‖ brasé; soudé. / ~-on flange ‖ aufgeschweißter Flansch m. ‖ bride f. brasée.

brazer ‖ Hartlot n. ‖ brasure f.; soudure f. forte.

brazier ‖ Hartlöter m. ‖ braseur m. / ~ (Copper sheet) ‖ Kupferschied m. ‖ chaudronnier m. / ~ (Found) ‖ Gelbgießer m. ‖ fondeur m. de bronze ou en cuivre. / ~ (Tin plate) ‖ Blechschmied m.; Klempner m. ‖ ferblantier m. / ~'s ware ‖ Messingware f. ‖ dinanderie f. / ~'s work ‖ Klempnerarbeit f. ‖ ouvrage m. du ferblantier.

braziery see brazier's ware.

brazil see Brazil-wood.

Brazil nut ‖ Paranuß f. ‖ noix f. du Brésil.

Brazil-wood ‖ Brasilholz n.; Fernambukholz n. ‖ bois m. de Brésil ou de Fernambouc; brésil m.; brésillet m.

brazing ‖ Lötstelle f.; Löten n.; Hartlötung f. ‖ brasure f.; soudure f. / ~ furnace ‖ Lötofen m. ‖ four m. à souder.

breach, to ‖ Bresche f. legen ‖ battre en brèche f.

breach of confidence ‖ Vertrauensbruch m. ‖ abus m. de confiance.

breach of contract ‖ Vertragsbruch m. ‖ violation f. de contrat. / to commit a ~ by refusal of service ‖ vertragswidrig den Dienst m. verweigern ‖ refuser le service contrairement au contrat. / to induce someone to commit ~ ‖ jemanden zum Vertragsbruch m. verleiten ‖ induire quelqu'un à la rupture d'un contrat.

breach of a wall ‖ Mauerbruch m.; Bresche f.; Mauersprung m. ‖ brèche f.

bread ‖ Brot n. ‖ pain m. / brown ~ ‖ Schwarzbrot n. ‖ pain m. bis. ou noir. / leavened ~ ‖ gesäuertes Brot n. ‖ pain m. levé. / sea ~ ‖ Schiffszwieback m. ‖ biscuit m. / unleavened ~ ‖ ungesäuertes Brot n. ‖ pain m. mollet. / Vienna ~ maker ‖ Wienerbrotbäcker m. ‖ viennois m. / Westphalian rye ~ ‖ Pumpernickel m. ‖ pain m. noir de Westphalie.

bread bakery ‖ Brotbäckerei f.; Bäckerei f. ‖ boulangerie f. / ~ baking ‖ Brotbacken n.; Backen n. ‖ cuisson f. du pain. / ~ basket ‖ Brotkorb m. ‖ panier m. à pain. / ~ box ‖ Brotkasten m. ‖ boîte f. à pain. / ~ carrier ‖ Gebäckträger m.; Gebäckausträger m. ‖ livreur m. ou porteur m. de pain. / ~ checker ‖ Brotzähler m. ‖ compteur m. de pains. / ~ corn ‖ Mengekorn n. ‖ mouture f.; blé m. mêlé. / ~ cutting machine ‖ Brotschneider m.: Brotschneidemaschine f. ‖ coupeuse f. à pain; machine f. à trancher du pain. / ~ deliveryman ‖ Gebäckträger m.; Gebäckausträger m. ‖ livreur m. ou porteur m. de pain. / ~ factory ‖ Brotfabrik f.: Brotherstellungsanlage f. ‖ installation f. de fabrication du pain; installation f. de panification. / fermentation of ~ ‖ Brotgärung f. ‖ fermentation f. panaire. / ~ maker ‖ Bäcker m.; Brotbäcker m. ‖ boulanger m. / ~ manufacture ‖ Brotfabrikation f. ‖ fabrication f. de pain; panification f. / ~ manufacturing plant see ~ factory. / ~ store ‖ Brotproviant m. ‖ vivres-pain mpl.

breadth ‖ Breite f. ‖ largeur f. / ~ of cloth ‖ Zeugbreite f. ‖ lé m. du drap. / ~ of cross-cut ‖ querschlägige Breite f. des Grubenfeldes ‖ largeur f. du champ d'exploitation. / ~ of the day (Build) ‖ Lichtenbreite f. ‖ largeur f. du jour. / extreme ~ see maximum ~. / maximum ~ ‖ größte Breite f. ‖ la plus grande largeur f.; largeur f. au fort. / moulded ~ (Shipb) ‖ größte Breite f. ‖ largeur f. maxima. / ~ over all (Shipb) ‖ Breite f. über alles ‖ largeur f. hors-tout.

bread toaster ‖ Brotröster m. ‖ grille-pain m.

break, to (Chem; Mach) ‖ zerschlagen; zerkleinern; mahlen ‖ concasser; briser. / ~ (Electr) ‖ ausschalten ‖ déconnecter; interrompre. / ~ (Flax) se al so tobrake ‖ brechen; braken ‖ macquer; broyer; briser; teiller. / ~ coal (Mine) ‖ die Kohle hauen ‖ haver ou couper le charbon. / a contact (Electr) ‖ einen Kontakt m. öffnen ‖ ouvrir un contact m. / ~ the cover (Bookb) ‖ den Umschlag m. biegen ‖ plier la couverture. / ~ down (Chem) ‖ zerfallen ‖ se décomposer. / ~ down (Mach) ‖ abbrechen; zu Bruch m. gehen ‖ renverser; rompre; ébouler. / ~ down the press (Print) ‖ die Presse abschlagen ‖ démonter la presse. / ~ the joint (Mas) ‖ den Verband m. verwerfen ‖ perdre la liaison. / ~ the line (Print) ‖ die Zeile abbrechen ‖ couper l'alinéa m. / ~ off ‖ abbrechen ‖ rompre. / ~ off (Chem) ‖ spalten ‖ couper. / liable to break off in splinters (Wood) ‖ splintrissig ‖ écailleux. / ~ open ‖ aufbrechen ‖ rompre. / ~ through a wall ‖ das Mauerwerk durchbrechen ‖ couper ou percer la maçonnerie. / ~ up (Chem) ‖ aufschließen; sich spalten ‖ désagréger; se couper. / ~ up (Mach) ‖ abbrechen; abmontieren; auseinandernehmen ‖ démonter; démolir. / ~ up a battery (Electr) ‖ eine Batterie auseinandernehmen ‖ démonter une pile. / ~ up a bridge ‖ eine Brücke f. abbrechen; abbrücken ‖ enlever ou démonter un pont. / ~ up the lump ‖ gar aufbrechen ‖ avaler la loupe. / ~ up a ship ‖ ein Schiff n. abbrechen ‖ démolir un vaisseau m.

break see also breakage ‖ Brechen n.; Zerbrechen n.; Bruch m. ‖ bris m.; rupture f.; cassure f.; fracture f. / ~ (Braking device) see brake. / ~ (Coachm) ‖ Radzirkel m.; Speichenmesser m. ‖ temple m. / ~ (Electr) ‖ Unterbrechung f. ‖ rupture f.; interruption f. / ~ (Electr; Breaking device) see breaker. / ~ (Flax; Hemp) see also brake ‖ Breche f.; Flachsbreche f.; Hanfbreche f.; Brake f. ‖ broie f. / ~ (Mach) ‖ Abbruch m.; Abbrechen f.; Auseinandernehmen n. / ~ (Met) ‖ Bruch m.; Abbruch m. ‖ casse f.; rompure f.; cassure f. / ~ (Mine) ‖ Schicht f. ‖ temps m. de repos. / ~ (Print) ‖ Ausgang m.; Absatz m. ‖ alinéa m.; terminaison f. / ~ (Trade) ‖ Preissturz m. ‖ baisse f. soudaine. / ~ of banks ‖ Uferabbruch m. ‖ corrosion f. des rives. ‖ mechanical make and ~ (Motor) ‖ Abreißvorrichtung f. ‖ dispositif m. de rupture mécanique. / quick ~ (Electr) ‖ Momentunterbrechung f. ‖ rupture f. brusque. / ~ of the types ‖ Anguß m. der gegossenen Typen ‖ jet m. des caractères d'imprimerie. / ~ of vision of a bifocal glass ‖ Bildsprung m. bei einer Zweistärkenbrille ‖ saut m. de l'image à la lunette à double foyer. / absence of any

~ of vision in a bifocal glass ‖ Unmerklichkeit f. des Bildsprungs bei einer Zweistärkenbrille ‖ suppression f. de la discontinuité de l'image à la lunette à double foyer. / ~ of wall ‖ vertieftes Mauerfeld n.; Blinde f. oder Nische f. oder Knick m. einer Mauer ‖ renfoncement m. dans le nu d'un mur; brisure f. d'une muraille.

breakable ‖ zerbrechlich ‖ fragile; brisable; cassable.

breakage ‖ Brechen n.; Zerbrechen n.; Bruch m. ‖ bris m.; rupture f.; cassure f.; fracture f. / ~ of the glass box (Acc) ‖ Bruch m. des Glasgefäßes ‖ bris m. du récipient en verre. / the point stands up well to ~ ‖ die Spitze ist widerstandsfähig gegen Bruch ‖ la pointe est résistante à la rupture. / risk of ~ ‖ Bruchgefahr f. ‖ risque m. de rupture.

breakdown (Auto) ‖ Panne f. ‖ panne f. / ~ attributable to bad workmanship ‖ Bruch m. infolge fehlerhafter Arbeit ‖ avarie f. par suite d'un travail défectueux. ‖ engine ~ ‖ Motorstörung f.; Versagen n. des Motors ‖ panne f. de moteur. / ~ of the mechanism ‖ Versagen n. des Mechanismus ‖ avarie f. du mécanisme. / owing to ~ of machinery ‖ wegen Betriebsstörung f. der Maschinen ‖ à cause de dérangements aux machines. / ~ of the piston rod ‖ Kolbenstangenbruch m. ‖ rupture f. de la tige de piston. /

breakdown lorry ‖ Abschleppwagen m. ‖ dépanneuse f. / ~ test ‖ Durchschlagprobe f. ‖ épreuve f. de percement.

breaker (Electr) ‖ Ausschalter m.; Unterbrecher m. ‖ interrupteur m. / ~ (Mar) ‖ Sturzsee f.; Brandungswelle f.; Brecher m. ‖ vague f. de ressac; gros coup m. de mer embarqué. / ~ (Mine) ‖ Abkohler m.; Häuer m. ‖ abatteur m.; rabatteur m. / ~ (Ore Dress) ‖ Brecher m. ‖ concasseur m. / ~ (Spinn) ‖ Reißkrämpel f.; Reißwolf m. ‖ briseur m.; carde f. en gros loup m. / additional ~ ‖ Nachbrecher m. ‖ re-concasseur m. / automatic circuit ~ (Electr) ‖ Selbstunterbrecher m. ‖ interrupteur m. automatique. / conical ~ ‖ Kegelbrecher m. ‖ concasseur m. à cônes. / cork ~ ‖ Korkbrecher m. ‖ concasseur m. de liège. / giratory ~ ‖ Rundbrecher m. ‖ concasseur m. à cône. / ~ with mantle of armoured steel plates ‖ Panzerbrecher m. ‖ concasseur m. avec enveloppe en acier à plaques de blindage. / ore ~ ‖ Erzbrecher m. ‖ concasseur m. de minerais. / pig ~ ‖ Masselbrecher m. ‖ casseur m. de gueuse. / salt ~ ‖ Salzbrecher m. ‖ concasseur m. de sel. / second ~ (Spinn) ‖ Pelzkrempel f. ‖ carde f. à tambour. / ship ~ ‖ Schiffsausschlachter m. ‖ démolisseur m. de bateaux.

breaker bolt ‖ Zerreißbolzen m. ‖ boulon m. de rupture. / ~ card ‖ Vorkarde f.; Grobkarde f.; Rauhkarde f.; Reißkrempel f. ‖ carde f. briseuse; briseur m. / ~ jaw suitably corrugated ‖ Brechbacke f. mit besonderer Riffelung ‖ mâchoire f. du concasseur d'une cannelure spéciale. / ~ mouth ‖ Brechmaul n. ‖ gueule f. du concasseur. / ~ shell ‖ Brechmantel m. ‖ enveloppe f. ou bâti m. du concasseur. / ~ strip of tyre (Auto) ‖ Leinwandschicht f. des Laufbandes ‖ toile f. de la bande de roulement.

breaking see also breakage ‖ Brechen n.; Zerbrechen n.; Bruch m. ‖ rupture f.;

bris m.; cassure f. / ~ (Electr) ‖ Unterbrechung f. ‖ rupture f.; interruption f.; coupure f. / ~ (Mill; Ore dress) ‖ Schroten n.; Brechen n. ‖ broyage m. / ~ of the axle ‖ Achsbruch m. ‖ rupture f. d'essieu. / ~ of circuit ‖ Stromunterbrechung f. ‖ interruption f. du courant. / ~ of a dike ‖ Deichbruch m. ‖ rupture f. d'une digue. / ~ of local calls for trunk calls ‖ Fernamtstrennung f. ‖ coupure f. par l'inter-urbain. / ~ by overload ‖ Bruch m. bei Überlastung ‖ rupture f. par suite de décharge ou en cas de surcharge. / ~ into small pieces ‖ Zerkleinerung f.; Zerstückelung f. ‖ concassage m. en morceaux. / preliminary ~ ‖ Vorbrechen n. ‖ avant-broyage m. / ~ of rails ‖ Schienenbruch m. ‖ rupture f. de rails. / sufficient security against ~ ‖ genügende Bruchsicherheit f. ‖ résistance f. suffisante contre ruptures. / sparkless ~ ‖ funkenlose Unterbrechung f. ‖ interruption f. sans étincelles. / ~ of the tube ‖ Rohrbruch m.; Zerreißen n. des Rohres ‖ rupture f. du tube. / ~ of the warpthread ‖ Schußfadenbruch m. ‖ rupture f. de la duite.

breaking card ‖ Vorkarde f.; Grobkarde; Reißkrempel f. ‖ carde f. briseuse; briseur m. / ~ cone ‖ Brechkegel m. ‖ cône m. de concassage. / ~ current ‖ Öffnungsinduktionsstrom m. ‖ courant m. d'induction d'ouverture. / ~ dilatation ‖ Bruchdehnung f. ‖ dilatation f. de rupture.

breaking-down ‖ Abbrechen n.; Abbruch m. ‖ démolition f. / ~ (Chem) ‖ Zerfall m. ‖ décomposition f. / ~ (Min) ‖ Zusammenbruch m.; Zubruchgehen n. ‖ éboulement m. / ~ mill (Met) ‖ Blockbrecher m. ‖ casseur m. de lingots. / ~ point ‖ Bruchfestigkeitsgrenze f. ‖ point m. de rupture. / ~ pressure (Electr) ‖ Durchschlagspannung f. ‖ tension f. au percement disruptif. / ~ temperature (Chem) ‖ Abbautemperatur f. ‖ température f. de dégradation ou de désagrégation ou de peptonisation. / ~ test (Electr) ‖ Durchschlagsversuch m. ‖ essai m. de percement disruptif.

breaking drum ‖ Brechtrommel f. ‖ tambour m. broyeur. / ~ load ‖ Bruchlast f.; Bruchfestigkeit f. ‖ charge f. de rupture. / ~ machine (Spinn) ‖ Schneidmaschine f. ‖ machine f. à couper le lin. / ~ mouth ‖ Brechmaul n. ‖ ouverture f. du concasseur. / ~ -off (Build) ‖ Abbruch m.; Abbruchsarbeit f. ‖ démolition f. / ~ place ‖ Bruchstelle f. ‖ point m. de rupture. / ~ plant ‖ Brechanlage f. ‖ installation f. de concassage. / ~ point ‖ Bruchgrenze f.; Festigkeitsgrenze f. ‖ limite f. de rupture. / ~ ring ‖ Brechring m.; Mahlring m. ‖ anneau m. concasseur; couronne f. de concassage. / ~ shaft ‖ Brechachse f. ‖ arbre m. de concassage. / ~ strength ‖ Bruchfestigkeit f.; Bruchwiderstand m. ‖ résistance f. à la rupture. / ~ stress ‖ Beanspruchung f. auf Bruch ‖ travail m. à la rupture; effort m. de rupture. / ~ test ‖ Bruchprobe f. ‖ essai m. de rupture. / ~ -up of a ship ‖ Abbruch m. eines Schiffes ‖ démolition f. d'un bateau. / ~ weight ‖ Bruchgewicht n. ‖ poids m. de rupture.

break key ‖ Unterbrechungstaste f. ‖ touche f. d'interruption. / ~ roller mill ‖ Schrotwalzenstuhl m.; Schrotmühle f. ‖ broyeur m.

break-sheer, to (Nav) ‖ vor Anker gieren ‖ roder.

break signalling system ‖ Pausensignalanlage f. ‖ installation f. pour émission de signaux horaires. / ~ spark ‖ Öffnungsfunke m. ‖ étincelle f. de rupture. / ~ tailings pl. (Mill) ‖ Schrotübergang m. ‖ refus m. de broyage.

breakwater ‖ Wellenbrecher m.; Rißbank f. ‖ brise-lames m.; brise-flots m.; éperon m.; risban m.

breast ‖ Brust f. ‖ poitrine f. / ~ (Build) see also breast wall ‖ Brüstung f.; Brustwehr f. ‖ parapet m.; appui m. / ~ (Plough) ‖ Streichbrett n. ‖ versoir m. / ~ of slipway (Shipb) ‖ Vorhelling f. ‖ avant-cale f.

breast beam (Weav) ‖ Brustbaum m.; Stiftbaum m. ‖ ensouple m. de devant. / ~ borer see ~ drill. / ~ button ‖ Westenknopf m. ‖ bouton m. de gilet. / ~ collar (Saddl) ‖ Brustriemen m.; Brustgurt m. ‖ poitrail m. / ~ drill ‖ Brustleier f.; Faustleier f.; Brustbohrer m. ‖ vilebrequin m. / ~ feather ‖ Brustfeder f.; Flaumfeder f. ‖ duvet m. / ~ height ‖ Brusthöhe f. ‖ hauteur f. d'homme. / ~ height (Build) see breast wall. / ~ hook (Shipb) ‖ Bugband n. ‖ écharpe f.

breasting (Build) see breast wall. / ~ (Pap) ‖ Kropf m.; Sattel m.; Berg m. ‖ gorge f.

breast lead ‖ Taucherherz n. ‖ plomb m. de poitrine. / ~ line (Bridge) ‖ Spanntau n. ‖ amarre f.; traversière f. / ~ moulding of a window ‖ Brüstungsgesims n. eines Fensters ‖ allège f.; tablette f. d'une fenêtre. / ~ pin ‖ Busennadel f.; Vorstecknadel f.; Brosche f. ‖ broche f. / ~ plate (Saddl) see ~ collar. / ~ plough ‖ Abstechpflug m.; Rasenpflug m. ‖ tranche-gazon m. / ~ pump ‖ Milchpumpe f.; Melkapparat m. ‖ tire-lait m.; pompe f. à lait. / ~ rail (Shipb) see breastwork (Shipb).

breast strap (Saddl) ‖ Kumtriemen m. ‖ coupliere f. / ~ ring (Saddl) ‖ Kumtfeder f.; Schake f. ‖ maille f. des attelles.

breastsummer see bressummer.

breast support ‖ Bruststütze f. ‖ support m. de poitrine.

breast wall ‖ Brüstungsmauer f.; Brüstung f.; Brustwehr f.; Schutzwehr f. ‖ appui m.; mur m. d'appui; parapet m. / ~ (Balustrade) ‖ Brustlehne f.; Geländer n.; Brustwehr f. ‖ balustrade f.; garde-fou m.; garde-corps m.

breast-wheel ‖ Kropfrad n.; mittelschlächtiges Wasserrad n. ‖ roue f. hydraulique de côté.

breastwork (Build) see breast wall. / ~ (Shipb) ‖ Reling f.; Schanzkleid n. ‖ lice f. ou lisse f. d'appui; bastingage m.; bastingue f.; pavois m.

breather ‖ Schnüffelventil n.; Schnarchventil n. ‖ reniflard m. / ~ (Pipe) ‖ Luftschlauch m. ‖ reniflard m.

breathing apparatus ‖ Atemgerät n.; Atmungsapparat m.; Respirationsapparat m.; Sauerstoffapparat m. ‖ appareil m. de respiration; masque m. respiratoire. / ~ appliance see ~ apparatus.

breech (Arm) ‖ Schwanzschraube f. ‖ culasse f. / false ~ (Arm) ‖ Scheibe f. oder Basküle f. am Gewehrschaft ‖ bascule f.; fausse-culasse f.

breechblock of howitzer ‖ Haubitzkopf m. ‖ tête f. d'obusier.

breeches pl. ‖ Hose f.; Beinkleid n. ‖ pantalon m.; culotte f. / ~ maker ‖ Hosenschneider m. ‖ culottier m.

breeching loop ‖ Öse f. ‖ anneau m. de brague.

breechings pl. of wool ‖ Kotspitzen fpl. der Wolle ‖ crottins mpl. de la laine.

breech loader ‖ Hinterlader m. ‖ arme f. à feu se chargeant par la culasse.

breeder ‖ Tierzüchter m.; Viehzüchter ‖ éleveur m.; nourrisseur m.

breeding ‖ Tierzucht f. ‖ élevage m. / ~ apparatus for fowl ‖ Zuchtapparat m. für Geflügel ‖ éleveuse f. pour volaille. / ~ cage ‖ Nistkasten m. ‖ couveuse f. pour oiseaux.

breeze (Coal) ‖ Kohlenklein n.; Grus m.; Kohlenlösche f. ‖ braise f.; poussière f. de charbon. / ~ (Mar) ‖ Brise f.; Kühlte f. ‖ brise f.; fraîcheur f.; petit vent m. / coke ~ ‖ Koksklein n. ‖ coke m. en grains menus. / commanding ~ ‖ günstige Brise f. ‖ brise f. maniable. / fresh ~ ‖ frischer Wind m.; frische Brise f. ‖ bonne ou jolie brise f. / gentle ~ ‖ schwacher Wind m.; leichte Brise f. ‖ légère ou petite ou faible brise f.; brise f. molle; vent m. faible ou mou. / land ~ ‖ Landbrise f. ‖ brise f. de terre. / light ~ ‖ leichter Wind m.; leichte Brise f. ‖ légère ou faible brise f.; brise f. molle; vent m. faible ou mou. / moderate ~ ‖ mäßiger Wind m.; mäßige Brise f. ‖ petite brise f.; brise f. modérée. / mountain ~ ‖ Bergwind m. ‖ brise f. ou vent m. de montagne. / sea ~ ‖ Seebrise f. ‖ brise f. du large. / stiff ~ ‖ steife Brise f. ‖ frais m.; grand frais m. / strong ~ ‖ starker Wind m.; steife Brise f. ‖ bonne brise f.; brise f. carabinée.

breithauptite ‖ Antimonnickel n. ‖ breithauptite f.

Bremen blue ‖ Bremerblau n. ‖ bleu m. de Brême.

bressommer see bressummer.

bressummer (Build) ‖ Saumschwelle f.; Oberschwelle f.; Trägerschwelle f. ‖ sommier m.

brevier (Print) ‖ Jungfernschrift f.; Petitschrift f. ‖ petit-texte m.

brew, to ‖ brauen ‖ brasser.

brew ‖ Gebräu n. ‖ brassin m.

brewage see brew.

brewer ‖ Brauer m. ‖ brasseur m.

brewer's barley ‖ Braugerste f. ‖ orge f. de brasserie. / ~ car ‖ Brauereiwagen m.; Bier(transport)wagen m. ‖ camion m. de brasseur. / ~ copper ‖ Braupfanne f. ‖ chaudière f. de brasserie ou à brasser. / ~ dray see ~ lorry. / ~ gimblet ‖ Handbohrer m. für Brauer ‖ vrille f. de brasseur. / ~ grains pl. ‖ Brauereitreber pl. ‖ drèche f. de brasserie. / ~ lorry ‖ Brauerwagen m.; Biertransportwagen m. ‖ haquet m. ou camion m. de brasseur. / ~ machinery plant ‖ Bierbrauereieinrichtung f. ‖ installation f. pour brasseries. / ~ malt ‖ Braumalz n. ‖ malt m. pour brasseries. / ~ pitch ‖ Brauerpech n. ‖ poix f. pour brasseurs.

brewery ‖ Brauerei f. ‖ brasserie f. / large ~ ‖ Großbrauerei f. ‖ grande brasserie f. / vinegar ~ ‖ Essigfabrik f. ‖ vinaigrerie f.

brewery articles pl. ‖ Brauereibedarfsartikel mpl. ‖ articles mpl. de brasserie. / ~ device for breweries ‖ Brauereiapparat m.; Brauereigerät n. ‖ appareil m. pour brasseries. / ~ installation ‖ Braue-

reieinrichtung f. ‖ installation f. pour brasseries. / ~ plant ‖ Brauereianlage f. ‖ installation f. de brasserie. / stirring device for breweries ‖ Brauereirührwerk n. ‖ agitateur m. de brasserie. / ~ yeast ‖ Brauereihefe f. ‖ levure f. de brasserie.

brewhouse ‖ Brauhaus n.; Brauerei f.; Sudhaus n. ‖ brasserie f. / ~ yield ‖ Sudhausausbeute f. ‖ rendement m. en chaudières ou de brasserie.

brewing ‖ Bierbrauen n.; Brauen n.; Brauerei f.; brassage m. / ~ (Matter) ‖ Gebräu n.; Sud m. ‖ brassin m. / ~ apparatus ‖ Brauereiapparat m. ‖ appareil m. pour brasseries. / ~ house see also brewhouse ‖ Sudhaus n.; Brauhaus n. ‖ salle f. de chaudières à brasser; brasserie f. / ~ machinery ‖ Brauereimaschine f. ‖ machine f. pour brasseries. / ~ master ‖ Braumeister m. ‖ maître m. brasseur. / ~ pan ‖ Braupfanne f.; Sudkessel m. ‖ chaudière f. à brasser. / ~ pan bottom ‖ Braupfannenboden m. ‖ fond m. de chaudière à brasser.

briar tooth (Saw) ‖ Wolfszahn m. ‖ dent-de-loup m. / ~ wood ‖ Bruyereholz n. ‖ bois m. de bruyère.

bribery ‖ Bestechung f. ‖ subornation f.; corruption f.

brick ‖ gebrannter Stein m.; Mauerziegel m.; Mauerstein m.; Backstein m.; Ziegel m. ‖ brique f. / acid-proof ~ ‖ säurefester Stein m. ‖ brique f. résistant aux acides. / air ~ ‖ Hohlziegel m. ‖ brique f. perforée. / air-dried ~ ‖ Luftziegel m.; Lehmstein m.; ungebrannter Ziegel m. ‖ brique f. crue ou séchée à l'air. / broken ~s pl. ‖ Ziegelbrocken mpl. ‖ débris mpl. de briques. / to burn the ~s pl. ‖ die Ziegel mpl. brennen ‖ cuire les briques fpl. / burnt ~ ‖ Backstein m.; gebrannter Mauerstein m. oder Ziegel m. ‖ brique f. cuite. / ~ burnt too little ‖ ungarer Ziegel m. ‖ brique f. mal cuite. / chromite ~ ‖ Chromitziegel m. ‖ brique f. en chromite. / clay ~ ‖ Luftziegel m. ‖ brique f. de limon. / clear-colored ~ ‖ hellfarbiger Mauerstein m. ‖ brique f. de couleur claire. / compass ~ ‖ Krummziegel m. ‖ brique f. courbée. / concave ~ see compass ~. / Dutch ~ ‖ Klinker m. ‖ brique f. hollandaise; biscuit m. / to eject the ~ ‖ den Preßling m. ausheben ‖ démouler la brique. / feather-edged ~ ‖ Keilziegel m. ‖ brique f. en coin; clef f. / fire ~ ‖ Schamottestein m.; Feuerziegel m. ‖ brique f. réfractaire. / fire-clay ~ see fire ~. / flat laid ~ ‖ auf die flache Kante gestellter Ziegel·m. ‖ brique f. posée de plat. / Flemish ~ see Dutch ~. / floating ~ ‖ Schwammziegel m.; Schwimmziegel m. ‖ brique f. flottante. / ~s pl. formed in moulds ‖ Handstrichziegel mpl. ‖ briques fpl. tassées dans des moules. / glazed ~ ‖ Verblendstein m. ‖ brique f. émaillée; pierre f. de revêtement. / hard-burnt ~ ‖ Hartbrandstein m. ‖ brique f. hollandaise. / hollow ~ ‖ hohler Ziegel m.; Hohlstein m.; Hohlziegel m. ‖ brique f. creuse. / ~ laid on edge ‖ auf die hohe Kante gestellter Ziegel m. ‖ brique f. posée de champ. / light ~ ‖ Leichtziegel m. ‖ brique f. légère. / machine-made ~ ‖ Maschinenziegel m. ‖ brique f. moulée à la machine. / to mould ~s pl. ‖ Ziegel mpl. streichen ‖ mouler des briques fpl. / ~ for paving-floors ‖ Pflasterziegel m. ‖ brique f. à paver; carreau m. à paver.

/ perforated ~ || Lochstein m. || brique f. perforée. / porous ~ || poröser Ziegel m. || brique f. poreuse. / porous hollow ~ || poriger Lochziegel m. || brique f. creuse poreuse. / porous ventilated ~ see porous hollow ~. / profilated ~ || Formziegel m. || brique f. profilée. / ~ of pumice and lime || Mauerstein m. aus Bimsstein und Kalk || brique f. en pierre ponce et chaux. / refractory ~ see fire ~. / shaped ~ || Formstein m. || brique f. moulée ou à façon. / silicious ~ || kieselhaltiger Mauerstein m. || brique f. siliceuse. / slag ~ || Schlacken(mauer)stein m. || brique f. de scorie ou de laitier. / slag ~s manufacturing plant || Schlackensteinherstellungsanlage f. || installation f. pour la fabrication des pierres en scorie. / solid ~ || Vollziegel m. || brique f. pleine. / solid porous ~ || poriger Vollziegel m. || brique f. pleine poreuse. / stock ~ || Hartbrand m.; Klinker m.; gebrannter Ziegel m. || brique f. la plus dure. / thin ~ || Ofenziegel m.; Kanalziegel m.; chantignolle f.; brique f. mince; demi-brique f. / tubular ~ || Röhrenziegel m.; Hohlziegel m. || brique f. tubulaire. / unburnt ~ see airdried ~. / vitrified ~ || Glasurstein m. || brique f. gobetée. / to wall the ~s pl. || die Ziegel mpl. auf Hage setzen || mettre les briques fpl. en haie. / wedge-shaped ~ see feather-edged ~. / white ~ || weißer Mauerstein m. || brique f. blanche.

brick and tile machine || Maschine f. zur Ziegelherstellung || machine f. pour la fabrication des briques et tuiles. / batch of ~s || Ziegelbrand m.; ein Brand m. Ziegel || cuite f. de briques. / ~ bats pl. || Ziegelschutt m. || briquaillons mpl. / box of ~s || Steinbaukasten m. || boîte f. de constructions en pierres. / ~ burner || Ziegelbrenner m. || briquetier m.; cuiseur m. de briques. / ~ car with water tank for tile works || Streichtischwagen m. mit Wasserbehälter für Ziegeleien || wagonnet m. à briques pour tuileries avec réservoir à eau. / ~ chimney || gemauerter Schornstein m. || cheminée f. maçonnée de briques. / ~ clay || Ziegelerde f.; Ziegelton m. || terre f. ou argile f. à briques. / ~ concrete || Ziegelbeton m. || béton m. de briquailles. / ~ course || Ziegelschicht f. || couche f. de briques. / ~ course laid on edge || Rollschicht f. || assise f. de champ; assise f. de briques posées de champ; roulage m. / ~ drying press || Ziegeltrockenpresse f. || presse f. à sécher les tuiles. / ~ dust || Ziegelmehl n. || poussière f. ou farine f. de briques. / ~ facing || Verkleidung f. mit Blendsteinen || revêtement m.; faux parement m. / ~ field || Feldziegelei f. || briqueterie f. en plein air. / ~ kiln || Backsteinofen m.; Ziegelofen m.; Ziegelei f. || four m. à briques; briqueterie f. / ~ layer || Maurer m.; Backsteinmaurer m. || briqueteur m.; maçon m. / ~ laying (Work) || Maurerarbeit f.; Ziegelarbeit f.; Mauerung f. || maçonnage m. ou murage m. en briques. / ~ laying (Masonry) see masonry. / ~ machinery || Ziegelmaschine f.; Mauersteinmaschine f. || machine f. à briques. / ~ maker || Ziegelformer m.; Ziegelbrenner m. || briquetier m. / ~ maker's wagon with water tank || Streichtischwagen m. mit Wasserbehälter || wagonnet m. pour tuileries avec réservoir à eau. / ~ making machine ||

Ziegelsteinformmaschine f. || machine f. à briques. / ~ masonry see also brickwork || Ziegelmauerwerk n.; Backsteinmauerung f. || maçonnerie f. ou murage m. en briques; briquetage m. / ~ mould || Ziegelsteinform f. || moule m. à briques.

brick moulding machine || Ziegelpresse f.; Ziegelformmaschine f. || presse f. à façonner les briques; machine f. à briques. / long-stringed ~ || Strangziegelpresse f. || presse f. pour le façonnage mécanique des briques à la filière.

brick nogging || Ziegelausmauerung f. einer Fachwand || posage m. de briques dans la cloison en charpente. / ~ pavement || Klinkerpflaster n. || pavé m. en brique. / ~ press || Ziegel(stein)presse f.; Steinpresse f. || presse f. à (faire les) briques ou à briqueter. / ~ road || Klinkerstraße f. || chaussée f. de briques; route f. pavée en briques. / standard ~ size || Normalziegelformat n. || dimensions fpl. normales des briques. / ~ steam works pl. || Dampfziegelei f. || briqueterie f. à vapeur. / ~ stone || Putzstein m.; Mauerstein m. || brique f. / ~ stone masonry see ~ masonry. / ~ wall || Ziegelmauer f. || mur m. de brique.

brickwork || Ziegel(stein)mauerwerk n.; Backsteinbau m.; Backsteinmauerung f. || maçonnerie f. ou muraillement m. ou murage m. en briques; briquetage m. / counterfeit ~ || nachgeahmter Backsteinbau m. || briquetage m. feint. / latticed ~ || gitterförmiges Mauerwerk n. || maçonnerie f. grillagée. / visible ~ || Ziegelrohbau m. || briquetage m. / ~ casing || Ziegelsteinverkleidung f. || muraillement m. en briques.

brick works pl. || Ziegelei f. || briqueterie f. / hand ~ || Handziegelei f. || briqueterie f. à la main. / mechanical ~ || Maschinenziegelei f. || briqueterie f. mécanique. / implements pl. for ~ || Ziegeleibedarfsartikel mpl. || articles mpl. pour tuileries. / ~ plant || Ziegeleieinrichtung f. || équipement m. de briqueterie.

bridal wreath || Myrtenkranz m.; Brautkranz m. || couronne f. en myrtes.

bridge, to ~ over || einen Fluß überbrücken || jeter un pont sur une rivière.

bridge || Brücke f. || pont m. / ~ (Boring machine) || Oberpfanne f. || chapiteau m. / ~ (Dentistry) || Brücke f. || couronne f. de dentier. / ~ (Spectacles) || Brücke f. || arcade f. / arched ~ || Bogenbrücke f. || pont m. arqué ou en arc. / bascule ~ || Klappbrücke f.; Hubbrücke f. || pont m. à bascule; pont-levis m. à trappe. / ~ of boiler furnace || Feuerbrücke f. || autel m. de foyer. / ~ of boards (Build) || Laufbrücke f.; Auflauf m.; Bahn f.; Pritsche f. || pont m. d'échafaudage. / cast-iron ~ || gußeiserne Brücke f. || pont m. en fonte. / charge ~ || Gichtbrücke f. || pont m. de chargement; pont m. de gueulard. / concrete ~ || Betonbrücke f. || pont m. de béton. / connecting ~ || Verbindungsbrücke f. || pont m. de communication. / ~ for conveying materials on the furnace top see charge ~. / diagonally braced girder ~ || Fachwerkbrücke f. mit geraden Trägern || pont m. à poutres droites composées. / floating ~ || schwimmende Brücke f. || pont m. ou bac m. volant. / flying ~ || fliegende Fähre f. || pont m. ou bac m. volant. / furnace ~ || Feuerbrücke f. || autel m. / furnace

~ cooled by air || luftgekühlte Feuerbrücke f. || autel m. refroidi à l'air. / ~ of gabions || Schanzkorbbrücke f. || pont m. de gabions. / hanging ~ || Hängebrücke f. || pont m. suspendu (à armatures). / iron ~ || Eisenbrücke f. || pont m. en fer. / lifting ~ || Zugbrücke f. || pont-levis m. / loading ~ || Verladebrücke f. || pont m. de chargement. / ~ of the nose (Spectacles) || Nasensteg m.; Nasenrücken m. || pont m.; nez m. / pendant ~ see hanging ~. / ~ on piles || Hochbrücke f.; Pfahlbrücke f. || pont m. de pilotis. / ~ with plug contacts (Electr) || Stöpselmeßbrücke f. || pont m. de mesure à fiches ou à chevilles. / provisory ~ || Notbrücke f.; Behelfsbrücke f.; provisorische Brücke f. || pont m. provisoire ou de service ou de circonstances. / ~ on rafts || Floßbrücke f. || pont m. de radeaux. / ~ upon rafts of inflated skins || Schlauchbrücke f. || pont m. d'outres de peaux de bouc. / rail ~ || Gleisbrücke f. || pont m. de voie. / railway ~ || Eisenbahnbrücke f. || pont m. de chemin de fer. / roofed ~ || überdachte Brücke f. || pont m. couvert. / rope ~ || Taubrücke f. || pont m. de cordes. / ~ of round wood || Rundholzbrücke f.; Stangenbrücke f. || pont m. de rondins. / simple ~ (Tel) || Einfachbrücke f. || pont m. simple. / small ~ || Steg m.; Brückchen n. || planche f.; passerelle f. / steel ~ see iron ~. / stone ~ || steinerne Brücke f. || pont m. en maçonnerie ou en pierre. / strutframe ~ || Spreng(werk)brücke f. || pont m. à contre-fiches. / suspension ~ || Hängebrücke f. || pont m. suspendu. / swing ~ || Drehbrücke f. || pont m. pivotant ou tournant. / temporary ~ see provisory ~. / thrust-block ~ (Shipb) || Drucklagerbügel m. || anneau m. de palier de butée. / timber ~ || hölzerne Brücke f. || pont m. en bois. / timber spandril ~ || hölzerne Bogenfachwerkbrücke f. || pont m. supporté par des cintres. / truss ~ || Gitterbrücke f.; Fachwerkbrücke f. || pont m. à poutres armées. / tubular ~ || Tunnelbrücke f.; Röhrenbrücke f. || pont m. tubulaire ou en tube. / ~ of a violin || Steg m. einer Geige || chevalet m. d'un violon. / wire (cable) suspension ~ || Drahtseilbrücke f. || pont m. suspendu à câbles d'acier. / wrought-iron ~ || schmiedeiserne Brücke f. || pont m. en fer forgé.

bridge arch || Brückenbogen m. || arc m. ou arche f. ou voûte f. de pont.

bridge-building || Brückenbau m. || construction f. de ponts. / ~ plant || Brückenbauanstalt f. || atelier m. pour la construction de ponts.

bridge connection (Electr) || Brückenschaltung f. || montage m. ou couplage m. en pont. / ~ connector (Electr) || Batterieklemme f.; Überbrückungsklemme f. || borne f. de batterie; pince f. de court-circuit ou de raccordement. / ~ construction see ~ building. / ~ crane || Brückenkran m.; Hochbahnkran m. || grue f. portique sur voie surélevée ou sur piliers; grue f. portique sur chevalets roulants. / ~ crossing (Electr line) || Brückenüberführung f. || traversée f. de pont. / ~ duplex-connection (Tel) || Brückengegensprechschaltung f. || couplage m. duplex à pont. / ~ feeding (Tel) || Brückenspeisung f. || alimentation f. par ponts.

/ ~ fuse (Electr) ∥ Brückensicherung f. ∥ coupe-circuit m. à pont. / ~ girder ∥ Brückenträger m. ∥ poutre f. de pont. / ~ head ∥ Brückenkopf m. ∥ tête f. de pont. / ~ method (Electr) ∥ Brückenmethode f. ∥ méthode f. du pont. / ~ opening ∥ Brückenöffnung f. ∥ ouverture f. de pont. / ~ pier ∥ Brückenpfeiler m. ∥ pied-droit m. ou pilier m. ou massif m. ou pile f. d'un pont. / ~ pile ∥ Brückengrundpfahl m. ∥ pieu m. ou pilot m. d'un pont. / ~ pole (Tel) ∥ Brückengestänge n. appui m. sur pont. / ~ raft ∥ Brückenfloß n. ∥ support m. flottant. / ~ rail ∥ Brückenschiene f.; Hohlschiene f. ∥ rail m. en V-inverse; rail m. Brunel. / ~ resistance (Electr) ∥ Brückenwiderstand m. ∥ résistance f. à pont. / ~ road ∥ Fahrbahn f. der Brücke ∥ tablier m. du pont. / ~ span ∥ Brückenspannung f. ∥ portée f. du pont. / static test of a ~ ∥ Belastungsprobe f. einer Brücke ∥ épreuve f. statique d'un pont. / ~ tree (Mill) ∥ Stellzeug n. ∥ trempure f.

bridge-type air hammer ∥ Brückenlufthammer m. ∥ marteau-pilon m. pneumatique à portique.

bridge wire with slide (Electr) ∥ Brückendraht m. mit Gleitschieber ∥ fil m. de pont avec contact glissant.

bridging beam (Carp) ∥ Zange f. ∥ poutre f. traversière; traversière f.; entrait m. / ~ coil (Tel) ∥ Abzweigspule f. ∥ bobine f. de bifurcation. / ~ contact ∥ Brückenkontakt m. ∥ contact m. à pont. / ~ joist (Join) ∥ Dielenlager n.; Polsterholz n. ∥ soliveau m.

bridging-over terminal (Acc) ∥ Überbrückungsklemme f. ∥ borne f. de raccordement.

bridgings pl. (Carp) ∥ Schwartenbretter npl.; Schallatten fpl.; Schalbretter npl. bois m. de couchis; planches fpl. de boisage.

bridle ∥ Zügel m.; Zaum m. ∥ bride f. / ~ chains pl. (Mar) ∥ Kettenhahnpot n. der Vertäuungsbojen ∥ pattes fpl. d'oie des corps morts. / ~ path ∥ Reitweg m. ∥ route f. muletière. / ~ rod (Steam eng) ∥ Leitstange f.; Gegenlenker m. ∥ manivelle f. ou bride f. du parallélogramme. / ~ rods pl. ∥ Lenkerpaar n. ∥ contre-balanciers mpl. du parallélogramme articulé.

bridoon button ∥ Trensenknopf m. ∥ olive f. du bridon.

brier see briar.

brig (Shipb) ∥ Brigg f. ∥ brick m. / hermaphrodite ~ ∥ Briggschoner m.; Marssegelschoner m. ∥ brick-goëlette f.

brigade, mechanized fighting ∥ mechanisierte Kampfbrigade f. ∥ brigade f. de combat mécanisée.

brigantine ∥ Schonerbrigg f. ∥ brigantin m.

brig-cutter ∥ Briggkutter m. ∥ cutter m. ou côtre m. gréé en brigantine.

bright ∥ glänzend ∥ luisant. / ~ (Liquid) ∥ klar ∥ claire; limpide. / to be ~ ∥ glänzen ∥ briller

bright adaptation (Opt) ∥ Hellanpassung f. ∥ adaptation f. à la lumière. / ~ coffee ∥ blanker Kaffee m. ∥ café m. luisant. / ~-drawn steel ∥ blank gezogener Stahl m. ∥ acier m. étiré à brillant. / ~ rice ∥ polierter Reis m. ∥ riz m. glacé. / ~ rim ∥ Lichtrand m. ∥ bord m. lumineux. / ~ surface ∥ glänzende Oberfläche f. ∥ surface f. polie ou brillante / ~ wire ∥ blanker Draht ∥ fil m. poli ou à surface lisse.

brighten, to (Glass) ∥ glätten; die Spiegelfolie polieren ∥ aviver la feuille d'étain. / ~ the colour ∥ eine Farbe f. auffrischen oder aufhellen ∥ éclaircir la couleur. / ~ a picture ∥ ein Bild aufhellen ∥ égayer un tableau.

brightener (Jewel) ∥ Glätter m. ∥ aviveur m. / ~ (Dyer) ∥ Abklärer m. ∥ aviveur m.

brightening (Opt) ∥ Erhellung f. ∥ éclat m. / to cease ~ (Met) ∥ abblicken ∥ ternir / cesser de faire l'éclair. / ~ of silver ∥ Silberblick m. ∥ éclair m. de l'argent.

brightening field observation (Opt) ∥ Hellfeldbeobachtung f. ∥ observation f. en fond clair. / ~ line on a dark ground ∥ heller Faden m. auf dunklem Grunde ∥ fil m. clair sur fond noir.

brightness ∥ Helligkeit f.; Glanz m. ∥ clarté f.; éclat m. / ~ of the cross lines (Opt) ∥ Fadenhelligkeit f. ∥ éclairement m. des fils. / ~ of the sun ∥ Sonnenhelligkeit f. ∥ clarté f. du soleil. / degree of ~ ∥ Helligkeitsstufe f. ∥ degré m. de clarté.

brightning see brightening.

brilliancy ∥ Helligkeit f. ∥ clarté f. / ~ of a beer ∥ Glanz m. oder Feuer n. eines Bieres ∥ brillant m. ou limpidité f. d'une bière.

brilliant ∥ glänzend ∥ brillant. / ~ lisle ∥ Glanzflor m. ∥ fil m. d'Écosse brillant. / ~ starch for linen ∥ Wäscheglanzstärke f. ∥ amidon m. brillant pour le linge.

brilliant ∥ Brillant m. ∥ brillant m.

brilliantine ∥ Haarpomade f. ∥ brilliantine f.; pommade f. pour les cheveux.

brim of a hat ∥ Krempe f. oder Rand m. eines Hutes ∥ bords mpl. d'un chapeau.

brimstone ∥ Schwefel m. ∥ soufre m. / to dip in ~ ∥ schwefeln ∥ soufrer. / roll ~ ∥ Stangenschwefel m. ∥ soufre m. en canon.

brimstone impression ∥ Schwefelabdruck m.; Schwefelpaste f. ∥ empreinte f. ou ectype f. en soufre. / ~ match ∥ Schwefelhölzchen n. ∥ allumette f. / ~ medal see ~ impression. / print in ~ see ~ impression. / ~ yellow ∥ Schwefelgelb n. ∥ jaune m. de soufre.

brine ∥ Sole f.; Salzsole f.; Salzlake; f. Lake f. ∥ saumure f.; eau f. salée. / chilled ~ ∥ kalte Sole f. ∥ saumure f. froide. / graduated ~ ∥ gradierte Sole f. ∥ eau f. graduée. / weak ~ ∥ schwache Lauge f. ∥ eaux fpl. faibles ou de lessivage ou petites.

brine bath ∥ Kochsalzbad n. ∥ bain m. d'eau salée. / ~ cooling ∥ Solekühlung f. ∥ refroidissement m. d'eau salée. / ~ ditch ∥ Solkanal m. ∥ brassour m.; brassoure f. / ~ evaporating pan ∥ Salzpfanne f. ∥ poêle f. à saumure. / ~ gauge ∥ Solwage f. ∥ pèse-sel m. / ~ inlet ∥ Soleeintritt m. ∥ entrée f. de la saumure.

Brinell aparatus ∥ Brinellkugeldruckgerät n.; Brinellapparat m.; Kugeldruckhärteprüfgerät n. ∥ appareil m. à bille (de) Brinell. / ~ hardness ∥ Brinellhärte f. ∥ dureté f. ou preinte f. Brinell. / ~ hardness test see ~ test (of hardness). / ~ machine see Brinell apparatus. / ~ process see ~ test (of hardness). / ~ test (of hardness) ∥ Kugeldruckprobe f. (nach Brinell) ∥ épreuve f. de pression à la Brinell; essai m. à la bille (Brinell). / ~ testing apparatus see Brinell apparatus.

brine marsh ∥ Salzsumpf m. ∥ marais m. salant. / ~ outlet ∥ Soleaustritt m. ∥ sortie f. de la saumure. / ~ pipe ∥ Salzablaßrohr n. ∥ tuyau m. de sortie de la saumure.

brine pump ∥ Solepumpe f. ∥ pompe f. à saumure. / ~ for rapid pickling ∥ Lakepumpe f. zum Schnellpökeln ∥ pompe f. à saumure pour saler ou mariner rapidement.

brine reservoir ∥ Salzwasserbehälter m. ∥ réservoir m. à eau salée. / ~ salt for hygienic and healing purposes ∥ Quellsalz n. für hygienische und Heilzwecke ∥ sel m. de source pour usages hygiéniques et curatifs. / ~ spring ∥ Solquelle f. ∥ source f. salée.

bring, to ~ down the roof (Mine) ∥ zu Bruche m. bauen ∥ faire ébouler les débris mpl. / ~ out ∥ herausgeben ∥ publier: éditer; faire paraître. / ~ to an anchor ∥ ein Schiff n. vor Anker legen ∥ mettre à l'ancre f. / ~ together ∥ in Kontakt m. bringen oder setzen ∥ mettre en contact m. / ~ up the crucibles ∥ die Schmelztiegel rotglühend machen ∥ chauffer les creusets au rouge cerise. / ~ up the materials (Met) ∥ den Ofen m. beschicken charger le fourneau. / ~ up a ship ∥ ein Schiff n. aufbringen ∥ capturer un bateau.

bringing-in the rotor of the turbo generator ∥ Einfahren n. des Turbogeneratorläufers ∥ essai m. de marche du rotor du turbo-alternateur.

briquette ∥ Brikett n.; Preßkohle f. ∥ briquette f.; aggloméré m. / brown coal ~ ∥ Braunkohlenbrikett n. ∥ lignite m. aggloméré. / wet-pressed ~ ∥ Naßbrikett n. briquette f. faite ou aggloméré m. fait par voie humide.

briquette making machine ∥ Brikettmaschine f. ∥ machine f. à faire les briquettes de charbon.

briquetting plant ∥ Brikettierungsanlage f. ∥ installation f. de briquettage ou pour la fabrication des agglomérés. / ~ for ore and blast furnace dust ∥ Anlage f. zur Brikettierung von Erz- und Gichtstaub installation f. pour briquetter les poussières des minerais.

briquetting press ∥ Brikettpresse f. ∥ presse f. à briquettes.

brisk ∥ lebhaft; heftig; rege ∥ vive; actif; animé. / ~ (Trade) ∥ schwunghaft ∥ florissant. / to boil ~ly ∥ lebhaft kochen bouillir vivement.

brisk demand ∥ rege Kauflust f. ∥ forte demande f. ou envie f. d'acheter. / ~ inquiry ∥ Zudrang m.; reger Zuspruch m. ∥ affluence f.; presse f.

bristle ∥ Schweinsborste f.; Borste f. ∥ soie f. ou poil m. de cochon; robe f. de soie. / basal ~ ∥ Basalborste f. ∥ brosse f. ou soie f. dorsale. / ~s pl. of the brush Borsten fpl. der Bürste ∥ soies fpl. de la brosse.

bristle brush ∥ Borstenbürste f.; Borstenpinsel m. ∥ brosse f. en poil; pinceau m. en soie. / ~ dyer ∥ Borstenfärber m. teinturier m. en soies. / ~ finishing works pl. ∥ Borstenzurichterei f. ∥ usine f. d'apprêt des soies de porcs. / ~ puller ∥ Borstenhaarausrupfer m. ∥ éjarreur m.

Bristol board ∥ Bristolpappe f. ∥ carton m. bristol. / ~ paper ∥ Bristolpapier n.; Isabeypapier m. ∥ papier m. Bristol.

brisure of a wall ∥ Knick m. einer Mauer ∥ brisure f. d'une muraille.

brittle ∥ morsch; zerbrechlich ∥ pourri; vermoulu; fragile. / ~ (Met) ∥ spröde ∥ paillé:

sec. / ~ (Mining) ‖ gebrech; bröcklig ‖ cassant; fragile. / ~ (Phys) ‖ spröde ‖ cassant; rouverin. / ~ when cold ‖ kaltbrüchig ‖ cassant à froid. / ~ when redhot ‖ warmbrüchig ‖ cassant à chaud. / for these objects the steel was too ~ ‖ für diesen Zweck war der Stahl zu spröde ‖ pour cet usage l'acier était trop cassant.

brittle hardness ‖ spröde Härte f. ‖ dureté f. d'un caractère cassant. / ~ iron / brüchiges Eisen n. ‖ fer m. cassant. / ~ metal ‖ Rotguß m.; Tombak m.; Rotmessing n. ‖ tombac m.; laiton m. rouge; bronze m. ‖ ~ silver ore ‖ Melanglanz m.; Stephanit m.; Schwarzgültigerz n. ‖ argent m. noir; argent sulfuré fragile, argent m. antimonié sulfuré noir. / ~ sulphide of silver see ~ silver ore.

brittleness ‖ Zerbrechlichkeit f. ‖ fragilité f. / ~ (Met) ‖ Sprödigkeit f. ‖ aigreur f. / ~ of iron while cold ‖ Kaltbruch m. von Eisen ‖ qualité f. d'un fer cassant à froid.

broach, to (Cask) ‖ anbohren ‖ mettre en perce f. / ~ a hole ‖ ein Loch n. aufräumen ‖ aléser un trou. / ~ in one operation with one broach ‖ mit einem Werkzeug in einem Zuge räumen ‖ brocher avec une broche en une seule passe.

broach ‖ Reibahle f.; Räumahle f.; Räumnadel f. ‖ broche f.; équarrissoir m.; alésoir m.; élargisseur m. / ~ (Build) ‖ Turmspitze f. ‖ aiguille f. ou pointe f. ou flèche f. d'une tour. / four-square ~ see square ~. / half-round ~ ‖ halbrunde Reibahle f. ‖ broche f. mi-ronde; alésoir m. ou équarrissoir m. mi-rond. / round ~ ‖ Polierahle f. ‖ alésoir m. rond. / six-square ~ ‖ sechseckige Reibahle f. ‖ alésoir m. à six pans. / square ~ ‖ Vierkantreibahle f.; viereckige Reibahle f. ‖ alésoir m. carré.

broach collecting grid ‖ Dornsammelrost m. ‖ grille f. collectrice des mandrins.

broaching (Beer) ‖ Anstich m. ‖ mise f. en perce. / ~ (Reaming) ‖ Räumen n. ‖ équarrissage m.; alésage m.

broaching machine ‖ Räummaschine f.; Nutenziehmaschine f. ‖ machine f. à aléser ou à brocher ou à dégorger. / ~ with hydraulic pump ‖ Räummaschine f. mit Flüssigkeitsgetriebe ‖ machine f. à aléser à commande hydraulique. / ~ drawhead ‖ Zugorgan n. der Räummaschine ‖ dispositif m. de traction de la machine à dégorger. / ~ stroke ‖ Ziehlänge f. der Räummaschine ‖ longueur m. de traction d'une machine à dégorger.

broaching press for tubes ‖ Rohraufweitepresse f. ‖ presse f. à élargir les tubes.

broachings pl. ‖ Schabspäne mpl. ‖ alésures fpl.

broach-post (Carp) ‖ Helmstange f. ‖ poinçon m.; aiguille f. d'une flèche en bois.

broad ‖ breit ‖ large; étendu. / ~ awl (Saddl) ‖ flache Ahle f. ‖ alène f. plate. / ~ axe (Coop) ‖ Lenkbeil n.; Breithacke f. ‖ doloire f. de tonnelier. / ~ bean ‖ Saubohne f. ‖ fève f. de marais. / ~ bran ‖ grobe Kleie f. ‖ gros son m. / ~ glass ‖ Fensterglas n.; Tafelglas n. ‖ verre m. à vitres. / ~ lath (Tiler) ‖ Schiefer(dach)-latte f. ‖ latte f. volige. / ~ thrashing machine ‖ Breitdreschmaschine f. ‖ batteuse f. large.

broadcasting ‖ Rundfunk m. ‖ broadcasting m.; radiodiffusion f. / ~ copyright ‖ Funkurheberrecht n. ‖ droit m. d'auteur en matière de radiodiffusion. /

~ disturbance ‖ Rundfunkstörung f. ‖ dérangement m. de la radiodiffusion. / ~ engineering ‖ Rundfunktechnik f. ‖ technique f. de la radiodiffusion. / ~ main transmitter ‖ Rundfunkhauptsender m. ‖ émetteur m. principal de radiodiffusion. / ~ microphone ‖ Mikrofon n. für Rundfunk; Rundfunkmikrofon n. ‖ microphone m. de radiodiffusion. / ~ receiver ‖ Rundfunkempfänger m. ‖ récepteur-broadcasting m. / ~ relay transmitter ‖ Rundfunkzwischensender m. ‖ émetteur m. accessoire de radiodiffusion. / ~ station ‖ Rundfunkstation f. ‖ station f. de radiodiffusion; poste m. transmetteur. / ~ transmission cable ‖ Rundfunkkabel n. ‖ câble m. pour la transmission radiophonique. / ~ transmitter ‖ Rundfunksender m. ‖ émetteur m. de radiodiffusion. / ~ wave ‖ Rundfunkwelle f. ‖ émission f. du broadcasting.

broad-drawing equalizing machine ‖ Breitstreckegalisiermaschine f. ‖ étireuse f. à égaliser en largeur.

broad-gauge (Railw) ‖ Breitspur f. ‖ voie f. ou écartement m. large. / ~ locomotive ‖ Breitspurlokomotive f. ‖ locomotive f. à voie large. / ~ railway ‖ Breitspurbahn f. ‖ chemin m. de fer à voie large.

broad-leaved garlic oil ‖ Bärlauchöl n. ‖ essence f. d'ail des ours.

broad-sheet (Size) ‖ Querformat n. ‖ format m. oblong. / ~ (Placard) ‖ Plakat n.; Anschlagzettel m. ‖ placard m.; affiche f.

broadside ‖ Breitseite f. ‖ côté m. large; flanc m. / ~ (Size) ‖ broad-sheet. / ~ (Ship) ‖ Breitseite f. ‖ bordée f.

broadside-on ‖ breitseits ‖ par le travers.

broadside port ‖ Breitseitpforte f. ‖ sabord m. de côté. / ~ sea ‖ Dwarssee f. ‖ mer f. de travers.

broadstone ‖ Quader m.; Quaderstein m. ‖ pierre f. carrée; carreau m.; moellon m. d'appareil. / ~ which forms a wagonload ‖ einführiger Quader m. ‖ quartier m. de voie ou de pierre. / ~ two to the load ‖ zweiführiger Quader m. ‖ carreau m.

broadwise ‖ der Breite f. nach; in der Breite ‖ en large; dans le sens de la largeur.

brocade ‖ Brokat m. ‖ brocart m.

broché silk ‖ broschierter Seidenstoff m. ‖ soie f. broché. / ~ stuff ‖ broschierter Stoff m. ‖ tissu m. broché.

broken (Chem) ‖ in Stücken npl. ‖ en morceaux mpl. / ~ (Coal; Ore; Stones) ‖ als Bruch m. ‖ cassé. / ~ (Topo) ‖ durchschnitten ‖ coupé; difficile; fourré; accidenté. / ~ (Trade) ‖ ruiniert ‖ être flambé ou frit ou ruiné. / to be ~ (Print) ‖ abfallen ‖ se coucher. / finely ~ (Chem) ‖ fein zerrieben ‖ finement broyé. / the vessel is ~ ‖ das Schiff hat einen Katzenrücken ‖ le navire est arqué ou éreinté ou goreté.

broken cassia lignea ‖ gestoßener Zimt m. ‖ cannelle f. concassé. / ~ coal ‖ Bruchkohle f.; Brechkohle f. ‖ déchets mpl. de charbon; charbon m. concassé. / ~ iron ‖ Eisenabfälle mpl. ‖ ferraille f.; débris mpl. de fer; riblons mpl. / ~ number (Math) ‖ Bruch m. ‖ fraction f.; nombre m. rompu. / ~-off ‖ abgebrochen ‖ détaché; abandonné. / ~ rock ‖ Schotter m.; Steinschlag m. ‖ cailloutis m.; pierres fpl. cassées ou con-

cassées. / ~-space saw ‖ Lochsäge f. ‖ scie f. à guichet; égohine f. / ~ test-piece ‖ durchgeschlagener Probestab m. ‖ éprouvette f. cassée en deux.

broken-stone ‖ Steinschlag m.; Schotter m.; Bruchstein m. ‖ pierraille f.; moellon m.; cailloutis m. / ~ industry ‖ Schotterindustrie f. ‖ industrie f. du caillloutis. / ~ manufacturing plant ‖ Schotteranlage f. ‖ installation f. de cailloutage. / washery for ~ (manufacturing) plants ‖ Wäsche f. für Schotteranlagen ‖ laveur m. pour installations de cailloutis. / ~ road ‖ Steinstraße f. ‖ route f. empierrée.

broker ‖ Makler m.; Unterhändler m. ‖ courtier m. / produce ~ ‖ Produktenmakler m. ‖ courtier m. en denrées.

brokerage ‖ Unterhändlergebühr f.; Maklergebühr f. ‖ courtage m. / free of ~ ‖ ohne jede Maklergebühr f. ‖ sans courtage m. / including ~ ‖ einschließlich Maklergebühr f. ‖ y inclus courtage m. / ~ of... ‖ Maklergebühr f. von... ‖ courtage m. de... / usual ~ ‖ übliche Maklergebühr f. ‖ courtage m. d'habitude.

broker's commission ‖ Maklerprovision f. ‖ commission f. de courtier. / ~ contract ‖ Schlußzettel m. ‖ bordereau m. de courtier. / ~ note see ~ contract.

bromate ‖ bromsaures Salz n. ‖ bromate m.

bromic acid ‖ Bromsäure f. ‖ acide m. bromique. / ~ silver ‖ Bromsilber n. ‖ bromure m. d'argent; bromyrite f.

bromide ‖ Bromid n.; Bromsalz n. ‖ bromide m.; bromure m. / ~ of ammonia ‖ Ammoniumbromid n. ‖ bromure m. d'ammonium. / ~ of arsenic ‖ Arsenbromid n. ‖ bromure m. d'arsenic. / ~ of calcium ‖ Bromkalk m.; Kalziumbromid n. ‖ bromure m. de calcium. / ~ of iodine ‖ Jodbromid n. ‖ bromure m. d'iode. / ~ of lime ‖ Bromkali n. ‖ bromure m. de chaux. / ~ of magnesium ‖ Brommagnesium n. ‖ bromure m. de magnésium: hydrobromate m. de magnésie. / ~ of mercury ‖ Quecksilberbromid n. ‖ bromure m. de mercure. / ~ of nickel ‖ Nickelbromid n. ‖ bromure m. de nickel. / ~ of potassium ‖ Bromkalium n.; Kaliumbromid n. ‖ bromure m. de potassium; hydrobromate m. de potasse. / ~ of radium ‖ Radiumbromid n. ‖ bromure m. de radium. / ~ of silver ‖ Bromsilber n. ‖ bromure m. d'argent.

bromide postcard ‖ Bromsilberpostkarte f. ‖ carte f. postale au bromure.

bromine ‖ Brom n. ‖ brome m. / ~-containing ‖ bromhaltig ‖ bromé. / ~ manufacturing plant ‖ Bromfabrikeinrichtung f. ‖ installation f. à produire le brome.

bromite ‖ Bromsilber n. ‖ bromure m. d'argent.

bromoform ‖ Bromoform n. ‖ bromoforme m.

bromyrite ‖ Bromsilber n. ‖ bromyrite f.; bromure m. argentique.

bronze, to ‖ bronzieren ‖ bronzer.

bronze ‖ Gelbguß m.; Bronze f. ‖ bronze m. / aluminium ~ ‖ Aluminiumbronze f. ‖ bronze m. d'aluminium. / forgeable ~ ‖ schmiedbare Bronze f. ‖ bronze m. forgeable. / genuine ~ ‖ echte Bronze f. ‖ bronze m. véritable. / ~ machine ‖ Maschinenbronze f. ‖ bronze m. pour machines. / malleable ~ ‖ schmiedbare Bronze f. ‖ bronze m. malléable. / mangane ~ ‖ Manganbronze f. ‖ manganbronze f. / real gold ~ ‖ Goldbronze f. ‖ bronze m. d'or. / steel ~

| Stahlbronze f. || bronze m. d'acier. / varnished ~ || unechte Bronzewaren fpl. || bronze m. verni.

bronze band || Bronzeband n. || bronze m. en bande. / ~ blue || Bronzeblau n. || bleu m. de bronze. / ~ casting || Bronzeguß m. || fonte f. en bronze; bronze m. fondu. / ~ chaser || Bronzeziselör m. || ciseleur m. sur bronze. / ~ colour || Bronzefarbe f. || couleur f. de bronze. / ~ colourer || Bronzör m. || bronzeur m. sur métaux. / ~ cyanide bath || Bronzebad n. || bain m. de bronzage. / ~ disk || Bronzescheibe f. || disque m. en bronze. / ~ fitter || Bronzearbeiter m. || ajusteur m. de bronze. / ~ forging || Bronzeschmiedestück n. || pièce f. de forge en bronce. / ~ foundry || Bronzegießerei f. || fonderie f. de bronze. / ~ ink || Bronzefarbe f. || couleur f. de bronze. / ~ lighting article || Beleuchtungskörper m. aus Bronze || article m. d'éclairage en bronze. / ~ paint || Bronzefarbe f. || couleur f. de bronze. / ~ pen || Bronzeschreibfeder f. || plume f. de bronze. / ~ pipe || Bronzeröhre f. || tuyau m. en bronze. / ~ powder || Bronzepulver n.; Bronzefarbe f. || bronze m. moulu; bronze m. en poudre. / ~ printing || Bronzedruck m. || impression f. bronzée.

bronzer || Bronzierer m. || bronzeur m.

bronze rod || Bronzestange f. || barre f. en bronce.

bronzes pl. || Bronzewaren fpl. || bronzes mpl.

bronze statue foundry || Bildgießerei f. || fonderie f. de bronzes d'art. / ~ tincture || Bronzetinktur f. || teinture f. au bronze. / ~ tincture gum || Bronzetinkturgummi m. || gomme f. pour teinture au bronze. / ~ varnish || Bronzelack m. || vernis m. au bronze. / ~ ware || Bronzewaren fpl. || articles mpl. en bronze. / ~ wire || Bronzedraht m. || fil m. de bronze; bronze m. en fil.

bronzing || Bronzieren n. || bronzage m. / ~ liquid see bronze tincture. / ~ machine || Bronzierapparat m. || machine f. à bronzer. / flat ~ machine (Print) || Flachbronziermaschine f. || machine f. à bronzer à plat. / ~ pickle (Met) || Brünierbeize f. || mordant m.

brooch || Brosche f.; Vorstecknadel f.; Brustnadel f. || broche f.; agrafe f. / painting || einfarbiges oder gemmenartiges Gemälde n.; Monochromie f. || camaïeu m.; peinture f. en camaïeu. / ~ plate || Broschenplatte f. || plaque f. de broche.

brood (Mine) || Gangart f. || gangue f.

brooding, artificial ~ establishment || Brutanstalt f. || couvoir m.; couveuse f.; incubateur m.

broom (Bot) || Ginster m. || genêt m. / ~ (Househ) || Kehrbesen m. || balai m. / birch ~ || Reisigbesen m. || balai m. de bouleau. / ~ of bristles || Borstenbesen m. || balai m. de soies de porc. / dog grass ~ || Queckenbesen m. || balai m. en chiendent. / feather ~ || Federbesen m. || plumeau m. / hair ~ || Haarbesen m. || balai m. de crin. / wood ~ || Strauchbesen m. || balai m. de brindilles de bois.

broom binder || Besenbinder m. || lieur m. de balais. / ~ factory || Besenfabrik f. || fabrique f. de balais. / ~ flower || Ginsterblüte f. || fleur f. de genêt. / ~ holder || Besenhalter m. || porte-balai m. / ~ maker || Besenmacher m. || fabricant m. de

balais. / ~ stick || Besenstiel m. || manche m. à balai.

broth, meat ~ || Fleischsaft m. || extrait m. ou jus m. de viande; bouillon m.

brow (Anatomy) || Stirn f. || front m. / ~ of an enclosure wall || Mauerabdeckung f. oder Abdach n. einer Einfriedigungsmauer || larmier m. d'un mur de clôture.

brown || braun || brun.

brown || Braun n. || brun m. / velvet ~ || Samtbraun n. || brun m. velours.

brown beer || Braunbier n. || bière f. brune. / double ~ || Doppelbraunbier n. || bière f. brune double.

brown coal || Braunkohle f. || lignite m.; houille f. brune ou terreuse. / earthy ~ || erdige Braunkohle f. || lignite m. terreux. / extraction of pulverulent ~ || Gewinnung mulmiger Braunkohle || extraction f. de lignite pulvérulent.

brown coal briquette || Braunkohlenbrikett n. || briquette f. de lignite; lignite m. aggloméré. / ~ firing || Braunkohlenfeuerung f. || chauffage m. au lignite. / ~ furnace for ~ || Feuerung f. für Braunkohle || foyer m. à lignite. / ~ gas || Braunkohlengas n. || gaz m. de lignite. / ~ mine || Braunkohlenbergwerk n. || mine f. de lignite. / ~ tar || Braunkohlenteer m. || goudron m. de lignite.

browner (Arm) || Brünierer m.; Braunmacher m. || brunisseur m.

brown hematite mine || Brauneisensteingrube f. || mine f. d'hématite brune.

browning || Zuckercoulör f. || teinture f. de caramel; couleurs fpl. de sucre.

brown iron ore || Brauneisenstein m. || limonite f.; hématite f. brune; fer m. oxydé hydraté.

brown ore || Braunerz n.; Blauerz n. || minerai m. brun; vivianite f.; fer m. spatique mûr.

brownstone || Braunstein m. || bioxyde m. de manganèse. / ~ mine || Braunsteingrube f. || mine f. de bioxyde de manganèse.

brown sugar || Rohzucker m. || sucre m. brut.

brow post (Build) || Querbalken m. || poutre f. transversale; traverse f.

bruise, to || knicken || briser; fêler; accabler. / ~ (Mill; Brew) || schroten || égruger.

bruise || Quetschung f. || contusion f.

bruised barley || Gerstenschrot n. || mouture f. d'orge.

bruiser (Glass) || konkave Schleifschale f. || bassin m.

bruising || Getreideschroterei f. || mouture f.

bruising mill || Quetschmühle f. || Schrotmühle f. || moulin m. écraseur ou concasseur; concasseur écraseur m. / artificial stone ~ || Kunststeinschrotmühle f. || moulin m. concasseur à meules artificielles. / malt ~ || Malzschrotmühle f. || moulin m. à concasser le malt.

brush, to || bürsten; abbürsten || brosser. / ~ off (Sugar) || plamotieren; abhaken || plamoter. / ~ off the form (Print) || die Form ausbürsten || brosser la forme.

brush || Bürste f. || brosse f. / ~ (Electr) || Bürste f.; Stromabnehmer m. || balai m.; brosse f. / ~ (Found) || Gußputzbürste f. || brosse f. à nettoyer. / ~ (Forg) || Löschwedel m.; Sprengwedel m. || goupillon m. / ~ (Mas) || Netzpinsel m.; Annässer m. || Annetzer m.; Quast m.; Maurerpinsel m.; Maurerbürste f. || brosse f.; brossette f.; balai m. / ~ (Mar) || Quast

m.; Pinsel m. || pinceau m.; guipon m.; broche f. / ~ (Paint) || Tünchpinsel m.; Quast m.; Bürste f. || brosse f. / barrel ~ || Faßbürste f. || brosse f. pour fûts. / blacking ~ (Mar) || Schwarzquast m. || brosse f. de barbouilleur. / black-leading ~ || Graphitierbürste f. || brosse f. à graphiter. / ~ for boiler tubes || Rohrbürste f. für Kesselrohre || brosse f. ou écouvillon m. pour tubes de chaudières. / bottle ~ || Flaschenbürste f. || goupillon m. / carbon ~ || Kohlenbürste f. || balai m. en charbon. / circular ~ || Radbürste f. || brosse f. à couronne circulaire. / cleaning ~ || Reinigungsbürste f.; Ausputzkratze f. || brosse f. à nettoyer. / clothes-~ || Kleiderbürste f. || brosse f. à habits. / cocoanut fibre ~ || Kokosfaserbürste f. || brosse f. en fibre de coco. / coppergauze ~ || Kupfergazebürste f. || balai m. en toile de cuivre. / cylindrical ~ || Rohrbürste f. || brosse f. cylindrique. / flat ~ || Flachpinsel m. || pinceau m. plat. / hair ~ || Haarbürste f. || brosse f. à cheveux. / hard ~ see scrubbing ~. / hat ~ || Hutbürste f. || brosse f. à chapeau. / heath ~ || Heidebürste f. || brosse f. de bruyère ou en buisson. / ~ for industrial use || Industriebürste f. || brosse f. industrielle. / laminated ~ || lamellierte Bürste f.; Blätterbürste f.; Folienbürste f. || balai m. en feuilles ou en clinquant. / long ~ || langer Pinsel m. || pinceau m. long / long-handled ~ (Mas) || Quast m. || brossette f. / metalized ~ || metallisierte Kohlenbürste f. || balai m. métallisé. / metallic ~ || Metallbürste f. || balai m. métallique. / movable ~ || verstellbare Bürste f. || balai m. mobile ou réglable. / nail ~ || Nagelbürste f. || brosse f. à ongles. / paint ~ see painter's ~. / painter's ~ || Malerpinsel m.; Pinsel m. || pinceau m. (de peintre). / ~ for painting see painter's ~. / ~ in permanent contact || dauernd aufliegende Bürste f. || balai m. toujours en contact. / piassava ~ || Piassavabürste f. || brosse f. en piazzava. / pressing ~ (Weav) || Andrückbürste f. || brosse f. conductrice. / rotating ~ || rotierende Bürste f. || balai m. tournant. / round ~ || Rundbürste f. || brosse f. circulaire; pinceau m. rond. / scrubbing ~ || Scheuerbürste f. || brosse f. à frotter ou de frottement. / shaving ~ || Rasierpinsel m. || blaireau m. / shining ~ || Glanzbürste f. || polissoire f. / soft ~ || weiche Bürste f. || brosse f. douce. / steel-wire ~ || Stahldrahtbürste f. || brosse f. en fil d'acier. / stiff ~ || harte Bürste f. || brosse f. dure. / tooth ~ || Zahnbürste f. || brosse f. à dents. / type-inking ~ || Bürste f. zum Färben der Typen || brosse f. pour l'encrage des types. / ~ for washing fermenting tuns || Gärbottichbürste f. || brosse f. pour le nettoyage des cuves de fermentation. / wire ~ || Drahtbürste f. || brosse f. métallique; grate-brosse f.

brush accessories pl. || Bürstenzubehör n. || accessoires mpl. de brosses. / ~ and filing machine || Bürst- und Florteilmaschine f. || machine f. à brosser et diviser le poil. / ~ backs pl. for ~es || Bürstenholz n. || bois m. de brosses. / ~ bolt (Electr) || Bürstenbolzen m. || tourillon m. de porte-balai. / ~ bolter (Mill) || Mehlbürstmaschine f. || bluterie f. à brosses. / ~ borer || Bürstenboh-

rer m. ‖ perceur m. de brosses. / ~ bundle ‖ Reisigbund n. ‖ fascine f. / ~ carriage (Electr) ‖ Bürstenschlitten m. ‖ chariot m. porte-balais. / ~ carrier (Electr) ‖ Bürstenführungsstange f. ‖ arbre m. porte-balais; porte-frotteur m. / ~ contact pressure ‖ Bürstenauflagedruck m. ‖ pression f. des balais. / ~ contact resistance ‖ Bürstenübergangswiderstand m. ‖ résistance f. de contact des balais. / ~ coupling ‖ Bürstenkupplung f. ‖ accouplement m. à brosses. / ~ detacher ‖ Bürstendetaschör m. ‖ détacheur m. à brosses. / ~ discharge ‖ elektrische Glimmentladung f.; dunkle Entladung f.; Bürstenentladung f. ‖ effluve f.; décharge f. en aigrette ou en brosse. / ~ displacement (Electr) ‖ Bürstenverstellung f. ‖ décalage m. des balais. / ~ displacement angle ‖ Bürstenverstellungswinkel m. ‖ angle m. de décalage des balais.

brusher (Mining) ‖ Abkohler m.; Hauer m. ‖ abatteur m.; rabatteur m. / ~ (Pap) ‖ Striegler m. ‖ brosseur m.

brushes pl. ‖ Bürstenwaren fpl.; Bürstenbinderwaren fpl. ‖ brosserie f. / bonebacked ~ ‖ Bürstenbinderwaren fpl. in Knochen ‖ brosserie f. montée sur os. / ~ of hog's bristles ‖ Borstenbürstenbinderwaren fpl. ‖ brosserie f. garni de poils. / ivory-backed ~ ‖ Bürstenbinderwaren fpl. in Elfenbein ‖ brosserie f. montée sur ivoire. / metal-backed ~ ‖ Bürstenbinderwaren fpl. auf Metall ‖ brosserie f. sur métal. / wood-backed ~ ‖ in Holz gebundene Bürstenbinderwaren fpl. ‖ brosserie f. montée sur bois.

brush fitter ‖ Bürsteneinzieher m. ‖ monteur m. de brosses. / ~ friction ‖ Bürstenreibung f. ‖ frottement m. des balais. / ~ friction loss ‖ Bürstenreibungsverlust m. ‖ perte f. par frottement des balais.

brush-holder (For painting brushes) ‖ Pinselkapsel f. ‖ virole f. de pinceaux. / ~ (Electr) ‖ Bürstenhalter m. ‖ porte-balai m.

brush-holder body ‖ Bürstenhalterblock m. ‖ corps m. du porte-balai. / ~ key ‖ Bürstenschlüssel m.; Bürstenhalterschlüssel m. ‖ clef f. de porte-balai. / ~ rod ‖ Bürstenstift m.; Bürstenhalterstift m. ‖ tige f. de porte-balai. / ~ star ‖ Bürstenhalterstern m. ‖ collier m. porte-balai.

brushing ‖ Abkohlen n. ‖ rabatage m.; abatage m. / ~ of cloth ‖ Bürsten n. des Tuches ‖ brossage m. de drap.

brushing device ‖ Bürstvorrichtung f. ‖ brossage m. / pile ~ device ‖ Strichbürsteinrichtung f. ‖ brosse f. pour l'effet de poil.

brushing machine ‖ Bürstmaschine f. ‖ machine f. à brosser; brosseuse f. / bottle ~ ‖ Flaschenbürstmaschine f. ‖ machine f. à brosser ou à nettoyer les bouteilles. / ~ with one cylinder ‖ Bürstmaschine f. mit einem Tambour ‖ machine f. à brosser avec un tambour. / ~ for sheets and paper in reels ‖ Bürstmaschine f. für Bogenpapier und Rollenpapier ‖ machine f. à brosser les papiers en feuilles et en rouleaux. / steam ~ ‖ Dampfbürstmaschine f. ‖ brosseuse f. à vapeur. / ~ for textiles ‖ Bürstmaschine f. für Textilien ‖ machine f. à brosser des textiles.

brushing mill see brushing machine.

brushing-off (Sugar) ‖ Plamotieren n.; Abhaken n. ‖ plamotage m.

brushing roller ‖ Rauhzylinder m. ‖ cylindre m. à lainer. / ~ worm ‖ Bürstenschnecke f. ‖ vis-brosse f.

brush key (Electr) ‖ Bürstenschlüssel m. ‖ clef f. à balais. / ~ lead (Electr) ‖ Bürstenstellung f. ‖ calage m. ou décalage m. des balais. / ~ lifting device ‖ Bürstenabhebevorrichtung f. ‖ relève-balais m.; dispositif m. de revelage des balais. / ~ light (Electr) ‖ Büschellicht n. ‖ aigrette f. lumineuse. / ~ maker ‖ Bürstenbinder m.; Bürstenmacher m. ‖ brossier m. / ~ making ‖ Bürstenbinderei f. ‖ brosserie f. / ~ making machine ‖ Bürstenherstellungsmaschine f. ‖ machine f. pour la fabrication des brosses. / ~ polisher ‖ Bürstenpolierer m. ‖ polisseur m. de bois de brosses. / ~ position ‖ Bürstenstellung f. ‖ position f. ou calage m. des balais. / ~ proof ‖ Bürstenabzug m. ‖ épreuve f. à la brosse. / ~ rocker ‖ Bürstenbrille f.; Bürstenbrücke f. ‖ joug m. ou lunette f. de porte-balais. / ~ roll ‖ Bürstenwalze f. ‖ brosse f. cylindrique. / ~ scraper ‖ Bürstenabstreicher m. ‖ racleur m. à brosses. / ~ selector (Electr) ‖ Bürstenwähler m. ‖ sélecteur m. de balais. / ~ spring (Electr) ‖ Bürstenfeder f. ‖ ressort m. de balai. / ~ tube paste ‖ Quetschtube f. mit Pinsel ‖ tube m. à matière pâteuse avec pinceau. / ~ cutter of veneer for ~es ‖ Bürstenfurniermacher m. ‖ débiteur m. de placages pour brosses. / ~ ware ‖ Bürsten(binder)waren fpl. ‖ ouvrages mpl. de brosserie; brosserie f. / ~ wheel ‖ Friktionsrad n.; Reibrad n. ‖ roue f. de friction.

brushwood ‖ Bürstenholz n. ‖ bois m. pour brosses. / ~ (Bot) ‖ Gesträuch n. ‖ broussailles fpl. / ~ (Hydr arch) ‖ Faschinenholz n.; Faschinenreisig n.; Buschholz n.; Reisholz n.; Strauchholz n. ‖ brins mpl.; branches fpl.; branchage m. / finished ~ ‖ fertig zugerichtetes Bürstenholz n. ‖ bois m. fini ou préparé pour brosses. / roughed ~ ‖ vorgearbeitetes Bürstenholz n. ‖ bois m. ébauché pour brosses. / ~ for timbering a shaft (Mine) ‖ Reisig n. zur Schachtzimmerung ‖ dosses fpl. dans un puits de mine.

brushwood borer ‖ Bürstenholzbohrer m. ‖ perceur m. de bois de brosses. / ~ cooling-stack ‖ Reisiggradierwerk n. ‖ hangar m. de graduation à fascines. / ~ cutter ‖ Reisigholzschneider m. ‖ coupeur m. de ramilles ou de broutilles ou de menu bois. / ~ cutter (Brushmaking) ‖ Bürstenbrettmacher m. ‖ débiteur m. de bois de brosses. / ~ dresser ‖ Bürstenholzzurichter m. ‖ façonneur m. de bois de brosses. / ~ revetment (Build) ‖ Reisigbett n. ‖ fagotaille f.

brush yoke (Electr) ‖ Bürstenjoch n. ‖ collier m. de porte-balais.

Brussels carpet ‖ Brüsseler Teppich m. ‖ moquette f. frisée ou bouclée. / ~ lace ‖ Brüsseler Spitze f. ‖ dentelle f. de Bruxelles.

bruyère pipe ‖ Bruyerepfeife f. ‖ pipe f. en bruyère. / ~ root ‖ Bruyerewurzel f. ‖ racine f. de bruyère.

bubble, to ~ through ‖ einblasen ‖ insuffler. / ~ up ‖ brausen ‖ bouillonner.

bubble ‖ Blase f.; Bläschen n. ‖ bulle f.; ampoule f.; bouillon m. / air ~ ‖ Luftblase f. ‖ bulle f. d'air. / the ~ is central ‖

die Wasserwage spielt ein ‖ la bulle d'air oscille. / the ~ is out of centre ‖ die Wasserwage schlägt aus ‖ la bulle d'air s'écarte. / to see the ~ without parallax through prisms ‖ die Libelle parallaxenfrei durch Prismen beobachten ‖ observer le niveau (d'eau) sans parallaxe par l'intermédiaire de prismes. / steam ~ ‖ Dampfblase f. ‖ bulle f. de vapeur. / the steel has the tendency to enclose many gas ~s of varying size in the fluid mass ‖ der Stahl hat die Neigung, in flüssigem Zustande viele Gasblasen in ungleichmäßiger Größe einzuschließen ‖ l'acier m. à l'état fluide a la tendance de renfermer de nombreuses bulles de gaz de grandeur variable.

bubble casing ‖ Libellengehäuse n. ‖ boîte f. du niveau d'eau. / ~ fermentation ‖ Blasengärung f. ‖ fermentation f. à bulles. / formation of ~s ‖ Blasenbildung f. ‖ formation f. de bulles. / ~ level ‖ Libelle f.; Wasserwage f. ‖ niveau m. d'eau; niveau m. à bulle d'air. / pair of ~s at right angles to each other ‖ Kreuzwasserwage f. ‖ niveaux mpl. d'eau croisés. / ~ sextant ‖ Libellensextant m. ‖ sextant m. à bulle d'air. / ~ test (Sug) ‖ Blasprobe f. ‖ preuve f. au soufflé.

bubbling ‖ Aufwallen n. ‖ bouillonnement m.

buchu leaves pl. ‖ Buccoblätter npl. ‖ feuilles fpl. de bucco.

buck, to ‖ laugen ‖ lessiver.

bucked ore ‖ Scheiderz n. ‖ minerai m. riche; minerai de scheidage.

bucker ‖ Erzpocher m. ‖ bocardeur m. or broyeur m. or pileur m. de minerai.

bucket ‖ Eimer m.; Kübel m.; Pütze f. ‖ seau m.; seilleau m.; bac m.; baquet m.; seillot m.; godet m.; benne f. / ~ (Coop) ‖ Gelte f.; Schöpfgefäß; hölzerner Eimer ‖ seau m. en bois; broc m.; manuel m. / ~ (Min) ‖ Schachtfördergefäß n.; Förderkübel m. ‖ tonne f.; tine f.; baquet m. / ~ (Pump) ‖ Kolben m. ‖ piston m. à clapet. / ~ (Water-wheel ‖ Zelle f. ‖ auge f. / ~ for blast charging furnaces ‖ Kübel m. für Hochofenbegichtung ‖ seau m. ou godet m. pour le remplissage des hauts-fourneaux. / coal ~ ‖ Kohleneimer m. ‖ seau m. à charbon. / dragging ~ ‖ Schleppschaufel f. ‖ benne f. traînante ou d'entraînement. / dredging ~ ‖ Baggereimer m. ‖ godet m. du cure-môle. / flower ~ ‖ Pflanzenkübel m. ‖ bac m. à fleurs. / guide ~ (Turbine) ‖ Leitschaufel f. ‖ aube f. directrice. / hinged ~ ‖ Klappkübel m. ‖ benne f. ouvrante. / ~ of a barrel (Shipb) ‖ Racktonne f. ‖ anneau m. ou baril m. de racage. / reversing ~ (Turbine) ‖ Umkehrschaufel f. ‖ aube f. d'inversion. / revolving ~ ‖ Drehkübel m. ‖ benne f. basculante. / sinking ~ ‖ Abteufkübel m. / benne f. de creusement. / slag ~ ‖ Schlackenkübel m. ‖ baquet m. à scories. / tin-plate ~ ‖ Blecheimer m. ‖ seau m. en fer-blanc. / tipping ~ ‖ Klappkübel m. ‖ benne f. basculante. / wooden ~ ‖ Holzeimer m. ‖ seau m. en bois.

bucket attemperator ‖ Eisschwimmer m. ‖ nageur m. ou plongeur m. à glace. / ~ capacity ‖ Becherinhalt m. ‖ contenance f. du godet. / ~ chain ‖ Baggerkette f.; Eimerkette f. ‖ Becherkette f.; Förderkette f. ‖ chaîne m. ou transporteur m. à godets; patenôtre f.; convoyeur m. / ~ chain conveyor ‖ Becherkettenförderer

m. ‖ transporteur m. *ou* élévateur m. à chaîne à godets; convoyeur m.; chapelet m.; noria f. / shaft ~ conveyance ‖ Schachtgefäßförderung f. ‖ transport m. de cages de montée. / ~ conveyor ‖ Kettenförderer m.; Becherwerk n. ‖ transporteur m. à chaîne *ou* à godets. / ~ crane ‖ Greiferkran m. ‖ grue f. à benne dragueuse *ou* à pelle automatique. / distance of ~s ‖ Becherabstand m. ‖ écartement m. des godets. / ~ dredger ‖ Eimerkettenbagger m.; Eimerbagger m. ‖ drague f. *ou* dragueur m. à godets.

bucket elevator ‖ Becherwerk n.; Paternosterwerk n. ‖ patenôtre f.; transporteur m. *ou* élévateur m. à augets *ou* à godets; chapelet m.; noria f. / inclined ~ ‖ Schrägbecherwerk m. ‖ noria f. inclinée. / vertical ~ ‖ Becherwerk n. für senkrechte Förderung ‖ élévateur m. à godets verticaux; noria f. verticale.

bucket excavator ‖ Eimerleiterbagger m. ‖ drague f. sèche à godets. / endless chain multi-~ ‖ Eimerkettenbagger m. ‖ excavateur m. à chaîne à godets.

bucket leather ‖ Lederstulpe f. ‖ manchette f. en cuir. / ~ lift (Min) ‖ Saugsatz m. in einem Pumpenschacht ‖ pompe f. aspirante dans une colonne élévatoire. / ~ truck ‖ Klappkübelwagen m. ‖ wagon m. à bennes ouvrantes. / ~ wheel ‖ Schöpfrad n. mit Eimern ‖ roue f. élévatoire; roue f. à baquets *ou* à seaux.

bucking of ores ‖ Scheidung f. der Erze ‖ scheidage m. et cassage m. des minerais.

bucking cloth (Bleach) ‖ Laugentuch n. ‖ charrier m. / ~ ore ‖ Scheiderz n. ‖ minerai m. riche; minerai de scheidage.

buckle, to ~ on ‖ anschnallen ‖ boucler; attacher.

buckle ‖ Schnalle f. ‖ boucle f. / closing ~ ‖ Verschlußschnalle f. ‖ boucle-fermoir m. / ~ of spring ‖ Federbund m. ‖ bride f. de ressort. / ~ of strap ‖ Riemenschnalle f. ‖ boucle f. de courroie. / ~ for trousers ‖ Hosenschnalle f. ‖ boucle f. de pantalons. / turn ~ ‖ Spannschloß n. ‖ tendeur m. / ~ for waistcoats ‖ Westenschnalle f. ‖ boucle f. de gilets.

buckled plate ‖ Buckelblech n. ‖ tôle f. bombée.

buckle maker ‖ Schnallenmacher m. ‖ fabricant m. de boucles.

buckles pl. and small fittings pl. ‖ Nadlerwaren pl. ‖ aiguilles fpl.; épingles fpl.

buckling ‖ Knickung f. ‖ flambage m. / ~ (Acc) ‖ Krümmen n.; Verziehen n. ‖ gondolement m.; distorsion f. / ~ load ‖ Knicklast f. ‖ charge f. au flambage. / ~ strength ‖ Knickfestigkeit f. ‖ résistance f. au flambage. / ~ stress ‖ Knickspannung f.; Knickbeanspruchung f. ‖ tension f. de flambage; effort m. de compression axiale. / ~ test machine ‖ Knickprüfmaschine f. ‖ machine f. à essayer les matériaux au flambage.

buckram (Weav) ‖ Steifleinen n.; Schetterleinen n.; Starrleinen n. ‖ bougran m.

buckshot ‖ Rehposten mpl.; Posten; Postenschuß m. ‖ chevrotines fpl.; postes fpl.

buckskin ‖ Damhirschfell n. ‖ peau f. de daim. / ~ (Weav) ‖ Buckskin m. ‖ buckskin m.; cuir m. de laine. / ~ weaving mill ‖ Buckskinweberei f. ‖ tissage m. de buckskin.

buckthorn ‖ Kreuzdornholz n.; Wegdornholz n. ‖ nerprun m.

buckthorn berry ‖ Kreuzdornbeere f. ‖ baie f. de nerprun. / extract of buckthorn berries ‖ Kreuzbeerextrakt m. ‖ extrait m. de baies de nerprun.

buckwheat ‖ Buchweizen m. ‖ blé m. noir *ou* sarrasin; sarrazin m. / ~ seed *see* buckwheat.

bud, to (Gard) ‖ okulieren ‖ écussonner.; greffer.

bud ‖ Knospe f. ‖ bouton m.

budding ‖ Blühen n. ‖ floraison f. / ~ knife ‖ Pfropfmesser n.; Okuliermesser n. ‖ écussonnoir m.; greffoir m.; couteau m. à greffer.

buddle ‖ Herd m. ‖ table f.; buddle. / round ~ ‖ Kegelherd m.; Rundherd m. ‖ table f. conique. / square ~ (Min) ‖ Schlämmgraben m. ‖ caisse f. *ou* table f. allemande; caisson m.; table f. servante au lavage des sables. / ~ for stamped ore *see* square ~.

buddling of ores ‖ Schlämmung f. der Erze ‖ lavage m. des minerais.

budget ‖ Budget n. ‖ budget m.

bud-shaped ‖ gemmenartig ‖ gemmiforme.

buff, to ‖ polieren ‖ polir; aviver.

buff *see* buffalo.

buffalo ‖ Büffel m. ‖ buffle m. / ~ (Hide) ‖ Büffelhaut f. ‖ peau f. de buffle. / ~ (Leather) ‖ Büffelleder n. ‖ cuir m. de buffle. ~ horn ‖ Büffelhorn n. ‖ corne f. de buffle. / ~ leather *see* buffalo (Leather). / ~ strip ‖ Büffellederriemen m. ‖ lanière f. en cuir de buffle. / worker in ~ ‖ Büffellederarbeiter m. ‖ bufifletier m.

buffer ‖ Puffer m. ‖ tampon m.; tampon m. de choc; buttoir m.; amortisseur m. / flat iron ~ ‖ Flacheisenpuffer m. ‖ tampon m. en fer plat. / india-rubber ~ ‖ Kautschukpuffer m. ‖ tampon m. à rondelles de caoutchouc. / movable ~ ‖ beweglicher Prellbock m. ‖ heurtoir m. mobile. / rigid ~ ‖ fester Puffer m. ‖ tampon m. rigide *ou* fixe. / round plate ~ ‖ bogenförmiger Puffer m. ‖ tampon m. à plaque à surface bombée. / rubber ~ ‖ Gummipuffer m. ‖ tampon m. de caoutchouc. / side ~ ‖ Seitenpuffer m. ‖ tampon m. latéral. / spring ~ ‖ Federpuffer m. ‖ tampon m. à ressort. / stationary ~ ‖ Prellbock m. ‖ heurtoir m. / wooden ~ ‖ Holzpuffer m. ‖ tampon m. en bois.

buffer action ‖ Pufferwirkung f. ‖ action f. amortisseuse.

buffer bar ‖ Stoßfänger m.; Stoßstange f. ‖ (tige f.) pare-chocs m. / squared shank of the ~ ‖ vierkantige Pufferstange f. ‖ tige f. carrée de tampon.

buffer battery ‖ Pufferbatterie f. ‖ batterie f. tampon. / system of ~ ‖ Pufferbetrieb m. ‖ système m. de batterie-tampon.

buffer block ‖ Abstützblock m. ‖ support m. d'entretoise. / ~ casing ‖ Pufferhülse f.; Pufferkreuz n.; Puffergehäuse n. ‖ boisseau m. de tampon. / ~ casing press ‖ Hülsenpufferpresse f. ‖ presse f. pour fabrication des tampons à boisseau. / ~ cross-beam ‖ Pufferquerriegel m. ‖ entretoise f. pour tampons. / ~ disc ‖ Pufferteller m.; Pufferscheibe f. ‖ disque m. de tampon. / ~ gear ‖ Stoßvorrichtung f. ‖ appareil m. de choc. / ~ head ‖ Pufferkopf m. ‖ tête f. de tampon. / ~ piston ‖ Pufferkolben m. ‖ contrepiston m. / ~ plate ‖ Pufferplatte f. ‖ plaque f. de tampon. / ~ plunger ‖ Pufferstange f. ‖ tige f. de tampon. / ~ shank *see* ~ plunger.

buffer spring (Railw) ‖ Stoßfeder f. ‖ ressort-amortisseur m. / ~ ‖ Pufferfeder f. ‖ ressort m. de tampon. / iron-disk of the ~ ‖ eiserne Pufferfederscheibe f. ‖ plaque f. en fer pour le ressort de tampon.

buffer stop (Railw) ‖ Prellbock m. ‖ buttoir m.; heurtoir m. / ~ (Iron constr) ‖ Prellträger m. ‖ poutre f. à taquet.

buffet ‖ Schenktisch m. ‖ buffet m.

buffing wheel ‖ Polierscheibe f. ‖ aviveuse f.; disque m. à polir.

buff leather ‖ Polierleder n. ‖ cuir m. à polir. / ~ wheel ‖ Schwabbelrad n. ‖ disque m. à multiple épaisseurs de cuir *ou* de drap.

buggy man ‖ Fördermann m. ‖ rouleur m. de wagonnets.

bugle ‖ Signalhorn n. ‖ clairon m. / ~ (Glass) ‖ Glaskoralle f. ‖ grain m. de verre noir.

buhl saw ‖ Laubsäge f.; kleine Schweifsäge f. ‖ scie f. à contourner *ou* à vider *ou* à découper *ou* à chantourner; scie f. à marqueterie *ou* d'horloger.

build, to ‖ erbauen; aufführen ‖ bâtir; construire; édifier. / ~ (Mas) ‖ maçonner. / ~ with baking ‖ mit Füllsteinen mpl. mauern ‖ bloquer. / ~-in ‖ einbauen ‖ monter. / ~ in day work ‖ in Regie f. bauen ‖ travailler en régie f. / ~ with loam and straw ‖ mit Lehm m. und Stroh n. mauern; wellern ‖ torcher; bousiller. / ~ slovenly ‖ fluchtlos bauen ‖ bâtir par épaulées. / ~-up ‖ aufbauen ‖ construire.

builder ‖ Erbauer m.; Konstruktör m.: Architekt m.; Baumeister m.; Bauunternehmer m. ‖ constructeur m.; architecte m.; entrepreneur m. en bâtiments. / iron structure ~ ‖ Eisenkonstruktör m. ‖ charpentier m. en fer. / ~ of sod work ‖ Rasenleger m. ‖ gazonneur m.

builder's certificate (Shipb) ‖ Beilbrief m. ‖ certificat m. de construction. / ~ hardware ‖ Bauutensilien pl. ‖ quincaillerie f. de bâtiment. / ~ requisites pl. ‖ Baubedarfsartikel mpl. ‖ fournitures fpl. pour constructions. / ~ trade ‖ Bauhandwerk n. ‖ métier m. de constructeur. / ~ winch or windlass ‖ Bauwinde f. ‖ treuil m. pour constructions en bâtiment.

builder-up (Min) ‖ Versatzarbeiter m.; remblayeur m.; restapleur m.; releveur m. de terres.

building ‖ Bau m.; Gebäude n.; Bauwerk n. ‖ bâtiment m.; édifice m.; construction f. / additional ~ ‖ Nebengebäude n.; Seitengebäude n.; Nebenbau m.; Seitenbau m. ‖ bâtiment m. accessoire *ou* additionnel; annexe m. / annexed ~ *see* additional ~. / ~ of automobile roads ‖ Automobilstraßenbau m. ‖ construction f. de routes pour automobiles. / ~ in day work ‖ Regiearbeit f. ‖ travaux mpl. par régie. / to demolish a ~ *see* to pull down a ~. / ~ above ground ‖ Hochbau m.; Oberbau m. ‖ construction f. au-dessus du sol; superstructure f. / iron ~ ‖ Eisenbauwerk n.; Eisenkonstruktion f. ‖ charpente f. en fer. / ~ of locomotives ‖ Lokomotivbau m. ‖ construction f. de locomotives. / ~ of motor cars ‖ Kraftwagenbau m. ‖ construction f. d'automobiles. / new ~ ‖ Neubau m. ‖ nouvelle construction f. / to pull down a ~ ‖ ein Gebäude n. abbrechen *oder* abreißen

|| démolir *ou* desceller un bâtiment. / square-framed ~ (Carp) || Riegelwerk n. || clayonnage m. *ou* colombage m. de charpente; cloisonnage m. de bois. / two-ailed ~ || Gebäude n. mit zwei Flügeln || bâtiment m. double *ou* à deux ailes. / ~ in wood || Holzbau m. || construction f. en bois.

building berth || Helling f. || cale f. / ~ made of concrete || in Beton aufgeführte Helling f. || cale f. construite en béton. / ~ for large vessels || Großhelling f. || cale f. pour grands bâtiments. / roofed-over ~ || überdachte Helling f. || cale f. couverte.

building box || Baukasten m. || boîte f. de construction. / ~ company || Baugesellschaft f. || société f. de construction. / ~ contract (Agreement) || Bauvertrag m. || contrat m. d'entreprise. / ~ contract (Piece work) || Bauakkord m. || marché m. d'ouvrage. / ~ contractor || Bauunternehmer m. || entrepreneur m. de bâtiments. / ~ crane || Baukran m.; Hochbaukran m. || grue f. de chantier *ou* (pour travaux) de construction. / ~ elevator || Bauaufzug m. || engin m. de levage; élévateur m. *ou* monte-charge m. pour constructions. / ~ enterprise || Bauunternehmung f. || entreprise f. de constructions. / ~ fittings pl. || Baubeschläge mpl. || ferrures fpl. de bâtiment; garniture f. de construction. / ~ garages || Garagenbau m. || construction f. de garages. / ~ ground || Baugrund m. || terrain m. à bâtir; sol m. de fondation. / ~ implements pl. || Baugerätschaften fpl.; Baugeräte npl. || ustensiles mpl. *ou* équipage m. de construction. / ~ iron || Baueisen n. || fer m. de construction. / ~ joiner || Bautischler m. || menuisier m. de bâtiment. / jolting of ~ || Gebäudeerschütterung f. || vibration f. *ou* trépidation f. au bâtiment. / ~ line || Bauflucht f. || alignement m. de bâtiments. / ~ machine || Baumaschine f. || machine f. de construction. / ~ manager || Bauführer m. || chef m. de chantier. / ~ material || Baustoff m.; Baumaterial n. || matière f. *ou* matériel m. de construction. / ~ material for overground construction || Baustoffe mpl. für Hochbauten || matériaux mpl. pour constructions au-dessus du sol. / ~ office || Baubüro n. || bureau m. d'ingénieur. / to desist from projected ~ operation || eine geplante Bebauung f. unterlassen || renoncer à des constructions projetées. / ornament for ~s || Baubeschlag m. || ferrure f. de bâtiment. / ~ permission || Baubewilligung f. || permission f. à bâtir. / plan for a ~ || Bauriß m. || plan m. d'un bâtiment. / ~ pump || Baupumpe f. || pompe f. de construction *ou* d'épuisement. / ~ regulations pl. || Bauvorschriften fpl. || instructions fpl. pour la construction. / shock to ~ *see* jolting of ~. / to lose its value as a ~ site / die Eigenschaft als Bauplatz verlieren || perdre sa qualité de terrain à bâtir. / ~ slip *see* building berth.

building stone || Mauerstein m.; Baustein m.; Quader m. || pierre f. à bâtir *ou* de construction *ou* de taille; brique f. de construction. / ~ squared by splitting *or* with the pointed chisel || durch Spalten *oder* mit dem Spitzmeißel vierkantig zugerichteter Baustein m. || pierre f. de construction équarri par clivage *ou* à la

pointe. / unhewn ~ || roher Baustein m. || pierre f. de construction brute.

building stone hewing || Bausteinbehauerei f. || taillerie f. de pierres à bâtir. / ~ quarry || Werksteinbruch m. || carrière f. de pierre à bâtir.

building, tax on ~ || Gebäudesteuer f. || impôt m. sur la propriété bâtie. / ~ timber || Bauholz n. || bois m. de charpente *ou* de construction.

building trade || Baugewerbe n. || industrie f. des bâtiments. / drying oven for the ~ || Trockenofen m. für Bauten; Bauaustrockner m. || four m. à sécher les bâtiments; braséro m.

building-up (Chem) || Synthese f. || synthèse f.

building work || Bauarbeit f. || travail m. de bâtiment.

building yard / Baustelle f. || pied m. d'œuvre. / ~ for motor boats || Motorbootwerft f. || chantier m. de construction de bateaux automobiles.

built by ... || gebaut von ... || construit par...; fabriqué par...; monté par ....

built-in filter || eingebautes Filter n. || filtre m. monté.

built-up || zusammengesetzt || monté; rapporté. / ~ crossing piece || Schienenkreuzungsstück n. || pièce f. de croisement composée de rails. / ~ frog || Schienenherzstück n. || cœur m. composé de rails. / ~ section || zusammengesetzter Träger m. || poutre f. composée.

bulb (Botany) || Knolle f.; Zwiebel f. || bulbe f. / ~ (Chem) || Kolben m.; Birne f. || ampoule f. / ~ (Electr) || Glühbirne f.; Birne f.; Glühlampe f. || ampoule f. (électrique); lampe f. incandescente. / glass ~ || Glasbirne f. || ampoule f. en verre. / pocket lamp ~ || Taschenlampenbirne f. || ampoule f. de lampe de poche. / ~ of a thermometer || Kugel f. eines Thermometers || cuvette f. d'un thermomètre.

bulb angle || Wulsteisen n. || fer m. à boudin. / ~ angle iron || Wulstwinkeleisen n. || cornière f. à boudin en fer. / ~ barometer || Gefäßbarometer n. || baromètre m. à cuvette. / ~ grower || Blumenzwiebelzüchter m. || jardinier m. producteur d'oignons à fleurs. / ~ iron || Wulsteisen n. || fer m. à boudin; barre f. en fer à boudin.

bulge, to (Mas) || ausbauchen || (se) bomber; forjeter.

bulge of a cask || Bauch m. eines Fasses || bouge m. d'un fût. / having a little ~ || leicht gekrümmt || bombé.

bulged (Pulley) || ballig gedreht || bombé.

bulge electrode || Bauchelektrode f. || électrode f. ventrale.

bulging device || Drückvorrichtung f. || dispositif à repousser les métaux. / ~ lathe || Drückbank f. || tour m. à repousser les métaux.

bulk || Größe f.; Masse f.; Volumen n. || volume m. / ~ (Mar) || Schiffsladung f.; cargaison f.; chargement m. / to break the ~ || mit dem Löschen n. (eines Schiffes) beginnen || entrer en déchargement.

bulk articles pl. || Massenartikel mpl. || articles mpl. faits en masse. / ~ beer || Faßbier n. || bière f. en fût. / ~ goods pl. || Massengüter npl. || marchandises fpl. en grandes masses. / ~ goods pl. (Mar) || Schüttgut n.; Stürzgut n. || marchandises fpl. en vrac.

bulkhead (Shipb) || Schott n. || cloison f.; cloisonnage m.; entourage m.; séparation f. / armour ~ || Panzerschott n. || cloison cuirassée. / batten and space ~ || Gitterschott n.; Lattenschott n.; Traljenschott n. || cloison f. à claire-voie *ou* à jour. / collision ~ || Kollisionsschott n. || cloison f. d'abordage. / cross ~ || Querschott n. || cloison f. transversale. / fireproof ~ || Brandschott n.; Feuerschott n.; || cloison f. pare-feu *ou* à localiser l'incendie. / longitudinal ~ || Längsschott n. || cloison f. longitudinale. / ~ of the manger || Klüsenschott n. || cloison f. de la gatte. / middle ~ || Gebeling f.; Längsschott zur Verhütung des Übergehens der Ladung ||cloison f. longitudinale. / partial ~ || Halbschott n. || cloison f. intermédiaire *ou* partielle. / ~ of the screw shaft || Stopfbüchsenschott n. || cloison f. au presse-étoupe du manchon de l'arbre de l'hélice. / temporary ~ || fliegendes Schott n. || cloison f. temporaire *ou* volante. / transversal *or* transverse ~ || Querschott n. || cloison f. transversale. / watertight ~ || wasserdichtes Schott n. || cloison f. étanche. / wing-passage ~ || Wallgangschott n. || cloison f. latérale.

bulkhead closing device || Schotten(tür)-schließvorrichtung f. || dispositif m. de fermeture pour cloisons étanches. / ~ door || Schott(en)tür f.; Schott(en)verschluß m. || porte f. de cloison. / ~ frame || Schottspant n. || cadre m. de cloison (étanche). / signal installation for ~s || Schottensignalanlage f. || installation f. de signal d'alarme pour la fermeture des portes de cloisons.

bulk load *see* bulk goods (Mar). / ~ production || Massenfertigung f.; Massenerzeugung f. || production f. en masse. / ~ tariff || Bauschtarif m. || tarif m. à forfait.

bulky || sperrig || encombrant. / ~ goods pl. (Railw) || sperrige Waren fpl; Sperrgut n. || pièces fpl. encombrantes *ou* volumineuses *ou* de dimensions démesurées. / ~ test-piece || sperrige Probe f. || spécimen m. de pièce encombrante.

bulldog (Met) || Bulldoggschlacke f.; Saigerschlacke f.; Puddelschlacke f. || bulldog m.; scorie f. de puddlage.

bullet || Kugel f. || balle f. / boat-tail ~ || Stromliniengeschoß n.; Geschoß n. mit verjüngtem Bodenteil || balle f. fuselée; obus m. à culot rétreint. / lead ~ || Bleikugel f. || balle f. de plomb. / tracer ~ || Rauchspurgeschoß n. || projectile m. traceur.

bullet mould || Kugelform f. || moule m. à balles.

bullet-proof || kugelfest; schußsicher || à l'épreuve f. des balles; invulnérable par des balles. / ~ plating || gewehrschußsicheres Blech n. || tôle f. à l'épreuve des balles *ou* du feu de fusil. / ~ shield || gewehrschußsicherer Schild m. || bouclier m. à l'épreuve des balles de fusil.

bull-headed rail || Doppelkopfschiene f. || rail m. à double champignon.

bullion (Glassm) || Ochsenauge n. eines Mondglases || boudine f.; œil m. de bœuf; noyau m. central. / ~ (Wiredr) || Kantille f.; Bouillon m. || cannetille f.; bouillon m. / ~ scale (Assay) || Kornwage f. || balance f. d'essai; bouillonscale m.

bull's-eye (Glassm) ‖ Butzen m. ‖ nœud m. d'un rond de verre. / ~ (Mar) ‖ hölzerne Kausch f. ‖ cosse f. de bois. / ~ (Shipb) ‖ Ochsenauge n.; Bullauge n. ‖ œil m. de bœuf; hublot m. / ~glass ‖ Butzenscheibe f. ‖ cul m. de bouteille; rond m. de verre.

bulwark (Shipb) ‖ Schanzkleid n. ‖ pavois m. / top-gallant ~ ‖ Oberschanzkleid n. ‖ fargue f.

bump, to ‖ glucksen ‖ soubresauter.

bumper (Railway wagon) see also buffer ‖ Puffer m. ‖ tampon m. / ~ (Railway construction) ‖ Prellbock m. ‖ heurtoir m. / ~ (Shipb) ‖ Eisfender m. ‖ pare-glace m. / leather ~ ‖ Lederpuffer m. ‖ amortisseur m. en cuir. / pneumatic ~ (Aero) ‖ Gummifederer m.; Luftpuffer m. ‖ extenseur m.; tampon m. de nacelle. / ~ of shock ‖ Stoßdämpfer m. ‖ tampon m. ou amortisseur m. de choc.

bumper rod ‖ Stoßstange f. ‖ pare-chocs m.

bumpiness ‖ Böigkeit f. ‖ air m. agité.

bumping bag ‖ Landungspuffer m. ‖ tampon m. ou amortisseur m. d'atterrissage. / ~ post ‖ Prellbock m. ‖ heurtoir m.; buttoir m.

bumpy road ‖ holperiger Weg m. ‖ route f. cahotante.

bunch, to ~ up (Textile) ‖ aufhocken ‖ rebrousser.

bunch ‖ Bündel n.; Büschel n. ‖ faisceau m. / ~ (Glass) ‖ Glasblase f. ‖ bosse f. / ~ of ore (Mine) ‖ Gangstock m. ‖ filon m. en forme d'amas. / ~ of ore (Miner) ‖ Erznest n.; Butzen m.; Putzen m. ‖ nid m. de minerai.

bunching-out (Join) ‖ Ausbauchung f. ‖ bombement m.

bundle ‖ Bündel n. ‖ faisceau m. / ~ (Pap) ‖ doppeltes Ries n. ‖ paquet m. de deux rames. / ~ (Ropem) ‖ Hanfbund n.; Loppe f. ‖ peignon m. / ~ (Trade) ‖ Kollo n.; Stückgut n. ‖ colli(s) m. / ~ of hoops (Coop) ‖ Bund m. Reifen ‖ paquet m. de cercles. / ~ of incoming trunks (Tel) ‖ ankommendes Leitungsbündel n. ‖ groupe f. de lignes d'arrivée. / ~ of outgoing trunks (Tel) ‖ abgehendes Leitungsbündel n. ‖ groupe f. de lignes de départ. / ~ of skeins (Silk) ‖ Bund m. Seide; Docke f. ‖ matteau m.; bouin m. / ~ of steel ‖ Bündel n. Stahl ‖ botte f. d'acier. / ~ of wire ‖ Drahtbund m. ‖ couronne f. de fil. / ~ of wood ‖ Reisigbündel n. ‖ faisceau m. de menu bois.

bundle file ‖ Rundfeile f. ‖ lime f. en paquet. / ~ press ‖ Bündelpresse f.; Garnpresse f.; Packpresse f. ‖ presse f. à empaqueter.

bundler (Spinn) ‖ Falzer m.; Bündler m. ‖ empaqueteur m. ou ployeur m. ou plieur m. de filés; panteur m. / ~ (Roll mill) ‖ Paketbinder m. ‖ botteleur m.

bundling machine, bar iron ‖ Stabeisenbündelmaschine f. ‖ machine f. à mettre les fers en barres en faisceaux. / scrap ~ ‖ Schrottbündelmaschine f. ‖ machine f. à emballer la mitraille.

bundling press ‖ Bündelpresse f.; Garnpresse f.; Packpresse f. ‖ presse f. à botteler ou à empaqueter.

bung, to (Coop) ‖ verspunden ‖ bondonner.

bung ‖ Spund m.; Spundzapfen m. ‖ bondon m.; (bonde f.); tape f. / ~ of a cask ‖ Faßspund m. ‖ bondon m. de fût. / drawing off ~ (Brew) ‖ Abfüllspund m. ‖ bondon m. de soutirage. / ~ of saggers (Porcel) ‖ Kapselstoß m. ‖

pile f. de cassettes. / wooden ~ ‖ Holzspund m. ‖ bondon m. en bois.

bung borer ‖ Spundbohrer m. ‖ bondonnière. / ~ closure ‖ Spundverschluß m. ‖ fermeture f. de bonde ou avec bondon. / ~ extracting ‖ Entspunden n. ‖ extraction f. ou enlèvement m. de bondons. / ~ extracting machine ‖ Entspundungsmaschine f. ‖ machine f. à enlever les bondons. / ~ extractor ‖ Entspundungsapparat m. ‖ appareil m. à enlever les bondons; tire-bonde m. / ~ flap ‖ Spundlappen m. ‖ pièce f. de drap pour bondons. / ~ fraiser see ~ milling machine.

bung hole ‖ Spund m.; Spundloch n.; Zapfloch n. ‖ bonde f.; (bondon m.); trou m. de bondon ou de mise en perce. / ~ boring machine ‖ Spundlochbohrmaschine f. ‖ machine f. à percer les bondes. / ~ fermentation (Brew) ‖ Spundguhr f. ‖ fermentation f. par la bonde.

bungle, to ‖ verpfuschen ‖ gâter; gâcher. / ~ ‖ pfuschen ‖ bousiller; tricher.

bungler ‖ Stümper m. ‖ bousilleur m. / ~ (Print) ‖ Papiersudler m. ‖ mâchurat m.

bung lifter ‖ Faßentspunder m. ‖ tire-bonde m. / ~ milling machine ‖ Spundfräsmaschine f. ‖ fraiseuse f. ou fraisoir m. pour bondons. / ~ planks pl. ‖ Spundwand f. ‖ file f. de palplanches; cloison f. en palplanches. / ~ tin ‖ Spundblech n. ‖ plaque f. métallique pour bondes. / ~ turner ‖ Spunddreher m. ‖ tourneur m. de bondons.

bunker ‖ Bunker m. ‖ soute f.; trémie f. / coal ~ ‖ Kohlenbunker m. ‖ soute f. à charbon. / cross ~ ‖ Querkohlenbunker m. ‖ soute f. à charbon transversale. / elevated ~ ‖ Hochbunker m. ‖ soute f. surélevée. / loading ~ ‖ Verladebunker m. ‖ trémie f. de chargement. / overhead storage ~ above boilers ‖ über den Kesseln gelagerter Bunker m. ‖ magasin m. de charbon à trémies au-dessus des chaudières. / trimmed in ~s pl. ‖ in den Bunkern gestaut ‖ arrimé dans les soutes fpl. / wing ~ ‖ Seitenkohlenbunker m. ‖ soute f. latérale.

bunker capacity ‖ Bunkerrauminhalt m. ‖ capacité f. de soute. / ~ coal ‖ Bunkerkohle f. ‖ charbon m. de soute.

bunkered ‖ gebunkert ‖ mis en soute f.

bunkering ‖ bunkernd ‖ mettant en soute f.

bunker scale ‖ Bunkerwage f. ‖ balance f. pour soutes.

bunny of ore ‖ Putzen m.; Butzen m.; Erznest n. ‖ nid m. de minerai.

Bunsen burner ‖ Bunsenbrenner m. ‖ brûleur m. (de) Bunsen. / acetylene ~ ‖ Azetylenbunsenbrenner m. ‖ brûleur m. de Bunsen à l'acétylène with yellow coloured flame ‖ Bunsenbrenner m. mit gelber Salzflamme ‖ bec m. Bunsen avec une flamme sodique jaune. / ~ stand ‖ Bunsenbrennerstativ n. ‖ support m. pour becs Bunsen.

Bunsen cell ‖ Bunsenelement n. ‖ élément m. Bunsen. / ~ flame ‖ Bunsenflamme f. ‖ flamme f. d'un bec Bunsen. / ~ flame yellow coloured with common salt ‖ mit Kochsalz gelb gefärbte Bunsenflamme f. ‖ flamme f. d'un bec Bunsen colorée en jaune par du sel de cuisine.

bunt of a sail ‖ Buk m. eines festgemachten Segels ‖ fonds mpl. d'une voile.

bunting ‖ Kongreßstoff m. ‖ étamine f.

bunt line (Mar) ‖ Gording f. ‖ cargue f.

buoy, to ‖ das Fahrwasser abbaken; betonnen ‖ baliser; metéer des balises.

buoy ‖ Ankerboje f.; Boje f.; Seetonne f. ‖ bouée f.; amarque f. / anchor ~ ‖ Ankerboje f. ‖ bouée f. d'ancre. / bell~ ‖ Glockenboje f. ‖ bouée f. sonnante ou à sonnerie. / fairway ~ ‖ Ansegelungsboje f. ‖ bouée f. d'atterrage. / life ~ ‖ Rettungsboje f. ‖ bouée f. de sauvetage. / light ~ ‖ Leuchtboje f.; Leuchttonne f. ‖ bouée f. lumineuse ou éclairée ou porte-feu. / lighted ~ see light ~. / sounding ~ see bell ~.

buoyage ‖ Betonnung f. ‖ balisage m.

buoyancy ‖ Schwimmfähigkeit f. ‖ flottabilité f. / ~ (Aero) ‖ statische Steigkraft f. ‖ force f. ascensionnelle statique. / defect of ~ ‖ statische Sinkkraft f. ‖ force f. descensionnelle statique.

buoyancy proceeding ‖ Schwimmaufbereitungsverfahren n. ‖ procédé m. de flottaison.

buoyant ‖ diastasereich ‖ riche en diastase f. / ~ gas ‖ Traggas n. ‖ gaz m. de gonflement ou de sustentation. / ~ power of a bridge ‖ Tragfähigkeit f. einer Brücke ‖ force f. d'un pont.

bur see burr.

burbling point ‖ Unstetigkeitsstelle f. point m. tourbillonnaire ou de discontinuité.

burden, to ~ the trade balance ‖ die Handelsbilanz f. belasten ‖ grever la balance commerciale.

burden ‖ Ladung f.; Last f.; Traglast f. charge f. / ~ (Met) ‖ Charge f.; Gicht f.; Satz m. ‖ charge f. du fourneau. / ~ (Ship) ‖ Lastigkeit f.; Tragfähigkeit f. ‖ tonnage m.; port m. / additional ~ (Steam) ‖ Überdruck m. ‖ surcharge m.

burdock oil ‖ Klettenwurzelöl n. ‖ huile f. de racines de bardane. / ~ root ‖ Klettenwurzel f. ‖ racine f. de bardane.

burette (Chem) ‖ Bürette f. ‖ burette f. / ~ pincer ‖ Bürettenklemme f. ‖ pince f. à burettes. / stand for ~s ‖ Bürettenhalter m. ‖ support m. à burettes.

burglar alarm ‖ Einbrecherglocke f. ‖ sonnerie f. d'alarme contre l'effraction. / electric ~ ‖ Türkontakt m.; Fensterkontakt m. ‖ contact m. va-et-vient.

burglar-proof window ‖ einbruchsicheres Fenster n. ‖ croisée f. à l'abri de l'effraction.

burglary policy ‖ Einbruchspolice f. ‖ police f. d'assurance contre le vol.

burgomaster ‖ Bürgermeister m. ‖ bourgmestre m.

burial ‖ Begräbnis n.; Beerdigung f. ‖ enterrement m.

burin ‖ Stichel m.; Grabstichel m. ‖ burin m. (de graveur).

burl, to ‖ noppen; belesen ‖ épincer; épinceler; épinceter; éplucher.

burlap ‖ Packleinwand f. ‖ toile f. d'emballage.

burler ‖ Nopper m. ‖ épinceteur m.

burling (Clothm) ‖ Noppen n.; Belesen n. ‖ épinçage m.; épincetage m. / ~ (Weav) ‖ Säubern n. ‖ énouage m. / cloth ~ ‖ Noppen n. des Tuches ‖ épetissage m. ou nopage m. de drap.

burling frame ‖ Nopprahmen m. ‖ pupitre m. à épinceter. / ~ iron ‖ Noppeisen n. ‖ épincette f.; brucelles fpl. / ~ machine ‖ Noppmaschine f.; Zupfmaschine f. ‖ épinceteuse f. mécanique.

burn, to ‖ brennen; verbrennen ‖ brûler. / ~ (Slowly) ‖ brennen; ‖ schwelen ‖ brûler

lentement *ou* sans flamme. / ~ a blue light (Nav) ‖ ein Blaufeuer abbrennen ‖ brûler un feu de Bengale. / ~ charcoal ‖ kohlen; verkohlen ‖ carboniser le bois. / ~ down ‖ vollständig verbrennen; niederbrennen ‖ brûler entièrement. / ~ lime ‖ den Kalk m. brennen ‖ cuire la chaux. / ~ on ‖ löten ‖ souder. / ~ out a cask (Coop) ‖ ein Faß ausfeuern ‖ chauffer une futaille montée. / ~ without flame ‖ schwelen ‖ brûler sans flamme.

burn ‖ Brandwunde f. ‖ brûlure f. / ointment for ~s and scalds ‖ Brandsalbe f. ‖ onguent m. contre les brûlures.

burnable residue ‖ brennbarer Rückstand m. ‖ résidu m. combustible.

burned-off ‖ vollständig gar ‖ parfaitement cuit *ou* distillé.

burner ‖ Brenner m. ‖ brûleur m.; bec m.; chalumeau m. / ~ (Charc) ‖ Köhler m. ‖ charbonnier m. / acetylene ~ ‖ Azetylenbrenner m. ‖ bec m. à acétylène. / autogenous fusing ~ ‖ Autogenschneidbrenner m. ‖ brûleur m. à coupure autogène. / autogenous welding ~ ‖ Autogenschweißbrenner m. ‖ brûleur m. à soudure autogène. / Bunsen ~ ‖ Bunsenbrenner m. ‖ brûleur m. Bunsen. / cutting ~ ‖ Schneidbrenner m. ‖ brûleur m. à découper. / ~ fitted with a fan blower mixer ‖ Brenner m. mit Schleuderradmischer ‖ brûleur m. avec mélangeur centrifuge. / gas ~ ‖ Gasbrenner m. ‖ brûleur m. pour chauffage à gaz; brûleur m. à gaz. / high-pressure gas ~ ‖ Preßgasbrenner m. ‖ brûleur m. pour gaz surpressé. / ~ fitted with mixer with orifice for the entering medium ‖ Brenner m. mit Strahldüsenmischer ‖ brûleur m. avec mélangeur à buse à jet. / oil ~ ‖ Ölbrenner m. ‖ brûleur m. à huile. / oil-gas ~ ‖ Ölgasbrenner m. ‖ brûleur m. à gaz d'huile. / oil lamp ~ ‖ Petroleumlampenbrenner m.‖ bec m. pour lampes à pétrole. / petroleum ~ ‖ Petroleumbrenner m. ‖ bec m. au pétrole. / ~ provided with a pilot burner ‖ Sparbrenner m. ‖ brûleur m. à veilleuse. / ~ provided with a ring ‖ Kronenbrenner m. ‖ brûleur m. à couronne. / round ~ ‖ Rundbrenner m.‖ bec m. rond. / split ~ ‖ Schlitzbrenner m.; Schnittbrenner m. ‖ bec m. fendu. / welding ~ ‖ Schweißbrenner m. ‖ brûleur m. à souder.

burner head ‖ Brennerdüse f. ‖ buse f. du brûleur.

burning ‖ Brand m. ‖ combustion f. / ~ (Ores) ‖ Rösten n.; Zubrennen n.; Brennen n. ‖ rôtissage m.; grillage m.; calcination f.; rouissage m. / ~ (Pott) ‖ Brennen n. ‖ cuisson f. / ~ of the boiler plate ‖ Überhitzung f. der Kesselwand ‖ coup m. de feu.

burning dust ‖ Kehrichtverbrennung f.; incinération f. d'ordures. / ~ fuze (Min) ‖ Minenzünder m.; Sprengkapsel f. ‖ détonateur m. / ~ glass ‖ Brennglas n. ‖ lentille f. convergente; verre m. ardent. / ~-hours discount ‖ Brennstundenrabatt m. ‖ rabais m. en rapport avec le nombre d'heures d'utilisation. / ~-in (Glassm) ‖ Einbrennen n. ‖ cuisson f. / ~ lamp for shoemakers ‖ Schuhmacherbrennzeuglampe f. ‖ lampe f. de cordonnier. / ~ machine ‖ Brennmaschine f. ‖ machine f. à marquer au fer rouge. / ~-off ‖ plötzliche Verbrennung f. ‖ déflagration f. / ~-out of the transformer

‖ Durchschlagen n. des Transformators ‖ claquage m. du transformateur. / ~ stamp ‖ Brennstempel m. ‖ fer m. à marquer *ou* marquoir m. au feu. / ~ stamp for wood printing ‖ Brennstempel m. für Holzdruck ‖ fer m. à marquer les bois au feu / ~ tongs pl. ‖ Lötzange f. ‖ pincette f. à souder. / ~ tools pl. for saddlers and shoemakers ‖ Brennzeug n. für Sattler und Schuhmacher ‖ fer m. à marquer au feu pour selliers et cordonniers.

burnish, to ‖ schleifen; polieren ‖ polir; adoucir; lustrer. / ~ (Gild) ‖ mit Blutstein polieren ‖ brunir. / ~ (Glassm) ‖ die Spiegelfolie polieren ‖ aviver la feuille d'étain.

burnisher ‖ Polierstein m. ‖ agate f.; brunissoir m.; pierre f. à brunir; pierre f. sanguine. / ~ (Curr) ‖ Gerbeisen n. ‖ écharnoir m.; dragoir m. / ~ (Engr) ‖ Mattpunze f. ‖ matoir m. / ~ (Watchm) ‖ Polierfeile f. ‖ brunissoir m.; carrelette f.

burnishing ‖ Glätten n.; Polieren n. mit dem Polierstahl *oder* Gerbstahl ‖ brunissage m. / ~ (Shaft) ‖ Prägepolieren n. ‖ galetage m. / ~ file ‖ Polierstahl m. ‖ brunissoir m. / ~ mill ‖ Polierstein m.; meule f. polissoire. / ~ stone ‖ Polierstein m. ‖ agate f.; brunissoir m.; pierre f. à brunir; pierre f. sanguine.

burn ointment ‖ Brandsalbe f. ‖ onguent m. contre brûlures. / ~ salve *see* burn ointment

burnt (Chem) (Met) ‖ gebrannt ‖ calciné. / ~ (Met) ‖ faulbrüchig ‖ brûlé. / ~ (Steel) ‖ übergar ‖ brûlé. / ~ black ‖ gebrannter Kienruß m. ‖ noir m. de fumée calciné. / ~ clay ‖ gebrannter Ton m. ‖ argile f. calcinée. / ~ gases pl. ‖ Abzugsgas n. ‖ gaz m. perdu. / the ~ gases pl. escape ‖ die Abgase npl. entweichen ‖ les gaz mpl. de la combustion s'échappent. / ~ gypsum ‖ gebrannter Gips m. ‖ plâtre m. cuit. / ~ insulation ‖ eingefressene Isolation f. ‖ isolement m. corrodé. / ~ lime ‖ Ätzkalk m.; gebrannter Kalk m. ‖ chaux f. cuite *ou* éteinte *ou* hydratée; oxyde m. de calcium. / ~ ore ‖ Kieselabbrand m. ‖ pyrite f. traitée. / ~ spot ‖ Brandstelle f. ‖ brûlure f. / ~ wood ‖ Brandholz n. ‖ bois m. arsin.

bur oil ‖ Klettensamenöl n. ‖ huile f. de bardane.

burr, to ‖ abgraten ‖ ébarber. / ~ (Clothm) ‖ krempeln ‖ chardonner; ratiner.

burr (Mach) ‖ Grat m. ‖ barbe f.; barbure f. / ~ (Found) ‖ Grat m.; Gußnaht f. ‖ bavure f.; ébarbure f. / to clean off ~s pl. ‖ den Grat entfernen; entgraten ‖ ébarber; ébavurer. / ~ made by drill ‖ Bohrgrat m. ‖ barbe f. de forage. / press for taking off the ~ ‖ Abgratpresse f. ‖ presse f. d'ébarbage.

burring willow (Spinn) ‖ Klettenwolf m. ‖ églouteronneuse f.; échardonneuse f. / ~ machine ‖ Abgratmaschine f. ‖ machine f. à ébarber.

burrow, to (Min) ‖ schürfen ‖ faire des recherches; fouiller; creuser.

burrow ‖ Grubenhalde f. ‖ halde f.

burr-picking machine ‖ Wollzupfmaschine f. ‖ machine f. échardonneuse.

burr-removing machine ‖ Abgratmaschine f. ‖ machine f. à ébarber (les bavures) / ébarbeuse f. / nut ~ ‖ Mutternabgratmaschine f. ‖ machine f. à ébarber les écrous.

burst, to ‖ explodieren; platzen; springen ‖ crever; éclater; faire explosion; se briser. / ~ (To split) ‖ aufreißen; rissig werden ‖ se fendiller.

burst ‖ Bruch m.; Riß m. ‖ rupture f. / ~ due to frost ‖ Frostsprengung f. ‖ rupture f. par le gel.

bursting ‖ Springen n.; Zerspringen n. ‖ éclatement m.; casse f. / ~ (Boil) ‖ Platzen n. ‖ explosion f. / ~ effect ‖ Sprengwirkung f. ‖ effet m. d'éclatement *ou* d'explosion. / ~ stress ‖ Bruchfestigkeit f. ‖ résistance f. à la rupture.

burton ‖ Stagtalje f. ‖ bredindin m.

bury, to (Conducts) ‖ in die Erde verlegen ‖ enterrer.

bus ‖ Omnibus m.; Kraftomnibus m.; Autobus m. ‖ omnibus m.; autobus m.; omnibus m. automobile. / double-decked ~ *see* double-deck motor. ~ / double-deck motor ~ ‖ Doppeldeckautobus m.; Decksitz(auto)omnibus m. ‖ omnibus m. *ou* autobus m. à impériale. / double-deck top-covered ~ ‖ geschlossener Doppeldeckautobus m. ‖ autobus m. à impériale américain. / inter-urban ~ ‖ Landomnibus m. ‖ autobus m. inter-urban. / long-distance ~ ‖ Überlandomnibus m. ‖ omnibus m. à grand rayon d'action. / one-man ~ ‖ Ein-Mann-Omnibus m. ‖ omnibus m. avec la conduite et la surveillance par le même employé. / single-deck ~ ‖ Autobus m. ohne Oberdeck ‖ autobus m.

bus bar (Electr) ‖ Sammelschiene f. ‖ barre f. omnibus. / distributing ~ ‖ Verteilungsschiene f. ‖ barre f. omnibus de distribution. / ~ room in the high-tension switch house ‖ Sammelschienenraum m. der Hochspannungsschaltanlage ‖ salle f. des barres du tableau de couplage haute tension.

bush, to ‖ ausbuchsen; füttern ‖ garnir; emboîter; revêtir (d'une douille); mettre une douille. / ~ (Mas) ‖ schellen ‖ boucharder.

bush (Bearing) *see also* bushing ‖ Lagerfutter n.; Lagerschale f.; Lagerbuchse f.; Buchse f.; Metallfutter n. ‖ coussinet m.; coquille f. de palier. / ~ (Pulley) ‖ Buchse f.; Hülse f. ‖ manchon m.; douille f. / ~ (Stuffing box) ‖ Grundbuchse f. ‖ grain m.; bague f. de fond. / ~ (Wheel) ‖ Radbuchse f. ‖ boîte f. de roue. / flanged ~ ‖ Flanschenbuchse f. ‖ douille f. à bride. / guide ~ ‖ Führungsbuchse f. ‖ boîte f. *ou* douille f. de guidage. / loose ~ ‖ Leerlaufbuchse f. ‖ douille f. pour la marche à vide. / ~ of a piston rod ‖ Hülse f. einer Kolbenstangenführung ‖ douille f. de la tige de piston. / spherical ~ ‖ Kugellagergehäuse n.; Kugellagerschale f. ‖ cage f. de roulement à billes; coussinet m. sphérique.

bush bundle ‖ Reisbund n.; Reisigbündel n. ‖ bourrée f.; cotret m.; fagot m.; faisceau m.; fascine f. / casting device for ~es ‖ Lagerschalengießvorrichtung f. ‖ dispositif m. à couler les coussinets.

bushel ‖ Scheffel m. ‖ boisseau m. / ~ maker ‖ Scheffelmacher m. ‖ boisselier m. / ~ making ‖ Weißküferei f. ‖ boissellerie f.

bush grinding machine ‖ Buchsenschleifmaschine f. ‖ machine f. à rectifier les douilles.

bush-hammer, to (Stone) ‖ krönln; stocken ‖ bretteler; bretter.

bushing *see also* bush || Lagerschale f.; Lagerbuchse f.; Buchse f.; Lagerhülse f. || coussinet m.; coquille f. / babbit ~ || Weißmetallagerschale f. || coussinet m. en métal blanc. / connecting rod ~ || Pleuelstangenlager n. || coussinet m. de tête de bielle. / crankshaft bearing ~ || Kurbelwellenlagerschale f. || coussinet m. de palier de vilebrequin. / no-load ~ || Leerlaufbuchse f. || douille f. pour marche à vide. / piston ~ || Kolbenbuchse f. || coussinet m. du piston. / piston pin ~ || Kolbenbolzenbuchse f. || coussinet m. de pied de bielle. / split ~ || geteilte Lagerbuchse f. || coussinet m. fendu *ou* en deux parties.

bushing current transformer || Stützenstromwandler m. || transformateur m. d'intensité type à isolateur pilier.

bush pile || Kantenpfahl m. || pieu m. de haie. / ~ wrench || Ausschraubwerkzeug n. für Spundringe || outillage m. à dévisser les anneaux de bondes métalliques.

busiest hour (Tel) || Hauptverkehrsstunde f. || heure f. la plus chargée.

business || Geschäft n.; Handel m. || affaire f.; opération f. commerciale. / ~ (Profession) || Beruf m. || profession f. / ~ agreed upon || vereinbartes Geschäft n. || affaire f. convenue *ou* conclue. / to close a ~ || ein Geschäft n. abschließen || traiter une affaire. / course of ~ || Geschäftsgang m. || marche f. des affaires. / giving-up ~ || Geschäftsaufgabe f. || cessation f. d'un commerce. / important ~ || wichtiges Geschäft n. || affaire f. considérable. / ~ on the instalment system || Abzahlungsgeschäft n. || maison f. de vente à crédit *ou* à terme; maison f. de vente par acompte. / knowledge of ~ || Geschäftskenntnis f. || connaissance f. des affaires. / local ~ || Platzgeschäft n. || affaire f. de la place. / no ~ || kein Geschäft n. || aucune affaire f.; pas d'affaire f. / opening of a ~ || Geschäftseröffnung f. || ouverture f. d'une maison. / pressure of ~ || Geschäftsdrang m. || presse f. des affaires. / relating to ~ || geschäftlich || d'affaires; pour affaires; commercial. / to retire from ~ || das Geschäft aufgeben || cesser le commerce. / retiring from ~ || Aufgabe f. des Geschäftes || cessation f. du commerce. / scope of ~ || Geschäftskreis m. || sphère f. d'activité. / versed in ~ || geschäftskundig || versé dans les affaires fpl.

business affair || Geschäftsangelegenheit f. || affaire f. commerciale. / ~ books pl. || Geschäftsbücher npl. || livres mpl. de commerce. / ~ clothes pl. || Berufskleidung f. || vêtements mpl. de travail *ou* d'ouvriers. / ~ clothing *see* ~ clothes.

business connexion || Geschäftsbeziehung f. || relation f. d'affaires; relation f. commerciale. / to form ~s pl. || Geschäftsverbindungen fpl. anknüpfen || entamer *ou* nouer *ou* ouvrir des relations fpl. d'affaires. / regular ~ || regelmäßige Geschäftsverbindung f. || relations fpl. suivies avec une maison.

business dealings pl. || Geschäftsverkehr m. || relations fpl. commerciales. / ~ die || Firmenstempel m. || timbre m. de la maison *ou* à firme. / ~ equipment || Geschäftseinrichtung f. || installation f. pour maison de commerce. / ~ hour || Geschäftsstunde f. || heure f. de bureau. / ~ house || Geschäftshaus n. || maison f. / ~ letter ||

Geschäftsbrief m. || lettre f. de commerce *ou* d'affaires. / ~ matter || Geschäftsangelegenheit f. || affaire f. / ~ prospects pl. || Geschäftsaussichten fpl. || possibilités fpl. *ou* attentes fpl. d'affaires.

business relations pl. *see also* business connexion. || Geschäftsverbindung f. || relation f. d'affaires. / to enter into ~ || in Geschäftsverbindung f. treten || entrer en relations fpl. commerciales.

business report || Geschäftsbericht m. || rapport m. de gestion. / ~ station (Tel) || Geschäftsanschluß m. || poste m. d'abonné dans un bureau. / ~ year || Geschäftsjahr n.; Rechnungsjahr n.; Betriebsjahr n. || année f. commerciale; exercice m.

busk || Blankscheit n. || busc m.

buss (Fish) || Heringsbüse f. || buse f.

bust || Büste f. || buste m. / ~ for shop windows || Schaufensterbüste f. || buste f. pour étalages.

bust holder || Büstenhalter m. || soutien-gorge m. / ~ support *see* ~ holder.

busy || tätig || actif. / to be ~ || zu tun haben || avoir de l'occupation f. *ou* à faire. / very ~ || vielbeschäftigt || très occupé.

busy back tone (Tel) || Besetztzeichen n. || signal m. de "pas libre". / ~ hours pl. || verkehrsstarke Zeit f. || heures fpl. de fort trafic. / ~ lamp || Besetztlampe f. || lampe f. de test. / ~ magnet || Besetztmagnet m. || électro m. d'occupation. / ~ period || Hauptgesprächszeit f. || heure f. la plus chargée. / ~ response || Besetztmeldung f. || signal m. d'occupation. / ~ test || Besetztprüfung f. || test m. de ligne occupée. / ~ tone || Besetztzeichen n. || signal m. d'occupation *ou* de „pas libre".

butcher, to || schlachten || abattre; égorger; tuer.

butcher || Schlächter m.; Fleischer m. || boucher m. / ~'s knife || Schlachtmesser n.; Schlächtermesser n. || couteau m. de bouchers. / ~'s saw || Metzgersäge f. || scie f. à bouchers. / ~'s shop || Schlächterei f. || boucherie f. / ~'s shop fitting || Fleischerladeneinrichtung f. || installation f. pour débit de boucheries. / ~'s stall || Fleischbank f. || étal m. pour bouchers. / ~'s stand || Fleischbank f. || étal m. / ~'s tool || Schlächtereiwerkzeug n. || outil m. de boucherie.

butchery || Schlächterei f.; Fleischerei f. || boucherie f. / wholesale ~ || Großschlächterei f. || boucherie f. en gros.

butchery machine || Schlächtereimaschine f. || machine f. de boucherie.

butler's implements pl. || Kellereiartikel mpl. || articles mpl. pour caves.

butment (Build) || Kämpfer m.; Widerlager n. || sommier m.; voussoir m.; aboutissement m.; naissance f. / intermediate ~ || Mittelpfeiler m. || pieddroit m. intermédiaire. / ~ on shore (Bridge) || Landbrücke f. || culée f. d'un pont.

butt (Arm) || Geschoßfang m.; Kugelfang m. || butte f. du polygone. / ~ (Curr) || Pfundleder n.; Schwerleder n. || cuir m. fort *ou* nerveux; gros cuir m. / ~ of a musket || Kolben m. eines Gewehres || crosse f. de fusil. / ~ of plates (Shipb) || Plattenstoß m. || about m. de tôles. / set of ~s for the manufacture of cardboard || Büttengarnitur f. für die Pappenfertigung || garniture f. de tonneaux à cartonneries.

butt end (Shipb) || Plankenende n. || bout m. *ou* tête f. de bordage. / ~ of a tree stem ||

Wurzelende n. eines Baumstammes || gros bout m. *ou* souche f. *ou* pied m. d'un (tronc d')arbre.

butter || Butter f. || beurre m. / ~ of antimony || Antimonbutter f. || beurre m. d'antimoine. / artificial ~ || Kunstbutter f. || beurre m. artificiel. / fresh ~ || frische Butter f. || beurre m. frais. / melted ~ || geschmolzene Butter f. || beurre m. fondu. / rich ~ || Schmierbutter f. || beurre m. gras. / salt ~ || gesalzene Butter f. || beurre m. salé. / salted ~ *see* salt ~. / ~ in tubs || Butter f. in Fässern || beurre m. en fûts. / vegetable ~ || Pflanzenbutter f. || beurre m. végétal.

butter blending machine || Butterknetmaschine f. || malaxeur m. de beurre. / ~ colouring || Butterfarbe f. || colorant m. pour beurre. / ~ and cheese colouring || Butter- und Käsefarbe f. || colorant m. pour beurre et fromage. / ~ cooler || Butterkühler m. || rafraîchissoir m. pour beurre. / ~ cooling room || Butterkühlraum m. || chambre f. froide à beurre. / ~ cutting wire || Butterschneidedraht m. || fil m. à couper le beurre. / ~ fat || Butterfett n. || beurre m. de vache; butyrine f.

butterfly cock || Flügelhahn m. || robinet m. papillon. / ~ damper || Rauchklappe f. || registre m. pivotant. / ~ nut || Flügelmutter f. || écrou m. à oreilles.

buttering machine || Buttermaschine f.; Butterfaß n. || baratteuse f.; baratte f.

butter maker || Buttermann m.; Butterhändler m. || beurrier m.; buturier m. / ~ manufactory || Tafelbutterfabrik f. || beurrerie f. / ~ melter || Buttersieder m. || fondeur m. de beurre.

buttermilk || Buttermilch f. || lait m. battu.

butter mould || Butterform f. || beurre m. / ~ moulder || Butterformerin f. || mouleuse f. de beurre. / ~ refractometer || Butterrefraktometer n. || réfractomètre m. à beurre. / ~ salter || Buttereinsalzer m. || saleur m. de beurre. / ~ salting || Buttereinsalzen n. || salaison f. de beurre. / tub for ~ || Butterfaß n. || tine f. de beurre; tinette f. à beurre.

butt hinge || Scharnier n.; Fischband n.; Einsatzband n. || charnière f.; fiche f. à vase. / ~ howel (Coop) || krumme Dechsel f.; Mollenhaue f. || asseau m. des tonneliers.

butt-joint, to || stumpf (aneinander)fügen; stumpf verbinden || abouter; mater; assembler bout à bout.

butt joint (Carp; Join) || Stoßfuge f.; Hirnfuge f.; Stumpfstoß m.; gerader oder stumpfer Stoß m. || joint m. abouté *ou* d'about; jointure f. à plat *ou* bout à bout; bout-à-bout m. / ~ (Carp; Join) (Joining) || Stoßverbindung f.; Endverbindung f. || assemblage m. à plat (joint) *ou* d'about *ou* bout à bout; aboutement m. / ~ (Rivetting) || Stoß m. mit Lasche; Überlaschung f. || joint m. *ou* jointure f. à couvre-joint. / ~ (Rivetting) (Joining) || Laschenverbindung f.; Laschennietung f.; Überlaschungsnietung f. || assemblage m. *ou* rivure m. à franc-bord *ou* à couvre-joint. / ~ with butt strap *see* ~ (Rivetting). / double-rivetted ~ || Laschennietung f.; Überlaschungsnietung f. || rivure f. à couvre-joint. / double-rivetted ~ || zweireihige Laschennietung f. || rivure f. à couvre-joint à deux rangs. / double-chain-rivetted ~ || zweireihige parallele

Laschennietung f. ‖ rivure f. à couvre-joint à deux rangs parallèles. / double-zigzag-rivetted ~ ‖ zweireihige Zickzack-laschennietung f. ‖ rivure f. à couvre-joint à deux rangs en quinconce. / single-rivetted ~ ‖ einreihige Laschennietung f. ‖ rivure f. à couvre-joint à un rang. / double-row ~ see double-rivetted ~. / single-row ~ see single-rivetted ~. / strap ~ ‖ Laschennietung f.; Überlaschnungsnietung f. ‖ rivure f. à couvre-joint. / double-strap ~ ‖ Doppellaschennietung f.; Nietung f. mit doppelter Laschung ‖ rivure f. à double couvre-joint. / single-strap ~ ‖ einseitige Laschennietung f.; Nietung f. mit einfacher Laschung ‖ rivure f. à couvre-joint simple. / ~ with one weld see single-strap ~. / ~ with two welds see double-strap ~.

butt-jointing see butt joint (Joining).

button ‖ Knopf m. ‖ bouton m. / ~ (Window) ‖ Wirbel m.; Fensterwirbel m. ‖ tourniquet m.; crémone f. / bone ~ ‖ Beinknopf m. ‖ bouton m. en os. / brace ~ ‖ Blechknopf m. ohne Öhr ‖ bouton m. découpé sans queue. / ~ for clothings ‖ Knopf m. für Kleidungsstücke ‖ bouton m. pour vêtements. / ~ of corozo ‖ Steinnußknopf m. ‖ bouton m. en corozo. / covered ~ ‖ überzogener Knopf m. ‖ bouton m. couvert. / ~ covered by hand ‖ überzogener Knopf m. ‖ bouton m. cousu. / ~ covered with cloth ‖ Zeugknopf m. ‖ bouton m. en étoffe. / fashion ~ ‖ Modeknopf m. ‖ bouton m. de mode. / ~ for gloves ‖ Handschuhknopf ‖ bouton m. pour gants. / gold ~ ‖ goldener Knopf m. ‖ bouton m. or. / ~ with holes ‖ Knopf m. mit Löchern ‖ bouton m. à trous. / horn ~ ‖ Hornknopf m. ‖ bouton m. en corne. / irised ~ ‖ Irisknopf m. ‖ bouton m. irisé. / lead ~ ‖ Bleiknopf m. ‖ bouton m. en plomb. / linen ~ ‖ Wäscheknopf m. ‖ bouton m. pour lingerie. / long ~ ‖ Knopf m. mit Öhr ‖ bouton m. à queue. / metal ~ ‖ Metallknopf m. ‖ bouton m. en métal. / military ~ ‖ Uniformknopf m. ‖ bouton m. d'uniforme. / paper ~ ‖ Papierknopf m. ‖ bouton m. en papier. / pearl ~ ‖ Perlmutterknopf m. ‖ bouton m. de nacre. / porcelain ~ ‖ Porzellanknopf m. ‖ bouton m. de porcelaine. / push ~ ‖ Druckknopf m. ‖ bouton m. de pression. / rotating ~ ‖ drehbarer Knopf m. ‖ bouton m. rotatif. / shell ~ ‖ hohler Blechknopf m. ‖ bouton m. à coquille. / silver ~ ‖ silberner Knopf m. ‖ bouton m. argent. / trimming ~ ‖ Besatzknopf m. ‖ bouton m. de lingerie ou de garniture. / uniform ~ ‖ Uniformknopf m. ‖ bouton m. pour uniformes. / ~ of a window sash ‖ Fensterknopf m. ‖ bouton m. ou tiroir m. ou olive f. de fenêtre. / wood ~ ‖ Holzknopf m. ‖ bouton m. en bois.

button borer ‖ Knopfbohrer m. ‖ perceur m. de boutons. / ~ card ‖ Knopfkarte f. ‖ carte m. pour encartage de boutons. / ~ cleaner ‖ Knopfgabel f. ‖ patience f.; patience-planchette f. / ~ cutter ‖ Knopfausschneider m. ‖ détacheur m. de boutons. / ~ dyer ‖ Knopffärber m. ‖ teinturier m. en boutons. / ~ founder ‖ Knopfgießer m. ‖ fondeur m. de boutons. / ~ hand-sewer ‖ Knopfannäher m. ‖ couseur m. de boutons à la main.

button hole finisher (Shoem) ‖ Knopflochvernäher m. ‖ brideur m. de bouton-

nières. / ~ machine ‖ Knopflochmaschine f. ‖ machine f. à faire les boutonnières. / ~ machine hand ‖ Knopflochmaschinennäher m. ‖ couseur m. de boutonnières à la machine. / ~ marker ‖ Knopflochanmerker m. ‖ marqueur m. de boutonnières. / ~ scissors pl. ‖ Knopflochschere f. ‖ ciseaux mpl. à boutonnière. / ~ sewing machine ‖ Knopflochnähmaschine f. ‖ machine f. à coudre les boutonnières. / ~ worker (Shoem) ‖ Knopflochnäher m. ‖ boutonniériste m.

button maker ‖ Knopfmacher m. ‖ boutonnier m. / ~ making machine ‖ Knopfherstellungsmaschine f. ‖ machine f. à fabriquer les boutons. / ~ mould ‖ Knopfform f. ‖ moule m. de bouton. / ~ polisher ‖ Knopfpolierer m. ‖ polisseur m. de boutons. / ~ scale ‖ Kornwage f. ‖ balance f. d'essai; bouillon-scale m. / ~ sewer-on (Shoem) ‖ Knopfannäher m. ‖ poseur m. de boutons. / ~ shank ‖ Knopföse f. ‖ queue f. de bouton. / ~ sorter ‖ Knopfsortiererin f. ‖ trieuse f. de boutons. / ~ stick ‖ Knopfgabel f. ‖ patience f. / ~ tracer ‖ Knopfanzeichner m. ‖ traceur m. en boutons. / ~ turner ‖ Knopfdreher m. ‖ tourneur m. de boutons.

butt pier ‖ Eckpfeiler m. ‖ pilastre m. cornier; cornière f. / ~ plate (Bridge) ‖ Stoßplatte f. ‖ couvre-joint m.

buttress, to ‖ stützen ‖ éperonner; arc-bouter. / ~ a wall by an arch ‖ eine Mauer durch einen Bogen stützen ‖ arc-bouter un mur.

buttress (Pillar) ‖ Pfeiler m. ‖ culée f.; massif m.; pied-droit m.; pilier m. / ~ (Strong pillar) ‖ Strebepfeiler f. ‖ contrefort m. / ~ (Farrier) ‖ Wirkeisen n.; Wirkmesser n. ‖ paroir m. / arched ~ ‖ Schwibbogen m. ‖ arc-boutant m.

butt-rivetting see butt joint (Rivetting).

butt strap (Rivetting) ‖ Lasche f. ‖ couvre-joint m.; bande f. de recouvrement; franc-bord m. / to place a ~ over a joint ‖ überlaschen ‖ poser un couvre-joint.

butt strap, attachment of a ~ ‖ Überlaschen n.; Überlaschung f. ‖ pose f. de couvre-joint. / ~ joint see butt joint, strap. / width of ~ ‖ Laschenbreite f. ‖ largeur f. du couvre-joint.

butt strip see butt strap.

butt-weld, to ‖ stumpf (aneinander-) schweißen ‖ souder à rapprochement. / the ends pl. are butt-welded to each other ‖ die Enden npl. sind stumpf aneinandergeschweißt ‖ les extrémités fpl. sont soudées par rapprochement.

butt-welded tube ‖ stumpf geschweißtes Rohr n. ‖ tube m. soudé à rapprochement.

butt-welding ‖ Stoßverschweißung f. ‖ soudure f. bout à bout; soudage m. abouté ou à rapprochement.

butty (Min) ‖ Grubenpächter m. ‖ forfaitier m.

butyl alcohol ‖ Butylalkohol m. ‖ alcool m. butylique.

butyric acid ‖ Buttersäure f. ‖ acide m. butyrique. / ~ acid fermentation ‖ Buttersäuregärung f. ‖ fermentation f. butyrique. / ~ ether ‖ Butteräther m. ‖ éther m. butyrique.

butyrine ‖ Butterfett n. ‖ beurre m. de vache; butyrine f.

buy, to ‖ einkaufen ‖ acheter. / ~ second-hand ‖ antiquarisch kaufen ‖ acheter d'occasion f. / ~ up (To forestall) ‖ auf-

kaufen ‖ acheter. / ~ up (To purchase) ‖ auskaufen ‖ acheter en totalité; enlever.

buyer ‖ Besteller m.; Abnehmer m.; Auftraggeber m. ‖ client m.; client-acheteur m.

buying, not thinking of ~ it ‖ nicht mehr darauf reflektieren ‖ renoncer à.

buying commission ‖ Einkaufsprovision f. ‖ commission f. d'achat. / ~ department ‖ Einkauf(s)büro n.; Einkaufabteilung f. ‖ service m. d'acquisition; section f. d'achat. / ~ power ‖ Kaufkraft f. ‖ pouvoir m. d'achat. / ~ rate (Exchange) ‖ Geldkurs m. ‖ taux m. d'achat. / ~ speculator ‖ Aufkäufer m. ‖ acheteur m.; accapareur m.; ramasseur m. / ~-up ‖ Aufkauf m. ‖ achat m.

buzzer (Tel) ‖ Summer m.; Schnarrsummer m.; Schnarrwecker m. ‖ ronfleur m.; sonnerie f. ronflante. / ~ for busy tone ‖ Summer m. für Besetztzeichen ‖ ronfleur m. d'occupation. / double magnet ~ ‖ Doppelmagnetsummer m. ‖ vibrateur m. à aimant double. / high-frequency ~ ‖ Töner m. ‖ vibreur-redresseur m. pour hautes fréquences. / ~ for ringing tone (Aut tel) ‖ Summer m. für frei ‖ ronfleur m. de sonnerie. / rhythmic ~ ‖ Unterbrechersummer m. ‖ ronfleur m. à interruption. / shunted ~ ‖ Nebenschlußsummer m. ‖ vibrateur m. à dérivation. / tuned ~ ‖ abgestimmter Summer m. ‖ vibrateur m. syntonisé.

buzzer indicator ‖ Summerschauzeichen n. ‖ voyant m. à courant de vibrateur. / ~ machine (Tel) ‖ Summermaschine f. ‖ trompette f. électrique.

buzzing noise ‖ Brummen n. ‖ bourdonnement m.

by-box for desk set (Tel) ‖ Beikasten m. für Tischgehäuse ‖ caisse f. pour l'appareil portatif.

by-key ‖ Nachschlüssel m. ‖ fausse clef f.

by-lane ‖ Seitenstraße f.; Seitenweg m.; Nebenweg m. ‖ chemin m. latéral ou retiré; voie f. détournée; rue f. latérale.

by-pass apparatus ‖ Umlaufapparat m. ‖ appareil m. de dérivation. / ~ system (Aut tel) ‖ Kreislaufsystem n. ‖ système m. circulaire.

by-product ‖ Nebenprodukt n.; Nebenerzeugnis n.; Abfallstoff m. ‖ sous-produit m.; dérivé m. / installation with ~ plant ‖ Anlage f. mit Gewinnung von Nebenerzeugnissen ‖ usine f. avec récupération des sous-produits. / plant with the recovery of ~s see installation with ~ plant. / ~ producing plant ‖ Nebenproduktengewinnungsanlage f. ‖ installation f. pour l'extraction des produits accessoires. / ~ recovery ‖ Nebenproduktgewinnung f. ‖ récupération f. des sous-produits. / ~ recovery plant ‖ Anlage f. zur Rückgewinnung von Nebenprodukten ‖ installation f. pour la récupération des sous-produits.

byrewoman ‖ Kuhmagd f. ‖ vachère f.

byroad see by-lane.

byssus ‖ Muschelseide f. ‖ bysse m.; byssus m.

byssus silk ‖ Byssusseide f. ‖ soie f. de byssus. / spinning of ~ ‖ Muschelseidenspinnerei f. ‖ filature f. de soie marine.

by-wash (Hydr arch) ‖ Seitengerinne n.; Umlaufgraben m.; Leerlauf m. ‖ canal m. de dérivation; déversoir m.; fossé m. de dérivation ou de circulation.

byway ‖ Nebenchaussee f.; Nebenweg m. ‖ petite route f.; chemin m. détourné.

C

cab (Carriage) ‖ Kab n.; Droschke f. ‖ cab m. / ~ (Loc) ‖ Schutzdach n. ‖ toit m. de protection f.

cabane ‖ Spannturm m. ‖ cabane f. ou pylône m. de haubannage.

cabbage ‖ Kohl m. ‖ chou m. / ~ compressor ‖ Krautpresser m. ‖ presseur m. de choux. / ~ cutter ‖ Krautschneidemaschine f.; Krauthobel m. ‖ coupe-choux m.; coupeur m. de choux. / ~ grower ‖ Krautgärtner m. ‖ planteur m. de choux. / ~ knife ‖ Krauthobel m. ‖ couteau m. à choucroute. / ~-lettuce ‖ Kopfsalat m. ‖ laitue f.

cabin (Build) ‖ Schuppen m.; Baracke f.; Bude f.; Hütte f.; échoppe f.; hangar m.; hutte f.; loge f. / ~ (Shipb) ‖ Koje f.; Kajüte f.; Kammer f. ‖ cajute f. / ~ for the examination of the safety-lamps ‖ Lampenstube f. einer Zeche ‖ lampisterie f. d'une houillère.

cabin bag ‖ Wäschesack m. ‖ sac m. à linge. / ~ door ‖ Kabinentür f. ‖ porte f. de cabine.

cabinet ‖ Kabinett n.; kleines Zimmer n. ‖ cabinet m. / ~ (Cupboard) ‖ Schrank m. coffret m.; armoire f. / ~ for documents ‖ Aktenschrank m. ‖ armoire f. à dossiers.

cabinet file ‖ Wälzfeile f. ‖ lime f. cabinette ou à arrondir. / ~ maker ‖ Kunsttischler m.; Kunstschreiner m.; Möbeltischler m.; Möbelschreiner m. ‖ ébéniste m. / ~ maker and decorator ‖ Möbelschreiner und -tapezierer m. ‖ ébéniste-tapissier m. / ~ making ‖ Kunstschreinerei f.; Möbelschreinerei f.; Kunsttischlerei f.; Möbeltischlerei f. ‖ ébénisterie f.; marqueterie f. / wood for ~ making ‖ Kunsttischlereiholz n. ‖ bois m. d'ébénisterie. / ~ varnish ‖ Möbellack m.; Möbelfirnis m. ‖ vernis m. à meubles. / ~ work ‖ Kunsttischlerarbeit f. ‖ ébénisterie f. / ~ worker see ~ maker.

cabin passenger ‖ Kajütenpassagier m. ‖ passager m. de cabine. / ~ stove ‖ Kajütenofen m. ‖ poêle m. de cabine. / ~ window ‖ Kabinenfenster n. ‖ fenêtre f. de cabine.

cable, to ‖ kabeln ‖ câbler.

cable ‖ Seil n.; Tau n.; Kabel n. ‖ câble m.; corde f. / ~ (Bridge) ‖ Drahtseil n. ‖ câble m. métallique. / ~ (Electr) ‖ Kabel n. ‖ câble m. / ~ (Mar) ‖ Tau n. ‖ câble m. / ~ (Ropem) ‖ Hanfseil n. ‖ corde f. ou cordage m. de chanvre; câble m. / aerial ~ ‖ oberirdisches Kabel n.; Luftkabel n. ‖ câble m. aérien. / suspender for aerial ~s ‖ Traghaken m. für Luftkabel ‖ crochet m. de suspension pour câbles aériens. / air ~ see aerial ~. / air-spaced paper core ~ ‖ Papierluftraumkabel n. ‖ câble m. sous papier et couche d'air. / armoured ~ ‖ Panzerkabel n.; armiertes Kabel n. ‖ câble m. armé. / armoured ~ for war purposes ‖ Panzerfeldkabel n. ‖ câble m. armé de campagne. / army ~ ‖ Feldkabel n. ‖ câble m. de campagne. / artificial ~ ‖ künstliches Kabel n. ‖ câble m. arti-

ficiel. / asphalted ~ ‖ Asphaltkabel n. ‖ câble m. bitumé. / bare ~ ‖ blankes Kabel n. ‖ câble m. nu. / bare-lead-sheathed ~ ‖ Bleikabel n. ‖ câble m. sous plomb nu. / bracing ~ ‖ Fangkabel n.; Spannkabel n. ‖ hauban m. inférieur de sustentation; câble m. tendeur. / braided ~ ‖ umklöppeltes Kabel n. ‖ câble m. sous tressage. / buried ~ ‖ Erdkabel n. ‖ câble m. souterrain. / check ~ ‖ Lenkkabel n. ‖ câble m. de commande de la roue arrière orientable. / coastal ~ ‖ Küstenkabel n. ‖ câble m. côtier. / coil-loaded ~ (Tel) ‖ Pupinkabel n. ‖ câble m. Pupin. / combined ~ ‖ gemischtpaariges oder kombiniertes Kabel n. ‖ câble m. combiné. / conduit ~ ‖ Röhrenkabel n.; Leitungskabel n. ‖ câble m. à tuyau; câble m. conducteur. / cotton and silk ~ ‖ Baumwollseidenkabel n. ‖ câble m. sous coton et soie. / covered ~ ‖ bewehrtes oder umhülltes Kabel n. ‖ câble m. guipé ou armé. / deep-sea ~ ‖ Tiefseekabel n. ‖ câble m. de haute mer. / desiccating of moist ~s ‖ Trocknen n. von feuchten Kabeln ‖ séchage m. ou desséchement m. de câbles humides. / distributing ~ ‖ Fernsprechverteilungskabel n. ‖ câble m. de distribution. / to draw a ~ ‖ ein Kabel n. ziehen ‖ poser un câble m. / drawing of ~s into pipes ‖ Einziehen n. von Leitungen in Rohre ‖ pose f. des câbles conducteurs dans les tubes. / duct-laid ~ ‖ in Röhren verlegtes Kabel n. ‖ câble m. posé en canalisations ou en tuyaux. / electric light ~ ‖ Beleuchtungskabel n.; Lichtkabel n. ‖ câble m. à éclairage. / electric power ~ ‖ Kraftübertragungskabel n.; Kraftkabel n. ‖ câble m. à transmission de force. / feeder ~ ‖ Speisekabel n. ‖ câble m. alimentaire ou d'alimentation. / fibre-covered ~ ‖ Faserstoffkabel n. ‖ câble m. sous fibre. / field subfluvial ~ ‖ Feldflußkabel n. ‖ câble m. fluvial militaire. / flexible ~ ‖ biegsames Kabel n. ‖ câble m. souple. / guttapercha ~ ‖ Guttaperchakabel n. ‖ câble m. sous gutta. / haulage ~ ‖ Förderseil n. ‖ câble m. d'extraction. / hauling-down ~ (Aero) ‖ Halteau n. ‖ câble m. de halage. / ~ covered with tarred hemp ‖ Kabel n. mit geteerter Hanfumspinnung ‖ câble m. sous guipage de chanvre goudronné. / ignition ~ ‖ Zündkabel n. ‖ câble m. d'allumage. / to imbed a ~ ‖ ein Kabel n. einbetten ‖ noyer un câble. / indoor ~ ‖ Hausleiterkabel n. ‖ câble m. d'intérieur. / influence of temperature in long distance ~s ‖ Temperatureinfluß m. bei Fernkabeln ‖ influence f. de la température sur câbles à grande distance. / insulated ~ ‖ isoliertes Kabel n. ‖ câble m. isolé. / intermediate ~ ‖ Kabelzwischenstück n. ‖ câble m. intermédiaire. / iron-clad ~ ‖ Panzerkabel n. ‖ câble m. cuirassé. / ~ insulated by jute ‖ Jutekabel n. ‖ câble m. sous jute ou à isolement de jute. / kind of ~ ‖ Kabelart f. ‖ espèce f. de

câble. / lac paper ~ ‖ Lackpapierkabel n. ‖ câble m. sous papier verni; câble m. émaillé sous papier. / lac paper shell ~ ‖ Lackpapiermantelkabel n. ‖ câble m. à gaîne de plomb avec isolement de laque et papier. / to lay a ~ ‖ ein Kabel n. verlegen ‖ poser un câble m. / to lay a submarine ~ ‖ ein Seekabel n. legen ‖ mouiller un câble m. / laying of ~ ‖ Kabelverlegung f. ‖ pose f. de câble. / laying an oil-filled ~ ‖ Verlegung f. eines Ölkabels ‖ pose f. d'un câble à huile. / lead ~ ‖ Bleikabel n. ‖ câble m. en plomb. / lead-covered ~ ‖ Bleikabel n.; Kabel n. mit Bleiummantelung ‖ câble m. sous plomb ou armé de plomb. / lead-covered rubber ~ ‖ Gummibleikabel n. ‖ câble m. au caoutchouc sous plomb. / leading ~ ‖ Leitkabel n. ‖ câble m. de guidage. / lighting ~ ‖ Lichtkabel n. ‖ câble m. d'éclairage. / to locate ~s ‖ Kabel npl. aufsuchen ‖ chercher à trouver la position de câbles. / equipment for locating ~s ‖ Gerät n. zum Aufsuchen von Kabeln ‖ appareil m. permettant de trouver la position de câbles. / long distance ~ ‖ Fernkabel n. ‖ câble m. à grande distance. / influence of temperature in long distance ~s ‖ Temperatureinfluß m. bei Fernkabeln ‖ influence f. de la température sur câbles à grande distance. / long-distance submarine ~ ‖ transatlantisches Kabel n. ‖ câble m. sous-marin à grande distance; câble m. d'outre-mer ou transocéanique. / long-distance telegraph ~ ‖ Telegrafenfernkabel n. ‖ câble m. télégraphique à grande distance. / long-distance telephone ~ ‖ Telefonfernkabel n. ‖ câble m. téléphonique à grande distance. / main telephone ~ ‖ Fernsprechhauptkabel n. ‖ câble m. téléphonique de transport. / measuring ~ ‖ Prüfkabel n. ‖ câble m. de mesure. / mining ~ ‖ Grubenkabel n. ‖ câble m. minier ou de mine. / desiccating of moist ~s ‖ Trocknen n. von feuchten Kabeln ‖ séchage m. ou desséchement m. de câbles humides. / ~ for no-delay service ‖ Schnellverkehrskabel n. ‖ câble m. pour le trafic rapide. / non-loaded ~ ‖ Thomsonkabel n. ‖ câble m. sans inductivité. / office ~ ‖ Amtskabel n. ‖ câble m. de bureau. / oil-filled ~ ‖ Ölkabel n.; Kabel n. mit Ölfüllung ‖ câble m. à huile. / laying an oil-filled ~ ‖ Verlegung f. eines Ölkabels ‖ pose f. d'un câble à huile. / outer ~ ‖ Außenkabel n. ‖ câble m. extérieur. / overhead ~ ‖ Luftkabel n.; Freileitung f. ‖ câble m. aérien. / paper ~ see paper-insulated ~. / paper and cotton-covered ~ ‖ Papierbaumwollkabel n. ‖ câble m. sous papier et coton. / paper-insulated ~ ‖ Papierkabel n. ‖ câble m. sous papier; câble m. isolé au papier. / paper and lead-covered ~ ‖ Papierbleikabel n. ‖ câble m. sous plomb et papier. / paper terminal ~ ‖ Papierabschlußkabel n. ‖ câble m. de fermeture sous papier. / to pay-out a ~ ‖ ein Kabel

n. abrollen *oder* auslegen ‖ dérouler un câble. / plain ~ ‖ Kabel n. ohne Bleimantel ‖ câble m. sans gaîne de plomb. / power ~ ‖ Kraftkabel n. ‖ câble m. pour le transport de force. / power-current ~ ‖ Starkstromkabel n. ‖ câble m. pour courant fort. / protecting ~ ‖ Schutzseil n. ‖ câble m. protecteur. / pull back ~ ‖ Rückzugsseil n. ‖ câble m. tracteur en arrière. / return ~ ‖ Rückleitungskabel n. ‖ câble m. de retour. / return feeder ~ ‖ Rückspeisekabel n. ‖ artère f. de retour. / river ~ ‖ Flußkabel n. ‖ câble m. sousfluvial. / rubber ~ *see* rubber-covered ~. / rubber-covered ~ ‖ Gummikabel n. ‖ câble m. sous caoutchouc. / shallow-water ~ ‖ Küstenkabel n. ‖ câble m. côtier. / small ~ (Mar) ‖ Treidelleine f. ‖ grelin m.; remorque f. / to split a ~ ‖ ein Kabel n. spleißen ‖ épisser un câble m. / starter ~ ‖ Anlaßkabel n. ‖ câble m. de démarrage. / subaqueous ~s ‖ Unterwasserkabel n. ‖ câble m. sous eau. / subfluvial ~ *see* river ~. / submarine ~ ‖ Unterseekabel n.; Überseekabel n.; Seekabel n. ‖ câble m. sous-marin. / to lay a submarine ~ ‖ ein Seekabel n. legen ‖ mouiller un câble m. / long-distance submarine ~ ‖ transatlantisches Kabel n. ‖ câble m. sous-marin à grande distance; câble m. d'outre-mer *ou* transocéanique. / protection of submarine ~s ‖ Schutzmaßnahmen fpl. für transatlantische Kabel ‖ protection f. des câbles sous-marins. / provisions pl. concerning the protection of submarine ~s ‖ Schutzrecht n. für transatlantische Kabel ‖ règlements mpl. concernant la protection des câbles sous-marins. / submarine telegraph ~ ‖ Seetelegrafenkabel n. ‖ câble m. télégraphique sous-marin. / submarine telephone ~ ‖ Seefernsprechkabel n. ‖ câble m. téléphonique sous-marin. / subscriber's ~ ‖ Teilnehmerkabel n. ‖ câble m. d'abonnés. / subterranean ~ ‖ unterirdisches Kabel n.; Landkabel n. ‖ câble m. souterrain *ou* sous-terre *ou* terrestre. / super-tension ~ ‖ Höchstspannungskabel n. ‖ câble m. très haute tension. / telegraph ~ ‖ Telegrafenkabel n. ‖ câble m. télégraphique. / long-distance telegraph ~ ‖ Telegrafenfernkabel n. ‖ câble m. télégraphique à grande distance. / submarine telegraph ~ ‖ Seetelegrafenkabel n. ‖ câble m. télégraphique sous-marin. / telephone ~ ‖ Telefonkabel n. ‖ câble m. téléphonique. / long-distance telephone ~ ‖ Telefonfernkabel n. ‖ câble m. téléphonique à grande distance. / main telephone ~ ‖ Fernsprechhauptkabel n. ‖ câble m. téléphonique principal. / submarine telephone ~ ‖ Seefernsprechkabel n. ‖ câble m. téléphonique sous-marin. / telephone end ~ ‖ Fernsprechabschlußkabel n. ‖ câble m. téléphonique de fermeture. / influence of temperature in long-distance ~ ‖ Temperatureinfluß m. bei Fernkabeln ‖ influence f. de la température sur câbles à grande distance. / ~ with three cores ‖ Dreileiterkabel n. ‖ câble m. à trois âmes. / transatlantic ~ ‖ transatlantisches Kabel n. ‖ câble m. d'outre-mer. / transmarine ~ *see* submarine ~. / underground ~ *see* subterranean ~. / water ~ *see* submarine ~. / weak-current ~ ‖ Schwachstromkabel n.; Fernmeldekabel

n. ‖ câble m. à faible intensité. / winding ~ ‖ Förderseil n. ‖ câble m. de trainage. / wire ~ ‖ Drahtseil n. ‖ câble m. en fil d'acier.

cable, armouring of ~s ‖ Kabelbewehrung f.; Kabelarmierung f.; Kabelarmatur f. ‖ armure f. *ou* armature f. de câbles.

cable armouring machine ‖ Kabelarmierungsmaschine f.; Kabelbewehrungsmaschine f. ‖ machine f. à armer les câbles. / asphalt box for ~ ‖ Asphaltkasten m. für Kabelarmierungsmaschinen ‖ bac m. *ou* récipient m. à asphalte pour machines à armer les câbles électriques.

cable balancing network ‖ Kabelnachbildung f. ‖ dispositif m. d'équilibrage d'un câble. / ~ basket ‖ Kabelkorb m. ‖ benne f. à câble. / ~ beacon ‖ Kabelbake f. ‖ balise f. de câble. / ~ bearer ‖ Kabelhalter m.; Kabelträger m. ‖ support m. de câble; porte-câble m.

cable box ‖ Kabelkasten m. ‖ boîte f. de câble. / underground ~ ‖ Landkabelmuffe f. ‖ manchon m. pour câble terrestre.

cable braiding ‖ Kabelumklöpplung f. ‖ tressage m. de câble. / ~ brake ‖ Seilbremse f. ‖ frein m. du câble. / ~ branching box ‖ Kabelverzweigungsmuffe f. ‖ manchon m. de branchement de câbles. / ~ buoy ‖ Kabelboje f. ‖ bouée f. de câble. / ~ car with traction engine ‖ Kabelwagen m. mit Zugmaschine ‖ voiture f. à câbles avec machine de traction *ou* avec tracteur. / ~ cellar ‖ Kabelkeller m. ‖ cave f. des câbles. / ~ chain shackle ‖ Ankerkettenschäkel m. ‖ manille f. d'un câble-chaîne. / ~ channel ‖ Kabelgraben m. ‖ caniveau m. de câble. / ~ circuit ‖ Kabelschaltung f. ‖ circuit m. pour les câbles. / ~ clamp ‖ Kabelklemme f. ‖ attache f. pour câbles. / ~ clip ‖ Kabelschuh m. ‖ raccord m. de câble. / ~ code ‖ Kabelschrift f. ‖ câble-code m. / three-key perforator for ~ code ‖ Handlocher m. für Kabelbetrieb ‖ perforateur m. / ~ compound ‖ Kabelausgußmasse f.; Kabelasphalt m. ‖ matériel m. *ou* mastix m. de remplissage de câbles; asphalte m. pour câbles. / ~ conduit ‖ Kabelkanal m. ‖ conduite f. de câble; fourreau m. porte-câbles. / ~ connecting box ‖ Kabelverbindungsmuffe f. ‖ manchon m. de jonction pour câbles. / core of the ~ ‖ Kabelkern m.; Kabelseele f. ‖ âme f. du câble. / core-conductor of the ~ ‖ Kabelader f. ‖ âme f. *ou* conducteur m. du câble. / core-maker ‖ Kabelspulmaschinenarbeiter m. ‖ bobineur m. de câbles. / ~ coupling ‖ Kabelkupplung f. ‖ joint m. de câbles. / ~ covering machine ‖ Kabelisoliermaschine f. ‖ machine f. à chemiser les câbles électriques. / ~ crab ‖ Seilaufkatze f.‖ porte-palan m. / ~ distribution box ‖ Kabelüberführungsendverschluß m.; Kabelüberführungskasten m. ‖ guérite f. de raccordement de câbles; chambre f. de concentration de câbles. / ~ distribution coupling box ‖ Kabelverzweigungsmuffe f. ‖ manchon m. de branchement de câbles. / ~ distribution plug ‖ Aufteilungsmuffe f. ‖ pièce f. de division. / ~ distribution point ‖ Kabelauführung f. ‖ point m. de distribution des lignes de câble. / ~ distribution pole ‖ Kabelüberführungssäule f. ‖ guérite f. de raccordement. / drawing ~s into pipes ‖ Einziehen n. von Kabeln in

Rohre ‖ pose f. des câbles dans des tubes. / ~ drum ‖ Kabeltrommel f. ‖ tambour m. à câble. / ~ drying ‖ Kabeltrocknung f. ‖ séchage m. des câbles. / ~ drying apparatus ‖ Kabeltrockenapparat m. ‖ séchoir m. pour câbles. / ~ drying shelf ‖ Kabeltrockenschrank m. ‖ étagère f. de séchage des câbles. / ~ drying stove ‖ Kabeltrockenofen m. ‖ four m. séchoir pour câbles. / drying tank for ~s ‖ Trockenkessel m. für Kabel ‖ bassin m. de séchage pour câbles. / ~ duct ‖ Kabelaufnahmeröhre f. ‖ tuyau m. à câble. / sealed ~ end ‖ Kabelstumpf m. ‖ capote f. de câble. / eye of the ~ ‖ Kabelschuh m.; Kabelklemme f. ‖ attache-câble m.; cosse f. de câble. / ~ factory ‖ Kabelfabrik f. ‖ fabrique f. de câbles; câblerie f. / ~ fault ‖ Kabelfehler f. ‖ défaut m. de câble. / filling-up ~s ‖ Ausgießen n. der Kabel ‖ remplissage m. des câbles. / ~ filling yarn ‖ Kabelfüllgarn n. ‖ matière f. fibreuse à remplir les câbles. / ~ finder *see* ~ locator. / ~ fittings pl. ‖ Kabelgarnitur f. ‖ accessoire m. pour câbles. / ~ and cord fittings pl. ‖ Tau- und Seilbeschläge mpl. ‖ armatures fpl. pour cordes de touage. / ~ form ‖ Kabelform f. ‖ peigne m. de câble. / ~ former ‖ Kabelformbrett n. ‖ gabarit m. pour peigne de câble. / ~ grease ‖ Kabelfett n.; Kabelschmiere f. ‖ graisse f. de câble. / ~ grip split ‖ Nachziehschlauch m. ‖ grip m. double. / ~ guard ‖ Absperrgestell n. ‖ garde-corps m. / ~ hanger ‖ Kabelhalter m. ‖ porte-câble m. / ~ haulage ‖ Seilförderung f. ‖ extraction f. par câble. / ~ head ‖ Kabelabschluß m. ‖ tête f. de câble. / impregnation of ~s ‖ Tränkung f. *oder* Imprägnierung f. von Kabeln ‖ imprégnation f. de câbles. / ~ impregnating tank ‖ Kabeltränkkessel m. ‖ vaisseau m. *ou* bac m. *ou* réservoir m. d'imprégnation des câbles. / ~ inlet ‖ Kabeleinführung f. ‖ entrée f. de câble. / ~ joint ‖ Spleißstelle f. ‖ raccordement m. du câble. / kind of ~ ‖ Kabelart f. ‖ espèce f. de câble. / laying (of) ~s ‖ Kabelverlegung f. ‖ pose f. de câbles. / ~ laying implement ‖ Kabeleinziehgerät n. ‖ ustensile m. pour la canalisation des câbles. / ~ laying machine ‖ Kabelverlegemaschine f. ‖ machine f. à poser les câbles électriques. / plan for laying ~s ‖ Kabellegeplan m. ‖ plan m. de pose des câbles. / ~ lead-in ‖ Kabeleinführung f. ‖ entrée f. de câbles. / ~ ('s) length ‖ Kabellänge f. ‖ encâblure f.; longueur f. *ou* tronçon m. de câble. / ~ letter telegram ‖ Kabelbrief m. ‖ câble-lettre-télégramme m. / ~ line ‖ Kabelleitung f. ‖ ligne f. de câbles. / ~ location equipment *see* ~ locator. / ~ locator ‖ Kabelsucher m.; Gerät n. zum Auffinden von Kabeln ‖ appareil m. permettant de trouver la position de câbles. / ~ machine *see* ~ making machine. / ~ maker ‖ Seilschläger m. ‖ câblier m.; câbleur m. / ~ making machine ‖ Maschine f. zur Herstellung von Kabeln ‖ machine f. à fabriquer les câbles. / ~ manufacturing ‖ Kabelherstellung f. ‖ fabrication f. des câbles. / ~ measurement ‖ Kabelmessung f. ‖ mesurage m. de câbles. / ~ measuring set ‖ tragbares Kabelmeßgerät n. ‖ appareil m. portatif de mesures de câbles. / ~ mortgaging act ‖ Kabelpfandgesetz n. ‖ loi f. sur le nantisse-

ment des câbles. / ~ paper ‖ Kabelpapier n. ‖ papier m. pour câbles. / ~ paying-out machine ‖ Kabelauslegemaschine f. ‖ machine f. à dérouler le câble. / pit for ~s ‖ Kabelkanal m.; Kabelschacht m. ‖ puits m. d'aboutissement des câbles. / ~ plaiter ‖ Kabelklöppelmaschinenarbeiter m. ‖ tresseur m. de câbles. / general plan of a ~ plant ‖ Lageplan m. einer Kabelanlage ‖ plan m. ou croquis m. d'une ligne de câbles. / ~ plough ‖ Seilpflugmaschine f. ‖ charrue f. à treuil. / position of a ~ ‖ Lage f. eines Kabels ‖ position f. d'un câble. / ~ press ‖ Kabelpresse f. ‖ presse f. à câbles. / lead ~ press ‖ Bleikabelpresse f. ‖ presse f. pour câbles sous plomb. / protection of ~s ‖ Schutzmaßnahmen fpl. für Kabel ‖ protection f. des câbles. / convention concerning the protection of ~s ‖ Kabelschutzvertrag m. ‖ convention f. pour la protection des câbles. / provisions pl. concerning the protection of ~s ‖ Kabelschutzrecht n. ‖ règlements mpl. concernant la protection des câbles. / protective covering of ~s ‖ Kabelschutzbekleidung f.; Kabelumhüllung f. ‖ enveloppe f. protectrice des câbles. / ~ pulley ‖ Seilrolle f. ‖ poulie f. à câble. / ~ rack ‖ Kabelgestell n. ‖ châssis m. de câble. / ~ railway ‖ Drahtseilbahn f. ‖ funiculaire m. / ~ reel ‖ Kabeltrommel f. ‖ tambour m. de câble. / ~ ring ‖ Tragring m. für Luftkabel ‖ bague f. de suspension. / ~ sheath ‖ Kabelmantel m. ‖ enveloppe f. de câble. / ~ sheath current ‖ Kabelmantelstrom m. ‖ courant m. d'enveloppe de câble. / ~ shelf ‖ Kabelrost m. ‖ étagère f. à câble. / ~ ship ‖ Kabelschiff n.; Kabelleger m. ‖ navire m. câblier; câblier m.; mouilleur m. de câbles. / ~ sleeve ‖ Kabelmuffe f. ‖ manchon m. de câble. / ~ soldering sleeve ‖ Kabellötmuffe f. ‖ manchon m. de soudage de câbles. / ~ sound ‖ Kabelsonde f. ‖ sonde f. pour l'immersion d'un câble. / ~ stage ‖ Kabelgatt n. ‖ soute f. aux câbles. / ~ standard ‖ Kabelnorm f. ‖ norme f. ou règle f. pour les câbles. / ~ steamer ‖ Kabeldampfer m. ‖ câblier m. / ~ stopper (Shipb) ‖ Kettenstopper m. ‖ bosse f. de câble. / ~ stranding machine ‖ Kabelumflechtmaschine f.; Kabelverseilmaschine f. ‖ machine f. à tresser le guipage du câble; toronneuse f. / strand-wire for ~s ‖ Kabellitze f. ‖ cordon m. conducteur pour câbles. / suspender for aerial ~s ‖ Tragband n. oder Traghaken m. für Luftkabel ‖ bague f. ou bride f. ou crochet m. de suspension pour câbles aériens. / taking-up ~s ‖ Kabelhochführung f. ‖ pose f. surélevée de câbles. / ~ tank ‖ Kabeltank m. ‖ tank m. à câbles. / ~ test with compressed air ‖ Druckluftprüfung f. der Kabel ‖ épreuve f. des câbles à l'air comprimé. / ~ test earth ‖ Kabelmeßerde f. ‖ terre f. pour les mesures électriques des câbles. / ~ tester ‖ Kabelprüfer m. ‖ essayeur m. de câbles. / ~ testing car ‖ Kabelmeßwagen m. ‖ voiture-laboratoire f. pour l'essai des câbles. / ~ testing outfit ‖ Kabelprüfeinrichtung f. ‖ équipement m. pour l'essai de câbles. / ~ testing set see ~ testing outfit. / ~ tramway ‖ Drahtseilbahn f. ‖ tramway m. à câbles. / ~ trolley ‖ Kabel(transport)wagen m. ‖ chariot m. de transport des câbles. /

~ trough ‖ Kabelkasten m. ‖ gaine f. pour ligne de câbles. / twisting of ~s ‖ Kabelverseilung f. ‖ câblage m. / ~ van see ~ trolley. / ~ wagon ‖ Kabelwagen m. ‖ chariot m. ou wagon m. à câbles. / vertical ~ wall duct ‖ Kabelhochführungsschacht m. ‖ cheminée f. d'ascension de câble. / ~ wax ‖ Kabelwachs n. ‖ cire f. à câbles.

cableway ‖ Drahtseilbahn f. ‖ transport m. par câble; transporteur m. à câble; chemin m. de fer aérien par câble; aerial ~ ‖ Drahtseilbahn f. ‖ funiculaire m.; chemin m. de fer funiculaire; transport m. par câble.

cableway car ‖ Seilbahnwagen m.‖ wagonnet m. de transport aérien. / ~ crane ‖ Kabelkran m. ‖ grue f. à câbles.

cable winch ‖ Kabelwinde f. ‖ treuil m. à câble. / ~ works pl. ‖ Kabelwerk n. ‖ câblerie f.; fabrique f. de câbles.

cabling (Arch) ‖ verstäbte Kannelierung f.; Stäbeausfüllung f. ‖ embâtonnage m.; rudenture f. / ~ (Tel) ‖ Verkabelung f. ‖ mise f. sous câble. / ~ diagram (Electr) ‖ Montagebild n. ‖ schéma m. de câblage.

cabotage (Geogr) ‖ Küstenkenntnis f. ‖ cabotage m.; expérience f. d'une côte. / ~ (Nav) ‖ Küstenschiffahrt f. ‖ cabotage m. / ~ (Trade) ‖ Küstenhandel m. ‖ cabotage m.

cabriolet ‖ Kabriolett n. ‖ cabriolet m.

ca' canny ‖ passiver Widerstand m. ‖ résistance f. passive.

cacao ‖ Kakao m. ‖ cacao m. / ~ nut oil ‖ Kakaobutter f. ‖ beurre m. de cacao.

cachou ‖ Cachou n. ‖ cachou m.

cactus fig ‖ Kaktusfeige f. ‖ figue f. de cactus.

cadastral survey ‖ Katastervermessung f. ‖ levé m. du cadastre. / ~ by the polar method ‖ Katastervermessung f. nach der Polarmethode ‖ levé m. du cadastre par la méthode polaire.

cadaver utilization plant ‖ Kadaververwertungsanlage f. ‖ installation f. d'utilisation de cadavres.

cadmia ‖ Galmei m. ‖ calamine f.; zinc m. carbonaté.

cadmiferous (Met) ‖ kadmiumhaltig ‖ cadmifère.

cadmium ‖ Kadmium n. ‖ cadmium m. / ~ chloride ‖ Chlorkadmium n. ‖ chlorure m. de cadmium. / ~ electrode ‖ Kadmiumelektrode f. ‖ électrode f. en cadmium. / measurement with ~ (Acc) ‖ Kadmiummessung f. ‖ mesure f. au cadmium. / ~ plating ‖ Kadmiumüberzug m. ‖ doublé m. de cadmium ou de cadmiumage. / ~ plating plant ‖ Verkadmiumierungsanlage f. ‖ installation f. de cadmiumage. / ~ salt ‖ Kadmiumsalz n. ‖ sel m. de cadmium. / ~ sulphide ‖ Schwefelkadmium n. ‖ sulfure m. de cadmium. / ~ yellow ‖ Kadmiumgelb n. ‖ jaune m. de cadmium.

caffeic acid ‖ Kaffeesäure f. ‖ acide m. caféique.

caffeine ‖ Koffein n.; Kaffein n.; Guaranin n. ‖ caféine f.; guaranine f.; théine f. / ~ salt ‖ Koffeinsalz n. ‖ sel m. de caféine.

cage ‖ Käfig m.; Korb m. ‖ cage f. / ~ (Mine) ‖ Förderkorb m.; Förderschale f. ‖ cage f. d'extraction. / ~ of delivery valve ‖ Druckventileinsatz m. ‖ corps m. de la soupape de compression. / hoisting ~ (Mine) ‖ Förderkorb m. ‖ cage f. d'extraction. / wire ~ ‖ Drahtkäfig m. ‖ cage

f. en fil de fer. / wooden ~ ‖ Holzkäfig m. ‖ cage f. en bois.

cage coil ‖ Käfigspule f. ‖ bobine f. à cage. / ~ loader (Met) ‖ Schrägaufzugarbeiter m. ‖ encageur m. / ~ maker ‖ Käfigmacher m. ‖ fabricant m. de cages. / ~ safety apparatus (Mine) ‖ Fallschutzvorrichtung f. für Förderkörbe ‖ parachute m. pour cages d'extraction.

caisson (Hydr arch) ‖ Senkkasten m. ‖ caisson m. (de fonçage). / ~ (Shipb) ‖ Verschlußponton m. ‖ bateau-porte m. / ~ (War impl) ‖ Munitionswagen m. ‖ voiture-caisson f.; caisson m. (à munition); fourgon m. / iron ~ ‖ eiserner Senkkasten m. ‖ caisson m. en tôle. / motor ~ ‖ Munitionskraftwagen m. ‖ voiture-caisson f. automotrice. / observatory ~ ‖ Beobachtungsmunitionswagen m. ‖ caisson-observatoire m. / tractor ~ ‖ Schleppermunitionswagen m. ‖ caisson-tracteur m. / trailer ~ ‖ Anhängermunitionswagen m. ‖ caisson-remorque m.

caisson gate (Hydr arch) ‖ Pontontor n. ‖ bateau-porte m.

cajeput oil ‖ Kajeputöl n. ‖ essence f. de cajéput.

cake ‖ Kuchen m. ‖ gâteau m. / ~ (Biscuit) ‖ Keks m.; Kek m.; Kake m. ‖ biscuit m. anglais. / ~ of cinder ‖ Schlackenkuchen m. ‖ gâteau m. ou plaque f. de scories. / ~ of clinker see ~ of cinder. / ~ of coke ‖ Kokskuchen m. ‖ tourteau m. de coke. / ~ of colour ‖ Tuschfarbe f.; Stück n. Tusche ‖ bâton m. de couleur. / fancy ~ ‖ Torte f. ‖ tarte f. / fancy ~ cutter ‖ Tortenheber m. ‖ pelle f. à gâteau. / ~ of ice ‖ Eisscholle f. ‖ glaçon m. / oil ~ ‖ Ölkuchen m. ‖ gâteau m. à l'huile ou d'huile; tourteau m. / oil ~ crusher ‖ Ölkuchenbrecher m. ‖ broyeur m. de gâteaux d'huile. / ~ of resin ‖ Harzkuchen m. ‖ disque m. de résine. / ~ of slag see ~ of cinder.

cake baker ‖ Feinbäcker m. ‖ pâtissier m.; biscuitier m. / ~ factory ‖ Keksfabrik f. ‖ fabrique f. de cakes. / ~ form ‖ Kuchenform f. ‖ forme f. à gâteaux.

cake mixer for beating eggs ‖ Schneeschlagmaschine f. ‖ batteuse f. pour battre le blanc d'œuf en neige. / the ~ beats the paste and the dough ‖ die Schlagmaschine f. schlägt die Masse und den Teig ‖ la batteuse bat toutes les qualités de pâtes. / ~ for mixing pastes ‖ Massenrührmaschine f. ‖ batteuse f. pour malaxer les pâtes.

cake pan ‖ Kuchenpfanne f. ‖ poêle f. à gâteaux. / ~ thickness (Filter) ‖ Kuchenstärke f. ‖ épaisseur f. de tourteau ou de gâteau. / ~ top strewing machine ‖ Streuselmaschine f. ‖ machine f. à morceler la pâte.

caking of the coal ‖ Backen n. der Kohle ‖ agglutination f. de la houille. / ~ coal ‖ Backkohle f. ‖ houille f. grasse.

Calabar bean ‖ Calabarbohne f. ‖ fève f. de Calabar.

calabash kernel ‖ Flaschenkürbiskern m. ‖ pépin m. de caldebasse.

calambac ‖ Kalambak n.; Adlerholz n. ‖ bois m. d'aigle; calambac m.

calamine ‖ Zinkspat m.; kohlensaures Zinkoxyd n.; Zinkkarbonat n.; Galmei m. ‖ carbonate m. de zinc; calamine f.; zinc m. carbonaté. / cupriferous ~ ‖ Kupferschaum m.; Tyrolit m. ‖ tyrolite

f.; kupaphrite f. / electric ~ ‖ Kieselgalmei m.; Zinkglas n. ‖ zinc m. oxydé silicifère; calamine f. électrique. / siliceous ~ ‖ Kieselzinkerz n.; Kieselzinkspat m.; Kieselgalmei m.; Hemimorphit m. ‖ hydro-silicate m. de zinc; zinc m. oxydé silicifère.

calamine brass ‖ Galmeimessing n. ‖ laiton m. à la calamine.

calamite ‖ Kalamit m. ‖ calamite f.

calamus oil ‖ Kalmusöl n. ‖ huile f. de calamus.

calash ‖ Kalesche f. ‖ calèche f.

calcar (Met) ‖ Kalzinierofen m.; Brennofen m. ‖ fourneau m. de calcinage; fourneau m. à calciner. / ~ arch (Glassm) ‖ Frittofen m. ‖ arche f. à fritter.

calcareous ‖ kalkartig ‖ calcaire. / ~ earth ‖ Kalkerde f. ‖ terre f. calcaire. / ~ gravel ‖ Kalkkies m. ‖ gravier m. calcaire. / ~ ironstone mine ‖ Minettegrube f. ‖ mine f. de minette ou de fer oolithique calcifère. / ~ mica schist ‖ Kalkglimmerschiefer m. ‖ schiste m. micacé calcaire. / ~ sandstone ‖ Kalksandstein m. ‖ pierre f. chaux-grès. / ~ sandstone making plant ‖ Kalksandstein-Herstellungsanlage f. ‖ installation f. pour fabriquer de briques silico-calcaires. / ~ spar ‖ Kalkspat m. ‖ spath m. calcaire. / ~ shale quarry ‖ Kalkschieferbruch m. ‖ carrière f. de schiste calcaire. / ~ sinter ‖ Kalksinter m.; Tropfstein m.; Stalaktit m. ‖ stalactite f. / ~ spar ‖ Kalkspat m. ‖ spath m. calcaire. / ~ tuff or tufa ‖ Kalktuff m.; Kalksinter m. ‖ tuf m. calcaire.

calcic ‖ kalziumhaltig ‖ calcique. / ~ hydrate ‖ Kalkhydrat n. ‖ hydrate m. de chaux; chaux f. hydratée. / ~ phosphate ‖ Kalziumphosphat n. ‖ phosphate m. de chaux. / precipitated ~ phosphate ‖ präzipitiertes Kalziumphosphat n. ‖ phosphate m. de chaux précipité. / ~ superphosphate ‖ Kalziumsuperphosphat n. ‖ superphosphate m. de chaux.

calcination ‖ Glühen n.; Kalzinieren n. ‖ calcination f. / ~ of gypsum ‖ Gipsbrennerei f. ‖ cuite f. du plâtre. / ~ of the ores ‖ Rösten n. der Erze ‖ grillage m. des minerais. / assay of ~ ‖ Röstprobe f. ‖ essai m. de grillage.

calcination pot ‖ Kalziniertopf m. ‖ vase m. ou bac m. à calcination.

calcine, to (Met) ‖ ausglühen; brennen; kalzinieren ‖ calciner; griller.

calcined ‖ gebrannt ‖ calciné.

calciner (Met) ‖ Röstofen m.; Kalzinierofen m. ‖ fourneau m. de calcinage ou à calciner. / ~ of gypsum ‖ Gipsbrenner m. ‖ plâtrier m. / iron ore ~ ‖ Eisensteinröster m. ‖ calcinateur m. de minerai de fer. / sulphur ~ ‖ Schwefelbrennofen m. ‖ cubilot m. à soufre.

calcining ‖ Kalzinieren n. ‖ calcinage m. / ~ furnace ‖ Kalzinierofen m.; Röstofen m. ‖ four m. à calciner ou de calcinage ou pour la cuisson. / ~ rod ‖ Röstspatel m. ‖ tige f. à calciner. / ~ test ‖ Röstscherben m.; Ansiedescherben m. ‖ têt m. à rôtir.

calcite ‖ Kalkspat m. ‖ calcite f.

calcium ‖ Kalzium n. ‖ calcium m. / ~ acetate ‖ essigsaurer Kalk m. ‖ acétate m. de chaux. / ~ carbide ‖ Karbid n.; Kalziumkarbid n. ‖ carbure m. de calcium. / ~ chloride ‖ Calciumchlorid n. ‖ chlorure m. de calcium. / ~ chloride cylinder ‖ Chlorkalziumzylinder m. ‖ cylindre m. pour chlorure de calcium. / ~ cyanamid

‖ Kalziumcyanamid n.; Kalkstickstoff m. ‖ cyanamide m. calcique; chaux-azote f. / ~ fertilizer ‖ Kalkdünger m. ‖ engrais m. à la chaux. / ~ fluoride ‖ Fluorkalzium n. ‖ fluorure m. de calcium. / ~ light ‖ Drummond'sches Kalklicht n. ‖ lumière f. de Drummond. / monosulphide of ~ ‖ Schwefelkalzium n. ‖ sulfure m. de calcium. / oxide of ~ ‖ Ätzkalk m. ‖ chaux f. vive ou caustique ou calcinée. / ~ peroxide ‖ Kalziumsuperoxyd n. ‖ peroxyde m. de calcium. / ~ phenylate ‖ karbolsaures Kalzium n. ‖ phénate m. de calcium. / ~ plumbate ‖ bleisaures Kalzium n. ‖ plombate m. de calcium. / ~ polysulphide ‖ Kalkschwefelleber f. polysulfure m. de calcium. / ~ soda electrolysis ‖ Kalzium-Natriumelektrolyse f. ‖ électrolyse f. de calcium et de sodium. / ~ sulphate ‖ Gips m. ‖ gypse m.; plâtre m. / ~ sulphide ‖ Schwefelkalzium n. ‖ sulfure m. de calcium.

calculagraph ‖ Kalkulagraf m. ‖ calculagraphe m.

calculate, to ‖ ausrechnen; berechnen ‖ calculer; compter. / ~ the volume of a solid ‖ die Masse f. kubizieren ‖ cuber la masse f.

calculated speed ‖ errechnete Geschwindigkeit f. ‖ vitesse f. donnée par le calcul; vitesse f. calculée ou prévue ou théorétique.

calculating device ‖ Rechenvorrichtung f. ‖ appareil m. à calculer. / ~ instrument ‖ Recheninstrument n. ‖ instrument m. de calcul ou à calculer. / ~ machine ‖ Rechenmaschine f. ‖ calculateur m.; machine f. à calculer. / ~ typewriter ‖ rechnende Schreibmaschine f. ‖ machine f. à écrire combinée avec machine à calculer.

calculation ‖ Ausrechnung f.; Berechnung f.; Rechnung f. ‖ calcul m.; compte m. / ~ of areas ‖ Flächenberechnung f. ‖ calcul m. des aires. / beyond ~ ‖ der Berechnung f. entzogen ‖ échappé au calcul m. / ~ of the current consumption (Electr) ‖ Strombedarfsberechnung f. ‖ calcul m. de la consommation de courant / ~ of economy ‖ Wirtschaftlichkeitsberechnung f. ‖ calcul m. de l'économie. / ~ of errors ‖ Fehlerrechnung f. ‖ calcul m. des erreurs. / to be far out in the ~s ‖ sich in der Rechnung f. irren ‖ être dans l'erreur m. de calcul. / ~ involving fractions ‖ Bruchrechnung f. ‖ calcul m. des fractions. / ~ of interest ‖ Zinsenberechnung f. ‖ calcul m. des intérêts. / ~ of probability ‖ Wahrscheinlichkeitsberechnung f. ‖ calcul m. de probabilité. / ~ of prospective profits ‖ Rentabilitätsberechnung f. ‖ calcul m. de rendement financier. / re-~ ‖ Neuberechnung f. ‖ calcul m. fait à nouveau / ~ of rents and annuities (Math) ‖ Rentenrechnung f. ‖ calcul m. des rentes et annuités. / ~ of stability ‖ Stabilitätsberechnung f. ‖ calcul m. de stabilité. / static ~ ‖ statische Berechnung f. ‖ calcul m. statique ou de stabilité. / ~ of strength ‖ Festigkeits(be)rechnung f. ‖ calcul m. de la résistance. / ~ of a traverse ‖ Polygonberechnung f. ‖ calcul m. d'un polygone. / ~ of the yield ‖ Ausbeuteberechnung f. ‖ calcul m. du rendement.

calculation speed see calculated speed.

calculator ‖ Rechenmaschine f. ‖ calculateur m.; machine f. à calculer.

calculus ‖ Rechnen n.; Rechnung f. ‖ calcul m.

caldron ‖ kleiner Kessel m.; Kupferkessel m. ‖ petite chaudière f.; chaudron m. en cuivre.

caledonite ‖ Kaledonit m. ‖ calédonite f.; plomb m. sulfaté-carbonaté cuprifère.

calefaction ‖ Erwärmung f. ‖ caléfaction f.

calendar ‖ Kalender m. ‖ calendrier m.; almanach m. / ~ (Watchm) ‖ Datumwerk n. ‖ quantième m. / tear-off ~ ‖ Abreißkalender m. ‖ calendrier m. ou bloc m. à feuilles détachables. / wall ~ ‖ Wandkalender m. ‖ calendrier m. de mur. / back of wall ~ ‖ Kalenderwand f. ‖ dos m. de calendrier. / writing table ~ ‖ Schreibtischkalender m. ‖ calendrier m. de table de bureau.

calendar back ‖ Kalenderrückwand f. ‖ dos m. de calendrier. / ~ block ‖ Kalenderblock m. ‖ bloc m. de calendrier. / replacing ~ block ‖ Kalenderersatzblock m. ‖ bloc m. de remplacement à calendrier. / special machine for making ~s ‖ Kalenderspezialmaschine f. ‖ machine f. spéciale pour la fabrication de calendriers. / ~ socle ‖ Kalenderuntersatz m. ‖ support m. de blocs calendriers ou de calendrier. / ~ watch ‖ Uhr f. mit Taganzeiger ‖ montre f. à quantième.

calender, to (Cloth) ‖ mangeln; kalandern; glätten ‖ calandrer; cylindrer; lustrer. / ~ (Pap) ‖ satinieren; kalandern ‖ calandrer; satiner.

calender ‖ Kalander m.; Glättmaschine f.; Mangel f. ‖ calandre f. / friction ~ ‖ Glanzkalander m.; Glättkalander m. ‖ calandre f. à lustrer. / glazing ~ ‖ Seidenglanzkalander m. ‖ calandre f. de satinage soyeux. / india-rubber ~ ‖ Gummikalander m. ‖ calandre f. à caoutchouc. / ~ for paper and cardboard ‖ Kalander m. für Papier und Pappe ‖ calandre f. à papier et carton. / ~ with rollers arranged alternatively for paper in rolls ‖ Kalander m. mit versetzter Walzenanordnung für Papier in Rollen ‖ calandre f. à disposition en quinconce des cylindres pour papier en rouleaux. / textile ~ ‖ Textilkalander m. ‖ calandre f. à textiles. / wet ~ ‖ Naßkalander m. ‖ calandre m. au mouillé. / x-roll twin ~ for glazing paper sheets (working width of rolls y mm) fitted with automatic sheet elevators and sheet transferring device ‖ x-walziger Doppelbogenkalander m. von y mm Walzenbreite mit selbsttätiger Bogen-, Hoch- und Überführung ‖ calandre-jumelle f. pour papier en feuilles (à x cylindres d'une largeur utile de y mm) et avec ascenseurs et transporteurs automatiques de feuilles.

calender bowl ‖ Kalanderwalze f. ‖ cylindre m. de calandres.

calendering ‖ Kalandern n. ‖ calandrage m. / ~ machine see calender.

calender man ‖ Kalanderführer m. ‖ calandreur m.; lamineur m. de papier. / part for ~ ‖ Kalanderteil m. ‖ pièce f. détachée de calandres. / ~ roller press ‖ Kalanderwalzenpresse f. ‖ presse f. à rouleaux.

calf (Anatomy) ‖ Wade f. ‖ mollet m. / ~ (Curr) ‖ Kalbleder n. ‖ veau m.; cuir m. de veau. / ~ (Mar) ‖ Eisscholle f. ‖ glaçon m. / bound in ~ (Bookb) ‖ in Kalbleder n. gebunden ‖ relié en veau m.

calf binding ‖ Ganzlederband m.; Franzband m. ‖ reliure f. en cuir plein *ou* en veau. / ~ bound volume *see* ~ binding. / ~ leather (Curr) ‖ Kalbleder n. ‖ veau m.; cuir m. de veau.

calf's foot dresser ‖ Kalbfußbereiter m. ‖ préparateur m. de pieds de veau. / ~ hair spinner ‖ Kälberhaarspinner m. ‖ fileur m. de poils de veau.

calf skin ‖ Kalbfell n.; Kalbleder n. ‖ peau f. *ou* cuir m. de veau.

caliber *see also* caliper ‖ Kalibriermaß n.; Kaliber n. ‖ calibre m. / ~ (Letter-found) Kernmaß n. ‖ prototype m.; typomètre m. / ~ (Mach) ‖ Schlüsselweite f. ‖ calibre m. / ~ (Print) ‖ Kaliber n. ‖ chape f. / ~ for polishing screw-knobs ‖ Schleifapparat m. für Schraubenköpfe ‖ moulin m. à polir les têtes de vis. / roll ~ ‖ Walzenkaliber n. ‖ calibre m. de cylindre de laminoir.

caliber gauge *see* caliber.

calibrate, to ‖ eichen ‖ graduer; calibrer.

calibrated chain ‖ kalibrierte Kette f. ‖ chaîne f. calibre. / non-~ chain ‖ unkalibrierte Kette f. ‖ chaîne f. non calibrée. / ~ link chain ‖ kalibrierte Gliederkette f. ‖ chaîne f. calibrée à maillons.

calibrating machine ‖ Kalibriermaschine f. ‖ machine f. à calibrer. / ~ press ‖ Kalibrierpresse f. ‖ presse f. à calibrer. / ~ resistance ‖ Eichwiderstand m. ‖ résistance f. d'étalonnage. / ~ room for electricity meters ‖ Zählereichsaal m. ‖ salle f. d'étalonnage *ou* de vérification de compteurs.

calibration ‖ Eichung f.; Gradeinteilung f. étalonnage m.; graduation f. / ~ of a spectroscope in terms of wave lengths ‖ Eichung f. eines Spektroskopes nach Wellenlängen ‖ étalonnage m. des échelles de longueurs d'ondes d'un spectroscope.

calibration cable (Electr) ‖ Eichkabel n. ‖ câble m. étalon. / ~ circuit (Tel) ‖ Eichleitung f. ‖ ligne f. artificielle étalon. / ~ coil (Tel) ‖ Eichspule f. ‖ bobine f. d'étalonnage. / ~ condenser ‖ Eichkondensator m. ‖ condensateur m. étalon. / ~ current regulator (Electr) ‖ Eichstromregler m. ‖ régulateur m. d'intensité d'étalonnage. / ~ curve ‖ Eichkurve f. ‖ courbe f. d'étalonnage. / ~ instrument (Electr) ‖ Eichmaß n. ‖ instrument m. étalon. / ~ potential regulator (Electr) ‖ Eichspannungsregler m. ‖ régulateur m. de tension d'étalonnage. / ~ shunt ‖ Eichnebenschluß m. ‖ résistance f. étalon. / ~ table ‖ Eichtabelle f. ‖ table f. d'étalonnage.

calibre *see* caliber.

calico ‖ Kaliko m.; Kattun m. ‖ toile-coton m.; calicot m. / printed ~ ‖ Zitz m. ‖ indienne f.

calico printer ‖ Kattundrucker m. ‖ indienneur m. / ~ printing ‖ Kattundruck m. ‖ impression f. de tissus de coton; impression f. d'indiennes. / ~ printing machine ‖ Kattundruckmaschine f. ‖ machine f. à imprimer le coton.

caligrapher *see* calligrapher.

caliper *see also* caliber ‖ Kalibriermaß n.; Kaliber n. ‖ calibre m. / ~ (Cylinder) ‖ Zylinderstichmaß n. ‖ jauge f. à coulisses. / ~ (Shaft) ‖ Tasterlehre f.; Rachenlehre f. ‖ calibre m. à mâchoire; calibre-mâchoire m. / lumber ~ ‖ Holzmeßgerät n. ‖ compas m. forestier. /

sliding ~ ‖ Schublehre f.; Schiebelehre f. ‖ calibre m. *ou* pied m. à coulisse; calibre m. coulant.

caliper gauge *see* caliper.

calipering instrument ‖ Kalibrierinstrument n. ‖ instrument m. de calibrage.

caliper rule ‖ Schraubenlehre f. ‖ calibre m. à vis.

calipers pl. ‖ Taster m.; Greifzirkel m.; Tasterzirkel m.; Kaliberzirkel m. ‖ compas m. / back ~ ‖ Taster m. mit Zahnbogen ‖ compas m. d'épaisseur à crémaillère. / egg ~ ‖ Ellipsenzirkel m. ‖ compas m. double calibre. / egg ~ with slider ‖ Ellipsenzirkel m. mit Gleitführung ‖ compas m. double calibre à glissière. / globe ~ ‖ Kugeltaster m. ‖ compas m. d'épaisseur pour billes. / inside ~ ‖ Lochtaster m.; Innentaster m. ‖ compas m. d'intérieur. / inside and outside ~ ‖ Innen- und Außentaster m. ‖ compas m. maître de danse *ou* d'intérieur et d'extérieur. / outside ~ ‖ Außentaster m. ‖ compas m. d'épaisseur *ou* d'extérieur. / ~ with regulating screw ‖ Taster m. mit Feinstellschraube ‖ compas m. de précision *ou* à vis de rappel de précision. / ~ with set screw ‖ Taster m. mit Einstellschraube ‖ compas m. à vis de rappel. / sliding ~ *see* sliding caliper. / spring ~ ‖ Federtaster m. ‖ compas m. à ressort. / thread ~ ‖ Gewindetaster m. ‖ compas m. d'épaisseur à vis.

calk . . . (Boil; Shipb) *see also* caulk . . .

calk, to (Drawing) ‖ pausen; durchzeichnen; durchpausen ‖ calquer. / ~ (Boil; Iron shipb) ‖ stemmen; verstemmen ‖ mater. / ~ (Wood shipb) ‖ kalfatern; dichten ‖ calfater. / ~ in ‖ einstemmen ‖ mater.

calk (Horse-shoe) ‖ Stollen m.; Eisgriff m. ‖ crampon m. / ~ with screws for horse-shoes ‖ Hufschraubstollen m. ‖ crampon m. à vis pour fers à cheval.

calked seam ‖ Stemmnaht f. ‖ joint m. maté; couture f. *ou* soudure f. matée.

calker (Shipb) ‖ Kalfaterer m. ‖ calfateur m.

calkin *see* calk.

calking (Drawing) ‖ Kopie f.; Pause f. ‖ poncis m.; poncif m.; calque m. / ~ (Shipb) ‖ Kalfaterung f. ‖ calfat m.; radoub m. / pneumatic ~ ‖ Stemmen n. mit Druckluft ‖ matage m. pneumatique. / ~ colours pl. ‖ Pausfarben fpl. ‖ couleurs fpl. à calquer.

call, to ~ for funds pl. ‖ Kapital n. einfordern ‖ appeler du capital. / ~ for tenders ‖ verdingen ‖ mettre en adjudication. / ~ in ‖ einkassieren ‖ encaisser. / ~ a person ‖ bei jemandem vorsprechen ‖ passer chez quelqu'un. / ~ up (Tel) ‖ anrufen ‖ téléphoner; appeler au téléphone; attaquer.

call ‖ Ruf m. ‖ appel m. / ~ (Hunting) ‖ Lockpfeife f. ‖ appeau m. / ~ (Mar) ‖ Kommando n. ‖ commandement m. / ~ (Tel) ‖ Anruf m. ‖ appel m. / ~ (Tel; Communication) ‖ Gespräch n. ‖ conversation f. / ~ (Trade) ‖ Abruf m. ‖ rappel m. / ~ for air traffic ‖ Luftgespräch n. ‖ communication f. de service des lignes d'aviation. / at ~ ‖ auf Abruf m. ‖ à convenance f. / audible ~ ‖ akustisches Signal n. ‖ appel m. phonique. / automatic ~ (Tel) ‖ selbsttätiger Anruf m. ‖ appel m. automatique. / avis d'appel (Tel) ‖ XP-Gespräch n. ‖ communication

f. avec avis d'appel. / ~ booked by prearrangement (Tel) ‖ Daueranmeldung f. ‖ communication f. demandée par listes préalables. / ~ booked the previous day (Tel) ‖ Vortagsanmeldung f. ‖ demande f. de communication déposée la veille. / fixed-time ~ (Tel) ‖ Festzeitgespräch n. ‖ communication f. fortuite à heure fixe. / government ~ (Tel) ‖ Staatsgespräch n. ‖ conversation f. d'état. / ticket for incoming ~s (Tel) ‖ Blatt n. zur Aufzeichnung eingehender Anrufe ‖ fiche f. d'arrivée d'appels. / ordinary ~ (Tel) ‖ gewöhnliches Gespräch n. ‖ conversation f. ordinaire. / preadvice ~ (Tel) ‖ Gespräch n. mit Voranmeldung ‖ conversation f. avec préavis. / semi-automatic ~ (Tel) ‖ halbselbsttätiger Anruf m. ‖ appel m. semiautomatique.

call bell (Tel) ‖ Melderufklingel f. ‖ sonnerie f. d'appel. / (telephone) ~ box ‖ Fernsprechzelle f. ‖ cabine f. téléphonique. / (mechanical) ~ counter ‖ Leistungszähler m. ‖ compteur m. statistique. / ~ counter key ‖ Leistungszähltaste f. ‖ bouton m. de compteur statistique. / ~ distributing system (Tel) ‖ Anrufverteilungssystem n. ‖ système f. à distributeur d'appel. / duration of a ~ (Tel) ‖ Gesprächsdauer f. ‖ durée f. de conversation.

called subscriber (Tel) ‖ angerufener Teilnehmer m. ‖ abonné m. demandé.

caller, to be ~ (Aut tel) ‖ besetzt sein durch eigenen Anruf ‖ être appelant m.

call finder (Tel) ‖ Anrufsucher m. ‖ chercheur m. d'appels. / frequency of ~s (Tel) ‖ Gesprächsdichte f. ‖ densité f. du trafic téléphonique.

calligrapher ‖ Kalligraph m. ‖ calligraphe m.

call indicator (Tel) ‖ Nummernanzeiger m. ‖ indicateur m. d'appel. / ~ indicator traffic (Tel) ‖ optischer B-Verkehr m. ‖ trafic m. avec des indicateurs d'appel.

calling ‖ Anrufen n.; Rufen n. ‖ appel m.; invocation f. / ~ circuit (Tel) ‖ Rufstromkreis m. ‖ circuit m. d'appel. / ~ current blocking condenser (Tel) ‖ Rufsperrkondensator m. ‖ condensateur m. d'arrêt de courant d'appel. / ~ key ‖ Wecktaste f. ‖ touche f. d'appel.

calling lamp (Tel) ‖ Anruflampe f. ‖ lampe f. d'appel. / ~ repeating (Tel) ‖ Anrufwiederholung f. ‖ répétition f. des lampes d'appel. / ~ strip (Tel) ‖ Anruflampenstreifen m. ‖ réglette f. de lampe d'appel.

calling line (Tel) ‖ Anrufleitung f. ‖ ligne f. d'appel. / ~ magneto ‖ Anrufinduktor m. ‖ inducteur m. d'appel. / ~ plug (Tel) ‖ Verbindungsstöpsel m.; Rufstöpsel m. ‖ fiche f. d'appel. / ~ relay (Tel) ‖ Anrufrelais n. ‖ relais m. d'appel. / ~ signal (Tel) ‖ Anrufzeichen n. ‖ signal m. d'appel. / ~ subscriber (Tel) ‖ anrufender Teilnehmer m. ‖ abonné m. demandeur. / ~ wave (Tel) ‖ Anrufwelle f. ‖ onde f. d'appel.

calliper *see* caliper.

callipers pl. *see* calipers.

call, loss of ~s (Aut tel) ‖ Verlustziffer f. im Fernsprechbetrieb ‖ perte f. des appels. / ~ metering (Tel) ‖ Gesprächszählung f. ‖ comptage m. des communications. / ~ number (Tel) ‖ Rufnummer f.; Telefonnummer f. ‖ numéro m. téléphonique. / order of ~s (Tel) ‖ Reihenfolge f. der Gesprächsverbindungen ‖ ordre m.

des communications. / ~ point (Fire station) ‖ Feuermeldestelle f. ‖ poste m. d'appel. / ~ receiving switch (Aut tel) ‖ Anrufverteiler m. ‖ répartiteur m. d'appels. / ~ relay for cable working (Tel) ‖ Anrufrelais n. für Kabelbetrieb ‖ relais m. d'appel pour le service des câbles. / ~ signal (Tel) ‖ Rufzeichen n. ‖ indicatif m. d'appel. / ~ signal apparatus (Tel) ‖ Anrufapparat m. ‖ appareil m. d'appel. / ~ signal band (Tel) ‖ Rufzeichenband n. ‖ bande f. des indicatifs d'appel. / storer-up for ~s (Aut tel) ‖ Wartefeld n.; Anrufsammler m. ‖ enregistreur m. des appels. / supervision of ~s (Tel) ‖ Gesprächsüberwachung f. ‖ contrôle m. des communications. / ticket for incoming ~s (Tel) ‖ Blatt n. zur Aufzeichnung eingehender Anrufe ‖ fiche f. d'arrivée d'appels.

calm (Sea) ‖ ruhig ‖ calme. / to become ~ ‖ sich beruhigen ‖ calmier; se calmer; mollir.

calm ‖ Windstille f. ‖ calme m. / dead ~ ‖ vollständige Windstille f. ‖ calme m. plat.

Calmon gummed tape (Tel) ‖ Calmonsches Gummiband n. ‖ ruban m. gommé Calmon.

calomel ‖ Quecksilberchlorür n.; Kalomel n. ‖ chlorure m. mercureux; protochlorure m. de mercure.

caloric capacity ‖ kalorische Leistung ‖ débit m. calorifique. / ~ engine ‖ Heißluftmotor m. ‖ moteur m. à air chaud. / ~ unit ‖ Wärmeeinheit f. ‖ unité f. de chaleur.

calorie ‖ Wärmeeinheit f.; Kalorie f. ‖ calorie f. / ~s pl. per hour ‖ Kalorien fpl. je Stunde ‖ colories fpl. en heure.

calorific power see calorific value.

calorific value ‖ Heizwert m. ‖ valeur f. ou pouvoir m. ou puissance f. calorifique. / ~ of the coal ‖ Heizwert m. oder Brennwert m. der Kohle ‖ valeur f. calorifique du charbon. / gas of high ~ ‖ heizkräftiges Gas n. ‖ gaz m. (doué) d'un grand pouvoir calorifique.

calorifier ‖ Heizkörper m.; Ofen m. ‖ calorifère m.; poêle m.

calorimeter ‖ Kalorimeter n.; Wärmemesser m.; Heizwertmesser m. ‖ calorimètre m. / ice ~ ‖ Eiskalorimeter n. ‖ calorimètre m. à glace. / water ~ ‖ Mischungskalorimeter n. ‖ calorimètre m. à eau.

calorimetric bomb ‖ kalorimetrische Bombe f. ‖ bombe f. calorimétrique. / ~ determination ‖ kalorimetrische Bestimmung f. ‖ détermination f. calorimétrique. / ~ measurement ‖ kalorimetrische Messung f. ‖ mesure f. au calorimètre. / ~ plant ‖ kalorimetrische Einrichtung f. ‖ installation f. calorimétrique. / ~ value see calorific value.

calorimetry ‖ Wärmemengenmessung f. ‖ calorimétrie f.

calorized steel ‖ alitierter Stahl m. ‖ acier m. calorisé.

calory see calorie.

calotte (Arch) ‖ Haube f.; Haubenlager n. ‖ calotte f.; chapeau m.; bonnet m.; chape f. / ~ (Geom) ‖ Kalotte f.; Kappe f. ‖ calotte f. / depressed ~ (Arch) ‖ gedrückte Kappe f. ‖ calotte f. surbaissé. / ~ for hats ‖ Hutkappe f. ‖ calotte f. pour chapeaux.

calx viva ‖ ungelöschter oder gebrannter Kalk m. ‖ chaux f. vive.

calyx (Bot) ‖ Kelch m. ‖ calice m.

cam (Mach) ‖ Knagge f.; Daumen m.; Nocken m.; Steuerdaumen m.; Nase f.; Kamm m. ‖ came f.; taquet m.; camme f. / adjustable ~ ‖ verstellbarer Nocken m. ‖ came f. réglable. / admission ~ ‖ Einlaßnocken m. ‖ came f. d'admission. / ~ of an arbor ‖ Hebedaumen m.; Wellendaumen m. ‖ mentonnet m. de l'arbre. / exhaust ~ ‖ Auslaßnocken m. ‖ came f. d'échappement. / exhaust valve ~ ‖ Auslaßventilnocken m. ‖ came f. de soupape d'échappement. / inlet ~ ‖ Einlaßnocken m. ‖ came f. d'admission. / ~ keyed on ‖ aufgekeilter Nocken m. ‖ came f. clavetée (sur). / knocking-over ~ (Weav) ‖ Abschlagexzenter n. ‖ came f. d'abattage. / ~ lead ~ (Tool) ‖ Leitkurve f. ‖ came f. de chariotage. / ~ milled-out of the solid ‖ mit der Welle aus einem Stück gefräster Nocken m. ‖ came f. et arbre m. fraisés d'une seule pièce. / ~ of the shaft ‖ Wellendaumen m.; Nocken m. ‖ mentonnet m. ou came f. de l'arbre. / sliding ~ ‖ verschiebbarer Nocken m. ‖ came f. à déplacement.

camber (Shoem) ‖ Hohlungsstück n. ‖ cambrure f. / ~ of beam (Shipb) ‖ Decksbalkenbucht f. ‖ bouge m. de barrot.

cambered axle ‖ gestürzte Achse f. ‖ essieu m. arqué. / ~ wing ‖ gewölbter Flügel m. ‖ aile f. cambrée.

cambering attachment ‖ Balligschleifeinrichtung f. ‖ dispositif m. à rectifier le bombé.

cam block, leather ‖ Lederklotz m. ‖ sabot m. en cuir. / ~ brake ‖ Kniehebelbremse f. ‖ frein m. à came ou à genouillère ou à levier articulé.

Cambrian system ‖ kambrisches System n. ‖ système m. cambrien.

cambric (Weav) ‖ Batist m. ‖ batiste f. / coarse linen ~ ‖ Batistleinwand f. ‖ grosse batiste f. / ~s pl. for wound dressing ‖ Verbandstoff m. ‖ matière f. pour pansements. / ~ weaver ‖ Batistweber m. ‖ mulquinier m.; tisseur m. de linon ou de batiste.

cam contour ‖ Nockengestalt f. ‖ profil m. de came. / ~ disc ‖ Schlagrad n. ‖ roue f. à came.

camel ‖ Kamel n. ‖ chameau m.

camel-hair ‖ Kamelhaar n. ‖ poils mpl. de chameau. / ~ belt ‖ Kamelhaartreibriemen m. ‖ courroie f. de poils de chameaux. / ~ cover ‖ Kamelhaardecke f. ‖ couverture f. en poils de chameau. / ~ fabric ‖ Kamelhaarstoff m. ‖ étoffe f. en poils de chameau. / ~ loden (cloth) ‖ Kamelhaarstrichloden m. ‖ lodier m. en poils de chameau. / ~ rug see ~ cover. / ~ spinner ‖ Kamelhaarspinner m. ‖ fileur m. de poils de chameau. / ~ tissue ‖ Kamelhaargewebe n. ‖ tissu m. en poils de chameau.

camellia seed ‖ Kameliensame m. ‖ graine f. de cameline.

cameo ‖ Gemme f.; geschnittener Stein m. ‖ gemme f.; camée f.

camera ‖ fotografischer Apparat m.; Kamera f. ‖ appareil m. photographique; chambre f. noire. / aerial ~ ‖ Luftbildkamera f. ‖ appareil m. photographique aérien. / aerial mapping ~ ‖ Luftbildmeßkamera f. ‖ chambre f. aéro-photogrammétrique. / air survey ~ see aerial mapping ~. / astro ~ see astrophotographic ~. / astrophotographic ~ ‖ astro-

fotografische Kamera f.; Astrokamera f. ‖ chambre f. astrophotographique. / cinematographic ~ ‖ Filmaufnahmekamera f.; Aufnahmeapparat m. ‖ chambre f. de prise de vues (pour cinémas). / ~ for copying work ‖ Reproduktionskamera f.; Wiedergabekamera f. ‖ chambre f. noire pour la reproduction. / double ~ ‖ Doppelkamera f. ‖ double-chambre f. noire. / enlarging ~ ‖ Vergrößerungskamera f. ‖ chambre f. noire pour l'agrandissement. / ~ with very long extension ‖ Kamera f. von sehr großer Brennweite ‖ chambre f. noire de très haute distance focale. / falling plate ~ ‖ Fallplattenkamera f. ‖ appareil m. photographique à plaques rabattantes. / ~ set to infinity ‖ auf unendlich eingestellte Kamera f. ‖ chambre f. noire reglée sur les lointains. / magazine ~ ‖ Magazinkamera f. ‖ appareil m. photographique magasin. / mirror reflex ~ ‖ Spiegelreflexkammer f. ‖ chambre f. à miroir de visée. / moon ~ ‖ Mondkamera f. ‖ chambre f. noire pour la lune. / motion or moving-picture ~ see cinematograph ~. / photo-micrographic ~ ‖ mikrofotografischer Apparat m. ‖ appareil ~ de microphotographie. / retinal ~ ‖ Netzhautkamera f. ‖ chambre f. rétinienne. / solar ~ ‖ Sonnenkamera f. ‖ chambre f. noire pour le soleil. / ~ of special design ‖ Spezialkamera f. ‖ chambre f. noire spéciale. / spectroscopic ~ see spectrum ~. / spectrum ~ ‖ Spektralkamera f.; spektrografische Kamera f. ‖ chambre f. spectroscopique. / sun and moon ~ ‖ Sonnemondkamera f. ‖ chambre f. noire pour le soleil et la lune. / ~ for the tropics ‖ Tropenkamera f. ‖ caméra f. pour pays tropics.

camera adjustment ‖ Kameraeinstellung f. ‖ mise f. au point de la chambre noire. / ~ attachment ‖ Kameraansatz m. ‖ chambre-rallonge f. / ~ lens ‖ Kameraobjektiv n. ‖ objectif m. pour chambres noires. / ~ mounting ‖ Kameragerüst n. ‖ cadre m. de la chambre photographique.

camera-obscura ‖ Dunkelkammer f. ‖ caméra f. obscura; chambre f. noire.

cam gear ‖ Nockensteuerung f.; Daumensteuerung f. ‖ commande f. par came; distribution f. à came. / ~ grinding machine ‖ Nockenschleifmaschine f. ‖ machine f. à rectifier les cames.

camlet ‖ Kamelott m. ‖ camelot m.

cam lever ‖ Nockenhebel m. ‖ levier m. à came. / ~ milling machine ‖ Nockenfräsmaschine f. ‖ fraiseuse f. à cames.

camomile ‖ Kamille f. ‖ camomille f. / ~ blossom ‖ Kamillenblüte f. ‖ fleur f. de camomille. / ~ oil ‖ Kamillenöl n. ‖ essence f. de camomilles.

camouflage, to ‖ tarnen ‖ camoufler. / ~ instruction ‖ Tarnvorschrift f. ‖ instruction f. sur le camouflage.

camouflet ‖ Quetschmine f. ‖ fourneau m. sous-chargé; camouflet m.

camp ‖ Lager n.; Feldlager n. ‖ camp m. / ~ (Agr) ‖ Miete f. (für Rüben oder Kartoffeln) ‖ silo m.

campaign ‖ Kampagne f. ‖ campagne f.

camp-bed ‖ Feldbett n. ‖ lit m. de camp(agne); lit m. brisé; brigantin m. / iron ~ ‖ eisernes Feldbett n.; eiserne Bettstelle f. ‖ lit m. de camp en fer; châlit m. en fer.

campeachy-wood ‖ Kampescheholz n. ‖ bois m. de Campêche. / ~ **extract** ‖ Kampescheholzauszug m. ‖ extrait m. de bois de Campêche.

camphor ‖ Kampfer m. ‖ camphre m.

camphorated spirit of wine ‖ Kampferspiritus m. ‖ alcool m. camphré.

camphoric acid ‖ Kampfersäure f. ‖ acide m. camphorique.

camphorine egg ‖ Kampferkugel f. ‖ boule f. de camphre; œuf m. de camphorine.

camphor oil ‖ Kampferöl n. ‖ essence f. de camphre. / ~ **refinery** ‖ Kampferraffinerie f. ‖ raffinerie f. de camphre.

cam-plate ‖ Hubscheibe f. ‖ disque m. à came.

camp sheathing ‖ Bollwerk n. ‖ palée f. de berge. / ~ **stool** ‖ Feldstuhl m.; Klappstuhl m. ‖ siège m. pliant ou de campagne.

camshaft ‖ Steuerwelle f.; Nockenwelle f.; Hubscheibenwelle f. ‖ arbre m. à cames ou de distribution. / **exhaust** ~ ‖ Auslaßnockenwelle f. ‖ arbre m. à cames d'échappement. / **movable** ~ ‖ verschiebbare Nockenwelle f. ‖ arbre m. à cames baladeur. / ~ **provided with two sets of cams** ‖ Steuerwelle f. mit einem doppelten Satz von Steuernocken ‖ arbre m. de distribution portant un double jeu de cames.

camshaft bearing ‖ Nockenwellenlager n.; Steuerwellenlager n. ‖ coussinet m. ou palier m. d'arbre à cames. / ~ **grinding machine** ‖ Nockenwellenschleifmaschine f. ‖ machine f. à rectifier les arbres à cames. / ~ **milling and copying machine** ‖ Exzenterwellenkopierfräsmaschine f. ‖ machine f. à fraiser les arbres à cames avec dispositif à copier. / ~ **side** ‖ Nockenwellenseite f. ‖ côté m. de distribution. / ~ **timing gear** ‖ Nockenwellenantriebsrad n. ‖ roue f. de l'arbre à cames.

cam shape ‖ Nockenform f. ‖ forme f. de came. / ~ **wheel** ‖ Knaggenrad n. ‖ roue f. à cames.

can, to ~ (Preserves) ‖ einmachen ‖ confire; conserver.

can ‖ Kanne f. ‖ bidon m.; burette f. / **ice** ~ ‖ Eiszelle f. ‖ mouleau m. à glace. / ~ **for motor cars** ‖ Kraftwagenkanister m.; Kanister m. ‖ bidon m. pour automobiles. / **nickel-plated** ~ ‖ vernickelte Kanne f. ‖ burette f. nickelée. / **varnish** ~ ‖ Lackkanne f. ‖ bidon m. à laque.

Canada balsam ‖ Kanadabalsam m. ‖ baume f. de Canada.

Canadian axe ‖ amerikanische Axt f. ‖ hache f. façon Canada. / ~ **poplar** ‖ kanadische Pappel f. ‖ peuplier m. du Canada.

canal ‖ Schiffskanal m.; Kanal m. ‖ canal m. navigable ou de navigation. / ~ **for irrigation** ‖ Bewässerungskanal m. ‖ canal m. d'irrigation. / **lateral** ~ ‖ Seitenkanal m.; Parallelkanal m. ‖ canal m. de dérivation. / ~ **with summit** ‖ Scheitelkanal m. ‖ canal m. à point de partage.

canal boat ‖ Pinasse f. ‖ péniche f.; pinasse f. / ~ **bottom** ‖ Kanalsohle f. ‖ lit m. du canal. / ~ **building** ‖ Kanalbau m. ‖ construction f. de canaux. / ~ **cleaning outfit** ‖ Kanalreinigungsgerät n. ‖ instrument m. pour le nettoyage des canaux. / ~ **closure** ‖ Schiffssperre f. auf einem Kanal ‖ chômage m. sur un canal. / ~ **dredger** ‖ Kanalbagger m. ‖ révoyeur m. / ~ **harbour** ‖ Kanalhafen m. ‖ garage m. ou gare f. d'un canal.

canalization ‖ Kanalisation f. ‖ canalisation f. / ~ **decanting plant** ‖ Kläranlage f. für Kanalisation ‖ installation f. de clarification pour canalisation. / ~ **plant** ‖ Kanalisationsanlage f. ‖ installation f. de canalisation. / ~ **works** pl. ‖ Kanalisationsanstalt f. ‖ entreprise f. de canalisations.

canalize, to ‖ kanalisieren ‖ canaliser.

canal lock ‖ Kanalschleuse f. ‖ écluse f. d'un canal. / ~ **mouth** ‖ Kanalmündung f. ‖ embouchure f. d'un canal. / ~ **navigation** ‖ Kanalschiffahrt f. ‖ navigation f. sur canaux. / ~ **navigation service** ‖ Kanalschiffahrtsverkehr m. ‖ exploitation de canaux mpl. de navigation. / ~ **piling** ‖ Kanalspundwand f. ‖ palplanche f. / ~ **rays** pl. (Electr) ‖ Kanalstrahlen mpl. ‖ rayons mpl. canaux. / ~ **regulator** ‖ Speiseschleuse f. oder Einlaßöffnung f. eines Kanals ‖ prise f. d'eau d'un canal. / ~ **slide-valve** ‖ Kanalschieber m. ‖ tiroir m. ou vanne f. de canal. / ~ **system** ‖ Kanalnetz n. ‖ réseau m. de canaux. / ~ **tug** ‖ Kanalschlepper m. ‖ remorqueur m. de canal.

Cananga oil ‖ Kanangaöl n. ‖ essence f. de Cananga. / ~ **water** ‖ Kanangawasser n. ‖ eau f. de Cananga.

canard-type airplane ‖ Entenflugzeug n. ‖ avion-canard m.

can-buoy ‖ Kegelboje f. ‖ bouée f. conique.

cancel, to (A contract etc.) ‖ ungültig machen; gerichtlich auflösen; kassieren; rückgängig machen ‖ casser; annuler; destituer; resilier. / ~ (Bookkeep) ‖ stornieren; ristornieren ‖ contre-passer; ristorner. / ~ (Stamps) ‖ entwerten ‖ oblitérer. / ~ **a purchase** ‖ einen Kauf m. rückgängig machen oder aufkündigen ‖ révoquer ou annuler un achat.

cancellation (Of a contract etc.) ‖ Nichtigkeitserklärung f.; Ungültigmachung f.; Streichung f. ‖ annulation f.; cassation f. / ~ (Of an order) ‖ Abbestellung f. ‖ annulation f. d'une commande ou d'un ordre; décommande f. / contre-ordre m. / ~ **suit** ‖ Nichtigkeitsklage f. ‖ demande f. en nullité.

cancelled ‖ gegenstandslos; nichtig ‖ annulé.

cancelling key (Electr) ‖ Rücknahmetaste f. ‖ touche f. de la remise en position.

candelabrum ‖ Kandelaber m.; Armleuchter m. ‖ candélabre m.

candelilla wax ‖ Kandelillawachs n. ‖ cire f. de candélille.

candidate ‖ Bewerber m.; Anwärter m.; Kandidat m. ‖ candidat m. / **to propose as** ~ ‖ als Kandidaten m. aufstellen ‖ porter ou présenter comme candidat m.

candied fruits pl. ‖ Konfitüre f.; kandierte Früchte fpl. ‖ confitures fpl.; fruits mpl. candis; candis mpl. / ~ **lemon peel** ‖ Zitronat n. ‖ citronnat m.

candle ‖ Kerze f.; Licht n. ‖ bougie f.; chandelle f.; cierge f. / **adorned** ~ ‖ verzierte Kerze f. ‖ bougie f. ornée. / **ceresin(e)** ~ ‖ Zeresinlicht n. ‖ chandelle f. en cérésine. / **coloured** ~ ‖ farbige Kerze f. ‖ bougie f. colorée. / **dipped** ~ ‖ gezogenes Licht n. ‖ chandelle f. plongée ou à la baguette. / **dyed** ~ ‖ gefärbte Kerze f. ‖ bougie f. colorée ou en couleur. / **fancy** ~ ‖ Luxuskerze f. ‖ bougie f. de luxe. / ~ **de luxe** see fancy ~. / **mould** ~ ‖ gegossenes Licht n. ‖ chandelle f. moulée ou au moule. / **paraffin** ~ ‖ Paraffinlicht n.; Paraffinkerze f. ‖ chandelle f. en paraffine. / ~ **with plaited wick** ‖ Licht n. mit geflochtenem Docht ‖ chandelle f. à mèche tressée. / **small coloured** ~ ‖ kleines farbiges Licht n. ‖ petite f. bougie colorée. / **standard** ~ (Opt) ‖ Normalkerze f. ‖ bougie f. normale. / **stearin** ~ ‖ Stearinlicht n. ‖ chandelle f. en stéarine. / **tallow** ~ ‖ Talglicht n. ‖ chandelle f. en suif. / ~ **with twisted wick** ‖ Licht n. mit gedrehtem Docht ‖ chandelle f. à mèche moulinée. / **wax** ~ ‖ Wachslicht n. ‖ chandelle f. en cire. / ~ **with woven wick** ‖ Licht n. mit gewebtem Docht ‖ chandelle f. à mèche tissée.

candle coal see cannel coal. / ~ **dipper** ‖ Lichtzieher m. ‖ chandelier m. / ~ **holder** see also candlestick ‖ Kerzenhalter m.; Handleuchter m. ‖ porte-bougie m.; bougeoir m.; chandelier m. / ~ **hour** ‖ Kerzenstunde f. ‖ bougie-heure f. / ~ **lamp** (Electr) ‖ Kerzenlampe f. ‖ lampe f. bougie. / ~ **lamp holder** ‖ Kerzenfassung f. ‖ douille f. à bougie. / ~ **lamp socket** see ~ lamp holder. / ~ **manufacturing** ‖ Kerzenfertigung f. ‖ fabrication f. de bougies. / ~ **manufacturing machine** ‖ Kerzenherstellungsmaschine f. ‖ machine f. à fabriquer des bougies. / ~ **meter** (Opt) ‖ Meterkerze f. ‖ bougie-mètre f. / ~ **moulding machine** ‖ Kerzengießmaschine f. ‖ machine f. à mouler les bougies. / ~ **polisher** ‖ Kerzenglätter m. ‖ lisseur ou polisseur de bougies.

candle-power ‖ Lichtstärke f. ‖ intensité f. ou puissance f. lumineuse (en bougies). / **mean hemispherical** ~ ‖ mittlere hemisphärische Lichtstärke f. ‖ intensité f. hémisphérique moyenne mesurée en bougies. / **mean spherical** ~ ‖ mittlere sphärische Lichtstärke f. ‖ bougies fpl. moyennes sphériques.

candle-power curve ‖ Leuchtkraftkurve f. ‖ courbe f. de l'intensité lumineuse. / ~ **intensity** see candle-power.

candlestick ‖ Leuchter m.; Kerzenhalter m.; Handleuchter m. ‖ bougeoir m.; porte-bougie m.; chandelier m. / **branched** ~ ‖ Armleuchter m. ‖ chandelier m. à bras; applique f. / ~ **for pianos** ‖ Pianoleuchter m. ‖ chandelier m. de pianos.

candle wick ‖ Kerzendocht m. ‖ mèche f. à bougies ou à chandelles. / ~ **wood** ‖ Zitronenholz n. ‖ bois m. de citron; bois m. jaune ou chandelle.

candy, to ‖ überzuckern; mit Zucker überziehen; kandieren ‖ candir; saupoudrer de sucre.

candy (Candied fruits) ‖ Konfitüre f.; eingezuckerte Früchte fpl. ‖ confitures fpl.; fruits mpl. candis; candis mpl. / ~ (Confectionery) ‖ Zuckerwerk n.; Zuckergebäck n. ‖ sucreries fpl. / ~ (Sugar) see sugar candy. / **stick** ~ ‖ Zuckerstange f. ‖ sucre m. en bâtons. / **sugar** ~ ‖ Kandis m.; Kandiszucker m.; Kandelzucker m.; Zuckerkand m. ‖ sucre m. candi; candi m.

candy maker ‖ Kandiskocher m. ‖ cuisinier m. de candi. / ~ **manufactory** ‖ Kandiszuckerfabrik f. ‖ fabrique f. de sucre candi; candiserie f.

cane (Bot) ‖ Rohr n. ‖ canne f.; roseau m. / ~ (Bamboo) ‖ Bambusrohr n.; Bambusstock m. ‖ bambou m. / ~ (Reed) ‖ Schilfrohr n.; Schilf n. ‖ roseau m.; jonc

m.; chalumeau m. / ~ (Stick) ‖ Rohrstock m. ‖ canne f.; jonc m. / ~ (Sugar) ‖ Zuckerrohr n. ‖ canne f. à sucre. / ~ (Walking stick) ‖ Spazierstock m.; Stock m. ‖ canne f. / ~ for chairs ‖ Stuhlrohr n. ‖ canne f. pour chaises. / ravelled ~ ‖ gefasertes Rohr n. ‖ canne f. filée / sugar ~ see cane (Sugar). / sweet ~ ‖ Kalmus m. ‖ acore m.

cane articles pl. ‖ Rohrwaren fpl. ‖ articles mpl. en roseau ou en jonc. / ~ brim stone ‖ Stangenschwefel m. ‖ soufre m. en canons. / ~ furniture ‖ Rohrmöbel npl. ‖ meubles mpl. en rotin. / ~ goods pl. see ~ articles. / ~ gun ‖ Stockflinte f. ‖ canne-fusil m. / ~ juice ‖ Zuckersaft m.; Rohrzuckersaft m. ‖ jus m. de canne à sucre; vesou m. / ~ seat ‖ Rohrsitz m. ‖ siège m. canné.

cane-sugar ‖ Rohrzucker m. ‖ sucre m. de canne; sucre m. colonial. / ~ fabrication ‖ Rohrzuckererzeugung f. ‖ fabrication f. de sucre colonial. / ~ producing plant ‖ Rohrzuckererzeugungsanlage f. ‖ installation f. de sucrerie pour canne à sucre.

cane worker ‖ Rohrarbeiter m. ‖ roselier m.

can-hook ‖ Faßhaken m. ‖ patte f. d'élingue.

canister ‖ Blechbüchse f. ‖ boîte f. métallique.

cankerous growth ‖ Holzkrebs m. ‖ tuméfaction f. de l'écorce.

canned fruits pl. ‖ Obstkonserve f.; eingemachte Früchte fpl. ‖ fruits mpl. conservés ou confits.

canned goods pl. ‖ Konserven fpl. ‖ conserves fpl. / ~ factory ‖ Konservenfabrik f. ‖ fabrique f. de conserves.

canned ham ‖ Schinkenkonserve f. ‖ jambon m. en conserve. / ~ meat ‖ Büchsenfleisch n.; Fleischkonserve f. ‖ conserve f. de viande. / ~ vegetables pl. ‖ Büchsengemüse n. ‖ légumes mpl. conservés.

cannel coal ‖ Fettkohle f.; Kannelkohle f. ‖ houille f. grasse; charbon m. gras; cannel-coal m.

canneled, double ~ ‖ zweiballig ‖ à deux biseaux mpl.

cannery ‖ Konservenfabrik f. ‖ fabrique f. de conserves.

cannon ‖ Geschütz n.; Kanone f. ‖ pièce f. (d'artillerie ou de canon); canon m. / ~ ball ‖ Kanonenkugel f. ‖ boulet m. / ~ foundry ‖ Stückgießerei f.; Kanonengießerei f. ‖ fonderie f. de canons. / ~ howitzer ‖ Kanonenhaubitze f. ‖ canon-obusier m. / ~ lock ‖ Perkussionsschloß n. ‖ percuteur m. / ~ metal ‖ Kanonengut n.; Kanonenbronze f.; Kanonenmetall n. ‖ bronze m. à canon. / ~ pinion (Watch) ‖ Viertelrohr m. ‖ chaussée f.

cannon-proof ‖ bombensicher; bombenfest ‖ à l'épreuve de la bombe.

cannon shot (Projectile) ‖ Geschoß n. einer Kanone ‖ projectile m. d'un canon. / ~ (Range) ‖ Schußweite f. oder Reichweite f. einer Kanone ‖ portée f. de canon.

cannula ‖ Kanüle f.; Wundröhrchen n. ‖ canule f.

canoe ‖ Kanu n. ‖ canot m.; pirogue f.

canon ‖ Regel f.; Richtschnur f.; Vorschrift f.; Kanon m. ‖ règle f.; prescription f. / ~ (Print) ‖ Kanon f.; Kanonschrift f.; Missal f. ‖ canon m.

cañon (Geol) ‖ Kañon n.; enge Schlucht f. ‖ ravin m.

---

canopy ‖ Betthimmel m.; Baldachin m. ‖ baldaquin m. / ~ (Lighting) ‖ Baldachin m. ‖ dais m. / ~ top ‖ Sonnendach n. ‖ capote f. à baldaquin.

cant, to (To incline) ‖ sich neigen; sich auf die Seite legen; schräg liegen ‖ s'incliner. / ~ (To set on edge) ‖ kanten; umkanten ‖ mettre sur la carne. / ~ while moving ‖ ecken beim Fahren ‖ coincer en travers de la voie. / ~ off ‖ abkanten ‖ écorner; émousser; équarrir. / ~ the wood ‖ das Holz kanten ‖ rouler le bois.

cant ‖ Schrägung f.; geneigte Fläche f. ‖ inclinaison f. / ~ of rails ‖ Schienenüberhöhung f. ‖ surélévation f. ou surhaussement m. du rail extérieur.

cant-chisel (Coachm) ‖ Wagnerbeitel m.; Kantbeitel m. ‖ ciseau m. en biseau ou de charron.

canted ‖ abgeschrägt; abgefast ‖ biseauté. / ~ timber (Carp) ‖ kantiger oder abgekanteter Balken m. ‖ bois m. écorné ou équarri.

canteen ‖ Werkskantine f.; Kantine f. ‖ cantine f. / ~ (Bottle) ‖ Feldflasche f. ‖ bidon m.

cant file ‖ Barettfeile f. ‖ barrette f.; lime f. à barrette. / ~ frame (Shipb) ‖ Kantspant m. ‖ couple m. dévoyé ou élancé.

cantharides pl. see cantharis.

cantharis ‖ Kantharide f.; spanische Fliege f. ‖ cantharide f.

cant-hook (Shipb) ‖ Kenterhaken m.; Kanthaken m.; Sethaken m. ‖ renard m.

cantilever ‖ freitragend ‖ encorbellé; portant en faux; en porte-à-faux. / ~ arm ‖ Auslegerbalken m.; freitragender Balken m. ‖ poutre f. en encorbellement. / ~ bridge ‖ Auslegerbrücke f. ‖ pont m. en encorbellement. / ~ crane ‖ Auslegerkran m. ‖ grue f. cantilever. / ~ monoplane ‖ freitragender Eindecker m.; monoplan m. en porte-à-faux. / ~ roof ‖ Kragendach n. ‖ toit m. en porte-à-faux. / ~ spring ‖ Auslegerfeder f. ‖ ressort m. à cantilever. / ~ trembler (Radio) ‖ Konsolenticker m. ‖ trembleur m. à console. / ~ wing (Aero) ‖ freitragender Flügel m. ‖ aile f. en porte-à-faux.

cantilla ‖ Kantille f. ‖ cannetille f.

canting of the shaft ‖ Ecken n. der Welle ‖ coincement m. de l'arbre.

cant timber (Shipb) see ~ frame.

canula needle (Med) ‖ Aderlaßhohlnadel f. ‖ aiguille f. trocart pour saignée.

canvas, to ‖ ausfüttern; mit Segeltuch n. überziehen ‖ doubler; revêtir de canevas m.

canvas (Weav) ‖ Kanevas m.; Baumwollstramin m.; baumwollene Gaze f. ‖ canevas m. / ~ (Mar) ‖ Segelleinwand f.; Segeltuch n. ‖ toile f. (à voile); toile f. canevas. / ~ (Packing-cloth) ‖ Packleinwand f. ‖ toile f. d'emballage; serpillière f. / ~ (Paint) ‖ Malerleinwand f.; Malertuch n. ‖ toile f. (de peinture). / ~ (Tent-cloth) ‖ Zeltleinwand f. ‖ toile f. (à voile). / decked out with all her ~ ‖ unter vollen Segeln ‖ à pleines voiles; toutes voiles dehors. / japanned ~ ‖ mit Lack getränkte Leinwand f.; mit Lack getränktes Segeltuch n. ‖ canevas m. ou toile f. enduit de vernis. / painted ~ ‖ geteerte oder gefirnißte Leinwand f. ‖ toile f. peinte ou goudronnée; prélat m. / painter's ~ ‖ grundierte Leinwand f.; Malerleinwand f. ‖ toile f. (de peinture); imprimature f.; imprimure f.; toile f. im-

---

primée. / painting ~ for theatrical purposes ‖ Malleinen n. für Theaterzwecke ‖ toile f. à decors de théâtre. / parcelling ~ (Mar) ‖ Schmartingleinwand f. ‖ fourrure f. / primed ~ see painter's ~. / to shorten ~ ‖ Segel npl. kürzen ‖ diminuer de voiles. / ~ of a survey ‖ Netz n. einer Vermessung ‖ réseau m. ou canevas m. d'une levée. / tarred ~ see painted ~. / ~ for trunks ‖ Koffersegeltuch n. ‖ toile f. d'emballage pour malles. / under ~ ‖ unter Segel n. ‖ sous voiles fpl. / varnished ~ see painted ~.

canvas bag ‖ Segeltuchfutteral n.; Leinwandtasche f. ‖ étui m. en grosse toile; sac m. en toile. / ~ blind ‖ Fenstervorhang m.; Sonnenblende f. ‖ marquise f.; rideau m. de fenêtre. / ~ case ‖ Segeltuchfutteral n. ‖ gaine f. en toile à voiles. / ~ cover ‖ Segeltuchüberzug m. ‖ gaine f. en toile à voile. / ~ hood ‖ „amerikanisches" Verdeck m.; Segeltuchverdeck n. ‖ capote f. en toile; capote f. américaine. / ~ hose ‖ Zeugschlauch m. ‖ conduite f. ou manche f. en toile. / ~ roller ‖ Tuchwalze f. ‖ rouleau m. à toile. / ~ shoe ‖ Leinenschuh m.; Segeltuchschuh m. ‖ chaussure f. de toile.

canvass, to (To debate) ‖ erörtern; debattieren; diskutieren ‖ débattre; discuter. / ~ (To investigate) ‖ prüfen; untersuchen; erforschen; sichten ‖ examiner; explorer; vérifier; scruter. / ~ for votes ‖ Stimmen fpl. werben ‖ solliciter les suffrages mpl.

canvass (Debate) ‖ Debatte f.; Erörterung f.; Diskussion f. ‖ débat m.; discussion f.

canvasser for advertisements ‖ Anzeigensammler m. ‖ courtier m. d'annonces.

caoutchouc see also rubber and india-rubber ‖ Kautschuk m.; Gummi n. ‖ caoutchouc m. / artificial ~ ‖ künstlicher Kautschuk m. ‖ caoutchouc m. artificiel. / ~ in balls ‖ Kautschuk m. in Kugeln ‖ caoutchouc m. en boules. / ~ in blocks ‖ Kautschuk m. in Blöcken ‖ caoutchouc m. en blocs. / ~ in breads ‖ Kautschuk m. in Brotform ‖ caoutchouc m. en pains. / ~ in cakes ‖ Kautschuk m. in Kuchen ‖ caoutchouc m. en nappes. / ~ in disks ‖ Kautschuk m. in Scheiben ‖ caoutchouc m. en disques. / factitious ~ ‖ Faktis n. ‖ caoutchouc m. factice. / hard(ened) ~ ‖ gehärteter Kautschuk m.; Hartgummi n.; Hartkautschuk m. ‖ caoutchouc m. durci; ébonite m. / turnery of hard ~ ‖ Kunstdrechslerwaren fpl. aus Hartkautschuk ‖ tabletterie f. en caoutchouc durci. / ~ in leaves ‖ Kautschuk m. in Blättern ‖ caoutchouc m. en feuilles. / ~ in liquid state ‖ Kautschuk m. in flüssigem Zustand ‖ caoutchouc m. à l'état liquide. / ~ in lumps ‖ Kautschuk m. in Klumpen ‖ caoutchouc m. en masses. / ~ in addition to metal ‖ Kautschuk m. in Verbindung mit Metall ‖ caoutchouc m. en addition de métal. / ~ mixed with other materials ‖ mit anderen Stoffen gemischter Kautschuk m. ‖ caoutchouc m. mélangé à d'autres matières. / natural ~ ‖ natürlicher Kautschuk m. ‖ caoutchouc m. naturel. / ~ in plates ‖ Kautschuk m. in Platten ‖ caoutchouc m. en plaques. / to purify ~ ‖ Kautschuk m. reinigen ‖ déchiqueter le caoutchouc. / raw ~ ‖ Rohkautschuk m.; unbearbeiteter oder

roher Kautschuk m. || caoutschouc m. rugueux *ou* brut. / recast ~ || umgeschmelzter Kautschuk m. || caoutchouc m. refondu. / regenerated ~ || regenerierter Kautschuk m. || caoutchouc m. régénéré. / rough ~ *see* raw ~. / soft ~ / Weichkautschuk m. || caoutschouc m. souple. / ~ in solution || Kautschuk m. in Lösung || caoutschouc m. en solution. / synthetic ~ || synthetischer Kautschuk m. || caoutchouc m. synthétique. / vulcanized ~ || vulkanisierter Kautschuk m. || caoutchouc m. vulcanisé.

caoutchouc breakage || Kautschukbruch m. | débris mpl. *ou* déchets mpl. *ou* chutes fpl. *ou* brisure f. de caoutchouc. / ~ cement || Kautschukkitt m. || ciment m. de caoutchouc. / ~ chips pl. || Kautschukschnitzel mpl. || rognures fpl. de caoutchouc. / ~ cipher || Kautschukbuchstabe m.; Kautschukziffer f. || chiffre m. *ou* caractère m. en caoutchouc. / ~ cloak || Gummimantel m. || manteau m. de caoutchouc. / ~ cover || Kautschuküberzug m. || enduit m. à base de caoutchouc. / ~ driving belt || Gummitreibriemen m.; Balatatreibriemen m. || courroie f. de commande en caoutchouc *ou* en balata. / ~ figure. || Kautschukziffer f. | chiffre m. en caoutchouc. / ~ filament | Kautschukfaden m. || fil m. de caoutchouc. / ~ glue || Kautschukleim m. || colle f. de caoutchouc. / ~ goods pl. || Kautschukwaren fpl. || ouvrages mpl. de caoutchouc. / ~ fragments pl. of old ~ goods || Bruchstücke npl. von alten Kautschukwaren || débris mpl. de vieux objets en caoutchouc. / ~hose || Kautschukschlauch m. || tuyau m. en caoutchouc.

caoutchoucin || Kautschuköl n. || essence f. de caoutchouc.

caoutchouc milk || Kautschukmilch f. || lait m. de caoutchouc. / vulcanized ~ paste for dentists || vulkanisierte Kautschukmasse f. für zahnärztliche Zwecke || pâte f. de caoutchouc vulcanisée pour dentistes. / ~ powder || Kautschukpulver n. || poudre f. de caoutchouc. / ~ sheets pl.|| Kautschukblätter npl.||feuilles fpl. de caoutchouc. / ~ sole (Shoe) || Kautschuksohle f.; Gummisohle f. || semelle f. en caouthouc. / ~ stopper || Kautschukstöpsel m. || bouchon m. de caoutchouc. / ~ tissue || Kautschukgewebe n. || tissu m. en caoutchouc. / machine for the ~ trade || Kautschukindustriemaschine f. || machine f. pour l'industrie du caoutchouc. / ~ tube || Kautschukröhre f. || tube m. en caoutchouc. / ~ waste || Kautschukabfall m. || déchet m. de caoutchouc.

cap, to ~ the piles pl. || die Pfähle mpl. beholmen || coiffer les pilotis mpl. de leurs chapeaux.

cap || Aufsatz m.; Kappe f.; Haube f.; Deckel m. || chapeau m. / ~ (Bearing) || Lagerdeckel m. || chapeau m. de palier. / ~ (Blasting) || Zündkapsel f. || détonateur m.; amorce f.; capsule f. de poudre fulminante. / ~ (Bridge) || Holm m.; Kappe f.; Kopfbalken m. || chapeau m.; chapiteau m.; chape f. / ~ (Carp) || Holm m.; Langschwelle f. || chapeau m.; longrine f. / ~ (Cloth) || Mütze f.; Kappe f.; Haube f. || casquette f.; bonnet m. / ~ (Pile) || Schlaghaube f. || chape f. *ou* casque m. de percussion; capuchon m. du pieu. / ~

(Watchm) || Staubdeckel m. || cuvette f. / bolted ~ || angeschraubter Deckel m. || couvercle m. boulonné. / bottle ~ || Flaschenkapsel f. || capsule f. à bouteilles. / exhaust valve ~ || Auslaßventilverschraubung f. || bouchon m. de soupape d'échappement. / filler ~ || Füllschraube f. || bouchon m. de remplissage. / inlet valve ~ || Einlaßventilverschraubung f. || bouchon m. de soupape d'admission. / linen ~ || leinene Haube f. || bonnet-linge m. / oil filler ~ || Ölfüllstutzen m. || bouchon m. de remplissage d'huile. / peaked ~ || Schirmmütze f. || casquette f. à visière. / screw ~ || Überwurfmutter f. || écrou m. creux *ou* à chapeau *ou* à raccord. / skull ~ || Käppchen n. || calotte f. / valve ~ || Ventilverschraubung f. || bouchon m. de soupape. / wheel hub ~ || Radkappe f. || chapeau m. de moyeu de roue.

capable || fähig; imstande; geeignet || capable. / ~ of being braked || bremsbar || freinable. / ~ of longitudinal motion || verschiebbar in der Längsrichtung || déplaçable en longueur.

capacious || geräumig; umfassend || spacieux.

capacitance (Electr) || kapazitive Impedanz f.; Scheinwiderstand m.; Wellenwiderstand m. || impédance f. capacitive.

capacitive (Electr) || kapazitiv || capacitif. / ~ load || kapazitive Belastung f. || charge f. capacitive.

capacity (Contents of a body) || Inhalt m.; Gehalt m.; Volumen n.; Fassungsraum m. || contenu m.; contenance f.; capacité f.; volume m.; teneur f. / ~ (Admissibility) || Fassungsvermögen n.; Aufnahmefähigkeit f.; Kapazität f. || capacité f.; admissibilité f. / ~ (Electr) || Kapazität f.; Aufspeicherungsvermögen n.; Fassungsvermögen n. || capacité f.; susceptibilité f. / ~ (Lifting power)||Tragfähigkeit f.; Tragkraft f.; Hebekraft f. || puissance f.; capacité f.; force f. portante *ou* de levage. / ~ (Productive power) || Leistungsfähigkeit f.; Leistung f. || puissance f.; productivité f.; capacité f.; rendement m.; débit m. / ~ (Privilege) || Befähigung f.; Befugnis f.; Qualifikation f. || qualification f.; capacité f. / absorptive ~ || Absorptionsfähigkeit f.; Aufnahmevermögen n. || pouvoir m. absorbant. / ~ of the boiler || Fassungsraum m. des Kessels || capacité f. d'une chaudière. / ~ of bucket (Dredger) || Löffelinhalt m. || capacité f. de la cuillère. / carrying ~ || Tragfähigkeit f. || charge f. admise. / cubic ~ || Rauminhalt m.; Kubikinhalt m. || capacité f. cube; cubage m. / distributed ~ || verteilte Kapazität f. || capacité f. répartie. / electric ~ || elektrische Kapazität f. || capacité f. électrique. / ~ of a furnace (Met) || Fassungsvermögen n. des Ofens || capacité f. du fourneau. / ~ of grab || Greiferinhalt m. || capacité f. de la pelle (preneuse) *ou* de la benne-drague. / ~ heatabsorptive || Wärmeaufnahmefähigkeit f. || capacité f. calorifique; pouvoir m. absorbant thermique. / internal ~ of a coil || Windungskapazität f. einer Spule || capacité f. entre les enroulements d'une bobine. / large ~ || große Kapazität f. || grande capacité f. / largest ~ || größtes Fassungsvermögen n. || contenance f. maximum. / legal ~ || Rechts-

fähigkeit f. || capacité f. légale. / lumped ~ || konzentrierte Kapazität f. || capacité f. concentrée. / for maximum ~ || für schwerste Leistungen fpl. || pour rendement maximum. / ~ to pay || Zahlungsfähigkeit f. || solvabilité f. / ~ of power station when completed || Ausbauleistung f. eines Kraftwerkes || puissance f. totale d'une usine complètement installée. / productive ~ || Ertragsfähigkeit f. || productivité f. / radiating ~ || Ausstrahlungsvermögen n. || pouvoir m. radiant. / rated ~ || Nennleistung f. || débit m. prévu. / ~ for reaction || Reaktionsfähigkeit f. || pouvoir m. *ou* capacité f. de réaction. / to loose the ~ for reaction || die Reaktionsfähigkeit f. verlieren || perdre le pouvoir *ou* la capacité de réaction. / reserve ~ || Reserveleistung f. || réserve f. de puissance. / small ~ || kleine Kapazität f. || faible capacité f. / small ~ meter || Kleinkapazitätsmesser m. || appareil m. pour la mesure de faibles capacités. / specific inductive ~ || spezifische Induktionskapazität f. || pouvoir inducteur m. spécifique. / to strain to the utmost limit of the ~ || bis zur Grenze f. der Leistungsfähigkeit ausnützen || utiliser jusqu'à la limite de la puissance. / ~ up to about . . . || Leistung f. bis etwa . . . || capacité f. jusqu'à environ . . . . / capacities pl. from x HP upwards || Leistungen fpl. von x PS aufwärts || capacités fpl. à partir de x CV. / ~ of a vessel || Fassungsvermögen n. eines Schiffes || capacité f. d'un navire. / water tower of x m³ ~ || Wasserturm m. von x cbm Fassung || château m. d'eau de x m³ de capacité. / working ~ || nutzbarer Inhalt m. || capacité f. utile. / ~ per working day || Tagesleistung f. || capacité f. de production par jour; débit m. journalier. / yearly ~ || Jahresleistungsfähigkeit f. || productivité f. annuelle.

capacity balancing in cables || elektrostatischer Induktionsschutz m. der Kabel || équilbrage m. de capacité des câbles. / ~ certificate || Befähigungszeugnis n. || certificat m. de capacité. / ~ curve || Kapazitätskurve f. || courbe f. de capacité. / ~ factor || Kapazitätsfaktor m. || facteur m. de capacité. / ~ increase || Steigerung f. der Leistungsfähigkeit || augmentation f. de la capacité. / ~ loss (Acc) || Kapazitätsschwund m. || perte f. de capacité. / ~ measure || Hohlmaß n. || mesure f. de capacité. / ~ measurement || Kapazitätsmessung f. || mesure f. de capacité. / ~ meter || Kapazitätsmesser m. || appareil m. pour la mesure des capacités. / ~ reactance || kapazitiver Blindwiderstand m. || réactance f. capacitive. / ~ test (Acc) || Kapazitätsprobe f. || essai m. de capacité. / ~ unbalance meter || Kopplungsmesser m.||appareil m. de mesure des déséquilibres de capacité.

caparison || Schabracke f.; Pferdedecke f.; Stalldecke f. || housse f.; caparaçon m.

cap bolt || Kopfschraube f. || boulon m. à tête *ou* à chapeau. / ~ clasp || Mützenschnalle f. || boucle f. à casquette.

cape (Geogr) || Kap n.; Vorgebirge n. || cap m.; promontoire m. / ~ cart-hood *see* ~ cart-top. / ~ cart-top || Segeltuchverdeck n.; amerikanisches Verdeck n. || capote f. en toile; capote f. américaine. / ~ chisel || Flachmeißel m. || burin m. bédane; bédane m.

7*

capel, to (Chem) ‖ kupellieren ‖ coupeller.

capers pl. ‖ Kapern fpl. ‖ câpres fpl.

capillarity ‖ Haarkraft f.; Kapillarität f. ‖ capillarité f.

capillary ‖ haarförmig; haarfein ‖ capillaire. / ~ action ‖ Kapillarwirkung f. ‖ action f. capillaire. / ~ attraction see capillarity. / ~ bottle ‖ Kapillarflasche f. ‖ bouteille f. capillaire. / ~ current ‖ Kapillaritätsstrom m. ‖ courant m. de capillarité. / ~ electrometer ‖ Kapillarelektrometer m.; Haarelektromesser m. ‖ électromètre m. capillaire. / ~ native copper ‖ Haarkupfer n. ‖ cuivre m. vierge capillaire. / ~ pressure ‖ Kapillardruck m. ‖ pression f. capillaire. / ~ pyrites pl. ‖ Haarkies m.; Schwefelnickelkies m. ‖ pyrite f. capillaire; nickel m. sulfuré; millérite f. / ~ reading microscope ‖ Kapillarenmeßmikroskop n. ‖ microscope m. de mesure capillaire. / ~ silver ‖ Haarsilber n. ‖ argent m. vierge capillaire. / ~ tube see capillary. / ~ working see ~ action.

capillary ‖ Haarrohr n.; Kapillare f.; Kapillargefäß n.; Haarröhrchen n. ‖ tube m. capillaire.

caping of an enclosure wall ‖ Abdeckung f. oder Abdach n. einer Einfriedigungsmauer ‖ larmier m. d'un mur de clôture.

capital (Arch) ‖ Kapitäl n.; Säulenknauf m.; Säulenkopf m. ‖ chapiteau m. / ~ (Money) ‖ Kapital n.; Stammvermögen n.; Fonds m. ‖ capital m.; fonds mpl. / ~ (Print) ‖ großer Anfangsbuchstabe m.; Majuskel f. ‖ capitale f.; lettre f. capitale. / authorized ~ ‖ autorisiertes Kapital n. ‖ capital m. autorisé. / dead ~ ‖ totes Kapital n. ‖ capital m. oisif ou improductif ou mort. / idle ~ ‖ brachliegendes Kapital n. ‖ capital m. oisif. / to invest ~ ‖ Kapital n. anlegen oder investieren ‖ placer de l'argent; engager du capital. / to lock-up the ~ ‖ das Kapital n. festlegen ‖ immobiliser le capital. / new ~ ‖ Nachtragszahlung f. ‖ payement m. supplémentaire; nouveau versement m. / opening ~ ‖ Grundkapital n. ‖ capital m. d'apport. / possessing ample ~ ‖ kapitalkräftig ‖ disposant d'importants capitaux. / to recall ~ ‖ Kapital n. kündigen ‖ revoquer du capital. / registered ~ ‖ eingetragenes Kapital n. ‖ capital m. social ou nominal. / to release ~ ‖ Kapital n. flüssig machen ‖ mobiliser du capital m. / share ~ ‖ Aktienkapital n. ‖ capital-actions m.; fonds m. social. / the share ~ is almost entirely in the hands of ... ‖ das Aktienkapital n. ist fast völlig im Besitz von ... ‖ la presque totalité f. des actions appartient à ... / spare ~ ‖ flüssiges Kapital n. ‖ fonds mpl. disponibles. / stock ~ ‖ Stammkapital n. ‖ fonds mpl. stock m., / subscribed ~ ‖ gezeichnetes Kapital n. ‖ capital m. souscrit. / to touch the ~ ‖ das Kapital angreifen ‖ attaquer ou entamer le capital; toucher au capital. / uncalled ~ ‖ noch nicht eingezahltes Kapital n. ‖ capital m. non-appelé. / working ~ ‖ Betriebskapital n. ‖ fonds mpl. ou capital m. de roulement; capital m. d'exploitation. / to write-off ~ ‖ Aktienkapital n. zusammenlegen ‖ réduire le capital.

capital account ‖ Kapitalkonto n. ‖ compte m. capital. / ~ cost of equipment ‖ Anlagekapital n.; Anlagekosten pl. ‖ prix m. coûtant d'équipement ou d'installation.

capitalization ‖ Kapitalisierung f. ‖ capitalisation f.

capitalize, to ‖ kapitalisieren ‖ capitaliser.

capitalized ‖ kapitalisiert ‖ capitalisé.

capital letter see capital (Print). / ~ share ‖ Kapitaleinlage f. ‖ apport m. / ~ ship ‖ Großkampfschiff n. ‖ cuirassé m. de première classe; bâtiment m. du type Dreadnought.

cap-jewel (Watchm) ‖ Deckstein m. ‖ contre-pivot m.

capmaker ‖ Mützenmacher m. ‖ casquettier m. / ladies' ~ ‖ Haubennäherin f. ‖ chapeleuse f.

cap materials pl. ‖ Mützenfurnitur f. ‖ fournitures fpl. pour casquettes. / ~ nut ‖ Überwurfmutter f.; Kapselmutter f. ‖ écrou m. à chape ou à chapeau ou à raccord.

cap-peak ‖ Mützenschirm m. ‖ visière f. de casquette. / cutter of ~s pl. ‖ Mützenschirmschneider m. ‖ coupeur m. de visières de casquettes. / to stitch the ~s pl. ‖ die Mützenschirme mpl. steppen ‖ piquer les visières fpl. de casquettes.

capped nut see cap nut.

capper (Tobacco) ‖ Deckblattleger m. ‖ capeur m.

cap-piece (Arch) ‖ Sturz m. ‖ linteau m. / ~ of a bearing ‖ Lagerdeckel m. ‖ chapeau m. de palier.

capping see cap.

capping brick ‖ Kappenziegel m.; Deckziegel m. ‖ brique f. à chaperon; mitron m.; dalle f. de brique. / ~ machine for springs ‖ Federkapselmaschine f. ‖ capsuleuse f. à ressorts. / ~ piece see cap-piece.

cap plug ‖ Hohlstopfen m. ‖ bouchon m. creux. / ~ pot (Glassm) ‖ bedeckter Hafen m. ‖ creuset m. couvert ou à la moufle. / ~ rock ‖ Deckgebirge n. ‖ terrains mpl. morts; couvert m. d'une couche.

caps pl. (Min) ‖ Käps pl.; Hinterstützen fpl. ‖ supports mpl. à charnières.

cap screw ‖ Kopfschraube f. ‖ vis f. à tête. / ~ sewer ‖ Haubennäher m. ‖ couseur m. de bonnets. / ~ sill of a gallery-frame ‖ Kappe f. eines Türstocks ‖ chapeau m. du châssis d'une galerie.

capsizable ‖ kenterbar; kippbar ‖ chavirable. / non ~ ‖ nicht kenterbar; nicht kenternd ‖ inchavirable.

capsize, to (Mar) ‖ kentern; umschlagen ‖ chavirer; sombrer; faire capot; capoter.

capstan (Min) ‖ Schachtwinde f. ‖ treuil m.; cabestan m. / ~ (Railw) ‖ Seilrangierwinde f. ‖ cabestan m. / ~ (Shipb) ‖ Spill n.; Gangspill n.; Schiffswinde f.; Ankerwinde f. ‖ cabestan m.; tourniquet m.; guindeau m.; vindas m. / ~ bar ‖ Handspill n. ‖ cabestan m. à bras. / cable ~ ‖ Kabelwinde f. ‖ cabestan m. ou treuil m. à câble. / double-headed ~ ‖ zweistufiges Spill n. ‖ cabestan m. à cloche bi-étagée. / gear ~ ‖ Verholgangspill n. ‖ petit cabestan m. / horse ~ ‖ Göpel m. ‖ manège m. / patent ~ (Shipb) ‖ Patentgangspill n. ‖ cabestan m. double. / ~ fitted with sproket wheel ‖ Gangspill n. mit Kettentrommel ‖ cabestan m. à couronne à empreinte.

capstan bar ‖ Gangspillspake f. ‖ barre f. du cabestan. / ~ barrel ‖ Spilltrommel f. ‖ cloche f. de cabestan. / ~ drive ‖ Göpel-antrieb m. ‖ commande f. par manège. / ~ drum head ‖ Gangspillkopf m. ‖ chapeau m. ou tête f. du cabestan. / ~ engine ‖ Ankerlichtmaschine f. ‖ machine f. pour cabestan. / ~ head (Shipb) ‖ Spillkopf m. ‖ cloche f. de cabestan. / ~ head (Turn) ‖ Revolverkopf m. ‖ tourelle f. revolver; porte-outil m. revolver. / ~ holes pl. (Shipb) ‖ Spillgatten npl. ‖ mortaises fpl. du vindas ou du cabestan; amelottes fpl. / ~ installation ‖ Spillanlage f. ‖ installation f. de cabestan.

capstan-lathe ‖ Revolverdrehbank f. ‖ tour m. à revolver. / ~ worker ‖ Fassondreher m. ‖ tourneur m. à façon; décolleteur m. / ~ working ‖ Fassondreherei f. ‖ décolletage m. de métaux.

cap strip (Aero) ‖ Holmgurt m. ‖ bride f. de longeron.

capsule ‖ Kapsel f. ‖ capsule f. / ~ of a bottle ‖ Flaschenkapsel f. ‖ capsule f. d'une bouteille. / enclosed in a ~ ‖ eingekapselt; gekapselt ‖ capsulé. / evaporating ~ ‖ Abdampfschale f.; Abdampfgefäß n. ‖ capsule f. évaporatoire ou à évaporation. / in ~ ‖ in einer Kapsel f.; in einem Futteral n. ‖ en étui m. / wooden ~ ‖ Holzkapsel f. ‖ étui m. en bois.

capsule fixer ‖ Pfropfenverkapseler m. ‖ capsuleur m.

capsuling machine ‖ Verkapselungsmaschine f.; Kapselmaschine f. ‖ machine f. à capsuler. / bottle ~ ‖ Flaschenverkapselmaschine f. ‖ capsulatrice f. de bouteilles.

captain (Mar) ‖ Kapitän m.; Kommandant m. ‖ capitaine m. (d'un bateau). / ~ (Mining) ‖ Steiger m.; Obersteiger m. ‖ maître-mineur m.; porion m. / ~ dresser (Dress ore) ‖ Pochsteiger m. ‖ maître-bocadeur m.

cap-tin ‖ Hutzinn n. ‖ étain m. en chapeaux.

captive-balloon ‖ Fesselballon m. ‖ ballon m. captif. / ~ station (Radio) ‖ Ballonstation f.; Fesselballonstation f. ‖ station f. de ballons captifs.

capture, to ~ a ship ‖ ein Schiff n. aufbringen oder kapern ‖ capturer un bateau.

caput mortuum (Chem) ‖ Phlegma n.; Caput-mortuum n. ‖ colcotar m.; caput-mortuum m.; résidu m.

car ‖ Wagen m.; Fahrzeug n. ‖ voiture f.; char m.; wagon m. / ~ (Aero) ‖ Gondel f. ‖ nacelle f. / ~ (Cableway) ‖ Wagen m. ‖ wagonnet m. / ~ (Railw) ‖ Eisenbahnwagen m.; Waggon m. ‖ wagon m. (de chemin de fer). / armoured ~ ‖ Panzerwagen m. ‖ automobile f. blindée; wagon m. blindé. / ~ of the balloon ‖ Ballonkorb m.; Gondel f. ‖ nacelle f. de ballon. / battery ~ (Railw) ‖ Akkumulatorenwagen m. ‖ automotrice f. à accumulateurs. / ~ for beer ‖ Bierwagen m. ‖ haquet m. de brasseur. / casting ~ ‖ Gießpfannenwagen m. ‖ chariot m. porte-poche ou de coulée; transporteur m. de la poche de coulée. / char-à-bancs ~ ‖ Gesellschaftswagen m. ‖ char m. à bancs automobile. / compartment ~ ‖ Abteilwagen m. ‖ voiture f. à compartiments. / corridor ~ ‖ Durchgangswagen m. ‖ voiture f. à couloir. / cross-country ~ (War impl) ‖ Querfeldeinwagen m. ‖ char m. à travers champ. / delivery ~ (Mine) ‖ Förderwagen m. ‖ berline f. / delivery ~ (Trade) ‖ Lieferwagen m. ‖ voi-

ture f. de livraison; camionette f. / demonstration ~ (Auto) ‖ Vorführungswagen m. ‖ voiture f. de démonstration. / double-deck tram ~ ‖ Decksitzwagen m. ‖ voiture f. à impériale. / drawing-room ~ ‖ Salonwagen m.; Luxuswagen m. ‖ voiture f. salon ou de luxe. / dump ~ ‖ Kippwagen m. ‖ wagon m. basculeur. / electric motor ~ ‖ elektrischer Kraftwagen m.; Elektromobil n. ‖ voiture f. électrique. / emergency ~ ‖ Rettungswagen m. ‖ voiture f. de secours. / freight ~ ‖ Güterwagen m. ‖ wagon m. à marchandises. / goods motor ~ ‖ Lastwagen m. ‖ autocamion m. / gondola ~ ‖ offener Güterwagen m. ‖ wagon m. découvert. / high-wheeled ~ ‖ hochrädriger Wagen m. ‖ voiture f. à grandes roues. / inspection ~ ‖ Revisionswagen m. ‖ voiture f. de génie. / iron box (freight) ~ ‖ normalspuriger eiserner Kastenwagen m. ‖ wagon m. à caisse en tôle à écartement normal. / ladle ~ ‖ Pfannenwagen m.; Gießwagen m. ‖ chariot m. porte-poche. / light ~ ‖ leichter Wagen m. ‖ voiture f. légère. / light railway ~ ‖ Kleinbahnwagen m. ‖ voiture f. à voie étroite. / ~ for transporting long beams ‖ Langholzwagen m. ‖ voiture f. à longs bois. / ~ for transporting long iron bars ‖ Langeisenwagen m. ‖ voiture f. à longs barres métalliques. / luggage ~ ‖ Gepäckwagen m. ‖ wagon m. à bagages. / ~ de luxe ‖ Luxuswagen m. ‖ voiture f. de luxe. / miner's ~ ‖ Grubenwagen m. ‖ wagonnet m. pour mines; berline f. (de mine); chien m. / motor ~ ‖ Kraftwagen m.; Automobil n.; Motorwagen m. ‖ automobile f. / observation ~ (Aero) ‖ Fesselballonkorb m. ‖ nacelle f. d'observation. / observation ~ (Railw) ‖ Aussichtswagen m. ‖ wagon m. vitré; voiture f. point de vue. / open ~ ‖ offener Wagen m. ‖ voiture f. ouverte. / passenger ~ ‖ Personenwagen m. ‖ voiture f. de voyageurs. / petrol ~ ‖ Benzinwagen m. ‖ automobile f. à essence. / platform ~ ‖ Plattformwagen m. ‖ wagon m. plat ou à plate-forme. / ~ with or without play in the axlebox guides ‖ Wagen m. mit oder ohne Spielraum in den Achsbuchsführungen ‖ voiture f. avec ou sans jeu dans les plaques de garde des boîtes à huile. / racing ~ (Auto) ‖ Rennwagen m. ‖ voiture f. de course. / rail ~ ‖ Schienenfahrzeug n.; Triebwagen m. ‖ voiture f. sur rails; automotrice f. sur rails. / runabout ~ ‖ Kleinauto n. ‖ voiturette f. / scoop-type tipping ~ ‖ Schnabelrundkipper m. ‖ wagonnet m. avec caisse en pelle basculante dans toutes les directions. / second-hand ~ ‖ gebrauchter Wagen m. ‖ voiture f. d'occasion. / side ~ ‖ Beiwagen m.; Seitenwagen m. ‖ cyclecar m.; voiturette f. à remorque latérale. / sight-seeing ~ ‖ Aussichtswagen m. ‖ voiture f. d'excursion. / slag ~ ‖ Schlackenwagen m. ‖ chariot m. à laitier. / sleeping ~ ‖ Schlafwagen m. ‖ wagon-lits m. / small ~ ‖ kleiner Wagen m. ‖ voiturette f. / steam ~ ‖ Dampfwagen m. ‖ voiture f. à vapeur. / stock ~ ‖ Serienwagen m. ‖ voiture f. de série. / tank ~ ‖ Tankwagen m. ‖ wagon-citerne m. / tipping ~ ‖ Kippwagen m. ‖ wagonnet m. basculant ou basculeur. /

touring ~ ‖ Tourenwagen m.; Reisewagen m. ‖ voiture f. de tourisme. / town ~ ‖ Stadtwagen m. ‖ voiture f. de ville. / tram-~ ‖ Straßenbahnwagen m. ‖ voiture f. de tramway. / transport ~ ‖ Transportwagen m. ‖ chariot m. de transport. / V-dump ~ with automatic locking device ‖ Muldenkipper m. mit selbsttätiger Muldenfeststellung ‖ basculeur m. à auge avec dispositif de fixation automatique de l'auge. / vestibule ~ ‖ Wagen m. mit Vorraum ‖ voiture f. à vestibule.

caramel ‖ Karamelle f. ‖ caramel m. / ~ maker ‖ Karamellenfabrikant m. ‖ fabricant m. de caramels. / ~ malt ‖ Karamellmalz n. ‖ malt m. de caramel; caramelmalt m.

caramelized sirup ‖ karamelisierter Sirup m. ‖ sirop m. caramélisé.

carapace ‖ Schildkrötenschale f. ‖ carapace f.

carat ‖ Karat n. ‖ carat m.

caravan ‖ Wohnwagen m. ‖ voiture f. d'habitation; roulotte f. / motor ~ ‖ Wohnmotorwagen m. ‖ automobile f. d'habitation; maison f. mobile.

caravan trade ‖ Karawanenhandel m. ‖ commerce m. par caravanes.

caraway seed ‖ Kümmel m. ‖ cumin m.; carvi m. / ~ oil ‖ Kümmelöl n. ‖ essence f. de carvi.

car axle ‖ Wagenachse f. ‖ essieu m. de voiture. / ~ barn ‖ Wagenschuppen m. ‖ remise f.; garage m.

carbide ‖ Karbid n. ‖ carbure m. / ~ of calcium ‖ Kalziumkarbid n. ‖ carbure m. de calcium. / ~ of iron ‖ Eisenkarbid n. ‖ carbure m. de fer.

carbide furnace ‖ Karbidofen m. ‖ four m. à carbure. / ~ lamp ‖ Karbidlampe f. ‖ lampe f. à carbure. / ~ pressing ‖ Preßkarbid n. ‖ carbure m. comprimé. / ~ works pl. ‖ Karbidfabrik f. ‖ fabrique f. de carbure.

carbine ‖ Karabiner m. ‖ carabine f.; mousqueton m.

car-body (Auto) ‖ Aufbau m.; Karosserie f. ‖ carrosserie f. / armoured ~ ‖ gepanzerter Wagenaufbau m. ‖ caisse f. cuirassée du char.

carbohydrate ‖ Kohlenhydrat n. ‖ hydrate m. de carbone.

carbolated cotton ‖ Karbolwatte f. ‖ ouate f. phéniquée. / ~ gauze ‖ Karbolgaze f. ‖ gaze f. phéniquée. / ~ soap ‖ Karbolseife f. ‖ savon m. (à l'acide) carbolique.

carbolic acid ‖ Karbolsäure f.; Karbol n. ‖ acide m. phénique ou carbolique. / ~ lime ‖ Karbolkalk m. ‖ chaux f. carbolique. / ~ soap ‖ Karbolseife f. ‖ savon m. (à l'acide) carbolique.

carbolineum ‖ Karbolineum n. ‖ carbolinéum m. / ~ coat of ~ ‖ Teerölanstrich m. ‖ peinture f. au carbolinéum.

carbon ‖ Kohlenstoff m. ‖ carbone m. / ~ (Arc lamp) see ~ for arc lamps. / to add ~ ‖ mit Kohlenstoff m. anreichern ‖ soumettre à l'action f. du carbone. / ~ for arc lamps ‖ Bogenlampenkohle f. ‖ crayon m. ou charbon m. de lampe à arc. / ~ contained in the coal ‖ Kohlenstoff m. der Kohle ‖ carbone m. contenu dans le charbon. / contents pl. of ~ ‖ Kohlenstoffgehalt m. ‖ teneur f. en carbone. / coppered ~ ‖ verkupferte Kohle f. ‖ charbon m. cuivré. / cored ~ ‖ Dochtkohle f. ‖ crayon m. à mèche. /

electric-lighting ~ ‖ Lichtkohle f. ‖ crayon m. à lumière. / ~ for elements ‖ Elementkohle f. ‖ charbon m. pour piles. / flame ~ ‖ Effektkohle f.; Flammenbogenkohle f. ‖ charbon m. flamme ou imprégné; charbon m. d'arc flammant. / flame-arc ~ see flame ~. / graphitic ~ ‖ graphitische Kohle f. ‖ charbon m. graphitique. / impregnated ~ see flame ~. / lighting ~ ‖ Beleuchtungskohle f. ‖ charbon m. à lumière. / to remove the ~ from ‖ frischen ‖ décarburer. / rich in ~ ‖ kohlenstoffreich ‖ riche en carbone m. / search-light ~ ‖ Scheinwerferkohle f. ‖ charbon m. pour projecteurs. / soldering ~ ‖ Lötkohle f. ‖ charbon m. à souder. / solid ~ ‖ Homogenkohle f. ‖ charbon m. homogène.

carbona of manganese ‖ Manganspat m. ‖ carbonate m. de manganèse.

carbonaceous see also carboniferous ‖ kohlenstoffhaltig ‖ carboné; carbonifère. / ~ coating of a crucible ‖ Futter n. eines Kohlentiegels ‖ brasque f. d'un creuset. / ~ lining of a crucible see ~ coating of a crucible.

carbon anode ‖ Kohleanode f. ‖ anode f. de charbon.

carbonatation man (Sugar) ‖ Saturationsarbeiter m. ‖ carbonateur m.

carbonate, to ‖ mit Kohlensäure f. behandeln oder sättigen ‖ saturer d'acide carbonique; carbonater.

carbonate ‖ kohlensaures Salz n.; Karbonat n. ‖ carbonate m. / acid ~ ‖ zweifach kohlensaures Salz n.; Bikarbonat n. ‖ bicarbonate m. / commercial ~ of ammonia ‖ Hirschhornsalz n. ‖ sel m. de corne de cerf; carbonate m. d'ammonium. / ~ of barium ‖ kohlensaurer Baryt m. ‖ carbonate m. de baryte. / crystallized sodium ~ ‖ Kristallsoda f. ‖ carbonate m. de sodium cristallisé. / earthy ~ of magnesia (Miner) ‖ Meerschaum m. ‖ écume f. de mer. / ~ of lead (Miner) ‖ Schwarzbleierz n.; Zerusit m. ‖ plomb m. carbonate noir; cérusite f. / ~ of lime ‖ Muschelkalkstein m.; Kalkspat m. ‖ chaux f. carbonatée. / ~ of magnesia ‖ Magnesiumkarbonat n.; kohlensaure Magnesia f. ‖ carbonate m. de magnésie; magnésie f. carbonatée. / ~ of manganese ‖ Manganspat m.; roter Braunstein m. ‖ manganèse m. carbonaté; carbonate m. de manganèse. / ~ of nickel ‖ kohlensaures Nickeloxydul n.; Nickelkarbonat n. ‖ carbonate m. de protoxyde de nickel. / ~ of potash ‖ Pottasche f.; kohlensaures Kalium n. ‖ potasse f.; carbonate m. de potasse. / ~ of silver ‖ Grausilber n. ‖ carbonate m. d'argent. / ~ of soda ‖ kohlensaures Natrium n.; Soda f. ‖ soude f. carbonatée; carbonate m. de soude. / ~ of zinc ‖ kohlensaures Zinkoxyd n.; Zinkspat m.; Zinkkarbonat n. ‖ carbonate m. de zinc; calamine m.

carbonated ‖ karbonathaltig ‖ carbonaté. / ~ beer ‖ mit Kohlensäure gesättigtes Bier n. ‖ bière f. artificiellement saturée d'acide carbonique.

carbon bisulphide ‖ Schwefelkohlenstoff m. ‖ sulfocarbure m.; sulfure m. de carbone. / ~ black ‖ Kohlenschwarz n.; Ruß m. ‖ noir m. de charbon ou de fumée. / ~ break-switch ‖ Kohlekontaktschalter m. ‖ interrupteur m. à contacts de charbon. / ~ brick ‖ Kohlenstoffstein

m. ‖ brique f. de carbone. / ~ brush ‖ Kohlenbürste f. ‖ balai m. en charbon. / ~ chloride ‖ Chlorkohlenstoff m. ‖ chlorure m. de carbone. / consumption of ~s (Arc lamp) ‖ Kohlenabbrand m. ‖ usure f. des charbons. / contact ‖ Kohlekontakt m. ‖ touche f. de charbon. / contents pl. of ~ (in steel) ‖ Kohlenstoffgehalt m. (des Stahls) ‖ teneur f. en carbone (de l'acier). / high contents pl. of ~ ‖ hoher Kohlenstoffgehalt m. ‖ teneur f. élevée en carbone; haute teneur f. en carbone. / low contents pl. of ~ ‖ niedriger Kohlenstoffgehalt m. ‖ teneur f. abaissée en carbone; basse teneur f. en carbone. / the content of ~ varies ‖ der Kohlenstoffgehalt m. schwankt ‖ la teneur en carbone varie. / ~ copy ‖ Durchschlag m. ‖ copie f. (au papier carbone); duplicata m.

carbon dioxide see also carbonic acid ‖ Kohlensäure f. ‖ acide m. carbonique. / ~ separator ‖ Kohlensäureabscheider m. ‖ séparateur m. d'acide carbonique. / ~ snow ‖ Kohlensäureschnee m. ‖ neige f. d'acide carbonique.

carbon disulphide ‖ Schwefelkohlenstoff m. ‖ sulfure m. de carbone; sulfocarbure m. / ~ electrode ‖ Kohlenelektrode f.; Kohlenbeutelelektrode f. ‖ électrode f. en charbon ou à sachet de charbon. / feeding of the ~s ‖ Kohlenvorschub m. ‖ avancement m. des charbons.

carbon filament ‖ Kohle(n)faden m. ‖ filament m. de charbon. / ~ lamp ‖ Kohlenfadenlampe f. ‖ lampe f. à filament de charbon. / twin ~ lamp ‖ Doppelkohlenfadenlampe f. ‖ lampe f. à double filament de carbone.

carbon holder ‖ Kohlenhalter m. ‖ portecharbon m. / magneto ~ ‖ Kohlenhalter m. des Magneten ‖ porte-charbon m. de magnéto.

carbonic acid see also carbon dioxide ‖ Kohlensäure f. ‖ acide m. carbonique. / liquid ~ ‖ flüssige Kohlensäure f. ‖ acide m. carbonique liquide.

carbonic acid compressor ‖ Kohlensäurekompressor m. ‖ compresseur m. d'acide carbonique. / ~ cooling machine ‖ Kohlensäurekältemaschine f. ‖ machine f. frigorifique à acide carbonique. / ~ determination ‖ Kohlensäurebestimmung f. ‖ dosage m. de l'acide carbonique. / ~ fitting ‖ Kohlensäurearmatur f. ‖ garniture f. pour acide carbonique. / ~ ice machine ‖ Kohlensäureeismaschine f. ‖ machine f. à glace à acide carbonique. / ~ impregnation apparatus ‖ Kohlensäureimprägnierapparat m. ‖ appareil m. pour l'imprégnation à acide carbonique. / ~ liquefying plant ‖ Kohlensäureverflüssigungsanlage f. ‖ installation f. de liquéfaction d'acide carbonique. / ~ producing plant ‖ Kohlensäureherstellungsanlage f. ‖ installation f. à fabriquer l'acide carbonique. / ~ washer ‖ Kohlensäurewäsche f. ‖ laveur m. d'acide carbonique.

carbonic oxide ‖ Kohlenoxyd n. ‖ oxyde m. de carbone. / ~ paper ‖ Kohlepapier n. ‖ papier m. carbone.

carboniferous see also carbonaceous ‖ kohlenstoffhaltig ‖ carbonifère. / ~ formation ‖ Kohlenformation f. ‖ formation f. houillère. / ~ rock ‖ Steinkohlengebirge n. ‖ terrain m. carbonifère. / ~ system

‖ karbonisches System n. ‖ système m. carbonifère.

carbonization ‖ Auskohlung f.; Verkohlung f.; Karbonisation f.; Karbonisierung f. ‖ carbonisage m.; carbonisation f. / ~ of bituminous coal ‖ Steinkohlenschwelung f. ‖ distillation f. lente de la houille. / ~ with circulation gases ‖ Spülgasschwelung f. ‖ distillation f. lente par moyen des gaz de balayage. / ~ of coal ‖ Kohleschwelung f. ‖ distillation f. lente des charbons. / dry ~ ‖ trockenes Auskohlen n. ‖ carbonisation f. à sec. / ~ of lignite ‖ Braunkohlenschwelung f. ‖ distillation f. lente du lignite. / ~ of oilshale ‖ Schwelung f. von Ölschiefergestein ‖ distillation f. lente des schistes huileux. / ~ of peat ‖ Verkohlung f. des Torfes ‖ carbonisation f. de la tourbe. / ~ in piles ‖ Haufenverkohlung f. ‖ carbonisation f. en tas. / ~ of pit-coal ‖ Verkokung f. ‖ cokéfaction f.; cokéfication f. / ~ of the silk rags ‖ Auskohlen n. der Seidenlumpen ‖ carbonisation f. des chiffons de soie. / ~ of solid fuels ‖ Brennstoffschwelung f. ‖ distillation f. lente de combustibles. / wet ~ ‖ nasses Auskohlen n. ‖ carbonisation f. humide. / ~ of wood ‖ Holzverkohlung f. ‖ carbonisation f. de bois. / ~ of woollen clothes ‖ Karbonisierung f. von Wollgeweben ‖ épaillage m. chimique des tissus de laine.

carbonization gas ‖ Schwelgas n. ‖ gaz m. dégagé; gaz m. de distillation lente. / ~ plant ‖ Holzverkohlungsanlage f. ‖ installation f. de carbonisation du bois. / ~ process ‖ Schwelvorgang m. ‖ procédé m. de distillation lente. ~ tar ‖ Schwelteer m. ‖ goudron m. de distillation lente. / ~ works pl. ‖ Schwelwerk n. ‖ installation ou usine f. de distillation lente.

carbonize, to ‖ auskohlen; verkohlen; karbonisieren ‖ carboniser; charbonner.

carbonized wool ‖ karbonisierte Wolle f. ‖ laine f. carbonisée.

carbonizer ‖ Karbonisierarbeiter m. ‖ carboniseur m. de laine.

carbonizing of pistons and valves (Mot) ‖ Verunreinigung f. von Kolben und Ventilen ‖ encrassage m. de pistons et soupapes.

carbonizing apparatus (Spinn) ‖ Karbonisierapparat m. ‖ appareil m. à carboniser. / ~ furnace (Spinn) ‖ Karbonisierofen m. ‖ four m. à carboniser. / ~ period (Mine) ‖ Garungsdauer f. ‖ durée f. de la distillation. / ~ plant ‖ Schwelanlage f. ‖ installation f. pour la distillation sèche. / ~ stove see ~ furnace. / ~ workshop ‖ Karbonisieranstalt f. ‖ atelier m. de carbonisation.

carbon lamp see carbon filament lamp. / ~ lightning arrester ‖ Kohlenblitzableiter m. ‖ paratonnere m. à charbon. / ~ maker (Lighting) ‖ Kohlenschleifer m. ‖ tailleur m. de charbon. / ~ monoxide see ~ oxide. / ~ oxide ‖ Kohlenoxyd n. ‖ oxyde m. de carbone. / ~ oxide gas ‖ Kohlenoxydgas n. ‖ gaz m. d'oxyde de carbone.

carbon paper ‖ Kohlepapier n. ‖ papier m. charbon; papier m. carbone. / ~ sheet of ~ ‖ Kohlepapierbogen m. ‖ feuille f. de papier carbone.

carbon pencil ‖ Kohlestift m.; Zeichenkohle f. ‖ fusain m. / percentage of ~ see contents of ~. / ~ pile ‖ Kohlenelement n. ‖ pile f. à charbon. / ~ print

Kohledruck m.; Pigmentdruck m. ‖ impression f. au charbon; héliogravure f. sur papier au charbon. / ~ recorder ‖ Rußschreiber m. ‖ récepteur m. à noir de fumée. / ~ steel ‖ Kohlenstoffstahl m. ‖ acier m. au carbone. / ~ terminal ‖ Kohlenklemme f. ‖ presse f. à charbon. / ~ tetrachloride ‖ Kohlenstofftetrachlorid n.; Tetrachlorkohlenstoff m. ‖ tétrachlorure m. de carbone. / ~ transmitter ‖ Kohlenmikrophon n. ‖ transmetteur m. à charbon.

carbonyl chloride ‖ Karbonylchlorid n.; Phosgen n. ‖ oxychlorure m. de carbone; phosgène m. / ~ composition ‖ Karbonylverbindung f. ‖ composé m. à carbonyle.

carborundum ‖ Karborundum n. ‖ carborundum m. / ~ paper ‖ Karborundumpapier n. ‖ papier m. au carborundum. / ~ resistance ‖ Karborundumwiderstand m. ‖ résistance f. en carborundum. / ~ rod ‖ Karborundumstab m. ‖ baguette f. en carborundum. / ~ wheel ‖ Karborundumscheibe f. ‖ meule f. en carborundum.

carbosulphide ‖ Schwefelkohlenstoff m. ‖ sulfure m. de carbone; sulfocarbure m.

carboy ‖ Säureballon m.; Korbflasche f.; Glasballon m. ‖ dame-jeanne f.; ballon m.; bonbonne f.; tourie f. à panier. / interlaced ~ ‖ eingeflochtene Transportflasche f. ‖ tourie f. à panier. / ~ put in baskets of metal or osier ‖ in Metall- oder Weidebehälter eingelassene Transportflasche f. ‖ tourie f. renfermée dans des paniers en métal ou en osier.

car brake ‖ Wagenbremse f. ‖ frein m. de voiture. / ~ builder (Railw) ‖ Eisenbahnwagenbauer m. ‖ constructeur m. de wagons.

carburate, to see to carburet.

carburet, to (Mot) ‖ vergasen ‖ carburer.

carburet ‖ Kohlenstoffverbindung f. ‖ carbure m. / ~ of calcium ‖ Kalziumkarbid n. ‖ carbure m. de calcium.

carbureting (Mot) ‖ Vergasung f. ‖ carburation f.

carburetor see carburettor.

carburettor ‖ Vergaser m. ‖ carburateur m. / alcohol ~ ‖ Spiritusvergaser m. ‖ carburateur m. à alcool. / atomizing ~ ‖ Einspritzvergaser m. ‖ carburateur m. à giclage. / constant-level ~ ‖ Vergaser m. mit Schwimmvorrichtung ‖ carburateur m. à niveau constant. / duplex ~ ‖ Doppelvergaser m. ‖ carburateur m. jumelé. / exhaust-jacketed ~ ‖ Vergaser m. mit Auspuffgasheizmantel ‖ carburateur m. réchauffé par l'échappement. / jet ~ ‖ Düsenvergaser m. ‖ carburateur m. à gicleur. / multiple-jet ~ ‖ Mehrdüsenvergaser m. ‖ carburateur m. à plusieurs gicleurs. / paraffin ~ ‖ Petroleumvergaser m ‖ carburateur m. à pétrole. / pressure-fed ~ ‖ Vergaser m. mit Druckförderung ‖ carburateur m. alimenté sous pression. / spray ~ ‖ Einspritzvergaser m.; Zerstäubungsvergaser m. ‖ carburateur m. à pulvérisation; atomisateur m. / surface ~ ‖ Oberflächenvergaser m. ‖ carburateur m. par surface ou à léchage. / vapourizing ~ ‖ Verdampfungsvergaser m. ‖ carburateur m. à évaporation. / water-jacketed ~ ‖ Vergaser m. mit Wasserheizmantel ‖ carburateur m. à chemise d'eau. / wick ~ ‖ Dochtvergaser m. ‖ carburateur m. à mèche.

carburettor adjustment ‖ Vergaserregelung f. ‖ réglage m. du carburateur. / ~ choke ‖ Mischrohr n. ‖ tuyau m. à mélange. / ~ control ‖ Vergasergestänge n. ‖ commandes fpl. du carburateur. / ~ float ‖ Vergaserschwimmer m. ‖ flotteur m. du carburateur. / ~ hot-air pipe ‖ Luftwärmer m. ‖ prise f. d'air chaud. / ~ lid ‖ Vergaserdeckel m. ‖ couvercle m. du carburateur. / ~ needle valve ‖ Kraftstoffnadelventil n. ‖ soupape f. à pointeau m. d'essence. / ~ primer ‖ Tipper m. des Vergasers ‖ poussoir m. de carburateur.

carburization ‖ Kohlung f. ‖ carburation f. / ~ by solid carbon ‖ Kohlung f. durch festen Kohlenstoff ‖ carburation f. par carbone solide.

carburize, to (Metal) ‖ mit Kohlenstoff m. anreichern; kohlen; karburieren ‖ carburer; cémenter avec du charbon.

carburized iron ‖ gekohltes Eisen n.; mit Kohlenstoff angereichertes Eisen n. ‖ fer m. carburé. / ~ layer ‖ kohlenstoffreiche Schicht f. ‖ couche f. carburée. / ~ surface ‖ kohlenstoffreiche oder einsatzgehärtete Oberfläche f. ‖ surface f. carburée.'

carburizing furnace ‖ Einsatzofen m. ‖ four m. à cémenter aux gaz carburants. / electric ~ plant ‖ elektrische Ofenanlage f. für Einsatzhärtung ‖ installation f. de fours électriques pour trempe de surface.

carburizing machine see carburizing furnace.

carcass (Corpse) ‖ Kadaver m.; Leichnam m.; Aas n. ‖ cadavre m.; charogne f. / ~ (Frame-work) ‖ Gerippe n.; Gestell n.; Rumpf m. ‖ carcasse f.; charpente f. / ~ (Fire-ball) ‖ Brandgeschoß n.; Brandgranate f.; Brandkugel f. ‖ projectile m. ou obus m. ou boulet m. incendiaire; carcasse f. / round ~ ‖ Brandbombe f. ‖ bombe f. incendiaire. / ~ of a building ‖ Zimmerwerk n. eines Gebäudes ‖ carcasse f. ou charpente f. d'un bâtiment. / ~ of a vessel ‖ Gerippe n. eines Schiffes ‖ carcasse f. ou membrure f. d'un navire.

carcass utilization plant ‖ Kadaververarbeitungsanlage f.; Kadaververwertungsanlage f. ‖ installation f. de préparation industrielle des cadavres d'animaux.

Carcel lamp ‖ Carcellampe f. ‖ lampe f. Carcel.

car, construction of ~s ‖ Wagenbau m.; Fahrzeugbau m. ‖ construction f. de voitures; carrosserie f.; charronnage m. / automatic ~ coupling (Railw) ‖ selbsttätige Wagenkupplung f. ‖ attelage m. automatique de wagons.

card, to ‖ krempeln; kratzen; streichen ‖ carder.

card (Pap) ‖ Karte f.; carte f. / ~ (Spinn; Weav) ‖ Wollkratze f.; Krempel f.; Karde f.; Kardätsche f. ‖ carde f. / ~ of admission ‖ Eintrittskarte f.; Einlaßkarte f. ‖ billet m. d'entrée; ticket m. / breaker ~ (Spinn) ‖ Reißkrempel f. ‖ briseur m.; carde f. briseuse ou en gros. / ~ for card registers ‖ Karteikarte f. ‖ carte f. de carthotèque. / ~ of a compass ‖ Kompaßzifferblatt n. ‖ rose f. des vents. / ~ of congratulation ‖ Glückwunschkarte f. ‖ carte f. de félicitation(s). / fillet ~ (Spinn) ‖ Deckelkarde f. ‖ carde f. à feuilles. / finishing ~ ‖ Wollkrempel f. ‖ finisseuse f. / flat ~ (Spinn) ‖ Deckelkarde f. ‖ carde f. à chapeaux. / hand ~ (Spinn) ‖ Kratze

f. ‖ carde f. à main; drousette f. / ~ for inner soles ‖ Brandsohlenpappe f. ‖ carton m. pour la seconde semelle. / ~ of invitation ‖ Einladungskarte f. ‖ billet m. d'invitation. / Jacquard ~ ‖ Webekartenpappe f. ‖ carton m. pour métier Jacquard. / ~ of needles and pins ‖ Nadelbrief m. ‖ paquet m. d'épingles. / roller ~ ‖ Walzenkarde f. ‖ carde f. à cylindres. / scribbler ~ see breaker ~. / ~ for spinning machines ‖ Karde f. für Spinnereimaschinen ‖ carde f. pour machines de filatures. / to take a ~ (Steam eng) ‖ indizieren ‖ relever un tracé m. d'indicateur.

cardamom ‖ Kardamom m. ‖ cardamome m. / ~ oil ‖ Kardamomöl n. ‖ essence f. de cardamome.

Cardan axle; cardan axle see Cardan shaft.

Cardanic suspension ‖ kardanische Aufhängung f. ‖ suspension f. de Cardan ou à la Cardan.

Cardan joint ‖ Kreuzgelenk n.; Kardangelenk n.; kardanisches Gelenk n. ‖ joint m. ou articulation f. de Cardan; cardan m. / ~ of the ring type ‖ ringförmiges Kardangelenk n. ‖ joint m. de Cardan sphérique.

Cardan shaft ‖ Kardanwelle f.; Gelenkwelle f. ‖ arbre m. de Cardan; arbre m. à cardan. / binding by means of ~s ‖ Gelenkwellenverbindung f. ‖ assemblage m. par arbres de Cardan.

cardboard ‖ Pappe f.; Karton m. ‖ carton m. / albuminized ~ ‖ mit Albumin überzogene Pappe f. ‖ carton m. albuminé. / asphalted ~ ‖ Teerpappe f. ‖ carton m. goudronné. / ~ for book-binding ‖ Buchbinderpappe f. ‖ carton m. de relieur. / ~ impregnated with coaltar ‖ mit Steinkohlenteer getränkte Pappe f. ‖ carton m. imprégné de goudron de houille. / common ~ ‖ gewöhnliche Pappe f. ‖ carton m. commun. / corrugated ~ ‖ Wellpappe f. ‖ carton m. ondulé. / corrugated ~ for packing ‖ gewellte Pappe f. für Verpackungszwecke ‖ carton m. d'emballage ondulé. / couched ~ for packing ‖ gegautschte Pappe f. für Verpackungszwecke ‖ carton m. d'emballage gaufré. / ~ for cutting-out dolls ‖ Ankleidepuppenbogen m. ‖ feuille f. avec poupées à habiller. / fine ~ ‖ Kartonpapier n. ‖ papier-carton m. / flinted ~ ‖ mit Feuersteinpulver bestreute Pappe f. ‖ carton m. silexé. / gelatinized ~ ‖ mit Gelatine überzogene Pappe f. ‖ carton m. gélatiné. / glassed ~ ‖ mit Glaspulver bestreute Pappe f. ‖ carton m. verré. / glazed ~ ‖ Kreidekarton m. ‖ carte f. couchée. / ivory ~ ‖ Elfenbeinkarton m. ‖ carton-ivoire m. / ondulated ~ ‖ Wellpappe f. ‖ carton m. ondulé. / ~ for packing (purposes) ‖ Packpappe f.; Pappe f. für Verpackungszwecke ‖ carton m. d'emballage. / ~ for packing strengthened by a paper sheet ‖ Pappe f. für Verpackungszwecke, durch einen Papierbogen verstärkt ‖ carton m. d'emballage renforcé par une (autre) feuille de papier. / ~ for packing with insertion of tissue ‖ Pappe f. für Verpackungszwecke mit Zwischenlage aus Gewebe ‖ carton m. d'emballage avec intercalation de tissu. / paraffined ~ ‖ mit Paraffin überzogene Pappe f. ‖ carton m. paraffiné. / photographic ~ ‖ fotografischer Karton m. ‖ carton m. pour la photographie. / pleated ~ for pack-

ing ‖ gefältete Pappe f. für Verpackungszwecke ‖ carton m. d'emballage plissé. / pressed ~ ‖ gepreßte Pappe f. ‖ carton m. comprimé. / sanded ~ ‖ mit Sand bestreute Pappe f. ‖ carton m. sablé. / stearin-coated ~ ‖ mit Stearin überzogene Pappe f. ‖ carton m. stéariné. / ~ strengthened by tissue ‖ durch Gewebe verstärkte Pappe f. ‖ carton m. renforcé par du tissu. / wrinkled for packing ~ ‖ gefältelte oder gerunzelte Pappe f. für Verpackungszwecke ‖ carton m. d'emballage froncé.

cardboard articles pl. ‖ Kartonagenwaren fpl. ‖ cartonnages mpl. / ~ bending machine ‖ Biegemaschine f. für Kartonagen ‖ machine f. à plier le carton.

cardboard box see also card box ‖ Pappschachtel f.; Pappkarton m. ‖ boîte f. en carton. / folded ~ ‖ Faltschachtel f. ‖ boîte f. pliante. / ~ for sending ‖ Versandschachtel f. ‖ boîte f. d'expédition en carton.

cardboard boxes pl. ‖ Kartonagen fpl. ‖ boîtes fpl. en carton; cartonnages mpl. / embossed ~ ‖ geprägte Kartonagen fpl. ‖ cartonnages mpl. estampés. / ~ for shoe manufacture ‖ Schuhkartonagen fpl. ‖ cartonnages mpl. pour chaussures.

cardboard boxes pl., fittings pl. for ‖ Kartonagenbeschläge mpl. ‖ ferrures fpl. d'articles en carton; garnitures fpl. pour cartonnages. / manufacture of ~ ‖ Kartonagenfertigung f.; Kartonagenfabrikation f. ‖ fabrication f. de cartonnages. / machine for the manufacture of ~ ‖ Maschine f. zur Herstellung von Pappwaren ‖ machine f. pour la fabrication de cartonnages.

cardboard case ‖ Pappkasten m. ‖ boîte f. en carton. / ~ cutter ‖ Pappschere f. ‖ cisaille f. à carton. / ~ factory ‖ Pappenfabrik f. ‖ fabrique f. de carton; cartonnerie f. / fancy articles pl. in ~ ‖ Pappgalanteriewaren fpl. ‖ articles mpl. de fantaisie en carton. / ~ frame ‖ Papprahmen m. ‖ cadre m. en carton. / ~ goods pl. ‖ Pappwaren fpl. ‖ cartonnages mpl. / manufacture of ~ goods pl. ‖ Kartonagenfabrikation f.; Kartonagenfertigung f.; Kartonagenherstellung f. ‖ fabrication f. de cartonnages. / ~ machine ‖ Kartonagenmaschine f. ‖ machine f. pour la fabrication des cartonnages. / ~ manufacture ‖ Pappenherstellung f.; Pappenfertigung f. ‖ manufacture f. de carton; cartonnerie f. / set of butts for the manufacture of ~ ‖ Büttengarnitur f. zur Pappenfertigung ‖ garniture f. de tonneaux pour cartonneries. / ~ manufacturing machine ‖ Kartonherstellungsmaschine f. ‖ machine f. à fabriquer le carton. / ~ matrice striking ‖ Pappmatrizenprägung f. ‖ empreinte f. des flans. / ~ plate ‖ Pappteller m. ‖ assiette f. en carton. / ~ toys making machine ‖ Pappspielwarenherstellungsmaschine f. ‖ machine f. à fabriquer des jouets en carton. / ~ works pl. ‖ Pappenfabrik f. ‖ cartonnerie f.

card box ‖ Pappschachtel f. ‖ boîte f. en carton. / ~ cutter ‖ Pappausschneider m. ‖ découpeur m. de carton. / ~ factory ‖ Schachtelfabrik f. ‖ fabrique f. de boîtes en carton. / ~ gluer ‖ Kartonagenkleber m. ‖ colleur m. de cartonnages. / ~ industry ‖ Kartonindustrie f. ‖ industrie f. du carton. / ~ machine ‖ Kartonagenmaschine f. ‖ machine f. à fabriquer les

cartonnages. / folded ~ machine || Faltschachtelmaschine f. || machine f. pour faire des boîtes pliantes. / ~ manufacturing || Herstellung f. von Pappschachteln || fabrication f. de boîtes en carton. / ~ wire stitcher || Kartonagendrahthefter m. || agrafeur m. de cartonnages.

card brusher (Spinn) || Kardätschenputzer m. || nettoyeur m. de cardes.

cardcase || Besuchskartentäschchen n. || porte-cartes m.

card cloth (Spinn) || Kratzentuch n. || tissu m. pour garniture de carde. / ~ clothier (Spinn) || Krempelbezieher m. || habilleur m. de cardes. / ~ clothing || Kratzenbeschlag m. || garniture f. de cardes. / ~ cutter (Weav) || Musterschläger m.; Kartenschläger m. || liseur m. au semple; piqueur m. de dessins.

carded cotton dyeing || Wattefärberei f. || teinture f. de coton cardé. / ~ wool yarn || Streichgarn n. || fil m. cardé. / ~ yarn || Halbkammgarn n.; Streichgarn n. || fil m. de laine cardée. / ~ yarn cloth || Streichgarnstoff m. || étoffe f. en laine cardée. / ~ yarn spinning machine || Streichgarnspinnereimaschine f. || machine f. à filer la laine cardée.

carder || Kardätscher m.; Krempler m. || cardeur m.; ouvrier m. de carderie; alimenteur m. de cardes.

card feeder || Krempler m. || cardeur m.; ouvrier m. de carderie; alimenteur m. de cardes. / ~ fillet || Kratzenband n. || ruban m. de cardes. / ~ fitting || Kardenbeschlag m. || garniture f. de cardes. / ~ fixer || Kardensetzer m. || régleur m. de cardes. / ~ girl (Weav) || Musteraufkleberin f. || encarteuse f. / ~ grinder || Kardenschleifer m.; Kardennadelrichter m. || aiguiseur m. de cardes. / ~ grinding machine || Krempelschleifmaschine f. || machine f. à affûter les cardes.

cardiac sounds pl. || Herztöne mpl. || bruits mpl. de souffles cardiaques.

cardinal advantage (Trade) || Hauptvorteil m. || avantage m. principal. / ~ importance || ausschlaggebende Bedeutung f. || importance f. capitale. / ~ point (Astron) || Kardinalpunkt m.; Himmelsgegend f. || point m. cardinal.

card-index || Kartei f. || casier m.; fichier m. / ~ box || Karteikasten m. || casier m. de cartothèques.

carding (Spinn) || Krempeln n.; Kratzen n.; Streichen n. || cardage m. / ~ of silk waste || Kardätschen n. des Seidenabfalls || peignerie f. de la bourre de soie. / ~ of wool || Krempeln n. der Wolle || cardage m. ou droussage m. de la laine.

carding machine (Spinn) || Krempelmaschine f. || machine f. à cardes. / ~ master (Spinn) || Krempelmeister m.; Kardenmeister m. || contremaître m. de cardes. / ~ wool spinner || Streichgarnspinner m. || fileur m. de laine cardée. / ~ wool spinning mill || Streichgarnspinnerei f. || filature f. de laine cardée. / ~ wool weaver || Streichgarnweber m. || tisseur m. de laine cardée.

cardioid || Kardioide f.; Herzkurve f. || cardioïde f. / ~ condenser || Kardioidkondensor m. || condensateur m. cardioïde. / ~ ultra-microscope || Kardioidultramikroskop n. || ultramicroscope m. cardioïde. / ~ ultra-microscope || Kardioidultramikroskop n. || ultramicroscope m. cardioïde.

---

card lacing (Weav) || Kartenverbinden n. || enlaçage m. de cartons. / ~ leather || Kratzenleder n. || cuir m. pour carde.

card paper || kartenstarkes Papier n. || papier m. fort pour cartes. / punched ~ for embroidery || Papierstramin m. || carton m. à broder.

card punching by hand (Weav) || Kartenschlägerei f. im Handbetrieb || piquage m. de cartons à la main. / ~ register || Kartei f. || casier m. / ~ repeating (Weav) || Kartenkopieren n. || repiquage m. de cartons. / ~ room (Spinn) || Krempelei f.; Karderie f. || carderie f. / ~ setter (Spinn) || Kardensetzer m. || débouteur m. / ~ sticker (Pap) || Kartenkleber m. || assembleur m. de cartes. / ~ tenter (Spinn) || Kardätschenführer m. || conducteur m. de cardes. / ~ winder (Spinn) || Garnaufkarter m. || encarteur m. de fils. / ~ writing feature || Postkartenschreibvorrichtung f. || dispositif m. pour écrire les cartes postales.

care || Aufsicht f.; Pflege f.; Wartung f. || soins mpl.; tenue f.; conservation f. / ~ of domestic animals || Haustierpflege f.; Tierpflege f. || soins mpl. aux animaux domestiques. / not to exercise due ~ || es an der erforderlichen Sorgfalt f. fehlen lassen || manquer de soins mpl. nécessaires.

care of || per Adresse f. || chez; aux bons soins mpl. de.

care, to take ~ to || Obacht f. geben auf || faire attention f. à. / with ~ || Vorsicht f. || attention.

careen, to ~ a vessel || ein Schiff n. kielholen || caréner un vaisseau; mettre un navire à la bande.

careen || Kielholen n. || abatage m.

carefully || sorgfältig || avec soin.

careful treatment || schonende Behandlung f. || traitement m. avec ménagement.

carelessness || Unachtsamkeit f. || inattention f.

car, end of the ~ || Kopfwand f. || paroi f. frontale ou de bout. / ~ frame || Laufgestell n.; Fahrgestell n. || bâti m. de chariot.

cargo || Schiffsladung f.; Fracht f. || cargaison f.; chargement m.; charge f. / general ~ || Stückgut n. || cargaison f. mixte. / ~ in grain || Schüttgüter npl.; Stürzgüter npl. || marchandises fpl. chargées en grenier. / homeward ~ || Rückladung f. || cargaison f. de retour. / ~ has shifted || die Ladung hat sich losgerissen || la cargaison s'est dérangée. / ~ badly stowed || schlecht verstaute Ladung f. || cargaison f. mal arrimée. / ~ of timber || Holzladung f. || charge f. de bois. / ~ of a vessel || Schiffsladung f. || chargement m.; cargaison f.

cargo block || Ladeblock m. || poulie f. de charge. / ~ boat || Frachtschiff n. || bâtiment m. de transport; cargo-boat m. / ~ discharging plant || Schiffslöscheinrichtung f. || installation f. de déchargement de navires. / ~ ship || Frachtschiff n. || cargo m. / ~ steamer || Frachtdampfer m. || vapeur m. de transport. / ~ steamer traffic || Frachtdampferverkehr m. || trafic m. des vapeurs marchands. / ~ vessel see ~ ship.

car headlight || Wagenlaterne f. || phare m. de voiture. / ~ kilometer || Wagenkilometer n. || kilomètre-voiture m.

---

carline-thistle root || Eberwurzel f. || racine f. de carline acaule.

carling || Kielschwein n.; Deckszwischenbalken m. || carlingue f.

carload || Wagenladung f. || wagon m. complet; charge f. de wagon. / ~ rate || Waggonladungsfrachtsatz m. || tarif m. par wagon complet.

carmine || Karminrot n. || carmin m. / ~ lac || Karminlack m. || laque f. carminée. / ~ spar || Karminspat m. || spath m. carmine; carminite f.

carminite see carmine spar.

carnation || Fleischfarbe f. || couleur f. de chair.

carnauba wax || Karnaubawachs n. || cire f. de carnauba. / ~ substitute || Karnaubawachsersatz m. || cire f. de carnauba factice.

carob || Johannisbrot n. || caroube f.; carouge f.

carotter (Fur) || Beizer m. || scréteur m. de peaux.

carpenter (Build) || Zimmermann m. || charpentier m. / ~ (Shipb) || Schiffszimmermann m. || matelot m. charpentier. / house ~ || Bauzimmerer m. || charpentier m. de bâtiment. / master ~ || Zimmermeister m. || maître charpentier m. / mill ~ || Mühlzimmermann m. || charpentier m. de moulin.

carpenter foreman || Zimmerpolier m. || charpentier m. appareilleur.

carpenter's bench || Hobelbank f. || établi m. ou banc m. de charpentier. / ~ bench clamp || Hobelbankeisen n. || valet m. d'établi de charpentier. / ~ crew (Mar) || Zimmermannsgasten mpl. || ouvriers mpl. charpentiers. / ~ gauge || Reißmaß n. || trusquin m. / ~ hammer axe || Zimmermannshammerbeil m. || hacherau m. de charpentier. / ~ hatchet || Zimmermannsbeil n. || hachette f. de charpentier. / ~ labourer || Zimmergeselle m. || compagnon m. charpentier. / ~ line || Schlagleine f. || ligne f. de charpentier. / ~ machine || Zimmereimaschine f. || machine f. pour charpentiers. / ~ pencil || Zimmermannsbleistift m. || crayon m. de charpentier. / ~ rule || Schmiege f.; Kluft f. || échelle f. pliante. / ~ tools pl. || Zimmermannsgerät n. || outils mpl. de charpentier. / ~ wood || Nutzholz n. || bois m. de construction. / ~ work || Zimmermannsarbeit f. || pièce f. de charpente. / ~ workshop || Zimmereiwerkstätte f. || atelier m. de charpentier.

carpentry || Zimmerei f. || charpenterie f. / house ~ || Bautischlerei f. || menuiserie f. de bâtiment.

carpet || Teppich m. || tapis m. / Axminster ~ || Axminsterteppich m. || tapis-chenille m. / Brussels ~ || Brüsseler Teppich m. || moquette f. frisée ou bouclée. / ~ of coconut fibre || Kokosteppich m. || tapis m. en fibre de coco. / felt ~ || Filzteppich m. || tapis m. en feutre. / jute ~ || Juteteppich m. || tapis m. en jute. / long-pile ~ || langhaariger Teppich m. || tapis m. velouté haute-laine. / plush ~ || Plüschteppich m.; Samtteppich m. || tapis m. velouté. / printed warp ~ || Teppich m. mit gedruckter Kette || moquette f. imprimée. / stair ~ || Läufer m. || tap is m. d'escalier. / strip of ~ || Läufer m. || chemin m. de velours. / Turkey ~ || Smyrnateppich m. || tapis m. à point noué ou d'Orient ou de Smyrne. / varie-

gated ~ ‖ geflammter Teppich m. ‖ tapis m. jaspé. / Wilton ~ ‖ Mokette f. ‖ tapis m. moquette. / woollen ~ ‖ Wollteppich m. ‖ tapis m. de laine.

carpet bag ‖ Reisesack m. ‖ sac m. de voyage. / ~ beater ‖ Teppichklopfer m. ‖ batteur m. de tapis. / ~ beating machine ‖ Teppichklopfmaschine f. ‖ batteuse-dépoussiéreuse f. pour tapis ; machine f. à battre les tapis. / ~ cleaning machine ‖ Teppichkehrmaschine f. ‖ machine f. à brosser les tapis. / ~ cover ‖ Teppichschoner m. ‖ pare-tapis m. / ~ darner ‖ Teppichstopfer m. ‖ rentrayeur m. de tapis. / ~ fringer ‖ Teppichfranser m. ‖ frangeur m. de tapis. / ~ fringe knotter ‖ Teppichfransenknüpfer m. ‖ noueur m. de franges de tapis. / ~ fringe sewer ‖ Teppichfransennäher m. ‖ couseur m. de franges de tapis. / ~ knife ‖ Fadenschneider m. ‖ tranche-fil m. / ~ loom see ~ weaver's loom. / ~ maker ‖ Teppichweber m. ‖ carpettier m. / ~ needle ‖ Teppichnadel f. ‖ aiguille f. de tapis. / ~ picker ‖ Teppichnopper m. ‖ noppeur m. de tapis. / ~ roller ‖ Teppichroller m. ‖ enrouleur m. de tapis. / ~ shearer ‖ Teppichscherer m. ‖ tondeur m. de tapis. / ~ shearing machine ‖ Teppichschermaschine f. ‖ tondeuse f. de tapis. / ~ sweeper ‖ Teppichkehrmaschine f. ‖ épousseteur m. pour les tapis. / ~ weaving ‖ Teppichweberei f. ‖ tissage m. des tapis. / ~ weaver's loom ‖ Teppichwebstuhl m. ‖ métier m. à tapis.

carrag(h)een ‖ Karragheenmoos n. ‖ chondre m. crispé.

Carrara marble ‖ Carraramarmor m. ‖ carrare m.

carriage ‖ Wagen m.; Fuhrwerk n. ‖ voiture f.; char m.; chariot m.; véhicule m. / ~ (Tel; Post) ‖ Bestellgebühr f.; Bestellgeld n. ‖ factage m. / ~ (Trade) ‖ Frachtgeld n.; Fuhrlohn m.; Rollgeld ‖ frais mpl. de transport; camionage m. / ~ (Transport) ‖ Anfuhr f. ‖ charroi m.; charriage m. / ~ (Typewr) ‖ Wagen m.; Schlitten m. ‖ chariot m. / ~ with accessories ‖ Wagen m. mit Zubehör ‖ chariot m. avec accessoires. / ~ for agriculture ‖ Wagen m. für die Landwirtschaft ‖ voiture f. d'agriculture. / ~ of a boring-machine ‖ Schlitten m. einer Bohrmaschine ‖ chariot m. d'une machine à percer. / boxtrail ~ (War impl) ‖ Kastenlafette f. ‖ affût m. rigide ou monoflèche. / by ~ ‖ per Achse f. ‖ par voiture f. / crab ~ ‖ Katzenwagen m. ‖ treuil m. roulant. / to detach a ~ ‖ einen Wagen m. abkuppeln ‖ découpler une voiture. / discharging ~ ‖ Abwurfwagen m. ‖ chariot m. de déchargement. / fire-brigade ~ ‖ Feuerwehrfahrzeug n. ‖ voiture f. de pompiers. / ~ for flying machines ‖ Flugzeugfahrgestell n. ‖ chariot m. de lancement pour machines volantes. / gun ~ ‖ Lafette f. ‖ affût m. / interchangeable ~ (Typewr) ‖ auswechselbarer Wagen m. ‖ chariot m. interchangeable. / land ~ ‖ Transport m. zu Lande ‖ transport m. par voie de terre. / long ~ ‖ Langschlitten m. ‖ chariot m. long. / mixed ~ (Railw) ‖ Wagen m. mit verschiedenen Klassen ‖ voiture f. mixte de voyageurs. / narrow-gauge ~ ‖ Schmalspurwagen m. ‖ wagon m. à voie étroite. / part of ~s ‖ Waggonbauteil m. ‖ pièce f. de construction du wagon. / passenger ~

Personenwagen m. ‖ voiture f. à voyageurs. / passenger ~ (Railw) ‖ Personenwagen m. ‖ wagon m. à voyageurs. / ~ for patients ‖ Krankenfahrzeug n. ‖ voiture f. pour malades. / ~ of printing press ‖ Karren m. einer Druckpresse ‖ coffre m. ou berceau m. d'une presse à imprimer. / splendid ~ ‖ Luxuswagen m. ‖ carrosse f. ou voiture f. de luxe. / split-trail ~ (War impl) ‖ Spreizlafette f. ‖ affût m. biflèche ou à flèches ouvrantes. / starting ~ (Aero) ‖ Anlaufgestell n. ‖ chariot n. de lancement. / ~ with tank ‖ Behälterwagen m. ‖ voiture f. (avec) réservoir. / travelling ~ (Typewr) ‖ beweglicher Schlitten m. ‖ chariot m. mobile. / under ~ (Aero) ‖ Fahrgestell n. ‖ châssis m. d'atterrissage. / universal gun ~ (War impl) ‖ Rundumlafette f. ‖ affût m. universel. / water ~ ‖ Transport m. zu Wasser ‖ transport m. par eau.

carriage beam ‖ Deichsel f. ‖ timon m. / ~ blacksmith ‖ Wagenschmied m. ‖ forgeron m. en voitures. / ~ body (Auto) ‖ Karosserie f. ‖ carrosserie f.; caisse f. d'auto. / ~ body (Coachm) ‖ Wagenkasten m. ‖ caisse f. de voiture. / ~ bridge ‖ Wagenbrücke f. ‖ pont m. de voiture. / ~ builder ‖ Wagenbauer m. ‖ carrossier m. / ~ building ‖ Wagenbau m. ‖ carrosserie f. ‖ ~ building locksmith's shop ‖ Wagenschlosserei f. ‖ serrurerie f. de voitures. / ~ cask ‖ Transportfaß n. ‖ Fuhrfaß n. ‖ fût m. d'expédition; foudre m. ou tonneau m. de transport; barrique f. / ~ cask cock ‖ Fuhrfaßhahn m. ‖ robinet m. de foudre de transport. / ~ clock ‖ Wagenuhr f. ‖ horloge f. de voiture. / ~ and coach trimming and lace ‖ Wagenposamenten pl. ‖ passementerie f. pour voitures. / ~ entrance ‖ Einfahrt f.; Torweg m. ‖ porte-cochère f. / ~ fee ‖ Abfuhrgebühr f. ‖ factage m. / ~ feed ‖ Transportwagenschaltwerk n. ‖ commande f. par chariot transporteur. / ~ fork ‖ Schlittengabel f. ‖ fourche f. du chariot. / ~ grease ‖ Wagenfett n.; Wagenschmiere f. ‖ vieux oing m.; graisse f. à voitures. / ~ guide ‖ Schlittenführung f. ‖ guidage m. du chariot. / ~ heating ‖ Wagenheizung f. ‖ chauffage m. des voitures. / ~ ironwork ‖ Wagenbeschlagteile mpl. ‖ ferrures fpl. de voitures. / ~ lantern ‖ Wagenlaterne f. ‖ lanterne f. de voitures. / ~ lever (Typewr) ‖ Wagenhebel m. ‖ levier m. de commande du chariot. / ~ lighting ‖ Wagenbeleuchtung f. ‖ éclairage m. de voiture. / ~ lock lever (Typewr) ‖ Wagenfeststeller m. ‖ levier m. de fixation du chariot. / ~ manufacture ‖ Wagenbauanstalt f. ‖ atelier m. de charronnage. / ~ movement (Typewr) ‖ Wagenbewegung f. ‖ mouvement m. du chariot. / ~ paid ‖ frachtfrei; Fracht f. bezahlt ‖ franc de fret ou de voiture; port m. payé. / ~ rail (Typewr) ‖ Gleitschiene f. des Wagens ‖ glissière f. du chariot. / ~ release key (Typewr) ‖ Wagenauslösetaste f. ‖ touche f. de libération du chariot. / ~ roller ‖ Wagenrolle f. ‖ galet m. du chariot. / ~ rolling-stock ‖ Wagenpark m. ‖ parc m. de voitures. / ~ rug ‖ Wagendecke f. ‖ bâche f. / ~ saddler ‖ Wagensattler m. ‖ sellier m. en voitures. / ~ spring ‖ Wagenfeder f. ‖ ressort m. du chariot ou de voiture. / ~ trimmings pl. ‖ Wagenbesatzartikel mpl. ‖ passementerie f. pour voitures. /

~ umbrella ‖ Wagenschirm m. ‖ ombrelle f. de voiture. / ~ wheel ‖ Wagenrad n. ‖ roue f. de voiture.

carried away by current ‖ durch den Strom abgetrieben ‖ drossé par le courant. / ~ by tide see ~ by current.

carrier (Chem) ‖ Überträger m. ‖ véhiculeur m. / ~ (Errand boy) ‖ Austräger m.; Bote m. ‖ porteur m. / ~ (Mach tool) ‖ Drehbankherz n.; Drehherz n. ‖ toc m. d'entraînement. / ~ (Mine) ‖ Fördergefäß n. ‖ benne f. / ~ (Railw) ‖ Gepäckträger m. ‖ porte-bagage m. / ~ (Traffic) ‖ Zubringer m. ‖ chariot m. / ~ (Transport) ‖ Rollfuhrmann m.; Rollfuhrunternehmer m. ‖ camionneur m. / ~ (Vehicle) ‖ Gepäckdreirad n. ‖ tri-porteur m. / differential ~ ‖ Differentiallagerkasten m. ‖ carter m. du différentiel. / electricity ~ ‖ Elektrizitätsträger m. ‖ conducteur m. d'électricité. / ~ for light cartridges ‖ Leuchtpatronenträger m. ‖ porte-fusées-éclairantes m. / spare-wheel ~ ‖ Ersatzradhalter m. ‖ porte-roue m. de secours. / tool ~ ‖ Werkzeughalter m.; Stichelhaus n. ‖ porte-outil m.

carrier current (Tel) ‖ Trägerstrom m. ‖ courant m. porteur. / ~ frequency (Tel) ‖ Trägerfrequenz f. ‖ fréquence f. du courant porteur. / ~ line (Tel) ‖ Trägerleitung f. ‖ ligne f. porteuse. / ~ pigeon ‖ Brieftaube f. ‖ pigeon m. voyageur. / ~ pigeon breeder ‖ Brieftaubenzüchter m. ‖ éleveur m. de pigeons-voyageurs. / ~ ring for roof space (Wind mill) ‖ Spannring m. ‖ cercle m. de tension. / ~ telegraphy and telephony with high frequencies ‖ Drahtfunk m. ; Hochfrequenztelegrafie f. und -telefonie f. auf Leitungen ‖ télégraphie f. et téléphonie f. à haute fréquence sur fils. / ~ telegraphy and telephony on lines ‖ Trägerstromtelegrafie f. und -telefonie f. ‖ télégraphie f. et téléphonie f. à courants porteurs. / ~ tricycle ‖ Gepäckdreirad n. ‖ tricycle m. porteur. / high frequency ~ wave ‖ Hochfrequenzübertragerwelle f. ‖ onde f. porteuse de haute fréquence. / ~ wave frequency ‖ Grundwellenfrequenz f. ‖ fréquence f. d'onde fondamentale.

carrigeen ‖ Karragheenmoos n. ‖ chondre m. crispé.

carrot ‖ Möhre f.; Karotte f. ‖ carotte f.

carry, to (Chem) ‖ übertragen ‖ véhiculer. / ~ off ‖ aufräumen ‖ enlever; écarter. / ~ off the current to earth ‖ den Strom m. in die Erde ableiten ‖ faire écouler le courant à la terre. / ~ off the heat ‖ die Wärme ableiten ‖ enlever la chaleur. / ~ on solid frames pl. ‖ in fester Stuhlung f. lagern ‖ monter sur un bâti fixe. / ~ out ‖ ausführen ‖ faire; effectuer; accomplir; produire. / completion of the extensions (which are) being carried out now ‖ Durchführung f. der im Gange befindlichen Ausbauarbeiten ‖ achèvement m. des constructions en cours. / ~ over ‖ stornieren ‖ contre-passer; résilier; ristorner. / ~ threads pl. (Textile) ‖ einziehen ‖ enfiler. / easy to be carried ‖ leicht zu transportieren ‖ facilement transportable.

carrying, wire ~ an electric current ‖ von elektrischem Strom durchflossener Draht ‖ fil m. parcouru par un courant.

carrying ‖ Abrollen n. ‖ camionnage m. / ~ away ‖ Abfuhr f. ‖ transport m. (en chariot). / ~ the goods ‖ Abfahren n. der

Güter ‖ enlèvement m. des marchandises. / ~ of miscellaneous goods for long distances ‖ Beförderung f. von Stückgütern über große Strecken ‖ transport m. de colis sur grandes distances.

carrying apparatus ‖ Transportgerät n. ‖ appareil m. de transport. / ~ axle ‖ Laufachse f. ‖ essieu m. porteur.

carrying capacity ‖ Tragfähigkeit f. ‖ capacité f. de charge. / ~ (Shipb) ‖ Tragfähigkeit f.; Ladefähigkeit f. ‖ port m. en lourd; capacité f. de chargement. / ~ (Tel) ‖ Belastungsfähigkeit f. einer Leitung ‖ densité f. de courant admissible. / ~ of track ‖ Tragfähigkeit f. des Gleises ‖ résistance f. de la voie.

carrying fork (Mach tool) ‖ Mitnehmergabel f. ‖ fourchette f. d'entraînement. / ~ rod ‖ Tragstange f. ‖ barre f. de support. / ~ roller ‖ Führungsrolle f. ‖ poulie f. conductrice. / ~ rope ‖ Tragseil n. ‖ câble m. porteur. / ~ sling (Acc) ‖ Tragbügel m. ‖ étrier m. de suspension. / ~ trade (Trade) ‖ Zwischenhandel m. ‖ commerce m. d'entrepôt ou intermédiaire. / ~ trade (Traffic) ‖ Transportgeschäft n. ‖ affaire f. ou commerce m. de transports.

car shed (Auto) ‖ Garage f. ‖ garage m. / ~ (Vehicle) ‖ Wagenschuppen m. ‖ dépôt m. des voitures; remise f.

car spring (Railw) ‖ Eisenbahnwagenfeder f. ‖ ressort m. pour wagons. / ~ (Vehicle) ‖ Wagenfeder f. ‖ ressort m. pour voitures ou wagons ou carrosseries.

cart, to ~ the soil ‖ die Erde abkarren ‖ brouetter de la terre.

cart (Hand) ‖ zweirädriger Karren m.; Karre f. ‖ charrette f. / ~ (Horse) ‖ Fuhrwerk n. ‖ véhicule m.; chariot m.; char m. / ~ covered ~ ‖ gedeckter Wagen m. ‖ carriole f. ‖ voiture f. couverte. / ~ doll's ~ ‖ Puppenwagen m. ‖ voiture f. de poupée. / hand ~ ‖ Handkarren m. ‖ charrette f. à bras. / open ~ ‖ Sturzkarren m. ‖ haquet m.; tombereau m. / ~ for steel bottles ‖ Transportkarren m. für Stahlflaschen ‖ voiture f. de transport pour bouteilles en acier. / timber ~ ‖ Langholzwagen m. ‖ char m. à bois de construction. / tipping ~ ‖ Kippkarren m. ‖ tombereau m. à benne basculante; charrette f. basculante. / ~ for transportation ‖ Transportkarre f. ‖ chariot m. de transport.

cartage (Post) ‖ Zustellungsgebühr f. ‖ exprès m. / ~ (Railway) ‖ Fuhrlohn m.; Rollgeld n. ‖ frais mpl. de transport ou de voiture; tarif m. de camionnage. / ~ (Transport) ‖ Anfuhr f. ‖ charroi m.; charriage m.

cart axle ‖ Wagenachse f. ‖ essieu m. de chariot.

carter (Mine) ‖ Fuhrmann m.; Kohlenfuhrmann m. ‖ charretier m.; varlet m. / ~'s office ‖ Rollfuhrwerkunternehmung f. ‖ entreprise f. de camionnage m.

carthamin (Chem) ‖ vegetabilisches Rot n.; Karthamin n. ‖ carthamine f.; rouge m. ou rose m. végétal.

Carthamus ‖ Färberdistel f.; falscher Safran m. ‖ carthame m.; safran m. bâtard.

cartilage ‖ Knorpel m. ‖ cartilage m.

cartilaginous ‖ knorpelig ‖ cartilagineux.

cart load ‖ Wagenladung f.; Fuhre f. ‖ charretée f.; charroi m. / ~ manual fire engine ‖ Abprotzspritze f. ‖ pompe f. à incendie sur charrette.

cartographer ‖ Kartograf m. ‖ cartographe m.

cartographic work ‖ kartografisches Werk n. ‖ ouvrage m. cartographique.

cartography ‖ Kartografie f.; Kartenzeichenkunst f. ‖ cartographie f.

carton, loose in ~s ‖ lose in Kartons mpl. ‖ en vrac.

carton pierre ‖ Steinpappe f. ‖ carton m. pierre.

car transfer platform ‖ Wagenschiebebühne f. ‖ transbordeur m. pour wagons; chariot m. roulant pour wagons.

cartridge (Mot) ‖ Zündstift m. ‖ cheville f. d'allumage. / ~ (War impl) ‖ Patrone f. ‖ cartouche f. / blank ~ ‖ Platzpatrone f. ‖ cartouche f. blanche. / fuse ~ (Electr) ‖ Sicherungspatrone f. ‖ cartouche f. de fusibles. / primer ~ (Mine) ‖ oberste Patrone f. ‖ cartouche f. d'amorce. / resorcinate explosive ~ ‖ Resorzinatsprengkapsel f. ‖ capsule f. explosive à la résorcine.

cartridge case ‖ Patronenhülse f. ‖ étui m. à cartouches; douille f. / ~ fuse (Electr) ‖ Sicherungspatrone f.; Patronensicherung f. ‖ cartouche f. de coupe-circuit. / ~ heated apparatus ‖ patronenbeheiztes Gerät n. ‖ appareil m. chauffé par des cartouches. / ~ heating member (Electr) ‖ Patronenheizkörper m. ‖ radiateur m. forme cartouche. / ~ maker ‖ Patronenmacher m. ‖ cartouchier m. / ~ making machine ‖ Maschine f. zur Herstellung von Patronen ‖ machine f. à fabriquer les cartouches. / ~ wad ‖ Patronenfilzpfropfen m. ‖ bourre f. de cartouche. / ~ works pl. ‖ Patronenfabrik f. ‖ cartoucherie f.; fabrique f. de cartouches.

car truck ‖ Wagenuntergestell n. ‖ châssis m. de voiture. / ~ frame (Railw) ‖ Untergestell n. des Wagens ‖ châssis m. de wagon.

cartwright ‖ Stellmacher m.; Rademacher m.; Wagner m. ‖ charron m. / ~'s work ‖ Stellmacherarbeit f. ‖ charronnage m.

cart type station (Radio) ‖ Karrenstation f.; fahrbare Station f. ‖ station f. du type sur chariot.

car type ‖ Wagenbauart f. ‖ type m. de véhicule.

car varnisher ‖ Wagenlackierer m. ‖ vernisseur m. en voitures.

carve, to ‖ ziselieren; schnitzen; ausarbeiten ‖ ciseler; sculpter.

carved, fit to be ~ ‖ zum Schnitzen n. geeignet ‖ susceptible d'être taillé. / ~ face (Mine) ‖ verschrämter Stoß m. ‖ paroi f. entaillée. / ~ weir ‖ durchbrochenes Wehr n. ‖ écluse f. à pertuis. / ~ wood ‖ Holzschnitzwaren fpl. ‖ bois mpl. sculptés. / ~ work ‖ Schnitzwerk n. aus Holz; Holzschnitzerei f. ‖ découpure f.; sculpture f. de bois.

carvel-built boat ‖ Kraweelboot n. ‖ canot m. bordé à joints ouverts ou carrées.

carvel work (Shipb) ‖ Kraweelbau m. ‖ bordé m. à plat-joint; bordé exécuté à franc bord.

carver (Print) ‖ Holzgravör m.; Holzschneider m. ‖ xylographe m. / ~ (Sculpt) ‖ Holzschnitzer m.; Holzbildhauer m. ‖ sculpteur m. en bois. / ~ of figures on wood ‖ Holzbildschnitzer m. ‖ sculpteur m. de figurines en bois. / hand ~ ‖ Handschnitzer m. ‖ sculpteur m. à la main. / image ~ ‖ Bildschnitzer m. ‖ imagier-tailleur m. / machine ~ ‖ Maschinen-

schnitzer m. ‖ sculpteur m. à la machine. / wood ~ (Print) ‖ Holzschneider m. ‖ graveur m. sur bois. / wood ~ (Sculpt) ‖ Holzschnitzer m. ‖ sculpteur m. sur bois.

carving ‖ Schnitzarbeit f.; Schnitzwerk n. aus Holz ‖ sculpture f. en bois; découpure f. / articles pl. from materials for ~ ‖ Schnitzstoffwaren fpl. ‖ ouvrages mpl. en matières à tailler. / geometrical ~ (Arch) ‖ Maßwerk n. ‖ broderie f.; tracé m.; découpure f.; réseau m.

carving chisel, spoon bit ~ ‖ Schaufelbetelchen n. für Bildschnitzer ‖ fermoir m. à spatule. / strong ~ ‖ Grobbeitel m. ‖ fermoir m. renforcé à dégrossir.

carving knife ‖ Vorlegemesser n.; Tranchiermesser n. ‖ couteau m. à découper. / ~ knife and fork ‖ Tranchierbesteck n. ‖ couvert m. à découper. / ~ material ‖ Schnitzstoff m. ‖ matière f. à tailler.

car, weight of the ~ ‖ Wagengewicht n.; Gewicht n. des Wagens; Leergewicht n. ‖ poids m. de la voiture.

car-wheel diameter ‖ Laufraddurchmesser m. ‖ diamètre m. de roue motrice.

caryatid ‖ Karyatide f. ‖ caryatide f.

cascade ‖ Wasserfall m.; Kaskade f. ‖ cascade f. / ~ amplification ‖ Kaskadenverstärkung f. ‖ amplification f. en cascade. / ~ connection ‖ Kaskadenschaltung f. ‖ montage m. en tandem; couplage m. en cascade.

cascades pl., artificial (Pyrot) ‖ Feuerregen m. ‖ cascades fpl. artificielles.

cascade voltage transformer ‖ Kaskadenspannungswandler m. ‖ transformateur m. de potentiel à cascade.

case, to ~ the glass (Glassm) ‖ überfangen ‖ plaquer le verre.

case ‖ Behälter m.; Kiste f.; Kasten m. ‖ caisse f.; boîte f. / ~ Schachtel f.; Etui n. ‖ emballage m.; étui m. / ~ Futteral n. ‖ gaine f. / ~ (Found) ‖ Mantel m. (einer Lehmform) ‖ chape f.; manteau m.; moulechape m.; surmoule m.; surtout m. / ~ (Loc, Boil) ‖ Kesselbekleidung f.; Mantel m. des Kessels ‖ chemise f. ou enveloppe f. d'une chaudière. / ~ (Locksm) ‖ Schloßkasten m.; boîte f.; palastre m. ou palâtre f. d'une serrure à palâtre. / ~ (Mach) ‖ Kapsel f.; Gehäuse n. ‖ enveloppe f.; carter m. / ~ (Jacket) ‖ Mantel m.; Umhüllung f. ‖ chemise f.; enveloppe f. / ~ (Met) ‖ Schachtscheider m. ‖ paroi f. de séparation; séparation f. / ~ (Pap) ‖ Haube f. (des Holländers) ‖ chapeau m.; chapiteau m. / ~ (Print) ‖ Setzkasten m.; Schriftkasten m. ‖ casse f. / bank-note ~ ‖ Geldscheintasche f. ‖ porte-feuille m. / ~ of chimney ‖ Schornsteinmantel m. ‖ chemise f. de cheminée. / ~ clock ~ ‖ Uhrgehäuse n. ‖ boîte f. de pendules. / ~ in court ‖ Rechtsfall m. ‖ cas m. litigieux. / crank ~ ‖ Kurbelgehäuse n. ‖ carter m. de la manivelle. / differential ~ ‖ Ausgleichgehäuse n. ‖ carter m. de différentiel. / ~ for documents ‖ Aktenmappe f. ‖ portefeuille m. (de bureau). / Dutch ~ ‖ holländischer Rahmen m. ‖ châssis m. hollandais. / engine ~ ‖ Motorgehäuse n. ‖ carter m. de moteur. / eyeglass ~ ‖ Pinzenezfutteral n. ‖ étui m. à pince-nez. / ~ of felt ‖ Filzfutteral n. ‖ étui m. en feutre. / glazed ~ ‖ Vitrine f. ‖ vitrine f. / ~ at issue ‖ Streitfall m.; Streitfrage f. ‖ point m. de controverse; différend m. / ~ for

knives, forks and spoons ‖ Behälter m. für Bestecke ‖ étui m. pour couverts de table. / ~ of a knife ‖ Scheide f. eines Messers ‖ gaîne f. ou étui m. d'un couteau. / leaden ~ ‖ Bleimantel m. ‖ enveloppe f. en plomb. / ~ fitted with lock and key ‖ verschließbarer Schrank m. ‖ armoire f. fermant à clé. / machine for the manufacture of ~s ‖ Kistenherstellungsmaschine f. ‖ machine f. à fabriquer des caisses. / ~ of mathematical instruments ‖ Reißzeug n. ‖ boîte f. à compas; étui m. de mathématique. / monocle ~ ‖ Einglasfutteral n. ‖ étui m. à monocles. / ~ for musical instruments ‖ Etui n. für Musikinstrumente ‖ étui m. pour instruments de musique. / ~ of necessity ‖ Notfall m. ‖ cas m. de besoin. / in ~ of need ‖ im Notfalle m. ‖ au besoin. / nominal diameter of ~ ‖ Gehäusenenndurchmesser m. ‖ diamètre m. nominal de la boîte. / ~ of osier ‖ Weidenfutteral n. ‖ étui m. en osier. / packing ~ ‖ Packkiste f. ‖ caisse f. d'emballage. / ~ for pendulum clocks ‖ Gehäuse n. für Penduluhren ‖ cage f. pour pendules. / ~ for photographic apparatus ‖ Hülle f. für fotografische Apparate ‖ étui m. pour appareils photographiques. / portable ~ ‖ Tragkasten m. ‖ caisse f. portative. / ~ for quick composition (Print) ‖ Schnellsetzkasten m. ‖ case f. tachéotype. / in ~ of repetition ‖ im Wiederholungsfalle m. ‖ en cas de récidive. / ~ for scent bottles ‖ Hülle f. für Riechfläschchen ‖ étui m. pour flacons. / sheet-iron ~ ‖ Blechkasten m. ‖ caisse f. en tôle. / shipping ~ for beer ‖ Bierversandkiste f. ‖ caisse f. à expédition pour la bière. / ~ for shooting weapons ‖ Hülle f. für Jagdwaffen ‖ étui m. pour armes de chasse. / small ~ ‖ Kästchen ‖ coffret m. / snap spectacle ~ ‖ Klappdeckelbrillenfutteral n. ‖ étui m. de lunettes à bascule. / spectacle ~ ‖ Brillenfutteral n. ‖ étui m. à lunettes. / ~ for tail pieces (Print) ‖ Leistenkasten m. ‖ caseau m.; bardeau m. / umbrella ~ ‖ Schirmfutteral n. ‖ fourreau m. de parapluie. / walnut ~ ‖ Nußbaumkasten m. ‖ boîte f. en noyer. / watch ~ ‖ Uhrgehäuse n. ‖ boîtier m. ou boîte f. de montre. / water-tight ~ ‖ wasserdichtes Gehäuse n. ‖ caisse f. hermétique. / ~ of a window ‖ Fensterfutter n.; Fensterzarge m. ‖ dormant m. d'un châssis; cadre m. de croisée. / wooden ~ ‖ Holzetui n.; Holzfutteral n. ‖ étui m. en bois. / ~ of yarn ‖ Garnkiste f. ‖ caisse f. de fil.

case bay (Carp) ‖ Fach n. oder Feld n. oder Raum m. zwischen zwei Balken ‖ claire-voie f.; travée f. / ~ bottle ‖ Feldflasche f. ‖ bidon m. / ~ casting ‖ Schalenguß m.; Hartguß m. ‖ fonderie f. ou fonte f. en coquilles. / ~ closing apparatus ‖ Kistenverschlußapparat m. ‖ appareil m. à fermer les caisses. / ~ closing appliance ‖ Kistenverschlußeinrichtung f. ‖ dispositif m. à fermer les caisses. / ~ fittings pl. ‖ Etuibeschläge mpl. ‖ ferrures fpl. pour étuis. / ~-full (Print) ‖ Schriftkasten m. voll Lettern ‖ cassettée f.

case-harden, to ‖ im Einsatz m. härten ‖ tremper en coquille; aciérer; cémenter.

case-hardened ‖ schalenhart; hart gegossen ‖ coulé en coquille. / ~ casting

Schalenguß m.; Kokillenguß m. ‖ coulage m. ou fonte f. en coquille. / ~ castiron ‖ Schalenguß m. ‖ fer m. fondu en coquille. / ~ crossing ‖ Hartgußkreuzung f.; Hartgußweiche f. Hartgußherzstück n. ‖ croisement m. en fonte en coquille. / ~ cylinder ‖ Hartgußwalze f. ‖ cylindre m. en fonte en coquille. / ~ frog see ~ crossing.

case-hardening ‖ Hartguß m.; Schalenguß m. ‖ fonte f. en coquilles; fonderie f. en coquilles. / ~ ‖ Einsatzhärtung f.; Oberflächenhärtung f. ‖ trempe f. en paquet ou en coquille ou de surface; cémentation f.; aciération f. / ~ ‖ Einsetzen n. ‖ aciérage m. / ~ box ‖ Einsatzkasten m. ‖ boîte f. à cémenter.

case-hardening machine, gas ‖ Druckgas- und Karbonisierofen m. ‖ four m. à cémenter aux gaz carburants.

case-hardening pot ‖ Einsetztiegel m. ‖ coffre m. de cémentation. / ~ steel ‖ Einsatzstahl m. ‖ acier m. à cémenter. / ~ work in a salt bath ‖ Zementation f. ‖ cémentation f. au bain de sel.

casein(e) ‖ Kasein n.; Käsestoff m. ‖ caséine f.; caséum m. / vegetable ~ ‖ Legumin n.; Pflanzenkasein n.; Pflanzenkäsestoff m. ‖ caséine f. végétale; légumine f.

casein cold-glue powder ‖ Kaseinkaltleimpulver n. ‖ colle f. froide de caséine en poudre. / ~ glue ‖ Kaseinleim m. ‖ colle f. de caséine. / ~ paint ‖ Kaseinfarbstoff m. ‖ couleur f. de caséine.

case lock ‖ Etuiverschluß m. ‖ fermeture f. pour étuis. / ~ lock ‖ Kastenschloß n. ‖ serrure f. à palastre ou à palâtre. / ~ maker (Bookb) ‖ Einhänger m. ‖ emboîteur m. / ~ making machine ‖ Etuisherstellungsmaschine f. ‖ machine f. à fabriquer des étuis.

casemate ‖ Kasematte f. ‖ casemate f.

casement ‖ Fensterrahmen m. ‖ châssis m. à verre ou à carreaux; croisée f.; cadre m. de croisée. / (French) ~ ‖ Flügelfensterrahmen m.; Futterrahmen m. ‖ croisée f. à vantaux; cadre m. de croisée à battants. / ~ of window ‖ Fensterflügel m. ‖ battant m. ou vantail m. de fenêtre.

casement staple ‖ Fensterkrampe f. ‖ crampon m. de fenêtre. / ~ stay ‖ Sturmhaken m. ‖ crochet m. à fenêtres ouvertes. / ~ window ‖ Flügelfenster n. ‖ croisée f. à vantaux.

case mill ‖ Kochermühle f. ‖ moulin m. de support ou à colonne creuse. / ~ nailing machine ‖ Kistennagelmaschine f. ‖ machine f. à clouer les caisses.

caseous ‖ kaseinartig; käsig ‖ caséeux.

case protector ‖ Kistenschoner m. ‖ protecteur m. de caisse. / ~ shot ‖ Kartätsche f. ‖ boîte f. à mitraille. / ~ spring (Watchm) ‖ Gehäusefeder f. ‖ secret m.

caseum ‖ Kasein n. ‖ caséine f.; caséum m. / ~ glue ‖ Käseleim m.; Kaseinleim m. ‖ colle f. caséine.

cash, to ~ a bill of exchange ‖ einen Wechsel m. auslösen ‖ acquitter ou racheter une lettre de change.

cash ‖ bares Geld n. ‖ argent m. comptant. / ~ (Cash register) ‖ Barverkauf m. ‖ comptant m. / ~ on delivery ‖ Zahlung f. gegen Lieferung; per Kasse ‖ paiement m. au comptant contre livraison; au comptant. / ~ on delivery ‖ Nachnahme f. ‖ remboursement m. / ~ against docu-

ments ‖ Kassa f. gegen Dokumente ‖ paiement m. contre documents. / for ~ ‖ gegen bar ‖ argent m. comptant; au comptant. / goods pl. sold for ~ ‖ gegen bar verkaufte Ware f. ‖ marchandise f. vendue au comptant. / ~ in hand ‖ verfügbares Geld n. ‖ fonds mpl. disponibles ou mobilisables; disponibilités fpl. / ~ in hand ‖ Kassenbestand m. ‖ encaisse f.; solde m. en caisse. / to send for ~ on delivery ‖ nachnehmen ‖ faire suivre en remboursement.

cash balance ‖ Kassenbestand m. ‖ montant m. de l'encaisse. / ~ ‖ Kassenabschluß m. ‖ clôture f. de la caisse.

cash block ‖ Kassenblock m. ‖ bloc m. de caisse. / ~ clip ‖ Kassenblockklammer f. ‖ pince f. pour blocs de caisse. / ~ rotary printing machine ‖ Kassenblockrotationsdruckmaschine f. ‖ machine f. rotative à imprimer les blocs de caisse.

cash-book ‖ Ausgabenbuch n. ‖ livre m. des dépenses. / ~ ‖ Kassenbuch n. ‖ livre m. de caisse.

cash box ‖ Kassette f. ‖ cassette f. / ~ ‖ Kassenschrank m. ‖ caisse f. / ~ lock ‖ Kastenschloß n.; Kofferschloß n.; Schrankschloß n. ‖ serrure f. d'armoire ou de coffre.

cash closed (Cash register) ‖ Kasse f. abgeschlossen ‖ caisse f. fermée. / ~ disbursements pl. ‖ Barausgang m. ‖ paiements mpl. effectués par la caisse. / ~ discount ‖ Rabatt m. gegen bar ‖ escompte m. au comptant. / ~ drawer ‖ Geldschublade f. ‖ tiroir m. de monnaie.

cashew nut ‖ Mahagoninuß f. ‖ pomme f. d'acajou.

cashier ‖ Kassierer m. ‖ caissier m.

cashing ‖ Inkasso n. ‖ encaissement m.; recouvrement m. / ~ of a bill ‖ Wechseleinlösung f. ‖ payement m. d'une lettre de change. / ~ up ‖ Kassenabschluß m. ‖ clôture f. de la caisse.

cashmere ‖ Kaschmir m. ‖ cachemir(e) m.

Cashmere goat's hair ‖ Kaschmirziegenhaar n. ‖ poil m. de chèvre cachemire. / ~ shawl ‖ Kaschmirschal m. ‖ châle m. de cachemire. / ~ wool ‖ Kaschmirwolle f. ‖ laine f. de cachemire.

cashoa ‖ Katechu n. ‖ cachou m.

cash order ‖ Kassenanweisung f.; Kassenschein m. ‖ bon m. de caisse; billet m. de banque. / ~ ‖ Postnachnahme f. ‖ remboursement m. postal.

cash receipts pl. ‖ Bareingang m. ‖ rentrées fpl. au comptant.

cash register ‖ Registrierkasse f.; Kontrollkasse f. ‖ caisse f. enregistreuse. / ~ with adding mechanism ‖ Registrierkasse f. mit Addierwerk ‖ caisse f. enregistreuse avec totalisateur. / ~ for checks ‖ Scheckdrucker m. ‖ totalisatrice f. à tickets. / ~ electric ‖ Registrierkasse f. mit elektrischem Antrieb ‖ caisse f. enregistreuse à commande électrique. / ~ with key action ‖ Tastenregistrierkasse f. ‖ caisse f. enregistreuse à touches. / ~ with key action for delivering tickets ‖ Tastenkasse f. für Scheckdruck ‖ caisse f. enregistreuse à touches imprimant sur ticket / ~ with lever action ‖ Hebelregistrierkasse f. ‖ caisse f. enregistreuse à leviers. / ~ with lever action for printing receipt on sale slips ‖ Hebelregistrierkasse f. für Quittungsdruck ‖ caisse f. enregistreuse à leviers imprimant sur

fiche. / multiple lever ~ ‖ Mehrzähler-hebelkontrollkasse f. ‖ totalisatrice f. multiple à leviers. / ~ operated by electric motor ‖ Registrierkasse für elektrischen Antrieb ‖ caisse f. enregistreuse à commande électrique. / ~ operated by hand ‖ Registrierkasse f. für Handbetrieb ‖ caisse f. enregistreuse à commande à main. / ~ with several adding mechanisms ‖ Registrierkasse f. mit mehreren Addierwerken ‖ caisse f. enregistreuse comportant plusieurs totalisateurs. / ~ for sale slips ‖ Zahlkassenquittungsdrucker m. ‖ totalisatrice f. à fiches. / single lever ~ ‖ Einzählerhebelkontrollkasse f. ‖ totalisatrice f. simple à leviers. / total adding ~ ‖ Totaladdierregistrierkasse f. ‖ caisse f. enregistreuse indiquant les totaux. / ~ in working ‖ Zahlkasse f. in Arbeitsstellung ‖ caisse f. enregistreuse prête au fonctionnement.

cash register lock ‖ Kassensperrschloß n. ‖ serrure f. de caisse.

cash remittance ‖ Geldsendung f. ‖ envoi m. d'argent *ou* d'espèces; remise f. / ~ sale ‖ Barverkauf m. ‖ vente f. au comptant. / ~ till ‖ Ladenkasse f. ‖ caisse f. de magasin *ou* de boutique.

casing ‖ Gehäuse n.; Einkapselung f. ‖ carter m.; cage f.; encastrement m. / ~ (Electr) ‖ Nutleiste f. ‖ planche f. à rainure. / ~ (Instrument) ‖ Fassung f. ‖ monture f. / ~ (Build) ‖ Schalung f. ‖ coffrage m. / ~ (Mach) ‖ Mantel m.; Umhüllung f. ‖ chemise f.; enveloppe f. / ~ (Textile) ‖ Futter n. ‖ doublure f. / ~ (Mine) ‖ Schachtausbau m. ‖ boisage m. *ou* cuvelage m. et muraillement m. d'un puits. / ~ of a boiler ‖ Mantel m. des Kessels; Kesselmantel m. ‖ enveloppe f. *ou* chemise f. d'une chaudière. / ~ of a chimney ‖ Mantel m. eines Schornsteins; Rauchfangmantel m. ‖ chemise f. de la cheminée. / ~ of the clock movement ‖ Einpassen n. des Uhrwerks in das Gehäuse ‖ emboîtage m. du mouvement. / ~ of column ‖ Säulenummantelung f. ‖ revêtement m. de la colonne. / ~ of a door ‖ Türfutter n. ‖ fourrure f. de porte. / double-groove ~ ‖ doppelt gerillte Holzleiste f. ‖ moulure f. en bois à deux rainures. / half of ~ ‖ Gehäusehälfte f. ‖ moitié f. de boîte. / inside wooden ~ ‖ Holzauskleidung f. ‖ coffrage m. en charpente. / leather ~ ‖ Lederfutteral n. ‖ étui m. en cuir. / light-proof ~ ‖ lichtdichtes Gehäuse n. ‖ boîte f. étanche à la lumière. / lower part of ~ ‖ Gehäuseunterteil m. ‖ partie f. inférieure de l'enveloppe. / machine for bending ~s pl. ‖ Mantelbiegemaschine f. ‖ machine f. à cintrer des viroles. / outer ~ (Blast-furnace) ‖ Mantel m. ‖ enveloppe f.; manteau m. / ~ of a pipe ‖ Rohrbekleidung f.; Rohrumkleidung f. ‖ enveloppe f. d'un tuyau. / ~ and pivot is a one-piece casting ‖ der Zapfen m. ist mit dem Trog in einem Stück gegossen ‖ pivot m. central venu de fonte avec la cuve; pivot faisant corps avec la cuve; le pivot est moulé avec la cuve en une seule pièce. / ~ of pressed material ‖ Preßstoffgehäuse n. ‖ boîte f. en matière comprimée. / single-groove ~ ‖ einfach gerillte Holzleiste f. ‖ moulure f. en bois à une rainure. / steel ~ ‖ Stahlgehäuse n. ‖ coffre m. d'acier. / tire ~ ‖ Mantel m. *oder* Laufdecke des Luftreifens ‖ bandage

m. pneumatique; enveloppe f. de roulement. / ~ for travelling crab ‖ Laufkatzengehäuse n. ‖ bâti m. de chariot roulant. / ~ of the turn-table ‖ Trog m. der Drehscheibe ‖ cuve f. de plaque tournante. / upper part of the ~ ‖ Gehäuseoberteil m. ‖ partie f. supérieure de l'enveloppe.

casing factory ‖ Futteralfabrik f. ‖ gainerie f. / bottle ~ machine ‖ Flaschenverkapselmaschine f. ‖ capsuleuse f. de bouteilles.

cask ‖ Faß n. ‖ baril m.; fût m.; futaille f.; tonneau m. / ~ (Coop) ‖ Gebinde n. ‖ futaille f.; barrique f. / to clean ~s pl. ‖ faßschwenken ‖ rincer les fûts mpl. / cleaning ~s pl. ‖ Faßschwenken n. ‖ rinçage m. des fûts. / to cleanse ~s pl. *see* to clean ~s. / to drive the hoops pl. of ~s ‖ Fässer npl. antreiben ‖ serrer les cercles mpl. de fûts. / entering ~s pl. through the manhole for washing purposes ‖ Faßschlupfen n. ‖ nettoyage m. intérieur des foudres. / export ~ ‖ Ausfuhrfaß n. ‖ fût m. d'exportation. / head of a ~ ‖ Faßboden n. ‖ fonçailles fpl. / to hoop a ~ ‖ die Reifen mpl. auf das Faß aufschlagen; ein Faß n. bereifen *oder* binden ‖ cercler un fût. / hoop of ~ ‖ Faßband n.; Faßreif m. ‖ cercle m. de fût. / to illuminate a ~ ‖ ein Faß n. ausleuchten ‖ visiter l'interieur m. d'un fût à la lumière. / iron ~ ‖ eisernes Faß n. ‖ tonneau m. en fer. / ~ for lixiviating ‖ Laugenfaß n. ‖ cuvier m. / machine for joining the ~s and driving on the hoops ‖ Faßzusammensetz- und Reifenaufziehmaschine f. ‖ machine f. à monter les fûts et à serrer les cercles. / machine for making ~s ‖ Faßmaschine f. ‖ machine f. pour la fabrication des tonneaux. / to rinse ~s pl. *see* to clean ~s. / rinsing ~s pl. *see* cleaning ~s. / small ~ ‖ Fäßchen n. ‖ barillet m. / to steam a ~ ‖ ein Faß n. dämpfen ‖ étuver un fût m. à la vapeur. / storage ~ ‖ Lagerfaß n. ‖ foudre m. (de garde). / to truss a ~ ‖ ein Faß n. zusammenziehen ‖ trousser un fût m. / to truss up the hoops pl. of ~s ‖ Fässer npl. antreiben ‖ serrer les cercles mpl. de fûts. / to turn up and dress a ~ ‖ ein Faß n. abdrehen *oder* behobeln ‖ raboter un fût m.

cask ausbrenner ‖ Faßentpichmaschine f. ‖ machine f. à dégoudronner les fûts. / ~ bridge ‖ Faßbrücke f.; Tonnenbrücke f. ‖ pont m. de tonneaux. / ~ brush ‖ Faßbürste f. ‖ brosse f. pour fûts / ~ buoy ‖ Tonnenboje f. ‖ bouée f. en baril. / ~ bung ‖ Faßspund m. ‖ bonde f. de fût; bondon m. / ~ bush ‖ Faßspundbüchse f. ‖ boisseau m.; douille f. de bonde. / ~ cellarage machine ‖ Faßkellereimaschine f. ‖ machine f. pour caves de magasinage du ... en tonneaux. / ~ clan ‖ Aufziehhaken m. ‖ griffe f. à foudres. / ~ clasp ‖ Faßspange f. ‖ penture f. à tonneau. / ~ clasp holder ‖ Faßspangenhalter m. ‖ attache f. de penture à tonneau. / ~ claws pl. ‖ Faßklauen fpl. ‖ griffes fpl. à foudres. / cleaning ~s pl. ‖ Faßschwenken n. ‖ rinçage m. de fûts. / ~ cleaning machine ‖ Faßreinigungsmaschine f. ‖ machine f. à laver les fûts. / ~ cleaning-off machine ‖ Faßabhobelmaschine f.; Faßdrehbank f. ‖ machine f. à blanchir *ou* à raboter les fûts. / ~ conveyor ‖ Faßtransportör m. ‖ transporteur m. de fûts. / ~ corking machine ‖ Faßverspundmaschine f. ‖ machine

f. à boucher les fûts; boucheuse f. pour fûts. / ~ croze-cutting machine ‖ Faßkrösemaschine f. ‖ machine f. à jâbler les douves. / ~ crozing machine ‖ Faßkrösemaschine f. ‖ machine f. à jâbler les douves. / ~ decanting apparatus ‖ Faßabfüllapparat m. ‖ appareil m. de soutirage pour tonneaux. / ~ deposit ‖ Faßgeläger n.; Faßhefe f. ‖ dépôt m. *ou* fonds mpl. *ou* lie f. de foudre. / ~ door ‖ Faßtürchen n. ‖ portière f. de foudre. / ~ door screw ‖ Faßtürchenschraube f. ‖ vis f. de portière de foudre. / ~ elevator ‖ Faßelevator m. ‖ élévateur m. à fûts. / ~ emptying apparatus ‖ Faßabfüllapparat m. ‖ soutireuse f. pour fûts.

casket ‖ Kassette f. ‖ cassette f. / ~ (Jewels) ‖ Schmuckkästchen n. ‖ écrin m.

casket fittings pl. ‖ Schatullenbeschläge mpl. ‖ ferrures fpl. de boîtes à nécessaires.

cask factory, equipment for cask factories ‖ Faßfabrikeinrichtung f. ‖ installation f. pour fabriques de tonneaux.

cask fermentation ‖ Faßgärung f. ‖ fermentation f. en fûts *ou* en tonneaux. / ~ filling apparatus ‖ Faßfüllapparat m. ‖ soutireuse f. *ou* appareil m. de soutirage pour fûts. / ~ funnel ‖ Faßtrichter m. ‖ entonnoir m. pour fûts. / ~ gauge ‖ Faßeiche f. ‖ pithomètre m. / ~ gauging apparatus ‖ Faßeichapparat m. ‖ appareil m. à litrer les fûts; litreuse f. pour fûts. / ~ gimblet ‖ Faßbohrer m. ‖ vrille f. à tonneau. / ~ groove-cutting machine ‖ Faßkrösemaschine f. ‖ machine f. à jâbler les douves. / ~ grooving machine *see* ~ groove-cutting machine. / ~ head ‖ Faßboden m. ‖ fond m. de fût; fonçailles fpl. / ~ head-rounding machine ‖ Faßbodendrehbank f.; Faßbodenschneidmaschine f. ‖ machine f. à arrondir les fonds de fûts. / ~ hoist *see* ~ windlass. / ~ hoop ‖ Faßband n.; Faßreifen m. ‖ cercle m. de fût; cerceau f.

casking ‖ Verpackung f. in Fässer ‖ encaquement m.; embarrillage m.

cask making machine ‖ Maschine f. zur Herstellung von Fässern ‖ machine f. à faire les tonneaux. / ~ manufactory ‖ Faßfabrik f. ‖ tonnellerie f. / ~ pitch ‖ Faßpech n. ‖ poix f. à fûts. / ~ pitching apparatus ‖ Faßpichapparat m. ‖ appareil m. à goudronner les fûts. / ~ planing machine ‖ Faßabhobelmaschine f.; Faßegalisiermaschine f. ‖ machine f. à blanchir *ou* à raboter les fûts. / ~ raising apparatus ‖ Faßaufsatzform f. ‖ forme f. d'assemblage pour douves. / ~ rinser ‖ Faßspülgerät n. ‖ rinceuse f. à fûts. / ~ rinsing machine ‖ Faßspülmaschine f. ‖ machine f. à laver les fûts; rinceuse f. à fûts. / ~ rolling apparatus ‖ Faßrollmaschine f. ‖ machine f. à rouler les fûts; rouleuse f. pour fûts. / ~ screw bung ‖ Faßverschluß m. ‖ bonde f. à vis. / ~ scrubber ‖ Faßbürstmaschine f. ‖ brosseuse f. pour fûts; machine f. à brosser les fûts. / ~ scrubbing machine ‖ Faßputzmaschine f. ‖ machine f. à brosser les fûts; brosseuse f. pour fûts. / ~ scrubbing and cleaning machine ‖ Faßputz- und Schwenkmaschine f. ‖ machine f. à brosser et à rincer les fûts. / ~ sediment *see* ~ deposit. / ~ setting up apparatus *see* ~ raising apparatus. / ~ sprinkler ‖ Faßausspritzer m. ‖ injecteur m. pour fûts. / ~ stave ‖ Faßdaube f.; Faßstab m.

‖ douelle f. *ou* douve f. de fût. / ~ support ‖ Faßlagergerüst n. ‖ chaix m. / ~ and vat support ‖ Ganter m. ‖ chantier m. / ~ test pump ‖ Faßprobierpumpe f. ‖ pompe f. à essayer les fûts à la pression. / ~ and barrel wagon ‖ Faßtransportwagen m. ‖ wagonnet m. à transporter des barils. / ~ washer ‖ Faßwaschapparat m. ‖ appareil m. à laver les fûts. / ~ windlass ‖ Faßwinde f.; Faßzug m. ‖ monte-fûts m.; treuil m. à fûts. / ~ wood ‖ Faßholz n. ‖ merrain m.

Cassel brown ‖ Kasselerbraun n. ‖ brun m. *ou* terre f. de Cassel. / ~ earth *see* ~ brown. / ~ yellow ‖ Kasselergelb n.; Mineralgelb n. ‖ jaune m. de Cassel.

cassia bark ‖ Xylokassia f.; Kassiarinde f.; casse f. ligneuse; cannelle f. de Malabar. / ~ kernels pl. ‖ Kassiakörner f. ‖ graines fpl. de cassie.

cassia lignea ‖ Zimtkassia f. ‖ cassia lignea m. / broken ~ ‖ gestoßener Zimt m. ‖ cannelle f. concassée.

cassia pod ‖ Kassiaschote f. ‖ casse f.

cassiterite ‖ Kassiterit m.; Zinnerz n. ‖ cassitérite f.

cassonade ‖ indischer Rohzucker m.; Farin m.; Kochzucker m. ‖ cassonade f.; castonade f.

cast, to ‖ gießen ‖ couler; fondre; jeter en fonte *ou* en moule; verser. / ~ (Wood) ‖ sich krumm ziehen ‖ se gauchir; se déjeter; se voiler. / ~ adrift (Mar) ‖ loslassen ‖ larguer; lâcher; relâcher. / ~ from the bottom (Found) ‖ mit dem Steigrohr n. gießen ‖ couler à cale *ou* à *ou* en siphon. / ~ cold ‖ kalt gießen ‖ couler à froid. / ~ on a core *see* ~ hollow. / ~ hollow (Found) ‖ über den Kern m. gießen; hohl gießen ‖ couler creux *ou* à noyau. / ~ in disks pl. (Chand) ‖ in Scheiben fpl. gießen ‖ écouler en disques. / ~-in the guide blades pl. ‖ die Leitschaufeln fpl. eingießen ‖ couler les aubes fpl. directrices dans la roue. / ~ in open sand ‖ im Herd m. gießen ‖ couler au découvert. / ~ iron ‖ Eisen n. schmelzen ‖ fondre le fer. / ~ loose (Arm) ‖ loslassen ‖ larguer; lâcher; relâcher. / ~ plate glass ‖ Spiegelglas n. gießen ‖ couler les glaces fpl. / ~ sideways in tempering (Steel) ‖ krumm werden beim Härten ‖ se voiler à la trempe. / ~ solid ‖ massiv gießen ‖ couler plein. / ~-up ‖ kalkulieren ‖ calculer. / ~ without core ‖ ohne Kern m. gießen ‖ couler en moule m. sans noyau; fondre dans le creux et à renversé.

cast, sand ‖ in Sand gegossen ‖ coulé au sable. / vertically ~ ‖ stehend gegossen ‖ coulé debout.

cast ‖ Guß m.; Gußstück n. ‖ pièce f. coulée *ou* moulée *ou* de fonte. / ~ (Wood engr) ‖ Abklatsch m.; Polytypie f.; Vieldruck m. ‖ polytypage m. / ~ (Engr) ‖ Abdruck m. ‖ empreinte f. / ~ en bloc ‖ im Block m. gegossen ‖ fondu monobloc. / die ~ ‖ Schalenguß m. ‖ coulée f. en coquille. / hollow ~ ‖ Hohlguß m. ‖ fonte f. creuse. / open ~ (Mine) ‖ Tagebau m. ‖ travail m. à ciel ouvert. / rough ~ (Build) ‖ Anwurf m.; Bewurf m.; Putz m.; Spritzwurf m. ‖ crépi m.; crépi m. et enduit m.; première couche f. d'enduit.

cast-away (Mar) ‖ Schiffbrüchiger m. ‖ naufragé m. / to be ~ ‖ scheitern; Schiffbruch leiden ‖ naufrager; crever; se boiser sur la côte.

cast brass ‖ Gußmessing n. ‖ laiton m. de fonte. / ~ concrete ‖ Gußbeton m. ‖ béton m. moulé. / ~ double jacket ‖ Gußdoppelmantel m. ‖ doubles parois fpl. en fonte.

castellated nut ‖ Kronenmutter f. ‖ écrou m. à entailles *ou* crénelé *ou* cannellé.

castel nut *see* castellated nut.

caster ‖ Gießer m. ‖ couleur m. / china ~ ‖ Porzellangießer m. ‖ mouleur m. en porcelaine. / ~ for furniture ‖ Möbelrolle f. ‖ rouleau m. pour meubles. / horn ~ ‖ Hornrolle f. ‖ roulette f. en corne. / porcelain ~ *see* china ~ / type ~ (Print) ‖ Schriftgießer m. ‖ fondeur m. de caractère *ou* de lettres. / wooden ~ ‖ Holzrolle f. ‖ roulette f. en bois.

caster wheel ‖ Lenkrad n. ‖ roue f. de guide.

cast goods pl., artistic ‖ Kunstguß m. ‖ articles mpl. artistiques en fonte.

cast hole ‖ Gußloch n. ‖ soufflure f. de fonte.

casting (Cast iron) ‖ (durch Umschmelzen von Roheisen erzeugtes) Gußeisen n. ‖ fer m. coulé *ou* fondu *ou* de fonte; fonte f.; fonte f. moulée. / ~ (Raw) ‖ gegossener Rohling m. ‖ pièce f. brute coulée *ou* moulée; pièce f. venue de fonte *ou* de fonderie. / ~ (Piece) *see also* castings ‖ Guß m.; Gußstück n. ‖ pièce f. coulée *ou* de fonte; moulage; objet m. coulé. / ~ (Cast; Copy) ‖ Abguß m. ‖ jet m. en moule; moulage m. / acid-proof ~ ‖ säurefester Guß m. ‖ fonte f. résistante aux acides. / ~ for automobiles ‖ Autoguß m. ‖ fonte f. d'automobiles. / the ~s are bolted direct to the foundation ‖ die Gußstücke npl. sind mit der Unterlage verbolzt ‖ les pièces fpl. sont boulonnées sur le socle. / case-hardened ~ ‖ Hartguß m.; Schalenguß m. ‖ fonte f. à la volée *ou* en coquille; fonte blanchie *ou* durcie; fonte f. coulée en coquille. / centrifugal ~ ‖ Zentrifugalguß m. ‖ fonte f. centrifuge. / chilled ~ ‖ Hartguß m. ‖ fonte f. moulée en coquille. / cleaning plant for ~s ‖ Gußputzanlage f. ‖ installation f. d'ébarbage de fonte. / ~ on a core ‖ Kernguß m.; Hohlguß m. ‖ fonte f. en creux; coulage m. en fonte à noyau. / ~ from the cupola ‖ Kupolofenguß m. ‖ fonte f. de deuxième fusion; fonte au cubilot. / dead-mould ~ ‖ Formguß m. ‖ coulage m. à creux perdu. / die ~ ‖ Spritzguß m. ‖ moulage m. mécanique *ou* sous pression; matricé m. / to fettle a ~ ‖ ein Gußstück n. putzen ‖ ébarber la fonte. / fire-proof ~ ‖ feuerfester Guß m. ‖ moulage m. à l'épreuve du feu. / flask ~ ‖ Kastenguß m. ‖ coulage m. en châssis. / formed in the ~ ‖ in Stahl m. gegossen ‖ moulé en acier m. / heavy ~ ‖ schweres Formstück m. ‖ lourde pièce coulée. / heavy iron ~ ‖ schweres eisernes Gußstück n. ‖ grosse pièce f. moulée en fonte. / hollow ~ ‖ Kernguß m.; Hohlguß m. ‖ fonte f. en creux; coulage m. en fonte à noyau. / hollow ~ (Piece) ‖ Hohlgußstück n. ‖ moulage m. creux; pièce f. coulée creuse. / horizontal ~ ‖ liegender Guß m. ‖ fonte f. coulée horizontalement. / ~ in iron moulds ‖ Schalenguß m.; Hartguß m. ‖ fonderie f. en coquilles; fonte f. en coquilles. / ~ of iron pots ‖ Topfgießerei f. ‖ fonte f. de marmites. / light iron ~ ‖ leichtes eisernes Gußstück n. ‖ petit moulage m. en fonte.

/ loam ~ ‖ Lehmguß m. ‖ coulage m. en argile. / ~ of a low frequency induction furnace ‖ Abgußgewicht n. eines Niederfrequenzinduktionsofens ‖ poids m. de coulée d'un four à induction à basse fréquence. / machine ~ (Casting for machine parts) ‖ Maschinenguß m. ‖ fonte f. pour pièces de machines. / machine ~ (Casting by machines) ‖ Maschinenguß m. ‖ coulée f. à la machine. / malleable ~ ‖ Temperguß m. ‖ fonte f. malléable. / malleable iron ~ ‖ Gußstück n. aus Temperguß ‖ moulage m. en fonte malléable. / ~ of metals ‖ Gießen n. der Metalle; Metallguß m. ‖ coulage m. *ou* fonte f. des métaux. / mould ~ ‖ Formguß m. ‖ moulage m. en châssis. / neat ~ ‖ sauberes Gußstück n. ‖ moulage m. d'exécution soigneuse. / ~ of pig ‖ Gießen n. des Roheisens ‖ coulage m. de la fonte brute. / ~ of plate glass ‖ Guß m. des Spiegelglases ‖ coulée f. des glaces. / ~ in rising stream ‖ Gießen n. in steigendem Strom ‖ coulée f. en source. / rough ~ ‖ Rohguß m. ‖ moulage m. brut. / ~ in open sand-moulds ‖ Herdguß m.; Herdgießen n. ‖ fonderie f. à découvert. / ~ with absence of contraction cracks ‖ Gußstück n. ohne Schrumpfriß ‖ pièce f. sans fissures par le retrait du métal. / sprinkled ~ ‖ Spritzguß m. ‖ moulage m. sous pression. / steel ~ ‖ Stahlgußstück n. ‖ moulage m. d'acier. / template ~ ‖ Schablonenguß m. ‖ coulage m. à la trusse. / ~ of thin sections ‖ dünnwandiges Stück n. ‖ pièce f. à parois de faibles épaisseurs / upright ~ ‖ stehender Guß m. ‖ fonte f. coulée verticalement. / ~ of works of art ‖ Kunstguß m. ‖ moulage m. d'objets d'art.

casting bay ‖ Gießhalle f. ‖ hall m. de coulée. / ~ bed ‖ Gießbett n. ‖ lit m. de coulée. / ~ bed crane ‖ Gießbettkran m. ‖ grue f. pour lits de coulée. / ~ box ‖ Gießlade f.; Gießkasten m.; Formkasten m. ‖ châssis m. de moulage. / ~ car ‖ Gießpfannenwagen m. ‖ chariot m. portepoche *ou* de coulée; transporteur m. de la poche de coulée.. / ~ carriage ‖ Gießwagen m. ‖ poche f. (roulante) de coulée. / ~ cleaner ‖ Gußschleifer m. ‖ ébarbeur m. à la meule.

casting cleaning equipment ‖ Gußputzeinrichtung f. ‖ installation f. à ébarber la fonte. / ~ hammer ‖ Gußputzhammer m. ‖ marteau m. à ébarber la fonte. / ~ cleaning machine ‖ Gußputzmaschine f. ‖ ébarbeuse f. / ~ tool ‖ Gußputzwerkzeug n. ‖ outil m. à ébarber la fonte.

casting crucible ‖ Schöpfherd m. ‖ creuset-puisard m. / ~ device for bushes ‖ Lagerschalengießvorrichtung f. ‖ dispositif m. à couler les coussinets. / ~ drum ‖ Gießtrommel f. ‖ tambour m. de coulée. / flask for ~ ‖ Formkasten m. ‖ châssis m. de moulage. / ~ gutter ‖ Einguß m.; Gukgerinne n. ‖ chenal m.; échenal m. / ~ house (Works) ‖ Schmelzhütte f.; Gießerei f. ‖ fonderie f. / ~ house (Hall) ‖ Gießhalle f.; halle f. de coulée. / ~ implements pl. for the graphic industry ‖ Gießanlage f. für das graphische Gewerbe ‖ installation f. de coulée pour les arts graphiques. / ~ ladle ‖ Gießkelle f.; Gießpfanne f. ‖ poche f. à fonte *ou* à couler.

casting machine (Letter-found) ‖ Gießmaschine f. ‖ machine f. à fondre; fondeuse f. / ~ for curved plates ‖ Gießmaschine f. für runde Stereoplatten ‖

fondeuse f. pour clichés ronds. / pasteboard ~ ‖ Pappengußmaschine f. ‖ machine f. à couler le carton.

casting model ‖ Gußmodell n. ‖ modèle f. de fonte. / ~ mould ‖ Gießform f.; Gußform f. ‖ moule m. à fonte; creux m. / ~ mould (Met) ‖ Metallgießform f. ‖ lingotière f. / ~ net ‖ Wurfnetz n. ‖ épervier m.; ressaut m. / ~ pig ‖ Gans f.; Massel f.; Flosse f. ‖ gueuse f.; gueuset m.; saumon m. / ~ pit ‖ Gießgrube f. ‖ fosse f. de coulée. / skin of ~ ‖ Gußrinde f. ‖ croûte f. de la fonte.

castings pl. *see also* casting (Piece) ‖ Guß m.; Gußstück n. ‖ moulage m.; fonte f.; fonte f. coulée *ou* moulée. / chilled ~ ‖ Kokillenguß m. ‖ moulage m. en coquille. / chilled ~ posses a superficial hard layer ‖ der Schalenguß m. besitzt eine harte Oberflächenschicht ‖ la fonte f. en coquille est recouverte d'une couche dure superficielle. / chilled iron ~ ‖ Schalenguß m. ‖ fonte f. en coquille. / die ~ ‖ Spritzguß m. ‖ pièces fpl. coulées sous pression. / dry sand ~ ‖ Masseguß m. ‖ moulage m. en terre. / grey iron ~ ‖ Grauguß m. ‖ moulage m. en fonte grise. / hard steel ~ ‖ harter Stahlguß m. ‖ moulage m. d'acier dur. / heat-resisting ~ ‖ hitzebeständiger Guß m. ‖ moulages mpl. de résistance à hautes températures. / heavy ~ ‖ schwerer Guß m. ‖ grosses pièces fpl. coulées. / loam ~ ‖ Lehmguß m. ‖ moulage m. en argile. / malleable ~ ‖ schmiedbares Gußeisen n.; Temperguß m. ‖ fonte f. malléable. / ~ for objects of arts ‖ Kunstguß m. ‖ objets mpl. d'art en fonte moulée. / ~ out of open sand-moulds (Found) ‖ Herdguß m. ‖ produits mpl. de fonte à découvert; fonte f. coulée à découvert. / pig-iron ~ ‖ Eisenguß m. ‖ fonte f. moulée. / sand ~ ‖ Sandguß m. ‖ coulage m. en sable. / soft ~ ‖ weicher Grauguß m. ‖ fonte f. douce. / special fire-proof and acid-resisting ~ ‖ feuer- und säurebeständiger Sonderguß m. ‖ fonte f. moulée speciale résistante à l'influence du feu et aux effets des acides. / steel ~ ‖ Stahlformguß m. ‖ moulage m. en acier. / ~ for structures ‖ Bauguß m. ‖ fonte f. moulée pour constructions de bâtiments.

casting shop ‖ Gießerei f. ‖ fonderie f. / ~ shovel ‖ Wurfschaufel f. ‖ pelle f. à cheminée. / ~ skin ‖ Gußhaut f. ‖ croûte f. de la fonte. / ~ slab (Glassm) ‖ Gießtafel f. ‖ table f. à couler. / ~ strain see ~ stress. / ~ stress ‖ Gußspannung f. ‖ tension f. de coulée. / ~ table see ~ slab.

cast-iron (Met) ‖ Gußeisen n.; Gießereiroheisen n. ‖ fonte f.; fer m. cru; fonte f. moulée *ou* de moulage. / case-hardened ~ ‖ Schalenguß m.; Kapselguß m.; Hartguß m. ‖ fonde f. coulée en coquille *ou* en moules en fonte. / malleable ~ ‖ schmiedbares Gußeisen n.; Temperguß m. ‖ fonte f. malléable. / non-malleable ~ ‖ nicht schmiedbarer Guß m. ‖ fonte f. non malléable. / of ~ ‖ gußeisern ‖ en fonte. / open-hearth ~ ‖ Herdguß m. ‖ produits mpl. de fonte coulés à découvert; fer m. fondu à découvert. / pure ~ ‖ geläutertes Roheisen n. ‖ fonte f. raffinée. / refined ~ ‖ gefeintes Roheisen n. ‖ metal m. fin; fonte f. raffinée; fonte f. mazée. / specular ~ ‖ Spiegeleisen n. ‖ fonte f. miroitante.

cast-iron box ‖ gußeiserne Muffe f. ‖ manchon m. de fonte. / ~ casing ‖ Gußeisengehäuse n. ‖ boîte f. *ou* bâti m. en fer fondu *ou* en fonte (de fer). / ~ castings pl. ‖ Grauguß m. ‖ fonte f. moulée; pièce f. en fonte; moulage m. en fonte grise. / ~ chilled frog ‖ Schalengußherzstück n. ‖ croisement m. en fer fondu en coquille. / ~ chilled wheel ‖ Schalengußrad n. ‖ roue f. pleine en fonte coulée en coquille. / ~ chips pl. ‖ Gußeisenspäne pl. ‖ riblons mpl. de fer fondu. / ~ construction exposed to great changes of temperature ‖ Gußeisenkonstruktion f., die starkem Temperaturwechsel ausgesetzt ist ‖ construction f. en fonte soumise à fortes variations de la température. / ~ frame ‖ Gußeisenrahmen m. ‖ cadre m. *ou* bâti en fonte. / ~ goods pl. ‖ Eisengußwaren fpl. ‖ pièces fpl. de fonte moulée. / ~ mould ‖ eiserne Gußschale f; Kokille f. ‖ moule m. en fonte; coquille f. / ~ pipe ‖ gußeisernes Rohr n.; Gußrohr n. ‖ tuyau m. en fonte. / ~ pulley ‖ Graugußriemenscheibe f. ‖ poulie f. en fonte. / ~ railing ‖ Geländer n. aus Gußeisen ‖ balustrade f. en fonte. / ~ stove ‖ Gußofen m.; Ofen m. aus Gußeisen ‖ poêle f. en fonte. / ~ scrap ‖ Gußbruch m. ‖ bocage m. de fonte. / ~ testing machine ‖ Gußeisenprüfmaschine f. ‖ machine f. à essayer la fonte. / ~ ware ‖ Gußwaren fpl.; Eisengußwaren fpl. ‖ fers mpl. coulés; fonte f. moulée; moulles fpl.; articles mpl. en fonte; quicailleries fpl. / ~ ware (Househ) ‖ Gußeisengeschirr n. ‖ poterie f. en fonte.

castle ‖ Schloß n.; Burg f. ‖ château m.; palais m.

cast-mould of sulphur ‖ Schwefelform f. ‖ moule m. à soufre.

cast-off burr (Weav) ‖ Abschlagrad n. ‖ roue f. d'abattage.

cast-on ‖ angegossen ‖ venu de fonte. / ~ test-piece ‖ angegossener Probestab m. ‖ eprouvette f. venue de fonte avec la pièce moulée; éprouvette f. attenant à la pièce. / ~ tyre ‖ angegossener Radreifen m. ‖ bandage m. venu de fonte avec la roue.

castor (Roller) ‖ Möbelrolle f.; Rollrädchen n. ‖ roulette f. / ~ (Miner) ‖ Kastor m. ‖ castor m.; castorite f. / ~ bean ‖ Rizinus m. ‖ ricin m.

castoreum ‖ Bibergeil n. ‖ castoréum m.

castor hair ‖ Biberhaar n. ‖ poil m. de castor. / ~ oil ‖ Rizinusöl n. ‖ huile f. de ricin.

cast part ‖ Gußstück n. ‖ pièce f. de fonte. / ~ plate ‖ Gußplatte f. ‖ plaque f. de fonte. / ~ rolls pl. ‖ Gußwalzen fpl. ‖ cylindres mpl. de fonte. / ~ scrap ‖ Gußschrott m. ‖ mitraille f. *ou* bocage m. de fonte. / ~ seam ‖ Gußnaht f. ‖ bavure f.. / ~ skin ‖ Gußhaut f. ‖ croûte f. de fonte.

cast steel ‖ Gußstahl m. ‖ acier m. fondu *ou* au creuset. / burnt ~ ‖ verbrannter Gußstahl m. ‖ acier m. fondu brûlé. / crucible ~ ‖ Tiegelgußstahl m. ‖ acier m. au creuset. / forged ~ ‖ geschmiedeter Gußstahl m. ‖ acier m. fondu forgé. / harsh ~ ‖ unschweißbarer Gußstahl m. ‖ acier m. fondu non soudable. / overheated ~ ‖ verbrannter Gußstahl m. ‖ acier m. fondu brûlé. / ~ one-piece wheel ‖ gegossenes Vollrad n. ‖ roue f. d'une seule pièce en acier moulé. / skilled ~ ‖

Tiegelgußstahl m. ‖ acier m. fondu au creuset. / soft ~ ‖ schweißbarer Gußstahl m. ‖ acier m. fondu soudable. / weldable ~ ‖ schweißbarer Gußstahl m. ‖ acier m. fondu soudable.

cast steel bell ‖ Stahlgußglocke f. ‖ cloche f. en acier coulé. / ~ block ‖ Gußstahlblock m. ‖ lingot m. d'acier. / ~ bullet ‖ Gußstahlkugel f. ‖ bille f. en acier fondu. / ~ crossing (Railw) ‖ Gußstahlherzstück n.; Gleiskreuzung f. ‖ croisement m. en acier fondu. / ~ flange ‖ Stahlgußflansch m. ‖ bride f. en acier coulé. / ~ frog *see* ~ crossing. / ~ hoop driver ‖ Gußstahlvollsetze f. ‖ chasse f. pleine en acier fondu. / ~ looking glass (Met) ‖ Gußstahlspiegel m. ‖ miroir m. en acier fondu. / ~ plate ‖ Gußstahlblech n. ‖ tôle f. d'acier fondu. / ~ spoke wheel ‖ Stahlgußspeichenrad n. ‖ roue f. à rayons en acier moulé. / ~ spring ‖ Gußstahlfeder f. ‖ ressort m. en acier fondu. / ~ stiffening ring (Boil) ‖ Eckring m. aus Gußstahl ‖ anneau m. d'angle en acier fondu. / ~ wire ‖ Gußstahldraht m. ‖ fil m. d'acier fondu. / ~ works pl. ‖ Gußstahlfabrik f.; Gußstahlwerk n. ‖ aciérie f.; fabrique f. d'acier coulé.

cast tubes pl. ‖ Gußröhren fpl. ‖ tuyaux mpl. moulés. / ~ window ‖ gußeisernes Fenster n. ‖ fenêtre f. en fonte. / ~ works pl. of art ‖ Kunstguß m.; Eisengußwaren fpl. ‖ objets m. d'art en fonte; ouvrages mpl. en fonte.

cat, to ~ the anchor ‖ den Anker katten ‖ caponner l'ancre m.

cat (Mar) ‖ Katt f. ‖ capon m.

catadioptric ‖ katadioptrisch ‖ catadioptrique.

catadioptrics pl. ‖ Katadioptrik f. ‖ catadioptrique f.

Catalan furnace *see* ~ hearth.

Catalan hearth ‖ Rennfeuer n. ‖ bas-foyer m.; forge f. catalane; feu m. catalan.

catalogue ‖ Preisliste f.; Katalog m.; Warenverzeichnis n. ‖ catalogue m.; brochure f. / ~ composition (Print) ‖ Katalogsatz n. ‖ composition de catalogues. / ~ cover ‖ Preislistenumschlag m. ‖ couverture f. de catalogue. / ~ price ‖ Katalogpreis m.; Listenpreis m. ‖ prix m. figurant au catalogue.

catalysis ‖ Katalyse f. ‖ catalyse f.

catalyst *see* catalyzer.

catalytic ‖ katalytisch ‖ catalytique. / ~ combustion ‖ katalytische Verbrennung f. ‖ combustion f. catalytique.

catalyze, to ‖ katalysieren ‖ catalyser.

catalyzer ‖ Katalysator m. ‖ catalyseur m.

catapult ‖ Katapult m.; Schleuder f.; Wurfmaschine f. ‖ catapulte f. / airimpulse ~ ‖ preßluftgetriebener Katapult m. ‖ catapulte f. de lancement à air comprimé. / powder-impulse ~ ‖ pulvergetriebener Katapult m. ‖ catapulte f. de lancement à poudre.

catapult launching gear (Aero) ‖ Schleudervorrichtung f.; Katapultvorrichtung f. ‖ dispositif m. de lancement.

cataract (Waterfall) ‖ Stromschnelle f.; Katarakt m. ‖ catarracte f. / ~ (Mach) ‖ Katarakt m. ‖ catarracte f. / ~ (Opt) ‖ Star m. ‖ catarracte f. / ~ blindness ‖ Starblindheit f. ‖ cécité f. qui résulte d'une catarracte. / ~ glass ‖ Starglas n. ‖ verre m. correcteur pour opérés de catarracte. / ~ knife (Med) ‖ Starmesser n. ‖ couteau m. à catarracte. / ~ reading glass ‖ Star-

leseglas n. ∥ verre m. pour la lecture destiné aux opérés de cataracte.

cat-block (Mar) ∥ Kattblock m. ∥ poulie f. de capon.

catch, to ∥ auffangen; fangen ∥ saisir; recueillir. / ~ (Lock) ∥ einschnappen ∥ se fermer à ressort *ou* au loquet. / ~ (Tooth wheel) ∥ ineinandergreifen; eingreifen ∥ engrener. / ~ fire ∥ anbrennen; Feuer fangen ∥ s'allumer; brûler.

catch (Fish) ∥ Fischfang m.; Fischzug m. ∥ pêche. f. / ~ (Door) ∥ Klinke f.; Drücker m.; Türdrücker m. ∥ loquet m.; clenche f. / ~ (Lock) ∥ Gesperre n.; Sperrhaken m.; Schließhaken m.; Arretierungsfalle f. ∥ fermoir m.; mentonnet m.; nappe f.; crochet m. *ou* cheville f. d'arrêt. / ~ (Spring lock) ∥ Schnapper m. ∥ loqueteau m. / ~ (Mach tool) ∥ Nase f.; Knagge f.; Daumen m.; Mitnehmer m. ∥ toc m.; taquet m.; mentonnet m.; / ~ for a bolt (Lock) ∥ Riegelhaken m.; Riegelhaspe f. ∥ verterelle f. de verrou. / boxed ~ (Lock) ∥ Schließkappe f.; überbauter Schließhaken m. ∥ gâche f. / cased ~ see boxed ~. / eccentric ~ ∥ Sektoren mpl. der Hebelsteuerung ∥ buttoirs mpl. *ou* heurtoirs mpl. à encliquetage.

catch-arrangement ∥ Fangvorrichtung f. ∥ dispositif m. d'arrêt. / ~ for elevator cages ∥ Fangvorrichtung f. für Förderkörbe ∥ dispositif m. d'arrêt pour cages de montée. / pneumatic ~ ∥ Druckluftfangvorrichtung f. ∥ dispositif m. d'arrêt à air comprimé.

catch-bolt ∥ Riegel m. mit einer Feder ∥ verrou m. à ressort.

catcher (Fish) ∥ Fischnetz n. ∥ filet m. de pêche; échiquier m. / ~ (Dredger) ∥ Greifkorb m. ∥ benne f. preneuse *ou* à griffe. / amalgam ~ ∥ Amalgamfänger m. ∥ collecteur m. de mercure.

catcher plate ∥ Fängerteller m. ∥ assiette f. du récepteur.

catch-fitted clip (Railw) ∥ Nasenklemmplatte f. ∥ plaque f. de serrage à talon.

catch hook ∥ Fanghaken m. ∥ crochet m. d'attache.

catching (Mach) ∥ Eingriff m. ∥ quottement m.

catch iron (Hunt) ∥ Fangeisen n. ∥ chaussetrape f. / ~ line (Print) ∥ Schlußzeile f. ∥ ligne f. perdue.

catchment area (Hydr arch) ∥ Einzugsgebiet n.; Sammelgebiet n.; Niederschlagsgebiet n. ∥ bassin m. hydrologique *ou* de réception. / ~ of a barrage ∥ Einzugsgebiet n. einer Talsperre ∥ superficie f. des terrains contribuants d'un barrage; domaine f. d'affluence d'un barrage. / ~ of a river ∥ Flußgebiet n. ∥ bassin m. d'un fleuve.

catchment basin see catchment area.

catch pit ∥ Wasserabzugsgraben m. ∥ fossé m. d'écoulement; drain m. / ~ spring ∥ Einschnappfeder f. ∥ ressort m. d'encliquetage. / ~ water drain ∥ Sickerkanal m. ∥ pierrée f. rigolée.

catechu ∥ Katechu n.; Kaschu n. ∥ cachou m. / ~ extract ∥ Katechuauszug m. ∥ extrait m. de catechou.

catenarian curve see catenary.

catenary ∥ Kettenlinie f. ∥ chaînette f.; ligne f. de chaînette. / ~ of equal resistance ∥ Kettenlinie f. von gleichem Widerstande ∥ chaînette f. d'égale résistance. / ~ of the strongest form see ~ of equal resistance.

catering ∥ Verpflegungswesen n.; Speisung f. ∥ service m. alimentaire.

caterpillar (Zool; Techn) ∥ Raupe f. ∥ chenille f. / ~ artillery ∥ Raupenartillerie f. ∥ artillerie f. sur chenilles. / ~ car ∥ Raupenwagen m. ∥ auto-chenille f. / ~ glue ∥ Raupenleim m. ∥ colle f. antichénillique. / ~ gun mount ∥ Raupenlafette f. ∥ affût m. sur chenilles. / ~ traction ∥ Raupenzug m. ∥ traction f. à chenilles. / ~ tractor ∥ Raupenschlepper m. ∥ tracteur m. sur chenilles. / ~ truck ∥ Raupenfahrwerk n.; Raupenunterwagen m. ∥ chariot m. sur chenilles. / ~ wheel ∥ Raupenrad n.; Raupe f. ∥ roue f. à chenilles; chenille f.

cat fall (Mar) ∥ Kattfall n.; Kattläufer m. ∥ garant m. du capon. / ~ gold ∥ Goldglimmer m.; Katzengold n. ∥ faux-or m.; mica m. jaune.

catgut ∥ Darmsaite f.; Catgut n. ∥ corde f. en boyaux; catgut m. / plated ~ ∥ übersponnene Saite f. ∥ corde f. filée.

catgut blower ∥ Darmbläser m. ∥ souffleur m. de boyaux. / ~ string see catgut.

cat-head (Shipb) ∥ Kranbalken m.; Ankerkran m. ∥ bossoir m.; bosseur m. / ~ stopper ∥ Porteurleine f.; Kattstopper m. ∥ bosse f. du bossoir. / ~ stopper chain ∥ Porteurleinkette f. ∥ bosse-debout f. en chaîne.

cathedral glass ∥ Kathedralglas n. ∥ verre m. cathédrale.

cathode ∥ Kathode f.; negativer Pol m. ∥ cathode f.; pôle m. négatif. / concave mirror ~ ∥ Hohlspiegelkathode f. ∥ cathode f. à miroir concave.

cathode-ray ∥ Kathodenstrahl m. ∥ rayon m. cathodique. / ~ oscillograph ∥ Kathodenoszillograf m.; Kathodenstrahloszillograf m. ∥ oscillographe m. cathodique. / ~ relay (Tel) ∥ Kathodenstrahlenrelais n. ∥ relais m. à rayons cathodiques.

cathodic separation of gold ∥ kathodische Abscheidung f. von Gold ∥ séparation f. cathodique de l'or.

cat hole (Ship) ∥ Klüse f.; Ankerklüse f. ∥ écubier m. / ~ hook ∥ Katthaken m.; Kattblockhaken m. ∥ croc m. du capon.

cation (Electr) ∥ Kation n. ∥ cation m. / ~s pl. ∥ Kationen npl. ∥ cations mpl.; produits mpl. dégagés au pôle négatif de la pile.

catkin (Bot) ∥ Dolde f. ∥ cône m.; ombelle f.

catoptrics pl. ∥ Katoptrik f. ∥ catoptrique f.

cat rake ∥ Bohrknarre f. ∥ Ratsche f. ∥ perçoir m. à rochet; drille f. à levier. / ~ rope (Mar) ∥ Kattläufer m. ∥ garant m. du capon.

cat's-eye (Miner) ∥ Katzenauge n. ∥ œil m. de chat. / ~ gum ∥ Dammarharz n. ∥ résine f. de dammara; dammara m.

cat's-foot flowers pl. ∥ (rote) Katzenpfötchenblüten fpl. ∥ fleurs fpl. de pied de chat.

catskin cover ∥ Katzenfellüberzug m. ∥ taie f. en peau de chat.

cat tackle (Mar) ∥ Kattakel n.; Kattgien f. ∥ palan m. de capon.

cattle ∥ Rindvieh n.; Hornvieh n. ∥ race f. *ou* espèce f. bovine.

cattle blood ∥ Viehblut n.; Rinderblut n. ∥ sang m. de bétail. / dried ~ ∥ getrocknetes Viehblut n. ∥ sang m. de bétail desséché. / liquid ~ ∥ flüssiges Viehblut n. ∥ sang m. de bétail liquide.

cattle bone ∥ Tierknochen m. ∥ os m. de bestiaux. / ~ breeder ∥ Rindviehzüchter m. ∥ éleveur m. de bêtes à cornes. / ~

breeding ∥ Rindviehzucht f.; Viehzucht f. ∥ élève f. des bestiaux. / ~ car ∥ Viehwagen m. ∥ wagon m. à bestiaux. / ~ draught ∥ Rindviehbespannung f. ∥ attelage m. à bœufs. / ~ feeder ∥ Stallschweizer m.; Viehfütterer m. ∥ laitiernourrisseur m. / ~ fodder ∥ Viehfutter n. ∥ fourrage m. / ~ fodderer see ~ feeder. / ~ fodder steamer ∥ Viehfutterdämpfer m. ∥ étuve. f. à pâture. / ~ food see ~ fodder. / ~ foot oil ∥ Klauenfett n.; Huffett n. ∥ huile f. de pied de bœuf. / ~ hair ∥ Haar n. vom Rindvieh ∥ poil m. de bêtes bovines. / ~ hoof ∥ Viehhuf m. ∥ sabot m. de bestiaux. / ~ horns pl. ∥ Viehhörner npl. ∥ cornes fpl. de bestiaux. / spotted ~ horn ∥ geflecktes Viehhorn n. ∥ corne f. bigarrée de bestiaux. / ~ house ∥ Viehstall m. ∥ étable f. pour bestiaux. / ~ keeping ∥ Viehhaltung f. ∥ garde f. d'animaux de ferme. / ~ loader ∥ Viehverlader m. ∥ embarqueur m. de bestiaux. / ~ mast powder ∥ Viehmastpulver n. ∥ engrais m. en poudre pour les bestiaux. / ~ muzzle ∥ Viehmaulkorb m. ∥ muselière f. pour bestiaux. / ~ salt ∥ Viehsalz n. ∥ sel m. pour bétail. / ~ shears pl. ∥ Viehschere f. ∥ ciseaux mpl. pour animaux. / ~ stunning apparatus ∥ Schlachtviehbetäubungsvorrichtung f. ∥ appareil m. à engourdir les bestiaux à abattre. / ~ steamer (Ship) ∥ Viehdampfer m. ∥ vapeur m. à bétail. / ~ train (Railw) ∥ Viehzug m. ∥ train m. de bestiaux. / ~ truck ∥ Viehwagen m. ∥ wagon m. à bestiaux; voiture f. pour le transport du betail. / ~ van see ~ truck. / ~ wagon see ~ truck. / ~ weighing machine ∥ Viehwage f. ∥ balance f. à bestiaux.

cauk . . . see also cog . . . (Carp; Join).

cauking joint (Join; Carp) ∥ Verkämmung f. ∥ assemblage m. à entailles; entaillure f.

caul (Veneer) ∥ Zulage f. ∥ cale f.

cauliflower ∥ Blumenkohl m. ∥ chou-fleur m.

caulk . . . see also calk . . .

caulk, to (Boil; Iron shipb) ∥ verstemmen ∥ mater. / ~ (Wood shipb) ∥ kalfatern; dicht machen ∥ calfater; étancher; imperméabiliser; radouber.

caulked edge ∥ Stemmkante f. ∥ bord m. maté. / ~ joint ∥ Stemmfuge f. ∥ joint m. maté. / ~ seam ∥ Stemmnaht f. ∥ joint m. maté; couture f. *ou* soudure f. matée.

caulker (Boil; Iron shipb) ∥ Stemmer m. mateur m. / ~ (Wood shipb) ∥ Kalfaterer m. ∥ calfateur m.; calfat m.; radoubeur m. / ~'s trift ∥ Treibeisen n. ∥ repoussoir m.

caulking (Boil; Iron shipb) ∥ Verstemmung f. ∥ matage m. / ~ (Wood shipb) ∥ Kalfatern n. ∥ calfatage m. / ~ the edges beyond the rim ∥ Verstemmen n. vorspringender Kanten ∥ matage m. des bords en saillie. / pneumatic ~ ∥ Stemmen n. mit Druckluft ∥ matage m. pneumatique.

caulking chisel ∥ Stemmeißel m. ∥ matoir m. / ~ edge ∥ Stemmkante f. ∥ bord m. maté. / cutting the ~ edges ∥ Anschneiden n. von Stemmkanten ∥ couper les bords mpl. matés. / ~ hammer ∥ Stemmhammer m. ∥ marteau m. à mater. / ~ iron (Boil; Iron shipb) ∥ Stemmeißel m. ∥ matoir m. / ~ iron (Wood shipb) ∥ Kalfateisen n.; Dichteisen n. ∥ fer m. à calfat; ciseau m. de

calfat. / ~ machine for boiler pipes ‖ Siederohrdichtmaschine f. ‖ machine f. à mater les tuyaux de chaudières. / ~ machine for pipes ‖ Rohrdichtmaschine f. ‖ machine f. à mater les tuyaux. / ~ mallet ‖ Kalfathammer m.; Dichthammer m. ‖ maillet m. de matage; marteau m. de calfat. / ~ piece ‖ Verstemmstück n. ‖ pièce f. de matage. / ~ seam ‖ Dichtungsnaht f. ‖ about m. pour le calfatage. / ~ tool ‖ Stemmsetze f. ‖ outil m. à mater; matoir m. / ~ tub ‖ Kalfatbutte f. ‖ caisse f. des calfateurs; sellette f. à calfat.

caulk weld ‖ Stemmnaht f. ‖ joint m. maté; couture f. *ou* soudure f. matée. / ~ welding ‖ Dichtungsschweißung f. ‖ soudure f. étanche.

cause ‖ Ursache f. ‖ cause f.

causeway ‖ Kunststraße f.; Heerstraße f.; Hauptstraße f.; Landstraße f.; Chaussee f. ‖ chaussée f.; grande route f. / ~ of fascines ‖ Faschinendamm m. ‖ pont m. *ou* digue f. de fascines.

causey ‖ Dammweg m.; Landstraße f. ‖ remblai m.; chaussée f.; jetée f.

caustic ‖ ätzend; kaustisch ‖ caustique; corrodant.

caustic (Chem) ‖ Ätzmittel n. ‖ caustique m.; agent m. corrosif. / ~ (Phys) ‖ kaustische Kurve f. ‖ courbe f. caustique.

caustic agent ‖ Ätzmittel n. ‖ caustique m. / ~ alkaline solution ‖ ätzalkalische Lösung f. ‖ solution f. alcaline caustique. / ~ ammonia ‖ wässeriges Ammoniak n.; Salmiakgeist m.; Salmiakspiritus m. ‖ esprit m. de sel ammoniac.

causticity ‖ Ätzkraft f. ‖ pouvoir m. corrosif; causticité f.

caustic lime ‖ Ätzkalk m. ‖ chaux f. vive. / ~ lye ‖ Ätzlauge f. ‖ lessive f. caustique. / ~ lye of soda ‖ Ätznatronlauge f. ‖ lessive f. caustique de soude.

caustic potash ‖ Ätzkali n. ‖ potasse f. caustique. / ~ in bars ‖ Ätzkali n. in Stangen ‖ potasse f. caustique en bâtonnets. / crude ~ ‖ rohes Ätzkali n. ‖ potasse f. à la chaux.

caustic process ‖ Ätzverfahren n. ‖ procédé m. au caustique. / ~ salt ‖ Beizsalz n. ‖ sel m. caustique.

caustic soda ‖ Ätznatron n.; Natronlauge f. ‖ lessive f. de soude; lessive f. soudique; soude f. caustique; hydrate m. de sodium. / ~ producing plant ‖ Ätznatronfabrikanlage f. ‖ installation f. à fabriquer la soude caustique. / ~ solution (Chem) ‖ Natronlauge f. ‖ soude f. caustique liquide.

caustic stick *see* ~ stone.

causticstone ‖ Höllenstein m.; Ätzstift m. ‖ pierre f. infernale *ou* à cautère; nitrate m. d'argent. / ~ holder (Chem app) ‖ Höllensteinhalter m. ‖ porte-nitrate m.

caustification ‖ Kaustizierung f. ‖ caustification f.

cautery ‖ Kauter m.; Brennstift m. ‖ cautère m.

caution ‖ Vorsicht f.; Vorsichtsmaßregel f.; Warnung f. ‖ attention f. / ~ ! ‖ Achtung! ‖ attention! / ~ board ‖ Warnungsschild n. ‖ tableau m. avertisseur.

cautiously ‖ vorsichtig ‖ avec précaution.

cave ‖ Höhle f. ‖ caverne f.; grotte f. / ~ (Cellar) ‖ Keller m. ‖ cave f.; cellier m. / exploration of ~s / Höhlenforschung f. ‖ exploration f. de cavernes.

cavern *see* cave.

cavernous ‖ höhlenartig ‖ caverneux. / ~ structure ‖ lückiges Gefüge n. ‖ structure f. caverneuse.

cavesson rein (Riding) ‖ Longe f.; Leine f. ‖ longe f.

caviar(e) ‖ Kaviar m. ‖ caviar m. / ~ dresser ‖ Kaviarzubereiter m. ‖ préparateur m. de caviar. / ~ substitute ‖ Kaviarersatzstoff m. ‖ succédané m. de caviar.

cavitation ‖ Hohlraumbildung f. ‖ cavitation f.

cavity ‖ Aushöhlung f.; Höhlung f. ‖ cavité f.; creux m. / ~ (Met) ‖ Lunker m. ‖ retassement m. / ~ (Mining) ‖ Mulde f. ‖ excavation f.; cavité f.; enfoncement m.

cease, to ‖ aufhören; ablassen; nachlassen ‖ cesser. / ~ running (Railw etc.) ‖ den Verkehr m. einstellen ‖ cesser de circuler. / ~ work ‖ die Arbeit niederlegen ‖ cesser *ou* abandonner le travail.

cedar ‖ Zeder f.; Zedernholz n. ‖ cèdre m. / rock red ~ ‖ Rotzeder f. ‖ genévrier m. savinier. / Virginia ~ ‖ Weichzeder f. ‖ genévrier m. de Virginie.

cedar-oil *see* cedar-wood oil.

cedar-wood ‖ Zedernholz n. ‖ bois m. de cèdre; cèdre m. / ~ (For pencils) ‖ Bleistiftholz n. ‖ bois m. à crayon. / ~ oil ‖ Zedernholzöl n. ‖ essence f. de bois de cèdre.

cede, to ‖ sich geben; nachgeben; weichen ‖ céder; plier; se relâcher; se détendre. / ~ a claim ‖ eine Forderung f. abtreten *oder* zedieren ‖ céder une créance.

cedrate ‖ Zedratfrucht f. ‖ cédrat m.

ceiling (Aero) ‖ Gipfelhöhe f. ‖ plafond m. / ~ (Build) ‖ Decke f. ‖ plafond m. / ~ (Shipb) ‖ Wegerung f.; Innenbeplankung f. ‖ vaigre f.; vaigrage m.; bordage m. du revêtement intérieur. / ~s pl. (Shipb) ‖ Binnenhäute fpl. ‖ vaigres fpl. / boarded ~ ‖ getäfelte Decke f. ‖ plafond m. lambrissé. / coffered ~ (Build) ‖ Felderdecke f.; Kassettendecke f. ‖ plafond m. à caissons. / inserted ~ (Build) ‖ blinde Decke f. ‖ faux plancher m. / lathed and plastered ~ ‖ Stuckdecke f. ‖ plafond m. de plâtre. / planked ~ ‖ Schaldecke f. ‖ plafond m. cloisonné. / pumiceconcrete ~ ‖ Bimsbetondecke f. ‖ plafond m. en béton de pierre-ponce. / ribbed ~ ‖ gerippte Decke f. ‖ plafond m. à nervures. / ~ of timbers ‖ Balkendecke f. ‖ plafond m. enfoncé; lambris m.

ceiling batten (Shipb) ‖ Wegerungsplanke f. ‖ bordage m. du revêtement intérieur; lambris m.; vaigre f. / ~ construction ‖ Deckenkonstruktion f. ‖ construction f. de plafond; plafonnage m. / ~ crab ‖ Deckenlaufkran m. ‖ chariot m. à poutre de plafond. / ~ duct ‖ Deckendurchführung f. ‖ traversée f. de plafond. / ~ fan ‖ Deckenventilator m.; Deckenwindflügel m.; Deckenfächer m. ‖ ventilateur m. plafonnier *ou* de plafond. / ~ hinge ‖ Deckenhaken m. ‖ crampon m. de plafond. / ~ illumination ‖ Deckenbeleuchtung f. ‖ éclairage m. de plafond. / ~ lamp ‖ Deckenlampe f. ‖ lampe f. plafonnière. / ~ lamp holder ‖ Deckenfassung f. ‖ douille f. de plafond. / ~ light fitting ‖ Deckenbeleuchtungskörper m. ‖ plafonnier m. / ~ picture ‖ Deckengemälde n. ‖ tableau m. de plafond. / ~ piece *see* ~ picture. / ~ plaster ‖ Deckenputz m. ‖ enduit m. de plafond. / ~ reflector ‖ Deckenreflektor m.; Deckenlichtspiegler m. ‖ réflecteur m. de plafond. / ~ rosette ‖ Decken-

rosette f. ‖ rosace f. de plafond. / ~ sink water-trap ‖ Deckensinkkasten m. ‖ siphon m. de décantation de plafond.

celery oil ‖ Sellerieöl n. ‖ essence f. de céleri. / ~ seed ‖ Selleriesamen m. ‖ semence f. de céleri.

celestial chart ‖ Himmelskarte f. ‖ carte f. céleste. / ~ globe ‖ Himmelsglobus m.; Himmelskugel f. ‖ globe m. céleste. / ~ vault ‖ Himmelswölbung f.; Himmelsgewölbe n. ‖ voûte f. céleste.

cell ‖ Zelle f. ‖ cellule f. / ~ (Electr) ‖ Element n.; Zelle f. ‖ élément m. (galvanique); pile f.; couple f. / ~ (Prison) ‖ Zelle f. ‖ cabanon m. / ~ (Small vessel) ‖ Küvette f. ‖ cuve f. / agglomerate ~ ‖ Brikettelement n. ‖ pile f. d'aggloméré. / bichromate ~ ‖ Chromsäureelement n. ‖ pile f. au bichromate. / closed ~ ‖ geschlossene Zelle f. ‖ élément m. fermé. / constant ~ ‖ konstantes Element n. ‖ élément m. constant. / copper oxide ~ ‖ Kupronelement n. ‖ élément m. à oxyde cuivrique. / doublé-fluid ~ ‖ Element n. mit zwei Flüssigkeiten ‖ pile f. à deux liquides. / dry ~ ‖ Trockenelement n. ‖ pile f. sèche. / electric ~ ‖ elektrisches Element n. ‖ élément m. *ou* pile f. électrique. / galvanic ~ ‖ galvanisches Element n. ‖ élément m. galvanique. / ~ of a known thickness ‖ Küvette f. mit bekannter Schichtdicke ‖ cuve f. d'épaisseur connue. / Meidinger ~ ‖ Ballonelement n. ‖ élément m. à ballon. / microphone ~ ‖ Mikrofonelement n. ‖ élément m. de microphone. / non-reversible ~ ‖ nicht umkehrbares Element n. ‖ pile f. non réversible. / open ~ ‖ offene Zelle f. ‖ élément m. ouvert. / photo-electric ~ ‖ fotoelektrische Zelle f. ‖ pile f. photoélectrique. / porous ~ ‖ poröse Zelle f.; Tonzelle f. ‖ vase m. poreux. / primary ~ ‖ Primärelement n. ‖ élément m. primaire. / primary electric ~ *see* galvanic ~. / regulating ~ (Acc) ‖ Gegenzelle f. ‖ élément m. de réduction. / reversible ~ ‖ umkehrbares Element n. ‖ pile f. réversible. / secondary ~ ‖ Sekundärelement n. ‖ élément m. secondaire. / single-fluid ~ ‖ Element n. mit einer Flüssigkeit ‖ pile f. à un seul liquide. / standard ~ ‖ Normalelement n. ‖ pile f. étalon. / wet ~ ‖ nasses Element n. ‖ pile f. humide; élément m. hydroélectrique.

cellar ‖ Keller m. ‖ cave f.; cellier m.; sous-sol m. / ~ for bottled beer ‖ Flaschenbierkeller m. ‖ cave f. à bouteilles à bière; entrepôt m. de bière en bouteilles. / ~ cleansing (Brew) ‖ Abfüllkeller m. ‖ cave f. de soutirage; soutirage m. / ~ cut out in a rock ‖ Felsenkeller m. ‖ cave f. taillée dans le roc. / to clear a ~ ‖ auskellern ‖ décaver. / fermentation ~ ‖ Gärkeller m. ‖ cave f. de fermentation. / filling ~ ‖ Abfüllkeller m. ‖ cave f. de soutirage.

cellarage ‖ Kellerei f. ‖ sommellerie f.

cellarage machine (Wine) ‖ Kellereimaschine f.; Weinkellereimaschine f. ‖ machine f. pour caves (de vin). / bottle ~ ‖ Flaschenkellereimaschine f. ‖ machine f. pour caves à bouteilles. / cask ~ ‖ Faßkellereimaschine f. ‖ machine f. pour caves de magasinage (du vin) en tonneaux. / champagne ~ ‖ Schaumweinkellereimaschine f.; Sektkellereimaschine f. ‖ machine f. pour caves de champagnes.

cellar-damp matrix ‖ kellerfeuchte Matrize f. ‖ flan m. humide.

cellarer see cellarman.

cellarman ‖ Kellereiarbeiter m.; Kellerbursche m.; Weinküfer m. ‖ ouvrier m. ou garçon m. de cave; caviste m.; garde-vins m.

cellar sink water-trap ‖ Kellersinkkasten m. ‖ siphon m. de décantation de cave. / ~ skylight ‖ liegendes Kellerfenster n. ‖ vue f. de terre. / ~ vault ‖ Kellergewölbe n. ‖ voûte f. de cave. / ~ wall ‖ Kellermauer f. ‖ mur m. de cave.

cell-destroying (Chem) ‖ zellwandzerstörend ‖ qui détruit les parois des cellules.

cell holder (Opt) ‖ Kammerhalter m. ‖ porte-chambre m. / ~ inspection lamp (Electr) ‖ Untersäurelampe f. für Bleisammler ‖ lampe f. d'examination des accumulateurs. / ~ jar ‖ Elementgefäß n.; Elementbehälter m. ‖ bac m. ou vase m. ou récipient m. de l'élément.

cello (Mus) ‖ Schello n.; Cello n. ‖ violoncelle m.

cellone ‖ Zellon n. ‖ cellone f. / ~ film ‖ Zellonfilm m. ‖ film m. de cellon. / metallized ~ film ‖ metallisierter Zellonfilm m. ‖ film m. de cellon métallisé. / ~ varnish ‖ Zellonlack m. ‖ vernis m. à la cellone.

cell switch ‖ Zellenschalter m. ‖ réducteur m. (pour accumulateur). / automatic ~ ‖ selbsttätiger Zellenschalter m. ‖ réducteur-adjoncteur m. automatique. / ~ for hand actuation ‖ Zellenschalter m. für Handbetrieb ‖ réducteur-adjoncteur m. à main. / ~ leads to ~ ‖ Zellenschalterleitung f. ‖ conducteur m. de réducteur.

cell terminal ‖ Polklemme f. ‖ borne f. d'élément. / ~ tester ‖ Elementprüfer m. ‖ essayeur m. de piles.

cellular arrangement ‖ Zelleneinbau m. ‖ construction f. cellulaire. / ~ pasteboard ‖ Holzpappe f. ‖ carton m. de bois. / ~ tissue ‖ Zellengewebe n.; Zellgewebe n. ‖ tissu m. cellulaire. / ~ wheel ‖ Kastenrad n.; Schöpfrad n. ‖ roue f. à godets ou à seaux.

celluloid ‖ Zellhorn n.; Zelluloid n. ‖ celluloïde m.; celluloïd m. / ~ articles pl. ‖ Zelluloidwaren fpl. ‖ objets mpl. en celluloïd. / ~ ball ‖ Zelluloidball m. ‖ balle f. en celluloïd. / ~ comb ‖ Zelluloidkamm m. ‖ peigne m. en celluloïd. / ~ facing ‖ Zelluloidauskleidung f. ‖ revêtement m. en celluloïde. / ~ goggles pl. ‖ Schutzbrille f. mit Zelluloidscheiben ‖ lunettes fpl. de protection en celluloïd. / ~ goods pl. see ~ articles pl. / ~ horn ‖ Zellhorn n. ‖ corne f. celluloïde. / ~ machine see ~ producing machine. / ~ paper ‖ Zelluloidpapier n. ‖ papier m. (de) celluloïde. / ~ press ‖ Zelluloidpresse f. ‖ presse f. pour celluloïd. / ~ producing machine ‖ Zelluloidherstellungsmaschine f.; Zellhornherstellungsmaschine f. ‖ machine f. pour la fabrication du celluloïd. / ~ sheet ‖ Zelluloidplatte f. ‖ plaque f. de celluloïd. / ~ solvent ‖ Zelluloidlösungsmittel n. ‖ dissolvant m. ou résolvant m. de celluloïd. / ~ spectacles pl. ‖ Zelluloidbrille f. ‖ lunettes fpl. en celluloïd. / ~ toys pl. ‖ Zelluloidspielwaren fpl. ‖ jouets mpl. en celluloïd.

cellulose ‖ Holzstoff m.; Zellulose f.; Zellstoff m.; Lignose f. ‖ pâte f. de bois; cellulose f. / ~ of sulphide ‖ Sulfitzellstoff m. ‖ cellulose f. sulfatée.

cellulose acetate ‖ Zelluloseazetat n. ‖ acétate m. de cellulose. / ~ acetate silk ‖ Zelluloseazetatseide f. ‖ soie f. à l'acétate de cellulose. / ~ extract ‖ Zellstoffauszug m.; Zelluloseextrakt m. ‖ extrait m. de cellulose. / ~ factory ‖ Zellstofffabrik f. ‖ fabrique f. de cellulose. / ~ filament ‖ Zellstofffaden m.; Zellulosefaden m. ‖ filament m. de cellulose. / ~ finish ‖ Zellstofflack m.; Zelluloselack m. ‖ émail m. à la cellulose; peinture f. cellulosique. / ~ lacquer see ~ finish. / ~ lye ‖ Zellstofflauge f.; Zelluloselauge f. ‖ lessive f. de cellulose.

cellulose-paper ‖ Zellstoffpapier n.; Zellulosepapier n. ‖ papier m. de cellulose. / ~ with satin finish ‖ geglättetes oder satiniertes Zellstoffpapier n. ‖ papier m. de cellulose satiné. / ~ smooth on one side ‖ einseitig glattes Zellstoffpapier n. ‖ papier m. de cellulose satiné d'un côté.

cellulose plate for filtering pulp ‖ Zellstoffplatte f. für Filtermasse ‖ plaque f. en pâte de cellulose pour masse filtrante. / ~ solvent ‖ Zellstofflösungsmittel n. ‖ dissolvant m. de cellulose. / ~ trade ‖ Zellstoffindustrie f. ‖ industrie f. de la cellulose. / ~ wadding ‖ Zellstoffwatte f. ‖ ouate f. de cellulose. / ~ yarn ‖ Zellstoffgarn n. ‖ fil m. en cellulose.

cement, to (Mason) ‖ mit Zement m. verputzen ‖ cimenter. / ~ (Met) ‖ zementieren ‖ cémenter. / ~ in ‖ mit Zement m. ausgießen ‖ couler du ciment dans les joints.

cement ‖ Zement m. ‖ ciment m. / ~ (Mortar) ‖ Mörtel m.; Kalkmörtel m. ‖ mortier m.; ciment m. / to inject ~ ‖ mit Zement m. torkretieren ‖ faire injection du ciment. / to mix ~ ‖ Zement m. anmischen ‖ doser et mélanger le ciment. / acidproof ~ ‖ säurefester Mörtel m. ‖ ciment m. résistant aux acides. / alcoholproof ~ ‖ alkoholfester Kitt n. ‖ mastic m. résistant à l'action de l'alcool. / artificial ~ ‖ künstlicher Zement m. ‖ ciment m. artificiel. / asphaltic ~ ‖ Asphaltkitt m. ‖ mastic m. d'asphalte. / bituminous ~ ‖ bituminöser Kitt m. ‖ mastic m. bitumineux. / black ~ see bituminous ~. / blast furnace ~ ‖ Hochofenzement m. ‖ ciment m. de haut-fourneau. / blast-furnace slag ~ ‖ Hochofenschlackenzement m. ‖ ciment m. de laitier de haut-fourneau. / calcareous ~ see hydraulic ~. / fire-proof ~ ‖ feuerfester Kitt m. ‖ mastic m. réfractaire. / gutta-percha ~ ‖ Guttaperchakitt m. ‖ mastic m. à la gutta-percha. / high-class ~ ‖ hochwertiger Zement m. ‖ ciment m. de haute valeur. / hydraulic ~ ‖ Zementmörtel m.; hydraulischer Mörtel m. ‖ ciment m. hydraulique; ciment m. romain. / iron Portland ~ ‖ Eisenportlandzement m. ‖ ciment m. de laitier Portland. / iron rust ~ ‖ Eisenkitt m. ‖ pouzzolane f. artificielle; mastic m. de fer. / joiner's ~ ‖ Leimkitt m. ‖ mastic m. à la colle. / melted ~ ‖ Schmelzzement m. ‖ ciment m. fondu. / natural ~ ‖ Naturzement m. ‖ ciment m. naturel. / ~ of plaster ‖ Gipsmörtel m. ‖ mortier m. au plâtre. / Portland ~ ‖ Portlandzement m. ‖ ciment m. Portland. / quickly taking ~ see quicksetting ~. / quick-setting ~ ‖ schnell abbindender Zement m. ‖ ciment m. à prise rapide. / resinous ~ ‖ Harzkitt m. ‖ mastic m. résineux. / Roman ~ ‖ Romanzement m. ‖ ciment m. romain. / rubber ~ ‖ Kautschukkitt m. ‖ mastic m. au caoutchouc. / slag ~ see iron Portland ~. / slow-setting ~ ‖ langsam abbindender Zement m. ‖ ciment m. à prise lente. / water-proofing ~ ‖ wasserbeständiger Zement m. ‖ ciment m. imperméable.

cement articles pl. ‖ Zementwaren fpl. ‖ objets mpl. en ciment.

cementation of malleable iron ‖ Glühfrischen n. ‖ cémentation f. de fer malléable.

cementation furnace ‖ Temperofen m.; Härteofen m.; Zementierofen m. ‖ four m. à cémenter. / ~ powder ‖ Härtepulver n. ‖ poudre f. à cémenter. / ~ process ‖ Einsatzhärtung f.; Zementierung f. ‖ cémentation f. / ~ water (Forg) ‖ Löschwasser n. ‖ pacquet m.

cement black ‖ Zementschwarz n. ‖ noir m. de ciment. / ~ bunker ‖ Zementsilo m. ‖ silo m. à ciment. / ~ colours pl. ‖ Farben fpl. für Zement ‖ peintures fpl. pour ciment. / ~ concrete ‖ Zementbeton m. ‖ béton m. de ciment. / ~ conduit see ~ pipe. / ~ deal ‖ Zementdiele f. ‖ planche f. en ciment. / ~ diaphragm ‖ Zementdiaphragma n. ‖ diaphragme m. en ciment.

cemented glass ‖ verkittetes Glas n. ‖ verre m. collé. / ~ lenses pl. ‖ verkittete Linsen fpl. ‖ lentilles fpl. collées; lentille f. constituée par verres accolés. / ~ steel ‖ Zementstahl m. ‖ acier m. de cémentation.

cementer ‖ Zementierer m. ‖ cimentier m.

cement flag ‖ Zementfliese f. ‖ carreau m. de ciment. / ~ floor ‖ Zementestrich m. ‖ aire f. en ciment. / ~ foundation ‖ Zementsockel m. ‖ socle m. en ciment. / ~ furnace (Build) ‖ Zementbrennofen m. ‖ four m. à ciment. / ~ glue ‖ Zementleim m. ‖ colle f. de ciment.

cementing (Build; Met) ‖ Zementierung f. ‖ cémentation. / ~ by jet ‖ Zementierung f. im Spritzverfahren ‖ cimentation f. par pistolet pulvérisateur.

cementing powder ‖ Einsatzpulver n.; Härtepulver n.; Zementierpulver n. ‖ poudre f. à cémenter. / action of the ~ ‖ Einwirkung f. des Härtepulvers ‖ influence f. de la poudre à cémenter. / the ~ must lie close to the surfaces to be cemented ‖ das Härtepulver n. muß überall fest anliegen ‖ la poudre à cémenter doit coller à toutes les surfaces à cémenter. / those parts which are carburetted by the ~ get a surface as hard as glass ‖ durch das Einsatzpulver gekohlte Stellen fpl. werden glashart ‖ aux endroits qui sont exposés à l'action de la poudre de cémentation, la surface devient dure comme le verre.

cementing process (Met) ‖ Einsatzhärtung f.; Zementierungsprozeß m. ‖ procédé m. de cémentation. / ~ duration ‖ Zementationsdauer f.; Zementierungsdauer f. ‖ durée f. de procédé de cémentation.

cement layer ‖ Zementschicht f. ‖ couche f. de ciment. / ~ laying-on machine ‖ Zementauftragemaschine f. ‖ machine f. à étaler le ciment. / ~ lime-concrete ‖ Zementkalkbeton m. ‖ béton m. de ciment et de chaux. / ~ lime-mortar ‖ Zementkalkmörtel m. ‖ mortier m. au ciment et à la chaux. / ~ maker ‖ Zementarbeiter m. ‖ cimentier m. / ~ manufactory ‖ Zementfabrik f.; Zementwerk n.;

Zementbrennerei f. ‖ fabrique f. de ciment. / ~ manufacture ‖ Zementerzeugung f. ‖ fabrication f. de ciment.

cement mill ‖ Zementmühle f. ‖ moulin m. à ciment. / ~ drive ‖ Zementmühlenantrieb m. ‖ dispositif m. de commande de moulins à ciment. / ~ plant ‖ Zementmahlanlage f. ‖ installation f. pour la fabrication du ciment.

cement mixer ‖ Zementmischmaschine f. ‖ betonnière f. / ~ mortar ‖ Zementmörtel m. ‖ mortier m. de ciment. / ~ paste ‖ Zementbrei m. ‖ lait m. de ciment. / ~ pipe ‖ Zementrohr n.; Zementröhre f. ‖ tuyau m. ou conduite f. en ciment. / ~ plant see ~ manufactory. / ~ plastering ‖ Zementverputz m. ‖ crépi m. de ciment. / ~ roofing-tile ‖ Zementdachziegel m. ‖ tuile f. de ciment. / ~ spatula ‖ Zementspatel m. ‖ spatule f. à ciment. / ~ steel ‖ Zementstahl m.; Blasenstahl m. ‖ acier m. cémenté (à soufflures).

cement-stone ‖ Zementstein m. ‖ pierre f. en ciment. / artificial ~ ‖ Zementkunststein m. ‖ pierre f. artificielle de ciment.

cement-stone quarry ‖ Zementmergelgrube f. ‖ carrière f. de marne à ciment.

cement testing press ‖ Zementprüfpresse f. ‖ presse f. à essayer le ciment. / ~ throwing jet ‖ Zementspritzvorrichtung f. ‖ projecteur-pulvérisateur m. pour ciment liquide. / ~ tile ‖ Zementdachziegel m. ‖ tuile f. en ciment. / ~ trassconcrete ‖ Zementtraßbeton m. ‖ béton m. de ciment et de trass.

cementware ‖ Zementwaren fpl. ‖ articles mpl. ou marchandises fpl. ou ouvrages mpl. en ciment. / ~ industry ‖ Zementwarenindustrie f. ‖ industrie f. des produits en ciment. / ~ making machine ‖ Zementwarenherstellungsmaschine f. ‖ machine f. pour la fabrication de produits en ciment.

cement work ‖ Zementieren n.; Zementierarbeit f. ‖ travaux en ciment.

cement works pl. see also ~ manufactory ‖ Zementwerk n.; Zementfabrik f. ‖ fabrique f. de ciment. / ~ equipment ‖ Zementwerkseinrichtung f. ‖ accessoires mpl. pour la fabrication du ciment. / ~ outfit see ~ equipment. / ~ plant ‖ Zementwerksanlage f. ‖ établissement m. pour la fabrication du ciment.

cenotine ‖ Ytterspat m. ‖ cénotime m.; phosphate m. d'yttrium.

censure, to ~ ‖ rügen ‖ blâmer.

censure ‖ Rüge f. ‖ blâme m.

census ‖ Volkszählung f. ‖ recensement m. (de la population). / ~ work ‖ Volkszählungsarbeiten fpl. ‖ travaux mpl. de recensement.

centaury herb ‖ Tausendgüldenkraut n. ‖ herbe f. de centaurée.

center, to ~ see to centre.

center see centre.

centering ‖ mittend; zentrierend ‖ centrant. / self-~ ‖ selbstmittend; selbstzentrierend ‖ à centrage m. automatique.

centering ‖ Einmitten n.; Mitten n.; Zentrieren n. ‖ centrage m. / ~ (Build) ‖ Lehrgerüst n.; Bogengerüst n.; Wölbgerüst n.; Bockverstellung f. ‖ armement m. de voûte; cintres mpl. d'une voûte; cintre m. de charpente. / ~ (Projectile) ‖ Führung f. ‖ forcement m.; guide m. / automatic ~ ‖ Zwangseinmittung f. ‖ centrage m. automatique. / machine for ~ the tyres ‖ Maschine f. zum Zentrieren

der Radreifen ‖ machine f. à centrer les bandages.

centering device ‖ Zentriervorrichtung f. ‖ dispositif m. de centrage ou à centrer. / ~ for circular nibbling works ‖ Zentriervorrichtung f. für Rundschnitte ‖ dispositif m. à centrer pour découper des cercles.

centering diaphragm ‖ Zentrierblende f. ‖ diaphragme m. de centrage. / ~ disc ‖ Mittungsscheibe f. ‖ plateau m. de centrage. / ~ gauge ‖ Mittelpunktlehre f. ‖ calibre m. de centrage. / ~ lens ‖ Mittungslinse f.; Zentrierlinse f. ‖ lentille f. de centrage. / ~ lens with ruled cross ‖ Zentrierglas n. oder Mittungsglas n. mit Strichkreuz ‖ verre m. de centrage muni d'un réticule.

centering machine ‖ Mittungsmaschine f.; Zentriermaschine f. ‖ machine f. à centrer. / multiple-spindle ~ ‖ mehrspindlige Zentriermaschine f. ‖ machine f. à centrer à plusieurs broches. / one-spindle ~ ‖ einspindlige Zentriermaschine f. ‖ machine f. à centrer à une broche. / two-spindle ~ ‖ zweispindelige Zentriermaschine f. ‖ machine f. à centrer à deux broches.

centering moulding machine ‖ Ausrichtmaschine f.; Zentriermaschine f.; Ankörnmaschine f. ‖ presse f. de précision; presse f. à centrer les chassis. / ~ pin ‖ Mittungsstift m. ‖ goupille f. ou ergot m. de centrage. / ~ press ‖ Zentrierpresse f. ‖ presse f. à centrer. / hydraulic ~ press ‖ hydraulisch angetriebene Zentrierpresse f. ‖ presse f. à centrer (actionnée d'après le système) hydraulique. / ~ ring ‖ Zentrierring m. ‖ anneau m. ou bague f. de centrage. / machined ~ ring ‖ gedrehter Zentrierring m. ‖ bague f. de centrage tournée. / ~ screw see also ~ pin ‖ Zentrierschraube f. ‖ vis f. de centrage. / ~ slide condenser (Opt) ‖ zentrierbarer Schlittenkondensor m. ‖ condenseur m. coulissant centrable. / ~ telescope ‖ Zentrierfernrohr n. ‖ lunette f. de centrage.

centesimal ‖ hundertteilig; prozentig ‖ centésimal. / ~ balance see ~ weighing machine. / ~ division ‖ Hundertteilung f. ‖ division f. centésimale. / ~ weighing machine ‖ Zentesimalwage f. ‖ bascule f. centésimale.

centiare ‖ Quadratmeter n.; Meter n. im Geviert ‖ mètre m. carré; centiare m.

centigrade ‖ hundertgradig ‖ centigrade.

centigrade ‖ Grad Celsius ‖ centigrade.

centigram ‖ Zentigramm n. ‖ centigramme m.

centiliter ‖ Zentiliter n. ‖ centilitre m.

centimeter ‖ Zentimeter n. ‖ centimètre m. / square ~ ‖ Quadratzentimeter m.; Zentimeter im Geviert ‖ centimètre m. carré.

centimeter division on a staff ‖ Zentimeterfeldeinteilung f. einer Latte ‖ division f. en centimètres d'une mire.

central adjusting (Mach) ‖ Zentralanstellung f. ‖ réglage m. central. / ~ arresting device ‖ Mittelstellarretiervorrichtung f. ‖ dispositif m. d'arrêt de la position centrale. / ~ bridge of the spectacles ‖ Mittelsteg m. der Brille ‖ pont m. des lunettes.

central-clock installation (Tel) ‖ Zentraluhrenanlage f. ‖ installation f. d'horloge centrale. / ~ network ‖ Zentraluhrennetz n. ‖ réseau m. d'horloges centrales. /

~ system ‖ Zentraluhrensystem n. ‖ système m. d'horloges centrales.

central condensation plant ‖ Zentralkondensationsanlage f. ‖ installation f. de condensation centrale. / ~ corridor ‖ Mittelgang m. ‖ couloir m. médian. / ~ depot ‖ Sammelstelle f. ‖ dépôt m. central. / ~ engine ‖ Mittelmotor m. ‖ moteur m. central.

Central European time ‖ mitteleuropäische Zeit f. ‖ heure f. de l'Europe centrale.

central grease lubrication ‖ zentrale Fettschmierung f. ‖ lubrification f. centrale à la graisse. / ~ guide ‖ Zentralführung f. ‖ guide m. central.

central-heating ‖ Zentralheizung f. ‖ chauffage m. central. / ~ apparatus ‖ Zentralheizungsvorrichtung f. ‖ appareil m. de chauffage central. / ~ fittings pl. ‖ Zentralheizungsarmaturen fpl. ‖ armatures fpl. pour le chauffage central. / ~ form-piece ‖ Zentralheizungsformstück n. ‖ pièce f. d'ajustage pour le chauffage central. / ~ plant ‖ Zentralheizungsanlage f. ‖ installation f. de chauffage central.

central hot-water supply ‖ Zentralwarmwasserversorgung f. ‖ chauffage m. central à eau chaude.

centralize, to ~ in one place ‖ in einem Punkt m. vereinigen ‖ centraliser; réunir au même endroit m.

central nervous system ‖ Zentralnervensystem n. ‖ système m. nerveux central. / ~ office of measurement ‖ Meßzentrale f. ‖ centrale f. de mesure. / ~ ophthalmoscope method ‖ zentrische Ophthalmoskopie f. ‖ ophtalmoscopie f. centrée. / ~ perspective ‖ zentralperspektivisch ‖ à perspective f. centrale. / ~ pivot ‖ Königssäule f.; Königszapfen m. ‖ pivot m. central. / ~ point see centre. / ~ position ‖ Mittellage f. ‖ position f. centrale. / ~ quad of a cable ‖ Kabelkern m.; Kernkabel n. ‖ quart m. central. / ~ rail ‖ Mittelschiene f. ‖ rail m. central. / ~ rail (Of a railway switch) ‖ Mittelschiene f. ‖ rail m. du milieu. / ~ station ‖ Hauptbahnhof m. ‖ gare f. centrale. / ~ tapping of the primary windings (Electr) ‖ Mittenanzapfung f. der Primärwicklung ‖ prise f. médiane de l'enroulement primaire. / ~ water-works pl. ‖ Hauptwasserwerk n.; Zentralwasserstation f. ‖ station f. à distribution centrale des eaux.

centration see centering.

centre, to ‖ mitten; zentrieren; einstellen ‖ centrer.

centre ‖ Mittelpunkt m.; Mitte f. ‖ centre m. / ~ (Carp) ‖ Lehrbogen m.; Bogenlehre f. ‖ cintre m. de charpente; tambour m. à voûter. / ~ (Tool) ‖ Körner m. ‖ centre m. / ~ (Turning lathe) ‖ Spitze f. ‖ pointe f. / axle ~ ‖ Achsschaft m.; Mittelachse f. ‖ corps m. de l'essieu. / back ~ (Mach tool) ‖ Reitstockspitze f. ‖ contre-pointe f. / ~ of curvature (Geom) ‖ Krümmungsmittelpunkt m. ‖ centre m. de courbure. / dead ~ ‖ Totpunkt m.; toter Punkt m. ‖ point m. mort. / ~ of distribution ‖ Verteilungspunkt m. ‖ centre m. de distribution. / from ~ to ~ ‖ von Mitte bis Mitte ‖ d'axe en axe; de centre en centre. / measured from ~ to ~ ‖ von Mitte zu Mitte gemessen ‖ mesuré de centre en centre. / ~ of fulcrum of a balance ‖ Mittelpunkt m. der Bewegung einer Wage ‖ centre m. de mou-

vement *ou* point m. d'appui d'une balance.

centre of gravity ‖ Schwerpunkt m. ‖ centre
m. *ou* point m. de gravité. / low position
of ~ ‖ Tieflage f. des Schwerpunktes ‖
position f. basse du centre de gravité. /
~ of the gyro *or* gyroscope ‖ Kreiselschwerpunkt m. ‖ centre m. de gravité
du gyroscope. / to place the ~ low ‖ den
Schwerpunkt m. tief legen ‖ placer en
bas de centre de gravité. / position of
the ~ ‖ Schwerpunktlage f. ‖ position f.
du centre de gravité. / ~ of a ship ‖
Schiffsschwerpunkt m. ‖ centre m. de
gravité du navire.

centre of gyration ‖ Drehungsmittelpunkt
m. ‖ centre m. de rotation. / instantaneous ~ ‖ Momentanzentrum n. ‖ centre
m. instantané. / ~ of mass (Mech) ‖ Mas-
senmittelpunkt m. ‖ centre m. de masse.
/ ~ of motion ‖ Drehpunkt m. ‖ point m.
d'appui *ou* de mouvement. / ~ of motion
of a balance *see* ~ of fulcrum of a balance.
/ ~ of oscillation ‖ Schwingungsmittel-
punkt m. ‖ centre m. d'oscillation. / ~
of parallel forces ‖ Mittelpunkt m. paral-
leler Kräfte ‖ centre m. des forces parallèles. / ~ of percussion ‖ Stoßmittelpunkt
m.; Stoßpunkt m. ‖ centre m. de
poussée. / ~ of pressure ‖ Druckmittelpunkt m.; Drucklinie f. ‖ centre m. de
pression; axe m. des centres de poussée. /
~ of rotation ‖ Drehungsmittelpunkt m. ‖
centre m. de rotation. / ~ of a surface ‖
Mittelpunkt m. einer Fläche ‖ centre m.
d'une surface. / ~ of traffic ‖ Verkehrs-
knotenpunkt m. ‖ centre m. de la circulation *ou* du trafic. / wheel ~ ‖ Radstern
m. ‖ étoile f. de la roue; étoile f. du moyeu.

centre-bit ‖ Zentrumsbohrer m. ‖ foret m.
ou mèche f. à centre; foret m. à trois
pointes.

centre-board (Shipb) ‖ Schwert n. ‖ aile
f. de dérive; semelle f. / ~ boat ‖ Schwert-
boot n. ‖ dériveur m.; bateau m. à dérive.

centre-boss ‖ Verstärkung f. in der Mitte ‖
renflement m. central. / ~ of a paddle
wheel (Shipb) ‖ Nabe f. eines Schaufelrades ‖ moyeu m. d'une roue à aubes.

centre cap (Mach) ‖ Druckhaupt n. ‖ chapeau m. de pivot. / ~ cock bit ‖ Zapfenbohrer m. ‖ mèche f. à bonde.

centred ‖ gemittet; zentriert; zentrisch
eingestellt ‖ placé au centre; centré.

centre-height (Lathe) ‖ Spitzenhöhe f. ‖
hauteur f. des pointes.

centreing *see* centering.

centre-lathe ‖ Spitzendrehbank f. ‖ tour m.
à pointes. / double ~ ‖ Schreinerdreh-
bank f. ‖ tour m. à deux pointes.

centre-line (Geom) ‖ Mittellinie f. ‖ axe
m. / distance between ~s ‖ Abstand m.
von Mitte zu Mitte ‖ distance f. de centre
en centre *ou* d'axe en axe.

centre-mark ‖ Körnermarke f.; Körner m.;
Körnerpunkt m. ‖ coup m. de pointeau. /
bringing down the punch on ~ ‖ Niederstellen n. des Stempels auf Körnermarke ‖ amenage m. du poinçon sur le
coup de pointeau.

centre-piece for table ‖ Tafelaufsatz m. ‖
sourtout m. de table.

centre-pin of the compass ‖ Pinne f. des
Kompasses ‖ pivot m. du compas.

centre-plate (Railw) ‖ Drehzapfenlager n.;
Drehgestellzapfenlager n. ‖ crapaudine
f. de pivot du bogie. / ~ on ball bearings ‖

Drehplatte f. auf Kugeln ‖ plaque f.
tournante sur billes.

centre-point (Mech) ‖ Drehpunkt m. ‖ centre-point m. de rotation. / ~ (Tool) *see*
~ punch.

centre punch ‖ Ankörner m.; Körner m. ‖
pointeau m.; amorçoir m. / ~ punching
apparatus ‖ Mittelsucher m.; Mittel-
punktsucher m. ‖ outil m. à centrer;
amorçoir m.; centreur m. / ~ support
(Crane) ‖ Drehstuhl m.; Königsstuhl m. ‖
support m. de la crapaudine. / ~ wheel
(Watchm) ‖ Minutenrad n.; großes Bodenrad n. ‖ roue f. de minute; roue f. de
longue tige.

centric setting of the instrument ‖ mittige
Aufstellung f. des Instrumentes ‖ disposition f. centrique de l'instrument.

centrifugal ‖ zentrifugal ‖ centrifuge.

centrifugal ‖ Zentrifuge f.; Schleudermaschine f.; Schleuder f. ‖ appareil m.
centrifuge. / accessories pl. of ~s ‖ Zentri-
fugenzubehör n. ‖ accessoire m. pour centrifuges. / ~ for cleaning oil ‖ Ölreinigungszentrifuge f. ‖ épurateur m. centrifuge d'huile. / hanging ~ ‖ Hänge-
zentrifuge f. ‖ centrifuge f. suspendue. / ~
with sieve jacket ‖ Schleudermaschine f.
mit Siebmantel ‖ essoreuse f. à panier
perforé. / ~ with solid jacket ‖ Schleudermaschine f. mit Vollmantel ‖ essoreuse f.
à panier plein.

centrifugal acceleration ‖ Fliehkraftbeschleunigung f. ‖ accélération f. centrifuge. / ~ beater ‖ Fliehkraftsichtmaschi-
nenflügel m. ‖ batteur m. *ou* aile f. d'une
bluterie centrifuge. / ~ bottom ‖ Zentri-
fugenboden m. ‖ fond m. de centrifuge. /
~ brake ‖ Fliehkraftbremse f. ‖ régulateur m. à force centrifuge; frein m.
centrifuge. / ~ casting ‖ Schleuderguß
m.; Zentrifugalguß m. ‖ coulée f. centrifuge. / ~ casting mould ‖ Schleuderguß-
form f. ‖ moule m. à coulée centrifuge. /
~ dressing machine (Mill) ‖ Fliehkraft-
sichtmaschine f.; Zentrifugalsichtma-
schine f. ‖ bluterie f. centrifuge. / ~
drum ‖ Schleudergefäß n.; Schleudertrommel f. ‖ tambour m. *ou* bol m. *ou*
récipient m. centrifuge. / ~ drum made
of pure nickel ‖ Schleudertrommel f. *oder*
Zentrifugentrommel f. aus Reinnickel ‖
tambour m. centrifuge en nickel pur. /
~ drying machine ‖ Trockenschleuder-
maschine f. ‖ hydro-extracteur m. / ~
engine *see* centrifugal. / ~ force ‖ Fliehkraft f.; Zentrifugalkraft f.; Schleuder-
kraft f. ‖ force f. centrifuge. / ~ governor ‖
Fliehkraftregler m.; Zentrifugalregler m. ‖
régulateur m. centrifuge. / ~ machine
see centrifugal. / ~ mill ‖ Schleudermühle f. ‖ broyeur m. centrifuge; désintégrateur m. / ~ oil purifier ‖ Ölreinigungszentrifuge f. ‖ épurateur m. centrifuge d'huile. / ~ pump ‖ Kreiselpumpe
f.; Zentrifugalpumpe f. ‖ pompe f. centrifuge. / ~ sifter ‖ Fliehkraftsichter m.;
Zentrifugalsichter m. ‖ bluterie f. centrifuge. / ~ strength *see* ~ force. / ~ weight
‖ Schwunggewicht n. ‖ poids m. de
lancement. / ~ wringing machine ‖ Flieh-
kraftwringmaschine f.; Zentrifugalwring-
maschine f. ‖ essoreuse f. centrifuge.

centrifuge, to ‖ schleudern ‖ centrifuger.

centrifuge *see* centrifugal.

centrifuging (To a certain degree of dryness) ‖ Trockenschleudern n. ‖ essorage

m. à sec. / ~ period ‖ Schleuderperiode f.;
Schleuderzeit f. ‖ période f. d'essorage.

centring *see* centering.

centripetal ‖ zentripetal ‖ centripète. / ~
force ‖ Zustrebekraft f.; Zentripetalkraft
f. ‖ force f. centripète.

century ‖ Jahrhundert n. ‖ siècle m.

ceramic ‖ keramisch ‖ céramique.

ceramic ‖ keramisches Erzeugnis n.; Keramik f.; (künstlerische) Töpferware f. ‖
céramique f. / dental ~ ‖ zahnärztliche
Keramik f. ‖ céramique m. dentaire. /
fine ~ ‖ Feinkeramik f. ‖ céramique f.
fine. / machine for fine ~s ‖ feinkera-
mische Maschine f. ‖ machine f. pour la
fabrication de céramique fine. / ordinary ~ ‖ Grobkeramik f. ‖ grosse céramique f.

ceramic art ‖ Keramik f.; Töpferkunst f.;
Kunstkeramik f. ‖ art m. céramique. / ~
industry ‖ keramische Industrie f. ‖ industrie f. céramique. / ~ kiln ‖ keramischer Brennofen m. ‖ four m. céramique. / ~ machine ‖ Keramikmaschine
f. ‖ machine f. pour la céramique. / ~
pavement slab ‖ keramische Fußbodenplatte f. ‖ carreau m. de carrelage céramique. / ~ wall slab ‖ keramische Wandplatte f. ‖ carreau m. de revêtement
céramique.

cereal *see* cereals.

cerealine ‖ Zerealin n. ‖ céréaline f.

cereals pl. ‖ Getreide n. ‖ céréales fpl.;
grains mpl.

cere cloth ‖ Wachstuch n. ‖ toile f. cirée.

ceresine ‖ Zeresin n.; gereinigtes Bergwachs n.; Mineralwachs n. ‖ cérésine f.;
cire f. fossile *ou* minérale. / ~ candle ‖
Zeresinlicht n. ‖ chandelle f. en cérésine. /
~ paper ‖ Zeresinpapier n. ‖ papier m. à
la cérésine. / ~ work equipment ‖ Zere-
sinfabrikeinrichtung f. ‖ installation f.
de fabrique de cérésine.

cerine (Chem) *see* ceresine. / ~ (Miner) *see*
cerium.

cerite earth ‖ Zeriterde f. ‖ terre f. de
cérite.

cerium ‖ Zerium n.; Orthit m.; Zerin n.;
Zer n. ‖ cérium m.; allanite f.; orthite f.
/ ~ metal ‖ Zermetall n. ‖ métal m. de
cérium. / ~ salt ‖ Zeriumsalz n. ‖ sel m.
de cérium.

cerotine colour ‖ Zerotinfarbe f. ‖ couleur
f. de cérotine.

certain ‖ zuverlässig ‖ éprouvé; certain.

certainty ‖ Zuverlässigkeit f. ‖ certitude f.;
authenticité f.; sûreté f.; solidité f.

certificate ‖ Bescheinigung f.; Zeugnis n.;
Diplom n.; Zertifikat n. ‖ certificat m.;
diplôme m. / ~ of capacity *see* ~ of proficiency. / ~ of (good) cha-
racter ‖ Leumundszeugnis n. ‖ certificat
m. de bonne conduite. / ~ of classification
‖ Klassifizierungsschein m. ‖ certificat
m. de classification. / ~ of delivery ‖
Einlieferungsschein m. ‖ bulletin m. de
dépôt; reçu m. / doctor's ~ ‖ ärztliche
Bescheinigung f.; ärztliches Attest n. ‖
certificat m. du médecin. / ~ of fitness *see*
~ of proficiency. / ~ of health (Mar) ‖
Schiffspaß m.; Gesundheitszeugnis n. ‖
certificat m. de santé. / ~ of insurance ‖
Versicherungsschein m. ‖ certificat m.
d'assurance. / ~ of origin ‖ Ursprungs-
zeugnis n. ‖ certificat m. d'origine *ou*
de provenance. / ~ of permission to
work a mine ‖ Mutungsschein m. ‖ brevet
m. de concession d'une mine. / ~ of

proficiency ‖ Befähigungszeugnis n. ‖ certificat m. de capacité. / ~ of registry ‖ Schiffszertifikat n.; Beilbrief m. ‖ certificat m. de navire *ou* de construction. / ~ of tonnage ‖ Meßbrief m. ‖ certificat m. de jaugeage *ou* du tonnage. / ~ of warranty ‖ Garantieschein m. ‖ billet m. de cautionnement m.; promesse f. de garantie.

certificated engineer ‖ Diplomingeniör m. ‖ ingénieur m. diplômé.

certify, to ‖ bescheinigen; attestieren ‖ certifier; attester.

ceruleo-sulphate ‖ indigschwefelsaures Salz n. ‖ céruléosulfate m.; sulfindigotate m.

ceruse (Chem) ‖ Bleiweiß n. ‖ céruse f.

cerusite ‖ Bleierde f.; Bleiglimmer m.; Zerusit m. ‖ cérusite f.

cervantite ‖ Antimonocker m. ‖ cervantite f.

cess ‖ Abort m. ‖ cabinet m. (d'aisance); latrines fpl.

cessation ‖ Stillstand m.; Ruhe f.; Aufhören n. ‖ cessation f.; arrêt m.

cession of the property ‖ Abtretung f. des Grundstückes ‖ cession f. du fonds.

cess pipe ‖ Abtrittschlotte f. ‖ chausse f. d'aisance; tuyau m. de chute. / ~ pool ‖ Senkgrube f.; Abortgrube f.; Kloake f.; Abzugschleuse f. ‖ creux m.; trappe f. à nettoyage; égout m.; cloaque m.; fosse f. d'aisance.

cetacea pl. (Zool) ‖ Waltiere npl. ‖ cétacés mpl.

chafe, to (Mar) ‖ schamfielen ‖ s'érailler; raguer.

chafed (Mar) ‖ schamfielt ‖ écaillé; ragué.

chafe rod (El line) ‖ Scheuerbock m.; Scheuerpfahl m. ‖ pieu m. protecteur.

chaff ‖ Häcksel m. ‖ paille f. hachée; hachée f. / ~ ‖ Kaff n. ‖ bourrier m. / ~ of corn ‖ Spreu f. von Getreide ‖ balle f. de céréales. / ~ cutter ‖ Häckselmaschine f. ‖ hache-pailles f. / ~ cutting machine *see* ~ cutter.

chaffering ‖ Schacher m. ‖ trafic m. sordide.

chafing board (Shipb) ‖ Schamfielungsplanke f. ‖ balustrade m. / ~ mat ‖ Schamfielungsmatte f. ‖ paillet m. de portage.

chain, to (Surv) ‖ mit der Kette messen ‖ chaîner.

chain ‖ Kette f. ‖ chaîne f. / ~ (Electr) ‖ Spannungsreihe f. ‖ chaîne f. / ~ (Weav) ‖ Aufzug m. ‖ chaîne f. / ~ (Carp) ‖ Meßkette f. ‖ chaîne f.; chaîne f. d'arpenteur. / anchor ~ ‖ Ankerkette f. ‖ chaîne f. d'ancre. / auxiliary ~ ‖ Notkette f.; Sicherheitskette f. ‖ chaîne f. de sûreté. / ~ of bicycle ‖ Fahrradkette f. ‖ chaîne f. de bicyclette. / block ~ ‖ Blockkette f. ‖ chaîne f. à blocs. / brass ~ ‖ Messingkette f. ‖ chaîne f. en laiton. / ~ of buckets ‖ Eimerkette f. ‖ chaîne f. de drague *ou* à godets. / calibrated ~ ‖ kalibrierte Kette f. ‖ chaîne f. calibre. / check ~ *see* auxiliary ~. / clear ~s pl. (Mar) ‖ klare Ketten fpl. ‖ chaînes fpl. dégagées. / copper ~ ‖ Kupferkette f. ‖ chaîne f. en cuivre. / coupling ~ (Railw) ‖ Kuppelkette f. ‖ chaîne f. d'attelage. / detachable (link) ~ ‖ zerlegbare Kette f. ‖ chaîne f. détachable. / double-roller ~ ‖ Doppelrollenkette f. ‖ chaîne f. à doubles rouleaux. / driving ~ ‖ Antriebskette f. ‖ chaîne f. de commande. / ~ of egg insulators ‖ Eierkette f. ‖ isolateurs mpl. de chaîne. / endless ~ ‖ endlose Kette f. ‖ chaîne f. sans fin. / eyeglass ~ ‖ Pin-

cenezkettchen n. ‖ chaînette f. pour pince-nez. / fan-driving ~ ‖ Ventilatorantriebskette f. ‖ chaîne f. de commande du ventilateur. / flat-link articulated ~ ‖ Gelenkkette f. ‖ chaîne f. à joints articulés. / flat-link articulated Gall's ~ ‖ Gallsche Kette f. ‖ chaîne f. de Galle. / forming a ~ ‖ kettenförmig ‖ en chaîne. / foul ~s pl. (Mar) ‖ unklare Ketten fpl. ‖ chaînes fpl. avec des tours. / Gall's ~ ‖ Gall'sche Kette f. ‖ chaîne f. Gall. / gold ~ ‖ Goldkette f. ‖ chaîne f. en or. / ~ of hills ‖ Hügelkette f. ‖ chaînon m.; chaîne f. de collines. / hook-link ~ ‖ Hakenkette f. ‖ chaîne f. à crochets. / immerged ~ (Hydr arch) ‖ Tauereikette f. ‖ chaîne f. noyée. / iron ~ ‖ eiserne Kette f. ‖ chaîne f. de fer. / lift ~ (Mill) ‖ Pansterkette f. ‖ chaîne f. à élever *ou* à baisser la roue à aubes. / long-link ~ ‖ langgliedrige Kette f. ‖ chaîne f. à maillons longs. / lubricating ~ ‖ Schmierkette f. ‖ chaîne f. de graissage. / ~ for measuring ‖ Meßkette f. ‖ chaîne f. d'arpenteurs. / motor car ~ ‖ Kraftwagenkette f. ‖ chaîne f. pour automobiles. / non-skid ~ ‖ Gleitschutzkette f. ‖ chaîne f. antidérapante; antidérapant m. à chaîne. / open-link ~ ‖ Schakenkette f. ‖ chaîne f. à maillons. / ~ with paddles (Shipb) ‖ Schaufelkette f. ‖ chaîne f. à aubes. / ~ to prevent expansion of store casks ‖ Faßspannkette f. ‖ chaîne-tendeur m. pour foudres. / roller ~ ‖ Rollenkette f. ‖ chaîne f. à rouleaux. / roundlink ~ ‖ Rundgliederkette f. ‖ chaîne m. à maillons ronds. / safety ~ *see* auxiliary ~. / short-link ~ ‖ kurzgliedrige Kette f. ‖ chaîne f. à maillons courts. / side ~ *see* auxiliary ~. / silent ~ ‖ geräuschlose Kette f. ‖ chaîne f. sourde *ou* silencieuse. / silver ~ ‖ Silberkette f. ‖ chaîne f. en argent. / ~ with spike hooks ‖ Kette f. mit Einschlagkeilen ‖ chaîne f. avec crampons. / sprocket ~ ‖ Gelenkkette f.; Laschenkette f. ‖ chaîne f. à fuseaux *ou* articulée. / striping ~ ‖ Ringelkette f. ‖ chaîne f. pour rayures horizontales. / stud link ~ ‖ Stegkette f. ‖ chaîne f. à étançons. / surveyor's ~ ‖ Meßkette f. ‖ chaîne f. d'arpenteur. / suspension ~ for clothes ‖ Rockaufhänger m. ‖ chaînette f. de suspension de robes. / tested ~ ‖ kalibrierte Kette f. ‖ chaîne f. calibre. / tyre ~ ‖ Schneekette f. ‖ chaîne f. à neige. / watch ~ ‖ Uhrkette f. ‖ chaîne f. de montre. / welded ~ ‖ geschweißte Kette f. ‖ chaîne f. à maillons soudés. / wire ~ ‖ Schuppenkette f. ‖ chaîne f. colonne.

chain and sprocket wheel drive ‖ Kettenantrieb m. ‖ entraînement m. par pignons et chaîne. / ~ barrier ‖ Kettenschranke f. ‖ barrière f. à chaîne. / ~ belt ‖ Kettenriemen m. ‖ courroie f. articulée. / ~ bracelet ‖ Kettenarmband n. ‖ bracelet m. de chaîne. / ~ brake ‖ Kettenbremse f. ‖ frein m. à chaîne. / ~ bridge ‖ Kettenbrücke f. ‖ Hängebrücke f. ‖ pont m. suspendu à chaînes; pont m. à chaînes. / ~ builder (Weav) ‖ Kettenbäumer m. ‖ monteur m. de chaîne. / ~ cable (Mar) ‖ Ankerkette f. ‖ câble-chaîne m.; chaîne f. de l'ancre. / ~ case ‖ Kettenkasten m. ‖ carter m. de chaînes. / ~ control (Mar) ‖ Kettensteuerung f. ‖ manœuvre f. par chaîne. / ~ conveyance ‖ Kettenförderung f. ‖ traînage m. par chaîne.

chain conveyor ‖ Kettenförderer m. ‖ transporteur m. à chaîne *ou* à godets. / bucket ~ ‖ Becherkettenförderer m. ‖ transporteur m. à chaîne à godets; convoyeur m. / ~s pl. ‖ Kettenbahn f. ‖ voie f. à chaîne flottante.

chain coupling ‖ Kettenkupplung f. ‖ attelage m. par chaînes. / ~ cutter ‖ Fräskette f. ‖ chaîne f. fraise. / ~ dredger ‖ Eimerbagger m. ‖ dragueur m. à godets.

chain drive ‖ Kettenantrieb m. ‖ transmission f. *ou* commande par chaînes. / double ~ ‖ Antrieb m. mit zwei Ketten ‖ transmission f. par deux chaînes. / single ~ ‖ Antrieb m. mit einer Kette ‖ transmission f. par chaîne unique.

chain-driven magneto ‖ Magnet m. mit Kettenantrieb ‖ magnéto m. à commande par chaîne.

chain drum ‖ Kettentrommel f. ‖ tambour m. à chaîne. / ~ ‖ Kettenstern m. ‖ étoile f. à chaîne.

chain dummy coupling ‖ Kettenleerkupplung f. ‖ faux accouplement m. avec chaînette. / ~ elevator ‖ Kettenbecherwerk n. ‖ élévateur m. à chaîne.

chainer (Min) ‖ Anschläger m. ‖ homme m. de chaîne; accrocheur m. à la chaîne.

chain follower (Surv) ‖ hinterster Kettenzieher m. ‖ premier aide m. du géomètre; porte-chaîne m. / ~ gear ‖ Kettengetriebe n. ‖ entraînement m. par chaîne; transmission f. à chaîne. / ~ gear case ‖ Kettenschutzkasten m. ‖ carter m. de chaîne.

chain grate ‖ Kettenrost m. ‖ grille f. à chaînes. / surface of ~ ‖ Wanderrostfläche f. ‖ surface f. de la grille mobile. / ~ stoker ‖ Kettenrostfeuerung f. ‖ foyer m. avec grille à chaîne.

chain guard ‖ Kettenschützer m.; Kettenschutz m. ‖ carter m. de la chaîne; gardechaîne m. / ~ haulage ‖ Kettenförderung f. ‖ transport m. par chaîne flottante. / ~ hook ‖ Kettenhaken m. ‖ crochet m. de chaîne. / ~ hook (Mar) ‖ Kettenhaken m. ‖ croc m. de câble.

chaining (Surv) ‖ Messen n. mit der Kette ‖ chaînage m.

chain insulator ‖ Kettenisolator m. ‖ chaîne f. d'isolateurs. / ~ iron ‖Ketteneisen n. ‖ fer m. de chaîne. / ~ jack ‖ Kettenwinde f. ‖ cric m. à noix. / ~ jack with two hooks ‖ Doppelklauenwinde f. ‖ cric m. à deux pattes à double noix. / ~ knot (Pont; Mar) ‖ Kettenstich m. ‖ nœud m. de chaîne. / ~ leader (Surv) ‖ vorderster Kettenzieher m. ‖ deuxième aide m. du géomètre; porte-chaîne m.

chainless ‖ kettenlos ‖ sans chaîne f.

chainlet ‖ Kettchen n. ‖ chaînette f.

chain link ‖ Kettenlasche f.; Kettenglied n. ‖ maillon m. *ou* flasque m. de chaîne. / two-fold ~ ‖ zweifache Kettenlasche f. ‖ maillon m. double.

chain locker (Shipb) ‖ Kettenkasten m. ‖ puits m. aux chaînes.

chain loom (Weav) ‖ Kettenstuhl m. ‖ métier m. chaîne. / high-speed ~ ‖ Schnellläuferkettenstuhl m. ‖ métier m. chaîne à grande vitesse. / Jacquard ~ ‖ Jacquardkettenstuhl m. ‖ métier m. chaîne Jacquard.

chain maker ‖ Kettenschmied m. ‖ chaînier m. / fancy ~ ‖ Kettenmacher m. ‖ chaîniste m.

chain making machine ‖ Kettenschmiedemaschine f. ‖ machine f. à forger les chaînes.

chainman (Min) ‖ Markscheidergehilfe m. ‖ ‖ jambot-niveleur m.; porteur m. de chaîne.

chain measuring ‖ Vermessung f. mit der Kette ‖ chaînage m. / ~ oiler ‖ Kettenöler m. ‖ graisseur m. à chaîne. / ~ pipe (Shipb) ‖ Kettenklüse f. ‖ écubier m. du câble. / ~ pitch ‖ Kettenteilung f. ‖ pas m. de chaîne. / ~ plate (Shipb) ‖ Rüsteisen n. ‖ chaîne f. ou cadène f. de haubans.

chain printing mill ‖ Kettendruckmaschine f. ‖ machine f. à chaîner ou à imprimer des chaînes.

chain pulley ‖ Kettenrolle f. ‖ poulie f. à chaîne. / ~ (Wheel) ‖ Kettenrad n. ‖ roue f. à chaîne. / ~ block ‖ Kettenflaschenzug m. ‖ palan m. épicycloïdal.

chain pump ‖ Kettenpumpe f. ‖ pompe f. à chapelet. / ~ (Work) ‖ Paternosterwerk n. ‖ patenôtre f.; chapelet m; noria m.

chain railway ‖ Kettenbahn f. ‖ transporteur m. à chaîne. / ~ built on the ground ‖ bodenständige Kettenbahn f. ‖ funiculaire m. fixe à chaîne.

chain ring ‖ Kettenkranz m. ‖ roue f. de chaîne. / ~ rivet ‖ Kettenniet n. ‖ rivet m. de chaîne. / ~ rivetting ‖ Kettennietung f. ‖ rivure f. en carré ou à chaîne. / ~ roller ‖ Kettenrolle f. ‖ poulie f. ou rouleau m. à chaîne. / ~ rule (Math) ‖ Kettenregel f. ‖ règle f. conjointe. / ~ runner (Min) ‖ Anschläger m. ‖ homme m. de chaîne; accrocheur m. à la chaîne. / ~ saw ‖ Kettensäge f. ‖ scie f. à chaîne. / ~ sheave ‖ Kettenscheibe f. ‖ rouet m. à chaîne. / ~ slip-stopper (Mar) ‖ Schlippstopper m. ‖ bosse f. à échappement. / ~ sprocket wheel ‖ Kettennuß f. ‖ noix f. ou pignon m. à chaîne. / stand clear of the ~! (Mar) ‖ Achtung! Kette! ‖ défiez la chaîne! / ~ steam tug ‖ Kettendampfer m. ‖ toueur m. à chaîne. / ~ stitch (Sewing mach) ‖ Kettenstich m. einer Nähmaschine ‖ point m. de chaînette. / ~ stitching ‖ Kettenstickerei f. ‖ broderie f. au crochet. / ~ stopper (Mar) ‖ Kettenstopper m. ‖ bosse f. de câble. / ~ stretching-device ‖ Kettenspanner m. ‖ tendeur m. de chaîne. / ~ surveying see ~ measuring. / ~ tension ‖ Kettenspannung f. ‖ tension f. de chaîne. / ~ testing machine ‖ Kettenprüfmaschine f. ‖ machine f. à essayer les chaînes. / ~ testing machine with sliding weight balance ‖ Kettenprüfmaschine f. mit Laufgewichtwage ‖ machine f. à éprouver les chaînes avec balance à poids curseur. / ~ and rope testing machine ‖ Ketten- und Seilprüfmaschine f. ‖ machine f. à essayer les chaînes et les câbles. / ~ tightener ‖ Kettenspanner m. ‖ tendeur m. de chaîne. / ~ toggel (Shipb) ‖ Katzenkopf m.; Normann m. ‖ trésillon m.; normand m. / ~ tool ‖ Kettenaufleger m. ‖ outil m. pour remettre la chaîne. / ~ towing ‖ Kettenschleppschiffahrt f. ‖ remorquage m. par chaîne. / ~ tractor ‖ Kettenschlepper m. ‖ tracteur m. à chaîne. / ~ transmission ‖ Kettenübertragung f. ‖ transmission f. par chaîne. / ~ tread ‖ Kettenlauffläche f. ‖ surface f. de roulement de chaîne. / ~ twisting (Weav) ‖ Kettenanknoten n. ‖ nouage m. de chaînes. / ~ wale (Shipb) ‖ Rüst f. ‖ porte-haubans m. / ~ way ‖ Kettenbahn f. ‖ transport m. par chaîne. / ~ weight (Textile) ‖ Kettengewicht n. ‖ poids m.

de chaîne. / ~ welding machine ‖ Kettenschweißmaschine f. ‖ machine f. à souder les chaînes. / ~ well (Shipb) ‖ Kettenkasten m. ‖ puits m. aux chaînes. / ~ wheel (Mach) ‖ Kettenrad n. ‖ roue f. à chaîne ou à empreintes. / ~ wheel (Cycle) ‖ Kettenrad n. ‖ roue f. de chaîne; hérisson m. / ~ windlass (Shipb) ‖ Kettenspill n. ‖ vireveau m.

chair ‖ Sessel m. ‖ Stuhl m. ‖ fauteuil m.; siège m.; chaise f. / ~ (Railw) ‖ Schienenstuhl m. ‖ chair m.; coussinet m. / ~ (Glassm) ‖ Glasmacherstuhl m. ‖ banc m. du verrier. / ~ (Min) ‖ Schachtfördergefäß n. ‖ tonne f.; tine f. / ~ of bent wood ‖ gebogener Stuhl m. ‖ chaise f. cintrée. / children's ~ ‖ Kinderstuhl m. ‖ siège m. pour enfants. / double ~ (Railw) ‖ Kreuzungsstuhl m.; Doppelstuhl m. ‖ double coussinet m.; coussinet m. du croisement. / fastening the ~s to the sleepers (Railw) ‖ Aufsetzen n. der Stühle auf den Querschwellen ‖ sabotage m. des traverses. / heel ~ of tongue (Railw) ‖ Drehstuhl m. ‖ coussinet m. de talon. / invalid wheel ~ ‖ Krankenfahrstuhl m. ‖ fauteuil m. roulant pour malades. / mechanical ~ ‖ verstellbarer Lehnstuhl m. ‖ fauteuil m. mécanique. / ~ for sick persons ‖ Krankenstuhl m. ‖ chaise f. de malades. / ~ of a wheel ‖ Radbuchse f. ‖ boîte f. de roue.

chair back ‖ Stuhllehne f. ‖ dossier m. / ~ cane for seats ‖ Stuhlflechtrohr n. ‖ jonc m. à sièges. / ~ frame ‖ Stuhlgestell n. ‖ châssis m. de chaise. / ~ leg ‖ Stuhlbein n. ‖ pied m. de (la) chaise. / ~ maker ‖ Stuhlmacher m.; Sitzmöbeltischler m. ‖ chaisier m.; ébéniste m. en sièges.

chairman ‖ Obmann m. ‖ chef m.; président m.

chair mender ‖ Stuhlsitzflechter m. ‖ empailleur m. de chaises.

chairs pl. ‖ Sitzmöbel npl. ‖ chaises fpl.

chair saw ‖ Schweifsäge f.; Stellsäge f. ‖ scie f. à tourner ou tourne-fond ou à chantourner ou à échancrer ou à évider. / ~ seat ‖ Stuhlsitz m. ‖ siège m. de chaise. / ~ spring ‖ Stuhlfeder f. ‖ ressort m. de chaise. / ~ turner ‖ Stuhlstabdrechsler m. ‖ tourneur m. en chaises.

chalcanthite ‖ Kupfervitriol n. ‖ chalcanthite f.

chalcocite ‖ Kupferglanz m. ‖ chalcosite f.

chalcographer ‖ Kupferstecher m. ‖ graveur m. en taille-douce ou au burin; chalcographe m.

chalcography ‖ Kupferstechkunst f. ‖ chalcographie f.

chalcopyrite ‖ Kupferkies m.; Gelferz n.; Chalkopyrit m. ‖ pyrite f. cuivreuse; cuivre m. pyriteux; chalcopyrite f.

chalcotrichite ‖ haarförmiges Rotkupfererz n.; Kupferblüte f.; Chalkotrichit m. ‖ cuivre m. oxydulé capillaire; ziguéline f. capillaire.

chalk, to ~ out ‖ skizzieren; entwerfen ‖ essquisser; ébaucher; crayonner.

chalk ‖ Kreide f. ‖ craie f.; chaux f. carbonatée crayeuse. / ~ (Chalky stone) ‖ Kalkstein m. ‖ pierre f. calcaire; castine f. / ~ (Lime) ‖ Kalk m. ‖ chaux f. / billiard ~ ‖ Billardkreide f. ‖ craie f. de billard. / black ~ ‖ schwarze Kreide f.; Zeichenschiefer m. ‖ craie f. noire; schiste m. graphique; crayon m. noir; ampélite f. graphique. / coloured ~ ‖

Buntstift m.; Farbstift m. ‖ crayon m. de couleur. / drawing ~ ‖ Zeichenkreide f. ‖ craie f. à dessiner. / floated ~ ‖ Schlämmkreide f. ‖ craie f. lavée; blanc m. lavé. / French ~ ‖ Schneiderkreide f. ‖ craie f. pour tailleurs. / layer of white ~ containing flints ‖ weiße Kreideschicht f. mit Feuersteinen ‖ cornus m. / litho ~ ‖ lithografische Kreide f. ‖ crayon m. ou craie f. lithographique. / lithographic(al) ~ see litho ~. / mountain ~ ‖ Bergkreide f. ‖ craie f. de montagne. / precipitated ~ ‖ Schlämmkreide f. ‖ blanc m. de Meudon. / prepared ~ see floated ~. / school ~ ‖ Schulkreide f. ‖ craie f. pour écoles. / silicious ~ ‖ Kieselkreide f. ‖ craie f. silicieuse. / Spanish ~ ‖ Talk m.; spanische Kreide f. ‖ craie f. de Briançon ou d'Espagne; talc m. / tailors' ~ ‖ Schneiderkreide f. ‖ craie f. pour tailleurs. / Vienna ~ ‖ Wiener Putzkalk m. ‖ chaux f. de Vienne. / white ~ ‖ weiße Kreide f.: Schreibkreide f. ‖ craie f. blanche. / writing ~ ‖ Schreibkreide f. ‖ craie f. à écrire.

chalk bench (Galv) ‖ Abkalktisch m. ‖ table f. de déchaulage. / ~ breaker ‖ Kreidebrecher m. ‖ concasseur m. à craie. / ~ digger ‖ Kreidegräber m. ‖ extracteur m. de craie. / ~ dressing ‖ Kreidebereitung f. ‖ apprêt m. de craie. / ~ line (Mas; Carp) ‖ Maurerschnur f.; Schlagleine f.; Zimmerschnur f. ‖ cordeau m.; fouet m.; ligne f. / ~ mill ‖ Kreidemühle f. ‖ moulin m. à craie.

chalkolite ‖ Kupferuranglimmer m. ‖ chalcolite f.

chalk pencil ‖ Kreidestift m. ‖ crayon m. à la craie. / ~ quarry ‖ Kreidebruch m. ‖ carrière f. de craie. / ~ works pl. ‖ Kreidewerk n. ‖ usine f. à craie.

chalky ‖ kalkhaltig ‖ calcifère. / ~ ‖ kalkig calcaire. / ~ sandstone ‖ Kalksandstein m. ‖ grès m. argilo-calcaire. / ~ soil ‖ Wiesenkalk m. ‖ chaux f. de prairie.

chamber ‖ Kammer f. ‖ chambre f. / ~ (Coachm) ‖ Schmierkammer f. ‖ dégagement m. ou évasement m. des boîtes. / ~ of accounts ‖ Rechnungshof m. ‖ cour f. des comptes. / ~ of a blast-furnace ‖ Oberschacht m. eines Hochofens ‖ cuve f. ou vide m. d'un haut-fourneau. / ~ of commerce ‖ Handelskammer f. ‖ chambre f. de commerce. / compressed-air ~ ‖ Druckluftkessel m. ‖ récipient m. à air comprimé.

Chamber of Deputies ‖ Abgeordnetenhaus n. ‖ chambre f. des députés.

chamber of a lock (Hydr arch) ‖ Kammer f. einer Schleuse ‖ chambre f. d'écluse. / ~ of mine ‖ Minenkammer f. ‖ fourneau m. de mine. / mixing ~ ‖ Mischkammer f. ‖ chambre f. de mélange. / sluice ~ ‖ Schleusenkammer f. ‖ chambre f. d'écluse.

chamber carpet ‖ Zimmerteppich m. ‖ tapis m. pour chambres. / ~ chair for prisons ‖ Leibstuhl m. für Gefängnisse ‖ chaise f. percée pour prisons. / ~ depth micrometer ‖ Kammertiefenmesser m. ‖ appareil m. pour mesurer la profondeur de la chambre. / ~ fan ‖ Zimmerventilator m. ‖ ventilateur m. à chambre.

chamber filter press ‖ Kammerfilterpresse f. ‖ filtre-presse m. à chambres. / ~ with double perfect upper lixiviation and exclusion of air ‖ Kammerfilterpresse f. mit oberer doppeltperfekter Auslaugung

unter Luftabschluß ‖ filtre-presse m. à chambres de lavage double parfait supérieur et à l'abri d'air. / ~ without lixiviation with aeration ‖ Kammerfilterpresse f. mit einfacher Auslaugung ‖ filtre-presse m. à chambres sans lavage avec aération.

chamber furnace ‖ Kammerofen m. ‖ fourneau m. à chambre. / discharging machine for ~s ‖ Entlademaschine f. für Kammeröfen ‖ machine f. de déchargement pour fours à chambre.

chamfer, to (To flute) ‖ einkehlen ‖ canneler. / ~ (To bevel) ‖ abkanten; abfasen; abschrägen ‖ délarder; écorner; émousser; tailler en chanfrein; dresser en biais; chanfreiner. / ~ (Carp) ‖ verjüngen (ein Holzstück) ‖ délarder en biseau. / ~ the edge ‖ die Kante brechen ‖ faire le chanfrein. / ~ the letters pl. ‖ die Lettern fpl. am Kopf schräg hobeln ‖ dégager les caractères mpl.

chamfer ‖ Schrägkante f.; abgeschrägte Kante f.; Abschrägung f.; Zuschärfung f. ‖ chanfrein m.; biseau m. / ~ (Arch) ‖ Fase f. ‖ facette f. / hollow ~ ‖ Hohlkehle f. ‖ chanfrein m. creux. / ~ for picture frames ‖ Bilderleiste f. ‖ baguette f. d'encadrement.

chamfered (Fluted) ‖ geriffelt ‖ cannelé. / ~ (Bevelled) ‖ abgeschrägt ‖ biseauté. / ~ edge ‖ abgeschrägte Kante f. ‖ chanfrein m.; biseau m.

chamfering ‖ Abschrägung f. ‖ biaisement m.; biseautage m.; chanfrein m.

chamfering machine, gear tooth ‖ Abrundmaschine f. für Zahnräder ‖ machine f. à arrondir les dentures d'engrenage. / ~ for pasteboard ‖ Pappenabschärfmaschine f. ‖ machine f. à affiler le carton.

chamois, to ‖ sämisch gerben ‖ chamoiser; passer en chamois.

chamois ‖ Sämischgarleder n.; Sämischleder n.; sämisches Leder n. ‖ cuir m. chamoisé; peau f. à la chamois *ou* chamoisée.

chamois-dressed leather ‖ sämischgares Leder n. ‖ peau f. chamoisée.

chamois dresser ‖ Sämischgerber m.; Weißgerber m. ‖ chamoiseur m. / ~ dressing ‖ Sämischgerben n.; Sämischgerberei f.; Weißgerberei f. ‖ chamoisage m.

chamoiser ‖ Sämischgerber m.; Weißgerber m. ‖ chamoiseur m.

chamoising *see* chamois dressing.

chamois leather ‖ Sämischleder n. ‖ peau f. chamoisée; chamois m.; cuir m. chamoisé. / ~ (Shammy) ‖ Gemsleder n.; Gemshaut f. ‖ peau f. de chamois; chamois m. / ~ (Wash leather) ‖ Waschleder n. ‖ peau f. chamoisée; cuir m. qui se lave.

chamot ‖ Schamotte f. ‖ chamotte f. / ~ breaker ‖ Schamottebrecher m. ‖ concasseur m. de chamotte. / ~ ware ‖ Schamottewaren fpl. ‖ articles mpl. en terre réfractaire. / ~ work plant ‖ Schamottefabrikanlage f. ‖ installation f. de fabriques de terre réfractaire.

champagne ‖ Schaumwein m. ‖ champagne m.; vin m. mousseux. / still ~ ‖ nicht moussierender Champagner m. ‖ champagne m. non mousseux.

champagne cellarage machine ‖ Sektkellereimaschine f. ‖ machine f. pour caves de champagnes. / ~ factory ‖ Schaumweinfabrik f. ‖ fabrique f. de vins mousseux. / disgorging plant in a ~ factory ‖ De-

gorgieranlage f. in einer Schaumwein kellerei ‖ installation f. de dégorgement dans une fabrique de vins mousseux. / ~ maker ‖ Champagnerküfer m. ‖ ouvrier m. en vins de Champagne. / ~ nippers pl. ‖ Champagnerzange f. ‖ pince f. à champagne. / ~ wine ‖ Champagnerwein m. ‖ vin m. de Champagne.

champaign ‖ Ebene f.; flaches Land n. ‖ campagne m.

champignon bed ‖ Champignonbeet n. ‖ champignonnière f.

chance, by ‖ durch Zufall m. ‖ par hasard m. / ~ of a court of judicature ‖ Schranke f. im Gerichtssaal ‖ parquet m. de justice.

Chancellor of Exchequer ‖ Finanzminister m. ‖ Ministre m. des Finances.

chandelier ‖ Kronleuchter m.; Kandelaber m.; Armleuchter m.; Lüster m. ‖ lustre m.; candélabre m.; girandole f.

change, to ‖ wechseln; umwechseln; auswechseln ‖ changer; échanger. / ~ colour ‖ verschießen ‖ se déteindre. / ~ the direction ‖ die Richtung verändern ‖ changer de direction. / ~ gears pl. (Auto) ‖ schalten ‖ changer de vitesse. / ~ one or the other part ‖ einzelne Teile npl. auswechseln ‖ remplacer certains organes mpl.

change ‖ Wechsel m.; Änderung f. ‖ changement m. / ~ (Coin) ‖ Scheidemünze f. ‖ billon m.; monnaie f. de billon; petite monnaie; menues espèces fpl. / ~ (Chem) ‖ Umformung f.; Verwandlung f. ‖ transformation f. / ~ of blades ‖ Messerwechsel m. ‖ changement m. des lames. / to bring about a ~ ‖ Wandel m. schaffen ‖ amener une reforme. / ~ of colours (Miner) ‖ Farbenwandlung f.; Labradorisieren n. ‖ chatoiement m. / ~ of connections for receiving (Radio) ‖ Umschaltung f. auf Empfang ‖ commutation f. pour la réception. / ~ of connections for transmitting (Radio) ‖ Umschaltung f. auf Senden ‖ commutation f. pour transmission. / converse ~ (Chem) ‖ umgekehrte Reaktion f. ‖ réaction f. inverse. / ~ of direction ‖ Richtungswechsel m. ‖ changement m. de direction *ou* de marche. / easy ~ ‖ schnelle Umstellung f. ‖ changement m. rapide. / ~ of feed by sliding gear box ‖ Vorschubänderung f. durch Schalträderkasten ‖ changement m. des avances par boîte d'engrenages. / ~ of gauge (Railw) ‖ Spurveränderung f. ‖ changement m. de l'écartement. / ~ of impulse ‖ Impulsveränderung f. ‖ variation f. de la quantité de mouvement. / ~ of length ‖ Längenänderung f. ‖ changement m. de longueur. / ~ of load ‖ Belastungsschwankung f. ‖ fluctuation f. *ou* variation f. de charge. / ~ of prices (Comm) ‖ Preisschwankung f. ‖ changement m. de prix. / quick ~ ‖ schnelle Umstellung f. ‖ changement m. rapide. / reverse ~ *see* converse ~. / ~ of register ‖ Kassenleergang m. ‖ change m. de caisse. / ~ of speed ‖ Gangwechsel m. ‖ change m. de marche. / ~ of stroke ‖ Hubwechsel m. ‖ changement m. de course. / ~ of tendency (Comm) ‖ Stimmungswechsel m. ‖ revirement m. de l'atmosphère. / ~ of the tide ‖ Gezeitenwechsel m.; Rückkehr f. der Ebbe und Flut; Widerzeit f. ‖ changement m. *ou* retour m. de la marée. / ~ of tone *see* ~ of tendency. / ~ of trim (Mar) ‖ Trimmänderung f. ‖ changement m. d'assiette. / sudden ~ of

weather ‖ plötzlicher Witterungsumschlag m. ‖ changement m. subit du temps *ou* vite du temps. / ~ of work people ‖ Belegschaftswechsel m. ‖ changement m. du personnel.

changeable ‖ auswechselbar ‖ interchangeable. / ~ (Textile) ‖ schillernd; changierend ‖ changeant; glacé.

change-box motion (Weav) ‖ Schützenwechsel m. ‖ appareil m. change-navette.

change gear ‖ Wechselrad n. ‖ roue f. de rechange. / ~s pl. to suit the number of teeth on the workpiece ‖ der Zähnezahl des Werkstückes entsprechende Wechselräder npl. ‖ roues fpl. de rechange combinées suivant le nombre de dents de la pièce.

change-over condenser ‖ Wechselkondensor m. ‖ condensateur m. interchangeable *ou* alternatif. / ~ lever ‖ Umschalthebel m. ‖ levier m. de changement de marche. / ~ relay ‖ Umschalterelais n. ‖ relais m. commutateur.

changer bar ‖ Wechseltaste f. ‖ barre f. de change.

change speed actuating shaft ‖ Umsteuerungswelle f. ‖ arbre m. de commande du changement de vitesse. / ~ gear ‖ Wechselgetriebe n. ‖ changement m. de vitesse. / ~ gear box ‖ Getriebekasten m. ‖ boîte f. des vitesses. / ~ lever ‖ Umsteuerungshebel m. ‖ levier m. de changement de vitesse.

change spur gear ‖ Wechselgetriebe n. ‖ engrenage m. de changement de vitesse. / ~ tune switch ‖ Wellenumschalter m. zum Abstimmen ‖ commutateur m. de longueurs d'onde. / ~ wheel ‖ Wechselrad n. ‖ roue f. de rechange.

changing, gear ‖ Schalten n. der Gänge ‖ changement m. de vitesses. / ~ of rails ‖ Einziehen n. von neuen Schienen ‖ changement m. de rails. / the ~ of the tools is effected in x minutes ‖ das Auswechseln n. der Werkzeuge erfolgt in x Minuten ‖ le changement des outils s'opère en x minutes.

changing attachment (Telescope) ‖ Wechselvorrichtung f. ‖ adaptateur m. / ~ collar (Telescope) ‖ Wechselring m. ‖ bague f. adaptatrice. / ~ device (Telescope) ‖ Wechselvorrichtung f. ‖ adaptateur m. / ~ device (Mech) ‖ Wechselvorrichtung f. ‖ dispositif-changeur m. / ~ over ‖ Umschaltung f. ‖ commutation f.

channel, to ‖ kannelieren; auskehlen; einkehlen; kehlen; riefeln ‖ canneler; rainer.

channel ‖ Kanal m. ‖ canal m. / ~ (Arch) ‖ Hohlkehle f.; Flächenrinne f. ‖ chenal m.; rainure f. / ~ (Build) ‖ Kehlleiste f. ‖ talon m. / ~ (Techn) ‖ Riefe f. ‖ rainure f.; cannelure f. / ~ (Iron) ‖ U-Eisen n. ‖ fer m. à U. / ~ (Join) ‖ Falz m. ‖ coulisse f. / ~ (Hydr Arch) ‖ Rinnsal n. ‖ chenal m.; cours m. d'eau; lit m. / Gerinne n.; oben offene Wasserleitung ‖ rigole f.; auge f. / ~ (Harbour) ‖ Fahrwasser n.; Fahrrinne f.; Hafeneinfahrt f. ‖ passe f.; chenal m.; passage m. / ~ (River) ‖ Flußbett n. ‖ lit m. d'un fleuve *ou* d'une rivière. / ~ (Mill) ‖ Mühlengerinne n. ‖ auge f.; bief m.; biez m.; coursier m.; coursière f. / ~ (Shipb) *see also* channels ‖ Rüst f. ‖ porte-haubans m. / ~ (Geol) ‖ Gesteinsgang m. ‖ crain m.; faille f.; filon m. de roche *ou* stérile. / ~ (Min) ‖ Schram m.; Rinne f.; Ritz m. ‖ entaille f.

/ ~ (Min: air escape) || Luftröhre f.; Wetterlutte f. || buse f.; buse f. d'airage. / ~ (Needl) || Kerbe f. (einer Nähnadel) || cannelure f. (d'une aiguille à coudre). / ~ formed by breakers || Brandungskehle f. || creux m. du ressac. / broad ~ (Arch) || Halskehle f.; stehende Hohlkehle f.; Einziehung f. || gorge f. / circular ~ || gekrümmtes Gerinne n.; Kropfgerinne n. || coursier m. ou coursière f. circulaire. / conveying ~ (Mill; Hydr arch) || Obergerinne n.; Vorarche f. || bief m.; bief m. d'amont. / fore ~ see conveying ~. / narrow ~ (Nav) || Priel n.; Rille f.; Balje f. || petit chenal m.; passe f. / navigable ~ of a river || Fahrrinne f. oder schiffbare Rinne f. eines Flusses || chenal m. d'une rivière. / outlet ~ (Factory) || Abführungskanal m. || canal m. de fuite. / ~ of a pulley || Rille f. einer Rolle || gorge f. d'une poulie. / ring-~ (Hydr arch) || Abfangekanal m. || canal m. de dérivation. / small ~ see narrow ~. / straight ~ || gerades Gerinne n.; Schußgerinne n. || auge f. droite; coursier m. rectiligne; coursière f. droite.

channel brick || Ofenziegel m.; Kanalziegel m. || chantignolle f.; brique f. mince; demi-brique f. / ~ brush || Kanalbürste f. || brosse f. à canaux.

channeled plate || Riffelblech n. || tôle f. striée ou gaufrée. / ~ sleeper (Railw) || Rillenschwelle f. || traverse f. à gorge.

channeling (Arch) || Kannelierung f.; Schaftrinnen fpl. || cannelures fpl.

channel iron || U-Eisen n. || fer m. à côtés ou à U. / ~ iron of a gutter || Haken m. einer Dachrinne || ferrement m. d'une gouttière. / ~ narrow (Geogr) || Meerenge f.; Straße f. || détroit m. / ~ rail || Rillenschiene f. || rail m. à gorge. / ~ rib || Fachwerkrippe f. || nervure f. en profilé.

channels pl. of a column see channeling. / fore ~ (Shipb) || Fockrüst f.; Vorrüst f. || petit porte-haubans m. / main ~ || Großrüst f. || grand porte-haubans m. / mizen ~ || Besamrüst f.; Kreuzrüst f. || portehaubans m. d'artimon.

channel support (Shipb) || Rüstenstütze f. || équerre f. de porte-haubans.

chantlate (Carp) || Saumlade f.; Saumlatte f.; Staublade f.; Staublatte f. || chanlatte f.

chantry altar (Build) || Meßaltar m. || chantrerie f.

chap, to || aufreißen; reißen; Risse bekommen; sich spalten || se crévasser; se fendre.

chap (Pincers) || Maul n. || mors m.; bouche f.

chapel || Kapelle f. || chapelle f.

chapiter || Kapital n.; Kapitell n.; Säulenknauf m.; Säulenkopf m. || chapiteau m.

chapiterel see chapiter.

chaplet (Found) || Kernbock m. || support m. de noyau. / ~ (Arch) || Paternoster n.; Perlenschnur f.; Perlenstab m.; Rosenkranz m.; beperlter Rundstab m. || chapelet m.; collier m.; fusarolle f.; patenôtre f.; perles fpl.; rosaire m. / bead for ~s || Kugel f. für Rosenkränze || perle f. à chapelets. / ~ hinge (Locksm) || Paternosterband n. || fiche f. à chapelet. / ~ maker || Rosenkranzmacher m. || chapeletier m.

chapped || gerissen; gespalten; rissig || lézardé; fendu. / ~ (Curr) || narbenbrüchig || cicatricé; crevassé.

chaptrell see chapiter.

char, to || verkohlen || charbonner.

char || tierische Kohle f. || noir animal m.

character (Print) || Type f.; Buchstabe m; || lettre f.; caractère m. d'imprimerie-type m. / arrow-headed ~s pl. || Keilschrift f. || caractères mpl. cunéiformes. / cuneiform ~s pl. see arrow-headed ~s. / ~ of the wind || Windcharakter m. || caractère m. du vent.

character doll || Charakterpuppe f. || poupée f. caractère.

characteristic (Elektr) || Merkmal n. || marque f.; caractère m. / ~ Charakteristik f. || caractéristique f. / ~ (Math) || Kennziffer f. || caractéristique f.

characteristic curve || Charakteristik f. || caractéristique f. / ~ curve of iron filament ballast lamps || Kennlinie f. von Eisenwiderständen || courbe f. caractéristique des ballasts. / ~ feature || Unterscheidungsmerkmal n. || marque f. distinctive. / ~ impedance || Wellenwiderstand m. || impédance f. caractéristique. / ~ line || Charakteristik f.; Kennlinie f. || courbe f. ou ligne f. caractéristique.

characterization || Kenntlichmachung f. || caractérisation f.

characterize, to || kennzeichnen || caractériser.

charcoal || Holzkohle f. || charbon m. de bois; fusain m. / ~ (Vegetable) || Pflanzenkohle f.; vegetabilische Kohle f. || charbon m. végétal. / animal ~ || tierische Kohle f.; Knochenkohle f. || noir m. animal; charbon m. animal. / black ~ || Schwarzkohle f. || charbon m. de bois noir ou ordinaire. / blood ~ || Blutkohle f. || charbon m. de sang. / bone ~ || Knochenkohle f. || charbon m. d'os. / to burn ~ || verkohlen; kohlen || carboniser le bois. / cylinder ~ || Retortenkohle f. || charbon m. de cornue. / ground ~ || Kohlenstaub m. || poussière f. ou poudre f. de charbon; charbon m. pilé. / hydrogenous ~ || Rotholz n.; Rotkohle f. || charbon m. roux. / meat ~ || Fleischkohle f. || charbon m. de viande. / medical ~ || Kohle f. für medizinische Zwecke || charbon m. médicinal. / ~ from near the chimney of the pile || Quandelkohle f. || charbon m. du côté de la cheminée d'une meule; cœur m. / powdered ~ || Holzkohlenpulver n.; Holzkohlenstaub m. || poudre f. ou poussière f. de charbon. / prepared ~ || Glühstoff m. || charbon m. composé. / quenched ~ || Löschkohle f.; abgedämpfte Kohle f. || charbon m. de braise. / red ~ see hydrogenous ~. / ~ of soft wood used for grinding || Schleifkohle f. zum Polieren von Metall || charbon m. pour adoucir. / sulphuric ~ || Krappkohle f. || charbon m. sulfurique de garance. / vegetable ~ || vegetabilische Kohle f.; Pflanzenkohle f. || charbon m. végétal. / wood ~ (For burning) || Holzkohle f. || charbon m. de bois.

charcoal bed || Löschboden m. || fond m. brasqué; sole f. en charbon. / ~ blast iron || Holzkohlenroheisen n. || fonte f. au charbon de bois. / ~ bottom see ~ bed. / ~ briquet || Holzkohlenbrikett n. || briquette f. de charbon de bois. / ~

briquette see ~ briquet. / ~ burner || Köhler m. || charbonnier m.

charcoal burning (Procedure) || Köhlerei f. || carbonisation f. du bois. / ~ (Place) || Köhlerei f.; Meiler m. || charbonnière f. / ~ in pits || Grubenverkohlung f. || carbonisation f. dans des fosses ou en fosses.

charcoal drawing || Kohlezeichnung f. || fusain m.; charbonnée f.

charcoal dust || Holzkohlenstaub m. || poussier m. de charbon de bois; charbon m. de bois en poudre; charbon m. de bois pilé. / ~ (mixed with ashes and earth) || Holzkohlenlösche f. || fraisil m.; frasil m.; frasin m. / ~ mixed with clay || Holzkohlengestübbe n. || fraisil m. à terre glace. / ~ mixture of ~ and clay (Met) || Kohlengestübbe n. für Öfen und Tiegel || brasque f. / mixture of ~ and soil || Kohlengestübbe n. bei der Köhlerei || fraisil m.

charcoal filter || Holzkohlenfilter n. || filtre m. à charbon de bois. / ~ fire || Holzkohlenfeuer n. || feu m. de braise ou de charbon de bois. / ~ hearth || Löschfeuer n. || feu m. brasqué. / ~ hearth steel || Holzkohlenfrischstahl m. || acier m. affiné au charbon de bois.

charcoal kiln || Köhlerei f. || charbonnière f. / ~ (Furnace) || Verkohlungsofen m. || four m. de carbonisation.

charcoal lining || Futter n. eines Kohlentiegels || brasque f. d'un creuset. / ~ lump || Holzkohle f. in Stücken || charbon m. de bois en morceaux. / ~ mound burner || Meilerbrenner m. || cuiseur m. de meules de bois. / ~ oven for turf || Torfkohlenofen m. || four m. pour la carbonisation de la tourbe. / ~ pencil || Zeichenkohle f. || charbon m. à dessin. / ~ pig || Holzkohlenroheisen n. || fonte f. crue au charbon de bois.

charcoal pile || Holzkohlenmeiler m. || meule f. de charbon de bois; charbonnière f. / large ~ || großer Meiler m.; Haufen m. || tas m.; grande meule f. / small ~ || kleiner Meiler m.; Kohlendocke f. || petite meule f. / the ~ sweats || der Meiler schwitzt || la meule exsude. / ~ dresser || Meilerbauer m. || dresseur m. de meules.

charcoal powder see ~ dust. / ~ stack see small ~ pile. / ~ steel || Holzkohlenstahl m. || acier m. au charbon de bois.

chare, to ~ an ashlar || einen Stein scharrieren || charruer une pierre.

charge, to ~ (Judical) || anklagen || accuser; dénoncer. / ~ (Trade) || in Anrechnung f. bringen || mettre en compte m. / ~ (puddling furnace) || einsetzen || charger. / ~ (Met) || beschicken; aufgeben; aufgichten; chargieren || charger; délivrer; alimenter. / ~ to the account || anrechnen || mettre ou passer en compte m.; compter. / ~ an accumulator || einen Akkumulator m. laden oder aufladen || charger un accumulateur. / ~ a blasthole || mit Sprengstoff m. besetzen || charger un trou de mine. / ~ extra || nachfordern || demander en sus. / ~ a furnace (Met) || einen Ofen besetzen oder laden oder beschicken || charger un fourneau. / ~ with (Chem) || beladen mit || charger de.

charge (Judical) || Anklage f. || accusation f.; prévention f.; incrimination f. / ~ (Cash register) || Kredit m. || crédit m. / ~ (Electr) || Ladung f. || charge f. / ~ (Met) || Einsatz m.; Charge f.; Einsatz-

material n.; Beschickung f. ‖ charge f.; fournée f.; matières fpl. chargées ou à fondre. / atomic ~ ‖ Elementarquantum n. ‖ charge f. atomique. / bound ~ (Electr) ‖ gebundene Ladung f. ‖ charge f. latente. / bound residual ~ (Electr) ‖ gebundener Ladungsrückstand m. ‖ charge f. résiduelle latente. / bursting ~ ‖ Sprengladung f. ‖ charge f. explosive. / ~ of charcoal (Met) ‖ Kohlengicht f. ‖ charge f. de charbon. / counter ~ ‖ Gegenklage f. ‖ reconvention f. / ~ for delivery ‖ Bestellgebühr f.; Bestellgeld n. ‖ factage m. / duration of ~ (Acc) ‖ Ladedauer f. ‖ durée f. de chargement. / exemption from ~ ‖ Gebührenfreiheit f. ‖ franchise f. de taxe. / extra ~ ‖ Nachforderung f. ‖ demande f. supplémentaire ou en sus. / extra ~ ‖ Mehrpreis m. ‖ supplément m. de prix. / extra ~ ‖ Aufgeld n. ‖ change m.; agio m. / final voltage on ~ (Acc) ‖ Endspannung f. der Ladung ‖ tension f. à la fin de la charge. / fixed ~ (Tel) ‖ Grundgebühr f. ‖ redevance f. fondamentale / free of ~ ‖ spesenfrei ‖ sans frais m. / the flames pl. melted the ~ ‖ die Flammen fpl. brachten den Einsatz zum Schmelzen ‖ les flammes fpl. provoquent la fusion du contenu du four. / gassing after the ~ (Acc) ‖ Nachkochen n. ‖ bouillonnement m. au repos. / initial voltage on ~ (Acc) ‖ Anfangsspannung f. der Ladung ‖ tension f. initiale de la charge. / loss of ~ (Acc) ‖ Ladeverlust m. ‖ perte f. de charge. / to make up the ~ (Met) ‖ gattieren ‖ préparer le lit de fusion. / method of making ~s ‖ Tarifpolitik f. ‖ régime f. des tarifs. / ~ making the metal bath not too tough ‖ Beschickung f., die das Metall nicht zu strengflüssig macht ‖ charge f. qui ne rende pas le métal difficilement fusible. / ~ of mine ‖ Minenladung f. ‖ charge f. de mine. / ~ of pig (Met) ‖ Heiße f.; Heize f. ‖ charge f. ou mise f. de fonte. / ~ of powder (for blasting) ‖ Pulverladung f.; Ladung f. ‖ charge f. du pétard. / ~ upon a realty ‖ dingliche Belastung f. ‖ charge f. réelle. / reloading ~s pl. ‖ Umladegebühr f. ‖ frais mpl. de transbordement. / residual ~ (Chem) ‖ Residuum n. ‖ charge f. résiduaire. / ~ for unloading ‖ Ausladegebühr f. ‖ frais mpl. de déchargement.

chargeable time (Tel) ‖ Gebührenminute f. ‖ minute f. comptable.

charge bogie ‖ Beschickungswagen m. ‖ wagon m. de chargement. / ~ bridge ‖ Gichtbrücke f. ‖ pont m. de chargement. ~ capacity (Acc) ‖ Kapazität f. ‖ capacité f. de charge. / ~ coke ‖ Schmelzkoks m. ‖ coke m. d'usine.

charged, to become positively ~ ‖ positiv elektrisch werden ‖ s'électriser positivement. / ~ paper ‖ beschwertes Papier n. ‖ papier m. chargé.

charge-hand ‖ Vorarbeiter m. ‖ chef m. d'équipe.

charger (Met) ‖ Gichtmann m. ‖ chargeur m.; arqueur m. / ~ (Met; Machine) ‖ Beschickmaschine f. ‖ enfourneuse f.

charges pl. ‖ Spesen pl.; Unkosten pl.; Kosten pl. ‖ frais mpl. / ~ for acquisition of land ‖ Grunderwerbskosten f. ‖ frais m. d'acquisition du terrain. / additional ~ ‖ Nebenkosten pl.; Nebenausgaben fpl.; außerordentliche Kosten pl. ‖ frais mpl. accessoires; faux-frais

mpl. / extraordinary ~ see additional ~. / incidental ~ see additional ~. / legal ~ ‖ Gerichtskosten pl. ‖ frais mpl. judiciaires. / petty ~ see additional ~. / ~ for protesting ‖ Protestkosten pl. ‖ frais mpl. de protêt; coût m. de l'acte. / ~ for salvage (Nav) ‖ Bergegeld n. ‖ droit m. de sauvetage. / ~ for storage ‖ Lagergeld n. ‖ frais mpl. de dépôt; magasinage m.

charges forward ‖ Spesennachnahme f. ‖ remboursement m. des frais.

charge wagon (Met) ‖ Beschickungswagen m. ‖ wagon m. ou benne f. de chargement.

charging (Acc) ‖ Ladung f. ‖ chargement m. / ~ (Met) ‖ Begichtung f.; Gichtung f. ‖ chargement m. / ~ of a battery ‖ Laden n. einer Batterie. ‖ charge f. d'une batterie. / device for ~ blast-furnaces ‖ Vorrichtung f. zum Beschicken von Hochöfen ‖ appareil m. de chargement pour hauts fourneaux. / ~ the oven (Pott) ‖ Einsetzen n. ‖ enfournage m.; enfournement m. / periodic supplementary ~ (Acc) ‖ Ladung f. mit Ruhepausen ‖ chargement m. suivi de repos. / ~ of storage cells ‖ Ladung f. von Sammlern ‖ charge f. des accumulateurs.

charging apparatus ‖ Beschick(ungs)vorrichtung f.; Begichtungsapparat m.; Chargiervorrichtung f. ‖ appareil m. de chargement ou d'alimentation; appareil-chargeur m. / ~ (Met) ‖ Gichtaufzug m. ‖ montecharges m. / ~ for the dressing industry ‖ Beschickungsapparat m. für die Aufbereitungsindustrie ‖ appareil m. de chargement pour l'industrie de préparation des matières premières.

charging barrow (Met) ‖ Begichtungswagen m. ‖ wagonnet m. de chargement. / ~ box ‖ Ladenmulde f. ‖ recipient m. de chargement. / ~ carriage (Met) ‖ Beschickungswagen m. ‖ chariot m. enfourneur. / ~ chute ‖ Verladerutsche f. ‖ glissière f. de chargement; couloir m. / ~ cone ‖ Aufgabetrichter m. ‖ trémie f. de chargement. / ~ connections pl. ‖ Ladeschaltung f. ‖ raccordement m. pour la charge.

charging crane ‖ Beschickungskran m.; Chargierkran m. ‖ grue f. de chargement; pont-roulant m. chargeur. / ~ (Met) ‖ Einsetzkran m. ‖ grue f. enfourneuse. / through ~ ‖ Muldenbeschickkran m. ‖ grue f. d'enfournement.

charging current (Acc) ‖ Ladestrom m. ‖ courant m. de charge(ment).

charging device ‖ Beschickungsapparat m.; Aufgabevorrichtung f. ‖ appareil m. ou dispositif m. de chargement. / ~ (Roll) ‖ Einsetzvorrichtung f. ‖ chargeur m. mécanique. / automatic ~ ‖ selbsttätige Aufgabevorrichtung f. ‖ mécanisme m. de chargement automatique. / ~ of blast-furnace ‖ Beschickungsvorrichtung f. des Hochofens ‖ appareil m. de chargement du haut-fourneau. / electrically moved ~ ‖ elektrisch betriebene Begichtungsvorrichtung f. ‖ appareil m. électrique de chargement des gueulards. / ~ for elevators ‖ Aufgabevorrichtung f. für Becherwerke ‖ dispositif m. de chargement pour élevateurs à godets. / ~ for railway carriages ‖ Beladevorrichtung f. für Eisenbahnwagen ‖ dispositif m. de chargement pour wagons de chemin de fer.

charging diagram ‖ Fülldiagramm n. ‖ diagramme m. de remplissage. / ~ door

Beschickungstür f. ‖ clapet m. d'alimentation. / ~ duration ‖ Ladedauer f. ‖ durée f. de charge. / ~ elevator ‖ Aufgabebecherwerk n. ‖ noria f. de chargement. / ~ floor (Met) ‖ Begichtungsbühne f.; Gichtbühne f. ‖ plate-forme f. / ~ gallery see ~ floor. / ~ gauge (Met) ‖ Gichtmaß n.; Gichtmesser m. ‖ sonde f.; bécasse f. / ~ generator (Acc) ‖ Lademaschine f. ‖ dynamo f. de charge. / ~ hoist for blast-furnaces ‖ Hochofenbegichtungsanlage f. ‖ dispositif m. de chargement de gueulard pour hauts-fourneaux. / ~ hoist with stationary winding house ‖ Begichtungsanlage f. mit festem Windenhaus ‖ dispositif m. de chargement de gueulard avec cabine de treuil fixe. / ~ hole ‖ Einschüttöffnung f. ‖ ouverture f. d'introduction. / ~ hopper ‖ Einfülltrichter m. ‖ trémie f. de chargement; remplisseur m. / ~ installation (Met) ‖ Begichtungsanlage f. ‖ installation f. de chargement de gueulard.

charging machine ‖ Einsetzmaschine f. ‖ pelle f. d'enfournement. / ~ (Acc) see charging generator. / ~ (Met) ‖ Beschickungsmaschine f.; Chargiermaschine f. ‖ enfourneuse f.; machine f. à charger ou à alimenter. / ~ for retorts ‖ Lademaschine f. für Retorten ‖ machine f. de chargement pour cornues.

charging man (Met) ‖ Gichtmann m. ‖ chargeur m.; arqueur m. / ~ outfit (Acc) ‖ Ladeeinrichtung f. ‖ installation f. de charge.

charging plant ‖ Beschickungsanlage f. ‖ installation f. de charge. ~ for accumulators ‖ Ladeeinrichtung f. für Sammler ‖ installation f. de charge pour accumulateurs. / ~ for cupola-furnaces ‖ Begichtungsanlage f. für Kuppelöfen ‖ installation f. de charge pour cubilots.

charging platform ‖ Beschickungsbühne f. ‖ plateforme f. de chargement. / ~ plug (Electr) ‖ Ladestöpsel m. ‖ bouchon m. de charge. / ~ position (Acc) ‖ Ladestellung f. ‖ position f. de charge. / ~ rectifier (Acc) ‖ Ladegleichrichter m. ‖ redresseur m. de charge. ~ resistance ‖ Ladewiderstand m. ‖ résistance f. de chargement. / ~ set (Acc) ‖ Ladeaggregat n.; Ladesatz m. ‖ agrégat m. ou groupe de charge. / clear ~ space inside cross bearers (Railw) ‖ freie Ladefläche f. zwischen den Drehgestellen ‖ espace m. de chargement libre — sans traverses — entre les bogies. / ~ station (Acc) ‖ Ladestation f.; Füllstation f. station f. de charge(ment). ~ stove ‖ Füllofen m. ‖ calorifère m. à chauffage constant. / ~ switch (Acc) ‖ Ladeschalter m. ‖ interrupteur m. ou commutateur m. de charge. / ~ switchboard ‖ Ladeschalttafel f. ‖ tableau m. de charge. / ~ tower ‖ Beschickungsturm m.; Gichtturm m. ‖ tour f. de chargement. / ~ trough ‖ Beschickungsmulde f. ‖ auge f. d'enfournement. / ~ valve ‖ Ladeventil n. ‖ soupape f. de charge. / ~ voltage ‖ Ladespannung f. ‖ voltage m. ou tension f. de charge.

charging wagon ‖ Möllerwagen m. ‖ benne f. de chargement. / coke-oven ~ ‖ Koksofenfüllwagen m. ‖ wagon m. de remplissage pour fours à coke.

charing chisel (Stone cutter) ‖ Scharriereisen n. ‖ ciseau m. à la charrue.

chariot (Tel) ‖ Läufer m.; Schlitten m. ‖ chariot m. / erecting ~ ‖ Montagewagen m. ‖ chariot m. de montage.

charlock destroying powder ‖ Hederichvernichtungspulver n. ‖ poudre f. à détruire herbe au chantre.

charnel house ‖ Totenkapelle f. ‖ charnier m.

charring in pits (Met) ‖ Grubenverkohlung f. ‖ carbonisation f. dans des fosses ou en fosses.

charring of wood ‖ Holzverkohlung f. ‖ carbonisation f. de bois. / ~ in heaps ‖ Meilerverkohlung f.; Meilerverfahren n. ‖ procédé m. des meules; carbonisation f. en meules. / ~ in large piles ‖ Haufenverkohlung f. ‖ carbonisation f. en tas ou aux tas; carbonisation en grandes meules.

charring pit ‖ Grube f. für Holzverkohlung ‖ fosse f. de carbonisation. / ~ place ‖ Meilerstelle f.; Meilerstätte f.; Köhlerei f. ‖ plancher m.; charbonnière f. / ~ wood - plant ‖ Holzverkohlungsanlage f. ‖ installation f. à carboniser le bois.

chart (Nav) ‖ Seekarte f.; Paßkarte f.; hydrografische Karte f. ‖ carte f. marine ou nautique ou hydrographique. / astronomical ~ ‖ astronomische Karte f.; Himmelskarte f. ‖ carte f. céleste ou astronomique. / ~ with the atomic weight of the elements ‖ Atomgewichtstabelle f. ‖ tableau m. des poids atomiques des éléments. / celestial ~ see astronomical ~. / Mercator's ~ ‖ Mercators Karte f. ‖ carte f. réduite. / plane ~ ‖ gleichgradige Karte f. ‖ carte f. plate ou plane. / sea ~ see chart. / track ~ (Nav) ‖ Segelkarte f. ‖ routier m.

chart-bearing drum ‖ Indikatortrommel f.; Diagrammtrommel f. ‖ tambour m. du diagramme.

charter, to ~ a vessel ‖ ein Schiff n. befrachten ‖ affréter un vaisseau.

charter by weight (Mar) ‖ Verfrachtung f. nach Gewicht ‖ affrètement m. au quintal.

charter master (Min) ‖ Grubenpächter m. ‖ forfaitier m. / ~ party ‖ Frachtvertrag m. ‖ contrat m. d'affrètement ou de nolissement.

chartographer ‖ Kartograf m. ‖ cartographe m.

chase, to ‖ punzen; treiben; ziselieren ‖ repousser (à l'aide de poinçons); ciseler. / ~ a screw thread ‖ ein Gewinde n. nachschneiden ‖ aviver ou repasser un filet.

chase ‖ Hetzjagd f. ‖ chasse f. à la courre. / ~ (Print) ‖ Rahmen m.; Formrahmen m. ‖ châssis m. / ~ for board sides (Print) ‖ Keilrahmen m. ‖ châssis m. à coin. / ~ without loose cross (Print) ‖ Rahmen m. ohne Mittelsteg ‖ ramette f.; châssis m. sans barre. / small ~ (Print) ‖ kleiner Rahmen m. ‖ ramette f.; petit châssis m.

chase-bar (Print) ‖ Rahmeisen n. ‖ règle f. de fer.

chased ‖ gehämmert ‖ martelé. / ~ work ‖ Ziselörarbeit f. ‖ ouvrage m. au maillet; travail m. repoussé.

chaser (Tool) ‖ Gewindestahl m.; Gewindesträhler m.; Strähler m. ‖ peigne m. / ~ (Found) ‖ Ziselör m. ‖ ciseleur m.; ciseleur réparateur m. / gold and silver ~ (Jewel) ‖ Gold- und Silbertreiber m. ‖ repousseur m. en or et argent. / metal ~ ‖ Metallschneider m. ‖ ciseleur m. sur pièce.

chaser grinding machine ‖ Gewindestrählerschleifmaschine f. ‖ machine f. à affûter les peignes à fileter.

chaser plane ‖ Kampfflieger m. ‖ avion m. de chasse.

chasing hammer ‖ Treibhammer m. ‖ marteau m. à bosseler ou à emboutir ou à bouge. / ~ machine ‖ Punzmaschine f. ‖ machine f. à repousser; repousseuse f. / ~ stake (Locksm) ‖ Treibstöckchen n. ‖ tasseau m. / ~ tool see chaser (Tool).

chassis (Auto) ‖ Rahmen m.; Untergestell n.; Fahrgestell n. ‖ châssis m. / low built ~ ‖ Niederrahmenfahrgestell n. ‖ châssis m. surbaissé. / truck ~ ‖ Lastkraftwagenrahmen m. ‖ châssis m. de camion automobile.

chasuble ‖ Meßgewand n. ‖ chasuble f. / ~ maker ‖ Meßgewandschneider m. ‖ chasublier m.

chatelaine bag ‖ Gürteltasche f. ‖ aumônière f.

chatoyant (Miner) ‖ schillernd ‖ chatoyant.

chat roller (Mine) ‖ Quetschwerk n. ‖ broyeur m.; broyeuse f.; machine f. à broyer.

chatter, to ‖ klappern ‖ cliqueter. / ~ (Contact springs) ‖ prellen ‖ rebondir.

chatterton (Electr) ‖ Chattertoncompound n. ‖ (mastic m.) chatterton m.

chauffeur ‖ Maschinenführer m. ‖ chauffeur-conducteur m. ou mécanicien m. / ~ (Auto) ‖ Kraftwagenführer m.; Schofför m. ‖ chauffeur m. / ~ equipment ‖ Kraftwagenführerausrüstung f. ‖ équipement m. de chauffeur.

cheap ‖ preiswert; wohlfeil ‖ économique; à bon marché. / ~ chimney ‖ Sparofen m. ‖ poêle m. économique. / ~ coffee ‖ Sparkaffee m. ‖ café m. économique.

cheapen, to ‖ verbilligen ‖ baisser le prix. / ~ the work to a considerable degree ‖ die Arbeit bedeutend verbilligen ‖ réduire considérablement le prix d'un travail.

cheapest ‖ billigst ‖ au plus bas prix m. / ~ contractor ‖ Mindestfordernder m. ‖ le moins offrant ou demandant. / ~ terms pl. ‖ billigste Bedingungen fpl. ‖ les meilleures conditions fpl.

cheap goods pl. ‖ Ramschware f. ‖ marchandise f. de rebut. / ~ yack ‖ Preis verderber m. ‖ Schleuderer m. ‖ gâtemétier m.; gâcheur m.

cheapness ‖ Billigkeit f. ‖ bon marché m.

check, to ‖ nachprüfen; nachzählen ‖ vérifier; contrôler; faire la revision. / ~ accounts ‖ nachrechnen ‖ recalculer; supputer. / ~ a bore ‖ eine Bohrung f. messen ‖ mesurer un alésage m. / ~ the roundness and parallelism ‖ auf Zylindrizität f. prüfen ‖ vérifier l'ovalisation f. et la cylindricité f. / ~ the ship's position ‖ den Schiffsort m. nachprüfen ‖ vérifier ou contrôler la position du navire.

check see also cheque.

check (Cash register) ‖ Scheck m.; Kontrollmarke f. ‖ Kontrolle f. ‖ ticket m.; plaque f. de contrôle; contrôle m. / back of the ~ ‖ Rückseite f. des Schecks ‖ verso m. du chèque. / cash register for ~s ‖ Scheckdrucker m. ‖ totalisatrice f. à tickets. / issue of ~s ‖ Scheckausgabe f. ‖ émission f. des tickets. / the register issues duplicate ~s ‖ die Kasse gibt den Scheck mit Doppelaufdruck aus ‖ la caisse émet ticket double.

check bore-hole ‖ Kontrollbohrloch n. ‖ sondage m. de contrôle. / ~ cable (Aero) ‖ Lenkkabel n.; Begrenzungsseil n.; Abfangseil n. ‖ câble m. de commande de la roue arrière orientable; câble m. limiteur de course. / ~ calculation ‖ Rechenprobe f. ‖ preuve f. du calcul. / ~ chain (Railw) ‖ Sicherheitskette f. ‖ chaîne f. de sûreté.

checked see checkered.

check endorser ‖ Scheckindossierapparat m. ‖ appareil m. à endosser les chèques.

checker (Bookb) ‖ Kollationnierer m. ‖ collationneur m. / ~ (Spinn) ‖ Fadenprüfer m. ‖ échantillonneur m. / ~ for petrol outlet ‖ Benzinausflußregler m. ‖ dosateur m. de sortie de l'essence. / wagon ~ (Railw) ‖ Wagenkontrollör m. ‖ contrôleur m. de wagons.

checkered see also chequered ‖ karriert; gewürfelt ‖ quadrillé; à carreaux. / ~ fabric ‖ gewürfelter Stoff m. ‖ étoffe f. à carreaux.

checkering (Crossing stripes) ‖ Fischhaut f. (Griffekerbung) ‖ quadrillage m.; face f. carrelée.

checking ‖ Nachprüfung f. ‖ vérification f. / ~ strip ‖ Justierleiste f. ‖ listel m. de réglage.

check instrument (Tel) ‖ Kontrollapparat m. ‖ appareil m. de contrôle. / ~ nut ‖ Gegenmutter f. ‖ contre-écrou m. / ~ rail ‖ Radlenker m.; Gegenschiene f.; Zwangschiene f. ‖ contre-rail m.

checks pl. (Coin) ‖ Rändeleisen npl. eines Rändelwerkes ‖ molettes fpl. d'une machine à molette. / ~ (Weav) ‖ gewürfeltes Zeug n. ‖ étoffe f. à petits carreaux.

check survey ‖ Kontrollvermessung f. ‖ arpentage m. de contrôle. / ~ switch (Tel) ‖ Kontrollumschalter m. ‖ commutateur m. de contrôle. / ~ test ‖ Nachprüfung f. ‖ épreuve f. de vérification.

check valve ‖ Rückschlagventil n. ‖ soupape f. de retenue. / ball ~ ‖ Kugelrückschlagventil n. ‖ soupape f. de retenue à boulet. / swing ~ ‖ Klappenrückschlagventil n. ‖ soupape f. de retenue à charnière.

checkweigher (Mine) ‖ Arbeitervertreter m. ‖ représentant m. des mineurs.

checkwriting machine ‖ Scheckschreibemaschine f. ‖ machine f. à écrire les chèques.

cheditte ‖ Cheddit n. ‖ chéddite f.

cheek (Met) ‖ Seigerblech n. ‖ paroi f. / ~ (Mar) ‖ Scheibenklampe f. ‖ poulie f. plate. / ~ ‖ Backe f. ‖ jotterau m. / ~ of beakhead ‖ Galionsstütze f. ‖ courbaton m. de l'éperon. / ~ of brake ‖ Bremsklotz m. ‖ mâchoire f. de frein. / crank ~ with hoops shrunk on ‖ Kurbelarm m. mit Schrumpfbändern ‖ bras m. de manivelle muni de frettes. / rounded ~ ‖ abgerundete Kurbelwange f. ‖ flasque m. arrondi. / ~ of shears ‖ Scherblatt n. ‖ lame f.; tranchant m.; mâchoire f.

cheeks pl. (Mach tool) ‖ Wangen fpl. ‖ montants mpl. / ~ of the balance ‖ Schere f. der Wage ‖ châsse f. de balance. / ~ of a press (Print) ‖ Preßwände fpl. ‖ Wangen fpl. einer Presse ‖ jumelles fpl. ou montants mpl. d'une presse.

cheek sluice ‖ Drempelschleuse f.; Schlagschleuse f.; Schleuse f. mit Stemmtoren ‖ écluse f. busquée ou en éperon

cheekstone of a kennel in paving ‖ Bordstein m. der Gosse ‖ pierre f. de bordure d'une rigole en pavé.

cheerless ‖ trostlos ‖ désolé; désolant.

cheese ‖ Käse m. ‖ fromage m. / artificial ~ ‖ Kunstkäse m. ‖ fromage m. artificiel. / fermented ~ ‖ gegorener Käse m. ‖ fromage m. fermenté. / fresh ~ ‖ frischer Käse m. ‖ fromage m. frais.

cheese baker ‖ Käsebäcker m. ‖ cuiseur m. de fromage. / ~ cooling room ‖ Käsekühlraum m. ‖ chambre f. froide à fromage. / equipment for ~ factories ‖ Käsereieinrichtung f. ‖ installation f. de fromagerie. / ~ glue ‖ Kaseinleim m. ‖ colle f. à la caséine. / ~ head (Screw) ‖ runder Kopf m. ‖ tête f. ronde. / ~ knife ‖ Käsemesser n. ‖ couteau m. à fromage. / ~ maker ‖ Milchkäser m. ‖ fromager m.; marcaire m. / ~ making ‖ Käserei f. ‖ fromagerie f. / ~ moulder ‖ Käseformer m. ‖ mouleur m. de fromage. / ~ salter ‖ Käsesalzer m. ‖ saleur m. de fromages. / ~ sieve ‖ Käseform f. ‖ caseret m. / ~ sieve maker ‖ Käseformmacher m. ‖ caserotier m. / ~ tryer ‖ Käseprüfnadel f. ‖ sonde f. à fromage. / ~ works pl. ‖ Sennerei f. ‖ laiterie-fromagerie f.

chemical ‖ chemisch ‖ chimique. / ~ action ‖ chemische Wirkung f. ‖ action f. chimique. / ~ analysis ‖ chemische Analyse f. ‖ analyse f. chimique. / ~ apparatus ‖ chemischer Apparat m. ‖ appareil m. chimique. / ~ arm ‖ chemische Waffe f. ‖ arme f. chimique. / ~ bleaching ‖ Schnellbleiche f. ‖ blanchissage m. chimique. / ~ bomb ‖ Gasbombe f. ‖ bombe f. à gaz. / ~ constitution ‖ chemische Zusammensetzung f. ‖ constitution f. chimique. / ~ decomposition ‖ chemische Zersetzung f. ‖ décomposition f. chimique. / gas mask with ~ filter ‖ Gasmaske f. mit Chemikalfilter ‖ masque m. à gaz à filtres chimiques. / ~ industry ‖ chemische Industrie f. ‖ industrie f. chimique. / apparatus for the ~ industry ‖ Apparat m. für die chemische Industrie ‖ appareil m. pour l'industrie des produits chimiques. / plant for the ~ industry ‖ Anlage f. für die chemische Industrie ‖ établissement m. pour l'industrie des produits chimiques. / ~ laboratory ‖ chemisches Laboratorium n. ‖ laboratoire m. de recherches chimiques. / ~ laboratory to control the current manufacture ‖ chemisches Laboratorium n. zur Überwachung der laufenden Fertigung ‖ laboratoire m. d'analyses chimiques pour contrôler la fabrication courante. / ~ and physical laboratory for scientific and technical research ‖ chemisch-physikalische Versuchsanstalt f. ‖ laboratoire m. de recherches chimiques-physiques. / plant for ~ manure manufacturing ‖ Gewinnungsanlage f. für künstliche Düngemittel ‖ installation f. pour la fabrication d'engrais chimique. / ~ mortar ‖ Gaswerfer m. ‖ projector m. de gaz. / ~ notation ‖ chemische Formel f. ‖ formule f. chimique. / ~ plant ‖ chemische Anlage f. ‖ installation f. pour produits chimiques. / ~ process ‖ chemisches Verfahren n. ‖ procédé m. chimique. / ~ product for photographic purposes ‖ chemisches Erzeugnis n. für fotografische Zwecke ‖ produit m. chimique à l'usage photographique. / ~ projectile ‖ Gasgeschoß n. ‖ projectile m. à gaz; obus m. à gaz. / ~ reaction ‖ che-mischer Vorgang m. ‖ réaction f. chimique. / ~ sedimentary rock ‖ chemisches Sedimentgestein n. ‖ roche f. sédimentaire chimique. / ~ structure ‖ chemische Zusammensetzung f. ‖ constitution f. chimique. / ~ transformation ‖ chemische Umsetzung f. ‖ transformation f. chimique. / ~ treatment ‖ chemische Behandlung f. ‖ traitement m. chimique. / ~ warfare agent ‖ chemisches Kriegsmittel n. ‖ agent m. de guerre chimique. / ~ works pl. ‖ chemische Fabrik f. ‖ fabrique f. de produits chimiques.

chemically combined ‖ chemisch gebunden ‖ chimiquement combiné. / ~ engraved metal board ‖ chemisch graviertes Metallschild n. ‖ plaque f. en métal gravée chimiquement. / ~ prepared paper ‖ chemisch präpariertes Papier n. ‖ papier m. chimique. / ~ pure ‖ chemisch rein ‖ chimiquement pur.

chemicals pl. ‖ Chemikalien fpl.; chemische Erzeugnisse npl. ‖ articles mpl. ou produits mpl. chimiques. / attaining a lower consumption of ~ ‖ geringerer Verbrauch m. an Chemikalien ‖ réduction f. de la consommation de produits chimiques. / mill for ~ ‖ Chemikalienmühle f. ‖ moulin m. pour produits chimiques.

chemicals pl. and dyes pl. ‖ Apothekerwaren fpl. ‖ articles mpl. pharmaceutiques.

chemisette ‖ Chemisett n.; Vorhemd n. ‖ chemisette f.

chemist ‖ Chemiker m.; Drogist m. ‖ chimiste m. / ~ to the law courts ‖ Gerichtschemiker m. ‖ chimiste m. assermenté par les tribunaux. / metallurgical ~ ‖ Metallurgiechemiker m. ‖ chimiste m. métallurgique.

chemistry ‖ Chemie f. ‖ chimie f. / agricultural ~ ‖ Agrikulturchemie f. ‖ chimie f. appliquée à l'agriculture. / analytical ~ ‖ analytische Chemie f. ‖ chimie f. analytique. / applied ~ ‖ angewandte Chemie f. ‖ chimie f. appliquée. / ~ of fermentation ‖ Gärungschemie f. ‖ chimie f. de la fermentation. / forensic ~ ‖ Gerichtschemie f. ‖ chimie f. légale. / inorganic ~ ‖ anorganische Chemie f. ‖ chimie f. anorganique ou inorganique. / organic ~ ‖ organische Chemie f. ‖ chimie f. organique. / philosophical ~ ‖ theoretische Chemie f. ‖ chimie f. théorique ou philosophique. / physical ~ ‖ physikalische Chemie f. ‖ chimie f. physique. / practical ~ ‖ angewandte Chemie f. ‖ chimie f. appliquée. / synthetical ~ ‖ synthetische Chemie f. ‖ chimie f. synthétique. / technical ~ ‖ technische Chemie f. ‖ chimie f. manufacturière ou technologique ou industrielle. / theoretical ~ ‖ theoretische Chemie f. ‖ chimie f. théorique. / vegetable ~ ‖ Pflanzenchemie f. ‖ chimie f. végétale; phytochimie f.

chemist's shop ‖ Apotheke f. ‖ pharmacie f. / ~ ‖ Drogenhandlung f. ‖ droguerie f. / ~ fittings pl. and requisites pl. ‖ Apothekenzubehör n. ‖ accessoires mpl. de pharmacies.

chemist's utensils pl. ‖ Apothekerutensilien pl. ‖ ustensiles mpl. de pharmaciens.

chemitypy, auxiliary printing machine for ~ ‖ Druckhilfsmaschine f. für Chemigrafie ‖ machine f. auxiliaire d'impression pour chimiographie.

chemograph ‖ elektrochemischer Bildschreiber m.; Chemograf m. ‖ chémographe m.

chenille machine ‖ Chenilliermaschine f. ‖ machine f. à cheniller. / ~ toys pl. ‖ Spielwaren fpl. aus Chenille ‖ jouets mpl. en chenille. / ~ weft cutter ‖ Chenilleschneiderin f. ‖ chenilleuse f. en tapis.

cheque ‖ Scheck m. ‖ chèque m. / crossed ~ ‖ Verrechnungsscheck m. ‖ chèque m. barré. / issue of a duplicate ~ ‖ Ausstellen n. eines Duplikatschecks ‖ émission f. d'un duplicata de chèque. / open ~ ‖ offener Scheck m. ‖ chèque m. ouvert. / payment has been stopped on this ~ ‖ der Scheck ist gesperrt ‖ ce chèque à été frappé d'opposition. / post-dated ~ ‖ vorausdatierter Scheck m. ‖ chèque m. post-daté. / to reject a ~ ‖ einen Scheck m. zurückweisen ‖ refuser un chèque f. / stale ~ ‖ verjährter Scheck m. ‖ chèque m. prescrit. / to stop the payment of a ~ ‖ einen Scheck m. nicht einlösen ‖ suspendre le paiement d'un chèque. / to write out a ~ ‖ einen Scheck m. ausstellen ‖ émettre un chèque m.

cheque blank ‖ Scheckformular n. ‖ chèque m. en blanc. / ~ perforating device ‖ Scheckdurchlochvorrichtung f. ‖ dispositif m. pour la perforation des chèques.

chequered cover ‖ Deckel m. mit Riffelung ‖ couvercle m. strié. / ~ jaw ‖ geriffelte Brechbacke f. ‖ mâchoire f. à crans. / ~ plate ‖ Riffelblech n. ‖ tôle f. gaufrée ou striée. / ~ plate covering ‖ Riffelblechbelag m. ‖ recouvrement m. en tôle striée. / wagon with ~ plate covering ‖ Wagen m. mit Riffelblechbelag ‖ wagon m. recouvert de tôle striée.

cheque typewriter ‖ Scheckschreibmaschine f. ‖ machine f. à écrire les chèques. / ~ writing device ‖ Scheckschreibvorrichtung f. ‖ dispositif m. à écrire les chèques.

cherry (Bot) ‖ Kirsche f. ‖ cerise f. / ~ (Locksm) ‖ Kugelfräser m.; Kugelsenker m. ‖ fraise f. sphérique. / ~ brandy ‖ Kirschbranntwein m.; Kirschwasser n. ‖ kirsch m. / ~ coal ‖ Sinterkohle f.; Grobkohle f. ‖ houille f. demi-grasse à longue flamme; charbon m. vif. / baking ~ coal ‖ backende Sinterkohle f. ‖ houille f. maigre collante. / ~ distillery ‖ Kirschwasserbrennerei f. ‖ distillerie f. de kirsch. / ~ gum ‖ Kirschbaumgummi m.; Kirschharz m. ‖ gomme f. de cerisier. / ~ laurel oil ‖ Kirschlorbeeröl n. ‖ essence f. de laurier-cerise.

cherry-red ‖ volle Rotglut f.; Kirschglut f.; Kirschglühhitze f. ‖ rouge m. cerise; cerise m. complet. / bright ~ ‖ helle Rotglut f. ‖ rouge m. cerise clair. / ~ heat ‖ Kirschrotglut f. ‖ chaude f. rouge-cerise.

cherry stalk ‖ Kirschenstiel m. ‖ queue f. de cerise. / ~ stone ‖ Kirschkern m. ‖ noyau m. de cerise. / ~ stoning machine ‖ Kirschenentkernmaschine f. ‖ dénoyauteuse f. de cerises. / ~ tree ‖ Kirschbaum m.; Weichselbaum m. ‖ cerisier m. / ~ wood ‖ Kirschbaumholz n. ‖ (bois m. de) cerisier m.

chess ‖ Schachspiel n. ‖ jeu m. d'échecs; échec m. / ~ (Pont) ‖ Belagbrett n. ‖ madrier m. / ~ board ‖ Schachbrett n. ‖ échiquier m.

chessmen pl. ‖ Schachfiguren fpl. ‖ pièces fpl. pour jeux d'échecs.

chessy copper ‖ Kupferlasur f.; Azurit m.; Chessylit m. ‖ azurite f.; cuivre m. carbonaté bleu; azur m. de cuivre.

chest ‖ Lade f.; Kiste f. ‖ caisse f.; coffre m.; tiroir m. / ~ of drawers ‖ Kommode f. ‖ commode f. / iron ~ ‖ Kassette f. ‖ cassette f. / ~ for provisions ‖ Proviantkiste f. ‖ caisson m. / ~ for rags (Pap) ‖ Lumpenbehälter m.; Lumpensortierkasten m. ‖ cassot m. / steam ~ ‖ Schieberkasten m. ‖ boîte f. à vapeur *ou* de distribution. / thrust-block ~ (Shipb) ‖ Drucklagergehäuse n. ‖ cage f. de palier de butée. / tool ~ ‖ Werkzeugschrank m. ‖ coffret m. à outils.

chest bellows pl. *see* ~ blowing machine. / ~ blowing machine ‖ Kastengebläse n. ‖ soufflets mpl. à caisse. / ~ lock ‖ Kastenschloß n. ‖ serrure f. à palâtre.

chestnut ‖ Kastanie f. ‖ châtaigne f. / edible ~ ‖ eßbare Kastanie f.; Marone f. ‖ marron m. / meal of ~s pl. ‖ Maronenmehl n. ‖ farine f. de marrons.

chestnut bark ‖ Kastanienrinde f. ‖ écorce f. du châtaignier. / ~ brown ‖ kastanienbraun ‖ brun châtain. / ~ extract ‖ Kastanienholzextrakt m. ‖ extrait m. de châtaignier. / ~ grinding mill ‖ Kastanienmühle f. ‖ moulin m. à châtaignes. / ~ meal ‖ Kastanienmehl n. ‖ farine f. de châtaignes. / ~ peeler ‖ Maronenklauberin f. ‖ éplucheuse f. de marrons. / ~ sorter ‖ Maronensortiererin f. ‖ trieuse f. de marrons. / ~ tree ‖ Kastanienbaum m. ‖ châtaignier m. / horse ~ tree ‖ Roßkastanienbaum m.; wilder Kastanienbaum m. ‖ châtaignier m. des Indes. / ~ wood extract ‖ Kastanienholzextrakt m. ‖ extrait m. de bois de châtaignier.

Chevalier barley ‖ Chevaliergerste f. ‖ orge f. Chevalier.

cheviot ‖ Cheviot m. ‖ cheviot m.

chew ‖ Priem m. ‖ chique f.

chewing tobacco ‖ Kautabak m.; Priemtabak m. ‖ tabac m. à mâcher *ou* à chiquer.

chiastolite ‖ Hohlspat m.; Chiastolit m. ‖ chiastolithe f.

chicory ‖ Zichorie f. ‖ chicorée f. / curled ~ ‖ krause Zichorie f. ‖ chicorée f. frisée.

chicory coffee ‖ Zichorienkaffee m. ‖ café m. de chicorée. / ~ drying ‖ Zichoriendarre f. ‖ cossetterie f.; sécherie f. de chicorée. / ~ manufacture ‖ Zichorienfabrik f. ‖ chicoraterie f.; fabrique f. de chicorée à café. / ~ mill ‖ Zichorienmühle f. ‖ moulin m. à chicorée. / ~ packer ‖ Zichorienpackerin f. ‖ paqueteuse f. de chicorée. / ~ roaster ‖ Zichorienbrenner m. ‖ torréfacteur m. de chicorée. / ~ root ‖ Zichorienwurzel f. ‖ racine f. de chicorée.

chief ‖ hauptsächlich ‖ principal; essentiel. / ~ administrative building ‖ Hauptverwaltungsgebäude n. ‖ bâtiment m. *ou* bureaux mpl. de l'administration centrale. / ~ arch (Bridge) ‖ Hauptbogen m. einer Brücke ‖ maîtresse f. arche. / ~ beam (Carp) ‖ Hauptbalken m.; Binderbalken m. ‖ maîtresse f. poutre. / ~ boatswain ‖ Oberbootsmann m. ‖ premier maître m. de manœuvre. / ~ engineer ‖ Hauptmaschinist m.; Maschinenmeister m. ‖ mécanicien m. en chef; maître-mécanicien m. / ~ engineer ‖ Oberingeniör m. ‖ ingénieur m. en chef. / ~ erector *see* ~ fitter. / ~ fitter ‖ Obermontör m. ‖ chef-monteur m. / ~ frame ‖ Hauptspant n.; Richtspant n. ‖ couple m. de levée. / ~ officer (Mar) ‖ Obersteuermann m. ‖

second lieutenant m. du navire; premier pilote m.; second m. / ~ partner ‖ Hauptteilhaber m. ‖ associé m. en chef. / ~ reaction ‖ Hauptreaktion f.; Hauptrückwirkung f. ‖ réaction f. principale. / ~ station (Railw) ‖ Hauptbahnhof m. ‖ station f. centrale. / ~ thing ‖ Hauptsache f. ‖ principal m.; essentiel m. / ~ wall (Build) ‖ Hauptmauer f. ‖ mur m. principal *ou* maîtresse. / ~ warden (Coin) ‖ Oberwardein m. ‖ essayeur m. en chef.

chiffonnier ‖ hohe Kommode f. ‖ chiffonnière f.

chignon ‖ Nackenzopf m. ‖ chignon m. / ~ maker ‖ Nackenzopfmacher m. ‖ chignonnier m.

chilblain ointment ‖ Frostsalbe f. ‖ onguent m. pour les engelures.

child's light ‖ Nachtlicht n. ‖ veilleuse f.

childrens' bag ‖ Kindertäschchen n. ‖ sac m. à main pour enfants. / ~ boots pl. and shoes pl. ‖ Kinderschuhwaren fpl. ‖ chaussures fpl. d'enfants. / ~ counting frame ‖ Kinderrechenmaschine f. ‖ machine f. à calculer pour enfants. / ~ food stuffs pl. ‖ Kindernahrung f. ‖ aliments mpl. pour enfants. / ~ furniture ‖ Kindermöbel pl. ‖ meubles mpl. pour enfants. / ~ hose ‖ Kinderstrumpf m. ‖ bas m. d'enfant. / ~ kitchen ‖ Kinderküche f. ‖ cuisine f. d'enfants. / ~ kitchen utensils pl. ‖ Kinderküchengeschirr n. ‖ vaisselles fpl. d'enfants. / ~ printing box ‖ Kinderdruckerei f. ‖ boîte f. d'imprimerie pour enfants. / ~ shoe ‖ Kinderschuh m. ‖ chaussure f. pour enfants. / ~ suit ‖ Kinderanzug m. ‖ costume m. pour enfants. / ~ toilet ‖ Kinderabort m. ‖ chaise f. percée pour enfants. / ~ toy ‖ Kinderspielzeug n. ‖ jouet m. pour enfants.

Chile nitre *see* ~ saltpetre. / ~ saltpetre ‖ Chilisalpeter m.; salpetersaures Natron n. ‖ salpêtre m. du Chili; azotate m. de soude.

chili ‖ Paprika m. ‖ paprica m.

chill, to (Brew) ‖ abkühlen ‖ réfrigérer; refroidir. / ~ (Found) ‖ in Koquillen fpl. gießen ‖ fondre en coquille f. / ~ (Met) ‖ abschrecken ‖ tremper. / ~ the cast iron ‖ das Roheisen abschrecken ‖ tremper la fonte.

chill (Found) ‖ Kapsel f.; Schale f. ‖ coquille f. / ~ casting (Found) ‖ Hartguß m.; Schalenguß m. ‖ fonte f. en coquille.

chilled (Metal) ‖ schalenhart; hart gegossen ‖ durci à la surface. / ~ castings pl. ‖ Kokillenguß m. ‖ fonte f. *ou* moulage m. en coquille. / ~ cast-iron frog (Railw) ‖ Schalengußherzstück n. ‖ cœur m. de fonte en coquille. / ~ cast-iron part ‖ Hartgußkörper m. ‖ élément m. en fonte durcie. / ~ goods pl. ‖ Kühlgut n. ‖ marchandise f. refroidie. / ~ iron castings pl. *see* ~ castings. / ~ iron runner ‖ Hartgußläufer m. ‖ meule f. de fonte en coquille. / ~ iron wheel ‖ Schalengußrad n. ‖ roue f. de fonte en coquille. / ~ roll ‖ Hartgußwalze f. ‖ cylindre m. en fonte durcie. / ~ wheel ‖ Hartgußrad n. ‖ roue f. en fonte durcie. / ~ work (Found) ‖ Schalenguß m.; Hartguß m. ‖ fonte f. coulée en coquille *ou* en moules de fonte.

chill gold ‖ Schalengold n. ‖ or m. de coquille.

chilli *see* chili.

chilling (Found) ‖ Hartguß m.; Schalenguß m. ‖ fonte f. durcie *ou* en coquille.

chill mould ‖ Kokille f. ‖ coquille f. / moule m. en fonte. / ~ room ‖ Kältelagerraum m.; Kühlraum m. ‖ chambre f. frigorifique.

chimb (Coop) ‖ Kimme f.; Gergel m. ‖ jable m.; échantignole f. / to make the ~s pl. of a cask ‖ ein Faß n. gargeln *oder* gergeln ‖ jabler un tonneau m. / ~ hoop of a cask (Coop) ‖ Schlußreif m. eines Fasses ‖ dernier cerceau m. *ou* sommier m. d'un tonneau.

chime (Mus) ‖ Glockenspiel n. ‖ carillon m. / ~ (Coop) *see* chimb. / sledge ~ ‖ Schlittengeläut n. ‖ clochette f. pour traineaux.

chimmer of gold-sand ‖ Goldwäscher m. ‖ orpailleur m.

chimney (Build; Loc) ‖ Rauchrohr n.; Schornstein m. ‖ cheminée f.; tuyau m. de cheminée. / ~ (Forg) ‖ Schmiedeesse f. ‖ forge f.; chaufferie f. / ~ (Gas) ‖ Schnittbrenner m. ‖ bec m. fendu. / ~ air ~ ‖ Ansaugschlot m. ‖ cheminée f. d'aspiration. / base of ~ ‖ Schornsteinsockel m. ‖ embase f. de cheminée. / ~ in brickwork ‖ Ziegelschornstein m. ‖ cheminée f. en briquetage. / ~ to catch the arsenic ‖ Giftfang m.; Giftturm m. ‖ cheminée f. pour l'arsenic. / ~ of a circular charcoal-pile ‖ Quandel m. eines Meilers ‖ cheminée f. d'une meule de carbonisation. / the draught of all furnaces was obtained by one ~ ‖ die Öfen mpl. erhalten ihren Zug sämtlich durch einen Kamin ‖ les fours mpl. ont un tirage commun par une cheminée. / fire-brick-lined ~ ‖ Schornstein m. mit Schamotteausfütterung ‖ cheminée f. à revêtement en briques réfractaires. / insulated ~ ‖ freistehender Kamin m. ‖ cheminée f. isolée. / isolated ~ *see* insulated ~. / octagonal ~ ‖ achteckiger Kamin m. ‖ cheminée f. de forme octogonale. / ~ of a power station ‖ Schornstein m. eines Kraftwerkes ‖ cheminée f. d'une usine génératrice. / steel-plate ~ ‖ Blechschornstein m. ‖ cheminée f. en tôle d'acier. / wrought-iron ~ ‖ Schornstein m. aus Schmiedeeisen ‖ cheminée f. en fer forgé.

chimney base ‖ Schornsteinsockel m. ‖ socle m. de cheminée. / ~ builder (Arch; Boil) ‖ Schornsteinbauer m. ‖ constructeur m. de cheminées. / ~ builder (Build) ‖ Ofensetzer m. ‖ fumiste m. / ~ cap ‖ Schornsteinhaube f. ‖ lanterne f. de cheminée. / ~ capping ‖ Schornsteinkopf m. ‖ chapiteau m. de cheminée. / ~ case ‖ Schornsteinmantel m. ‖ chemise f. de cheminée. / ~ construction ‖ Schornsteinbau m. ‖ construction f. de cheminées. / ~ cooler ‖ Kaminkühler m. ‖ réfrigérant m. à cheminée. / ~ corner (Build) ‖ Ofennische f.; Hölle f. ‖ ruelle f. / ~ damper ‖ Rauchschieber m. ‖ régistre m. de cheminée. / ~ draught ‖ Kaminzug m. ‖ tirage m. de cheminée. / plant for artificial ~ draught ‖ Schornsteinanlage f. mit künstlichem Zug ‖ installation f. de tirage artificiel pour les cheminées. / ~ erection ‖ Schornsteinbau m. ‖ construction f. de cheminées. / ~ flue ‖ Rauchkanal m.; Fuchs m. ‖ carneau m. de cheminée. / ~ funnel ‖ Schlot m.; cheminée f. / ~ head ‖ Schornsteinaufsatz m.; Schornsteinkappe f.; Schornsteinhaube f. ‖ chapeau m. de cheminée. / ~ hood ‖ Rauchfang m.; Rauchmantel m.;

Schurz m. ‖ hotte f. *ou* manteau m. de cheminée. / ~ jambs pl. ‖ Kamingewände n.; Kamineinfassung f. ‖ jambages mpl. de cheminée. / ~ labourer (Mine) ‖ Arbeiter m. an der Sturzgasse ‖ monteur m. de cheminées. / ~ mantle ‖ Schornsteinmantel m. ‖ enveloppe f. de cheminée. / ~ piece ‖ Kamin m. ‖ cheminée f. / ~ pipe ‖ Kaminrohr n. ‖ tuyau m. de cheminée. / ~ pot ‖ Tonschornsteinröhre f. ‖ tuyau m. de cheminées en terre cuite. / rotating ~ pot ‖ drehbarer Schornsteinaufsatz m. ‖ gueule f. de loup. / ~ soot ‖ Schornsteinruß m. ‖ suie f. de cheminée. / ~ spring ‖ Kaminschnäpper m. ‖ ressort m. de cheminée. / ~ sweeper ‖ Kaminfeger m.; Schornsteinfeger m. ‖ ramoneur. / ~ sweeper's scraper ‖ Kaminfegerscharre f. ‖ grappin m. de ramoneur. / ~ sweeping ‖ Schornsteinfegen n. ‖ ramonage m. / ~ top (Build) ‖ Schornsteinaufsatz m.; Schornsteinhaube f.; Schornsteinkappe f. ‖ cage f. de cheminée. / ~ ventilator (Build) ‖ Windklappe f. ‖ clapet m. de cheminée; éolipyle m.

chin ‖ Kinn n. ‖ menton m.

china ‖ Porzellan n. ‖ porcelaine f. / English ~ ‖ englisches Porzellan n. ‖ porcelaine f. tendre anglaise. / Meissen ~ ‖ Meißener Porzellan n. ‖ porcelaine f. de Meissen.

china articles pl. for chemical and technical purposes ‖ Porzellan n. für chemisch-technische Zwecke ‖ porcelaine f. à l'usage chimique et technique. / ~ blue ‖ Englischblau n. ‖ bleu m. de faïence *ou* de Chine *ou* anglais. / ~ borer ‖ Porzellanbohrer m. ‖ perceur m. de porcelaine. / ~ caster ‖ Porzellangießer m. ‖ mouleur m. en porcelaine. / ~ clay ‖ Porzellanerde f.; Kaolin n.; Steinmark n. ‖ terre f. à porcelaine; kaolin m.; lithomarge f. / ~ clay quarry ‖ Kaolingrube f. ‖ carrière f. de terre à porcelaine. / ~ clay washing ‖ Kaolinschlämmerei f. ‖ laverie f. de kaolin. / ~ decoration ‖ Porzellanverzierung f. ‖ décoration f. de porcelaine. / ~ grass ‖ Ramie f. ‖ ramie f. / ~ grass spinning ‖ Ramiespinnerei f. ‖ filature f. de ramie. / ~ ink ‖ chinesische Tusche f. ‖ encre f. de Chine. / ~ insulator ‖ Porzellanisolator m. ‖ isolateur m. en porcelaine. / ~ mat ‖ chinesische Matte f. ‖ natte f. de Chine. / ~ ornamentation ‖ Auftragen n. von Reliefs auf Porzellan ‖ pastillage m. / ~ painter ‖ Porzellanmaler m. ‖ peintre m. sur porcelaine. / ~ painting ‖ Porzellanmalerei f.; Steingutmalerei f. ‖ peinture f. sur porcelaine *ou* sur faïence. / ~ thrower ‖ Porzellandreher m. ‖ tourneur m. en porcelaine.

chinaware ‖ Porzellan n. ‖ porcelaines fpl. / ~ for household ‖ Gebrauchsporzellan n. ‖ porcelaine f. à l'usage courant. / tender ~ ‖ weiches Porzellan n. ‖ porcelaine f. tendre.

chinchona wine ‖ Chinawein m. ‖ vin m. au quinquina.

chine ‖ Kimm n. ‖ rebord m. en saillie.

chink, to (Material) ‖ aufreißen; reißen; Risse mpl. bekommen ‖ se crévasser; se fendre. / ~ (Mus) ‖ auf den Klang prüfen ‖ examiner par le son.

chink (Pott) ‖ Haarriß m. ‖ fêlure f.; gerçure f. / ~ (Wood) ‖ Spalt m. ‖ fente f. / ~ of a wall ‖ Mauerspalte f.; Mauersprung m.; Riß m. im Mauerwerk ‖ lézarde f. *ou* fente f. d'un mur.

chinked (Wood) ‖ gerissen; gespalten; rissig ‖ fendu; fissuré.

chin-rest ‖ Kinnstütze f. ‖ mentonnière f.

chinsing iron ‖ Stemmeisen n. ‖ burin m. à sertir.

chintz (Calico) ‖ Zitz m. ‖ indienne f.

chip, to ‖ meißeln ‖ buriner. / ~-off the scale ‖ den Kesselstein abkratzen *oder* abklopfen ‖ écailler m. *ou* détartrer.

chip (Jewel) ‖ Splitter m. ‖ éclat m. / ~ (Planer) ‖ Span m.; Hobelspan m. ‖ copeau m. / ~ (Turn) ‖ Span m.; Drehspan m. ‖ copeau m.; tournure f. / to cut ~s pl. ‖ Späne mpl. schneiden ‖ faire des copeaux mpl. / ~ of wood ‖ Holzspan m. ‖ éclat m. *ou* copeau m. de bois.

chip axe ‖ Breitaxt f.; Breitbeil n.; Schlichtbeil n. ‖ épaule f. / ~ basket ‖ Spankorb m. ‖ panier m. en éclats de bois. / ~ box ‖ Spanschachtel f. ‖ boîte f. de copeaux. / ~ carved ware ‖ Kerbschnittarbeit f. ‖ travail m. à l'encoche. / ~ deflector ‖ Spanlenker m. ‖ guide-copeaux m. / ~ hat ‖ Basthut m. ‖ chapeau m. d'écorce.

chipped ‖ gespänt ‖ sur copeaux mpl.

chipper (Tool) ‖ Abklopfhammer m. ‖ marteau m. détartreur. / ~ (Workman) ‖ Meißelarbeiter m. ‖ burineur m.

chipping (Iron; Wood) ‖ Span m. ‖ copeau m. / ~ of stone ‖ Schotter m.; Steinschlag m. ‖ cailloutis m.; pierres fpl. cassées *ou* concassées.

chipping chisel ‖ Hartmeißel m.; Schrotmeißel m. ‖ burin m.

chippings pl. ‖ Kleinschlag m.; Steinbrocken mpl. ‖ gros gravier m.; pierres fpl. concassées.

chips pl. (Boring mach) ‖ Bohrspäne mpl. ‖ bûchilles fpl.; alésures fpl.; copeaux mpl. de foret. / ~ (Lathe) ‖ Drehspäne mpl. ‖ tournure f.; copeaux mpl. / ~ (Stone cutter) ‖ Arbeitszoll m.; Steinsplitter mpl. ‖ écailles fpl.; décombres mpl.

chip, strip of ~s ‖ Spanband n. ‖ bande f. de copeaux. / thickness of ~ ‖ Spanstärke f. ‖ épaisseur f. de la tournure. / ~ tray ‖ Spänepfanne f. ‖ bac m. à copeaux. / width of ~ ‖ Spanbreite f. ‖ largeur f. du copeau *ou* de coupe.

chisel, to (Metal) ‖ meißeln ‖ ciseler; buriner. / ~ (Wood) ‖ stemmen ‖ tailler au ciseau m. / ~ off ‖ abmeißeln ‖ enlever au burin; buriner. / ~ out ‖ auskreuzen ‖ buriner avec le bédane.

chisel (Forg) ‖ Schrotmeißel m.; Abschrot m.; Setzeisen n. ‖ tranche f. d'enclume; burin m.; bédane m. / ~ (Joiner) ‖ Reitel m.; Stecheisen n. ‖ ciseau m.; gouge f. / ~ (Letter-found) ‖ Grundeisen n. ‖ ciseau m. / ~ (Locksm) ‖ Meißel f. ‖ burin m.; ciseau m. / ~ (Mine) ‖ Bohrmeißel m.; trepan m. de sondage; burin m. / bolt ~ ‖ Kreuzmeißel m. ‖ bédane m.; burin m. d'âne. / boring ~ (Mine) ‖ Bohrmeißel m. ‖ trepan m. de sondage; burin m. / cape ~ ‖ Flachmeißel m. ‖ burin m. plat. / caulking ~ ‖ Stemmeißel m.; Stemmsetze f. ‖ matoir m. / chipping ~ ‖ gerader Meißel m. ‖ burin m. / cold ~ ‖ Nuteisen n.; Kaltmeißel m.; Bankmeißel m. ‖ ciseau m. à froid *ou* d'établi; bec m. d'âne. / ~ for cold metal ‖ Kaltmeißel m.; Kaltschrotmeißel m. ‖ ciseau m. *ou* tranche f. à froid. / cope ~ ‖ Nuteisen n. ‖ ciseau m. à rainurer. / corner ~ (Joiner) ‖ Geißfuß m. ‖ gouge f. triangulaire. /

cross-cutting ~ ‖ Kreuzmeißel m. ‖ bec m. d'âne; bédane m. / cross-mouthed ~ ‖ Kronenbohrer m. ‖ pistolet m. à pointe carrée; bonnet m. de prêtre. / to cut away with the ~ ‖ wegmeißeln ‖ couper au ciseau m. / ~ for cutting iron when heated ‖ Schrotmeißel m. ‖ ciseau m. à chaud; tranche f. / ~ dented ~ (Sculpt) ‖ Gradierbetel m.; Gradiereisen n. ‖ ciseau m. gradin; gradine f. / driving ~ ‖ Stemmeisen n. ‖ bédane m. / drop ~ ‖ Kuttermeißel m. ‖ burin m. brise-rocs. / file ~ ‖ Feilenmeißel m. ‖ étoile f. du tailleur de limes. / firmer ~ ‖ Stechbeitel m. ‖ ciseau m. fort. / flat ~ ‖ Flachmeißel m. ‖ burin m. plat. / granulated chasing ~ (Engr) ‖ Grainpunze f. ‖ égrenoir m.; grenoir m. / great ~ ‖ Schroteisen n. ‖ ébarboir m.; ébauchoir m. / hand cold ~ ‖ Handmeißel m. ‖ ciseau m. à main. / hollow ~ ‖ Hohlmeißel m. ‖ gouge f. / hot ~ ‖ Schrotmeißel m. ‖ tranche f. à chaud. / joiner's ~ ‖ Tischlermeißel m. ‖ ciseau m. de menuisier. / mortise ~ ‖ Lochbeitel m. ‖ bédane m. / nailsmith's ~ ‖ Hauer m.; Stockmeißel m.; Blockmeißel m. ‖ tranchet m. / plugging ~ ‖ Locheisen n. ‖ perce-meule m. / pneumatic ~ ‖ Druckluftmeißel m. ‖ marteau-burineur m. pneumatique; burin m. à air comprimé. / pointed ~ (Stone-cutter) ‖ Spitzmeißel m.; Grabstichel m.; Spitzeisen n. ‖ ciseau m. pointu *ou* conique; aiguille f.; onglet m. / pointed ~ with hexagon shank ‖ Spitzmeißel m. mit sechskantigem Einsteckende ‖ aiguille f. à queue hexagonale. / pointed ~ with round shank ‖ Spitzmeißel m. mit rundem Einsteckende ‖ aiguille f. à queue cylindrique. / print cutter's ~ (Engr) ‖ Gravierbetel m. ‖ ciseau m. pour graveurs. / print cutter's nose ~ (Engr) ‖ Gravierstemmbetel ‖ fermoir m. pour graveurs. / print cutter's spoon bit ~ (Engr) ‖ Gravierschaufelbetel m. ‖ fermoir m. à spatule pour graveurs. / to remove the rivet with cross ~ ‖ den Nietkopf m. auskreuzen ‖ couper la tête du rivet. / ripping ~ *see* firmer ~. / rock ~ ‖ Felsmeißel m. ‖ Kuttermeißel m. ‖ burin m. brise-roc. / rod ~ ‖ Stielmeißel m. ‖ tranche f. à manche. / spade-type ~ ‖ Schaufelmeißel m. ‖ bêche f. / stone ~ ‖ Steinmeißel m. ‖ ciseau m. à pierres. / twice bevelled ~ ‖ Stemmeisen n. ‖ fermoir m. / wall ~ ‖ Steinbohrer m. ‖ perce-meule m.; bonnet m. de prêtre. / ~ for warm metal ‖ Warmmeißel m.; Warmschrotmeißel m. ‖ tranche f. à chaud. / ~ for working cold metal ‖ Hartmeißel m. ‖ ciseau m. *ou* tranche f. à froid. / from a ~ only y inches were worn away ‖ der Meißel zeigte nur eine Abnutzung von x m ‖ le burin ne montrait qu'une usure de x mètres.

chisel hammer ‖ Meißelhammer m. ‖ marteau m. à ciseler. / ~ shank ‖ Bohrereinsteckende n. ‖ queue f. de fleuret. / ~ steel ‖ Meißelstahl m. ‖ acier m. à ciseau. / ~ work ‖ Ziselörarbeit f. ‖ ouvrage m. au maillet; travail m. repoussé.

chit, to (Grain) ‖ äugeln; guzen ‖ piquer.

chitting (Grain) ‖ Äugeln n. ‖ piquage m.

chloracetic acid ‖ Chloressigsäure f. ‖ acide m. chloracétique.

chloral ‖ Chloral n. ‖ chloral m. / hydrate of ~ ‖ Chloralhydrat n. ‖ hydrate m. de chloral.

chlorate ‖ chlorsaures Salz n.; Chlorat n. ‖ chlorate m. / ~ of potassium ‖ chlorsaures Kali n. ‖ chlorate m. de potasse.
chlorauric acid ‖ Aurichloridwasserstoffsäure f. ‖ acide m. chloraurique.
chloric, detonating ~ gas ‖ Chlorknallgas n. ‖ chlore m. gazeux fulminant.
chloride ‖ Chlorid n. ‖ chlorure m. / ~ of antimony ‖ Chlorantimon n. ‖ trichlorure m. d'antimoine. / ~ of calcium ‖ Chlorkalzium n. ‖ chlorure m. de calcium; chaux f. chlorurée. / lead ~ ‖ Kotunnit m.; Chlorblei n. ‖ cotunnite f.; chlorure m. de plombe. / ~ of lime ‖ Chlorkalk m. ‖ chlorure m. de chaux. / ~ of magnesium ‖ salzsaure Magnesia f.; Chlormagnesium n. ‖ chlorure m. de magnésium. / ~ of manganese ‖ Manganchlorür n. ‖ chlorure m. (protochlorure) de manganèse. / mercurous ~ ‖ Kalomel n.; Quecksilberchlorür n. ‖ calomel m.; chlorure m. mercureux. / ~ of nickel ‖ Chlornickel n. ‖ chlorure m. de nickel. / ~ of nitrogen ‖ Chlorstickstoff m. ‖ chlorure m. de nitrogène. / ~ of potassium ‖ Chlorkalium n. ‖ chlorure m. de potassium. / ~ of potassium and platinum ‖ Kaliumplatinchlorür n. ‖ chlorure m. de potasse et de platine. / ~ of potassium producing plant ‖ Chlorkaliumfabrikanlage f. ‖ installation f. à fabriquer le chlorure de potassium. / ~ of silver ‖ Chlorsilber n.; Hornsilber n. ‖ chlorure m. d'argent; argent m. corné. / ~ of tin ‖ Chlorzinn n. ‖ chlorure m. d'étain. / ~ of zinc ‖ Chlorzink n. ‖ chlorure m. de zinc.
chloride-containing ‖ chloridhaltig ‖ chloruré. / ~ gas apparatus for sterilization ‖ Chlorgasentkeimungsapparat m. ‖ appareil m. de stérilisation au chlore.
chlorinate, to ‖ chlorieren ‖ chlorer.
chlorination ‖ Verchlorung f. ‖ chlorination f. / ~ plant for gold and silver ores ‖ Verchlorungsanlage f. für Gold- und Silbererze ‖ installation f. de chloruration pour les minerais d'or et d'argent.
chlorine ‖ Chlor n.; Bleichsäure f. ‖ chlore m.; chlorine f. / liquid ~ ‖ verflüssigtes Chlor n. ‖ chlore m. liquide.
chlorine developer ‖ Chlorentwickler m. ‖ appareil m. à développer le chlore. / ~ gas ‖ Chlorgas n. ‖ gaz m. de chlore. / ~ liquifying plant ‖ Chlorverflüssigungsanlage f. ‖ installation f. de liquéfaction de chlore. / ~ water ‖ wässeriges Chlor n. ‖ eau f. de chlore.
chloring (Bleach) ‖ Chloren n. ‖ chlorage m.
chlorite ‖ chlorigsaures Salz n. ‖ chlorite m.
chloroaurate of sodium ‖ Goldsalz n. ‖ sel m. d'or; chloreaurate m. de soude.
chloroform ‖ Chloroform n. ‖ chloroforme m. / ~ producing plant ‖ Chloroformfabrikeinrichtung f. ‖ installation f. de fabrique de chloroforme.
chlorophane ‖ Pyrosmaragd m. ‖ chlorophane m.
chlorophyll ‖ Chlorophyll n.; Blattgrün n. ‖ chlorophylle f.
chlorous acid ‖ chlorige Säure f. ‖ acide m. chloreux.
chock, to (Wheel) ‖ hemmen (durch Keil) ‖ enrayer; caler.
chock ‖ Bremskeil m. ‖ cale f. / ~ (Shipb) ‖ Klampe f. ‖ taquet m.
chocked (Print) ‖ klecksig ‖ pâteux.
chocolate ‖ Schokolade f. ‖ chocolat m. / ~ cake ‖ Schokoladetafel f. ‖ tablette f. de chocolat. / ~ candy ‖ Pralinen npl. ‖

pralinés fpl. / ~ coating plant ‖ Schokoladenüberziehanlage f. ‖ installation f. à dresser le chocolat. / ~ cooling cupboard ‖ Schokoladenkühlschrank m. ‖ armoire f. frigorifique à chocolat. / ~ drop ‖ Schokoladenplätzchen n. ‖ pastille f. de chocolat. / ~ drops moulding machine ‖ Schokoladenplätzchenmaschine f. ‖ machine f. à faire les croquettes de chocolat. / ~ factory ‖ Schokoladenfabrik f. ‖ chocolaterie f. / ~ factory hand ‖ Schokoladenfabrikarbeiter m. ‖ ouvrier m. chocolatier. / ~ machine ‖ Schokoladenfabrikmaschine f. ‖ machine f. pour chocolaterie. / ~ maker ‖ Schokoladenfabrikant ‖ chocolatier m. / ~ manufacture ‖ Schokoladenwarenfabrik f. ‖ chocolaterie f. / ~ mould ‖ Schokoladenform f. ‖ forme f. à chocolats. / roller of ~ paste ‖ Schokoladenteigwalzer m. ‖ lamineur m. de pâte de chocolat. / ~ powder ‖ Schokoladenpulver n. ‖ chocolat m. en poudre. / ~ tablet delivery apparatus ‖ Schokoladenautomat m. ‖ distributeur m. de tablettes de chocolat.
choice ‖ trefflich; auserlesen ‖ excellent.
choice ‖ Auswahl f.; Auslese f. ‖ choix m.
choke, to (Mach) ‖ abdrosseln ‖ étrangler; amortir; supprimer. / ~ the teeth of a file ‖ eine Feile f. verschmieren ‖ empâter la lime.
choke (Electr) ‖ Drosselspule f. ‖ bobine f. autoinductive ou d'impédance ou de réactance. / air-core protecting ~ (Radio) ‖ Impedanzspule f. mit Luftkern für hohe Frequenz ‖ bobine f. de réactance sans noyau de fer. / carburettor ~ (Motor) ‖ Mischrohr n. ‖ tuyau m. de mélange. / smoothing ~ ‖ Abflachungsdrossel f. ‖ self m. d'aplatissement; bobine f. d'étouffement de bruits.
choke coil ‖ Schutzspule f. ‖ bobine f. de protection ou de réactance.
choked ‖ verstopft ‖ noyé; obstrué; bouché.
choke damp (Mine) ‖ Schwaden m.; böses Wetter n. ‖ mofette f.; gaz mpl. délétères. / ~ indicator ‖ Schlagwetteranzeiger m. ‖ indicateur m. de grisou. / ~ proof ‖ schlagwettersicher ‖ antidéflagrant.
choking ‖ Verstopfung f. ‖ obstruction f.
choking coil (Electr) ‖ Drosselspule f. ‖ bobine f. autoinductive ou d'impédance ou de réactance. / ~ for higher harmonics ‖ Oberwellendrossel f. ‖ bobine f. de réactance pour les ondes supérieures. / protection by ~ ‖ magnetischer Kopplungsschutz m. ‖ protection f. par des bobines de décharge.
choking resistance ‖ Drosselwiderstand m. ‖ résistance f. à réactance. / ~ tube (Tel) ‖ Drosselröhre f. ‖ tube m. d'étranglement. / ~ up ‖ Verstopfung f. ‖ engorgement m.; obstruction f.
choose, to ‖ aussuchen; auswählen ‖ choisir.
chop, to ~ off (Iron) ‖ abschroten ‖ couper à la tranche; trancher. / ~ (Wood) ‖ abhacken ‖ couper; abattre à coups mpl. de hache.
chop hammer ‖ Schrothammer m.; Setzeisen n. ‖ tranche f. à manche.
chopped ‖ gespalten ‖ fendu. / ~ and fermented cabbage ‖ Sauerkohl m. ‖ choucroute f. / ~ straw ‖ Häcksel m. ‖ paille f. hachée.
chopper (Agr) ‖ Häckselmesser n. ‖ couperet m. / ~ (Aut tel) ‖ Zerhacker m. ‖ hacheur m. / kitchen ~ ‖ Hackbeil n. ‖

couperet m. de cuisine. / slotting machine for ~s ‖ Bestoßmaschine f. für Schnitzelmesser ‖ machine f. à écorner les couteaux à rognures.
chopping (Sea) ‖ Kabbelung f.; Scholken n.; stoßweises Schlagen n. der Wellen ‖ clapotage m.; clapotis m. / ~ bench ‖ Futterschneidemaschine f. ‖ hache-fourage f. / ~ blade ‖ Futterklinge f.; Häckselmesser n. ‖ couteau m. ou lame f. du hache-paille. / ~ block ‖ Hackblock m. ‖ billot m. / straw ~ factory ‖ Häckselfabrik f. ‖ fabrique f. de paille hachée. / ~ knife ‖ Hackmesser n.; Wiegemesser n. ‖ couperet m.
chopping machine (Butch) ‖ Hackmaschine f. ‖ hachoir m. / ~ (Match) ‖ Abschlagemaschine f. ‖ machine f. à couper. / fat ~ ‖ Fettzerkleinerungsmaschine f. ‖ machine f. à dépiécer la graisse.
chord (Geom) ‖ Sehne f. eines Bogens oder Winkels ‖ corde f.; soustendante f. / ~ (Mus) ‖ Saite f. ‖ corde f. / ~ of an arch (Arch) ‖ Kämpferlinie f. eines Bogens ‖ corde f. d'un arc. / lower ~ (Beam) ‖ Untergurt m. ‖ bride f. inférieure. / ~ for pianos ‖ Klaviersaite f. ‖ corde f. à piano. / ~ of spar (Aero) ‖ Holmgurt m. ‖ bride f. de longeron. / ~ of a truss (Bridge) ‖ Gurt m. oder Gurtung f. eines Trägers ‖ semelle f. d'une poutre. / upper ~ ‖ Obergurt m. ‖ bride f. ou semelle f. supérieure.
chord wire (Aero) ‖ Spanndraht m. ‖ tirant m. en corde; corde f. de tension. / ~ zither ‖ Akkordzither f. ‖ cithare f. d'accord.
choroid membrane ‖ Aderhaut f. ‖ membrane f. choroïde.
christening of a vessel (Act of naming) ‖ Schiffstaufe f. ‖ baptême m. d'un bâtiment.
christianite ‖ Kalkharmotom m.; Phillipsit m. ‖ christianite; harmotome m. calcaire; phillipsite m.
Christmas card ‖ Weihnachtskarte f. ‖ carte f. de Noël.
Christmas tree adornment of wadding ‖ Christbaumschmuck m. aus Watte ‖ décoration f. en ouate pour arbres de Noël. / ~ decoration ‖ Christbaumschmuck m. ‖ décoration f. ou ornement pour arbres de Noël. / ~ illumination ‖ Christbaumbeleuchtung f. ‖ lampes fpl. pour arbres de Noël. / ~ ornaments pl. ‖ Christbaumschmuck m. ‖ ornements mpl. ou articles mpl. pour arbres de Noël. / ~ ornaments making machine ‖ Christbaumschmuckherstellungsmaschine f. ‖ machine f. à fabriquer des ornements pour arbres de Noël.
chromate ‖ Chromat n.; chromsaures Salz n. ‖ chromate m. / red ~ of lead ‖ Rotbleierz n.; chromsaures Bleioxyd n.; Krokoit n. ‖ plomb m. chromaté rouge; minium m. / ~ of potassium ‖ chromsaures Kali n. ‖ Kaliumchromat n. ‖ chromate m. de potassium.
chromate bath ‖ Chrombad n. ‖ bain m. de chromage. / ~ plant ‖ Chromsalzanlage f. ‖ installation f. de chromate.
chromatic (Opt) ‖ farbig; chromatisch ‖ chromatique.
chromatic (Opt) ‖ Chromat m.; Chromatlinse f. ‖ chromate m. / two-lens ~ ‖ zweiteiliger Chromat m. ‖ chromate m. à deux lentilles.

chromatic aberration || Farbenabweichung f. || aberration f. chromatique. / ~ correction || chromatische Korrektion f. || correction f. chromatique *ou* des aberrations chromatiques. / ~ deviation from a mean principal focus || chromatische Abweichung f. von einem mittleren Brennpunkt || aberration f. chromatique d'un foyer moyen. / ~ difference of the spherical aberrations || chromatische Differenz f. der sphärischen Abweichungen || différence f. chromatique des aberrations de sphéricité. / ~ identity || Farbengleichheit f. || identité f. chromatique. / ~ purity || Farbenreinheit f. || pureté f. des teintes. / ~ quartz lens || Quarzchromatlinse f. || objectif m. chromatique en quartz. / ~ sensation || Farbenempfindung f. || sensation f. chromatique.

chromatics pl. || Farbenlehre f. || chromatique f.

chrome || Chrom n. || chrome m. / ~ alum || Chromalaun m. || alun m. de chrome. / ~ bath || Chrombad n. || bain m. chromaté. / ~ battery || Chromelement n. || pile f. au bichromate de potasse. / ~ colour || Chromfarbe f. || couleur f. de chrome.

chromed neat's leather || Chromrindleder n. || cuir m. de vache chromé.

chrome green || chromgrün || vert de chrome m. / ~ iron ore || Chromeisenerz n. || minerai m. de fer chromaté. / ~ leather || Chromleder n. || cuir m. chromé. / ~ leather belt || Chromlederriemen m. || courroie f. en cuir chromé. / ~ mine || Chromeisensteingrube f. || mine f. de chrome. / ~ nickel || Chromnickel n. || nickel-chrome m. / ~ nickel steel || Chromnickelstahl m. || acier m. chrome-nickel. / ~ nickel steel with a high percentage of chromium || Chromnickelstahl m. mit hohem Chromgehalt || acier m. au nickel-chrome de haute teneur en chrome. / ~ nickel wire || Chromnickeldraht m. || fil m. de nickel-chrome. / ~ orange || chromorange || chrome-orange. / ~ ore || Chromerz n. || minerai m. de chrome. / ~ ore brick || Chromerzstein m. || brique m. en minerai de chrome. / ~ patent leather || Chromrindlackleder n. || cuir m. chromé verni pour chaussures. / ~-plated || verchromt || chromé. / ~ plating plant || Verchromungsanlage f. || installation f. de chromage. / ~ red || Chromrot n. || rouge m. de chrome. / ~ steel || Chromstahl m. || acier m. chromé *ou* au chrome. / ~ steel plate || Chromstahlplatte f. || plaque f. en acier chromé. / ~ tanning || Chromgerbung f. || tannage m. au chrome. / ~ yellow || Chromgelb n. || jaune m. de chrome.

chromic acid || Chromsäure f. || acide m. chromique. / ~ acid cell || Chromsäureelement n. || élément m. au bichromate de potasse. / ~ anhydride || Chromsäure f. || acide m. chromique.

chromite || Chromeisenerz n. || chromite f.; sidérochrome m. / ~ brick || Chromitziegel m. || brique f. en chromite.

chromium || Chrom n. || chrome m.

chromium acetate || essigsaures Chrom n. || acétate m. de chrome. / ~ acidol colour || Chromazidolfarbe f. || couleur f. acidol au chrome. / ~ chloride || Chlorchrom n. || chlorure m. de chrome. / ~ fluoride || Fluorchrom n. || fluorure m. de chrome. / ~ nickel steel || Chromnickelstahl m. || acier m. au chrome et nickel. /

~ nitrate || salpetersaures Chrom n. || nitrate m. de chrome. / ~ oxide || Chromoxyd n. || oxyde m. de chrome. / ~ phosphate || phosphorsaures Chrom n. || phosphate m. de chrome. / ~ plating || Verchromungsanlage f. || installation f. de chromage. / ~ salt || Chromsalz n. || sel m. de chrome. / ~ special steel || Chromspezialstahl m. || acier m. spécial au chrome. / ~ steel || Chromstahl m. || acier m. au chrome. / ~ sulphate || schwefelsaures Chrom n. || sulfate m. de chrome. / ~ tanning extract || Chromgerbextrakt m. || extrait m. de chrome tannant *ou* tannique.

chromolithographic printer || Chromolithograf m. || chromolithographe m. / ~ printing works pl. || Farbensteindruckerei f. || imprimerie f. chromolithographique.

chromolithography || Chromolithografie f.; lithografischer Farbendruck m. || chromolithographie f.

chromo-paper || Chromopapier n. || papier m. chromé.

chromophotolithography || Chromophotolithografie f. || chromophotolithographie f.

chromotype || Chromotypie f.; Farbdruck m. || chromotype m.; chromo m.

chromotypography || Chromotypografie f.; Farbenbuchdruck m. || chromotypographie f.; impression f. en couleurs.

chromotypogravure, three-colour ~ || Dreifarbenätzung f. || chromotypogravure f. aux trois couleurs.

chromyl chloride || Chlorchromsäure f. || chlorure m. de chromyle.

chronic || chronisch || chronique.

chronograph || Chronograf m.; Zeitschreiber m. || chronographe m. / ~ watch || Uhr f. mit Beobachtungssekunden || montre f. chronographe.

chronometer || Chronometer n.; Zeitmesser m. || chronomètre m.; montre f. marine. / marine ~ || Schiffsuhr f. || chronomètre m. de marine. / observer's ~ || Beobachtungsuhr f. || montre f. *ou* horloge f. pour les observations. / pocket ~ || Taschenuhr f. || chronomètre m. de poche. / standard ~ || Regelchronometer n. || chronomètre-étalon m.

chronometer test calibration || Uhrprüfung f. || vérification f. de la montre; contrôle m. du chronomètre.

chrysalis, to suffocate the ~ shortly before setting-off to spin || die Larve erst kurz vor dem Abspinnen töten || étouffer la chrysalide peu de temps avant le filage.

chrysoberyl || Goldberyll m. || chrysobéril m.; cymophane f.

chrysocolla || Kieselkupfer n.; Kupfergrün n.; Pechkupfer n.; Kieselmalachit m.; Berggrün n. || chrysocolle f.; cuivre m. hydro-siliceux; vert m. de montagne.

chrysolite || Olivin m.; Chrysolit m.; Peridot m. || péridot m.; olivine f.; chrysolithe f.

chuck || Spannfutter n.; Klemmfutter n. || chuck m.; mandrin m. de serrage. / air-operated ~ || Druckluftfutter n. || mandrin m. pneumatique. / ball turning ~ || Kugelspannfutter n. || mandrin m. creux pour boules. / bell ~ || Schraubenfutter n. || chuck m. *ou* mandrin m. à vis. / clamping ~ || Klemmfutter n. || mandrin m. de serrage. / elastic ~ || Klemmfutter n. || mandrin m. brisé. / electromagnetic ~ || elektromagnetische Aufspannplatte

f. || plaque f. de serrage électromagnétique. / ~ with holdfasts || Stachelfutter n. || mandrin m. à pointes. / ~ with four jaws || Vierbackenfutter n. || chuck m. à quatre mâchoires. / jaw ~ || Backenfutter n. || chuck m. *ou* mandrin m. à mâchoires. / lathe ~ || Drehbankfutter n.; Spannfutter n. || manchon m. pour tours; chuck m.; mandrin m. de serrage. / magnetic ~ || magnetisches Futter n. || mandrin m. magnétique. / oval ~ || Futter m. zum Ovaldrehen || chuck m. à tourner en oval; mandrin m. du tour à oval. / patent lever ~ || Patenthebelfutter n. || chuck m. à serrage instantané par levier. / rotary magnetic ~ || Magnetfutter n.; magnetisches Spannfutter n. || mandrin m. électromagnétique. / rotary screwing ~ || umlaufender Schneidkopf m. || filière f. rotative. / screwing ~ || Schneidkopf m. || cage f. de filière. / screwing ~ with automatic release || selbstauslösender Schneidkopf m. || cage f. de filière automatique; filière f. à ouverture automatique. / screw point ~ || Schraubenfutter n. || mandrin m. à vis. / scroll ~ || Spiralfutter n. || mandrin m. à spirale plate. / self-centring ~ || Zentrierfutter n. || mandrin m. à centrage automatique. / socket ~ || Spundfutter n.; Einschlagfutter n. || mandrin m. creux. / split-socket ~ || Kreuzfutter n. || mandrin m. creux en quatre sections. / split-screw ~ || geteilter Gewindeschneidkopf m. || mandrin m. (à) gueule-de-loup. / three-jawed ~ || Dreibackenfutter n. || mandrin m. à trois griffes. / two-jawed ~ || Zweibackenfutter n.; Klemmfutter n. mit zwei Backen || mandrin m. à deux mâchoires. / universal ~ || Universalfutter n. || mandrin m. universel.

chucking by compressed-air || Preßluftfestspannung f. || dispositif m. de serrage à air comprimé. / ~ of the work piece || Einspannen n. des Arbeitsstückes || serrage m. de la pièce à usiner. / semi-automatic ~ machine || Futterhalbautomat m. || machine f. semi-automatique pour le travail *ou* mandrin.

chuck job, turning lathe for ~s || Drehbank f. für Futterarbeiten || tour m. à décolleter au mandrin.

chuck lathe || Kopfdrehbank f.; Planscheibendrehbank f. || tour m. en l'air *ou* à plateau. / semi-automatic machine for ~ work || Halbautomat m. für Futterarbeit || machine f. semi-automatique pour le décolletage au mandrin.

church banner || Kirchenfahne f. || oriflamme f. / ~ bell || Kirchenglocke f. || cloche f. d'église. / ~ brocade || Kirchenbrokat m. || brocard m. d'église. / ~ bronzes pl. || Bronzewaren fpl. für Kirchen || bronzes mpl. d'église. / ~ candle || Kirchenkerze f. || cierge m. d'église. / ~ clock || Turmuhr f. || horloge f. / ~ damask || Kirchendamast m. || damas m. d'église. / ~ furniture || Kirchenmöbel pl. || meubles mpl. d'église; mobilier m. religieux. / ~ heating || Kirchenbeheizung f. || chauffage m. d'église. / ~ implements pl. || Kirchengerät n. || vases mpl. sacrés d'église. / ~ organ || Kirchenorgel f. || orgue m. d'église. / ~ ornaments pl. || Kirchenparamente mpl. || ornements mpl. et vêtements mpl. sacerdotaux. / ~ pendant || Kirchenwimpel m. || flamme f. de

la messe. / ~ tower ‖ Kirchturm m. ‖ clocher m. / ~ window ‖ Kirchenfenster n. ‖ vitrail m. d'église. / painted ~ window ‖ bemaltes Kirchenfenster n. ‖ vitrail m. peint. / ~ woven ornament ‖ gewebter Kirchenschmuck m. ‖ ornement m. d'église en tissu.

churchyard gardener ‖ Friedhofsgärtner m. ‖ jardinier m. de cimetière.

churn, to ‖ buttern ‖ battre le beurre ou le lait; baratter.

churn ‖ Butterfaß n. ‖ baratte f.; baratteuse f.; tine f. de beurre. / motor-driven ~ ‖ Buttermaschine f. mit Motorantrieb ‖ baratteuse f. (mue) à moteur. / power-driven ~ ‖ Butterfaß n. für Kraftbetrieb ‖ baratteuse f. (à commande) mécanique.

churner ‖ Butterfaßarbeiter m. ‖ baratteur m.

chute ‖ Gerinne n.; Rutsche f. ‖ couloir m.; gouttière f.; chéneau m.; chute f. / coal ~ ‖ Kohlenrutsche f. ‖ plan m. incliné à charbon. / conveyor ~ ‖ Förderrutsche f. ‖ plan m. incliné de transport. / ~ of the hopper (Wind mill) ‖ Schuh m. des Rumpfes ‖ auget m. de la trémie. / oscillating ~ ‖ Schwingrinne f. ‖ gouttière f. oscillante.

chute adjustment ‖ Rutscheneinstellung f. ‖ réglage m. du plan incliné. / ~ trap ‖ Schüttklappe f. ‖ trappe f. de déchargement.

cider ‖ Apfelwein m. ‖ cidre m. / ~ manufactory ‖ Apfelweinkelterei f. ‖ cidrerie f.; brasserie f. de cidre. / ~ press ‖ Apfelweinpresse f.; Apfelkelter f. ‖ presse f. ou pressoir m. à cidre.

cigar ‖ Zigarre f. ‖ cigare m. / well-seasoned ~ ‖ gut abgelagerte Zigarre f. ‖ cigare m. bien sec.

cigar box ‖ Zigarrenkiste f. ‖ boîte f. ou caisse f. à cigares. / ~ case ‖ Zigarrenetui n.; Zigarrenbehälter m. ‖ étui m. à cigares. / ~ cutter ‖ Zigarrenabschneider m. ‖ coupe-cigares m.

cigarette ‖ Zigarette f. ‖ cigarette f. / anti-asthmatic ~ ‖ Asthmazigarette f. ‖ cigarette f. antiasthmatique. / perfumed ~ ‖ parfümierte Zigarette f. ‖ cigarette f. parfumée.

cigarette book ‖ Zigarettenpapierheft n. ‖ cahier m. de papier à cigarettes. / ~ box ‖ Zigarettenschachtel f. ‖ boîte f. pour cigarettes. / ~ cardboard box ‖ Zigarettenkarton m. ‖ carton m. pour cigarettes. / ~ case ‖ Zigarettenetui n.; Zigarettenbehälter m. ‖ étui m. à cigarettes. / ~ holder ‖ Zigarettenspitze f. ‖ fume-cigarettes m. / ~ husk ‖ Zigarettenhülse f. ‖ enveloppe f. de cigarette. / pneumatic mixing plant for ~ industry ‖ pneumatische Mischanlage f. für die Zigarettenindustrie ‖ installation f. de mélangeur pneumatique pour l'industrie des cigarettes. / ~ machine ‖ Zigarettenmaschine f. ‖ machine f. à cigarettes. / ~ machine ribbon ‖ Zigarettenmaschinenband n. ‖ ruban m. pour machines à cigarettes. / ~ maker ‖ Zigarettenarbeiterin f. ‖ cigarettière f. / ~ making machine ‖ Zigarettenherstellungsmaschine f. ‖ machine f. à fabriquer des cigarettes. / ~ manufacture ‖ Zigarettenfabrik f. ‖ fabrique f. de cigarettes. / ~ packing machine ‖ Zigarettenpaketiermaschine f. ‖ machine f. à empaqueter les cigarettes. / ~ paper ‖ Zigarettenpapier n. ‖ papier m. à cigarettes. / ~ paper in tubes ‖ Zigarettenpapier n. in Hülsen ‖ papier

m. à cigarettes en tubes. / tobacco for ~s ‖ Zigarettentabak m. ‖ tabac m. à cigarettes. / ~ tube ‖ Zigarettenhülse f. ‖ tube m. ou enveloppe f. de cigarette. / ~ tube machine ‖ Zigarettenhülsenmaschine f. ‖ machine f. à tubes à cigarettes.

cigar holder ‖ Zigarrenspitze f. ‖ fume-cigares m. / ~ lighter ‖ Zigarrenanzünder m. ‖ allumeur m. pour cigares; allume-cigares m. / ~ maker ‖ Zigarrenmacher m. ‖ cigarier m. / ~ making ‖ Zigarrenfabrikation f. ‖ manufacture f. de cigares. / ~ making machine ‖ Zigarrenherstellungsmaschine f. ‖ machine f. à fabriquer des cigares. / ~ mould ‖ Zigarrenpreßform f. ‖ moule m. à cigares. / ~ roller ‖ Zigarrenroller m. ‖ cigarier m. / ~ sorter ‖ Zigarrensortierer m. ‖ trieur m. de cigares.

cilia forceps pl. ‖ Zilienpinzette f. ‖ pince f. à épiler les cils.

cill ‖ Fenstersohle f.; Fensterschwelle f. Fensterbank f. ‖ seuil m.; banquette f.

cinchona ‖ Chinin n. ‖ quinine f. / ~ bark ‖ Chinarinde f. ‖ quinquina m.; quina m.

cinder ‖ Kohlenlösche f.; Lösche f.; Schlacke f. ‖ escarbille f.; scorie f. / blast-furnace ~ ‖ Schlacke f. eines Hochofens ‖ laitier m. d'un haut-fourneau. / smith's ~s pl. ‖ Schmiedekohle f. ‖ houille f. pour feux de forge; charbon m. de forge; houille f. maréchale; fine-forge f.

cinder bed ‖ Schlackenbett n. ‖ lit m. de scorie. / ~ box ‖ Schlackenkasten m.; Schlackenbehälter m. ‖ caisse f. à laitiers. / ~ concrete ‖ Löschebeton m. ‖ béton m. d'escarbille. / ~ fall ‖ Schlackentrift f. ‖ pissée f.; voie f. de déchargement de laitier ou des scories. / ~ frame (Loc) ‖ Funkenrost m.; Funkensieb n. ‖ grille f. de flammèches. / ~ hair ‖ Schlackenwolle f. ‖ poil m. de laitier ou de scorie. / ~ hole of the blast-furnace ‖ Schlackenloch n. des Hochofens ‖ trou m. de laitier d'un haut-fourneau. / ~ hook ‖ Schlackenhaken m. ‖ ringard m. / ~ pit-man ‖ Schlackengrubenreiniger m.; Pitsreiniger m. ‖ nettoyeur m. de pits. / ~ stone ‖ Aschenstein m.; Schlackenstein m. ‖ pierre f. de cendre ou de laitier ou de scorie. / ~ tip ‖ Schlackenhalde f. ‖ crassier m. / ~ tub ‖ Schlackenwagen m. ‖ wagon m. à laitier ou à crasses.

cinema ‖ Lichtspieltheater n.; Kino n. ‖ cinéma m. / accessories pl. for ~s ‖ Kinozubehör n. ‖ accessoires mpl. pour cinémas.

cinema film ‖ Kinofilm m. ‖ film m. pour cinémas. / ~ lending institute ‖ Filmverleihanstalt f. ‖ institut m. de louage de films.

cinema lamp ‖ Kinolampe f. ‖ lampe f. cinéma. / ~ poster ‖ Filmplakat n. ‖ placard m. de cinéma. / ~ projector ‖ Kinovorführungsapparat m. ‖ appareil m. pour représentations cinématographiques. / ~ spotter ‖ Kinoflecker m. ‖ instrument m. cinématographique d'observation du tir.

cinematic spotting device ‖ Kinoflecker m ‖ instrument m. cinématographique d'observation du tir.

cinematograph ‖ Kinematograf m. ‖ cinématographe m. / rolls pl. for ~s ‖ Rollen fpl. oder Streifen für Kinematografen ‖ rouleaux mpl. ou bandes pour ciné-

matographes. / stripes pl. for ~s see rolls pl. for ~s.

cinematograph camera ‖ Filmaufnahmekamera f. ‖ chambre f. de prise de vues. / ~ film spool ‖ Kinofilmrolle f. ‖ rouleau m. de film cinématographe.

cinematographical apparatus ‖ Kinematografenapparat m. ‖ appareil m. cinématographique. / ~ instrument ‖ kinematografischer Apparat m. ‖ instrument m. cinématographique. / ~ picture ‖ kinematografisches Bild n. ‖ vue f. cinématographique. / ~ raw film ‖ Kinorohfilm m. ‖ film m. brut cinématographique.

cinematography ‖ Kinematografie f. ‖ cinématographie f.

cinerary urn ‖ Aschenurne f. ‖ urne f. sépulcrale.

cinnabar ‖ Schwefelquecksilber n.; Zinnober m. ‖ mercure m. sulfuré; cinabre m. / artificial ~ ‖ Quecksilbersulfid n.; künstl. Zinnober m. ‖ cinabre m. artificiel; sulfure m. de mercure. / hepatic ~ ‖ Kohlenzinnober m.; Quecksilberlebererz n. ‖ mercure m. sulfuré compact; mercure m. hépatique.

cinnabar red ‖ Zinnoberrot n. ‖ cinabre m. artificiel.

cinnamic acid ‖ Zimtsäure f. ‖ acide m. cinnamique. / ~ aldehyde ‖ Zimtaldehyd n. ‖ aldéhyde m. cinnamique.

cinnamon ‖ Zimt m. ‖ cannelle f. / Chinese ~ ‖ chinesischer Zimt m. ‖ cannelle f. de Chine. / ~ bark ‖ Zimtrinde f. ‖ écorce f. de cannelle. / ~ flower ‖ Zimtblüte f. ‖ fleur f. de cannelier. / ~ oil ‖ Zimtöl n. ‖ essence f. de cannelle. / ~ stone ‖ Kaneelstein m.; Hessonit m. ‖ essonite f. / ~ tincture ‖ Zimttinktur f. ‖ teinture f. de cannelle.

cipher, to ‖ chiffrieren ‖ chiffrer.

cipher ‖ Chiffre f. ‖ chiffre m. / art of ~ code-writing ‖ Chiffrierkunst f. ‖ art m. cryptographique. / ~ code typewriter ‖ Geheimschreibmaschine f. ‖ machine f. à écrire l'écriture chiffrée.

ciphered writing ‖ chiffrierte Schrift f. ‖ écriture f. chiffrée. / writing of ~ documents ‖ Abfassung f. chiffrierter Schriftstücke ‖ rédaction f. de documents chiffrés.

cipher language ‖ chiffrierte Sprache f. ‖ langage m. chiffré. / ~ roll for counters ‖ Zahlenrolle f. für Zählwerke ‖ rouleau m. à chiffres pour compteurs. / ~ telegram ‖ Chiffretelegramm n. ‖ dépêche f. chiffrée.

circle ‖ Kreis m.; Kreislinie f. ‖ cercle m. ‖ Kreisfläche f.; Kreisinhalt m.; aire f. du cercle. / antarctic ~ ‖ antarktischer oder südlicher Polarkreis m. ‖ cercle m. antarctique ou méridional. / arctic ~ ‖ arktischer oder nördlicher Polarkreis m. ‖ cercle m. polaire arctique. / area of a ~ ‖ Kreisfläche f.; Kreisinhalt m. ‖ cercle m.; aire f. du cercle. / astronomical ~ ‖ astronomischer Kreis m. ‖ cercle m. d'astronomie. / complete ~ ‖ Vollkreis m. ‖ cercle m. entier. / concentric ~ ‖ konzentrischer Kreis m. ‖ cercle m. concentrique. / ~ of curvature ‖ Krümmungskreis m. einer Kurve ‖ cercle m. osculateur; cercle de courbure. / described ~ (Ship) ‖ Drehkreis m. ‖ cercle m. décrit. / ~ divided ‖ Teilkreis m.; Einstellkreis m. ‖ cercle m. divisé ou de calage. / eccentric ~ ‖ exzentrischer Kreis m. ‖ cercle m. excentrique. / great ~ of a sphere ‖ größter Kreis m. einer Kugel ‖

grand cercle m. d'une sphère. / horary ~ || Stundenkreis m. || cercle m. horaire. / ~ of illumination || Lichtkreis m. || cercle m. d'illumination. / to move in a ~ || kreisen || tourner dans le cercle; circuler. / osculatory ~ || Krümmungskreis m. || cercle m. de courbure; cercle m. osculateur. / parallel ~ || konzentrischer Kreis m. || cercle m. concentrique. / pitch ~ || Teilkreis m. eines Zahnrades || cercle primitif d'une roue dentée. / ~ of railway || Ringgleis n. || réseau m. circulaire. / reflecting ~ || Spiegelkreis m. || cercle m. de réflexion.

circle cutting machine || Kreisschere f. || cisaille f. circulaire pour découper les disques. / ~ diagram || Kreisdiagramm n. || diagramme m. circulaire.

circonium || Zirkon n. || zirconium m.

circuit, to ~ see also to circulate || zirkulieren; umlaufen; fließen || circuler. / to short-circuit || kurzschließen || court-circuiter.

circuit (Electr) || Schaltung f.; Stromkreis m. || circuit m. / ~ (Hydr arch) || Serpentine f.; Schlängelung f. || virage m. / all-busy ~ (Aut tel) || Abschaltung f. || coupure f. / aperiodic ~ (Radio) || aperiodischer Kreis m. || circuit m. apériodique. / brake ~ || Bremsstromkreis m. || circuit m. de freinage. / branch ~ || Verzweigungsschaltung f. || circuit m. de bifurcation. / calling ~ || Rufstromkreis m. || circuit m. d'appel. / to close the ~ || den Stromkreis m. schließen || fermer le circuit. / closed ~ with intermediate receiving-stations || Schleifenschaltung f. || embrochage m. / closed oscillating ~ || geschlossener Erregerkreis m. || circuit m. oscillant fermé. / combined ~ || Vierersprechkreis m. || circuit m. de conversation quadruple ou combiné. / coupled ~ (Radio) || gekoppeltes System n. || circuit m. accouplé. / the ~ is cut || der Stromkreis m. ist unterbrochen || le circuit m. est coupé. / dead-level ~ (Electr) || Leerstromkreis m. || courant m. en circuit ouvert. / derived ~ || Nebenschlußstromkreis m.; Abzweigstromkreis m. || circuit m. dérivé. / driving ~ || Antriebsstromkreis m. || circuit m. d'entrainement. / earth-return ~ || Erdrückleitung f. || retour m. par la terre. / enquiry ~ || Rückfragestromkreis m. || circuit m. de demande. / galvanic ~ || galvanischer Strom m. || courant m. galvanique. / grounded simplex ~ || Einfachleitung f. mit Erde || ligne f. simple avec la terre. / in ~ (Electr) || angeschlossen || en circuit m. / incoming ~ || ankommende Leitung f. || circuit m. d'arrivée. / input ~ (Electr) || Eingangsstromkreis m. || circuit m. d'entrée. / input ~ (Radio) || Empfangsstromkreis m. || circuit m. de réception. / intermediate ~ (Radio) || Zwischenkreis m. || circuit m. intermédiaire. / joining in the ~ || Einschaltung f. || mise f. en circuit. / local ~ || Ortsstromkreis m. || circuit m. local. / main ~ || Hauptstromkreis m. || circuit m. principal. / marking ~ (Tel) || Anreizkreis m. || circuit m. de marquage. / meshed and tuned combination ~ || abgestimmter Schwingungskreis m. || circuit m. oscillatoire syntonisé. / metallic ~ || Drahtleitung f. || circuit m. protecteur. / multiple ~ || Mehrfachstromkreis m. || circuit m. bifurqué. / open radiating ~ || offener Strahlungs-

kreis m. || circuit m. radiant ouvert. / open wire ~s || Freileitung f. || ligne f. aérienne. / oscillating transmitting ~ (Radio) || Sendererregerkreis m. || circuit m. oscillant. / outgoing ~ || abgehende Leitung f. || circuit m. de départ. / output ~ || Verbrauchszweig m. || circuit m. d'utilisation. / plate-to-filament ~ (Radio) || Anodendrahtstromkreis m. || circuit m. filament plaque. / power ~ || Kraftleitung f. || ligne f. de force motrice. / primary ~ || Primärstromkreis m. || circuit m. inducteur ou primaire. / principal ~ || Hauptstromkreis m. || circuit m. principal. / to put in ~ || einschalten || mettre en circuit m. / regenerative ~ || Rückkopplung f. || couplage m. en arrière; rétroaction f. / repeatered ~ || mit Verstärkern versehene Leitung f. || circuit m. muni d'amplificateurs. / secondary ~ || sekundärer Stromkreis m.; Nebenstromkreis m. || circuit m. secondaire. / short ~ || Kurzschluß m. || court-circuit m. / to short ~ || kurzschalten; kurzschließen || court-circuiter. / side ~ || Stammleitung f. || circuit m. reel; ligne f. de base. / single-wire ~ || Einzelleitung f. || circuit m. à simple fil. / superposed ~ || Vierersprechkreis m. || circuit m. de conversation quadruple ou combiné.

circuit breaker (Elektr) || selbsttätiger Ausschalter m. || disjoncteur m. / ~ (Hydr arch) || Stromschütz n. || interrupteur m. de courant. / automatic ~ (Electr) || Selbstausschalter m. || déclencheur m. automatique. / hand ~ || Handausschalter m. || interrupteur m. à main. / no-voltage ~ || Nullspannungsausschalter m. || interrupteur m. à manque d'intensité. / overload ~ || Maximalausschalter m. || interrupteur m. à maximum.

circuit breaker and closer || Stromunterbrecher m. und Stromschließer m. || disjoncteur m. et conjoncteur m.

circuit changer (Electr) || Kommutator m. || commutateur m. / ~ closer || Schalter m. || conjoncteur m. / ~ closing position (Electr) || Stromschließstellung f. || position f. de fermeture du circuit. / ~ crossing (El line) || Platzwechsel m. || croisement m. des circuits; rotation f. / luminous ~ diagram || Leuchtschaltbild n. || diagramme m. lumineux. / ~ equilizer (Tel) || Stromreiniger m. || filtreur m.

circuiting, short ~ device || Kurzschließer m. || dispositif m. de mise en court-circuit

circuit studies pl. || Schaltungslehre f. || étude f. des circuits.

circular || Rundschreiben n.; Zirkular n. || lettre f. circulaire; circulaire f.

circular || kreisrund; kreisförmig || circulaire. / division of a ~ arc || Kreisteilung f. || graduation f. circulaire. / ~ brush || Radbürste f. || brosse f. à couronne circulaire. / ~ bubble || Dosenlibelle f. || nivelle f. sphérique; niveau m. rond. / ~ convex concentrator (Ore dress) || Rundherd m. || table f. tournante. / ~ conveyor || Kreisförderer m. || transporteur m. circulaire. / ~ disc || kreisförmiges Kurbelblatt n. || flasque f. circulaire de manivelle. / ~ dividing apparatus for toothed wheels || Kreisteilapparat m. für Zahnräder || appareil m. diviseur pour roues dentées. / ~ fabric || Rundware f. || tricot m. circulaire. / ~ flanging press || Kümpelpresse f. || presse f. à emboutir les fonds de

chaudière. / ~ form tool holder || Rundstahlhalter m. || porte-outil m. pour outil circulaire. / ~ frame || Rundstrickstuhl m. || métier m. circulaire. / ~ function || Kreisfunktion f.; Winkelfunktion f. || fonction f. circulaire. / ~ graduating machine || Kreisteilmaschine f. || machine f. à graduer les cercles. / ~ grinding machine || Rundschleifmaschine f. || machine f. à rectifier les pièces rondes. / ~ groove (Mach) || ringförmige Nute f. || rainure f. annulaire. / ~ knit goods pl. || Rundwirkwaren fpl. || tricot m. circulaire. / ~ knitter || Rundstrickmaschine f. || machine f. à tricoter circulaire. / Jacquard ~ knitting machine || Jacquardrundstrickmaschine f. || machine f. à tricoter circulaire Jacquard. / ~ letter || Rundschreiben n. || circulaire m.; lettre f. circulaire. / ~ line || Kreislinie f. || ligne f. circulaire. / ~ linking machine || Rundkettelmaschine f. || remmailleuse f. circulaire. / ~ micrometer || Kreismikrometer n. || micromètre m. à cercles. / ~ milling machine || Rundfräsmaschine f. || fraiseuse f. circulaire. / ~ motion || Kreisbewegung f. || mouvement m. circulaire. / ~ nut || Stellmutter f. || écrou n. à trous ou de fixage. / ~ profiling apparatus || Zylinderkurvenfräsapparat m. || appareil m. à fraiser les cames cylindriques. / ~ railway || Ringbahn f. || chemin m. de fer de ceinture. / ~ wooden ~ rib || Holzring m. || bague f. en bois.

circular saw || Kreissäge f. || scie f. circulaire ou ronde. / metal ~ || Metallkreissäge f. || scie f. circulaire à métaux. / ~ for mines || Grubenkreissäge f. || scie f. circulaire pour mines. / ~ for roof shingles || Schindelmaschine f. || scie f. circulaire à bardeaux. / ~ bench for cutting metal || Tischkreissäge f. für Metall || scie f. circulaire d'établi à métaux. / ~ sharpening machine || Schärfmaschine f. für Metallkreissägeblätter || affûteuse f. pour lames de scie circulaire à métal.

circular scale || Kreisskala f. || cercle m. gradué. / ~ section || kreisrunder Querschnitt m. || section f. circulaire. / ~ shape of the crank web || kreisförmiges Kurbelblatt n. || flasque f. circulaire de manivelle. / ~ shaping attachment || Rundhobelvorrichtung f. || dispositif m. à raboter circulaire. / ~ shears pl. || Kreisschere f. || cisaille f. circulaire ou cylindrique. / ~ slide || Rundschieber m. || tiroir m. cylindrique. / ~ spirit level || Dosenlibelle f. || nivelle f. sphérique; niveau m. rond. / ~ spring needle machine || Rundwirkmaschine f. || métier m. circulaire à tricot. / ~ stone saw (Techn) || Steinkreissäge f. || scie f. circulaire à tailler des pierres. / ~ subsidence || Kesselsenkung f. || affaissement m. en forme de cuvette. / ~ suspension of the type bars || kreisförmige Aufhängung f. der Typenhebel || suspension f. circulaire des tiges à caractères. / ~ table (Slotting mach) || Rundtisch m. || table f. circulaire. / ~ work (Weav) || Schlauch m. || tricot m. tube.

circulate, to || umlaufen; zirkulieren || circuler.

circulated, widely || weitverbreitet || très répandu (dans le public).

circulating library || Leihbibliothek f. || cabinet m. de lecture. / ~ library for the

workmen and staff ‖ Bücherhalle f. für Arbeiter und Beamte ‖ bibliothèque f. pour les ouvriers et employés. / ~ pump ‖ Umlaufpumpe f. ‖ pompe f. de circulation.

circulation of air ‖ Luftzirkulation f. ‖ circulation f. d'air. / ~ of bank notes ‖ Banknotenumlauf m. ‖ circulation f. des billets de banque. / ~ of bills of exchange ‖ Wechselverkehr m. ‖ circulation f. des traites. / force ~ ‖ Druckumlauf m. ‖ circulation f. forcée. / in ~ ‖ kursfähig ‖ coté. / to be in ~ ‖ kursieren ‖ circuler; courir. / ~ of money ‖ Geldumlauf m. ‖ circulation f. monétaire. / to recall from ~ ‖ außer Kurs m. setzen ‖ mettre hors (de) circulation f. / water ~ ‖ Wasserumlauf m. ‖ circulation f. d'eau. / supervising the water ~ ‖ Überwachung f. des Wasserkreislaufes ‖ surveillance f. de la circulation d'eau.

circulation constant ‖ Umlaufgröße f. ‖ constante f. de circulation. / ~ heating ‖ Umlaufheizung f. ‖ chauffage m. à circulation. / ~ pipe ‖ Umlaufrohr n. ‖ tube m. de circulation f. / ~ pump ‖ Zirkulationspumpe f. ‖ pompe f. de circulation.

circumference ‖ Peripherie f.; Umfang m. ‖ circonférence f.; périphérie f. / ~ of a circle ‖ Kreisumfang m.; Kreisperipherie f. ‖ circonférence f. du cercle. / ~ of the wheel ‖ Umfang m. des Rades ‖ circonférence f. de la roue.

circumferential seam welding machine ‖ Rundnahtschweißmaschine f. ‖ machine f. à souder les joints circonférentiels ou les joints ronds.

circumferentor ‖ Hängekompaß m. ‖ poche f. de mineur; boussole f.

circumscribe, to (Opt) ‖ abgrenzen ‖ délimiter.

circumstances pl. ‖ Sachlage f.; besondere Verhältnisse npl. ‖ circonstances fpl.

cistern ‖ Zisterne f. ‖ citerne f. / ~ (Mach) ‖ Wasserkasten m. ‖ citerne f.; réservoir m. à eau / ~ of barometer ‖ Gefäß n. eines Barometers ‖ cuvette f. d'un baromètre. / sheet-iron ~ ‖ Eisentank m. ‖ bâche f. en tôle. / wooden ~ ‖ Bottich m. ‖ cuve f. en bois.

cistern barometer ‖ Gefäßbarometer n. ‖ baromètre m. à cuvette.

citrate ‖ zitronensaures Salz n.; Zitrat n. ‖ citrate m.

citric acid ‖ Zitronensäure f. ‖ acide m. citrique. / ~ producing plant ‖ Zitronensäureherstellungsanlage f. ‖ installation f. à fabriquer l'acide citrique.

citron ‖ Zitrone f. ‖ citron m.

citronella oil ‖ Zitronellöl n. ‖ essence f. de citronelle.

city exchange line (Tel) ‖ Postleitung f. ‖ ligne f. au réseau. / ~ fog ‖ Stadtnebel m. ‖ brouillard m. de ville. / ~ map ‖ Stadtplan m. ‖ plan m. de ville. / ~ water ‖ Wasserleitungswasser n. ‖ eau f. de la conduite.

civary of vaulting (Arch) ‖ Gewölbfach n.; Feld zwischen Gewölbrippen ‖ pan m. de voûte.

civet ‖ Zibet n. ‖ civette f.

civil aeroplane ‖ Zivilflugzeug n. ‖ avion m. civile. / ~ engineer ‖ Bauingenieur m. ‖ ingénieur m. des ponts et chaussées ou des constructions civiles. / ~ servant ‖ Beamte m. ‖ fonctionnaire m.; employé m. / ~ service duty ‖ Zivildienstpflicht f.; Arbeitsdienstpflicht f. ‖ devoir m. civique.

civilized, force of public opinion of the ~ world ‖ Macht f. der öffentlichen Meinung der zivilisierten Welt ‖ force f. d'opinion publique des nations civilisées.

clack see also ~ valve ‖ Klappe f.; Ventilklappe f. ‖ clapet m. / ~ of the pump ‖ Pumpenventil n. ‖ clapet m. de pompe.

clack box ‖ Ventilgehäuse n. ‖ boîte f. à clapet; corps m. de soupape.

clack seat ‖ Ventilsitz m. ‖ siège m. du clapet. / ~ ring ‖ Ventilsitzring m. ‖ anneau m. de siège de clapet.

clack valve ‖ Klappenventil n.; Windklappe f. ‖ clapet m.; soupape f. à clapet ou à charnière.

claim, to ‖ fordern; beanspruchen; Anspruch erheben ‖ demander; réclamer. / ~ (Mining) ‖ muten ‖ demander la concession d'une mine.

claim (Demand) ‖ Anspruch m.; Rechtsanspruch m. ‖ prétention f.; exigence f. / ~ (Right) ‖ Anrecht n. ‖ droit m.; titre m. / ~ (Refusal of goods) ‖ Beanstandung f.; Reklamation f. ‖ réclamation f. / ~ (Min share) ‖ Kux m.; Anteil m. ‖ part f. de mine; valeur f. minière. / to abandon all ~s ‖ alle Ansprüche mpl. fallen lassen ‖ abandonner toutes réclamations fpl. / ~ based on a bill of exchange ‖ Wechselforderung f. ‖ créance f. fondée sur lettre de change. / ~ for compensation for damage see ~ for damages. / ~ for damages ‖ Anspruch m. auf Schadenersatz; Entschädigungsanspruch m. ‖ réclamation f. (en réparation de dommage). / to give proof of a ~ for damages ‖ einen Schadensersatzanspruch m. glaubhaft machen ‖ justifier une réclamation en réparation de dommage. / ~ to delivery ‖ Bezugsrecht n. ‖ droit m. d'émission. / legitimacy of a ~ ‖ Rechtmäßigkeit f. einer Forderung ‖ légitimité f. d'une prétention. / the ~ of the patent is clearly defined ‖ der Patentanspruch ist klar ausgedrückt ‖ les points principaux qui sont protégés par le brevet résultent clairement de la demande. / to raise a ~ ‖ reklamieren ‖ réclamer. / to satisfy all ~s pl. ‖ allen Ansprüchen mpl. genügen ‖ satisfaire toutes les exigences fpl. / ~ of a technical character ‖ Reklamation f. oder Beanstandung f. technischer Art ‖ réclamation f. de nature technique.

claimant (Mining) see claimholder.

claimholder (Mining) ‖ Muter m. ‖ aspirant m. d'une mine.

claim validity ‖ Rechtsgültigkeit f. einer Forderung ‖ validité f. d'une créance.

clamp, to ‖ festklemmen; klemmen ‖ serrer; caler; brider. / ~ (Mach) ‖ arretieren; festklemmen ‖ bloquer; serrer à bloc. / ~ a board ‖ ein Brett mit Hirnleisten versehen ‖ emboîter une planche. / ~ the bricks ‖ den Satz machen ‖ mettre les briques fpl. en haies. / ~ by means of a lever ‖ durch einen Hebel m. festklemmen ‖ serrer au moyen d'un levier. / ~ the tool ‖ den Drehstahl m. einspannen ‖ serrer l'outil m.

clamp ‖ Zwinge f.; Klammer f.; Krampe f. ‖ crampon m.; agrafe f.; pince f. / ~ (Join) ‖ Hirnleiste f. ‖ listel m. de traverse; emboîture f. / ~ (Locksm) ‖ Haspe f. ‖ happe f.; crampon m. / ~ adjustable ‖ Schraubzwinge f. ‖ serre-joints m.; presse f. à main. / bench ~ ‖ Bankzwinge f.; Hobelbankeisen n. ‖ presse f. d'établi. / ~ of bricks ‖ Hag m.; geschichteter Ziegelhaufen m. ‖ haie f. de briques. / ~ for coarse adjustment (Telescope etc.) ‖ Klemmung f. der Grobverstellung ‖ blocage m. du mouvement rapide. / crossing-over ~ ‖ Kreuzklemme f. ‖ pince f. de croisement. / exhaust manifold ~ ‖ Auspuffrohrschelle f. ‖ collier m. de tuyau d'échappement. / ~ for eyepiece draw tube ‖ Klemmung f. des Okularauszuges ‖ vis f. immobilisant le tube-tirage. / ~ for fine adjustment (Telescope etc.) ‖ Klemmung f. der Feinverstellung ‖ blocage m. du mouvement lent. / ~ for focussing attachment ‖ Klemmung f. der Fokussierung ‖ blocage m. de la mise au point. / gripping ~ ‖ Greifhaken m. ‖ crochet m. prenant. / lockfiler's ~s pl. ‖ Reifkloben m.; Reifkluppe f. ‖ mordache f. à chanfrein. / splicing ~ ‖ Verbindungsklammer f. ‖ agrafe f. de joint. / straining ~ ‖ Abspannklemme f. ‖ borne f. de suspension ou d'arrêt. / terminal ~ ‖ Anschlußklemme f. ‖ pince f. de bout. / twist ~ ‖ Würgeklemme f. ‖ pince f. à torsade. / vice ~s pl. ‖ Spannkluppe f. mordache f.; crampon m. / wire ~ (Electr) ‖ Drahthalter m.; Drahtklemme f. ‖ serre-fil m.; borne f. / wooden ~ (Clothespeg) ‖ Holzklammer f.; hölzerne Wäscheklammer f. ‖ fichoir m. pince f. à linge.

clamp connection ‖ Klemmverbindung f. ‖ accouplement m. à serrage / ~ connector ‖ Bindeklammer f. ‖ serre-joints m. / ~ dog ‖ Spannkloben m. ‖ griffe f. de serrage; étau m. à main.

clamper ‖ Furnierpresser m. ‖ plaqueur m. à la presse.

clamping apparatus (Electr) ‖ Aufspannvorrichtung f. ‖ appareil m. de serrage et de fixation. / ~ bolt ‖ Druckbolzen m.; Klemmbolzen m. ‖ boulon m. de serrage. / ~ chuck ‖ Klemmfutter n.; Spannfutter n. ‖ mandrin m. de serrage. / ~ collar ‖ Klemmring m. ‖ bague f. de fixation. / ~ device for sheets ‖ Blechspannvorrichtung f. ‖ mécanisme m. à serrer les tôles. / ~ eye ‖ Klemmauge n. ‖ œil m. de blocage. / ~ lever ‖ Klemmhebel m. ‖ manette f. de serrage a. / ~ plate ‖ Klemmplatte f. ‖ plaque f. de serrage. / electromagnetic ~ plate ‖ elektromagnetische Aufspannplatte f. ‖ plaque f. électromagnétique de fixation. / ~ ring ‖ Klemmring m.; Druckring m.; Eisenzwinge f. ‖ anneau m. ou bague f. ou rondelle f. de pression; collier m. de fer; virole f. ou collier m. de serrage. / ~ screw ‖ Klemmschraube f. ‖ vis f. d'arrêt ou de serrage; serre-fil m. / to release the ~ screw ‖ die Klemmschraube f. lösen ‖ desserrer la vis de pression. / ~ sleeve and nut ‖ Spannhülse f. mit Mutter ‖ manchon m. de serrage avec écrou. / ~ surface on slide ‖ Stößelspannfläche f. ‖ surface f. de la semelle du coulisseau. / ~ tool ‖ Spannwerkzeug n.; Aufspannwerkzeug n.; Einspannwerkzeug n. ‖ outil m. à serrer ou de serrage.

clamp iron for walls ‖ Mauerklammer f. ‖ crampon m. de maçon. / ~ jaw ‖ Klemmbacke f. ‖ coussinet m. ou mâchoire f. ou mordache f. de serrage. / displaceable ~ lever ‖ Klemmverschiebehebel m. ‖ levier m. à pince déplaçable. / ~ ring see clamping ring. / ~ roller (El line) ‖ Klemmrolle f. ‖ bobine f. à borne.

clamps pl. see clamp.

clapper (Bell) ‖ Klöppel m. ‖ battant m. / ~ (Mill) ‖ Rührnagel m.; Rüttelstock m. ‖ battant m. / ~ (Pump) *see* clack valve.

clap-sill ‖ Schlagschwelle f.; Schleusenschwelle f. ‖ seuil m. du busc; seuillet m. d'écluse; busc m.

clarification (Brew) ‖ Abläuterung f.; filtration f. / ~ (Chem) ‖ Klärung f. ‖ clarification f. / ~ of beer ‖ Klären n. des Biers ‖ collage m. de la bière. / ~ of the juice ‖ Saftreinigung f. ‖ épuration f. de jus sucrés. / ~ of wine ‖ Klärung f. des Weines ‖ collage m. de vin.

clarifier ‖ Klärpfanne f. ‖ chaudière f. à défécation *ou* à clarification; claire f.; poisonnière f.

clarifierman (Brew) ‖ Klärmeister m. ‖ clarifieur m.

clarify, to (Brew) ‖ abläutern; läutern ‖ filtrer. / ~ (Chem) ‖ (sich) abklären; (sich) klären ‖ (se) clarifier.

clarifying (Brew) ‖ Schönen n.; Läuterung f. ‖ clarification f.; collage m. / ~ (Sugar) ‖ Scheidung f.; Klärung f.; Läuterung f. ‖ défécation f. première; clarification f. / ~ agent ‖ Klärmittel n. ‖ clarifiant m. / ~ filter ‖ Klärfilter m. ‖ filtre m. clarificateur. / ~ plant ‖ Kläranlage f. ‖ installation f. de décantation. / ~ tank ‖ Klärbecken n.; Läuterungsbecken n. ‖ bassin m. de décantation.

clarinet ‖ Klarinette f.; Gellflöte f. ‖ clarinette f. / ~ mouthpiece ‖ Klarinettenmundstück n. ‖ bec m. de clarinette.

clarion (Mus instr) ‖ Zinke f. ‖ clairon m.

clarionet *see* clarinet.

clash gears pl. ‖ Zahnradwechselgetriebe n. ‖ changement m. de vitesse par engrenages.

clasp, to ‖ einhaken; klammern ‖ enclencher; cramponner; accrocher.

clasp ‖ Agraffe f.; Schnalle f.; Spange f. ‖ agrafe f.; boucle f.; fermoir m. / ~ (Bookb) ‖ Buchschloß n. ‖ fermoir m. / ~ (Dentistry) ‖ Klammer f.; Zahnklammer f.; Gebißklammer f. ‖ crochet m. / ~ (Locksm) ‖ Riegelhaken m.; Riegelhaspe f. ‖ verterelle f. de verrou. / ~ and eye ‖ Haken m. und Öse f. ‖ crochet m. et œillet m. / apron ~ ‖ Schürzenhaken m. ‖ agrafe f. de tablier. / cask ~ ‖ Faßspange f. ‖ penture f. / ~ for iron hoops ‖ Verschluß m. für Bandeisen ‖ boucle f. pour feuillard de fer. / ~ for trousers ‖ Hosenschnalle f. ‖ boucle f. de pantalons. / ~ for waistcoats ‖ Westenschnalle f. ‖ boucle f. de gilets.

clasp knife ‖ Taschenmesser n.; Klappmesser n.; Federmesser n. ‖ canif m. / ~ nail (Carp) ‖ Schindelnagel m. ‖ clou m. à bardeaux.

class, to ~ *see* to classify.

class ‖ Klasse f.; Kategorie f. ‖ classe f.; catégorie f. / ~ (Nav) ‖ Klasse f. ‖ division f. / ~ (Course) ‖ Kursus m.; Lehrgang m. ‖ cours m. / ~ for instruction in first-aid work ‖ Lehrgang m. für erste Hilfe bei Unglücksfällen ‖ cours m. d'instruction de premiers secours. / ~ of workers ‖ Arbeiterkategorie f. ‖ catégorie f. d'ouvriers.

classification ‖ Klasseneinteilung f.; Einteilung f. (in Klassen); Klassifikation f.; Klassenordnung f. ‖ classification f. / ~ (Dress ore) ‖ Sortierung f. ‖ classification f.; calibrage m. / ~ of ore ‖ Erzbezeichnung f. ‖ désignation f. du minerai. / ~ certificate (Mar) ‖ Klassi-

fizierungsschein m. ‖ certificat m. de classification.

classifier (Dress ore) ‖ Setzmaschine f. ‖ classificateur m.; classeur m.; classeurtrieur m. / hydraulic ~ ‖ Spitzlutte f. ‖ spitz-luttle m.

classify, to ‖ in Klassen fpl. einordnen; klassifizieren; einteilen; sichten ‖ classifier; classer; mettre à part; trier. / ~ (Dress ore) ‖ scheiden; ausklauben; klassieren ‖ classer; trier; séparer. / ~ the powder ‖ Pulver n. sortieren ‖ égaliser la poudre.

classifying apparatus (Dress ore) ‖ Klassiervorrichtung f. ‖ appareil-classeur m. / ~ drum ‖ Klassiertrommel f. ‖ tambourclasseur m. / ~ grate ‖ Klassierrost m. ‖ grille-classeur f. / ~ screen ‖ Klassiersieb n. ‖ crible-classeur m.

class microscope ‖ Lehrmikroskop n.; Kursusmikroskop n. ‖ microscope m. de travaux pratiques *ou* d'instruction.

clause ‖ Klausel f.; (einschränkende) Nebenbestimmung f. ‖ clause f. / agreement ~ ‖ Vertragsklausel f. ‖ clause f. de contrat. / most favoured nation ~ ‖ Meistbegünstigungsklausel f. ‖ clause f. de la nation la plus favorisée.

clausthalite ‖ Clausthalit m.; Selenblei n. ‖ claustalite f.; plomb m. sélénié.

claw ‖ Klaue f.; Haken m.; Kralle f. ‖ griffe f. / ~ (Mach) ‖ Klaue f.; Kralle f. ‖ griffe f. / double ~ of the rack of a jack ‖ Gabel f. der Zahnstange einer Winde ‖ griffes fpl. de la crémaillère d'un cric. / ~ of the hammer ‖ Hammerklaue f.; Splitt m.; gespaltene Finne f. des Hammers ‖ panne f. fendue du marteau.

claw coupling ‖ Klauenkupplung f. ‖ embrayage m. *ou* accouplement m. à griffes. / ~ crane ‖ Pratzenkran m. ‖ grue f. à griffes. / ~ hammer ‖ Hammer m. mit gespaltener Finne ‖ Klauenhammer m.; Splitthammer m. ‖ marteau m. à panne fendue. / ~ hook ‖ Klauenhaken m. ‖ crochet m. à griffes. / ~ wrench ‖ Nagelzieher m.; Nagelheber m.; Nagelauszieher m. ‖ arrache-clou m.; tire-clou m.

clay ‖ Ton m.; Letten m. ‖ terre f. glaise; glaise f.; argile f. / to bake the ~ ‖ Ton m. brennen ‖ cuire l'argile. / to mould the ~ (Brick) ‖ Ton m. streichen ‖ mouler l'argile. / to tread ~ ‖ den Ton m. treten ‖ piétiner l'argile. / blue ~ ‖ blauer Ton m. ‖ argile f. bleue. / burnt ~ ‖ gebrannter Ton m. ‖ argile f. cuite. / china ~ ‖ Porzellanerde f. ‖ terre f. à porcelaine. / coloured ~ ‖ Farberde f. ‖ engobe m. / common ~ *see also* loam ‖ Lehm m.; sandiger Ton m. ‖ limon m.; argile f.; terre f. franche ou limoneuse. *ou* à briques. / fine ~ *see* china ~. / fire ~ ‖ Schamotte f.; feuerfester Ton m. ‖ argile f. *ou* terre f. réfractaire. / fusible ~ ‖ schmelzbarer Ton m. ‖ argile f. fusible. / meager ~ ‖ magerer Ton m. ‖ argile f. maigre. / ochrey ~ ‖ ockriger Ton m. ‖ argile f. ocreuse. / pipe ~ ‖ Pfeifenton m.; plastischer *oder* bildfähiger Ton m.; Töpferton m. ‖ argile f. plastique; glaise f. / plastic ~ *see* pipe ~. / potter's ~ *see* pipe ~. / rich ~ ‖ fetter Ton m. ‖ argile f. grasse. / saliferous ~ ‖ Salzton m. ‖ argile f. salifère. / schistous ~ ‖ Schieferton m. ‖ argile f. schisteuse. / ~ mixed with silver ‖ Silberletten m. ‖ argile f. mêlée d'argent. / tamping ~ ‖ Stampf-

masse f. ‖ pisé m. réfractaire damé. / unctuous ~ *see* rich ~.

clay bank (Geol) ‖ Tonschicht f. ‖ banc m. ou couche f. d'argile. / ~ bead ‖ Tonperle f. ‖ perle f. d'argile. / ~ brick ‖ Luftziegel m.; Lehmstein m.; ungebrannter Ziegel m. ‖ brique f. crue; brique f. séchée à l'air; brique f. de limon. / ~ composition ‖ Tonmasse f. ‖ composition f. à base d'argile. / ~ crucible ‖ Tontiegel m. ‖ creuset m. en terre réfractaire. / ~ cutter (Pott) ‖ Tonschneide f.; Tonschneider m. ‖ épée f.; coupe-argiles m.; machine f. à découper l'argile. / ~ cutting device *see* ~ cutter. / ~ dust ‖ Tonmehl n. ‖ argile f. en poudre.

clayey (Miner) ‖ tonig; lettig ‖ argileux; glaiseux. / ~ marl ‖ Tonmergel m. ‖ marne f. argileuse. / ~ soil ‖ Lehmboden m. ‖ terre f. glaiseuse; sol m. argileux.

clay galls pl. ‖ Tongalle f. ‖ taches fpl. d'argile. / ~ industry ‖ Tonindustrie f. ‖ céramique f.; industrie f. céramique.

claying of the sugar ‖ Decken n. *oder* Weißen n. des Zuckers ‖ terrage m. du sucre. / ~ of a bore-hole ‖ Verletten n. eines Bohrlochs ‖ glaisage m. d'un trou de sonde.

claying apparatus (Chem) ‖ Deckapparat m.; Deckvorrichtung f. ‖ appareil m. de couverture. / ~ bar (Min) ‖ Lettenbohrer m.; Trockenbohrer m. ‖ pilon m.; boulon m. à terre glaise.

clay kneader ‖ Lehmkneter m. ‖ malaxeur m. d'argile. / ~ kneading machine ‖ Tonknetmaschine f. ‖ machine f. à pétrir la glaise. / ~ layer *see* ~ bank. / ~ maker *see* ~ mixer. / ~ mass ‖ Tonmasse f. ‖ pâte céramique f. / ~ mill ‖ Tonmühle f. ‖ moulin m. à préparer l'argile. / ~ miner ‖ Lehmgräber m. ‖ argileur m. / ~ mixer ‖ Tonmischer m. ‖ malaxeur m. d'argile. / ~ mixing machine ‖ Tonmischmaschine f. ‖ machine f. à mélanger la glaise. / ~ mortar ‖ Lehmmörtel m. ‖ mortier m. de terre. / ~ pigeon ‖ Tontaube f. ‖ pigeon m. d'argile. / ~ pipe ‖ Tonpfeife f. ‖ pipe f. en terre glaise. / ~ pit ‖ Tongrube f.; Lehmgrube f. ‖ glaisière f. / ~ plate press ‖ Tonplattenpresse f. ‖ presse f. pour plaques en argile. / ~ plug (Found) ‖ Lehmpfropf m.; Stichpfropf m. ‖ tampon m. / ~ press ‖ Tonpresse f. ‖ presse f. à argile. / ~ retort ‖ Tonretorte f. ‖ cornue f. en terre réfractaire. / ~ rollers pl. (Brick) ‖ Tonwalzwerk n. ‖ laminoir m. à l'argile. / ~ slate ‖ Tonschiefer m. ‖ schiste m. argileux. / ~ sod (Hydr arch) ‖ Kleisode f. ‖ gazon m. pour digues. / ~ stove ‖ Tonofen m. ‖ poêle m. de faïence. / ~ toys pl. ‖ Tonspielwaren fpl. ‖ jouets mpl. en argile. / ~ tube ‖ Tonröhre f. ‖ tuyau m. en terre cuite. / ~ ware ‖ Tonwaren fpl. ‖ poterie f. produits mpl. d'argile. / ~ wetting (Brick) ‖ Einsumpfen n. des Tones ‖ détrempage m. de l'argile. / ~ worker ‖ Kunsttöpfer m. ‖ céramiste m. / ~ working machine ‖ Tonbearbeitungsmaschine f. ‖ machine f. à travailler la terre glaise. / ~ works pl. ‖ Tonwerk n.; Tongrube f. ‖ usines fpl. à argile.

cleading of a boiler ‖ Mantel m. des Kessels ‖ enveloppe f. *ou* chemise f. d'une chaudière.

clean ‖ sauber; rein ‖ propre; net. / ~ coal ‖ reine Kohle f. ‖ charbon m. pur. / ~ coffee ‖ rein schmeckender Kaffee m. ‖

café m. droit de goût. / ~ cut ‖ sauberer Schnitt m. ‖ coupe f. bien franche; coupe f. nette. / ~ ore ‖ reines Erz n. ‖ minerai m. pure. / ~ proof (Print) ‖ Revisionsbogen m.; Ansichtsbogen m. ‖ seconde f.; seconde épreuve f. / ~ sheet ‖ Aushängebogen m. ‖ bonne feuille f.

clean, to ‖ putzen; reinigen; säubern ‖ nettoyer; purifier. / ~ (Cotton) ‖ auskörnen ‖ égrener. / ~ (Hydr arch) ‖ schlämmen ‖ débourber. / ~ (Mining) ‖ décombrer. / ~ (Wood) ‖ schlichten ‖ planer; replanir le bois. / ~ a beam with an axe ‖ einen Balken nachbeilen ‖ raviver une poutre. / ~ casks pl. ‖ faßschwanken ‖ rincer les fûts mpl. / ~ the castings pl. ‖ Guß m. putzen ‖ ébarber ou nettoyer la fonte. / ~ the cloth ‖ das Tuch reinigen oder säubern ‖ épautier le drap. / ~ a file ‖ eine Feile reinigen ‖ dégraisser une lime. / ~ raw soda ‖ die Soda reinigen ‖ touiller la soude. / ~ a road ‖ eine Straße reinigen ‖ débourber ou nettoyer une rue. / ~ from rust ‖ rostfrei machen ‖ dérouiller. / ~ the stones from mortar ‖ die alten Steine mpl. vom Mörtel reinigen ‖ décrotter les vieilles pierres ou les vieux carreaux. / ~ the type (Print) ‖ die Schrift waschen ‖ brosser les lettres.

cleaned, electro-chemically ‖ elektrochemisch gereinigt ‖ nettoyé électro-chimiquement.

cleaned coal ‖ Schlämmkohle f. ‖ charbon m. débourbé.

cleaner ‖ Putzer m. ‖ nettoyeur m. / ~ (Coop) ‖ Putzmesser n. ‖ plane f. curette. / ~ (Agent) see cleaning material. / ~ (Met) ‖ Beizer m. ‖ décapeur m. / ~ (Mill) ‖ Reinigungsmaschine f. ‖ nettoyeur m. / ~ (Mining) ‖ Ausräumer m. ‖ vidangeur m. / ~ (Of a mowing machine) ‖ Abstreifer m. ‖ tringle f. de débourrage. / amalgam ~ ‖ Reinigungspfanne f. ‖ cuve f. de purification. / ~ for boiler pipes ‖ Kesselrohrreiniger m. ‖ nettoyeur m. de tubes de chaudières. / locomotive ~ ‖ Lokomotivputzer m. ‖ nettoyeur m. de locomotives. / ~ for pots ‖ Topfreiniger m. ‖ nettoyeur m. de pots. / dry ~ ‖ Trockenreiniger m. ‖ appareil m. épurateur à sec. / quantity of dust deposited in the dry ~s ‖ in den Trockenreinigern sich absetzende Staubmassen fpl. ‖ masses fpl. de poussière qui se déposent dans les appareils épurateurs. / vacuum ~ ‖ Staubsauger m. ‖ aspirateur m. de poussière.

cleaner-up ‖ Gußputzer m.; Reiniger m. ‖ ébarbeur m.; nettoyeur m.

cleaning ‖ Reinigung f. ‖ nettoyage m. / ~ (Found) ‖ Verputzen n. ‖ ébarbage m.; nettoyage m. / ~ (Sweeping) ‖ Kehren n.; Auskehren n.; Wegfegen n. ‖ balayage m. / water motor for ~ bottles ‖ Wassermotor m. zur Flaschenreinigung ‖ moteur m. hydraulique pour le rinçage des bouteilles. / ~ of brass-work ‖ Metallreinigung f. ‖ fourbissage f. / ~ of casks pl. ‖ Faßschwanken n. ‖ rinçage m. des fûts. / ~ of the chain (Weav) ‖ Putzen n. der Kette ‖ remondage m. / ~ of the cisterns (Glassm) ‖ Reinigen n. der Gießhäfen ‖ curage m. des cuvettes. / ~ of the gas ‖ Reinigung f. des Gases ‖ épuration f. du gaz. / mechanical ~ ‖ selbsttätige Reinigung f. ‖ nettoyage m. automatique. / ~ the pieces ‖ Putzen n. von Werkstücken ‖ nettoyage m. des pièces. /

preliminary ~ ‖ Vorreinigung f. ‖ épuration f. préalable. / ~ of sheets ‖ Reinigen n. oder Säubern n. des Bleches ‖ récurage m. de la tôle. / ~ of the ship ‖ Schiffsreinigung f. ‖ nettoyage m. du navire. / thorough ~ ‖ gründliche Reinigung f. ‖ nettoyage m. à fond. / ~ and working with lime-cream ‖ Schwöden n. ‖ enchaussenage m.

cleaning apparatus for castings ‖ Gußputztrommel f. ‖ appareil m. d'ébarbage pour moulages. / ~ for hoses ‖ Schlauchreinigungsapparat m. ‖ appareil m. à nettoyer les tuyaux flexibles. / ~ for the packing space of boiler tube closures ‖ Dichtungsflächenreiniger f. für Siederohrverschlüsse ‖ nettoyeur m. de surfaces des joints des tubes bouilleurs.

cleaning brush ‖ Reinigungsbürste f.; Ausputzkratze f. ‖ brosse f. à nettoyer ou de nettoyage. / ~ (Print) ‖ Waschbürste f. ‖ brosse f. à laver.

cleaning card ‖ Putzkratze f. ‖ carde f. de nettoyage. / ~ cellar (Brew) ‖ Abfüllraum m.; Abfüllkeller m. ‖ cave f. de soutirage; soutirage m. / ~ device for castings ‖ Gußputzanlage f. ‖ installation f. pour l'ébarbage de la fonte. / ~ door (Met) ‖ Reinigungsklappe f. ‖ porte f. de nettoyage. / ~ drum ‖ Scheuerglocke f.; Scheuertrommel f. ‖ cloche f. ou tambour m. de rinçage. / ~ drum (Dress ore) ‖ Putztrommel f. ‖ trommel m. de nettoyage. / ~ hole ‖ Putzloch n. ‖ trou m. de nettoyage. / ~ installation for blast-furnace gas ‖ Gichtgasreinigungsanlage f. ‖ installation f. pour l'épuration des gaz de hauts fourneaux. / ~ knife see cleaner (Coop).

cleaning machine ‖ Putzmaschine f.; Reinigungsmaschine f. ‖ machine f. à nettoyer ou de nettoyage; appareil m. nettoyeur. / ~ for bands ‖ Putzmaschine f. für Bandmaterial ‖ nettoyeur à feuillards. / ~ for carpets ‖ Teppichkehrmaschine f. ‖ machine f. à brosser les tapis. / ~ for corn ‖ Reinigungsmaschine f. für Getreide ‖ machine f. à nettoyer les (grains de) céréales. / ~ for ore delivery wagons ‖ Maschine f. zum Reinigen von Erzförderwagen ‖ machine f. de nettoyage pour wagonnets à minerais. / ~ for rags ‖ Lumpenreinigungsmaschine f. ‖ machine f. à nettoyer les chiffons. / ~ for seeds ‖ Reinigungsmaschine f. für Samen ‖ machine f. à nettoyer les grains de semences. / tram ~ ‖ Förderwagenreinigungsmaschine f. ‖ machine f. à nettoyer les wagonnets de transport. / wool ~ ‖ Wollreinigungsmaschine f. ‖ machine f. de nettoyage de la laine.

cleaning and scrubbing machine for casks ‖ Faßputz- und Schwenkmaschine f. ‖ machine f. à brosser et à rincer les fûts.

cleaning material ‖ Putzmittel n.; Reinigungsmittel n. ‖ moyen m. ou produit m. à nettoyer; matière f. pour nettoyer. / ~ mill ‖ Schrotmühle f. ‖ moulin m. à décortiquer.

cleaning-off machine for casks ‖ Faßabhobelmaschine f. ‖ machine f. à blanchir ou à raboter les fûts.

cleaning plant ‖ Reinigungsanlage f. ‖ installation f. de nettoyage. / bottle ~ ‖ Flaschenreinigungsanlage f. ‖ installation f. pour le rinçage des bouteilles. / malt ~ ‖ Malzputzereianlage f. ‖ installation f. de nettoyage du malt. / waste paper

~ ‖ Altpapierreinigungsanlage f. ‖ installation f. de nettoyage des vieux papiers. / biological ~ for waste water ‖ biologische Kläranlage f. für Abwasser; biologische Abwasserkläranlage f. ‖ installation f. de nettoyage biologique des eaux résiduaires.

cleaning preparation (Shoem) ‖ Ausputzpräparat n. ‖ préparation f. de nettoyage. / ~ preparation of ores ‖ Erzaufbereitung f. ‖ préparation f. mécanique des minerais. / ~ rag ‖ Putzlappen m.; Putztuch n. ‖ toile f. à nettoyer; chiffon m. de nettoyage. / ~ rod (Gun) ‖ Reinigungsstock m. ‖ baguette f. à nettoyer. / ~ roller (Weav) ‖ Sauberkeitswalze f. ‖ rouleau m. de propreté.

cleaning and dressing shop for castings ‖ Gußputzerei f. ‖ atelier m. d'ébarbage.

cleaning table (Found) ‖ Putzbank f. ‖ table f. de dessablage. / ~ tool ‖ Reinigungsgerät n. ‖ outil m. à nettoyer. / ~ tool for nozzle ‖ Düsenreiniger m. ‖ nettoyeur m. du gicleur. / ~ wool ‖ Putzwolle f. ‖ étoupe f. à nettoyer.

cleanse, to ~ see to clean.

cleanser (Mining) see also cleaner ‖ Krätzer m.; Löffelräumer m.; Raumlöffel m. ‖ curette f. de mineur; tire-sable m.

cleansing see cleaning.

clear ‖ hell; klar; blank ‖ clair. / ~ (Easy to survey) ‖ übersichtlich ‖ sommaire; clair. / ~ (Meteor) ‖ heiter; klar ‖ clair. / ~ (Railw) ‖ entblockt ‖ débloqué. / ~ (Trade) ‖ ohne Abzug m.; netto ‖ net. / ~ (Wood) ‖ astrein ‖ dépourvu de branches. / ~ cutting ‖ Waldabtrieb m.; Kahlschlag m. ‖ coupe f. définitive ou à blanc étoc. / ~ felling see ~ cutting. / ~ freight ‖ Nettofracht f. ‖ fret m. net. / ~ gain ‖ Reingewinn m. ‖ bénéfice m. net; profit m. tout clair. / ~ glass globe ‖ Klarglasglocke f. ‖ globe m. en verre clair. / ~ glass screen ‖ Blankscheibe f. ‖ glace f. transparente. / ~ head-room (Build) ‖ Lichtenhöhe f. ‖ hauteur f. du jour; hauteur f. libre. / ~ profit ‖ Reingewinn m.; Nettogewinn m. ‖ bénéfice m. net; profit m. tout clair. / ~ span (Build) ‖ Lichtenweite f. einer Öffnung ‖ échappé f. du jour. / ~ weather ‖ sichtiges oder klares Wetter n. ‖ temps m. clair.

clear, to ~ (A house etc.) ‖ leeren; räumen ‖ vider; évacuer. / ~ (A street etc.) ‖ frei machen ‖ dégager. / ~ (Bookkeep) ‖ saldieren ‖ balancer; solder; arrêter un compte. / ~ (Chem) see also to clarify ‖ klären; läutern; rektifizieren ‖ laisser déposer; clarifier. / ~ (Duty) ‖ verzollen; freimachen; ausklarieren ‖ dédouaner. / ~ (Forest) ‖ lichten ‖ élaguer. / ~ (Nav) ‖ freikommen ‖ échapper; parer; éviter. / ~ (Phot) ‖ abschwächen ‖ affaiblir. / ~ an adit (Mining) ‖ einen Stollen aufräumen ‖ saigner une galerie. / ~ the attle (Mining) ‖ einen alten Schacht aufsäubern ‖ décombrer un vieux puits de mine. / ~ the bars (Boiler) ‖ den Rost von Schlacken reinigen ‖ décrasser la grille. / ~ the cellar ‖ auskellern ‖ décaver. / ~ for coming to anchor ‖ klarmachen zum Ankern ‖ faire péneau. / ~ from customs ‖ verzollen ‖ dédouaner. / ~ a furnace ‖ einen Ofen brechen ‖ nettoyer un fourneau. / ~-in (Mar) ‖ einklarieren ‖ acquitter à la douane. / ~-off old stock ‖ das Lager n.

räumen ‖ vider *ou* déblayer le magasin. / ~ pools from mud ‖ Teiche ausschlämmen ‖ débourber des étangs. / ~ a rope (Nav) ‖ ein Tau klarlegen ‖ parer un câble. / ~ a tackle ‖ ein Takel n. klaren ‖ détordre un palan; défaire les tours mpl. d'un palan. / ~-up (Meteor) ‖ aufheitern; aufklaren ‖ s'éclaircir. / ~-up (Nav) ‖ abwehen ‖ s'appaiser.

clear (Build) ‖ Lichtes n.; lichter *oder* innerer Raum m. ‖ jour m. / ~ (Sugar) ‖ Klärsel n. ‖ clairce f.; clairée f. / in the ~ ‖ im Lichten n.; licht ‖ à l'intérieur; dans l'œuvre.

clearance (Clear space) ‖ Lichtraumprofil n.; lichter Raum m.; Raum m. im Lichten ‖ espace m. libre; jour m. / ~ (Room to move) ‖ Spielraum m. ‖ jeu m.; chasse f. / ~ (Duty) ‖ zollamtliche Abfertigung f.; Zollbelastung f. ‖ expédition f. en douane; taxation f. / ~ (Trade) ‖ Vertrieb m. ‖ débit m.; placement m.; écoulement m.; vente f. / ~ of goods (Duty) ‖ Verzollung f. von Waren ‖ dédouanement m. de marchandises. / noxious ~ of a cylinder ‖ schädlicher Raum m. eines Dampfzylinders ‖ espace m. mort *ou* nuisible d'un cylindre. / ~ of the piston ‖ Kolbenspielraum m. ‖ jeu m. du piston. / side ~ of the piston ‖ seitliche Luft f. des Kolbens ‖ jeu m. latéral de piston. / sufficient ~ provided in the rest ‖ genügender Spielraum m. in der Auflage ‖ jeu m. suffisant dans l'appui.

clearance angle ‖ Ansatzwinkel m. ‖ angle m. de dépouille. / ~ angle of the turning steel ‖ Anstellwinkel m. des Drehmeißels ‖ angle m. de coupe de l'outil de tour. / ~ gauge for rolling stock ‖ Umgrenzungslinie f. für Eisenbahnfahrzeuge ‖ gabarit m. des wagons de chemin de fer. / ~ height with jib lowered (Crane) ‖ Durchfahrtshöhe f. bei gesenktem Ausleger ‖ hauteur f. de passage avec flèche baissée. / ~ limit (Railw) ‖ Ladelehre f.; Lademaß n. ‖ gabarit m. (de chargement). / ~ limit of running-through ‖ Durchfahrtprofil n. ‖ gabarit m. de libre passage. / ~ method (Duty) ‖ Verzollungsmethode f. ‖ méthode f. de dédouanement. / ~ paper ‖ Klarierungsschein m. ‖ acquit m. de douane. / ~ sale ‖ Ausverkauf m.; Räumungsausverkauf m. ‖ vente f. totale; liquidation f. générale; débarras m. / ~ space ‖ Kompressionsraum m. ‖ chambre f. de compression. / ~ volume of cylinder ‖ Kompressionsvolumen n. ‖ volume m. de la chambre de compression.

clearer (Mining) ‖ Hauer m. ‖ coupeur m.; piqueur m.

clearing (Bank) ‖ Skontro n. ‖ arrêté m. de compte. / ~ (Chem) ‖ Klärung f.; Läuterung f. ‖ clairçage m.; défécation f.; clarification f. / ~ (Of ores) ‖ Läuterung f. ‖ débourbage m. / ~ (Of toothed wheel) ‖ Zahnlücke f.; creux m. / ~ of kräusen (Brew) ‖ Durchfallen n. der Kräusen ‖ chute f. des kräusen. / ~ in private branch exchanges (Tel) ‖ Schlußzeichengebung f. in Nebenstellenanlagen ‖ signalisation f. de fin de conversation pour les installations supplémentaires. / ~ of silo ‖ Siloentleerung f. ‖ vidange m. de silos. / ~ from wood ‖ Abholzen n. déboisement.

clearing apparatus ‖ Ausräumapparat m. ‖ appareil m. de curage. / ~ attachment (Textile) ‖ Ausfaserapparat m. ‖ appareil m. à brosser. / ~ block mechanism (Railw) ‖ Freigabeblock m. ‖ bloc m. de libération. / ~ cistern ‖ Klärbecken n. ‖ réservoir m. à clarifier. / ~ hole (Loc) ‖ Spülloch n. ‖ trou m. de lavage.

clearinghouse ‖ Abrechnungsstelle f.; Clearinghouse n. ‖ comptoir m. de règlement. / ~ for Foreign currency ‖ Abrechnungsstelle f. *oder* Clearinghouse n. für Valuta ‖ caisse f. de liquidations pour affaires en change. / ~ business ‖ Giroverkehr m. ‖ opération f. de virement.

clearing iron ‖ Formstecher m. ‖ stoqueur m. / ~ lamp repeating (Tel) ‖ Schlußzeichenübertragung f. ‖ transmission f. de la lampe de clôture. / ~ locomotive (Mining) ‖ Abraumlokomotive f. ‖ locomotive f. de déblai. / ~ pan ‖ Klärpfanne f. ‖ chaudière f. à défécation *ou* à clarification; claire f.; poissonnière f. / ~ pitman ‖ Abraumarbeiter m. ‖ déblayeur m. / ~ relay (Tel) ‖ Schlußrelais n.; Schlußschütz n. ‖ relais m. de fin de conversation. / ~ section (Tel) ‖ Abrückabschnitt m. ‖ canton m. débloqueur. / ~ signal (Tel) ‖ Schlußzeichen n. ‖ signal m. de fin.

clearness (Opt) ‖ Schärfe f. ‖ clarté f. / ~ of modulation (Radio) ‖ Abstimmschärfe f. ‖ sélectivité f.

cleat (Mar) ‖ Kreuzholz n.; Klampe f. ‖ taquet m. / belaying ~ ‖ Kreuzklampe f.; Belegklampe f. ‖ taquet m. d'amarrage. / porcelain insulating ~ ‖ Porzellanklemme f. ‖ taquet m. en porcelaine. / stern ~ ‖ Heckklampe f. ‖ taquet m. de poupe.

cleat insulator ‖ Klemmisolator m. ‖ isolateur m. à collier.

cleavable ‖ spaltbar ‖ clivable. / ~ wood ‖ spaltbares Holz n. ‖ bois m. fendable.

cleavage (Chem) ‖ Spaltung f. ‖ décomposition f. / ~ (Miner) ‖ Spaltung f.; Spaltbarkeit f.; Blätterbruch m. ‖ clivage m. / ~ (Geol) ‖ falsche *oder* sekundäre *oder* transversale Schieferung f. ‖ clivage m.

cleave, to (Chem) ‖ spalten ‖ cliver; écafer; rocter. / ~ (Wood) ‖ klöben; spalten; schlitzen ‖ fendre; refendre. / ~ hard rocks with quoins (Mine) ‖ ketzern ‖ fendre la roche.

cleaver ‖ Baumaxt f.; Holzhaueraxt f. ‖ cognée f. du bûcheron; marlin m. / ~ (Butch) ‖ Bankmesser n. ‖ couperet m.

cleaving grain (Build) ‖ Lager n.; Bruchlager n. ‖ lit m. de carrière. / lower ~ (Build) ‖ unteres Lager n.; Unterlager n.; Lagerfläche f. ‖ lit m. de dessous. / upper ~ (Build) ‖ oberes Lager n.; Oberlager n.; Haupt n. eines Steines ‖ lit m. de dessus.

cleaving chisel for roofing slate ‖ Spaltmeißel m. für Dachschiefer ‖ rabattoir m. pour ardoises. / ~ tool (Basketm) ‖ Schmaler m. ‖ fer m. à écaffer. / ~ wedge ‖ Spaltkeil m. ‖ coin m.

cleft ‖ klüftig ‖ crevassé. / ~ (Wall) ‖ gerissen; gespalten; rissig ‖ lézardé. / ~ stave ‖ gespaltene Faßdaube f. ‖ douve f. fendue. / ~ wood ‖ Langholz n. ‖ bois m. de fil.

cleft ‖ Riß m.; Spalte f.; Spalt m. ‖ déchirure f.; fendille f.; fissure f.; gerçure f.; fêlure f.; crevasse f.

clench, to ‖ einhaken ‖ enclencher; accrocher.

clerestory ‖ Dachaufsatz m. ‖ lanterneau m.

clerical robe ‖ Priesterkleidung f. ‖ vêtements mpl. sacerdotaux.

clerk ‖ Fabrikbeamter m. ‖ employé m. de fabrique. / desk of ~ in charge (Tel) ‖ Aufsichtsplatz m. ‖ place f. de surveillance.

clevelandite ‖ Natronfeldspat m.; Albit m. ‖ albite f.; clevelandite f.

clever utilization ‖ geschickte Ausnutzung f. ‖ utilisation f. habile.

clevis (Mine) ‖ Haken m. am Förderseil ‖ croc m. / ~ (Mach) ‖ gabelförmige Zugstange f. ‖ tirant m. *ou* bielle f. en fourche.

clevy see clevis.

clew to ~ down, a yard (Mar) ‖ herunterholen; niederholen ‖ haler bas; peser sur *ou* dessus.

clew ‖ Knäuel n. ‖ peloton m. / ~ (Mar) ‖ Schothorn n. ‖ point m. d'écoute. / ~ of a course (Mar) ‖ Schothorn n. der Untersegel ‖ point m. d'amure d'une basse voile. / ~ of yarn ‖ Garnknäul m. ‖ pelote f. *ou* peloton m. de fil.

clewed up (Mar) ‖ in der Gei f. hängen; aufgegeit ‖ cargué.

clew-garnet block ‖ Geitaublock m. ‖ poulie f. de cargue-point. / ~ line block see clew-garnet block.

cliché ‖ Klischee n. ‖ cliché m. / ~ manufacturing ‖ Klischeeherstellung f. ‖ fabrication f. de clichés.

click ‖ Schaltklinke f.; Sperrklinke f. ‖ doigt m. d'encliquetage; cliquet m. / ~ (Of a door) ‖ Türklopfer m. ‖ heurtoir m. / ~ (Tel) ‖ Knackton m. ‖ toc m.

clicker (Shoem) ‖ Ausstanzer m. ‖ brocheur m.

click lever ‖ Klingenhebel m. ‖ levier m. à cliquet. / ~ steel (Watchm) ‖ Sperrkegelstahl m. ‖ acier m. à cliquets. / ~ testing (Tel) ‖ Knackprüfung f. ‖ test m. par clic. / ~ and ratchet wheel ‖ Gesperre n. ‖ encliquetage m.

clickwork see click and ratchet wheel.

client ‖ Kunde m. ‖ client m.

cliff ‖ steile Felsküste f. ‖ falaise f. / waves pl. dashing against ~s ‖ Klippenbrandung f. ‖ ressac m. contre les roches.

climate, in any ‖ in jedem Klima n. ‖ sous tous les climats mpl. / continental ~ ‖ Binnenklima n. ‖ climat m. continental. / temperate ~ ‖ gemäßigtes Klima n. ‖ climat m. tempéré.

climatic condition ‖ klimatisches Verhältnis n. ‖ condition f. climatérique.

climatology ‖ Klimalehre f. ‖ climatologie f.

climb, to ‖ bergauf fahren; aufsteigen; klettern ‖ monter. / ~-up a shaft (Mine) ‖ einen Schacht hinauffahren ‖ remonter un puits.

climb ‖ Aufstieg m. ‖ montée f.

climber (El Line) ‖ Kletterschuh m.; Steigeisen n. ‖ étrier m. à grimper; griffe f.

climbing, ramp for ~ ‖ Auflaufzunge f. ‖ aiguille f. en rampe. / ~ ability ‖ Bergsteigefähigkeit f. ‖ tenue f. en côte. / ~ with high ~ capacity (Aero) ‖ mit großem Steigvermögen n. ‖ susceptible de monter très rapidement. / ~ iron ‖ Steigeisen n. ‖ griffe f. / ~ power ‖ Kletterfähigkeit f. ‖ aptitude f. de grimper. / ~ shaft ‖ Fahrschacht m.; Steigschacht m. ‖ puits m. aux échelles. / ~ turntable ‖ Kletterdrehscheibe f. ‖ plaque f. en saillie.

clinch, to ‖ kalt nieten ‖ river à froid. / ~ a rivet head ‖ eine Niete f. stauchen ‖ river une tête.

clinch (Locksm) ‖ Haspen m.; Haspe f. ‖ harpon m.; crochet m.; picolet m.; verterelle f.

clinched and rivetted ‖ niet- und nagelfest ‖ tenant à fer et à clou.

clincher ‖ Klammer f.; Krampe f.; Kramme f.; Haspe f. ‖ crampon m.; clameau m.; crampe f.; agrafe f. / square ∼ ‖ Klammerhaken m.; Zulagklammer f. ‖ lien m. d'assemblage; clameau m. à deux faces; bride f.

clincher rim ‖ Wulstfelge f. ‖ jante f. à talon. / ∼ tyre ‖ Wulstreifen m. ‖ pneu m. à talons.

cling, to ‖ anhaften ‖ se coller. / ∼ to ‖ haften an . . . ‖ être fixé; tenir.

clink, to (Met) ‖ reißen; beim Härten rissig werden ‖ criquer.

clink (Locksm) ‖ Haspe f.; Klammer f.; Krampe f. ‖ happe f.; crampe f.; crampon m.

clinker ‖ Klinker m. ‖ brique f. hollandaise *ou* à four; chantignole f. / ∼ (Slag) ‖ Schlacke f. ‖ scorie f.; mâchefer m.; laitier m. / acidproof ∼ ‖ säurefester Klinker m. ‖ brique f. hollandaise à l'épreuve d'acides. / basic ∼ ‖ basische Schlacke f. ‖ scorie f. basique. / cement ∼ ‖ Zementschlacke f. ‖ laitier m. de ciment. / to form ∼ ‖ sintern ‖ se crasser. / hard ∼ ‖ harter Klinker m. ‖ brique f. dure hollandaise.

clinker brick ‖ Klinker m.; Schlackenstein m. ‖ brique f. hollandaise *ou* de scories. / ∼ built ‖ geklinkert ‖ bâti *ou* bordé à clin.

clinker grate ‖ Schlackenrost m. ‖ grille f. à scories.

clinkering ‖ Backen n. des Kokses ‖ agglutination f. *ou* cuisson f. du coke. / self-∼ ‖ Selbstabschlackung f. ‖ décrassage m. automatique.

clinkering coal ‖ schlackende Kohle f. ‖ charbon m. scorifère.

clinker pit ‖ Schlackenfall m. ‖ puits m. à scories de foyer. / ∼ utilization machine ‖ Schlackenverwertungsmaschine f. ‖ machine f. à utiliser le mâchefer.

clinking (Met) ‖ Reißen n. des Stahls beim Härten ‖ déchirure f.; crique f.

clinkstone ‖ Phonolith m.; Klingstein m. ‖ phonolithe m.

clinometer ‖ Bergwage f.; Gefällsmesser m.; Neigungsmesser m. ‖ éclimètre m.; indicateur m. de pente. / ∼ pendulum ‖ Krängungspendel n. ‖ pendule m. à clinomètre.

clip, to (Pap) ‖ beschneiden ‖ ébarber; rogner. / ∼ (Forg; Locksm) ‖ abschroten ‖ couper à la tranche; trancher.

clip (Farr) ‖ Schnebbe f. *oder* Schneppe f. *oder* Stoß m. *oder* Vorschuh m. am Hufeisengriff ‖ pinçon m. / ∼ (Railw) ‖ Lasche f.; Stoßlasche f. ‖ éclisse f. / ∼ (Weav) ‖ Kluppe f. ‖ tenaille f. / ∼ (Tube) ‖ Schelle f. ‖ collier m.; bride f. / ∼ (Tel) ‖ Bügel m. ‖ étrier m. / ∼ (Plate) ‖ Klemmplatte f. ‖ crapaud m.; plaque f. de serrage. / binding ∼ (Electr) ‖ Befestigungsklemme f. ‖ borne f. de fixation. / brass ∼ (Office) ‖ Messingbriefklammer f. ‖ attache-feuilles f. en laiton. / ∼ for cash blocks ‖ Kassenblockklammer f. ‖ pince f. pour blocs de caisse. / catch-fitted ∼ (Railw) ‖ Nasenklemmplatte f. ‖ plaque f. de serrage à talon. / ∼ with corrugated sides (Office) ‖ Briefklammer f. mit geriffelten Schenkeln ‖ attache-feuilles f. à branches ondulées. / ∼ for fixing light rails ‖ Klemmplatte f. ‖ étrier m. / laundry ∼ ‖ Wäsche-

klammer f. ‖ fichoir m.; épingle f. à linge. / ∼ with looped ends (Office) ‖ Briefklammer f. mit umgebogenen Enden ‖ attache-feuilles f. à pointes recourbées. / nickel-plated ∼ ‖ vernickelte Briefklammer f. ‖ attache f. nickelée. / paper ∼ ‖ Briefklammer f.; Büroklammer f. ‖ attache-lettres f.; attache-feuilles f. / screw rail ∼ ‖ Schraubenklemme f. ‖ crampon m. à vis de serrage. / special ∼ ‖ Spezialklemme f. ‖ pince f. spéciale. / spring ∼ ‖ Federbügel m. ‖ bride f. de ressort. / ∼ with straight ends (Office) ‖ Briefklammer f. mit geraden Enden ‖ attache-feuilles f. à extrémités droites.

clip binder (Office) ‖ Klammerhefter m. ‖ classeur m. à pince.

clipper (Shipb) ‖ Klipper m. ‖ clipper m.

clippers pl. ‖ Haarschneidemaschine f. ‖ tondeuse f. / ∼ for animals ‖ Haarschneidemaschine f. für Tiere ‖ tondeuse f. pour animaux. / hair ∼ ‖ Haarschneidemaschine f. ‖ tondeuse f.

clipping machine for curtain fabrics ‖ Ausschneidemaschine f. für Gardinenstoffe ‖ machine f. à découper les rideaux.

clippings pl. (Coin) ‖ Münzgekrätz n. ‖ cisailles fpl. / ∼ (Gloves) ‖ Abfall m. ‖ rognure f. / ∼ of iron ‖ Eisenabfälle mpl. ‖ ferraille f.; débris mpl. de fer; riblons mpl.

clip plate ‖ Spannplatte f.; Klemmplatte f. ‖ plaque f. de serrage *ou* de fixation. / ∼ spring switch ‖ Federschalter m. ‖ interrupteur m. à ressort.

cloaca ‖ Kloake f.; Abzugschleuse f. ‖ égout m.; cloaque m. / ∼ system ‖ Schwemmsystem n. für Abfallstoffe ‖ système m. d'égouts *ou* de cloaques.

cloak ‖ Mantel m. ‖ manteau m. / caoutchouc ∼ ‖ Gummimantel m. ‖ manteau m. de caoutchouc. / woven material for ∼s ‖ Mantelstoff m. ‖ étoffe f. à manteaux.

cloak bag ‖ Mantelsack m. ‖ porte-manteau m. / leather ∼ ‖ lederne Reisetasche f. ‖ valise f. en cuir.

cloak room ‖ Kleiderablage f. ‖ vestiaire m. / ∼ strap (Saddl) ‖ Mantelpackriemen m.; Mantelriemen m. ‖ courroie f. de manteau.

clock ‖ Uhr f. ‖ montre f.; horloge f. / ∼ (Stocking) ‖ Zwickel m. ‖ grisotte f. / alarm ∼ ‖ Weckeruhr f. ‖ réveille-matin m. / ∼ wound annually ‖ Jahresuhr f. ‖ horloge f. marchant une année. / astronomical ∼ ‖ astronomische Uhr f. ‖ horloge f. astronomique. / ∼ for automatic staircase lighting ‖ Uhr f. für selbsttätige Treppenbeleuchtung ‖ interrupteur m. d'éclairage électrique automatique d'escalier. / auxiliary master ∼ ‖ Unterhauptuhr f. ‖ horloge f. principale secondaire. / ∼ for buildings ‖ Uhr f. für Gebäude ‖ horloge f. d'édifice. / control ∼ ‖ Kontrolluhr f. ‖ horloge-contrôle f. / cuckoo ∼ ‖ Kuckucksuhr f. ‖ pendule à coucou. / electric ∼ ‖ elektrische Uhr f. ‖ horloge f. électrique. / electrically controlled ∼s pl. ‖ elektrische Uhrenanlage f. ‖ horloges fpl. électriques. / ∼ for equatorial telescope mounting ‖ Uhrwerk n. für parallaktische Fernrohrmontierung ‖ mouvement m. d'horlogerie pour lunettes parallactiques. / hanging ∼ ‖ Wanduhr f. ‖ pendule f. suspendue. / large ∼ ‖ Großuhr f. ‖ grande horloge f. / master ∼ ‖ Normal-

uhr f. ‖ horloge f. régulatrice. / monumental ∼ ‖ Prachtuhr f. ‖ horloge f. monumentale. / ∼ for motion in right ascension (Astro) ‖ Uhrwerk n. für den Stundenantrieb ‖ horloge f. pour actionner le mouvement horaire. / pendulum ∼ ‖ Penduluhr f. ‖ pendule f. / pneumatic ∼ ‖ pneumatische Uhr f. ‖ horloge f. pneumatique. / secondary ∼ ‖ Nebenuhr f.; sympathische Uhr f. ‖ horloge f. secondaire. / spring ∼ ‖ Federuhr f. ‖ horloge f. à ressort. / spring-driven ∼ ‖ Uhrwerk n. mit Federantrieb ‖ mouvement m. d'horlogerie à ressort. / synchronous electric ∼ system ‖ elektrisches Synchronuhrensystem n. ‖ régime m. d'horloges électriques synchronisées. / the ∼s pl. synchronize ‖ die Uhren fpl. laufen im Gang überein ‖ les horloges fpl. marchent synchroniquement. / table ∼ ‖ Standuhr f. ‖ pendule f. de cheminée. / to take a ∼ to pieces ‖ eine Uhr auseinandernehmen ‖ démonter une horloge. / telephone ∼ ‖ Telefonuhr f. ‖ chronomètre m. pour téléphones. / tower ∼ ‖ Turmuhr f. ‖ horloge f. de clocher. / travelling ∼ ‖ Reiseuhr f. ‖ pendule f. de voyage. / weight-driven ∼ ‖ Uhrwerk n. mit Gewichtsantrieb ‖ mouvement m. d'horlogerie à poids. / wooden ∼ ‖ hölzerne Uhr f. ‖ horloge f. en bois. / x-day ∼ ‖ x-Tageuhr f. ‖ montre f. x jours.

clock bell ‖ Uhrglocke f. ‖ cloche f. d'horloges. / ∼ box ‖ Uhrgehäuse n. für Großuhren ‖ caisse f. d'horloges. / ∼ case see ∼ box. / ∼ case for pendulum ∼s ‖ Gehäuse n. für Penduluhren ‖ cage f. pour pendules. / ∼ circle (Astro) ‖ Uhrkreis m. ‖ roue f. hélicoïdale pour entraîner l'ensemble des axes de rotation. / ∼ dial ‖ Zifferblatt n. ‖ cadran m. de montres. / ∼ factory ‖ Uhrenfabrik f. ‖ fabrique f. d'horloges; horlogerie f. / ∼ glass ‖ Uhrglas n. ‖ verre de montre. / ∼ hand ‖ Uhrzeiger m. ‖ aiguille f. de cadran.

clock installation ‖ Uhrenanlage f. ‖ installation f. d'horloge. / central ∼ ‖ Zentraluhrenanlage f. ‖ installation f. centrale d'horloges. / secondary clock of an electric ∼ ‖ Nebenuhr f. einer elektrischen Uhrenanlage ‖ horloge f. secondaire d'une installation d'horloges électriques.

clock key ‖ Uhrschlüssel m. ‖ clef f. de pendules. / ∼ machine (Stocking) ‖ Zwickelmaschine f. ‖ machine f. à faire la grisotte.

clockmaker ‖ Uhrmacher m. ‖ horloger m. / ∼'s file ‖ Uhrmacherfeile f. ‖ lime f. d'horloger.

clock materials pl. ‖ Uhrzubehör n. ‖ fournitures fpl. d'horlogerie.

clock movement ‖ Uhrwerk n. ‖ mouvement m. d'une horloge. / casing of the ∼ ‖ Einpassen n. des Uhrwerks in das Gehäuse ‖ emboîtage m. du mouvement. / ∼ factory ‖ Uhrwerkfabrik f. ‖ fabrique f. de mouvements de montres.

clock sub-station ‖ Uhrenunterzentrale f. ‖ centrale f. d'horloges à relais. / ∼ test ‖ Uhrprüfung f. ‖ vérification f. de la montre; contrôle m. du chronomètre. / ∼ weight ‖ Uhrgewicht n. ‖ poids m. d'horloge.

clockwise ‖ Uhrzeigersinn m. ‖ sens m. des aiguilles d'une montre. / in counter-direction ‖ entgegengesetzt dem Uhrzeigersinn ‖ en sens m. inverse des aiguil-

les d'une montre. / in ~ direction ‖ im Uhrzeigersinn m. ‖ dans le sens m. des aiguilles d'une montre. / rotation in ~ direction ‖ Drehung f. im Sinne des Uhrzeigers ‖ tournant m. dans le sens des aiguilles d'une montre. / ~ rotation ‖ Rechtsdrehung f. ‖ rotation f. à droite.

clockwork ‖ Uhrwerk n. ‖ mouvement m. d'une horloge. / ~ arc-lamp ‖ Uhrwerkbogenlampe f. ‖ microlampe f. à arc munie d'un mouvement d'horlogerie. / ~-driven disk ‖ durch Uhrwerk angetriebene Scheibe f. ‖ disque m. entraîné par mouvement d'horlogerie. / ~ factory ‖ Uhrwerkfabrik f. ‖ fabrique f. de mouvements de montres. / automatic ~ regulating ‖ selbsttätige Regelung f. des Uhrwerks ‖ réglage m. automatique du mouvement d'horlogerie.

clod of earth ‖ Erdballen m. ‖ motte f. de terre.

clod beetle ‖ Schollenschlägel m. ‖ émottoir m. / ~ coal ‖ Sinterkohle f.; Grobkohle f. ‖ houille f. demi-grasse à longue flamme; charbon m. vif; houille f. grossière ou en mottes. / ~ crusher ‖ Ackerschleife f.; Schollenbrecher m.; Schollenzerteiler m. ‖ rouleau m. brise-mottes; traînoir m.

cloddish ‖ klumpig ‖ motteux.

clog, to ‖ verstopft werden; sich verstopfen ‖ se boucher; s'obstruer.

clog ‖ Holzschuh m.; Pantine f. ‖ sabot m.; socque m. / ~ (Ram) ‖ Knebel m. ‖ garrot m.

clogged ‖ verstopft ‖ bouché; obstrué.

clogging (Mill) ‖ Verkleisterung f. ‖ gommage m. / precocious ~ of the separating apparatus ‖ vorzeitiges Verstopfen n. der Trennungsapparate ‖ obstruction f. précoce des appareils séparateurs.

clog lace ‖ Holzschuhriemen m. ‖ bride f. de sabot.

close ‖ geschlossen ‖ fermé. / ~ (Print) ‖ kompreß; dicht; undurchschossen ‖ serré. / ~ coupling (Radio) ‖ feste Kopplung f. ‖ accouplement m. fort; couplage m. serré. / ~ goods pl. ‖ dichte Ware f. ‖ tricot m. serré.

close, to ‖ schließen; zuschließen; verschließen; abschließen ‖ fermer; serrer. / ~ a business ‖ ein Geschäft n. abschließen ‖ conclure ou terminer une affaire. / ~ a circuit ‖ einen Stromkreis schließen ‖ fermer un circuit. / ~ the open cover sheets (Bookb) ‖ die offenen Umschlagseiten fpl. umlegen ‖ refermer les deux parties fpl. ouvertes de la couverture.

close of mail ‖ Postschluß m. ‖ heure f. du courrier.

closed ‖ abgeschlossen ‖ terminé. / ~ (Electr) ‖ eingeschaltet ‖ fermé. / ~ ashpit furnace ‖ Unterwindfeuerung f. ‖ foyer m. à soufflage sous grille. / ~ cell (Electr) ‖ geschlossene Zelle f. ‖ élément m. fermé.

closed-circuit connections pl. ‖ Ruhestromschaltung f. ‖ couplage m. à courant permanent. / ~ current ‖ Ruhestrom m. ‖ courant m. permanent. / ~ working ‖ Ruhestrombetrieb m. ‖ exploitation f. à circuit fermé.

closed-coil armature ‖ Anker m. mit geschlossener Wicklung ‖ induit m. à circuit fermé. / ~ armature dynamo ‖ Dynamo f. mit geschlossener Ankerwicklung ‖ dynamo f. avec induit à bobinage fermé.

closed-end impedance ‖ Kurzschlußwiderstand m. ‖ impédance f. en court-circuit.

closed type (Mach) ‖ geschlossene Bauart f. ‖ construction f. fermée.

close-range conveyor ‖ Nahförderer m. ‖ transporteur m. à courte distance.

closer, door ‖ Türschließer m. ‖ fermeporte m.

close-support artillery ‖ direkte Unterstützungsartillerie f. ‖ artillerie f. d'appui direct. / ~ tank (War impl) ‖ Nahkampf- oder Unterstützungskampfwagen m. ‖ char m. d'accompagnement.

closet ‖ Spülabort m.; Klosett n. ‖ cabinet m. d'aisance; latrines fpl. / water ~ ‖ Wasserklosett n. ‖ water-closet m.; latrine f. à l'anglaise; W. C.

closet hopper ‖ Aborttrichter m. ‖ cuvette f. du water-closet. / ~ rinsing box ‖ Abortspülkasten m. ‖ caisse f. à rincer les closets. / ~ seat ‖ Klosettsitz m. ‖ siège m. de cabinet. / ~ spring (Locksm) ‖ Schrankschnäpper m. ‖ ressort m. de placard.

close-wall ‖ Einfriedigungsmauer f. ‖ mur m. de clôture ou d'enceinte.

closing ‖ Verschluß m. ‖ fermeture f. / airtight ~ ‖ luftdichter Verschluß m. ‖ fermeture f. hermétique. / automatic ~ ‖ selbsttätiger Verschluß m. ‖ fermeture f. automatique. / ~ of books ‖ Rechnungsabschluß m. ‖ arrête m. ou règlement m. de compte. / ~ of the circuit ‖ Schließung f. oder Schluß m. des Stromkreises ‖ fermeture f. d'un circuit. / frame-worked ~ (Carp) ‖ Bundwand f.; Fachwand f.; Riegelwand f. ‖ cloison f. en charpente; pans mpl. de bois; paroi f. en clayonnage ou en colombage. / ~ for preserve jars ‖ Einkochverschluß m. ‖ fermeture f. pour bocaux à conserves.

closing arc (Electr) ‖ Schließungsbogen m. ‖ arc m. de clôture. / ~ capsule ‖ Verschlußkapsel f. ‖ capsule f. de fermeture. / ~ clasp (Locksm) ‖ Überfall m. ‖ garniture f. à serrer.

closing device ‖ Abschlußvorrichtung f.; Schließvorrichtung f. ‖ dispositif m. de fermeture; fermeture f. / ~ for bags ‖ Taschenbügel m. ‖ fermoir m. pour sacs à main. / ~ for bottles ‖ Flaschenverschluß m. ‖ agrafe f. pour bouteilles. / bulkhead ~ ‖ Schottentürschließvorrichtung f. ‖ dispositif m. de fermeture pour cloisons étanches.

closing flap ‖ Abschlußklappe f. ‖ clapet m. de fermeture. / parts pl. of ~ mechanism ‖ Verschlußteile mpl. ‖ pièces fpl. de fermeture. / ~ rope ‖ Schließseil m. ‖ câble m. de fermeture. / ~ trap of storage hopper (Mach) ‖ Füllrumpfverschluß m. ‖ fermeture f. de trémie.

closing-up stone (Glassm) ‖ Rauchkuchen m.; tönerner Vorsetzer m. zum Schließen der Arbeitslöcher ‖ taraison f.

closing window of a thermo-couple ‖ Abschlußfenster n. eines Thermoelementes ‖ fenêtre f. de fermeture pour un élément thermo-électrique.

closure ‖ Verschluß m. ‖ fermeture f.; clôture f. / ~ (Locksm) ‖ Schließbeschläge mpl. ‖ fermeture f. / emergency ~ ‖ Notverschluß m. ‖ bâtardeau m.

cloth (Wool) ‖ Tuch n. ‖ drap m. / ~ (Linen) ‖ Leinwand f. ‖ toile f. / ~ (Stuff for clothing) ‖ Gewebe n.; Stoff m.; Zeug n. ‖ étoffe f.; tissu m. / army ~ ‖ Militärtuch n. ‖ drap m. de troupe. / bolting ~ ‖ Müllergaze f. ‖ tissu-gaze f. pour blu-

toirs. / ~ for book-covers ‖ Bucheinbandstoff m. ‖ étoffe f. de reliure. / ~ for civilian dress ‖ Ziviltuch n. ‖ drap m. civil. / felted ~ ‖ Filztuch n. ‖ drap m. feutré ou feutre. / friction ~ ‖ gummierte Leinwand f. ‖ toile f. caoutchoutée. / grass ~ ‖ Grasleinen n. ‖ batiste f. de Canton; drap m. d'herbe. / gummed ~ ‖ gummierter Stoff m. ‖ étoffe f. gommée. / ~ with heavy selvedge ‖ Tuch n. mit starken Leisten ‖ drap m. à fortes lisières. / lined ~ ‖ Futtertuch n. ‖ drap m. fourré. / ~ for lining ‖ Futterstoff m. ‖ (étoffe f. à) doublure f. / mixed ~ ‖ meliertes Tuch n. ‖ drap m. melangé. / packing ~ ‖ Packleinwand f. ‖ toile f. ou canevas m. d'emballage. / ~ which prevents the spattering ‖ Schutzlappen m. am Schleifstein ‖ rabat d'eau. / rough ~ ‖ Loden m. ‖ drap m. brut; drap m. non foulé. / rubberized ~ ‖ gummierte Leinwand f. / toile f. caoutchoutée. / ~ of the sail ‖ Kleid n. des Segels ‖ laize f. de la voile. ‖ silk ~ ‖ Seidentuch n. ‖ drap m. de soie. / ~ for tents ‖ Zeltleinwand f. ‖ toile f. à voile. / tracing ~ ‖ Pausleinwand f.; Kopierleinwand f. ‖ toile f. à calquer; papier m. toile. / tweeled ~ with two faces ‖ zweiseitiger, beidrechter Köper m. ‖ étoffe f. croisée à double face; batavia m. / tweeled woollen ~ ‖ Köpertuch n. ‖ drap m. de Berry. / ~ for uniforms ‖ Uniformtuch n. ‖ drap m. d'uniforme. / unmilled ~ see rough ~. / varnished ~ ‖ gefirnißter Stoff m. ‖ tissu m. vernis. / waterproof ~ ‖ wasserdichter Stoff m. ‖ tissu m. imperméable. / woollen ~ ‖ (wollenes) Tuch n. ‖ drap m. / ~ for workmen ‖ Berufskleidung f. ‖ vêtement m. d'ouvriers ou de travail.

cloth back (Bookb) ‖ Leinenrücken m. ‖ dos m. en toile. / ~ beam (Weav) ‖ Zeugbaum m. ‖ ensouple f. / ~ belt ‖ Zeuggürtel m. ‖ ceinture f. en tissu. / ~ brush see clothes brush. / ~ brusher ‖ Tuchbürster m. ‖ brosseur m. de drap. / ~ brushing ‖ Bürsten n. des Tuches ‖ brossage m. de drap. / ~ burling ‖ Noppen n. des Tuches ‖ épetissage m. ou noppage m. de drap. / ~ button ‖ Stoffknopf m. ‖ bouton m. d'étoffe. / ~ carbonizing ‖ Tuchkarbonisation f. ‖ carbonisation f. des draps. / ~ covering ‖ Stoffbespannung f. ‖ entoilage m. / ~ cutter ‖ Zuschneider m. ‖ découpeur m. (d'étoffes). / ~ cutting ‖ Zuschneiden n. des Zeuges ‖ découpage m. d'étoffes. / ~ doll ‖ Stoffpuppe f. ‖ poupée f. en étoffe. / ~ envelope ‖ Stoffhülle f. ‖ enveloppe f. en tissu ou en étoffe.

clothe, to ‖ mit Stoff m. beziehen; mit Zeug n. ausschlagen ‖ garnir de tissu.

clothes pl. ‖ Bekleidung f. ‖ habillement m. / ready-made ~ ‖ Fertigkleidung f.; Konfektion f. ‖ confection f. / sporting ~ ‖ Sport(be)kleidung f. ‖ vêtement m. de sport.

clothes bag ‖ Reisesack m. ‖ sac m. de voyage. / ~ brush ‖ Kleiderbürste f. ‖ brosse à habits. / ~ cleaner ‖ Kleiderreiniger m. ‖ nettoyeur m. d'habits. / hanging chain for ~ ‖ Garderobenkette f. ‖ chaîne f. pour accrocher les vêtements. / ~ line ‖ Wäscheleinen n. ‖ corde f. à linge. / ~ line winder ‖ Wäscheleinenwickler m. ‖ bobine f. pour les cordes à linge.

clothes-peg (Linen) ‖ Wäscheklammer f. ‖ pince f. ou fichoir m. à linge. / ~ (Dress-

ings) ‖ Kleiderhalter m.; Kleiderhaken m. ‖ porte-manteaux m.; patère f.

clothes pin *see* clothes-peg. / ~ **rack** ‖ Kleiderleiste f. ‖ porte-vêtement m. / ~ **stand** ‖ Kleiderständer m. ‖ porte-habit m. / ~ **tongs** pl. ‖ Wäschezange f. ‖ tenailles fpl. à linge.

cloth filter ‖ Seihetuch n.; Filtriertuch n. ‖ couloir m. *ou* filtre m. en toile. / ~ **finishing** (Weav) ‖ Ausrüsten n. der Gewebe ‖ finissage m. de tissus. / ~ **finishing establishment** ‖ Appreturanstalt f. ‖ atelier m. d'apprêtage de draps. / ~ **frame tenter** ‖ Tuchrahmer m. ‖ rameur m. de drap. / ~ **friezing** ‖ Kräuseln n. des Tuches ‖ frisage m. de drap. / ~ **fulling** ‖ Walke f. *oder* Walken n. des Tuches ‖ foulonnage m. du drap. / ~ **gaiter** ‖ Stoffgamasche f.; Zeuggamasche f. ‖ guêtre f. en étoffe. / ~ **gilding** ‖ Stoffvergoldung f. ‖ dorure f. sur étoffes. / ~ **glazing** ‖ Glanzieren n. von Geweben ‖ glaçage m. d'étoffes. / ~ **glove** ‖ Stoffhandschuh m. ‖ gant m. en étoffe. / ~ **glove sewing** ‖ Stoffhandschuhnäherei f. ‖ cousage m. de ganterie d'étoffe.

clothier ‖ Tuchmacher m. ‖ drapier m. / ~'s **cutter** ‖ Zuschneider m. ‖ coupeur-tailleur m.

clothing ‖ Bekleidung f.; Kleidung f.; Kleidungsstücke npl. ‖ confection f.; vêtements mpl.; habillement m. / ~ **of a boiler** ‖ Mantel m. des Kessels ‖ enveloppe f. *ou* chemise f. d'une chaudière. / ~ **for boys** ‖ Knabenkleidung f. ‖ vêtement m. pour garçonnets. / ~ **for children** ‖ Kinderkleidung f. ‖ vêtement m. pour enfants. / ~ **for girls** ‖ Mädchenkleidung f. ‖ vêtement m. pour fillettes. / ~ **of the man-hole** ‖ Mannlochverkleidung f. ‖ chemise f. du trou d'homme. / ~ **for men** ‖ Männerkleidung f. ‖ vêtement m. pour hommes. / ~ **professional** ‖ Berufskleidung f. ‖ vêtements mpl. de travail. / ~ **water-proof** ‖ wasserdichte Kleidungsstücke npl.; wasserdichte Kleidung f. ‖ vêtement m. imperméable. / ~ **for women** ‖ Frauenkleidung f. ‖ vêtement m. pour femmes. / ~ **safety** ~ **for workmen** ‖ Arbeiterschutzkleidung f. ‖ vêtements mpl. de protection pour ouvriers. / ~ **for young men** ‖ Jünglingskleidung f. ‖ vêtement m. pour jeunes gens.

clothing accessories pl. ‖ Kleidungszubehör n. ‖ accessoires mpl. du vêtement. / ~ **factory** ‖ Konfektionsgeschäft n.; Kleiderfabrik f. ‖ confection f. de vêtements. / ~ **industry** ‖ Bekleidungsindustrie f. ‖ industrie f. des vêtements. / ~ **material** ‖ Kleiderstoff m. ‖ étoffe f. pour habits.

cloth looker (Weav) ‖ Warenbeschauer m. ‖ réceptionnaire m. / ~ **manufactory** ‖ Tuchfabrik f. ‖ fabrique f. de drap. / ~ **manufacture** ‖ Tuchfertigung f. ‖ manufacture f. des draps. / ~ **marking** ‖ Markieren n. des Tuches ‖ marquage m. de drap. / ~ **measuring machine** ‖ Stoffmeßmaschine f. ‖ machine f. à mesurer les étoffes. / ~ **mop** (Galv) ‖ Lappenscheibe f. ‖ disque m. en drap. / ~ **mounted paper** ‖ Leinenpapier n. ‖ papier m. entoilé. / ~ **oiler** ‖ Wachstuchlackierer m. ‖ cireur m. de taffetas. / ~ **paper** ‖ Tuchpapier n. ‖ papier m. imitation-toile. / ~ **peg** *see* clothes peg. / ~ **presser** ‖ Tuchplätter m. ‖ presseur m. de drap. / ~ **pressing** ‖ Pressen n. von Geweben ‖ pressage m. d'étoffes. / ~ **printing** ‖ Zeugdruckerei f. ‖

impression f. sur étoffes. / ~ **printing machine** ‖ Zeugdruckmaschine f. ‖ machine f. à imprimer les tissus *ou* d'impression sur étoffes. / ~ **printing plant** ‖ Gaufrieranstalt f. ‖ atelier m. à gaufrer. / ~ **prover** ‖ Leinwandmikroskop n.; Leinwandprober m.; Fadenzähler m. ‖ compte-fil m.; loupe f. du tisserand. / ~ **purse** ‖ Stoffhandtasche f. ‖ bourse f. en étoffe. / ~ **renterer** ‖ Tuchstopfer m. ‖ rentrayeur m. de drap. / ~ **rentering** ‖ Stopfen n. des Tuches ‖ stoppage m. de drap. / ~ **sawyer** (Tailor) ‖ Zeugzuschneider m. ‖ découpeur m. d'étoffes. / ~ **scraper** (Mill) ‖ Tuchabstreifer m. ‖ racloir m. en drap. / ~ **shearer** ‖ Tuchscherer m. ‖ tondeur m. de drap. / ~ **shears** pl. ‖ Tuchschere f. ‖ forces fpl. à tondre les draps. / ~ **shirt** ‖ Tuchhemd n. ‖ chemise f. de drap (bleu). / ~ **shrinker** ‖ Dekatierer m.; Tuchdekatierer m. ‖ décatisseur m. de drap. / ~ **singer** ‖ Tuchsenger m. ‖ grilleur m. de drap. / ~ **valve** ‖ Stoffventil n. ‖ soupape f. en étoffe. / ~ **washer** ‖ Tuchwäscher m. ‖ laveur m. de drap. / ~ **washing machine** ‖ Tuchwaschmaschine f. ‖ machine f. à laver le drap. / ~ **weaver** ‖ Tuchweber m.; Tuchmacher m. ‖ tisseur m. de drap; tisserand m. de drap. / ~ **weaving** ‖ Tuchweberei f. ‖ tissage m. des draps.

clotted ‖ klumpig ‖ grumeleux.

clottings pl. of the wool ‖ Kotspitzen fpl. der Schafwolle ‖ crottins mpl. de la laine.

cloud, to (Weav) ‖ moirieren; wässern ‖ moirer; tabiser.

cloud ‖ Wolke f. ‖ nuage m. / ~ **in the marble** ‖ Marmorader f. ‖ fil m. dans le marbre. / ~ **of poisonous gas** ‖ Giftgaswolke f. ‖ vague f. de gaz toxique. / ~ **of smoke** ‖ Rauchwolke f. ‖ nuage m. de fumée. / ~ **coloured** ~ **of smoke** ‖ farbige Rauchwolke f. ‖ nuage m. d'éclatement coloré. / **staken** ~s pl. ‖ Haufenwolken fpl. ‖ nuages mpl. en groupes. / **stratified** ~ ‖ Schichtenwolke f. ‖ nuage m. stratifié. / **without** ~s ‖ wolkenlos ‖ sans nuages; serein.

clouded stuff ‖ Moiree m. ‖ moiré m.; moirée f.; étoffe f. moirée.

cloud bank ‖ Wolkenbank f. ‖ banc m. de nuages. / ~ **electricity** ‖ Wolkenelektrizität f. ‖ électricité f. des nuages. / ~ **gas attack** ‖ Gasangriff m. ‖ attaque f. par vagues gazeuses.

cloudiness, bacterial ‖ Bakterientrübung f. ‖ trouble m. de bactéries.

cloudless ‖ wolkenlos; wolkenfrei ‖ sans nuages; serein. / ~ **zone** ‖ wolkenfreie Zone f. ‖ zone f. sans nuages.

cloud measurement ‖ Wolkenmessung f. ‖ mesure f. des nuages. / ~ **motion** ‖ Wolkenzug m. ‖ marche f. *ou* mouvement m. des nuages. / ~ **procession** *see* ~ motion. / ~ **theodolite** ‖ Wolkentheodolit m. ‖ théodolite m. de nuages *ou* pour la mesure de la hauteur des nuages. / ~ **train** *see* ~ motion. / ~ **velocity** ‖ Wolkengeschwindigkeit f. ‖ vitesse f. des nuages. / ~ **zone** ‖ Wolkenzone f. ‖ zone f. de nuages; troposphère f.

clough-arch ‖ Gerinne n. einer Schleuse ‖ auge f. *ou* conduit m. d'une écluse.

clove ‖ Gewürznelke f.; girofle f. / ~ **bark** ‖ Nelkenzimt m. ‖ cannelle f. giroflée. / ~ **oil** ‖ Gewürznelkenöl n. ‖ essence f. de girofles.

clover ‖ Klee m. ‖ trèfle m.

club ‖ Verein m. ‖ société f. / ~ (Railw) ‖ Bremsknüppel m. ‖ barre f. d'enrayement. / **articles** pl. for social ~s ‖ Vereinsartikel mpl. ‖ articles mpl. de tombola pour sociétés. / **workmen's** ~ ‖ Arbeiterverein m. ‖ union f. d'ouvriers; syndicat m. ouvrier.

club chair ‖ Klubsessel m. ‖ siège m. de club. / ~ **furniture** ‖ Klubmöbel npl. ‖ ameublement m. de club. / ~ **handle** ‖ Keulengriff m. ‖ poignée f. en forme de massue. / ~ **ribbon** ‖ Vereinsband n. ‖ ruban m. pour sociétés. / ~ **rooms** pl. **for foremen** ‖ Werkmeisterkasino n. ‖ casino m. de contre-maîtres.

clue (Mar) ‖ Schothorn n. ‖ point m. d'écoute.

clumsy ‖ schwerfällig ‖ lourd.

cluster (Bot) ‖ Dolde f. ‖ cône m.; ombelle f. / ~ **of crystals** ‖ Kristalldruse f. ‖ géode f. cristallifère; groupe m. de cristaux.

clustered ‖ gebündelt ‖ fasciculé. / ~ **teeth** pl. (Saw) ‖ Gruppenzahnung f. ‖ denture f. en groupes.

clutch ‖ Klaue f.; Haken m. ‖ griffe f. / ~ (Coupling) ‖ Kupplung f. ‖ accouplement m.; embrayage m. / **bayonet** ~ ‖ Bajonettgriff m. ‖ manette f. à douille de baïonette. / **claw** ~ ‖ Klauenkupplung f. ‖ embrayage m. à griffes; manchon m. d'embrayage à dents. / **coil** ~ ‖ Spiralfederkupplung f.; Federbandkupplung f. ‖ embrayage m. à spirale *ou* à ruban. / **cone** ~ ‖ Kegelkupplung f. ‖ embrayage m. à cône. / ~ **of a coupling-box** ‖ Klaue f. einer Kupplungsmuffe ‖ griffe f. d'un manchon d'accouplement. / **direct driving** ~ ‖ Klaue f. für unmittelbaren Gang ‖ griffe f. de prise directe. / **disk** ~ ‖ Lamellenkupplung f. ‖ embrayage m. à disques. / **double-cone** ~ ‖ Doppelkonuskupplung f. ‖ embrayage m. à double cône. / **expanding** ~ ‖ Ausdehnungskupplung f. ‖ embrayage m. extensible. / **fierce** ~ ‖ harte Kupplung f. ‖ embrayage m. trop brusque. / **friction** ~ ‖ Reibungskupplung f. ‖ embrayage m. à friction. / **friction disc** ~ ‖ Reibungslamellenkupplung f. ‖ embrayage m. à friction à disques. / **high-speed** ~ ‖ Kupplung f. für unmittelbaren Gang ‖ embrayage m. de prise directe. / **loose** ~ ‖ lösbare Kupplungsmuffe f. ‖ manchon m. d'embrayage et de débrayage. / **magnetic** ~ ‖ magnetelektrische Kupplung f. ‖ embrayage m. magnétique. / **multiple disc** ~ ‖ Lamellenkupplung f. ‖ embrayage m. à disques. / **overrunning** ~ ‖ Freilaufkupplung f. ‖ embrayage m. à roue libre. / **plate** ~ ‖ Scheibenkupplung f. ‖ embrayage m. par plateau. / **safety** ~ ‖ Fangvorrichtung f. ‖ frein m. à mâchoires. / **self-acting instantaneous gear** ~ ‖ selbsttätig ausrückende Momentklauenkupplung f. ‖ accouplement m. à mâchoires et à débrayage automatique instantané. / **self-actuating** ~ ‖ selbsteinrückende Kupplung f. ‖ accouplement m. automatique. / **sliding** ~ ‖ Rutschkupplung f. ‖ accouplement m. à coulisse. / **spring band** ~ ‖ Federbandkupplung f. ‖ embrayage m. à ruban. / ~ **for starting handle** ‖ Andrehkurbelklaue f.; Ankurbelungsklaue f. ‖ griffe f. de démarreur.

clutch bearing, radial ‖ Kupplungslager n. ‖ roulement m. annulaire d'embrayage. / ~ **brake** ‖ Kupplungsbremse f. ‖

frein m. d'embrayage. / ~ cam ‖ Kupplungsdaumen m. ‖ came f. d'embrayage. / ~ case ‖ Kupplungsgehäuse n. ‖ carter m. d'embrayage. / ~ cone ‖ Kupplungskegel m. ‖ cône m. d'embrayage. / ~ coupling ‖ Klauenkupplung f. ‖ accouplement m. ou manchon m. à griffes. / ~ coupling-box ‖ lösbare Kupplungsmuffe f. ‖ manchon m. d'embrayage et de débrayage. / ~ disk ‖ Kupplungsscheibe f. ‖ disque m. d'embrayage. / ~ facing ‖ Kupplungsbelag m. ‖ garniture f. d'embrayage. / ~ fork ‖ Kupplungsgabel f. ‖ fourchette f. de débrayage. / ~ operating device ‖ Kupplungsgestänge n. ‖ tringle f. de commande de l'embrayage. / ~ pedal ‖ Kupplungspedal n.; Kupplungsfußhebel m. ‖ pédale f. de débrayage. / ~ plate ‖ Kupplungsscheibe f. ‖ plateau m. d'embrayage. / ~ release sleeve ‖ Kupplungsausrückmuffe f. ‖ épaulement m. de débrayage; manchon m. d'accouplement. / ~ shaft ‖ Mitnehmerwelle f.; Antriebswelle f. ‖ essieu m. d'entraînement; arbre m. primaire. / ~ spindle ‖ Kupplungswelle f. ‖ arbre m. d'embrayage. / ~ spring ‖ Kupplungsfeder f. ‖ ressort m. d'embrayage. / ~ taper ‖ Kuppelkonus m. ‖ cône m. d'embrayage.

clydnograph (Electr) ‖ Klydonograf m.; Wellenschreiber m. ‖ appareil m. régistreur d'ondes de surtension.

CO₂ tester ‖ Rauchgasprüfer m. ‖ indicateur m. de CO_2.

coach ‖ Wagen m. ‖ voiture f.; wagon m. / motor ~ (Railw) ‖ Triebwagen m. ‖ automotrice f.

coach building ‖ Wagenbau m. ‖ carrosserie f. / ~ house ‖ Wagenremise f. ‖ remise f. / ~ leather ‖ Verdeckleder n. ‖ cuir m. à capot. / ~ letting-out ‖ Lohnfuhrwerk n.; Wagenvermietung f. ‖ location f. de voitures.

coachmaker ‖ Wagenbauer m. ‖ carrossier m.

coachman ‖ Kutscher m. ‖ cocher m. / ~'s seat ‖ Kutschersitz m. ‖ siège m. du cocher.

coach painting ‖ Anstreichen n. von Wagen ‖ peinture f. de voitures. / ~ screw (Railw) ‖ Schienenschraube f.; Schwellenschraube f. ‖ tirefond m. / ~ wrench ‖ Engländer m.; Franzose m.; englischer Schraubenschlüssel m. ‖ clef f. anglaise.

coagulate, to ‖ gerinnen; käsig ausscheiden; verdicken ‖ se cailler; se coaguler; se figer; se prendre.

coagulation ‖ Gerinnung f. ‖ coagulation f.

coagulum ‖ Gerinnsel n. ‖ coagulum m.

coaking coal see coking coal.

coal, to ‖ Kohlen fpl. einnehmen ‖ faire du charbon.

coal ‖ Kohle f. ‖ charbon m.; houille f. / agglomerated ~ ‖ zusammengepreßte Kohle f. ‖ charbon m. aggloméré. / artificial ~ ‖ Kunstkohle f. ‖ charbon m. artificiel. / baked ~ ‖ zusammengebackene Kohle f. ‖ charbon m. cuit. / baking ~ ‖ backende Kohle f. ‖ charbon m. collant. / best quality ~ ‖ Kohle f. bester Qualität ‖ charbon m. de la meilleure qualité. / bituminous ~ ‖ bituminöse Kohle f.; Fettkohle f. ‖ charbon m. bitumineux ou gras. / black ~ ‖ Schwarzkohle f.; Steinkohle f. ‖ houille f.; charbon m. de terre. / blind ~ ‖ magere Steinkohle f.; Kohlenblende f. ‖ charbon m.

sec. / bright ~ see glance ~. / broken ~ ‖ Bruchkohle f.; Brechkohle f. ‖ déchets mpl. de charbon; charbon m. concassé. / brown ~ ‖ Braunkohle f. ‖ lignite m.; houille f. brune. / bunker ~ ‖ Bunkerkohle f. ‖ charbon m. de soute. / caking ~ ‖ Backkohle f.; Fettkohle f. ‖ houille f. grasse. / cannel ~ ‖ Fettkohle f.; Pechkohle f.; Kannelkohle f. ‖ houille f. grasse ou compacte ou flambante; charbon m. gras. / carbonized ~ ‖ verkokte Steinkohle f. ‖ houille f. carbonisée. / cleaned ~ ‖ Schlämmkohle f. ‖ charbon m. débourbé. / clinkering ~ ‖ schlackende Kohle f. ‖ charbon m. scorifère. / close-burning ~ ‖ Sandkohle f.; magere Steinkohle f. ‖ houille f. sèche ou maigre ou noncollante. / coking ~ ‖ Kokerkohle f. ‖ houille f. à coke. / columnar ~ ‖ stengeliger Anthrazit m. ‖ houille f. bacillaire. / decolourizing ~ ‖ Entfärbungskohle f. ‖ charbon m. décolorant. / dressed ~ ‖ aufbereitete oder gewaschene Kohle f. ‖ charbon m. lavé ou préparé. / dull ~ ‖ Mattkohle f. ‖ charbon m. mat. / fat ~ ‖ fette Steinkohle f. ‖ houille f. grasse. / fine ~ rich in tar ‖ teerreiche Feinkohle f. ‖ charbon m. fin riche en goudron. / first-class ~ ‖ Kohle f. bester Sorte ‖ charbon m. de première qualité. / flaming ~ ‖ Flammförderkohle f. ‖ charbon m. flambant. / foliated ~ ‖ Schieferkohle f.; Blätterkohle f. ‖ houille f. schisteuse. / forge ~ ‖ Eßkohle f.; Schmiedekohle f. ‖ charbon m. de forge. / free-burning ~ see close-burning ~. / gas ~ ‖ Fettkohle f.; gasreiche Kohle f. ‖ houille f. grasse; charbon m. à gaz. / gas motor ~ ‖ Gasmotorkohle f. ‖ charbon m. pour moteur à gaz. / ~ for gas works ‖ Generatorkohle f. ‖ charbon m. de générateur. / glance ~ ‖ Glanzkohle f.; Kohlenblende f. ‖ houille f. luisante ou sèche. / green ~ ‖ frische Kohle f. ‖ charbon m. vierge. / imported ~ ‖ eingeführte Kohle f. ‖ charbon m. d'importation. / large ~ ‖ Stückkohle f. ‖ gaillettes fpl. / layer of ~ ‖ Kohlenschicht f. ‖ couche f. de charbon. / long-flame ~ ‖ langflammige Kohle f. ‖ houille f. flambante. / lustrous ~ ‖ Glanzkohle f. ‖ charbon m. luisant. / mineral ~ ‖ Schwarzkohle f.; Steinkohle f. ‖ houille f.; charbon m. de terre. / mixing of ~ ‖ Kohlenmischung f. ‖ mélange m. de charbon. / non-baking ~ ‖ Magerkohle f. ‖ houille f. maigre. / non-bituminous ~ ‖ Magerkohle f. ‖ houille f. maigre. / non-caking ~ see close-burning ~. / non-gaseous ~ ‖ gasarme Kohle f. ‖ charbon m. pauvre en gaz. / obtaining the ~ in the open workings ‖ Kohlengewinnung f. im Tagebau ‖ production f. du charbon à ciel ouvert. / open burning ~ ‖ Gasflammkohle f.; Sinterkohle f.; Grobkohle f. ‖ charbon m. flambant; houille f. en gros morceaux. / peat ~ ‖ Torfkohle f. ‖ charbon m. de tourbe. / pit ~ ‖ Schwarzkohle f.; Steinkohle f. ‖ houille f.; charbon m. de terre. / pitchy ~ ‖ Pechkohle f. ‖ houille f. poisseuse. / at pithead ‖ Kohle f. an der Zeche ‖ charbon m. au puits. / plastic ~ ‖ Formkohle f. ‖ charbon m. plastique. / powdered ~ ‖ Kohlenstaub m. ‖ charbon m. pulvérisé; poussière f. de charbon. / pressed ~ ‖ gepreßte Kohle f. ‖ charbon m. comprimé. / pulverized ~ see powdered ~. / refuse ~ ‖ Kohlenabfall m. ‖ déchet m. de char-

bon. / rich ~ see bituminous ~. / rough ~ ‖ rohe Förderkohle f. ‖ houille f. crue ou tout-venante; tout-venant m. / run-of-mine ~ ‖ Förderkohle f. ‖ tout-venant m. / screened ~ ‖ gesiebte Kohle f. ‖ charbon m. criblé. / semi-bituminous ~ ‖ halbfette Kohle f. ‖ houille f. demi-grasse. / combined pieces of ~ and shale ‖ verwachsene Kohle f. ‖ charbon m. barré. / short-flame ~ ‖ kurzflammige Kohle f. ‖ houille f. à courte flamme. / sifted ~ ‖ Siebkohle f. ‖ charbon m. criblé. / size of ~ (Min) ‖ Kohlengröße f. ‖ grandeur f. ou grosseur f. du charbon. / slack ~ ‖ Gruskohle f.; Kohlenklein n. ‖ fin m.; houille f. menue; menu m.; poussier m. / ~ free of slag ‖ schlackenreine Kohle f. ‖ charbon m. exempt de scories. / slaty ~ ‖ schieferige Steinkohle f. ‖ houille f. schisteuse. / small ~ ‖ Feinkohle f.; Kohlenklein n.; Grus m.; Gruskohle f. ‖ charbon m. fin ou pulvérisé ou menu; fines fpl. / smith's ~ ‖ Schmiedekohle f. ‖ fine f. forge; houille f. pour feux de forge. / smokeless ~ ‖ rauchlose Kohle f. ‖ charbon m. sans fumée. / transformation into ~ ‖ Inkohlung f. ‖ carburation f. / uninflammable ~ see close-burning ~. / unmined ~ ‖ unverritztes Kohlenflöz n. ‖ veine f. vierge. / vegetable ~ ‖ Pflanzenkohle f.; vegetabilische Kohle f. ‖ charbon m. végétal. / vein of ~ ‖ Kohlenflöz n. ‖ couche f. de charbon ou de houille. / washed ~ ‖ gewaschene Kohle f. ‖ charbon m. lavé.

coal analysis ‖ Kohlenanalyse f. ‖ analyse f. du charbon. / ~ ascent ‖ Kohlenladeplatz m. ‖ quai m. à houille. / ~ bag ‖ Kohlensack m. ‖ sac m. à charbon. / automatic ~ balance ‖ selbsttätige Kohlenwage f. ‖ bascule f. automatique à charbon. / ~ balls pl. ‖ Peras fpl. ‖ peras fpl. / ~ basin ‖ Kohlenbecken n. ‖ bassin m. houiller. / ~ basket ‖ Kohlenkorb m. ‖ corbeille f.; rasse f.; benne f.; panier m. à charbon. / ~-bearing sandstone ‖ Kohlensandstein m. ‖ grès m. houiller. / ~-bearing strata pl. ‖ flözführendes Steinkohlengebirge n. ‖ terrain m. carbonifère ou contenant des couches de charbon. / ~ borer ‖ Kohlenbohrer m. ‖ foreur m. en veine. / ~ box ‖ Kohlenkasten m.; Kohlenfüller m. ‖ caisse f. à charbon. / ~ box (Room) ‖ Kohlenraum m. ‖ soute f. à charbon. / ~ breaker ‖ Kohlenbrecher m. ‖ concasseur m. de charbon. / ~ breaking ‖ Kohlenzerkleinerung f. ‖ concassage m. de charbon. / bringing-up the ~ by rail and discharging on to elevated bunkers ‖ Kohlenzufuhr f. durch eine Standbahn auf die Hochbunker ‖ transport m. du charbon par voie fixe aux soutes surélevées. / ~ briquette ‖ gepreßte Steinkohle f.; Steinkohlenbrikett n. ‖ aggloméré m. de houille. / ~ briquetting plant ‖ Kohlebrikettierungsanlage f. ‖ installation f. à agglomérer le charbon. / ~ bucket ‖ Kohleneimer m. ‖ seau m. à charbon. / ~ bunker ‖ Kohlensilo m. ‖ silo m. à charbon. / ~ bunker (Shipb) ‖ Kohlenbunker m. ‖ soute f.; tremie f. à charbon. / ~ bunker grating ‖ Kohlenlochgräting f. ‖ grille f. ou rosace f. à jour de charbon. / ~ cake ‖ Preßkohle f.; Brikett n. ‖ aggloméré m.; briquette f. / ~ cake pressman ‖ Brikettpresser m. ‖ presseur m. de briquettes. / ~ cargo ‖ Kohlenladung f. ‖

cargaison f. de charbon. / ~ carrier ‖ Kohlenschlepper m. ‖ brouetteur m. *ou* brouettier m. de charbon. / ~ cartman ‖ Fuhrmann m.; Kohlenfuhrmann m. ‖ charretier m.; varlet m. / ~ charging plant ‖ Bekohlungsanlage f. ‖ installation f. de chargement de charbon. / ~ chute ‖ Kohlenrutsche f. ‖ couloir m. incliné pour charbons. / ~ consumption ‖ Kohlenverbrauch m. ‖ consommation f. de charbon. / ~ consumption per kilowatt hour ‖ Kohlenverbrauch m. je Kilowattstunde ‖ consommation f. de charbon par kilowatt-heure. / total ~ consumption ‖ Gesamtkohlenverbrauch m. ‖ dépense f. de charbon totale. / ~ contract ‖ Kohlenkontrakt m. ‖ contrat m. pour charbon. / ~ conversion industry *see* ~ gasification industry. / ~ conveying plant ‖ Bekohlungsanlage f.; Kohlenförderanlage f. ‖ approvisionnement m. en charbon; installation f. de chargeur mécanique de charbon *ou* pour le chargement des charbons. / ~ cooking works ‖ Verkokungsanstalt f. ‖ usine f. de carbonisation de la houille. / ~ crucible ‖ Kohlentiegel m. ‖ creuset m. en charbon. / ~ crusher ‖ Kohlenbrecher m. ‖ casseur m. *ou* broyeur m. *ou* concasseur m. à charbon. / ~ and coke crusher ‖ Kohlen- und Koksbrecher m. ‖ broyeur m. à charbon et à coke. / ~ crushing ‖ Kohlenzerkleinerungsanlage f. ‖ concassage m. de charbon. / ~ cutting machine ‖ Schrämmaschine f. ‖ machine f. à entailler; haveuse f.; houilleuse f.; déhouilleuse f. / ~ deposits pl. which can be extracted ‖ gewinnbares Kohlenvorkommen n. ‖ charbon m. reconnu récupérable. / ~ deposits pl. which are at present undeveloped ‖ noch unaufgeschlossenes Kohlenfeld n. ‖ gisement m. houiller pas encore mis en exploitation. / ~ distillation ‖ Kohlenvergasung f. ‖ gazéification f. du charbon. / ~ dressing ‖ Kohlenaufbereitung f. ‖ préparation f. mécanique de charbon. / ~ drying plant ‖ Kohlentrocknungsanlage f. ‖ installation f. de séchage du charbon. / ~ drying plant for coal grinding ‖ Kohlentrocknungsanlage f. für die Kohlenvermahlung ‖ installation f. de séchage du charbon à pulvériser.

coal-dust ‖ Kohlenstaub m. ‖ poussier m. de charbon; charbon m. en poussière. / ~ brick ‖ Brikett n. ‖ briquette f. de houille *ou* de lignite. / ~ burning ‖ Kohlenstaubfeuerung f. ‖ chauffage m. à la poussière de charbon. / ~ cake ‖ Kohlenstaubziegel m. ‖ briquette f. de poussier de charbon. / ~ firing ‖ Kohlenstaubfeuerung f. ‖ chauffage m. à la poussière de charbon. / boiler for firing ~ ‖ Kohlenstaubkessel m. ‖ chaudière f. chauffée au poussier de charbon. / ~ furnace ‖ Kohlenstaubfeuerung f. ‖ foyer m. à chauffage de charbon pulvérisé. / ~ mill ‖ Kohlenstaubmühle f. ‖ pulvérisateur m. de charbon. / ~ plant ‖ Kohlenstaubanlage f. ‖ installation f. de poussier de charbon. / ~ wagon ‖ Kohlenstaubwagen m. ‖ wagon m. à poussier de charbon.

coal elevator, floating ‖ schwimmender Kohlenheber m. ‖ élévateur m. flottant à charbon.

coalesce, to ‖ zusammenfließen ‖ se combiner.

coal, electric ~ face lighting in a coal mine ‖ elektrische Abbaubeleuchtung f. in einer Kohlengrube ‖ éclairage m. électrique de l'abattage de charbon dans une mine. / ~ field ‖ Kohlenfeld n. ‖ terrain m. carbonifère. / ~ field not yet „won" ‖ noch unaufgeschlossenes Kohlenfeld n. ‖ gisement m. houiller pas encore mis en exploitation. / ~ filter ‖ Kohlenfilter n. ‖ filtre m. au charbon. / ~ firing ‖ Kohlenfeuerung f. ‖ chauffage m. au charbon. / ~ fork ‖ Koksgabel f. ‖ fourche f. à coke. / ~ forming fermentation ‖ Kohlegärung f. ‖ -fermentation f. carbonique. / ~ free of slag ‖ schlackenreine Kohle f. ‖ charbon m. exempt de scories. / ~ gas ‖ Kohlengas n.; Leuchtgas n. ‖ gaz m. de houille. / ~ gasification industry ‖ Kohleumwandlungstechnik f. ‖ industrie f. de transformation du charbon. / ~ getter *see* ~ hewer. / ~ grinding ‖ Kohlenvermahlung f. ‖ pulvérisation f. de charbon. / ~ grinding plant ‖ Kohlenmahlanlage f. ‖ installation f. de broyage de charbon. / ~ grit ‖ Kohlensandstein m. ‖ carrache m.; grès m. houiller. / ~ handling platform ‖ Ladegerüst n. für Kohlen ‖ pont m. de chargement de charbons. / ~ heaver ‖ Kohlenlader m. ‖ charbonnier m. / ~ hewer ‖ Kohlenhauer m. ‖ houilleur m.; piqueur m. de houille. / ~ hewer (Overhand stooping) ‖ Steigortshauer m. ‖ monteur m. en veine. / ~ hewer (Underhand stooping) ‖ Hauer m. im Fallort ‖ avaleur m. en veine. / ~ hoist ‖ Kohlenladeplatz m. ‖ estacade f. *ou* fosse f. à houille. / ~ hole ‖ Kohlenraum m. ‖ soute f. à charbon; charbonnière f.; charbonnerie f. / ~ holing machine ‖ Kohlenschrämmaschine f. ‖ machine f. à entailler les couches de charbon. / ~ hopper ‖ Kohlentrichter m. ‖ entonnier m. *ou* trémie f. à charbon. / ~ hutch ‖ Kohlenwagen m. ‖ benne f. piocheuse *ou* à charbon; chien m.

coaling crane ‖ Kohlenladekran m.; Bekohlungskran m. ‖ grue f. à charger le charbon; grue f. charbonnière. / ~ for railway stations ‖ Bekohlungskran m. für Bahnhöfe ‖ grue f. à charger le charbon aux gares.

coaling plant ‖ Bekohlungsanlage f. ‖ installation f. de chargeur mécanique de charbon. / boiler ~ ‖ Kesselbekohlungsanlage f. ‖ installation f. à charger le charbon dans les chaudières. / ~ for locomotives ‖ Lokomotivbekohlungsanlage f. ‖ installation f. de chargement de charbon pour locomotives.

coaling screen ‖ Kohlenpersenning f. ‖ masque m. pour l'embarquement du charbon.

coaling scuttle ‖ Kohlenloch n. ‖ trou m. à charbon. / ~ cover ‖ Kohlenlochdeckel m. ‖ plaque f. du trou à charbon.

coaling station (Mar) ‖ Kohlenstation f. ‖ station f. à charger le charbon pour navires.

coal kindler ‖ Feueranzünder m. ‖ allume-feu m. / ~ layer of ~ ‖ Kohlenschicht f. ‖ couche f. de charbon. / ~ liquefaction ‖ Kohlenverflüssigung f. ‖ liquéfaction f. du charbon. / ~ loading plant ‖ Kohlenverladeanstalt f. ‖ installation f. de chargement de charbon. / ~ loading wagon ‖ Kohlenfüllwagen m. ‖ wagon-chargeur m. de charbon. / ~ measure ‖ Füllfaß n.; Füllkorb m. ‖ seau m. à puiser; panier m. à charger; corbeille f.

/ ~ measures (Geol) ‖ Kohlengebirge n. ‖ terrain m. houiller. / ~ measurer ‖ Kohlenverwieger m. ‖ mesureur m. de charbon.

coalmerchant, wholesale ‖ Kohlengroßhändler m. ‖ négociant m. en gros de charbons.

coal mill ‖ Kohlenstaubmühle f. ‖ moulin m. à charbon. / ~ mine ‖ Kohlenzeche f.; Kohlenbergwerk n. ‖ mine f. de houille; houillère f.; charbonnage m. / ~ mine district ‖ Kohlenbecken n. ‖ bassin houiller m. / ~ mine plant ‖ Kohlengrubenanlage f. ‖ établissement m. houillier. / ~ miner ‖ Kohlenbergmann m. ‖ houilleur m. / ~ mixing plant ‖ Kohlenmischturm m.; Kohlenmischanlage f. ‖ tour m. de mélangeurs à houille; installation f. à mélanger le charbon. / ~ mud ‖ Kohlenschlamm m. ‖ schlamm m. *ou* boue f. de charbon. / conveying bridge for ~ open working ‖ Abraumförderbrücke f. ‖ pont m. de transport pour travaux de déblaiement. / ~ outcrop ‖ Kohlenausbiß m. ‖ affleurement m. de charbon. / ~ picker on the heap ‖ Kohlenausklauber m. auf der Bergehalde ‖ ramasseur m. de gailletteries. / ~ pile ‖ Kohlenmeiler m.; Kohlenhalde f. ‖ pile f. à charbon; feu m. *ou* foyer m. à coke. / ~ pit ‖ Kohlenzeche f. ‖ mine f. de houille *ou* de charbon. / exhausted ~ pit ‖ abgebaute Kohlenzeche f. ‖ houillère f. épuisée. / ~ poker ‖ Kohlenkrücke f.; Schüreisen n. ‖ ringard m.; fourgon m.; attisoir m. / ~ preparation ‖ Kohlepräparat n. ‖ préparation f. de charbon. / ~ preparing plant ‖ Kohlenaufbereitungsanlage f. ‖ installation f. pour la préparation du charbon. / ~ protection (Shipb) ‖ Kohlenschutz m. ‖ protection f. au moyen du charbon. / ~ raising costs pl. ‖ Gestehungspreis m. ‖ prix m. coûtant *ou* de revient. / ~ rake ‖ Kohlenkrahle f.; Schüreisen n. ‖ ringard m.; fourgon m.; perce-fournaise m. / ~ screen ‖ Kohlensieb n. ‖ crible m. à charbon. / ~ screening plant ‖ Siebanlage f. für Kohlen ‖ installation f. de tamisage de la houille. / ~ seam ‖ Kohlenflöz n. ‖ couche f. de charbon *ou* de houille. / ~ separation plant ‖ Kohlenseparation f. ‖ installation f. pour la séparation de la houille. / ~ shed ‖ Kohlenschuppen m. ‖ hangar m. à charbon. / ~ ship ‖ Kohlenschiff n. ‖ charbonnier m. / ~ shortage ‖ Kohlenknappheit f. ‖ disette f. de charbons. / ~ shovel ‖ Kohlenschaufel f. ‖ pelle f. à charbon. / ~ shoveller ‖ Kohlenschaufler m. ‖ bouteur m. / ~ sifting plant ‖ Kohlensieberei f. ‖ installation f. à tamiser le charbon. / size of ~ ‖ Kohlengröße f. ‖ grandeur f. *ou* grosseur f. du grain de charbon. / ~ skip ‖ Füllfaß n.; Füllkorb m. ‖ seau m. à puiser; panier m. à charger; corbeille f. / ~ slate ‖ Brandschiefer m. ‖ pyroschiste m.; schiste m. bitumineux. / ~ slime ‖ Schlammkohle f. ‖ schlammkohle f. / ~ sludge ‖ Kohlenschlamm m. ‖ schlamm m. *ou* boue f. de charbon. / ~ sorter ‖ Kohlenklauber m. ‖ classeur m. de charbon m. / ~ stamper ‖ Kohlenstampfer m. ‖ pileur m. de charbon. / ~ stamping plant ‖ Kohlenstampfanlage f. ‖ pilonnage m. de charbon.

coal storage ‖ Kohlenlagerplatz m. ‖ dépôt m. de charbon. / ~ hopper ‖ Kohlenvorratstrichter m. ‖ trémie f. d'emmaga-

sinage de charbon. / ~ room ‖ Kohlen-vorratsbehälter m. ‖ récipient m. à emmagasiner le charbon. / ~ tower ‖ Kohlenturm m. ‖ tour f. à charbon.

coal stores pl. ‖ Kohlenlager n. ‖ dépôt m. de charbon. / ~ stove for bath ‖ Kohlen-badeofen m. ‖ poêle f. à charbon pour bain. / ~ supplies pl. of recent formation ‖ jüngere Kohlenvorkommen npl. ‖ char-bon m. provenant de couches de forma-tion récente. / ~ supply ‖ Kohlenzufuhr f. ‖ arrivage m. de charbon.

coal tar ‖ Steinkohlenteer m. ‖ goudron m. de houille. / ~ burner ‖ Kohlenteerschwe-ler m. ‖ distillateur m. de goudron. / ~ co-lour ‖ Teerfarbe f. ‖ colorant m. dérivé des goudrons. / ~ distillation ‖ Steinkohlen-teerdestillation f. ‖ distillation f. de gou-dron de houille. / ~ dye ‖ Kohlenteerfarbe f. ‖ matière f. colorante dérivée des gou-drons de houille. / ~ oil ‖ Steinkohlenteer-öl n. ‖ huile f. de goudron de houille. / ~ pitch ‖ Steinkohlenteerpech n. ‖ brai m. de houille. / ~ produce ‖ Steinkohlen-teerpräparat n. ‖ produit m. dérivé du goudron de houille.

coal tip ‖ Kohlenkippvorrichtung f. ‖ appareil m. basculeur à décharger les charbons; estacade f. à houille. / ~ tip wagon ‖ Kohlenkippwagen m. ‖ wa-gonnet m. basculant à charbon. / ~ tower ‖ Kohlenturm m. ‖ tour m. ou tré-mie f. à charbon; tour m. d'emmagasine-ment des charbons. / ~ train ‖ Kohlen-zug m. ‖ train m. de charbon. / ~ trans-port car ‖ Kohlentransportwagen m. ‖ wagon m. à transporter la houille. / ~ trimmer ‖ Kohlentrimmer m. ‖ soutier-charbonnier m. / ~ truck (Railw) ‖ Koh-lenwagen m. ‖ wagon m. à charbon. / vein of ~ ‖ Kohlenflöz n. ‖ couche f. de charbon ou de houille. / ~ wagon ‖ Koh-lenwagen m. ‖ wagon m. à houille. / ~ washer ‖ Kohlenwäsche f. ‖ laveur m. de charbon. / ~ washery ‖ Koh-lenwäsche f. ‖ installation f. de la-vage du charbon. / ~ washing plant ‖ Kohlenwäsche f. ‖ installation f. de la-vage du charbon; lavoir m. à charbon. / ~ washings pl. ‖ Kohlenschlamm m. ‖ boue f. de charbon; schlamms pl. / ~ wharf ‖ Kohlenlöschplatz m. ‖ quai m. à houille. / ~ wheeler ‖ Kohlenkarrer m. ‖ rouleur m. de charbon.

coarse (Coal dress) ‖ grob gepreßt; rösch ‖ légèrement bocardé. / with ~ cut ‖ mit grobem Schnitt m. ‖ enlevant de gros copeaux mpl.

coarse adjustment ‖ Grobeinstellung f. ‖ dégrossissage m. / clamp for ~ ‖ Klem-mung f. der Grobverstellung ‖ blocage m. du mouvement rapide.

coarse bruising of grain ‖ Grobschroten n. von Getreide ‖ concassage f. grossier de grains. / ~ cloth ‖ Loden m. ‖ drap m. brut; loden m. / ~ copper ‖ Rohkupfer n.; Gelbkupfer; Schwarzkupfer ‖ cuivre m. noir; cuivre brut. / ~ crushing ‖ Grob-zerkleinerung f. ‖ broyage m. grossier. / ~ grain ‖ grobes Korn n.; grobe Körnung f. ‖ gros grain m. / ~ grain washer ‖ Grobkornsetzmaschine f. ‖ machine-la-veuse f. de gros. / ~ gravel ‖ Grobschotter m. ‖ gros gravier m. / ~ grinding ‖ Grobzerkleinerung f.; Schroten n. ‖ broyage m. grossier. / ~ limestone ‖ Grobkalk m. ‖ chaux f. grossière.

coarsely, to grind ~ ‖ grob zerkleinern ‖ moudre grossièrement. / ~ grained (Met) ‖ grobkörnig ‖ à gros grain. / ~ grained (Wood) ‖ grobfaserig ‖ à fibre f. grossière. / ~ ground grist ‖ Grobschrot n. ‖ farine f. grossière; blé m. égrugé. / ~ powdered ‖ grob gepulvert ‖ grossièrement pulvérisé. / ~ ringed (Wood) ‖ grobjährig ‖ aux cernes mpl. larges.

coarse metal ‖ Rohstein m.; Kupferrohstein m. ‖ matte f. brute de cuivre; métal m. brut de cuivre. / ~ metal slag ‖ Roh-steinschlacke f. ‖ scorie f. de la matte brute. / ~ motion ‖ Grobverstellung f. ‖ mouvement m. rapide. / ~ motion of a telescope in altitude ‖ Grobverstellung f. eines Fernrohrs in Höhe ‖ tournement m. vertical de la lunette à la main.

coarseness ‖ Grobkörnigkeit f. ‖ granula-tion f. grossière.

coarse plaster ‖ Berapp m.; Bewurf m.; Spritzwurf m. ‖ enduit m. hourdé; hour-dage m.; ravalement m. / ~ plate ‖ Grob-blech n. ‖ grosse tôle f. / ~ roving (Spinn) ‖ Vordergespinst n.; Lunte f. ‖ boudin m.; mèche f. / ~ silk glove ‖ Bourrettehand-schuh m. ‖ gant m. de filoselle. / ~ silk spinning ‖ Bourrettegarnspinnerei f. ‖ filature f. de bourette. / ~ sugar ‖ Roh-zucker m. ‖ moscouade f.; sucre m. brut. / ~ thread ‖ grober Bindfaden m. ‖ ficelle f. grossière. / thick ~ wool ‖ Schwanzwolle f. ‖ coaille f.; écuaillé f.

coast, to (Mar) ‖ längs der Küste f. fahren ‖ côtoyer une terre; descendre une côte. / ~ (Trade) ‖ Küstenhandel m. treiben ‖ caboter.

coast ‖ Küste f. ‖ côte f. / along the ~ ‖ an der Küste f. entlang ‖ le long de la côte. / bold ~ ‖ steile Küste f. ‖ côte f. accore ou élevée. / clear ~ ‖ reine Küste f. ‖ côte f. saine. / flat ~ ‖ flache oder niedrige Küste f. ‖ côte f. basse. / high ~ ‖ gefährliche Küste f. ‖ côte f. malsaine. / high ~ ‖ hohe Küste f.; hohes Land n. ‖ terre f. grosse ou haute. / iron-bound ~ see bold ~. / off the ~ ‖ in der Nähe f. der Küste ‖ au large de la côte. / rocky ~ ‖ felsige Küste f. ‖ côte f. rocheuse ou garnie de roches.

coastal cable ‖ Küstenkabel n. ‖ câble m. côtier. / ~ lowlands pl. ‖ Küstentiefland n. ‖ pays m. bas côtier. / ~ radio station ‖ Küstenfunkstelle f. ‖ radio-station f. côtière. / ~ river ‖ Küstenfluß m. ‖ fleuve m. côtier. / ~ survey ‖ Küstenaufnahme f.; Küstenvermessung f. ‖ levé m. d'une côte. / ~ terrace ‖ Küstenterrasse f. ‖ plate-forme f. littorale.

coaster (Mar) ‖ Küstenfahrer m.; Küsten-schiffer m. ‖ caboteur m. / ~ (Ship) ‖ Küstenfahrzeug n. ‖ cabotier m.

coast-guard vessel (Duty) ‖ Zollwachtschiff n. ‖ douanier m. / ~ (Warship) ‖ Küsten-wachtschiff n. ‖ garde-côte m.

coasting ‖ Küstenfahrt f. ‖ cabotage m. / ~ fisherman ‖ Küstenfischer m. ‖ pêcheur m. côtier. / ~ navigation ‖ Küstenschiff-fahrt f. ‖ navigation f. côtière; cabotage m. / ~ pilot ‖ Küstenlotse m. ‖ côtier m.; pilote-côtier m. / ~ trade ‖ Küstenhandel m. ‖ cabotage m.; commerce m. côtier. / ~ vessel ‖ Küstenfahrzeug n. ‖ cabotier m.

coast line ‖ Brandungslinie f. ‖ ligne f. de démarcation des eaux sur les côtes. / displacement of the ~ ‖ Strandverschie-bung f. ‖ déplacement m. du littoral.

coat, to ‖ einhüllen; einkleiden ‖ enrober; envelopper. / ~ (Paint) ‖ überstreichen ‖ enduire. / ~ the concrete ‖ den Beton m. verkleiden ‖ revêtir le bèton. / ~ the wood ‖ den Stamm verkleiden ‖ revêtir le bois.

coat ‖ Rock m.; Mantel m. ‖ habit m.; robe f.; manteau m. / ~ (Found) ‖ Form-mantel m.; Überform f. ‖ surmoule m.; surtout m. / ~ (Paint) ‖ Überzug m.; Anstrich m.; Schicht f. ‖ couche f. / ~ (Shipb) ‖ Kragen m. (von Segeltuch der Masten und Pumpen) ‖ braie f.; embre-lure f. ou toile f. goudronnée. / anti-acid ~ ‖ säurebeständiger Anstrich m. ‖ couche f. anti-acide. / finishing ~ (Paint) ‖ Deck-anstrich m. ‖ deuxième couche f. / first ~ (Paint) ‖ Grundfarbe f.; Grund-anstrich m. ‖ couleur f. de fond; pre-mière couche f. de peinture. / first ~ (Build) ‖ Anwurf m.; Bewurf m.; Putz m.; Spritzwurf m. ‖ crépi m.; crépi et enduit m.; première couche f. d'enduit. / ground ~ (Paint) ‖ Grundanstrich m. ‖ première couche f. de peinture. / ~ of ice ‖ Eishülle f. ‖ couche f. de glace. / light short ~ ‖ Joppe f. ‖ vareuse f.; casaquin m. / ~ of loam (Mould) ‖ Lehmauftrag m.; Lehmaufschlag m. ‖ couche f. d'argile ou de terre. / ~ of mail ‖ Panzerhemd n. ‖ cotte f. de mailles. / ~ of paint ‖ Farb-anstrich m. ‖ couche f. de peinture. / ~ of red lead ‖ Menniganstrich m. ‖ couche f. (de peinture) au minium. / second ~ (Build) ‖ zweite Lage f.; Putz m.; Tün-cherschicht f. ‖ chemise f.; enduit m.

coat baster ‖ Kleiderhefter m. ‖ faufileur m. d'habits. / ~ hanger ‖ Kleiderbügel m. ‖ cintre m.; porte-vêtements m.; cerceau m.

coated, with handle ~ with oil ‖ mit ge-öltem Stiel m. ‖ avec manche m. enduit de vernis. / ~ paper ‖ gestrichenes Pa-pier n. ‖ papier m. couché.

coat hook ‖ Kleiderhaken m. ‖ porte-vête-ments m.; patère f.

coating (Cast iron) ‖ Gußhaut f.; croûte f. / ~ (Chem) ‖ Beschlag m. ‖ enduit m. / ~ (Clothm) ‖ Fries m.; Flaus m. ‖ frise f. / ~ (Furnace) ‖ Futter n.; Ausfütterung f. ‖ revêtement m. / ~ (Mas) ‖ Gipsbewurf m. ‖ crépi m. de plâtre. / ~ (Metal) ‖ Über-zug m.; Schicht f. ‖ couche f. / ~ brass ~ ‖ Messingbeschlag m. ‖ laitonnage m. / ~ with broken stones (Road) ‖ Aufschütten n. des Steinschlags ‖ cailloutage m.; em-pierrement m. / ~ chocolate ~ plant in a cake-factory ‖ Schokoladenüberziehan-lage f. in einer Keksfabrik ‖ installation f. de dressage au chocolat dans une fabrique de cakes. / fireproof ~ ‖ feuer-fester Anstrich m. ‖ peinture f. ignifuge. / ~ of gravel (Road) ‖ Lage f. Kies ‖ couche f. de gravier. / ~ of ice ‖ Eisbelag m. ‖ couche f. de glace. / ~ of japan ‖ Lack-schicht f. ‖ couche f. de vernis. / ~ of paint ‖ Fertiganstrich m. ‖ peinture f. définitive. / non-corrosive ~ of paint ‖ Rostschutzanstrich m. ‖ couche f. de peinture antirouille. / to give a ~ of paint ‖ anstreichen ‖ peindre; enduire de couleur f. / ~ of a projectile ‖ Geschoß-mantel m. ‖ chemise f. du projectile. / to protect by a ~ of colourless lacquer ‖ mit farblosem Schutzlack m. überziehen ‖ recouvrir d'un vernis protecteur inco-lore. / protective ~ to counteract the ef-fects of the weather ‖ Schutzanstrich m.

gegen Witterungseinflüsse ‖ couche f. de peinture pour protéger contre les influences atmosphériques. / protective metallic ~ ‖ metallischer Schutzüberzug m. ‖ revêtement m. protecteur métallique. / red lead ~ ‖ Menniganstrich m. ‖ couche f. de minium. / ~ of silver ‖ Silberschicht f. ‖ argenture f. / wire ~ apparatus ‖ Drahtumspulapparat m. ‖ appareil m. à enrouler dans du fil métallique.

coat pocket ‖ Rocktasche f. ‖ poche f. du veston. / ~ rack ‖ Mantelhaken m. ‖ patère f.; porte-manteaux. m. / ~ stitcher ‖ Fantasiewirker m. ‖ tricoteur m. en fantaisie.

cob, to ~ ores (Mine) ‖ handscheiden ‖ trier.

cobalt ‖ Kobalt m. ‖ cobalt m. / bright white ~ ‖ Glanzkobalt m.; Kobaltglanz m.; Kobaltit m. ‖ cobalt m. gris; cobaltine f. / chloride of ~ ‖ Kobaltchlorür n. ‖ chlorure m. de cobalt; hydro-chlorate m. d'oxyde de cobalt. / containing ~ ‖ kobalthaltig ‖ cobaltique. / earthy ~ ‖ Rußkobalt m.; schwarzer Erdkobalt m. ‖ cobalt m. oxydé noir. / resembling ~ ‖ kobalthaltig ‖ cobaltique. / silver-white ~ ‖ Glanzkobalt m.; Kobaltglanz m.; Kobaltit m. ‖ cobalt m. gris; cobaltine f. / tin-withe ~ ‖ Smaltin m.; Speisekobalt m. ‖ smaltine f.; cobalt m. arsénical.

cobalt bloom ‖ Kobaltblüte f.; roter Erdkobalt m.; arseniksaurer Kobalt m.; Erythrin n. ‖ cobalt arséniaté; fleur f. de cobalt; erythrine f. / ~ blue ‖ Kobaltblau n. ‖ bleu m. de cobalt. / ~ bottle ‖ Kobaltflasche f. ‖ bouteille f. à cobalt. / ~ colouring ‖ Kobaltierung f. ‖ cobaltinage m. / ~ crust ‖ Kobaltbeschlag m.; erdige Kobaltblüte f. ‖ arséniate m. de cobalt terreux; efflorescence f. de cobalt. / ~ drier ‖ Kobalttrockenpräparat n. ‖ siccatif m. de cobalt. / ~ glance ‖ Glanzkobalt m.; Kobaltglanz m.; Kobaltit m. ‖ cobalt m. gris; cobaltine f. / ~ green ‖ Kobaltgrün n. ‖ vert m. de cobalt.

cobaltic ‖ kobalthaltig ‖ cobaltique. / ~ bath ‖ Kobaltbad n. ‖ bain m. de cobaltage. / ~ oxide ‖ Kobaltoxyd n. ‖ sesquioxyde m. de cobalt.

cobaltiferous ‖ kobalthaltig ‖ cobaltifère.

cobaltine see cobalt glance.

cobalt mica (Miner) see ~ bloom. / ~ mine ‖ Kobaltbergwerk n. ‖ mine f. de cobalt.

cobalt ochre, black ~ ‖ Kobaltschwärze f.; Kobaltmulm m.; Kobaltruß m. ‖ cobalt m. terreux noir; cobalt m. oxydé noir. / red ~ see cobalt bloom.

cobalt ore ‖ Kobalterz n. ‖ minerai m. de cobalt. / black-earthy ~ see cobalt ochre, black.

cobaltous oxide ‖ Kobaltoxydul n. ‖ protoxyde m. de cobalt.

cobald pyrites pl. ‖ Kobaltkies m. ‖ cobalt sulfuré; linnéite f.; koboldine f. / ~ regulus ‖ Kobaltspeise f. ‖ speiss m. de cobalt; arséniure m. de cobalt. / ~ speiss see ~ regulus.

cobbler (Shoem) ‖ Schnellsohler m. ‖ ressemeleur m. de chaussures. / ~'s tool ‖ Schusterhandwerkzeug n. ‖ crépin m. / ~'s wax ‖ Schuhmacherwachs n.; Schusterpech n. ‖ cire f. de cordonnerie.

cobbles pl. ‖ Füllkohle f.; Würfelkohle f. ‖ gailettes fpl.; gailletins mpl.; grélats mpl.

cobble stone ‖ Feldstein m. ‖ caillou m. tout venant.

cob brick ‖ Luftziegel m.; Lehmstein m.; ungebrannter Ziegel m. ‖ brique f. crue ou séchée à l'air. / ~ wall ‖ Lehmziegelmauer f. ‖ mur m. de briques crues.

cob mortar ‖ Lehmmörtel m. ‖ mortier m. d'argile. / ~ wall ‖ Lehmwand f.; Wellerwand f. ‖ mur m. de bousillage. / ~ work (Build) ‖ Lehmstampfwerk n. ‖ manière f. de bâtir en pisé; ouvrage m. pisé.

cocaine ‖ Kokain n. ‖ cocaïne f.

coca leaves pl. ‖ Kokablätter npl. ‖ feuilles fpl. de coca.

coccolith ‖ Kokkolith m. ‖ coccolithe f.

Cochin copra ‖ Cochinkopra f. ‖ coprah m. cochinchinois.

cochineal ‖ Koschenillefarbe f. ‖ cochenille f. / carmine of ~ ‖ Koschenillefarbe f. ‖ cochenille f.; carmin m. de cochenille. / ~ red ‖ Koschenillerot n. ‖ rouge m. de cochenille.

cock, to ~ a hat ‖ einen Hut m. aufkrämpen ‖ retrousser ou retaper un chapeau.

cock ‖ Hahn m. ‖ robinet m. / angle ~ ‖ Winkelhahn m. ‖ robinet m. d'angle. / ball ~ ‖ Schwimmerhahn m. ‖ robinet-flotteur m. / bibb ~ ‖ Niederschraubhahn m. ‖ robinet m. à vis. / blow-off ~ ‖ Abblasehahn m. ‖ robinet m. d'évacuation ou de purge. / body of ~ ‖ Hahngehäuse n. ‖ boisseau m. de robinet. / cylinder grease ‖ ~Zylinderschmierhahn m. ‖ robinet m. graisseur du cylindre. / discharge ~ ‖ Ablaßhahn m. ‖ robinet m. de vidange. / draining ~ ‖ Ablaßhahn m. ‖ robinet m. de vidange. / draining ~ for condensed water ‖ Kondenswasserablaßhahn m. ‖ robinet m. de vidange pour l'eau de condensation. / draw-off ~ ‖ Entwässerungshahn m. ‖ robinet m. de purge. / feed ~ ‖ Speisehahn m. ‖ robinet m. d'alimentation. / four-way ~ ‖ Vierweghahn m.; Kreuzhahn m. ‖ robinet m. à quatre voies. / gas ~ ‖ Gashahn m. ‖ robinet m. à gaz. / gauge ~ ‖ Wasserstandshahn m. ‖ robinet m. du niveau d'eau. / to grind in a ~ ‖ einen Hahn m. einschleifen ‖ roder un robinet m. / to grind in ~ plug ‖ ein Hahnküken n. einschleifen ‖ ajuster une noix de robinet par rodage. / ~ for guard's van ‖ Schaffnerbremshahn m. ‖ robinet m. du gardefrein. / ~ of a gun ‖ Hahn m. eines Gewehres ‖ chien m. d'un fusil. / head of ~ ‖ Hahnkopf m. ‖ tête f. de robinet. / injection ~ ‖ Einspritzhahn m. ‖ robinet m. d'injection. / mixing ~ ‖ Mischhahn m. ‖ robinet m. de mélange. / mud ~ ‖ Schmutzhahn m. ‖ robinet m. d'ébouage. / to open the ~ ‖ den Hahn m. aufdrehen ‖ ouvrir le robinet. / ~ with long pipe to strain off store casks ‖ Abseihhahn m.; Abseihwechsel m. ‖ robinet m. pour soutirer les fonds de foudres. / plug of a ~ ‖ Hahnkegel m.; Hahnküken n. ‖ noix f. ou clef f. d'un robinet. / plug of two-way ~ ‖ Durchgangshahnküken n. ‖ noix f. pour robinet droit. / ~ of pressure gauge with oval test flange ‖ Manometerhahn m. mit ovalem Prüfflansch ‖ robinet m. de manomètre à bride de contrôle ovale. / priming ~ ‖ Kompressionshahn m. ‖ robinet m. de compression. / radiator drain ~ (Auto) ‖ Kühlerablaßhahn m. ‖ robinet m. de vidange du radiateur. / shut-off ~ ‖ Absperrhahn m. ‖ robinet m. d'arrêt. / to shut-off a ~ ‖ einen Hahn m. schließen oder zudrehen ‖ fer-

mer un robinet m. / stop ~ see shut-off ~. / stop ~ of pressure gauge with air discharge ‖ Manometerabsperrhahn m. mit Entlüftung ‖ robinet m. d'arrêt de manomètre avec évacuation d'air. / straight-way ~ ‖ Durchgangshahn m. ‖ robinet m. droit ou ordinaire. / stuffing box ~ ‖ Stopfbuchsenhahn m. ‖ robinet m. avec presse-étoupe. / surface blow-off ~ (Boiler) ‖ Salzablaßhahn m.; Abschäumhahn m. ‖ robinet m. d'extraction à hauteur de niveau. / testing ~ ‖ Probhahn m. ‖ robinet m. de jauge. / three-way ~ ‖ Dreiweghahn m. ‖ robinet m. à trois voies. / try ~ see testing ~. / two-way ~ ‖ Durchgangshahn m. ‖ robinet m. à deux orifices. / two-way ~ without stuffing box ‖ Durchgangshahn m. ohne Stopfbuchse ‖ robinet m. droit sans boîte de bourrage. / valve ~ ‖ Ventilhahn m. ‖ robinet-valve m. / water ~ ‖ Wasserhahn m. ‖ robinet m. d'eau. / water-gauge ~ see gauge ~. / way of ~ ‖ Hahnöffnung f. ‖ passage m. du robinet. / wood ~ ‖ Holzhahn m. ‖ robinet m. en bois.

cock bead (Arch) ‖ Rundstab m. ‖ boudin m.; bosel m.; baguette f.

cocket center (Build) ‖ gesprengtes Lehrgerüst n. ‖ cintre m. retroussé.

cockle (Flower) ‖ Kornrade f. ‖ nielle f. / ~ (Grain) ‖ Kokkelskorn n. ‖ coque f. du Levant. / ~ (Mussel) ‖ Muschel f. ‖ coquille f. / ~ gatherer ‖ Muschelfischer m. ‖ pêcheur m. de coquilles ou de coquillages ou de moules. / ~ shell (Mar) ‖ Seelenverkäufer m.; Nußschale f. (Boot) ‖ périssoire m.; coque f. de noix. / ~ stairs pl. ‖ Wendeltreppe f. ‖ escalier m. en limaçon ou à vis.

cockling sea ‖ Kabbelsee f.; kabbelige See f. ‖ clapotage m.; clapotis m.

cockpit, pilot's ~ (Aero) ‖ Führerraum m. oder Führersitz m. im Flugzeug ‖ cabine f. ou poste m. ou siège m. du pilote.

cockpit heater (Aero) ‖ Sitzraumerwärmer m.; Sitzraumheizung f. ‖ réchauffeur m. de carlingue.

cock plug ‖ Hahnküken n. ‖ noix f. de robinet. / ~ spanner ‖ Hahnschlüssel m. ‖ clef f. de robinet. / ~ spring ‖ Winkelfeder f. ‖ ressort m. à chien. / ~ wrench see ~ spanner.

coco ‖ Kokospalme f. ‖ cocotier m.

cocoa ‖ Kakao m. ‖ cacao m. / ~ in beans ‖ Kakao m. in Bohnen ‖ cacao m. en fèves. / ~ bean crusher ‖ Kakaobohnenquetscher m. ‖ broyeur m. de fèves de cacao. / cleaner of raw ~ beans ‖ Kakaobohnensortierer m. ‖ trieur m. de fèves de cacao. / presser of ~ beans ‖ Kakaobohnenpresser m. ‖ presseur m. de fèves de cacao. / broken ~ ‖ Kakaobruch m. ‖ brisure f. de cacao. / ~ in powder ‖ gemahlener Kakao m. ‖ cacao m. en poudre. / preparation of ~ ‖ Kakaozubereitung f. ‖ préparation f. de cacao.

cocoa bean ‖ Kakaobohne f. ‖ fève f. de cacao. / ~ butter ‖ Kakaobutter f. ‖ beurre m. de cacao. / ~ grinder ‖ Kakaomühle f. ‖ moulin m. à cacao. / ~ hull ‖ Kakaohülse f. ‖ pelure f. de cacao. ‖ ~ husk ‖ Kakaoschale f. ‖ coque f. de cacao. / ~ meal ‖ Kakaopulver m. ‖ cacao m. en poudre. / ~ nut see coconut. / ~ oil soap ‖ Kokosölseife f. ‖ savon m. de coco. / ~ paste ‖ Kakaopaste f. ‖ pâte f. de cacao. / ~ pellicle ‖ Kakao-

häutchen n. ‖ pellicule f. de cacao. / ~ powder ‖ Kakaopulver n. ‖ poudre f. de cacao. / ~ preparing machine ‖ Kakaomaschine f. ‖ machine f. à préparer le cacao. / ~ roaster ‖ Kakaobrenner m. ‖ grilleur m. de cacao. / ~ shell ‖ Kakaoschale f. ‖ coque f. *ou* pelure f. de cacao.

coco fibre ‖ Kokosfaser f. ‖ fibre f. de coco. / ~ matting ‖ Kokosmatte f. ‖ natte f. de coco.

coconut ‖ Kokosnuß f. ‖ noix f. de coco; coco m. / ~ butter ‖ Kokosnußbutter f. ‖ beurre m. de coco; cocose f. / ~ fibre ‖ Kokosfaser f. ‖ brou m. de coco; fibre f. des noix de coco. / ~ fibre brush ‖ Kokosbürste f. ‖ brosse f. en fibre de coco. / ~ fibre ware ‖ Kokosfaserwaren fpl. ‖ articles mpl. en fibres de coco. / ~ oil ‖ Kokosnußöl n. ‖ huile f. de coco. / ~ palm ‖ Kokospalme f. ‖ cocotier m. / ~ shell ‖ Kokosnußschale f. ‖ coque f. de noix de coco. / ~ tree ‖ Kokosnußbaum m. ‖ cocotier m.

cocoon ‖ Kokon m.; Seidengehäuse n. ‖ cocon m.; soie f. en cocon. / without any damage to the ~s pl. ‖ ohne Beschädigung f. der Kokons ‖ sans endommagement m. des cocons. / to extend the working-up of the ~s over a longer lapse of time ‖ Verarbeiten n. der Kokons auf größere Zeiträume verteilen ‖ répartir le traitement des cocons sur un laps de temps plus étendu.

cocoon fibre ‖ Kokonfaden m. ‖ fil m. de cocon. / ~ opener ‖ Kokonöffner m. ‖ coconneuse f. / ~ reeler ‖ Kokonhaspeler m. ‖ dévideur m. de cocons. / ~ reeling ‖ Kokonwinderei f. ‖ dévidage m. de cocons. / ~ sorter ‖ Kokonsortierer m. ‖ trieur m. de cocons. / ~ spinner ‖ Kokonspinner m. ‖ fileur m. de cocons.

cod ‖ Kabeljau m.; Dorsch m. ‖ morue f. (fraîche); cabillaud m.; dorsch m. / dried ~ ‖ Stockfisch m.; getrockneter Kabeljau m. ‖ morue f. sèche *ou* séchée; merluche f.; stockfisch m.

code ‖ Telegraphencode m.; Telegrammschlüssel m. ‖ code m.; private ~ ‖ Privattelegrammschlüssel m. ‖ code m. personnel. / ~ of signals ‖ Signalbuch n. ‖ code m. des signaux. / telegraphic ~ ‖ Codeschlüssel m. ‖ code m. télégraphique.

coded ‖ chiffriert ‖ codifié; chiffré.

codeine ‖ Kodein n. ‖ codéine f.

code language ‖ verabredete Sprache f. ‖ langage m. convenu. / ~ list ‖ Schiffsliste f.; Schiffsregister n. ‖ liste f. des bâtiments et leur numéros officiels dans le Code international. / ~ name of stations ‖ Rufzeichen n. einer Telegrafenstation ‖ indicatif m. / ~ number ‖ Kennziffer f. ‖ numéro m. indicatif. / ~ signalling system in fire-alarm installations ‖ Morsesystem n. in Feuermeldeanlagen ‖ système m. Morse dans les installations d'avertisseurs d'incendie. / ~ telegram ‖ Chiffretelegramm n. ‖ dépêche f. chiffrée. / ~ typewriter ‖ Chiffriermaschine f. ‖ machine f. cryptographique. / ~ word ‖ Telegrammwort n.; Drahtungswort n.; Schlüsselwort n. ‖ désignation f. *ou* code m. télégraphique.

codfish *see* cod.

cod fisherman ‖ Kabeljaufischer m.; Stockfischfänger m. ‖ pêcheur m. de morues; morutier m. / ~ liver oil ‖ Lebertran m.; Dorschtran m. ‖ huile f. de foie de morue.

coe (Mine) ‖ Kaue f. ‖ hangar m.; abri m.

coefficient ‖ Koeffizient m.; Beiwert m.; Beizahl f. ‖ coefficient m. / ~ of coupling (Radio) ‖ Kopplungsgrad m. ‖ coefficient m. d'accouplement. / ~ of charge (Electr) ‖ Ladungskoeffizient m. ‖ coefficient m. de charge. / ~ of cyclic variation (Mach) ‖ Ungleichförmigkeitsgrad m. ‖ degré m. d'irrégularité. / ~ of elongation ‖ Dehnungszahl f. ‖ coefficient m. d'allongement. / ~ of expansion ‖ Ausdehnungskoeffizient m.; Wärmeausdehnungskoeffizient m. ‖ coefficient m. d'expansion *ou* de dilatation. / ~ of friction ‖ Reibungskoeffizient m.; Reibungszahl f.; Reibungsziffer f. ‖ coefficient m. de frottement. / ~ of linear expansion ‖ linearer Ausdehnungskoeffizient m. ‖ coefficient m. de dilatation linéaire. / quality ~ ‖ Gütekoeffizient m. ‖ coefficient m. qualitatif *ou* de qualité. / ~ of selfinduction ‖ Selbstinduktionskoeffizient m. ‖ coefficient m. de self. / temperature ~ ‖ Temperaturkoeffizient m. ‖ coefficient m. de température. / ~ of thermal expansion ‖ Wärmeausdehnungskoeffizient m. ‖ coefficient m. de dilatation thermique.

coelostat ‖ Coelostat m. ‖ cœlostat m.

coercitive force *see* coercive force.

coercive force ‖ Koerzitivkraft f. ‖ force f. coercitive. / ~ measures pl. ‖ Zwangsverfahren n. ‖ procédure f. coercitive.

coe-steads pl. (Mine) ‖ Kaue f. ‖ hangar m. du puits.

coffee ‖ Kaffee m. ‖ café m. / artificial ~ ‖ künstlicher Kaffee m. ‖ café m. artificiel. / ~ in beans ‖ Kaffee m. in Bohnen ‖ café m. en fèves. / bright ~ ‖ blanker Kaffee m. ‖ café m. luisant. / clean ~ ‖ reinschmeckender Kaffee m. ‖ café m. droit de goût. / fine and good ~ ‖ guter Kaffee m. ‖ bon café m. / good middling ~ ‖ mittelguter Kaffee m. ‖ café m. bon marchand. / green ~ ‖ grüner Kaffee m. ‖ café m. vert. / ground ~ ‖ gemahlener Kaffee m. ‖ café m. moulu. / large-berried ~ ‖ großbeeriger Kaffee m. ‖ café m. à gros grains. / pale ~ ‖ blasser Kaffee m. ‖ café m. pale. / raw ~ ‖ ungebrannter Kaffee m. ‖ café m. vert. / roasted ~ ‖ gerösteter Kaffee m. ‖ café m. torréfié. / ~ in shells or husks ‖ Kaffee m. in Pergament oder in Samenhülsen ‖ café m. en parchemin ou en crosses. / small-berried ~ ‖ kleinbohniger Kaffee m. ‖ café m. à petits grains. / sorting and cleaning machine for ~ and ~ surrogates ‖ Sortier- und Reinigungsmaschine f. für Kaffee und Kaffeeersatz ‖ machine f. pour le triage et le nettoyage du café et des succédanés du café. / unclean ~ ‖ unreiner Kaffee m. ‖ café m. gras. / unroasted ~ ‖ ungebrannter Kaffee m. ‖ café m. vert.

coffee boiling apparatus ‖ Kaffeemaschine f. ‖ machine f. à café; appareil m. à préparer le café. / ~ cup ‖ Kaffeetasse f. ‖ tasse f. à café. / ~ dressing machine ‖ Kaffeebearbeitungsmaschine f. ‖ machine f. pour le traitement du café. / ~ essence ‖ Kaffeeessenz f. ‖ essence f. de café. / ~ extract ‖ Kaffeeauszug m. ‖ extrait m. de café. / ~ glazing ‖ Kaffeeglasur f. ‖ glaçure f. pour café. / ~ machine ‖ Kaffeemaschine f. ‖ machine f. à préparer le café. / ~ making plant ‖ Kaffeereinigungs- und -sortieranlage f. ‖ machines fpl. à

dresser et conditionner le café. / ~ mill ‖ Kaffeemühle f. ‖ moulin m. à café. / ~ mill for attaching to wall ‖ Wandkaffeemühle f. ‖ moulin m. à café mural. / ~ percolator ‖ Kaffeemaschine f. ‖ machine f. à préparer le café. / ~ pot ‖ Kaffeekanne f. ‖ cafetière f. / ~ pot with flow-over system ‖ Kaffeekanne f. mit Überkochung ‖ cafetière f. à fontaine. / ~ preparing machine ‖ Kaffeeaufbereitungsmaschine f. ‖ machine f. de préparation du café. / ~ range ‖ Kaffeeherd m. ‖ cuisinière-cafetière f. / ~ roaster ‖ Kaffeeröstmaschine f.; Kaffeebrenner m.; Kaffeetrommel f. ‖ brûloir m. *ou* grilleur m. de café; tambour m. à griller le café. / ~ roasting ‖ Kaffeerösterei f. ‖ brûlerie f. *ou* torréfaction f. *ou* grillage m. de café. / ~ roasting machine ‖ Kaffeeröstmaschine f. ‖ torréfacteur m. à café. / ~ service ‖ Kaffeegeschirr n. ‖ service m. à café. / ~ set ‖ Kaffeegeschirr n. ‖ service m. à café. / ~ set in coloured enamel ‖ Kaffeegeschirr n. in farbigen Glasuren ‖ service m. à café en émail coloré. / ~ shrub ‖ Kaffeebaum m. ‖ cafier m. / ~ sorter ‖ Kaffeesortierer m. ‖ trieur m. de café. / ~ sorting ‖ Kaffeeauslesen n. ‖ triage m. de café. / ~ stall ‖ Kaffeeschenke f. ‖ cantine f. à café. / ~ strainer ‖ Kaffeetrichter m. ‖ entonnoir m. à café. / ~ substitute ‖ Kaffee-Ersatzstoff m.; Kaffeesurrogat n. ‖ succédané m. du café.

coffer ‖ Koffer m. ‖ coffre m. / ~ dam ‖ Fangdamm m. ‖ batardeau m. / ~ wall ‖ Füllmauer f. ‖ mur m. de remplage *ou* de blocage *ou* rempli de hourdage. / ~ work ‖ Füllmauerwerk n. ‖ maçonnerie f. en blocage *ou* de remplage *ou* remplie de hourdage; murage m. bloqué. / ~ work of loam earth (Build) ‖ Lehmstampfbau m. ‖ manière f. de bâtir en pisé; construction f. en pisé.

coffin ‖ Sarg m. ‖ cercueil m. / ~ (Print) ‖ Fundament n. (einer Schnellpresse) ‖ châssis m. du coffre. / lead ~ ‖ Bleisarg m. ‖ cercueil m. en plomb. / metallic ~ ‖ Metallsarg m. ‖ cercueil m. métallique. / zinc ~ ‖ Zinksarg m. ‖ cercueil m. en zinc.

coffin fittings of cardboard ‖ Sargbeschläge mpl. aus Pappe ‖ garnitures fpl. de cercueil en carton. / ~ fringe ‖ Sargfranse f. ‖ frange f. de cercueil. / ~ handle ‖ Sarggriff m. ‖ anse f. pour cercueils. / ~ lid ‖ Sargdeckel m. ‖ couvercle m. de cercueil. / ~ maker ‖ Sargtischler m. ‖ fabricant m. de cercueils. / ~ metal furniture ‖ Sargbeschläge mpl. aus Blech ‖ ferrures fpl. en tôle pour cercueils. / ~ ornament ‖ Sargverzierung f. ‖ ornement m. pour cercueils. / ~ screw ‖ Sargschraube f. ‖ vis f. pour cercueils. / ~ slab ‖ Sargdeckel m. von Stein ‖ dalle f. tumulaire.

cog, to (Carp) ‖ aufkämmen ‖ assembler à entailles. / ~ (Mach) ‖ mit Zähnen mpl. versehen ‖ endenter. / ~ (Mine) ‖ mit Bergen versetzen ‖ remblayer. / ~ (Wheels) ‖ kämmen ‖ engrener.

cog (Mach) ‖ Nase f. ‖ prisonnier m. / ~ (Mill) ‖ Holzzahn m.; Kamm m. ‖ alluchon m.; dent f. de bois. / ~ (Mine) ‖ Versatzpfeiler m. ‖ pilier m. de remblais. / ~ of wheel ‖ Radzahn m. ‖ dent f. de roue.

cogent ‖ triftig ‖ fondé; valable.

cogged (Mach) ‖ mit Zähnen mpl. versehen ‖ denté.

cogger (Metal) ‖ Zwischenwalzer m. ‖ lamineur m. au train intermédiaire. / ~ (Mine) ‖ Versatzarbeiter m. ‖ remblayeur m.; restapleur m.; releveur m. de terres.

cogging ‖ Verzahnung f. ‖ denture f. / ~ joint (Carp) ‖ Überkämmen n. ‖ assemblage m. à entailles. / ~ mill (Roll mill) ‖ Vorstraße f.; Blockwalzwerk n. ‖ train m. dégrossisseur; laminoir m. à blooms.

cogman (Mine) ‖ Grubenmaurer m. ‖ maçon m. de puits.

cognac distiller ‖ Schnapsbrenner m. ‖ distillateur m. de vin; brandevinier m.

cog railway ‖ Zahnradbahn f. ‖ chemin m. de fer à crémaillère. / ~ tooth ‖ Holzzahn m. ‖ alluchon m.; dent f. de bois.

cog wheel ‖ Kammrad n. ‖ roue f. à dents. / ~ of a jack ‖ Stirnrad n. einer Wagenwinde ‖ roue f. dentée d'un cric.

cog wheel escapement ‖ Zahnradauslösung f. ‖ échappement f. à roue dentée. / ~ transmission ‖ Zahnradübersetzung f. ‖ transmission f. par engrenage.

cohere, to ‖ zusammenhaften ‖ cohérer.

coherence ‖ Frittung f. ‖ cohérence f.

coherency see coherence.

coherer ‖ Fritter m.; Kohärer m. ‖ cohéreur m. / ~ with adaptable sensitiveness ‖ Fritter m. mit einstellbarer Empfindlichkeit ‖ cohéreur m. à sensibilité variable. / granular ~ ‖ Körnerfritter m. ‖ cohéreur m. à grenaille.

coherer current ‖ Fritterstrom m. ‖ courant m. du cohéreur. / ~ pole ‖ Fritterpol m. ‖ pôle m. de cohéreur. / ~ protector ‖ Frittersicherung f. ‖ cohéreur m. protecteur. / ~ resistance ‖ Fritterwiderstand m. ‖ résistance f de cohéreur / ~ terminal ‖ Fritterklemme f ‖ borne f. de cohéreur. / ~ tester ‖ Fritterprüfer m. ‖ essayeur m. de cohéreur.

cohesion ‖ Kohäsion f. ‖ cohésion f.

cohesive attraction ‖ Kohäsion f. ‖ cohésion f. / ~ hold ‖ Kohäsion f.; Anhaften n. ‖ prise f. cohésive.

coif ‖ Frauenhaube f. ‖ coiffe f. de femme.

coil, to ‖ wickeln ‖ bobiner. / ~ round ‖ wickeln ‖ rouler; enrouler. / ~ up ‖ wickeln ‖ rouler; enrouler. / ~ up a rope in a roll ‖ ein Tau n. (in einer Scheibe) aufrollen ‖ glèner; faire une glène de filin.

coil (Electr) ‖ Rolle f.; Spule f. ‖ bobine f.; rouleau m. / ~ (Mach) ‖ Rohrschlange f.; serpentin m. / ~ (Mar) ‖ Scheibe f. (von Tauwerk‖ glène f.; glène de filin. / basket ~ ‖ Korbspule f. ‖ bobine f. type fond de panier. / choking ~ (Electr) ‖ Drosselspule f. ‖ self m. / cylindrical ~ ‖ Zylinderspule f. ‖ bobine f. cylindrique. / disc ~ ‖ Flachspule f. ‖ bobine f. plate. / double wound ~ ‖ Spule f. mit doppelter Drahtwindung ‖ bobine f. bifilaire. / drainage ~ (Electr) ‖ Ableitungsspule f. ‖ bobine f. de drainage. / exciting ~ ‖ Erregerspule f. ‖ bobine f. inductrice. / field ~ (Electr) ‖ Magnetspule f. ‖ bobine f. à aimants. / heating ~ ‖ Heizschlange f. ‖ serpentin m. réchauffeur. / honeycomb ~ ‖ Honigwabenspule f. ‖ bobine f. en nid d'abeille. / ignition ~ ‖ Zündspule f. ‖ bobine f. d'allumage. / iron-powder core ~ ‖ Massekernspule f. ‖ bobine f. à noyau en poudre de fer comprimé. / ~ of low capacity ‖ kapazitätsarme Spule f. ‖ bobine f. à capacité réduite. / ~ of magnet

‖ Magnetspule f. ‖ bobine f. d'électroaimant. / new winding of the ~ (Electr) ‖ Neuwicklung f. der Spule ‖ rébobinage m. de la bobine. / ~ in opposite direction ‖ gegenläufige Windung f. ‖ enroulement m. en sens inverse. / pressed dust core ~ ‖ Massekernspule f. ‖ bobine f. à noyau (formé d'un aggloméré) de limaille de fer. / quenching ~ ‖ Löschdrossel f. ‖ extinction f. à réactance. / radiator ~ ‖ Schlangenkühler m. ‖ radiateur m. à serpentin. / Ruhmkorff's ~ ‖ Induktorium n. ‖ bobine f. de Ruhmkorff. / ~s pl. per slot ‖ Spulen fpl. je Nut ‖ bobines fpl. par rainure. / syntonizing ~ (Radio) ‖ Abstimmspule f. ‖ bobine f. de syntonisation. / ~ in three sections ‖ Dreifachspule f. ‖ bobine triple. / triple ~ (Electr) ‖ Dreifachspule f. ‖ bobine f. triple. / tripping ~ ‖ Auslösespule f. ‖ bobine f. d'amorce. / ~ of wire ‖ Drahtrolle f.; Drahtbund m.; Drahtring m. ‖ couronne f. ou botte f. ou torche f. de fil.

coil antenna ‖ Spulenantenne f. ‖ antenne f. en boucle. / ~ box ‖ Spulenkasten m. ‖ boîte f. à selfs. / ~ clutch ‖ Federbandkupplung f. ‖ embrayage m. à ressort ou à ruban ou à spirale. / ~ coupler ‖ Spulenkoppler m. ‖ coupleur m. de selfs.

coiled disc wheel ‖ Wickelrad n. ‖ roue f. à disque composé de bandes de fer enroulées; roue f. enroulée. / ~ spring ‖ Keilfeder f. ‖ ressort m. à languette. / ~ -up paper ribbon ‖ Papierschlange f. ‖ serpentin m. de papier.

coil end (Electr) ‖ Spulenkopf m. ‖ tête f. de bobine.

coiler ‖ Drehtopf m. ‖ pot m. tournant. / hemp ~ (Mach) ‖ Hanfflechte f.; Hanfzopf m. ‖ tresse f. de chanvre.

coiling machine, automatic ~ ‖ selbsttätige Wickelmaschine f. ‖ machine f. d'enroulement automatique. / insulating tube ~ ‖ Isolierrohrwickelmaschine f. ‖ enrouleuse f. de tubes isolants. / scrap ~ ‖ Schrotwickelmaschine f. ‖ machine f. à enrouler les riblons. / strip ~ ‖ Rundbiegemaschine f. für Bänder ‖ machine f. à cintrer les feuillards. / wire ~ ‖ Drahtaufrollmaschine f. ‖ machine f. à enrouler le fil.

coil loaded cable ‖ Pupinkabel n. ‖ câble m. Pupin. / ~ line with musical loading (Tel) ‖ musikpupinisierte Leitung f. ‖ ligne f. pupinisée à charge musicale.

coil loading ‖ Pupinverfahren n. ‖ pupinisation f. / ~ of cables ‖ Pupinisierung f. der Kabel ‖ pupinisation f. des câbles. / first section of ~ (tel) ‖ Anlauflänge f. in Pupinleitungen ‖ première section f. de pupinisation.

coil sleeve ‖ Spulenmuffe f. ‖ manchon m. à bobine. / ~ space ‖ Spulenabstand m. ‖ espacement m. des bobines. / ~ spring ‖ zylindrische Schraubenfeder f. ‖ ressort m. à boudin ou en spirale à boudin ou hélicoïdal. / ~ stand ‖ Spulenständer m. ‖ support-bobine m. / ~ winder (Electr) ‖ Wickler m. ‖ bobineur m.; bobinier m.

coil-winding ‖ Spulenwickelung f. ‖ enroulement m. à bobines. / small ~ ‖ Kleinspulerei f. ‖ fabrication f. de petites bobines.

coil winding bench see ~ machine. / ~ machine ‖ Spulenwickelmaschine f. ‖ bobineuse f.; machine f. à bobiner.

coin, to ‖ prägen ‖ estamper; frapper; battre. / ~ (Coin) ‖ Münzen fpl. schlagen ‖ monnayer. / ~ between the ferrel ‖ ringprägen ‖ frapper à virole. / ~ money ‖ münzen ‖ frapper des monnaies fpl.; monnayer.

coin ‖ Geld n. ‖ argent m.; monnaie f. / ~ (Piece) ‖ Geldstück n. ‖ pièce f. d'argent ou de monnaie. / ~ (Arch) ‖ Ecke f. ‖ encoignure f. / bad ~ ‖ geringhaltige Münze f. ‖ monnaie f. de mauvais aloi. / copper ~ ‖ Kupfermünze f. ‖ monnaie f. de cuivre. / counterfeit ~ see false ~. / false ~ ‖ falsche Münze f. ‖ fausse-monnaie f. / gold ~ ‖ Goldmünze f. ‖ monnaie f. d'or. / hardened ~ ‖ stempelharte Münze f. ‖ monnaie f. écrouie. / light ~ ‖ geringhaltige Münze f. ‖ monnaie f. de mauvais aloi. / nickel ~ ‖ Nickelmünze f. ‖ monnaie f. de nickel. / rimmed ~ ‖ gerändelte Münze f. ‖ monnaie f. cordonnée. / silver ~ ‖ Silbermünze f. ‖ monnaie f. d'argent. / small ~ ‖ Scheidemünze f. ‖ monnaie f. d'appoint; billon m.; petite monnaie f.

coinage ‖ Münzwesen n. ‖ monnaie f.; monnayage m. / ~ ‖ Ausmünzung f.; Münzprägung f. ‖ frappe f.

coin bag ‖ Geldsack m. ‖ sachet m. à monnaie. / ~ box (Tel) ‖ Münzfernsprecher m. ‖ poste m. à prépaiement.

coincide, to (Phys) ‖ zusammenfallen ‖ coincider.

coincidence telemeter ‖ Schnittbildentfernungsmesser m.; Koinzidenzentfernungsmesser m. ‖ télémètre m. à coincidence.

coin counter ‖ Münzenzähler m. ‖ appareil m. à compter les pièces de monnaie. / ~ cup ‖ Geldschale f. ‖ coupe f. pour pièces de monnaie. / ~ engraver ‖ Münzgravör m ‖ graveur m. en monnaies / ~ engraving ‖ Münzprägegravierung f ‖ gravure f. en monnaies.

coiner ‖ Münzschläger m. ‖ frappeur m. de monnaies.

coining (Coin) ‖ Münzen n. ‖ monnayage m.; monnéage m. / ~ (Print) ‖ Prägung f. ‖ empreinte f. / ~ die ‖ Münzstempel m. ‖ coin m.; poinçon m.

coining ferrule ‖ Prägring m. ‖ virole f. / channelled ~ ‖ gekerbter Prägring m. ‖ virole f. cannelée. / closed ~ ‖ ganzer oder ungeteilter Prägring m. ‖ virole f. pleine. / non-channelled ~ ‖ glatter Prägring m. ‖ virole f. lisse. / open ~ made of three parts ‖ dreiteiliger, gebrochener Prägring m. ‖ virole f. brisée.

coining hammer ‖ Münzschlaghammer m. ‖ bouard m. / ~ machine ‖ Münzprägemaschine f. ‖ machine f. à frapper les monnaies. / ~ machine for gold ‖ Goldprägepresse f. ‖ presse f. pour frapper l'or. / ~ press see ~ machine.

coin manufacturing ‖ Münzprägeanstalt f. ‖ fabrique f. de monnaies. / ~ plate ‖ Münzplatte f.; Scheibe f. ‖ flan m. / ~ sorter ‖ Münzensortierapparat m. ‖ appareil m. pour trier les pièces de monnaie.

coir ‖ Kokosfaser f.; Kokosbast m. ‖ fibre f. de coco; coir m. / ~ rope ‖ Basttau n. ‖ filin m. en coir. / ~ yarn ‖ Kokosgarn n. ‖ fil m. de brou du coco.

coke, to ‖ verkoken ‖ cokefiér. / ~ the coal ‖ die Steinkohle verkoken ‖ cokéfier la houille f. / ~ well ‖ einen guten Gang m. des Koksofens haben ‖ produire une belle gueule de four à coke.

coke ‖ Koks m. ‖ coke m.; coak m.; charbon m. de houille. / broken ~ ‖ Bruchkoks m. ‖ coke m. concassé; coke m. moyen. / ~ of brown coal ‖ Grudekoks m. ‖ coke m. de lignite. / cake of ~ ‖ Kokskuchen m. ‖ tourteau m. de coke. / crushed ~ ‖ Knabbelkoks m.; Brechkoks m. ‖ coke m. concassé. / glowing ~ ‖ glühender Koks m. ‖ coke m. ardent. / hard lumpy ~ ‖ fester stückiger Koks m. ‖ coke m. solide en morceaux. / ~ of a high class ‖ Koks m. von hochwertiger Beschaffenheit ‖ coke m. de meilleure qualité. / large ~ ‖ Großkoks m. ‖ coke m. en morceaux; gros coke m. / lean ~ ‖ Magerkoks m. ‖ coke m. maigre. / machine-broken ~ see broken ~. / to remove deposited ~ residues pl. ‖ von anhaftenden Koksrückständen mpl. reinigen ‖ nettoyer les résidus mpl. de coke adhérents. / semi-~ ‖ Halbkoks m. ‖ semi-coke m. / sifted ~ ‖ gesiebter Koks m. ‖ coke m. criblé ou tamisé. / small ~ ‖ Koksklein n.; Kokslösche f. ‖ coke m. menu; braise f. de coke.

coke barrow ‖ Kokskarren m. ‖ wagonnet m. à coke. / ~ basket ‖ Kokskorb m. ‖ corbeille f. à coke. / inclined ~ bench ‖ Koksrampe f. ‖ plan m. incliné de four à coke. / ~ blast-furnace ‖ Hochofen m. für Koks ‖ haut-fourneau m. au coke. / ~ breaker ‖ Koksbrecher m. ‖ broyeur m. ou concasseur m. de coke. / ~ breeze ‖ Koksklein n. ‖ braise f. de coke; coke m. en grains menus. / ~ burner ‖ Koksarbeiter m. ‖ cokeur m. / ~ burning of pit-coal ‖ Verkokung f. der Steinkohle ‖ carbonisation f. de la houille. / cake of ~ ‖ Kokskuchen m. ‖ tourteau m. de coke. / ~ charging machine ‖ Kokseinsetzmaschine f. ‖ machine-enfourneuse f. à coke. / ~ cinder ‖ Kokslösche f. ‖ fraisil m. de coke. / ~ cooler ‖ Koksablöscher m. ‖ extincteur m. de coke. / ~ cooling plant ‖ Kokskühlungsanlage f. ‖ installation f. à refroidir le coke. / ~ crusher ‖ Koksbrecher m. ‖ concasseur m. de coke.

coke crushing ‖ Koksbrechen n. ‖ concassage m. de coke. / ~ plant ‖ Koksbrechanlage f. ‖ installation f. de concassage de coke. / ~ rolls pl. ‖ Koksmühle f. ‖ rouleaux mpl. concasseurs à coke. / ~ and sizing plant ‖ Kokszerkleinerungs- und sortieranlage f. ‖ concassage m. et triage m. des cokes.

coke dressing plant ‖ Koksaufbereitungsanlage f. ‖ installation f. de préparation de coke.

coke dust ‖ Koksstaub m.; Kokslösche f. ‖ poussière f. de coke. / ~ firing ‖ Koksstaubfeuerung f. ‖ foyer m. à poussière de coke.

coke filler ‖ Koks(auf)lader m. ‖ chargeur m. de coke. / ~ furnace (Met) ‖ Hochofen m. für Koks ‖ haut-fourneau m. au coke. / ~ kiln ‖ Verkokungsofen m. ‖ four m. à coke.

coke-like residue of distillation ‖ koksähnlicher Rückstand m. von der Destillation ‖ résidu m. similaire au coke provenant de la distillation.

coke loading plant ‖ Koksverladeanlage f. ‖ installation f. à charger le coke.

coke oven ‖ Koksofen m. ‖ four m. à coke. / closed ~ ‖ geschlossener Koksofen m. ‖ four m. à coke fermé. / battery of coke ovens ‖ Koksofengruppe f. ‖ batterie f. de fours à coke. / ~ charging wagon ‖ Koksofenfüllwagen m. ‖ wagonnet-chargeur m. de four à coke; wagon m. de remplissage pour fours à coke. / ~ equipment ‖ Koksofenausrüstung f. ‖ équipement m. de four à coke. / ~ gas ‖ Koksofengas n.; Kokereigas n. ‖ gaz m. de four à coke. / ~ industry ‖ Kokereiindustrie f. ‖ industrie f. du coke. / ~ plant ‖ Kokereianlage f. ‖ installation f. pour la fabrication du coke. / set of coke ovens see battery of coke ovens. / ~ stoker ‖ Koksarbeiter m. ‖ cokeur m.

coke pusher ‖ Koksausdrückmaschine f. ‖ défourneuse f. de coke. / ~ pushing machine see ~ pusher.

coke quenching car ‖ Kokslöschwagen m. ‖ chariot m. pour l'extinction du coke. / ~ and conveying car ‖ Kokslösch- und -transportwagen m. ‖ wagon m. d'extinction et de transport du coke. / ~ tower ‖ Kokslöschturm m. ‖ tour f. d'extinction de coke. / ~ wagon see ~ car.

coke riddler ‖ Koksaussieber m. ‖ cribleur m. de coke. / ~ screener ‖ Kokssieber m. ‖ cribleur m. de coke. / ~ screening ‖ Kokssieberei f. ‖ criblage m. du coke. / ~ separation ‖ Kokstrennung f. ‖ séparation f. de coke. / ~ shovel with sheet head handle ‖ Koksschaufel f. mit Griffstiel ‖ pelle f. à charger le coke avec manche à poignée. / ~ sifting plant ‖ Kokssieberei f. ‖ installation f. de tamisage de coke. / ~ sifting wagon ‖ Kokssiebwagen m. ‖ wagon m. pour le tamisage du coke. / ~ sorting plant ‖ Kokssortierungsanlage f. ‖ installation f. de triage de coke. / ~ tar ‖ Zechenteer m. ‖ goudron m. de coke. / ~ waste ‖ Koksabfall m. ‖ déchets mpl. de coke. / ~ works pl. ‖ Kokerei f. ‖ fabrique f. de coke; cokerie f.

cokify, to ‖ verkoken ‖ cokéfier.

coking of bituminous coal ‖ Steinkohlenverkokung f. ‖ distillation f. de la houille. / ~ in closed ovens (Met) ‖ Ofenverkokung f.; Verkoken n. in geschlossenen Öfen ‖ carbonisation f. de la houille dans des fours fermés. / ~ of the coke ‖ Backen n. des Kokses ‖ agglutination f. ou cuisson f. du coke. / ~ in heaps ‖ Meilerverkokung f. ‖ carbonisation f. en meules. / ~ in open kilns ‖ Verkoken n. in offenen Öfen ‖ carbonisation f. en fours ouverts. / ~ in piles ‖ Verkoken n. in Meilern ‖ carbonisation f. en meules. / ~ in pits ‖ Verkokung f. in Gruben ‖ carbonisation f. en fosses. / ~ under pressure ‖ Druckverkokung f. ‖ cokéfaction f. ou cokéfication f. sous pression.

coking capacity ‖ Verkokungsfähigkeit f. ‖ pouvoir m. cokéfiant. / ~ coal ‖ Kokskohle f.; Kokerkohle f. ‖ charbon m. ou houille f. ou menue f. à coke. / ~ coal charging wagon ‖ Koksofenfüllwagen m. ‖ wagonnet m. de chargement de fours à coke. / ~ duff ‖ Koksgrus m. ‖ poussier m. de coke. / ~ machine ‖ Kokereimaschine f. ‖ machine f. pour la fabrication du coke. / ~ plant ‖ Kokereieinrichtung f.; Kokereianlage f. ‖ installation f. de cokerie; cokerie f. / ~ process of ~ ‖ Vorgang m. beim Verkoken ‖ marche f. de la carbonisation. / ~ process ‖ Kohlenvergasung f. ‖ gazéification f. du charbon. / ~ small ‖ Koksfeinkohle f. ‖ fines fpl. à coke.

colander ‖ Durchschlag m. ‖ crible m.

colarin ‖ Säulenhals m. ‖ col m. ou gorge f. de colonne.

colation ‖ Durchseihen n. ‖ colature f.

colature see colation.

colchicum corn ‖ Herbstzeitlosewurzel f. ‖ racine f. de colchique. / ~ seed ‖ Herbstzeitlosesamen m. ‖ semence f. de colchique.

colcothar ‖ Caput mortuum n.; Englischrot n. ‖ colcothar m.; rouge m. anglais ou d'Angleterre.

cold ‖ kalt ‖ froid.

cold ‖ Kälte f. ‖ froid m. / ~ due to evaporation ‖ Verdunstungskälte f. ‖ froid m. dû à l'évaporation.

cold-air (Meteor) ‖ Kaltluft f. ‖ air m. froid. / ~ current ‖ Kaltluftstrom m.; kalter Luftstrom m. ‖ courant m. d'air froid. / ~ layer ‖ Kaltluftschicht f.; kalte Luftschicht f. ‖ couche f. d'air froid. / ~ machine ‖ Kaltluftmachine f. ‖ machine f. à air froid. / ~ mass ‖ Kaltluftmasse f. ‖ masse f. d'air froid. / ~ stratum see cold-air layer.

cold-bend ‖ kaltgebogen ‖ plié à froid. / ~-bend iron ‖ kaltgebogenes Eisen n. ‖ fer m. plié à froid. / ~-bend test ‖ Kaltbiegeprobe f. ‖ essai m. de pliage à froid. / to ~-bend ‖ kaltbiegen ‖ plier ou courber à froid. / ~-bending ‖ Kaltbiegen n. ‖ pliage m. à froid.

cold-blast (Metal) ‖ Kaltluft f.; kalter Wind ‖ vent m. ou air m. froid. / ~ iron ‖ kalterblasenes oder kaltgeblasenes Eisen n. ‖ fonte f. ou fer m. à l'air froid. / ~ pig see ~ iron. / ~ slide ‖ Kaltwindschieber m. ‖ vanne f. à air froid. / ~ valve see ~ slide.

cold-break ‖ Kaltbruch m. ‖ cassure f. à froid. / ~-brittle (Met) ‖ kaltbrüchig ‖ cassant à froid. / ~-brittle iron ‖ kaltbrüchiges Eisen n. ‖ fer m. cassant à froid. / ~-brittleness ‖ Kaltbrüchigkeit f. ‖ qualité f. (d'un fer) cassant à froid. / ~ chisel ‖ Hartmeißel m.; Schrotmeißel m.; Kaltmeißel m. ‖ ciseau m. ou burin m. ou tranche f. à froid. / ~ chisel (For Grooves) ‖ Nuteisen n.; Kreuzmeißel m. ‖ bédane m.; bec-d'âne m. / ~ circular saw ‖ Kaltkreissäge f. ‖ scie f. circulaire à froid. / ~ cream ‖ Coldkrem m.; Cold-creme f.; Cold-cream m. ‖ cold-cream m.; crème f. froide. / ~-cutting of rolled iron ‖ Kaltschneiden n. von Walzeisen ‖ coupe f. à froid de fers laminés. / to ~-draw ‖ kaltziehen ‖ étirer à froid; écrouir. / ~-drawn ‖ kaltgezogen ‖ étiré à froid. / ~-drawn steel wire ‖ kaltgezogener Eisendraht m. ‖ fil m. d'acier étiré à froid. / to ~-dress ‖ kaltrichten ‖ dresser à froid. / ~-dressing ‖ Kaltrichten n. ‖ dressage m. à froid. / ~ flash ‖ kalter Blitzschlag m. ‖ coup m. de foudre froid. / ~ glue ‖ Kaltleim m. ‖ colle f. à froid. / ~ green house ‖ Kalthaus n. ‖ serre f. froide. / to ~-hammer ‖ kalthämmern; kaltstrecken; hartschlagen; federhart machen; kaltschmieden ‖ battre ou marteler à froid; écrouir. / ~-hammered ‖ federhart; kaltgeschlagen; kaltgestreckt ‖ écroui. / ~-hammered iron ‖ kaltgeschmiedetes Eisen n. ‖ fer m. écroui. / ~-hammering ‖ Kaltschmieden n.; Hartschlagen n. ‖ écrouissage m.; écrouissement m. / ~ insulator ‖ Kälteisoliermittel n. ‖ isolation f. frigorifique; matière f. (isolante) calorifuge.

coldness ‖ Kälte f. ‖ froid m.; froideur f.

cold nut press ‖ Kaltmutternpresse f. ‖ presse f. à froid pour la fabrication d'écrous. / to ~-press ‖ kaltpressen ‖ presser *ou* comprimer à froid. / ~ press ‖ Kaltpresse f. ‖ presse f. à froid. / ~ press for forming and upsetting rivets. ‖ Kaltpresse f. zum Formen und Stauchen von Nieten ‖ presse f. à froid pour former et fouler des rivets. / ~-pressing ‖ Kaltpressen n. ‖ pressée f. à froid. / ~ protective ‖ Kälteschutzmittel n. ‖ préservatif m. contre le froid. / ~-resisting property ‖ Frostbeständigkeit f. ‖ résistance f. à la gelée. / to ~-rivet ‖ kaltnieten ‖ river à froid. / ~-rivetted boiler sheet ‖ kaltgenietetes Kesselblech n. ‖ tôle f. de chaudière rivé à froid. / ~-rivetting ‖ Kaltnietung f.; Kaltnieten n.; kalte Nietung f. ‖ rivetage m. à froid; rivure f. (faite) à froid. / to ~-roll ‖ kaltwalzen ‖ laminer *ou* cylindrer à froid. / ~ roll ‖ Kaltwalze f. ‖ cylindre m. de laminoir (pour laminage) à froid. / ~-rolled ‖ kaltgewalzt ‖ laminé à froid. / ~-rolling ‖ Kaltwalzen n. ‖ laminage m. à froid. / ~ rolling mill ‖ Kaltwalzwerk n. ‖ laminoir m. à froid. / ~ saw ‖ Kaltsäge f. ‖ scie f. à froid. / ~ saw for cutting the risers ‖ Trichterkaltsäge f. ‖ scie f. à froid pour tronçonnage des masselottes. / ~ sawing machine for metals ‖ Kaltsäge f. für Metalle ‖ machine f. à scier les métaux à froid. / to ~-shear ‖ kaltscheren; kaltschneiden ‖ cisailler à froid. / ~-sheared ‖ kaltgeschnitten ‖ cisaillé *ou* découpé à froid. / ~-shearing ‖ Kaltscheren n. ‖ cisaillement m. à froid. / ~-short (Met) ‖ kaltbrüchig ‖ cassant à froid. / ~-shortness ‖ Kaltbrüchigkeit f. ‖ qualité f. (d'un fer) cassant à froid. / ~-shot (Found) ‖ unvollkommen *oder* unscharf gegossen ‖ coulé non précisément. / ~-spots fermentation ‖ Gärung f. mit kahlen Stellen ‖ fermentation f. à places chauves. / ~-stirred soap ‖ kaltgerührte Seife f. ‖ savon m. touillé à froid.

cold-storage ‖ Kühlwerk n. ‖ usine f. frigorifique. / ~ house ‖ Kühlhaus n. ‖ entrepôt m. frigorifique. / ~ room ‖ Kühlraum m. ‖ chambre f. froide *ou* de rafraîchissement. / ~ wagon ‖ Kühlwagen m. ‖ voiture f. frigorifique.

cold store ‖ Kühlhalle f. ‖ entrepôt m. frigorifique *ou* de refroidissement. / to ~-straighten ‖ kaltrichten ‖ redresser à froid. / ~-straightened ‖ kaltgerichtet ‖ redressé à froid. / ~ stroke ‖ kalter Blitzschlag m. ‖ coup m. de foudre froid. / ~ test (Oil) ‖ Kälteprobe f. ‖ essai m. à froid. / ~ thread rolling machine ‖ Kaltgewindewalze f. ‖ machine f. à laminer à froid les filets de vis. / ~-upsetting machine ‖ Kaltstauchmaschine f. ‖ machine f. à refouler à froid.

cold-water pump ‖ Kaltwasserpumpe f. ‖ pompe f. à eau froide. / ~ test ‖ Kaltwasserprobe f. ‖ essai m. à l'eau froide. / ~ trough ‖ Kaltwassertrog m. ‖ cuve f. à eau froide.

cold wave ‖ Kältewelle f. ‖ onde f. *ou* vague f. de froid. / to ~-work ‖ kaltrecken ‖ écrouir; autofretter. / ~-worked (Gun) ‖ kaltgereckt; autofrettiert ‖ frotté. / ~-worked blank ‖ kaltgereckter Rohling m. ‖ ébauche f. autofrettée. / ~-working (Met) ‖ Kaltrecken n.; Kalt-

verformung f. ‖ écrouissage m. / ~-working (Blast-furnace) ‖ kalter Gang m. ‖ allure f. froide.

coleseed ‖ Kolzasame m. ‖ graine f. de colza.
collaborator ‖ Mitarbeiter m. ‖ collaborateur m.; compagnon m. de travail.
collapse, to ‖ zusammenstürzen ‖ s'écrouler. / ~ (To buckle) ‖ knicken ‖ flamber.
collapse ‖ Knick m. ‖ flambage m. / ~ of kraeusen head (Brew) ‖ Durchbruch m. der Kräusen ‖ chute f. des kraeusen. / ~ of prices ‖ Sturz m. der Kurse ‖ débâcle f.
collapsible ‖ auseinandernehmbar ‖ démontable. / ~ (Lens) ‖ zusammenlegbar ‖ à manche f. pliante. / ~ boat ‖ zusammenklappbares Boot n.; Faltboot n. ‖ canot n. *ou* bateau m. pliable *ou* pliant *ou* démontable. / ~ box ‖ Faltschachtel f. ‖ boîte f. pliante. / ~ dinghy ‖ Faltboot n. ‖ canot m. repliable. / ~ erecting crane ‖ zerlegbarer Rüstkran m. ‖ pont m. roulant démontable pour le montage. / ~ head ‖ Klappverdeck n. ‖ capote f. pliable.
collapsible tube ‖ Tube f.; Quetschtube f. ‖ tube m. / ~s closing machine ‖ Tubenschließmaschine f. ‖ machine f. à fermer les tubes. / ~ filling machine ‖ Tubenfüllmaschine f. ‖ machine f. à remplir les tubes.
collapsing boat *see* collapsible boat.
collar (Linen) ‖ Kragen m.; Halskragen m. ‖ collet m.; faux-col m. / ~ (Mach) ‖ Kragen m.; Rand m.; Reifen m.; Reif m.; Bund m. ‖ collier m.; cercle m.; virole f.; cerceau m. / ~ (Forg) ‖ Hammerhülse f. ‖ bogue f.; hulse f.; hurasse f. du gros marteau de forge. / ~ (Hydr Arch) ‖ Halsband n. des Schleusentores ‖ collier m. / ~ (Screw coupling) ‖ Bundring m. ‖ collet m. / axle ~ ‖ Achsbund m. ‖ collet m. de l'essieu. / axle ~ forged from the solid ‖ aus dem Vollen n. geschmiedeter Achsbund m. ‖ collet m. de l'essieu venu de forge. / clamping ~ ‖ Klemmring m. ‖ bague f. de fixation. / everclean ~s pl. and stuffs pl. ‖ Dauerwäsche f. ‖ linge m. économique *ou* durable. / ~ of harness (Saddl) ‖ Kumt n. ‖ collier m. de cheval. / jumping ~s on wheel spokes ‖ Anstauchen n. der Bunde an Radspeichen ‖ refoulement m. des embases aux rayons de roues. / ~ of the mandrel ‖ Dornansatz m. ‖ collet m. du mandrin. / ~ of a mast ‖ Spielkragen m. eines Mastes ‖ écoutillon m. d'un mât. / ~ of nearhorse (Saddl) ‖ Sattelkummet n. ‖ collier m. du porteur. / ~ of paper ‖ Papierkragen m. ‖ collet m. en papier. / ~ of a shaft (Min) ‖ Schachtkranz m. ‖ garniture f. d'un puits. / shirt ~ ‖ Hemdkragen m. ‖ col m. (de chemise); faux-col m. / shrunk-on ~ ‖ Schrumpfring m. ‖ frette f. calée à chaud; collet m. rapporté. / turn down ~ ‖ Umlegekragen m. ‖ col m. rabattu. / ~ of volatilizing tube ‖ Riechrohrschelle f. ‖ collier m. pour tuyaux à substances volatiles. / wood ~ for horses ‖ Holzkummet n. ‖ collier m. en bois de cheval.
collar beam (Carp) ‖ Kehlbalken m. ‖ entrait m. supérieur; poutre f. de noue. / ~ bearing (Mach) ‖ Halslager n. ‖ palier m. à collets. / ~ board (Weav) ‖ Halsbrett n.; Platinenbrett n. ‖ planche f. de collets. / ~ bone ‖ Schlüsselbein n. ‖ clavicule f. / ~ button ‖ Kragenknopf m. ‖ bouton m. à faux-col. / ~ chain ‖ Halfterkette

f. ‖ chaîne f. de licou. / ~ cutter (Tail) ‖ Kragenzuschneider m. ‖ coupeur m. de cols.
collarer ‖ Kragenmacher m. ‖ confectionneur m. de cols.
collarette guide (Machine) ‖ Bortennäher m. ‖ appareil m. à coudre les collarettes.
collar harness (Saddl) ‖ Kumtgeschirr n. ‖ harnais m. à colliers. / housing of the ~ (Saddl) ‖ Kumtdeckel m.; Kumtkappe f. ‖ coiffe f. du collier. / ~ machinist ‖ Kragennäherin f. ‖ piqueuse f. de faux-cols. / ~ needle (Saddl) ‖ Packnadel f.; Schneidnadel f. ‖ aiguille f. à réguiller; carrelet m. / ~ nut ‖ Bundmutter f.; Achsmutter f. ‖ écrou m. à chapeau.
collar pad (Saddl) ‖ Kumtkissen n. ‖ corps m. du collier. / to stuff the ~s pl. ‖ die Kumtkissen npl. stopfen ‖ rembourrer les corps mpl. du collier.
collar plate (Turn) ‖ Lünette f.; Brille f. ‖ (poupée f. à) lunette f. / ~ protector ‖ Kragenschoner m. ‖ cache-col m. / ~ rein ‖ Halfterzügel m. ‖ longe f. du licou *ou* bouclée; corde f. du licou. / ~ ring ‖ Halfterring m. ‖ anneau m. du licou; porte-barres mpl. / ~ step bearing ‖ Ringspurlager n. ‖ crapaudine f. annulaire. / ~ stud *see* button. / ~ thrust bearing ‖ Kammlager n.; Drucklager n. ‖ palier m. à cannelures *ou* de butée.
collate, to ‖ kollationieren ‖ collationner.
collator ‖ Kollationierer m. ‖ collationneur m.
colleague ‖ Mitarbeiter m. ‖ collaborateur m.; compagnon m. (de travail).
collect, to (Chem) ‖ sich ansammeln ‖ s'accumuler. / ~ (Trade) ‖ einziehen ‖ encaisser.
collecting (Oil; Residues) ‖ Abfangen n. ‖ récupération f.; recoupement m. / ~ business ‖ Inkassogeschäft n. ‖ affaire f. d'encaissement. / ~ pipe ‖ Sammelrohr n.; Sammelröhre f. ‖ tuyau-collecteur m. / ~ plate ‖ Sammelplatte f. ‖ plaque f. collectrice. / ~ shaft ‖ Sammelschacht m. ‖ puits-collecteur m. / ~ tank ‖ Sammelbecken n. ‖ réservoir collecteur m. / ~ trough ‖ Sammelmulde f. ‖ auge f. collectrice. / ~ worm ‖ Sammelschnecke f. ‖ vis f. sans fin collectrice.
collection ‖ Sammlung f. ‖ collection f. / ~ (Assemblage) ‖ Zusammenstellung f. ‖ rapprochement m.; combinaison f.; association f. / ~ (Trade) ‖ Inkasso n. ‖ encaissement m.; recouvrement m. / ~ of copper plate engraving ‖ Kupferstichkabinett n. ‖ cabinet m. de gravures en taille-douce. / ~ of samples ‖ Musterkollektion f.; Musterzusammenstellung f. ‖ collection f. d'échantillons. / ~ object of ~ ‖ Sammlungsgegenstand m. ‖ objet m. de collection.
collective number (Tel) ‖ Sammelnummer f. ‖ numéro m. collectif. / ~ weather radiogram ‖ Wettersammelfunkspruch m. ‖ radiogramme m. météorologique collectif.
collector (Trade) ‖ Sammler m. ‖ collectionneur m.; quêteur m. / ~ (Electr) ‖ Kollektor m. ‖ collecteur m. / ~ bow (Electr) ‖ Bügel m. eines Stromabnehmers ‖ archet m. de prise de courant. / ~ ring ‖ Bürstensammelring m. ‖ anneau m. collecteur. / ~ slip ring ‖ Kollektorschleifring m. ‖ bague f. collectrice *ou* de contact.
college ‖ Lehranstalt f. ‖ établissement m. d'instruction; école f.; institution f.

/ army industrial ~ ‖ Heereswirtschafts-hochschule f. ‖ école f. industrielle de l'armée. / veterinary ~ ‖ tierärztliche Hochschule f. ‖ école f. vétérinaire.

collet ‖ Zwinge f. ‖ virole f. / ~ (Jewel) ‖ Kalotte f. *oder* Kalette des Brillanten ‖ collet m.; calotte f. / ~ chuck ‖ Spannzange f. ‖ pince f. de serrage. / ~ chuck attachment ‖ Spannpatroneneinrichtung f. ‖ dispositif m. de serrage par pinces.

Colleton, cylindrical wooden pontoon of ~ ‖ Faßponton m. ‖ ponton m. cylindrique *ou* ponton-tonneau m. selon Colleton. / ~ cylinder pontoon bridge ‖ Faßpontonbrücke f. (nach Colleton) ‖ pont m. de tonneaux à l'anglaise; pont m. à la Colleton.

collier (Mine) ‖ Bergmann m. ‖ mineur m. / ~ (Hewer) ‖ Häuer m. ‖ coupeur m.; piqueur m. / ~ (Shipb) ‖ Kohlenschiff n. ‖ charbonnier m. / steam ~ ‖ Kohlentransportdampfer m. ‖ charbonnier m. à vapeur / ~'s boy (Mine) ‖ Lehrhauer m. ‖ galibot m.

colliery ‖ Kohlenzeche f.; Zeche f.; Steinkohlengrube f. ‖ houillère f.; mine f. de charbon; charbonnage m. / ~ branch-line ‖ Kohlenflügelbahn f. ‖ embranchement m. de chemin de fer de mines. / ~ labourer ‖ Grubenarbeiter m. ‖ ouvrier m. du fond; mineur m. / ~ railway ‖ Zechenbahn f. ‖ chemin m. de fer de mine.

collimator ‖ Kollimator m.; Visiervorrichtung f. ‖ collimateur m. / ~ ‖ Spaltrohr n. ‖ collimateur m. / ~ with focussing mechanism ‖ Kollimatorrohr m. mit Fokussierung ‖ collimateur m. muni d'un dispositif de mise au point. / ~ mirror ‖ Kollimatorspiegel m. ‖ miroir-collimateur m. / ~ telescope ‖ Kollimatorfernrohr n. ‖ lunette f. collimatrice.

collision ‖ Zusammenstoß m. ‖ collision f. / ~ with other types ‖ Zusammentreffen n. mehrerer Typen ‖ collision f. avec d'autres caractères. / insurance against ~ ‖ Unfallversicherung f. ‖ assurance f. contre les accidences causés par des tiers.

collodion ‖ Kollodium n. ‖ collodion m.; collode m. / ~ cotton ‖ Schießbaumwolle f.; Nitrozellulose f. ‖ coton m. explosif. / ~ cotton ‖ Kollodium(baum)wolle f. ‖ coton m. collodion. / ~ emulsion ‖ Kollodiumemulsion f. ‖ émulsion f. au collodion. / ~ layer ‖ Kollodiumschicht f. ‖ couche m. de collodion. / ~ proceeding ‖ Kollodiumverfahren n. ‖ procédé m. au collodion. / ~ silk ‖ Kollodiumseide f. ‖ soie f. de collodion.

colloid ‖ Kolloid n. ‖ colloide m.

colloidal ‖ gallertartig; kolloidal ‖ gélatineux; colloïdal. / ~ silver ‖ kolloidales Silber n. ‖ argent m. colloïdal. / ~ solution ‖ kolloidale Lösung f. ‖ solution f. colloïdale. / ~ state ‖ Kolloidzustand m. ‖ état m. colloïdal. / in a ~ state ‖ im Kolloidalzustand m. ‖ à l'état m. colloïdal. / ~ substance ‖ Kolloidalsubstanz f. ‖ matière f. colloïdale.

collotype ‖ Farbenlichtdruck m. ‖ impression f. héliographique; héliographie. f

collyrite ‖ Kollyrit m. ‖ collyrite f.

collyrium ‖ Augenwasser n. ‖ collyre m. liquide.

colmation ‖ Kolmation f.; Aufschlämmung f. ‖ colmation f.

colocynth ‖ Koloquinte f. ‖ coloquinte f. / ~ apple *see* colocynth.

Cologne water ‖ Kölnisches Wasser n. ‖ eau f. de Cologne.

colon (Print) ‖ Kolon n.; Doppelpunkt m. ‖ deux-points m.; deux points mpl.

colonial flour mill ‖ Exportmühle f. ‖ moulin m. pour exportation. / ~ goods pl. ‖ Kolonialwaren fpl. ‖ denrées fpl. coloniales. / ~ helmet ‖ Tropenhelm m.; Kolonialhelm m. ‖ casque m. colonial. / ~ machine ‖ Kolonialmaschine f. ‖ machine f. coloniale. / ~ oil mill ‖ Kolonialölmühle f. ‖ moulin m. à huile pour les colonies. / ~ products pl. *see* ~ goods. / ~ railway ‖ Kolonialbahn f. ‖ ligne f. coloniale.

colonist ‖ Ansiedler m. ‖ colon m.

colony, bacteria ‖ Bakterienkolonie f. ‖ colonie f. de bactéries. / workmen's ~ ‖ Arbeiterkolonie f. ‖ cité f. ouvrière.

colophonite ‖ Kolophonit m.; Pechgranat m. ‖ colophonite m.

colophony ‖ Geigenharz n.; Kolophonium n. ‖ arcanson m.; brai m. sec; colophane f. / pulverised ~ ‖ Kolophoniumpulver n. ‖ colophane f. pulvérisée. / ~ burner ‖ Kolophoniumgießer m. ‖ fondeur m. de colophane. / ~ solder ‖ Kolophoniumlötzinn n. ‖ soudure f. à colophane.

color *see* colour.

coloration ‖ Färbung f. ‖ coloration f. / ~ of the keys (Typewr) ‖ Färbung f. *oder* Farbe f. der Tasten ‖ coloration f. des touches. / degree of ~ ‖ Färbungsgrad m. ‖ degré m. de coloration. / ~ test ‖ kolorimetrische Probe f. ‖ essai m. par colorimétrie; colorimétrie f.

colorimeter ‖ Farbenmeßapparat m.; Farbenmesser m. ‖ colorimètre m.

colorimetric determination ‖ kolorimetrische Bestimmung f. ‖ détermination f. colorimétrique.

colour, to ‖ färben ‖ teindre. / ~ a plan ‖ einen Riß m. farbig anlegen ‖ mettre un plan en couleurs.

colour *see also* colours ‖ Farbe f. ‖ couleur f. / ~ (Build; Join etc.) ‖ Farbe f.; Anstrich m. ‖ peinture f. / ~ (Chem) *see also* colouring matter ‖ Farbe f.; Farbstoff m. ‖ teinture f.; colorant m. / ~ (Coloration) *see also* coloration ‖ Farbengebung f.; Färbung f. ‖ coloris m.; coloration f. / ~ (Paint) ‖ Farbe f.; Mischfarbe f.; teinte f. ‖ couleur f. / ~ (Print) ‖ Farbe f. ‖ encre f. (d'imprimerie). / ~ alcoholsoluble ~ ‖ spritlösliche Farbe f. ‖ couleur f. soluble en alcool. / ~ of animal origin ‖ Farbe f. tierischen Ursprungs ‖ couleur f. d'origine animale. / ~ for artificial flowers ‖ Farbe f. für künstliche Blumen ‖ couleur f. pour feuillages et fleurs artificielles. / artist's ~ ‖ Künstlerfarbe f. ‖ couleur m. pour peintre-artistes. / body ~ ‖ Deckfarbe f. ‖ couleur f. non transparente; couleur f. opaque / to brighten the ~ ‖ die Farbe auffrischen ‖ rafraîchir la couleur. / ~ for cement ‖ Zementfarbe f. ‖ couleur f. pour ciment. / chemical ~ ‖ chemische Farbe f. ‖ couleur f. chimique. / dry or pasty ~ of coal tar ‖ trockene oder teigförmige Steinkohlenteerfarbe f. ‖ teinture f. dérivée du goudron de houille sec ou en pâte. / earthy ~s pl. ‖ Erdfarben fpl. ‖ couleurs fpl. minérales. / fast ~ ‖ echte *oder* beständige Farbe f. ‖ couleur f. stable *ou* solide; grand teint m. / ~ is faulty ‖ mißfarbig ‖ de mauvaise teinte / first print ~ ‖ Vordruckfarbe f. ‖ couleur f.

de première impression. / fugitive ~ ‖ unechte *oder* unbeständige Farbe f. ‖ couleur f. fugitive ou instable. / glazing ~ ‖ durchscheinende Farbe f. ‖ couleur f. transparente. / glue ~ ‖ Leimfarbe f. ‖ couleur f. à la colle. / ~ of gold ‖ Goldfarbe f. ‖ couleur f. à dorer / ground ~ ‖ Grundfarbe f. ‖ première couche f. / ~ interference ~ ‖ Interferenzfarbe f. ‖ couleur f. d'interférence. / lasting. ~ *see* fast ~. / mineral ~s pl. ‖ Erdfarben fpl. ‖ couleurs fpl. minérales. / mixed ~ ‖ Mischfarbe f. ‖ couleur f. mélangée. / ~ for motor cars ‖ Kraftwagenfarbe f. ‖ couleur f. pour automobiles. / non-poisonous ~ ‖ giftfreie Farbe f. ‖ couleur f. exempt du poison ou non vénéneuse ou inoffensive. / not fast ~ *see* fugitive ~. / ~ prepared with oil ~ ‖ mit Öl zubereitete Farbe f. ‖ couleur f. préparée à l'huile. / opaque ~ ‖ Deckfarbe f.; Körperfarbe f. ‖ couleur f. couvrante. / passing ~ (Chem) ‖ Übergangsfarbe f. ‖ couleur f. de variation de teinte / paste ~ ‖ Teigfarbe f. ‖ couleur f. en pâte. / permanent ~ *see* fast ~. / pigment ~ *see* opaque ~. / powdered ~ ‖ Farbe f. in Pulverform ‖ couleur f. en poudre. / prime ~ ‖ Grundfarbe f. ‖ couleur f. de fond. / priming ~ ‖ Grundierfarbe f. ‖ fond m. ou première couche de couleur. / primitive ~ ‖ Hauptfarbe f. ‖ couleur f. simple ou originaire ou primitive. / ~ for printing on textiles ‖ Zeugdruckfarbe f. ‖ couleur f. pour impression sur étoffes. / secondary ~ ‖ zusammengesetzte Farbe f. ‖ couleur f. secondaire ou composée ou hétérogène. / sharp ~ ‖ hervorstechende Farbe f. ‖ couleur f. vive. / shell-gold ~ ‖ Malergoldfarbe f. ‖ couleur f. d'or moulu. / transparent ~ ‖ durchscheinende Farbe f.; Lasurfarbe f. / couleur f. transparente. / ~ prepared with turpentine ‖ mit Terpentinöl zubereitete Farbe f. ‖ couleur f. préparée à l'essencede terpentine. / of vast ~ ‖ farbecht ‖ de couleur f. résistante. / vegetable ~ ‖ vegetabilische Farbe f.; Pflanzenfarbe f. ‖ couleur f. végétale. / vitrifiable ~ ‖ Schmelzfarbe. ‖ couleur f. vitrifiable. / water ~ ‖ Aquarellfarbe f. ‖ couleur f. à l'aquarelle.

colour, addition of ‖ Farbzusatz m. ‖ addition f. de matière colorante. / ~ agglutinant ‖ Farbenbindemittel m. ‖ liant m. pour couleurs. / ~ apparatus (Lug) ‖ Couleurapparat m. ‖ appareil m. à couleur. / ~ band ‖ Farbensaum m. ‖ frange f. colorée. / ~ blind ‖ farbenblind ‖ daltonien. / ~ blindness ‖ Farbenblindheit f. ‖ achromatopsie f.; daltonisme m. / ~ box ‖ Tuschkasten m. ‖ boîte f. à lavis. / ~ box filler ‖ Farbkasteneinleger m. ‖ metteur m. de couleurs en boîtes. / ~ coke for black colours ‖ Farbkoks f. für Schwarzfarben ‖ coke m. colorant pour couleurs noires. / ~ comparator ‖ Farbenkomparator m. ‖ comparateur m. de teintes. / ~ defect ‖ Farbfehler m. ‖ défaut m. de chromatisme. / ~ defect of the objective ‖ Farbenfehler m. des Objektivs ‖ aberration f. chromatique de l'objectif. / depth of ~ ‖ Farbtiefe f. ‖ intensité f. de la coloration. n. ~ determination ‖ Farbbestimmung f. ‖ détermination f. de la coloration. / revolving ~ disc (Opt) ‖ Farbglasrevolver m. ‖ revolver m. à verres colorés. / ~ dispersion ‖ Farbenzerstreuung f. ‖ dispersion

f. des couleurs. / ~ earths pl. || Farberden fpl. || terres fpl. colorantes.

coloured || gefärbt || coloré. / ~ || eigenfarbig || coloré par soi-même. / ~ || farbig || de couleur f. / ~ crayon || farbiger Schreibstift m.; Farbstift m.; Buntstift m. || crayon m. de couleur. / ~ earth || Ockerfarbe f. || ocre f.; terre f. d'ombre. / ~ embroidery || Buntstickerei f. || broderie f. en couleurs. / ~ glaring glass || farbiges Blendglas n. || verre m. coloré antiéblouissant. / ~ glass || Farbglas n. || verre m. de couleur. / ~ goods pl. || Buntwaren fpl. || rouenneries fpl. / ~ impression || Farbabdruck m. || impression f. colorée. / ~ paper || Buntpapier n. || papier m. de couleur ou peint. / ~ paper making plant || Buntpapierherstellungsanlage f. || installation f. à fabriquer les papiers peints. / ~ pencil || Farbstift m. || crayon m. de couleur. / ~ plate || Farbenkunstdruck m. || image f. imprimée en couleurs. / ~ printing ink || Druckfarbe f. || encre f. (de couleur) d'imprimerie. / ~ rings pl. || Farbenringe mpl. || anneaux mpl. colorés. / ~ table cover || bunte Tischdecke f. || nappe f. de table en couleur. / ~ throughout || durchgefärbt || coloré dans la masse. / ~ wrapping paper || farbiges Umschlagpapier n. || papier m. coloré à envelopper.

colourer || Kolorist m. || coloriste m.

colour filter || Farbfilter m. || écran m. coloré. / ~ on quartz plate || Farbfilter m. auf Quarzplatte || écran m. coloré monté sur plaque en quartz.

colour glass disc || Farbgläserscheibe f. || disque m. de verres colorés. / ~ grinder || Farbenreiber m. || broyeur m. / ~ grinding || Farbenreiben n. || broyage m. de couleurs. / ~ grinding machine || Farbenreibmaschine f. || machine f. à broyer les couleurs; broyeur m. à couleurs. / ~ grinding mill || Farbenmühle f. || moulin m. à couleurs. / impression of ~s || Farbendruck m. || impression f. en couleurs.

colouring (Colour) || Anstrich m.; Farbe f. || peinture f. / ~ (Coloration) || Färbung f. || coloration f. / ~ (Paint) || Kolorieren n. || coloriage m. / absorption ~ || Färbung f. durch Einsaugung; Einsaugungsfärbung f. || coloration f. par absorption. / ~ in oil || Ölanstrich m.; Ölfarbenanstrich m. || peinture f. à l'huile. / wood ~ || Holzfärben n. || coloration f. du bois.

colouring agent || Färbemittel n. || agent m. de coloration.

colouring bodies pl. absorption band of the ~ of blood || Absorptionsbande f. von Blutfarbstoffen || bande f. d'absorption des colorants du sang. / ~ for enamel works || Farbkörper mpl. für die Emaille-Industrie || corps mpl. colorants pour émailleries.

colouring brush || Tuschpinsel m. || pinceau m. / degree of ~ || Farbtönung f. || degré m. de coloration. / ~ earth || Farberde f. || terre f. colorante. / ~ extract || Farbstoffauszug m. || extrait m. tinctorial.

colouring matter || Farbstoff m.; Farbsubstanz f. || colorant m.; matière f. ou substance f. colorante. / ~ for the industry of ceramics || Farbstoff m. für die keramische Industrie || matière f. colorante pour l'industrie céramique. / ~ of vegetable origin || Farbstoff m.

pflanzlichen Ursprungs || matière f. colorante d'origine végétale.

colouring power || Färbekraft f. || pouvoir m. colorant. / ~ value || Färbekraft f. || pouvoir m. colorant. / ~ workshop || Kolorieranstalt f. || atelier m. de coloriage.

colourist || Kolorist m. || coloriste m.

colour, laying-on of ~s || Farbenauftrag m.; Farbenlage f. || couche f. de couleur.

colourless (Opt) || farbenrein || pur. / ~ (Chem) || farblos || incolore.

colourlessness || Farblosigkeit f. || absence f. de colorations.

colour maker || Farbenanmacher m. || préparateur m. de couleurs. / ~ malt || Farbmalz n. || farbmalz m.; malt m. grillé ou torréfié. / ~ manufactory || Farbenfabrik f. || fabrique f. de couleurs. / ~ measurer || Farbenmeßapparat m. || mesureur m. de couleurs. / ~ mill for oil colours || Ölfarbmühle f. || moulin m. à broyer les couleurs à l'huile.

colour mixer (Dyer) || Farbenmischer m. || mélangeur m. de couleurs. / ~ (Machine) || Farbmischmaschine f. || mélangeur m. à couleurs.

colour painting, water || Aquarellmalerei f.; Wasserfarbenmalerei f. || peinture f. à gouache ou à l'aquarelle ou en détrempe.

colour pencil || Farbstift m. || crayon m. de couleur. / ~ photography || Chromofotografie f. || photographie f. en couleurs. / ~ plant || Farbenfabrikeinrichtung f. || installation f. à fabriquer les couleurs. / ~ poster works pl. || Plakatdruckerei f. || imprimerie f. d'affiches en couleurs. / ~ printer || Farbendrucker m. || imprimeur m. en couleurs.

colour printing || Buntdruck m.; Farbendruck m. || chromotyp(ograph)ie f.; chromo m.; lithocromie f.; impression f. en couleurs. / ~ machine || Farbendruckmaschine f. || machine f. pour l'impression en couleurs. / ~ works pl. || Farbendruckerei f. || imprimerie f. en couleurs.

colour reaction || Farbenreaktion f. || réaction f. des couleurs.

colours pl. (Mar) || Flagge f. || pavillon m.; enseigne f. / ~ for merchant-ships || Handelsflagge f. || pavillon m. marchand.

colours pl. (Chem) || Farben fpl. || couleurs fpl. / alcohol-soluble ~ || spritlösliche Farben fpl. || couleurs fpl. solubles en alcool. / ~ for artificial flowers || Farben fpl. für künstliche Blumen || couleurs fpl. pour fleurs artificielles. / ~ for artificial leather || Farben fpl. für Kunstleder || couleurs fpl. pour cuir artificiel. / ~ for artificial stones || Farben fpl. für Kunststeine || couleurs fpl. pour pierres artificielles. / ~ for asbestos slate || Farben fpl. für Asbestschiefer || couleurs fpl. pour schiste d'amiante. / ~ for bootpolish || Farben fpl. für Schuhkrem || couleurs fpl. pour crème à chaussures. / ~ for enamel works || Farben fpl. für Emaillierwerke || couleurs fpl. pour émailleries. / harmless ~ || giftfreie Farben fpl. || colorants mpl. inoffensifs. / ~ for lincrusta || Farben fpl. für Linkrusta || couleurs fpl. pour lincrusta. / ~ for linoleum || Farben fpl. für Linoleum || couleurs fpl. pour linoléum. / ~ for liqueurs || Farben fpl. für Liköre || colorants mpl. pour liqueurs. / ~ for machinery || Farben fpl. für Eisenanstrich || couleurs fpl. pour la peinture du fer. / ~ for mineral waters

|| Farben fpl. für Mineralwässer || couleurs fpl. pour eaux gazeuses. / oil-soluble ~ || fettlösliche Farben fpl. || colorants mpl. au gras. / ~ for paper stainers || Farben fpl. für Papierfabrikation || couleurs fpl. pour papier. / ~ for painting on china || Scharffeuerfarben fpl. für Porzellan || couleurs fpl. au grand feu pour porcelaine. / ~ for polishing inks || Farben fpl. für Kaltpoliertinten || couleurs fpl. pour encres à polir. / potter's ~ || keramische Farben fpl. || couleurs fpl. céramiques. / primitive ~ || Grundfarben fpl. || couleurs fpl. simples ou originaires ou primitives. / ~ for rubber goods || Farben fpl. für Gummiwaren || couleurs fpl. pour articles en caoutchouc. / ~ for schools || Schulfarben fpl. || couleurs fpl. pour écoles. / ~ for sealing-wax || Farben fpl. für Siegellack || couleurs fpl. pour cire à cacheter. / secondary ~ || Nebenfarben fpl. || couleurs fpl. secondaires ou composées ou hétérogènes. / vitrifiable ~ || Schmelzfarben fpl. || couleurs fpl. vitrifiables.

colour scale || Farbenskale f. || échelle f. des couleurs. / ~ screen (Phot) || Farbschirm m.; Farbscheibe f. || écran m. coloré. / ~ sprayer || Farbenzerstäuber m. || appareil m. pulvérisateur pour couleurs. / ~ stamping pad || Dauerstempelfarbkissen n. || coussin m. permanent d'encre à timbrer. / ~ strength || Farbtiefe f. || intensité f. de la coloration. / ~ test || kolorimetrische Probe f. || colorimétrie f.; essai m. par colométrie. / ~ tube || Farbentube f. || tube m. de couleur. / ~ weaving || Buntweberei f. || tissage m. en couleurs.

colter of a plough || Pflugschar f. || coutre m. de charrue.

columbarium || Rüstloch n. || trou m. de boulin; trou m. de traverse.

columbite || Kolumbit m. || columbite f.

Columbo root || Kolumbowurzel f. || racine f. de colombo.

column || Säule f. || colonne f.; pilier m.; montant m.; poteau m. / ~ (Math) || Reihe f. || colonne f. / ~ (Print) || Spalte f.; Kolonne f. || colonne f. / barometric ~ || Quecksilbersäule f. || colonne f. barométrique. / base of ~ || Säulenfuß m. || embase f. de colonne. / ~ of a bridge || Brückenpfeiler m. || pilier m. de pont. / iron ~ with concrete casing || eiserne Säule f. mit Betonummantelung || colonne f. en fer avec enveloppe en béton. / diminished ~ || geradlinig verjüngte Säule f. || colonne f. diminuée. / in ~s pl. || gespalten; in Spalten fpl. || en colonnes fpl. / ~ for iron structures || Säule f. für Eisenhochbauten || colonne f. pour constructions en fer. / rotating ~ || drehbarer Ständer m. || montant m. pivotant. / steering ~ || Lenksäule f. || colonne f. de direction. / with two ~s (Print) || zweispaltig || à deux colonnes fpl.

column apparatus (Chem) || Kolonnenapparat m. || appareil m. colonne.

columnar || säulenförmig || en forme f. de colonne. / ~ argillaceous iron || Schindelnagel m. || argile f. ferrugineuse en tiges. / ~ basalt || Säulenbasalt m. || basalte m. en colonnes. / ~ red iron-ore || Nagelerz n.; Schindelerz n.; stengliger roter Toneisenstein m. || argile f. ferrugineuse en tiges. / ~ structure (Geol) || Säulenform f. || structure f. en colonnes.

column casing ‖ Säulenverschalung f. ‖ coffrage m. en colonne. / ~ milling machine ‖ Ständerfräsmaschine f. ‖ fraiseuse f. à montant. / ~ press ‖ Säulenpresse f. ‖ presse f. à colonnes.

colza mill ‖ Rapsmühle f. ‖ moulin m. à colza. / ~ oil ‖ Rapsöl n.; Rüböl n. ‖ huile f. de colza.

comb, to (Wool) ‖ auskämmen ‖ peigner.

comb ‖ Kamm n. ‖ peigne m. / ~ (Flax) ‖ Hechel f.; Hechelkamm m. ‖ peigne m.; sérançoir m.; peigne-séran m. / ~ (Spinn) ‖ Kämmaschine f. ‖ peigneuse f. / ~ (Warping mach) ‖ Rietblatt n.; Scherblatt n. ‖ peigne m. / ~ (Worsted spinn) ‖ Nadelkamm m. ‖ peigne m.; barette f. à aiguilles. / artificial horn ~ ‖ Kunsthornkamm m. ‖ peigne m. en corne artificielle. / bone ~ ‖ Knochenkamm m. ‖ peigne m. en os. / ~ of a book ‖ gewölbter Rücken m. eines Buches ‖ endos m. d'un livre. / celluloid ~ ‖ Zelluloidkamm m. ‖ peigne m. en celluloïde. / curry ~ ‖ Mähnenkamm m. ‖ étrille f. / fancy ~ ‖ Galanteriekamm m. ‖ peigne m. fantaisie. / horn ~ ‖ Hornkamm m. ‖ peigne m. en corne. / horse ~ ‖ Striegel m. ‖ étrille f. ‖ / ivory ~ ‖ Elfenbeinkamm m. ‖ peigne m. en ivoire. / knocking-over ~ (Weav) ‖ Abschlagkamm m. ‖ peigne m. d'abattage. / metallic ~ ‖ Metallkamm m. ‖ peigne m. métallique. / tortoise shell ~ ‖ Schildpattkamm m. ‖ peigne m. en. écaille. / wooden ~ ‖ Holzkamm m. ‖ peigne m. en bois.

combat airplane ‖ Schlachtflugzeug n. ‖ avion m. de bataille.

comb blade (Spinn) ‖ Hackerblatt n. ‖ lame f. de peigne. / ~ cleaning apparatus ‖ Kammreinigungsapparat m. ‖ appareil m. à nettoyer les peignes. / ~ cutter's web ‖ Kammsäge f. ‖ scie f. pour peignes. / ~ decorator ‖ Kammverzierer m. ‖ ornementier m. en peignes.

combed material ‖ Kammzug m. ‖ peigné m.; trait m. / ~ wool ‖ gekämmte Wolle f. ‖ laine f. peignée.

comber (Spinn) ‖ Kämmer m. ‖ peigneur m. / ~ fly ‖ Kammflug m. ‖ duvet m. / ~ lap machine ‖ Bandwickelapparat m. ‖ réunisseuse f. de rubans.

comb finisher ‖ Kammschleifer m. ‖ arrondisseur m. de peignes.

combination ‖ Verbindung f.; Zusammensetzung f. ‖ union f.; jonction f. / ~ (Assemblage) ‖ Zusammenstellung f.; Kombination f. ‖ rapprochement m.; combinaison f.; association f. / ~ (Mine) ‖ Grubenkonzern m. ‖ communauté f. d'intérêts de mines. / chemical ~ ‖ chemische Verbindung f. ‖ combinaison f. chimique; composé m. / doctrine of ~ (Math) ‖ Kombinatorik f. ‖ théorie f. des combinaisons. / to enter into ~ (Chem) ‖ Verbindung f. eingehen ‖ entrer en combinaison f. / ~ of interests ‖ Interessengemeinschaft f. ‖ convention f.; syndicat m.; communauté f. d'intérêts. / ~ of metals ‖ Legierung f. ‖ alliage m.; combinaison f. de métaux. / ~ of protoxide and oxide ‖ Oxydoxydul n. ‖ combination f. de protoxyde et de sesquioxyde; oxyde intermédiaire entre le protoxyde et le sesquioxyde.

combination apparatus ‖ Kombinationsgerät n. ‖ récepteur m. à montages de combinaison. / ~ baking steam oven ‖

Verbunddampfbackofen m. ‖ four m. de boulangerie combiné à vapeur. / ~ die (Tool) ‖ Kompoundschnitt m. ‖ matrice f. combinée. / ~ fuze ‖ kombinierter Zünder m. ‖ fusée f. combinée. / ~ letter padlock ‖ Kombinationsschloß n.; Vexierschloß n. ‖ cadenas m. à combinaison à lettres; serrure f. à secret. / ~ telescope for astronomical and terrestrial observation ‖ Fernrohr n. für Himmel und Erde ‖ lunette f. pour les observations célestes et terrestres.

combinator ‖ Kombinator m. ‖ combinateur m.

combine, to ‖ vereinigen ‖ combiner. / ~ additively ‖ addieren ‖ additionner. / ~ in order to obtain better wage conditions ‖ sich zur Erlangung besserer Lohnbedingungen zusammenschließen ‖ se coaliser pour obtenir de meilleures conditions de salaire. / ~ sources pl. of power together to form one closed system ‖ Kraftquellen fpl. zu einem System zusammenfassen ‖ réunir des sources fpl. d'énergie dans un seul système.

combine ‖ Interessengemeinschaft f.; convention f.; syndicat m.; communauté f. d'intérêts. / ~ (Agr) ‖ Mähdrescher m. ‖ moissonneuse-batteuse f. / prohibition to ~ ‖ Koalitionsverbot n. ‖ interdiction f. de coalition.

combined ‖ kombiniert ‖ combiné. / ~ circuit (Tel) ‖ Vierersprechkreis m. ‖ circuit m. de conversation quadruple ou combiné. / ~ effect ‖ Verbundwirkung f. ‖ effet m. compound. / ~ fire-alarm and clock system ‖ kombinierte Feuermelde- und Uhrenanlage f. ‖ installation f. combinée d'avertisseurs d'incendie et d'horloges électriques. / ~ fire-alarms and fire station alarms with automatic indication of locality ‖ kombinierte Feuermelde- und Alarmanlage f. mit selbsttätiger Ortsangabe ‖ installation f. combinée d'avertisseurs d'incendie et d'alarme avec indication automatique du lieu. / ~ fire-alarm and wachtman's control system ‖ kombinierte Feuermelde- und Wächterkontrollanlage f. ‖ installation f. combinée d'avertisseurs d'incendie et de contrôleurs de rondes. / ~ Morse safety circuit ‖ kombinierte Morse-Sicherheitsschaltung f. ‖ montage m. Morse combiné de sécurité. / ~ movements pl. ‖ zusammengesetzte Bewegungen fpl. ‖ mouvements mpl. conjugués. / ~ parts pl. ‖ miteinander verbundene Teile ‖ pièces fpl. assemblées ou combinées. / ~ switch and fuse ‖ Schalter m. mit Sicherung ‖ interrupteur m. avec coupe-circuit. / ~ steam baking oven ‖ Kombinationsdampfbackofen m. ‖ four m. de boulangerie à vapeur combiné. / ~ with (Chem) ‖ gebunden an ‖ combiné à.

combing (Spinn) ‖ Kämmen n. ‖ peignage m. / ~ (Spinn) ‖ Kämmling m. ‖ peignon m. / ~ of wool ‖ Kämmen n. der Wolle ‖ peignage m. de la laine. / ~ machine for wool ‖ Wollkämmaschine f.; Kämmmaschine f. für Wolle ‖ machine f. à peigner la laine; peigneuse f. / ~ segment for ~ machines (Spinn) ‖ Segment n. für Krempelmaschinen ‖ segment m. pour peigneuses. / ~ waste from ~ (Wool) ‖ Abfall m. vom Kämmen ‖ déchet m. de peignage.

combings pl. (Spinn) ‖ Kammabfall m. ‖ bourre f.

combing works pl. (Spinn) ‖ Kämmerei f. ‖ peignerie f.

comb maker ‖ Kammacher m. ‖ fabricant m. de peignes. / ~ pliers pl. ‖ Hechlerzange f. ‖ pince f. de peigneur. / ~ polisher ‖ Kammschleifer m. ‖ polisseur m. de peignes. / ~ saw ‖ Kammsäge f. ‖ scie f. pour peignes. / ~ tenter (Spinn) ‖ Kammstuhlarbeiter m. ‖ peigneur m. à la machine.

combustible ‖ brennbar; verbrennbar ‖ combustible. / of easy ~ nature ‖ leichtentzündlich ‖ de combustibilité f. facile. / non-~ ‖ unbrennbar ‖ incombustible.

combustible gas ‖ brennbares Gas n. ‖ gaz m. combustible ou inflammable. / ~ mixture ‖ brennbares Gemisch n. ‖ mélange m. combustible.

combustible ‖ Brennstoff m.; Heizstoff m. ‖ combustible m. / artificial ~ ‖ künstlicher Brennstoff m. ‖ combustible m. artificiel ou aggloméré.

combustion ‖ Verbrennung f. ‖ combustion f. / accelerated ~ / beschleunigte Verbrennung f. ‖ combustion f. accélérée. / air for ~ ‖ Verbrennungsluft f. ‖ air m. comburant. / catalytic ~ ‖ katalytische Verbrennung f. ‖ combustion f. catalytique. / ~ of coal ‖ Kohlen(ver)feuerung f. ‖ combustion f. de la houille. / complete ~ ‖ vollkommene Verbrennung f. ‖ combustion f. complète ou parfaite. / ~ by explosion ‖ Explosivverbrennung f. ‖ combustion f. par explosion. / imperfect ~ ‖ unvollkommene Verbrennung f. ‖ combustion f. imparfaite. / perfect ~ ‖ vollkommene Verbrennung f. ‖ combustion f. complète. / product of ~ ‖ Verbrennungsprodukt n. ‖ produit m. de combustion. / rapid ~ ‖ lebhafte Verbrennung f. ‖ combustion f. vive. / slow ~ ‖ langsame Verbrennung f. ‖ combustion f. lente. / spontaneous ~ ‖ Selbstentzündung f. ‖ inflammation f. spontanée. / ~ of turf ‖ Torffeuerung f. ‖ combustion f. de la tourbe. / ~ of wood ‖ Holzfeuerung f. ‖ combustion f. du bois.

combustion analysis ‖ Verbrennungsanalyse f. ‖ analyse f. par combustion. / ~ block ‖ Brennerstein m. ‖ brique f. réfractaire. / ~ chamber (Boil) ‖ Verbrennungskammer f.; Verbrennungsraum m. ‖ chambre f. de combustion. / ~ curve ‖ Verbrennungslinie f. ‖ courbe f. de combustion.

combustion engine, crane with ~ ‖ Kran m. mit Brennkraftmotor ‖ grue f. à moteur à combustion. / internal ~ ‖ Verbrennungsmotor m. ‖ moteur m. à combustion. / valveless air-cooled internal ~ operating on the two-stroke cycle ‖ ventilloser luftgekühlter Zweitaktverbrennungsmotor m. ‖ moteur m. à explosion fonctionnant suivant le cycle à deux temps et dépourvu de soupapes.

combustion furnace ‖ Verbrennungsofen m. ‖ four m. à combustion; grille f. à analyse. / ~ gases pl. ‖ Feuergase npl. ‖ gaz mpl. de combustion. / ~ heat ‖ Verbrennungswärme f. ‖ chaleur f. de combustion. / ~ motor ‖ Verbrennungskraftmaschine f. ‖ moteur m. à combustion. / ~ product ‖ Verbrennungsprodukt n. ‖ produit m. de combustion. / ~ recorder ‖ registrierende Verbrennungskontrollvorrichtung f. ‖ appareil m. enregistreur de contrôle de combustion. / ~ space ‖ Verbrennungsraum m. ‖ chambre f. de com-

bustion. / ~ tubing ‖ Verbrennungsrohr
n. ‖ tube m. de combustion.
come, to ‖ kommen ‖ venir. / ~ alongside
(Mar) ‖ längsseits kommen ‖ venir accoster. / ~ at anchor ‖ ein Schiff vor
Anker legen ‖ mettre à l'ancre. / ~ down
(Chem) ‖ ausfällen ‖ précipiter. / ~ to
light ‖ zutagekommen; zutagetreten ‖ se
manifester. / ~ off (Chem) ‖ sich entwickeln ‖ se dégager. / ~ to shore ‖ landen; anlanden ‖ prendre terre f.
comet finder (Opt) ‖ Kometensucher m. ‖
chercheur m. de comètes. / altazimuth ~
‖ azimutaler Kometensucher m. ‖ chercheur m. de comètes azimutale. / equatorial ~ ‖ parallaktischer Kometensucher
m. ‖ chercheur m. de comètes parallactique.
comfort ‖ Komfort m.; Bequemlichkeit f.
‖ confort m.; commodité f.
comfortable ‖ wohnlich ‖ confortable; commode.
comma ‖ Komma n. ‖ virgule f. / inverted
~ ‖ Anführungszeichen n. ‖ guillemet m.
command, to (Trade) ‖ bestellen ‖ commettre.
command ‖ Kommando n. ‖ commandement m. / ~ of the sea ‖ Seeherrschaft f.
‖ maîtrise f. des mers. / steering ~ ‖
Ruderkommando n. ‖ commandement
m. à la barre.
commander ‖ Kommandant m. ‖ commandant m.; capitaine m.
commanding officer's compartment ‖ Kom-
mandantenraum m. ‖ poste m. du commandant.
commendable ‖ ratsam ‖ prudent; à propos.
commendation ‖ Anpreisung f. ‖ réclame
f.; boniment m.
commentation ‖ Marginale f.; Randbemerkung f. ‖ glosse f. ou note f. marginale;
manchette f.
commerce ‖ Handelsverkehr m.; Handel
m.; Geschäft n. ‖ commerce m. / ~ raider
(Mar) ‖ Handelszerstörer m. ‖ croiseur m.
corsaire.
commercial ‖ wirtschaftlich ‖ économique. /
~ aeroplane ‖ Verkehrsflugzeug n. ‖ avion
m. de transport commercial. / ~ book ‖
Geschäftsbuch n. ‖ livre m. de commerce.
/ ~ book factory ‖ Geschäftsbücherfabrik
f. ‖ fabrique f. de livres commerciaux. /
~ career ‖ kaufmännische Laufbahn f. ‖
carrière f. commerciale. / ~ casting ‖
Handelsguß m. ‖ fonte f. marchande
/ ~ clerk ‖ Handlungsbeflissener m. ‖
employé m. de commerce; commis m. /
~ code ‖ Handelsgesetzbuch n. ‖ code m.
de commerce. / ~ college ‖ Handels-
akademie f. ‖ académie f. commerciale;
école f. des hautes études commerciales.
/ ~ court ‖ Handelsgericht n. ‖ tribunal
m. de commerce. / ~ directory ‖ Han-
delsadreßbuch n. ‖ annuaire m. commercial. / ~ efficiency ‖ wirtschaftlicher
Nutzen m. ‖ rendement m. économique.
/ ~ enterprise ‖ Geschäftsunternehmen
n.; Geschäftsunternehmung f. ‖ entreprise commerciale. / ~ firm ‖ Geschäfts-
haus n. ‖ maison f. de commerce. / ~ guide
‖ Firmenbuch n.; Firmenregister n. ‖ dictionnaire m. des adresses; annuaire m. /
~ house ‖ Geschäft n. ‖ maison f. de commerce; établissement m. / ~ intercourse ‖
Handelsverkehr m. ‖ relations fpl. commerciales; marché m. des affaires. / ~ law
‖ Handelsgesetz n. ‖ loi f. relative au
commerce. / ~ ledger ‖ Geschäftsbuch n.

‖ livre m. de commerce. / to divide into
~ lengths ‖ Zerteilen n. auf handelsübliche Längen ‖ débiter en longueurs fpl.
commerciales. / ~ paper ‖ Geschäftspapier
n. ‖ papier m. d'affaire. / ~ parts pl. ‖
handelsübliche Teile mpl. ‖ éléments
mpl. du type commercial. / ~ right ‖
Handelsrecht n. ‖ droit m. commercial.
/ ~ section (Bank) ‖ Handelsabteilung f.
‖ service m. commercial. / ~ sign ‖
Firmenschild n. ‖ enseigne f. commerciale. / ~ speed ‖ Verkehrsgeschwindigkeit f. ‖ vitesse f. commerciale ou d'utilisation. / ~ statistics pl. ‖ Handelsstatistik f. ‖ statistique f. commerciale. / ~
steel ‖ handelsüblicher Stahl m. ‖ acier
m. du type commercial. / ~ treaty ‖ Handelsvertrag m. ‖ traité m. de commerce.
/ ~ value ‖ Handelswert m. ‖ valeur f.
commerciale. / ~ world ‖ Geschäftswelt
f. ‖ monde m. commercial.
commission ‖ Auftrag m. ‖ commission f.;
ordre m. / ~ (Trade) ‖ Provision f. ‖ provision f.; commission f. / ~ (Mar) ‖ Indienststellung f. ‖ armement m. / to accord a ~ ‖ eine Vergütung f. gewähren ‖
allouer une commission f. / does not include ~ ‖ schließt keine Provision f. ein
‖ ne comprend pas de commission f. / ~
for examination of the communication
services ‖ verkehrstechnische Prüfungs-
kommission f. ‖ commission f. de vérification des moyens techniques de trafic. / free of ~ ‖ frei von Provision f. ‖
sans commission f. / including ~ ‖ einschließlich Provision f. ‖ commission f.
comprise. / on ~ ‖ in Kommission f. ‖
en dépôt m.
commission account ‖ Provisionskonto n. ‖
compte m. de commission. / ~ agency ‖
Kommissionsgeschäft n. ‖ maison f. de
commission. / ~ agent ‖ Zwischenhändler
m.; Kommissionär m. ‖ intermédiaire
m.; commissionaire m. / ~ business ‖
Kommissionsgeschäft n. ‖ affaire f. en
commission.
commissioner ‖ Kommissionär m. ‖ commissionaire m.; facteur m.
commissioning of a ship ‖ Indienststellung
f. eines Schiffes ‖ mise f. en armement
d'un navire.
commission statement ‖ Vergütungs-
abrechnungsbeleg m. ‖ bulletin m. de
commission. / ~ trade ‖ Kommissions-
handel m. ‖ commerce m. de commission.
/ ~ weaver ‖ Webwarenfabrikant m. ‖
fabricant m. de tissus.
commissure ‖ Fuge f. ‖ joint m.
committee ‖ Ausschuß m. ‖ comité m. /
economic ~ ‖ Wirtschaftskomitee n. ‖
comité m. économique. / to refer to a ~ ‖
einem Ausschuß m. überweisen ‖ renvoyer à une commission f.
committer ‖ Besteller m. ‖ client m.; client-
acheteur m.
commodities pl. ‖ Gebrauchsgüter npl. ‖
objets mpl. d'utilité.
common battery working ‖ Zentralbatterie-
betrieb m. ‖ exploitation f. à batterie
centrale. ~ black pitch ‖ Schusterpech
n. ‖ poix f. de cordonnier. / ~ design ‖
gewöhnliche Ausführung f. ‖ construction f. normale. / ~ feldspar powdered
or calcined ‖ gemeiner feldspat m., gepulvert oder gebrannt ‖ Feldspath m.
commun pulvérisé ou calciné. / ~ law ‖
Landrecht n. ‖ droit m. civil ou commun

ou coutumier. / ~ letters pl. ‖ gewöhnliche Buchstabenschrift f. ‖ caractères
mpl. d'écriture ordinaire. / ~ maple ‖
Feldahorn m. ‖ érable m. champêtre. /
~ salt ‖ Kochsalz n.; Chlornatrium n. ‖
chlorure m. de sodium; sel m. de cuisine. /
~ workman ‖ Lohnarbeiter m. ‖ manœuvre m.; manouvrier m.
communal forest ‖ Gemeindewald m. ‖
forêt f. communale.
communicating ‖ kommunizierend ‖ com-
municant.
communication (Tel) ‖ Verbindung f. ‖
interconnexion f.; communication f.;
conversation f. / ~ of motion ‖ Fortpflanzung f. der Bewegung ‖ communication
f. de mouvement. / ~ by railway ‖ Eisenbahnverbindung f. ‖ communication f.
par chemin de fer. / ~s pl. per minute
(Tel) ‖ Verbindungen fpl. je Minute ‖
communications fpl. par minute.
communication cord ‖ Notleine f. ‖ corde
f. d'alarme. / power circuit interference
with ~ line ‖ Beeinflussung f. der Fernmeldeleitung durch Starkstromanlagen ‖
influence f. perturbatrice engendrée dans
la ligne de communication par les installations d'énergie électrique.
communicator ‖ Fernmelder m. ‖ com-
municateur m.
community automatic exchange ‖ Landfernsprechnetz n. für Selbstanschlußbetrieb ‖ téléphone m. rural.
commutated current ‖ kommutierter Strom
m. ‖ courant m. commuté.
commutating field ‖ Kommutierungsfeld
n. ‖ champ m. de commutation. / ~ pole
‖ Wendepol m. ‖ pôle m. de commutation.
commutation current ‖ Kommutierungs-
strom m. ‖ courant m. de commutation.
commutator ‖ Stromwender m. ‖ commutateur m. / oval ~ ‖ unrunder Kollektor m.
‖ collecteur m. ovalisé. / testing ~ ‖ Meß-
umschalter m. ‖ clef f. de mesure. /
vertical ~ ‖ Stirnkollektor m. ‖ collecteur
m. frontal.
commutator balsam in bars (Electr) ‖
Kollektorbalsam m. in Stangen ‖ lubrifiant m. pour collecteurs en bâtonnets.
/ ~ bar (Electr) ‖ Kommutatorseg-
ment n. ‖ lame f. ou touche f. du collecteur. / ~ bush ‖ Kollektorbuchse f. ‖
douille f. de collecteur. / ~ casing ‖
Kollektorgehäuse n. ‖ cage f. de collecteur. / ~ clamping ring ‖ Kollektorklemm-
ring m. ‖ bague f. de calage du collecteur.
/ ~ composition ‖ Kollektorfett n. ‖ apprêt m. pour collecteurs. / ~ ears pl. ‖
Kommutatoransätze mpl. ‖ talons mpl.
du collecteur. / ~ end core-head ‖ Ankerflansch m. der Kommutatorseite ‖ flasque m. de l'induit du côté collecteur. /
~ grease ‖ Kommutatorfett n. ‖ lubrifiant m. pour collecteur. / ~ lubricant ‖
Kollektorschmiere f. ‖ lubrifiant m. pour
collecteur. / ~ lug ‖ Kollektorfahne f. ‖
talon m. du collecteur. / ~ machining
device ‖ Kollektorabdrehvorrichtung f. ‖
appareil m. pour dresser le collecteur au
tour. / ~ motor with auxiliary slip-rings ‖
Kollektormotor m. mit Hilfsschleifrin-
gen ‖ moteur m. à collecteur à bagues
collectrices auxiliaires. / ~ pitch ‖ Kol-
lektorteilung f. ‖ pas m. du collecteur. /
~ polishing paper ‖ Kollektorschleifpa-
pier n. ‖ papier m. à polir le collecteur. /
rectifier ‖ Kommutatorgleichrichter m. ‖

permutatrice f. / ~ segment || Kollektor-lamelle f. || lame f. du collecteur. / ~ shaft || Kommutatorwelle f. || arbre m. de commutateur. / ~ sleeve || Kollektor-hülse f. || manchon m. du collecteur. / ~ switch || Fahrrichtungsschalter m. || commutateur m. de sens de marche. / ~ truing-up device || Vorrichtung f. zum Abdrehen des Kollektors || rectifieuse f. pour collecteurs.

compact (Geol) || massives Gestein n. || compacte m.

compact || gedrängt || compact. / ~ construction || gedrängte Bauart f. || construction f. serrée.

compact design (Mach) || gedrängte Bau-art f. || construction f. compacte. / motor of ~ || Motor m. gedrängter Bau-art || moteur m. de construction com-pacte.

companion (Shipb) || Niedergangskappe f. || capot m. d'échelle. / ~ ladder || Kajüts-treppe f. || échelle m. de la chambre.

company || Firma f. || maison f. (de commerce); raison f. commerciale. / ~ || Handelsgesellschaft f. || société f. (de commerce); compagnie f. / joint stock ~ || Aktiengesellschaft f. || société f. anonyme ou par actions. / ~ limited by shares || Aktiengesellschaft f. || société f. anonyme ou par actions. / ship's ~ || Mannschaft f. || équipage m.

company law || Gesellschaftsgesetz n. || lois fpl. sur les sociétés. / ~'s report || Gesellschaftsbericht m. || rapport m. de la compagnie.

comparative firing || Vergleichsschießen n. || tir m. comparatif. / ~ value || Vergleichs-wert m. || valeur f. comparative.

comparator || Komparator m.; Rundlauf-lehre f. || comparateur m. / colour ~ || Farbenkomparator m. || comparateur m. de teintes. / spectro ~ || Spektrokompara-tor m. || spectrocomparateur m.

compare, to || vergleichen || comparer.

comparison, numerical || zahlenmäßiger Vergleich m. || comparaison f. numéri-que. / device for ~ of spectrums con-sisting of a wave-length scale and a ~ spectrum || Einrichtung f. zur Verglei-chung des Spektrums mit einer Wellen-längenskale und einem Vergleichsspek-trum || dispositif m. à comparer le spectre avec une échelle de longueurs d'onde et un autre spectre.

comparison electrode || Vergleichselektrode f. || électrode f. de comparation. / ~ eye-piece || Vergleichsokular n. || oculaire m. de comparaison. / ~ method || Vergleichs-methode f.; Vergleichsverfahren n. || mé-thode f. de comparaison. / ~ microscope for spectra || Vergleichsmikroskop n. für Spektren || microscope m. pour la com-paraison des spectres. / ~ prism || Ver-gleichsprisma n. || prisme m. de compa-raison. / ~ spectroscope || Vergleichs-spektroskop n. || spectroscope m. de com-paraison. / ~ spectrum || Vergleichsspek-trum n. || spectre m. de comparaison. / ~ star scale || Vergleichssternskala f. || échelle f. de comparaison des astres.

compartment (Railw) || Abteil n. || com-partiment m.; coupé m.

compass || Kompaß m. || boussole f. / auxiliary ~ || Tochterkompaß m. || bous-sole f. filiale ou secondaire ou comman-dée. / azimuth ~ || Azimuthkompaß m. || compas m. azimutal. / barrel ~ || Trom-

melkompaß m. || compas-tambour m. / bearing ~ || Peilkompaß m. || compas m-de relèvement. / boat's ~ || Bootskom.paß m. || volet m.; compas m. d'embar-cation. / cabin ~ || Kajütskompaß m.; Hängekompaß m. || compas m. suspendu ou de chambre. / compensator of the ~ || Kompensationsmagnet m. || compensa-teur m. magnétique du compas. / ~ with deviation table || Kompaß m. mit De-viationstafel || boussole f. avec table des déclinaisons. / division of the ~ || Kom-paßeinteilung f. || division f. de la rose des vents. / drum ~ || Trommelkompaß m. || compas-tambour m. / dry ~ || Trocken-kompaß m. || compas m. sec. / fluid ~ || Fluidkompaß m. || compas m. à liquide. / gyro ~ || Kreiselkompaß m. || compas m. ou boussole f. gyroscopique. / hanging ~ see cabin ~. / immersed ~ || Flüssigkeits-kompaß m. || compas m. à liquide. / li-quid ~ || Fluidkompaß || compas m. à li-quide. / magnetic ~ || Magnetkompaß m. || compas m. magnétique. / marine ~ || Schiffskompaß m. || boussole f. marine; compas m. de navires. / mariner's ~ || Marinekompaß m. || boussole f. marine. / miner's ~ || Grubenkompaß m. || Hänge-kompaß m. || boussole f. ou poche f. de mineur. / nautical ~ see marine ~. / pe-destal of the ~ || Kompaßsäule f. || piédes-tal m. du compas. / pin of the ~ || Kom-paßpinne f.; Kompaßspitze f. || pivot m. du compas. / pocket ~ || Taschenkompaß m. || boussole f. de poche. / point of the ~ || Kompaßstrich m. || aire f. ou rumb m. de vent de la rose. / radio ~ || Radiokom-paß m.; Richtungsfinder m. || radio-bous-sole] f.; boussole f. sans fil. / repeater ~ see auxiliary ~. / to reverse the ~ || den Kompaß m. umsetzen || retourner le compas ou la boussole. / rhumb of the ~ see point of the ~. / sea ~ || Seekompaß m. || compas m. de mer ou de route; bous-sole f. / standard ~ || Regelkompaß m. || compas m. étalon. / steering ~ || Steuer-kompaß m. || compas m. de route ou d'habitacle. / surveying ~ || Feldmesser-kompaß m. || boussole f. d'arpenteur. / surveyor's ~ see surveying ~. / travelling ~ || Reisekompaß m. || compas m. de route. / wheel ~ see steering ~. / wireless ~ || Radiokompaß m.; Richtungsfinder m. || radio-boussole m.; boussole f. sans fil.

compass bearing || Kompaßpeilung f. || re-lèvement m. au compas. / ~ board (Weav) || Harnischbrett m.; Löcherbrett n.; Schnürbrett n. || planche f. d'arcades ou à trous. / ~ bowl || Kompaßgehäuse n. || cuvette f. du compas. / ~ box || Kom-paßdose f.; Kompaßbüchse f. || boîte f. du compas. / ~ card || Kompaßrose f.; Windrose f. || rose f. des vents ou du compas. / ~ direction see ~ course. / compensation of ~ error || Kompaßbe-richtigung f. || compensation f. des er-reurs de la boussole.

compasses pl. || Zirkel m. || compas m. / ~ (Dividers) || Handzirkel m.; Stechzirkel m. || compas m. de mesure; compas m. à pointes sèches. / beam ~ || Stangenzirkel m. || compas m. à verge. / bow ~ see spring bow ~. / box of ~ || Reißzeug n. || boîte f. à compas. / ~ with detachable legs || Einsatzzirkel m. || compas m. à rallonges. / ~ with fixed needle points || Stechzirkel m. mit festen Spitzen || com-

pas m. à pointes sèches fixes. / foot of ~ || Zirkelfuß m. || pied m. de compas. / handle of ~ || Zirkelkopf m. || tête f. du compas. / lead ~ || Bleistiftzirkel m. || compas m. à crayon. / leg of ~ || Zirkel-schenkel m. || jambe f. de compas. / ~ with movable points || Einsatzzirkel m. || compas m. à pointes rapportées. / pair of ~ || Zirkel m.; Scharnierzirkel m. || compas m. (à charnière). / point of ~ || Zirkelspitze f. || pointe f. du compas. / proportional ~ see reducing ~. / ~ with quick return || Schnellspannzirkel m. || compas m. à détente rapide. / reducing ~ || Reduktionszirkel m. || compas m. de réduction ou de proportion. / ~ with regulating screw || Haarzirkel m. || com-pas m. de précision. / ~ with removable points || Zirkel m. mit auswechselbaren Spitzen || compas m. à pointes de re-change. / round leg ~ || Zirkel m. mit runden Schenkeln || compas m. droit à branches rondes. / ~ with screwed wing || Zirkel m. mit auswechselbarem Füh-rungsbügel || compas m. avec quart de cercle rapporté. / set of ~ see box of ~. / scribing ~ || Parallelzirkel m.; Kreis-reißer m. || compas m. à point réglable. / ~ with shifting legs || Zirkel m. mit ein-setzbaren Spitzen || compas m. à pointes de rechange. / ~ with solid wing || Zirkel m. mit festem Führungsbügel || compas m. avec quart de cercle fixe. / spring bow ~ || Nullenzirkel m. || compas m. à pompe. / straight ~ (Mar) || grader Passer m. || compas m. droit. / ~ with wing || Zirkel m. mit Führungsbügel || compas m. droit avec quart de cercle.

compass kettle || Kompaßkessel m. || caisse f. du compas. / ~ key || Zirkelschlüssel m. || clef f. de compas. / ~ lens || Kompaß-glas n. || verre m. pour boussoles. / ~ liquid || Kompaßflüssigkeit f. || liquide m. pour compas. / ~ maker || Reißzeug-macher m. || fabricant m. de compas. / ~ needle || Kompaßnadel f.; Magnetnadel f. || aiguille f. aimantée ou de compas ou de boussole. / socket of the ~ needle || Dop m.; Kompaßhütchen n. || chape f. de la rose des vents. / ~ plane || Schiff-hobel m. || rabot m. rond ou cintré. / ~ saw || Lochsäge f.; Stichsäge f.; Spitz-säge f. || scie f. à guichet; scie f. égoïne. / ~ stand || Kompaßstativ n. || trépied m. d'un compas.

compensate, to || kompensieren; ausglei-chen || compenser. / ~ the error in obser-vation || den Beobachtungsfehler m. aus-gleichen || compenser l'erreur m. d'ob-servation. / ~ for || entschädigen || dé-dommager; indemniser. / ~ for loss of gas || den Gasverlust m. ausgleichen || compenser la perte ou la fuite de gaz. / ~ for loss of profit || einen entgangenen Gewinn m. ersetzen || compenser une perte de bénéfice. / to over-compensate ~ the compass || den Kompaß m. über-kompensieren || surcompenser le compas. / ~ by periodic payments || durch Zah-lung einer Rente f. entschädigen || in-demniser par le payement d'une rente. / ~ a person (for a thing) || jemanden entschädigen || indemniser quelqu'un. / to under-compensate the compass || den Kompaß m. unterkompensieren || sous-compenser le compas.

compensating battery (Electr) || Ausgleich-batterie f. || pile f. de compensation.

/ ~ beam ‖ Ausgleichhebel m. ‖ balancier m. / ~ current ‖ Ausgleichstrom m. ‖ courant m. compensateur. / ~ eyepiece ‖ Kompensationsokular n. ‖ oculaire m. compensateur. / ~ jet ‖ Ausgleichdüse f. ‖ gicleur m. de compensation. / ~ lever for brake ‖ Bremsausgleichhebel m. ‖ levier m. compensateur de frein. / ~ magnet ‖ Ausgleichmagnet m. ‖ aimant m. correcteur *ou* de compensation. / magnetic ~ plate ‖ magnetischer Kompensator m. ‖ compensateur m. magnétique. / ~ set ‖ Kompensationsapparat m. ‖ appareil m. compensateur. / ~ sheave ‖ Ausgleichrolle f. ‖ galet m. de compensation. / ~ spring ‖ Ausgleichfeder f. ‖ ressort m. compensateur.

compensation (Mar) ‖ Kompensation f. ‖ compensation f. / ~ (Trade) ‖ Schadloshaltung f.; Entschädigung f. ‖ dédommagement m.; indemnisation f.; indemnité f.; rémuneration f. / amount of ~ ‖ Abfindungssumme f. ‖ indemnité f.; payement m. par compensation. / appropriate (A) ~ *see* improvised ~. / automatic ~ ‖ selbsttätiger Ausgleich m. ‖ compensation f. automatique. / to come to an amicable settlement for ~ ‖ sich über eine Entschädigung f. gütlich einigen ‖ s'arranger amiablement pour une indemnité. / ~ of compass error ‖ Kompaßberichtigung f. ‖ compensation f. des erreurs de la boussole. / to bring an action for ~ for damage ‖ Klage f. auf Schadensersatz erheben ‖ porter une plainte pour réparation d'un dommage. / claim for ~ for damage ‖ Schadensersatzanspruch m. ‖ réclamation f. en réparation de dommage. / to give proof of a claim for ~ for damage ‖ einen Schadenersatzanspruch m. glaubhaft machen ‖ justifier une réclamation en réparation de dommage. / to be liable to make ~ for damage ‖ entschädigungspflichtig sein ‖ être tenu à dédommagement. / to determine a ~ ‖ eine Entschädigung f. festsetzen ‖ fixer un dédommagement. / determination of ~ ‖ Entschädigungsfestsetzung f. ‖ fixation f. des indemnités. / ~ of errors ‖ Fehlerausgleichung f. ‖ compensation f. des erreurs. / improvised ~ ‖ behelfsmäßige Kompaßausgleichung f. ‖ compensation f. de la boussole par des moyens de fortune. / to pay ~ in advance ‖ eine Entschädigung f. im voraus zahlen ‖ verser l'indemnité d'avance. / to pay annual ~ ‖ eine jährliche Entschädigung f. leisten ‖ accorder une indemnité annuelle. / to pay ~ in cash ‖ in Geld n. entschädigen ‖ indemniser en argent m.

compensation apparatus ‖ Kompensationsapparat m. ‖ appareil m. de compensation. / ~ circuit (Electr) ‖ Ausgleichskreis m. ‖ circuit m. d'équilibre. / ~ engine (Electr) ‖ Ausgleichmaschine f. ‖ compensatrice f. / ~ line (Electr) ‖ Ausgleichsleitung f. ‖ ligne f. de compensation. / ~ magnet ‖ Kompensationsmagnet m. ‖ aimant-compensateur m. / ~ method ‖ Ausgleichsverfahren n. ‖ méthode f. de compensation. / ~ pendulum ‖ Kompensationspendel n. ‖ pendule m. compensateur *ou* à compensation. / ~ set (Electr) ‖ Ausgleichsatz m. ‖ groupe m. de compensation. / ~ test apparatus ‖ Kompensationsapparat m. ‖ appareil m. à mesures de compensation.

compensator ‖ Kompensator m. ‖ compensateur m. / tube ~ ‖ Rohrausgleicher m. ‖ compensateur m. à tuyaux.

compense, to ‖ kompensieren; ausgleichen ‖ compenser.

compete, able to ‖ konkurrenzfähig ‖ capable de soutenir la concurrence.

competent ‖ maßgebend ‖ compétent. / ~ court ‖ Gerichtsstand m. ‖ tribunal m. compétent. / ~ judge ‖ Sachverständiger m. ‖ expert m.

competing ‖ konkurrierend ‖ en concurrence.

competition ‖ Wettbewerb m.; Konkurrenz f. ‖ concurrence f.; concours m. / dishonest ~ ‖ unlauterer Wettbewerb m. ‖ concurrence f. déloyale. / to enter into ~ ‖ in Wettbewerb m. treten ‖ entrer en concurrence f. / international ~ ‖ internationaler Wettbewerb m. ‖ concours m. international. / severe ~ ‖ scharfe Konkurrenz f. ‖ forte concurrence f. / unfair ~ ‖ unlauterer Wettbewerb m. ‖ concurrence f. déloyale. / without ~ ‖ konkurrenzlos; wettbewerbslos ‖ sans concurrence f.

competitive ‖ konkurrenzfähig ‖ capable de soutenir la concurrence.

competitor ‖ Konkurrent m. ‖ concurrent m.; compétiteur m.

compile, to ‖ zusammenstellen ‖ assembler; grouper; compiler.

complain, to ‖ reklamieren; beanstanden ‖ réclamer.

complaint ‖ Reklamation f. ‖ plainte f.; réclamation f. / to bring ~s pl. to notice ‖ Beanstandungen fpl. geltend machen ‖ faire réclamations fpl. / to make a ~ ‖ eine Klage f. einreichen ‖ déposer une plainte.

complaint desk (Tel) ‖ Beschwerdestelle f. ‖ table f. de réclamation.

complement of an angle (Geom) ‖ Komplement n. *oder* Ergänzung f. eines Winkels ‖ complément m. d'un angle. / ~ of an arc (Geom) ‖ Komplement n. *oder* Ergänzung f. eines Bogens ‖ complément m. d'un arc.

complementary ‖ komplementär ‖ complémentaire. / ~ colours pl. ‖ Komplementärfarben fpl. ‖ couleurs fpl. complémentaires. / ~ ending ‖ Abschiedsformel f. ‖ formule f. de politesse. / ~ strip for the foreign type wheel ‖ Ergänzungsblatt n. für das fremdsprachliche Typenrad ‖ bande f. servant à completer le jeu de caractères de la roue à caractères en langue étrangère.

complete ‖ vollwertig; komplet ‖ complet; parfait. / ~ damping ‖ vollkommene Dämpfung f. ‖ amortissement m. parfait. / ~ revolution ‖ ganze Umdrehung f. ‖ tour m. complet.

complete, to ‖ ausbauen; vervollständigen ‖ achever; développer. / ~ a circuit ‖ einen Stromkreis m. herstellen ‖ compléter un circuit m.

completely ‖ vollständig ‖ entièrement; tout à fait.

completion ‖ Ausbau m. ‖ achèvement m. / ~ of the extensions which are being carried out now ‖ Durchführung f. der im Gang befindlichen Ausbauarbeiten ‖ achèvement m. des constructions en cours. / ~ of interior rooms ‖ Innenausbau m. ‖ achèvement m. des intérieurs.

complex number ‖ komplexe Zahl f. ‖ expression f. complexe.

compliance ‖ Einwilligung f. ‖ consentement m.

complicated, part of ~ shape ‖ kompliziert geformtes Stück n. ‖ élément m. de formes compliquées.

complications pl. ‖ Weiterungen fpl. ‖ difficultés fpl.; formalités fpl.

comply, to ~ with ‖ übereinstimmen ‖ accorder.

component (Chem) ‖ Bestandteil n. ‖ composant m. / ~ (Mech) ‖ Seitenkraft f.; Komponente f. ‖ composante f. / ~ of the air ‖ Luftbestandteil m. ‖ composant m. de l'air; élément m. constituant de l'air. / automobile ~s ‖ Automobilteile mpl. ‖ pièces fpl. d'automobiles. / field ~ ‖ Feldkomponent m. ‖ composante f. du champ. / horizontal ~ ‖ wagerechte Komponente f. ‖ composante f. horizontale. / ~ of a power (Mech) ‖ Seitenkraft f. ‖ composante f. d'une force. / ~ of velocity ‖ Komponente f. der Geschwindigkeit ‖ composante f. de vitesse. / vertical ~ ‖ senkrechte Komponente f. ‖ composante f. verticale.

component force ‖ Teilkraft f.; Seitenkraft ‖ force f. composante. / ~ magnification (Opt) ‖ Einzelvergrößerung f. ‖ grossissement m. propre *ou* partiel. / ~ optician ‖ Fachoptiker m. ‖ opticien m. spécialiste. / ~ part ‖ Bestandteil m. ‖ partie f. constituante.

compose, to (Mech) ‖ zusammensetzen ‖ composer. / ~ (Print) ‖ setzen ‖ composer. / ~ closely (Print) ‖ eng setzen ‖ composer serré. / ~ in columns (Print) ‖ spaltenweise setzen ‖ composer par colonnes fpl. / ~ in companionship (Print) ‖ stückweise setzen ‖ travailler en galée f.

composed motion ‖ zusammengesetzte Bewegung f. ‖ mouvement m. composé. / ~ of ‖ bestehend aus ‖ composé de.

composer ‖ Komponist m. ‖ compositeur m. de musique.

composing, correctly ~ the charge ‖ richtiges Einsetzen n. des Metalls ‖ manière irréprochable de charger le métal dans le creuset.

composing, rapid ~ apparatus (Print) ‖ Schnellsetzer m. ‖ appareil m. à composer rapide. / ~ device (Print) ‖ Setzvorrichtung f. ‖ appareil m. à composer. / ~ galley ‖ Setzschiff n.; Schiffchen n. ‖ galée f. / ~ stick (Locksm) ‖ Scheinecke f.; Winkelband n.; Eckschiene f. ‖ équerre f. de fer. / ~ stick (Print) ‖ Winkelhaken m.; Setzwinkel m. ‖ composteur m. / ~ tool (Print) ‖ Setzwerkzeug n. ‖ outil m. à composer.

composite building (Shipb) ‖ Kompositbau m. ‖ construction f. en bois et en fer. / ~ line (Tel) ‖ zusammengesetzte Leitung f. ‖ ligne f. composée. / ~ tool ‖ aufgeschweißter Stahl m. ‖ outil m. terminé par soudure de la mise. / ~ vessel ‖ Kompositschiff n. ‖ navire m. en bois et en fer.

compositing table (Print) ‖ Setztisch m. ‖ table f. de composition.

composition (Build) ‖ Kreidepaste f.; Masse f. ‖ gros-blanc m. / ~ (Chem) ‖ Zusammensetzung f. ‖ constitution f. / ~ (Mech) ‖ Zusammensetzung f. ‖ composition f. / ~ (Print) ‖ Setzen n.; Schriftsetzen n.; Satz m. ‖ composition f. / ~ of catalogues ‖ Katalogsatz m. ‖ composition f. de catalogues. / chemical ~ ‖ chemisches Präparat n. ‖ produit m.

chimique. / complicated ~ || schwieriger *oder* komplizierter Satz m. || composition f. compliquée *ou* difficile. / to distribute a ~ (Print) || einen Satz m. ablegen || distribuer une composition. / ~ of forces || Zusammensetzung f. von Kräften || composition f. de forces. / hand ~ (Print) || Handsatz m. || composition f. à main. / ~ of metals || Metalllegierung f. || alliage m. métallique. / mixed ~ (Print) || gemischter Satz m. || composition mixte. / narrow ~ || enger Satz m. || composition f. serrée. / preparing of ~ || Satzansetzen n. || mélange m. des matières. / ~ of packets (Print) || Stücksatz m.; Paketsatz m. || paquetage m. / ~ for powder || Pulversatz m. || composition f. de la poudre à canon. / ~ for rockets || Raketensatz m.; Treibsatz m. || matière f. fusante. / rules of ~ (Print) || Satzanweisung f. || marche f. typographique. / ~ of slips (Print) || Paketsatz m.; Stücksatz m. || paquetage m. / slow-burning ~ || fauler Satz || composition lente. / uniform ~ || gleichmäßiger Satz m. || composition f. régulière. / wide ~ || weiter Satz m. || composition f. espacée.

composition metal || Legierung f. || alliage m. / ~ part of rockets || Satzsäule f. || cylindre m. de composition. / ~ sieve || Trommelsieb n. || tamis m. à tambour.

compositor (Print) || Schriftsetzer m.; Setzer m. || compositeur m. d'imprimerie. / ~ of the companionship (Print) || Stücksetzer m.; Paketsetzer m. || paquetier m.; piècier m. / ~ who makes omissions (Print) || schlechter Setzer m.; bourdonneur m. / ~ and printer (Print) || Schweizerdegen m. || amphibie m. / ~'s scale of prices || Setzertarif m. || tarif m. des compositeurs.

compound || zusammengesetzt; Verbund-composé; compound. / to work ~ || nach dem Verbundsystem n. arbeiten || fonctionner en compound.

compound (Chem) || Verbindung f. || combinaison f.; composé m. / ~ (Substance) || Gemisch n.; Masse f. || alliage m.; mélange m.; matière f. / additive ~ || Anlagerungsverbindung f. || produit m. *ou* combinaison f. d'addition. / ~ cable || Kabelisoliermasse f.; Kabelvergußmasse f. || matière f. de remplissage (de la boîte de jonction) pour câbles. / chemical ~ || chemische Verbindung f. || combinaison f. chimique; composé m.

compound arrangement || Verbundanordnung f. || disposition f. compound. / ~ dryer || Verbundtrockner m. || sécheur m. système compound. / ~ dynamo || Doppelschlußmaschine f. || dynamo f. compound. / ~ engine || Verbundmaschine f. || machine f. compound. / ~ excitation || Kompounderregung f. || excitation f. composée. / ~ interest || Zinseszins m. || intérêt m. composé. / ~ locomotive || Verbundlokomotive f. || locomotive f. compound. / six-coupled ~ locomotive || ^{3}/$_3$ gekuppelte Verbundlokomotive f. || locomotive f. compound à trois essieux couplés. / ~ mill || Verbundmühle f. || moulin m. combiné *ou* compound. / ~ motion || zusammengesetzte Bewegung f. || mouvement m. composé. / ~ slide rest || Kreuzsupport m. || support m. à chariots croisés. / ~ slides pl. || Kreuzschlitten m. || chariot m. à mouvements croisés. / ~

spring hammer || Verbundfederhammer m. || marteau m. à ressort compound. / ~ steam blowing engine || Verbunddampfgebläsemaschine f. || machine-soufflante f. à vapeur compound. / ~ steam engine || Verbunddampfmaschine f.; Kompounddampfmaschine f. || machine f. à vapeur compound *ou* du système compound. / ~ table || Kreuztisch m. || table f. composée. / ~ tank for cable manufacturing || Massebehälter m. für die Kabelherstellung || récipient m. de masse isolante pour la fabrication des câbles. / ~ tube mill || Verbundrohrmühle f. || moulin m. tubulaire combiné. / ~ winding (Electr) || Kompoundwicklung f. || enroulement m. compound. / ~-wound dynamo || Verbunddynamomaschine f. || dynamo f. compound.

compound, additive (Chem) || Anlagerungsverbindung f. || produit m. d'addition. / chemical ~ || chemische Verbindung f. || combinaison f. chimique.

compounding || Kompoundierung f. || compoundage m.

compound tank for cable manufacturing || Massebehälter m. für die Kabelherstellung || récipient m. de masse isolante pour la fabrication des câbles.

comprehensive || umfassend || étendu; vaste.

compress, to || zusammendrücken; zusammenpressen; komprimieren || comprimer.

compressed-air || Preßluft f.; Druckluft f. || air m. comprimé. / hoop actuated by ~ for overhead contact || Druckluftbügelbetätigung f. für Stromabnehmer || prise f. de courant à archet actionnée à l'air comprimé.

compressed-air accumulator || Druckluftsammler m. || accumulateur m. à air comprimé. / ~ accumulator for firings with travelling grates || Druckluftstauer m. für Wanderrostfeuerungen || chargeur m. à air comprimé pour foyers à grille mobile. / ~ brake || Druckluftbremse f. || frein m. pneumatique *ou* à air comprimé. / ~ chamber || Druckluftkessel m. || récipient m. à air comprimé. / ~ chamber for torpedoes || Torpedoluftkessel m. || réservoir m. à air comprimé pour torpilles. / ~-driven conveying plant || durch Druckluft betriebene Förderanlage f. || installation f. d'engin de transport fonctionnant au moyen de l'air comprimé. / ~ conveyor || Druckluftförderer m. || transporteur m. pneumatique. / ~ drilling machine || Preßluftbohrmaschine f. || perceuse f. à air comprimé. / ~ engine || Preßluftmotor m. || moteur m. à air comprimé. / ~ fittings pl. || Preßluftarmatur f. || armature f. pour air comprimé. / ~ hammer || Preßlufthammer m. || marteau m. à air comprimé. / ~ impulse catapult || preßluftgetriebener Katapult m. || catapulte f. de lancement à air comprimé. / ~ injection || Drucklufteinspritzung f. || injection f. pneumatique. / ~ jet apparatus || Druckluftstrahlapparat m. || appareil m. à jet d'air comprimé. / ~ locomotive || Druckluftlokomotive f. || locomotive f. à air comprimé. / ~ machine || Druckluftmaschine f. || moteur m. à air comprimé. / ~ manometer || geschlossenes Luftmanometer n. || manomètre m. à air comprimé. / ~ meter || Preßluftmesser m. || appareil m. à mesurer la consommation d'air comprimé. / ~ motor || Preßluftmotor m.

|| moteur m. à air comprimé. / ~ oil separator || Preßluftentöler m. || déshuileur m. à air comprimé. / ~ piping || Druckluftleitung f. || conduite f. d'air comprimé. / ~ plant || Druckluftanlage f. || installation f. d'air comprimé. / ~ receiver || Druckluftbehälter m. || réservoir m. à air comprimé. / ~ receiver for starting || Druckluftanlaßgefäß n. || réservoir m. d'air comprimé pour mise en marche. / ~ receiver for ticket distributing systems || Druckluftempfänger m. für Zettelrohrposten || récepteur m. d'air comprimé pour le transport des fiches. / ~ reservoir || Druckluftbehälter m. || réservoir m. à air comprimé. / ~ rock drilling machine || Preßluftgesteinbohrmaschine f. || perceuse f. de roches à air comprimé. / ~ sender for ticket distributing systems || Druckluftsender m. für Zettelrohrposten || transmetteur m. d'air comprimé pour le transport des fiches. / ~ service || Druckluftbetrieb m. || service m. à air comprimé. / ~-driven sledge hammer || Preßluftschmiedehammer m. || marteau m. de forge à air comprimé. / ~ starter || Preßluftanlasser m. || démarreur m. à air comprimé. / ~ switch || Druckluftschalter m. || interrupteur m. à air comprimé. / ~ tank || Druckluftkessel m. || réservoir m. à air comprimé. / ~ transmission || Druckluftübertragung f. || transmission f. par air comprimé. / ~ turbine || Preßluftturbine f. || turbine f. à air comprimé. / ~ water separator || Preßluftwasserabscheider m. || séparateur m. d'eau à air comprimé.

compressed concrete articles pl. || Zementgußwaren fpl. || objets mpl. en béton moulé. / ~ gas || komprimiertes Gas m. || gaz m. comprimé. / ~ paper || Hartpapier n. || papier m. durci. / ~ rubble || Preßgipsplatte f. || hourdis m. comprimé. / ~ tube (Techn) || gepreßtes Rohr n. || tuyau m. étiré sans soudure; tuyau m. repoussé. / ~ turf making || Preßtorffabrikation f. || fabrication f. de tourbe comprimée. / ~ vegetables pl. || komprimiertes Gemüse n. || légume m. comprimé. / ~ wadding || Verbandwatte f. || ouate f. à pansements. / ~ water reservoir || Druckwasserbehälter m. || réservoir m. à eau comprimée. / ~ yeast || Preßhefe f. || levure f. pressée.

compressibility || Zusammendrückbarkeit f. || compressibilité f. / ~ of the air || Zusammendrückbarkeit f. der Luft || compressibilité f. de l'air.

compressible || zusammendrückbar || compressible. / ~ fluid || zusammendrückbare Flüssigkeit f. || liquide m. compressible. / ~ soil || sich setzender Boden m. || sol m. *ou* terrain m. compressible.

compressing of industrial gases || Verdichtung f. von Industriegasen || compression f. des gaz de fabriques.

compression || Kompression f.; Verdichtung f.; Zusammendrückung f. || compression f. / in ~ || auf Druck m. beansprucht || sous compression. / ~ mechanic ~ || mechanische Verdichtung f. || compression f. mécanique. / ratio of ~ || Verdichtungsverhältnis n.; Kompressionsverhältnis n. || taux m. *ou* rapport m. de compression. / single-stage ~ || einstufige Verdichtung f. || compression f. mono-étagée. / two-stage ~ || zweistufige Verdichtung f. || compression f. bi-étagée. / work done under ~ || Ver-

dichtungsarbeit f. || travail m. de compression.

compression cooling machine || Kompressionskühlmaschine f. || machine f. frigorifique à compression. / ~ curve || Kompressionskurve f. || courbe f. de compression. / ~ cylinder || Preßzylinder m. || cylindre m. de pression. / elasticity of ~ || Druckelastizität f. || élasticité f. de compression. / ~ loss || Kompressionsverlust m. || perte f. par compression. / ammonia ~ machine || Ammoniakkompressionsmaschine f. || machine f. à compression d'ammoniaque. / pressure reached at end of ~ || Kompressionsenddruck m. || degré m. de compression final. / ~ pump || Kompressionspumpe f. || pompe f. de compression. / ~ ratio || Verdichtungsverhältnis n. || rapport m. *ou* taux m. *ou* degré m. de compression. / ~ release || Verdichtungsminderer m. || décompresseur m. / ~ rib of the wing (Aero) || Hauptrippe f. des Flügels f. || nervure f. principale d'aile; nervure f. de compression d'aile. / ~ spring || Druckfeder f. || ressort m. de pression. / ~ stop cock || Quetschhahn m. zum Verschließen von Kautschukröhren || pince f. à fermer les tubes en caoutchouc. / ~ strain || Druckbeanspruchung f. || effort m. pe compression. / ~ strength || Druckfestigkeit f. || résistance f. à la compression. / ~ stroke || Kompressionshub m. || course f. de compression. / ~ test || Druckprobe f. || épreuve f. de compression.

compressive force || Druckkraft f. || force f. de compression. / ~ protection spiral || Druckschutzspirale f. || spirale f. de protection contre la compression. / ~ strength || Druckfestigkeit f. || résistance f. à la compression. / minimum ~ strength || Mindestdruckfestigkeit f. || résistance f. minimum à la compression. / ~ stress || Druckspannung f. || tension f. de compression.

compressor || Verdichter m.; Kompressor m. | compresseur m. / ~ (Shipb) || Kettenkneifer m.; Kettenstopper m. || col m. de cygne. / air ~ || Luftkompressor m. || compresseur m. d'air. / air-injection ~ || Einblaseluftpumpe f. || compresseur m. d'injection d'air. / ammonia ~ || Ammoniakkompressor m. || compresseur m. à ammoniac. / axle ~ (Railw) || Achskompressor || compresseur m. d'essieux. / carbonic acid ~ || Kohlensäurekompressor m. || compresseur m. d'acide carbonique. / double ~ || Doppelkompressor m. double. / double carbonic acid ~ of one single stage || einstufiger Kohlensäuredoppelverdichter m. || compresseur m. double à acide carbonique à un seul étage. / enclosed ~ || Kapselkompressor m. || compresseur m. blindé. / ~ actuated by fall of water || Hydrokompressor m. || compresseur m. à chute d'eau. / gas ~ || Gaskompressor m. || compresseur m. à gaz. / high-pressure air ~ || Hochdruckluftkompressor m. || compresseur m. d'air à haute pression. / injection air ~ || Einblaseluftpumpe f. || compresseur m. d'air d'injection. / metallic part of ~s || Kompressorenmetallteil m. || pièce f. métallique pour compresseurs. / rotary ~ || Rotationskompressor m. || compresseur m. rotatif. / small piston ~ || Kleinkolbenkompressor m. || compresseur m. à piston de faibles

dimensions. / three-stage ~ || dreistufiger Kompressor m. || compresseur m. à trois étages. / two-stage ~ || zweistufiger Verdichter m. *oder* Kompressor m. || compresseur m. à deux étages.

compressorless Diesel engine || kompressorloser Dieselmotor m. || moteur m. Diesel sans compresseur. / eight cylinder direct reversible ~ Diesel engine || umsteuerbarer kompressorloser Achtzylinderdieselmotor m. || moteur m. Diesel réversible à injection mécanique à huit cylindres. / ~ marine Diesel engine || kompressorloser Schiffsdieselmotor m. || moteur m. Diesel marin à injection mécanique. / ~ single-acting four-stroke engine with trunk pistons ||kompressorlose einfachwirkende Viertaktmaschine f. mit Tauchkolben || machine f. à quatre temps et à simple effet à injection mécanique et à pistons plongeurs.

compressor oil || Kompressoröl n. || huile f. de compresseur.

compriming engine || Komprimiermaschine f. || machine f. à comprimer.

compromise || Übereinkommen n.; Übereinkunft; Vergleich m. || arrangement m.; convention f. / suggested ~ || Vermittlungsvorschlag m. || proposition f. de conciliation *ou* d'arrangement.

compromised || gütlich geordnet || arrangé; transigé.

compulsion || Zwang m. || contrainte f. / to liquidate under ~ || zwangsweise auflösen || liquider par contrainte f. / ~ to organize || Organisationszwang m. || contrainte f. au syndicat. / under ~ || notgedrungen || contraint par la nécessité.

compulsory administration || Zwangsverwaltung f. || administration f. judiciaire. / ~ alienation || Enteignung f. || expropriation f. / ~ passport system || Paßzwang m. || obligation f. de se munir d'un passeport. / ~ sale by auction || Zwangssteigerung f. || vente f. forcée.

compute, to || schätzen || évaluer.

computer, firing data ~ || Richtwerteberechner m.; Schußwerteberechner m. || compteur m. des données du tir.

computing machine || Rechenmaschine f. || machine f. à calculer.

concave || hohl; konkav || concave. / gilt ~ glass mirror || vergoldeter Glashohlspiegel m. || réflecteur m. concave en verre doré. / ~ lens || Konkavlinse f.; Zerstreuungslinse f. || verre m. divergent *ou* concave.

concave mirror || Hohlspiegel m.; Konkavspiegel m. || miroir m. concave. / gilt ~ || vergoldeter Hohlspiegel m. || réflecteur m. concave doré. / glass ~ || Glashohlspiegel m. || réflecteur m. concave en verre. / parabolic ~ || parabolischer Hohlspiegel m. || miroir m. parabolique concave. / ~ of short focal length || Hohlspiegel m. von kurzer Brennweite || miroir m. concave de foyer court. / ~ of speculum metal || Hohlspiegel m. aus Spiegelmetall || miroir m. concave en métal spéculaire.

concave mirror cathode || Hohlspiegelkathode f. || cathode f. à miroir concave. / ~ spectrometer || Hohlspiegelspektrometer n. || spectromètre m. à miroirs concaves. / ~ spectroscope || Hohlspiegelspektroskop n. || spectroscope m. à miroirs concaves.

concave ophthalmoscope || hohler Augenspiegel m. || ophtalmoscope m. concave.

/ ~ shaping attachment || Konkavhobelvorrichtung f. || dispositif m. à raboter concave. / ~ sound || Hohlsonde f. || sonde f. cannelée.

concavity || Konkavität f. || concavité f.

concede, to || konzessionieren || concéder.

concentrate, to (Chem) || einkochen; eindampfen; eindicken; konzentrieren || évaporer; concentrer. / ~ (Met) || anreichern || concentrer.

concentrated (Chem) || konzentriert; stark concentré. / highly ~ (Chem) || hochgradig à haut degré. / ~ acid || konzentrierte Säure f. || acide m. concentré.

concentrates pl. (Mine) || Schlich m. || schlich m.

concentrating || Eindampfen n. || concentration f. / ~ percussion table (Metal) || Stoßherd m. || table f. à secousses. / ~ table (Tel) || Anruftisch m. || table f. d'appels.

concentration (Acc) || Konzentration f. || concentration f. / ~ (Chem) || Grädigkeit f. || degré m. / ~ (Hydr arch) || Einengung f. || rétrécissement m. / ~ (Ore dress) || Anreicherung f. || enrichissement m. / at the maximum of ~ (Chem) || höchstprozentig || au maximum m. de concentration. / ~ of ore by flotation process || Naßaufbereitung f. der Erze || enrichissement m. des minerais par flottaison. / ~ of trunk lines (Tel) || Zusammenlegung f. von Fernleitungen || concentration f. des circuits interurbains.

concentration machinery || Konzentrationsapparat m. || appareil m. de concentration. / ~ plant || Konzentrationsanlage f. || installation f. de concentration. / ~ plant for sulphuric acid || Schwefelsäurekonzentrationsanlage f. || installation f. de concentration d'acide sulfurique. / ~ position (Tel) || Sammelplatz m. || position f. de concentration. / ~ smelting (Metal) || Konzentrationsschmelzen n. || fondage m. de concentration.

concentrator (Tel) || Anrufschrank m. || tableau m. d'appel. / circular convex ~ (Ore dress) || Rundherd m. || table f. tournante.

concentric || konzentrisch || concentrique. / ~ light diffuser || Lichtverteilungsschirm m. || diffuseur m. de la lumière concentrée. / ~ scale || konzentrische Skale f. || échelle f. concentrique.

concentricity || Rundheit f. || concentricité f.

concern || Konzern m. || groupe m. / ~ for renting films || Filmverleih m. || location f. de films.

concerning || bezüglich || concernant; touchant.

concertina || Konzertino n. || concertino m. / ~ fittings pl. || Harmonikabeschläge mpl. || garnitures fpl. d'harmonicas.

concession || Konzession f.; Genehmigung f. || concession f. / adjoining ~ (Bergb) || Nachbarfeld n. || champ m. voisin. / to alienate the right to a ~ || das Recht auf Mutung veräußern || aliéner le droit de demande. / to apply for a mining ~ || muten || demander la concession. / to apply for several minerals in one ~ || mehrere Mineralien npl. in einer Mutung begehren || viser plusieurs minéraux dans une seule demande. / to decline to grant a ~ || eine Mutung f. abweisen || rejeter une demande de concession. / deed of ~ || Mutungsurkunde f. || acte m. de demande de concession. / extended ~ ||

Geviertfeld n. ‖ champ m. d'exploitation quadraturé *ou* élargé. / to grant ~ ‖ konzessionieren ‖ concéder. / holder of a ~ ‖ Konzessionsinhaber m. ‖ concessionnaire m. / invalidity of the ~ ‖ Ungültigkeit f. der Mutung ‖ invalidité f. de la concession. / longitudinally extended ~ (Min) ‖ gestrecktes Feld n. ‖ champ m. longitudinal. / to make ~ ‖ konzessionieren ‖ concéder. / ~ of a mine ‖ Verleihung f. eines Bergwerks ‖ concession f. d'une mine. / perimeter of a ~ (Mine) ‖ Feld n.; Schlagkreis m. ‖ cercle *ou* périmètre m. embrassé par la (demande de) concession. / plan of the ~ (Mine) ‖ Situationsriß m. der Mutung ‖ plan m. de la demande de concession. / right to a ~ ‖ Mutungsrecht n. ‖ droit m. du requérant. / to withdraw an application for a ~ ‖ eine Mutung f. zurückziehen ‖ retirer une demande de concession.

concessionaire, to oppose the claim of a ~ by means of another concession ‖ dem Anspruch des Muters eine andere Mutung f. entgegenhalten ‖ opposer une autre demande à la prétention du requérant. / prior right of the ~ ‖ Vorzugsrecht n. des Muters ‖ droit m. de priorité du demandeur.

concession plan ‖ Mutungskarte f. ‖ carte f. des demandes de concession. / ~ syndicate ‖ Mutungsgemeinschaft f. ‖ communauté f. pour une demande de concession.

conchoidal ‖ muschelig ‖ conchoïdal.

conclude, to ‖ beschließen ‖ résoudre; décider. / ~ an agreement ‖ einen Vertrag m. schließen ‖ contracter un accord.

conclusion ‖ Beschluß m. ‖ résolution f.; conclusion f. / ~ of contract ‖ Abschluß m. des Vertrages ‖ passation f. du contrat. / false ~ ‖ Trugschluß m. ‖ fausse conclusion f.

concordances pl. (Print) ‖ Konkordanzen fpl. ‖ concordances fpl.

concrete ‖ dicht; fest; kompakt ‖ concret. / ~ number (Math) ‖ benannte Zahl f. ‖ nombre m. concret.

concrete, to (Build) ‖ betonieren ‖ bétonner. / ~ (Cristals) ‖ anschießen; ansetzen ‖ cristalliser.

concrete (Build) ‖ Gußmörtel m.; Beton ‖ béton m. / armoured ~ *see* reinforced ~. / asbestos ~ ‖ Asbestbeton m. ‖ béton m. d'amiante. / ash ~ ‖ Löschebeton m. ‖ béton m. de cendre. / asphalt ~ ‖ Asphaltbeton m. ‖ béton m. d'asphalte. / brick ~ ‖ Ziegelbeton m. ‖ béton m. de briquailles. / broken ~ ‖ Betonbruch m. ‖ béton m. concassé. / broken stone ~ ‖ Steinschlagbeton m. ‖ béton m. de pierraille. / cast ~ *see* floated ~. / cement ~ ‖ Zementbeton m. ‖ béton m. de ciment. / slowsetting cement ~ ‖ Zementkalkbeton m.; verlängerter Zementbeton m. ‖ béton m. de ciment et de chaux. / compressed ~ *see* rammed ~. / dry ~ ‖ wasserarmer Beton m. ‖ béton m. peu mouillé. / to encase with ~ *see* to imbed in ~. / floated ~ ‖ Gußbeton m. ‖ béton m. coulé. / granite ~ ‖ Granitbeton m. ‖ béton m. de granite. / gravel ~ ‖ Kiesbeton m. ‖ béton m. de gravier. / gypsum ~ ‖ Gipsbeton m. ‖ béton m. de plâtre. / heaped ~ ‖ Schüttbeton m. ‖ béton m. coulé. / to imbed in ~ ‖ in Beton m. einbetten ‖ enrober *ou* encastrer dans le béton. / to lay ~ under

water ‖ Beton m. unter Wasser schütten ‖ couler du béton m. sous l'eau. / to let into ~ *see* to imbed in ~. / lime ~ ‖ Kalkbeton m. ‖ béton m. de chaux. / cement ~ diluted with lime *see* slowsetting cement ~. / meagre ~ ‖ magerer Beton m. ‖ béton m. maigre. / moist ~ ‖ erdfeuchter Beton m. ‖ béton m. à consistance de terre humide. / plaster ~ ‖ Gipsbeton m. ‖ béton m. de plâtre. / preparation of ~ ‖ Betonbereitung f. ‖ préparation f. du béton. / articles pl. of pressed ~ ‖ Waren fpl. aus gepreßtem Beton ‖ ouvrages mpl. en béton comprimé. / pumice-stone ~ ‖ Bimsbeton m. ‖ béton m. de pierre ponce. / to ram with ~ ‖ mit Beton m. ausstampfen ‖ damer avec du béton m. / rammed ~ ‖ gestampfter Beton m. ‖ béton m. comprimé *ou* damé. / reinforced ~ ‖ Eisenbeton m. ‖ béton m. armé. / skeleton of reinforced ~ ‖ Betongerippe n. ‖ ossature f. en béton. / rendering ~ ‖ Spritzbeton m. ‖ crépi m. en béton. / rich ~ ‖ fetter Beton m. ‖ béton m. gras. / to set in ~ ‖ in Beton m. einbetten ‖ encastrer en béton m. / slag ~ ‖ Schlackenbeton m. ‖ béton m. de scorie *ou* de laitier. / steel ~ ‖ Stahlbeton m. ‖ béton m. en acier. / stone ~ ‖ Steinbeton m. ‖ béton m. de pierre *ou* de cailloux. / wet ~ ‖ wasserreicher Beton m. ‖ béton m. très mouillé. / to work ~ ‖ betonieren ‖ bétonner. / to work ~ in frosty weather ‖ bei Frost m. betonieren ‖ bétonner pendant la gelée.

concrete arch ‖ Betongewölbe n. ‖ voûte f. en béton. / ~ articles pl. ‖ Betonwaren fpl. ‖ objets mpl. en béton. / ~ bed ‖ Betonbett n.; Betonunterlage f. ‖ lit m. *ou* couche f. de béton. / ~ block ‖ Betonblock m. ‖ bloc m. de béton. / ~ bridge ‖ Betonbrücke f. ‖ pont m. de béton. / ~ conduit ‖ Betonrohr n. ‖ tuyau m. en béton. / ~ construction ‖ Betonbau m. ‖ construction f. en béton. / ~ foundation ‖ Betonfundament n.; Betongründung f.; Zementsockel m. ‖ fondation f. en béton; bétonnage m.; socle m. en béton. / ~ hinge ‖ Betongelenk n. ‖ articulation f. en béton. / ~ iron ‖ Betoneisen n. ‖ fer m. rond *ou* à béton. / ~ iron construction ‖ Betoneisenkonstruktion f. ‖ construction f. de béton-fer. / ~ kneading machine ‖ Betonknetmaschine f. ‖ machine f. à pétrir le béton. / ~ layer ‖ Betonschüttung f. ‖ coulage m. de béton; remblai m. en béton. / ~-made furniture ‖ Betonmöbel n. ‖ meuble m. en béton. / ~ mixer ‖ Betonmischer m.; Betonmischmaschine f. ‖ machine f. à mélanger le béton; malaxeur m. à béton; bétonnière f. / ~ mixing machine *see* ~ mixer. / ~ paving ‖ Betonstraßendecke f. ‖ pavage m. en béton. / ~ pile ‖ Betonpfahl m. ‖ pilier m. *ou* pieu m. en béton. / ~ pillar *see* ~ pile. / ~ pipe ‖ Betonrohr n. ‖ tuyau m. en béton. / ~ plant ‖ Betonbereitungsanlage f. ‖ installation f. pour préparer du béton. / ~ pole (El line) ‖ Betonmast m., Schleudermast m. ‖ poteau m. en béton; pylône m. en ciment armé.

concreter ‖ Betonarbeiter m. ‖ bétonneur m.

concrete road ‖ Betonstraße f. ‖ route f. bétonnée. / surface of the ~ cover ‖ Oberschicht f. der Betonstraßendecke ‖ couche f. supérieure du revêtement de la

route en béton. / machine for the construction of ~s ‖ Maschine f. für den Betonstraßenbau ‖ machine f. pour la construction des routes en béton. / ~ finisher ‖ Straßendeckenfertiger m. ‖ finisseur m. pour revêtements de routes. / ~ surface padded down ‖ gedichtete Betonstraßendecke f. ‖ revêtement m. de route en béton damé et comprimé. / ~ surface true to profile ‖ profilgerechte Betonstraßendecke f. ‖ revêtement m. de route en béton d'un profil bien dressé. / ~ surface without undulations ‖ wellenfreie Betonstraßendecke f. ‖ revêtement m. de route en béton exempt d'ondulations.

concrete safe ‖ Geldschrank m. aus Beton ‖ coffre-fort m. en béton. / ~ slab ‖ Betonplatte f. ‖ plaque f. en béton. / hollow ~ slab ‖ Zementhohldiele f. ‖ dalle f. creuse en ciment. / ~ stairs pl. ‖ betonierte Treppe f. ‖ escalier m. bétonné. / ~ steel ‖ Betoneisen n. ‖ fer m. à béton. / ~ stone ‖ Betonstein m. ‖ pierre f. en béton. / hollow ~ stone ‖ Betonhohlstein m. ‖ pierre f. creuse en béton. / ~ testing press for compression, buckling and bending tests ‖ Betonprüfpresse f. für Druck-, Knick- und Biegeversuche ‖ presse f. à essayer le béton à la compression, à la brisure et à la flexion. / ~ vault ‖ Betongewölbe n. ‖ voûte f. en béton. / ~ window frame ‖ Fensterrahmen m. aus Beton ‖ châssis m. de fenêtre en béton. / ~ work ‖ Betonieren n. ‖ travail m. en béton.

concretion (Geol) ‖ Konkretion f.; Zusammenwachsen n. ‖ concrétion f. / spheroidal ~ (Miner) ‖ Niere f. ‖ rognon m.; globule m. oblong. / spheroidal ~ of marl ‖ Mergelniere f. ‖ marne f. sphéroïdale *ou* cloisonnée.

concussion ‖ Aufprall m.; Erschütterung f. / secousse f.; choc m.; trépidation f. / ~ spring ‖ Stoßdämpfer m. ‖ ressort m. amortisseur. / ~ test ‖ Schlagversuch m. ‖ essai m. au choc.

condemn, to ~ a ship ‖ ein Schiff n. kondemnieren ‖ condamner un navire.

condensability (Chem) ‖ Verdichtbarkeit f. ‖ condensabilité f.

condensable ‖ verdichtbar ‖ condensable.

condensate ‖ Kondensat n. ‖ condensé m. / ~ container ‖ Kondensationsgefäß n. ‖ récipient m. de condensation.

condensation ‖ Verdichtung f.; Verflüssigung f.; Kondensation ‖ condensation f. / counter-flow ~ ‖ Gegenstromkondensation f. ‖ condensation f. à courants de sens contraire; condensation f. à contrecourant. / effect of the ~ ‖ Kondensationsleistung f. ‖ effet m. de la condensation. / to increase the effect of the ~ ‖ die Leistung der Kondensation erhöhen ‖ augmenter l'effet de la condensation. / heat of ~ ‖ Kondensationswärme f. ‖ chaleur f. de condensation. / ~ by injection ‖ Einspritzkondensation f. ‖ condensation f. par injection *ou* par mélange. / jet ~ ‖ Einspritzkondensation f. ‖ condensation f. par mélange *ou* par injection. / liquid product of ~ ‖ flüssiges Kondensationsergebnis n. ‖ produit m. liquide de condensation; liquide m. résultant de la condensation. / loss due to ~ ‖ Kondensationsverlust m. ‖ perte f. par condensation. / ~ by mixing ‖ Mischkondensation f. ‖ condensation f. par mé-

lange. / parallel-flow ~ ‖ Gleichstromkondensation f. ‖ condensation f. à courants de même sens. / saving of steam due to ~ ‖ Dampfersparnis f. durch Kondensation ‖ économie f. de vapeur par condensation. / surface ~ ‖ Oberflächenkondensation f. ‖ condensation f. à surface. / water of ~ ‖ Kondenswasser n. ‖ eau f. condensée ou de condensation.

condensation chamber ‖ Abscheidekammer f.; Kondensationskammer f. ‖ chambre f. de condensation. / ~ plant ‖ Kondensationsanlage f. ‖ installation f. de condensation. / central ~ plant ‖ Zentralkondensationsanlage f. ‖ installation f. de condensation centrale. / ~ product ‖ Kondensat n. ‖ produit m. de condensation. / ~ space ‖ Kondensationsraum m.; Niederschlagraum m. ‖ chambre f. de condensation.

condensation water ‖ Kondenswasser n. ‖ eau f. de condensation. / ~ filter ‖ Kondenswasserfilter n. ‖ filtre m. à eau de condensation. / ~ remover ‖ Kondenswasserableiter m. ‖ conduite f. ou purgeur m. d'eau de condensation.

condense, to ‖ kondensieren; verdichten ‖ condenser. / ~ (Chem) ‖ sich verdichten ‖ se condenser. / ~ water vapour ‖ Wasserdampf m. ausscheiden ‖ condenser la vapeur d'eau.

condensed milk ‖ kondensierte Milch f.; Büchsenmilch f. ‖ lait m. condensé. / ~ manufactory ‖ Milchkondensieranstalt f. ‖ fabrique f. de lait condensé.

condensed steam pump ‖ Kondensatpumpe f. ‖ pompe f. à vapeur condensée.

condenser (Chem) ‖ Kühler m. ‖ réfrigérant m. / ~ (Electr; Mach) ‖ Kondensator m.; Verflüssiger m.; Verdichter m. ‖ condenseur m.; condensateur m. / adjustable ~ (Radio) ‖ variabler Kondensator m.; Drehkondensator m. ‖ condensateur m. réglable. / adjustable ~ with delicate regulation (Radio) ‖ Drehkondensator m. mit Feineinstellung ‖ condensateur m. variable à air avec réglage micrométrique. / air ~ ‖ Luftkühler m. ‖ réfrigérant m. à air. / aspiration of the ~ ‖ Saugkraft f. des Kondensators ‖ force f. d'aspiration du condenseur. / atmospheric ~ with horizontal pipes ‖ Berieselungsverflüssiger m. mit liegenden Rohren ‖ condenseur m. de ruissellement à tuyaux horizontaux. / atmospheric ~ with vertical tubes ‖ Berieselungsverflüssiger m. mit stehenden Rohren ‖ condenseur m. de ruissellement à tuyaux verticaux. / auxiliary ~ ‖ Hilfskondensator m. ‖ condenseur m. auxiliaire. / back-flow ~ ‖ Rückflußkühler m. ‖ réfrigérant m. à reflux. / bulb ~ ‖ Kugelkühler m. ‖ réfrigérant m. à boules. / centrifugal ~ ‖ Kreiselkondensator m. ‖ condenseur m. centrifuge. / closed surface ~ ‖ geschlossener Oberflächenkondensator m.; Oberflächengefäßkondensator m. ‖ condenseur m. à surface fermé. / ~ combined with the turbine ‖ in die Turbine eingebauter Kondensator m. ‖ condenseur m. monté dans la turbine. / ~ by contact ‖ Röhrenkondensator m. ‖ condenseur m. tubulaire. / counter-current ~ ‖ Gegenstromkondensator m. ‖ condenseur m. à courants de sens contraire; condenseur m. à contre-courant. / double-pipe ~ ‖ Doppelrohrverflüssiger m. ‖ condenseur m. à doubles tubes. /

ejector ~ ‖ Strahlkondensator m. ‖ condenseur m. à jet. / external ~ ‖ Röhrenkondensator m. ‖ condenseur m. tubulaire. / ~ independent of turbine ‖ von der Turbine getrennter Kondensator m. ‖ condenseur m. séparé de la turbine. / injection ~ see jet ~. / interchangeable ~ ‖ Wechselkondensor m. ‖ condensateur m. interchangeable ou alternatif. / intermediate circuit ~ (Radio) ‖ Kondensator m. im Zwischenkreis ‖ condensateur m. du circuit intermédiaire. / jet ~ ‖ Einspritzkondensator m. ‖ condenseur m. à injection ou par mélange ou à jet. / luminous spot ring ~ ‖ Leuchtbildkondensor m. ‖ condensateur m. à images lumineuses. / mixing ~ ‖ Mischkondensator m. ‖ condenseur m. à mélange. / parallel current ~ ‖ Gleichstromkondensator m. ‖ condenseur m. à courants de même sens. / plate ~ ‖ Plattenkondensator m. ‖ condensateur m. à plateaux. / preliminary ~ ‖ Vorkondensator m. ‖ condenseur m. préalable. / quenching ~ ‖ Löschkondensator m. ‖ condensateur m. étouffant. / reflux ~ ‖ Rückflußkühler m. ‖ réfrigérant m. à reflux. / secondary ~ (Mach) ‖ Nachkondensator m. ‖ condenseur m. auxiliaire. / secondary circuit ~ (Radio) ‖ Kondensator m. im Sekundärkreis ‖ condensateur m. du circuit secondaire. / shorting ~ (Radio) ‖ Verkürzungskondensator m. ‖ condensateur m. de raccourcissement. / spherical ~ ‖ Kugelkondensator m. ‖ condensateur m. sphérique. / ~ without steps ‖ stufenloser Kondensator m. ‖ condensateur m. sans étages. / submerged ~ ‖ Tauchkondensator m. ‖ condenseur m. à immersion. / standpipe for ~ suction ‖ Kondensatorsaugstutzen m. ‖ tubulure f. d'aspiration de condenseur. / surface ~ (Brew) ‖ Flächenberieselungskondensator m. ‖ condenseur m. à ruissellement à surface. / surface ~ (Steam) ‖ Oberflächenkondensator m. ‖ condenseur m. tubulaire. / swing-out ~ ‖ ausklappbarer Kondenser m. ‖ condensateur m. s'écartant. / tuning ~ (Radio) ‖ Abstimmkondensator m. ‖ condensateur m. de syntonisation. / twin-coupled ~ ‖ doppeltgeschalteter Kondensator m. ‖ condensateur m. jumelé. / two-lens ~ ‖ zweilinsiger Kondensator m. ‖ condensateur m. à deux lentilles. / tubular ~ ‖ Röhrenkondensator m. ‖ condenseur m. tubulaire. / universal ~ ‖ Universalkondensator m. ‖ condenseur m. universel. / variable ~ (Radio) ‖ variabler Kondensator m. ‖ condensateur m. variable ou réglable. / wet ~ ‖ nasser Kondensator m. ‖ condenseur m. humide. / wound ~ ‖ Wickelkondensator m. ‖ condensateur m. enroulé (de feuilles d'étain).

condenser agitator ‖ Kondensatorrührwerk n. ‖ agitateur m. du condenseur. / air pump ‖ Kondensatorluftpumpe f. ‖ pompe f. à air du condenseur. / ~ armature ‖ Kondensatorbelegung f. ‖ armature f. de condensateur. / ~ bed ‖ Kondensatorbett n. ‖ lit m. du condensateur. / ~ body ‖ Kondensatorkörper m. ‖ corps m. du condenseur. / ~ capacity ‖ Kondensatorkapazität f. ‖ capacité f. du condensateur. / ~ chamber ‖ Kondensatorkammer f. ‖ chambre f. de condenseur. / circuit ‖ Kondensatorkreis m. ‖ circuit m. du condensateur. / ~ cock ‖ Konden-

satorhahn m. ‖ robinet m. du condenseur. / ~ foundation ‖ Kondensatorfundament n. ‖ fondation f. de condenseur. / ~ lens ‖ Kondensorlinse f.; Beleuchtungslinse f. ‖ lentille f. de condensateur. / ~ lightning-arrester ‖ Kondensatorblitzschutzvorrichtung f. ‖ parafoudre m. du condensateur. / ~ mirror ‖ Kondensorspiegel m. ‖ miroir-condensateur m. / ~ pipe ‖ Kondensatorrohr n. ‖ tuyau m. de condenseur. / ~ piping ‖ Kondensatorrohrleitung f. ‖ conduite f. du condenseur. / ~ plant ‖ Verflüssigeranlage f. ‖ installation f. de condensation. / ~ plant of the turbo set ‖ Kondensationsanlage f. zum Turbosatz ‖ installation f. de condensation pour le groupe turbogénérateur. / ~ pressure ‖ Kondensatordruck m. ‖ pression f. dans le condenseur. / ~ receiver ‖ Kondensatorkasten m.; Kondensationsgefäß n. ‖ récipient m. de condenseur. / ~ reel for switching device (Tel) ‖ Anschlußrolle f. des Anschaltgeräts ‖ condensateur m. de prise de courant pour l'appareil d'écoute. / ~ section (Tel) ‖ Kondensatorabschnitt m. ‖ section f. de condensateur. / ~ system ‖ Kondensatorsystem n. ‖ système m. de condensateurs. / ~ tank ‖ Kondensatorgefäß n. ‖ réservoir m. du condenseur. / ~ temperature ‖ Kondensatortemperatur f. ‖ température f. du condenseur. / ~ test set ‖ Kondensatorprüfer m. ‖ indicateur m. de rendement pour condenseurs. / ~ transmitter ‖ Kondensatormikrofon n. ‖ microphone m. à condensateur. / ~ tube ‖ Kühlrohr n. ‖ tube m. de condenseur.

condensing of a text (Print) ‖ Zusammenziehung f. eines Textes ‖ condensation f. d'un texte. / to work ~ ‖ auf Kondensation f. arbeiten ‖ marcher à condensation f. / to work non-~ ‖ mit Auspuff m. arbeiten ‖ marcher à échappement libre.

condensing box ‖ Kondenstopf m. ‖ séparateur m. d'eau de condensation. / ~ capacity ‖ Kondensiervermögen n. ‖ pouvoir m. condensatif. / ~ chamber ‖ Kondensationsraum m. ‖ chambre f. de condensation. / ~ coil ‖ Kühlschlange f. ‖ serpentin m. refroidisseur. / ~ force ‖ Niederschlagsvermögen n. ‖ force f. condensante. / ~ lens ‖ Sammellinse f.; Konvexlinse f. ‖ lentille f. convergente. / ~ plant ‖ Kondensationsanlage f. ‖ installation f. de condensation. / ~ plant with air jet pump ‖ Kondensationsanlage f. mit Strahlluftpumpe ‖ appareil m. condensateur avec pompe centrifuge de circulation et éjecteur condenseur. / ~ steam engine ‖ Kondensationsdampfmaschine f. ‖ machine f. à vapeur à condensation. / ~ tower ‖ Kondensationsturm m. ‖ tour f. de condensation. / ~ tub ‖ Kondensationsgefäß n. ‖ bonbonne f. de condensation / two cylinder ~ turbine ‖ zweigehäusige Kondensationsturbine f. ‖ turbine f. à condensation à deux corps. / ~ vessel ‖ Kondensationsgefäß n.; Niederschlaggefäß n. ‖ récipient m. ou vaisseau m. de condensation. / ~ water see also condensation water ‖ Kondenswasser n. ‖ eau f. de condensation. / ~ water reconductor ‖ Kondenswasserrückleiter m. ‖ reconducteur m. d'eau de condensation. / apparatus for returning the ~ water ‖

Kondenswasserrückspeiseanlage f. ‖ installation f. alimentaire à recouvrement de l'eau de condensation. / ~ worm ‖ Kühlschlange f. ‖ serpentin m. réfrigérant.

condiment ‖ Genußmittel n.; Würzmittel n. ‖ denrée f. alimentaire; condiment m.

condition ‖ Bedingung f. ‖ condition f. / ~s pl. of acceptance ‖ Übernahmebedingungen fpl. ‖ conditions fpl. de l'acceptation / the ~s are adverse ‖ die Bedingungen fpl. sind ungünstig ‖ les conditions fpl. sont défavorables. / ~ of aggregation ‖ Aggregatzustand m. ‖ état m. d'agrégation. / ~s pl. of auction ‖ Versteigerungsbedingungen fpl. ‖ conditions fpl. de la vente aux enchères. / bad ~ ‖ Schadhaftigkeit f. ‖ mauvais état m. / change of ~ ‖ Zustandsänderung f. ‖ changement m. d'état. / ~ for connecting ‖ Anschlußbedingung f. ‖ condition f. de jonction. / ~s pl. of contract ‖ Vertragsbedingungen fpl. ‖ conditions fpl. de contrat. / ~s pl. of delivery ‖ Bezugsbedingungen fpl. ‖ conditions fpl. de livraison. / ~ of equilibrium ‖ Gleichgewichtsbedingung f. ‖ condition f. d'équilibre. / ~ of a furnace (Metal) ‖ Ofengang m.; Gang m. eines Ofens ‖ marche f. ou allure f. d'un fourneau. / ~ of the grain ‖ Getreidebeschaffenheit f. ‖ composition f. des grains. / to make ~s ‖ Bedingungen fpl. aufstellen ‖ établir des conditions fpl. / ~ of operation ‖ Betriebsbedingung f. ‖ condition f. de fonctionnement. / ~ of payment ‖ Zahlungsbedingung f. ‖ condition f. de paiement. / ~ of sale ‖ Verkaufsbedingung f. ‖ condition f. de vente. / ~ of test ‖ Prüfungsvorschrift f. ‖ condition f. d'essai.

conditioner, wheat ~ ‖ Weizenvorbereiter m. ‖ conditionneur m. à blé.

conditioning apparatus ‖ Konditionierapparat m. ‖ appareil m. de conditionnement.

condolatory card ‖ Beileidskarte f. ‖ carte f. de condoléance.

conduct, to ‖ leiten ‖ conduire. / ~ the current (Electr) ‖ Strom m. führen ‖ être sous tension f. / ~ heat ‖ Wärme f. leiten ‖ conduire la chaleur.

conduct of affairs ‖ Geschäftsführung f. ‖ conduite f. des affaires. / ~ of fermentation ‖ Gärführung f. ‖ conduite f. de la fermentation.

conductance (Electr) ‖ Wirkleitwert m.; Leitfähigkeit f. ‖ conductance f.; conductivité f.

conductibility ‖ Leitfähigkeit f.; Leitvermögen n. ‖ conductibilité f. / electric ~ ‖ elektrische Leitfähigkeit f. ‖ conductibilité f. électrique. / ~ of the ground ‖ Erdleitungsvermögen n. ‖ conductibilité f. du sol. / heat ~ ‖ Wärmeleitung f. ‖ conductibilité f. calorifique ou de chaleur.

conducting, to render ~ ‖ leitend machen ‖ rendre conducteur. / ~ cord ‖ Leitungslitze f. ‖ cordon m. conducteur. / ~ salt ‖ Leitsalz n. ‖ sel m. conducteur. / ~ surface ‖ Leitungsfläche f. ‖ surface f. conductrice. / ~ wire ‖ Drahtleitung f.; elektrischer Leitungsdraht m. ‖ fil m. conducteur.

conduction (Electr) ‖ Übertragung f. ‖ conduction f. / ~ current (Electr) ‖ Leitungsstrom m. ‖ courant m. de conduction; courant conduit.

conductivity ‖ Leitfähigkeit f.; Leitungsfähigkeit f. ‖ conductivité f.; conductibilité f. / low ~ ‖ geringe Leitfähigkeit f. ‖ faible conductivité f. / thermal ~ ‖ Wärmeleitfähigkeit f. ‖ conductibilité f. calorifique. / unilateral ~ ‖ einseitige Leitfähigkeit f. ‖ conductance f. unilatérale.

conductivity vessel ‖ Leitfähigkeitsgefäß n. ‖ boîte f. de conductibilité. / ~ water ‖ Leitfähigkeitswasser n. ‖ eau f. de conductivité.

conductor (Elektr) ‖ Leiter m. ‖ conducteur m. / acidproof ~ ‖ säurefeste Leitung f. ‖ ligne f. résistant aux acides. / ~ of the balloon ‖ Ballonführer m. ‖ pilote m. du dirigeable. / bare ~ ‖ blanker Leiter m. ‖ conducteur m. nu. / battery switch ~ (Acc) ‖ Zellenschalterleitung f. ‖ conducteur m. de réducteur-adjoncteur. / ~ of the cable ‖ Kabelseele f. ‖ âme m. du câble. / ~ carrying current ‖ stromführender Leiter m. ‖ conducteur m. parcouru par un courant. / charged ~ ‖ stromführender Leiter m. ‖ conducteur m. chargé. / copper ~ ‖ Kupferleitung f. ‖ conducteur m. en cuivre. / ~ of electricity ‖ Elektrizitätsleiter m. ‖ conducteur m. d'électricité. / first-class ~ ‖ Leiter m. erster Klasse ‖ conducteur n. de premier chaix. / ~ of heat ‖ Wärmeleiter m. ‖ conducteur m. de la chaleur. / ~ for moist places ‖ Feuchtraumleitung f. ‖ fil m. conducteur pour locaux humides. / poor ~ ‖ schlechter Leiter m. ‖ corps m. de faible conductivité. / return ~ ‖ Rückleiter m.; Rückleitung f. ‖ conducteur m. ou ligne f. de retour. / rubber-insulated ~ ‖ Gummiader f.; Gummiaderleiter m. ‖ fil m. isolé au acoutchouc. / second-class ~ ‖ Leiter m. zweiter Klasse ‖ conducteur m. de deuxième ou de seconde classe. / single ~ ‖ Einfachleitung f. ‖ ligne f. simple. / solid ~ ‖ körperlicher Leiter m. ‖ conducteur m. solide. / solid copper ~ ‖ massiver Kupferstableiter m. ‖ conducteur m. en cuivre plein. / triple ~ ‖ Dreifachleitung f.; Drillingsleiter m. ‖ ligne f. ou conducteur m. triple. / twin ~ ‖ Doppelleiter m.; Doppelleitung f. ‖ conducteur m. jumelé; ligne f. double. / varnished ~ ‖ Lackader f. ‖ conducteur m. verni. / weatherproof ~ ‖ wetterfeste Leitung f. ‖ ligne f. résistant aux intempéries.

conductor rail ‖ Leitungsschiene f. ‖ rail m. conducteur. / ~ wire ‖ Leitungsdraht m. ‖ fil m. conducteur.

conduit ‖ Röhre f.; Rohrleitung f. ‖ tube m.; conduite f. / ~ (Electr) ‖ Leitungsrohr n. ‖ tube m. / ~ (Hydr arch) ‖ Rohrnetz n. ‖ conduite f. de tuyaux; tuyautage m. / ~ (Insulating tube) ‖ Isolierrohr n. ‖ tube m. isolant. / air ~ (Mine) ‖ Lutte f. ‖ buse f. d'aérage. / armoured insulated ~ ‖ Isolierrohr n. mit Eisenarmierung ‖ tube m. isolateur armé d'acier ou de fer. / brass ~ ‖ Messingisolierrohr n. ‖ tube m. isolant de laiton. / butt-jointed ~ ‖ stumpfgestoßenes Isolierrohr n. ‖ tube m. isolant maté. / cement ~ ‖ Zementrohr n. ‖ tube m. en ciment. / charging ~ ‖ Schüttkanal m. ‖ conduit m. à charbon ou d'alimentation ou de chargement. / ~ with closed seam ‖ Isolierrohr n. mit geschlossener Naht ‖ tube m. isolant à couture fermée. / cooling ~ ‖ Kühlgraben m. ‖ caniveau m. réfrigérant. / ~ of earthenware troughing ‖ Steinzeugrinne f. ‖ caniveau m. en

argile cuite. / fireproof ~ ‖ feuerbeständiges Isolierrohr n. ‖ tube m. isolant réfractaire. / flexible seamless ~ with socket ‖ biegsames nahtloses Rohr n. mit Muffe ‖ tube m. flexible sans soudure avec manchon. / guide ~ ‖ Umführungskanal m. ‖ conduit m. ou canal m. annullaire. / insulating ~ ‖ Isolationskanal m. ‖ conduit m. ou canal m. isolant. / ~ with lead-covered iron lining ‖ Isolierrohr n. mit verbleitem Eisenüberzug ‖ tube m. isolant armé de tôle plombée. / ~ for leading-in cables (Tel) ‖ Einführungskanal m. für unterirdische Hauseinführungen ‖ conduite f. pour l'introduction de câbles dans un immeuble. / ~ with open seam ‖ Isolierrohr m. mit offener Naht ‖ tube m. isolant à fente. / paper ~ ‖ Papierisolierrohr n. ‖ tube m. isolant en papier. / ~ of pipes ‖ Rohrleitung f.; Röhrenleitung f. ‖ conduite f. de tuyaux. / ~ under pressure ‖ Druckleitung f. ‖ conduite f. forcée ou de refoulement. / screwed ~ ‖ Rohrverschraubung f. ‖ assemblage m. par tube fileté. / steel-armoured ~ ‖ Stahlpanzerrohr n. ‖ tube m. isolant armé d'acier; tube m. protecteur en acier. / ~ for ventilation ‖ Lüftungskanal m. ‖ canal m. de ventilation. / ~ of water ‖ Wasserleitung f. ‖ conduite f. d'eau.

conduit box, angle ~ ‖ Winkelabzweigdose f. ‖ boîte f. de dérivation d'angle ou d'équerre. / cast iron ~ ‖ Abzweigdose f. aus Gußeisen ‖ boîte f. de dérivation en fonte. / porcelain ~ ‖ Abzweigdose f. aus Porzellan ‖ boîte f. de dérivation en porcelaine. / watertight ~ ‖ wasserdichte Abzweigdose f. ‖ boîte f. de dérivation étanche d'eau.

conduit cable ‖ Röhrenkabel n. ‖ câble m. à tuyau. / ~ fittings pl. ‖ Isolierrohrzubehör m. ‖ garnitures fpl. de tubes isolants. / ~ pipe ‖ Leitungsrohr n. ‖ tuyau m. de conduite. / ~ system ‖ Isolierrohrsystem n. ‖ système m. de tubes isolants. / ~ tube ‖ Isolierrohr n. ‖ tube m. isolant. / ~ wire ‖ Rohrdraht m. ‖ fil m. dans tubes.

condurango root ‖ Kondurangorinde f. ‖ racine f. de condurango.

cone (Bot) ‖ Dolde f. ‖ cône m.; ombelle f. / ~ (Geol) ‖ Gesteinskuppe f.; dôme m. / ~ (Math) ‖ Kegel m. ‖ cône m. / adjustable ~ ‖ verstellbare Düse f. ‖ tuyère f. réglable ou ajustable. / alluvial ~ ‖ Schuttkegel .m; Schutthalde f. ‖ cône m. de déjection. / axis of the ~ ‖ Kegelachse f. ‖ axe m. du cône. / blast furnace ~ ‖ Gichtglocke f. ‖ cloche f. de hautfourneau. / breaking ~ ‖ Brechkegel m. ‖ cône m. de concassage. / clutch ~ ‖ Kupplungskegel m. ‖ cône m. d'accouplement ou d'embrayage. / corrugated grinding ~ ‖ geriffelter Mahlkonus m. ‖ cône m. broyeur cannelé. / ~ of debris (Geol) ‖ Aufschüttungskegel m. ‖ cône m. de débris. / delivery ~ ‖ Druckdüse f. ‖ tuyère f. de refoulement. / discharge ~ see delivery ~. / ~ of light ‖ Lichtkegel m. ‖ cône m. de lumière. / ~ of light rays ‖ Lichtkegel m. ‖ cône m. lumineux. / melting ~ ‖ Brennkegel m.; Schmelzkegel m. ‖ cône m. pyrométrique. / oblique ~ ‖ schiefer Kegel m. ‖ cône m. oblique. / ~ of rays ‖ Strahlenbündel n.; Strahlenkegel m. ‖ cône m. de rayons. / right ~ ‖ gerader Kegel m. ‖ cône m.

droit *ou* vertical. / scalene ~ ‖ schiefer
Kegel m. ‖ cône m. scalène *ou* oblique. /
Seger's ~ ‖ Segerkegel m. ‖ montre f. de
Seger. / shape of ~ ‖ kegelige Gestalt f. ‖
forme f. conique. / spindle ~ ‖ Achsen-
kegel m. ‖ cône m. d'axe. / steam ~ ‖
Dampfdüse f. ‖ tuyère f. à vapeur. / taper
of the ~ ‖ Anzug m. *oder* Steigung f. des
Kegels; Konizität f. ‖ conicité f.; fruit m.
du cône. / truncated ~ ‖ Kegelstumpf m.
‖ cône m. tronqué. / upright ~ *see*
right ~. / ~ of valve ‖ Ventilkegel m. ‖
cône m. de soupape. / volcanic ~ ‖ vulkanischer Kegel m. ‖ cône m. volcanique.
cone bearing ‖ Kegellager n. ‖ coussinet m.
à cône. / ~ block (Mach) ‖ Ausricht-
keil m. ‖ cale f. de réglage. / ~ brake ‖
Kegelbremse f. ‖ frein m. à cône de
friction. / ~ clutch ‖ Kegelkupplung f. ‖
embrayage m. à cône. / ~ coupling ‖
Kegelkupplung f. ‖ accouplement m.
par cônes. / ~ mill ‖ Glockenmühle f.;
Konusmühle f. ‖ moulin m. à cône;
broyeur m. à cloche. / ~ pulley ‖ Stufenscheibe f. ‖ poulie f. de transmission à
cônes. / ~ pulley drive ‖ Antrieb m. durch
Stufenscheibe ‖ commande f. par cône.
/ ~ stem (Bot) ‖ Doldenstiel m. ‖ tige f.
de cloche. / ~ strig (Bot) ‖ Dolden-
spindel f. ‖ pédoncule m.; pétiole m.
/ ~ winder (Weav) ‖ Weftgarnspuler m. ‖
épeuleur m.
confection (Cloth) ‖ Konfektionsware f. ‖
confection f. / ~ (Sugar) ‖ Konfitüre f. ‖
confiture f.
confectioner ‖ Konditor m. ‖ confiseur m.
/ ~'s oven ‖ Konditoreibackofen m. ‖
four m. de pâtissier. / machine for ~'s
shop ‖ Maschine f. für Konditoreien ‖
machine f. pour pâtissiers. / ~'s supplies
pl. ‖ Konditoreibedarfsartikel mpl. ‖
fournitures fpl. pour confiseurs.
confectionery (Work) ‖ Konditorei f.; Bon-
bonfabrik f. ‖ confiserie f. / ~ (Ware) ‖
Zuckerwaren fpl.; Konditorwaren fpl.;
Konfekt n. ‖ confiserie f.; sucrerie f.;
produits mpl. de la pâtisserie. / ~ colour
‖ Konditorfarbe f. ‖ couleur f. pour la
confiserie. / ~ machine ‖ Konditoreimaschine f. ‖ machine f. pour confiseries
ou pour pâtisseries. / ~ manufacture ‖
Zuckerwarenfabrik f. ‖ fabrication f.
de confiserie en gros. / ~ sole ‖ Konditorherd m. ‖ four m. pâtissier.
confer, to ‖ konferieren ‖ conférer / ~
a favour ‖ eine Vergünstigung f. erweisen ‖ accorder un avantage m. / ~
power of attorney ‖ Vollmacht f. erteilen
‖ donner pouvoir m.
conference ‖ Konferenz f. ‖ conférence f.
/ disarmament ~ ‖ Abrüstungskonferenz
f. ‖ conférence f. du désarmement.
confetti ‖ Konfetti n. ‖ confetti m.
confidence, to misuse ~ ‖ das Vertrauen n.
mißbrauchen ‖ abuser de la confiance.
confident ‖ zuversichtlich ‖ confiant.
configuration of coast ‖ Küstengestaltung
f. ‖ configuration f. de la côte. / ~ of
flow ‖ Stromlinienbild n. ‖ image f. *ou*
représentation f. des filets fluides. / ~s
pl. of flow ‖ Strömungsgebilde n. ‖ image
f. d'écoulement; lignes fpl. de courant.
confine, to ‖ begrenzen; einengen ‖ confiner; limiter; resserrer. / ~ by dikes ‖
eindämmen ‖ diguer; enfermer d'une
digue.
confined space liable to contain explosive
mixtures ‖ Betriebsraum m. zur Unter-

bringung von Sprengstoffen ‖ local m.
industriel où l'on peut loger des matières explosives.
confirm, to ~ a report ‖ eine Nachricht f.
bestätigen ‖ confirmer une nouvelle.
confirmation of private rights ‖ Privat-
berechtigung f. ‖ autorisation f. accordée à un particulier.
confiscate, to ‖ konfiszieren ‖ confisquer;
saisir.
confiscation ‖ Beschlagnahme f. ‖ confiscation f.; saisie f.
confiture ‖ Konfitüre f. ‖ confiture f.
conflagration ‖ Brand m. ‖ conflagration
f.; incendie m.; embrasement m.
conflict ‖ Konflikt m.; Widerstreit m. ‖
conflit m.
confluence ‖ Zusammenfluß m. ‖ confluent m.
confluent ‖ Nebenfluß m. ‖ confluent m.
conform, to ‖ in Übereinstimmung f. sein;
sich richten ‖ se conformer.
conformability of strata (Geol) ‖ Konkor-
danz f. ‖ concordance f. des couches.
conformation of the soil ‖ Bodenbildung f.
‖ formation f. du terrain.
conformity ‖ Übereinstimmung f. ‖ concordance f.; accord m.; conformité f. / to
be in ~ with (Electr) ‖ der Belastung f.
entsprechen ‖ correspondre à la charge.
confusion ‖ Wirrwarr m. ‖ confusion f.
congeal, to ‖ zusammenfrieren; frieren ‖
congeler; glacer. / ~ ‖ gefrieren; erstarren; gerinnen ‖ se congeler; se figer;
se solidifier; se prendre.
congealable ‖ gefrierbar; gerinnbar ‖ susceptible de congélation; congélable.
congealed ‖ fest geworden; solidifié. / ~ ‖
gefroren ‖ congélé.
conger fisherman ‖ Aalfischer m. ‖ pêcheur
m. d'anguilles *ou* de congres.
congeries pl. ‖ Komplex m. ‖ ensemble m.;
assemblage m.
conglomerate (Geol) ‖ Trümmergestein n.;
Konglomerat n. ‖ conglomérat m. / ~
mixing machine (Glassm) ‖ Gemengemischmaschine f. ‖ machine f. à mélanger.
Congo paper ‖ Congopapier n. ‖ papier m.
Congo.
congratulation card ‖ Glückwunschkarte
f. ‖ carte f. de félicitation.
congratulatory card *see* congratulation
card.
congress ‖ Kongreß m. ‖ congrès m.
congreve match ‖ Reibzündhölzchen n. ‖
allumette f. à friction; congrève f.
congruence (Math) ‖ Deckung f.; Kon-
gruenz f. ‖ coïncidence f.
conical ‖ kegelförmig; konisch ‖ en cone;
conique. / ~ bending arrangement ‖
Konischbiegevorrichtung f. ‖ dispositif
m. de cintrage conique. / ~ boiler flask ‖
Becherkolben m. ‖ fiole f. conique à
ébullition. / ~ breaker ‖ Kegelbrecher m.
‖ concasseur m. à cônes. / ~ course ‖
konischer Schuß m. ‖ virole f. conique.
/ ~ cylinder cover ‖ konischer Zylinderdeckel m. ‖ plateau m. *ou* couvercle m.
de cylindre conique. / ~ drum ‖ kegelförmige Trommel f. ‖ tambour m. conique. / ~ expansion ‖ konische Aufwei-
tung f. ‖ évasement m. conique. / ~ face ‖
kegelförmige Gleitfläche f.; konische
Lauffläche f. ‖ surface f. frottante conique. / ~ form of the wheel tyres ‖ Konizität f. der Radreifen ‖ conicité f. des
bandages de roues. / rivetting die for ~

head ‖ Nietstempel m. für spitzen Nietkopf ‖ bouterolle f. (pour tête de rivet)
conique. / ~ pendulum ‖ Schwungkugel-
regler m. ‖ régulateur m. à force centrifuge. / ~ ring ‖ Keilring m. ‖ bague f.
conique. / ~ rivet head ‖ spitzer Nietkopf m. ‖ tête f. de rivet conique *ou*
en pointe de diamant. / ~ rivetting ‖
Spitzkopfnietung f. ‖ rivure f. conique
ou à point de diamant. / ~ shape ‖
Kegelgestalt f. ‖ forme f. conique. / ~
shell ring ‖ konischer Schuß m. ‖ virole f.
conique. / ~ seat ‖ geneigte Sitzfläche f. ‖
surface f. inclinée du siège. / ~ socket
‖ Kegelhülse f. ‖ douille f. conique.
conifer oil ‖ Koniferenöl n. ‖ essence f. de
conifères.
coniferous wood ‖ Nadelholz n. ‖ bois m.
conifère.
conifers pl. ‖ Nadelhölzer mpl. ‖ bois mpl.
conifères.
coniform *see* conical.
conjugate point ‖ konjugierter Punkt m. ‖
point m. conjugué.
conjunctiva ‖ Bindehaut f. ‖ conjonctive f.
connect, to (Chem) ‖ ketten ‖ lier. / ~
(Electr) ‖ koppeln; schalten ‖ coupler. /
~ (Tel) ‖ anschließen ‖ donner la communication. / ~ in parallel ‖ parallel schalten
‖ coupler en parallèle. / ~ in series ‖
hintereinander schalten ‖ réunir en série
f. / ~ through (Tel) ‖ durchschalten ‖ relier par. / ~ with ‖ beziehen auf ‖ rapporter à.
connected (Mech) ‖ gekoppelt; verbunden ‖
conjugué. / ~ across a resistance ‖ über
einen Widerstand m. geschaltet ‖ bran-
ché sur la résistance. / ~ load of an electric annealing furnace ‖ Anschlußwert
m. eines elektrischen Glühofens ‖ puissance f. installée d'un four électrique
à recuire.
connecting band ‖ Verbindungsschiene f. ‖
bande f. de jonction. / ~ bar (Acc) ‖ Blei-
leiste f. ‖ barre f. de jonction. / ~ box of
a cable ‖ Verbindungsmuffe f. eines Kabels ‖ manchon m. de jonction d'un
câble. / house ~ box ‖ Hausanschluß-
kasten m. ‖ boîte f. de raccordement
pour maisons. / ~ branch ‖ Einsatzstutzen m. ‖ tubulure f. de jonction / ~
branch stamped out of one plate without
seam (Boil) ‖ Stutzen m. aus einem Blech
ohne Naht gepreßt ‖ tubulure f. emboutie
d'une seule tôle sans soudure. / ~ cord
(Tel) ‖ Verbindungsschnur f. ‖ cordon m.
de raccordement. / ~ flange ‖ Anschlußflansch m. ‖ bride f. de raccordement. / ~
knee ‖ Verbindungsknie n. ‖ coude f. de
jonction. / ~ link ‖ Zwischenglied n. ‖
membre m. intermédiaire. / ~ organ
(Tel) ‖ Verbindungsorgan n. ‖ organe m.
de raccordement. / ~ piece ‖ Zwischenstück n.; Verbindungsstück n. ‖ pièce f.
intermédiaire *ou* de raccord. / ~ piece
(Boil) ‖ Verbindungsstutzen m. ‖ tubulure f. de communication. / ~ pipe ‖ Anschlußleitung f. ‖ conduite f. *ou* tuyauterie f. de jonction. / ~ plate ‖ Verbin-
dungsblech n. ‖ plaque f. de raccord. / ~
platform (Wagon) ‖ Übergangsbrücke f. ‖
passerelle f. / ~ plug ‖ Kontaktstöpsel m.
‖ cheville f. de contact.
connecting rod ‖ Schubstange f.; Pleuelstange f. ‖ bielle f. / ~ for conveying
troughs ‖ Förderrinnenschubstange f. ‖
bielle f. pour transporteurs à auges. /
forked ~ ‖ gegabelte Schubstange f. ‖

bielle f. à fourche. / offset ~ || versetzte Pleuelstange f. || bielle f. déportée.

connecting rod bearing || Pleuelstangenlager n.; Treibstangenlager n. || palier m. *ou* coussinet m. de la bielle. / big end of ~ || Pleuelstangenkopf m. || tête f. de bielle. / ~ bushing || Pleuelstangenlager n. || coussinet m. de tête de bielle. / ~ fork || Schubstangengabel f. || fourche f. de la bielle / ~ motion || Schubstangenbewegung f. || mouvement m. de la bielle. / ~ tip || Kolbenstangenkopf m. || tête f. de bielle.

connecting shaft || Verbindungswelle f.; Transmissionswelle f. || arbre m. de transmission. / ~ spiral || Verbindungsschnecke f. || hélice f. de jonction. / ~ spring || Kontaktfeder f. || ressort m. de connexion. / ~ terminal || Anschlußklemme f. || borne f. de raccord. / ~ tube || Verbindungsrohr n.; Ansatzrohr n. || tube m. de jonction *ou* de connexion.

connection || Verbindung f. || jonction f.; connexion f. / ~ (Electr) || Anschluß m. || raccordement m.; raccord m. / ~ (Electr Coupling) || Schaltung f. || couplage m.; montage m.; circuit m. / ~ (Tel) || Gesprächsverbindung f. || communication f. / ~ (Trade) || Kundschaft f. || pratique f.; clientèle f. / ~ of adjoining rails || Schienenverbindung f. || assemblage m. *ou* éclissage m. des rails. / back ~ (Electr) || hinterer Anschluß m. || prise f. arrière. / to cut a ~ (Tel) || eine Verbindung f. trennen || couper une communication; déconnecter. / diagram of ~s || Schaltungsschema n. || schéma m. de connexion. / ~ for electric cooking consumption || Wärmeanschluß m. einer Elektroküche || raccordement m. pour usages calorifiques d'une cuisine électrique. / frame ~ (Auto) || Rahmenverbindung f. || raccord m. de châssis. / ~ in groups (Acc) || Gruppenverbindung f. || réunion f. par groupes. / house service ~ (Electr) || Hausanschluß m. || branchement m. d'abonné. / ~ of incandescent lamps || Glühlampenschaltung f. || connection f. de lampes à incandescence. / ~ for measurement || Meßschaltung f. || montage m. de mesure. / to miss the ~ || den Anschluß m. versäumen || manquer la correspondance. / multiple ~ (Electr) || Parallelschaltung f. || montage m. en parallèle. / ~ in opposition || Gegenschaltung f. || installation f. par opposition; montage m. en opposition. / ~ of pipes || Anschluß m. von Rohren || raccordement m. pour tubes. / private ~ || Privatanschluß m. || connexion f. privée. / ~ in quantity || Nebeneinanderschaltung f. || montage m. en quantité. / ~ by railway || Eisenbahnverbindung f. || communication f. de chemin m. de fer. / rear ~ || rückseitiger Anschluß m. || branchement m. arrière. / series ~ || Reihenschaltung f.; Hintereinanderschaltung f. || connexion f. en série f. / to set up a ~ || eine Verbindung f. herstellen || établir une communication; faire un raccord. / star ~ || Sternschaltung f. || branché en étoile. / in star ~ || in Sternschaltung f. || en étoile f. / stardelta ~ || Sterndreieckschaltung f. || montage m. en étoile-triangle. / starting-up ~ || Anlaßschaltung f. || raccordement m. de démarrage. / tachometer ~ || Tachometeranschluß m. || prise f. de tachymètre. / temporary ~ || Aushilfsschal-

tung f. || raccordement m. provisoire. / total ~s pl. || Gesamtanschluß m. || nombre m. global d'abonnés. / trunk ~ || Fernanschluß m. || raccordement m. interurbain. / ~ for welding and fusion burner || Anschluß m. für Schweiß- und Schneidbrenner || raccord m. pour brûleurs à souder et à découper. / wrong ~ (Electr) || falsche Schaltung f. || montage m. incorrect. / wrong ~ (Tel) || falsche Verbindung f. || faux raccordement.

connection box || Anschlußkasten m. || boîte f. de jonction. / ~ cable || Anschlußkabel n. || câble m. de raccordement. / ~ cord || Anschlußschnur f. || cordon m. / ~ cube || Anschlußwürfel m. || cube m. à prises. / ~ lever || Verbindungshebel m. || levier m. intermédiaire. / to develop a network of high capacity transmission and ~ lines || ein Netz n. leistungsfähiger Übertragungsleitungen und Kupp(e)lungsleitungen ausbauen || installer un réseau m. puissant de lignes de diffusion et d'interconnections. / ~ link || Verbindungsglied n. || membre m. de raccord. / ~ material || Anschlußmaterial n. || matériel m. de raccordement. / ~ organ || Anschlußorgan n. || organe m. de raccordement. / ~ scheme || Schaltungsschema n. || schéma m. des connexions. / ~ sleeve || Verbindungsmuffe f. || manchon m. de raccordement. / ~ terminal || Verbindungsklemme f. || borne f. de raccordement; serre-fils m. / ~ terminal board || Verbindungsklemmbrett n. || tableau m. à bornes de jonction. / ~ value || Anschlußwert m. || wattage m.

connector (Electr) || Kontaktfeder f. || ressort m. de connexion. / ~ (Tel aut) || Leitungswähler m. || connecteur m. / line ~ || Leitungswähler m. || sélecteur m. final. / ~ for local and long distance traffic (Aut tel) || Leitungswähler m. für Orts- und Fernverkehr || connecteur m. combiné. / plug ~ (Tel) || Verbindungsstöpsel m. || fiche f. de raccordement. / ~ for testing points (Tel) || Untersuchungsklemme f. || borne f. de coupure.

connector box (Electr) || Abzweigdose f. für Isolierrohr || boîte f. de distribution. / ~ clip || Lüsterklemme f. || agrafe f.

connexion *see* connection.

conning bridge || Kommandobrücke f. || passerelle f. de navigation. / ~ tower || Kommandoturm m. || tourelle f. de commandement.

connoisseur || Sachkenner m.; Sachkundiger m. || expert m.

conoid || Konoid n. || conoïde m. / elliptic ~ || Rotationsellipsoid n.; Sphäroid n. || ellipsoïde m. de révolution *ou* de rotation.

conoscopic || konoskopisch || conoscopique.

consecution controller (Aut tel) || Anrufordner m. || classeur m. d'appels.

consecutive number || laufende Nummer f. || numéro m. d'ordre.

consent || Zustimmung f. || consentement m.; approbation f.; assentiment m. / ~ (Trade) || Zusage f. || assentiment m.; promesse f.

consequent pole || Folgepol m. || pôle m. *ou* point m. conséquent.

conservation || Haltbarmachen n. || conservation f. / ~ of energy || Erhaltung f. der Energie || conservation f. de l'énergie.

conservatory (Gard) || Gewächshaus n. || serre f.

conserve tin box || Konservenbüchse f. || boîte f. à conserves.

consider, to || berücksichtigen || considérer; prendre en considération f.

considerable || ansehnlich || considérable.

consideration, after mature ~ || nach reiflicher Überlegung f. || réflexion f. faite; après mûre réflexion f. / first ~ || Hauptsache f. || principal m.; essentiel m. / in ~ of || mit Rücksicht f. auf || en considération f. de; par égard m. pour. / to take into ~ || in Anschlag m. bringen || mettre en ligne f. de compte.

consignee, address of the ~ || Versandanschrift f. || adresse f. pour l'expédition.

consignment, collective ~ || Sammelladung f. || envoi m. en groupage. / ~ by the lorry load || Sammelgut n. || expédition f. collective *ou* en cueillette. / ~ of valuables || Wertsendung f. || valeur f.

consistence || Konsistenz f.; Dichtigkeit f.; Dichte f. || consistance f.; densité f.

consistency || Dickflüssigkeit f. || consistance f.

consistent || konsistent; dick || consistant. / ~ grease || konsistentes Fett n. || graisse f. consistante.

console || Konsole f. || console f. / ~ of stone (Arch) || Kragstein m.; Konsole f.; Tragstein m. || corbeau m.; console f. en pierre; ancone f.

consolidate, to || konsolidieren || consolider; fonder.

consolidated loan || Konsolidierungsanleihe f. || emprunt m. de consolidation.

consolidation || Konsolidation f. || consolidation f. / lateral ~ of poles || Seitenbefestigung f. der Telegraphenstangen || consolidation f. des poteaux.

consolidation pole (El line) || Stützpfosten m. || soutien m.

consonant (Mus) || wohlklingend || consonnant.

constancy of volume || Raumbeständigkeit f. || constance f. de volume.

constant || stetig || continu; constant. / ~ly active bell || Fortschellklingel f. || sonnerie f. continue. / ~ current dynamo || Dynamo f. für konstante Stromstärke || dynamo f. á intensité constante. / ~ current modulation (Tel) || Anodenspannungsmodulation f. || modulation f. à courant constant. / ~ level carburettor || Vergaser m. mit gleichbleibendem Stand || carburateur m. à niveau constant. / ~ pressure dynamo || Dynamo f. für konstante Spannung || dynamo f. à tension constante. / ~ speed || unveränderliche Umlaufzahl f. || vitesse f. constante.

constant || konstante Größe f.; Konstante f.; Festwert m. || constante f.; quantité f. constante; valeur f. fixe. / galvanometer ~ || Galvanometerkonstante f. || constante f. d'un galvanomètre. / keeping ~ || Konstanthaltung f. || maintien m. de la constante.

constantan || Konstantan n. || constantan m.

constituent || Bestandteil m. || composant m. / ~ of the air || Luftbestandteil m. || élément m. constituant de l'air.

constitution (Chem) || Bau m.; Beschaffenheit f. || constitution f. / chemical ~ of the steel || chemische Beschaffenheit f. des Stahles || constitution f. chimique de l'acier.

constriction ‖ Einschnürung f. ‖ étranglement m.; contraction f.; gorge f.

construct, to (Build) ‖ erbauen; ausführen ‖ bâtir; construire; édifier. / ~ a bridge ‖ eine Brücke f. bauen oder schlagen ‖ construire *ou* établir *ou* jeter un pont. / ~ a railway ‖ eine Eisenbahn f. bauen ‖ construire un chemin de fer.

constructed, well ~ machine ‖ durchkonstruierte Maschine f. ‖ machine f. de construction tout à fait perfectionnée.

constructer ‖ Erbauer m.; Konstruktör m. ‖ constructeur m.

construction ‖ Konstruktion f.; Bauart f. ‖ construction f. / boiler ~ ‖ Bau m. des Kessels; Kesselbau m. ‖ construction f. de chaudières; travail m. de chaudronnerie. / boiler in process of ~ ‖ im Bau begriffener Kessel m. ‖ chaudière f. en cours de construction. / ~ of bridges ‖ Brückenbau m. ‖ construction f. de ponts. / ~ of corrugated iron ‖ Wellblechbau m. ‖ construction f. en tôles ondulées. / in course of ~ ‖ im Bau m. befindlich ‖ en voie f. de construction. / piece of ~ in dismounted state ‖ Konstruktionsstück n. in zerlegtem Zustand ‖ pièce f. de construction à l'état démonté. / extra heavy ~ ‖ besonders schwere Ausführung f. ‖ construction f. extra robuste. / fault of ~ ‖ Konstruktionsfehler m. ‖ défaut m. de construction. / in general ~ ‖ in allgemeiner Ausführung f. ‖ en construction f. générale. / glazed iron-frame ~s pl. ‖ verglaste in Eisenkonstruktion gehaltene Hallen fpl. ‖ halls mpl. en charpente métallique vitrée. / imitated ~ ‖ nachgeahmte Bauart f. ‖ construction f. imitée. / in ~ ‖ im Bau m. ‖ en construction f. / iron ~ ‖ Eisenkonstruktion f. ‖ construction f. de fer. / latest ~ ‖ neuste Konstruktion f. ‖ dernière construction f. / low ~ ‖ niedriger Bau m. ‖ construction f. basse. ‖ ~ of machines ‖ Maschinenbau m. ‖ construction f. de machines. / mixed ~ ‖ Gemischtbau m. ‖ construction f. mixte. / modern ~ ‖ Neubau m. ‖ construction f. moderne. / new ~ ‖ Neubau m. ‖ nouvelle construction f. / open-face ~ ‖ offene Bauart f. ‖ construction f. ouverte. / particularly strong ~ ‖ besonders kräftige Bauart f. ‖ construction f. particulièrement robuste. / robust ~ ‖ unverwüstliche Ausführung f. ‖ exécution f. robuste. / self-contained ~ ‖ geschlossener Aufbau m. ‖ construction f. serrée. / the ~ is of the simplest kind imaginable ‖ die Konstruktion ist die denkbar einfachste ‖ la construction est la plus simple possible. / solid ~ ‖ stabile Konstruktion f. ‖ construction f. robuste. / special heavy ~ ‖ besonders schwere Bauart f. ‖ construction f. particulièrement robuste. / steel ~ ‖ Eisenbauwerk n.; Eisenkonstruktion f. ‖ charpente f. métallique. / ~ of steel castings pl. ‖ Stahlgußausführung f. ‖ exécution f. en acier coulé. / ~ of a timber work ‖ Holzverband m.; Zimmerverband m. ‖ assemblage m. des bois. / under ~ ‖ im Bau m. ‖ en construction f. / vertical ~ ‖ stehende Ausführung f. ‖ exécution f. verticale.

construction accuracy ‖ Herstellungsgenauigkeit f. ‖ exactitude f. d'exécution.

constructional drawing ‖ Konstruktionszeichnung f. ‖ dessin m. de construction.

/ ~ element of the aircraft ‖ Flugzeugbauelement n. ‖ élément m. de construction d'avions. / power driven machine for the ~ engineering ‖ Eisenbearbeitungsmaschine f. für Kraftbetrieb ‖ machine à force motrice à travailler les fers profilés. / ~ iron ‖ Baueisen n. ‖ fer m. de construction. / where larger dimensions are inconvenient for ~ reasons pl. ‖ wo größere Abmessungen fpl. aus konstruktiven Rücksichten unbequem sind ‖ dans le cas où pour des raisons techniques on ne peut choisir des dimensions plus grandes. / ~ sketch ‖ Konstruktionsskizze f. ‖ croquis m. de construction. / producing ~ steels pl. from a cold charge ‖ Herstellung f. von Konstruktionsstählen aus kaltem Einsatz ‖ fabrication f. d'acier de construction à partir de la charge froide.

construction car for telephone ‖ Fernsprechbauwagen m. ‖ voiture f. pour constructions téléphoniques. / ~ circuit (Tel) ‖ Bauschaltung f. ‖ circuit m. de construction. / ~ cost ‖ Anlagekosten pl. ‖ frais m. d'installation *ou* d'établissement. / ~ drawing ‖ Konstruktionszeichnung f. ‖ plan m. de construction. / ~ method ‖ Baugrundsatz m. ‖ méthode f. de construction. / ~ unit gang (Tel) ‖ Telegraphenbautrupp m. ‖ équipe f. d'ouvriers de constructions télégraphiques.

constructive ‖ konstruktiv ‖ constructif. / ~ elements pl. for which a minimum of wear and tear is desirable and which at the same time demand the greatest possible safety against breakage ‖ Konstruktionsteile mpl., die höchsten Widerstand gegen Verschleiß und größte Sicherheit gegen Bruch bieten müssen ‖ pièces fpl. qui doivent présenter à la fois un minimum d'usure et un maximum de résistance à la rupture.

constructor ‖ Erbauer m. ‖ constructeur m. / naval ~ ‖ Schiffbaumeister m. ‖ ingénieur m. de constructions navales *ou* de la marine. / ~ of roads ‖ Wegebaumeister m.; Bauingeniör m. ‖ ingénieur m. des ponts et des chaussées.

consular, certificate ‖ konsularische Beglaubigung f. ‖ certificat m. consulaire.

consulate ‖ Konsulat n. ‖ consulat m.

consul general ‖ Generalkonsul m. ‖ consul m. général.

consult, to ‖ um Rat fragen; zu Rate ziehen ‖ consulter.

consultation ‖ Rücksprache f. ‖ pourparler m.

consulting engineer ‖ beratender Ingeniör m. ‖ ingénieur-conseil m. / ~'s fees pl. ‖ Honorar n. des beratenden Ingeniörs ‖ honoraires mpl. de l'ingénieur-conseil.

consulting office ‖ Beratungsstelle f. ‖ bureau m. de conseil.

consumable load ‖ Verbrauchslast f. ‖ charge f. de consommation. / ~ stores pl. ‖ Verbrauchsgegenstände mpl.; objets mpl. de consommation.

consume, to ‖ konsumieren; verbrauchen; aufbrauchen ‖ consommer; épuiser.

consumer ‖ Konsument m.; Verbraucher m. ‖ consommateur m.; abonné m. / ~'s ware ‖ Verzehrungsgegenstand m. ‖ produit m. de consommation.

consumption ‖ Konsum m.; Verbrauch m. ‖ consommation f.; dépense f. / annual ~ *see* yearly ~. / average ~ ‖ durchschnitt-

licher Verbrauch m. ‖ consommation f. moyenne. / ~ of carbons (Arc lamp) ‖ Kohlenabbrand m. ‖ usure f. des charbons. / ~ of coal ‖ Kohlenverbrauch m. ‖ consommation f. de charbon. / ~ of current ‖ Stromentnahme f. ‖ consommation f. de courant. / daily ~ of coal ‖ täglicher Kohlenverbrauch m. ‖ consommation f. journalière de houille. / ~ of energy (Electr) ‖ Energieverbrauch m. ‖ consommation f. d'énergie; wattage m. / ~ of energy (Mach) ‖ Kraftverbrauch m. ‖ dépense f. de force. / ~ of free air ‖ Verbrauch m. an atmosphärischer Luft ‖ consommation f. d'air libre. / ~ of fuel ‖ Brennstoffverbrauch m. ‖ consommation f. de combustible. / gasoline ~ ‖ Benzinverbrauch m. ‖ consommation f. de gasoline. / ~ per horse-power hour ‖ Verbrauch m. für die Pferdestärke und Stunde ‖ consommation f. au cheval-heure. / ~ of hot water ‖ Heißwasserverbrauch m. ‖ consommation f. d'eau chaude. / internal ~ ‖ Eigenverbrauch m. ‖ consommation f. parasite. / low ~ of power ‖ geringer Kraftbedarf m. ‖ force f. absorbée minime. / ~ of materials ‖ Materialbedarf m. ‖ consommation f. de matériaux. / minimum specific ~ ‖ kleinster spezifischer Verbrauch m. ‖ consommation f. spécifique minima. / ~ of petrol per brake horse-power hour ‖ Benzinverbrauch m. je Bremspferdestärke und Stunde ‖ consommation f. d'essence par cheval-heure effectif. / ~ of raw material ‖ Verbrauch m. an Werkstoffen ‖ consommation f. de matière brute. / ~ of steam ‖ Dampfverbrauch m. ‖ dépense f. de vapeur. / yearly ~ ‖ Jahresverbrauch m. ‖ consommation f. annuelle.

consumption characteristic ‖ Verbrauchskennlinie f. ‖ courbe f. caractéristique de consommation. / good for ~ ‖ Verzehrungsgegenstand m. ‖ produit m. de consommation. / place of ~ ‖ Verbrauchsstelle f. ‖ lieu m. de consommation. / point of ~ ‖ Verbraucherbezirk m. ‖ région f. d'utilisation. / ~ test ‖ Verbrauchsprüfung f. ‖ essai m. de consommation.

contact ‖ Kontakt m.; Berührung f. ‖ contact m. / area of ~ ‖ Kontaktfläche f. ‖ surface f. de contact. / back ~ ‖ Ruhekontakt m. ‖ contact m. de repos. / bad ~ ‖ schlechter Kontakt m. ‖ contact m. imparfait. / bow ~ ‖ Bügelkontakt m. ‖ contact m. d'archet. / to break a ~ ‖ einen Kontakt m. öffnen ‖ ouvrir un contact. / to bring in ~ ‖ in Berührung f. bringen ‖ mettre en contact m. / brush type of ~ ‖ Bürstenkontakt m. ‖ contact m. à balai. / carbon ~ ‖ Kohlekontakt m. ‖ contact m. à charbon. / ~ controllable from one point ‖ von einem Punkt aus regelbare Anstrichverstellung f. ‖ contact m. réglable d'un seul endroit. / dead ~ *see* vacant ~. / dummy ~ ‖ Leerkontakt m. ‖ contact m. inoccupé. / fixed ~ between rails ‖ fester Kontakt m. in der Gleismitte ‖ crocodile m. / indicator for earthen electric ~ ‖ Erdschlußanzeiger m. ‖ indicateur m. de contact électrique terrestre. / to keep from ~ with the air ‖ unter Luftausschluß m. aufbewahren ‖ conserver à l'abri m. de l'air. / loosening of the ~ ‖ Kontaktlockerung f. ‖ desserrage m. des contacts.

/ to make ~ || in Kontakt m. bringen || mettre en contact m. / material for ~s || Kontaktmaterial n. || matériel m. pour les contacts. / mechanical ~ || Kopfkontakt m. || contact m. supérieur. / medium ~ || mittlerer Kontakt m. || contact m. médian. / off-normal ~ || Kopfkontakt m. || contact m. de travail. / period of ~ || Eingriffsdauer f. || durée f. d'engrènement. / rail ~ || Schienenkontakt m. || contact m. de voie. / rest ~ || Ruhekontakt m. || contact m. de repos. / rubbing ~ || Streichkontakt m. || contact m. à frottement. / secondary off-normal~ || Wellenkontakt m. || contact m. de repos secondaire. / seconds ~ with synchronizing devices. || Sekundenkontakt m. mit Synchronisierungseinrichtungen || contact m. régulateur avec installations de synchronisation. / sliding ~ || Gleitkontakt m.; Schleifkontakt m. || collecteur m. courseur; contact m. frotteur./ treading ~ || Tretkontakt m. || contact m. à pédale. / vacant ~ || Leerkontakt m. || plot m. mort; contact m. vacant. / ~ of wires || Drahtberührungsstelle f. || contact m. de fils.

contact bank || Kontaktbank f. || banc m. des contacts. / suspended ~ box || Hängeanschlußdose f. || prise f. de courant à suspension. / ~ breaker spring || Unterbrecherfeder f. || ressort m. de trembleur. / ~ bush || Kontaktbüchse f. || boîte f. de contact. / ~ button || Kontaktknopf m. || bouton m. de contact. / ~ carrier || Kontaktträger m. || porte-contact m. / ~ carrying member || kontaktführender Teil m. || plot m. de contact. / ~ cathode || Berührungskathode f. || cathode f. de contact. / ~ chariot (Tel) || Kontaktschlitten m. || chariot m. de contact. / ~ clip || Kontaktschelle f. || collier m. de contact. / ~ detector || Kontaktdetektor m. || détecteur m. de contact. / ~ device see ~ mechanism. / ~ dryer || Berührungstrockner m. || sécheur m. à contact. / ~ fault || Kontaktfehler m. || contact m. défectueux. / ~ finger || Kontaktfinger m. || touche f. / ~ lamp || Kontaktglühlampe f. || lampe f. à contact. / ~ mechanism for watermeters || Kontaktwerk n. für Wassermesser || mécanisme m. de contact pour compteurs d'eau. / ~ panel || Kontaktfeld n. || panneau m. des contacts. / ~ plate || Kontaktplatte f. || plaque f. de contact. / ~ plate at rest || Ruheschiene f. || plaque f. de contact de repos. / ~ plug || Stecker m. || prise f. de courant à fiches. / ~ point || Kontaktstift m. || pointe f. de contact; goujon m. / ~ point electrode || Punkt(schweiß)elektrode f. || électrode f. pontiforme. / ~ pressure || Kontaktdruck m. || pression f. des contacts. / ~ process || Kontaktverfahren n. || procédé m. par contact. / ~ regulating switch || Regelwiderstand m. || résistance f. réglable; rhéostat m. / ~ ring || Kontaktring m. || bague f. de contact. / ~ rod || Abnehmerstange f. || tige f. de contact. / ~ roller || Rollenelektrode f. || électrode f. à rouleau. / ~ screw (Electr) || Kontaktschraube f. || vis f. de contact. / ~ screw (Mech) || Meßschraube f. || vis f. de mesure. / ~ segment || Kontaktscheibe f. || disque m. de contact. / ~ skate || Kontaktschlitten m. || frotteur m. de prise. / ~ spring || Unterbrecherfeder f. || ressort m. de contact. /

~ stud || Kontaktknopf m. || bouton m. de contact; goutte f. de suif. / ~ surface || Kontaktfläche f. || surface f. du contact. / ~ theory || Kontakttheorie f. || théorie f. du contact. / ~ treadle of mercury || Quecksilberschienenkontakt m. || contact m. de rail à mercure. / ~ trolley || Kontaktrolle f. || trolley m. / mercury ~ tube || Quecksilberschaltröhre f. || interrupteur m. à ampoule à mercure. / ~ wire insulator || Fahrdrahtisolator m. || isolateur m. de fil de contact.

contagion || Ansteckung f. || contagion f.

contain, to || fassen || contenir.

container || Behältnis n. || récipient m.; étui m. / ~ for fire arms || Schußwaffenbehälter m. || caisse f. pour les armes à feu. / ~ bottom || Behälterboden m. || fond m. de réservoir.

containing oxygen || sauerstoffhaltig || oxygénifère; oxygéné; oxydé.

contaminate, to || verunreinigen || souiller.

contamination, bacteria ~ || Bakterieninfektion f. || infection f. bactérienne.

content see contents.

contentious || streitig || contentieux; litigieux.

contentiously pleaded || streitig verhandelt || débattu contradictoirement.

contents pl. || Inhalt m. || contenu m.; capacité f.; teneur f. / ~ (Print) || Register n.; Inhaltsverzeichnis n.; Inhalt m. || index m.; table f. des matières. / ~ of ashes || Aschengehalt m. || teneur m. en cendres. / ~ of a body || Rauminhalt m. eines Körpers || capacité f. ou volume m. d'un corps. / ~ cubical ~ || Rauminhalt m.; Volumen n. || volume m. / ~ of gas || Gasgehalt m. || teneur f. en gaz. / low ~ || niedriger Gehalt m. || faible teneur f. / ~ of slate || Schiefergehalt m. || teneur f. en schiste. / small ~ || schwacher Gehalt m. || faible teneur f. / visible ~ || sichtbarer Inhalt m. || contenu m. visible.

contiguous angle || Nebenwinkel m. || angle m. adjacent ou contigu.

continental climate || Binnenklima n. || climat m. continental.

contingencies pl. || unvorhergesehene Ausgaben fpl. || faux frais mpl. divers.

continual || stetig || continu; constant.

continuance || Bestand m. || permanence f.

continuation || Weiterführung f. || continuation f.

continue, to ~ service relations pl. || das Arbeitsverhältnis fortsetzen || prolonger le contrat de travail.

continuity || Kontinuität f.; Stetigkeit f. || continuité f.

continuous || stetig; fortlaufend; kontinuierlich || continu. / ~ carbonizing plant for clothes || kontinuierliche Karbonisieranlage f. für Tuche || installation f. continue à carboniser les draps. / ~ control shaft || durchgehende Welle f. || arbre m. de commande traversant. / ~ conveyor || stetiger Förderer m. || transporteur m. continu.

continuous current || Gleichstrom m. || courant m. continu. / ~ alternating-current convertor || Gleichstromwechselstromeinankerumformer m. || commutatrice f. à courant continu en courant alternatif. / ~alternating-current dynamotor || Gleichstromwechselstromumformer m. || dynamoteur m. à courant continu en courant alternatif. / ~ continuous current dynamotor || Gleichstromgleichstromumfor-

mer m. || dynamoteur m. à courant continu en courant continu. / ~ dynamo || Gleichstromdynamo f. || génératrice f. à courant continu. / ~ field || Gleichstromfeld n. || champ m. continu. / ~ generator || Gleichstromdynamomaschine f. || dynamo f. à courant continu. / ~ motor (Electr) || Gleichstrommotor m. || électromoteur m. à courant continu. / ~ rotary-current convertor || Gleichstromdrehstromeinankerumformer m. || commutatrice f. à courant continu en courant triphasé. / ~ rotary-current dynamotor || Gleichstromdrehstromumformer m. ||dynamoteur m. à courant continu en courant triphasé. / ~ side || Gleichstromseite f. || côté m. du courant continu. / ~ transformer || Gleichstromtransformator m. || transformateur m. à courant continu.

continuous drive shaft see ~ control shaft. / ~ finishing machine || kontinuierliche Dekatiermaschine f.; Duffmaschine f. || machine f. à décatir à la continue. / fitting flat irons pl. on the ~ flow system || Bandmontage f. von Bügeleisen || montage m. à la chaîne de fers à repasser. / ~ girder || durchgehender Träger m. || poutre f. continue. / ~ heating furnace || Rollofen m. || four m. roulant. / ~ helio spectrograph for recording the ultraviolet spectrum of the sun || Apparat m. zur Dauerfotografie des ultravioletten Sonnenspektrums || appareil m. à photographier continuellement le spectre solaire ultraviolet. / ~ impost || um einen Pfeiler herum geführter Kämpfer m. || imposte f. cintrée. / ~ production installation || Fließarbeitseinrichtung f. || dispositif m. pour le travail continu. / ~ rolling mill || kontinuierliches Walzwerk n. || laminoir m. à mouvement continu. / ~ slow motion || endlose Feinbewegung f. || mouvement m. lent sans fin. / ~ spectrum || kontinuierliches Spektrum n. || spectre m. continu. / ~ supply || ununterbrochene Zufuhr f. || arrivée f. ininterrompue. / ~ wave receiver (Radio) || Empfänger m. für ungedämpfte Wellen || récepteur m. d'ondes non amorties. / ~ working || Dauerbetrieb m. || service m. continu. / ~ working kiln || Ofen m. mit Dauerbetrieb || four m. continu.

contour || Kontur f. || contour m. / ~ of wing || Flügelumriß m. || contour m. d'aile. / ~ line || Horizontalkurve f. || ligne f. de niveau. / ~ map || Schichtlinienplan m. || plan m. des courbes de niveau; carte f. en courbes de niveau.

contraband || Schleichhandel m. || contrebande f.; commerce m. interlope.

contract, to || verengern; zusammenziehen || rétrécir; se contracter. / ~ (Delivery) || akkordieren || accorder. / ~ (Trade) || einen Vertrag m. abschließen || conclure un marché ou un traité. / ~ mould || schimmeln || chancir; se chancir; moisir; se moisir.

contract || Abschluß m.; Vertrag m.; Kontrakt m. || contrat m.; forfait m. / according to ~ || vertragsmäßig || selon le contrat. / to be under ~ || kontraktmäßig oder vertragsmäßig gebunden || être lié par contrat m. / breach of ~ || Vertragsbruch m. || violation f. du contrat. / to break a ~ || vertragsbrüchig werden || rompre un contrat. / by ~ || kontraktlich; vertragsmäßig || contractuel. / to con-

clude a ~ ‖ einen Vertrag m. abschließen ‖ contracter. / conclusion of ~ ‖ Abschließung f. des Vertrages ‖ passation f. du contrat. / to draw up a ~ ‖ einen Vertrag m. entwerfen ‖ rédiger un contrat. / ~ between employer and employee ‖ Dienstvertrag m. ‖ contrat m. de service. / ~ of employment ‖ Arbeitsvertrag m. ‖ contrat m. de travail. / to decide rights and privileges by the ~ of employment ‖ die Rechte npl. und Pflichten fpl. aus dem Arbeitsvertrag regeln ‖ régler les droits m. et les devoirs mpl. d'après le contrat de travail. / to enter into a ~ ‖ einen Vertrag m. schließen ‖ passer un contrat. / freedom of ~ ‖ Vertragsfreiheit f. ‖ liberté f. de contrat. / ~ for future delivery ‖ Lieferungsvertrag m. ‖ contrat m. de livraison. / to make a ~ ‖ kontrahieren; einen Vertrag m. schließen ‖ contracter. / a period of ~ expires ‖ eine Vertragszeit f. läuft ab ‖ une période f. de contrat expire. / ~ of purchase ‖ Kaufvertrag m. ‖ contrat m. de vente. / to put out to ~ ‖ auf dem Submissionswege m. vergeben ‖ mettre en adjudication f. / ~ for service ‖ Dienstvertrag m.; Arbeitsverhältnis n. ‖ contrat m. de louage *ou* de service *ou* de travail. / to conceal from the employer the existence of another ~ for service ‖ dem Arbeitgeber das Bestehen eines anderen gleichzeitigen Arbeitsverhältnisses verschweigen ‖ cacher à l'employeur l'existence d'un autre contrat de travail. / sum of the ~ ‖ Verdingungssumme f. ‖ montant m. du forfait. / term of ~ ‖ Vertragsbedingung f. ‖ terme m. du contrat. / to terminate the ~ for service ‖ das Vertragsverhältnis n. lösen ‖ annuler le contrat de travail. / to work by ~ ‖ in Akkord m. arbeiten ‖ travailler à la pièce *ou* à la tâche. / yearly ~ ‖ Jahreskontrakt m. ‖ contrat m. annuel.

contract date ‖ Liefertermin m. ‖ date f. de livraison. / ~ fulfilling place ‖ Erfüllungsort m. ‖ lieu m. de payement; élection f. de domicile.

contracting the iris diaphragm ‖ Verengern n. der Irisblende ‖ réduction f. de l'ouverture du diaphragme-iris. / ~ party ‖ Kontrahent m. ‖ contractant m.

contraction ‖ Schrumpfung f.; Schwindung f.; Einschnürung f.; Verengung f. ‖ étranglement m.; contraction f.; gorge f.; rétrécissement m. / ~ of a casting ‖ Schwindmaß n. eines Gußstückes ‖ retrait m. du moulage. / coefficient of ~ ‖ Kontraktionskoeffizient m. ‖ coefficient m. de contraction. / ~ of the concrete during setting ‖ Schwinden n. des Betons beim Erhärten ‖ retrait m. du béton à la prise. / ~ of diameter ‖ Querkontraktion f. ‖ contraction f. transversale. / imperfect ~ ‖ unvollkommene Kontraktion f. ‖ contraction f. imparfaite. / incomplete ~ ‖ unvollständige Kontraktion f. ‖ contraction f. incomplète. / ~ of metals ‖ Schwinden n. der Metalle ‖ retrait m. des métaux. / ~ of the pupil ‖ Pupillenverengerung f. ‖ rétrécissement m. de la pupille. / ~ of a seam ‖ Verdrückung f. eines Flözes ‖ étranglement m. d'une couche sédimentaire. / ~ of volume ‖ Volumenkontraktion f. ‖ contraction f. de volume.

contraction cavity (Found) ‖ Lunker m. ‖ retassement m.; retassure f. / ~ crack

(Found) ‖ Schrumpfriß m. ‖ fissure f. de retrait. / the casting shows absence of ~ cracks ‖ Gußstück n. ohne Schrumpfriß ‖ pièce f. de fonte sans fissures par le retrait du métal. / ~ rule ‖ Schwindmaßstab m. ‖ mesure f. *ou* règle f. à retrait.

contract note ‖ Schlußschein m. ‖ bordereau m.

contractor ‖ Kontrahent m. ‖ contractant m. / ~ ‖ Unternehmer m. ‖ entrepreneur m. de travaux. / ~ (Build) ‖ Bauunternehmer m. ‖ entrepreneur m. de bâtiments. / ~ (Delivery) ‖ Lieferant m. ‖ pourvoyeur m.; fournisseur m. / ~ (Mine) ‖ Grubenpächter m. ‖ forfaitier m. / ~ of carrying ‖ Frachtunternehmer m. ‖ entrepreneur m. de transport. / ~ for farm labour ‖ Unternehmer m. landwirtschaftlicher Arbeiten ‖ entrepreneur m. de travaux agricoles. / ~ for public works ‖ Unternehmer m. für öffentliche Bauten ‖ entrepreneur m. de travaux publics.

contractor's crossing ‖ Schleppkreuzung f. ‖ croisement m. à rails pivotants. / ~ locomotive ‖ Baulokomotive f. ‖ locomotive f. d'entrepreneur. / ~ points ‖ Schleppweiche f. ‖ aiguille f. déplaçable. / ~ pump ‖ Baupumpe f. ‖ pompe f. de construction *ou* d'épuisement. / ~ railway ‖ Feldbahn f. ‖ chemin m. de fer portatif; Decauville m.; voie f. portative. / ~ railway track ‖ verlegbares Gleis n. ‖ voie f. portative. / ~ switch ‖ Schleppweiche f. ‖ changement m. à rails mobiles. / ~ work ‖ Unternehmerarbeit f. ‖ travaux mpl. à l'entreprise.

contract price ‖ Gedingepreis m. ‖ prix m. convenu. ~ / work ‖ Akkordarbeit f. ‖ travail m. à la tâche *ou* à forfait.

contra-flow ‖ Gegenstrom m. ‖ contre-courant m.

contrary ‖ Gegenteil n. ‖ contraire m. / in ~ direction ‖ in entgegengesetztem Sinne m. ‖ en contre-sens m. / on the ~ ‖ gegenteilig ‖ contraire.

contrary direction ‖ Gegenbewegung f. ‖ contre-mouvement m.; mouvement m. opposé *ou* contraire. / ~ wind ‖ Gegenwind m. ‖ vent m. contraire.

contrast ‖ Gegensatz m. ‖ contraste m. / to accentuate ~s pl. ‖ die Kontrastwirkung f. erhöhen ‖ obtenir des contrastes mpl. plus vifs. / ~ micrometer ‖ Kontrastmikrometer n. ‖ micromètre m. à contraste.

contrate wheel (Watchm) ‖ Kronrad n. ‖ roue f. à dents de côté *ou* à couronne.

contribution ‖ Geldbeitrag m. ‖ contribution f.; cotisation f.; fournissement m.

contrivance ‖ Einrichtung f. ‖ combinaison f.; dispositif m. / ~ for blowing out barrels ‖ Faßausblasevorrichtung f. ‖ appareil m. à nettoyer les tonneaux. / wedge-closing ~ ‖ Keilverschluß m. ‖ fermeture f. à coin.

control, to ‖ überwachen; prüfen ‖ contrôler; surveiller. / ~ a relay ‖ ein Relais n. steuern ‖ contrôler un relais.

control ‖ Kontrolle f. ‖ contrôle m. / ~ (Aero) ‖ Steuerung f. ‖ commande f. / ~ centralized in one place ‖ an einem Punkt vereinigte Bedienung f. ‖ service m. réuni au même endroit. / electric ~ ‖ elektrische Steuerung f. ‖ commande f. électrique. / ~ of manufacturing opera-

tions ‖ Betriebskontrolle f. ‖ contrôle m. de fabrication. / radio ~ of moving objects ‖ Fernlenkung f. ‖ commande f. à grande distance des objets mouvants par télégraphie sans fil. / reversive ~ (Aut tel) ‖ Rückkontrolle f. ‖ contrôle m. réversible. / shop ~ ‖ instrument ‖ Betriebsüberwachungsgerät n. ‖ appareil m. pour la surveillance d'exploitation. / stoke-hole ~ ‖ Überwachung f. der Feuerung ‖ surveillance f. du foyer. / system of ~ ‖ Steuerungsart f. ‖ système f. de commande. / ~ for training purposes ‖ Schulsteuerung f. ‖ commande f. de l'élève.

control apparatus ‖ Kontrollapparat m. ‖ appareil m. de contrôle. / ~ monthly balance (Bookkeep) ‖ monatliche Kontrollbilanz f. ‖ balance f. mensuelle vérificative. / ~ battery (Hydr press) ‖ Steuerbatterie f. ‖ batterie f. de distributeurs. / thermal ~ board for supervising the circulation of water ‖ Wärmewarte f. zur Überwachung des Wasserkreislaufes ‖ poste m. de contrôle thermique destiné à surveiller la circulation d'eau. / ~ board for turbine plant ‖ Turbinenüberwachungstafel f. ‖ tableau m. de contrôle d'une installation de turbines. / ~ cabin ‖ Führerstand m. ‖ cabine f. de manœuvre. / ~ cable (Aero) ‖ Steuerseil n. ‖ câble m. de commande. / ~ car ‖ Führergondel f. ‖ nacelle f. de pilotage. / ~ cash register ‖ Kontrollregistrierkasse f. ‖ caisse f. enregistreuse de contrôle. / ~ clock ‖ Kontrolluhr f. ‖ horloge-contrôle f. / ~ current ‖ Kuppelstrom m. ‖ circuit m. répétiteur d'itinéraire. / ~ cylinder (Electr) ‖ Steuerwalze f. ‖ hélice f. de controlleur. / ~ device for side movement of cylinder for cloth with heavy selvedge ‖ Einrichtung f. zum seitlichen Verstellen der Mulde für Tuche mit starken Leisten ‖ dispositif m. pour déplacement latéral de la cuvette pour draps à fortes lisières. / ~ drive (Aero) ‖ Steuerwerk n. ‖ commande f. de l'avion. / ~ equipment for fire-alarm systems ‖ Kontrolleinrichtung f. in Feuermeldeanlagen ‖ installation f. de contrôle dans le service des avertisseurs d'incendie. / ~ experiment ‖ Gegenversuch m. ‖ contre-essai m. / ~ flange ‖ Kontrollflansch m. ‖ bride f. de contrôle. / ~ gauge ‖ Kontrollmanometer n. ‖ manomètre m. de contrôle. / ~ gear (Hydr press) ‖ Steuerorgan n. ‖ organe m. de distribution. / ~ gear (Mach) ‖ Kontrollanlage f. ‖ installation f. de contrôle. / elevator ~ gear ‖ Aufzugsteuerung f. ‖ dispositif m. de commande pour ascenseurs. / ~ -heavy (Aero, Nav) ‖ steuerlastig ‖ lourd sur la queue. / ~ instrument ‖ Kontrollinstrument n. ‖ instrument m. de contrôle.

controller ‖ Aufseher m.; Kontrollör m. ‖ contrôleur m. / ~ (Electr) ‖ Fahrschalter m.; Steuerschalter m. ‖ contrôleur m. / ~ (Watchm) ‖ Wächteruhr f. ‖ contrôleur m. de rondes. / derricking ~ ‖ Hubsteuerschalter m. ‖ contrôleur m. pour appareils de levage.

controller cylinder ‖ Schaltwalze f. ‖ cylindre m. de contrôleur. / ~ drum ‖ Fahrschaltertrommel f. ‖ cylindre m. de combinateur. / ~ finger ‖ Kontaktfinger m. der Schaltwalze ‖ doigt m. de contact du contrôleur.

control level ‖ Prüflibelle f. ‖ niveau m. de contrôle. / ~ lever ‖ Betätigungshebel m.; Schalthebel m. ‖ levier m. de manœuvre; palonnier m.

controlling, apparatus for ~ the generation of power ‖ Apparat m. zur Überwachung der Energieerzeugung ‖ appareil m. pour le contrôle de la production d'énergie. / device for ~ the distribution of power ‖ Einrichtung f. zur Überwachung der Energieverteilung ‖ installation f. pour le contrôle de la répartition d'énergie. / ~ the flowering season and the growth of plants ‖ Regelung f. der Blütezeit und des Pflanzenwachstums ‖ réglage m. de la mise en fleurs et de la croissance de plantes. / ~ machines ‖ Maschinenkontrolle f. ‖ contrôle m. des machines.

controlling apparatus ‖ Kontrollapparat m.; Steuerapparat m. ‖ appareil m. de contrôle. / ~ clock ‖ Kontrolluhr f. ‖ montre f. ou horloge f. de contrôle. / ~ device ‖ Kontrollvorrichtung f. ‖ dispositif m. de contrôle. / electric ~ device ‖ elektrischer Kontrollapparat m. ‖ contrôleur m. électrique. / ~ grid (Radio) ‖ Steuergitter n. ‖ grille f. régulatrice. / ~ spring for locomotive trucks ‖ Rückstellfeder f. für Lokomotivdrehgestelle ‖ ressort m. de rappel pour bogies de locomotives. / ~ switch (Electr) ‖ Steuerschalter m. ‖ appareil m. de commande; contrôleur m. / ~ valve ‖ Steuerventil n.; Umsteuerschieber m. ‖ valve f. ou soupape f. de manœuvre.

control lock ‖ Sperrschloß n. ‖ verrou m. de blocage. / ~ period see ~ reaction time. / ~ pillar for the speedy and distinct transmission of instructions ‖ Kommandosäule f. zur schnellen und eindeutigen Übermitt(e)lung von Befehlen ‖ colonne f. de commande pour la transmission rapide et irréprochable des ordres. / remote ~ plant (Electr) ‖ Fernsteueranlage f. ‖ installation f. de télécommande. / ~ reaction time ‖ Steuerzeit f. ‖ temps m. d'action de gouvernail. / ~ room (Aero) ‖ Kommandoraum m. ‖ poste m. de commandement. / ~ section (Tel) ‖ Kontrollamt n. ‖ bureau m. de contrôle. / ~ sleeve ‖ Kontrollmuffe f. ‖ manchon m. de contrôle. / ~ station of the engine ‖ Bedienungsstand m. der Maschine ‖ poste m. de manœuvre de la machine. / ~ strip ‖ Kontrollstreifen m. ‖ bande f. de contrôle. / ~ system ‖ Kontrollsystem n. ‖ système m. de contrôle. / ~ test ‖ Prüfmessung f. ‖ mesure f. de contrôle. / ~ transmitter (Radio) ‖ Steuersender m. ‖ transmetteur m. de commande. / automatic ~ valve for pig iron mixers ‖ selbsttätige Mischersteuerung f. für Roheisenmischer ‖ distributeur m. automatique pour mélangeurs de fonte.

contusion ‖ Quetschung f. ‖ contusion f.

convalescents' home ‖ Erholungshaus n. für Genesende ‖ maison f. de convalescents.

convection, electrolytic ~ ‖ elektrolytische Konvektion f. ‖ convection f. électrolytique. / ~ current (Electr) ‖ Konvektionsstrom m. ‖ courant m. de convection.

convenience ‖ Geeignetheit f.; Eignung f. ‖ convenance f.

convenient ‖ geeignet ‖ convenable. / ~ operation ‖ bequeme Bedienung f. ‖ manipulation f. simple.

convention for meteorological service ‖ internationales Wetterdienstabkommen n. ‖ convention f. internationale pour le service météorologique.

converge, to ‖ konvergieren; zusammenlaufen ‖ converger.

convergence ‖ Konvergenz f. ‖ convergence f. / ~ angle ‖ Konvergenzwinkel m. ‖ angle m. de convergence. / ~ angle gauge ‖ Konvergenzwinkelmesser m. ‖ appareil m. pour mesurer l'angle de convergence.

converging cylinder lens ‖ sammelnde Zylinderlinse f. ‖ lentille f. cylindrique convergente. / ~ lens ‖ Sammellinse f.; Konvexlinse f. ‖ lentille f. convergente.

conversation ‖ Rücksprache f.; Besprechung f. ‖ pourparler m. / installation for secret ~s (Tel) ‖ Geheimsprecheinrichtung f. ‖ installation f. pour conversations secrètes. / meter for duration of ~ ‖ Gesprächszeitmesser m. ‖ compteur m. du temps de conversation. / ~ by subscription (Tel) ‖ Abonnementsgespräch n.; Monatsgespräch n. ‖ conversation f. par abonnement.

conversation counter ‖ Gesprächszähler m. ‖ compteur m. de conversations. / ~ meter see ~ counter.

conversion (Chem) ‖ Umsetzung f. ‖ conversion f.; transformation f. / ~ (Electr) ‖ Umwandlung f. ‖ changement m. / ~ (Math) ‖ Umkehrung f. ‖ renversement m. / ~ from alternating current to direct current ‖ Umformung f. von Wechselstrom in Gleichstrom ‖ transformation f. de courant alternatif en courant continu. / factory plan for the ~ from peace to war footing ‖ Plan m. zur Werksumstellung auf Kriegsfertigung ‖ plan m. de l'usine à transformer le travail de la paix à celui de la guerre. / ~ into salt (Chem) ‖ Salzbildung f. ‖ salification f.

convert, to ‖ umwandeln ‖ convertir; transformer. / ~ alternating current into continuous current ‖ Wechselstrom m. in Gleichstrom m. umwandeln ‖ convertir le courant alternatif en courant continu. / ~ into ‖ umrechnen ‖ réduire; calculer. / ~ mechanical energy into electrical energy ‖ mechanische Energie f. in elektrische verwandeln ‖ transformer l'énergie mécanique en énergie électrique. / ~ into power ‖ in Energie f. umsetzen ‖ transformer en énergie.

converted, to be ~ into (Chem) ‖ sich umwandeln in ‖ se transformer en. / ~ steel ‖ Blasenstahl m.; Zementstahl m. ‖ acier m. cémenté; acier m. de cémentation à soufflures.

converter (Electr) ‖ Umformer m. ‖ convertisseur m. / ~ (Metal) ‖ Konverter m.; Birne f. ‖ convertisseur m.; cornue f. / acid lining of the ~ ‖ saures Konverterfutter n. ‖ garnissage m. acide de la cornue. / alternating-current-continuous-current ~ ‖ Wechselstromgleichstromeinankerumformer m. ‖ commutatrice f. à courant alternatif en courant continu. / cascade ~ ‖ Kaskadenumformer m. ‖ convertisseur m. en cascade. / continuous-current-alternating-current ~ ‖ Gleichstromwechselstromeinankerumformer m. ‖ commutatrice f. à courant continu en courant alternatif. / continuous-current-rotary-current ~ ‖ Gleichstromdrehstromeinankerumformer m. ‖ commutatrice f. à courant continu en courant triphasé. / electrolytic ~ ‖ elektrolytischer Gleichrichter m. ‖ convertisseur m. électrolytique. / flywheel ~ ‖ Schwungrad-

umformer m. ‖ transformateur m. à volant. / mercury vapour static ~ ‖ Quecksilberdampfgleichrichter m. ‖ convertisseur m. statique à vapeur de mercure. / pear-shaped ~ (Met) ‖ birnenförmiger Konverter m. ‖ convertisseur m. en forme de poire. / plug ramming machine for ~s ‖ Bodenstampfmaschine f. für Konverter ‖ machine f. à damer les fonds de convertisseurs. / rotary ~ ‖ Einankerumformer m.; Drehumformer m. ‖ commutatrice f.; convertisseur m. / rotary-current-continuous-current ~ ‖ Drehstromgleichstromeinankerumformer m. ‖ commutatrice f. à courant triphasé en courant continu. / ~ with self-starter ‖ Umformer m. mit Selbstanlauf ‖ convertisseur m. avec auto-démarrage. / single-phase voltage ~ ‖ einphasiger Spannungswandler m. ‖ transformateur m. monophasé.

converter bottom ‖ Konverterboden m. ‖ fond m. de convertisseur. / machine for stamping ~ bottoms ‖ Konverterbodenstampfmaschine f. ‖ machine f. à damer les fonds de convertisseurs. / ~ leverman ‖ Konverterkipper m. ‖ opérateur m. du convertisseur. / ~ plant (Electr) ‖ Umformeranlage f. ‖ installation f. de commutation. / ~ process (Metal) ‖ Birnenverfahren n. ‖ procédé m. au convertisseur. / ~ ring ‖ Konverterring m. ‖ anneau m. de convertisseur. / ~ shop ‖ Konverterhalle f. ‖ hall m. des convertisseurs. / underground ~ station ‖ unterirdisches Umformerwerk n. ‖ poste m. de conversion souterrain. / ~ switch board ‖ Umformerschalttafel f. ‖ tableau m. de distribution de transformateur. / truck for ~s ‖ Bodeneinsatzwagen m. für Konverter ‖ wagonnet m. de transport pour fonds de convertisseurs. / ~ trunnion ‖ Schildzapfen m. des Konverters ‖ tourillon m. du convertisseur.

convertible ‖ verwandelbar; umwandelbar ‖ convertible.

converting (Electr) ‖ Umwandlung f. ‖ transformation f. / ~ (Metal) ‖ Windfrischen n. ‖ affinage m. au vent ou par soufflage. / half ~ ‖ Einsatzhärtung f.; Oberflächenhärtung f. ‖ trempe f. en paquet ou en coquille ou de la surface. / ~ process ‖ Windfrischverfahren n. ‖ procédé m. d'affinage au vent.

convertor see converter.

convex ‖ bauchig; erhaben ‖ convexe. / silvered ~ glass reflector ‖ versilberter Glaskonvexspiegel m. ‖ miroir m. convexe en verre argenté. / ~ lens ‖ Konvexlinse f.; Sammellinse f. ‖ verre m. convergent ou convexe. / ~ shaping attachment ‖ Konvexhobelvorrichtung f. ‖ dispositif m. à raboter convexe. / ~ turning attachment ‖ Einrichtung f. zum Balligdrehen ‖ dispositif m. à tourner bombé.

convexly turned ‖ ballig gedreht ‖ bombé.

convey, to ‖ befördern ‖ transférer; transporter; conduire.

conveyance (Jur) ‖ Übertragung f. ‖ cession f. / ~ (Mine) ‖ Förderung f. ‖ extraction f. / ~ (Trade) ‖ Beförderung f. ‖ transport m.; charriage m. / ~ in bulk see ~ in mass. / deed of ~ ‖ Überschreibungsurkunde f.; Übertragungsurkunde f. ‖ acte m. translatif de propriété; acte m. de cession. / ~ of earth (Railw) ‖ Erdtransport m. ‖ mouvement m. de

terre. / ~ of goods || Gütertransport m. || messagerie f. / ~ in mass || Massenförderung f. || déplacement m. ou extraction f. ou transport m. en masse. / means pl. of ~ || Beförderungsmittel n. || moyen m. de transport. / to modify the ~ of the grant || die Verleihungsurkunde f. ändern || modifier l'acte m. de concession. / ~ of parcels || Paketbeförderung f. || transport m. de colis. / ~ by post || Postbeförderung f. || transport m. postal. / ~ of sludge || Schlammförderung f. || transport m. de schlamms. / vertical ~ || Vertikaltransport m. || transport m. entre niveaux différents.

conveyance band || Transportband n. || ruban m. de transport. / ~ worm || Transportschnecke f. || hélice f. transporteuse.

conveyer || Förderer m.; Transportör m. || transporteur m. / ~ (Mill) || Transportschnecke f. || vis f. sans fin. / aerial ~ || elektrische Seilbahn f. || transporteur m. électrique aérien. / aerial ~ of telpheric type || Hängebahn f. || transporteur m. telphérique. / ash ~ || Aschetransportanlage f. || installation f. de transport des cendres. / band ~ || Bandtransportör m. || transporteur m. à courroie. / belt ~ || Gurtförderer m. || transporteur m. à courroie ou à bande. / bucket ~ || Becherwerk n. || transporteur m. à godets. / bucket chain ~ || Becherkettenförderer m. || transporteur m. à chaîne de godets; convoyeur m. / chain ~ || Kettenförderer m. || transporteur m. à chaîne. / circular ~ || Kreisförderer m. || transporteur m. circulaire. / close-range ~ || Nahförderer m. || transporteur m. à courte distance. / compressed-air ~ || Druckluftförderer m. || transporteur m. pneumatique. / continuous belt ~ || laufendes Band n. || table f. roulante; transporteur à courroie sans fin. / dipping bucket ~ || schaufelndes Becherwerk n. || élévateur m. à augets piocheurs. / flour worm ~ || Mehlförderschnecke f. || vis f. sans fin pour le transport de farine. / mechanical ~ for the precipitations of sand || mechanische Fördereinrichtung f. für Bergeversatz || engin m. mécanique de transport pour éboulis à remblayage. / rake ~ || Rechenförderer m. || transporteur m. à râteaux. / screw ~ || Schneckenförderer m. || transporteur m. à vis sans fin. / short-distance ~ || Nahförderer m. || transporteur m. à petite distance. / suction air ~ || Saugluftförderer m. || transporteur m. à aspiration d'air. / worm ~ see screw ~.

conveyer belt || Förderband n.; Transportband n. || bande f. ou courroie f. transporteuse. / ~ chain || Becherkette f.; Förderkette f. || transporteur m. ou chaîne f. à godets; convoyeur m. / mechanically driven ~ chute || mechanisch bewegte Förderrutsche f. || plan m. incliné de transport à actionnement mécanique. / ~ pipe line || Förderleitung f. || tuyauterie f. de transport. / ~ spiral || Förderschnecke f. || vis f. (sans fin) transporteuse. / ~ trough (Mine) || Förderrinne f. || gouttière f. à secousses. / mechanically driven ~ trough || mechanisch bewegte Förderrinne f. || gouttière f. de transport à actionnement mécanique. / mechanically driven ~ tube || mechanisch bewegtes Förderrohr n. || tuyau m. de

transport à actionnement mécanique. / ~ worm || Förderschnecke f. || vis f. transporteuse.

conveying || Förderung f. || transport m.; extraction f.; charriage m. / ~ the spring water || Quellfassung f. || captation f. de source.

conveying appliances pl. || Fördergerät n. || accessoires mpl. de transport. / ~ band link || Transportbandgelenk n. || articulation f. de la bande de transport. / ~ batten || Transportleiste f. || listeau m. transporteur. / ~ belt || Förderband n. || courroie f. de transport; bande f. transporteuse. / ~ bridge for coal open working || Abraumförderbrücke f. || pont m. de transport pour travaux de déblaiement. / ~ bucket || Transportbecher m. || godet m. de transport. / ~ channel (Hydr arch) || Obergerinne n.; Vorarche f. || bief m. / oscillating ~ channel || Schwingförderrinne f. || gouttière f. de transport oscillante. / ~ device || Fördervorrichtung f. || transporteur m. / ~ equipment || Fördermittel n. || engin m. de transport. / ~ equipment for wholesale goods || Transportmittel n. für Massengüter || dispositif m. de transport de marchandises en masses. / ~ implement || Transportgerät n. || ustensile m. de transport. / fecal ~ implement || Fäkalienabfuhrgerät n. || ustensile m. pour l'enlèvement des matières fécales. / ~ installation || Fördereinrichtung f. || installation f. de transport. / harvest ~ means pl. || Erntetransportmittel n. || dispositif m. de transport de récoltes. / ~ medium || Transportmittel n. || moyen m. de transport.

conveying plant || Förderanlage f. || installation f. de transport. / coal ~ || Bekohlungsanlage f. || installation de chargeur mécanique de charbon. / ~ driven by compressed-air || durch Druckluft betriebene Förderanlage f. || installation f. (d'engin) de transport fonctionnant au moyen de l'air comprimé. / ~ driven by suction air || durch Saugluft betriebene Förderanlage f. || installation f. (d'engin) de transport fonctionnant au moyen de l'air aspiré. / fixed ~ || ortsfeste Förderanlage f. || installation f. d'extraction fixe. / oil ~ || Ölfördereinrichtung f. || installation f. pour le transport de l'huile. / pneumatic ~ for stone powder || Druckluftgesteinsstaubförderanlage f. || installation f. pour le transport pneumatique de la poussière de pierres. / portable ~ || fahrbare Förderanlage f. || installation f. d'extraction mobile. / structural steelwork for ~s || eiserne Gerüstbauten mpl. für Transportanlagen || charpentes fpl. métalliques pour installations de transport. / ~ for wholesale goods || Förderanlage f. für Massengut || installation f. (d'engin) de transport pour marchandises en masses.

conveying, spent grains ~ screw || Austreberschnecke f. || vis f. sans fin pour l'évacuation des drèches. / ~ tank || Fördergefäß n. || benne f. transporteuse; cuve f. ou cage f. d'extraction. / ~ trough || Förderrinne f.; Kastenrinne f. || gouttière f. transporteuse; chéneau m. de transport en bois; transporteur m. à caisse. / connecting rod for ~ troughs || Förderrinnenschubstange f. || bielle f. pour transporteurs à auge. / ~ truck

see ~ tank. / ~ tube || Förderrohr n. || tuyau m. de transport.

conveyor see conveyer.

convincing || überzeugend; triftig || valable; frappant; démonstratif; fondé.

convolution of the winding || Umlauf m. der Wicklung || circonvolution f. de l'enroulement.

cook, to || kochen || cuire.

cooker (Pap) || Holzkocher m. || cuiseur m.; bouiller. / self-~ || Kochkiste f. || fourréchaud m.

cooking, electric || elektrisches Kochen n. || cuisson f. électrique. / ~ of herbs || Kräuterkochen n. || cuisson f. d'herbes. / ~ and house-keeping school || Koch- und Haushaltungsschule f. || école f. ménagère.

cooking apparatus || Kochapparat m.; Kocher m.; Kochmaschine f.; Kochherd m. || appareil m. ou fourneau m. à cuire ou de cuisine. / economical ~ || Sparkocher m. || fourneau m. économique. / electric ~ || elektrischer Kochapparat m. || appareil m. électrique pour la cuisine; fourneau m. électrique de cuisine. / petrol ~ || Petroleumkocher m. || appareil m. à cuire au pétrole.

cooking box || Kochkiste f. || caisse f. garde-chaleur de cuisine; caisse f. à cuire. / connection for electric ~ consumption || Wärmeanschluß m. einer Elektroküche || raccordement m. pour usages calorifiques d'une cuisine électrique. / bone coal ~ drum || Knochenkohlekochbottich m. || tonneau m. de cuisson de charbon d'os. / ~, frying and baking apparatus || Koch-, Brat- und Backapparat m. || appareil m. à cuire, rôtir et frire. / ~ installation for ships || Schiffsküche f. || cuisine f. de bord. / ~ knife || Küchenmesser n. || couteau m. de cuisine. / ~ plant || Kochanlage f. || cuisine f. / ~ plate || Kochplatte f. || réchaud m. (de cuisine). / electrically heated ~ plate || elektrisch beheizte Kochplatte f. || réchaud m. électrique. / ~ point || Kochstelle f. || feu m. / ~ pot || Kochtopf m. || marmite f.; casserole f. / ~ range || Kochofen m. || poêle f. de cuisine. / electric ~ range || elektrischer Kochherd m. || appareil m. électrique à cuire. / four-holed ~ range || Vierstellenkochherd m. || cuisinière f. à quatre trous. / ~ and house-keeping school || Koch- und Haushaltungsschule f. || école f. ménagère. / ~ stove || Kochherd m. || fourneau m. de cuisine. / ~ utensil || Kochgeschirr n. || casserole f. / electrically heated ~ vessel || elektrisch beheiztes Kochgefäß n. || marmite f. chauffée électriquement.

cookroom (Shipb) || Schiffsküche f.; Kombüse f. || cuisine f. de l'équipage.

cool || kühl || frais; froid. / to get ~ || kalt werden || se refroidir. / keep in ~ place || kühl aufbewahren || garder en lieu m. frais.

cool, to || kühlen || réfrigérer; refroidir. / ~ || erkalten; kalt werden || refroidir. / ~ down || abkühlen || refroidir; réfrigérer. / ~ down from x⁰ to y⁰ || von x⁰ auf y⁰ abkühlen || refroidir de x⁰ à y⁰. / the pieces pl. should be cooled down carefully under warm ashes || man läßt die Stücke unter warmer Asche langsam erkalten || on laisse refroidir avec précaution les pièces fpl. sous de la cendre chaude. / ~ the gelatine || Gelatine f. kühlen || refroidir la gélatine.

/ ~ immediately ‖ abschrecken ‖ refroidir brusquement. / ~ -off ‖ sich abkühlen ‖ se rafraîchir; se refroidir.

cool ‖ Kühle f. ‖ fraîcheur f.

cooled, air-~ ‖ luftgekühlt ‖ refroidi par air. / water-~ ‖ wassergekühlt ‖ à refroidissement d'eau.

cooled area ‖ gekühlte Bodenfläche f. ‖ surface f. refroidie. / ~ injection nozzle ‖ gekühlte Einspritzdüse f. ‖ gicleur m. refroidi. / ~ inspection cell in a mortuary ‖ gekühlte Schauzelle f. in einem Leichenschauhaus ‖ cellule f. refroidie dans une morgue.

cooler ‖ Kühler m.; Kühlapparat m. ‖ réfrigérant m.; réfroidisseur m.; rafraîchissoir m. / ~ (Auto) ‖ Kühler m. ‖ radiateur m. / ~ (Brew) ‖ Kühlschiff n. ‖ bac m. refroidissoir ou rafraîchissoir. / ~ of closed type with forced draught ‖ Kühlwerk n. mit künstlichem Zug ‖ réfrigérant m. clos à tirage forcé. / ~ for the compressor ‖ Kompressorkühler m. ‖ réfrigérant m. du compresseur. / counter-current ~ ‖ Gegenstromkühler m. ‖ réfrigérateur m. à contre-courant. / ~ of faggot type ‖ Reisiggradierwerk n. ‖ réfrigérant m. à fascinage. / ~ with forced draught ‖ Kühlturm m. mit künstlichem Zug ‖ réfrigérant m. à tirage forcé. / gilled ~ see ribbed ~. / high-pressure pipe ~ ‖ Hochdruckröhrenkühlapparat m. ‖ appareil m. frigorifique tubulaire à haute pressions / ~ of lath type ‖ Kühlturm m. aus Lattenwerk ‖ réfrigérant m. à persiennes. / lye ~ ‖ Laugenkühler m. ‖ rafraîchissoir m. à lessives. / ~ with natural draught ‖ selbstlüftender Kaminkühler m. ‖ réfrigérant m. à tirage naturel. / oil ~ ‖ Ölkühler m. ‖ refroidisseur m. d'huile. / ~ in the open air with spraying nozzles ‖ im Freien aufgestellter Rieselkühler m. mit Düsen ‖ réfrigérant m. à air libre à tuyères de pulvérisation. / ~ of open type ‖ offener Kaminkühler m. ‖ réfrigérant m. ouvert. / rain air ~ ‖ Regenluftkühler m. ‖ frigorifère m. à pluie. / ribbed ~ ‖ Rippenkühler m. ‖ radiateur m. ou réfrigérant m. à ailettes. / ~ with stages ‖ Zonenkühler m. ‖ réfrigérant m. à gradins. / surface evaporative ~ ‖ Berieselungskondensator m. ‖ condenseur m. à ruissellement. / tube air ~ ‖ Röhrenluftkühler m. ‖ frigorifère m. à tuyaux. / tubular ~ ‖ Batteriekühler m. ‖ réfrigérant m. tubulaire.

coolerman ‖ Zuckerkocher m. ‖ évaporeur m.

cooler turbine ‖ Kühlerturbine f. ‖ turbine f. de réfrigérant.

cool-hammer, to ~ the iron ‖ das Eisen kaltschmieden ‖ battre le fer à froid; écrouir le fer.

cool-hammered iron ‖ kaltgeschmiedetes Eisen n. ‖ fer m. écroui.

cooling ‖ kühlend ‖ réfrigérant.

cooling ‖ Kühlung f.; Abkühlung f. ‖ refroidissement m.; réfrigération f. / air ~ ‖ Luftkühlung f. ‖ refroidissement m. par air. / ~ of the air in dwelling and working rooms in the tropics ‖ Luftkühlung f. bei Wohn- und Arbeitsräumen in den Tropen ‖ réfrigération f. de l'air aux lieux habités et d'ateliers dans les pays tropicaux. / ~ by automatic circulation ‖ Kühlung f. durch selbsttätigen Umlauf ‖ refroidissement m.

par thermo-siphon. / ~ by means of circulating water ‖ Wasserumlaufkühlung f. ‖ refroidissement m. par circulation d'eau. / ~ by conduction ‖ Abkühlung f. durch Leitung ‖ refroidissement m. par conduite. / ~ by evaporation ‖ Verdunstungskühlung f. ‖ refroidissement m. par évaporation. / forced draught ~ ‖ Druckluftkühlung f. ‖ refroidissement m. à courant d'air forcé. / ~ by gills ‖ Rippenrohrkühlung f. ‖ refroidissement m. par ailettes. / intermediate ~ ‖ Zwischenkühlung f. ‖ refroidissement m. intermédiaire. / loss due to ~ ‖ Wärmeverlust m. ‖ perte f. par refroidissement. / ~ of malt floors ‖ Kühlung f. der Malztennen ‖ refroidissement m. des germoirs. / ~ by melting ‖ Abkühlung f. durch Verflüssigung ‖ refroidissement m. par fusion. / natural circulation water ~ ‖ Thermosiphonkühlung f. ‖ refroidissement m. par thermosiphon. / ~ by outspread surface ‖ Flächenkühlung f. ‖ refroidissement m. à surface déployée. / period of ~ ‖ Abkühlungszeit f. ‖ période f. de refroidissement. / ~ by radiation ‖ Abkühlung f. durch Strahlung ‖ refroidissement m. par rayonnement. / ~ by vaporization see ~ by evaporisation. / water ~ ‖ Wasserkühlung f. ‖ refroidissement m. à l'eau.

cooling action ‖ Kühlwirkung f. ‖ action f. refroidissante. / ~ agent ‖ Kälteträger m. ‖ agent m. frigorifique. / highly volatile ~ agent ‖ hochflüchtiger Kälteträger m. ‖ agent m. frigorifique très volatil. / ~ arch (Glassm) ‖ Kühlofen m. ‖ carcaise f. / ~ band ‖ Kühlband n. ‖ bande-refroidisseur m. / ~ bed ‖ Kühlbett n. ‖ refroidissoir m. / continuous ~ bed ‖ kontinuierliches Warmbett n. ‖ lit m. refroidisseur continu. / roller ~ bed ‖ Rollenkühlbett n. ‖ refroidissoir m. à rouleaux. / ~ cell ‖ Kühlzelle f. ‖ case f. froide. / ~ chamber ‖ Kühlraum m. ‖ chambre f. de refroidissement. / ~ coil ‖ Kühlschlange f. ‖ serpentin m. de refroidissement. / ~ cupboard ‖ Kühlschrank m. ‖ armoire f. frigorifique. / chocolate ~ cupboard ‖ Schokoladenkühlschrank m. ‖ armoire f. frigorifique à chocolat. / ~ curve ‖ Abkühlungskurve f. ‖ courbe f. de refroidissement. / steam ~ device ‖ Dampfkühler m. ‖ réfrigérant m. à vapeur. / ~-down ‖ Erkaltung f. ‖ refroidissement m. / ~ floor (Brew) ‖ Kühlschiff n. ‖ bac m. refroidissoir ou rafraîchissoir. / blast ~ furnace for bottle industry ‖ Zugkühlofen m. für die Flaschenindustrie ‖ four m. de refroidissement à tirage pour l'industrie des bouteilles. / ~ jacket ‖ Kühlmantel m. ‖ enveloppe f. réfrigérante. / ~ jacket motor ‖ Kühlmantelmotor m. ‖ moteur m. à enveloppe de refroidissement. / ~ liquid ‖ Kühlflüssigkeit f. ‖ liquide m. de refroidissement.

cooling machine, absorption ~ ‖ Absorptionskühlmaschine f. ‖ machine f. à froid à absorption. / ammonia absorption ~ ‖ Ammoniakabsorptionsmaschine f. ‖ machine f. à ammoniaque à absorption. / carbonic acid ~ ‖ Kohlensäurekältemaschine f. ‖ machine f. frigorifique à acide carbonique. / compression ~ ‖ Kompressionskühlmaschine f. ‖ machine f. frigorifique à compression.

cooling pipe ‖ Kühlleitung f. ‖ conduite f. de réfrigération. / coiled ~ pipe ‖ Kühlschlange f. ‖ serpentin m. réfrigérateur. / ~ pipes pl. ‖ Kühlrohrsystem n. ‖ tuyaux mpl. réfrigérants.

cooling plant ‖ Kühlanlage f. ‖ installation f. frigorifique. / air ~ ‖ Luftkühlanlage f. ‖ installation f. de rafraîchissement d'air. / fittings pl. for ~s ‖ Armatur f. für Kühlanlagen ‖ armature f. pour installations frigorifiques. / gas ~ ‖ Gaskühlanlage f. ‖ installation f. de réfrigération des gaz. / ~ for low temperature ‖ Tiefkühlanlage f. ‖ installation f. frigorifique à basse température. / milk ~ ‖ Milchkühlanlage f. ‖ installation f. frigorifique dans une laiterie. / soap slab ~ ‖ Seifenplattenkühlanlage f. ‖ refroidisseur m. de plaques de savon.

cooling pump ‖ Kühlpumpe f. ‖ pompe f. de refroidissement. / ~ rack (Roll mill) ‖ Kühlbett n. ‖ lit m. refroidisseur. / ~ rib ‖ Kühlrippe f. ‖ ailette f. de refroidissement. / ~ ring for blast furnaces ‖ Kühlring m. für Hochöfen ‖ anneau m. refroidisseur pour hauts-fourneaux.

cooling room, butter ~ ‖ Butterkühlraum m. ‖ chambre f. froide à beurre. / cheese ~ ‖ Käsekühlraum m. ‖ chambre f. froide à fromage. / egg ~ ‖ Eierkühlraum m. ‖ chambre f. froide pour œufs. / fish ~ ‖ Fischkühlraum m. ‖ chambre f. froide à poissons. / fixed ~ ‖ fester Kühlraum m. ‖ chambre f. froide fixe. / fruit ~ ‖ Obstkühlraum m. ‖ chambre f. froide fruits. / meat ~ ‖ Fleischkühlraum m. ‖ chambre f. froide à viande. / milk ~ ‖ Milchkühlraum m. ‖ chambre f. froide à lait. / poultry ~ ‖ Geflügelkühlraum m. ‖ chambre f. froide à volaille. / ~ for raw furs ‖ Kühlraum m. für Rohpelze ‖ chambre f. froide pour fourures brutes. / ~ for ready-made furs ‖ Kühlraum m. für Konfektionspelze ‖ chambre f. froide pour fourrures confectionnées. / removable ~ ‖ transportabler Kühlraum m. ‖ chambre f. froide transportable. / ~ for seeds of lily of the valley ‖ Kühlraum m. für Maiblumenkeime ‖ chambre f. froide pour des germes de muguet.

cooling ship ‖ Kühlschiff n. ‖ navire m. frigorifique. / ~ stack ‖ Gradierwerk n. ‖ bassin m. de graduation. / ~ surface ‖ Kühlfläche f. ‖ surface f. de refroidissement ou de réfrigération. / ~ system ‖ Kühlsystem n. ‖ système m. de refroidissement. / ~ tank ‖ Kühlapparat m. ‖ appareil m. réfrigérant ou refroidisseur. / ~ tower ‖ Kühlturm m. ‖ tour f. refroidissante; réfrigérant m. à cheminée. / ~ tower (Saltworks) ‖ Gradierwerk n. ‖ réfrigérant m. à fascines. / chimney type ~ tower ‖ Kühlturm m. ‖ réfrigérant m. à cheminée. / hourly capacity of a ~ tower ‖ Stundenleistung f. eines Kühlturmes ‖ puissance f. horaire d'un tour de réfrigération. / ~ train ‖ Kühlzug m. ‖ train m. frigorifique. / ~ trough ‖ Kühlbett n.; Kühltrog m. ‖ lit m. refroidisseur; auge m. de rafraîchissement. / ~ tube ‖ Kühlwanne f. ‖ cuve f. réfrigérante. / ~ tube (Brew) ‖ Berieselungsrohr n. ‖ tube m. rafraîchisseur. / ~ wagon ‖ Kühlwaggon m. ‖ wagon m. frigorifique.

cooling water ‖ Kühlwasser n. ‖ eau f. réfrigérante ou de refroidissement. / quantity of ~ required ‖ Kühlwasserbedarf m.

|| quantité f. d'eau nécessaire au refroidissement.

cooling water admission || Kühlwassereintritt m. || entrée f. de l'eau refroidissante. / ~ circulation || Kühlwasserumlauf m. || circulation f. d'eau de refroidissement. / ~ discharge || Kühlwasseraustritt m. || sortie f. de l'eau refroidissante. / ~ feed pipe || Kühlwasserzuleitung f. || conduite f. d'amenée de l'eau de refroidissement. / ~ inlet || Kühlwassereintritt m. || entrée f. de l'eau de refroidissement. / ~ outlet || Kühlwasseraustritt m. || sortie f. de l'eau de refroidissement. / ~ pipe line || Kühlwasserleitung f. || conduite f. de l'eau réfrigérante. / ~ pump || Kühlwasserpumpe f. || pompe f. à eau froide. / ~ space || Kühlwasserraum m. || enveloppe f. de circulation d'eau froide. / ~ supply || Kühlwasserbeschaffung f. || alimentation f. en eau réfrigérante. / ~ tenter || Kühlwasserwärter m. || régleur d'eau de refroidissement.

cooling worm || Kühlschnecke f. || hélice-refroidisseur f.

cool-shear, to || kaltscheren || cisailler à froid.

coom || Achsschmiere f.; verdickte Schmiere f. || cambouis m.

coop || Bottich m.; Wanne f. || cuve f.; cuveau m. / ~ (Mine) || Kaue f. || hangar m.; abri m.

cooper || Böttcher m.; Küfer m. || tonnelier m.; futailleur m.; foudrier m.

cooperage || Böttcherei f.; Küferei f. || tonnellerie f.; barillage m.

co-operate, to || zusammenarbeiten || coopérer.

co-operation || Zusammenarbeit f. auf gemeinnütziger Grundlage || cooperation f.

co-operative association || Gegenseitigkeitsgesellschaft f. || société f. mutuelle. / ~ association || Erwerbsgenossenschaft f. || société f. coopérative d'achats. / ~ bank || Genossenschaftsbank f. || banque f. coopérative. / ~ society || Konsumverein m. || société f. ou coopérative f. de consommation. / ~ trading system || Genossenschaftswesen n. || associations fpl.; syndicalisme m.

cooper's heading-knife || Schroppmesser n. für Küfer || plane f. à foncer pour tonneliers. / ~ jointer || Fügebank f. || colombe f.; colombe f. à joindre. / ~ plane || Fügebank f.; Fügblock m. || colombe f.; joindeux m. / double iron for ~ plane || Fügblockdoppeleisen n. || fer m. avec contrefer à colombe. / ~ plane iron || Fügblockeisen n. || fer m. de colombe. / ~ tools pl. || Böttcherwerkzeug n. || outils mpl. de tonneliers. / ~ wood || Faßholz n. || bois m. à tonneaux.

coop repairer || Büttenflicker m. || tonnelier-réparateur m.

co-ordinate || Koordinate f. || coordonnée f. / Cartesian ~s pl. || Punktkoordinaten fpl.; kartesische Koordinaten fpl. || coordonnées fpl. cartésiennes. / correction of ~s || Koordinatenverbesserung f. || rectification f. des coordonnées. / oblique-angled ~s pl. || schiefwinklige Koordinaten fpl. || coordonnées fpl. obliques. / origin of ~s || Koordinatenanfang m. || origine f. des coordonnées. / parallel shift of systems of ~s || Parallelverschiebung f. der Koordinatensysteme || déplacement m. parallélique des systèmes de coordonnées. / polar ~s pl. || Polarkoordinaten fpl. || coor-

données fpl. polaires. / rectangular ~s pl. || rechtwinklige Koordinaten fpl. || coordonnées fpl. cartésiennes. / plotting by rectangular ~s || Aufnahme f. nach dem Koordinatenverfahren || levé m. de plan par la méthode des coordonnées. / right-angled ~s pl. || rechtwinklige Koordinaten fpl. || coordonnées fpl. rectangulaires. / system of ~s || Achsenkreuz n.; Koordinatensystem n. || système m. de coordonnées. / tangential ~s pl. || Tangentenkoordinaten fpl. || coordonnées fpl. tangentielles. / transformation of ~s || Umrechnung f. von Koordinaten || transformation f. des coordonnées. / trilinear ~s pl. || Dreilinienkoordinaten fpl. || coordonnées fpl. triangulaires.

co-ordinate axes || Koordinatenachsen fpl. || axes mpl. des coordonnées. / ~ measuring apparatus || Koordinatenmeßapparat m. || appareil m. pour la mesure des coordonnées. / ~ measuring instrument for astronomical purposes || Koordinatenmeßapparat m. für astronomische Zwecke || appareil m. pour la mesure des coordonnées destiné aux travaux astronomiques. / ~ reading microscope for measuring watch parts || Koordinatenmeßmikroskop n. für Taschenuhrbestandteile || microscope m. pour la mesure de pièces d'horlogie.

cop, to ~ ores || Erze scheiden || scheider ou séparer les minerais.

cop || Garnkötzer m. || cannette f.

copaiba balsam || Kopaivabalsam m. || baume f. copaiba.

copal || Kopal n.; Kopalharz n. || résine f. copal; copal m.

copaline || Kopalin m. || copal m. fossile; copaline f.

copal oil || Kopalöl n. || huile f. de copal. / ~ resin see copal. / ~ varnish || Kopallack m. || vernis m. au copal. / ~ washing || Kopalwäscherei f. || laverie f. de copal.

copartnership ltd. || Gesellschaft f. m. b. H. || société f. à responsabilité limitée.

cope, to ~ with the peak consumption || den Spitzenbedarf m. decken || couvrir la pointe.

cope (Build) || Gewölbbogen m.; Mauerabdeckung f. || voûte f.; chaperon m. / ~ (Cloth) || Chorrock m. || chape f. / ~ (Found) || Oberkasten m.; Mantel einer Lehmform || châssis m. supérieur; manteau m.

copied || abschriftlich || en copie f.

copier of notes || Notenschreiber m. || copiste m. de musique; noteur m.

coping (Build) || Mauerabdeckung f.; Mauerkappe f. || chaperon m.; couronnement m. / convex ~ || konvexe Mauerabdeckung f. || chaperon m. en bahut. / ~ of an enclosure wall || Mauerabdeckung f. oder Abdach n. einer Einfriedigungsmauer || larmier m. d'un mur de clôture. / segmental ~ see convex ~.

coping brick || Kappenziegel m.; Deckziegel m. || brique f. à chaperon; mitron m.; dalle f. de brique. / ~ piece see ~ plate. / ~ plate (Carp) || Wandrahmen m.; Oberschwelle f. || raineau m.; chapeau m. de cloison. / ~ stone (Road) || Abdeckplatte f. || pierre f. à chaperon; plateau m. de couverture.

cop lath || Kötzerständer m. || support m. des bobines.

copper || kupfern || de cuivre.

copper, to || verkupfern || cuivrer. / ~ a ship || ein Schiff n. kupfern || doubler en cuivre un bateau.

copper || Kupfer n. || cuivre m. / ~ (Brew) || Braupfanne f. || chaudière f. de brasserie ou à brasser. / ~ (Coin) || Kupfermünze f. || monnaie f. de cuivre. / ~ (Dyer) || Küpe f. || cuve f. de teinture. / ~ (Househ) || kleiner Kessel m. || chaudron m.; petite chaudière f. / ammonico-muriatic ~ || Kupferchloridammoniak n. || cuivre m. ammonico-muriatique. / ~ amount of ~ used || Kupferverbrauch m. || cuivre m. employé. / ~ in bars || Stangenkupfer n. || cuivre m. en barres. / bean shot ~ || granuliertes Kupfer n. || cuivre m. en grains ou en dragées. / black ~ || Schwarzkupfer n.; Rohkupfer n. || cuivre m. noir; cuivre m. brut. / black oxide of ~ || Kupferoxyd n. || protoxyde m. de cuivre. / ~ in blocks || Kupfer n. in Blöcken || cuivre m. en blocs. / blue ~ || Kupferindig(o) m. || cuivre m. sulfureux; covellite f. / blue carbonate of ~ || Kupferlasur f.; Azurit m.; Chessylit m. || azurite f.; cuivre m. carbonaté bleu; azur m. de cuivre. / capillary red oxide of ~ || haarförmiges Rotkupfererz n.; Kupferblüte f.; Chalkotrichit m. || cuivre m. oxydulé capillaire; zigueline f. capillaire. / carbonate of ~ || kohlensaures Kupferoxyd n. || carbonate m. de cuivre; cuivre m. carbonaté. / cast ~ || gegossenes Kupfer n. || cuivre m. coulé. / coarse ~ see black ~. / dry ~ || übergares oder kaltbrüchiges Kupfer n. || cuivre m. cassant ou sec. / electrolytic ~ || elektrolytisches Kupfer n. || cuivre m. électrolytique. / electrotype ~ || galvanisch gefälltes Kupfer n. || cuivre m. galvanique ou de la pile. / feathered shot ~ see bean shot ~. / ferruginous red oxide of ~ || Kupferbraun n.; Ziegelerz n. || cuivre m. oxydulé ferrifère ou oxydulé terreux. / first refined ~ || rohgares Kupfer n. || cuivre m. demi-fin ou raffiné une fois. / flat bar ~ || Flachkupfer n. || cuivre m. plat. / to free from ~ || entkupfern || décuivrer. / to get coarse ~ || Kupfer n. schwarz machen || fondre le cuivre noir. / gilt ~ || vergoldetes Kupfer n. || cuivre m. doré. / granulated ~ || granuliertes Kupfer n. (in groben Körnern) || cuivre m. en grains ou en dragées. / gray ~ see grey ~. / grey ~ || Kupferpfahlerz n. || cuivre m. gris. / hard ~ || Hartkupfer n. || métal m. dur. / hard-drawn ~ || hart gezogenes Kupfer n. || cuivre m. écroui; cuivre étiré à froid. / hemichloride of ~ || Kupferchlorür n. || chlorure m. cuivreux ou de cuivre. / hemioxyde of ~ || Kupferoxydul n. || oxydule m. de cuivre; cuivre m. oxydulé. / ~ in ingots || Kupfer n. in Barren || cuivre m. en lingots. / micaceous ~ || Kupferglimmer m. || cuivre m. micacé ou mica. / native ~ || gediegenes oder natürliches Kupfer n. || cuivre m. natif. / of ~ || kupfern || de cuivre. / over-poled ~ || überpoltes Kupfer n. || cuivre m. surraffiné. / oxide of ~ || Kupferoxyd n. || protoxyde m. de cuivre. / phosphate of ~ || phosphorsaures Kupferoxyd n. || cuivre m. phosphaté. / ~ in pigs || Kupfer n. in Mulden || cuivre m. en saumons. / plated ~ || versilbertes Kupfer n. || cuivre m. argenté. / to pole ~ || Kupfer n. polen || travailler le cuivre avec la perche. / protochloride of ~ || Kupferchlorid n. ||

chlorure m. cuivrique; chlorure m. *ou* protochlorure de cuivre. / protoxyde of ~ ‖ Kupferoxyd n. ‖ protoxyde m. de cuivre. / raw ~ ‖ rohes Kupfer n. ‖ cuivre m. brut. / recovery of electrolytic ~ from iron lye ‖ Gewinnung f. von Elektrolytkupfer aus Erzlauge ‖ extraction f. du cuivre électrolytique des lessives métalliques. / red ~ ‖ Rotkupfer n. ‖ cuivre m. rouge. / red oxide of ~ ‖ Kupferoxydul n. ‖ oxydule m. de cuivre; cuivre m. oxydulé. / to refine ~ ‖ Kupfer n. hammergar machen ‖ raffiner le cuivre. / refined ~ ‖ hammergares Kupfer n.; Raffinatkupfer n. ‖ cuivre m. fin *ou* raffiné. / to revive ~ ‖ Kupfer n. frischen ‖ rafraîchir le cuivre. / the ~ rises ‖ das Kupfer steigt (Raffinieren) ‖ le cuivre monte *ou* s'élève. / rhomboidal arseniate of ~ ‖ Kupferglimmer m. ‖ cuivre m. arséniaté lamelliforme. / ~ in rods ‖ Stangenkupfer n. ‖ cuivre m. en barres. / rolled ~ ‖ Walzkupfer n. ‖ cuivre m. laminé. / ~ in rolls ‖ Rollenkupfer n. ‖ cuivre m. en rouleaux. / rosette ~ ‖ Rosettenkupfer n. ‖ cuivre m. en rosettes; gâteaux mpl. de rosette; plaques fpl. de cuivre. / round ~ ‖ Rundkupfer n. ‖ cuivre m. rond. / scale of ~ ‖ Kupferglühspan m. ‖ écailles fpl. de cuivre. / scale oxide of ~ ‖ Kupferasche f.; Kupferhammerschlag m. ‖ paille f. *ou* cendres fpl. de cuivre; battiture f. / ~ in sheets ‖ Kupfer n. in Platten ‖ cuivre m. en planches. / ~ in solid blanks ‖ Kupfer n. in Masseln ‖ cuivre m. en lingots. / subacetate of ~ (Chem) ‖ Kupfergrün n. ‖ acétate m. de cuivre; verdit m. / suboxide of ~ ‖ Kupferoxydul n. ‖ oxydule m. de cuivre; cuivre m. oxydulé. / sulphate of ~ ‖ schwefelsaures Kupferoxyd m.; Kupfervitriol n.; Kupfersulphat n. ‖ sulfate de cuivre; cuivre m. sulfaté *ou* vitriolé; vitriol m. bleu; couperose f. bleue. / sulphide of ~ ‖ Kupferglanz m.; Kupferglas n.; Schwefelkupfer n. ‖ cuivre m. sulfuré *ou* vitreux. / under-poled ~ ‖ übergares *oder* kaltbrüchiges Kupfer n. ‖ cuivre m. cassant *ou* sec. / vitreous ~ *see* sulphide of ~. / washing ~ ‖ Waschkessel m. ‖ chaudron m. à bouillir le linge; chaudière f. de buanderie. / work of forged ~ ‖ Kupferschmiedearbeit f. ‖ travail m. de chaudronnerie. / yellow ~ (Metal) ‖ Gelbguß m.; Messing n. ‖ laiton m.; cuivre m. jaune. / yellow ~ (Miner) ‖ Gelbkupfererz n.; Kupferkies m. ‖ pyrite f. cuivreuse; cuivre m. pyriteux.

copper acetate ‖ Grünspan m.; essigsaures Kupfer n. ‖ vert-de-gris m.; acétate m. de cuivre. / ~ and ammonia sulphate ‖ Kupferammoniumsulfat n. ‖ sulfate m. de cuivre ammoniacal. / ~ apparatus ‖ Kupferapparat m. ‖ appareil m. en cuivre. / ~ arsenite ‖ arsenigsaures Kupfer n. ‖ arsénite m. de cuivre. / ~ articles pl. ‖ Kupferwaren fpl. ‖ objets mpl. en cuivre.

copperas ‖ Vitriol n. ‖ couperose f. / blue ~ ‖ blaues Vitriol n.; Kupfervitriol n. ‖ vitriol m. bleu *ou* de cuivre; couperose f. bleue. / green ~ ‖ Eisenvitriol m. ‖ vitriol m. vert *ou* de fer; couperose f. verte.

copper ashes pl. ‖ Kupferasche f.; Kupferhammerschlag m. ‖ paille f. *ou* cendres fpl. de cuivre; battiture f. / ~ assay ‖ Kupferprobe f. ‖ essai m. de cuivre. / ~ band ‖ Kupferband n. ‖ cuivre m. en

bande. / ~ bar ‖ Kupferstange f. ‖ barre f. *ou* tringle m. en cuivre. / ~ bars for stay-bolts and rivets ‖ Kupferstangen fpl. für Stehbolzen und Niete ‖ barres fpl. en cuivre rouge pour entretoises et rivets. / ~ bath ‖ Kupferbad n. ‖ bain m. de cuivre. / ~ beach ‖ Blutbuche f. ‖ hêtre m. rouge. / ~ bearing ‖ kupferführend ‖ cuprifère. / to roll the ~ billet to bars ‖ den Kupferbarren m. zu Stangen auswalzen ‖ laminer la barre f. de cuivre en barreaux. / ~ bit ‖ Lötkolben m. ‖ soudoir m.; fer m. *ou* barre f. à souder. / ~ bit with an edge ‖ Hammerlötkolben m. ‖ fer m. à souder en marteau. / ~ bit with a point ‖ Spitzlötkolben m. ‖ fer m. à souder pointu. / ~ blow pipe ‖ kupfernes Lötrohr n. ‖ chalumeau m. à souder en cuivre. / ~ borate ‖ borsaures Kupfer n. ‖ borate m. de cuivre. / ~ bottom ‖ Kupferboden m. ‖ fond m. de cuivre. / ~ bottom (Shipb) ‖ Kupferbeschlag m.; Kupferhaut f. ‖ doublage m. en cuivre. / ~-bottomed (Shipb) ‖ kupferbodig ‖ doublé en cuivre. / ~-bottomed vessel ‖ Gefäß n. mit Kupferboden ‖ récipient m. à fond de cuivre. / ~ cake ‖ Kupferkuchen m. ‖ saumon m. de cuivre. / ~ cap for sport shooting ‖ Zündhütchen n. für Schieß(sport)zwecke ‖ capsule f. de poudre fulminante de tir. / ~ carbonate ‖ kohlensaures Kupfer n. ‖ carbonate m. de cuivre. / ~ carbon brush ‖ Kupferkohlenbürste f. ‖ balai m. charbon-cuivré. / ~ chain ‖ Kupferkette f. ‖ chaîne f. en cuivre. / ~ chloride ‖ Chlorkupfer n. ‖ chlorure m. de cuivre. / ~ chromate ‖ chromsaures Kupfer n. ‖ chromate m. de cuivre. / ~ coating ‖ Verkupferung f. ‖ couche f. de cuivre. / ~ coin ‖ Kupfermünze f. ‖ pièce f. *ou* monnaie f. de cuivre. / ~-containing ‖ kupferhaltig ‖ cuprique. / ~ coverer ‖ Kupferdecker m. ‖ couvreur m. en cuivre. / ~ covering of roofs ‖ Kupferbedachung f. ‖ couverture f. en cuivre des toits. / ~ cyanide ‖ Kupferzyanür n.; Zyankupfer n. ‖ cyanure m. de cuivre. / ~ damping (Electr) ‖ Kupferdämpfung f. ‖ amortissement m. en cuivre. / ~ disk ‖ Kupferscheibe f. ‖ rosette f. de cuivre. / ~ disks pl. ‖ Rosettenkupfer n. ‖ plaque f. de cuivre; rosette f. / ~ double salt ‖ Kupferdoppelsalz n. ‖ double sel m. de cuivre. / ~ embossing ‖ Kupferprägerei f. ‖ repoussage m. du cuivre. / ~ engraver ‖ Kupferstecher m. ‖ graveur m. sur cuivre.

copperer ‖ Verkupferer m. ‖ cuivreur m.

copper fire box ‖ kupferne Feuerbuchse f. *oder* Feuerkiste f. ‖ foyer m. en cuivre. / ~ fittings pl. ‖ Kupferbeschlag m. ‖ ferrure f. en cuivre. / ~ foil ‖ Kupferfolie f. ‖ feuille f. *ou* paillon m. de cuivre. / ~ forging ‖ Kupferschmiedestück n. ‖ pièce f. en cuivre forgé. / ~ founder ‖ Kupfergießer m. ‖ fondeur m. en cuivre. / ~ foundry ‖ Kupfergießerei f. ‖ fonderie f. de cuivre. / ~ froth ‖ Kupferschaum m.; Tyrolit m. ‖ tyrolite f.; kupaphrite f. / ~ fur (Salt) ‖ Kesselstein m.; Pfannenstein m.; Schaben fpl. ‖ écailles fpl. / ~ galvanoplastic plant ‖ Kupfergalvanoplastikanlage f. ‖ installation f. de galvanoplastic au cuivre. / ~ gauze brush ‖ Kupfergewebebürste f. ‖ balai m. en toile de cuivre. / ~ glance ‖ Kupferglanz m.; Kupferglas n.; Schwefelkupfer n. ‖ cuivre m. sulfuré *ou* vitreux. / ~ green ‖ Kiesel-

kupfer n.; Kupfergrün n.; Kieselmalachit m. ‖ chrysocolle f.; cuivre m. hydrosiliceux. / ~ hammer ‖ Kupferhammer m. ‖ masse f. en cuivre. / ~ handle ‖ Kupfergriff m. ‖ poignée f. en cuivre. / ~ hinge ‖ Scharnier n. aus Kupfer ‖ charnière f. en cuivre.

coppering ‖ Verkupfern n. ‖ cuivrage m. / ~ bath ‖ Kupferbad n. ‖ bain m. de cuivrage.

copper and iron pyrites ‖ Kupfereisenkies m. ‖ pyrites fpl. de cuivre et de fer. / ~ kettle ‖ Kupferkessel m.; kupferner Kochkessel m. ‖ chaudron m. *ou* bouilloire f. en cuivre. / ~ linoleate ‖ leinölsaures Kupfer n. ‖ linoléate m. de cuivre. / ~ loss (Electr) ‖ Kupferverlust m. ‖ perte f. dans le cuivre. / ~ manufacture ‖ Kupfergewinnung f. ‖ métallurgie f. du cuivre. / ~ matt *see* copper metal.

copper metal (Miner) ‖ Kupferstein m. ‖ matte f. cuivreuse *ou* de cuivre. / blue ~ ‖ blauer Kupferstein m. ‖ matte f. de cuivre bleue. / calcined ~ ‖ gerösteter Kupferstein m. ‖ matte f. de cuivre grillée *ou* calcinée. / close ~ ‖ dichter Kupferstein m. ‖ matte f. de cuivre serrée. / concentrated ~ ‖ Spurstein m.; konzentrierter Kupferstein m. ‖ matte f. de cuivre concentrée *ou* enrichie. / enriched ~ *see* concentrated ~. / granulated ~ ‖ Kupfersteingranalien pl.; granulierter Kupferstein m. ‖ matte f. de cuivre granulée. / roasted ~ *see* calcined ~. / white ~ ‖ weißer Kupferstein m.; Konzentrationsstein m. ‖ matte f. blanche de cuivre. / ~ calciner ‖ Kupfersteinröstofen m. ‖ four m. de grillage des mattes de cuivre.

copper mica ‖ Kupferglimmer m. ‖ cuivre m. micacé *ou* mica. / ~ mill ‖ Kupferhammer m. ‖ forge f. de cuivre. / ~ mine ‖ Kupferbergwerk n. ‖ mine f. de cuivre. / ~ money ‖ Kupfergeld n. ‖ monnaie f. de cuivre. / ~ nail ‖ Kupfernagel m.; Kupferspieker m. ‖ clou m. de cuivre. / ~ nickel ‖ Kupfernickel m.; Rotnickelkies m. ‖ kupfernickel m.; nickeline f.; nickel m. arsenical. / ~ nitrate ‖ salpetersaures Kupfer n. ‖ nitrate m. de cuivre.

copper ore ‖ Kupfererz n. ‖ minerai m. de cuivre. / argentiferous grey ~ ‖ Silberfahlerz n. ‖ cuivre m. gris argentifère. / green ~ ‖ Malachit m. ‖ malachite f. / red ~ ‖ Rotkupfererz n.; Kuprit m. ‖ cuivre m. oxydulé *ou* oxydé rouge. / velvet ~ ‖ Kupfersamterz n. ‖ cuivre m. velouté. / yellow ~ ‖ Gelbkupfererz n.; Kupferkies m. ‖ pyrite f. cuivreuse; cuivre m. pyriteux. / assay of ~ ‖ Kupfererzprobe f. ‖ essai m. des minerais de cuivre.

copper oxide ‖ Kupferoxyd n. ‖ oxyde m. de cuivre.

copper plate ‖ Kupferplatte f.; Kupferblech n. ‖ plaque f. de cuivre. / ~ (Engr) ‖ Kupferstich m.; Kupferstichplatte f. ‖ estampe f. *ou* gravure f. en taille-douce; gravure f. en cuivre. / etched ~ (Print) ‖ Strichätzung f. ‖ phototypographie f. au trait. / impressed ~ ‖ Kupfertiefdruck m. ‖ impression f. en taille-douce. / ~ for stencilling ‖ Schablonenkupferblech n. ‖ feuille f. de cuivre pour patrons.

copper plate engraver ‖ Kupferstecher m. ‖ graveur m. sur cuivre *ou* en taille-

douce. / ~ engraving ‖ Kupferstich m. ‖ gravure f. sur cuivre. / ~ paper ‖ Kupferdruckpapier n. ‖ papier m. à estampes. / ~ press ‖ Kupferdruckpresse f. ‖ presse f. d'imprimerie en taille-douce. / ~ print ‖ Kupferdruck m. ‖ gravure f. / ~ printer ‖ Kupferdrucker m. ‖ imprimeur m. en taille-douce. / ~ printing ‖ Tiefdruck m.; Kupferdruck m. ‖ impression f. en creux ou en taille-douce. / ~ printing machine ‖ Tiefdruckmaschine f. ‖ machine f. à impression en creux. / ~ printing press ‖ Tiefdruckpresse f.; Kupferdruckpresse f. ‖ presse f. à impression en taille-douce. / ~ rectifier ‖ Plattengleichrichter m. ‖ redresseur m. à plaques de cuivre.

copper plating, electrolytic ‖ galvanisches Verkupfern n. ‖ cuivrage m. galvanique. / ~ plant ‖ Verkupferungsanlage f. ‖ installation f. de cuivrage.

copper point ‖ Kupferspitze f. ‖ pointe f. en cuivre. / ~ powder ‖ Kupferpulver n. ‖ cuivre m. en poudre. / ~ printer ‖ Kupferdrucker m. ‖ imprimeur m. en taille-douce. / ~ printing works pl. ‖ Kupferdruckerei f. ‖ imprimerie f. en taille-douce. / ~ product ‖ Kupferfabrikat n.; Kupfererzeugnis n. ‖ produit m. en cuivre. / ~ punt (Mar) ‖ Scheuerprahm m. ‖ plate f. / ~ pyrites pl. ‖ Kupferkies m.; Chalkopyrit m. ‖ pyrite f. cuivreuse; cuivre m. pyriteux; chalcopyrite f. / ~ rain ‖ Kupferregen m.; Sprühkupfer n. ‖ pluie f. de cuivre. / ~ recovery plant (Galv) ‖ Kupferrückgewinnungsanlage f. ‖ installation f. de récupération de cuivre. / ~ refining ‖ Kupferraffinieren n. ‖ raffinage m. du cuivre. / ~ refining hearth ‖ Kupfergarherd m. ‖ foyer m. à affiner le cuivre. / ~ refining slag ‖ Kupfergarschlacke f. ‖ scorie f. de cuivre raffiné. / ~ resinate ‖ harzsaures Kupfer n. ‖ résinate m. de cuivre. / ~ rivet ‖ Kupferniet n. ‖ rivet m. en cuivre. / ~ rolling mill ‖ Kupferwalzwerk n. ‖ usine f. de laminage de cuivre; laminoir m. à cuivre. / ~ salt ‖ Kupfersalz n. ‖ sel m. de cuivre. / ~ scales pl. ‖ Kupferasche f.; Kupferhammerschlag m.; Kupferglühspan m. ‖ paille f. ou cendres fpl. ou écailles fpl. ou battiture f. de cuivre. / ~ schist ‖ Kupferschiefer m. ‖ schiste m. cuivreux. / ~ setting ‖ Kupferheftung f. ‖ sertissage m. en cuivre. / ~ sheathing (Shipb) ‖ Kupferbeschlag m. ‖ doublage m. en cuivre.

copper sheet ‖ Kupferblech n. ‖ planche f. en cuivre rouge; tôle f. de cuivre. / gold-plated ~ ‖ goldplattiertes Kupferblech n. ‖ cuivre m. plaqué d'or ou doublé d'or. / plated ~ ‖ plattiertes Kupferblech n. ‖ cuivre m. plaqué ou doublé. / silver-plated ~ ‖ silberplattiertes Kupferblech n. ‖ cuivre m. plaqué d'argent ou doublé d'argent.

copper shell ‖ Kupferniederschlag m. ‖ coquille f. galvanoplastique.

copper slag ‖ Kupferschlacke f. ‖ crasse f. de cuivre. / blister ~ ‖ Schwarzkupferschlacke f. ‖ scorie f. du cuivre brut; scorie f. grillée. / coarse ~ see copper slag, blister.

copper sleeve ‖ Kupferhülse f. ‖ manchon m. en cuivre. / ~ smelting ‖ Kupferverhüttung f. ‖ fonte f. du cuivre. / ~ smelting in blast-furnaces ‖ Schachtofenkupferarbeit f. ‖ fonte f. du cuivre dans les fourneaux à cuve.

coppersmith ‖ Kupferschmied m. ‖ chaudronnier m. en cuivre. / ~'s shop ‖ Kupferschmiede f. ‖ chaudronnerie f. de cuivre. / ~'s ware ‖ Kupferschmiedewaren fpl. ‖ petite chaudronnerie f.; cuivrerie f. / ~'s works pl. ‖ Kesselschmiede f. ‖ petite chaudronnerie f.

copper smoke ‖ Kupferrauch m. ‖ fumée f. de cuivre. / ~ soda ‖ Kupfersoda f. ‖ soude f. de cuivre. / ~ solder ‖ Kupferlot n. ‖ soudure f. de cuivre. / ~ sponge ‖ Kupferschwamm m. ‖ cuivre. m. spongieux. / ~ standard ‖ Kupfernorm f. ‖ norme f. ou règle f. pour le cuivre. / ~ suboxide ‖ Kupferoxydul n. ‖ protoxyde m. de cuivre. / ~ sulphate ‖ Kupfervitriol n. ‖ vitriol m. bleu; sulfate m. de cuivre. / ~ sulphate plant ‖ Kupfervitriolanlage f. ‖ installation f. de sulfate de cuivre. / ~ sulphide ‖ Schwefelkupfer n. ‖ sulfure m. de cuivre. / ~ test ‖ Kupferprobe f. ‖ essai m. de cuivre / ~ thorns pl. ‖ Kupferdorn m. ‖ épines fpl. de ressuage. / ~ tube ‖ Kupferrohr n.; Kupferröhre f. ‖ tuyau m. ou tube m. en cuivre. / ~ turnings pl. ‖ Kupferdrehspäne mpl. ‖ tournure f. de cuivre. / ~ type ‖ Kupferbuchstabe m. ‖ lettre f. en cuivre. / ~ vessel ‖ Kupferkessel m. ‖ récipient m. en cuivre. / ~ vitriol ‖ Kupfervitriol n. ‖ sulfate m. de cuivre. / ~ ware ‖ Kupferwaren fpl. ‖ articles mpl. en cuivre. / ~ washer ‖ Kupferunterlegscheibe f. ‖ rondelle en cuivre. / ~ weight ‖ Messinggewicht n. ‖ poids m. en cuivre ou en laiton.

copper wire ‖ Kupferdraht m. ‖ fil m. de cuivre. / to anneal the ~ ‖ den Kupferdraht m. ausglühen ‖ recuire le fil de cuivre. / cemented ~ ‖ zementierter Kupferdraht m. ‖ trait m. de cuivre jaune ou cémenté. / hard-drawn ~ ‖ Hartkupferdraht m. ‖ fil m. de cuivre écroui. / non-insulated ~ ‖ unisolierter Kupferdraht m. ‖ fil m. de cuivre non isolé. / stranded ~ ‖ Kupferseil n. ‖ câble m. en cuivre.

copper works pl. (Found) ‖ Kupferwerk n. ‖ fonderie f. (de minerai) de cuivre; cuivrerie f. / ~ (Ore dress) ‖ Kupferhammer m.; Kupferhütte f. ‖ forge f. de cuivre; cuivrerie f.

copper worm ‖ Kupferschlange f. ‖ serpentin m. en cuivre.

coppice cutter ‖ Reisigholzschneider m. ‖ coupeur m. de ramilles ou de broutilles ou de menu bois. / ~ wood ‖ Unterholz n. ‖ taillis m.

copra ‖ Kopra f. ‖ coprah m. / Cochin ~ ‖ Cochinkopra f. ‖ coprah m. cochinchinois. / ~ dryer ‖ Kopratrockner m. ‖ séchoir m. à coprah. / ~ oil ‖ Koprahöl n. ‖ huile f . de coprah.

copse cutter see coppice cutter.

cop winding machine ‖ Kötzerspulmaschine f. ‖ cannetière f.; bobinoir m. à cannette.

copy, to ‖ nachmachen; kopieren ‖ copier. / ~ from life ‖ nach dem Leben n. oder nach der Natur f. zeichnen ‖ dessiner d'après nature f. / ~ a picture ‖ ein Gemälde n. nachzeichnen ‖ prendre le trait.

copy ‖ Kopie f.; Abschrift f. ‖ copie f. / ~ (Drawing) ‖ Pause f.; Kopie f. ‖ calque m. / ~ (Print) ‖ Exemplar n. ‖ exemplaire m. / ~(Typewr) ‖ Durchschlag m. ‖ copie f.; duplicata m. / ~ for the ac-

countancy department ‖ Buchhaltungsdurchschlag m. ‖ copie f. pour le service de la comptabilité. / attested ~ ‖ beglaubigte Abschrift f. ‖ copie f. certifiée. / ~ of a bill ‖ Wechselkopie f. ‖ copie de traite. / certified ~ ‖ beglaubigte Abschrift f. ‖ copie f. vérifiée ou légalisée ou certifiée. / ~ of contract ‖ Vertragsabschrift f. ‖ copie f. de contrat. / ~ of invoice ‖ Fakturabschrift f. ‖ copie f. de facture. / legalized ~ ‖ rechtsgültige Abschrift f. ‖ copie f. valide. / legally attested ~ ‖ beglaubigte Abschrift f. ‖ copie f. vérifiée ou légalisée. / to make a fair ~ ‖ ins reine schreiben ‖ mettre au net m. / printer's ~ ‖ Manuskript n. ‖ manuscrit m. / to set a ~ ‖ ein Manuskript n. aufsetzen ‖ composer une copie. / ~ for signing ‖ Mundum n.; Reinschrift f. ‖ expédition f. à signer. / true ~ ‖ gleichlautende Abschrift f. ‖ copie f. conforme. / by way of ~ ‖ abschriftlich ‖ en copie f.

cop yarn ‖ Kötzergarn n. ‖ fil m. en cannette.

copy book ‖ Schreibheft n. ‖ cahier m. / ~ (Office) ‖ Kopierbuch n. ‖ livre m. à copier les lettres; copie f. de lettres; copie-lettres m. / ~ press ‖ Briefkopierpresse f. ‖ presse f. à copier les lettres.

copy cutter ‖ Manuskriptverteiler m. ‖ découpeur m. de copie. / ~ holder (Print) ‖ Manuskripthalter m. ‖ porte-copie m. / ~ holder (Typewr) ‖ Konzepthalter m. ‖ porte-sténogramme m.

copying apparatus ‖ Kopierapparat m. ‖ autocopiste m.; appareil m. multiplicateur. / ~ apparatus (Drawing) ‖ Lichtpausapparat m. ‖ appareil m. à calquer. / ~ arc lamp ‖ Kopierbogenlampe f. ‖ lampe f. à arc pour travaux de copie. / ~ bath ‖ Kopierbad n. ‖ bain m. à copier. / ~ book ‖ Kopierbuch n. ‖ livre m. de copies. / ~ cloth ‖ Kopiertuch n. ‖ drap m. à copier. / ~ cloth (Drawing) ‖ Kopierleinwand f.; Pausleinewand f. ‖ toile f. à calquer. / ~ fees pl. ‖ Schreibgebühr f. ‖ frais mpl. de copie. / ~ ink ‖ Kopiertinte f.; Hektografentinte f. ‖ encre f. à copier ou à autocopier. / ~ ink for printing ‖ Kopierfarbe f. ‖ encre f. à copier. / ~ ink pencil ‖ Tintenstift m. ‖ crayon m. aniline ou encre. / ~ lamp ‖ Kopierlampe f. ‖ lampe f. à photocalquer. / ~ lathe ‖ Kopierdrehbank f. ‖ tour m. à copier.

copying machine ‖ Kopiermaschine f. ‖ machine f. à copier. / ~ for cinematographic films ‖ Kinofilmkopiermaschine f. ‖ machine f. à copier les films cinématographiques. / photo ~ ‖ Lichtpausapparat m. ‖ appareil m. à tirer les bleus. / relief ~ ‖ Reliefkopiermaschine f. ‖ machine f. à copier en relief.

copying mine ‖ Kopiermine f. ‖ mine f. à copier. / ~ paper ‖ Kopierpapier n. ‖ papier m. autocopiste ou à copier. / ~ paste ‖ Hektografenmasse f. ‖ pâte f. à autocopier. / ~ pencil ‖ Kopierstift m. ‖ crayon m. encre ou à copier. / ~ press ‖ Kopierpresse f. ‖ presse f. à copier. / rotative ~ press ‖ Kopierdrehpresse f. ‖ presse f. rotative à copier. / ~ property ‖ Kopierfähigkeit f. ‖ propriété f. copiante. / ~ telegraph ‖ Kopiertelegraf m. ‖ télégraphe m. autographique. / ~ work ‖ Kopierarbeiten fpl. ‖ travaux mpl. de copie. / ~ print ‖ Abdruck m. ‖ impression f.

copyright ‖ Verlagsrecht n. ‖ droit m. d'impression.

copyright name ‖ gesetzlich geschützter Name m. ‖ nom m. déposé.

coque ‖ schalenförmig gewundener Bandknoten m. ‖ coque f.

coral ‖ Koralle f. ‖ corail m. / ~ in branches ‖ Zweigkoralle f. ‖ corail m. en branches. / ~ in fragments ‖ Bruchkoralle f. ‖ corail m. en morceaux. / imitation ~ ‖ künstliche Korallenwaren fpl. ‖ objets mpl. en corail artificiel. / natural ~ ‖ natürliche Koralle f. / corail m. naturel.

coral beads pl. ‖ Korallenhalsband n. ‖ collier m. de corail. / branch of ~ ‖ Korallenzinke f. ‖ branche f. de corail. / ~ diver see fisherman. / ~ fisherman ‖ Korallenfischer m. ‖ pêcheur m. de corail; corailleur m. / ~ fishery ‖ Korallenfischerei f. ‖ pêche f. du corail. / ~ island ‖ Koralleninsel f. ‖ attolle m.; île f. de coraux. / ~ limestone see ~ rag. / ~ ore ‖ Korallenerz n. ‖ mercure m. sulfuré bitumineux. / ~ ornament ‖ Korallenschmuck m. ‖ parure f, en corail. / ~ rag ‖ Korallenkalk m. ‖ calcaire m. corallien; marbre m. coralloïde. / ~ reef ‖ Korallenriff n. ‖ récif m. de coraux. / ~ worker ‖ Korallenzubereiter m. ‖ préparateur m. de coraux; corailleur m.

corbel, to (Build) ‖ auskragen ‖ encorbeller.

corbel (Build) ‖ Konsole f.; Kragstück n. ‖ corbeau m.; console f.; ancone f. / ~ of a capital (Arch) ‖ Kern m. oder Glocke f. eines Kapitells ‖ vase m. ou corbeille f. ou tambour m. de chapiteau. / ~ of a chimney mantle ‖ Schornsteinmantelknagge f. ‖ courge f. de manteau d'une cheminée. / iron ~ ‖ eiserner Konsolträger m. ‖ console f. en fer. / stone ~ (Build) ‖ Kragstein m.; Tragstein m. ‖ corbeau m. ou console f. en pierre. / ~ under the purlins (Carp) ‖ Pfettenknagge f. ‖ chantignole f. / ~ under a window-jamb ‖ Fensterkonsole f.; Kragstein m. unter dem Fenstergewände ‖ allége f.

corbelled-out ‖ freitragend; vorgekragt ‖ encorbellé; en porte-à-faux. / to be ~ ‖ auf Kragsteinen mpl. ruhen; vorgekragt sein ‖ porter à faux; porter en saillie f.

corbelling out ‖ Überkragung f. ‖ encorbellement m. / ~ of bricks ‖ Auskragen n. des Mauerwerks ‖ encorbellement n. de la maçonnerie.

corbel piece (Carp) ‖ Sattelholz n.; Schirrbalken m. ‖ corbeau m.; racinal m.; sous-poutre f.; sous-longueron m. / ~ tree (Carp) ‖ Kraftbalken m.; Notbalken m. ‖ force f. de solivure; poutre f. de force.

cord, to ‖ zuschnüren ‖ ficeler; lacer. / ~ a book ‖ einen Bücherrücken m. rippen ‖ nerver un livre.

cord ‖ Leine f.; Schnur f.; Strick m.; Tau n.; Seil n. ‖ corde f.; ligne f.; cordelière f.; câble m. / ~ (Electr) ‖ Leitungsschnur f. ‖ corde f. / ~ (Surv) ‖ Meßschnur f. ‖ cordeau m. / ~ (Textile) ‖ Manchester m. ‖ cordelet m.; velours m. à côtes. / ~ (Wood) ‖ Klafter f. ‖ corde f.; stère m. / endless ~ ‖ Seil n. ohne Ende ‖ corde f. sans fin. / flexible ~ ‖ biegsame Leitung f. ‖ conducteur m. souple. / flexible ~ for telegraphs ‖ Telegrafenschnur f. ‖ cordon m. pour télégraphie. / flexible ~ for telephone ‖ Telefonschnur f. ‖ cordon m. pour téléphonie. / having ~s pl. ‖ streifig ‖ cordé. / ~ of hemp ‖ Hanfseil n. ‖ corde f.

de chanvre. / to knot ~s ‖ Schnüre fpl. knüpfen ‖ nouer des lacets mpl. / ~ for lorgnettes ‖ Lorgnettenschnur f. ‖ cordon m. pour faces à main. / neck ~ (Weav) ‖ Platinenschnur f. ‖ collet m. / pair of ~s (Tel) ‖ Schnurpaar n. ‖ dicorde f. / to splice ~s pl. ‖ Schnüre fpl. flechten ‖ épisser des lacets mpl.

cordage ‖ Seilerwaren fpl. ‖ cordages mpl. / ~ (Shipb) ‖ Tauwerk n. ‖ cordage m. / ~ making machine ‖ Tauwerkherstellungsmaschine f. ‖ machine f. pour la fabrication des cordages. / ~ reel (Shipb) ‖ Taurolle f. ‖ tourniquet m. pour amarres.

cord circuit ‖ Schnurstromkreis m. ‖ circuit m. de cordon. / ~ circuit repeater ‖ Schnurverstärker m. ‖ répéteur m. sur cordon. / ~ eye ‖ Kausche f.; Öse f. ‖ passe-fils m.; œillet m.

cordial (Distill) ‖ Likör m. ‖ liqueur f.

cordierite ‖ Peliom m.; Kordierit m.; Dichroit m. ‖ cordiérite m.; dichroïte m.

cording (Weav) ‖ Schnürung f.; Harnischstechen n. ‖ encordage m.; empoutage m.

cordovan leather ‖ Korduanleder n. ‖ cordouan m.

cord packing ‖ Schnurpackung f. ‖ garniture f. en cordon. / ~ pendant (Electr) ‖ Schnurpendel n. ‖ suspension f. à cordon. / ~ plaiter ‖ Senkelmacher m. ‖ tresseur m. de cordons. / ~ repeater (Tel) ‖ Schnurverstärker m. ‖ répéteur m. sur cordon. / ~ repeater station (Tel) ‖ Schnurverstärkeramt n. ‖ station f. de répéteurs sur cordon. / ~ roller ‖ Schnurrolle f. ‖ poulie f. à corde.

cordwain see cordovan leather.

cordwainer ‖ Schuhmacher m.; Schuster m. ‖ bottier m.; cordonnier m.

core (Found) ‖ Formkern m.; Kern m. ‖ marron m.; noyau m. du moule. / ~ of apple ‖ Kerngehäuse n. vom Apfel ‖ trognon m. de pomme. / ~ of the cable ‖ Kabelseele f.; Kabelkern m. ‖ âme f. du câble. / ~ for casting shells ‖ Geschoßkern m. ‖ noyau m. pour la fonte des obus. / ~ of the distribution network ‖ Schwerpunkt m. des Verteilungsnetzes ‖ centre m. de distribution. / false ~ (Found) ‖ Kernstück n. ‖ pièce f. rapportée. / the ~ shows a fibrous tough texture ‖ der Kern zeigt eine zähe, sehnige Beschaffenheit ‖ les piéces fpl. sont tenaces et nerveuses au noyau. / grained ~ ‖ körnige Bruchfläche f. ‖ cassure f. à grain grossier ordinaire. / ~ of leading coils (Electr) ‖ Spulenkern m. ‖ noyau m. de bobines. / magnetic ~ ‖ Weicheisenkern m. ‖ barreau m. de fer magnétique. / ~ of a rope ‖ Seele eines Seiles ‖ âme f. ou mèche f. d'une corde. / hempen ~ of a rope ‖ Hanfseele f. eines Seiles ‖ âme f. en chanvre d'un cordage. / ~ of a screw ‖ Schraubenkern m. ‖ noyau m. d'une vis. / ~ of a section ‖ Kern m. eines Querschnittes ‖ noyau m. central d'une section. / soft iron ~ (Electr) ‖ Weicheisenkern m. ‖ noyau m. de fer doux. / the whole ~ was taken out of the shaft after it was bored out ‖ der Kern m. wurde nach dem Bohren ganz herausgehoben ‖ le noyau fut enlevé en entier après le forage.

core bar (Found) ‖ Kernspindel f.; Kernstange f. ‖ arbre m. (en fer) à noyau. / ~ binding material (Found) ‖ Kernbindemittel n. ‖ agglomérant m. pour noyaux.

/ ~ board (Found) ‖ Kernbrett n. ‖ échantillon m. de noyau. / ~ box (Found) ‖ Kernkasten m. ‖ boîte f. à noyau. / pressed dust ~ coil ‖ Massekernspule f. ‖ bobine f. à noyau en limaille de fer. / ~ conductor of the cable ‖ Kabelader f. ‖ conducteur m. du câble.

cored ‖ hohl ‖ creux. / ~ carbon ‖ Dochtkohle f. ‖ charbon m. à mèche.

core drying stove ‖ Kerntrockenofen m. ‖ four m. à sécher les noyaux; étuve f. à noyaux. / ~ frame ‖ Kernstütze f. ‖ support m. de noyau. / ~ grinder (Found) ‖ Kernschleifmaschine f. ‖ machine f. à rectifier les noyaux. / ~ hammer drill ‖ Kernbohrhammer m. ‖ marteau m. à retirer les noyaux. / ~ iron (Electr) ‖ Kerneisen n. ‖ fer m. à noyau. / ~ lathe ‖ Kerndrehbank f. ‖ tour m. à noyaux. / ~ losses pl. (Electr) ‖ Eisenkernverluste mpl. ‖ pertes fpl. dans le noyau. / ~ maker (Mould) ‖ Kernmacher m. ‖ ouvrier m. noyauteur; noyauteur m. / ~ making machine see ~ moulding machine. / ~ mark ‖ Kernmarke f. ‖ portée f. de noyau. / ~ moulding machine ‖ Kernformmaschine f. ‖ machine f. à mouler les noyaux ou à noyauter. / ~ moulding shop ‖ Kernmacherei f. ‖ atelier m. à noyaux. / ~ press ‖ Kernpresse f. ‖ presse f. à noyaux. / ~ print see ~ mark. / ~ sand ‖ Kernsand m. ‖ sable m. à noyau. / ~ sheet ‖ Dynamoblech n. ‖ tôle f. pour dynamos. / ~ skeleton ‖ Kerngerippe n. ‖ armature f. ou cage f. du noyau. / ~ spindle ‖ Kernspindel f. ‖ arbre m. à noyau. / ~ spindle pipe ‖ Kernspindelrohr n. ‖ tube-broche f. de noyau. / ~ strickle ‖ Kernschablone f. ‖ trousse f. à noyau. / ~ templet see ~ strickle. / ~ transformer ‖ Kerntransformator m. ‖ transformateur m. à noyau.

corf (Mine) ‖ Kübel m. ‖ seau m.; tonne f.; tine f.

coriander ‖ Koriander m. ‖ coriandre f.

cork, to ‖ korken ‖ boucher.

cork ‖ Korkstopfen m.; Korken m.; Korkpfropfen m. ‖ bouchon m. de liège. / ~ (Bot) ‖ Kork m.; Korkholz n.; Korkrinde f. ‖ liège m. / ~ (Tree) ‖ Korkeiche f. ‖ chêne-liège m. / agglomerated ~ ‖ zusammengepreßter Kork m. ‖ liège m. aggloméré. / articles pl. of ~ ‖ Korkwaren fpl. ‖ ouvrages mpl. en liège. / artificial ~ ‖ Kunstkork m. ‖ liège m. factice. / ~ of barrel ‖ Faßkork m. ‖ bouchon m. de tonneau. / ~ of bottle ‖ Flaschenkork m. ‖ bouchon m. de bouteille. / crushed ~ ‖ zerstoßener Kork m. ‖ liège m. concassé. / cut ~ ‖ zerschnittener Kork m. ‖ liège m. découpé. / to extract ~s ‖ entkorken ‖ déboucher. / extracting of ~s ‖ Entkorken n. ‖ débouchage m. / ground ~ ‖ gemahlener Kork m. ‖ liège m. moulu. / piece of ~ ‖ Korkstück n. ‖ morceau m. de liège. / ~ in plates ‖ Kork m. in Tafeln ‖ liège m. en planches. / prepared ~ ‖ zugerichteter Kork m. ‖ liège m. préparé. / pulverized ~ ‖ gepulverter Kork m. ‖ liège m. pulvérisé. / raw ~ ‖ roher Kork m. ‖ liège m. brut. / rubber ~ ‖ Kautschukstopfen m. ‖ bouchon m. de caoutchouc. / scraped ~ ‖ abgekratzter Kork m. ‖ liège m. raclé. / smoothed ~ ‖ geglätteter Kork m. ‖ liège m. lissé.

cork articles pl. ‖ Korkwaren fpl. ‖ objets mpl. en liège. / ~ bark ‖ Korkrinde f. ‖ liège m. / ~ borer ‖ Korkbohrer m. ‖

perce-bouchon m. / ~ breaker || Korkbrecher m. || concasseur m. de liège. / ~ brick || Korkstein m. || brique f. en liège. / agglomerated ~ brick || Korkstein m. || liège m. aggloméré. / ~ buoy || Korkboje f. || bouée f. de liège. / ~ chips pl. || Korkabfälle mpl. || déchets mpl. de liège. / ~ composition || Korkmasse f. || enveloppe f. linogomme. / ~ cutter || Korkschneider m.; Pfropfenschneider m. || bouchonnier m. / ~ disk || Korkscheibe f. || disque m. en liège. / ~ dust || Korkmehl n. || poudre f. de liège.

corker || Flaschenkorker m. || boucheur m. de bouteilles.

cork extractor || Kork(en)zieher m.; Pfropfenzieher m.; Entkorker m. || tire-bouchon m. appareil m. à retirer les bouchons. / ~ fastener || Flaschenverkorkmaschine f. || machine f. à boucher les bouteilles. / ~ float || Korkschwimmer m. || macaron m. / ~ flour || Korkmehl n. || poudre f. de liège.

corking machine || Verkorkmaschine f. || machine f. à boucher; bouchonneuse f. / bottle ~ || Flaschenkorkmaschine f. || machine f. à boucher les bouteilles. / cask ~ || Faßkorkmaschine f. || machine f. à boucher les fûts; boucheuse f. pour fûts. / ~ man || Flaschenkorker m. || boucheur m. de bouteilles.

cork insulation || Korkisolation f. || isolement m. de liège. / ~ insulator for heat || Wärmeschutzschale f. aus Kork || enveloppe f. calorifuge en liège. / ~ jacket || Schwimmweste f. || gilet m. en liège; ceinture f. de sauvetage. / ~ mill || Korkmühle f.; Korkmüllerei f. || moulin m. à liège. / ~ moulds pl. || Korkrohrbekleidung f. || coquilles fpl. en liège. / ~ oak || Korkeiche f. || chêne-liège m. / ~ paper || Korkpapier n. || papier-liège m. / ~ plate || Korkplatte f. || carreau m. en liège. / ~ plate factory || Korkplattenfabrik f. || fabrique f. de plaques de liège. / ~ powder || Korkmehl n. || poudre f. ou sciures fpl. de liège. / ~ presser || Korkpresse f. || mâche-bouchon m. / ~ remover || Korkzieher m. || tire-bouchon m. / ~ ring || Korkring m. || rondelle f. en liège. / ~ screw || Kork(en)zieher m. || tire-bouchon m. / ~ sculpture || Phelloplastik f.; Korkmodellierkunst f. || phelloplastique f. / ~ shavings pl. || Korkschnitzel mpl. || miettes fpl. de liège. / ~ sheet || Korkplatte f. || planche f. de liège. / ~ sole || Korksohle f. || semelle f. en liège. / ~ stopper || Korkstopfen m. || bouchon m. en liège. / ~ stopple || Korkstöpsel m. || bouchon m. de liège. / ~ tip for cigarettes || Zigarettenkorkmundstück n. || bout m. en liège pour cigarettes. / ~ tree see ~ oak. / ~ ware || Korkwaren fpl. || objets mpl. en liège. / ~ waste || Korkabfall m. || déchet m. de liège. / ~ wheel || Korkrad n. || roue f. en liège. / ~ wirer || Drahtverschnürer m. || metteur m. de fil de fer. / ~ wood || Korkholz n. || bois m. de liège. / ~ worker || Korkschneider m. || découpeur m. de liège.

Corliss valve gear || Corlißsteuerung f. || distribution f. Corliss.

corn, to (Met) || granulieren; körnen || grainer, granuler.

corn (Agr) || Getreide n.; Korn n. || blé m.; grains mpl. / (Anatomy) || Hühnerauge n. || cor m. / blasted ~ || brandiges Korn n.; Mutterkorn n. || blé m. cornu ou ergoté; ergot m.; ébrun m. / ~ in ears || Getreide n. in Ähren || céréales fpl. en épi. / ~ laid-down by rain || durch Regen umgelegtes Getreide n. || récolte f. versée par la pluie. / ~ in sheaves || Getreide n. in Garben || céréales fpl. en gerbes. / smutted ~ see blasted ~. / sorting machine for ~ || Getreidesortiermaschine f. || machine f. à trier les grains de céréales.

corn brandy distillery || Getreidebrennerei f. || distillerie f. de grains. / ~ chandler || Getreidehändler m.; Mehlhändler m. || marchand m. de blé ou de farine; minotier m. / ~ cleaner || Getreidereiniger m. || nettoyeur m. de blé. / cleaning of ~ || Getreidereinigung f.; Kornputzerei f. || nettoyage m. des grains. / ~ cleaning machine || Getreidereinigungsmaschine f. || machine f. à nettoyer les céréales / ~ coffee || Kornkaffee m. || café m. de grains. / ~ crusher || Getreidezerkleinerungsmaschine f. || broyeur m. de grain. / ~ cure || Hühneraugenmittel n. || remède m. contre les cors; anti-cors m. / ~ dealer || Getreidehändler m. || marchand m. de blé. / ~ distiller || Getreidebrenner m. || distillateur m. de grains. / ~ drill || Sämaschine f. || semoir m. / ~ duty || Getreidezoll m. || droit m. sur les céréales.

cornea (Eye) || Hornhaut f. || cornée f. / curved profile of the ~ || Hornhautkrümmung f. || courbure f. de la cornée.

corneal microscope, binocular || binokulares Hornhautmikroskop n. || microscope m. cornéen binoculaire. / image erecting ~ || bildaufrichtendes Hornhautmikroskop n. || microscope m. cornéen redresseur.

corneal nerve || Hornhautnerv m. || nerf m. de la cornée.

corned (Curr) || gekörnt; genarbt || grainé. / ~ (Metal) || körnig || granulaire; grenu. / ~ beef || Büchsenfleisch n.; Corned beef n. || conserve f. de viande.

corn elevator || Getreideelevator m. || élévateur m. ou transporteur m. de blé. / floating ~ || schwimmender Getreideheber m. || élévateur m. flottant à grains.

cornelian || Karneol m. || carnéole f.; cornaline f.

cornel wood || Kornelkirschenholz n. || cornouiller m.

corner, to ~ an article || eine Ware f. (spekulativ) aufkaufen || accaparer un article.

corner || Ecke f.; Winkel m. || angle m.; coin m. / blunted ~ (Build) || abgestumpfte Ecke f. || entrecoupé f. / broken ~ (Stone cutter) || abgestoßene Kante f. || écornure f. / ~ of the eye || Augenwinkel m. || angle m. de l'œil. / sharp interior ~ || scharfe inwendige Ecke f. || vive arête f. intérieure. / ~ angle || Eckwinkel m. || cornière f. d'angle / ~ box || Proszeniumsloge f. || baignoire f. d'avant-scène. / ~ column || Ecksäule f. || colonne f. d'angle. / ~ connection (Build) || Eckverband m. || assemblage m. d'angle / ~ iron || cramp Winkelband n.; Eckschiene f. || ferrure f. angulaire. / ~ cupboard || Eckschrank m. || écoinçon m.; encoignure f.

cornered (Build) || eckig; winkelig || angulaire; angulé. / ~ (Trade) || spekulativ aufgekauft || accaparé.

corner fillet || Eckleiste f. || baguette f. de coin. / ~ house || Eckhaus n. || maison f. du coin. / ~ iron || Winkeleisen n.; Eisenklammer f. || cornière f. / ~ jamb-stone || Eckpfeiler m. || jambe f. d'encoignure. / ~ joint || Winkelstoß m. || joint m. angulaire. / ~ pilaster || Eckpilaster m. || pilastre m. cornier. / ~ pillar || Eckpfeiler m. || pilastre m. cornier; cornière f. / ~ plate (Railw) || Eckband n. || équerre f. / ~ post || Eckpfosten m.; Ecksäule f. || poteau m. cornier; colonne f. d'angle. / ~ room || Eckzimmer n. || chambre f. du coin. / ~ stiffening || Eckversteifung f. || renforcement m. des angles. / ~ stone || Kropfstein m.; Eckstein m. || pierre f. d'encoignure. / ~ tile || Kehlziegel m.; Hohlziegel m. || noue f.; tuile f. imbricée ou cornière. / ~ window || Eckfenster n. || fenêtre f. du coin / ~ working machine (Bookb) || Eckenbearbeitungsmaschine f. || machine f. à travailler les coins.

cornet || Klapphorn n. || cornet m. à piston.

corn flour manufacturer || Getreidemühlenbesitzer m. || minotier m. / ~ grinding || Müllerei f. || minoterie f. / ~ house || Kornspeicher m.; Kornboden m. || grenier m.

cornice || Gesims n.; Mauerbrüstung f. || moulure f.; faîte m. de mur. / ~ flashing || Simsabdeckung f. || chaperon m. de l'entablement. / ~ plane || Karnieshobel m. || rabot m. à doucine; doucine f.

corn industry || Getreideindustrie f. || industrie f. des céréales.

Cornish boiler || Einflammrohrkessel m.; Cornwallkessel m. || chaudière f. à un seul tube-foyer.

corn knife || Hühneraugenmesser n. || coupe-cors m. / ~ loft || Getreidespeicher m.; Kornboden m. || magasin m. à blé; grenier m. / ~ magazine see ~ loft. / ~ market || Getreidemarkt m. || marché m. des blés. / ~ mill || Getreidemühle f. || moulin m. à céréales ou à farine ou à grains ou à blé. / ~ miller || Müller m. || meunier m. / ~ mower || Getreidemähmaschine f. || moissonneuse-javeleuse f.; moissonneuse f. / ~ oil || Maisöl n. || huile f. de maïs. / ~ rake || Getreiderechen m. || rateau m. à blé. / ~ shovel || Kornschaufel f. || pelle f. à grains ou à blé. / ~ sieve || Kornsieb n.; Rätter m. || crible m. à blé. / ~ sorting machine || Getreidesortiermaschine f. || machine f. à trier les grains (de céréales). / ~ trade || Getreidehandel m.; Mehlhandel m. || commerce m. des blés; minoterie f. / ~ transshipping plant. / Getreideumschlaganlage f. || installation f. de déchargement des grains. / ~ van || Kornschwinge f. || van m. émotteur. / ~ washerman || Getreidewäscher m. || laveur m. de grains.

corona (Arch) || Kranzleiste f. || larmier m. / ~ (Astron) || Hof m. (um Sonne oder Mond) || couronne f.; halo m. / ~ (Electr) || Glimmentladung f. || couronne f.; phénomène m. de fluorescence. / ~ effect || Koronaeffekt m. || effet m. couronne. / ~ nut || Steinnuß f. || corozo m.; noix f. de corozo.

corozo, button of ~ || Steinnußknopf m. || bouton m. en corozo.

corporate body || juristische Person f. || personne f. civile. / ~ obliged to maintain the public roads || Wegeunterhaltungspflichtiger m. || personne f. civile obligée d'entretenir les chemins publics.

corporation ‖ Innung f. ‖ corps m. de métier; corporation f.

corpse ‖ Leiche f. ‖ corps m. / to preserve ~s ‖ Leichen fpl. aufbewahren ‖ conserver des cadavres mpl.

corpuscle ‖ Massenteilchen n. / corpuscule m. / blood ~s pl. ‖ Blutkörperchen npl. ‖ globules mpl. du sang.

corpuscular ‖ korpuskular ‖ corpusculaire.

corrasion (Geol) ‖ Ausnagung f. ‖ corrasion f.

correct ‖ regelrecht ‖ normal; correct; en règle. / if found ~ ‖ nach Richtigbefund m. ‖ à bientrouvé m.

correct, to ‖ berichtigen; korrigieren ‖ corriger. / ~ (Print) ‖ Korrektur lesen ‖ lire ou relire une épreuve. / ~ an error ‖ einen Fehler m. ausschalten ‖ éliminer une erreur f. / ~ the instrumental errors pl. ‖ die Instrumentalfehler mpl. berichtigen ‖ corriger les erreurs fpl. instrumentales. / ~ the map ‖ die Karte berichtigen ‖ corriger la carte. / ~ parallax ‖ die Parallaxe f. beseitigen ‖ écarter la parallaxe. / ~ type (Print) ‖ Korrekturen fpl. auf dem Blei lesen ‖ corriger sur le plomb.

correcting a river ‖ Kanalisierung f. eines Flusses ‖ correction f. d'une rivière.

correction ‖ Richtigstellung f. ‖ rectification f.; misc f. au point. / ~ (Opt) ‖ Berichtigung f.; Korrektion f. ‖ correction f. / ~ (Phys) ‖ Fehlerverbesserung f. ‖ compensation f. / ~ (Print) ‖ Korrektur f. ‖ correction f. / chromatic ~ ‖ chromatische Korrektion f. ‖ correction f. chromatique. / full ~ ‖ Vollberichtigung f. ‖ correction f. complète. / ~ for hysteresis ‖ Nachwirkungsberichtigung f. ‖ correction f. d'effet résiduel ou de fatigue. / ~ for lag see ~ for hysteresis. / ~ of an instrument ‖ Berichtigung f. eines Instrumentes ‖ correction f. d'un instrument. / ~ of the map ‖ Kartenverbesserung f. ‖ correction f. de la carte; correction f. apportée à la carte. / ~ of rivers ‖ Flußregulierung f. ‖ canalisation f. d'une rivière. / spherical ~ ‖ sphärische Korrektion f. ‖ correction f. sphérique; correction f. de l'aberration de sphéricité. / ~ for temperature ‖ Wärmegradberichtigung f. ‖ correction f. thermométrique ou de température.

correction key ‖ Verbesserungstaste f. ‖ touche f. de correction. / ~ mark ‖ Berichtigungsmarke f. ‖ repère m. d'ajustage. / ~ mount ‖ Korrektionsfassung f. ‖ monture f. à correction. / ~ sheet ‖ Deckblatt n. ‖ feuille f. rectificative.

correctness ‖ Richtigkeit f. ‖ exactitude f.; régularité f.

corrector (Print) ‖ Hauskorrektor m.; Korrektor m. ‖ compositeur-corriger m.; correcteur m.

correlation ‖ Korrelation f. ‖ corrélation f.

correspondence ‖ Schriftverkehr m.; Schriftwechsel m.; Briefwechsel m. ‖ correspondance f.; échange m. de lettres. / ~ in foreign languages ‖ fremdsprachiger Schriftwechsel m. ‖ correspondance f. en langues étrangères.

correspondent (Math) ‖ gleichnamig ‖ correspondant; homonyme.

correspondent ‖ Geschäftsfreund m.; Berichterstatter m. ‖ correspondant m.

corresponding to ‖ entsprechend ‖ correspondant à . . .

corresponding train ‖ Anschlußzug m. ‖ train m. de correspondance.

corridor ‖ Gang m.; Korridor m. ‖ couloir m.; corridor m. / private ~ ‖ geheimer Gang m.; Nebengang m. ‖ passage m. dégagé. / ~ carriage (Railw) ‖ Durchgangswagen m. ‖ voiture f. à couloir.

corrode, to ‖ korrodieren; anfressen ‖ corroder. / the steel will not corrode under the action of acids ‖ der Stahl ist widerstandsfähig gegen den Angriff von Säure ‖ l'acier m. résiste à l'influence d'acides. / ~-off ‖ abbeizen ‖ toucher à l'eau forte; faire disparaître à l'aide d'un corrosif.

corroded ‖ abgefressen ‖ corrodé.

corrodible see corrosive.

corroding agent ‖ Korrosionsmittel n. ‖ corrodant m.

corrosion ‖ Lochfraß m.; Ätzung f.; Zerfressung f.; Korrosion f. ‖ corrosion f. / ~ of boiler plates ‖ Ausfressung f. der Kesselbleche ‖ corrosion f. des tôles de chaudière. / electrolytic ~ ‖ elektrolytische Zerstörung f. ‖ corrosion f. électrolytique. / machine elements pl. impervious to ~ ‖ Maschinenteile mpl., die Widerstandsfähigkeit gegen Korrosion verlangen ‖ pièces fpl. de machines d'une grande résistance à l'action corrosive. / owing to ~ ‖ infolge von Korrosion f. ‖ dû à la corrosion. / terminal ~ (Electr) ‖ Anfressen n. der Klemmen ‖ corrosion f. des bornes.

corrosive ‖ fressend ‖ corrosif. / ~ (Iron) ‖ rostig ‖ rouillé. / non-~ ‖ rostsicher ‖ antirouille.

corrosive gas ‖ ätzendes Gas n. ‖ gaz m. corrosif. / ~ stick ‖ Ätzstift m. ‖ crayon m. corrosif. / ~ sublimate ‖ Sublimat n. ‖ sublimé m. corrosif. / ~ sublimat gauze ‖ Sublimatgaze f. ‖ gaze f. au sublimé.

corrosive ‖ Korrosionsmittel n. ‖ corrodant m.

corrugate, to ‖ riffeln; wellen ‖ canneler; onduler.

corrugated ‖ wellenförmig; gewellt; gerieft; geriffelt ‖ ondulé; cannelé; ridé. / ~ cardboard ‖ Wellpappe f. ‖ carton m. ondulé. / ~ covering ‖ Wellblechbeplankung f. ‖ revêtement m. en tôle ondulée. / ~ flue ‖ gewelltes Flammrohr n.; Wellrohr n. ‖ tube-foyer m. ondulé; tube m. ondulé. / ~ flue with tube plate welded on ‖ Wellrohr m. mit angeschweißter Rohrwand ‖ tube-foyer m. ondulé avec plaque tubulaire raccordée par soudage.

corrugated iron ‖ Wellblech n. ‖ tôle f. ondulée. / construction of ~ ‖ Wellblechbau m. ‖ construction f. en tôles ondulées. / decomposable ~ barrack ‖ zerlegbare Wellblechbaracke f. ‖ baraque f. de tôle ondulée démontable. / ~ covering ‖ Wellblecheindeckung f. ‖ couverture f. en tôle ondulée. / ~ press for ~ ‖ Wellblechpresse f. ‖ presse f. à tôle ondulée.

corrugated paper ‖ Wellpapier n. ‖ papier m. ondulé. / ~ sheet (iron) ‖ Wellblech n. ‖ tôle f. ondulée. / ~ sheet iron for roof ‖ Wellblech n. für Dachdeckung ‖ tôle f. ondulée pour toitures. / ~ sheet iron structure ‖ Wellblechbau m. ‖ bâtiment m. en tôle ondulée. / ~ sheet metal ‖ Riffelblech n. ‖ tôle f. ridée. / ~ tube ‖ Wellrohr n. ‖ tuyau m. ondulé. / rolling mill for ~ tubes ‖ Wellrohrwalzwerk n. ‖ laminoir m. à tubes ondulés.

corrugation ‖ Riffeln n.; Riffelung f. ‖ cannelage m.; cannelure f. / ~s on rails ‖ Wellenbildung f. auf den Schienen ‖

usure f. ondulatoire des rails. / ~ of rocky strata ‖ Faltung f. der Schichten ‖ plissement m. des couches.

corsair warfare ‖ Piratenkrieg m. ‖ guerre f. de corsairs.

corset ‖ Korsett n. ‖ corset m. / woven ~ ‖ gewebtes Korsett n. ‖ corset m. tissé. / machine for ~ hooks ‖ Korsettösenmaschine f. ‖ machine f. pour la fabrication des agrafes pour corsets. / ~ material ‖ Korsettstoff m. ‖ étoffe f. à corsets. / ~ spring ‖ Korsettfeder f. ‖ ressort m. à corsets.

corundum ‖ Korund m. ‖ corindon m. / artificial ~ ‖ künstlicher Korund m. ‖ corindon m. artificiel. / common ~ ‖ gemeiner Korund m.; Diamantspat m. ‖ corindon m. harmophane; spathe m. adamantin. / precious ~ ‖ Saphir m.; edler Korund m.; Rubin m. ‖ corindon m. hyalin; rubis m.; saphir m. / ~ brick ‖ Korundstein m. ‖ brique f. de corindon.

corver (Mine) ‖ Wagenausbesserer m. ‖ réparateur m. de berlines.

corvette ‖ Korvette f. ‖ corvette f.

cosecant ‖ Kosekante f. ‖ cosécante f.

cosine ‖ Kosinus m. ‖ cosinus m. / series of the ~ ‖ Kosinusreihe f. ‖ série f. du cosinus.

cosmetic ‖ verschönernd ‖ cosmétique. / ~ article ‖ kosmetischer Artikel m. ‖ article m. cosmétique. / ~ box ‖ Dose f. für kosmetische Mittel ‖ boîte f. à cosmétiques. / ~ pharmaceutical industry ‖ kosmetisch pharmazeutische Industrie f. ‖ industrie f. cosmétique-pharmaceutique. / ~ preparation ‖ kosmetisches Präparat n. oder Erzeugnis n. ‖ préparation f. cosmétique; produit m. cosmétique.

cosmetic ‖ Mittel n. zur Körperpflege; Schönheitsmittel n.; Kosmetik f. ‖ cosmétique m.

cosmic event ‖ kosmischer Vorgang m. ‖ fait m. cosmique.

cosmographical instrument ‖ kosmografisches Instrument n. ‖ instrument m. de cosmographie.

cosmonautics pl. ‖ Fahrt f. in den Weltenraum; Raumflug m.; Kosmonautik f. ‖ super-aviation f.; navigation f. extra-atmosphérique.

cost ‖ Unkosten pl. ‖ dépense f.; frais mpl. / ~s pl. (Trade) ‖ Spesen pl. ‖ frais mpl.; déboursé m. / additional ~ ‖ Mehrkosten pl. ‖ frais mpl. additionnels. / ~s of administration ‖ Verwaltungskosten pl. ‖ frais mpl. d'administration. / at ~ ‖ zum Selbstkostenpreis m. ‖ à prix m. coûtant. / at the ~ of the purchaser ‖ die Kosten pl. gehen zu Lasten des Bestellers ‖ les frais mpl. sont à la charge du client. / ~ to build ‖ Baukosten pl. ‖ prix m. de construction / small ~ of buying ‖ geringe Anschaffungskosten pl. ‖ frais mpl. bas d'achat. / ~ of construction ‖ Anlagekosten pl. ‖ frais mpl. d'installation ou d'établissement. / ~ of conversion ‖ Umwandlungskosten pl. ‖ frais mpl. de transformation. / extra ~ ‖ Aufschlag m. ‖ hausse f.; augmentation f. / extremely low first ~ ‖ außergewöhnlich niedrige Anlagekosten pl. ‖ des frais mpl. d'installation très modérés. / ~ of fuel ‖ Brennstoffkosten pl. ‖ prix m. du combustible. / ~ of housing ‖ Unterbringungskosten pl. ‖ frais mpl. de logement ou d'emménagement. / the

labour ~ is fixed beforehand mutually by the firm and the hands || die Arbeit wird zu einem bestimmten Lohn ausgeführt, der vorher zwischen Betrieb und Arbeitern festgesetzt wird || travail m. exécuté à prix fixe convenu entre la société *ou* le patron et les ouvriers. / law ~ || Gerichtskosten pl. || dépens mpl. / ~ for making up (Text) || Aufmachungsksoten pl. || frais mpl. de la confection. / minimal operating ~ || geringste Betriebskosten pl. || des frais mpl. d'exploitation très avantageux. / prime ~ || Anlagekosten pl. || frais mpl. d'établissement. / prime ~ of installation || Baukosten pl. || frais mpl. d'installation. / ~ of production || Macherlohn m. || main-d'œuvre f.; façon f. / ~ of production || Gestehungspreis m. || prix m. coûtant de la production; prix m. de revient. / ~ of removing || Umzugskosten pl. || frais mpl. de déménagement. / ~ of renovation || Wiederherstellungskosten pl. || frais mpl. de réparation. / goods pl. are forwarded at the ~ and risk of the consignee || der Versand der Waren erfolgt auf Rechnung und Gefahr des Empfängers || les marchandises fpl. voyagent aux frais, risques et perils du destinataire. / ~ of sales || Verkaufsunkosten pl. || frais mpl. de vente. / ~ of smelting || Schmelzkosten pl. || coût m. de fusion. / ~ of telegram || Telegrammspesen pl. || frais mpl. de dépêche. / ~ of testing || Prüfungskosten pl. || frais mpl. d'épreuve.

cost accounting department || Selbstkostenberechnungsabteilung f. || service m. de détermination des prix de revient / ~book mining company || Gewerkschaft f. || société f. d'exploitation. / ~ price || Selbstkosten pl.; Gestehungspreis m. || prix m. coûtant *ou* de revient. / ~ sheet || Kostenaufstellung f. || bordereau m. du prix de revient.

costs pl. *see* cost.

costume || Kostüm n.; Tracht f. || costume m.; mode f.; charge f.

cot (Mar) || Hängematte f. mit Rahmen | cadre m.; hamac m. à cadre. / finger ~ || Gummifinger m. || doigtier m. / thatched ~ || Strohhütte f. || chaumière f.

cotangent || Kotangente f. || cotangente f.

cotillon articles pl. || Kotillonartikel mpl. || articles mpl. de cotillon.

cottager || Häusler m.; Kossäte m. || journalier m.

cottar *see* cotter.

cotter, to ~ a bolt || einen Bolzen m. versplinten || goupiller un boulon.

cotter (Carp) || Keil m. || coin m.; clef f. / ~ (Mach) || Splint m.; Vorsteckstift; Schließkeil m. || clavette f.; goupille f.; clavette f. de serrage. / to tighten the ~ || den Keil m. anziehen || serrer la clavette. / inlet valve ~ || Einlaßventilkeil m. || clavette f. de soupape d'admission.

cotter bolt || Schließbolzen m.; Keilbolzen m. || boulon m. à clavette *ou* en forme de coin. / ~ drift || Keiltreiber m. || chasse-clavette m. / ~ file || dickflache Feile f. || carrelet m. plat; lime f. à pilier. / ~ key || Keiltreiber m. || chasse-clavette m. / ~ pin || Vorsteckkeil m. || clavette f.; goupille f. / ~ slot || Keilloch n. || encoche f. de clavette.

cotton || Baumwolle f. || coton m. / artificial ~ || Kunstbaumwolle f. || coton m. artificiel. / artificial ~ making machine ||

Kunstbaumwollefertigungsmaschine f. || machine f. de fabrication du coton artificiel. / ginned ~ || entkernte Baumwolle f. || coton m. égrené. / long-stapled ~ || langstapelige Baumwolle f. || coton m. longue soie. / picked ~ || gezupfte Baumwolle f. || coton m. épluché. / pressing plant for ~ || Baumwollpreßanlage f. || installation f. à comprimer le coton. / raw ~ || Rohbaumwolle f. || coton m. brut. / seedy ~ || nicht entkernte Baumwolle f. || coton m. non égrené. / shortstapled ~ || kurzstapelige Baumwolle f. || coton m. courte-soie. / waste ~ || Putzbaumwolle f.; Twist m. || déchet m. de coton.

cotton bale breaker || Baumwollballenaufmacher m. || déballeur m. de coton; déchireur m. de balles de coton. / ~ bale picker hand || Baumwollballenöffner m. || démêleur m. de balles de coton. / ~ baling press || Baumwollballenpresse f. || presse f. à mettre le coton en balles. / ~ bandage || Baumwollbinde f. || bande f. de coton. / ~ bed tick || Bettbarchent m. || coutil m. de coton pour literie. / ~ belt || Baumwollriemen m. || courroie f. en coton *ou* Balata. / ~ blanket || Baumwolldecke f. || couverture f. en coton. / ~ bleaching || Baumwollbleichung f. || blanchiment m. de coton. / ~-braided covering || Baumwollumklöppelung f. || revêtement m. en coton tressé. / ~ cambric || Kambrik m. || batiste m. de coton. / ~ carder || Baumwollkrempler m. || cardeur m. de coton. / ~ carding || Baumwollkrempelei f. || cardage m. de coton. / ~ cleaner || Baumwollreiniger m. || nettoyeur m. de coton m. / waste ~ cleaner || Baumwollabfallreinigungsmaschine f. || machine f. à nettoyer les déchets de coton. / ~ cloth || Baumwollgewebe n. || toile f. de coton. / ~ comber || Baumwollkämmer m. || peigneur m. de coton. / ~ combing || Baumwollkämmerei f. || peignage m. de coton. / ~ cord || Baumwollschnur f. || ganse f. de coton. / ~ covering || Baumwollumspinnung f. || guipage m. de coton. / ~ crop || Baumwollernte f. || récolte f. de coton. / ~ deviller || Baumwollwolfer m. || diableur m. *ou* ouvrier m. de wellow de coton. / ~ doubler || Baumwolldublierer m. || doubleur m. *ou* réunisseur m. de coton. / ~ drawing-frame tenter || Baumwollstrecker m. || étireur m. de coton. / ~ dressing machine || Baumwollvorbereitungsmaschine f. || machine f. à préparer le coton. / ~ drill || Baumwolldrillich m.; Baumwolldrell m. || treillis m. de coton. / ~ dyer || Baumwollfärber m. || teinturier m. sur coton. / ~ dyeing plant || Baumwollfärberei f. || teinturerie f. de coton. / ~ enucleation machine || Baumwollentkernungsmaschine f. || machine f. à égrener le coton. / ~ finisher machine-minder || Baumwollvollender m. || finisseur m. de coton. / ~ gin || Baumwollentkörnungsmaschine f. || égreneuse f. de coton. / ~ glove || Baumwollhandschuh m. || gant m. de coton. / ~ goods pl. || Baumwollwaren fpl. || articles mpl. *ou* tissus m. de coton. / ~ hard waste breaker tenter || Baumwollreißer m. || effilocheur m. de coton m. / ~ hard waste breaking || Baumwollabfallreißerei f. || effilochage m. de coton. / ~ hose || Baumwollenschlauch

m. || tuyau m. en coton. / ~ hosiery || Baumwollstrickerei f. || bonneterie f. de coton. / ~ intermediate tenter || Baumwollmittelstrecker m. || étireur m. intermédiaire de coton. / ~ lap machine minder *see* ~ lapper. / ~ lapper || Baumwollausbreiter m. || étaleur m. *ou* nappeur m. de coton. / ~ lining || Baumwollfutterstoff m. || doublure f. en coton. / ~-made || baumwollen || de coton. / ~ manufacture || Baumwollindustrie f. || industrie f. cotonnière. / ~ mill || Baumwollspinnerei f. || filature f. de coton. / ~ mixer || Baumwollmischraumarbeiter m. || mélangeur m. de coton. / ~ oil || Baumwollsamenöl n. || huile f. de coton. / ~ operative || Baumwollarbeiter m. || cotonnier m. / ~ packing (Mach) || Baumwollpackung f. || tresses fpl. en coton. / ~ picker || Baumwollpflückmaschine f. || éplucheuse f. de coton. / ~ picker tenter || Baumwollrupper m. || éplucheur m. de coton. / ~ plug || Wattebausch m. || tampon m. de ouate. / ~ plush || Baumwollplüsch m. || peluche f. de coton. / ~ powder || Schießbaumwolle f. || fulmi-coton m. / ~ press || Baumwollballenpresse f. || presse f. à coton. / ~ rag paper || Baumwollpapier n. || papier m. de coton. / ~ reeler || Baumwollweifer m. || dévideur m. de coton. / ~ ribbon hand || Baumwolldublierer m. || doubleur m. *ou* réunisseur m. de coton. / ~ rope || Baumwollseil n. || câble m. *ou* corde f. de coton. / ~ roving-frame tenter || Baumwollvorgarnspinner m. || boudineur m. de coton. / ~ seed || Baumwollsamen m. || coton m. en cogne. / ~ seed cake || Baumwollsamenölkuchen m. || tourteau m. de graine de coton. / ~ seed oil || Baumwollsamenöl n. || huile f. de coton. / ~ seed peeling machine || Baumwollsamenschälmaschine f. || broyeuse f. de graines de coton. / ~ seed residue || Baumwollsamenrückstand m. || résidu m. de graines de coton. / ~ serge || baumwollener Sergestoff m. || serge f. en coton. / ~ sewing thread || Baumwollnähgarn n. || fil m. de coton. / ~ shag || baumwollener Felbel m. || panne f. de coton. / ~ shaker || Baumwollschlagmaschinenarbeiter m. || batteur m. *ou* briseur m. de coton. / ~ silk cable || Baumwollseidenkabel n. || câble m. sous coton et soie. / ~ speeder tenter || Baumwollfeinstrecker m. || étireur m. en fin de coton. / ~ spinner || Baumwollspinner m. || fileur m. de coton. / ~ spinning || Baumwollspinnerei f. || filature f. de coton. / waste ~ spinning || Baumwollabfallspinnerei f. || filature f. de déchets de coton. / ~ spinning machine || Baumwollspinnereimaschine f. || machine f. à filer le coton. / ~ spinning master || Baumwollspinnereimeister m. || contremaître m. de filature de coton. / owner of a ~ spinning mill || Baumwollspinnereibesitzer m. || propriétaire m. d'une filature de coton. / ~ spinning railway head tenter || Baumwollspinnmaschinendublierer m. || soigneur m. réunisseur de filature de coton. / ~ steamer || Baumwolldämpfer m. || vaporisateur m. de coton. / brake band *ou* strap || Bremsband n. aus Baumwollgurten || ruban m. à frein en ceinture de coton. / ~ supplier || Baumwollspeiser m. || alimenteur m. de coton. / uncalendered ~ tape || ungeglättetes Baumwollband n. || ruban m. de coton

non calandré. / ~ textiles pl. || Baumwollgewebe n. || tissus mpl. de coton. / ~ thread || Baumwollzwirn || fil m. de coton. / ~ thread cabler || Baumwollfadendreher m. || câbleur m. de coton. / ~ threader || Baumwollüberspinner m. || guipeur m. de coton. / ~ tissue || Baumwollgewebe n. || tissu m. de coton. / ~ tree || Baumwollstaude f. || cotonnier m. / ~ tress (Mach) || Baumwollzopf m. || garniture f. en coton. / ~ twist || Baumwollzwirn m. || fil m. retors de coton. / ~ twister || Baumwollzwister m.; Baumwollzwirner m. || retordeur m. de coton. / ~ twist mill || Baumwollzwirnerei f. || retordage m. de coton. / ~ velvet || Baumwollsamt m. || velours m. de coton. / ~ wadding || Baumwollwatte f. || ouate f. de coton. / ~ warp || Baumwollkette f. || chaîne f. de coton. / ~ washer || Baumwollwäscher m. || laveur m. de coton.

cotton waste || Baumwollabfall m. || déchets mpl. de coton. / ~ for cleaning || Putzbaumwolle f. || déchets mpl. de coton. / ~ bleaching || Baumwollabfallbleichung f. || blanchiment m. de déchets de coton. / ~ hand || Baumwollabfallsortierer m. || trieur m. de déchets de coton. / ~ washer / Baumwollabfallwäscher m. || laveur m. de déchets de coton. / ~ yarn || Baumwollabfallgarn n. || fil m. de déchets en coton.

cotton weaver || Baumwollweber m. || tisseur m. de coton. / ~ weaving || Baumwollweberei f. || tissage m. de coton. / ~ wick || Baumwolldocht m. || mèche f. de coton. / ~ willower || Baumwollwolfer m. || diableur m. ou ouvrier m. de wellow de coton. / ~ winding || Baumwollhaspelei f. || dévidage m. de coton. / ~ wood || Baumwollholz n. || bois m. de cotonier. / ~ wool || Watte f. || ouate f. / ~ wool ball || Wattekugel f. || boule f. en ouate. / ~ wool figure || Wattefigur f. || figure f. en ouate . / ~ wool probe (Med) || Watteträger m. || portecotons m. / ~ yarn || Baumwollfaden m. || fil m. de coton. / ~ yarn made up for the retail || Baumwollgarn n. in Aufmachung für den Kleinverkauf || fil m. de coton conditionné pour la vente au détail. / ~ yarn bleaching || Baumwollgarnbleichung f. || blanchiment m. de fils de coton.

cottrel see cotter.

cotunnite || Kotunnit n.; Chlorblei n. || cotunnite f.

couch, to ~ the leaves || die Blätter npl. gautschen || coucher les feuilles fpl. / ~ the paper-sheets || das Papier gautschen / coucher les feuilles fpl. de papier.

couch || Ruhebett n. || lit m. de repos. / ~ (Bot) || Quecke f. || chiendent. / ~ (Brew) || Beet n.; Schicht f. (Gerste) || couche f. / ~ of paper || Lage f. oder Packen m. Papier || liasse f. de papier; jetée f.

coucher (Pap) || Papierstreicher m.; Gautscher m. || coucheur m.

couching (Pap) || Gautschen n. || couchage m.

cough || Husten m. || toux f. / ~ drop || Hustenbonbon m. || bonbon m. contre la toux.

coughing spell || Hustenanfall m. || accès m. de toux.

couleur (Sugar) || Zuckercouleur f. || teinture f. de caramel; couleurs fpl. de sucre.

coulter || Kolter n.; Pflugeisen n.; Sech n. || coutre m.; sep m.

coumarone resin || Kumaronharz n. || résine f. de coumarone.

coumarou, leaf of || Coumaroublatt n. || feuille f. de coumarou. / ~ bean || Coumaroubohne f. || fève f. de coumarou.

council || Kollegium n.; Ratsversammlung f.; Rat m. || conseil m.; corps m. enseignant; collège m.

counsel, to || beraten || conseiller.

counsel || Ratschlag m. || conseil m. / ~ (Legal Advise) || Rechtsbeistand m. || conseil m. judiciaire; avocat m. / ~ 's opinion || Gutachten n. || avis m.; opinion f.; rapport m.; expertise f.

count, to || zählen || compter. / ~ clockwise || rechtsläufig zählen || compter dans le sens des aiguilles d'une montre. / ~ contra-clockwise || linksläufig zählen || compter dans le sens opposé des aiguilles d'une montre. / ~ from left to right || rechtsläufig zählen || compter dans le sens des aiguilles d'une montre. / ~ paper || Papier n. abzählen || compter les feuilles fpl. de papier. / ~ from right to left || linksläufig zählen || compter dans le sens opposé des aiguilles d'une montre. / ~-up || aufzählen || compter; énumérer.

counter (Game) || Spielmarke f. || jeton m. / ~ (Mach) || Zählwerk n.; Zähler m. || mécanisme m. compteur; compteur m. / ~ (Print) || Punzen m. || poinçon m. / ~ (Shipb) || Gillung f. des Hinterschiffes || voûte f. d'arrière. / ~ (Sign) || Marke f. || marque f. / ~ (Trade) || Ladentisch m. || comptoir m. / conversation ~ || Gesprächszähler m. || compteur m. de conversation. / ~ of customers (cash register) || Kundenzähler m. || compteur m. de clients. / ~ electrically driven ~ || elektrisch aufziehbares Zählwerk n. || minuterie f. à commande électrique. / ~ of a meter || Zählwerk n. || mécanisme m. de compteur; minuterie f. / pocket ~ || Taschenzähler m. || compteur m. de poche. / ~ of sheets (Print) || Bogenzähler m. || compteur m. de feuilles. / ~ of transactions (Cash register) || Postenzähler m. || compteur m. de ventes.

counteract, to || gegenwirken || contrebalancer.

counteracting stirring mechanism || gegeneinander arbeitendes Rührwerk n. || agitateur m. à mouvement contraire. / ~ forces pl. (Phys) || Gegenkräfte fpl. || forces fpl. antagonistes ou opposées.

counterbalance, to || ausgleichen; ausbalanzieren; auswuchten || équilibrer; égaliser; balancer.

counterbalance || Ausgleichgewicht n. || contrepoids m. (d'équilibre). / ~ of weight || Gewichtsausgleich m. || équilibre m. de poids. / ~ weight || Ausgleichgewicht n. || contrepoids m.

counterbass || Kontrabaß m. || contrebasse f.

counterbill || Gegenwechsel m. || rechange m.; retraite f.; retour m.

counterbore || Senker m.; Krauskopf m. || fraise f. conique.

counter calibrating room || Zählereichsaal m. || salle f. de vérification de compteurs.

counter-chain wheel || Gegenkettenrad n. || contre-roue f. à chaîne. / ~-charge || Gegenklage f. || reconvention f. / ~-

claim || Gegenforderung f.; Gegenschuld f. || créance f. en compensation; demande f. reconventionnelle; dette f. passive.

counter-contact connection || Gegentaktschaltung f. || système m. en va-etvient. / ~ strengthener || Gegentaktverstärker m. || amplificateur m. pushpull. / ~ switching see connection.

counter-current || Rückstrom m.; Gegenstrom m. || contre-courant m. / ~ apparatus || Gegenstromapparat m. || appareil m. à contre-courant. / ~ boiler || Gegenstromkessel m. || chaudière f. système contre-courant; générateur m. de vapeur à contre-courant. / ~ condenser || Gegenstromkondensator m. || condenseur m. à contre-courant. / ~ condenser of the multi-pass type || Elementenbündelverflüssiger m. || condenseur m. à contrecourant du type multitubulaire.

counterdraft || Gegenwechsel m. || rechange m.; retraite f.; traite f. de retour.

counterdraw, to (Drawing) || durchpausen || calquer. / ~ (Print) || Gegenabdruck m. machen || décalquer; contre-tirer.

counterdrawing || Pause f.; Kopie f. || calque m.

counter-electromotive || gegenelektromotorisch || contre-électromotif.

counterfeit || künstlich; nachgemacht || artificiel; factice; imité.

counterfeit, to (Print) || unerlaubt nachdrucken || contrefaire.

counterfeit (Print) || Nachdruck m. || contrefaçon f.; contrefaction f.

counterfeiter (Print) || Urheber m. eines unberechtigten Nachdruckes || contrefacteur m. / ~ (Money) || Falschmünzer m. || faux monnayeur m.

counter-flange || Gegenflansch m. || contrebride f. / ~-floor || Blendboden m.; Blindboden m. || fausse-aire f.; fauxparquet m. / ~-flow || Gegenströmung f. || contre-courant m. / ~-flow condenser || Gegenstromkühler m. || condenseur m. à contre-courant. / ~-force || Gegenkraft f. || contre-force f. / ~-fort || Gewölbpfeiler m. || appui m.; contrefort m. / ~-impression || Gegenabdruck m.; Gegenabzug m. || contre-épreuve f. / ~-item || Gegenposten m. || contre-partie f.

countermand, to || absagen; widerrufen || contremander; décommander.

countermark || Gegenzeichnung f. || contresignature f.

countermine || Gegenmine f. || contre-mine f.; mine f. défensive.

counter-motion || Gegenbewegung f. || contre-mouvement m.; mouvement m. réactionnaire.

counter-movement see counter-motion.

counter-obligation || Gegenverpflichtung f. || obligation f. réciproque.

counter-offer || Gegenvorschlag m.; Gegenangebot n. || contre-proposition f.

counter-order || Gegenauftrag m.; Gegenorder f. || contre-ordre m.

counterpane || Bettdecke f. || courtepointe f. / cotton ~ || baumwollene Bettdecke f. || couvre-lit m. coton.

counterpart || Vertragsgegner m. || contrepartie f.

counterparty || Gegenpartei f. || partie f. adverse.

counter pedometer || Schrittmesser m. || compte-pas m.

counterpoise (Mach) || Gegengewicht n. || contre-poids m. / ~ (Radio) || ausglei-

chender Erdschluß m.; Gegengewicht n. || mise f. à la terre compensée; capacité f. d'équilibre. / ~ of a draw-bridge || Schwengel m. *oder* Wippe f. einer Aufzugbrücke || bascule f. *ou* flèche f. d'un pont-levis.

counter-pressure || Gegendruck m. || contrepression f. / ~ filling apparatus || Gegendruckfüllapparat m. || appareil m. de soutirage à contre-pression. / ~ racker *see* ~ racking apparatus. / ~ racking apparatus || Gegendruckfaßfüllapparat m. || soutireuse f. à contre-pression pour fûts. / ~ turbine || Gegendruckturbine f. || turbine f. à contre-pression.

counter-proof (Print) || Gegenabdruck m.; Gegenabzug m. || contre-épreuve f.

counter-punch || Gegenpunzen m. || contrepoinçon m.

counter-rail (Railw) || Gegenschiene f.; Leitzwangschiene f. || contre-rail m.

counter-reckoning *see* counter-claim.

counter-remittance || Gegenrimesse f. || retour m.

counter-sample || Gegenmuster n. || échantillon m. de contrôle.

countershaft (Intermediate Shaft) || Vorgelegewelle f. || arbre m. de renvoi; arbre m. secondaire. / ~ (Intermediate Gearing) || Vorgelege n. || engrenage m. *ou* transmission f. intermédiaire; renvoi m. (de mouvement); contre-arbre m. / loose ~ pulley || Losscheibe f. am Deckenvorgelege n. || poulie f. folle du renvoi.

countersign, to || gegenzeichnen || contresigner.

countersink, to || ausfräsen; versenken || noyer. / ~ a screw head || einen Schraubenkopf m. versenken || noyer la tête d'une vis.

countersink (Mach) || Senker m.; Versenker m.; Krauskopf m. || foret m. à fraiser; fraise f. conique. / ~ (Opt) || Einschliff m. || excavation f. / cone ~ || konischer Senker m. || fraise f. à tête conique *ou* en forme de cône. / rose-head ~ || Krauskopf m. || fraise f. champignon.

countersinking drilling machine || Bohr- und Versenkmaschine f. || perceuse f. à noyer les trous.

counterspring || Gegenfeder f. || contreressort m.

countersteam || Gegendampf m. || contrevapeur f. / ~ brake || Gegendampfbremse f. || frein m. à contre-vapeur.

countersunk head || versenkter Kopf m. || tête f. noyée. / bolt screw with ~ || Bolzen m. mit versenktem Kopf || boulon m. à tête noyé.

countersunk rivet || versenktes Niet n. || rivet m. noyé *ou* à tête perdue *ou* à tête fraisée. / ~ screw || Senkschraube f. || vis f. à tête fraisée; vis f. noyée.

counter-support (Mach tool) || Gegenständer m. || montant m. à supports.

counter-value || Gegenwert m. || contrevaleur f.; équivalent m.

counter-veneer, to || auf beiden Seiten furnieren; gegenfurnieren || contre-plaquer.

counter-voltage conductor || Gegenspannungsdraht m. || fil m. de contretension.

counterweight || Gegengewicht n. || contrepoids m. / balanced by a ~ || durch ein Gegengewicht n. ausbalanziert || équilibré par un contre-poids. / ~ box (Mach) || Gewichtskasten m.; Gegengewichtskasten m. || caisse f. à contrepoids. /

~ brake || Wurfhebelbremse f. || frein m. à levier à contrepoids.

counterweighted || durch Gegengewicht n. ausgeglichen || équilibré.

counterweight fitting || Schnurzugaufhängung f. || suspension f. à contre-poids.

counting apparatus || Zähleinrichtung f.; Zähler m.; Zählwerk n. || (appareil m.) compteur m. / ~ apparatus for miners' trucks || Förderwagenzähleinrichtung f. || compteur m. pour wagonnets de mine. / ~ chamber || Zählkammer f. || cellule f. à compter. / ~ device || Zählapparat m. || appareil m. à compter; compteur m.

counting machine || Rechenmaschine f. || machine f. à calcul. / money ~ || Geldzählmaschine f. || compteur m. d'argent. / ~ for tickets || Fahrkartenzählmaschine f. || machine f. compteuse de billets.

counting mechanism || Zählwerk n. || mécanisme m. compteur. / ~ with jumping figures || Zählwerk n. mit springenden Zahlen || mécanisme m. compteur à chiffres sautants.

countless || zahllos || innombrable; sans nombre.

country || Gegend f. || pays m.; terrain m. / ~ (Mine) || Gebirge n.; Nebengestein n. || terrain m. / cross-~ || querfeldein || à travers champ m. / ~ which is poor in bituminous coal || steinkohlenarmes Gebiet n. || pays m. pauvre en houille. / ~ that is poor in oils || ölarmes Land n. || pays m. sans sources naturelles d'huiles minérales. / up to ~ || landeinwärts || vers l'intérieur m.

country baker || Landbrotbäcker m. || boulanger m. rural. / ~ fog || Landnebel m. || brouillard m. de campagne. / ~ house || Landhaus n. || maison f. de campagne; villa f. / ~-man || Bauer m. || paysan m. / ~ mill || Bauernmühle f. || moulin m. agricole. / ~ rock (Mine) || Nebengestein n. || roche f. encaissante *ou* de parois. / ~ smith || Dorfschmied m. || forgeron m. de village.

county court || Amtsgericht n. || tribunal m. cantonal de première instance; justice f. de paix.

coupé || Halbkutsche f.; geschlossener Zweisitzer m.; Coupé n. || coupé m. / ~ (Railw) || Eisenbahnabteil n. || coupé m.; compartiment m. d'un wagon.

couple, to (Auto) || kuppeln; schalten || embrayer. / ~ (Chem) || koppeln || copuler; coupler. / ~ (Mach) || einrücken || embrayer. / ~ (Radio) || koppeln || accoupler. / ~ (Railw) || ankuppeln || atteler. / ~ back (Radio) || rückkoppeln || rétrocoupler; monter en réaction f. / ~ wagons (Railw) || Wagen mpl. anhängen || accrocher des wagons mpl.

couple || Paar n. || couple m. / ~ of forces || Kräftepaar n. || couple m. de forces. / principal ~ || Dachbinder m. || maîtresse f. ferme. / righting ~ (Shipb) || aufrichtendes Kräftepaar n. || couple m. de rappel. / ~ of a roof || Gebinde n. eines Dachstuhls || ferme f. de comble. / ~ of wheels || Radsatz m. || paire f. de roues.

couple-close || Sparrwerk n. || chevrons mpl. d'un comble.

coupled || gekuppelt || accouplé. / directly ~ || unmittelbar gekuppelt || couplé directement. / ~ axle || Kuppelachse f. || essieu m. couplé. / ~ circuit (Radio) || gekoppeltes System n. || circuit m. accouplé. / ~ oscillations pl. || gekoppelte Schwingun-

gen fpl. || oscillations fpl. liées. / ~ pole (Electr) || Doppelständer m. || poteau m. moisé.

coupled wheel || Kuppelrad n. || roue f. accouplée. / ~ with lateral play || verschiebbares Kuppelrad n. || roue f. couplée à jeu latéral. / ~ set || Kuppelradsatz m. || essieu m. accouplé monté.

coupler (Railw) || Wagenkuppler m.; Rangierer m. || accrocheur m. / automatic car ~ (Railw) || selbsttätige Kupplung f. || attelage m. automatique. / reversible pilot ~ (Railw) || Hornschwenkkopf m. || attelage m. à griffe pivotante. / ~ head (Railw) || Kuppelkopf m. || tête f. d'attelage. / ~ jaw (Railw) || Kuppelklaue f. || griffe f. de la tête d'attelage.

coupler-on (Mine) || Wagenkuppler m. || coupleur m. de berlines.

coupling (Chem) || Kopplung f. || copulation f. / ~ (Mach) || Kupplung f. || accouplement m.; embrayage m. / ~ (Radio) || Kopplung f. || couplage m.; accouplement m. / ~ (Railw) || Wagenkupplung f.; Kupplung f. || attelage m. / ~ (Tel) || Platzzusammenschaltung f. || groupement m. de position. / articulated ~ || Gelenkkupplung f. || attelage m. articulé. / automatic ~ (Railw) || selbsttätige Kupplung f. || attelage m. automatique. / band ~ || Bandkupplung f. || accouplement m. à ruban. / bar ~ || Stangenkupplung f. || accouplement m. à barres. / ~ of brake type || Bremskupplung f. || accouplement m. à frein. / brush ~ (Mach) || Bürstenkupplung f. || accouplement m. à brosses. / claw ~ || Klauenkupplung f. || embrayage m. *ou* accouplement m. à griffes. / close ~ (Radio) || feste Kopplung f. || accouplement m. rigide; couplage m. serré. / clutch ~ || Klauenkupplung f. || accouplement m. à griffes. / coefficient of ~ (Radio) || Kopplungsgrad m. || coefficient m. d'accouplement. / the combined ~s pl. are arranged to swing about a vertical bolt || die Kupplungen fpl. sind um einen gemeinsamen senkrechten Bolzen schwenkbar angeordnet || les attelages mpl. combinés pivotent autour d'un boulon vertical commun. / conical ~ || Konuskupplung f. || accouplement m. à cône. / direct ~ || unmittelbare Kupplung f. || couplage m. direct. / disengaging ~ || Ausrückvorrichtung f. || mécanisme m. de débrayage. / disk ~ || Scheibenkupplung f. || embrayage m. à plateaux. / dog ~ || Klauenkupplung f. || accouplement m. à griffes. / elastic ~ || nachgiebige Kupplung f. || accouplement m. élastique *ou* flexible. / elastic spring ~ || elastische Federkupplung f. || accouplement m. à ressort élastique. / electromagnetic ~ || Elektromagnetkupplung f. || accouplement m. électromagnétique. / emergency ~ (Railw) || Notkupplung f. || attelage m. de sûreté. / engaging and disengaging ~ || lösbare Kupplung f. || accouplement m. à embrayage et débrayage. / expansion ~ || Ausdehnungskupp(e)lung f. || accouplement m. extensible. / flange ~ || Flanschenkupplung f. || accouplement m. à plateau. / flange of ~ (Railw) || Kupplungsflansch m. || bride f. d'accouplement / flexible ~ || nachgiebige Kupplung f. || accouplement m. élastique *ou* flexible. / friction ~ || Reibungskupplung f. || accouplement m. à friction. / hose ~ (Railw) || Schlauch-

kupplung f. ‖ accouplement m. *ou* raccordement m. à tuyaux flexibles. / hydraulic ~ ‖ hydraulische Kupplung f. ‖ accouplement m. hydraulique. / inductive ~ (Radio) ‖ induktive Kopplung f. ‖ accouplement m. inductif. / jaw ~ ‖ Klauenkupplung f. ‖ attelage m.à griffe. / jointed ~ ‖ Gelenkkupplung f. ‖ accouplement m. à articulation. / leather ~ ‖ Lederkupplung f. ‖ accouplement m. en cuir. / loose ~ (Radio) ‖ lose *oder* schlaffe Kopplung f. ‖ couplage m. lâche. / magneto ~ ‖ Magnetkupplung f. ‖ accouplement m. de magnéto. / percentage of ~ (Radio) ‖ prozentuale Kopplung f. ‖ expression f. procentuelle du couplage. / permutation ~ (Railw) ‖ Umsteckkupplung f. ‖ attelage m. à griffes et à vis amovible. / rigid ~ ‖ starre Kupplung f. ‖ accouplement m. fixe. / safety ~ (Railw) ‖ Sicherheitskupplung f. ‖ attelage m. de sûreté. / screw ~ (Railw) ‖ Schraubenkupplung f. ‖ attelage m. à vis. / ~ operated from side (Railw) ‖ seitlich bedienbare Kupplung f. ‖ attelage m. latéral *ou* à commande latérale. / slipping ~ ‖ Rutschkupplung f. ‖ accouplement m. à coulisse; friction f. glissante. / steel bolt ~ ‖ Stahlbolzenkupplung f. ‖ embrayage m. à clavettes en acier. / steel plate ~ ‖ Stahlblattkupplung f. ‖ accouplement m. à lames en acier. / to throw the ~ out ‖ entkuppeln ‖ débrayer. / transition ~ (Railw) ‖ Übergangskupplung f. ‖ attelage m. temporaire. / ~ for tube conduit ‖ Rohrverschraubung f. ‖ manchon m. de tube; raccord m. de tuyaux. / universal ~ ‖ Kreuzgelenkkupplung f. ‖ joint m. brisé.
coupling bar ‖ Kupplungsstange f. ‖ barre f. d'attelage. / ~ bolt ‖ Kupplungsbolzen m. ‖ boulon m. d'attelage.
coupling box (Electr) ‖ Kabelmuffe f. ‖ manchon m. d'assemblage. / ~ (Mach) ‖ Muffenhülse f. ‖ manchon m. d'accouplement. / screw ~ ‖ Schraubenkupplungsmuffe f. ‖ manchon m. d'accouplement à vis. / suspension ~ ‖ Hängekupplungsdose f. ‖ boîte f. d'accouplement suspendue.
coupling chain (Railw) ‖ Zugkette f.; Kuppelkette f. ‖ chaîne f. d'attelage. / ~ changer (Radio) ‖ Kopplungswechsler m. ‖ commutateur m. de couplage. / ~ cock (Railw) ‖ Kupplungshahn m. ‖ robinet m. d'accouplement. / ~ coefficient of ~ (Radio) ‖ Kopplungsgrad m. ‖ coefficient m. de couplage. / ~ coil (Radio) ‖ Kopplungsspule f. ‖ bobine f. de couplage. / ~ locomotive ~ device ‖ Lokomotivkuppelvorrichtung f. ‖ dispositif m. d'attelage pour locomotives. / ~ disk ‖ Kupplungsscheibe f. ‖ plateau m. d'accouplement. / ~ factor ‖ Kopplungsgrad m. ‖ coefficient m. de couplage. / ~ flange ‖ Kupplungsflansch m. ‖ bride f. d'accouplement. / ~ counter-balance weight fastened to the ~ flange ‖ am Kupplungsflansch angebrachtes Gegengewicht n. ‖ contrepoids m. rapporté à la bride d'assemblage. / ~ fork ‖ Ausrückgabel f. ‖ fourche f. de débrayage. / ~ hook (Railw) ‖ Kuppelhaken m. ‖ crochet m. d'attelage. / ~ hose ‖ Kupplungsschlauch m. ‖ tube m. coupleur flexible; boyau m. d'attelage. / ~ lever ‖ Kupplungshebel m. ‖ levier m. d'embrayage et de débrayage. / ~ and uncoupling-lever ‖ Ein- und Ausrückhebel m. ‖ levier m.

d'embrayage et de débrayage./ ~ lining ‖ Kupplungsbelag m. ‖ garniture f. d'embrayage. / ~ nut ‖ Kupplungsmutter f.; Spannmutter f. ‖ écrou m. de tendeur. / ~ pin (Railw) ‖ Kuppelzapfen m. ‖ bouton m. de manivelle d'accouplement. / ~ plate ‖ Achszwinge f. ‖ bride f. d'étrier d'essieu.
coupling rod ‖ Kuppelstange f. ‖ bielle f. d'accouplement. / front ~ (Loc) ‖ vordere Kuppelstange f. ‖ bielle f. d'attelage avant. / rear ~ ‖ hintere Kuppelstange f. ‖ bielle f. d'attelage arrière.
coupling screw (Railw) ‖ Kupplungsspindel f. ‖ tendeur m. d'attelage. / ~ shackle (Railw) ‖ Kupplungsbügel m. ‖ maille f. de tendeur. / ~ spindle thread for railway work ‖ Eisenbahnkupplungsspindelgewinde n. ‖ filet m. des tiges d'attelage pour chemin de fer. / ~ strap (Locksm) ‖ Schelle f. ‖ agrafe f. de serrage. / ~ strap (Saddl) ‖ Kumtstrippe f.; Kumtgurtriemen m. ‖ courroie f.; couplière f. / ~ tappet ‖ Auslösungsknagge f. ‖ taquet m. d'embrayage et de débrayage. / ~ value (Radio) ‖ Kopplungszahl f. ‖ valeur f. de couplage.
coupon ‖ Kupon m.; Zinsschein m. ‖ coupon m. ‖ Zinsschein m. / outstanding ~ ‖ notleidender Kupon m. ‖ coupon m. en souffrance. / ~ sheet ‖ Kuponbogen m. ‖ feuille f. de coupons.
course (Mas) ‖ Lage f. *oder* Schicht f. von Steinen ‖ assise f. / ~ (Mine) ‖ Erzgang m. ‖ filon m.; veine f. / ~ (Nav) ·‖ Kurs m. ‖ route f. / ~ (Travel) ‖ Fahrt f. ‖ passage m.; trajet m. / ~ (Sail) ‖ Untersegel n. ‖ basse voile f. / altering ~ (Nav) ‖ Kursänderung f. ‖ changement m. de route. / ~ of business ‖ Geschäftsgang m. ‖ marche f. des affaires; opération f. / compound ~ (Nav) ‖ Koppelkurs m. ‖ journée f.; route f. compliquée. / to correct the ~ (Nav) ‖ den Kurs m. verbessern ‖ corriger la route. / ~ of a curve ‖ Verlauf m. einer Kurve ‖ allure f. d'une courbe. / curved ~ ‖ bogenförmiger Verlauf m. ‖ allure f. en forme d'arc. / to deviate from the ~ ‖ vom Fahrweg m. abkommen ‖ s'écarter de la route. / direct ~ ‖ direkter Kurs m. ‖ route f. directe. / ~ of fascines ‖ Faschinenschicht f. ‖ couche f. de fascines. / first ~ ‖ Vorkost f. ‖ comestibles mpl. / first ~ of a file ‖ Unterhieb m. einer Feile ‖ première taille f. d'une lime. / ~ of flight ‖ Flugweg m. ‖ voie f. aérienne; cours m. de l'avion. / forced ~ of exchange ‖ Zwangskurs m. ‖ cours m. forcé du change. / fore ~ (Mar) ‖ Focksegel n. ‖ misaine f. / ~ of headers (Mas) ‖ Binderschicht f.; Kopfschicht f.; Kopfstückenschicht f. ‖ assise f. par boutisses *ou* en demi-boutisse. / to keep ~ ‖ Kurs m. steuern ‖ faire route f. / ~ of large stones at the base of a foundation ‖ Grundschicht f. ‖ assise f. de fondation en pierres dures. / to loose the ~ ‖ vom Fahrweg m. abkommen ‖ s'écarter de la route. / main ~ (Mar) ‖ Großsegel n. ‖ grande voile f. / ~ of manufacture ‖ Arbeitsgang m. ‖ cours m. de la fabrication; phase f. d'élaboration *ou* de travail. / mizen ~ (Mar) ‖ Kreuzsegel n. ‖ basse voile f. d'artimon. / ~ of purlins (Carp) ‖ Pfettenlage f. ‖ cours m. de pannes. / ~ of the rays ‖ Verlauf m. der Strahlen ‖ marche f. des rayons. / ~ of a river ‖ Flußlauf m.; Lauf m. eines Flus-

ses ‖ cours m. d'une rivière. / second ~ of a file ‖ Kreuzhieb m. *oder* Oberhieb m. einer Feile ‖ seconde taille f. d'une lime. / to shape the ~ for ... (Nav) ‖ Kurs m. setzen auf ... ‖ tracer la route pour ... / steered ~ (Nav) ‖ gesteuerter Kurs m. ‖ route f. suivie. / ~ of stretchers (Build) ‖ Läuferschicht f.; Laufschicht f.; Streckerschicht f. ‖ assise f. en panneresse *ou* en parement *ou* par carreaux. / upright ~ (Mas) ‖ Rollschicht f. ‖ assise f. de champ; roulage m.; assise f. de briques posées de champ. / ~ of the wind ‖ Gang m. des Windes ‖ marche f. du vent.
course angle ‖ Kurswinkel m. ‖ angle m. de route. / ~ recorder ‖ Kursschreiber m. ‖ navigraphe m.; enregistreur m. de route. / ~ signals pl. ‖ Kurssignale npl.; Fahrtzeichen npl. ‖ signaux mpl. de route.
coursing ‖ Hetzjagd f. ‖ chasse f. à la courre.
court ‖ Hof m.; Hofraum m. ‖ cour f.; préau m.; enclos m. / ~ (Blind alley) ‖ Sackgasse f. ‖ cul-de-sac m.; impasse f. / ~ (Law court; Building) ‖ Gerichtsgebäude n. ‖ maison m. *ou* palais m. de justice; tribunal m. / ~ (Palace) ‖ Palast m. ‖ palais m.; hôtel m. / ~ (Play ground) ‖ Spielplatz m. ‖ jeu m.; place f. des jeux; préau m. / ~ (Session of a court) ‖ Gerichtssitzung f. ‖ séance f. d'une cour; audience f. / ~ (Session hall) ‖ Gerichtssaal n. ‖ salle f. d'audience. / ~ (Tribunal) ‖ Gericht n.; Gerichtshof m. ‖ tribunal m.; cour f.
Court ‖ Gericht n.; Gerichtshof m. ‖ tribunal m.; cour f. / ~ of Appeal ‖ Berufungsgericht n. ‖ cour m. d'appel. / ~ of Arbitration ‖ Schiedsgericht n. ‖ tribunal m. arbitral; conseil m. des prud'hommes. / ~ of Cassation ‖ Kassationsgericht n. ‖ cour m. de cassation. / ~ of Common Pleas ‖ Zivilgerichtshof m. ‖ tribunal m. de première instance. / Competent ~ ‖ Gerichtsstand m. ‖ tribunal m. compétent. / ~ of Justice ‖ Gericht n. ‖ tribunal m.; cour f. de justice juridiction f. / ~ of Law *see* ~ of Justice. / High ~ of Parliament ‖ Parlament n. ‖ parlement m.
court plaster ‖ englisches Heftpflaster n. ‖ taffetas m. anglais; emplâtre m. adhésif anglais. / ~ yard ‖ Hof m.; Hofraum m. ‖ cour f.; enclos m.
cousinet (Mas) ‖ Kämpferschicht f. ‖ imposte f.; coussinet m.
covelline ‖ Kupferindigo m. ‖ cuivre m. sulfureux; covellite f.
cover, to ‖ abdecken ‖ couvrir; recouvrir. / ~ (Colour) ‖ decken ‖ couvrir. / ~ bombproof ‖ bombensicher eindecken ‖ mettre à l'abri *ou* à l'épreuve de bombes. / ~ the expenses pl. ‖ die Kosten pl. decken ‖ couvrir les frais mpl. / ~ glass ‖ Glas n. plattieren *oder* überfangen ‖ doubler le verre. / ~ a roof ‖ das Dach eindecken ‖ poser la couverture. / ~ with tapestry ‖ tapezieren ‖ tendre la tapisserie. / ~ the tympan (Print) ‖ den Deckel m. der Handpresse überziehen ‖ coller le tympan.
cover (Bookb) ‖ Einbanddecke f.; Einband m. ‖ couverture f.; reliure f. / ~ (Hole) ‖ Abdeckung f. ‖ couverture f.; couvercle m. / ~ (Househ) ‖ Tischbesteck n. ‖ couvert m. / ~ (Letter) ‖ Briefumschlag m. ‖ enveloppe f. à lettres. / ~ (Textile) ‖ Decke f. ‖ couverture f. / ~ (Vessel) ‖ Deckel m. ‖ cou-

vercle m. / ~ (Wrapper) ‖ Hülle f. ‖ enveloppe f. / acidproof composition rubber ~ ‖ säurefester Gummideckel m. ‖ couvercle m. en caoutchouc inattaquable aux acides. / air-tight ~ ‖ luftdichter Deckel m. ‖ couvercle m. hermétique. / ~ for books ‖ Buchdeckel m. ‖ couverture f. de livres. / to wrap the ~ to a booklet (Bookb) ‖ den Umschlag m. an eine Broschüre einhängen ‖ appliquer la couverture à la brochure. / camel hair ~ ‖ Kamelhaardecke f. ‖ couverture f. en poil de chameau. / ~ of caoutchouc ‖ Überzug m. aus Kautschuk ‖ enduit m. à base de caoutchouc. / ~ of chimney ‖ Kamindeckel m. ‖ capuchon m. de la cheminée. / conical ~ of steam cylinder ‖ konischer Boden m. des Dampfzylinders ‖ fond m. conique du cylindre à vapeur. / back ~ of cylinder ‖ hinterer Zylinderdeckel m. ‖ couvercle m. arrière de cylindre. / front ~ of cylinder ‖ vorderer Zylinderdeckel m. ‖ couvercle m. avant de cylindre. / ~ of a dam-wall ‖ Rücken m. eines Stauwehres ‖ cape f. d'un batardeau. / hard rubber ~ ‖ Hartgummideckel m. ‖ couvercle m. en ébonite. / hinged ~ ‖ Scharnierdeckel m.; Klappenverschluß m. ‖ couvercle m. à charnière; fermeture f. à clapet. / ~ with inscription ‖ Schriftdeckel m. ‖ couvercle m. avec inscription. / inserted ~ ‖ Einsatzdeckel m. ‖ couvercle m. inséré. / ~ for legal documents ‖ Aktendeckel m. ‖ chemise f. dun dossier. / non-skid ~ (Auto) ‖ Gleitschutz m. ‖ enveloppe f. antidérapante. / non-skid ~ with studded leather band (Auto) ‖ Gleitschutzdecke f. mit aufgenietetem Ledermantel ‖ enveloppe f. antidérapante. / non skid ~ with studs vulcanized-in (Auto) ‖ Gleitschutzdecke f. mit einvulkanisierten Nieten ‖ enveloppe f. antidérapante avec rivets vulcanisés dans la masse. / ~ of oil ‖ Überzug m. aus Öl ‖ enduit m. à base d'huile. / overhead ~ (Mine) ‖ Schutzdach n. ‖ toit m. de défense. / ~ of a pile (Charc) ‖ Meilerdecke f. ‖ couverture f. d'une meule. / quilted ~ ‖ Steppdecke f. ‖ couverture f. ouatée ou piquée; courte-pointe f. / revolving ‖ Aufklappdeckel m. ‖ couvercle m. basculant. / ~ for saltpetre boiler ‖ Salpeterkesseldeckel m. ‖ couvercle m. de chaudières à salpêtre. / sheet metal ~ ‖ Blechmantel m. ‖ revêtement m. en tôle. / table ~ ‖ Tischbesteck n. ‖ couvert m. de table. / ~ of wheel gearing ‖ Räderverdeckung f. ‖ couvre-engrenage m. / ~ containing window ‖ Deckel m. mit Beobachtungsfenster ‖ couvercle m. à regard.

covercoating ‖ Paletotstoff m. ‖ étoffe f. à manteau.

cover design ‖ Umschlagzeichnung f. ‖ dessin m. de couverture.

covered, the berths are ~ throughout their lengths with glazed iron-frame constructions ‖ die Hellinge fpl. sind ihrer ganzen Ausdehnung nach mit verglasten, in Eisenfachwerk ausgeführten Hallen überdacht ‖ sur les cales fpl. s'élèvent des halls en charpente métallique vitrée, qui abritent toute la longueur des bassins. / rubber ~ ‖ mit Gummi m. überzogen ‖ à revêtement m. en caoutchouc. / ~ by timber planking ‖ durch Holz-

belag m. abgedeckt ‖ garni d'une couverture en madriers. / the turntable is ~ by a timber planking (Railw) ‖ die Drehscheibe ist durch einen Bohlenbelag abgedeckt ‖ la plaque tournante est garnie d'une couverture en madriers.

covered goods wagon ‖ bedeckter Güterwagen m. ‖ wagon m. à marchandises couvert ou fermé. / ~-in all round ‖ allseitig geschlossen ‖ fermé de toutes parts. / ~ market ‖ Markthalle f. ‖ halle f. de marché. / ~ truck see ~ goods wagon / ~ van ‖ Planwagen m. ‖ voiture f. à bâche; wagon m. bâché. / ~ wagon see ~ goods wagon. / ~ fil m. guipé. / ~ wire ‖ umsponnener Draht

cover fabric ‖ Deckenstoff m. ‖ drap m. à couvertures. / ~ glass ‖ Deckglas n. ‖ couvre-objet m. / ~ glass gauge ‖ Deckglastaster m. ‖ calibre m. pour lamelles couvre-objet. / ~ hook ‖ Schrankhaken m. ‖ patère f.

covering ‖ Abdeckung f.; Beplankung f.; Decke f. ‖ recouvrement m.; couverture f. / ~ (Aero) ‖ Bespannung f. ‖ entoilage m. / ~ (Build) ‖ Dachdeckerarbeit f. ‖ couverture f. d'un toit. / ~ (Metal) ‖ Schutzdecke f. (über flüssigem Metall) ‖ couverture f. / ~ with asbestos ‖ Asbestbekleidung f. ‖ revêtement m. en asbeste. / ~ of the cable ‖ Kabelhülle f. ‖ enveloppe f. de câble. / chequered plate ~ ‖ Riffelblechbelag m. ‖ recouvrement m. en tôle striée. / ~ of the concrete ‖ Bedeckung f. des Betons ‖ chevauchement m. du béton. / ~ cotton ‖ Baumwollumspinnung f. ‖ guipage m. de coton. / ~ of earth's surface ‖ Bodenbedeckung f. ‖ revêtement m. du sol. / framework for ~ (Aero) ‖ Bespannungsgerüst n. ‖ ossature f. pour le revêtement. / ~ of the hats ‖ Plattieren n. oder Überziehen n. der Hüte ‖ dorage m. des chapeaux. / hemp ~ ‖ Hanfumspinnung f. ‖ guipage m. de chanvre. / ~ with insulating material ‖ Abdecken m. mit Isolierstoff ‖ application f. d'un isolant. / jute ~ ‖ Juteumspinnung f. ‖ guipage m. de jute. / ~ of a mirror ‖ Spiegelbelag m. ‖ étamure f.; tain m.; feuille f. d'étain d'un miroir. / ~ of snow ‖ Schneedecke f. ‖ couverture f. ou couche f. de neige. / ~ of the soil ‖ Bodenbedeckung f. ‖ recouvrement m. du sol. / ~ in sheet steel ‖ Stahlblechmantel m. ‖ enveloppe f. en tôle d'acier. / ~ for steps ‖ Stufenbelag m. ‖ recouvrement m. de marche. / ~ for walls ‖ Wandbekleidungsstoff m. ‖ étoffe f. de revêtement de murs. / yarn ~ ‖ Zwirnumspinnung f. ‖ guipage m. en fils retors.

covering device ‖ Abdeckvorrichtung f. ‖ dispositif m. de recouvrement. / ~ glass ‖ Deckgläschen n. ‖ couvre-objet m.; borderau m. / ~ letter ‖ Begleitbrief m. ‖ lettre f. d'envoi. / ~ machine (Spinn) ‖ Umspinnmaschine f.; Bespinnmaschine f. ‖ métier m. à guiper; guipeuse f. / ~ needle (Textile) ‖ Decknadel f. ‖ poinçon m. / ~ plate (Road) ‖ Abdeckplatte f. plaque f. ‖ de recouvrement. / ~ platerivetting ‖ Laschennietung f. ‖ rivure f. à couvre-joint. / ~ power (Colour) ‖ Deckkraft f. ‖ opacité f.; propriété f. couvrante. / ~ strip ‖ Eckkappe f. ‖ couvre-joint m. d'angle.

coverlet ‖ Bettdecke f.; Oberbett n. ‖ édredon m.; couverture f. de lit. / ~

quilter ‖ Deckenstepperin f. ‖ piqueuse f. de couvertures.

cover marbler (Bookb) ‖ Buchdeckelmarmorierer m. ‖ racineur m. en reliure.

cover plate (Boil) ‖ Lasche f. ‖ couvrejoint m. / ~ (Rivetting) ‖ Lasche f. ‖ couvre-joint m.; bande f. de recouvrement; franc-bord m. / ~ (Road) ‖ Abdeckplatte f. ‖ plaque f. de recouvrement; pierre f. à chaperon; plateau m. de couverture. / ~ (Vessel) ‖ Abschlußdeckel m. ‖ couvercle m. de fermeture. / ~ of a lock ‖ Schloßdeckel m. ‖ couverture f. d'une serrure. / perforated ~ ‖ gelochte Abdeckplatte f. ‖ plaque f. de recouvrement perforée.

cover printing unit ‖ Umschlagwerk n. ‖ groupe m. d'impression de couverture. / ~ protector (Auto) ‖ Reifenschutz m. ‖ protecteur m. de pneu. / ~ remover (Brew) ‖ Abschöpflöffel m. ‖ écumoire f.

co-versed sine (Math) ‖ Kosinusversus m. ‖ cosinus m. verse.

cover slide ‖ Abschlußschieber m.; Bodenklappe f. ‖ tiroir m. de fermeture; trappe f. de fond. / ~ spring ‖ Deckelfeder f. ‖ ressort m. de couvercle. / ~ strap ‖ Deckelriemen m. ‖ cordon m. de couvercle. / ~ strip ‖ Abschlußkappe f. ‖ chapeau m. arêtier. / ~ strip of tailplane tip (Aero) ‖ Flossenendkappe f. ‖ chapeau m. de rive du stabilisateur.

cow bell ‖ Kuhglocke f. ‖ clochette f. pour vaches. / ~ boy ‖ Kuhknecht m. ‖ gardeur m. de vaches. / ~ breeder ‖ Kuhzüchter m. ‖ éleveur m. de vaches. / ~ catcher (Railw) ‖ Schienenräumer m. ‖ chasse-bestiaux m. / ~ dung ‖ Kuhmist m. ‖ bouse f. de vache. / ~ dung bath (Dyer) ‖ Kuhmistbad n. ‖ bousage m. des indiennes. / ~ hair cloth ‖ Kuhhaargewebe n. ‖ thibaude f. / ~ hide ‖ Rindsleder n. ‖ cuir m. de vache. / ~ horn ‖ Kuhhorn n. ‖ corne f. de vache.

cowl (Build) ‖ drehbare Schornsteinkappe f. aus Blech ‖ chapeau m. en tôle d'une cheminée. / ~ (Shipb) ‖ Windhaube f. ‖ chapeau m. aspirateur. / ~ (Wind mill) ‖ Klappe f.; Haube f. ‖ toit m. du moulin. / ~ of a forge ‖ Rauchfang m. einer Schmiede ‖ chaperon m. de forge. / movable ~ ‖ bewegliche Schornsteinkappe f. ‖ hotte f. ou champignon m. mobile.

cowling (Mach) ‖ Haube f. ‖ capot m. / engine ~ ‖ Motorhaubenblech n. ‖ tôle f. frontale de capotage du moteur.

Cowper hot-blast apparatus see Cowper stove.

Cowper stove for heating the blast ‖ Cowper-Apparat m. für Winderhitzung ‖ appareil m. Cowper pour le chauffage du vent.

cow shed ‖ Kuhstall m. ‖ vacherie f.

crab (Build) ‖ Bockwinde f. ‖ treuil m. à bâti. / ~ (Crane) ‖ Laufkatze f. ‖ chariot m. ou treuil m. roulant / ~ (Mach) ‖ Haken m. (am Förderseil) ‖ croc m. / ~ (Mine) ‖ Schachtwinde f. ‖ treuil m. de mine / ~ (Shipb) ‖ Spill n. ‖ cabestan m. / ~ (Zool) ‖ Krabbe f. ‖ crabe m. / crane ~ ‖ Kranlaufkatze f. ‖ chariot m. de pont roulant. / ~ fitted with driver's stand ‖ Führerstandlaufkatze f. ‖ chariot m. roulant à poste de conducteur. / inside running ~ ‖ innenlaufende Katze f. ‖ chariot m. circulant à l'intérieur. / rope-drawn ~ ‖ Seilzugkatze f. ‖ chariot m. mu par câbles. / travelling ~ ‖ Lauf-

katze f.; Hängekatze f. ‖ chariot m. roulant *ou* suspendu; treuil m. roulant. **crabbing machine,** boiling and ~ (Weav) ‖ Koch- und Fixiermaschine f. ‖ machine f. à cuire et à fixer.

crab bolt ‖ durchgehender Ankerbolzen m. ‖ boulon m. d'ancrage travers. / ~ carriage ‖ Katzenwagen m. ‖ treuil m. roulant. / ~ fisherman ‖ Krabbenfischer m. ‖ pêcheur m. de crabes. / ~ crane with ~ gear ‖ Greiferkran m. ‖ grue f. à benne dragueuse. / ~ traversing ‖ katzfahren ‖ déplacer le chariot. / ~ tree ‖ Holzapfelbaum m. ‖ pommier m. sauvage. / ~ winch ‖ Katzenwinde f. ‖ treuil m. de chariot roulant.

crack, to ‖ springen; platzen; rissig werden ‖ (se) fendre; fissurer; craquer. / ~ (Glass) rissig werden ‖ fêler. / ~ (Mas) ‖ aufreißen ‖ se crevasser. / ~ (Metal) ‖ rissig werden ‖ se fendre; se gercer; se fendiller; criquer. / ~ (Pott) ‖ Risse mpl. bekommen ‖ se fendiller. / ~ (Wood) ‖ spalten; schlitzen ‖ fendre; refendre. / the steel cracks ‖ der Stahl wird rissig ‖ l'acier m. se fendille.

crack ‖ Riß m.; Härteriß m.; Sprung m.; Borste f. ‖ crevasse f.; fissure f.; fente f.; déchirure f.; gerçure f.; crique f. / ~ (Glassm) ‖ Knick m. ‖ fêlure f. / ~ (Pott) ‖ Haarriß m.; Trockenriß m. ‖ fissure f.; gerçure f. / ~ (Wood) ‖ Windriß m.; Trockenriß m. ‖ gerçure f. / bar which while cold was bent without showing any ~s (Met) ‖ Stab m., der im kalten Zustande verbogen wurde, ohne Risse zu bekommen ‖ barreau m. déformé à froid: il ne s'est pas produit de fentes. / the ~s take rise in the cheeks ‖ die Risse mpl. gehen von den Kurbelblättern aus ‖ les fentes fpl. prennent leur origine dans les disques de la manivelle. / ~s in the direction of the grain (Wood) ‖ Spiegelklüfte fpl. ‖ fissures fpl. dans le sens du fil du bois fendu. / the ~s pl. start in the discs and extend into the pin (Met) ‖ die Risse mpl. gehen von den Kurbelblättern aus und setzen sich in den Zapfen hinein fort ‖ les fissures fpl. partent des disques de la manivelle et se prolonge dans le tourillon. / the ingots showed ~s pl. during forging ‖ die Barren mpl. bekamen Risse beim Schmieden ‖ les lingots mpl. d'acier fondu montraient des fissures au forgeage. / the ~s pl. extend into the pin ‖ die Risse mpl. erstrecken sich in den Zapfen hinein ‖ les fentes fpl. se prolongent dans le tourillon. / starting point of the ~s ‖ Stelle f., wo in der Regel die Risse entstehen ‖ endroit m. où les fissures se produisent ordinairement. / tyre bent together while cold without showing any ~s ‖ im kalten Zustande ohne Rißbildung f. zusammengebogener Reifen m. recourbé à froid sans présenter la moindre fissure. / the wheel got a ~ ‖ das Rad bekam einen Sprung ‖ la roue s'était fissurée.

crack of a conducting-pipe ‖ Ritze f. *oder* Spalte f. in einer Wasserröhre ‖ renard m. d'une conduite. / ~ due to contraction ‖ Schwindriß m. ‖ crevasse f. dû au retrait. / ~ due to expansion ‖ Treibriß m. ‖ fissure f. due à la poussée. / formation of ~s ‖ Rißbildung f. ‖ formation f. de fissures. / longitudinal ~ ‖ Längsriß m. ‖ fissure f. longitudinale. / season ~ in

metal ‖ Altersriß m. im Metall ‖ crique f. de vieillissement dans les métaux. / surface ~ (Metal) ‖ Hautriß m. ‖ crevasse f. superficielle. / ~ in steel ‖ Härteriß m. im Stahl ‖ gerçure f. *ou* crevasse f. *ou* crique f. de l'acier. / ~ from tempering ‖ Härteriß m. ‖ fissure f. de trempe. / transverse ~ ‖ Querriß m. ‖ fissure f. transversale. / ~ of a wall ‖ Mauerspalte f.; Mauersprung m.; Riß m. im Mauerwerk ‖ lézarde f. *ou* fente f. d'une muraille.

cracked ‖ gespalten; rissig ‖ crévassé; gercé; lézardé; fissuré. / during the forging process the steel gets ~ ‖ der Stahl wird beim Schmieden rissig ‖ l'acier m. se fendille sous la forge. / ~ gasoline ‖ durch Crackprozeß m. gewonnenes Benzin n. ‖ essence f. de cracking.

cracker (Roll mill) ‖ Brechwalze f. ‖ cylindre m. broyeur. / ~ (Sugar) ‖ Knallbonbon m. ‖ bonbon m. fulminant.

cracking (Metal) ‖ Reißen n. ‖ déchirure f. / ~ (Tel) ‖ Knallgeräusch n. ‖ crépitement m.; friture f. / to convert tar by ~ into motor spirit and Diesel oil ‖ Teer m. durch das Krackverfahren in Treiböle überführen ‖ traiter le goudron par cracking de manière à le transformer en essences pour moteurs. / ~ of the hot-bulb ‖ Reißen n. des Glühkopfes ‖ rupture f. de la bulbe chaud *ou* de la calotte chaude. / safety against ~ and fracturing ‖ Sicherheit f. gegen Bildung von Rissen und Brüchen ‖ sécurité f. contre les crevasses et les ruptures. / season ~ ‖ Altersriß m. ‖ craquelure f. saisonnière. / ~ of the wood ‖ Reißen n. des Holzes ‖ gercement m. du bois.

cracking coal (Glassm) ‖ Sprengkohle f. ‖ charbon m. à détacher. / ~ process ‖ Crackprozeß m. ‖ procédé m. de fractionnement. / ~ ring (Glassm) ‖ Sprengeisen n. ‖ anneau m. à détacher.

crackling of tin ‖ Knirschen n. des Zinnes; Zinnschrei m. ‖ cric m. *ou* cri m. de l'étain.

cracknel biscuit maker ‖ Kringelbäcker m. ‖ fabricant m. de craquelins.

cradle ‖ Gestell n. ‖ berceau m. / ~ (Metal) ‖ Röstkratze f. ‖ grattoir m. / ~ (Shipb) ‖ Schlitten m.; Ablaufgerüst n. ‖ ber m.; berceau m. / ~ (Rail) ‖ Schienenstuhl m. ‖ chair m. *ou* coussinet m. de rail. / ~ (Railw) ‖ Abrollwiege f. ‖ courbe f. de glissement. / ~ assuming the shape of a sleeve ‖ muffenartige Wiege f. ‖ berceau m. à moufles. / boiler ~ ‖ Kesselrollbock m. ‖ chevalet m. à chaudières; tréteau m. à galets pour chaudières. / cast-steel ~ ‖ Stahlgußwiege f. ‖ berceau m. en acier moulé. / starting ~ (Aero) ‖ Anlaufgestell n. ‖ chariot n. de lancement. / telescope ~ ‖ Fernrohrwiege f. ‖ berceau m. de lunette. / tipping ~ ‖ Kippwiege f. ‖ berceau m. basculant.

cradle tipping hopper ‖ Wiegenkipper m. ‖ wagonnet m. basculant par glissement sur un plan incliné. / ~ tip wagon see cradle tipping hopper.

craft ‖ Handwerk n. ‖ métier m.; profession f. manuelle. / ~ (Mar) ‖ Fahrzeug n. ‖ bâtiment m. / river ~ ‖ Fahrzeug n. der Binnenschiffahrt ‖ navire m. pour la navigation intérieure. / sea-going ~ ‖ Fahrzeug n. der Seeschiffahrt ‖ navire m. pour la navigation maritime.

craftsman ‖ Handwerker m. ‖ homme m. de métier; ouvrier m.; artisan m. / ~ (Build) ‖ Handlanger m. ‖ manœuvrier m.; journalier m. / ~ (Trade) ‖ Gewerbetreibender m. ‖ industriel m.

cram, to ‖ mästen; engraisser.

cramp, to ‖ klammern; ankrampen ‖ cramponner; acclamper.

cramp (Carp) ‖ Klammer f.; Krampe f.; Kramme f.; Haspe f. ‖ crampon m.; clameau m.; crampe f.; serre f. / ~ (Join) ‖ Schraubzwinge f.; Leimzwinge f. ‖ serre-joint m.; presse f. à main. / ~s pl. (Med) ‖ Krämpfe mpl. ‖ crampes fpl. / ~ (Upholsterer) ‖ Tapezierstift m. ‖ crampon m. / to fix the ~s ‖ die Klammern fpl. einsetzen ‖ enclaver les crampons mpl. / to run-in the ~s (Mas) ‖ die Klammern fpl. vergießen ‖ couler *ou* sceller les crampons mpl. / ~ for slippers ‖ Pantoffelklammer f. ‖ crampon m. à pantoufles. / ~ of square iron for fixing rails on sleepers ‖ Klammer f. aus Vierkanteisen zur Befestigung von Schienen auf Schwellen ‖ crampon m. en fer carré pour fixer les rails aux traverses.

cramp bark ‖ Viburnumrinde ‖ écorce f. de viburnum. / ~ frame (Join; Carp) ‖ Leimzwinge f.; Schraubzwinge ‖ presse f. à main; serre-joint m. / ~ gauge (Railw) ‖ Spurlehre f. bei Stuhlschienen ‖ jauge f. pour la vérification du sabotage. / ~ hole ‖ Klammerloch n. ‖ trou m. de crampon.

cramp iron ‖ Kramme f.; Krampe f.; Haspe f. ‖ crampon m.; clameau m.; crampe f.; happe f.; agrafe f. / ~ for fastening the jamb-stones on the wall (Build) ‖ Wandanker m. ‖ patte f. des lancis *ou* en plâtre. / ~ for fastening wainscots (Join) ‖ Täfelwerksklammer f. ‖ patte f. à lambris. / ~ for wood ‖ Bankeisen n.; Holzkrampe f. ‖ patte f. d'ancrage; patte-fiche f.

cramps pl. ‖ Krämpfe mpl. ‖ crampes fpl.

cranage ‖ Krangeld n. ‖ droits mpl. de grue.

cranberry ‖ Preißelbeere f. ‖ airelle f.

crane ‖ Kran m. ‖ grue f. / ~ (Mine) ‖ Schachtwinde f.; Hebezeug n. ‖ treuil m. de mine; cabestan m. / ~ (Railw) ‖ Ladekran m. ‖ grue f. de chargement. / all movements pl. of the ~ can be performed with the greatest exactness ‖ der Kran m. gestattet in all seinen Bewegungen die größte Genauigkeit ‖ la grue permet d'exécuter tous les mouvements avec la plus grande précision. / angle ~ ‖ Winkelkran m. ‖ grue f. à support triangulaire. / area served by a ~ ‖ Reichweite f. eines Kranes ‖ rayon m. desservi par une grue. / automatic grab ~ ‖ Greiferkran m. ‖ grue f. à pelle automatique. / benzol ~ ‖ Benzolkran m. ‖ grue f. à benzol. / bicycle ~ ‖ Velozipedkran m. ‖ grue-vélocipède f. / boat's ~ ‖ Bootskran m. ‖ grue f. d'embarcation. / bracket ~ ‖ Konsolkran m. ‖ grue f. à console. / bridge ~ ‖ Hochbahnkran m. ‖ grue f. sur voie surélevée. / building ~ ‖ Baukran m.; Hochbaukran m. ‖ grue f. de construction. / cableway ~ ‖ Kabelkran m. ‖ grue f. à câbles. / elevated cableway ~ ‖ Kabelhochbahnkran m. ‖ grue f. à câbles aériens; Blondin m. / casting bed ~ ‖ Gießbettkran m. ‖ grue f pour lits

de coulée. / casting ~ || Gieß(erei)kran m. || grue f. de coulée *ou* de fonderie. / charging ~ || Beschickungskran m.; Einsetzkran m.; Chargierkran m. || grue f. enfourneuse *ou* de chargement; pont-roulant m. chargeur. / claw ~ || Pratzenkran m. || grue f. à griffes. / coaling ~ || Bekohlungskran m.; Kohlenladekran m. || grue f. à charger le charbon. / ~ with combustion engine || Kran m. mit Brennkraftmotor || grue f. à moteur à combustion. / ~ controlled from floor level || vom Boden aus betätigter Kran m. || pont m. commandé du sol. / ~ with crab gear || Greiferkran m. || grue f. à benne preneuse. / derrick ~ || Derrick(kran) m.; Dreifußkran m. || grue f. derrick *ou* de chevalement. / derrick wagon ~ || Waggonkran m.; Eisenbahndrehkran m. || grue f. sur chariot; grue f. montée sur wagon. / Diesel ~ || Dieselmotorkran m. || grue f. à moteur Diesel. / dock ~ || Werftkran m. || grue f. de chantier de constructions navales. / equipment ~ || Ausrüstungskran m. || grue f. d'armement. / erecting ~ || Montagekran m. || pont m. roulant de montage. / fixed stile ~ || Kran m. mit feststehender Säule || grue f. à pivot fixe. / floating ~ || Schwimmkran m. || grue f. flottante. / forge ~ || Schmiedekran m. || grue f. de forge. / foundry ~ || Gießereikran m.; Gießkran m. || grue f. de fonderies *ou* pour la coulée. / foundry travelling ~ || Gießereilaufkran m. || pont m. roulant pour fonderie. / frame ~ || Bockkran m. || grue f. portique. / gantry ~ || Bockkran m.; Portalkran m. || grue f. à chevalet *ou* à portique. / goliath ~ || Schwerlastkran m. || grue f. titan *ou* géante. / hammer-head ~ || Hammerkran m. || grue-marteau f. / handling ~ || Verladekran m. || grue f. de chargement. / handling ~s for the unloading of the raw materials || Ausladebrücken fpl. für das Entladen der Rohstoffe || élévateur-transporteur m. pour le déchargement des matières premières. / hand travelling ~ || Laufkran m. für Handbetrieb || pont m. à commande à main. / harbour ~ || Hafenkran m. || grue f. de port. / hatch ~ || Lukenkran m. || grue f. de lucarne. / helmet ~ || Helmkran m. || grue f. à casque *ou* à volée variable. / ~ with hinged jib || Kran m. mit beweglichem Ausleger || grue f. à flèche mobile. / hydraulic ~ || hydraulischer Kran m. || grue hydraulique. / ~ for ice producers || Eisgeneratorenkran m. || grue f. pour générateurs de glace. / ladle-handling ~ || Gießkran m. || grue f. pour poche de coulée. / locomotive ~ || Lokomotivkran m.; Kranlokomotive f. || grue f. à locomotive; grue-locomotive f. / locomotive lifting ~ || Lokomotivhebekran m. || pont m. roulant pour le transport des locomotives. / ~ for long timber || Langholzkran m. || grue f. pour bois de grandes longueurs. / magnet ~ || Magnetkran m. || grue f. à aimant. / mast ~ || Mastenkran m. || grue f. à poteau. / ~ for metallurgical works || Hüttenwerkskran m. || pont m. roulant d'usines métalliques. / movable ~ || Fahrkran m. || grue f. mobile *ou* sur chariot. / ore loading ~ || Erzverladekran m. || grue f. de chargement de minerais. / outreach of a ~ || Ausladung f. eines Krans ||

portée f. d'une grue. / piece-goods ~ || Stückgutkran m. || grue f. pour colis. / pillar ~ || Säulendrehkran m. || grue f. à colonne. / isolated pillar ~ || freistehender Säulendrehkran m. || grue f. pivotante isolée. / pivot slewing ~ || Zapfendrehkran m. || grue f. tournante sur pivot. / portable slewing ~ || fahrbarer Drehkran m. || grue f. pivotante transportable. / portal ~ || Portalkran m. || grue f. à portique. / portal revolving ~ || Vollportaldrehkran m. || grue f. pivotante à portique entier *ou* tournante à plein portique. / quay ~ || Uferkran m. || grue f. de quai. / radius of a ~ || Ausladung f. eines Krans || portée f. d'une grue. / ram ~ || Fallwerkskran m. || grue f. à bélier. / revolving ~ || Drehkran m. || grue f. pivotante. / hand-worked revolving ~ || von Hand getätigter Drehkran m. || grue f. à bras *ou* pivotante à main. / revolving wagon ~ || Wagendrehkran m. || grue f. pivotante sur wagon. / self-propelled Diesel-engine-driven revolving wagon ~ || selbstfahrender Wagendrehkran m. mit Dieselmotorenantrieb || grue f. automotrice pivotante sur wagon actionnée par moteur Diesel. / rolling mill ~ || Walzwerkkran m. || grue f. de laminoir. / rotary ~ || Drehkran m. || grue f. pivotante *ou* tournante. / semi-portal ~ || Halbportalkran m. || semi-portal crane. / semi-portal gantry ~ || Halbportalkran m. || grue f. à demi-portique. / semi-portal revolving ~ || Halbportaldrehkran m. || grue f. pivotante à semi-portique *ou* tournante à demi-portique. / sheet iron ~ || Blechtransportkran m. || grue f. à transporter les tôles. / shifting ~ || Versatzkran m. || grue f. à déplacer. / shipbuilding ~ || Hellingkran m. || grue f. de cale sèche. / slewing ~ || schwenkbarer Kran m. || Drehkran m. || grue f. tournante *ou* pivotante. / slewing tower ~ || Turmdrehkran m. || grue f. à pylône pivotante. / slipway ~ || Hellingdrehkran m. || grue f. pivotante pour cales sèches. / smelting house ~ || Hüttenwerkskran m. || grue f. pour usines métallurgiques. / soaking pit ~ || Tiefofenkran m. || pont m. roulant pour four pit. / stationary ~ || ortsfester Kran m. || grue f. fixe *ou* stationnaire. / stationary steam-driven revolving ~ || feststehender Dampfdrehkran m. || grue f. à vapeur pivotante et fixe. / stationary tower revolving ~ || feststehender Turmdrehkran m. || grue f. pivotante géante et fixe. / stationary wharf revolving ~ || feststehender Hafendrehkran m. || grue f. fixe pivotante de port. / steam ~ || Dampfkran m. || grue f. à vapeur. / steam-driven floating ~ || Dampfschwimmkran m. || grue f. flottante à vapeur. / steam-driven locomotive slewing ~ || Dampflokomotivdrehkran m. || grue f. locomotive pivotante à vapeur. / steam-driven revolving ~ || Dampf- drehkran m. || grue f. à vapeur pivotante. / ~ for steelworks || Stahlwerkskran m. || pont m. roulant d'aciéries. / store ~ || Lagerkran m. || grue f. de magasin *ou* de dépôt. / stripping ~ || Abstreifkran m. || grue f. à démouler. / ~ with suction frame for the transport of plate glass || Saugrahmenkran m. für Spiegelglastransporte || grue f. à cadre suçoir pour le transport des glaces. / three-

legged ~ || Dreibeinkran m. || grue f. de chevalement. / tower ~ || Turmkran m. || grue f. à pylône. / tower revolving ~ || Turmdrehkran m. || grue f. tournante sur pylône. / transport ~ || Versatzkran m. || grue f. de transbordement. / travelling ~ || Laufkran m. || pont m. roulant; grue f. roulante. / travelling ~ arranged under the roofs of the shops || unter dem Hallendach angebrachter Laufkran m. || pont m. roulant disposé sous la toiture des halls. / the bay is equipped with heavy travelling ~s to handle the rolls || die Halle f. ist mit schweren Kranen zum Auswechseln der Walzen versehen || le hall est desservi par de puissants ponts-roulants pour le changement des cylindres. / travelling gantry ~ || Portalkran m. || grue-portique f. roulante. / travelling ~ with underslung jib || Laufkran m. mit drehbarer Laufkatze || pont m. canard à flèche pivotante suspendue. / travelling ~ operated by hand || Handlaufkran m. || pont m. roulant actionné à bras. / travelling revolving ~ || fah barer Drehkran m. || grue f. pivotante mobile. / triple ~ || Dreifachkran m. || grue f. triple. / trough-charging ~ || Muldenbeschickkran m. || grue f. d'enfournement. / unloading ~ || Umladekran m. || grue f. de transbordement. / the ~ has a useful radius of x meters || der Kran m. hat eine nutzbare Ausladung von x m. || la grue f. a une portée utile de x mètres. / visor ~ *see* helmet ~. / wall ~ || Wandkran m.; Konsolkran m. || grue-murale f.; grue f. à console. / wall slewing ~ || Wanddrehkran m. || grue f. murale pivotante. / warehouse ~ || Speicherkran m.; Schuppenkran m. || grue f. de grenier *ou* de dépôt. / weight of ~ || Krangewicht n. || poids m. d'une grue. / workshop ~ || Werkstattkran m. || pont m. roulant d'atelier. / x ton semiportal revolving ~ with luffing jib, y m radius || Halbportaldrehkran m. mit Wippausleger, x t Tragfähigkeit, y m Ausladung || grue f. pivotante à semiportique avec flèche relevable de x t de puissance, y m de portée. / yard ~ || Hofkran m. || grue f. de cour.

crane arm || Kranausleger m. || flèche f. *ou* bras m. *ou* fauconneau m. de grue. / ~ arm of lattice type || gitterförmiger Ausleger m. || flèche f. en treillis. / ~ balks pl. || Laufbahn f. || voie f. de roulement. / ~ beam || Kranausleger m. || fauconneau m. de grue. / ~ bridge || Kranbrücke f. || pont m. de la grue. / ~ car || Kranwagen m. || wagon m. dépanneur *ou* grue. / ~ carriage || Laufkatze f. || chariot m. roulant. / ~ crab || Kranlaufkatze f. || chariot m. de pont roulant. / ~ driver || Kranführer m. || conducteur de grue f. / ~ dues pl. || Krangeld n. || droits mpl. de grue. / ~ engineer || Laufkranführer m. || conducteur m. de échafaudage m. frame || Krangerüst n. || charpente f. de grue. / locomotive fitted up with ~ gear || Lokomotive f. mit Kranausrüstung || locomotive f. à grue. / ~ installation || Krananlage f. || installation f. de grue. / ~ jib || Kranausleger m.; Kranarm m. || flèche f. *ou* bras m. de grue. / ~ locomotive || Lokomotive f. mit Kranausrüstung; Kranlokomotive f. || locomotive f. à grue. / ~ magnet || Kran-

magnet m.; Lastmagnet m. ‖ aimant m. à monter la charge; (électro-)aimant m. de grue. / ~ man ‖ Kranführer m. ‖ conducteur m. de grue. / outreach of a ~ see radius of a ~. / ~ pontoon ‖ Kranprahm m. ‖ ponton m. de grue. / ~ post ‖ Kranbaum m. ‖ pylône m. / ~ post of iron framework ‖ Kransäule f. in Eisenfachwerk ‖ pylône m. en charpente métallique. / radius of a ~ ‖ Ausladung f. oder Reichweite f. eines Krans ‖ portée f. d'une grue. / ~ rail ‖ Kranschiene f. ‖ rail m. pour grues. / ~ runner see ~ man. / ~ runway girder ‖ Kranbahnträger m. ‖ poutre f. de pont roulant. / salvage ~ ship ‖ Bergungskranschiff n. ‖ bateau m. à grue de sauvetage. / ~ tip for wagons ‖ Eisenbahnwagenkrankipper m. ‖ basculeur m. à grue pour wagons. / ~ tipping device ‖ Aufzugkipper m.; Krankipper m. ‖ basculeur m. à grue. / ~ track see ~ balks. / ~ truck ‖ Kranwagen m. ‖ wagon-grue m. / ~ truck for cleaning the sinkwater traps (Road) ‖ Kranwagen m. zum Reinigen der Sinkkästen ‖ voiture-grue f. à nettoyer les siphons de décantations. / motor ~ truck ‖ Motorkranwagen m. ‖ grue f. automobile. / ~ way ‖ Kranbahn f. ‖ chemin m. de roulement. / weight of ~ ‖ Krangewicht n. ‖ poids m. d'une grue.

crank (Shipb) ‖ rank; leicht kenterbar ‖ faible du côté.

crank, to ‖ kröpfen ‖ couder. / ~ (Auto) ‖ ankurbeln ‖ démarrer.

crank ‖ Kurbel f.; Maschinenkurbel ‖ manivelle f. / angle of rotation of the ~ ‖ Kurbeldrehwinkel m. ‖ angle m. de rotation de la manivelle. / balanced ~ ‖ Kurbel f. mit Gegengewichten ‖ manivelle f. à plateaux équilibrateurs. / bent ~ with handle ‖ gekröpfte Handkurbel f. ‖ manivelle f. à main coudée. / ~ brake ‖ Bremskurbel f. ‖ manivelle f. de serrage. / the ~ is in two parts bolted together ‖ die Kurbel f. ist aus zwei Stücken zusammengebaut ‖ la manivelle f. est assemblée de deux pièces. / ~ built up of two steel-cast discs shrunk on having between them a forged crank pin ‖ durch aufgeschrumpfte Stahlgußkurbelblätter und einen geschmiedeten Kurbelzapfen gebildete Kurbel f. ‖ manivelle f. formée de deux plateaux en acier fondu serrés à chaud et sur un tourillon forgé. / ~ with counter-balance weight ‖ Kurbel f. mit Gegengewicht ‖ manivelle f. à contrepoids. / to force the ~ by pressure ‖ die Kurbel aufpressen ‖ caler la manivelle à la presse. / forged ~ ‖ geschmiedete Kurbel f. ‖ manivelle f. forgée. / ~ forged out of one massive piece ‖ in einem Stück massiv geschmiedete Kurbel f. ‖ manivelle f. forgée d'une seule pièce massive. / high-pressure ~ ‖ Hochdruckkurbel f. ‖ manivelle f. à haute pression. / inside ~ ‖ Innenkurbel f. ‖ manivelle f. intérieure. / inside ~s which necessitate a cranked axle ‖ Innenkurbeln fpl., die Achskröpfung erfordern ‖ manivelles fpl. intérieures lesquelles exigent que l'essieu soit coudé. / to key on the ~ ‖ die Kurbel aufkeilen ‖ claveter la manivelle sur l'arbre. / knocking in the ~ ‖ Kurbelschlag m. ‖ cogne m. dans la manivelle. / lift of ~ ‖ Erhebung f. der Kurbel ‖ levée f. de la manivelle. / low-pressure ~ ‖ Niederdruckkurbel f. ‖

manivelle f. à basse pression. / machine for making ~s ‖ Kurbelherstellungsmaschine f. ‖ machine f. à fabriquer des manivelles. / opposite ~s pl. ‖ gegenläufige Kurbeln fpl. ‖ manivelles fpl. contraires. / ~ outside ‖ Außenkurbel f. ‖ manivelle f. extérieure. / return ~ ‖ Gegenkurbel f. ‖ contre-manivelle f. / ~s pl. at right angles ‖ im rechten Winkel versetzte Kurbeln fpl. ‖ manivelles fpl. à 90°. / ~ of a shaft ‖ Wellenkröpfung f. ‖ coude m. de l'arbre. / to shrink-on the ~ ‖ die Kurbel warm aufziehen ‖ emmancher la manivelle à chaud sur l'arbre. / spherical ~ with handle ‖ Kugelhandkurbel f. ‖ manivelle f. à main à boule. / straight ~ with handle ‖ gerade Handkurbel f. ‖ manivelle f. à main droite. / to turn the ~ to the starting position ‖ die Kurbel in die Anlaßstellung einstellen ‖ placer la manivelle dans la position de mise en marche. / working by ~ ‖ Kurbelantrieb m. ‖ commande f. à manivelle.

crank arm ‖ Kurbelarm m. ‖ flasque m. de manivelle. / ~ arrangement ‖ Kurbelgestänge n. ‖ manivelles fpl. / ~ axle see crankshaft. / ~ bearing ‖ Kurbellager n. ‖ palier m. de l'arbre à manivelle. / ~ bearing pedestal ‖ Kurbellagerbock m. ‖ chevalet m. de palier d'arbre à manivelle. / ~ brace ‖ Brustleier f.; Brustbohrer m.; Faustleier f. ‖ vilebrequin m.; virebrequin m.; drille m. à arçon.

crank case ‖ Kurbelgehäuse n.; Kurbelkasten m. ‖ boîte f. de la manivelle; carter m. des manivelles. / barrel-type ~ ‖ Tunnelkurbelgehäuse n. ‖ carter m. tubulaire. / split-type ~ ‖ geteilter Kurbelkasten m. ‖ carter m. en deux moitiés.

crank case lower half ‖ Kurbelgehäuseunterteil m.; untere Kurbelkastenhälfte f. ‖ cuvette f. inférieure du carter. / upper half ‖ Kurbelgehäuseoberteil m.; obere Kurbelkastenhälfte f. ‖ cuvette f. supérieure du carter.

crank cheek with hoops shrunk on ‖ Kurbelarm m. mit Schrumpfbändern ‖ bras m. de manivelle muni de frettes. / forged-on ~ cheek ‖ angeschmiedete Kurbelwange f. ‖ flasque m. forgé. / ~ circle ‖ Kurbelkreis m. ‖ cercle m. de manivelle. / ~ coupling ‖ Kurbelkupplung f. ‖ accouplement m. de manivelles.

crank disc ‖ Kurbelblatt n.; Kurbelscheibe f. ‖ flasque m. ou plateau m. de manivelle. / opening provided in the ~ ‖ ausgespartes Kurbelblatt n. ‖ plateau m. évidé; évidement m. des flasques de manivelle. / solid ~ ‖ volles Kurbelblatt n. ‖ flasque m. plein.

crank drawing press ‖ Kurbelziehpresse f. ‖ presse f. excentrique à étirer. / ~-driven rivetting machine ‖ Kurbelnietmaschine f. ‖ riveuse f. à manivelle.

cranked ‖ gekröpft ‖ coudé. / inside cranks which necessitate a ~ axle ‖ Innenkurbeln fpl., die Achskröpfung erfordern ‖ manivelles fpl. intérieures lesquelles exigent que l'essieu soit coudé. / ~ dynamo ‖ Kurbeldynamo f. ‖ dynamo m. manuel. / ~ fish plate (Railw) ‖ Übergangslasche f. ‖ éclisse f. de raccordement. / ~ roughing tool ‖ gekröpfter Schruppstahl m. ‖ outil m. d'ébauche coudé. / ~ lathe tool ‖ gekröpfter Stichel m. oder Drehstahl m. ‖ outil m. de tour coudé; lame f. coudée.

crank fulling mill (Textile) ‖ Druckwalke f. ‖ foulon m. à ressort. / oscillating ~ gear ‖ schwingende Kurbelschleife f. ‖ coulisse-manivelle f. oscillante. / ~ guillotine shearing machine ‖ Parallelkurbelblech-(tafel)schere f. ‖ cisailles fpl. à guillotine pour tôles. / ~ hammer ‖ durch Kurbel f. angetriebener Hammer ‖ marteau m. à manivelle. / ~ handle ‖ Handhabe f. einer Kurbel; Kurbel(hand)griff m. ‖ manche m. ou poignée f. de manivelle. / knocking in the ~ ‖ Kurbelschlag m. ‖ cogne m. dans la manivelle. / ~ lever ‖ Kurbelhebel m.; Winkelhebel m. ‖ levier m. coudé. / ~ lever press ‖ Kniehebelpresse f. ‖ presse f. à genouillère. / lift of ~ ‖ Erhebung f. der Kurbel ‖ levée f. de la manivelle. / ~ making machine ‖ Kurbelherstellungsmaschine f. ‖ machine f. à fabriquer les manivelles. / ~ mechanism ‖ Kurbelgetriebe n.; Kurbelmechanismus m. ‖ mécanisme m. à manivelle.

crankness ‖ Rankheit f. ‖ jalousie f.

crank pin ‖ Kurbelzapfen m.; Treibzapfen m. ‖ bouton m. ou tourillon m. de manivelle; maneton m. / inserted ~ ‖ eingesetzter Kurbelzapfen m. ‖ bouton m. de manivelle emmanché. / outside ~ ‖ äußerer Kurbelzapfen m. ‖ bouton m. de manivelle extérieur.

crank pin brass ‖ Kurbelzapfenlagerschale f. ‖ coussinet m. du maneton. / ~ bearing ‖ Kurbelzapfenlager n. ‖ palier m. de tête de bielle; coussinet m. du bouton de manivelle. / ~ lathe ‖ Kurbelzapfendrehbank f. ‖ tour m. pour les manetons (d'arbres à vilebrequin).

crank press ‖ Kurbelpresse f. ‖ presse f. à manivelle. / inclinable ~ ‖ neigbare Exzenterpresse f. ‖ presse f. à excentrique inclinable.

crank rod ‖ Kurbelstange f. ‖ tige f. de manivelle.

crankshaft ‖ Kurbelwelle f.; Kurbelachse f.; Kropfachswelle f.; gekröpfte Welle f. ‖ arbre m. coudé ou à manivelle ou de manivelle; essieu m. coudé; vilebrequin m. / ball bearing ~ ‖ Kurbelwelle f. mit Kugellagern ‖ arbre-manivelle m. ou vilebrequin m. à roulements à billes. / x bearing ~ ‖ x-mal gelagerte Kurbelwelle f. ‖ vilebrequin m. à x paliers. / built-up ~ ‖ zusammengebaute Kurbelwelle f. ‖ arbre m. à manivelle en plusieurs pièces. / counterbalanced ~ ‖ Kurbelwelle f. mit Gegengewicht ‖ vilebrequin m. à flasques équilibrées. / to draw-off the ~ ‖ von der Kurbelwelle abziehen ‖ enlever de l'arbre m. à manivelle. / four-throw ~ of four built-up and hole-bored cranks ‖ vierfache Kurbelwelle f. aus vier zusammengebauten durchbohrten Kurbelwellenstücken ‖ arbre m. à quatre coudes composé de quatre tronçons perforés chacun assemblé de plusieurs pièces. / the ~ was fractured at the crank by a shock ‖ die Kropfachswelle f. erlitt bei einem heftigen Schlag einen Bruch in der Kurbel ‖ l'essieu m. coudé essuya un choc violent et se rompit à la manivelle. / locomotive ~ ‖ Lokomotivkurbelachse f.; Triebradachse f. ‖ essieu m. coudé pour locomotives. / oblique ~ ‖ schrägschenklige Kurbelwelle f. ‖ essieu m. coudé à flasque oblique. / seven-bearing ~ ‖ siebenmal gelagerte Kurbelwelle f. ‖ vilebrequin m. à sept paliers. / shoul-

der on ~ || Kurbelwellenbund m. || collet m. du vilebrequin *ou* d'arbre à manivelle. / single-throw ~ || einfach gekröpfte Kurbelwelle f. || arbre m. coudé à une seule manivelle. / solid ~ || aus einem Stück geschmiedete Kurbelwelle f. || coude m. faisant corps avec l'arbre. / to start in any position of the ~ || in jeder Kurbelstellung f. anspringen || pouvoir être lancé dans n'importe quelle position de la manivelle. / three-bearing ~ || dreimal gelagerte Kurbelwelle f. || vilebrequin m. à trois paliers. / x-throw ~ || x-mal gekröpfte Kurbelwelle f. || vilebrequin m. à x coudes. / throw of ~ || Kröpfung f. der Kurbelwelle || coude m. du vilebrequin. / two-throw ~ || doppelt gekröpfte Kurbelwelle f. || arbre m. à deux coudes *ou* double coudé.

crankshaft bearing || Kurbelwellenlager n. || palier m. de manivelle *ou* de vilebrequin. / ~ bushing || Kurbelwellenlagerschale f. || coussinet m. de palier de vilebrequin. / ~ cap || Kurbelwellenlagerdeckel m. || chapeau m. de palier de vilebrequin. / ~ grinding machine || Kurbelwellenschleifmaschine f. || machine f. à rectifier les arbres vilebrequins. / ~ lathe || Kurbelwellendrehbank f. || tour m. à tourner les arbres coudés. / shoulder on ~ || Kurbelwellenbund m. || collet m. de l'arbre à manivelle. / ~ starting clutch || Andrehklaue f. der Kurbelwelle || griffe f. de mise en marche de vilebrequin. / throw of ~ || Kröpfung f. der Kurbelwelle || coude m. de l'arbre à manivelle.

crank shears pl. || Kurbelschere f. || cisailles fpl. à guillotine. / ~ slotting machine || Kurbelstoßmaschine f. || mortaiseuse f. à manivelle. / ~ turning moment || Kurbeldrehmoment n. || moment m. de rotation de la manivelle.

crankweb || Kurbelarm m.; Kurbelwange f. || bras m. de manivelle; flasque f. de vilebrequin. / circular shape of the ~ || kreisförmiges Kurbelblatt n. || flasque m. circulaire de manivelle. / ~ with hoops shrunk-on || Kurbelarm m. mit Schrumpfbändern || bras m. de manivelle muni de frettes.

crank wheel drive || Kurbelradantrieb m. || commande f. par manivelle.

crape, to (Weav) || krausen; kreppen || crêper.

crape || Krepp m. || crêpe m. / ~-like tissue || kreppartiges Gewebe n. || crépon m. / ~ paper || Kreppapier n. || papier m. crêpé. / ~ paper making machine || Kreppapierherstellungsmaschine f. || machine f. à fabriquer le papier crêpé.

craping machine || Kreppmaschine f. || machine f. à crêper.

crash || Absturz m. || chute f. / ~ on the exchange || Börsenkrach m. || débâcle f. à la bourse; krach m. / ~ linen || Drillichleinen n. || toile f. treillis.

crate || Lattenkiste f. || caisse f. à claire-voie.

crater || Krater m. || cratère m. / ~ of the arc || Krater m. des Lichtbogens || cratère m. de l'arc. / ~ of a mine || Minentrichter m. || entonnoir m. de mine.

crater-lake || Kratersee m.; Maar n. || cratère-lac m.; maar m.

cravat || Krawatte f.; Binder m.; Schlips m. || cravate m. / ~ maker || Krawattennäherin f. || cravatière f.

crawler || Raupenkette f. || chenille f.

crayfish || Krebs m. || écrevisse f. / ~ bow net || Krebskorb m.; Krebsreuse f. || nasse f. à écrevisses. / ~ catcher || Krebsfänger m. || pêcheur m. d'écrevisses.

crayon || Farbstift m.; Pastellstift m. || crayon m. de couleur; pastel m. / ~ board || Zeichenkarton m. || carton m. à dessins. / ~ drawing || Kreidezeichnung f. || dessin m. à la craie *ou* au crayon. / ~ painting || Pastellmalerei f.; Trockenmalerei f. || peinture f. en pastel. / ~ paper || Zeichenpapier n. || papier m. à dessins. / ~ pencil || Pastell n.; Pastellstift m.; Pastellfarbe f.; farbiger Zeichenstift m. || pastel m.; crayon m. à pastel; couleur f. à pastel.

craze, to (Pott) || rissig werden || se fendiller; se fissurer.

craze || Haarriß m. || fendille f.

cream, to || rahmen || écrémer.

cream || Sahne f.; Rahm m. || crème f. (de lait). / formation of ~ || Rahmbildung f. || formation f. de la crème. / preserved ~ || konservierter Rahm m. || crème f. conservée. / ~ of tartar || gereinigter Weinstein m.; Cremortartari m. || crème f. de tartre.

cream chocolate || Kremschokolade f. || chocolat m. à la crème. / ~ colour || Kremfarbe f. || couleur f. crème.

creamery || Molkerei f.; laiterie f. / ~ worker || Milcharbeiter m. || laitier m.

cream heater || Rahm(vor)wärmer m. || réchauffeur m. pour crème. / ~ ripener || Rahmreifer m. || appareil m. à fermentation de la crème. / ~ separator || Milchentrahmer m.; Milchschleuder f. || écrémeuse f. centrifuge; écrémoir m.

crease, to (Tinm) || sieken || soyer; suager; ourler. / ~ the cover (Bookb) || den Umschlag m. nuten || rainer la couverture.

creasing die (Tinm) || Siekenform f.; Siekeneisen n. || bille f. à moulures. / ~ hammer || Siekenhammer m. || marteau m. à suage *ou* à soyer. / ~ tool (Tinm) || Siekenstock m. || suage m.

credentials pl. || Beglaubigungsschreiben n. || lettres fpl. de créance. / ~ (Bank) || Akkreditiv n. || lettre f. de crédit; accréditif m.

credit (Bookkeep) || Haben n. || avoir m. / ~ (Banking) || Kredit m. || crédit m. / ~ (Trade) || Ruf m. || réputation f.; renom m. / additional ~ || zusätzlicher Kredit m. || crédit m. additionnel. / ~ for bills || Wechselkredit m. || crédit m. en banque. / ~ in blank || Blankokredit m. || crédit m. à découvert. / to buy on ~ || auf Kredit m. kaufen || acheter à crédit. / circular letter of ~ || Zirkularkreditbrief m. || lettre f. circulaire de crédit. / extension of a letter of ~ || Verlängerung f. eines Akkreditifs || prolongation f. d'un accréditif. / to give ~ || kreditieren || créditer. / goods pl. sold on ~ || auf Ziel verkaufte Ware f. || marchandise f. vendue à terme. / letter of ~ || Kreditbrief m.; Akkreditiv n. || lettre f. de crédit; accréditif m. / limited ~ || beschränkter Kredit m. || crédit m. limité. / open ~ *see* ~ in blank. / to open a ~ || akkreditieren || accréditer. / opened ~ *see* letter of ~. / permanent ~ || permanentes Akkreditiv n. || accréditif m. permanent. / to place a thing to the ~ of a person || jemandem etwas gutschreiben *oder* kreditieren || créditer quelque chose à quelqu'un. / real estate ~ association || Bodenkredit-

bank f. || crédit-foncier m. / restricted ~ || beschränkter Kredit m. || crédit m. restreint. / secured ~ || sichergestellter Gläubiger m. || créancier m. garanti. / unconfirmed ~ || widerruflicher Kredit m. || crédit m. non-confirmé. / unlimited ~ *see* ~ in blank.

crediting || Gutschrift f. || créance f.; dette f. active.

creditor || Gläubiger m.; Kreditor m. || créditeur m.; créancier m. / ~'s meeting || Gläubigerversammlung f. || réunion f. des créanciers.

creditors pl. *see* credit side.

credit posting (Bookkeep) || Kreditkolonne f. || colonne f. des crédits. / ~ sale || Kreditverkauf m. || vente f. à crédit. / ~ side || Haben n.; Habenseite f. || côté m. du crédit. / ~ terms pl. || Kreditbedingungen fpl. || termes mpl. de crédit. / ~ transaction || Termingeschäft n.; Terminhandel m. || affaire f. à terme.

creek || Schlupfhafen m. || calangue f.; calanque f.; crique f.

creel || Weidenkorb m. || panier m. d'osier. / ~ (Spinn) || Aufsteckgatter m. || râtelier m. / ~ frame (Spinn) || Aufsteckrahmen m. || porte-râtelier m.

creeling the bobbins || Aufstecken n. der Spulen || embrochement m. des bobineaux.

creep, to || wandern; kriechen || cheminer; grimper.

creep (Mine) || Sohlenauftrieb m. || boursouflement m. du sol. / ~ of track (Railw) || Verschieben n. des Gleises in der Längsrichtung || déplacement m. des rails en sens longitudinal.

creepage of current along the surface of the porcelain (Electr) || Kriechen n. des Stromes über Porzellan || glissement m. des étincelles sur la superficie de la porcelaine.

creeper || Transportschnecke f. || vis f. sans fin. / ~ blade || Kratzerschaufel f. || raclette f.

creeping of rails || Wandern n. der Schienen || cheminement m. des rails. / angle stop preventing ~ || Stemmwinkel m. gegen das Wandern der Schienen || équerre f. d'arrêt contre le cheminement des rails. / splice bar to prevent ~ || Stemmlasche f. gegen Schienenwandern || éclisse f. épaulée pour empêcher le cheminement des rails.

creeping current (Electr) || Kriechstrom m. || courant m. circulant à la surface de l'isolant.

cremate, to || einäschern || incinérer.

cremation || Feuerbestattung f.; Leichenverbrennung f. || crémation f.

cremato(rium) *see* crematory.

crematory || Feuerbestattungsanstalt f. || crématorium m. / ~ (stove) || Einäscherungsofen m. || four m. crématoire.

cremnitz white || Kremserweiß n. || blanc m. d'argent.

creosote, to || mit Kreosot n. tränken || créosoter.

creosote || Kreosot n. || créosote f. / to impregnate with ~ || mit Kreosot n. tränken || imprégner à la créosote; créosoter.

creosote carbonate || Kreosotkarbonat n.; kohlensaures Kreosot n. || carbonate m. de créosote.

creosoted wood || mit Kreosot getränktes Holz n. || bois m. créosoté.

creosote oil ‖ Kreosotöl n.; Teeröl n. ‖ huile f. de créosote. / ~ testing ‖ Teerölprüfung f. ‖ essai m. du créosote.

creosoting of wood ‖ Imprägnieren n. von Holz; Kreosotierung f. des Holzes ‖ créosotage m. de bois.

crêpe (Caoutchouc) ‖ Krepp m.; Kreppgummi n. ‖ crêpe m. / ~ rubber ‖ Schaumgummi n. ‖ crêpe m. de latex.

crescent and French roll shaping machine (Bak) ‖ Hörnchenwickelmaschine f. ‖ machine f. à faire les croissants. / ~ moon ‖ zunehmender Mond m. ‖ lune f. croissante. / ~ shape ‖ Sichelgestalt f. ‖ forme f. de croissant. / ~-shaped ‖ sichelförmig ‖ en croissant.

cresol ‖ Kresol n. ‖ crésol m.

cresotic acid ‖ Kresotinsäure f. ‖ acide m. crésotique.

crest ‖ Gipfel m.; Scheitel m.; Grat m. ‖ crête f. / ~ (Arch) ‖ Krone f.; Firstkamm m. ‖ crête f. / ~ of a wave (Mar) ‖ Kamm m. einer Welle; Wellenkopf m. ‖ couronne f. d'une lame; sommet m. de vague.

crest factor ‖ Scheitelfaktor m. ‖ facteur m. d'amplitude. / ~ tile ‖ Kaminziegel m. ‖ tuile f. de crête.

cretaceous system ‖ Kreidesystem n. ‖ système m. crétacé.

cretonne (Weav) ‖ Doppelshirting m. ‖ cretonne f.

crevasse of glacier ‖ Gletscherspalte f. ‖ crevasse f. d'un glacier.

crevice ‖ Sprung m.; Riß m.; Spalt m. ‖ crevasse f.; fissure f. fêlure f. / ~ of a wall ‖ Mauerspalte f.; Mauersprung m.; Riß m. im Mauerwerk ‖ lézarde f. ou fente f. d'une muraille. / circular ~ in wood ‖ Kernschäle f.; Ringkluft f. ‖ roulure f.

crew (Mar) ‖ Besatzung f.; Mannschaft f.; Gasten mpl. ‖ équipage m.; équipe f. / carpenter's ~ (Mar) ‖ Zimmermannsgasten mpl. ‖ ouvriers mpl. charpentiers. / cooper's ~ (Mar) ‖ Böttchersgasten mpl. ‖ ouvriers mpl. tonneliers. / handling ~ (Aero) ‖ Haltemannschaft m. ‖ équipe f. de manœuvre. / ~ of a mine ‖ Belegschaft f. (einer Grube) ‖ personnel m. d'une mine. / to pay off the ~ (Mar) ‖ die Schiffsmannschaft abmustern ‖ congédier l'équipage. / to ship the ~ ‖ die Schiffsmannschaft anmustern ‖ engager l'équipage.

crewspace ‖ Mannschaftsraum m. ‖ poste m. de l'équipage.

crib (Mine) ‖ Kranz m. (von Holz oder Gußeisen beim Schachtausbau) ‖ rouet m.; couronne f. / ~ tubbing ‖ wasserdichter Schachtausbau m. durch Holzringe ‖ cuvelage m. en bois circulaire; tubage m. en bois.

cricket ‖ Kricketspiel n. ‖ jeu m. de cricquet.

crimp ‖ Faltenstoff m. ‖ crêpé m.

crimson ‖ Karmoisin n. ‖ cramoisi m. / ~ red ‖ Berlinerrot n. ‖ rouge m. cramoisi.

crimper (Glassw) ‖ Randschleifer m. ‖ fletteur m.

crimping machine for wire netting ‖ Krippmaschine f. für die Drahtweberei ‖ machine f. à plier le gros fil pour les buts du tissage.

cringle (Mar) ‖ Legel m. ‖ patte f.

crinkled paper ‖ Wellpapier n. ‖ papier m. plissé.

crinoidean ‖ Krinoide f. ‖ crinoïde m.

crinoline ‖ Roßhaarstoff m. für Unterröcke ‖ crinoline f.

cripper see crippling board.

cripple, to (Curr) ‖ krispeln ‖ rebrousser; crêpir.

cripple timber ‖ Halbholz m. ‖ bois m. mi-plat.

crippling board (Curr) ‖ Krispelholz n. ‖ paumelle f.; grènetoir m. / ~ machine ‖ Krispelmaschine f. ‖ machine f. à rebrousser.

crisis ‖ Wendepunkt m.; Krise f. ‖ crise f.; moment m. critique. / economic ~ ‖ Wirtschaftskrise f. ‖ crise f. économique. / ~ in the money market ‖ Geldkrise f.; Krise f. auf dem Geldmarkt ‖ crise f. financière ou monétaire.

crisp almond ‖ gebrannte Mandel f. ‖ praline f.

crisper (Cloth) ‖ Kräusler m. ‖ crêpeur m.

crisping (Cloth) ‖ Kräuseln n. ‖ crêpage m. / ~ (Mach) ‖ Fräsung f. ‖ fraisure f.

cristite ‖ Cristite n. ‖ cristite m.

critic ‖ Kritiker m.; Kunstrichter m. ‖ critique m.; censeur m.

critical ‖ kritisch ‖ critique. / ~ resistance ‖ kritischer Widerstand m. ‖ résistance f. critique. / ~ speed ‖ kritische Geschwindigkeit f. ‖ vitesse f. critique. / ~ temperature ‖ kritische Temperatur ‖ température f. critique.

criticism ‖ Kritik f. ‖ critique f.

criticize, to ‖ kritisieren; beurteilen; rezensieren ‖ faire la critique; critiquer.

crochet, to ‖ häkeln ‖ broder ou travailler au crochet. / ~ by hand ‖ mit der Hand f. häkeln ‖ crocheter à la main.

crocheter ‖ Häkler m. ‖ tricoteur m. au crochet.

crochet hook ‖ Häkelhaken m. ‖ crochet m. à crocheter. / ~ needle ‖ Häkelnadel f. ‖ crochet m. / ~ pin ‖ Häkelnadel f. ‖ crochet m. / artistic ~ work ‖ künstlerische Häkelarbeit f. ‖ ouvrage m. artistique au crochet.

crocidolite ‖ Krokydolith m.; Blaueisenstein m. ‖ crocidolite f.

crockery ‖ irdenes Geschirr n.; Steingut n.; Fayence f. ‖ vaisselle f. pour ménage; faïence f. / table ~ ‖ Tischgeschirr n. ‖ vaisselle f. / utility ~ ‖ Gebrauchsgeschirr n. ‖ vaisselle f. d'usage courant.

crockery earth extraction ‖ Fayenceerdegrube f. ‖ carrière f. de terre à faïence. / ~ maker ‖ Steingutarbeiter m. ‖ faïencier m. / ~ ware ‖ Tonzeugwaren fpl. ‖ poterie f. / ~ washing machine ‖ Geschirrwaschmaschine f. ‖ machine f. à laver la vaisselle.

crocket of a tile ‖ Nase f. eines Dachziegels ‖ crochet m. de tuile.

crocoite ‖ Rotbleierz n.; chromsaures Bleioxyd n.; Krokoit m. ‖ plomb m. chromaté.

crocus ‖ safrangelb ‖ safrané.

crocus (Bot) ‖ Krokus m.; Safran m. ‖ safran m. / ~ (Chem) ‖ Polierrot n. ‖ rouge m. à polir.

crook, to ‖ krümmen ‖ plier; couder; courber.

crook ‖ Haken n. ‖ crochet m.

crooked ‖ gekrümmt ‖ courbé. / ~ grown tree ‖ krummgewachsenes Holz n. ‖ arbre m. déjeté ou contourné. / ~ lever ‖ Winkelhebel m. ‖ levier m. coudé. / ~ timber ‖ Krummholz n. ‖ bois m. déjeté.

crookedness ‖ Krümmung f. ‖ courbe f.; courbure f.; coude m.

crop, to ‖ scheren ‖ tondre. / ~ out (Mine) ‖ anstehen; zu Tage ausgehen ‖ paraître à fleur f. de terre; se montrer au jour; affleurer.

crop (Agr) ‖ Ernte f. ‖ moisson f. / ~ (Cereals) ‖ Getreide n. auf dem Halm ‖ grains mpl. sur pied. / ~ (Dyer) ‖ beraubter Krapp m. ‖ garance f. robée. / ~ (Mine) ‖ Ausgehendes n. ‖ affleurement m. / ~ (Miner) ‖ Scheiderz n. ‖ minerai m. riche ou de scheidage. / stand ~ see ~ (Cereals).

crop butt see ~ hide. / harvesting machine for ~s ‖ Getreideerntemaschine f. ‖ machine f. de récolte pour céréales. / ~ hide (Curr) ‖ Pfundleder n.; Schwerleder n. ‖ cuir m. fort ou nerveux; gros cuir m.

cropper ‖ Baumausschneider m. ‖ élagueur m.; émondeur m.

cropping (Clothm) ‖ Scheren n. des Tuches ‖ tondage m.; tonte f.; tonture f. / ~ machine ‖ Tuchschermaschine f. ‖ tondeuse f. / ~-out of the bed (Mine) ‖ Zutagetreten n. der Schicht ‖ affleurement m. de la couche. / ~ waste ‖ Scherflocken fpl. ‖ déchêts mpl. de tondeuse.

crop report ‖ Ernteberícht m. ‖ rapport m. de récolte. / ~ shear for iron ‖ Eisenschere f. ‖ cisaille f. à dresser les bouts. / ~ transporting implements pl. ‖ Erntetransportmittel npl. ‖ ustensiles mpl. à transporter la récolte.

croquet ‖ Krocketspiel n. ‖ jeu m. de croquet.

cross ‖ quer ‖ transversal.

cross, to ‖ kreuzen ‖ croiser; traverser; passer. / ~ a lode (Mine) ‖ einen Gang m. überfahren ‖ traverser un filon. / ~ paper ‖ das Papier schränken ‖ croiser le papier en le comptant. / ~ a rope lashing (Mar) ‖ Bändsel oder Taue kreuzen ‖ brider. / ~ a seizing see ~ a rope lashing. / to ~ the warp threads (Weav) ‖ die Fäden mpl. verkreuzen oder ins Kreuz legen ‖ enverger ou encroiser les fils. / ~ wires pl. ‖ Telegrafendrähte mpl. kreuzen ‖ croiser des fils mpl. télégraphiques.

cross ‖ Kreuz n. ‖ croix f. / ~ (Anchor) ‖ Kreuz n. ‖ croisée f.; crosse f.; diamant m. ~ (Opt) ‖ Strichkreuz n.; Fadenkreuz n. ‖ réticule m. / ~ (Print) ‖ Kreuzzeichen n. ‖ Kreuzstern m. ‖ croix f.; étoile f. croisée. / ~ (Rope) ‖ halber Schlag m. ‖ croix f. / ~ (Surv) ‖ Kreuzmaß n. ‖ équerre f. d'arpenteur. / beating ~ ‖ Schlagkreuz n. ‖ croisillon m. portant des marteaux. / four-way ~ (Railw) ‖ Kreuzstück n. ‖ pièce f. en croix. / long ~ (Print) ‖ Mittelsteg m. ‖ barre f. du compositeur. / Scotch ~ (Draff) ‖ Anschwänzer m.; Anschwänzkreuz n. ‖ croix f. écossaise. / short ~ (Print) ‖ Kreuzsteg m. ‖ lingots mpl. ou garnitures fpl. à croisillons. / St. Andrew's ~ ‖ Andreaskreuz n. ‖ croix f. de Saint-André. / upright ~ ‖ Stehkreuz n. ‖ crucifix m. à pied.

cross action ‖ Gegenklage f. ‖ contre-accusation f.; reconvention f.

cross arm ‖ Querträger m. ‖ traverse f.; bras m. croisé. / ~ of a fly-press ‖ Schwunghebel m. einer Schraubenpresse ‖ balancier m. ou verge f. du balancier à vis. / ~ plate (El line) ‖ Vorlegestück n. für Querträger ‖ semelle f.

cross bar (Carp) ‖ Querholz n.; Querriegel m. ‖ traverse f. ou entretoise f. en bois.

/ ~ (Mach) ‖ Querriegel m.; Traverse f. ‖ traverse f. / ~ (Railw) ‖ Querstrebe f.; Querträger m. ‖ traverse f.; entretoise f. transversale. / ~ (Weav) ‖ Querstock m.; Spannbalken m. ‖ traversier m.; traverse f. de tension. / ~ of a bay-work ‖ Riegel m. oder Querholz n. einer Fachwand; Wandriegel m. ‖ entretoise f. de cloison. / ~ of the man-hole ‖ Mannlochbügel m. ‖ étrier m. de trou d'homme. / ~ between the panes of glass ‖ Fenstersprosse f. ‖ croisillon m. de croisée. / ~ of a trellis ‖ Querstange f. eines Gitters ‖ traverse f. d'une grille.

cross bar micrometer ‖ Kreuzstabmikrometer n. ‖ micromètre m. à barres croisées.

cross-barred end (Railw) ‖ Gitterkopfwand f. ‖ paroi f. frontale à grille ou de bout à grille.

cross bar switch (Aut tel) ‖ Koordinatenwähler m. ‖ sélecteur m. coordinateur. / swaged ~ traction eye ‖ gesenkgeschmiedete Kreuzlappenzugöse m. ‖ anneau m. d'attelage à joint en croix forgé en matrice.

cross beam (Aero) ‖ Holm m. ‖ sommier m. / ~ (Bridge) ‖ Querträger m. ‖ entretoise f. / ~ (Build) ‖ Querbalken m. ‖ poutre f. transversale; traverse f. / ~ (Carp) ‖ Querholz n.; Querriegel m. ‖ traverse f. ou entretoise f. en bois. / ~ (Carp) ‖ Unterzug m. ‖ solive f. / ~ (Ship) ‖ Querbalken m.; Dwarsbalken traversin m. / ~ (Windmill) ‖ Schwertbalken m. ‖ poutre f. pivotante. / ~ of a cask ‖ Faßriegel m. ‖ verrou m. de portière de foudre.

cross bearer of a furnace-grate ‖ Rostträger m. oder Rostbalken m. eines Feuerrostes ‖ sommier m. ou traverse f. d'une grille de foyer.

cross bearing (Nav) ‖ Kreuzpeilung f. ‖ relèvement m. croisé.

cross beater mill ‖ Schlagkreuzmühle f. ‖ moulin m. à croisillon percuteur; broyeur m. à marteaux frappeurs ou de percussion à crouse.

cross beating ‖ Hirnleiste f. ‖ emboîture f.

cross bedding (Mine) ‖ Kreuzschichtung f. ‖ stratification f. croisée.

cross belt ‖ Quertransportband n. ‖ courroie f. croisée; bande f. transversale.

cross bill ‖ Rückwechsel m. ‖ rechange m.; retraite f.

cross blade for sectioning the burr (Plate shear) ‖ Quermesser n. zum Zerteilen des Grates ‖ lame f. d'équerre à cisailler la bavure.

cross board (Mine) ‖ Pfeilerdurchhieb m.; Pfeilerort m. ‖ traverse f. dans un pilier.

cross bond see cross brace (Build).

cross brace (Build) ‖ Kreuzverband m.; Strebe f. ‖ liaison f. croisée ou diagonale; diagonale f.; tringle f. / ~ (El line) ‖ Mittelriegel m. ‖ entretoise f.

cross bracket (El line) ‖ Querriegel m. ‖ entretoise f.

cross bulkhead ‖ Querschott n. ‖ cloison f. transversale. / ~ bunker ‖ Querbunker m. ‖ soute f. de travers. / ~-chap hand vice ‖ Feilkloben m. mit breitem Maul ‖ étau m. à main à grande ouverture des mâchoires. / ~ coil aerial ‖ Kreuzrahmenantenne f. ‖ antenne f. à cadre double. / ~ composition (Print) ‖ Quersatz m. ‖ composition f. en traverse. / ~ connection (Tel) ‖ Querverbindung f. ‖ connexion f. transversale.

cross-country ‖ querfeldein ‖ à travers champ. / ~ car (Arm) ‖ Querfeldeinwagen m. ‖ char m. à travers champ. / ~ flight ‖ Überlandflug m. ‖ survol m. / ~ road ‖ Querstraße f.; Seitenstraße f.; Nebenstraße; Nebenweg m. ‖ chemin m. de traverse.

cross cracks pl. in the iron ‖ Querrisse mpl. im Eisen ‖ criques fpl. transversales dans le fer. / ~ current ‖ Querstrom m. ‖ courant m. transversal.

cross-cut, to ‖ quer durchschneiden ‖ couper ou scier en travers. / ~ a stone ‖ einem Stein m. den Kreuzhieb geben ‖ traverser une pierre. / ~ wood ‖ das Holz über Hirn sägen ‖ couper ou scier le bois en travers.

cross-cut ‖ Querhieb m. ‖ coupe f. en travers. / ~ (File) ‖ Kreuzhieb m. ‖ taille f. croisée. / ~ (Mine) ‖ Querschlag m. ‖ galerie f. à travers banc. / ~ (Saw) ‖ Quersäge f. ‖ scie f. à tronçonner ou à couper en travers. / breadth of ~ ‖ querschlägige Breite f. des Grubenfeldes ‖ largeur f. du champ d'exploitation.

cross-cut chisel ‖ Kreuzmeißel m. ‖ bec m. d'âne; bédane m. / turner's ~ ‖ Stechbeitel m. für Drechsler ‖ bédane m. pour tourneurs.

cross-cut end (Wood) ‖ Hirnfläche f. ‖ surface f. de bout. / ~ saw ‖ Schrotsäge f. ‖ scie f. passe-partout. / ~ saw for logs ‖ Abkürzsäge f.; Steifsäge ‖ scie f. à tronçonner. / ~ wood ‖ Stirnholz n.; Querholz n.; Hirnholz n. ‖ bois m. de bout.

cross-cutter (Mine) ‖ Schrämmaschine f. ‖ haveuse f.

cross-cutting machine (Pap) ‖ Querschneider m. ‖ coupeuse f. transversale.

cross-demand ‖ Gegenforderung f. ‖ créance f. en compensation; demande f. reconventionnelle.

cross-dike ‖ Querdamm m. ‖ duit m.

crossed ‖ gekreuzt; geschränkt ‖ croisé. / ~ cheque ‖ Verrechnungsscheck m. ‖ chèque m. barré. / ~ currents pl. ‖ gekreuzte Ströme mpl. ‖ courants mpl. croisés.

cross-entry ‖ Gegenposten m. ‖ contrepartie f.

cross-eyed, to be ~ ‖ schielen ‖ loucher.

cross feed (Tool mach) ‖ Planvorschub m. ‖ avance f. transversale. / ~ fibre ‖ Querfaser f. ‖ fibre f. transversale. / ~ file ‖ Vogelzunge f. ‖ lime f. ovale ou double demironde. / ~ fissure ‖ Querriß m. ‖ fissure f. transversale. / ~ flucan ‖ Lettenquerkluft f. ‖ faille f. glaiseuse. / ~ frog (Railw) ‖ Herzstück n. einer Kreuzung ‖ croisement m. de traversée. / ~ gallery ‖ Querschlag m. ‖ galerie f. à travers banc. / ~ girder ‖ Querbalken m.; Querträger m. ‖ poutre f. transversale; traverse f. / ~-grained slate ‖ Griffelschiefer m. ‖ ardoise f. à écrire. / ~-grained wood ‖ Hirnholz n. ‖ bois m. taillé contre le fil; bois m. de bout. / ~ grooving machine ‖ Quernutenmaschine f. ‖ machine f. à faire les rainures transversales. / ~ table with ~ guide ledges ‖ Drehscheibe f. mit Kreuzspurleisten ‖ plaque f. tournante avec railsguides entrecroisés.

cross hair diopter ‖ Fadendiopter n. ‖ pinnule f. à fils.

cross hairs pl. ‖ Fadenkreuz n. ‖ réticule m.

cross-handle of an earth-borer (Mine) ‖ Kopfstück n. eines Erdbohrers ‖ manivelle f. d'un trépan.

crosshead ‖ Kreuzkopf m. ‖ tête f. ou crosse f. de piston; cross(ett)e f. / closed type of ~ ‖ geschlossener Kreuzkopf m. ‖ crosse f. fermée. / forked type of ~ ‖ gabelförmiger Kreuzkopf m. ‖ crosse f. bifurquée. / ~ of a press ‖ Preßholm m. ‖ sommier m. de presse. / slotted ~ ‖ Langlochkreuzkopf m. ‖ crossette f. à coulisse. / ~ with upper and lower slippers ‖ Kreuzkopf m. mit oberer und unterer Führung ‖ guidage m. de crossette à deux savates.

crosshead bearing ‖ Kreuzkopflager n. ‖ coussinet m. de la crosse. / ~ block ‖ Gleitbacke f.; Schlitten m.; Gleitklotz m. ‖ coulisseau m. de crosse; glissoir m. / ~ body see ~ block. / ~ cap ‖ Kreuzkopflagerdeckel m. ‖ chapeau m. de la crosse. / ~ gib see ~ shoe. / ~ gudgeon ‖ Kreuzkopfbolzen m. ‖ tourillon m. de la crosse. / ~ guide ‖ Kreuzkopfgleitbahn f.; Kreuzkopfführung f. ‖ glissière f. de la crosse; guide f. de la tête de piston. / ~ pin ‖ Kreuzkopfzapfen m. ‖ tourillon m. de crosse. / ~ shoe ‖ Kreuzkopfschuh m. ‖ patin m. ou semelle f. de crosse. / ~ slipper see ~ shoe.

cross-hole nut ‖ Kreuzlochmutter f. ‖ écrou m. percé.

crossing (Cross road) ‖ Kreuzweg m. ‖ carrefour m. / ~ (El line) ‖ Überquerung f. ‖ traversée f. / ~ (Railw) ‖ Kreuzung f.; Kreuzungsstelle f. ‖ croisement m. de voie. / ~ (Railway frog) ‖ Kreuzung f.; Herzstück n.; Kreuzungsstück n. ‖ cœur m. (de croisement); croisement m.; pièce f. de croisement. / ~ (Road) ‖ Straßenkreuzung f. ‖ croisement m. / ~ (Road and railway) ‖ Schienenübergang m.; Straßenübergang m. ‖ passage m. / ~ (Two railway tracks crossing one another) ‖ Kreuzung f.; Kreuzungsstelle f. ‖ traversée f. ou croisement m. de voie. / ~ (Zool) ‖ Kreuzung f. ‖ croisement m. / ~ (Weav) ‖ Fadenkreuz n.; Kreuzung f. ‖ croisure f. / ~ at acute angles (Railw) ‖ spitzwinklige Kreuzung f. ‖ croisement m. de voie oblique ou à angle aigu. / aerial ~ (El line) ‖ Luftkreuzung f. ‖ croisement m. aérien. / angle of ~ (Railw) ‖ Herzstückneigung f.; Kreuzungsverhältnis n.; Kreuzungswinkel m. ‖ déviation f. du cœur; angle m. de croisement. / contractors' ~ ‖ Schleppkreuzung f. ‖ croisement m. à rails pivotants. / curve ~ (Railw) ‖ Kurvenkreuzung f. ‖ croisement m. en courbe. / diamond ~ (Railw) ‖ Kreuzungsstück n.; Doppelherzstück n. ‖ croisement m. ou cœur m. double; pièce f. de croisement. / ~ with forged point ‖ Kreuzungsstück n. mit geschmiedeter Spitze ‖ pièce f. de croisement avec pointe forgée. / ~ of lines (Railw) ‖ Gleiskreuzung f. ‖ croisement m. de voie. / ~ between a main track and a narrow gauge line ‖ Kreuzung f. zwischen einer Hauptbahn und einer Kleinbahn ‖ croisement m. entre une ligne normale et une ligne étroite. / manganese steel ~ (Railw) ‖ Kreuzung f. aus Manganstahl ‖ croisement m. en acier manganèse. / ~ between narrow gauge railways ‖ Kreuzung f. zwischen Kleinbahnen unter sich ‖ croisement m. entre deux voies étroites. / oblique ~ (Railw) ‖ schräge Bahnkreuzung f. ‖ traversée f. oblique. / parabolic ~ ‖ Parabelweiche f. ‖ croisement m.

parabolique. / ~ for pedestrians ‖ Fahrdammübergang m. für Fußgänger ‖ passage m. pour piétons. / ~ of rails ‖ Gleiskreuzung f. ‖ croisement m. de voies. / ~ over the railway ‖ Eisenbahnübergang m. ‖ passage m. au dessus du chemin de fer. ‖ ~ reversible ~ (Railw) ‖ umwendbare Kreuzung f. ‖ croisement m. à retournement. / ~ at right angle (Railw) ‖ rechtwinklige Kreuzung f. ‖ croisement m. de voie à angle droit; traversée f. rectangulaire. / ~ subway ~ ‖ Wegunterführung f. ‖ passage m. souterrain; subterraneau m. / ~ with wheel flange ramp ‖ Herzstück n. mit Flanschen ‖ cœur m. avec semelle guide-bourrelet.

crossing, angle of the ‖ Kreuzungswinkel m.; Weichenwinkel m. ‖ angle m. du changement *ou* de croisement. / ~ nose *see* crossing point.

crossing piece ‖ Herzstück n.; Kreuzung f.; Kreuzungsstück n. ‖ cœur m. (de croisement); croisement m.; pièce f. de croisement. / built-up ~ ‖ Schienenkreuzungsstück n.; aus Schienen zusammengestelltes Herzstück n. ‖ pièce f. de croisement composée de rails. / rectangular ~ ‖ rechtwinkliges Kreuzungsstück n. ‖ pièce f. de croisement rectangulaire.

crossing place in the line (Railw) ‖ Ausweichung f. ‖ voie f. de garage. / ~ point ‖ Herzstückspitze f. ‖ pointe f. de cœur (de croisement). / ~ pole ‖ Kreuzungsmast m.; Kreuzungsstange f. ‖ support m. de traversée. / scheme of ~s (El line) ‖ Kreuzungsschema n.; Kreuzungsfolge f. ‖ schéma m. de croisements *ou* de transpositions. / ~ sign ‖ Warnungstafel f. ‖ écriteau m. *ou* plaque f. d'avertissement. / ~ switch ‖ Kreuzungsweiche f. ‖ traversée-jonction f. ‖ ~ track ‖ Kreuzgleis n. ‖ voie f. de croisement. ‖ ~ vein (Mine) ‖ Kreuzgang m.; Quergang m. ‖ filon m. croiseur.

cross iron ‖ Kreuzeisen n. ‖ fer m. en croix. / ~ jack ‖ Kreuzsegel n. ‖ basse voile f. d'artimon. / ~ key ‖ Querkeil m. ‖ clavette f. transversale. / ~ lever ‖ Gestängekreuz n. ‖ levier m. en croix.

cross line (Print) ‖ Bruchstrich m.; Querstrich m. ‖ ligne f. transversale; barre f. de fraction. / ~ eyepiece ‖ Strichkreuzokular n. ‖ oculaire m. à réticule. / ~ illuminating attachment ‖ Fadenkreuzbeleuchtungseinrichtung f. ‖ dispositif m. à éclairer le réticule. / ~ micrometer ‖ Netzmikrometer n. ‖ micromètre m. à réseau.

cross lines pl. ‖ Fadenkreuz n. ‖ réticule m. / ~ etched on glass ‖ auf Glas geätztes Fadenkreuz n. ‖ réticule m. gravé sur verre. / ~ on glass plate ‖ Fadenkreuz n. auf Glasplatte ‖ réticule m. sur (plaque de) verre. / ~ photographed-on ‖ fotografisch aufgetragenes Fadenkreuz n. ‖ réticule m. reproduit par la photographie. / brightness of the ~ ‖ Fadenhelligkeit f. ‖ clarté f. des fils.

cross lode ‖ Quergang m.; Kreuzgang m. ‖ filon m. croiseur. / ~ loop ‖ Kreuzschlaufe f. ‖ nœud m. croisé. / ~magnetization ‖ Quermagnetisierung f. ‖ aimantation f. transversale. / ~ member ‖ Querträger m. ‖ entretoise f. / ~-mouthed chisel ‖ Kreuzbohrer m. ‖ fleuret m. à double tranchant en croix *ou* en bonnet de prêtre. / ~ movement ‖ Kreuzbewegung f.; Querverschiebung f. ‖

mouvement m. croisé; déplacement m. transversal. / ~ movement of the table (Mach tool) ‖ Querverschiebung f. des Tisches ‖ déplacement m. transversal de la table.

crossover (Railw) ‖ Kreuzungsweiche f. ‖ traversée-jonction f. / ~ road ‖ Verbindungsgleis n. ‖ voie f. de jonction.

crosspiece ‖ Querstück; Querverband m.; Querhaupt n. ‖ entretoisement m.; traverse f. / ~ (Shipb) ‖ Querbalken m.; Dwarsbalken m. ‖ traversin m. / ~ of the framing of a lock-gate ‖ Riegel m. einer Schleusenzarge ‖ entretoise f. du cadre d'une porte d'écluse à vannes. / middle ~ ‖ Mittelriegel m. einer Schleuse ‖ entretoise f. *ou* traverse f. moyenne. / ~ of the slide ‖ Querschlitten m. ‖ traverse f. du chariot. / wooden ~ ‖ Querholz n.; Querriegel m. ‖ traverse f. *ou* entretoise f. en bois. / ~ die for windows ‖ Fensterkreuzsprossenstanze f. ‖ matrice f. à estamper les croisées de fenêtres.

cross rafter ‖ Quersparren m. ‖ linçoir m.

cross rail ‖ Querbalken m. ‖ traverse f. / adjustable ~ ‖ verstellbarer Querbalken m. ‖ traverse f. mobile. / fixed ~ ‖ feststehender Querbalken m. ‖ traverse f. fixe. / ~ motion ‖ Querbalkenbewegung f. ‖ mouvement m. de la traverse.

cross ring ‖ Kreuzring m. ‖ bague f. *ou* anneau m. à deux axes en croix.

crossroad ‖ Kreuzweg m. ‖ carrefour m.

cross sea ‖ Kreuzsee f.; kreuzweis laufende See f. ‖ mer f. contraire *ou* creuse.

cross section ‖ Querprofil n.; Querschnitt m. ‖ coupe f. transversale *ou* en travers. / perfectly close-grained ~ ‖ vollkommen dichter Querschnitt f. ‖ section f. transversale absolument homogène.

cross shaft ‖ Querwelle f. ‖ arbre m. transversal.

cross-shearing machine ‖ Querschermaschine f. ‖ tondeuse f. transversale.

cross sleeper (Railw) ‖ Querschwelle f. ‖ traverse f. / ~ of a grating ‖ Querschwelle f. eines Rostes ‖ traversine f. *ou* racinal m. de palée.

cross springer (Arch) ‖ Gratbogen m.; Kreuzbogen m. ‖ arc m. arêtier *ou* diagonal; croisée f. d'ogive. / ~ staff (Surv) ‖ Kreuzmaß n. ‖ équerre f. d'arpenteur. / ~ stitch ‖ Kreuzstich m. ‖ point m. croisé. / ~ stitcher ‖ Kreuzstepper m. ‖ piqueur m. en croix. / ~ stone ‖ Kreuzstein m.; Harmotom m. ‖ pierre f. de croix; harmotome f. ‖ ~ stud ‖ Kreuzstrebe f. ‖ jambe f. de force croisée. / ~-tail butt ‖ Gabelstück n. der Pleuelstange ‖ bielle f. latérale du grand té. / ~-tail strap *see* ~-tail butt.

crosstalk, to (Tel) ‖ übersprechen ‖ diaphoner.

crosstalk (Tel) ‖ Nebensprechen n.; Übersprechen n. ‖ diaphonie f.; mélange m. de conversation. / ~ meter (Tel) ‖ Übersprechmesser m. ‖ diaphonomètre m.

cross tie (Railw) ‖ Querschwelle f. ‖ traverse f.; entretoise f. d'écartement. / automatic ~ traverse of the table ‖ selbsttätige Querbewegung f. des Tisches ‖ course f. transversale automatique de la table. / ~ tree (Mar) ‖ Querspaling f. ‖ barre f. de travers *ou* traversière; traversin m. / ~ tube boiler ‖ Quersieder m. ‖ générateur m. transversal. / ~

vaulting ‖ Kreuzgewölbe n. ‖ voûte f. croisée. / ~ vein (Mine) ‖ Kreuzgang m.; Quergang m. ‖ filon m. croiseur. / ~ wall (Build) ‖ Quermauer f. ‖ mur m. en traverse. / ~ way ‖ Kreuzweg m. ‖ carrefour m. / ~ way of the grain ‖ Hirnseite f. des Holzes ‖ côté m. de la moelle; coupe f. transversale. / ~ weaving ‖ Weben n. mit gekreuzter Kette ‖ tissage m. à chaîne croisée. / ~ web being spider's web stretched over a plate (Opt) ‖ Fadenkreuz n. aus Spinnenfäden mit Fadenplatte ‖ réticule m. en fils d'araignée avec plaque porte-réticule. / ~ winding ‖ Kreuzwicklung f. ‖ bobinage m. en forme de croix.

cross wires pl. ‖ Fadenkreuz n. ‖ réticule m. / the ~ are not in the focal plane ‖ das Fadenkreuz n. liegt nicht in der Bildebene ‖ le réticule m. n'est pas dans le plan de l'image.

cross wire sight ‖ Fadendiopter n. ‖ pinnule f. à fils.

crosswise ‖ quer ‖ transversale(ment); en travers.

cross work ‖ Fensterkreuz n. ‖ croisillon m.; croix f. de croisée. / ~ working (Mine) ‖ Querbau m. ‖ ouvrage m. en travers.

crotch ‖ Gabelholz n.; Gabelung f. der Zweige ‖ gibelot m.

crotchets pl. (Print) ‖ eckige Klammern fpl. ‖ crochets mpl.

croton leaf ‖ Färberkrotonblatt n. ‖ feuille f. de tournesol. / ~ oil ‖ Krotonöl n. ‖ huile f. de croton. / ~ seeds pl. ‖ Krotonsaat f. ‖ semence f. de croton.

crowbar ‖ Brecheisen n.; Brechstange f.; Kuhfuß m. ‖ levier m. de fer; pied m. de biche *ou* de chèvre. / crooked ~ (Build) ‖ Kuhfuß m.; Geißfuß m. ‖ pince f. à levier. / crooked ~ (Locksm) ‖ Spitzzange f. ‖ bec m. à corbin. / clow-ended ~ (Mine) ‖ Brecheisen mit gespaltener Klaue ‖ pied-de-biche m.

crowd ‖ Zudrang m. ‖ affluence f.; presse f.

crowfoot ‖ Hahnpot f. ‖ araignée f.

crown ‖ Krone f. ‖ couronne f. / ~ (Anchor) ‖ Kreuz m. ‖ croisée f.; crosse f.; diamant m. / ~ (Mar) ‖ Hahnpot ‖ tête f. de mort; couronne f. / ~ (Ropem) ‖ Hakenkrone f. ‖ croisille f. / ~ (Watch) ‖ Aufzugskrone f. ‖ couronne f. / ~ of an arch ‖ Bogenscheitel m. ‖ sommet m. *ou* vertex m. *ou* apex m. d'un arc. / ~ of the case (Watch) ‖ Gehäusekrone f. ‖ couronne f. de la boîte. / ~ of the dam ‖ Krone f. der Sperrmauer ‖ crête f. du mur de barrage. / ~ of a hat ‖ Rand m. eines Hutes ‖ bords mpl. d'un chapeau. / ~ of pearls ‖ Perlenkranz m. ‖ couronne f. de perles. / ~ of a sluice ‖ Haupt n. einer Schleuse ‖ tête f. d'écluse. / tooth ~ ‖ Zahnkrone f. ‖ couronne f. de dents.

crown bar ‖ Deckenträger m. ‖ ferme f. de ciel. / ~ bit ‖ Kronenbohrer m. ‖ foret m. à couronne. / ~ cork ‖ Kronenkorkverschluß m. ‖ bouchon-couronne m.; fermeture f. à bouchon-couronne. / ~ drill ‖ Kronenbohrer m. ‖ perçoir m. à couronne. / ~ escapement (Clockm) ‖ Spindelhemmung f. ‖ échappement m. à verge. / ~ gate of a canal lock ‖ oberes Schleusentor n.; Fluttor n. ‖ porte f. d'amont d'une écluse de canal. / ~ glass (Opt) ‖ Kronglas n. ‖ crown-glass m. / ~ glass lens ‖ Kronglaslinse f. ‖ lentille f. en crown-glass.

crowning (Arch) ‖ Krone f. ‖ crête f. ~ (Pulley) ‖ Wölbung f. ‖ bombage m.; flèche f.

crown knot ‖ Hahnpot f. ‖ tête f. de mort. / ~ plate rivetting apparatus ‖ Deckennietapparat m. ‖ appareil m. à riveter le ciel de foyer. / ~ post roof ‖ Hängewerk n. mit einer Säule ‖ simple arbalète f. / ~ ring with rollers ‖ Rollenkranz m. ‖ couronne f. de rouleaux. / ~ saw ‖ Kronsäge f.; Ringsäge f. ‖ scie f. cylindrique. / ~ stiffening rib ‖ Verstärkungsbalken m. in der Decke ‖ nervure f. de renforcement du ciel. / ~ wheel (Mach) ‖ Kettenkranz m. ‖ couronne f. pour chaîne. / ~ wheel (Watchm) ‖ Kronrad n. ‖ roue f. à dents de côte *ou* à couronne.

crow's nest (Shipb) ‖ Mastkorb m. ‖ échanguette f.

croze (Coop) ‖ Kröse f. ‖ jabloire f. / cask ~ cutting machine ‖ Faßkrösemaschine f. ‖ machine f. à jabler les douves. / ~ iron ‖ Kröseisen n. ‖ peigne m. à jable.

crozer ‖ Kimmhobel m.; Gergelmesser n. ‖ jabloire f.

crucible (Chem; Metal) ‖ Tiegel m.; Schmelztiegel m. ‖ creuset m. (à fondre). / black-lead ~ ‖ Graphittiegel m. ‖ creuset m. de plombagine. / ~ composed of two parts of graphite and one part of clay ‖ Tiegel m. aus zwei Teilen Graphit und einem Teil Ton ‖ creuset m. formé de deux parties de graphite et d'une partie de terre argileuse. / empty ~ ‖ leerer Tiegel m. ‖ creuset m. vide. / filtering ~ ‖ Filtertiegel m. ‖ creuset m. à filtrer. / graphite ~ ‖ Graphittiegel m. ‖ creuset m. en graphite. / porcelain ~ ‖ Schmelztiegel m. aus Porzellan ‖ creuset m. en porcelaine. / portable ~ ‖ fahrbarer Schmelztiegel m. ‖ creuset m. sur roues. / pressed steel ~ ‖ Tiegel m. aus Preßstahl m. ‖ creuset m. en acier embouti.

crucible filler ‖ Tiegelfüller m. ‖ remplisseur m. de creusets.

crucible furnace ‖ Schmelzofen m.; Tiegelofen m. ‖ four m. à creusets. / electric ~ ‖ Elektrotiegelofen m. ‖ four m. électrique à creuset. / ~ for gas ‖ Tiegelofen m. für Gasfeuerung ‖ four m. à creuset à chauffage au gaz. / multiple ~ ‖ mehrtiegliger Ofen ‖ four m. à plusieurs creusets. / ~ with preheater ‖ Tiegelofen m. mit Vorwärmer ‖ four m. à creuset avec avant-chambre de chauffe. / single ~ ‖ eintiegliger Ofen m. ‖ four m. à un seul creuset.

crucible making shop ‖ Tiegelkammer f. ‖ atelier m. pour la fabrication de creusets. / ~ man ‖ Tiegelformer m. ‖ creusetier m. / manufacture of ~s ‖ Tiegelherstellung f. ‖ confection f. de creusets. / ~ material ‖ Tiegelmasse f. ‖ matière f. pour creusets. / ~ melting furnace *see* ~ furnace. / ~ pliers pl. with olive nose ‖ Tiegelzange f. ‖ pince f. à creuset à olive. / ~ puller-out (Met) ‖ Tiegelzieher m. ‖ tireur m. de creusets. / ~ stand (Metal) ‖ Käse m.; Untersatz m. ‖ fromage m.; tourte f.

crucible steel ‖ Tiegelstahl m.; Tiegelguß m. ‖ acier m. au creuset. / natural hard ~ ‖ naturharter Tiegelstahl m. ‖ acier m. au creuset de dureté naturelle.

crucible steel foundry ‖ Tiegelstahlwerk n. ‖ fonderie f. d'acier au creuset. / ~ furnace ‖ Tiegelofen m. ‖ four m. à creuset. / ~ manufacture ‖ Tiegelstahl-

erzeugung f. ‖ fabrication f. de l'acier au creuset. / ~ plant ‖ Tiegelstahlwerk n. ‖ aciérie f. aux creusets. / ~ process ‖ Tiegelstahldarstellung f. ‖ fabrication f. de l'acier au creuset. / ~ works pl. ‖ Tiegelstahlhütte f. ‖ aciérie f. (pour la fabrication) au creuset.

crucible tongs pl. ‖ Schmelztiegelzange f. ‖ pince f. *ou* tenailles fpl. à creusets. / ~ works pl. ‖ Tiegelfabrik f. ‖ atelier m. pour la fabrication des creusets.

crucifix ‖ Kruzifix n. ‖ crucifix m.

cruciform girder ‖ Kreuzträger m. ‖ poutre f. en croix.

crucite ‖ Chiastolit m.; Hohlspat m. ‖ chiastolithe f.

crude (Apparatus) ‖ roh; einfach; primitiv ‖ simple; primitif. / ~ (Materials) ‖ roh ‖ cru; brut. / ~ gas ‖ Rohgas n. ‖ gaz m. brut. / ~ helium ‖ Rohhelium n. ‖ hélium m. brut. / ~ iron ‖ Roheisen n. ‖ fer m. cru; fonte f. crue. / ~ lignite ‖ Rohbraunkohle n. ‖ lignite m. cru. / ~ metal ‖ Rohmetall n. ‖ métal m. cru. / ~ mining methods pl. ‖ rohe *oder* primitive Arten fpl. des Bergbaues ‖ méthodes fpl. primitives d'exploitation des mines. / ~ naphtha ‖ Rohnaphta n. ‖ naphte m. brut.

crude oil ‖ Rohöl n. ‖ huile f. lourde. / ~ engine ‖ Rohölmotor m. ‖ moteur m. à pétrole brut. / portable ~ engine ‖ Rohölmotorlokomobile f. ‖ locomobile f. à huile brute. / mixture of gas oil and ~ ‖ Mischung f. von Gasöl mit Rohöl ‖ mélange m. de gasoil avec de l'huile lourde. / ~ motor locomotive ‖ Rohölmotorlokomotive f. ‖ locomotive f. à moteur à huile lourde.

crude ore ‖ Roherz n. ‖ minerai m. brut. / ~ silk ‖ Rohseide f. ‖ soie f. écrue. / ~ steel ‖ Rennstahl m. ‖ acier m. naturel. / ~ steel ingot ‖ Rohstahlblock m. ‖ lingot m. d'acier brut.

cruise, to (Nav) ‖ kreuzen; lavieren ‖ croiser. / ~ along the coast ‖ an der Küste f. kreuzen ‖ croiser en vue de terre *ou* sur une côte. / ~ off ‖ auf der Höhe von . . . kreuzen ‖ croiser à la hauteur de.

cruiser (Shipb) ‖ Kreuzer m. ‖ croiseur m. / armoured ~ ‖ Panzerkreuzer m. ‖ croiseur m. cuirassé. / auxiliary ~ ‖ Hilfskreuzer m. ‖ croiseur m. auxiliaire. / battle ~ ‖ Linienkreuzer m. ‖ grand croiseur m. de bataille; cuirassé m. rapide. / belted ~ ‖ armierter Panzerkreuzer m. ‖ croiseur m. à ceinture cuirassée. / light ~ ‖ kleiner Kreuzer m. ‖ petit croiseur m. / protected ~ ‖ geschützter Kreuzer m. ‖ croiseur m. protégé. / scout ~ ‖ Aufklärungskreuzer m. ‖ éclaireur m. / small ~ ‖ kleiner Kreuzer m. ‖ petit croiseur m. / submarine ~ ‖ Unterseekreuzer m. ‖ croiseur m. submersible.

cruising altitude (Aero) ‖ Fahrhöhe f. ‖ hauteur f. de voyage. / ~ speed ‖ Reisegeschwindigkeit f. ‖ vitesse f. commerciale *ou* de croisière. / ~ turbine ‖ Vorwärtsturbine f. ‖ turbine f. de marche avant.

crumble, to (Chem) ‖ zerstäuben ‖ réduire en poussière f. / ~ (Glassm) ‖ abkröseln ‖ gréser; grésiller. / ~ to pieces ‖ zerbröckeln ‖ réduire en petits fragments mpl. ; émie(tte)r.

crumbling ‖ Abbröcklung f.; Abbröckeln n. ‖ écaillement m.; émiettement m.; désagrégation f. / ~-away ‖ Abbröckeln n. ‖ écaillement m. / ~ of brickwork ‖

Abbröckeln n. des Mauerwerkes ‖ désagrégation f. de maçonneries de briques. / ~ of a rock ‖ Abbröckeln n. eines Felsens ‖ émiettement m. *ou* effritement d'une roche.

crumbling iron (Glassm) ‖ Kröseleisen n.; Fügeeisen n. ‖ grugeoir m.; grésoir m.

crumped ‖ gekröpft ‖ cambré à gorge; coudé.

crumple, to ‖ zerknittern ‖ chiffonner.

crunching of sand ‖ Knirschen n. des Sandes ‖ crissement m. du sable.

crupper (Saddl) ‖ Schwanzriemen m. ‖ croupière f.; coulière f. / ~ loop (Saddl) ‖ Schwanzriemenöse f. ‖ chape f. de croupière.

crush, to ‖ zerquetschen; zerkleinern; brechen ‖ concasser; broyer; écraser. / ~ and wash (Coal dress) ‖ mahlen und waschen ‖ broyer et laver. / ~ ores pl. ‖ Erze npl. quetschen ‖ broyer les minerais mpl.

crushed ‖ zerquetscht ‖ broyé. / slightely ~ (Mine) ‖ grob gepreßt; rösch ‖ légèrement bocardé. / ~ oats pl. ‖ Haferflocken fpl. ‖ avoine f. broyée.

crusher ‖ Vorbrecher m.; Brecher m.; Zerkleinerungsmaschine f. ‖ concasseur m.; broyeur m. / ball type of ~ ‖ Kugelmühle f. ‖ broyeur m. à boulets. / bone ~ ‖ Knochenbrecher m. ‖ concasseur m. pour os. / centrifugal ~ ‖ Kreiselbrecher m. ‖ broyeur m. centrifuge. / claw of a ~ ‖ Klaue f. eines Brechers ‖ griffe f. de concasseur. / clod ~ ‖ Ackerschleife f. ‖ traînoir m. / coal ~ ‖ Kohlenbrecher m. ‖ concasseur m. à charbon. / coke ~ ‖ Koksbrecher m. ‖ concasseur m. à coke. / gypsum ~ ‖ Gipsbrecher m. ‖ concasseur m. à plâtre. / ~ for olives ‖ Olivenquetsche f. ‖ moulin m. à olives. / rotary ~ ‖ Rundbrecher m. ‖ concasseur m. à cône oscillant.

crusher roll ‖ Brechwalze f. ‖ cylindre m. broyeur. / ~ worm ‖ Brechschnecke f. ‖ vis f. concasseuse.

crushing ‖ Zerkleinerung f.; Brechen n. ‖ broyage m.; concassage m. / coarse ~ ‖ grobkörnige Zerkleinerung f. ‖ broyage m. grossier. / ~ of hard materials ‖ Hartzerkleinerung f. ‖ concassage m. de matières dures. / resistance to ~ ‖ Druckfestigkeit f. ‖ résistance f. à la compression. / scrap iron ~ ‖ Verschrotung f. ‖ broyage m. de ferrailles.

crushing ball ‖ Mahlkugel f. ‖ boulet m. de broyage. / ~ cylinder ‖ Quetschwalze f. ‖ cylindre m. broyeur. / ~ cylinder (Caoutchouc) manufacture ‖ Reinigungswalzen fpl. ‖ machine f. à déchiqueter. / ~ drum ‖ Mahltrommel f. ‖ tambour m. de broyage. / ~ head ‖ Brechkopf m.; Brechkegel m. ‖ cône m. broyeur. / ~ jaw ‖ Brechbacke f. ‖ mâchoire f. broyeuse.

crushing machine ‖ Zerkleinerungsmaschine f.; Brechmaschine f. ‖ concasseur m.; broyeur m.; machine f. à broyer. / ice ~ ‖ Eiszerkleinerungsmaschine f. ‖ machine f. à désagréger la glace. / stone ~ ‖ Steinspaltmaschine f. ‖ machine f. à fendre lä pierre.

crushing mill ‖ Walzenbrecher m.; Quetschwerk n.; Kollergang m. ‖ broyeuse f.; concasseur m.; broyeur m. à meules. / ~ for cured malt ‖ Quetsche f. für Darrmalz ‖ concasseur à malt touraillé. / ore ~ ‖ Erzquetsche f. ‖ broyeur m. pour minerais.

crushing plant ‖ Zerkleinerungsanlage f. ‖ installation f. de broyage. / coke ~ and sizing plant ‖ Kokszerkleinerungs- und -sortieranlage f. ‖ installation f. de concassage et de triage du coke. / ~ for hard materials ‖ Hartzerkleinerungsanlage f. ‖ broyeur m. à matières dures. / ~ for hydraulic packing material (Mine) ‖ Versatzaufbereitung f. zur Zerkleinerung von Spülmaterial ‖ installation f. pour le broyage des remblais hydrauliques. / lime-stone ~ ‖ Kalksteinzerkleinerungsanlage f. ‖ installation f. de broyage de calcaire.

crushing ring ‖ Brechring m. ‖ anneau m. broyeur. / ~ surface ‖ Mahlfläche f. ‖ surface f. travaillante d'un broyeur. / ~ strength ‖ Druckfestigkeit f. ‖ résistance f. à l'écrasement. / ~ test ‖ Schlagprobe f. ‖ essai m. à l'écrasement; épreuve f. d'aplatissement.

crush rock ‖ Quetschling m. ‖ roche f. comprimée.

crust (Chem) ‖ Haut f. ‖ croûte f. / ~ (Found) ‖ Gußrinde f. ‖ croûte f. de la fonte. / ~ (Metal) ‖ Zunder m. ‖ battiture f. / ~ of bread ‖ Kruste f. oder Rinde f. des Brotes ‖ croûte f. de pain. / ~ of cement ‖ Zementkruste f. ‖ croûte f. de ciment. / crystal ~ ‖ Kristallhaut f. ‖ croûte f. cristalline. / ~ of the earth ‖ Erdrinde f. ‖ écorce f. ou croûte f. terrestre. / character of the earth's ~ ‖ Oberflächenbeschaffenheit f. des Bodens ‖ constitution f. de la surface du sol. / ~ of iron ‖ Hammerschlag m. ‖ battitures fpl. de fer. / ~ separated from the crumb (Bak) ‖ abgebackene Rinde f. ‖ écaille f. de pain. / ~ of weathered material ‖ Verwitterungskruste f. ‖ croûte f. altérée par l'action atmosphérique.

crustacean ‖ Schaltier n. ‖ crustacé m.

crutch ‖ Krücke f. ‖ béquille f. / ~ (Boat) ‖ Dolle f. ‖ tolat m. à fourches; dame f. / ~ (Chem) ‖ Krückwerk n. ‖ agitateur m. / ~ (Shipb) ‖ Piekstück n. ‖ fourcat m.; fourque f. / ~ of the rack of a jack ‖ Gabel f. der Zahnstange einer Winde ‖ corne f. de la crémaillère d'un cric.

crutcher (Soapm) ‖ Forke f. ‖ brassoir m. en fourche.

cry of tin ‖ Zinnschrei m. ‖ cri m. de l'étain.

cryolite ‖ Kryolith m. ‖ cryolithe f. / ~ glass ‖ Milchglas n. ‖ verre m. en cryolithe.

crystal ‖ Kristall n.; Bleiglas n. ‖ cristal m.; verre m. plombifère. / ~ (Miner) ‖ Kristall m. ‖ cristal m. / ~ (Metal) ‖ Korn n. ‖ grain m.; cristal m. / biaxial ~ ‖ zweiachsiger Kristall m. ‖ cristal m. biaxe. / device for the examination of defective surfaces of ~s ‖ Einrichtung f. für die Untersuchung mangelhafter Kristallflächen ‖ dispositif m. pour l'examen de surfaces cristallines défectueuses. / homohedral ~s pl. ‖ homoedrische Kristalle mpl. ‖ solides mpl. homoèdres. / hopper-shaped ~s pl. ‖ Kastendrusen fpl. ‖ cristaux mpl. en forme d'escalier. / image of the ~ ‖ Kristallbild n. ‖ image f. du cristal. / leaden chamber's ~ ‖ Bleikammerkristall m. ‖ cristal m. de chambre de plomb. / mixed ~ ‖ Mischkristall m. ‖ cristal m. mixte. / pseudomorphous ~ ‖ Pseudomorphose f. ‖ pseudomorphose f. / uniaxial ~ ‖ einachsiger Kristall m. ‖ cristal m. à axe unique.

crystal axis ‖ Kristallachse f. ‖ axe m. de cristal. / ~ control (Radio) ‖ Quarzsteuerung f. ‖ contrôle m. d'onde à quartz. / ~ cutter ‖ Glasschleifer m. ‖ tailleur m. de cristaux. / ~ cutting ‖ Kristallschleifen n. ‖ taille f. de cristaux. / ~ cymometer ‖ Quarzkristallwellenmesser m. ‖ ondemètre m. à quartz. / ~ detector (Radio) ‖ Kristalldetektor m. ‖ détecteur m. à cristal ou à galène. / balanced ~ detector (Radio) ‖ ausbalanzierter Kristallwellenanzeiger m. ‖ détecteur m. à cristal équilibré. / ~ engraving ‖ Kristallgravierung f. ‖ gravure f. sur cristaux.

crystal glass ‖ Kristallglas n.; Bleiglas n. ‖ anglésite f.; cristal m.; flint-glass m.; flint m. / ~ for mirrors ‖ Kristallspiegelglas n. ‖ cristal m. pour miroirs. / ~ works pl. ‖ Kristallglasfabrik f. ‖ cristallerie f.

crystal goniometer ‖ Kristallgoniometer n. ‖ goniomètre m. à cristaux. / ~ with one circle ‖ einkreisiges Kristallgoniometer n. ‖ goniomètre m. à cristaux muni d'un cercle. / ~ with two circles ‖ zweikreisiges Kristallgoniometer n. ‖ goniomètre m. à cristaux muni de deux cercles.

crystal house man ‖ Kristallsodaarbeiter m. ‖ ouvrier m. de cristaux de soude.

crystalline ‖ kristallhell; kristallinisch ‖ cristallin. / ~ fracture (Metal) ‖ körniger Bruch m. des Eisens ‖ texture f. de fer grenue ou cristalline; grainure f. / ~ ice ‖ Destillateis n. ‖ glace f. cristalline. / ~ iron ‖ Feinkorneisen n. ‖ fer m. aciéreux; fer m. dur; fer m. à texture grenue. / in the ~ state ‖ in Kristallform f. ‖ à l'état m. cristallin.

crystallizable ‖ kristallisierbar ‖ cristallisable.

crystallization ‖ Kristallisieren n.; Kristallbildung f. ‖ cristallisation f. / disturbed ~ ‖ gestörte Kristallisation f. ‖ cristallisation f. troublée. / ~ by electric currents ‖ Kristallisation f. mittels elektrischen Ströme ‖ cristallisation f. par courants électriques. / fractional ~ ‖ fraktionierte Kristallisation f. ‖ cristallisation f. fractionnée. / grainlike ~ ‖ körnige Kristallisation f. ‖ cristallisation f. à grains fins. / ~ of iron in a magnetic field ‖ Kristallisation f. des Eisens in einem magnetischen Felde ‖ cristallisation f. du fer dans un champ magnétique. / liquid of ~ ‖ Kristallflüssigkeit f. ‖ liquide m. de cristallisation. / ~ of monohydrates of sulphuric acid ‖ Auskristallisieren n. von Schwefelsäuremonohydraten ‖ cristallisation f. d'acide sulphurique monohydraté. / ~ of soda ‖ Auskristallisieren n. von Soda ‖ cristallisation f. de soude. / water of ~ ‖ Kristallwasser n. ‖ eau f. de cristallisation.

crystallize, to ‖ kristallisieren ‖ cristalliser. / ~ out ‖ auskristallisieren ‖ se dissocier en cristaux mpl.; cristalliser; se cristalliser.

crystallizer ‖ Kristallisationsgefäß n. ‖ cristallisoir m.

crystallizing ‖ Kristallbildung f.; Kristallisierung f. ‖ cristallisation f. / ~ apparatus ‖ Kristallisierapparat n. ‖ appareil m. de cristallisation. / ~ dish ‖ Kristallisationsschale f. ‖ cristallisoir m. / ~ pan ‖ Kristallisiergefäß n. ‖ chauderon m. de cristallisation.

crystallography ‖ Kristallografie f. ‖ cristallographie f.

crystal paper ‖ Kristallpapier n. ‖ papier m. cristal. / hemisphere ~ refractometer ‖ Halbkugelkristallrefraktometer n. ‖ réfractomètre m. demi-boule à cristaux. / ~s pl. ‖ Glasbehang m.; Leuchtergehänge ‖ applique f. de verre; pendeloque f. / ~ sand ‖ Kristallsand m. ‖ sable m. cristallin. / ~ set ‖ Detektorempfänger m. ‖ récepteur m. à galène. / ~ skeleton ‖ Kristallskelett n. ‖ squelette m. de cristal; armature f. cristalline. / ~ structure ‖ Kristallstruktur f. ‖ structure f. de cristal ou cristalline. / irregular ~ surfaces pl. ‖ unregelmäßige Kristallflächen fpl. ‖ surfaces fpl. cristallines irrégulières. / ~ system ‖ Kristallsystem n. ‖ système m. cristallin.

cubage see cubature.

cubanite ‖ Weißkupfererz n. ‖ cubane f.; cubanite f.

cubature ‖ Kubikberechnung f. ‖ cubature f.

Cuba yellow wood ‖ kubanisches Gelbholz n. ‖ bois m. jaune de Cuba.

cube, to ‖ kubieren ‖ cuber; élever au cube m.

cube (Geom) ‖ Kubus m.; Würfel m. ‖ cube m.; hexaèdre m. / ~ (Math) see number. / ~ having a length of side of x meters ‖ Würfel m. von x m Kantenlänge ‖ cube m. d'une longueur d'arête de x mètres. / ~ of a number ‖ dritte Potenz f.; Kubus m. einer Zahl ‖ cube m.; troisième puissance. / ~ standard ‖ Normalwürfel m. ‖ cube m. normal.

cube game ‖ Würfelspiel n. ‖ jeu m. de cubes. / ~ number ‖ Kubikzahl f.; Kubus m.; dritte Potenz f. ‖ cube m. d'un nombre; nombre m. cube; troisième puissance f. / ~ root ‖ Kubikwurzel f.; dritte Wurzel f. ‖ racine f. cubique. / ~ shape ‖ Würfelform f. ‖ forme f. ou moule m. cubique.

cubic (Geom) ‖ würfelförmig ‖ cubique. / ~ (Math) ‖ kubisch ‖ du troisième degré. / ~ capacity ‖ Kubikinhalt m. ‖ capacité f. cube. / ~ centimeter ‖ Kubikzentimeter n. ‖ centimètre m. cube. / ~ content ‖ Kubikinhalt m.; Fassungsvermögen n. ‖ cubage m.; volume m.; capacité f. cube. / ~ decimeter ‖ Kubikdezimeter n. ‖ décimètre m. cube. / ~ foot ‖ Kubikfuß m. ‖ pied m. cube. / ~ inch ‖ Kubikzoll m. ‖ pouce m. cube. / ~ measure ‖ Kubikinhalt m.; Fassungsvermögen n. ‖ mesure f. de volume. / ~ measure of packing ‖ Rauminhalt m. der Verpackung ‖ volume m. de l'emballage. / ~ meter ‖ Kubikmeter n.; Raummeter n. ‖ mètre m. cube. / ~ number see cube number. / ~ system ‖ Kubiksystem n. ‖ système m. cubique ou régulier. / ~ yard ‖ Kubikyard n. ‖ yard m. cube.

cubicle (Electr) ‖ Schaltzelle f. ‖ habitacle m.

cubing ‖ Kubatur f.; Raummessung f.; Körpermessung f. ‖ cubature f.; cubage m. / ~ formula ‖ Kubierungsformel f. ‖ formule f. cubique.

cuckoo clock ‖ Kuckucksuhr f. ‖ pendule f. à coucou.

cucumber ‖ Gurke f. ‖ concombre m. / pickled ~ ‖ Salzgurke f. ‖ concombre m. confit au sel. / ~ pip ‖ Gurkenkern m. ‖ pépin m. de concombre.

cucurbit (Bot) ‖ Kürbis m. ‖ courge f. / ~ (Chem) ‖ Glaskolben m.; Destilliergefäß n. ‖ matras m.

cudbear ‖ Persio m. ‖ orseille f. violette. / ~ derivatives pl. ‖ Orseillepräparate npl. ‖ dérivés mpl. d'orseille.

cuff (Cloth) ‖ Manschette f.; Aufschlag m. ‖ manchette f.; revers m. / leather ~ ‖ Ledermanschette f. ‖ manchette f. en cuir. / paper ~ ‖ Papiermanschette f. ‖ manchette f. en papier.

cuff link see ~ stud. / ~ knitter ‖ Kunststricker m. ‖ tricoteur m. de parements. / ~ sewer (Tail) ‖ Rändernäher m. ‖ couseur m. ou surjeteur m. de parements. / ~ stud ‖ Manschettenknopf m. ‖ bouton m. de manchette.

cul-de-sac ‖ Sackgasse f. ‖ cul-de-sac m.; impasse f.

cull, to (Cloth) ‖ noppen; belesen ‖ épincer; épinceler; épinceter. / ~ (Wool) ‖ auszupfen ‖ égrateronner.

cullet (Glassm) ‖ Bruchglas n.; Glasbruch m. ‖ cassure f. de verre.

culm ‖ Kohlenlösche f.; Staubkohle f.; Grus m.; Steinkohlenklein n. ‖ fraisil m.; charbon m. poussiéreux; menu m. / ~ coke ‖ Feinkoks m.; Perlkoks m. ‖ grésillon m.

culminate, to ‖ den Höhepunkt m. erreichen; kulminieren; durch den Meridian m. gehen ‖ être au point m. culminant; culminer.

culminating edge ‖ Scheitelkante f. ‖ arête f. culminante. / ~ point ‖ Wurfhöhe f. ‖ sommet m.

culmination ‖ Kulmination f.; Gipfelung f. ‖ culmination f.; passage m. au méridien. / point of ~ ‖ Kulminationspunkt m.; Gipfelpunkt m. ‖ point m. de culmination; point m. culminant.

cultivate, to ‖ den Boden m. bearbeiten; anbauen ‖ cultiver. / ~ the land ‖ das Land n. bestellen ‖ cultiver la terre.

cultivated plant ‖ Kulturpflanze f. ‖ plante f. cultivée.

cultivating lands pl. and forests pl. ‖ Urbarmachung f. von Ländereien und Forsten ‖ cultivation f. des fonds de terre et des forêts. / seed ~ plant ‖ Saatveredelungsanlage f. ‖ installation f. d'amélioration des semences.

cultivation ‖ Pflanzung f.; Anbau m. ‖ plantation f.; culture f. / ~ of corn ‖ Getreidebau m. ‖ culture f. des grains. / ~ of flowers ‖ Blumenzucht f. ‖ floriculture f. / ~ of plants ‖ Pflanzenpflege f. ‖ cultivation f. des plantes. / ~ of the soil ‖ Bodenbearbeitung f. ‖ culture f. ou cultivation f. du sol. / ~ of wood ‖ Holzpflanzung f. ‖ culture f. de bois.

cultivator ‖ Kultivator m. ‖ cultivateur m. / ~ tooth ‖ Kultivatorzinke f. ‖ dent m. de cultivateur.

culture ‖ Kultur f.; Ackerbau m. ‖ agriculture f. / bacterial ~ ‖ Bakterienkultur f. ‖ culture f. de bactéries. / pure ~ ‖ Reinkultur f. ‖ culture f. pure.

culvert ‖ Düker m.; Entwässerungsstollen m.; Bachdurchlaß m.; Rigole f. ‖ petit aqueduc; ponceau m.; rigole f. transversale. / ~ (Build) ‖ Rinnstein m. ‖ culière f.; caniveau m.; rigole f. / arched ~ ‖ überwölbter Wasserabzug m. ‖ aqueduc m. voûté. / discharging ~ in a dyke ‖ Pumpsiel n. ‖ canal m. colateur. / open ~ ‖ offener Wasserlauf m. ‖ canal m. à ciel ouvert. / ~ covered by slabs ‖ Kanalisation f. mit Plattenüberdeckung. ‖ aquéduc m. couvert par dalles. / underground ~ ‖ unterirdischer Wasserlauf m. ‖ canal m. en galerie. / underground brick ~ ‖ Straßenkanalisation f. ‖ canalisation f. en briquetage. / wooden ~ ‖ Holzrinne f. ‖ caniveau m. en planches.

cuneiform characters pl. ‖ Keilschrift f. ‖ caractères mpl. cunéiformes.

cup ‖ Schale f. ‖ cuvette f. / ~ (Chem) ‖ Schälchen n.; Näpfchen n. ‖ coupelle f.; godet m. / ~ (Drinking vessel) ‖ Becher m. ‖ timbale f. / ~ (Househ) ‖ Tasse f. ‖ tasse f. / head ~ (Rivetting) ‖ Gegenhalter m. ‖ contre-bouterolle f. / hopper ~ ‖ Trichter m. ‖ trémie f.; entonnoir m. / leather ~ ‖ Ledersturp m. ‖ manchon m. en cuir embouti. / ~ for mocca ‖ Mokkatasse f. ‖ tasse f. à mocca. / replenishing ~ ‖ Einfülltopf m. ‖ pot m. de remplissage.

cup anemometer type of speed indicator ‖ Schalenkreuzfahrtmesser m. ‖ indicateur m. de vitesse à coquilles. / ~ barometer ‖ Gefäßbarometer n. ‖ baromètre m. à cuvette. / ~ and cone bearing ‖ Kegelkugellager n. ‖ roulement m. à billes par cône et cuvettes.

cupboard ‖ Schrank m. ‖ armoire f. / ~ (Househ) ‖ Speiseschrank m.; Tellerschrank m.; Silberschrank m. ‖ buffet m. / chocolate cooling ~ ‖ Schokoladenkühlschrank m. ‖ armoire f. frigorifique à chocolat. / cooling ~ ‖ Kühlschrank m. ‖ armoire f. frigorifique. / freezing ~ ‖ Gefrierschrank m. ‖ armoire f. de congélation. / freezing ~ for temperatures down to-x degrees centigrade ‖ Tiefgefrierschrank m. mit -x⁰ C ‖ armoire f. de congélation pour températures basses jusqu'à -x centigrades. / fume ~ (Chem) ‖ Abzugschrank m. ‖ hotte f. fermée. / ~ for spices ‖ Gewürzschrank m. ‖ armoire f. pour épices. / ~ for typewriting machines ‖ Schreibmaschinenschrank m. ‖ armoire f. pour machines à écrire.

cupboard fitting ‖ Schrankbeschlag m. ‖ garniture f. d'armoire.

cupel, to ‖ kupellieren ‖ coupeller.

cupel ‖ Probtiegel m. ‖ coupelle f. / assaying by the ~ ‖ Kupellieren n.; Kupellation f. ‖ coupellation f. en petit.

cupel assay (Chem) ‖ Kapellenprobe f. ‖ essai m. de coupelle. / ~ dust ‖ Kapellenasche f.; Kläre f. ‖ cendre f. de coupelle; claire f. / ~ furnace ‖ Kapellenofen m. ‖ four m. à coupelles.

cupellation ‖ Kupellieren n. ‖ coupellation f. / ~ of silver ‖ Silberscheidung f. ‖ affinage m. ou coupellation f. de l'argent.

cupel pyrometer ‖ Legierungspyrometer n. ‖ pyromètre m. à la coupelle ou de Princep. / ~ tongs pl. ‖ Kapellenkluft f.; Kapellenzange f. ‖ pince f. ou tenailles fpl. à coupelle.

cup insulator ‖ Glockenisolator m. ‖ isolateur m. à simple cloche. / ~ leather ‖ Manschettenleder n.; getriebenes Leder n. ‖ cuir m. embouti.

cupola (Arch) ‖ Kuppelgewölbe n. ‖ coupole f.; dôme m.; voûte f. sphérique ou en dôme. / ~ (Metal) see ~ furnace. / casting from ~ ‖ Kuppelofenguß m. ‖ coulage m. du cubilot.

cupola dome ‖ Kuppelgewölbe n. ‖ voûte f. en coupole.

cupola furnace ‖ Schachtofen m.; Kupolofen m. ‖ cubilot m. / charging plant for ~s ‖ Begichtungsanlage f. für Kupolöfen ‖ installation f. de charge pour cubilots. / ~ for melting pig-iron of x tons daily output ‖ Kupolofen m. zum Schmelzen des Roheisens mit einer täglichen Leistung von x t ‖ cubilot m. pour la fusion de la fonte d'une capacité de x t par jour.

cupola keeper ‖ Kupolofenwärter m. ‖ fondeur m. au cubilot. / ~ tenter ‖ Kupolofenschmelzer m. ‖ fondeur m. au cubilot.

cupped plate end ‖ vertiefter Tellerboden m. ‖ fond m. à plateau bombé.

cupping glass ‖ Schröpfkopf m. ‖ ventouse f.

cuprammonium silk ‖ Kupferseide f. ‖ soie f. au cuivre.

cupric carbonate ‖ kohlensaures Kupferoxyd n. ‖ carbonate m. de cuivre; cuivre m. carbonaté. / ~ chloride ‖ Kupferchlorid n. ‖ chlorure m. cuivrique ou de cuivre. / ~ oxide ‖ Kupferoxyd n. ‖ protoxyde m. de cuivre. / ~ sulphosteatite ‖ Kaliumkupfersulfat n. ‖ sulfostéatite m. cuprique.

cup ring for gas holders ‖ Tassenring m. für Gasbehälter ‖ anneau m. à bord recourbé en U pour récipients à gaz.

cuprite ‖ Kupferglas n.; Rotkupfererz n. ‖ cuprite f.; cuivre m. vitreux rouge.

cuproid ‖ Pyramidentetraeder n.; Trigondodekaeder n. ‖ tétraèdre m. pyramidé; hémiicositétraèdre m.

cuprous chloride ‖ Kupferchlorür n. ‖ chlorure m. cuivreux ou de cuivre. / ~ oxide ‖ Kupferoxydul n. ‖ cuivre m. oxydulé. / ~ pickling bath ‖ Kuprodekupierbad n. ‖ bain m. de décuivrage.

cup-shaped die ‖ Schellhammer m.; Aufsatzhammer m. ‖ bouterolle f.; chasserivet m.

cup spring ‖ Tellerfeder f.; Bellevillefeder f. ‖ ressort m. ou rondelle f. Belleville. / ~ valve ‖ Tellerventil n. ‖ soupape f. en chapeau. / ~ wheel ‖ Topfscheibe f. ‖ meule-boisseau m.

curaçao peel oil ‖ Curaçaoschalenöl n. ‖ essence f. d'écorces de curaçao.

curb (Build) ‖ Einfassung f.; clôture f.; enceinte f.; falère m. / ~ (Mine) ‖ Schachtausbaukranz m. ‖ rouet m.; couronne f.; cadre m. porteur. / ~ (Saddl) see ~ chain. / ~ (Tel) ‖ Kurb n. ‖ curb m. / ~ of a well ‖ Brunneneinfassung f.; Brunnenbrüstung f. ‖ margelle f. de puits; puisard m.

curb beam of a timber-bridge ‖ Saumholz m. einer Holzbrücke; Brückenschwelle f. ‖ garde-pavé m. d'un pont en bois. / ~ chain (Saddl) ‖ Kinnkette f. ‖ gourmette f. / link of the ~ chain ‖ Kinnkettenglied n. ‖ maille f. de la gourmette. / ~ hook (Saddl) ‖ Kinnhaken m. ‖ touret m. de chaînette. / near-side ~ hook ‖ Kinnkettenhaken m. ‖ crochet m. de gourmette. / ~ plate ‖ Mauerlatte f. oder Spannring m. eines runden Daches ‖ taflement m.; semelle f. à courbe. / ~ rafter (Build) ‖ Obersparren m. ‖ chevron m. du faux comble. / ~ stone ‖ Bordschwelle f.; Bürgersteigbordstein m.; Abweisstein m. ‖ parement m.; bordure f. de trottoirs; garde-roue m.

curcuma ‖ Gelbwurzel f.; Kurcume f. ‖ safran m. ou souchet m. des Indes; curcuma m. / ~ paper ‖ gelbes Reagenspapier n.; Kurkumapapier n. ‖ papier m. de curcuma. / ~ root ‖ Kurkumawurzel

f.; gelber Ingwer m. ‖ racine f. de curcuma.

curdle, to ‖ gerinnen lassen ‖ cailler; coaguler. / ~ ‖ gerinnen; käsig ausscheiden ‖ se cailler; se coaguler; se figer; se prendre.

curds pl. ‖ Quark m. ‖ caillebotte f. / ~ (Chem) ‖ käsiger Niederschlag m.; Gerinnsel n. ‖ caillé m.; coagulum m.

curdy ‖ käsig ‖ caséeux; caillebotté.

cure, to ‖ Fehler mpl. beseitigen; korrigieren ‖ corriger. / ~ fish ‖ Fische mpl. räuchern ‖ fumer les poissons mpl. / ~ hay ‖ Heu n. trocknen ‖ sécher le foin. / ~ malt ‖ das Malz darren ‖ dessécher le grain germé. / ~ meat ‖ Fleisch n. einpökeln ‖ saler la viande.

cured malt crushing mill ‖ Quetsche f. für Darrmalz ‖ concasseur m. à malt touraillé.

curette ‖ Kürette f. ‖ curette f.

curing of skins ‖ Häutekonservierung f. ‖ conservation f. des peaux brutes. / ~ house (Sugar) ‖ Tropfhaus n.; Zuckerraffinerie f. ‖ purgerie f. / ~ room ‖ Räucherei f. ‖ chambre f. à fumer.

curiosity ‖ Sehenswürdigkeit f. ‖ curiosités fpl.

curious manuscript ‖ seltenes Manuskript n. ‖ manuscrit m. curieux.

curl ‖ Wirbel m. ‖ tournoiement m.; tourbillon m.; tourbillonnement m. / ~ (Vector) ‖ Rotation f. ‖ rotation f. / ~ of smoke ‖ Rauchwirbel m. ‖ tourbillon m. de fumée. / ~ in the speckled wood ‖ Flader m. im Maserholz ‖ ronce f. dans la madrure. / ~ cloud ‖ Haarwolke f. ‖ nuage m. bouclé.

curled border ‖ gefräster Rand m. ‖ fraisure f. / ~ wood ‖ Maser f.; Maserholz n.; Flader m. ‖ madrure f.; bois m. madré; bois tapiré.

curling, apparatus for ~ feathers and feather boas ‖ Apparat m. zum Kräuseln von Federn und Federboas ‖ appareil m. pour friser les plumes et boas en plumes.

curling iron ‖ Frisiereisen n.; Brenneisen n. ‖ fer m. à friser. / ~ heater ‖ Brennscherenwärmer m. ‖ chauffe-fer à friser m.

curling machine for animal hairs ‖ Tierhaarkräuselmaschine f. ‖ ratineuse f. pour poils d'animaux. / ~ for vegetable fibres ‖ Pflanzenfaserkräuselmaschine f. ‖ ratineuse f. pour filaments végétaux. / ~ for vegetable fibres and animal hairs ‖ Kräuselmaschine f. für Pflanzenfasern und Tierhaare ‖ ratineuse f. pour filaments végétaux et poils d'animaux.

curling pin ‖ Lockennadel f.; Frisiernadel f. ‖ épingle f. à friser. / ~ stuff see curled wood. / ~ tongs pl. ‖ Frisierschere f.; Brennschere f. ‖ fer m. à friser.

currant ‖ Johannisbeere f. ‖ groseille f. / ~ growing ‖ Johannisbeerkultur f. ‖ culture f. de groseilles. / ~ wine ‖ Johannisbeerwein m. ‖ vin m. de groseilles.

currency see also current ‖ Strom m.; Strömung f. ‖ courant m. / ~ (Circulation) ‖ Kursieren n.; Umlauf m.; Zirkulation f. ‖ circulation f. / ~ (Standard of coinage) ‖ Währung f. ‖ valeur f. monétaire. / ~ (Money) ‖ umlaufendes Geld n.; Umlaufsmittel n.; Kurant n. ‖ argent m. en circulation. / ~ (Trade) ‖ Wert m.; Kurs m. ‖ cours m. / ~ conversion table ‖ Währungsumrechnungstabelle f. ‖ table f. de conversion en monnaies

étrangères. / ~ paper ‖ Papiergeld n.; Banknoten fpl. ‖ papier-monnaie m.

current ‖ laufend; fließend ‖ courant. / ~ (Coin) ‖ umlaufend; kursierend ‖ en cours m. / ~ (Exchange) ‖ kursfähig ‖ coté. / ~ (Math) ‖ fortlaufend ‖ continu; formant une série. / to be ~ ‖ kursieren ‖ circuler; courir.

current (Electr) ‖ Strom m. ‖ courant m. / ~ (River) ‖ Strömung f. ‖ courant m. / absence of ~ (Electr) ‖ Stromlosigkeit f. ‖ absence f. de courant. / ~ of air ‖ Zug m.; Zugluft f.; Luftströmung f. ‖ courant m. d'air. / alternating ~ ‖ Wechselstrom m. ‖ courant m. alternatif. / alternating ~ ammeter ‖ Wechselstrommesser m. ‖ ampèremètre m. pour courant alternatif. / blowing ~ (Electr) ‖ Abschmelzstromstärke f. ‖ intensité f. (de courant) de fusion. / to branch a ~ (Electr) ‖ einen Strom m. abzweigen ‖ dériver un courant. / branched ~s pl. (Electr) ‖ verzweigte Ströme mpl. ‖ courants mpl. bifurqués. / ~ induced at break ‖ Unterbrechungsstrom m.; Öffnungsstrom m. ‖ courant m. de coupure; extra-courant m. de rupture. / to bring about a close union between the generation and the distribution of ~ ‖ eine starke Verflechtung f. zwischen Stromerzeugung und Stromverteilung herstellen ‖ établir un rapprochement entre la production du courant et de sa distribution. / charging ~ (Acc) ‖ Ladestrom m. ‖ courant m. de chargement. / ~ along the coast ‖ Küstenströmung f. ‖ courant m. côtier. / continuous ~ ‖ Gleichstrom m.; Ruhestrom m. ‖ courant m. continu. / continuous ~ machine (Electr) ‖ Gleichstromgenerator m. ‖ machine f. à courant continu. / crossed ~s pl. ‖ gekreuzte Ströme mpl. ‖ courants mpl. croisés. / decrease of ~ ‖ Stromabnahme f. ‖ diminution f. du courant. / to deliver the ~ economically and reliably to the supply area ‖ Strom m. wirtschaftlich und betriebssicher den Verbrauchsgebieten zuführen ‖ conduire le courant m. jusqu'aux centres de consommation avec la plus grande économie et sûreté possibles. / to deliver ~ to... ‖ mit Strom m. beliefern ‖ alimenter de courant m. / descending ~ ‖ Abwind m. ‖ vent m. descendant; courant m. d'air descendant. / direct ~ (Electr) ‖ Gleichstrom m. ‖ courant m. continu. / direct ~ ammeter ‖ Gleichstrommesser m. ‖ ampèremètre m. pour courant continu. / direction of the ~ ‖ Stromrichtung f. ‖ sens m. du courant. / downward ~ ‖ Abwind m. ‖ vent m. descendant; courant m. d'air descendant. / ~ to earth ‖ Erdschlußstrom m. ‖ courant m. de perte à la terre. / electric ~ ‖ elektrischer Strom m. ‖ courant m. électrique. / excess ~ (Electr) ‖ Überstrom m.; surintensité f. / external ~ (Tel) ‖ Außenstrom m. ‖ courant m. parasite. / feed ~ ‖ Anodenruhestrom m. ‖ courant m. permanent de plaque. / filament ~ (Radio) ‖ Heizstrom m. ‖ courant m. de chauffage. / device for flattening rectifier ~s ‖ Glättungseinrichtung f. ‖ dispositif m. pour lisser les courants redressés. / forming ~ (Acc) ‖ Formationsstrom m. ‖ courant m. de formation. / eddy ~ ‖ Wirbelstrom m. ‖ courant m. de Foucault. / ~ which is generated close to the source of power ‖ unmittelbar auf der Energiegrundlage

erzeugter Strom m. ‖ courant m. obtenu sur les lieux-mêmes des sources d'énergie. / generation of ~ (Electr) ‖ Stromerzeugung f. ‖ production f. de courant. / heavy ~ ‖ Starkstrom m. ‖ courant m. de haute intensité. / high-frequency ~ ‖ Hochfrequenzstrom m. ‖ courant m. à haute fréquence. / high-tension ~ ‖ hochgespannter Strom m. ‖ courant m. à haute tension. / high-voltage ~ see high-tension ~. / holding ~ ‖ Haltestrom m. ‖ courant m. de collage. / idle ~ see wattless ~. / incoming audible ~ (Tel) ‖ ankommender Hörstrom m. ‖ courant m. de transmission arrivant. / induced ~ ‖ Sekundärstrom m. ‖ courant m. induit. / inducing ~ ‖ Primärstrom m. ‖ courant m. inducteur. / inductive ~ ‖ Induktionsstrom m. ‖ courant m. induit. / intensity of ~ ‖ Stromstärke f. ‖ intensité f. de courant. / interlinked ~ ‖ verketteter Strom m. ‖ courant m. accouplé. / inverse ~ ‖ Schließungsstrom m. ‖ courant m. de fermeture. / kind of ~ ‖ Stromgattung f. ‖ nature f. du courant. / leakage ~ ‖ Sickerstrom m. ‖ courant m. d'infiltration. / low-frequency ~ ‖ Niederfrequenzstrom m. ‖ courant m. à basse fréquence. / low-tension ~ ‖ niedrig gespannter Strom m. ‖ courant m. à basse tension. / low-voltage ~ see low-tension ~. / luminous ~ ‖ Glimmstrom m. ‖ courant m. de luminescence; effluves mpl. / ~ induced at make ‖ Schließungsstrom m. ‖ courant m. de fermeture. / minute ~ ‖ Schwachstrom m. ‖ courant m. faible. / multiphase ~ ‖ Mehrphasenstrom m. ‖ courant m. polyphasé. / oceanic ~ ‖ Meeresströmung f. ‖ courant m. marin. / path of the ~ ‖ Stromweg m. ‖ parcours m. du courant. / permanent ~ ‖ Dauerstrom m. ‖ courant m. permanent. / ~ in phase ‖ Phasenstrom m. ‖ courant m. en phase. / ~ of polarization ‖ Polarisationsstrom m. ‖ courant m. de polarisation. / pulsating ~ see pulsatory ~. / pulsatory ~ ‖ pulsierender Strom m. ‖ courant m. pulsatoire. / ~ for the pump ‖ Pumpstrom m. ‖ courant m. de pompage. / quantity of ~ ‖ Strommenge f. ‖ quantité f. du courant. / rectified ~ ‖ gleichgerichteter Strom m. ‖ courant m. redressé. / redressed ~ see rectified ~. / retaining ~ ‖ Haltestrom m. ‖ courant m. de collage. / reverse ~ (Acc) ‖ Gegenstrom m. ‖ courant m. contraire. / rush of ~ ‖ Stromstoß m. ‖ coup m. de courant. / single phase ~ ‖ Einphasenstrom m. ‖ courant m. monophasé. / space discharge ~ ‖ Raumentladestrom m. ‖ courant m. de décharge de l'espace. / splitting-up of ~ ‖ Stromteilung f. ‖ division f. de courant. / strong ~ ‖ Starkstrom m. ‖ courant m. de haute intensité. / supply of ~ ‖ Stromlieferung f. ‖ fourniture f. de courant. / to supply demand for ~ ‖ mit Strom m. versorgen ‖ suppléer aux demandes fpl. de courant. / stray ~ see vagrant ~. / taking of ~ ‖ Stromabnahme f. ‖ captation f. du courant. / thermoelectric ~ ‖ Thermostrom m. ‖ courant m. thermoélectrique. / three-phase ~ ‖ Drehstrom m.; Dreiphasen(wechsel)strom m. ‖ courant m. (alternatif) triphasé. / voltaic ~ ‖ galvanischer Strom m. ‖ courant m. galvanique. / vagrant ~ ‖ vagabondierender oder herumirrender Strom m. ‖ courant m. vagabond. / wattless ~ ‖

wattloser Strom m. ‖ courant m. dé-watté. / weak ~ ‖ Schwachstrom m. ‖ courant m. faible; courant m. de basse intensité. / without ~ ‖ stromlos ‖ sans courant m. / working with any kind of ~ ‖ für jede Stromart f. geeignet ‖ tournant sur tout courant m.
current acceptivity ‖ Stromaufnahme f. ‖ absorption f. de courant. / ~ account ‖ Girokonto n. ‖ compte m. de virement. / ~ balance ‖ Stromwippe f. ‖ commutateur m. de courant. / ~ capacity ‖ Stromkapazität f. ‖ capacité f. de courant.
current-carrying ‖ stromführend ‖ parcouru par le courant. / ~ capacity of a cable in regard to heating ‖ Belastungsfähigkeit f. eines Kabels in bezug auf Erwärmung ‖ capacité f. d'un cable limitée par rapport à l'échauffement. / ~ lug (Acc) ‖ stromführende Fahne f. ‖ queue f. conductrice. / ~ parts pl. ‖ stromführende Teile mpl. ‖ pièces fpl. sous tension. / ~ spring ‖ Stromzuführungsfeder f. ‖ ressort m. d'amenée du courant.
current coil ‖ Stromspule f. ‖ bobine f. d'intensité. / ~ collecting rail ‖ Stromschiene f. ‖ rail m. de contact. / ~ consumption ‖ Stromverbrauch m. ‖ consommation f. de courant. / ~ converter ‖ Stromumformer m.; Stromwandler m. ‖ convertisseur m. de courant; réducteur m. de courant. / ~ curve ‖ Stromkurve f. ‖ courbe f. d'intensité ou de courant. / decrease of ~ ‖ Stromabnahme f. ‖ diminution f. du courant. / ~ density ‖ Stromdichte f. ‖ densité f. de courant. / admissible ~ density ‖ zulässige Stromdichte f. ‖ densité f. de courant admissible. / direction of ~ ‖ Stromrichtung f. ‖ sens m. du courant. / ~ direction indicator ‖ Stromrichtungsanzeiger m. ‖ indicateur m. de sens de courant. / ~ displacement ‖ Stromverschiebung f. ‖ déplacement m. de courant. / extension of the ~ distributing system ‖ Erweiterung f. der Stromverteilungsanlage ‖ augmentation f. de l'installation de distribution de courant électrique. / secondary ~ distributing system ‖ Unterverteilung f. von Strom ‖ sous-distribution f. de courant. / ~ flow indicator ‖ Stromrichtungsanzeiger m. ‖ indicateur m. de sens de courant. / ~ fluctuation (Electr) ‖ Stromschwankung f. ‖ variation f. de courant. / extension of the ~ generating system ‖ Erweiterung f. der Stromerzeugungsanlage ‖ agrandissement m. de l'installation de production de courant électrique. / generation of ~ ‖ Stromerzeugung f. ‖ production f. du courant. / plant for ~ generation ‖ Stromerzeugungsanlage f. ‖ installation f. génératrice de courant. / undertaking engaged in ~ generation ‖ Unternehmung f. für Stromerzeugung ‖ entreprise f. de production de courant électrique. / ~ impulse ‖ Stromstoß m. ‖ impulsion f. de courant. / increase of ~ ‖ Stromzunahme f. ‖ augmentation f. du courant. / ~ integral ‖ Stromintegral n. ‖ intégrale f. de courant. / ~ intensity ‖ (elektrische) Stromstärke f. ‖ intensité f. de courant. / ~ intensity at make ‖ Einschaltestromstärke f. ‖ intensité f. de courant lors de la fermeture du circuit. / ~ interrupter ‖ Stromunterbrecher m. ‖ interrupteur m. de courant.

currentless (Electr) ‖ stromlos ‖ sans courant m.
current limiter ‖ Strombegrenzer m. ‖ limiteur m. de courant ou d'intensité. / ~ load diagram ‖ Stromverbrauchskurve f. ‖ courbe f. de courant. / ~ loop (Radio) ‖ Stromschleife f. ‖ boucle f. de courant. / ~ meter (Electr) ‖ Stromzähler m. ‖ compteur m. de courant. / ~ meter (Hydr arch) ‖ Stromgeschwindigkeitsmesser m. ‖ moulinet m. / ~ node ‖ Stromknoten m. ‖ nœud m. d'intensité du courant. / ~ number ‖ laufende Nummer f. ‖ numéro m. d'ordre ou de série. / ~ path ‖ Leitungsweg m.; Stromweg m. ‖ parcours m. ou passage m. du courant. / ~ price see ~ rate. / quantity of ~ ‖ Strommenge f. ‖ quantité f. du courant. / ~ rate (Trade) ‖ Börsenkurs m. ‖ cours m. de la bourse. / ~ rate of exchange ‖ Tageskurs m. ‖ change m. du jour. / ~ rectifier ‖ Stromgleichrichter m. ‖ redresseur m. de courant. / ~ reverser (Electr) ‖ Stromwender m. ‖ inverseur m. de courant. / rush of ~ ‖ Stromstoß m. ‖ coup m. de courant. / source of ~ ‖ Stromquelle f. ‖ source f. de courant. / ringing ~ source ‖ Rufstromquelle f. ‖ source f. de courant d'appel. / space ~ source ‖ Raumstromquelle f. ‖ source f. de courant spatial. / splitting-up of ~ ‖ Stromteilung f. ‖ division f. du courant. / ~ strength ‖ Stromstärke f. ‖ intensité f. de courant.
current supply ‖ Stromlieferung f. ‖ fourniture f. de courant. / ~ to extension telephones ‖ Nebenstellenspeisung f. ‖ alimentation f. des postes supplémentaires. / to draw up a plan for the ~ of the whole area ‖ einen Stromversorgungsplan m. für das Gesamtgebiet aufstellen ‖ établir un plan m. de distribution de courant pour le rayon entier. / ~ from primary batteries ‖ Stromversorgung f. aus Primärelementen ‖ distribution f. de courant par piles primaires. / rational ~ organization ‖ rationelle Gestaltung f. des Stromabsatzes ‖ organisation f. rationelle de la vente du courant. / fusion of ~ systems ‖ Zusammenfassung f. der Stromversorgung ‖ concentration f. de réseaux.
current transformer ‖ Stromwandler m. ‖ transformateur m. d'intensité du courant. / bushing ~ ‖ Stützerstromwandler m. ‖ transformateur m. d'intensité type à isolateur pilier. / service voltage of a ~ ‖ Betriebsspannung f. eines Stromwandlers ‖ tension f. de service d'un transformateur d'intensité. / spreader ~ in an open air transformer station ‖ Stützerstromwandler m. in einer Freiluftanlage ‖ transformateur m. d'intensité à support installé dans un poste à l'air libre. / testing device for ~ ‖ Stromwandlerprüfeinrichtung f. ‖ installation f. de vérification de transformateurs d'intensité. / test voltage of a ~ ‖ Prüfspannung f. eines Stromwandlers ‖ tension f. d'essai d'un transformateur d'intensité.
current triangle ‖ Stromdreieck n. ‖ triangle m. de courant. / ~ tube ‖ Stromfaden m.; Stromröhre f. ‖ tube m. de courant. / ~ yield ‖ Stromausbeute f. ‖ rendement m. en courant.
currier ‖ Gerber m. ‖ chamoiseur m.; mégissier m.; tanneur m. / goat skin ~

‖ Ziegenledergerber m. ‖ corroyeur-chevrier m.
currier's knife ‖ Gerbermesser n. ‖ couteau m. de corroyeur. / ~ works pl. ‖ Lederzurichtungswerkstatt f. ‖ corroirie f.
curriery ‖ Lederfabrik f. ‖ corroirie f.
curry, to ‖ gerben ‖ corroyer; tanner.
curry comb ‖ Striegel m.; Mähnenkamm m. ‖ étrille f.
currying of leather ‖ Lederzurichtung f. ‖ corroyage m. de cuir.
cursor of a mathematical instrument ‖ Schieber m. eines mathematischen Instrumentes ‖ curseur m. d'un instrument de mathématiques.
curtail, to ~ expenses pl. ‖ die Ausgaben fpl. einschränken ‖ rétrécir ou diminuer ou réduire les dépenses.
curtailing (Expenses) ‖ Einschränkung f. ‖ rétrécissement m.; diminution f.
curtain ‖ Vorhang m.; Gardine f. ‖ rideau m. / fireproof ~ ‖ Asbestvorhang m. ‖ rideau m. de sûreté. / guipure ~ ‖ Gipürgardine f. ‖ rideau-guipure m. / iron ~ ‖ eiserner Vorhang m. ‖ courtine f. / machine-embroidered ~ ‖ gestickte Gardine f. ‖ rideau m. brodé. / ~ before the stage ‖ Theatervorhang m. ‖ toile f.; rideau m. / white ~ ‖ weiße Gardine f. ‖ rideau m. blanc.
curtain clasp ‖ Vorhanghalter m. ‖ embrasse f. à rideau. / ~ drawing device ‖ Gardinenzugvorrichtung f. ‖ dispositif m. pour le tirage des rideaux. / ~ embroiderer ‖ Vorhangsticker m. ‖ brodeur m. de rideaux. / ~ embroidery ‖ Gardinenstickerei f. ‖ broderie f. de rideaux. / ~ fabric ‖ Gardinenstoff m. ‖ étoffe f. à rideau. / ~ framing ‖ Gardinenwaschanstalt f. ‖ blanchissage m. de rideaux. / ~ hanger ‖ Gardinenaufmacher m. ‖ poseur m. de rideaux. / ~ hook ‖ Gardinenhaken m. ‖ crochet m. à rideaux. / ~ lace machine ‖ Gardinenwebstuhl m.; Gardinenmaschine f. ‖ métier m. à rideaux. / ~ lace machine see ~ lace machine. / ~ net ‖ Gardinentüll m. ‖ tulle m. pour stores. / ~ pole see ~ rod. / ~ ring ‖ Gardinenring m. ‖ anneau m. pour rideau. / ~ rod ‖ Gardinenstange f. ‖ tringle f. à rideaux. / ~ stretcher ‖ Gardinenspanner m. ‖ étire-rideaux m. / ~ stuff ‖ Vorhangstoff m. ‖ étoffe f. pour stores.
curvature ‖ Bogen m.; Krümmung f. ‖ courbe f.; courbure f. / ~ (Geol) ‖ Falte f.; Biegung f. ‖ pli m. / ~ (Geom) ‖ Bogenlinie f. ‖ arceau m.; courbure f. / ~ (Plane) ‖ Wölbung f. ‖ voussure f.; bombement m.; curvature f. / ~ (Shipb) ‖ Bucht f. ‖ courbure f.; bouge m.; tonture f. / double ~ ‖ doppelte Krümmung f. ‖ double courbure f. / radius of ~ ‖ Krümmungshalbmesser m. ‖ rayon m. de courbure. / radius of ~ of the concave mirror ‖ Krümmungshalbmesser m. des Hohlspiegels ‖ rayon m. de courbure du miroir concave. / ~ of rim of a wheel ‖ Krümmung f. einer Radfelge ‖ courbure f. de la jante d'une roue. / ~ of a road ‖ Krümmung f. eines Weges ‖ courbe f. d'un chemin. / ~ of track ‖ Gleiskrümmung f. ‖ courbe f. de voie.
curve, to (Join) ‖ schweifen ‖ échancrer; évider; cambrer; arquer. / ~ (Mine) ‖ schrämen ‖ entailler. / ~ (Roll) ‖ krümmen; biegen ‖ courber; cintrer; plier.
curve ‖ Krümmung f.; Kurve f. ‖ courbure f.; courbe f. / ~ (Math) ‖ Kurve f. ‖

courbe f. / ~ of adjustment (Railw) ‖ Übergangskurve f. ‖ courbe f. de raccordement. / ~ after one hour's run ‖ Kurve f. nach einstündigem Betrieb ‖ courbe f. au bout d'une marche d'une heure. / algebraical ~ ‖ algebraische Kurve f. ‖ courbe f. algébrique. / ~ of an arc ‖ Gewölbebogen m. ‖ cintre m. d'une voûte. / arrival ~ ‖ Empfangskurve f. ‖ courbe f. à l'arrivée. / caustic ~ ‖ kaustische Kurve f. ‖ courbe f. caustique. / to conform to a ~ ‖ sich in einer Kurve f. bewegen ‖ s'inscrire dans la courbe. / ~ of displacement (Shipb) ‖ Lastenmaßstab m. ‖ échelle f. de solidité. / elastic ~ ‖ elastische Linie f. ‖ courbe f. élastique. / expansion ~ ‖ Expansionskurve f. ‖ courbe f. de détente. / ~ of extinction of rolling (Shipb) ‖ Ausschwingungskurve f. eines Schiffes ‖ courbe f. de l'extinction des oscillations de roulis. / harmonic ~ ‖ harmonische Kurve f. ‖ courbe f. harmonique. / ~ showing iron losses (Electr) ‖ Eisenverlustkurve f. ‖ courbe f. des pertes dans le fer. / irregular ~ ‖ Kurvenlineal n. ‖ pistolet m. / junction ~ ‖ Verbindungskurve f. ‖ courbe f. de raccordement. / ~ of kilowatts ‖ Kilowattkurve f. ‖ courbe f. de puissance en kilowatts. / large-radius ~ (Railw) ‖ flache Kurve f. ‖ courbe f. de grand rayon. / left-hand ~ ‖ Linkskurve f. ‖ courbe f. à gauche. / ~ of line ‖ Gleiskrümmung f. ‖ courbe f. de voie. / logarithmic ~ ‖ logarithmische Linie f. ‖ logarithmique f. / magnetic ~ ‖ magnetische Kurve f. ‖ courbe f. magnétique. / ~ of magnetization ‖ Magnetisierungskurve f. ‖ courbe f. d'aimantation. / metacentric ~ ‖ metazentrische Kurve f. ‖ courbe f. métacentrique. / ~ of the second, third order ‖ Kurve f. zweiten, dritten Grades ‖ courbe f. du second, du troisième degré. / parabolic ~ ‖ Parabel f. höherer Ordnung ‖ parabole f. d'ordre supérieur. / peak of the ~ ‖ Kurvenscheitelpunkt m. ‖ apogée m. de la courbe. / plane ~ ‖ einfach gekrümmte Linie f. ‖ courbe f. plane. / ranging ~s (Railw; Road) ‖ Kurvenabsteckung f. ‖ tracé m. de courbes. / ~ of resistance (Shipb) ‖ Widerstandskurve f. ‖ courbe f. de résistance. / reversed ~ ‖ Gegenkrümmung f. ‖ contre-courbe f. / right-hand ~ ‖ Rechtskurve f. ‖ courbe f. à droite. / road ~ ‖ Wegkurve f. ‖ virage m. / setting out ~s see ranging ~s. / sharp ~ (Railw) ‖ scharfe Gleiskrümmung f. oder Kurve f. ‖ courbe f. prononcée ou accentuée ou raide. / to run through a sharp ~ ‖ eine scharfe Kurve f. durchfahren ‖ franchir une courbe à faible rayon. / short-radius ~ ‖ stark gekrümmte Kurve f. ‖ courbe f. de faible rayon. / slope of the ~ ‖ Steilheit f. der Kurve ‖ pente f. de la courbe. / ~ of stability (Shipb) ‖ Stabilitätskurve f. ‖ courbe f. de la stabilité. / statistical ~ ‖ statistische Kurve f. ‖ courbe f. statistique. / transcendental ~ ‖ transzendente Kurve f. ‖ courbe f. transcendante.

curve crossing (Railw) ‖ Kurvenkreuzung f. ‖ croisement m. en courbe. / ~ cutting machine ‖ Kurvenschere f. ‖ cisaille f. circulaire à découper les courbes.

curved ‖ gebogen; gekrümmt ‖ courbe; courbé. / ~ (Arch) ‖ gewölbt ‖ cintré. /

~ (Join) ‖ geschweift ‖ relevé. / slightly ~ ‖ leicht gekrümmt ‖ bombé.

curved chisel ‖ Halbmondmeißel m. ‖ gouge f. / ~ form ‖ geschweifte Form f. ‖ forme f. cambrée. / ~ forceps pl. ‖ gebogene Pinzette f. ‖ pince f. coudée. / ~ furrow ‖ gekrümmte Fläche f. ‖ rayon m. courbé. / ~ lancet ‖ gebogene Lanzette f. ‖ couteau m. lancéolaire coudé. / ~ piece ‖ Kurvenstück n. ‖ pièce f. courbée. / ~ profile of the cornea ‖ Hornhautkrümmung f. ‖ courbure f. de la cornée. / ~ sail (Wind mill) ‖ gewölbter Flügel m. ‖ palette f. cintrée. / ~ sheet ‖ gebogenes Blech n. ‖ tôle f. cintrée. / ~ surface ‖ krumme oder gekrümmte Fläche f. ‖ surface f. courbe ou bombée.

curve car (El line) ‖ Leitungsklemme f. für Kurven ‖ joint-manchon m. pour courbes. / entering the ~ ‖ Einfahren n. in die Kurve ‖ entrée f. en courbe. / ~ frame (Railw) ‖ Kurvenrahmen m. ‖ châssis m. de voie courbe. / ~ line ‖ Kurve f. ‖ courbe f.; ligne f. courbe. / ~ milling machine see curves milling machine. / peak of the ~ ‖ Kurvenscheitelpunkt m. ‖ apogée m. de la courbe. / ~ pen ‖ Kurvenziehfeder f. ‖ tire-curviligne m. / ~ rail (Railw) ‖ Kurvenschiene f. ‖ rail m. pour courbes.

curves pl. ‖ Kurvenlineal n. ‖ pistolet m.; règle f. courbe. / ~ milling machine ‖ Kurvenfräsmaschine f. ‖ machine f. à fraiser en courbes.

curve section (Railw) ‖ Kurvenjoch n. ‖ tronçon m. de voie courbe. / slope of the ~ ‖ Steilheit f. de Kurve ‖ pente f. de la courbe. / ~ table ‖ Kurventafel f. ‖ tableau m. des courbes. / ~ templet see curves.

curvilineal see curvilinear.

curvilinear ‖ krummlinig ‖ curviligne. / ~ saw ‖ Kronsäge f.; Ringsäge f. ‖ scie f. cylindrique.

curving ‖ Krümmung f. ‖ courbage m.; courbe f.; courbure f. / ~ (Join) ‖ Schweifung f. ‖ échancrure f.; cambrure f.; chantournement m. / ~ of rails ‖ Biegung f. der Schienen ‖ courbure f. des rails. / ~ saw ‖ Schweifsäge f.; Stellsäge f. ‖ scie f. à tourner ou à chantourner ou à échancrer.

curvometer ‖ Krümmungsmesser m. ‖ curvimètre m.

cushion, to (Steam engine) ‖ Stöße mpl. auffangen ‖ amortir les chocs. / ~ with steam buffering ‖ mit Dampf m. bremsen ‖ donner contre-vapeur f.

cushion ‖ Kissen n.; Polster n. ‖ coussin m. / ~ (Arch) ‖ Kämpferschicht f.; Anfall m. eines Gewölbes; Polsterkapitell n. ‖ imposte f.; coussinet m. / ~ (Electr) ‖ Reibkissen n. (einer Elektrisiermaschine) ‖ coussin m.; frottoir m. / air ~ ‖ Luftkissen n. ‖ coussin m. à air. / rubber ~ ‖ Gummikissen n. ‖ coussin m. en caoutchouc. / ~ of steam in the cylinder ‖ Dampfpolster n. im Zylinder ‖ matelas m. de vapeur dans le cylindre.

cushioning ‖ Prellvorrichtung f. ‖ dispositif m. amortisseur. / ~ spring ‖ Pufferfeder f. ‖ ressort m. de choc.

cushion seat ‖ Polstersitz m. ‖ siège m. rembourré. / ~ tyre ‖ Vollgummireifen m. ‖ bandage m. plein.

cusp of a curve ‖ Scheitelpunkt m. einer Kurve ‖ point m. de rebroussement d'une courbe.

custom ‖ Gewohnheit f.; Sitte f.; Brauch m. ‖ coutume f.; usage m.; habitude f. / ~ (Trade) ‖ Kundschaft f. ‖ pratique f.; clientèle f. / ~ of craftsmen ‖ Handwerksbrauch m. ‖ usage m. des gens du métier; usage du métier.

customary ‖ üblich ‖ usuel; habituel. / ~ in a country ‖ landesüblich ‖ à l'usage du pays. / ~ in a place ‖ ortsüblich ‖ conforme à l'usage local.

customer ‖ Mandant m. ‖ commettant m. / ~ (Trade) ‖ Kunde m.; Kundschaft f. ‖ client m.; chalandise f. / to introduce ~s pl. ‖ Kunden mpl. zuweisen ‖ procurer des clients mpl. / counter of ~s (Cash register) ‖ Kundenzähler m. ‖ compteur m. de clients. / ~'s receipt printer ‖ Quittungsdruckvorrichtung f. ‖ dispositif m. imprimant les reçus des clients.

custom house ‖ Zollamt n. ‖ douane f. / ~ duty ‖ Zollabgabe f. ‖ droit m. de douane. / ~ officer ‖ Zollbeamter m. ‖ employé m. des douanes; douanier m.

custom law ‖ Zollgesetz n. ‖ loi f. de douane.

customs pl. ‖ Zoll m.; Steuer f. ‖ douane f. / to clear from ~ ‖ verzollen ‖ dédouaner. / administration of ~ ‖ Zollverwaltung f. ‖ administration f. douanière. / ~ airport ‖ Zollflughafen m. ‖ aérodrome m. douanier. / ~ duty ‖ Einfuhrzoll m. ‖ droit m. d'importation ou d'entrée. / ~ entry ‖ zollamtliche Abfertigung f. ‖ expédition f. en douane. / ~ examination ‖ Zolluntersuchung f. ‖ visite f. douanière. / ~ frontier ‖ Zollgrenze f. ‖ frontière f. douanière. / ~ launch ‖ Zollkreuzer m. ‖ croiseur m. de douane. / ~ nomenclature ‖ Zollnomenklatur f. ‖ nomenclature f. douanière. / ~ seal ‖ Zollverschluß m. ‖ scellés mpl. douaniers. / ~ tariff ‖ Zolltarif m. ‖ tarif m. douanier. / ~ tariff rate ‖ Zollsatz m. ‖ classe f. du tarif douanier. / ~ warehouse ‖ Zollspeicher m. ‖ entrepôt m. de douane. / ~ wharf ‖ Zollmole f. ‖ môle m. de la douane. / ~ yard ‖ Zollhof m. ‖ entrepôt m. de douane.

cut ‖ geschnitten ‖ coupé. / ~ across the grain ‖ quergeschnitten ‖ tranché à travers le fibre. / ~ from the solid steel ‖ aus dem vollen Stahl m. geschnitten ‖ taillé de l'acier m. massif. / ~ glass ‖ facettiertes Glas n. ‖ verre m. taillé. / ~ glass cup ‖ Schale f. aus facettiertem Glas ‖ coupe f. en verre taillé. / ~ goods pl. (Textile) ‖ geschnittene Ware f. ‖ marchandise f. coupée. / ~ hemp ‖ Schnitthanf m. ‖ chanvre m. coupé. / ~ nail ‖ Polsternagel m. ‖ clou m. découpé. / ~ plush ‖ Schneidplüsch m. ‖ peluche f. coupée. / ~ stone ‖ Haustein m. ‖ pierre f. taillée ou de taille. / ~ tooth ‖ geschnittener oder gefräster Zahn m. ‖ dent f. taillée à la fraise. / ~ velvet ‖ geschorener Samt m. ‖ velours m. coupé.

cut, to ‖ schneiden ‖ tailler; cisailler; couper; découper; trancher. / ~ (Agr) ‖ mähen ‖ faucher. / ~ (Sheep) ‖ scheren ‖ tondre. / ~ across ‖ quer zur Faser f. schneiden ‖ couper à travers le fibre. / ~ across (Mine) ‖ das Gebirge durchörtern ‖ percer le terrain. / ~ across the grain (Join) ‖ querschneiden ‖ couper ou trancher à travers le fibre. / ~ again see ~ anew. / ~ anew (File) ‖ aufhauen ‖ retailler. / ~-away a mast ‖ einen Mast m. kappen ‖ couper un mât. / ~-away metal ‖ hinterdrehen ‖ détalonner;

dégager la surface en arrière. / the bearing cuts ‖ das Lager frißt ‖ le palier grippe. / ~ bricks pl. ‖ Ziegel mpl. zuhauen ‖ couper les briques fpl. à la règle. / ~ the cable ‖ das Ankertau kappen ‖ couper *ou* tailler le câble. / ~ chips pl. ‖ Späne mpl. schneiden ‖ faire des copeaux mpl. / ~ the clay (Pott) ‖ den Ton m. schneiden ‖ couper l'argile f. / ~ cloth ‖ das Tuch scheren ‖ tondre le drap. / ~ coal ‖ Kohlen fpl. hauen ‖ haver *ou* couper le charbon. / ~ a connection (Tel) ‖ eine Verbindung f. trennen ‖ couper une communication; déconnecter. / ~ current off the motor ‖ einen Motor m. abschalten ‖ couper le courant du moteur. / ~-down a forest ‖ abholzen ‖ ~-down a ship ‖ ein Schiff n. abwracken ‖ démolir un navire. / ~-down timber ‖ Holz n. fällen ‖ couper le bois. / ~-down trees ‖ Bäume mpl. fällen ‖ abattre *ou* couper les arbres mpl. / ~-down the voltage (Electr) ‖ die Spannung erniedrigen ‖ réduire la tension. / ~ each other (Geom) ‖ einander *oder* sich schneiden ‖ s'entre-couper. / ~ an edge into a bevelling ‖ abkanten ‖ délarder; écorner; chanfreiner. / ~ files ‖ Feilen fpl. hauen ‖ tailler les limes fpl. / ~ flax ‖ den Flachs m. schneiden ‖ couper la lin. / ~ glass ‖ Glas n. schneiden ‖ tailler le verre. / ~ grooves pl. ‖ Nuten fpl. stoßen ‖ rainurer. / ~ the grooves in a roll ‖ eine Walze f. kalibrieren ‖ tourner les cannelures d'un cylindre; canneler *ou* calibrer un cylindre. / ~-in automatically ‖ selbsttätig einschalten ‖ s'intercaler automatiquement. / ~ into a brilliant ‖ den Brillanten m. schleifen ‖ tailler le diamant. / ~ into lengths ‖ ablängen ‖ couper de longueur f. / ~ into pieces ‖ stückeln ‖ morceler. / ~ the kiln ‖ den Gasofen m. durchstoßen ‖ nettoyer le fourneau. / ~ lengthwise ‖ der Länge f. nach schneiden ‖ couper de longueur f. / ~ the mast ‖ den Mast m. kappen ‖ couper *ou* tailler le mât. / ~ and notch the cross-sleepers (Carp) ‖ die Schwellen fpl. zuschneiden und auskämmen ‖ tailler et entailler les traversines fpl. / ~ and notch the joists (Carp) ‖ die Balken mpl. zuschneiden und verkämmen ‖ débiter et entailler les solives fpl. / ~-off (Axe) ‖ abhacken ‖ couper; abattre à coups mpl. de hache. / ~-off (Parting tool) ‖ abstechen ‖ tronçonner. / ~-off (Scissors) ‖ abschneiden ‖ couper. / ~-off the branches of timber ‖ das Stammholz abästen ‖ découper les branches fpl. du bois. / ~-off the grain ‖ das Fell abnarben ‖ effleurer la peau. / ~-off the steam ‖ den Dampf m. absperren ‖ couper la vapeur. / ~-off the top of a pile ‖ einen Pfahl m. abschneiden ‖ couper les têtes d'un pieu. / ~-off velvet ‖ den Sammet m. glattschneiden ‖ raser le velours m. / ~-out ‖ ausschneiden ‖ découper. / ~-out resistance ‖ den Widerstand m. ausschalten ‖ mettre la résistance hors circuit. / ~-out the rivets ‖ Nieten npl. losschlagen; entnieten ‖ enlever les rivets mpl.; dériver. / ~-out with the saw ‖ mit der Säge f. ausschneiden ‖ dégauchir à la scie. / ~-out sheet metal ‖ Blech n. stanzen ‖ estamper la tôle. / ~-out a strake of plank etc. (Shipb) ‖ einen Plankengang m. heraushauen ‖ déli-

vrer une virure de bordé. / ~-out test specimens pl. ‖ Probestäbe mpl. entnehmen ‖ prélever des éprouvettes fpl. / ~ the piston rod is cut ‖ die Kolbenstange ist riefig ‖ la tige de piston est rayée. / ~ the rags pl. (Pap) ‖ die Lumpen mpl. schneiden ‖ déromper les chiffons mpl. / ~ the roots pl. ‖ die Wurzeln fpl. abhauen ‖ couper les racines fpl. / ~ screws pl. ‖ Gewinde n. schneiden ‖ fileter; tarauder. / ~ screws with the chaser ‖ Schrauben fpl. mit dem Drehstahl schneiden ‖ fileter au tour. / ~ screws with a die ‖ Schrauben fpl. mit Gewindeeisen schneiden ‖ fileter à la filière. / ~ screws by hand ‖ Gewinde n. aus freier Hand schneiden ‖ fileter *ou* tarauder à la volée. / ~ to size ‖ zuschneiden ‖ couper à dimension f. / ~ sods pl. ‖ abrasen ‖ peler la terre. / ~ a solid core out of the ingot in one piece (Met) ‖ aus dem Rohrblock einen massiven Kern m. herausschneiden ‖ enlever le métal autour d'un noyau solide. / ~ steep down ‖ steil abböschen ‖ escarper. / ~ stones pl. ‖ Steine mpl. behauen ‖ tailler des pierres fpl. / ~ stones with a saw ‖ die Steine zersägen *oder* zerschneiden ‖ débiter la pierre à la scie. / ~ the teeth of a wheel (Mach) ‖ die Radzähne mpl. einschneiden ‖ tailler les dents fpl. d'une roue dentée. / ~ the teeth of a saw ‖ Sägezähne mpl. hauen ‖ faire *ou* tailler les dents de scie. / ~ a thread ‖ ein Gewinde n. schneiden ‖ fileter une vis / ~ through ‖ durchschneiden ‖ couper en deux. / ~ timber (Carp) ‖ Holz n. der Länge nach sägen ‖ refendre *ou* tailler le bois; scier le bois de longueur. / ~ the timber (Forest) ‖ Holz n. fällen *oder* hauen ‖ abattre le bois. / ~-up ‖ in Stücke npl. zerschneiden ‖ subdiviser. / ~ the velvet ‖ den Samt m. reißen *oder* schneiden ‖ ciseler *ou* couper le velours. / ~ veneers pl. ‖ Furnierholz n. schneiden ‖ couper des feuilles fpl. à plaquer. / ~ wood ‖ Holz n. ausschneiden ‖ découper le bois. / ~ the wood into curved surfaces with the sweep- or bow-saw (Join) ‖ ausschweifen ‖ chantourner.

cut *see also* cutting ‖ Schnitt m.; Schneiden n.; Ausschneiden n. ‖ coupe f.; taille f.; coupage m. / ~ (Carp; Join; Mach) ‖ Ausschnitt m.; Einschnitt m.; Kerbe f. ‖ taille f.; entaille f.; coupure f.; échancrure f.; encoche f. / ~ (Fashion) ‖ Schnitt m. ‖ taille f. / ~ (File) ‖ Hieb m.; Feilenhieb m. ‖ taille f. / ~ (Hydr arch) ‖ kleiner Graben m.; Krecke f. ‖ crique f. / ~ (Mine) ‖ Rinne f.; Ritz m.; Schramm m. ‖ entaille f. / ~ (Print) ‖ Druckstock m.; Bildstock m.; Klischee n. ‖ cliché m. / ~ (Print) ‖ Abbildung f. ‖ gravure f. / ~ (Terrain) ‖ Einschnitt m. ‖ tranchée f.; déblai m. / ~ (Wood) ‖ Schnittfläche f. ‖ coupe f. / bastard ~ (File) ‖ Bastardhieb m. ‖ taille f. bâtarde. / ~ of the cloth *see* ~ of the fabric. / with coarse ~ ‖ mit grobem Schnitt m. ‖ enlevant de gros copeaux mpl. / ~ of cotton yarn ‖ Gebinde n. Baumwollgarn ‖ échevette f. de fil de coton. / cross-grained ~ ‖ Hirnschnitt m. im Holz ‖ coupe f. transversale. / ~ of a dike ‖ Durchfahrt f. in einem Deich ‖ coupure f. d'une digue. / end-grained ~ *see*

cross-grained ~. / ~ of the fabric ‖ Stoffzuschnitt m. ‖ coupe f. du drap. / ~ of file ‖ Feilenhieb m. ‖ taille f. de lime. / fine ~ (File) ‖ Schlichthieb m. ‖ taille f. douce. / finishing ~ ‖ Fertigschnitt m. ‖ coupe f. à terminer. / to make a finishing ~ ‖ sauber fertig arbeiten ‖ donner la dernière coupe. / to take the finishing ~ ‖ nachdrehen ‖ repasser au tour m. / first ~ (File) ‖ Grundhieb m.; Unterhieb m. ‖ première taille f. / to make a first ~ ‖ vorschneiden ‖ donner la première coupe. / ~ with the grain (Carp) ‖ Längenschnitt m. ‖ coupe f. longitudinale; coupe f. de fil. / length of ~ ‖ Schnittlänge f. ‖ longueur f. de coupe. / ~ of a letter ‖ Schnitt m. einer Schrift ‖ gravure f. d'une lettre. / mid ~ (File) ‖ Halbschlichthieb m. ‖ taille f. demidouce. / ~ of a roller ‖ Riefe f. einer Walze ‖ entaille f. d'un cylindre-lamineur. / rough ~ (File) ‖ grober Hieb m. ‖ grosse taille f. / rough ~ (Turn) ‖ Schruppschnitt m. ‖ coupe f. à dégrossir. / smooth ~ (File) ‖ feiner Hieb m. ‖ fine *ou* douce taille f. / specially smooth ~ ‖ besonders glatter Schnitt m. ‖ coupe f. spécialement lisse. / ~ of timber for deals ‖ Trennschnitt m. ‖ sciage m. de long. / ~ in wages ‖ Lohnabbau m. ‖ réduction f. des salaires. / ~ through the woods ‖ Durchschlag m. im Walde ‖ percée f.; trouée f. / with coarse ~ ‖ mit grobem Schnitt m. ‖ enlevant de gros coupeaux mpl.

cutaneous eruption ‖ Hautausschlag m. ‖ éruption f. cutanée.
cut capacity ‖ Schnittleistung f. ‖ rendement m. de coupe.
cutch ‖ Katechu n. ‖ catéchou m.
cutlass ‖ Jagdmesser n.; Hirschfänger m. ‖ couteau m. de chasse; coutelas m.
cutler ‖ Messerschmied m. ‖ coutelier m. / ~'s trade ‖ Messerschmiede f.; Messerschmiederei f. ‖ coutellerie f.
cutlery ‖ Messerschmiedewaren fpl. ‖ coutellerie f. / ~ for tourists ‖ Touristenbesteck n. ‖ couvert m. de touristes.
cutlery factory ‖ Besteckfabrik f. ‖ fabrique f. de couverts. / ~ ware ‖ Messerwaren fpl. ‖ coutellerie f.
cut looker (Weav) ‖ Schaumeister m.; Stückbeschauer m. ‖ inspecteur m. des étoffes tissues.
cut-off (Electr) ‖ abgeschaltet isolé; déconnecté.
cut-off, to ~ *see under* to cut.
cut-off (Steam engine) ‖ Füllung f.; Füllungsgrad m. ‖ admission f.; degré m. d'admission. / maximum ~ (Steam engine) ‖ Maximalfüllung f. ‖ admission f. maximum.
cut-off curve ‖ Expansionskurve f. ‖ courbe f. de détente. / ~ jack (Tel) ‖ Abschalteklinke f. ‖ jack m. de déconnexion. / ~ key (Tel) ‖ Trenntaste f. ‖ clef f. de rupture. / ~ magnet ‖ Abschaltemagnet m. ‖ électro m. de déconnexion. / ~ relay (Tel) ‖ Trennrelais n. ‖ relais m. de coupure. / ~ saw ‖ Ablängsäge f. ‖ scie f. à tronçonner. / ~ signal ‖ Trennzeichen n. ‖ signal m. de coupure.
cut-out (Electr) ‖ Sicherung f. ‖ coupecircuit f.; fusible m. de sûreté. / automatic ~ (Electr) ‖ Selbstunterbrecher m. ‖ disjoncteur m. *ou* interrupteur m. automatique. / ~ cartridge ‖ Patronenausschalter m. ‖ cartouche f. fusible. / double ~ ‖ Doppelausschalter m. ‖ in-

terrupteur m. *ou* disjoncteur m. double. / exhaust ~ ‖ Auspuffklappe f. ‖ clapet m. d'échappement. / ~ with free release ‖ Freiauslösung f. ‖ déclenchement m. libre. / high-tension ~ ‖ Hochspannungssicherung f. ‖ coupe-circuit m. à fusible pour haute tension. / lead ~ ‖ Bleisicherung f. ‖ coupe-circuit m. à lame de plomb. / low-tension ~ ‖ Niederspannungssicherung f. ‖ coupe-circuit m. à fusible pour basse tension. / maximum ~ ‖ Maximalausschalter m.; Überstromschalter m. ‖ interrupteur m. à maximum. / no-voltage ~ ‖ Nullspannungsschalter m. ‖ disjoncteur m. à minimum; interrupteur m. automatique de tension nulle. / reverse current ~ ‖ selbsttätiger Ausschalter m. ‖ conjoncteur-disjoncteur m. / safety ~ ‖ Sicherheitsausschalter m. ‖ interrupteur m. de sûreté.

cut-out device ‖ Stillsetzvorrichtung f. ‖ dispositif m. d'arrêt. / ~ section ‖ Sichtausschnitt m. ‖ regard m. *ou* échancrure f. à inspection.

cutter ‖ Schneidapparat m.; Abschneider m.; Schneidmaschine f. ‖ coupoir m.; coupeur m.; coupeuse f. / ~ (Bookb) ‖ Beschneider m. ‖ ébarbeur m.; rogneur m. / ~ (Boring) ‖ Zentrumsbohrer m. ‖ foret m. *ou* mèche f. à centre; foret m. à trois pointes. / ~ (Carp) ‖ Löffelbohrer ‖ cuiller f. / ~ (Coin) ‖ Münzschere f. ‖ coupoir m.; emporte-pièce m. / ~ (Mach) ‖ Keil m.; Schlüssel m.; Splint m. ‖ clavette f.; clef f. / ~ (Milling cutter) ‖ Fräser m.; Fräsmesser n.; Schneidzahn m. ‖ fraise f.; lame f. / ~ (Mine) ‖ Häuer m.; Gesteinshauer m. ‖ coupeur m.; enfonceur m.; piqueur m.; bouveleur m.; haveur m.; abatteur m. / ~ (Planing mach) ‖ Stichel m.; Meißel m.; Stahl m. ‖ outil m.; burin m. / ~ (Shears) ‖ Parallelschere f. ‖ cisaille f. parallèle. / ~ (Shipb) ‖ Kutter m. ‖ cotre m.; cutter m. / ~ (Tool) ‖ Schneidwerkzeug n. ‖ outil m. tranchant. / adjustable ~ ‖ einsetzbare Schneidzunge f. ‖ languette f. rapportable. / adjustment of top ~ ‖ Obermesserverstellung f. ‖ réglage m. de la lame supérieure. / angular ~ ‖ Winkelfräser m. ‖ fraise f. conique. / backed-off ~ ‖ hinterdrehter Fräser m. ‖ fraise f. avec profil invariable. / bar iron ~ ‖ Stabeisenschere f. ‖ cisailles fpl. pour fers en barres. / ~ of a boring-machine ‖ Messer n. *oder* Schneide f. einer Bohrmaschine ‖ lame f. *ou* couteau m. d'une machine à aléser. / bread ~ ‖ Brotschneidemaschine f. ‖ machine f. à couper le pain. / chain ~ ‖ Fräserkette f. ‖ chaîne f. dentée à fraiser; chaîne f. porte-couteaux. / to change the ~s ‖ die Messer npl. auswechseln ‖ échanger les couteaux mpl. / ~ of chopped straw cutting machine ‖ Häckselmesser n. ‖ couteau m. de hachoir. / ~ for cucumbers ‖ Gurkenhobel m. ‖ coupe-concombres m. / cylindrical ~ ‖ Walzenfräser m. ‖ fraise f. cylindrique. / damage to ~ ‖ Fräserbruch m. ‖ rupture f. de fraise. / facing ~ ‖ Planfräser m. ‖ fraise f. de fraiseuse-raboteuse. / file ~ ‖ Feilenaufhauer m. ‖ retailleur m. de limes. / finishing ~ ‖ Nachfräser m. ‖ fraise f. finisseuse. / formed ~ ‖ Fassonfräser m. ‖ fraise f. profilée. / ~ for gear wheels ‖ Zahnradfräser m. ‖ fraise f. à engrenages. / helicoidal ~s pl. of a cylindrical shearing

machine ‖ Messer npl. einer Zylinderschermaschine ‖ couteaux mpl. d'une tondeuse hélicoïde. / high-speed ~ ‖ Schnellfräser m. ‖ fraise f. rapide. / hole-boring ~ ‖ Lochfräser m. ‖ fraise f. à aléser. / ~ with horizontal adjustment ‖ horizontal verstellbarer Fräser m. ‖ fraise f. avec déplacement horizontal. / interchangeable ~ ‖ auswechselbares Messer n. ‖ lame f. remplaçable. / key way ~ ‖ Nutstahl m. ‖ outil m. à raboter les rainures. / milling ~ ‖ Fräser m. ‖ fraise f. / moulding ~ ‖ Kehlmesser m. ‖ fer m. à moulurer. / pipe ~ ‖ Rohrabschneider m. ‖ coupe-tubes m. / profiling ~ ‖ Profilfräser m. ‖ fraise f. profilée. / rebating ~ ‖ Falzfräser m. ‖ fraise f. à feuillures. / removable ~ ‖ verschiebbare Schneide f. ‖ lame f. remplaçable. / screw slotting ~ ‖ Schraubenschlitzfräser m. ‖ fraise f. à entailler les têtes de vis. / sectional iron ~ ‖ Formeisenschere f. ‖ cisailles fpl. pour fers profilés. / semicircular ~ ‖ Pilzfräser m. ‖ fraise f. à champignon. / shank ~ ‖ Messerkopfzapfen m. ‖ tourillon m. sur lequel est fixé le porte-outil. / side milling ~ ‖ Scheibenfräser m. ‖ fraise f. en disque. / slot milling ~ ‖ Nutenfräser m. ‖ fraise f. à rainures *ou* à cannelures. / solid ~ in one piece (Plane iron) ‖ Messer n. aus einem Stück ‖ lame f. *ou* couteau m. d'une seule pièce. / stocking ~ ‖ Vorfräser m. ‖ fraise f. ébaucheuse. / thread milling ~ ‖ Gewindefräser m. ‖ fraise f. à fileter. / ~ for vegetables ‖ Gemüsehobel m. ‖ coupe-légumes m. / wheel ~ *see* ~ for gear wheels. / tillage ~ ‖ Ackerfräser m. ‖ fraiseuse f. de labour.

cutter adjusting screw ‖ Messerstellschraube f. ‖ vis f. de réglage du coupoir. / ~ arbor ‖ Fräsbolzen m. ‖ mandrin m. porte-pièce.

cutter bar ‖ Bohrstange f. ‖ tige f. porte-foret. / ~ (Reaper) ‖ Fingerschneidbalken m.; Schneidbalken m. ‖ barre f. coupeuse. / ~ of a boring-machine ‖ Bohrspindel f. einer Bohrmaschine ‖ arbre m. d'une machine à aléser. / medium pitch ~ ‖ Schneidbalken m. für Mittelschnitt ‖ barre f. coupeuse pour coupe moyenne. / narrow pitch ~ ‖ Schneidbalken m. für Tiefschnitt ‖ barre f. coupeuse pour coupe profonde. / standard pitch ~ ‖ Schneidbalken m. für Normalschnitt ‖ barre f. coupeuse pour coupe normale.

cutter clamp (Mover) ‖ Messerklemme f. ‖ écrou m. à fixer le coupoir. / ~ disc ‖ Messerscheibe f. ‖ disque m. à couteaux. / automatic ~ grinder ‖ Fräserschärfautomat m. ‖ machine f. automatique à affûter et à creuser les fraises. / ~ grinding machine ‖ Messerschleifmaschine f. ‖ machine f. à affûter les couteaux. / ~ hand (Pap) ‖ Papierschneider m. ‖ coupeur m.

cutter head (Milling cutter) ‖ Fräskopf m. ‖ tête f. de fraisage. / ~ (Wood working) ‖ Messerkopf m. ‖ tête f. porte-outil; porte-lames m. / ~ grinder ‖ Messerkopfschleifmaschine f. ‖ machine f. à affûter les fraises à dents rapportées *ou* à émoudre les têtes à couteaux. / ~ grinding machine *see* ~ grinder.

cutter holder (Mover) ‖ Messerhalter m. ‖ porte-coupoir m. ‖ support (Mover) ‖ Messerbock m. ‖ chevalet m. du porte-coupoir.

cutter making machine ‖ Fräserherstellungsmaschine f. ‖ machine f. à tailler les fraises.

cutter-out (Weav) ‖ Ausschneider m. ‖ refendeur m. de tissus.

cutter rig ‖ Kuttertakelage f. ‖ gréement m. en cotre.

cutter's fee ‖ Hauerlohn m. ‖ salaire m. du haveur; équarrissage m.

cutter spindle (Milling cutter) ‖ Frässpindel f. ‖ arbre m. porte-fraise. / ~ spindle (Wood working) ‖ Messerwelle f. ‖ arbre m. porte-outils. / ~ spindle speed ‖ Frässpindelgeschwindigkeit f. ‖ vitesse f. de l'arbre porte-fraise. / springing upwards of ~ ‖ Ausweichen n. des Fräsers nach oben ‖ décalage m. vertical de la fraise.

cutter's wages pl. *see* cutter's fee.

cutting (Tool) ‖ schneidend; scharf ‖ tranchant; coupant.

cutting ‖ Schneiden n.; Ausschneiden n. ‖ coupe f.; coupage m.; découpage m. / ~ (Carp; Join; Mach) ‖ Ausschnitt m.; Einschnitt m. ‖ coupure f.; entaille f.; échancrure f. / ~ (Build; Railw; Road) ‖ Einschnitt m.; Durchstich m.; Abgrabung f. ‖ tranchée f.; déblai m. / ~ (Forest) ‖ Holzschlag m.; Schlag m.; Fällen n. ‖ abatage m.; abattage m.; abatis m. / ~ (Milling) ‖ Fräsen n. ‖ fraisage m. / ~ (Mine) ‖ Kerb m.; Schlitz m. ‖ échancrure f. latérale. / ~ (Textile) ‖ Scheren n. ‖ tondage m.; coupe f.; tonture f. / ~ (Tool) ‖ Span m. ‖ copeau m. / ~ (Turn) ‖ Drehspan m. ‖ copeau m.; tournure f. / autogenous ~ ‖ autogenes Schneiden n. ‖ coupure f. autogène. / autogenous ~ shop ‖ Werk n. für autogenes Schneiden ‖ atelier m. de coupure autogène. / ~ of caulking edges ‖ Stemmkantenschneiden n. ‖ coupe f. de rebords chanfreinés. / ~ chamfer edges ‖ Stemmkantenschneiden n. ‖ coupe f. des bords en chanfrein. / ~ of clay ‖ Schneiden n. des Tons ‖ coupage m. de l'argile. / clean ~ (Reaper) ‖ sauberer Schnitt m. ‖ coupe f. nette. / cold ~ of rolled iron ‖ Kaltschneiden n. von Walzeisen ‖ coupe f. à froid de fers laminés. / electric-arc ~ ‖ elektrisches Schneiden n. ‖ coupe f. à l'arc électrique. / ~ of files ‖ Feilenhauen n. ‖ taille f. *ou* taillage m. des limes. / file ~ anew ‖ Feilenaufhauen n. ‖ retaillage m. de limes. / the frame has ~s pl. for the reception of the axle boxes ‖ der Rahmen hat Ausschnitte für die Achslager ‖ le châssis est dépourvu d'échancrures à recevoir les boîtes d'essieu. / hand ~ (Textile) ‖ Handschnitt m. ‖ coupage m. à la main. / ~ of the hands (Watchm) ‖ Ausstanzen n. der Zeiger ‖ découpage m. d'aiguilles. / hot ~ of rolled iron ‖ Warmschneiden n. von Walzeisen ‖ coupe f. à chaud de fers laminés. / machine ~ (Textile) ‖ Maschinenschnitt m. ‖ coupage m. à la machine. / to make a ~ (Railw; Road) ‖ einen Einschnitt m. stechen ‖ percer une tranchée f. / ~ of a seam ‖ Abbau m. eines Flözes ‖ ouvrage m. *ou* abatage m. d'une couche. / ~ of sheets ‖ Trennen n. von Blechtafeln ‖ cisaillement m. de tôles. / ~ trenches ‖ Ziehen n. von Gräben ‖ mise f. de tranchées. / under water ~ ‖ Schneiden n. unter Wasser ‖ découpage m. sous l'eau. / ~ of velvet ‖ Reißen n. *oder* Schneiden n. des Samts ‖ ciselage m.

du velours. / ~ welding edges on angle-iron rings ‖ Beschneiden n. von Winkelringen auf Stemmkante ‖ chanfreinage m. des anneaux de récipients. / ~ the welding-edge on irregular-shaped vessel plates ‖ Beschneiden n. der Schweißkante an unregelmäßig geformten Behälterblechen ‖ chanfreinage m. des tôles de récipients d'une forme irrégulière. / ~ welding edges on plain shell plates ‖ Anschneiden n. von Stemmkanten an geraden Mantelblechen ‖ chanfreinage m. des tôles planes de revêtement.

cutting-across (Mine) ‖ Durchschlag m.; Durchörterung f. ‖ percement m. (souterrain). / ~ angle ‖ Schneidwinkel m.; Schnittwinkel m. ‖ angle m. de coupe.

cutting apparatus ‖ Schneidvorrichtung f. ‖ cisailleuse f.; appareil m. à découper. / cross-~ (Bookb) ‖ Querschneidapparat m. ‖ appareil m. à coupe transversale. / round-iron ~ ‖ Betoneisenschneider m. ‖ appareil m. à trancher le fer rond.

cutting appliance see ~ apparatus. / internal ~ attachment for automatic spur gear cutting machines ‖ Innenfräsvorrichtung f. für selbsttätige Stirnradfräsmaschinen ‖ dispositif m. de taillage intérieur pour machines alternatives automatiques à tailler les engrenages droits. / ~ block of files ‖ Hauamboß m. für Feilen ‖ enclume f. à entailler les limes; tas m. / ~ board (Househ) ‖ Schneidbrett n.; Hackbrett n. ‖ tailloir m. / ~ board (Tail) ‖ Zuschneidetisch m. ‖ établi m. / ~ burner ‖ Schneidbrenner m. ‖ brûleur m. à découper. / autogenous ~ burner ‖ Brenner m. für autogenes Schneiden; Autogenschneidbrenner m. ‖ brûleur m. ou chalumeau m. à coupure autogène. / under water ~ burner ‖ Unterwasserschneidbrenner m. ‖ brûleur m. à découper au-dessous de l'eau. / ~ capacity ‖ Schneidfähigkeit f. ‖ capacité f. de coupe. / ~ chisel (Carp) ‖ Abstechmeißel m. ‖ ébarboir m. / ~ chisel (Forg) ‖ Aufhauer m. ‖ langue f. de carpe. / ~ compasses pl. ‖ Schneidzirkel m. ‖ coupe-cercle m. / ~ depth by means of finely divided scale ‖ Schnittiefe f. nach Millimeterskale ‖ profondeur f. de coupe d'après la graduation millimétrique. / ~ detachment of a ~ ‖ Spanabhebung f. ‖ détachement m. d'un copeau. / ~ device ‖ Abschneidapparat m. ‖ dispositif m. coupeur. / ~ die ‖ Schneidbacke f. ‖ coussinet m. de filière. / ~ direction (Reaper) ‖ Schnittrichtung f. ‖ direction f. de coupe. / ~-down a wood ‖ Schlagen n. eines Waldes ‖ coupe f. d'un bois. / heaviest ~ duty ‖ höchste Schnittleistung f. ‖ débit m. maximum de coupe. / ~ edge of tool ‖ Schneide f. oder Schneidekante f. eines Werkzeuges ‖ tranchant m. ou face f. de l'outil.

cutting effect, of great ~ ‖ schneidhaltig ‖ résistance f. du tranchant. / hardness of great ~ ‖ schneidhaltende Härte m. ‖ dureté f. tranchante ou d'outil coupant.

cutting, example of ~ work ‖ Schnittprobe f. ‖ exemple m. de découpage. / ~ flame ‖ Schneidflamme f. ‖ flamme f. découpante. / ~ formation of a ~ ‖ Spanbildung f. ‖ formation f. d'un copeau. / ~ frame ‖ Reißwerk n. ‖ traçoir m. / ~ hardness ‖ Schneidhärte f. ‖ trempe f. active ou d'outil coupant. / ~ height ‖ Schnitthöhe f. ‖ hauteur f. de coupe.

/ ~ height (Dredger) ‖ Reichhöhe f. ‖ hauteur f. d'attaque. / ~ iron (Agr) ‖ Sech n. eines Pfluges ‖ coutre m. de charrure; sep m. / ~ knife (Saddl) ‖ Werkmesser n.; Halbmondmesser n. ‖ couteau m. à pied. / layer of ~s ‖ Spanschicht f. ‖ strie f. de copeaux. / ~ line (Print) ‖ Abschnittlinie f.; Schnittlinie f. ‖ marque f. à couper. / ~ line pointer ‖ Schnittlinienanzeiger m. ‖ indicateur m. des lignes de coupe.

cutting machine ‖ Schneid(e)maschine f. ‖ coupeuse f.; machine f. à couper. / ~ (Bookb) ‖ Beschneidmaschine f. ‖ rogneuse f. / ~ (Spinn) ‖ Schneidemaschine f. ‖ machine f. à couper. / ~ (Textile) ‖ Schermaschine f. ‖ machine f. à tondre; tondeuse f. / automatic spur gear ~ with horizontal work arbor ‖ selbsttätige Stirnradfräsmaschine f. mit horizontalem Aufspannbolzen ‖ machine f. alternative horizontale à tailler les engrenages droits avec mandrin porte-pièce horizontal. / bevel gear ~ ‖ Kegelräderfräsmaschine f. ‖ machine f. à tailler les roues coniques. / bevel edge ~ for welding ‖ Schweißkantenschere f. ‖ cisaille f. circulaire à chanfreiner les bords des tôles pour soudure. / bolt-screw ~ ‖ Bolzenschneidmaschine f. ‖ machine f. à tarauder les boulons. / ~ for botanical drugs ‖ Kräuterschneidmaschine f. ‖ machine f. à couper les herbes. / chaff ~ ‖ Häckselmaschine f. ‖ hache-paille f. / circle ~ ‖ Kreisschere f. ‖ cisaille f. circulaire pour découper les disques. / cross-~ ‖ Querschneidmaschine f. ‖ coupeuse f. en travers. / curve ~ ‖ Kurvenschere f. ‖ cisaille f. circulaire pour découper les courbes. / figure ~ ‖ Figurenschere f. ‖ cisaille f. circulaire pour découper les formes. / herringbone gear ~ ‖ Pfeilräderfräsmaschine f. ‖ machine f. à fraiser les roues à denture à chevrons. / long-~ ‖ Langschneidmaschine f. ‖ coupeuse f. en long. / mitre ~ ‖ Gehrungsschere f. ‖ cisaille f. à onglets. / ~ for rubber working ‖ Schneidemaschine f. für die Gummibearbeitung ‖ machine f. à trancher pour la préparation du caoutchouc. / rack ~ ‖ Zahnstangenfräsmaschine f. ‖ machine f. à fraiser les crémaillères. / strip ~ ‖ Streifenschere f. ‖ cisaille f. circulaire à découper les bandes. / ~ for strips ‖ Abschneidmaschine f. für Streifen ‖ machine f. à couper des bandes.

cutting material ‖ Schnittstoff m. ‖ matière f. à tailler ou à sculpter. / ~ mechanism ‖ Schneidvorrichtung f. ‖ appareil m. coupeur. / ~ nippers pl. ‖ Beißzange f.; Vorschneider m.; Schneidezange f. ‖ pince f. ou tenaille f. coupante.

cutting-off ‖ Abschneiden n. ‖ découpage m. / ~ (Mach; Turn) ‖ Abstechen n. ‖ tronçonnage m. / ~ (Steam) ‖ Absperren n. ‖ arrêt m. / ~ (Turn) ‖ Fassondrehen n. ‖ décolletage m. / ~ the steam ‖ Dampfabsperrung f. ‖ arrêt m. de la vapeur. / ~ blade holder ‖ Abstechstahlhalter m. ‖ porte-outil m. à tronçonner. / ~ device ‖ Abschneidegerät n. ‖ appareil m. à decouper; coupeur m.

cutting-off machine ‖ Abstechmaschine f. ‖ machine f. à tronçonner. / ~ with rotating tool ‖ Abstechmaschine f. mit umlaufendem Werkzeug ‖ machine f. à tronçonner à outil tournant. / ~ with rotating work piece ‖ Abstechmaschine f.

mit umlaufendem Werkstück ‖ machine f. à tronçonner à pièce à travailler tournante. / wire ~ ‖ Drahtabschneidemaschine f. ‖ machine f. à couper le fil.

cutting-off shears pl. ‖ Trennschere f. ‖ cisaille f. à sectionner. / ~ slide ‖ Abstechsupport m. ‖ chariot m. de tronçonnage. / ~ tool ‖ Einstechstahl m. ‖ outil m. à saigner. / ~ tubes pl. ‖ Rohrabschneiden n. ‖ tronçonnage m. des tubes.

cutting-out ‖ Ausstanzen n.; Stanzen n. ‖ découpage m. / ~ (Textile) ‖ Ausschnitt m. ‖ entaille f. / ~ window openings ‖ Ausschneiden n. von Fensteröffnungen ‖ découpage m. des ouvertures de fenêtre. / ~ board ‖ Zuschneidetisch m. ‖ écofrai m.; établi m. / ~ machine (Coin) ‖ Münzschere f. ‖ coupoir m.; emporte-pièce m. / ~ machine for clothes ‖ Zuschneidemaschine f. für Stoffe ‖ machine f. à découper les étoffes. / ~ machine for sacks ‖ Sackzuschneidmaschine f. ‖ machine f. à découper les sacs. / ~ scissors pl. ‖ Ausschneideschere f. ‖ ciseaux mpl. de découpage.

cutting-over traffic to telephone exchanges ‖ Betriebsüberleitung f. auf neue Fernsprechvermittlungsämter ‖ passage m. du service aux centrales téléphoniques.

cutting plant ‖ Schneidanlage f. ‖ installation f. de découpage. / autogeneous ~ ‖ autogene Schneidanlage f. ‖ installation f. de coupe autogène. / waste paper ~ ‖ Altpapierschneideanlage f. ‖ installation f. pour le découpage des vieux papiers.

cutting pliers pl. ‖ Drahtabschneider m. ‖ pince f. coupe-fils. / ~ point ‖ Reißer m. ‖ couteau m.; tracelet m. / ~ power ‖ Schnittkraft f. ‖ force f. de coupe. / steel retaining its ~ power ‖ schneidhaltiger Stahl m. ‖ acier m. à grande résistance f. du tranchant. / ~ press (Bookb) ‖ Beschneidepresse f. ‖ presse f. à rogner. / ~ pressure (Plate shear) ‖ Schnittdruck m. ‖ pression f. de coupe. / iron ~ saw ‖ Metallsäge f. ‖ scie f. à métaux. / shoot for ~s ‖ Spanfall m. ‖ bac m. à copeaux. / ~ speed ‖ Schnittgeschwindigkeit f. ‖ vitesse f. de coupe. / ~ stroke ‖ Schnittlänge f. ‖ longueur f. de coupe. / succession of ~s (Forest) ‖ Hiebfolge f. ‖ succession f. des coupes. / ~ table ‖ Schneidtisch m. ‖ dérompoir m. / ~ thickness of chip ‖ Spanstärke f. ‖ épaisseur f. de la tournure.

cutting tool ‖ Schneidewerkzeug n. ‖ outil m. tranchant. / a set of ~s ‖ ein Satz m. Hobelstähle ‖ un jeu m. d'outils à raboter. / ~ alloy ‖ Schneidlegierung f. ‖ alliage m. pour outils tranchants. / ~ metal ‖ Schneidmetall n. ‖ métal m. pour outils tranchants.

cutting tooth of a broach ‖ Schneidzahn m. d'une broche. / ~ torch ‖ Schneidbrenner m. ‖ brûleur m. à découper.

cutting-up ‖ Abdecken n. ‖ équarrissage m.

cutting width (Agr) ‖ Schnittbreite f. ‖ largeur f. de coupe. / ~ (Turn) ‖ Spanbreite f. ‖ largeur f. du copeau ou de coupe.

cuttle bone ‖ Blackfischbein n. ‖ os m. de sèche. / ~ fish ‖ Sepia f.; Tintenfisch m. ‖ sèche f.; cornet m.; calmar m.

cut-work printing ‖ Illustrationsdruck m. ‖ impression f. des illustrations.

cyanic acid ‖ Zyansäure f. ‖ acide m. cyanique.

cyanide ‖ Zyansalz n.; Zyanür n. ‖ cyanure m.; prussiate m.; cyanide m. / ~ of barium ‖ Zyanbarium n. ‖ cyanure m. de baryum. / ~ of copper ‖ Zyankupfer n. ‖ cyanure m. de cuivre. / ~ of copper bath (Met) ‖ zyankalisches Kupferbad n. ‖ bain m. de cyanure de cuivre. / ~ of hydrogen ‖ Blausäure f. ‖ acide m. cyanhydrique. / ~ of mercureous potassium see ~ of potassium and mercury. / ~ of potassium ‖ Kaliumzyanid n.; Zyankalium n.; blausaures Kalium n. ‖ cyanure m. de potassium; prussiate m. de potasse. / ~ of potassium and copper ‖ Kaliumkupferzyanür n. ‖ cyanure m. de potassium et de cuivre. / ~ of potassium and mercury ‖ Kaliumquecksilberzyanur n. ‖ cyanure m. de potassium et de mercure. / ~ of silver ‖ Zyansilber n. ‖ cyanure m. d'argent. / ~ of silvery potassium ‖ Zyansilberkalium n. ‖ cyanure m. d'argent et de potassium. / ~ of sodium ‖ Zyannatrium n. ‖ cyanure m. de soude. / ~ of zinc ‖ Zyanzink n. ‖ cyanure m. de zinc. / ~ of zinky potassium ‖ Zyanzinkkalium n. ‖ cyanure m. de zinc et de potassium. / ~ of zinc and potassium see ~ of zinky potassium.

cyanogen ‖ Zyan n. ‖ cyanogène m. / ~ gas ‖ Zyangas n. ‖ cyanogène m.; azoture m. de carbone.

cycle ‖ Fahrrad n. ‖ cycle m.; bicyclette f.; vélocipède m.; vélo m. / ~ (Phys; Electr) ‖ Zykel m.; Kreisprozeß m. ‖ cycle m. / complete ~ ‖ geschlossener Kreisprozeß m. ‖ cycle m. parfait. / closed ~ ‖ geschlossener Kreisprozeß m. ‖ cycle m. fermé. / four-stroke ~ ‖ Viertakt m.; Viertaktverfahren n. ‖ cycle m. à quatre temps. / high ~ ‖ Hochrad n. ‖ bicycle m. / ~ for invalids. ‖ Invalidenfahrrad n. ‖ bicyclette f. pour infirmes. / lunar ~ ‖ Mondzyklus m. ‖ cycle m. lunaire. / metonic ~ see lunar ~. / ~ of the moon see lunar ~. / ~ motor ‖ Motorrad n. ‖ motocyclette f. / non-reversible ~ ‖ nicht umkehrbarer Kreisprozeß m. ‖ cycle m. non-réversible. / ~ of operations ‖ Arbeitsgang m. ‖ cours m. de la fabrication; phase f. d'élaboration ou de travail. / ~ of operations (Phys) ‖ Kreisprozeß m. ‖ cycle m. / ~ pl. per second ‖ Perioden fpl. je Sekunde ‖ cycles mpl. par seconde. / two-stroke ~ ‖ Zweitakt m.; Zweitaktverfahren n. ‖ cycle m. à deux temps. / thermal ~ ‖ thermischer Kreisprozeß m. ‖ évolution f. ou cycle m. thermique.

cycle accessories pl. ‖ Fahrradzubehörteile mpl. ‖ accessoires mpl. pour vélocipèdes. / ~ bag ‖ Fahrradtasche f. ‖ poche f. ou sacoche f. pour bicyclette. / ~ bell ‖ Fahrradglocke f. ‖ timbre m. pour cycles. / ~ brake ‖ Fahrradbremse f. ‖ frein m. pour bicyclettes. / ~ car ‖ Kleinauto n. ‖ cycle-car. / ~ driving gear ‖ Fahrradantrieb m. ‖ commande f. pour bicyclettes. / ~ engineer ‖ Fahrradmechaniker m. ‖ mécanicien m. pour vélocipèdes. / ~ felloe ‖ Fahrradfelge f. ‖ jante f. de bicyclette. / ~ fitter ‖ Fahrradmonteur m. ‖ monteur m. de vélos. / ~ fittings pl. ‖ Fahrradbestandteile mpl. ‖ accessoires mpl. pour bicyclettes. / ~ fork ‖ Radgabel f. ‖ fourche f. / ~ frame ‖ Fahrradgestell n. ‖ cadre m. de bicyclette.

/ ~ free-wheel ‖ Fahrradfreilauf m. ‖ axe f. ou roue f. libre pour vélocipèdes. / ~ hub ‖ Fahrradnabe f. ‖ moyeu m. de bicyclette. / ~ lamp ‖ Fahrradlaterne f. ‖ lanterne f. de cycle. / ~ motor ‖ Fahrrad(einbau)motor m. ‖ moteur m. pour bicyclettes. / ~ oil ‖ Fahrradöl n. ‖ huile f. pour bicyclettes. / ~ part ‖ Fahrradersatzteil m. ‖ piece f. détachée pour vélocipèdes. / ~ parts pl. for ~s ‖ Fahrradzubehör n. ‖ accessoires mpl. de bicyclette. / ~ path ‖ Radfahrweg m. ‖ accotement m. cyclable. / ~ pedal ‖ Fahrradpedal n. ‖ pédale f. pour bicyclettes. / ~ repair workshop ‖ Fahrräderausbesserungswerkstätte f. ‖ atelier m. de réparation de vélocipèdes. / ~ spoke ‖ Fahrradspeiche f. ‖ rayon m. de bicyclette. / ~ support ‖ Fahrradständer m. ‖ support m. de bicyclettes. / ~ tread crank ‖ Fahrradtretkurbel f. ‖ manivelle f. à pédaler pour bicyclettes. / ~ wheel fitter ‖ Radeinsetzer m. ‖ monteur m. de roues pour vélocipèdes.

cyclic ‖ zyklisch; ringförmig ‖ cyclique. / ~ constant ‖ Umlaufgröße f. ‖ constante f. de circulation. / ~ flow ‖ Kreisströmung f. ‖ courant m. circulaire. / ~ variation ‖ Ungleichförmigkeit f. ‖ variation f. cyclique.

cycling road ‖ Radfahrweg m. ‖ route f. de vélo; accotement m. cyclable.

cycloid ‖ Zykloide f. ‖ cycloïde f. / ~ toothing ‖ Zykloidenverzahnung f. ‖ denture f. cycloïdale.

cyclometry (Math) ‖ Kreismessung f. ‖ cyclométrie f.

cyclone (Meteor) ‖ Zyklon m. ‖ tempête f. tournante; cyclone m. / ~ (Mill) ‖ Windsichter m. ‖ séparateur m. à vent.

cylinder (Engine; Math) ‖ Zylinder m. ‖ cylindre m. / ~ (Chem) ‖ Stahlflasche f.; Gaszylinder m. ‖ bombe f.; obus m. / ~ (Mach) ‖ Rolle f.; Walze f.; Zylinder m. ‖ rouleau m.; cylindre m. / ~ (Pap) ‖ Holländer m.; Stoffmühle f. ‖ moulin m. à cylindre; cylindre m. / ~ (Pott) ‖ Mangelholz n.; Rollholz n. ‖ billette f.; rouleau m. / ~ (Pump) ‖ Zylinder m. ‖ cylindre m.; corps m.; barillet m. / air cooled ~ ‖ luftgekühlter Zylinder m. ‖ cylindre m. à refroidissement par air. / to bore a ~ ‖ einen Zylinder m. ausbohren ‖ aléser un cylindre m. / bore of ~ ‖ Zylinderbohrung f. ‖ alésage m. du cylindre. / ~ cast in one piece ‖ in einem Stück gegossener Zylinder m. ‖ monobloc m. / clearing ~ (Ore dress) ‖ Läutertrommel f. ‖ débourbeur m.; tambour m. débourbeur. / compression ~ ‖ Preßzylinder m. ‖ cylindre m. de pression. / ~ with detachable head ‖ Zylinder m. mit abnehmbarem Kopf ‖ cylindre m. à culasse amovible. / ~ with double walls ‖ Zylinder m. mit doppelter Wandung ‖ cylindre m. à double paroi. / eight ~ motor ‖ Achtzylindermotor m. ‖ moteur m. à huit cylindres. / engraving ~ (Print) ‖ Bildwalze f. ‖ cylindre m. gravé. / four ~ motor ‖ Vierzylindermotor m. ‖ moteur m. à quatre cylindres. / front cover of ~ ‖ Zylindervorderdeckel m. ‖ couvercle m. avant de cylindre. / graduated ~ (chem) ‖ Meßzylinder m. ‖ éprouvette f. graduée. / head of ~ ‖ Zylinderkopf m. ‖ culasse f. de cylindre. / high-pressure ~ ‖ Hochdruckzylinder m. ‖ cylindre m. à haute

pression. / horizontal ~ ‖ liegend angeordneter Zylinder m. ‖ cylindre m. couché. / jacketed ~ ‖ Zylinder m. mit Mantel ‖ cylindre m. à chemise. / low-pressure ~ ‖ Niederdruckzylinder m. ‖ cylindre m. à basse pression. / main ~ ‖ Hauptwalze f. ‖ cylindre m. principal. / measuring ~ ‖ Meßzylinder m. ‖ éprouvette f. graduée. / oscillating ~ ‖ schwingender oder oszillierender Zylinder m. ‖ cylindre m. oscillant. / ~ of the rag-engine (Pap) ‖ Holländerwalze f. ‖ cylindre m. de la pile défileuse. / reverse ~ ‖ Umkehrwalze f. ‖ cylindre m. de renversement de marche. / revolving ~s pl. ‖ umlaufende Zylinder mpl. ‖ cylindres mpl. rotatifs. / ribbed ~ ‖ Zylinder m. mit Kühlrippen ‖ cylindre m. à ailettes. / separately cast ~ ‖ einzelstehender Zylinder m. ‖ cylindre m. séparé. / sheet metal ~ ‖ Blechzylinder m. ‖ cylindre m. de tôle. / six-~ motor ‖ Sechszylindermotor m. ‖ moteur m. à six cylindres. / steam ~ ‖ Dampfzylinder m. ‖ cylindre m. à vapeur. / ~ of steam ‖ Dampfinhalt m. des Zylinders; Zylinder m. voll Dampf ‖ cylindrée f. de vapeur. / surface of a ~ ‖ Zylinderfläche f. ‖ surface f. cylindrique. / switch ~ (Electr) ‖ Schaltwalze f. ‖ tambour m. à segments. / triple-~ engine ‖ Dreizylindermaschine f. ‖ machine f. à trois cylindres. / tubular watercooled ~ open at both ends ‖ beiderseitig offener rohrförmiger wassergekühlter Zylinder m. ‖ cylindre m. en forme de tube à refroidissement par eau et ouvert des deux côtés. / turbine ~ ‖ Turbinengehäuse n. ‖ enveloppe f. de turbine. / twelve-~ motor ‖ Zwölfzylindermotor m. ‖ moteur m. à doux cylindres. / twin-~ engine ‖ Zweizylindermaschine f. ‖ machine f. à deux cylindres. / vibrating ~ see oscillating ~. / washing ~ see clearing ~. / water-cooled ~ ‖ wassergekühlter Zylinder ‖ cylindre m. à refroidissement d'eau. / working surface of ~ ‖ Zylinderlauffläche f. ‖ portée f. du cylindre. / ~ worn out of truth ‖ unrunder Zylinder m. ‖ cylindre m. ovalisé. / the ~ is worn out ‖ der Zylinder ist ausgelaufen ‖ le cylindre s'est ovalisé; le cylindre est usé.

cylinder, adjustment of the distance between ~ and types ‖ Einstellung f. des Abstandes zwischen Walze und Typen ‖ réglage m. d'écartement entre le cylindre et les caractères. / ~ back end ‖ Hinterdeckel m. ‖ couvercle m. arrière. / ~ barrel ‖ Zylindermantel m. ‖ corps m. du cylindre. / ~ base flange ‖ Zylinderunterflansch m. ‖ semelle f. de cylindre. / ~ bearing ‖ Walzenlager n. ‖ palier m. à rouleaux. / ~ block ‖ Zylinderblock m. ‖ bloc m. de cylindres. / ~ blowing engine ‖ Zylindergebläsemaschine f. ‖ machine f. soufflante à piston. / ~ body ‖ Zylinderkörper m. ‖ corps m. du cylindre. / ~ boiler ‖ Zylinderkessel m. ‖ chaudière f. cylindrique. / ~ bore ‖ Zylinderbohrung f. ‖ alésage m. du cylindre. / ~ boring apparatus ‖ Zylinderbohrapparat m. ‖ appareil m. à aléser les cylindres. / ~ double-spindle boring machine ‖ zweispindlige Zylinderbohrmaschine f. ‖ machine f. double à aléser les cylindres. / ~ burning ‖ Retortenverkohlung f. ‖ carbonisation f.

dans des cylindres. / ~ capacity ‖ Zylinderinhalt m.; Hubvolumen n. ‖ cylindrée f. / ~ casing ‖ Zylindermantel m. ‖ enveloppe f. du cylindre. / ~ casting ‖ Zylinderguß m. ‖ fonte f. à cylindres. / ~ charring see ~ burning. / ~ clearance / schädlicher Raum m. ‖ espace m. mort. ~ cleating ‖ Zylinderverkleidung f. ‖ enveloppe f. ou revêtement m. du cylindre. / ~ cloth pressing machine with firmly fixed cylinder ‖ Muldenpresse f. mit festgelagertem Zylinder ‖ presse f. cylindrique à une cuvette avec cylindre fixe. / ~ clothing ‖ Zylinderbekleidung f. ‖ enveloppe f. du cylindre. / ~ coking ‖ Retortenverkokung f. ‖ carbonisation f. par distillation. / ~ content ‖ Zylinderinhalt m. ‖ cylindrée f. / ~ cover ‖ Zylinderdeckel m. ‖ couvercle m. de cylindre. / ~ cover bolt ‖ Zylinderdeckelschraube f. ‖ boulon m. du couvercle de cylindre. / ~ diagram ‖ Volumendiagramm n. ‖ diagramme m. des cylindrées. / ~ diameter ‖ Zylinderdurchmesser m. ‖ diamètre m. ou alésage m. de cylindre. / triple-~ engine ‖ Dreizylindermaschine f. ‖ machine f. à trois cylindres. / ~ engraving works pl. ‖ Walzengravieranstalt f. ‖ atelier m. pour graver des cylindres. / ~ fittings pl. ‖ Zylinderausrüstung f. ‖ garnitures fpl. du cylindre. / ~ flange ‖ Zylinderflansch m. ‖ bride f. de cylindre. / ~ foot ‖ Zylindergestell n. ‖ pied m. de cylindre. / ~ fulling machine see ~ milling machine. / ~ gauge ‖ Zylinderkaliber n. ‖ calibre m. ou vérificateur m. à cylindre. / ~ glass ‖ Zylinderglas n. ‖ verre m. cylindrique. / ~ grinder ‖ Zylinderschleifmaschine f. ‖ rectifieuse f. pour cylindres. / ~ grinding machine ‖ Zylinderschleifmaschine f. ‖ machine f. à rectifier l'intériéur des cylindres. / ~ head ‖ Zylinderkopf m. ‖ culasse f. de cylindre. / detachable ~ head ‖ abnehmbarer Zylinderkopf m. ‖ culasse f. amovible ou rapportée. / integral ~ head ‖ angegossener Zylinderkopf m. ‖ culasse f. fondu d'un seul morceau de cylindre. / ~ head gasket ‖ Zylinderkopfdichtung f. ‖ joint m. de culasse. / ~ lagging see ~ cleating. / ~ length ‖ Zylinderlänge f. ‖ longueur f. du cylindre.

cylinder lens ‖ Zylinderlinse f. ‖ lentille f. cylindrique. / converging ~ ‖ sammelnde Zylinderlinse f. ‖ lentille cylindrique convergente. / dispersing ~ ‖ zerstreuende Zylinderlinse f. ‖ lentille f. cylindrique divergente. / ~ in mount ‖ Zylinderlinse f. in Fassung ‖ lentille f. cylindrique avec monture. / ~es pl. of different powers ‖ Zylinderlinsen fpl. von verschiedener Stärke ‖ lentilles fpl. cylindriques de puissance différente.

cylinder liner / Zylindereinsatz m.; Zylinderlaufbuchse f. ‖ corps m. intérieur ou fourreau m. du cylindre; chemise f. / ~ lubrication ‖ Zylinderschmierung f. ‖ lubrification f. ou graissage m. du cylindre. / ~ lubricator ‖ Zylinderschmiergefäß n. ‖ graisseur m. du cylindre. / ~ measuring instrument ‖ Zylindermeßgerät n. ‖ appareil m. pour la mesure des cylindres. / ~ mill ‖ Walzenmühle f. ‖ moulin m. à cylindres. / four-~ motor ‖ Vierzylindermotor m. ‖ moteur m. à quatre cylindres. / ~ needle ‖ Zylindernadel f. ‖ aiguille f. de cylindre. / ~ oil ‖ Zylinderöl n. ‖ huile f. à cylindre. / ~ oiler ‖ Zylinderschmierapparat m. ‖ graisseur m. de cylindre. / ~ packing ring ‖ Zylinderdichtungsring m. ‖ anneau m. ou bague f. de garniture de cylindre. / ~ polishing machine ‖ Zylinderpoliermaschine f. ‖ machine f. à polir l'intérieur des cylindres. / ~ pontoon ‖ Faßponton m. ‖ ponton m. cylindrique. / ~ port ‖ Öffnung f. für Ein- und Austritt des Dampfes ‖ orifice m. du cylindre.

cylinder press with felt wrapper control ‖ Muldenpresse f. mit Mitläuferfilzeinrichtung ‖ presse f. cylindrique avec feutre sans fin. / ~ with liftable cylinder ‖ Muldenpresse f. mit abhebbarem Zylinder ‖ presse f. cylindrique avec cylindre relevable. / ~ with long cloth take-off ‖ Muldenpresse f. mit langem Warenabzug ‖ presse f. cylindrique avec plieuse longue.

cylinder printing ‖ Walzendruck m. ‖ impression f. au rouleau. / ~ machine ‖ Walzendruckmaschine f. ‖ machine f. à imprimer au rouleau.

cylinder protecting device ‖ Walzenschutzvorrichtung f. ‖ dispositif m. pour la protection des cylindres.

cylinders pl. (Mill) ‖ Läufer mpl. ‖ meules fpl. roulantes. / ~ arranged in fan shape ‖ fächerförmig angeordnete Zylinder mpl. ‖ cylindres mpl. disposés en éventail. / ~ cast in block ‖ Zylinderblock m. ‖ cylindres mpl. en bloc. / ~ cast in pairs ‖ paarweise zusammengegossene Zylinder mpl.; Zweizylinderblock m. ‖ cylindres mpl. fondus en paires; bloc m. de cylindres jumelés. / revolving ~ ‖ umlaufende Zylinder mpl. ‖ cylindres mpl. rotatifs.

cylinder safety valve ‖ Zylindersicherheitsventil n. ‖ soupape f. de sûreté du cylindre. / ~ shaving machine ‖ Walzenabschleifmaschine f. ‖ machine f. à repasser les cylindres. / ~ shavings pl. ‖ Walzenspäne mpl. ‖ copeaux mpl. du cylindre. / ~ stiffening piece ‖ Zylinderverstrebung f. ‖ renforcement m. de cylindre. / ~ stuffing box ‖ Zylinderstopf-

buchse f. ‖ boîte f. à bourrage de cylindre. / ~ support ‖ Walzenstuhl m. ‖ cage f. à cylindres. / surface of the ~ ‖ Zylinderfläche f. ‖ surface f. cylindrique. / ~ trip (Print) ‖ Druckabsteller m. ‖ dispositif m. de soulèvement du cylindre d'impression. / ~ volume ‖ Zylinderinhalt m.; Zylindervolumen n. ‖ volume m. du cylindre; cylindrée f. / ~ wall ‖ Zylinderwand f. ‖ paroi f. du cylindre. / cooling of ~ walls ‖ Kühlung f. der Zylinderwandung ‖ refroidissement m. des parois du cylindre. / working surface of ~ ‖ Zylinderlauffläche f. ‖ portée f. du cylindre.

cylindrical ‖ zylindrisch; walzenförmig ‖ cylindrique. / ~ bending of boiler shells ‖ zylindrisches Einrollen n. von Kesselschüssen ‖ roulage m. cylindrique des tôles de chaudières. / ~ boiler ‖ Zylinderkessel m. ‖ chaudière f. cylindrique. / ~ cutter ‖ Walzenfräser m. ‖ fraise f. cylindrique. / ~ grinding machine ‖ Rundschleifmaschine f. ‖ machine f. à rectifier (extérieurement) les pièces cylindriques. / ~ key ‖ Rundkeil m. ‖ clavette f. ronde. / ~ milling machine ‖ Zylinderwalke f. ‖ fouleuse f. cylindrique. / ~ part of a boiler ‖ Langkessel m. ‖ corps m. cylindrique de la chaudière. / ~ pin ‖ Zylinderstift m. ‖ goupille f. cylindrique. / ~ ring armature ‖ Zylinderringanker m. ‖ induit m. en anneau cylindrique. / ~ screw with rounded-off top edge ‖ Zylinderschraube f. mit runder oberer Kopfkante f. ‖ vis f. à tête cylindrique avec arête ronde supérieure. / ~ screw with sharp top edge ‖ Zylinderschraube f. mit scharfer oberer Kopfkante f. ‖ vis f. à tête cylindrique avec arête aigue supérieure. / ~ spiral spring ‖ zylindrische Spiralfeder f. ‖ ressort m. à boudin cylindrique. / ~ surface ‖ Zylinderfläche f. ‖ surface f. cylindrique.

cyma reversa (Arch) ‖ Kehlstoß m.; Kehlleiste f. ‖ talon m.; gueule f. renversée; cymaise f. lesbienne.

cymbal ‖ Zimbel f.; Schallbecken n. ‖ cymbale f.

cymometer (Radio) ‖ Wellenmesser m. ‖ cymomètre m.

cymoscope (Radio) ‖ Zymoskop n. ‖ cymoscope m.

cypress ‖ Zypresse f. ‖ cyprès m. / ~ grass ‖ Zypergras n. ‖ souchet m. / ~ wood ‖ Zypressenholz n. ‖ bois m. de cyprès.

Cyprus wine ‖ Zyperwein m. ‖ vin m. de Chypre.

cystotome ‖ Zystotom n. ‖ kystotome m.

cytidine phosphate ‖ Zytidinphosphat n. ‖ phosphate m. de cytidine.

cytogenesis (Biol) ‖ Zellbildung f. ‖ formation f. de cellules.

D

D slide valve ‖ Muschelschieber m.; D-Schieber m. ‖ tiroir m. à coquille *ou* en D.
dab, to ‖ abklatschen; klischieren ‖ clicher.
dab of the compass needle ‖ Nut m. der Kompaßnadel ‖ chape f. *ou* chaperon m. de l'aiguille du compas.
dabbed drawing ‖ gewischte Zeichnung f. ‖ dessin m. à l'estompe.
dabbing machine ‖ Klischiermaschine f.; Abklatschmaschine f. ‖ machine f. à clicher.
dabber (Spinn) ‖ Einschlagbürste f. ‖ brosse f. enfonceuse.
dagger ‖ Dolch m. ‖ poignard m. / ~ (Print) ‖ Kreuzzeichen n. ‖ croix f.
dagging the yarn ‖ Besprengen n. des Garnes mit Wasser ‖ arrosage m. du fil avec de l'eau.
dag wool ‖ Kotspitzen fpl. der Wolle ‖ crottins mpl. de la laine.
daily allowance ‖ Tagegelder npl. ‖ salaire m. journalier. / ~ output ‖ Tagesleistung f.; Tageserzeugung f. ‖ capacité f. de production par jour; débit m. journalier; production f. journalière. / ~ output up to x tons ‖ Tagesleistung f. bis zu x ts ‖ rendement m. allant jusqu'à x tonnes par jour. / ~ paper ‖ Tageszeitung f. ‖ feuille f. quotidienne. / ~ turnout *see* ~ output. / ~ wages pl. ‖ Tagelohn m. ‖ journée f.; salaire m. d'une journée. / ~ wages pl. (Mine) ‖ Schichtlohn m. ‖ salaire m. fixe.
dainty provisions pl. ‖ Feinkostware f. ‖ comestibles mpl. fins.
dairy ‖ Milchwirtschaft f.; Molkerei f.; Meierei f. ‖ laiterie f. / accessory products for dairies ‖ Molkereihilfsstoffe mpl. ‖ produits mpl. accessoirs pour laiteries.
dairy carter ‖ Milchwagenkutscher m. ‖ charretier m. de laiterie; cocher-laitier m. / ~ equipment ‖ Molkereieinrichtung f. ‖ installation f. de laiterie. / ~ farm ‖ Meierei f.; Pachtgut n. ‖ métairie f.; ferme f. / ~ farmer ‖ Milchpächter m. ‖ laitier-fermier m. / ~ implements pl. ‖ Molkereigerät n. ‖ instrument m. de laiterie. / ~ machine ‖ Molkereimaschine f. ‖ machine f. de laiterie.
dairymaid ‖ Milchmädchen n. ‖ laitière f.
dairyman ‖ Milchmann m. ‖ laitier m.
dairy manager ‖ Molkereiverwalter m. ‖ gérant m. de laiterie. / ~ owner ‖ Molkereibesitzer m. ‖ propriétaire m. de laiterie. / ~ plant ‖ Molkereianlage f. ‖ laiterie f. / ~ product ‖ Molkereierzeugnis n. ‖ produit m. de laiterie.
Daltonism ‖ Farbenblindheit f.; Rotblindheit f. ‖ achromatopsie f.; daltonisme m.
dam, to ‖ stauen ‖ endiguer; barrer. / ~ in ‖ eindämmen ‖ diguer; enfermer d'une digue. / ~ up ‖ eindämmen; abdämmen ‖ diguer; enfermer d'une digue; détourner par une digue. / ~ up a water course ‖ einen Wasserlauf m. stauen ‖ barrer un cours d'eau.
dam ‖ Damm m.; Staudamm m. ‖ digue f. / ~ across a valley ‖ Talsperre f. ‖

barrage-réservoir m.; digue f. de vallée. / auxiliary ~ ‖ Vorsperre f. ‖ barrage m. de sûreté. / building of dikes and ~s across a valley ‖ Talsperrenbau m. ‖ construction f. de digues et de barrages de vallée. / ~ of brickwork (Mine) ‖ Mauerdamm m. ‖ barrage m. en maçonnerie. / ~ of a harbour ‖ Hafendamm m.; Mole f.; Wellenbrecher m. ‖ jetée f. *ou* digue f. d'un port; môle m. de port; brise-lames m. / ~ in loose stuff ‖ Sickerdamm m. ‖ barrage m. criblant. / movable ~ ‖ bewegliches Wehr n. ‖ barrage m. mobile. / movable ~ with stores ‖ Rolltafelwehr n. ‖ barrage m. à rideaux. / permanent ~ ‖ festes Wehr n. ‖ barrage m. fixe. / ~ of a pond ‖ Teichdamm m. ‖ bachasse f. / ~ of quarry stone masonry ‖ Sperrmauer f. in Bruchsteinmauerung ‖ mur m. de barrage en pierre de carrière *ou* en moellons. / segment ~ ‖ Segmentwehr n. ‖ barrage m. à secteur *ou* à segment. / stone ~ ‖ massives Wehr n. ‖ barrage m. en maçonnerie. / wooden ~ ‖ hölzernes Wehr n. ‖ barrage m. en bois.
damage, to ‖ beschädigen ‖ endommager; avarier; détériorer.
damage ‖ Schaden m. ‖ dommage m.; dégât m.; avarie f.; préjudice m. / ~ amounting to ‖ Schaden m. in Höhe von / dégâts mpl. s'élèvant à. / appraisal of ~ ‖ Schadensabschätzung f. ‖ taxation f. du dommage. / ~ assessed at ‖ Schaden m. festgesetzt auf ‖ dégâts mpl. évalués à. / to call-in experts to determine the ~ ‖ zur Ermittlung eines Schadens einen Sachverständigen m. heranziehen ‖ appeler des experts pour la reconnaissance *ou* l'évaluation des dommages. / ~ caused by ‖ Schaden m. verursacht durch ‖ dégâts mpl. causés par. / ~ caused by fire ‖ Feuerschaden m. ‖ dommage m. causé par le feu; sinistre m. / ~ caused by hail ‖ Hagelschaden m.; Hagelschlag m. ‖ dommage m. causé par la grêle; coup m. de grêle. / to claim ~s pl. ‖ Entschädigung f. fordern; Schadenersatz m. beanspruchen ‖ réclamer des dommages mpl. *ou* la réparation d'un dommage. / claim of ~s ‖ Entschädigungsanspruch m. ‖ réclamation f. / ~ to cutter ‖ Fräserbruch m. ‖ rupture f. de fraise. / deliberate and unlawful ~ ‖ vorsätzliche und rechtswidrige Sachbeschädigung. ‖ dommage m. prémédité et illégal causé aux choses. / to determine the amount of ~ ‖ den Umfang m. eines Schadens feststellen ‖ fixer l'étendue f. d'un dommage. / to do ~ ‖ Schaden m. anrichten ‖ occasionner. / duty to compensate for ~ ‖ Schadensersatzpflicht f. ‖ obligation f. de dédommagement. / ~ estimated at ‖ Schaden m. geschätzt auf ‖ dégâts mpl. estimés. à / to incur ~ ‖ eine Beschädigung f. verschulden ‖ être responsable d'un dommage. / mechanical ~ ‖ mechanische Beschädigung f. ‖ dégât m. *ou* endommagement m. mécanique. / ~ done by moths ‖ Mottenfraß m. ‖ mangeure

f. de teignes. / to be responsible for ~ ‖ schadenersatzpflichtig sein ‖ être responsable du dommage causé. / ~ by sea water ‖ Seeschaden m. ‖ avarie f. / serious ~ ‖ ernstlicher Schaden m. ‖ dégâts mpl. sérieux. / to settle for ~ ‖ einen Schaden m. ersetzen ‖ régler un dommage; dédommager / slight ~ ‖ geringer Schaden m. ‖ dégâts mpl. légers. / ~ by sea ‖ Havarie f.; Seeschaden m. ‖ avarie f.
damaged ‖ schadhaft ‖ endommagé; défectueux; avarié. / arrived ~ ‖ beschädigt angekommen ‖ arrivé endommagé *ou* avarié. / ~ by ‖ beschädigt durch ‖ endommagé par. / ~ goods pl. ‖ Ausschußware f.; schadhafte Ware f. ‖ marchandise f. de rebut; pacotille f.; camelote f.; marchandises fpl. avariées. / ~ part ‖ Knick m. ‖ fêlure f.; brissure f.; faux pli m.; coude m.
damages pl. ‖ Schaden m. ‖ avaries fpl.; dommage m.; dégâts mpl. / ~ (Jur) ‖ Schadenersatz m. ‖ dommages mpl. intérêts.
damaging ‖ Beschädigung f. ‖ endommagement m.
damask, to ‖ damastartig weben ‖ damasser.
damask ‖ Damast m.; Damaststoff m. ‖ damas m.; linge m. damassé; damassé m. / furniture ~ ‖ Möbeldamast m. ‖ tissu m. damassé pour meubles. / half ~ ‖ Halbdamast m. ‖ damas m. cafard. / manufactory of ~s ‖ Damastweberei f. ‖ damasserie f. / silk ~ ‖ Seidendamast m. ‖ damas m. de soie.
damask blade ‖ Damaszener Klinge f. ‖ lame f. damasquinée; damas m. / ~ linen ‖ Damastleinwand f. ‖ damassé m. / ~ machine ‖ Damastmaschine f. ‖ machine f. pour damassé. / ~ paper ‖ Damastpapier n. ‖ papier-linge m.; papier-étoffe m. / ~ pattern (Weav) ‖ Damastmuster n. ‖ damassure f. / ~ steel ‖ Damaszenerstahl m. ‖ lame f. damasquinée; damas m. / ~ table-linen ‖ Leinendamast m.; Damasttischzeug m. ‖ linge m. damassé. / ~ towel ‖ Damasthandtuch n. ‖ essuie-main m. en damassé. / ~ weaver ‖ Damastweber m. ‖ damasseur m.
dam beam seal ‖ Dammbalkenverschluß m. ‖ fermeture f. de digue à poutres en acier. / ~ building ‖ Dammbau m.; Deichbau m. ‖ construction f. de digue. / ~ door for mining plants ‖ Dammtür f. für Bergwerksbetriebe ‖ porte f. de barrage pour services miniers. / ~ drawing water ‖ Schöpfbuhne f. ‖ épi m. à puiser.
dame's violet ‖ Nachtviole f. ‖ julienne f.
dam keeper ‖ Deichmeister m. ‖ garde-chaussée m. privé.
dammar gum ‖ Dammarharz n. ‖ gomme f. de dammar. / ~ lac ‖ Dammarlack m. ‖ laque f. de dammar. / ~ resin ‖ Dammarharz n. ‖ résine f. de dammar. / ~ substitute ‖ Dammarersatz m. ‖ dammar m. artificiel.
damming of torrents ‖ Wildbacheindämmung f. ‖ correction f. des torrents; dé-

fenses fpl. contre les torrents. / sluice for ~ plants ‖ Schütz n. für Stauanlagen ‖ vanne f. de barrage.

damp ‖ feucht; naß ‖ humide. / ~ steam ‖ nasser Dampf m. ‖ vapeur f. humide.

damp, to ‖ anfeuchten ‖ humecter; humidifier; mouiller. / ~ (Phys) ‖ dämpfen ‖ amortir. / ~ the clockwork ‖ das Uhrwerk n. bremsen ‖ freiner le mouvement m. d'horlogerie. / ~ the shocks ‖ die Stöße mpl. dämpfen ‖ amortir les chocs mpl.

damp ‖ Dunst m.; Feuchtigkeit f. ‖ humidité f. / ~ (Mine) ‖ Schwaden m. ‖ mofette f. / after ~ ‖ Nachschwaden m. ‖ gaz mpl. délétères. / black ~ (Mine) ‖ schlagende Wetter npl. ‖ grisou m. / choke ~ see after ~. / fire ~ see black ~. / fulminating ~ see black ~.

dampener, glass-roll ~ ‖ Glaswalzenanfeuchter m. ‖ cylindre m. de verre d'humecteur.

damper (Boil) ‖ Zugregler m. ‖ registre m. / ~ (Mill) ‖ Feuchter m.; Netzer m. ‖ mouilleur m. / ~ (Phys) ‖ Dämpfer m. ‖ amortisseur m.; sourdine f. / air ~ ‖ Lufteinlaßklappe f. ‖ clapet m. d'entrée d'air. / ash ~ ‖ Aschfallklappe f. ‖ porte f. du cendrier. / sound ~ ‖ Dämpfungsmittel n. ‖ sourdine f. / ~ of a stove ‖ Ofenrohrklappe f. ‖ clef f. d'un tuyau de poêle. / ~ of superheater ‖ Überhitzerdämpfer m. ‖ étouffoir m. de surchauffeur. / vibration ~ ‖ Schwingungsdämpfer m. ‖ amortisseur m. de vibration.

damper plate ‖ Luftschieber m. ‖ registre m.

damping (Phys) ‖ Dämpfung f.; Abdämpfung f. ‖ amortissement m.; amortissage m. / ~ of the arc ‖ Dämpfung f. des Lichtbogens ‖ amortissement m. de l'arc. / complete ~ ‖ vollkommene Dämpfung f. ‖ amortissement m. parfait. / degree of ~ ‖ Dämpfungsgrad m. ‖ degré m. d'amortissement. / ~ down of the coke ‖ Löschen n. des Kokses ‖ extinction f. du coke. / high ~ (Radio) ‖ große Dämpfung f. ‖ amortissement m. élevé. / ~ of the oscillation ‖ Abklingen n. der Schwingung ‖ décroissement m. ou amortissement m. de l'oscillation.

damping circuit (Electr) ‖ Dämpfungskreis m. ‖ circuit m. amortisseur. / ~ current ‖ Dämpfungsstrom m. ‖ courant m. amortisseur. / ~ curve ‖ Dämpfungskurve f. ‖ courbe f. d'amortissement. / ~ device ‖ Dämpfungsvorrichtung f. ‖ dispositif m. amortisseur. / ~ factor ‖ Dämpfungsfaktor m. ‖ facteur m. d'amortissement. / ~ measurement ‖ Dämpfungsmessung f. ‖ mesure f. de l'amortissement. / ratio of ~ ‖ Dämpfungsverhältnis m. ‖ rapport m. d'amortissement. / ~ reduction ‖ Dämpfungsverminderung f. ‖ réduction f. de l'amortissement. / ~ resistance ‖ Dämpfwiderstand m. ‖ résistance f. à l'amortissement.

dam plate (Metal) ‖ Wallplatte f.; Schlackenblech n. ‖ plaque f. de dame.

dampness ‖ Feuchtigkeit f. ‖ humidité f.

damp-proof ‖ gegen Feuchtigkeit f. beständig ‖ imperméable à l'humidité f.

damps pl. in mines ‖ Schlagwetter npl. ‖ gaz mpl. délétères dans les mines.

damps pl. in mines ‖ Grubenwetter n. ‖ air m. ou airage m. dans les mines.

dancing pump see ~ shoe. / ~ room ‖ Tanzsaal m.; Ballsaal m. ‖ salle f. de danse. / ~

shoe ‖ Ballschuh m.; Tanzschuh m. ‖ soulier m. de bal ou de soirée.

dandelion root ‖ Löwenzahnwurzel f. ‖ racine f. de pissenlit.

danger ‖ Gefahr f. ‖ danger m.; péril m. / ~ of back-firing of the flame ‖ Rückschlaggefahr f. der Flamme ‖ danger m. de retour de flamme. / ~ of explosion ‖ Explosionsgefahr f. ‖ danger m. d'explosion. / ~ from fall of rock ‖ Steinschlaggefahr f. ‖ danger m. de chute de pierres. / ~ of fire ‖ Feuersgefahr f. ‖ danger m. d'incendie. / ~ of life ‖ Lebensgefahr f. ‖ danger m. de mort.

danger, absence of ~ ‖ Gefahrlosigkeit f. ‖ absence f. de danger. / ~ alarm system ‖ Gefahrmeldeanlage f. ‖ installation f. d'avertisseurs de danger. / ~ card ‖ Warnungstafel f. ‖ tableau m. d'avis affichée.

dangerous section ‖ gefährlicher Querschnitt m. ‖ section f. dangereuse. / ~ stresses pl. in the neck of the pin ‖ gefährliche Beanspruchung f. in der Hohlkehle des Zapfens ‖ effort m. dangereux au congé du tourillon.

danger report ‖ Gefahrenmeldung f. ‖ avis m. ou annonce f. de danger. / ~ signal ‖ Alarmsignal n. ‖ signal m. d'alarme. / ~ signal (Railw) ‖ Notsignal n. ‖ signal m. de détresse ou d'arrêt. / warning of ~ ‖ Gefahrenmeldung f. ‖ avis m. ou avertissement m. de danger. / to carry to a point outside the ~ zone ‖ bis außerhalb der Gefahrenzone f. führen ‖ conduire jusqu'en dehors de la zone dangereuse.

Daniell's cell ‖ Daniell-Element n. ‖ élément m. de Daniell.

dark ‖ dunkel ‖ noir. / ~ adaptation ‖ Dunkelanpassung f. ‖ adaptation f. à l'obscurité. / ~ beer extract ‖ Farbbierauszug m. ‖ extrait m. concentré de colorant de bière. / ~ blue ‖ dunkelblau ‖ bleu foncé. / ~ body ‖ dunkler Körper m. ‖ corps m. de couleur sombre ou foncée. / ~ brown ‖ dunkelbraun ‖ brun foncé. / ~ coloured ‖ dunkelfarbig ‖ de couleur f. foncée. / ~ discharge ‖ dunkle Entladung f. ‖ décharge f. obscure.

darken, to ‖ dunkeln; nachdunkeln ‖ se foncer.

darkening device for rooms ‖ Saalverdunkler m. ‖ interrupteur m. de courant pour salles. / ~ plant ‖ Verdunkelungsanlage f. ‖ installation f. d'obscurcissement.

dark-field condenser ‖ Dunkelfeldkondensor m. ‖ condensateur m. à fond noir. / ~ equipment ‖ Dunkelfeldeinrichtung f. ‖ dispositif m. pour l'observation en fond noir. / ~ observation ‖ Dunkelfeldbeobachtung f. ‖ observation f. sur fond noir.

dark glass ‖ Sonnenblendglas n. ‖ verre m. sombre. / ~ gray ‖ dunkelgrau ‖ gris brun; gris foncé. / ~ green ‖ dunkelgrün ‖ vert foncé. / ~ grey see ~ gray. / bright line on a ~ ground ‖ heller Faden m. auf dunklem Grunde ‖ fil m. clair sur fond noir. / ~ ground illumination attachment ‖ Dunkelfeldbeleuchtungseinrichtung f. ‖ dispositif m. d'éclairage à fond noir. / ~ ground stop ‖ Dunkelfeldblende f. ‖ diaphragme m. à fond noir.

darkness ‖ Dunkelheit f. ‖ obscurité f.

dark red heat ‖ Dunkelrotglut f. ‖ chaude f. sombre; chaude f. rouge naissant. / the forging must be finished at ~ ‖ das Schmieden muß bei Dunkelrotglut be-

endet sein ‖ le forgeage doit être fini à la chaleur rouge-sombre.

dark-room ‖ Dunkelkammer f. ‖ chambre noire ou obscure. / ~-lamp ‖ Dunkelkammerlampe f. ‖ lampe f. à chambre obscure.

dark-slide, double ~ (Photo) ‖ Doppelschlittenkassette f. ‖ châssis-double m. à coulisse; châssis m. à deux chariots. / ~ loaded with a plate ‖ mit einer Platte geladene Kassette f. ‖ châssis m. chargé d'une plaque. / ~ carriage ‖ Kassettenschlitten m. ‖ chariot m. du châssis.

dark violet ‖ dunkelviolett ‖ violet foncé.

darn, to ~ stockings pl. ‖ Strümpfe mpl. stopfen ‖ ravauder ou repriser des bas mpl.

darner ‖ Stopferin f. ‖ rentrayeuse f.

darning, fine ‖ Kunststopfen n. ‖ stoppage m. / ~ needle ‖ Stopfnadel f. ‖ aiguille f. à repriser ou à ravauder. / cotton ~ yarn ‖ baumwollenes Stopfgarn n. ‖ fil m. de coton à repriser.

dartre ointment ‖ Flechtensalbe f. ‖ onguent m. ou pommade f. antidartreux.

dash, to ‖ besprengen; bespritzen ‖ arroser.

dash (Mach) ‖ Riß m. oder Vorzeichnung f. auf dem Werkstück ‖ trait m. de repère. / ~ (Print) ‖ Gedankenstrich m. ‖ moins m.; trait m. suspensif; tiret m.

dashboard ‖ Spritzbrett n. ‖ garde-crotte m.; tablier m.

dash counting (Tel) ‖ Strichzählung f. ‖ comptage m. exprimé par des traits.

dashpot ‖ Stoßdämpfer m.; Bremszylinder m. ‖ amortisseur m. à piston; dash-pot m. / adjustable ~ ‖ regelbarer Stoßdämpfer m. ‖ amortisseur m. à cylindre réglable. / ~ type shock absorber ‖ hydraulischer Stoßfänger m. ‖ amortisseur m. à liquide.

data pl. see datum. / to correct the ~ ‖ die Angaben fpl. verbessern ‖ corriger les indications fpl.

dataller ‖ Tagelöhner m. ‖ journalier m.

date, to ‖ datieren ‖ dater. / ~ forward ‖ vordatieren ‖ postdater.

date ‖ Datum n. ‖ date f.; quantième m. / ~ (Bot) ‖ Dattel f. ‖ datte f. / ~ of delivery ‖ Lieferfrist f. ‖ délai m. ou terme m. de livraison. / due ~ see ~ of maturity. / ~ of maturity ‖ Fälligkeitstag m.; Verfalltag m. ‖ échéance f.; terme m. de l'échéance. / ~ of payment ‖ Zahlungstermin m. ‖ date m. de paiement. / ~ of receipt of tenders ‖ Termin m. für Submissionsangebote ‖ terme m. ou délai m. pour les enchères.

date block ‖ Abreißblock m.; Merkblock m.; Kalenderblock m.; Notizblock m. ‖ bloc m. notes; éphéméride f.; bloc m. pour calendrier. / ~ indicating watch ‖ Datumanzeigeruhr f. ‖ montre-quantième f. / ~ meal ‖ Dattelmehl n. ‖ farine f. de dattes.

dater ‖ Datumstempel m. ‖ dateur m.

dates pl., critical ~ ‖ kritische Daten pl. ‖ constantes fpl. critiques. / ~ of delivery should not be regarded as binding upon somebody ‖ Lieferfristangaben fpl. gelten als unverbindlich ‖ les délais mpl. de livraisons s'entendent sans engagement.

date stamp (Office) ‖ Tagesstempel m. ‖ timbre m. dateur. / ~ (Print) ‖ Datumstempel m. ‖ dateur m.-vitesse m.

dating device ‖ Datiervorrichtung f. ‖ dispositif m. dateur.

datum ‖ Gegebene n.; gegebene Tatsache f.; Angabe f. ‖ donnée f. / incorrect ~ ‖ fehlerhafte Angabe f. ‖ donnée f. incorrecte / ~ level see ~ line. / ~ line (Surv) ‖ Standebene f.; Grundebene f. ‖ plan m. de niveau. / ~ plane see ~ line. / ~ point (Surv) ‖ Normalfixpunkt m. ‖ point m. de repère. / ~ surface ‖ Normalnull f.; Normalhorizont m. ‖ niveau m. vraie ou moyen d'une mer.

daub, to ~ a foundry ladle ‖ einen Gießlöffel m. ausschmieren ‖ garnir à l'intérieur une poche de fonderie.

dauber (Brew) ‖ Faßpicher m. ‖ enduiseur m. de tonneaux. / ~ of paper (Print) ‖ Papiersudler m. ‖ mâchurat m.

davit ‖ Davit m. ‖ davier m. / quarter ~ ‖ Seitendavit m. ‖ bossoir m. ou portemanteau m. des embarcations de côté. / ship's ~ ‖ Schiffsdavit m. ‖ davier m. de bateau.

Davy-man (Mine) ‖ Lampenwärter m. ‖ lampiste m.

Davy's safety lamp ‖ Davysche Sicherheitslampe f. ‖ lampe f. de sûreté de Davy.

dawn ‖ Morgendämmerung f. ‖ crépuscule m. du matin.

day ‖ Tag m. ‖ journée f.; jour m. / ~ (Build) ‖ Öffnung f.; Lichte n. ‖ jour m.; ouverture f. / by the ~ ‖ am Tage m. ‖ pendant le jour. / capacity per working ~ ‖ Tagesleistung f. ‖ capacité f. de production journalière; débit m. journalier. / for ~s ‖ tagelang ‖ des jours mpl. entiers. / on the ~ after ‖ tags darauf ‖ le lendemain. / on the ~ previous ‖ tags zuvor ‖ la veille. / ~ of rest ‖ Ruhetag m. ‖ jour m. de repos. / ~ of settlement ‖ Vergleichstag m.; Vergleichstermin m. ‖ jour m. de conciliation. / ~ without sunshine ‖ sonnenloser Tag m. ‖ jour m. sans soleil. / ~ of a window ‖ Lichte n. eines Fensters ‖ jour m. de fenêtre.

day-bomber ‖ Tagbombenflugzeug n. ‖ avion m. de bombardement de jour. / ~ bombing aeroplane see ~ -bomber.

daybook ‖ Tagebuch n. ‖ agenda m. ‖ Kladde f. ‖ brouillon m.; brouillard m.

daybreak ‖ Tagesanbruch m. ‖ pointe f. du jour.

dayflight ‖ Tagflug m. ‖ vol m. de jour. / ~ height (Build) ‖ Lichtenhöhe f. ‖ hauteur f. libre ou du jour. / ~ labourer ‖ Tagelöhner m.; Handlanger m. ‖ journalier m.; manœuvrier m.

daylight ‖ Tageslicht n. ‖ lumière f. du jour; plein jour m. / artificial ~ ‖ künstliches Tageslicht n. ‖ lumière f. du jour artificielle.

daylight; fluctuation of ~ ‖ Schwankung f. des Tageslichtes ‖ fluctuation f. de la lumière du jour. / ~ lamp ‖ Tageslichtlampe f. ‖ lampe f. à lumière solaire. / ~ photograph ‖ Tagesaufnahme f. ‖ prise f. de vue diurne ou pendant le jour.

dayman ‖ Tagelöhner m. ‖ journalier m.

days pl. of grace ‖ Respekttage mpl. ‖ jours mpl. de répit ou de grâce. / ~ of respite see ~ of grace.

day shift ‖ Tagschicht f. ‖ équipe f. ou poste f. de jour. / ~ shirt ‖ Oberhemd n. ‖ chemise f. de jour.

dayspring ‖ Tagesanbruch m. ‖ jour m. naissant; aube f.; pointe f. du jour.

day-tariff ‖ Tagestarif m. ‖ tarif m. de jour. / ~-thunderstorm ‖ Tagesgewitter

n. ‖ orage m. de jour. / ~-wages pl. ‖ Tagelohn m. ‖ salaire m. à la journée. / ~-wage man ‖ Tagelohnarbeiter m. ‖ ouvrier m. payé par salaire journalier.

daywork ‖ Tagarbeit f. ‖ travail m. de jour ou à la journée.

dazzling ‖ blendend ‖ éblouissant.

dead ‖ matt; glanzlos ‖ mat. / ~ (Arch) ‖ falsch; nachgeahmt ‖ faux; imité. / ~ (Electr) ‖ spannungslos ‖ sans tension f. / ~ before the wind ‖ platt vor dem Winde m. ‖ droit de l'arrière m. / to make ~ (Electr) ‖ stromlos machen ‖ couper le courant; déconnecter.

dead bacteria ‖ Bakterienleiche f. ‖ cadavres mpl. de bactéries. / ~ beat (Phys) ‖ aperiodisch; gedämpft ‖ apériodique. / ~ calm ‖ vollständige Windstille f. ‖ calme m. plat.

dead centre (Mach) ‖ Totpunkt m. ‖ point m. mort. / ~ (Mach tool) ‖ Körnerspitze f. ‖ contre-pointe f.

dead conductor ‖ stromloser Leiter m. ‖ conducteur m. sans courant. / ~ dipping of brass ‖ Mattieren n. des Messingblechs ‖ décapage m. du laiton.

deaden, to ‖ mattieren; färben ‖ mater; donner le mat; mettre au mat.

dead end binding (El line) ‖ Endbund m. ‖ ligature f. à l'isolateur d'arrêt. / ~ end sliding ‖ totes Gleis n. ‖ voie f. en cul-de-sac ou en impasse. / ~ ending (Aerial lines) ‖ Abspannbund m. ‖ ligature f. à l'isolateur de tension.

deadening ‖ Mattieren n. ‖ mise f. au mat; matage m. / ~ matter ‖ Mattfarbe f. ‖ mat m. / ~ vat ‖ Mattiertonne f. ‖ tonneau m. au mat.

deadeye (Shipb) ‖ Jungfernblock m. ‖ cap m. de mouton.

dead face (Arch) ‖ blinde Fassade f.; Blendfassade f. ‖ façade f. feinte. / ~ floor (Carp) ‖ Blendboden m.; Blindboden m. ‖ fausse-aire f.; faux-parquet m. / ~ gilding ‖ Mattvergoldung f. ‖ dorage m. mat. / ~ gold ‖ mattes Gold n. ‖ or m. mat. / ~-grinded (Glass) ‖ matt geschliffen ‖ dépoli.

deadhead (Found) ‖ Gießkopf m.; verlorener Kopf m. ‖ masselotte f.

dead hole ‖ blindes Loch n. ‖ trou m. aveugle.

deadlight ‖ Lukendeckel m.; Verschlußdeckel m. für Kajütfenster ‖ couvercle m. d'un hublot ou d'une écoutille.

dead line (Electr) ‖ Indifferenzlinie f.; ligne f. neutre. / ~ load ‖ Totlast f. ‖ poids m. mort.

deadlock (Locksm) ‖ Einriegelschloß n. ‖ serrure f. à pêne dormant.

deadman's eye ‖ kleines rundes Fenster n. ‖ œil-de-bœuf m.

dead mould (Found) ‖ verlorene Gießform f. ‖ moule m. perdu. / ~ nettle ‖ Taubnessel f. ‖ ortie f. blanche. / ~ plate of a furnace ‖ Feuerplatte f. ‖ sole f. ou table f. de foyer. / ~ point ‖ Totpunkt m. ‖ point m. mort. / the engine has no ~ points ‖ die Maschine hat keinen toten Punkt ‖ la machine ne possède pas de point mort. / inner ~ point position ‖ innere Totpunktlage f. ‖ position f. de point mort inférieur. / ~ reckoning ‖ Gissung f. ‖ gisement m. / ~ rock (Mine) ‖ Quergestein n. ‖ roche f. transversale ou des parois.

deads pl. (Mine) ‖ taubes Gestein n. ‖ gangue f.; matière f. ou roche f. stérile.

dead-smooth file ‖ Doppelschlichtfeile f.; Feinschlichtfeile f. ‖ lime f. superfine ou très douce. / ~ standing tree ‖ dürres Holz n. ‖ bois m. mort en estant. / ~ steam ‖ Abdampf m. ‖ vapeur f. épuisée ou passive. / ~-stroke hammer ‖ Federhammer m. ‖ marteau m. à ressort. / ~-time (War impl) ‖ Verzugszeit f.; Ladeverzug m.; Befehlsverzug m. ‖ temps m. mort. / ~ wall ‖ blinde oder fensterlose Mauer f. ‖ mur m. orbe. / ~ water ‖ Kielwasser n.; Sog m. ‖ remous m. / ~ weight ‖ Verpackungsgewicht n.; Tara f. ‖ poids m. mort; tare f.

deadwood (Forest) ‖ abgestorbenes Holz n.; Reisig n. ‖ bois m. mort. / ~ (Shipb) ‖ Totholz n. ‖ bois m. mort.

dead work (Shipb) ‖ totes Werk n. ‖ œuvres fpl. mortes.

deaf and dumb ‖ taubstumm ‖ sourd et muet; sourd-muet. / ~ alphabet ‖ Taubstummenalphabet n. ‖ alphabet m. des sourds-muets.

deaf, apparatus for the ~ ‖ Hörgerät n. ‖ appareil m. acoustique.

deafness ‖ Taubheit f. ‖ surdité f.

deal, to ~ individually ‖ individuell behandeln ‖ traiter d'une façon individuelle.

deal ‖ Menge f.; Teil m. ‖ quantité f. / ~ (Bot) ‖ Tannenholz n. ‖ bois m. de sapin. / ~ (Carp) ‖ Holzbohle f.; Diele f. ‖ bois m. blanc; planche f.; ais m. / dull-edged ~ ‖ ungesäumte Diele f. ‖ planche f. flacheuse. / other side in a ~ ‖ Vertragsgegner m. ‖ contre-partie f.

dealer ‖ Händler m. ‖ négociant m. / ~ in hides ‖ Fellhändler m. ‖ pelletier m. / ~ in stocks ‖ Aktienhändler m. ‖ agioteur m.; courtier m. / ~ in wood ‖ Holzhändler m. ‖ marchand m. de bois.

deal level circuit ‖ Leerstromkreis m. ‖ courant m. en circuit ouvert. / ~ wood ‖ Tannenholz n. ‖ bois m. de sapin.

dear ‖ teuer; wertvoll ‖ cher.

dearness of provisions ‖ Teuerung f. ‖ cherté f.; disette f.; renchérissement m.

death ‖ Tod m. ‖ mort f. / ~ linen ‖ Totenwäsche f. ‖ linge m. mortuaire. / ~ rate ‖ Sterblichkeitsziffer f. ‖ mortalité f.

debacle of the ice ‖ Eisgang m.; Eisbruch m. ‖ débâcle f. des glaces.

debenture ‖ Schuldschein m.; Verschreibung f. ‖ reconnaissance f.; obligation f. / issue of ~s ‖ Ausgabe f. von Obligationen ‖ émission f. d'obligations.

debenture bond ‖ Pfandbrief m. ‖ lettre f. de gage; cédule f. hypothécaire.

debit, to ‖ in Anrechnung f. bringen ‖ mettre en compte m. / ~ to one's account ‖ in Rechnung f. stellen ‖ mettre en ligne de compte m.

debit ‖ Debet n.; Soll n. ‖ débit m.; doit m. / to the ~ of ‖ zu Lasten fpl. von ‖ à la charge f. de. / to pass to the ~ ‖ anrechnen ‖ compter; mettre ou passer en compte m. / ~ and credit ‖ Soll n. und Haben n. ‖ doit m. et avoir m.

debit note ‖ Belastungsanzeige f.; Debetnote f. ‖ note f. de débit. / ~ posting ‖ Debetkolonne f. ‖ colonne f. des débits. / ~ side ‖ Soll n. ‖ doit m.

deblooming agent ‖ Entscheinungsmittel n. ‖ préparation f. déluisante. / ~ colour ‖ Entscheinungsfarbe f. ‖ couleur f. déluisante.

debouchure ‖ Mündung f. ‖ bouche f.

debrassing bath ‖ Entmessingungsbad n. ‖ bain m. de délaitonnage.

debris ‖ Trümmer pl. ‖ débris m.

debrominate, to ‖ entbromen ‖ débromer.

debt, barred ~ ‖ verjährte Schuld f. ‖ dette f. caduque. / **deeply involved in** ~ ‖ überschuldet ‖ surchargé ou criblé de dettes fpl. / **to discharge a** ~ ‖ eine Schuld f. ablösen ‖ amortir une dette. / **floating** ~ ‖ schwebende Schuld f. ‖ dette f. flottante. / **outstanding** ~s pl. ‖ Außenstände pl. ‖ actifs mpl. créances; sommes fpl. à recouvrer. / **reciprocal** ~ ‖ Gegenschuld f. ‖ dette f. passive. / **to work off a** ~ ‖ eine Schuld f. abarbeiten ‖ acquitter une dette par son travail.

debtor ‖ Debitor m. ‖ débiteur m. / ~ (Bookkeep) ‖ Debet n.; Debetseite f. ‖ débit m.; doit m. / ~ **and creditor** ‖ Soll n. und Haben n. ‖ doit m. et avoir m.

decade ‖ Dekade f.; Reihe f. oder Satz m. von 10 Stück ‖ décade f. / **three** ~ **resistance** (Aut tel) ‖ Dreidekadenstufenwiderstand m. ‖ résistance f. de 3 (trois) décades. / ~ **system** (Tel) ‖ Dekadensystem n. ‖ système m. décimal.

decagon ‖ Zehneck n. ‖ décagone m.

decalage ‖ Schränkung f. ‖ interinclinaison f.

decalcomania picture ‖ Abziehbild n. ‖ décalcomanie f.

decalescence ‖ Abschreckung f. ‖ décalescence f.

decant, to (Brew) ‖ abresten; abseihen ‖ tirer les fonds mpl. de foudres. / ~ (Chem) ‖ abfüllen; dekantieren; umfüllen ‖ décanter; filtrer; transvaser.

decantation ‖ Dekantieren n. ‖ décantation f.; transvasement m.

decanted ‖ abgefüllt ‖ décanté.

decanter ‖ Abklärgefäß n.; Dekantierungsgefäß n. ‖ décanteur m.

decanting (Brew) ‖ Abresten n. ‖ action f. de tirer les fonds de foudres. / ~ **apparatus for analyses** ‖ Abfüllapparat m. oder Schlämmapparat m. für Analysen ‖ appareil m. de décantage pour analyses. / **cask** ~ **apparatus** ‖ Faßabfüllapparat m. ‖ appareil m. de soutirage pour tonneaux / ~ **cellar** ‖ Abfüllkeller m. ‖ cave f. de soutirage. / **kaolin** ~ **machine** ‖ Kaolinschlämmaschine f. ‖ machine f. à laver le kaolin. / ~ **plant for canalisation** ‖ Kläranlage f. für Kanalisation ‖ installation f. de clarification pour canalisation. / ~ **plant for water supply works** ‖ Wasserwerkkläranlage f. ‖ installation f. de clarification de canalisations d'eau. / ~ **tank** ‖ Dekantationsbehälter m. ‖ cuve f. à décanter. / ~ **vessel** ‖ Dekantiergefäß n. ‖ réservoir m. de decantation.

decapper ‖ Zündhütchenzange f. ‖ amorceur m.

decarbonization ‖ Entkohlung f. ‖ décarburation f.

decarbonize, to ‖ entkohlen ‖ décarburer.

decarbonizing of the steel ‖ Entkohlen n. des Stahls ‖ décarbonisation f. de l'acier.

decarburate, to ‖ entkohlen ‖ décarburer.

decarburization in the open-hearth furnace ‖ Herdfrischen n. ‖ affinage m. au basfoyer.

decarburize, to ‖ frischen; entkohlen ‖ décarburer.

decare ‖ Dekare f.; Dekar n. ‖ décare m.

decatizing, removable ~ **cylinder** ‖ aushebbarer Dekatierzylinder m. ‖ cylindre m. à décatir amovible. / ~ **machine** ‖ Kesseldekatierapparat m. ‖ appareil m. de décatissage. / **wet** ~ **machine** ‖ Naßdekatiermaschine f. ‖ décatisseuse f. au mouillé.

/ ~ **roller** ‖ Dekatierwalze f. ‖ cylindre m. à décatir.

decay, to ‖ verfallen; verderben ‖ pourrir; carier. / ~ **in the open air** ‖ vermodern ‖ pourrir à l'air m.

decay ‖ Verfall m. ‖ décadence f.; ruine f. / ~ (Build) ‖ Baufälligkeit f. ‖ caducité f. / ~ (Wood) ‖ Vermoderung f. ‖ croupissement m. / **to fall into** ~ (Build) ‖ verfallen ‖ se dégrader; tomber en ruines fpl. / ~ **of a wall by efflorescence** ‖ Mauerfraß m. ‖ carie f. des murailles.

decayed ‖ morsch ‖ pourri; vermoulu. / ~ **walling** ‖ verfallenes Mauerwerk n. ‖ masure f. / ~ **wood** ‖ angefaultes Holz n. ‖ bois m. piqué.

deceitful ‖ trügerisch ‖ trompeur; illusoire.

decelerate, to ‖ die Geschwindigkeit f. mindern ‖ ralentir.

deceleration ‖ Geschwindigkeitsabnahme f. ‖ retardation f.

deception ‖ Täuschung f. ‖ tromperie f.; illusion f.

deceptive see deceitful.

decide, to ‖ den Ausschlag m. geben; entscheiden ‖ décider; emporter la balance.

decigram ‖ Dezigramm n. ‖ décigramme m.

decimal ‖ Dezimale f. ‖ décimale f. / **to approximate** ~ (Math) ‖ eine Dezimale f. aufrunden ‖ forcer une décimale. / ~ **balance** ‖ Dezimalwage f. ‖ balance f. ou bascule f. décimale. / ~ **figure** ‖ Dezimalstelle f. ‖ décimale f. / ~ **fraction** ‖ Dezimalbruch m. ‖ fraction f. décimale. / ~ **numbering** ‖ dekadische Bezifferung f. ‖ numérotation f. décimale. / ~ **resistance** ‖ Dekadenwiderstand m. ‖ résistance f. en décades. / ~ **resistance-box** ‖ Dekadenwiderstandskasten m. ‖ boîte f. de résistance en décades. / ~ **rheostat** ‖ Dekadenrheostat m. ‖ rhéostat m. en décades. / ~ **system** ‖ Dezimalsystem n. ‖ système m. décimal. / ~ **weighing machine** ‖ Dezimalwage f. ‖ bascule f. au dixième.

decimeter ‖ Dezimeter n. ‖ décimètre m.

decipher, to ‖ dechiffrieren; entziffern ‖ déchiffrer.

deciphering-device ‖ Dechiffriervorrichtung f. ‖ dispositif m. à déchiffrer.

decision ‖ Beschluß m. ‖ résolution f.; conclusion f. / ~ **of the court** ‖ richterliche Entscheidung f.; Gerichtsbeschluß m. ‖ décision f. du tribunal.

decisive ‖ ausschlaggebend ‖ décidant; décisif.

deck (Aero) ‖ Tragfläche f. ‖ plan m. / ~ (Shipb) ‖ Deck n.; Verdeck n. ‖ pont m. / **between** ~ ‖ Zwischendeck n. ‖ entre-pont m. / ~ **between the paddle-boxes** see **flying** ~. / **flying** ~ ‖ Raddeck n. ‖ passerelle f. entre les tambours. / **promenade** ~ ‖ Promenadendeck n. ‖ pont-promenade m. / ~ **of a ship** ‖ Deck n. eines Schiffes ‖ pont m. d'un navire. / **upper** ~ ‖ Oberdeck n. ‖ pont m. des gaillards ou supérieur.

deck beam ‖ Deckbalken m. ‖ barrot m. de pont. / ~ **boy** ‖ Leichtmatrose m. ‖ novice m. / ~ **cargo** ‖ Deckladung f. ‖ pontée f.; cargaison f. de pont.

decked (Shipb) ‖ mit Verdeck n.; gedeckt ‖ ponté.

deck, height between ~s ‖ Deckshöhe f. ‖ hauteur f. des ponts.

deckhouse ‖ Decksalon m. ‖ kiosque m. sur le pont.

deck load ‖ Deckslast f. ‖ chargement m. sur le pont. / ~ **passenger** ‖ Deckpassagier m. ‖ passager m. de pont. / ~ **plank** ‖ Decksplanke f. ‖ bordage m. des ponts. / ~ **planking** ‖ Deckbeplankung f. ‖ bordages mpl. des ponts. / ~ **plate** (Shipb) ‖ Deckplatte f. ‖ plaque f. de recouvrement du pont. / ~ **seam** (Shipb) ‖ Decksnaht f. ‖ couture f. de pont. / ~ **sheer** ‖ Deckssprung m. ‖ tonture f. d'un pont. / ~ **stopper** (Shipb) ‖ Kettenstopper m. ‖ bosse f. de chaîne. / ~ **strip** ‖ Deckstreifen m. ‖ bande f. de recouvrement. / ~ **tackle** ‖ Decktakel n. ‖ palan m. à main.

declaration (Duty) ‖ Inhaltsangabe f.; Angabe f. ‖ déclaration f. (du contenu). / ~ **of bankruptcy** ‖ Insolvenzerklärung f.; Konkursanmeldung f. ‖ déclaration f. de faillite. / ~ **of value** ‖ Wertangabe f. ‖ déclaration f. de valeur.

declination (Phys) ‖ Abweichung f.; Deklination f. ‖ déclinaison f. / **angle of** ~ ‖ Abweichungswinkel m. ‖ angle m. de déclinaison. / ~ **axis** ‖ Deklinationsachse f. ‖ axe m. de déclinaison. / ~ **circle graduated in minutes of arc** ‖ in Bogenminuten geteilter Deklinationskreis m. ‖ cercle m. de déclinaison divisé en minutes d'arc. / ~ **circle reading to one minute** ‖ auf eine Minute ablesbarer Deklinationskreis m. ‖ cercle m. de déclinaison permettant de lire la minute. / ~ **circle with vernier** ‖ Deklinationskreis m. mit Noniusablesung ‖ cercle m. de déclinaison à vernier. / ~ **compass** ‖ Deklinationsbussole f. ‖ boussole f. de déclinaison. / ~ **instrument** ‖ Abweichungskompaß m. ‖ boussole f. déclinatoire. / ~ **needle** ‖ Deklinatorium n. ‖ boussole f. de déclinaison. / **table of** ~ ‖ Abweichungstafel f. ‖ table f. de déclinaison.

declivity ‖ Gefäll n.; Abdachung f.; Fall m.; Neigung f.; Böschung f. ‖ déclivité f.; pente f.; penchant m.; inclinaison f.

decoction ‖ Abkochung f.; Dekoktion f. ‖ décoction f. / ~ (Brew) ‖ Absud m. ‖ décoction f. / ~ **method** ‖ Dekoktionsverfahren n. ‖ méthode f. de décoction.

decocting medium ‖ Abkochmittel n. ‖ produit m. à décoction.

decohere, to ‖ entfritten ‖ décohérer.

decoherence ‖ Entfrittung f. ‖ décohésion f.

decoherer (Radio) ‖ Klopfer m.; Entfritter m. ‖ frappeur m.; décohéreur m.

decolour, to ‖ entfärben ‖ décolorer.

decoloured ‖ abgefärbt ‖ decoloré.

decolouring ‖ Entfärben n. ‖ décoloration f.

decolourize, to ‖ entfärben ‖ décolorer.

decolourizing ‖ bleichend; entfärbend ‖ décolorant. / ~ **agent** ‖ Bleichmittel n. ‖ décolorant m. / ~ **coal** ‖ Entfärbungskohle f. ‖ charbon m. décolorant; noire f. décolorante.

decomposable ‖ zerlegbar; zersetzbar ‖ décomposable. / ~ **barrack of corrugated iron** ‖ zerlegbare Wellblechbaracke f. ‖ baraque f. de tôle ondulée démontable.

decompose, to ‖ zerlegen ‖ décomposer. / ~ **the force** ‖ die Kraft zerlegen ‖ décomposer la force. / ~ **the sunlight** ‖ das Sonnenlicht n. zerlegen ‖ décomposer la lumière du soleil.

decomposition (Chem) ‖ Zersetzung f. ‖ décomposition f. / ~ (Miner) ‖ Verwitterung f. ‖ désagrégation f.; effritement m. / ~ (Phys) ‖ Zerlegung f. ‖ décomposition f. / **deep-**

13*

seated ~ || Tiefenzersetzung f. || décomposition f. dans les profondeurs. / electrolytic ~ || elektrolytische Spaltung f. || décomposition f. électrolytique. / ~ of forces || Zerlegung f. der Kräfte || décomposition f. des forces. / ~ of gas mixtures || Zerlegung f. von Gasgemischen || décomposition f. de mélanges de gaz. / plant for ~ of gas compounds || Anlage f. zur Zerlegung von Gasgemischen || installation f. de décomposition des mélanges gazeux. / plant for the ~ of water || Wasserzersetzungsanlage f. || installation f. pour la décomposition de l'eau.

decomposition chamber || Zersetzungskammer f. || chambre f. de réaction *ou* de décomposition.

decopperize, to || entkupfern || décuivrer.

decorate, to || verzieren || orner; décorer.

decorating, interior ~ || Innenarchitektur f. || décoration f. d'intérieurs. / inside ~ article || Innendekorationsartikel m. || article m. décoratif pour l'intérieur. / ~ flower || Dekorationsblume f. || fleur f. pour la décoration.

decoration || Abzeichen n. || marque f.; insigne m. / ~ of glass and crystal || Glasveredelung f. || décoration f. de verres et de cristaux. / ~ plush || Dekorationsplüsch m. || peluche f. pour décoration. / ~ ribbon || Ordensband n. || ruban m. pour ordres. / ~ stuff || Dekorationsstoff m. || étoffe f. de décoration. / ~ varnish || Dekorationslack m. || vernis m. à décorations.

decorative sheet iron || Zierblech n. || tôle f. d'ornamentation.

decorator (Furniture) || Möbelverzierer m.; Dekoratör m. || décorateur m. en meubles. / ~ (Paint) || Dekorationsmaler m. || peintre-décorateur m.

decorticate, to || entrinden || écorcer; décortiquer.

decorticator || Schälmaschine f. || décortiqueuse f.; machine f. à décortiquer.

decoy (Hunting) || Lockvogel m. || appelant m.

decrease, to || kleiner werden || diminuer.

decrease || Verminderung f. || décroissement m.; diminution f.; amoindrissement m. / ~ (Exchange) || Abnahme f. || baisse f. / ~ (Metal) || Abbrand m. || perte f. / ~ (Phys) || Abfall m. || chute f. / ~ (Valuables) || Verschlechterung f. || détérioration f.; dépréciation f. / ~ of current || Sinken n. des Stromes || décroissement m. du courant. / ~ of lift || Auftriebsverminderung f. || diminution f. de la force ascensionnelle *ou* de poussée. / ~ of pressure || Druckabfall m. || chute f. *ou* diminution f. de pression. / ~ in receipts || Mindereinnahme f. || déficit m. / rate of ~ in temperature || Wärmegefälle n. || chute f. de température; gradient m. de la température. / ~ of value || Wertverminderung f.; Wertverringerung f. || diminution f. de valeur; dépréciation f.; moins-value f. / to compensate for the ~ in value || den Minderwert m. ersetzen || compenser la moins-value.

decreased || vermindert || diminué.

decreasing motion || verzögerte Bewegung f. || mouvement m. retardé. / ~ tariff || nach unten abgestufter Tarif m. || tarif m. dégressif.

decree, to || beschließen || résoudre; décider.

decree of the court || Gerichtsbeschluß m || décision f. du tribunal.

decrement (Phys) || Dekrement n.; Verminderung f.; Abnahme f. || décrément m. / logarithmic ~ || logarithmisches Dekrement n. || décrément m. logarithmique.

decremeter || Dämpfungsmesser m. || appareil m. de mesure d'amortissement; décrémètre m.

deduct, to (Price) || ablassen; in Abzug m. bringen; abziehen || rabattre; défalquer.

deduction (Price) || Abzug m. || déduction f.

deed of assignment || Zessionsurkunde f.; Abtretungsurkunde f. || acte m. d'abandon *ou* de délaissement *ou* de transfert. / ~ of conveyance || Übertragungsurkunde f.; Überschreibungsurkunde f. || acte m. de cession; acte m. translatif de propriété. / ~ of partnership || Gesellschaftsvertrag m. || contrat m. de société; acte m. constitutif. / ~ of sale || Verkaufsurkunde f. || acte m. de vente. / ~ of transfer *see* ~ of assignment.

de-energize, to (Electr) || aberregen; außer Strom m. setzen || désexciter; mettre hors circuit m.; désamorcer.

deep || tief || profond. / ~ gap || tiefgehende Ausladung f. || col m. de cygne très profond. / ~ greenhouse || halbunterirdisches Gewächshaus n.; Erdgewächshaus n. || serre f. profonde. / ~ plate || tiefer Teller m. || assiette f. creuse. / ~ water channel || Tiefwasserrinne f. || chenal m. d'eau profonde.

deep || Tiefe f. || profondeur f. / ~-boring (Mine) || Tiefbohrung f. || sondage m. *ou* forage m. à grande profondeur. / ~-boring tools pl. || Tiefbohrgerät n.; Tiefbohrwerkzeuge npl. || outils mpl. à forer *ou* de sondage en grande profondeur. / sheet-iron for ~-drawing || Tiefziehblech n.; Tiefstanzblech n. || tôle f. à estampage et à emboutir.

deepen, to || vertiefen || approfondir; creuser. / ~ (Colours) || verdunkeln || assombrir. / ~ (Print) || abdunkeln || donner une nuance plus foncée.

deepened || tiefer gemacht || approfondi.

deepening || Austiefung f. || approfondissement m.; creusement m.

deep foundation || Tiefgründung f. || fondation f. à grande profondeur. / ~ mine working || Tiefbaubetrieb m. || exploitation f. souterraine; concession f. avec excavation souterraine. / ~ mining *see* ~ mine working.

deep-sea cable || Tiefseekabel n. || câble m. océanique *ou* de mer profonde. / ~ lead || Tieflot n. || plomb m. pour les grandes sondes. / ~ lead-line || Tieflotleine f. || grande ligne f. de sonde. / ~ sounding || Tiefseelotung f. || sondage m. des grandes profondeurs.

deep-seated decomposition || Tiefenzersetzung f. || décomposition f. dans les profondeurs. / ~-webbed U-iron || hochstegiges U-Eisen n. || fer m. en U avec âme de grande hauteur.

deerskin imitation || Wildlederimitation f. || imitation f. de peau de daim.

defalcate, to (Math) || abziehen || défalquer.

default || Mangel m. || défaut m. / in ~ of || mangels; à défaut m. de; faute f. de. / notice of ~ || Inverzugsetzung f. || mise f. en retard m.

defaulter || Restant m.; säumiger Zahler m. || reliquataire m.

defecate, to (Chem) || reinigen; abklären || désinfecter.

defecation || Desinfektion f. || désinfection f.

defect || Fehler m. || défaut m.; défectuosité f.; vice m. / ~ (Metal) || Fehlstelle f. || vice m. / bent without showing any ~s || verbogen, ohne daß sich schadhafte Stellen zeigten || déformé sans qu'il y eût des dégats. / ~ of construction || Konstruktionsfehler m. || défaut m. de construction. / ~ of fabrication || Herstellungsfehler m. || défaut m. de fabrication. / having ~s || mit Mängeln mpl. behaftet || comportant d'imperfections fpl. / ~ in insulation || Isolationsfehler m. || défaut m. d'isolement. / owing to ~s || wegen Mängel mpl. || à cause des défauts mpl. / ~ of sight || Fehlsichtigkeit f. || amétropie f. / ~s pl. of timber || Holzfehler mpl. || vices mpl. de bois.

defective || fehlerhaft; schadhaft || défectueux; endommagé. / ~ coil || schadhafte Spule f. || bobine f. défectueuse. / ~ insulation || schadhafte Isolation f. || isolement m. défectueux. / ~ pieces pl. of faulty workmanship are excluded from further manufacture || fehlerhafte Stücke npl. sind von der Weiterverarbeitung ausgeschlossen || les pièces fpl. présentant des défauts quelconques sont rébutées de la fabrication ultérieure. / ~ tightness || schlechte Abdichtung f. || étanchéité f. défectueuse. / ~ trunk (Tel) || gestörte Verbindungsleitung f. || ligne f. de jonction dérangée. / ~ working || fehlerhaftes Arbeiten n. || fonctionnement m. irrégulier. / ~ workmanship || mangelhafte Arbeit f. || travail m. défectueux *ou* mal fait.

defectively sighted eye || fehlsichtiges Auge n. || œil m. amétrope.

defence || Schutz m.; Verteidigung f. || défense f. / ~ of downs || Dünenschutzwerk n. || défense f. des dunes. / national ~ act || nationales Verteidigungsgesetz n. || loi f. de défense nationale.

defendant || Angeklägter m.; Beklagter m. || prévenu m.; accusé m.; defendeur m.

defensive arm || Verteidigungswaffe f. || arme f. de défense.

deferred call (Tel) || Zurückstellung f. einer Gesprächsanmeldung || communication f. différée. / ~ share || Genußschein m. || action f. de jouissance.

deferrization || Enteisenung f. || déferrisation f.

deficiency || Mangel m. || manque m. / ~ (Trade) || Unterbilanz f. || bilan m. passif; déficit m. / ~ of air || Luftmangel m. || manque m. d'air. / ~ in receipts || Mindereinnahme f. || déficit m. / ~ of sight || Fehlsichtigkeit f. || amétropie f.

deficit || Ausfall m.; Mindereinnahme f.; Defizit n.; Unterbilanz f. || déficit m.; perte m.; manque m.; bilan m. passif. / ~ of vapour for saturation || Sättigungsunterschuß m. des Dampfes || déficit m. de saturation de la vapeur.

defile || Klamm f.; Schlucht f.; Gebirgspaß m. || ravin m.; défilé m.

define, to || bestimmen; definieren || définir.

defined, well ~ || eindeutig || parfaitement déterminé.

definible, statically ~ || statisch bestimmbar || déterminable par la statique.

definition || Begriffsbestimmung f. || définition f. / ~ of image || Bildschärfe f. || netteté f. de l'image.

deflagrate, to ‖ explosionsartig verbrennen ‖ déflagrer; fuser.

deflagration (Chem) ‖ explosionsartige Verbrennung f. ‖ déflagration f.

deflate, to ‖ entleeren; herauslassen ‖ vidanger; vider; dégonfler.

deflating sleeve ‖ Entleerungsschlauch m. ‖ manche f. de dégonflement.

deflect, to ‖ ablenken ‖ dévier; défléchir. / ~ backwards ‖ nach rückwärts ausschlagen ‖ dévier vers l'arrière m. / ~ the flow ‖ die Strömung f. ablenken ‖ défléchir ou dévier les filets fluides. / ~ forwards ‖ nach vorwärts ausschlagen ‖ dévier vers l'avant m. / ~ a magnetic needle ‖ eine Magnetnadel f. ablenken ‖ dévier une aiguille aimantée.

deflectable ‖ lenkbar ‖ déviable.

deflecting force ‖ Ablenkungskraft f.; ablenkende Kraft f. ‖ force f. déviante ou de déviation.

deflection (Electr; Phys) ‖ Ausschlag m.; Ablenkung f. ‖ déviation f.; déflexion f. / ~ (Material) ‖ Durchbiegung f. ‖ flexion f.; flèche f. / average ~ ‖ mittlerer Ausschlag m. ‖ déviation f. moyenne. / ~ in either direction ‖ doppelseitiger Zeigerausschlag m. ‖ déviation f. bilatérale. / ~ of a girder ‖ Durchbiegung f. eines Balkens ‖ flèche f. de la poutre. / ~ to the left ‖ Linksablenkung f. ‖ déviation f. vers la gauche. / ~ of the lines of force ‖ Kraftlinienablenkung f. ‖ déviation f. des lignes de force. / ~ of the load stipulated ‖ Durchbiegung f. bei vorgeschriebener Belastung ‖ flèche f. à la charge prescrite. / ~ of plate spring ‖ Aufbiegung f. der Blattfeder ‖ flèche f. de ressort à lame ou à feuille. / ~ of the plumb line ‖ Lotabweichung f. ‖ déviation f. de la verticale. / ~ of pointer ‖ Zeigerausschlag m. ‖ déviation f. de l'aiguille. / ~ to the right ‖ Rechtsablenkung f. ‖ déviation f. vers la droite. / steady ~ ‖ gleichmäßiger Ausschlag m. ‖ déviation f. soutenue. / ~ of tyre ‖ Plattwerden n. des Radreifens ‖ aplatissement m. du bandage. / ~ of the wind by mountains ‖ Windablenkung f. durch das Gebirge ‖ déviation f. du vent par les montagnes.

deflection angle ‖ Ausschlagwinkel m. ‖ angle m. d'écartement. / ~ method ‖ Ausschlagmethode f. ‖ méthode f. de déviation. / ~ minimum ‖ geringste Durchfederung f. ‖ minimum m. de flexion. / ~ test ‖ Durchbiegeversuch m. ‖ essai m. de flexion. / ~ test by pressure on the hub ‖ Nabendurchbiegeprobe f. ‖ essai m. de défonçage du moyeu.

deflectometer ‖ Ablenkungsmesser m. ‖ déclinomètre m.; flectomètre m.

deflector ‖ Ablenker m. ‖ déflecteur m.

deforestation ‖ Entwaldung f. ‖ déboisement m.

deform, to ‖ deformieren ‖ déformer.

deformation ‖ Deformation f.; Formänderung f. ‖ déformation f. / permanent ~ ‖ bleibende Formänderung f. ‖ déformation f. permanente. / slight ~ of the axle ‖ geringe Verbiegung f. der Achse ‖ légère déformation f. de l'essieu.

deformation test ‖ Deformationsprobe f. ‖ essai m. de déformation. / ~ by loads without a permanent set ‖ Belastungsprobe f. ohne bleibende Durchbiegung f. ‖ essai m. de flexion par charge morte sans déformation permanente.

deformed wave ‖ verzerrte Welle f. ‖ onde f. faussée.

degasifying ‖ Entgasung f. ‖ dégazation f.; dégazage m.; dégagement m. de gaz.

degelation ‖ Auftauen n. ‖ dégèlement m.

degeneration ‖ Degeneration f.; Verschlechterung f. ‖ dégénérescence f. / ~ of the yeast ‖ Ausarten n. der Hefe ‖ dégénération f. de la levûre.

degerminate, to ‖ entkeimen ‖ dégermer.

degermination ‖ Entkeimung f. ‖ dégermination f.; stérilisation f.

degerminator ‖ Entkeimer m. ‖ dégermeur m.

degradation (Geol) ‖ Verwitterung f. ‖ dégradation f.; décomposition f.; effrittement m.; désagrégation f.

degras ‖ Degras n.; Gerberfett n.; Weißbrühe f. ‖ dégras m.

degreasing, benzol for ~ ‖ Benzol n. für Fettextraktion ‖ benzol m. de dégraissage.

degreasing bath ‖ Entfettungsbad n. ‖ bain m. de dégraissage. / electrolytical ~ ‖ elektrolytisches Entfettungsbad n. ‖ bain m. de dégraissage électrolytique.

degreasing bench ‖ Entfettungstisch m. ‖ table f. de dégraissage. / ~ composition ‖ Entfettungsmasse f. ‖ composition f. à dégraisser. / ~ plant ‖ Entfettungsanlage f. ‖ installation f. de dégraissage.

degree (Astron; Math; Phys) ‖ Grad m. ‖ degré m.

degree of accuracy ‖ Genauigkeitsgrad m. ‖ degré m. de précision. / ~ of accuracy with which readings can be taken ‖ Ablesegenauigkeit f. ‖ précision f. des lectures. / ~ of admission ‖ Füllgrad m. ‖ degré m. d'admission. / ~ of approximation ‖ Annäherungsgrad m. ‖ degré m. d'approximation. / ~ Baumé ‖ Grad m. Beaumé ‖ degré m. Beaumé. / ~ of damping ‖ Dämpfungsgrad m. ‖ degré m. d'amortissement. / ~ of density ‖ Dichtigkeitsgrad m. ‖ degré m. de compacité. / to divide into ~s ‖ in Grade mpl. einteilen ‖ diviser en degrés mpl.; graduer. / ~ of expansion ‖ Expansionsgrad m. ‖ degré m. de détente ou d'expansion. / ~ of feeding (Mach) ‖ Füllungsgrad m. ‖ degré m. de détente ou d'expansion. / ~ of final attenuation ‖ Endvergärungsgrad m. ‖ degré m. d'atténuation limite. / ~ of fineness ‖ Feinheitsgrad m. ‖ degré m. de finesse. / ~ of freedom (Phys) ‖ Freiheitsgrad m. ‖ degré m. de liberté. / ~ of hardness ‖ Härtegrad m. ‖ degré m. de dureté. / ~ of heat ‖ Hitzegrad m. ‖ température f. ou degré m. de chaleur. / ~ of irregularity ‖ Ungleichförmigkeitsgrad m. ‖ degré m. d'irrégularité. / ~ of luminosity ‖ Helligkeitsstufe f. ‖ degré m. de clarté. / ~ of pulverization ‖ Feinheitsgrad m. ‖ degré m. de finesse. / ~ of spectacles ‖ Brillennummer f. ‖ numéro m. des lunettes. / ~ of superheat ‖ Überhitzungsgrad m. ‖ degré m. de surchauffe.

degrees pl. centigrade ‖ Grade mpl. Celsius ‖ degrés mpl. centigrades. / ~ Fahrenheit ‖ Grade mpl. Fahrenheit ‖ degrés mpl. Fahrenheit. / ~ of frost ‖ Kältegrade mpl. ‖ degrés mpl. de froid. / ~ Réaumur ‖ Grade mpl. Réaumur ‖ degrés mpl. Réaumur.

dehiscence (Bot) ‖ Aufspringen n. ‖ déhiscence f.

dehydrate, to ‖ wasserentziehen ‖ déshydrater.

dehydrating agent ‖ wasserentziehendes Mittel n. ‖ déshydratant m. / ~ oil ‖ Ölentwässerung f. ‖ déshydratation f. de l'huile.

de-ironing, water ~ ‖ Wasserenteisenung f. ‖ déferrisation f. de l'eau.

delay, to ‖ verschleppen; verschieben ‖ retarder.

delay ‖ Verzögerung f.; Verzug m.; Aufschub m. ‖ retard m. / ~ (Mine) ‖ Fristung f.; Stillstandsfrist f.; Frist f. ‖ délai m.; sursis m. / ~ in the arrival of the machinery ‖ verspätete Anlieferung f. der Maschinenanlage ‖ retard m. dans l'arrivée des machines. / ~ in delivery ‖ Lieferungsverzögerung f. ‖ retard m. de la livraison. / ~ without ~ ‖ ohne Verzögerung f. ‖ sans délai m.

del-credere ‖ Delkredere n. ‖ ducroire m. / ~ fund ‖ Delkrederefonds m. ‖ fonds m. de ducroire.

delegate ‖ Abgeordneter m. ‖ député m.; délégué m.

delessite ‖ Eisenchlorit m. ‖ delessite f.

delftware ‖ Delfter Porzellan n. ‖ poterie f. de Delft.

deliberate, to ‖ konferieren; beratschlagen ‖ conférer.

delicate balance ‖ empfindliche Wage f. ‖ balance f. sensible.

deliquesce, to ‖ zerfließen ‖ tomber en déliquescence.

deliquescent ‖ zerfließend ‖ déliquescent.

deliver, to (Chem) ‖ ausfließen ‖ débiter. / ~ (Metal) ‖ aufgeben ‖ charger; delivrer; alimenter. / ~ (Trade) ‖ liefern; aushändigen ‖ livrer; délivrer; fournir; alimenter. / ~ subsequently ‖ nachliefern ‖ livrer plus tard. / ~ a thing to a person ‖ jemandem etwas zustellen ‖ remettre ou faire tenir quelque chose à quelqu'un.

delivered at building site ‖ frei Baustelle f. ‖ marchandise f. rendue sur chantier. / ~ at site ‖ frei Verwendungsstelle f. ‖ franco à pied d'œuvre. / to be ~ ‖ lieferbar ‖ livrable; à livrer. / when ~ ‖ im Anlieferungszustand m.; bei Anlieferung f. ‖ dans l'état m. de livraison.

delivering bowl (Weav) ‖ Abzugswalze f. ‖ rouleau m. de tirage ou d'appel.

delivery ‖ Lieferung f.; Abgabe f. ‖ livraison f.; fourniture f. / ~ of current ‖ Stromabgabe f. ‖ fourniture f. de courant. / ~ of energy ‖ Energiebelieferung f. ‖ envoi m. d'énergie. / final ~ ‖ Restlieferung f.; Schlußlieferung f. ‖ livraison f. finale. / front ~ (Print) ‖ Frontbogenausleger m. ‖ sortie f. frontale. / ~ of fuel ‖ Brennstofflieferung f. ‖ livraison f. du combustible. / half-monthly ~ ‖ Halbmonatslieferung f. ‖ livraison f. par quinze jours ou par une quinzaine. / for immediate ~ ‖ zur sofortigen Lieferung f. ‖ à livraison f. immédiate. / ~ by instalments ‖ Teilzahlungslieferung f. ‖ livraison f. à crédit. / machine with a ~ of x cubic metres per minute of water ‖ Maschine f. mit einer minutlichen Leistung von x Kubikmetern Wasser ‖ machine f. refoulant x mètres cubes d'eau par minute. / to observe the time of ~ ‖ die Lieferzeit f. innehalten ‖ respecter le délai de livraison. / open sheet ~ (Print) ‖ Planoausleger m.; Planobogenausgang m. ‖ sortie f. de feuilles à plat. / ~ per hour ‖ stündliche Liefermenge f. ‖ débit m. par heure. / ~ per minute ‖ Liefermenge f. in der Minute ‖ débit

m. par minnte. / pile ~ (Print) ‖ Stapelausleger m. ‖ receveur m. *ou* sortie f. de feuilles à pile. / place for ~, payment and for legal questions is N. N. ‖ Erfüllungsort m. für Lieferung und Zahlung sowie Gerichtsstand ist N. N. lieu m. de livraison, payement et tribunal compétent est N. N. / ~ of a pump ‖ Förderwassermenge f. einer Pumpe ‖ eau f. refoulée par pompe; refoulement m. *ou* débit m. d'une pompe. / refused ~ ‖ Lieferung f. verweigert ‖ livraison f. refusée / reversible ~ (Print) ‖ umsteuerbarer Ausleger m. ‖ sortie f. de feuilles réversible. / ~ of telegrams at destination ‖ Telegrammzustellung f. am Bestimmungsort ‖ remise f. des télégrammes à destination. / total amount of ~ of a pump ‖ Gesamtliefermenge f. einer Pumpe ‖ rendement m. *ou* débit m. total en eau d'une pompe.

delivery apparatus, automatic ~ ‖ Verkaufsautomat m. ‖ distributeur m. automatique. / chocolate tablet ~ ‖ Schokoladenautomat m. ‖ distributeur m. de tablettes de chocolat. / drink ~ ‖ Getränkeautomat m. ‖ distributeur m. de boissons. / pastry ~ ‖ Speiseautomat m. ‖ distributeur m. de pâtisseries. / postage stamp ~ ‖ Briefmarkenautomat m. ‖ distributeur m. de timbres-poste. / railway ticket ~ ‖ Fahrkartenautomat m. ‖ distributeur m. de tickets de chemin de fer.

delivery, bill of ‖ Ablieferungsschein m. ‖ bulletin m. *ou* bon m. *ou* ordre m. de livraison; reçu m. / ~ car ‖ Lieferwagen m. ‖ voiture f. de livraison; camionette f. / ~ car (Mine) ‖ Förderwagen m. ‖ berline f. / ~ certificate ‖ Einlieferungsschein m. ‖ bulletin m. de dépôt; reçu m. / ~ charge ‖ Bestellgebühr f.; Bestellgeld n. ‖ factage m. / ~ chute ‖ Aufgaberutsche f.; Ausschüttrinne f. ‖ glissoir m. de chargement *ou* de déversement. / ~ conduit ‖ Druckleitung f. ‖ conduite f. de refoulement. / ~ cylinder (Print) ‖ Übergabezylinder m. ‖ cylindre m. de transmission. / ~ hose ‖ Druckschlauch m. ‖ manche f. foulante *ou* de refoulement. / automatic ~ machine ‖ Verkaufsautomat m. ‖ distributeur m. automatique. / ~ note ‖ Lieferschein m. ‖ bulletin m. de livraison. / ~ order ‖ Bezugsschein m. ‖ quittance f. d'abonnement; bon m. / ~ pipe (Pump) ‖ Ausflußrohr n.; Druckrohr n. ‖ tuyau m. de refoulement. / ~ pipe (Steam) ‖ Dampfausströmungsrohr n. ‖ tuyau m. d'échappement de vapeur. / ~ platform ‖ ‖ Abzughängebank f. ‖ palier m. d'exploitation. / ~ receipt ‖ Aufgabeschein m. ‖ récépissé m.; reçu m. / ~ roll ‖ Lieferwalze f. ‖ cylindre m. de décharge. / ~ side of the pump ‖ Ausflußseite f. der Pumpe ‖ côté m. de refoulement de la pompe. / term of ~ ‖ Lieferfrist f. ‖ délai m. *ou* terme m. de livraison. / ~ time ‖ Ablieferungszeit f. ‖ terme m. de livraison. / ~ tube (chem) ‖ Ableitungsrohr n. ‖ tube m. de dégagement. / ~ valve ‖ Druckventil n. ‖ soupape f. de refoulement. / ~ valve cage ‖ Druckventileinsatz m. ‖ corps m. de la soupape de compression *ou* de refoulement. / ~ valve spring ‖ Druckventilfeder f. ‖ ressort m. de soupape de refoulement.

delivery wagon ‖ Lieferwagen m. ‖ voiture f. de livraison. / cleaning machine for ore ~ ‖ Förderwagenreinigungsmaschine f. ‖ machine f. de nettoyage pour wagonnets à minerais. / fitting of ore ~ ‖ Förderwagenbeschlagteil m. ‖ ferrure f. de chien. / automatic ~ circulation ‖ selbsttätiger Wagenumlauf m. ‖ installation f. automatique de circulation pour wagonnets.

delta ‖ Flußdelta n.; Delta n. ‖ delta m. / ~ connected ‖ in Dreieckschaltung f. ‖ couplé en triangle m. / ~ connection ‖ Dreieckschaltung f. ‖ couplage m. en triangle. / ~ metal ‖ Deltametall n. ‖ métal m. delta. / ~ star connection ‖ Sterndreieckschaltung f. ‖ raccordement m. étoile-triangle.

delusion ‖ Täuschung f. ‖ tromperie f.; illusion f.

demagnetization ‖ Entmagnetisierung f. ‖ démagnétisation f.; désaimantation f.

demagnetize, to ‖ entmagnetisieren ‖ désaimanter.

demagnetizer ‖ Entmagnetisierungsapparat m. ‖ appareil m. de désaimantation.

demagnetizing field ‖ entmagnetisierendes Feld n. ‖ champ m. démagnétisant.

demand ‖ Bedarf m. ‖ demande f. / active ~ for ‖ lebhafte Nachfrage f. nach ‖ demande f. active pour. / ~ for concession without previous discovery ‖ blinde Mutung f. ‖ demande f. (de concession) à l'aveugle *ou* sans découverte préalable. / to be in great ~ ‖ großen Zulauf m. haben ‖ être en vogue f. / growing ~ ‖ wachsende Anforderung f. ‖ demande f. croissante. / ~ for money ‖ Geldnachfrage f. ‖ marché m. monétaire. / ~ for a patent ‖ Patentgesuch n. ‖ demande f. de brevet. / on special ~ ‖ auf besonderen Wunsch m. ‖ sur demande f. spéciale. / ~ for water ‖ Wasserbedarf m. ‖ quantité f. d'eau nécessaire.

demanganizing, water ~ ‖ Wasserentmanganung f. ‖ séparation f. de manganèse d'eau.

demanganization ‖ Entmanganung f. ‖ démanganisation f.

demarcation ‖ Grenzreglung f. ‖ délimitation f.

demeloir (Spinn) ‖ Stapelzugmaschine f. ‖ démêloir m.

demethylate, to ‖ entmethylieren ‖ déméthyler.

demijohn ‖ Korbflasche f. ‖ dame-jeanne f.; bouteille f. clissée.

demilune ‖ Ravelin n.; Halbmondschanze f. ‖ demilune f.; ravelin m.

demolish, to ~ a building ‖ ein Gebäude n. abbrechen *oder* niederreißen ‖ démolir *ou* raser un bâtiment.

demolition ‖ Abbruch m. ‖ démolition f. / ~ bomb ‖ Zerstörungsbombe f. ‖ bombe f. à démolir.

demonetize, to ‖ entwerten; außer Kurs m. setzen ‖ démonétiser.

demonstration apparatus ‖ Vorführungsapparat m. ‖ appareil m. pour démonstration. / ~ car (Auto) ‖ Vorführungswagen m. ‖ voiture f. de démonstration. / ~ eye ‖ Demonstrationsauge n. ‖ œil m. de démonstration. / ~ eyepiece ‖ Demonstrationsokular n. ‖ oculaire m. de démonstration. / ~ instrument ‖ Demonstrationsinstrument n. ‖ instrument m. de démonstration. / ~ object ‖ Demon-

strationspräparat n. ‖ préparation f. de démonstration. / ~ ophthalmoscope ‖ Demonstrationsspiegel m.; Demonstrationsophthalmoskop n. ‖ ophtalmoscope m. de démonstration. / ~ purpose ‖ Lehrzweck m. ‖ but m. d'enseignement.

demurrage ‖ Überliegegeld n. ‖ frais mpl. de surestarie.

demurrer ‖ Rechtseinwand m. ‖ exception f. péremptoire; dénégation f.

denaturate, to *see* denature, to.

denaturating agent ‖ Vergällungsmittel n.; Denaturierungsmittel n. ‖ dénaturant m.

denaturation ‖ Denaturieren n.; Vergällung f. ‖ dénaturation f.

denature, to ‖ vergällen; denaturieren ‖ dénaturer.

denatured alcohol ‖ denaturierter Spiritus m. ‖ alcool m. dénaturé.

denaturing preparation ‖ Denaturierungsmittel n. ‖ préparation f. dénaturante. / ~ wood naphtha ‖ Vergällungsholzgeist m. ‖ alcool m. méthylique dénaturant.

denickelling bath ‖ Entnickelungsbad n. ‖ bain m. de dénickelage.

denitrating plant ‖ Denitrieranlage f. ‖ installation f. de dénitrification.

denitrify, to ‖ denitrieren ‖ dénitrer.

denomination, exact ~ ‖ genaue Benennung f. ‖ dénomination f. exacte.

denominator (Math) ‖ Nenner m. ‖ dénominateur m.

dense ‖ dicht ‖ dense. / very ~ (Phys) ‖ spezifisch schwer ‖ très dense.

densimeter ‖ Dichtigkeitsmesser m. ‖ densimètre m.

density ‖ Dichtigkeit f.; Dichte f. ‖ densité f. / ~ of acid ‖ Säuredichte f. ‖ densité f. de l'acide. / varying ~ of the air ‖ wechselnde Dichte f. der Luft ‖ densité f. variable de l'air. / ~ of the earth ‖ Erddichte f. ‖ densité f. de la terre. / electric ~ ‖ elektrische Dichte f. ‖ densité f. électrique. / ~ of a gas ‖ Gasdichte f. ‖ densité f. d'un gaz. / ~ by surface ‖ Flächendichte f. ‖ densité f. superficielle. / ~ of traffic ‖ Verkehrsdichte f. ‖ intensité f. du traffic. / ~ by volume ‖ Raumdichte f. ‖ densité f. en volume. / ~ of wind ‖ Winddichte f. ‖ densité f. du vent.

dent, to ‖ einbeulen ‖ se cabosser. / ~ (Join) ‖ zahnen ‖ bretteler.

dent (Mach) ‖ Zahn m. ‖ dent m. / ~ (Mech) ‖ Kugellehre f.; Kugelkaliber n. ‖ logement m. du boulet.

dental fillings pl. ‖ Füllmaterialien npl. für Zahnärzte ‖ matériaux mpl. de remplissage pour dentistes. / ~ forceps ‖ Zahnarztpinzette f. ‖ précelle f. de dentiste. / ~ guttapercha ‖ Zahnkautschuk m. ‖ guttapercha f. dentaire. / ~ instrument ‖ zahnärztliches Instrument n. ‖ instrument m. de dentisterie. / ~ paste ‖ Zahnpasta f. ‖ pâte f. dentifrice. / ~ powder ‖ Zahnpulver n. ‖ poudre f. dentifrice. / ~ preparation ‖ zahnärztliches Präparat m. ‖ préparation f. dentale. / base plate for ~ prosthesis ‖ Gebißplatte f. ‖ plaque f. base pour dentiers. / instrument for ~ surgery ‖ zahnärztliches Instrument n. ‖ instrument m. de chirurgie dentaire. / ~ water ‖ Zahnwasser n. ‖ eau f. dentifrice.

denticulation (Mach) ‖ Bezahnung f. ‖ denture f.

dentifrice *see* dental paste *or* powder *or* water.

dentistry, requisites pl. for ~ ‖ zahntechnische Bedarfsgegenstände mpl. ‖ articles mpl. pour la chirurgie dentaire.

dentist's clinique ‖ Zahnklinik f. ‖ clinique f. dentaire. / ~ material ‖ zahnärztliche Bedarfsartikel mpl. ‖ fournitures fpl. pour dentistes.

denture, artificial ~ ‖ Kunstgebiß n. ‖ dentier m.; prothèse f. dentaire. / rubber ~ ‖ Kautschukgebiß n. ‖ dentier m. en caoutchouc. / reinforcing wire and wire gauze strengthener for rubber ~s ‖ Verstärkungseinlage aus Draht und Drahtgaze für Kautschukgebisse ‖ renfort m. en fil et grillages pour dentiers en caoutchouc.

denture-spring ‖ Gebißfeder f. ‖ ressort m. de dentier. / ~ carrier ‖ Gebißfederträger m. ‖ porte-ressort m. de dentier.

denudation of the earth-crust ‖ Abtragung f. der Erdoberfläche ‖ dénudation f. du sol. / ~ by water ‖ Auswaschung f. ‖ érosion f. par lavage.

denude, to (Electr) ‖ die Isolation entfernen ‖ dénuder.

denuded area ‖ Kahlschlagfläche f. ‖ surface f. dénudée.

denunciation ‖ Anzeige f. ‖ dénonciation f.; citation f.; signification f.

deodorization ‖ Geruchlosmachen n. ‖ désodorisation f. / ~ of oils and fats ‖ Desodorisierung f. von Ölen und Fetten ‖ désodorisation f. des huiles et graisses. / high-vacuum ~ plant ‖ Hochvakuumdesodorisierungsanlage f. ‖ installation f. de désodorisation sous haute vide.

deodorize, to ‖ geruchlos machen ‖ désodoriser.

deodorizing preparation ‖ Desodorisierungsmittel n. ‖ préparation f. à désodoriser.

deoxidate, to see deoxidize, to.

deoxidation ‖ Desoxydation f. ‖ désoxydation f.; désoxygénation f.

deoxidize, to ‖ desoxydieren ‖ désoxyder; désoxygéner.

deoxygenate, to see deoxidize, to.

deoxygenation see deoxidation.

Department of Agriculture ‖ Ackerbauministerium n. ‖ département m. agricole.

department of business ‖ Geschäftszweig m. ‖ branche f. de commerce; spécialité f. / ~ store ‖ Warenhausverkaufsraum m. ‖ salle f. de vente d'un grand magasin.

departure ‖ Abgang m.; Abfahrt f. ‖ départ m. / station of ~ ‖ Abgangsamt n. ‖ bureau m. de départ.

depend, to ~ upon ‖ abhängig sein von; sich verlassen auf ‖ dépendre de; compter sur; s'attendre à; avoir confiance f. en. /, the price depends on the goods ‖ der Preis richtet sich nach der Ware ‖ le prix se règle sur la marchandise.

dependence ‖ Abhängigkeit f. ‖ dépendance f. / ~ of attenuation on frequency ‖ Dämpfungsverlauf m. ‖ dépendence f. de l'affaiblissement des fréquences.

dependent ‖ abhängig ‖ dépendant. / ~ variable ‖ abhängige Veränderliche f. ‖ variable f. dépendante.

deperdition of heat ‖ Wärmeverlust m. ‖ déperdition f. de la chaleur.

dephosphoration ‖ Entphosphorung f. ‖ déphosphoration f.

dephosphorize, to ‖ entphosphoren ‖ déphosphoriser.

depilate, to ‖ abpälen; enthaaren ‖ épiler. / ~ the hide ‖ die Haut enthaaren oder

abpälen ‖ débourrer ou épiler les peaux fpl.

depilating agent ‖ Enthaarungsmittel n. ‖ épilant m.

depitching machine with oil-firing ‖ Entpichmaschine f. für Ölfeuerung ‖ dégoudronneur m. à l'huile. / ~ station (Brew) ‖ Entpichstation f. ‖ atelier m. de dégoudronnage.

depolarization ‖ Depolarisation f. ‖ dépolarisation f.

depolarize, to ‖ depolarisieren ‖ dépolariser.

depose, to ‖ eidlich aussagen ‖ déposer; déclarer.

deposit, to (Chem) ‖ niederschlagen ‖ déposer. / ~ (Chem) ‖ sich absetzen ‖ se déposer. / ~ the earth ‖ Erde f. aufschütten ‖ déposer a terre.

deposit (Acc) ‖ Schlammablagerung f. ‖ dépôt m. de boue. / ~ (Boil) ‖ Kesselstein m. ‖ incrustations fpl.; sédiments mpl.; dépôts mpl.; vidanges fpl.; tartres mpl. / ~ (Chem) ‖ Niederschlag m. ‖ dépôt m. / ~ (Mine) ‖ Ablagerung f.; Lagerstätte f. ‖ gisement m.; gîte m. / ~ (Trade) ‖ Angeld n. ‖ arrhes fpl. / ~ of carbon ‖ Rußansatz m. ‖ dépôt m. de suie. / ~ of coke ‖ Koksansatz m. ‖ résidu m. de coke. / disturbed ~ ‖ gestörtes Gebirge n. ‖ terrain m. disloqué. / galvanoplastic ~ ‖ galvanoplastischer Niederschlag m. ‖ dépôt m. galvanique. / insoluble ~ ‖ unlöslicher Bodensatz m. ‖ dépôt m. insoluble. / to investigate the width of a ~ ‖ die Lagerstätte auf ihre Mächtigkeit untersuchen ‖ explorer le gisement sous le rapport de sa puissance. / ~ of iron ore ‖ Eisenerzlager n.; Eisensteinlager n. ‖ gisement m. de fer. / littoral ~ ‖ Küstenablagerung f. ‖ sédiment m. côtier. / ~ of minerals ‖ Mineralvorkommen n. ‖ gisements mpl. de minerais. / probable extent of the ~ ‖ mutmaßliche Erstreckung f. der Lagerstätte ‖ extension f. présumée des gisements. / unstratified ~ ‖ Geschiebeformation f. ‖ terrain m. de transport.

deposit bank ‖ Depositenbank f. ‖ banque f. de dépôt. / ~ business ‖ Lombardgeschäft n. ‖ prêt m. ou avance f. sur gages. / ~ case ‖ Depositenfach n. ‖ casier m. pour dépôts de banque. / ~ department ‖ Depositenkasse f. ‖ caisse f. des dépôts.

deposited ‖ abgelagert ‖ déposé.

deposition of moisture ‖ Feuchtigkeitsniederschlag m. ‖ dépôt m. de buées. / electrolytic ~ of silver ‖ elektrolytischer Silberniederschlag m. ‖ dépôt m. électrolytique d'argent.

depository ‖ Aufbewahrungsort m.; Aufbewahrungsraum m. ‖ dépôt m.

deposits pl. ‖ Depositengelder npl. ‖ fonds mpl. en dépôt.

depossess, to ‖ enteignen ‖ exproprier.

depossession ‖ Enteignung f. ‖ expropriation f.

depot ‖ Niederlage f.; Lagerhaus n.; Warenlager n. ‖ dépôt m.; magasin m.; entrepôt m. / freight ~ see, goods ~ . / goods ~ ‖ Güterschuppen m. ‖ dépôt m. des marchandises; hangar m. à marchandises. / submarine ~ ship ‖ Hilfsschiff n. für Unterseeboote ‖ bâtiment m. de sauvetage de sous-marins.

depreciate, to ‖ an Wert m. verlieren; entwerten ‖ déprécier.

depreciated ‖ entwertet ‖ déprécié.

depreciation ‖ Wertverlust m.; Minderwert m. ‖ dépréciation f.; moins-value f. / allowance for ~ ‖ Abzug m. für Entwertung ‖ remise f. pour dépréciation. / annual ~ ‖ jährliche Abschreibung f. ‖ amortissement m. annuel. / ~ on machinery and buildings ‖ Abschreibung f. an Maschinen und Gebäuden usw. ‖ amortissement m. ou dépréciation f. de la valeur des machines et des bâtiments. / ~ of value ‖ Wertverminderung f. ‖ dépréciation f.

depress, to ‖ niederdrücken; erniedrigen ‖ déprimer. / ~ several keys at one time (Typewr) ‖ mehrere Tasten zu gleicher Zeit herunterdrücken ‖ abaisser plusieurs touches à la fois. / the market is depressed ‖ der Markt ist gedrückt ‖ le marché est lourd.

depressed arch ‖ abgeflachter Bogen m. ‖ arc m. surbaissé. / ~ calotte ‖ gedrückte Kappe f. ‖ calotte f. surbaissée.

depression (Aero) ‖ Senkung f. ‖ abaissement m. / ~ (Chem) ‖ Erniedrigung f. ‖ abaissement m. / ~ (Phys) ‖ Unterdruck m. ‖ dépression f. / ~ telemeter ‖ Entfernungsmesser m. mit senkrechter Basis; Depressionsentfernungsmesser m. ‖ télémètre m. de dépression.

depressive ‖ niederdrückend ‖ dépressif.

deprive, to ~ of air ‖ Luft f. entziehen; entlüften ‖ évacuer l'air m.

depth ‖ Tiefe f. ‖ profondeur f. / ~ (Mine) ‖ Teufe f. ‖ profondeur f. / ~ (River) ‖ Fahrtiefe f. ‖ profondeur f. navigable / ~ of bore ‖ Bohrtiefe f. ‖ profondeur f. de forage. / ~ to the centre of the earth see unlimited ~ / ~ of cut ‖ Schnittiefe f. ‖ profondeur f. de coupe. / ~ of dipping ‖ Eintauchtiefe f. ‖ profondeur f. d'immersion. / ~ of dishing ‖ Tiefe f. der Kümpelung ‖ flèche f. du cintrage. / changeable ~ of drop ‖ veränderliche Sturzhöhe f. ‖ hauteur f. de déversement variable. / ~ of focus ‖ Fokustiefe f. ‖ profondeur f. de foyer. / great ~ ‖ tiefer Grund m. ‖ grand fond m. / ~ of hardening ‖ Härtetiefe f. ‖ épaisseur f. de trempe ou de la couche durcie. / ~ of hole ‖ Bohrtiefe f. ‖ profondeur f. de forage. / ~ of immersion see ~ of dipping. / ~ of jaws (Vice) ‖ Backenhöhe f. ‖ hauteur f. des mors. / ~ of key way ‖ Nuttiefe f. ‖ profondeur f. de la rainure. / ~ of a letter ‖ Schriftkegel m. ‖ corps m. de lettre; force f. de corps d'un caractère. / ~ to which a pile is driven into the earth ‖ Rammtiefe f. eines Pfahles ‖ fichée f. d'un pieu. / greatest ~ of pit ‖ größte Förderteufe f. ‖ profondeur f. maximum d'un puits. / ~ of rainfall ‖ Niederschlagshöhe f. ‖ hauteur f. de pluie. / registered ~ ‖ Vermessungstiefe f. ‖ creux m. du registre. / ~ of a sail ‖ Heiß m. eines Segels ‖ chûte f. au mât d'une voile. / ~ of a shaft ‖ Schachtteufe f. ‖ plomb m. de bure. / ~ of a ship ‖ Tiefe f. eines Schiffes ‖ creux m. d'un vaisseau. / ~ in tank ‖ Flüssigkeitsstand m. im Behälter ‖ hauteur f. de pige. / ~ of throat ‖ Maultiefe f. ‖ profondeur f. de gorge. / ~ for tonnage ‖ Vermessungstiefe f. ‖ creux m. de la cale pour le jaugeage. / ~ of tooth ‖ Zahntiefe f. ‖ hauteur f. du dent. / unlimited ~ ‖ ewige Teufe f. ‖ profondeur f. insondable ou abyssale. / ~ of water (Hydr arch) ‖ Wasserstand m. ‖ hauteur f. d'eau.

·/~ of water (Nav) || Tiefe f. des Wassers || profondeur f. d'eau. / ~ of water in the fairway || Fahrwassertiefe f. || profondeur f. d'eau navigable.

depth, difference in ~ || Tiefenunterschied m. || différence f. de relief. / ~ gauge (Hydr arch) || Pegel m. || échelle f. fluviale. / ~ gauge (Mach) || Tiefentaster m.; Tiefenmesser m. || calibre m. ou pied m. à profondeur. / ~ indicator (Mine) || Teufenanzeiger m. || indicateur m. de position des cages.

depthing tool (Watchm) || Eingriffzirkel m. || compas m. aux engrenages.

depth rudder || Höhensteuer n. || gouvernail m. de profondeur. / ~ stop (Drilling mach) || Tiefenanschlag m. || butée f. permettant de régler la profondeur des trous à percer.

deputy (Mine) || Steiger m. || délégué mineur m. / ~ overman of timber setters (Mine) || Zimmerersteiger m. || chef-boiseur m.

derail, to || entgleisen || dérailler.

derailment || Entgleisung f. || déraillement m.

derange, to || in Unordnung f. bringen || déranger; fausser.

derangement || Unordnung f. || dérangement m.

derelict (Mar) || Seetrift f.; herrenloses Gut n.; treibendes Wrack n. || debri m. ou épave f. de mer.

deresinify, to (chem) || entharzen || dérésinifier.

derivable || ableitbar || dérivable.

derivate || Derivat n. || dérivé m.

derivation (Electr) || Stromableitung f.; Abzweigung f. || dérivation f.; branchement m.; bifurcation f. / ~ (Projectile) || Seitenablenkung f. || dérivation f.

derivative (Chem) || Abkömmling m. || dérivée f. / ~ (Math) || Ableitung f.; Differentialquotient m. || dérivée f.

derive, to (Chem) || abstammen || dériver. / ~ (Electr) || verzweigen || dériver. / ~ (Math) || ableiten || dériver.

derived function || abgeleitete Funktion f. || fonction f. dérivée. / ~ unit || abgeleitete Einheit f. || unité f. dérivée.

dermatoscope || Dermatoskop n. || dermatoscope m.

derrick || beweglicher Ausleger m. Ladebaum m. || derrick m.; potence f.; chevalement m. / ~ (Mine) || Bohrturm m. || échafaud m. à forer. / ~ (Shipb) || Ladebaum m. || mât m. de charge. / ~ car || Kranwagen m. || wagon m. dépanneur; wagon-grue m. / ~ crane || Derrickkran m.; Dreifußkran m. || grue f. derrick;

derricking controller || Hubsteuerschalter m.; Hubfahrschalter m. || contrôleur m. pour appareils de levage.

derrick wagon-crane Waggonkran m.; Eisenbahnwagendrehkran m. grue f. || montée sur wagon.

desactivate, to || entaktivieren || désactiver.

desactivation || Entaktivierung f. || désactivation f.

descend, to ~ into a mine || einfahren; in die Grube fahren || descendre dans une mine. / ~ into a shaft (Mine) || einen Schacht m. befahren || descendre dans un puits. / ~ a slope (Mine) || hinunterfahren || descendre une pente.

descending current || Abwind m. || vent m. descendant; courant m. (d'air) descendant. / region of ~ currents || Abwind-

feld n. || champ m. de vent descendant. / static ~ force || statische Sinkkraft f. || force f. descensionnelle statique.

descent (Mine) || Einfahrt f. || descente f. / ~ (Road; River) || Gefälle n. || inclinaison f.; pente f.; chute f. / ~ (Surface) || Neigung f. || descente f.; inclinaison f.; pente f. / ~ of the charges of a blast furnace || Niedersinken n. der Gichten eines Hochofens || descente f. des charges d'un haut-fourneau. / irregular ~ of the furnace-charges || Kippen n. oder Rücken n. der Gichten || éboulement m. ou descente f. irrégulière des charges d'un haut-fourneau. / ~ of the keys || Tastenniedergang m. || plongée f. des touches.

description || Schilderung f. || description f. / detailed ~ || ausführliche Beschreibung f. || description f. détaillée. / detailed ~ of a building || genaue Baubeschreibung f. || description f. détaillée de la construction d'un bâtiment. / of every ~ || jeder Art f. || de toute espèce f.

descriptive geometry || darstellende Geometrie f. || géométrie f. descriptive.

desemulsify, to || entemulgieren || désémulsionner.

desert sand || Wüstensand m. || sable m. du désert.

deserving of credit || kreditwürdig || digne de crédit.

dessicate, to || austrocknen || dessécher.

desiccating || trocknend || desséchant. / ~ machine for wood pulp || Entwässerungsmaschine f. für Zellulosebrei || installation f. pour la déshydratation de la pâte de cellulose.

desiccation || Trocknen n. || dessiccation f.; séchage m. / to achieve the rapid and uniform ~ of the emulsion || rasches und gleichmäßiges Erstarren n. der Emulsion bewirken || provoquer la solidification rapide et uniforme de l'émulsion. / ~ of wood || Holztrocknung f. || dessiccation f. ou séchage m. du bois.

desiccator || Trockenmittel n. || agent-dessiccateur m.

design, to || projektieren; entwerfen; aufzeichnen || projeter; dessiner. / ~ (Weav) || patronieren; ausnehmen; absetzen || mettre en carte f.

design || Zeichnung f. || dessin m. / ~ (Weav) || Muster n.; Aufzeichnung f. || dessin m. / common ~ || see standard ~ / compact ~ || (Mach) gedrängte Bauart f. || construction f. compendieuse ou compacte. / ~ in full size || Detailzeichnung f. || dessin m. de détail ou détaillé / ~ of a microscope || Mikroskopbau m. || construction f. d'un microscope. / registered ~ || Gebrauchsmuster n. || brevet m. pour modèles d'utilité; modèle m. déposé. / ~ of a self-supporting overhead cable || Stufenmuster n. eines Luftkabels || construction f. d'un câble aérien à auto-support. / simple ~ || einfache Bauart f. || construction f. simple. / standard ~ || gewöhnliche Ausführung f. || constructions fpl. normales. / strong ~ || widerstandsfähige Ausführung f. || exécution f. résistante ou robuste.

designate, to || bezeichnen || indiquer.

designation || Bezeichnung f. || désignation f. / ~ consisting of a letter with a number affixed || Bezeichnung f., die aus einem Buchstaben und einer Nummer besteht || désignation f. par une lettre suivie d'un chiffre.

design-cutting machine for indentation || Musterschneidemaschine f. für Zackenschnitt || machine f. à denteler les échantillons.

designer || Musterzeichner m. || dessinateur m. industriel. / ~ (Weav) || Musterausnehmer m. || metteur m. en cartes pour tissus; encarteur m. de tissus.

designing for fabrics || Musterzeichnen n. || piquage m. de cartes de dessins. / ~ machine || Schablonen(stech)maschine f. || machine f. à piquer.

design painter (Weav) || Musterkolorierer m. || coloriste m. de tissus. / ~ paper (Drawing) || Zeichenpapier n. || papier m. à dessin ou à dessiner. / ~ paper (Weav) || Patronenpapier n.; Tupfpapier n.; Musterpapier n. || papier m. à patron; papier m. quadrille; carte f. / ~ sheet || Konstruktionsblatt n. || feuille f. de construction.

desilvering bath || Entsilberungsbad n. || bain m. de désargenture.

desilverize, to || entsilbern || extraire l'argent m.; désargenter.

desintering mill || Entzunderungswalzwerk n. || laminoir m. à enlever les batitures. / ~ plant || Entsinterungsanlage f. || installation f. d'appareils à enlever les batitures.

desizing agent || Entschlichtungsmittel n. || produit m. de dégommage.

desk || Pult n. || pupitre m. / bookkeeper's ~ || Stehpult n. || bureau m. de comptable. / ~ for children || Kinderschreibpult n. || pupitre m. pour enfants. / double ~ || Doppelpult n. || pupitre m. double / roll-top ~ || Schreibtisch m. mit Rollverschluß || bureau m. à rideaux.

desk blotter pad || Schreibunterlage f. mit Löschpapier || sous-main m. avec buvard. / ~ form || Pultform f. || forme f. de pupitre. / folding ~ pad || aufklappbare Schreibunterlage f. || sous-main m. pliable. / ~ set || Schreibgarnitur f. || garniture f. de bureau. / ~ telephone set || Tischfernsprecher m. || poste m. de téléphone mobile. / ~ tray || Briefablegekasten m. || corbeille f. à correspondance pour le bureau. / ~ type board || Schaltpult n. || tableau m. de manœuvre à pupitre.

despatch see dispatch.

desquamation || Abschuppung f. || desquamation f.

dessert knife || Nachtischmesser n. || couteau m. à dessert. / ~ knife and fork || Dessertbesteck n. || couvert m. à dessert. / ~ plate || Obstteller m. || assiette f. à dessert. / ~ wine || Likörwein m.: Dessertwein m. || vin m. de liqueur.

dessinier rolling mill || Dessinierwalzwerk n. || laminoir m. à dessins.

destitute || leer; entblößt || dépourvu.

destroy, to || zerstören || détruire.

destroyed || abgetötet; zerstört || détruit.

destroyer, tank ~ (War impl) || Kampfwagenjäger m.; Tankzerstörer m. || contre-tank m.

destroying, paper ~ machine || Papiervernichtungsmaschine f. || machine f. pour la destruction du papier.

destructibleness || Zerstörbarkeit f. || destructibilité f.

destructibility (Chem) || Zersetzbarkeit f. || destructibilité f.

destruction of the bacteria || Bakterienvernichtung f.; Keimtötung f. || destruc-

tion f. des microbes. / ~ of the molecule ‖ Zertrümmerung f. des Moleküls ‖ destruction f. de la molecule. / ~ of a wall ‖ Mauereinsturz m. ‖ rupture f. ou écroulement m. d'un mur.

destructive ‖ vernichtend; zerstörend ‖ destructif. / ~ distillation ‖ trockene Destillation f. ‖ distillation f. sèche. / ~ lightning ‖ Schadenblitz m. ‖ foudre f. ou éclair m. produisant des dégâts.

desulphurate, to ‖ entschwefeln ‖ dessoufrer.

desulphuration ‖ Entschwefelung f. ‖ dessoufrage m.

desurgarization of molasses ‖ Melasseentzuckerung f. ‖ sucraterie f.

detach, to ~ a carriage ‖ einen Wagen m. abkuppeln ‖ découpler une voiture. / ~ the stock ‖ das Mahlgut auflösen ‖ désagréger le produit.

detachable ‖ abnehmbar ‖ amovible; mobile; rapporté. / ~ ball journal bearing ‖ Schulterkugellager n. ‖ roulement m. à billes démontable. / ~ crank ‖ abnehmbare Kurbel f. ‖ manivelle f. détachable.

detached revetment ‖ freistehende Mauer f. ‖ revêtement m. ou mur m. détaché.

detacher (Mill) ‖ Detaschör m. ‖ détacheur m.

detaching mill ‖ Lösewalzwerk n. ‖ laminoir m. pour décoller le tube du mandrin.

detachment of the retina ‖ Netzhautablösung f. ‖ décollement m. de la rétine.

detail ‖ Einzelteil n. ‖ détail m.; pièce f. détachée. / fine ~s pl. ‖ Feinheiten fpl. ‖ fins détails mpl. / ~ drawing ‖ Detailzeichnung f. ‖ dessin m. de détails.

detailed ‖ ausführlich ‖ détaillé. / ~ description ‖ ausführliche Beschreibung f. ‖ description f. détaillée. / ~ scheme ‖ Angabe f. ‖ instruction f.; dispositif m. / ~ strip ‖ Kontrollstreifen m. ‖ bordereau m. de contrôle.

detan, to ‖ entgerben ‖ détanner.

detar, to ‖ entteeren ‖ dégoudronner.

detect, to ‖ nachweisen; (qualitativ) bestimmen ‖ déceler; rechercher.

detectable (Chem) ‖ nachweisbar ‖ décelable.

detecting simulated amblyopia ‖ Feststellung f. vorgetäuschter Sehschwäche ‖ décèlement m. de l'amblyopie simulée.

detection (Chem) ‖ Nachweis m.; (qualitative) Bestimmung f. ‖ décèlement m.

detector ‖ Detektor m. ‖ détecteur m. / balanced crystal ~ (Radio) ‖ ausbalanzierter Kristallwellenanzeiger m. ‖ détecteur m. à cristal équilibré. / contact ~ ‖ Kontaktdetektor m. ‖ détecteur m. de contact. / double ~ ‖ Gegendetektor m. ‖ détecteur m. double. / electrolytic ~ ‖ elektrolytischer Detektor m. ‖ détecteur m. électrolytique. / ~ of fire damp (Mine) ‖ Anzeiger m. für schlagende Wetter ‖ indicateur m. de grisou. / galena ~ ‖ Bleiglanzdetektor m. ‖ détecteur m. de galène. / magnetic ~ ‖ magnetischer Detektor m. ‖ détecteur m. magnétique. / pyrite ~ ‖ Pyritdetektor m. ‖ détecteur m. à pyrite. / recording ~ ‖ registrierender Detektor m. ‖ détecteur enregistreur m. / thermoelectric ~ ‖ thermoelektrischer Detektor m. ‖ détecteur m. thermoélectrique. / watchman's time-~ ‖ Kontrolluhr f.; Wächteruhr f. ‖ contrôleur m. des rondes. / wave ~ (Radio) ‖ Wellenanzeiger m. ‖ détecteur m. d'onde.

detector bar ‖ Fühlschiene f.; Druckschiene f. ‖ pédale f. de calage ou de sûreté. / ~ bar stop ‖ Anschlag m. der Druckschiene ‖ taquet m. d'arrêt de la pédale. / ~ receiver (Radio) ‖ Detektorempfänger m. ‖ récepteur m. à détection sur galène. / ~ receiving set ‖ Detektorapparat m. ‖ appareil m. détecteur. / ~ tube (Radio) ‖ Detektorröhre f. ‖ lampe f. détectrice. / ~ valve ‖ Detektorröhre f. ‖ tube m. détecteur. / ~ voltmeter ‖ Detektorvoltmesser m. ‖ voltmètre m. à détecteur.

detent (Clockm; Tel) ‖ Auslösung f. ‖ échappement m. / ~ (Mach) ‖ Arretierung f.; Sperrklinke f. ‖ détente f.; arrêtage m.; dispositif m. d'arrêt. / ~ gear of an indicator ‖ Anhaltevorrichtung f. eines Indikators ‖ dispositif m. d'arrêt d'un indicateur. / ~ pin (Locksm) ‖ Sperrstift m. oder Vorstecker m. am Bolzen der Schloßfeder ‖ étoquiau m.; goupille f. d'arrêt.

deteriorate, to ‖ verschlechtern; verderben ‖ détériorer; avarier. / ~ ‖ schlechter werden ‖ se gâter; se détériorer.

deterioration ‖ Verschlechterung f. ‖ détérioration f.; dépréciation f.

determinant of the line equations ‖ Gleichungsdeterminante f. für Leitungen ‖ déterminant m. des équations de ligne.

determination of the age of strata ‖ Altersbestimmung f. der geologischen Schichten ‖ détermination f. de l'âge des couches. / ~ of the alcohol ‖ Alkoholbestimmung f. ‖ dosage m. de l'alcool. / ~ of the amids ‖ Amidenbestimmung f. ‖ dosage m. des amides. / ~ of ashes ‖ Aschenbestimmung f. ‖ détermination f. des cendres. / ~ of the attenuation ‖ Vergärungsgradbestimmung f. ‖ détermination f. de l'atténuation. / ~ of calorific value ‖ Heizwertbestimmung f. ‖ détermination f. de la puissance calorifique. / calorimetric ~ ‖ kalorimetrische Bestimmung f. ‖ détermination f. calorimétrique. / ~ of compensation ‖ Entschädigungsfestsetzung f. ‖ fixation f. des indemnités. / ~ of earth-thrust ‖ Erddruckermittlung f. ‖ détermination f. de la poussée des terres. / ~ of the heating value ‖ Heizwertbestimmung f. ‖ détermination f. de la valeur calorifique. / ~ of the metal contents of an ore ‖ Bestimmung f. des Metallgehaltes eines Erzes ‖ détermination f. de la teneur en métal d'un minerai. / ~ of a point ‖ Punktbestimmung f. ‖ fixation f. d'un point. / ~ of position by measurement of star's altitude ‖ Ortsbestimmung f. durch Gestirnhöhenmessung ‖ détermination f. astronomique du point. / ~ of position by one base line ‖ Ortsbestimmung f. mit einer Standlinie ‖ détermination f. de la position d'un lieu au moyen d'une ligne de base. / ~ of the value of fuels ‖ Brennstoffwertbestimmung f. ‖ détermination f. de la valeur calorifique des combustibles. / ~ of the viscosity ‖ Viskositätsbestimmung f. ‖ détermination f. de la viscosité. / ~ of yield ‖ Ertragsfeststellung f.; Etatsfestsetzung f. ‖ calcul m. de rendement; fixation f. de la possibilité.

determine, to ‖ beschließen ‖ résoudre; décider. / ~ (Chem) ‖ feststellen ‖ déterminer; fixer. / ~ the position of a point ‖ die Lage f. eines Ortes bestimmen ‖ fixer

la position d'un lieu. / ~ the relative positions of various points to each other ‖ verschiedene Punkte mpl. zueinander festlegen ‖ déterminer différents points les uns par rapport aux autres.

determining the working resistance ‖ Bestimmung f. des Arbeitswiderstandes ‖ détermination f. de la résistance des matériaux.

detonate, to (Chem) ‖ verpuffen ‖ détoner.

detonating ball ‖ Knallerbse f. ‖ pois m. fulminant. / ~ cap ‖ Sprengkapsel f. ‖ détonateur m. / ~ composition ‖ Zündsatz m.; Zündmasse f. ‖ composition f. fulminante; matière f. fulminante. / ~ signal ‖ Knallsignal n. ‖ pétard m.

detonation ‖ Knall m. ‖ détonation f.

detonator (Mine) ‖ Sprengkapsel f.: Zünder m. ‖ détonateur m. / ~ (Railw) ‖ Knallsignal n. ‖ pétard m.

detour ‖ Umweg m. ‖ détour m.

detriment ‖ Schädigung f.; Verlust m. ‖ détriment m.; tort m.; préjudice m.

detrimental ‖ schädlich ‖ nuisible: préjudicielle.

detrital deposit ‖ Trümmerlagerstätte f. ‖ gisement m. secondaire.

detrition apparatus (Malt) ‖ Abreibeapparat m. ‖ appareil m. à frottement.

detritus (Geol) ‖ Geröllmasse f. ‖ éboulis m.; détritus mpl. / ~ (Hydr arch) ‖ Sinkstoffe mpl. ‖ détritus mpl.

develop, to ‖ enthüllen; entfalten ‖ développer. / ~ (Math) ‖ entwickeln ‖ développer. / ~ (Phot) ‖ entwickeln ‖ développer; révéler.

developer (Phot) ‖ Entwickler m. ‖ bain m. de développement; révélateur m.: développeur m.

developing of photos ‖ Entwickeln n. von fotografischen Aufnahmen ‖ développement m. des images photographiques. / ~ bath (Phot) ‖ Entwicklungsbad n. ‖ bain m. révélateur ou de développement. / photographic ~ device ‖ fotografischer Entwicklungsapparat m. ‖ appareil m. de développement photographique.

development ‖ Entwicklung f. ‖ développement m. / ~ of the city ‖ Entwicklung f. der Stadt ‖ développement m. urbain ou de la ville. / ~ of gas (Acc) ‖ Gasentwicklung f. ‖ formation f. de gaz. / ~ into series (Math) ‖ Reihenentwicklung f. ‖ développement m. en série. / ~ of the surface (Acc) ‖ Oberflächenentwicklung f. ‖ développement m. de surface. / ~ of the traffic ‖ Entfaltung f. des Verkehrs ‖ développement m. du trafic.

development paper (Phot) ‖ Entwicklungspapier n. ‖ papier m. de développement. / ~ study of a telephone network ‖ Bauplan m. für die Erweiterung eines Fernsprechnetzes ‖ étude f. d'organisation d'un réseau téléphonique.

deviate, to ‖ abweichen; abirren ‖ dévier; obliquer; s'écarter. / ~ from the course ‖ vom Fahrweg m. abkommen ‖ s'écarter de la route. / ~ a magnetic needle ‖ eine Magnetnadel f. ablenken ‖ dévier une aiguille aimantée.

deviation ‖ Abweichung f.; Deviation f. ‖ déviation f. / ~ (Aero) ‖ Abtrift m.: Abtrieb m.; Abdrängung f. ‖ dérive f. / ~ (Tel) ‖ Umleitung f. ‖ détournement m. / ~ of the compass ‖ Deviation f. des Kompasses ‖ déviation f. de la boussole. / left-hand ~ ‖ Linksabweichung f. ‖ déviation f. à gauche. / local ~ from the

regular variation of declination || örtliche Abweichung f. vom regelmäßigen Verlauf der Deklinationsänderung || écart m. local par rapport au cours régulier de la variation de la déclinaison. / ~ of the magnetic needle || Mißweisung f. der Magnetnadel || déviation f. ou déclinaison f. de l'aiguille aimantée. / ~ of x millimeters || Abweichung f. von x Millimetern || écart m. de x millimètres. / prismatic ~ || prismatische Ablenkung f. || déviation f. prismatique. / ~ in the quality || Ungleichmäßigkeit f. in der Herstellung f. || manque f. d'uniformité. / ~ from the straight line || Abweichung f. von der Geraden || déviation f. de la ligne droite. / ~ roller || Ablenkungsrolle f. || galet m. de déviation.

device || Vorrichtung f. || dispositif m. / adjusting ~ || Verstellvorrichtung f. || dispositif m. de réglage. / ~ for adjusting the tension of the belt || Riemenspanner m. || tendeur m. de courroie. / automatic recording ~ || selbsttätig registrierende Schreibvorrichtung f. || dispositif m. enregistreur automatique. / ~ for controlling the distribution of power || Einrichtung f. zur Überwachung der Energieverteilung || installation f. pour le contrôle de la répartition de l'énergie. / ~ for controlling the generation of power || Einrichtung f. zur Überwachung der Energieerzeugung || installation f. pour le contrôle de la production de l'énergie. / indicating ~ || Anzeigevorrichtung f. || dispositif m. indicateur. / make-and-break ~ || Unterbrecherscheibe f. || disrupteur m. / mechanical ~ || mechanische Vorrichtung f. || dispositif m. mécanique. / ~ for narrowing (Textile) || Deckvorrichtung f. || mécanique f. à diminuer. / ~ for writing pay envelopes || Schreibvorrichtung f. für Lohnbeutel || dispositif m. pour enveloppe de paye.

devil (Spinn) || Reißwolf m. || diable m.; loup m. / ~ (Winding cable) || Haken m. || croc m. / ~'s claw (Build) || Kropfeisen n.; Steinklaue f.; Teufelsklaue f. || louve f.; renard m.

devitrify, to || entglasen || dévitrifier.
devotional articles pl. || Devotionalien fpl. || articles mpl. de piété.
dew, to || betauen || couvrir de rosée.
dew (Meteor) || Tau m. || rosée f. / to become covered with ~ || betauen || se couvrir de rosée. / ~ cap of a telescope || Taukappe f. eines Fernrohres || tube m. pare-rosée d'une lunette. / hinged lens cap in ~ mount || aufklappbarer Objektivdeckel m. in der Taukappe || couvercle m. à clapet de l'objectif dans le tube pare-rosée. / ~ point || Taupunkt m. || point m. de rosée. / ~ retting (Spinn) || Taurotte f. || rouissage m. sur terre.
dextrine || Dextrin n. || dextrine f.
dextrose || Dextrose f. || dextrose f.
diacetylmorphine || Diazetylmorphin n. || diacéthylmorphine f.
diacid || zweibasisch || bibasique.
diadochite || Phosphoreisensinter m.; Diadochit n. || diadochite f.
diagnosis || Diagnose f. || diagnostic m. / x-ray ~ || Röntgenuntersuchung f. || radiodiagnostic m.; diagnostic m. par les rayons X.
diagonal || Diagonale f. || diagonale f. / ~ brace || Diagonalstrebe f. || entretoise f. diagonale. / ~ bracing || Diagonalverstre-

bung f.; Kreuzverspannung f. || croisillons mpl. d'incidence; entretoisement m. en diagonale. / ~ drilling machine || Diagonalbohrmaschine f. || machine f. à percer en diagonale. / ~ members pl. || Strebenkreuz n. || croix f. de contrefiches ou de jambes de force; contrefiches fpl. ou jambes fpl. de force en croix. / ~ rail || Diagonalschiene f. || rail m. diagonal. / ~ rib || Diagonalrippe f. || nervure f. diagonale. / ~ stay see ~ brace. / ~ trussing || Querstrebe f. || renforcement m. diagonal.

diagram || Diagramm n. || diagramme m. / accompanying ~ || nebenstehendes Diagramm n. || schéma m. ci-contre. / ~ of connections (Electr) || Schaltungsanordnung f.; Stromlaufschema n. || schéma m. des connexions; plan m. de montage. / ~ of connections for railroad stations (Tel) || Blockschaltplan m. || schéma m. des connexions aux gares. / ~ of forces || Kräfteplan m. || diagramme m. der forces. / to planimeter a ~ || ein Diagramm n. planimetrieren || planimétrer un diagramme. / ~ of the slide-valve || Schieberdiagramm n. || diagramme m. de la distribution de la vapeur.
diagram paper || Indikatorpapier n. || papier m. de l'indicateur. / ~ recording || Diagrammaufzeichnung f. || traçage m. à diagramme.
diagraph || Storchschnabel m. || diagraphe m.
dial, to (Aut tel) || die Nummer wählen || combiner le numéro. / ~ underground (Mine) || markscheiden || lever ou tracer des plans de mine.
dial || Zifferblatt n. || cadran m. / clock ~ || Zifferblatt n. || cadran m. de montres. / ~ of the compass || Rose f. des Kompasses || rose f. des vents de la boussole. / ~ with concentric scale || Zifferblatt n. mit konzentrischer Skale || cadran m. à échelle concentrique. / ~ with eccentric scale || Zifferblatt n. mit exzentrischer Skale || cadran m. à échelle excentrique. / enamelled ~ || emailliertes Zifferblatt n.; Emailzifferblatt n. || cadran m. émaillé. / interchangeable ~ || auswechselbares Zifferblatt n. || cadran m. interchangeable. / luminous ~ || leuchtendes Zifferblatt n. || cadran m. lumineux. / metallic ~ || metallisches Zifferblatt n. || cadran m. métallique. / ~ printed in two colours || zweifarbiges Zifferblatt n. || cadran m. en deux couleurs. / slanting ~ || schräges Zifferblatt n. || cadran m. oblique. / surveyor's ~ || Feldkompaß m. || boussole f. d'arpenteur.
dial apparatus system for fire alarm installations || Zeigerapparatsystem n. in Feuermeldeanlagen || système m. de poste à cadran dans les installations d'avertisseurs d'incendie. / ~ balance || Zeigerwage f. || bascule f. à aiguille. / ~ card || Windrose f. || rose f. des vents ou au compas.
dialing (Mine) || Markscheiderzug m. || levé m. ou tracé m. à la boussole. / ~ tone (Aut tel) || Aufforderungssignal n.; Amtszeichen n. || bruit m. de réseau; signal m. de cadran. / ~ tone coil (Tel) || Signalspule f. || bobine f. d'alarme.
diallogite || roter Braunstein m.; Diallogit m.; Manganspat m. || carbonate m. de manganèse; diallogite f.

dial painter || Zifferblättermaler m. || peintre m. en cadran. / ~ plate || Teilscheibe f. || plateau-diviseur m. / ~ plate (Watchm) || Zifferblatt n. || cadran m.; limbe m. / ~ speed tester (Aut tel) || Prüfeinrichtung f. für Nummernscheiben || appareil m. pour éprouver les cadrans. / ~ switch (Aut tel) || Nummernscheibe f. || cadran m. / ~ telegraph || Zeigertelegraf m. || télégraphe m. à cadran. / ~ tone (Tel) || Amtszeichen n. || ton m. de cadran. / ~ work (Watchm) || Vorlegewerk n. || cadrature f.
dialysis || Dialyse f. || dialyse f.
dialyze, to || dialysieren || dialyser.
dialyzer || Dialysator m. || dialyseur m.
diamagnetic || diamagnetisch || diamagnétique.
diamagnetism || Diamagnetismus m. || diamagnétisme m.
diameter || Durchmesser m. || diamètre m. / ~ of an axle-tree || Stärke f. einer Welle || diamètre m. d'un arbre. / ~ of disc || Scheibendurchmesser m. || diamètre m. du disque. / inside ~ || Innendurchmesser m. || diamètre m. intérieur. / lens ~ || Linsendurchmesser m. || diamètre m. de la lentille. / outside ~ || Außendurchmesser m. || diamètre m. extérieur. / principal ~ of the parabola || Parabelachse f. || axe m. de la parabole. / ~ and gauged length of test bars || Probestababmessungen fpl. || dimensions fpl. de l'éprouvette.
diameter class (Wood) || Stärkeklasse f. || catégorie f. de grosseur. / ~ growth (Wood) || Dickenwachstum n. || croissance f. en diamètre.
diamond (Geom) || Raute f.; Rhombus m. || losange m.; rhombe m. / ~ (Miner) || Diamant m. || diamant m. / ~ (Print) || Diamant m. || Sédanoise f. / ~ for glass-cutting || Schneiddiamant m. zum Glasschneiden || diamant m. pour couper le verre. / glazier's ~ || Glaserdiamant m. || diamant m. de vitrier ou à couper le verre. / ~ for rock drills || Diamant m. für Tiefbohrungen || diamant m. pour forages. / rose ~ (Jewel) || Rautenstein m.; Rose f.; Rosette f. || diamant-rose m.; rose f. / rough ~ || roher Diamant m. || diamant m. brut.
diamond bearing || diamanthaltig || diamantifère. / ~ black || Diamantschwarz n. || noir m. diamant. / ~ boring crown see ~ rock drill crown. / ~ cleaver || Diamantspalter m. || cliveur m. de diamant. / ~ crossing (Railw) || Kreuzungsstück n.; Doppelherzstück n. || croisement m. ou cœur m. double; pièce f. de croisement. / ~ cutter || Diamantschneider m.; Diamantschleifer m. || tailleur m. de diamants. / ~ cutting shop || Diamantschneiderei f. || taillerie f. de diamants. / ~ draught (Weav) || Pointieren n. || passage m. en pointe. / ~ drill || Diamantbohrer m.; Diamantbohrmaschine f. || foret m. ou perforateur m. à diamants. / ~ dust || Diamantine f.; Diamantpulver n.; Diamantstaub m. || poudre f. ou poussière f. de diamant; égrisée f. / ~ fabric (Weav) || Atlastrikot m. || satin-fil m. / ~ gray || Diamantgrau n. || gris m. diamant. / ~ mine || Diamantengrube f. || mine f. de diamant. / ~ pencil || Glaserdiamant m. || diamant m. pour couper le verre; rabot m. à diamant. / ~ polishing factory || Diamantschleiferei f. || taillerie f. de diamant. / ~ powder see

~ dust. / ~ rock drill ‖ Diamantbohrmaschine f. ‖ machine f. à percer au diamant. / ~ rock drill crown ‖ Diamantbohrkrone f. ‖ couronne f. diamantée à percer. / ~ truing device for grinding wheels ‖ Schleifscheibenabrichtvorrichtung f. ‖ dispositif m. à redresser pour meules. / ~ white ‖ Diamantweiß n. ‖ blanc m. diamant.

diapered (Build) ‖ jaspiert ‖ diapré; gaufré. ~ (Weav) ‖ gemustert ‖ façonné.

diaphane ‖ Diaphanie f. ‖ diaphanie f.; vitrauphanie f.

diaphanoscope ‖ Durchleuchtungsapparat m.; Diaphanoskop n. ‖ diaphanoscope m.

diaphanous ‖ durchscheinend ‖ diaphane.

diaphragm, to ‖ abblenden ‖ diaphragmer.

diaphragm (Masch) ‖ Diaphragma n.; Querwand f.; Scheidewand f. ‖ diaphragme m. / ~ (Med) ‖ Zwerchfell n. ‖ diaphragme m. / ~ (Opt) ‖ Blende f. ‖ diaphragme f. / ~ (Pump) ‖ Federplatte f. ‖ diaphragme f. / ~ (Tel) ‖ Schallplatte f.; Membran f. ‖ diaphragme m. / adjustable ~ ‖ einstellbare Blende f. ‖ obturateur m. réglable. / azimuth ~ ‖ Azimutblende f. ‖ diaphragme f. azimutal. / centring ~ ‖ Zentrierblende f. ‖ diaphragme m. de centrage. / cylindrical ~ ‖ Zylinderblende f. ‖ diaphragmecylindre m. / eyepiece ~ ‖ Okularblende f. ‖ diaphragme m. oculaire. / five-point ~ ‖ Fünfpunktblende f. ‖ diaphragme m. à cinq points. / front ~ ‖ Abschlußblende f. ‖ diaphragme m. de fermeture. / ~ of x grades ‖ x-teilige Stufenblende f. ‖ diaphragme m. à x parties. / inset ~ ‖ Einhängeblende f. ‖ diaphragme m. à suspendre. / interchangeable ~ ‖ Einsatzblende f. ‖ diaphragme m. interchangeable. / iris ~ ‖ Irisblende f. ‖ diaphragme m. en forme d'iris. / measuring ~ ‖ Meßmembran f. ‖ membrane f. d'organe de mesure. / to open the ~ ‖ die Blende f. öffnen ‖ ouvrier le diaphragme. / sliding ~ ‖ Blendschieber m. ‖ diaphragme m. coulissant. / slip-on ~ ‖ Aufsteckblende f. ‖ diaphragme m. s'adaptant à la monture de l'objectif. / to shut the ~ ‖ die Blende schließen ‖ fermer le diaphragme. / sliding ~ ‖ Schiebeblende f. ‖ diaphragme m. coulissant. / ~ in soft iron (Tel) ‖ Membran f. aus weichem Eisen ‖ membrane f. en fer doux.

diaphragm aperture (Opt) ‖ Blendenöffnung f. ‖ ouverture f. du diaphragme. / ~ arrangement capable of rotation about the axis of the telescope ‖ eine um die Fernrohrachse drehbare Blendvorrichtung f. ‖ diaphragme m. tournant autour de l'axe de la lunette. / ~ cap (Tel) ‖ Membrandeckel m. ‖ couvercle m. du diaphragme ou de la membrane. / ~ carrier ‖ Diaphragmenträger m. ‖ portediaphragme m. / ~ case ‖ Federplattengehäuse n. ‖ boîte f. de diaphragme. / ~ current ‖ Diaphragmenstrom m. ‖ courant m. de diaphragme. / ~ pressure gauge ‖ Plattenfedermanometer m. ‖ manomètre m. à diaphragme ondulé ou à plaque. / ~ process ‖ Diaphragmenverfahren n. ‖ procédé m. à diaphragmes. ~ pump ‖ Membranpumpe f.; Diaphragmapumpe f. ‖ pompe f. à diaphragme. / ~ slider furnished with stops arranged side by side ‖ mit stufenförmig nebeneinander angeordneten Blenden versehener Blendenschieber m. ‖ coulisse

f. munie de diaphragmes échelonnés juxtaposés.

diapositive ‖ Diapositiv n. ‖ diapositive f. / ~ glass ‖ Diapositivglas n. ‖ verre m. diapositif.

diary ‖ Tagebuch n. ‖ agenda m.

diastase ‖ Diastase f.; Malzbildner m. ‖ diastase f. / translocation ~ ‖ Translokationsdiastase f. ‖ diastase f. de translocation.

diastatic ‖ diastatisch ‖ diastatique. / ~ action ‖ Diastasewirkung f. ‖ action f. diastatique.

diathermic apparatus ‖ Diathermieapparat m. ‖ appareil m. pour la diathermie.

diatoric ‖ Lochzahn m. ‖ dent f. diatorique.

dice (pl. of die) ‖ Würfel mpl. ‖ dés mpl. / game of ~ ‖ Würfelspiel n. ‖ jeu m. de dés. / ~ box of cardboard ‖ Würfelbecher m. aus Pappmaché ‖ gobelet m. à dés en carton pâte.

dichloræthylen ‖ Dichloräthylen n. ‖ ethylène m. dichlorique.

dichloride of platinum ‖ Platinchlorür n.; Platindichlorid n. ‖ chlorure m. de platine.

dichotomic ‖ dichotomisch ‖ dichotomique.

dichroism ‖ Dichroismus m. ‖ dichroïsme m.

dichroite ‖ Peliom m.; Cordierit m.; Dichroit m. ‖ cordiérite m.; dichroïte m.

dicky ‖ Vorhemd n. ‖ devant m. de chemise.

dictating machine see dictograph. / ~ tube (Office) ‖ Sprechschlauch m. ‖ tube m. pour la dictée.

dictograph ‖ Diktiermaschine f. ‖ dictographe m.; machine f. à dicter.

die (Coin) ‖ Prägestempel m. ‖ coin m. / ~ (Forg) ‖ Gesenk n. ‖ étampe f. / ~ (Game) (pl.: dice) ‖ Würfel m. ‖ dé m. / ~ (Mach tool) ‖ Schnitt m. ‖ étampe f. / ~ (Press) ‖ Ziehring m.; Preßring m. ‖ perçoir m. / ~ (Print) ‖ Prägeplatte f. ‖ plaque f. à gaufrer. / ~ (Punch) ‖ Matrize f. ‖ matrice. / ~ (Screw cutting) see also dies ‖ Schneidbacke f. ‖ coussinet m. de filière. / ~ (Wiredrawing) ‖ Kaliber n. ‖ matrice f. à étirer. / the changing of the ~s is effected in x minutes ‖ das Auswechseln n. der Werkzeuge erfolgt in x Minuten ‖ le changement des outils s'opère en x minutes. / the ~ coined without any appreciable wear ‖ der Prägestempel prägte ohne wesentliche Abnutzung ‖ le coin frappa sans une trace d'usure apparente. / coining ~ ‖ Münzstempel m. ‖ coin m. / engraver's ~ ‖ Gravörstempel m. ‖ coin m. pour graveurs. / first operation ~ (Press) ‖ Werkzeug n. zum Vorziehen ‖ outil m. pour l'emboutissage en première passe. / to forge in ~s ‖ im Gesenk n. schmieden ‖ matricer; forger en matrices fpl. / lower ~ ‖ Untergesenk n. ‖ matrice f. / milled tapped lappedthreading ~ ‖ gefräst-gebohrt-geläppte Schneidbacke f. ‖ peigne m. à fileter fraisé, taraudé et rodé. / multiple ~ ‖ Mehrfachgesenk n. ‖ étampé f. multiple. / over-~ ‖ Obergesenk n.; Oberstempel m. ‖ étampe f. supérieure. / upper-~ see over-~.

die cast ‖ Schalenguß m. ‖ coulée f. en coquille. / ~ casting ‖ Spritzguß m. ‖ fonte f. lancée; pièce f. coulée sous pression. / ~ engraving ‖ Gaufriergravierung f. ‖ gravure f. de gaufrois. / ~ forged point of frog (Railw) ‖ im Gesenk

geschmiedete Herzstückspitze f. ‖ pointe f. de cœur matricée. / ~ forging ‖ Gesenkschmiedstück n. ‖ pièce f. matricée. / ~ grinding attachment for threading dies ‖ Gewindeschneidbackenschleifvorrichtung f. ‖ dispositif m. à affûter les coussinets à fileter. / ~ hammer ‖ Nummerschlägel m. ‖ marteau m. compteur. / ~ hammered forging ‖ Gesenkschmiedestück n. ‖ pièce f. forgée en matrice.

die head ‖ Gewindeschneidkopf m. ‖ cage f. de filière. / self-opening ~ ‖ selbstöffnender Gewindeschneidkopf m. ‖ filière f. automatique. / ~ of a threading machine ‖ Schneidkopf m. einer Gewindeschneidmaschine ‖ filière f. d'une machine à fileter.

djelectric ‖ nichtleitend; dielektrisch ‖ diélectrique.

dielectric ‖ Nichtleiter m.; Dielektrikum n. ‖ diélectrique m.; corps m. isolateur. / ~ after-effect ‖ dielektrische Nachwirkung f. ‖ effet m. ultérieur diélectrique. / ~ constant ‖ Dielektrizitätskonstante f. ‖ constante f. diélectrique. / ~ displacement ‖ dielektrische Verschiebung f. ‖ déplacement m. diélectrique. / ~ fatigue ‖ dielektrische Nachwirkung f. ‖ fatigue f. diélectrique. / ~ hysteresis ‖ Dielektrizitätshysterese f. ‖ hystérèse f. diélectrique. / ~ hysteretic constant ‖ Dielektrizitätshysteresiskonstante f. ‖ constante f. d'hystérèse diélectrique. / ~ medium ‖ dielektrisches Medium n. ‖ milieu m. diélectrique. / ~ rigidity see ~ strength. / ~ space ‖ Dielektrikum n. ‖ champ m. diélectrique. / ~ strength (Electr) ‖ dielektrische Festigkeit f. ‖ rigidité f. diélectrique. / ~ stress ‖ dielektrische Beanspruchung f. ‖ effort m. diélectrique.

die press ‖ Gesenkpresse f. ‖ presse f. à matricer ou à estamper.

dies pl. (Locksm) ‖ Schneidbacken fpl. ‖ coins mpl.; coins mpl. à vis; coussinets mpl. à fileter. / opening of the ~ ‖ Klemmbackenöffnung f. ‖ ouverture f. des mâchoires.

Diesel crane ‖ Dieselmotorkran m. ‖ grue f. à moteur Diesel.

Diesel engine ‖ Dieselmaschine f.; Dieselmotor m. ‖ moteur m. Diesel. / air injection type ~ ‖ Dieselmotor m. mit Drucklufteinspritzung ‖ moteur m. Diesel à injection pneumatique. / airless injection ~ ‖ kompressorloser Dieselmotor m. ‖ moteur m. Diesel sans compresseur. / ~ with airless injection of the fuel ‖ Dieselmotor m. mit luftloser Brennstoffeinführung ‖ moteur m. Diesel muni d'injection sans air comprimé. / compressorless ~ ‖ kompressorloser Dieselmotor m. ‖ moteur m. Diesel sans compresseur. / directly reversible ~ ‖ direkt umsteuerbarer Dieselmotor m. ‖ moteur m. Diesel directement réversible. / marine ~ ‖ Schiffsdieselmotor m. ‖ moteur m. Diesel marin. / opposed-piston ~ ‖ Doppelkolbendieselmotor m. ‖ moteur m. Diesel à double piston. / reversible marine ~ ‖ umsteuerbarer Schiffsdieselmotor m. ‖ moteur m. marin Diesel directement réversible. / six-cylinder ~ ‖ Sechszylinderdieselmotor m. ‖ moteur m. Diesel à six cylindres. / six-cylinder double-acting two-stroke marine ~ ‖ doppeltwirkender Zweitaktsechszylinderschiffsdieselmotor m. ‖ moteur m. Diesel marin à deux temps à double effet à six cylin-

dres. / six-cylinder single-acting two-stroke marine ~ ‖ einfachwirkender Zweitaktsechszylinderschiffsdieselmotor m. ‖ moteur m. Diesel marin à simple effet à deux temps à six cylindres. / three-cylinder ~ ‖ Dreizylinderdieselmotor m. ‖ moteur m. Diesel à trois cylindres. / three-cylinder airless injection stationary four-stroke ‖ kompressorloser ortsfester Viertaktdreizylinderdieselmotor m. ‖ moteur m. Diesel fixe sans compresseur à trois cylindres de cycle à quatre temps. / vessel with marine ~s as main engines ‖ Schiff n. mit Dieselmotoren als Hauptmaschinen ‖ navire m. propulsé par moteurs Diesel marins.

Diesel engined sailing vessel ‖ Dieselmotorsegelschiff n. ‖ voilier m. à moteur auxiliaire Diesel. / ~ generator with exciter ‖ Dieselgenerator m. mit angebauter Erregermaschine ‖ groupe m. Diesel-électrogène avec excitatrice en bout d'arbre. / ~ locomotive ‖ Diesellokomotive f. ‖ locomotive f. Diesel ou à moteur Diesel. / ~ motor see ~ engine. / three-wheel ~ motor roller ‖ Dieselmotordreiradwalze f. ‖ rouleau m. compresseur à trois rouleaux à moteur Diesel. / ~ oil ‖ Dieselöl n. ‖ huile f. Diesel; Dieseloil f. / ~ ship ‖ Dieselschiff n.; Dieselmotorschiff n. ‖ navire m. à moteur Diesel; bateau m. Diesel.

die shaper ‖ Kopierstoßmaschine f. ‖ machine f. à mortaiser suivant gabarit; mortaiseuse f. à copier. / ~ square ‖ scharfkantig ‖ à vive arête f. / ~ stamp ‖ Prägepresse f. ‖ presse f. à empreindre. / ~ stamping ‖ Prägedruck m. ‖ gravure f. en creux.

diestock ‖ Schneidkluppe f. ‖ filière f. à coussinets; porte-filière m. / ~ for gas pipes ‖ Gasgewindeschneidkluppe f. ‖ filière f. pour tarauder les tuyaux à gaz.

dietary preparation ‖ diätetisches Präparat n. ‖ préparation f. diététique.

differ, to ‖ unterscheiden ‖ différencier. / ~ fundamentally ‖ grundlegend unterscheiden ‖ distinguer d'une façon fondamentale.

difference ‖ Unterschied m. ‖ différence f. / ~ at the assay ‖ Ungleichheit f. in der Probe ‖ tressaut m. / ~ in the focal adjustment ‖ Einstellungsdifferenz f. ‖ différence f. de réglage. / ~ of level between two points ‖ Höhenunterschied m. zweier Punkte ‖ différence f. de niveau entre deux points. / ~ of longitude ‖ Längenunterschied m. ‖ différence f. en longitude. / ~ of phases ‖ Phasendifferenz f. ‖ différence f. des phases. / ~ of potentials (Electr) ‖ Potentialdifferenz f. ‖ différence f. des potentiels. / ~ in price ‖ Preisunterschied m. ‖ différence f. de prix. / to split the ~ ‖ sich den Unterschied m. teilen ‖ partager la différence. / ~ in time ‖ Zeitunterschied m. ‖ différence f. de temps. / ~ of value ‖ Wertunterschied m. ‖ différence f. de valeur. / ~ of vergency ‖ Neigungsdifferenz f. ‖ différence f. d'inclinaison. / ~ in weight ‖ Gewichtsunterschied m. ‖ différence f. de poids.

differential (Auto; Math) ‖ Differential n. ‖ différentiel m. / ~ alarm ‖ Differentialmelder m. ‖ avertisseur m. différentiel. / ~ arc lamp ‖ Differentialbogenlampe f. ‖ lampe f. à arc différentielle. / ~ brake ‖ Differentialbremse f. ‖ frein m. de différentiel. / ~ bridge ‖ Differentialbrücke f. ‖ pont m. différentiel. / ~ calculus ‖ Differentialrechnung f. ‖ calcul m. différentiel. / ~ case ‖ Ausgleichgehäuse n. ‖ carter m. de différentiel. / ~ casing ‖ Differentialgehäuse n. ‖ boîte f. de différentiel. / ~ compound dynamo ‖ Antikompounddynamo f. ‖ dynamo m. anti-compound. / ~ compound winding ‖ Gegenwickelung f.; Antikompoundwicklung f. ‖ contre-enroulement m.; enroulement m. anticompound. / ~ counting chart ‖ Differentialzähltafel f. ‖ tableau m. différentiel pour la numération. / ~ draught gauge ‖ Differentialzugmesser m. ‖ indicateur m. différentiel de tirage. / ~ equation ‖ Differentialgleichung f. ‖ équation f. différentielle. / ~ feed (Boring mach) ‖ Differentialvorschub m. ‖ avance f. par différentiel. / ~ galvanometer ‖ Differentialgalvanometer n. ‖ galvanomètre m. différentiel. / ~ galvanoscope ‖ Differentialgalvanoskop n. ‖ galvanoscope m. différentiel. / ~ gear (Auto) ‖ Ausgleichgetriebe n. ‖ engrenage m. différentiel. / ~ manometer ‖ Differentialmanometer n. ‖ manomètre m. différentiel. / ~ mechanism ‖ Differentialgetriebe n. ‖ mécanisme m. différentiel. / ~ piston ‖ Differentialkolben m.; Stufenkolben m. ‖ piston m. différentiel. / ~ pulley ‖ Differentialflaschenzug m. ‖ palan m. différentiel. / ~ pupilloscope ‖ Differentialpupilloskop n. ‖ pupilloscope m. différentiel. / ~ purchase see ~ tackle. / ~ relay ‖ Differentialrelais n. ‖ relais m. différentiel. / ~ screw ‖ Differentialschraube f. ‖ vis f. différentielle. / ~ shaft ‖ Differentialwelle f. ‖ arbre m. du différentiel. / ~ spider (Auto) ‖ Ausgleichstern m. ‖ axe m. des satellites du différentiel. / ~ system ‖ Differentialschaltung f. ‖ méthode f. différentielle. / ~ tackle ‖ Differentialflaschenzug m. ‖ palan m. différentiel. / ~ windlass ‖ Differentialwinde f. ‖ treuil m. différentiel.

differentiate, to (Math) ‖ differenzieren; ableiten ‖ différencier; prendre la dérivée.

differentiation (Math) ‖ Differentiation f. ‖ différentiation f.

difficult to sell ‖ schwerverkäuflich ‖ de vente f. difficile. / part ~ to shape ‖ schwierig herzustellendes Stück n. ‖ pièce f. de main-d'œuvre difficile.

difficulties pl. ‖ Schwierigkeiten fpl.; Weiterungen fpl. ‖ difficultés fpl.; formalités fpl. / to find one's self in ~ owing to the failure of a bank ‖ durch den Zusammenbruch m. einer Bank sich in Schwierigkeiten befinden ‖ se trouver aux prises avec des difficultés résultant d'un krach de banque.

difficult manufacture ‖ schwierige Verarbeitung f. ‖ difficultés fpl. que présentent le façonnage et l'usinage.

difficulty, obtained without any serious ~ ‖ ohne besondere Schwierigkeit f. ‖ obtenu sans difficultés sérieuses.

diffract, to (Phys) ‖ beugen ‖ diffracter.

diffraction ‖ Diffraktion f.; Beugung f. ‖ diffraction f. / ray deflected by ~ ‖ durch Beugung abgelenkter Strahl m. ‖ rayon m. dévié par diffraction.

diffraction grating ‖ Beugungsgitter n. ‖ réseau m. à diffraction. / ~ spectroscope ‖ Gitterspektroskop n. ‖ spectroscope m. à réseau.

diffraction hand grating spectroscope ‖ Gitterhandspektroskop n. ‖ spectroscope m. à réseau à main. / ~ spectrum ‖ Beugungsspektrum n.; Gitterspektrum n. ‖ spectre m. de réseau ou de diffraction.

diffuse (Opt) ‖ zerstreut ‖ diffus.

diffuse, to (Opt) ‖ diffundieren ‖ diffuser.

diffused light ‖ zerstreutes Licht n. ‖ lumière f. diffuse.

diffuser ‖ Zerstäuber m. ‖ diffuseur m. / water spray ~ ‖ Wasserzerstäubungsdüse f. ‖ irrigateur m. d'eau.

diffuser bottom ‖ Diffusorboden m. ‖ fond m. de diffuseur. / ~ top ‖ Diffusorhaube f. ‖ chapeau m. de diffuseur.

diffusibility ‖ Diffusionsvermögen n. ‖ diffusibilité f.

diffusion (Phys; Chem) ‖ Diffusion f. ‖ diffusion f. / speed of ~ ‖ Diffusionsgeschwindigkeit f. ‖ vitesse f. de diffusion.

diffusion apparatus ‖ Diffusionsapparat m. ‖ appareil m. de diffusion. / ~ pump ‖ Diffusorpumpe f. ‖ pompe f. à diffusion. / ~ worker (Sugar) ‖ Diffusionsarbeiter m. ‖ diffuseur m.

dig, to ‖ graben ‖ fouiller; creuser. / ~ a ditch ‖ einen Graben m. ziehen ‖ faire ou creuser un fossé. / ~ the earth ‖ die Erde ausheben; den Boden m. ausgraben ‖ creuser ou extraire la terre; déblayer le terrain. / ~ the ground ‖ die Erde abtragen ‖ déblayer un terrain. / ~ out (Mine) ‖ (Erze, Kohlen) gewinnen ‖ extraire. / ~ out (Soil) ‖ ausgraben ‖ creuser la terre. / ~ peat ‖ Torf m. stechen ‖ extraire la tourbe. / ~ up the earth ‖ den Boden m. aufgraben ‖ fouiller le terrain.

digester ‖ Kocher m.; Digestor m.; Extraktionsapparat m. ‖ extracteur m.; digesteur m.; autoclave m.; marmite f. autoclave.

digger ‖ Erdarbeiter m. ‖ terrassier m. / ~'s shovel ‖ Erdarbeiterschaufel f. ‖ pelle f. de terrassier.

digging (Mine) ‖ Schurf m.; Schürfung f. ‖ fouille f. de recherche. / ~ depth (Shovel) ‖ Reichtiefe f. ‖ profondeur f. de fouille. / ~ machine ‖ Grabemaschine f.; Bagger m. ‖ excavateur m.; drague f.

digitalis leaves pl. ‖ Fingerhutblätter npl. ‖ feuilles fpl. de digitale.

dike ‖ Flußdamm m.; Kai m.; Pier m.; Deich m. ‖ quai m.; digue f. / ~ across a valley ‖ Talsperre f. ‖ digue f. ou barrage m. de vallée. / building of ~s and dams across a valley ‖ Talsperrenbau m. ‖ construction f. de digues et de barrages de vallée. / encircling ~ ‖ Ringdamm m.; Ringdeich m. ‖ digue f. de ceinture. / half-moon ~ ‖ Kesseldeich m. ‖ digue f. en demi-lune. / main ~ ‖ Winterdeich m. ‖ digue f. insubmersible. / mat ~ ‖ Mattendamm m. ‖ digue f. mattée. / provisional ~ ‖ Pinnplanke f. ‖ digue f. provisoire. / ~ built of stone ‖ Steindamm m. ‖ digue f. de pierre. / stone ~ ‖ Ziegelsteindeich m. ‖ digue f. maçonnée.

dike building ‖ Deichbau m. ‖ construction f. de digues. / ~ dam ‖ Buhne f.; Schlange f.; Schlechte f. ‖ clayonnage m.; crèche f.; éperon m. / ~ drain ‖ Deichschleuse f. ‖ écluse f. pratiquée dans une digue. / ~ lock see ~ drain. / ~ master ‖ Deichmeister m. ‖ surintendant m. des digues; dyck-grave m. / ~ path see ~ way.

/ ~ sluice ‖ Deichschleuse f. ‖ retenue f. de chasse; vanne f. de digue. / ~ union ‖ Deichverband m. ‖ société f. chargée de la construction et de l'entretien d'une digue. / ~ way ‖ Deichweg m. ‖ chemin m. le long d'une digue; chemin m. de halage.

diking ‖ Deichbau m. ‖ travail m. aux digues. / ~ in ‖ Eindeichung f. ‖ endiguement m.

dilapidate, to ‖ verfallen lassen ‖ délabrer.

dilapidated ‖ baufällig ‖ caduc; délabré.

dilapidation ‖ Baufälligkeit f. ‖ délabrement m.

dilatability ‖ Ausdehnbarkeit f. ‖ expansibilité f.; dilatabilité f.

dilatable (Phys) ‖ dehnbar ‖ dilatable; expansible.

dilatation ‖ Expansion f.; Ausdehnung f. ‖ expansion f.; étendue f.; dilatation f. / ~ of length ‖ Längendehnung f. ‖ dilatation f. linéaire.

dilate, to ‖ (sich) dehnen ‖ (s')étendre; (se) dilater; (s')allonger.

dilatometer ‖ Dilatometer n. ‖ dilatomètre m.

dilatory ‖ säumig ‖ retardataire.

diligent ‖ arbeitsam ‖ laborieux; travailleux.

dill seed ‖ Dillsamen m. ‖ semence f. d'aneth.

diluent (Chem) ‖ Streckungsmittel n.; Verdünnungsmittel n. ‖ diluant m.

dilute, to ‖ verdünnen ‖ diluer. / ~ the acid ‖ die Säure verdünnen ‖ diluer l'acide m.

diluted ‖ verdünnt ‖ étendu; dilué. / ~ acid ‖ schwache *oder* verdünnte Säure f. ‖ acide m. dilué *ou* étendu *ou* faible.

diluting agent ‖ Verdünnungsmittel n. ‖ diluant m.

dilution ‖ Verdünnung f. ‖ dilution f. / extent of ~ ‖ Verdünnungsgrad m. ‖ degré m. de dilution. / ~ heat ‖ Verdünnungswärme f. ‖ chaleur m. de dilution.

diluvial ‖ diluvial; angeschwemmt ‖ diluvial. / ~ ore ‖ Seifenerz n.; Wascherz n. ‖ minerai m. d'alluvion *ou* de lavage.

dim ‖ trüb ‖ trouble.

dim, to ~ the head lamps ‖ die Scheinwerfer mpl. abblenden ‖ baisser les phares mpl.

dimasted ‖ mastlos ‖ démâté.

dimension ‖ Dimension f.; Abmessung f. ‖ dimension f.; cote f. / to draw the ~s into a design ‖ Maße npl. einschreiben ‖ coter un dessin. / figured ~ ‖ eingeschriebenes Maß n. ‖ cote f. d'un dessin. / general ~ ‖ Hauptabmessung f. ‖ dimension f. principale. / larger ~s are inconvenient for constructional reasons ‖ größere Abmessungen fpl. sind aus baulichen Rücksichten unangebracht ‖ pour des raisons techniques on ne peut choisir des dimensions plus fortes. / to write the ~s into a design ‖ die Maße npl. in eine Zeichnung einschreiben ‖ coter un dessin.

dimensioned sketch ‖ Maßskizze f. ‖ esquisse f. cotée; croquis m. coté.

dimethyl aniline ‖ Dimethylanilin n. ‖ diméthylaniline f. / ~ sulphate ‖ Dimethylsulfat n. ‖ sulfate m. de diméthyl.

diminish, to ‖ vermindern; verkleinern; herabsetzen ‖ diminuer; réduire. / ~ a piece of wood ‖ ein Holzstück n. schwächen oder verjüngen ‖ délarder une pièce de bois; démaigrir en biseau. / ~ a wall ‖ eine Mauer f. verschwächen ‖ allégir *ou* affaiblir un mur.

diminished ‖ vermindert ‖ réduit. / in ~ proportions pl. ‖ in verjüngtem Maßstabe m. ‖ à échelle f. de réduction.

diminishing plank ‖ Verjüngungsplanke f. ‖ bordé m. de diminution.

diminution ‖ Schmälerung f. ‖ rétrécissement m.; diminution f.; amoindrissement m. / ~ (Metal) ‖ Abbrand m. ‖ perte f. / ~ of influenced tension by exceeding wires (Tel) ‖ Spannungssenkung f. durch überschießende Leitungsstrecken ‖ diminution f. de la tension influencée par des tronçons de ligne excédants.

dimmer, headlight ~ ‖ Blendschutzvorrichtung f. bei Scheinwerfern ‖ dispositif m. antiéblouissant aux phares.

dimmer switch ‖ Dunkelschalter m. ‖ interrupteur m. à résistance réglable.

dinas brick ‖ Dinasstein m. ‖ brique f. dinas. / ~ works pl. ‖ Dinaswerk n. ‖ usine f. de brique dinas.

dinghy ‖ Jolle f.; kleines Boot n. ‖ barge f.; chaloupe f.; nacelle f. / collapsible ~ ‖ Faltboot n. ‖ canot m. repliable.

dining hall to take meals brought from home ‖ Speisesaal m. zur Einnahme der mitgebrachten Mahlzeit ‖ réfectoire m. pour prendre les repas apportés. / ~ house ‖ Speisehaus n. ‖ établissement-réfectoire m. / ~ room ‖ Speisesaal m. ‖ salle f. à manger.

dinner napkin ‖ Mundtuch n. ‖ serviette f. / ~ service ‖ Tafelgerät n. ‖ service m. de table. / ~ set ‖ Eßgeschirr n. ‖ vaisselle f. pour dîner. / porcelain ~ set ‖ Tafelgeschirr n. aus Porzellan ‖ service m. de table en porcelaine.

dint ‖ Beule f.; Striemen m. ‖ enfonçure f.; bosse f.; raie f.

dioptase ‖ Kupfersmaragd m.; Dioptas m. ‖ dioptase f.; cuivre m. dioptase.

diopter ‖ Diopter n. ‖ dioptre f.; viseur m.; pinnule f. / cross hair ~ ‖ Fadendiopter n. ‖ pinnule f. à fils. / eyepiece adjustable in terms of ~s ‖ Okular n. mit Dioptrieneinstellung f. ‖ oculaire m. muni d'un dispositif de mise au point gradué en dioptries.

diopter focusing mount ‖ Dioptrieneinstellung f. ‖ mise f. au point à dioptries. / ~ scale ‖ Dioptrienskale f. ‖ échelle f. des dioptries; graduation f. en dioptries.

dioptric ‖ dioptrisch ‖ dioptrique. / ~ disturbance ‖ dioptrische Störung f. ‖ dérangement m. *ou* trouble m. dioptrique.

dioxide ‖ Dioxyd n. ‖ bioxyde m.

dip, to ‖ eintauchen; einsenken ‖ plonger. / ~ (Dyer) ‖ eintauchen ‖ immerger. / ~ (Geol) ‖ einfallen ‖ plonger; s'incliner. / ~ (Metal) ‖ eintränken; eintauchen ‖ imbiber; plonger; tremper. / ~ (Pap) ‖ schöpfen ‖ ouvrer; plonger. / ~ the brass ‖ das Messing abbrennen; gelbbrennen ‖ relever la couleur du laiton; dérocher *ou* décaper le laiton.

dip (El line) ‖ Durchhang m.; Leitungshang m.; Pfeilhöhe f. ‖ flèche f. / ~ (Geol) ‖ Einfallen n. ‖ inclinaison f.; plongement m.; pendage m. / ~ (Candle) ‖ gezogenes Licht n. ‖ chandelle f. plongée *ou* à la baguette. / to give a ~ ‖ nach unten neigen ‖ incliner vers le bas. / ~ of the horizon ‖ Kimmtiefe f. ‖ abaissement m. de l'horizon. / ~ of the paddle-wheels of a paddle steamer ‖ Tauchung f. der Räder eines Raddampfers ‖ im-

mersion f. des roues à aubes d'un vapeur à roues.

dip, angle of ~ (Mine) ‖ Fallwinkel m. ‖ angle m. de pendage. / direction of full ~ of the mineral deposit ‖ Fallinie f. der Lagerstätte ‖ pendage m. du gisement. / to determine the direction of full ~ of a mineral deposit ‖ die Linie f. des stärksten Fallens einer Lagerstätte bestimmen ‖ déterminer la ligne de la plus grande pente d'un gisement. / height of the ~ (Build) ‖ Abdachungshöhe f. ‖ hauteur f. de plongée. / the line of strike and direction of ~ vary (Geol) ‖ Streichrichtung f. und Fallrichtung f. wechseln ‖ la direction f. et l'inclinaison f. changent.

diphenylarsinchloride ‖ Blaukreuzkampfstoff m. ‖ chlorure m. de diphénylarsine; cyanure m. de diphénylarrine.

diphosgene ‖ Perstoff m.; Grünkreuzkampfstoff m. ‖ croix f. verte; surpalite m.; chloroformiate m. de méthyle trichloré.

diplex working (Tel) ‖ Diplexbetrieb m. ‖ transmission f. diplex.

diplococcus ‖ Diplokokkus m. ‖ diplocoque m.

diploma ‖ Diplom n. ‖ diplôme m.; brevet m.

dip needle ‖ Inklinationsnadel f. ‖ boussole f. d'inclinaison.

dipped ‖ getaucht ‖ plongé.

dipper (Mar) ‖ Schöpfer m. (des Kochs) ‖ petit seau m. / ~ (Pap) ‖ Büttgeselle m.; Schöpfer m. ‖ plongeur m.; puiseur m. / biscuit ware ~ (Pott) ‖ Glasierer m. ‖ trempeur m. de biscuit. / wood ~ ‖ Holzfärber m. ‖ trempeur m. de bois.

dipper arm ‖ Löffelstiel m. ‖ bras m. du godet. / ~ capacity ‖ Löffelinhalt m. ‖ capacité f. du godet. / ~ discharging ‖ Löffelentleerung f. ‖ déchargement m. du godet. / ~ displacement ‖ Löffelverschiebung f. ‖ déplacement m. du godet. / ~ relay ‖ Tauchrelais n. ‖ relais m. à plongeurs. / ~ slide ‖ Löffelklappe f. ‖ trappe f. du godet. / ~ tooth ‖ Löffelzahn m. ‖ dent m. du godet. / ~ width ‖ Tieflöffelbreite f. ‖ largeur f. du godet en fouille.

dipping (Metal) ‖ Abbeizen n. ‖ décapage m.; dérochage m. / ~ of liquids in vessels ‖ Abstechen n. *oder* Abziehen n. von Flüssigkeiten in Gefäßen ‖ soutirage m. de liquides dans des réservoirs. / ~ the matches into the inflammable compound ‖ Betupfen n. *oder* Eintauchen n. der Zündhölzchen ‖ chimicage m. des allumettes chimiques.

dipping depth ‖ Eintauchtiefe f. ‖ profondeur f. d'immersion. / ~ electrode ‖ Tauchelektrode f. ‖ électrode f. plongeante. / ~ frame (Dyer) ‖ Küpenrahmen m. ‖ cadre m. / ~ motion (Ship) ‖ Tauchbewegung f. ‖ mouvement m. d'immersion. / ~ needle ‖ Inklinationsnadel f. ‖ aiguille f. *ou* boussole f. d'inclinaison; inclinatoire m. magnétique. / ~ plant for oil linen ‖ Tauchanlage f. für Ölleinen ‖ installation f. d'immersion de la toile huilée. / ~ plant for oil silk ‖ Tauchanlage f. für Ölseide ‖ installation f. d'immersion de la soie huilée. / ~ refractometer ‖ Eintauchrefraktometer n. ‖ réfractomètre m. à immersion. / ~ rod (Candle) ‖ Dochtspieß m. ‖ baguette f. / ~ tube ‖ Tauchrohr n. ‖ tuyau-plongeur m.

dip regulation table ‖ Durchhangstafel f. ‖ tableau m. des flèches. / ~ rod ‖ Ölmeßstab m. ‖ réglette-jauge m.

dipteral (Arch) ‖ doppelflügelig ‖ diptère.
dipteros *see* dipteral.
direct ‖ gerade ‖ direct. / ~ communication ‖ Direktsprechen n. ‖ communication f. directe. / ~ coupling ‖ unmittelbare Kupplung f. ‖ couplage m. direct.
direct-current ammeter ‖ Gleichstromamperemeter n. ‖ ampèremètre m. pour courant continu. / ~ arc ‖ Gleichstrombogenlampe f. ‖ lamqe f. à arc à courant continu. / ~ circuit ‖ Gleichstromleitung f. ‖ circuit m. de courant continu. / ~ generator ‖ Gleichstromdynamo f. ‖ dynamo f. à courant continu. / ~ measuring-instrument ‖ Gleichstrommeßinstrument n. ‖ instrument m. de mesure pour courant continu. / ~ motor ‖ Gleichstrommotor m. ‖ moteur m. à courant continu. / ~ network ‖ Gleichstromnetz n. ‖ réseau m. à courant continu. / ~ side ‖ Gleichstromseite f. ‖ côté m. continu. / ~ three-wire plant ‖ Gleichstrom-Dreileiteranlage f. ‖ installation f. à courant continu à trois conducteurs. / ~ traction ‖ Gleichstromfahrbetrieb m. ‖ traction f. à courant continu. / ~ two-wire plant ‖ Gleichstromzweileiteranlage f. ‖ installation f. à courant continu à deux conducteurs. / ~ voltmeter ‖ Gleichstromvoltmesser m. ‖ voltmètre m. pour courant continu.
direct drive ‖ unmittelbarer Antrieb m. ‖ commande f. directe. / ~ electric drive ‖ unmittelbarer elektrischer Antrieb m. ‖ commande f. électrique directe.
directed, as ~ ‖ wie vorgeschrieben ‖ comme indiqué.
direct expansion cooling system ‖ direktes Expansionssystem n. für Kühlung ‖ système m. de réfrigération par détente directe. / ~ firing ‖ unmittelbare Feuerung f. ‖ feu m. nu *ou* direct.
directing force (Electr) ‖ Richtkraft f.; Richtvermögen n. ‖ force f. directrice. / ~ gun ‖ Leitgeschütz n. ‖ pièce f. directrice; pièce-guide f. / ~ picket (Hydr arch) ‖ Lehrpfahl m. ‖ piquet-directeur m. / ~ staff (Surv) ‖ Absteckstange f.; Meßstange f. ‖ jalon m. / ~ wheel (Windmill) ‖ Stellrad n. ‖ roue f. d'orientation.
direction (Aero) ‖ Richtung f. ‖ direction f. / ~ (Auto) ‖ Steuerung f. ‖ direction f. / ~ (For use) ‖ Anweisung f. ‖ instructions fpl. de service; indication f.; ordre m. / ~ (Geol) ‖ Streichen n. ‖ direction f. / ~ (Trade) ‖ Direktion f. ‖ direction f.; directoire m. / to change ~ ‖ umsteuern ‖ changer de direction *ou* de sens de marche. / ~ of the cloud's motion ‖ Zugrichtung f. der Wolken ‖ direction f. du mouvement *ou* de la marche des nuages. / ~ of current (Electr) ‖ Stromrichtung. f. ‖ sens m. du courant. / to set itself in the ~ of the field ‖ sich in die Feldrichtung einstellen ‖ se placer dans la direction du champ. / in the ~ of flight ‖ längs der Flugrichtung f. ‖ dans le sens du vol. / ~ of flow ‖ Strömungsrichtung f. ‖ direction f. du flux. / ~ of the force ‖ Kraftrichtung f.; Kraftsinn m. ‖ direction f. de la force. / ~ of the hands of a watch ‖ Uhrzeigerbewegung f. ‖ sens m. des aiguilles d'une montre. / ~ of motion ‖ Fahrtrichtung f. ‖ direction f. de marche *ou* de translation. / ~ of motion (Mech) ‖ Bewegungsrichtung f. ‖ sens m. de mouvement. / in opposite ~ ‖ in entgegengesetzter Richtung f. ‖ dans la di-

rection opposée. / ~ of relative wind ‖ Flugwindrichtung f. ‖ direction f. du vent relatif. / in reversed ~ ‖ in umgekehrter Richtung f. ‖ dans le sens inverse. / across the ~ of rolling ‖ quer zur Walzrichtung f. ‖ transversalement au laminage. / ~ of rotation ‖ Drehrichtung f. ‖ direction f. *ou* sens m. de rotation. / ~ of running ‖ Laufrichtung f.; Drehrichtung f. ‖ direction f. de la marche; sens m. de rotation. / general ~ of strike (Mine) ‖ Hauptstreichrichtung f. ‖ ligne f. de direction principale. / ~ of view ‖ Blickrichtung f. ‖ direction f. du regard. / ~ of the wind ‖ Windrichtung f. ‖ direction f. du vent. / ~ of works ‖ Bauleitung f. ‖ administration f. des travaux.
directional, uni-~ ‖ in einer Richtung f. wirkend ‖ agissant à sens m. unique.
directional aerial (Radio) ‖ gerichtete Antenne f. ‖ aérien m. à ondes dirigées. / ~ operation (Tel) ‖ Richtungsbetrieb m. ‖ exploitation f. à direction déterminée. / ~ telescope ‖ Richtfernrohr n. ‖ lunette f. de direction.
direction finder (Radio) ‖ Peilempfänger m. ‖ récepteur m. radiogoniomètre. / radio ~ ‖ Radioortungsgerät n. ‖ radio-chercheur m. de position. / ~ station (Radio) ‖ Peilstation f. ‖ station f. radiogoniométrique.
direction finding ‖ Standortsbestimmung f. ‖ orientation f. / ~ by radiogoniometry (Aero) ‖ radiogoniometrische Ortung f. ‖ repérage m. *ou* localisation f. radiogoniométrique. / ~ service (Radio) ‖ Peildienst m. ‖ service m. goniométrique.
direction indicator ‖ Richtungszeiger m. ‖ indicateur m. de direction. / gyroscopic ~ ‖ Kreiselfeuerleitanzeiger m. ‖ indicateur m. gyroscopique.
direction line (Print) ‖ Normzeile f. ‖ ligne f. de pied. / ~ post ‖ Wegweiser m. ‖ poteau-guide m.
directions pl. ‖ Verhaltungsmaßregel f. ‖ instruction f. / ~ for use ‖ Gebrauchsanweisung f.; Gebrauchsvorschrift f. ‖ mode m. d'emploi.
directive antenna ‖ gerichtete Antenne f. ‖ antenne f. directrice.
directiveness ‖ Lenkbarkeit f.; Richtbarkeit f. ‖ dirigeabilité f.
directive power (Antenna) ‖ Richtvermögen n. ‖ pouvoir m. directif.
director ‖ Direktor m. ‖ directeur m. / ~ (Aut tel) ‖ Umrechner m. ‖ directeur m. / managing ~ ‖ Betriebsleiter m. ‖ directeur m. technique. / managing ~ (Bank) ‖ Verwaltungsrat m. ‖ administrateur m.; conseil m. d'administration.
directorate, member of the ~ ‖ Mitglied n. des Direktoriums ‖ membre m. du comité de direction.
director, board of ~s ‖ Aufsichtsrat m. ‖ conseil m. d'administration.
directory ‖ Adreßbuch n. ‖ annuaire m. *ou* répertoire m. d'adresses. / commercial ~ ‖ kaufmännisches Adreßbuch n. ‖ annuaire m. du commerce.
directrix (Geom) ‖ Directrix f.; Leitlinie f. ‖ directrice f. / ~ (Mach) ‖ Leitschaufel f. ‖ aube f. directrice.
direct-vision prism ‖ Prisma n. mit gerader Durchsicht ‖ prisme m. à vision directe. / ~ set of prisms ‖ geradsichtiger Prismensatz m. ‖ jeu m. de prismes à vision directe.
dirigible ‖ Lenkluftschiff n.; lenkbares Luftschiff n. ‖ dirigeable m.; ballon m.

dirigeable. / semi-rigid ~ ‖ halbstarres Luftschiff n. ‖ dirigeable m. semi-rigide.
dirt ‖ Schmutz m. ‖ crasse f. / adhering ~ ‖ anhaftender Schmutz m. ‖ crasse f. adhérente. / ~ cage (Centrifuge) ‖ Flügeleinsatz m. ‖ cage f. à boues. / ~ carrier (Mine) ‖ Waschbergeabzieher m. ‖ chargeur m. de résidues.
dirty (Mach) ‖ verstopft ‖ encrassé. / ~ linen ‖ schmutzige Wäsche f. ‖ linge m. sale.
disability insurance ‖ Invalidenversicherung f. ‖ assurance f. contre l'invalidité.
disabled ship ‖ seeunfähiges Schiff n. ‖ navire m. désemparé *ou* en pantenne. / ~ workman ‖ invalider Arbeiter m. ‖ ouvrier m. invalide.
disablement insurance *see* disability insurance.
disadvantage ‖ Nachteil m. ‖ désavantage m.; préjudice m.
disadvantageous ‖ ungünstig; nachteilig ‖ défavorable; désavantageux.
disaggregation (Brew) ‖ Auflösung f. ‖ désagrégation f.; dissolution f.; friabilité f.
disappointment ‖ Enttäuschung f.; Mißlingen n.; üble Lage f. ‖ contretemps m.
disarm, to ‖ abrüsten ‖ désarmer.
disarmament ‖ Abrüstung f. ‖ désarmement m. / ~ conference ‖ Abrüstungskonferenz f. ‖ conférence f. du désarmement.
disassembled machine ‖ zerlegte Maschine f. ‖ machine f. démontée.
disbark, to ‖ entrinden; abborken ‖ écorcer; décortiquer.
disbarked wood ‖ geschältes Holz n. ‖ bois m. écorcé.
disbursement ‖ Geldausgabe f. ‖ argent m. dépensé; dépense f. / ~ (Mine) ‖ Zubuße f. ‖ cotisation f.
disc *see also* disk ‖ Scheibe f. ‖ disque m.; plateau m.
discentered, with rolls arranged in a ~ order ‖ mit versetzter Walzenanordnung f. ‖ avec rouleaux mpl. placés en position alterne.
discerning ‖ scharfsinning ‖ perspicace; sagace.
discharge, to (Car) ‖ abladen; entladen ‖ décharger; débarquer. / ~ (Debt) ‖ tilgen ‖ amortir. / ~ (Electr) ‖ entladen ‖ décharger. / ~ (Liquid) ‖ ablassen ‖ faire écouler. / ~ (Ship) ‖ löschen; entladen ‖ décharger; débarquer. / ~ (Workman) ‖ entlassen ‖ renvoyer; congédier. / ~ an accumulator ‖ einen Akkumulator m. entladen ‖ décharger un accumulateur. / ~ the battery ‖ die Batterie entladen ‖ décharger la batterie. / ~ a debt ‖ eine Schuld f. ablösen ‖ amortir une dette. / ~ a kiln (Malt) ‖ eine Darre f. abräumen ‖ décharger une touraille. / ~ a ship ‖ ein Schiff n. löschen ‖ décharger un navire.
discharge (Car) ‖ Ausladen n. ‖ décharge f. / ~ (Debt) ‖ Tilgung f. ‖ amortissement m. / ~ (Electr) ‖ Entladung f. ‖ décharge f. / ~ (Liquid) ‖ Ablauf m.; Abfluß m. ‖ décharge f.; épanchoir m. / ~ (Ship) ‖ Entladen n.; Löschen n. ‖ déchargement m.; débarquement m. / ~ at anode ‖ Anodenentladung f. ‖ décharge f. de l'anode. / ~ of a coil ‖ Spulenentladung f. ‖ surtension f. de rupture. / ~ of condenser ‖ Entladung f. eines Kondensators ‖ décharge f. d'un condensateur. / continuous ~ ‖ kontinuierlicher Abfluß m. ‖ écoulement m. continu. / corona ~

(Electr) || Sprühen n. || effluve m. en couronne. / dark~||dunkle Entladung f.|| décharge f. obscure; effluve m. / disruptive ~ || plötzliche Entladung f.; Funkenentladung f. || décharge f. disruptive. / ~ in gas || Gasentladung f. || décharge f. dans le gaz. / gradual ~ || allmähliche Entladung f. || décharge f. graduelle. / intermittent ~ || intermittierende Entladung f. || décharge f. intermittente. / lateral ~ || Nebenentladung f. || décharge f. latérale. / oscillatory ~ || oszillierende Entladung f. || décharge m. oscillatoire. / ~ on pointed conductors || Spitzenentladung f. || décharge f. par pointes. / ~ by points *see* ~ on pointed conductors. / ~ of a river || Abflußmenge f. eines Flusses || quantité f. de l'écoulement d'une rivière; débit m. d'une rivière. / silent ~ *see* dark ~. / slow ~ || langsame Entladung f. || décharge f. lente. / surface ~ || Oberflächenentladung f. || glissement m. superficiel. / ~ of water || Wasserauslauf m. || sortie f. de l'eau. / to utilize the ~ of water day and night || die anfallenden Wassermengen fpl. dauernd ausnutzen || utiliser d'une façon continue les eaux des rivières.

discharge capacity || Entladefähigkeit f. || capacité f. de décharge. / ~ cock || Entleerungshahn m. || robinet m. de décharge. / ~ current || Entladungsstrom m. || courant m. de décharge.

discharged valve regulation || entlastete Ventilsteuerung f. || régulation f. par soupapes déchargées.

discharge device || Entladevorrichtung f.; Entladungsvorrichtung f. || dispositif m. de décharge. / duration of ~ || Entladedauer f.; Entladezeit || durée f. ou temps m. de déchargement. / final voltage on ~ (Acc) || Endspannung f. der Entladung || tension f. à la fin de la décharge. / initial voltage on ~ (Acc) || Anfangsspannung f. der Entladung || tension f. initiale de la décharge. / ~ key || Entladetaste f.||clé f. de décharge. / ~ opening || Entleerungsöffnung f. || orifice m. de vidange. / ~ paper || Abkehrschein m. || congé m. / phenomenon of ~ || Entladungserscheinung f. || phénomène m. de décharge. / ~ pipe || Schüttrohr n. || tuyau m. de décharge. / ~ process (Acc) || Entladungsvorgang m. || procédé m. de déchargement. / ~ quantity || Abflußmenge f. || volume m. débité.

discharger (Car) || Entladungsvorrichtung f. || déchargeur m. / ~ (Electr) || Funkenzieher m. || excitateur m. / ~ (Radio) || Entlader m. || éclateur m. / automatic ~ || Selbstentleerer m. || déchargeur m. automatique. / high-speed disc ~ || schnellumlaufende Scheibenfunkenstrecke f. || éclateur m. à disque à grande vitesse. / micrometric spark ~ (Radio) || Mikrometerfunkenstrecke f. || éclateur m. à intervalle micrométrique. / rapid ~ leaving no remainder || restlos entleerender Selbstentlader m. || déchargeur m. rapide à déchargement sans restants.

discharge rate (Ship) || Löschungsgeschwindigkeit f. || quantité f. à décharger. / ~ rope || Entleerungsseil n. || câble m. de vidange. / ~ switch || Entladungsschalter m. || interrupteur m. de décharge. / ~ test (Acc) || Entladeprobe f. || essai m. de déchargement. / ~ ticket || Entlassungs-

schein m. || billet m. de congé. / ~ tube (Electr) || Entladungsröhre f. || tube m. de décharge. / ~ valve || Auslaßventil n. || soupape f. d'émission. / ~ velocity || Abflußgeschwindigkeit f. || vitesse f. d'écoulement.

discharging of accumulators || Entladen n. der Sammler || décharge f. des accumulateurs. / ~ of the cargo (Mar) || Löschen n. der Ladung || débarquement m. ou déchargement m. de la cargaison. / ~ of a kiln (Malt) || Abladen n. *oder* Abräumen n. einer Darre || déchargement m. d'une touraille. / ~ of load || Entlastung f. || déchargement m. / ~ of a river || Einmündung f. eines Flusses || débouchement m. ou embouchure f. d'une rivière.

discharging, carboy ~ apparatus || Ballonentleerungsapparat m. || appareil m. de vidange ou de décharge de bonbonnes. / ~ basin || Brunnensumpf m. || fond m. du puits ou de la pompe. / ~ bucket elevator || Entladebecherwerk n. || noria f. de déchargement. / ~ carriage || Abwurfwagen m. || chariot m. de déchargement. / ~ coil || Entladespule f. || bobine f. de décharge. / ~ culvert || Hauptsiel n. || pertuis m. / ~ current || Entladungsstrom m. || courant m. de décharge. / ~ device || Entleerungsvorrichtung f. || dispositif m. de vidange. / ~ device for railway carriages || Entladevorrichtung f. für Eisenbahnwagen || dispositif m. de déchargement pour wagons de chemin de fer. / ~ hole (Build) || Ausmündung f. am Ende des Fallrohres; Gußsteinauslauf m. || écouloir m.; dégorgement m.; dégorgeoir m. / ~ hopper || Sturzkasten m. || trémie f. de déversement. / ~ machine for chamber furnaces || Entlademaschine f. für Kammeröfen || machine f. de déchargement pour fours à chambre. / ~ pipe (Loc) || Blasrohr n.; Ausblasrohr n. || tuyau m. d'échappement de la vapeur. / ~ point || Entladestelle f. || lieu m. de déchargement. / ~ port || Löschungshafen m. || port m. de déchargement. / ~ stamp (Briquette press) || Ausstoßstempel m. || tampon m. d'expulsion; piston m. de chasse-briquettes. / ~ station || Entladestation f. || station f. de déchargement. / ~ tackles (Mar) || Löschgeschirr n. || palan m. de débarquement. / ~ trough || Brunnensumpf m. || fond m. du puits de la pompe. / ~ syrup ~ valve || Sirupablaufventil n. || soupape f. de vidange pour sirop. / ~ vault (Build) || Laibungsbogen m. || arrière-voussure f.

discharging wagon || Entladewagen m. || chariot m. de déchargement. / either-side ~ (Railw) || zweiseitiger Seitenentleerer m. || wagon m. se déchargeant de deux côtés. / one-side ~ (Railw) || einseitiger Seitenentleerer m. || wagon m. de décharge latérale; wagonnet m. se déchargeant d'un seul côté. / self-~ || Selbstentlader m. || wagon m. autodéchargeur ou à déchargement automatique.

discharging wharf || Löschplatz m. || débarcadère m.; quai m. de déchargement.

disc mill || Scheibenmühle f. || moulin m. à disque.

discolour, to || entfärben || décolorer.

discolouring || Entfärbung f. || décoloration f. / ~ agent || Entfärbungsmittel n. || agent m. décolorant.

disconnect, to (Auto motor) || auskuppeln || débrayer; désembrayer. / ~ (Electr) || abschalten || mettre hors circuit m. / ~ (Railw) || entkuppeln || déclencher; désassembler. / ~ (Tel) || trennen||couper; déconnecter. / to ~ the blast || den Wind m. abstellen || arrêter le vent. / to ~ the locomotive from the train || die Lokomotive vom Zug abhängen || détacher ou découpler la machine du train.

disconnected (Electr) || abgeschaltet || déconnecté.

disconnecting, automatic || Selbstauslösung f. || débrayage m. automatique. / ~ double throw switch (Tel) || Trennumschalter m. || sectionneur-inverseur m. / ~ fork || Ausrückgabel f. || fourche f. de débrayage. / ~ insulator (Tel) || Trennisolator m. || isolateur m. de coupure. / ~ key (Tel) || Trenntaste f. || clef f. de rupture. / ~ magnet || Abschaltemagnet m. || électro m. de déconnexion. / ~ relay || Abschalterelais n. || relais m. de déconnexion. / ~ station (Tel) || Trennstelle f. || station f. de coupure. / ~ strip (Tel) || Trennstreifen m. || réglette f. de coupure. / ~ switch (Electr) || Trennschalter m. || interrupteur m.

disconnection (Electr) || Abschaltung f. || mise f. hors circuit. / ~ (Mach) || Ausrückung f. || débrayage m. / ~ (El line) || Unterbrechung f. von Leitungen || disconnection. / ~ (Tel) || Fernsprechsperre f. || suspension f. / forced ~ (Aut tel) || zwangsläufige Freigabe f. || déconnexion f. forcée. / ~ by reversing || ruckweises Ausschalten n. || mise f. hors de circuit ou de fonctionnement par retrait. / safety ~ || Notausrücker m. || débrayeur m. supplémentaire.

discontinue, to (Electr) || unterbrechen || cesser; discontinuer.

discontinuity || Unstetigkeit f. || discontinuité f. / point of ~ || Unstetigkeitsstelle f. || point m. tourbillonnaire ou de discontinuité.

discontinuous (Mach) || aussetzend || intermittent.

discontinuous (Math; Phys) || unstetig; diskontinuierlich || discontinu.

discount, to || diskontieren || escompter.

discount (Bank) || Diskont m. || escompte m. / ~ (Trade) || Nachlaß m.; Preisermäßigung f. || remise f.; rabais m.; diminution f. / to allow ~ || Rabatt m. geben; einen Skonto m. gewähren || accorder une remise ou une escompte; vendre au rabais. / ~ for cash || Rabatt m. bei Barzahlung || rabais m. en cas de payement comptant. / to lower the ~ || den Diskont m. herabsetzen || baisser l'escompte m. / to raise the ~ || den Diskont m. erhöhen || majorer l'escompte m. / special ~ || Sonderrabatt m. || escompte m. de faveur.

discount bank || Diskontbank f. || banque f. d'escompte. / bank rate of ~ || Bankzinsfuß m. || escompte m. officiel.

discounted bill || diskontierter Wechsel m. || effet m. escompté.

discount house || Wechselbank f. || banque f. de change.

discounting of bills || Diskontierung f. von Wechseln || escompte m. d'effets.

discount market || Diskontmarkt m. || marché m. d'escompte. / ~ rate || Diskontsatz m. || taux m. de l'escompte. / ~ stamp || Rabattmarke f. || marque f.

de rabais. / ~ tariff ‖ Rabattarif m. ‖ tarif m. à rabais. / ~ ticket ‖ Rabattmarke f. ‖ timbre-escompte m.

discoupling ‖ Entkopplung f. ‖ désaccouplement m. / ~ hammer (Railw) ‖ Entkupplungshammer m. ‖ marteau m. de débrayage.

discover, to (Mine) ‖ erschürfen ‖ découvrir en creusant. / ~ layers by digging ‖ Lagerstätten fpl. erschürfen ‖ découvrir des mines fpl.

discoverer (Mine) ‖ Schürfer m. ‖ découvreur m.

discovery ‖ Fund m. ‖ découverte f.; trouvaille f. / ~ (Mine) ‖ Aufschluß m. ‖ ouverture f. d'un gisement. / proof of the ~ ‖ Nachweis m. der Fündigkeit ‖ preuve f. de la découverte. / voyage of ~ ‖ Forschungsreise f. ‖ voyage m. de découverte.

discrepancy, to adjust a ~ ‖ eine Differenz f. regeln ‖ régler une différence.

discriminating gear ‖ Uhrwerkschalter m. ‖ dispositif m. à section sélective. / ~ relay ‖ Selektivschutz m. ‖ protection f. par relais discriminateur. / ~ tone (Tel) ‖ Unterscheidungston m. ‖ son m. de distinction.

discrimination ‖ Scharfsinn m. ‖ perspicacité f.; sagacité f.

disease, to ‖ erkranken ‖ tomber malade.

disease ‖ Leiden n. ‖ maladie f. / ~ of timber ‖ Holzkrankheit f. ‖ maladie f. ou vice m. de bois. / ~ of trees see ~ of timber. / ~ of wood see ~ of timber.

disembark, to ‖ landen; ausschiffen ‖ débarquer.

disemboguement of a canal ‖ Ausmündung f. eines Kanals ‖ débouquement m. d'un canal.

disengage, to (Bank) ‖ entlasten ‖ dégager. / ~ (Mach) ‖ entkuppeln; ausrücken ‖ découpler; décrocher; dételer; désembrayer. / ~ (Railw) ‖ loskuppeln ‖ désembrayer; découpler.

disengaged line (Tel) ‖ freie Leitung f. ‖ jonction f. disponible.

disengagement of table drive ‖ Stillsetzen n. des Tisches ‖ arrêt m. de la table. / ~ lever spindle ‖ Ausrückhebelwelle f. ‖ arbre m. du levier de débrayage.

disengaging bar ‖ Ausrückschiene f. ‖ tige f. modificatrice. / ~ bracket under the stanchions ‖ Auslöseknagge f. unter den Rungen ‖ support m. de déclenchement sous les ranchers. / ~ coupling ‖ Ausrückvorrichtung f. ‖ accouplement m. de débrayage. / ~ fork ‖ Ausrückgabel f. ‖ fourche f. de débrayage. / ~ gear ‖ Ausrückvorrichtung f. ‖ appareil m. de débrayage; dispositif m. de désaccouplement. / ~ pawl ‖ ausrückbare Sperrklinke f. ‖ cliquet m. débrayable. / ~ rod ‖ Ausrückstange f. ‖ tige f. de débrayage. / ~ shaft ‖ Ausrückwelle f. ‖ arbre m. de débrayage. / ~ strap ‖ Ausrückbügel m. ‖ bride f. de débrayage.

disgorging plant ‖ Degorgieranlage f. ‖ installation f. de dégorgement.

dish, to ~ the plate ‖ Blech n. kümpeln ‖ cintrer ou emboutir la tôle.

dish ‖ Schale f.; Napf m. ‖ écuelle f.; cuvette f. / ~ of asbestos ‖ Asbestschale f. ‖ capsule f. d'amiante. / ~ of cardboard for the penholder ‖ Federhalterschale f. aus Pappmaché ‖ plumier m. en carton-pâte. / evaporating ~ ‖ Abdampfschale f.; Abdampfgefäß n. ‖ capsule f. évapo-

ratoire ou à évaporation. / flat ~ ‖ flache Schale f. ‖ coupelle f. plate. / perforated ~ ‖ Durchbruchgeschirr n. ‖ batterie f. de cuisine perforée.

dishcloth ‖ Geschirrtuch n. ‖ essuie-main m.

dished and flanged parts ‖ Kümpelteile mpl. ‖ pièces fpl. embouties. / ~ boiler end ‖ gewölbter Kesselboden m. ‖ fond m. bombé. / ~ plate ‖ Buckelblech n.; gewölbtes Blech n. ‖ tôle f. bombée ou à voûte ou emboutie à panneau.

dishing (Metal) ‖ Kümpelarbeit f. ‖ travail m. d'emboutissage. / ~ (Wood) ‖ Wölbung f.; Schweifung f. ‖ bombement m. / depth of ~ ‖ Tiefe f. der Kümpelung ‖ flèche f. du cintrage.

dishing press ‖ Kümpelpresse f.; Bördelpresse f. ‖ presse f. à emboutir. / ~ radius ‖ Wölbungsradius m. ‖ rayon m. du bombement ou des fonds bombés.

dish mill ‖ Tellermühle f. ‖ moulin m. à plateaux.

dishonest ‖ unredlich; schmutzig ‖ malhonnête. / ~ competition ‖ unlauterer Wettbewerb m. ‖ concurrence f. déloyale.

dishwarmer ‖ Tellerwärmer m. ‖ chauffe-plat m.

dish washing table ‖ Geschirraufwaschtisch m. ‖ table f. à laver la vaisselle.

dishwater ‖ Spülwasser n. ‖ lavure f. de vaisselle.

disincrustant ‖ Kesselsteinlösemittel n. ‖ désincrustant m.; tartrifuge f.

disinfect, to ‖ desinfizieren ‖ désinfecter.

disinfectant ‖ desinfizierend ‖ désinfectant.

disinfectant ‖ Desinfektionsmittel n. ‖ désinfectant m.

disinfecting ‖ desinfizierend ‖ désinfectant. / ~ apparatus ‖ Desinfektionsapparat m. ‖ appareil m. de désinfection. / ~ composition ‖ Desinfektionsmittel n. ‖ désinfectant m. / ~ plant ‖ Desinfektionseinrichtung f. ‖ installation f. de désinfection.

disinfection ‖ Desinfektion f. ‖ désinfection f. / ~ of bedding ‖ Desinfektion f. von Bettzeug ‖ désinfection f. de literie. / ~ plant ‖ Entseuchungsanstalt f.; Desinfektionsanlage f. ‖ installation f. de désinfection. / ~ plant for linen ‖ Wäschedesinfektionsanlage f. ‖ installation f. de désinfection du linge.

disinflate, to (Tire) ‖ die Luft f. entweichen lassen ‖ dégonfler.

disintegrate, to ‖ zerstückeln ‖ concasser.

disintegrated, simultaneous existence of a number of ~ net-works ‖ Nebeneinanderbestehen n. kleiner vielfach zersplitterter Versorgungsnetze ‖ un grand nombre m. de petits réseaux existants les uns à côté des autres.

disintegrating and dressing machine ‖ Erzaufbereitungsmaschine f. ‖ machine f. à désagréger et à préparer. / ~ machinery (Mill) ‖ Desintegrator m. ‖ désagrégeur m.

disintegration ‖ Zersetzung f.; Zerfall n.; Zerkleinerung f.; Verwitterung f. ‖ désintégration f.; désagrégation f.; décomposition f. / to prevent the chemical ~ of the ammunition ‖ chemische Zersetzungsprozesse mpl. in der Munition verhindern ‖ prévenir la décomposition chimique de la munition. / product of ~ ‖ Zerfallgebilde n.; Verwitterungsgebilde n. ‖ produit m. de désagrégation ou de décomposition.

disintegrator ‖ Schleudermühle f.; Desintegrator m. ‖ broyeur m. centrifuge; des intégrateur m.; désagrégeur m.

disjoin, to ‖ zerlegen ‖ démonter; désassembler. / the hull of the boat can be disjoined into three parts ‖ der Rumpf des Bootes ist in drei Teile zerlegbar ‖ la coque de canot se peut désassembler en trois pièces.

disjointing table stand ‖ zerlegbares Tischstativ n. ‖ pied m. de table démontable.

disk ‖ Scheibe f. ‖ disque m.; plateau. / ~ for centering a plummet wire ‖ Lotteller m. ‖ plateau m. de plomb. / circular ~ ‖ kreisförmiges Kurbelblatt n. ‖ flasque m. circulaire de manivelle. / ~ for constructing a column (Arch) ‖ Säulentrommel f.; Trommelstein m. ‖ tambour m. de colonne. / elliptic ~ ‖ längliche Scheibe f. ‖ disque m. ovale. / friction ~ ‖ Reibscheibe f. ‖ plateau m. de friction. / ~ with furnace openings pressed outwards (Boil) ‖ Boden m. mit ausgepreßten Rohrlöchern ‖ disque m. avec bride à tube obtenue par emboutissage. / grooved ~ ‖ Kanalscheibe f. ‖ disque m. à gorges. / hard rubber ~ ‖ Hartgummischeibe f. ‖ disque m. en caoutchouc durci. / intermediate ~ (Turbine) ‖ Zwischenboden m. ‖ cloison m. ou disque m. intermédiaire. / ~ for stopping ‖ Scheibe f. für Verschlußzwecke ‖ rondelle f. de bouchage. / ~ of tin for bottle corks ‖ Auflegeplatte f. aus Zinn für Flaschenkorke ‖ rondelle f. en tôle d'étain pour bouchons. / toothed ~ ‖ gezahnte Scheibe f. ‖ disque m. denté. / wheel ~ (Auto) ‖ Radverblendscheibe f. ‖ couvre-rayons m. de roue. / ~ of the wheel represented in section an undulating line ‖ Radscheibe f. von wellenförmigem Querschnitt ‖ roue f. en forme de disque dont la section transversale présentait une ligne à ondulation.

disk armature ‖ Scheibenanker m. ‖ induit m. en disque. / ~ -type arrester ‖ Scheibenblitzableiter m. ‖ parafoudre m. à disques. / ~ bearing ‖ Scheibenlager n. ‖ palier m. à disque. / ~ brake ‖ Scheibenbremse f. ‖ frein m. à disque. / ~ buffer ‖ Stoßpuffer m. ‖ tampon m. de choc. / ~ chuck ‖ Scheibenfutter n. ‖ mandrin m. à disque. / ~ clutch ‖ Lamellenkupplung f. ‖ embrayage m. à disques. / ~ coil ‖ Flachspule f. ‖ bobine f. plate. / ~ coil variometer ‖ Flachspulenvariometer n. ‖ variomètre m. à bobines plates. / ~ condenser ‖ Drehkondensator m. ‖ condensateur m. réglable à plaques. / ~ coupling ‖ Scheibenkupplung f. ‖ embrayage m. à plateaux. / diameter of ~ ‖ Scheibendurchmesser m. ‖ diamètre m. du disque. / ~ diaphragm (Opt) ‖ Blendscheibe f. ‖ disque-diaphragme m. / ~ discharger ‖ Scheibenfunkenstrecke f. ‖ éclateur m. à disque / studded ~ discharger (Radio) ‖ rotierende Scheibenfunkenstrecke f. mit Zähnen ‖ éclateur m. à disque muni de prisonniers latéraux. / ~ feeder ‖ Abstreichteller m.; Abstreichtisch m. ‖ table f. de distribution; table f. doseur. / ~ friction gear ‖ Planscheibengetriebe n. ‖ transmission f. par plateaux de friction. / ~ gap transmitter ‖ Plattenfunkenstrecke f. ‖ éclateur m. à plaques. / ~ generator ‖ Scheibenankerdynamo f. ‖ génératrice f. à induit en disque. / ~ pile (Railw) ‖ Scheibenpfahl m. ‖ pieu m. à disque. / perforated ~ powder ‖ Würfel-

pulver n. ‖ poudre f. prismatique *ou* en dés. / ~ saw ‖ Kreissäge f. ‖ scie f. circulaire *ou* ronde. / hand ~ signal (Railw) ‖ Handscheibensignal n. ‖ disque m. manœuvré à la main. / ~ spring ‖ Scheibenfeder f. ‖ ressort m. Belleville. / ~ wheel ‖ Scheibenrad n. ‖ roue f. à disque. / ~ wheel with boss and rim all in one piece ‖ Scheibenrad n. mit Nabe und Kranz in einem Stück ‖ roue f. pleine avec moyeu et couronne d'une pièce. / rolled mild steel ~ wheel ‖ gewalztes flußeisernes Scheibenrad n. ‖ roue f. à disque laminée en acier doux. / ~ wheel centre ‖ Radscheibe f. ‖ centre m. de roue à disque. / rolled ~ wheel centre ‖ gewalzte Radscheibe f. ‖ centre m. de roue à disque laminé.

dislocation of strata ‖ Schichtenstörung f. ‖ dislocation f. des couches.

dismantle, to (Build) ‖ abtragen; abreißen ‖ démanteler; démolir; raser. / ~ (Fortification) ‖ schleifen ‖ démolir; démanteler. / ~ (Mach) ‖ auseinandernehmen ‖ démonter. / ~ a bridge ‖ eine Brücke abbrechen; abbrücken ‖ enlever *ou* replier un pont.

dismantling (Mach) ‖ Auseinanderbau m. ‖ démontage m. / ~ (Ship) ‖ Abtakelung f. ‖ dégréement m.; dégréage m. / ~ the engine ‖ Zerlegung f. des Motors ‖ démontage m. du moteur.

dismembrator ‖ Dismembrator m. ‖ démembreur m.

dismiss, to ‖ den Dienst m. kündigen ‖ dénoncer le contrat de service. / ~ workmen ‖ Arbeiter mpl. entlassen ‖ licencier *ou* renvoyer *ou* congédier les ouvriers. / ~ a workman without notice ‖ einen Arbeiter m. ohne Kündigung entlassen ‖ congédier un ouvrier sans préavis.

dismissal ‖ Aufkündigung f. ‖ résiliation f.; dénonciation f. / ~ from service ‖ Dienstentlassung f. ‖ congé m. / ~ of workmen ‖ Entlassung f. der Arbeiter ‖ licenciement m. *ou* renvoi m. des ouvriers.

dismissal, legal period of notice of ~ ‖ gesetzliche Aufkündigungsfrist f. ‖ délai m. de congédiement *ou* de renvoi légal. / x-days notice of ~ ‖ x-tägige Kündigungsfrist f. ‖ délai m. de congédiement *ou* de renvoi de x jours. / reason for ~ ‖ Kündigungsgrund m. ‖ motif m. de congédiement. / right of ~ ‖ Kündigungsrecht n. ‖ droit m. de congédiement *ou* de renvoi.

dismissed worker ‖ abkehrender Arbeiter m. ‖ ouvrier m. congédié.

dismount, to ~ a building ‖ ein Gebäude n. abbrechen *oder* niederreißen ‖ démolir *ou* raser un bâtiment. / ~ by parbuckle (Coop) ‖ ein Faß n. schroten ‖ descendre un tonneau à l'aide d'une trévire.

dismounting machines pl. ‖ Abbau m. von Maschinen ‖ démontage m. de machines.

disobedience, to be guilty of gross ~ ‖ sich groben Ungehorsams m. schuldig machen ‖ se rendre coupable de grossière désobéissance f.

disorder, to get into ‖ in Unordnung f. geraten ‖ se détraquer.

dispart, to ‖ kalibrieren ‖ calibrer.

dispatch ‖ Versand m. ‖ expédition f. / icc ~ ‖ Eisabgabe f. ‖ expedition f. de la glace. / ~ of traffic ‖ Abwicklung f. des Verkehrs ‖ écoulement m. du trafic.

dispatcher ‖ Versender m. ‖ expéditeur m.

dispatch goods ‖ Eilgut n. ‖ marchandises fpl. en grande vitesse. / ~ jar ‖ Versandgefäß n. ‖ vase m. d'expédition. / ~ vessel (Mar) ‖ Aviso m.; Depeschenboot n. ‖ aviso m.

dispense, to (Beer) ‖ ausschänken ‖ débiter.

dispensing bottle ‖ Medizinflasche f. ‖ bouteille f. à médicaments. / ~ glass ‖ Medizinglas n. ‖ verre m. à médicaments.

disperse, to (Opt) ‖ zerstreuen ‖ disperger.

dispersing cone ‖ Streuungskegel m. ‖ cône m. d'émission. / ~ cylinder lens ‖ zerstreuende Zylinderlinse f. ‖ lentille f. cylindrique divergente. / ~ lens ‖ zerstreuende Linse f.; Zerstreuungslinse f. ‖ lentille f. divergente. / ~ prism ‖ Dispersionsprisma n. ‖ prisme m. à dispersion.

dispersion ‖ Streuung f. ‖ fuites fpl. / colour ~ (Opt) ‖ Farbenzerstreuung f. ‖ dispersion f. des couleurs.

dispersive power (Opt) ‖ Zerstreuungsvermögen n. ‖ pouvoir m. dispersif.

displace, to ‖ verschieben; verlegen ‖ déplacer. / ~ the air ‖ entlüften; die Luft verdrängen ‖ déplacer *ou* chasser l'air m. quantity of water displaced by a vessel ‖ die durch das Schiff verdrängte Wassermenge ‖ volume m. d'eau déplacé par la coque d'un navire.

displaceable clamp lever ‖ Klemmverschiebehebel m. ‖ levier m. à pince déplaçable. / ~ sleeve ‖ Verschiebemuffe f. ‖ manchon m. déplaçable.

displacement ‖ Verschiebung f.; Verstellung f. ‖ déplacement m. / ~ (Shipb) ‖ Verdrang m.; Wasserverdrängung f.; Deplazement n. ‖ déplacement m. / angular ~ ‖ Winkelverschiebung f. ‖ déplacement m. angulair. / ~ of the body of the car ‖ Verschiebung f. des Wagenkastens ‖ déplacement m. de la caisse de voiture. / ~ of the centre of pressure ‖ Druckpunktwanderung f. ‖ déplacement m. du centre de pression. / dielectric ~ ‖ dielektrische Verschiebung f. ‖ déplacement m. diélectrique. / ~ fully loaded ‖ Wasserverdrängung f. bei voller Ladung ‖ déplacement m. en charge. / ~ of phase ‖ Phasenverschiebung f. ‖ déphasage m. / piston ~ ‖ Kolbenverdrängung f. ‖ déplacement m. de piston. / apparatus for measuring ~ of spectrum lines ‖ Apparat m. zur Messung der Linienverschiebung im Spektrum ‖ appareil m. pour mesurer le déplacement des raies du spectre.

displacing the iris diaphragm ‖ Verschieben n. der Irisblende ‖ déplacement m. du diaphragme-iris. / ~ device ‖ Verfahreinrichtung f. ‖ dispositif m. de déplacement.

display, to ‖ zur Schau f. stellen ‖ exposer.

display (Ware) ‖ Aushang m. ‖ devanture f.; étalage m. / ~ of types ‖ Schriftanordnung f. ‖ arrangement m. des types.

disposal ‖ Absatz m.; Verkauf m. ‖ débit m. / right of ~ by the owner of the soil ‖ Verfügungsrecht n. des Grundeigentümers ‖ droit m. du propriétaire de disposer du fonds.

dispose, to ~ of (Trade) ‖ absetzen ‖ placer.

disposition of the step-grooves ‖ Nuteneinteilung f. an der Treppe ‖ disposition f. des commarchures.

disproportion ‖ Mißverhältnis n. ‖ disproportion f.

dispute ‖ Streitsache f. ‖ différend m.

disruptive ‖ disruptiv; Bruch m. bewirkend ‖ disruptif. / ~ discharge ‖ plötzliche Entladung f.; Funkenentladung f. ‖ décharge f. disruptive *ou* à étincelle. / ~ potential ‖ Funkenpotential n. ‖ potentiel m. disruptif. / ~ strength ‖ Durchschlagfestigkeit f. ‖ rigidité f. diélectrique. / ~ voltage ‖ Durchschlagspannung f. ‖ tension f. disruptive.

diss ‖ Diß n. ‖ diss m.

dissect, to ‖ zergliedern ‖ démembrer; disséquer.

dissecting instruments pl. ‖ Präparierbesteck n. ‖ trousse f. à dissection. / ~ magnifier ‖ Präparierlupe f. ‖ loupe f. à dissection. / stereoscopic ~ microscope ‖ stereoskopisches Präpariermikroskop n. ‖ microscope m. à dissection stéréoscopique. / ~ needle ‖ Präpariernadel f. ‖ aiguille f. à dissection. / ~ stand ‖ Präparierstativ n. ‖ statif m. à dissection. / ~ table ‖ Präpariertisch m. ‖ table f. *ou* platine f. à dissection.

dissection ‖ Sektion f. ‖ ouverture f. / ~ forceps pl. ‖ anatomische Pinzette f. ‖ pince f. à dissection.

dissembling ‖ blind; nachgeahmt ‖ dissimulé; feint; imité.

dissimulated electricity ‖ gebundene Elektrizität f. ‖ électricité f. dissimulée.

dissipate, to ~ the excess water vapour ‖ den überschüssigen Wasserdampf m. ausscheiden ‖ séparer *ou* éliminer la vapeur d'eau en excès.

dissipation of energy ‖ Energievergeudung f. ‖ dissipation f. d'énergie. / ~ up to ... ‖ Abdrosselung f. (Verlust) bis auf ... ‖ étranglement m. jusqu'au ...

dissociate, to ‖ dissoziieren ‖ dissocier.

dissociation ‖ Entmischung f.; Dissoziation f. ‖ dissociation f. / ~ of fatty matters ‖ Spaltung f. von Fettstoffen ‖ dissociation f. des corps gras. / ~ chamber ‖ Zersetzungskammer f. ‖ chambre f. de réaction *ou* de décomposition.

dissolution (Chem) ‖ Abbau m.; Lösung f. ‖ dissolution f.; solubilisation f.

dissolve, to (Chem) ‖ auflösen ‖ dissoudre; faire fondre. / ~ partnership ‖ eine Genossenschaft f. aufheben ‖ dissoudre une association coopérative.

dissolvent (Chem) ‖ Lösungsmittel n. ‖ dissolvant m. / ~ power ‖ Auflösungsvermögen n. ‖ pouvoir m. dissolvant. / ~ producing factory ‖ Lösungsmittelfabrik f. ‖ fabrique f. de dissolvants.

dissolver, rapid ~ ‖ Schnellauflöser m. ‖ appareil m. destiné à la dissolution rapide.

dissolving ‖ Auslaugung f.; Auflösung f. ‖ dissolution f. / ~ of rocks ‖ Auslaugung f. des Gesteins ‖ solution f. de la roche.

dissolving apparatus ‖ Auflöseapparat m. ‖ appareil m. de dissolution. / apparatus for ~ lyes ‖ Laugeauflöseapparat m. ‖ appareil m. à dissoudre la lessive. / ~ machinery ‖ Aufschließmaschine f. ‖ mélangeur-dissolveur m. / ~ vat ‖ Laugfaß n. ‖ cuve f. de dissolution.

dissonance ‖ Dissonanz f. ‖ dissonance f.

dissonant ‖ mißklingend ‖ dissonant.

dissymmetry ‖ Unsymmetrie f. ‖ asymétrie f.

distaff ‖ Spinnrocken m.; Rocken m. ‖ quenouille f.

distance ‖ Entfernung f.; Abstand m. ‖ distance f. / ~ between centres ‖ Abstand m. von Mitte zu Mitte ‖ distance f. de centre en centre *ou* d'axe en axe. / ~ centre of ... to centre of ... ‖ Entfernung f. von Mitte ... bis Mitte ... ‖

distance f. de l'axe de ... à l'axe de ... /
~ between centre-lines ‖ Abstand m. von
Mitte zu Mitte ‖ écartement m. d'axe en
axe. / ~ between centre line of stern
post and centre of propeller shaft ‖ Entfernung f. von Mitte Hintersteven bis
Mitte Schraubenwelle ‖ écartement m. de
centre d'étambot en centre d'arbre porte-
hélice. / ~ covered (Aero; Nav) ‖ durch-
laufene Strecke f. ‖ distance f. parcourue. / to determine the ~ ‖ den Abstand
m. bestimmen ‖ déterminer la distance. /
~ of distinct vision ‖ deutliche Sehweite
f. ‖ distance f. de la vision distincte. /
equal ~s pl. ‖ gleiche Abstände mpl. ‖
distances fpl. égales. / ~ between the
eyes ‖ Augenabstand m. ‖ distance f. ou
écartement m. des yeux. / horizontal ~
between two points ‖ Wagerechtentfer-
nung f. zweier Punkte ‖ distance f. horizontale entre deux points. / interocular ~
‖ Augenabstand m. ‖ distance f. des yeux.
/ ~ between letters ‖ Buchstabenabstand
m. ‖ écartement m. entre les caractères. /
~ of level ‖ Niveauabstand m. ‖ distance
f. du niveau. / ~ between lines ‖ Zeilen-
abstand m. ‖ écartement m. entre les
lignes. / to measure the ~ with compasses
‖ die Entfernung mit dem Zirkel abgreifen
‖ mesurer la distance avec le compas. /
observation ~ ‖ Beobachtungsabstand m.
‖ distance f. d'observation. / to pace a ~
‖ eine Strecke f. abschreiten ‖ mesurer
une distance au pas. / ~ between the
pupils ‖ Pupillenabstand m. ‖ écartement m. pupillaire. / reading ~ ‖ Leseabstand m. ‖ distance f. de lecture. / ~
sailed see ~ covered. / to span a ~ ‖
Strecke f. überbrücken ‖ franchir une
distance. / sparking ~ ‖ Funkenstrecke f.
‖ éclateur m.; contact m. tournant;
distance f. explosive. / to stake off the
~ see to measure ~ / stopping ~ ‖ Bremsweg m. ‖ distance f. parcourue après
freinage. / at a sufficient ~ ‖ in genü-
gendem Abstand m. ‖ à une distance
suffisante. / take-off ~ (Aero) ‖ Startlänge f. ‖ distance f. parcourue pour
envol. / visual ~ ‖ Sehweite f.; Weite f.
des deutlichen Sehens ‖ distance f. de
la vue distincte; portée f. visuelle. /
~ of wheel-centres (Railw) ‖ Radstand
m. ‖ écartement m. des essieux. /
working ~ ‖ Arbeitsabstand m. ‖ distance f. frontale.
distance control for ensuring reliable co-
operation between power station and
sub-stations ‖ Fahrplanfernregler m. für
Kraftwerke ‖ régulateur m. à distance
permettant le fonctionnement d'ensemble entre l'usine d'énergie et les sous-
stations. / ~ control instrument ‖ Fernmeßgerät n. ‖ appareil m. de mesure à
distance. / ~ glass ‖ Fernglas n. ‖ lunette
f. / exact optical ~ measurement ‖ Prä-
zisionsdistanzmessung f. auf optischem
Wege ‖ mesure f. optique précise des
distances. / ~ piece ‖ Einsatzstück n.;
Zwischenstück n.; Abstandhülse f. ‖
pièce f. intercalaire ou intermédiaire. /
~ pipe ‖ Distanzrohr n. ‖ tuyau m.
d'écartement. / ~ plate ‖ Distanzplatte
f. ‖ plaque f. de distance. / ~ spectacles
pl. ‖ Fernbrille f. ‖ lunette f. pour les
lointains ou destinée aux lointains. /
~ switch ‖ Fernschalter m. ‖ téléinterrupteur m. / ~ thermometer ‖ Fernthermometer n. ‖ téléthermomètre m. /

~ tube piece ‖ Distanzrohrstück n. ‖
tube m. d'espacement.
distant action theory ‖ Fernwirkungs-
theorie f. ‖ théorie f. d'action par distance. / ~ control switch ‖ Schalter m.
mit Fernsteuerung ‖ téléinterrupteur m.
/ ~ effect ‖ Fernwirkung f. ‖ effet m. à
distance.
distantly controlled ‖ ferngesteuert ‖ commandé à distance. / ~ controlled relay ‖
ferngesteuertes Relais n. ‖ relais m. à
distance.
distant recorder ‖ Fernzähler m. ‖ compteur m. à distance. / ~ regulation ‖ Fern-
reg(e)lung f. ‖ télérégulation f. / ~ wave
zone ‖ Fernbereich m. ‖ zône f. d'action
lointaine.
distemper ‖ Stubenweißer m. ‖ peintre m.
en détrempe. / ~ painting ‖ Tempera-
malerei f. ‖ peinture f. en détrempe.
distil, to see distill, to.
distill, to ‖ destillieren ‖ distiller. / ~ off ‖
abdestillieren ‖ chasser par distillation. /
~ over ‖ überdestillieren ‖ passer. / ~ with
steam ‖ mit Wasserdampf m. destillieren
‖ entraîner à la vapeur.
distillability ‖ Destillierbarkeit f. ‖ distilla-
bilité f.
distillable ‖ destillierbar ‖ distillable.
distillate, to ‖ abdestillieren ‖ distiller.
distillate ‖ Destillat n. ‖ distillat m. / aromatic ~ ‖ aromatisches Destillat n. ‖
eau f. distillée aromatique.
distillation ‖ Destillation f. ‖ distillation f. /
~ of coal in closed retorts ‖ Destillation
f. von Kohlen in geschlossenen Retorten
‖ distillation f. de la houille en vases
clos. / ~ of coal tar ‖ Destillation f. von
Steinkohlenteer ‖ distillation f. de goudron de houille. / destructive ~ see dry ~. /
dry ~ ‖ trockene Destillation f.; Trocken-
destillation f. ‖ distillation f. sèche. / ~
of dwarf-pine oil ‖ Latschenkiefernöl-
destillation f. ‖ distillerie f. d'huile de
pin nain. / fractional ~ ‖ teilweise oder
fraktionierte Destillation f. ‖ distillation
f. fractionnée. / ~ of mineral fuels ‖
Destillation f. von mineralischen Brennstoffen ‖ distillation f. des combustibles
minéraux. / resin ~ ‖ Harzdestillation f. ‖
distillation f. de résine. / ~ and refinery
of tar ‖ Teerzerlegung f. ‖ traitement m.
du goudron. / wood ~ ‖ Holzdestillation
f. ‖ distillation f. du bois.
distillation apparatus ‖ Destillierapparat
m. ‖ appareil m. de distillation. / ~ plant
‖ Destillationsanlage f. ‖ installation f. de
distillation. / ~ plant for the removal of
free fatty acids from edible oil ‖ Destillationsanlage f. zur Entfettung der freien
Fettsäure aus Speiseöl ‖ installation f.
pour l'élimination par distillation des
acides gras libres dans les huiles alimentaires. / ~ pot ‖ Destilliertopf m. ‖
récipient m. de distillation. / ~ process ‖
Destillationsverfahren n. ‖ procédé m.
par distillation. / ~ very valuable ~ product ‖ hochwertiges Destillationserzeug-
nis n. ‖ produit m. de distillation de
grande valeur. / ~ vessel see ~ pot.
distilled beverage ‖ abgezogenes Getränk
n. ‖ boisson f. distillée. / ~ water ‖ destilliertes Wasser n. ‖ eau f. distillée.
distiller (Apparatus) ‖ Destillierblase f. ‖
alambic m. / ~ (Person) ‖ Brenner m. ‖
distillateur m. / ~'s grains pl. ‖ Brennerei-
treber pl. ‖ drèche f. de distillerie. / furnace for ~'s wash ‖ Schlempeofen m. ‖

four m. à vinasse. / ~'s wash drying apparatus ‖ Schlempetrockner m. ‖ séchoir m. à vinasse.
distillery ‖ Brennerei f. ‖ distillerie f. / ~
of lees' spirit ‖ Weinhefebrennerei f. ‖
distillerie f. ou brûlerie f. de lie de vin. /
~ installation for liqueur distilleries ‖
Likörfabrikeinrichtung f. ‖ équipement
m. à distiller des liqueurs. / ~ of spirits ‖ Branntweinbrennerei f. ‖ distillerie
f.; brûlerie f. / ~ plant ‖ Brennereianlage
f. ‖ installation f. de distillerie. / ~
workman ‖ Brennereiarbeiter m. ‖ ouvrier
m. de distillerie. / ~ yeast ‖ Brennerei-
hefe f. ‖ levure f. de distillerie.
distilling apparatus ‖ Destillierapparat m.
‖ appareil m. de distillerie ou à distiller. / ~ flask ‖ Destillationskolben
m. ‖ ballon m. à distiller; matras m. à
distillation. / ~ off the substances pl.
possessing smell ‖ Abdestillieren n. der
Duftstoffe ‖ distiller les matières fpl.
odorantes. / ~ plant ‖ Destillationsanlage
f. ‖ installation f. de distillation. / wood
~ plant ‖ Holzdestillationsanlage f.
‖ appareil m. de distillation du bois. /
~ tube ‖ Destillationsrohr n.; Destillier-
rohr n. ‖ tube m. distillatoire. / ~ vessel ‖
Destilliergefäß n. ‖ vase m. distillatoire;
alambic m.; cucurbite f.
distinctness (Opt) ‖ Deutlichkeit f. ‖ netteté f.
distinguishing mark ‖ Kennzeichen n. ‖
marque f. distinctive; signe m. caractéristique. / ~ (Telegraph pole) ‖ Bezeich-
nungsnagel m. ‖ clou m. estampillé; marque f. distinctive.
distorsion (Tel) ‖ Verzerrung f. im Fernsprechbetrieb ‖ distorsion f.
distort, to (Material) ‖ sich krumm ziehen;
sich verziehen; sich werfen ‖ se gauchir;
se déjeter; se voiler. / ~ (Tel) ‖ verzerren ‖ présenter de déformations fpl. /
~ (Mach) ‖ verdrehen ‖ tordre. / ~ (Opt)
‖ verzerren; verzeichnen ‖ distordre;
défigurer; tordre. / the axle was distorted to an angle of about x° ‖ die
Achse war um ungefähr x° verdreht f. ‖
l'essieu m. fut tordu d'un angle de x°
environ.
distortion (Aero) ‖ Verwindung f. ‖ distorsion f. / ~ (Opt) ‖ Formänderung f.; Ver-
zeichnung f. ‖ déformation f.; distorsion
f. / ~ (Tel) ‖ Verzerrung f. ‖ déformation
f. / ~ due to hardening ‖ Härteverzug m.;
beim Härten auftretender Verzug m. ‖
déformation f. qui se produit lors de la
trempe. / total ~ (El line) ‖ elektrisches
Längenmaß n. ‖ longueur f. électrique. /
correcting of ~ (Repeater) ‖ Entzerrung
f. ‖ correction f. de distorsion; anti-
distorsion f. / ~ freedom from ~ of the
magnifier ‖ Verzeichnungsfreiheit f. der
Lupe ‖ absence f. de distorsion de la
loupe.
distortionless (Tel) ‖ verzerrungsfrei ‖ sans
distorsion; dénué de distorsion. / ~ line
(Tel) ‖ verzerrungsfreie Leitung f. ‖ ligne
f. sans distorsion.
distortion regenerative (Radio) ‖ Rück-
kopplungsverzerrung f. ‖ distorsion f.
par réaction.
distrain, to ‖ auspfänden ‖ saisir.
distraint ‖ Auspfändung f. ‖ saisie f.; saisie-
exécution f.
distress at sea ‖ Seenot f. ‖ sinistre m. de
mer. / ~ call ‖ Seenotanruf m. ‖ appel m.
de détresse. / ~ message ‖ Seenottele-

gramm n. ‖ message m. de détresse. / ~ service (Radio) ‖ Seenotmeldedienst m. ‖ ‖ service m. de détresse. / ~ signal ‖ Notsignal n. ‖ signal m. de secours. / ~ traffic ‖ Seenotverkehr m. ‖ trafic m. de détresse.

distribute, to ‖ austeilen; verteilen ‖ distribuer. / ~ (Print) ‖ ablegen ‖ distribuer. / ~ the error ‖ den Fehler m. ausgleichen ‖ compenser l'erreur f. / ~ the ink (Print) ‖ die Farbe *oder* Schwärze auf die Form auftragen ‖ encrer les lettres fpl.; toucher la forme. / ~ the ink over the roller (Print) ‖ die Walze f. einfärben ‖ toucher le rouleau. / ~ the loads pl. ‖ die Lasten fpl. verteilen ‖ répartir les charges fpl. / ~ the letters into wrong boxes (Print) ‖ die Buchstaben mpl. in falsche Fächer legen; die Buchstaben mpl. falsch ablegen ‖ faire des coquilles fpl.

distributing apparatus (Print) ‖ Ablegeapparat m. ‖ appareil m. distributeur. / ~ for lemon juice and Seltzer's water ‖ Limonaden- und Seltersausschankapparat m. ‖ appareil m. pour débits de limonades et d'eau de Seltz.

distributing basin (Hydr arch) ‖ Hauptbehälter m. ‖ réservoir m. principal. / ~ board (Electr) ‖ Batterieverteilungstafel f. ‖ tableau m. des piles *ou* de distribution. / ~ box (Electr) ‖ Abzweigkasten m.; Verteilungsdose f. ‖ boîte f. de branchement *ou* de distribution. / ~ box (Steam) ‖ Schieberkasten; Ventilkasten m. ‖ boîte f. de distribution d'une machine à vapeur. / ~ bus-bar ‖ Verteilungsschiene f. ‖ barre f. omnibus de distribution. / ~ cable (Tel) ‖ Verteilungskabel n. ‖ artère f. de distribution. / intermediate ~ frame (Tel) ‖ Zwischenverteiler m. ‖ répartiteur m. intermédiaire. / ~ fuse ‖ Verteilungssicherung f. ‖ fusible m. *ou* coupe-circuit m. de distribution. / ~ gear controlling the starting air ‖ Anlaßluftsteuerung f. ‖ distribution f. d'air de mise en marche. / ~ network (Electr) ‖ Verteilungsnetz n. ‖ réseau m. de distribution. / ~ pipes pl. (Hydr arch) ‖ Rohrnetz n. ‖ conduite f. de tuyaux; tuyautage m. / ~ pole (El line) ‖ Verteilungsgestänge n. ‖ appui m. de répartition. / ~ reservoir (Hydr arch) ‖ Hochbehälter m. ‖ réservoir m. à distribution d'eau; chateau m. d'eau / ~ table ‖ Verteilertisch m. ‖ table f. doseuse. / ~ valve ‖ Verteilungsschieber m. ‖ tiroir m. distributeur. / ~ valve controlling starting air ‖ Anlaßluftsteuerventil n. ‖ soupape f. de distribution de l'air de mise en marche. / ~ worm ‖ Verteilungsschnecke f. ‖ vis f. sans fin d'alimentation.

distribution ‖ Austeilung f.; Verteilung f. ‖ distribution f. / ~ of charge (Electr) ‖ Ladungsverteilung f. ‖ distribution f. de la charge. / ~ of circulation ‖ Umlaufverteilung f. ‖ répartition f. *ou* distribution f. de la circulation. / ~ of dividends ‖ Verteilung f. der Dividende ‖ répartition f. du dividende. / ~ of electricity ‖ Elektrizitätsverteilung f. ‖ distribution f. d'électricité. / ~ of energy ‖ Energieverteilung f. ‖ distribution f. d'énergie. / ~ of energy in sunlight ‖ Energieverteilung f. im Sonnenlicht ‖ répartition f. des radiations solaires. / ~ of errors ‖ Fehlerausgleichung f. ‖ compensation f. des erreurs. / ~ of incoming exchange calls to the stations of a private automatic tele-

phone installation ‖ Vermittlung f. von ankommenden Amtsgesprächen nach den Apparaten einer selbsttätigen Hausfernsprechanlage ‖ écoulement m. des conversations de service d'arrivée vers les appareils d'une installation téléphonique privée automatique. / ~ of ink ‖ Farbenverreibung f. ‖ broyage m. de l'encre. / the ~ of power device for controlling ‖ Einrichtung f. zur Überwachung der Energieverteilung ‖ installation f. pour le contrôle de la répartition de l'énergie. / ~ of precipitation ‖ Niederschlagsverteilung f. ‖ distribution f. de la pluie. / ~ of pressure ‖ Druckverteilung f. ‖ répartition f. des pressions. / radial ~ (Tel) ‖ offene Verteilung f. ‖ distribution f. radiale. / ~ of rainfall ‖ Niederschlagsverteilung f. ‖ distribution f. de la pluie. / ring-like ~ (El line) ‖ ringförmige Verteilung f. ‖ distribution f. circulaire. / ~ of tools ‖ Werkzeugausgabe f. ‖ distribution f. des outils. / ~ of the stresses ‖ Verteilung f. der Spannungen ‖ distribution f. des tensions. / ~ by tubular slide valve ‖ Rohrschiebersteuerung f. ‖ distribution f. par valve tubulaire. / ~ of type (Print) ‖ Ablegen n. des Satzes ‖ distribution f. de la composition. / plant for the ~ of water ‖ Wasserverteilungsanlage f. ‖ installation f. pour la distribution d'eau.

distribution band ‖ Verteilungsband n. ‖ bande f. de distribution. / ~ board ‖ Verteilungstafel f. ‖ tableau m. de distribution. / ~ box (El line) ‖ Verzweiger m. ‖ chambre f. de distribution. / ~ conduit (Cable) ‖ Verteilungskanal m. ‖ conduit m. de distribution. / ~ network ‖ Verteilungsnetz n. ‖ réseau m. de distribution. / core of the ~ network ‖ Schwerpunkt m. des Verteilungsnetzes ‖ centre m. de distribution. / ~ plate ‖ Streuteller m. ‖ assiette f. de répartition.

distributor (Mach) ‖ Verteiler m. ‖ distributeur m. / ~ (Print) ‖ Ableger m. ‖ distributeur m. / ~ (Tel) ‖ Mehrfachverteiler m. ‖ distributeur m. / ignition ~ ‖ Zündungsverteiler m. ‖ distributeur m. d'allumage. / ~ of ladles ‖ Gießmeister m. ‖ chef m. de coulée. / magneto ~ ‖ Verteiler m. des Magnetapparates ‖ distributeur m. du magnéto. / manure ~ ‖ Düngerstreumaschine f. ‖ sémoir m. d'engrais. / sole ~ ‖ Alleinvertreter m. ‖ représentant m. exclusif.

distributor shaft ‖ Verteilerwelle f. ‖ arbre m. de distribution.

district ‖ Distrikt m.; Bezirk m. ‖ district m.; région f. / aseismic ~ ‖ erdbebenfreies Gebiet n. ‖ domaine m. asméismique. / ~ of guard (Forest) ‖ Schutzbezirk m. ‖ triage m.; district m. *ou* périmètre m. de protection. / ~ mining ‖ Grubengebiet n.; Kohlenrevier n. ‖ district m. minier; région f. minière. / seismic ~ ‖ Erdbebengebiet n. ‖ domaine m. séismique.

district cable (Electr) ‖ Bezirkskabel n. ‖ câble m. de district. / ~ court ‖ Amtsgericht n. ‖ tribunal m. cantonal de première instance; justice f. de paix. / ~ traffic (Tel) ‖ Bezirksverkehr m. ‖ trafic m. régional.

disturb, to ‖ stören ‖ troubler; déranger.

disturbance ‖ Störung f. ‖ perturbation f.; dérangement m.; tapage m.; trouble m. / ~ by breaking of the wire ‖ Störung f. durch Drahtbruch ‖ dérangement m. dû à une

rupture de fil. / ~ of broadcast reception ‖ Rundfunkempfangsstörung f. ‖ perturbation f. de radio réception. / ~ of directional force ‖ Richtkraftstörung f. ‖ perturbation f. de force directrice. / ~ by earth (Electr) ‖ Störung f. durch Erdschluß ‖ dérangement m. dû à une mise à la terre. / ~ of equilibrium ‖ Gleichgewichtsstörung f. ‖ perturbation f. d'équilibre. / magnetic ~ ‖ magnetische Störung f. ‖ perturbation f. magnétique. / ~ in the movement of the ribbon (Typewr) ‖ Farbbandstörung f. ‖ perturbation f. dans la course du ruban encré. / partial ~ ‖ Teilstörung f. ‖ perturbation f. partielle; composante f. de perturbation. / ~ of service ‖ Betriebsstörung f. ‖ dérangement m. de service. / ~ of traffic ‖ Verkehrsstörung f. ‖ perturbation f. du trafic. / ~ of work ‖ Baustörung f. ‖ dérangement m. des travaux.

disturbance current (Tel) ‖ Störstrom m. ‖ courant m. perturbateur. / ~ elimination ‖ Beseitigung f. einer Störung ‖ relève f. d'un dérangement.

disturbed ‖ gestört ‖ troublé. / ~ deposit (Mine) ‖ gestörtes Gebirge n. ‖ terrain m. disloqué.

disturbing admixture ‖ störende Beimengung f. ‖ constituant m. perturbateur. / ~ current (Tel) ‖ Störstrom m. ‖ courant m. perturbateur. / ~ noise ‖ Störgeräusch n. ‖ bruit m. perturbateur. / ~ unbalance measuring set ‖ Geräuschunsymmetriemesser m. ‖ appareil m. de mesure du déséquilibre des circuits.

disuint, to ‖ entfetten ‖ dessuinter; dégraisser.

ditch ‖ Graben m. ‖ fossé m.; canal m. rigole f. / ~ (Agr) ‖ Abzugsgraben m.; Entwässerungsanlage f. ‖ fossé m. d'écoulement. / ~ (Hydr arch) ‖ Sickergraben m. ‖ perré m. / ~ boundary — *see* partition ~. / to dig a ~ ‖ einen Graben m. ziehen ‖ creuser un fossé. / to dig the ~es for foundation (Build) ‖ den Grund m. graben; die Grundgräben mpl. ziehen; die Fundamentgräben mpl. ausheben ‖ creuser *ou* fouiller les fondations mpl. / ~ for foundation (Build) ‖ Grundgraben m.; Grundgrube f. ‖ creux m.; fondation m. / partition ~ (Build; Hydr arch) ‖ Grenzgraben m. ‖ fossé m. limitrophe *ou* mitoyen.

ditch canal ‖ Sackkanal m. ‖ canal m. à niveau.

ditcher (Mach) ‖ Tieflöffelbagger m. ‖ drague f. de pelle fouille. / ~ (Person) ‖ Deicharbeiter m. ‖ travailleur m. aux digues.

divan ‖ Diwan m. ‖ divan m. / ~ carpet ‖ Diwandecke f. ‖ tapis m. de divan. / ~ cover ‖ Diwanteppich f. ‖ couverture f. pour canapés.

dive, to (Aero) ‖ abstürzen ‖ piquer. / ~ (Hydr arch) ‖ versenken ‖ plonger. / ~ (Mar) ‖ tauchen ‖ plonger.

dive (Aero) ‖ Absturz m.; steiler Abflug m. ‖ chute f.; vol m. piqué. / vertical ~ (Aero) ‖ Sturzflug m. ‖ vol m. piqué, piqué m.

diver ‖ Taucher m. ‖ scaphandrier m.; plongeur m.

diverge, to ‖ auseinandergehen; abweichen; divergieren ‖ diverger.

divergence ‖ Divergenz f. ‖ divergence f. / ~ of the balls ‖ Streuung f. der Kugeln ‖ dispersion f. des balles.

14*

divergency see divergence.

divergent ‖ divergent ‖ divergent. / ~ **lens** ‖ Konkavlinse f.; Zerstreuungslinse f. ‖ verre m. divergent ou concave; lentille f. divergente.

diverging see divergent.

diversion of a river-channel ‖ Geradleitung f. oder Rektifikation f. oder Streckung f. eines Flusses ‖ rectification f. du lit d'un fleuve.

diver's work ‖ Taucherarbeit m. ‖ travail m. de scaphandrier.

divert, to ‖ die Richtung f. ändern ‖ dévier; détourner. / ~ **a water course** ‖ einen Wasserlauf m. verändern ‖ détourner la direction d'un cours d'eau.

divide, to (Geom) ‖ teilen ‖ partager. / ~ (Math) ‖ dividieren ‖ diviser; faire la division. / ~ **into degrees** ‖ in Grade mpl. einteilen ‖ diviser en degrés mpl. / ~ **an ingot up into several slices** ‖ Gußblöcke mpl. in einzelne Stücke zerteilen ‖ tronçonner les lingots mpl. en disques. / the ingots are divided up into discs on special lathes by means of chisels ‖ die Blöcke mpl. werden durch Einstechmeißel auf Spezialbänken in Scheiben abgeteilt ‖ sur les tours speciaux les lingots mpl. sont tronçonnés par des outils à couper dans des disques. / ~ **a word into syllables** ‖ ein Wort n. in Silben abbrechen oder abteilen ‖ diviser un mot en syllabes. / ~ **into commercial lengths** ‖ auf handelsübliche Längen zerteilen ‖ débiter en longueurs commerciales. / ~ **up** ‖ stückeln ‖ morceler.

divided battery (Electr) ‖ unterteilte Batterie f. ‖ batterie f. sectionnée. / ~ **beam of a balance** ‖ mit Teilstrichen versehener Wageaalken m. ‖ réglette f. d'une balance. / ~ **circle** (Opt) ‖ Einstellkreis m. ‖ cercle m. gradué ou de calage. / ~ **into** ‖ eingeteilt in ‖ partagé ou divisé en. / ~ **in two** ‖ in zwei Teile gespalten ‖ partagé; coupé en deux.

dividend ‖ Dividende f. ‖ dividende m. / **final** ~ ‖ Schlußdividende f. ‖ solde m. du dividende. / **to fix the** ~ ‖ die Dividende bestimmen ‖ fixer le dividende. / **interim** ~ ‖ Abschlagsdividende f.; Interimsdividende f. ‖ dividende m. provisoire; acompte m. sur le dividende. / **to pay a** ~ ‖ die Dividende verteilen ‖ payer le dividende. / **sham** ~ ‖ fiktive Dividende f. ‖ dividende m. fictif. / **special** ~ ‖ Superdividende f. ‖ boni m. / **surplus** ~ ‖ außerordentliche Dividende f. ‖ superdividende m. / **unclaimed** ~ ‖ nicht erhobene Dividende f. ‖ dividende m. non réclamé.

dividend, distribution of the ~ ‖ Verteilung f. der Dividende ‖ répartition f. du dividende. / ~ **fixing** ‖ Festsetzung f. der Dividende ‖ fixation f. du dividende. / ~ **payment** ‖ Dividendenausschüttung f. ‖ distribution f. du dividende. / ~ **warrant** ‖ Dividendenschein m.; Kupon m. ‖ coupon m.

divider ‖ Teilgerät n. ‖ diviseur m.

dividers pl. ‖ Handzirkel m.; Stechzirkel m. ‖ compas m. de mesure ou à pointes sèches. / **proportional** ~ ‖ Reduktionszirkel m. ‖ compas m. de réduction.

dividing apparatus ‖ Teilapparat m. ‖ appareil m. à diviser. / **circular** ~ **apparatus for toothed wheels** ‖ Kreisteilapparat m. für Zahnräder ‖ appareil m. diviseur pour roues dentées / **special** ~ **attach-**ment (Mach tool) ‖ Einzelteileinrichtung f. ‖ dispositif m. diviseur à la main.

dividing device ‖ Teilvorrichtung f. ‖ dispositif m. à diviser. / **horizontal** ~ ‖ horizontaler Teilapparat m. ‖ diviseur m. horizontal. / **vertical** ~ ‖ vertikaler Teilapparat m. ‖ diviseur m. vertical.

dividing head ‖ Teilkopf m. ‖ poupée f. diviseur. / ~ **machine** ‖ Teilmaschine f. ‖ machine f. à diviser ou à doser. / ~ **machine for lengths** ‖ Längenteilmaschine f. ‖ machine f. à diviser des longueurs. / ~ **plate** ‖ Teilscheibe f. ‖ plateau m. diviseur. / ~ **spindle** ‖ Teilspindel f. ‖ broche f. de la poupée à diviser. / ~ **spindle bore** ‖ Teilspindelbohrung f. ‖ alésage m. de broche de la poupée à diviser. / **strip shear for** ~ **strips and plates** ‖ Streifenschere f. zum Teilen von Streifen und Platten ‖ cisailles fpl. à bandes pour diviser les bandes et les plaques. / ~ **taster** ‖ Teilzirkel m. ‖ compas m. à diviser. / ~ **wheel** ‖ Teilrad n. ‖ roue f. à diviser. / ~ **worm wheel** ‖ Teilschneckenrad n. ‖ roue f. hélicoïdale diviseur.

dividivi pod ‖ Dividivischote f. ‖ gousse f. de divi-divi.

diving air pump ‖ Taucherluftpumpe f. ‖ pompe f. à air de scaphandrier. / ~ **apparatus** ‖ Taucherapparat m. ‖ appareil m. de plongeurs ou de scaphandriers; scaphandre m. / ~ **bell** ‖ Taucherglocke f. ‖ cloche f. à plongeur. / ~ **boat** ‖ Taucherboot n. ‖ bateau m. plongeur. / ~ **depth** ‖ Tauchtiefe f. ‖ profondeur f. d'immersion. / ~ **dress** ‖ Taucheranzug m. ‖ vêtement m. de scaphandriers. / ~ **helmet** ‖ Tauchhelm m. ‖ casque m. de plongeur. / ~ **material** ‖ Tauchergerät n. ‖ matériel m. de scaphandriers.

divining rod ‖ Wünschelrute f. ‖ baguette f. divinatoire ou de sourcier. / ~ **movement** ‖ Wünschelrutenausschlag m. ‖ mouvement m. ou réaction f. de la baguette divinatoire.

divisibility ‖ Teilbarkeit f. ‖ divisibilité f.

divisible ‖ teilbar ‖ divisible.

division (Forest) ‖ Waldabteilung f. ‖ parcelle f. de bois. / ~ (Math) ‖ Division f. ‖ division f. / ~ (Phys) ‖ Teilstrich m. ‖ trait m. de division. / ~ (Ship register) ‖ Klasse f. ‖ division f. / ~ **of the circle into 60⁰** ‖ Sechzigerteilung f. ‖ division f. sexagésimale. / ~ **of a circular arc** ‖ Kreisteilung f. ‖ graduation f. circulaire. / **with double set of** ~ ‖ mit doppelter Teilung f. ‖ à deux échelles fpl. / ~ **graduated** ‖ Gradteilung f. ‖ division f. en degrés. / ~ **into hours** ‖ Stundenteilung f. ‖ division f. horaire. / ~ **of load** ‖ Belastungsverteilung f. ‖ répartition f. de charge. / ~ **of a shaft** (Mine) ‖ Schachttrumm m.; Schachtabteilung f. ‖ compartiment m. d'un puits. / **with single** ~ ‖ mit einfacher Teilung f. ‖ division f. à une échelle. / ~ **in squares** ‖ Feldereinteilung f. ‖ division f. par compartiments. / ~ **in squares like a chessboard** ‖ Schachbrettfeldereinteilung f. ‖ division f. par compartiments en damier. / ~ **of a straight line** ‖ g(e)radlinige Teilung f. ‖ graduation f. rectiligne. / ~ **of work** ‖ Arbeitseinteilung f. ‖ répartition f. du travail.

division line ‖ Teilstrich m. ‖ graduation f. / ~ **mark** ‖ Strichmarke f. ‖ trait m. de repère. / ~ **plate** ‖ Teilscheibe f. ‖ plateforme f. à diviser; plateau m. diviseur.

divisions pl. **inside the box** ‖ im Inneren des Kastens angebrachte Abteile npl. ‖ casiers mpl. pratiqués à l'intérieur du coffre.

divisor (Math) ‖ Teiler m.; Divisor m. ‖ diviseur m.

do, to ‖ leisten ‖ faire; effectuer; accomplir; produire. / **the wheels** pl. **have done up to x kilometers** ‖ die Räder npl. haben ungefähr x Kilometer zurückgelegt ‖ les roues fpl. ont parcourues une voie de x kilomètres.

doatiness ‖ Spreufleckigkeit f. ‖ taches fpl. pailletées.

dobby card punching (Weav) ‖ mechanische Kartenschlägerei f. ‖ perçage m. au métier; piquage m. des cartes.

docimastic art see docimasy.

docimasy (Chem) ‖ Probierkunst f. ‖ docimasie f.; docimastique f.; art m. d'essayeur.

dock, to ~ **a ship** ‖ ein Schiff n. docken ‖ amener un vaisseau dans un dock ou dans une cale; entrer un navire dans un bassin; mettre un navire au bassin.

dock (Hydr arch) ‖ Dockhafen m.; Binnenhafen m. ‖ bassin m. à flot. / ~ (Mar) ‖ Dock n.; Trockendock n. ‖ dock m.; bassin m. de radoub. / ~ (Trade) ‖ Lagerhof m. ‖ dock m.; entrepôt m. / **dry** ~ ‖ Trockendock m. ‖ cale f. sèche; bassin m. de radoub. / **floating** ~ ‖ Schwimmdock n. ‖ dock m. flottant; cale f. flottante. / **graving** ~ see dry / **to take a ship in** ~ see **to dock a ship**.

dockage ‖ Dockgebühren fpl. ‖ droit m. de bassin.

dock apron ‖ Dockboden m. ‖ radier m. du bassin. / ~ **dues** pl. see dockage.

docked ‖ gedockt ‖ entré au bassin m. ou au dock m.

docker ‖ Hafenarbeiter m.; Dockarbeiter m. ‖ débardeur m.; docker m.

docket ‖ Faktur f. ‖ facture f.

dock gate ‖ Schleusentür f. ‖ porte f. de bassin.

docking gear (Aero) ‖ Haltegestell n. ‖ dispositif m. de retenue.

dockman ‖ Dockarbeiter m. ‖ docker m.

dock plant ‖ Dockanlage f. ‖ installation f. de bassin de radoub. / ~ **pontoon** ‖ Pontonverschluß m. für Docks ‖ ponton m. à dock. / ~ **pump** ‖ Dockpumpe f. ‖ pompe f. de docks ou à cales. / ~ **warrant** ‖ Lagerschein m. ‖ bulletin m. de dépôt; warrant m.

dockyard ‖ Schiffswerft f. ‖ chantier m. / **naval** ~ ‖ Seearsenal n. ‖ arsenal m. maritime.

dockyard crane ‖ Werftkran m. ‖ grue f. de chantier naval.

doctor (Med) ‖ Arzt m. ‖ médecin m. / ~ (Pap) ‖ Schaber m. ‖ docteur m. / ~ (Text print) ‖ Rakel f. ‖ Abstreichmesser n. ‖ racle f.; doctor m. / ~ **blade** (Text print) ‖ Rakelmesser n. ‖ lame f. de racle. / ~**'s certificate** ‖ ärztliches Attest n. ‖ certificat m. du médecin. / ~ **rule** (Text print) ‖ Rakellineal n. ‖ règle f. de la racle. / ~ **web** (Pap) ‖ Papiermaschinenschaber m. ‖ docteur m.; lame f. de docteur pour papeteries.

doctrine of equilibrium ‖ Lehre f. vom Gleichgewicht ‖ pondération f.

documentary analysis ‖ Beweisanalyse f. ‖ analyse f. documentaire. / ~ **confirmation** see ~ **proof**. / ~ **proof** ‖ Beweis

stück n.; Beleg m. ‖ pièce f. justificative; document m. justificatif.

document case ‖ Aktenmappe f. ‖ serviette f. d'avocat *ou* à dossiers / ~ conveying the grant ‖ Verleihungsurkunde f. ‖ acte m. de concession. / ~ cover ‖ Aktendeckel m. ‖ couverture f. de dossiers. / ~ envelope ‖ Aktenumschlag m. ‖ enveloppe f. à documents. / ~ file ‖ Aktenhefter m. ‖ classeur m.

dodecahedron ‖ Dodekaeder n. ‖ dodécaèdre m. / pentagonal ~ ‖ Pyritoeder n.; Pentagondodekaeder n. ‖ hémihexatétraèdre m.; dodécaèdre m. pentagonal. / rhombic ~ ‖ Rhombendodekaeder n. ‖ dodécaèdre m. rhomboïdal; rhombododécaèdre m. / trigonal ~ ‖ Pyramidentetraeder n.; Trigondodekaeder n. ‖ tétraèdre m. pyramidé *ou* pyramidal; tétratrièdre m.; hémiicositétraèdre m.

dodge, mechanical ~ ‖ Handwerkskniff m. ‖ coup m. de main; secret m.; truc m.

dodges pl. ‖ Winkelzüge mpl. ‖ détours mpl.; subterfuges mpl.

doffer (Spinn) ‖ Peignör m.; Filet m. ‖ peigneur m.

doffing cylinder (Spinn) ‖ Kammwalze f.; Abnehmer m. ‖ peigneur m. enleveur.

dog ‖ Hund m. ‖ chien m. / ~ (Build) ‖ Kropfeisen n.; Steinklaue f. ‖ louve f.; renard m. / ~ (Claw coupling) ‖ Klauenkörper m. ‖ griffe f. / ~ (Lathe) ‖ Drehherz n. ‖ toc m. de tour. / ~ (Lift) ‖ Hebezwinge f. ‖ griffe f. / ~ (Pawl) ‖ Sperrklinkenzahn m. ‖ cliquet m. / adjustable ~ (Mach tool) ‖ verstellbarer Anschlag m.; Umsteuerknagge f. ‖ taquet m. ajustable. / brake ~ (Railw) ‖ Bremssperrklinke f. ‖ cliquet m. d'arrêt du frein. / cant ~ (Wood rolling) ‖ Wendehaken m.; Kanthaken m. ‖ crochet m. pour rouler les troncs. / catching ~ (Wagon) ‖ Fangfrosch m. ‖ bloc m. d'arrêt sur rail. / clamp ~ (Tool) ‖ Spannkloben m. ‖ bloc m. *ou* griffe f. de serrage. / ~ of the coupling ‖ Kupplungsklaue f. ‖ griffe f. d'accouplement. / engaging ~ ‖ Mitnehmerklaue f. ‖ griffe f. d'entraînement. / rolling ~ see cant ~.

dog biscuit ‖ Hundekuchen m. ‖ biscuit m. *ou* gâteau m. pour chiens. / ~ cake *see* ~ biscuit. / ~ clutch (Coupling) ‖ Ausrückmuffe f. ‖ lösbare Kupplungsmuffe f. ‖ manchon m. de débrayage. / ~ collar ‖ Hundehalsband n. ‖ collier m. de chiens. / grass broom ‖ Queckenbesen m. ‖ balai m. en chiendent. / ~ hair (Spinn) ‖ Stichelhaare pl. ‖ jarre m. / ~ headed spike ‖ Schienennagel m.; Hakennagel m. ‖ crampon m. de rail. / ~ hook ‖ Klammerhaken m. ‖ griffe f. de serrage. / ~ house ‖ Hundehütte f. ‖ niche f. à chien. / ~ nail (Coachm) ‖ Kuppennagel m. ‖ caboche f. / ~ pint (Mar) ‖ Hundepünt f.; zugespitztes Tauende ‖ queue f. de rat.

dogs pl. (Nippers) ‖ Ziehzange f. ‖ pince f. à étirer. / box for ~ (Railw) ‖ Hundekasten m.; Hundekotter m. ‖ niche f. à chien.

dog spike (Railw) ‖ Schienennagel m. ‖ crampon m. de rail. / ~ sports article ‖ Hundesportartikel m. ‖ article m. de sport de chiens. / ~ stay of fire box ‖ Feuerbolzen m. ‖ cavalier m. du ciel de boîte à feu.

dogwood ‖ Kornelholz n. ‖ bois m. de cornouiller.

dokimasy *see* docimasy.

doll ‖ Puppe f. ‖ poupée f. / ~ (Railw) ‖ kurzer Signalmast m.; Weichenpfahl m. ‖ poteau m. à signal bas. / artistic ~ ‖ Künstlerpuppe f. ‖ poupée f. artistique. / celluloid ~ ‖ Zelluloidpuppe f. ‖ poupée f. en celluloïd. / china ~ ‖ Porzellanpuppe f. ‖ poupée f. en porcelaine. / cloth ~ ‖ Stoffpuppe f. ‖ poupée f. en étoffe. / ~ for cutting out of cardboard ‖ Ankleidepuppenbogen m. ‖ feuille f. avec poupées pour habiller. / dressed ~ ‖ gekleidete Puppe f. ‖ poupée f. habillée. / ~ for dressing ‖ Ankleidepuppe f. ‖ poupée f. pour habiller. / undressed ~ ‖ ungekleidete Puppe f. ‖ poupée f. non habillée.

doll dresser ‖ Puppenkleiderarbeiterin f. ‖ habilleuse f. de poupées. / ~ factory ‖ Puppenfabrik f. ‖ fabrique f. de poupées. / ~ fitter ‖ Puppenzusammensetzer m. ‖ assembleur m. en poupées. / ~ maker ‖ Puppenmacher m. ‖ fabricant m. de poupées.

doll's article ‖ Puppenartikel m. ‖ article m. pour poupées. / ~ bedstead ‖ Puppenbett n. ‖ lit m. pour poupées. / ~ carriage ‖ Puppenwagen m. ‖ voiture f. de poupée. / ~ cart see ~ carriage. / ~ eyes pl. ‖ Puppenaugen npl. ‖ yeux mpl. pour poupées. / ~ furniture ‖ Puppenmöbel n. ‖ meuble m. de poupées. / ~ head ‖ Puppenkopf m. ‖ tête f. de poupée. / ~ house ‖ Puppenhaus n. ‖ maison f. de poupées. / ~ house lamp ‖ Puppenstubenlampe f. ‖ lampe f. pour chambres de poupée. / ~ kitchen ‖ Puppenküche f. ‖ cuisine f. de poupée. / ~ outfit see ~ trousseau. / ~ perambulator see ~ carriage. / ~ room ‖ Puppenstube f. ‖ chambre f. de poupées. / ~ set ‖ Puppengarnitur f. ‖ garniture f. pour poupées. / ~ trousseau ‖ Puppenausstattung f. ‖ trousseau m. de poupées. / ~ trunk ‖ Puppenkoffer m. ‖ coffre m. de poupée. / ~ wig ‖ Puppenperücke f. ‖ perruque f. de poupées. / ~ wigger ‖ Puppenfriseur m. ‖ perruquier m. pour poupées.

dolly (Riveting) ‖ Nietkloben m.; Vorhalter m. ‖ mandrin m. d'abattage; appuyeur m.; contre-bouterolle f. / ~ tub (Mine) ‖ Rührfaß n.; Schlämmfaß n. ‖ cuve f. à délayer.

dolomite ‖ Dolomit m.; Bitterkalk m. ‖ dolomite f.; dolomie f.; chaux f. carbonatée magnésifère. / ~ brick ‖ Dolomitstein m. ‖ brique f. dolomitique *ou* de dolomie. / ~ mill ‖ Dolomitmühle f. ‖ moulin m. à dolomite. / ~ plant ‖ Dolomitanlage f. ‖ installation f. de dolomie. / ~ powder ‖ Dolomitmehl n. ‖ poudre m. de dolomite.

dome (Arch) ‖ Kuppel f.; Kuppelgewölbe n. ‖ dôme m.; voûte f. sphérique. / ~ (Boil) ‖ Dampfdom m. ‖ dôme m. de chaudière à vapeur. / ~ (Watchm) ‖ Staubdeckel m. ‖ cuvette f. / low ~ ‖ Flachkuppel f. ‖ voûte f. en cul-de-four surbaissée. / ~ of the mill. ‖ Mühlenhaube f. ‖ calotte f. du moulin / one-piece pressed ~ ‖ aus einem Blech gepreßte Domhaube f. ‖ calotte f. du dôme embouti d'une seule tôle. / ~ with reinforced flange welded-on ‖ Dom m. mit angeschweißtem verstärktem Ring ‖ dôme m. avec bride de renforcement soudée. / the ~ is secured to the boiler by rivetting through a heavy flanged collar ‖ der Dom m. ist auf den Kessel mit Hilfe eines schweren Flanschenringes

aufgenietet ‖ le dôme est rivé à la chaudière au moyen d'un solide anneau bridé. / ~ with thickened flange ‖ Haube f. mit verstärktem Flansch ‖ calotte f. avec bride renforcée. / truncated ~ (Build) ‖ oben abgebrochenes Kuppelgewölbe n. ‖ voûte en bonnet de prêtre.

dome base angle-ring ‖ Domwinkelring m. ‖ collerette f. emboutie du dôme. / welded ~ body ‖ geschweißter Dommantel m. ‖ virole f. de dome soudée. / pressed ~ cover ‖ gepreßter Domdeckel m. ‖ fond m. de dôme embouti. / ~ joint grinding machine ‖ Schleifmaschine f. für Dampfdomdichtungsflächen ‖ machine f. à rectifier les surfaces de joints des dômes de vapeur. / ~ lamp ‖ Deckenlampe f. ‖ lampe f. de plafond. / ~ nozzle ‖ Domstutzen m. ‖ tubulure f. de dôme. / ~ rivetter ‖ Domnietmaschine f. ‖ riveuse f. pour dômes. / ~ seating (Boil) ‖ Domsitz m. ‖ siège m. de dôme. / ~ -shaped summit ‖ Bergkuppe f. ‖ sommet m. en forme de dôme.

domestic ‖ inländisch ‖ indigène; intérieur. / care of ~ animals ‖ Haustierpflege f. ‖ soins mpl. à animaux domestiques. / ~ coal ‖ Hausbrandkohle f. ‖ charbon m. domestique *ou* de ménage. / ~ fuel ‖ Hausbrand m.; Hausbrandstoff m. ‖ combustible m. domestique. / ~ industry ‖ Hausindustrie f. ‖ industrie f. à domicile. / ~ machine ‖ Haushaltmaschine f. ‖ appareil m. domestique. / ~ plant ‖ Haushaltanlage f. ‖ installation f. pour usages domestiques. / ~ (servant) ‖ Dienstbote m. ‖ domestique m. / ~ smoothing-iron ‖ Haushaltbügeleisen n. ‖ fer m. à repasser de ménage. / ~ utensils pl. ‖ Hausgerät n. ‖ ustensile m. domestique. / ~ watering installation ‖ Hauswasseranlage f. ‖ installation f. d'eaux de service dans l'appartement. / ~ water pump ‖ Hauswasserpumpe f. ‖ pompe f. de cuisine. / ~ water supply ‖ Hauswasserversorgung f. ‖ adduction f. des eaux à la maison.

dome stiffening ring ‖ Domversteifungsring m. ‖ anneau m. de renforcement du dôme.

domeykite ‖ Arsenikkupfer n. ‖ domeykite f.

domiciled ‖ ansässig; wohnhaft ‖ établi; domicilié.

dominoes pl. (Game) ‖ Domino n. ‖ domino m.

donation ‖ Schenkung f. ‖ don m.; donation f.

done (Bak) ‖ gar ‖ assez cuit.

donkey (Crane) ‖ Laufkatze f. ‖ chariot m.; treuil m. roulant. / ~ (Zool) ‖ Esel m. ‖ âne m. / ~ boiler ‖ Hilfskessel m. ‖ chaudière f. auxiliaire. / ~ engine see ~ pump. / ~ pump ‖ Kesselspeisepumpe f. ‖ pompe f. alimentaire; machine f. d'alimentation; petit cheval alimentaire. / ~ winch ‖ Hilfsdampfwinde f. ‖ treuil m. à vapeur supplémentaire *ou* de secours.

doomage for retarded delivery ‖ Konventionalstrafe f. für verspätete Lieferung ‖ retenue f. à cause de retard à la livraison.

door ‖ Tür f. ‖ porte f. / ~ (Coach) ‖ Wagenschlag m.; Kutschenschlag m. ‖ portière f. de voiture. / ~ armour (Shipb) ‖ Panzertür f. ‖ porte f. blindée *ou* cuirassée. / ~ automatically closing ~ ‖ selbstschließende Tür f. ‖ porte f. fermant automatiquement. / cast iron ~ ‖ gußeiserne Tür f. ‖ porte f. en fonte. / ~ for chimneys ‖ Kamintür f. ‖

porte f. de cheminées. / dam ~ (Mine) ||
Dammtür f. || porte f. de digue *ou* de bar-
rage. / dead ~ (Arch) || blinde Tür f. || fausse
porte f. / fire-proof ~ || feuersichere Tür f.
|| porte f. à l'épreuve du feu. / four-panel-
led ~ || Kreuztür f.; Vierfüllungstür f. ||
porte f. à quatre panneaux. / framed ~ ||
eingestemmte Tür f. || porte f. encadrée. /
side ~ || Innentür f. || porte f. intérieure. /
inledged ~ || Tür f. mit aufgenagelten
Leisten || porte f. avec emboîtures
clouées. / planed ~ || gehobelte Tür f. ||
porte f. pleine *ou* rabotée. / planed and
clamped ~ || gehobelte Tür f. mit ein-
geschobenen Leisten || porte f. rabotée
ou pleine emboîtée. / ploughed and ton-
gued ~ || gespundete Tür f. || porte f.
emboîtée à rainures et languettes. / roll-
ing~ || Rolltür f. || porte f. coulissante.
/ side ~ || Seitentür f. || entrée f. latérale.
/ ~ with two leaves || Flügeltür f. || porte
f. à deux battants. / water-jacketed ~ ||
wassergekühlte Tür f. || porte f. refroidie
par circulation d'eau. / watertight ~
(Shipb) || wasserdichte Tür f. || porte f.
étanche. / wooden ~ || Holztür f. || porte
f. en bois.
door bay (Build) || Türjoch n. || baie f. de
porte. / ~ bolt || Türriegel m. || verrou m.
de porte. / ~ bolting installation || Tür-
verriegelungsanlage f. || installation f.
de verrouillage de portes. / ~ boy
(Metal) || Türzieher m. || leveur m. de por-
tes. / ~ boy (Mine) || Wettertürenwärter
m. || fermeur m. de porte. / ~ case *see*
~ casing. / ~ casing || Türfutter n.; Tür-
verkleidung f. || fourrure f.; jambage m.;
dormant m. de porte. / ~ closer || Tür-
schließer m. || ferme-porte m. / compres-
sed air ~ closer || pneumatischer Tür-
schließer m. || ferme-porte m. à air com-
primé. / pneumatic ~ closing device ||
Druckluftürschließeinrichtung f. || dis-
positif m. de ferme-porte à air com-
primé. / ~ contact interrupter || Türkon-
taktschalter m. || contact m. de porte. / ~
cushion || Türpolster n. || bourrelet m.
pour porte. / ~ fastener || Türschließer m.
|| ferme-portes m. / ~ fittings pl. || Tür-
beschläge mpl. || ferrures fpl. de portes. /
~ frame || Türrahmen m. || bâti m. de
porte. / ~-handle || Türgriff m.; Tür-
drücker m. || poignée f. de porte; clenche f.
pour portes. / ~ hole || Guckloch n.; Schau-
loch n. || ouvreau m.; espion m. / ~ iron
fitting || eiserner Türbeschlag m. || ferrure
f. de portes. / ~-keeper's room || Pfört-
nergemach n. || loge f. de concierge. /
knob || Türgriff m. || bouton m. de portes.
/ ~ knob turning machine || Türknopfdreh-
bank f. || tour m. pour boutons de porte.
/ ~ latch || Türklinke f. || clenche f. / ~
leaf || Torflügel m. || vantail m. de porte. /
~ lintel (Build) || Türkappe f.; Türsturz
m. || linteau m. de porte. / ~ lock || Tür-
schloß n. || serrure f. de porte. / ~ lock
safety device || Türschloßsicherung f. ||
dispositif m. de sûreté pour serrures de
portes. / ~ mat || Fußmatte f. || paillas-
son m. / ~ matcher || Türstocksetzer m. ||
monteur m. de portes. / ~ mountings
pl. *see* ~ fittings. / electric ~ opener ||
elektrischer Türöffner m. || ouvre-porte
m. électrique. / ~ opening || Türöffnung f.
|| baie f. de porte. / ~ panel || Türverklei-
dung f. || panneau m. de porte. / ~ plate ||
Firmenschild n. || enseigne f. / ~ post ||
Türpfosten m. || poteau m. d'huisserie;

montant m. de porte. / ~ protector ||
Türschoner m. || plaque f. de propreté
pour portes. / ~ safety appliance || Tür-
sicherung f. || bec m. de cane pour por-
tes. / ~ scraper || Fußkratze f. || décrottoir
m. / ~ sill || Türschwelle f. || seuil m. de
porte. / ~ tender (Mine) || Wettertür-
wärter m. || ouvrier m. à la manœuvre
des portes d'aérage. / ~ unlocker || Tür-
öffner m. || ouvre-portes m. / ~-way ||
Torweg m. || porte f. cochère.
dormant (Build) || Schwelle f.; Grund-
schwelle f.; Bodenschwelle f.; Grund-
balken m. || racinal m.; dormant m.;
semelle f. / ~ of a ground-floor || Lager-
schwelle f. *oder* Unterzug m. eines Fuß-
bodens || racinal m.; sole f. de plancher. /
~ bolt (Locksm) || hebende Falle f. ||
pêne m. dormant. / ~ tree of a window ||
Kämpfer m. eines Fensters || dormant m.
de croisée.
dormer *see* dormant. / ~ window || Dach-
fenster n.; Bodenfenster n. || lucarne f.
dormitory || Schlafsaal m. || dortoir m.
dorsal (Textile) || Rückenteppich m.; Wand-
behang m. || dossier m.
dose || Dosis f. || dose f.
dosed preparation || dosiertes Präparat n. ||
préparation f. dosée.
dosel *see* dorsal.
dosing machine || Dosiermaschine f.; Ab-
wiegemaschine f. || machine f. à doser. /
plant || Dosieranlage f. || installation f. de
dosage.
dosser || Kiepe f.; Tragkorb m. || hotte f.
dot, to || punktieren || pointiller.
dotted || punktiert || pointillé. / ~ line ||
punktierte Linie f. || ligne f. pointillée;
ligne f. en traits interrompus; trait m.
pointillé.
dotting (Coppersm) || Punktiermanier f. ||
gravure f. pointillée. / ~ needle (Engr) ||
Punktiernadel f.; Radiernadel f. || échoppe
f.; aiguille f. à pointiller. / ~ pen || Punk-
tierfeder f. || plume f. à pointiller. / ~
wheel || Punktierrädchen n. || roue f. à
pointiller.
double; to || verdoppeln || doubler. / ~ over
|| umkrempeln || tomber le bord.
double || doppelt || double.
double-acting || doppelwirkend || à double
effet m. / ~ brake || doppeltwirkende
Bremse f. || frein m. à double effet. / ~
compound pumping engine || doppelt-
wirkende Verbundpumpmaschine f. ||
machine f. d'épuisement compound à
double effet. / ~ engine || doppeltwirken-
der Motor m. || moteur m. à double effet.
/ ~ four-stroke cycle tandem large gas
engine || doppeltwirkende Viertaktgroß-
gasmaschine f. in Reihenanordnung ||
gros moteur à gaz à quatre temps à
double effet disposé en série.
double air cock || Doppellufthahn m. || ro-
binet m. à air double. / ~ anchor plate
for T-head bolts for sole plates || Doppel-
ankerplatte f. für Fundamentschrauben
zu Sohlplatten || double plaque f. d'ancrage
de fondation pour boulons. / ~ angle
fish plate (Railw) || Doppelwinkellasche
f. || éclisse f. cornière double. / ~ barrel
infantry accompanying gun || doppel-
rohriges Infanteriegeschütz n. || canon m.
d'infanterie à double tube. / ~ battery
switch (Acc) || Doppelzellenschalter m. ||
réducteur-adjoncteur m. double. / ~
beam frame || Doppelbalkenrahmen m. ||
bâti m. à deux paliers. / ~ beer || Doppel-

bier n. || bière f. double. / ~-bladed knife
switch || Doppelschalter m.; Doppel-
messerschalter m. || interrupteur m. bi-
polaire à lames. / ~ blow cold upsetting
machine || Doppeldruckkaltpresse f. ||
presse f. à froid à double effet. / ~ bottom
|| Doppelboden m. || double fond m. / ~
brewing plant || Doppelsudwerk n. || in-
stallation f. double de brassage. / ~
brown beer || Doppelbraunbier n. || bière
f. brune double. / ~ cableway || Doppel-
seilbahn f. || chemin m. de fer aérien
dédoublé. / ~ camera || Doppelkamera f. ||
double-chambre f. noire. / ~ card tenter
(Spinn) || Doppelkrempelführer m. || con-
ducteur m. de double-carde. / ~ cell-
switch || Doppelzellenschalter m. || ré-
ducteur m. double.
double chain || Doppelkette f. || chaîne f.
double. / ~ stitch sewing machine ||
Doppelkettenstichnähmaschine f. || ma-
chine f. à coudre à point de chaînette
double.
double cleat || Doppelklemme f. || borne f.
double *ou* à deux vis. / ~ cloth (Weav) ||
Doppelgewebe n. || matelassé n.; étoffe f.
matelassée. / ~ column press with upper
plate which can be swung outwards ||
Zweisäulenpresse f. mit ausschwenkba-
rem Oberholm || presse f. à deux colonnes
et traverse pivotante. / ~ column vertical
boring mill || Zweiständerkarusselldreh-
bank f. || tour m. vertical à deux mon-
tants. / ~ commutator || Doppelumschal-
ter m. || commutateur m. double. / ~
compressor || Doppelkompressor m. ||
compresseur m. double. / ~ cone aerial ||
Doppelkonusantenne f. || antenne f. bi-
conique. / ~ cone spring || Doppelkegel-
feder f. || ressort m. biconique.
double-conical || doppelkonisch || biconique.
/ ~ runner || doppelkonischer Läufer m. ||
meule f. biconique. / ~ connection ||
Doppelverbindung f. || connection f.
double. / ~ cord operation || Zweischnur-
betrieb m. || exploitation f. par dicorde. /
~ cover || Doppelasche f. || double couvre-
joints mpl. / ~ cross-coil aerial || Doppel-
kreuzrahmenantenne f. || double-cadre
antenne f.
double current key || Doppelstromtaste f. ||
touche f. *ou* manipulateur m. pour deux
pôles. / ~ repeater || Doppelstromüber-
tragung f. || translation f. à deux cou-
rants. / ~ working || Doppelstrombetrieb
m. || télégraphie f. à deux courants.
double curve points pl. || Kurvenweiche f.
|| changement m. en courbe. / ~ cut
(File) || Kreuzhieb m. || taille f. croisée. /
~ cut-out || Doppelausschalter m. || inter-
rupteur m. *ou* disjoncteur m. double. /
~ cutting drill || zweischneidiger Bohrer
m. || mèche f. à deux tranches.
double-cylinder engine || Zwillingsmaschine
f. || machine f. bicylindrique. / ~ teasel
rod gig (Weav) || Doppelrauhmaschine f.
für Strichrauherei || laineuse f. double
pour poil long et couché.
double dark slide (Phot) || Doppelschlitten-
kassette f. || châssis-double m. à coulisse;
châssis m. à deux chariots. / ~ diagonal
roller mill || Doppeldiagonalwalzenstuhl
m. || moulin m. à cylindres diagonale
double.
double-ended boiler || Doppelender m.;
Doppelkessel m. || chaudière f. à foyers
aux deux extrémités. / ~ spanner with
equal width of jaws || Doppelschrauben-

schlüssel m. mit gleichen Maulweiten ǁ clé f. double avec ouvertures semblables. / ~ spanner with unequal width of jaws ǁ Doppelschraubenschlüssel m. mit ungleichen Maulweiten ǁ clé f. double avec différentes ouvertures.

double-expansion engine ǁ Zweifachexpansionsmaschine f. ǁ machine f. à double expansion.

double exposure ǁ Doppelbelichtung f. ǁ exposition f. ou pose f. double.

double-faced (Textile) ǁ doppelseitig ǁ à deux faces fpl. / ~ winding machine ǁ doppelseitige Spulmaschine f. ǁ bobinoir m. à deux faces.

double flange wheel ǁ Rad n. mit zwei Spurkränzen; zweiflanschiges Rad n. ǁ roue f. à deux boudins. / ~ frequency meter (Electr) ǁ Doppelfrequenzmesser m. ǁ fréquencemètre m. à deux systèmes. / ~ frog (Railw) ǁ Doppelherzstück n. ǁ cœur m. de rail double. / ~ gain-controller ǁ Doppelschwächungswiderstand m. ǁ régulateur m. d'amplification en double. / ~ grate ǁ Doppelrost m. ǁ grille f. de foyer double. / ~ grid lamp ǁ Zweigitterröhre f. ǁ lampe f. à deux grilles. / ~ grid valve (Radio) ǁ Doppelgitterröhre f. ǁ lampe f. bigrille; tube m. à vide à deux grilles. / ~ halfround file ǁ Vogelzunge f. ǁ feuille f. de sauge. / ~-handled cheese knife ǁ Doppelgriffkäsemesser n. ǁ couteau m. à fromage à deux mains. / ~ haulage (Mine) ǁ Doppelförderung f. ǁ extraction f. double. / ~-headed rail (Railw) ǁ Doppelkopfschiene f. ǁ rail m. à champignon double. / ~ helical tooth ǁ Winkelzahn m. ǁ dent m. à chevron. / ~ high pass filter (Tel) ǁ Doppelspulenkette f. ǁ filtre m. passe haut en double. / ~ ignition ǁ Doppelzündung f. ǁ allumage m. double.

double image ǁ Doppelbild n. ǁ image f. double. / ~ micrometer ǁ Doppelbildmikrometer n. ǁ micromètre m. à image double.

double iron for cooper's plane ǁ Fügblockdoppeleisen n. ǁ fer m. avec contre-fer à colombe. / ~ jack (Tel) ǁ Doppelklinke f. ǁ jack m. double. / ~ lead-covering ǁ Doppelbleimantel m. ǁ double gaine f. en plomb. / ~ line (Electr) ǁ Doppelleitung f. ǁ ligne f. double. / ~ link ǁ Doppelbandgelenk n. ǁ maillon m. double. / ~ lock stitch ǁ Doppelsteppstich m. ǁ point m. redoublé. / ~ magazine composing machine (Print) ǁ Doppelmagazinsetzmaschine f. ǁ machine f. à composer à deux magasins. / ~ magnifier ǁ Doppellupe f. ǁ loupe f. double.

double-milling machine ǁ Doppelfräsmaschine f. ǁ fraiseuse f. à deux broches horizontales. / ~ with spindles adjustable horizontally and vertically ǁ Doppelfräsmaschine f. mit horizontal und vertikal verstellbaren Spindelstöcken ǁ machine f. à fraiser double avec réglage horizontal et vertical des arbres porte-fraises. / ~ with two main pairs of rollers ǁ Doppelzylinderwalke f. mit zwei Hauptwalzenpaaren ǁ foulon m. cylindrique double avec deux pairs de cylindres à laver. /

double packing rings pl. ǁ Doppelringpakkung f. ǁ anneaux mpl. doubles de garniture. / ~ pin (El line) ǁ Doppelstütze f. ǁ console f. double. / ~ building of a ~ pit plant ǁ Bau m. einer Doppelschachtanlage ǁ construction f. d'une installation à deux

puits. / ~ plane ǁ Doppelhobel m. ǁ rabot m. à double fer. / ~ plate wheel ǁ Doppelscheibenrad n. ǁ roue f. à double disque. / ~ plush ǁ Doppelplüsch m. ǁ double peluche f. / ~ pole ǁ Doppelmast m. ǁ poteaux mpl. jumelés. / ~ pole switch ǁ zweipoliger Schalter m. ǁ interrupteur m. bipolaire. / ~ prism with Landolts graduation ǁ Doppelprisma n. mit Landoltscher Teilung ǁ prisme m. double muni de la graduation de Landolt.

doubler (Frame) ǁ Zwirnmaschine f. ǁ métier m. à retordre. / ~ (Spinn) ǁ Zwirner m.; Twister m. ǁ retordeur m. / ~ (Worker) ǁ Dublierer m.; Facher m. ǁ doubleur m.; réunisseur m. / ~ and twister ǁ Doppelzwirner m. ǁ câbleur m. / piece ~ (Weav) ǁ Warendoppler m. ǁ dosseur m.

double-railed ǁ doppelgleisig ǁ à double voie. **double-rail line** ǁ Doppelschienengleis n. ǁ voie f. à rails doubles. / ~ rate meter ǁ Doppeltarifzähler m. ǁ compteur m. à double tarif. / ~ reception (Radio) ǁ Doppelempfang m. ǁ réception f. double. / ~ refracting crystal ǁ doppeltbrechender Kristal m. ǁ cristall m. biréfringent. / ~ refracting prism ǁ doppeltbrechendes Prisma n. ǁ prisme m. biréfringent. / ~ refraction ǁ Doppelbrechung f. ǁ réfraction f. double. / ~ seat valve ǁ Doppelsitzventil n. ǁ soupape f. à double siège. / ~ shaking screen ǁ Doppelstoßschwingsieb n. ǁ double tamis m. à secousse. / ~ shear rivet joint ǁ doppelschnittige Nietung f. ǁ rivure f. bicisaillée / ~ shell boiler ǁ doppelwandiger Kessel m. ǁ chaudière f. à double paroi. / ~ shutting lock ǁ Doppelschloß n. ǁ serrure f. bénarde.

double-sided eccentric press ǁ Doppelständerexzenterpresse f. ǁ presse f. à excentrique à double montant. / ~ overhang eccentric press ǁ ausladende Doppelständerexzenterpresse f. ǁ presse f. à excentrique à deux montants avec bâti à col de cygne.

double-sliding door ǁ Doppelflügeltür f. ǁ porte f. à deux battants.

double-sluice ǁ Doppelschütz n. ǁ double vanne f. / ~ weir ǁ Doppelschützenwehr n. ǁ barrage m. à vannes doubles.

double-sole ǁ Doppelsohle f. ǁ semelle f. double. / ~ attachment ǁ Doppelsohlenvorrichtung f. ǁ mécanisme m. à semelles doubles.

double-spoke wheel ǁ Doppelspeichenrad n. ǁ roue f. à rayons doubles. / ~ stage ǁ zweistufig ǁ à deux étages. / ~ standard plate shear ǁ Zweiständerblechtafelschere f. ǁ cisaille f. à tôles à double montant. / ~ stitch ǁ Doppelmasche f. ǁ maille f. gardée. / ~ storied boiler ǁ Etagenkessel m. ǁ chaudière f. à étages. / ~ stroke bell ǁ Doppelschlagwecker m. ǁ sonnerie f. à deux battants.

double-switch ǁ Doppelweiche f. ǁ changement m. de voie à double aiguille. / symmetrical ~ ǁ symmetrische Doppelweiche f. ǁ changement m. double à voies symétriques.

doublet (Opt) ǁ Doppellinie f. ǁ doublet m. **double-tariff** ǁ Doppeltarif m. ǁ tarif m. double. / ~ thread ǁ doppelgängiges Gewinde n. ǁ vis f. à double pas. / ~ thrilling of the wires ǁ doppelte Verdrehung f. des Drahtes ǁ rotation f. double des fils. / ~ throw switch ǁ zweipoliger Umschal-

ter m. ǁ commutateur m. bipolaire à deux directions. / ~ tyres pl. ǁ Zwillingsbereifung f. ǁ pneus mpl. jumelés / ~-T-iron ǁ Doppel-T-Eisen ǁ fer m. en T double. / ~ tool post ǁ doppelter Support m. ǁ support m. porte-outil double. / ~ track ǁ Doppelgleis n. ǁ double voie f. / ~ tube ǁ Doppelröhre f. ǁ lampe f. à deux systèmes. / ~ tube telescope ǁ Doppelfernrohr n. ǁ jumelle f. / ~-walled ǁ doppelwandig ǁ à double paroi f. / ~ way (Railw) ǁ doppelspurige Eisenbahn f. ǁ chemin m. de fer à deux voies. / ~ winch ǁ Doppelwinde f. ǁ treuil m. double. / ~ window ǁ Doppelfenster n. ǁ doublefenêtre f.; contre-fenêtre f. / ~ wire circuit (Tel) ǁ Doppelleitung f. ǁ boucle f. / ~ working repeater ǁ doppeltwirkende Übertragung f. ǁ translation f. à deux directions.

doubling (Shipb) ǁ Dopplung f. ǁ (Spinn) ǁ Dublieren n. ǁ doublage m.; réunissage m.; doublage m. / ~ of planks (Shipb) ǁ Plankenverdopplung f. ǁ soufflage m. / ~ of plates (Shipb) ǁ Plattenverdopplung f. ǁ doublure f. de tôles. / ~ of a sail ǁ Saum m. oder Saumstreifen m. des Segels ǁ gaine f. de voile.

doubling machine ǁ Dubliermaschine f. ǁ machine f. à doubler. / plate ~ ǁ Blechdoppler m. ǁ plieuse f. de tôles.

douche ǁ Brausebad n. ǁ douche f.

dough, to ~ in (Brew) ǁ einteigen; einmaischen ǁ empâter.

dough ǁ Teig m. ǁ pâte f. / the ~ raises (Bak) ǁ der Teig geht auf ǁ la pâte se lève. / silky, light and well-formed ~ ǁ lockerer luftiger Teig m. ǁ pâte f. légère et volumineuse. / to work out the ~ ǁ den Teig m. verarbeiten ǁ travailler la pâte f.

dough beating machine ǁ Teigschlagmaschine f. ǁ machine f. à battre la pâte. / ~ dividing machine ǁ Teigteilmaschine f. ǁ machine f. à diviser la pâte; diviseur m. de pâte. / ~ dividing and moulding machine ǁ Teigteil- und Teigwirkmaschine f. ǁ diviseuse-rouleuse f. ou diviseuse-batteuse f. de pâtes.

doughing-in (Brew) ǁ Einteigen n.; Einmaischen n. ǁ empâtage m. / ~ temperature ǁ Einmaischtemperatur f. ǁ température f. d'empâtage.

dough kneading machine ǁ Teigknetmaschine f. ǁ machine f. à pétrir la pâte.

doughmaker ǁ Kneter m. ǁ pétrisseur m.; gindre m.; mitron m.

dough mixer ǁ Knetmaschine f. ǁ pétrin m. mécanique. / ~ mixing machine ǁ Teigmischmaschine f. ǁ machine f. à malaxer ou mélanger la pâte. / ~ moulding machine ǁ Teigwirkmaschine f. ǁ machine f. à rouler la pâte; rouleur m. de la pâte. / ~ roller ǁ Teigwalze f. ǁ rouleau m. à pâte. / ~ stirring machine ǁ Teigrührmaschine f. ǁ machine f. à remuer la pâte.

doughy ǁ teigig ǁ pâteux. / ~ mass ǁ teigige Masse f. ǁ pâte f.

douse, to ~ a sail ǁ ein Segel n. streichen ǁ amener une voile.

dovetail, to ǁ zinken ǁ assembler à queue d'aronde.

dovetail ǁ Schwalbenschwanz m. ǁ queue f. d'aronde. / common ~ ǁ offener Schwalbenschwanz m. ǁ queue f. d'aronde percée. / false ~ ǁ umgekehrter Schwalbenschwanz m. ǁ contre-queue f. d'aronde. / lapped ~ ǁ verdeckter Schwalbenschwanz m. ǁ queue f. d'aronde recouverte.

/mitred~‖versenkter Schwalbenschwanz m. ‖ queue f. d'aronde perdue. / ordinary ~ see common ~.

dovetailed ‖ schwalbenschwanzförmig ‖ à queue f. d'aronde. / ~ groove ‖ schwalbenschwanzförmige Nut f. ‖ rainure f. à queue d'aronde. / ~ guide ‖ Schwalbenschwanzführung f. ‖ glissière f. à queue d'aronde.

dovetail file ‖ Schwalbenschwanzfeile f. ‖ lime f. à queue d'aronde. / ~ hole ‖ Schwalbenschwanzeinschnitt m. ‖ entaille f. en queue d'aronde; aronde f. / ~ indent ‖ schwalbenschwanzförmiger Zahn m. ‖ adent m. à queue d'aronde.

dovetailing ‖Schwalbenschwanzverzapfung f.; Zusammenzinken n. ‖ assemblage m. à queue d'aronde. / common ~ ‖ offene Schwalbenschwanzverbindung f. ‖ assemblage m. à queue d'aronde ordinaire. / lapped ~ ‖ verdeckte Schwalbenschwanzverbindung f. ‖ assemblage m. à queue d'aronde recouverte. / mitred ~ ‖ versenkte Schwalbenschwanzverbindung f. ‖ assemblage m. à queue d'aronde perdue.

dovetailing machine ‖ Zinkenmaschine f. ‖ machine f. à faire les tenons en queue d'aronde. / ~ plane ‖ Zinkenhobel m. ‖ bouvet m. à languette.

dovetail jag ‖ Schwalbenschwanzblatt n. ‖ entaille f. d'aronde. / ~ slot ‖ schwalbenschwanzförmige Nute f. ‖ entaille f. en queue d'aronde. / ~ wire ‖ Schwalbenschwanzdraht m. ‖ fil m. à queue d'aronde.

dowel, to ‖ zusammendübeln; verdübeln ‖ cheviller; goujonner.

dowel ‖ Dübel m.; Holzpflock m. ‖ tenon m.; goujon m.; cheville f. / ~ of beech wood ‖ Buchenholzdübel m. ‖ tampon m. de hêtre. / iron ~ ‖ Eisendübel m. ‖ goujon m. de fer. / knob ~ ‖ Knopfdübel m. ‖ goujon m. à bouton. / lead ~ ‖ Bleidübel m. ‖ goujon m. de plomb. / spiral ~ ‖ Spiraldübel m. ‖ goujon m. à spirale. / steel ~ ‖ Stahldübel m. ‖ goujon m. d'acier. / wall ~ ‖ Mauerdübel m. ‖ goujon m. mural. / wooden ~ ‖ Holzdübel m. ‖ cheville f. en bois.

dowel borer ‖ Dübelbohrer m. ‖ tarière f. à goujon.

doweled ‖ gedübelt ‖ chevillé; assemblé par chevilles.

dowel hole ‖ Dübelloch n. ‖ trou m. de goujon.

dowel(l)ing ‖ Verdübeln n.; Verdübelung f. ‖ assemblage m. avec clefs ou goujons; chevillage m. / ~ pin ‖ Dübel m.; Holzdübel m.; Zylinderzapfen m. ‖ goujon m.; tampon m. cylindrique.

dowel machine ‖ Dübelmaschine f. ‖ machine f. à faire les chevilles. / ~ pin (Mach) ‖ Prisonstift m. ‖ cheville f. de répérage. / ~ saw ‖ Zapfensäge f. ‖ scie f. à chevilles. / ~ socket hammer ‖ Dübellochhammer m. ‖ marteau m. pour trous à cheville.

dowlas (Weav) ‖ Kreas m.; Lederleinwand f. ‖ crès f.; crée f.

down ‖ hinab ‖ vers le bas. / breaking ~ (Mine) ‖ Zubruchgehen n. ‖ éboulement m. / up and ~ ‖ auf und ab ‖ mouvement m. montant et descendant.

down stroke of the piston ‖ Kolbenniedergang m. ‖ descente f. du piston. / ~ train ‖ abwärts fahrender Zug m. ‖ train m. descendant. / ~-and up-train ‖ Pendelzug m. ‖ train m. d'aller et retour.

down (Bird) ‖ Daune f. ‖ duvet m. / ~ (Geol) ‖ Düne f. ‖ dune f. / ~ for beds ‖ Bettdaune f. ‖ duvet m. de lit. / ~ glued upon tissue ‖ auf Gewebe aufgeklebtes Flaumhaar n. ‖ duvet m. collé sur tissus. / vegetable ~ ‖ Pflanzendaune f. ‖ duvet m. végétal.

downcast of a mine ‖ einziehender Schacht m. ‖ puits m. descendant.

downcomer (Boil) ‖ Fallröhre f. ‖ tube m. descendant.

down crops pl. ‖ auf dem Halm durch Unwetter umgelegtes Getreide n. ‖ récolte f. versée. / ~ defence ‖ Dünenschutzwerk n. ‖ défense f. des dunes. / ~ feather ‖ Flaumfeder f. ‖ plume f. à duvet. / ~ grade ‖ Gefälle n. ‖ pente f. / ~ hair ‖ Flaumhaar n. ‖ poil m. follet. / ~ picker ‖ Federreißer m. ‖ trieur m. de plumes. / ~ pipe ‖ Abfallrohr n. ‖ tuyau m. de descente ou de gouttières.

downpour ‖ Platzregen m. ‖ averse f.

downward current ‖ Abwind m. ‖ vent m. descendant; courant m. d'air descendant. / region of ~ currents ‖ Abwindfeld n. ‖ champ m. de vent descendant. / dynamic ~ force ‖ dynamische Sinkkraft f. ‖ force f. descensionnelle dynamique. / ~ journey ‖ Niederfahrt f. ‖ descente f. / ~ movement ‖ Rückgang m. ‖ ralentissement m. / ~ voyage ‖ Talfahrt f. ‖ voyage m. en aval.

downwards ‖ abwärts ‖ vers le bas; en aval.

downwash ‖ Abwind m. ‖ vent m. descendant; courant m. d'air descendant.

dowser, metal ‖ Wünschelrute f. aus Metall ‖ baguette f. divinatoire en métal.

draff ‖ Brauereitreber pl. ‖ drèche f. de brasserie. / to rerake the ~ ‖ Treber pl. aufhacken ‖ piocher ou piquer les drèches fpl. / to sparge the ~ ‖ Treber pl. anschwänzen ‖ arroser ou laver les drèches fpl.

draft see draught.

drag, to ‖ schleifen; schleppen ‖ traîner. / ~ the collector (Electr) ‖ den Stromabnehmer m. nachschleppen ‖ remorquer le dispositif de prise de courant. / the brake blocks drag on the wheels ‖ die Bremsklötze mpl. schleifen an den Rädern ‖ les sabots mpl. frottent sur les roues.

drag (Aero) ‖ Rücktrieb m. ‖ dérive f. / ~ (Carr) ‖ Hemmzeug n.; Hemmvorrichtung f. ‖ enrayement m.; enrayage m.; enrayure f.; dispositif m. de freinage. / ~ (Dredger) ‖ Baggerschaufel f. ‖ drague f. / ~ (Sawn) ‖ Klotzwagen m. ‖ chariot m. à bois. / hand ~ ‖ Handbagger m. ‖ drague f. à main.

dragbar (A) (Agr mach) ‖ Deichsel f. ‖ faux-timon m. / ~ (Loc) ‖ Kupplungsstange f. ‖ barre f. d'attelage. / ~ (Bridge) ‖ Hängestange f. ‖ tige f. de suspension.

drag bolt ‖ Kupplungsbolzen m. ‖ boulon m. d'attelage. / ~ chain ‖ Kupplungskette f. ‖ chaîne f. d'attelage. / ~ chain (Carr) ‖ Hemmkette f. ‖ chaîne f. d'enrayure ou d'enrayage. / ~ coefficient ‖ Widerstandskoeffizient m. ‖ coefficient m. de traînée. / ~ flask (Found) ‖ Unterkasten m. einer Gießform ‖ châssis m. inférieur.

dragger-out (Roll mill) ‖ Paketierofenmann m. ‖ paqueteur m.

dragging bucket ‖ Schleppschaufel f. ‖ benne f. traîneuse ou piocheuse ou pre-

neuse. / capacity of ~ ‖ Schleppschaufelinhalt m. ‖ capacité f. de la benne traîneuse ou piocheuse ou preneuse.

drag hook ‖ Kuppelhaken m. ‖ crochet m. de la barre d'attelage; crochet m. d'attelage.

drag line ‖ Schleppleine f. ‖ câble m. ou cordage m. ou aussière f. de remorque. / ~ (Dredger) ‖ Schleppschaufelbagger m.; Dragline f. ‖ pelle-dragline f. / ~ bucket excavator ‖ Schürfkübelkabelbagger m. ‖ drague f. sèche à benne et à câble. / ~ equipment ‖ Greifereinrichtung f.; Schleppschaufeleinrichtung f. ‖ équipement m. de benne preneuse ou de drag line. / winch of the ~ for pulling in the ropes ‖ Einziehwinde f. des Schleppschaufelbaggers ‖ treuil m. de halage de la pelle-drag line.

drag net ‖ Schleppnetz n.; Kratzgarn n. ‖ drège f.

dragon piece (Carp) ‖ Stichbalken m. ‖ blochet m.

dragon's blood ‖ Drachenblut n. ‖ sang-dragon m.

drag rope ‖ Schleppseil n. ‖ guide-rope m.; amarre f. de touage. / ~ slide cable ‖ Schleppseil n. ‖ câble m. entraîné. / ~ spring ‖ Kupplungsstangenfeder f. ‖ ressort m. de la barre d'attelage. / ~ staff of a cart ‖ Bergstütze f.; Hemmstütze f. ‖ servante f. d'un chariot. / ~ turf ‖ Baggertorf m. ‖ tourbe f. extraite à la drague.

drain, to ‖ entwässern ‖ drainer; dessécher; égoutter. / ~ a cylinder ‖ einen Zylinder entwässern ‖ purger un cylindre. / ~ a mine by means of an adit ‖ die Wasser npl. einer Grube durch einen Stollen lösen ‖ assécher une mine par une galerie d'écoulement. / ~ off ‖ entleeren ‖ vidanger; vider; dégonfler. / to let mine water drain off ‖ das Grubenwasser n. versickern lassen ‖ laisser les eaux fpl. de mine se perdre par infiltration. / ~ off swamps ‖ entsumpfen ‖ dessécher des marais mpl. / ~ a piece of swampy or bog-land ‖ ein versumpftes Grundstück n. entwässern ‖ assécher un terrain marécageux. / ~ a pit ‖ einen Schacht m. trockenlegen ‖ mettre un puits à sec. / ~ a pond ‖ einen Teich m. austrocknen ‖ mettre un étang à sec. / ~ a seam ‖ ein Flöz n. lösen ‖ saigner ou démerger une veine.

drain (Agr) ‖ Drain m. ‖ drain m.; rigole f. souterraine. / ~ (Build) ‖ Abzugskanal m.; Gerinne n. ‖ drain m.; égout m.; rigole f. / ~ (Found) ‖ Rinne f. ‖ canal m. de coulage; chenal m.; écheneau m. / ~ (Hydr arch) ‖ Entwässerungsrohr n.; Sickerrohr n. ‖ drain m. / ~ (Mine) ‖ Senkgrube f.; Senkloch n. ‖ puisard m.; puits m. absorbant; égougeoir m. / ~ (Money) ‖ Abfluß m. ‖ sortie f. / ~ (Road) ‖ Straßenrinne f.; Gosse f. ‖ rigole f. ou canal m. de pavé; ruisseau m. ou égout m. de rue. / catch-water ~ (Build) ‖ Sickerkanal m. ‖ canal m. découlement ou d'infiltration. / main-~ ‖ Hauptentwässerungsrohr n. ‖ égout m. collecteur. / side ~ (Railw) ‖ Bahngraben m. ‖ contre-fossé m. / ~ covered by slabs ‖ Plattendurchlaß m. ‖ aquéduc m. couvert de dalles.

drainage (Agr) ‖ Dränage f.; Entwässerung f.; Trockenlegung f. ‖ vidange f.; égouttage m.; drainage m. / ~ (Brew) ‖

Abläuterung f. ‖ filtration f. / ~ (Hydr arch) ‖ Trockenlegung f. ‖ saignée f.; desséchement m. / ~ (Pipe) ‖ Entwässerung f. ‖ purge f. / ~ of cable conduits ‖ Dränrohranlage f. an Kabelkanälen ‖ drainage m. des conduites. / ~ of surface (Railw) ‖ Trockenlegung f. des Planums ‖ assainissement m. de plate-forme. / ~ of the trench ‖ Trockenlegung f. der Baugrube ‖ asséchement m. de la fouille.

drainage area ‖ Niederschlagsgebiet n. ‖ bassin m. / ~ area of a river ‖ Flußgebiet n. ‖ bassin m. d'un fleuve. / ~ canal ‖ Hauptentwässerungsgraben m. ‖ fossé m. principal de desséchement ou d'assainissement. / ~ channel ‖ Entwässerungskanal m. ‖ canal m. de desséchement. / ~ coil (Radio) ‖ Entladespule f. ‖ bobine f. de décharge. / to disturb existing ~ conditions ‖ die Vorflut f. stören ‖ arrêter la dérivation naturelle de l'eau. / ~ ditch ‖ Entwässerungsgraben m. ‖ fossé m. d'écoulement. / ~ pipe ‖ Dränröhre f.; Entwässerungsröhre f. ‖ tuyau m. de drainage. / ~ sieve ‖ Entwässerungssieb n. ‖ tamis m. de drainage ou d'égouttage. / ~ sluice ‖ Entwässerungsschleuse f. ‖ écluse f. d'assèchement. / ~ system ‖ Entwässerungsanlage f. ‖ installation f. de drainage; fossé m. d'écoulement. / ~ and irrigating-works pl. ‖ Meliorierung f.; Ent- und Bewässerung f. ‖ amélioration f.

drain channel ‖ Abflußkanal m.; Abzugkanal m. ‖ canal m. d'écoulement. / ~ implement for ~ cleansing ‖ Kanalreinigungsgerät n. ‖ ustensile m. pour le nettoyage des conduits de drainage.

drain cock ‖ Ablaßhahn m.; Entleerungshahn m. ‖ robinet m. de vidange ou de purge. / ~ for condensed water ‖ Kondenswasserablaßhahn m. ‖ robinet m. de vidange d'eau de condensation. / ~ for emptying heaters ‖ Ablaßhahn m. für die Entleerung der Heizanlage ‖ robinet m. de décharge pour évacuation de chauffage. / oil-pan ~ (Motor) ‖ Ölablaßhahn m. ‖ robinet m. de vidange du carter.

drained ‖ ausgeleert ‖ drainé.

drainerman (Pap) ‖ Bleichergehilfe m. ‖ homme m. de fosse.

draining see also drainage.

draining ‖ Dränarbeit f. ‖ entreprise f. de drainage. / ~ (Chem) ‖ Abnutschen n.; Absaugen n. ‖ essorage m. / ~ of the foundation ground ‖ Entwässerung f. des Baugrundes ‖ drainage m. du terrain à bâtir. / ~ of mines (Mine) ‖ Wasserhaltung f. ‖ épuisement m. des eaux.

draining apparatus ‖ Entwässerungsapparat m. ‖ appareil m. d'égouttage. / ~ box for bottles ‖ Flaschenauslaufkasten m. ‖ panier-égouttoir m. de bouteilles. / ~ channel ‖ Entwässerungskanal m. ‖ couloir m. de drainage. / ~ dish ‖ Abtropfschale f. ‖ passoire f.; vase f. à égoutter; égouttoir m. / ~ ditch ‖ Entwässerungsgraben m.; Abzugsgraben m. ‖ fossé m. d'écoulement. / ~ elevator ‖ Entwässerungsbecherwerk n. ‖ noria f. d'épuisement. / ~ engine ‖ Entwässerungsmotor m. ‖ moteur m. d'épuisement. / ~ machine (Chem) ‖ Ausschleudermaschine f. ‖ essoreuse f. / ~ pipe ‖ Dränröhre f. ‖ tuyau m. de drainage ou de desséchement ou de conduite. / soil ~ plant ‖ Bodenentwässerungsanlage f. ‖ instal-

lation f. de drainage des terrains. / ~ plough ‖ Rinnenpflug m. ‖ charrue f. à rigole. / ~ screen ‖ Tropfsieb n. ‖ crible m. d'égouttage. / ~ sieve ‖ Entwässerungssieb n. ‖ tamis m. d'égouttage. / ~ tank ‖ Entwässerungsbehälter m. ‖ réservoir m. d'égouttage. / ~ tower ‖ Entwässerungsturm m. ‖ tour f. d'égouttage. / ~ truck for bottles ‖ Flaschenauslaufgestell n. ‖ égouttoir m. pour bouteilles. / ~ well (Build) ‖ Senkgrube f.; Senkloch n. ‖ puisard m.; puits m. absorbant; égougeoir m.

drain metal (Found) ‖ Gerinnstücke npl.; Metallrückstände in der Gußrinne ‖ échenaux mpl.; résidu m. de métal.

drainpipe ‖ Dränröhre f.; Entwässerungsröhre f. ‖ tuyau m. de drainage ou de dessèchement.

drain plug ‖ Ablaßstopfen m. ‖ bouchon m. de vidange. / oil pan ~ (Motor) ‖ Ölablaßstopfen m. ‖ bouchon m. de vidange du carter inférieur.

drain rinsing apparatus ‖ Kanalspüler m. ‖ appareil m. à rincer les conduits de drainage. / ~ tile ‖ Abzugsziegel m.; Dränziegel m. ‖ brique f. à drainage. / ~ trap ‖ Abflußrohrkniestück n.; Traps m. ‖ siphon m. / ~ valve ‖ Ablaßventil n.; Entwässerungsventil n. ‖ purgeur m.; soupape f. de vidange. / acidulous ~ water ‖ säurehaltiges Abwasser n. ‖ eau f. vanne acide.

draisine ‖ Draisine f. ‖ draisienne f.

draper ‖ Weißwarenhändler m. ‖ linger m.

drapery ‖ Modewaren fpl. ‖ modes fpl.; nouveautés fpl.

draught (Air) ‖ Luftzug m. ‖ vent m.; courant m. d'air. / ~ (Bill) ‖ Tratte f. ‖ traite f.; disposition f. / ~ (Carp) ‖ volle Kante f. ‖ arête f. vive. / ~ (Furnace) ‖ Zug m. ‖ tirage m.; appel m.; vent m.; air m. / ~ (Haulage) ‖ Zug m. ‖ traction f. / ~ (Payment) ‖ Anweisung f. ‖ assignation f.; délégation f. / ~ (Shipb) ‖ Tiefgang m. ‖ tirant m. d'eau. / ~ (Weav) ‖ Passage f.; Fadeneinzug m. ‖ passage m.; remettage m. / artificial ~ ‖ künstlicher Zug m. ‖ tirage m. artificiel. / bad ~ ‖ schlechter Zug m. ‖ mauvais tirage m. / ~ in a chimney ‖ Schornsteinzug m. ‖ tirage m. d'une cheminée. / diamond ~ (Weav) ‖ pointierte Passage f. ‖ passage m. à point; remettage m. à retour. / to draw a ~ ‖ einen Wechsel m. ausstellen ‖ émettre une traite. / first ~ ‖ Urschrift f. ‖ original m. / forced ~ ‖ künstlicher Zug m. ‖ tirage m. forcé. / ~ of furrows ‖ Zug m. der Furchen oder Rillen oder Nuten ‖ excentricité f. ou chasse f. des rayons. / geometrical ~ ‖ Grundriß m.; Plan m.; Riß m. ‖ plan m.; tracé m.; projection f. horizontale. / induced ~ ‖ Saugzug m. ‖ tirage m. induit ou par aspiration. / light ~ (Mar) ‖ Tiefgang m. des leeren Schiffes ‖ tirant m. d'eau lège. / with light ~ ‖ leichtzügig ‖ à traction f. légère; de traction f. douce. / load ~ (Mar) ‖ Tiefgang m. des beladenen Schiffes ‖ tirant m. d'eau en charge. / mean ~ (Mar) ‖ mittlerer Tiefgang m. ‖ tirant m. d'eau moyen. / natural ~ ‖ natürlicher Luftzug m. ‖ tirage m. naturel; courant m. d'air naturel. / on ~ ‖ vom Faß n. ‖ en fût m.; en perce f. / to put on ~ (Beer) ‖ ausschenken ‖ débiter m. / ~ of a ship ‖ Tiefgang m. eines Schiffes ‖ tirant m. d'eau d'un navire. / ~ of a stove ‖ Luft-

zug m. oder Zug m. eines Ofens ‖ appel m. ou tirage m. d'un foyer. / ~ by steam jet ‖ durch Dampfstrahlgebläse n. erzeugter Zug ‖ tirage m. à jet de vapeur. / ~ of traverse ‖ Polygonzug m. ‖ tracé m. polygonal ou de polygones. / underground ~ (Survey) ‖ Zug m. unter Tage ‖ levé m. de plan au fond.

draught agreement ‖ Vertragsentwurf m. ‖ projet m. de contrat. / export ~ beer ‖ Ausfuhrfaßbier n. ‖ bière f. d'exportation en fûts. / ~ board ‖ Dambrett n. ‖ damier m. / ~ cattle ‖ Zugtier n. ‖ bête f. de trait. / ~ check plate ‖ Zugabsperrklappe f. ‖ registre m. d'arrêt de tirage. / forced ~ cooling ‖ Druckluftkühlung f. ‖ refroidissement m. à courant d'air forcé. / induced ~ fan ‖ Saugzugventilator m. ‖ ventilateuraspirateur m. pour tirage induit. / ~ form ‖ Wechselformular n. ‖ formulaire m. de traite. / ~ furnace ‖ Zugofen m.; Windofen m. ‖ fourneau m. à vent. / forced ~ furnace ‖ Unterwindfeuerung f. ‖ foyer m. à soufflage sous grille. / ~ gauge ‖ Zugmesser m.; Flutometer n. ‖ indicateur m. de tirage; flutomètre m. / recording ~ gauge ‖ registrierender Zugmesser m. ‖ enregistreur m. de tirage. / ~ horse ‖ Zugpferd n. ‖ cheval m. de trait. / ~ indicator see ~ gauge.

draughting machine ‖ Zeichenmaschine f. ‖ machine f. à dessiner.

draught machine for the soil-tilling ‖ Bodenbearbeitungszuggerät n. ‖ machines fpl. à attelage pour cultiver le sol. / ~ mark ‖ Gedingezeichen n. ‖ témoin m. de travail. / ~ plough ‖ Gespannpflug m. ‖ charrue f. attelée. / ~ preventer ‖ Fensterdichter m. ‖ bourrelet m. de fenêtre. / ~ regulator ‖ Zugregler m. ‖ régulateur m. de tirage.

draughts pl. of a ship ‖ Schiffspläne mpl. ‖ plans mpl. d'un navire.

draughtsman ‖ Zeichner m. ‖ dessinateur m. / ~ (Weav) ‖ Musterzeichner m.; Webereizeichner m. ‖ dessinateur m. sur tissus. / professional ~ ‖ Berufszeichner m. ‖ dessinateur m. professionnel.

draw, to (Design) ‖ zeichnen; aufzeichnen ‖ dessiner. / ~ (Forg) ‖ strecken ‖ marteler; étirer. / ~ (Metal) ‖ ziehen ‖ étirer; tréfiler. / ~ (Mine) ‖ fördern; gewinnen ‖ extraire. / ~ (Textile) ‖ verziehen; strecken ‖ étirer; laminer. / ~ a bill ‖ eine Tratte f. ausstellen ‖ émettre une traite. / ~ a cable ‖ ein Kabel n. ziehen ‖ tirer ou poser un câble. / to cold-draw ‖ kaltziehen ‖ étirer à froid. / to cold-draw with the hammer ‖ kaltschmieden ‖ battre à froid; écrouir. / ~ the dimensions into a design ‖ die Maße npl. in eine Zeichnung einschreiben ‖ coter un dessin. / ~ down iron (Forg) ‖ strecken; recken ‖ étirer le fer sous le marteau. / ~ down while cold ‖ kaltstrecken ‖ marteler à froid. / ~ a draft or draught ‖ einen Wechsel m. ausstellen ‖ émettre une traite. / ~ the fires (Boil) ‖ die Feuer npl. herausziehen ‖ mettre bas les feux mpl. / ~ in ‖ einzeichnen ‖ dessiner dans; marquer. / ~ with Indian ink ‖ mit Tusche f. zeichnen; dessiner au lavis m. / ~ on litho-stone ‖ lithografieren ‖ lithographier. / ~ off (Air) ‖ ansaugen ‖ aspirer. / ~ off (Brew) ‖ abfüllen ‖ soutirer. / ~ off (Mas) ‖ abziehen ‖ aplanir. / ~ off (Textile) ‖ abziehen ‖ tirer. / ~ off (Water) ‖ absaugen ‖ épuiser; essorer; aspirer; pomper.

/ ~ off the crank shaft || von der Kurbelwelle f. abziehen || enlever de l'arbre m. de manivelle. / ~ off the deposit || den Schlamm m. ablassen || décharger la boue des chaudières. / ~ off the oil || das Öl ablassen || faire écouler l'huile. / ~ out (Forg) || ausschmieden; recken || allonger; battre; forger; marteler. / ~ out (Mine) || fördern || extraire. / ~ out the furnace || den Hochofen m. auskratzen || décrasser le haut-fourneau. / ~ out gravel || Kies m. baggern || draguer le gravier. / ~ out iron by rolls || Eisen n. auswalzen || étirer le fer aux cylindres de laminoir. / ~ a perpendicular line || ein Lot n. fällen || abaisser une perpendiculaire. / ~ a pile || einen Pfahl m. ausziehen || retirer un pieu. / ~ a pond || einen Teich m. austrocknen || mettre à sec un étang. / ~ to scale || maßstäblich zeichnen || dessiner à l'échelle f. / ~ by sight || nach dem Augenmaß n. zeichnen || dessiner à vue f. / ~ tubes || Rohre ziehen || étirer des tuyaux mpl. / ~ up || ausfertigen; entwerfen || dresser; rédiger. / ~ up an account || eine Rechnung f. ausschreiben || dresser un compte. / ~ up a contract || einen Vertrag m. entwerfen || élaborer ou rédiger un contrat. / ~ up a telegram || ein Telegramm n. abfassen || rédiger un télégramme. / ~ up water || Wasser n. schöpfen || puiser de l'eau f. / ~ water (Mar) || Tiefgang m. haben || avoir un tirant d'eau. / ~ wire || Draht m. ziehen || tréfiler; étirer le fil.

draw, normal ~ and buffer appliances || normale Zug- und Stoßvorrichtung f. || appareil m. de choc et de traction du type normal.

drawback || Übelstand m. || inconvénient m.; désavantage m. / ~ (Found) || Kernstück n.; Keilstück n. || pièce f. rapportée ou de rapport; rapport m.; tiroir m. / ~ (Trade) || Rückzoll m. || draw-back m.; prime f. de réexportation.

drawbarrier || Zugbarriere f. || barrière f. à distance.

drawbar (Agr mach) || Deichsel f. || fauxtimon m. / ~ (Haulage) || Zugstange f. || barre f. d'attelage ou de traction. / ~ (Loc) || Kupplungsstange f. || barre f. d'attelage. / continuous ~ (Railw) || durchgehende Zugstange f. || barre f. d'attelage continue.

drawbeam (Hydr arch) || Schwengel m.; Hebebalken m. || bascule f.; flèche f.; balancier m. / ~ (Mach) || Haspelwelle f. || arbre m. d'un guindeau ou d'un treuil. / ~ of a drawbridge || Schwengel m. oder Wippe f. einer Aufzugbrücke || bascule f. à fléau; flèche f. / ~ of a well || Brunnenschwengel m. || levier m. ou balancier m. d'une pompe.

drawbench (Wiredr) || Drahtziehbank f. || banc m. à tréfiler ou à étirer; argue f. / ~ for rods || Stangenziehbank f. || banc m. à étirer les barres. / ~ for wire drawing || Drahtzug m. || banc m. à tréfiler; banc m. de tréfilerie.

drawbolt || Kupplungsbolzen m. || boulon m. d'attelage.

drawboy (Mine) || Hundsläufer m. || rouleur m. de chiens. / ~ (Weav) || Aufzieher m. || tireur m.

drawbridge || Zugbrücke f. || pont-levis m. / ~ with chains || Zugbrücke f. mit Ketten || pont-levis m. à chaînes.

draw-cord for opening lens cap || Zugschnur f. zum Aufklappen des Objektivdeckels || cordon m. de tirage pour relever le couvercle de l'objectif.

drawer (Design) || Zeichner m. || dessinateur m. / ~ (Join) || Schubfach n.; Schublade f. || tiroir m.; layette f. / ~ (Metal) || Zieher m. || étireur m. tréfileur m. / ~ (Spinn) || Strecker m. || étireur m.; étirageur m. / ~ (Trade) || Trassant m. || tireur m.; émetteur m. / ~ (Weav) || Musterzeichner m. || dessinateur m. industriel. / each of these adding devices is combined with a ~ || jedes dieser Addierwerke npl. ist mit einer Schublade verbunden || chacun de ces totalisateurs est combiné avec un tiroir. / automatic ~ (Cash reg) || selbsttätige Schublade f. || tiroir m. automatique. / working at open ~ (Cash reg) || Arbeiten n. bei offener Schublade || marche f. à tiroir ouvert. / spike ~ (Railw) || Nagelklaue f. || pince f. à pied-de-biche.

drawer pull || Muschelgriff m. || coquille f.

drawers pl. || Unterhose f. || caleçon m.

drawgear || Zugvorrichtung f. || appareil m. de traction. / ~ lying at the same level with the buffers (Railw) || die Zugvorrichtung f. liegt in gleicher Höhe wie die Puffer || l'appareil m. de traction est disposé à la même hauteur que les tampons. / rigid ~ || feste Zugvorrichtung f. || attelage m. rigide.

drawhead (Railw) || Zugstangenkopf m. || tête f. de la barre d'attelage. / ~ of broaching machine (Mine) || Zugorgan n. der Räummaschine || pièce f. d'attelage de la machine à brocher.

drawhole (Metal) || Ziehloch n. || trou m. de filière.

drawhook (Railw) || Zughaken m. || crochet m. de traction ou d'attelage. / elastic ~ (Railw) || federnder Zughaken m. || crochet m. de traction à ressort. / rigid ~ (Railw) || fester Zughaken m. || crochet m. de traction rigide. / the ~ is arranged to turn about a vertical bolt (Railw) || der Zughaken ist um einen senkrechten Bolzen drehbar || le crochet de traction pivote autour d'un boulon vertical.

drawhook guide (Railw) || Zughakenführung f. || guide m. de crochet de traction.

drawing (Design) || Zeichnung f. || dessin m. / ~ (Mine) || Förderung f. || extraction f. / ~ (Textile) || Strecken n.; Verziehen n.; étirage m.; laminage m. / ~ (Weav) || Musterzeichnung f. || dessin m. / additional ~s (Trade) || Mehreinnahme f. || excédent m. de recette. / art ~ || künstlerische Zeichnung f. || dessin m. artistique. / ~ of cables into pipes || Einziehen n. von Leitungen in Rohre || tirage m. de conducteurs dans les tubes. / ~ in a chimney || Schornsteinzug m. || tirage m. d'une cheminée. / China-ink ~ || Tuschzeichnung f. || dessin m. lavé. / constructional ~ || Konstruktionszeichnung f. || plan m. de construction. / copying ~s || Kopieren n. von Zeichnungen || copie f. des dessins mpl. / ~ cylindrically || Zylindrischziehen n. || étirage m. cylindrique. / dabbed ~ || gewischte Zeichnung f. || dessin m. à l'estompe. / detail ~ || Zeichnung f. von Einzelteilen; Detailzeichnung f. || dessin m. de détails; tracé m. en détail. / dimensioned ~ || Maßzeichnung f. || plan m. coté. / enlarging ~s || Vergrößerung f. von Zeichnungen || agrandissement m. de dessins. / erection ~ || Montagezeichnung f. || dessin m. de montage. / general ~ || Zusammenstellungszeichnung f.; Übersichtszeichnung f. || croquis m. d'assemblage; dessin m. d'ensemble. / geometrical ~ || geometrische Zeichnung f. || dessin m. géométrique. / hatched ~ || schraffierte Zeichnung f. || dessin m. hachuré. / to ink-in a ~ || eine Zeichnung f. ausziehen || passer un dessin à l'encre. / ~ of irregularly formed hollow parts || Ziehen n. unregelmäßig geformter Hohlkörper || étirage m. des corps creux de forme irrégulière. / lead-pencil ~ || Bleistiftzeichnung f. || dessin m. au crayon. / lithographic ~ || lithografierte Zeichnung f. || dessin m. lithographié. / marginal ~s || Randzeichnungen fpl. || dessins mpl. sur le bord. / mechanical ~ || technische Zeichnung f. || dessin m. technique. / metal ~ || Metallzieherei f. || étirage m. de métaux. / outline ~ || Strichzeichnung f. || dessin m. à la plume. / pencil ~ || Handzeichnung f. || dessin m. à la plume. / perspective ~ || perspektivische Zeichnung f. || dessin m. perspectif. / precision ~ || Präzisionszieherei f. || étirage m. de précision. / preliminary ~ || Entwurfszeichnung f. || croquis m. / pricked ~ (Paint) || Schablone f.; durchstochenes Muster n. || poncis m.; poncif m. / propaganda ~ || Werbezeichnung f. || dessin m. de réclame. / reducing ~s || Verkleinerung f. von Zeichnungen || réduction f. des dessins. / ~ upon stone || Steinzeichnung f. || dessin m. lithographié. / washed ~ || getuschte Zeichnung f. || dessin m. au lavis; feuille f. lavée. / ~ in water colours see washed ~. / working ~ || Werkzeichnung f. || dessin m. d'atelier.

drawing apparatus || Zeichenapparat m. || appareil m. à dessiner. / ~ awl (Saddl) || Riemenahle f. || alêne f. à bredir ou à passer les lanières; passe-corde m. / ~ back || Fuß m. oder Grundlinie f. einer Böschung || reculement m. d'un talus.

drawing bench || Drahtziehbank f.; Drahtzug m. || banc m. à étirer; banc m. de tréfilerie; argue f. / bar ~ || Stangenziehbank f. || banc m. à étirer les barres. / ~ for thick wires || Grobzug m. || banc m. de tréfilerie à gros fil.

drawing board || Zeichenbrett n.; Reißbrett n. || planche f. ou planchette f. à dessin. / ~ book || Zeichenbuch n. || livre m. à dessiner.

drawing chalk || Zeichenkreide f. || craie f. à dessiner. / coloured ~ || farbige Zeichenkreide f. || craie f. à dessin de couleur. / white ~ || weiße Zeichenkreide f. || craie f. à dessin blanche.

drawing charcoal || Zeichenkohle f. || fusain m.; charbon m. à dessin. / ~ compasses pl. || Reißzirkel m.; Steckzirkel m. || compas m. à pointes de rechange. / ~ depth || Ziehtiefe f. || profondeur f. d'emboutissage. / pneumatic ~ device || pneumatischer Ziehapparat m. || appareil m. d'emboutissage pneumatique. / ~ down of iron || Recken n. des Eisens || étirage m. au marteau. / ~ engine (Mine) || Fördermaschine f. || machine f. ou moteur m. d'extraction. / ~ form for cardboard || Ziehform f. für Pappe || forme f. d'étirage pour carton.

drawing frame (Mine) || Förderkorb m.; Fördergestell n.; Förderschale f. || cage f. d'extraction. / ~ (Spinn) || Streckwerk n.; Vorspinnmaschine f. || banc m. d'é-

tirage. / ~ with bobbins (Spinn) || Spulen-strecke f. || banc m. d'étirage à bobines.

drawing-in (Cable) || Einziehen n. || tirage m. / cables ~ || Einziehen n. von Kabeln || tirage m. des câbles. / warp ~ (Weav) || Ketteneinziehen n. || remettage m. de chaînes.

drawing-in box || Einziehdose f. || boîte f. de tirage. / ~ rope || Zugseil n. || corde f. de tirage. / ~ wire || Ziehdraht m. || fil m. de tirage.

drawing ink || Zeichentinte f.; Tusche f. || encre f. à dessiner ou de Chine. / liquid ~ || Ausziehtusche f.; flüssige Tusche f. || encre f. de Chine; encre f. liquide à dessin.

drawing instrument || Zeicheninstrument n.; Zeichengerät n. || instrument m. à dessin. / case of ~s || Reißzeugkasten m. || étui m. de mathématiques. / ~s pl. in wood || Zeichenutensilien fpl. aus Holz || instruments mpl. de dessin en bois.

drawing knife (Coop) || Reifmesser m.; Ziehmesser n.; Küfermesser n. || plane f. / ~ knife (Farrier) || Hufmesser n. || rogne-pied m.; couteau m. du maréchal. / ~ loft (Shipb) || Schnürboden m. || salle f. de gabarits. / ~ machine || Zeichenmaschine f. || machine f. à dessiner. / machine for reproduction of ~s || Zeichnungskopiermaschine f. || machine f. à reproduire les plans. / ~ machine for wires || Ziehmaschine f. für Drähte || tréfilerie f. pour fils métalliques. / ~ master || Zeichenlehrer m. || maître m. à dessiner. / ~ material || Zeichenmaterial n. || ustensiles mpl. pour le dessin. / ~ mill || Drahtzieherei f. || tréfilerie f.; tirerie f. / mistake in the ~ || Konstruktionsfehler m. || vice f. de construction.

drawing-off bung (Brew) || Abfüllspund m. || bondon m. à écoulement. / ~ cock || Abzapfhahn m. || robinet m. de soutirage. / ~ funnel || Abfülltrichter m. || entonnoir m. de soutirage. / ~ gas || Abgas n. || gaz m. d'échappement. / ~ machine (Brew) || Abfüllmaschine f. || machine f. de soutirage ou à écoulement ou à soutirer; machine f. à remplir les bouteilles.

drawing office || Zeichenbüro n.; Zeichensaal m. || bureau m. de dessin.

drawing-out (Mine) || Förderung f. || extraction f. / ~ blow || Reckschlag m. || coup m. pour travail d'étirage. / ~ device (Roll mill) || Auszieher m. || défourneur m.; extracteur m.

drawing paper || Zeichenpapier n. || papier m. à dessin. / ~ in rolls || Zeichenpapier n. in Rollen || papier m. à dessin en rouleaux.

drawing pattern || Zeichenvorlage f. || modèle m. de dessin. / ~ pen || Zeichenfeder f.; Reißfeder f. || plume f. à dessin; tire-ligne m. / ~ pencil || Zeichenstift m. || crayon m. à dessin. / ~ pin || Reißnagel m.; Reißstift m.; Heftzwecke f. || punaise f. / ~ plant || Ziehanlage f. || installation d'étirage. / ~ plate (Wiredr) || Zieheisen n. || filière f. (à étirer); plaque f. / ~ pliers pl. (Wiredr) || Schleppzange f. || tenaille f. continue. / ~ point || Reißnadel f. || pointe f. à tracer. / ~ prism || Zeichenprisma n. || prisme m. à dessiner. / ~ press || Ziehpresse f. || presse f. à étirer. / ~ rod || Zugstange f. || tringle f. de traction / ~ roller (Spinn) || Streckwalze f.; Verzugswalze f. || étireur-laminoir m.; cylindre-étireur m. / ~ -room see ~ office / ~rope || Manntausendseil n. || corde f. de main

douce. / ~ rule || Reißschiene f. || règle f. ou T m à dessiner.

drawings pl. **of fermentation** || Abseihbier n. || fonds mpl. de cuves ou de foudres.

drawing school || Zeichenschule f. || école f. de dessin. / ~ scraper (Coop) || Zugschaber m. || racloir m. à deux manches. / ~ set || Reißzeug n. || boîte f. de compas. / ~ shaft || Förderschacht m. || puits m. d'extraction. / ~ stage || Zeichentisch m. || pupitre m. à dessiner. / ~ stand || Zeichengestell n. || chevalet m. à dessin. / wire ~ stone || Drahtziehstein m. || pierre f. à tréfiler. / ~ table || Zeichentisch m. || table f. à dessiner. / ~ table with board || Zeichentisch m. mit Brett || table f. à dessin avec planche. / ~ tongs pl. || Kniehebelklemme f. || patte f. de grenouille. / ~ tool || Ziehwerkzeug n. || outil m. à étirer ou à tréfiler. / ~ up (Clockm) || Aufzug m. || remontage m. / ~ up (Mine) || Förderung f. || extraction f. / ~ up (Weav) || Einzug m. || rentrée f. du chariot. / ~ utensils pl. || Zeichenutensilien npl. || articles mpl. de dessin. / ~ wheel || Schöpfrad n. || roue f. de puisage. / accurate wire ~ works pl. || Genaudrahtzieherei f.; Präzisionsdrahtzieherei f. || tréfilerie f. de précision.

draw-in roller || Einzugswalze f. || rouleau m. inférieur.

drawknife (Coop) || Ziehmesser n.; Zugmesser n.; Schnitzmesser n. || plane f.; couteau m. à deux manches.

drawline Ziehleine f. einer Ramme || tiraude f.

drawn (Wire) || gezogen || étiré. / hollow ~ article || gezogener Hohlkörper m. || pièce f. creuse étiré sans soudure. / ~ axle || gezogene Achse f. || essieu m. en acier étiré. / seamless ~ tube || nahtlos gezogene Röhre f. || tube m. étiré sans soudure. / solid ~ axle || gezogene Achse f. || essieu m. étiré.

drawn-in system (Electr) || Verlegung f. in Isolierrohr || système m. sous tubes.

drawn iron wire || gezogener Eisendraht m. || fil m. de fer tréfilé. / ~ piece || Ziehteil m. || pièce f. étirée. / ~ steel wire || gezogener Stahldraht m. || fil m. d'acier tréfilé. / ~ tube || gezogene Röhre f. || tube m. étiré sans soudure / ~ wire || gezogener Draht m. || fil m. tréfilé ou étiré. / ~ work installation || Ziehwerksanlage f. || installation f. d'atelier d'étirage.

drawnet || Hamen m.; Senke f. || ableret m.; ablerette f.

draw-off roll || Abzugwelle f. || roulette f. de tirage.

draw-out radiator || einziehbarer Kühler m. || radiateur m. amovible ou escamotable.

drawplate || Drahtzieheisen n. || filière f. de tréfilerie ou étirage. / ~ steam baking oven || Auszugsdampfbackofen m. || four m. à vapeur amovible ou avec chariot sortant.

drawshave see drawknife.

drawspring || Zugfeder f. || ressort m. de traction.

drawtube (Opt) || Ausziehtubus m. || tube m. à tirage.

drawvice || Spannschraube f. || vis f. de tension.

draw well || Ziehbrunnen m. || puits m. à roue. / ~ wire barrier || von der Station betätigte Zugschranke f. || barrière f. à bascule manœuvrée à distance.

dray || Rollwagen m. || camion m.; chariot m. à bords. / ~ ladder || Schrotleiter f. || poulain m.

dreadnaught (Cloth) || Fries m.; Flaus m. || frise f.

dredge, to || baggern || draguer. / ~ a harbour || einen Hafen ausbaggern || draguer un port.

dredge (Agr) || Mengkorn n. || méteil m. / ~ (Fish) || Schleppnetz n. || drague f. / ~ boat || Baggerschiff n. || bateau-dragueur m.; bateau-rabot m. / ~ bucket || Baggereimer m.; godet m. de drague. / ~ bucket ladder || Baggereimerleiter f. || élinde f. de drague. / ~ captain || Baggerschiffsmeister m. || capitaine m. de bateau-drague.

dredge chain || Baggerkette f. || chaîne f. dragueuse.

dredged peat || Baggertorf m. || tourbe f. draguée. / ~ net || Schleppnetz n.; || drège f.

dredger || Bagger m.; Baggermaschine f. || drague f. / ~ see dredgerman. / bucket ~ || Eimerbagger m. || drague f. à godets / dry ~ || Trockenbagger m. || drague f. sèche. / ~ using the guide roller system || mit Führungsrollen versehener Bagger m. || excavateur m. pourvu de galets de guidage. / sand pump ~ || Sandpumpenbagger m. || dragueur m. à pompes de sable. / steam ~ Dampfbagger m. || drague f. à vapeur. / suction ~ for emptying barges || Schutensauger m. || aspirateur m. pour gabares. / ~ transporting the ballast || Prahmbagger m. || dragueurporteur m.

dredger bucket || Baggereimer m. || godet m. de drague. / ~ drum || Turas m. || tambour m. de drague. / ~ joint pin || Baggerbolzen m. || boulon m. de drague. / ~ ladle || Baggerlöffel m. || cuiller f. de drague.

dredgerman || Baggerarbeiter m.; Baggerer m. || dragueur m.; ouvrier m. dragueur.

dredger tooth || Baggerzahn m. || griffe f. pour dragues.

dredges pl. || Dragees fpl.; Zuckerwerk n. || dragées fpl.

dredging || Baggerung f.; Baggerei f. || curage m.; dragage m. / ~ boat || Baggerschiff n. || bateau m. dragueur. / ring for ~ bolt || Baggerbolzenring m. || anneau m. pour boulons à dragues. / ring for ~ bolts forged of manganese steel || aus Manganstahl geschmiedeter Ring m. für Baggerbolzen || anneau m. pour boulons à dragues forgé en acier au manganèse. / ~ bucket see dredger bucket. / knife for ~ bucket || Baggereimermesser n. || bec m. de godet. / ~ machine || Bagger m.; Baggermaschine f. || machine f. à draguer; drague f. / ~ roll || Turas m. || rouleau m. à drague. / ~ service || Baggerbetrieb m. || service m. de dragues. / ~ tumbler || Baggertrommel f.; Kettentreibscheibe f.; Turas m. || tambour m. de drague.

dregs pl. || Bodensatz m.; Bodenhefe f. || lie f. / ~ of oil || Ölhefe f. || lie f. d'huile. / ~ of tar || Teersatz m. || lie f. ou sédiment m. de goudron.

dress, to (Curr) || gerben || tanner; corroyer. / ~ (Mach) || ausrichten; richten || dresser; redresser. / ~ (Weav) || zurichten || appareiller. / ~ cloth || das Tuch rauhen || garnir; lainer. / ~ flax || Flachs m. hecheln || racler le lin. / ~ hides pl. || gerben || corroyer. / ~ leather || das Leder zurichten || corroyer le cuir. / ~ ores ||

Erz n. aufbereiten ‖ traiter un minerai. / ~ the outer surface ‖ die Außenfläche verputzen ‖ enduire la surface extérieure. / ~ the points (Print) ‖ Punktur f. zurichten ‖ poser les pointures fpl. / ~ a quarrystone ‖ einen Bruchstein m. bossieren *oder* behauen ‖ piquer un moellon. / ~ skins by the application of oil ‖ sämisch gerben ‖ chamoiser; passer en chamois. / ~ a timber with the twibil ‖ einen Balken m. mit der Queraxt abputzen ‖ dresser et aviver une poutre avec la besaiguë. / ~ the warp (Weav) ‖ die Kette schlichten ‖ parer *ou* encoller la chaîne.

dress ‖ Kleidung f. ‖ habit m.; habillement m.; confection f. / fancy ~ ‖ Maskenkostüm n. ‖ costume m. de bal masqué. / ~ of huntsmen ‖ Jägerkleidung f. ‖ habillement m. de chasseurs. / night ~ ‖ Nachtkleid n.; Nachtrock m. ‖ robe f. de nuit. / workmen's ~es pl. ‖ Arbeiterkleidung f. ‖ habillements mpl. d'ouvrier.

dress designer ‖ Kostümzeichner m. ‖ dessinateur m. de costumes.

dressed products pl. ‖ aufbereitetes Gut n. ‖ matières fpl. préparées.

dresser (Furniture) ‖ Anrichte f. ‖ dressoir m. / ~ (Mas) ‖ Putzer m. ‖ pareur m. / ~ (Ore dress) ‖ Pocharbeiter m. ‖ bocardeur m.

dress guard ‖ Kleiderschützer m. ‖ gardejupe m. / ~ hook ‖ Kleiderhaken m. ‖ porte-manteaux m.

dressing (Build) ‖ Verkleidung f. ‖ revêtement m. / ~ (Clothm) ‖ Appretieren n. ‖ apprêt m. / ~ (Metal) ‖ Aufbereitung f. ‖ préparation f. mécanique; traitement m.; manutention f. / ~ (Millstone) ‖ Schärfung f. ‖ rhabillage m. / ~ (Sheet iron) ‖ Richten n. ‖ dressage m. / ~ (Weav) ‖ Schlichte f. ‖ parement m.; encollage m. / door ~ ‖ Türbekleidung f. ‖ chambranle m. / fermentation ~ ‖ Gärungsbeschleunigungsmittel n. ‖ addition f. faite pour accélérer une fermentation / ~ of floor ‖ Abhobeln n. des Parketts ‖ aplanissage m. de parquets. / ~ ores ‖ Erzaufbereitung f.; Aufbereitung f. von Erzen ‖ préparation f. *ou* traitement m. des minerais. / ~ of ores on the wet method ‖ nasses Verfahren n. bei der Erzaufbereitung ‖ traitement m. des minerais par voie humide. / pit coal ~ ‖ Steinkohlenaufbereitung f. ‖ préparation f. de la houille. / ~ of skins ‖ Fellzurichterei f.; Zurichten n. von Fellen ‖ peausserie f. / ~ with slabs see ~ with tables. / ~ with tables (Build) ‖ Verblendung f. *oder* Vertäfelung f. mit Platten ‖ lambrissage m. / warp ~ (Weav) ‖ Kettenschlichten n. ‖ dressage m. de chaînes. / wet ~ ‖ nasse Aufbereitung f. ‖ traitement m. par voie humide.

dressing apparatus ‖ Schlichtapparat m. ‖ pareur m. mécanique. / ~ brush (Weav) ‖ Schlichtbürste f. ‖ brosse f. à parer. / ~ case ‖ Necessairetäschchen n.; Reisenecessaire n. ‖ sac m. de toilette. / ~ case (Med) ‖ Verbandskasten m. ‖ armoire f. pour les premiers secours *ou* à pansement. / ~ chisel (Carp) ‖ Balleneisen n.; Schrotmeißel m. ‖ ébauchoir m.; fermoir m. / ~ glue ‖ Appreturleim m. ‖ colle f. pour apprêts. / ~ gown ‖ Schlafrock m. ‖ robe f. de chambre. / ~ hammer (Metal) ‖ Pritschhammer m.; Abrichthammer m. ‖ marteau m. de parage. / ~ hammer (Pav) ‖ Zurichtehammer m. ‖ épinçoir m. / ~

industry ‖ Aufbereitungsindustrie f. ‖ industrie f. de préparation des matières premières.

dressing machine (Chem) ‖ Aufbereitungsmaschine f. ‖ machine f. à préparer. / ~ (Clothm) ‖ Appreturmaschine f. ‖ machine f. d'apprêtage. / ~ (Mill) ‖ Beutel m.; Beutelei f. ‖ bluterie f.; bluteau m.; blutoir m. / ~ (Ores) ‖ Aufbereitungsmaschine f. ‖ machine f. pour le traitement des minerais. / ~ (Weav) ‖ Leimmaschine f. ‖ encolleuse f.; colloir m. / cotton ~ ‖ Baumwollvorbereitungsmaschine f. ‖ machine f. à préparer le coton. / ~ for permanent way materials ‖ Aufbereitungsmaschine f. für Eisenbahnoberbaustoffe ‖ machine f. à préparer les matériaux de superstructure de chemin de fer. / wool ~ ‖ Wollaufbereitungsmaschine f. ‖ machine f. à préparer de la laine.

dressing and sizing machine (Weav) ‖ Kettenschlichtmaschine f. ‖ pareuse-encolleuse f.

dressing plant (Ores) ‖ Aufbereitungsanlage f. ‖ installation f. de préparation *ou* de triage. / coal ~ ‖ Kohleaufbereitungsanlage f. ‖ installation f. de préparation de charbon. / lignite ~ ‖ Braunkohlenaufbereitungsanlage f. ‖ installation f. de préparation de lignite. / moulding sand ~ ‖ Formsandaufbereitungsanlage f. ‖ installation f. de préparation de sable de moulage. / ore ~ ‖ Erzaufbereitungsanlage f. ‖ installation f. de préparation de minerais. / rag ~ (Pap) ‖ Holländer m. ‖ installation f. pour la préparation des chiffons. / sand ~ ‖ Sandaufbereitungsanlage f. ‖ installation f. de préparation du sable. / silicon ~ ‖ Silikaaufbereitungsanlage f. ‖ installation f. de préparation de la silice.

dressing plate (Railw) ‖ Schienenrichtplatte f. ‖ table f. en fonte pour le dressage des rails. / ~ room ‖ Ankleidezimmer n. ‖ garde-robe f. / cleaning and ~ shop for castings ‖ Gußputzerei f. ‖ atelier m. de nettoyage et d'ébarbage. / ~ size ‖ Appretur f. ‖ Klebestoff m. ‖ encollage m.; parement m.; apprêt m. / ~ stake (Tinm) ‖ Spannstock m. ‖ tas m. à dresser. / ~ style ‖ Tracht f. ‖ costume m.; mode f.; charge f. / ~ table ‖ Putztisch m. ‖ toilette f. / ~ technic ‖ Aufbereitungstechnik f. ‖ technique f. du traitement. / ~ tool ‖ Richtwerkzeug n.; Abrichtwerkzeug n. ‖ outil m. à dresser.

dressmaker's scissors ‖ Schneiderinnenschere f. ‖ ciseaux mpl. de couturière. / ~ smoothing-iron ‖ Konfektionsbügeleisen n. ‖ fer m. à repasser pour la confection.

dressmaking ‖ Bekleidungsgewerbe n. ‖ industrie f. du vêtement.

dress manufacture of ready-made articles ‖ Manufakturwaren fpl. ‖ articles mpl. manufacturés. / ~ material ‖ Kleiderstoff m. ‖ étoffe f. pour costumes. / ~ material embroidery ‖ Stickerei f. für Kostüme ‖ broderie f. de costumes. / ~ preserver ‖ Schweißblatt n. ‖ sous-bras m.; dessous m. de bras. / ~ shield see ~ preserver. / ~ shirt ‖ Oberhemd n. ‖ chemise f. de jour.

dried getrocknet ‖ étuvé; séché. / ~ blood ‖ getrocknetes Blut n.; Blutmehl n. ‖ sang m. séché. / ~ blood for manure ‖ Blutdünger m. ‖ sang m. desséché pour

engrais. / ~ fruits pl. ‖ Dörrobst n. ‖ fruits mpl. desséchés. / ~ vegetables ‖ Dörrgemüse n. ‖ légume m. sec *ou* séché.

drier (Mach) ‖ Trockengerät n. ‖ dessécheuse f.; séchoir m.; sécheur m. / ~ (Paint) ‖ Sikkativ n. ‖ siccatif m. / ~ (Spinn) ‖ Aufhänger m. ‖ empercheur m. / centrifugal ~ ‖ Trockenschleuder f. ‖ essoreuse f. / paint and varnish ~ ‖ Sikkativ n. ‖ siccatif m. / rotary ~ ‖ Trockentrommel f. ‖ séchoir m. rotatif. / ~ with single roller ‖ Einwalzentrockner m. ‖ sécheur m. à cylindre unique.

drift, to (Mach) ‖ aufreiben ‖ étamper. / ~ (Nav) ‖ triften; abgetrieben werden; versetzen ‖ flotter; drosser. / ~ a hole (Locksm) ‖ ein Loch n. ausdornen *oder* aufreiben ‖ étamper *ou* aléser un trou.

drift (Aero; Nav) ‖ Abtrift f.; Abtrieb m.; Abdrängung f. ‖ dérive f. / ~ (Agr) ‖ Trift f. ‖ pâturage m. / ~ (Forg) ‖ Lochhammer m. ‖ chasse f. à percer; poinçon m. / ~ (Locksm) ‖ Reibahle f.; Dorn m. ‖ étampe f.; mandrin m.; alésoir m. / ~ (Mine) ‖ Sohlenstrecke f. ‖ galerie f. d'allongement; chassage m. / ~ (Pyrot) ‖ Setzer m.; Treibstock m. ‖ baguette f. à charger. / ~ of an arch ‖ Seitenschub m. eines Bogens ‖ poussée f. horizontale d'un arc. / to enlarge holes pl. with a ~ ‖ Löcher npl. aufdornen ‖ mandriner des trous mpl. / hollow ~ ‖ Hohlsetzer m.; Hohlstempel m. ‖ baguette f. creuse *ou* percée. / horizontal ~ of a vault ‖ Seitenschub m. eines Gewölbes ‖ effort m. *ou* poussée f. d'une voûte. / in the ~ (Mine) ‖ vor Ort m. ‖ à la veine; à front m. de taille. / water level ~ (Mine) ‖ Grundstrecke f. ‖ chasse f. *ou* voie f. de fond.

drift current ‖ Driftströmung f. ‖ dérive f. de la mer.

drifter (Mine) ‖ Querschlaghauer m. ‖ bacneur m.; bouveleur m.

drift fishery see ~ net fishing. / ~ ice ‖ Drifteis n. ‖ glace f. flottante; glaces fpl. en dérive. / ~ indicator ‖ Abdrängungsmesser m.; Abdriftmesser m. ‖ dérivomètre m.

drifting of snow ‖ Schneetreiben n. ‖ tourmente f. *ou* tourbillons mpl. de neige.

drift keel ‖ Schlingerkiel m.; Kimmkiel m.; Seitenkiel m. ‖ quille f. latérale *ou* auxiliaire *ou* de dérive. / ~ mining (Mine) ‖ Stollenbetrieb m. ‖ exploitation f. par galeries. / ~ net ‖ Treibnetz n. ‖ filet m. traînant; caurantille f. / ~ net fishing ‖ Treibnetzfischerei f. ‖ pêche f. aux filets trainants *ou* aux caurantilles. / ~ sail ‖ Treibsegel n. ‖ voile f. flottante. / ~ sand ‖ Flugsand m. ‖ sable m. emporté par le vent; sable m. mouvant. / ~ wood ‖ Treibholz n. ‖ bois m. flottant.

drill, to (Agr) ‖ säen ‖ semer. / ~ (Mach) ‖ bohren ‖ forer; percer. / ~ finished holes ‖ Löcher npl. nachbohren ‖ reforer *ou* repercer des trous.

drill (Agr) ‖ Furche f.; Rille f. ‖ rainure f.; rigole f. / ~ (Agr mach) ‖ Sämaschine f. ‖ machine f. à semer; semoir m.; sembrador m. / ~ (Clothm) ‖ Drell m.; Drillich m. ‖ treillis m. / ~ (Mach) ‖ Drillbohrer m. ‖ mèche f.; foret m. / core hammer ~ ‖ Kernbohrhammer m. ‖ marteau m. à retirer les noyaux. / corn ~ ‖ Sämaschine f. ‖ semoir m. / diamond ~ ‖ Diamantbohrmaschine f. ‖ perforateur m. à diamants. / double-cutting ~ ‖ zweischneidiger Bohrer m. ‖ mèche f. à deux tran-

chants. / electric hand ~ ‖ Elektrohandbohrmaschine f. ‖ perceuse f. électrique à main. / guano ~ see manure ~. / manure ~ ‖ Düngerstreumaschine f. ‖ machine f. à épandre les engrais ou le guano. / metal ~ ‖ Metallbohrer m. ‖ foret m. à métaux. / multiple-spindle ~ ‖ mehrspindlige Bohrmaschine f. ‖ machine f. à percer à broches multiples. / one-meter ~ ‖ Einmeterbohrer m. ‖ foret m. d'un mètre de long. / pin ~ ‖ Zapfenbohrer m. ‖ tarière f.; mèche f. à tenon. / pneumatic ~ ‖ Preßluftbohrhammer m. ‖ marteau-foreur m. à air comprimé. / portable ~ ‖ fahrbare Bohrmaschine f. ‖ machine f. à percer roulante. / radial ~ ‖ Radialbohrmaschine f. ‖ machine f. à percer radiale. / rock ~ ‖ Gesteinsbohrer m. ‖ perforateur m. de roche. / ~ for rock drilling ‖ Bohrstahl m. für Gesteinsbohrungen ‖ acier m. de tarière pour le forage de roches. / sensitive ~ ‖ Schnellbohrmaschine f. ‖ perceuse f. rapide. / spiral ~ ‖ Schneckenbohrer m. ‖ tarière f. à spirale ou en hélice; tarière f. tordue ou torse. / twist ~ ‖ Spiralbohrer m. ‖ foret m. hélicoïdale. / wall ~ ‖ Wandbohrmaschine f. ‖ foreuse f. murale.

drill borer ‖ Drillbohrer m. ‖ drille f. / ~ bow ‖ Drillbogen m. ‖ archet m.; archelet m. / ~ box (Agr) ‖ Saatkasten m. der Sämaschine ‖ boîte f. du semoir. / ~ carriage ‖ Bohrschlitten m. ‖ chariot m. porte-broche. / ~ charger (Mine) ‖ Lademeister m. ‖ chargeur m. de mines. / ~ chuck ‖ Bohrfutter n. ‖ manchon m. porte-mèches.

drilled ‖ gebohrt ‖ foré.

driller (Mach) ‖ Bohrmaschine f. ‖ machine f. à percer ou à forer; perceuse f. / ~ (Workman) ‖ Bohrer m. ‖ perceur m.

drillet ‖ Knopper f.; Gallapfel m. ‖ gallon m.

drill gauge ‖ Bohrlehre f. ‖ calibre m. de perçage. / ~ grinding machine ‖ Bohrerschleifmaschine f. ‖ machine f. à affûter les forets. / rock ~ hammer (Mine) ‖ Bohrhammer m. ‖ marteau-perforateur m. / multiple-spindle ~ head ‖ Mehrspindelbohrkopf m. ‖ mandrin m. d'alésage à forets multiples.

drilling of the bore hole out of the hard steel blocks ‖ Ausbohren n. der Löcher aus harten Stahlblöcken ‖ forage m. de trous dans un bloc d'acier d'une grande dureté. / heavy duty ~ work ‖ schwere Bohrarbeit f. ‖ gros travail m. de perçage.

drilling apparatus ‖ Bohrapparat m. ‖ appareil m. à percer; perceuse f. / ~ capacity ‖ Bohrleistung f. ‖ capacité f. de perçage. / ~ frame ‖ Bohrgerüst n. ‖ chevalet m. à forer.

drilling hammer ‖ Bohrhammer m. ‖ marteau-foreur m. / pneumatic ~ ‖ Preßluftbohrhammer m. ‖ marteau-foreur m. à air comprimé. / rock ~ ‖ Gesteinsbohrhammer m. ‖ marteau-foreur m. pour roches.

drilling machine ‖ Bohrmaschine f. ‖ machine f. à percer ou à forer; foreuse f.; perceuse f. / ~ for boilers ‖ Kesselbohrmaschine f. ‖ machine f. à percer les tôles be chaudières. / compressed air ~ ‖ Preßluftbohrmaschine f. ‖ perceuse f. à air comprimé. / five-spindle ~ ‖ Fünfspindelbohrmaschine f. ‖ perceuse f. à cinq broches. / high-speed ~ ‖ Schnellbohrmaschine f. ‖ machine f. à percer

rapide. / ~ for metal working ‖ Bohrmaschine f. für die Metallbearbeitung ‖ perceuse f. pour métaux. / multiple-~ ‖ Vielspindelbohrmaschine f. ‖ perceuse f. multiple ou à forets multiples. / percussion ~ ‖ Stoßbohrmaschine f. ‖ perforatrice f. mortaiseuse. / ~ for driving pin holes ‖ Mitnehmerlöcherbohrmaschine f. ‖ perceuse f. pour trous d'entraînement. / precision high-speed ~ ‖ Genauigkeitsschnellbohrmaschine f. ‖ perceuse f. rapide de précision. / radial ~ ‖ Auslegerbohrmaschine f. ‖ machine f. à percer radiale à broche centrale; perceuse f. radiale. / rock ~ ‖ Gesteinsbohrmaschine f. ‖ foreuse f. pour roches. / rotary coal ~ ‖ Kohlendrehbohrmaschine f. ‖ perceuse f. tournante à houille. / single purpose ~ ‖ Bohrmaschine f. für Sonderzwecke ‖ machine f. à percer spéciale. / two-column ~ ‖ Doppelsäulenbohrmaschine f. ‖ perceuse f. à deux montants. / upright ~ ‖ Säulenbohrmaschine f. ‖ machine f. à percer verticale. / ~ for wood ‖ Bohrmaschine f. für Holz ‖ perceuse f. pour bois.

drilling machine table for testing ‖ Versuchsbohrtisch m. ‖ table f. pour essais de perçage sur les foreuses. / ~ worker ‖ Bohrmaschinenarbeiter m. ‖ perceur m. à la machine.

drilling plant ‖ Bohranlage f. ‖ installation f. de foreuses. / ~ tools pl. ‖ Bohrzeug n. ‖ outils mpl. de forage.

drill machine (Agr) ‖ Drillmaschine f.; Sämaschine f. ‖ semoir m. en lignes. / ~ plough see ~ machine. / ~ press ‖ Säulenbohrmaschine f. ‖ foreuse f. à colonne.

drills pl. (Weav) ‖ Drell m. ‖ treillis m.; linge m. ouvré.

drill, set of ~s ‖ Satz m. Bohrer ‖ jeu m. de mèches. / ~ sharpening machine ‖ Bohrerschärfmaschine f. ‖ machine f. à aiguiser les forets. / rock ~ sharpening machine ‖ Gesteinsbohrerschärfmaschine f. ‖ machine f. à aiguiser les forets à roches. / ~ smith ‖ Bohrschmied m. ‖ forgeur m. de mèches.

drill spindle, distance from centre of ~ to column ‖ Bohrspindelausladung f. ‖ distance f. de l'axe de la broche au bâti. / travel of ~ ‖ Bohrspindelhub m. ‖ course f. de la broche. / ~ guide ‖ Bohrspindelführung f. ‖ guide m. de la broche de perçage.

drill upsetting machine ‖ Bohrerstauchmaschine f. ‖ machine f. à refouler les forets. / rock ~ ‖ Gesteinsbohrerstauchmaschine f. ‖ machine f. à refouler les forets à roches.

drink ‖ Getränk n. ‖ boisson f. / fermented ~ ‖ gegorenes Getränk n. ‖ boisson f. fermentée. / non-alcoholic ~ ‖ alkoholfreies Getränk n. ‖ boisson f. non-alcoolique ou anti-alcoolique ou sans alcool. / not spirituous ~ see non-alcoholic ~.

drinkable ‖ trinkbar ‖ potable. / ~ water ‖ Trinkwasser n. ‖ eau f. potable ou buvable.

drink delivery apparatus ‖ Getränkeautomat m. ‖ distributeur m. de boissons.

drinking glass ‖ Trinkglas n. ‖ verre m. à boire. / ~ pot ‖ Bierkrug m. ‖ cruche f. à bière. / ~ set ‖ Trinkgerät n. ‖ vaisselle f. pour boire. / ~ straw ‖ Trinkhalm m. ‖ chalumeau m. / ~ vessel ‖ Trinkgefäß n. ‖ gobelet m. / ~ water supply ‖ Trinkwasserversorgung f. ‖ distribution f. d'eau potable.

drip, to ‖ herabtropfen; träufeln ‖ dégoutter. / the white-metal drips out of the bearing ‖ das Lagermetall läuft aus ‖ le métal blanc s'écoule des coussinets.

drip cock ‖ Entwässerungshahn m. ‖ purgeur m.; robinet m. purgeur. / ~ cover ‖ Abtropfschale f. ‖ passoire f.; vase m. à égoutter; égouttoir m. / ~ cup ‖ Tropfbecher m. ‖ poche f. de vidange. / ~ edge ‖ Abtropfkante f. ‖ bord m. pour dégoutter. / ~ feed ‖ Tropfölschmierung f. ‖ graissage m. à compte-gouttes. / ~ feed lubricator ‖ Tropfölschmiergefäß n. ‖ graisseur m. compte-gouttes. / ~ nozzle ‖ Tropfdüse f. ‖ ajutage m. ou gicleur m. à gouttes. / ~ oiler ‖ Tropföler m. ‖ graisseur m. à compte-gouttes. / ~ oil lubrication ‖ Rieselölschmierung f. ‖ distribution f. d'huile par gouttes.

dripping board ‖ Abtropfbrett n. ‖ égouttoir m. / ~ plant ‖ Rieselanlage f. ‖ installation f. d'irrigation. / ~ tube (Med) ‖ Tropfpipette f. ‖ pipette f. compte-gouttes.

drip plate ‖ Tropfblech n. ‖ tôle f. d'égouttement.

drips pl. (Sugar) ‖ Nachlauf m.; Tröpfel m. ‖ sirop m. d'égout.

dripstone (Arch) ‖ Kranzleiste f.; Rinnleiste f. ‖ larmier m. / ~ (Geol) ‖ Filterstein m.; Filtersandstein m. ‖ Filterkalkstein m. ‖ pierre f. filtrante ou à filtre; grès m. fitrant.

drip tray ‖ Tropfschale f. ‖ gouttière f.

drive, to ~ down a river ‖ einen Fluß m. hinunter treiben ‖ descendre une rivière. / ~ a gallery of a mine ‖ eine Strecke f. treiben ‖ construire ou pratiquer une galerie de mine. / ~ a gallery to the hade of a seam ‖ eine einfallende Strecke f. treiben; einen Abbau m. machen ‖ faire une descente. / ~ a heating through (Mine) ‖ das Gebirge durchörtern ‖ percer le terrain. / ~ the hoops pl. of casks ‖ Fässer npl. antreiben ‖ serrer les cercles mpl. de fûts. / ~ in ‖ eintreiben; einschlagen; einrammen ‖ chasser; enfoncer; battre. / ~ in the coins (Print) ‖ die Form einkeilen ‖ arrêter ou assujettir les caractères avec des coins. / ~ a cotter ‖ einen Keil m. eintreiben ‖ serrer une clavette. / ~ in piles ‖ Pfähle mpl. eintreiben ‖ mettre en fiche ou enfoncer des pieux mpl. / ~ in the wedges pl. ‖ die Keile mpl. antreiben ‖ enfoncer les cales fpl. / ~ a level (Mine) ‖ auffahren ‖ avancer ou percer une galerie. / ~ the lines (Print) ‖ die Zeilen fpl. enger machen ‖ serrer les lignes fpl. / ~ oakum into the seams ‖ das Werg in die Nähte treiben ‖ chasser l'étoupe f. dans les coutures. / ~ on a gallery ‖ einen Stollen m. vortreiben ‖ construire ou pratiquer une galerie. / ~ on the plane iron ‖ das Hobeleisen vortreiben ‖ donner du fer au rabot. / ~ on the rim ‖ auf der Felge f. fahren ‖ rouler dégouflé. / ~ out (Bolt) ‖ heraustreiben ‖ retirer; repousser. / ~ the piles ‖ die Pfähle einrammen oder einschlagen oder eintreiben ‖ enfoncer les pilotis mpl. par le mouton. / ~ with the tide ‖ mit der Flut f. treiben ‖ aller avec le jusant. / ~ up the coins (Print) ‖ die Form einkeilen ‖ arrêter ou assujettir avec des coins. / ~ a wedge ‖ den Keil m. antreiben ‖ serrer ou enfoncer le coin.

drive (Arch) ‖ Anfahrt f. ‖ approche f.; rampe f. / ~ (Aut tel) ‖ Wählerantrieb m.

|| entraînement m. / ~ (Carriage) || Fahrt f. || voyage m.; course f.; excursion f. / ~ (Mach) || Antrieb m.; Betrieb m.; Trieb m. || commande f. / bell crank ~ || Winkelantrieb m. || commande f. à levier coudé. / level gear ~ || Kegelradantrieb m. || transmission f. à pignon conique. / chain ~ || Kettenantrieb m. || commande f. à chaîne. / direct ~ (Auto) || unmittelbarer Gang m. || prise f. directe. / direct ~ (Mach) || unmittelbarer Antrieb m. || commande f. directe. / direct electric ~ || unmittelbarer elektrischer Antrieb m. || commande f. électrique directe. / electric ~ || elektrischer Antrieb m. || commande f. électrique. / electric multimotor ~ of a paper making machine || elektrischer Mehrmotorenantrieb m. einer Papiermaschine || commande f. électrique multiple d'une machine à papier. / flywheel ~ || Schwungradantrieb m. || commande f. sur le volant. / free-swinging ~ || freischwingender Antrieb m. || commande f. autobalanceuse. / friction ~ || Reib(ungs)antrieb m. || commande f. à friction. / final ~ || Hinterachsantrieb m. || transmission f. aux roues arrières; commande f. de l'essieu arrière. / front wheel ~ || Vorderradantrieb m. || transmission f. sur la roue avant. / group ~ || Gruppenantrieb m. || commande f. par groupe. / lateral ~ || seitlicher Antrieb m. || commande f. latérale. / magneto ~ || Magnetantrieb m. || commande f. de la magnéto. / mechanical ~ || mechanischer Antrieb m. || commande f. mécanique. / normal ~ || Normalantrieb m. || commande f. normale. / pump ~ || Pumpenantrieb m. || commande f. de pompe. / rear ~ || rückwärts liegender Antrieb m. || commande f. à l'arrière. / rear wheel ~ || Hinterradantrieb m. || transmission f. sur la roue arrière. / separate ~ || Einzelantrieb m. || commande f. individuelle. / shaft ~ || Kardanantrieb m. || transmission f. à cardan. / single ~ || Einzelantrieb m. || commande f. individuelle. / ~ by single pulley || Einscheibenantrieb m. || commande f. par monopoulie. / tachometer ~ || Tachometerantrieb m. || commande f. du tachymètre. / worm ~ || Schneckenantrieb m. || transmission f. par vis sans fin.

driven || angetrieben || commandé. / ~ by compressed air || Preßluftantrieb m. || commande f. par air comprimé. / electrically ~ || elektrisch betrieben || mu électriquement. / ~ by electric motor || elektromotorisch angetrieben || commandé par un moteur électrique. / hand ~ || für Handbetieb m. || à commande f. à main. / the machine tools are ~ by electromotors || die Bearbeitungsmaschinen fpl. werden durch elektrische Motoren angetrieben || la commande des machinesoutils se fait à l'aide de moteurs électriques. / power ~ || für Kraftbetrieb m. || à commande f. mécanique. / ~ on the rise (Mine) || schwebend || montant; rampant.

driver (Auto) || Wagenführer m. || chauffeur m. / ~ (Brew) || Bierfahrer m. || cocherlivreur m. / ~ (Cart) || Kutscher m. || cocher m. / ~ (Lathe) || Mitnehmer m. || toc m. / ~ (Mach) || Mitnehmer m.; Knaggen m.; Nase f. || toc m.; taquet m. / ~ of a ribbon-loom || Rechen m. oder Treiber m. eines Bandwebstuhles || chasse

navettes m. d'un métier à rubans. / spike ~ || Nagelhammer m. || marteau m. de cloueur.

drivers pl. of the cradle (Shipb) || Schlittenständer mpl. || colombiers mpl.

driver's brake valve || Führerbremsventil n. || robinet m. de frein du mécanicien. / ~ cab (Loc) || Führerhaus n.; Führerstand m. || abri m. du mécanicien. / ~ cabin || Kranführerhaus n. || guérite f. de manœuvre. / ~ cage || Führerhäuschen n. || cabine f. du conducteur. / ~ seat || Führersitz m. || siège m. de conducteur. / ~ stand || Maschinenstand m. || plateforme f. du mécanicien. / fixed ~ stand alongside (Crane) || seitlich fester Führerstand m. || poste m. latéral fixe du conducteur.

drive shaft || Antriebswelle f.; Steuerwelle f. || arbre m. de commande. / ~ housing || Gelenkwellenrohr n. || jambe f. tubulaire de l'arbre à cardan.

drive side || Antriebseite f. || côté m. de la commande. / ~ wheel || Treibrad n. || roue f. motrice.

driving (Mach) || Antrieb m. || commande f.; transmission f. / ~ (Mine) || Niederhauen n. || descente f. / ~ of ice || Eisgang m. || débâcle f. ou charriage m. de glaçons. / ready for ~ || fahrfertig || prêt à conduire.

driving anchor (Mar) || Treibanker m. || ancre f. flottante. / ~ axle || Motorachse f.; Treibachswelle f. || essieu m. moteur.

driving belt || Treibriemen m. || courroie f. de transmission. / balata ~ || Balatatreibriemen m. || courroie f. de transmission en balata. / endless leather ~ || endloser Ledertreibriemen m. || courroie f. de transmission sans fin en cuir. / leather ~ || Ledertreibriemen m. || courroie f. de transmission en cuir. / width of ~ || Breite f. des Antriebriemens || largeur f. de la courroie de commande.

driving belt pulley || Riemenscheibe f. || poulie f. motrice.

driving car || Triebwagen m. || voiture f. automotrice. / ~ chain || Treibkette f.; Antriebskette f. || chaîne f. motrice ou de commande. / ~ circuit (Electr) || Antriebsstromkreis m. || circuit m. d'entraînement.

driving clock of an instrument || Uhrwerkantrieb m. eines Instrumentes || mouvement m. d'horlogerie d'un instrument. / ~ for the motion in right ascension of large refractors || Antriebsvorrichtung f. für die Stundenbewegung großer Refraktoren || moteur m. pour le mouvement horaire des grands réfracteurs.

driving cog wheel || Antriebszahnrad n. || pignon m. de commande. / ~ cone pulley || Antriebstufenscheibe f. || poulie f. de commande à gradins.

driving crank || Treibkurbel f. || manivelle f. motrice. / ~ with eccentric || Treibkurbel f. mit Exzenter || manivelle f. motrice avec excentrique. / ~ with inserted coupling pin || Treibkurbel f. mit eingesetztem Steuerzapfen || manivelle f. motrice avec bouton d'excentrique emmanché. / ~ with inserted return crank || Treibkurbel f. mit eingesetzter Gegenkurbel || manivelle f. motrice avec contremanivelle emmanchée.

driving device || Triebwerk n. || dispositifmoteur m.; mécanisme m. d'entraînement. / ~ direct-current motor || An

triebsgleichstrommotor m. || électromoteur m. de commande à courant continu. / ~ down of piles || Rammen n. oder Einrammen n. der Pfähle || pilotage m. / ~ edge of the lands in bore of gun || Führungskante f. der Felder in der Geschützseele || flanc m. d'appui des cloisons dans l'âme d'une bouche à feu. / ~ engine || Antriebsmaschine f. || machine f. motrice. / ~ force of the wind || treibende Wirkung f. des Windes || poussée f. du vent.

driving gear || Antriebsvorrichtung f.; Triebwerk n. || mécanisme m. de commande; mouvement m.; commande f. / ~ (Mill) || Gangwerk n.; Gangzeug n. || moulage m. / ~ above || mit oberem Antrieb m. || à commande f. supérieure. / ~ below || mit unterem Antrieb m. || à commande f. inférieure. / elastic ~ || elastisches Vorgelege n. || transmission f. élastique.

driving gear part || Triebwerksteil m. || pièce f. ou organe m. du mécanisme moteur.

driving head-ways pl. (Mine) || Vorrichtungsarbeit f. beim Grubenbau || travail m. préparatoire. / ~ license || Fahrschein m.; Führerschein m. || permis m. de conduire.

driving machine (Coop) || Antreibmaschine f. || machine f. à serrer; serreuse f. / cone truss hoop ~ (Coop) || Arbeitsreifenanziehmaschine f. mit Konus || machine f. à serrer les cercles de travail avec cône. / hoop ~ || Faßreifenauftreibmaschine f. || machine f. à serrer les cercles de fûts; serreuse f. pour cercles de fûts. / truss hoop ~ (Coop) || Arbeitsreifenanziehmaschine f. || machine f. à serrer les cercles de travail.

driving magnet (Electr) || Antriebsmagnet m. || électro m. d'entraînement. / ~ motor || Antriebsmotor m. || moteur m. de commande. / ~ off the gases || Entgasung f. || dégazation f.; dégazage m.; dégagement m. de gaz. / ~ out (Print) || weiter Satz m.; Überschreitung f. || enjambage m.; composition f. large. / ~ output || Antriebsleistung f. || puissance f. de propulsion. / ~ pin || Treibzapfen m. || bouton m. moteur ou de manivelle.

driving pinion || Ritzel m.; Antriebszahnrad n. || pignon m. ou engrenage m. de commande. / staggered ~s pl. || gegeneinander versetzte Zahnscheiben fpl. || disques mpl. à dents placés alternativement.

driving plate || Mitnehmerscheibe f. || plateau-toc m. / ~ power || Betriebskraft f. || force f. motrice. / ~ pulley || Scheibenschwungrad n. || poulie-volant f. / ~ ring || Mitnehmerring m. || bague f. d'entraînement. / ~ rod || Treibstange f. || bielle f. motrice. / ~ roller || Transportrolle f. || galet m. de transport. / ~ rope || Treibseil m. || câble f. de transmission. / ~ seat || Kutschersitz m. || siège m. du cocher. / ~ shaft || Triebwelle f. || arbre m. à manivelle. / ~ shield || Vortriebsschild m. || bouclier m. / ~ shield for tunnels || Vortriebsschild m. für Tunnel || bouclier m. d'avancement pour le fonçage de tunnels. / ~ side || Antriebseite f. || côté m. de la commande. / ~ snow || Schneegestöber n. || turbillons mpl. ou tourmente f. de neige. / ~ spindle || Antriebsspindel f. || bielle f. de commande. / ~ water (Mill) ||

Aufschlagwasser n. ‖ eau f. motrice. / ~ weight ‖ Zuggewicht n. ‖ contrepoids m. d'entraînement; poids-entraîneur m. / ~ weight for clock ‖ Antriebsgewicht n. für das Uhrwerk ‖ poids m. moteur pour le mouvement d'horlogerie. / ~ wheel ‖ Ritzel n.; Treibrad ‖ pignon m.; roue f. dentée motrice. / ~ wheel for flyers (Spinn) ‖ Fleierantriebsrad n. ‖ roue f. de commande à ailettes.

drizzle ‖ Staubregen m. ‖ bruine f.

drop, to (Cask) ‖ auslaufen; rinnen ‖ couler; fuir. / do not drop! ‖ nicht fallen lassen! ‖ laissez pas tomber!; gare aux chutes fpl.! / ~ down with the current (Mar) ‖ mit dem Strom m. treiben ‖ dériver avec le courant. / ~ down a river ‖ einen Fluß m. hinunter treiben ‖ descendre une rivière. / ~ forging ‖ im Gesenk n. schmieden ‖ matricer; forger en matrices. / ~ in ‖ zutropfen lassen ‖ faire couler goutte f. à goutte. / not to be dropped! ‖ nicht stürzen! ‖ ne pas renverser! / ~ out of step (Electr) ‖ außer Tritt m. fallen ‖ tomber hors de phase f. / the pressure drops ‖ der Druck m. sinkt ‖ la pression f. tombe.

drop ‖ Tropfen m. ‖ goutte f. / ~ (Curve) ‖ Abfall m. (Kurve) ‖ chute f. (courbe). / ~ (El line) ‖ Abfall m. ‖ chute f.; descente f. / ~ (Sugar) ‖ Zuckerbonbon m. ‖ bonbon m. de sucre. / ~ (Tel) ‖ Klappe f. ‖ annonciateur m.; indicateur m. téléphonique. / by ~s pl. ‖ tropfenweise ‖ par gouttes fpl. / ~ by drop ‖ tropfenweise ‖ goutte f. à goutte. / changeable depth of ~ ‖ veränderliche Sturzhöhe f. ‖ hauteur f. de déversement variable. / potential ~ (Electr) ‖ Spannungsabfall m. ‖ chute f. de tension. / ~ in temperature ‖ Wärmegefälle n. ‖ chute f. de température; gradient m. de la température. / ~ in voltage ‖ Spannungsabfall m. ‖ chute f. de tension.

drop base rim ‖ Tiefbettfelge f. ‖ jante f. à base creuse. / ~ box slay (Weav) ‖ Wechsellade f. ‖ battant m. à plusieurs navettes. / ~ chisel ‖ Kuttermeißel m. ‖ burin m. brise-rocs. / ~ conduit ‖ Tropfleitung f. ‖ conduite f. à gouttes. / ~ counter ‖ Tropfenzähler m. ‖ compte-gouttes m.

drop door ‖ Falltür f. ‖ porte f. bascule ou à guillotine. / ~ in the wagon floor (Railw) ‖ Bodenklappe f. ‖ trappe f. de fond. / ~ girder (Railw) ‖ Bodenklappenträger m. ‖ longeron m. de la trappe du fond.

drop-forging (Action) ‖ Gesenkschmieden n. ‖ forgeage m. mécanique; estampage m. ou matriçage m. / ~ (Piece) ‖ Gesenkschmiedestück n. ‖ pièce f. matricée. / ~ industry ‖ Gesenkschmiedeindustrie f. ‖ industrie f. de l'estampage. / ~ machine ‖ Gesenkschmiedemaschine f. ‖ machine f. à estamper ou à matricer.

drop gate (Sluice) ‖ Klapptor n. ‖ porte f. à trappe. / ~ hammer ‖ Fallhammer m. ‖ marteau-pilon m. (à friction); mouton m. à chute libre. / ~ indicator (Tel) ‖ Fallscheibe f. ‖ indicateur m. à volets. / ~ lake ‖ Lackfarbe f. ‖ laque m. / ~ oiler ‖ Tropföler m. ‖ graisseur m. à compte-gouttes.

dropper (Med) ‖ Tropfenzähler m. ‖ compte-gouttes m.; pipette f. / ~ (Pap) ‖ Eguttör m. ‖ égoutteur m.

dropping of speed ‖ Abfallen n. der Geschwindigkeit ‖ diminution f. de la vitesse. / ~ board ‖ Abtropfbrett n. ‖ égouttoir m. / ~ bottle ‖ Tropfflasche f. ‖ flacon m. compte-goutte. / ~ glass ‖ Bürette f. ‖ burette f. / ~ out (Vessel) ‖ Lecken n.; Rinnen n.; Schwitzen n. ‖ fuite f.

drop pipe ‖ Sickerrohr n. ‖ tube m. stillatoire. / ~ scene ‖ Zwischenaktvorhang m. ‖ toile f. pour entr'actes. / ~ shutter (Tel) ‖ Einfallscheibe f. ‖ disque m. de cliquetage. / ~ shutter for „Full" and „Empty" indication (Tel) ‖ Voll- und Leermelder m. ‖ indicateur m. à clapets pour „Plein" et „Vide". / ~ side (Railw) ‖ umlegbare Wand f. ‖ paroi f. à rabattement. / ~ switchboard (Tel) ‖ Klappenschrank m. ‖ tableau m. commutateur à volets. / ~ tap ‖ Tropfenhahn m. ‖ robinet m. à goutte.

drop test ‖ Schlagprobe f.; Schlagversuch m. ‖ essai m. au choc. / during the ~ the wheel was so placed that the whole rim was uniformly supported ‖ bei der Schlagprobe f. ruhte das Rad am ganzen Umfange gleichmäßig auf ‖ durant l'essai m. de choc la roue reposait avec toute la circonférence sur son assiette. / horizontal ~ ‖ horizontale Schlagprobe f. ‖ essai m. de choc en position horizontale. / vertical ~ ‖ vertikale Schlagprobe f. ‖ essai m. au choc en position verticale.

drop weight ‖ Tropfengewicht n. ‖ poids m. de goutte. / ~ window ‖ herablaßbares Fenster n. ‖ fenêtre f. coulissante. / ~ wire (Tel) ‖ Einführungsdraht m. ‖ fil m. d'entrée. / ~ work ‖ Fallwerk n. ‖ installation f. à chute libre.

drosometer (Phys) ‖ Taumesser m. ‖ drosomètre m.

dross (Coal) ‖ gewaschene Feinkohle f. ‖ fines fpl. lavées. / ~ (Metal) ‖ Schlacke f. ‖ crasse f.; laitier m.; scorie f. / goldsmith's ~ ‖ Goldkrätze f. ‖ cendres fpl. ou lavure f. d'or. / ~ for the mould ‖ Schlacke f. für die Halde ‖ crasse f. à haler. / ~ of sulphur ‖ Schwefelschlacke f. ‖ crasse f. de soufre.

drosses pl., skimmings pl. and ~ ‖ Metallabfall m. ‖ scraps mpl.; fer m. de riquette; ferraille f.

dross heap ‖ Schlackenhalde f. ‖ crassier m.

drossy ‖ schlackig; schlackenartig ‖ crasseux. / ~ iron ‖ schlackenreiches Eisen n. ‖ fer m. gras.

drought ‖ Dürre f.; Trockenheit f. ‖ sécheresse f.

drowned (Mine) ‖ ersoffen ‖ noyé; submergé.

drowsiness ‖ Schläfrigkeit f. ‖ somnolence f.

drudge, to ‖ schuften ‖ bûcher; piocher.

drug ‖ Droge f. ‖ drogue f. / ~ (Med) ‖ Arznei f. ‖ médicine f. / compressed ~ ‖ gepreßtes Medikament n. ‖ médicament m. comprimé. / botanical ~s pl. ‖ Vegetabilien fpl. ‖ herboristeries fpl.

drug colour ‖ Drogistenfarbe f. ‖ couleur f. de droguerie. / ~ cutting machine ‖ Drogenschneidmaschine f. ‖ coupeuse f. de drogues.

druggist ‖ Drogenhändler m.; Materialist m. ‖ droguiste m.

drug grinder ‖ Drogenmüller m. ‖ pileur m. de drogues. / ~ mill ‖ Drogenmühle f. ‖ moulin m. à drogues.

drugs pl. ‖ Drogenwaren fpl. ‖ drogues fpl.; droguerie f. / ~ ‖ Apothekerwaren fpl. ‖ articles mpl. pharmacologiques.

drum ‖ Trommel f. ‖ tambour m. / ~ (Agr) ‖ Trommel f. ‖ crible m. rotatif. / ~ (Mine) ‖ Seilkorb m.; Seiltrommel f. ‖ tambour m. / ~ (Wiredr) ‖ Scheibe f. ‖ bobine f. d'une filière; casette f. / ahead ~ ‖ Vorwärtstrommel f. ‖ tambour m. à marche avant. / amalgam ~ ‖ Amalgamtrommel f. ‖ tambour m. à amalgame. / astern ~ ‖ Rückwärtstrommel f. ‖ tambour m. à marche arrière. / breaking ~ ‖ Brechtrommel f. ‖ tambour m. broyeur. / cable ~ ‖ Kabeltrommel f. ‖ tambour m. à câble. / ~ of a capital (Arch) ‖ Kern m. oder Glocke f. eines Kapitells ‖ vase m.; corbeille f.; tambour m. de chapiteau. / chain ~ ‖ Kettentrommel f. ‖ tambour m. à chaîne. / ~ of a crane ‖ Trommel f. eines Kranes ‖ treuil m. d'une grue. / drying ~ ‖ Trockentrommel f. ‖ tambour m. de séchage. / ~ for film ribbon ‖ Trommel f. zur Auflage des Filmbandes ‖ tambour m. sur lequel s'applique le film. / graduated ~ ‖ Meßtrommel f. ‖ tambour m. divisé ou à mesure. / iron ~ ‖ eisernes Faß n. ‖ tonneau m. en fer. / luffing ~ ‖ Einziehtrommel f. ‖ tambour m. à enrouler. / metallic ~ see iron ~. / picking ~ ‖ Läutertrommel f. ‖ tambour m. de triage. / quenching ~ ‖ Löschtrommel f. ‖ tambour m. extincteur. / recording ~ ‖ Schreibtrommel f. ‖ tambour-enregistreur m. / revolving ~ ‖ drehbare Trommel f. ‖ tambour m. rotatif. / rotary ~ ‖ drehbare Trommel f. ‖ tambour m. rotatif. / screening ~ (Mine) ‖ Siebtrommel f. ‖ tambour m. de tamisage. / semi-circular ~ ‖ halbkreisförmige Trommel f. ‖ tambour m. semi-circulaire. / sifting ~ ‖ Rundsieb n. ‖ tamis m. circulaire. / sizing ~ ‖ Sortiertrommel f. ‖ tambour m. cribleur. / ~ for steaming (Dyer) ‖ Faß n. zum Dämpfen; Küpe f. ‖ cuve f. ou tonneau m. de vaporisage. / tin ~ ‖ Blechtrommel f. ‖ tambour m. en fer-blanc. / with upper ~ laterally arranged (Boil) ‖ mit einseitig liegendem Oberkessel m. ‖ avec collecteur m. supérieur disposé sur le côté. / varnishing ~ ‖ Lackiertrommel f. ‖ tambour m. à laquer. / washing ~ ‖ Waschtrommel f. ‖ tambour m. de lavage. / wooden ~ ‖ Würtel m. ‖ bobine f. en bois.

drum apparatus ‖ Trommelapparat m. ‖ appareil m. à tambour. / ~ armature (Electr) ‖ Trommelanker m. ‖ induit m. en tambour. / ~ bench (Wiredr) ‖ Rollenbank f.; Scheibenziehbank f. ‖ filière f. à bobine. / ~ brake ‖ Trommelbremse f. ‖ frein m. à tambour. / ~ case ‖ Trommelgehäuse n. ‖ caisse f. de tambour. / ~ compass ‖ Trommelkompaß m. ‖ compas-tambour m. / back ~ disc ‖ hintere Trommelscheibe f. ‖ disque m. arrière de tambour. / front ~ disc ‖ vordere Trommelscheibe f. ‖ disque m. avant de tambour. / ~ frame ‖ Trommelgerüst n. ‖ carcasse f. du trommel.

drumhead of capstan (Shipb) ‖ Spillkopf m. ‖ chapeau m. de cabestan.

drum house ‖ Trommelhaus n. ‖ atelier m. de tambours extincteurs. / ~ mill ‖ Trommelmühle f. ‖ moulin m. ou broyeur m. à tambour. / ~ mixer ‖ Trommelmischer m.; Trommelmischmaschine f. ‖ mélangeur-tambour m.

Drummond's lime-light ‖ Drummondsches Kalklicht n. ‖ lumière f. de Drummond.

drum net ‖ Fischreuse f.; Reuse f. ‖ bire f.; nasse f.; panier m. / ~ nickel-plating ‖ Trommelvernickelung f. ‖ nickelage m. au tambour. / ~ parchment ‖ Trommelleder n. ‖ peau f. pour tambours. / ~ process ‖ Trommelverfahren n. ‖ procédé m. par tambour. / ~ receiver (Pneumatic tube) ‖ Walzenempfänger m. ‖ poste m. récepteur à cylindre. / ~ saw ‖ Kronsäge f.; Ringsäge f. ‖ scie f. cylindrique. / ~ shaft ‖ Trommelwelle f. ‖ arbre m. à tambour. / ~ starter ‖ Walzenanlasser m.; Schaltwalzenanlasser m. ‖ démarreur m. à cylindre; rhéostat m. à tambour tournant. / ~ washer ‖ Trommelwascher m. ‖ laveur m. à tambour. / ~ weir ‖ Trommelwehr n. ‖ barrage m. à tambour.
druss ‖ Grus m.; Kohlengrus ‖ charbon m. fin; fouailles fpl.; menu m.
dry ‖ trocken ‖ sec. / ~ (Meteor) ‖ niederschlagsarm ‖ pauvre en condensations *ou* en pluies; à faibles condensations. / to keep ~ ‖ trocken halten ‖ protéger contre humidité f. / keep ~! ‖ vor Nässe f. zu schützen!; trocken aufzubewahren! ‖ à préserver de l'humidité!
dry accumulator ‖ Trockenakkumulator m. ‖ accumulateur m. sec. / ~ battery ‖ Trockenbatterie f. ‖ batterie f. sèche. / ~ bulb thermometer ‖ trockenes Thermometer n. ‖ thermomètre m. sec. / ~ cask ‖ Packfaß n. ‖ boucaut m. / ~ cell ‖ Trockenelement n. ‖ pile f. sèche. / ~ closet cleaning ‖ Tonnenabfuhrwesen n. ‖ entreprise f. de fosses mobiles. / ~ compass ‖ Trockenkompaß m. ‖ compas m. sec. / working on ~ compression ‖ Überhitzungsverfahren n. ‖ régime m. *ou* marche f. en surchauffement. / ~ concrete ‖ wasserarmer Beton m. ‖ béton m. peu humide. / ~ condenser ‖ Trockenkondensor m. ‖ condensateur m. fonctionnant à sec. / ~ crop ‖ Trockenfutter n. ‖ fourrage m. sec. / ~ crushing ‖ Trockenmahlung f. ‖ broyage m. par voie sèche. / ~ distillation ‖ trockene Destillation f. ‖ distillation f. sèche. / ~ dock ‖ Trockendock n. ‖ cale f. sèche; bassin m. de radoub. / ~ dock gate ‖ Trockendocktor n. ‖ portail m. de cale sèche. / ~ dock plant ‖ Trockendockanlage f. ‖ installation f. de cale sèche. / ~ dredger ‖ Trockenbagger m. ‖ drague f. sèche. / ~ element (Electr) ‖ Trockenelement n. ‖ élément m. sec; pile f. sèche. / ~ extract ‖ Trockenextrakt m. ‖ extrait m. sec. / ~ finishing machine ‖ Trockenappreturmaschine f. ‖ machine f. pour l'apprêt à sec. / ~ frame (Spinn) ‖ Trockenspinnmaschine f. ‖ métier m. à sec. / ~ fruit manufacturer ‖ Obstdörrer m. ‖ fabricant m. de fruits séchés. / ~ gas meter ‖ trockener Gasmesser m. ‖ compteur m. à gaz sec. / ~ goods ‖ Schnittwaren fpl. ‖ mercerie f. / ~ goods elevator ‖ Trockenelevator m. ‖ élévateur m. pour matières sèches. / ~ grinding (Mach) ‖ Trockenschliff m. ‖ affûtage m. *ou* rectification f. à sec. / ~ grinding (Mine) ‖ Trockenvermahlung f. ‖ broyage m. à voie sèche. / ~ grinding (Needl) ‖ Trockenschleifen n. ‖ polissage m. à sec. / ~ insulation ‖ Trockenisolierung f. ‖ isolement m. sec. / ~ kiln man (Sawmill) ‖ Trockenofenwärter m. ‖ sécheur m. de planches au four. / ~ lenses pl. (Opt) ‖ Trockensystem n. ‖ objectifs mpl. à sec. / ~ measure of volume ‖ Raummaß n. ‖ mesure f. de capacité. / ~

mixture ‖ Trockenmischung f. ‖ mélange m. sec. / ~ oil (Paint) ‖ Trockenöl n. ‖ siccatif m. / ~ packing (Mine) ‖ Versatz m. ‖ remblais mpl. / ~ pile (Electr) ‖ Trockenelement n. ‖ élément m. sec; pile f. sèche. / ~ plate (Phot) ‖ Trockenplatte f. ‖ plaque f. sèche. / ~ preparation ‖ Trockenpräparat n. ‖ préparation f. à sec. / ~ press ‖ Trockenpresse f. ‖ presse f. à sec. / ~ process ‖ Trockenverfahren n. ‖ procédé m. par voie sèche. / ~ products packing machine ‖ Packmaschine f. für Trockenerzeugnisse ‖ machine f. à empaqueter les produits secs. / ~ purification ‖ Trockenreinigung f. ‖ épuration f. par voie sèche. / ~ rot ‖ trockene Fäulnis f.; Trockenfäule f.; Holzschwamm m. ‖ pourriture f. *ou* putréfaction f. sèche; mérule m.; champignon m. des maisons *ou* du bois. / ~ sand (Mould) ‖ Formmasse f. ‖ masse f. à mouler. / ~ sand casting ‖ Masseguß m. ‖ moulage m. en sable gras *ou* en terre. / ~ sand mould ‖ Masseform f. ‖ moule m. en sable gras. / ~ sand moulding ‖ Masseformerei f.; Formerei f. mit fettem Sand ‖ moulage m. en sable gras. / ~ separation ‖ trockene Scheidung f. ‖ séparation f. à sec. / ~ spinning ‖ Trockenspinnen n. ‖ filage m. au sec. / ~ spraying clothes with insect powder ‖ Einstäuben n. von Kleidungsstücken mit Insektenpulver ‖ dispersion f. de la poudre insecticide pour la préservation de vêtements. / ~ state of the air ‖ Trockenheit f. des Luftmeeres ‖ état m. de sécheresse de l'atmosphère. / ~ steaming ‖ Trockendekatur f. ‖ décatissage m. par vapeur sèche. / ~ treatment ‖ Trockenverfahren n.; trocknes Verfahren n. ‖ voie f. sèche. / ~ vapour ‖ trockener Dampf m. ‖ vapeur f. sèche. / ~ washing ‖ trockene Waschung f. ‖ lavage m. par voie sèche. / ~ weight ‖ Trockengewicht n. ‖ poids m. sec. / ~ wine ‖ abgelagerter Wein m. ‖ vin m. mûr *ou* reposé. / ~ yeast ‖ Trockenhefe f. ‖ levain m. sec. / ~ zone ‖ Trockenzone f.; regenarmes Gebiet n. ‖ zone f. sèche; région f. de pluies rares.
dry, to ‖ trocknen ‖ sécher. / ~ (Bleach) ‖ trocknen (an der Luft) ‖ essorer. / ~ (Brew) ‖ darren ‖ touraîller. / ~ fruits pl. ‖ Früchte fpl. dörren ‖ torréfier des fruits mpl. / ~ malt ‖ das Malz darren ‖ dessécher le grain germé. / ~ with a mob (Mar) ‖ schwabbern; feudeln ‖ fauberter. / ~ off (Malt) ‖ abdarren ‖ maintenir la température finale. / ~ in the open air ‖ an der Luft f. trocknen ‖ sécher. / ~ paper by pressing ‖ das Papier trocken pressen ‖ écacher le papier. / ~ up ‖ austrocknen ‖ dessécher.
dryer *see also* drier ‖ Trockenmaschine f. ‖ machine f. à sécher; sécheuse f.; dessécheuse f.; séchoir m.
drying ‖ Austrocknung f. ‖ séchage m.; dessication f. / artificial ~ ‖ künstliches Trocknen n. ‖ séchage m. artificiel. / ~ of cable joint ‖ Trocknen n. der Lötstelle ‖ dessèchement m. de l'épissure. / marsh ~ ‖ Feldbereinigung f. ‖ dessèchement m. de marais. / preliminary ~ ‖ Vortrocknung f. ‖ préséchage m. / ~ of skins ‖ Trocknung f. von Häuten ‖ séchage m. de peaux. / ~ of wood ‖ Trocknung f. von Holz ‖ séchage m. *ou* dessication f. des bois.
drying agent ‖ Trockenmittel n.; Trocken-

substanz f. ‖ desséchant m.; déshydratant m.
drying apparatus ‖ Dörrapparat m.; Trockenapparat m. ‖ séchoir m.; étuve f. à sécher; appareil m. de séchage. / distiller's wash ~ ‖ Schlempetrockner m. ‖ séchoir m. à vinasse. / ~ with evaporation ‖ Eindampfapparat m. ‖ séchoir m. par évaporation. / soap ~ ‖ Seifentrockenapparat m. ‖ séchoir m. à savon.
drying battery ‖ Trockenbatterie f. ‖ batterie f. de séchage. / ~ board ‖ Trockengestell n. ‖ glacis m. / ~ cabinet ‖ Trockenschrank m. ‖ étuve f. à sécher. / ~ chamber ‖ Trockenkammer f.; Trockenraum m. ‖ chambre f. de séchage *ou* de dessiccation; étendoir m. / ~ chamber carriage ‖ Trockenkammerwagen m. ‖ chariot m. d'étuve. / ~ cylinder ‖ Trockencylinder m. ‖ cylindre m. sécheur.
drying device, blood ~ ‖ Bluttrockner m. ‖ appareil m. à sécher le sang. / gas ~ ‖ Gastrockner m. ‖ sécheur m. à gaz. / ~ for rotors ‖ Ankertrocknungsapparat m. ‖ appareil m. de séchage d'induits.
drying drum ‖ Trockentrommel f. ‖ sécheur m. rotatif; tambour m. de séchage. / ~ fabrics pl. ‖ Entnässen n. von Geweben ‖ essorage m. de tissus. / ~ flask ‖ Trockenflasche f. ‖ flacon m. sécheur. / ~ frame ‖ Trockengestell n. ‖ cadre m. de séchage. / ~ ground ‖ Trockenfeld n. ‖ champ m. de haies. / ~ house ‖ Trockenhaus n. ‖ sécherie f. / ~ hurdle ‖ Darrhorde f.; Darrboden m.; Trockendarre f. ‖ claie f. à sécher; plateau m. de touraille; étuve f. / ~ installation ‖ Trockenvorrichtung f. ‖ installation f. à sécher. / ~ kiln ‖ Darre f. ‖ étuve f. séchoire; touraille f. / ~ machine ‖ Trockenmaschine f. ‖ machine f. à sécher; sécheuse f. / ~ twin cylinder ~ machine ‖ Zweiwalzentrockner m. ‖ séchoir m. à cylindres conjugués. / ~ method ‖ Eintrocknungsmethode f. ‖ méthode f. de dessiccation. / ~ oil ‖ Trockenöl n.; Sikkativ n. ‖ huile f. siccative; siccatif m. / ~ oven ‖ Trockenschrank m.; étuve f. / ~ oven for the building trades ‖ Bauaustrockner m. ‖ appareil m. à sécher les bâtiments. / ~ place (Blech) ‖ Aufhängeboden; Trockenboden m. ‖ séchoir m.; étandage m.
drying plant ‖ Trockenanlage f. ‖ installation f. de séchage; séchoir m. / ~ for chemical works ‖ Trocknungsanlage f. für chemische Fabriken ‖ installation f. de séchage pour fabriques de produits chimiques. / coal ~ ‖ Kohlentrocknungsanlage f. ‖ installation f. de séchage du charbon. / pulp ~ ‖ Pülpetrocknungsanlage f. ‖ installation f. de séchage de pulpes.
drying plate (Weav) ‖ Trockenplatte f. ‖ plaque f. à sécher. / ~ press (Pap) ‖ Trockenpresse f. ‖ presse f. à sécher. / ~ process ‖ Trockenverfahren n. ‖ procédé m. de séchage. / ~ room (Bleach) ‖ Aufhängeboden m.; Trockenboden m. ‖ séchoir m.; étandage m. / ~ room (Found) ‖ Trockenkammer f. ‖ séchoir m. / ~ shed ‖ Trockengerüst n. ‖ séchoir m. / ~ sheet ‖ Trockenblech n. ‖ tôle f. sécheuse. / ~ stove (Found; Phot) ‖ Trockenofen m.; Trockenkammer f. ‖ four m. à sécher; étuve f. séchoire. / ~ stove (Brew) ‖ Darrofen m. ‖ fourneau m. à sécher. / ~ tower ‖ Trockenturm m. ‖ colonne f. sécheuse.

dryness ‖ Trockenheit f. ‖ sécheresse f. / ~ (Chem) ‖ Trockenheit f. ‖ siccité f. / to evaporate to ~ (Chem) ‖ abrauchen ‖ évaporer à siccité f. / to reach a state of ~ ‖ eintrocknen ‖ arriver à siccité.

dub, to ~ the timber ‖ das Holz dächseln ‖ dresser le bois à l'herminette.

dubbing ‖ Schlichten n.; Abschlichten n. ‖ parage m.

duck (Weav) ‖ Bramtuch n.; leichtes Segeltuch n. ‖ toile f. à voiles légère; toile f. mélis ou d'olonne. / ~ stone ‖ Duckstein m. ‖ trass m.; tuf m. calcaire. / ~ type aeroplane ‖ Entenflugzeug n. ‖ avion-canard m.

duct (Mach) ‖ Rohr n.; Röhre f. ‖ tuyau m. / ~ for change of type (Cable) ‖ Übergangsformstück n. ‖ tuyau m. intermédiaire. / oil ~ ‖ Ölleitung f. ‖ canalisation f. d'huile.

duct bolt ‖ Durchführungsbolzen m. ‖ boulon m. de traversée.

ductibility ‖ Geschmeidigkeit f. ‖ ductibilité f.

ductile (Metal) ‖ schmiedbar; streckbar ‖ ductile; malléable. / ~ (Paste) ‖ knetbar, plastisch ‖ plastique. / ~ (Phys) ‖ dehnbar ‖ ductile.

ductility ‖ Dehnbarkeit f.; Ziehbarkeit f. ‖ ductilité f.

ductor knife (Print) ‖ Abstreichmesser n.; Farbmesser n. ‖ couteau m. à encre. / ~ roller (Print) ‖ Farbzylinder m. ‖ cylindre m. à l'encre.

due ‖ fällig ‖ payable; échéant.

due, mine ~s ‖ Bergwerksabgabe f. ‖ contribution f. minière. / ~ appertaining to the State ‖ dem Staat zufließende Abgabe f. ‖ impôt m. revenant à l'Etat. / ~ date ‖ Verfalltag m.; Fälligkeitstag m. ‖ échéance f. / ~ weight (Coin) ‖ Schrot n. der Münzen ‖ poids m. total.

duff (Mine) ‖ Nußgruskohle f.; gewaschene Feinkohle f. ‖ menus mpl. grenus; fines fpl. lavées.

duffer ‖ Stümper m. ‖ bousilleur m.

dug peat ‖ Stichtorf m. ‖ tourbe f. extraite à la bêche. / ~ sand ‖ Grubensand m. ‖ sable m. fouille ou de fouille.

dulcifying material ‖ Süßstoff m. ‖ dulcifiant m. / ~ substance ‖ Süßholz n. ‖ bois m. de réglisse.

dulcein ‖ Dulcin n. ‖ dulcine f.

dulcite ‖ Düngemittel n. ‖ dulcite f.

dull (Metal) ‖ matt; glanzlos ‖ mat. / ~ (Trade) ‖ träge ‖ languissant. / to grow ~ (Metal) ‖ abblicken ‖ cesser d'être luisant ou poli.

dull, to (Tool) ‖ abstumpfen; stumpf machen ‖ émousser; perdre le tranchant.

dull black ‖ mattschwarz ‖ noir terne. / ~ coal ‖ Mattkohle f. ‖ charbon m. mat.

dulled ‖ mattiert; matt ‖ dépoli; mat.

dull edged deal ‖ ungesäumte Diele f. ‖ planche f. flacheuse. / ~ timber ‖ wankantiges Holz n. ‖ bois m. flacheux.

dullness of business ‖ Stillstand m. der Geschäfte ‖ stagnation f. des affaires.

duly ‖ ordnungsgemäß ‖ dûment.

dumb aerial (Radio) ‖ verstimmte Antenne f. ‖ antenne f. muette. / ~-bell ‖ Hantel m. ‖ haltère m. / ~ chalder (Shipb) ‖ Ruderträger m. ‖ fausse-penture f. / ~-craft ‖ Fußwinde f. mit Haken; Hebelade f. ‖ cric m. à crochet. / ~ iron (Auto) ‖ Federhand f.; Federarm m. ‖ main f. du châssis. / ~ piece (Coin) ‖

klanglose Münzplatte f. ‖ flan m. cendreux. / ~ match (Shipb) ‖ Lippklampe f. ‖ taquet m. à manche. / ~-waiter ‖ Speisenaufzug m. ‖ monte-plat m.

dummy contact (Electr) ‖ Leerkontakt m. ‖ contact m. inoccupé. / ~ coupling ‖ Leerkupplung f. ‖ faux accouplement m. / ~ tank ‖ Scheinkampfwagen m. ‖ char m. de combat simulé; faux char m. de combat.

dump, to (Material) ‖ stürzen; abladen ‖ déverser. / ~ (Trade) ‖ schleudern ‖ vendre à vil prix; gâcher la marchandise. / ~ the ore into the bins without any rehandling ‖ das Erz ohne Umladen in die Vorratsräume stürzen ‖ verser ou vider le minerai dans les poches sans déchargement. / ~ the raw materials into the storage bins ‖ Erz n. in die Vorratsräume abstürzen ‖ verser des minerais aux dépôts. / the ore is dumped into railway wagons ‖ das Erz wird in die Eisenbahnwagen abgestürzt ‖ le minerai se vide dans des wagons. / the ores pl. are dumped on the storing place ‖ die Erze npl. werden auf die Lagerplätze abgestürzt ‖ les minerais mpl. sont versés sur les dépôts.

dump (Miner) ‖ Schlackenhalde f. ‖ halde f.; crassier m.; monceau m.

dump car ‖ Kippwagen m. ‖ wagon m. basculeur. / ~ tipping to either side ‖ zweiseitiger Kippwagen m. ‖ wagonnet m. à double bascule. / V-~ with automatic locking device ‖ Muldenkipper m. mit selbsttätiger Muldenfeststellung ‖ basculeur m. à auge avec dispositif de fixation automatique de l'auge.

dumper ‖ Gichtmann m. ‖ chargeur m.

dumping ‖ Dumping n. ‖ dumping m. / ~ ground (Dredger) ‖ Absturzplatz m. ‖ terrain m. de décharge. / ~ height (Dredger) ‖ Ausschütthöhe f. ‖ hauteur f. de déchargement ou de déversement. / invariable ~ height ‖ unveränderliche Sturzhöhe f. ‖ hauteur f. de chute constante. / ~ kiln (Brew) ‖ Klapphorde f. ‖ touraille f. à jalousie. / ~ radius (Dredger) ‖ Ausschüttweite f. ‖ portée f. de déversement ou de déchargement. / ~ truck see dump car. / ~ valve (Aero) ‖ Bodenventil m. ‖ soupape f. de fond.

dun, to ‖ zur Zahlung f. auffordern ‖ inviter à payer.

dune, isolated ‖ Einzeldüne f. ‖ dune f. isolée. / shifting ~ ‖ Wanderdüne f. ‖ dune f. mobile. / steep coast covered with ~s ‖ Dünensteilküste f. ‖ côte f. escarpée couverte de dunes. / protection for ~s by planting ‖ Dünenbepflanzung f. ‖ plantation f. sur les dunes. / protection for ~s by tree planting ‖ Dünenaufforstung f. ‖ plantation f. sylvestre sur les dunes. / ~ protection works pl. ‖ Dünenbau m. ‖ art m. de fixation des dunes. / ~ sand ‖ Dünensand m. ‖ sable m. des dunes.

dung, to ‖ düngen ‖ engraisser les terres.

dung ‖ Dünger m. ‖ engrais m.; fiente f; fumier m.; crottin m. / ~ bath (Dyer) ‖ Mistbad n. ‖ bain de fiente; bousage m. des indiennes. / ~ cart ‖ Mistwagen m. ‖ voiture f. à fumier. / ~ fork ‖ Düngerforke f. ‖ fourche f. à fumier. / ~ yard ‖ Dungstätte f. ‖ pailler m.; fosse f. à fumier.

duo-rolls pl. ‖ Zwillingswalzen fpl. ‖ cylindres mpl. duo.

duplex lathe ‖ Duplexdrehbank f. ‖ tour m. à double outil. / ~ repeater (Tel) ‖ Doppelstromgegensprechübertragung f. ‖ translation f. duplex. / ~ system (Tel) ‖ Doppelbetrieb m. ‖ exploitation f. en duplex. / ~ telegraphy ‖ Gegensprechtelegrafie f.; Duplextelegrafie f. ‖ télégraphie f. duplex. / ~ telephony ‖ Gegensprechtelefonie f.; Duplextelefonie f. ‖ téléphonie f. duplex. / ~ working (Tel) ‖ Gegensprechbetrieb m.; Duplexbetrieb m. ‖ transmission f. duplex.

duplicate (Typewriter) ‖ Durchschlag m. ‖ copie f.; duplicate m. / in ~ ‖ abschriftlich ‖ en copie f. / by way of ~ see ~, in.

duplicate lens (Opt) ‖ Ersatzglas n. ‖ verre m. de rechange. / ~ part ‖ Austauschstück n. ‖ pièce f. de rechange.

duplicating book ‖ Durchschreibebuch n. ‖ livre m. à copier; cahier m. à écrire les duplicatas. / ~ machine ‖ Vervielfältigungsapparat m. ‖ appareil m. duplicateur. / ~ paper ‖ Kopierpapier n.; Durchschreibepapier n.; Kohlepapier n. ‖ papier m. à copier; papier-carbone m.

durability ‖ Dauerhaftigkeit f.; Haltbarkeit f. ‖ durabilité f.; solidité f.

durable ‖ haltbar ‖ durable. / ~ linen ‖ Dauerwäsche f. ‖ linge m. durable ou permanent ou économique.

durableness ‖ Dauerhaftigkeit f. ‖ durabilité f.; solidité f.; stabilité f.

dural profile ‖ Duralprofil n. ‖ profil m. de duralumin.

duralumin ‖ Duralumin n. ‖ duralumin m.

duration ‖ Bestand m. ‖ permanence f. / ~ of action ‖ Einwirkungsdauer f. ‖ durée f. d'action. / ~ of a call (Tel) ‖ Gesprächsdauer f. ‖ durée f. de conversation. / ~ of charge (Acc) ‖ Ladedauer f. ‖ durée f. de chargement. / ~ of discharge (Acc) ‖ Entladedauer f. ‖ durée f. de déchargement. / ~ of drying ‖ Trockendauer f. ‖ durée f. de séchage. / ~ of fermentation ‖ Gärdauer f. ‖ durée f. de fermentation. / ~ of filling ‖ Füllungsdauer f. ‖ durée f. de remplissage. / ~ of flight ‖ Flugzeit f.; Flugdauer f. ‖ durée f. de vol. / ~ of setting ‖ Erhärtungsdauer f. ‖ durée f. de prise. / ~ of test ‖ Versuchsdauer f. ‖ durée f. de l'essai.

durra ‖ Dari n. ‖ dari m.

dust, to (Chem) ‖ zerstäuben ‖ vaporiser; réduire en poussière f. / ~ the rags (Pap) ‖ die Lumpen mpl. sieben ‖ trier les chiffons mpl. / ~ a road ‖ eine Straße f. reinigen ‖ débourber ou nettoyer la rue.

dust (Chem) ‖ Staub m. ‖ poudre f.; poussière f. / ~ (Found) ‖ Staub m. ‖ portée f. / ~ (Mine) ‖ Kohlenlösche f.; Gestübbe n. ‖ fraisil m.; poussière f. de charbon. / charcoal ~ see coal ~. / coal ~ ‖ Kohlenstaub m. ‖ poussière f. de charbon. / metallic ~ ‖ metallischer Staub m. ‖ poussière f. métallique. / to reduce the development of ~ to a minimum ‖ die Staubentwicklung f. auf ein Mindestmaß beschränken ‖ réduire au minimum la formation de poussières. / ~ of roasted ore ‖ Röststaub m. ‖ poussier m. de grillage. / the ~ is worked up into briquettes for further use in the blast furnaces ‖ die Staubmassen fpl. werden für die Wiederverwertung im Hochofen in Brikette verwandelt ‖ les masses fpl. de poussières sont traitées dans une usine à briquetter pour être réutilisées dans le haut-fourneau.

dust bag (Found) ‖ Staubbeutel m. ‖ poncis m. / sheet iron ~ bin ‖ Müllkasten m. aus Eisenblech ‖ caisse f. à ordures en tôle. / ~ box ‖ Müllkasten m. ‖ panier m. aux balayures; poubelle f.; boîte f. à ordures. / ~ brush ‖ Staubbürste f.; Handfeger m. ‖ brosse f. à épousseter; balayette f. / ~ bucket ‖ Mülleimer m. ‖ seau m. à ordures. / ~ cap ‖ Staubkappe f.; Staubdeckel m. ‖ chapeau m. à poussière; couvercle m. antipoussiéreux. / ~ cart ‖ Müllwagen m.; Müllabfuhrwagen m. ‖ tombereau m. à déchets ménagers; tombereau m. de transport des ordures / motor ~ cart ‖ Motormüllwagen m. ‖ chariot m. automobile pour le transport des ordures. / ~ casing ‖ Staubgehäuse n. ‖ enveloppe f. à poussière. / ~ catcher ‖ Staubfänger m. ‖ collecteur m. à poussière; attrappe-poussière f. / flue ~ catcher ‖ Flugaschefänger m. ‖ capuchon m. à cendres. / ~ chamber ‖ Staubhaube f. ‖ chapeau m. à poussière. / ~ chamber (Mill) ‖ Staubkammer f. ‖ chambre f. à poussière. / ~ cloth ‖ Staubtuch n. ‖ torchon m. à épousseter. / ~ coal ‖ Grießkohle f. ‖ charbon m. menu ou poussiéreux. / ~ collector ‖ Staubsammler m. ‖collecteur m. à poussière. /~ counter ‖ Staubzähler m. ‖ compteur m. de poussière. / ~ cover ‖ Staubdecke f. ‖ couverture f. protectrice contre la poussière. / ~ development ‖ Staubentwicklung f. ‖ dégagement m. de poussière. / ~ dry ‖ staubtrocken ‖ sec comme de la poussière.

duster ‖ Lappen m. ‖ chiffon m.; loque f.; guenille f.; lobe m. / ~ (Dusting cloth) ‖ Wischtuch n.; Staubtuch n.; torchon m. / ~ (Dust brush) ‖ Handfeger m.; balayette f. / ~ (Pap) ‖ Siebmaschine f.; Stäuber m. ‖ blutoir m.

dust exhausting plant ‖ Staubabsauganlage f. ‖ installation f. d'aspiration de poussière. / ~ extracting installation see ~ extracting plant. / ~ extracting method ‖ Entstaubungsverfahren n. ‖ procédé m. de dépoussiérage. / ~ extracting plant ‖ Entstaubungsanlage f. ‖ installation f. de dépoussiérage. / ~ extraction ‖ Staubabzug m.; Entstaubung f. ‖ aspiration f. de la poussière; dépoussiérage m. / ~ extractor ‖ Ausblaseapparat m. ‖ appareil m. pneumatique pour le nettoyage. / ~ filter ‖ Staubfilter n.; Staubreiniger m. ‖ filtre m. à poussière; soufflet m. / ~ filter with straight mouthpiece ‖ Staubreiniger m. mit geradem Mundstück ‖ soufflet m. à buse droite. / formation of ~ ‖ Staubentwicklung f. ‖ développement m. de poussière. / ~ gold ‖ Staubgold n. ‖ or m. en poudre impalpable. / ~ haze ‖ Staubnebel m. ‖ brouillard m. de poussière. / ~ helmet ‖ Staubschutzhelm m. ‖ casque m. de protection contre la poussière. / ~ hood (Auto) ‖ Verdeckhülle f. ‖ capote f. de voiture.

dusting arrangement ‖ Streuvorrichtung f. ‖ dispositif m. d'épandage. / ~ brush ‖ Staubpinsel m.; Abstaubpinsel m. ‖ époussetoir m.; pinceau m. à épousseter. / ~ machine ‖ Abstaubmaschine f. ‖ machine f. à épousseter. / bag ~ machine ‖ Sackausstäuber m. ‖ batteuse f. de sacs. / ~ material ‖ Streugut n. ‖ produit m. à épandre.

dust-laden air ‖ staubhaltige Luft f. ‖ air m. chargé de poussière.

dust-laying washing ‖ Berieselung f. ‖ ruissellement m.

dustless air ‖ staubfreie Luft f. ‖ air m. sans poussière.

dust mass ‖ Staubmasse f. ‖ masse f. de poussière. / coal ~ mill ‖ Kohlenstaubmühle f. ‖ pulvérisateur m. à charbon. / ~ oil ‖ staubbindendes Öl n.; Stauböl n. ‖ huile f. antipoussière. / ~ ore ‖ Mulm m. ‖ mine f. pulvérulente.

dustpan ‖ Kehrichtschaufel f. ‖ pelle f. à ordures.

dust penetration ‖ Eintritt m. von Staub ‖ entrée f. de la poussière. / ~ pipe (Watchm) ‖ Hut m. ‖ chapeau m. / ~ pipe factory (Watchm) ‖ Hütefabrik f. ‖ fabrique f. de chapeaux. / ~ pocket ‖ Staubsack m. ‖ poche f. de la poussière. / ~ preserving shell ‖ Staubmantel m. ‖ enveloppe f. protectrice contre la poussière.

dustproof ‖ staubdicht ‖ imperméable à la poussière.

dust, protected from ~ ‖ vor Staub m. geschützt ‖ protégé de la poussière. / quantity of ~ deposited in the dry cleaners ‖ in den Trockenreinigern absetzende Staubmassen fpl. ‖ les masses fpl. de poussière qui se déposent dans les appareils épurateurs à sec. / ~ removal plant see ~ removing plant.

dust removing from carpets ‖ Entstauben n. von Teppichen ‖ époussetage m. de tapis.

dust removing device by means of centrifuges ‖ Entstaubungsvorrichtung f. durch Zentrifugen ‖ appareil m. pour la séparation de poussière par centrifuges. / ~ by electric precipitation ‖ Entstaubungsvorrichtung f. durch elektrischen Niederschlag ‖ appareil m. pour la séparation de poussière par extinction électrique. / ~ by filters ‖ Entstaubungsvorrichtung f. durch Filter ‖ appareil m. pour la séparation de poussière par filtres.

dust removing plant ‖ Entstaubungsanlage f. ‖ installation f. de dépoussiérage. / ~ screen ‖ Staubschützer m. ‖ garde-poussière f. / ~ separation ‖ Staubabscheidung f. ‖ dépoussiérage m. / ~ separator ‖ Staubabscheider m. ‖ séparateur m. de poussière. / ~ settling ‖ Staubniederschlagung f. ‖ captation f. de la poussière. / ~ shell ‖ Staubsieb n. ‖ tamis m. à poussière. / ~ shield ‖ Staubschutz m.; Staubschutzschild n. ‖ pare-poussière m.; plaque f. anti-poussière. / ~ shot ‖ Vogeldunst m.; cendre f. de plomb; cendrée f.; plomb m. menu; menuise f.; menuisaille f. / ~ sieve see ~ shell. / ~ sucking apparatus ‖ Staubsaugvorrichtung f. ‖ appareil m. aspiratoire de poussière; dispositif m. d'aspiration ou aspirateur m. de poussier. / ~ van see ~ cart.

Dutch brick ‖ Klinker m. ‖ brique f. hollandaise; chantignole f. / ~ clinker see ~ brick. / ~ drop (Glassm) ‖ Glasträne f. ‖ goutte f. de verre; larme f. batavique. / ~ drops pl. ‖ Haarlemer Öl n. ‖ huile f. de Haarlem. / ~ flight ‖ Fleutschiff n. ‖ flûte f. / ~ gin ‖ Genever m.; feiner Wacholderbranntwein ‖ genièvre m. fin. / ~ gold ‖ Goldschaum m. ‖ or m. faux en feuilles. / ~ linen ‖ holländische Leinwand f. ‖ toile f. de Hollande. / ~ pink ‖ Schüttgelb n. ‖ stil-de-grain m. / ~ scoop (Hydr arch) ‖ Schöpfschaufel f.; Schwungschaufel f. ‖ pelle f. hollandaise. / ~ tile ‖ Fliese f.; Kachel f. ‖ carreau m. céramique. / cement ~ tile ‖ Zementfliese f. ‖ carreau m. de ciment. / ~ tongs pl. ‖ Froschklemme f. ‖ tendeur-grenouille m. / ~ windmill ‖ holländische Windmühle f. ‖ moulin m. à vent hollandais.

dutiable ‖ zollpflichtig ‖ soumis aux droits.

duty ‖ Abgabe f.; Zollabgabe f. ‖ impôt m.; droit m.; taxe f. / ~ (Mach) ‖ Nutzleistung f. ‖ effet m. ou travail m. utile. / ~ ad valorem ‖ Wertzoll m. ‖ droit m. ad valorem ou d'après la valeur. / ~ of anchorage ‖ Ankergeld n.; Hafengeld n. ‖ droit m. d'ancrage ou du port. / ~ on barley ‖ Gerstensteuer f. ‖ impôt m. sur l'orge. / ~ on board-ship ‖ Borddienst m. ‖ service m. à bord. / ~ to compensate for damage ‖ Schadenersatzpflicht f. ‖ obligation f. de dédommagement. / contrary to ~ ‖ pflichtwidrig ‖ déloyal. / entrance ~ ‖ Einfuhrzoll m. ‖ droits mpl. d'entrée. / export ~ ‖ Ausfuhrzoll m. ‖ droits mpl. d'exportation ou de sortie. / import ~ ‖ Eingangszoll m. ‖ droit m. d'entrée. / ~ on luxuries ‖ Luxussteuer f. ‖ impôt m. sur le luxe. / ~ on malt ‖ Malzaufschlag m. ‖ droit m. de fabrication sur le malt. / to persist in a refusal to carry out duties ‖ sich weigern, den obliegenden Verpflichtungen fpl. nachzukommen ‖ se refuser de remplir ses obligations fpl. / ~ free ‖ zollfrei ‖ exempt de droits mpl.; en franchise f. / ~ stamp ‖ Zollstempel m. ‖ décharge f. / ~ unpaid ‖ unverzollt ‖ sans payer la douane.

dwarf elder root ‖ Attichwurzel f. ‖ racine f. de hièble.

dwarf pine oil ‖ Latschenkiefernöl n. ‖ essence f. de pin de montagne.

dwelling, workmen's ~s ‖ Werksiedlung f. ‖ colonie f. ouvrière. / workman's ~ in the colonies ‖ Arbeiterwohnhaus n. in den Siedlungen ‖ maison f. ouvrière des colonies.

dwelling car ‖ Wohnwagen m. ‖ roulotte f.; voiture f. d'habitation. / ~ carriage see ~ car. / ~ house ‖ Wohnhaus n. ‖ maison f. d'habitation.

dye ‖ färbend ‖ tinctorial.

dye, to ‖ färben ‖ teindre. / ~ parchment yellow ‖ das Pergament gilben ‖ jaunir ou teindre en jaune le parchemin. / ~ in the wool ‖ in der Wolle f. färben ‖ teindre en poil m.

dye ‖ Färbung f. ‖ teint m. / acid ~ ‖ Säurefarbstoff m. ‖ colorant m. acide. / ~ in balls ‖ Farbkugel f. ‖ boule f. de teinture. / fast ~ ‖ echte Färbung f. ‖ grand ou bon teint m. / fugitive ~ ‖ unechte Färbung f. ‖ petit ou mauvais teint m.; biscuit m. / indigoïd ~ ‖ indigoartiger Farbstoff m. ‖ colorant m. indigoïde. / to give the last ~ (Text print) ‖ ausfärben ‖ parachever la teinture. / permanent ~ see fast ~. / sulphur ~ ‖ Schwefelfarbstoff m. ‖ colorant m. au soufre.

dyed in the grain ‖ waschecht ‖ bon teint m. / ~ of one colour ‖ einfarbig ‖ uni. ~ wood ‖ gefärbtes Holz n. ‖ bois m. teint.

dye hand ‖ Färbereiarbeiter m. ‖ ouvrier m. de teinturerie.

dyeing ‖ Färben n.; Färberei f. ‖ teinture f. / ~ of fibres before spinning ‖ Textilfasernfärberei f. vor dem Spinnen ‖ teinture f. de fibres textiles avant filature. / fur ~ ‖ Färben n. von Pelzen ‖ teinture f.

de peaux en poils. / mending of defects in ~ ‖ Ausbesserung f. von Farbfehlern ‖ débarrage m. d'étoffes. / piece ~ ‖ Stückfärberei f. ‖ teinture f. d'étoffes. / straw ~ ‖ Strohfärben n. ‖ teinture f. de paille. / Turkey-red ~ ‖ Türkischrotfärberei f. ‖ teinture f. en rouge turc. / wood ~ ‖ Holzbeizen n. ‖ teinture f. du bois.

dyeing articles pl. ‖ Färbereiartikel mpl. ‖ articles mpl. de teinturerie. / ~ house ‖ Färberei f. ‖ teinturerie f. / ~ industry ‖ Färbereiindustrie f. ‖ industrie f. des teintureries. / ~ machine ‖ Färbereimaschine f. ‖ machine f. de teinturerie. / wire netting ~ machine ‖ Drahtgewebefärbmaschine f. ‖ machine f. à peindre les toiles métalliques. / materials pl. for ~ ‖ Rohstoffe mpl. zum Färben ‖ matières fpl. premières de teinture. / wool ~ plant ‖ Wollfärberei f. ‖ teinturerie f. de laine. / ~ vat ‖ Färbfaß n.; Küpe f. ‖ cuve f. de teinture.

dye maker ‖ Farbenfabrikant m. ‖ fabricant m. de colorants chimiques.

dyer ‖ Färber m. ‖ teinturier m. / woollen ~ ‖ Wollfärber m. ‖ teinturier m. sur laine.

dyer's bath ‖ Färberflotte f. ‖ bain m. de teinture. / ~ broom flower ‖ Färberginsterblüte f. ‖ fleur f. de genestrolle. / ~ moss ‖ Färberflechte f. ‖ lichen m. tinctorial. / ~ spirit ‖ Zinnkomposition f. ‖ composition f. d'étain.

dyes pl. for home use ‖ Farben fpl. für den Haushalt ‖ teintures fpl. pour le ménage.

dye sprayer ‖ Farbenspritzapparat m. ‖ appareil m. pulvérisateur de couleurs.

dyestuff ‖ Farbstoff m. ‖ matière f. colorante. / artificial ~ ‖ künstlicher Farbstoff m. ‖ matière f. colorante artificielle. / natural ~ ‖ natürlicher Farbstoff m. ‖ matière f. colorante naturelle.

dyestuff factory ‖ Farbenfabrik f. ‖ industrie f. des matières colorantes. / ~ industry ‖ Farbstoffindustrie f. ‖ industrie f. des colorants.

dye trade machine ‖ Farbenindustriemaschine f. ‖ machine f. pour l'industrie des couleurs.

dyewood ‖ Farbholz n. ‖ bois m. colorant *ou* de teinture. / red ~ ‖ Rotholz n. ‖ bois m. rouge *ou* d'huile.

dyewood cutter ‖ Farbholzmüller m. ‖ râpeur m. de bois de teinture. / ~ extract ‖ Farbholzauszug m.; Farbholzextrakt m. ‖ extrait m. de bois colorant *ou* de teinture. / ~ grinding ‖ Farbholzmühle f. ‖ trituration f. de bois de teinture.

dye works pl. ‖ Färberei f. ‖ teinturerie f.

dying out of the oscillation ‖ Abklingen n. der Schwingung ‖ décroissement m. *ou* amortissement m. de l'oscillation.

dyke *see* dike.

dynameter ‖ Dynameter n. ‖ dynamètre m.

dynamic ‖ dynamisch ‖ dynamique. / ~ equilibrium ‖ dynamisches Gleichgewicht n. ‖ équilibre m. dynamique. / ~ experiments pl. ‖ dynamische Untersuchung f. ‖ vérification f. dynamique. / ~ lift ‖ dynamischer Auftrieb m. ‖ poussée f. dynamique. / ~ pressure ‖ Staudruck n. ‖ pression f. dynamique. / ~ strength ‖ Schwingungsfestigkeit f.; dynamische Festigkeit f. ‖ résistance f. à l'oscillation. / ~ stresses pl. ‖ Schwingungsbeanspruchung f.; dynamische Beanspruchung f. ‖ effort m. d'oscillation. / ~ unit ‖ Krafteinheit f. ‖ unité f. de force.

dynamical *see* dynamic.

dynamically balanced ‖ dynamisch ausgewuchtet ‖ équilibré dynamiquement. / ~ stable ‖ dynamisch stabil ‖ dynamiquement stable.

dynamics pl. ‖ Dynamik f. ‖ dynamique f.

dynamite ‖ Dynamit n. ‖ dynamite f. / ~ cartridge ‖ Dynamitpatrone f. ‖ cartouche f. de dynamite. / ~ works pl. ‖ Dynamitfabrik f. ‖ fabrique f. de dynamite; dynamiterie f.

dynamo ‖ Dynamomaschine f. ‖ machine f. dynamo-électrique; dynamo f. / belt-driven ~ ‖ Dynamo f. für Riemenantrieb ‖ génératrice f. à commande par courroie. / brake ~ ‖ Bremsdynamo f. ‖ dynamo-frein m. / ~ for buffer battery ‖ Dynamomaschine f. für Pufferbetrieb ‖ dynamo f. pour les batteries-tampon. / closed-coil armature ~ ‖ Dynamo f. mit geschlossener Ankerwicklung ‖ dynamo f. avec induit à bobinage fermé. / compound-wound ~ ‖ Verbunddynamomaschine f. ‖ dynamo f. compound. / constant-current ~ ‖ Dynamo f. für konstante Stromstärke ‖ dynamo f. à intensité constante. / constant-pressure ~ ‖ Dynamo f. für konstante Spannung ‖ dynamo f. à tension constante. / constant-voltage ~ ‖ Dynamo f. mit konstanter Spannung ‖ dynamo f. à voltage constant. / continuous-current ~ ‖ Gleichstromdynamo f. ‖ génératrice f. à courant continu. / cranked ~ ‖ Kurbeldynamo n. ‖ dynamo m. manuel. / direct driven ~ ‖ direkt gekuppelte Dynamo f. ‖ dynamo f. à couplage direct. / flat-ring ~ ‖ Flachringdynamo f. ‖ génératrice f. à anneau plat. / gas ~ ‖ Gasdynamomaschine f. ‖ dynamo f. accouplée à un moteur à gaz. / lighting ~ ‖ Lichtdynamo f. ‖ dynamo f. d'éclairage. / low-tension ~ ‖ Niedrigspannungsdynamo f. ‖ dynamo f. à basse tension. / motorcycle ~ ‖ Motorfahrraddynamo f. ‖ dynamo f. de motocyclette. / ~ with salient poles ‖ Außenpoldynamo f. ‖ dynamo f. à pôles extérieurs. / selfexcited ~ ‖ selbsterregte Dynamo f. ‖ dynamo f. auto-excitatrice. / self-regu-

lating ~ ‖ selbstregelnde Dynamo f. ‖ dynamo f. auto-régulatrice. / separately-excited ~ ‖ fremderregte Dynamo f. ‖ génératrice f. à excitation indépendante. / series ~ ‖ Hauptstrommaschine f. ‖ dynamo m. en série. / shunt wound ~ ‖ Nebenschlußdynamo f. ‖ dynamo f. shunt. / train-lighting ~ ‖ Zugbeleuchtungsdynamo f. ‖ dynamo f. pour l'éclairage des trains. / unipolar ~ ‖ Unipolardynamo f. ‖ dynamo f. unipolaire.

dynamo brush ‖ Dynamobürste f. ‖ brosse f. à dynamos; balai m. de génératrice. / ~ electric machine ‖ Dynamomaschine f. ‖ machine f. dynamo-électrique; dynamo f. / ~ fitter ‖ Dynamomonteur m. ‖ monteur m. de dynamos. / ~ machine *see* dynamo.

dynamometer ‖ Kraftmesser m.; Dynamometer n. ‖ dynamomètre m.

dynamometrical ‖ dynamometrisch ‖ dynamométrique. / ~ brake ‖ Bremsdynamometer n. ‖ frein m. dynamométrique.

dynamo oil ‖ Dynamoöl n. ‖ huile f. pour dynamo. / plate pasting machine for ~ plates ‖ Blechbeklebemaschine f. für Dynamobleche ‖ machine à coller du papier sur les tôles de dynamo. / ~ shaft ‖ Dynamowelle f. ‖ arbre m. de dynamo. / ~ sheet ‖ Dynamoblech n. ‖ tôle f. à dynamos. / ~ steel ‖ Dynamostahl m. ‖ acier m. à dynamos. / ~ steel casting ‖ Dynamostahlguß m. ‖ moulage m. en acier à dynamos.

dynamotor ‖ Umformer m. ‖ dynamoteur m. / alternating-current continuous-current ~ ‖ Wechselstromgleichstromumformer m. ‖ dynamoteur m. à courant alternatif en courant continu. / continuous-current alternating-current ~ ‖ Gleichstromwechselstromumformer m. ‖ dynamoteur m. à courant continu en courant alternatif. / continuous-current continuous-current ~ ‖ Gleichstromgleichstromumformer m. ‖ dynamoteur m. à courant continu en courant continu. / continuous-current rotary-current ~ ‖ Gleichstromdrehstromumformer m. ‖ dynamoteur m. à courant continu en courant triphasé. / rotary-current continuous-current ~ ‖ Drehstromgleichstromumformer m. ‖ dynamoteur m. à courant triphasé en courant continu.

dynamo wagon ‖ Dynamowagen ‖ voiture f. porte générateur.

dynatron ‖ Dynatron n. ‖ dynatron m.

dyne ‖ Dyne f. ‖ dyne f.

dyscrasite ‖ Antimonsilber n. ‖ discrase f.

dysentery ‖ Ruhr f. ‖ dysenterie f.

dysluite ‖ Gahnit m.; Dysluit m.; Zinkspinell m. ‖ gahnite f.; dysluite f.

E

eagle wood ‖ Adlerholz n. ‖ bois m. d'aigle.

ear ‖ Ohr n. ‖ oreille f. / ~ (Corn) ‖ Ähre f. ‖ épi m. / ~ (Hammer) ‖ Feder f. ‖ languette f. / ~ (Vessel) ‖ Henkel m.; Griff m. ‖ anse f. / ~ of a cord ‖ Seilöhr n.; Tauöse f. ‖ élingue f. d'un cordage. / ~ of a rope see ~ of a cord. / small ~ ‖ Ährchen n. ‖ épillet m.; jeune épi m. / young ~ see small ~.

ear-ache ‖ Ohrenschmerzen mpl.; Ohrenweh n. ‖ mal m. d'oreille.

ear cap see ear shield. / ~ chain ‖ Ohrkettchen n. ‖ chaînette f. pour l'oreille. / ~ cleaner ‖ Ohrenreiniger m. ‖ cure-oreille m.

earing (Mar) ‖ Nockhorn n. ‖ empointure f.

ear loop ‖ Ohrspange f. ‖ arc m. auriculaire.

early hop ‖ Augusthopfen m. ‖ houblon m. hâtif ou précoce.

earnest money ‖ Angeld n. ‖ arrhes fpl.

earnings pl. ‖ Gewinn m. ‖ gain m.; bénéfice m.; profit m. / average ~ ‖ Durchschnittsverdienst m. ‖ recettes fpl. moyennes. / gross ~ ‖ Rohgewinn m. ‖ recettes fpl. brutes. / net ~ ‖ Nettogewinn m. ‖ recettes fpl. nettes.

ear protection ‖ Gehörschutz m. ‖ protection f. contre les chocs acoustiques. / ~ ring ‖ Ohrring m. ‖ boucle f. d'oreilles.

ears pl., empty ‖ Abschöpfgerste f. ‖ écume f.; orge f. légère; petite orge f.; orgette f.

ear shell ‖ Irismuschel f. ‖ oreille f. de mer. / ~ shield ‖ Ohrenschützer m. ‖ protège-oreille m.; oreillet m.

earshot, within ~ (Tel) ‖ in Hörbereich m. ‖ en intervalle f. audible; à distance d'oreille.

ears-lifter ‖ Ährenheber m. ‖ releveur m. d'épis.

ear spindle (Corn) ‖ Ährenspindel f. ‖ tige f. d'épi. / ~ tab see ~ shield.

earth ‖ Erde m. ‖ terre f. / ~ (Build) ‖ Erdboden m. ‖ terre f.; terrain m. / ~ (Electr) ‖ Erdschluß m. ‖ terre f. accidentelle; mise f. à la terre. / ~ (El line) ‖ Erdverbindung f. ‖ connexion f. de ou à la terre. / ~ (Motor) ‖ Erdung f. ‖ masse f. / ~ for manufacturing ceramic products ‖ Erde f. zur Herstellung von keramischen Erzeugnissen ‖ terre f. réfractaire pour la fabrication de produits céramiques. / to deposit the ~ ‖ den Boden m. aufschütten ‖ recharger ou relever le terrain. / to excavate the ~ ‖ den Boden m. ausgraben ‖ creuser ou extraire la terre; déblayer le terrain. / fire-proof ~ ‖ feuerfeste Erde f. ‖ terre f. réfractaire. / green ~ ‖ Grünerde f.; Seladonit m. ‖ céladonite f. / ~ in natural state ‖ gewachsener Boden m. ‖ terrain m. naturel; terre f. vierge. / rare ~s pl. ‖ seltene Erden fpl. ‖ terres fpl. rares. / rotten ~ ‖ Modererde f. ‖ terre f. pourrie. / x-million cubic meters of ~ had to be shifted ‖ die Bodenbewegung f. war x Millionen Kubikmeter ‖ le déblayage de terres atteignit un volume de x millions de mètres cubes. / ~ obtained from the soil ‖

aus dem Boden gewonnene Erde f. ‖ terre f. extraite du sol. / vegetable ~ ‖ Pflanzenerde f. ‖ terreau m.

earth antenna ‖ Erdantenne f. ‖ antenne f. basse. / ~ arrester (Radio) ‖ Erdverbindungskurzschließer m. ‖ court-circuiteur m. de mise à la terre. / ~ auger ‖ Tellerbohrer m.; Erdbohrer m. ‖ tarière f. à large spire. / ~ bank ‖ Schüttdamm m.; Erddamm m. ‖ levée f. ou barrage m. de terre; remblai m. / ~ borer ‖ Erdbohrer m. ‖ sonde f.; tarière f. / ~ boring auger ‖ Erdbohrer m. ‖ sonde f.; trépan m.; tarière f. à sonder. / ~ brace (El line) ‖ Unterriegel m. ‖ entretoise f. souterraine. / ~ capacity difference meter ‖ Erdkapazitätsdifferenzmesser m. ‖ appareil m. de mesure de différence de capacité par rapport à la terre. / ~ clip (Tel) ‖ Erdschelle f. ‖ borne f. de mise à la terre. / ~ coal ‖ bituminöse Holzerde f. ‖ bois m. bitumineux terreux; lignite m. terreux.

earth connection ‖ Erdableitung f.; Erdverbindung f. ‖ dérivation f. à la terre; connexion f. de terre. / power plant's ~ (Electr) ‖ Erdleitung f. für Stromversorgungsanlagen ‖ mise f. à la terre d'installations d'énergie. / ~ of roof poles ‖ Rohrständererdleitung f. ‖ conducteur m. de mise à la terre du chevalet en tubes. / ~ for terminal poles (Electr) ‖ Erdleitung f. für Kabelüberführungen ‖ terre f. des boîtes de raccordement.

earth contact ‖ Erdkontakt m. ‖ contact m. terrestre. / ~ control for fire-alarm systems ‖ Erdschlußkontrolle f. in Feuermeldeanlagen ‖ contrôle m. de mise à la terre aux avertisseurs d'incendie.

earth current (Electr) ‖ Erdstrom m.; Erdrückstrom m. ‖ courant m. dans la terre; courant m. tellurique ou de retour par terre. / metallic circuit against ~ ‖ Erdstromschleife f. ‖ compensation f. du courant tellurique. / ~ equalizer ‖ Erdstromausgleicher m. ‖ compensateur m. de courant tellurique.

earth density ‖ Erddichte f. ‖ densité f. de la terre. / ~ direction finder ‖ Erdpeilgerät n. ‖ dispositif m. de recherche de la direction par le sol. / ~ dissymmetry of a loop ‖ Erdungsunsymmetrie f. einer Doppelleitung ‖ déséquilibre m. d'une boucle par rapport à la terre. / ~ drill ‖ Erdbohrer m. ‖ trépan m.

earthed circuit ‖ geerdeter Stromkreis m. ‖ circuit m. à la terre. / ~ iron strap ‖ Erdungsbügel m. ‖ arceau m. protecteur de mise à la terre.

earthen ‖ irden ‖ en terre. / ~ floor ‖ Lehmestrich m.; Lehmtenne f. ‖ aire f. en argile ou de repous. / ~ pipe ‖ Tonröhre f. ‖ tuyau m. en poterie.

earthenware ‖ Steingutwaren fpl.; Tonwaren fpl.; irdenes Geschirr n. ‖ articles mpl. en faïence; poterie f. / burnt ~ ‖ gebrannter Ton m. ‖ terre f. cuite. / glazed ~ ‖ verglaste Tonwaren fpl. ‖ faïence f.

earthenware decorator ‖ Steingutmaler m. ‖ peintre m. sur faïence. / ~ disk for clocks and watches ‖ Ziffernblatt n. aus Steingut ‖ cadran m. en faïence. / ~ fill-in plate for mounting metal and wood ‖ Steinguteinlageplatte f. für Metall- und Holzmontage ‖ plaque f. en faïence pour montures en métal et en bois. / ~ glazing ‖ Steingutglasur f. ‖ glaçure f. de ou à faïence. / ~ jar ‖ Tonzelle f. ‖ récipient m. en grès; vase m. en terre cuite. / ~ manufacturing plant ‖ Steingutfabrikeinrichtung f. ‖ équipement m. de fabrique de faïences. / ~ pipe ‖ Tonröhre f. ‖ tuyau m. en poterie. / ~ sieve ‖ Steinzeugsieb n. ‖ passoire f. en grès. / ~ slab ‖ Tonplatte f. ‖ carreau m. de faïence. / ~ stove ‖ Tonofen m. ‖ poêle f. en faïence. / ~ tank ‖ Steingutwanne f. ‖ cuve f. en grès. / ~ vat ‖ Steingutbottich m. ‖ cuve f. en grès.

earth flat ‖ Asbest n. ‖ amiante m.; filasse f. de montagne. / ~ hoe ‖ Erdhacke f.; Haue f. ‖ hoyau m.; pic m.; binette f. / ~ indicator for clock systems ‖ Erdschlußanzeiger m. in Uhrenanlagen ‖ indicateur m. de mise à la terre accidentelle de centrales d'horloges. / ~ inductor compass ‖ Erdinduktionskompaß m. ‖ compas m. d'induction terrestre.

earthing (Electr) ‖ Erdung f. ‖ mise f. à la terre. / unexceptionable ~ ‖ einwandfreie Erdung f. ‖ bon état m. de la mise à la terre.

earthing leakage resistance (Tel) ‖ Erdableitungswiderstand m. ‖ résistance f. d'écoulement de la mise à la terre. / ~ screw ‖ Erdungsschraube f. ‖ vis m. de mise à la terre.

earth interior ‖ Erdinneres n. ‖ intérieur m. de la terre. / ~ leakage indicator ‖ Erdschlußanzeiger m. ‖ indicateur m. de contact électrique terrestre. / ~ leakage meter ‖ Erdschlußmesser m. ‖ indicateur m. de perte à la terre. / ~ loop (Tel) ‖ Erdfehlerschleife f. ‖ boucle f. de mise à la terre.

earth-magnetic declination ‖ erdmagnetische Deklination f. ‖ déclinaison f. magnétique terrestre. / ~ instrument ‖ erdmagnetisches Instrument n. ‖ instrument m. magnétique terrestre.

earth masses pl. ‖ Erdmassen fpl. ‖ terres fpl.; masses fpl. de terre. / ~ moving ‖ Erdbewegung f. ‖ mouvement m. des terres. / ~ noise (Tel) ‖ Erdgeräusch n. ‖ bruit m. dû à la terre. / ~ nut ‖ Erdnuß f. ‖ pistache f. de terre; arachide f. / ~ nut oil ‖ Erdnußöl n. ‖ huile f. d'arachides. / ~ oil ‖ Naphtha f.; Erdöl n. ‖ naphte f. / ~ pipe ‖ Erdleitungsrohr n. ‖ tube m. de mise à la terre. / ~ pitch ‖ Erdpech n.; Asphalt m. ‖ bitume m. solide; asphalte m. / ~ plate (Electr) ‖ Erder m. ‖ prise f. ou plaque f. de terre. / ~ pressure ‖ Erddruck m. ‖ pression f. ou poussée f. des terres.

earthquake ‖ Erdbeben n. ‖ tremblement m. de terre; séisme m. / ~ wave ‖ Erd-

bebenwelle f. ‖ onde f. produite par un tremblement de terre.

earth rammer ‖ Handramme f.; Erdstampfe f. ‖ dame f.; demoiselle f.; hie f. à main. / measurement of ~ resistance ‖ Erdleitungsmessung f. ‖ mesure f. de la resistance d'une mise à la terre. / ~ return circuit ‖ Erdrückleitung f. ‖ retour m. par terre. / ~ roof (Mine) ‖ Abraum m. ‖ couche f.; lit m. de terre; lit m. de décombres *ou* de déblai.

earths pl., rare ‖ Edelerden fpl. ‖ terres fpl. rares.

earthshake due to tunnelling ‖ Tunnelbeben n. ‖ tremblement m. de terre causé par le percement d'un tunnel. / ~ due to underground collapse ‖ Einsturzbeben n. ‖ séisme m. dû à un effondrement.

earth shifting *see* ~ moving. / ~ short-circuit ‖ Erdkurzschluß m. ‖ court circuit m. de mise à la terre. / ~ stay (El line) ‖ Fußanker m. ‖ hauban m. de poteau. / ~ subsidence ‖ Erdsenkung f. ‖ affaissement m. du sol. / ~ surface ‖ Erdoberfläche f. ‖ surface f. de la terre. / ~ telegraphy ‖ Erdtelegrafie f. ‖ télégraphie f. par la terre. / ~ terminal arrester (Tel) ‖ unterbrochener Erdanschluß m. ‖ éclateur m. de mise à la terre. / ~ tester ‖ Erdungsmesser m. ‖ mesureur m. de la mise à la terre. / ~ traverse (El line) ‖ Unterriegel m. ‖ entretoise f. souterraine *ou* inférieure. / ~ wall ‖ Erdwall m. ‖ digue f.; rempart m. de terre. / ~ wax ‖ Ozokerit m.; Erdwachs n.; Bergwachs n. ‖ ozokérite m. / ~ wire for poles (El line) ‖ Stangenerdleitung f. ‖ conducteur m. de terre pour poteaux. / ~ wire clamp ‖ Untersuchungsklemme f. für Blitzableiterseile ‖ borne f. de coupure pour conducteurs de paratonnerre. / ~ wireless telegraphy ‖ Erdfunkerei f. ‖ radiotélégraphie f. par la terre.

earthwork ‖ Erdarbeit f.; Bodenbewegung f. ‖ terrassement m.; transport m. des terres. / ~ (Railw) ‖ Unterbau m. ‖ infrastructure f. de la voie. / slipping ~ ‖ rutschender Boden m. ‖ terrain m. mouvant.

earthwork contractor ‖ Erdbauunternehmer m. ‖ entrepreneur m. de terrassements. / ~ device ‖ Vorrichtung f. für Erdarbeiten ‖ appareil m. pour travaux de terrassement. / ~ foreman ‖ Schachtmeister m. ‖ maître-terrassier m. / ~ labourer ‖ Erdarbeiter m. ‖ terrassier m.

earthy ‖ erdig; mulmig ‖ terreux; friable. / ~ brown coal (Geol) ‖ mulmige Braunkohle f. ‖ lignite m. terreux. / ~ colours pl. ‖ Erdfarben fpl. ‖ couleurs fpl. minérales. / ~ gypsum ‖ Gipserde f. ‖ terre f. gypseuse. / ~ peat ‖ erdiger Torf m. ‖ tourbe f. terreuse. / ~ talc ‖ Talk m. ‖ talc m. terreux.

ease, to (Build) ‖ entlasten ‖ soulager. / ~ down ‖ sacken lassen; herunterfieren ‖ mollir; larguer *ou* laisser descendre. / ~ off the sheets (Mar) ‖ die Schoten fpl. abfieren ‖ filer les écoutes fpl.

ease (Anti difficulty) ‖ Leichtigkeit f. ‖ légèreté f.; facilité f. / ~ (Room) ‖ Komfort m. ‖ confort m.

easel ‖ Staffelei f. ‖ chevalet m. d'un peintre.

easiness *see* ease.

east ‖ Ost m.; Osten m. ‖ est m.

Easter article ‖ Osterartikel m. ‖ cadeau m. de Pâques. / ~ card ‖ Osterkarte f. ‖

carte f. pour fêtes de Pâques. / ~ fair ‖ Ostermesse f. ‖ foire f. de Pâques.

east gale ‖ heftiger Ostwind m. ‖ coup m. de vent d'est. / ~ longitude ‖ östliche Länge f. ‖ longitude f. est *ou* orientale.

eastward ‖ östlich; ostwärts ‖ à l'est; dans l'est; vers l'est.

easy ‖ leicht ‖ facile. / ~ access ‖ leicht zugänglich ‖ d'un accès facile. / ~ management ‖ große Handlichkeit f. ‖ maniement m. *ou* service m. facile. / ~ to read ‖ leicht leserlich ‖ facile à lire.

eaves pl. ‖ Dachtraufe f. ‖ égout m.; larmier m. / ~ course ‖ Traufziegelreihe f. ‖ battellement m. / ~ lead ‖ Traufplatte f. ‖ bavette f. d'égout. / ~ mouldings pl. ‖ Dachgesims n. ‖ corniche f. au pied du toit *ou* de toiture. / ~ trough hanger ‖ Haken m. einer Dachrinne ‖ crochet m. de gouttière.

ebb ‖ Ebbe f. ‖ basse-marée f.; jusant m.; reflux m. / half ~ ‖ halbe Ebbe f. ‖ mijusant m.; demi-marée f. baissante.

ebb beginning ‖ Vorebbe f. ‖ commencement m. de jusant. / first quarter of the ~ ‖ Vorebbe f. ‖ commencement m. du jusant. / ~ stream ‖ Ebbstrom m. ‖ courant m. de jusant; courant m. de sortie. / ~ tide ‖ Ebbe f. ‖ marée descendante f.

ebonist ‖ Kunsttischler m.; Kunstschreiner m. ‖ ébéniste m.

ebonite ‖ Ebonit n.; Hartgummi m. ‖ ébonite f.; caoutchouc m. durci. / ~ cell ‖ Ebonitzelle f. ‖ élément m. avec bac en ébonite. / ~ guard insulator ‖ Ebonitschutzglocke f. ‖ protecteur m. d'isolateur en ébonite. / ~ plate ‖ Ebonitscheibe f. ‖ disque m. en ébonite. / ~ pulley ‖ Hartgummischeibe f. ‖ poulie f. en ébonite. / ~ ring ‖ Hartgummiring m. ‖ bague f. *ou* anneau m. en ébonite. / ~ tube ‖ Ebonitrohr n. ‖ tube m. en ébonite. / ~ ware ‖ Hartgummiwaren fpl. ‖ articles mpl. en caoutchouc durci.

ebonize, to ‖ schwarz beizen ‖ ébéner.

ebony ‖ schwarzes Ebenholz n. ‖ ébène f.; bois m. d'ébène. / German ~ ‖ Eibenholz n. ‖ ébène f. d'Allemagne. / red ~ ‖ Rotebenholz n.; rotes Ebenholz n. ‖ ébène f. rouge.

ebony oil ‖ Ebenholzöl n. ‖ huile f. *ou* essence f. de bois d'ébène. / ~ tree ‖ Ebenholzbaum m. ‖ ébénier m. / ~ wood *see* ebony.

ebullioscope ‖ Ebullioskop n.; Siedemesser m. ‖ ébullioscope m.

ebullition ‖ Kochen n.; Sieden n. ‖ ébullition f.

eccentric ‖ exzentrisch ‖ excentrique.

eccentric ‖ Exzenter m. n. ‖ excentrique m. / ~ drive ‖ Exzenterantrieb m. ‖ commande f. par excentrique. / ~ gab ‖ Exzentergabel f. ‖ fourche f. excentrique. / ~ gitting press ‖ Exzenterabkneifpresse f. ‖ presse f. à excentrique pour entailler; coupe-jet m. / ~ hoop ‖ Exzenterbügel m. ‖ collier m. / ~ lever ‖ Exzenterhebel m.; exzentrischer Hebel m. ‖ levier m. excentrique. / ~ movement ‖ Exzenterbewegung f. ‖ mouvement m. par excentrique. / ~ ophtalmoscope method ‖ azentrische Ophtalmoskopie f. ‖ ophtalmoscopie f. excentrique. / ~ pin ‖ Exzenterzapfen m. ‖ bouton m. excentré. / ~ press ‖ Exzenterpresse f. ‖ presse f. à excentrique. / ~ pressure ‖ Exzenterdruck m. ‖ pression f. d'excentrique. /

~ ring ‖ Exzenterring m. ‖ collier m. d'excentrique. / ~ rod ‖ Schwingenstange f.; Exzenterstange f. ‖ tige f. d'excentrique. / ~ scale ‖ exzentrische Skale f. ‖ échelle f. excentrique. / ~ setting of the instrument ‖ außermittige Aufstellung f. des Instrumentes ‖ disposition f. excentrique de l'instrument. / ~ shaft ‖ Exzenterwelle f. ‖ arbre m. d'excentrique. / ~ sheave ‖ Exzenterscheibe f.; Hubscheibe f. ‖ plateau m. *ou* disque m. d'excentrique. / ~ sheave half ‖ Hubscheibenhälfte f. ‖ moitié f. de disque d'excentrique. / ~ strap ‖ Exzenterbügel m.; Hubscheibenring m. ‖ collier m. *ou* bride f. d'excentrique. / ~ trimming press ‖ Exzenterabgratpresse f. ‖ presse f. excentrique à ébarber.

eccentricity ‖ Exzentrizität f. ‖ excentricité f. / ~ error ‖ Exzentrizitätsfehler m. ‖ erreur f. d'excentricité.

echelon working (Tel) ‖ Staffelbetrieb m. ‖ service m. échelonné.

echo ‖ Echo n.; Widerhall m. ‖ écho m. / ~ attenuation measuring set ‖ Echomesser m. ‖ appareil m. à mesurer l'affaiblissement de l'écho; échomètre m. / ~ current (Tel) ‖ Echostrom m. ‖ courant m. de l'écho. / ~ effect ‖ Echowirkung f. ‖ effet m. d'echo.

echometer ‖ Schallmesser m.; Tonmesser m. ‖ sonomètre m.; échomètre.

echo microphone ‖ Echomikrofon n. ‖ microphone m. d'écho. / ~ receiver ‖ Echoempfänger m. ‖ récepteur m. d'écho. / ~ sounding ‖ Echolotung f. ‖ sondage m. par son *ou* par écho. / ~ suppresser ‖ Echosperre f. ‖ suppresseur m. d'écho.

eclipse of the light of a light-house ‖ Verfinsterung f. des Lichtes eines Leuchtfeuers ‖ éclipse f. de la lumière d'un phare. / ~ system ‖ Eklipsenbauart f. ‖ système m. à éclipse.

economic ‖ sparsam; wirtschaftlich; ökonomisch ‖ économique. / ~ committee ‖ Wirtschaftskomitee n. ‖ comité m. économique. / the varied ~ conditions pl. of the different countries ‖ die verschiedenartigen Wirtschaftsverhältnisse npl. der einzelnen Länder ‖ les conditions fpl. économiques qui varient d'un pays à l'autre. / ~ conference ‖ Wirtschaftskonferenz f. ‖ conférence f. économique. / ~ fluctuation ‖ wirtschaftliche Schwankung f. ‖ variation f. économique. / ~ limit ‖ Wirtschaftlichkeitsgrenze f. ‖ limite f. économique. / ~ oiling apparatus ‖ Ölsparapparat m. ‖ graisseur m. économique. / ~ plan ‖ Wirtschaftsplan m. ‖ plan m. d'économie. / ~ point of view ‖ wirtschaftlicher Gesichtspunkt m. ‖ point m. de vue économique. / ~ policy ‖ Wirtschaftspolitik f. ‖ politique f. économique. / ~ production ‖ wirtschaftliche Herstellung f. ‖ fabrication f. économique. / to exploit something for private ~ purposes ‖ etwas privatwirtschaftlich ausbeuten ‖ exploiter quelque chose à titre d'entreprise privée. / ~ work ‖ rationelle Arbeitsweise f. ‖ mode m. de travail économique; production rationnelle.

economical *see* economic.

economics pl. *see also* economy ‖ Wirtschaft f.; Wirtschaftslehre f. ‖ économie f. / ~ of a mine ‖ Grubenhaushalt m. ‖ économie f. de la mine.

economist, political ~ ‖ Volkswirt m.; Nationalökonom m. ‖ économiste m.

economizer ‖ Ekonomiser m.; Vorwärmer m. ‖ économiseur m. / ~ press ‖ Ekonomiserpresse f. ‖ presse f. à assembler les économiseurs.

economizing furnace ‖ Sparfeuerung f. ‖ foyer-économiseur m.

economy ‖ Wirtschaftlichkeit f.; Sparsamkeit f. ‖ économie f. / national ~ see political ~. / political ~ ‖ Staatswirtschaft f.; Volkswirtschaft f.; National-ökonomie f.; Volkswirtschaftslehre f. ‖ économie f. politique. / ~ in space ‖ Raumersparnis f. ‖ économie f. de place. / ~ of time and wages ‖ Ersparnis f. an Zeit und Arbeitslohn ‖ économie f. de temps et de salaire. / ~ in running the works ‖ Betriebswirtschaft f. ‖ économie f. d'exploitation.

ecrasite ‖ Ekrasit n. ‖ écrasite m.

ectype (Coin) ‖ Abdruck m. einer Münze ‖ ectype m.

eddy ‖ Neer f.; Wasserstrudel m.; Wasserwirbel m. ‖ remous m.; tournant m. / free from eddies pl. ‖ wirbellos ‖ sans remous mpl.; sans tourbillons mpl.

eddy current (Electr) ‖ Wirbelstrom m. ‖ courant m. parasite ou de Foucault. / ~ brake ‖ Wirbelstrombremse f. ‖ frein m. à courants parasites. / ~ loss ‖ Wirbelstromverlust m. ‖ perte f. par courants de Foucault.

edge, to (Millstone) ‖ schärfen; picken ‖ repiquer; rhabiller. / ~ (Pap) ‖ beschneiden ‖ ébarber; rogner. / ~ (Tinm) ‖ bördeln ‖ border. / ~ (Textile) ‖ einfassen ‖ border; ourler. / ~ the paper ‖ das Papier beschneiden ‖ rogner le papier. / ~ the timber ‖ das Holz abschwarten ‖ couper les flaches fpl.

edge (Curve) ‖ Knick m. ‖ angle m. / ~ (Geom; Miner) ‖ Kante f. ‖ arête f. / ~ (Join) ‖ Kante f.; Fase f.; Rand m. ‖ carne f.; arête f.; tranche f. / ~ (Millstone) ‖ Schärfe f. ‖ riblage m. / ~ (Shipb) ‖ Naht f. ‖ couture f. / ~ (Tool) ‖ Schärfe f.; Schneide f. ‖ mèche f.; fil m.; coupant m.; tranchant m.; face f. de l'outil.; airscrew blade leading ~ ‖ Leitkante f. der Luftschraube ‖ coupant m. de la pale d'hélice. / ~ going beyond the piece ‖ vorspringende Kante f. ‖ saillie f.; rebord m. en saillie. / blunt ~ ‖ stumpfe Schneide f. ‖ tranchant m. émoussé. / ~ of a board ‖ hohe Kante f. oder schmale Seite f. eines Brettes ‖ carne f. ou champ m. ou tranche f. d'une planche. / ~ of a book ‖ Schnitt m. an einem Buche ‖ tranche f. d'un livre. / ~ of a brick ‖ hohe Kante f. eines Ziegels ‖ tranche f. ou carne f. ou champ m. d'une brique. / ~ of a coin ‖ Rand m. einer Münze ‖ tranche f. d'une monnaie. / ~ of a crystal ‖ Kristallkante f. ‖ arête f. d'un cristal. / droved ~ ‖ volle Kante f. ‖ arête f. vive. / feather ~ ‖ dünne Kante f. ‖ chanfrein m.; biseau m. / ~ of flange ‖ Bordkante f. ‖ arête f. de tôle. / full ~ ‖ scharfe Kante f. ‖ arête f. vive. / ~ of a hammer ‖ Finne f. des Hammers ‖ panne f. du marteau. / ~ hammered down (Tinm) ‖ Falz m.; Fugenfalz m. ‖ avisure f. / ~ of hole ‖ Lochkante f. ‖ arête f. de trou. / hollowground ~ ‖ eingeschliffene Hohlkehle f. ‖ cannelure f. obtenue à la meule. / inner ~ of the horseshoe ‖ innerer Rand m. des Hufeisens ‖ rive f. ou bordure f. intérieure d'un fer à cheval. / to machine the ~s ‖ Kanten fpl. bearbeiten oder be-

stoßen ‖ dresser les arêtes fpl. / milled ~ ‖ geriefter Rand m. ‖ bord m. moleté. / on ~ ‖ hochkant; auf der schmalen oder auf der hohen Kante f. ‖ de champ m. / outer ~ of the horseshoe ‖ äußerer Rand m. des Hufeisens ‖ rive f. ou bordure f. extérieure du fer à cheval. / ~ of a pendulum ‖ Schneide f. eines Pendels ‖ couteau m. d'un pendule. / ~ of plate ‖ Blechrand m. ‖ bord m. de tôle. / with polished ~s pl. ‖ mit polierten Kanten fpl. ‖ à bords mpl. polis. / ~ of regression (Geom) ‖ Rückkehrlinie f. ‖ arête f. de rebroussement. / rough ~ ‖ Grat m. ‖ barbe f.; morfil m. / with rounded ~s ‖ mit abgerundeten Kanten fpl. ‖ à coins mpl. arrondis. / ~ of the screwblade ‖ Kante f. des Schiffsschraubenflügels ‖ arête f. d'une aile d'hélice. / sharp ~ ‖ scharfe Kante f. ‖ arête f. vive. / with sharp ~s pl. ‖ mit scharfen Kanten fpl. ‖ à vives arêtes fpl.; présentant d'angles vifs mpl. / ~ of slope ‖ Böschungsrand m. ‖ bord m. du talus. / to take off the ~s pl. ‖ die Kanten fpl. brechen ‖ chanfreiner. / terminal ~ (Crystal) ‖ Polkante f. ‖ arête f. culminante. / trimmed ~ ‖ bestoßene Kante f. ‖ arête f. écornée ou taillée. / up to the ~ ‖ bis zum Rand m. ‖ jusqu'au bord m. / upper ~ of the tooth ‖ Zahnoberkante f. ‖ arête f. extérieure de dent. / ~ of a wedge ‖ Schärfe f. oder Schneide f. eines Keils ‖ tranchant m. d'un coin ou d'une cale. / the width of ~ can be made as required ‖ die Breite f. des Randes kann beliebig bemessen werden ‖ les tôles fpl. sont fournis avec bords de largeurs quelconques. / window ~ ‖ Fensterecke f. ‖ coin m. de fenêtre. / wire ~ ‖ Grat m. ‖ morfil m.; barbe f.

edge bolt ‖ Augbolzen m.; Ringbolzen m.; Ringnagel m. ‖ cheville f. à boucle; piton m.; piton m. à anneau. / ~ bond ‖ Eckverband m. ‖ assemblage m. angulaire. / ~ colourer (Bookb) ‖ Schnittausmaler m. ‖ colorieur m. sur tranches.

edged, keen ‖ mit scharfen Karten fpl.; mit scharfer Schneide f. ‖ à vives arêtes fpl.; présentant d'angles vifs mpl. / ~ surface ‖ kantige Oberfläche f. ‖ surface f. anguleuse.

edge fracture ‖ Kantenriß m. ‖ crevasse f. du bord. / ~ joint ‖ Eckverzapfung f.; Eckverband m. ‖ assemblage m. d'angle. / ~ joint file ‖ Scharnierfeile f. ‖ lime f. plate à coulisse ou à charnière. / ~ mill ‖ Kollergang m. ‖ broyeur m. à meules; (tordoir m.). / ~ milling machine ‖ Kantenfräsmaschine f. ‖ chanfreineuse f.

edger (Metal) ‖ Beschneider m. ‖ rogneur m.

edge rail ‖ Kantenschiene f.; ornière f.; bande f. saillante; rail m. à rebord.

edge runner ‖ Kollergang m. ‖ broyeur m. à meules. / ~ with stationary and rotary grinding track ‖ Kollergang m. mit feststehender und umlaufender Mahlbahn ‖ broyeur m. à meules verticales avec plaque de roulement fixe ou tournant. / ~ for wet grinding ‖ Naßkollergang m. ‖ broyeur m. à meules par voie humide.

edge runner bandage ‖ Kollergangbandage f. ‖ bandage m. pour broyeurs à meules. / ~ mill ‖ Kollergang m. ‖ broyeur m. à meules verticales. / ~ plate ‖ Kollerplatte f. ‖ plaque f. de broyeur vertical. / ~ ring ‖ Kollerring m. ‖ anneau m. de broyeur vertical.

edge seam (Mine) ‖ rechtes oder stehendes Flöz n. ‖ dressant m.; droit m. / ~ stress ‖ Randspannung f. ‖ tension f. au bord.

edge tool ‖ Schneidwerkzeug n.; Schneidzeug n. ‖ outil m. tranchant; taillanderie f. / ~ maker ‖ Grobschmied m. ‖ taillandier m.

edgeways ‖ hochkant; auf der schmalen Seite f. ‖ de champ m. / flush mounted ~ pattern instrument ‖ eingebautes Profilinstrument n. ‖ instrument m. de profil encastré. / ~ wound ‖ hochkant gewickelt ‖ enroulé ou bobiné de champ m.

edgewise see edgeways.

edge work (Coin) ‖ Kräuselwerk n.; Rändelwerk n. ‖ machine f. à moleter ou à moletter ou à tranche.

edging of a sail ‖ Saum m. oder Saumstreifen m. des Segels ‖ gaine f. de voile. / ~ machine ‖ Abkantmaschine f. ‖ machine f. à plier la tôle.

edible fat ‖ Speisefett n. ‖ graisse f. alimentaire ou de table. / ~ oil ‖ Speiseöl n. ‖ huile f. alimentaire ou de table. / ~ oil industry ‖ Speiseölindustrie f. ‖ industrie f. des huiles alimentaires.

edifice ‖ Gebäude n.; Bauwerk n. ‖ bâtiment m.; édifice m.; construction f.

edify, to ‖ erbauen ‖ bâtir; construire; édifier.

Edison accumulator ‖ Edisonsammler m.; Nickeleisenakkumulator m. ‖ accumulateur m. Edison ou nickel-fer. / ~ (incandescent) lamp ‖ Edison(glüh)lampe f. ‖ lampe f. incandescente Edison. / ~ lamp holder with suspension hook ‖ Edisonfassung f. mit Aufhängehaken ‖ douille f. Edison à crochet de suspension. / ~ screw cap ‖ Edisonsockel m. ‖ culot m. Edison.

edit, to ‖ herausgeben ‖ publier; éditer; faire paraître.

edition of a booklet ‖ Auflage f. einer Broschüre ‖ édition f. d'une brochure.

editor ‖ Herausgeber m. ‖ éditeur m.; gérant m. / ~ (Newspapers) ‖ Schriftleiter m. ‖ rédacteur m.

education of apprentices ‖ Lehrlingsausbildung f. ‖ éducation f. d'apprentis. / additional ~ both manual and mental given to apprentices ‖ Fortbildung f. der Lehrlinge ‖ instruction f. ultérieure des apprentis. / technical ~ ‖ Fachbildung f. ‖ éducation f. professionnelle.

educational appliance ‖ Lehrmittel n. ‖ article m. pour l'enseignement. / ~ establishment ‖ Lehranstalt f. ‖ établissement m. d'instruction; école f.; institution f. / proof of the necessary ~ training ‖ Nachweis m. der erforderlichen Vorbildung ‖ preuve f. de la préparation nécessaire.

eduction of the steam ‖ Dampfausströmung f. ‖ échappement m. de la vapeur.

eduction canal ‖ Abflußgraben m.; Abzugsgraben m. ‖ rigole f.; canal m. d'écoulement; colateur m. / main ~ canal ‖ Hauptabzugsgraben m. ‖ canal m. principal d'écoulement. / ~ channel see eduction canal. / ~ pipe ‖ Abzugsrohr n. ‖ tuyau m. d'émission.

eel dam ‖ Aalwehr n.; Aalsprung m. ‖ anguillère f.; gord m. / ~ fisherman ‖ Aalfischer m. ‖ pêcheur m. d'anguilles. / ~ ladder ‖ Aalrinne f.; Aaltreppe f. ‖ échelle f. aux anguilles; écrille f. / ~ smoker ‖ Aalräucherer m. ‖ fumeur m. d'anguilles. / ~ weir see ~ dam.

efface, to ‖ auswischen; austilgen ‖ essuyer; effacer.

effect, to ‖ leisten ‖ faire; effectuer; accomplir; produire.

effect ‖ Effekt m. ‖ effet m.; rendement m. / absorbing ~ ‖ absorbierende Wirkung f. ‖ action f. absorbante. / final ~ ‖ Endergebnis n. ‖ effet m. final; résultat m. / magnifying ~ (Opt) ‖ vergrößernde Wirkung f. ‖ action f. grossissante; effet m. grossissant. / ~ of a machine ‖ Leistung f. einer Maschine ‖ rendement m. d'une machine. / strongest ~ ‖ stärkste Wirkung f. ‖ effet m. maximum. / ~ of suction ‖ Saugwirkung f. ‖ effet m. d'aspiration. / ultimate ~ ‖ Nachwirkung f. ‖ effet m. ultérieur; suites fpl. / useful ~ ‖ Nutzeffekt m. ‖ rendement m. industriel; effet m. utile. / weakest ~ ‖ schwächste Wirkung f. ‖ effet m. minimum.

effect film ‖ Geräuschfilm m. ‖ film m. bruyant.

effective ‖ wirksam ‖ effectif. / ~ component (Electr) ‖ Wirkwert m. ‖ composante f. effective ou réelle. / ~ horse power ‖ Nutzpferdestärke f.; effektive Pferdestärke f. ‖ cheval m. effectif. / ~ impedance (Electr) ‖ Wirkwiderstand m. ‖ résistance f. effective. / ~ power ‖ Nutzkraft f. ‖ force f. effective ou utile. / ~ value ‖ Effektivwert m. ‖ valeur f. efficace ou virtuelle.

effect yarn, twisted ~ ‖ Effektzwirn m.; Musterzwirn m.; Zierzwirn m. ‖ fil m. fantaisie retordu.

effervesce, to (Chem) ‖ aufbrausen; aufwallen ‖ faire effervescence f.; bouillonner.

effervescence (Chem) ‖ Schäumen n.; Aufbrausen n.; Wallen n. ‖ effervescence f.

effervescent ‖ aufbrausend ‖ effervescent. / ~ salt ‖ Brausesalz n. ‖ sel m. effervescent.

efficiency ‖ Wirkungsgrad m.; Nutzeffekt m.; Gütegrad m. ‖ rendement m. / ~ of boiler ‖ Wirkungsgrad m. des Kessels ‖ rendement m. de la chaudière. / commercial ~ ‖ Nutzeffekt m. ‖ rendement m. industriel; effet m. utile. / ~ of delivery of the blower ‖ Förderleistung f. des Gebläses ‖ débit m. du ventilateur. / ~ of engines ‖ Maschinenstärke f.; Maschinenleistung f. ‖ puissance f. des machines, / ~ including gear losses ‖ Wirkungsgrad m. einschließlich Zahnradverlust ‖ rendement m. y compris la perte par friction de l'engrenage. / high thermal ~ ‖ guter thermischer Wirkungsgrad m. ‖ rendement m. thermique élevé. / ideal ~ ‖ idealer Wirkungsgrad m. ‖ rendement m. idéal. / mechanical ~ ‖ mechanischer Wirkungsgrad m. ‖ rendement m. mécanique. / thermal ~ ‖ thermischer Wirkungsgrad m. ‖ rendement m. thermique. / ~ of transmitters ‖ Wirkungsgrad m. von Mikrofonen ‖ rendement m. de microphones. / ~ of working ‖ Arbeitsgrad m. ‖ efficacité f. de travail.

efficiency factor ‖ Gütegrad m. ‖ coefficient m. de qualité. / increase of ~ by x to y% ‖ Verbesserung f. des Wirkungsgrades um x bis y% ‖ augmentation f. du rendement de x à y%. / ~ test ‖ Wirkungsgradbestimmung f. ‖ détermination f. de rendement.

efficient ‖ leistungsfähig ‖ capable; suffisant. / ~ height ‖ Nutzhöhe f. ‖ hauteur f. utile.

effloresce, to ‖ verwittern ‖ s'effleurer.

efflorescence ‖ Auswitterung; Effloreszenz f.; Ausblühung f. ‖ efflorescence f.; effritement m. / to be in ~ ‖ ausblühen ‖ s'effleurer. / ~ of saltpetre flower ‖ Ausblühen n. des Salpeters ‖ efflorescence f. du sel de nitre. / ~ of wall ‖ Mauerfraß m. ‖ carie f. de la maçonnerie.

efflux ‖ Abfluß m.; Ausfließen n. ‖ sortie f.; dépense f.

effort ‖ Anstrengung f.; Mühe f. ‖ effort m. / to make ~s ‖ sich anstrengen ‖ s'efforcer; faire des efforts mpl. / without special ~ ‖ ohne besondere Anstrengung f. ‖ sans effort f. apparent.

effort current (Electr) ‖ Wirkstrom m. ‖ courant m. watté.

egg ‖ Ei n. / œuf m. / eatable ~ ‖ eßbares Ei n. ‖ œuf m. comestible. / ~ freed from the shell ‖ von der Schale befreites Ei n. ‖ œuf m. sans coque.

egg albumin (Chem) ‖ Eialbumin n. ‖ albumine f. d'œuf. / ~ beater ‖ Schneebesen m. ‖ batteuse f. d'œufs. / ~ boiling device ‖ Eierkocher m. ‖ bouilloire f. à œufs. / ~ calipers pl. ‖ Ellipsenzirkel m. ‖ compas m. double calibre. / ~ calipers pl. with slider ‖ Ellipsenzirkel m. mit Gleitführung ‖ compas m. double calibre à coulisse. / ~ coal ‖ Eierbrikett n. ‖ boulet m. / ~ cooling room ‖ Eierkühlraum m. ‖ chambre f. froide pour œufs. / ~ cup ‖ Eierbecher m. ‖ coquetier m. / ~ glass ‖ Eieruhr f. ‖ sablier m. / ~ insulator ‖ Eiisolator m. ‖ isolateur m. ovoïde. / ~ oil ‖ Eieröl n. ‖ huile f. d'œufs. / ~ package ‖ Eierpackung f. ‖ emballage m. pour (des) œufs. / ~ preparation ‖ Eipräparat n. ‖ préparation f. aux œufs. / ~ preservative ‖ Eierkonservierungsmittel n. ‖ matière f. conservatrice pour œufs. / ~ preserve ‖ Eikonserve f. ‖ conserve f. d'œufs. / ~-shaped ‖ eiförmig ‖ ovale; oviforme; ovoïde. / white of ~ ‖ Eiweiß n. ‖ blanc m. d'œuf; albumine f. / ~ yolk ‖ Eigelb n.; Eidotter m. ‖ jaune m. d'œuf.

egoutteur ‖ Eguttör m. ‖ égoutteur m.

Egyptian type ‖ Blockschrift f. ‖ égyptienne f.

eider down ‖ Eiderdaune f. ‖ édredon m.

eight-fold twisting (Cable) ‖ Achterverseilung f. ‖ câblage m. en huit fils.

eight-hours' day ‖ Achtstundentag m. ‖ journée f. de huit heures.

eight-lock knitting machine ‖ Achtschloßstrickmaschine f. ‖ machine f. à tricoter à huit serrures.

eight-shoe, truck with ~ screw brake ‖ Drehgestell n. mit achtklotziger Spindelbremse ‖ bogie f. avec frein à vis à huit sabots.

eight-sided ‖ achtseitig ‖ à huit pans mpl.

eikonogen ‖ Eikonogen n. ‖ iconogène m.

eject, to ~ the brick ‖ den Preßling m. ausheben ‖ démouler la brique.

ejecting cylinder ‖ Ausstoßzylinder m. ‖ cylindre m. éjecteur. / ~ device ‖ Auswurfvorrichtung f. ‖ dispositif m. éjecteur. / ~ motion ‖ Auswurfbewegung f. ‖ mouvement m. d'éjection. / ~ piston ‖ Ausstoßkolben m. ‖ piston-éjecteur m.

ejector (Arm) ‖ Auswerfer m. ‖ éjecteur m. / ~ (Mach) ‖ Stahlsaugapparat m. ‖ aspirateur m. ou exhausteur m. à jet. / ash ~ ‖ Aschenejektor m.; Aschenauswerfer m. ‖ éjecteur m. de cendres. / cartridge ~ ‖ Patronenauswerfer m. ‖ boutoir m. / ~ with relapse ‖ Auswerfer

m. mit Rückfall ‖ éjecteur m. avec retour.

ejector condensor ‖ Strahlkondensator m. ‖ éjecteur-condenseur m.; condenseur m. à jet.

elæolite ‖ Nephelin m.; Eläolith m.; Davyn m.; Fettstein m. ‖ néphéline f.; elaeolite f.; pierre f. grasse.

elastic ‖ elastisch; dehnbar ‖ élastique. / ~ band ‖ elastisches Band n. ‖ Gummiband n. ‖ ruban m. en caoutchouc; cordon m. élastique. / ~ band weaver ‖ Gummizeugweber m. ‖ tisseur m. d'élastiques. / ~ chuck ‖ mandrin m. de serrage. / ~ collar ‖ federnder Ring m. ‖ bague f. faisant ressort. / ~ deformation ‖ elastische Formänderung f. ‖ déformation f. élastique. / ~ force ‖ Federkraft f. ‖ force f. élastique. / ~ force tester ‖ elastischer Kraftprüfer m. ‖ machine f. élastique pour l'essai des forces. / ~ gum ‖ Kautschuk n.; Gummielastikum n. ‖ caoutchouc m.; gomme f. élastique. / ~ hysteresis ‖ Elastizitätshysteresis f. ‖ hystérésis f. élastique. / ~ limit ‖ Elastizitätsgrenze f. ‖ limite f. d'élasticité. / ~ spring ‖ Springfeder f.; Sprungfeder f. ‖ ressort m. élastique. / ~ spring coupling ‖ elastische Federkupplung f. ‖ accouplement m. à ressort élastique. / ~ stocking ‖ Gummistrumpf m. ‖ bas m. élastique. / ~ stocking maker ‖ Gummistrumpfstricker m. ‖ tricoteur m. de bas élastiques. / ~ suspension ‖ federnde Aufhängung f. ‖ suspension f. élastique. / ~ web ‖ elastisches Gewebe n. ‖ tissu m. élastique.

elastic ‖ Gummiband n.; Gummizug m. ‖ cordon m. élastique; élastique m. / round ~ ‖ Gummifaden m. ‖ fil m. élastique ou en caoutchouc. / ~s pl. ‖ Strumpfbänder npl. ‖ jarretières fpl.

elasticity ‖ Elastizität f.; Federkraft f. ‖ élasticité f.; force f. élastique. / ~ of compression ‖ Druckelastizität f. ‖ élasticité f. de compression. / ~ of extension ‖ Zugelastizität f. ‖ élasticité f. de traction. / ~ of flexure ‖ Biegungselastizität f. ‖ élasticité f. de flexion. / ~ of tank tracks ‖ elastische Nachgiebigkeit f. der Kampfwagenraupen ‖ souplesse f. des chenilles de char de combat. / apparatus for testing ~ of yarn ‖ Garnelastizitätsmesser m. ‖ indicateur m. d'élasticité des fils.

elasticity, limit of ~ ‖ Elastizitätsgrenze f. ‖ limite f. d'élasticité. / modulus of ~ ‖ Elastizitätsmodul m.; Elastizitätszahl f. ‖ module m. d'élasticité. / modulus of extension ~ ‖ Zugelastizitätsmodul n. ‖ module m. d'élasticité à la traction. / modulus of transverse ~ ‖ Gleitmodul m. ‖ module m. d'élasticité au cisaillement.

elastin ‖ Elastin n. ‖ élasticine f.

elaterite ‖ elastisches Erdpech n. ‖ caoutchouc m. minéral.

elbow ‖ Ellbogen m. ‖ coude m. / ~ (Mach) ‖ Knie n.; Kniestück; Muffenkrümmer m. ‖ coude m. / ~ bend (Pipe) ‖ Kniestück n. ‖ coude m. / ~ board (Window) ‖ Fensterbrett n.; Latteibrett n. ‖ accoudoir m.; planche f. d'appui. / ~ cushion ‖ Armlehne f. ‖ appui-bras m. ou accoud oir m. / ~ joint ‖ Knieverbindung f. ‖ jointure f. en L. / ~ lever ‖ Winkelhebel m. ‖ levier m. coudé. / ~ place (Build) ‖

Brüstung f. || appui m.; parapet m. / ~ rail see ~ cushion.

elder || Holunderholz n. || sureau m. / ~ berry || Holunderbeere f. || baie f. de sureau. / ~ blossom || Holunderblüte f. || fleur f. de sureau. / ~ pith || Holundermark n. || moelle f. de sureau. / ~ pith ball || Holundermarkkugel f. || balle f. de moelle de sureau. / ~ wood || Holunderholz n. || sureau m.; bois m. du sureau.

electric || elektrisch || électrique. / ~ accumulator || (elektrischer) Sammler m.; Akkumulator m. || accumulateur m. (électrique). / ~ advertising || Lichtreklame f. durch leuchtende Zeichen || avertissement m. par enseignes lumineuses.

electric apparatus || elektrischer Apparat m. || appareil m. électrique. / ~ for domestic use || elektrotechnischer Apparat m. für den Hausgebrauch || appareil m. électrotechnique à usage domestique. / ~ for laboratory || elektrischer Apparat m. für Laboratorium || appareil m. électrique pour laboratoire.

electric baking-oven for the household || elektrischer Hausbackofen m. || four m. à cuire électrique à l'usage domestique. / ~ bell contact for door plates || elektrischer Klingelkontakt m. für Türschilder || contact m. de sonnerie électrique pour plaques de porte. / ~ car || Elektromobil n. || voiture f. électrique. / ~ carbon || galvanische Kohle f. || charbon m. électrique. / ~ central station || elektrische Kraftstation f.; Elektrizitätswerk n. || centrale f. d'électricité; usine f. d'électricité. / ~ company || Elektrizitätsgesellschaft f.; Elektrizitätswerk n. || société f. d'électricité. / ~ control || elektrische Steuerung f. || commande f. électrique.

electric current || elektrischer Strom m. || courant m. électrique. / wire carrying an ~ || von elektrischem Strom durchflossener Draht || fil m. parcouru par un courant électrique. / ~ supply system || elektrisches Stromverteilungsnetz n. || réseau m. de distribution du courant électrique.

electric discharge device || elektrische Entladevorrichtung f. || dispositif m. de décharge électrique. / ~ door opener || elektrischer Türöffner m. || ouvreur m. de porte électrique. / ~ drag || Elektroschlepper m. || tracteur m. électrique. / ~ engine see ~ machine. / ~ engineer || Elektroingeniör m.; Elektrotechniker m. || ingénieur-électricien m. / ~ engineering || Elektrotechnik f. || électrotechnique f. / ~ equipment || elektrische Einrichtung f. oder Ausrüstung f. || équipement m. électrique. / ~ fan || elektrischer Lüfter m. || ventilateur m. électrique. / ~ field || elektrisches Feld n. || champ m. électrique. / ~ field of the air || luftelektrisches Feld n. || champ m. électrique de l'air ou de l'atmosphère. / ~ fitter || Elektromonteur m. || monteur-électricien m. / ~ fitting || elektrischer Apparat m. || appareillage m. pour l'électricité. / ~ furnace || elektrischer Ofen m. || four m. électrique. / ~ generating station see ~ central station. / ~ generator || Stromerzeuger m. || génératrice f. électrique. / ~ hearth || Elektroofen m. || four m. électrique. / ~ heating machine || elektrische Erwärmungsmaschine f. || machine f. à rechauffer électriquement. /

~ hoisting gear || Elektrohubwerk n; Elektrowinde f. || treuil m. électrique. / ~ horn || elektrische Sirene f. || sirène f. électrique. / ~ image of a conductor as to earth || elektrostatisches Spiegelbild n. eines Leiters || image m. électrique d'un conducteur par rapport à la terre. / ~ insulating materials pl. || Isolierstoffe pl. für Elektrotechnik || matières fpl. isolantes pour l'électricité. / ~ lamp || elektrische Lampe f. || lampe f. électrique. / ~ lamp for lighting off || elektrische Ableuchtlampe f. oder Ausleuchtlampe || lampe f. électrique à éclairer l'intérieur. / ~ lifting magnet || Elektrohebemagnet m. || électroaimant m. de levage. / ~ light || elektrisches Licht n. || lumière f. électrique. / yellow ~ light || gelbes elektrisches Licht n. || lumière f. électrique jaune-rouge. / ~ light cable || Lichtkabel n. || câble m. pour lumière électrique. / ~ lighting || elektrische Beleuchtung f. || éclairage m. électrique. / ~ lighting for trains || elektrische Zugbeleuchtung f. || éclairage m. électrique des trains. / ~ locomotive || elektrische Lokomotive f. || locomotive f. électrique. / ~ machine || elektrische Maschine f. || machine f. électrique. / construction of ~ machines || Bau m. elektrischer Maschinen || construction f. de machines électriques. / ~ main || Hauptleitung f. || conduite f. principale. / auto car for ~ measurement || Meßkraftwagen m. || automobile m. pour mesure électrique. / ~ measuring device || elektrische Meßvorrichtung f. || appareil m. de mesure électrique. / ~ measuring instrument || elektrisches Meßinstrument n. || instrument m. de mesures électriques. / ~ meter || elektrischer Zähler m.; Elektrizitätszähler m. || compteur m. électrique ou d'électricité. / ~ motor || Elektromotor m. || moteur m. électrique. / ~ motor car || Elektromobil n. || automobile m. électrique. / ~ muffle furnace || elektrischer Muffelofen m. || four m. électrique à moufle. / ~ operating of signals and switches || elektrische Signal- und Weichenstellung f. || commande f. électrique des signaux et aiguilles. / ~ operating of switches || elektrische Weichenstellung f. || commande f. électrique des aiguilles. / ~ oscillation || elektrische Schwingung f. || oscillation f. électrique. / ~ overhead railway || Elektrohängebahn f. || transporteur m. aérien ou chemin m. de fer suspendu électrique; électro-chemin m. suspendu. / steam power ~ plant || Elektrizitätswerk n. mit Dampfkraft; Dampfelektrizitätswerk n. || usine f. électrique à vapeur. / ~ pocket lamp lens || Linse f. für elektrische Taschenlampen || lentille f. lampes électriques de poche. / ~ power cable || Kraftübertragungskabel n. || câble m. à transmission de force. / ~ power house || elektrische Zentrale f. || centrale f. d'électricité. / ~ power station || Elektrizitätswerk n. || usine f. électrique. / wind driven ~ power station || Windelektrizitätswerk n. || centrale f. aéroélectrique. / dust removing device by ~ precipitation || Entstaubungsvorrichtung f. durch elektrischen Niederschlag || appareil m. pour la séparation de poussière par extinction électrique. / ~ precipitation of tar mists || elektrische Niederschlagung f. der Teernebel || précipitation f. électrique des

brouillards de goudron. / ~ pulley block || Elektroflaschenzug m. || palan m. électrique. / ~ quantity || Elektrizitätsmenge f. || quantité f. d'électricité. / ~ railway || elektrische Bahn f. || chemin m. de fer électrique; tramway m. électrique. / suspended ~ railway see ~ overhead railway. / ~ regulating device || elektrische Drehzahleinstellvorrichtung f. || régulateur m. de vitesse électrique. / ~ shock || elektrischer Schlag m. || commotion f. ou choc m. électrique. / ~ sign || elektrische Lichtreklame f. || enseigne f. lumineuse. / ~ smelting bath || elektrisches Schmelzbad n. || bain m. de fusion électrique. / ~ soldering tools pl. || elektrisches Lötwerkzeug n. || fers mpl. à souder électriques. / ~ spark || elektrischer Funken m. || étincelle f. électrique. / ~ steam boiler || Elektrodampfkessel m. || chaudière f. chauffée à l'électricité. / ~ steel || Elektrostahl m. || acier m. électrique. / ~ steel in ingots || Elektrostahl m. in Blöcken || acier m. au four électrique en lingots. / ~ steel furnace || Elektrostahlofen m. || four m. électrique à acier. / ~ steel plant see ~ steel works || steel works pl. || Elektrostahlwerk n. || aciérie f. électrique; fonderie f. d'acier électrique. / ~ suspension railway || Elektrohängebahn f. || transporteur m. aérien électrique; chemin m. de fer électrique à suspension; électrochemin m. suspendu. / ~ thermometer || Fernthermometer n. || téléthermomètre m.; thermomètre m. électrique. / ~ tool || Elektrowerkzeug n. || outil m. électrique. / ~ tractor || Elektroschlepper m. || remorqueur m. électrique. / ~ tramway || elektrische Straßenbahn f. || tramway m. électrique. / ~ transformer || elektrischer Transformator m. || transformateur m. électrique. / ~ traversing gear || Elektrofahrwerk n. || commande f. électrique de déplacement. / ~ trolley || Elektrokarren m. || charrette f. électrique. / ~ truck || Elektrokarren m. || chariot m. électrique. / ~ tube || elektrische Röhre f. || tube m. électrique. / ~ welding machine || elektrische Schweißmaschine f. || machine f. à souder électrique. / ~ welding shop || Elektroschweißwerk n. || atelier m. de soudage électrique. / ~ wire || Draht m. für elektrische Leitungen || fil m. de canalisation électrique.

electrical see electric.

electrically controlled clock || elektrische Uhrenanlage f. || horloge f. électrique. / ~ driven || elektrisch betrieben || mu électriquement. / ~ driven clock || elektrisch betriebene Turmuhr f. || horloge f. à mouvement électrique. / ~ driven pendulum hydro-extractor || Elektropendelzentrifuge f. || essoreuse f. oscillante électrique. / ~ heated furnace (Met) || elektrisch beheizter Ofen m. || four m. électrique pour le traitement thermique. / ~ heated laboratory apparatus || elektrisch beheiztes Laboratoriumsgerät n. || appareil m. de laboratoire à chauffage électrique.

electrician || Elektrotechniker m. || électricien m.

electricity || Elektrizität f. || électricité f. / atmospheric ~ || atmosphärische Elektrizität f.; Luftelektrizität f. || électricité f. atmosphérique. / bound ~ || gebundene

Elektrizität f. ‖ électricité f. dissimulée *ou* latente. / contact ~ ‖ Berührungselektrizität f. ‖ électricité f. par contact. / dissimulated ~ *see* bound ~. / frictional ~ ‖ Reibungselektrizität f. ‖ électricité f. de frottement. / induced ~ ‖ Induktionselektrizität f. ‖ électricité f. d'inductive; électricité induite *ou* par influence. / ~ of precipitations ‖ Niederschlagselektrizität f. ‖ électricité f. des précipitations. / radiating ~ ‖ strahlende Elektrizität f. ‖ électricité f. rayonnante. / resinous ~ ‖ Harzelektrizität f. ‖ électricité f. résineuse *ou* négative. / static ~ ‖ statische Elektrizität f. ‖ électricité f. statique. / ~ of thunder and lightning ‖ Gewitterelektrizität f. ‖ électricité f. d'un temps orageux. / vitreous ~ ‖ Glaselektrizität f. ‖ électricité f. vitrée *ou* de verre *ou* positive.

electricity distribution ‖ Elektrizitätsverteilung f. ‖ distribution f. d'électricité. / ~ leakage ‖ Elektrizitätsverlust m. ‖ perte f. d'électricité. / ~ measurement ‖ Elektrizitätsmessung f. ‖ mesure f. électrique. ~ meter ‖ Elektrizitätszähler m. ‖ compteur m. d'électricité. / ~ quantity ‖ Elektrizitätsmenge f. ‖ quantité f. d'électricité. / ~ source ‖ Elektrizitätsquelle f. ‖ source f. d'électricité.

electricity supply ‖ Elektrizitätsversorgung f. ‖ distribution f. d'électricité. / influence which a company is able to exert on the public ~ ‖ Einflußnahme f. einer Gesellschaft auf die Elektrizitätswirtschaft ‖ prise f. d'influence d'une société sur l'économie électrique. / ~ engineering ‖ Elektrizitätswirtschaft f. ‖ économie f. électrique. / ~ enterprise ‖ Elektrizitätslieferungsunternehmen n. ‖ entreprise f. de distribution d'électricité. / rational amalgamation of the existing public ~ systems ‖ rationelle Zusammenfassung f. der bestehenden elektrowirtschaftlichen Anlagen ‖ exploitation f. rationnelle d'installations électriques déjà existantes.

electricity works pl. ‖ Elektrizitätswerk n. ‖ station f. centrale d'électricité; usine f. d'électricité.

electrification ‖ Elektrisierung f.; Elektrifikation f. ‖ électrisation f.; électrification f.

electrified ‖ elektrisiert; elektrifiziert ‖ électrisé; électrifié. / ~ railway ‖ elektrische Vollbahn f. ‖ chemin m. de fer électrifié.

electrify, to ‖ elektrisieren ‖ électriser.

electrization *see* electrification.

electro (Print) ‖ Druckstock n.; Bildstock m.; Klischee n.; Galvano n. ‖ cliché m.; galvano m.

electro-acoustics pl. ‖ Elektroakustik f. ‖ électroacoustique f.

electrobus ‖ elektrscher Omnibus m. ‖ omnibus m. électrique.

electro-capillary current ‖ Kapillaritätsstrom m. ‖ courant m. électro-capillaire.

electro-cement ‖ Elektrozement m. ‖ électrociment.

electro-chemical ‖ elektrochemisch ‖ électrochimique. / ~ apparatus ‖ elektrochemischer Apparat m. ‖ appareil m. électrochimique. / ~ plant ‖ elektrochemische Anlage f. ‖ installation f. électrochimique.

electro-chemically cleaned ‖ elektrochemisch gereinigt ‖ nettoyé électro-chimiquement.

electro-chemist ‖ Elektrochemiker m. ‖ électrochimiste m.

electro-chemistry ‖ Elektrochemie f. ‖ électrochimie f.

electro-chilled iron foundry ‖ Elektrohartgußgießerei f. ‖ fonderie f. électrique de fonte durcie.

electro-chromic rings pl. ‖ Farbenringe m pl. ‖ anneaux m pl. colorés.

electrode ‖ Elektrode f. ‖ électrode f. / bare ~ ‖ nackte Elektrode f. ‖ électrode f. nue. / carbon ~ ‖ Kohlenbeutelelektrode f. ‖ électrode f. à sachet de charbon. / contact-point ~ *see* spot ~. / dipping ~ ‖ Tauchelektrcde f. ‖ électrode f. plongeante. / ~ for incandescent lamps manufacture ‖ Elektrode f. für die Glühlampenfertigung ‖ électrode f. pour la fabrication des lampes à incandescence. / negative ~ ‖ Kathode f.; negativer Pol m. ‖ cathode m.; pôle m. négatif. / net-shaped ~ ‖ Netzelektrode f. ‖ électrode f. réticulaire. / outward ~ ‖ Außenelektrode f. ‖ électrode f. extérieur. / roller ~ ‖ Rollenelektrode f. ‖ électrode f. à rouleau. / silver ~ ‖ Silberelektrode f. ‖ électrode f. d'argent. / spot ~ ‖ Punkt-(schweiß)elektrode f. ‖ électrode f. punctiforme. / tip ~ *see* spot ~. / wire-gauze ~ ‖ Netzelektrode f. ‖ électrode f. en toile métallique.

electrode carbon ‖ Elektrodenkohle f. ‖ charbon m. à électrodes. / consumption of ~s ‖ Abbrand m. der Elektroden ‖ usure f. des électrodes. / ~ frame ‖ Elektrodenrahmen m. ‖ cadre m. d'électrodes. / ~ holder ‖ Elektrodenhalter m. ‖ porte-électrode m. / life of ~s ‖ Lebensdauer f. von Elektroden ‖ durée f. des électrodes.

electro-dental apparatus ‖ Elektrodendentalapparat m. ‖ appareil m. électrique de chirurgie dentaire.

electro-deposit ‖ galvanischer Niederschlag m. ‖ précipité m. galvanique.

electro-deposited ‖ galvanisch gefällt ‖ galvanique; galvanique de la pile; déposé par la pile.

electro-deposition ‖ Galvanoplastik f. ‖ galvanoplastie f.

electrode potential ‖ Elektrodenpotential n. ‖ potentiel m. des électrodes. / ~ press ‖ Elektrodenpresse f. ‖ presse f. à électrodes. / ~ terminal ‖ Polklemme f. ‖ borne f. d'élément. / water circulation jacket surrounding the ~s ‖ Elektrodenwasserkühlmantel m. ‖ enveloppe f. à circulation d'eau entourant les électrodes.

electro-diagnosis ‖ Elektrodiagnose f. ‖ électrodiagnose f.

electro-dynamic(al) ‖ elektrodynamisch ‖ électrodynamique.

electro-dynamometer ‖ Elektrodynamometer n. ‖ électrodynamomètre m.

electro-economic interest ‖ elektrowirtschaftliche Beteiligung f. ‖ participation f. électro-économique.

electro-furnace ‖ Elektroofen m. ‖ four m. électrique.

electro-gild, to ‖ galvanisch vergolden ‖ dorer galvanique.

electro-gilding ‖ galvanische Vergoldung f. ‖ dorure f. galvanique.

electro-hydraulic press ‖ elektrohydraulische Presse f. ‖ presse f. électro-hydraulique. / ~ riveting machine ‖ elektrohydraulische Nietmaschine f. ‖ riveuse f. électro-hydraulique.

electrolier ‖ elektrischer Kronleuchter m. ‖ lustre m. électrique.

electrolysis ‖ Elektrolyse f. ‖ électrolyse f. / ~ in the dry-way ‖ schmelzflüssige Elektrolyse f. ‖ électrolyse par fusion.

electrolyte ‖ Elektrolyt m.; Füllsäure f. ‖ électrolyte m.; acide m. de remplissage *ou* pour accumulateurs.

electrolytic(al) ‖ elektrolytisch ‖ électrolytique. / ~ analysis ‖ elektrolytische Analyse f. ‖ analyse f. électrolytique. ~ apparatus ‖ elektrolytischer Apparat m. ‖ appareil m. électrolytique. / ~ bronzing ‖ galvanisches Bronzieren n. ‖ bronzage m. galvanique. / ~ copper ‖ Elektrolytkupfer n. ‖ cuivre m. électrolytique. / ~ copper plating ‖ galvanisches Verkupfern n. ‖ cuivrage m. galvanique. / ~ decomposition ‖ elektrolytische Spaltung f. ‖ décomposition f. électrolytique. / ~ detector ‖ elektrolytischer Detektor m. ‖ déceleur m. électrolytique. / ~ efficiency ‖ Stromausbeute f. ‖ rendement m. électrolytique en courant. / ~ generation of hydrogen ‖ elektrolytische Herstellung f. von Wasserstoffgas ‖ production f. électrolytique de l'hydrogène. / ~ interrupter ‖ elektrolytischer Unterbrecher m.; Wehneltunterbrecher m. ‖ interrupteur m. électrolytique. / ~ proceeding ‖ elektrolytisches Verfahren n. ‖ procédé m. électrolytique. / ~ rectifier ‖ Polarisationszelle f. ‖ élément m. de polarisation. / ~ reproductions pl. ‖ galvanoplastische Waren f pl. ‖ objets m pl. en galvanoplastie.

electrolyze, to ‖ elektrolysieren ‖ électrolyser.

electro-magnet ‖ Elektromagnet m. ‖ électro-aimant m.; électro m. / tubulated ~ ‖ Glockenmagnet m. ‖ aimant m. campanulé.

electro-magnetic(al) ‖ elektromagnetisch ‖ électromagnétique. / ~ coupling ‖ Elektromagnetkupplung f. ‖ accouplement m. électromagnétique. / ~ field ‖ elektromagnetisches Feld n. ‖ champ m. électromagnétique. / ~ oscillator ‖ Magnetsummer m.; Blattfedersummer m. ‖ oscillateur m. électromagnétique. / ~ theory of light ‖ elektromagnetische Lichttheorie f. ‖ théorie f. électromagnétique de la lumière.

electro-magnetism ‖ Elektromagnetismus m. ‖ électromagnétisme m.

electro-mechanics pl. ‖ Elektromechanik f. ‖ électromecanique f.

electro-medical apparatus ‖ elektromedizinischer Apparat m. ‖ appareil m. électromédical *ou* d'électricité médicale.

electro-medicinal *see* electro-medical.

electro-metallurgy ‖ Elektrometallurgie f. ‖ électro-métallurgie f.

electrometer ‖ Elektrometer n. ‖ électromètre m. / capillary ~ ‖ Kapillarelektrometer n. ‖ électromètre m. capillaire. / gold leaf ~ ‖ Goldblattelektrometer n. ‖ électromètre m. à feuilles d'or. / quadrant ~ ‖ Quadrantelektrometer n. ‖ électromètre m. à quadrants.

electrometer gauge ‖ Prüfelektrometer n. ‖ jauge f. électrométrique.

electromobile ‖ Elektromobil n. ‖ électromobile m.; automobile f. électrique.

electromotive ‖ elektromotorisch ‖ électromoteur. / ~ force ‖ elektromotorische Kraft f. ‖ force f. électromotrice. / back ~ force ‖ rückelektromotorische Kraft f. ‖

force f. électromotrice inverse. / counter ~ force || elektromotorische Gegenkraft f. || force f. contre-électromotrice.

electromotive || elektrische Lokomotive f. || locomotive f. électrique.

electromotograph || Elektromotograf m. || électromotographe m.

electromotor || Elektromotor m. || moteur m. électrique; électromoteur m. / small power ~ || Kleinelektromotor m. || petit électromoteur m.

electron (Electr) || Elektron n. || électron m. / ~ (Met) || Elektrometall n.; Elektron n. || métal m. électron; électron m. / ~ current || Elektronenstrom m. || courant m. électronique.

electro-negative || elektronegativ || électronégatif.

electron emission || Elektronenemission f. || émission f. électronique.

electronic || elektronisch || électronique.

electron metal || Elektronmetall n.; Elektron n. || métal m. électron; électron m. / ~ relay || Elektronenrelais n. || relais m. à gaz ionisé. / ~ theory || Elektronentheorie f. || théorie f. des électrons. / ~ tube generator || Röhrensummer m. || oscillateur m. à lampe amplificatrice.

electro-optics pl. || Elektrooptik f. || électro-optique f.

electrophorus || Elektrophor m. || électrophore m.

electro-physiology || Elektrophysiologie f. || électrophysiologie f.

electroplate, to || galvanisch versilbern; auf galvanischem *oder* elektrolytischem Wege mit einem Metallüberzug versehen || recouvrir d'un dépôt métallique par voie électrolytique; métalliser *ou* argenter galvaniquement.

electroplate || galvanisch plattierte Ware f. || plaqué m. galvanique. / ~ on wood-base || Galvano n. auf Holzfuß *oder* auf Holzunterlage || galvano m. sur bois.

electroplated wares pl. || galvanisch versilberte Waren fpl. || orfèvrerie f. argentée.

electroplating || Galvanostegie f.; Elektroplattierung f.; galvanische Metallüberziehung f. || galvanostégie f.; galvanisation f.; métallisation f. galvanique. / ~ bath || galvanoplastisches Bad n. || bain m. galvanoplastique. / ~ plant || galvanoplastische Anlage f.; Galvanisierungsanlage f. || installation f. de galvanoplastie *ou* de galvanisation. / wholesale ~ plant || Massengalvanisierapparat m. || appareil m. de grand débit pour dépôts galvaniques. / ~ practice || Galvanotechnik f. || galvanotechnique f.

electro-pneumatic hammer || elektropneumatischer Hammer m. || marteau m. électro-pneumatique. / ~ rock drill(ing machine) || elektropneumatische Gesteinbohrmaschine f. || marteau m. piqueur électro-pneumatique.

electro-positive || elektropositiv || électropositif.

electro pulley block || Elektroflaschenzug m. || moufle f. *ou* palan m. électrique.

electro-railway || elektrische Bahn f. *oder* Eisenbahn f. || chemin m. de fer électrique. / suspended ~ || Elektrohängebahn f. || électro-chemin m. suspendu; funiculaire f. à commande électrique; installation f. de transport aérien électrique; chemin m. de fer funiculaire électrique.

electroscope || Elektroskop n. || électroscope m. / gold leaf ~ || Goldblattelektroskop n. || électroscope m. à feuilles d'or.

electroscopic || elektroskopisch || électroscopique.

electro-silvering || galvanische Versilberung f. || argenture f. galvanique.

electro-smelt, to || elektrisch schmelzen || fondre au four électrique.

electro-smelting || elektrisches Schmelzen n. || fusion f. électrique.

electrostatic(al) || elektrostatisch || électrostatique. / ~ induction || Influenz f. || induction f. électrostatique. / ~ machine || Elektrisiermaschine f. || machine f. électrostatique. / ~ relay || elektrostatisches Relais n. || relais m. électrostatique.

electrostatics pl. || Elektrostatik f. || électrostatique f.

electro-steel || Elektrostahl m. || acier m. électrique.

electrotechnic(al) || elektrotechnisch || électrotechnique.

electrotechnics pl. || Elektrotechnik f. || électrotechnique f.

electro-therapeutics pl. || Elektrotherapie f. || électro-thérapeutique f.

electrothermic || elektrothermisch || électrothermique. / ~ recovery || elektrothermische Gewinnung || extraction f. électrothermique. / ~ treatment apparatus || Apparat m. für Hochfrequenz- und Diathermiebehandlung || appareil m. pour le traitement à haute fréquence et pour l'électro-thermothérapie.

electrotype || galvanisch || galvanique.

electrotype, to || klischieren || clicher.

electrotyper || Galvanoplastiker m. || électrotypeur m.; galvanoplasticien m.

electrotyping || Elektrotypie f.; Galvanoplastik f. || électrotypie f.; galvanoplastie f. / auxiliary printing machine for ~ || Druckhilfsmaschine f. für Galvanoplastik || machine f. auxiliaire d'impression pour galvanoplastie.

electrotypy *see* electrotyping.

electuary || Latwerge f. || électuaire m.

element || Element n. || élément m. / ~ (Chem) || Element n.; einfacher Körper m.; Grundstoff m. || élément m.; substance f. élémentaire; corps m. simple. / ~ (Electr) *see also* cell || Zelle f.; Element n.; pile f.; élément m. / ~ of arc || Bogenelement n. || élément m. d'arc. / ~ with a balloon (Electr) || Ballonelement n. || pile f. à ballon. / metallic ~ || metallisches Element n. || élément m. métallique. / non-metallic ~ || Metalloid n. || métalloïde m. / ~ with a papier-mâché diaphragm || Pappelement n. || élément m. à diaphragme de papier-mâché. / single ~ || Einzelelement m. || élément m. individuel. / ~s pl. with large number of spectrum lines || linienreiche Elemente npl. || éléments mpl. riches en raies. / ~ of surface || Flächenelement n. || élément m. de surface. / voltaic ~ || Voltaelement n. || pile f. voltaïque.

elementary || elementar || élémentaire. / ~ analysis || Elementaranalyse f. || analyse f. élémentaire *ou* ultime. / ~ coil of the armature || Ankerglied n. || bobine f. élémentaire de l'induit. / ~ geometry || elementare Geometrie f. || géométrie f. élémentaire. / ~ mass || Massenpunkt m. || quantité f. élémentaire. / ~ substance ||

einfacher Körper m. || élément m.; substance f. élémentaire.

elemi gum || Elemiharz n. || gomme f. élémi.

elephant boiler || Elefantenkessel m. || chaudière f, à bouilleurs.

elevated || hochliegend || surélevé. / ~ cableway crane || Kabelhochbahnkran m. || grue f. à câbles aériens; Blondin m. / ~ discharging track || Sturzbahn f. || voie f. surélevée de déchargement. / ~ reservoir || *see* ~ tank. / ~ tank || Hochbehälter m. || réservoir m. surélevé.

elevating device attached to the tripod || Stativ n. mit Hochstellvorrichtung || trépied m. avec dispositif de levage. / ~ gear of a telescope tripod || Hochstellvorrichtung f. eines Fernrohrstativs || dispositif m. élévateur d'un pied de lunette. / ~ plant || Elevatoranlage f.; Hebeanlage f. || installation f. d'élévateur *ou* de levage.

elevation (Build) || Aufriß m.; Voderansicht f. || élévation f.; vue f. en coupe verticale. / ~ (Geol) || Bodenerhebung f. || soulèvement m. du sol. / ~ (Surv) || Erhebung f.; Höhe f. || élévation f.; altitude f.; hauteur f. / ~ of the pole || Polhöhe f. || élévation f. *ou* hauteur f. du pôle; latitude f.

elevation, angle of ~ (Surv) || Erhöhungswinkel m.; Erhebungswinkel m. || angle m. d'élévation *ou* de hauteur. / ~ plan || Höhenplan m. || plan m. d'élévations.

elevator (Aero) || Höhensteuer n.; Tiefenruder n. || gouvernail m. de profondeur *ou* d'altitude *ou* de hauteur *ou* horizontal. / ~ (Lift) || Elevator m.; Lift m.; Fahrstuhl m.; Aufzug m. || élévateur m.; ascenseur m.; monte-charge m. / blast-furnace ~ || Gichtaufzug m.; Hochofenaufzug m. || élévateur m. *ou* monte-charge m. de haut-fourneau. / bucket ~ || Becherwerk n. || élévateur m. de *ou* à godets. / ~ with bucket-tipping device || Kippbecherwerk n. || élévateur m. à godets basculants. / chain ~ || Kettenelevator m. || élévateur m. à chaîne. / charging ~ || Aufgabebecherwerk n. || noria f. de chargement. / discharging bucket ~ || Entladebecherwerk n. || noria f. de déchargement. / dry goods ~ || Trockenelevator m. || élévateur m. pour matières sèches. / feeding ~ *see* charging ~. / freight ~ || Lastaufzug m. || monte-charge m. / ~ by hand || Handaufzug m. || treuil m. à main. / hydraulic ~ || senkrechte Hebung f.; bewegliche Schleuse f. || ascenseur m. hydraulique. / inclined ~ || schräger Aufzug m.; Becherwerk n. für schräge Förderung || monte-charge m. incliné; élévateur m. à godets obliques. / paternoster ~ || Becherwerk n.; Paternosteraufzug m. || élévateur m. à godets; chapelet m. || noria f.; monte-charge m. à chaîne sans fin. / vertical ~ || senkrechter Aufzug m. || monte-charge m. vertical; élévateur-transporteur m. vertical.

elevator belt || Elevatorgurt m. || toile f. d'élévateur. / ~ cable || Tiefenruderkabel n. || câble m. de commande du gouvernail de profondeur. / ~ cage || Förderkorb m. || cage f. de montée. / catch device for ~ cages || Fangvorrichtung f. für Förderkörbe || dispositif m. d'arrêt pour cages de montée. / holding apparatus for ~ cages || Aufsetzvorrichtung f. für Förderkörbe || dispositif m. de chargement pour cages de montée. / ~ canvas || Ele-

vatorgurt m. ǁ bande f. *ou* courroie f. d'élévateur. / ~ control gear ǁ Aufzugsteuerung f. ǁ dispositif m. de commande pour ascenseurs. / ~ foot ǁ Becherwerkfuß m. ǁ partie f. inférieure de la noria. / ~ frame ǁ Fördergerüst n.; ·Becherwerkgerüst n. ǁ charpente f. de monte-charge *ou* de noria. / ~ gear ǁ Förderhaspel m. ǁ treuil m. / ~ head ǁ Becherwerkkopf m. ǁ partie f. supérieure de noria. / ~ helmsman (Aero) ǁ Höhensteuermann m. ǁ timonier m. d'altitude. / ~ pit ǁ Becherwerkgrube f. ǁ fosse f. de *ou* à noria. / ~ push and pull rod (Aero) ǁ Höhenruderstoßstange f. ǁ biellette f. de la commande de profondeur. / ~ rocking shaft ǁ Höhenruderwelle f. ǁ arbre m. oscillant de la commande de profondeur. / ~ tower ǁ Elevatorenturm m. ǁ tour f. élévatoire. / ~ trunking ǁ Elevatorschlotte f. ǁ corps m. *ou* manche m. d'élévateur. / ~ tube ǁ Elevatorschlauch m. ǁ tube f. d'élévateur.

eliminate, to ǁ ausscheiden; ausstoßen; herauslösen ǁ éliminer; supprimer. / ~ the acid ǁ entsäuern ǁ désacidifier. / ~ heat ǁ Wärme f. abführen ǁ éliminer la chaleur.

elimination ǁ Elimination f.; Absonderung f. ǁ élimination f. / with ~ of ǁ unter Austritt m. von ǁ avec élimination f. de. / ~ race ǁ Ausscheidungsrennen n. ǁ éliminatoire f.

eliquate, to ǁ abseigern ǁ liquater; ressuer; achever de ressuer.

eliquation ǁ Seigerprozeß m.; Seigerung f. ǁ liquation f.; ressuage m. / to separate by ~ *see* to eliquate. / ~ hearth ǁ Seigerherd m. ǁ four m. de liquation *ou* de ressuage.

ellipse ǁ Ellipse f. ǁ ellipse f.

ellipsograph ǁ Ellipsenzirkel m. ǁ ellipsographe m.

ellipsoid *see* ellipsoidal.

ellipsoid ǁ Ellipsoid n. ǁ ellipsoïde m. / oblate ~ ǁ flaches Rotationsellipsoid n. ǁ ellipsoïde m. de rotation aplati. / oblong ~ *see* prolate ~. / prolate ~ ǁ verlängertes Rotationsellipsoid n. ǁ ellipsoïde m. de révolution prolongé. / ~ of revolution ǁ Rotationsellipsoid n.; Sphäroid n. ǁ ellipsoïde m. de révolution *ou* de rotation.

ellipsoidal ǁ ellipsoidförmig; ellipsoidisch ǁ ellipsoïdale; ellipsoïde.

elliptical ǁ elliptisch ǁ elliptique. / ~ compasses pl. ǁ Ellipsenzirkel m.; Ovalzirkel m. ǁ compas m. à ellipse. / ~ reflector ǁ elliptischer Reflexionsspiegel m. ǁ miroir m. plan elliptique. / ~ spring ǁ Elliptikfeder f. ǁ ressort m. elliptique. / three-coupled ~ spring ǁ dreifach gekuppelte Doppelfeder f. ǁ ressort m. elliptique couplé par trois.

elliptograph *see* ellipsograph.

elm ǁ Ulme f. ǁ orme m. / Scotch ~ ǁ Bergulme f. ǁ orme m. de montagne. / ~ wood ǁ Ulmenholz n. ǁ bois m. d'orme.

elongate, to ǁ strecken; verlängern ǁ étirer; allonger.

elongated ǁ verlängert ǁ allongé. / ~ hole ǁ längliches Loch n. ǁ trou m. longitudinal.

elongation ǁ Verlängerung f.; Längenausdehnung f. ǁ dilatation f. linéaire; allongement m. / permanent ~ ǁ bleibende Dehnung f. ǁ allongement m. permanent. / ~ at rupture ǁ Bruchdehnung f. ǁ allongement m. de rupture.

elongation cord ǁ Verlängerungsschnur f. ǁ cordon m. de prolongement. / ~ speed ǁ Dehnungsgeschwindigkeit f. ǁ vitesse f. d'allongement.

elucidate, to ǁ verdeutlichen; erläutern ǁ rendre clair; élucider; **expliquer.**

elutriation ǁ Schlämmung f.; Abschlämmen n. ǁ élutriation f.

elvan ǁ quarzführender Porphyr m.; Felsit m. ǁ porphyre m. quartzifère.

emanate, to ǁ ausgehen; ausfließen; ausströmen ǁ sortir; s'épuiser; émaner; provenir de; partir de.

emanation ǁ Ausströmung f.; Emanation f. ǁ émanation f.

embank, to ǁ eindämmen; eindeichen ǁ remblayer; endiguer; terrasser.

embankment ǁ Erdwall m.; Erddamm m.; Wall m.; Damm m. ǁ digue f.; rempart m. *ou* levée f. de terre; barrage m.; remblai m. / ~ (Railw) ǁ Erddamm m. ǁ remblai m. / loose stone ~ ǁ Steinschüttung f. ǁ enrochement m.; couche f. de cailloux *ou* de pierres concassées. / the material required for the ~ work ǁ das zu Aufschüttungen fpl. notwendige Material ǁ les matériaux nécessaires aux travaux de terrassement en remblai.

embargo ǁ Hafensperre f. ǁ embargo m.

embark, to ǁ sich einschiffen ǁ s'embarquer.

embarrass, to ǁ verwirren; in Verlegenheit f. setzen ǁ embarrasser; gêner.

embarrassing situation ǁ Zwangslage f. ǁ état m. de contrainte.

embed, to ~ *see also* to imbed ǁ einbetten; lagern ǁ sceller; encastrer. / ~ a frame in a wall ǁ einen Fundamentrahmen m. in eine Mauer einbetten ǁ sceller un châssis dans un mur.

embedded ǁ eingebettet ǁ scellé.

embedding ǁ Einmauerung f. ǁ scellement m. dans la maçonnerie.

embolite ǁ Chlorbromsilber n. ǁ embolite f.

emboss, to ǁ bossieren; erhaben ausarbeiten; treiben ǁ emboutir; bosseler; repousser; gaufrer.

embossed globe ǁ Reliefglobus m. ǁ globe m. en relief. / ~ metal articles pl. ǁ Metallprägewaren fpl. ǁ articles mpl. en métal repoussés. / ~ printing for blind ǁ Blindendruck m. ǁ impression f. en relief pour aveugles. / ~ writing picture telegraphy ǁ Reliefbildtelegrafie f. ǁ téléphotographie f. à écriture en relief.

embosser (Build) ǁ Bossierer m. ǁ piqueur m. de moellons. / ~ (Tel) ǁ Stiftsschreiber m. ǁ appareil m. Morse à couteau.

embossing ǁ Aufprägung f. ǁ gaufrage m. / leather ~ ǁ Lederpressen n. ǁ gaufrage m. de cuir; coréoplastie f.

embossing batten (Weav) ǁ Wirklatte f. ǁ battant-brocheur m. / ~ iron ǁ Bossiereisen n. ǁ talon m.; ébauchoir m. en fer. / ~ machine (Pap) ǁ Gaufriermaschine f. ǁ machine f. à gaufrer. / ~ machine (Print) ǁ Prägepresse f. ǁ presse f. à empreindre. / ~ press for hot process ǁ Heißprägepresse f. ǁ presse f. à empreindre à chaud.

embossment ǁ Bossierarbeit f. ǁ ouvrage m. en basse; bosselage m. / ~ of reliefs ǁ Auftreibung f. ǁ degré m. du relief. / ~ machine ǁ Wulstmaschine f. ǁ machine f. à bossage.

embouchure of a river ǁ Mündung f. eines Flusses ǁ embouchure f. d'une rivière.

embrasure of a window ǁ Fensterschmiege f. *oder* Ausschrägung f. der Laibung des Fensters ǁ embrasement m. d'une fenêtre.

embrace, to ǁ umfassen; umschließen ǁ embrasser.

embroider, to ǁ sticken; brodieren ǁ broder.

embroidered flag ǁ bestickte Fahne f. ǁ drapeau m. brodé. / ~ table linen ǁ gestickte Tischwäsche f. ǁ lingerie f. de table brodée. / ~ trimming ǁ Besatzstickerei f. ǁ garniture f. brodée.

embroiderer ǁ Sticker m. ǁ brodeur m. / gold thread ~ ǁ Golddrahtsticker m. ǁ brodeur m. en filets d'or. / hand ~ ǁ Handstickerin f. ǁ brodeuse f. à la main. / machine ~ ǁ Maschinensticker m. ǁ brodeur m. à la mécanique. / name ~ ǁ Namensticker m. ǁ brodeuse f. du nom. / pantograph ~ ǁ Pantografsticker m. ǁ pantographiste-brodeur m. / silver thread ~ ǁ Silberdrahtsticker m. ǁ brodeur m. en filets d'argent. / tapestry ~ ǁ Teppichsticker m. ǁ brodeur m. en tapisserie.

embroidering apparatus ǁ Stickapparat m. ǁ machine f. à broder. / ~ awl ǁ Stecher m. zum Sticken ǁ poinçon m. à broder. / ~ frame ǁ Stickrahmen m. ǁ tambour m. à broder.

embroidering machine ǁ Stickmaschine f. ǁ machine f. à broder; brodeuse f.; couso-brodeur m. / cop winder of ~ ǁ Stickmaschinenspulerin f. ǁ naveteuse f. de machine à broder. / ~ needle ǁ Stickmaschinennadel f. ǁ aiguille f. pour métiers à broder. / needle filler of ~ ǁ Stickmaschineneinfädlerin f. ǁ enfileuse f. d'aiguilles de machines à broder.

embroidering needle ǁ Sticknadel f. ǁ aiguille f. à broder. / ~ wool ǁ Stickwolle f. ǁ laine f. à broder.

embroidery ǁ Stickerei f. ǁ broderie f. / artistic ~ ǁ Kunststickerei f. ǁ broderie f. artistique. / ~ on cambric ǁ Batiststickerei f. ǁ broderie f. sur batiste. / cotton ~ ǁ Stickerei f. auf Baumwolle ǁ broderie f. en coton. / flat-stitch ~ ǁ Plattstichstickerei f. ǁ broderie f. hachebachée. / hand ~ ǁ Handstickerei f. ǁ broderie f. à la main. / ~ on linen ǁ Stickerei f. auf Leinen ǁ broderie f. sur tissu de lin. / machine ~ ǁ Maschinenstickerei f. ǁ broderie f. mécanique. / machine-made coloured ~ ǁ Maschinenbuntstickerei f. ǁ broderie f. mécanique de couleur. / machine-made white ~ / Maschinenweißstickerei f. ǁ broderie f. blanche à la mécanique. / ~ on net ǁ Tüllstickerei f. ǁ broderie f. sur tulle. / trimming stitch ~ *see* flat-stitch ~.

embroidery bleacher ǁ Stickereibleicher m. ǁ blanchisseur m. de broderies. / ~ cotton ǁ Stickbaumwolle f. ǁ coton m. à broder. / ~ designer ǁ Stickereizeichner m. ǁ dessinateur m. pour broderies. / ~ finishing ǁ Stickereiappretur f. ǁ apprêt m. de broderies. / ~ frame ǁ Stickrahmen m. ǁ métier m. à broder. / ~ machine ǁ Stickmaschine f. ǁ métier m. *ou* machine f. à broder; brodeuse f. / ~ needle ǁ Sticknadel f. ǁ aiguille f. à broder. / ~ pattern ǁ Stickmuster n. ǁ patron m. pour broderies. / ~ printer ǁ Drucker m. von Stickereimustern ǁ imprimeur m. de broderies. / ~ tracer ǁ Stickereimusterpauser m. ǁ ponceur-dessinateur m. en broderies. / ~ yarn ǁ Stickgarn n. ǁ fil m. à broderie *ou* à broder.

emerald ǁ Smaragd m. ǁ émeraude f. / ~ copper ǁ Kupfersmaragd m.; Dioptas m. ǁ dioptase f.; cuivre m. dioptase.

/ ~ green ‖ smaragdgrün ‖ vert émeraude. / ~ nickel ‖ Nickelsmaragd m.; Zaratit m. ‖ émeraude f. de nickel.

emergency brake ‖ Notbremse f. ‖ frein m. de secours. / ~ apparatus ‖ Notbremseinrichtung f. ‖ appareils mpl. du frein de secours des voyageurs. / ~ valve ‖ Notbremsventil n. ‖ valve f. du frein de secours.

emergency call ‖ Notsignal n. ‖ signal m. de détresse. / ~ car ‖ Rettungswagen m.; Hilfswagen m. ‖ voiture f. d'ambulance. / ~ closure ‖ Notverschluß m. ‖ bâtardeau m. / ~ coupling ‖ Notkupplung f. ‖ attelage m. de sûreté. / ~ exit ‖ Notausgangstür f. ‖ porte f. de réserve. / ~ lighting ‖ Notbeleuchtung f. ‖ éclairage m. de secours ou de réserve. / automatically controlled ~ lighting set ‖ selbsttätig sich einschaltendes Notbeleuchtungsaggregat n. ‖ groupe m. d'éclairage de secours à enclenchement automatique. / ~ outlet of a dam ‖ Notablaß m. einer Sperrmauer ‖ vanne f. de secours d'un mur du barrage. / ~ power station ‖ Aushilfskraftwerk n. ‖ usine f. auxiliaire. / ~ repair ‖ Notausbesserung f. ‖ réparation f. improvisée. / ~ set ‖ Notschaltung f. ‖ connexion f. de secours. / ~ switch ‖ Notausschalter m. ‖ interrupteur m. de sécours. / ~ tank ‖ Nottank m. ‖ réservoir m. de secours. / ~ transmitter (Radio) ‖ Notsender m. ‖ transmetteur m. de secours.

emery ‖ Schmirgel m. ‖ émeri m. / to lay on with ~ ‖ mit Schmirgel m. belegen ‖ émeriser. / levigated ~ ‖ geschlämmter Schmirgel m. ‖ potée f. d'émeri.

emery cloth ‖ Schmirgelleinwand f.; Schmirgelleinen n. ‖ toile f. d'émeri. / ~ cloth making plant ‖ Schmirgelleinenherstellungsanlage f. ‖ installation f. à fabriquer la toile à l'émeri. / ~ disc ‖ Schmirgelscheibe f. ‖ meule f. d'émeri. / ~ dust ‖ Schmirgelpulver n. ‖ poudre f. d'émeri. / ~ goods pl. ‖ Schmirgelwaren fpl. ‖ articles mpl. en émeri. / ~ grinding machine ‖ Schmirgelschleifmaschine f. ‖ machine f. à polir à l'émeri. / ~ mill ‖ Schmirgelwerk n. ‖ fabrique f. d'émeri. / ~ paper ‖ Schmirgelpapier n. ‖ papier-émeri m.; papier m. émerisé. / ~ paper making plant ‖ Schmirgelpapierherstellungsanlage f. ‖ installation f. à fabriquer du papier à l'émeri. / ~ plate ‖ Schmirgelscheibe f. ‖ meule f. en émeri. / ~ powder ‖ Schmirgelpulver n. ‖ émeri m. en poudre. / ~ quarry ‖ Schmirgelbruch m. ‖ carrière f. d'émeri. / ~ stick ‖ Schmirgelfeile f.; Schmirgelholz n. ‖ polissoir m.; rodoir m. à l'émeri. / ~ stone ‖ Schmirgelstein m. ‖ pierre f. d'émeri. / ~ wheel ‖ Schmirgelscheibe f. ‖ meule f. d'émeri. / ~ wheel man ‖ Schmirgelschleifer m. ‖ meuleur m. émeri. / ~ work plant ‖ Schmirgelwerkeinrichtung f. ‖ équipement m. à travailler l'émeri.

emetic ‖ Brechmittel n. ‖ émétique m. / ~ tartar ‖ Brechweinstein m. ‖ tartre m. émétique.

emigrant ship ‖ Auswandererschiff n. ‖ navire m. d'émigrants.

emigrate, to ‖ auswandern ‖ émigrer.

emigration ‖ Auswanderung f. ‖ émigration f.

emission ‖ Emission f. ‖ émission f. / secondary ~ (Electr) ‖ Sekundärausstrahlung n. ‖ émission f. secondaire.

emission spectrum ‖ Emissionsspektrum n. ‖ spectre m. d'émission. / ~ theory ‖ Emissionstheorie f. ‖ théorie f. de l'émission.

emit, to ‖ emittieren; aussenden ‖ émettre. / ~ rays pl. ‖ ausstrahlen ‖ rayonner.

emitter ‖ Sender m. ‖ poste m. transmetteur; transmetteur m. ‖ émetteur m.

emitting power ‖ Emissionsvermögen n. ‖ pouvoir m. émissif.

emmetropic eye ‖ rechtsichtiges Auge n. ‖ œil m. emmétrope.

emolument, to appoint with fixed ~s ‖ gegen feste Bezüge mpl. anstellen ‖ engager à gage m. fixe.

empirical ‖ empirisch; erfahrungsmäßig ‖ empirique. / ~ formula ‖ empirische Formel f. ‖ formule f. empirique ou brute.

emplectite ‖ Kupferwismutglanz m.; Kupferwismuterz n.; Emplektit m. ‖ bismuth m. sulfuré cuprifère; emplectite f.

employ, to ‖ aufwenden; anwenden ‖ employer; dépenser; utiliser. / ~ (Person) ‖ beschäftigen ‖ donner du travail m. à …

employed as ‖ beschäftigt als ‖ employé comme. / ~ by a firm ‖ bei einer Firma f. tätig ‖ employé dans une maison.

employe see employee.

employee ‖ Arbeitnehmer m.; Angestellter m. ‖ employé m. / ~ of the firm ‖ Werksangehöriger m. ‖ personne f. occupée par des établissements. / total of ~s ‖ Gesamtbelegschaft f. ‖ ensemble m. du personnel ouvrier ou occupé. / the whole of the ~s of a concern ‖ gesamte Belegschaft f. eines Konzerns ‖ personnel m. entier occupé par un groupe.

employer ‖ Arbeitgeber m. ‖ employeur m.; patron m.

employer's liability insurance association ‖ Berufsgenossenschaft f. ‖ association f. professionnelle. / ~ policy ‖ Gewerbeunfallversicherungspolice f. ‖ police f. d'assurance contre la responsabilité des patrons.

employment ‖ Stellung f.; Beschäftigung f.; Tätigkeit f. ‖ emploi m.; place f. / to terminate an ~ unlawfully ‖ das Arbeitsverhältnis unberechtigt lösen ‖ rompre sans justification le contrat de travail.

employment agency ‖ Stellenvermittlung f. ‖ bureau m. de placement. / ~ bureau ‖ Arbeitsnachweis m. ‖ bureau m. de placement. / ~ contract ‖ Arbeitsvertrag m. ‖ contrat m. de travail. / ~ field ‖ Anwendungsgebiet n. ‖ domaine m. d'emploi ou d'application ou de travail ou d'occupation. / ~ office ‖ Arbeiterannahmestelle f. ‖ bureau m. d'embauchage. / ~ records pl. ‖ Arbeitsstatistik f. ‖ statistique f. se rapportant au travail.

empower, to ‖ bevollmächtigen; ermächtigen ‖ autoriser; donner pouvoir m.

emptier ‖ Entleerer m.; Ausleerer m. ‖ videur m. / bottle ~ ‖ Flaschenausgießer m. ‖ bouchon-verseur m. pour bouteilles.

emptiness ‖ Leere f. ‖ vide m. / ~ (Soil) ‖ Taubheit f. ‖ stérilité f.

empties pl. ‖ leere Emballagen fpl. ‖ emballages mpl. vides.

empty ‖ leer ‖ vide; vidé. / when ~ return to ‖ leer zurück nach ‖ vide retour à. / ~ crucible ‖ leerer Tiegel m. ‖ creuset m. vide. / ~ ears pl. (Agr) ‖ Abschöpfgerste f. ‖ orge f. légère; écume f.; petite orge f.; orgette f.

empty, to ‖ leeren; entleeren; ablassen; ausschütten ‖ vider; évacuer; déverser; culbuter; dégonfler.

emptying ‖ Entleerung f. ‖ vidange f. / suction dredger for ~ barges ‖ Schutensauger m. ‖ aspirateur m. pour gabares. / ~ of the shaft furnace ‖ Schachtofenentleerung f. ‖ vidange f. ou défournement m. des fours à puits.

emptying brake ‖ Entleerungsbremse f. ‖ frein m. d'ouverture.

empyreumatic (al) ‖ brenzlich; empyreumatisch ‖ empyreumatique.

emulation ‖ Wetteifer m.; Nacheiferung f. ‖ émulation f.

emulsifier see emulsifying machine.

emulsify, to ‖ emulgieren ‖ émulsionner.

emulsifying machine ‖ Emulgiermaschine f. ‖ émulsionneuse f.

emulsion ‖ Emulsion f. ‖ émulsion f. / to achieve the rapid and uniform desiccation of the ~ ‖ rasches und gleichmäßiges Erstarren n. der Emulsion bewirken ‖ effectuer la solidification rapide et uniforme de l'émulsion.

emulsion coater ‖ Emulsionaufstreicher m. ‖ émulsionneur m. / ~ gelatine ‖ Emulsionsgelatine f. ‖ gélatine f. d'émulsion.

emulsioning plant ‖ Emulsionsanlage f. ‖ installation f. à émulsionner.

emulsion trough ‖ Emulsionstrog m. ‖ auget m. d'émulsion.

enamel, to ‖ emaillieren ‖ émailler.

enamel ‖ Email n.; Emaille f. ‖ émail m. / ~ (Paint) ‖ Emaillefarbe f. ‖ couleur f. émail; peinture f. vernissante. / acid-proof ~ ‖ säurefestes Email n. ‖ émail m. résistant aux acides. / black ~ ‖ Schwarzschmelz m. ‖ nielle f.; émail m. noir. / black ~ finish ‖ schwarze Lackierung f. ‖ achèvement m. en noir émaillé. / coloured ~ ‖ farbige Glasur f. ‖ émail m. coloré. / crude ~ ‖ rohe Emaille f.; rohes Email n. ‖ émail m. brut. / embossed ~ ‖ Reliefemail n. ‖ émail m. en haute taille. / to inlay with black ~ ‖ niellieren ‖ nieller. / ivory ~ finish ‖ elfenbeinfarbige Lackierung f. ‖ achèvement m. en couleur d'ivoire émaillé. / painter's ~ ‖ Maleremail n. ‖ émail m. de peinture. / ~ ready for use ‖ gebrauchsfertige Emaille f.; gebrauchsfertiges Email n. ‖ émail m. préparé.

enamel colour ‖ Emailfarbe f. ‖ couleur f. émail. / with ~ facing ‖ mit Emailüberzug m. ‖ émaillé. / ~ finish ‖ Lackierung f.; Emaillierung f. ‖ achèvement m. ou accomplissement m. en émail. / ~ foil(ing) ‖ Emailbelag m. ‖ couche f. d'émail. / ~ foil(ing) of the watch case bottom ‖ Emailbelag m. des Uhrgehäusedeckels ‖ couche f. d'émail sur le fond de la boîte d'une montre. / ~ furnace ‖ Emaillierofen m. ‖ four m. à émailler.

enamelled ‖ emailliert ‖ émaillé. / white ~ ‖ weiß emailliert ‖ en émail blanc.

enamelled dial ‖ Emailzifferblatt n. ‖ cadran m. émaillé. / ~ metal plate ‖ Emailleschild n. ‖ plaque f. métallique ou enseigne f. émaillée. / ~ name plate ‖ Emailschild n. ‖ enseigne f. émaillée. / ~ plate ‖ Emailplatte f. ‖ plaque f. émaillée. / ~ sign see ~ plate. / ~ stoneware ‖ emailliertes Steinzeug n. ‖ grès m. émaillé. / ~ utensils pl. ‖ emailliertes Geschirr n. ‖ batteries fpl. de cuisine émaillées. / ~ ware ‖ Emailwaren fpl. ‖

articles mpl. émaillés. / ~ wire ‖ Emaildraht m. ‖ fil m. émaillé.

enameller ‖ Emaillierer m. ‖ émailleur m. / metal ~ ‖ Metallemaillierer m. ‖ émailleur m. sur métaux.

enamelling ‖ Emaillieren n. ‖ émaillage m. / ~ of casting ‖ Emaillierung f. von Guß ‖ émaillage m. de fonte. / ~ at the lamp ‖ Emaillieren n. vor der Lampe ‖ émaillure f. à la lampe. / ~ on precious metals ‖ Emaillierung f. edler Metalle ‖ émaillage m. sur métaux précieux.

enamelling equipment ‖ Emailliereinrichtung f. ‖ installation f. d'émaillage. / ~ furnace ‖ Emaillierofen m. ‖ four m. à émailler. / ~ machine ‖ Emaillierwerkmaschine f. ‖ machine f. pour äteliers d'émaillage. / ~ manufactory ‖ Emaillierwerk n. ‖ émaillerie f. industrielle.

enamel mill ‖ Emailmühle f. ‖ moulin m. à émail. / ~ paint *see* enamel colour. / ~ painter ‖ Emailmaler m. ‖ peintre m. sur émail.

enamel painting ‖ Emailmalerei f. ‖ peinture f. sur émail. / ~ on a deep ground ‖ Schmelzmalerei f. mit vertieftem Grund ‖ émail m. champlevé. / ~ with deepened out-lines ‖ Schmelzmalerei f. mit vertieften Konturen ‖ émail m. translucide. / ~ with inlaid metal-lines ‖ Schmelzmalerei f. mit eingelegten Metallstreifen ‖ émail m. cloisonné.

enamel pitch ‖ Emailpech n. ‖ poix m. émail.

enamels pl. *see* enamel ware.

enamel varnish ‖ Emaillack m. ‖ laque-émail f. / ~ ware ‖ Emailwaren fpl. ‖ articles mpl. émaillés; émaux mpl.

encase, to ‖ umhüllen; einschalen; ineinanderschieben ‖ encaisser; enrober; s'emboîter.

encased water-tank ‖ Wasserbehälter m. mit Verkleidung ‖ réservoir m. d'eau recouvert.

encashment ‖ Inkasso n. ‖ encaissement m.; recouvrement m.

encaustic tile ‖ farbig glasierter Ziegel m. ‖ tuile f. vernie (hollandaise).

encaustic ‖ Bohnerwachs n. ‖ encaustique f. / ~ for furniture ‖ Möbelpolitur f. ‖ encaustique f. pour meubles.

enchase, to ‖ einspannen ‖ tendre sur; tendre dans. / ~ (Goldsm) ‖ fassen; einfassen ‖ enchasser; entourer d'ornements mpl. / ~ (Jewel) ‖ fassen ‖ enchâsser; sertir. / ~ (Wood) ‖ einlegen ‖ incustrer.

encircle, to ‖ umfassen; umschließen; umschnüren ‖ fretter; entourer; ceinturer.

encircling line ‖ Ringbahn f. ‖ ligne f. de ceinture.

enclose, to ‖ umschließen; einfriedigen ‖ enfermer; entourer. / ~ (Letter) ‖ anschließen; beischließen ‖ ajouter. / ~ a machine ‖ eine Maschine f. einfriedigen ‖ entourer une machine.

enclosed (In letter) ‖ anliegend; inliegend ‖ ci-inclus; ci-joint; sous ce pli m. / ~ in a capsule ‖ gekapselt ‖ capsulé. / ~ entirely ‖ vollkommen gekapselt ‖ entièrement blindé *ou* fermé. / fully ~ *see* entirely ~.

enclosed arc ‖ eingeschlossener Lichtbogen m. ‖ arc m. enfermé. / ~ arc lamp ‖ Dauerbrandbogenlampe f. ‖ lampe f. à arc enfermé *ou* à arc de longue durée. / ~ casing (Turbine) ‖ geschlossenes Gehäuse n. ‖ enveloppe f. *ou* corps m. de turbine. / ~ compressor ‖ Kapselkom-

pressor m. ‖ compresseur m. blindé. / ~ gear ‖ gekapseltes Getriebe n. ‖ engrenage m. enfermé dans des carters.

enclosing of the fly-wheel ‖ Umwehrung f. des Schwungrades ‖ entourage m. du volant.

enclosure ‖ Gehege n.; Einfriedigung f. ‖ enclos m.; enceinte f. / ~ (Build) ‖ Wall m. ‖ terrasse f. / ~ (Gard) ‖ lebende Hecke f. ‖ haie f. vive. / plaited wirework for ~s ‖ Drahtgeflecht n. zur Einfriedigung ‖ treillis m. en fil métallique d'enceinte. / ~ wall ‖ Einfriedigungsmauer f. ‖ mur m. de clôture *ou* d'enceinte.

encroachment ‖ Übergriff m. ‖ empiètement m. / ~ (Tel) ‖ Spielraum m. ‖ empiètement m.

encumber, to (Mine) ‖ verschütten ‖ encombrer.

end, to ‖ endigen; aufhören ‖ finir; achever; terminer.

end ‖ Ende n.; Spitze f.; Ziel n. ‖ bout m.; extrémité f. / ~ (Season) ‖ Ende n.; Schluß m. ‖ fin f. / ~ of an adit (Mine) ‖ Stollenort m. ‖ extrémité f. d'une galerie. / boiler ~ ‖ Kesselboden m. ‖ fond m. de chaudière. / the ~s of the boiler are supplied with machined edges ‖ die Kesselböden mpl. werden mit abgedrehten Kanten geliefert ‖ les fonds mpl. de la chaudière sont fournis avec arêtes passées au tour. / ~ of car (Railw) ‖ Kopfwand f.; Stirnwand f. ‖ paroi f. frontale *ou* de bout. / ~ of the ebb ‖ Hinterebbe f.; Achterebbe f.; letzte Ebbe f. ‖ fin f. du jusant. / flanged ~ ‖ gebördelter Boden m. ‖ fond m. bridé *ou* à bride. / flanged boiler ~ ‖ umgezogener Kesselboden m. ‖ fond m. de chaudière à brides. / ~ of a gallery (Mine) ‖ Feldort m. ‖ lieu m. de travail dans une galerie; fin f. d'une galerie. / the ~s of the girders are kept very low ‖ die Trägerenden npl. liegen tief ‖ les extrémités fpl. des longerons sont placées bas. / ~ of a level *see* ~ of a gallery. / ~ of the line ‖ Zeilenschluß m. ‖ fin f. de ligne. / on ~ ‖ aufrecht ‖ debout. / ~ of a perpendicular ‖ Fußpunkt m. eines Lotes ‖ pied m. d'un perpendiculaire. / ~ of a rail ‖ Schienenende n. ‖ about. du rail. / ~ of a rope (Mar) ‖ Pünt f. ‖ bout m. / staved ~ ‖ angestauchter Rand m. ‖ bord m. refoulé. / tapered ~ ‖ verjüngtes Ende n. ‖ rétrécissement m. de l'extrémité. / ~ of thread ‖ Gewindeauslauf m. ‖ fin m. de filetage. / ~ of tunnel (Mine) ‖ Stollenmundloch n. ‖ orifice m. du tunnel.

end abutment (Bridge) ‖ Landpfeiler m.; Widerlager n. ‖ pile f. culée.

endanger, to ‖ gefährden; in Gefahr f. bringen ‖ mettre en danger m.

endangering, indirect ~ ‖ mittelbare Gefährdung f. ‖ danger m. indirect. / ~ clamp ‖ Hirnleiste f.; Türfries m. ‖ listel m. de traverse; emboîture f. / ~ cut-out switch ‖ Endausschalter m. ‖ interrupteur m. de fin de course. / ~ face (Crystal) ‖ Kristallendfläche f. ‖ base f. de crystal. / ~ frame ‖ Stirnbügel m. ‖ cadre m. d'avant. / ~ grain ‖ Hirnholz n. ‖ bois m. de bout. / ~ hoop (Coachm) ‖ Achsring m. ‖ anneau m. de bout d'essieu.

endive ‖ Endivie f. ‖ endive f.

end journal bearing ‖ Stirnlager n. ‖ palier m. frontal.

endless ‖ endlos; ohne Ende n. ‖ sans fin f. / ~ band ‖ endloses Band n. ‖ bande f. sans fin. / ~ belt ‖ endloser Riemen m. ‖ courroie f. sans fin. / ~ chain ‖ endlose Kette f. ‖ chaîne f. sans fin. / ~ chain multi-bucket excavator ‖ Eimerkettenbagger m. ‖ excavateur m. à chaîne de godets. / ~ cord ‖ Seil n. ohne Ende ‖ corde f. sans fin. / ~ ribbon ‖ endloses Band n. ‖ ruban m. sans fin. / ~ rope ‖ Seil n. ohne Ende ‖ corde f. sans fin. / ~ saw ‖ Bandsäge f. ‖ scie f. à lame continue *ou* à ruban. / ~ saw blade ‖ Bandsägeblatt n. ‖ lame f. de scie à ruban. / ~ screw ‖ Schnecke f.; Schraube f. ohne Ende ‖ vis f. sans fin. / ~ strip ‖ endloser Streifen m. ‖ bande f. sans fin. / ~ thread ‖ endloser Faden m. ‖ fil m. continu. / ~ wire rope way ‖ Drahtseilbahn f. ‖ chemin m. de fer funiculaire.

end mill ‖ Walzenstirnfräser m.; Fingerfräser m. ‖ fraise f. cylindrique en bout; fraise f. à queue.

endogenous ‖ endogen ‖ endogène.

endorse, to ‖ girieren; indossieren ‖ endosser; virer.

endorsee ‖ Indossat m.; Indossatar m. ‖ endossé m.

endorsement ‖ Giro n.; Indossament n. ‖ endossement m. / ~ in blank ‖ Blankoindossament n. ‖ endossement m. en blanc.

endorser ‖ Girant m.; Indossant m. ‖ endosseur m.

endosmose ‖ Endosmose f. ‖ endosmose f. / electric ~ ‖ kataphorische Wirkung f. ‖ endosmose f. électrique.

endosmosis *see* endosmose.

endosperm ‖ Endosperm n. ‖ endosperme m.

endospore ‖ Endospore f. ‖ endospore f.

endosporic ‖ endospor ‖ endosporé.

endosporium ‖ Endosporium n. ‖ endospore f.

endothermic ‖ endotherm ‖ endothermique.

endowment ‖ Stiftung f. ‖ fondation f.; établissement m. / ~ fund ‖ Unterstützungskasse f. ‖ caisse f. de secours.

end paper (Bookb) ‖ Vorsatzpapier n.; Vorsatzblatt n. ‖ papier m. *ou* feuillet m. de garde. / ~ piece ‖ Kopfträger m. ‖ traverse f. de bout. / ~ plank (Railw) ‖ Kopfwand f.; Stirnwand f. ‖ paroi f. de bout. / ~ plate (Acc) ‖ Endplatte f. ‖ plaque f. extrême. / front ~ plate of boiler ‖ Kesselstirnwand f. ‖ fond m. avant de chaudière. / ~ platform ‖ Endplattform f. ‖ plateforme f. extrême de derrière. / ~ play ‖ leerer *oder* toter Gang m. ‖ jeu m. / ~ play of a screw ‖ toter Gang m. einer Schraube ‖ jeu m. d'un filet. / ~ product ‖ Endprodukt n. ‖ produit m. final. / ~ screw of joiner's bench ‖ Hinterzange f. einer Hobelbank ‖ presse f. de derrière d'un établi du menuisier. / ~ and side tip wagon ‖ Hinter- und Seitenkipper m. ‖ wagon m. se vidant de côté et par le bout. / to prevent the ~ sleepers from becoming disconnected (Railw) ‖ eine Verschiebung f. der Endschwellen verhindern ‖ empêcher le déplacement des traverses terminales. / ~ stanchion (Railw) ‖ Eckrunge f.; Kopfrunge f. ‖ rancher m. cornier *ou* de bout. / ~ stirrup of lead ‖ Endbleibügel m. ‖ étrier m. de calage en plomb. / ~ stone (Watchm) ‖ Deckstein m. ‖ contre-pivot m. / ~ straining ‖ Endverankerung f. ‖ ancrage m. dans le sol.

/ ~ support (Car) ‖ Kopfquerträger m. ‖ traverse f. extrême. / ~ terminal ‖ Endpolklemme f. ‖ borne f. d'attache. / ~ tile ‖ Schlußziegel m. ‖ tuile f. recourbée. / ~ tipper ‖ Hinterkipper m. ‖ camion m. basculant à l'arrière. / ~ tube (Tel) ‖ Endröhre f. ‖ lampe f. de puissance. / ~ turns pl. of armature ‖ Schlußwindungen fpl. des Ankers ‖ tours mpl. d'aboutissement de l'induit.

endurance ‖ Haltbarkeit f. ‖ durabilité f.; endurance f. / ~ flight ‖ Dauerflug m. ‖ vol m. de durée. / ~ test ‖ Dauerversuch m. ‖ essai m. de longue durée. / ~ test machine ‖ Dauerprüfmaschine f. ‖ machine f. pour essais de longue durée.

end way of the grain ‖ Hirnseite f. des Holzes ‖ côté m. de la moelle; coupe f. transversale.

endwise tipping truck ‖ Hinterkipper m. ‖ wagon m. à déchargement par le bout.

energetic(al) ‖ tatkräftig; energisch ‖ énergique.

energize, to (Electr) ‖ erregen ‖ amorcer.

energy ‖ Energie f. ‖ énergie f. / to accumulate ~ ‖ Energie f. aufspeichern ‖ accumuler de l'énergie f. / available ~ ‖ verfügbare Energie f. ‖ énergie f. disponible. / electro-magnetic ~ ‖ elektromagnetische Energie f. ‖ énergie f. électromagnétique. / kinetic ~ ‖ kinetische Energie f. ‖ énergie f. cinétique. / magnetic ~ ‖ magnetische Energie f. ‖ énergie f. magnétique. / mechanical ~ ‖ mechanische Energie f. ‖ énergie f. mécanique. / to convert mechanical ~ into electrical energy ‖ mechanische Energie f. in elektrische verwandeln ‖ transformer l'énergie mécanique en énergie électrique. / potential ~ ‖ potentielle Energie f. ‖ énergie f. potentielle. / radiant ~ ‖ strahlende Energie f. ‖ énergie f. rayonnante. / ~ radiated by the aerial ‖ Antennenstrahlung f. ‖ énergie f. rayonnée par l'aérien. / ~ of radiation in sunlight ‖ Strahlungsenergie f. des Sonnenlichtes ‖ radiation f. solaire. / wasted ~ ‖ Leerlaufarbeit f. ‖ dépense f. à vide.

energy, conservation of ~ ‖ Erhaltung f. der Energie ‖ conservation f. de l'énergie. / consumption of ~ ‖ Kraftverbrauch m. ‖ dépense f. de force. / delivery of ~ ‖ Energiebelieferung f. ‖ fourniture f. d'énergie. / ~ destroyer ‖ Energievernichter m. ‖ destructeur m. d'énergie. / distribution of ~ ‖ Energieverteilung f. ‖ distribution f. d'énergie. / flow of ~ ‖ Energiestrom m. ‖ flux m. d'énergie. / ~ flux see flow of ~. / loss of ~ ‖ Energieverlust m. ‖ perte f. d'énergie. / small loss of ~ by hysteresis ‖ geringer Energieverlust m. durch Hysteresis ‖ petite perte f. d'énergie par hystérésis. / mutation of ~ ‖ Energieumwandlung f. ‖ mutation f. de l'énergie. / storage of ~ ‖ Energieaufspeicherung f. ‖ accumulation f. d'énergie. / transformation of ~ ‖ Energieumwandlung f. ‖ transformation f. d'énergie. / transmission of ~ ‖ Energieleitung f.; Kraftübertragung f. ‖ transmission f. de l'énergie. / cost of transmission of ~ ‖ Energieleitungskosten pl. ‖ frais mpl. de transmission de l'énergie. / waste of ~ ‖ Energieverschwendung f. ‖ gaspillage m. d'énergie.

enfeebled ‖ matt; kraftlos ‖ faible.

enforced ‖ verstärkt ‖ armé.

engage, to (Mach) ‖ einrücken; kuppeln ‖ embrayer. / ~ (Workmen) ‖ anstellen ‖ assigner une poste; engager; embaucher; prendre. / ~ together (Bricks) ‖ in Verband m. mauern ‖ enlier; placer les briques fpl. en liaison. / ~ with ... (Gear) ‖ ineinandergreifen ‖ engrener sur ... / ~ a workman ‖ einen Arbeiter m. einstellen ‖ engager ou embaucher un ouvrier.

engaged (Gear) ‖ im Eingriff m. ‖ engrené; être en prise. / to be ~ (Aut tel) ‖ besetzt sein (durch fremden Anruf) ‖ être occupé.

engagement (Gearing) ‖ Verzahnung f.; Eingriff m. ‖ engrènement m.; engrenage m.; prise f. / ~ (Obligation) ‖ Verpflichtung f.; Verbindlichkeit f. ‖ engagement m. / ~ (Tel) ‖ Besetztfall m. ‖ cas m. d'occupation. / to terminate an ~ ‖ das Arbeitsverhältnis beendigen ‖ terminer ou mettre fin à un engagement de travail.

engaging and disengaging ‖ Ein- und Ausschalten n. ‖ embrayage m. et débrayage m. / instantaneous ~ ‖ augenblickliches Einrücken n. und Ausrücken n. ‖ embrayage m. et débrayage m. instantané. / ~ coupling box ‖ Auslösungskupplung f. ‖ accouplement m. de débrayage. / ~ gear see ~ coupling box.

engaging and release see engaging and disengaging.

engaging clutch ‖ Einrückklaue f. ‖ dispositif m. d'embrayage. / ~ dog ‖ Mitnehmerklaue f. ‖ griffe f. d'entraînement.

engine ‖ Maschine f.; Motor m.; Kraftmaschine f. ‖ machine f.; moteur m. / ~ (Pap) ‖ Holländer m.; Stoffmühle f. ‖ pile f. ou moulin m. à cylindre. / ~ (Railw) ‖ Lokomotive f. ‖ locomotive f.; machine f. locomotive. / ~ for air injection of the fuel ‖ motorisch angetriebene Einblasevorrichtung f. ‖ moteur m. du type à injection pneumatique. / ~ with airless injection of the fuel ‖ Einspritzmotor m. ‖ moteur m. à injection mécanique. / ~ has broken down ‖ Maschine f. ist betriebsunfähig ‖ la machine est en panne. / crude oil ~ ‖ Rohölmotor m. ‖ moteur m. à huile lourde. / driving ~ ‖ Antriebsmaschine f. ‖ machine f. motrice. / eight-cylinder ~ ‖ Achtzylindermotor m. ‖ machine f. à huit cylindres. / four-cylinder ~ ‖ Vierzylindermotor m. ‖ machine f. à quatre cylindres. / four-stroke cycle ~ ‖ Viertaktmotor m. ‖ moteur m. à quatre temps. / ~ with four times see four-stroke cycle ~. / horizontal ~ ‖ liegender Motor m. ‖ moteur m. horizontal. / horse-power ~ ‖ Maschine f. von x Pferdestärken ‖ machine f. de x chevaux. / hydraulic ~ ‖ hydraulischer Motor m. ‖ moteur m. hydraulique. / internal combustion ~ ‖ Verbrennungsmaschine f. ‖ moteur m. à combustion interne. / internal explosion ~ ‖ Explosionsmotor m. ‖ moteur m. à explosion. / left-hand ~ ‖ Linksmaschine f. ‖ machine f. à gauche. / one-cylinder ~ ‖ Einzylindermotor m. ‖ machine f. à un cylindre ou monocylindrique. / petrol ~ ‖ Bezinmotor m. ‖ moteur m. à essence. / portable ~ ‖ Lokomobile f. ‖ locomobile f. / quadruple-expansion ~ ‖ Vierfachexpansionsmaschine f. ‖ machine f. à quadruple détente. / right-hand ~ ‖ Rechtsmaschine f. ‖ machine f. à droite.

/ rose ~ ‖ Guillochiermaschine f. ‖ machine f. à guillocher; tour m. à guillocher ou à rosettes. / semiportable ~ ‖ halbortsfeste Antriebsmaschine f. ‖ machine f. demi-fixe. / single-cylinder ~ ‖ Einzylindermaschine f. ‖ machine f. monocylindrique. / six-cylinder ~ ‖ Sechszylindermotor m. ‖ machine f. à six cylindres. / steam ~ ‖ Dampfmaschine f. ‖ machine f. à vapeur. / fixed steam ~ ‖ ortsfeste Dampfmaschine f. ‖ machine f. à vapeur fixe ou stationnaire. / highpressure steam ~ ‖ Hochdruckdampfmaschine f. ‖ machine f. à haute pression. / land steam ~ see fixed steam ~. / stationary steam ~ see fixed steam ~. / supercharged ~ ‖ überverdichtender Motor m. ‖ moteur m. suralimenté. / superheated ~ ‖ Heißdampfmaschine f. ‖ machine f. à vapeur surchauffée. / vertical steam ~ ‖ stehende Dampfmaschine f. ‖ machine f. à vapeur verticale. / ~ with tender ‖ Lokomotive f. mit Schlepptender ‖ locomotive f. avec tendeur. / triple-expansion ~ ‖ Dreifachexpansionsmaschine f. ‖ machine f. à triple détente. / twelve-cylinder ~ ‖ Zwölfzylindermotor m. ‖ machine f. à douze cylindres. / two-cylinder ~ ‖ Zweizylindermotor m. ‖ machine f. à deux cylindres. / water-cooled ~ ‖ wassergekühlter Motor m. ‖ moteur m. à refroidissement d'eau ou par l'eau.

engine bed ‖ Motorbock m. ‖ support m. de moteur. / ~ bed plate ‖ Maschinenrahmen m. ‖ bâti m. de machine. / ~ belt ‖ Maschinenriemen m. ‖ courroie f. de machines. / ~ breakdown ‖ Motorstörung f. ‖ panne f. de moteur. / ~ builder ‖ Maschinenbauer m. ‖ mécanicien m.; constructeur-mécanicien m. / ~ building ‖ Maschinenbau m. ‖ construction f. de machines. / ~ car (Aero) ‖ Maschinengondel f. ‖ nacelle-moteur m. / ~ case ‖ Motorgehäuse n. ‖ carter m. de moteur. / ~ cowling ‖ Motorverkleidung f. ‖ capote f. du moteur. / ~ drive ‖ Lokomotivbetrieb m. ‖ traction f. par locomotive. / ~ driver (Mach) see engineer (Engineman). / ~ driver (Railw) ‖ Lokomotivführer m.; Maschinenführer m.; Maschinist m. ‖ mécanicien m.; conducteur m. de la locomotive. / ~ driving ‖ Führung f. der Maschine ‖ conduite f. de machine.

engineer ‖ Ingeniör m. ‖ ingénieur m. / ~ (Engineman) see also engineman ‖ Maschinist m.; Maschinenführer m.; Maschinenwärter m.; Maschinenmeister m. ‖ machiniste m.; mecanicien m. / certificated ~ ‖ Diplomingeniör m. ‖ ingénieur m. diplômé. / chief ~ ‖ Oberingeniör m. ‖ ingénieur m. en chef. / civil ~ ‖ Bauingeniör m. ‖ ingénieur m. des ponts et chaussées. / consulting ~ ‖ beratender Ingeniör m. ‖ ingénieur m. conseil. / ~ of a district ‖ Bezirksingeniör m. ‖ ingénieur m. de district. / ~ for drainage ‖ Kulturtechniker m. ‖ ingénieur m. agricole. / government's ~ ‖ Bauinspektor m. ‖ inspecteur m. des travaux publics; ingénieur m. ou architecte m. du gouvernement. / ~ of illumination ‖ Beleuchtungsingeniör m. ‖ ingénieur m. éclairagiste. / ~-in-chief see chief ~. / ~ of lighting see ~ of illumination. / mechanical ~ ‖ Maschineningeniör m. ‖ ingénieur-mécanicien m. / mining ~ ‖ Bergbauingeniör m. ‖ ingénieur m. des mines. / ~ for subterraneous work ‖ Tiefbauingeniör m. ‖ in-

génieur m. de travaux souterrains *ou* en dessous du sol.

engineer board || Ingeniörabteilung f. || comité m. des ingénieurs.

engineer's hammer || Maschinenbauerhammer m. || marteau m. d'ajusteur.

engineering || Ingeniörwesen n. || science f. de l'ingenieur; génie m. civil. / canal-, harbour- and river-~ || Hydrotechnik f. || service m. hydraulique; service m. des canaux, des rivières et des ports maritimes. / general ~ || allgemeiner Maschinenbau m. || construction f. mécanique générale. / hydraulic ~ || Wasserbaukunst f. || hydrotechnique f. / mechanical ~ || Maschinenbau m. || construction f. de machines.

engineering department || Maschinenbauabteilung f.; Konstruktionsbüro n. || département m. de la construction mécanique; salle f. de construction *ou* de dessins *ou* bureau des études. / ~ drawing || Maschinenzeichnung f. || dessin m. de machine. / ~ practice || Technik f. technique f.

engine fitter || Monteur m. || monteurmécanicien m. / ~ frame || Maschinenrahmen m. || bâti m. de machine. / ~ head lamp || Lokomotivlaterne f. || lanterne f. de locomotive. / ~ house || Maschinenhaus n. || hall m. aux machines *ou* salle f. des machines. / ~ house (Railw) || Maschinenhaus n.; Lokomotivschuppen m. || dépôt m. de machines; remise f. à locomotives.

engineman *see also* engineer || Maschinist m.; Maschinenwärter m. || machiniste m.; mécanicien m. / ~ (Railw) || Maschinenführer m.; Maschinist m.; Lokomotivführer m. || mécanicien m.; conducteur m. de locomotive.

engine oil || Maschinenöl n. || huile f. de machines. / ~ parts pl. || Maschinenelemente npl.; Maschinenteile mpl. || organes mpl. de machines. / ~ pit || Pumpenschacht m. || puits m. d'épuisement. / ~ pit (Railw) || Reinigungsgrube f. || fosse f. à piquer. / ~ platform || Maschinenplattform f. || plateforme f. de la machine. / ~ power || Maschinenkraft f. || force f. de la machine. / ~ repair || Maschinenreparatur f. || réparation f. de la machine.

engine room || Maschinenraum m. || chambre f. *ou* emplacement m. *ou* salle f. des machines. / ~ of the ship || Schiffsmaschinenraum m. || compartiment m. de machines du navire.

engine room complement (Mar) || Maschinenpersonal n. || personnel m. de la machine. / hatch to ~ (Shipb) || Maschinenluke f. || écoutille f. de la machine. / ~ skylight (Shipb) || Maschinenraumoberlicht n. || claire-voie f. de la chambre des machines.

engine shaft (Mine) || Pumpenschacht m.; Wasserhaltungsschacht m. || puits m. d'épuisement *ou* d'exhaure. / ~ system || Motorsystem n. || système m. de moteur. / ~ tender (Railw) || Tender m. || tender m. / ~ truck || Drehgestell n. (der Lokomotive) || bogie m. *ou* truck m. (de locomotive). / ~ and tender turntable || Drehscheibe f. mit großem Durchmesser || pont m. tournant pour locomotive et tender. / ~ varnish || Maschinenlack m. || vernis m. pour machines. / ~ works pl. ||

Maschinenfabrik f. || atelier m. de construction de machines.

English blue || Englischblau n. || bleu m. de faïence *ou* de Chine *ou* anglais. / ~ china || englisches Porzellan n. || porcelaine f. tendre anglaise. / ~ cord machine || englische Schnurmaschine f. || mécanique f. anglaise à cordons. / ~ pitch lead screw || Zollleitspindel f. || vis-mère f. au pas de pouces. / ~ red || Caput mortuum; Englischrot n. || colcothar m.; rouge m. anglais. / ~ sauce || englische Soße f. *oder* Tunke || sauce f. anglaise. / ~ standard screw || englisches Schraubengewinde n. || filet m. au pas de vis anglais.

engrailed ring round a coin || Kräuselung f. einer Münze; Rand m. einer Münze || cordon m. d'une monnaie.

engrave, to || gravieren || graver. / ~ in wood || in Holz n. schneiden || graver sur bois m.

engraved || eingeätzt || gravé. / ~ printing roller || Gravurdruckwalze f. || cylindre m. d'imprimerie gravé.

engraver (Person) || Gravör m.; Ziselör m. || graveur m.; ciseleur m. réparateur. / ~ (Tool) || Grabstichel m.; Stichel m.; Zeiger m. || burin m. du graveur; matoire f. / ~ on copper || Kupferstecher m. || graveur m. en taille-douce *ou* au burin; chalcographe m. / metal ~ || Metallgravör m.; Gravör m. || graveur m. sur métal. / ~ of music || Notenstecher m. || graveur m. en musique. / ~ in steel || Stahlstecher m. || graveur m. sur acier. / ~ on wood || Holzgravör m. || xylographe m.

engraver's die || Gravörstempel m. || coin m. pour graveurs.

engraving || Gravierung f.; Stich m. || gravure f. / ~ (For handles) || Fischhaut f. (Griffekerbung) || quadrillage m.; face f. carrelée. / ~ artistic || künstlerischer Stich m. || gravure f. artistique. / coin ~ || Münzprägegravierung f. || gravure f. en monnaies. / ~ on copper || Kupferstich m.; Kupferstichplatte f. || gravure f. sur cuivre; estampe f. *ou* gravure f. en taille-douce. / die ~ || Gaufriergravierung f. || gravure f. de gaufroirs. / ~ on marble || Marmorgravierung f. || gravure f. sur marbre. / medal ~ || Medaillengravierung f. || gravure f. de médailles. / ~ on metal || Metallschneidekunst f.; Metallgravierung f. || gravure f. sur métaux. / ~ of music || Notenstechen n. || gravure f. de la musique. / ~ for plates || Gravierung f. für Platten || gravure f. de plaques. / precious stone ~ || Edelsteingravierung f. || gravure f. en pierres fines. / ~ of the rolls || Gravur f. der Walzen || gravure f. de cylindres. / ~ on steel || Stahlstich m. || gravure f. sur *ou* en acier; estampe f. sur acier. / ~ on stone || Steingravierung f. || gravure f. sur pierre. / ~ stroke ~ || Stichelschneiden n. || gravure f. au burin. / wood ~ || Holzschnitt m. || gravure f. sur bois.

engraving cylinder (Print) || Bildzylinder m.; Bildwalze f. || cylindre m. gravé. / ~ establishment || Gravieranstalt f. || établissement m. *ou* atelier m. de gravure. / ~ machine || Graviermaschine f. || machine f. à graver. / ~ machine worker || Maschinengravör m. || graveur m. à la machine. / ~ needle || Graviernadel f. || pointe f. pour taille-douce. / ~ plate || Druckplatte f. || estampe f.; gravure f.

enhanced permeability to ultra-violet rays of the glass || gesteigerte Durchlässigkeit f. im Ultraviolett des Glasmaterials || transparence f. plus grande dans l'ultraviolet de coulées de verre. / ~ view plastic || erhöhte Plastik f. des Bildes || relief m. rehaussé de l'image.

enlarge, to || vergrößern || agrandir. / ~ holes pl. with a drift || Löcher npl. aufdornen || mandriner des trous mpl. / ~ a hole with the rimer || ein Loch mit der Reibahle aufreiben || agrandir un trou à l'alésoir. / ~ the mouth of a tube || die Mündung einer Röhre erweitern || élargir le bout d'un tuyau.

enlarged || vergrößert; erweitert || agrandi.

enlargement || Ausbreitung f.; Ausdehnung f.; Vergrößerung f.; Erweiterung f. || extension f.; propagation f.; agrandissement m. / ~ (Phot) || Vergrößerung f. || agrandissement m. / ~ (Mine) || Keller m.; Kreuzpunkt m. mehrerer Strecken || case f.; carre-four m.; point m. de croisement d'une mine.

enlarging || Vergrößern n.; Erweitern n. || agrandissement m.; élargissement m. / ~ drawings || Vergrößerung f. von Zeichnungen || agrandissement m. des dessins. / ~ old shops || Erweiterung f. bestehender Betriebe || élargissement m. des exploitations existantes.

enlarging apparatus || Vergrößerungsapparat m. || appareil m. d'agrandissement. / ~ camera || Vergrößerungskamera f. || chambre f. noire d'agrandissement. / ~ hammer || Streckhammer m. || marteau m. à dégrossir. / ~ process || Vergrößerungsverfahren n. || procédé m. d'agrandissement.

enliven, to ~ a picture (Paint) || ein Bild n. aufhellen || aviver un tableau.

enneagon || Neuneck n. || ennéagone m.

enormous || übermäßig || démesuré; exagéré; exorbitant; énorme.

enosmite || Erdharz n.; Kampferharz n. || énosmite f.

enquire, to (Tel) || abfragen || demander.

enquiry || Anfrage f. || demande f. / ~ (Tel) || Abfragen n. || demande f.; réponse f. / local ~ (Tel) || Rückfrage f. || demande f.

enquiry button (Tel) || Rückruftaste f. || clé f. de rappel. / ~ circuit || Rückfragestromkreis m. || circuit m. de demande. / ~ line (Tel) || Bescheidleitung f. || ligne f. de référence. / ~ position (Tel) || Bescheidstelle f. || table f. de référence. / ~ ticket (Tel) || Auskunftsblatt n. || fiche f. de renseignement.

enrange, to (Build; Surv) || einfluchten || dresser à la ligne; dresser d'alignement; enligner; aligner.

enrich, to (Chem) || anreichern || enrichir.

enriched (Metal) || angereichert || enrichi.

enriching (Metal) || Anreicherungsvorgang m. || enrichissement m.

enrichment (Metal) || Veredelung f.; Anreicherung f. || enrichissement m.

enroll, to ~ the crew (Mar) || die Mannschaft anmustern || engager l'équipage m.

ensiform || schwertförmig || ensiforme.

ensign || Abzeichen n.; Kennzeichen n. || enseigne m.; marque f. / ~ (Mar) || Flagge f. || pavillon m.; enseigne f. / ~ halliard || Flaggenfall n.; Flaggenleine f. || drisse f. du pavillon. / ~ staff || Flaggenstock m. || épart m.; mât m. *ou* bâton m. de pavillon.

entablature ‖ Gebälk n. ‖ entablement m.

entanglement, barbed-wire ~ ‖ Stacheldrahtverhau m. ‖ abatis m. de fil barbelé.

entangling ‖ Verflechtung f. ‖ enlacement m.

enter, to (Trade) ‖ vormerken ‖ prendre note f. de. / ~ into combination (Chem) ‖ Verbindung f. eingehen ‖ entrer en combinaison. / ~ into force ‖ in Kraft f. treten ‖ entrer en vigueur f. / ~ into negociations ‖ in Unterhandlung f. treten ‖ entrer en pourparler m.

entering of the water (Mine) ‖ Wasserdurchbruch m. ‖ venue f. ou brèche f. d'eau. / ~ arrangement for fabric ‖ Gewebeeinführapparat m. ‖ appareil m. pour l'introduction du tissu.

enterprise ‖ Unternehmen n.; Unternehmung f. ‖ entreprise f. / commercial ~ ‖ kaufmännisches Unternehmen n. ‖ entreprise f. commerciale. / ~ of hydraulic structure ‖ Wasserbauunternehmung f. ‖ entreprise f. de construction hydraulique. / industrial ~ ‖ industrielles Unternehmen n. ‖ entreprise f. industrielle. / joint ~ ‖ gemeinschaftliches Unternehmen n. ‖ entreprise f. en participation. / for facilitating public intercourse ‖ öffentliche Verkehrsanstalt f. ‖ entreprise f. des communications publiques. / ~ of turnpike engineering ‖ Straßenbauunternehmung f. ‖ entreprise f. de construction de routes. / ~ of underground workings ‖ Tiefbauunternehmung f. ‖ entreprise f. de construction au-dessous du sol. / aim of the ~ ‖ Gegenstand m. des Unternehmens ‖ but m. de l'entreprise.

entitle, to ‖ benennen; betiteln ‖ dénommer; qualifier. / to be entitled to a testimonial ‖ Anspruch m. auf ein Zeugnis haben ‖ avoir droit m. à un certificat.

entrance ‖ Einfahrt f.; Eintritt m.; Eingang m. ‖ entrée f. / ~ into the harbour ‖ Einfahrt f. eines Schiffes ‖ entrée f. dans un port. / ~ of the port ‖ Hafeneinfahrt f. ‖ entrée f. du port. / ~ of underground subscriber's cables ‖ Einführung f. von unterirdischen Fernsprechleitungen ‖ entrée f. souterraine de postes.

entrance box for lines ‖ Einführungskasten m. für Leitungen ‖ boîte f. d'entrée des fils. / ~ door ‖ Eingangstür f. ‖ porte f. d'entrée. / ~ duty ‖ Einfuhrzoll m. ‖ droit m. d'entrée. / ~ fee ‖ Eintrittsgeld n. ‖ droit m. d'entrée. / ~ gate ‖ Eingangstor n. ‖ porte f. d'entrée. / ~ gate of iron ‖ eisernes Tor n. ‖ porte-cochère m. en fer. / ~ hall ‖ Vorhalle f. ‖ vestibule m. / ~ line (Railw) ‖ Einfahrtsgleis n. ‖ voie f. d'entrée.

entresol (Build) ‖ Halbgeschoß n.; Zwischenstock m. ‖ entre-sol m.; mezzanine f.

entropy ‖ Entropie f. ‖ entropie f.

entrust, to ‖ anvertrauen ‖ confier.

entry ‖ Eintreten n.; Eintritt m. ‖ entrée f. / ~ (Book-keep) ‖ Eintragung f.; Buchung f. ‖ inscription f.; écriture f. / ~ (Door) see entry door. / ~ (Duty) ‖ Wertangabe f.; Deklaration f. ‖ déclaration f. / ~ (Mine) ‖ Strecke f.; galerie f. / ~ (Money) ‖ Eingang m. ‖ rentrée f.; recouvrement m. ‖ arrivée f.; entrée f.

entry door ‖ Haustür f.; Eingang m.; Eingangstür f.; Zugang m. ‖ porte f. d'entrée ou de rue ou de la maison; entrée f.

entwine, to ‖ verflechten ‖ enlacer.

enucleation machine, cotton ~ ‖ Baumwollentkernungsmaschine f. ‖ machine f. à égrener le coton.

enumerate, to ‖ aufzählen ‖ compter; énumérer.

enumeration ‖ Aufzählung f. ‖ énumération f.

envelope ‖ Briefumschlag m.; Briefhülle f. ‖ enveloppe f. à lettres. / cloth-lined ~ ‖ gefütterter Leinenumschlag m. ‖ enveloppe f. fourrée. / loose ~ (Aero) ‖ schlappe Hülle f. ‖ enveloppe f. flasque ou détendue. / outlook ~ see window ~. / to put in an ~ ‖ kuvertieren ‖ envelopper. / slack ~ see loose ~. / transparent ~ see window ~. / window ~ ‖ Fensterbriefumschlag m. ‖ enveloppe f. à guichet. / wooden ~ ‖ Holzhülle f. ‖ enveloppe f. en bois.

envelope, folding machine for ~s ‖ Briefumschlagfaltmaschine f. ‖ plieuse f. pour enveloppes à lettres. / gumming machine for ~s ‖ Briefumschlagklebemaschine f. ‖ colleuse f. pour enveloppes à lettres. / ~ (manufacturing) machine ‖ Briefumschlagmaschine f. ‖ machine f. à fabriquer des enveloppes à lettres. / manufactory of ~s ‖ Kuvertfabrik f.; Briefumschlagfabrik f. ‖ fabrique f. d'enveloppes. / ~ opener ‖ Brieföffner m. ‖ ouvre-lettres m. / ~ sealer ‖ Briefschließer m. ‖ machine f. à fermer les lettres. / ~ sealer and stamp affixer ‖ Briefschließ- und Freimachungsmaschine f. ‖ machine f. à fermer les enveloppes et à affranchir.

enveloping machine ‖ Einwickelmaschine f. ‖ machine f. à envelopper.

envelopment ‖ Umhüllung f.; Hülle f.; Verkleidung f.; revêtement m. / ~ for heating bodies ‖ Heizkörperverkleidung f. ‖ revêtement m. de radiateurs.

enzym(e) ‖ Enzym n. ‖ enzyme m. / oxidizing ~ ‖ oxydierendes Enzym n. ‖ enzyme m. oxydant. / ~ action ‖ Enzymwirkung f. ‖ action f. enzymatique.

enzymology ‖ Enzymologie f. ‖ enzymologie f.

eosin ‖ Eosin n. ‖ éosine f.

epaulet(te) ‖ Epaulette f. ‖ épaulette f.

epicyclic gear ‖ Planetengetriebe n.; Umlaufgetriebe n. ‖ engrenage m. épicycloidal ou planétaire ou à satellite.

epicycloid ‖ Epizykloide f. ‖ épicycloïde f. / interior ~ ‖ Hypozykloide f. ‖ épicycloide f. intérieure ou inférieure; hypocycloïde f. / internal ~ see interior ~.

epicycloidal wheel ‖ Epizykloidenrad n. ‖ roue f. épicycloïdale.

epidemic ‖ Epidemie f. ‖ épidémie f. / ~ antidote ‖ Seuchenbekämpfungsmittel n. ‖ antidote m. contre epidémies.

epidermis ‖ Epidermis f.; Oberhaut f. ‖ épiderme m.

epidiascope ‖ Epidiaskop n. ‖ épidiascope m.

epidote, manganesian ~ ‖ Manganepidot m.; piemontesischer Braunstein m. ‖ épidote m. manganésifère.

epigenite ‖ Arsenwismutkupfererz n. ‖ épigénite f.

episcope ‖ Episkop n. ‖ épiscope m.

epistolary style ‖ Briefstil m. ‖ style m. épistolaire.

epistyle ‖ Architrav m.; Hauptbalken m. ‖ architrave f.; épistyle m.

epithelium ‖ Epithel n. ‖ épithélium m.

epoch ‖ Epoche f. ‖ époque f.

eprouvette (Metal) ‖ Problöffel m.; Probestab m. ‖ éprouvette f.

epsomite extraction ‖ Bittersalzgewinnung f. ‖ extraction f. de sulfate de magnésie.

Epsom salts ‖ Bittersalz n.; Magnesiumsulfat n. ‖ sulfate m. de magnésie; sel m. d'Epsom ou amer. / ~ plant ‖ Bittersalzanlage f. ‖ installation f. de sulfate de magnésie.

equal (Math) ‖ gleich ‖ égal. / on an ~ footing ‖ paritätisch ‖ égalitaire.

equal-armed beam scale ‖ gleicharmige Balkenwage f. ‖ balance f. à fléaux égaux.

equality (Trade) ‖ Parität f. ‖ parité f. / ~ (Math) ‖ Gleichheit f.; Gleichmäßigkeit f. ‖ égalité f.

equalization ‖ Gleichmachen n. ‖ égalisation f. / ~ of earthwork ‖ Massennivellement n. ‖ compensation f. des déblais et des remblais. / ~ of energy ‖ energiewirtschaftlicher Ausgleich m. ‖ échange m. d'énergie. / ~ in series (Tel) ‖ Längsentzerrung f. ‖ compensation f. de la distorsion en série. / ~ of temperature ‖ Temperaturausgleich m. ‖ compensation f. de température.

equalize, to ‖ ausgleichen ‖ équilibrer; égaliser; compenser.

equalizer (Electr) ‖ Ausgleichsmaschine f. ‖ compensatrice f.; égalisatrice f. / ~ (Mach) ‖ Ausgleichhebel m. ‖ balancier m. (compensateur). / ~ (Pressure) ‖ Ausgleicher m. ‖ égalisateur m. / ~ (Tel) ‖ Entzerrer m. ‖ dispositif m. compensateur de distorsion; correcteur m. / longitudinal ~ (Mach) ‖ Längsausgleichhebel m. ‖ balancier m. compensateur longitudinal. / traverse ~ (Mach) ‖ Querausgleichhebel m. ‖ balancier m. transversal.

equalizer connection (Electr) ‖ Ausgleichverbindung f. ‖ connexion f. égalisatrice ou équipotentielle. / ~ spring ‖ Balanzierfeder f.; Einstellfeder f. ‖ ressort m. de balancier ou de rappel ou de compensation.

equalizing basin for a pump storage station ‖ Ausgleichbecken n. für ein Pumpspeicherwerk ‖ bassin m. compensateur d'une usine d'accumulation par pompage. / ~ current ‖ Ausgleichstrom m. ‖ courant m. égalisateur. / ~ flow ‖ Ausgleichströmung f. ‖ écoulement m. ou courant m. de compensation. / broad drawing ~ machine ‖ Breitstreckegalisiermaschine f. ‖ étireuse f. à égaliser en largeur. / ~ pond ‖ Ausgleichweiher m. ‖ étang m. compensateur. / ~ rolling mill ‖ Egalisierwalzmaschine f. ‖ machine f. à cylindres pour égaliser. / ~ wire (Electr) ‖ Ausgleichleitung f. ‖ circuit m. d'équilibrage.

equalling file ‖ Flachstumpffeile f. ‖ lime f. rectangulaire.

equally sided ‖ gleichseitig ‖ équilatéral.

equal mark ‖ Gleichheitszeichen n. ‖ signe m. d'égalité.

equation ‖ Gleichung f. ‖ équation f. / algebraical ~ ‖ algebraische Gleichung f. ‖ équation f. algébrique. / ~ of the centre ‖ Mittelpunktsgleichung f. ‖ équation f. du centre. / ~ of condition ‖ Bedingungsgleichung f. ‖ équation f. de condition. / ~ of condition of perfect gases ‖ Zustandsgleichung f. der vollkommenen Gase ‖ équation f. caractéristique des gaz parfaits. / ~ of continuity ‖ Konti-

equation 241 erinite

nuitätsgleichung f. || équation f. de continuité. / ~ of a curve in polar coordinates || Polargleichung f. einer Kurve || équation f. polaire. / to form an ~ || eine Gleichung f. ansetzen || mettre les quantités fpl. en équation. / fundamental ~ || Grundgleichung f.; Fundamentalgleichung f. || équation f. fondamentale. / personal ~ || persönliche Gleichung f. || équation f. personnelle. / ~ of state || Zustandsgleichung f. || équation f. d'état. / ~ of time || Zeitgleichung f. || équation f. du temps.

equation formula || Gleichungsformel f. || formule f. d'équation.

equator || Äquator m. || équateur m. / ~ of a magnet || neutrale Linie f. eines Magneten || ligne f. neutre d'un aimant.

equatorial || äquatorial || équatorial. / ~ axis || parallaktische Achse f. || axe m. parallactique. / ~ comet finder || parallaktischer Kometensucher m. || chercheur m. de comètes parallactique. / ~ directing system of the telescope mounting || parallaktisches Führungssystem n. der Fernrohrmontierung || système-conducteur m. parallactique de la monture de lunette. / ~ head with hour circle and declination circle || parallaktisches Achsensystem n. mit Stundenkreis und Deklinationskreis || système m. d'axes parallactiques avec cercle horaire et cercle de déclinaison. / ~ mounting of a comet finder || parallaktische Montierung f. eines Kometensuchers || monture f. parallactique d'un chercheur de comètes. / ~ section || Äquatorialschnitt m. || section f. équatoriale. / ~ system of carriers of the telescope mounting || parallaktisches Tragsystem n. der Fernrohrmontierung || système-support m. parallactique de la monture de lunette. / ~ telescope || parallaktisches Fernrohr n. || lunette f. parallactique. / ~ telescope mounting || parallaktische Fernrohrmontierung f. || monture f. parallactique de lunette.

equatorial || Äquatorial n. || équatorial m.

equatorially mounted telescope || parallaktisch montiertes Fernrohr n. || lunette f. à monture parallactique.

equiangular || gleichwinklig || équiangle.

equilateral || gleichseitig || équilatéral; équilatère. / ~ triangle || gleichseitiges Dreieck n. || triangle m. équilatéral.

equilibrate, to || abgleichen || balancer; équilibrer.

equilibrator, pneumatic ~ (Gun) || Druckluftausgleicher m. || équilibreur m. à air.

equilibrium || Gleichgewicht n. || équilibre m. / atmospheric ~ || atmosphärisches Gleichgewicht n. || équilibre m. atmosphérique. / to disturb the ~ || das Gleichgewicht stören || troubler ou déranger ou rompre l'équilibre m. / dynamical ~ || dynamisches Gleichgewicht n. || équilibre m. dynamique. / ~ of forces || Kräftegleichgewicht n. || équilibre m. des forces. / indifferent ~ || indifferentes Gleichgewicht n. || équilibre m. indifférent. / ~ of moments || Momentenausgleich m. || compensation f. des moments. / neutral ~ see indifferent ~. / to restore the ~ || das Gleichgewicht wiederherstellen || rétablir l'équilibre m. / stable ~ || stabiles Gleichgewicht n. || équilibre m. stable. / unstable ~ || labiles Gleichgewicht n. || équilibre m. instable.

equilibrium, condition of ~ || Gleichgewichtsbedingung f. || condition f. d'équilibre. / disturbance of ~ || Gleichgewichtsstörung f. || perturbation f. ou dérangement m. d'équilibre. / position of ~ || Gleichgewichtslage f. || position f. d'équilibre. / state of ~ see position of ~.

equilibrize, to || ins Gleichgewicht bringen || mettre en équilibre m.; équilibrer.

equinoctial || Himmelsäquator m. || équateur m. céleste.

equip, to || ausstatten || garnir. / ~ with ... || ausrüsten mit ... || munir de ...

equipment || Werksgerät n.; Ausrüstung f. || outillage m.; équipement m. / ~ for producing aluminium || Aluminiumgewinnungseinrichtung f. || installation f. de production de l'aluminium. / ~ for boiler house || Kesselhauseinrichtung f. || installation f. de la chaufferie. / electric ~ || elektrische Ausrüstung f. || équipement m. électrique. / ~ of an electric lighting plant || elektrische Beleuchtungseinrichtung f. || appareils mpl. à éclairage électrique. / military ~ || Militäreffekten pl. || équipements mpl. militaires. / optical ~ || optische Ausrüstung f. || équipement m. optique. / ~ of a repeater valve with a socket || Sockelung f. einer Lampe || équipement m. d'une lampe amplificatrice de socle. / ~ for round calls || Rundgesprächseinrichtung f. || installation f. pour l'appel et les conversations simultanées d'un groupe d'abonnés. / single motor ~ || Einmotorenantrieb m. || commande f. par un seul moteur. / ~ for soldiers || Armeeausrüstung f. || équipement m. militaire. / ~ of the stand || Stativausstattung f. || équipement m. du statif. / ~ for stereotyping || Stereotypieeinrichtung f. || installation f. de stéréotypie. / ~ for tropical climates || Tropenausrüstung f. || vêtement m. pour les tropics. / ~ for xylolithe factory || Steinholzfabrikeinrichtung f. || installation f. pour fabriques de xylolithe.

equipment article for ships || Schiffsausrüstungsgegenstand m. || objet m. d'armement de bateaux. / ~ crane || Ausrüstungskran m. || grue f. d'armement.

equipoise see equilibrium.

equipotential || isoelektrisch || équipotentiel. / ~ surface || Niveaufläche f. oder Gleichgewichtsfläche f. || surface f. équipotentielle.

equipping machine for finishing plants || Ausrüstungsmaschine f. für Appreturanstalten || machine d'équipment pour ateliers d'apprêtage.

equisetum || Schachtelhalm m. || tige f. de prèle.

equivalence || Gleichwertigkeit f.; Äquivalenz f. || équivalence f. / sign of ~ || Kongruenzzeichen n. || signe m. d'équivalence.

equivalent (Bank) || Gegenwert m. || contrevaleur f.; équivalent m. / ~ (Phys) || Äquivalent n. || équivalent m. / ~ (Trade) || Geldeswert m. || valeur f. marchande ou en argent. / chemical ~ in weight || chemisches Äquivalentgewicht n. || poids m. équivalent chimique. / electro-chemical ~ || elektrochemisches Äquivalent n. || équivalent m. électrochimique. / mechanical ~ of heat || mechanisches Wärmeäquivalent n. || équivalent m. mécanique de la chaleur.

equivalent circuit || Ersatzstromkreis m. || circuit m. équivalent.

erase, to || radieren || gratter; effacer.

eraser (Gum) || Radiergummi m. || gomme-grattoir m.; gomme f. à effacer. / ~ (Knife) || Radiermesser n. || grattoir m. / combination ink and pencil ~ || Blei- und Tintengummi m. || gomme f. à crayon et à encre. / steel ~ || Radiermesser n. || grattoir m. / typewriter ~ || Schreibmaschinengummi m. || gomme f. pour machine à écrire.

eraser cartridge || Radiergummihülse f. || porte-gomme m. / grinding cylinder for ~s || Schleiftrommel f. für Radiergummi || tambour m. à polir la gomme à effacer.

erasing knife || Radiermesser n. || grattoir m. / ~ shield || Radierschablone f. || plaque f. à échancrures à effacer. / ~ water || Radierwasser n. || eau f. grattoire ou à effacer.

erasure || Rasur f. || effaçage m.; rature f.

erbium preparation || Erbiumpräparat n. || préparation f. d'erbium.

erect || aufrecht || droit; debout. / ~ image || aufrechtes Bild n. || image f. droite ou debout.

erect, to || erbauen; errichten || bâtir; construire; édifier; ériger. / ~ the image of the prism-glass by multiple reflection of the rays || das Bild n. im Feldstecher durch mehrfache Spiegelung des Lichtstrahles wieder aufrichten || redresser l'image m. dans la jumelle à prismes par plusieurs réflexions successives du rayons lumineux. / ~ a machine || eine Maschine montieren oder aufstellen || monter une machine.

erecting || Montage f.; Zusammenbau m.; Aufstellung f. || montage m. / ~ the image || Bildaufrichtung f. || redressement m. de l'image.

erecting crane || Montagekran m. || grue f. de montage. / collapsible ~ crane || zerlegbarer Rüstkran m. || grue f. démontable de montage. / expenses pl. for ~ || Montagekosten pl. || frais mpl. de montage. / the hall is provided with x ~ pits || die Halle f. ist mit x Montageschächten ausgestattet || le hall est muni de x puits de montage. / ~ scaffolding || Montagegerüst n. || échafaudage m. de montage. / ~ shop || Aufstellungswerkstatt f.; Montageraum m. || atelier m. de montage. / ~ shop of a dynamo work || Montagehalle f. eines Dynamowerkes || hall m. de montage d'une usine des dynamos. / movable ~ system of lenses || verstellbares bildumkehrendes Linsensystem n. || système m. de lentilles redresseur réglable. / ~ tool || Montagewerkzeug n. || outil m. de montage. / ~ wagon || Montagewagen m. || voiture f. ou wagon m. de montage.

erection see also erecting || Aufstellung f.; Errichtung f.; Montage f. || montage m.; érection f. / ~ of the poles || Aufstellung f. der Stützen || plantation f. des poteaux de lignes. / ~ drawing || Montagezeichnung f. || dessin m. de montage.

erector (Mach) || Monteur m. || monteur m.

erg || Erg n. || erg m.

ergot (of rye) (Agr) || Mutterkorn n. || seigle m. ergoté; blé m. cornu ou ergoté; ergot m. / ~ extract || Mutterkornextrakt m. || extrait m. de seigle ergoté.

ericite || Heidenstein m. || éricite f.

erinite || Erinit m. || érinite f.

eriometer ‖ Eriometer n. ‖ ériomètre m.

erosion by rain ‖ Auswaschung f. durch Regen ‖ érosion f. *ou* accrousement m. par la pluie.

err, to ‖ sich versehen; sich täuschen; sich irren ‖ se tromper; faire une bévue.

errand boy ‖ Laufbursche m. ‖ garçon m. de course *ou* de magasin; galopin m.; garçon-messager m. / ~ **girl** ‖ Laufmädchen n. ‖ trottin m.

errata pl. (Print) ‖ Angabe f. von Druckfehlern; Druckfehlerverzeichnis n. ‖ errata mpl.

erratic block *see* ~ **boulder.** / ~ **boulder** ‖ Findling m.; Geschiebeblock m. ‖ bloc m. erratique; moraine f.

erratum (Print) ‖ Druckfehler m. ‖ faute f. d'impression.

erroneous ‖ irrtümlich ‖ erroné; faux.

error ‖ Fehler m. ‖ erreur f. / ~ (Print) ‖ Druckfehler m. ‖ faute f. d'impression. / ~ (Trade) ‖ Mißgriff m. ‖ méprise f. / ~ **of calibration** ‖ Nennwertfehler m. ‖ erreur f. d'étalonnage. / ~ **of the compositor** (Print) ‖ Setzerfehler m. ‖ faute f. de composition. / ~ **of division** ‖ Teilungsfehler m. ‖ erreur f. de graduation. / ~ **from external sources** (Tel) ‖ Fremdfehler m. ‖ erreur f. par des causes étrangères. / ~ **in fabrication** ‖ Herstellungsfehler m. ‖ défaut m. *ou* faute f. de fabrication. / ~ **in indication** ‖ Falschweisung f. ‖ instruction f. inexacte *ou* fausse. / ~ **of observation** ‖ Beobachtungsfehler m. ‖ erreur f. d'observation. / **to compensate the** ~ **in observation** ‖ den Beobachtungsfehler m. ausgleichen ‖ compenser l'erreur m. d'observation. / **partial** ~ ‖ Teilfehler m. ‖ erreur f. partielle. / ~ **in reading off** ‖ Ablesungsfehler m. ‖ erreur f. de lecture. / ~ **of sighting** ‖ Visierfehler m. ‖ faute f. *ou* méprise f. de visée. / **small** ~ ‖ geringfügiger Fehler m. ‖ erreur f. peu importante.

error, compensation of the ~**s** ‖ Fehlerausgleichung f. ‖ compensation f. des erreurs. / **determination of** ~ ‖ Fehlerbestimmung f. ‖ détermination f. d'erreur. / ~ **function** ‖ Fehlerintegral n. ‖ intégrale f. de Laplace. / ~ **law** ‖ Gaußsches Fehlergesetz n. ‖ loi f. de Gauß. / **limit of** ~ ‖ Fehlergrenze f. ‖ limite f. d'erreur. / **to remain below the limit of** ~ ‖ unter der Fehlergrenze f. bleiben ‖ rester en deçà de la limite d'erreur. / **to exceed the limit of** ~ ‖ die Fehlergrenze f. überschreiten ‖ dépasser la limite d'erreur. / **source of** ~ ‖ Fehlerquelle f. ‖ cause f. d'erreur. / **source of** ~ **resulting from the influence of heat** ‖ durch den Einfluß der Wärme entstehende Fehlerquelle ‖ source f. d'erreurs causée par l'influence de la chaleur.

eruginous copper ‖ mit Grünspan beschlagenes Kupfer n. ‖ cuivre m. érugineux.

eruption ‖ Ausbruch m. ‖ éruption f.

eruptive rock ‖ Eruptivgestein n. ‖ roche f. éruptive. / ~ **stones** pl. ‖ Eruptivgestein n. ‖ pierres fpl. d'éruption.

erythrine ‖ Erythrin m. ‖ cobalt m. arséniaté; érythrine f.

escape, to (Gas) ‖ entweichen; ausströmen ‖ écouler; s'échapper.

escape ‖ Ausströmen n.; Entweichen n. ‖ fuite f.; échappement m.

escapement (Clockm) ‖ Hemmung f. ‖ échappement m. / ~ (Typewr) ‖ Schaltung f. ‖ échappement m. / **cog wheel** ~ ‖

Zahnradauslösung f. ‖ échappement f. à roue dentée. / **cylinder** ~ (Clockm) ‖ Zylinderhemmung f. ‖ échappement m. à cylindre. / **dead-beat** ~ (Clockm) ‖ ruhende *oder* schleifende Hemmung f. ‖ échappement m. à l'épine *ou* à repos. / **recoil** ~ (Clockm) ‖ zurückspringende Hemmung f. ‖ échappement m. à recul. / **repose** ~ (Clockm) *see* **dead-beat** ~.

escapement setting (Watchm) ‖ Gangsetzen n. ‖ plantage m. / ~ **wheel** (Clockm) ‖ Hemmungsrad n.; Steigrad n. ‖ roue f. de rencontre. / ~ **wheel** (Typewr) ‖ Schaltrad n. ‖ roue f. d'échappement.

escape pipe ‖ Dampfausströmungsrohr n. ‖ tuyau m. d'échappement de la vapeur. / ~ **steam** ‖ Auspuffdampf m. ‖ vapeur f. d'échappement. / ~ **valve** ‖ Auslaßventil n.; Sicherheitsventil n. ‖ soupape f. de trop-plein *ou* de sûreté.

escarp, to ‖ böschen ‖ taluter.

escarpment ‖ Abdachung f.; Böschung f.; steiler Abhang m. ‖ escarpement m.; talus m.

escutcheon (Lock) ‖ Schild n. ‖ entrée f. / ~ **of vaulting** (Arch) ‖ Gewölbfach n.; Feld zwischen den Gewölbrippen ‖ pan m. de voûte.

espalier ‖ Spalier n. ‖ espalier m.

esparto ‖ Esparto n.; Spartgras n. ‖ sparte m. / ~ **plait** ‖ Espartogeflecht n. ‖ tresse f. de sparte. / ~ **spinning** ‖ Spartgrasspinnerei f. ‖ filature f. d'alfa.

esplanade (Build) ‖ freier Platz m.; Graswiese f. ‖ esplanade f.

especially ‖ hauptsächlich ‖ principal; essentiel.

essay, to ‖ prüfen; versuchen; erproben; einen Versuch durchführen *oder* machen ‖ essayer.

essay ‖ Versuch m. ‖ essai m.; épreuve f.; recherche m.

essayer's tongs ‖ Probzange f. ‖ pince f. d'essayeur.

essence ‖ Essenz f. ‖ essence f. / **artificial** ~ ‖ künstliche Essenz f. ‖ essence f. artificielle. / ~ **for liquors** ‖ Liköressenz f. ‖ essence f. pour liqueurs. / ~ **of pearls** ‖ Perlenessenz f. ‖ essence f. des écailles de l'ablette *ou* d'orient. / ~ **for perfumers** ‖ Parfümerieessenz f. ‖ essence f. pour parfumeurs. / ~ **vegetable** ~ ‖ pflanzliche Essenz f. ‖ essence f. végétale.

essential ‖ ätherisch ‖ essentiel; éthéré. / ~ **oil** ‖ ätherisches Öl n. ‖ huile f. éthérée *ou* essentielle. / **factory of** ~ **oils** ‖ Fabrik f. für ätherische Öle ‖ fabrique f. *ou* extraction f. d'huiles éthérées.

essential ‖ wesentlich; wichtig ‖ essentiel; considérable. / ~ **point** ‖ Anhaltspunkt m. ‖ point m. d'appui.

essonite ‖ Kaneelstein m.; Hessonit m. ‖ essonite f.

establish, to ‖ aufstellen ‖ mettre; placer.

establishment ‖ Anstalt f.; Institut n.; Etablissement n.; Einrichtung f.; Niederlassung f.; Fabrik f. ‖ établissement m. / ~ (Met) ‖ Hütte f. ‖ usine f.; établissement m. / ~ (Trade) ‖ Gründung f.; Einrichtung f. ‖ fondation f.; établissement m.; création f. / ~ **for artistic equipment and decoration of rooms** ‖ Werkstätte f. für Innenkunst ‖ atelier m. d'équipement et décors artistiques de logements *ou* d'appartements. / ~ **for building purposes** ‖ Einrichtung f. einer Baustelle ‖ installation f. du chantier. / ~ **of the connections** ‖ Verbindungsauf-

bau m. im Fernsprechbetrieb ‖ schéma m. des connections du réseau téléphonique. / ~ **in co-operation with somebody** ‖ Gemeinschaftsgründung f. ‖ fondation f. en commun avec quelqu'un. / ~ **for interior art decoration** ‖ Werkstätte f. für Innendekoration ‖ atelier m. de décors artistiques des intérieurs. / ~ **for light treatment** ‖ Lichtheilstation f. ‖ établissement m. de bains de lumière. / ~ **for washing wool** ‖ Wollwäscherei f. ‖ lavoir m. de laines.

establishment charges pl. ‖ Einrichtungskosten pl. ‖ frais mpl. d'établissement.

estamin ‖ Etamin n. ‖ étamine f.

estate ‖ Landgut n. ‖ terre f.; domaine m. / **real** ~ ‖ Grundeigentum n. ‖ biens-fonds mpl.

estate tenant ‖ Gutspächter m. ‖ fermier m. de terre *ou* de domaine rural.

ester ‖ Ester m. ‖ éther m. / **hydrogen** ~ ‖ Estersäure f. ‖ éther-acide m.

estimable ‖ schätzbar ‖ estimable. / ~ (Chem) ‖ bestimmbar ‖ dosable.

estimate, to ‖ abschätzen; veranschlagen; taxieren ‖ estimer; évaluer; taxer. / ~ (Chem) ‖ (quantitativ) bestimmen ‖ doser. / ~ **volumetrically** ‖ titrimetrisch feststellen ‖ doser volumétriquement; titrer.

estimate ‖ Schätzung f.; Veranschlagung f.; Anschlag m.; Abschätzung f. ‖ devis m.; estimation f.; évaluation f. / **approximate** ~ ‖ ungefährer Anschlag m. ‖ évaluation f. approximative. / **builder's** ~ ‖ Überschlag m. ‖ calcul m. estimatif. / ~ **of construction** ‖ Bauanschlag m. ‖ devis m. de construction. / ~ **of cost** ‖ Kostenanschlag m. ‖ devis m. (estimatif); évaluation f. des frais. / **to make an** ~ ‖ einen Überschlag m. machen ‖ faire un calcul approximatif. / **rough** ~ ‖ Kostenvoranschlag m. ‖ devis m. estimatif *ou* approximatif.

estimated ‖ veranschlagt; bewertet ‖ estimé; évalué. / **over** ~ ‖ zu hoch veranschlagt ‖ surfait; tayé *ou* évalué trop cher. / **under** ~ ‖ zu niedrig veranschlagt ‖ évalué en-dessous de la valeur.

estimation (Trade) *see* **estimate.** / ~ (Chem) ‖ (quantitative) Bestimmung f. ‖ dosage m.

estuary of a river ‖ Einmündung f. eines Flusses ‖ embouchure f. d'une rivière.

etamine ‖ Etamin n. ‖ étamine f.

etch, to ‖ ätzen; anätzen; radieren ‖ graver à l'eau-forte *ou* à la pointe sèche; décaper; ronger.

etchant ‖ Ätzmittel n. ‖ rongeur m.; morsure m.

etched copper plate ‖ Radierung f. ‖ gravure f. à la pointe sèche. / ~ **figure** ‖ Ätzfigur f. ‖ figure f. obtenue au moyen de corrosifs; figure f. de corrosion. / ~ **lines** pl. ‖ eingeätzte Linien fpl. ‖ traits mpl. gravés à l'eau forte.

etcher ‖ Radierer m.; Ätzer m. ‖ graveur m. à l'eau-forte *ou* à la pointe sèche.

etching ‖ Radierung f.; Radierkunst f. ‖ gravure f. à l'eau-forte. / **line** ~ ‖ Strichätzung f. ‖ gravure f. au trait. / **art of** ~ ‖ Ätzkunst f. ‖ art m. de graver à l'eauforte.

etching basin ‖ Ätzschale f. ‖ cuvette f. à mordant. / ~ **figure** ‖ Ätzfigur f. ‖ figure f. de corrosion. / ~ **ground** ‖ Ätzgrund m. ‖ couche f. de vernis. / ~ **ink** ‖ Ätzfarbe f. ‖ encre f. mordante; mordant m. / ~ **material** ‖ Ätzmittel n. ‖ agent m.

corrosif; caustique m. / ~ needle || Radier-
nadel f.; Ätznadel f. || pointe f. sèche;
échoppe f. / ~ powder || Ätzpulver n. ||
poudre f. corrosive. / ~ solution for zinc ||
Zinkätze f. || eau-forte f. pour planches
de zinc. / ~ trough || Ätztrog m. || baquet
m. à mordant.

ethane, machine working on ~ || Äthan-
maschine f. || machine f. à éthane.

ether || Äther m. || éther m. / acetic ~ ||
Essigäther m.; Essigsäureäthylester m.
|| éther m. acétique. / ethylic ~ || Schwefel-
äther m.; Äthyläther m. || éther m. sul-
furique. / methylacetic ~ || Essigsäure-
methylester m. || éther m. méthylacéti-
que. / nitrous ~ || Salpetersäureäthyl-
ester m. || éther m. nitrique *ou* azoti-
que. / ordinary ~ || gewöhnlicher Äther
m. || éther m. ordinaire. / measuring the
proportion of ~ in air || Messung f. des
Äthergehaltes in der Luft || dosage m. de
l'éther dans l'air. / sulphuric ~ || Schwe-
feläther m. || éther m. sulfurique.

ether distiller || Ätherbrenner m. || distilla-
teur m. d'éther.

ethereal || ätherisch || essentiel; éthéré.

etherification || Ätherbildung f. || éthérifica-
tion f.

ether producing plant || Ätherfabrikein-
richtung f. || installation f. de distillerie
d'éther. / ~ refrigerating machine ||
Ätherkühlmaschine f. || machine f. fri-
gorifique à éther. / ~ smelling || ätherisch
riechend || d'odeur f. éthérée. / ~ steam
engine || Ätherdampfmaschine f. || ma-
chine f. à vapeur d'éther. / ~ tester ||
Ätherprober m. || essayeur m. d'éther. /
~ theory || Äthertheorie f. || théorie f.
de l'éther.

ethiops, antimonial ~ || Spießglanzmohr m.
|| éthiops m. antimonial.

ethnographic(al) || ethnografisch || ethno-
graphique.

ethyl || Äthyl n. || éthyle m. / ~ acetate ||
Äthylazetat n.; Essigester m. || acétate
m. d'éthyle. / ~ alcohol || Äthylalkohol
m.; Gärungsalkohol m. || alcool m. éthyli-
que *ou* de fermentation. / ~amidophenol
|| Äthylamidophenol n. || éthylamidophé-
nol m. / ~ benzoate || Benzoesäureäther
m. || benzoate m. d'éthyle. / ~ bromide ||
Bromäthyl n. || bromure m. d'éthyle. /
~ chloride || Chloräthyl n. || chlorure m.
d'éthyle.

ethylene || Äthylen n.; schweres Kohlen-
wasserstoffgas n. || éthylène m.; gaz n.
oléfiant; gaz m. lourd d'hydrogène car-
buré. / ~ chloride || Chloräthylen n. ||
chlorure m. d'éthylène. / ~ oxide || Äthyl-
oxyd n. || oxyde m. d'éthyle.

ethyl ether || Äthyläther m. || oxyde m.
d'éthyle. / ~ iodide || Jodäthyl n. || iodure
m. d'éthyle. / ~ nitrate || Äthylnitrat n. ||
nitrate m. d'éthyle. / ~ preparation ||
Äthylpräparat n. || préparation f. d'éthyle.
/ ~tartaric acid || Äthylweinsäure f. ||
acide m. éthylotartrique.

etna || Spirituskocher m. || réchaud m. à
alcool.

eucalyptol || Eukalyptol n. || eucalyptol m.

eucalyptus leaves pl. || Eukalyptusblätter
npl. || feuilles fpl. d'eucalyptus. / ~ oil ||
Eukalyptusöl n. || essence f. d'eucalyp-
tus.

euchlorine || Euchlorin n. || euchlorine f.

euchroite || Euchroit m. || euchroïte f.

euclase || Euklas m. || euclase f.

eudialyte || Eudialyt m. || eudialyte f.

eudiometer || Eudiometer n.; Gasprüfer m.
|| eudiomètre m.

eulytite || Wismutblende f. || eulytite f.

euphonious || wohlklingend || harmonieux;
sonore; euphonique.

euphorbium resin || Euphorbiumharz n. ||
gomme-résine f. d'euphorbe.

euphrasy || Augentrost m. || euphraise f.

European and extra-European system (Tel)
|| europäischer und außereuropäischer
Vorschriftenbereich m. || régime m. euro-
péen et extra-européen.

eutectic || eutektisch || eutectique.

eutectic || Eutektikum n. || eutectique m.

euxenite || Euxenit m. || euxénite f.

evacuate, to || luftleer machen; evakuieren;
entleeren || évacuer.

evacuated || verdünnt || évacué. / highly ~
|| hochevakuiert || évacué fortement. /
~ glass || luftleeres Glas n. || récipient m.
en verre évacué.

evacuating, by ~ || durch Evakuieren n. ||
en faisant le vide.

evacuation || Auspumpung f.; Verdünnung
f.; Evakuierung f. || évacuation f.; pom-
page m.

evaluate, to *see also* to estimate || ab-
schätzen; schätzen; berechnen; veran-
schlagen; den Wert m. bestimmen ||
évaluer.

evaluating grating spectroscope || Gitter-
meßspektroskop n. || spectromètre m. à
réseau.

evaluation *see also* estimate || Schätzung f.;
Anschlag m.; Berechnung f. || évaluation
f.; estimation f.

evaporate, to || verdampfen; eindampfen;
abdampfen; verdunsten || évaporer; con-
centrer. / ~ (Cable joint) || abdämpfen ||
faire évaporer. / to allow ~ (Chem) || ab-
dampfen lassen || faire évaporer. / ~ to
dryness (Chem) || abrauchen; bis zur
Trockenheit verdampfen || évaporer à
siccité f. *ou* à sec. / ~ the juice || den
Saft m. einkochen || cuire *ou* cuire
le jus. / to make evaporate || abdampf-
fen lassen || faire évaporer. / ~ off || ab-
dampfen || chasser par évaporation f.

evaporating apparatus || Abdampfappa-
rat m. || appareil m. d'évaporation. / ~
basin || Abdampfschale f.; Abdampf-
gefäß n. || cuve f. *ou* capsule f. évapora-
toire *ou* à évaporation. / ~ boiler || Ab-
dampfkessel m. || chaudière f. évapora-
toire *ou* d'évaporation. / ~ capsule *see*
~ basin. / ~ dish *see* ~ basin. / ~ flask ||
Abdampfkolben m. || flacon m. d'évapo-
ration. / ~ kettle (Brew) || Abdampf-
pfanne f. || chaudière f. à évaporer. / ~
pan *see* ~ basin. / ~ vessel *see* ~ basin.

evaporation || Verdampfung f. || évapora-
tion f.; vaporisation f. / ~ of saltpetre ||
Salpetersud m. || cuite f. *ou* évaporation
f. du salpêtre. / ~ of water || Verdampfung
f. des Wassers || évaporation f. de l'eau.

evaporation condenser || Verdampfungs-
kondensator m. || condenseur m. à éva-
poration. / ~ dish || Abdampfschale f. ||
capsule f. d'évaporation. / ~ hopper
cooling || Verdampfungskühlung f. || re-
froidissement m. par évaporation. / ~
meter || Verdunstungsmesser m. || éva-
porimètre m.; atmomètre m. / ~ tem-
perature || Verdampfungstemperatur f. ||
température f. d'évaporation.

evaporative capacity || Verdampfungsfä-
higkeit f. || capacité f. de vaporisation. /
surface ~ cooler || Berieselungskondensa-

tor m. || condenseur m. à ruissellement. /
~ cooling || Heißkühlung f.; Verdun-
stungskühlung f. || refroidissement m.
par évaporisation. / ~ duty of boiler ||
Verdampfungsvermögen n. des Kessels ||
débit m. évaporatif de la chaudière. /
~ efficiency || Verdampfungsfähigkeit f.
|| puissance f. d'évaporation.

evaporator || Eindampfapparat m.; Ver-
dampfer m. || appareil m. à ébullition;
évaporateur m. / untreated water ~ ||
Rohwasserverdampfer m. || vaporisa-
teur m. d'eau brute. / ~ of the vertical
tube type || Steilrohrverdampfer m. ||
évaporateur m. à éléments tubulaires
verticaux.

evaporimeter || Verdunstungsmesser m. ||
évaporimètre m.; atmomètre m.

evasive reply || ausweichende Antwort f. ||
réponse f. évasive.

even || glatt || lisse. / ~ (Math) || gerade
pair. / upon an ~ keel (Shipb) || gleich-
lastig || sans différence de tirant d'eau. /
~ page (Print) || gerade Seite f. || page f.
paire; verso m. / ~ surface || glatte Fläche
f. || surface f. lisse.

even, to || planieren; abgleichen; einebnen
|| aplanir; niveler; égaler; égaliser; égalir.

evening robe || Abendkleid n. || robe f. de
soirée.

evenly-distributed load || gleichmäßige Ge-
wichtsverteilung f. || charge f. uniformé-
ment répartie.

everlasting crown || Immortellenkranz m. ||
couronne f. en immortelles.

evidence || Beweisaufnahme f.; Zeugen-
aussage f. || audition f. des preuves *ou*
des témoins.

evident || augenfällig || évident. / to be ~ ||
sich herausstellen; zutageliegen || être
évident.

evil || Übelstand m. || inconvénient m.

evolute || Evolute f. || développée f.

evolution (Math) || Radizieren n.; Wurzel-
ziehen n. || extraction f. d'une racine.

evolve, to (Chem) || entwickeln || dégager.

evolvent (Math) || Evolvente f. || dévelop-
pante f.

ewe || Mutterschaf n. || brebis f.

ewer || Wasserkanne f. || aiguière f.; pot m.
à eau.

exact || pünktlich || ponctuel; exact. / ~
optical distance measurement || Präzi-
sionsdistanzmessung f. auf optischem
Wege || mesure f. optique précise des
distances.

exactly cast || sauber gegossen || coulé pro-
prement et exactement.

exactness || Genauigkeit f. || exactitude f.;
précision f.

exaggerate, to || übertreiben; übertrieben
darstellen || exagérer.

exaggerated || übertrieben || outré; exorbi-
tant; exagéré.

examination || Prüfung f.; Besichtigung f.;
inspection f.; visite f.; examen m. / ~
of blood || Blutuntersuchung f. || examen
m. du sang. / ~ of the car || Untersuchung
f. des Wagens || visite f. de la voiture. /
careful ~ || sorgfältige Untersuchung f. ||
examen m. consciencieux. / ~ by experts
|| Expertise f.; Sachverständigenbegut-
achtung f.; Sachverständigengutachten
n. || expertise f. / magnifier for the ~ of
the eyes || Lupe f. für die Augenuntersu-
chung || loupe f. pour l'examen des yeux.
/ ~ of the soil || Bodenuntersuchung f. ||
exploration f. *ou* examen m. du sol.

examine, to ‖ prüfen; besichtigen ‖ examiner; inspecter; visiter; vérifier. / ~ again ‖ nachprüfen ‖ vérifier. / ~ the ground ‖ den Grund m. auspeilen ‖ sonder le fond. / ~ by squaring (Build) ‖ nachwinkeln; auf den Winkel m. prüfen ‖ équerrer; examiner par l'équerre f.

examiner (Metal) ‖ Sortierer m. ‖ visiteur m. / ~ of plates (Phot) ‖ Plattenbeschauer m. ‖ visiteur m. des plaques.

example ‖ Beispiel n. ‖ exemple m. / ~ of cutting work ‖ Schnittprobe f. ‖ exemple m. de découpage. / for ~ ‖ beispielsweise; zum Beispiel n. ‖ par exemple. / ~ of work ‖ Arbeitsbeispiel n. ‖ exemple m. de travail.

excandescence (Forg) ‖ Weißglühhitze f. ‖ chaude f. blanche.

excavate, to ‖ ausschachten; ausgraben ‖ excaver; foncer; fouiller. / ~ the earth ‖ Erde f. ausheben ‖ creuser ou extraire la terre; déblayer le terrain.

excavated material ‖ Baggergut n. ‖ curure f.; déblai m. de fouille ou de dragage.

excavating apparatus ‖ Baggergerät n. ‖ appareil m. d'excavation. / ~ depth ‖ Baggertiefe f. ‖ profondeur f. de fouille. / ~ machine- ‖ Grabemaschine f. ‖ excavateur m. / ~ power of the skimmer ‖ Reißkraft f. des Planierlöffels ‖ force f. d'arrachement du godet à niveler.

excavation ‖ Aushöhlung f.; Höhlung f. ‖ cavité f.; creux m. / ~ (Ground) ‖ Ausgrabung f. ‖ déterrement m.; fouille f. / ~ (Railw; Road) ‖ Einschnitt m.; Durchstich m. ‖ déblai m.; tranchée f.; creux m.; fosse f.; fouille m. / ~ (Salt) ‖ Kammerbau m. ‖ exploitation f. en chambres ou halles. / ~ of the foundations ‖ Fundamentaushub m. ‖ excavation f. des fondations.

excavator ‖ Exkavator m.; Bagger m. ‖ excavateur m.; drague f. / drag line bucket ~ ‖ Schürfkübelkabelbagger m. ‖ drague f. sèche à benne et à câble. / endless chain multi-bucket ~ ‖ Eimerkettenbagger m. ‖ excavateur m. ou drague f. à chaîne de godets. / ~ for lignite mines ‖ Braunkohlenbagger m. ‖ excavateur m. pour mines de lignite. / ~ for underground work ‖ Tiefbaubagger m. ‖ excavateur m. pour travaux souterrains.

exceed, to ‖ überschreiten ‖ excéder; dépasser. / ~ the speed limit ‖ die zulässige Geschwindigkeit überschreiten ‖ excéder ou dépasser la vitesse permise ou autorisée.

exceeded ‖ überschritten ‖ dépassé; excédé.

excel, to ‖ übertreffen ‖ surpasser.

excellent ‖ trefflich ‖ excellent.

excentric see eccentric.

except ‖ ausgenommen ‖ excepté.

exception ‖ Ausnahme f. ‖ exception f. / by way of ~ ‖ ausnahmsweise ‖ exceptionnellement.

exceptional ‖ außergewöhnlich ‖ exceptionnel. / ~ offer ‖ Sonderangebot n. ‖ offre f. de faveur.

exceptionally clear type of prism crown glass ‖ besonders klares Prismenkronglas n. ‖ verre m. crown à prismes parfaitement transparent.

excess ‖ Überschuß m.; Übermaß n. ‖ excès m.; excédent m. / ~ of air ‖ Luftüberschuß m. ‖ excès m. de vent. / ~ of gas ‖ Gasüberschuß m. ‖ gaz m. en excès. / in ~

‖ überschüssig; überzählig ‖ en excès m.; de trop. / ~ of oil ‖ Überschuß m. an Öl ‖ excès m. d'huile. / ~ of steam ‖ überschüssiger Wasserdampf m. ‖ vapeur f. (d'eau) en excédent.

excess current (Electr) ‖ Überstrom m. ‖ surintensité f. / ~ cut-out ‖ Überstromausschalter m. ‖ disjoncteur m. de surcharge. / ~ switch ‖ Überstromschalter m. ‖ interrupteur m. de surcharge. / ~ zero voltage cut-out switch ‖ Überstromnullspannungsausschalter m. ‖ disjoncteur m. à voltage nul de surcharge.

excess pressure ‖ Überdruck m. ‖ excès m. de pression; surtension f. / ~ pressure valve ‖ Überdruckventil n. ‖ soupape f. de sûreté ou de surpression. / ~ voltage ‖ Überspannung f. ‖ tension f. supplémentaire. / ~ voltage wave ‖ Überspannungswelle f. ‖ onde f. de surtension. / ~ weight ‖ Mehrgewicht n. ‖ surpoids m.; excédent m. de poids. / ~ weight to be strictly observed ‖ bestimmt einzuhaltendes Übergewicht n. ‖ surpoids m. exact.

exchange, to ‖ eintauschen; austauschen; tauschen ‖ échanger; troquer. / ~ (Money) ‖ wechseln; einwechseln; umwechseln ‖ changer; échanger. / ~ salutes pl. (Mar) ‖ Salute mpl. austauschen oder wechseln ‖ échanger des saluts mpl.

exchange ‖ Tausch m.; Umtausch m.; Austausch m. ‖ échange m.; troc m. / ~ (Bank) ‖ Wechselagio n. ‖ change m. / ~ (Bullion trade) ‖ Geldhandel m. ‖ trafic m. d'argent; change m. / ~ (Build) ‖ Börse f.; Börsengebäude n. ‖ bourse f. / ~ (Tel) ‖ Vermittlungsstelle f. ‖ bureau m. des communications. / ~ of air ‖ Luftaustausch m. ‖ échange m. d'air. / automatic ~ for up to 1000 subscribers operating on the preselector system ‖ Selbstanschlußzentrale f. nach dem 1000er Vorwählersystem ‖ commutateur m. téléphonique pour 1000 abonnés maximum d'après le système à présélecteurs. / to give in ~ ‖ eintauschen; austauschen; in Tausch m. geben ‖ échanger; changer. / local ~ (Tel) ‖ Amt n. ‖ bureau m. central. / main ~ (Tel) ‖ Hauptamt n. ‖ bureau m. principal. / manual ~ (Tel) ‖ Handamt n. ‖ bureau m. central manuel. / ~ of old shares for new ones ‖ Umtausch m. alter Aktien gegen neue ‖ échange m. de vieilles actions contre des nouvelles. / originating ~ (Tel) ‖ Anmeldeanstalt f.; Ursprungsanstalt f. ‖ bureau m. de départ ou d'origine. / printed ~ ‖ Kurszettel m. ‖ bulletin m. de la bourse. / semi-automatic ~ (Tel) ‖ halbselbsttätiges Amt n. ‖ bureau m. demi-automatique.

exchangeable ‖ vertauschbar; austauschbar ‖ échangeable. / ~ ball race of a ball bearing ‖ auswechselbarer Laufring m. eines Kugellagers ‖ anneau m. de fond ou bague f. extérieure interchangeable d'un roulement à billes.

exchange broker ‖ Wechselmakler m. ‖ courtier m. de change. / ~ call (Tel) ‖ Amtsanruf m. ‖ appel m. du bureau central. / distribution of incoming ~ calls to the stations of a private automatic telephone installation ‖ Vermittlung f. von ankommenden Amtsgesprächen nach den Apparaten einer selbsttätigen Hausfernsprechanlage ‖ écoulement m. de conversations de service d'arrivée vers les appareils d'une installa-

tion téléphonique privée automatique. / first of ~ ‖ Primawechsel m. ‖ première f. de change. / fluctuation of ~ ‖ Kursschwankung f. ‖ fluctuation f. des changes. / law of ~ ‖ Wechselrecht n. ‖ droit m. cambial. / ~ line (Tel) ‖ Amtsleitung f. ‖ ligne f. de jonction. / branch of the ~ line (Tel) ‖ Amtsleitungszweig m. ‖ branche f. de la ligne de jonction. / ~ list ‖ Geldkurszettel m. ‖ cote f. de l'agio. / ~ maintenance work (Tel) ‖ Amtspflege f. ‖ entretien m. de l'installation du bureau. / ~ mill ‖ Kundenmühle f. ‖ moulin m. à façon. / panic on the ~ ‖ Börsenkrach m. ‖ débâcle f.; krach m. / ~ report ‖ Börsenbericht m. ‖ bulletin m. de la bourse. / stock ~ rules pl. ‖ Börsenordnung f. ‖ règlement m. de la bourse. / second of ~ ‖ Sekundawechsel m. ‖ deuxième f. de change. / ~ stamp tax ‖ Wechselstempelsteuer f. ‖ impôt m. du timbre des effets de commerce. / ~ switch (Tel) ‖ Amtsschalter m. ‖ commutateur m. du bureau.

Exchequer ‖ Fiskus m.; Finanzamt n. ‖ fisc m.

excise ‖ Akzise f.; städtische Verzehrssteuer f. ‖ accise f.; octroi m.; contributions fpl. indirectes.

excitation (Electr) ‖ Erregung f. ‖ excitation f. / separate ~ (Electr) ‖ Fremderregung f. ‖ excitation f. indépendante.

excitation output ‖ Erregerleistung f. ‖ puissance f. d'excitation. / ~ voltage ‖ Erregerspannung f. ‖ tension f. d'excitation.

excite, to ‖ reizen ‖ exciter. / ~ (Electr) ‖ erregen ‖ exciter. / ~ a field ‖ ein Feld n. erregen ‖ exciter un champ. / ~ the plate (Phot) ‖ die Platte empfindlich machen ‖ exciter la plaque.

excitement ‖ Reiz m.; Anregung f. ‖ excitation f.

exciter (Electr) ‖ Erregerdynamomaschine f. ‖ dynamo m. d'excitation. / directcoupled ~ (Electr) ‖ unmittelbar gekuppelte Erregermaschine f. ‖ génératrice f. d'excitation directement accouplée.

exciting bobbin ‖ Erregerspule f. ‖ bobine f. excitatrice. / ~ circuit ‖ Erregerstromkreis m. ‖ circuit m. d'excitation. / ~ coil ‖ Erregerspule f. ‖ bobine f. inductrice. / ~ current ‖ Erregerstrom m.; Magnetisierungsstrom m. ‖ courant m. d'excitation ou d'aimantation; courant m. excitateur. / ~ current strength ‖ Erregerstromstärke f. ‖ intensité f. du courant d'excitation. / ~ fluid (Acc) ‖ Erregerflüssigkeit f. ‖ liquide m. excitateur. / ~ salt ‖ Erregersalz n. ‖ sel m. excitateur.

exclusion ‖ Ausschluß m. ‖ exclusion f.

exclusive ‖ ausschließlich ‖ exclusif. / ~ interest ‖ Sonderinteresse n. ‖ intérêt m. séparé ou particulier. / ~ property ‖ Alleinbesitz m. ‖ propriété f. / ~ right ‖ Vorrecht n. ‖ privilège m. / ~ sale ‖ Alleinverkauf m. ‖ vente f. exclusive.

excorticated ‖ abgerindet ‖ écorcé.

excrements pl. ‖ Fäkalstoffe mpl. ‖ fèces fpl.; matières fpl. fécales.

excrescence ‖ unregelmäßiger Auswuchs m. ‖ excroissance f.

excreta pl. see excrements.

excursion train ‖ Ausflüglerzug m. ‖ train m. de plaisir. / ~ vehicle ‖ Aussichtswagen m. ‖ voiture f. d'excursion.

excuse || Rechtfertigungsgrund m.; Ausrede f.; Entschuldigung f. || excuse f.; raison f. justificative.

execute, to || ausführen; durchführen || exécuter; effectuer; remplir. / ~ in economy || in eigener Verwaltung f. ausführen || exécuter en régie f.

execution || Ausführung f.; Durchführung f. || exécution f. / ~ (Judicature) || Auspfändung f. || saisie f.; saisie-exécution f. / ~ in wood || hölzerne Ausführung f. || exécution f. en bois.

exemption || Befreiung f.; Freisein n. || exemption f.; dispense f. / ~ from charge || Gebührenfreiheit f. || en franchise f. de taxe. / ~ from fee see ~ from charge.

exercise || Ausübung f. || exercise f.; pratique f. / ~ paper || Konzeptpapier n. || papier m. à minutes.

exertion || Anspannung f. || tension f.

exfoliate, to (Metal) || (sich) abblättern || s'exfolier; s'effeuiller.

exfoliation || Abblättern n. || effeuillage m. / ~ of the rail || Abblätterung f. der Schiene || exfoliation f. du rail.

exhalation || Exhalation f.; Verdunstung f. || exhalaison f.

exhaust, to (Chem) || absaugen || faire le vide. / ~ (Mot) || auspuffen || échapper.

exhaust || Exhaustor m. || aspirateur m. / ~ (Mot) || Auspuff m. || échappement m. / the colour of ~ deteriorates || die Auspuffarbe verschlechtert sich || les fumées fpl. d'échappement deviennent moins claires. / open ~ || freier Auspuff m. || échappement m. libre. / water cooled ~ manifold || wassergekühlte Auspuffleitung f. || tubulure f. d'échappement à chemise d'eau.

exhaust alarm || Auspuffpfeife f. || sifflet m. sur l'échappement. / ~ bend || Auspuffkrümmer m. || coude m. d'échappement. / ~ box || Auspuffkasten m. || boîte f. d'échappement. / ~ cam || Auslaßnocken m. || came f. d'échappement. / ~ camshaft || Auslaßnockenwelle f. || arbre m. à cames d'échappement. / ~ collector || Auspuffsammler m. || pot m. d'échappement. / ~ cut-out || Auspuffklappe f. || disjoncteur ou clapet m. d'échappement.

exhausted (Goods) || vergriffen || vendu. / ~ (Mine) || abgebaut || exploité à fond. / the mines are still only partially ~ || die Gruben fpl. sind erst teilweise abgebaut || les mines fpl. ne sont encore exploitées que partiellement. / ~ coal pit || abgebaute Kohlenzeche f. || houillière f. épuisée.

exhauster || Sauglüfter m.; Exhaustor m. || exhausteur m.; aspirateur m. / ~ installation || Saugzuganlage f. || installation f. de tirage forcé.

exhaust gas || Abgas n.; Auspuffgas n. || gaz m. brulé ou d'échappement. / heat of ~es || Abgaswärme f. || chaleur f. des gaz perdus. / leading of the ~es || Auspuffführung f. || canalisation f. ou conduite f. des gaz d'échappement. / ~ pressure || Auspuffgasdruck m. || pression f. des gaz d'échappement. / ~ trunk || Gasentlüftungsschacht m. || cheminée f. d'évacuation du gaz. / ~ utilization || Abgasverwertung f. || utilisation f. des gaz brûlés ou d'échappement.

exhausting to atmosphere || Auspuffen n. || échappement m. à l'air libre. / ~ machine for ventilating mines || Grubenventila-

tor m.; Wettermaschine f.; Wettertrommel f. || machine f. aspirante ou pneumatique pour l'aération des mines. / ~ plant for smoke || Rauchabsaugeanlage f. || installation f. d'aspiration des fumées. / ~ plant for steam || Absaugeanlage f. für Dämpfe || installation f. d'aspiration des vapeurs. / ~ valve (Steam) || Einlaßventil n. || soupape f. d'aspiration.

exhaustion || Erschöpfung f.; Abbau m. || épuisement m. / ~ of the ores of a mine || Abbauen. eines Feldes || épuisement m. d'une mine.

exhaust lap of the slide-valve || innere Überdeckung f. des Dampfschiebers || recouvrement m. intérieur ou à l'échappement. / ~ lever (Mot) || Auslaßhebel m. || levier m. d'échappement.

exhaust manifold || Auspuffleitung f. || tubulure f. d'échappement. / water-cooled ~ || wassergekühlte Auspuffleitung f. || tubulure f. d'échappement à chemise d'eau. / ~ clamp || Auspuffrohrschelle f. || collier m. de tuyau d'échappement.

exhaust nipple || Ausblasemundstück n. || raccord m. fileté d'échappement. / ~ passage || Dampfausströmungskanal m. || canal m. d'échappement de la vapeur. / ~ pipe || Auspuffrohr n. || tuyau m. d'échappement. / ~ port || Auspuffkanal m. || lumière f. d'échappement.

exhaust steam || Ausströmungsdampf m. || vapeur f. d'échappement ou de décharge. / ~ main || Abdampfleitung f. || conduite f. de vapeur d'échappement. / ~ oil separator || Abdampfentöler m. || déshuileur m. de vapeur d'échappement. / ~ pipe || Abdampfrohr n. || tuyau m. de vapeur d'échappement. / ~ preheater || Abdampfvorwärmer m. || réchauffeur m. à vapeur d'échappement. / ~ separator || Abdampfkondenswasserabscheider m. || séparateur m. d'eau de condensation ou d'échappement. / ~ turbine || Abdampfturbine f. || turbine f. à vapeur d'échappement. / ~ turbine for utilizing the exhaust steam of other engines || Abdampfturbine f. zur Verwertung des Abdampfes anderer Maschinen || turbine f. à vapeur d'échappement alimentée par la vapeur d'échappement d'autres machines. / ~ utilizing plant || Abdampfverwertungsanlage f. || installation f. d'utilisation des vapeurs d'échappement. / ~ valve || Abblaseventil n || soupape f. de purge ou d'échappement.

exhaust strainer || Auspuffseiher m. || crépine f. ou filtre m. d'échappement. / ~ stub || Auspuffstutzen m. || raccord m. d'échappement. / ~ tank with hydraulic water seal || Auspufftopf m. mit Wasserabschluß || pot m. d'échappement à fermeture hydraulique. / ~ throttle disc || Auspuffdrosselklappe f. || bouchon m. de réglage d'échappement.

exhaust valve || Auslaßventil n.; Auspuffventil n. || soupape f. ou clapet m. d'échappement. / ~ overhead ~ || hängendes Auslaßventil n. || soupape f. d'échappement commandée par le haut.

exhaust valve box || Abblaseventilkammer f. || chambre f. de soupape d'échappement. / ~ cam || Auslaßventilnocken m. || came f. de soupape d'échappement. / ~ cap || Auslaßventilverschraubung f. || bouchon m. de soupape d'échappement. / ~ cone || Auslaßventilkegel m. || cône

m. de soupape d'échappement. / ~ lifter || Auslaßventilheber m. || toc m. de soupape d'échappement; lève-soupape f. d'échappement. / ~ lift rod || Auspuffventilheberstange f. || tige f. de poussoir de la soupape d'échappement. / ~ stem || Auspuffventilstange f. || tige f. de soupape d'échappement.

exhaust way (Build) || Ausflußkanal m. || canal m. de déversement. / ~ (Mach) || Dampfausströmungskanal m. || canal m. d'échappement de la vapeur.

exhibit, to || eine Ausstellung f. beschicken; ausstellen || exposer.

exhibition || Ausstellung f.; Schau f. || vue f.; inspection f.; revue f.; exposition f.; étalage m. / ~ international ~ || Weltausstellung f. || exposition f. universelle.

exhibition case || Schaukasten m. || vitrine f. / ~ hall || Ausstellungshalle f. || halle f. d'exposition. / ~ machine || Ausstellungsmaschine f. || machine-modèle m. d'exposition.

exist, to || existieren; vorhanden sein || exister.

existence || Bestand m.; Existenz f. || existence f.

exit || Ausgang m. || sortie f.

exit pupil (Opt) || Austrittspupille f. || pupille f. de sortie; cercle m. ou anneau m. oculaire. / ~ of a microscope || Austrittspupille f. eines Mikroskopes || pupille f. de sortie d'un microscope. / ~ of a telescope || Austrittspupille f. eines Fernrohres || pupille f. d'émergence d'un télescope; anneau m. oculaire d'un télescope.

exophthalmometer || Exophthalmometer n. || exophtalmomètre m. / reflecting ~ || Spiegelophthalmometer n. || exophtalmomètre m. à miroirs.

exorbitant || übermäßig; übertrieben || démesuré; exagéré; exorbitant.

exosmose || Exosmose f. || exosmose f.

exospore || Exospore f. || exospore f.

exosporic || exospor || exosporé. / ~ bacterium || exospores Bakterium n. || bactérie f. exosporée.

exosporium || Exosporium n. || exospore f.

exothermic || exothermisch || exothermique.

exotic || exotisch || exotique. / ~ timber || Überseeholz n. || bois m. d'outre-mer ou exotique ou des colonies.

expand, to (Mach) || weiten; aufweiten; erweitern || élargir. / ~ (Gases) || entspannen; nachlassen || détendre. / ~ (Phys) || sich dehnen; (sich) ausdehnen || se dilater. / ~ a boiler tube || Kesselrohr n. aufwalzen || mandriner ou renfler ou élargir un tube de chaudière.

expanded metal || Streckmetall n. || métal m. déployé.

expander, tube ~ || Rohraufweiter m. || cylindre m. à mandriner les tubes.

expanding clutch || Ausdehnungskupplung f. || accouplement extensible. / ~ pulley || Ausdehnungsriemenscheibe f. || poulie f. extensible. / ~ roller (Mach) || Spannrolle f. || galet-tendeur m. / ~ vessel || Expansionsgefäß n. || pot m. à détente.

expansibility || Ausdehnbarkeit f. || expansibilité f.; dilatabilité f.

expansible || ausdehnbar || expansif.

expansion (Mach) || Aufweitung f. || élargissement m. / ~ (Phys) || Expansion f.; Ausdehnung f.; Entspannung f. || expansion f.; détente f.; dilatation f. / ~ of cast-iron || Quellen n. des Guß-

eisens || expansion f. de la fonte. / mean coefficient of cubical ~ || mittlerer kubischer Ausdehnungskoeffizient m. || coefficient m. moyen de dilatation cubique. / linear ~ || Längenausdehnung f. || dilatation f. linéaire. / radial ~ || radiale Ausdehnung f. || expansion f. radiale. / radial ~ (Gun) || Kaltreckung f.; radiale Expansion f. || autofrettage m. / ~ of steam || Ausdehnung f. des Dampfes || expansion f. de la vapeur.

expansion appearance || Treiberscheinung f. || phénomène m. de gonflement. / ~ back (Bookb) || dehnbarer Rücken m. || reliure f. extensible. / ~ coefficient || Ausdehnungskoeffizient m. || coefficient m. de dilatation. / ~ coupling || Ausdehnungskupplung f. || accouplement m. extensible; embrayage m. à segments extensibles. / ~ curve || Expansionskurve f. || courbe f. de détente. / ~ degree || Expansionsgrad m. || degré m. de détente ou d'expansion. / ~ joint (Build) || Ausdehnungsfuge f. || joint m. de dilatation. / ~ joint (Mach) || Ausdehnungsverbindung f.; Kompensator m. (in Rohrleitungen) || joint m. coulissant; compensateur m. / ~ pipe || Dehnungsrohr n. || tube m. de compensation. / ~ piston (Steam) || Spannungskolben m. || expansif m. / ~ rate || Füllungsgrad m. || degré m. de détente ou d'expansion. / ~ slide valve || Expansionsschieber m. || tiroir m. à expansion ou de détente. / ~ sliding block || Expansionsgleitbacke f. || coulisseau m. de détente. / ~ steam engine || Expansionsdampfmaschine f. || machine f. à vapeur à détente. / ~ stroke (Mot) || Explosionshub m. || course f. d'explosion. / ~ stuffing box || Ausgleichstopfbuchse f. || boîte f. à bourrage compensatrice; presse-étoupe m. compensateur. / ~ valve || Regelventil n. || vanne f. de réglage. / ~ valve gear || Expansionssteuerung f. || distribution f. à détente.

expansive force || Ausdehnungskraft f. || force f. expansive. / ~ power see expansive force.

expansiveness || Ausdehnungsvermögen n. || pouvoir m. expansif.

expect, to || erwarten; rechnen oder hoffen oder zählen auf ... || attendre; compter; espérer.

expectation value || voraussichtlicher Ertragswert m. || valeur f. d'attente.

expel, to (Motor) || ausstoßen; herausspülen || expulser; chasser; éliminer; évacuer. / ~ the burnt gases pl. || die Verbrennungsgase npl. herausspülen || chasser les gaz mpl. brûlés.

expelling device (Office) || Auswerfvorrichtung f. || éjecteur m.

expenditure see expense.

expense || Geldausgabe f.; Aufwand m.; Auslage f. || argent m. dépensé; dépense f.; déboursé m. / ~ distribution || Unkostenverteilung f. || répartition f. des frais accessoires.

expenses pl. || Kosten pl.; Unkosten pl.; Spesen pl. || frais mpl.; dépense f. / to add the ~ to the goods || die Unkosten pl. auf die Waren schlagen || se prévaloir ou charger les frais sur la marchandise. / additional ~ see extraordinary ~. / the ~ amount to || die Kosten pl. belaufen sich auf || les frais mpl. s'élèvent à. / to clap the ~ to the goods || die Unkosten pl.

auf die Waren schlagen || se prévaloir ou charger les frais sur la marchandise. / to cover the ~ pl. || die Kosten pl. decken || couvrir les frais mpl. / ~ for erecting || Montagekosten pl. || frais mpl. de montage. / extraordinary ~ || außergewöhnliche Ausgaben fpl.; außerordentliche Kosten pl. || faux-frais mpl. / legal ~ || Rechtsunkosten pl.; Gerichtskosten pl. || dépens mpl. ou frais mpl. de justice. / all ~ paid || spesenfrei || frais mpl. déduits; sans frais; net de tous frais. / out of pocket ~ || Barauslagen pl. || ménues dépenses fpl. / travelling ~ || Reiseunkosten pl. || frais mpl. de voyage. / ~ of upkeep || Unterhaltungskosten pl. || prix m. d'entretien.

expensive, to be very ~ || mit großen Unkosten pl. verknüpft sein || être très coûteux.

experience, to || experimentieren || expérimenter; éprouver; essayer.

experience || Erfahrung f. || expérience f. / ~ of a coast || Küstenkenntnis f. || cabotage m.; expérience f. d'une côte. / long ~ || langjährige Erfahrung f. || expérience f. longue. / practical ~ || praktische Erfahrung f. || expérience f. pratique.

experienced || geschäftskundig || versé dans les affaires.

experiment || Experiment n.; Probe f.; Versuch m. || expérience f.; essai m.; épreuve f. / wearisome ~ || mühevoller Versuch m. || expérience f. difficile. / station for ~s || Versuchsanstalt f. || station f. d'essai.

experimental || experimentell || expérimental. / ~ area (Agr) || Versuchsfläche f. || place f. ou surface f. d'essai. / ~ boiler || Versuchskessel m. || chaudière f. d'essai. / ~ flight || Versuchsflug m. || vol m. d'essai. / ~ model || Versuchsmodell n. || modèle m. d'épreuves. / ~ section of the signal corps || Versuchsabteilung f. der Verkehrstruppen || section f. d'essais des troupes des communications. / ~ station in the heavy current laboratory || Experimentieranlage f. im Starkstromlaboratorium || installation f. d'expériences au laboratoire pour courants intenses ou forts. / ~ station for ore treatment || Erzaufbereitungsversuchsanstalt f. || atelier m. d'essai du traitement des minerais. / ~ work || Versuch m. || essai m.

expert || Sachverständiger m.; Fachmann m. || expert m.; spécialiste m.; homme m. de métier. / to submit to an ~ || begutachten lassen || faire émettre son avis; demander une expertise. / examination by ~s || Expertise f.; Sachverständigenbegutachtung f.; Sachverständigengutachten n. || expertise f.

expert's opinion || Sachverständigengutachten n. || avis m. d'expert; expertise f. / ~ report || Sachverständigenbericht m. || rapport m. d'expert; expertise f.

expiration (Bill) || Fälligkeit f.; Verfall m. || échéance f.; exigibilité f.

explain, to || auseinandersetzen; erklären || expliquer.

explanation || Erklärung f. || explication f.

explode, to || explodieren; platzen; zerspringen || exploser; faire explosion.

exploder || Explosionsmittel n.; Sprengmittel n. || explosif m. / incandescent ~ || Glühzünder m. || exploseur m. à incandescence.

exploit || ausbeuten; ausnutzen || exploiter. / to ~ mining areas for private economic purposes || Felder npl. privatwirtschaftlich ausbeuten || exploiter des terrains miniers au titre d'entreprise privée. / to be entitled to exploit (Min) || gewinnungsberechtigt sein || être autorisé à exploiter.

exploitation || Wirtschaftsführung f. || exploitation f.

exploiter || Ausbeuter m. || exploiteur m.

explorating shaft (Mine) || Versuchsschacht m. || puits m. de recherche.

exploration || Erforschung f.; Untersuchung f. || exploration f. / ~ paper (Chem) || Reagenspapier n. || papier m. réactif. / ~ work || Forschungsarbeiten pl. || travaux mpl. d'exploration ou de recherche.

explore, to || erforschen; untersuchen || explorer; rechercher. / ~ (Mine) || aufschließen || ouvrir. / ~ the ground by bore-holes || ein Terrain n. abbohren || faire des recherches fpl. dans un terrain; sonder un terrain.

explored || untersucht || exploré.

explosibility || Explodierbarkeit f. || explosibilité f.

explosible || explodierbar || explosible.

explosion || Explosion f.; Plotz m.; Sprengschlag m.; Entladung f. || explosion f. / ~ of a boiler || Bersten n. eines Kessels || explosion f. d'une chaudière. / delayed ~ || verzögerte Explosion f. || explosion f. retardée. / muffler ~ || Knallen n. im Auspufftopf || pétarade f. au pot d'échappement.

explosion chamber || Explosionsraum m. || chambre f. d'explosion. / ~ danger || Explosionsgefahr f. || danger m. d'explosion. / ~ door (Metal) || Explosionsklappe f. || soupape f. de protection contre les explosions. / ~ motor || Explosionsmotor m. || moteur m. à explosion.

explosion-proof heating member || explosionsgesicherter Heizkörper m. || corps m. de chauffe inexplosible. / ~ plant || explosionssichere Anlage f. || installation f. inexplosible. / ~ tank see ~ vessel. / ~ vessel || explosionssicheres Gefäß n. || vase m. à l'épreuve d'explosion; récipient m. inexplosible.

explosive || explodierend; explosiv || explosif. / ~ agent see explosive. / ~ ball || Explosionskugel f. || balle f. explosible. / ~ cartridge || Sprengkapsel f. || capsule f. fulminante. / resorcinate ~ cartridge || Resorzinatsprengkapsel f. || capsule f. explosive ou détonateur m. à la résorcine. / ~ cotton || Schießbaumwolle f. || fulmicoton m.; coton m. azotique. / ~ field (Electr) || Entladestärke f. || champ m. explosif. / ~ gas || Grubengas n. || grisou m. / ~ mixture || explosibles oder explosionsfähiges Gemisch n. || mélange m. explosif ou détonnant. / confined space liable to contain ~ mixtures || Betriebsraum m. zur Lagerung von Sprengstoffen || local m. permettant l'emmagasinage de produits explosifs. / ~ strength || Sprengwirkung f. || puissance f. explosive. / ~ substance see explosive.

explosive || Sprengstoff m.; Sprengmittel n. || explosif m.; matière f. explosive; composé m. explosif. / safety ~ || Sicherheitssprengstoff m. || explosif m. de sûreté ou difficilement inflammable.

explosives manufacture || Sprengstoffherstellung f. || fabrication f. de matières

explosives. / ~ manufacturing plant ‖ Sprengstofffertigungseinrichtung f. ‖ équipement m. *ou* installation f. de fabriques d'explosifs. / ~ testing apparatus ‖ Sprengstoffuntersuchungsapparat m. ‖ appareil m. à analyser *ou* à essayer les explosifs.

exponent (Math) ‖ Exponent m. ‖ exposant m.

export, to ‖ exportieren; ausführen ‖ exporter.

export *see also* exportation ‖ Export m.; Ausfuhr f. ‖ exportation f.

export article ‖ Ausfuhrgegenstand m.; Exportartikel m. ‖ article m. d'exportation.

exportation ‖ Ausfuhr f.; Export m. ‖ exportation f. / ~ of birds ‖ Vogelexport m. ‖ exportation f. d'oiseaux. / ~ of grain ‖ Getreideausfuhr f. ‖ exportation f. des blés. / total ~s pl. ‖ Gesamtausfuhr f. ‖ total m. de l'exportation. / wood ~ ‖ Holzausfuhr f. ‖ exportation f. de bois.

exportation, permit of ~ ‖ Ausfuhrerlaubnis f. ‖ permis m. d'exportation. / prohibition of ~ ‖ Ausfuhrverbot n. ‖ défense f. d'exporter. / to remove a prohibition of ~ ‖ ein Ausfuhrverbot n. aufheben ‖ abroger une défense d'exporter.

export beer ‖ Exportbier n.; Ausfuhrbier n. ‖ bière f. d'exportation. / bottled ~ beer ‖ Ausfuhrflaschenbier n.; Exportflaschenbier n. ‖ bière f. d'exportation en bouteilles. / draught ~ beer ‖ Ausfuhrfaßbier n.; Exportfaßbier n. ‖ bière f. d'exportation en fûts. / ~ cask ‖ Ausfuhrfaß n.; Exportfaß n. ‖ fût m. d'exportation. / ~ concern ‖ Exportgeschäft n. ‖ maison f. d'exportation. / ~ department ‖ Ausfuhrabteilung f.; Exportabteilung f. ‖ département m. d'exportation. / ~ duty ‖ Ausfuhrzoll m. ‖ droits mpl. d'exportation.

exporter ‖ Exportör m.; Ausführer m. ‖ exportateur m.

export merchant ‖ Ausfuhrhändler m. ‖ négociant m. exportateur. / ~ mill ‖ Exportmühle f. ‖ moulin m. pour l'exportation.

exports pl. ‖ Ausfuhrwaren fpl.; Exportartikel mpl. ‖ marchandises fpl. *ou* articles mpl. d'exportation.

export trade ‖ Ausfuhrhandel m. ‖ commerce m. d'exportation.

expose, to (Phot) ‖ belichten; exponieren ‖ exposer (à la lumière); poser. / ~ (Goods) *see also* to exhibit ‖ ausstellen; auslegen ‖ exposer. / ~ to the action of rays ‖ bestrahlen ‖ exposer à l'action f. de rayons. / ~ to the air (Dyer) ‖ lüften ‖ aérer. / ~ goods pl. for sale ‖ Waren fpl. zum Verkauf aufstellen ‖ étaler des marchandises fpl. pour la vente. / ~ to the light *see* to expose.

exposed (Phot) ‖ belichtet ‖ exposé.

exposing position (Phot) ‖ Belichtungsstellung f. ‖ position f. d'exposition. / ~ time (Phot) ‖ Belichtungszeit f.; Belichtungsdauer f. ‖ durée f. *ou* temps m. de pose.

exposition ‖ Erklärung f.; Darlegung f. ‖ exposition f.; exposé m.; explication f. / ~ (Exhibition) ‖ Ausstellung f. ‖ exposition f.

exposure (Phot) ‖ Belichtung f. ‖ exposition f. (à la lumière); pose f. / ~ to the action of rays ‖ Bestrahlung f. ‖ exposition f. à l'action de rayons. / ~ to the light *see* to exposure. / ~ of the plate in the

camera ‖ Belichten n. der Platte in der Kamera ‖ exposition f. de la plaque dans la chambre noire.

exposure chamber (Phot) ‖ Belichtungskammer f. ‖ chambre f. d'exposition. / ~ meter (Phot) ‖ Belichtungsmesser m. ‖ pose-mètre m. / ~ time (Phot) ‖ Belichtungsdauer f.; Aufnahmezeit f. ‖ durée f. de l'exposition à la lumière; temps m. de pose.

express, to ‖ ausdrücken ‖ exprimer. / ~ an opinion ‖ ein Urteil n. abgeben ‖ manifester une opinion. / ~ the wind velocity in meters per second ‖ die Windgeschwindigkeit in Metern je Sekunde angeben ‖ exprimer la vitesse du vent en mètres par seconde.

express ‖ ausdrücklich; bestimmt; deutlich ‖ exprès. / ~ express; Schnell...; Eil... ‖ express; à grande vitesse. / ~ cargo boat ‖ Eilgüterboot n. ‖ chaland m. *ou* vapeur m. *ou* cargo m. pour service accéléré. / ~ forwarding ‖ Eilbeförderung f. ‖ transport m. en grande vitesse. / ~ freighter ‖ Eilfrachtschiff n. ‖ vapeur m. *ou* cargo m. pour service accéléré. / ~ goods pl. ‖ Eilgut n.; Eilfracht f.; Expreßgut n. ‖ grande vitesse f.; marchandises fpl. expédiées par grande vitesse. / ~ goods tariff ‖ Eilguttarif m.; Eilgutgebührensatz m. ‖ tarif m. de grande vitesse. / ~ letter ‖ Eilbrief m. ‖ lettre f. expresse. / ~ locomotive ‖ Schnellzuglokomotive f. ‖ locomotive f. d'express. / ~ motor coach ‖ Schnelltriebwagen m. ‖ voiture f. motrice rapide. / ~ movement ‖ Eilbewegung f. ‖ marche f. rapide / ~ passenger locomotive *see* ~ locomotive. / ~ rate *see* ~ goods tariff. / ~ steamer ‖ Schnelldampfer m. ‖ vapeur m. rapide. / ~ train *see* express.

express ‖ Schnellzug m.; Eilzug m.; Expreßzug m.; Expreß m.; Blitzzug m.; (F-)D-Zug m. ‖ train m. à grande vitesse; express m.; train m. express *ou* accordéon. / ~ paid ‖ Eilbote m. bezahlt ‖ exprès m. payé.

expression ‖ Ausdruck m. ‖ expression f.

expropriate, to ‖ enteignen ‖ exproprier. / to carry out the decision to expropriate ‖ den Enteignungsbeschluß m. vollstrecken ‖ exécuter l'ordonnance f. d'expropriation.

expropriation ‖ Enteignung f. ‖ expropriation f. / decision to ~ ‖ Enteignungsbeschluß m. ‖ ordonnance f. d'expropriation. / proceedings pl. of ~ ‖ Enteignungsverfahren n. ‖ procédure f. d'expropriation.

expunged ‖ ausgestrichen ‖ effacé.

exsiccate, to ‖ trocknen ‖ dessécher.

extend, to ‖ ausbreiten; ausdehnen; auseinanderbiegen ‖ étendre. / ~ (To spread) ‖ sich ausbreiten; sich ausdehnen ‖ se répandre; s'étendre. / ~ to infinity ‖ sich ins Unendliche n. erstrecken ‖ s'étendre à l'infini. / ~ the iron ‖ das Eisen strecken *oder* recken ‖ étirer le fer sous le marteau.

extended ‖ verlängert ‖ prolongé. / ~ to ... ‖ verlängert bis ... ‖ prolongé jusqu'à ... / can be ~ to ... ‖ kann verlängert werden bis ... ‖ peut être prolongé jusqu'à ... / widely ~ ‖ weitverzweigt ‖ très étendu *ou* répandu.

extended concession ‖ ·Geviertfeld n. ‖ champ m. d'exploitation quadraturé *ou*

élargé. / ~ ⊤-shaped antenna ‖ verlängerter ⊤-Luftleiter m.; verlängerte ⊤-Antenne f. ‖ antenne f. en ⊤ à branches horizontales prolongées.

extensibility ‖ Ausdehnbarkeit f. ‖ expansibilité f.; dilatabilité f.

extensible ‖ ausdehnbar ‖ extensible.

extension ‖ Ausbreitung f.; Erweiterung f.; Vergrößerung f.; Ausdehnung f. ‖ extension f.; propagation f.; élargissement m.; accroissement m.; agrandissement m.; dilatation f. / ~ (Phys) ‖ Dehnung f.; Ausdehnung f. ‖ extension f.; dilatation f.; allongement m. / ~ (Trade) ‖ Verlängerung f. ‖ prolongation f. / ~ of building ‖ Erweiterungsbau m. ‖ agrandissement m. des ateliers. / completion of the ~s which are being carried out now ‖ Durchführung f. der im Gange befindlichen Ausbauarbeiten ‖ achèvement m. des constructions supplémentaires en cours. / ~ of the current distributing system ‖ Erweiterung f. der Stromverteilungsanlage ‖ augmentation f. de l'installation de distribution de courant électrique. / ~ of the heating and power station ‖ Erweiterung f. des Heizwerkes und Kraftwerkes ‖ agrandissement m. de la centrale de chauffage et d'énergie. / lateral ~ ‖ seitliche Auslaung f. ‖ porte f. à faux latérale. / ~ of a line (Tel) ‖ Leitungsverlängerung f. ‖ prolongement m. de ligne. / ~ of a plant ‖ Erweiterung f. einer Anlage ‖ agrandissement m. d'une installation. / ~ of a power station ‖ Erweiterungsbau m. eines Kraftwerkes ‖ agrandissement m. d'une usine génératrice. / with regard to the ~s to come ‖ im Hinblick auf künftige Erweiterungen fpl. ‖ en vue des extensions fpl. à venir. / ~ of time ‖ Fristverlängerung f.; Fristgewährung f. ‖ délai m.; sursis m.; prorogation f. / ~ of the track of a railway ‖ Eisenbahngleisverlängerung f. ‖ allongement m. d'une voie de chemin de fer.

extension arm ‖ Ausleger m. ‖ console f. *ou* support m. supplémentaire. / ~ bar (El line) ‖ Verlängerungsschiene f. ‖ barre f. d'allongement. / ~ board for the distribution of incoming exchange calls to the stations of a private automatic telephone installation ‖ Nebenstellenzentrale f. zur Vermittlung von ankommenden Amtsgesprächen nach den Apparaten einer selbsttätigen Hausfernsprechanlage ‖ poste m. supplémentaire pour l'écoulement de conversations de service d'arrivée vers les appareils d'une installation téléphonique privée automatique. / ~ cord ‖ Anschlußschnur f. ‖ corde f. de raccord. / ~ line (Tel) ‖ Nebenanschlußleitung f. ‖ ligne f. de poste auxiliaire; communication f. téléphonique secondaire. / ~ multiple section (Tel) ‖ Ansatzfeld n. ‖ tableau m. d'extension. / ~ rod for the tripod ‖ Verlängerungsstange f. für das Stativ ‖ tige f. de rallonge du trépied.

extension station (Tel) ‖ Nebenanschluß m. ‖ poste m. supplémentaire. / outside ~ (Tel) ‖ Außennebenstelle f. ‖ poste m. auxiliaire *ou* supplémentaire extérieur. / ~ without exchange facilities (Tel) ‖ Hausstelle f. ‖ poste m. privé. / ~ service (Tel) ‖ Nebenstellenbetrieb m. ‖ service m. des postes supplémentaires.

extension strength ‖ Zugfestigkeit f. ‖ résistance f. à la traction. / ~ subscriber's

station (Tel) *see* outside extension station.

extensive ‖ vielseitig; umfassend ‖ à plusieurs faces; étendu; varié. / ~ network of standard and narrow-gauge lines of railroads ‖ weitverzweigtes Netz n. normal- und schmalspuriger Schienenstränge ‖ réseau m. étendu de voies normales et étroites. / ~ use ‖ ausgedehnte Verwendung f. ‖ emploi m. vaste.

extent ‖ Ausdehnung f.; Umfang m. ‖ étendue f. / ~ of dilution ‖ Verdünnungsgrad m. ‖ degré m. de dilution.

exterior ‖ äußerer; äußerlich ‖ extérieur. / ~ flange ‖ Außenflansch m. ‖ bride f. extérieure. / ~ plate (Shipb) ‖ Außenhautplatte f. ‖ plaque f. de bardé extérieur. / ~ resistance ‖ äußerer Widerstand m. ‖ résistance f. extérieure.

exterminate, to ‖ vertilgen; ausrotten ‖ exterminer.

extermination of noxious animal and plant sicknesses ‖ Schädlingsbekämpfung f. ‖ extermination f. d'animaux nuisibles et de maladies des plantes.

external crack ‖ Außenriß m. ‖ fente f. extérieure. / ~ dimension ‖ Außenabmessung f. ‖ dimension f. extérieure. / ~ form ‖ äußere Form f. ‖ présentation f.

external pole ‖ Außenpol m. ‖ pôle m. extérieur. / ~ armature (Electr) ‖ Außenpolanker m. ‖ induit m. à pôles extérieurs. / ~ dynamo ‖ Außenpoldynamo f. ‖ dynamo f. à pôles extérieurs.

external screwing machine ‖ Außengewindeschneidmaschine f. ‖ machine f. de filetage extérieur. / ~ taper grinding ‖ Außenschliff m. konischer Rundflächen ‖ rectification f. extérieure de surfaces circulaires coniques. / for ~ use (Med) ‖ zum äußerlichen Gebrauch m. ‖ à l'usage m. externe. / ~ wall ‖ Außenwand f. ‖ paroi f. extérieure.

externally (Med) ‖ äußerlich ‖ à l'usage m. externe. / ~ and internally ‖ außen und innen ‖ au dehors et en dedans.

extinction ‖ Ablöschung f.; Auslöschung f. ‖ extinction f. / ~ of the arc ‖ Auslöschen n. des Lichtbogens ‖ extinction f. de l'arc. / ~ fittings pl. ‖ Feuerlöscharmatur f. ‖ armature f. d'appareils extincteurs d'incendie.

extinguish, to ‖ löschen; auslöschen ‖ éteindre; s'éteindre. / ~ the fire ‖ Feuer n. löschen ‖ éteindre l'incendie m.

extinguisher ‖ Löschhütchen n. ‖ éteignoir m. / ~ for distance ‖ Fernlöscher m. ‖ extincteur m. à distance. / fire ~ ‖ Feuerlöschgerät n. ‖ appareil extincteur d'incendies.

extinguishing apparatus ‖ Löschgerät n.; Löschapparat m. ‖ appareil m. d'extinction. / ~ plant for coking coal ‖ Kokslöschanlage f. ‖ installation f. d'extinction de coke.

extirpator ‖ Scharegge f. ‖ extirpateur m.

extra-axial aberration ‖ Aberration f. außerhalb der Axe ‖ aberration f. en dehors de l'axe. / ~ charge ‖ Aufgeld m.; Agio n.; Mehrpreis m. ‖ change m.; agio m.; supplément m. de prix. / ~ cost ‖ Aufschlag m. ‖ hausse f.; augmentation f.; renchérissement m. / ~ current (Electr) ‖ Extrastrom m. ‖ courant m. supplémentaire. / ~ expense ‖ Extraausgabe f. ‖ dépense f. extraordinaire; extra m. / ~ hand ‖ Hilfsarbeiter m. ‖ ouvrier m. supplémen-

taire *ou* auxiliaire. / ~ high-tension plant ‖ Höchstspannungsanlage f. ‖ installation f. à très haute tension. / ~-judicial ‖ außergerichtlich ‖ extra-judiciaire. / ~-weft figured fabric ‖ lanzierter Stoff m. ‖ tissu m. lancé. / ~ weight ‖ Übergewicht m.; surpoids m.; poids m. supplémentaire *ou* excédent m. de poids. / ~ work ‖ Überarbeit f.; Nacharbeit f. ‖ travail m. supplémentaire.

extract, to (Chem) ‖ extrahieren; auslaugen; ausziehen; entziehen ‖ extraire; épuiser. / ~ (Mine) ‖ gewinnen ‖ extraire. / ~ bungs pl. ‖ entspunden ‖ tirer des bondes fpl. / ~ corks pl. ‖ entkorken ‖ déboucher. / ~ gold ‖ entgolden ‖ extraire l'or m. / ~ a root (Math) ‖ eine Wurzel f. ziehen ‖ extraire une racine.

extract (Chem) ‖ Extrakt m.; Auszug m. ‖ extrait m.; débouilli m. / ~ of account ‖ Rechnungsauszug m. ‖ extrait m. *ou* relevé m. de compte; bordereau m. / apparent ~ ‖ scheinbarer Extraktgehalt m. ‖ teneur f. apparente en extrait. / colouring ~ ‖ Farbstoffauszug m. ‖ extrait m. tinctorial. / dry ~ ‖ Trockenextrakt m. ‖ extrait m. sec. / ~ of dye wood ‖ Farbholzauszug m. ‖ extrait m. de bois de teinture. / ~ of gall-nuts ‖ Galläpfelauszug m. ‖ extrait m. de noix de galle. / ~ of green malt ‖ Grünmalzauszug m. ‖ extrait m. de malt vert. / in ~s pl. ‖ auszugsweise ‖ par extrait m. / ~ of meat ‖ Fleischextrakt m. ‖ extrait m. de viande. / original ~ ‖ ursprünglicher Extraktgehalt m. ‖ teneur f. originale en extrait. / real ~ ‖ wirklicher Extraktgehalt m. ‖ teneur f. réelle en extrait. / vegetable ~ ‖ Pflanzenauszug m. ‖ extrait m. végétal.

extractable radiator ‖ ausziehbarer Kühler m. ‖ radiateur m. amovible *ou* escamotable.

extract determination ‖ Extraktbestimmung f. ‖ détermination f. de l'extrait.

extracting *see also* extraction ‖ Extraktion f.; Gewinnung f. ‖ extraction f. / ~ of corks pl. ‖ Entkorken n. ‖ débouchage m. / ~ fuels pl. from coal ‖ Brennstoffgewinnung f. aus Kohle ‖ obtention f. du combustible par extraction du charbon.

extracting plant *see* extraction plant. / ~ process (Office) ‖ Wachspapierverfahren n. ‖ procédé m. de polycopie au papier ciré. / ~ produce *see* extraction agent. / ~ winch (Mine) ‖ Förderwinde f. ‖ treuil m. d'extraction.

extraction (Chem) ‖ Ausziehen n.; Gewinnung f.; Extraktion f. ‖ extraction f. / direct ~ of iron in a charcoal-hearth ‖ Luppenfrischarbeit f. ‖ extraction f. directe du fer dans les fours d'affinage. / direct ~ of iron from the ore ‖ Rennarbeit f. ‖ extraction f. directe du fer de son minerai. / ~ of the juice *see* ~ of the syrup. / ~ of a root (Math) ‖ Wurzelziehen n. ‖ extraction f. d'une racine. / ~ of sugar ‖ Entzuckerung f. ‖ extraction f. du sucre. / ~ of the syrup (Sugar) ‖ Gewinnung f. des Saftes; Saftgewinnung f. ‖ extraction f. du jus *ou* des jus de diffusion.

extraction agent ‖ Extraktionsmittel n. ‖ produit m. *ou* agent m. *ou* moyen m. d'extraction. / ~ apparatus *see* extractor. / ~ contrivances pl. for ~ *see* extraction

agent. / ~ flask ‖ Extraktionskolben m. ‖ ballon m. à extraction.

extraction plant ‖ Extraktionsanlage f. ‖ installation f. d'extraction. / ~ for cotton waste for cleaning ‖ Extraktionsanlage f. für Putzwolle ‖ installation f. d'extraction de déchets de coton. / oil ~ ‖ Ölextraktionsanlage f. ‖ installation f. d'extracteurs d'huile. / ~ for oil cakes ‖ Extraktionsanlage f. für Ölpreßkuchen ‖ installation f. d'extraction de tourteaux. / ~ for oil seed ‖ Extraktionsanlage f. für Ölsaaten ‖ installation f. d'extraction pour semences oléagineuses.

extraction process ‖ Gewinnungsverfahren n. ‖ procédé m. d'extraction. / ~ vessel ‖ Extraktionsgefäß n. ‖ chaudron m. d'extraction. / ~ works pl. ‖ Laugeanstalt f. ‖ usine f. d'extraction.

extractive ‖ Extraktstoff m.; Extrakt m. ‖ extractif m. / ~ matter ‖ Extraktivstoff m. ‖ matière f. extractive; extractif m.

extract measurer ‖ Extraktbemesser m. ‖ pèse-extrait m.

extractor ‖ Extraktionsapparat m. ‖ appareil m. à extraction *ou* à lessive; extracteur m.; digesteur m.

extract table ‖ Extrakttabelle f. ‖ table f. indiquante l'extrait. / ~ yield ‖ Extraktausbeute f. ‖ rendement m. en extrait. / ~ yielding capacity ‖ Extraktergiebigkeit f. ‖ richesse f. en extrait.

extreme ‖ übermäßig ‖ démesuré; exagéré; exorbitant.

extremes pl. (Math) ‖ äußere Glieder npl. einer Proportion ‖ termes mpl. extrêmes.

extrude, to ‖ ausstoßen ‖ expulser; rejeter.

extruded spar (Aero) ‖ gezogener Holm m. ‖ longeron m. filé à chaud à la presse.

extruding press *see* extrusion press.

extrusion press ‖ Ziehpresse f. ‖ presse f. à étirer. / ~ for metal bars ‖ Metallstangenpresse f. ‖ presse f. à filer à chaud les barres métalliques. / ~ for tubes ‖ Rohrziehpresse f. ‖ presse f. à filer à chaud les tubes métalliques.

eye ‖ Auge n. ‖ œil m. / ~ (Mar; Pont) ‖ Schleife f.; Schlinge f.; Öse f. ‖ anneau m.; ganse f.; œil m. / ~ (Needl) ‖ Nadelöhr n. ‖ œil m.; trou m.; chas m. / ~ (Techn) ‖ Öse f.; Auge n. ‖ œillet m. / artificial ~ ‖ künstliches Auge n. ‖ œil m. artificiel. / astigmatic ~ ‖ astigmatisches Auge n. ‖ œil m. astigmate. / axially symmetrical ~ ‖ achsensymmetrisches Auge n. ‖ œil m. symétrique par rapport à l'axe. / ~ of a bolt ‖ Auge n. eines Bolzens ‖ œillet m. *ou* œil m. d'un boulon. / bull's ~ ‖ Schiffsfenster n. ‖ hublot m. / ~ of the crank ‖ Auge n. der Kurbelwelle ‖ œil m. de l'arbre à manivelle. / ~ of the crank-pin ‖ Kurbelzapfenauge n. ‖ œil m. de bouton de la manivelle. / defectively sighted ~ ‖ fehlsichtiges Auge n. ‖ œil m. amétrope. / emmetropic ~ ‖ rechtsichtiges Auge n. ‖ œil m. emmétrope. / ~ of a furnace (Metal) ‖ Ofenauge n. ‖ évent m. de fourneau. / ~ of a hatchet ‖ Helmloch *oder* Stielloch n. eines Beils ‖ douille f. *ou* œil f. d'une cognée. / ~ accommodated for infinity ‖ auf Unendlich akkommodiertes Auge n. ‖ œil m. accommodé à l'infini. / long-sighted ~ ‖ übersichtiges Auge n. ‖ œil m. hypermétrope. / far point surface of a long-sighted ~ ‖ Fernpunktfläche f. eines übersichtigen Auges ‖ lieu m. des punctum remotum d'un œil hypermé-

trope. / ~ of a mill-stone (Mill) ‖ Loch n. eines Mühlsteins ‖ œillard m. d'une meule. / to the naked ~ ‖ mit freiem Auge n. ‖ à l'œil nu. / night-blind ~ ‖ nachtblindes Auge n. ‖ œil m. héméralope. / normal ~ ‖ normales Auge n. ‖ œil m. normal. / ~ of the rod (Mach) ‖ Stangenauge n. ‖ œillet m. de tige. / ~ of a spherical vault (Arch) ‖ Nabelöffnung f. einer Kuppel ‖ œil m. de dôme. / static ~ ‖ akkommodationsloses Auge n. ‖ œil m. en état d'accommodation relâchée. / ~ of the tuyere ‖ Formauge n. ‖ œil m. de la tuyère. / unaided ~ ‖ unbewaffnetes Auge n. ‖ œil m. nu.

eyeball ‖ Augapfel m. ‖ globe m. oculaire; pupille f.

eyebolt [Augenbolzen m.; Ösenbolzen m. ‖ boulon m. à œillet; piton m. / ~ with a forelock (Mar) ‖ Schotbolzen m. ‖ cheville f. à boucle et à goupille. / ~ and key ‖ Keilbolzen m.; Schließbolzen m. ‖ boulon m. à clavette ou à goupille.

eyebolt screw ‖ Ringschraube f. ‖ vis f. à œil ou piton.

eyebrow ‖ Augenbraue f. ‖ sourcil m. / ~ pencil ‖ Augenbrauenstift m. ‖ crayon m. pour les sourcils.

eye cap (Opt) ‖ Okulardeckel m. ‖ bonnette f. d'oculaire. / unscrewing ~ cap for astro eyepieces ‖ abschraubbarer Augendeckel m. der Astrookulare ‖ bonnette f. dévissable des oculaires astronomiques. / ~ clinic ‖ Augenheilanstalt f. ‖ clinique f. ophtalmologique. / ~ corner ‖ Augenwinkel m. ‖ angle m. de l'œil. / ~ cup see ~ douche. / ~ douche ‖ Augendusche f. ‖ œillère f. / distance between the ~s ‖ Augenabstand m. ‖ écartement m. des yeux. / scale for varying the distance between the ~s pl. ‖ Skale f. zur Einstellung des Augenabstandes ‖ échelle f. pour le réglage de l'écartement des yeux.

eye end of the reflector tube ‖ Okularende n. des Spiegelrohres ‖ extrémité f. oculaire du tube du réflecteur. / ~ of the refractor ‖ Okularende n. des Refraktors ‖ extrémité f. intérieure du réfracteur.

eye, wooden box for ~ examining lenses ‖ Augenprobgläserkasten m. aus Holz ‖ boîte f. de bois pour verres à examiner les yeux. / ~ front segment ‖ Augenvordergrund m. ‖ partie f. antérieure de l'œil. / fundus of the ~ ‖ Augenhintergrund m. ‖ fond m. d'œil. / image of the fundus of the ~ ‖ Augenhintergrundbild n. ‖ image f. du fond de l'œil.

eyeglass (Opt) ‖ Augenglas n. ‖ lunette f. / ~ (Phys) ‖ Okular n. ‖ verre m. oculaire; oculaire m. / single ~ ‖ Einglas n.; Lorgnon n. ‖ monocle m.; lorgnon m. / watchmaker's ~ ‖ Uhrmacherlupe f. ‖ loupe f. d'horloger.

eyeglass case ‖ Pincenezfutteral n. ‖ étui m. pour pince-nez. / ~ chain ‖ Pincenezkettchen n. ‖ chaînette f. de pince-nez.

eyeglasses pl. ‖ Klemmer m.; Kneifer m.; Pincenez n. ‖ pince-nez m.

eyeglass hook ‖ Pincenezhaken m. ‖ portepince-nez m. / ~ screw ‖ Pincenezschraube f. ‖ vis f. de pince-nez. / ~ spring ‖ Pincenezfeder f. ‖ ressort m. de pince-nez.

eyelens ‖ Augenlinse f.; Okular n. ‖ lentille f. en verre d'œil; oculaire m.

eyelet see also eye ‖ Öse f. ‖ œillet m. / ~ (Build) ‖ Dachluke f. ‖ lucarne f. / screw ~ ‖ Ösenschraube f. ‖ vis à œillet ou à piton.

eyelet hole (Build) ‖ Dachluke f. ‖ lucarne f.; œillet m.; œillette f. / ~ (Mar) ‖ Reffgatt n.; Reffbandloch n. im Segel; Gaatjen n. ‖ œillet m. des voiles; œil m. de pic.

eyelet punch pliers ‖ Ösenzange f. ‖ pince f. à œillet. / ~ row stitcher (Shoem) ‖ Ösenstepper m. ‖ piqueur m. de bordure d'œillets. / ~ stay stitcher (Shoem) ‖ Ösenriemenstepper m. ‖ piqueur m. de sous-œillets. / ~ tape ‖ Band n. mit Ösen ‖ bande f. garnie d'œillets.

eyeletter (Tail) ‖ Schnürlochmacher m. ‖ poseur m. d'œillets.

eyeletting machine ‖ Öseneinsetzmaschine f. ‖ machine f. à placer les œillets.

eye level ‖ Augenhöhe f. ‖ à hauteur f. de l'œil.

eyelid ‖ Augenlid n. ‖ paupière f. / ~ retractor ‖ Lidhalter m. ‖ releveur m. de paupière. / ~ retractor with clamp ‖ Lidsperre f. ‖ blépharostat m.

eye microscope ‖ Augenmikroskop n. ‖ microscope m. ophtalmologique. / operation on the ~ ‖ Augenoperation f. ‖ opération f. ophtalmologique ou de l'œil. / lamp for operations on the ~ ‖ augenärztliche Operationslampe f. ‖ lampe f. pour les opérations ophtalmologiques. / ~ pair ‖ Augenpaar n. ‖ les deux yeux.

eyepiece ‖ Okular n. ‖ oculaire m. / ~ adjustable in terms of diopters ‖ Okular m. mit Dioptrieneinstellung ‖ oculaire m. muni d'un dispositif de mise au point gradué en dioptries. / astronomical ~ ‖ astronomisches Okular n. ‖ oculaire m. astronomique. / ~ for astronomical observations ‖ Okular n. für astronomische Beobachtungen ‖ oculaire m. pour les observations astronomiques. / binocular ~ ‖ binokulares Okular n. ‖ oculaire m. binoculaire. / comparison ~ ‖ Vergleichsokular n. ‖ oculaire m. de comparaison. / compensating ~ ‖ Kompensationsokular n. ‖ oculaire m. compensateur. / crossline ~ ‖ Strichkreuzokular n. ‖ oculaire m. à réticule. / double ~ ‖ Doppelokular n. ‖ oculaire m. double. / ~ with double refracting prism screwed on ‖ Okular n. mit aufgeschraubtem doppeltbrechendem Prisma ‖ oculaire m. sur lequel un prisme biréfringent est vissée. / finder ~ ‖ Sucherokular n. ‖ oculaire-chercheur m. / focal length of the ~ ‖ Okularbrennweite f. ‖ distance f. focale de l'oculaire. / ~ free from colour ‖ farbenfreies Okular n. ‖ oculaire m. achromatique. / ~ free from distortion ‖ verzeichnungsfreies Okular n. ‖ oculaire m. exempt de distorsion. / ~ free from internal reflection ‖ reflexfreies Okular n. ‖ oculaire m. exempt de reflets. / ~ of long focal length ‖ langbrennweitiges Okular n. ‖ oculaire m. à long foyer. / micrometer ~ ‖ Mikrometerokular n. ‖ oculaire m. pour micromètres. / microspectroscopic ~ ‖ Mikrospektralokular n. ‖ oculaire m. microspectroscopique. / orthoscopic micrometer ~ ‖ orthoskopisches Mikrometerokular n. ‖ oculaire m. orthoscopique pour micromètres. / photographic ~ ‖ fotografisches Okular n. ‖ oculaire m. photographique. / pointer ~ ‖ Zeigerokular n. ‖ oculaire m. indicateur. / provision for ensuring that the ~ moves parallel to the axis of the telescope ‖ zentrischparallele Führung f.

des Okularrohres ‖ guidage m. centroparallèle du tube oculaire. / quartz ~ ‖ Quarzokular n. ‖ oculaire m. en quartz. / revolving ~ head ‖ Okularrevolver m. ‖ revolver-oculaire m. / revolving ~ head furnished with three ~s ‖ Okularrevolver m. mit drei Okularen ‖ revolver m. à trois oculaires. / to set the ~ to the observer's sight ‖ das Okular n. auf die Sehschärfe des Beobachters einstellen ‖ adapter la mise au point à la vue de l'observateur. / spectroscopic ~ ‖ Spektralokular n. ‖ oculaire m. spectroscopique. / terrestrial ~ ‖ terrestrisches Okular n. ‖ oculaire m. terrestre.

eyepiece cap ‖ Okulardeckel m. ‖ bonnette f. de l'oculaire. / ~ changer see ~ head. / ~ clamping device ‖ Okularklemmung f. ‖ serrage m. de l'oculaire. / ~ correcting lens (Opt) ‖ Okularaufsteckglas n. ‖ verre m. à emboîter sur l'oculaire. / ~ cross line micrometer ‖ Okularnetzmikrometer n. ‖ micromètre-oculaire m. à réseau. / ~ cup ‖ Okularmuschel f. ‖ bonnette f. oculaire. / shallow ~ cup for spectacle wearers ‖ flache Okularmuschel f. für Brillenträger ‖ bonnette f. plate pour les porteurs de lunettes. / ~ diaphragm ‖ Okularblende f. ‖ diaphragme m. oculaire. / distance between the ~s ‖ Okularabstand m. ‖ écartement m. des oculaires.

eyepiece draw tube ‖ Okularauszug m. ‖ tube-tirage m. oculaire; tirage m. oculaire. / ~ with rack and pinion ‖ Okularauszug m. mit Zahn und Trieb oder mit Zahnstange ‖ tube-tirage m. oculaire à crémaillère. / ~ with three-spindle mechanism ‖ Okularauszug m. mit Dreispindelantrieb ‖ tirage m. oculaire à trois tiges filetées.

eyepiece draw tube clamp ‖ Klemmung f. des Okularauszuges ‖ vis f. immobilisant le tube-tirage. / ~ extension ‖ Verstellung f. des Okularauszuges ‖ course f. du tirage oculaire.

eyepiece field of view ‖ Gesichtsfeld n. des Okulars ‖ champ m. visuel de l'oculaire. / ~ field stop ‖ Okularblendebene f. ‖ plan m. du diaphragme de l'oculaire. / revolving ~ head ‖ Okularrevolver m. ‖ revolver-oculaire m.

eyepiece microscope ‖ Okularmikroskop n. ‖ microscope m. oculaire. / ~ with variable magnification ‖ Okularmikroskop n. mit veränderlicher Vergrößerung ‖ microscope m. oculaire à grossissement variable.

eyepiece pair ‖ Okularpaar n. ‖ pair f. d'oculaires. / ~ prism ‖ Okularprisma n. ‖ prisme m. oculaire. / ~ revolver ‖ Okularrevolver m. ‖ revolver-oculaire m. ~ revolving head see ~ revolver. / ~ screw micrometer ‖ Okularschraubenmikrometer n. ‖ oculaire-micromètre m. à vis. / ~ sliding sleeve ‖ Okularsteckhülse f. ‖ douille f. porte-oculaire; douille f. pour les oculaires. / ~ spectroscope ‖ Okularspektroskop n. ‖ spectroscope m. oculaire.

eye-point ‖ Augenpunkt m. ‖ pupille f. de sortie.

eye-pointed needle ‖ Lochnadel f. ‖ aiguille f. percée.

eye preserver ‖ Schutzbrille f. ‖ lunette f. protectrice. / ~ protectors for workers

Arbeiterschutzbrille f. ‖ lunettes fpl. protectrices pour ouvriers.

eyer (Needl) ‖ Bohrer m. ‖ perceur m.

eye radiation apparatus ‖ Augenbestrahlungsapparat m. ‖ appareil m. pour l'irradiation de l'œil. / ~ salve ‖ Augensalbe f. ‖ collyre m. gras. / ~ screw ‖ Ösenschraube f. ‖ vis à œil *ou* à œillet; piton m. / ~ shade ‖ Augenschirm m. ‖ garde-vue m.; abat-jour m. / ~ slit of tank ‖ Sehschlitz m. am Kampfwagen ‖ épiscope m. du tank. / ~ splice (Sailm) ‖ Augsplissung f. ‖ épissure f. à œil. / ~ tube *see* eyepiece. / ~ wash ‖ Augenwasser n. ‖ collyre m. (liquide). / ~ water *see* ~ wash.

eye-witness ‖ Augenzeuge m. ‖ témoin m. oculaire.

F

fabric ‖ Gewebe n.; Stoff m. ‖ tissu m. / ~ for balloons ‖ Ballonstoff m. ‖ étoffe m. à ballon. / à jour ~ ‖ à jour Ware f. ‖ tissu m. à jour. / checkered ~ ‖ gewürfelter Stoff m. ‖ étoffe f. à carreaux. / device for alternative backward and foreward movement of the ~ (Weav) ‖ Einrichtung f. zum abwechselnden vor- und rückwärtslaufenden Wagengang ‖ dispositif m. permettant la marche alternative du tissu. / drying ~s pl. ‖ Entnässen n. von Geweben ‖ essorage m. de tissus. / entering arrangement for ~ ‖ Gewebeeinführapparat m. ‖ appareil m. pour l'introduction du tissu. / extra-weft figured ~ ‖ lanzierter Stoff m. ‖ tissu m. lancé. / knitted ~ ‖ gewirkter Stoff m. ‖ tissu m. par mailles. / ~ open on one side ‖ einseitig offener Stoff m. ‖ étoffe f. ouverte d'un côté. / ~ for ready-made clothes ‖ Konfektionsstoff m. ‖ étoffe f. pour la confection. / rubberized ~ ‖ gummierter Stoff m. ‖ tissu m. caoutchouté. / tubular ~ ‖ Schlauchware f. ‖ tricot m. tubulaire. / ~ of worsted yarn ‖ Kammgarnstoff m. ‖ tissu m. peigné.

fabricate, to ‖ herstellen; verfertigen; fabrizieren ‖ fabriquer.

fabrication *see also* manufacture ‖ Fabrikation f.; Herstellung f.; Anfertigung f. ‖ fabrication f. / interchangeable ~ ‖ Austauscharbeit f. ‖ fabrication f. interchangeable. / ~ plant ‖ Fabrikanlage f. ‖ installation f. d'une fabrique.

fabric envelope ‖ Stoffhülle f. ‖ enveloppe f. en tissu *ou* en étoffe. / ~ guard ‖ Stoffschoner m. ‖ garde m. d'étoffe. / ~ inspecting and measuring machine (Weav) ‖ Warenschau- und Meßmaschine f. ‖ machine f. à visiter et mesurer les tissus. / ~ valve ‖ Stoffventil n. ‖ soupape f. en étoffe.

façade of a building ‖ Vorderseite f. eines Gebäudes; Fassade f. ‖ façade f. d'un édifice. / ~ washer ‖ Fassadenputzer m. ‖ ravaleur m.

face, to (Build) ‖ verkleiden; verblenden; belegen ‖ revêtir. / ~ (Mach) ‖ aufstellen; montieren; adjustieren ‖ dresser; ajuster. / ~ (Turn) ‖ flachdrehen; plandrehen ‖ dresser (un plateau); surfacer au tour. / ~ a baywork with bricks ‖ die Fachwand verblenden ‖ revêter les pans en brique.

face ‖ Gesicht n. ‖ visage m. / ~ (Arch) ‖ Front f.; Außenseite f.; Fassade f. ‖ façade f.; face f. / ~ (Clockm) ‖ Zifferblatt n. ‖ cadran m. / ~ (Letter-found) ‖ Gesicht n.; Kopf m. ‖ face f.; œil m. / ~ (Miner) ‖ Kristallfläche ‖ face f.; facette f.; plan m. / anvil ~ ‖ Amboßbahn f. ‖ face f. *ou* table f. de l'enclume. / at the ~ (Mine) ‖ vor Ort m. ‖ à la veine; à front de taille. / bold ~ (Print) ‖ fette Schrift f. ‖ type m. épais. / carved ~ (Mine) ‖ verschrämter Stoß m. ‖ paroi f. entaillée. / ~ of cleavage (Miner) ‖ Spaltungsfläche f. ‖ face f. de clivage. / ~ of the compass ‖ Rose f. des Kompasses ‖ rose f. de vent. / fat ~ (Print) ‖ fetter Schriftkegel m. ‖ corps m. gros. / lateral ~ of a wedge ‖ Seitenfläche f. eines Keils ‖ face f. latérale *ou* côté m. d'un coin. / ~ of a plane ‖ Sohle f. *oder* Bahn eines Hobels ‖ plan m. *ou* semelle f. d'un rabot. / ~ of a shaft (Mine) ‖ Schachtstoß m. ‖ face f. *ou* paroi f. d'un puits. / slide main ~ ‖ Schieberspiegel m. ‖ table f. de tiroir. / ~ of a solid ‖ Seitenfläche f. eines Körpers ‖ face f. d'un corps. / ~ of a stone ‖ Kopfseite f. eines Steines ‖ panneau m. de tête *ou* parement m. d'une pierre. / ~ of a tooth of a wheel ‖ Stirnfläche f. eines Radzahnes ‖ face f. de la dent d'une roue. / ~ of a wall ‖ Mauerfront f.; Außenseite f. einer Mauer ‖ parement m. *ou* côté m. frontal d'un mur. / ~ of working (Mine) ‖ Abbaustoß m. ‖ front m. *ou* fond m. de taille.

face and circular grinding machine ‖ Plan- und Rundschleifmaschine f. ‖ machine f. à rectifier les surfaces circulaires et planes. / ~ cam operating the long turning slides ‖ Mantelkurve f. für den Langvorschub ‖ came f. périphérique pour l'avance de chariotage.

faced ‖ verkleidet ‖ recouvert; revêti.

face lathe ‖ Scheibendrehbank f. ‖ tour m. à plateau. / ~ mould (Build) ‖ Lehre f. ‖ panneau m. / ~ plan (Build) ‖ Hauptansicht f. ‖ élévation f. principale. / ~ plate ‖ Planscheibe f. ‖ plateau m. de tour. / ~ powder ‖ Puder m.; Gesichtspuder m. ‖ poudre f. de riz. / ~ powder press ‖ Puderpresse f. ‖ presse f. à poudre de riz. / ~ shield ‖ Schutzschild m. ‖ protecteur m.; plaque f. protectrice; bouclier m. / ~ shield (Mask) ‖ Gesichtsmaske f.; Schutzmaske f. ‖ masque m. protecteur *ou* de sûreté.

facet ‖ Facette f.; Rautenfläche f.; Schleifseite f. an Glas und Steinen ‖ facette f.

face wheel ‖ Kronrad n. ‖ roue f. à dents de côté *ou* à couronne.

facility ‖ Leichtigkeit f. ‖ légèreté f.; facilité f.

facing (Turn) ‖ Plandrehen n. ‖ dressagem. d'un plateau. / ~ (Grinding) ‖ Planflächenschliff m. ‖ rectification f. de surfaces circulaires planes. / ~ in bricks ‖ Mauerwerkverblen dung f. ‖ parement m. en briques. / ~ of brake ‖ Bremsfutter m. ‖ garniture f. du frein. / clutch ~ ‖ Kupplungsbelag m. ‖ garniture f. d'embrayage. / ~ with cup wheel ‖ Planflächenschliff m. mit Schleiftopf ‖ rectification f. de surfaces circulaires planes avec meule boisseau. / ~ with disk wheel ‖ Stirnflächenschliff m. mit Schleifrad ‖ rectification f. de surfaces circulaires planes avec meule ordinaire. / ~ with fascines (Hydr arch) ‖ Berauhwehrung f.; Rauchwehr f. ‖ fascinage m. et clayonnage m. / finish ~ ‖ Planschlichten n. ‖ dressage-finissage m.

facing board ‖ Blendholz n. ‖ bois m. de placage. / ~ brick ‖ Verblendstein m.; Blendziegel m. ‖ brique f. de parement *ou* de revêtement. / glazed ~ brick ‖ glasierter Verblender m. ‖ brique f. glacée pour parement. / ~ cutter (Shoem) ‖ Stulpenzuschneider m. ‖ coupeur m. de revers. / ~ head ‖ Planfräskopf m. ‖ tête-fraise f. à surfacer. / ~ lathe ‖ Plandrehbank f. ‖ tour m. en l'air. / ~ marble ‖ Verblendmarmor m. ‖ marbre m. de parement. / ~ plastering ‖ Edelputz m. ‖ enduit m. imperméable. / ~ sand (Mould) ‖ feingesiebter Formsand m. ‖ sable m. fin de moulage. / ~ sewer (Tail) ‖ Besetzer m. ‖ attacheur m. / ~ stitcher (Shoem) ‖ Stulpenstepper m. ‖ piqueur m. de revers. / ~ stone ‖ Verblendstein m.; Fassadestein m. ‖ pierre f. de parement *ou* d'ornementation. / ~ wall ‖ Stirnmauer f. ‖ mur m. frontal.

factitious ‖ künstlich; nachgemacht ‖ artificiel; factice. / ~ gem ‖ künstlicher Edelstein m. ‖ pierre f. précieuse artificielle.

factor (Math) ‖ Faktor m. ‖ facteur m. / ~ of capacity ‖ Kapazitätsfaktor m. ‖ facteur m. de capacité. / ~ of safety ‖ Sicherheitsfaktor m. ‖ coefficient m. de sécurité.

factorial magnification ‖ Lupenvergrößerung f. ‖ grossissement m. à la loupe.

Factories and Workshops Act ‖ Gewerbeordnung f. ‖ loi f. sur le travail industriel.

factory ‖ Fabrik f.; Betrieb m. ‖ fabrique f.; usine f. / ~ (Trade) ‖ Handelsniederlassung f. ‖ factorerie f. / artificial stone ~ ‖ Kunststeinfabrik f. ‖ fabrique f. de pierres artificielles. / ~ of blank movements (Watchm) ‖ Rohuhrwerkfabrik f. ‖ fabrique f. d'horlogerie *ou* de mouvements en blanc; blantier m. / ~ of canned goods ‖ Konservenfabrik f. ‖ fabrique f. de conserves. / ~ of church organs ‖ Kirchenorgelbauanstalt f. ‖ fabrique f. d'orgues pour églises. / cork plate ~ ‖ Korkplattenfabrik f. ‖ fabrique f. de plaques de liège. / india rubber ~ ‖ Gummifabrik f. ‖ fabrique f. de caoutchouc. / ~ for preparing materials to be woven at home ‖ Webefaktorei f. für Hausweberei ‖ atelier m. de préparation pour tissage à domicile. / ~ for printing colours ‖ Druckfarbenfabrik f. ‖ fabrique f. de couleurs d'impression.

factory building ∥ Fabrikgebäude n. ∥ bâtiment m. d'usine. / ~ chimney ∥ Fabrikschornstein m. ∥ cheminée f. d'usine. / / ~ cost ∥ Selbstkosten pl. ∥ prix m. de revient. / ~ hand ∥ Fabrikarbeiter m. ∥ ouvrier m. d'usine ou de fabrique ou de manufacture. / ~ inspector ∥ Fabrikinspektor m. ∥ inspecteur m. de travail. / professional ~ inspector ∥ Betriebskontrollör m. ∥ contrôleur m. de fabrication. / ~ installation ∥ Fabrikeinrichtung f. ∥ installation f. d'usines. / ~ labourer ∥ Fabrikarbeiter m.· ∥ ouvrier m. d'usine ou de fabrique ou de manufacture. / ~ ladder ∥ Fabrikleiter f. ∥ échelle f. de fabrique. / ~ length ∥ Werklänge f.; Fabrikationslänge f. ∥ longueur f. de fabrication. / ~ number ∥ Fabrikationsnummer f. ∥ numéro m. de fabrication. / ~ plan for the conversion from peace to war footing ∥ Plan m. zur Werksumstellung auf Kriegsfertigung ∥ plan m. de l'usine pour le renversement du travail de paix en celui de guerre. / ~ plant see fabrication plant. / ~ price ∥ Fabrikpreis m. ∥ prix m. de fabrique. / ~ regulation ∥ Arbeitsordnung f. ∥ règlement m. d'atelier.

faculty of vision, apparatus for ascertaining the ~ ∥ Apparat m. zur Sehschärfenprüfung ∥ appareil m. pour la détermination de l'acuité visuelle.

fade, to ∥ den Glanz m. verlieren; verblassen; verschießen ∥ perdre son éclat; se ternir; se faner.

fading (Opt) ∥ Überblendung f. ∥ obturation f. / ~ (Radio) ∥ Schwunderscheinung f.; Fading n.; Fadingeffekt m. ∥ fading m.; effet m. de fading. / ~ control equipment ∥ Überblendungseinrichtung f. ∥ dispositif m. d'obturation. / ~ control unit (Sound film) ∥ Übergangsschaltung f. ∥ système m. de commutation. / ~ effect see fading (Radio).

faeces pl. ∥ Fäkalstoffe mpl. ∥ fèces fpl.; matières fpl. fécales.

fag-end (Mar) ∥ Hundeende n. ∥ œillet m. ou tistre m.

faggot ∥ Ried m.; Riedgras n. ∥ fagot m. / ~ (Metal) ∥ Schweißpaket n. ∥ paquet m. de fer à souder. / ~ (Wood) ∥ Holzbündel n.; Reisigbündel n. ∥ fagot m.; bourrée f. / ~ of iron bars ∥ Paket n. Eisenstäbe ∥ paquet m. ou trousse f. de fer en barres / magnetized ~ ∥ Magnetbündel n. ∥ faisceau m. aimanté. / ~ of old iron ∥ Paket n. Eisenabfälle ∥ fagot m. ou ramasse f. ou paquet de mitrailles. / small ~ ∥ kleines Reisigbündel n. ∥ margotin m. / ~ of steel ∥ Zange f. oder Paket von Stahlstäben ∥ trousse f. ou fagot m. ou paquet m. d'acier. / ~ of wood ∥ Paket n. von Reisholz ∥ paquet m. ou faisceau m. de menu bois.

faggot bundling press ∥ Bündelpresse f. ∥ presse f. à botteler. / ~ maker ∥ Holzbündelmacher m. ∥ fagotier m. / ~ sticks pl. ∥ Knüppelholz n.; Rundholz n. ∥ rondins mpl. / ~ trolley ∥ Spreißelwagen m. ∥ wagonnet m. pour transporter les fagots.

fagot see faggot.

fahlerz ∥ Fahlerz n. ∥ fahlerz m.

fahlunite ∥ Fahlunit m. ∥ fahlunite f.

Fahrenheit scale ∥ Gradeinteilung f. nach Fahrenheit ∥ échelle f. Fahrenheit. / ~ thermometer ∥ Fahrenheitthermometer n. ∥ thermomètre m. Fahrenheit.

faïence ∥ Fayence f.; Halbporzellan n. ∥ faïence f. / common ~ ∥ gemeine Fayence f. ∥ faïence f. commune.

fail, to ∥ mißlingen ∥ ne pas réussir; échouer. / ~ (Trade) ∥ in Konkurs m. geraten; fallieren ∥ tomber en faillite f.; faillir.

failed piece ∥ mißratenes Stück n. ∥ pièce f. manquée.

failing ∥ mangels ∥ à défaut de; faute de.

failure ∥ Mißerfolg m. ∥ insuccès m. / ~ (Trade) ∥ Konkurs m. ∥ faillite f. / to find one's self in difficulties owing to the ~ of a bank ∥ durch den Zusammenbruch m. einer Bank sich in Schwierigkeiten befinden ∥ se trouver aux prises avec des difficultés résultant d'un krach de banque. / ~ of a firm ∥ Zusammenbruch m. einer Firma ∥ faillite f. d'une maison de commerce. / ~ of a wall ∥ Mauereinsturz m. ∥ rupture f. ou écroulement m. d'un mur.

fainting ∥ Ohnmacht f. ∥ évanouissement m.

faints pl., weak ~ (Distill) ∥ Nachlauf m. ∥ après-coulant m.; eau-de-vie f. dernière; repasses fpl.

fair ∥ Jahrmarkt m. ∥ foire f. / annual ~ ∥ Jahresmesse f. ∥ foire f. annuelle. / sample ~ ∥ Mustermesse f. ∥ foire f. aux échantillons (spécimens). / technical ~ ∥ technische Messe f. ∥ foire f. technique.

fair copy ∥ Reinschrift f. ∥ copie f. au net; expédition f. à signer. / ~ dealing ∥ Reellität f. ∥ solidité f.; honnêteté f. / ~ leader (Shipb) ∥ Scherbrett n. ∥ râteau m.; conduit m.

fairing plate ∥ Verkleidungsblech n. ∥ tôle f. de carénage.

fairness ∥ Ehrlichkeit f. ∥ honnêteté f.

fair price ∥ annehmbarer Preis m. ∥ prix m. raisonnable.

fairway ∥ Fahrwasser n.; Fahrrinne f. ∥ passe f.; chenal m.; passage m. / ~ arch (Bridge) ∥ Schiffsöffnung f. ∥ arche f. marinière. / ~ buoy ∥ Ansegelungsboje f.; Fahrwassertonne f. ∥ bouée f. d'atterrage ou de direction.

fall, to ∥ fallen ∥ tomber; descendre. / ~ (Build) ∥ einstürzen ∥ s'écrouler; ébouler. / ~ back (Brew) ∥ durchfallen ∥ tomber. / ~ to the bottom (Chem) ∥ sich setzen ∥ se déposer. / ~ down (Chem) ∥ niederfallen ∥ se déposer; être déposé. / ~ down to the bottom ∥ sich abscheiden; sich absetzen ∥ se déposer; être déposé. / ~ down a river with the tide (Nav) ∥ absacken; sich vom Strome treiben lassen ∥ descendre une rivière en dérivant. / ~ in (Build; Mine) ∥ einfallen ∥ s'écrouler; s'enfoncer; s'ébouler. / ~ in (Locksm) ∥ einschnappen ∥ encliqueter. / to cause the roof of a mine to ~ in ∥ den Bruch m. niedergehen lassen oder einstürzen ∥ ébouler ou faire ébouler le débris. / ~ to leeward (Nav) ∥ leewärts abtreiben ∥ tomber sous le vent. / ~ out of step (Electr) ∥ außer Tritt m. fallen ∥ perdre le synchronisme.

fall ∥ Gefälle n.; Neigung f.; Steigung f. ∥ pente f.; remont f.; déclivité f. / ~ (Aero) ∥ Absturz m. ∥ chute f.; vol m. piqué. / ~ (Phys) ∥ Fall m. ∥ chute f. ∥ electric ~ ∥ Gefälle n.; Spannungsunterschied m. ∥ chute f. électrique. / free ~ ∥ der freie Fall ∥ chute f. libre. / height of ~ ∥ Fallhöhe f. ∥ hauteur f. de chute. / ~ of the main-top-gallant halliard ∥ Großbramfall n. ∥ garant m. de la drisse

du grand perroquet. / ~ of potential ∥ Potentialabfall m. ∥ chute f. de potentiel. / ~ of pressure ∥ Druckabfall m. ∥ chute f. ou diminution de pression. / ~ of a river ∥ Stromgefälle n. ∥ pente f. d'un fleuve. / ~ of the rock (Mine) ∥ Hereinbrechen n. des Gebirges ∥ dévers m. / steep ~ ∥ steiler Abfall m. ∥ descente f. précipitée. / ~ of temperature ∥ Temperaturabfall m. ∥ chute f. de température. / ~ of a wall ∥ Mauereinsturz m. ∥ rupture f. ou écroulement m. d'un mur. / compressor working by ~ of water ∥ Hydrokompressor m. ∥ compresseur m. à chute d'eau.

fallacy ∥ Trugschluß m. ∥ fausse conclusion f.

falling back of kräusen (Brew) ∥ Durchfallen n. der Kräusen ∥ chute f. des kräusen. / ~-down ∥ Zusammensturz m. ∥ éboulement m. / ~ gradient ∥ Gefälle n. ∥ pente f. / ~-in (Min) ∥ zu Bruche gehen ∥ éboulement m. / ~ latch ∥ Fallklinke f.; Fallriegel m. ∥ loqueteau m.; loquet m. / ~ stone ∥ Meteorstein m. ∥ aérolithe m.; météorolithe m.; pierre f. météorique; météorite f.

fall proof ∥ Schlagprobe f. ∥ épreuve f. au choc. / ~ rate ∥ Fallgeschwindigkeit f. ∥ vitesse f. de chute.

false (Arch) ∥ falsch; nachgeahmt ∥ faux; imité; feint; orbe; simulé. / ~ ceiling ∥ Fehlboden m.; Einschub m.; Schragboden m. ∥ plancher m. de remplissage. / ~ collar ∥ Hemdkragen m. ∥ col m. de chemise; faux-col m. / ~ core ∥ Kernstück n. ∥ pièce f. rapportée. / ~ flames pl. ∥ bengalisches Feuer n. ∥ feux mpl. de bengale. / ~ keel ∥ Loskiel m. ∥ fausse quille f. / ~ key ∥ Dietrich m.; Nachschlüssel m. ∥ fausse clef f. / ~ ogive type of projectile ∥ Haubengeschoß m.; Geschoß n. mit falscher Bogenspitze ∥ projectile m. à fausse ogive. / ~ pearl ∥ unechte Perle f. ∥ perle f. fausse. / ~ zero (Electr) ∥ falscher Nullpunkt m. ∥ faux zéro m.

famatinite ∥ Famatinit m. ∥ famatinite f.

family accommodation ∥ Familienwohnung f. ∥ habitation f. de famille. / ~ allowance ∥ Familienzulage f. ∥ allocation f. de famille.

famine ∥ Hungersnot f. ∥ famine f.

fan ∥ Fächer m. ∥ éventail m. / ~ (Agr) ∥ Kornschwinge f.; Schwingwanne f.; Staubmühle f.; Wurfmaschine f. ∥ van m. émotteur; sabbat m. / ~ (Mach) see also ventilator ∥ Windflügel m.; Ventilator m. ∥ ventilateur m. / belt-driven ~ ∥ Ventilator m. mit Riemenantrieb ∥ ventilateur m. commandé par courroie. / blast ~ see blowing ~. / blowing ~ ∥ Blasventilator m.; Gebläse n. ∥ ventilateur m. soufflant. / centrifugal ~ ∥ Zentrifugalventilator m. ∥ ventilateur m. centrifuge. / electric ~ ∥ Elektroventilator m.; elektrischer Ventilator m. ∥ ventilateur m. électrique. / flywheel ~ ∥ Schwungradventilator m. ∥ ventilateur m. à volant. / gear-driven ~ ∥ Ventilator m. mit Zahnradantrieb ∥ ventilateur m. commandé par engrenage. / helicoidal ~ ∥ Schraubenventilator m. ∥ ventilateur m. hélicoïdal. / induced draught ~ ∥ Saugzugventilator m. ∥ aspirateur m. pour tirage induit. / mine ~ ∥ Wettermaschine f. ∥ ventilateur m. de mine / non-folding ~ ∥ unzusammenklappbarer Fächer m. ∥ écran m. à main. / propeller ~ ∥ Schrauben-

ventilator m. ǁ ventilateur m. hélicoïdal. / radiator ~ ǁ Kühlerventilator m. ǁ ventilateur m. du radiateur. / rotary ~ / Rotationsventilator m. ǁ ventilateur m. rotatif. / ~ acting by suction with a delivery per minute of x cubic metres ǁ saugend wirkender Ventilator m. mit x cbm Leistung in der Minute ǁ ventilateur m. aspirant avec un rendement de x m⁰ par minute.
fan aerial ǁ Fächerantenne f. ǁ antenne f. en éventail. / ~ belt ǁ Ventilatorriemen m.; Windflügelriemen m. ǁ courroie f. du ventilateur. / ~ belt fastener ǁ Ventilatorriemenschloß n. ǁ agrafe f. de courroie du ventilateur. / ~ blade ǁ Windflügelblech n. ǁ ailette f. de ventilateur. / ~ blower ǁ Flügelgebläse n.; Ventilator m. ǁ ventilateur m. / ~ blower mixer ǁ Schleuderradmischer m. ǁ mélangeur m. centrifuge. / ~ bracket ǁ Ventilatorstütze f. ǁ support m. du ventilateur. / ~ brake ǁ Flügelbremse f. ǁ frein m. à moulinet.
fancet plug ǁ Ansteckspund m. ǁ bonde-certaine f.; certaine f.
fancy (Weav) ǁ gemustert ǁ façonné.
fancy ǁ Neuheit f. ǁ nouveauté f. / ~ article ǁ Fantasiegegenstand m. ǁ article m. de fantaisie. / ~ button ǁ Fantasieknopf m. ǁ bouton m. fantaisie. / ~ cake cutter ǁ Tortenheber m. ǁ pelle f. à gâteau. / ~ cardboard-box ǁ Luxuskartonnage f. ǁ cartonnage m. de luxe. / ~ comb ǁ Galanteriekamm m. ǁ peigne m. fantaisie. / ~ cord ǁ Schnürsenkel m. ǁ cordelière f.; lacet m. / ~ dash (Print) ǁ Zierlinie f. ǁ filet m. orné. / ~ dress ǁ Maskenkostüm n. ǁ costume m. de bal masqué. / ~ feather ǁ Schmuckfeder f. ǁ plume f. d'ornement ou de parure. / dressed ~ feather ǁ zugerichtete Schmuckfeder f. ǁ plume f. de parure apprêtée. / ~ frame (Print) ǁ Fantasieeinfassung f. ǁ cadre m. de fantaisie. / ~ furniture ǁ Luxusmöbel n. ǁ meuble m. de luxe. / ~ glassware ǁ Zierglaswaren pl. ǁ verrerie f. de décoration.
fancy goods pl. ǁ Galanteriewaren fpl.; articles mpl. de fantaisie. / ~ ǁ Modewaren fpl. ǁ modes fpl.; nouveautés fpl. / ~ ǁ Nippsachen fpl. ǁ bibelots mpl. / ~ turner ǁ Kunstdrechsler m. ǁ tourneur m. en tabletterie.
fancy hat ǁ Fantasiehut m. ǁ chapeau m. fantaisie. / ~ hosiery ǁ Fantasiestrumpfwaren fpl. ǁ bonneterie f. fantaisie. / ~ knob ǁ Zierkugel f. ǁ boule f. d'ornementation. / ~ leather goods pl. ǁ Ledergalanteriewaren fpl. ǁ articles mpl. de fantaisie en cuir. / ~ letters pl. ǁ Zierschrift f. ǁ caractères mpl. de fantaisie. / ~ nail ǁ Dekorationsnagel m.; Ziernagel m. ǁ clou m. de fantaisie. / ~ needle ǁ Schmucknadel f. ǁ aiguille f. de fantaisie.
fancy paper ǁ Fantasiepapier n. ǁ papier m. de fantaisie. / ~ decoration ǁ Fantasiepapierausstattung f. ǁ décoration f. par papier de fantaisie. / ~ making machine ǁ Luxuspapierherstellungsmaschine f. ǁ machine f. à fabriquer du papier de luxe.
fancy porcelain ǁ Luxusporzellan n. ǁ porcelaine f. de luxe. / ~ roller (Spinn) ǁ Schnellwalze f.; Trommelputzwalze f. ǁ volant m. / ~ scent ǁ Blütenöl n. ǁ essence f. de fleurs. / ~ shop ǁ Galanteriewarenhandlung f. ǁ magasin m.

d'articles de luxe. / ~ silk spinning ǁ Seidenschoddyspinnerei f. ǁ filature f. de soie fantaisie. / ~ sock ǁ Fantasiesocke f.; gemusterte Socke f. ǁ chaussette f. fantaisie. / ~ stitcher (Shoem) ǁ Ornamentstepper m. ǁ piqueur m. de fantaisie. / ~ stitcher (Weav) ǁ Fantasiewirker m. ǁ tricoteur m. en fantaisie. / ~ straw articles pl. ǁ Fantasiewaren fpl. aus Stroh ǁ agréments mpl. en paille. / ~ threader ǁ Drahtumspinner m. ǁ guipeur m. de fil. / ~ twill ǁ verzierter Köper m. ǁ croisé m. façonné ou de fantaisie. / ~ weaving ǁ Bildweberei f.; Musterweberei f. ǁ tissage m. d'étoffes façonnées. ǁ ~ work frame ǁ Stickrahmen m. ǁ cadre m. pour broderie. / ~ yarn ǁ Kunstgarn n.; Zierfaden m. ǁ fil m. fantaisie.
fan delivery ǁ geförderte Luftmenge f. ǁ rendement m. en air soufflé. / ~ driving ǁ Ventilatorantrieb m. ǁ commande f. du ventilateur. / ~ driving chain ǁ Ventilatorantriebskette f. ǁ chaîne f. de commande du ventilateur. / ~ driving pulley ǁ Ventilatorriemenscheibe f. ǁ poulie f. de ventilateur. / ~ embroiderer ǁ Fächerstickerin f. ǁ brodeuse f. sur éventails. / ~ engine man (Mine) ǁ Ventilatorwärter m. ǁ conducteur m. de ventilateur m. / ~ fitter ǁ Fächerzurichter m. ǁ monteur m. d'éventails. / ~ frame ǁ Fächergestell n. ǁ monture f. d'éventail.
fanbolt ǁ durchgehender Ankerbolzen m. ǁ boulon m. d'ancrage traversant ou de scellement ordinaire.
fan hub ǁ Windflügelnabe f. ǁ moyeu m. de ventilateur. / ~ light ǁ Guckfenster m. ǁ vasistas m. / ~ maker ǁ Fächerfabrikant m. ǁ éventaillier m. / ~ motor ǁ Ventilator m. ǁ ventilateur m. / radial ~ motor ǁ Radialfächermotor m. ǁ moteur m. pour ventilateur radial.
fanner (Agr) ǁ Kornreinigungsmaschine f.; Kornschwinge f.; Staubmühle f. ǁ émotteur m.; sabbat m. / ~ (Mine) ǁ Grubenventilator m.; Wettermaschine f.; Wettertrommel f. ǁ machine f. aspirante ou ventilateur m. pour l'aérage des mines; ventilateur m.
fanning machine (Agr) see fanner.
fan-shaped ǁ fächerförmig ǁ en forme f. d'éventail. / ~ antenna ǁ Fächerluftleiter m. ǁ antenne f. en éventail.
fan support ǁ Ventilatorstütze f. ǁ support m. de ventilateur.
fan wheel ǁ Flügelrad n.; Ventilatorrad n. ǁ tourniquet m. de ventilateur. / ~ for mine ventilators ǁ Flügelrad n. für Grubenventilatoren ǁ moulinet m. pour ventilateurs à mines. / ~ anemograph ǁ Windradanemograf m. ǁ anémographe m. à moulinet.
farad ǁ Farad n. ǁ farad m.
Faraday's disc ǁ Faraday's Scheibe f. ǁ disque m. de Faraday.
faradic ǁ faradisch ǁ faradique.
fare ǁ Fahrgeld n. ǁ prix m. de passage.
farina ǁ Kartoffelmehl n.; Kartoffelstärke f. ǁ fécule f.
farm ǁ Pachtgut n.; Meierei f. ǁ ferme f.; métairie f. / ~ rented for half the products ǁ Halbpachtwirtschaft f. ǁ métairie f.; métayage f. / ~ administrator ǁ Gutsverwalter m. ǁ administrateur m. ou régisseur m. d'une ferme.
farmer ǁ Bauer m. ǁ paysan m. / ~ ǁ Landwirt m. ǁ cultivateur m. / ~ ǁ Pächter m.

ǁ fermier m. / ~ line (Tel) ǁ Farmerleitung f. ǁ ligne f. rurale.
farmer's apprentice ǁ Landwirtschaftsstudent m. ǁ élève m. agriculteur. / ~ workshop ǁ Farmerwerkstatt f. ǁ atelier m. de fermier.
farm hand ǁ Landarbeiter m. ǁ ouvrier m. agricole.
farming ǁ Ackerbau m.; Landbau m.; Landwirtschaft f.; landwirtschaftlicher Betrieb m. ǁ agriculture f.; économie f. rurale; exploitation f. agricole. / ~ building ǁ Wirtschaftsgebäude n. ǁ bâtiment m. d'exploitation rurale. / ~ implements pl. ǁ Ackergerät n. ǁ outils mpl. aratoires ou de labourage. / ~ machine ǁ landwirtschaftliche Maschine f. ǁ machine f. agricole.
farm irrigation ǁ Farmbewässerung f. ǁ irrigation f. pour ferme. / ~ land ǁ Ackerboden m. ǁ sol m.; terroir m.; terre f. arable ou labourable ou végétable. / ~ manager ǁ Gutsverwalter m. ǁ administrateur m. ou régisseur m. d'une ferme. / ~ owner ǁ Grundbesitzer m. ǁ propriétaire m. foncier ou terrien. / ~ school ǁ Ackerbauschule f. ǁ ferme-école f. / ~ servant ǁ Ackerknecht m.; Knecht m. ǁ valet m. ou garçon m. de ferme. / ~ tractor ǁ Zugmaschine f. für die Landwirtschaft ǁ tracteur m. agricole. / ~ worker ǁ Landarbeiter m. ǁ ouvrier m. agricole. / ~ yard ǁ Pachthof m.; Meierhof m.; Wirtschaftshof m. ǁ ferme f.; métairie f.
faröelite ǁ Faröelith m. ǁ mésole f.; féroëlite f.
far-point ǁ Fernpunkt m. ǁ punctum m. remotum. / ~ finder ǁ Fernpunktsucher m. ǁ chercheur m. du punctum remotum. / ~ surface of a long-sighted eye ǁ Fernpunktfläche f. eines übersichtigen Auges ǁ lieu m. du punctum remotum d'un œil hypermétrope.
farrier ǁ Hufschmied m. ǁ maréchal-ferrant m.; maître-ferrant.
farrier's pincers pl. ǁ Hufzange f. ǁ tricoises fpl. / ~ school ǁ Lehrschmiede f. ǁ école f. de maréchalerie.
farriery ǁ Hufschmiede f. ǁ maréchalerie f.; forge f. maréchale.
far-sighted ǁ weitblickend ǁ prévoyant. / ~ (Opt) ǁ weitsichtig ǁ presbyte.
farthest point ǁ Endpunkt m. ǁ extrémité f.; point m. terminus.
fascine ǁ Faschine f. ǁ fascine f. / headed ~ ǁ Kopffaschine f. ǁ fascine f. de retraite. / pitched ~ ǁ Pechfaschine f. ǁ fascine f. goudronnée. / causeway of ~s ǁ Faschinendamm m. ǁ pont m. ou digue f. de fascines. / layers of ~s ǁ Packfaschinen fpl. ǁ fascinages mpl.
fascine choker (Hydr arch) ǁ Reitel m.; Würgetau n. ǁ cabestan m. / ~ dike ǁ Faschinendamm m, ǁ pont m. ou digue f. de fascines. / ~ revetment ǁ Faschinenbekleidung f. ǁ revêtement m. en fascines. / ~ wood ǁ Reisholz n.; Faschinenholz n. ǁ bois m. de fascinage. / ~ work ǁ Faschinenwerk n. ǁ fascinage m.
fash of seams (Found) ǁ Grat m.; Guß naht f. ǁ ébarbure f.
fashion, to ǁ formen; gestalten ǁ façonner; former; modeler.
fashion ǁ Tracht f. ǁ costume m.; mode f.; charge f.
fashionable ǁ modern ǁ moderne; à la mode.

fashion art ‖ Modekunst f. ‖ art m. de mode. / ~ articles pl. ‖ Modeartikel mpl. ‖ articles mpl. de mode. / ~ button ‖ Modeknopf m. ‖ bouton m. de mode.

fashion piece (Shipb) ‖ Randsomholz n. ‖ estain m. / top-timber of the ~ (Shipb) ‖ Heckseitenstütze f.; Heckstütze f. ‖ cornière f.; allonge f. de cornière.

fassaite ‖ Fassait m. ‖ fassaïte f.

fast ‖ fest ‖ fixe; ferme. / ~ ‖ schnell ‖ rapide. / ~ (Mach) ‖ ortsfest; stationär ‖ stationnaire. / ~ (Textile) ‖ waschecht ‖ bon teint. / ~ cargo vessel ‖ Eilgüterschiff n. ‖ bateau m. pour service accéléré. / ~ countershaft pulley ‖ Festscheibe f. am Deckenvorgelege ‖ poulie f. fixe du renvoi. / ~ goods traffic ‖ Eilgutverkehr m. ‖ trafic m. des marchandises à grande vitesse. / ~ pulley ‖ Festscheibe f. ‖ poulie f. fixe. / ~ recorder ‖ Schnellschreiber m. ‖ transmetteur m. à grande vitesse. / ~ red ‖ Echtrot n. ‖ rouge m. solide. / ~ roll ‖ festgelagerte Walze f. ‖ cylindre m. fixe. / ~ sailer ‖ Schnellsegler m. ‖ fin voilier m.; navire m. bon marcheur à voiles. / ~ sailing vessel see ~ sailer.

fasten, to ‖ anknüpfen ‖ attacher; nouer. / ~ (Join) ‖ anschlagen; befestigen ‖ affermir. / ~ (Mach) ‖ feststellen ‖ arrêter; fixer. / ~ (Mar) ‖ festmachen ‖ amarrer; encapeler (un cordage). / ~ (Rope) ‖ festbinden ‖ fixer à l'aide de cordes; serrer. / ~ with bolts ‖ verbolzen ‖ cheviller. / ~ with pales see ~ with stakes. / ~ with stakes ‖ pfählen; anpfählen; verpfählen ‖ échalasser. / ~ the tyre by spring ring ‖ den Radreifen m. mittels Sprengrings sichern ‖ agrafer le bandage. / ~ by wedges ‖ festkeilen ‖ claveter; caler; assujettir.

fastener (Locksm) ‖ Schubriegel m. ‖ barre f.; targette f.; verrou m. (glissant). / ~ (Window) ‖ Wirbel m. ‖ tourniquet m.; crémone f. / belt ~ ‖ Riemenschloß n. ‖ attache f. pour courroies. / hood ~ (Auto) ‖ Motorhaubenverschluß m. ‖ fermeture f. de capot. / label and ticket ~ ‖ Kollianhänger m. ‖ étiquette f. pour colis. / office ~ ‖ Büronadel f. ‖ épingle f. de bureau.

fastener setter (Shoem) ‖ Ösenschlager m. ‖ poseur m. d'œillets.

fastening ‖ Anknüpfung f. ‖ liaison f. / ~ (Build) ‖ Befestigungsorgan n. ‖ pièce f. d'attache. / ~ (Mach) ‖ Sicherung f. ‖ crampon m. / ~ fish plates to the ends of the rail ‖ Befestigung f. der Laschen an den Schienenenden ‖ attache f. des éclisses aux bouts des rails. / lateral ~ (Build; Mach) ‖ Querverbindung f. ‖ entretoise f. / manner of ~ ‖ Befestigungsart f. ‖ manière f. de fixation. / ~ for preserve glasses ‖ Konservenglasverschluß m. ‖ fermeture f. pour verres de conserves. / ~ the rails to the sleepers with bolt and spikes ‖ Nagelung f. der Schienen ‖ clouage m. des rails. / ~ of the springs ‖ Federbefestigung f. ‖ fixation f. des ressorts. / ~ of track ‖ Gleisbefestigung f. ‖ fixation f. des rails.

fastening iron ‖ Moniereisen n. ‖ fer m. de consolidation. / ~ key ‖ Befestigungskeil m. ‖ clavette f. de fixation. / ~ ring (Tyre) ‖ Schließring m.; Sprengring m. ‖ cercle m. à simple agrafage; agrafage m. annulaire.

fastenings pl. (Locksm) ‖ Schließbeschläge n. ‖ fermeture f.

fastening strip ‖ Befestigungsstreifen m. ‖ bande f. de fixation.

fastidious ‖ anspruchsvoll; wählerisch ‖ exigeant; prétentieux.

fat ‖ schlüpfrig; fettig ‖ adipeux. / ~ (Coal) ‖ fett ‖ gras; bitumineux.

fat ‖ Fett n. ‖ graisse f. / animal ~ ‖ tierisches Fett n. ‖ graisse f. (d'origine) animale. / bone ~ ‖ Knochenfett n. ‖ suif f. d'os. / ~ for cooking ‖ Speisefett n. ‖ graisse f. comestible. / edible ~ ‖ Speisefett n. ‖ graisse f. alimentaire ou de table. / fulling ~ ‖ Walkfett n. ‖ graisse f. de foulage. / to grow ~ (Soapm) ‖ zäh werden ‖ s'engraisser. / leather ~ ‖ Lederfett n. ‖ graisse f. pour cuir. / beef marrow ~ ‖ Rindermarkfett n. ‖ graisse f. de moelle de bœuf. / melted ~ ‖ Schmalz n. ‖ graisse f. fondue. / perfumed ~ ‖ wohlriechendes Fett n. ‖ graisse f. parfumée. / spraying ~ ‖ Spritzfett n. ‖ graisse f. à jet. / thick ~ ‖ konsistentes Fett n. ‖ graisse f. consistante. / vegetable ~ ‖ Pflanzenfett n. ‖ vegetabilisches Fett n. ‖ graisse f. végétale. / wagon ~ ‖ Wagenfett n. ‖ graisse f. pour voitures.

fat chopping machine ‖ Fettzerkleinerungsmaschine f. ‖ machine f. à couper la graisse. / ~ coal ‖ Fettkohle f. ‖ charbon m. gras. / ~ determination ‖ Fettbestimmung f. ‖ détermination f. ou dosage m. des matières grasses. / ~ drop ‖ Fettropfen m. ‖ gouttelette f. de graisse.

fat-faced (Print) ‖ fett ‖ plein. / ~ type (Print) ‖ fette Schrift f. ‖ caractère m. plein.

fat globule ‖ Fettkorn n. ‖ globule m. gras.

fathom (Mar) ‖ Faden m. ‖ brasse f. / ~ (Mine) ‖ Lachter m.; n. ‖ brasse f.; toise f.

fatigue ‖ Ermüdung f. ‖ fatigue f. / dielectric ~ ‖ dielektrische Nachwirkung f. ‖ fatigue f. diélectrique. / ~ of metal ‖ Ermüdung f. des Metalls ‖ fatigue f. du métal.

fat knife ‖ Fettstecher m. ‖ sonde f. pour graisses. / ~ letter ‖ fetter Buchstabe m. ‖ caractère m. plein. / ~ lime ‖ Fettkalk m. ‖ chaux f. grasse. / ~ soap ‖ Fettseife f. ‖ savon m. d'axonge. / ~ -splitting ‖ fettspaltend ‖ dédoublant les graisses.

fatten, to ‖ mästen ‖ engraisser.

fattening pasture (Agr) ‖ Mast f. ‖ engrais m. / ~ stable (Shipb) ‖ Mastkäfig m. ‖ épinette f.

fat-tight ‖ fettdicht ‖ imperméable aux graisses.

fat tracing paper for pencils ‖ Fettreproduktionspapier n. für Schreibstifte ‖ papier m. gras à décalquer pour crayons.

fatty ‖ fett; fettig; ölig ‖ gras; graisseux.

fatty acid ‖ Fettsäure f. ‖ acide m. gras. / ~ saponified by bases ‖ mittels Laugen verseifte Fettsäure f. ‖ acide m. gras saponifié au moyen d'alcalis.

fatty acid distillation plant ‖ Fettsäuredestillationsanlage f. ‖ installation f. de distillation des acides gras. / ~ distiller ‖ Fettsäuredestillierer m. ‖ distillateur m. d'acides gras.

fatty aspect ‖ Fettglanz m. ‖ aspect m. graisseux. / ~ goods pl. ‖ Fettwaren fpl. ‖ graisses fpl. / ~ matter see ~ substance. / ~ substance ‖ Fettbestandteil

m.; Fettstoff m. ‖ graisse f.; corps m. gras.

faucet ‖ Faßhahn m. ‖ robinet m.; chantepleure f. / ~ from pure nickel ‖ Hahn m. aus Reinnickel ‖ robinet m. en nickel pur. / ~ joint (Mach) ‖ eingelassener Rand m.; Randverbindung f. ‖ couture f. à clin. / ~ pipe ‖ Muffenrohr n. ‖ tuyau m. à emboîtement.

faujasite ‖ Faujasit m. ‖ faujasite f.

fault ‖ Fehler m.; Versehen n. ‖ faute f. / ~ (Geol) ‖ Verwerfung f. ‖ rejet m. / ~ (Mine) ‖ Gangspalte f.; Spalte f. ‖ fente f.; fissure f. / ~ of balancing (Tel) ‖ Nachbildungsfehler m. ‖ défaut m. d'équilibrage. / ~ of construction ‖ Konstruktionsfehler m. ‖ défaut m. de construction. / downcast ~ (Mine) ‖ Verwerfung f. in die Tiefe ‖ renfoncement m. / there was this ~ inherent in the piece that it broke easily ‖ das Stück hatte den Fehler, leicht zu reißen ‖ la pièce avait le défaut de casser facilement. / ~ due to torsion (Mine) ‖ Drehverwerfung f. ‖ rejet m. par torsion. / upcast ~ (Mine) ‖ Verwerfung f. in die Höhe ‖ relèvement m. / ~ in weaving ‖ Webfehler m. ‖ défaut m. de tissage.

fault block (Geol) ‖ Horst m. ‖ horst m. / ~ detector ‖ Erdschlußsucher m. ‖ indicateur m. de pertes à la terre. / ~ fissure ‖ Bewegungsspalt m. ‖ fente f. par mouvement du sol.

faultless ‖ fehlerfrei; tadelfrei; tadellos ‖ irréprochable; sans défauts mpl.

fault, localisation of ~ (Electr) ‖ Fehlerortsbestimmung f. ‖ localisation f. du dérangement. / office for removal of ~s (Tel) ‖ Störungsstelle f. ‖ bureau m. pour la réparation des dérangements.

faultsman (Tel) ‖ Störungssucher m. ‖ surveillant m.

fault trough (Geol) ‖ Graben m. ‖ renforçage m.

faulty ‖ fehlerhaft; schadhaft; mangelhaft ‖ endommagé; défectueux; avarié. / ~ piece ‖ fehlerhaftes Stück n. ‖ pièce f. défectueuse. / ~ wire (Tel) ‖ Fehlerader f. ‖ fil m. défectueux.

fauserite ‖ Fauserit m. ‖ fauserite f.

favor see favour.

favour ‖ Vergünstigung f. ‖ faveur f. / ~ (Article) ‖ Ansteckartikel m.; Geschenkartikel m.; Reklamezugabeartikel m. ‖ fleur f. ou insigne m. à mettre à la boutonnière; article m. de prime. / to confer a ~ ‖ eine Vergünstigung f. erweisen ‖ accorder un avantage m. / in ~ of ‖ zu Gunsten von ‖ en faveur f. de.

fawn coloured ‖ rehbraun ‖ fauve.

fayalite ‖ Fayalit m.; Eisenchrysolit m. ‖ fayalite f.; péridot m. ferrugineux.

faying surface (Mach) ‖ Dichtungsfläche f. ‖ surface f. de contact.

fearnaught (Weav) ‖ Fries m.; Flaus m. ‖ frise f.; drap m. frisé; montagnac m.

feasibility ‖ Durchführbarkeit f.; Ausführbarkeit f. ‖ possibilité f. d'exécution.

feasible ‖ tunlich; ausführbar ‖ faisable; réalisable; exécutable.

feather, to ~ the oars (Mar) ‖ die Riemen mpl. platt werfen ‖ dévirer les avirons mpl.

feather ‖ Feder f. ‖‖ plume f. / ~ (Carp) ‖ dünner Spund m. ‖ languette f. / ~ (Mach) ‖ Rippe f. ‖ nervure f. / ~ (Mech) ‖ Federkeil m. ‖ clavette f. linguiforme ou à rainure. / bed ~ manufacturing

machine ‖ Bettfedernverarbeitungsmaschine f. ‖ machine f. à apprêter des duvets de lit. / ~ for carnivals ‖ Karnevalfeder f. ‖ plume f. de carneval. / down ~ ‖ Flaumfeder f. ‖ plume f. à duvet. / fancy ~ ‖ Fantasiefeder f.; Schmuckfeder f. ‖ plume-fantaisie f.; plume f. d'ornement. / dressed fancy ~ ‖ zugerichtete Schmuckfeder f. ‖ plume f. de parure apprêtée. / ~ glued upon tissue ‖ auf Gewebe aufgeklebte Feder f. ‖ plume f. collée sur tissu. / ~ for ladies' hats ‖ Damenhutfeder f. ‖ plume f. pour chapeaux de dames. / ostrich ~ ‖ Straußfeder f. ‖ plume f. d'autruche. / plain ~ ‖ geradstirnige Paßfeder f. ‖ clavette f. plate encastrée. / quill ~ ‖ Kielfeder f. ‖ plume f. à tige. / ~ including quill ‖ Feder f. einschließlich Federkiel ‖ plume f. y compris le tuyau de plume. / round-ended ~ ‖ rundstirnige Paßfeder f. oder Gleitfeder f. ‖ clavette f. coulissante ou plate encastrée.

feather alum ‖ Federalaun m. ‖ alun m. de plume. / ~ articles pl. ‖ Waren fpl. aus Federn ‖ articles mpl. en plumes. / ~ beard ‖ Federfahne f. ‖ barbe f. de plume. / ~ bleaching ‖ Putzfederbleicherei f. ‖ décoloration f. de plumes pour parures. / ~ broom ‖ Federbesen m. ‖ plumeau m. / ~ broom stick ‖ Federbesenstiel m. ‖ manche m. de plumeau. / ~ brush see ~ broom. / ~ cleaner ‖ Federreiniger m. ‖ épureur m. de plumes. / ~ cloud ‖ Federwolke f. ‖ nuage m. très léger et floconneux. / ~ curler ‖ Federkrauslerin f. ‖ friseuse f. de plumes. / ~ duster see ~ broom.

feathered shot (Metal) ‖ Federkupfergranalien fpl. ‖ cuivre m. en plumes.

feather fastener ‖ Federkleberin f. ‖ colleuse f. de plumes. / ~ furs pl. ‖ Federpelzwaren fpl. ‖ fourrures fpl. en plumes. / ~ sewer ‖ Federnäherin f. ‖ couseuse f. de plumes. / ~ shag ‖ Felbel m. ‖ peluche f. long-poil. / silk ~ shag ‖ seidener oder echter Felbel m. ‖ peluche f. long-poil de soie. / woollen ~ shag ‖ wollener Felbel m. ‖ peluche f. long-poil de coton; panne f. de laine. / ~ tuft ‖ Federbusch m. ‖ panache m. de plumes. / ~ whale bone ‖ Federfischbein n. ‖ baleine f. de plume. / ~ whisk ‖ Staubwedel m. ‖ plumeau m. / ~ wind mill ‖ Federwindmühle f. ‖ moulinet m. en plumes. / ~ zeolite ‖ Mesotyp m.; Natrolit m.; Faserzeolit m. ‖ mésotype m.; natrolithe m.

feature ‖ äußere Erscheinung f.; Charakteristik f. ‖ trait m.; caractéristique f. / leading ~ ‖ Hauptmerkmal n. ‖ caractéristique f. principale.

feaze, to (Cable) ‖ ausfasern; sich aufdrehen ‖ effiler; éfaufiler.

feazing cylinder ‖ Lumpenzupfer m. ‖ effilocheur m.

fecal see feces.

feces pl. ‖ Fäkalien pl. ‖ fèces fpl.; matières fpl. fécales; vidanges fpl. / removal of ~ ‖ Fäkalienabfuhr f. ‖ enlèvement m. des matières fécales. / ~ conveying implement ‖ Fäkalienabfuhrgerät n. ‖ instrument m. pour l'enlèvement des matières fécales.

fecula ‖ Kartoffelmehl n.; Stärkemehl n. ‖ fécule f. amylacée ou de pommes de terre. / ~ brandy distillery ‖ Stärkebrennerei f. ‖ distillerie f. d'eau de vie

ou d'alcool de fécule. / ~ works pl. ‖ Stärkemehlfabrik f. ‖ féculerie f.

fee ‖ Honorar n.; Gebühren fpl. ‖ honoraires mpl. / additional ~ ‖ Zuschlaggebühr f. ‖ surtaxe f. / carriage ~ ‖ Abfuhrgebühr f. ‖ factage m. / cutter's ~ (Stone-cutter) ‖ Hauerlohn m. ‖ équarrissage m. / ~ for delivery ‖ Bestellgebühr f.; Bestellgeld n. ‖ factage m. / exemption from ~ ‖ Gebührenfreiheit f. ‖ franchise f. de taxe. / ~ for inspection and report ‖ Gebühr f. für Besichtigung und Bericht ‖ honoraires mpl. pour inspection et rapport. / ~ for writing ‖ Schreibgebühr f. ‖ frais mpl. de copie.

feeble ‖ matt; kraftlos ‖ faible. / ~ and fickle (Meteor) ‖ schwach und unbeständig ‖ échars. / ~ constructed machine ‖ leichtgebaute Maschine f. ‖ machine f. de construction légère. / ~ current ‖ Schwachstrom m. ‖ courant m. faible.

feed, to (Boil) ‖ speisen ‖ alimenter. / ~ (Metal) ‖ aufgeben ‖ charger; délivrer; alimenter. / ~ back (Radio) ‖ rückkoppeln ‖ rétrocoupler; monter en réaction f. / ~ the furnace ‖ den Ofen m. beschicken ‖ charger le fourneau.

feed (Boil) ‖ Speisung f. ‖ alimentation f. / ~ (Mach tool) ‖ Vorschub m. ‖ avance f.; déplacement m.; serrage m. / ~ (Metal) ‖ Beschickungsmenge f. ‖ charge f. / automatic ~ ‖ selbsttätiger Vorschub m. ‖ avance f. automatique. / bar ~ ‖ Stangenvorschub m. ‖ avance f. des barres. / ~ of boring spindle ‖ Verschiebung f. der Bohrspindel ‖ course f. axiale de la broche à percer. / cattle ~ ‖ Viehfutter n. ‖ fourrage m. / cross ~ ‖ Planvorschub m. ‖ avance f. de planage. / ~ of drilling spindle see ~ of boring spindle. / drip ~ ‖ Tropfölschmierung f. ‖ graissage m. à compte-gouttes. / fully automatic ~ ‖ ganz selbsttätige Zuführung f. ‖ alimentation f. absolument automatique. / intermittent ~ ‖ Sprungvorschub m. ‖ avance f. par intermittence. / longitudinal ~ ‖ Langvorschub m. ‖ avance f. de chariotage. / ~ of material ‖ Werkstoffzuführung f. ‖ amenage m. de matériel. / self-acting disengaging of the ~ ‖ selbsttätige Auslösung f. des Vorschubs ‖ débrayage m. automatique de l'avancement. / semi-automatic ~ ‖ halbselbsttätige Zuführung f. ‖ alimentation f. semi-automatique. / ~ of tool ‖ Werkzeugvorschub m. ‖ avance f. des outils. / ~ for the turn table ‖ Revolvertellerzuführung f. ‖ amenage m. par plateau révolver.

feed adjuster ‖ Vorschubregelung f. ‖ réglage m. de l'avance. / ~ apparatus ‖ Füllapparat m. ‖ appareil m. d'alimentation. / ~ back ‖ Rückkopplung f. ‖ réaction f. / roller ~ bed ‖ Auflaufrollgang m. ‖ train m. de rouleaux d'amenée. / ~ boiler ‖ Speisekessel m. ‖ chaudière f. alimentaire. / ~ box (Found) ‖ Eingußkasten m. ‖ bassin m. de coulée. / ~ cam ‖ Vorschubkurve f. ‖ came f. d'avance. / ~ change (Mach tool) ‖ Vorschubwechsel m. ‖ changement m. de l'avance. / ~ change gear ‖ Vorschubwechselrad n. ‖ roue f. de rechange pour l'avance. / ~ chuck ‖ Vorschubpatrone f. ‖ pince f. d'avance. / master ~ chuck with jaws ‖ Vorschubpatronenkörper m. mit Einsätzen ‖ corps m. de pinces d'avance avec mors. / ~ cock ‖ Speisehahn m. ‖

robinet m. d'alimentation. / ~ current ‖ Anodenruhestrom m. ‖ courant m. permanent de plaque. / ~ cylinder ‖ Speisezylinder m. ‖ cylindre m. alimentaire. / vacuum fuel ~ device (Motor) ‖ Unterdruckförderer m. ‖ élévateur m. d'essence. / ~ eccentric ‖ Vorschubexzenter m. ‖ excentrique m. d'entraînement.

feeder (Electr) ‖ Speiseleitung f. ‖ feeder m.; câble m. ou ligne f. d'alimentation; artère m. / ~ (Hydr arch) ‖ Speisegraben m.; Bewässerungsgraben m. ‖ rigole f. d'alimentation; fossé m. d'irrigation. / ~ (Metal) ‖ Aufgabevorrichtung f. ‖ appareil m. chargeur; feeder m. / ~ (Print) ‖ Anleger m. ‖ margeur m. / ~ (Textile) ‖ Kinderlatz m. ‖ bavette f. / ~ (Weav) ‖ Vorschieber m. ‖ mécanisme m. alimentaire. / ~ of a fire engine ‖ Zubringer m. einer Feuerspritze ‖ tuyau m. alimentaire d'une pompe à incendie. / main ~ ‖ Hauptspeiseleitung f. ‖ feeder m. principal. / percussion ~ ‖ Stoßaufgabevorrichtung f. ‖ appareil m. chargeur à choc. / piston ~ ‖ Kolbenaufgabevorrichtung f. ‖ feeder m. à piston. / push ~ ‖ Schubaufgabevorrichtung f. ‖ feeder m. à poussée. / roller ~ ‖ Walzenaufgabevorrichtung f. ‖ feeder m. à cylindres. / shaking ~ ‖ Schüttelaufgabevorrichtung f. ‖ feeder m. à secousses. / worm ~ ‖ Schneckenaufgabevorrichtung f. ‖ appareil m. chargeur à vis.

feeder bar ‖ Speiseschiene f. ‖ barre f. des artères. / ~ line ‖ Zubringerlinie f. ‖ ligne f. auxiliaire. / ~ line aeroplane ‖ Zubringerflugzeug n. ‖ avion m. de ligne auxiliaire. / ~ plug-box on the ground ‖ Erdanschlußkasten m. ‖ boîte f. de prise de courant disposée sur le bord.

feed gear box ‖ Vorschubräderkasten m. ‖ boîte f. d'engrenages des avances. / ~ heater ‖ Vorwärmer m. ‖ réchauffeur m. / ~ hopper (Mill) ‖ Einlauftrichter m.; Beschickungstrichter m. ‖ entonnoir m. d'entrée; trémie f. de chargement.

feeding ‖ Speisung f. ‖ alimentation f. / ~ of the boiler ‖ Kesselspeisung f. ‖ alimentation f. de la chaudière. / hand ~ ‖ Handzuführung f. ‖ alimentation f. à main. / ~ the loose sheets (Bookb) ‖ Einlegen n. der losen Bogen ‖ marge f. des feuilles détachées. / ~ simultaneously two sides (Electr) ‖ gleichzeitige zweiseitige Speisung f. ‖ alimentation f. simultanée des deux côtés.

feeding apparatus (Metal) ‖ Aufgabevorrichtung f. ‖ appareil-chargeur m.; appareil d'alimentation; feeder m. / ~ (Print) ‖ Anlegeapparat m. ‖ rectificateur m. de marge. / ~ (Typewr) ‖ Bogenanleger m. ‖ margeur m. automatique. / oscillating ~ ‖ Schüttelspeiseapparat m. ‖ trémie f. d'alimentation à secousses. / sheet ~ ‖ Bogenanlegeapparat m. ‖ appareil m. margeur.

feeding arrangement (Print) ‖ Anlegevorrichtung f. ‖ taquet m. automatique. / pneumatic ~ ‖ pneumatische Bogenführung f. ‖ transport m. pneumatique de feuilles; transporteur m. de feuilles à aspiration.

feeding barley ‖ Futtergerste f. ‖ orge f. fourragère. / ~ board (Print) ‖ Anlegebrett n. ‖ table f. à pile. / ~ bottle ‖ Milchflasche f. ‖ bouteille f. à lait. / to cut out the ~ bridge (Tel) ‖ die Speisebrücke f. abschalten ‖ rompre le circuit

d'alimentation. / ~ crane (Railw) ‖ Speisewasserkran m. ‖ grue f. d'alimentation. / ~ disc ‖ Aufgabetisch m. ‖ table f. de distribution. / ~ elevator ‖ Aufgabebecherwerk n. ‖ noria f. de chargement. / ~ gear (Arc lamp) ‖ Nachschubvorrichtung f. ‖ mécanisme m. d'avance. / ~ head (Found) ‖ Anguß m.; Gießkopf m. ‖ jet m.; masselotte f.; saumon m. / ~ pipe ‖ Speiserohr n. ‖ tuyau m. alimentaire. / ~ roller (Spinn) ‖ Aufgabewalze f. ‖ cylindre m. rotatif d'alimentation. / ~ shoe (Mill) ‖ Beimischer m. ‖ distributeur m. à secousses. / ~ stuff ‖ Futtermittel n. ‖ produit m. fourrage; fourrage m. / ~ wick ‖ Farbdocht m. ‖ mèche f. d'encrage.

feed lime ‖ Futterkalk m. ‖ chaux f. pour nourriture ou à affourrager. / ~ motion ‖ Schaltbewegung f. ‖ mouvement m. de déplacement. / ~ pipe ‖ Speiserohr n. ‖ tube m. d'alimentation. / ~ pump ‖ Speisepumpe f. ‖ pompe f. alimentaire ou d'alimentation. / ~ automatic ~ regulator ‖ Wasserstandsregler m. ‖ régulateur m. du niveau d'eau. / ~ retardation coil (Tel) ‖ Speisebrücke f. ‖ pont m. d'alimentation. / ~ roll (Mach tool) ‖ Zugwalze f.; Zuführungswalze f. ‖ rouleau m. entraîneur. / ~ roll (Typewr) ‖ Andrückwalze f. ‖ cylindre m. presse-papier. / ~ salt ‖ Nährsalz n. ‖ sel m. nutritif. / ~ screw ‖ Förderschnecke f. ‖ vis f. transporteuse. / ~ tank ‖ Speisewasserbehälter m. ‖ réservoir m. d'alimentation. / ~ valve ‖ Kesselspeiseventil n. ‖ robinet m. d'alimentation. / regulating ~ valve ‖ Schieberdruckregler m. ‖ soupape f. d'alimentation automatique.

feed water ‖ Speisewasser n. ‖ eau f. d'alimentation. / ~ consumption ‖ Speisewasserverbrauch m. ‖ consommation f. en eau d'alimentation. / ~ heater ‖ Speisewasservorwärmer m. ‖ réchauffeur m. d'eau d'alimentation. / ~ plant ‖ Speisewasseranlage f. ‖ installation f. d'eau d'alimentation. / ~ preheater see ~ heater. / ~ purifier ‖ Speisewasserreiniger m. ‖ épurateur m. d'eau alimentaire. / ~ purifying ‖ Speisewasserreinigung f. ‖ épuration f. d'eau d'alimentation. / ~ regulator ‖ Speisewasserregler m. ‖ régulateur m. d'eau alimentaire.

feeler (Loc) ‖ Schienenräumer m. ‖ chassepierres m.; garde f. d'une locomotive. / ~ gauge ‖ Spion m. ‖ calibre m. à lames; jauge f. d'épaisseur. / ~ lever ‖ Tasthebel m. ‖ levier m. à touche. / ~ point ‖ Tasterspitze f. ‖ pointe f. du calibre. / ~ scale ‖ Tasterskale f. ‖ échelle f. du calibre.

feigned (Arch) ‖ blind; nachgeahmt ‖ dissimulé; feint; imité.

feldspar ‖ Feldspat m. ‖ feldspath m. / common ~ ‖ rechtwinkliger Feldspat m.; Orthoklas m. ‖ orthoclase f. / compact ~ ‖ dichter Feldspat m.; Feldstein m. ‖ feldspath m. compacte; pétrosilex m. / glassy ~ ‖ glasiger Feldspat m.; Eisspat m.; Sanidin m. ‖ feldspath m. vitreux; sanidine m.

feldspar-free rocks pl. ‖ feldspatfreies Gestein n. ‖ roches fpl. sans feldspaths; roches fpl. afeldspathiques.

feldspar porphyry ‖ Euritporphyr m. ‖ porphyre m. euritique; eurite m. porphyroïde. / ~ quarry ‖ Feldspatgrube f.

‖ carrière f. de feldspath. / ~ screen ‖ Feldspathsieb n. ‖ crible m. à feldspath. / ~ washer ‖ Feldspatsetzmaschine f. ‖ lavoir m. ou laveuse f. à feldspath.

felicitation telegram ‖ Glückwunschtelegramm n. ‖ télégramme m. de félicitation.

fell, to ~ timber ‖ Holz n. hauen ‖ abattre du bois m.

felled timber ‖ geschlagenes Holz n. ‖ bois m. abattu.

feller ‖ Holzfäller m. ‖ abatteur m. de bois; bûcheron m. / ~ foreman ‖ Haumeister m. ‖ maître-bûcheron m.

fellies pl. see also felloe ‖ Felgenkranz m.; Radkranz m. ‖ jante f.; anneau m. ou couronne f. circulaire d'une roue. / ~ assembling machine ‖ Felgenbank f. ‖ jantière f.

felling ‖ Fällen n.; Holzschlag m.; Abhieb m. des Holzes ‖ abat(t)age m.; coupe f. / final ~ ‖ gänzlicher Abhieb m. ‖ coupe f. définitive. / first ~ ‖ Anhieb m. ‖ mise f. en coupe. / secondary ~ (Forest) ‖ Nachhieb m. ‖ coupe f. secondaire. / ~ trees ‖ Holzfällen n. ‖ abatage f. du bois ou des arbres. / ~ a wood ‖ Schlagen n. eines Waldes ‖ coupe f. d'un bois.

felling area (Forest) ‖ Hiebfläche f. ‖ parterre m. des coupes. / ~ axe ‖ Holzhaueraxt f.; Baumaxt f. ‖ cognée f. ou hache f. de bûcheron. / ~ brush (Hatt) ‖ Walkbürste f. ‖ lustre m. / ~ registration ‖ Holzaufnahme f. ‖ inventaire m. de la coupe. / ~ saw ‖ Waldsäge f. ‖ scie f. ventrue ou passe-partout; passe-partout m. / ~ season ‖ Fällzeit f.; Wadelzeit f. ‖ époque f. de l'abattage.

felloe ‖ Felge f. ‖ jante f. / ~ made of bended wood ‖ Felge f. aus gebogenem Holz ‖ jante f. en bois courbé. / ~ of bicycle ‖ Fahrradfelge f. ‖ jante f. de bicyclette. / round ~ ‖ Rundfelge f. ‖ jante f. ronde. / ~ for solid tyres ‖ Felge f. für Massivreifen ‖ jante f. à bandages pleins. / steel ~ ‖ Stahlfelge f. ‖ jante f. en acier. / wheel ~ ‖ Grundfelge f. ‖ jante f. de roue. / wood ~ for cycles ‖ Holzfelge f. für Fahrräder ‖ jante f. en bois pour vélocipèdes.

felloe brake ‖ Felgenbremse f. ‖ frein m. sur jante. / ~ driver ‖ Radfelgenmontör m. ‖ monteur m. de jantes. / ~ mortising machine hand ‖ Felgenbohrer m. ‖ mortaiseur m. de jantes. / ~ pattern ‖ Felgenmodell n. ‖ jeumérante f. / ~ stabilizing machine ‖ Felgenstabilisiermaschine f. ‖ machine f. à équilibrer les jantes. / ~ wood ‖ Felgenholz n. ‖ bois m. pour jante.

fellow ‖ Geselle m. ‖ compagnon m.; ouvrier-compagnon m. / ~ workman ‖ Arbeitskollege m.; Kollege m. ‖ compagnon m. ou camarade m. de travail; confrère m.

felly see felloe.

felsobanyite ‖ Felsobanyit m. ‖ felsobanyite f.

felspar see feldspar.

felspath see feldspar.

felt, to ‖ verfilzen ‖ feutrer. / ~ (Grain heap) ‖ greifen ‖ prendre. / proper to felt ‖ verfilzungsfähig ‖ feutrable.

felt ‖ Filz m. ‖ feutre m. / ~ for grinding ‖ Schleiffilz m. ‖ feutre m. à polir. / ~ for paper mills ‖ Filz m. für Papierfabrikation ‖ feutre m. pour papéteries. / ~ for polishing ‖ Polierfilz m. ‖ feutre m. à

polir. / roofing ~ ‖ Dachfilz m. ‖ feutre m. pour toiture. / ~ for technical purposes ‖ Filz m. für technische Zwecke ‖ feutre m. pour l'industrie. / wool ~ ‖ Wollfilz m. ‖ feutre m. de laine.

felt cardboard ‖ Filzpappe f. ‖ cartonfeutre m. / ~ carpet ‖ Filzteppich m. ‖ tapis m. en feutre. / ~ case ‖ Filzfutteral n. ‖ étui m. en feutre. / ~ covering (Build) ‖ Filzdach n. ‖ toiture f. en feutre. / ~ covering (Mach) ‖ Filzumkleidung f. ‖ feutrage m. / ~ cushion (Office) ‖ Filzkissen n. ‖ coussin m. en feutre. / ~ cushion (Typewr) ‖ Filzplatte f. ‖ tapis m. de feutre. / ~ disk ‖ Filzscheibe f. ‖ disque m. en feutre.

felted ‖ filzig; verfilzt ‖ feutré. / ~ cloth ‖ Filztuch n. ‖ drap-feutre m.

felter, hat ~ ‖ Hutwalker m. ‖ fouleur m. en chapeaux.

felt fuller ‖ Filzwalker m. ‖ fouleur m. de feutre. / ~ goods pl. ‖ Filzwaren fpl. ‖ articles mpl. de feutre. / ~ hat ‖ Filzhut m. ‖ chapeau m. de feutre.

felting (Grain heap) ‖ Greifen n. ‖ prise f. / ~ machine ‖ Filzmaschine f. ‖ machine f. à feutrer. / extractor of grease from ~ water ‖ Walkwasserextraktör m. ‖ extracteur m. de graisse des eaux de foulonnage.

felt ink pad ‖ Filzfarbkissen n. ‖ coussin-encreur m. en feutre. / ~ joint ‖ Filzdichtung f. ‖ joint m. de feutre. / ~-like ‖ filzartig ‖ feutré. / ~ maker ‖ Filzmacher m. ‖ feutrier m. / ~ mop ‖ Filzscheibe f. ‖ disque m. en feutre. / ~ packing see ~ joint. / ~ pad ‖ Filzplatte f.; Filzunterlage f. ‖ plaque f. ou coussin m. de feutre. / ~ polishing wheel ‖ Filzpolierscheibe f. ‖ meule f. à polir en feutre. / ~ post ‖ befilzter Pauscht m. ‖ porse-feutre m.; porse-flotre f. / ~ ring ‖ Filzring m. ‖ rondelle f. de feutre. / ~ roll ‖ Wickelwalze f. ‖ cylindre m. en feutre. / ~ rollers press (Pap) ‖ Naßpresse f. ‖ presse f. humide. / ~ shoe ‖ Filzschuh m.; Filzsocke f. ‖ chausson m. ou chaussure f. de feutre. / ~ shoulder blade (Tail) ‖ Schulterblatt n. aus Filz ‖ épaulette f. ou omoplate f. en feutre. / ~ shoulder strap see ~ shoulder blade. / ~ sole ‖ Filzsohle f. ‖ semelle f. en feutre. / ~ washer see ~ ring. / ~ working machine ‖ Filzbearbeitungsmaschine f. ‖ machine f. à travailler le feutre.

felucca ‖ Felucke f. ‖ felouque f.

female ‖ weiblich ‖ femelle. / ~ screw ‖ Schraubenmutter f.; Mutter f. einer Schraube ‖ écrou m.; vis f. femelle ou creuse. / ~ tank (War mat) ‖ weiblicher Kampfwagen m.; weiblicher Tank m. ‖ char m. mitrailleuse; tank m. femelle. / ~ thread ‖ Innengewinde n. ‖ filet m. intérieur ou femelle. / ~ worker ‖ Arbeiterin f. ‖ ouvrière f.

fence, to ‖ einfriedigen ‖ enclore. / ~-in (Build) ‖ einzäunen ‖ échalasser; clôturer; enceindre. / ~ with pales ‖ einpfählen ‖ palisader; entourer de palis mpl.

fence (Build) ‖ Einfriedigung f.; Einzäunung f.; Zaun m. ‖ clôture f.; enceinte f.; enclos m.; haie f. / ~ (Mach) ‖ Galerie f. ‖ galerie f.; garde-corps m. / ~ of the bank (Hydr arch) ‖ Uferschutz m. ‖ défense f. de la rive. / boarding ~ ‖ Bretterzaun m. ‖ clôture f. en planches. / ~ of pales ‖ Lattenzaun m. ‖

clôture f. en lattis; palissade f. / ~ of a
plane ‖ verstellbarer Hobelanschlag m.
‖ joue f. mobile. / protective ~ ‖ Fangzaun m. ‖ haie f. ou palissade f. d'arrêt.
/ ~ of rods ‖ Rutenzaun m. ‖ haie f. morte.
/ ~ of sleepers ‖ Schwellenzaun m. ‖
clôture f. en vieilles traverses. / stone ~ ‖
Steineinfriedigung f. ‖ clôture f. en
pierres. / top ~ (Railw) ‖ Gitteraufsatz
m. ‖ treillage m. de surhaussement. /
wire ~ ‖ Drahtzaun m. ‖ clôture f. métallique.
fenced with pales (Build) ‖ eingezäunt ‖
entouré de palis mpl.
fence lath ‖ Spalierlatte f. ‖ échalas m. ou
latte f. à lattis. / ~ pale ‖ Zaunpfahl m.
‖ palis m. / ~ wire ‖ Zaundraht m. ‖ fil
m. de fer pour clôture.
fencing ‖ Einfriedigung f.; Einzäunung f.
‖ clôture f.; enceinte f. / ~ of a station ‖
Bahnhofseinfriedigung f. ‖ clôture f.
d'une gare. / wire net ~ ‖ Drahtzaun m.;
Drahtgewebeeinfriedigung f. ‖ clôture f.
en toile métallique.
fencing, gymnastic and ~ club ‖ Turn- und
Fechtklub m. ‖ club m. de gymnastique
et d'escrime. / ~ equipment ‖ Fecht-
ausrüstung f. ‖ équipement m. pour
escrime. / ~ goods pl. ‖ Fechtartikel mpl.
‖ articles mpl. pour l'escrime. / ~ mask
‖ Fechtmaske f. ‖ masque m. d'escrime.
fender (Auto) ‖ Kotflügel m. ‖ garde-
boue m. / ~ (Fire place) ‖ Ofenvorsetzer
m.; Kaminvorsetzer m. ‖ garde-cendre
m. / ~ (Mar) ‖ Fender m. ‖ défense f. /
~ (Mill) ‖ Griesholm m. ‖ traverse f. de
poteaux. / ~ (Road) ‖ Prellpfahl m. /
~ (Road) ‖ Prellstein m. ‖ Prellstein m. ‖ borne f. chasse-roue. / ~
(Shipb) ‖ Scheuerleiste f.; Reibholz n. ‖
défense f. en bois. / boat's ~ ‖ Boots-
fender m. ‖ défense f. d'embarcation. /
cork ~ ‖ Korkfender m. ‖ défense f. en
liége. / hanging ~ see wood ~. / ~ wood
‖ Holzfender m. ‖ défense f. en bois.
fender beam of an ice breaker ‖ Eisbalken
m. eines Eisbrechers ‖ chapeau m. incliné d'un brise-glace. / inclined ~ beam
of an ice breaker ‖ schräger Holm m.
eines Eisbrechers; Eispfahl m. ‖ chapeau
m. incliné d'un brise-glace. / ~ pile
(Hydr arch) ‖ Prellpfahl m. ‖ repoussoir
m. / ~ support ‖ Kotflügelstütze f. ‖ support m. de pare-boue. / ~ wales pl.
(Hydr arch) ‖ Prellzangen fpl. ‖ défenses fpl. / ~ wall (Hydr arch) ‖ Schutzwand f. ‖ mur m. de garde ou de barrage.
fen land farmer ‖ Marschbauer m.; Moorbauer m. ‖ habitant m. d'un pays marécageux; exploitant m. de polders m. ou
de watringues m.
fennel ‖ Fenchel m. ‖ fenouil m. / ~ honey
‖ Fenchelhonig m. ‖ miel m. de fenouil.
/ ~ oil (Chem) ‖ Fenchelöl n. ‖ essence f.
de fenouil. / common ~ oil ‖ Bitter-
fenchelöl n. ‖ essence f. de fenouil amer.
fenugreek ‖ Bockshornklee m. ‖ fenugrec
m.
fergusonite ‖ Fergusonit m. ‖ fergusonite f.
ferment, to ‖ gären ‖ fermenter.
ferment ‖ Gärungsstoff m.; Ferment n. ‖
ferment m. / alcoholic ~ ‖ Alkohol-
gärungspilz m. ‖ ferment m. alcoolique.
/ hydrolytic ~ ‖ hydrolytisches Ferment
n. ‖ ferment m. hydrolyte. / inverting ~
‖ invertierendes Ferment n. ‖ ferment m.
invertissant.

fermentability ‖ Gärungsfähigkeit f.; Gär-
barkeit f. ‖ fermentescibilité f.
fermentable ‖ gärungsfähig; gärbar ‖ fermentescible; fermentatif. / ~ liquid ‖
Gärflüssigkeit f. ‖ liquide m. fermentescible.
fermentation ‖ Gärung f.; Gur f.; Gärvorgang m. ‖ fermentation f.; procédé m.
de fermentation. / accelerated ~ ‖ Schnell-
gärung f. ‖ fermentation f. rapide. /
acetic acid ~ ‖ Essigsäuregärung f. ‖
fermentation f. acétique. / acid ~ ‖ saure
Gärung f. ‖ fermentation f. acide. /
after ~ ‖ Nachgärung f. ‖ fermentation f.
secondaire ou insensible. / alcoholic ~ ‖
Alkoholgärung f. ‖ fermentation f. alcoolique. / to animate the ~ ‖ die Gärung
beleben ‖ activer la fermentation; battre
la guilloire. / arrested ~ ‖ Rastgärung f.
‖ fermentation f. d'arrêt ou d'interruption. / bare spots ~ ‖ Gärung f. mit
kahlen Stellen ‖ fermentation f. à places
chauves. / ~ from below ‖ Untergärung f.
‖ fermentation f. avec dépôt. / boiling ~
‖ stürmische oder kochende Gärung f. ‖
fermentation f. tumulteuse. / ~ of bread
‖ Brotgärung f. ‖ fermentation f. panaire.
/ bung hole ~ ‖ Spundgärung f. ‖ fermentation f. par la bonde. / cask ~ ‖
Faßgärung f. ‖ fermentation f. en fûts
ou en tonnes. / coal-forming ~ ‖ Kohlegärung f. ‖ fermentation f. carbonique. /
cold spots ~ ‖ Gärung f. mit kahlen
Stellen ‖ fermentation f. à places chauves.
/ fiery ~ see boiling ~. / fission fungi ~ ‖
Spaltpilzgärung f. ‖ fermentation f. pro-
voquée par les bactéries schizomycètes.
/ insensible ~ ‖ Nachgärung f.; guillage
m.; fermentation f. insensible. / main ~
‖ Hauptgärung f. ‖ fermentation f. principale. / oxidation ~ ‖ Oxydationsgärung f. ‖ fermentation f. d'oxydation.
/ ~ under pressure ‖ Druckgärung f. ‖
fermentation f. sous pression. / to prevent ~ ‖ Gärungsprozesse mpl. vermeiden ‖ empêcher la fermentation. / putrid
~ ‖ faulende Gärung f. ‖ fermentation f.
putride. / putrifactive ~ see putrid ~. /
rest ~ see arrested ~. / rim ~ ‖ Randgärung f. ‖ fermentation f. périphérique.
/ ropy ~ see slimy ~. / secondary ~ see
after ~. / sedimentary ~ see ~ from
below. / slimy ~ ‖ schleimige Gärung f.
‖ fermentation f. visqueuse. / sluggish ~
‖ träge Gärung f. ‖ fermentation f. lente
ou paresseuse. / sour ~ ‖ saure Gärung f.
‖ fermentation f. acide. / spontaneous ~
‖ spontane Gärung f. ‖ fermentation f.
spontanée. / standing ~ ‖ Standgur f.
‖ fermentation f. en cuves. / surface ~ ‖
Obergärung f. ‖ fermentation f. ordinaire.
/ symbiotic ~ ‖ symbiotische Gärung f. ‖
fermentation f. symbiotique. / upper ~
see surface ~. / vinous ~ ‖ geistige
Gärung f. ‖ fermentation f. vineuse ou
spiritueuse. / viscous ~ see ropy ~. /
water-forming ~ ‖ Wassergärung f. ‖
fermentation f. produisant de l'eau. /
anomaly during the ~ ‖ Gärungserschei-
nung f. ‖ anomalie f. en cours de fermentation.
fermentation bacterium ‖ Gärungsbakte-
rium n. ‖ bactérie f. de fermentation. /
beginning of ~ of the wort ‖ Ankommen
n. der Würze ‖ entrée f. en fermentation
du moût. / ~ cellar ‖ Gärkeller m. ‖ cave
f. de fermentation. / ~ cellar with cooled
air circulation for x hl of must in alu-

minium tanks ‖ Gärkeller m. für x hl
Weinmost mit Aluminiumbottichen und
Ventilationskühlung ‖ cave f. de fermentation, desservie par un frigorifère
d'air, pour x hl de moût en tanks d'aluminium. / ~ chemistry ‖ Gärungschemie
f. ‖ chimie f. de la fermentation. / ~ conduct ‖ Gärführung f. ‖ conduite f. de la
fermentation. / ~ cylinder ‖ Gärzylinder
m. ‖ cylindre m. à fermentation. / ~
degree ‖ Vergärungsgrad m. ‖ degré m.
de fermentation. / ~ duration ‖ Gärdauer f. ‖ durée f. de fermentation. / ~
heat ‖ Gärungswärme f. ‖ chaleur f.
produite durant la fermentation. / ~ industry ‖ Gärungsindustrie f. ‖ industrie f.
de la fermentation. / ~ organism ‖ Gä-
rungsorganismus m. ‖ organisme m. de
fermentation. / ~ phenomenon ‖ Gärungs-
phänomen n. ‖ phénomène m. de fermentation. / ~ process ‖ Gärverfahren n. ‖
procédé m. de fermentation. / ~ product
‖ Gärerzeugnis n. ‖ produit m. de la
fermentation. / ~ temperature ‖ Gär-
temperatur f.; Gärwärmegrad m. ‖ température f. de fermentation. / ~ theory
‖ Gärungstheorie f. ‖ théorie f. de la
fermentation.
fermentative ‖ gärfähig; fermentativ ‖
fermentescible; fermentatif. / ~ activity
‖ Gärtätigkeit f. ‖ activité f. fermentative. / ~ energy ‖ Gärvermögen n. ‖
énergie f. fermentative. / ~ power ‖
Gärkraft f. ‖ force f. fermentative; pouvoir m. fermentatif. / ~ test ‖ Gärkraft-
bestimmung f. ‖ détermination f. du
pouvoir fermentatif.
fermented ‖ gegoren ‖ fermenté. / ~ from
below ‖ untergärig ‖ à fermentation
basse. / ~ from top ‖ obergärig ‖ à fermentation f. haute. / ~ beverage ‖ gegorenes Getränk n. ‖ boisson f. fermentée.
fermenter ‖ Gärbottich m.; Gärgeschirr
n.; Gärbütte f.; Gärgefäß n. ‖ cuve f.
de fermentation.
fermentescible ‖ gärfähig ‖ fermentescible.
fermenting ‖ Gärung f. ‖ fermentation f. /
~ board ‖ Gärschrank m. ‖ armoire f.
de fermentation. / ~ cellar ‖ Gärkeller m.
‖ cave f. de fermentation. / ~ cellar man
‖ Gärkellerarbeiter m. ‖ fermenteur m. /
~ cylinder ‖ see fermentation cylinder. /
~ man ‖ Gärführer m. ‖ fermenteur m. /
~ meal ‖ Triebmehl n. ‖ farine f. fermentatif. / ~ plant ‖ Gäranlage f. ‖
installation f. de fermentation. / ~ power
‖ Fermentativvermögen n.; Gärkraft f. ‖
pouvoir m. fermentatif. / ~ process ‖
Gärverfahren n.; Gärvorgang m. ‖ procédé m. de fermentation. / vacuum ~
process ‖ Vakuumgärverfahren n. ‖ procédé m. de fermentation dans le vide. /
~ room ‖ Gärraum m. ‖ cave f. de fermentation. / ~ room for the rum distillery ‖ Gärraum m. für die Rumfabrika-
tion ‖ rhummerie f.; guildive f. / ~ room
man ‖ Gärkellerbursche m. ‖ garçon m.
de cave de fermentation. / ~ round see
~ tun. / ~ tub see ~ tun. / ~ tube ‖
Gärungsröhrchen n. ‖ tube f. à fermentation.
fermenting tun ‖ Gärbottich m.; Gärbütte
f. ‖ cuve f. guilloire ou de fermentation. /
attemperator coil for ~s ‖ Gärbottich-
kühlschlange f. ‖ serpentin m. réfrigérant pour cuves de fermentation. / ~
beer ‖ Gärbottichbier n.; Jungbier n. ‖

bière f. provenant des cuves de fermentation; bière f. jeune. / ~ cock ‖ Gärbottichhahn m. ‖ robinet m. de cuve de fermentation. / brush for washing ~s ‖ Gärbottichbürste f. ‖ brosse f. pour le nettoyage des cuves de fermentation. / float for ~s ‖ Gärbottichschwimmer m. ‖ flotteur m. réfrigérant pour cuves de fermentation. / swimmer for ~s see float for ~s. / varnish for ~s ‖ Gärbottichglasur f. ‖ vernis m. pour cuves de fermentation.

fermenting vessel ‖ Gärgefäß n.; Gärgeschirr n. ‖ tonneau m. à fermenter ou servant à la fermentation. / closed ~ ‖ geschlossenes Gärgefäß n. ‖ tonneau m. clos pour la fermentation. / enamelled ~ ‖ emaillierter Gärbottich m. ‖ cuve f. de fermentation émaillée. / glass ~ ‖ gläserner Gärbottich m. ‖ cuve f. de fermentation en verre.

fern extract ‖ Farnkrautextrakt m. ‖ extrait m. de fougère. / ~ gatherer ‖ Farnkrautschneider m. ‖ ramasseur m. de fougère. / ~ root ‖ Farnkrautwurzel f. ‖ racine f. de fougère.

ferrel (Carp; Forg) ‖ Frette f.; Zwinge f. ‖ frette f.; virole f.

ferret silk ‖ Flockseide f.; ungezwirnte Seide f. ‖ effiloches fpl.

ferric chloride ‖ Eisenchlorid n. ‖ perchlorure m. de fer; chlorure m. ferrique. / ~ concrete sleeper ‖ Betoneisenschwelle f. ‖ traverse f. en béton armé. / ~ ferrocyanide ‖ Ferrozyaneisen n. ‖ ferrocyanide m. de fer. / ~ oxide ‖ Eisenoxyd n. ‖ oxyde m. ferrique. / ~ sulphate ‖ Ferrisulphat n.; Eisenvitriol n. ‖ sulfate m. ferrique.

ferricyanide ‖ Ferrizyanid n.; Eisenzyanid n. ‖ ferricyanure m. / ~ of potassium ‖ rotes Blutlaugensalz n. ‖ cyanoferride m. de potassium; cyanure m. rouge de potassium et de fer; ferricyanure m. de potassium.

ferriferous ‖ eisenhaltig ‖ ferrugineux; ferrifère.

ferril ‖ Ringbeschlag m.; Stockzwinge f.; Zwinge f. ‖ virole f.

ferro-alloy ‖ Ferrolegierung f. ‖ ferroalliage m.; alliage m. ferrique ou de fer.

ferroarsenite water ‖ Eisenarsenwasser n. ‖ eau m. à l'arséniate de fer.

ferrochromium ‖ Ferrochrom n. ‖ ferrochrome m.

ferro-concrete ‖ Eisenbeton m. ‖ béton m. armé. / ~ beam ‖ Eisenbetonbalken m. ‖ poutre f. en béton armé. / ~ pipe ‖ Eisenbetonrohr n. ‖ tuyau m. en béton armé.

ferrocyanic acid ‖ Ferrozyanwasserstoffsäure f. ‖ acide m. ferro-cyanique.

ferrocyanide ‖ Ferrozyanid n.; Eisenzyanür n. ‖ ferrocyanure m. / ~ of hydrogen ‖ Ferrozyanwasserstoffsäure f. ‖ acide m. ferro-cyanique. / ~ of iron ‖ Ferrozyaneisen n. ‖ ferrocyanide m. de fer. / ~ of potassium ‖ gelbes Blutlaugensalz n. ‖ prussiate m. jaune de potasse; ferrocyanure m. de potassium; cyanoferrure m. de potassium jaune; lessive f. du sang.

ferrogallic ink ‖ Eisengallustinte f. ‖ encre f. gallique ferrée.

ferromagnetism ‖ Ferromagnetismus m. ‖ ferro-magnétisme m.

ferromanganese ‖ Ferromangan n. ‖ ferromanganèse m.

ferrometallic alloy ‖ Ferrometallegierung f. ‖ alliage m. ferro-métallique.

ferromolybdenum ‖ Ferromolybdän n. ‖ ferromolybdène m.

ferronickel ‖ Ferronickel n.; Nickeleisen n. ‖ ferronickel m.

ferrooxide ‖ Eisenoxydul n. ‖ oxyde m. ferreux; protoxyde m. de fer.

ferrophosphorus ‖ Ferrophosphor m.; Eisenphosphor m. ‖ ferrophosphore. m.

ferroprussic acid ‖ Ferrozyanwasserstoffsäure f.; Eisenblausäure f. ‖ acide m. ferro-cyanique.

ferrosilicon ‖ Siliziumeisen n.; Ferrosilizium n. ‖ ferro-silicium. / acid-resisting ~ castings pl. ‖ säurebeständiger Siliziumeisenguß m. ‖ moulages mpl. en ferrosilicium résistants aux (effets des) acides.

ferrosoferric ‖ ferroferri ‖ ferrosoferrique; ferrous ferri.

ferrotherm steel ‖ Ferrothermstahl m. ‖ acier m. ferrotherm.

ferrotin ‖ Ferrozinn n. ‖ ferro-étain m.

ferrotitanium ‖ Ferrotitan n. ‖ ferrotitane m.

ferrotungsten ‖ Ferrowolfram n. ‖ ferrotungstène m.

ferrous chloride ‖ Eisenchlorür n. ‖ protochlorure m. de fer; chlorure m. ferreux. / ~ cyanide ‖ Eisenzyanür n. ‖ ferrocyanure m. / ~ oxide ‖ Eisenoxydul n. ‖ protoxyde m. de fer; fer m. oxydulé. / ~ sulphate ‖ Eisenvitriol n. ‖ sulfate m. ferreux.

ferrovanadium ‖ Eisenvanadium n.; Ferrovanadium n.; Vanadiumeisen n. ‖ ferrovanadium m.

ferruginous ‖ eisenhaltig ‖ ferrugineux; ferrifère. / ~ quartz ‖ Eisenkiesel m. ‖ quartz m. ferrugineux. / ~ sand ‖ Eisensand m. ‖ sable m. ferrugineux. / ~ water ‖ Stahlbrunnen m. ‖ eau f. ferrugineuse.

ferrule, to ‖ mit einer Zwinge f. versehen ‖ emboutir.

ferrule ‖ Ringbeschlag m.; Stockzwinge f.; Zwinge f.; virole f.; ferrure f. annulaire. / ~ for boiler tubes (Loc) ‖ Feuerrohrring m. ‖ bague f. pour tubes de chaudières. / grooved ~ (Coin) ‖ gekerbter Prägring m. ‖ virole f. cannelée. / smooth ~ (Coin) ‖ glatter Prägring m. ‖ virole f. lisse. / ~ of a tool-handle ‖ Heftzwinge f. eines Werkzeuges ‖ virole f. d'un manche d'outil.

ferry ‖ Fähre f. ‖ bac m.; bateau m. de passage. / ~ ‖ Fährschiff n. ‖ bateau m. transbordeur. / moving ~ ‖ Trajekt m. ‖ bac m. / railway ~ ‖ Eisenbahnfähre f. ‖ bac m. porte-train; ferry-boat m. / suspension ~ ‖ Schwebefähre f. ‖ bac m. suspendu. / train ~ see railway ~.

ferry boat ‖ Fährboot n. ‖ bateau m. de passage; bac m. / ~ (Railw) see railway ferry. / ~ for horses ‖ Fährboot n. für Pferde; Pferdefähre f. ‖ barguette f.; passe-cheval m. / steam ~ (Railw) ‖ Fährdampfer m.; Trajektschiff n. ‖ vapeur m. de passage; bac m. à vapeur pour trains.

ferry bridge ‖ Fährbrücke f. ‖ pont m. transbordeur.

fertility ‖ Ergiebigkeit f. ‖ fertilité f. / grade of ~ (Forest) ‖ Standortsgüte f. ‖ classe f. de fertilité.

fertilized ‖ gedüngt ‖ fertilisé.

fertilizer ‖ Düngemittel n. ‖ engrais m. / ~ factory ‖ Düngemittelfabrik f. ‖ fabrique f. d'engrais. / ~ salt ‖ Düngesalz n. ‖

sel m. d'engrais. / ~ producing plant ‖ Düngemittelherstellungsanlage f. ‖ installation f. pour la fabrication des engrais.

festoon ‖ Blumengehänge n.; Fruchtgehänge n.; Laubgehänge n.; Girlande f. ‖ guirlande f. / ~ embroiderer ‖ Festonstickerin f. ‖ festonneuse f.

fetch, to ~ up ‖ aufholen ‖ ramasser. / ~ way (Mach) ‖ Spiel n. oder Spielraum m. haben ‖ jouer; avoir du jeu.

fetching ‖ Abholgebühr f. ‖ prise f. à domicile.

fetid ‖ faulig ‖ fétid. / ~ shale ‖ Stinkschiefer m. ‖ schiste m. suant.

fettling ‖ Abschroten n. ‖ ébarbage m. / ~ of earthenware ‖ Nachputzen n. der Tonwaren ‖ évidage m. des pièces céramiques moulées. / ~ machine (Grinder) ‖ Entgratmaschine f. ‖ machine f. à ébarber; ébarbeuse f. / ~ shop (Found) ‖ Gußputzerei f. ‖ atelier m. d'ébarbage.

fibre ‖ Fiber f.; Faser f. ‖ fibre f.; fil m.; filament m. / ~ (Metal) ‖ Sehne f. ‖ nerf m. / to attain a better preservation of the ~ ‖ bessere Erhaltung f. der Faser erzielen ‖ conserver mieux la fibre. / to destroy the torsion of the ~ ‖ einen Faden m. austordieren ‖ éliminer la torsion du fil. / ~ of hemp ‖ Hanffaser f. ‖ brin m. de chanvre. / ~ leaf ‖ Blattfaser f. ‖ fibre f. extraite de la feuille. / nerve ~ ‖ Nervenfaser f. ‖ fibre f. nerveuse. / neutral ~ ‖ neutrale Faser f. ‖ fibre f. neutre ou invariable. / vegetable ~ ‖ Pflanzenfaser f. ‖ fibre f. végétale. / vulcanized ~ ‖ Vulkanfiber f. ‖ fibre f. vulcanisée. / ~ of wood ‖ Holzfaser f. ‖ fil m. du bois.

fibre brush ‖ Fiberbürste f. ‖ brosse f. en fibre. / ~ bundle ‖ Fadenbündel n. ‖ faisceau m. de fils. / ~ dressing machine ‖ Faseraufbereitungsmaschine f. ‖ machine f. à préparer les filaments. / ~ electrometer ‖ Saitengalvanometer n. ‖ électromètre m. à fil. / ~ gear ‖ Fiberrad n. ‖ roue f. en f bre. / ~ material ‖ (Pflanzen)Faserstoff m. ‖ matière f. de fibre. / ~ packing ‖ Fiberdichtung f. ‖ joint m. de fibre. / ~ pinion ‖ Fiberzahnrad n. ‖ pignon m. de fibre. / ~ producing machine ‖ Fasergewinnungsmaschine f. ‖ machine f. à extraire des filaments. / ~ ring ‖ Fiberring m. ‖ bague f. en fibre. / ~ suspension ‖ Fadenaufhängung f. ‖ suspension f. à fil. / ~ tension ‖ Faserspannung f. ‖ tension f. de fibres.

fibrillation, auricular ~ ‖ Vorhofflimmern n. ‖ fibrillation f. auriculaire. / ~ of the lens ‖ Faserung f. der Linse ‖ fibres fpl. du cristallin.

fibrine ‖ Faserstoff m.; Fibrin n. ‖ fibrine f. / vegetable ~ ‖ Pflanzenfibrin n. ‖ fibrine f. végétale.

fibrocement ‖ Fibrozement m. ‖ fibrociment m.

fibroferrite ‖ Fibroferrit m. ‖ fibroferrite f.

fibroin ‖ Fibroin n. ‖ fibroïne f.

fibrolite ‖ Fibrolit m. ‖ fibrolite f.

fibrous ‖ faserig ‖ fibreux. / ~ (Glassm) ‖ krätzig; zaserig, zellig ‖ galeux. / ~ (Marble) ‖ faserig; aderig ‖ filandreux. / ~ (Metal) ‖ sehnig ‖ fibreux; nerveux. / ~ calcite ‖ Faserkalk m. ‖ calcite f. fibreuse. / tough and ~ fracture (Metal) ‖ sehniger und zäher Anbruch m. ‖ crique f. nerveuse et tenace. / ~ gneiss ‖

Holzgneis m. || gneiss m. fibreux. / ~ material || Faserstoff m. || matière f. fibreuse. / ~ peat || Fasertorf m. || tourbe f. fibreuse *ou* de racines. / ~ steel || sehniger Stahl m. || acier m. nerveux. / ~ texture || faseriges Gefüge n. || structure f. fibreuse. / a rather violent operation, which, however, the ~ texture of the steel can very well stand || gewaltsame Behandlung, die der sehnige Stahl sehr gut verträgt || ce traitement m. est passablement violent, mais un acier nerveux le supporte fort bien.

fichtelite || Fichtelit m. || fichtelite f.

fictitious bargain || Scheingeschäft n. || opération f. fictive. / ~ purchase || Scheinkauf m. || achat m. simulé.

fid, to (Shipb) || das Schloßholz einsetzen || passer la clef.

fid (Shipb) || Schloßholz n.; Schlußkeil m. || clef f. *ou* clé f. d'un mât. / lever ~ (Shipb) || Schloßholz n. mit Hebel || clef f. à levier *ou* à bascule. / ~ of a top gallant mast || Bramstängeschloßholz n. || clef f. d'un mât de perroque. / ~ of a topmast || Stängeschloßholz n. || clef f. d'un mât de hune.

fiddle case || Geigenkasten m. || étui m. *ou* boîte f. à violon. / ~ stick || Geigenbogen m. || archet m.

fid hole (Shipb) || Schloßholzloch n.; Gat n. || trou m. pour la clef de mât. / ~ lanyard (Shipb) || Schloßholzreep n. || aiguilette f. de clef de mât.

fiducial, illumination of ~ lines (Opt) || Fadenbeleuchtung f. || éclairage m. des fils.

field || Feld n. || champ m. / to extend the ~ of activities of a company || das Tätigkeitsfeld n. einer Gesellschaft erweitern || élargir la zône de travail d'une société. / alternative ~ || Wechselfeld n. || champ m. alternatif. / ~ of attraction (Magnet) || anziehender Bereich m. || zône f. *ou* région f. d'attraction. / commutating ~ || Kommutierungsfeld n. || champ m. de commutation. / ~ of danger || gefährliches Feld n. || champ m. dangereux. / electric ~ || elektrisches Feld n. || champ m. électrique. / electric ~ between parallel conductors || elektrisches Feld n. von parallelen Leitungen || champ m. électrique entre conducteurs parallèles. / electromagnetic ~ || elektromagnetisches Feld n. || champ m. électromagnétique. / ~ of employment || Anwendungsgebiet n. || domaine m. d'application *ou* d'emploi. / to excite a ~ || ein Feld n. erregen || exciter un champ. / ~ of force || Kraftfeld n. || champ m. de force. / gravitational ~ || Gravitationsfeld n.; Schwerefeld n. || champ m. de gravitation. / magnetic ~ || magnetisches Feld n. || champ m. magnétique. / ~ of magnetic force || magnetisches Kraftlinienfeld n. || champ m. magnétique. / ~ of opposite polarity || ungleichnamiges Feld n. || champ m. de polarité contraire. / rotating ~ || Drehfeld n.; rotierendes Feld n. || champ m. tournant.

field of view || Gesichtsfeld n. || champ m. visuel. / enlarged ~ || erweitertes Sehfeld n. || grand champ m. visuel. / ~ of the eyepiece || Gesichtsfeld n. des Okulars || champ m. de l'oculaire. / apparent ~ of the eyepiece || scheinbares Gesichtsfeld n. des Okulars || champ m. apparent de l'oculaire. / ~ of a telescope

|| Sehfeld n. eines Fernrohres || champ m. d'une lunette. / edge of the ~ || Rand m. des Sehfeldes || bord m. du champ de vision. / number of the ~ || Sehfeldzahl f. || coefficient m. de champ.

field of vision || Blickfeld n. || champ m. visuel. / ~ in angular measure || Sehfeld n. im Winkelmaß || champ m. réel en degrés.

field book (Surv) || Notizbuch n.; Manual n. || livre m. de notes; carnet m. / ~ for calculations (Surv) || Winkelbuch n. || carnet m. de calculs d'arpentage.

field break switch || Magnetausschalter m. || interrupteur m. de l'excitation. / ~ cable joint || Feldkabelverbindung f. || jonction f. de câbles militaires. / ~ carriage || Ackerwagen m. || voiture f. agricole. / ~ coil || Magnetspule f. || bobine f. pour aimants. / ~ diaphragm || Sehfeldblende f. || diaphragme m. de champ. / ~ distortion || Feldverzerrung f. || distorsion f. du champ. / ~ excitation || Felderregung f. || excitation f. de champ. / ~ forge || Feldschmiede f. || forge f. portative.

field glass || Feldstecher m.; Fernrohr n. || jumelle f.; longue-vue f. / ~ fitted with central twin focussing device || mit Mitteltrieb ausgestatteter Feldstecher m. || jumelle f. à molette centrale. / ~ high-power ~ for long distances || stark vergrößernder Feldstecher m. für große Entfernungen || jumelle f. à fort grossissement pour observations à grande distance. / ~ of great light transmitting capacity || lichtstarker Feldstecher m. || jumelle f. lumineuse. / miniature ~ || Miniaturfeldstecher m. || jumelle f. de poche. / wide-angle ~ || Weitwinkelfeldstecher m. || jumelle f. grand-angulaire.

field glass magnifier || Feldstecherlupe f. || jumelle-loupe f. / rain-guard of the ~ || Regenschutzdeckel m. des Feldstechers || pare-pluie m. de la jumelle.

field gray || Feldgrau n. || gris m. de champ. / ~ guard || Feldhüter m. || garde-champêtre m.; garde-chasse m. / ~ kiln (Brick) || Feldbrand m. || four m. de campagne. / ~ kitchen || Feldküche f. || cuisine f. portable. / ~ lens || Kollektivlinse f.; Vorderlinse f. || verre m. de champ; lentille f. avant. / ~ line construction (Tel) || Feldleitungsbau m. || construction f. de lignes militaires. / ~ lineman (Tel) || Streckenläufer m. || surveillant m. de ligne. / ~ magnet || Feldmagnet m. || inducteur m. / ~ punching (Electr) || Feldblech n. || tôle f. de champ. / ~ railway || Feldbahn f. || voie f. étroite *ou* militaire; decauville m. / ~ roller || Ackerwalze f. || rouleau m. pour le travail des champs. / ~ set (Tel) || Feldstation f. || station f. de campagne.

field stop (Opt) || Gesichtsfeldblende f. || diaphragme m. de champ. / ~ of the eyepiece || Gesichtsfeldblende f. des Okulars || diaphragme m. de champ de l'oculaire.

field telegraph construction || Feldtelegrafenbau m. || construction f. des télégraphes militaires.

field tripod || Feldstativ n. || pied m. de campagne. / ~ with adjustable stays || Feldstativ n. mit verstellbaren Spreizen || trépied m. de campagne avec allong écartant les trois pieds. / folding ~ || zusammenklappbares Feldstativ n. || pied

m. de campagne pliant. / steel ~ || Stahlrohrfeldstativ n. || pied m. de campagne en tubes d'acier.

field variations pl. (Electr) || Feldschwankungen fpl. || fluctuations fpl. de champ. / ~ winch || Ackerwinde f. || treuil m. de labourage. / ~ wire for cavalry || Kavalleriedraht m. || fil m. isolé pour la cavallerie.

fiery (Mine) || mit schlagenden Wettern npl. behaftet || grisouteux. / ~ atmospheres pl. (Mine) || schlagende Wetter npl. || grisou m. / ~ fermentation || stürmische Gärung f. || fermentation f. tumultueuse. / ~ vapours pl. *see* ~ atmospheres.

fifteen-reel unit rotary printing machine in line arrangement || Fünfzehnrollenrotationsdruckmaschine f. in Reihenanordnung || machine f. rotative à 15 unités d'impression en groupe.

fig || Feige f. || figue f. / ~ coffee || Feigenkaffee m. || café m. de figues. / ~ meal || Feigenmehl n. || farine f. de figues.

fighting, track ~ machine || Raupenkampfmaschine f. || engin m. de combat muni de chenilles.

figuline || Töpferton m. || argile f. figuline; terre f. à poteries.

figure, to || formen; gestalten || façonner; former; modeler; bosseler. / ~ a graduation || eine Teilung f. beziffern || porter des chiffres mpl. sur une graduation.

figure || Figur f.; Form f. || figure f.; forme f. / ~ (Math) || Ziffer f. || chiffre m. / as is shown in the ~ annexed || wie nebenstehende Abbildung f. zeigt || comme on le voit sur la figure ci-contre. / etched ~ || Ätzfigur f. || figure f. obtenue par l'attaque des réactifs; figure f. de corrosion. / ~ formed in plaster || Gipsabguß m. || plâtre m.; moule m. en plâtre. / magnetic ~ || magnetisches Kraftlinienbild n. || spectre m. magnétique.

figure blank (Tel) || Zifferweiß n. || blanc m. de chiffres. / ~ key || Zifferblanktaste f. || touche f. du blanc de chiffres.

figure cutting machine || Figurenschere f. || cisaille f. circulaire pour découper les formes.

figured || gemustert || façonné. / ~ linen weaver || Bildweber m. || tisseur m. de toiles à figurines. / ~ weaving || Gebildweberei f. || tissage m. au Jacquard *ou* des étoffes grand façonnées. / ~ wire || Fassondraht || fil m. gaufré *ou* façonné.

figure stamp || Zahlenpunze f. || poinçon m. à chiffre.

figuring of the graduation || Bezifferung f. der Teilung || chiffrage m. de la graduation. / ~ machine (Weav) || Leinwandmaschine f. || mécanique f. à cylindre.

filament || Faden m.; Faser f. || filament m.; fil m.; fibre f. / ~ (Electr) || Glühfaden m. || filament m. / bamboo ~ || Bambusfaser f. || fibre f. de bambou. / a ~ produced by means of the bobbin procedure || ein im Spulenverfahren ersponnener Faden m. || un fil m. préparé dans la méthode de filature avec bobines. / ~ of carbon || Kohlenfaden m. || filament m. de charbon. / economical ~ (Electr) || Sparfaden m. || filament m. à faible consommation. / ~ of incandescent lamp || Glühlampenfaden m. || filament m. de lampe à incandescence. / twisted ~ || gezwirnter Faden m. || fil m. organsiné. / ~ of valve (Electr) || Heizkathode f. || cathode f. de chauffage.

filament battery ‖ Heizbatterie f. ‖ batterie f. de chauffage. / ~ current key (Electr) ‖ Zündtaste f. ‖ touche f. de courant de chauffage. / three ~ lamp (Electr) ‖ Dreifadenlampe f. ‖ lampe f. à trois filaments. / ~ screening grid ‖ Raumladungsgitter n. ‖ grille f. auxiliaire ou intérieure. / ~ temperature ‖ Glühfadentemperatur f. ‖ température f. du filament. / ~ voltage (Radio) ‖ Heizspannung f. ‖ tension f. de chauffage.

filar micrometer, position ~ ‖ Positionsfadenmikrometer n. ‖ micromètre m. de position.

file, to (Bookkeep) ‖ registrieren ‖ enregistrer. / ~ (Mach) ‖ feilen; abfeilen; befeilen ‖ limer. / ~ across ‖ überfeilen; querfeilen ‖ limer en travers. / ~ by hand ‖ im Feilkloben m. feilen ‖ limer à main. / ~ off ‖ abfeilen ‖ enlever à la lime; limer. / ~ over ‖ überfeilen ‖ passer la lime sur. / ~ out ‖ ausfeilen ‖ limer. / ~ roughly ‖ aus dem Groben n. feilen ‖ ébaucher à la lime. / ~ smooth ‖ schlichtfeilen ‖ finir à la lime.

file (Mach) ‖ Feile f. ‖ lime f. / ~ (Pap) ‖ Aktenbündel n.; Aktenstoß m.; Stoß m. Papier ‖ liasse f. / ~ (Row of persons etc.) ‖ Reihe f.; rang m.; file f. / adjusting ~ (Coin) ‖ Justierfeile f. ‖ lime f. à ajuster. / arm ~ ‖ Armfeile f. ‖ lime f. à bras; carreau m. / bastard ~ ‖ Bastardfeile f. ‖ lime f. bâtarde. / to choke the teeth of a ~ ‖ eine Feile f. verschmieren ‖ empâter la lime. / clockmaker's ~ ‖ Uhrmacherfeile f. ‖ lime f. d'horloger. / coarse ~ ‖ Grobfeile f. ‖ lime f. grosse ou rude. / to cut ~s pl. ‖ Feilen fpl. hauen ‖ tailler les limes fpl. / dead-smooth ~ ‖ Feinschlichtfeile f. ‖ lime f. superfine. / double-cut ~ ‖ zweihiebige Feile f. ‖ lime f. à taille croisée ou à taille double ou à contre-taille. / double half-round ~ ‖ Vogelzunge f. ‖ feuille f. de sauge. / dovetail ~ ‖ Schwalbenschwanzfeile f. ‖ lime f. à queue d'aronde. / entering ~ ‖ schmale Spitzfeile f.; Lochfeile f. ‖ lime f. d'entrée. / equalling ~ ‖ Zahnfeile f.; Ausstreichfeile f. ‖ lime f. à égalir. / fine-toothed ~ ‖ Mittelfeile f. ‖ lime f. moyenne. / flat ~ ‖ flache Feile f.; Ansatzfeile f. ‖ lime f. plate; plate f. à main. / hack ~ see knife ~. / half-round ~ ‖ halbrunde Feile f. ‖ lime f. demi-ronde. / hand ~ ‖ Handfeile f. ‖ lime f. plate. / handling ~s pl. ‖ Feilenbearbeitung f. ‖ travail m. des limes. / to harden ~s pl. ‖ Feilen fpl. härten ‖ tremper des limes fpl. / key ~ ‖ Schlüsselfeile f. ‖ lime f. à clef. / knife ~ ‖ Messerfeile f. ‖ lime f. à couteau. / knife ~ for cross-cut saws ‖ Schrotsägenschärffeile f. ‖ lime f. à couteau pour passepartout. / nail ~ ‖ Nagelfeile f. ‖ lime f. à ongles. / oval ~ ‖ Ovalfeile f. ‖ lime f. ovale. / pinion ~ ‖ Triebfeile f. ‖ lime f. à pignon. / to place on ~s pl. ‖ den Akten fpl. einverleiben ‖ ajouter aux archives fpl. / rapid letter ~ ‖ Schnellhefter m. ‖ classeur m. rapide. / rasping ~ ‖ Raspel f. ‖ râpe f. / to re-cut ~s pl. ‖ Feilen fpl. aufhauen ‖ retailler les limes fpl. / rough ~ ‖ Grobfeile f.; Strohfeile f. ‖ lime f. grosse ou rude ou en paille. / round ~ ‖ Rundfeile f. ‖ lime f. ronde. / saw ~ ‖ Sägenfeile f. ‖ lime f. à scies. / second-cut ~ ‖ Halbschlichtfeile f. ‖ lime f. demi-douce. / sharp ~ ‖ Stoßfeile f. ‖ lime f. à bouter. / single-cut ~ ‖ einhiebige Feile f. ‖ lime f. à taille simple. / small ~ ‖ Handfeile f. ‖ lime f. à main; limatule f. / smooth ~ ‖ Schlichtfeile f.; Abziehfeile f. ‖ lime f. douce. / ~ with spattle ‖ Spatelfeile f. ‖ lime f. à palette. / square ~ ‖ Vierkantfeile f. ‖ lime f. carrée. / standard ~ ‖ Normalfeile f. ‖ lime f. normale. / super-fine ~ ‖ Doppelchlichtfeile f. ‖ lime f. superfine. / ~ for talking machine records ‖ Schallplattenordner m. ‖ classeur m. pour disques de machines parlantes. / thinning ~ ‖ Flankierfeile f. ‖ lime f. à efflanquer. / three-square ~ ‖ Dreikantfeile f. ‖ lime f. triangulaire. / triangular ~ ‖ dreikantige Feile f. ‖ lime m. triangulaire.

file bench ‖ Feilbank f. ‖ établi m. ou banc m. à limer. / ~ cut ‖ Feilenhieb m. ‖ taille f. de lime. / ~ cutter ‖ Feilenaufhauer m. ‖ retailleur m. de limes. / ~ cutter's chisel ‖ Feilenmeißel m. ‖ étoile f. du tailleur de limes.

file cutting anew ‖ Feilenaufhauen n. ‖ retaillage m. de limes. / ~ anvil ‖ Hauamboß m. für Feilen ‖ enclume f. à retailler les limes. / ~ machine ‖ Feilenhaumaschine f. ‖ machine f. à tailler les limes.

filed (Patent) ‖ eingereicht ‖ déposé.

file dust ‖ Feilspäne mpl.; Feilstaub m. ‖ limaille f.; limature f. / ~ forger ‖ Feilenschmied m. ‖ forgeur m. de limes. / ~ grinder ‖ Feilenschleifer m. ‖ émeuleur m. de limes. / ~ handle ‖ Feilenheft n.; Feilengriff m. ‖ manche m. ou poignée f. de limes. / ~ hardener ‖ Feilenhärter m. ‖ trempeur m. de limes. / ~ hardening furnace ‖ Feilenhärteofen m. ‖ four m. à tremper les limes. / ~ maker ‖ Feilenmacher m. ‖ ouvrier m. en limes. / ~ manufacturing machine ‖ Feilenherstellungsmaschine f. ‖ machine f. à faire les limes. / ~ manufactory ‖ Feilenfabrik f. ‖ fabrique f. de limes. / ~ rasp ‖ Raspelfeile f. ‖ râpe-lime f. / ~ recutting ‖ Feilenhauen n. ‖ retaillage m. de limes. / ~ resharpening see ~ recutting.

filer, saw ~ ‖ Sägenfeiler m. ‖ limeur m. de scies.

file scratch see ~ stroke. / ~ steel ‖ Feilenstahl m. ‖ acier m. à limes. / ~ stroke ‖ Feilstrich m. ‖ trait m. ou coup m. de lime.

filet apparatus ‖ Filetrahmen m. ‖ appareil m. à jour.

file testing machine ‖ Feilenprüfmaschine f. ‖ machine f. à essayer les limes. / ~ working machine ‖ Feilenbearbeitungsmaschine f. ‖ machine f. à travailler les limes.

filigrane see filigree.

filigree ‖ Drahtarbeit f.; Filigranarbeit f. ‖ filigrane f.

filigreed paper ‖ Papier n. mit Wasserzeichen ‖ papier m. filigrané.

filigree worker ‖ Filigranarbeiter m. ‖ filigraniste m.

filing ‖ Feilen n. ‖ limure f. / ~ block ‖ Feilholz n.; Feilfutter n. ‖ bois m. à limer. / ~ board see ~ block. / ~ machine ‖ Feilmaschine f. ‖ limeuse f.; machine f. à limer. / ~ and sawing machine ‖ Feil- und Sägemaschine f. ‖ machine f. à limer et à scier.

filings pl. ‖ Feilspäne mpl. ‖ limaille f.

filing vice ‖ Feilkloben m. ‖ étau m. à limer.

fill, to ‖ füllen; laden; beschicken ‖ remplir; charger; emplir. / ~ a battery ‖ eine Batterie f. füllen. ‖ remplir une batterie. / ~ and fasten the kibble (Mine) ‖ anschlagen ‖ charger et attacher les barriquets. / ~-in the joints with mortar (Build) ‖ die Fugen fpl. mit Mörtel vergießen ‖ sceller les joints mpl. en mortier. / ~ the sails ‖ (ab)brassen ‖ brasser; décharger les voiles. / ~ up (Brew) ‖ abfüllen ‖ mettre en fûts mpl.; soutirer. / ~ up (Liquid) ‖ vollgießen; ausgießen; ausfüllen ‖ sceller. / ~ up (Soil) ‖ anschütten; aufschütten; ausfüllen ‖ remblayer; remplir. / ~ up a ditch with rubbish ‖ einen Graben mit Schutt ausfüllen ‖ remblayer un fossé. / ~ up a hollow tooth ‖ einen Zahn m. plombieren ‖ plomber une dent. / ~ up the joints with garretings ‖ verzwicken ‖ garnir les joints de cales.

filled section rail (Railw) ‖ Blockschiene f.; Vollschiene f.; Zungenschiene f. ‖ rail m. de section pleine ou en U renversé. / ~ with point of blade welded-on (Railw) ‖ Blockschiene f. mit angeschweißtem Zungenende ‖ rail m. de section pleine avec pointe d'aiguille raccordée par soudage.

filled up with hot liquid material ‖ mit flüssiger heißer Masse vergossen ‖ scellé d'une masse liquide à chaud.

filler (Chem) ‖ Grundierlack m. ‖ laque f. de première couche. / ~ (Furniture) ‖ Verkitter m. ‖ mastiqueur m. / ~ (Person) ‖ Beschicker m. ‖ chargeur m.; enfourneur m. / ~ (Saddl) ‖ Polsterholz n.; Stopfholz n. ‖ rembourroir m. / coke ~ ‖ Kokslader m. ‖ chargeur m. de coke. / ore ~ ‖ Erzlader m. ‖ chargeur m. de minerai. / wagon ~ (Mine) ‖ Kohlenwagenfüller m. ‖ chargeur m. de wagonnets ou berlines.

filler cap ‖ Füllschraube f.; Füllöffnung f. ‖ bouchon m. ou orifice m. de remplissage. / ~ tank (Auto) ‖ Einfüllstutzen m. des Behälters ‖ bouchon m. de réservoir.

filler plug of a lubricator ‖ Füllschraube f. eines Schmierapparates ‖ bouchon m. de remplissage d'un graisseur. / ~ stripper (Cigar) ‖ Einlageripper m. ‖ écôteur m. de tripe.

fillet (Arch) ‖ Leistchen n.; Plättchen n.; Riemchen n.; Saum m.; Steg m. ‖ bandelette f.; fillet m.; listel m. carré; réglet m. / ~ (Mach) ‖ Anlauf m.; Ausrundung f. ‖ congé m. / ~ of a screw ‖ Gewinde f. einer Schraube ‖ filet m. ou pas m. d'une vis.

fillet card (Weav) ‖ Deckelkarde f. ‖ carde f. à feuilles. / ~ plane ‖ Leistenhobel m. ‖ tarabiscot m.; bouvet m. à doucine; grain m. d'orge.

fill furnace ‖ Füllofen m. ‖ four m. au gonflement.

filling ‖ Füllung f. ‖ remplissage m. / ~ (Material) ‖ Füllmasse f. ‖ matière f. de remplissage. / ~ (Weav) ‖ Einschlag m.; Einschußgarn n. ‖ trame f.; fil m. pour trame. / ~ of empty spaces in velvet ‖ Ausbesserung f. fehlerhafter Stellen im Samt ‖ resarcissage m. de velours. / attachment for ~ into sacks ‖ Absackvorrichtung f. ‖ dispositif m. d'ensachage ou à remplir les sacs. / ~ with sand (Hydr arch) ‖ Versandung f. ‖ ensablement m. / ~ into wagons (Railw) ‖ Aufladen n. in Waggons ‖ chargement m. ou mise f. en wagon. / ~ of a wall (Build) ‖ Füllmund m. einer

Füllmauer ‖ remplage m. *ou* emplage m. d'un mur.

filling apparatus ‖ Füllapparat m. ‖ appareil m. de remplissage. / ~ apparatus (Brew) ‖ Abfüllapparat m. ‖ soutireuse f. / ~ bottle for travellers ‖ Reisetintenflasche f. ‖ bouteille f. à encre pour le voyage. / ~ chest process ‖ Füllkastenverfahren n. ‖ procédé m. à boîte de remplissage. / ~ chute ‖ Füllkanal m. ‖ conduit m. réservoir. / ~ compound ‖ Ausgußmasse f.; ‖ pâte f. de remplissage. / ~ device (Ice manufacture) ‖ Eiszellenfüller m. ‖ appareil m. de remplissage; remplisseur-jaugeur m. / ~ duration ‖ Füllungsdauer f. ‖ durée f. de remplissage. / ~ earth ‖ Füllerde f. ‖ terre f. à remblai. / ~ frame (Found) ‖ Aufsetzrahmen m. ‖ cadre m. supérieur. / ~ frame (Shipb) ‖ Füllspant n. ‖ couple m. de remplissage. / ~ height ‖ Füllhöhe f. ‖ hauteur f. de remplissage. / ~ hole ‖ Einfüllöffnung f. ‖ encoche f. d'entrée; trou m. de remplissage. / ~ hole plug ‖ Füllochschraube f. ‖ vis f. du trou de remplissage. / ~ hopper ‖ Fülltrichter m. ‖ trémie f. de chargement. / ~-in of a wall ‖ Füllmund m. einer Füllmauer ‖ remplage m. *ou* emplage m. d'un mur. / ~-in work (Build) ‖ Füllmauerwerk n. ‖ remplage m.; muraille f. de remplage. / ~ machine ‖ Füllmaschine f. ‖ mécanisme m. de remplissage. / ~ and corking machine ‖ Füll- und Korkmaschine f. ‖ tireuse-boucheuse f. / ~ material ‖ Füllmaterial f. ‖ matériel m. de remplissage. / ~ paste (Acc) ‖ Füllmasse f. ‖ pâte f. de remplissage. / ~ pencil ‖ Füllbleistift m. ‖ crayon m. à mines de rechange. / ~ piece (Mach) ‖ Futter n. ‖ garniture f. / ~ piece fitted across the opening between the pedestal legs ‖ quer über dem Rahmenausschnitt zwischen den Achsstreben angeordneter Steg m. ‖ pièce f. de remplissage qui se trouve en travers de la découpure entre les jambes. / ~ pipe ‖ Füllrohr n. ‖ tuyau m. de remplissage. / ~ plant ‖ Füllanlage f. ‖ installation f. de remplissage. / ~ plate of locomotive frame ‖ Lokomotivrahmenfutterstück n. ‖ fourrure f. de châssis de locomotive. / ~ rack ‖ Abfüllstation f. ‖ station f. de mise en bouteilles. / ~ hydraulic press for ~ rolls ‖ hydraulische Walzenpresse f. ‖ presse f. hydraulique à cylindres. / ~ room ‖ Füllkammer f.; Füllstube f. ‖ empli m. / ~ sleeve ‖ Füllansatz m. ‖ manche f. de gonflement *ou* de remplissage *ou* d'apprentice. / ~ slot *see* ~ hole. / ~ stone ‖ Füllstein m. ‖ blocaille f. / ~ substance ‖ Füllstoff m. ‖ matière f. inerte. / ~ tap (Brew) ‖ Abfüllhahn m. ‖ robinet m. de soutirage. / ~ timber *see* ~ frame. / ~ tube (Brew) ‖ Flaschenfüllrohr n. ‖ canule f.

filling-up (Build) ‖ Anschüttung f. ‖ remblai m. / ~ material ‖ Spülversatz m. ‖ remblayage m. hydraulique. / ~ pipe (Brew) ‖ Abfüllschlauch m. ‖ tuyau m. de soutirage.

fillister ‖ Falzhobel m. ‖ feuilleret m. / side ~ ‖ Plattbank f.; Plattenhobel m. ‖ feuilleret m. à plates-bandes. / ~ head (Screw) ‖ runder Kopf m. ‖ tête f. ronde. / ~ screw head ‖ runder Schraubenkopf ‖ tête f. de vis ronde.

fill-up valve ‖ Füllventil n. ‖ soupape f. de remplissage.

film (Chem) ‖ Häutchen n.; Überzug m.; Schicht f. ‖ pellicule f.; couche f. / ~ (Phot) ‖ Film m.; Filmband n. ‖ film m.; pellicule f. (photographique *ou* cinématographique). / ~ of acid ‖ Säureschicht f. ‖ couche f. d'acide. / black and white ~ ‖ Schwarzweißfilm m. ‖ film m. *ou* image m. noir-blanc. / brush contact ~ (Electr) ‖ Bürstenübergangsschicht f. ‖ pellicule f. de contact des balais. / celluloid ~ ‖ Zelluloidfilm m. ‖ film m. en celluloïde. / to change the ~ ‖ den Filmstreifen m. auswechseln ‖ remplacer la pellicule du film. / chloride of silver celluloid ~ ‖ Chlorsilberzelluloidfilm m. ‖ pellicule f. en celluloïde au chlorure d'argent. / cinematographic ~ ‖ Film m.; Kinofilm m. ‖ film m. cinématographique. / cinematographic raw ~ ‖ Kinorohfilm m. ‖ film m. brut cinématographique. / coloured ~ ‖ kolorierter *oder* bunter Film m.; Farbenfilm m. ‖ film m. coloré. / to fix a ~ ‖ einen Film m. fixieren *oder* lichtunempfindlich machen ‖ rendre une pellicule inaltérable à la lumière; fixer une pellicule. / graphitic ~ ‖ Graphitüberzug m. ‖ pellicule f. de graphite. / to harden the ~ ‖ den Film m. härten ‖ durcir la pellicule. / music ~ ‖ Musikfilm m. ‖ film m. musical. / non-inflammable ~ ‖ nicht entzündlicher Film ‖ filme m. ininflammable. / ~ of oxide ‖ Oxydschicht f.; Lufthaut f. ‖ couche f. d'oxyde. / to remove the ~ of oxide ‖ die Oxydschicht f. entfernen ‖ enlever la couche f. d'oxyde. / ~ with photographed sound ‖ Lichttonfilm m.; Film m. mit fotografiertem Ton ‖ film m. à sons photographiés. / photographic ~ ‖ Film m.; Rollfilm m. ‖ film m. photographique. / portrait ~ ‖ Porträtfilm m. ‖ portrait-film m. / raw ~ ‖ Rohfilm m. ‖ film m. brut. / Röntgen ~ *see* X-ray ~. / roll ~ ‖ Rollfilm m. ‖ roll-film m. / ~ of rust ‖ Rosthaut f. ‖ pellicule f. de rouille. / selenium ~ ‖ Selenschicht f. ‖ couche f. de sélénium. / sensitized ~ ‖ lichtempfindlich gemachter Film m. ‖ pellicule f. sensibilisée. / sound ~ ‖ Tonfilm m.; Klangfilm m. ‖ film m. parlant. / speaking ~ *see* talking. / talking ~ ‖ sprechender Film m.; Sprechfilm m.; Klangfilm m.; Tonfilm m. ‖ film m. parlant. / white ~ ‖ Schimmel m. (Pilz) ‖ chancissure f.; moisi m.; moisissure f. / X-ray ~ ‖ Röntgenfilm m. ‖ film m. radiographique. / zellite ~ ‖ Zellitfilm m. ‖ film m. en zellite.

film band ‖ Filmband n.; Filmstreifen m.; pellicule f. / ~ camera ‖ Filmkamera f.; Filmaufnahmeapparat m. ‖ chambre f. *ou* appareil m. à pellicules. / ~ car ‖ Filmwagen m. ‖ voiture f. à films. / ~ cup ‖ Filmschale f. ‖ coupe f. pour films. / ~ cut-out ‖ Durchschlagsicherung f. ‖ dispositif m. de protection contre les claquages. / ~ demonstration ‖ Filmvorführung f.; Filmwiedergabe f. ‖ présentation f. du film. / ~ drum ‖ Filmtrommel f. ‖ tambour m. à film *ou* à pellicule. / roller ~ guide ‖ Filmkurve f.; Führungsschiene f. ‖ glissière f. pour films. / ~ lending institute ‖ Filmverleih m. ‖ institut m. de louage de films. / ~ manufacture ‖ Filmfabrikation f.; Filmfertigung f.; Filmherstellung f. ‖ fabrication f. de films. / ~ material ‖ Filmwerkstoff m. ‖ matériel m. de film. / ~ pack ‖ Packfilm m.; Filmpack m. ‖ film-pack m.; block-film m. / ~ producing machine ‖ Filmherstellungsmaschine f. ‖ machine f. à fabriquer des films. / ~ reel *see* ~ spool. / concern for renting ~s ‖ Filmverleih m. ‖ location f. de films. / ~ reproduction ‖ Filmaufnahme f. ‖ réproduction f. d'une scène sur un film. / drum for ~ ribbon ‖ Trommel f. zur Auflage des Filmbandes ‖ tambour m. à enrouler le film. / ~ roll ‖ Filmrolle f. ‖ rouleau m. de film. / ~ roll cutting machine ‖ Filmrollenschneidmaschine f. ‖ machine f. à découper les rouleaux de film. / ~ roller ‖ Filmwalze f. ‖ cylindre m. pour films. / ~ spool ‖ Filmspule f. ‖ bobine f. de film *ou* de pellicule.

film studio ‖ Filmatelier n.; Aufnahmeatelier n.; Aufnahmeraum m. ‖ studio m. / effect lamp for ~ ‖ Kinofilmeffektlampe f.; Effektaufnahmelampe f. ‖ lampe f. destinée à produire certains effets de lumière dans les prises de vues cinématographiques. / lamp for ~ ‖ Atelieraufnahmelampe f.; Filmaufnahmelampe f.; Aufnahmelampe f. ‖ lampe f. pour prises de vues cinématographiques; lampe f. de prise de vue dans les studios.

film travel mechanism ‖ Filmtransportvorrichtung f. ‖ mécanisme m. d'avance du film. / ~ waste ‖ Filmabfall m. ‖ déchet m. de films. / ~ winding mechanism ‖ Filmaufwickelvorrichtung f. ‖ mécanisme m. d'enroulement de la pellicule. / ~ works pl. ‖ Filmfabrik f. ‖ fabrique f. de films.

filmy ‖ häutig; mit dünnem Häutchen n. bedeckt ‖ pelliculé.

filter, to ‖ filtern; abklären ‖ filtrer; décanter; suinter. / ~ hot ‖ heiß filtern ‖ filtrer chaud.

filter (Chem) ‖ Filter n.; Seiher m. ‖ filtre m.; couloire f. / ~ (Tel) ‖ Kettenleiter m.; Siebkette f.; Sperrkette f. ‖ système m. iteratif; filtre m. / ~ air ‖ Luftfilter n.; Luftreiniger m. ‖ filtre m. à air. / amalgam ~ ‖ Amalgamfilter n. ‖ filtre m. pour amalgame. / charcoal ~ ‖ Holzkohlenfilter n. ‖ filtre m. en charbon de bois. / gas mask with chemical ~ ‖ Gasmaske f. mit Chemikalfilter ‖ masque m. à gaz à filtres chimiques. / ~ for chromophotography ‖ Filter n. für Farbenfotografie ‖ filtre m. pour photographie en couleurs. / coke ~ ‖ Koksfilter n. ‖ filtre m. à coke. / colour ~ ‖ Farbfilter n. ‖ écran m. coloré. / dust ~ ‖ Staubfilter n. ‖ filtre m. à poussière. / ~ of the fresh water condensing apparatus ‖ Destillierapparatfilter n. ‖ filtre m. du condenseur à eau douce. / gravel ~ ‖ Kiesfilter n. ‖ filtre m. à gravier. / green ~ ‖ Grünfilter n. ‖ écran m. vert. / high-pass ~ (Tel) ‖ Kondensatorleiter f. ‖ filtre m. passe-haut. / to load a ~ ‖ ein Filter n. einlegen ‖ charger un filtre. / low-pass ~ ‖ Spulenleiter m.; Niederfrequenzsiebkette f. ‖ filtre m. passe-bas. / oil ~ ‖ Ölfilter n. ‖ filtre m. à huile. / press ~ ‖ Druckfilter n. ‖ filtre m. à pression. / rapid ~ ‖ Schnellfilter n. ‖ filtre m. rapide. / sand ~ ‖ Sandfilter n. ‖ filtre m. au sable. / sewage ~ ‖ Klärteich m. ‖ bassin m. de décantation. / suction cell ~ ‖ Saugzellenfilter n. ‖ filtre m. à cellules d'aspiration. / unfolded ~ ‖ Rundfilter n. ‖ filtre m. sans plis. / water ~ ‖ Wasserfilter n. ‖ filtre m. à eau.

filter beaker ‖ Filterbecher m. ‖ gobelet m. de filtration. / ~ bed ‖ Filterbett n.; Filtrationsschicht f.; Kläranlage f. ‖ couche f. filtrante ou de filtrage. / ~ box ‖ Filterkasten m. ‖ boîte f. filtrante. / ~ box for inks ‖ Farbenfilter n. ‖ filtre m. à encres. / ~ casing ‖ Filtergehäuse n. ‖ boîte f. de filtration. / ~ charcoal ‖ Filterkohle f. ‖ charbon m. à filtres. / ~ chamber ‖ Filterkammer f. ‖ chambre f. filtrante ou de filtration ou de filtre.

filter cloth ‖ Filterstoff m.; Filtertuch n.; Seihetuch n. ‖ étoffe f. ou étamine f. à filtrer; blanchet m.; couloire f.; filtre m. en toile; toile f. filtrante. / ~ washing machine ‖ Filtertuchwaschmaschine f. ‖ machine f. à laver les tissus filtrants.

filtered ‖ gefiltert ‖ filtré. / ~ arc light ‖ gefiltertes Bogenlicht n. ‖ lumière f. fournie par un arc électrique muni d'un filtre.

filter frame ‖ Filterrahmen m. ‖ cadre m. filtrant ou de filtre. / ~ gauze ‖ Filtergaze f. ‖ gaze f. à filtrer. / ~ housing ‖ Filtergehäuse n. ‖ carter m. du filtre.

filtering ‖ Filterung f.; Filtern n. ‖ filtration f.; filtrage m.

filtering ‖ filternd ‖ filtrant. / ~ apparatus ‖ Filterapparat m.; Filtrierapparat m. ‖ appareil m. à filtrer. / ~ bag ‖ Filtersack m. ‖ sac m. filtrant. / ~ bag of trellis ‖ Seihesack m. von Zwillich ‖ chausse f. en toile de treillis pour filtrer. / ~ basin ‖ Filterbecken n.; Klärbecken n. ‖ bassin m. de décantation. / ~ capacity ‖ Seihvermögen n. ‖ pouvoir m. de filtration. / ~ charcoal ‖ Filterkohle f. ‖ charbon m. à filtrer. / ~ cone ‖ Filterkonus m. ‖ cône m. à filtrer. / ~ crucible ‖ Filtertiegel m. ‖ creuset m. à filtration. / ~ funnel ‖ Seihetrichter m. ‖ entonnoir m. à filtre; filtre m. / ~ gravel. ‖ Filterkies m. ‖ gravier m. à filtrer. / ~ mass washing apparatus ‖ Filtermassewaschapparat m. ‖ appareil m. de lavage de la masse à filtrer. / ~ material ‖ Filtermaterial n. ‖ matériel m. de filtrage. / ~ paper ‖ Filterpapier n. ‖ papier-filtre m.; papier m. à filtrer. / ~ pipette ‖ Filterpipette f. ‖ pipette f. à filtration. / ~ plant ‖ Filteranlage f. ‖ installation f. de filtres ou de filtrage ou de filtration. / ~ press ‖ Filterpresse f. ‖ presse-filtre f. / ~ pulp ‖ Filtermasse f. ‖ masse f. filtrante. / ~ stone ‖ Filterstein m.; Filtersandstein m.; Filterkalkstein m. ‖ pierre f. filtrante ou à filtrer; grès m. filtrant. / ~ tank see ~ basin. / ~ tube ‖ Filterrohr n. ‖ tube-filtre m. / ~ tube filled with asbestos ‖ Asbestfilterrohr n. ‖ tube-filtre m. d'amiante.

filter installation ‖ Filteranlage f. ‖ installation f. de filtration. / ~ layer ‖ Filterbett n.; Filtrationsschicht f. ‖ couche f. filtrante. ~ lid ‖ Filterdeckel m. ‖ couvercle m. de filtre. / ~ man ‖ Filtrierer m. ‖ filtreur m. / ~ mass ‖ Filtermasse f. ‖ masse f. filtrante. / ~ paper ‖ Filterpapier n. ‖ papier m. filtrant. / ~ pipe ‖ Filterrohr n. ‖ tube-filtre m.; tuyau-filtre m.

filter plate ‖ Filterplatte f. ‖ plaque f. filtrante. / finely channelled ~ ‖ fein kannellierte Filterplatte f. ‖ plaque f. filtrante à cannelures fines.

filter press ‖ Filterpresse f. ‖ filtre-presse f. / experimental ~ ‖ Versuchsfilterpresse f. ‖ filtre-presse m. à essais. / laboratory ~

‖ Laboratoriumsfilterpresse f. ‖ filtrepresse m. pour laboratoire. / long ~ ‖ langgebaute Filterpresse f. ‖ filtre-presse m. de grande dimension.

filter-press cloth ‖ Filterpressentuch n. ‖ toile f. de filtre-presse. / ~ pump ‖ Filterpreßpumpe f. ‖ pompe f. de pression à filtre.

filter pulp ‖ Filtermasse f. ‖ masse f. filtrante. / ~ stand ‖ Filterstativ n.; Filtergestell n. ‖ support m. de filtre ou d'entonnoir.

filter stuff ‖ Filtermasse f. ‖ masse f. filtrante. / ~ press ‖ Filtermassenpresse f. ‖ presse f. à masse filtrante. / ~ washer ‖ Filtermassenwaschapparat m. ‖ appareil m. à laver la masse filtrante. / ~ washing and sterilizing apparatus ‖ Filtermassenwasch- und -sterilisierapparat m. ‖ appareil m. à laver et à stériliser la masse filtrante.

filter surface ‖ Filterfläche f. ‖ surface f. filtrante. / ~ transformer ‖ Filtertransformator m. ‖ transformateur m. filtre. / ~ wadding ‖ Filterwatte f. ‖ ouate f. à filtrer. / ~ washer ‖ Filterwäscher m. ‖ laveur m. de filtres.

filth in cloth ‖ Schmutzknötchen n. im Tuch ‖ épouti m. dans le drap.

filtrate, to ‖ durchfiltern ‖ filtrer. / ~ off ‖ abfiltern ‖ séparer par filtration f.

filtrate ‖ Filtrat n. ‖ filtrat m.; filtratum m.; produit m. filtré.

filtration ‖ Filtern n. ‖ filtration f. / ~ (Brew) ‖ Abläuterung f. ‖ filtration f. / ~ chamber ‖ Filterkammer f. ‖ chambre f. de filtration. / ~ flask ‖ Filterflasche f. ‖ flacon m. à filtrer. / ~ paper ‖ Filterpapier n. ‖ papier m. à filtrer. / ~ plant with pump service ‖ Filteranlage f. mit Pumpenbetrieb ‖ installation f. de filtrage avec pompe. / ~ vat ‖ Filterschale f. ‖ bac m. de filtration; cuve f. filtrante.

fin (Aero) ‖ Flosse f.; Kielflosse f.; Leitfläche f. ‖ dérive f.; empennage m.; plan m. de quille.

final ‖ endgültig ‖ définitif. / ~ account ‖ Abschlußrechnung f. ‖ compte m. en définitive. / ~ amplification (Tel) ‖ Endverstärkung f. ‖ amplification f. finale. / ~ attenuation ‖ Endvergärung f. ‖ atténuation f. finale. / degree of ~ attenuation ‖ Endvergärungsgrad m. ‖ degré m. d'atténuation limite. / ~ declaration of a concessionaire in respect of his concession ‖ Schlußerklärung f. (Mutung) ‖ déclaration f. définitive d'un demandeur concernant son gîte. / ~ delivery ‖ Restlieferung f.; Schlußlieferung f. ‖ livraison f. finale. / ~ dividend ‖ Schlußdividende f. ‖ solde m. du dividende. / ~ effect ‖ Endergebnis n. ‖ effet m. final; résultat m. / ~ felling ‖ gänzlicher Abhieb m. ‖ coupe f. définitive. / the rolls pl. get a ~ grinding ‖ Walzen fpl. nachschleifen ‖ finir les cylindres mpl. à la meule. / ~ instalment ‖ Restzahlung f. ‖ payement m. pour solde. / ~ kilning temperature (Malt) ‖ Abdarrtemperatur f. ‖ température f. finale de touraillage. / ~ mashing heat ‖ Abmaischtemperatur f. ‖ température f. finale de saccharification. / ~ project ‖ endgültiger Entwurf m. ‖ projet m. définitif. / ~ result ‖ Endergebnis n. ‖ résultat m. final. / ~ result of the calculation ‖ Endergebnis n. der Rechnung ‖ résultat m. final du calcul. / ~ selector

(Aut tel) ‖ Linienwähler m. ‖ sélecteur m. final ou de ligne. / ~ stage (Forest) ‖ Schlag m. für den Abtrieb ‖ partie f. pour la coupe. / ~ velocity ‖ Endgeschwindigkeit f. ‖ vitesse f. finale. / ~ voltage on charge ‖ Endspannung f. der Ladung ‖ tension f. à la fin de la charge. / ~ voltage on discharge ‖ Endspannung f. der Entladung ‖ tension f. à la fin de la décharge.

finance, to ‖ finanzieren ‖ commanditer.

finance ‖ Geldwesen n.; Finanzwesen n.; Finanzwirtschaft f. ‖ finances fpl. / ~ stamp ‖ Effektenstempel m. ‖ timbre m. d'effets.

financial ‖ finanziell ‖ financier. / ~ business see finance. / ~ concerns pl. see finance. / ~ embarrassment ‖ Geldverlegenheit f. ‖ embarras m. financier ou d'argent; gêne f. / ~ enterprise ‖ Geldgeschäft n. ‖ affaire f. d'argent ou de banque. / ~ records filing department (Bank) ‖ finanzstatistische Abteilung f. ‖ service m. de documentation financière. / ~ situation ‖ Vermögenslage f.; Finanzlage f. ‖ situation f. financière. / ~ statement ‖ Finanzbericht m. ‖ état m. financier. / ~ world ‖ Finanzwelt f. ‖ monde m. financier; (haute) finance f. / ~ year ‖ Betriebsjahr n.; Etatsjahr n.; Finanzjahr n. ‖ exercice m.; année f. financière.

financially powerful see ~ strong. / ~ strong ‖ kapitalkräftig ‖ disposant d'importants capitaux.

financier ‖ Finanzmann m. ‖ financier m.

financiers pl. ‖ Geldleute pl. ‖ gens mpl. de finance.

find, to ‖ finden ‖ trouver. / ~ by boring (Mine) ‖ erbohren ‖ découvrir par la sonde. / ~ a fault ‖ einen Fehler m. feststellen ‖ relever une erreur. / ~ one's way about ‖ sich zurechtfinden ‖ s'orienter.

finder ‖ Entdecker m. ‖ découvreur m. / ~ (Duty) ‖ Zollvisitator m. ‖ inspecteur m. des douanes; douanier m. / ~ (Opt) ‖ Suchglas n.; Sucher m. ‖ chercheur m. trouveur m.; viseur m. / comet ~ ‖ Kometensucher m. ‖ chercheur m. de comètes. / ~ of a telescope ‖ Sucher m. eines Fernrohres ‖ chercheur m. d'une lunette. / x-times ~ (Opt) ‖ x-facher Sucher m. ‖ chercheur m. x fois.

finder circle (Spectroscope) ‖ Finderteilkreis m. ‖ cercle m. de repère. / ~ eyepiece ‖ Sucherokular n. ‖ oculaire-chercheur m. / ~ telescope ‖ Richtfernrohr n. ‖ lunette f. de direction ou de repère.

finding ‖ Auffinden n. ‖ découverte f. / ~ (Object) ‖ Fund m. ‖ objet m. trouvé. / ~ the centre of gravity of a body ‖ Bestimmung f. des Schwerpunkts eines Körpers; Schwerpunktsbestimmung f. ‖ recherche f. du centre de gravité d'un corps. / ~ the ship's place ‖ Ortsbestimmung f. eines Schiffes ‖ détermination f. du point du navire.

fine, to (Met) ‖ frischen ‖ affiner; raffiner; décarburer; mazer. / ~ the iron ‖ das Eisen frischen ‖ affiner la fonte.

fine ‖ fein ‖ fin. / ~ adjusting screw ‖ Feineinstellschraube f. ‖ vis f. micrométrique.

fine adjustment ‖ Feineinstellung f. ‖ mise f. au point juste; mise f. au point précise. / clamp for ~ ‖ Klemmvorrichtung f. der Feinverstellung ‖ dispositif m. de blocage du mouvement lent. / ~ screw ‖ Fein(ein)stellschraube f. ‖ vis f. de fin calage.

fine-art forge ‖ Kunstschmiede f. ‖ forge f. artistique. / ~ object ‖ Kunstgegenstand m.; kunstgewerblicher Gegenstand m. / objet m. d'art.

fine baker's ware ‖ Feinbackware f. ‖ boulangerie-pâtisserie f. / ~ carding of jute ‖ Feinkrempeln n. der Jute ‖ cardage m. en fin du jute. / ~ ceramic ‖ Feinkeramik f. ‖ céramique f. fine. / ~ chain (Jewel) ‖ Kette f. mit feinen Ringen ‖ chaîne f. en jaseran.

fine coal ‖ Feinkohle f. ‖ fines fpl.; menu charbon m.; charbonnaille f. / ~ sump ‖ Feinkohlensumpf m. ‖ fosse f. à fines.

fine compasses pl. ‖ Genauigkeitszirkel m. ‖ compas m. droit fin. / ~-count mill ‖ Feinspinnerei f. ‖ filature f. de fin numéros. / ~ crusher (Ore dress) ‖ Feinmühle f. ‖ broyeur m. des fins. / ~ crushing ‖ Feinmahlung f.; Feinzerkleinerung f. ‖ broyage m. fin. / ~ cut ‖ Schlichthieb m. ‖ taille f. douce. / ~ darning ‖ Kunststopfen n. ‖ stoppage m. / ~ drawer (Spinn) ‖ Feinstrecker m. ‖ étireur m. en fin. / ~ emery cloth ‖ feines Schmirgelleinen n. ‖ toile-émeri f. fine; toile f. à l'émeri fine; toile f. émerisée fine. / ~ filter for fuel ‖ Brennstoffeinfilter m. ‖ filtre m. fin du combustible. / ~ flax ‖ feiner Flachs m. ‖ lin m. fin. / ~ fluting ‖ feine Riffelung f. ‖ cannelures fpl. fines. / ~ gold ‖ Feingold n. ‖ or m. fin. / ~ and good coffee ‖ guter Kaffee m. ‖ café m. bon. / ~ grain ‖ Feinkorn n.; feines Korn n. ‖ grain m. fin.

fine-grained (Metal) ‖ feinkörnig ‖ à grain m. fin. / ~ (Wood) ‖ feinfaserig ‖ à fibre f. fine. / ~ carbon ‖ feinkörnige Kohle f. ‖ charbon m. à fin grain. / ~ grinding wheel ‖ feinkörnige Schleifscheibe f. ‖ meule f. à grain fin. / ~ iron ‖ feinkörniges Eisen n. ‖ fer m. à grain fin. / ~ powder ‖ feinkörniges Pulver n. ‖ poudre f. à grains fins. / ~ semi-coke ‖ feinkörniger Halbkoks m. ‖ semi-coke m. sous forme de grains très fins. / ~ texture ‖ feinkörniges Gefüge n. ‖ structure f. à grains fins.

fine grain washing machine ‖ Feinkornsetzmaschine f. ‖ machine f. laveuse de fins. / ~-granular see ~-grained. / ~ gravel ‖ Feinkies m. ‖ menu gravier m. / ~ grinding see ~ crushing. / ~ grist ‖ Feinschrot n. ‖ mouture f. fine.

fine iron ‖ Kleineisen n. ‖ petit-fer m. / ~ (Metal) ‖ Feineisen n. ‖ mazée f.; fonte f. mazée; fer m. fin ou raffiné. / ~ mill ‖ Feineisenstraße f. ‖ laminoir m. ou train m. à fers fins. / ~ structure ‖ Feineisenkonstruktion f. ‖ construction f. en fer fin. / ~ train see ~ mill.

fine liquor ‖ Klärsel n. ‖ clairce f.; clairée f. / ~ mechanics pl. ‖ Feinmechanik f. ‖ mécanique f. fine ou de précision. / ~ metal ‖ gefeintes Roheisen n.; Feineisen n. ‖ métal m. fin; fonte f. raffinée; fonte f. mazée. / ~ paper making machine ‖ Feinpapierherstellungsmaschine f. ‖ machine f. à fabriquer le papier fin. / ~ pitch spur wheel ‖ fein geteiltes Stirnrädchen n. ‖ roue f. droite à dents fines. / ~ polishing wheel ‖ Glanzscheibe f. ‖ meule f. (en) émeri douce ou à fin grain. / ~ rain ‖ Sprühregen m. ‖ pluie f. fine. / ~ regulation of the primary current ‖ Feinregelung f. des Primärstromes ‖ réglage m. précis du circuit primaire. / ~ semi-coke made from bituminous coal ‖ feinkörniger Steinkohlenhalbkoks m. ‖

semi-coke m. de houille à grains fins. / ~ silver ‖ Feinsilber n. ‖ argent m. fin. / ~ thread ‖ Feingewinde n. ‖ filet m. fin. / ~ thread silk ‖ feinfädige oder feine Rohseide f. ‖ soie f. grège à brin fin. / ~ wire clothing (Spinn) ‖ feiner Kratzenbeschlag m. ‖ garniture f. de carde à dents fines. / ~ wire rope ‖ feindrähtiges Seil n. ‖ câble m. métallique en fils minces. / ~ yarn ‖ feines Garn n. ‖ fil m. fin.

fine ‖ Geldstrafe f. ‖ amende f. / ~ (Trade) ‖ Reugeld n. ‖ dédit m.; folle enchère f. / ~ for retarded delivery ‖ Konventionalstrafe f. für verspätete Lieferung ‖ retenue f. du chef des retards.

finely powdered (Chem) ‖ feinverteilt ‖ finement divisé.

fineness of bow lines ‖ Feinheit f. oder Schärfe f. des Vorschiffes ‖ finesse f. de la proue. / ~ of cocoon filament ‖ Feinheit f. des Kokonfadens ‖ finesse f. du filament de cocon. / coefficient of ~ (Shipb) ‖ Völligkeitszahl f. ‖ coefficient m. de finesse. / degree of ~ ‖ Feinheitsgrad m. ‖ degré m. de finesse. / ~ for and aft (Shipb) ‖ Feinheit f. oder Schärfe f. der Schiffsenden ‖ finesse f. des extrémités du bateau. / ~ of gold ‖ Feinheit f. des Goldes ‖ finesse f. ou titre m. d'or. / ~ of grain ‖ Korngröße f. ‖ grosseur f. de grain. / ~ of grinding ‖ Mahlfeinheit f. ‖ finesse f. de mouture. / ~ of grist ‖ Mehlfeinheit f. ‖ degré m. ou finesse f. de mouture. / ~ of lines ‖ Feinheit f. der Linien ‖ finesse f. des lignes. / ~ of the screw ‖ Feinheit f. des Gewindes / pas m. fin f. du filet. / ~ of the stern ‖ Feinheit f. oder Schärfe f. des Hinterschiffes ‖ finesse f. de la poupe.

fineness ratio of the strut ‖ Schlankheitsverhältnis n. der Strebe ‖ allongement m. du mât.

finer (Metal) ‖ Frischer m. ‖ affineur m.; mazeur m.

finery (Metal) ‖ Frischwerk n. ‖ finerie f.; mazerie f.; usine f. d'affinage. / ~ cinders pl. ‖ Frischschlacke f. ‖ scorie f. des feux d'affinerie; laitier m. de forge. / ~ forge see ~ furnace. / ~ furnace ‖ Frischfeuer n.; Frischherd m. ‖ forge f.; feu m. de forge ou d'affinerie; fourneau m. d'affinage. / ~ iron ‖ Frischeisen n. ‖ fer m. affiné.

fines pl. see fine coal.

finger ‖ Finger m. ‖ doigt m. / adjustable ~ ‖ (ver)stellbarer Finger m. ‖ doigt m. réglable. / catch ~ ‖ Klinke f. ‖ cliquet m. / centering ~ ‖ Rückstelldaumen m. ‖ doigt m. de rappel. / change ~ ‖ Anstoßstange f.; Auslösestift m. ‖ tringle f. de butée. / distributing ~ ‖ Steuerfinger m.; Steuerstift m. ‖ doigt m. ou goupille f. de distribution. / fixed ~ ‖ feststehender Finger ‖ doigt m. fixe. / guide ~ ‖ Führungsfinger m. ‖ doigt-guide m. / indicator ~ ‖ Drehungszeiger m. ‖ doigt m. indicateur. / lifting ~ ‖ Hebedaumen m. ‖ came f. de soulèvement. / locking ~ ‖ Anschlag m. ‖ doigt m. de butée.

finger arm ‖ Kontakthebel m. ‖ manette f. du contrôleur. / ~ board ‖ Klaviatur f.; Griffbrett n.; Tastenbrett n. ‖ clavier m. / ~ cot see ~ stall. / ~ disk ‖ Nummernscheibe f. ‖ disque m. d'appel. / ~ end ‖ Fingerspitze f. ‖ bout m. de doigt. / ~ grip (Mine) ‖ Geißfuß m. ‖ repertoire m.; caracole f. / ~ guard ‖ Fingerabweiser m. ‖ joue f. de protection

pour les doigts. / ~ holder ‖ Fingerhalter m. ‖ appui-doigt m. / ~ machine (Textile) ‖ Fingermaschine f. ‖ tricoteuse f. pour doigt de gants. / ~ mark ‖ Fingerabdruck m. ‖ empreinte f. digitale ou du doigt. / ~ pad ‖ Fingerhut m. ‖ doigtier m.; dé m. (à coudre). / ~ plate ‖ Schutzscheibe f.; Türschoner m. ‖ plaque f. de propreté. / ~ post ‖ Wegweiser m. ‖ poteau m. indicateur. / ~ protector see ~ stall. / ~ ring ‖ Fingerring m. ‖ bague f.; anneau m. / ~ stall ‖ Fingerling m.; Gummifinger m. ‖ doigtier m. / ~ tip see ~ end.

fining (Brew) ‖ Schönen n.; Klären n. ‖ clarification f.; collage f. / ~ (Chem) ‖ Klärungsmittel n. ‖ clarifiant f. / ~ (Metal) ‖ Frischen n. ‖ affinage m.; raffinage m.; mazéage m. / ~ in a charcoal-bed ‖ Löscharbeit f. ‖ affinage m. dans le feu brasqué. / ~ gray iron ‖ Rohfrischen n. ‖ affinage m. imparfait; premier affinage m. / ~ of iron ‖ Frischen n. des Eisens ‖ affinage m. de la fonte. / ~ of pig ‖ Frischen n. des Roheisens ‖ affinage m. de la fonte crue. / ~ of pig in a reverberatory furnace while constantly stirring it with iron rakes ‖ Frischen n. des Roheisens im Flammofen unter beständigem Umrühren (Puddeln) mit eisernen Haken ‖ affinage m. de la fonte au four à réverbère et brassage (puddlage) constant au moyen de ringards en fer. / ~ of pig iron see ~ of pig. / white iron ~ ‖ Garfrischen n. ‖ affinage m. complet.

fining effect, the ore had a ~ on the pig ‖ das Erz wirkte frischend auf das Roheisen ‖ le minerai affinait la fonte.

fining forge see finery furnace. / ~ furnace see finery furnace.

fining process ‖ Frischarbeit f.; Frischprozeß m. ‖ affinage m. / charcoal ~ see fining in a charcoal bed. / ~ of iron see fining of iron.

fining slag (Forg) ‖ Schmiedesinter m. ‖ laitier m. de forge. / ~ (Metal) ‖ Frischschlacke f. ‖ scorie f. des feux d'affinerie.

finish, to ‖ beenden; vollenden ‖ finir; achever. / ~ (Build) ‖ ausbauen ‖ achever; développer. / ~ (Pap) ‖ feinmahlen ‖ raffiner. / ~ the hole ‖ die Bohrung f. schlichten ‖ finir l'alésage. / ~-off (Join; Locksm) ‖ schlichten ‖ planer; replanir. / ~-off a planed surface of wood ‖ das Holz nach dem Hobeln abschlichten oder nachputzen ‖ replanir le bois.

finish ‖ Ende n.; Schluß m. ‖ fini m. / ~ (Build) ‖ Ausführung f.; Bauart f.; Ausstattung f. ‖ construction f.; exécution f. / artistic ~ ‖ künstlerische Ausführung f. ‖ modèle m. artistique. / buffed ~ ‖ polierte Ausstattung f. oder Ausführung f. ‖ exécution f. luisante.

finished ‖ beendet ‖ fini. / after having ~ ‖ nach Fertigstellung f. ‖ après avoir terminé. / partly ~ articles pl. ‖ Zwischenerzeugnisse npl. ‖ demi-produits mpl. / ~ axle ‖ fertig bearbeitete Achse f. ‖ essieu m. fini. / ~ forging ‖ fertiges Schmiedestück n. ‖ pièce f. de forge finie. / the machined surfaces pl. are highly ~ ‖ die Flächen fpl. sind sehr sauber bearbeitet ‖ les surfaces fpl. sont proprement finies. / ~ iron ‖ Handelseisen n. ‖ fer m. marchand. / ~ marking tool ‖ gravierter Stempel m. ‖ poin-

çon m. gravé. / not ~ (Weav) ∥ rauh; hart ∥ rude; dur au toucher. / ~ product ∥ Fertigerzeugnis n. ∥ produit m. fini.

finisher ∥ Fertigbearbeiter m.; Fertigmacher m. ∥ acheveur m.; finisseur m. / ~ (Clothm) ∥ Appretör m. ∥ apprêteur m. / ~ (Pap) ∥ Ganz(zeug)holländer m.; Feinzeugholländer m. ∥ cylindre m. affineur ou raffineur ou broyeur; pile f. raffineuse. / ~ (Roll mill) ∥ Fertigwalzer m. ∥ lamineur m. finisseur. / ~ (Spinn) see finishing box minder. / ~ (Weav) ∥ Ausrüster m.; Ausführer m. ∥ apprêteur-finisseur m.

finish facing ∥ Planschlichten n. ∥ dressage-finissage m.

finish grinding (Mach) ∥ Fertigschleifen n. ∥ finissage m. à la meule ou d'affûtage. / ~ of grain ∥ Feinmahlen n. von Getreide ∥ mouture f. fine de grains.

finishing ∥ Fertigstellen n.; Fertigmachen n.; Vollenden n.; Ausarbeitung f. ∥ finissage m.; achèvement m. / ~ (Clothm) ∥ Appretur f.; Appretieren n. ∥ apprêt m. / ~ (Grinding) see finish grinding. / ~ (Railw) ∥ Planierung f.; Einebnung f. ∥ régalage m. / cloth ~ ∥ Appretur f. oder Zurichtung f. von Stoffen ∥ apprêt m. d'étoffes. / hosiery ~ ∥ Strickwarenappretur f. ∥ apprêt m. de bonneterie. / leather ~ ∥ Lederappretur f. ∥ apprêt m. de cuir. / ~ of tissues ∥ Appretieren n. von Geweben ∥ apprêt m. de tissus. / white ~ ∥ Weißappretur f. ∥ apprêt m. en blanchiment. / ~ a working piece ∥ Fertigbearbeitung f. eines Werkstückes ∥ finissage m. d'une pièce. / yarn ~ ∥ Garnappretur f. ∥ apprêt m. de fils.

finishing bit (Mach) ∥ Kaliberbohrer m.; Schlichtbohrer m. ∥ alésoir m.; polissoir m. / ~ blow (Pav) ∥ Schlichtschlag m. ∥ coup m. (pour l'opération) de parage. / ~ box minder (Spinn) ∥ Wollfeinstrecker m. ∥ finisseur m. de laine. / ~ coat (Paint) ∥ Deckanstrich m. ∥ deuxième couche f.

finishing cut (Mach) ∥ Fertigschnitt m. ∥ coupe f. finisseuse. / to take the ~ ∥ nachdrehen ∥ repasser au tour.

finishing cutter ∥ Nachfräser m. ∥ fraise f. finisseuse. / ~ draft (Spinn) ∥ Endverzug m. ∥ étirage m. finisseur. / ~ drum (Ore dress) ∥ Tratschtrommel f.; Fertigwaschtrommel f. ∥ trommel m. finisseur. / ~ fly frame (Spinn) ∥ Feinfleier m. ∥ banc m. (à broches) en fin. / ~ groove ∥ Fertigkaliber n. ∥ finisseur m.; cannelure f. finisseuse. / ~ hammer ∥ Polierhammer m. ∥ marteau m. finisseur. / ~ head (Spinn) ∥ letzte Spulmaschine f. ∥ bobinoir m. finisseur. / ~ heat (Malt) see ~ temperature.

finishing machine (Clothm) ∥ Appreturmaschine f. ∥ machine f. à apprêter ou d'apprêtage. / ~ (Pap) ∥ Feinmühle f. ∥ raffineur m. / ~ (Spinn) ∥ Feinspinnmaschine f. ∥ machine f. à filer en fin; métier m. en fin. / dry ~ (Clothm) ∥ Trockenappreturmaschine f. ∥ machine f. pour l'apprêt à sec. / ~ for wax cloth ∥ Wachstuchfertigungsmaschine f. ∥ machine f. pour le finissage de toile cirée.

finishing and decatizing machine (Weav) ∥ Finish- und Dekatiermaschine f. ∥ machine f. à finir et décatir. / ~ material (Clothm) ∥ Appreturzusatzmittel n. ∥ apprêt m. / ~ mill see rolling mill. / ~

mortar ∥ Putzmörtel m. ∥ mortier m. d'achèvement.

finishing oven (Textile) ∥ Formofen m. ∥ four m. d'apprêt. / ~ for gloves ∥ Handschuhformofen m. ∥ four m. d'apprêt pour gants. / ~ for stockings ∥ Strumpfformofen m. ∥ four m. d'apprêt pour bas.

finishing preparation ∥ Appreturmittel n. ∥ encollage m. pour tissus; apprêt m. / ~ roasting (Metal) ∥ Garrösten n. ∥ grillage m. définitif. / ~ roll ∥ Fertigwalze f. ∥ cylindre m. finisseur. / ~ roller see ~ roll. / ~ rolling mill ∥ Fertigwalzwerk n.; Justierwalzwerk n.; Nachwalzwerk n.; Fertigstraße f.; Fertigstrecke f. ∥ laminoir m. finisseur ou à calibrer; train m. finisseur; équipage m. finisseur. / ~ rolls see ~ rolling mill. / ~ shop ∥ Werkstatt f. zur Fertigbearbeitung ∥ atelier m. de parachèvement. / ~ temperature (Malt) ∥ Abdarrtemperatur f. ∥ température f. finale de touraillage. / ~ tool (Turn) ∥ Schlichtwerkzeug n.; Schlichtmeißel m. ∥ outil m. à repasser. / ~ yarn ∥ Auszwirn m. ∥ retors m. final.

finlike ∥ flossenförmig ∥ pinniforme.

fir ∥ Tanne f. ∥ sapin m.

fir board ∥ Tannenbrett n. ∥ planche f. en sapin. / thick ~ ∥ tannene Bohle f. ∥ tavaillon m.

fire, to ∥ anzünden; anstecken ∥ allumer. / ~ (To catch fire) ∥ (sich) entzünden; Feuer n. fangen ∥ enflammer; s'enflammer; s'embraser. / ~ (Arm) ∥ schießen ∥ tirer. / ~ (Boiler; Stove) ∥ heizen ∥ chauffer. / ~ (Loc) ∥ anheizen; anzünden ∥ commencer à chauffer; mettre en feu. / ~ with ball cartridges ∥ scharf schießen ∥ tirer à balle ou à boulet ou à projectiles. / ~ with full service-charge (Arm) ∥ mit voller Ladung f. schießen ∥ tirer à charge pleine. / ~ a mine ∥ eine Mine f. zünden ∥ mettre le feu à un fourneau. / ~ with reduced charge (Arm) ∥ mit schwacher Ladung schießen ∥ tirer à charge faible. / ~ a round (Arm) see ~ with ball cartridges. / ~ a shot ∥ einen Schuß m. lösen ∥ tirer un coup. / ~ standing ∥ freihändig schießen ∥ tirer à bras francs. / ~-up see ~ (Boiler; Stove).

fire ∥ Feuer n. ∥ feu m. / ~ (Conflagration) ∥ Brand m. ∥ incendie m.; embrasement m. / ~ (Forg) ∥ Feuer n.; Hitze f. ∥ chaude f. / ~ (Metal) ∥ Herd m. ∥ foyer m.; feu m.; four m. / to catch ~ ∥ anbrennen; sich entzünden ∥ s'allumer; brûler; s'enflammer. / charcoal ~ ∥ Holzkohlenfeuer n. ∥ feu m. de braise ou de charbon de bois. / chimney ~ ∥ Schornsteinbrand m. ∥ feu m. de cheminée. / to damp the ~ ∥ die Feuerung dämpfen ∥ modérer le feu. / to draw the ~s ∥ das Feuer n. herauswerfen ∥ mettre bas les feux mpl. / to extinguish the ~ ∥ Feuer n. löschen ∥ éteindre l'incendie m. / great ~ ∥ Feuersbrunst f. ∥ incendie m. / to kindle a ~ see to set ~. / to light a ~ see to set ~. / to set ~ ∥ Feuer n. anmachen oder anzünden ∥ allumer les feux mpl. / to set on ~ ∥ entzünden; anzünden; anstecken ∥ mettre en feu m.; allumer; mettre le feu à . . . / to set ~ to . . . see to set on ~. / to slacken the ~ ∥ einen Ofen m. kühlen; das Feuer dämpfen ∥ modérer le feu. / to take ~ see to catch ~. / ~ in the timbering of the roof ∥ Dachstuhlbrand m. ∥ incendie m. dans les combles.

fire alarm ∥ Feueralarm m. ∥ appel m. d'incendie. / ~ (Device) ∥ Feuermelder m. ∥ avertisseur m. d'incendie. / automatic ~ ∥ selbsttätiger Feuermelder m. ∥ avertisseur m. d'incendie automatique. / indoor ~ ∥ Innenfeuermelder m. ∥ avertisseur m. d'incendie d'intérieur. / transmitter of a combined ~ and watchman control installation ∥ Melder m. einer vereinigten Feuermeldeanlage und Wächterkontrollanlage ∥ avertisseur m. d'une installation combinée d'avertissement d'incendie et de contrôle de rondes.

fire alarm box see ~ (Device). / ~ indicator see ~ (Device). / ~ installation ∥ Feuermeldeanlage f. ∥ installation f. d'avertisseurs d'incendie.

fire alarm system ∥ Feuermeldesystem n. ∥ système m. d'avertisseurs d'incendie. / ~ for ships ∥ Feuermeldeanlage f. für Schiffe ∥ installation f. d'avertisseurs d'incendie pour navires.

fire and salvage boat ∥ Spritzendampfer m. ∥ bateau-pompe m. / ~ arch ∥ Frittofen m. ∥ arche f. à fritter.

firearm ∥ Schußwaffe f.; Feuerwaffe f. ∥ arme f. à feu; arme f. de tir. / portable ~ ∥ Handfeuerwaffe f. ∥ arme f. à feu portative.

fire arm container ∥ Schußwaffenbehälter m. ∥ caisse f. pour les armes à feu.

fire ball (Astron) ∥ Meteor m.; Feuerball m.; Feuerkugel f. ∥ météore m.; aérolithe m.; bolide m.

fire bar ∥ Roststab m. ∥ barreau m. de grille; barre f. de grille ou de foyer ou de fourneau. / ~ bearer ∥ Rostträger m.; Rostbalken m.; Rostschwelle f. ∥ sommier m. ou chevalet m. ou châssis m. ou traverse f. d'une grille; support m. des barreaux de la grille. / ~ frame ∥ Rostlager n. ∥ cadre m. ou support m. de la grille.

fire blende ∥ Feuerblende f. ∥ feuer-blende f. / ~ board (Mine) ∥ Schlagwetterwarnungstafel f. ∥ écriteau m. signalant un quartier grisouteux. / ~ boiling ∥ Feuerkochung f. ∥ cuisson f. au feu.

fire box (Loc) ∥ Feuerbuchse f. ∥ boîte f. à feu; foyer m. / copper ~ ∥ kupferne Feuerbuchse f. ∥ boîte f. à feu en cuivre. / ~ with corrugated flue welded on ∥ Feuerbuchse f. mit angeschweißtem Wellrohr ∥ boîte f. à feu avec tube ondulé raccordé par soudage. / inside ~ ∥ innere Feuerbuchse f. ∥ foyer m.; boîte f. à feu intérieure. / iron ~ ∥ eiserne Feuerbuchse f. ∥ foyer m. en fer. / mild steel ~ ∥ flußeiserne Feuerbuchse f. ∥ boîte f. à feu en tôle d'acier. / outside ~ ∥ äußere Feuerbuchse f. ∥ boîte f. à feu; boîte f. à feu extérieure. / wall of the ~ ∥ Feuerbuchswand f. ∥ paroi f. du foyer.

fire box back plate ∥ Feuerbuchsrückwand f. ∥ paroi f. postérieure ou face f. arrière du foyer. / ~ boiler ∥ Feuerbuchskessel m.; Lokomotivkessel m. ∥ chaudière f. de locomotive. / ~ bottom flange ∥ Feuerbuchsbodenring m.; Feuerbuchs(fuß)ring m. ∥ cadre m. intérieur d'un foyer; bague f. de la boîte à feu.

fire box crown ∥ Feuerbuchsdecke f. ∥ ciel m. ou plafond m. du foyer. / ~ plate ∥ Feuerbuchsdeckenblech n. ∥ tôle f. pour le ciel de foyer.

fire box door (Loc) ∥ Feuerbuchstür f. ∥ porte f. du foyer. / ~ (Metal) ∥ Heiztür f.; Feuerungstür f. ∥ porte f. du foyer ou de

la chauffe. / ~ plate ‖ Feuerbuchstürwand f. ‖ paroi f. de la porte du foyer.

fire box drilling machine ‖ Feuerbuchsbohrmaschine f. ‖ machine f. à percer les foyers. / ~ foundation ring *see* ~ bottom flange. / ~ front plate ‖ Feuerbuchsvorderwand f. ‖ paroi f. antérieure *ou* face f. avant de la boîte à feu. / ~ outside plate *see* ~ shell. / ~ plate ‖ Feuerbuchswand f. ‖ paroi f. de la boîte à feu; plaque f. *ou* tôle f. de foyer. / ~ ring *see* ~ bottom flange. / ~ sheet *see* ~ plate. / ~ shell ‖ Feuerbuchsmantel m. ‖ enveloppe f. du foyer *ou* de la boîte à feu. / ~ shell plate *see* ~ side plate. / ~ side plate ‖ Feuerbuchsseitenwand f. ‖ paroi f. latérale *ou* flanc m. de la boîte à feu. / ~ sides pl. ‖ Feuerbuchs(seiten)platten fpl. ‖ parois fpl. latérales de la boîte à feu. / ~ stay (Loc) ‖ kupferner Stehbolzen m. ‖ entretoise f. en cuivre rouge. / ~ wall ‖ Feuerbuchswand f. ‖ paroi f. de la boîte à feu *ou* du foyer.

fire boy *see* fireman.

fire brick ‖ feuerfester Stein m. *oder* Ziegel m. *oder* Mauerstein m.; Feuerziegel m.; Ofenziegel m.; Schamottestein m. ‖ brique f. réfractaire. / ~ plant ‖ Schamotte(stein)fabrik f. ‖ fabrique f. de briques réfractaires.

fire brickwork ‖ Schamottemauerwerk n. ‖ maçonnerie f. de briques réfractaires. / ~ bridge ‖ Feuerbock m.; Feuerbrücke f. ‖ autel m. *ou* marche f. (de foyer); pont m. de chauffe.

fire brigade ‖ Feuerwehr f. ‖ pompiers mpl.; corps m. *ou* équipe f. de (sapeurs-)pompiers. / ~ autocar *see* ~ motor car. / ~ carriage ‖ Feuerwehrfahrzeug n. ‖ voiture f. pour pompiers *ou* à incendie. / ~ equipment ‖ Feuerwehrausrüstung f. ‖ équipement m. de pompiers. / ~ motor car ‖ Feuerwehrmotorwagen m.; Feuerwehrkraftfahrzeug n. ‖ voiture f. automobile pour pompiers; automobile f. pour le service d'incendie.

fire bucket ‖ Feuereimer m. ‖ seau m. à incendie. / ~ chest (Loc) ‖ Feuerkammer f.; Feuerung f. ‖ caisse f. à feu.

fire clay ‖ feuerfester Ton m.; Schamotte(erde) f.; Feuerton m. ‖ argile f. *ou* terre f. réfractaire. / ~ crucible ‖ Tontiegel m. ‖ creuset m. en terre réfractaire. / ~ extraction ‖ Gewinnung f. feuerfester Erde ‖ extraction f. de terre réfractaire. / ~ goods pl. ‖ Schamottewaren fpl.; Schamotteerzeugnisse npl. ‖ articles mpl. en terre réfractaire; produits mpl. réfractaires. / ~ ware *see* ~ goods.

fire cock ‖ Feuerhahn m.; Hydrant m. ‖ bouche f. d'incendie; robinet m. à feu.

fire company *see* fire brigade.

fire control (Arm) ‖ Feuerleitung f. ‖ direction f. *ou* conduite f. du tir. / ~ director ‖ Kommandogerät n. ‖ téléindicateur m. / ~ method ‖ Feuerleitverfahren n. ‖ système m. *ou* méthode f. de conduite du tir.

fire damp (Chem) ‖ Grubengas n.; leichtes Kohlenwasserstoffgas n.; Sumpfgas n. ‖ gaz m. léger d'hydrogène carburé; gaz m. des marais. / ~ (Mine) ‖ schlagende Wetter npl. ‖ grisou m.; brisou m.; feu m. terrou; mofette f. / ~ tester ‖ Grubengasinterferometer n.; Grubengasmesser m. ‖ grisoumètre m. (interférentiel).

fire door (Loc) ‖ Feuertür f. ‖ porte f. de foyer. / air inlet in the ~ ‖ Lufteinlaß-

öffnung f. in der Feuertür ‖ appel m. d'air par la porte du foyer. / to close the ~ ‖ die Feuertür schließen ‖ fermer la porte du foyer. / flap ~ ‖ Klappfeuertür f. ‖ porte f. de foyer à rabattement. / to open the ~ ‖ die Feuertür öffnen ‖ ouvrir la porte du foyer.

fire door frame ‖ Heiztürrahmen m.; Feuergeschränkrahmen m. ‖ cadre m. de la devanture de foyer. / ~ handle ‖ Heiztürgriff m. ‖ poignée f. de porte de foyer. / ~ hole ring ‖ Feuertürlochring m. ‖ cadre m. de la porte du foyer.

fire engine ‖ Feuerspritze f. ‖ pompe f. à incendie. / cart manual ~ ‖ Abprotzspritze f. ‖ pompe f. à incendie sur charrette. / manual ~ ‖ Handspritze f. ‖ pompe f. à main. / motor ~ ‖ Motor(feuer)spritze f. ‖ pompe f. automobile à incendie. / portable ~ ‖ tragbare Feuerspritze f. ‖ pompe f. à incendie portative. / small hand ~ ‖ Handfeuerspritze f. ‖ pompe f. à incendie à main. / steam ~ ‖ Dampffeuerspritze f. ‖ pompe f. à incendie à vapeur. / steam ~ with implement wagon ‖ Dampffeuerspritze f. mit Gerätewagen ‖ pompe f. à incendie à vapeur avec chariot d'agrès.

fire engine hose ‖ Feuerspritzenschlauch m. ‖ tuyau m. flexible pour pompes à incendie. / ~ steam boiler ‖ Feuerspritzenkessel m. ‖ chaudière f. pour pompe à incendie.

fire escape ‖ Rettungsleiter f. ‖ de sauvetage.

fire extinction, organisation of ~ ‖ Feuerlöschwesen n. ‖ service m. d'incendie. / ~ fittings pl. ‖ Feuerlöscharmatur f. ‖ armature f. pour l'extinction du feu.

fire extinguisher ‖ Feuerlöscher m. ‖ extincteur m. d'incendie. / hand ~ ‖ Handfeuerlöscher m. ‖ extincteur m. manuel *ou* à main.

fire extinguishing apparatus *see* fire extinguisher. / ~ lever ‖ Brandhahn m. ‖ robinet m. d'incendie. / ~ material ‖ Feuerlöschmittel n. ‖ matière f. d'extinction du feu. / ~ mountings pl. ‖ Feuerlöscharmaturen pl. ‖ garnitures fpl. d'extinction des incendies. / ~ rose ‖ Feuerlöschbrause f. ‖ crépine f. d'incendie.

fire fan ‖ Feuerungsventilator m. ‖ ventilateur m. du foyer. / flash of ~ ‖ Feuergarbe f. ‖ gerbe f. de feu. / ~ gilding ‖ Feuervergoldung f. ‖ dorure f. au feu.

fire grate ‖ Feuerrost m. ‖ grille f. du foyer. / double-inclined ~ ‖ Sattelrost m.; Schweinerücken m. ‖ grille f. à deux plaines.

fire grate bar ‖ Feuerroststab m. ‖ barreau m. de grille du foyer.

fire hole ‖ Feuerloch n.; Heizloch n.; Schürloch n. ‖ chauffe f.; embrasure f.; ouverture f. de la chauffe; porte f. de chauffe; taquerie f. / ~ ring ‖ Feuerlochring m. ‖ cadre m. circulaire *ou* circulaire m. de la porte du foyer.

fire hook ‖ Feuerhaken m.; Feuerspieß m.; Rühreisen n.; Schüreisen n.; Schürhaken m. ‖ attisoir m.; attisonnoir m.; perce-fournaise m.; ringard m.; tisonnier m.

fire insurance ‖ Feuerversicherung f.; Brandassekuranz f. ‖ assurance f. contre l'incendie. / ~ company ‖ Feuerversicherungsgesellschaft f. ‖ compagnie f. d'assurance contre l'incendie. / ~ office ‖ Brandkasse f. ‖ caisse f. d'assurance contre l'incendie. / ~ policy ‖ Feuer-

versicherungspolice f. ‖ police f. d'assurance contre l'incendie.

fire iron ‖ Schüreisen n. ‖ lance f. à feu; ringard m. / ~ irons pl. ‖ Ofengeräte npl.; Kamingeräte npl. ‖ garniture f. de feu. / ~ ladder ‖ Feuerleiter f. ‖ échelle f. à incendie.

fireless locomotive ‖ feuerlose Lokomotive f. ‖ locomotive f. sans foyer.

firelight ‖ Feueranzündholz n. ‖ tolinet m.; bûche f.

fire lighter ‖ Feueranzünder m.; Kohlenanzünder m. ‖ allume-feu f.; briquet m. allumeur. / ignition material for ~s ‖ Feuerzeuglunte f. ‖ mèche f. pour briquets.

fireman (Boiler) ‖ Heizer m. ‖ chauffeur m.; homme m. de four. / ~ (Member of fire company) ‖ Feuerwehrmann m. ‖ pompier m. / ~ (Metal; Pott) ‖ Brenner m. ‖ chauffeur m. / locomotive ~ ‖ Lokomotivheizer m. ‖ chauffeur m. de locomotive.

fireman's equipment ‖ Feuerwehrausrüstung f. ‖ matériel m. *ou* équipement m. pour pompiers. / ~ glasses pl. *see* ~ goggles. / ~ goggles pl. ‖ Rauchschutzbrille f. ‖ lunettes fpl. protectrices pour pompiers. / ~ hatchet ‖ Feuerwehrbeil n. ‖ hachette f. de pompier. / ~ helmet ‖ Feuerwehrhelm m. ‖ casque m. de pompiers. / ~ material *see* ~ equipment. / ~ pick hatchet ‖ Feuerwehrspitzbeil n. ‖ hachette f. de pompier à pointes.

fire opal ‖ Feueropal m. ‖ opale f. à flammes.

fire place ‖ Feuerherd m. ‖ foyer m. / ~ ‖ Kamin m. ‖ cheminée f. / ~ (Forg) ‖ Schmiedeherd m. ‖ feu m. de forge. / ~ of a caldron ‖ Kesselfeuerung f. ‖ âtre m. de chaudron. / electric ~ ‖ Elektrokamin m. ‖ foyer m. électrique. / ~ of the flame furnace ‖ Feuerraum m. des Flammofens ‖ foyer m. du four à réverbère. / ~ of a kettle ‖ Kesselfeuerung f. ‖ foyer m. de chaudière. / stove imitating a ~ ‖ Kaminofen m. ‖ cheminée f. à la prussienne. / ~ tergiform ~ (Metal) ‖ Schweinerücken m. ‖ grille f. en forme de toit.

fire place fitting ‖ Kaminausrüstung f. ‖ garniture f. de foyer.

fire plug ‖ Hydrant m. ‖ borne-fontaine f.; bouche f. d'incendie. / ~ poker ‖ Feuerschürer m. ‖ attisoir m. / ~ policy ‖ Feuerversicherungspolice f. ‖ police f. d'assurance contre l'incendie.

fireproof ‖ feuersicher; unverbrennbar; feuerbeständig ‖ incombustible; à l'épreuve f. du feu; réfractaire. / ~ (Chem) ‖ flammensicher ‖ ignifugé. / ~ articles pl. ‖ feuerfeste Artikel mpl. ‖ articles mpl. réfractaires. / testing method for ~ building material ‖ Prüfverfahren n. für feuerfeste Baustoffe ‖ méthode f. d'essai pour matériaux de construction réfractaires. / ~ bulkhead ‖ Brandschott n. ‖ cloison f. pare-feu *ou* à localiser l'incendie. / ~ casing ‖ feuerfeste Auskleidung f. ‖ brasque f.; manteau m. intérieur réfractaire. / ~ castings pl. ‖ feuerfester Guß m. ‖ moulage m. à l'épreuve du feu. / ~ cement ‖ feuerfester Kitt m. ‖ mastic m. réfractaire. / ~ cement (Build) ‖ Schamottemörtel m. ‖ ciment m. réfractaire. / ~ conduit ‖ feuerbeständiges Isolierrohr n. ‖ tube m. incombustible. / ~ door ‖ feuersichere Tür f. ‖ porte f. à l'épreuve du feu. / ~ earth ‖ feuerfeste Erde f. ‖

terre f. réfractaire. / ~ goods pl. formed from argillaceous mass || aus toniger Masse geformte feuerfeste Waren fpl. || ouvrages mpl. réfractaires moulés en masse argileuse. / ~ impregnation || Feuerschutzmittel n. || masse f. d'imprégnation réfractaire. / ~ masonry || feuerfeste Ausmauerung f. || maçonnerie f. réfractaire. / ~ material || feuersicheres Material n. || matière f. à l'épreuve du feu. / method of testing ~ material || Prüfverfahren n. für feuerfeste Baustoffe || méthode f. d'essai pour matériaux réfractaires. / ~ mortar || feuerfester Mörtel m. || mortier m. réfractaire. / ~ pottery || Feuertonware f. || objet m. en terre réfractaire. / ~ roofing || feuersicheres Eindecken n. || couverture f. incombustible. / ~ stone || feuerfester Stein m. || pierre f. réfractaire. / ~ ware || feuerfeste Waren fpl. || ouvrages mpl. réfractaires.

fire protection, means for ~ || Feuerschutz m. || protection f. contre l'incendie. / ~ organisation || Feuerlöschwesen n. || service m. d'incendie.

firer || Heizer m. || chauffeur m.; homme m. de four.

fire, rain of ~ (Pyrot) || Feuerregen m. || cascades fpl. de feux d'artifice.

fire-resisting || feuerbeständig || réfractaire. / ~ bulkhead || Feuerschott n.; Brandschott n. || cloison f. transversale d'incendie. / ~ casting || feuerbeständiger Guß f. || fonte f. allant au feu. / ~ duct's closing || feuersicherer Abschluß m. für Kanäle || obturation f. des conduites par matières incombustibles.

fire roll || Feuerlärm m. || signal m. d'incendie. / ~ room (High-furnace) || Schacht m.; Seele f. || cuve f. ou cheminée f. intérieure. / ~ room (Shipb) || Heizraum m. || chambre f. de chauffe. / ~ screen || Ofenschirm m.; Kaminschirm m. || garde-feu m.; écran m. de poêle. / ~ ship || Feuerschiff n. || bateau-feu m. / ~ shovel || Kohlenschaufel f. || pelle f. à charbon. / ~ side || Kamin m. || cheminée f. / ~ silvering || Feuerversilberung f. || argenture f. au feu. / ~ slice || Schüreisen n. || lance f. à feu; ringard m.

firestone || Feuerstein m. || pierre f. ou silex m. à feu.

fire tender || Heizer m. || chauffeur m.; homme m. de four. / ~ test || Brandprobe f. || épreuve f. par le feu; essai m. de résistance à l'incendie.

fire-tinned wire || feuerverzinnter Draht m. || fil m. étamé au feu.

fire tongs pl. || Feuerzange f. || pince f. à feu.

fire tube (Boil) || Siederohr n.; Rauchrohr n. || tube m. de fumée ou de chaudière. / ~ (Build) || Feuerkanal m.; Feuerzug m. || carneau m. / ~ boiler || Feuerrohrkessel m. || chaudière f. multitubulaire à tubes de fumée.

fire tug || Feuerlöschboot n. || bateau-pompe m. à incendie. / ~ wall || Brandmauer f. || mur m. réfractaire ou massif ou mitoyen. / ~ watch || Feuerwache f. || quart m. aux feux.

firewood || Brennholz n. || bois m. de chauffage ou à brûler. / splitted ~ || gespaltenes Brennholz n. || bois m. à brûler refendu.

firewood sawing || Sägen n. von Brennholz || sciage m. de bois de chauffage. / ~ splitter || Brennholzspalter m. || fendeur m. de bois à brûler. / ~ splitting || Spalten

n. von Brennholz || fendage m. de bois de chauffage. / ~ splitting machine || Brennholzmaschine f. || machine f. à débiter le bois de chauffage.

firework maker || Kunstfeuerwerker m. || artificier m.

fireworks pl. || Feuerwerkskörper m. || feu m. d'artifice; artifice m.; objet m. pyrotechnique. / composition for ~ || Feuerwerkssatz m. || composition f. d'artifices.

firing (Metal) || Feuerung f. || feu m.; four m. / ~ (Stoking) || Heizung f. || chauffage m. / adjustable ~ || regelbare Feuerung f. || foyer m. réglable. / brown coal ~ || Braunkohlenfeuerung f. || chauffage m. au lignite. / direct ~ || unmittelbare Feuerung f. || feu m. nu. / gas ~ || Gasfeuerung f. || chauffage m. au gaz. / grate ~ || Rostfeuerung f. || foyer m. à grille. / pulverized coal ~ || Staubkohlenfeuerung f. || chauffage m. au charbon pulvérisé. / ~ with travelling grate || Wanderrostfeuerung f. || foyer m. à grille mobile.

firing circuit || Zündleitung f. || circuit m. d'inflammation. / ~ data computer || Richtwerteberechner m.; Schußwerteberechner m. || compteur m. des données du tir. / ~ data transmission device (Arm) || Richtwerteübertragungsgerät n. || transmetteur m. de données du tir. / anti-aircraft ~ director || Flugabwehrfeuerkommandogerät n. || instrument m. de direction du tir antiaérien. / fittings pl. of ~ for breweries || Feuerungsarmatur f. für Brauereien || armature f. de foyer pour brasseries. / ~ key || Minenzünder m. || exploseur m. magnétique. / ~ order see ~ point. / ~ plant || Feuerungsanlage f. || installation f. de foyer. / ~ point || Zündzeitpunkt m. || point m. d'allumage.

firing table, anti-aircraft ~ || Flugabwehrschußtafel f. || table f. de tir antiaérien. / mechanical ~ in form of slide rule || mechanische Schußtafel f. in Form eines Rechenschiebers || table f. de tir mécanique sous forme de règle à calcul.

firing tape || Zündschnur f. || cordon m. détonnant; étoupille f.; mèche f.

fir leaf oil || Tannennadelöl n. || essence f. de sapin.

firm || fest || fixe; ferme. / ~ (Mach) || fest; stationär || stationnaire. / ~ (Mine) || klemmig || dur; compact; serré. / to make ~ (Join; Locksm) || anschlagen; befestigen || assembler; fixer.

firm || Firma f. || maison f. (de commerce); raison f. commerciale. / ~ || Handelsgesellschaft f. || société f. (de commerce); compagnie f. / of good standing || solides Haus n.; solide Firma f. || une maison f. sérieuse ou solide. / long ~ || Schwindelfirma f. || maison f. de filous. / to sign for the ~ || firmieren || signer sous la raison de.

firmer's chisel || Stechbeitel m. || bédane m.; ciseau m. fort. / ~ compasses pl. || Bildhauerzirkel m. || compas m. de sculpteur. / ~ gouge || Gutsche f. für Bildhauer || gouge f. pour sculpteurs.

firm, position of a ~ || Geschäftslage f. einer Firma || situation f. des affaires d'une maison. / ~'s stamp || Firmenstempel m. || timbre m. de la maison ou à firme.

fir oil || Kiefernadelöl n. || essence f. d'aiguilles de (sa)pin. / ~ plank || tannene Bohle f. || tavaillon m.

first, in the ~ place || erstens || en premier lieu m. / ~ of exchange || Primawechsel m. || première f. de change.

first-aid station in case of accidents || Verbandstelle f. für erste Hilfe in Unglücksfällen || poste m. de premier secours en cas d'accidents. / ~ treatment in case of accidents || erste Hilfe f. bei Unglücksfällen || premier secours m. en cas d'accidents. / ~ work || erste Hilfe f. (bei Unglücksfällen) || premiers secours mpl.

first breaking-up (Metal) || Rohaufbrechen n. || soulèvement m.; désornage m.

first-class || prima || de première qualité. / ~ workshop || leistungsfähige Werkstatt f. || important atelier m.

first consideration || Hauptsache f. || principal m.; essentiel m. / reducing of ~ costs pl. || Verbilligung f. der Anschaffungskosten || réduction f. des frais d'achat. / ~ drawer (Spinn) || Grobstrecker m. || étireur m. en gros. / ~ edition || Erstausgabe f. || première édition f. / ~ felling (Wood) || Anhieb m. || mise f. en coupe. / ~ finisher (Glassm) || Fertigmacher m. || souffleur m.; ouvreur m. / ~ frame of a shaft (Mine) || Füllbaum m. || premier cadre m. d'un puits. / ~-hand || aus erster Hand f. || de première main f. / ~ instalment || Angeld n.; Anzahlung f. || arrhes fpl.; acompte m. / ~ lump (Metal) || Sauer m.; Urdeul m. || principe m. du massé. / ~ mate || Maat m. || capitaine m. en second. / ~-rate || erstklassig || de premier ordre m. / ~ runnings product || Anfangserzeugnis n.; Anfangsergebnis n. || produit m. initial.

firsts pl. (Spinn) || Reinhanf m. || brin m.; chanvre m. net ou sérancé.

first stuff (Pap) || Halbzeug n. || pâte f. effilochée; demi-pâte f.; défilé m.

fir wood || Kiefernholz n.; Tannenholz n.; Föhrenholz n. || bois m. de pinastre ou de sapin. / ~ sleeper (Railw) || Tannenschwelle f. || traverse f. en sapin.

fisc || Fiskus m. || fisc m.

fiscal year || Geschäftsjahr n.; Rechnungsjahr n. || exercice m.

fiscus see fisc.

fish, to (Railw) || verlaschen || éclisser. / ~ the anchor || den Anker m. fischen || traverser l'ancre f. / ~ with the line || angeln || pêcher à la ligne. / ~ a mast || Schalen fpl. um einen Mast legen || acclamper ou jumeller un mât. / ~ a piece of timber || durch Anblattung verstärken || armer une pièce de charpente par fourrures.

fish || Fisch m. || poisson m. / ~ (Railw) || Lasche f. || éclisse f. / ~ (Shipb) || Schale f.; Backe f. || jumelle f. / angular ~ || Winkellasche f. || éclisse f. à cornières. / fresh-water ~ || Süßwasserfisch m. || poisson m. d'eau douce. / ~ of a made-mast || Seitenstück n. eines gebauten Mastes || armure f. d'un mât composé. / ~ of an oar || Riemendollstück n. || galaverne f. d'aviron. / salt-water ~ || Seefisch m. || poisson m. de mer.

fish basket || Reuse f. || nasse f. / ~ bolt || Laschenbolzen m. || boulon m. d'éclisse.

fishbone || Fischbein n. || baleine f.

fish bow net || Fischreuse f. || nasse f. à poissons. / ~ plaiter || Fischreusenflechter m. || tresseur m. de nasses à poissons.

fish breeder || Fischzüchter m. || pisciculteur m. / ~ breeding || Fischzucht f. || pisciculture f.; élevage f. de poissons. / ~ breeding house || Fischzuchtanstalt f. || établissement m. de pisciculture. / ~ canning in cans || Fischkonserven fpl. in Dosen || conserves fpl. de poissons en boîtes.

/ ~ chair (Railw) || Stuhllasche f. || coussinetéclisse m. / ~ channel || Fischgraben m. || étier m. / ~ chopper || Fischköpferin f. || étêteuse f. de poissons. / ~ cooling room || Fischkühlraum m. || chambre f. froide à poissons. / ~ curer || Fischzubereiter m. || fabricant m. de conserves de poissons. / ~ curing || Fischkonservenfabrik f. || fabrique f. de conserves de poissons. / motor ~ cutter || Motorfischkutter m. || cutter m. pêcheur à moteur. / ~ davit || Fischdavit m. || bossoir m. de traversière. / ~ drying || Fischtrocknerei f. || séchage m. de poissons.

fished (Railw) || verlascht || éclissé. / ~ joint || verlaschter Schienenstoß m. || joint m. éclissé. / ~ mast || gefischter Mast m. || mât m. réclampé ou jumelé ou renforcé.

fish egg || Fischei n. || œuf m. de poisson.

fisher see fisherman.

fisherman || Fischer m. || pêcheur m. / open sea ~ || Hochseefischer m. || pêcheur m. en haute mer.

fishery || Fischerei f. || pêche f.; pêcherie f. / great ~ || Großfischerei f. || grande pêche f. / inshore ~ || Küstenfischerei f.; Strandfischerei f. || pêche f. côtière. / little ~ || kleine Fischerei f. || petite pêche f. / on the open sea || Hochseefischerei f. || pêche f. en haute mer; pêche f. hauturière. / lessee of ~ || Fischereipächter m. || fermier m. de pêche. / master mariner of the ~ || Fischmeister m. || patron m. pêcheur.

fish flour || Fischmehl n. || farine f. de poissons. / ~ fork || Fischbesteck n. || couvert m. à poisson. / ~ garth || Fischwehr n. || écrille f. / ~ gig || Harpune f. || harpon m.; foène f.; fouane f. / ~ glue || Fischleim m. || colle f. de poisson; ichtyocolle f. / ~ grease || Fischfett n. || graisse f. de poisson. / ~ grease for shammy leather || Gerberfett n. || dégras m. / ~ guano || Fischguano m. || guano m. de poisson. / ~ hatching see ~ breeding. / ~ header see ~ chopper.

fishhook (Fish) || Angelhaken m. || hameçon m. / ~ (Mar) || Fischhaken m. || croc m. de candelette; cantonnière f. de l'ancre. / ~ (Shipb) || Penterhaken m.; Fischhaken m. || croc m. de traversière. / ~ maker || Angelzeugmacher m. || hameçonnier m.

fishing || Fischfang m.; Fischerei f. || pêche f. / ~ (Railw) || Laschenverbindung f. || éclissage m. / ~ accessory || Angelgerät n. || ustensile m. de pêche. / ~ basket || Fischkorb m. aus Draht || nasse f. pour la pêche en treillis. / ~ boat || Fischerboot n. || barque f. de pêche; canot m. de pêcheur. / ~ boat master || Fischerbootmeister m. || patron m. de barque ou bateau de pêche. / ~ fleet || Fischerflotte f. || flotte f. de pêcheurs. / ~ ground || Fischgrund m. || fond m. poissonneux. / ~ line || Angel f. || ligne f. à pêche. / ~ net || Fischnetz n. || filet m. ou rets m. de pêche. / ~ plate (Railw) || Lasche f.; Schienenlasche f. || éclisse f.; éclisse f. plate. / ~ product || Fischereierzeugnis n. || produit m. de la pêche. / ~ rod || Angelrute f. || canne f. à pêche. / ~ tackle || Fischereigerät n. || article m. de pêche; engin m. de pêche à la ligne. / ~ vessel || Fischereifahrzeug n. || bateau m. de pêcheur; chalutier m.

fish joint (Railw) || Stoßverbindung f. || assemblage m. des éclisses avec les rails. / ~ keeper || Fischzüchter m. || piscicul-teur m. / ~ ladder || Fischpaß m.; Fischleiter f. || échelle f. à poissons. / ~ manure || Fischdünger m. || engrais m. de poisson. / ~ meal || Fischmehl n. || farine f. de poisson. / ~ net || Fischnetz n. || filet m. de pêche.

fish oil || Fischtran m. || huile f. de poisson. / ~ refining plant || Tranraffinerieanlage f. || installation f. de raffinerie d'huile de poisson.

fishplate || Schienenlasche f. || éclisse f.; couvre-joint m. / angle ~ || Winkellasche f. || éclisse f. cornière. / cranked ~ || Übergangslasche f. || éclisse f. de raccordement. / double-angle ~ || Doppelwinkellasche f. || éclisse f. double cornière. / ~ fastening ~s to the ends of the rail || Befestigung f. der Laschen an den Schienenenden || attache f. des éclisses aux bouts des rails. / flat ~ || Flachlasche f. || éclisse f. plate. / ~ beneath the foot of the rail || Fußlasche f. || coussinet-éclisse m. / hooked ~ || Hakenlasche f. || éclisse f. à crampon. / inner ~ with round holes || Innenlasche f. mit runden Löchern || éclisse f. intérieure avec trous ronds. / ~ to join the rails || Lasche f. zur Verbindung der Schienen untereinander || éclisse f. pour l'assemblage des rails. / joint with inside rivetted ~s for portable lines || Stoßverbindung f. mit innen angenieteten Stecklaschen || joint m. par éclisses Spalding rivées intérieurement. / the ~ was notched || die Stoßlasche wurde ausgeklinkt || les éclisses fpl. des joints reçurent des encoches. / outer ~ with elongated holes || Außenlasche f. mit Langlöchern || éclisse f. extérieure avec trous oblongs. / rail ~ || Schienenlasche f. || éclisse f. de rails. / shoe-shaped ~ || Schuhwinkellasche f. || éclisse f. cornière à sabots. / straight ~ || gerade Lasche f. || éclisse f. plate. / Z ~ || Z-Lasche f. || éclisse f. Z.

fishplate, designation of ~ || Laschenbezeichnung f. || désignation f. des éclisses. / flange of the ~ || Laschenschenkel m. || cornière f. de l'éclisse. / ~ hole in the ~ for the bolt || Laschenloch n. || trou m. pour le boulon d'éclisse. / notching of the ~ || Laschenausklinkung f. || encochement m. ou entaille f. de l'éclisse. / ~ punching-machine || Laschenlochmaschine f. || machine f. à percer les éclisses. / ~ rail || Laschenschiene f. || rail m. éclisse. / ~ section || Laschenprofil n. || profil m. d'éclisse. / ~ weight of ~ per meter run || Laschengewicht n. je laufenden Meter || poids m. de l éclisse par mètre courant.

fish-pond || Fischteich m. || vivier m.; étang m. / ~ owner || Fischteichbesitzer m. || propriétaire m. d'un étang.

fish pot || Reuse f. || nasse f. / ~ roe || Fischrogen m. || œufs mpl. de poisson. / ~ salter || Fischsalzer m. || saleur m. de poissons. / ~ salting || Fischsalzerei f. || salage m. de poissons. / ~ shop || Fischhandlung f. || magasin m. de poissons. / ~ skin || Fischhaut f. || peau f. de poisson. / ~ sluice || Fischschleuse f. || écluse f. à poissons. / ~ smoker || Fischräucherer m. || enfumeur m. de poissons. / ~ smoking || Fischräucherei f. || fumage m. de poissons. / ~ spear || Harpune f. || harpon m. / ~ tackle || Fischtakel n. || traversière f. / ~ tail burner || Flachbrenner m. || bec m. plat. / ~ torpedo || Fischtorpedo m. || torpille-poisson f. / ~ utilization plant || Fischverwertungsanlage f. || installation f. d'utilisation de poissons.

fisk see fisc.

fissile || spaltbar || fissile.

fissility || Spaltbarkeit f. || fissilité f.

fission fungi fermentation || Spaltpilzgärung f. || fermentation f. provoquée par les bactéries schizomycètes.

fissure, to || spalten || fissurer; fendiller.

fissure (Geol) || Kluft f. || fente f.; gerçure f. / ~ (Metal) || Riß m.; Härteriß m.; Borste f. || fissure f.; fente f.; déchirure f.; crevasse f.; gerçure f.; crique f.; fêlure f. / ~ (Wood) || Frostspalte f.; Kaltriß m. || gélivure f. / ~ of a conducting-pipe || Ritze f. oder Spalte f. in einer Wasserröhre || renard m. / ~ of overthrust (Geol) || Überschiebungsspalte f. || fissure f. du pli-faille. / ~ due to subsidence || Einbruchspalte f. || fente f. par éboulement. / ~ vein (Mine) || echter Gang m. || vrai filon m.; filon m. de fracture.

fist || Faust f. || poing m. / ~ size || Faustdicke f. || grosseur f. du poing.

fistula || Fistel f. || fistule f.

fit || tauglich || utile. / ~ to be carved || zum Schnitzen n. geeignet || susceptible d'être taillé.

fit, to || zusammenpassen; aufeinanderpassen; einbauen || appareiller; adapter; triquer; monter. / ~-in || einpassen || ajuster. / ~ into one another || ineinanderstecken || s'emboîter les uns dans les autres. / ~-on || anproben || essayer. / ~-out || ausstatten || garnir. / ~-out a vessel || ein Schiff n. ausrüsten || équiper un navire. / the parts pl. fit together || die Teile mpl. passen ineinander || les pièces fpl. s'adaptent l'une à l'autre. / ~ the pieces together || die Stücke npl. zusammenpassen || ajuster les pièces fpl. l'une sur l'autre; raccorder les éléments mpl.; adapter les pièces fpl. l'une à l'autre. / the plates pl. fit to the rails || die Platten fpl. passen zu den Schienen || les selles fpl. conviennent aux rails. / ~ tightly (Cloth) || eng anschließen || (se) coller. / ~ together by grinding one on the other || aufeinander schleifen || roder l'un sur l'autre. / ~-up a machine || eine Maschine f. montieren || monter une machine.

fitch handle and penholder turning machine || Drehbank f. für Pinselgriffe und Federhalter || tour m. pour manches de pinceaux et porteplumes.

fitness || Qualifikation f.; Eignung f. || qualification f.; capacité f. / ~ for his work || Arbeitstüchtigkeit f. || aptitude f. au travail. / certificate of ~ || Befähigungszeugnis n. || certificat m. de capacité.

fitted || eingebaut || monté. / ready ~ (Mach) || fertig eingebaut || complètement monté. / ~ with || ausgerüstet mit || pourvu de. / ~ with lock and key || verschließbar || fermant à clef.

fitter || Herrichter m.; Zurichter m.; Ausstatter m. || dresseur m.; monteur m. / ~ (Engine) || Montör m. || monteur m. / ~ (Cloth) || Probiererin f. || essayeuse f. / ~ (Hardware) || Fertigarbeiter m. || monteur m. / ~ (Mach) || Schlosser m.; Maschinenbauer m. || serrurier m.; mécanicien m.; serrurier-ajusteur m. / ~ (Tail) || Zuschneider m. || découpeur m.

/ ~ (War mat) ‖ Schaftmontör m. ‖ équipeur m. / chief ~ ‖ Obermontör m. ‖ chef-monteur m. / ~ of dust pipes (Watchm) ‖ Kappensetzer m. ‖ poseur m. de chapeaux. / iron railing ~ ‖ Eisengeländermacher m. ‖ rampiste m. en fer. / metal ~ ‖ Maschinenschlosser m. ‖ ajusteur-mécanicien m. / tool ~ ‖ Einrichter m.; Werkzeugmontör m. ‖ monteur m. d'outils.

fitter's dress ‖ Montöranzug m. ‖ habit m. de monteur. / ~ hammer ‖ Schlosserhammer m.; Montörhammer m. ‖ marteau m. d'ajusteur. / ~ shop ‖ Schlosserwerkstatt f. ‖ serrurerie f. / ~ tools pl. ‖ Schlosserwerkzeug n. ‖ outil m. de serrurier.

fitting ‖ Montieren n.; Einrichtung f. ‖ ajustage m.; adaption f.; montage m. / ~ and drilling of the watch glasses ‖ Einsetzen n. und Durchbohren n. der Uhrgläser ‖ posage m. et perçage m. des verres de montres. / ~ flat irons on the continuous flow system ‖ Bandmontage f. von Bügeleisen ‖ montage m. à la chaîne de fers à repasser. / ~ of the spectacles by an optician ‖ Anpassung f. der Brille durch den Optiker ‖ adaption f. de la lunette par l'opticien. / ~ machine for armatures ‖ Maschine f. zur Herstellung von Armaturen ‖ machine f. spéciale pour fabriquer des armatures.

fitting-out of a wagon ‖ Ausrüstung f. eines Wagens ‖ équipement m. de wagon. / ~ basin (Shipb) ‖ Ausrüstungsbassin n. ‖ bassin m. d'équipement. / ~ dock ‖ Baudock n. ‖ chantier m. d'armement.

fitting piece ‖ Paßstück n. ‖ cale f. d'ajustage.

fittings pl. ‖ Zubehör n.; Zubehörteile npl.; Fittings npl.; Armaturen fpl. ‖ garnitures fpl.; accessoires mpl.; armatures fpl. / ~ (Locksm) ‖ (eiserne Beschläge) mpl. ‖ ferrures fpl. / ~ for acid pipe line ‖ Säureleitungsarmaturen fpl. ‖ accessoires mpl. pour conduites d'acide. / blast furnace ~ ‖ Hochofenarmatur f. ‖ armature f. pour hauts-fourneaux. / boiler ~ ‖ Kesselarmaturen fpl. ‖ accessoires mpl. de chaudières. / brass ~ ‖ Messingarmaturen fpl. ‖ laitonnerie f.; garnitures fpl. en cuivre. / buckles pl. and small ~ ‖ Nadlerwaren pl. ‖ aiguilles fpl.; épingles fpl. / compressed air ~ ‖ Preßluftarmatur f. ‖ armature f. pour air comprimé. / ~ for cooling plants ‖ Armatur f. für Kühlanlagen ‖ armature f. pour installations frigorifiques. / ~ for curtains ‖ Eisenbeschläge mpl. für Gardinen ‖ ferrures fpl. de rideaux. / engine ~ ‖ Armaturen pl. für Maschinen ‖ accessoires mpl. de moteurs. / fire extinction ~ ‖ Feuerlöscharmatur f. ‖ armature f. pour l'extinction du feu. / ~ of firing for breweries ‖ Feuerungsarmatur f. für Brauereien ‖ armature f. de foyer pour brasseries. / machine for making furniture ~ ‖ Möbelbeschlagherstellungsmaschine f. ‖ machine f. à fabriquer des garnitures de meubles. / high-pressure ~ ‖ Hochdruckarmatur f. ‖ armature f. à haute pression. / ~ for horse harnesses and coaches ‖ Reit- und Fahrgeschirrbeschläge mpl. ‖ garnitures fpl. pour harnais de chevaux et de voitures. / ~ for hose pipes ‖ Schlaucharmatur f. ‖ accessoire m. pour tuyaux flexibles. / iron ~ ‖ Eisenbeschlag m. ‖ ferrure f. / iron ~ for

building purposes ‖ Baubeschlag m. ‖ ferrure f. de bâtiment. / little iron ~ of the permanent-way ‖ Schienenbefestigungsstücke npl. ‖ petit matériel m. de la voie. / office ~ ‖ Büromöbel pl. ‖ meubles mpl. de bureau. / ~ of ore delivery wagon ‖ Förderwagenbeschlagteil m. ‖ ferrure f. de chien de mine. / overhead line ~ ‖ Freileitungsarmatur f. ‖ accessoire m. pour ligne aérienne. / pump ~ ‖ Pumpenarmatur f. ‖ armature f. pour pompes. / ~ for railway carriages ‖ Eisenbahnwagenarmatur f. ‖ armature f. pour wagons de chemin de fer. / ~ for revolving shutters ‖ Rolladenbeschläge mpl. ‖ ferrures fpl. de jalousies. / ~ for sanitary plants ‖ Armatur f. für sanitäre Anlagen ‖ garniture f. pour installations sanitaires. / ~ for shoe machines ‖ Schuhmaschinenarmatur f. ‖ accessoires mpl. pour machines à chaussures. / shop ~ ‖ Ladenmöbel pl. ‖ meubles mpl. de magasin. / ~ for shop window stands ‖ Schaufenstergestellbeschläge mpl. ‖ ferrures fpl. de supports pour étalages. / small iron ~ ‖ Kleineisenzeug n. ‖ accessoires mpl. de voie. / steam boiler ~ ‖ Dampfkesselarmaturen fpl. ‖ accessoires mpl. de chaudières à vapeur. / machine for making structural ~ ‖ Baubeschlagherstellungsmaschine f. ‖ machine f. à fabriquer les ferrures de bâtiment / ~ for travellers' bags ‖ Reisetaschenbeschläge mpl. ‖ ferrures fpl. pour sacs à main. / ~ for walking sticks and for umbrellas ‖ Stock- und Schirmbeschläge mpl. ‖ ferrures fpl. de cannes et de parapluies. / ~ for water piping ‖ Wasserleitungsarmaturen fpl. ‖ robinetterie f. de conduites d'eau. / ~ for wooden boxes and cases ‖ Kistenbeschläge mpl. ‖ ferrures fpl. pour caisses en bois.

fitting shop ‖ Montagehalle f. ‖ halle f. de montage.

fittings smith (Coachm) ‖ Geschirrbeschläger m. ‖ ferreur m. d'accessoires.

fitting-up ‖ Aufstellung f.; Montierung f. ‖ montage m.

fitting work ‖ Paßarbeit f. ‖ ajustage m. / ~ workshop for farmers ‖ Schlosserwerkstatt f. für Farmer ‖ atelier m. de serrurerie pour fermiers.

five-electrode tube (Radio) ‖ Fünfelektrodenröhre f. ‖ lampe f. à cinq électrodes.

five-engined ‖ fünfmotorig ‖ pentamoteur.

five-mast bark ‖ Fünfmastbark f. ‖ cinq-mât barque f.

five-point diaphragm ‖ Fünfpunktblende f. ‖ diaphragme m. à cinq points.

five-unit code ‖ Fünferalphabet n. ‖ code m. à cinq émissions.

five-wire network ‖ Fünfleiternetz n. ‖ réseau m. à cinq fils. / ~ system ‖ Fünfleitersystem n. ‖ système m. à cinq conducteurs ou fils.

fix, to ‖ festmachen ‖ fixer. / ~ (Mach) ‖ feststellen ‖ arrêter; fixer. / ~ (Mach tool) ‖ einspannen ‖ serrer; fixer. / ~ (Phot) ‖ fixieren ‖ fixer. / ~ the chairs on the sleepers ‖ die Schwellen fpl. aufstuhlen ‖ saboter les traverses fpl. / ~ the cramps into the ashlar ‖ die Klammern fpl. in den Stein einsetzen ‖ enclaver les crampons mpl. / ~ with a grapnel (Carp) ‖ aufklauen ‖ empâter ou fixer avec le grappin. / ~ with hemp (Insulator) ‖ aufhanfen ‖ garnir de chanvre m. / ~ a nut with a splitpin ‖ eine Mutter f. versplinten ‖ goupiller un écrou. / ~ a position

(Aero) ‖ franzen ‖ repérer. / ~ struts pl. ‖ ausstreben ‖ entretoiser. / ~ with a toggle ‖ knebeln ‖ garotter. / ~ work-pieces ‖ Werkstücke npl. aufspannen ‖ fixer des pièces à travailler.

fixation forceps pl. ‖ Fixierpinzette f.; Feststellpinzette f. ‖ pince f. à fixer.

fixative bath (Phot) ‖ Fixierbad n. ‖ bain m. fixateur ou de fixage. / ~ salt ‖ Fixiersalz n. ‖ sel m. de fixage ou fixateur ou fixative.

fixed ‖ fest ‖ fixe; ferme. / ~ (Mach) ‖ fest; stationär ‖ stationnaire. / ~ at right angles to ... ‖ rechtwinklig angesetzt ‖ disposé perpendiculairement à ... / ~ beam ‖ eingespannter Balken m. ‖ poutre f. encastrée. / ~ charge and message rate tariff ‖ Grund- und Gesprächsgebühr f. ‖ régime m. de la redevance fondamentale et de la conversation taxée. / ~ point ‖ fester Punkt m. ‖ point m. fixe. / ~ price ‖ fester Preis m. ‖ prix m. fixe ou fait. / ~ pulley ‖ Festscheibe f. ‖ poulie f. fixe.

fixer (Phot) ‖ Fixiermittel n. ‖ fixateur m.

fixing ‖ Fixierung f. ‖ fixage m. / ~ of the blades ‖ Schaufelbefestigung f. ‖ montage m. des aubes. / ~ of dividends ‖ Festsetzung f. der Dividende ‖ fixation f. du dividende. / ~ of photos ‖ Fixieren n. von fotografischen Aufnahmen ‖ fixation f. des images photographiques. / ~ on the wall ‖ Wandbefestigung f. ‖ fixage m. au mur.

fixing agent ‖ Fixiermittel n.; Bindemittel n. ‖ fixateur m. / ~ band ‖ Befestigungsstreifen m. ‖ bande f. de fixation. / ~ bar ‖ Befestigungsschiene f. ‖ rail m. de fixation. / ~ bath ‖ Fixierbad n. ‖ bain m. fixateur ou de fixage. / ~ bolt ‖ Stellriegel m. ‖ verrou m. de fixation. / ~ catch ‖ Befestigungsriegel m. ‖ barre f. ou verrou m. de fixation. / ~ clip ‖ Befestigungsschelle f.; Feststellklammer f. ‖ collier m. ou pince f. de fixation. / ~ device ‖ Aufspannvorrichtung f.; Feststellvorrichtung f. ‖ dispositif m. de fixation. / ~ key ‖ Verbindungskeil m. ‖ clavette f. d'assemblage. / ~ muff ‖ Verbindungsmuffe f. ‖ manchon m. d'assemblage. / ~ plate ‖ Befestigungsplatte f.; Aufspannplatte f. ‖ plaque f. de fixation. / ~ point ‖ Einspannstelle f. ‖ point m. de serrage. / ~ salt ‖ Fixiersalz n. ‖ fixateur m. / ~ screw ‖ Befestigungsschraube f. ‖ vis f. de fixation; boulon m. d'attache ou d'ancrage. / ~ slot ‖ Aufspannschlitz m. ‖ rainure f. de fixation. / ~ tape ‖ Befestigungsband n. ‖ bride f. de fixation.

fixture ‖ Befestigung f. ‖ pièce f. fixe.

fizgig ‖ Harpune f. ‖ harpon m.; foène f.; fouane f.

fizzing ‖ Aufzischen n. ‖ sifflement m.

flabby ballonet ‖ unpralles Ballonett n. ‖ ballonnet m. flasque.

flacon see flagon.

flag, to ‖ die Fliesen fpl. legen; mit Fliesen fpl. pflastern ‖ poser les carreaux mpl.; carreler.

flag ‖ Fahne f. ‖ drapeau m. / ~ (Build) ‖ Fliese f. ‖ dalle f. / ~ (Mar) ‖ Flagge f. ‖ pavillon m.; enseigne f. / ~s pl. in diamond pavement ‖ Fliesen fpl. in Schlagverband ‖ carreaux mpl. rangés obliquement. / hexagonal ~ ‖ sechseckige Fliese f. ‖ carreau m. à six pans. / mosaic ~ ‖ Mosaikfliese f. ‖ carreau m.

en mosaïque. / ~ for motor cars ∥ Kraftwagenflagge f. ∥ pavillon m. pour automobiles. / national ~ ∥ Nationalflagge f. ∥ pavillon m. national. / paper ~ ∥ Papierfahne f. ∥ drapeau m. en papier. / ~ for sighting on ∥ Signalfahne f. ∥ fanion-signal m. / signal ~ ∥ Signalflagge f. ∥ pavillon m. à signaux. / surveyor's ~ ∥ Absteckfähnchen n. ∥ guidon m. d'arpenteur. / ~ of truce ∥ Friedensflagge f. ∥ pavillon m. parlementaire *ou* de paix *ou* blanc. / ~ with a waft flag ∥ Notflagge f. ∥ pavillon m. de détresse. / ~ for wall covering ∥ Wandbekleidungsplatte f. ∥ carreau m. de revêtement.

flagellum (Brew) ∥ Geißel m. ∥ cil m.

flag embroiderer ∥ Fahnensticker m. ∥ brodeur m. de drapeaux.

flageolet ∥ Flageolett n. ∥ flageolet m.

flag fittings pl. ∥ Fahnenbeschläge mpl. ∥ garnitures fpl. de drapeaux.

flagger ∥ Fliesenleger m. ∥ poseur m. de dalles.

flagging ∥ Fliesenlegen n. ∥ dallage m.; pose f. de dalles; carrelage m.

flag layer for foot-way ∥ Bürgersteigleger m. ∥ poseur m. de dalles de trottoir.

flagon ∥ Flakon n. ∥ flacon m.

flag pavement ∥ Fliesenfußboden m.; Backsteinpflaster n. ∥ pavé m. carrelé *ou* de carreaux. / ~ root ∥ Kalmuswurzel f. ∥ racine f. de calamus. / ~ staff ∥ Flaggenstock m. ∥ hampe f.; lance f. de pavillon.

flagstone ∥ Fliese f.; Kachel f. ∥ dalle f. / ~ pavement ∥ Plattenpflaster n. ∥ carrelage m.

flag waif ∥ Notflagge f. ∥ pavillon m. de détresse.

flail ∥ Dreschflegel m. ∥ fléau m. à battre.

flake, to ∥ abblättern ∥ s'effeuiller.

flake (Chem) ∥ Flocke f. ∥ flocon m. / white ~ ∥ Schieferweiß n. ∥ blanc m. de plomb feuilleté; céruse f. en lamelles.

flaked graphite ∥ Flockengraphit m. ∥ graphite m. en flocons.

flaky ∥ schuppig; flockig ∥ floconneux.

flame ∥ Flamme f. ∥ flamme f. / lighting ~ ∥ Zündflamme f. ∥ flamme f. d'allumage. / reducing ~ ∥ Reduktionsflamme f. ∥ flamme f. réductrice. / small pointed ~ ∥ Stichflamme f. ∥ petite flamme f. pointue; jet m. de flamme.

flame arc ∥ Flammenbogen m. ∥ arc m. (à) flamme. / ~ arch ∥ Feuergewölbe n. ∥ voûte f. d'inflammation. / ~ bridge ∥ Feuerbrücke f. ∥ pont m. de chauffe; autel m.; marche f. / ~ furnace ∥ Flammofen m. ∥ fourneau m. à réverbère *ou* dormant. / ~ guard ∥ Flammenschutz m. ∥ pare-flamme m. / ~ killer ∥ Mittel n. gegen Mündungsfeuer ∥ antilueurs mpl.

flame-proof ∥ feuersicher ∥ ignifuge.

flame thrower ∥ Flammenwerfer m. ∥ lance-flammes m. / ~ tube ∥ Flammrohr n. ∥ tuyau m. de flamme; tube-foyer m.

flaming ∥ Fensterlaibung f.; Anschlagmauer f. ∥ embrasure f. / ~ of the arc ∥ Aufflammen n. des Lichtbogens ∥ flamboiement m. de l'arc. / ~ coal (Min) ∥ Flammförderkohle f. ∥ charbon m. flambant.

flang ∥ Doppelkeilhaue f. ∥ marteau m. à (deux) pointes; pic m. à deux branches.

flange, to ∥ flanschen; umbördeln; umkrempeln ∥ brider; mettre une bride; tomber le bord.

flange (Girder) ∥ Gurtung f. ∥ membrure f.; aile f. / ~ (Grooved roll) ∥ Rand m.

(einer Kaliberwalze) ∥ rebord m.; collier m. / ~ (Mach) ∥ Flansch m. ∥ bride f. / ~ (Railway rail) ∥ Spurkranz m. ∥ patin m. / ~ (Tube) ∥ umgezogener Bord m. ∥ bord m.; collet m. / ~ (Wheel) ∥ Radkranz m.; Radflansch m. ∥ bourrelet m.; mentonnet m.; rebord m. / angle ring ~ ∥ Winkelflansch m. ∥ bride f. à cornières. / blank ~ ∥ Deckelflansch m.; Blindflansch m. ∥ bride f. aveugle; fausse bride f. / ~ of boiler end ∥ Krempe f. des Kesselbodens ∥ bord m. du fond. / boiler plate ~ ∥ Kesselblechflansch m. ∥ bride f. d'une tôle de chaudière. / ~ of the boss ∥ Nabenflansch m. ∥ contre-disque m. de moyeu. / bottom ~ (Bridge) ∥ Gurtungsblech m. ∥ plate-bande f.; semelle f. / bottom ~ (Mach) ∥ Fußflansch m. ∥ aile f. inférieure. / ~ of brass bearings ∥ Lagerschalenbund m. ∥ rebord m. du coussinet. / brazed-on ~ ∥ aufgelöteter Flansch m. ∥ bride f. soudée. / caststeel ~ ∥ Stahlgußflansch m. ∥ couronne f. à bride en acier coulé. / connecting ~ ∥ Anschlußflansch m. ∥ bride f. de raccordement. / core head ~ (Electr) ∥ Ankerflansch m. ∥ flasque m. fixe de l'induit. / ~ of coupling ∥ Kupplungsflansch m. ∥ bride f. d'accouplement. / cylinder ~ ∥ Zylinderflansch m. ∥ bride f. de cylindre. / ~ with feet ∥ Fußflansch m. ∥ couronne f. à pieds. / ~ of the fish plate (Railw) ∥ Laschenschenkel m. ∥ cornière f. de l'éclisse. / ~ of girder ∥ Trägerflansch m. ∥ aile f. de la poutre. / intermediate ~ ∥ Zwischenflansch m. ∥ bride f. intermédiaire. / loose ~ ∥ loser Bord m. ∥ bride f. folle. / lower ~ ∥ Unterflansch m. ∥ semelle f. inférieure. / ~ of a pipe ∥ Flansch m. eines Rohres ∥ bourrelet m. *ou* bride f. d'un tuyau. / ~ of rail base (Railw) ∥ Flansch m. des Schienenfußes ∥ aile f. du patin de rail. / with rivetted-on ~ ∥ mit angenietetem Flansch m. ∥ à bride f. rivetée. / with ~ rivetted to the piece ∥ mit angenietetem Flansch m. ∥ à bride f. rivetée. / screwed-on ~ ∥ aufgeschraubter Flansch m. ∥ bride f. vissée. / shaft coupling ~ ∥ Kurbelwellenkupplungsflansch m. ∥ plateau m. d'accouplement du vilebrequin. / steam boiler dome ~ rolled-on ∥ angewalzter Flansch m. am Kesseldom ∥ collerette f. laminée au dôme de chaudière. / top ~ (Bridge) ∥ Obergurt m.; obere Gurtung ∥ semelle f. supérieure. / top ~ (Mach) ∥ Oberflansch m. ∥ aile f. supérieure. / ~ of a truss ∥ Gurt m.; die Gurtung ∥ semelle f. / wheel ~ ∥ Radflansch m.; Spurkranz m. ∥ rebord m. *ou* boudin m. d'une roue.

flange angle ∥ Gurtungswinkel m. ∥ cornière f. de la membrure. / ~ coupling ∥ Flanschenkupplung f. ∥ accouplement m. à plateaux.

flanged ∥ geflanscht ∥ à bords mpl. tombés. / ~ and dished parts pl. ∥ Kümpelteile mpl. ∥ pièces fpl. bombées. / ~ inwards ∥ boiler end ∥ eingehalster Kesselboden m. ∥ fond m. de chaudière à collet intérieur. / machine ~ boiler end ∥ mittels maschineller Vorrichtung umgezogener Kesselboden m. ∥ fond m. de chaudière bridé à la machine. / ~ outwards ~ boiler end ∥ ausgehalster Kesselboden m. ∥ fond m. à collet extérieur. / ~ at right angles ∥ im rechten Winkel angeflanscht ∥ bridé en angle droit. / ~ boiler end ∥ um-

gezogener Kesselboden m. ∥ fond m. de chaudière à collet. / ~ bush ∥ Flanschenbuchse f. ∥ douille f. à bride. / ~ edge joint ∥ Stirnstoß m. ∥ joint m. à arête bridée. / ~ end ∥ gebördelter Boden m. ∥ fond m. bridé *ou* à bride. / ~ fire box ∥ umgebördelte Feuerbuchse f. ∥ boîte f. à feu bridée. / ~ pipe ∥ Flanschenrohr n. ∥ tuyau m. à brides. / ~ plate ∥ Kümpelblech n. ∥ tôle f. à bord tombé; plaque f. bridée. / ~ rim ∥ umgekrempelter Rand m. ∥ bord m. bridé. / ~ ring ∥ Flanschenring m. ∥ anneau m. à bride. / ~ shaft ∥ Planscheibenwelle f. ∥ arbre m. à plateaux. / ~ sheet work ∥ Bördelarbeit f. ∥ pièce f. emboutie. / ~ wheel ∥ geflanschtes Rad n. ∥ roue f. à boudin.

flange groove (Railw) ∥ Spurrille f. ∥ ornière f. de passage. / ~ hub ∥ Flanschnabe f. ∥ moyeu m. à flasques. / ~ joint ∥ Flanschverbindung f. ∥ joint m. à brides *ou* à boudin.

flangeless wheel ∥ Rad n. ohne Spurkranz ∥ roue f. à surface cylindrique *ou* sans boudin.

flange motor ∥ Flanschmotor m. ∥ moteur m. à bride. / ~ pipe ∥ Flanschenrohr n. ∥ tuyau m. à bride. / ~ planing and trimming machine ∥ Flanschendrehbank f. ∥ tour m. à brides. / ~ rail ∥ Breitfußschiene f.; Vignoleschiene f. ∥ rail m. Vignole *ou* à patin. / ~ stiffening ∥ Gurtversteifung f. ∥ entretoisement m. des membrures. / ~ tie plate ∥ Krempenplatte f. ∥ plaque f. à rebord; selle f. à crochet. / ~ turning jig ∥ Flanschendrehsupport m. ∥ dispositif m. à dresser les brides. / ~-type cock with bent outlet ∥ Flanschenhahn m. mit gebogenem Auslauf ∥ robinet m. à bride avec sortie recourbée. / ~-type cock with cap screw ∥ Flanschenhahn m. mit Überwurfmutter ∥ robinet m. à bride avec écrou à chapeau.

flangeway clearance ∥ Zungenrille f. ∥ ornière f. des boudins.

flange working machine ∥ Flanschenbearbeitungsmaschine f. ∥ machine f. à travailler les brides. / ~ works pl. ∥ Flanschfabrik f. ∥ fabrique f. de brides.

flanging ∥ Flanschen n.; Kümpelung f. ∥ bordage m.; emboutissage m.; rabattement m. des bords. / ~ the boiler ends ∥ Bördeln n. von Kesselböden ∥ bridage m. de fonds de chaudières.

flanging attachment ∥ Bördeleinrichtung f. ∥ dispositif m. à border. / ~ machine ∥ Bördelmaschine f. ∥ machine f. à border; machine f. à tomber *ou* à brider les bords.

flanging press ∥ Kümpelpresse f.; Bördelpresse f. ∥ presse f. à border *ou* à emboutir. / hydraulic circular ~ ∥ hydraulische Kümpelpresse f. ∥ presse f. hydraulique à emboutir les fonds de chaudières.

flank ∥ Flanke f. ∥ flanc m. / concave ~ ∥ hohle Flanke f. ∥ flanc m. creux. / right-lined ~ ∥ gerade Flanke f. ∥ flanc m. droit. / tooth ~ ∥ Zahnflanke f. ∥ flanc m. de dent. / ~ of a tooth of a wheel ∥ Zahnflanke f. ∥ flanc m. d'une dent de roue.

flank front ∥ Seitenfront f. ∥ façade f. de côté.

flannel ∥ Flanell m. ∥ flanelle f. / treble-milled ~ ∥ Wollmolton m. ∥ molleton m. de laine.

flannel bandage ∥ Flanellbinde f. ∥ bande f. de flanelle. / ~ shirt ∥ Flanellhemd n. ∥ chemise f. de flanelle. / ~ waistcoat ∥

Flanellunterjacke f. ‖ gilet m. de flanelle.

flap, to (Sail) ‖ schlagen; flattern ‖ battre.

flap ‖ Klappe f.; Falltür f. ‖ abattant m.; trappe f. / bottom ~ (Wagon) ‖ Bodenklappe f. ‖ trappe f. de fond; fond m. à rabattement. / gummed ~ ‖ gummierte Briefumschlagklappe f. ‖ patte f. d'enveloppe gommée. / hinged ~ ‖ um ein Gelenk drehbare Klappe f. ‖ clapet m. à charnières. / non-return ~ ‖ Rückschlagklappe f. ‖ clapet m. de retenue. / pneumatic operation of ~s for self-tippings ‖ Druckluftklappenbetätigung f. für Selbstentlader ‖ clapet m. pour déchargeurs automatiques commandé à l'air comprimé. / self-opening ~ ‖ selbsttätige Klappenöffnung f. ‖ ouverture f. automatique de clapet. / ventilation ~ ‖ Lüftungsflügel m. ‖ clapet m. de ventilation.

flap-covered oil cup ‖ Klappenöler m. ‖ graisseur m. à clapet. / ~ door ‖ Klapptür f. ‖ trappe f.; porte f. à rabattement.

flapping of the sails ‖ Schlagen n. der Segel ‖ battements mpl. des voiles.

flap seat ‖ Klappsitz m. ‖ siège m. à rabattement; strapontin m. / ~ valve ‖ Klappenventil n. ‖ clapet m.; soupape f. à clapet ou à charnière.

flare, to ~ up ‖ sich entzünden ‖ s'enflammer.

flare ‖ heller Lichtschein m.; flackerndes Licht n. ‖ éclat m. / light ~ ‖ Leuchtrakete f. ‖ fusée f. éclairante. / parachute ~ ‖ Fallschirmrakete f. ‖ fusée f. à parachute.

flare light, acetylene ‖ Azetylenleuchtfeuer n. ‖ phare m. à acétylène. / ~ pistol ‖ Leuchtpistole f. ‖ pistolet m. éclairant ou pour fusées d'éclairage. / ~ signal ‖ Raketenzeichen n. ‖ signal m. de fusée.

flash, to (Lamp) ‖ flackern ‖ vaciller.

flash ‖ Aufleuchten n.; Blitz m. ‖ éclair m. / cold ~ ‖ kalter Blitzschlag m. ‖ coup m. de foudre froid. / ~ of fire ‖ Feuergarbe f. ‖ gerbe f. de feu. / ~ of lightning ‖ Blitzschlag m. ‖ coup m. de foudre.

flash apparatus see flashing apparatus.

flashback of the flame ‖ Rückschlag m. oder Rückschlagen n. der Flamme ‖ retour m. de la flamme.

flash cistern ‖ Klosettspülgerät n. ‖ appareil m. de chasse pour W. C.

flashed glass (Glassm) ‖ Überfangglas n. ‖ verre m. à deux couches.

flashing ‖ Kehlblech n. ‖ noquet m. / cornice ~ ‖ Simsabdeckung f. ‖ chaperon m. de l'entablement.

flashing apparatus ‖ Blinkgerät n. ‖ appareil m. miroiteur; miroiteur m. / ~ furnace (Glassm) ‖ Auslaufofen m. ‖ four m. par l'embrasure. / ~ light ‖ Blinkfeuer n. ‖ phare m. à éclats. / ~ light signal ‖ Blinksignal n. ‖ signal m. à éclats. / ~ point ‖ Flammpunkt m. ‖ point m. d'inflammation. / ~ signal (Mar) ‖ Lichtblitzsignal n. ‖ signal m. à éclats. / ~ signal (Tel) ‖ Flackerzeichen n. ‖ signal m. de scintillement.

flash instrument (Tel) ‖ Blink m. ‖ appareil m. miroiteur. / ~ light (Electr) ‖ Scheinwerfer m. ‖ phare m.; projecteur m. / ~ light (Phot) ‖ Blitzlicht n. ‖ poudre f. fulminante· lumière f. au magnésium.

flash-over test ‖ Überschlagsversuch m. ‖ essai m. de décharge par arc. / ~ voltage (Electr) ‖ Überspannung f. ‖ surtension f.

flash point ‖ Flammpunkt m. ‖ point m. d'inflammation. / ~ tester ‖ Flammpunktprüfer m. ‖ essayeur m. du point d'inflammation. / ~ tester for oil ‖ Flammpunktprüfer m. für Öl ‖ appareil m. à déterminer le point d'inflammation des huiles.

flash ranging ‖ Lichtmessen n. ‖ repérage photométrique. / ~ troop ‖ Lichtmeßtrupp m. ‖ section f. de repérage optique sur terre.

flash signal gear ‖ Lichtsignalgerät n. ‖ appareil m. de signalisation lumineuse. / ~ test ‖ Entflammungsprobe f. ‖ épreuve f. d'inflammation.

flask ‖ Fläschchen n.; Flakon m. ‖ petite bouteille f.; flacon m.; fiole f. / ~ (Chem) ‖ Glaskolben m. ‖ matras m. / ~ (Mould) ‖ Formkasten m. ‖ châssis m. de moulage. / ~ with flat bottom ‖ Stehkolben m. ‖ ballon m. à fond plat. / glass-stoppered ~ ‖ Stöpselflasche f. ‖ flacon m. bouché à l'émeri. / graduated ~ ‖ Meßkolben m.; Meßflasche f. ‖ ballon m. jaugé; fiole f. jaugée. / middle ~ (Mould) ‖ Mittelkasten m. ‖ châssis m. de milieu. / round ~ ‖ Rundkolben m. ‖ ballon m. à fond rond. / spherical ~ ‖ Glaskugel f. ‖ ballon m. de verre. / washing ~ ‖ Spritzflasche f. ‖ pissette f.

flask board (Mould) ‖ Formbrett n.; Modellbrett n. ‖ planche f. de fond; couche f. / ~ mould ‖ Kastenform f. ‖ moule m. en châssis. / ~ moulding ‖ Kastenformerei f. ‖ moulage m. en châssis.

flat ‖ platt; flach ‖ plat. / ~ (Trade) ‖ still ‖ tranquille; inactif; calme. / with ~ bottom ‖ flachrund ‖ à fond m. plat.

flat (Build) ‖ Altan m.; flaches Dach n.; Plattform f. ‖ comble m. plat; plateforme f. / ~ (Mar) ‖ Untiefe f.; Flach n.; Watt n. ‖ estuaire f.; basse f.; hautfond m.

flat American spade ‖ amerikanischer Flachspaten m. ‖ bêche f. américaine plate.

flat bar (Metal) ‖ Platine f. ‖ larget m. / ~copper ‖ Flachkupfer n. ‖ cuivre m. plat. / ~ iron ‖ Flacheisen n. ‖ méplat m. (en fer); plat m. à faible épaisseur.

flat-bed printing machine for job work ‖ Akzidenzschnellpresse f. ‖ machine f. à imprimer en blanc. / ~ rotary machine (Print) ‖ Flachdruckrotationsmaschine f. ‖ machine f. rotative à formes plates. / ~ web machine see ~ rotary machine.

flat billet ‖ Breiteisen n. ‖ large-plat m. / ~ blade holder ‖ Flachstahlhalter m. ‖ porte-outil m. ordinaire. / ~ bottom ‖ Flachboden m. ‖ fond m. plat. / ~ bottomed ‖ flachbodig ‖ à fond m. plat. / ~ bottom rail ‖ Breitfußschiene f.; Vignoleschiene f. ‖ rail m. Vignoll ou à patin. / ~ bottom self-discharging car to be dumped to one side ‖ Flachbodenselbstentlader m. für Entladung nach einer Seite ‖ wagon m. auto-déchargeur à fond plat se vidant par un côté. / ~ burner ‖ Flachbrenner m. ‖ bec m. plat. / ~ chisel ‖ Flachmeißel m. ‖ burin m. plat. / ~ clock pliers pl. ‖ Flachzange f. für Uhrmacher ‖ pince f. plate d'horloger. / ~ coal ‖ Flaches n.; Plattes n. ‖ plat m.; plateur f. / ~ coil ‖ Flachspule f. ‖ galette f. / ~ copper bar ‖ Flachkupfer n. ‖ cuivre m. plat. / ~ cut ‖ flache Strichätzung f. ‖ gravure f. en trait. / ~ em-

broidery ‖ Plattstickerei f. ‖ broderie f. plate. / set for ~ feet ‖ Plattfußeinlage f. ‖ semelle f. pour pieds plats. / ~ file Flachfeile f. ‖ lime f. plate. / ~ fish plate (Railw) ‖ Flachlasche f. ‖ éclisse f. plate. / ~ front of wave (Electr) ‖ flach ansteigende Welle f. ‖ onde f. à front aplati. / ~ furnace mixer ‖ Flachherdmischer m. ‖ mélangeur m. à sole plate. / ~ groove ‖ Flachkaliber n. ‖ cannelure f. plate.

flathammer, to ‖ glatthämmern; nachhämmern ‖ repasser la tôle.

flat hammerer ‖ Blechrichter m. ‖ planeur m. sur métaux. / ~ handle ‖ flaches Werkzeugheft n. ‖ manche m. plat. / ~ indent ‖ Kerb m.; Kerbe f. ‖ entaille f.; encoche f.

flat-iron (Metal) ‖ Flacheisen n. ‖ larget m.; fer m. plat ou méplat. / ~ (Tail) ‖ Bügeleisen n.; Plätteisen n. ‖ fer m. à repasser. / electric ~ ‖ elektrisches Bügeleisen n. oder Plätteisen n. ‖ fer m. à repasser électrique. / gas ~ ‖ Gasbügeleisen n. ‖ fer m. à repasser chauffé au gaz. / household ~ ‖ Haushaltbügeleisen n. ‖ fer m. à repasser pour ménages.

flat-iron bar ‖ Flacheisenstab m. ‖ barre f. en fer plat. / ~ grate ‖ Flacheisenrost m. ‖ grille f. à fers plats.

flat-iron buffer ‖ Flacheisenpuffer m. ‖ tampon m. à fer plat. / ~ rod ‖ Flacheisenstab m. ‖ barre f. en fer plat. / ~ shearing machine ‖ Flacheisenschere f. ‖ cisaille f. à fers plats. / ~ stirrup ‖ Flacheisenbügel m. ‖ étrier m. en fer plat. / gas ~ stove ‖ Gasbügelofen m. ‖ four m. à gaz pour fers à repasser. / ~ train (Roll mill) ‖ Bandeisenstraße f. ‖ laminoir m. à fers plats.

flat key ‖ Flachkeil m. ‖ clavette f. conique sur méplat. / ~ knitting machine ‖ Flachstrickmaschine f. ‖ machine f. à tricoter rectiligne. / ~-link chain ‖ Gelenkkette f. ‖ chaîne f. d'articulations; chaîne f. de maillons. / ~-lying mineral deposit ‖ flache Lagerstätte f. ‖ gisement m. horizontal ou en plateure.

flatness (Trade) ‖ Stille f. ‖ inaction f.; stagnation f.

flat-nosed and cutting nippers ‖ Flachzange f. mit Seitenschneider ‖ pince f. plate et coupante. / ~ pliers pl. ‖ Flachzange f.; Drahtzange f. mit flachen Backen ‖ pince f. plate.

flat oval lens ‖ flachovales Brillenglas n. ‖ verre m. à ovale aplati. / ~ part ‖ flächiges Werkstück n. ‖ pièce f. plate. / ~ pick ‖ Kreuzhacke f.; Spitz- und Flachhacke f. ‖ pioche f. / ~ piece of clay (Porcel) ‖ Schwarte f. ‖ croûte f. d'argile. / ~ pipe principle ‖ Flachrohrprinzip n. ‖ principe m. des tubes aplatis. / ~ pliers pl. ‖ Flachzange f. ‖ pince f. ou pincette f. ou tenaille f. plate. / ~ rail ‖ Flachschiene f. ‖ rail m. plat. / ~ rammer (Mould) ‖ Stampfer m. ‖ batte f.; cogneuse m. / ~ rate ‖ Pauschgebühr f. ‖ abonnement m. forfaitaire.

flat-ring armature ‖ Flachringanker m. ‖ induit m. à anneau aplati. / ~ dynamo ‖ Flachringdynamo f. ‖ génératrice f. à anneau plat.

flat rods pl. (Mine) ‖ Feldgestänge n. ‖ piston m.; tirant m. / ~ roof ‖ flaches Dach n. ‖ toit m. plat; toiture f. plate. / ~ rubber collar ‖ Flachgummiring m. ‖ anneau m. plat ou rondelle f. plate

en caoutchouc. / ~ rubber ring *see* ~ rubber collar.

flats pl. (Mine) ‖ Feldgestänge n. ‖ piston m.; tirant m. / ~ (Writing pen) ‖ Plättchen npl.; Stahlblech n. ‖ flans mpl. *ou* lamettes fpl. d'acier.

flat section instrument ‖ Flachprofilinstrument n. ‖ instrument m. de mesure de profil plat./ ~ shale ‖ Flachschiefer m. ‖ schistes mpl. plats. / ~-shaped ‖ flach aplati. / ~ sieve ‖ Flachsieb n. ‖ tamis m. plat. / ~ sliding bolt ‖ Vierkantschubriegel m. ‖ targette f. pêne plat. / ~ spherical vault ‖ Flachkuppel f. ‖ voûte f. en cul-de-four surbaissée. / ~ spiral spring ‖ Schloßfeder f. ‖ ressort m. en spirale plate. / with a ~ surface ‖ ebenflächig ‖ plan; planièdre.

flatten, to (Glassm) ‖ glattstreichen ‖ estriquer. / ~ (Locksm) ‖ glatthämmern; nachhämmern ‖ repasser la tôle *ou* le fer-blanc. / ~ (Metal) ‖ abflachen; platt drücken ‖ aplatir. / ~ the eyes (Needl) ‖ pflöcken; das Öhr breit schlagen ‖ palmer. / ~ iron ‖ strecken; recken (das Eisen) ‖ étirer le fer sous le marteau. / ~ out (Aero) ‖ im Gleitflug m. heruntergehen ‖ descendre en vol m. plané. / ~ out (Metal) ‖ ausschlagen ‖ battre. / ~ the planks (Coin) ‖ die Schrötlinge mpl. ausschlagen *oder* plätten ‖ flatir les flans mpl. / ~ the wire ‖ den Draht m. plätten ‖ aplatir *ou* écacher le fil.

flattened ‖ abgeflacht; abgeplattet ‖ aplati.

flattener (Forg) ‖ Setzhammer m.; Flachhammer m. ‖ chasse f. à parer; paroir m. / ~ (Needl) ‖ Breitschläger m. ‖ aplatisseur m. / ~ (Railw) ‖ Schienenrichtplatte f. ‖ table f. en fonte pour le dressage des rails. / ~ (Wiredr) ‖ Plättwalze f.; Streckwalze f. ‖ aplatisseur m.

flattening ‖ Abflachung f. ‖ aplatissement m. / ~ of iron ‖ Recken n. des Eisens ‖ étirage m. au marteau.

flattening arrangement (Tel) ‖ Abflachschaltung f. ‖ dispositif m. de compensation. / ~ furnace (Glassm) ‖ Streckofen m. ‖ fourneau m. d'étendage. / ~ hammer (Coin) ‖ Plätthammer m.; Schrötlingshammer m. ‖ flatoir m. / ~ stone (Glassm) ‖ Lager n.; Streckstein m. ‖ plaque f. à étendre. / ~ tool ‖ Streckeisen n. ‖ polis m.; rabot m. / ~ wood (Glassm) ‖ Filzstock m.; befilztes Polierholz n. ‖ estrique f.; lustroir m.; moëllette f.

flatter *see* flattener.

flat tile ‖ Flachziegel m.; Plattziegel m. ‖ tuile f. plate *ou* à crochet.

flatting (Locksm) ‖ Platthämmern n. ‖ aplatissement m. / ~ (Paint) ‖ Ölanstrich m. ohne Lackierung; matter Ölanstrich m. ‖ vernis m. à l'huile mat. / ~ mill ‖ Plättmaschine f.; Plättwerk n. ‖ aplatissoire f.

flat tool ‖ Schlichtstahl m. ‖ outil m. finisseur. / ~ tuning (Radio) ‖ unscharfes Abstimmen n. ‖ syntonisation f. non aiguë. / ~ valve ‖ Flachschieber m. ‖ tiroir m. plan. / ~ wagon ‖ Plateauwagen m. ‖ wagon m. à plateau. / ~ wire ‖ Flachdraht m. ‖ fil m. plat.

flaw (Cloth) ‖ Platte f. ‖ rupture f. d'une maille. / ~ (Found) ‖ Gußblase f. ‖ soufflure f. de fonte. / ~ (Gem) ‖ Fleck m. ‖ glace f.; paille f.; fumée f. / ~ (Material) ‖ Riß m.; Fehlstelle f.; Fehler m. ‖ fissure f.; fente f.; déchirure f.; fêlure f.; paille f.; tache f.; défectuosité f. / ~

(Wood) ‖ Windriß m. ‖ gerçure f. / ~ in the casting ‖ Gußfehler m.; Galle f. ‖ défaut m. de coulée; paille f. / ~ in the glass ‖ Glasfehler m. ‖ défaut m. du verre. / ~ from tempering ‖ Härteriß m. ‖ fissure f. de trempe.

flawed ‖ gespalten; rissig ‖ crévassé; gercé; lézardé.

flawless ‖ fehlerfrei ‖ sans défaut m.

flax ‖ Flachs m. ‖ lin m. / broken ~ ‖ gebrochener Flachs m. ‖ lin m. broyé. / clean ~ ‖ reiner Flachs m. ‖ lin m. propre. / common ~ ‖ gemeiner Flachs m. *oder* Lein m. ‖ lin m. commun *ou* cultivé. / cottoned ~ ‖ Flachsbaumwolle f.; Flachswolle f. ‖ lin-coton m. / cut ~ ‖ geschnittener Flachs m.; Kurzflachs m. ‖ lin m. coupé. / dew-retted ~ ‖ im Tau gerotteter Flachs m. ‖ lin m. roui à la rosée. / ~ in fibres ‖ Flachs m. in Fasern ‖ lin m. en fibres. / fine ~ ‖ feiner Flachs m. ‖ lin m. fin. / hackled ~ ‖ Hechelflachs m.; Kernflachs m.; gekämmter Flachs m. ‖ lin m. peigné; filasse f. *ou* brin m. de lin peigné. / heckled ~ *see* hackled ~. / late ~ ‖ Spätlein m.; Spätflachs m. ‖ lin m. froid *ou* tardif. / long ~ ‖ Langflachs m. ‖ lin m. long. / New Zealand ~ ‖ neuseeländischer Flachs m. ‖ chanvre m. de la Nouvelle Zélande. / raw ~ ‖ roher Flachs m. ‖ lin m. cru. / retted ~ ‖ gerösteter Flachs m. ‖ lin m. roui. / scutched ~ ‖ Schwingflachs m.; Reinflachs m. ‖ lin m. teillé *ou* en filasse. / steeped ~ ‖ Rösteflachs m.; Rotteflachs m. ‖ lin m. naisé *ou* de rouissage. / ~ in stems ‖ Flachs m. in Stengeln ‖ lin m. en tiges. / swingled ~ *see* scutched ~. / toad ~ ‖ gemeiner Lein m. ‖ lin m. sauvage; linaire f. / undressed ~ ‖ roher Flachs m. ‖ lin m. cru. / uniform ~ ‖ gleichmäßiger Flachs m. ‖ lin m. homogène. / unsorted ~ ‖ ungesichteter Flachs m. ‖ lin m. non trié *ou* non classé. / water-retted ~ ‖ Wasserflachs m. ‖ lin m. roui à l'eau. / weak ~ ‖ schwacher Flachs m. ‖ lin m. faible.

flax bale ‖ Flachsballen m. ‖ balle f. de lin. / ~ bast ‖ Flachsbast m.; Flachsfaser f. ‖ filasse f. de lin. / ~ binder ‖ Flachsbinder m. ‖ botteleur m. de lin. / ~ bleaching ‖ Flachsbleichung f. ‖ blanchiment m. de lin. / ~ brake ‖ Flachsbreche f. ‖ broie f.; macque f.; brisoir m. / ~ breaker ‖ Flachsbrecher m. ‖ broyeur m. de lin. / ~ breaking ‖ Flachsknickerei f. ‖ broyage m. de lin. / ~ breaking machine ‖ Flachsbrechmaschine f. ‖ machine f. à broyer le lin. / ~ buyer ‖ Flachskäufer m. ‖ acheteur m. de lin. / ~ comb ‖ Hechel f. ‖ séran m.; peigne m.; sérin m.; sérançoir m. / ~ comb cleaner ‖ Flachshechelmaschinenputzer m. ‖ débourreur m. de peigneuses. / ~ comber ‖ Flachshechler m. ‖ peigneur m. *ou* séranceur m. de lin. / ~ cotton ‖ Flachsbaumwolle f.; Flachswolle f. ‖ lin-coton m. / ~ cutter ‖ Flachsschneider m. ‖ coupeur m. de lin. / ~ cutting machine ‖ Flachsschneidemaschine f.; Flachsreißmaschine f. ‖ machine f. à couper le lin; coupeuse f. de lin. / ~ dealer ‖ Flachshändler m. ‖ marchand m. de lin; négociant m. en lin. / ~ dewretting ‖ Flachstauröste f. ‖ rouissage m. de lin à la rosée. / ~ drawer ‖ Flachsstrecker m. ‖ étireur m. de lin. / ~ drawing frame ‖ Flachsbandmaschine f.; Strecke f. für Flachs ‖ machine f. à étirer de lin; étirage m. de lin. / ~ drawing ma-

chine *see* ~ drawing frame. / ~ dresser ‖ Flachsbereiter m. ‖ préparateur de lin. / ~ dresser's knife ‖ Ribbemesser n. (zum Flachsribben) ‖ racloir m. / ~ dressing ‖ Zubereitung f. von Flachs ‖ préparation f. du lin. / ~ dressing machine ‖ Flachsaufbereitungsmaschine f. ‖ machine f. à préparer le lin. / ~ drier ‖ Flachstrockner m. ‖ sécheur m. de lin. / ~ dry spinner ‖ Flachstrockenspinner m. ‖ fileur m. de lin au sec. / ~ dri spinning ‖ Flachstrockenspinnerei f. ‖ filature f. de lin au sec.

flaxen ‖ flachsen ‖ de lin.

flaxen linen ‖ Flachsleinwand f. ‖ toile f. de lin. / ~ bleaching ‖ Leinenwarenbleichung f. ‖ blanchiment m. des toiles de lin. / ~ weaving ‖ Flachsleinwandweberei f. ‖ tissage m. de toiles de lin.

flax fibre *see* ~ bast. / ~ finisher ‖ Flachssichter m.; Flachsreiniger m. ‖ finisseur m. *ou* repasseur m. de lin. / ~ ginning machine ‖ Flachsentkern(ungs)maschine f. ‖ machine f. à égrener le lin. / ~ grower ‖ Flachsbauer m. ‖ cultivateur m. *ou* producteur m. de lin. / ~ hackler ‖ Flachshechler m. ‖ peigneur m. *ou* séranceur m. de lin. / ~ hackling ‖ Flachshechelei f. ‖ peignage m. *ou* sérançage m. de lin. / ~ hackling machine ‖ Flachshechelmaschine f. ‖ peigneuse f. mécanique du lin. / ~ hand combing ‖ Flachshandhechelei f. ‖ peignage m. de lin à la main. / ~ harl *see* ~ bast. / ~ hards pl. ‖ Flachswerg n.; Flachshede f. ‖ étoupe f. de lin. / ~ land retting ‖ Flachsfeldröste f. ‖ rouissage m. de lin sur pré. / ~ linen ‖ Flachsleinwand f. ‖ toile f. de lin. / ~ line yarn ‖ Langflachsgarn n. ‖ fil m. de lin. / ~ machine combing ‖ Flachsmaschinenhechelei f. ‖ peignage m. de lin à la machine. / ~ machine hackler ‖ Flachsmaschinenhechler m. ‖ peigneur m. de lin à la machine. / ~ merchant *see* ~ dealer. / ~ mill ‖ Flachsspinnerei f. ‖ filature f. de lin. / ~ plant ‖ Lein m.; Leinpflanze f. ‖ lin m. / ~ preparing plant ‖ Flachsaufbereitungsanstalt f. ‖ installation f. à préparer du lin. / ~ purchaser ‖ Flachseinkäufer m. ‖ agent m. acheteur de lin. / ~ reeler ‖ Flachshaspler m. ‖ dévideur m. de lin. / ~ retter ‖ Flachsröster m. ‖ rouisseur m. de lin. / ~ retting ‖ Flachsröste f. ‖ rouissage m. de lin. / ~ retting tank ‖ Flachsröstegrube f. ‖ routoir m. de lin. / ~ ribbon ‖ Flachsfaserband n. ‖ ruban m. de lin. / ~ rippler ‖ Flachsriffler m. ‖ drégeur m. de lin. / ~ rippling ‖ Flachsriffeln n. ‖ égrenage m. *ou* drégeage m. de lin. / ~ rougher ‖ Flachsvorhechler m. ‖ émoucheur m. *ou* éplucheur m. *ou* débloqueur m. de lin. / ~ roving ‖ Flachsvorgarn n. ‖ mèche f. de préparation de lin. / ~ scutching ‖ Flachsschwingen n. ‖ teillage m. de lin. / ~ seed ‖ Leinsamen m. ‖ graine f. de lin. / ~ sliver *see* ~ ribbon. / ~ sorter ‖ Flachssortierer m. ‖ braqueur m. de lin. / ~ spinner ‖ Flachsspinner m. ‖ fileur m. de lin. / ~ spinning ‖ Flachsspinnerei f. ‖ filage m. de lin. / ~ spinning machine ‖ Flachsspinnereimaschine f. ‖ machine f. à filer le lin. / ~ spinning mill ‖ Flachsspinnerei f. ‖ filature f. de lin. / ~ spreading machine for ~ spinning ‖ Anlegemaschine f. für Flachsspinnereien ‖ machine f. à étaleur pour le filage du lin. / ~ spreader (Spinn) ‖ Flachsaufleger m. ‖ étaleur m. de lin. / ~ straw ‖ Rohflachs m.; Stroh-

flachs m.; Flachsstroh n. ‖ lin m. en paille *ou* en bois *ou* en chaume; paille f. de lin. / ~ tow ‖ Flachshede f. ‖ étoupe f. de lin. / ~ tow spinning ‖ Flachswergspinnerei f. ‖ filature f. d'étoupes de lin. / ~ tow yarn ‖ Flachswerggespinst n. ‖ fil m. d'étoupe et de lin. / ~ twine ‖ Flachsfaden m. ‖ ficelle f. de lin. / ~ twisting (mill) ‖ Flachszwirnerei f. ‖ retorderie f. de lin. / ~ warm water retting ‖ Flachsdampfrotte f. ‖ rouissage m. de lin à la vapeur. / ~ water retting ‖ Flachswasserröste f. ‖ rouissage m. de lin à l'eau. / ~ wax ‖ Flachswachs n. ‖ cire f. de lin. / ~ weaving ‖ Leinenweberei f. ‖ tissage m. du lin. / ~ wet spinner ‖ Flachsnaßspinner m. ‖ fileur m. de lin au mouillé. / ~ wet spinning ‖ Flachsnaßspinnerei f. ‖ filature f. de lin au mouillé. / ~ wholesale trade ‖ Flachsgroßhandel m. ‖ commerce m. de lin en gros. / ~ winder *see* ~ reeler. / ~ winding ‖ Flachshaspelei f. ‖ dévidage m. de lin. / ~ wool *see* ~ cotton. / ~ worker ‖ Flachsarbeiter m. ‖ linier m.; ouvrier m. en lin. / machine for working ~ ‖ Flachsaufbereitungsmaschine f. ‖ machine f. à travailler le lin. / ~ yarn ‖ Flachsgarn n. ‖ fil m. de lin. ‖ yarn weaver ‖ Flachsgarnweber m. ‖ tisseur m. de lin.

flayer ‖ Abdecker m.; Schinder m. ‖ équarrisseur m.; écorcheur m.

flaying ‖ Schindern n. ‖ écorchage m. / ~ ground *see* ~ place. / ~ machine ‖ Abdeckereimaschine f. ‖ machine f. d'équarrissage. / ~ place ‖ Abdeckerei m. ‖ voirie f.

flayman ‖ Kornsortierer m. ‖ trieur m. de grains.

fley-bane wood (Ioin) ‖ Leistenholz n. ‖ bois m. de Guillaume.

fleece, to ‖ rupfen ‖ arracher.

fleece (Spinn) ‖ Vlies n.; Pelz m.; Schurwolle f. ‖ toison f.; nappe f.; couche f. de voiles. / ~ dyeing ‖ Vliesfärberei f. ‖ teinture f. de laine en vrac.

fleeced goods pl. (Textile) ‖ gerauhte Ware f. ‖ marchandise f. cardée.

fleece machine ‖ Vliesmaschine f. ‖ nappeuse f. / ~ roller ‖ Aufroller m.; Vliestrommel f.; Felltrommel f. ‖ cylindre m. *ou* tambour m. à nappe. / ~ scribbler ‖ Fellmaschine f. ‖ droussette f. à nappe. / ~ washing ‖ Pelzwäsche f.; Rückenwäsche f. der Wolle ‖ lavage m. à dos. / ~ washing machine ‖ Vlieswaschmaschine f. ‖ laveuse f. mécanique pour toison entière. / ~ wool ‖ Schurwolle f. ‖ laine f. de toison.

fleecing-machine *see* fleece machine.

fleecings pl. *see* fleece wool.

fleecy ‖ wollig ‖ laineux.

fleet ‖ Flotte f. ‖ flotte f.; armée f. navale. / ~ of merchant-men ‖ Handelsflotte f. ‖ flotte f. marchande. / small ~ ‖ Flottille f. ‖ flottille f.

Flemish window ‖ Halbgeschoßfenster n. ‖ fenêtre f. mezzanine.

flesh charcoal ‖ Fleischkohle f. ‖ charbon m. de viande. / ~ colour ‖ Fleischfarbe f. ‖ couleur f. de chair. / ~-coloured ‖ fleischfarbig ‖ de couleur chair.

flesher ‖ Fleischer m. ‖ boucher m. / ~-out (Curr) ‖ Ausfleischer m. ‖ écharneur m.

fleshing knife ‖ Schabmesser n.; Streicheisen n.; Ausfleischeisen n.; Schabeisen n.; Ausschneidemesser n. ‖ couteau m. rond; fer m. à écharner; écharnoir m.

---

flesh juice ‖ Fleischsaft m. ‖ jus m. de viande. / ~ meal ‖ Fleischmehl n. ‖ viande f. moulue. / ~ mixing machine ‖ Fleischmengmaschine f.; Fleischmischmaschine f. ‖ malaxeur m. de viande. / ~ shaving machine ‖ Häutehobelmaschine f. ‖ machine f. à écharner. / ~ side ‖ Aasseite f.; Fleischseite f. ‖ côté m. de la chair; chair f. de la peau.

Flettner rudder ‖ Flettnerruder n. ‖ gouvernail m. Flettner.

flexibility ‖ Biegsamkeit f. ‖ flexibilité f.; souplesse f. / ~ of the connection ‖ Unstarrheit f. der Verbindung ‖ non-rigidité f. de la liaison *ou* du couplage.

flexible ‖ biegsam ‖ flexible; souple. / ~ axle frame for motor cars ‖ Lenkachsuntergestell n. für Motorwagen ‖ châssis m. à essieux articulés pour voitures motrices. / ~ conductor ‖ biegsamer Leiter m. ‖ conducteur m. souple. / ~ cord ‖ biegsame Leitung f.; Leitungsschnur f. ‖ conducteur m. souple. / ~ gas pipe ‖ Gasschlauch m. ‖ tuyau m. flexible à gaz. / ~ metal hose ‖ biegsamer Metallschlauch m. ‖ manche f. métallique flexible. / ~ metal tube *see* ~ metal hose. / ~ pipe ‖ Schlauch m. ‖ tuyau m. flexible. / ~ shaft ‖ biegsame Welle f. ‖ arbre m. flexible. / ~ stalk of plant ‖ biegsamer Pflanzenstengel m. ‖ tige f. végétale flexible.

flexion ‖ Biegung f. ‖ flexion f. / moment of ~ ‖ Biegungsmoment n. ‖ moment m. de flexion. / number of ~s ‖ Zahl f. der Biegungen ‖ nombre m. des flexions.

flexure (Geol) ‖ Falte f.; Biegung f. ‖ pli m. / ~ (Mech) ‖ Biegung f.; Durchbiegung f. ‖ flexion f.; courbure f.

flicker, to ‖ flackern ‖ vaciller.

flicker microscope ‖ Blinkmikroskop n. ‖ microscope m. à éclipses. / ~ photometer ‖ Flimmerfotometer n.; Flackerfotometer n. ‖ photomètre m. à éclats.

flier *see* flyer.

flight (Aero) ‖ Flug m.; Aufschwung m. ‖ vol m. / ~ (Build) ‖ Treppenarm m.; Treppe f. ‖ rampe f.; volée f. / altitude ~ ‖ Höhenflug m. ‖ vol m. d'altitude. / climbing ~ ‖ Aufwärtsflug m. ‖ vol m. ascendant. / day ~ ‖ Tagflug m. ‖ vol m. de jour. / gliding ~ ‖ Gleitflug m. ‖ vol m. plané *ou* à voile. / ~ of locks ‖ Schleusentreppe f. ‖ escalier m. d'écluses. / long-distance ~ ‖ Fernflug m. ‖ raid m. / non-stop ~ ‖ Flug m. ohne Halt ‖ vol m. sans escale. / at right angles to the direction of ~ ‖ quer zur Flugrichtung f. ‖ dans un sens perpendiculaire à la direction du vol. / solo ~ ‖ Einzelflug m. ‖ vol m. seul. / ~ of steps ‖ Treppenflucht ‖ rampe f.

flight altitude ‖ Flughöhe f. ‖ altitude f. de vol. / attitude of ~ ‖ Fluglage f. ‖ position f. *ou* régime m. en vol. / course of ~ ‖ Flugweg m. ‖ voie f. aérienne; cours m. de l'avion. / duration of ~ ‖ Flugzeit f. ‖ durée f. de vol. / ~ path ‖ Flugbahn f. ‖ trajectoire f. du vol. / position of ~ *see* attitude of ~. / ~ speed (Aero) ‖ Fahrtgeschwindigkeit f. ‖ vitesse f. d'avancement. / time of ~ *see* duration of ~.

flimsy ‖ schwach ‖ léger.

flint ‖ Flintstein m.; Feuerstein m. ‖ flint m.; caillou m.; silex m. pyromaque *ou* à feu. / ~ cloth ‖ Feuersteinleinen n. ‖ toile f. silexée. / ~ crystal ‖ englisches Kristallglas n.; Flintglas n. ‖ verre m. de cristal. / ~ digger ‖ Steinarbeiter m.

‖ tireur m. de cailloux. / gatherer of ~s pl. ‖ Steinfischer m. ‖ ramasseur m. de galets.

flint glass ‖ Flintglas n.; Bleiglas n. ‖ flint m.; verre m. plombifère *ou* de cristal. / heavy ~ ‖ Schwerflintglas n. ‖ verre m. en flint lourd.

flint gun ‖ Steinschloßgewehr n. ‖ fusil m. à silex.

flintiness ‖ Glasigkeit f. ‖ état m. vitreux; glaçage m.; vitrification f.

flint paper ‖ Kieselpapier n.; Feuersteinpapier n. ‖ papier m. silexé *ou* de silex. / ~ pebble for grinding mills ‖ Flintstein m. für Kugelmühlen ‖ silex m. rond *ou* galet m. sphérique pour broyeurs.

flinty ‖ kieselig; glasig; muschelig ‖ caillouteux; vitreux; conchoïdal. / ~ earth ‖ Kieselerde f. ‖ terre f. siliceuse. / ~ ground ‖ Kiesgrund m.; Keigrund; Kegelgrund m. ‖ fond m. de cailloutage. / ~ malt ‖ Glasmalz n. ‖ malt m. vitreux.

flivver (War mat) ‖ Querfeldeinwagen m. ‖ char m. à travers champ.

float, to ‖ schwimmen ‖ nager; flotter. / ~ (Log) ‖ flößen ‖ flotter. / ~ and set (Build) ‖ glatt *oder* fertig putzen ‖ enduire. / ~ a wall ‖ eine Wand f. schlämmen ‖ flotter un mur.

float (Log) ‖ Floß n. ‖ radeau m.; train m. de bois. / ~ (Mach) ‖ Schwimmer m. ‖ flotteur m. / ~ (Shipb) ‖ Schaufel f.; Radschaufel f. ‖ aube f.; pale f.; palette f. ball ~ *see* spherical ~. / cork ~ ‖ Korkschwimmer m. ‖ macaron m. / feathering ~ ‖ bewegliche Schaufel f. ‖ aube f. articulée *ou* mobile. / ~ for fermenting tuns ‖ Gärbottichschwimmer m. ‖ flotteur m. réfrigérant pour cuves de fermentation. / ~ of a fishing net ‖ Flott n. eines Fischernetzes ‖ flotte f. d'un filet de pêche. / ~ of masons ‖ Reibebrett n. ‖ aplanissoire f. / narrow ~ ‖ Schleusenfloß n. ‖ écluse f.; demi-train m. de bois. / outboard ~ ‖ Stützschwimmer m. ‖ flotteur m. de bout d'aile. / ~ of paddle wheel ‖ Radschaufel f. ‖ aube f. *ou* pale f. de la roue. / punctured ~ ‖ durchlochter Schwimmer m. ‖ flotteur m. percé. / spherical ~ ‖ Kugelschwimmer m. ‖ flotteur m. sphérique. / wing ~ (Aero) ‖ Gleitflosse f. ‖ nageoire f. latérale.

floatable (River) ‖ flößbar ‖ flottable.

floatation plant ‖ Flotationsanlage f. ‖ installation de triage par le flottage.

floatboard (Paddle of water wheel) ‖ Schaufel f. ‖ aile f.; aube f.; palette f.; aileron m.; alichon m. / ~ wheel ‖ Schaufelrad n. ‖ roue f. à aubes.

float chamber ‖ Schwimmergehäuse n. ‖ chambre f. du flotteur; cuve f. à niveau constant.

floated concrete ‖ Gußbeton m. ‖ béton m. coulé. / ~ timber yard ‖ Floßplatz m. ‖ chantier m. de bois flotté.

floater ‖ Schwimmer m. ‖ flotteur m. / ~ (Hydr arch) ‖ Pegel n.; Peil m. ‖ échelle f. d'eau *ou* fluviale; marque m. d'eau. / ~ with eyelet ‖ Schwimmer m. mit Öse ‖ flotteur m. avec crochet. / ~ switch with contact device ‖ Schwimmerschalter m. mit Kontaktvorrichtung ‖ interrupteur m. à flotteur avec dispositif d'établissement de contacts.

floating ‖ Trift f. ‖ flottage m. / ~ of ice ‖ Eisgang m. ‖ débâcle f.

floating ‖ schwimmend ‖ flottant. / ~ aerodrome ‖ Flugzeugmutterschiff n.; Flug-

zeugträger m. || ~ navire m. porte-avions; bâtiment m. porte-aéronefs. / ~ bath soap || Schwimmbadeseife f. || savon m. léger. / ~ body || Schwimmkugel f.; Schwimmer m. || flotteur m. / ~ bridge || Schiffsbrücke f.; Schwimmbrücke f. || pont flottant *ou* de bateaux. / ~ capital || Betriebskapital n. || fonds m. de roulement. / ~ coal elevator || schwimmender Kohlenheber m. || élévateur m. flottant à charbon. / ~ corn elevator || schwimmender Getreideheber m. || élévateur m. flottant à grains. / ~ crane || Schwimmkran m. || grue f. flottante; grue-ponton f. / ~ debt || schwebende Schuld f. || dette f. flottante. / ~ dock || Schwimmdock n. || dock m. flottant; cale f. forme flottante. / ~ dredger || Schwimmbagger m. || drague f. flottante. / ~ gate || Schwimmtor n.; Verschlußponton m. || porte f. flottante; bateau-porte m. / ~ grain elevator || schwimmender Getreideheber m. || élévateur m. flottant à grains. / ~ level luffing crane || Schwimmwippkran m. || grue f. flottante à volée variable. / ~ line || Wasserlinie f. || ligne f. de flottaison *ou* d'eau. / ~ loading plant || schwimmende Verladeanlage f. || installation f. flottante de chargement. / automatic ~ pump || selbsttätige Schwimmerpumpe f. || pompe f. flottante automatique. / ~ support || Brückenfloß n. || support m. flottant. / ~ system || Schwimmerwerk n. || système m. flottant. / ~ wharf of a flying bridge || Landungssteg m. einer fliegenden Brücke || culée f. mobile d'un pont volant.

float light || Nachtlicht n. || veilleuse f. / ~ planking || Schwimmerhaut f. || recouvrement m. du flotteur. / ~ section of ~ || Schwimmerausschnitt m. || section f. de flotteur. / ~ skin || Schwimmerhaut f. || recouvrement m. du flotteur. / ~ spindle || Schwimmerstange f. || tige f. de flotteur. / ~ switch for pumps || Schwimmerschalter m. für Pumpen || interrupteur m. à flotteur pour pompes. / ~ valve || Schwimmerventil n. || soupape f. à flotteur.

flock paper || Samttapete f.; Samtpapier n. || papier m. velouté. / ~ silk || Flockseide f.; Abseide f. || bourre f.; frison m.; capiton m.

flocky || flockig || floconneux.

flocculation || Flockung f. || floculation f.

flood, to || überschwemmen || noyer; inonder; submerger.

flood (Inundation) || Überschwemmung f. || inondation f.; déluge m. / ~ (Tide) || Flut f. || flot m.; marée f. montante; flux m. / first quarter ~ || Viertelflut f. || quart m. de flot. / half ~ || halbe Flut f.; mi-flot m. / last quarter ~ || Dreiviertelflut f. || trois quarts mpl. de flot. / ~ of a river || Hochwasser n. eines Flusses || crue f. *ou* haute-eau f. d'une rivière.

flood anchor || Flutanker m. || ancre f. de flot. / ~ arch || Flutbrücke f. || avant-pont m. / beginning of the ~ || Vorflut f. || montant m. de marée.

flooded || überschwemmt || inondé. / ~ lands pl. || Überschwemmungsgebiet n. || plaine f. submersible.

flood gate of a lock || Fluttor n. *oder* Obertor n. *oder* Vordertor n. einer Schleuse || porte f. d'amont *ou* de tête. / to open the ~s pl. || die Schützen npl. ziehen || lever les vannes fpl. / to stop the ~s pl. || die

Schützen npl. einstellen || vantiller; mettre les vannes fpl.

flooding || Überschwemmung f. || inondation f. / ~ (Hydr arch) || Versumpfung f. || paludification f. / vine yard ~ || Rebenbewässerung f. || arrosage m. des vignes.

flood-light || Scheinwerfer m.; Lichtstrahler m. || projecteur m. / ~ landing || Landebahnleuchte f. || projecteur m. d'atterrissage.

flood-lightning || Flutlichtbeleuchtung f. || illumination f. par projecteurs à flots de lumière.

flood-light projector || Flutlichtleuchte f. || projecteur m. à flots de lumière.

flood mark || Flutzeichen n. || hauteur f. de marée marquée. / ~ plain of a river || Überschwemmungsfläche f. eines Flusses || plaine f. d'inondation d'une rivière. / ~ span || Überbau m. für Flutöffnungen; Flutöffnung f. || avant-pont m. / ~ stream || Flutstrom m. || courant m. d'entrée *ou* de flot. / ~ tide *see* flood (Tide).

flookan (Geol) || Schmerkluft f. || crevasse f. remplie d'argile. / ~ (Mine) || Letten m. || glaise f.; terre f. glaise. / ~ course || Lettensalband n. || éponte f. glaiseuse.

floor, to (Brew) || ausweichen || décuver. / ~ (Build) || den Fußboden m. legen || planchéier; parqueter; poser le plancher; carreler.

floor || Fußboden m. || plancher m.; aire f. / ~ (Acc) || Laufboden m. || plancher m. / ~ (Build) || Geschoß n.; Stockwerk n. || étage m. / ~ (Car) || Wagenboden m. || plancher m. / ~ (Mine) || liegende Schicht f.; Liegendes n. || sol m.; mur m.; couche f. inférieure. / asphalt ~ || Asphaltestrich m. || aire f. en asphalte. / boarded ~ || gedielter Fußboden m.; Bretterfußboden m. || plancher m.; sol m. planchéié. / boarded ~ of a turn table || Abdeckung f. einer Drehscheibe || plancher m. d'une plaque tournante. / ~ of bridge || Brückenbahn f.; Fahrbahn f. einer Brücke || tablier m. *ou* chaussée f. d'un pont. / caking ~ (Brew) || Greifhaufen m. || couche f. qui prend. / cased ~ || Halbparkett n.; Friesfußboden m. || demi-parquet m.; plancher m. à compartiments. / cement ~ || Zementestrich m. || aire f. en ciment; sol m. cimenté. / clamped ~ *see* cased ~. / clay ~ || Lehmestrich m. || aire f. en argile. / concrete ~ || Betonfußboden m. || pavage m. bétonné. / counter ~ || Blindboden m.; Blendboden m. || faux-parquet m. / dead ~ *see* counter ~. / fireproof ~ || feuerfester Fußboden m. || plancher m. incombustible. / folded ~ || gespundeter Fußboden m. || plancher m. de planches jointes à rainure et languette. / framed ~ || Friesfußboden m. || plancher m. à compartiments *ou* en frises. / ~ of fuselage || Rumpfboden m. || fond m. du fuselage. / ground ~ || Erdgeschoß n. || rez-de-chaussée m. / inlaid ~ || Parkettboden m.; Stabfußboden m. || parquet m. de plancher; parqueterie f. / iron ~ || eiserner Fußboden m. || plancher m. en fer. / ~ without joints || fugenlose Fußböden mpl. || plancher m. sans joints. / to lay down the ~ || den Fußboden m. legen || poser le plancher; faire l'aire f. / to let in flush with the ~ || in den Fußboden m. einlassen || mettre au niveau du sol. / ~ of lime-grains || Kalkrumpenestrich m. || aire f. à repous. / ~ of a lock (Hydr arch) || Kammerböden m. || busc

m. d'une écluse. / ~ of the machine house || Maschinenhausfußboden m. || seuil m. de la salle des machines. / mosaic ~ || Mosaikfußboden m. || parquet m. mosaïque. / ~ in the moulding room || Herd m. in der Formerei || sole f. au moulage. / old ~ || Althaufen m. || vieille couche f. / parquet ~ || Parkett(fuß)boden m. || parquet m. / paved ~ || gepflasterter Fußboden m. || pavé m.; pavage m. / ~ of the reverberatory furnace || Herd m. des Flammofens || foyer m. du four à réverbère. / rough ~ *see* caking ~. / ~ of a seam || Liegendes n. || base f. d'une couche *ou* d'un gisement. / ~ of slabs || Plattenbelag m. || carrelage m.; dallage m. / sound-proof ~ || Fehlboden m. || entrevous m. / ~ of stone wood || Steinholzfußboden m. || plancher m. en xylolithe. / straight-joint ~ || stumpfgefügter Fußboden m. || plancher m. à joint plat. / terra-cotta ~ || Terrakottafußboden m. || dallage m. en terre cuite. / ~ of a tunnel || Tunnelsohle f. || base f. d'un tunnel. / wooden ~ || Holzfußboden m. || plancher m. en bois.

floor beam || Fußbodenbalken m. || solive f. / ~ board || Bodenbrett n. || planche f. de fond. / ~ boarding || Dielung f. || voligeage m. / ~ boards pl. || Dielung f. || planchéiage m. / ~ carpet || Fußbodenteppich m. || tapis m. (de plancher). / ~ casing || Deckenschalung f. || coffrage m. de plafond. / ~ clearer (Malt) || Abräumer m.; Darresel m. || déchargeur m. mécanique de touraille. / ~ contact || Fußbodenkontakt m. || contact m. de parquet. / ~ covering || Fußbodenbelag m. || revêtement m. du plancher; enduit m. de plancher. / ~ dressing || Parkettabhobeln n. || rabattage m. de parquets / ~ flogger || Parketthobler m. || raboteur m. de parquets. / ~ girders pl. and lateral bracing (Bridge) || Querverbindung f.; Querversteifung f.; Windverkreuzung f. || contreventement m.; cours m. d'entretoises; pièces fpl. de pont. / ~ grid || Rostfußboden m.; Lattenrost m. || plancher m. en grille; grillage m. en lattes. / ~ heads pl. || Kimm f.; Bilge f. || fleurs fpl.; petits fonds mpl. / ~ heating installation || Etagenheizungsanlage f. || installation f. de chauffage d'étage.

flooring || Fußboden m. || plancher m. / ~ (Brew) || Ausweichen n. || décuvage m. / ~ (Build) || Fußbodenbelag m. || pavement m.; planchéiage m.; parquetage m.; carrelage m. / ~ of a bridge || Brückenbelag m. || platelage m. d'un pont. / planed ~ || Hobeldiele f. || plancher m. raboté. / to put down the ~ || den Fußboden m. legen || poser le plancher; to remove the ~ || den Fußboden m. aufreißen || relever le plancher.

flooring inlayer || Parkettfußbodenleger m. || parqueteur-menuisier m. / ~ inlayer on bitumen || Alphaltfußbodenverleger m. || parqueteur m. sur bitume. / ~ machine || Fußbodenmaschine f. (zum Zurichten der Fußbodendielen) || machine f. à bouveter. / ~ oil || Fußbodenöl n. || huile f. pour parquets. / ~ plate || Belagblech n. || tôle f. de revêtement. / ~ slab || Fußbodenplatte f. || dalle f. à carrelage.

floor mat || Fußbodenmatte f. || natte f. de plancher. / ~ moulder || Dammgrubenformer m. || mouleur m. en fosse.

/ ~ paints pl. ‖ Farben fpl.für Fußböden ‖ couleurs fpl. pour planchers. / ~ plank ‖ Querbohle f. ‖ madrier m. transversal. / ~ plate (Build) ‖ Fußbodenplatte f. ‖ carreau m. de plancher. / ~ plate (Shipb) ‖ Flurplatte f. ‖ tôle f. du parquet. / ~ plate with planed slots ‖ Aufspannplatte f. mit gehobelten Schlitzen ‖ taque f à rainures de fixation rabotées. / ~ polish ‖ Bohnerwachs n. ‖ encaustique f. / ~ polisher ‖ Bohnermaschine f. ‖ cireuse f. / ~ polishing paste ‖ Bohnermasse f. ‖ encaustique f. pour parquets; cire f. à parquets. / ~ preserving preparation ‖ Fußbodenpflegemittel n. ‖ enduit m. pour l'entretien des planchers. / ~ scrubber ‖ Parkettbohner m. ‖ cireur m. de parquets.

floor space (Build) ‖ Grundfläche f. ‖ surface f. / ~ (Wagon) ‖ Grundfläche f.; Bodenfläche f. ‖ surface f. du plancher. / effective ~ ‖ bebaute Fläche f. ‖ superficie f. couverte. / required ~ ‖ erforderliche Bodenfläche f. ‖ encombrement m. requis. / ~ required with ram in extreme outer position ‖ Raumbedarf m. bei größtem Stößelhub ‖ encombrement m. la course maximum du coulisseau comprise.

floor stone see ~ tile. / ~ tile ‖ Fliese f. ‖ carreau m. / ~ tiler ‖ Fliesenleger m.; Plattenleger m. ‖ carreleur m. / ~ varnish ‖ Fußbodenlack m. ‖ vernis m. pour planchers. / ~ wax ‖ Bohnerwachs n. ‖ encaustique f.; cire f. à parquets.

flora ‖ Flora f. ‖ flore f. ‖

floral otto ‖ Blütenöl n. ‖ essence f. de fleurs.

Florence marble ‖ Ruinenmarmor m. ‖ marbre m. ruiniforme ou de Florence.

floret ‖ Florett m.; Florettseide f. ‖ bourre f. de soie de fleuret m.; fleuron m.

floret silk, spun ~ ‖ Schappegarn n. ‖ schappe f. / ~ spinning ‖ Florettseidenspinnerei f. ‖ filature f. de soie de fleuret.

floriculture ‖ Blumenzucht f.; Blumengärtnerei f. ‖ floriculture f.

floss hole (Metal) ‖ Schlackenauge n.; Schlackenloch n. ‖ chio m. ou trou m. de laiterol. / ~ silk ‖ Wattseide f. ‖ première bourre f.; bourreuse f.; blaise m.

flotation ‖ Schwimmen n. ‖ flottage m. / ~ axis ‖ Schwimmachse f. ‖ axe m. de flottaison. / ~ gear (Aero) ‖ Schwimmergestell n. ‖ appareil m. d'amerrissage. / ~ plant (Metal) ‖ Schwimmaufbereitungsanlage f. ‖ installation f. de triage par flottation.

flotilla ‖ Flottille f. ‖ flottille f. / ~ leader ‖ Flottillenführer m. ‖ conducteur m. de flottilles.

flour, to (Mill) ‖ mahlen; beuteln ‖ moudre; bluter.

flour ‖ Mehl n. ‖ farine f. / ~ for brown bread ‖ Schwarzmehl n. ‖ bisaille f. / containing ~ ‖ mehlhaltig ‖ farineux. / ~ of corn ‖ Mehl n. von Getreide ‖ farine f. de céréales. / ~ of emery ‖ Schmirgelstaub m. ‖ fleur f. ou potée f. d'émeri. / fossil ~ ‖ fossiles Mehl n. ‖ farine f. fossile. / superfine ~ ‖ feinstes Mehl n.; Auszug(s)mehl n. ‖ fleur f. de farine; farine f. bien blutée. / wheaten ~ ‖ Weizenmehl n. ‖ farine f. de froment.

flour and bran mixing plant ‖ Mehl- und Kleiemischanlage f. ‖ installation f. à mélanger la farine et le son. / ~ box ‖

Mehlkasten m. ‖ huche f. de moulin; farinière f. / ~ chest see ~ box. / ~ dealer see ~ merchant. / ~ gold ‖ goldführender Sand m. ‖ gravier m. aurifère. / ~ improving process ‖ Mehlveredlungsverfahren n. ‖ procédé m. pour l'amélioration des farines.

flourish (Print) ‖ Randverzierung f. ‖ bordure f.; vignette f.

flourished (Print) ‖ musiert ‖ orné; gris.

flourishing ‖ schwunghaft ‖ florissant.

flour merchant ‖ Mehlhändler m. ‖ marchand m. de farine; minotier m. / ~ mill ‖ Mehlmühle f. ‖ moulin m. à farine. / ~ mill for wheat and rye ‖ Getreidemühle f. ‖ moulin m. à froment et à seigle. / ~ milling ‖ Müllerei f. ‖ minoterie f. / ~ mixer ‖ Mehlmischmaschine f. ‖ mélangeuse f. à farine. / ~ packer ‖ Mehlpackmaschine f. ‖ ensacheuse f. à farines. / ~ packing machine ‖ Mehlverpackungsmaschine f. ‖ machine f. d'emballage de farine. / ~ paste ‖ Mehlkleister m. ‖ colle f. ou pâte f. de farine. / ~ scale ‖ Mehlwage f. ‖ balance f. à farine. / ~ sifter ‖ Mehlsichter m. ‖ crible m. à farine. / ~ sifting machine ‖ Mehlsiebmaschine f. ‖ machine f. à tamiser la farine. / ~ sifting plant ‖ Mehlsiebanlage f. ‖ installation f. de tamis à farine. / ~ sulphur ‖ Schwefelblumen fpl.; Schwefelblüten fpl. ‖ fleurs fpl. de soufre; soufre m. en fleurs. / ~ trade ‖ Mehlhandel n. ‖ minoterie f. / ~ tester ‖ Mehlprüfer m. ‖ sonde f. à farine. / ~ wafer ‖ Mehloblate f. ‖ hostie f. de farine. / ~ worm conveyer ‖ Mehlförderschnecke f. ‖ vis f. transporteuse à farine.

floury ‖ mehlig ‖ farineux.

flow, to ‖ fließen ‖ couler; s'écouler. / ~ in ‖ einströmen ‖ affluer dans; se déverser dans. / ~ out ‖ abfließen ‖ s'écouler. / ~ over ‖ überlaufen ‖ déborder.

flow (Chem) ‖ Fluß m. ‖ coulée f. / ~ (Hydr) ‖ Strommenge f. ‖ débit m.; portée f. / ~ (Phys) ‖ Strömung f. ‖ écoulement m.; cours. / ~ (Tide) ‖ Flut f. ‖ flot m.; marée f. montante; flux m. / configuration of ~ ‖ Stromlinienbild n. ‖ image f. ou représentation f. des filets fluides; lignes fpl. de courant. / ~ constant in time see ~ steady in time. / ~ of current (Electr) ‖ Stromlauf m. ‖ circulation f. du courant. / cyclic ~ ‖ Kreisströmung f. ‖ courant m. circulaire. / to deflect the ~ ‖ die Strömung f. ablenken ‖ défléchir ou dévier les filets mpl. fluides. / disturbed ~ ‖ gestörte Strömung f. ‖ écoulement m. perturbé. / ~ of energy ‖ Energiestrom m. ‖ flux m. d'énergie. / equalizing ~ ‖ Ausgleichsströmung f. ‖ écoulement m. ou courant m. de compensation. / maximum ~ at high water ‖ höchstes Hochwasser n. ‖ eau f. maximum. / ~ invariable with the time ‖ zeitlich gleichbleibende Strömung f. ‖ écoulement m. constant dans le temps. / ~ minimum ~ at low water ‖ kleinstes Niedrigwasser n. ‖ eau f. minima. / ~ of material fibres ‖ Verlauf m. der Materialfasern ‖ contexture f. des fibres de la matière. / parallel ~ ‖ Parallelströmung f. ‖ courant m. parallèle. / ~ steady in time ‖ zeitlich gleichbleibende Strömung f. ‖ écoulement m. constant dans le temps. / undisturbed ~ ‖ ungestörte Strömung f. ‖ écoulement m. non pertubé.

flower ‖ Blume f. ‖ fleur f. / artificial ~ ‖ künstliche Blume f.; Kunstblume f. ‖ fleur f. artificielle. / bead ~ ‖ Blume f. aus Perlen ‖ fleur f. en perles. / coloured ~ ‖ gefärbte Blüte f. ‖ fleur f. teinte. / ~ of gypsum ‖ Stuckgips m. ‖ fleur f. de plâtre. / luminous ~ ‖ Leuchtblume f. ‖ fleur f. lumineuse. / ~ for millinery ‖ Schmuckblume f. ‖ fleur f. de parure. / paper ~ ‖ Papierblume f. ‖ fleur f. en papier. / ~ of sulphur ‖ Schwefelblume f. ‖ fleur f. de soufre. / ~ of the vat (Dyer) ‖ Blume f. auf der Küpe ‖ fleurée f. de cuve.

flower box ‖ Blumenkasten m.; Pflanzenkasten m. ‖ caisse f. pour fleurs. / ~ bucket ‖ Pflanzenkübel m. ‖ bac m. à fleurs. / ~ bulb ‖ Blumenzwiebel f. ‖ oignon m. de fleur; bulbe m. / cultivation of ~s ‖ Blumenzucht f. ‖ floriculture f. / ~ cutter ‖ Blätterausstanzer m. ‖ découpeur m. de fleurs. / ~ dyer ‖ Blumenfärber m. ‖ teinturier m. de fleurs.

flowered ‖ geblümt ‖ fleuri.

flower gatherer ‖ Blumensammler m. ‖ ramasseur m. de fleurs. / ~ grower ‖ Blumenzüchter m. ‖ jardinier-fleuriste m. / cut ~ grower ‖ Schnittblumenzüchter m. ‖ jardinier m. cultivant des fleurs à couper. / ~ holder ‖ Blumenhalter m. ‖ porte-bouquets m. / ~ house ‖ Gärtnerei f. ‖ jardinage m. / ~ line ‖ Röschenzeile f. ‖ cordelière f. / ~ manure ‖ Blumendünger m. ‖ engrais m. pour fleurs. / ~ oil ‖ Blütenöl n. ‖ essence f. de fleurs.

flower pot ‖ Blumentopf m. ‖ pot m. à fleurs. / ~ cover ‖ Blumentopfhülle f. ‖ enveloppe f. de pots à fleurs. / ~ mould ‖ Form f. für Blumentöpfe ‖ moule m. pour pots à fleurs. / ~ press ‖ Blumentopfpresse f. ‖ presse f. pour pots à fleurs.

flower stand ‖ Blumenständer m.; Blumentisch m. ‖ jardinière f. / ~ trellis ‖ Blumengitter n. ‖ treillis m. pour fleurs. / ~ tub ‖ Blumenkübel m. ‖ cache-pot m. à fleurs. / ~ vase ‖ Blumenvase f. ‖ bouquetier m.; vase m. à fleurs. / ~ wire ‖ Blumendraht m. ‖ fil m. pour fleurs.

flow formation ‖ Strömungsgebilde n. ‖ image f. d'écoulement; lignes fpl. de courant. / ~ heater ‖ Durchflußerhitzer m. ‖ chauffe-eau m. pour eau coulante.

flowing ‖ fließend ‖ coulant; inondant; écoulant. / ~ furnace ‖ Floßofen m.; Flußofen m.; Blauofen m. ‖ fourneau m. à fonte.

flowing-out of a liquid ‖ Ausfluß m. einer Flüssigkeit ‖ écoulement m. d'un liquide.

flow limit ‖ Fließgrenze f. ‖ limite f. d'allongement. / ~ pattern ‖ Stromlinienbild n. ‖ image f. ou représentation f. des filets fluides. / ~ sheet ‖ Mühlendiagramm n. ‖ diagramme m. de mouture.

flow system, fitting flat irons pl. on the continuous ~ ‖ Bandmontage f. von Bügeleisen ‖ montage m. à la chaîne de fers à repasser. / ~ of small motor manufacture ‖ Fließfertigung f. von Kleinmotoren ‖ fabrication f. à la chaîne de petits moteurs électriques.

flow, time of ~ ‖ Durchflußzeit f. ‖ durée f. de passage. / ~ velocity ‖ Strömungsgeschwindigkeit f. ‖ vitesse f. du flux ou de passage. / ascensional velocity of ~ ‖ Strömungsgeschwindigkeit f. (aufwärts) ‖ vitesse f. ascentionnelle.

flucan ‖ Schmerkluft f. ‖ crevasse f. remplie d'argile.

fluctuate, to ‖ schwanken ‖ fluctuer; varier.
fluctuating ‖ schwankend ‖ fluctuant. / ~ current ‖ unregelmäßiger Strom m. ‖ courant m. d'intensité variable.
fluctuation ‖ Schwankung f. ‖ fluctuation f.; variation f.; oscillation f. / ~ of current ‖ Stromschwankung f. ‖ fluctuation f. du courant. / ~ of day light ‖ Schwankung f. des Tageslichtes ‖ fluctuation f. de la lumière du jour. / economical ~ ‖ wirtschaftliche Schwankung f. ‖ variation f. économique. / ~ of exchange ‖ Kursschwankung f. ‖ fluctuation f. du change. / ~ of load ‖ Belastungsschwankung f. ‖ fluctuation f. de charge. / ~ of pressure ‖ Druckschwankung f. ‖ fluctuation f. de pression. / ~ in prices ‖ Preisschwankung f. ‖ fluctuation f. des prix. / ~ of temperature ‖ Temperaturschwankung f. ‖ fluctuation f. de la température.
flue | Feuerzug m. ‖ carneau m. / ~ (Boil) ‖ Flammrohr n. ‖ tube-foyer m. / the ~s pl. are arranged in the boiler end ‖ die Rohre npl. sind in den Boden eingebaut ‖ les tubes mpl. sont montés dans le fond. / bottom ~ ‖ Unterzug m.; Grundzug m. ‖ carneau m. inférieur. / chimney ~ ‖ Rauchabzugrohr n. ‖ tuyau m. de fumée. / collecting ~ ‖ Sammelfuchs m. ‖ carneau m. collecteur. / corrugated ~ ‖ gewelltes Flammrohr n. ‖ tube-foyer m. ondulé. / external ~ see outer ~. / furnace ~ see flue. / inner ~ ‖ Innenzug m. ‖ carneau m. intérieur. / internal ~ see inner ~. / main ~ ‖ Fuchs m.; Fuchskanal m. ‖ carneau m.; renard m.; conduit m. de raccordement des carneaux à la cheminée. / masonry ~ ‖ gemauerter Zug m. ‖ carneau m. maçonné. / outer ~ ‖ Außenzug m. ‖ carneau m. extérieur. / overtop ~ see upper ~. / return ~ ‖ zurückkehrender Feuerzug m. ‖ carneau m. de retour. / side ~ ‖ Seitenzug m. ‖ carneau m. latéral. / smoke ~ see main ~. / smooth ~ ‖ glattes Flammrohr n. ‖ tube-foyer m. lisse ou uni. / straight ~ ‖ glattes Feuerrohr n. ‖ tube-foyer m. lisse. / telescopic ~ ‖ Rauchfang m. mit Rohrauszug ‖ cheminée f. téléscopique. / ~ with throttle valve ‖ Rauchfang m. mit Drosselklappe ‖ cheminée f. à papillon. / top ~ see upper ~. / upper ~ ‖ Oberzug m. ‖ carneau m. supérieur. / ventilating ~ ‖ Dunstrohr n.; Wrasenrohr n.; Ventilationsrohr n. ‖ tuyau m. aérique ou de fumée ou de ventilation.
flue ash ‖ Flugasche f. ‖ cendre f. volante. / ~ retainer ‖ Flugaschefänger m. ‖ boîte f. à recueillir les cendres volantes.
flue belt see flue ring.
flue boiler ‖ Flammrohrkessel m. ‖ chaudière f. à tube-foyer. / combination of ~ and smoke-tube boiler ‖ kombinierter Flammrohrrauchrohrkessel m. ‖ chaudière f. combinée à tube-foyer et à tubes de fumée. / one ~ ‖ Einflammrohrkessel m. ‖ chaudière f. à un seul tube-foyer ou à un seul foyer.
flue bridge ‖ Fuchsbrücke f. ‖ pont m. de rampant ou d'échappement; petit autel m. de foyer. / ~ brush ‖ Rohrwischer m. ‖ torchetubes m. / steam jet ~ cleaning apparatus ‖ Dampfstrahlrauchrohrreiniger m. ‖ purgeur m. de tubes à fumée à jet de vapeur. / ~ damper see ~ shutter. / accessor ~ door ‖ Einsteigtür f. ‖ porte f. d'accès.

flue dust ‖ Rauchkammerlösche f. ‖ poussier m. de charbon déposé dans les boîtes à fumée. / to blow the ~ from the tubes ‖ die Flugasche von den Röhren abblasen ‖ débarrasser les tubes mpl. des cendres volantes qui s'y sont déposées. / pocket for the collection of the ~ ‖ Vertiefung f. zum Ansammeln der Flugasche ‖ poche f. à cendres folles; enfoncement m. recevant les cendres folles. / settling of ~ ‖ Flugascheablagerung f. ‖ dépôt m. de cendres volantes. / ~ briquetting plant ‖ Gichtstaubbrikettierungsanlage f. ‖ installation f. pour la fabrication des agglomérés de poussière de gueulard. / ~ catcher ‖ Flugaschefänger m. ‖ capuchon m. à folles cendres collecter les.
flue furnace ‖ Flammrohrfeuerung f. ‖ foyer m. intérieur tubulaire.
flue gas ‖ Rauchgas n. ‖ gaz m. de fumée. / drawing off the ~es ‖ Absaugen m. der Rauchgase ‖ aspiration f. de la fumée ou des gaz de fumée. / ~ analyser ‖ Rauchgasprüferanlage f.; Rauchgasanalysator m. ‖ analyseur m. des gaz de fumée. / analysis of ~es ‖ Rauchgasanalyse f. ‖ analyse f. des gaz de carneau. / ~ explosion ‖ Rauchgasexplosion f.; Gasexplosion f. im Feuerzug ‖ explosion f. des gaz de fumée; explosion f. de gaz dans le carneau. / ~ preheater ‖ Rauchgasvorwärmer m. ‖ réchauffeur m. à gaz de fumée. / ~ pump ‖ Rauchgaspumpe f. ‖ pompe f. à fumée ou à gaz. / ~ tester see ~ testing apparatus. / ~ testing apparatus ‖ Rauchgasprüfer m. ‖ appareil m. d'analyse ou analyseur m. des gaz de fumée. / ~ utilization ‖ Abgasverwertung f. ‖ utilisation f. du gaz de fumée. / ~ utilizing plant ‖ Abgasverwertungsanlage f. ‖ installation f. pour l'utilisation des gaz de fumée.
flue heating surface ‖ Flammrohrheizfläche f. ‖ surface f. de chauffe des tubes-foyers.
fluellite ‖ Fluellit m. ‖ fluellite f.
flue loss ‖ Fuchsverlust m. ‖ perte f. dans le carneau.
flueman ‖ Kaminfeger m. ‖ nettoyeur m. de cheminées; ramoneur m.
flue measure ‖ Zugmesser m. ‖ appareil m. à mesurer le courant d'air. / ~ opening ‖ Schornsteinrohr n.; Schornsteinhöhlung f. ‖ creux m. ou conduit m. de la cheminée. / ~ pipe see flue (Boil). / ~ pyrometer ‖ Fuchspyrometer n. ‖ pyromètre m. de carneau.
flue ring ‖ Flammrohrschuß m. ‖ virole f. ou tronçon m. de tube-foyer. / flanged ~ ‖ gebördelter oder geflanschter Flammrohrschuß m. ‖ virole f. bridée ou à bride de tube-foyer. / rivetted ~s pl. ‖ zusammengenietete Flammrohrschüsse mpl. ‖ viroles fpl. de tubes-foyers rivetées ensemble. / welded ~s pl. ‖ zusammengeschweißte Flammrohrschüsse mpl. ‖ viroles fpl. de tubes-foyers soudées ensemble.
flue section see flue ring. / ~ shutter ‖ Zugabsperrklappe f. für Flammrohre ‖ clapet m. d'arrêt de tirage pour tubes-foyers. / ~ stay ‖ Flammrohrstütze f. ‖ soutien m. ou support m. de tube-foyer. / ~ stiffening ‖ Flammrohrversteifung f.; Flammrohrverstärkung f. ‖ renforcement m. ou consolidation f. de tube-foyer. / ~ support see ~ stay. / ~ tube see flue (Boil).
fluid ‖ flüssig ‖ fluide. / in a thick ~ state ‖ dickflüssig ‖ en état m. peu fluide.

fluid ‖ Flüssigkeit f.; Fluidum n. ‖ fluide m.; fluidité f. / absorbing ~ ‖ absorbierende Schicht f. ‖ couche f. absorbante. / aëriform ~ see elastic ~. / alcoholic ~ ‖ weingeisthaltige Flüssigkeit f. ‖ liquide m. alcoolique. / dipping ~ for incandescent mantles ‖ Tauchfluid n. für Glühstümpfe ‖ fluide m. à plonger les manchons à incandescence. / elastic ~ ‖ elastische Flüssigkeit f. ‖ fluide m. aériforme ou gazeux.
fluid accumulator ‖ Akkumulator m. mit Säurefüllung ‖ accumulateur m. à liquide.
fluidal structure ‖ Fluidalstruktur f. ‖ structure f. fluidale.
fluid cell ‖ nasses Element n. ‖ pile f. à liquide. / ~ friction ‖ Flüssigkeitsreibung f. ‖ frottement m. du liquide. / ~ gauge chamber ‖ hydraulische Meßdose f. ‖ boîte f. de mesure hydraulique. / inside plashing of the ~ ‖ Flüssigkeitsschlag m. ‖ impact m. du liquide.
fluidity ‖ Flüssigkeit f. ‖ fluide m.; fluidité f.
fluid metal ‖ Flußmetall n.; Flußeisen n. ‖ fer m. doux ou en lingot. / internal ~ motion ‖ Flüssigkeitsbewegung f. im Innern ‖ mouvement m. interne du liquide. / ~ pipe ‖ Flüssigkeitsleitung f. ‖ tuyauterie f. à liquides. / ~ stratum ‖ Flüssigkeitsschicht f. ‖ couche f. de liquides.
fluocerite ‖ Fluozerit m. ‖ fluocerine f.
fluor ‖ Flußspat ‖ fluorine f.; spath fluor. / earthy ~ ‖ Flußerde f.; erdiger Flußspat m. ‖ chaux f. fluatée terreuse. / ~ earth ‖ Flußerde f.; erdiger Flußspat m. ‖ chaux f. fluatée terreuse.
fluoresce, to ‖ fluoreszieren ‖ entrer en fluorescence f.
fluorescence ‖ Fluoreszenz f. ‖ fluorescence f. / glass ~ ‖ Glasfluoreszenz f. ‖ fluorescence f. du verre.
fluorescence radiation ‖ Fluoreszenzstrahlung f.; Eigenstrahlung f. ‖ rayonnement m. par fluorescence. / ~ spectrum ‖ Fluoreszenzspektrum n. ‖ spectre m. fluorescent ou de fluorescence. / ~ strip ‖ Fluoreszenzstreifen m. ‖ bande f. fluorescence.
fluorescent ‖ fluoreszierend ‖ fluorescent. / ~ glass ‖ fluoreszierendes Glas n. ‖ verre m. fluorescent. / ~ reflector ‖ Fluoreszenzreflektor m. ‖ réflecteur m. fluorescent. ~ screen ‖ Fluoreszenzschirm m. ‖ écran m. fluorescent.
fluorhydric acid ‖ Fluorwasserstoffsäure f.; Flußsäure f. ‖ acide m. fluorhydrique.
fluoric acid ‖ Flußsäure f. ‖ acide m. fluorhydrique. / ~ silicate ‖ Fluorsilikat n. ‖ fluorsilicate m.
fluoride ‖ Fluorid n.; Fluorsalz n. ‖ fluorure m.
fluorine ‖ Fluor n. ‖ fluor m. / ~ compound ‖ Fluorpräparat n. ‖ fluorure m. / ~ silicate ‖ Fluorsilikat n. ‖ fluorsilicate m.
fluorite ‖ Flußspat m.; Fluorit m. ‖ fluorine f. / ~ lens ‖ Flußspatlinse f. ‖ lentille f. en spath fluor. / ~ plate ‖ Flußspatplatte f. ‖ lame f. de fluorine.
fluoroscope ‖ Fluoroskop m.; Kryptoskop n. ‖ bonnette f. radioscopique.
fluoroscopic diaphragm ‖ Durchleuchtungsblende f. ‖ diaphragme m. radioscopique. / ~ picture ‖ Leuchtschirmbild n. ‖ image m. radioscopique (sur l'écran). / ~ screen examination ‖ Leuchtschirmuntersuchung f. ‖ examen m. pulmonaire par écran radioscopique.

fluoroscopy see also radioscopy ‖ Röntgendurchleuchtung f. ‖ radioscopie f. / room for ~ ‖ Durchleuchtungsraum m. ‖ chambre f. de radioscopie.

fluor-spar ‖ Flußspat m. ‖ spath m. fluor; fluorine f. / ~ quarry ‖ Flußspatgrube f. ‖ carrière f. de spath fluor.

fluosilicate ‖ Siliziumfluorwasserstoffsalz n. ‖ fluosilicate m.

fluosilicic acid ‖ Kieselfluorsäure f. ‖ acide m. hydrofluosilicique.

flush ‖ abgeglichen; eben; platt ‖ plain.

flush, to (Hydr arch) ‖ schlämmen; spülen ‖ débourber; curer. / ~ (To make even) ‖ bündig oder gleich machen; in gleiche Flucht f. bringen ‖ affleurer; niveler.

flush (Hydr) ‖ Aufwallung f. ‖ chasse f. d'eau. / ~ bolt of a door ‖ Kantenriegel m. ‖ verrou m. à coulisse. / ~ box (Tel) ‖ Untersuchungsbrunnen m. für unterirdische Leitungen ‖ boîte f. d'affleurement.

flushing see flush. / ~ tank ‖ Schwimmerbehälter m. ‖ réservoir m. flottant. / ~ water ‖ Schwemmwasser n. ‖ eau f. d'arrosage.

flush joint ‖ bündiger Stoß m. ‖ franc-bord m.; joint m. lisse. / ~ rivet ‖ versenktes Niet n. ‖ rivet m. noyé ou à tête perdue ou à tête fraisée. / ~ screw ‖ Senkschraube f. ‖ vis f. à tête fraisée; vis f. noyée. / ~ switch ‖ Unterputzschalter m.; Dosenschalter m. für Montage unter Putz ‖ interrupteur m. encastré. / ~ sluice ‖ Spülschleuse f. ‖ écluse f. de chasse. / ~ water pipe ‖ Spülrohr n. ‖ tuyau m. d'injection d'eau. / ~ wood screw ‖ Senkholzschraube f. ‖ vis f. à bois à tête fraisée.

flute, to ‖ riffeln; einkehlen ‖ canneler; évider; rainer.

flute (Groove) ‖ Hohlkehle f.; Riffel f. ‖ cannelure f.; rainure f. / ~ (Mus) ‖ Flöte f. ‖ flûte f.

fluted ‖ geriffelt ‖ ridé. / ~ roll ‖ Riffelzylinder m.; geriffelte Walze f. ‖ cylindre m. cannelé. / ~ roller see ~ roll.

flute-harmonium ‖ Harmoniflöte f. ‖ harmoni-flûte f.

fluting ‖ Kannelierung f. ‖ cannelure f.; évidage m.; rainurage m. / ~ gouge ‖ hohle Gutsche f. ‖ gouge f. creuse. / ~ machine ‖ Kanneliermaschine f.; Riffelmaschine f. ‖ machine f. à canneler. / ~ plane ‖ Kannelierhobel m.; Rinnenhobel m. ‖ guillaume m. à canneler. / ~ support ‖ Riffelsupport m. ‖ support m. de cannelage.

flutometer ‖ Flutometer n. ‖ flutomètre m.

flutter, to ‖ flattern ‖ flotter; battre.

flutter effect (Tel) ‖ Flatterwirkung f. ‖ effet m. vibratoire ou de battement.

fluttering of the arc ‖ Flackern n. des Lichtbogens ‖ vacillement m. de l'arc.

fluviatic hydraulics pl. ‖ Flußwasserbau m. ‖ hydraulique f. fluviale.

flux, to ‖ das Flußmittel zusetzen ‖ ajouter le fondant.

flux ‖ Fluß m. ‖ flux m. / ~ (Electr) ‖ Kraftlinienfluß m. ‖ flux m. de force. / ~ (Metal) ‖ Schmelzpulver n.; Flußmittel n.; Zuschlag m. ‖ flux m.; fondant m. / ~ (Tide) ‖ Flut f. ‖ flot m.; marée f. montante. / armature ~ (Electr) ‖ Ankerkraftfluß m. ‖ flux m. de l'induit. / black ~ (Metal) ‖ schwarzer Fluß m. ‖ flux m. noir. / compact ~ ‖ dichter Fluß m. ‖ chaux f. fluatée compacte. / ~ of force ‖

Kraftfluß m. ‖ flux m. de force. / ~ of light ‖ Lichtstrom m. ‖ flux m. lumineux. / siliceous ~ ‖ Kieselfluß m.; kieseliger Zuschlag m. ‖ fondant m. silicieux. / white ~ ‖ weißer Fluß m. ‖ flux m. blanc.

fluxing ‖ Zuschlag m. ‖ addition f. de fondant. / ~ ore ‖ Zuschlagerz n. ‖ minerai m. fondant ou à gangue fusible.

flux limestone ‖ Zuschlagkalkstein m. ‖ fondant m. calcaire. / ~ material ‖ Zuschlagmaterial n. ‖ fondant m. / ~ stone ‖ Flußkalkstein m. ‖ castine f.

fly, to ‖ fliegen ‖ voler. / to let fly the sheets pl. ‖ die Schoten fpl. fliegen lassen ‖ filer les écoutes fpl. / ~ with the wind ‖ mit dem Wind m. fliegen ‖ voler avec le vent.

fly (Aero) ‖ Flug m.; Fliegen n. ‖ vol m.; volée f. / ~ (Mach) ‖ Schwungrad n. ‖ volant m. / ~ (Spinn) ‖ Zwirnflügel m. ‖ ailette f.

fly ash see flying ashes.

fly ball ‖ Schwungkörper m.; Schwungkugel f.; Reglerkugel f. ‖ corps m. (agissant par force) centrifuge; boule f. de régulateur. / governor ~ ‖ Reglerkugel f.; Schwungkugel f. ‖ boule f. du régulateur. / ~ arm ‖ Schwungkugelarm m. ‖ bras m. de la boule du régulateur.

fly catcher ‖ Fliegenfänger m. ‖ attrapemouches m. / ~ cement ‖ Fliegenleim m. ‖ colle f. d'attrape-mouches.

flyer (Aero) ‖ Flieger m. ‖ aviateur m. / ~ (Spinn) ‖ Flügel m.; Spindelflügel m.; Fleier m. ‖ ailette f. / ~ and bobbin ‖ Flügel m. und Spule f. ‖ ailette f. et bobine f. / hollow ~ with spring finger (Spinn) ‖ hohler Flügel m. mit Preßfinger ‖ ailette f. creuse avec doigt comprimeur. / solid ~ (Spinn) ‖ voller Flügel m. ‖ ailette f. pleine.

flyer frame see fly frame. / ~ guide ‖ Flügelfadenführer m. ‖ guide-fil m. à ailettes. / ~ lathe (Weav) ‖ Federlade f. ‖ battant m. régulateur ou à claquette.

flyers pl. ‖ freitragende Treppe f.; Freitreppe f. ‖ escalier m. suspendu; perron m.

flyer spinning frame ‖ Flügelfeinspinnmaschine f.; Flügelspinner m. ‖ métier m. continu à ailettes; banc m. ou continu m. à ailettes. / ~ twister ‖ Flügelzwirnmaschine f. ‖ continu m. à ailettes à retordre. / ~ yarn ‖ Flügelgarn n. ‖ fil m. de continu à ailettes.

fly frame (Spinn) ‖ Fleier m.; Vorspinnmaschine f. ‖ banc m. à broches. / differential ~ ‖ Differentialflyer m. ‖ banc m. à broches à mouvement differentiel. / finishing ~ ‖ Feinfleier m. ‖ banc m. à broches en fin. / intermediate ~ ‖ Mittelfleier m. ‖ banc m. à broches intermédiaire.

fly frame helper ‖ Fleiergehilfe m. ‖ aidebancbrocheur m. / ~ spinner ‖ Flügelspinner m. ‖ fileur m. au continu à ailettes. / fine ~ tenter ‖ Feinfleier m. ‖ bancbrocheur m. en fin. / intermediate ~ tenter ‖ Mittelfleier m. ‖ bancbrocheur m. intermédiaire.

fly glue ‖ Fliegenleim m. ‖ colle f. aux mouches.

flying (Aero) ‖ Flug m.; Fliegen n. ‖ vol m.; volée f. / ~ on back see reverse ~. / formation ~ ‖ Geschwaderflug m. ‖ vol m. en formation ou d'escadre. / reverse ~ ‖ Rückenflug m. ‖ vol m. sur le dos. / trick ~ ‖ Kunstflug m. ‖ vol m. acrobatique.

flying ashes pl. ‖ Flugasche f. ‖ cendre f. volante ou mouvante ou volatile.

flying boat ‖ Flugboot n. ‖ hydravion m. à coque. / goods-carrying ~ ‖ Frachtflugboot n. ‖ hydravion m. à coque de frêt.

flying bridge ‖ fliegende Brücke f. oder Fähre f. ‖ pont m. volant. / ~ cadet ‖ Flugschüler m. ‖ élève-pilot m. / ~ hours pl. ‖ Flugstunden fpl. ‖ nombre m. d'heures en vol. / ~ jib ‖ Außenklüver m. ‖ clinfoc m. / ~ jib man ‖ Außenklüvergast m. ‖ gabier m. de clin-foc. / ~ machine ‖ Flugmaschine f.; Flugapparat m.; Flugzeug n. ‖ appareil m. d'aviation; machine f. volante; aéroplane m. / ~ school ‖ Fliegerschule f. ‖ école f. d'aviation. / ~ shears pl. ‖ Schere f. mit fliegender Lagerung ‖ cisailles fpl. à porte-à-faux. / ~ shuttle ‖ Schnellschützen m. ‖ navette f. volante. / ~ speed ‖ Fluggeschwindigkeit f. ‖ vitesse f. de vol. / ~ target ‖ Flugscheibe f. ‖ cible f. planeuse ou volante. / ~ weight ‖ Fluggewicht n. ‖ poids m. en ordre de vol. / ~ yeast ‖ Flughefe f. ‖ voltigeurs mpl.

fly leaf (Print) ‖ Allonge f. ‖ feuille f. attachée. / ~ lever ‖ Schwunghebel m.; levier m. oscillant. / ~ nut ‖ Flügelmutter f. ‖ écrou m. à oreilles. / ~ paper ‖ Fliegenpapier n. ‖ papier m. à mouches; papier m. tue-mouches. / ~ press ‖ Schraubenpresse f.; Stoßwerk n.; Prägwerk n. ‖ balancier m. ou presse f. à vis.

fly-proof cover ‖ Fliegenglocke f. ‖ cloche-moustiquaire f.

fly sheet ‖ Flugblatt n.; Flugschrift f. ‖ feuille f. volante; pamphlet m.

flywheel (Mach) ‖ Schwungrad n. ‖ volant m. / ~ (Spinn) ‖ Twistwirtel m. ‖ volant m. / auxiliary ~ ‖ Zusatzschwungrad n. ‖ deuxième volant f. / to balance the ~ ‖ das Schwungrad ausbalanzieren ‖ équilibrer le volant. / balanced ~ ‖ ausbalanziertes Schwungrad n. ‖ volant m. contre-pesé. / ~ between two bearings ‖ beiderseitig gestütztes Schwungrad n. ‖ volant m. entre deux paliers. / ~ elastically coupled to the generator shaft ‖ federnd mit der Generatorwelle gekuppeltes Schwungrad n. ‖ volant m. accouplé d'une manière élastique à l'arbre de l'alternateur. / engine with a huge ~ ‖ Maschine f. mit einem gewaltigen Schwungrade ‖ machine f. munie d'un puissant volant. / ~ consisting of two parts ‖ zweiteiliges Schwungrad n. ‖ volant m. en deux parties. / ~ divided between the arms ‖ zwischen den Armen geteiltes Schwungrad n. ‖ volant m. divisé entre les bras. / ~ divided through the spokes ‖ in den Armen geteiltes Schwungrad n. ‖ volant m. divisé dans les bras. / overhung ~ ‖ fliegendes Schwungrad n. ‖ volant m. monté à porte à faux. / the ~ runs true ‖ das Schwungrad läuft genau rund ‖ le volant tourne bien rond. / the ~ runs untrue ‖ das Schwungrad schlägt oder taumelt ‖ le volant ne tourne pas rond; le volant tourne à faux. / without ~ ‖ schwungradlos ‖ sans volant m. / the ~ wobbles see the ~ runs untrue.

flywheel action ‖ Schwungradwirkung f. ‖ action f. de volant. / arm of the ~ ‖ Schwungradarm m. ‖ bras m. du volant. / boss of the ~ ‖ Schwungradnabe f. ‖ moyeu m. de volant. / ~ brake ‖ Schwung-

radbremse f. ‖ frein m. sur le volant. / bursting of the ~ see ~ explosion. / ~ calculation ‖ Schwungradberechnung f. ‖ calcul m. du volant. / ~ cogging ‖ Schwungradverzahnung f. ‖ denture f. du volant. / ~ connection ‖ Schwungradschaltung f. ‖ connection f. à couplage volant. / ~ converter ‖ Schwungradumformer m. ‖ transformateur m. à volant. / ~ dowelling ‖ Schwungradverklammerung f. ‖ agrafage m. du volant. / ~ drive ‖ Schwungradantrieb m. ‖ commande f. sur le volant. / ~ effect ‖ Schwungradwirkung f. ‖ action f. ou effet m. de volant. / ~ engine ‖ Schwungradmaschine f. ‖ dynamo-volant f. / ~ explosion ‖ Schwungradexplosion f. ‖ éclatement m. ou explosion f. du volant. / ~ fan ‖ Schwungradventilator m. ‖ volant-ventilateur m. / ~ governor ‖ Schwungradregler m. ‖ régulateur-volant m. / ~ housing ‖ Schwungradgehäuse n. ‖ carter m. de volant. / ~ hub see boss of the ~. / ~ key ‖ Schwungradkeil m. ‖ clavette f. de volant. / ~ mass ‖ Schwungmasse f. ‖ masse f. mobile. / ~ moment ‖ Schwungmoment n. ‖ moment m. d'inertie. / ~ pit ‖ Schwungradgrube f. ‖ fosse f. de volant. / ~ pointer ‖ Schwungradzeiger m. ‖ index m. du volant. / ~ shaft ‖ Schwungradwelle f. ‖ arbre m. de volant. / ~ side (Mach) ‖ Schwungradseite f. ‖ côté m. du volant. / ~ toothing see ~ cogging. / ~-type generator ‖ Schwungraddynamomaschine f. ‖ dynamo-volant f. / ~ weight ‖ Schwunggewicht n. ‖ poids m. du volant.
foal (Mine) ‖ Fördermann; Wagenstößer m.; Schlepper m. ‖ galibot m.; aidegalibot m.
foam, to ‖ aufbrausen ‖ bouillonner.
foam ‖ Schaum m. ‖ écume f.; mousse f. / ~ silica ‖ Kieselkreide f. ‖ craie f. siliceuse.
fob ‖ Täschchen n. ‖ pochette f.
f. o. b. (free on board) ‖ frei an Bord m. ‖ franco à bord m.
focal ‖ den Brennpunkt m. betreffend ‖ focal. / difference in the ~ adjustment ‖ Einstellungsdifferenz f. ‖ différence f. de calage. / ~ distance ‖ Brennweite f. ‖ distance f. focale ou du foyer. / ~ intercept of a condenser ‖ Schnittweite f. eines Kondensors ‖ distance f. frontale d'un condensateur.
focal length ‖ Brennweite f. ‖ distance f. focale. / ~ of the eyepiece ‖ Okularbrennweite f. ‖ distance f. focale de l'oculaire. / with long ~ ‖ mit langer Brennweite f. ‖ avec distance f. focale longue; à long foyer m. / ratio of the aperture of the lens and the ~ ‖ Verhältnis n. der Objektivöffnung zur Brennweite ‖ rapport m. entre l'ouverture de l'objectif et la distance focale. / short ~ (Opt) ‖ kurze Brennweite f. ‖ foyer m. court. / standard value of ~ of object glasses ‖ Normalbrennweite f. der Objektive ‖ distance f. focale normale des objectifs.
focal line ‖ Brennlinie f. ‖ ligne f. focale.
focal plane ‖ Brennebene f. ‖ plan m. focal. / back ~ ‖ hintere Brennebene f. ‖ plan m. focal arrière. / front ~ ‖ vordere Brennebene f. ‖ plan m. focal avant.
focal plane shutter ‖ Fokalschlitzverschluß m. ‖ obturateur m. de rideau ou de plaque.

focal point ‖ Brennpunkt m. ‖ foyer m. / ~ ray ‖ Brennstrahl m. ‖ rayon m. focal. / ~ skin distance ‖ Fokushautabstand m. ‖ distance f. foyer-peau f. / ~ surface ‖ Fokalfläche f. ‖ surface f. focale.
focus, to (Opt; Phot) ‖ auf richtige Entfernung einstellen ‖ mettre au point m. / ~ independently either eyepiece of a field glass ‖ jedes Okular n. eines Feldstechers für sich einstellen ‖ mettre séparément au point chaque oculaire d'une jumelle. / ~ a picture ‖ das Bild auf richtige Entfernung einstellen ‖ mettre au point de l'image.
focus (Opt) ‖ Brennpunkt m. ‖ foyer m. / depth of ~ ‖ Fokustiefe f. ‖ profondeur f. de foyer. / with a long ~ ‖ mit langer Brennweite f. ‖ avec distance f. focale longue; à long foyer m. / ~ of microbes ‖ Mikrobenherd m. ‖ foyer m. microbien. / ~ of a mine ‖ Minenherd m. ‖ foyer m. de mine; endroit m. des feux; point m. du feu. / out of ~ image ‖ unscharfes Bild n. ‖ image f. manquante de netteté. / virtual ~ ‖ Zerstreuungspunkt m. ‖ foyer m. virtuel.
focusing see focussing.
focussed for infinity ‖ auf Unendlich n. eingestellt ‖ réglé à l'infini m.
focussing (Opt) ‖ Einstellung f. ‖ mise f. au point. / ~ the image on the ground glass screen ‖ Einstellung f. des Bildes auf der Mattscheibe ‖ mise f. de l'image au point sur le dépoli. / ~ for near a field glass ‖ Einstellen n. eines Feldstechers auf die Nähe ‖ mise f. au point sur les objets rapprochés d'une jumelle. / pinion head for ~ eyepiece upon cross lines ‖ Okularfokussierung f. auf dem Fadenkreuz ‖ mise f. au point de l'oculaire sur le réticule.
focussing adjustment ‖ Fokussierungseinstellung f. ‖ mise f. au foyer. / ~ attachment on eyepiece draw tube ‖ Fokussierung f. am Okularauszug ‖ mise f. au point du tirage oculaire. / clamp for ~ attachment ‖ Klemmung f. für Fokussierung ‖ blocage m. de la mise au point. / ~ arc lamp ‖ Fixpunktbogenlampe f. ‖ lampe f. à arc à point lumineux fixe. / ~ lens ‖ Einstellupe f. ‖ loupe f. à monture; loupe f. de mise. / non-spherical aplanatic ~ lens ‖ aplanatische asphärische Ophthalmoskoplinse f. ‖ lentille f. ophtalmoscopique aplanétique asphérique. / ~ magnifier ‖ Einstellupe f. ‖ loupe f. de mise au point. / ~ mechanism ‖ Fokussiervorrichtung f.; Einstellvorrichtung f. ‖ dispositif m. de mise au point. / ~ microscope ‖ Einstellmikroskop n.; Fokussierungsmikroskop n. ‖ microscope m. de mise au point. / ~ plane ‖ Einstellungsebene f. ‖ plan m. de mise au point. / ~ screen ‖ Mattscheibe f. ‖ verre f. dépoli.
fodder ‖ Viehfutter n. ‖ aliment m. pour le bétail; fourrage m.; pâture f. / ~ with molasses ‖ Futter n. mit Melassezusatz ‖ fourrage m. mélassé. / ~ barley see feeding barley. / ~ bin ‖ Futterbehälter m.; mangeoire f. (de fourrage). / ~ distributor ‖ Futterverteiler m. ‖ distributeur m. de fourrage. / ~ lime ‖ Futterkalk m. ‖ chaux f. pour nourriture ou à affourrager. / ~ making ‖ Futterzubereitung f. ‖ préparation f. du fourage. / ~ preparing ‖ Futterbereitung f. ‖ préparation f. de fourrage. / ~ steamer ‖ Futterdämpfer m. ‖ appareil m. à étuver le fourrage; étuve

f. pour fourrage. / ~ stewer see ~ steamer. / ~ stewing apparatus see ~ steamer.
fog ‖ Nebel m. ‖ brouillard m.; brume f. / ~ (Agr) ‖ Grumt n. ‖ refoin m.; regain m. / artificial ~ ‖ künstlicher Nebel m. ‖ brouillard m. artificiel. / damp ~ ‖ Nebelschwaden m. ‖ brouillard m. flottant. / land ~ ‖ Landnebel m. ‖ brouillard m. de campagne. / formation of ~ ‖ Nebelbildung f. ‖ formation f. de brouillard ou de brume.
foggy ‖ nebelig ‖ nébuleux. / ~ season ‖ nebelreiche Jahreszeit f. ‖ saison f. brumeuse ou à nombreux brouillards.
fog horn ‖ Nebelhorn n. ‖ trompe f. de brume. / ~ mist ‖ Dunstnebel m. ‖ brouillard m. de fumées ou de vapeurs légères. / ~ repeater ‖ Nebellichtsignal n. ‖ répétiteur m. lumineux. / ~ signal ‖ Nebelsignal n.; Nebelzeichen n. ‖ signal m. de brume. / ~ signal (Railw) ‖ Knallsignal n. ‖ pétard m. / ~ trumpet see horn. / wall of ~ ‖ Nebelschicht f. ‖ couche f. de brouillard. / ~ whistle ‖ Nebelpfeife f. ‖ sifflet m. de brume.
foil (Arm) ‖ Florett n. ‖ fleuret m. / ~ (Metal) ‖ Folie f.; Blattmetall n. ‖ feuille f. / brass ~ ‖ Messingfolie f. ‖ feuille f. de laiton. / ~ of a mirror ‖ Spiegelbelag m. ‖ étamure f.; tain m.; feuille f. d'étain d'un miroir. / tin ~ ‖ Stanniol n. ‖ feuille f. d'étain.
foil brush ‖ Blätterbürste f.; Folienbürste f. ‖ balai m. de clinquant.
foiling of a mirror see foil of a mirror.
foil loudspeaker ‖ Folienlautsprecher m. ‖ haut-parleur m. à feuille. / ~ metal ‖ Blattmetall n. ‖ métal m. en feuilles.
fold, to (Bookb) ‖ falten; zusammenlegen ‖ plier. / ~ (Tinm) ‖ falzen ‖ replier; agrafer. / ~ in ‖ einknicken ‖ plier. / ~ up a sail ‖ ein Segel n. auftuchen ‖ plier une voile f.
fold (Bookb) ‖ Falz m. ‖ pli m. / ~ (Geol) ‖ Falte f. ‖ pli m.; flexure f. / ~ (Tinm) ‖ Falz m. ‖ repli m.; agrafe f.
foldable ‖ zusammenlegbar ‖ qui peut être plié; pliable.
folded ‖ zusammengefaltet ‖ plié. / ~ card box machine ‖ Faltschachtelmaschine f. ‖ machine f. pour faire des boîtes pliantes. / ~ rocks pl. ‖ Faltengebirge n. ‖ montagne f. formée par plissement.
folder (Bookb; Pap; Print; Weav) ‖ Falzer m. ‖ plieur m. / ~ (Office; Bookb) ‖ Falzbein n. ‖ plioir m.; cambreur m. / ~ (Opt) ‖ Kneifer m. ‖ binocle m.; pincenez m. / ~ (Spinn) ‖ Falzer m. ‖ empaqueteur m. ou ployeur m. ou plieur m. de filés. / ~ (Tinm) ‖ Falzer m. ‖ agrafeur m. / ~ with few tapes ‖ bänderarmer Falzapparat m. ‖ plieuse f. à peu de cordons. / hand ~ with sticks ‖ Stoffleger m. mit Stäbchen ‖ plieur m. à la baguette. / ~ with metal blade ‖ Falzbein n. mit Metallklinge ‖ coupe-papier m. à lame de métal. / rimmed ~ ‖ Klemmer m. mit Randfassung ‖ pince-nez m. à chasse. / tape-less ~ ‖ bänderloser Falzapparat m. ‖ plieuse f. sans cordons. / ~ without points ‖ punkturloser Falzapparat m. ‖ plieuse f. sans pointures.
folding ‖ zusammenlegbar ‖ pliant. / ~ apparatus ‖ Falzgerät n. ‖ plieuse f.; appareil m. à plier. / ~ bed ‖ Klappbettstelle f. ‖ lit m. pliant. / ~ board ‖ Falzbrett n. ‖ ais m. à pliage. / ~ boat ‖ Faltboot n. ‖

canot m. démontable *ou* pliant. / ~ bone ‖ Falzbein n. ‖ plioir m.; coupe-papier m. / ~ border ‖ Falz m.; Fugenfalz m. ‖ avisure f. / ~ box ‖ Faltschachtel f. ‖ boîte f. pliante. / ~ cardboard-box ‖ Faltpappschachtel f. ‖ cartonnage m. pliant. / ~ chair ‖ Klappstuhl m. ‖ siège m. pliant. / ~ door ‖ Flügeltür f. ‖ porte f. à deux battants. / ~ dormer-window ‖ Dachklappe f. ‖ lucarne f. à rabattement. / ~ field tripod ‖ zusammenklappbares Feldstativ n. ‖ pied m. de campagne pliant. / ~ floor ‖ gespundeter Fußboden m. ‖ plancher m. de planches jointes à rainure et languette. / ~ funnel ‖ Falztrichter m. ‖ entonnoirplieur m. / ~ gate ‖ Flügeltor m. ‖ porte f. à deux battants. / ~ goggles pl. ‖ zusammenklappbare Brille f. ‖ lunettes fpl. pliantes. / ~ head ‖ Klappverdeck n. ‖ capot m. à soufflet. / ~ job ‖ Abkantarbeit f. ‖ travail m. d'émoussement. / ~ ladder ‖ Klappleiter f. ‖ échelle f. brisée *ou* pliante.

folding machine (Bookb) ‖ Falzmaschine f. ‖ plieuse f. mécanique; machine f. à plier. / ~ (Tinm) ‖ Abkantmaschine f. ‖ machine f. à border à vive arête; machine f. à replier. / ~ for boxes ‖ Schachtelfaltmaschine f. ‖ plieuse f. pour boîtes. / ~ for envelopes ‖ Briefumschlagfaltmaschine f. ‖ plieuse f. pour enveloppes à lettres.

folding and grooving machine ‖ Langfalz-, Biege- und Zudrückmaschine f. ‖ machine f. à plier et à agrafer. / universal ~, rounding and boxforming machine ‖ Universalabkant-, Rund- und Kastenbiegemaschine f. ‖ machine f. universelle à plier, rouler et former des boîtes.

folding magnifier ‖ Einschlaglupe f. ‖ loupe f. fermante *ou* pliante. / ~ mould ‖ Klappform f. ‖ moule m. à bascule. / ~ mount of a magnifier ‖ Einschlagfassung f. einer Lupe ‖ monture f. pliante d'une loupe; forme f. fermante d'une loupe. / ~ rule ‖ Zollstock m. ‖ mètre m. pliant; règle f. brisée; réglet m. / ~ sale ‖ Schmiege f.; Kluft f. ‖ équerre f. pliante. / ~ screen ‖ Wandschirm m. ‖ paravent m. / ~ screen embroiderer ‖ Wandschirmstickerin f. ‖ brodeuse f. sur paravents. / ~ seat ‖ Klappsitz m. ‖ strapontin m. / ~ stick *see* ~ bone. / ~ stool for theatres ‖ Theaterklappstuhl m. ‖ strapontin m. pour salles de théâtre. / ~ table ‖ Tisch m. zum Zusammenklappen ‖ table f. à rabattement. / ~ test ‖ Faltprobe f. ‖ essai m. de ployage. / ~ tongs pl. ‖ Falzzange f. ‖ tenaille f. de pliage. / ~ tool ‖ Deckschaufel f.; Schalleisen n. ‖ outil m. à replier. / ~ top (Auto) ‖ zusammenklappbares Verdeck n. ‖ capote f. ployante. / ~ under ‖ herunterklappbar ‖ rabattable; repliable vers le bas. / ~-up side (Wagon) ‖ Seitenwand f. zum Hochklappen ‖ paroi f. latérale se relevant en forme de clapet.

foliaceous ornaments ·pl. ‖ Laubwerk n. ‖ rinceaux mpl.

foliage ‖ Laubwerk n. ‖ feuillage m.; rinceaux mpl. / ~ plant ‖ Blattpflanze f. ‖ plante f. à feuillage.

foliated tellurium ‖ Blättererz n. ‖ argent m. vierge en lames; nagyagite f.

folio (Accountbook) ‖ Seitenzahl f. ‖ folio m. / ~ (Size) ‖ Folioformat n. ‖ folio m. / ~ volume ‖ Foliant m. ‖ in-folio m.; livre m. in-folio.

follow, to ‖ folgen ‖ suivre. / ~ a trade ‖ ein Gewerbe n. treiben ‖ exercer un métier.

follower (Mach) ‖ Stopfbuchsbrille f.; Stopfbuchsdeckel m. ‖ chapeau m. de boîte à bourrage. / ~ (Surv) ‖ Kettenzieher m. ‖ porte-chaîne m. / ~ of piston ‖ Kolbendeckel m. ‖ couvercle m. *ou* plateau m. du piston. / ~ plate *see* ~ of piston.

follow rest (Mach tool) ‖ laufende *oder* mitgehende Brille f. ‖ lunette f. mobile *ou* à suivre.

fondant ‖ Fondant m. ‖ fondant m.

food ‖ Nahrungsmittel n.; Nährmittel n.; Genußmittel n.; Lebensmittel n. ‖ produit m. *ou* denrée f. alimentaire; comestible m.; aliment m.; nourriture f. / ~ of high nutritive value ‖ hochwertiges Nährmittel n. ‖ comestible m. de haute valeur nutritive. / ~ for infants ‖ Kindernährmittel n. ‖ farine f. pour enfants. / perishable ~s ‖ leichtverderbliche Lebensmittel npl. ‖ comestibles mpl. périssables. / saccharine ~ ‖ zuckerhaltiges Nährmittel n. ‖ préparation f. alimentaire au sucre.

food industry ‖ Nahrungsmittelindustrie f. ‖ industrie f. de l'alimentation. / ~ paste ‖ getrocknete Mehlspeise f. ‖ pâte f. alimentaire. / ~ preparation ‖ Nährpräparat n. ‖ préparation f. alimentaire.

foods pl. *see* food *and* foodstuff.

foodstuff ‖ Nährstoff m.; Nährmittel n.; Nahrungsmittel n. ‖ produit m. alimentaire. / animal ~ ‖ Futtermittel n.; Viehfutter n. ‖ fourrage m.; nourriture f. pour bétail. / ~ for children ‖ Kindernahrung f. ‖ aliment m. pour enfants.

foodstuff factory ‖ Nährmittelfabrik f. ‖ fabrique f. alimentaire; fabrique f. de denrées comestibles. / ~ industry ‖ Nährmittelfabrik f. ‖ industrie f. de denrées comestibles.

flooproof apparatus ‖ gegen falsche Bedienung gesicherter Apparat m. ‖ appareil m. à l'épreuve de fausses manœuvres.

foot, to ~ on (Textile) ‖ anwirken; anstricken ‖ enter.

foot ‖ Fuß m. ‖ pied m. / ~ = 0,304794 m. ‖ englischer Fuß m. ‖ pied m. anglais.

foot of bearing ‖ Lagerfuß m. ‖ pied m. de palier. / ~ of casing ‖ Gehäusefuß m.; Anschlußflansch m. ‖ pied m. d'enveloppe; bride f. de raccordement. / cylinder ~ ‖ Zylinderstütze f. ‖ support m. du cylindre. / ~ of frame ‖ Rahmenfuß m. ‖ pied m. de bâti. / ~ of an organ pipe ‖ Orgelpfeifenboden m.; Orgelpfeifenkern m. ‖ noyau m. *ou* pied m. d'une orgue. / ~ of a page ‖ Fuß m. einer Druckseite ‖ bas m. de page. / ~ of the rest (Turn) ‖ Fuß m. des Supports ‖ semelle f. du support. / ~ of the rail ‖ Schienenfuß m. ‖ patin m. du rail. / ~ of a rail chair ‖ Fuß m. *oder* Fußplatte f. eines Schienenstuhls ‖ semelle f. d'un coussinet de rails.

football ‖ Fußball m. ‖ football m. / ~ pump ‖ Fußballpumpe f. ‖ pompe f. pour footballs.

footboard (Build) ‖ Trittbrett n. ‖ planchette f. / ~ (Mach) ‖ Galerie f. ‖ galerie f.; garde-corps m. / ~ (Wagon) ‖ Auftritt m.; Fußtritt m. ‖ marchepied m. / ~ bracket ‖ Trittbrettträger m. ‖ soutien m. de planchette.

foot brake ‖ Fußbremse f.; Tritthebelbremse f. ‖ frein m. à pied; frein m. (à levier) à pédale.

foot bridge ‖ Fußgängerbrücke f.; Laufbrücke f. ‖ pont m. pour piétons; passerelle f. / ~ passenger ~ ‖ Bahnsteigbrücke f.; Übergangssteg m.; Bahnhofssteg m. ‖ passerelle f. de gare; passage m. aérien *ou* par dessus.

foot castor of a stand ‖ Fußrolle f. eines Stativs ‖ roulette f. d'un pied. / ~ control ‖ Fußsteuerung f. ‖ commande f. à pied. / ~ delivery-car ‖ Lebensmittellieferwagen m. ‖ voiture f. de livraison pour denrées alimentaires. / ~ disease ‖ Klauenseuche f.; piétin m. / ~ ease ‖ Fußstreupulver n. ‖ poudre f. pour les pieds.

footer (Textile) ‖ Fußmaschine f. ‖ machine f. à pieds.

foothill ‖ Vorgebirge n. ‖ promontoire m.; cap m.

footing (Textile) ‖ Leinenspitze f. ‖ bisette f. / ~ of wall ‖ Mauerfuß m. ‖ pied m. de mur. / ~ of walls ‖ Grundschicht f. ‖ assise f. de fondement en pierres dures.

foot keys pl. (Organ) ‖ Pedal n. ‖ pédale f. / ~ lathe ‖ Fußdrehbank f. ‖ tour m. à pédale. / ~ lever ‖ Fußhebel m. ‖ pédale f. / ~ lights pl. ‖ Rampenbeleuchtung f. ‖ éclairage m. de rampe. / ~ muff ‖ Fußsack m. ‖ chancelière f. / ~ note ‖ Fußnote f. ‖ note f. courante. / ~-operated brake ‖ Fußtrittbremse f. ‖ frein m. commandé par pédale.

footpath ‖ Bürgersteig m. ‖ trottoir m.; estrade f. / ~ (Canal) ‖ Fußpfad m. ‖ chemin m. de contrehalage.

foot plank (Wagon) ‖ Trittbrett n. ‖ planche f. de marche-pied. / ~ plate of locomotive ‖ Lokomotivplattform f. ‖ tablier m. d'une locomotive. / ~ plate of pole ‖ Mastfußplatte f. ‖ plaque f. d'assise du poteau. / ~ pound ‖ Fußpfund n. ‖ livre-pied f. / ~ powder ‖ Fußstreupulver n. ‖ poudre f. pour les pieds. / ~ press (Print) ‖ Fußpresse f. ‖ presse f. à pédale. / ~ rail ‖ Fußschiene f. ‖ rail m. à patin. / ~ rest ‖ Fußraste f. ‖ repose-pied m.; dépose m. pied. / ~ ring for fowls ‖ Geflügelfußring m. ‖ anneau m. pour pattes de volailles. / ~ roller ‖ Fußrolle f. ‖ galet m. conducteur *ou* de soutien.

footrope ‖ Strickleiter f. ‖ échelle f. de corde. / ~ (Mar) ‖ Paard n.; Rahepferd n. ‖ marchepied m. de vergue.

foot rule ‖ Zollstock m. ‖ règle f. graduée en pouces; mètre m. pliant; réglet m. / ~ scraper ‖ Fußkratzeisen n. ‖ essuie-pieds m.; décrottoir m. / ~ screw of a stand ‖ Fußschraube f. eines Stativs ‖ vis f. calante d'un pied. / ~ sole ‖ Fußsohle f. ‖ plante f. du pied.

footstep (Join) ‖ Fußbank f.; Fußschemel m. ‖ escabeau m.; escabelle f. / ~ (Wagon) ‖ Tritt m. ‖ marche-pied m. / ~ (Mach) ‖ Drehpfanne f. ‖ grain m.; crapaudine f. / ~ bearing ‖ Spurlager m. ‖ crapaudine f.; porte-culot m. / ~ heating ‖ Schemelheizung f. ‖ chauffage m. des bancs de pieds. / ~ liner (Mill) ‖ Spurlatte f. ‖ plaque f. de butée. / ~ plate (Mill) *see* ~ liner.

footstick (Print) ‖ Fußsteg m. ‖ lingot m. en pied de page.

footstool *see* footstep (Join).

foot stove ‖ Fußwärmer m.; Feuerkieke f. ‖ chaufferette f. / ~ support ‖ Fußstütze f. ‖ appui-pied m. / ~ throttle ‖ Akzelera-

tor m. ‖ accélérateur m. / ~ valve ‖ Saugventil n.; Fußventil n.; Bodenventil n. ‖ soupape f. d'aspiration; clapet m. de pied.

footwall of a lode (Mine) ‖ Liegendes n. eines Ganges ‖ sol m. *ou* mur m. d'un filon.

foot warmer ‖ Fußwärmer m. ‖ chaufferette f.; chauffe-pied m.; bouillotte f. / ~ warming plate ‖ Fußwärmplatte f. ‖ banc m. chauffe-pied.

for instance ‖ beispielsweise; zum Beispiel n. ‖ par exemple m.

for once ‖ ausnahmsweise ‖ d'exception; exceptionnellement.

for sale ‖ zum Verkauf m.; zu verkaufen ‖ à vendre.

forage ‖ Futtermittel npl. ‖ produit m. fourrage; fourrage m.

forage cake ‖ Futterkuchen m. ‖ tourteau m. / ~ for cattle ‖ Futterkuchen m. zur Viehfütterung ‖ tourteau m. pour l'alimentation du bétail. / ~ from oil seed ‖ Futterkuchen m. aus ölhaltigen Samen ‖ tourteau m. de graines oléagineuses.

forage cap ‖ Schirmmütze f. ‖ casquette f. à visière. / ~ cutting machine ‖ Futterschneidemaschine f. ‖ machine f. à hacher le fourrage; hache-fourrage m.; hache-paille m. / ~ ladder (Carr) ‖ Schoßkelle f. ‖ fourragère f. / ~ plant ‖ Futterpflanze f. ‖ plante f. fourragère. / ~ straw ‖ Futterstroh n. ‖ paille f. fourrageuse.

force, to ‖ forzieren ‖ forcer. / ~ a blast of air through the molten metal ‖ einen Luftstrom m. durch das flüssige Roheisen pressen ‖ faire passer un courant d'air à travers la fonte liquide. / ~ the boiler ‖ den Kessel m. anstrengen *oder* forcieren ‖ forcer la chaudière. / ~ the crank by pressure ‖ die Kurbel aufpressen ‖ presser ou emmancher la manivelle sur l'arbre. / ~ down the commander (Hatt) ‖ das Formband n. herabtreiben ‖ avaler *ou* faire descendre la ficelle. / ~ the piston off the rod ‖ den Kolben m. von der Stange abdrücken ‖ décaler la tige de piston; déchasser le piston de sa tige.

force ‖ Kraft f. ‖ force f.; puissance f.; effort m. / the ~s pl. act in an opposite direction ‖ die Kräfte fpl. wirken in entgegengesetzter Richtung ‖ les forces fpl. agissent *ou* sont dirigées dans la direction opposée *ou* contraire. / active ~ ‖ Wucht f. ‖ force f. vive. / the ~ acts on a point ‖ die Kraft greift in einem Punkte an ‖ la force s'applique sur un point. / adhesive ~ ‖ Adhäsionskraft f. ‖ adhérence f.; adhésion f. / back electromotive ~ ‖ rückelektromotorische Kraft f. ‖ force f. électromotrice inverse. / ~ of the blow ‖ Schlagstärke f. ‖ puissance f. de frappe. / centrifugal ~ ‖ Fliehkraft f.; Zentrifugalkraft f. ‖ force f. centrifuge. / centripetal ~ ‖ Zustrebekraft f.; Zentripetalkraft f. ‖ force f. centripète. / coercive ~ ‖ Koerzitivkraft f. ‖ force f. coercitive. / component ~ ‖ Teilkraft f. ‖ force f. composante. / counter ~ ‖ Gegenkraft f. ‖ contre-force f. / counter electromotive ~ ‖ elektromotorische Gegenkraft f. ‖ force f. contre-électromotrice. / to decompose the ~ *see* to resolve the ~. / deflecting ~ ‖ ablenkende Kraft f. ‖ force f. déviante. / electromotive ~ ‖ elektromotorische Kraft f. ‖ force f. électromotrice. / to enter into ~ ‖ in

Kraft f. treten ‖ entrer en vigueur f. / ~ of gravity ‖ Erdschwere f.; Schwerkraft f.; Gravitationskraft f. ‖ force f. de gravitation; pesanteur f. / internal ~ ‖ widerstehende Kraft f. ‖ force f. intérieure. / magnetic ~ ‖ magnetische Kraft f. ‖ force f. magnétique. / magnetomotive ~ (Electr) ‖ magnetomotorische Kraft f. ‖ force f. magnétomotrice. / mechanized ~s pl. (Arm) ‖ mechanisierte Streitkräfte fpl. ‖ forces fpl. motorisées. / moving ~ ‖ treibende Kraft f. ‖ force f. motrice. / opposed ~ ‖ Gegenkraft f. ‖ contre-force f. / ~ of percussion (Mach) ‖ Durchschlagskraft f. ‖ force f. de pénétration. / portative ~ ‖ Tragkraft f. ‖ force f. portante. / ~ of public opinion of the civilized world ‖ Macht f. der öffentlichen Meinung der zivilisierten Welt ‖ force f. d'opinion publique des nations civilisées. / to resolve the ~ ‖ die Kraft f. zerlegen ‖ décomposer la force. / resisting ~ ‖ Widerstandskraft f. ‖ force f. résistante; résistance f. / ~s pl. in the same direction ‖ gleichgerichtete Kräfte fpl. ‖ forces fpl. dans la même direction. / shearing ~ ‖ Scherkraft f. ‖ force f. de cisaillement. / superior ~ ‖ höhere Gewalt f. ‖ force f. majeure. / tangential ~ ‖ Tangentialkraft f. ‖ force f. tangentielle. / tensile ~ ‖ Zugkraft f. ‖ force f. de traction. / thermo-electric electromotive ~ ‖ thermoelektrische Kraft f. ‖ force f. électromotrice thermoélectrique. / torsional ~ ‖ Drehkraft f.; Verdrehungskraft f.; Torsionskraft f. ‖ force f. de torsion. / tractive ~ ‖ Zugkraft f. ‖ force f. de traction. / ~ of wind ‖ Windstärke f. ‖ force f. de vent. / to measure the ~ of the wind ‖ die Windstärke f. messen ‖ mesurer la force *ou* l'intensité du vent.

force circulation ‖ Druckumlauf m. ‖ circulation f. forcée.

forced ‖ notgedrungen ‖ contraint par la nécessité. / ~ action guard lid ‖ zwangsweiser Schutzdeckel m. ‖ couvercle m. protecteur fonctionnant automatiquement. / ~ commutation (Electr) ‖ erzwungene Kommutierung f. ‖ commutation f. forcée. / ~ course of exchange ‖ Zwangskurs m. ‖ cours m. forcé du change.

forced draught ‖ Druckzug m. ‖ tirage m. forcé. / ~ cooling ‖ Druckluftkühlung f. ‖ refroidissement m. à courant d'air forcé. / ~ furnace ‖ Unterwindfeuerung f. ‖ foyer m. à soufflage sous grille.

forced-in key ‖ Treibkeil m. ‖ clavette f. chassée *ou* conique.

forced landing (Aero) ‖ Notlandung f. ‖ atterrissage m. forcé. / ~ loan ‖ Zwangsanleihe f. ‖ emprunt m. forcé. / ~ lubrication ‖ Preßschmierung f.; Druckschmierung f. ‖ graissage m. sous pression. / ~ oil air-cooled transformer with blower ‖ Öltransformator m. mit Ölumlauf und äußerer Selbstlüftung ‖transformateur m. à circulation d'huile sous pression et à refroidissement naturel. / ~ oscillation ‖ erzwungene Schwingung f. ‖ oscillation f. contrainte. / ~ proof ‖ Gewaltversuch m. ‖ essai m. à outrance. / ~ radial position of the fore-axle ‖ zwangläufige Radialstellung f. der Vorderachse ‖ orientation f. radiale commandée du premier essieu. / ~ ventilation ‖ Durchzuglüftung f. ‖ ventilation f. forcée.

force equation ‖ Kraftgleichung f. ‖ équation f. des forces. / origin of ~ ‖ Angriffspunkt m. einer Kraft ‖ point m. d'application d'une force. / ~ polygon ‖ Kräftepolygon n. ‖ polygone m. de forces. / ~ and lift pump ‖ Druck- und Saugpumpe f. ‖ pompe f. foulante et aspirante. / unit of ~ ‖ Krafteinheit f.; Einheit f. der Kraft ‖ unité f. de force.

forceps pl. ‖ Zange f.; Pinzette f. ‖ forceps m.; pince f. / abortion ~ ‖ Abortuszange f. ‖ pince f. à faux germes. / artery ~ ‖ Arterienklemme f. ‖ pince f. hémostatique. / assayer's ~ ‖ Probierzange f. ‖ pince f. d'essayeur. / curved ~ ‖ gebogene Pinzette f. ‖ pince f. coudée. / fixation ~ ‖ Feststellpinzette f. ‖ pince f. à fixer. / hooked ~ ‖ Hakenpinzette f. ‖ pince f. à griffes *ou* à crochets. / iris ~ ‖ Augenpinzette f.; Irispinzette f. ‖ pince f. à iris. / large ~ ‖ Kornzange f. ‖ grande pince f. / ~ with nibbed points ‖ Pinzette f. mit Querriefen ‖ pince f. à mors striés transversalement. / ~ of rustless steel ‖ Klemme f. *oder* Pinzette f. aus nichtrostendem Stahl ‖ pince f. en acier inoxydable. / ~ with serrated points ‖ geriefte Pinzette f. ‖ pince f. striée. / splinter ~ ‖ Splitterpinzette f. ‖ pince f. à écharde. / straight ~ ‖ gerade Pinzette f. ‖ pince f. droite. / surgical ~ ‖ chirurgische Pinzette f. ‖ pince f. de chirurgie. / ~ with teeth ‖ Hakenpinzette f.; gezahnte Pinzette f. ‖ pince f. à griffes *ou* à crochets.

forcer ‖ Stempel m.; Druckpumpenkolben m. ‖ piston m. de pompe foulante.

force, unit of ‖ Krafteinheit f.; Einheit f. der Kraft ‖ unité f. de force.

forcing (Projectile) ‖ Führung f. ‖ forcement m.; guide m. / ~ the edge of the tyre beyond the rim and hammering it down against it ‖ Verstemmen n. vorspringender Kanten am Radreifen ‖ martelage m. et matage m. des bords en saillie des bandages. / ~ pump ‖ Druckpumpe f. ‖ pompe f. foulante; refouleur m. / ~ screw ‖ Abdrückschraube f. ‖ boulon m. de détente.

ford ‖ Furt f. ‖ gué m.

fore-and-aft ‖ längsschiffs ‖ de l'avant à l'arrière. / ~ sails pl. ‖ Schrotsegel npl. ‖ voiles mpl. auriques. / ~ schooner ‖ Gaffelschoner m. ‖ goëlette f. franche.

fore-and-after (Mar) ‖ Gaffelschoner m. ‖ goëlette f. franche.

fore axle ‖ Leitachse f.; Vorderachse f. ‖ essieu m. avant.

forebay ‖ Wasserschloß n. ‖ château m. d'eau.

fore bolster ‖ Vorderachsschale f. ‖ sellette f. de devant. / ~ brace ‖ Fockbraß f. ‖ bras m. de la misaine. / ~ breast-work of the forecastle ‖ vorderes Schott n. der Back ‖ fronteau m. avant.

fore carriage ‖ Vorderwagen m. ‖ avant-train m. / ~ of a plough ‖ Vordergestell n. eines Pfluges ‖ avant-train m. de charrue.

forecast ‖ Voranschlag m. ‖ calcul m. préalable.

forecastle (Shipb) ‖ Back f.; Logis n.; Schlafplatz m. ‖ gaillard m.

fore-cellar ‖ Vorkeller m. ‖ avant-cave f.

fore-channel ‖ Obergerinne n.; Vorarche f. ‖ bief m. d'amont.

fore chill-room ‖ Vorkühlraum m. ‖ entrée f. de la salle frigorifique.

fore clew-garnet ‖ Fockgeitau n. ‖ carguepoint f. de misaine.

foreclosure, right of ~ by creditors ‖ Beschlagnahmerecht n. der Gläubiger ‖ droit m. de saisie des créanciers.

fore eccentric see forward eccentric.

fore edge of stem ‖ Vorstevenvorderkante f. ‖ arête f. avant de l'étrave.

fore-foot (Shipb) ‖ Stevenlauf m. oder Unterlauf m. des Kieles ‖ brion m.; marsouin m.; ringeau m.; ringeot m.

fore front ‖ Vorderseite f. ‖ façade f. de devant.

fore-ganger (Mar) ‖ Kettenvorlauf m. ‖ premier maillon m.; première longueur f.

forego, to ‖ Abstand m. nehmen von ‖ renoncer à.

foregoer see fore-ganger.

fore head (Mine) ‖ Streckenort m.; Abbaustoß m. ‖ fond m. d'une galerie; front m. de taille.

forehearth (Metal) ‖ Vorherd m. ‖ avantcreuset m.

foreign ‖ fremd; auswärtig; ausländisch ‖ étranger. / ~ bill (of exchange) ‖ Auslandwechsel m. ‖ lettre f. de change sur l'étranger. / ~ body ‖ Fremdkörper m. ‖ corps m. étranger. / ~ branch ‖ Auslandsfiliale f. ‖ succursale f. étrangère. / ~ country ‖ Ausland n. ‖ étranger m.

foreigner ‖ Ausländer m. ‖ étranger m.

foreign exchange broker ‖ Devisenhändler m. ‖ agent m. de change.

foreign language ‖ Fremdsprache f. ‖ langue f. étrangère. / correspondence in ~ languages ‖ fremdsprachiger Schriftwechsel m. ‖ correspondance f. en langues étrangères. / ~ matter ‖ Fremdkörper m. ‖ corps m. étranger.

Foreign Office ‖ Auswärtiges Amt n. ‖ ministère m. des affaires étrangères.

foreign post-paper ‖ Überseebriefpapier n. ‖ papier m. pelure. / ~ trade ‖ Außenhandel m.; Auslandhandel m. ‖ commerce m. extérieur. / ~ trade ‖ Überseehandel m. ‖ commerce m. d'outremer.

fore-lift (Mar) ‖ Focktoppwant f. ‖ balancine f. de misaine.

forelock ‖ Splint m. ‖ clavette f. / ~ bolt ‖ Keilbolzen m. ‖ boulon m. à clavette ou à goupille.

fore-maker (Forg) ‖ Schmiedemeister m. ‖ forgeron m.

foreman ‖ Werkmeister m. ‖ contremaître m.; chef m. d'atelier; maître-ouvrier m. / ~ (Carp) ‖ Polier m. ‖ appareilleur m.; maître-ouvrier m.; parlier m. / ~ (Factory) ‖ Aufseher m. ‖ piqueur m. / ~ (Forg) ‖ Schmiedemeister m. ‖ forgeron m. / ~ (Mar) ‖ Vormann m. ‖ brigadier m. / ~ (Mas) ‖ Maurerpolier m. ‖ maître-compagnon m. / ~ (Print) ‖ Faktor m. ‖ prote m. / ~ (Shipb) ‖ Meisterknecht m. ‖ second maître m. de construction. / ~ (Trade) ‖ Aufsichtsposten m. ‖ chef m. d'équipe. / carpenter ~ ‖ Zimmerpolier m. ‖ charpentier m. appareilleur. / ~ of mines ‖ Steiger m. ‖ maître-mineur m.; porion m. / ~ of the navvies ‖ Schachtmeister m. ‖ chef m. des ouvriers ou des terrassiers. / ~ of plate-layers (Railw) ‖ Kolonnenführer m. ‖ chef m. d'équipe. / ~ for roofing ‖ Dachdeckermeister m. ‖ maître m. couvreur. / ~ of the strikers (Forg) ‖ Zuschläger m.; Schmied m. ‖ forgeron m.

foreman bottler ‖ Flaschenkellermeister m. ‖ chef m. de canetterie. / club rooms pl.

---

for foremen ‖ Werkmeisterkasino n. ‖ casino m. pour contremaîtres. / ~ glass blower ‖ Glasbläsermeister m. ‖ chef m. de place.

fore-mast ‖ Fockmast m. ‖ mât m. de misaine.

foremost-frame (Shipb) ‖ Ohrspant n. ‖ couple m. de coltis.

fore part of a stone ‖ Kopfseite f. oder Hauptseite f. eines Steines ‖ parement m. d'une pierre.

fore platform ‖ Vorderplattform f. ‖ plateforme f. extrème de devant.

forerunner (Mar) ‖ Vorläufer m. der Logleine ‖ houache m. / ~ tackle ‖ Focktakel n. ‖ caliorne f. de misaine; candelette f.

foresail ‖ Fock f.; Focksegel n. ‖ misaine f.; voile f. de misaine. / ~ sheet ‖ Fockschot f. ‖ écoute f. de la misaine.

forescene ‖ Proszenium n. ‖ avant-scène f.

fore screen ‖ Vorsieb n. ‖ crible m. préparatoire. / ~ shoe (Farr) ‖ Vordereisen n. ‖ fer m. de devant.

fore-shrouds pl. ‖ Fockwanten fpl. ‖ haubans mpl. de misaine.

foresight ‖ Vorsorge f. ‖ prévoyance f. / moveable ~ (Arm) ‖ Klappkorn n. ‖ guidon m. à charnière.

forest, to ‖ beholzen ‖ boiser.

forest ‖ Forst m.; Wald m. ‖ forêt m. / communal ~ ‖ Gemeindewald m. ‖ forêt f. communale. / low ~ ‖ Niederwald m. ‖ taillis m. simple. / resinous ~ ‖ Nadelwald m. ‖ forêt f. résineuse.

forestal instruments pl. and utensils pl. ‖ forstwirtschaftliche Geräte npl. und Instrumente npl. ‖ instruments mpl. et ustensiles mpl. forestiers.

forestall, to ‖ zuvorkommen ‖ devancer; prévenir.

fore starling ‖ Kronpfeilerkopf m.; Pfeilervorhaupt n. ‖ avant-bec m. d'une pile; bec m. d'amont.

forestay ‖ Fockstag n. ‖ étai m. de misaine. / ~ sail ‖ Fockstagsegel n. ‖ petit foc m.; trinquette f.; tourmentin m.

forest college ‖ Forstakademie f. ‖ école f. forestière supérieure. / tools pl. for ~ economy ‖ Forstkulturgerät n. ‖ outils mpl. pour l'économie forestière.

forester (Person) ‖ Förster m. ‖ garde-forestier m. / ~ (Tree) ‖ Waldbaum m.; Forstbaum m. ‖ arbre m. forestier. / ~'s house ‖ Försterei f. ‖ maison f. forestière.

forest inspector ‖ Waldaufseher m. ‖ garde m. forestier. / ~ master ‖ Forstmeister m. ‖ inspecteur m. des forêts. / ~ officer ‖ Forstbeamter m. ‖ agent m. forestier. / to manage through ~ officers ‖ der Forst f. unterstellen ‖ soumettre au régime forestier. / ~ pasture ‖ Waldweide f. ‖ pâturage m. en forêt. / ~ plant ‖ Forstpflanze f. ‖ plante f. forestière. / ~ possessor ‖ Waldbesitzer m. ‖ propriétaire m. de forêts ou de bois. / ~ railway ‖ Forstbahn f. ‖ chemin m. de fer forestier. / ~ ranger ‖ Förster m. ‖ garde forestier m.

forestry ‖ Forstkultur f.; Forstwirtschaft f. ‖ exploitation f. des forêts; sylviculture f. / ~ association ‖ Waldgenossenschaft f. ‖ association f. de propriétaires forestiers.

forest shrub ‖ Forststrauch m. ‖ arbuste m. forestier. / ~ stand ‖ Waldbestand m. ‖ peuplement m. d'une forêt. / ~ survey ‖ Forstvermessungswesen n. ‖ arpentage

---

m. forestier. / ~ tree ‖ Forstbaum m.; Waldbaum m. ‖ arbre m. forestier.

fore studding-sail (Mar) ‖ Fockleesegel n. ‖ bonnette f. de misaine. / ~ halliard ‖ Fockleesegelfall m. ‖ drisse f. de la bonnette de misaine.

forest wool ‖ Waldwolle f. ‖ laine f. de pins ou de bois.

foretack ‖ Fockhals m. ‖ amure f. de misaine.

foretackle ‖ Focktakel n. ‖ caliorne f. de misaine; candelette f.

fore yard (Build) ‖ Vorhof m. ‖ avant-cour f. / ~ (Mar) ‖ Fockrahe f. ‖ vergue f. de misaine.

forfeit ‖ Abstandsgeld n.; Reugeld n. ‖ forfait m.; dédit m.; indemnité f.

forge, to ‖ schmieden; hämmern; treiben; schlagen ‖ forger; battre; marteler; martiner; faire au marteau. / the ingot was forged out under an hydraulic press (the actual forging required x heats) ‖ der Block m. wurde in x Hitzen unter einer hydraulischen Presse ausgeschmiedet ‖ le lingot m. fut forgé sous une presse hydraulique en x chauffes. / ~ coin ‖ falschmünzen ‖ contrefaire des monnaies. / ~ cold ‖ kaltschmieden ‖ forger à froid. / ~ in dies pl. ‖ im Gesenk n. schmieden ‖ matricer; forger en matrices fpl. / ~ hollow ‖ hohlschmieden ‖ forger à creux. / ~ hot ‖ warmschmieden ‖ forger à chaud. / ~ the iron ‖ das Eisen (aus)schmieden ‖ forger ou battre le fer. / ~ out ‖ ausschmieden; recken ‖ étirer sous le marteau. / ~ in the rough ‖ vorschmieden ‖ forger grossièrement. / ~ square ‖ viereckig schmieden ‖ équarrir.

forge ‖ Schmiede f.; Hammerschmiede f.; Hammerwerk n. ‖ forge f. / ~ (Hearth) ‖ Schmiedefeuer n.; Schmiedeherd m.; Schmiedeesse f.; Feuer n.; Herd m. ‖ forge f.; feu m. ou four m. (de forge); chaufferie f. / Biscayan ~ (Met) ‖ biskayisches Feuer n. ‖ feu m. biscayen. / bloomery ~ (Met) ‖ Luppenherd m.; Rennherd m.; Luppenfrischfeuer n.; Rennfeuer n. ‖ feu m. ou foyer m. ou forge f. à loupes; bas-foyer m. / boiler ~ ‖ Kesselschmiede f. ‖ chaudronnerie f.; atelier m. de chaudronnerie. / Catalan ~ (Met) ‖ katalonisches Feuer n. oder Frischfeuer n. ‖ feu m. catalan; forge f. (à la) catalane. / electric ~ ‖ elektrische Schmiedeesse f. ‖ forge f. électrique. / field ~ see portable ~. / fine art ~ ‖ Kunstschmiede f. ‖ forge f. artistique. / large ~ ‖ Dampfhammerschmiede f. ‖ grosse forge f. / portable ~ ‖ Feldschmiede f. ‖ forge f. portative ou volante ou de campagne. / travelling ~ see portable ~.

forgeable ‖ schmiedbar ‖ forgeable; malléable. / ~ while cold ‖ im kalten Zustand schmiedbar ‖ forgeable à froid.

forge bellows pl. see ~ blower. / ~ blower ‖ Schmiedefeuergebläse n. ‖ soufflet m. ou soufflante f. ou soufflerie f. de forge; vache f. / ~ blower with electric drive ‖ Schmiedefeuergebläse n. mit elektrischem Antrieb ‖ soufflante f. de forge à commande électrique. / ~ coal ‖ Schmiedekohle f.; Eßkohle f. ‖ charbon m. de forge; houille f. maréchale ou pour feux de forge; maréchale f. / ~ coal coke ‖ Eßkohlenkoks m. ‖ coke m. de charbon de forge. / ~ crane ‖ Schmiedekran m. ‖ grue f. de forge ou pour forges.

forged see also to forge ‖ geschmiedet ‖ forgé; venu de forge. / artistically ~ article ‖ Kunstschmiedeerzeugnis n. ‖ travail m. d'art en fer forgé. / hand ~ ‖ von Hand f. geschmiedet; handgeschmiedet ‖ forgé à (la) main. / ~ in one piece ‖ in einem Stück geschmiedet ‖ forgé d'une seule pièce. / rough ~ ‖ roh geschmiedet ‖ (venant) brut de forge. / ~ solidly ‖ massiv geschmiedet ‖ forgé massif. / ~ thoroughly ‖ durchgeschmiedet ‖ forgé jusqu'au cœur.

forged iron ‖ geschmiedetes Eisen n.; Schmiedeeisen n. ‖ fer m. forgé ou de forge. / ~ nail ‖ Schmiedenagel m. ‖ clou m. forgé. / ~ piece see forging (Piece). / ~ roll ‖ geschmiedete Walze f. ‖ cylindre m. de laminoir forgé.

forge hammer ‖ Schmiedehammer m. ‖ marteau m. de forge ou à forger; marteau-pilon m.; martinet m. / ~ hearth see forge (Hearth). / ~ heater ‖ Ofenmann m. ‖ chauffeur m. de forge.

forgeman ‖ Schmiedearbeiter m.; Schmied m.; Grobschmied m.; Hammerschmied m. ‖ ouvrier m. forgeur ou de forge; forgeur m.; forgeron m.

forge manager ‖ Betriebsleiter m. einer Schmiede ‖ chef m. de fabrication de la forge. / ~ master ‖ Schmiedemeister m. ‖ maître m. de forge; maître-forgeron m. / ~ pig (iron) ‖ Puddelroheisen n.; Frischereiroheisen n. ‖ fonte f. (crue) de puddlage ou d'affinage. / plant of ~s pl. ‖ Hammerwerk n.; Schmiede f. ‖ forge f.; martinet m.

forger ‖ Schmied m.; Grobschmied m.; Hammerschmied m. ‖ forgeur m.; forgeron m.

forge roll ‖ Schmiedewalze f. ‖ cylindre m. forgeur. / ~ rolling machine see forging rolls. / ~ rolling mill see ~ train.

forgery see forging (Action).

forge scales pl. ‖ Hammerschlag m.; Glühspan m.; Zunder m. ‖ pailles fpl. ou scories fpl. ou écailles fpl. ou battiture f. de fer; mâchefer m.; martelures fpl. / ~ shop ‖ Hammerwerk n.; Hammerschmiede f. ‖ atelier m. de forge. / ~ steel ‖ Schmiedestahl m.; Puddelstahl m. ‖ acier m. de forge; acier m. puddlé ou de puddlage.

forget, to ‖ vergessen ‖ oublier.

forgetful ‖ nachlässig ‖ négligent.

forge tongs pl. ‖ Schmiedezange f. ‖ pince f. ou tenailles fpl. du forgeron; tricoises fpl. / ~ train (Met) ‖ Puddelwalzwerk n.; Luppenwalzwerk n.; Rohschienenwalzwerk n. ‖ train m. ébaucheur ou forgeur; train de puddlage ou de loupes; laminoir m. ébaucheur. / ~ wagon see portable forge. / ~ welding ‖ Herdschweißen n. ‖ soudure f. au foyer. / ~ welding ‖ Hammerschweißung f. ‖ soudage m. au marteau.

forging (Action) ‖ Schmieden n.; Schmiedearbeit f.; Verschmieden n. ‖ forgeage m.; forgement m.; travail m. de forge; action f. de forger; ouvrage m. du forgeron; traitement m. au forgeage. / ~ (Piece) ‖ Schmiedestück n.; Schmiedling m. ‖ pièce f. forgée ou de forge ou à forger. / ~s pl. of any sizes ‖ Schmiedestücke mpl. von beliebig großen Abmessungen ‖ pièces fpl. de forge de dimensions quelconques. / artistic ~ ‖ Kunstschmiedestück n. ‖ pièce f. d'art en fer forgé. / die ~ (Hammered) ‖ Gesenk-

schmiedestück n. ‖ pièce f. matricée ou forgée en matrice. / electric ~ ‖ elektrisches Schmieden n. ‖ forgeage m. électrique. / rough ~ ‖ geschmiedeter Rohling m. ‖ pièce f. brute forgée; pièce f. venue de forge. / square ~ ‖ Viereckigschmieden n. ‖ équarrissage m. / thorough ~ of heavy masses ‖ durchdringendes Schmieden n. schwerer Massen ‖ forgement m. à fond des masses volumineuses. / ~ at too low a temperature ‖ Schmieden n. bei zu niedriger Temperatur ‖ forgeage m. à une température trop bas. / ulterior ~ of special steels already rough-forged ‖ Weiterverschmieden n. bereits vorgeschmiedeter Sonderstahle ‖ second procédé m. de forgeage des aciers spéciaux déjà forgés. / well finished ~s pl. of the best material without welding ‖ aus bestem Werkstoff sauber ohne Schweißung geschmiedete Stücke npl. ‖ pièces fpl. soigneusement forgées du meilleur métal sans soudure.

forging hammer see forge hammer. / ~ hearth see forge hearth. / ~ iron ‖ Schmiedeeisen n. ‖ fer m. de forge. / ~ machine ‖ Schmiedemaschine f. ‖ machine f. à forger.

forging-out the iron ‖ Recken n. des Eisens ‖ étirage m. du fer.

forging press ‖ Schmiedepresse f. ‖ presse f. à forger ou de forge. / electric ~ ‖ Elektroschmiedepresse f. ‖ presse f. à forger électrique. / hammer-type ~ ‖ Schmiedepresse f. in Hammerform ‖ presse f. à forger modèle pilon. / high-speed ~ ‖ Schnellschmiedepresse f. ‖ presse f. à forger à action rapide. / hydraulic ~ ‖ hydraulische Schmiedepresse f. ‖ presse f. à forger hydraulique. / steam ~ see steam-power ~. / steam-hydraulic ~ ‖ dampfhydraulische Schmiedepresse f. ‖ presse f. à forger vapohydraulique; presse f. hydraulique à vapeur à forger. / steam-power ~ ‖ Dampf(druck)schmiedepresse f. ‖ presse f. à vapeur à forger.

forging rolls pl. ‖ Schmiedewalze f.; Schmiedewalzmaschine f. ‖ laminoir m. à forger.

forgings manufactory see forge steel.

forging stamping press ‖ Gesenkschmiedepresse f. ‖ presse f. à matricer; presse f. à étamper. / ~ steel see forge steel. / ~ temperature ‖ Schmiedetemperatur f. ‖ température necessaire pour le forgeage. / ~ test ‖ Schmiedeprobe f. ‖ essai m. à la forge ou de forgeage ou au forgeage. / ~ tool ‖ Schmiedewerkzeug n. ‖ outil m. de forge ou du forgeron.

fork, to ~ a belt off ‖ einen Riemen m. herunterwerfen ‖ débrayer une courroie. / ~ a belt on ‖ einen Riemen m. auflegen ‖ embrayer une courroie.

fork ‖ Gabel f. ‖ fourche f.; fourchette f. / ~ of bicycle ‖ Fahrradgabel f. ‖ fourche f. de bicyclette. / dung ~ ‖ Düngerforke f. ‖ fourche f. à fumier. / front ~ with stay rods ‖ Doppelvordergabel f. ‖ double fourche f. d'avant / ~ of furnace ‖ Ofengabel f.; Forke f. ‖ fourchette f. de fourneau. / gear shift ~ ‖ Schaltgabel f. ‖ fourchette f. de tirette. / gold ~ ‖ Goldgabel f. ‖ fourchette f. en or. / hay ~ ‖ Heugabel f. ‖ fourche f. à foin. / ~ of a river ‖ Flußgabelung f. ‖ bifurcation f. d'un fleuve. / ~ of a road ‖ Weggabelung f. ‖ bifurcation f. d'une route. / ~ for running boards ‖ Gabelstütze f. für Laufbretter ‖ console f. à fourche pour marchepieds. / ~ for se-

parating malt ‖ Greifhaufenschüttelgabel f. ‖ râteau-secoueur m. pour démêler les mottes d'une couche de malt. / silver ~ ‖ Silbergabel f. ‖ fourchette f. en argent. / spring ~ ‖ Federgabel f. ‖ fourche f. à ressort. / table ~ ‖ Tischgabel f. ‖ fourchette f. de table. / tedding ~ ‖ Gabel f. zum Heuwenden ‖ fourche f. faneuse. / tuning ~ ‖ Stimmgabel f. ‖ diapason m. / vibrating tuning ~ see tuning ~. / ~ for wires ‖ Drahtgabel f. ‖ lance f. à fourche. / wooden ~ ‖ Holzgabel f. ‖ fourchette f. en bois.

fork and spoon ‖ Tischbesteck n. ‖ couvert m. / horn ~ ‖ Tischbesteck n. aus Horn ‖ couvert m. en corne. / ivory ~ ‖ Tischbesteck n. aus Elfenbein ‖ couvert m. en ivoire. / ~ for salad ‖ Salatbesteck n. ‖ couvert m. à salade.

fork arm ‖ Gabelarm m. ‖ bras m. porte-fourches. / three-pronged ~ centre (Mach tool) ‖ Dreizack m.; dreizackige Spindelspitze f. ‖ pointe f. triple d'arbre de tour; trident m.

forked ‖ gegabelt ‖ bifurqué. / ~ axle ‖ Gabelachse f. ‖ essieu m. chapé. / ~ bed (Mot) ‖ Gabelrahmen m. ‖ bâti m. en fourche. / ~ connecting rod ‖ gegabelte Schubstange f. ‖ bielle f. fourchue. / ~ connector ‖ Gabelverbindung f. ‖ raccordement m. ou connexion f. à fourche. / ~ end of the rack of a jack ‖ Gabel f. der Zahnstange einer Winde ‖ cornes fpl. de la crémaillère d'un cric. / ~ hoe ‖ Gabelhaue f.; zweizinkige Hacke f. ‖ bigot m.; houe f. fourchure. / ~ lightning ‖ Linienblitz m. ‖ éclair m. linéaire. / ~ projection ‖ gegabelte Nase f. ‖ appendice m. en forme de fourche. / ~ stanchion ‖ Gabelstütze f. ‖ support m. à fourche. / ~ terminal ‖ gabelförmiger Kabelschuh m. ‖ cosse f. à fourche pour câbles.

fork guide ‖ Gabelführung f. ‖ guidage m. à fourche. / ~ head ‖ Gabelkopf m. ‖ tête f. de fourche. / ~ joint ‖ Gabelgelenk n. ‖ chape f. / ~ lever ‖ Gabelhebel m. ‖ levier m. à fourche. / ~ link (Mach) ‖ Gabel f. ‖ enfourchement m.

forkman (Roll mill) ‖ Hebelmann m. ‖ fourcheur m.

fork rod ‖ Gabelstange f. ‖ tige f. à fourchette. / ~-shaped ‖ gabelförmig ‖ en forme de fourche. / ~-type tedder ‖ Gabelheuwender m. ‖ faneuse f. à fourches.

fork wrench ‖ Gabelschlüssel m. ‖ clef f. à fourche.

form, to ‖ formen; gestalten ‖ former; façonner; modeler; bosseler. / ~ (Chem) ‖ entstehen ‖ se former. / ~ an angle (Locksm) see ~ a knee. / ~ a bridge ‖ eine Brücke bauen oder schlagen ‖ construire ou établir ou jeter un pont. / ~ clinkers (Coal) ‖ backen ‖ s'agglutiner. / ~ an equation (Math) ‖ eine Gleichung f. ansetzen ‖ mettre les quantités fpl. en équation. / ~ a knee ‖ kröpfen ‖ couder; épauler. / ~ loops pl. ‖ Schlingen fpl. bilden ‖ former des serpentins mpl.

form ‖ Form f.; Gestalt f. ‖ forme f. / ~ (Mach) ‖ Schablone f. ‖ échantillon m.; calibre m. / ~ (Print) ‖ Druckform f. ‖ forme f. / ~ of application ‖ Antragsformular n. ‖ demande f.; proposition f. / ~ of arch ‖ Bogenform f. ‖ forme f. de l'arc. / to change the ~ ‖ die Gestalt f. ändern ‖ modifier la figure ou la forme.

/ ~ coefficient of ~ ‖ Völligkeitszahl f. ‖ coefficient m. de finesse; rapport m. de renflement. / concave ~ ‖ ausgebauchte Form f. ‖ forme f. concave. / ~ of crystal ‖ Kristallform f. ‖ forme f. cristalline. / crystalline ~ *see* ~ of crystal. / curved ~ ‖ geschweifte Form f. ‖ forme f. cintrée. / dished ~ *see* concave ~. / to distribute the letters of a ~ (Print) ‖ eine Form ablegen ‖ rompre une forme; distribuer les lettres. / first ~ (Print) ‖ Prime f.; Schöndruck m. ‖ prime f.; forme f. première. / ~ of fracture ‖ Bruchform f. ‖ forme f. de la fracture. / to gauge a ~ (Print) ‖ Format n. machen ‖ mettre la garniture. / inner ~ (Print) ‖ zweite Form f.; Widerdruck m. ‖ retiration f.; seconde f.; verso m. / to lay on the ~ (Print) ‖ die Form einheben ‖ mettre la forme sous presse. / to lift out the ~ (Print) ‖ die Form ausheben ‖ relever la forme. / ~ of load ‖ Belastungsweise f. ‖ mode m. de chargement. / long ~ ‖ längliche Form f. ‖ forme f. oblongue. / ~ of a pavement ‖ Planum n. einer zu pflasternden Straße ‖ aire f. d'un pavé; plate-forme f. / perfecting ~ ‖ Widerdruckform f. ‖ forme f. d'impression au verso. / ~ of rails ‖ Schienenform f.; Schienenprofil n. ‖ section f. du rail. / ~ in rolls ‖ Formular n. in Rollenform ‖ formulaire m. en forme de rouleaux. / ~ of roofing tile ‖ Dachziegelform f. ‖ forme f. de la tuile. / ~ of the stairs ‖ Treppenform f. ‖ forme f. de l'escalier. / stamping ~ ‖ Stampfform f. ‖ moule m. ou caisse f. de pilonnage. / tightly stretched ~ of balloon ‖ feste *oder* pralle Form f. des Ballons ‖ form f. invariable du ballon. / to turn-up a ~ (Print) ‖ ein Format n. umschlagen *oder* umstülpen ‖ renverser une forme. / to unlock a ~ (Print) ‖ ein Format n. aufschlagen *oder* abschlagen ‖ dégager les garnitures fpl.; desserrer une forme. / to untie a ~ *see* to unlock a ~.

formaldehyde ‖ Formaldehyd n. ‖ aldéhyde m. formique; formaldéhyde m.

formaline ‖ Formalin n. ‖ formaline f.

formate ‖ ameisensaures Salz n. ‖ formiate m.

formation (Acc) ‖ Formierung f. ‖ formation f. / ~ (Geol) ‖ Formation f.; Schichtenbildung f. ‖ formation f.; terrain m. / ~ of bubbles (Acc) ‖ Blasenbildung f. ‖ formation f. de bulles. / carboniferous ~ ‖ Kohlenformation f. ‖ formation f. houillère. / ~ of dust ‖ Staubentwicklung f. ‖ développement m. de poussière. / ~ of eddies ‖ Wirbelbildung f. ‖ formation f. de tourbillons. / to avoid the ~ of folds ‖ Faltenbildung f. vermeiden ‖ éviter le plissage. / ~ of gas ‖ Gasbildung f.; Gasentwicklung f. ‖ formation f. ou développement m. de gaz. / geological ~ ‖ geologische Formation f. ‖ formation f. géologique. / ~ of humus ‖ Bildung f. von Ackererde ‖ formation f. d'humus. / ~ of hydrogen ‖ Wasserstoffbildung f. ‖ dégagement m. d'hydrogène. / ~ of land ‖ Terrainverhältnisse npl. ‖ formation f. du terrain. / ~ of mould *see* ~ of humus. / ~ of oxygen ‖ Sauerstoffentwicklung f. ‖ formation f. d'oxygène. / ~ of oxy-hydrogen gas ‖ Knallgasentwicklung f. ‖ formation f. de gaz détonant. / ~ of pipes (Blast furnace) ‖ Trichterbildung f. ‖ formation f. d'entonnoir. / primitive ~ ‖ Urgebirge n.; Urfels m. ‖ roche f. primitive. / ~ of rock ‖ Gesteinsbildung f. ‖ for-

mation f. de roche. / ~ of rust ‖ Rostbildung f. ‖ formation f. de rouille. / ~ of scum ‖ Schaumbildung f. ‖ formation f. de mousse. / ~ of smoke ‖ Rauchbildung f. ‖ formation f. de fumée. / ~ of soot ‖ Rußbildung f. ‖ formation f. de suie. / ~ spark ~ ‖ Funkenbildung f. ‖ formation f. d'étincelles. / ~ of steam ‖ Dampfbildung f.; Dampfentwicklung f. ‖ formation f. de vapeur. / ~ of vortices *see* ~ of eddies.

formation flying ‖ Geschwaderflug m. ‖ vol m. en formation *ou* d'escadre. / ~ level ‖ Kronlinie f. der Erdarbeiten ‖ niveau m. des remblais. / ~ line ‖ Planumsohle f. ‖ niveau m. de l'assiette.

form cutter (Wood engr) ‖ Formschneider m. ‖ graveur m. en bois. / ~ cutting (Print) ‖ Schriftschneidekunst f. ‖ gravure f. de caractères. / ~ cylinder (Print) ‖ Formzylinder m. ‖ cylindre m. gravé.

formed in the casting ‖ in Stahl m. gegossen ‖ moulé en acier m. / ~ cutter ‖ Profilfräser m. ‖ fraise f. profilé *ou* de forme.

former (Bookb) ‖ Deckelbeschneider m. ‖ rabaisseur m. ~ (Join; Carp) ‖ Schroteisen n.; Stechbeitel m. ‖ ciseau m. à planche; fermoir m. / ~ (Pott) ‖ Former m.; Dreher m. ‖ mouleur m. / ~ roller ‖ Führungsrolle f.; Leitrolle f. ‖ roulette f. à gabarit; galet m. de guidage. / ~ 's tool ‖ Formerwerkzeug n. ‖ outils mpl. pour mouleurs. / ~ winding ‖ Schablonenwicklung f. ‖ enroulement m. sur gabarit.

formiate ‖ ameisensaures Salz n. ‖ formiate m.

formic acid ‖ Ameisensäure f. ‖ acide m. formique. / ~ aldehyde ‖ Formaldehyd n. ‖ aldéhyde m. formique. / ~ ether ‖ Ameisenäther m. ‖ éther m. formique.

forming (Acc) ‖ Formation f. ‖ formation f. / ~ (Porcel) ‖ Formerei f.; Formen n. ‖ moulage m. / ~ of an accumulator ‖ Formieren n. des Akkumulators ‖ formation f. de l'accumulateur. / ~ collars on wheel spokes ‖ Anstauchen n. der Bunde an Radspeichen ‖ refoulement m. des embases aux rayons de roues. / ~ a knee (Mach) ‖ gekröpft ‖ coudé. / ~ the knots ‖ Knotenbildung f.; Knotenherstellung f. ‖ formation f. de nœud. / ~ in plaster moulds ‖ Gipsformerei f. ‖ moulage m. à la housse. / ~ with sheets (Porc; Pott) ‖ Schwartenformerei f. ‖ moulage m. à la croûte.

forming acid (Acc) ‖ Formiersäure f. ‖ acide m. de formation. / ~ battery ‖ Formierbatterie f. ‖ batterie f. de formation. / ~ cam ‖ Formkurve f. ‖ came f. de façonnage. / ~ current (Acc) ‖ Formationsstrom m. ‖ courant m. de formation. / ~ die (Staves) ‖ Aufsatzform f. ‖ forme f. d'assemblage. / ~ machine (Spinn) ‖ Austreibemaschine f.; Ausziehwagen m. ‖ chariot m.; coureuse f. / sliver ~ machine (Spinn) ‖ Austückelmaschine f. ‖ machine f. pour former une mèche continue. / ~ pliers pl. ‖ Drückzange f. ‖ pince f. à emboutir. / ~ shop (Acc) ‖ Formationsraum m. ‖ salle f. de formation. / ~ tank (Acc) ‖ Formiergefäß n. ‖ récipient m. de formation.

form lock (Print) ‖ Formenhalter m. ‖ serre-forme m. / ~ piece (Letter found) ‖ Formstück n. ‖ platine f. / central heating ~ piece ‖ Zentralheizungsformstück n. ‖ pièce f. d'ajustage pour le chauffage central. / ~ powder ‖ Formpuder m. ‖

poudre f. de moulage. / ~ setter (Print) ‖ Formenausschießer m. ‖ metteur m. en forme.

formula ‖ Formel f. ‖ formule f. / empirical ~ ‖ empirische Formel f. ‖ formule f. empirique. / horse power ~ ‖ Leistungsformel f. ‖ formule f. de puissance. / horse power ~ for purposes of taxation ‖ Steuerformel f. ‖ formule f. de puissance pour l'impôt. / mathematical ~ ‖ mathematische Formel f. ‖ formule f. mathématique. / ~ of mathematics *see* mathematical ~. / physical and chemical ~ ‖ physikalisch-chemische Formel f. ‖ formule f. physique et chimique. / ~ for raising force ‖ Hubkraftformel f. ‖ formule f. de poussée. / ~ relating to rigid system ‖ Rahmenformel f. ‖ formule f. relative aux systèmes rigides. / ~ for transformation ‖ Umformungsformel f. ‖ formule f. de transformation.

formulary ‖ Formular n.; Vordruck m. ‖ formulaire f.

forril ‖ Pergament n. ‖ parchemin m.

forsterite ‖ Forsterit m.; Magnesiumolivin m. ‖ forstérite f.

fortified ‖ fest; befestigt ‖ fort; fortifié.

fortify, to ‖ befestigen ‖ fortifier.

fortress radio station ‖ Festungsfunkenstation f. ‖ station f. radiotélégraphique de place. / ~ telegraph ‖ Festungstelegraf m. ‖ télégraphe m. de forteresse.

forward, to ‖ expedieren; befördern ‖ expédier. / ~ a thing to a person ‖ jemandem etwas zustellen ‖ remettre *ou* faire tenir quelque chose à quelqu'un.

forwarder (Bookb) ‖ Rundpresser m.; Rundschläger m. ‖ arrondisseur m.

forward eccentric ‖ Vorwärtsexzenter n. ‖ excentrique m. de marche avant. / ~ guide surface ‖ Vorwärtsgleitbahn f. ‖ coulisse f. pour marche avant.

forwarding, in the absence of special instructions for ~ ‖ falls keine besonderen Versandvorschriften fpl. vorliegen ‖ à défaut d'indications spéciales d'expédition. / ~ of goods ‖ Güterbeförderung f. ‖ transport m. des marchandises. / ~ by railway ‖ Eisenbahntransport m. ‖ transport m. par chemin de fer.

forwarding agent ‖ Spediteur m. ‖ expéditeur m. / ~ 's motor car ‖ Speditionskraftwagen m. ‖ automobile d'expéditeur.

forwarding bill ‖ Speditionszettel m. ‖ bulletin m. d'expédition. / ~ business ‖ Transportgeschäft n. ‖ affaire f. *ou* commerce m. de transports. / ~ charge ‖ Speditionsgebühr f. ‖ frais mpl. d'expédition.

forward motion ‖ Vorwärtsbewegung f. ‖ mouvement m. en avant. / ~ piston stroke ‖ Kolbenhingang m. ‖ aller m. du piston.

forwards, going ~ ‖ Vorwärtsgang m. ‖ marche f. avant.

forward speed ‖ Vorwärtsgang m. ‖ vitesse f. avant. / ~ standard (Mach) ‖ Vorwärtsständer ‖ montant m. antérieur *ou* avant.

fossil ‖ Versteinerung f. ‖ fossile m. / ~ flour ‖ fossiles Mehl n. ‖ farine f. fossile. / ~ material ‖ fossiler Rohstoff m. ‖ matière f. fossile.

fossil meal ‖ Kieselgur f.; Infusorienerde f. ‖ kieselguhr m.; farine f. fossile; terre f. d'infusoires. / ~ brick ‖ Kieselgurstein ‖ brique f. en farine fossile. / ~ covering ‖ Isolierung f. aus Kieselgur ‖ revêtement m. en farine fossile.

fossil remains pl. ‖ Abdruck m. einer Versteinerung ‖ empreinte f. d'un fossil.

fother, to ~ a leak ‖ ein Leck n. stopfen *oder* futtern ‖ aveugler *ou* boucher une voie d'eau.

foul, to ~ water ‖ das Wasser n. verunreinigen ‖ salir l'eau f.

foul ‖ faul; verdorben; widerwärtig ‖ vicié; toxique. / **~ air** ‖ verdorbene Luft f. ‖ air m. vicié. / **~ air** (Mine) ‖ gebrauchte Wetter npl. ‖ air m. vicié.

foulard ‖ bunt bedruckter Taft m.; Foulard m. ‖ foulard m.

foul impression ‖ Fehldruck m. ‖ feuille f. mal venue *ou* mal imprimée. / **~ linen** ‖ schmutzige Wäsche f. ‖ linge m. sale. / **~ page** ‖ fehlerhafte Druckseite f. ‖ page f. mal venue.

found, to ~ ‖ gießen ‖ couler; fondre; jeter en fonte *ou* en moule; verser. / **~ ... into** ‖ mit ... ausgießen ‖ sceller de ... / **~ a wall** ‖ eine Mauer f. gründen ‖ établir *ou* fonder *ou* rendre stable un mur.

foundation (Build) ‖ Fundament n.; Unterbau m.; Gründung f.; Fundierung f. ‖ fondation f. / **~** (Trade) ‖ Gründung f.; fondation f.; établissement m.; création f. / the **~** of the anvil-block is excessively heavy and very steady in return ‖ Unterbau m. für den Amboß ist sehr schwer und erschütterungsfrei ‖ le massif de fondation pour l'enclume est excessivement pesant mais à l'abri de toute trépidation. / **~** on bearing piles ‖ Pfahlgründung f. ‖ parc m. *ou* radier m. de pilotis. / the **~** is blown up by water *see* the **~** is undermined. / boiler **~** ‖ Kesselfundament n. ‖ fondation f. de chaudière. / brick **~** ‖ Ziegelwerkfundament n. ‖ fondation f. en briques. / caisson **~** ‖ Senkkastengründung f. ‖ fondation f. par caisson. / to carry down the **~** to natural soil ‖ das Fundament bis auf gewachsenen Boden führen ‖ faire aller la fondation jusqu'au terrain naturel. / cement **~** ‖ Zementfundament n. ‖ fondation f. en ciment. / **~** of a company effected jointly with somebody ‖ Gründung f. einer Gesellschaft, die gemeinschaftlich mit jemandem erfolgte ‖ fondation f. d'une société qui se fit en commun avec quelqu'un. / compressed air **~** ‖ pneumatische Gründung f. ‖ fondation f. à l'air comprimé. / concrete **~** ‖ Betongründung f. ‖ fondation f. par immersion de béton. / **~** on concrete between sheet pilings ‖ Gründung f. auf Beton zwischen Spundwänden ‖ fondation f. sur béton entre palplanches. / condenser **~** ‖ Kondensatorfundament n. ‖ fondation f. du condenseur. / **~** between crib coffer-dams ‖ Gründung f. mit Fangdämmen ‖ fondation f. par l'emploi de bâtardeaux. / direct **~** ‖ direkte Gründung f. ‖ fondation f. sur le sol. / **~** by means of diving bells ‖ Taucherglockengründung f. ‖ fondation f. par cloche de plongeur. / **~** of a firm ‖ Geschäftsgründung f. ‖ fondation f. d'une maison. / **~** by means of freezing ‖ Gefriergründung f. ‖ fondation f. par congélation. / **~** on a grillage ‖ Pfahlrostgründung f. ‖ fondation f. sur pieux *ou* sur un radier de pilotis. / iron tubular **~** ‖ Röhrengründung f. ‖ fondation f. par cuvelage. / light **~** ‖ leichtes Fundament n. ‖ fondations fpl. légères. / masonry **~** ‖ gemauertes Fundament n. ‖ fondation f. en maçonnerie. / **~** on piles ‖

Pfahlgründung f. ‖ fondation f. sur pieux. / **~** by pit sinking ‖ Gründung f. mittels Schachtabteufung ‖ fondation f. par fonçage de puits. / pneumatic **~** ‖ Preßluftgründung f. ‖ fondation f. à l'air comprimé. / random stone **~** ‖ Gründung f. auf Steinschüttung ‖ fondation f. à pierres perdues *ou* par enrochements. / **~** of rock ‖ Felsengrund m. ‖ fondement m. de roc. / **~** between sheet piles ‖ Gründung f. zwischen Spundwänden ‖ fondation f. au moyen (de cours) de palplanches. / a special **~** is not needed ‖ ein besonderer Unterbau m. ist nicht erforderlich ‖ il n'est pas besoin de construire une fondation spéciale. / **~** of a stone pitching ‖ Grundböschung f. einer Uferpflasterung ‖ fondation f. d'une défence de rive. / **~** on sunken stone coffins ‖ Senkkastengründung f. ‖ encaissement m. *ou* fondation f. sur des coffres remplis de pierres. / sunk shaft **~** ‖ Senkröhrengründung f. ‖ fondation f. par tubage. / sunk well **~** ‖ Senkbrunnengründung f. ‖ fondation f. sur des puits foncés. / timber-and iron-cased concrete **~** ‖ Mantelgründung f. ‖ fondation f. par encaissement. / **~** by means of a timber crib without a bottom ‖ Senkkastengründung f. ‖ fondation f. par caissons. / **~** on timber platform ‖ Rostgründung f. ‖ fondation f. par grillages *ou* par patins de charpente. / **~** of a trench filled with sand ‖ Sandschüttung f. ‖ fondation f. sur une couche de sable. / turbine **~** ‖ Turbinenfundament n. ‖ fondation f. de la turbine. / the **~** is undermined by water ‖ der Grund m. ist unterspült *oder* unterwaschen ‖ le fondement m. est déchaussé *ou* affouillé par l'eau.

foundation block ‖ Fundamentklotz m. ‖ plot m. *ou* bloc m. de fondation. / **~ bolt** ‖ Fundamentanker m.; Fundamentbolzen m. ‖ ancre f. *ou* boulon m. de fondation. / **~ bolt hole** ‖ Ankerloch n. ‖ trou m. d'ancre. / **~ brickwork** ‖ Grundmauerwerk n. ‖ maçonnerie f. de fondation. / **~ cellar** ‖ Fundamentkeller m. ‖ cave f. de fondation. / **depth of ~** ‖ Fundamenttiefe f. ‖ profondeur f. de fondation. / **~ ditch** (Build) ‖ Schachtgrube f. ‖ fondement m. / **~ frame** (Mach) ‖ Grundrahmen m. ‖ châssis m. *ou* cadre m. de fondation. / **~ insulation** ‖ Fundamentisolierung f. ‖ isolement m. de la fondation. / **method of laying ~s** ‖ Gründungsverfahren n. ‖ méthode f. de fondation. / **~ masonry** ‖ Grundmauerwerk n. ‖ maçonnerie f. de fondement. / **~ pile** (Bridge) ‖ Grundpfeiler m. ‖ pilier m. de fondation. / **~ pile** (Build) ‖ Rostpfahl m. ‖ pilot m. de support; pilotis m. de grillage. / **~ piling** ‖ Grundpfählung f. ‖ palée-basse f. / **~ pillar** ‖ Grundpfeiler m. ‖ pilier m. de fondation. / **~ pit** ‖ Baugrube f. ‖ fouille f. de fondation. / **~ plate** (Roll) ‖ Grundplatte f. ‖ plaque f. de fondation. / **~ ring of firebox** ‖ Feuerbuchsring m. ‖ bague f. de boîte à feu. / **~ screw** ‖ Fundamentschraube f. ‖ boulon m. de fondation. / **~ sill** ‖ Grundschwelle f. ‖ longuerine f. de fondation. / **reinforced concrete ~ slab** ‖ Fundamentplatte f. aus Eisenbeton ‖ semelle f. de fondation en béton armé. / **~ slide rail** ‖ Fundamentschiene f. ‖ glissière f. de fondation. / **~ soil** (ground) ‖ Baugrund m.

terrain m. à bâtir; sol m. de fondation. / **~ stone** ‖ Fundamentstein m.; Grundstein m. ‖ pierre f. de fondement *ou* fondamentale. / **~ structures** pl. and equipments pl. ‖ Gründungen fpl. und Gründungsausrüstungen fpl. ‖ fondations fpl. sous eau et matériel pour ces ouvrages. / **~ tester** ‖ Baugrundprüfer m. ‖ vérificateur m. du sol de fondation. / **~ trench** ‖ Fundamentgraben m. ‖ tranchée f. de fondation. / **~ wall** ‖ Grundmauer f. ‖ mur m. de fondation. / **~ walling** ‖ Grundmauerwerk n. ‖ maçonnerie f. de fondement.

founder, to ~ (Ship) ‖ sinken; untergehen ‖ couler; couler bas; sancir; sombrer.

founder ‖ Gründer m. ‖ fondateur m. / **~** (Caster) ‖ Gießer m. ‖ fondeur m. / iron **~** ‖ Eisengießer m. ‖ ouvrier m. de fonderie. / letter **~** ‖ Schriftgießer m. ‖ fondeur m. de caractères *ou* de lettres. / metal **~** ‖ Metallgießer m. ‖ fondeur m. en métaux. / type **~** *see* letter **~**. / yellow-metal **~** ‖ Gelbgießer m. ‖ fondeur m. de bronze *ou* en cuivre.

founder's black ‖ Schlichte f. ‖ noir m. pour fonderie. / **~ cataract** ‖ Gießerstar m. ‖ cataracte f. des fondeurs. / **~ shaft** ‖ Fundgrube f. ‖ mine f. de découverte. / **~ share** ‖ Gründeraktie f. ‖ action f. de fondateur. / **~ truck** ‖ Schlitten m.; Schleppe f. ‖ chariot m. de transport.

foundery *see* foundry.

founding ‖ Guß m.; Abguß m. ‖ fonte f.; moulage m.; jet m. en moule. / **~** of metals ‖ Guß m.; Gießen n. der Metalle ‖ coulage m. *ou* fonte f. des métaux. / **~** of rollers ‖ Walzenguß m. ‖ fonte f. de rouleaux. / **~ furnace** ‖ Gießofen m. ‖ fourneau m. de fonderie.

foundry ‖ Gießerei f. ‖ fonderie f. / **art ~** ‖ Kunstgießerei f. ‖ fonderie f. d'art. / **artistic ~** ‖ Kunstgießerei f. ‖ fonderie f. artistique. / **bell ~** ‖ Glockengießerei f. ‖ fonderie f. de cloches. / **brass ~** ‖ Gelbgießerei f.; Messinggießerei f. ‖ fonderie f. de cuivre jaune *ou* de laiton. / **brass ~** (Bronze) ‖ Bronzegießerei f. ‖ fonderie f. de bronze. / **bronze statue ~** ‖ Bildgießerei f. ‖ fonderie f. de bronzes d'art. / **copper ~** ‖ Kupfergießerei f. ‖ fonderie f. de cuivre. / **crucible steel ~** ‖ Schmelzbau m. für Tiegelguß ‖ fonderie f. d'acier au creuset. / **iron ~** ‖ Eisengießerei f. ‖ fonderie f. de fer. / **lead ~** ‖ Bleigießerei f. ‖ fonderie f. de plomb. / **letter ~** ‖ Schriftgießerei f. ‖ fonderie f. de caractères *ou* de lettres. / **red copper ~** ‖ Rotgießerei f. ‖ fonderie f. de cuivre rouge. / **special ~** ‖ Sondergießerei f. ‖ fonderie f. spéciale. / **stereotype ~** ‖ Stereotypengießerei f. ‖ fonderie f. de stéréotypes. / **tin ~** ‖ Zinngießerei f. ‖ fonderie f. d'étain. / **type ~** ‖ Schriftgießerei f. ‖ fonderie f. typographique *ou* de caractères *ou* de letters. / **zinc ~** ‖ Zinkgießerei f. ‖ fonderie f. de zinc.

foundry blower ‖ Gießereigebläse n. ‖ ventilateur m. de fonderie. / **~ car** ‖ Gieß(erei)wagen m. ‖ chariot m. de coulée. / **~ coke** ‖ Gießereikoks m. ‖ coke m. de fonderie. / **~ crane** ‖ Gießkran m. ‖ grue f. pour la coulée *ou* de fonderie. / **~ equipment** ‖ Gießereieinrichtung f. ‖ accessoires mpl. d'une fonderie. / **~ foreman** ‖ Gießmeister m. ‖ chef m. de fosse *ou* de fonderie. / **~ goods** pl. ‖ Gußwaren fpl. ‖ fers mpl. coulés; fonte f. moulée;

moulées fpl.; ouvrages mpl. en fonte. / ~ implements pl. ‖ Gießereibedarfsartikel mpl. ‖ articles mpl. courants pour fonderies. / ~ iron ‖ Gießereiroheisen n. ‖ fonte f. de moulage. / ~ ladle ‖ Gießpfanne f. ‖ poche f. de coulée. / ~ machine ‖ Gießereimaschine f. ‖ machine f. pour fonderie.

foundryman ‖ Gießer m. ‖ fondeur m.; ouvrier m. fondeur.

foundry material ‖ Gießereibedarf m. ‖ matériel m. pour fonderies. / owner of a ~ ‖ Gießereibesitzer m. ‖ propriétaire m. d'une fonderie. / ~ pattern ‖ Gußmodell n. ‖ modèle m. de fonderie. / ~ pattern maker ‖ Gießereimodelltischler m. ‖ menuisirr-modeleur m. de fonderie. / ~ pig see ~ pig iron. / ~ pig iron ‖ Gießereiroheisen n. ‖ fonte f. crue de fonderie ou de moulage. / ~ pit ‖ Dammgrube f. ‖ fosse f. de coulée. / ~ plant ‖ Gießereianlage f. ‖ installation f. fonderie. / ~ reverberatory furnace ‖ Gießereiflammofen m. ‖ four m. à réverbère de fonderie. / ~ sand ‖ Gießereisand m. ‖ sable m. de fonderie ou de moulage. / ~ travelling crane ‖ Gießereilaufkran m. ‖ pont m. roulant de fonderie.

fount (Letter found) ‖ Guß m. ‖ fonte f.

fountain ‖ Fontäne f.; Springbrunnen m. ‖ fontaine f. / ~ maker ‖ Brunnenbauer m. ‖ constructeur m. de fontaines.

fountain pen ‖ Füllfederhalter m. ‖ plume f. fontaine; porte-plume m. réservoir; stylo(graphe) m. / cap of the ~ ‖ Füllfederhalterkappe f. ‖ chapeau m. du porte-plume réservoir ou du stylographe. / ~ ink ‖ Füllfedertinte f. ‖ encre f. pour plume-réservoir. / safety type ~ ‖ Sicherheitsfüllfederhalter m. ‖ stylographe m. ou porte-plume m. reservoir de sûreté.

fountain water ‖ Quellwasser n. ‖ eau f. de source.

fount case (Print) ‖ Leistenkasten m.; Vorratskasten m. ‖ caseau m.; bardeau m.

four, in ~ columns pl. (Print) ‖ vierspaltig ‖ à quatre colonnes fpl.

four-colour printing ‖ Vierfarbendruck m. ‖ impression f. en quatre couleurs.

four-column hydraulic press for building transformer cores ‖ Viersäulenpresse f. für den Transformatorenbau ‖ presse f. à quatre colonnes pour serrage des tôles de transformateurs.

four-cusped ‖ vierlappig ‖ quadrilobé.

four-cycle engine ‖ Viertaktmaschine f. ‖ machine f. à quatre temps.

four-cylinder motor ‖ Vierzylindermotor m. ‖ moteur m. à quatre cylindres.

four-electrode tube ‖ Vierelektrodenröhre f. ‖ lampe f. à quatre électrodes.

four-engined ‖ viermotorig ‖ quadrimoteur.

Fourier series pl. ‖ Fouriersche Reihe f. ‖ série f. de Fourier.

fourpole ‖ Vierpol m. ‖ quadripôle m.

four-roller plate bending machine ‖ Vierwalzenblechbiegemaschine f. ‖ machine f. à cintrer les tôles à quatre cylindres.

four-rowed barley ‖ vierzeilige Gerste f. ‖ orge f. à quatre rangs.

four-seater ‖ Viersitzer m. ‖ voiture f. à quatre sièges.

four-shoe brake ‖ Vierklotzbremse f. ‖ frein m. à quatre sabots.

four-sided ‖ vierseitig ‖ quadrilatéral; quadrilatère; à quatre pans mpl.

four-stroke cycle ‖ Viertakt m. ‖ cycle m. à quatre temps. / ~ engine ‖ Viertaktmo-

tor m. ‖ moteur m. à quatre temps. / ~ motor ‖ Viertaktmotor m. ‖ moteur m. à quatre temps.

four-stroke motor see four-stroke cycle motor.

four-way cock ‖ Vierwegehahn m. ‖ robinet m. à quatre voies.

four-wheel brake ‖ Vierradbremse f. ‖ frein m. sur quatre roues. / ~ vehicle ‖ Vierradfahrzeug n. ‖ véhicule m. à quatre roues.

four-wire circuit (Tel) ‖ Vierdrahtschaltung f. ‖ système à quatre fils. / ~ intermediate repeater ‖ Vierdrahtzwischenverstärker m. ‖ amplificateur m. intermédiaire à quatre fils. / ~ repeater ‖ Vierdrahtverstärker m. ‖ répéteur m. à quatre fils. / ~ termination ‖ Vierdrahtgabelschaltung f. ‖ termineur m. d'un circuit à quatre fils.

fowlerite ‖ Fowlerit m. ‖ fowlérite f.

fowling piece ‖ Jagdflinte f.; Jagdgewehr n. ‖ fusil m. de chasse.

fowl-yard ‖ Hühnerhof m.; Geflügelhof m. ‖ basse-cour f.

foxed (Pap) ‖ moderfleckig ‖ piqué.

foxing (Brew) ‖ Säuern n. der Würze ‖ acidification f. du moût de bière.

fox saw ‖ Fuchsschwanzsäge f.; Fuchsschwanz m. ‖ scie f. à main ou à manche.

foxtail-wedging ‖ Verkeilen n. eines Zapfens ‖ assemblage m. à contre-clavette.

foxy ‖ fuchsig ‖ fauve.

fraction ‖ Bruch m.; Bruchstück n. ‖ fraction f. / (Math) ‖ Bruch m. ‖ fraction f.; nombre m. rompu. / calculation involving ~s ‖ Bruchrechnung f. ‖ calcul m. des fractions. / compound ~ ‖ zusammengesetzter Bruch m. ‖ fraction f. de fraction. / continued ~ ‖ Kettenbruch m. ‖ fraction f. continue. / improper ~ ‖ unechter Bruch m. ‖ expression f. fractionnaire. / irrational ~ ‖ irrationaler Bruch m. ‖ fraction f. irrationnelle. / partial ~ ‖ Teilbruch m. ‖ fraction f. partielle. / proper ~ ‖ echter Bruch m. ‖ fraction f. proprement dite. / rational ~ ‖ rationaler Bruch m. ‖ fraction f. rationnelle. / ~ of a revolution ‖ Bruchteil m. einer Umdrehung ‖ fraction f. d'un tour. / simple ~ (Math) ‖ einfacher oder gemeiner Bruch m. ‖ fraction f. simple ou ordinaire. / vulgar ~ see simple ~.

fractional ‖ fraktioniert ‖ fractionné. / ~ currency ‖ Scheidemünze f. ‖ monnaie f. d'appoint; billon m.; petite monnaie f. / ~ distillation ‖ teilweise Destillation ‖ distillation f. fractionnée. / ~ pitch ‖ Teilschritt m. ‖ pas m. partiel.

fractionate, to ‖ fraktionieren ‖ fractionner.

fractionating column ‖ Destillieraufsatz m. ‖ colonne f. à distiller.

fractionation ‖ Fraktionierung f. ‖ fractionnement m.

fraction stroke ‖ Bruchstrich m.; Schrägstrich m. ‖ barre f. de fraction.

fracture, to ‖ brechen ‖ fracturer; casser; briser; rompre.

fracture ‖ Bruch m. ‖ fracture f.; rupture f.; cassure f. / the axle was distorted without showing any incipient ~ ‖ die Achse wurde verdreht, ohne einen Anbruch zu zeigen ‖ l'arbre m. fut tordu sans présenter une amorce de fissure. / ~ with coarse grains ‖ grobkörniger Bruch m. ‖ cassure f. à gros grains. / conchoidal ~ ‖ muscheliger Bruch m. ‖ cassure f. conchoïde. / crystalline ~ ‖

krystallinischer Bruch m. ‖ cassure f. crystalline. / ~ with fine grains ‖ feinkörniger Bruch m. ‖ cassure f. à grains fins. / the ~ showed a fine-grained texture ‖ die Bruchstelle f. wies ein feinkörniges Gefüge auf ‖ la cassure présentait une structure finement granulée. / granular ~ of cast-iron ‖ körniger Bruch m. des Gußeisens ‖ cassure f. grenue ou à grains de la fonte. / hackly ~ ‖ zackiger Bruch m. ‖ cassure f. crochue. / incipient ~ ‖ Anbruch m. ‖ amorce f. de fissure. / longitudinal ~ ‖ Längsbruch m. ‖ cassure f. longitudinale. / the ~s pl. occured in the pin ‖ die Brüche mpl. traten im Zapfen auf ‖ les ruptures fpl. se manifestaient dans le tourillon. / ~ of the skull ‖ Schädelbruch m. ‖ fracture f. du crâne. / splintery ~ ‖ splitteriger Bruch m. ‖ cassure f. à éclats. / ~ of spring ‖ Federbruch m. ‖ rupture f. ou bris m. de ressort. / tough and fibrous ~ ‖ sehniger und zäher Anbruch m. ‖ cassure f. nerveuse et tenace. / the ~ presented a very tough texture ‖ die Bruchfläche zeigte ein sehr zähes Gefüge ‖ la cassure montrait une texture très tenace.

fracture test ‖ Bruchprobe f. ‖ épreuve f. de cassure.

fracturing, safety against cracking and ~ ‖ Sicherheit f. gegen Bildung von Rissen und Brüchen ‖ sécurité f. contre les crevasses et les ruptures.

fragile (Mine) ‖ gebrech; bröcklig ‖ cassant; fragile. / ~ (Packing up) ‖ zerbrechlich ‖ fragile. / ~ (Wood) ‖ morsch ‖ pourri; vermoulu.

fragmentation bomb ‖ Splitterbombe f. ‖ bombe f. à fragmentation.

fragments pl. ‖ Trümmer pl. ‖ débris mpl. / ~ of old caoutchouc goods ‖ Bruchstücke npl. von alten Kautschukwaren ‖ débris mpl. de vieux objets de caoutchouc.

frame, to (Ornament) ‖ einfassen ‖ entourer d'ornements mpl. / ~ (Picture) ‖ einrahmen ‖ encadrer; emborurer; enchâsser dans une bordure. / ~ poles together ‖ Telegraphenstangen fpl. verkuppeln ‖ coupler des poteaux mpl.

frame (Arch) ‖ Leiste f. ‖ tringle f. / ~ (Build) ‖ Bock m.; Gerüst n.; Gestell n. ‖ chevalet m.; tréteau m. / ~ (Hide) ‖ Einspannrahmen m. ‖ herse f. / ~ (Loc) ‖ Rahmen m. ‖ longeron m. / ~ (Loom) ‖ Stuhlgestell n. ‖ bâti m. / ~ (Mach) ‖ Gestell n. ‖ bâti m. / ~ (Mot) ‖ Gehäuse n. ‖ carcasse f. / ~ (Piano) ‖ Rahmen m. ‖ barrage m.; châssis m. / ~ (Sawm) ‖ Gatter n.; Sägegatter n. ‖ châssis m. de scie; châssis porte-lames; porte-scie m. / ~ (Shipb) ‖ Spant n. ‖ couple m. / ~ (Vehicle) ‖ Rahmen m. ‖ châssis m. / after ~ (Shipb) ‖ Achterspant n. ‖ couple m. de l'arrière. / angular ~ of the wagon ‖ eckiger Rahmen m. des Wagens ‖ châssis m. rectangulaire du wagon. / automobile ~ ‖ Autorahmen m.; Autogestell n. ‖ châssis m. d'automobile. / ~ bent of one piece ‖ aus einem Stück gebogener Rahmen m. ‖ cadre m. courbé d'une seule pièce. / ~ carried on rollers (Typewr) ‖ auf Rollen laufender Rahmen m. ‖ cadre m. marchant sur galets. / to carry on solid ~s ‖ in fester Stuhlung f. lagern ‖ monter sur un bâti fixe. / chief ~ (Shipb) ‖ Richtspant n. ‖ couple

m. de levée. / ~ of cold-rolled steel ‖ Rahmen m. aus kalt gewalztem Stahl ‖ bâti m. en acier laminé à froid. / ~ of convenient working height ‖ gebrauchshoher Ständer m. ‖ montant m. en hauteur convenable. / ~ of the crane ‖ Krangerüst n. ‖ châssis m. de grue. / cross ~ ‖ Kreuzverband m. ‖ contrefiches fpl. diagonales. / ~ for crowns (Gard) ‖ Kranzreifen m. ‖ carcasse f. pour couronnes. / curve ~ ‖ Kurvenrahmen m. ‖ châssis m. de voie courbe. / curved ~ ‖ gebogener Rahmen m. ‖ châssis m. courbé. / the ~ has cut-outs for the reception of the axle boxes ‖ der Rahmen m. hat Ausschnitte für die Achslager ‖ le châssis m. est pourvu d'échancrures destinées à recevoir les boîtes d'essieu. / ~ of cycle ‖ Fahrradrahmen m. ‖ cadre m. de bicyclette. / the ~s pl. become disconnected ‖ die Gleisrahmen mpl. ziehen sich auseinander ‖ les châssis mpl. de voie se déplacent. / ~ of a door ‖ Rahmenwerk n. oder Einrahmung f. einer Tür ‖ bâti m. d'une porte. / double ~ ‖ Doppelrahmen m. ‖ châssis m. double. / double ~ (Shipb) ‖ doppeltes Spant n. ‖ membrure f. double. / double-rib ~ ‖ Fangstuhl m. ‖ métier m. à double fonture. / dropped ~ ‖ gekröpfter Rahmen m. ‖ châssis m. coudé. / dry ~ (Spinn) ‖ Trockenspinnmaschine f. ‖ métier m. à sec. / Dutch ~ (Spinn) ‖ holländischer Rahmen m. ‖ châssis m. hollandais. / elevator ~ ‖ Fördergerüst n. ‖ charpente f. de montecharge. / end ~ (Wagon) ‖ Stirnbügel m. ‖ cadre m. d'avant. / ~ of an engine ‖ Gestell n. einer Maschine usw. ‖ bâti m. ou cadre m. ou châssis m. d'une machine. / false ~ (Build) ‖ Hilfsjoch n.; Notjoch n. ‖ faux-cadre m. / false ~ (Mine) ‖ Hilfskranz m. ‖ faux cadre m. / fixed ~ ‖ fester Rahmen m. ‖ cadre m. fixe. / foremost ~ (Shipb) ‖ Ohrspant n.; vorderstes Spant n. ‖ couple m. de coltis; colts m.; colti m. / gallery ~ ‖ Rahmen m.; Türgerüst n. ‖ châssis m. d'une galerie de mine. / gold ~ ‖ Goldrahmen m. ‖ cadre m. doré. / ~ for grinding mill ‖ Schleifsteingestell n. ‖ bâti m. de meule. / hinged ~ (Typewr) ‖ aufklappbarer Rahmen m. ‖ cadre m. réversible. / immovable ~ see sash ~. / in ~s pl. (Shipb) ‖ in Spanten npl. ‖ en couples mpl. / inside ~ ‖ innerer Rahmen m. ‖ châssis m. intérieur. / intermediate ~ (Shipb) ‖ Zwischenspant n. ‖ couple m. intermédiaire. / intermediate ~ (Spinn) ‖ Mittelfleier m. ‖ banc m. à broches intermédiaires. / intermediate ~ (Tel) ‖ Zwischenverteiler m. ‖ répartiteur m. intermédiaire. / iron ~ (Iron building) ‖ Eisengerippe n. ‖ charpente f. de fer. / iron ~ (Shipb) ‖ eisernes Spant n. ‖ couple m. en fer. / japanned iron ~ ‖ Rahmen m. aus lackiertem Eisen ‖ châssis m. en fer laqué. / ~ for lifting the screw (Shipb) ‖ Heberahmen m. ‖ châssis m. ou cadre m. des hélices amovibles. / longitudinal ~ (Shipb) ‖ Längsspant n. ‖ membrure f. longitudinale; lisse f. / ~ of the machine ‖ Maschinengestell n. ‖ bâti m. de la machine. / main ~ (Mine) ‖ Hauptjoch n. ‖ cadre m. uni ordinaire. / midship ~ ‖ Hauptspant n.; Nullspant n. ‖ maîtrecouple m. / ~ of the mill ‖ Stuhl m. oder Bock m. der Mühle ‖ chaise f. ou pylone m. du moulin. / outside ~ ‖ äußerer

Rahmen m. ‖ châssis m. extérieur. / oval ~ ‖ Ovalrahmen m. ‖ cadre m. oval. / ~ of paper ‖ Papierrahmen m. ‖ cadre m. en papier. / permanent ~ (Build) ‖ Hauptjoch n. ‖ cadre m. uni ordinaire. / picture ~ ‖ Bilderrahmen m. ‖ cadre m. / pit head ~ (Mine) ‖ Fördergerüst n. ‖ charpente f. du puits. / pressed steel ~ ‖ gepreßter Stahlrahmen m.; Rahmen m. aus gepreßtem Stahlblech ‖ châssis m. en tôle d'acier emboutie. / principal ~ see chief ~. / reversed ~ (Shipb) ‖ Gegenspant n. ‖ couple m. renversé. / revolving ~ ‖ Getriebequerhaupt n. ‖ traverse f. tournante. / rivetted ~ ‖ genieteter Rahmen m. ‖ châssis m. riveté. / ~ of a saddle ‖ Sattelgestell n. ‖ chapuis m.; fût m. d'une selle. / sash ~ ‖ Fensterzarge f. ‖ croisée f. ou huisserie f. de fenêtre; châssis m. dormant de fenêtre. / ~ of a saw ‖ Sägegestell n. einer Spannsäge ‖ châssis m. ou monture f. d'une scie; porte-scie m.; arc m. ou arçon m. de scie. / ~ for scoop wheel ‖ Heberadgerüst n. ‖ charpente f. métallique pour roue élévatrice. / shaking ~ (Ore dress) ‖ Glauchherd m.; Kehrherd m. ‖ table f. dormante. / ~ of a ship ‖ Schiffspant n. ‖ couple m. d'un navire. / single ~ (Shipb) ‖ einfaches Spant n. ‖ couple m. simple. / ~ of the slide ‖ Rahmen m. des Schiebers ‖ guide m. du tiroir. / slubbing ~ (Spinn) ‖ Grobfleier m. ‖ banc m. à broches en gros. / square ~ (Shipb) ‖ Winkelspant n. ‖ couple m. carré. / ~ of the stern (Shipb) ‖ Heckspant n. ‖ montant m. ou allonge f. de poupe. / sub-~ ‖ Hilfsrahmen m. ‖ faux-châssis m. / ~ under a table-board ‖ Tischzarge f. ‖ châssis m. de tablette. / temporary ~ see false ~. / three-needle ~ (Textile) ‖ Dreinadelstuhl m. ‖ métier m. à trois aiguilles. / ~ of timber (Mine) ‖ Holzgeviere n. ‖ cadre m. de boisage. / transverse ~ (Ship) ‖ Querspant n. ‖ couple m. transversal. / travelling ~ ‖ fahrbares Gerüst n. ‖ charpente f. sur roues. / trial ~ for spectacles ‖ Probbrille f. ‖ lunettes fpl. d'essai. / trussed ~ (Auto) ‖ versteifter Rahmen m. ‖ châssis m. renforcé. / tubular ~ ‖ Rohrrahmen m. ‖ châssis m. tubulaire. / ~ of a turning lathe ‖ Drehbankgestell n. ‖ banc m. d'un tour. / ~ with valves ‖ Flügelfensterrahmen m. ‖ croisée f. à vantaux; cadre m. de croisée à battants. / washing ~ see shaking ~. / window ~ ‖ Fensterrahmen m. ‖ châssis m. de fenêtre. / wooden ~ ‖ Holzrahmen m. ‖ châssis m. en bois. / wood filled ~ ‖ armierter Holzrahmen m. ‖ châssis m. en bois armé.

framed (Shipb) ‖ in Spanten stehend ‖ monté en couples mpl. / ~ building ‖ Fachwerkbau m. ‖ bâtisse f. en treillis; construction f. en cloisonnage. / ~ spec-

frame aerial ‖ Rahmenantenne f. ‖ antenne f. à cadre. / ~ angle-iron (Shipb) ‖ Spantwinkeleisen n. ‖ cornière f. membrure. / ~ antenna ‖ Rahmenantenne f. ‖ collecteur m. sur cadre. / ~ bead (Join) ‖ Rahmleiste f. ‖ baguette f. de cadre. / ~ bending press ‖ Rahmenbiegepresse f. ‖ presse f. à cintrer les châssis. / ~ connection ‖ Rahmenverbindung f. ‖ raccord m. de châssis. / ~ crane ‖ Bockkran m. ‖ grue f. portique. / ~ cross member ‖ Querträger m. des Rahmens ‖ traverse f. de châssis.

tacles pl. ‖ gefaßte Brille f. ‖ lunettes fpl. à verres enchassés. / ~ whip-saw ‖ Örtersäge f. ‖ scie f. à débiter.

frame filter press ‖ Rahmenfilterpresse f. ‖ filtre-presse m. à cadres. / ~ with aeration and with exclusion of air ‖ Rahmenfilterpresse m. mit einfacher Auslaugung unter Luftabschluß ‖ filtre-presse m. à cadres avec aération et à l'abri d'air. / ~ with double perfect lateral lixiviation and with exclusion of air ‖ Rahmenfilterpresse m. mit seitlicher doppelperfekter Auslaugung unter Luftabschluß ‖ filtre-presse m. à cadres à lavage double parfait latéral et à l'abri d'air.

frame fixer (Spinn) ‖ Stuhlsetzer m. ‖ monteur m. de métiers. / ~ gilder ‖ Rahmenvergolder m. ‖ doreur m. de cadres. / half of ~ ‖ Rahmenhälfte f. ‖ moitié f. de châssis. / ~ handle (Textile) ‖ Daumendrücker m. ‖ pouce m. / ~ head (Print hand press) ‖ Krone f. ‖ chapeau m.; chapiteau m.; chaperon m. / ~ knitter ‖ Maschinenstricker m. ‖ tricoteur m. à la machine. / ~ knitting machine ‖ Kettelmaschine f. ‖ machine f. à mailler.

frameless spectacles pl. ‖ ungefaßte Brille f. ‖ lunettes fpl. à verres non enchassés.

frame maker ‖ Einrahmer m. ‖ encadreur m. / ~ maker (Piano) ‖ Rahmenmacher m. ‖ barragier m. / ~ opening ‖ Rahmenausschnitt m. ‖ découpure f. du châssis. / ~ part of pressed steel plate ‖ aus Blechen gepreßter Rahmenteil m. ‖ élément m. de châssis embouti en tôle. / ~ piece of a panel door ‖ Rahmenstück n. oder Fries m. an einer gestemmten Tür ‖ emboîture f. ou membrure f. d'une porte à panneaux.

frame plate (Acc) ‖ Rahmenplatte f. ‖ plaque f. de cadre. / ~ (Loc) ‖ Rahmenwange f.; Träger m. ‖ longeron m. / ~ (Wagon) ‖ Rahmenblech n. ‖ tôle f. pour châssis. / steel-cast ~ stay ‖ Rahmenversteifung f. in Stahlguß ‖ caissonnement m. en acier moulé.

frame press ‖ Rahmenpresse f. ‖ presse f. à cadres. / ~ (Loc) ‖ Rahmenpresse f. ‖ presse f. à longeron.

frame profile (Shipb) ‖ Spantprofil n. ‖ profilé m. de couple. / ~ receiving (Radio) ‖ Rahmenempfang m. ‖ réception f. sur cadre.

frame saw ‖ Gattersäge f. ‖ scie f. alternative ou à cadre. / ~ (Carp) ‖ Zuschneidesäge f.; Örtersäge f. ‖ scie f. à débiter. / ~ (Join) ‖ Klobsäge f.; Furniersäge f. ‖ scie f. à refendre ou à placage.

frame section (Shipb) ‖ Spantprofil n. ‖ profilé m. de couple. / ~ shears pl. ‖ Rahmenschere f. ‖ cisaille f. à guillotine. / ~ side member ‖ Längsträger m. des Rahmens ‖ longeron m. de châssis. / ~ slotting machine ‖ Rahmenstoßmaschine f. ‖ mortaiseuse f. pour longerons. / stiffening of the ~ ‖ Rahmenversteifung f. ‖ renforcement m. du châssis. / support of the ~ ‖ Rahmenstütze f. ‖ support m. du châssis. / ~-supported electric motor ‖ Elektromotor m. mit Rahmenaufhängung ‖ moteur m. électrique suspendu au châssis. / trussing of ~ ‖ Rahmenunterzug m. ‖ renforcement m. du châssis. / ~ tube ‖ Rahmenrohr n. ‖ tube m. de cadre. / ~ window ‖ Blendrahmenfenster n.; Zargenfenster n. ‖ fenêtre f. à châssis.

framework ‖ Fachwerk n.; Gebälk n. ‖ charpente f.; treillis m.; cloisonnage m. / ~ for blast furnaces ‖ Hochofengerüst n. ‖ charpente f. pour hauts-fourneaux. / ~ of body ‖ Karosserieholzgestell n. ‖ encadrement m. de menuiserie de carrosserie; carcasse f. de caisse. / ~ of a building ‖ Holzwerk n. eines Gebäudes ‖ charpente f. d'un bâtiment. / ~ for covering (Aero) ‖ Bespannungsgerüst n. ‖ ossature f. pour le revêtement. / iron ~ ‖ Eisenfachwerk n. ‖ charpente f. métallique. / iron ~ for roofs ‖ eiserne Dachkonstruktion f. ‖ charpente f. de toits en fer. / ~ of the sail (Wind mill) ‖ Flügelgerippe n. ‖ ossature f. de l'aile. / the shops pl. consist of iron ~ ‖ die Gebäude npl. sind aus Eisenfachwerk hergestellt ‖ les ateliers mpl. sont en charpente métallique. / supporting ~ ‖ Traggerüst n. ‖ charpente f. de support. / ~ of the washing machine ‖ Waschmaschinengestell n. ‖ bâti m. de la machine à laver. / wicker ~ ‖ rohrgeflochtenes Fachwerk n. ‖ treillis m. d'osier. / ~ of a window ‖ Fenstergerähme n. ‖ croisée f. d'une fenêtre; fenêtre-croisée f.

framework, construction of a ~ (Carp) ‖ Holzverband m.; Zimmerverband m. ‖ assemblage m. des bois. / ~ knitting ‖ Wirkerei f. ‖ bonneterie f. / ~ wall ‖ Fachwerkwand f. ‖ cloison m. en charpente.

framing ‖ Fassung f.; Einfassung f. ‖ cadre m.; châssis m. / ~ of a building ‖ Holzwerk n. eines Gebäudes ‖ charpente f. d'un bâtiment. / ~ for the carcass of a roof ‖ Dachbalkenlage f. ‖ plancher m. ou enrayure f. de comble. / cross ~ ‖ Querriegel m. ‖ entremise f. / diagonal ~ ‖ Kreuzverband m. ‖ membrure f. diagonale; liaisons fpl. diagonales. / ~ of a door ‖ Rahmenstuhl m. einer Tür ‖ cadre m. ou encadrement m. de porte; bâti m. des panneaux. / ~ of joists (Carp) ‖ Balkenlage f. ‖ solivure f. / ~ of a machine ‖ Maschinengerüst n.; Maschinengestell n. ‖ bâti m. ou bâti m. d'une machine. / ~ of a mill ‖ Mühlgerüst n. ‖ beffroi m. d'un moulin. / ~ of a panel ‖ Rahmholz n. ‖ bâti m. des panneaux. / ~ of timbers ‖ Holzverbindung f. ‖ assemblage m. des bois. / ~ (Mill) ‖ Trichter m. einer Mühle ‖ cône m. d'un moulin à farine. / of a vessel ‖ Gerippe n. eines Schiffes ‖ carcasse f. ou membrure f. d'un navire.

framing timber ‖ Rahmenholz n. ‖ bois m. pour châssis. / ~ (Carp) ‖ Verbandholz n.; Ausbindeholz n. ‖ bois m. d'assemblage.

Francis spiral turbine, set of double horizontal shaft ~s ‖ Doppelfrancisspiralturbinensatz m. in horizontaler Anordnung ‖ groupe m. de turbines Francis à spirale et à arbre horizontal.

Francis turbine ‖ Francisturbine f. ‖ turbine f. de Francis. / set of vertical shaft single-wheel ~s ‖ Franciseinradturbinensatz m. in vertikaler Anordnung ‖ groupe m. de turbines Francis à couronne unique et à arbre vertical.

Francis vertical shaft turbine, single-wheel ~ ‖ einkränzige Francisvertikalturbine f. ‖ turbine f. Francis à couronne unique et à arbre vertical.

francolite ‖ Frankolith m. ‖ francolite f.

Frankfort black ‖ Frankfurter Schwarz n.; Kupferdruckfarbe f. ‖ noir m. de Frankfort.

frankincense ‖ Weihrauch m. ‖ encens m.

franklinite ‖ Franklinit m.; Zinkeisenerz n. ‖ franklinite f.

fraudulent ‖ schwindelhaft ‖ vertigineux.

Fraunhofer lines pl. ‖ Fraunhofersche Linien fpl. ‖ raies fpl. de Fraunhofer.

free, to ~ from acid ‖ entsäuern ‖ désacidifier; neutraliser. / ~ from branches ‖ entästen ‖ émonder. / ~ from gas ‖ entgasen ‖ éliminer le gaz. / ~ the oil from all septic matter ‖ das Öl von allen fäulniserregenden Stoffen befreien ‖ débarrasser l'huile f. de toutes les matières septiques. / ~ a ship ‖ ein Schiff n. auspumpen oder lenzpumpen ‖ agréner ou affranchir un navire. / ~ spirit from amylic alcohol ‖ den Branntwein m. entfuseln ‖ défuseler ou débarrasser l'esprit-de-vin de l'alcool d'amyle.

free ‖ frei ‖ libre. / ~ (Trade) ‖ kostenlos ‖ gratuit. / ~ (Wind) ‖ raum ‖ largue. / ~ depot ‖ frei Bahnhof m. ‖ franco gare f. / ~ from ‖ frei von ‖ exempt de. / ~ from duty ‖ abgabenfrei ‖ exempt de droits mpl. / ~ from end play ‖ frei von totem Gang m. ‖ libre de tout jeu m. / ~ from parallax ‖ parallaxenfrei ‖ exempt de parallaxe. / ~ of charge ‖ spesenfrei; kostenfrei ‖ frais mpl. déduits; sans frais mpl.; net de tous frais mpl. / ~ of interest ‖ zinsfrei ‖ sans intérêt m. / ~ on board (f. o. b.) ‖ frei an Bord m. ‖ franco (à) bord m. / ~ on rail ‖ bahnfrei ‖ franco sur wagon m. / to set ~ ‖ freimachen ‖ mettre en liberté. / in the ~ state (Chem) ‖ in freiem Zustande m. ‖ à l'état m. libre. /

free adjudication ‖ freihändige Vergebung f. ‖ adjudication f. libre. / ~ alternating current ‖ freier Wechselstrom m. ‖ courant alternatif libre. / ~ antenna ‖ Freiantenne f. ‖ antenne f. haute. / ~ balloon ‖ Freiballon m. ‖ ballon m. libre.

freeboard (Shipb) ‖ Freibord m. ‖ francbord m. / vessel with low ~ ‖ Niederbordschiff n. ‖ navire m. bas de bord. / ~ mark ‖ Ladelinie f. ‖ ligne f. de charge.

freebooter ‖ Schmugglerschiff n. ‖ interlope m.; aventurier.

free copy ‖ Freiexemplar n. ‖ exemplaire m. gratuit.

freedom, to limit the ~ of contract ‖ die Vertragsfreiheit f. einschränken ‖ limiter la liberté de contrat. / ~ to mine ‖ Bergbaufreiheit f. ‖ liberté f. ou droit m. d'exploitation minière. / ~ of movement ‖ Freizügigkeit f. ‖ liberté f. de résidence. / ~ to organize ‖ Koalitionsfreiheit f. ‖ liberté f. à l'organisation. / ~ of the press ‖ Pressefreiheit f. ‖ liberté f. de la presse. / ~ to prospect (Min) ‖ Schürffreiheit f. ‖ liberté f. des recherches. / ~ to settle anywhere see ~ of movement. / ~ of trade ‖ Gewerbefreiheit f. ‖ liberté f. industrielle ou du commerce.

free fall ‖ Freifall m. ‖ chute f. libre. / ~ of the tup ‖ freier Fall m. des Bären ‖ chute f. libre du mouton.

free-hand drawing ‖ Freihandzeichnen n. ‖ dessin m. à main levée.

free-jet turbine ‖ Freistrahlturbine f. ‖ turbine f. à libre jet.

free lance ‖ freizügiger Arbeiter m. ‖ ouvrier m. ayant la liberté de résidence. / ~ luggage ‖ Freigepäck n. ‖ bagages mpl. en franchise.

freely supported ‖ frei aufliegend ‖ reposant librement.

free motion ‖ Freilauf m. ‖ course f. libre. / ~ oscillation ‖ freie Schwingung f. ‖ oscillation f. libre. / ~ pass (Railw) ‖ Freifahrkarte f. ‖ parcours m. libre.

freeport ‖ Freihafen m. ‖ port m. franc; entrepôt m. réel.

freestone ‖ Mauerstein m. ‖ moellon m.; pierre f. à bâtir ou de construction. / ~ ‖ Quader m.; Quaderstein m.; Haustein m.; Werkstein m. ‖ pierre f. carrée; carreau m.; moellon m. d'appareil ou de taille. / gritty ~ ‖ Sandstein m. ‖ grès m. / to set the ~s horizontally ‖ die Werksteine mpl. einwiegen ‖ poser les pierres fpl. à niveau.

freestone mason ‖ Hausteinmaurer m. ‖ maçon m. (qui travaille) en pierres de taille. / ~ masonry ‖ Quadermauerwerk n.; Hausteingemäuer n. ‖ maçonnerie f. en pierres de taille. / ~ walling see ~ masonry.

free-swinging drive ‖ freischwingender Antrieb m. ‖ commande f. autobalanceuse. / ~ plansifter ‖ freischwingender Plansichter m. ‖ plansichter m. autobalanceur.

free trade ‖ Freihandel m.; Handelsfreiheit f. ‖ commerce m. libre; libre-échange m.; liberté f. de commerce.

free-trading ‖ freihändlerisch ‖ libreéchangeale.

freewheel ‖ Freilaufrad n. ‖ roue f. libre. / ~ hub ‖ Freilaufnabe f. ‖ moyeu m. de roue libre.

free wind direction ‖ freie Windrichtung f. ‖ direction f. libre du vent.

freeze, to ‖ frieren ‖ congeler; glacer. / ~ (Chem) ‖ erstarren ‖ se congeler; se prendre en masse; se concréter. / ~ (Cloth) ‖ ratinieren ‖ friser; ratiner. / ~ together ‖ zusammenfrieren ‖ congeler.

freezer ‖ Eiserzeugungsapparat m. ‖ congélateur m.; glacière f.

freezing ‖ Gefrieren n. ‖ congélation f. / ~ apparatus for making ice ‖ Eiserzeugungsapparat m.; Gefrierapparat m.; Eismaschine f. ‖ appareil m. de congélation; glacière f. / ~ chamber ‖ Gefrierraum m. ‖ chambre f. de congélation. / ~ cupboard ‖ Gefrierschrank m. ‖ armoire f. de congélation. / ~ cupboard for temperatures down to — x degrees centigrade ‖ Tiefgefrierschrank m. mit —x⁰ Celsius ‖ armoire f. de congélation pour températures basses jusqu'à —x centigrades. / ~ machine (Cloth) ‖ Ratiniermaschine f. ‖ ratineuse f. / ~ mixture ‖ Kältemischung f. ‖ mélange m. réfrigérant ou frigorifique. / ~ plant ‖ Gefrieranlage f. ‖ installation f. de congélation. / ~ point ‖ Gefrierpunkt m. ‖ point m. de congélation. / ~ process ‖ Tiefkälteverfahren n. ‖ procédé m. de congélation à basse température. / meat ~ room ‖ Fleischgefrierraum m. ‖ chambre f. de congélation pour de la viande. / ~ shaft ‖ Gefrierschacht m. ‖ puits m. frigorifique. / ~ tank ‖ Eiserzeuger m. ‖ générateur m. à glace. / ~ tank for manufacture of plate ice ‖ Platteneiserzeuger m. ‖ générateur m. à glace en plaques. / ~ water ‖ Gefrierwasser n. ‖ eau f. à congeler.

freieslebenite ‖ Freieslebenit m.; Schilfglaserz n. ‖ freieslébénite f.

freight (Costs) ‖ Fracht f.; Frachtgeld n. ‖ frais mpl. de transport. / ~ (Load) ‖ Fracht f. ‖ charge f. / ~ (Nav) ‖ Heuer-

geld n. ∥ paye f. / ~ (Ship's load) ∥ Schiffsladung f. ∥ cargaison f.; chargement m. / additional ~ ∥ Frachtzuschlag m.; Überfracht f. ∥ frais mpl. additionnels; taxe f. supplémentaire. / back ~ ∥ Rückfracht f. ∥ port m. pour le retour. / direct ~ ∥ Durchfracht f. ∥ fret m. direct. / extra ~ ∥ Frachtzuschlag m. ∥ frais mpl. additionnels; taxe f. supplémentaire. / full ~ ∥ ganze oder volle Fracht f. ∥ fret m. entier. / gross ~ ∥ Bruttofracht f. ∥ fret m. brut. / homeward ~ ∥ Rückfracht f. ∥ fret m. de retour. / lowest ~ ∥ Minimalfacht f. ∥ minimum m. de fret. / ~ stipulated in a lump sum ∥ in Bausch und Bogen bedungene Fracht f. ∥ fret m. en bloc. / to make ~ ∥ die Fracht abschließen oder bedingen ∥ conclure le fret. / ~ by measure ∥ Verfrachtung f. nach Maß ∥ affrètement m. au tonneau. / normal ~ ∥ Normalfracht f. ∥ fret m. normal. / ~ per register ton ∥ Fracht f. für die Registertonne ∥ fret m. par tonneau d'encombrement. / through ~ ∥ Durchfracht f. ∥ fret m. de transit. / ~ per ton weight ∥ Fracht f. für die Gewichtstonne ∥ fret m. par tonneau (pesant). / ~ by water ∥ Wasserfracht f. ∥ fret m.

freight accounting ∥ Frachtbuchführung f. ∥ comptabilité f. des frais de transport. / ~ aeroplane ∥ Frachtflugzeug n. ∥ avion m. de fret.

freightage ∥ Frachtgeld n. ∥ affrétage m.

freight automobile ∥ Lastautomobil n.; Lastkraftwagen m. ∥ camion m. automobile. / ~ boat ∥ Frachtkahn m. ∥ chaland m.

freight car ∥ Lastwagen m. ∥ camion m.; voiture f. à marchandises. / ~ (Railw) ∥ Güterwagen m. ∥ wagon m. à marchandises. / motor ~ ∥ Lastkraftwagen m. ∥ camion m. automobile. / motor ~ train ∥ Kraftwagenlastzug m. ∥ convoi m. de camions automobiles. / box of ~ ∥ Wagenkasten m. ∥ caisse f. de voiture. / iron box ~ for standard gauge ∥ normalspuriger eiserner Kastenwagen m. ∥ wagon m. à caisse en fer à voie normale.

freight charges pl. ∥ Frachtspesen pl. ∥ frais mpl. de transport. / deferred payment of ~ ∥ Frachtstundung f. ∥ répit m. pour le paiement de la taxe.

freight elevator ∥ Lastaufzug m. ∥ montecharge m. / ~ engine ∥ Güterzuglokomotive f. ∥ locomotive f. à marchandises.

freighter (Person) ∥ Verfrachter m.; Schiffsbefrachter m. ∥ expéditeur m.; armateur m. / ~ (Shipb) ∥ Frachtschiff n. ∥ cargo m.

freight-free ∥ frachtfrei ∥ exempt de fret m. / ~ baggage ∥ Freigepäck n. ∥ bagages mpl. transportés gratuitement. / ~ luggage ∥ Freigepäck n. ∥ bagages mpl. transportés gratuitement.

freight goods traffic ∥ Frachtgutverkehr m. ∥ trafic m. des marchandises à petite vitesse. / ~ hold ∥ Frachtraum m. ∥ cale f. pour fret.

freighting ∥ Schiffsbefrachtung f. ∥ affrètement m.

freight lift ∥ Frachtaufzug m. ∥ montecharge m. / ~ locomotive see ~ engine. / ~ market ∥ Frachtmarkt m. ∥ marché m. du fret. / ~ motor car ∥ Lastkraftwagen m. ∥ camion-automobile m. / ~ note ∥ Frachtbrief m. ∥ liste f. de fret. / ~ rate ∥ Frachtsatz m. ∥ taux m. de fret; tarif m. fret. / ~ steamer ∥ Frachtdampfer m. ∥ cargo m. à vapeur. / ~ train ∥ Güterzug m. ∥ train m. de marchandises. / ~ wagon see ~ car.

French boiler ∥ Elefantenkessel m. ∥ bouilleur m. / ~ brandy ∥ Franzbranntwein m. ∥ eau-de-vie f. de France. / ~ casement ∥ Flügelfenster n. ∥ croisée f. à vantaux. / ~ chalk ∥ Talkum m. ∥ talc m. / ~ polish ∥ Schellackpolitur f. ∥ poli m. à gomme laque.

frequency ∥ Frequenz f. ∥ fréquence f. / acoustic ~ ∥ Hörfrequenz f. ∥ fréquence f. acoustique. / ~ of calls (Tel) ∥ Gesprächsdichte f. ∥ densité f. du trafic. / cut-off ~ ∥ Grenzfrequenz f. ∥ fréquence f. limite. / dot ~ ∥ Telegrafierfrequenz f. ∥ fréquence f. fondamentale. / harmonic ~ ∥ harmonische Frequenz f. ∥ fréquence f. harmonique. / high ~ ∥ Hochfrequenz f. ∥ haute fréquence f. / low ~ ∥ Niederfrequenz f. ∥ basse fréquence. / natural ~ ∥ Eigenschwingung f. ∥ oscillation f. naturelle. / tone ~ ∥ Tonfrequenz f. ∥ fréquence f. musicale.

frequency band ∥ Frequenzband n. ∥ bande f. de fréquences. / limitation of ~s ∥ Frequenzbegrenzung f. ∥ limitation f. des bandes de fréquences.

frequency changer ∥ Frequenzwandler m. ∥ transformateur m. de fréquence. / ~ converter ∥ Frequenzwandler m. ∥ changeur m. de fréquence. / ~ curve ∥ Häufigkeitskurve f. ∥ courbe f. de fréquence. / ~ dependency on ∥ Frequenzabhängigkeit f. ∥ dépendance f. de fréquence. / ~ indicator ∥ Frequenzanzeiger m. ∥ indicateur m. de fréquence. / ~ meter ∥ Frequenzmesser m. ∥ fréquencemètre m. / ~ multiplication ∥ Frequenzvervielfachung f.; Frequenzsteigerung f. ∥ multiplication f. de fréquences. / ~ polygon ∥ Häufigkeitspolygon n. ∥ polygone m. de fréquence. / ~ range ∥ Tonbereich m. ∥ bande f. de fréquence. / ~ recorder ∥ Frequenzschreiber m. ∥ fréquencemètre m. / ~ response (Tel) ∥ Frequenzabhängigkeit f. ∥ dépendance f. de la fréquence. / ~ scale ∥ Frequenzband n. ∥ bande f. de fréquences. / ~ transformer ∥ Frequenztransformator m. ∥ transformateur m. ou changeur m. de fréquence. / ~ variation in ∥ Frequenzänderung f. ∥ variation f. de la fréquence.

frequent ∥ oftmalig ∥ fréquent.

fresco ∥ Freske f.; Fresko n. ∥ fresque f.; peinture f. ou tableau m. à fresque. / ~ painting ∥ Freskomalerei f. ∥ fresque f.; peinture f. à la fresque.

fresh ∥ frisch ∥ frais. / ~ air ∥ Frischluft f. ∥ air m. frais. / ~ bacon ∥ frischer Speck m. ∥ lard m. frais. / ~ grouping of the supply area ∥ Neugliederung f. des Stromversorgungsgebietes ∥ réorganisation f. du centre de consommation. / ~ meat ∥ frisches Fleisch n. ∥ viande f. fraîche. / ~ snow ∥ Neuschnee m. ∥ neige f. tombée récemment. / in ~ state ∥ in frischem Zustand m. ∥ à l'état m. frais.

freshwater ∥ Frischwasser n.; Süßwasser n. ∥ eau f. fraîche ou pure. / ~ deposit (Geol) ∥ Süßwasserablagerung f. ∥ dépôt m. d'eau douce. / ~ fishing ∥ Binnenfischerei f. ∥ pêche f. en eau douce. / ~ lake ∥ Haff n. ∥ haff m. / ~ limestone ∥ Süßwasserkalk m. ∥ calcaire m. lacustre. / ~ pump ∥ Frischwasserpumpe f. ∥ pompe f. à eau fraîche.

fret ∥ Gärungsstoff m.; Ferment n. ∥ ferment m.; zumine f.

fret saw ∥ Dekupiersäge f.; Schweifsäge f.; Laubsäge f. ∥ scie f. à chantourner ou à découper.

fret sawing manufacture ∥ Laubsägearbeit f. ∥ chantournage m.

fretting ∥ stürmische Gärung f. ∥ fermentation f. tumultueuse.

fretwork (Arch) ∥ Maßwerk n.; ausgeschnittenes Motiv n. ∥ broderie f.; découpure f.; tracé m.; réseau m. / ~ (Glassm) ∥ Glasmosaik f.; Millefiori fpl. ∥ verre m. millefleurs ou mosaïque.

friability (Chem) ∥ Zerreibbarkeit f. ∥ déagrégation f.; friabilité f.

friable ∥ zerreibbar; mulmig ∥ friable. / ~ soil ∥ krümeliger Boden m. ∥ sol m. friable.

friableness (Techn) ∥ Zerreibbarkeit f. ∥ friabilité f.

friar (Print) ∥ Mönchsbogen m.; Mönch m. ∥ feinte f.; bouquet m. feuille f. venue par bouquets; moine m.

friction ∥ Friktion f.; Reibung f. ∥ frottement m. / air ~ ∥ Luftreibung f. ∥ frottement m. de l'air. / ~ of journal ∥ Zapfenreibung f. ∥ frottement m. du tourillon. - rolling ~ ∥ Schienenreibung f. ∥ friction f. des rails. / skin ~ ∥ Oberflächenreibung f. ∥ frottement m. superficiel. / sliding ~ ∥ gleitende Reibung f. ∥ frottement m. au glissement. / ~ between wheel and rail ∥ Reibung f. zwischen Rad und Schiene ∥ frottement m. entre les roues et les rails.

frictional grooved gearing ∥ Keilrädergetriebe n. ∥ engrenage m. à friction; transmission f. à friction par poulies à gorge. / ~ loss ∥ Reibungsverlust m. ∥ perte f. de frottement.

friction brake ∥ Reibungsbremse f. ∥ frein m. à friction. / ~ brush ∥ Frottierbürste f. ∥ brosse f. à friction. / ~ calender ∥ Friktionskalander m.; Glanzkalander m. ∥ calandre f. à friction ou à lustrer. / ~ change gear ∥ Reibungswendegetriebe n. ∥ changement m. de marche à friction. / ~ cloth ∥ gummierte Leinwand f. ∥ toile f. caoutchoutée. / ~ clutch ∥ Reibungskupplung f. ∥ embrayage m. ou accouplement m. à friction. / coefficient of ~ ∥ Reibungskoeffizient m. ∥ coefficient m. de frottement. / ~ cone ∥ Friktionskegel m. ∥ cône m. de friction. / ~ coupling ∥ Rutschkupplung f.; Reibungskupplung f. ∥ embrayage m. ou accouplement m. à friction. / adjustable ~ coupling ∥ nachstellbare Friktionskupplung f. ∥ embrayage m. à friction réglable. / longitudinal ~ device (Bak) ∥ Längsreiber m. ∥ broyeur m. longitudinal; conche f.

friction disk ∥ Reibscheibe f. ∥ plateau m. de friction. / ~ disc clutch ∥ Reibungslamellenkupplung f. ∥ embrayage m. à friction à disques. / ~ drive ∥ Reibräderantrieb m. ∥ commande f. par roues à friction. / ~ duck ∥ gummierte Leinwand f. ∥ toile f. frictionnée. / ~ effect ∥ Reibungseinfluß m. ∥ influence f. de friction. / ~ forging hammer ∥ Friktionsschmiedehammer m. ∥ marteau m. de forge à friction. / ~ forging press ∥ Friktionsschmiedepresse f. ∥ presse f. à forger à friction. / ~ gear ∥ Reibungsgetriebe n. ∥ engrenage m. à friction. / ~ gearing ∥

Friktionsantrieb m. ‖ commande f. par friction. / ~ governor ‖ Friktionsregler m. ‖ régulateur m. à friction. / ~ hammer ‖ Friktionshammer m. ‖ marteau m. *ou* pilon m. à friction. / ~ hoist ‖ Friktionsaufzug m. ‖ monte-charge m. commandé par friction. / brush ~ loss ‖ Bürstenreibungsverlust m. ‖ perte f. par frottement des balais. / ~ match ‖ Reibzündhölzchen n. ‖ allumette f. à friction; congrève f. / ~ pillar screw press ‖ Friktionssäulenpresse f. ‖ presse f. à friction à colonnes. / ~ plate ‖ Friktionsscheibe f.; Reibscheibe f. ‖ plaque f. de friction. / ~ press ‖ Friktionspresse f. ‖ presse f. à friction. / ~ resistance ‖ Reibungswiderstand m. ‖ résistance f. due à la friction. / ~ reversing gear ‖ Friktionswendegetriebe n. ‖ dispositif m. de renversement de marche à friction. / ~ roller ‖ Reibrolle n. ‖ galet m. de friction. / ~ screw press ‖ Friktionsspindelpresse f. ‖ presse f. à friction. / ~ spring hammer ‖ Friktionsfederhammer m. ‖ marteau m. à ressort à friction. / ~ starting clutch ‖ Antriebreibungskupplung f. ‖ accouplement m. à friction de commande. / ~ surface ‖ Reibungsfläche f. ‖ surface f. de friction. / ~ wheel ‖ Reibrad n. ‖ roue f. à friction. / grooved ~ wheel ‖ Keilrad n. ‖ roue f. à gorge *ou* à coin. / ~ winch ‖ Friktionswinde f. ‖ treuil m. à friction. / ~ winding-on machine (Weav) ‖ Friktionsaufwickelapparat m. ‖ enrouleuse f. à friction. / ~ windlass *see* ~ winch.

frieze, to ‖ kräuseln ‖ friser; crêper. / ~ (Clothm) ‖ aufkratzen; aufrauhen; kraus machen ‖ friser; carder; aplaner; aplaigner; égratigner; lainer.

frieze (Arch) ‖ Fries m.; Borte f. ‖ frise f.; plate-bande f. / ~ (Clothm) ‖ Fries m.; Flaus m. ‖ frise f.; drap m. frisé; montagnac m.

friezed cloth *see* frieze (Clothm).

frieze-lined tension disc (Weav) ‖ Friesscheibe f. zur Fadenbremsung ‖ disque m. garni de frise pour la tension du fil.

friezing ‖ Kräuseln n. ‖ frisage m. / cloth ~ ‖ Kräuseln n. *oder* Rauhen n. des Tuches ‖ frisage m. *ou* lainage m. du drap.

friezing cylinder (Cloth) ‖ Stachelwalze f. ‖ cylindre m. garni d'aiguilles. / ~ iron *see* frizzling iron. / ~ tool (Engr; Turn) ‖ Krauspunze f.; Körnchenpunze f.; grobe Mattpunze f. ‖ frisoir m.

frigate ‖ Fregatte f. ‖ frégate f.

frigid ‖ kalt; eisig; eiskalt ‖ glacé; froid. / ~ zone ‖ Kältezone f. ‖ zone f. glaciale.

frigidity ‖ Kälte f.; Eiskälte f. ‖ froid m. (glacial).

frigorific ‖ Kälte f. erzeugend ‖ frigorifique. /~ machine ‖ Kältemaschine f. ‖ machine f. frigorifique. / ~ mixture ‖ Kältemischung f. ‖ mélange m. frigorifique.

frill, to ‖ kräuseln ‖ crêper.

fringe, to ‖ mit Fransen fpl. versehen *oder* verzieren ‖ franger; border.

fringe ‖ Rand m.; Grenze f.; Einfassung f. ‖ bordure f.; lisière f. / ~ (Clothm; Lace-m) ‖ Franse f. ‖ frange f. / chromatic ~ (Opt) ‖ bunter Rand m. ‖ contours mpl. irrisés. / twisted ~ (Lace-m) ‖ gedrehte Franse f. ‖ frange f. tordue. / fringe knotter ‖ Fransenknüpfer m. ‖ noueur m. de franges. / ~ twister ‖ Fransendreherin f. ‖ guipeuse f. en franges.

fringing machine ‖ Fransenknüpfmaschine f. ‖ machine f. à franger. / ~ reef (Geol) ‖ Saumriff n. ‖ récif m. côtier *ou* bordure; récif frangeant.

frisket (Print) ‖ Rähmchen n.; Gesperr n. ‖ frisquette f. / to cover part of the ~ with paper ‖ den Satz m. beleimen ‖ masquer.

frisket griper *see* frisket.

frit, to (Glassm) ‖ fritten; (die Glasmasse) schmelzen ‖ fritter.

frit ‖ Fritte f.; Glassatz m. ‖ fritte f. / ~ furnace *see* fritting furnace.

frith (Mar) ‖ Meerenge f. ‖ détroit m. / ~ (Forest) ‖ Unterholz n.; Gestrüpp n. ‖ taillis m.

fritter, to ‖ zerstücke(l)n; in Stücke npl. (zer)schneiden *oder* (zer)teilen ‖ découper; mettre en morceaux mpl.

fritting bench (Glassm) ‖ Frittafel f.; Mengtafel f. ‖ fonceau m. / ~ furnace (Glassm) ‖ Frittofen m. ‖ four m. à fritter. / ~ table *see* ~ bench.

friz, to *see* to frieze *and* to frizzle.

frizz, to *see* to frieze *and* to frizzle.

frizzle, to ‖ kräuseln ‖ friser; crêper. / ~ ‖ sich kräuseln ‖ se crêper; friser. / ~ (Clothm) ‖ aufrauhen ‖ friser; lainer. / ~ (To roast) ‖ braten; rösten ‖ griller; rôtir.

frizzle ‖ Locke f. ‖ boucle f.

frizzler ‖ Haarbrenner m.; Haarkräus(e)ler m. ‖ coiffeur m.

frizzling iron ‖ Kräuseleisen n.; Brenneisen n. ‖ fer m. à boucler *ou* à friser.

frizzly ‖ kraus; gekräuselt ‖ frisé; bouclé; crépu.

frizzy *see* frizzly.

frog ‖ Frosch m. ‖ grenouille f. / ~ (Railw) ‖ Herzstück n. ‖ cœur m.; cœur m. de croisement; croisement m. (de voie). / ~ (Railway switch) ‖ Weiche f. ‖ changement m. (de voie); branchement m.; aiguillage m. / aerial ~ (Railw) ‖ Luftweiche f.; Luftkreuzung f. ‖ changement m. de voie *ou* croisement m. aérien. / cast-iron chilled ~ ‖ Schalengußherzstück n. ‖ croisement m. en fer fondu en coquille. / diagonal ~ ‖ Diagonalweiche f. ‖ aiguillage m. diagonal. / double ~ ‖ Doppelherzstück n.; Kreuzungsstück n. ‖ cœur m. double; pièce f. de croisement. / ~ with forged steel point ‖ Herzstück n. mit angeschmiedeter Flußstahlspitze ‖ cœur m. de croisement avec pointe forgée en acier. / ~ made by joining steel rails ‖ aus Stahlschienen zusammengesetztes Herzstück n. ‖ cœur m. composé de rails en acier. / manganese steel ~ ‖ Manganstahlherzstück n. ‖ cœur m. en acier manganèse. / overhead ~ ‖ Fahrdrahtweiche f. ‖ croisement m. *ou* changement m. de fil; changement m. de voie du fil de contact. / left-hand ~ ‖ Linksweiche f. ‖ branchement m. *ou* changement m. à gauche. / right-hand ~ ‖ Rechtsweiche f. ‖ branchement m. *ou* changement m. à droite. / spring rail ~ ‖ Herzstück n. mit beweglicher Flügelschiene ‖ croisement m. à patte de lièvre mobile. / steel-cast ~ ‖ Stahlgußherzstück n. ‖ cœur m. en acier moulé. / swing-rail ~ ‖ Drehschienenherzstück n. ‖ croisement m. avec bout de rail mobile. / two-way ~ ‖ zweifache Weiche f. ‖ double changement m. de voie; double aiguillage m.

frog, base plate for (Railw) ‖ Herzstückunterlagsplatte f. ‖ plaque f. de surhaussement pour cœurs. / body of ~ (Railw) ‖ Weichenkörper m. ‖ corps m. du changement de voie. / ~ catcher ‖ Froschfänger m. ‖ pêcheur m. de grenouilles. / ~ clamp (Railw) ‖ Weichenklemme f. ‖ pince f. d'aiguillage. / ~ guard (Railw) ‖ Weichenschutz m. ‖ traverse f. d'écartement de l'aiguillage. / hind-leg of a ~ ‖ Froschschenkel m. ‖ patte f. de grenouille. / ~ piece (Railw) ‖ Weichenstück n. ‖ pièce f. du changement de voie.

frog point (Railw) ‖ Herzstückspitze f. ‖ pointe f. du croisement *ou* de cœur. / die-forged ~ (Railw) ‖ im Gesenk geschmiedete Herzstückspitze f. ‖ pointe f. de cœur matricée.

frog spawn ‖ Froschlaich m. ‖ frai m. de grenouilles. / ~ wire (Railw) ‖ Weichendraht m. ‖ fil m. de changement de voie aérien.

front ‖ vorn ‖ en avant; devant.

front ‖ Vorderseite f. ‖ front m. / ~ (Arch) ‖ Fassade f.; Außenseite f. ‖ façade f.; face f. / ~ of the foot (Stocking) ‖ Fußdecke f. ‖ devant m. du pied. / from ~ to back ‖ von vorn nach hinten ‖ d'avant en arrière. / in ~ ‖ vorn ‖ par devant. / ~ of the rear of the fuselage (Aero) ‖ Rumpfspitze f. ‖ extrémité f. avant du fuselage.

frontage (Build) ‖ Mittelbau m. ‖ corps m. central du bâtiment.

frontal (Arch) ‖ Giebel m.; Ziergiebel m. ‖ fronton m. / ~ screen ‖ Vordersieb n. ‖ tamis m. de devant. / ~ side of a stone ‖ Kopfseite f. *oder* Stirnfläche f. *oder* Vorderfläche f. eines Steins ‖ parement m. d'une pierre.

front axle ‖ Vorderachse f. ‖ essieu m. avant. / ~ (Auto) ‖ Gabelachse f. ‖ essieu m. à chapes ouvertes. / ~ suspension ‖ Vorderachsaufhängung f. ‖ suspension f. d'essieu avant.

front building ‖ Vordergebäude n. ‖ avant-corps m. / ~ cart ‖ Vorderwagen m. ‖ avant-train m. de charrettes. / ~ connection ‖ vorderseitiger *oder* vorderer Anschluß m. ‖ branchement m. avant; raccordement m. de devant. / ~ contact ‖ Arbeitskontakt m. ‖ contact m. de travail. / ~ coupling rod ‖ vordere Kuppelstange ‖ bielle f. d'avant. / ~ cycle fork ‖ Vorderradgabel f. ‖ fourche f. avant. / ~ delivery (Print) ‖ Frontbogenausleger m. ‖ sortie f. frontale. / ~ diaphragm ‖ Abschlußblende f. ‖ diaphragme m. de fermeture. / ~ door ‖ Haustür f. ‖ porte f. de face *ou* de la maison. / ~ elevation (Drawing) ‖ Vorderansicht f. ‖ vue f. de face. / ~ elevator (Mach tool) ‖ einseitiger Hebetisch m. ‖ releveur m. unilatéral. / ~ end door (Boil) ‖ Rauchkammertür f. ‖ porte f. de boîte à fumée. / ~ end plate (Boil) ‖ Vorderboden m. ‖ fond m. avant. / ~ face (Arch) ‖ Frontseite f.; Hauptseite f.; Vorderseite f. ‖ face f. / ~ focal plane ‖ vordere Brennebene f. ‖ plan m. focal avant. / ~ frame ‖ Vorderrahmen m. ‖ châssis m. avant. / ~ gate ‖ Portal n. ‖ portail m. / ~ hammer (Techn) ‖ Stirnhammer m. ‖ marteau m. frontal.

frontier station ‖ Grenzstation f. ‖ station f. frontière.

frontispiece ‖ Giebel m.; Giebelseite f. ‖ frontispice m.

front lens ‖ Frontlinse f. ‖ lentille f. frontale *ou* additionnelle. / ~ operating as

an objective ‖ Objektivvorsatzlinse f. ‖ lentille f. additionnelle se plaçant devant l'objectif.

front plate ‖ Stirnplatte f.; Vorderwand f. ‖ panneau m. avant; volet m. antérieur. / ~ of a firebox ‖ Stehkesselvorderwand f. ‖ paroi f. *ou* face f. avant de la caisse f. à feu. / ~ for a locomotive firebox ‖ Vorderplatte f. einer Lokomotivfeuerbüchse ‖ face f. avant d'une boîte à feu de locomotive.

front radiator ‖ Stirnkühler m. ‖ radiateur m. frontal. / ~ rake ‖ Brustwinkel m. ‖ angle m. d'affûtage. / ~ rollers pl. (Spinn) ‖ vorderste Riffelwalzen fpl. ‖ premiers cylindres mpl. cannelés. / ~ seat ‖ Vordersitz m. ‖ siège m. avant. / ~ shield ‖ Windschutzscheibe f. ‖ paravent m.; pare-brise m. / ~ side ‖ Vorderseite f. ‖ côté f. de front. / ~ side (Arch) ‖ Fassade f. ‖ façade f. / ~ spring ‖ Vorderfeder f. ‖ ressort m. avant. / ~ spring bracket ‖ Vorderfederbock m. ‖ support m. de ressort avant. / ~ stage ‖ Proszenium n. ‖ avant-scène f. / ~ strut ‖ Vorderstrebe f. ‖ jambe f. de force avant. / ~ tipper ‖ Vorderkipper m. ‖ wagonnet m. basculant en bout. / ~ view ‖ Vorderansicht f. ‖ vue f. de face. / to face a ~ wall with bricks ‖ eine Fassade f. mit Ziegeln verblenden *oder* mit Blendziegeln verkleiden ‖ revêtir une façade de briques.

front wheel ‖ Vorderrad n. ‖ roue f. avant. / ~ brake ‖ Vorderradbremse f. ‖ frein m. sur roue avant. / ~ drive ‖ Vorderradantrieb m. ‖ transmission f. à roue avant. / ~ motor ‖ Vorderradmotor m. ‖ moteur m. attaquant l'essieu avant. / ~ pair ‖ Vorderradsatz m. ‖ train m. de roues avant. / ~ tyre brake ‖ Vorderradreifenbremse f. ‖ frein m. sur pneu de la roue avant.

front window ‖ Doppelfenster n. ‖ contrefenêtre f.; contre-châssis m.

frost, to ~ over ‖ bereifen ‖ couvrir de gelée blanche. / ~ with sand and water (Glassm) ‖ matt schleifen ‖ dépolir *ou* égriser.

frost ‖ Frost m.; Rauhreif m. ‖ gelée f.; givre m. / ~ in the cracks of rocks (Geol) ‖ Spaltenfrost m. ‖ congélation f. dans les crevesses. / late ~ ‖ Spätfrost m. ‖ gelée f. printanière.

frost action ‖ Frostwirkung f. ‖ action f. du froid. / burst due to ~ ‖ Frostsprengung f. ‖ rupture f. par le gel. / ~-cleft *see* ~-cracked. / ~-cracked (Wood) ‖ frostrissig; eisklüftig ‖ gélif; gélis.

frosted glass ‖ Mattglas n.; Eisglas n. ‖ verre m. mat *ou* craquelé *ou* dépoli. / ~ globe ‖ mattierte Glasglocke f. ‖ globe m. dépoli. / ~ lamp ‖ mattierte Glühlampe f. ‖ lampe f. à incandescence dépolie.

frosting (Glassm) ‖ Mattschleifen n. ‖ dépolissage m.; dépolissement m.

frost nail ‖ Eisnagel m. ‖ clou m. à glace. / ~ preservative ‖ Frostschutzpräparat n. ‖ préservatif m. contre la gelée.

frostproof ‖ frostbeständig ‖ résistant à la gelée. / to be ~ ‖ frostbeständig sein ‖ résister au gel m.

frost shake ‖ Frostriß m. ‖ gélivure f.

frosty weather ‖ Frostwetter n. ‖ temps m. de gelée.

froth, to ‖ schäumen ‖ mousser; rocher. / ~ up ‖ aufschäumen ‖ mousser. / ~ over ‖ überschäumen ‖ passer par-dessus.

froth ‖ Schaum m. ‖ écume f.; mousse f.

frozen ‖ gefroren ‖ congelé; glacé. / ~ materials pl. ‖ gefrorene Baustoffe f. ‖ matériaux m. gelés. / ~ meat ‖ gefrorenes Fleisch n.; Gefrierfleisch n. ‖ viande f. congelée *ou* frigorifiée. / ~ soil ‖ Eisboden m. ‖ sol m. congelé.

fructose ‖ Fruktose f.; Fruchtzucker m. ‖ fructose f.; sucre m. de fruit.

fruit ‖ Frucht f. ‖ fruit m. / artificial ~ ‖ künstliche Frucht f. ‖ fruit m. artificiel. / canned ~ *see* preserved ~. / ~ cultivated in a hothouse ‖ im Treibhaus gezogene Frucht f. ‖ fruit m. produit en serre. / divided ~ ‖ zerteilte Frucht f. ‖ fruit m. divisé. / dried ~ ‖ gedörrte Frucht f. ‖ fruit m. torréfié. / dried ~s pl. ‖ Dörrobst n. ‖ fruits mpl. séchés. / eatable ~ ‖ eßbare Frucht f. ‖ fruit m. comestible. / exotic ~ ‖ exotische Frucht f. ‖ fruit m. exotique. / indigenous ~ ‖ einheimische Frucht f. ‖ fruit m. indigène. / preserved ~ ‖ eingemachte Frucht f.; Fruchtkonserve f. ‖ fruit m. confit; conserve f. de fruits. / ~ preserved in brandy ‖ Fruchtkonserve f. in Branntwein ‖ conserve f. de fruits à l'eau-de-vie. / ~ preserved in salt ‖ Salzkonserve f. ‖ conserve f. au sel. / ~ preserved in spirits ‖ in Branntwein eingemachte Frucht f. ‖ fruit m. conservé à l'eau-de-vie. / ~ preserved in sugar ‖ eingezuckerte Frucht f. ‖ fruit m. confit. / ~ preserved with syrup ‖ Kompottfrucht f. ‖ conserve f. de fruits au sirop. / ~ preserved in vinegar ‖ Essigkonserve f. ‖ conserve f. au vinaigre. / stewed ~ ‖ Kompott n. ‖ compote f. de fruits. / whole ~ ‖ ganze Frucht f. ‖ fruit m. entier.

fruit coffee ‖ Fruchtkaffee m. ‖ café m. de fruits. / ~ cooling room ‖ Obstkühlraum m. ‖ chambre f. froide pour des fruits. / ~ culture ‖ Obstbaumzucht f. ‖ arboriculture f. fruitière. / ~ drier ‖ Obsttrockner m. ‖ sécheur m. de fruits. / ~ drying ‖ Dörren n. von Obst ‖ séchage m. de fruits. / ~ essence ‖ Fruchtessenz f. ‖ essence f. de fruits. / ~ ether ‖ Fruchtäther m. ‖ éther m. de fruits. / ~ fibre ‖ Fruchtfaser f. ‖ fibre f. extraite du fruit. / ~ grower ‖ Obstbaumzüchter m. ‖ pomiculteur m. / ~ hurdle ‖ Obsthorde f. ‖ claie f. pour fruits.

fruit juice ‖ Fruchtsaft m. ‖ jus m. de fruits. / concentrated ~ ‖ eingedickter Fruchtsaft m. ‖ jus m. concentré de fruits.

fruit mill ‖ Obstpresse f.; Fruchtmühle f. ‖ presse f. à fruits; moulin m. à fruits charnus. / ~ paste ‖ Fruchtpaste f. ‖ pâte f. de fruits; pulpe f. de fruits desséchés. / ~ peel ‖ Fruchtschale f. ‖ écorce f. de fruit. / ~ picking instrument ‖ Obstpflücker m. ‖ cueilloir m. / ~ preserves manufacturer ‖ Fruchtkonservenfabrikant m. ‖ fabricant m. de conserves de fruits. / ~ preserving ‖ Obstverwertung f. ‖ industrie f. du fruit. / ~ press ‖ Obstpresse f.; Obstfruchtpresse f. ‖ presse f. à fruits charnus. / ~ presser ‖ Fruchtkelterer m. ‖ pressureur m. de fruits. / ~ pressing ‖ Fruchtkeltern n.; Obstkeltern n. ‖ pressurage m. de fruits. / ~ pulp ‖ Obstmark m. ‖ pulpe f. de fruits. / ~ room hand ‖ Fruchtzubereiter m. ‖ préparateur m. de fruits. / ~ shrub ‖ Obststrauch m. ‖ arbuste m. fruitier. / ~ sugar ‖ Fruchtzucker m.; Fruktose f.

levulose f.; fructose f.; sucre m. de fruit. / ~ syrup ‖ Fruchtsirup m. ‖ sirop m. de fruits. / ~ tincture ‖ Fruchtäther m. ‖ éther m. *ou* essence f. de fruits. / ~ tree ‖ Obstbaum m. ‖ arbre m. fruitier. / ~ vegetables pl. ‖ Fruchtgemüse n. ‖ fruit-légume m. / ~ wine ‖ Fruchtwein m.; Obstwein m. ‖ vin m. de fruits.

frustum of a pyramid ‖ Pyramidenstumpf m.; abgestumpfte Pyramide f. ‖ pyramide f. tronquée; tronc m. de pyramide.

frying pan ‖ Bratpfanne f. ‖ poêle f. à frire; sauteuse f. / ~ tube ‖ Bratröhre f. ‖ petit four m. à rôtir.

fuchsia ‖ Fuchsia f. ‖ fuchsia m.

fuchsine (Chem) ‖ Fuchsin n. ‖ fuchsine f. / ~ solution ‖ Fuchsinlösung f. ‖ solution f. de fuchsine.

fuchsite (Miner) ‖ Fuchsit m. ‖ fuchsite f.

fuel, to ‖ mit Brennstoff versorgen; Brennstoff liefern ‖ fournir *ou* pourvoir de combustible. / ~ ‖ Brennstoff m. einnehmen; tanken ‖ faire du combustible.

fuel ‖ Brennstoff m.; Kraftstoff m.; Feuerungsmaterial n.; Heizmaterial n.; Betriebsstoff m. ‖ combustible m. / artificial ~ ‖ künstlicher Brennstoff m. ‖ combustible m. artificiel *ou* aggloméré. / ~ having a high asphalt content ‖ asphaltreicher Brennstoff m. ‖ combustible m. riche en bitume. / carbonized ~ ‖ verkohlter Brennstoff m. ‖ combustible m. carbonisé. / flaming ~ ‖ flammender Brennstoff m. ‖ combustible m. flambant. / ~ absolutely free from water ‖ einwandfrei entwässerter Brennstoff m. ‖ combustible m. complètement purgé d'eau. / gaseous ~ ‖ gasförmiger Brennstoff m. ‖ combustible m. gazeux. / heavy ~ ‖ Schwerkraftstoff m. ‖ combustible m. lourd. / high-grade ~ ‖ hochwertiger Brennstoff m. ‖ combustible m. de supérieure qualité. / to inject ~ by means of a nozzle ‖ Brennstoff m. durch eine Düse einspritzen ‖ injecter le combustible à l'aide d'un gicleur. / light ~ ‖ Leichtbrennstoff m. ‖ combustible m. léger. / liquid ~ ‖ flüssiger Kraftstoff m. ‖ combustible m. liquide. / low-grade ~ ‖ minderwertiger Brennstoff m. ‖ combustible m. de qualité inférieure. / mineral ~ ‖ mineralischer Brennstoff m. ‖ combustible m. minéral *ou* d'origine minérale. / pulverized ~ ‖ Kohlenstaub m. aspiure f.; poussier m.; poussière f. de charbon. / ~ of recent geological formation ‖ jüngerer Brennstoff m. ‖ combustible m. des couches de formation récente. / solid ~ ‖ fester Brennstoff m. ‖ combustible m. solide. / thick ~ ‖ dickflüssiger Brennstoff m. ‖ combustible m. visqueux. / ~ worth transporting ‖ versandwürdiger Brennstoff m. ‖ combustible m. susceptible d'être transporté.

fuel, atomizing (of) the ‖ Zerstäubung f. des Brennstoffs ‖ pulvérisation f. du combustible. / ~ bunker ‖ Brennstoffbunker m. ‖ soute f. à combustible. / determination of the calorific value of ~s ‖ Brennstoffwertbestimmung f.; Heizwertbestimmung f. von Brennstoffen ‖ détermination f. de la puissance calorifique des combustibles. / ~ cam (Mot) ‖ Brennstoffnocken m. ‖ came f. de pompe à combustible. / ~ capacity (Loc) ‖ Brennstoffaufnahmefähigkeit f. ‖ capacité f. de la soute à combustible. / carbonization of ~s ‖ Brennstoffschwelung f. ‖ distilla-

tion f. lente de combustibles. / ~ conditions pl. ‖ Brennstoffverhältnisse npl. ‖ conditions fpl. d'approvisionnement en combustible. / ~ conduit see ~ piping. / ~ consumption ‖ Brennstoffverbrauch m.; Kraftstoffverbrauch m. ‖ consommation f. de combustible. / cost of ~ see ~ expenses. / delivery of ~ ‖ Brennstoffzufuhr f. ‖ amenée f. du combustible. / the situation of the ~ district is an advantage from the point of view of transport ‖ die Lage des Brennstoffgebietes ist frachtgünstig ‖ le bassin m. du combustible est favorablement situé au point de vue du transport. / ~ economy ‖ Kraftstoffersparnis f. ‖ économie f. de combustible.

fuel expenses pl. ‖ Brennstoffkosten pl. ‖ prix m. de combustible. / low ~ ‖ niedrige oder geringe Brennstoffkosten pl. ‖ bas prix m. de revient du combustible.

fuel, extracting ~s pl. from coal ‖ Brennstoffgewinnung aus Kohle ‖ extraction f. du combustible par le traitement du charbon. / ~ feed (Mot) ‖ Kraftstoffzuführung f.; Brennstoffzuleitung f. ‖ alimentation f. de combustible. / ~ feed pipe ‖ Brennstoffzuleitungsrohr n. ‖ tuyau m. d'amenée ou tube m. d'arrivée de combustible. / to expropriate ~ fields pl. in favour of the public electricity supply ‖ Brennstoffelder npl. zugunsten der öffentlichen Elektroversorgung enteignen ‖ exproprier des gisements mpl. de combustible au profit de la distribution publique de l'électricité. / ~ filler cap (Mot) ‖ Brennstoffeinfüllstutzen m. ‖ orifice m. de remplissage d'essence. / fine filter for ~ ‖ Brennstofffeinfilter m. ‖ filtre m. fin pour combustible. / ~ gage see ~ gauge. / ~ gas ‖ Heizgas n.; Brenngas n. ‖ gaz m. combustible ou de combustion. / gasification of ~s ‖ Brennstoffvergasung f. ‖ gazéification f. de combustibles. / ~ gauge ‖ Kraftstoffuhr f.; Brennstoffvorratsmesser m. ‖ indicateur m. de niveau de combustible; indicateur m. volumétrique du carburant. / heating of the ~ ‖ Anwärmen n. des Brennstoffs ‖ réchauffage m. du combustible. / injecting ~ after compression ‖ Brennstoffeinspritzung f. nach der Kompression ‖ injection f. du combustible après la compression. / ~ injection ‖ Brennstoffeinspritzung f.; Brennstoffzuführung f. ‖ injection f. de combustible. / ~ inlet pipe see ~ feed pipe. / ~ level ‖ Kraftstoffstand m. ‖ niveau m. de combustible. / ~ level gauge see ~ gauge.

fuel oil ‖ Heizöl n.; Brennöl n.; Feuerungsöl n. ‖ huile f. à brûler ou à chauffer; huile f. ou pétrole m. combustible. / ~ for explosion motors ‖ Treiböl n. für Explosionsmotoren ‖ huile f. combustible pour moteurs à explosion.

fuel pipe ‖ Brennstoffleitung f.; Kraftstoffrohr n. ‖ tuyau m. de combustible. / ~ pipe union ‖ Anschlußstutzen m. der Brennstoffleitung ‖ raccord m. du tuyau de combustible. / ~ piping ‖ Brennstoffleitung f. ‖ conduite f. de combustible. / ~ preheater ‖ Brennstoffvorwärmer m. ‖ réchauffeur m. de combustible.

fuel pump ‖ Brennstoffpumpe f. ‖ pompe f. à combustible. / ~ casing ‖ Brennstoffpumpengehäuse n. ‖ carter m. de la pompe à combustible. / ~ liner ‖ Brenn-

stoffpumpenlaufbuchse f. ‖ fourreau m. de la pompe à combustible. / ~ piston ‖ Brennstoffpumpenkolben m. ‖ piston m. de pompe à combustible. / ~ tappet ‖ Brennstoffpumpenstößel m. ‖ tige f. poussoir de la pompe à combustible.

fuel reserve tank ‖ Kraftstoffhilfsbehälter m. ‖ réservoir m. à combustible de réserve. / ~ reservoir see ~ tank. / ~ saving see ~ economy. / ~ storage ‖ Brennstofflagerung f. ‖ stockage m. de combustible. / ~ strainer ‖ Kraftstoffreiniger m.; Brennstofffilter n. ‖ filtre m. de combustible. / ~ supply ‖ Brennstoffzufuhr f. ‖ admission f. ou arrivée f. de combustible. / ~ system diagram ‖ Brennstoffleitungsschema n. ‖ schéma m. de la circulation du combustible. / ~ tank ‖ Brennstoffbehälter m.; Kraftstoffbehälter m. ‖ réservoir m. à combustible. / ~ utilization ‖ Brennstoffausnutzung f. ‖ utilisation f. de combustible. / ~ valve ‖ Brennstoffventil n. ‖ soupape f. à combustible. / outlet velocity of the ~ vapour ‖ Ausströmungsgeschwindigkeit f. des Brennstoffdampfes oder der Brenngase ‖ vitesse f. d'écoulement de la vapeur de combustible. / ~ weight ‖ Brennstoffgewicht n. ‖ poids m. de combustible.

fugacious see fugitive.

fugitive (Dyer) ‖ unecht; flüchtig ‖ fugitif. / ~ colour ‖ unechte Farbe f. ‖ couleur f. fugitive.

fulcrum (Of a lever) ‖ Stützpunkt m.; Gelenkpunkt m.; Drehpunkt m. ‖ point m. d'appui ou d'articulation. / ~ of a lever with equal arms forming a notch ‖ als Kimme ausgebildeter Drehpunkt m. eines gleicharmigen Hebels ‖ centre m. de rotation d'un levier à bras égaux constitué par une encoche.

fulcrum bracket ‖ Hebelträger m. ‖ support m. de point fixe. / ~ pin ‖ Drehbolzen m. ‖ tourillon m.

fulfilment, place of ‖ Erfüllungsort m. ‖ lieu m. de payement; élection f. de domicile.

fulguration (Galv) ‖ Fulguration f. ‖ fulguration f.

fulgurite ‖ Blitzröhre f. ‖ tube m. à éclairs. / ~ (Miner) ‖ Fulgurit m. ‖ fulgurite f.

fuliginous ‖ rußig; rußfarbig ‖ fuligineux.

full, to (Clothm) ‖ walken ‖ fouler.

full ‖ voll ‖ plein; rempli. / ~ admission ‖ Vollfüllung f. ‖ pleine admission f.

full-centre vault (Arch) ‖ Rundbogengewölbe n. ‖ voûte f. en cintre ou en plein cintre.

full-edged ‖ vollkantig ‖ équarri m. à vives arêtes.

full-faced (Print) ‖ fett ‖ plein. / ~ type ‖ fette Schrift f. ‖ caractère m. plein.

full-grown (Forest) ‖ hochstämmig ‖ de haute futaie f. / ~ workman ‖ volljähriger Arbeiter m. ‖ ouvrier m. adulte.

fulled goods pl. ‖ Walkware f. ‖ articles mpl. ou tissus mpl. foulés. / ~ stuff (Weav) see ~ goods.

fuller (Clothm) ‖ Walkmüller m.; Tuchwalker m. ‖ foulonnier m.; foulon m.; fouleur m. / ~ (Forg) ‖ runder Setzhammer m. ‖ chasse f. ronde; dégorgeoir m. / ~ (Shipb) ‖ Füllungseisen n. ‖ matoir m.

fuller's earth ‖ Walkerde f.; Seifenerde f. ‖ terre f. à foulon; argile f. savonneuse. / ~ oil ‖ Walkfett n. ‖ huile f. à foulon. / ~ soap ‖ Walkseife f. ‖ savon m. à foulon.

fulling of cloth ‖ Tuchwalke f. ‖ foulonnage m. ou foulage m. du drap. / hosiery ~ ‖ Wirkwarenwalke f. ‖ foulage m. de bonneterie.

fulling club see ~ pestle. / ~ fat ‖ Walkfett n. ‖ graisse f. de foulage. / ~ grease see ~ fat. / ~ machine (Clothm) ‖ Walkmaschine f. ‖ machine f. à fouler. / ~ mill ‖ Walkmühle f.; Walkerei f.; Hammerwalke f. ‖ moulin m. à foulons; foulerie f.; foulon m. / ~ miller ‖ Walker m. ‖ fouleur m. / ~ pestle (Curr) ‖ Pumpkeule f. ‖ enfonçoir m. à tête. / ~ soap see fuller's soap. / ~ stock (Hammer) ‖ Walkhammer m. ‖ maille m. à fouler. / ~ stocks pl. ‖ Kurbel- und Hammerwalke f. ‖ foulon m. à manivelle. / ~ trough ‖ Walktrog m. ‖ fouloir m.

full-load ‖ Vollbelastung f.; Vollast f. ‖ pleine charge f. / to start under ~ ‖ unter Vollast f. anlaufen ‖ démarrer sous pleine charge f.

full-load current ‖ Vollaststrom m. ‖ courant m. à pleine charge. / ~ winding ‖ Wicklung f. für volle Last ‖ enroulement m. pour pleine charge.

full-loaded ‖ vollbelastet ‖ à pleine charge f.

full moon ‖ Vollmond m. ‖ pleine lune f.

fullness indicator (Balloon) ‖ Prallanzeiger m. ‖ indicateur m. de plénitude ou de remplissage du ballon.

full particulars pl., to give ~ of ... ‖ Auskunft f. erteilen über ... ‖ donner des renseignements mpl. sur ...

full power (of attorney) ‖ Vollmacht f. ‖ plein pouvoir m. / to furnish with ~s pl. ‖ mit Vollmacht f. ausstatten ‖ munir d'un pouvoir.

full-range ‖ volles Maß n. ‖ mesure f. comble. / ~ scale ‖ natürliche Größe f. ‖ vraie grandeur f. / ~ speed ‖ größte Geschwindigkeit f. ‖ vitesse f. maxima. / ~ trough ‖ Fülltrog m. ‖ auge f. à remplir.

fully-automatic machine ‖ vollselbsttätige oder vollautomatische oder ganzselbsttätige Maschine f. ‖ machine f. entièrement automatique. / multiple-spindle ~ ‖ Mehrspindelvollautomat m. ‖ machine f. automatique à broches multiples. / ~ for producing parts from bar-stock or for bar-work see fully-automatic bar machine.

fully automatic bar machine ‖ Stangenvollautomat m.; Vollautomat m. für Stangenarbeit ‖ machine f. automatique pour le travail de la barre.

fulminating gold ‖ Knallgold n.; Goldoxydammoniak n. ‖ or m. fulminant; aurate m. d'ammoniaque. / ~ mercury ‖ Knallquecksilber n. ‖ fulminate m. de mercure; mercure m. fulminant. / ~ powder ‖ Friktionspulver n. ‖ poudre m. fulminante. / ~ silver. ‖ Knallsilber n.; knallsaures Silberoxyd n. ‖ argent m. fulminant.

fulminic acid ‖ Knallsäure f. ‖ acide m. fulminique.

fume see also fumes ‖ Dunst m.; Dampf m. ‖ fumée f. / ~ cup-board (Chem) ‖ Abzugschrank m. ‖ hotte f. fermée. / ~ dispersion installation ‖ Entnebelungsanlage f. ‖ installation f. pour dissiper le brouillard.

fumeless ‖ dunstlos ‖ sans fumée.

fumes pl. see also smoke ‖ Rauch m.; Dampf m.; Dunst m. ‖ fumées fpl.

fumigate, to ~ a vessel ‖ ein Schiff zur Rattenvertilgung ausräuchern oder räu-

chern || fumiger *ou* parfumer un navire.

fumigating candle || Räucherkerze f. || pastille f. (odorante) à brûler. / ~ paper || Räucherpapier n. || papier m. d'Arménie. / ~ powder || Räucherpulver n. || poudre f. à parfumer.

fumigation || Räucherung f. || fumigation f. / ~ (Of a zinc-plate) || Anräucherung f. || fumigation f.

fuming || rauchend || fumant. / ~ nitric acid || rote rauchende Salpetersäure f. || acide m. azotique fumant. / ~ sulphuric acid || Oleum n. || acide m. sulfurique fumant.

fumitory (Botan) || Erdrauch m. || fumeterre f.

fumivorous || rauchverzehrend || fumivore.

function (Service) || Dienstleistung f. || service m. / ~ (Math) || Funktion f. || fonction f. / algebraical ~ || algebraische Funktion f. || fonction f. algébrique. / circular ~ || Winkelfunktion f.; trigonometrische Funktion f. || fonction f. circulaire *ou* trigonométrique. / composite ~ || zusammengesetzte Funktion f. || fonction f. composée. / compound ~ || Funktion f. von Funktionen || fonction f. de fonctions. / derived ~ || abgeleitete Funktion f. || fonction f. dérivée. / elliptic ~ || elliptische Funktion f. || fonction f. elliptique. / explicit ~ || explizite *oder* entwickelte Funktion f. || fonction f. explicite. / homogenous ~ || homogene Funktion f. || fonction f. homogène. / implicit ~ || implizite *oder* unentwickelte Funktion f. || fonction f. implicite. / inverse ~ || gegenseitige *oder* reziproke Funktion f. || fonction f. inverse. / simple ~ || einfache Funktion f. || fonction f. simple. / spherical ~ || Kugelfunktion f.; sphärische Funktion f. || fonction f. sphérique. / transcendental ~ || transzendente Funktion f. || fonction f. transcendante. / trigonometrical ~ || trigonometrische Funktion f. || fonction f. trigonométrique.

fund, to || in Kapital umwandeln; kapitalisieren; fundieren || fonder; consolider.

funds pl. || Betriebskapital n.; Grundkapital n.; Fonds m. || fonds mpl.; capital m. / to call for ~ || Kapital einfordern || appeler du capital. / ~ at one's disposal || verfügbares Kapital n. || fonds mpl. disponibles *ou* mobilisables; disponibilités fpl. / guarantee and del credere ~ || Delkredere- und Garantiefonds m. || fonds m. de ducroire et de garantie. / sinking ~ || Tilgungsfonds m. || fonds m. d'amortissement.

fundament (Build) || Gründung f.; Grundbau m. || Fundament n. || fondement m.

fundamental circuit || Grundschaltung f.; Grundstromkreis m. || schéma m. de principe; circuit m. fondamental. / ~ equation (Math) || Fundamentalgleichung f. || équation f. fondamentale. / ~ frequency (Electr) || Grundfrequenz f. || fréquence f. fondamentale. / ~ oscillation || Grundschwingung f. || oscillation f. fondamentale. / ~ tone (Tel) || Grundton m. || son m. fondamental. / ~ wave || Grundwelle f. || onde f. fondamentale. / ~ wave of aerial || Grundwelle f. der Antenne || onde f. fondamentale de l'antenne.

fundus of the eye || Augenhintergrund m. || fond m. de l'œil.

funeral carriage || Leichenwagen m.; Bestattungswagen m.; Begräbniswagen m.;

|| voiture f. funèbre. / ~ lowering device || Leichenversenkvorrichtung f. || dispositif m. pour descendre des cadavres. / ~ ornament of beads || Grabschmuck m. aus Perlen || ornement m. funéraire en perles. / ~ undertaking || Beerdigungsinstitut n. || entreprise f. de funérailles. / ~ vault || Grabgewölbe n. || caveau m.; sépulcre m.; sépulture f. / ~ wreath || Grabkranz m. || couronne f. funéraire.

fungicides pl. || Schwammtod m.; Schwammvernichtungsmittel n. || fungicides mpl.

fungoid growth || Pilzbildung f. || formation f. de champignons.

fungous growth || Hausschwamm m.; Holzschwamm m. || champignon m. du bois; bolet m. destructeur.

fungus || Pilz m.; Schwamm m. || champignon m.; fungus m.

funicular polygone (Mech) || Seilpolygon n. || polygone m. funiculaire.

funicular railway || Hängebahn f.; chemin m. de fer à suspension. / electric ~ || Elektrohängebahn f. || chemin m. de fer électrique à suspension.

funnel || Trichter m. || entonnoir m. / ~ (Found) || Gußtrichter m.; Gußloch n. || gueule f. *ou* jet m. *ou* ouverture f. de moule. / ~ (Build; Steam; Shipb) || Schornstein m.; Esse f.; Rauchrohr m.; Schlot m.; Kamin m. || cheminée f. / ~ (Outlet pipe) || Abzugsrohr n. || tuyau m. d'échappement. / to sweep the ~ || den Schornstein m. fegen || ramoner la cheminée. / acoustic ~ (dictating machine) || Schallbecher m. || pavillon m. / discharging ~ || Ablauftrichter m. || trémie f. de décharge. / drawing-off ~ || Abfülltrichter m. || trémie f. de vidange. / ~ for letting down (Shipb) || umklappbarer Schornstein m.; Schornstein m. zum Umklappen || cheminée f. à rabattement. / light-proof connecting ~ (Opt) || Lichtverschlußtrichter m. || cône m. à intercepter le faux jour. / ~ of a mine || Minentrichter m. || entonnoir m. *ou* excavation f. de mine. / ~ (Found) || Einguß m. || jet m.; trompe f. *ou* trou m. de coulage des moules. / ~ of sheet-iron || Blechschornstein m. || cheminée f. en tôle. / ~ telescope ~ (Of a steamship) || Teleskopschornstein m. || cheminée f. à coulisse.

funnel flange || Trichterflansch m. || bride f. d'entonnoir. / ~ holder || Trichterstativ n. || support m. d'entonnoir. / ~ pipe *see* ~ (Build etc). / ~ shroud (Shipb) || Schornsteinstag n. || étai m. *ou* hauban m. de cheminée. / ~ stag *see* ~ shroud. / ~ stand (Chem) || Trichtergestell n. || support m. d'entonnoir. / ~ tube || Trichterrohr n.; Einfülltrichter m. || tube m. d'entonnoir.

fur, to ~ the boiler || den Kessel m. ausklopfen || piquer *ou* désincruster *ou* détartrer la chaudière.

fur || Pelz m.; Pelzwerk n. || fourrure f.; pelleterie f. / ~ (Boil) || Kesselstein m. || couche f. de tartre; calcin m.; incrustations fpl.; tartre m. / boiler ~ *see* fur (Boil). / calcareous ~ || Kesselstein m.; Pfannenstein m. || écailles fpl. du sel. / raw ~ || Rohpelz m. || fourrure f. brûte. / cooling room for raw ~s pl. || Kühlraum m. für Rohpelze || chambre f. froide pour des fourrures brûtes. / ready-made ~ || Konfektionspelz m. || fourrure f. confectionnée. / cooling room for ready-made

~s pl. || Kühlraum m. für Konfektionspelze || chambre f. froide pour fourrures confectionnées.

furbish, to || schleifen; polieren || frotter; polir; adoucir; fourbir.

furbisher || Drahtscheurer m. || éclaircisseur m.

fur bleaching || Pelzbleichen n. || décoloration f. de fourrures. / ~ boot || Pelzstiefel m. || bottine f. fourrée. / ~ boots and shoes pl. || Pelzschuhwerk n. || chaussure f. en fourrure. / bower of ~s || Haarfacher m. || arçonneur m. de poils.

furca || Gabel f.; Forke f. || fourche f.

fur cap || Pelzmütze f. || casquette f. en fourrure.

furcation || Gabelteilung f.; Gabelung f. || fourchure f.; bifurcation f. / ~ of a river || Flußgabelung f. || bifurcation f. d'un fleuve.

fur collar || Pelzkragen m. || palatine f.

fur deposit || Kesselsteinablagerung f. || entartrage m. / ~ removing || Kesselsteinbeseitigung f. || détartrage m.

fur dyeing || Pelzfärben n. || teinture f. de fourrures. / ~ machine || Pelzfärbemaschine f. || machine f. à teindre les fourrures.

fur felt hat || Haarhut m. || chapeau m. en feutre de poil. / ~ glove || Pelzhandschuh m. || gant m. en pelleterie. / ~ goods pl. || Pelzwaren fpl. || articles mpl. de pelleterie. / industry of ~s || Rauchwarenindustrie f. || industrie f. de la fourrure.

furl, to ~ a flag (Mar) || eine Flagge f. aufrollen || ferler un pavillon. / ~ a sail || ein Segel n. auftuchen || plier une voile. / ~ the sails pl. (Mar) || die Segel npl. festmachen || serrer *ou* ferler les voiles.

fur lining || Pelzfutter n. || doublure f. de pelisse; fourrure f.

furlong = 201,167 m. || Achtelmeile f. || furlong m.

fur lustring || Pelzlustrieren n. || lustrage m. de fourrures. / ~ mantilla || Pelzmantille f. || pelisse f.

furnace || Ofen m. || four m.; fourneau m. / ~ (Boiler) || Feuerung f. || foyer m. / the acid ~s were converted into basic ones || die sauren Öfen mpl. wurden in basische umgewandelt || les fours mpl. à sole acide furent transformés en fours à sole basique. / amalgam distilling ~ || Amalgamdestillationsofen m. || four m. pour la distillation des amalgames. / annealing ~ || Glühofen m. || four m. à recuire. / to arm a ~ || einen Ofen m. beschlagen || armer un fourneau. / asphalt ~ || Asphaltofen m. || four m. à asphalte. / automatic(al) ~ || selbsttätige Feuerung f. || foyer m. automatique. / bar heating ~ || Vorwärmeofen m. für Stabeisen || four m. à réchauffer les barres d'acier. / blast ~ || Hochofen m. || haut-fourneau m. / to blow down the blast ~ || den Hochofen m. ausgehen lassen || mettre hors feu le haut-fourneau. / ~ of a boiler || Kesselfeuerung f. || foyer m. d'une chaudière. / ~ for brown coal || Feuerung f. für Rohbraunkohle || foyer m. à lignite. / calcining ~ || Brennofen m. || four m. à calciner. / carburizing ~ || Karbonisierofen m. || four m. à cémenter aux gaz carburants. / cementation ~ || Zementierofen m. || four m. à cémenter. / charcoal blast ~ || Holzkohlenhochofen m. (mit Holzkohlen betriebener Hochofen) || haut-fourneau m. au charbon de

bois. / to charge the ~ || den Ofen m. beschicken || alimenter le foyer. / chemical ~ || Retortenofen m. || fourneau m. à cornues. / ~ with closed breast || Ofen m. mit geschlossener Brust || fourneau m. à poitrine fermée. / coal-dust ~ || Kohlenstaubfeuerung f. || foyer m. à poussier de charbon. / cold blast ~ || Kaltgebläseofen m. || fourneau m. à vent froid. / ~ with automatic charging and discharging of the work for use in quantity production || Ofen m. mit selbsttätiger Beschickung und Abkühlung des Massengutes || four m. à alimentation et refroidissement automatiques pour grandes séries. / continuous heating ~ || Rollofen m. || four m. roulant. / crucible ~ || Tiegelofen m. || four m. à creuset. / ~ for the distillation of sulphur || Schwefelläuterofen m.; Galeerenofen zur Läuterung des Schwefels || fourneau m. de galère pour la distillation de soufre. / ~ for distiller's wash || Schlempeofen m. || four m. à vinasse. / ~ under draught || Windfeuerung f. || four m. à tirage forcé. / electrical ~ || elektrischer Ofen m. || fourneau m. électrique. / steel making in the electric ~ || Stahlgewinnung f. im Elektroofen || production f. de l'acier dans un four électrique. / electrically heated ~ || elektrisch beheizter Ofen m. || four m. à chauffage électrique. / electric arc ~ || Lichtbogenofen m. || four m. électrique à arc. / electric muffle ~ || elektrischer Muffelofen m. || fourneau m. à moufle électrique. / electric resistance ~ || elektrischer Widerstandsofen m. || four m. électrique à résistance. / external ~ || Außenfeuerung f. || foyer m. extérieur. / gas ~ || Gasfeuerung f. || foyer m. à gaz. / gas-fired ~ || Gasfeuerung f. || foyer m. à gaz. / ~ for gas works || Gaswerkofen m. || four m. pour usines à gaz. / hammer ~ || Schmiedeofen m. || four m. à forger. / heating ~ || Wärmofen m. || four m. à rechauffer. / heat-treatment ~ || Wärmebehandlungsofen m. || four m. pour le traitement thermique. / horizontal ~ || liegender Ofen m. || Flammofen m. || four m. dormant. / ~ with horizontal grate || Planrostfeuerung f. || foyer m. à grille horizontale. / inclined grate ~ || Schrägrostfeuerung f. || foyer m. à grille inclinée. / induction ~ || Induktionsofen m. || four m. à induction. / ~ for the industry || Industrieofen m. || four m. pour l'industrie. / internal ~ || Innenfeuerung f. || foyer m. intérieur. / light arc ~ || Lichtbogenofen m. || four m. à arc. / mechanical ~ || mechanische Feuerung f. || foyer m. mécanique. / melting ~ || Schmelzofen m. || four m. de fusion ou à fondre. / ~ with movable hearth || Ofen m. mit beweglichem Herd || four m. à sole mobile. / oil-firing ~ || Ölfeuerung f. || foyer m. à huile. / open-hearth ~ || Siemens-Martin-Ofen m. || four m. Siemens-Martin / ~ out of blast || ausgeblasener Ofen m. || fourneau m. arrêté. / the ~ is overcharged || der Ofen m. ist übersetzt || le fourneau m. est surchargé ou irrégulier. / ~ for peat || Torffeuerung f. || foyer m. à tourbe. / porcelain calcining ~ || Porzellanbrennofen m. || four m. à cuire la porcelaine. / pot ~ (Glassm) || Hafenofen m. || four m. à pots. / ~ for producing plates for sets of teeth || Ofen m. zur Herstellung von Gebißplatten || four m.

à fabriquer les plaques de dentiers. / pulverized coal ~ || Kohlenstaubfeuerung f. || foyer m. à charbon pulvérisé. / push-heating ~ || Stoßofen m. || four m. à secousses. / recuperative ~ || Rekuperativfeuerung f. || foyer m. de récupération de chaleur. / regenerative ~ || Regenerativofen m.; Regenerativofen m. || four m. à régénérateur au gaz. / reheating ~ || Wärmofen m. || four m. à rechauffer. / resistance ~ || Widerstandsofen m. || four m. à résistance. / reverberatory ~ || Flammofen m. || four m. à reverbère. / roasting ~ || Röstofen m. || four m. de grillage. / rolling mill ~ || Walzwerksofen m. || four m. de laminoirs. / rotary ~ || Drehherdofen m. || four m. tournant. / to set a ~ to work || einen Ofen m. in Betrieb setzen; einen Hochofen m. anblasen || mettre un fourneau en activité ou à feu. / smelting ~ || Schmelzofen m.; Kupolofen m. || four m. de fusion; cubilot m. / smoke-consuming ~ || rauchverzehrende Feuerung f. || foyer m. fumivore. / smokeless ~ || rauchfreie Feuerung f. || foyer m. fumivore. / steam boiler ~ || Dampfkesselfeuerung f. || foyer m. de chaudière. / ~ with stepped grate || Treppenrostfeuerung f. || foyer m. avec grille à gradins. / tempering ~ || Temperofen m. || four m. de cémentation. / three-phase current ~ || Drehstromofen m. || four m. à courant triphasé. / tilting ~ || Kippofen m. || four m. culbutant. / travelling grate ~ || Wanderrostfeuerung f.; Wanderfeuerung f. || foyer m. à grille mouvante ou mobile. / ~ for waste || Feuerung f. für Abfälle || foyer m. à déchets. / water jacket ~ || Wassermantelofen m.; wassergekühlter Ofen m. || four m. à chemise d'eau. / welding ~ || Schweißofen m. || four m. à souder. / ~ for wood || Holzfeuerung f. || foyer m. à bois. / the ~ works || der Ofen geht || le fourneau va ou marche.

furnace bar || Roststab m. || barre f. du gril; barreau m. de grille. / ~ blast || Hochofenwind m. || vent m. de haut-fourneau. / drying the ~ blast || Trocknung f. des Hochofenwindes || séchage m. de l'air des hauts-fourneaux. / ~ bridge || Hochofenbrücke f. || autel m. ou pont m. de hauts-fourneaux. / ~ cadmia || Ofenbruch m.; Gichtschwamm m.; Tutia f. || cadmie f. ou calamine f. de fourneau; tutie f.; (Belg:) kiess f.; spode m. / ~ calamine see ~ cadmia. / ~ casting || Ofengußstück n. || pièce f. de fonte du four. / ~ charge || Charge f.; Gicht f.; Satz m. || charge f. du fourneau. / ~ coal || Ofenkohle || charbon m. pour fours métallurgiques. / ~ coke || Hochofenkoks m. || coke m. métallurgique ou de hauts-fourneaux. / ~ construction || Feuerungsbauart f. || système m. de foyer. / ~ cover see ~ door. / ~ door || Ofentür f.; Ofendeckel m. || porte f. de fourneau. / the ~ doors pl. run on rollers for lateral displacement || die Ofentüren mpl. laufen auf Rollen mit seitlicher Verschiebung || les portes fpl. des fours se manœuvrent latéralement sur galets. / ~ door lifting device || Ofentürhebevorrichtung f. || dispositif m. de soulèvement pour portes de four. / ~ dust || Gichtstaub m. || poussière f. de haut-fourneau. / ~ fittings pl. || Ofenzubehör n.; Ofenarmatur f. || armature f. pour four; accessoires mpl. de four. / ~ fore-

man || Hochofenmeister m. || chef m. de four; premier fondeur m.; garde-feu m. / ~ fore-plate (Forge) || Schaffplatte f. || seuil m. du four. / ~ fork || Ofengabel f.; Forke f. || fourchette f. de fourneau. / ~ front || Feuergeschränk n. || devanture f. de foyer. / ~ gas || Gichtgas n. || gaz m. de gueulard. / ~ generator for heating the retorts || Ofengenerator m. für Heizung der Retorten || générateur m. de four servant au chauffage des cornues. / ~ grate || Ofenrost m. || grille f. de four. / ~ hoist || Gichtaufzug m. || chargeur m.; ou monte-charge m. ou élévateur m. de haut-fourneau. / ~ hole || Ofenloch n. || encadrement m. de fourneau. / ~ house || Ofenhaus n. || salle f. de fours. / ~ installation || Feuerungsanlage f. || installation f. de foyer. / iron shell of ~ || Blechmantel m. des Ofens || enveloppe f. en tôle du four. / ~ lead || Herdblei n. || plomb m. de foyer. / ~ lining || Ofenausmauerung f. || maçonnage m. ou revêtement m. du four. / ~ lining of silica bricks || saures Ofenfutter n. || revêtement m. acide du four.

furnaceman || Ofenarbeiter m. || enfourneur m. / ~ (Glassm) || Glasofenarbeiter m.; fliqueur m. / ~ (Found) || Gießer m.; Schmelzer m. || fondeur m.

furnace mantle of a blast-furnace || Rauhgemäuer n. eines Hochofens || maçonnerie f. extérieure d'un haut-fourneau. / ~ nickelplating || Herdvernickelung f. || nickelage m. pour foyers de cuisine. / flat disc with ~ opening stamped out || flacher Boden m. mit ausgepreßtem Rohrloch || disque m. plat avec bride à tube obtenue par emboutissage. / ~ plate || Ofenblech n. || tôle f. de foyer. / ~ shovel || Ofenschaufel f. || pelle f. de four. / ~ steel || Rohstahl m. || acier m. naturel ou brut. / ~ tongs pl. || Ofenzange f. || pinces fpl. pour fours. / ~ tools pl. || Schürgerät n.; Schürzeug n. || outils mpl. pour foyers. / ~ top || Gicht f. eines Hochofens || gueulard m.; gueule f.

fur nailer || Fellnagler m. || cloueur m. de peaux.

furnish, to ~ with || versehen mit; ausrüsten mit || garnir de; munir de. / ~ with leather (Mach) || beledern || garnir de cuir. / ~ with points || zacken || denteler; déchiqueter. / ~ with provisions || mit Proviant m. versehen oder versorgen || fournir en vivres mpl.; avitailler.

furnisher || Lieferant m. || fournisseur m.

furnishing for apothecary's shops || Apothekeneinrichtung f. || installation f. de pharmacies. / ~ with boards || Bebohlung f. || planchéiage m.

furniture || Wohnungseinrichtung f.; Zimmerausstattung f.; Möbel npl. || ameublement m.; meubles mpl. / ~ (Fittings) || Garnitur f. || garniture f.; monture f. / artistic ~ || Kunstmöbel n. || meuble m. artistique. / ~ of bamboo || Bambusmöbel n. || meuble m. en bambou. / bentwood ~ || Wiener Möbel npl. || meubles mpl. en bois ployé. / ~ of casement || Fensterbeschlag m.; Windeisen n. || armature f. de vitrage; ferrure f. de croisée. / ~ for children || Kindermöbel npl. || meubles mpl. pour enfants. / church ~ || Kirchenmöbel npl. || mobilier m. religieux; meubles mpl. d'église. / concrete-made ~ || Betonmöbel n. || meuble m. en béton. / fancy ~ || Luxusmöbel n. || meuble m.

de luxe. / fine-art upholstered ~ ‖ Luxus-polstermöbel n. ‖ meuble m. rembourré de luxe. / garden ~ ‖ Gartenmöbel npl. ‖ meubles mpl. de jardin. / iron ~ ‖ eisernes Möbel n. ‖ meuble m. en fer. / iron ~ for gardens ‖ eiserne Gartenmöbel npl. ‖ meubles mpl. de jardin en fer. / kitchen ~ ‖ Küchenmöbel npl. ‖ meubles mpl. de cuisine. / ~ for large kitchen ‖ Groß-kücheneinrichtung f. ‖ installation f. de grande cuisine. / leather ~ ‖ Ledermöbel npl. ‖ meubles mpl. de cuir. / ornamental ~ ‖ Luxusmöbel npl. ‖ meubles mpl. de luxe. / of osier ‖ Möbel n. aus Weide ‖ meuble m. en osier. / ~ from plaited osiers ‖ Möbel n. aus Flechtstoffen ‖ meuble m. en vannerie. / ~ from plaited twigs see ~ from plaited osiers. / polished and varnished ~ ‖ Schleiflackmöbel npl. ‖ meubles mpl. polis et laqués. / ~ of rush ‖ Rohrmöbel n. ‖ meuble m. en jonc. / school ~ ‖ Schulmöbel npl. ‖ mobilier m. scolaire. / small ~ ‖ Kleinmöbel n. ‖ petit meuble m. / small art ~ ‖ Klein-kunstmöbel npl. ‖ meubles mpl. mignons. / steel ~ ‖ Stahlmöbel n. ‖ meuble m. en acier. / to take-off the ~ (Print) ‖ das Format n. abschlagen ‖ abattre ou en-lever la garniture. / toy ~ ‖ Kindermöbel npl. ‖ meubles mpl. d'enfants. / up-holstered ~ ‖ Polstermöbel mpl. ‖ meub-les mpl. rembourrés. / ~ of wood ‖ Holz-möbel npl. ‖ meubles mpl. en bois.

furniture, beater for upholstered ~ ‖ Mö-belklopfer m. ‖ bat-meuble m. / ~ belt ‖ Möbelgurt m. ‖ sangle f. de meuble. / ~ brush ‖ Möbelpinsel m. ‖ pinceau m. pour meubles. / ~ cloth ‖ Möbelstoff m. ‖ étoffe f. pour meubles. / ~ copper fit-tings pl. ‖ kupferner Möbelbeschlag m. ‖ décor m. d'ameublement en cuivre. / ~ cover ‖ Staubdecke f. ‖ housse f. / ~ covering ‖ Möbelstoff m. ‖ étoffe f. pour meubles. / ~ damask ‖ Möbeldamast m. ‖ tissu m. damassé pour meubles. / ~ filler ‖ Möbelverkitter m. ‖ mastiqueur m. de meubles. / ~ fitter ‖ Beschlägearbeiter m. ‖ garnisseur m.

furniture fittings pl. ‖ Möbelbeschlag m. ‖ garniture f. pour meubles. / machine for making ~ ‖ Möbelbeschlagherstellungs-maschine f. ‖ machine f. à fabriquer des garnitures de meubles. / metal ~ ‖ Möbelbeschläge mpl. aus Metall ‖ gar-nitures fpl. métalliques pour meubles.

furniture gimp ‖ Gipürtüll m. ‖ guipure f. d'ameublement. / ~ leather ‖ Mö-belleder n. ‖ cuir m. à meubles. / ~ lock ‖ Möbelschloß n. ‖ serrure f. pour meub-les. / ~ maker ‖ Möbeltischler m. ‖ me-nuisier m. en meubles; ébéniste m. / art ~ making ‖ Kunsttischlerei f. ‖ menuiserie f. d'art. / ~ mountings pl. ‖ Möbelbe-schläge mpl. ‖ garnitures fpl. pour meub-les. / object of ~ ‖ Zimmerausstattungs-gegenstand m. ‖ objet m. d'ameuble-ment. / ornamental fittings for ~ ‖ Mö-belverzierung f. ‖ ornement m. pour meubles. / ~ painter ‖ Holzmaler m. ‖ peintre m. sur bois. / piece of ~ ‖ Möbel n. ‖ meuble m. / ~ plush ‖ Möbelplüsch m. ‖ peluche f. pour meubles. / ~ polish ‖ Möbelpolitur f. ‖ brillant m. ou encaus-tique m. pour meubles. / ~ polisher ‖ Mö-belpolierer m. ‖ polisseur m. de meubles. / ~ polisher ‖ Möbelwachser m. ‖ cireur m. en ébénisterie. / ~ removal by auto-mobiles ‖ Möbeltransport m. mit Kraft-

wagen ‖ transport m. de meubles par automobiles. / ~ removing ‖ Möbeltrans-port m. ‖ transport m. de meubles. / ~ trimmings pl. ‖ Möbelposamenten npl. ‖ passementerie f. pour meubles. / ~ van ‖ Möbelwagen m. ‖ voiture f. de déménage-ment; tapissière f. / ~ varnish ‖ Möbel-lack m. ‖ vernis m. pour meubles. / ~ varnisher ‖ Möbellackierer m. ‖ vernis-seur m. de meubles. / ~ varnishing ‖ Mö-belfirnissen n. ‖ vernissage m. de meub-les. / ~ velvet ‖ Möbelsamt m. ‖ velours m. d'ameublement.

furrier ‖ Kürschner m.; Fellhändler m. ‖ pelletier m. / ~'s knife ‖ Kürschnermesser n. ‖ lame f. ou couteau m. de pelletier. / ~'s pliers pl. ‖ Kürschnerzange f. ‖ pince f. de pelletier.

furriery ‖ Pelzwaren fpl.; Pelzwerk n. ‖ fourrures fpl.; pelleterie f. / artificial ~ ‖ künstliches Pelzwerk n. ‖ fourrure f. fac-tice.

furrow ‖ Riefe f. ‖ rainure f.; cannelure f. / ~ (Agr) ‖ Furche f.; Rille f. ‖ rainure f.; rigole f.; sillon m. / ~ (Join) ‖ Nut f. ‖ enrayure f.; rainure f. / ~ (Mine) ‖ Rinne f.; Ritz m.; Schramm m. ‖ entaille f. / curved ~ ‖ gekrümmte Rille f. ‖ rayon m. courbé. / master ~ (Millstone) ‖ Hauptfurche f. ‖ maître-rayon m. / ~ of a millstone ‖ Mühlsteinfurche f. ‖ silon m. d'une meule.

furs pl. ‖ Pelzwaren fpl.; Pelzwerk n. ‖ pel-leteries fpl.; fourrures fpl. / dressed ~ ‖ verarbeitetes Pelzwerk n. ‖ pelleterie f. ouvrée. / dried ~ ‖ getrocknetes Pelz-werk n. ‖ pelleterie f. séchée. / raw ~ ‖ rohes Pelzwerk n. ‖ pelleterie f. brute.

fur-stone, boiler ~ ‖ Kesselstein m. ‖ in-crustations fpl.; sédiments mpl.; dépôts mpl.

further, to ‖ Vorschub m. leisten ‖ aider; favoriser.

further particulars pl. ‖ weitere Angaben fpl. ‖ informations fpl. ultérieures.

fur trade ‖ Fellhandel m. ‖ pelleterie f. / ~ trimming ‖ Pelzbesatz m. ‖ garniture f. en pelleterie.

fuse, to ‖ schmelzen; raketenartig ver-brennen ‖ fuser. / ~ off ‖ ausschmelzen; abschmelzen ‖ séparer par fusion f. / ~ together ‖ verschmelzen ‖ fondre en-semble.

fuse (Electr) ‖ (elektrische) Sicherung f.; Stromsicherung f.; Schmelzeinsatz m.; Schmelzstreifen m. ‖ fusible m.; coupe-circuit m. / ~ (Mine) ‖ Zündschnur f. ‖ mèche f. d'allumage. / ~ (War mat) ‖ Zünder m. ‖ allumeur m. déton-ateur; fusée f. / the ~ blows (Electr) ‖ die Sicherung schmilzt aus ‖ le fusible fond. / combination ~ ‖ kombinierter Zünder m. ‖ fusée f. combinée. / com-bined ~ and switch (Electr) ‖ Schalter m. und Sicherungen kombiniert ‖ interrup-teur m. avec coupe-circuit. / double-acting ~ ‖ Doppelzünder m. ‖ fusée f. à double effet. / electric ~ ‖ elektrischer Zünder m. ‖ fusée f. électrique. / high-sensitive ~ ‖ Augenblickszünder m. ‖ fu-sée f. instantanée. / lead ~ ‖ Bleisiche-rung f. ‖ coupe-circuit m. à lame de plomb. / ~ for light and power currents ‖ Starkstromsicherung f. ‖ coupe-circuit m. pour courants de haute intensité. / main ~ (Electr) ‖ Hauptsicherung f. ‖ coupe-circuit m. principal. / ~ with re-tarded action ‖ Verzögerungszünder m. ‖

fusée f. à effet retardé. / strip ~ ‖ Strei-fensicherung f. ‖ fusible m. à lame.

fuse alarm ‖ Sicherungsüberwachungsein-richtung f. ‖ alarme f. pour les fusibles. / ~ block ‖ Klemmbrett n. für Schmelz-streifen ‖ plaque f. porte-fusibles. / ~ board ‖ Sicherungsgestell n. ‖ bâti m. à fusibles. / body of the ~ ‖ Satzstück n.; Zünderkörper m. ‖ corps m. de fusée. / ~ box (Cable) ‖ Sicherungskasten m. ‖ boîte f. à fusible; guérite f. / ~ composition ‖ Zünderfüllmasse f. ‖ composition f. à fusées.

fused ‖ geschmolzen ‖ fondu.

fusee (Clockm) ‖ Schnecke f. ‖ fusée f. / ~ arbor (Clockm) ‖ Schneckendrehstift m.; Schneckenabläufer m. ‖ arbre m. à fu-sées. / ~ chain (Clockm) ‖ Schnecken-kette f. ‖ chaîne f. à fusée. / ~ cutting machine (Clockm) ‖ Schneidzeug n. zur Herstellung der Kettenschnecke ‖ ma-chine f. à rayer les fusées ou à tailler les fusées. / ~ escapement (Clockm) ‖ Spin-delhemmung f. ‖ échappement m. à verge. / ~ notching arbor see ~ arbor.

fuse equipment ‖ Sicherungsanlage f. ‖ établissement m. des fusibles.

fusee wheel (Clockm) ‖ Schneckenrad n. ‖ roue f. de fusée.

fuse insulator ‖ Sicherungsisolator m. ‖ isolateur m. porte-fusible.

fusel ‖ Fusel m. ‖ mauvaise eau-de-vie f.

fuselage (Aero) ‖ Rumpf m. ‖ fuselage m. / ~ framework ‖ Rumpfgerippe n. ‖ car-casse f. du fuselage.

fuse lamp (Tel) ‖ Sicherungslampe f. (für Telegraphenbatterien) ‖ lampe f. de sû-reté.

fusel oil ‖ Fuselöl n. ‖ alcool m. d'amyle ou de queue; huile f. de pommes de terre; fusol m.

fuse manufacturing machinery ‖ Zünder-fertigungsmaschine f. ‖ machine f. pour la fabrication de détonateurs. / ~ ma-terial ‖ Zündmittel n. ‖ matière f. allu-mante. / ~ plug ‖ Sicherungsstöpsel m. ‖ cartouche f. / ~ rack for repeater stations (Tel) ‖ Sicherungsgestell n. für Verstär-kerämter ‖ bâti m. des organes d'alimen-tation des lampes amplificatrices.

fuses pl. ‖ Sicherungsmaterial n. ‖ coupe-circuits mpl.

fuse strip (Electr) ‖ Sicherungsschmelz-streifen m. ‖ lame f. fusible. / ~ supervi-sory relay ‖ Sicherungskontrollrelais n. ‖ relais m. pilote avertisseur pour coupe-circuit. / ~ table ‖ Sicherungstisch m. ‖ table f. des coupe-circuits. / ~ wire ‖ Ab-schmelzdraht m.; Schmelzsicherungs-draht m. ‖ fil m. fusible.

fusibility ‖ Schmelzbarkeit f. ‖ fusibilité f.

fusible ‖ schmelzbar ‖ fusible. / difficultly ~ ‖ schwer schmelzbar ‖ difficilement fu-sible. / easily ~ ‖ leicht schmelzbar; leicht-flüssig ‖ facilement fusible.

fusible alarm ‖ Schmelzlotmelder m. ‖ aver-tisseur m. à fusible. / ~ alloy ‖ Schnellot n. ‖ alliage m. fusible. / ~ conductor ‖ schmelzbarer Leiter m. ‖ conducteur m. fusible. / ~ cone ‖ Brennkegel m.; Schmelz-kegel m.; Segerkegel m. ‖ cône m. pyro-métrique. / ~ glass ‖ schmelzbares Glas n. ‖ verre m. fusible. / ~ metal see ~ alloy. / ~ piece ‖ Schmelzeinsatz m. ‖ pièce f. fusible. / ~ plug ‖ Schmelz-pfropfen m. ‖ bouchon m. fusible.

fusing agent ‖ Schmelzmittel n. ‖ agent m. de fusion. / ~ apparatus for wool spin-

ners ‖ Schmelzapparat m. für Wollspinnereien ‖ appareil m. de fusion pour filatures de laine. / ~ burner ‖ Schneidbrenner m. ‖ brûleur m. à découper. / acetylene-oxyhydrogen ~ burner ‖ Azetylensauerstoffschneidbrenner m. ‖ brûleur m. acétylène-oxyhydrique à découper. / autogenous ~ burner ‖ autogener Schneidbrenner m. ‖ brûleur m. autogène à découper. / ~ current ‖ Abschmelzstromstärke f. ‖ intensité f. de courant de fusion. / ~ point ‖ Schmelzpunkt m. ‖ point m. de fusion.

fusion (Found) ‖ Schmelzung f. ‖ fusion f. / ~ (Trade) ‖ Fusion f. oder Zusammenschluß m. von Gesellschaften ‖ fusion f. / complete ~ ‖ Schmelzfluß m. ‖ fusion f. complète. / ~ of current supply systems ‖ Zusammenfassung f. der Stromversorgung ‖ concentration f. de réseaux de courant électrique. / ~ of the local supply networks ‖ Zusammenfassung f. der lokalen oder örtlichen Versorgungsnetze ‖ groupement m. des petits réseaux d'alimentation locaux. / ~ of metals ‖ metallische Verbindung f. ‖ liaison f. de métaux. / ~ of rays ‖ Strahlenvereinigung f. ‖ concentration f. des rayons. / second ~ ‖ zweite Schmelzung f. ‖ refonte f.; deuxième fusion f.

fusion, heat of ~ ‖ Schmelzwärme f. ‖ chaleur f. de fusion. / time of ~ ‖ Abschmelzzeit f. ‖ durée f. de fusion. / ~ welding ‖ Schmelzschweißung f. ‖ soudage m. par fusion.

fustet ‖ Fisettholz n. ‖ bois m. de fustet; fustet m. / ~ bud ‖ Fisettknospe f. ‖ feuille f. de brindille de fustet.

fustian ‖ Baumwollsamt m.; unechter Samt m.; Manchester m.; Velvetin m. ‖ velours m. de coton; velvantine f.

fustic ‖ kubanisches Gelbholz n. ‖ bois m. jaune (de Cuba). / young ~ ‖ Fisettholz n. ‖ bois m. de fustet. / extract of ~ ‖ Gelbholzauszug m. ‖ extrait m. de bois jaune.

fusty ‖ schimmelig; muffig ‖ chanci; couvert de moisissure; moisi. / to turn ~ ‖ stocken; schimmeln ‖ se moisir; se chancir.

futtock (Shipb) ‖ Gabelholz n. ‖ gibelot m. / ~ shrouds pl. ‖ Püttings pl.; Püttingswanten fpl. ‖ gambes fpl. de hune; haubans mpl. de réserve.

fuze see fuse.

fuzee see fusee.

fuzz, to ‖ zerfasern ‖ effiler; effilocher.

fuzzy ‖ flockig; faserig; flaumig ‖ floconneux; filandreux; fibreux; duveté; cotonneux.

G

gabion ‖ Schanzkorb m. ‖ gabion m. / ~ bridge ‖ Schanzkorbbrücke f. ‖ pont m. de gabions.

gable ‖ Giebel m. ‖ pignon m. / ~ bottom ‖ Sattelboden m. ‖ fond m. en dos d'âne. / ~-bottomed car (Railw) ‖ Sattelwagen m.; Wagen m. mit Sattelboden ‖ wagon m. à fond en dos d'âne.

gabled dormer-window ‖ Giebelfenster n. ‖ lucarne f. faîtière.

gable end see ~ wall. / ~ wall ‖ Giebelmauer f.; Giebelwand f. ‖ mur m. ou pan m. de pignon. / ~ window ‖ Giebelfenster n. ‖ lucarne f. de pignon ou de mansarde.

gad (Mine) ‖ eiserner Keil m.; Fimmel m.; Bergeisen n. ‖ coin m. en fer; aiguille f. du mineur.

gadolinite ‖ Gadolinit m.; Ytterbyt m.; Ytterit m. ‖ gadolinite f.

gaff ‖ Gaffel f. ‖ vergue f. à corne.

gaffer (Glassm) ‖ Fertigmacher m. ‖ souffleur m.; ouvreur m.

gaff flag ‖ Gaffelflagge f. ‖ pavillon m. national à la corne. / ~ sail ‖ Gaffelsegel m. ‖ voile f. à corne.

gag ‖ Knebel m. ‖ garrot m.

gage see also gauge ‖ Lehre f.; Stichmaß n. ‖ calibre m.; jauge f.; gabarit m. / ~ cock ‖ Wasserstands(prob)hahn m. ‖ robinet m. (d'épreuve) de niveau d'eau.

gahnite ‖ Gahnit m.; Zinkspinell m.; Automolit m. ‖ spinelle m. zincifère; gahnite f.; automolithe f.

gain, to ‖ gewinnen; Vorteil m. erlangen ‖ gagner; profiter.

gain ‖ Gewinn m.; Profit m. ‖ gain m.; bénéfice m.; profit m. / ~ controller (Tel) ‖ Schwächungswiderstand m. ‖ régulateur m. d'amplification.

gainful ‖ gewinnbringend; einträglich ‖ productif; profitable; lucratif. / ~ly employed ‖ erwerbstätig ‖ industriel.

gain regulator see gain controller.

gaiter ‖ Gamasche f. ‖ guêtre f. / leather ~s pl. ‖ Ledergamaschen fpl. ‖ guêtres fpl. en cuir.

gaiter cutter ‖ Gamaschenschneider m. ‖ coupeur m. de guêtres. / ~ knitter ‖ Gamaschenwirker m. ‖ tricoteur m. de guêtres. / ~ maker ‖ Gamaschenmacher m. ‖ guêtrier m. / ~ quilter ‖ Gamaschenstepper m. ‖ piqueur m. de guêtres. / ~ weaver ‖ Gamaschenweber m. ‖ tisseur m. de guêtres.

galactite ‖ Galaktit m. ‖ galactite f.

galactometer ‖ Milchmesser m.; Milchwage f.; Galaktometer n.; Galaktoskop ‖ lactomètre m.; galactomètre m.; pèse-lait m.

galactose ‖ Galaktose f. ‖ galactose f.

galalite see galalith.

galalith ‖ Galalith m. ‖ galalithe f. / ~ handle ‖ Galalithgriff m. ‖ manche f. en galalithe. / ~ producing machine ‖ Galalithherstellungsmaschine f. ‖ machine f. à fabriquer la galalithe.

Galam gum ‖ Galamgummi m. ‖ gomme f. de Galam.

galangal root ‖ Galgantwurzel f. ‖ racine f. de galanga.

galban gum ‖ Galbanharz n. ‖ gomme f. galban.

gale ‖ Brise f.; Kühlte f. ‖ brise f. / fresh ~ ‖ stürmischer Wind m. ‖ vent m. frais. / moderate ~ ‖ harter Wind m. ‖ forte brise f. / strong ~ ‖ Sturm m. ‖ vent m. grand frais. / whole ~ ‖ starker Sturm m. ‖ coup m. de vent; bourrasque f.

galena ‖ Bleiglanz m.; Galenit m.; Grauerz n. ‖ galène f. / ~ detector ‖ Bleiglanzdetektor m. ‖ détecteur m. de galène. / ~ mine ‖ Bleierzbergwerk n. ‖ mine f. de plomb ou de galène.

galenical (Chem) ‖ galenisches Präparat n. ‖ préparation f. galénique.

galenite see galena.

galeopsis herb ‖ Hanfnesselkraut n.; Hohlzahnkraut n. ‖ herbe f. de galéopsis.

galipot ‖ Galipot n.; weißes Fichtenharz n. ‖ galipot m.

gall, to (Mar) ‖ schamfielen ‖ s'érailler; raguer.

gall (Anatomy) ‖ Galle f. ‖ bile f. / ~ (Botan) ‖ Gallapfel m. ‖ noix f. de galle. / Aleppo ~ see black ~. / black ~ grüner oder levantischer oder schwarzer Gallapfel m. ‖ galle f. noire ou verte. / blue ~ see black ~. / ~ of glass ‖ Glasgalle f. ‖ fiel m. ou sel m. de verre; suin m.

galled (Mar) ‖ schamfielt ‖ écaillé; ragué.

gallery ‖ Galerie f.; Laufgang m. ‖ galerie f. / ~ (Corridor) ‖ Korridor m. ‖ corridor m. / ~ (Mining) ‖ Stollen m.; Strecke f. ‖ galerie f. / ascending ~ (Mining) ‖ steigende Strecke f. ‖ galerie f. montante. / charging ~ (Met) ‖ Gichtbühne f. ‖ plateforme f. du gueulard. / descending ~ (Mining) ‖ einfallende Strecke f. ‖ galerie f. descendante ou en pente. / to drive a ~ to the hade of a seam ‖ eine einfallende Strecke treiben ‖ faire une descente. / to drive-on a ~ ‖ einen Stollen m. oder eine Strecke f. vortreiben ‖ construire ou pratiquer une galerie. / inclined ~ ; Flaches n. ‖ flach einfallende Förderstrecke f. ‖ descenderie f.; vallée f. ‖ toret m. / level ~ ‖ söhlige Strecke f. ‖ galerie f. horizontale. / ~ of a mine ‖ Stollen m.; Strecke f.; Minengang m. ‖ galerie f. ou conduit m. de mine. / ~ of paintings ‖ Gemäldegalerie f.; Bildergalerie f. ‖ galerie f. de tableaux. / ~ of a quarry ‖ Stollen m. einer Steingrube ‖ chemin m. d'une carrière.

gallery construction ‖ Stollenbau m. ‖ construction f. de galerie. / ~ entrance (Mine) ‖ Stolleneingang m.; Minenauge n. ‖ entrée f. de galerie. / ~ frame (Mine) ‖ Türgerüst n. ‖ châssis m. d'une galerie.

galley (Print) ‖ Satzschiff m. ‖ galée f. / ~ (Shipb) ‖ Schiffsküche f.; Kombüse f. ‖ cuisine f. de l'équipage. / ~ press (Print) ‖ Abziehpresse f. ‖ presse f. à tirer des épreuves. / ~ slice (Print) ‖ Schiffszunge f. ‖ coulisse f. de galée.

gallic acid ‖ Gallussäure f. ‖ acide m. gallique. / ~ ink see gallotannic ink.

galling (Dyer) ‖ Galläpfelfarbe f.; Galläpfelbeize f. ‖ engallage m.

galling leather (Saddl) ‖ Scheuerleder n. ‖ pièce f. de frottement; plaque f. d'appui.

gallnut ‖ Gallapfel m. ‖ noix f. de galle. / French ~ (Dyer) ‖ Puischgallapfel m.; französischer Gallapfel m. ‖ cassenole f. / extract of ~s ‖ Galläpfelauszug m. ‖ extrait m. de noix de galle.

gallonic acid ‖ Galloninsäure f. ‖ acide m. galloninique.

galloon ‖ Borte f.; Tresse f. ‖ galon m.; tresse f. / ~ **maker** ‖ Bortenwirker m. ‖ galonnier m.

gallotannic acid ‖ Gerbsäure f.; Tannin n.; Galläpfelsäure f. ‖ acide m. gallotannique ou gallique. / ~ **ink** ‖ Gallustinte f. ‖ encre f. gallo-tannique ou gallique.

gallows pl. ‖ Galgen m. ‖ potence f. / ~ (Print) ‖ Sattel m.; Galgen m.; Deckelstuhl m. ‖ chevalet m. du timpan.

Gall's chain ‖ Gall'sche Kette f.; Gelenkkette f. ‖ chaîne f. Gall; chaîne f. articulée ou de maillons.

gallsteep see galling (Dyer).

gall tannin ‖ Galläpfeltannin n. ‖ tannine f. de noix de galle.

gally see galley (Print).

galmey ‖ Galmei m. ‖ calamine f. / ~ **mine** ‖ Galmeigrube f. ‖ mine f. de calamine.

galosh ‖ Galosche f.; Überschuh m. ‖ galoche f. / rubber ~ ‖ Gummischuh m.; Gummigalosche f. ‖ galoche f. ou chaussure f. en caoutchouc.

galvanic ‖ galvanisch ‖ galvanique. / ~ **cell** ‖ galvanisches Element n. ‖ élément m. galvanique. / ~ **colouring of metals** ‖ Galvanochromie f.; galvanische Metallfärbung f. ‖ colorisation f. électrochimique des métaux. / ~ **coupling** (Radio) ‖ galvanische Kopplung f. ‖ couplage m. galvanique. / ~ **etching** ‖ Galvanokaustik f. ‖ galvano-caustic f.

galvanism ‖ Berührungselektrizität f. ‖ galvanisme m.

galvanization ‖ Galvanisierung f. ‖ galvanisation f. / ~ (Of iron etc.) ‖ Verzinkung f. ‖ galvanisation f.

galvanize, to ‖ galvanisieren ‖ galvaniser. / ~ (To zinkify) ‖ verzinken ‖ galvaniser.

galvanized ‖ galvanisiert; verzinkt ‖ galvanisé. / ~ **sheet** ‖ verzinktes Blech n. ‖ tôle f. galvanisée. / ~ **tube** ‖ verzinktes Rohr n. ‖ tuyau m. galvanisé. / ~ **wire** ‖ verzinkter Draht m. ‖ fil m. galvanisé.

galvanizer ‖ Verzinker m. ‖ galvaniseur m.

galvanizing of iron ‖ galvanische Verzinkung f. des Eisens ‖ zingage m. galvanique du fer; galvanisation f. / ~ **furnace** ‖ Verzinkungsofen m. ‖ four m. à zinguer. / ~ **plant** ‖ Verzinkerei f. ‖ installation f. de zingage.

galvano ‖ Klischee n. ‖ cliché m.

galvanometallurgy ‖ Galvanometallurgie f. ‖ galvanométallurgie f.

galvanometer ‖ Galvanometer n. ‖ galvanomètre m. / ballistic ~ ‖ ballistisches Galvanometer n. ‖ galvanomètre m. balistique. / ~ with ball-shaped shield ‖ Kugelpanzergalvanometer n. ‖ galvanomètre m. à cuirasse sphérique. / hot-wire ~ ‖ Hitzdrahtgalvanometer n. ‖ galvanomètre m. à fil chauffé. / mirror ~ ‖ Spiegelgalvanometer n. ‖ galvanomètre m. à miroir. / moving-coil ~ ‖ Spulengalvanometer n.; Drehspulgalvanometer n. ‖ galvanomètre m. à cadre mobile. / moving-magnet ~ ‖ Nadelgalvanometer n.; Drehmagnetgalvanometer n. ‖ galvanomètre m. à aimant mobile ou à aiguille. / portable ~ ‖ Taschengalvanometer n. ‖ galvanomètre m. de poche. / reflecting ~ ‖ Reflexionsgalvanometer n.; Spiegelgalvanometer n. ‖ galvanomètre m. à miroir. / registering ~ ‖ schreibendes Galvanometer n.; Registriergalvanometer m. ‖ galvanomètre m. enregistreur.

galvanometer constant ‖ Eichzahl f. eines Galvanometers; Galvanometerkonstante f. ‖ constante f. d'un galvanomètre.

galvanoplastic art see galvanoplastics. / ~ **bath** ‖ galvanoplastisches Bad n. ‖ bain m. galvanique. / ~ **copper bath** ‖ galvanisches Kupferbad n. ‖ bain m. de cuivrage galvanique. / ~ **deposit** ‖ galvanoplastischer Niederschlag m. ‖ dépôt m. galvanique. / ~ **machine** ‖ Galvanoplastikmaschine f. ‖ machine f. de galvanoplastie. / ~ **plant** ‖ Galvanoplastikanlage f. ‖ installation f. de galvanoplastie. / ~ **workshop** ‖ galvanoplastische Anstalt f. ‖ atelier m. de galvanoplastie.

galvanoplastics pl. ‖ Galvanoplastik f. ‖ galvanoplastie f.

galvanoscope ‖ Galvanoskop n. ‖ galvanoscope m.

galvanotyping ‖ Galvanotypie f. ‖ galvanotypie f.

gambier ‖ Gambir n. ‖ gambir m.

gambir see gambier.

gambling table ‖ Spieltisch m. ‖ table f. de jeu.

gamboge gum ‖ Gummigutt n. ‖ gomme-gutte f.

game ‖ Spiel n. ‖ jeu m. / ~ (Hunting) ‖ Wild n. ‖ gibier m. / feathered ~ ‖ Federwild n. ‖ gibier m. à plume.

game bag ‖ Jagdtasche f. ‖ gibecière f. / ~ **carrier** ‖ Wildknecht m. ‖ porte-carnier m.; porte-gibier m. / ~ **keeper** ‖ Feldhüter m. ‖ garde-champêtre m.; garde-chasse m. / ~ **pocket** see ~ bag. / ~ **shop** ‖ Wildhandlung f. ‖ magasin m. à gibier.

gang (Mine) ‖ Gebirge n.; Nebengestein n.; Ganggestein n.; Gangart f.; gangue f. / ~ (Weav) ‖ Gang m. (beim Scheren der Kette) ‖ portée f. / ~ of pitmen ‖ Schicht f.; Kameradschaft f. ‖ poste m.; équipe f. / ~ of workmen ‖ Arbeiterrotte f.; Arbeitergruppe f. ‖ équipe f. d'ouvriers.

gang-board ‖ Landsteg m.; Steg m.; Landplanke f.; Laufplanke f. ‖ planche f. à débarquer ou à passerelle.

gang cutter ‖ Satzfräser m. ‖ fraise f. d'un jeu combiné.

ganger ‖ Werkmeister m. ‖ contremaître m.; chef m. d'atelier; maître-ouvrier m.

gang foreman ‖ Bautruppführer m. ‖ brigadier-poseur m.

gangman (Agr) ‖ Sommerarbeiter m. ‖ bandier m.; solatier m.; ouvrier m. agricole. / ~ (Railw) ‖ Rangierer m. ‖ homme d'équipe m.

gangmaster see ganger.

gang press ‖ Stufenpresse f. ‖ presse f. à poinçons multiples. / ~ **punch operator** (Shoem) ‖ Ösenlochmacher m. ‖ perforeur m. d'œillets.

gangrene ‖ Holzbrand m. ‖ gangrène f.

gang saw ‖ Spaltgatter n.; Trenngatter n. ‖ scie f. verticale alternative à refendre. / ~ **sawyer** ‖ Gattersäger m. ‖ conducteur m. de scie à plusieurs lames. / ~ **slitting machine** ‖ Rollenschere f. ‖ cisaille f. circulaire multiple. / ~ **stone** (Geol) ‖ Gangstein m.; Gangmasse f.; taubes Gestein n. einer Lagerstätte ‖ gangue f.; matière f. stérile d'un gîte; filon m. sauvage. / ~ **trench** (Mine) ‖ Schram m.; Rinne f.; Ritz m. ‖ entaille f.

gangue see gang (Mine). / ~ **stone** ‖ Gangstein m. ‖ filons mpl. sauvages.

gangway (Way through) ‖ Gang m.; Durchgang m. ‖ allée f. / ~ (Running board) ‖ Laufbühne f. ‖ passerelle f.; marchepied m. / ~ (Shipb) ‖ Laufplanken fpl.; Laufbrücke f. ‖ passe-avant m. ou planche-passerelle f.

ganomatite ‖ Ganomatit m.; Gänsekötigerz n. ‖ ganomatite f.

gantry (For casks) ‖ Faßlager n. ‖ chantier m. à tonneaux. / ~ (Of a crane) ‖ Krangerüst n.; Portal n. ‖ portique m.; portail m.

gantry crane ‖ Portalkran m.; Bockkran m. ‖ grue f. (mobile) à portique ou à chevalet. / revolving one-ledged ~ ‖ schwenkbarer Halbportalkran m. ‖ grue f. tournante à demi-portique.

gantry pillar ‖ Pendelstütze f.; Portalstütze f. ‖ pilier-pendule m.; pilier m. de portique. / ~ **post** see ~ pillar.

gap ‖ Öffnung f.; Spalte f.; Ritze f.; Zwischenraum m. ‖ intervalle m.; vide m.; ouverture f.; espace m. / ~ (Space between the wings of an aeroplane) ‖ Flügelabstand m. ‖ entreplan m. / ~ (Forest) ‖ Blöße f. ‖ vide m. / ~ (Mine) ‖ Schmierkluft f. ‖ fente f. glaiseuse. / air ~ ‖ Luftzwischenraum m. ‖ entrefer m. / deep ~ (Mach tool) ‖ tiefgehende Ausladung f. ‖ col m. de cygne très profond. / extra deep ~ ‖ besonders große Ausladung f. ‖ col m. de cygne extraordinairement profond. / ~ at joint ‖ Stoßfuge f. ‖ jeu m. du joint. / having many ~s ‖ lückenhaft ‖ incomplet; défectueux. / rotating ~ (Electr) ‖ rotierende Funkenstrecke f. ‖ éclateur m. rotatif. / spark ~ ‖ Funkenstrecke f. ‖ éclateur m.; parafoudre m. à armorçage d'arc. / ~ of a wall ‖ Mauerspalte f.; Mauersprung m.; Riß m. im Mauerwerk ‖ lézarde f. ou fente f. d'une muraille.

gap depth (Mach tool) ‖ Ausladung f. ‖ portée f.; profondeur f. de gorge; col m. de cygne.

gape, to ‖ sich spalten ‖ se fendre.

garage ‖ Wagenhalle f.; Einstellhalle f. für Wagen; Wagenschuppen m.; Garage f. ‖ garage m. / ~ of corrugated sheet iron ‖ Garage f. oder Wagenschuppen m. aus Wellblech ‖ garage m. en tôle ondulée.

garage air pump ‖ Garagenluftpumpe f. ‖ pompe f. à air pour garages. / ~ **building** ‖ Wagenschuppenbau m.; Garagenbau m. ‖ construction f. de garages. / ~ **heating** ‖ Garagenheizung f.; Wagenschuppenheizung f. ‖ chauffage m. de garages. / ~ explosion-proof ~ stove ‖ explosionssicherer Garagenheizofen m. ‖ poêle m. de sûreté pour garages.

garance (Dyer) ‖ Krapp m. ‖ garance f.

garancine (Dyer) ‖ Garanzin n.; Krappkohle f. ‖ garancine f.

garbage ‖ Hausmüll n.; Küchenabfall m.; Küchenabfälle mpl. ‖ ordures fpl. ménagères ou de cuisine; issues fpl. / ~ and town-refuse burning furnace ‖ Müll- und Abfallverbrennungsofen m. ‖ four m. à brûler ou à incinérer les ordures et déchets. / ~ bin see ~ collector. / ~ collector ‖ Müllbehälter m. ‖ collecteur m. à ordures. / ~ removal truck ‖ Müllabfuhrwagen m. ‖ camion m. pour ordures ménagères; voiture f. de transport des ordures. / ~ truck see ~ removal truck.

garbling ‖ Aussieben n.; Sichten n.; sassement m.; tamisage m.; triage m.

garblings pl. (Mill) ‖ Siebsel n. ‖ criblure f.

garboard plank (Shipb) ‖ Kielgangplanke f. ‖ madrier m. de gabord. / ~ **strake** ‖ Kielgang m. ‖ gabord m.

garden ‖ Garten m. ‖ jardin m. / to lay-out a ~ ‖ einen Garten m. anlegen ‖ installer un jardin. / gardens pl. which are let-out to the workers ‖ Pachtgärten mpl. für Arbeiter ‖ jardins mpl. loués aux ouvriers.

garden bed ‖ Gartenbeet n. ‖ plate-bande f.; planche f. ou couche f. de jardin. / ~ chair ‖ Gartenstuhl m. ‖ chaise f. de jardin ou rustique. / ~ city ‖ Gartenstadt f. ‖ cité-jardin f. / ~ earth ‖ Gartenerde f. ‖ terreau m.; terre f. franche ou végétale.

gardener ‖ Gärtner m. ‖ jardinier m. / ~ (Day-labourer) ‖ Tagelöhner für Gartenarbeit f. ‖ jardinier-journalier m. / ~ (Helper) ‖ Gärtnergehilfe m. ‖ aide-jardinier m. / ~ (Labourer) ‖ Gartenarbeiter m. ‖ jardinier m.

gardener's bill ‖ Gärtnerhippe f.; gekrümmtes Gartenmesser n. ‖ serpe f. de jardinier. / ~ hoe ‖ Gartenhaue f. ‖ aisseau m. / ~ knife ‖ Gärtnermesser n. ‖ couteau m. de jardinier.

garden furniture ‖ Gartenmöbel pl. ‖ meubles mpl. de jardin. / ~ of iron ‖ eiserne Gartenmöbel pl. ‖ meubles mpl. de jardin en fer.

gardenia oil ‖ Gardeniaöl n. ‖ essence f. de gardénie.

gardening ‖ Gärtnerei f.; Gartenbau m.; Blumenzucht f. ‖ jardinage m. / ~ implements pl. ‖ Gartengeräte npl. ‖ outils mpl. de jardinage. / ~ knife see gardener's knife. / ~ owner ‖ Gärtnereibesitzer m. ‖ jardinier m. propriétaire. / ~ tool ‖ Gartengerät n.; Gartenwerkzeug n. ‖ instrument m. de jardinage. / ~ tools pl. see ~ implements. / ~ undertaker ‖ Gartenbauunternehmer m. ‖ entrepreneur m. de jardinage.

garden mould ‖ Gartenerde f. ‖ terreau m.; terre f. franche ou végétale. / ~ plant ‖ Gartenpflanze f. ‖ plante f. horticole. / ~ plot ‖ Gartenanlage f. ‖ parterre m.; jardin m. / ~ square m. / ~ roller ‖ Gartenwalze f. ‖ rouleau m. ou cylindre m. de jardin. / ~ screen ‖ Gartenschirm m. ‖ parasol m. pour jardins. / ~ seeds pl. ‖ Gärtnereisamen mpl. ‖ graines fpl. horticoles. / ~ shears pl. ‖ Gartenschere f. ‖ ciseaux mpl. de jardinier. / ~ sprayer see ~ syringe. / ~ syringe ‖ Gartenspritze f. ‖ seringue f. ou pompe f. de jardin; arrosoir m. / ~ tent ‖ Gartenzelt n. ‖ pavillon m.

gargle ‖ Gurgelwasser n.; Mundwasser n. ‖ gargarisme m.

gargoyle ‖ Wasserspeier m.; Abtraufe f. der Dachrinne ‖ gargouille f.; canon m. de gouttière.

garland ‖ Girlande f.; Blumengehänge n. ‖ guirlande f. / paper ~ ‖ Papiergirlande f. ‖ guirlande f. en papier.

garlic ‖ Knoblauch m. ‖ aulx mpl.; ail m. / ~-like ‖ knoblauchartig ‖ alliacé; d'odeur f. d'ail. / ~ oil ‖ Knoblauchöl n. ‖ essence f. d'ail.

garment ‖ Bekleidung f.; Kleidung f. ‖ confection f.; vêtements mpl. / men's ~ ‖ Herrenbekleidung f. ‖ vêtements mpl. d'hommes.

garment mending ‖ Kleiderausbesserung f. ‖ raccommodage m. de vêtements.

garnet (Miner) ‖ Granat m. ‖ grenat m. / double ~ (Locksm) ‖ Kreuzband n. ‖ té m. (simple). / ~ (Chem) ‖ Grossular m.; Kalkgranat m.; grüner Granat m. ‖ grenat m. calcarifère ou grossulaire; grossu-

laire m. / oriental ~ ‖ Almandin m.; orientalischer Granat m.; Karfunkel m. ‖ alabandine f.; grenat m. almandin ou oriental ou rouge. / precious ~ see oriental ~. / single ~ (Locksm) ‖ Scheinecke f.; Winkelband n.; Eckschiene f. ‖ équerre f. de fer.

garnets pl. (Mar) ‖ Ladetakel npl. ‖ bredindins mpl.; petits palans mpl. d'étai.

garnet cutter ‖ Granatschleifer m. ‖ polisseur m. de grenats. / ~ work ‖ Granatwaren fpl. ‖ articles mpl. de grenat.

garniture ‖ Beschlag(teil) m.; Garnitur f. ‖ garniture f.; ferrure f.; monture f. / ~ of a boiler ‖ Garnitur f. eines Dampfkessels ‖ garniture f. ou robinetterie f. d'une chaudière à vapeur.

garret ‖ Dachgeschoß n.; Bodengeschoß n.; Dachstube f.; Bodenkammer f. ‖ étage m. en galetas; mansarde f.; galetas m.; grenier m. / ~ door ‖ Bodentür f. ‖ porte f. de grenier. / ~ staircase ‖ Bodentreppe f. ‖ escalier m. du grenier. / ~ story see garret.

garter ‖ Sockenhalter m.; Strumpfband n. ‖ jarretelle f.; jarretière f.

gas, to (Acc) ‖ kochen ‖ cuire. / ~ (Clothm) ‖ mit Gas sengen; gasen ‖ gazer.

gas ‖ Gas n. ‖ gaz m. / the ~es pl. escape ‖ die Gase npl. entweichen ‖ les gaz mpl. s'échappent. / to remove the dust from the ~ ‖ das Gas von mitgerissenem Staub befreien ‖ débarasser le gaz des poussières entraînées. / to rid the ~ of dust see to remove the dust from the ~. / to shut-off the ~ ‖ Gas n. abdrosseln ‖ étrangler le gaz. / to store the ~ ‖ das Gas aufspeichern oder lagern ‖ stocker ou emmagasiner le gaz. / blast furnace ~ ‖ Hochofengas n.; Gichtgas n. ‖ gaz m. du gueulard ou de hauts-fourneaux. / buoyant ~ see lifting ~. / burnt ~es pl. ‖ Abzugsgas n.; Abgas n. ‖ gaz m. perdu ou de combustion. / clear ~ ‖ Klargas n. ‖ gaz m. clair. / coke oven ~ ‖ Kokereigas n.; Koksofengas n. ‖ gaz m. provenant des fours à coke; gaz m. de four à coke. / combustible ~ ‖ brennbares Gas n. ‖ gaz m. combustible ou inflammable. / ~es pl. of combustion (Chem) ‖ Verbrennungsgase npl. ‖ gaz mpl. de la combustion. / compressed ~ ‖ komprimiertes Gas n. ‖ gaz m. comprimé. / crude ~ ‖ Rohgas n. ‖ gaz m. brut. / drawing-off ~ see exhaust ~. / earth's ~ ‖ Erdgas n. ‖ gaz m. naturel. / exhaust ~ ‖ Abgas n. ‖ gaz m. d'échappement. / fat ~ ‖ Fettgas n. ‖ gaz m. gras. / flue ~ ‖ Feuergas n.; Rauchgas n. ‖ gaz m. du foyer ou de fumée. / generator ~ ‖ Sauggas n.; Generatorgas n. ‖ gaz m. de gazogène; gaz m. pauvre. / heating ~ ‖ Heizgas n. ‖ gaz m. de chauffage. / ~ of high calorific value ‖ heizkräftiges Gas n. ‖ gaz m. d'un grand pouvoir calorifique. / high-pressure ~ ‖ Preßgas n. ‖ gaz m. surpressé. / hydro-carbon ~ ‖ Kohlenwasserstoffgas n. ‖ gaz m. d'hydrogène carburé. / impure ~ ‖ unreines Gas n. ‖ gaz m. impur. / industrial ~ ‖ Industriegas n. ‖ gaz m. industriel. / inflammable ~ see combustible ~. / lifting ~ ‖ Traggas n. ‖ gaz m. de gonflement ou de sustentation. / lighting ~ ‖ Leuchtgas n. ‖ gaz m. d'éclairage. / liquefied ~ ‖ verflüssigtes oder flüssiges Gas n. ‖ gaz m. liquéfié. / natural ~ ‖ Erdgas n. ‖ gaz m. naturel. /

non-inflammable ~ ‖ unbrennbares Gas n. ‖ gaz m. non inflammable. / oil ~ ‖ Fettgas n.; Ölgas n. ‖ gaz m. d'huile ou de pétrole. / peat ~ ‖ Torfgas n. ‖ gaz m. de tourbe. / poor ~ ‖ Schwachgas n. ‖ gaz m. pauvre. / power ~ see also generator ~ ‖ Kraftgas n. ‖ gaz m. pour moteurs. / producer ~ see generator ~. / rare ~ ‖ Edelgas n. ‖ gaz m. rare ou noble. / reducing or reduction ~ ‖ reduzierendes Gas n. ‖ gaz m. réducteur ou réductif. / residual ~ ‖ Gasrest m. ‖ gaz m. résiduel. / resin ~ ‖ Harzgas n. ‖ gaz m. de résine. / suction ~ ‖ Sauggas n. ‖ gaz m. aspiré. / supporting ~ see lifting ~. / top ~ see blast-furnace ~. / unburnt ~es pl. ‖ unverbrannte Gase pl. ‖ gaz mpl. non brûlés. / waste ~ see blast-furnace ~. / water ~ ‖ Wassergas n. ‖ gaz m. à l'eau. / wood ~ ‖ Holzgas n. ‖ gaz m. de bois.

gas admission tubing ‖ Gaszuführungsrohr n. ‖ tuyau m. d'amenée du gaz. / ~ air mixer ‖ Gasluftmischer m. ‖ mélangeur m. d'air et de gaz. / ~ air mixture ‖ Gasluftgemisch n. ‖ mélange m. d'air et de gaz. / ~ analysis ‖ Gasuntersuchung f. ‖ analyse f. des gaz. / ~ apparatus ‖ Gasapparat m. ‖ appareil m. à gaz. / ~ automatic ~ apparatus ‖ Gasautomat m. ‖ distributeur m. automatique de gaz. / ~ attack ‖ Gasangriff m. ‖ attaque f. de gaz. / ~ bag (Of a balloon) ‖ Gashülle f. ‖ enveloppe f. à gaz. / ~ baking oven ‖ Gasbackofen m. ‖ four m. au gaz à faire cuire. / ~ balance ‖ Gaswage f. ‖ balance f. à gaz. / ~ balloon see ~ bag. / ~ balloon for children ‖ Kinderluftballon m. ‖ ballon m. à gaz pour enfants. / ~ bath oven ‖ Gasbadeofen m. ‖ four m. à gaz pour bains. / ~ battery ‖ Gasbatterie f. ‖ pile f. à gaz. / ~ black (Dyer) ‖ Gasruß m. ‖ noir m. de fumée. / ~ blastlamp ‖ Gasgebläselampe f. ‖ lampe f. à soufflerie f. à gaz. / ~ blow (Found) ‖ Gasblase f. ‖ soufflure f. / ~ blower see ~ blowing engine. / ~ blowing engine ‖ Gasgebläse n.; Gasmaschinengebläse n.; Gasgebläsemaschine f. ‖ soufflerie f. ou soufflante à gaz; machine f. soufflante à gaz. / ~ blowing engine for steelworks ‖ Stahlwerkgasgebläse n. ‖ machine f. soufflante à gaz pour aciéries. / ~ blowpipe ‖ Lötbrenner m.; Lötrohr n. ‖ chalumeau m. / ~ bottle ‖ Gasentbindungsflasche f. ‖ flacon m. générateur pour gaz. / ~ bracket ‖ Gasarm m. ‖ genouillère f. ou applique f. à gaz. / ~ buoy ‖ Leuchtboje f.; Gastonne f. ‖ bouée f. à gas. / ~ burette ‖ Gasburette f. ‖ burette f. à gaz.

gasburner ‖ Brenner m. für Gasfeuerung; Gasbrenner m. ‖ brûleur m. pour chauffage au gaz; brûleur m. à gaz; bec m. de ou à gaz. / acetylene ~ ‖ Azetylengasbrenner m. ‖ bec m. de gaz à l'acétylène. / high air pressure ~ for operation on low-pressure gas ‖ Brenner m. für Niederdruckgas mit Luftansaugung ‖ brûleur m. pour gaz à basse pression avec aspiration d'air. / ~ with mixer and gas governor ‖ Gasbrenner m. mit Mischer und Gasdruckregler ‖ brûleur m. à gaz équipé d'un mélangeur et d'un régulateur de pression de gaz. / ~ pliers pl. ‖ Gasbrennerzange f.; Brennerzange f. ‖ pince f. à brûleur à gaz.

gas burning ‖ Gasfeuerung f. ‖ chauffage m. au gaz. / ~ carbon (Chem) ‖ Retorten-

graphit m.; Retortenkohle f. ‖ charbon m. de cornue. / ~ carburizing furnace see ~ case hardening furnace. / ~ case-hardening ‖ Gaseinsatzhärtung f. ‖ cémentation f. aux gaz carburants. / ~ case-hardening furnace ‖ Gaseinsatzofen m.; Druckgas- und Karbonisierofen m. ‖ four m. à cémenter aux gaz carburants. / ~ case-hardening process ‖ Druckgaseinsetzverfahren n. ‖ procédé m. à cémenter aux gaz carburants. / ~ chandelier ‖ Gaskrone f. ‖ lustre m. à gaz. / ~ clean(s)ing ‖ Reinigung f. des Gases ‖ épuration f. du gaz.

gas cleaning plant ‖ Gasreinigungsanlage f. ‖ installation f. pour l'épuration de gaz. / ~ for dry purification ‖ Gasreinigungsanlage f. für Trockenreinigung ‖ installation f. d'épuration sèche de gaz. / ~ for electric purification ‖ Gasreinigungsanlage f. für elektrische Reinigung ‖ installation f. d'épuration électrique de gaz. / ~ for wet purification ‖ Gasreinigungsanlage f. für Naßreinigung ‖ installation f. d'épuration humide de gaz.

gas cloud (Of poisonous gas) ‖ Gaswolke f. ‖ vague f. de gaz. / ~ coal ‖ Gaskohle f.; Fettkohle f. ‖ houille f. grasse ou à gaz; charbon m. gras ou à gaz. / ~ coke ‖ Gaskoks m. ‖ coke m. de gaz; coke m. (provenant) des usines à gaz. / ~ compound ‖ Gasgemisch n. ‖ mélange m. gazeux. / plant for decomposition of ~ compounds ‖ Anlage f. zur Zerlegung von Gasgemischen ‖ installation f. de décomposition des mélanges gazeux. / ~ compressor ‖ Gasverdichter m.; Gaskompressor m. ‖ compresseur m. à gaz. / ~ conduit ‖ Gasleitung f. ‖ conduite f. de gaz. / ~ constituents pl. ‖ Gasbestandteile mpl. ‖ principes mpl. constituants du gaz. / ~ consumption ‖ Gasverbrauch m. ‖ consommation f. de gaz.

gas container see also gasometer ‖ Gasbehälter m. ‖ gazomètre m.; récipient m. à gaz. / ~ with floating cover ‖ Gasbehälter m. mit schwimmender Glocke ‖ gazomètre m. à cloche flottante. / ~ projector ‖ Gaswerfer m. ‖ projector m. de gaz. / ~ valve ‖ Gasflaschenventil n. ‖ soupape f. de la bouteille à gaz.

gas cooker ‖ Gaskocher m. ‖ fourneau m. à gaz. / ~ cooler ‖ Gaskühler m. ‖ réfrigérateur m. à gaz. / S-shaped ~ cooler (Met) ‖ S-Apparat m.; S-förmiger Gaskühler m. ‖ appareil m. réfrigérant à gaz en S. / ~ cooling plant ‖ Gaskühlanlage f. ‖ installation f. de réfrigération des gaz. / ~ density ‖ Gasdichte f. ‖ densité f. du gaz; densimètre m. à gaz. / ~ density gauge ‖ Gasdichtemesser m. ‖ appareil m. de mesure de la densité du gaz. / ~ density meter see ~ density gauge. / ~ desiccator see also ~ drying device (Acc) ‖ Gastrockner m. ‖ sécheur m. du gaz. / ~ detector (Tel) ‖ Gasdetektor m. ‖ détecteur m. à gaz. / ~ development (Acc) ‖ Gasentwicklung f. ‖ formation f. ou développement m. de bulles de gaz. / ~ drying device ‖ Gastrockner m. ‖ sécheur m. à gaz. / ~ dynamo ‖ Gasdynamomaschine f. ‖ dynamo f. à gaz.

gaselier see gas chandelier.

gas engine ‖ Gasmotor m.; Gasmaschine f.; Gaskraftmaschine f. ‖ moteur m. ou machine f. à gaz; gazomoteur m. / high-

power ~ ‖ Großgasmaschine f. ‖ moteur m. à gaz de grande puissance; gros moteur m. à gaz. / high-speed ~ ‖ schnelllaufende Gasmaschine f. ‖ moteur m. à gaz à grande vitesse. / large ~ see high power ~.

gas-engineer ‖ Gasingenieur m. ‖ ingénieur-gazier m.

gaseous ‖ gasförmig ‖ gazeux; aériforme. / ~ body ‖ gasförmiger Körper m. ‖ corps m. gazeux. / ~ mineral water ‖ Säuerling m. ‖ eau f. minérale gazeuse. / ~ mixture ‖ Gasmischung f.; Gasgemisch n. ‖ mélange m. gazeux.

gas excess ‖ Gasüberschuß m. ‖ excès m. de gaz. / ~ exhauster ‖ Gassauger m. ‖ aspirateur m. de gaz. / ~ exhausting plant ‖ Absaugeanlage f. für Gase ‖ installation f. d'aspiration des gaz. / ~ explosion ‖ Gasexplosion f. ‖ explosion f. de gaz.

gas-filled bomb ‖ Gasbombe f. ‖ bombe f. à gaz. / ~ lamp ‖ Glühlampe f. mit Gasfüllung ‖ lampe f. à atmosphère gazeuse.

gas-fired furnace ‖ gasgefeuerter Ofen m.; Gasfeuerung f. ‖ four m. pour le traitement thermique au gaz; chauffage m. ou foyer à gaz. / ~ furnace for hardening ‖ Gashärteofen m. ‖ four m. de trempe au gaz. / ~ muffle furnace ‖ gasgeheizter Muffelofen m. ‖ four m. à moufle au gaz.

gas firing ‖ Gasfeuerung f. ‖ chauffage m. ou foyer m. à gaz. / regenerative ~ ‖ Regenerativgasfeuerung f. ‖ régénérateur m. à gaz. / ~ installation see gas firing plant. / ~ plant ‖ Gasfeuerungsanlage f. ‖ installation f. de chauffage au gaz.

gas fitter ‖ Gasinstallatör m. ‖ installateur m. pour le gaz. / ~ fitting ‖ Gaseinrichtung f. ‖ installation f. de gaz. / ~ fittings pl. ‖ Gasarmatur f. ‖ armature f. pour gaz. / ~ flat-iron stove ‖ Gasbügelofen m. ‖ four m. à gaz pour fers à repasser. / ~ formation (Acc) ‖ Gasentwicklung f.; Gasbildung f. ‖ formation f. de gaz.

gas furnace see also gas firing ‖ Gasofen m. ‖ four m. à gaz ou au gaz. / regenerative ~ ‖ Regenerator m.; Regeneratorofen m.; Regenerativofen m. ‖ four m. à régénérateur au gaz. / ~ with regenerators see regenerative ~.

gas generation ‖ Gasgewinnung f. ‖ production f. de gaz.

gas generator ‖ Gaserzeuger m.; Gasgenerator m.; Gasentwicklungsgerät n.; Gasentwickler m. ‖ appareil m. gazogène ou à produire les gaz; générateur m. à gaz; gazogène m. / ~ with recovery of by-products ‖ Gaserzeuger m. mit Nebenerzeugnisgewinnung ‖ gazogène f. avec récupération de sousproduits. / revolving-grate ~ ‖ Drehrostgaserzeuger m. ‖ gazogène m. à grille tournante. / steam-generation ~ ‖ Gaserzeuger m. mit Dampferzeugung ‖ gazogène m. à production de vapeur.

gas generator furnace ‖ Gasgeneratorofen m. ‖ fourneau m. générateur (de gaz). / ~ plant ‖ Gaserzeugungsanlage f. ‖ installation f. de gazogène.

gas geyser ‖ Gasbadeofen m. ‖ poêle m. à gaz pour bains. / ~ governor ‖ Gasregler m. ‖ régulateur m. de gaz. / ~ hearth ‖ Gasherd m. ‖ foyer m. à gaz. / ~ heating ‖ Gasheizung f. ‖ chauffage m. au gaz.

gas holder see also gas container and gasometer ‖ Gasbehälter m.; Gasometer m. ‖ gazomètre m.; réservoir m. à gaz. / ~ volume ‖ Fassungsraum m. des Gasometers ‖ contenance f. d'un gazomètre.

gas hull (Of a balloon) ‖ Ballonhülle f. ‖ enveloppe f. du ballon.

gasifiability ‖ Vergasbarkeit f. ‖ gazéifiabilité f.

gasifiable ‖ vergasbar ‖ gazéifiable.

gasification ‖ Vergasung f. ‖ gazéification f. / ~ (Spraying) ‖ Zerstäubung f. ‖ pulvérisation. / ~ of coal ‖ Vergasung f. der Kohle ‖ transformation f. ou gazéification f. du charbon. / ~ of (solid) fuels ‖ Brennstoffvergasung f. ‖ gazéification f. de combustibles. / ~ process ‖ Vergasungsvorgang m.; Vergasungsprozeß m. ‖ procédé m. de gazéification.

gasiform see gaseous.

gasify, to ‖ vergasen ‖ gazéifier.

gas ignition ‖ Gasentzündung f. ‖ inflammation f. des gaz. / ~ indicator ‖ Gasanzeiger m. ‖ indicateur m. de gaz. / ~ inlet pipe ‖ Gaszuleitungsrohr n. ‖ tube m. d'arrivée du gaz. / ~ interferometer ‖ Gasinterferometer n. ‖ interféromètre m. à gaz.

gas jet (Flame) ‖ Gasflamme f. ‖ jet m. de gaz. / ~ (Burner) ‖ Gasbrenner m. ‖ brûleur m. à gaz; bec m. de gaz.

gasket ‖ Dichtung f.; Flanschendichtung f. ‖ étoupage m.; bourrage m.; garniture f. (de brides). / cylinder head ~ ‖ Zylinderkopfdichtung f. ‖ joint m. de culasse. / hemp ~ ‖ Hanfdichtung f. ‖ étoupage m. de chanvre. / spark(ing) plug ~ ‖ Zündkerzendichtung f. ‖ joint m. de bougie. / valve cap ~ ‖ Ventilverschraubungsdichtung f. ‖ joint m. de bouchon de soupape.

gas lamp ‖ Gaslampe f. ‖ lampe f. à gaz. / ~ leakage recorder ‖ Kontrollvorrichtung f. für Gasverlust; Gasverlustanzeiger m. ‖ appareil m. contrôleur des fuites de gaz. / ~ lever (Auto) ‖ Gashebel m. ‖ manette f. d'admission de gaz.

gaslight ‖ Gaslicht n. ‖ lumière f. du gaz. / incandescent ~ ‖ Gasglühlicht n. ‖ lumière f. à incandescence par le gaz.

gas lighter ‖ Gasanzünder m. ‖ allume-gaz m.; allumoir m. de gaz. / automatic ~ ‖ Gasselbstanzünder m. ‖ allumoir m. automatique pour gaz. / electric ~ ‖ elektrischer Gasanzünder m. ‖ allumoir m. électrique pour gaz. / self ~ see automatic ~.

gas lighting ‖ Gasbeleuchtung f. ‖ éclairage m. au gaz.

gaslight paper ‖ Gaslichtpapier n. ‖ papier m. à la lumière du gaz.

gas-like see gaseous.

gas liquefying plant ‖ Gasverflüssigungsanlage f.; Anlage f. zur Verflüssigung von Gasen ‖ installation f. de liquéfaction de gaz. / ~ maker (Met) ‖ Gasarbeiter m. ‖ gazier m. / ~ manometer ‖ Gasdruckmesser m. ‖ manomètre m. à gaz.

gas mask ‖ Gasmaske f. ‖ masque m. à gaz. / absorbing ~ ‖ Absorptionsgasmaske f. ‖ masque m. à gaz absorbant. / ~ with chemical filter ‖ Gasmaske f. mit Chemikalfilter ‖ masque m. à gaz à filtres chimiques.

gas meter ‖ Gasmesser m.; Gasuhr f.; Gaszähler m.; Gasometer m. ‖ compteur m. à gaz; gazomètre m. / ~ inspector ‖ Gasuhrableser m. ‖ contrôleur m. de compteurs à gaz. / ~ testing apparatus ‖ Gasmesserprüfvorrichtung f. ‖ appareil m. de contrôle des compteurs à gaz.

gas mixture ‖ Gasgemisch n. ‖ mélange m. gazeux ou de gaz. / ~ decomposition ‖

Zerlegung f. von Gasgemischen ‖ décomposition f. de mélanges de gaz. / ~ liquefaction ‖ Verflüssigung f. von Gasgemischen ‖ liquéfaction f. de mélanges de gaz.

gas motor ‖ Gasmotor m.; Gaskraftmaschine f. ‖ moteur m. à gaz. / ~ for lighting gas ‖ Leuchtgasmotor m. ‖ moteur m. à gaz d'éclairage. / ~ for marsh gas ‖ Erdgasmotor m. ‖ moteur m. à gaz naturel. / ~ for suction gas ‖ Sauggasmotor m. ‖ moteur m. à gaz aspiré.

gas motor coal ‖ Gasmotorkohle f. ‖ charbon m. pour moteur à gaz.

gas oil ‖ Gasöl n. ‖ gasoil m.; huile f. à gaz. / ~ from lignite ‖ Gasöl n. aus Braunkohle ‖ gasoil m. extrait du lignite. / mixture of ~ and crude oil ‖ Mischung f. von Gasöl mit Rückständen ‖ mélange m. de gasoil avec résidus de distillation.

gasolene see gasoline.

gasoline ‖ Gasolin n. ‖ gazoline f. / ~ (Petrol) ‖ Benzin n. ‖ benzine f.; essence f. / ~ gauge ‖ Benzinstandanzeiger m. ‖ indicateur m. de niveau d'essence. / ~ lamp ‖ Gasolinlampe f. ‖ lampe f. à gazoline. / ~ motor ‖ Benzinmotor m. ‖ moteur m. à essence. / ~ separator ‖ Benzinfilter n.; Benzinreiniger m. ‖ filtre m. à essence. / ~ storage ‖ Benzinlagerung f. ‖ entrepôt m. d'essence. / ~ strainer ~ separator. / ~ valve ‖ Benzinhahn m. ‖ robinet m. d'essence.

gasometer ‖ Gasbehälter m.; Gasometer m. ‖ gazomètre m.; réservoir m. à gaz. / ~ with floating cover ‖ Gasbehälter m. mit schwimmender Glocke ‖ gazomètre m. à cloche flottante. / ~ prover ‖ Gasbehälteraufseher m. ‖ surveillant m. de gazomètre.

gas pipe ‖ Gasrohr n. ‖ tuyau m. à gaz. / ~ fittings ‖ Zubehörteile npl. für Gasleitungen ‖ accessoires mpl. de canalisation pour le gaz. / ~ thread ‖ Gasgewinde n. ‖ filet m. des tuyaux à gaz.

gas piping ‖ Gasrohrleitung f. ‖ tuyauterie f. à gaz. / ~ and air mains pl. for blast furnaces ‖ Hochofengas- und -luftleitung f. ‖ conduite f. de gaz et d'air pour hauts-fourneaux.

gas plant for long distance supply ‖ Gasfernversorgungsanlage f. ‖ installation f. d'alimentation à distance avec gaz. / ~ pliers pl. ‖ Gaszange f.; Gasrohrzange f. ‖ pince f. à gaz. / ~ power engine see gas motor.

gas pressure ‖ Gasspannung f.; Gasdruck m. ‖ tension f. ou pression f. de gaz. / ~ igniter at distance ‖ Gasdruckfernzünder m. ‖ allumage m. à distance par la pression du gaz. / ~ indicator (balloon) ‖ Prallanzeiger m. ‖ indicateur m. de plénitude ou de remplissage du ballon. / ~ reducing valve ‖ Gasdruckminderer m.; Gasdruckminderventil n. ‖ soupape f. réductrice de gaz. / ~ regulator ‖ Gasdruckregler m. ‖ régulateur m. de pression du gaz.

gas producer see ~ generator. / ~ projectile ‖ Gasgeschoß n. ‖ projectile m. à gaz. / ~ puddling ‖ Gasflammofenfrischen n.; Gasfrischen n.; Gaspuddeln n. ‖ puddlage m. au gaz. / ~ purifier ‖ Gasreinigungsanlage f.; Gasreiniger m. ‖ épurateur m. ou purgeur m. de gaz. / ~ purifying material ‖ Gasreinigungsmasse f. ‖ matière f. à épurer le gaz. / ~ purifying plant ‖ Gasreinigungsanlage f. ‖ installation f.

à purifier le gaz. / ~ quantity ‖ Gasmenge f. ‖ quantité f. de gaz. / ~ rate collector ‖ Gasgebührenerheber m. ‖ receveur-gazier m. / reduction into ~ see gasification. / ~ regulator ‖ Gasregler m.; Gasdruckregler m. ‖ régulateur m. de pression du gaz. / ~ reservoir see gasometer. / ~ residue ‖ Gasrest m. ‖ résidu m. gazeux. / ~ retort ‖ Gasretorte f. ‖ cornue f. à gaz. / ~ rolling mill engine ‖ Gaswalzenzugmaschine f. ‖ moteur m. à gaz pour laminoirs. / ~ safety device ‖ Gassicherheitsapparat m. ‖ appareil m. de sûreté pour le gaz. / ~ sample ‖ Gasprobe f. ‖ prise f. de gaz. / ~ shell ‖ Gasgeschoß n. ‖ obus m. à gaz; projectile m. à gaz.

gassing (Acc) ‖ Gasentwicklung f. ‖ formation f. de gaz. / ~ after the charge (Acc) ‖ Nachkochen n. ‖ bouillonnement m. au repos.

gas singeing ‖ Sengen n. mit Gas ‖ grillage m. au gaz; gazage m. / ~ slack (coal) ‖ Gasfördergrus m. ‖ menu charbon m. à gaz. / ~ slide valve ‖ Gasschieber m. ‖ vanne f. à gaz. / ~ smoothing iron ‖ Gasplätteisen n. ‖ fer m. à repasser au gaz. / ~ soldering copper ‖ Gaslötkolben m. ‖ fer m. à souder au gaz. / ~ stove ‖ Gasofen m.; Gasheizofen m. ‖ four(neau) m. à gaz. / kitchen ~ stove ‖ Gasherd m.; Gaskochapparat m. ‖ fourneau m. de cuisine à gaz.

gas supply ‖ Gasversorgung f.; Gaszufuhr f. ‖ distribution f. de gaz. / long-distance ~ plant ‖ Gasfernversorgungsanlage f. ‖ installation f. de distribution de gaz à grandes distances. / long-distance ~ network ‖ Ferngasversorgungsnetz n. ‖ réseau m. de distribution de gaz à grande distance.

gassure, non ~ coal ‖ gasarme Kohle f. ‖ charbon m. pauvre en gaz.

gassy ‖ gashaltig ‖ gazeux. / ~ (Charged with fire damp) ‖ schlagwetterreich ‖ grisouteux.

gas take (Blast furnace) ‖ Gichtgasfang m. ‖ prise f. de gaz (de haut-fourneau). / ~ tank see also gasometer and ~ container ‖ Gasbehälter m. ‖ gazomètre m. / ~ tar ‖ Steinkohlenteer m.; Gasteer m. ‖ goudron m. de houille. / ~ testing apparatus ‖ Gasanalysenapparat m.; Gasmeßgerät n. ‖ appareil m. d'analyse de gaz. / ~ testing chamber ‖ Gasmeßkammer f. ‖ compartiment m. de mesure à gaz. / ~ thread pipe-stock ‖ Gasgewindeschneidkluppe f. ‖ filière f. au pas du gaz. / ~ tongs pl. see ~ pliers. / ~ turbine ‖ Gasturbine f. ‖ turbine f. à gaz. / ~ valve ‖ Gasventil n. ‖ soupape f. à gaz. / ~ vent ‖ Gasabzug m. ‖ évent m. de gaz. / ~ warfare ‖ Gaskrieg m.; Giftgaskrieg m. ‖ guerre f. aux gaz. / ~ washer ‖ Gaswascher m.; Skrubber m. ‖ scrubber m.; laveur m. à gaz. / ~ washing apparatus ‖ Gaswaschvorrichtung f. ‖ appareil m. à laver le gaz. / ~ washing bottle ‖ Wäscher m.; Gaswaschflasche f. ‖ laveur m.; barbotteur m.; flacon-laveur m. de gaz. / ~ washing tube ‖ Gaswaschaufsatz m. ‖ tube-laveur m. de gaz. / ~ welding ‖ Gas(schmelz)schweißung f. ‖ soudage m. au gaz. / ~ worker ‖ Gasarbeiter m. ‖ ouvrier-gazier m.

gasworks pl. ‖ Gasanstalt f.; Gaswerk n.; Gasfabrik f. ‖ usine f. à gaz. / ~ building ‖ Gaswerkbauanstalt f. ‖ entrepreneur m. pour la construction d'usines à gaz. / ~ construction ‖ Gaswerkbau m. ‖

construction f. d'usines à gaz. / ~ furnace ‖ Gaswerkofen m. ‖ four m. pour usines à gaz. / ~ machine ‖ Gaswerkmaschine f. ‖ machine f. pour usines à gaz.

gat (Nav) ‖ Fahrwasser n. zwischen den Bänken; Seegat n. ‖ passe f. entre les sables.

gate (Door) ‖ Tor n. ‖ porte f. / ~ (Passage) ‖ (enge oder schwierige) Durchfahrt f.; Einfahrt f.; Torweg m. ‖ porte f. cochère. / ~ (Found) ‖ Einguß m. ‖ jet m.; trompe m.; échenal m. / ~ (Mining) see also gateway ‖ Strecke f. ‖ galerie f. / ~ of iron ‖ eisernes Tor n. ‖ porte f. cochère en fer. / lock ~ (Hydr arch) ‖ Schleusentor n. ‖ porte f. d'écluse. / ~ of the mould see gate (Found). / ~ of railway ‖ Eisenbahnwegschranke f. ‖ barrière f. de chemin de fer. / safety ~ (Hydr arch) ‖ Nottor n.; Sicherheitstor n. ‖ écluse f. de garde. / sliding ~ ‖ Schiebetor n. ‖ porte f. coulissante.

gate hole see gate (Found).

gatehouse ‖ Pförtnerhaus n.; Wärterhaus n. ‖ maison f. du garde.

gatekeeper (Railw) ‖ Schrankenwärter m. ‖ garde-barrière m.

gate-like supports pl. (Crane) ‖ portalartiger Unterbau m. ‖ montants mpl. en forme de portail.

gate road see gateway (Mining).

gateway (Passage) see gate (Passage). / ~ (Mining) ‖ Förderstrecke f. ‖ galerie f. ou voie f. de roulage.

gather, to (Garden) ‖ pflücken; lesen ‖ cueillir. / ~ (Print) ‖ Lagen fpl. machen ‖ assembler le cahier.

gatherer ‖ Sammler m. ‖ collectionneur m.; quêteur m. / ~ (Bookb) ‖ Ausschießer m. ‖ assembleur m.

gathering (of harvest) ‖ Ernte f.; Ernten n. ‖ récolte f. / ~ (Print) ‖ Lage f. von gedruckten Bogen ‖ cahier m. de feuilles imprimées. / ~ of gold from the riversand ‖ Goldwaschen n. ‖ cueillette f. de l'or dans les sables des rivières. / to lay down a ~ see to gather (Print).

gathering ground of a river ‖ Flußgebiet n. ‖ bassin m. d'un fleuve.

gathering machine ‖ Erntemaschine f. für Hackfrüchte ‖ machine f. à récolter les fruits de labour. / potato ~ ‖ Kartoffelerntemaschine f. ‖ machine f. à récolter les pommes de terre. / turnip ~ ‖ Rübenerntemaschine f. ‖ machine f. à récolter les betteraves et navets.

gatway see gat (Nav).

gaudy green ‖ hellgrün ‖ vert clair.

gauffer ‖ Waffel f. ‖ gaufre f.

gauge, to ‖ eichen; kalibrieren; adjustieren ‖ étalonner; jauger; calibrer; ajuster. / ~ (Join) ‖ vorreißen ‖ trusquiner; pointer; poinçonner. / ~ a form (Print) ‖ Format n. machen ‖ mettre la garniture.

gauge ‖ Eichmaß n.; Urmaß n. ‖ échantillon m.; étalon m.; jauge f. / ~ (Measuring instrument) ‖ Lehre f.; Maßlatte f.; Stichmaß n. ‖ calibre m.; jauge f.; gabarit m. / ~ (Pattern) ‖ Schablone f. ‖ échantillon m.; gabarit m.; calibre m. / ~ (Pressure indicator) ‖ Manometer n. ‖ manomètre m.; indicateur m. / ~ (Distance between rails) ‖ Spurweite f. ‖ voie f.; écartement m. de voie. / ~ (Railw loading) ‖ Durchgangsprofil n.; Ladelehre f. ‖ gabarit m. (de chargement); profil-limite m. / ~ (Print) ‖ Längenlinie f.; Kolumnen-

maß n. ‖ réglette m. du longueur. / to inspect by a ~ ‖ mit einer Lehre f. nachprüfen ‖ vérifier (par un calibre). / adjusting ~ ‖ Einstellehre f. ‖ calibre m. de réglage. / air pressure ~ ‖ Luftdruckmesser m.; manomètre m. d'air. / angle ~ ‖ Winkelmesser m. ‖ goniomètre m. / clearance ~ for rolling stock ‖ Umgrenzungslinie f. für Eisenbahnfahrzeuge ‖ gabarit m. des wagons de chemin de fer. / depth ~ (Hydr arch) ‖ Pegel m. ‖ échelle f. fluviale. / depth ~ (Measuring) ‖ Tiefentaster m. ‖ calibre m. de profondeur. / glass ~ ‖ Standglas n. ‖ tube m. de niveau. / ~ of goods-wagons ‖ Lademaß n. (für offene Güterwagen) ‖ gabarit m. de chargement. / height ~ ‖ Höhenmaßstab m. ‖ pied m. à hauteur. / ~ of inclination of rails ‖ Spurlehre f. für die Schienenneigung ‖ gabarit m. d'inclinaison des rails / interpupillary distance ~ ‖ Augenabstandmesser m. ‖ appareil m. pour mesurer l'écartement des yeux. / ~ for the control of the height of teeth ‖ Zahnhöhenmeßlehre f. ‖ calibre m. à coulisse pour mesurer la hauteur des dents. / ~ of laths (Railw) ‖ Lattenprofil n. eines Dammes ‖ gabarit m. en lattes clouées; profil m. de terrassement. / limit ~ ‖ Grenzlehre f.; Toleranzlehre f. ‖ calibre m. de tolérance. / loading ~ see ~ (Railw loading). / ~ for looking glasses (Glassm) ‖ Spiegelmesser m. zum Messen der Glasdicke ‖ pochomètre m. / mercurial ~ ‖ Quecksilberdruckmesser m.; Quecksilbermanometer n. ‖ manomètre m. à mercure. / micrometer ~ ‖ Schraubenlehre f.; palmer m. / oil level ~ ‖ Ölstandzeiger m. ‖ indicateur m. d'huile. / pressure vacuum ~ ‖ Manovakuummesser m. ‖ manovacuomètre m. / quartering ~ ‖ Nachprüfvorrichtung f. ‖ dispositif m. de contrôle. / ~ for running true ‖ Rundlauflehre f. ‖ comparateur m. / setting ~ ‖ Einstellehre f. ‖ appareil m. de réglage. / standard ~ (Railw) ‖ Spurlehre f. ‖ gabarit m. / standard clearance ~ (Railw) ‖ Normalprofilschablone f. ‖ gabarit m. de profil normal. / steam ~ ‖ Manometer n.; Dampfdruckmesser m. ‖ manomètre m. / straddle ~ ‖ Reiterlehre f. ‖ calibre m. cavalier. / universal surface ~ ‖ Universalreißstock m. ‖ trusquin m. universal. / water ~ ‖ Wasserstandglas n. ‖ niveau m. d'eau à tube de verre. / ~ of way (Railw) ‖ Spurweite f. ‖ distance f. entre les rails; écartement m. des rails; largeur f. de la voie. / wooden ~ ‖ Holzlehre f.; Lehrbrett n. ‖ gabarit m. en bois; moulet m.

gauge board ‖ Manometertafel f. ‖ tableau m. de manomètre. / ~ change (Railw) ‖ Spurveränderung f. ‖ changement m. de l'écartement. / ~ cock ‖ Probierhahn m.; Wasserstandhahn m. ‖ robinet m. de jauge ou d'épreuve; robinet m. vérificateur; indicateur m. d'eau. / ~ cock (Of the pressure gauge) ‖ Manometerhahn m. ‖ robinet m. de manomètre.

gauge connexion, fixed ~ ‖ feste Manometeraufnahme f. ‖ raccordement m. du manomètre fixe. / ~ by fixed socket ‖ Manometeraufnahme f. durch feste Muffe ‖ raccordement m. du manomètre par raccord fixe. / movable ~ ‖ drehbare Manometeraufnahme f. ‖ raccordement m. du manomètre mobile. / ~ by movable socket ‖ Manometeraufnahme f. durch drehbare Muffe ‖ raccordement m. du manomètre par raccord mobile.

gauged ‖ geeicht ‖ jaugé. / ~ length ‖ Meßlänge f. ‖ longueur f. entre repères.

gauge glass ‖ Wasserstandzeiger m.; Wasserstandglas n. ‖ indicateur m. de niveau d'eau; tube m. de niveau d'eau; niveau m. à tube de verre. / ~ lamp (Loc) ‖ Laterne f. am Wasserstandzeiger ‖ lanterne f. de niveau d'eau. / ~ narrowing (Railw) ‖ Spurverengung f. ‖ rétrécissement m. de la voie. / ~ pipe ‖ Manometerleitung f. ‖ tuyauterie f. de manomètre. / ~ plate ‖ Ziehring m. für Röhren ‖ filière f. à tuyaux. / ~ point (Assay) ‖ Endmarke f. ‖ repère m.

gauger ‖ Eichinspektor m. ‖ inspecteur m. des poids et mesures. / ~ of the navvies (Railw construction) ‖ Schachtmeister m. ‖ chef m. des ouvriers ou des terrassiers.

gauger's fee ‖ Eichgebühr f. ‖ frais mpl. de jaugeage. / ~ stamp ‖ Eichstempel m. ‖ timbre m. de jaugeage: poinçon m. (des poids et mesures).

gauge ring (Electr) ‖ Paßring m.; Einsatzring m. ‖ bague f. de fixation. / ~ rod (Railw) ‖ Spurstange f. ‖ tringle f. d'écartement. / ~ rod of a pump (Mar) ‖ Pumpenpeilstock m. ‖ sonde f. de pompe. / ~ setting device (Railw) ‖ Spurrichter m. ‖ appareil m. à redresser ou à rectifier les alignements des rails. / ~ slide (Opt) ‖ Meßeinsatz m. ‖ cadre-mesure m. / ~ tap see ~ cock.

gauging apparatus ‖ Eichvorrichtung f.; Eichgerät n.; Eichapparat m. ‖ appareil m. à calibrer ou à jauger. / ~ for casks ‖ Faßeichvorrichtung f. ‖ appareil m. à litrer les fûts; litreuse f. pour fûts.

gauging instrument ‖ Meßwerkzeug n. ‖ instrument m. ou outil m. de mesure.

gauging office ‖ Eichamt n. ‖ service m. des poids et mesures.

gauze ‖ Gaze f.; Flor m. ‖ gaze f.; crêpe m. / book-binding ~ ‖ Buchbindergaze f. ‖ gaze f. pour la reliure. / common ~ ‖ glatte Gaze f. (ohne Muster) ‖ gaze f. unie (ni brochée ni façonnée). / cotton ~ ‖ Baumwollgaze f. ‖ gaze f. de coton. / ~ with damask-figures ‖ Damastgaze f. ‖ gaze f. damassée. / fancy ~ ‖ gemusterte Gaze f. ‖ gaze f. façonnée ou à dessin. / figured ~ ‖ geblumte oder gemodelte oder broschierte Gaze f. ‖ gaze f. brochée ou à fleurs. / metal ~ ‖ Metalltuch n. ‖ toile f. métallique. / plain ~ see common ~. / silk(en) ~ ‖ Seidengaze f. ‖ gaze f. de soie; canevas m. en soie. / tweeled ~ ‖ Köpergaze f. ‖ gaze f. croisée. / wire ~ ‖ Drahtgewebe n.; Drahtgaze f.; Metallgewebe n. ‖ toile f. ou tissu m. ou gaze f. métallique.

gauze bandage ‖ Gazebinde f. ‖ bande f. de gaze. / ~ bottom ‖ Gazeboden m. ‖ fond m. de gaze. / ~ loom (Weav) ‖ Gazegeschirr n.; Gazestuhl m. ‖ métier m. à gaze. / ~ ribbon ‖ Gazeband n.; Dünntuchband n. ‖ ruban m. gaze ou de gaze. / ~ sieve ‖ Drahtgewebesieb n.; Gazesieb n. ‖ tamis m. à gaze ou à toile métallique. / ~ weaver ‖ Gazeweber m. ‖ gazetier m.

Gay Lussac tower ‖ Gay-Lussac-Turm m. ‖ tour f. de Gay-Lussac.

gaylussite (Miner) ‖ Gaylussit m. ‖ gaylussite f.; hydrocarbonate m. de soude et de chaux.

gear, to ~ (into one another) ‖ ineinandergreifen ‖ s'engrener.

gear (Mach) see also gearing ‖ Getriebe n.; Rädervorgelege n.; Übersetzung f. ‖ engrenage m.; transmission f. / ~ (Mechanism) ‖ Vorrichtung f.; Mechanismus m. ‖ appareil m.; mécanisme m.; armature f. / ~ (Toothed wheel) ‖ Zahnrad n. ‖ roue f. dentée. / ~ (Auto) ‖ Antriebsrad n. ‖ roue f. de commande. / to cut ~s pl. ‖ Zahnräder npl. fräsen ‖ fraiser les gorges fpl. des dents. / wheels pl. which are constantly in ~ ‖ ständig im Eingriff stehende Räder npl. ‖ roues fpl. dentées continuellement en prise. / axle drive bevel ~ ‖ großes Differentialantriebskegelrad n. ‖ pignon m. du centre du différentiel. / bevel ~ ‖ Kegelrad(ausgleichs)getriebe n. ‖ engrenage m. différentiel à pignon conique; engrenage m. conique (d'angle). / bottom ~ (Auto) ‖ geringe Geschwindigkeit f. ‖ première vitesse f. / brake ~ ‖ Bremsvorrichtung f. ‖ timonerie f. de frein. / cam ~ ‖ Nockensteuerung f.; Daumensteuerung f. ‖ distribution f. ou commande f. à came. / camshaft timing ~ ‖ Nockenwellenantriebsrad n. ‖ roue f. de l'arbre à cames. / change speed ~ ‖ Wechselgetriebe n. ‖ boîte f. à vitesse; engrenage m.; changement m. de vitesse. / clash ~s pl. (Auto) ‖ Zahnradwechselgetriebe n. ‖ changement m. de vitesse par engrenages. / ~ pl. closed together ‖ Räderwerk n. in gedrängter Anordnung ‖ mécanisme m. d'engrenages très ramassé. / composite ~ ‖ Metallfiberzahnrad n. ‖ roue f. dentée en fibre et métal. / crankshaft timing ~ ‖ Kurbelwellenzahnrad n. ‖ roue f. dentée du vilebrequin. / differential ~ (Auto) ‖ Ausgleichgetriebe n. ‖ engrenage m. différentiel. / differential side ~ ‖ Hinterachswellenzahnrad n. ‖ roue f. dentée du différentiel. / disengaging ~ ‖ Ausrückvorrichtung f. ‖ appareil m. de débrayage. / docking ~ (Aero) ‖ Haltegestell n. ‖ dispositif m. de retenue. / double reduction ~ ‖ Reduktionsgetriebe n. ‖ engrenage m. de démultiplicateur. / driving ~ ‖ Antriebsvorrichtung f. ‖ mécanisme m. de commande. / epicyclic ~ ‖ Planetengetriebe n.; Umlaufgetriebe n. ‖ engrenage m. épicycloidal ou planétaire ou à sattelites. / fibre ~ ‖ Fiberzahnrad n. ‖ roue f. dentée en fibre. / fourth ~ ‖ vierter Gang m. ‖ quatrième vitesse f. / friction ~ ‖ Reibrädergetriebe n.; Reibungsgetriebe n.; Reibgetriebe n. ‖ engrenage m. à galet de friction ou à roues à friction; engrenage m. à friction. / herringbone ~ ‖ Pfeilradgetriebe n. ‖ engrenage m. hélicoïdal double. / hoisting ~ ‖ Hubwerk n. ‖ treuil m.; engins mpl. de levage. / electric hoisting ~ ‖ Elektrohubwerk n. ‖ treuil m. électrique. / hydraulic ~ ‖ Flüssigkeitsgetriebe n. ‖ mécanisme m. de commande hydraulique. / intermediate ~ ‖ Zwischenschaltung f. ‖ transmission f. intermédiaire. / internal ~ ‖ Innenverzahnung f. ‖ denture f. intérieure. / lantern ~ ‖ Triebstockkranz m. ‖ couronne f. à fuseaux. / luffing ~ ‖ Einziehwerk n. ‖ dispositif m. de relevage. / planetary ~ see epicyclic ~. / precision ~ ‖ Genauigkeitsverzahnung f. ‖ denture f. de précision. / rack ~ f ‖ Zahnstangenantrieb m. ‖ commande à crémaillère. / rawhide ~ ‖ Rohhaut-

getriebe n. ‖ engrenage m. en cuir vert. / reduction ~ see transmission ~. / relieving ~ ‖ Feststellvorrichtung f. ‖ appareil m. de calage. / reverse ~ (Aut) ‖ Rückwärtsgang m. ‖ marche f. arrière. / reverse idler ~ ‖ Rücklaufrad n. ‖ pignon m. de marche arrière. / reversing ~ ‖ Umsteuergetriebe n.; Räderwendegetriebe n.; Umsteuerung f. ‖ mécanisme m. de renversement ou de changement de marche; mécanisme m. de renversement de marche par engrenage. / second ~ (Auto) ‖ zweiter Gang m. ‖ deuxième vitesse f. / self-locking ~ ‖ selbstsperrendes Getriebe n. ‖ mécanisme m. d'encliquetage automatique. / ~ of a set of timber (Mining) ‖ Türstock m. bei der Streckenzimmerung ‖ poteau m. d'un cadre de boisage. / spur ~ ‖ Stirnradgetriebe n. ‖ engrenage m. droit. / starting ~ (Auto) ‖ erster Gang m. ‖ première vitesse f. / steering ~ ‖ Lenkeinrichtung f. ‖ mécanisme m. de direction. / striking ~ (Electr) ‖ Zündvorrichtung f. ‖ mécanisme m. d'amorçage. / third ~ (Auto) ‖ dritter Gang m. ‖ troisième vitesse f. / three-speed ~ (Auto) ‖ Dreiganggetriebe n. ‖ boîte f. à trois vitesses. / throwing-in ~ ‖ Einrückvorrichtung ‖ appareil m. d'embrayage. / toothed (wheel) ~ ‖ Zahnradgetriebe n. ‖ commande f. par engrenage. / top ~ ‖ hohe Übersetzung f. ‖ prise f. directe. / transmission ~ (Mach) ‖ Vorgelege n. ‖ engrenage m. ou transmission f. intermédiaire; renvoi m. (de mouvement); contre-arbre m. / traversing ~ ‖ Fahrwerksantrieb n.; Fahrwerk n. ‖ commande f. du déplacement du chariot. / electric traversing ~ ‖ Elektrofahrwerk n. ‖ commande f. électrique de déplacement. / hand traversing ~ ‖ Handfahrwerk n. ‖ déplacement m. commandé à la main. / trip ~ ‖ Ausklinkmechanismus m. ‖ mécanisme m. de déclanchement. / valve ~ ‖ Ventilantrieb m. ‖ commande f. de soupape. / worm ~ ‖ Schneckengetriebe n.; Wurmgetriebe n. ‖ dispositif m. ou engrenage m. à vis sans fin.

gear box ‖ Getriebekasten m.; Zahnradkasten m.; Räderkasten m. ‖ boîte f. ou carter m. des engrenages. / ~ (Of the change-speed gear) ‖ Wechselgetriebekasten m.; Getriebekasten m. ‖ boîte f. ou carter m. des vitesses; boîte f. de changement de vitesse.

gear case (Mach) see gear box. / ~ (Bicycle) ‖ Kettenschützer m. ‖ garde-chaîne m.

gear casing (Mach) see gear box. / ~ changing (Auto) ‖ Schalten n. der Gänge; Umschalten n. ‖ changement m. de vitesse. / selective system of ~ changing ‖ Verschiebeschaltung f. ‖ système m. sélectif de changement de vitesse. / ~ cutting machine ‖ Räderfräsmaschine f. ‖ machine f. à fraiser les roues ou à fraiser les engrenages.

gear-driven ‖ durch Zahnräder npl. angetrieben ‖ commandé par roues fpl. dentées. / ~ face plate ‖ Zahnkranzplanscheibe f. ‖ plateau m. à couronne dentée. / ~ pump ‖ Zahnradpumpe f. ‖ pompe f. à engrenage.

geared ‖ mit Zahnradgetriebe ‖ à commande par engrenage. / ~ lubricating pump ‖ Zahnradschmierölpumpe f. ‖ pompe f. à huile de graissage à commande par pignons.

gear grinding machine ‖ Zahnräderschleifmaschine f. ‖ machine f. à rectifier les engrenages.

gear hobbing machine ‖ nach dem Wälzverfahren arbeitende Räderfräsmaschine f.; Abwälzfräsmaschine f. ‖ machine f. à tailler les engrenages cylindriques par fraise-mère développante. / ~ with horizontal work arbor and hob arbor carriage on adjustable knee ‖ Abwälzräderfräsmaschine f. mit horizontalem Aufspannbolzen und verstellbarem Winkeltisch ‖ machine f. à tailler les engrenages cylindriques avec mandrin porte-pièce horizontal et console déplaçable. / ~ with vertical work arbor and adjustable circular work table ‖ Räderfräsmaschine f. mit vertikalem Aufspannbolzen und verschiebbarem Aufspanntisch ‖ machine f. à tailler les engrenages cylindriques avec mandrin porte-pièce vertical et table circulaire déplaçable.

gearing (Coming into gear) ‖ Eingriff m. ‖ engrènement m.; engrenage m.; quottement m. / ~ (Gear wheels) see also gear (Mach) ‖ Getriebe n.; Räderwerk n.; Zahnräderwerk n.; Rädervorgelege n.; Vorgelege n. ‖ engrenage m.; rouage m.; harnais m. d'engrenages. / ~ (Turning lathe: train of gear wheels) ‖ Antriebsräder npl. ‖ rouage m. / automobile ~ ‖ Automobilgetriebe n. ‖ engrenage m. d'automobile. / ~ of draw frame (Spinn) ‖ Streckentriebwerk n.; Streckengetriebe n. ‖ train m. d'engrenages du banc d'étirage. / intermediate ~ ‖ Vorgelege n. ‖ engrenage m. intermédiaire. / multiple ~ ‖ mehrfache Räderübersetzung f. ‖ harnais m. d'engrenages multiple. / reduction ~ see transmission ~. / spiral ~ ‖ Schraubenradgetriebe n. ‖ couple m. de roues hélicoïdales. / ~ of tooth- and cogwheels see ~ (Gear wheels). / transmission ~ (Mach) ‖ Vorgelege n. ‖ engrenage m. ou transmission f. intermédiaire; renvoi m. (de mouvement); contre-arbre m. / transmitting ~ ‖ Transmissionsgetriebe n.; Übersetzungsgetriebe n. ‖ engrenage m. de transmission. / worm ~ ‖ Schneckenradgetriebe n. ‖ couple m. de roue et vis tangente.

gearing ratio see gear ratio.

gear moulding machine ‖ Zahnradformmaschine f. ‖ machine f. à mouler les engrenages. / number of ~s (Auto) ‖ Gangzahl f. ‖ nombre m. des vitesses. / ~ pinion ‖ Getriebezahnrad n. ‖ pignon m. de boîte à vitesse. / ~ planing machine ‖ Räderhobelmaschine f. ‖ machine f. à raboter les engrenages. / ~ pump ‖ Zahnradpumpe f. ‖ pompe f. à engrenage. / ~ ratio ‖ Übersetzungsverhältnis n. ‖ rapport m. d'engrenage ou de transmission ou de réduction. / ~ shaft ‖ Transmissionswelle f. ‖ arbre m. de transmission. / ~ shift bar (Auto) ‖ Schaltstange f. ‖ tige f. de commande des fourchettes. / ~ shift fork (Auto) ‖ Schaltgabel f. ‖ fourchette f. d'engrenage. / ball and socket ~ shifting ‖ Kugelschaltung f. ‖ changement m. de vitesse par levier oscillant. / progressive system of ~ shifting ‖ Durchzugsschaltung f. ‖ système m. de changement progressif de vitesse. / selective system of ~ shifting ‖ Verschiebeschaltung f. ‖ système m. de changement coulissant de vitesse. / ~ shift lever ‖ Schalthebel m. ‖ levier m. de vitesse. / ~ shift lever shaft ‖

Schaltwelle f. ‖ arbre m. de levier de vitesse. / ~ slotting machine ‖ Räderstoßmaschine f. ‖ machine f. à mortaiser les engrenages. / ~ testing machine ‖ Zahnräderprüfmaschine f. ‖ machine f. à vérifier les engrenages. / ~ tooth chamfering machine ‖ Abrundmaschine f. für Radzähne ‖ machine f. à arrondir les dents d'engrenages.

gear wheel ‖ Getrieberad n.; Zahnrad n. ‖ roue f. d'engrenage. / ~ cutter see also gear cutting machine and gear hobbing machine ‖ Zahnradfräser m.; Zahnfräser m. ‖ fraise f. à engrenages. / ~ guard ‖ Zahnräderschutzkasten m. ‖ cage f. protectrice d'engrenage. / ~ making machine ‖ Zahnräderherstellungsmaschine f. ‖ machine f. à fabriquer les roues dentées. / ~ moulding machine ‖ Zahnräderformmaschine f. ‖ machine f. à mouler les roues d'engrenages. / ~ rim ‖ Zahnkranz m. ‖ couronne f. dentée.

gear working machine ‖ Zahnräderbearbeitungsmaschine f. ‖ machine f. à usiner ou à travailler les roues dentées.

gedrite (Miner) ‖ Gedrit m. ‖ gédrite f.

gehlenite (Miner) ‖ Gehlenit m. ‖ gehlénite f.

Geissler tube ‖ Geißler'sche Röhre f. ‖ tube m. de Geissler.

gelatine ‖ Gelatine f.; weißer Leim m.; Gallert m. ‖ gélatine f. / to cool the ~ ‖ die Gelatine f. kühlen ‖ refroidir la gélatine. / bichromatic ~ ‖ Chromgelatine f. ‖ gélatine f. chromatée. / ~ of bones ‖ Knochenleim m. ‖ colle f. ou gélatine f. d'os; ostéocolle f. / ~ for food purposes ‖ Gelatine f. für Genußzwecke ‖ gélatine f. alimentaire. / ~ in laminæ ‖ Gelatine f. in Blättchen ‖ gélatine f. en feuillets. / ~ in plates ‖ Tafelgelatine f. ‖ gélatine f. en plaques. / ~ in form of powder ‖ Gelatine f. in Pulverform ‖ gélatine f. en poudre. / sheet ~ (Chem) ‖ Gelatinefolien fpl.; Gelatine f. in Blättern ‖ gélatine f. en feuilles.

gelatine capsule ‖ Gelatinekapsel f. ‖ capsule f. gélatineuse ou de gélatine. / ~ capsule machinery ‖ Gelatinekapselmaschine f. ‖ machine f. pour la fabrication de capsules gélatineuses. / ~ culture ‖ Gelatinekultur f. ‖ culture f. sur gélatine. / ~ dynamite ‖ Gelatinedynamit n. ‖ dynamite-gélatine m. / ~ filling (Acc) ‖ Gelatinefüllung f. ‖ remplissage m. de gélatine. / ~ foil ‖ Gelatinefolie f. ‖ feuille f. de gélatine. / ~ glue ‖ Gelatineleim m. ‖ colle f. gélatineuse ou à la gélatine. / ~ manufactory ‖ Gelatinefabrikanlage f. ‖ installation f. de gélatine. / ~ paper ‖ Gelatinepapier n. ‖ papier m. gélatiné. / ~ picture ‖ Hauchbild n. aus Gelatine ‖ image f. en gélatine. / ~ plate see ~ foil. / ~ producing plant ‖ Gelatineherstellungsanlage f. ‖ installation f. à fabriquer la gélatine. / ~ spangle ‖ Gelatineflitter m. ‖ paillette f. en gélatine. / ~ sugar ‖ Leimzucker m. ‖ sucre m. de gélatine; glycocolle f. / ~ tracing paper ‖ Gelatinepauspapier n. ‖ papier-gélatine m. à calquer. / ~ worker ‖ Gelatinearbeiter m. ‖ gélatineur m. / ~ works pl. ‖ Gelatinewerk n. ‖ fabrique f. de gélatine.

gelatinize, to ‖ gelatinisieren ‖ gélatinifier; gélatin(is)er.

gelatinized acid (Acc) ‖ gelatinisierte Säure f. ‖ acide m. gélatiné.

gelatinizing agent ‖ Gelatinisierungsmittel n. ‖ moyen m. de gélatinisation.

gelatinous ‖ gelatinös; gallertartig ‖ gélatineux; mucilagineux. / ~ paste ‖ gelatinehaltige Masse f. ‖ pâte f. à base de gélatine.

gelsemium root ‖ Gelsemiumwurzel f. ‖ racine f. de gelsémium.

gem ‖ Gemme f.; Edelstein m.; Schmuckstein m. ‖ pierre f. gemme; pierre f. précieuse ou noble. / artificial ~ ‖ künstlicher oder unechter Edelstein m. ‖ imitation f. de pierre précieuse; pierre f. précieuse artificielle. / factitious ~ see artificial ~.

gem cutter ‖ Edelsteinschleifer m. ‖ polisseur m. de pierres précieuses. / ~ cutting ‖ Edelsteinschleiferei f. ‖ taillerie f. de pierres précieuses. / ~ engraver ‖ Edelsteingravör m. ‖ graveur m. en pierres fines. / ~ formation ‖ Gemmenbildung f. ‖ formation f. de gemmes. / ~ magnifier ‖ Edelsteinlupe f. ‖ loupe f. de bijouterie.

genealogical tree ‖ Stammbaum m. ‖ arbre m. généalogique.

genealogy ‖ Genealogie f.; Geschlechterkunde f. ‖ généalogie f.

general ‖ allgemein ‖ général. / ~ assembly see ~ meeting. / ~ cargo ‖ Stückgut n. ‖ cargaison f. mixte. / ~ direction of strike (Geol) ‖ Hauptstreichrichtung f. ‖ ligne f. de direction principale. / ~ hand ‖ ungelernter Arbeiter m. ‖ simple ouvrier m.; ouvrier m. sans métier.

generality ‖ Allgemeingültigkeit f. ‖ généralité f.

generalization ‖ Verallgemeinerung f. ‖ généralisation f.

generalize, to ‖ verallgemeinern ‖ généraliser.

general ledger ‖ Hauptbuch n. ‖ grand livre m. / ~ manager ‖ Generaldirektor m. ‖ directeur m. général. / ~ meeting ‖ Plenarsitzung f.; Vollversammlung f.; Generalversammlung f. ‖ séance f. publique ou plénière; assemblée f. générale. / ~ methods pl. of treatment ‖ allgemeine Richtlinien fpl. für die Verarbeitung ‖ directives mpl. généraux pour le traitement. / ~ plan ‖ Übersichtsplan m.; Lageplan m. ‖ plan m. fondamental ou de situation. / ~ power of attorney ‖ Generalvollmacht f. ‖ procuration f. générale; plein pouvoir m. / ~ rain ‖ Landregen m. ‖ pluie f. générale. / ~ view ‖ Überblick m.; Gesamtansicht f. ‖ résumé m.; coup m. d'œil; vue f. d'ensemble.

generate, to ‖ erzeugen ‖ produire; engendrer. / ~ steam ‖ Dampf m. erzeugen ‖ produire ou engendrer de la vapeur.

generating plant (Electr) see ~ station. / ~ set see generator set. / ~ station (Electr) ‖ Stromerzeugungsanlage f. ‖ station f. ou usine f. génératrice; usine f. centrale électrique. / linking-up of ~ stations into a closed system ‖ Ausbau m. von Stromerzeugungsanlagen zu einem geschlossenen System ‖ conjugaison f. des usines génératrices en vue de former un seul système.

generation of current (Electr) ‖ Stromerzeugung f. ‖ production f. de courant. / ~ of power (Electr) ‖ Energieerzeugung f. ‖ production f. d'énergie. / apparatus for controlling the ~ of power ‖ Apparat m. zur Überwachung der Energieerzeugung ‖ appareil m. pour le contrôle de la production d'énergie. / ~ of strong

current ‖ Erzeugung f. des Starkstroms ‖ production f. de courant fort. / ~ of weak current ‖ Erzeugung f. des Schwachstroms ‖ production f. de courant faible.

generator (Electr) ‖ Lichtmaschine f.; Generator m.; Dynamomaschine f. ‖ dynamo f.; générateur m.; génératrice f. / ~ (Gas) ‖ Gaserzeuger m.; Gasgenerator m. ‖ générateur m.; gazogène m. / ~ (Steam) ‖ Dampferzeuger m.; Dampfkessel m. ‖ chaudière f. à vapeur. / acetylene ~ ‖ Azetylenentwickler m. ‖ générateur m. d'acétylène. / alternating current ~ ‖ Wechselstromdynamomaschine f. ‖ dynamo f. à courant alternatif; alternateur m. / battery-lighting-ignition ~ ‖ Lichtbatteriezünder m. ‖ générateur m. et batterie f. à éclairage et à allumage. / ~ with boiler tubes ‖ Siederohrkessel m. ‖ chaudière f. à bouilleurs. / continuous current ~ ‖ Gleichstromdynamomaschine f. ‖ dynamo f. ou génératrice f. à courant continu. / direct current shunt-wound ~ ‖ Gleichstromnebenschlußgenerator m. ‖ dynamo f. excitée en dérivation pour courant continu. / double-wound ~ ‖ Doppelschlußgenerator m. ‖ génératrice f. à excitation mixte. / electric ~ ‖ Stromerzeuger m. ‖ génératrice f. électrique. / hand ~ (Electr) ‖ Kurbelinduktor m. ‖ magnéto m. à manivelle. / ignition-lighting ~ ‖ Lichtmagnetzünder m. ‖ génératrice f. à éclairage et à allumage. / railway ~ ‖ Bahndynamo f. ‖ dynamo f. de traction. / shunt-wound ~ ‖ Nebenschlußgenerator m. ‖ génératrice f. shunt; dynamo f. excitée en dérivation. / shunt-wound ~ with voltage divider ‖ Nebenschlußgenerator m. mit Spannungsteiler ‖ dynamo f. excitée en dérivation à trois conducteurs. / three-phase current flywheel ~ ‖ Drehstromschwungradgenerator m. ‖ dynamo-volant f. à courant triphasé. / three-phase current synchronous ~ ‖ Drehstromsynchrongenerator m. ‖ génératrice f. synchrone triphasée. / two-phase ~ ‖ Zweiphasengenerator m. ‖ générateur m. à deux phases.

generator circuit (Radio) ‖ Generatorkreis m. ‖ circuit m. générateur. / ~ fittings pl. ‖ Generatorarmaturen fpl. ‖ armatures fpl. pour générateurs. / ~ gas ‖ Generatorgas n. ‖ gaz m. de gazogène. / ~ set (Electr) ‖ Maschinensatz m.; Stromerzeugungsaggregat n. ‖ groupe m. électrogène. / ~ set for telegraphy ‖ Telegrafierstromgenerator m. ‖ groupe m. électrogène pour la télégraphie. / ~ tube (Radio) ‖ Generatorröhre f. ‖ lampe f. génératrice.

genetic ‖ genetisch ‖ génétique.

geneva ‖ Genever m.; Wacholderbranntwein m. ‖ genièvre m. fin.

Genoa cord ‖ Genuakord m. ‖ corde f. de Gênes.

gentian ‖ Enzian m. ‖ gentiane f. / ~ extract ‖ Enzianauszug m.; Enzianextrakt m. ‖ extrait m. de gentiane. / ~ root ‖ Enzianwurzel f. ‖ racine f. de gentiane.

gentleman farmer ‖ Grundbesitzer m. ‖ propriétaire m. foncier ou terrien.

gentlemen's body linen ‖ Herrenwäsche f. ‖ lingerie f. pour hommes. / ~ boots pl. and shoes pl. ‖ Herrenschuhwaren fpl. ‖ chaussures fpl. pour hommes. / ~ hat ‖ Herrenhut m. ‖ chapeau m. pour hommes. / ~ hatchet ‖ Haushaltsbeil n. ‖

hachette f. de ménage. / ~ linen ‖ Herrenwäsche f. ‖ linge m. pour hommes. / ~ tailor ‖ Herrenschneider m. ‖ tailleur m. pour hommes. / requisites pl. for ~ tailors ‖ Herrenschneiderartikel mpl. ‖ articles mpl. de tailleurs pour hommes.

gentle-slate ‖ Alaunschiefer m. ‖ ampélite f. alumineuse ou aluminifère; schiste m. alunifère.

genuine ‖ echt; rein ‖ naturel. / ~ (Authentic) ‖ authentisch ‖ authentique.

geocronite ‖ Geokronit m. ‖ géocronite f.; sulfure m. de plomb antimonifère et arsenifère.

geode (Geol) ‖ Mandel f. ‖ amande f.; géode f.

geodesy ‖ Geodäsie f.; höhere Vermessungskunde f.; Erdmessung f. ‖ géodésie f. / instrument of ~ ‖ Erdmeßinstrument n. ‖ instrument m. de géodésie.

geodetic head ‖ geodätisches Gefälle n. ‖ chûte f. géodésique. / ~ instrument ‖ geodätisches Instrument n. ‖ instrument m. géodésique ou de géodésie. / ~ measuring instrument ‖ geodätisches Meßinstrument n. ‖ instrument m. de mesure géodésique. / ~ survey see geodesy.

geodynamics pl. ‖ Dynamik f. fester Körper; Geodynamik f. ‖ dynamique f. des corps solides.

geographical engraver ‖ Kartenstecher m. ‖ graveur m. de cartes géographiques. / ~ latitude ‖ geografische Breite f. ‖ latitude f. géographique; (Mar:) hauteur f. / ~ map ‖ Landkarte f. ‖ carte f. de géographie; carte f. terrestre.

geography ‖ Erdkunde f.; Geografie f. ‖ géographie f. / mathematical ~ ‖ mathematische Erdkunde f. ‖ géographie f. mathématique.

geoid ‖ abgeplattetes Rotationsellipsoid n. ‖ ellipsoïde m. de révolution aplati; géoïde m.

geological examination see ~ research. / ~ map ‖ Formationskarte f.; geologische Karte f. ‖ carte f. des couches géologiques; carte f. géologique. / ~ report ‖ geologisches Gutachten n. ‖ expertise f. géologique. / ~ research ‖ geologische Untersuchung f. ‖ recherche f. géologique. / ~ section ‖ geologisches Profil n. ‖ profil m. géologique. / ~ specimen ‖ Handstück n. ‖ échantillon m. géologique. / ~ survey ‖ geologische Landesaufnahme f. ‖ relèvement m. géologique.

geologist ‖ Geologe m.; Geognost m. ‖ géologue m.; géognoste m. / ~'s hammer ‖ geologischer Hammer m. ‖ marteau m. de géologue.

geology ‖ Erdbildungskunde f.; Geologie f.; Erdgeschichte f. ‖ géologie f. / practical ~ ‖ angewandte Geologie f. ‖ géologie f. pratique ou appliquée.

geometer ‖ Feldmesser m. ‖ géomètre m. / ~ of mines ‖ Markscheider m. ‖ arpenteur m. des mines; géomètre m. souterrain.

geometral see geometrical.

geometrical ‖ geometrisch ‖ géométrique. / ~ drawing ‖ geometrische Zeichnung f.; geometrisches Zeichnen n. ‖ dessin m. géométrique. / ~ locus ‖ geometrischer Ort m. ‖ lieu m. géométrique. / ~ mean ‖ geometrisches Mittel n. ‖ moyenne f. géométrique. / ~ progression ‖ geometrische Progression f. ‖ progression f. géométrique.

geometry ‖ Geometrie f.; Raumlehre f. ‖ géométrie f. / analytic ~ ‖ analytische

Geometrie f. ‖ géométrie f. analytique. / analytic ~ of two dimensions ‖ analytische Geometrie f. in der Ebene ‖ géométrie f. analytique à deux dimensions. / analytic ~ of three dimensions ‖ analytische Geometrie f. im Raume ‖ géométrie f. analytique à trois dimensions. / common ~ see elementary ~. / descriptive ~ ‖ darstellende Geometrie f. ‖ géométrie f. descriptive. / elementary ~ ‖ elementare oder niedere Geometrie f. ‖ géométrie f. élémentaire. / higher ~ ‖ höhere Geometrie f. ‖ géométrie f. supérieure. / modern ~ ‖ neuere Geometrie f. ‖ méthodes fpl. modernes de géometrie. / sublime ~ see higher ~. / subterraneous ~ ‖ Markscheidekunst f. ‖ géometrie f. souterraine.

geomorphology ‖ Lehre f. von der Gestalt der Erdoberfläche ‖ géomorphologie f.

geophysics pl. ‖ Erdphysik f.; Geophysik f. ‖ géophysique f.

geranium oil ‖ Geraniumöl n. ‖ essence f. de géranium.

geranyl acetate ‖ essigsaures Geranyl n. ‖ acétate m. de géranyle.

germ (Botan) ‖ Keim m. ‖ germe m. / ~ (Bacterium) ‖ Keim m. ‖ bactérie f. / free from ~s pl. ‖ keimfrei ‖ exempt de bactéries.

German ocean ‖ Nordsee f. ‖ mer f. d'Allemagne ou du Nord.

German silver ‖ Neusilber n.; Argentan n. ‖ maillechort m.; alfénide m.; argentan m.; argent m. allemand. / ~ band ‖ Neusilberband n. ‖ bande f. en maillechort. / ~ goods pl. ‖ Neusilberwaren fpl.; Chinasilberwaren fpl. ‖ produits mpl. en maillechort. / ~ rolling mill ‖ Argentanwalzwerk n. ‖ laminage m. de maillechort. / ~ sheet ‖ Neusilberblech n. ‖ feuille f. en maillechort. / ~ ware see ~ goods. / ~ wire ‖ Neusilberdraht m. ‖ fil m. en maillechort.

German text see ~ type. / ~ tinder ‖ Feuerschwamm m. ‖ amadou m. préparé. / ~ type (Print) ‖ Frakturschrift f.; Fraktur f. ‖ caractère m. gothique ou allemand; écriture f. gothique. / ~ windmill ‖ deutsche Mühle f.; Bockmühle f.; Ständermühle f. ‖ moulin m. allemand ou à pile ou à pylone.

germinate, to ‖ keimen ‖ germer.

germinated ‖ gekeimt ‖ germé.

germinating power ‖ Keimkraft f. ‖ faculté f. germinative.

germination ‖ Keimung f. ‖ germination f.

gersdorffite (Miner) ‖ Arsennickelglanz m.; Gersdorffit m. ‖ gersdorffite f.

get, to ~ afloat (Mar) ‖ freikommen; flott werden ‖ renflouer. / ~ dirty ‖ sich verstopfen ‖ s'encrasser. / ~ hot ‖ heißwerden, heißlaufen ‖ s'échauffer. / ~ loose (Mach) ‖ Spiel n. oder Spielraum m. haben ‖ jouer; avoir du jeu. / ~ off (Mar) see to get afloat. / ~ off the rails ‖ entgleisen ‖ dérailler; sortir des rails. / ~ the offing (Mar) ‖ die offene See gewinnen ‖ gagner le large. / ~ ready ‖ fertigmachen ‖ apprêter. / ~ a set (Met) ‖ krumm werden (beim Härten) ‖ (se) courber ou se voiler (à la trempe).

getter (Min) ‖ Abkohler m.; Häuer m. ‖ abatteur m.; piqueur m. à la veine; rabatteur m.

getter-in (Weav) ‖ Fadenanleger m. ‖ aiderentreur m.

getting (Min) ‖ Abkohlen n. ‖ rabatage m.; abatage m. / ~ barren (Geol) ‖ Vertaubung f. ‖ appauvrissement m.

getting-hot (Masch) ‖ Warmlaufen n.; Heißlaufen n. ‖ chauffage m.; échauffement m.

getting-off the rails ‖ Entgleisung f. ‖ déraillement m.

get-up of a book ‖ Buchausstattung f. ‖ décoration f. du livre.

geyser ‖ Springquelle f.; Geiser m. ‖ geyser m. / gas ~ ‖ Gasbadeofen m. ‖ four m. à gaz pour (les) bains.

gherkin ‖ (kleine) Pfeffergurke f. ‖ cornichon m.

ghost line (Met) ‖ Schleifriß m. ‖ strie f. de polissage (de la plaquette polie).

giant aeroplane ‖ Riesenflugzeug n. ‖ avion m. géant. / ~ airship ‖ Riesenluftschiff n. ‖ dirigeable m. géant. / ~ loudspeaker ‖ Großlautsprecher m. ‖ haut-parleur m. géant.

gib (Of a crane) ‖ Kranbalken m.; Kranarm m.; Kranschnabel m.; Rollenholm m. ‖ volé f.; bec m.; fauconneau m.; flèche f. de grue. / ~ (Key) see gibheaded key. / ~ and cotter ‖ Keil m. und Lösekeil m. ‖ clavette f. et contreclavette.

gibbet ‖ Galgen m. ‖ potence f. / ~ (of a crane) see gib (of a crane).

gib head (Mach) ‖ Nase f. ‖ talon m. / height of ~ ‖ Nasenhöhe f. ‖ hauteur f. du talon.

gib-headed flat key ‖ Nasenflachkeil m. ‖ clavette f. à talon sur méplat. / ~ key ‖ Nasenkeil m.; Hakenkeil m. ‖ clavette f. à talon ou à mentonnet. / ~ saddle key ‖ Nasenhohlkeil m. ‖ clavette f. creuse à talon.

gib key see gib-headed key. / hollow ~ see gib-headed saddle key.

gid (Of sheep) ‖ Drehkrankheit f. ‖ tournis m.

giddiness ‖ Schwindel m. ‖ vertige m.; étourdissement m.

gid-rolls pl. ‖ Kurierwalzen fpl.; Schnellwalzen fpl. ‖ cylindres mpl. d'une grande vitesse; laminoir m. accéléré ou à marche rapide.

gift ‖ Schenkung f. ‖ don m.; donation f.

gig (Coachm) ‖ Jagdwagen m.; zweiräderiger Gabelwagen m.; Gig n. ‖ guigue f. / ~ (Mar) ‖ leichtes Ruderboot n.; Gig n. ‖ guigue f. / ~ (Clothm) ‖ Rauhmaschine f. ‖ laineuse f.; machine f. à lainer. / laying-down ~ see water ~. / raising ~ ‖ Verstreichrauhmaschine f. ‖ striqueuse f.

gigantolite (Miner) ‖ Gigantolith m. ‖ gigantolite f.

gig mill see gig (Clothm).

gild, to ‖ vergolden ‖ dorer.

gilder ‖ Vergolder m. ‖ doreur m. / ~ on carved wood ‖ Holzschnitzwarenvergolder m. ‖ doreur m. sur bois sculptés. / ~ of mouldings ‖ Leistenvergolder m. ‖ doreur m. de baguettes. / ~ on wood ‖ Holzvergolder m. ‖ doreur m. sur bois.

gilder's knife ‖ Vergoldermesser n. ‖ avivoir m. / ~ tip ‖ Vergolderpinsel m. ‖ doroir m. / ~ tongs pl. ‖ Vergolderzange f. ‖ moustache f. / ~ varnish ‖ Vergolderlack m. ‖ vernis m. pour dorure. / ~ wax ‖ Vergolderwachs n. ‖ cire f. à dorer.

gilding ‖ Vergoldung f. ‖ dorure f.; dorage m. / ~ by amalgamation ‖ Quecksilbervergoldung f. ‖ dorure f. au feu. / burnished ~ ‖ Leimvergoldung f.; Glanzvergoldung f.; Wasservergoldung f.; Ver-

goldung f. auf Leimgrund ‖ dorure f. en détrempe. / burnished ~ of metals ‖ Vergoldung f. mit Blattgold ‖ dorure f. des métaux avec de l'or en feuilles. / cold ~ ‖ kalte Vergoldung f. ‖ dorure f. au pouce. / dead ~ ‖ Mattvergoldung f. ‖ dorage m. mat. / ~ by dipping ‖ Eintauchvergoldung f. ‖ dorage m. par immersion. / ~ in distemper see burnished ~. / dry ~ see hot ~. / electro ~ ‖ galvanische Vergoldung f. ‖ dorure f. galvanique. / fire ~ see hot ~. / hot ~ ‖ Quecksilbervergoldung f.; Feuervergoldung f. ‖ dorure f. au feu. / ~ in oil ‖ Ölvergoldung f. ‖ dorure f. à l'huile. / rugged ~ ‖ rauhe Vergoldung f. ‖ or m. haché. / water ~ see burnished ~. / ~ on water-size see burnished ~. / ~ by a weak liquid amalgam ‖ Vergoldung f. durch Eintauchen in ein flüssiges Amalgam ‖ dorure f. au sauté. / wet ~ ‖ nasse Vergoldung f. ‖ dorure f. au trempe. / ~ of wood ‖ Holzvergoldung f. ‖ dorure f. sur bois.

gilding bench ‖ Vergoldetisch m. ‖ table f. de doreur. / ~ mender ‖ Ausbesserer m. für vergoldete Gegenstände ‖ réparateurdoreur m. / ~ press (Bookb) ‖ Vergoldepresse f. ‖ presse f. à dorer ou à tranchefiler. / ~ and embossing press ‖ Vergolde- und Prägepresse f. ‖ presse f. à dorer et à estamper. / ~ size ‖ Vergoldergrund m. ‖ assiette f.

gill box (Spinn) ‖ Nadelstabstreckwerk n. ‖ étirage m. à barrettes d'aiguilles; gill-box m. / ~ minder ‖ Rohstrecker m. ‖ gill-boxeur m.

gilt ‖ vergoldet; goldfarbig ‖ doré.

gilt ‖ Vergoldung f. ‖ dorure f. / to lose the ~ ‖ die Vergoldung verlieren ‖ se dédorer.

gilt brass-mounting (Opt) ‖ (fein)vergoldete Messingfassung f. ‖ monture f. cuivre doré. / ~ concave mirror ‖ vergoldeter Hohlspiegel m. ‖ réflecteur m. concave doré. / ~ copper ‖ vergoldetes Kupfer n. ‖ cuivre m. doré.

gilt-cornice varnish ‖ Goldleistenlack m. ‖ vernis m. pour tringles dorées.

gilt-edged (Bookb) ‖ mit Goldschnitt m. ‖ doré sur tranche f.; à tranche dorée.

gilt-edges pl. ‖ Goldschnitt m. ‖ tranche f. dorée. / ~ burnisher ‖ Goldschnittpolierer m. ‖ brunisseur m. ou doreur m. sur tranches. / ~ maker ‖ Goldschnittarbeiter m. ‖ doreur m. sur tranches.

gilt moulding ‖ Goldleiste f. ‖ baguette f. ou tringle f. dorée. / ~ silver ‖ Vermeil n.; vergoldete Silberware f. ‖ vermeil m.

gimbal mounting ‖ Kardanaufhängung f. ‖ suspension f. à cardan.

gimbals pl. ‖ Kompaßringe mpl. ‖ balanciers mpl. du compas.

gimblet see gimlet.

gimlet ‖ Holzbohrer m.; Nagelbohrer m.; Frettbohrer m.; Zimmermannsbohrer m.; Vorbohrer m. ‖ vrille f.; foret m. à bois; amorçoir m. / brewer's ~ ‖ Handbohrer m. für Brauer ‖ vrille f. de brasseur. / cask ~ ‖ Faßbohrer m. ‖ vrille f. à tonneau.

gin, to (Cotton) ‖ auskörnen; entkörnen ‖ égrener.

gin (Cotton) ‖ Entkörnemaschine f.; Egreniermaschine f. ‖ égreneuse f.; machine f. à égrener. / ~ (Dist) ‖ Genever m.; Wacholder(branntwein) m.; Gin m. ‖ genièvre m.; genévrette f.; gin m.

/ ~ (Hoisting gear) ‖ Hebebock m.; Hebezeug n.; Winde f. ‖ chèvre f.; engin m. /
~ (Mine) ‖ Göpel m.; Fördermaschine f.;
Fördergöpel m. ‖ machine f. ou manège
m. à molette; machine f. d'extraction. /
cotton ~ ‖ Baumwollentkörnungsma-
schine f. ‖ égreneuse f. de coton.
gin-fall (Hoisting) ‖ Hebezeugtau n. ‖ câble
m. de chèvre.
ginger ‖ Ingwer m. ‖ gingembre m.
gingerbread ‖ Honigkuchen m.; Ingwer-
kuchen m.; Pfefferkuchen m.; Lebkuchen
m. ‖ pain m. d'épice. / ~ maker ‖ Honig-
kuchenbäcker m. ‖ fabricant m. de pain
d'épices. / ~ manufacturing machine ‖
Lebkuchenfertigungsmaschine f. ‖ machine f. pour la fabrication de pain
d'épice.
gingergrass oil ‖ Gingergrasöl n. ‖ essence
f. de géranium des Indes.
ginger oil ‖ Ingweröl n. ‖ essence f. de
gingembre.
gingham ‖ Matratzendrell m.; englische
oder schottische Leinwand f. ‖ treillis m.
pour matelas; guingamp m.
ginging of a shaft ‖ wasserdichte Schachtausmauerung f. oder Schachtmauerung
f. ‖ cuvelage m. en maçonnerie; muraillement m. d'un puits.
ginning of the cotton ‖ Entkörnen n. der
Baumwolle ‖ égrenage m. du coton.
ginning machine for flax ‖ Flachsentsa-
mungsmaschine f. ‖ machine f. à égrener
le lin.
gin-saw for cotton cleaning ‖ Egrenier-
kreissäge f. ‖ scie f. à égrener le coton.
ginseng root ‖ Ginsengwurzel f. ‖ racine f.
de ginseng.
gips; gipsum see gypsum.
girasol (Miner) ‖ Feueropal m.; orientalischer Opal m. ‖ girasol m.
giratory breaker ‖ Walzenbrecher m.;
Rundbrecher m. ‖ concasseur m. giratoire ou à cône.
girder ‖ Träger m. ‖ poutrelle f.; fer m. à
plancher; longeron m.; support m. en fer.
/ arched ~ ‖ Bogenträger m. ‖ poutre f. cintrée. / braced ~ ‖ Parallelträger m. ‖
poutre f. droite. / broad-flanged ~ ‖
Breitflanschträger m. ‖ poutre f. à larges
semelles. / central ~ ‖ Mittelträger m.;
Mittelbalken m. ‖ poutre f. centrale ou
du milieu. / cruciform ~ ‖ Kreuzträger
m. ‖ poutre f. en croix. / inverted chain-
and-bow ~ ‖ Fischbauchträger m.; Linsenträger m. ‖ poutre f. lenticulaire. /
iron ~ ‖ eiserner Träger m.; Eisenträger
m. ‖ support m. en fer; poutrelle f. (en
fer). / lattice-work ~ ‖ Gitterträger m. ‖
poutre f. cloisonnée ou en treillis. / main
~ ‖ Hauptlangträger m.; Hauptträger m.
‖ longeron m. principal; poutre f. principale ou maîtresse. / outside ~ ‖ Außenträ-
ger m. ‖ poutrelle f. extérieure. / perforated ~ ‖ gelochter Träger m. ‖ poutre f.
perforée. / pressed ~ ‖ Preßträger m.; gepreßter Träger m. ‖ poutre m. embouti.
/ rivetted ~ ‖ genieteter Träger m. ‖
poutrelle f. rivetée. / rolled ~ ‖ gewalzter
Träger m. ‖ poutrelle f. laminée. / secondary ~ ‖ Zwischenträger m. ‖ poutre f.
intermédiaire. / T-~ ‖ T-Träger m. ‖
poutre f. en T. / double-T-~ ‖ Doppel-T-
Träger m. ‖ poutre f. en double T. /
transverse ~ ‖ Querträger m.; Binderquerträger m. ‖ entretoise f.; transversale f. dans les fermes. / trussed ~ ‖
armierter Träger m. ‖ poutre f. armée. /

tubular ~ ‖ vollwandiger Träger m. ‖
poutre f. pleine. / under ~ ‖ Unterzug m.
‖ sous-poutre f. / way-board ~ (Mining) ‖
Hängebankunterzug m. ‖ poutre f.
traverse de recette.
girder base ‖ Trägerfuß m. ‖ semelle f. d'une
poutrelle.
girder-bridge ‖ Balkenbrücke f. ‖ pont m.
à longerons ou à poutres. / fish-bellied ~
‖ Brücke f. mit Fischbauchträgern ‖ pont
m. à poutres en ventre de poisson.
girder casing ‖ Balkenschalung f. ‖ coffrage
m. de poutre. / deflection of a ~ ‖ Durchbiegung f. eines Balkens ‖ flèche f. de
la poutre. / ~ flange ‖ Trägerflansch m.
‖ aile f. de la poutre. / ~ iron ‖ Träger-
eisen n. ‖ fer m. à poutre.
girderless floor ‖ balkenlose Decke f. ‖
plancher m. sans nervures ou sans
poutres.
girder mill ‖ Trägerwalzwerk n. ‖ laminoir
m. à poutrelles. / ~ and section mill ‖
Formeisenwalzwerk n. ‖ laminoir m. à
fers profilés et cornières. / ~ mould ‖
Trägerschalung f. ‖ coffrage m. de poutre.
/ ~ pole ‖ Gittermast m. ‖ pylône m.;
poteau m. en treillis. / ~ rail ‖ Breitfußschiene f.; Vignoleschiene f. ‖ rail m.
Vignole ou à patin.
girdle, to ~ (Trees) ‖ ringeln ‖ anneler.
girdle see also girth ‖ Gurt m.; Gürtel m.
‖ sangle f.; ceinture f.
girdler ‖ Gürtler m. ‖ ceinturier m.; cro-
chetier m.
girth (Saddl) ‖ Hängegurt m.; Bauchgurt
m.; Sattelgurt m. ‖ ventrière f.; sangle
f. / ~ (Weav) ‖ Gurt m.; Gurte f.; dickes
bandförmiges Gewebe n. ‖ sangle f.
/ ~ (Of the printing roller) ‖ Gurte f.
oder Riemen m. an der Walze ‖ corde f. du
rouleau. / rein and saddle ~ ‖ Zügel- und
Sattelgurt m. ‖ sangle f. pour rênes et
selles.
girth buckle ‖ Gurtschnalle f. ‖ boucle f. de
sangle ou de ceinture. / ~ conveyor ‖
Gurtförderer m.; Gurttransportör m. ‖
transporteur m. à ruban. / ~ line ‖ Jollentau n. ‖ agui m.; cartahu m. / ~ line
block ‖ Jollblock m. ‖ poulie f. de cartahu. / ~ pulley ‖ Gurtscheibe f. ‖ poulie
f. à ruban.
gismondine (Miner) ‖ Gismondin m. ‖ gismondine f.
git (Found) see also gate (Found) ‖ Einguß
m.; Gußtrichter m. ‖ jet m.; gueule m.;
trou m. de coulée. / removing ~s from
brass castings ‖ Abkneifen n. von Messinggußtrichtern ‖ enlever les masselottes en
laiton.
gitting press (Found) ‖ Abkneifpresse f. ‖
presse f. pour entailler; coupe-jet m. /
eccentric ~ ‖ Exzenterabkneifpresse f. ‖
presse f. à excentrique pour entailler;
coupe-jet m. à excentrique.
give, to ‖ geben ‖ donner. / ~ batter ‖ abböschen ‖ taluter; mettre en talus. /
full particulars of ... / ~ Auskunft f. erteilen über ... ‖ donner des renseignements mpl. sur ... / ~ iron to the plane ‖
das Hobeleisen n. vortreiben ‖ donner du
fer au rabot. / ~-out heat ‖ Wärme f. abgeben oder ausstrahlen ‖ rayonner ou irradier de la chaleur. / ~-out in tender ‖
in Submission f. vergeben ‖ donner en soumission. / ~ up ‖ aufgeben; verzichten
auf ...; Abstand m. nehmen von ... ‖
renoncer à ... / ~-up a mine ‖ ein Bergwerk auflassen oder verlassen ‖ aban-

donner une mine. / ~ way (Mech) ‖ sich
geben; nachgeben; weichen ‖ céder; plier;
lâcher; se détendre.
Givet-glue ‖ Givetleim m. ‖ colle f. anglaise
ou blonde ou brillante de Givet.
giving of an order ‖ Vergebung f. eines
Auftrages; Auftragserteilung f. ‖ passation f. d'une commande.
giving-up of business ‖ Geschäftsaufgabe
f. ‖ cessation f. d'un commerce.
glacé (Weav) ‖ schillernd ‖ changeant;
glacé. / ~ glove ‖ Glanzlederhandschuh
m.; Glacéhandschuh m. ‖ gant m. glacé.
glacial acetic acid ‖ Eisessig m. ‖ acide m.
acétique glacial.
glance coal ‖ Glanzkohle f.; Anthrazit m. ‖
charbon m. luisant; houille f. luisante. / ~
copper ‖ Kupferglanz m.; Kupferglas n.;
Schwefelkupfer n. ‖ cuivre m. sulfuré ou
vitreux.
gland (Anatomy) ‖ Drüse f. ‖ glande f. / ~
(Bot) ‖ Eichel f. ‖ gland m. / ~ (Mach) ‖
Stopfbuchs(en)deckel m. ‖ presse-garni-
ture m.; chapeau m. d'un presse-étoupe.
/ adjustable ~ ‖ verstellbare Stopfbuchsbrille f. ‖ chapeau m. réglable de presse-
étoupe. / water ~ ‖ Flüssigkeitsabschluß
m. ‖ fermeture f. ou obturation f. hydraulique.
gland bolt ‖ Stopfbuchsschraube f. ‖ boulon
m. de boîte à bourrage. / ~ bush see ~
lining. / flange of ~ ‖ Stopfbuchsbrillen-
flansch m. ‖ bride f. de chapeau de
presse-étoupe. / ~ follower see gland
(Mach). / ~ lining ‖ Stopfbuchsbrillen-
futter m. ‖ fourrure f. de chapeau de
presse-étoupe.
glans (Bot) see gland (Bot).
glare, to ‖ blenden ‖ éblouir.
glare ‖ Glanz m.; Funkeln n.; blendendes
Licht n. ‖ éblouissement m.; éclat m. /
headlight ~ ‖ Blenden n. der Scheinwerfer ‖ éblouissement m. des phares.
glaring ‖ blendend; glänzend; funkelnd ‖
éblouissant.
glaring (Subst) see glare. / ~ glass ‖ Blendglas n. ‖ verre m. éblouissant. / coloured
~ glass ‖ farbiges Blendglas n. ‖ verre m.
coloré éblouissant. / ~ light of arc ‖ blendendes oder grelles Licht n. der Bogenlampe ‖ loueur f. éblouissante de l'arc. /
~ light of motor car headlights ‖ grelles
Licht n. der Autolampen ‖ lumière f.
éblouissante des phares d'automobile.
glaserite ‖ Glaserit m. ‖ glasérite f.
glass ‖ Glas n. ‖ verre m. / ~ (Barometer) ‖
Barometer n. ‖ baromètre m. / ~ (Cup) ‖
Glas n.; Trinkglas n. ‖ verre m. / ~
(Disk) ‖ Glasscheibe f. ‖ vitre f.; plaque
f. de verre. / ~ (Field glass) see glasses. /
~ (Magnifying glass) ‖ Lupe f.; Vergrößerungsglas n. ‖ loupe f. / ~ (Mar) ‖ Stundenglas n.; Sandglas n.; Sanduhr f. ‖
horloge f. de sable; sablier m.; ampoulette f. / ~ (Mirror) see looking ~. / ~
(Spectacles) see glasses. / ~ (Thermometer) ‖ Thermometer n. ‖ thermomètre
m. / armoured ~ ‖ Drahtglas n. ‖ verre
m. armé. / artistic ~ ‖ Kunstglas n. ‖
cristal m. d'art. / art-stained ~ for church
windows ‖ gemaltes Glas n. für Kirchenfenster ‖ vitrail m. d'art. / bevelled ~ ‖
Facettenglas n.; facettiertes Glas n. ‖
glace f. biseautée ou taillé. / to blow ~ ‖
Glas n. blasen ‖ souffler du verre. / blown
~ ‖ geblasenes Glas n. ‖ verre m. soufflé. /
blue ~ ‖ Blauglas n. ‖ verre m. bleu. / to
blunt ~ ‖ Glas n. blind machen ‖ émous-

ser *ou* dépolir le verre. / bottle ~ ‖ Flaschenglas n. ‖ verre m. à bouteilles. / broken ~ ‖ Glasbrocken mpl.; Glasbruch m.; Glasscherben fpl. ‖ calcin m.; grésil m.; groisil m.; rognure f. de verre; tessons mpl. / bull's-eye ~ ‖ Mondglas n. ‖ verre m. à boudines *ou* à vitres soufflé en plateaux ronds. / cast ~ ‖ gegossenes Glas n. ‖ verre m. coulé. / cast in sand ‖ in Sand gegossenes Glas n. ‖ verre m. sablé. / ~ cast on wire-cloth ‖ auf Metalltuch gegossenes Glas n. ‖ verre m. coulé sur toile métallique. / cataract ~ ‖ Starglas n. ‖ verre m. correcteur pour opérés de cataracte. / cataract reading ~ ‖ Starleseglas n. ‖ verre m. à lunettes destinées aux opérés de cataracte. / cathedral ~ ‖ Kathedralglas n. ‖ verre m. cathédrale. / cemented ~ ‖ verkittetes Glas n. ‖ verre m. mastiqué. / ~ for church windows ‖ Glas n. für Kirchenfenster ‖ verre m. pour vitraux. / clear ~ ‖ klares Glas n. ‖ verre m. clair. / coloured ~ ‖ farbiges Glas n.; Farbglas n. ‖ verre m. coloré; verre m. de couleur. / colourless ~ ‖ ungefärbtes Glas n. ‖ verre m. incolore. / concave ~ ‖ Hohlglas n. ‖ verre m. concave. / convex ~ ‖ gebogenes Glas n. ‖ verre m. bombé *ou* convexe. / corrugated ~ ‖ gestreiftes Glas n. ‖ verre m. strié. / cut ~ ‖ geschliffenes Glas n. ‖ verre m. poli *ou* taillé. / to cut ~ ‖ Glas n. schneiden ‖ tailler le verre. / devitrified ~ ‖ entglastes Glas n. ‖ verre m. dévitrifié. / diamonded ~ ‖ diamantiertes Glas n. ‖ verre m. diamanté. / distance ~ *see* field ~. / shop with double ~ ‖ Werkstatt f. mit Doppelverglasung ‖ hall m. à double vitrage. / drawn ~ ‖ gezogenes Glas n. ‖ verre m. étiré. / field ~ ‖ Feldstecher m.; Fernglas n. ‖ jumelles fpl. (de campagne); lunette f. / flashed ~ ‖ Überfangglas n. ‖ verre m. doublé *ou* à deux couches. / flint ~ ‖ Bleiglas n.; Flintglas n. ‖ verre m. plombifère; flint-glass m. / ~ which has flown into the hearth ‖ Herdglas n. ‖ picadil m. / frosted ~ ‖ Mattglas n.; Eisglas n. ‖ verre m. mat *ou* dépoli *ou* craquelé. / fusible ~ ‖ schmelzbares Glas n. ‖ verre m. fusible. / green claret ~ ‖ Römer m. ‖ verre m. à vin. / to grind the ~ ‖ Glas n. schleifen ‖ tailler le verre. / grooved ~ ‖ gerieftes *oder* kanneliertes Glas n. ‖ verre m. cannelé. / hammered ~ ‖ gehämmertes Glas n. ‖ verre m. martelé. / hard ~ ‖ Hartglas n. ‖ verre m. dur. / hardened ~ ‖ Hartglas n. ‖ verre m. trempé. / hardly fusible ~ ‖ schwer schmelzbares Glas n. ‖ verre m. difficilement fusible. / high-polish lead crystal ~ ‖ Hochglanzbleikristall n. ‖ verre m. plombifére brillant. / ~ for household use ‖ Wirtschaftsglas n. ‖ verre f. de ménage. / lead ~ ‖ Bleiglas n. ‖ verre m. plombifère. / lenticular ~ ‖ Glaslinse f. ‖ lentille f.; verre m. lenticulaire. / ~ for lighting purposes ‖ Beleuchtungsglas n.; Glas n. für Beleuchtungszwecke ‖ verre m. pour l'éclairage. / lime ~ ‖ Kalkglas n. ‖ verre m. à la chaux. / looking ~ ‖ Spiegel m.; Spiegelglas n. ‖ miroir m.; glace f. / lozenged ~ ‖ gerautetes Glas n. ‖ verre m. losangé. / magnifying ~ ‖ Vergrößerungsglas n. ‖ verre m. grossissant. / marbled ~ ‖ marmoriertes Glas n. ‖ verre-marbre m. / marine ~ ‖ Marineglas n. ‖ jumelles fpl. de marine. / milk ~ ‖ Milchglas n. ‖ verre m. opale. / moderating ~ ‖ Blendglas n. ‖ verre m. sombre. /

moulded ~ ‖ geformtes Glas n.; Preßglas n. ‖ verre m. moulé *ou* comprimé. / ~ offering great resistance to chemical influences ‖ hartes und chemisch widerstandsfähiges Glas n. ‖ verre m. dur résistant bien aux agents chimiques. / opera ~ ‖ Operngucker m. ‖ lorgnette f. de théâtre. / ~ required for optical parts ‖ Glasmaterial n. für optische Teile ‖ fontes fpl. pour les pièces optiques. / ornamental ~ ‖ Ornamentglas n. ‖ verre m. ornemental. / packing ~ ‖ Verpackungsglas n. ‖ verre m. pour emballage. / ~ for pendulum clocks ‖ Großuhrglas n. ‖ verre m. de pendule. / plain ~ ‖ glattes Glas n. ‖ verre m. uni. / powdered ~ ‖ Glasmehl n.; Glasstaub m. ‖ verre f. en poudre. / pressed ~ ‖ Preßglas n. ‖ verre m. comprimé. / pressed hard ~ for ship windows ‖ Preßhartglas n. für Schiffsfenster ‖ verre m. moulu et trempé à vitres du bateau. / printed ~ ‖ bedrucktes Glas n. ‖ verre m. imprimé. / prismatic ~ ‖ Prismenglas n. ‖ verre m. prismatique. / race ~ ‖ Rennglas n. ‖ jumelles fpl. de courses. / raw ~ ‖ rohes Glas n. ‖ verre m. brut. / refined ~ ‖ geläuterte Glasmasse f. ‖ verre m. affiné. / reticulated ~ ‖ Fadenglas n.; Filigranglas n. ‖ verre m. filigrané. / ribbed ~ ‖ geripptes Glas n.; Riffelglas n. ‖ verre m. rayé *ou* cannelé. / roofing ~ ‖ Dachglas n. ‖ verre m. de toiture. / scientific optic ~ ‖ wissenschaftliches optisches Glas n. ‖ verre m. d'optique scientifique. / ~ without secondary spectrum ‖ Glas n. ohne sekundäres Spektrum ‖ verre m. exempt de spectre secondaire. / semi-white ~ ‖ halbweißes Glas n. ‖ verre m. demi-blanc. / sheet ~ ‖ Tafelglas n. ‖ verre m. en plaques. / ~ in sheets ‖ Glas n. in Blättern ‖ verre m. en feuilles. / special optical ~ ‖ optisches Spezialglas n. ‖ verre m. optique spécial. / splinter-proof ~ ‖ splitterfreies *oder* splittersicheres Glas n. ‖ verre m. incassable. / articles pl. of spun ~ ‖ Waren fpl. aus gesponnenem Glas ‖ objets mpl. en verre filé. / stained ~ ‖ Kathedralglas n. ‖ verre m. cathédrale. / straw-coloured ~ ‖ halbweißes Glas n. ‖ verre m. demiblanc. / supplementary ~ ‖ Aufsteckglas n. ‖ verre m. additionnel. / tempered ~ ‖ Vulkanglas n. ‖ verre m. trempé. / tinted ~ ‖ Rauchglas n. ‖ verre m. fumé. / ~ of unusual transparency for ultra violet rays ‖ besonders gut ultraviolettdurchlässiges Glas n. ‖ verre m. particulièrement bon pour les radiations ultraviolettes. / volcanic ~ ‖ Glaslava f.; Obsidian m. ‖ laitier m. de volcan; lave f. vitreuse obsidienne; obsidiane f. / wired rolled ~ ‖ Drahtglas n. ‖ verre m. armé de fil de fer.

glass accumulator box *see* ~ accumulator jar. / ~ accumulator jar (Electr) ‖ Akkumulatorglas n. ‖ bac m. d'accumulateur en verre. / ~ advertising article ‖ Glasreklameartikel m.‖ article-réclame m. de verre. / ~ apparatus ‖ Glasapparat m. ‖ appareil m. en verre. / ~ articles pl. ‖ Glaswaren fpl. ‖ verrerie f. / surgical ~ article ‖ chirurgischer Glasartikel m. ‖ article m. de chirurgie en verre. / ~ ball blower ‖ Glaskugelbläser m. ‖ bossetier m. de ballons en verre. / ~ balloon ‖ Glasballon m. ‖ bonbonne f.; damejeanne f. ballon m. de verre. / ~ bead ‖ Glasperle f. ‖ perle f. artificielle. / ~ bell

‖ Glasglocke f. ‖ cloche f. en verre. / ~ bending workshop ‖ Glasbiegewerkstatt f. ‖ atelier m. à courber le verre. / ~ bevelling machine ‖ Facettenschliffapparat m. für Spiegelglas ‖ appareil m. à biseauter les glaces. / ~ bleaching material ‖ Glasentfärbungsmittel n. ‖ matière f. à décolorer le verre. / ~ block ‖ Glasblock m. ‖ bloc m. de verre. / to grind into a ~ block ‖ in einen Glasblock m. einschleifen ‖ tailler dans un bloc de verre.

glass blower ‖ Glasbläser m. ‖ souffleur-verrier m. / ~'s cataract ‖ Glasbläserstar m. ‖ cataracte f. de souffleurs de verre. / ~'s lamp ‖ Glasbläserlampe f.; Schmelzlampe f. ‖ lampe f. de souffleur de verre. / ~'s pipe ‖ Glasblaserohr n. ‖ tube m. de souffleur de verre.

glass blowing ‖ Glasbläserei f. ‖ soufflage m. de verre. / ~ at the lamp ‖ Glasblasen n. vor der Lampe ‖ soufflage m. à la lampe. / ~ machine ‖ Glasblasemaschine f. ‖ machine f. à souffler le verre. / ~ mould ‖ Glasblasform f. ‖ matrice f. pour la soufflerie du verre. / ~ plant ‖ Glasbläsereieinrichtung f. ‖ installation f. de soufflerie du verre.

glass bottle ‖ Glasflasche f. ‖ bouteille f. de verre. / ~ box (Acc) ‖ Akkumulatorenglas n.; Glasgefäß n. ‖ bac m. d'accumulateurs en verre. / breakage of the ~ box (Acc) ‖ Bruch m. des Glasgefäßes ‖ bris m. du récipient en verre. / ~ brick ‖ Glasbaustein m. ‖ brique f. de verre. / ~ brush ‖ Glasbürste f. ‖ brosse f. en verre. / ~ bulb ‖ Glasbirne f.; Glaskugel f. ‖ ampoule f.; boule f. de verre. / ~ button ‖ Glasknopf m. ‖ bouton m. de verre. / ~ canopy *see* ~ dome. / ~ carboy ‖ Glasballon m. ‖ dame-jeanne f. / ~ case ‖ Glaskasten m. ‖ caisse f. vitrée. / ~ chimney ‖ Lampenzylinder m. ‖ cheminée f. de lampe; verre m. à lampe. / ~ cleaning ‖ Fensterreinigung f. ‖ nettoyage m. de carreaux *ou* de fenêtres. / ~ cloth ‖ Glasleinen n. ‖ toile f. verrée. / ~ crucible (Glassm) ‖ Glashafen m.; Hafen m.; Schmelzhafen m. ‖ creuset m.; padelin m.; pot m. / ~ coating factory ‖ Glasbelegerei f. ‖ atelier m. de garniture en verre. / ~ cock ‖ Glashahn m. ‖ robinet m. en verre. / ~ colour ‖ Glasfarbe f. ‖ couleur f. pour verre. / ~ colouring ‖ Glasfärbung f. ‖ coloration f. de verre. / ~ colouring substance ‖ Glasfärbungsmittel n. ‖ matière f. à colorer le verre. / ~ cover ‖ Deckglas n. ‖ couvercle m. en verre; verre m. de couverture. / spherically curved ~ cover ‖ kugelförmig ge krümmter Glasdeckel m. ‖ couvercle m. de verre sphérique. / ~ cup ‖ Glasschale f. ‖ coupe f. en verre. / ~ cup for pins ‖ Glasbehälter m. für Briefklammern ‖ coupe f. en verre pour attaches métalliques. / ~ cutter ‖ Glasschneider m. ‖ coupeur m. de verre; coupe-verre m. / ~ cutting machine ‖ Glasschneidemaschine f. ‖ machine f. à couper le verre. / ~ ground disc ‖ Mattscheibe f. ‖ verre m. dépoli. / ~ dish ‖ Glasschale f. ‖ cuvette f. de verre. / moulded ~ dish ‖ gegossene Glasschale f. ‖ cuvette f. de verre coulé. ‖ ~ dome ‖ Glaskuppel f. ‖ globe m. de verre. / ~ drop ‖ plötzlich abgekühlter Glastropfen m.; Glasträne f. ‖ goutte f. de verre; larme f. batavique. / ~ emballing ‖ Glasemballage f. ‖ emballage m. en

verre. / ~ engraving ‖ Glasgravierung f. ‖ gravure sur verre m.

glasses pl. (Field glass) ‖ Fernglas n.; Opernglas n. ‖ jumelles fpl.; lunette f. / ~ (Spectacles) ‖ Brille f.; Augenglas n. ‖ lunettes fpl. / firemen's ~ ‖ Rauchschutzbrille f. ‖ lunettes fpl. de pompiers. / ~ for strabismus ‖ Schielglas n. ‖ verres mpl. louchettes.

glass etching workshop ‖ Glasätzerei f. ‖ atelier m. de gravure sur verre. / ~ eye ‖ Glasauge n. ‖ œil m. de verre. / ~ fancy ware ‖ Glasluxusware f. ‖ verrerie f. de luxe. / ~ fibre ‖ Glasgespinst n. ‖ verre m. filé. / ~ fittings pl. for metal goods factories ‖ Fassungsartikel mpl. für Metallwarenfabriken ‖ articles mpl. de montage en verre pour fabriques d'objets en métal. / flaw in the ~ ‖ Glasfehler m. ‖ défaut m. du verre. / fragments pl. of ~ ‖ Glassplitter mpl. ‖ débris mpl. de verre. / ~ furnace ‖ Glasofen m. ‖ four m. de verrerie ou de fusion. / ~ gall ‖ Glasgalle f. ‖ fiel m. ou sel m. de verre; suin m. / ~ gauge ‖ Standglas n. ‖ tube m. de niveau. / ~ gilder ‖ Glasvergolder m. ‖ doreur m. sur verre. / ~ gilding ‖ Glasvergoldung f. ‖ dorure f. sur verre. / ~ goblet ‖ Becherglas n.; Glasbecher m. ‖ goblet m. en verre. / ~ grain ‖ Glaskorn n. ‖ grain m. de verre. / ~ grinder ‖ Glasschleifer m. ‖ tailleur m. de verre. / ~ grinding workshop ‖ Glasschleiferei f. ‖ atelier m. de polissage de verre. / ~ guard of glow lamp ‖ Glühlampenschutzglas n. ‖ globe m. protecteur de lampe à incandescence.

glasshard ‖ glashart ‖ dur comme le verre.

glass hardness (Met) ‖ Glashärte f. ‖ trempe f. glacée.

glass-house (Gard) ‖ Gewächshaus n. ‖ serre f. / ~ (Glassm) ‖ Glashütte f. ‖ verrerie f.

glass industry ‖ Glasindustrie f. ‖ industrie f. du verre.

glassiness ‖ Glasigkeit f. ‖ état m. vitreux; glaçage m.; vitrification f.

glass inlaying ‖ Glasmosaik n. ‖ verre m. mosaïque. / ~ instrument ‖ Glasinstrument n. ‖ instrument m. en verre. / ~ insulator ‖ Glasisolator m. ‖ isolateur m. en verre.

glass jar ‖ Glasgefäß n.; Glas n. ‖ vase m. ou cuve en verre. / ~ for cells (Electr) ‖ Elementglas n. ‖ vase m. à piles en verre. / graduated ~ (Chem; Phys) ‖ graduierte oder kalibrierte Glasglocke f. ‖ cloche f. graduée. / ~ with stop-cock ‖ Glasglocke f. mit Hahnstück ‖ cloche f. à robinet en verre.

glass jewelry ‖ Glaskurzwaren fpl. ‖ objets mpl. en verre. / ~ ladle (Found) ‖ Glasgießlöffel m. ‖ poche f. de coulée à verre. / ~ letter ‖ Glasbuchstabe m. ‖ lettre f. en verre. / ~ light fittings pl. ‖ Glasbeleuchtungsgegenstände mpl. ‖ articles mpl. d'éclairage en verre. / ~ lozenge ‖ Rautenglas n. ‖ vitre f. rhomboïde; rhombe m. de vitre. / ~ maker ‖ Glasfabrikarbeiter m. ‖ ouvrier m. de verrerie. / ~ making ‖ Glasfertigung f.; Glasfabrikation f. ‖ verrerie f. / ~ manufacture ‖ Glasmanufaktur f. ‖ manufacture f. de verre. / ~ manufacturer see ~ maker. / ~ marble ‖ Glasperle f. ‖ bille f. de verre. / ~ mastic layer ‖ Glasverkitter m. ‖ mastiqueur m. de vitres. / ~ matrass ‖ Glaskolben m. ‖ matras m. ou ballon m. de verre. / ~ meal

‖ Glasmehl n. ‖ verre m. en poudre. / ~ melter ‖ Glasschmelzer m. ‖ fondeur m. de verre. / ~ melting-pot ‖ Glashafen m.; Hafen m.; Schmelzhafen m. ‖ creuset m.; padelin m.; pot m. / ~ metal ‖ Glasmasse f. ‖ verre m. en masse. / ~ mirror ‖ Glasspiegel m. ‖ miroir m. en verre. / silvered ~ mirror ‖ Glassilberspiegel m. ‖ miroir m. en verre argenté. / ~ mosaics pl. ‖ Glasmosaik f.; Millefiori fpl. ‖ verre m. mosaïque; verre m. mille-fleurs. / ~ mould ‖ Glasgießform f. ‖ presse f. à mouler les verres. / ~ muddying ‖ Glastrübung f. ‖ ternissure f. de verre. / ~ muddying substance ‖ Glastrübungsmittel n. ‖ matière f. à ternir le verre. / ~ oven ‖ Glasofen m. ‖ four m. de fusion ou de verrerie. / ~ painter ‖ Glasmaler m. ‖ peintre-verrier m. / ~ painting ‖ Glasmalerei f.; Glasmalereierzeugnis n. ‖ peinture f. sur verre. / ~ pane ‖ Glasscheibe f. ‖ vitre f. / ~ paper ‖ Glaspapier n. ‖ papier m. verre ou de verre. / ~ paper making machine ‖ Glaspapierherstellungsmaschine f. ‖ machine f. à fabriquer le papier de verre. / ~ partition (Build) ‖ Glasverschlag m.; Glaswand f. ‖ cloison f. en verre; vitrage m. / ~ paving ‖ Glaspflaster n. ‖ pavement m. en verre; dalle f. lumineuse; isoloir m. / ~ pearl ‖ Glasperle f. ‖ perle f. en verre. / ~ pearl powder for relief painting ‖ Streuperlen fpl. für Reliefmalerei ‖ perles fpl. pour la peinture en relief. / ~ pencil ‖ Glaspinsel m. ‖ pinceau m. en verre. / ~ photography ‖ Glasfotografie f. ‖ photographie f. sur verre. / ~ picture ‖ Glasbild n. ‖ peinture f. sur verre. / ~ placard ‖ Glasfirmenschild n.; Glasplakat n. ‖ plaque f. de réclame en verre; affiche f. en verre. / ~ plate see also ~ placard ‖ Glasplatte f. ‖ plaque f. de verre; panneau m. ou table f. en verre. / ~ point ‖ Glasspitze f. ‖ pointe f. de verre. / ~ (full) pole ‖ (volle) Glasstange f. ‖ tige f. (pleine) en verre. / ~ polisher ‖ Glaspolierer m. ‖ polisseur de verre m. / ~ polishing ‖ Polieren n. von Glas; Glaspolieren n. ‖ polissage du verre m. / ~ polishing machine ‖ Glasschleifereimaschine f. ‖ machine f. à tailler le verre. / ~ poster see ~ placard. / ~ powder ‖ Glaspulver n.; Glasmehl n. ‖ verre m. pulvérisé; poudre f. de verre. / ~ powder for decoration purposes ‖ Glasschnee m. für Dekorationszwecke ‖ poudre f. de verre à décorer. / ~ press ‖ Glaspresse f.; Glaspreßform f. ‖ moule m. à comprimer le verre. / ~ printing ‖ Glasdruck m. ‖ impression f. sur verre. / ~ prism ‖ Glasprisma n. ‖ prisme m. en verre. / ~ refining material ‖ Glasläuterungsmittel n. ‖ matière f. à raffiner le verre. / ~ parabolic reflector ‖ Glasparabolspiegel m. ‖ réflecteur m. parabolique en verre. / ~ retort ‖ Glasretorte f. ‖ rétorte f. en verre; cornue f. en verre. / ~ ring ‖ Glasring m. ‖ anneau m. de verre. / ~ rod (Chem) ‖ Glasstab m. ‖ baguette f. ou tige m. en verre. / little ~ rod ‖ Glasstäbchen n. ‖ petite baguette f. en verre. / ~ rolling machine ‖ Glaswalzmaschine f. ‖ laminoir m. à verre. / ~ roof ‖ Glasdach n. ‖ toit m. en verre. / non-puttied ~ roof ‖ kittloses Glasdach n. ‖ toit m. (à tuiles) en verre sans mastique. / trellis for ~ roof ‖ Glasdachsprosse f. ‖ treillage m. pour toits en verre. / ~

roofing ‖ Glasbedachung f. ‖ toiture f. en verre. / ~ roundle ‖ Butzenscheibe f. ‖ cul m. de bouteille; rond m. de verre. / ~ sand ‖ Glassand m. ‖ sable m. verrier; sable m. pour verrerie. / ~ scale ‖ Glasmaßstab m. ‖ division f. sur verre. / ~ shade ‖ Glasglocke f.; Glasstürze f. ‖ cloche f. en verre. / ~ side (wall) ‖ Glasseitenwand f. ‖ paroi f. latérale en verre. / ~ silverer ‖ Glasversilberer m. ‖ argenteur m. sur verre. / ~ silvering ‖ Glasversilberung f. ‖ argenture f. sur verre. / ~ slide ‖ Glasschieber m. ‖ pièce f. coulissante en verre. / ~ slide (Object glass) ‖ Objektträger m. ‖ porte-objets m. / species of ~ ‖ Glasart f. ‖ espèce f. de verre. / ~ spectrograph ‖ Glasspektrograf m. ‖ spectrographe m. à optique en verre. / ~ spectrograph of great light transmitting power ‖ lichtstarker Glasspektrograf m. ‖ spectrographe m. à optique en verre lumineux. / ~ sphere ‖ Glaskugel f. ‖ boule f. de verre. / ~ spinning ‖ Glasspinnerei f. ‖ filature f. de verre. / ~ spinning machine ‖ Glasspinnmaschine f. ‖ machine f. à filer le verre. / ~ square ‖ Fensterscheibe f. ‖ carreau m. de vitre; vitre f.; panneau m. de vitre ou de verre. / ~ staining ‖ eingebrannte Glasmalerei f. ‖ art m. de l'émailleur; peinture f. d'apprêt ou en apprêt. / ~ stirrer see ~ rod. / ~ stopper ‖ Glasstöpsel m. ‖ bouchon m. de verre ou à verre. / ground(-in) ~ stopper ‖ eingeschliffener Glasstöpsel m. ‖ bouchon m. de verre rodé ou en verre hermétique. / ~ stopple see ~ stopper. / ~ strip ‖ Glasstreifen m. ‖ bande f. en verre. / ~ syringe ‖ Glasspritze f. ‖ seringue f. en verre. / ~ table see ~ plate. / ~ tear ‖ Glasträne f.; (plötzlich abgekühlter) Glastropfen m. ‖ goutte f. de verre. / ~ thread ‖ Glasfaden m.; Glasgespinst n. ‖ fil m. de verre. / ~ tile (Building-stone) ‖ Glasbaustein m. ‖ carreau m. ou brique f. de verre. / ~ tile (Roofing tile) ‖ Glasdachziegel m. ‖ tuile f. en verre. / ~ tube ‖ Glasrohr n.; Glasröhre f. ‖ tube m. ou tuyau m. ou canon m. en verre. / ~ tube for lighting apparatus ‖ dicke Glasröhre f. für Beleuchtungsvorrichtungen ‖ verrine f. pour appareils d'éclairage. / ~ tubing see ~ tube. / ~ tulip ‖ Glastulpe f. ‖ tulipe f. / ~ utensil ‖ Glasgerät n. ‖ ustensile m. en verre. / ~ varnish ‖ Glaslack m. ‖ vernis m. pour verre. / ~ vase ‖ Glashafen m. ‖ vase m. en verre. / ~ vessel ‖ Glasgefäß n. ‖ vaisseau m. ou récipient m. en verre.

glassware ‖ Glaswaren fpl. ‖ objets mpl. ou articles mpl. en verre; verrerie f. / artistic ~ ‖ Kunstglaswaren fpl. ‖ verrerie f. artistique. / cut ~ ‖ geschliffene Glaswaren fpl. ‖ verrerie f. taillée. / hollow ~ ‖ Hohlglaswaren fpl.; Hohlglas n.; Hohlglasgefäße npl. ‖ gobeleterie f. / mounted ~ ‖ gefaßte Glaswaren fpl. ‖ verrerie f. garnie. / scientific ~ ‖ wissenschaftliche Glaswaren fpl. ‖ verreries fpl. scientifiques. / special ~ ‖ Spezialglaswaren fpl. ‖ verrerie f. spéciale. / utility ~ ‖ Glasgebrauchsgegenstände mpl. ‖ verrerie f. d'usage courant.

glass washer ‖ Plattenwäscher m. ‖ laveur m. de plaques. / ~ waste ‖ Glasabfall m. ‖ déchet m. de verre. / ~ wool ‖ Glaswolle f. ‖ laine f. ou coton m. de verre; verre m. filé. / ~ working machine ‖ Glasbearbeitungsmaschine f. ‖ ma-

chine f. pour le travail du verre *ou* à travailler le verre. / modern ~ working method ‖ neuzeitliches Glasbearbeitungsverfahren n. ‖ outillage m. de verre perfectionné.

glass-works pl. *see also* glass-house ‖ Glashütte f. ‖ verrerie f. / ~ plant ‖ Glasfabrikeinrichtung f. ‖ équipement m. pour verreries. / ~ product ‖ Glashüttenerzeugnis n. ‖ (produit m. de) verrerie f.

glassy ‖ glasig; glasartig ‖ vitreux. / ~ malt ‖ Glasmalz n. ‖ malt m. vitreux.

glauberite ‖ Glauberit n. ‖ glaubérite f.

Glauber's salt ‖ schwefelsaures Natron n.; Natriumsulfat n.; Glaubersalz n. ‖ sulfate m. de soude; sel m. de Glauber.

glaucodot (Miner) ‖ Glaukodot m.; Kobaltarsenikkies m. ‖ fer m. arsenical cobaltifère.

glaucoma (Eye disease) ‖ grüner Star m. ‖ cataracte f. glaucomateuse.

glauconite (Miner) ‖ Glaukonit m.; Grünerde f. ‖ glaukonite f.

glaucophane (Miner) ‖ Glaukophan m. ‖ glaucophane f.

glaze, to ‖ glasieren; mit Glasur f. überziehen ‖ vernisser; vernir; glacer. / ~ (Chem) ‖ verglasen ‖ verrer. / ~ (Pap) ‖ glätten; satinieren ‖ satiner; lisser; glacer. / ~ a window ‖ ein Fenster verglasen ‖ vitrer une fenêtre; poser des carreaux.

glaze ‖ Glasur f. ‖ glaçure f. / ~ (Protective coating) ‖ Schutzanstrichmittel n. ‖ enduit m.; peinture f. protectrice. / refractory ~ ‖ strengflüssige Glasur f. ‖ couche f. *ou* vernis m. qui ne se vitrifie que très-difficilement.

glaze baking ‖ Einbrennen n. der Glasur; Glasurbrand m. ‖ cuisson f. du vernis.

glazed (Glazier) ‖ verglast ‖ vitré. / ~ (Pott) ‖ glasiert ‖ vernissé. / ~ appearance ‖ glänzendes Aussehen n. ‖ apparence f. luisante. / ~ board ‖ Glanzpappe f. ‖ carton m. glacé *ou* lissé; carton m. porcelaine. / ~ facing brick ‖ glasierter Verblender m. *oder* Verblendstein m. ‖ brique f. glacée *ou* hollandaise pour parement. / ~ cardboard ‖ Kreidekarton m. ‖ carton m. couché. / ~ earthenware ‖ verglaste Ton waren fpl. ‖ faïence f. / ~ iron-frame constructions pl. ‖ verglaste in Eisenkonstruktion gehaltene Hallen fpl. ‖ halls mpl. en charpente métallique vitrée. / ~ linen ‖ Glanztuch n. ‖ toile f. gommée. / ~ paper ‖ Glanzpapier n.; geglättetes Papier n.; Satinpapier n. ‖ papier m. glace *ou* satiné *ou* lissé. / ~ paper making machine ‖ Glanzpapierherstellungsmaschine f. ‖ machine f. à fabriquer le papier glacé. / ~ stove ‖ Kachelofen m. ‖ poêle m. en faïence.

glaze grinder (Pott) ‖ Glasurarbeiter m. ‖ broyeur m. d'émail. / ~ kiln ‖ Glasurofen m. ‖ four m. à glaçure.

glazer (Cutl) ‖ Feinschleifer m. ‖ affileur m. / ~ (Emery wheel) ‖ Schmirgelscheibe f.; Polierscheibe f. ‖ meule f. d'émeri; polissoire f. / ~ for paper machines ‖ Papiermaschinenglättwerk n. ‖ lisseuse f. pour machines à papier. / ~ waves pl. (Fault in the glaze) ‖ Glanzwellen fpl. ‖ retirement m. / ~ wheel *see* glazer (Emery wheel).

glazier ‖ Glaser m. ‖ vitrier m.

glazier's diamond ‖ Glaserdiamant m. ‖ diamant m. de vitrier *ou* à couper le

verre. / ~ hammer ‖ Glaserhammer m. ‖ marteau m. de vitrier. / ~ lead ‖ Fensterblei n.; Glaserblei n. ‖ plomb m. à vitres *ou* du vitrier. / ~ material ‖ Glaserbedarfsartikel mpl. ‖ fournitures fpl. pour vitriers. / ~ pick-hammer ‖ Glaserhammer m. mit spitzer Finne ‖ besaiguë f. *ou* putty ‖ Glaserkitt m. ‖ mastic m. de vitrier. / ~ straper (Glassm) ‖ Ofenkrücke f. ‖ graton m. / ~ work ‖ Glaserarbeiten fpl. ‖ travaux mpl. de vitrier. / ~ workshop for buildings ‖ Bauglaserei f. ‖ atelier m. de vitriers en bâtiments.

glazing *see also* glaze ‖ Glasur f. ‖ glaçure f. / ~ (Clothm) ‖ Glanz m.; Wasserappretur f. ‖ apprêt m. à l'eau f. / ~ (Cutl) ‖ Schmirgeln n. (von Metallflächen); Feinschleifen n. ‖ polissage m. à l'émeri; rodage m.; affilage m. / ~ (Glazier's trade) ‖ Glaserei f.; Glaserhandwerk n. ‖ vitrerie f. / ~ (Paint) ‖ Glasur f.; Malerglasur f. ‖ glacis m. / ~ (Pap) ‖ Satinieren n. ‖ satinage m. / ~ for ceramics ‖ Glasur f. für Keramik ‖ vernissage m. pour la céramique. / cloth-piece ‖ Glanzieren n. von Geweben ‖ glaçage m. d'étoffes. / ~ by dipping ‖ Glasur f. durch Eintauchen ‖ glaçure f. par immersion *ou* par trempage. / easily fusible ~ ‖ leichtflüssige Glasur f. ‖ glaçure f. *ou* vernis m. facilement vitrescible. / ~ by pouring ‖ Glasur f. durch Begießen ‖ glaçure f. par arrosement. / ~ by powdering ‖ Glasur f. durch Bestäuben ‖ glaçure f. par saupoudration *ou* aspersion. / ~ by volatilization ‖ Glasur f. durch Verflüchtigung ‖ glaçure f. par volatilisation. / ~ by washing *see* ~ by pouring ‖ Glasur f. durch Begießen ‖ glaçure f. par arrosement. / white-lead ~ ‖ Zinnglasur f. ‖ glaçure f. stannifère. / ~ of a window ‖ Verglasung f. eines Fensters ‖ vitrage m. (d'une fenêtre). / ~ of yarn ‖ Glänzen n. des Zwirnes ‖ lustrage m. de fils.

glazing apparatus (Pap) ‖ Satinierapparat m.; Glättvorrichtung f. (für Papier) ‖ satineuse f. / ~ barrel ‖ Poliertrommel f. ‖ lissoir m.; tambour m. polisseur. / ~ calender (Clothm) ‖ Glanzkalander m.; Glättkalander m. ‖ calandre f. à lustrer. / ~ calender (Pap) ‖ Satinierkalander m.; Seidenglanzkalander m. ‖ calandre f. de satinage soyeux. / ~ colour ‖ durchscheinende Farbe f.; Lasurfarbe f. ‖ couleur f. transparente.

glazing-machine (Clothm) ‖ Glänzmaschine f.; Glättmaschine f. ‖ lissoir m. / ~ (Pap) ‖ Satiniermaschine f.; Satinierwalzwerk n. ‖ machine f. à satiner; satineuse f.; laminoir m. / ~ (Pott) ‖ Glasiermaschine f. ‖ machine f. à glacer.

glazing manufacturer's plant ‖ Glasurenfabrikseinrichtung f. ‖ installation f. de fabrique de glaçures. / ~ rolls pl. ‖ Satinierwalzen fpl. ‖ rouleaux mpl. satineurs. / ~ wheel *see* glazer (Emery wheel).

gleam ‖ Lichtstrahl m. ‖ rayon m. lumineux. / ~ of silver ‖ Silberblick m. ‖ fulguration f. d'argent.

glen ‖ Klamm f.; Schlucht f. ‖ ravin m.

gliadin ‖ Pflanzenleim m.; Indigoleim m. ‖ gliadine f.

glide, to ‖ gleiten ‖ glisser.

glider ‖ Segelflugzeug n.; Gleitflugzeug n.; motorloses Flugzeug n. ‖ avion m. du vol à voile; planeur m.

gliding ‖ Gleiten n. ‖ glissage m. / noiseless ~ ‖ geräuschloses Gleiten n. ‖ glissage m. silencieux.

gliding angle (Aero) ‖ Gleitwinkel m. ‖ angle m. (de vol) plané. / ~ boat ‖ Wassergleitboot n.; Gleitboot n. ‖ hydroglisseur m. / ~ flight ‖ Gleitflugzeug n. ‖ vol m. plané *ou* à voile.

glimmer (Miner) ‖ Glimmer m. ‖ mica m. / ~ in coarse disks ‖ Glimmer m. in groben Scheiben ‖ mica m. en disques grossiers. / yellow ~ ‖ Katzengold n.; Goldglimmer m. ‖ faux-or m.; mica m. jaune.

glimmer-light rectifier ‖ Glimmlichtgleichrichter m. ‖ lampe f. redresseuse à fluorescence.

glisten, to ‖ flimmern; glitzern; funkeln ‖ briller.

glistening ‖ flimmernd ‖ brillant.

glister ‖ Schein m.; Glanz m.; Funkeln n. ‖ éclat m.; scintillement m.; étincellement m.

globe ‖ Kugel f.; kugelförmiger Gegenstand m. ‖ globe m. / ~ (Geogr) ‖ Globus m. ‖ globe m. / celestial ~ ‖ Himmelsglobus m. ‖ globe m. céleste. / embossed ~ ‖ Reliefglobus m. ‖ globe m. en relief. / glass ~ ‖ Glasglocke f. ‖ globe m. en verre. / frosted ~ of glass ‖ mattierte Glasglocke f. ‖ globe m. en verre dépoli. / terrestrial ~ ‖ Erdglobus m. ‖ globe m. terrestre.

globigerina mud ‖ Globigerinenschlamm m. ‖ boue f. à globigérines.

globular chart ‖ Planiglob m.; (Erd- oder Himmels-)Kugelkarte f. ‖ planisphère m. / ~ lens ‖ kugelige Linse f. ‖ lentille f. à surface sphérique. / ~ structure (Geol) ‖ kugeliger Bau m.; kugelige Struktur f. ‖ structure f. globulaire.

globule ‖ Kügelchen n. ‖ globule m. / ~ (Met) ‖ Metallkorn n. ‖ grain m. métallique.

glockenspiel ‖ Glockenspiel n. ‖ carillon m.

glonoïne ‖ Nitroglyzerin n.; Glonoin n. ‖ nitroglycérine f.

gloss, to (Clothm) ‖ falschen Glanz m. geben; laudieren ‖ lustrer; donner un faux lustre. / ~ (Pott) *see* to glaze (Pott). / ~ the cloth ‖ das Tuch pressen ‖ catir le drap. / ~ over ‖ vertuschen; bemänteln ‖ dissimuler.

gloss (Clothm) ‖ Preßglanz m. ‖ cati m. / ~ (Pott) *see* glaze (Pott). / ~ of cloth ‖ Preßglanz m. eines Tuches ‖ cati m. d'une étoffe *ou* d'un drap. / marginal ~ (Print) ‖ Randbemerkung f.; Randglosse f.; Marginalie f. ‖ glosse f. *ou* note f. marginale; manchette f.

glossing (Clothm) ‖ Pressen n. ‖ catissage m.

gloss ink ‖ Glanzfarbe f. ‖ couleur f. de lustre.

glossy (Weav) ‖ glatt; geschoren ‖ ras.

glove ‖ Handschuh m. ‖ gant m. / coarse silk ~ ‖ Burettehandschuh m. ‖ gant m. de filoselle. / cloth ~ ‖ Stoffhandschuh m. ‖ gant m. en étoffe. / cotton ~ ‖ Baumwollhandschuh m. ‖ gant m. de coton. / ~ of deerskin ‖ Wildlederhandschuh m. ‖ gant m. en peau de daim. / elbow ~ ‖ langer Handschuh m. ‖ gant m. passe-coude. / glacé ~ ‖ Glanzlederhandschuh m.; Glacéhandschuh m. ‖ gant m. glacé. / knitted ~ ‖ Strickhandschuh m. ‖ gant m. tricoté. / lace ~ ‖ Filethandschuh m. ‖ gant m. de filet. / leather ~ ‖ Lederhandschuh m. ‖ gant m. de peau. / lined ~ ‖ gefütterter Handschuh m. ‖ gant m.

fourré. / rubber ~ ‖ Gummihandschuh m. ‖ gant m. en caoutchouc. / silk ~ ‖ Seidenhandschuh m. ‖ gant m. de soie. / thread ~ ‖ Zwirnhandschuh m. ‖ gant m. de fil. / wash-leather ~ ‖ Waschlederhandschuh m. ‖ gant m. de peau lavable. / worsted ~ ‖ Wollhandschuh m. ‖ gant m. de laine. / ~ of woven material see cloth ~.

glove button ‖ Handschuhknopf m. ‖ bouton m. pour gants. / ~ buttoner ‖ Handschuhknöpfer m. ‖ crochet m. à gants. / ~ cutter ‖ Handschuhschneider m. ‖ coupeur m. de gants. / ~ embroiderer ‖ Handschuhstickerin f. ‖ brodeuse f. sur gants. / ~ fabric ‖ Handschuhstoff m. ‖ tissu m. pour gants. / ~ finger ‖ Handschuhfinger m. ‖ doigt m. de gant. / ~ finger knitting machine ‖ Fingerstrickmaschine f. ‖ machine f. à tricoter les doigts. / ~ finishing oven ‖ Handschuhformofen m. ‖ four m. d'apprêt pour gants. / ~ folder ‖ Handschuhformer m. ‖ apprêteur m. de gants. / ~ knitter ‖ Handschuhstricker m. ‖ tricoteur m. de gants. / ~ leather ‖ Handschuhleder n. ‖ cuir m. pour gants. / tawed ~ leather ‖ weißgares Handschuhleder n. ‖ cuir m. mégissé pour gants. / ~ machine ‖ Handschuhmaschine f. ‖ machine f. de ganterie. / ~ maker ‖ Handschuhmacher m. ‖ gantier m. / ~ making ‖ Handschuhfertigung f.; Handschuhmacherei f. ‖ ganterie f. / ~ manufactory ‖ Handschuhfabrik f. ‖ ganterie f.

glover ‖ Handschuhmacher m. ‖ gantier m. **glove** scourer ‖ Handschuhreiniger m. ‖ nettoyeur m. de gants. / ~ sewer ‖ Handschuhnäherin f. ‖ couseuse f. de ganterie. / ~ sewing machine ‖ Handschuhnähmaschine f. ‖ machine à coudre les gants. / ~ stitcher ‖ Handschuhstepperin f. ‖ piqueuse f. en ganterie (de peau). / ~ stitching ‖ Handschuhstepperei f. ‖ piqûre f. de gants. / ~ trade see ~ making.

gloving leather see glove leather. / ~ machine ‖ Handschuhnähmaschine f. ‖ machine f. à coudre les gants.

glow, to ‖ verglühen; glühen; glimmen ‖ brûler sans flamme; être ardent. / to ~ the iron ‖ das Eisen glühen; dem Eisen eine Hitze geben ‖ chauffer ou faire rougir le fer; donner une chaude au fer.

glow ‖ Glühen n.; Glut f. ‖ lueur f. / ~ after discharge (Electr) ‖ Nachglühen n. oder Nachleuchten n. (eines Körpers) nach elektrischer Entladung ‖ lueur f. après une décharge. / blue ~ (Elektr) ‖ Glimmlicht n. ‖ lueur f. de décharge.

glow body ‖ Glühkörper m. ‖ corps m. à incandescence.

glow cathode ‖ Glühkathode f. ‖ cathode f. incandescente. / ~ rectifier ‖ Glühkathodengleichrichter m. ‖ lampe f. redresseuse à cathode incandescente.

glow cooking plate (Electr) ‖ Glühkochplatte f. ‖ réchaud m. à chauffage électrique. / ~ discharge ‖ Glimmentladung f. ‖ décharge f. à faible lueur.

glowing (Electr) ‖ Glühen n. ‖ incandescence f. / ~ (Met ect.) ‖ Glühen n. ‖ ardeur f. / ~ coke ‖ glühender Koks m. ‖ coke m. ardent. / ~ furnace for bone-coal glowing ‖ Knochenkohleglühofen m. ‖ four m. à calciner le charbon d'os. / ~ heat ‖ Glühhitze f. ‖ chaude f. vive.

glow lamp (Electr) ‖ Glühlampe f. ‖ lampe f. à incandescence. / pear-shaped ~ ‖ bir-

nenförmige Glühlampe f.; Glühbirne f. ‖ lampe f. à incandescence piriforme.

glow-lamp base ‖ Glühlampensockel m. ‖ culot m. de lampe à incandescence. / ~ cap see ~ base. / ~ filament ‖ Glühlampenfaden m. ‖ filament m. de lampe à incandescence. / ~ fitting see ~ holder. fittings pl. ‖ Glühlampenarmatur f. ‖ garnitures fpl. de lampe à incandescence. / ~ holder ‖ Glühlampenfassung f. ‖ douille f. de lampe à incandescence. / ~ photometer ‖ Glühlampenfotometer n. ‖ photomètre m. de lampes à incandescence. / ~ reflector ‖ Glühlampenreflektor m.; Glühlampenlichtspiegler m. ‖ réflecteur m. de lampe à incandescence. / ~ resistance ‖ Glühlampenwiderstand m. ‖ résistance f. à incandescence.

glow light ‖ Glimmlicht n. ‖ lueur f. de décharge. / ~ oscillograph ‖ Glimmlichtoszillograf m. ‖ oscillographe m. à décharge de faible lueur.

glow rod ‖ Glühstab m. ‖ crayon m. incandescent.

glow transmitter ‖ Glimmlichtsender m. ‖ émetteur m. à faible lueur.

glucina ‖ Beryllerde f. ‖ glucine f.

glucose ‖ Stärkezucker m.; Glukose f.; Traubenzucker m.; Glykose f. ‖ glucose f.; glycose f. / ~ factory ‖ Glukosefabrik f.; Stärkezuckerfabrik f. ‖ glucoserie f. / ~ manufactory see ~ factory.

glue, to ‖ leimen ‖ coller. / ~-in ‖ einleimen ‖ coller; encoller. / ~-on ‖ ankleben; anleimen ‖ coller. / ~ paper bags ‖ Tüten fpl. kleben ‖ coller des sacs mpl. en papier.

glue ‖ Leim m.; Tischlerleim m.; Klebstoff m.; Klebemittel n. ‖ colle f.; colle f. forte; gluten m. / to apply the ~ to ... ‖ den Leim m. auftragen ‖ enduire de colle f. / to give the ~ time to set ‖ dem Leim m. die Zeit zum Anziehen geben ‖ donner à la colle le temps de prendre. / animal ~ ‖ Tierleim m. ‖ colle f. à gélatine. / bird ~ ‖ Vogelleim m. ‖ glu m. / ~ of bones ‖ Knochenleim m. ‖ colle f. d'os; gélatine f. d'os; ostéocolle f. / boxmaker's ~ ‖ Kartonnagenleim m. ‖ colle f. pour cartonnages. / cheese ~ ‖ Käseleim m.; Kaseinleim m. ‖ colle f. caséine. / cold ~ ‖ Kaltleim m. ‖ colle f. à froid. / Cologne ~ ‖ kölnischer Leim m. ‖ colle f. de Cologne. / dressing ~ ‖ Appreturleim m. ‖ colle f. pour apprêts. / Flanders ~ ‖ flandrischer Leim m. ‖ colle f. de Flandre ou de Hollande. / ~ of gluten ‖ Kleberleim m.; Eiweißleim m. ‖ colle f. végétale ou albuminoïde; gluten m. / ~ for grinding ‖ Schleiferleim m. ‖ colle f. d'émouleur. / ~ from hides ‖ Leim m. aus Häuten ‖ colle f. de peaux. / ~ in form of jelly ‖ Leim m. in Gallertform ‖ colle f. en gelée. / joiner's ~ ‖ Tischlerleim m. ‖ colle f. forte. / leather ~ ‖ Lederleim m. ‖ colle f. de cuir. / ~ in leaves ‖ Leim m. in Blättern ‖ colle f. en feuilles. / liquid ~ ‖ flüssiger Leim m. ‖ colle f. (blanche) liquide. / ~ for paperhangings ‖ Tapetenleim m. ‖ colle f. pour papiers-peints. / ~ in form of paste ‖ Leim m. in Teigform ‖ colle f. en pâte. / ~ in plates ‖ Tafelleim m. ‖ colle f. en plaques. / powdered ~ ‖ Leimpulver n.; Pulverleim m.; Leim m. in Pulverform ‖ colle f. en poudre. / pulverized ~ see powdered ~. / ~ (made of the refuse) of sheep skins

(Join) ‖ Hammelfelleim m. ‖ poissonnure f. / ~ in tables see ~ in plates. / vegetable ~ ‖ Pflanzenleim m. ‖ colle f. végétale. / ~ of whale tendons ‖ Leim m. aus Walfischsehnen ‖ colle f. de tendons de baleine.

glue boiler ‖ Leimkocher m.; Leimsieder m.; Leimkochapparat m. ‖ cuiseur m. ou chaudière f. à colle forte; bouilleur m. de colle; appareil m. à cuire la colle. / ~ boiling apparatus see ~ boiler. / ~ brush ‖ Leimpinsel m. ‖ brosse f. à coller. / ~ colour ‖ Leimfarbe f. ‖ couleur f. à la colle ou en détrempe; détrempe f. / ~ cooking apparatus see ~ boiler. / ~ cutting machine ‖ Leimschneidemaschine f. ‖ machine f. à couper la colle.

glued on wood ‖ auf Holz n. geleimt ‖ collé sur bois m. / ~ back of the booklet ‖ beleimter Rücken m. der Broschüre ‖ dos m. de la brochure enduit de colle.

glue dryer ‖ Leimtrockner m. ‖ séchoir m. de colle. / ~ drying plant ‖ Leimtrockenanlage f. ‖ installation f. de séchage de colle. / ~ glass ‖ Leimglas n. ‖ verre m. à colle. / ~ heating apparatus ‖ Leimwärmeapparat m. ‖ appareil m. à chauffer la colle. / ~ heating pan ‖ Leimwärmpfanne f. ‖ cassole f.

glueing machine ‖ Leimauftragmaschine f. ‖ machine f. à étendre la colle. / book ~ ‖ Broschürenleimmaschine f. ‖ machine f. à coller les brochures.

glueing-on machine for reels of paper ‖ Anleimmaschine f. für Rollenpapier ‖ machine f. à coller le papier en rouleaux.

glueing strip (Bookb) ‖ Leimleiste f. ‖ lame f. de collage.

glue jelly ‖ Gallert m. ‖ gélatine f. / to solidify the ~ jelly ‖ den Gallert m. erstarren lassen ‖ solidifier la gélatine (de la colle). / ~ maker ‖ Leimsieder m. ‖ fabricant m. ou faiseur m. de colle (forte). / ~ manufacturer see ~ maker. / ~ manufacturing plant ‖ Leimfabrikeinrichtung f. ‖ installation f. à fabriquer la colle. / ~ materials pl. ‖ Leimgut n.; Leimleder n. ‖ colles fpl.; matières fpl. ou rognures fpl. des peaux. / ~ mould ‖ Leimform f. ‖ moule m. à colle. / ~ pot ‖ Leimtiegel m.; Leimtopf m. ‖ pot m. à colle. / ~ powder ‖ Leimpulver n. ‖ colle f. en poudre. / ~ press (Join) ‖ Leimzwinge f.; Schraubzwinge f. ‖ sergent m.; presse f. à main ou à serrer; serre-joint m. / ~ priming (Gild) ‖ Leimgrund m. ‖ encollage m. / ~ producing plant ‖ Leimherstellungsanlage f. ‖ installation f. de fabrication de la colle.

gluer (Bookb) ‖ Leimer m. ‖ encolleur m. **glue** spreading machine ‖ Leimauftragmaschine f. ‖ machine f. à étendre la colle. / ~ stamp (Bookb) ‖ Leimstempel m. ‖ tampon m. à colle. / ~ stock ‖ Leimleder n. ‖ oreillons mpl. / ~ water ‖ Leimwasser n. ‖ eau f. de colle. / ~ watercolour ‖ Leimfarbe f. ‖ détrempe f. ou peinture f. à la colle. / ~ water-paint see ~ water-colour. / ~ works pl. ‖ Leimfabrik f.; Leimwerk n. ‖ fabrique f. de colle.

glut, to ‖ überhäufen; überladen; übersättigen ‖ surcharger; (trade:) encombrer. **gluten** ‖ Kleber m.; Schuhmacherleim m. ‖ gluten m.; colle f. de gluten. / animal ~ see bone glue.

glutine ‖ Glutin n. ‖ glutine f. / ~ cloudiness ‖ Glutintrübung f. ‖ trouble m. de glu-

tine. / ~ turbid ‖ glutintrüb ‖ trouble de glutine. / ~ turbidity *see* ~ cloudiness.

glutted (Trade) ‖ überhäuft ‖ encombré.

glycerinated sulphuric acid ‖ Glyzerinschwefelsäure f. ‖ acide m. sulfurique glycérique.

glycerine ‖ Glyzerin n.; Ölsüß n. ‖ glycérine f. / ~ distilling plant ‖ Glyzerindestillationsanlage f. ‖ installation f. de distillation de glycérine. / ~ extracting plant ‖ Glyzeringewinnungsanlage f. ‖ installation f. d'extraction de glycérine. / ~ pitch ‖ Glyzerinpech n. ‖ poix f. de glycérine. / ~ producing plant *see* ~ extracting plant. / ~ refinery ‖ Glyzerinraffinerie f. ‖ raffinerie f. de glycérine. / ~ soap ‖ Glyzerinseife f. ‖ savon m. à la glycérine. / ~ substitute ‖ Glyzerinersatz m. ‖ succédané m. de glycérine.

glycerol *see* glycerine.

glycocoll ‖ Leimzucker m. ‖ glycocolle m.; sucre m. de gélatine.

glycogen ‖ tierisches Stärkemehl n.; Glykogen n. ‖ glycogène m.

glycolic acid ‖ Glykolsäure f. ‖ acide m. glycolique.

glycose *see* glucose.

glycosid ‖ Glykosid n. ‖ glycoside f.

glyptography ‖ Steinschneidekunst f.; Glyptographie f. ‖ glyptographie f.

glyptotheca ‖ Skulpturensammlung f.; Glyptothek f. ‖ glyptothèque f.

gmelinite ‖ Gmelinit m. ‖ gmélinite f.

gneiss (Miner) ‖ Gneis m. ‖ gneiss m.; granit m. stratifié *ou* veiné. / amphibolic ~ ‖ Hornblendegneis m. ‖ gneiss m. amphibolique. / fibrous ~ ‖ Holzgneis m. ‖ gneiss m. fibreux. / granitic ~ ‖ granitischer Gneis m. ‖ gneiss m. granitoïde. / porphyroid ~ ‖ porphyrartiger Gneis m. ‖ gneiss m. porphyroïde.

gneiss granite ‖ Gneisgranit m. ‖ granitgneiss m. / ~ quarry ‖ Gneisbruch m. ‖ carrière f. de gneiss.

gnomon (Astro) ‖ Sonnenzeiger m.; Zeiger m. der Sonnenuhr; Gnomon m. ‖ gnomon m.

go, to ‖ gehen ‖ aller. / ~ (Mach) ‖ im Gange m. *oder* in Betrieb m. sein; gehen; laufen ‖ marcher; fonctionner; être en marche. / ~ (Nav) ‖ laufen ‖ courir. / to set going (Mach) ‖ in Gang m. setzen ‖ mettre en marche. / ~ astern (Nav) ‖ über Steuer gehen ‖ culer. / ~ backward (Nav) ‖ rückwärtsgehen ‖ aller en arrière. / ~ down (Ship) ‖ zugrunde gehen; untergehen ‖ couler bas; couler à fond. / ~ down (Water) ‖ fallen ‖ baisser; descendre. / ~ forward ‖ vorwärts gehen ‖ aller en avant. / ~-out (Fire) ‖ ausgehen; verlöschen ‖ s'éteindre. / ~ to pieces pl. ‖ auseinanderfallen ‖ en Stücke npl. gehen ‖ tomber en morceaux mpl. / ~ to sea ‖ in See gehen; in See stechen ‖ mettre *ou* prendre la mer. / ~ under ground (Mining) ‖ einfahren ‖ descendre; entrer.

goaf (Mining) ‖ alter Mann m.; abgebautes Grubenfeld n. ‖ vieux travaux mpl.; vieux ouvrages mpl.

go-and-return line ‖ Hin- und Rückleitung f. ‖ ligne f. d'aller et de retour.

go-astern motion ‖ Rückwärtsbewegung f. ‖ marche f. en arrière.

goat ‖ Ziege f. ‖ chèvre f. / ~ breeder ‖ Ziegenzüchter m. ‖ éleveur m. de chèvres. / ~ herd ‖ Ziegenhirt m. ‖ chevrier m.

goat's hair ‖ Ziegenhaar n. ‖ poil m. de chèvre. / ~ spinner ‖ Ziegenhaarspinner m. ‖ fileur m. de poils de chèvre.

goat's horn ‖ Ziegenhorn n. ‖ corne f. de chèvre. / meal of ~ clover ‖ Mehl n. aus Bockshornklee ‖ farine f. de fenugrec.

goatskin ‖ Ziegenleder n. ‖ cuir m. de chèvre. / ~ currier ‖ Ziegenledergerber m. ‖ corroyeur-chevrier m.

gob (Mine) *see* goaf.

gobbing (Mine) *see* gob stuff.

gobelin ‖ Gobelin m.; ‖ Gobelintapete f.; Webbildtapete f. ‖ gobelin m. / ~ carpet *see* gobelin. / ~ tapestry *see* gobelin. / ~ weaver ‖ Gobelinweber m. ‖ hautlissier m.

gobile rib knitter ‖ Rippenstrickerin f. ‖ côtière f. à la gobile.

goblet ‖ Kelch m.; Becher m.; Kelchglas n.; Humpen m. ‖ coupe f.; gobelet m. / glass ~ ‖ Becherglas n.; Kelchglas n. ‖ goblet m. en verre.

goblet works pl. ‖ Glaswarenfabrik f. ‖ gobeleterie f.

gob stuff (Mine) ‖ Versatz m. ‖ remblais mpl.

goethite *see* göthite.

gofer *see also* wafer ‖ Waffeloblate f. ‖ gaufre f. / ~ baker ‖ Waffelbäcker m. ‖ fabricant m. de gaufres.

goffer, to ‖ kräuseln; gaufrieren ‖ gaufrer.

goffer *see* gofer.

goffered paper ‖ gaufriertes *oder* krauses *oder* gekraustes Papier n. ‖ papier m. gaufré. / ~ plate ‖ Waffelblech n. ‖ tôle f. gaufrée.

goffering iron ‖ Waffeleisen n. ‖ gaufrier m.

goggles pl. ‖ Schutzbrille f.; Staubbrille f. ‖ lunettes fpl. protectrices; automobile ~ ‖ Kraftfahrerbrille f.; Autobrille f. ‖ lunettes fpl. d'automobiliste. / folding ~ ‖ zusammenklappbare Brille f. ‖ lunettes fpl. pliantes. / motor ~ *see* automobile ~. / protective ~ *see* goggles. / ~ for workmen ‖ Arbeiterschutzbrille f. ‖ lunettes fpl. protectrices pour ouvriers.

going-down (Exchange) ‖ Abschwächung f. ‖ perte f. au change.

gold ‖ Gold n. ‖ or m. / ~ alloyed ~ ‖ legiertes Gold n.; Karatgold n. ‖ or m. allié. / alluvial ~ ‖ Waschgold n. ‖ or m. alluvionnaire *ou* d'alluvion. / bar ~ ‖ Barrengold n. ‖ or m. en barres *ou* en lingots. / beaten ~ ‖ Blattgold n.; geschlagenes Gold n. ‖ or m. battu *ou* en feuilles; feuilles fpl. d'or. / ~ in blocks ‖ Gold n. in Blöcken ‖ or m. en barres fpl. *ou* ingots mpl. / burnish ~ ‖ Glanzgold n. ‖ or m. imité. / ~ of x carats ‖ Gold n. mit einem Feingewicht von x Karat ‖ or m. à x carats. / ~ in cast bars ‖ Gold n. in gegossenen Barren ‖ or m. en barres coulées. / colloidal ~ ‖ wasserlösliches *oder* kolloidales Gold n. ‖ or m. colloïdal. / dead ~ ‖ mattes Gold n. ‖ or m. mat. / Dutch ~ ‖ Flittergold n.; Rauschgold n. ‖ clinquant m.; oripeau m. / fine ~ ‖ Feingold n. ‖ or m. fin. / fulminating ~ ‖ Knallgold n.; Goldoxydammoniak n. ‖ or m. fulminant; or m. d'ammoniaque. / graphic ~ ‖ Schrifterz n.; Schrifttellur m.; Sylvanit m.; Weißtellur n. ‖ or m. graphique; sylvane m.; tellure m. natif auro-argentifère. / ~ in ingots *see* bar ~. / native ~ ‖ gediegenes Gold n. ‖ or m. natif *ou* vierge. / painter's ~ ‖ Malergold n.; Muschelgold n. ‖ or m. d'applique. / ~ in form of powder ‖

Gold n. in Pulverform ‖ or m. en poudre. / rolled ~ ‖ Golddoublé n. ‖ doublé-or m. / shell ~ *see* painter's ~. / water ~ *see* painter's ~.

gold amalgam (Met) ‖ Goldamalgam n. ‖ amalgame m. d'or.

gold-and-silver chaser ‖ Gold- und Silbertreiber m. ‖ repousseur m. en or et en argent. / ~ embroidery ‖ Gold- und Silberstickerei f. ‖ broderie f. d'or et d'argent. / ~ filigree ‖ Gold- und Silbergespinst n. ‖ filigrane m. en or et argent. / ~ smith ‖ Goldschmied m. ‖ bijoutier m.

gold articles pl. for churches ‖ Goldwaren fpl. für Kirchen ‖ orfèvrerie f. religieuse. / ~ articles pl. for table ‖ Tafelartikel mpl. aus Gold ‖ orfèvrerie f. de table. / ~ ash ‖ Goldasche f. ‖ cendre f. d'orfèvre. / ~ ash washer *see* ~ washer. / ~ balance ‖ Goldwage f. ‖ balance f. d'essayeur; biquet m.; trébuchet m. / ~ bath ‖ Goldbad n. ‖ bain m. de dorure.

gold-bearing (Miner) ‖ goldführend ‖ aurifère. / ~ quartz deposit ‖ Goldquarzlager n. ‖ gisement m. de quartz aurifère.

gold-beater ‖ Goldschläger m. ‖ batteur m. d'or.

gold-beater's book ‖ Goldschlägerbuch n. ‖ livret m. de batteurs d'or. / ~ gut *see* ~ skin. / ~ mould ‖ Goldschlägerform f. ‖ moule m. pour le battage d'or. / ~ skin ‖ Goldschlägerhaut f. ‖ baudruche f.; peau f. pour batteurs d'or. / hygienic articles of ~ skin ‖ hygienische Artikel pl. aus Goldschlägerhäutchen ‖ objets mpl. hygiéniques en baudruche. / ~ skin envelope ‖ Goldschlägerhauthülle f. ‖ enveloppe f. en baudruche.

gold beating ‖ Goldschlägerei f.; Schlagen n. der Goldblättchen ‖ battage m. d'or; batte f. des feuilles d'or. / ~ blocking ‖ Blattvergoldung f. ‖ dorure f. en feuilles. / ~ brocade ‖ Goldstoff m.; Goldbrokat m. ‖ drap m. d'or. / ~ bronze ‖ Goldbronze f. ‖ bronze m. d'or. / powdered ~ bronze ‖ Goldbronzepulver n. ‖ bronze m. d'or en poudre. / ~ burnisher ‖ Goldglätter m. ‖ brunisseur m. en or. / ~ button ‖ goldener Knopf m. ‖ bouton m. d'or. / ~ case (Watch) ‖ Goldgehäuse n. ‖ boîte f. d'or. / ~ chain ‖ Goldkette f. ‖ chaîne f. en or. / ~ chips pl. ‖ Goldabfall m. ‖ bractéole f.; rognure f. (de feuilles) d'or. / ~ chloride ‖ Chlorgold n. ‖ chlorure m. d'or. / ~ coin ‖ Goldmünze f.; Goldstück n. ‖ pièce f. *ou* monnaie f. d'or. / ~ colour ‖ Goldfarbe f. ‖ couleur f. d'or.

gold-colour, to ‖ goldfärben; golden färben ‖ mettre en couleur d'or.

gold-containing ‖ goldhaltig ‖ aurique.

gold digger ‖ Goldgräber m. ‖ bêcheur m. d'or. / ~ dust ‖ Goldstaub m. ‖ poussière f. d'or; grains mpl. d'or. / ~ embroidery ‖ Goldstickerei f. ‖ broderie f. d'or.

golden ‖ golden; aus Gold; goldig ‖ d'or; doré. / ~ ochre (Chem) ‖ Goldocker m. ‖ ocre m. doré. / ~ pen ‖ Goldfeder f. ‖ plume f. en or. / ~ seal root ‖ Hydrastiswurzel f. ‖ racine f. d'hydrastis. / ~ sulphide of antimony ‖ Goldschwefel m. ‖ sulfure m. doré d'antimoine.

gold engraver ‖ Goldgravör m. ‖ graveur m. sur or. / ~ field ‖ Goldfeld n. ‖ région f. aurifère. / ~ filigree ‖ Goldgespinst n. ‖ filigrane m. en or. / ~ foil ‖ Goldfolie f.; Blattgold n.; Goldbelag m. ‖ feuille f. d'or. / false ~ foil ‖ unechtes Blattgold n. ‖ or m. faux battu en feuilles.

/~ fountain pen ‖ Goldfüllfeder f. ‖ plume f. à réservoir en or. /~ frame ‖ Goldrahmen m. ‖ cadre m. doré *ou* en or. / ~ ingot ‖ Goldbarren m. ‖ barre f. d'or. / ~ ink ‖ Goldtinte f. ‖ encre f. d'or. / ~ jewellery ‖ goldene Schmuckwaren fpl. ‖ bijouterie f. *ou* orfèvrerie f. fine. / ~ lace ‖ Goldborte f. ‖ galon m. d'or.

gold leaf ‖ Blattgold n.; Goldblatt n. ‖ or m. en feuilles; feuille f. d'or. / false ~ ‖ unechtes Blattgold n. ‖ or m. faux battu en feuilles. / genuine ~ ‖ echtes Blattgold n. ‖ or m. pur en feuilles.

gold-leaf electrometer ‖ Goldblattelektrometer m. ‖ électromètre m. à feuilles d'or. / ~ electroscope ‖ Goldblattelektroskop n. ‖ électroscope m. à feuilles d'or. / ~ grinding machine ‖ Blattgoldanreibemaschine f. ‖ machine f. à frotte *ou* broyer l'or en feuilles. / ~ sweeping machine ‖ Blattgoldabkehrmaschine f. ‖ machine f. à brosser l'or en feuilles.

gold mine ‖ Goldgrube f.; Goldmine f.; Goldbergwerk n. ‖ mine f. d'or. / ~ money ‖ Goldgeld n. ‖ monnaie f. d'or. / ~ nugget ‖ Goldklumpen m. ‖ pépite f. *ou* tas m. d'or. / ~ ochre *see* golden ochre. / ~ ore ‖ Golderz n. ‖ minerai m. d'or. / amalgamating device for ~ ores ‖ Amalgamiereinrichtung f. für Golderze ‖ installation f. pour amalgamer les minerais d'or. / ~ paper ‖ Goldpapier n. ‖ papier m. doré. / ~ pen ‖ Goldfeder f. ‖ plume f. en or. / ~ peroxide ‖ Goldoxyd n. ‖ oxyde m. d'or. / ~ pin tooth (Dentistry) ‖ Goldknopfzahn m. ‖ dent f. à pointe en or. / ~ plated ‖ goldplattiert ‖ doublé *ou* plaqué d'or. / ~ plating ‖ Goldplattierung f. ‖ doublé m. *ou* plaqué m. d'or. / electro ~ plating ‖ galvanische Vergoldung f. ‖ dorure f. galvanique. / plant for ~ plating ‖ Vergoldungsanlage f. ‖ installation f. de dorure. / ~ printer (Paint) ‖ Goldschreiber m. ‖ chrysographe m. / ~ printing ‖ Golddruck m. ‖ impression f. en or. / ~ purple ‖ Goldpurpur m. ‖ pourpre m. de Cassius. / ~ purse ‖ goldener Geldbeutel m. ‖ bourse f. en or. / ~ purse maker ‖ Goldwirker m. ‖ fabricant m. d'objets en mailles d'or. / ~ recovery plant ‖ Goldgewinnungsanlage f. ‖ installation f. d'extraction de l'or. / electrolytical ~ recovery plant ‖ elektrolytische Goldgewinnungsanlage ‖ installation f. d'extraction électrolytique de l'or. / ~ refiner ‖ Goldscheider m. ‖ affineur m. d'or. / ~ refining ‖ Goldraffination f.; Goldscheidung f. ‖ raffinage m. *ou* affinage m. de l'or. / ~ roller ‖ Goldwalzer m. ‖ lamineur m. d'or. / ~ rolling mill ‖ Goldwalzwerk n. ‖ laminoir m. à or. / ~ salts pl. ‖ Goldsalze npl. ‖ sels mpl. d'or. / ~ scales pl. *see* ~ balance. / ~ size (Gild) ‖ Leimgrund m. ‖ encollage m.; or m. couleur. / ~ sizer (Bookb) ‖ Vergoldungsgrundierer m. ‖ assietteur m. / ~ size varnish ‖ Goldgrundfirnis m. ‖ vernis m. à fond d'or. / ~ smelter ‖ Goldgießer m. ‖ fondeur m. d'or.

goldsmith ‖ Goldschmied m. ‖ orfèvre m. / ~'s work ‖ Goldschmiedearbeit f. ‖ orfèvrerie f. / ~'s work polishing ‖ Goldwarenpolierung f. ‖ polissage m. en orfèvrerie.

gold spangle ‖ Goldflitter m. ‖ paillette f. en or. / ~ spectacles pl. ‖ Goldbrille f. ‖ lunettes fpl. d'or. / ~ stamping (Bookb) ‖ Goldprägung f. ‖ empreinte f. dorée. /

~ stamping works pl. ‖ Goldprägeanstalt f. ‖ établissement m. d'estampage d'or. / ~ stampman ‖ Golderzpocher m. ‖ bocardeur m. *ou* broyeur m. de minerai aurifère. / ~ standard ‖ Goldwährung f. ‖ étalon m. d'or. / ~ thread ‖ Goldfaden m. ‖ fil m. d'or. / ~ threads pl. ‖ Goldgespinst n. ‖ filet m. d'or; or m. filé. / ~ tissue ‖ Goldgewebe n. ‖ tissu m. d'or. / ~ trimming ‖ Goldposamenten pl. ‖ passementerie f. d'or. / ~ trioxide ‖ Goldoxyd n. ‖ oxyde m. d'or. / ~ turner ‖ Golddreher m. ‖ tourneur m. en or. / ~ varnish ‖ Goldlack m. ‖ vernis m. d'or. / to spread the ~ varnish ‖ den Goldfirnis m. aufstreichen ‖ étendre le vernis d'or. / **goldware** ‖ Goldwaren fpl. ‖ orfèvrerie f.; articles mpl. en or. / ~ setting ‖ Goldfassung f. ‖ sertissage m. en or. **gold washer** ‖ Goldwäscher m. ‖ orpailleur m.; laveur m. de sables aurifères. / ~ washing ‖ Goldwaschen n. ‖ lavage m. des sables aurifères; cueillette f. de l'or dans les sables des rivières. / ~ watch ‖ goldene Uhr f. ‖ montre f. d'or. / ~ weight ‖ Goldgewicht n. ‖ poids m. d'or. / ~ wire ‖ Golddraht m. ‖ fil m. *ou* trait m. en or; or m. trait. / ~ wire drawing works pl. ‖ Golddrahtzieherei f. ‖ fabrique f. de fils d'or.

golf coat ‖ Golfjacke f. ‖ golf m. / ~ ball ‖ Golfball m. ‖ balle f. golf. / ~ hose ‖ Sportstrumpf m. ‖ bas m. de sport.

gome ‖ Wagenschmiere f. ‖ cambouis m.

gompholite (Geol) ‖ Nagelfluhe f. ‖ poudingue m. calcaire *ou* polygénique.

gondola ‖ Gondel f. ‖ gondole f. / ~ car (Railw) ‖ offener Güterwagen m. ‖ wagon m. découvert *ou* ouvert.

goniometer ‖ Goniometer n.; Winkelmesser m. ‖ goniomètre m. / crystal ~ with two circles ‖ zweikreisiges Kristallgoniometer n. ‖ goniomètre m. à cristaux muni de deux cercles. / reflective ~ ‖ Reflexionsgoniometer n. ‖ goniomètre m. à réflexion.

goniometric ‖ goniometrisch ‖ goniométrique.

goniometry ‖ Goniometrie f. ‖ goniométrie f.

good ‖ gut ‖ bon. / ~ (Serviceable) ‖ tauglich ‖ qui peut servir; utile. / ~ (Met) ‖ gar; fein ‖ bon; fin. / to make ~ ‖ entschädigen ‖ dédommager; indemniser.

good middling coffee ‖ mittelguter Kaffee m. ‖ café m. bon marchand. / ~ quality ‖ gute Qualität f. ‖ bonne qualité f. / ~ quality through (Coal) ‖ Melierte f. ‖ tout-venant m. (bonne composition).

goods pl. ‖ Waren fpl.; Güter npl.; Erzeugnisse npl. ‖ marchandises fpl.; objets mpl.; articles mpl.; produits mpl. / ~ (Load of a ship etc.) ‖ Ladung f.; Fracht f.; Frachtgut n. ‖ charge f. / ~ are forwarded at the cost and risk of the consignee ‖ der Versand m. der Waren erfolgt auf Rechnung und Gefahr des Empfängers ‖ les marchandises fpl. voyagent aux frais, risques et périls du destinataire. / ~ to be conveyed ‖ Beförderungsgut n.; Fördergut n.; Transportgut n. ‖ matières fpl. à transporter. / double-sided ~ ‖ doppelseitige Ware f. ‖ article m. à double face. / ~ to be dryed ‖ Trocknungsgut n. ‖ matières fpl. à sécher. / duty-paid ~ ‖ verzollte Waren fpl. ‖ marchandises fpl. taxées *ou* dédouanées. / half-finished ~ ‖ Halbzeug

n. ‖ demiproduits mpl. / ~ of high value ‖ hochwertige Waren fpl. ‖ marchandises fpl. de haute valeur. / ~ laden in bulk (Mar) ‖ Schüttgüter npl.; Stürzgüter npl. ‖ marchandises fpl. en vrac. / perishable ~ ‖ leichtverderbliche Waren fpl. ‖ marchandises fpl. périssables. / ~ of a ready sale ‖ gangbare Ware f. ‖ marchandise f. courante. / ~ sold for cash ‖ gegen bar verkaufte Ware f. ‖ marchandise f. vendue au comptant. / ~ to be transported *see* ~ to be conveyed.

goods-carrying ‖ frachtbefördernd ‖ transportant des marchandises. / ~ aeroplane ‖ Frachtflugzeug n. ‖ avion m. de fret. / ~ flying boat ‖ Frachtflugboot n. ‖ hydravion m. à coque de fret.

goods conveyance on roads ‖ Spedition f.; Frachtbeförderung f. auf dem Landwege ‖ transport m. de marchandises par voie de terre. / group of ~ ‖ Warengruppe f. ‖ catégorie f. de marchandises. / kind of ~ ‖ Warengattung f. ‖ nature f. des marchandises. / ~ lift ‖ Lastenaufzug m. ‖ monte-charges m. / list of ~ ‖ Warenverzeichnis n. ‖ tableau m. *ou* spécification des marchandises. / ~ locomotive ‖ Güterzuglokomotive f. ‖ locomotive f. de train à marchandises. / ~ platform ‖ Güterladeplatz m. ‖ quai m. à marchandises. / ~ shed ‖ Güterschuppen m. ‖ hangar m. à marchandises. / ~ station ‖ Güterbahnhof m. ‖ gare f. des marchandises. / ~ tank locomotive ‖ Güterzugtenderlokomotive f. ‖ locomotive f. tender de train à marchandises. / ~ traffic ‖ Güterverkehr m.; Frachtverkehr m. ‖ trafic m. des marchandises *ou* d'expédition; roulage m. / ~ train ‖ Güterzug m. ‖ train m. de marchandises. / ~ transport ‖ Güterbeförderung f.; Gütertransport m. ‖ transport m. des marchandises.

goods wagon ‖ Güterwagen m. ‖ wagon m. à marchandises. / covered ~ ‖ bedeckter Güterwagen m. ‖ wagon m. à marchandises couvert *ou* fermé. / open ~ ‖ offener Güterwagen m. ‖ wagon m. à marchandises découvert.

goodsyard ‖ Lagerplatz m. ‖ dépot m. de matériel.

goom *see* gome.

goose ‖ Gans f. ‖ oie f. / ~ (Tail) ‖ Bügeleisen n.; Plätteisen n. ‖ carreau m.; fer m. à repasser.

gooseberry ‖ Stachelbeere f. ‖ groseille f. verte. / ~-bush growing ‖ Stachelbeerkultur f. ‖ culture f. de groseilles à maqueraux. / ~ wine ‖ Stachelbeerwein m. ‖ vin m. de groseilles vertes.

goose-breast ‖ Gänsebrust f. ‖ poitrine f. d'oie. / ~ smoker ‖ Gänsebrusträucherer m. ‖ fumeur m. de poitrines d'oies.

goose-dung ore ‖ Ganomatit m.; Gänsekötigerz n. ‖ argent-merde m. d'oie; ganomatite f.

goose fat ‖ Gänsefett n. ‖ graisse f. d'oie.

goose-liver pie ‖ Gänseleberpastete f. ‖ pâté m. de foie gras.

gorge (Groove) ‖ Kehle f.; Rille f.; Rinne f.; Einschnitt m. ‖ cannelure f.; gorge f. / ~ (Ravine) ‖ Schlucht f.; Hohlweg m. ‖ ravin m.; gorge f.

goslarite ‖ Gallitzenstein m.; Goslarit m.; Zinkvitriol n. ‖ goslarite f.; zinc m. sulfaté.

gossan (Miner) ‖ eiserner Hut m. ‖ chapeau m. ferrugineux.

Gothic letter ‖ Grotesk f.; Gotisch f. ‖ caractère m. gothique. / ~ type see ~ letter.
göthite (Miner) ‖ Nadeleisenerz n.; Rubinglimmer m.; Eisenglimmer m.; Goethit m. ‖ gœthite f.; fer m. pourpré.
gouch ‖ Quecke f. ‖ chiendent m. / medicinal ~ ‖ Quecke f. zum Heilgebrauch ‖ chiendent m. officinal.
gouge (Join) ‖ Hohlbeitel m.; Hohleisen n.; Gutsche f.; Hohlmeißel m. ‖ gouge f. / ~ (Turn) see turning ~. / bent ~ ‖ gebogene Gutsche f. ‖ gouge f. coudée. / bent-neck ~ ‖ Rabenschnabel m.; gebogenes oder gekröpftes Hohleisen n. ‖ bec m. de corbin; gouge f. à bec de corbin. / crooked ~ see bent-neck ~. / dented ~ ‖ Gradiergutsche f. ‖ gouge f. gradine. / flat ~ ‖ flache Gutsche f. ‖ gouge f. plate. / fluting ~ ‖ hohle Gutsche f. ‖ gouge f. creuse. / middle ~ ‖ halbflache Gutsche f. ‖ gouge f. demi-plate. / print-cutter's ~ ‖ Graviergutsche f. ‖ gouge f. de graveurs. / round ~ ‖ rundes Hohleisen n. ‖ gouge f. ronde. / square ~ ‖ viereckiges Hohleisen n.; Viereisen n. ‖ gouge f. carrée. / turning ~ ‖ Schrotmeißel m.; Hohlmeißel m.; Röhrenmeißel m. ‖ gouge f. (à ébaucher). / veining ~ ‖ Kanneliergutsche f. ‖ gouge f. à bretter.
gouge bit ‖ Hohlbohrer m. ‖ mèche-cuiller f.; évidoir m.
goulet (Mar) ‖ enge Einfahrt f. ‖ goulet m.
gourd ‖ Kürbis m. ‖ courge f.; citrouille f. / ~ pip ‖ Kürbiskern m. ‖ pépin m. de citrouille.
govern, to (Mach) ‖ regeln ‖ régler. / ~ (Nav) ‖ steuern; lenken ‖ gouverner; diriger.
governing (Mach) ‖ Reg(e)lung f. ‖ réglage m. / perfect ~ ‖ einwandfreie Reglung f. ‖ réglage m. parfait.
Government ‖ Regierung f.; Regierungsbehörde f. ‖ gouvernement m. / ~ bond ‖ Staatsschuldverschreibung f. ‖ obligation f. d'État. / ~ grant ‖ Staatssubvention f.; staatliche Beihilfe f. ‖ subvention f. d'État. / ~ loan ‖ Staatsanleihe f. ‖ emprunt m. public. / ~ official ‖ staatlicher Beamter m.; Staatsbeamter m. ‖ fonctionnaire m. ou employé m. de l'État. / ~ regulation ‖ Regierungsverordnung f. ‖ ordonnance f. du gouvernement. / ~ security ‖ Staatspapier n. ‖ fonds m. public. / ~ stock ‖ staatliche Anleihepapiere npl.; Staatspapiere npl. ‖ titres mpl. d'emprunt public; effets mpl. publics. / ~ tax ‖ Staatsabgabe f.; impôt m. public. / ~ telegram ‖ Staatstelegramm n. ‖ télégramme m. d'État.
governor (Mach) ‖ Regler m.; Regulator m.; Moderator m. ‖ régulateur m.; modérateur m. / ~ (Tel) ‖ Richtungsmagnet n. ‖ aimant m. correcteur ou directeur. / centrifugal ~ (Mach) ‖ Fliehkraftregler m.; Zentrifugalregler m. ‖ régulateur m. centrifuge. / hydraulic ~ ‖ hydraulischer Regler m. ‖ régulateur m. hydraulique. / parabolic ~ ‖ parabolischer Regler m. ‖ régulateur m. parabolique. / pressure ~ ‖ Druckregler m. ‖ régulateur m. de pression. / speed ~ ‖ Geschwindigkeitsregler m. ‖ régulateur m. de vitesse.
governor and brake ‖ Brems- und Reguliervorrichtung f. ‖ dispositif m. de freinage et de régulation.
governor ball ‖ Reglergewicht n. ‖ boule f. de régulateur. / ~ fly ‖ Windfang m. ‖ ailette f. de régulateur. / ~ housing ‖ Reglergehäuse n. ‖ cage f. de régulateur. / ~

lever ‖ Reglerhebel m. ‖ levier m. de réglage ou de régulateur. / ~ pin (Mot) ‖ Regelspindel f. ‖ pointeau m. de réglage. / ~ rod ‖ Regelgestänge n. ‖ biellette f. de réglage; tige f. du régulateur. / ~ sleeve ‖ Reglermuffe f. ‖ manchon m. de régulateur. / ~ spring ‖ Reglerfeder f. ‖ ressort m. de régulateur. / ~ valve ‖ Regelventil n. ‖ soupape f. de réglage. / ~ valve case ‖ Regelventileinsatz m. ‖ boîte f. de soupape de réglage. / ~ weight (Mot) ‖ Reglergewicht n. ‖ masses fpl. centrifuges. / ~ wheel ‖ Reglerrad n. ‖ roue f. de régulateur.
grab (Dredger) ‖ Baggergreifer m.; Greifer m.; Greifer m. ‖ grappin m.; pelle f. (de chargement); benne f. (prenante). / the ~ works at an angle ‖ der Greifer m. setzt sich beim Graben schief auf den Boden ‖ le grappin m. se pose en biais en attaquant le sol. / automatic ~ ‖ Selbstgreifer m.; Greifkübel m. ‖ pelle f. automatique; benne f. à griffe. / ~ for bulk goods ‖ Greifer m. für Massengüter ‖ grue f. à griffes pour marchandises en vrac. / coal ~ ‖ Kohlengreifer m. ‖ grappin m. à charbon. / ~ loaded with spoil ‖ mit Baggergut gefüllter Greifer m. ‖ grappin m. rempli de matériel dragué. / loading ~ ‖ Verladegreifer m. ‖ benne f. ou pelle f. de chargement. / motor ~ ‖ Motorgreifer m. ‖ benne f. preneuse ou grappin m. à moteur. / polyp ~ ‖ Polypgreifer m. ‖ grappin m. polype. / ~ for round timber ‖ Greifer m. für Rundholz ‖ griffe f. pour rondin. / ~ for stones ‖ Steingreifer m. ‖ benne f. à ramasser les pierres et les cailloux.
grab capacity ‖ Greiferinhalt m. ‖ capacité f. du grappin. / ~ crane ‖ Greiferkran m. ‖ grue f. à benne (piocheuse). / ~ dredger ‖ Greifbagger m. ‖ grappin m.
graben (Geol) ‖ Graben m. ‖ renforçage m.
grab head ‖ Greiferkopf m. ‖ tête f. de benne. / ~ hook ‖ Greiferhaken m. ‖ crochet m. du grappin. / reach of ~ ‖ Grabweite f. ‖ portée f. de fouille. / winch for the ~ ‖ Greiferwinde f. ‖ treuil m. du grappin.
grade, to ‖ staffeln; abstufen ‖ échelonner; graduer. / ~ (To sort) ‖ sichten; klassieren ‖ classer.
grade (Degree of quality) ‖ Gehalt m.; Gütegrad m.; Güte f.; Qualität f. ‖ teneur f.; titre m.; classe f.; qualité f. / ~ (Gradient) see gradient. / down ~ ‖ Gefälle n. ‖ pente f.; inclinaison f. / ~ of fertility (Agr) ‖ Standortsgüte f.; Fruchtbarkeitsgrad m. (des Bodens) ‖ classe f. de fertilité. / high ~ ‖ gute Qualität f. ‖ bonne qualité f. / higher ~ ‖ bessere Qualität f. ‖ qualité f. supérieure. / low ~ ‖ geringe Qualität f. ‖ qualité f. inférieure. / lower ~ ‖ geringere Qualität f.; geringere Güte f. ‖ qualité f. plus inférieure. / medium ~ ‖ mittlere Sorte f.; mittlere Qualität f. ‖ qualité f. moyenne. / ~ of steel ‖ Stahlmarke f. ‖ marque f. d'acier. / up ~ ‖ Steigung f. ‖ montée f.; rampe f.; inclinaison f. / usual ~ ‖ gewöhnliche oder gebräuchliche Sorte f.; übliche Qualität f. ‖ qualité f. usuelle.
graded ‖ sortiert ‖ classé; trié.
gradient ‖ geneigte Fläche f.; schiefe Ebene f.; Gradient m. ‖ plan m. incliné; gradient m. / ~ (Rising) ‖ Steigung f.; Steigungsverhältnis n. ‖ rampe f.; inclinaison f.; montée f. / ~ (Falling) ‖ Gefälle n.; Neigung f.; Neigungsverhältnis n. ‖

pente f.; déclivité. / easy ~ ‖ schwache Steigung f. ‖ pente f. faible. / greatest ~ (Railw) ‖ Maximalgefälle n.; größtes Gefälle n. ‖ maximum m. des pentes. / ~ of the line (Railw) ‖ Neigung f. der Strecke ‖ déclivité f. de la voie. / ~ of a road ‖ Steigung f. einer Straße ‖ montée f. d'une route. / ~ of a slope ‖ Böschungswinkel m. ‖ inclinaison f. d'un talus. / steep ~ ‖ starke Steigung f. ‖ pente f. forte.
gradient indicator ‖ Steigungsmesser m. ‖ clinomètre m.; indicateur m. de pente. / ~ with water tube ‖ Steigungsmesser m. mit Wasserfüllung. ‖ clinomètre m. à tube d'eau.
gradient meter see gradient indicator. / percentage of ~ ‖ Steigungsverhältnis n. in Vomhundert ‖ taux m. de la rampe en %. / ~ post (Railw) ‖ Neigungszeiger m. ‖ indicateur m. de déclivité.
grading (Bringing to a level) ‖ Planierung f.; Einebnung f. ‖ régalage m.; nivellement m. / ~ (Tel) ‖ Staffelung f. ‖ échelonnage m. / ~ of lines ‖ Staffeln n. von Leitungen ‖ échellonement m. des lignes multiples.
grading machine ‖ Sortiermaschine f. ‖ trieuse f.
gradual ‖ stufenweise ‖ raduel; echelonné.
gradually ‖ allmählich; schrittweise; stufenweise; nach und nach ‖ peu à peu; graduellement.
gradual reduction (Mill) ‖ Stufenauflösung f. ‖ réduction f. graduelle.
graduate, to ‖ abstufen ‖ graduer. / ~ (in degrees) ‖ in Grade mpl. einteilen; mit Meßeinteilung f. versehen ‖ graduer; diviser en degrés mpl.; jauger. / ~ (Salt) ‖ (die Sole) gradieren ‖ graduer.
graduated arc ‖ Skalebogen m. ‖ arc m. gradué; échelle f. / ~ cylinder ‖ Meßzylinder m. ‖ éprouvette f. graduée. / ~ division ‖ Gradteilung f.; Gradeinteilung f. ‖ division f. en degrés. / ~ drum ‖ Meßtrommel f. ‖ tambour m. divisé. / ~ flask ‖ Meßflasche f.; Meßkolben m. ‖ fiole f. jaugée; ballon m. jaugé. / ~ instruments pl. (of glass or porcelain) ‖ Mensuren fpl. ‖ verreries fpl. ou porcelaines fpl. graduées. / ~ tariff ‖ Staffeltarif m. ‖ tarif m. échelonné. / ~ tube ‖ Meßrohr n.; Meßröhre f. ‖ tube m. gradué. / ~ vessel ‖ Meßgefäß n. ‖ vase m. jaugé.
graduating drum ‖ Meßwalze f. ‖ tambour m. de mesure. / ~ house see graduation house. / ~ machine ‖ Maßeinteilungsmaschine f.; Mengeneinteilungsmaschine f. ‖ machine f. de dosage. / ~ stem guide ‖ Graduierstangenführung f. ‖ bouchon m. guide de la tige de graduation. / ~ valve ‖ Abstufungsventil n. ‖ valve f. de graduation.
graduation ‖ Gradeinteilung f.; Teilung f. ‖ graduation f. (de l'échelle). / to figure a ~ ‖ eine Teilung f. beziffern ‖ porter des chiffres mpl. sur une graduation. / figuring of the ~ ‖ Bezifferung f. der Teilung ‖ chiffrage m. de la graduation. / coarse ~ ‖ grobe Einteilung f. ‖ graduation f. grossière. / fine ~ ‖ feine Teilung f. ‖ graduation f. fine. / ~ on scale ‖ Teilstrich m. ‖ marque f. de subdivision.
graduation house (Salt) ‖ Gradierwerk n. ‖ bâtiment m. de graduation. / ~ pan (Salt) ‖ Gradierpfanne f. ‖ poêle f. de graduation. / ~ photometer ‖ Stufenfotometer n. ‖ photomètre m. graduel.

graft, to (Garden) ‖ pfropfen ‖ enter; greffer. / ~-up ‖ **aufpropfen** ‖ enter.

graft ‖ Pfropfreis n. ‖ greffe f.

grafter ‖ Pfropfer m. ‖ greffeur m.

grafting-knife ‖ Pfropfmesser n.; Okuliermesser n. ‖ entoir m.; greffoir m.; écussonnoir m. / foldable ~ ‖ zusammenklappbares Okuliermesser n. ‖ greffoir m. fermant.

grafting saw ‖ Pfropfsäge f.; Gärtnersäge f. ‖ scie f. à enter ou à greffer; scie f. de jardinier. / ~ wax ‖ Baumwachs n. ‖ cire f. à greffer.

grain, to ‖ körnen ‖ grener. / ~ (Curr) ‖ narben; krispeln ‖ greneler; grener; donner les croisés du grain; rebrousser; crêpir.

grain (Particle) ‖ Korn n. ‖ grain m. / ~ (Botan) ‖ Samenkorn n.; Samen m. ‖ graine f.; semence f. / ~ (Cereals) ‖ Getreide n. ‖ blé m.; céréales f pl.; grains pl. / ~ (Fibre) ‖ Faser f. ‖ fibre f. / ~ (Fibre of wood) ‖ Längsfaser f. ‖ fil m. / ~ (Degree of coarseness or texture) ‖ Körnung f.; Kornart f.; grainure f.; grain m. / ~ (Troy weight) ‖ Gran n.; Korn n. ‖ grain m. / ~ (Mill) ‖ Mahlgut n. ‖ mouture f. / ~ (Pott) ‖ Krebs m. (im Ton) ‖ féramine f.

grain, across the ‖ quer zur Faser f. ‖ en travers aux fibres f pl.; perpendiculairement aux fibres. / ~ against the ~ ‖ gegen die Faser ‖ contre la fibre. / artificial ~ of morocco leather ‖ künstliche Narbe f. des Maroquins ‖ croisé m. du grain du maroquin. / bruised ~ ‖ Getreideschrot n.; geschrotetes Korn n. ‖ blé m. concassé. / cleaned ~ ‖ gereinigtes Getreide n. ‖ blé m. nettoyé. / coarse ~ ‖ grobes Korn n.; grobe Körnung f. ‖ gros grain m. / cross the ~ see across the ~. / ~ of culinary salt ‖ Salzkorn n. ‖ grain m. de sel. / to cut-off the ~ ‖ das Fell abnarben ‖ effleurer la peau. / ~ of dust ‖ Staubkorn n. ‖ grain m. de poussière. / hard ~ (Agr) ‖ harte Frucht f. ‖ blé m. dur. / hard ~ (In the marble) ‖ Kern m.; Krebs m. ‖ durillon m. / ~ with long beard ‖ Getreide n. mit langen Spelzen ‖ grains mpl. à longues glumes. / malted ~ ‖ gemälztes Getreide n. ‖ grain m. malté. / medium ~ ‖ mittleres Korn n.; mittlere Körnung f. ‖ grain m. moyen. / ~ of morocco leather ‖ Narbe f. des Saffians oder Maroquins ‖ grain m. du maroquin. / in the ~ mortar ‖ Kalkkern m.; Kalkkrumpe f. ‖ lopin m.; écrevisse f.; rigaud m. / ~ of sand ‖ Sandkorn n. ‖ grain m. de sable. / smooth ~ ‖ glattes Korn n.; glatte Körnung f. ‖ grain m. lisse. / soft ~ ‖ weiche Frucht f. ‖ blé m. tendre. / spent ~s pl. ‖ Treber pl. ‖ drèches f pl. / ~ of a stone ‖ Gefüge n. oder Struktur f. eines Steines ‖ structure f. ou grainure f. ou texture f. d'une pierre. / with the ~ ‖ mit der Faser ‖ dans le sens des fibres; en long. / ~ without beard ‖ Getreide n. ohne Spelzen ‖ grains mpl. sans glumes. / ~ of wood ‖ Faserung f. oder Maserung f. des Holzes; Holzfaser f.; Längsfaser f. im Holz ‖ fil m. de bois.

grain alcohol ‖ Gärungsalkohol m. ‖ alcool m. de fermentation. / ~ balance ‖ Getreidewage f. ‖ balance f. pèse-grains; balance f. à grains. / ~ binder ‖ Bindemäher m. ‖ moissonneuse-lieuse f. / ~ cleaning machine ‖ Getreidereinigungsmaschine f.; Triör m.; Getreideputzmaschine f. ‖ machine f. à nettoyer le blé; trieur m. / coarse bruising of ~ ‖ Grobschroten n. von Getreide ‖ concassage m. grossier de grains. / condition of the ~ ‖ Getreidebeschaffenheit f. ‖ composition f. des grains. / ~ counter ‖ Kornzähler m.; Granometer n. ‖ compte-grains m.; granomètre m. / ~ counting apparatus see ~ counter. / ~ crusher ‖ Schrotmühle f. ‖ concasseur m. de grains. / ~ drill ‖ Sämaschine f.; Drillmaschine f. ‖ semoir m. / ~ drier ‖ Korntrockner m. ‖ sécheur m. de grains. / ~ drying apparatus ‖ Trebertrockenvorrichtung f. ‖ appareil m. à sécher les drèches.

grained ‖ körnig ‖ granulaire; grenu. / close-~ (Met) ‖ feinkörnig ‖ à grain fin; finement granulé. / coarse-~ (Wood) ‖ grobfaserig ‖ à fibre f. grossière. / compact-~ see close-~. / end-~ wood ‖ Stirnholz n.; Querholz n.; Hirnholz n. ‖ bois m. de bout. / fine-~ (Met) see close-~. / fine-~ (Wood) ‖ feinfaserig ‖ à fibre f. fine.

grain elevator ‖ Getreideheber m.; Elevator m. ‖ élévateur m. à grains. / floating ~ ‖ schwimmender Getreideheber m. ‖ élévateur m. flottant à grains.

grainer of wood ‖ Holzfaserimitator m. ‖ graineur m. sur bois.

grain, finish-grinding of ‖ Feinmahlen n. von Getreide ‖ mouture f. fine de grains.

graining board (Curr) ‖ Krispelholz n. ‖ paumelle f.; pommelle f.; marguerite f. à la main; grènetoir m. / ~ machine (Curr) ‖ Krispelmaschine f. ‖ machine f. à rebrousser; marguerite f. mécanique. / ~ sand ‖ Kornsand m. ‖ sable m. à grainer.

grain lead ‖ Probierblei n.; Kornblei n. ‖ plomb m. d'essai; plomb m. en grains. / nature of ~ ‖ Kornbeschaffenheit f.; Körnungsbeschaffenheit f. ‖ nature f. du grain. / ~ reel ‖ Getreidehaspel f. ‖ treuil m. de blé.

grains pl. (Agr) ‖ Samen m.; Saat f. ‖ graines f pl.; semence f. / ~ (Silk) ‖ Grains pl.; Eier npl. der Seidenraupe ‖ grains mpl.; graines f pl. / spent ~ ‖ Treber pl. ‖ drèches f pl.

grain sampler ‖ Getreidestecher m. ‖ sonde f. à grain. / ~ saucing machine ‖ Getreidebeizmaschine f. ‖ machine f. à corroder le blé. / ~ scale see ~ balance. / ~ side (Curr) ‖ Narbenseite f.; Haarseite f. ‖ fleur f. ou grain m. de la peau. / ~ sieve ‖ Körnersieb n. ‖ tamis m. à grains. / ~ silo ‖ Getreidesilo m. ‖ silo m. à céréales. / ~ tester ‖ Getreideprober m. ‖ coupe-grains m. / ~ tin ‖ Zinnpulver n.; Körnerzinn n.; Kornzinn n. ‖ sable m. stannifère; étain m. en larmes ou en grains. / ~ warehouse ‖ Getreidelagerhaus n.; Getreidespeicher m. ‖ grenier m. ou silo m. à blé. / ~ wood ‖ Langholz n.; Aderholz n. ‖ bois m. en grume; bois m. de fil.

grainy ‖ körnig ‖ granulaire; grenu. / ~ (Glass) ‖ sandig; höckerig ‖ verre m. plein de pontis; verre m. avec pontis.

gram see also gramme ‖ Gramm n. ‖ gramme f.

graminæ pl. ‖ Gräser npl. ‖ graminées f pl.

gramineal plants pl. see graminæ.

grammatite ‖ Grammatit m.; Tremolit m. ‖ grammatite f.; trémolite f.

gramme ‖ Gramm n. ‖ gramme m. / ~ atome ‖ Grammatom n. ‖ atome-gramme m. / ~ calory ‖ Grammkalorie f.; kleine Kalorie f. ‖ petite calorie f. / ~ molecule ‖ Grammolekel n. ‖ molécule-gramme f.

gramophone ‖ Grammofon n.; Sprechapparat m. ‖ gramophone m. / ~ needle ‖ Grammofonnadel f. ‖ aiguille f. de gramophone. / ~ record ‖ Schallplatte f.; Grammofonplatte f. ‖ disque m. de gramophone. / ~ shell see ~ record.

granary (Agr) ‖ Kornspeicher m.; Kornboden m.; Getreidespeicher m. ‖ grenier m. (à blé). / ~ (Trade) ‖ Getreidelagerhaus n.; Getreidespeicher m. ‖ silo m. à blé. / ~ spray and hopper ‖ Rieselspeicher m. ‖ magasin m. à ruissellement.

grand-front (Build) ‖ Hauptfront f. ‖ façade f. principale.

grand-piano ‖ Flügel m. ‖ piano m. à queue.

grange ‖ Meierei f.; Pachtgut n. ‖ métairie f.; ferme f.

granillo ‖ Grenadillholz n.; rotes Ebenholz n. ‖ grenadille f.

granite ‖ Granit m. ‖ granit m. / Egyptian ~ see red ~. / graphic ~ ‖ Pegmatit m.; Schriftgranit m. ‖ pegmatite f.; granit m. graphique. / fine-grained ~ ‖ feinkörniger Granit m. ‖ granitelle f. / gray ~ ‖ grauer Granit m. ‖ granit m. gris. / porphyritic ~ see gray ~. / red ~ ‖ roter oder ägyptischer Granit m. ‖ granit m. rouge ou égyptien. / small ~ ‖ Kleingranit m. ‖ petit granit m.

granite axe ‖ Granithaue f. ‖ pioche f. à granit. / ~ concrete ‖ Granitbeton m. ‖ béton m. de granit. / ~ cutting ‖ Granithauen n. ‖ piquage m. de granit. / ~ feller ‖ Granithauer m. ‖ piqueur m. de granit. / ~ quarry ‖ Granitbruch m. ‖ carrière f. de granit. / ~ sand ‖ Granitsand m. ‖ sable m. granitique. / ~ ware ‖ Granitwaren f pl. ‖ objets mpl. en granit.

granitic ‖ granitartig ‖ granité.

granometer ‖ Kornzähler m.; Granometer n. ‖ compte-grains m.; granomètre m.

grant, to ‖ bewilligen ‖ accorder; concéder. / ~ mining rights for a mineral ‖ ein Mineral n. verleihen ‖ concéder le droit à l'extraction d'un minéral. / ~ a prior right ‖ ein Vorzugsrecht n. einräumen ‖ accorder un droit de préférence.

grant ‖ Bewilligung f.; Verleihung f. ‖ concession f.; vote m. / ~ (Mine) ‖ Schürfschein m. ‖ enseignement m. / to apply for the ~ of a concession (Mine) ‖ um Verleihung f. eines Feldes nachsuchen ‖ solliciter la concession d'un gîte minier. / document conveying the ~ ‖ Verleihungsurkunde f. ‖ acte m. de concession. / to modify the conveyance of the ~ ‖ die Verleihungsurkunde f. ändern ‖ modifier l'acte m. de concession. / to oppose a ~ ‖ gegen die Verleihung f. einsprechen ‖ faire opposition à la concession. / procedure to obtain a ~ ‖ Verleihungsverfahren n. ‖ procédure f. de concession. / to put an end to a ~ ‖ eine Verleihung f. aufheben ‖ retirer une concession. / to refuse the ~ ‖ die Verleihung f. versagen ‖ refuser la concession. / ~ of respite ‖ Fristgewährung f. ‖ délai m.; sursis m.; prorogation f.

granted (Patent) ‖ erteilt ‖ accordé.

granting see also grant. / ~ of railway ‖ Eisenbahnverleihung f. ‖ concession f. de chemin de fer.

granular ‖ körnig ‖ granuleux; granulaire; granulé; grenu. / ~ fracture (Met) ‖ körniger Bruch m. ‖ fracture f. granulé. / ~ ore ‖ Graupenerz n. ‖ mine f. en grains. / ~ structure (Met) ‖ körniges Gefüge n. ‖ texture f. *ou* structure f. granulée. / ~ texture *see* ~ structure. / ~ transmitter (Tel) ‖ Körnermikrofon n. ‖ microphone m. à granules sphériques.
granulate, to ‖ körnen; granulieren ‖ granuler; grainer; réduire en grains.
granulated *see* granular.
granulating crusher ‖ Granuliermühle f. ‖ broyeur-granulateur m. / ~ hammer ‖ Stockhammer m. ‖ boucharde f.; marteau m. à granuler. / ~ machine ‖ Körnmaschine f.; Granuliermaschine f. ‖ machine f. à granuler; grenoir m. / ~ plant for blast furnace slag ‖ Granulierungsanlage f. für Hochofenschlacke ‖ installation f. à granuler le laitier de haut-fourneau.
granulation ‖ Körnung f.; Granulierung f.; Körnen n. ‖ granulation f.; grainage m. / ~ of sugar ‖ Kristallisation f. *oder* Kornbildung f. des Zuckers ‖ cristallisation f. du sucre. / of varying ~s ‖ von verschiedener Körnung f. ‖ de grains mpl. de grosseur différente. / size of the ~s ‖ Korngröße f. ‖ grosseur f. des grains.
granulator ‖ Granulator m. ‖ granulateur m. / fine ~ ‖ feines Schrotsieb n. ‖ guillaume m. en fin. / first ~ ‖ Schrotsieb n. ‖ guillaume m.
granulite ‖ Granulit m.; Weißstein m. ‖ granulite f.
granulose ‖ Granulose f. ‖ granulose f.
grape ‖ Weintraube f. ‖ raisin m. / crushed ~ ‖ zerquetschte Weintraube f. ‖ raisin m. écrasé.
grape crusher ‖ Traubenquetsche f. ‖ égrugeoir m. pour égruger les grappes. / ~ fruit ‖ Pampelmuse f. ‖ pamplemousse f. / ~ gatherer ‖ Winzer m. ‖ vendangeur m. / ~ husks pl. ‖ Weintreber pl. ‖ marc m. du vin.
grape juice ‖ Traubensaft m. ‖ jus m. de raisin. / unfermented ~ ‖ ungegorener Traubensaft m. ‖ jus m. de raisin non fermenté.
grape mill ‖ Traubenquetsche f. ‖ pressoir m. à raisins. / ~ picker *see* ~ gatherer. / ~ pip ‖ Weintraubenkern m.; Traubenkern m. ‖ pépin m. de raisin. / ~ press ‖ Traubenpresse f. ‖ pressoir m. à raisins. / sack for ~s ‖ Traubensack m. ‖ sac m. à raisins. / ~ shot (War mat) ‖ Kartätsche f. ‖ boîte f. à mitraille. / ~ shovel ‖ Traubenschaufel f. ‖ pelle f. à raisins. / ~ sugar ‖ Traubenzucker m. ‖ sucre m. de raisin; glucose f. / ~ vine ‖ Weinrebe f. ‖ cep m. de vigne.
graphic *see* graphical.
graphical ‖ grafisch ‖ graphique. / ~ art institute ‖ grafische Kunstanstalt f. ‖ institut m. d'art graphique. / ~ calculation ‖ grafische Berechnung f. ‖ calcul m. graphique. / ~ establishment ‖ grafische Anstalt f. ‖ atelier m. d'art graphique. / ~ industry ‖ grafsiches Gewerbe n. ‖ art m. graphique. / ~ method ‖ grafisches Verfahren n. ‖ procédé m. graphique. / ~ product ‖ grafisches Erzeugnis n. ‖ produit m. d'art graphique. / ~ range table (Arm) ‖ grafische Schußtafel f. ‖ table f. graphique de tir. / ~ representation ‖ grafische Darstellung f. ‖ tracé m. graphique. / ~ solution ‖ gra-

fische Auswertung f. ‖ évaluation f. *ou* solution f. graphique. / ~ statics pl. ‖ Grafostatik f. ‖ statique f. graphique.
graphically determined ‖ zeichnerisch ermittelt ‖ déterminé graphiquement. / ~ represented ‖ grafisch dargestellt ‖ tracé graphiquement.
graphite ‖ Graphit m.; Pottlot n.; Reißblei n. ‖ graphite m.; plombagine f. / ~ (Met) ‖ Temperkohle f. ‖ graphite m.; carbone m. de recuit. / artificial ~ ‖ künstlicher Graphit m. ‖ graphite m. artificiel. / deflocculated ~ ‖ entflockter Graphit m. ‖ graphite m. défloculé. / flaked ~ ‖ Flockengraphit m. ‖ graphite m. en flocons. / ~ in rods ‖ Graphit m. in Stäbchen ‖ graphite m. en baguettes.
graphite brick ‖ Graphitstein m. ‖ brique f. en graphite. / ~ coating apparatus ‖ Graphitiervorrichtung f. ‖ appareil m. de graphitage. / ~ coating machine ‖ Graphitiermaschine f. ‖ graphiteuse f.; machine f. à graphiter. / ~ crucible ‖ Graphit(schmelz)tiegel m. ‖ creuset m. (de fusion) en graphite.
graphited ‖ graphitiert ‖ graphité. / ~ cord ‖ Graphitschnur f. ‖ cordon m. graphité.
graphite electrode ‖ Graphitelektrode f. ‖ électrode f. en graphite. / ~ flake ‖ Graphitschuppe f. ‖ écaille f. de graphite. / ~ grease *see* ~ lubricant. / ~ lubricant ‖ Graphitschmiere f. ‖ graisse f. graphitée *ou* de graphite. / ~ mill ‖ Graphitmühle f. ‖ moulin m. à graphite. / ~ mine ‖ Graphitbergwerk n. ‖ mine f. de graphite. / ~ mining ‖ Graphitgewinnung f. ‖ extraction f. de graphite. / ~ paste ‖ Graphitmasse f. ‖ pâte f. de graphite. / ~ schist ‖ Graphitschiefer m. ‖ gneiss m. graphitique. / segregation of ~ ‖ Graphitausscheidung f. ‖ ségrégation f. de graphite. / ~ works pl. ‖ Graphitwerk n.; Graphit ‖ fabrique f. de graphite.
graphitic (Met) ‖ graphitisch ‖ limailleux; graphitique. / ~ brush ‖ Graphitbürste f. ‖ balai m. graphitique.
graphiting ‖ Graphitieren n. ‖ graphitage m. / ~ (spray) device ‖ Grafitiervorrichtung f. ‖ appareil m. à graphiter.
graphitize, to ‖ graphitieren ‖ graphiter.
graphophone *see also* phonograph ‖ Grafofon n. ‖ graphophone m.
grapnel (Build) ‖ Kropfeisen n.; Steinklaue f.; Teufelsklaue f. ‖ louve f.; renard m. / ~ (Mar) ‖ Bootsanker m.; Suchanker m.; Dragge f.; Quirlanker m. ‖ grappin m. / boarding ~ (Mar) ‖ Enterhaken m. ‖ grappin d'abordage.
grapple, to ‖ verankern ‖ ancrer. / ~ the boiler ‖ den Kessel m. verankern ‖ ancrer la chaudière.
grapplers pl. ‖ Klettereisen npl. ‖ grappins mpl.
grappling of a submarine cable ‖ Hebung f. eines Unterwasserkabels ‖ relèvement m. d'un câble sousmarin.
grappling anchor *see* boarding grapnel. / ~ rope (Aero) ‖ Fangleine f. ‖ amarre f.
grass ‖ Gras m. ‖ herbe f. / ~ after cutting ‖ Mahd f. ‖ fauchaison f.; fauchage f.; foin m.; andain m. / China ~ (Spinn) ‖ chinesisches Gras n.; Chinagras n. ‖ rhéa m.
grass bleacher ‖ Naturbleicher m. ‖ blanchisseur m. sur pré. / ~ bleaching ‖ Grasbleiche f.; Rasenbleiche f. ‖ blanchiment m. au pré. / ~ cutting tool (Garden) ‖

Rasenmähgerät n.; Rasenmäher m. ‖ outil m. à faucher (l'herbe). / ~ keeper ‖ Wiesenwächter m. ‖ gardien m. d'herbages. / ~ land ‖ Grasboden m. ‖ terrain m. gazonné; sol m. herbeux. / ~ mower (Agr) ‖ Grasmäher m.; Grasmähmaschine f. ‖ faucheuse f. (d'herbe). / motor ~ mower ‖ Motorgrasmäher m. ‖ faucheuse f. à moteur. / ~ mowing machine *see* ~ mower. / ~ peat ‖ Rasentorf m. ‖ tourbe f. gazonneuse. / ~ plot ‖ Rasenplatz m. ‖ boulingrin m.; pelouse f. / ~ plot watering appliance ‖ Rasensprengvorrichtung f.; Rasensprenger m. ‖ appareil m. d'arrosage de pelouses. / ~ seed ‖ Grassaat f.; Grassamen mpl. ‖ graines fpl. de graminées. / ~ shears pl. ‖ Grasschere f. ‖ sécateur m.
grate, to ‖ kratzen; reiben; schaben ‖ gratter; gratteler; racler. / ~ (To provide with a grate) ‖ gittern; umgittern; mit einem Gitter versehen ‖ grillager. / ~ down ‖ abschaben ‖ gratter; racler.
grate ‖ Gitter n.; Gitterrost m.; Rost m. ‖ grille f. / ~ (Mine) ‖ Rätter m. ‖ grille f. / adjustable ~ ‖ verstellbarer Rost m. ‖ grille f. déplaçable. / admission of cold air over the ~ ‖ Zuführung f. kalter Luft über dem Rost ‖ amenée f. d'air froid par-dessus la grille. / making a ~ into a blast-furnace ‖ Rostschlagen n. beim Hochofen ‖ grillage m. des hauts-fourneaux. / clinker ~ ‖ Schlackenrost m. ‖ grille f à scories. / conical ~ ‖ Kegelrost m. ‖ grille f. conique. / double-inclined ~ ‖ Schweinerücken m.; doppelt geneigter Rost m. ‖ grille f. en forme d'un toit. / fire ~ ‖ Feuerrost m.; Rost m. einer Feuerstätte ‖ grille f. du foyer. / horizontal ~ (Met) ‖ Planrost m. ‖ grille f. horizontale. / inclined ~ ‖ schräger *oder* geneigter Rost m.; Schrägrost m. ‖ grille f. inclinée. / movable ~ ‖ beweglicher Rost m. ‖ grille f. mobile. / multi-stage ~ *see* step ~. / revolving ~ ‖ beweglicher *oder* drehbarer Rost m. ‖ grille f. mobile *ou* tournante. / rotary ~ *see* revolving ~. / smoke-consuming ~ ‖ rauchverzehrender Rost m. ‖ grille f. fumivore. / steep ~ ‖ geneigter *oder* schräger Rost m. ‖ grille f. inclinée. / step ~ ‖ Etagenrost m. ‖ grille f. à étages *ou* a gradins. / wooden ~ (To retain floated wood) ‖ Hauptfang m.; Fangrechen m. ‖ batardeau m. / wrought-iron ~ ‖ schmiedeeisernes Gitter n. ‖ grille f. en fer forgé.
grate area ‖ Rostfläche f. ‖ surface f. de grille. / ~ bar (Boil) ‖ Roststab m. ‖ barreau m. de grille; barre f. de foyer. / space between the ~ bars ‖ Rostspalte f. ‖ intervalle m. entre les barreaux de grille. / ~ firing ‖ Rostfeuerung f. ‖ foyer m. à grille. / ~ frame ‖ Rostgestell n.; Rostrahmen m. ‖ cadre m. *ou* support m. *ou* chassis m. de grille. / ~ pendulum ‖ Rostpendel n. ‖ pendule m. (de compensation) à grille.
grater ‖ Raspel f.; Reibeisen n. ‖ râpe f.; lime f. mordante; égrugeoir m. / ~ file ‖ Feinraspel f. ‖ écouane f. à bois.
grate-shaped ‖ rostartig ‖ en forme de grille.
grate surface *see* grate area.
graticule for estimating distances ‖ Strichplatte f. zum Entfernungschätzen ‖ micromètre m. à estimer la distance. / illuminated ~ ‖ beleuchtetes Fadenkreuz n. ‖ réticule m. éclairé.

grating (Build) ‖ Gatter n.; Gitter n.; Gitterwerk n. ‖ grille f.; treillis m. / ~ (Hydr arch) ‖ Flechtwerk n.; Rost m. ‖ clayonnage m. / ~ (Opt) ‖ Gitter n. ‖ réseau m. / ~ (Shipb) ‖ Gräting f.; Rostwerk n. ‖ caillebotis m.; caillebottis m.; grillage m.; grille f. / boat's ~ ‖ Bootsgräting f. ‖ caillebottis m. d'embarcations. / ~ for floating wood ‖ Floßrechen m. ‖ barrage m. pour le bois de flot. / ~ of timbers ‖ Balkenrost m. ‖ grille f. en bois. / ~ for tin-plates ‖ Schragen m. zum Abtropfen der Weißbleche ‖ grillage m. pour laisser dégoutter le fer-blanc. / ~ for ventilation and light wells ‖ Rost m. für Luft- und Lichtschächte ‖ grille f. pour prises de ventilation et de lumière.

grating beam (Build) ‖ Rostschwelle f.; Langschwelle f. *oder* Schwelle f. eines Rostes ‖ longrine f.; chapeau m.; sablière f.; plate-forme f. / ~ machine (Household) ‖ Reibmaschine f. ‖ machine f. à râper. / ~ machine for grating ointments to perfect fineness ‖ Maschine f. zum Feinzerreiben von Salben ‖ machine f. pour râper *ou* moudre des onguents. / ~ nippers pl. ‖ Käfigbauerzange f. ‖ pince f. de cagiste. / ~ plank ‖ Rostbohle f.; Rostdiele f. ‖ plate-forme f.; couchis m. d'un grillage. / silvered ~ replica ‖ versilberte Gitterkopie f. ‖ copie f. argentée d'un réseau. / ~ sill *see* ~ beam.

grating spectroscope ‖ Gitterspektroskop n. ‖ spectroscope m. à réseau. / autocollimating ~ ‖ Gitterspektroskop n. mit Autokollimation ‖ spectroscope m. à réseau autocollimateur. / (diffraction) hand ~ ‖ Gitterhandspektroskop n. ‖ spectroscope m. à réseau à main. / evaluating ~ ‖ Gittermeßspektroskop n. ‖ spectromètre m. à réseau à mesurer.

grating spectrum ‖ Gitterspektrum n. ‖ spectre m. de réseau.

gratuitous article ‖ Zugabegegenstand m.; Zugabeartikel m. ‖ article m. à donner par-dessus le marché.

grave, to ~ a ship ‖ den Boden eines Schiffes reinigen ‖ gratter le fond d'un navire.

grave ‖ schwerwiegend; wichtig ‖ grave.

grave ‖ Grab n. ‖ tombe f. / artificial flowers pl. for ~s ‖ Grabschmuck m. aus künstlichen Blumen ‖ garniture f. pour tombes en fleurs artificielles.

grave cleaning ‖ Grabenreinigung f. ‖ curage m. de fossés. / ~ digger ‖ Totengräber m. ‖ fossoyeur m.

gravel, to ~ the pavement ‖ das Pflaster bekiesen ‖ couvrir le pavé de gravier.

gravel ‖ Kies m.; grober Sand m.; Kiessand m.; Grand m. ‖ gravier m. / to draw-out ~ ‖ Kies m. baggern ‖ draguer le gravier. / ~ for blast-engines ‖ Gebläsekies m. ‖ sable m. à machines soufflantes. / calcareous ~ ‖ Kalkkies m.; gravier m. calcaire. / ~ for concrete ‖ Betonschotter m. ‖ gravier m. de béton. / excavated ~ ‖ Baggergut n. ‖ déblais mpl. / loamy ~ ‖ lehmhaltiger Kies m. ‖ gravier m. argileux. / ~ pit ‖ Grubenkies m. ‖ gravier m. de carrière. / washed ~ ‖ gewaschener Kies m. ‖ pierraille f. lavée.

gravel ballast (Road) ‖ Kiesbettung f. ‖ ballast m. de gravier. / ~ bank ‖ Kiesbank f. ‖ banc m. de gravier. / ~ bottom (Mar) ‖ Kieselgrund m.; Schingelgrund m. ‖ fond m. de gravier. / ~ car (Railw) ‖ Kieswagen m.; Schotterwagen m. ‖ wa-

gon m. d'ensablement. / ~ concrete ‖ Kiesbeton m. ‖ béton m. de gravier. / ~ dredging ‖ Sandbaggerei f. ‖ dragage m. de sable. / ~ extraction ‖ Kiesgewinnung f.; Kiesförderung f. ‖ extraction f. de gravier. / ~ getting installation ‖ Kiesgewinnungsanlage f. ‖ installation f. à extraire le gravier. / ~ ground ‖ Kiesboden m. ‖ terrain m. graveleux. / ~ labourer ‖ Kiesgräber m. ‖ grèvier m. / ~ layer ‖ Kiesschicht f. ‖ couche f. de gravier.

gravelling with broken rock (Road) ‖ Aufschotterung f.; Schotterbelag m. ‖ empierrement m.

gravelly ‖ sandig; kiesig ‖ sablonneux; graveleux. / ~ soil ‖ Kiesboden m. ‖ terrain m. graveleux.

gravel pit ‖ Kiesgrube f. ‖ ballastière f.; carrière f. de gravier. / ~ washing machine ‖ Kieswaschmaschine f. ‖ laveur m. à gravier; machine f. à laver le gravier. / ~ washing and sorting machine ‖ Kieswasch- und sortiermaschine f. ‖ machine f. à laver et à trier le gravier.

graver (Coin) ‖ Stempelschneider m. (Werkzeug) ‖ cisoire f. / ~ (Engr) ‖ Grabstichel m.; Stichel m. ‖ ciselet m.; matoire f.; burin m. de graveur. / ~ (Turn) ‖ Drehstichel m. ‖ burin m. du tourneur. / flat ~ ‖ platter Grabstichel m. ‖ burin m. plat; onglette f. / lozenge ~ ‖ Grabstichel m. mit rautenförmigem Querschnitt ‖ burin m. losange. / sharp ~ (Jewel) ‖ Grabeisen n. ‖ échoppe f. à champlever. / square ~ ‖ Grabstichel m. mit quadratischem Querschnitt ‖ burin m. carré. / twice-bent ~ ‖ gekröpfter Grabstichel m. ‖ burin m. coudé.

grave railing ‖ Grabgitter n. ‖ grille f. de tombe.

gravimetric ‖ gewichtsanalytisch ‖ gravimétrique. / ~ effect ‖ Schwerewirkung f. ‖ action f. de la pesanteur.

gravimetry ‖ Gewichtsanalyse f. ‖ analyse f. pondérale.

graving dock ‖ Trockendock n. ‖ bassin m. de radoub; cale f. sèche.

graving tool ‖ Stichel m.; Grabstichel m. ‖ burin m. (de graveur).

gravitation ‖ Schwerkraft f.; Gravitation f. ‖ gravitation f. / law of ~ ‖ Schweregesetz n.; Gravitationsgesetz n. ‖ loi f. de gravitation.

gravitational field ‖ Schwerefeld n.; Gravitationsfeld n. ‖ champ m. de gravitation. / ~ mass ‖ Schweremasse f.; Gravitationsmasse f. ‖ masse f. gravitationelle.

gravity ‖ Erdschwere f.; Schwere f. ‖ pesanteur f.; poids m. / ~ (Brew) ‖ Extraktgehalt m. ‖ teneur f. en extrait. / apparent ~ (Brew) ‖ scheinbarer Extraktgehalt m. ‖ teneur f. apparente en extrait. / original ~ (Brew) ‖ ursprünglicher Extraktgehalt m. ‖ teneur f. originale en extrait. / real ~ (Brew) ‖ wirklicher Extraktgehalt m. ‖ teneur f. réelle en extrait. / specific ~ (Phys) ‖ Volumgewicht n.; spezifisches Gewicht n. ‖ densité f.; gravité f. *ou* pesanteur f. *ou* poids spécifique.

gravity cable ‖ Riese f. ‖ glissoir m. à câble. / centre of ~ ‖ Schwerpunkt m. ‖ centre m. *ou* point m. de gravité. / center of ~ of the gyroscope ‖ Kreiselschwerpunkt m. ‖ centre m. de gravité du gyroscope. / force of ~ ‖ Schwerkraft f.; Erdschwere f.; Gravitationskraft f. ‖ force f. de gravitation; pesanteur f. /

measurement of the force of ~ ‖ Schweremessung f. ‖ mesure f. de la gravité. / ~ mixer (Concrete) ‖ Freifallmischer m. ‖ malaxeur m. à chute libre. / ~ purifier of semolina ‖ Maschine f. zum Sortieren von Grieß nach spezifischem Gewicht ‖ purifieur m. à gravité de la semoule. / ~ tank (Aero) ‖ Hochbehälter m.; Fallbehälter m.; Falltank m. ‖ réservoir m. en charge.

gray *see* grey.

gray-wacke ‖ Grauwacke f. ‖ grès m. schisteux; grauwacke m.

graze of a projectile ‖ Aufschlag m. eines Geschosses ‖ point m. de chute d'un projectile.

grazier ‖ Wiesenbauer m.; Wiesenwirt m. ‖ herbager m.

grease, to ~ (with fat) ‖ (mit Fett) schmieren; einfetten; einschmieren ‖ graisser. / ~ (with oil) ‖ ölen; (mit Öl) schmieren ‖ huiler; lubrifier (à l'huile). / ~ duly ‖ abschmieren ‖ graisser complètement. / ~ leather ‖ Leder n. abflammen ‖ donner le suif au cuir.

grease ‖ Fett n.; Fettstoff m. ‖ graisse f.; corps m. gras. / ~ (Mach) ‖ Schmiere f.; Schmiermittel n.; Schmierfett n. ‖ enduit m.; graisse f. / to remove ~ ‖ entfetten ‖ dessuinter; dégraisser. / adhesive ~ ‖ Adhäsionsfett n. ‖ graisse f. adhérente. / alimentary ~ ‖ Speisefett n. ‖ graisse f. alimentaire. / animal ~ ‖ tierischer Fettstoff m. ‖ corps m. gras animal. / antifriction ~ ‖ Heißachsenschmiere f. ‖ graisse f. pour essieux chauds. / artificial ~ ‖ Kunstfett n. ‖ graisse f. préparée. / ~ for automobiles ‖ Kraftwagenfett n.; Kraftwagenschmiermittel n. ‖ graisse f. pour automobiles. / belt ~ ‖ Treibriemenfett n.; Riemenfett n. ‖ graisse f. pour courroies. / consistent ~ ‖ konsistentes Fett n. ‖ graisse f. consistante. / graphite ~ ‖ graphithaltiges Fett n. ‖ graisse f. à graphite. / (melted) horse ~ ‖ Kammfett n. ‖ graisse f. de cheval. / lubricating ~ ‖ Schmierfett n. ‖ graisse f. (de graissage) *ou* à lubrifier. / machine ~ ‖ Maschinenfett n. ‖ graisse f. pour machines. / melted ~ ‖ Schmalz n. ‖ graisse f. fondue. / Stauffer ~ ‖ Staufferfett n.; konsistentes Fett n. ‖ graisse f. Stauffer; graisse f. consistante. / vegetable ~ ‖ pflanzlicher Fettstoff m. ‖ corps m. gras d'origine végétale. / watersoluble ~ ‖ wasserlösliches Fett n. ‖ graisse f. soluble dans l'eau. / ~ of the wool ‖ Schweiß m. in der Wolle ‖ suint m.

grease axle box ‖ Schmierbuchse f. ‖ boîte f. à graisse. / ~ axle box with white metal bushings ‖ Achsschmierbuchse f. mit Weißmetallager ‖ boîte f. à graisse avec coussinets d'alliage blanc. / ~ bag (Auto) ‖ Lederschutzhülle f. ‖ enveloppe f. en cuir. / ~ boiler ‖ Fettschmelzer m. ‖ fondeur m. de graisse. / ~ box ‖ Schmierbüchse f. ‖ boîte f. à graisse. / ~ chamber (Coachm) ‖ Schmierkammer f. ‖ dégagement m. *ou* évasement m. des boîtes pour recevoir la graisse. / ~ cock ‖ Schmierhahn m. ‖ robinet m. de graissage; robinet-graisseur m. / ~ cup (Railw) ‖ Schmierkapsel f. ‖ godet m. graisseur. / ~ extracting apparatus for extracting grease from bones ‖ Knochenentfettungsvorrichtung f. ‖ appareil m. de dégraissage des os. / ~ extracting plant ‖ Entfettungsanlage f. ‖ installation f. de

dégraissage *ou* d'extraction de graisse. / ~ filter for steam || Dampfentöler m. || déshuileur m. de vapeur; séparateur m. d'huile de la vapeur. / ~ gun || Fettspritze f. || pompe f. *ou* compresseur m. à graisse. / ~ industry || Fettindustrie f. || industrie f. de graisse. / ~ industry plant || Fettindustrieanlage f. || installation f. d'industrie de graisse. / ~ lubrication || Fettschmierung f. || lubrification f. à la graisse. || ~ paint || Schminke f. || fard m. / ~ pan / Fettschmelzpfanne f. || casserole f. à fondre la graisse.

grease-proof paper || Butterbrotpapier n.; fettdichtes Papier n. || papier m. pour tartines au beurre.

grease pump || Fettpumpe f. || pompe f. pour graisses.

greaser || Öler m.; Schmierer m. || graisseur m.; huileur m.

grease rag (Wire-dr) || Fettlappen m. || affile m. / ~ remover for metals || Metallentfettungsmittel n. || dégraisseur m. de métaux.

grease-spot || Fettfleck m. || tache f. de graisse. / ~ photometer || Fettfleckfotometer n.; Fettflecklichtstärkemesser m. || photomètre m. à tache d'huile.

grease syringe || Fettspritze f. || seringue f. de graissage. / ~ tap *see* ~ cock.

greasing (Mach) || Schmieren n.; Schmierung f. || graissage m.; lubrifaction f.; lubrifiage m. / ~ apparatus || Schmierapparat m. || appareil m. de graissage. / ~ substance || Schmiermittel n.; Schmiere f. || matière f. grasse *ou* lubrifiante.

greasy || fettig; ölig || graisseux; onctueux; adipeux. / to feel ~ || sich fettig anfühlen || être onctueux au toucher.

greasy wool || Schweißwolle f.; Fließwolle f.; Schmutzwolle f. || surge f.; laine f. en suint; laine f. surge.

great || groß || grand. / ~ calory || große Kalorie f.; Kilokalorie f. || grande calorie f.; kilocalorie f. / ~ cattle || Großvieh n. || gros bétail m. / ~ circle || größter Kreis m.; Orthodrome f. || orthodrome m.

green || grün || vert.

green || Grün n.; grüne Farbe f. || vert m. / Brunswick ~ || Braunschweiger Grün n.; grüne Kupferfarbe f. || vert m. de Brunsvic. / emerald ~ || Smaragdgrün n. || vert m. émeraude. / olive ~ || Olivgrün n. || vert m. olive. / Paris ~ || Schweinfurter Grün n. || vert m. de Paris *ou* de Schweinfurt. / permanent ~ || Permanentgrün n. || vert m. permanent. / Silesian ~ || Silesiagrün n. || vert m. Silésie. / Urania ~ || Uraniagrün n. || vert m. d'urane.

green coffee || grüner Kaffee m. || café m. vert. / ~ copperas || Eisenvitriol m.; grüner Vitriol m. || sulfate m. de fer. / ~ crop || Grünfutter n. || fourrage m. vert. / ~ cross (poisonous gas) || Grünkreuzkampfstoff m.; Perstoff m. || chloroformiate m. de méthyle trichloré; croix f. verte; surpalite m. / ~ earth || Grünerde f.; Seladonit m. || terre f. verte; céladonite f. / ~ field || Wiesenboden m. || sol m. de prairie. / ~ filter (Opt) || Grünfilter n. || écran m. vert. / ~ grocer || Gemüsehändler m. || fruitier m.; marchand m. des quatre saisons.

greenheart || Grünholz n. || bois m. laurier dur.

greenhouse || Gewächshaus n.; Treibhaus n. || serre f. / cold ~ || Kalthaus n. || serre f. froide. / iron ~ || eisernes Gewächs-

haus n. || serre f. en charpente métallique.

greenhouse construction || Gewächshausbau m. || construction f. de serres. / ~ plant || Treibhauspflanze f.; Treibhausgewächs n. || plante f. de serres.

green iron ore || Grüneisenerz n.; Grüneisenstein m. || fer m. phosphaté vert; dufrénite f.

greenish || grünlich || verdâtre. / ~ black || grünschwarz || noir verdâtre.

Greenlandman || Grönlandfahrer m. || groënlandier m.; marin m. groënlandier.

green light || grünes Licht n. || feu m. vert.

green-malt || Grünmalz n. || malt m. vert. / ~ extract || Grünmalzauszug m. || extrait m. de malt vert. / ~ moistener || Grünmalzbefeuchter m. || arroseur m. de malt vert. / ~ oar *see* ~ turning device. / ~ turning device || Grünmalzwender m. || retourneur m. mécanique de malt vert.

greenockite || Greenockit m. || greenockite f.; sulfure m. de cadmium.

green oil || Grünöl n. || huile f. verte. / ~ pasture || Grünfutter n. || fourrage m. vert. / ~ rot || Grünfäule f. || putréfaction f. due à la pezize. / ~ sand (Found) || magerer *oder* nasser *oder* grüner Formsand m. || sable m. vert *ou* maigre. / ~ sand (Geol) || Grünsand m. || sable m. *ou* grès m. vert; glauconite f. / ~ soap || grüne Seife f. || savon m. vert *ou* de potasse.

greenstone || Grünstein m.; Diabas m. || diabase f.; diorite f. / ~ quarry || Diabassteinbruch m. || carrière f. de diabase.

green-wood || Grünholz n. || cèdre m. vert; bois m. vert. / ~ saw || Grünholzsäge f. || scie f. à bois vert.

greisen || Greisen m.; Granit m. ohne Feldspat || hyalomicte f.; greisen m.

grenade || Granate f. || obus m.; grenade f. / hand ~ || Handgranate f. || grenade f. à main. / ~ thrower || Granatwerfer m. || lance-grenades m.

grenadillo || Grenadillholz n.; rotes Ebenholz n. || grenadille f.

grenadine (Textile) || Grenadin n. || grenadine f.

grengesite || Grengesit m. || grengésite f.

grey || grau || gris. / ~ castings pl. || Grauguß m. || fonte f. grise. / ~ mercurial ~ copper (Miner) || Graugültigerz n. || cuivre m. gris mercurifère. / ~ iron castings pl. *see* ~ castings. / ~ iron foundry || Graugießerei f. || fonderie f. à fonte grise.

greyish || gräulich || grisâtre. / ~ green || graugrün || gris-vert.

greywacke *see* graywacke.

grid || Schutzgitter n.; Gitter n. || grille f. (de protection). / auxiliary ~ (Radio) || Hilfsgitter n. || grille f. auxiliaire. / control ~ || Raumladungsgitter n. || grille f. de contrôle. / heating ~ || Heizgitter n. || grille f. de chauffage.

grid battery || Gitterbatterie f. || batterie f. de grille. / ~ circuit || Gitterkreis m. || circuit m. de grille. / ~ condenser || Gitterkondensator m. || condensateur m. de grille. / ~ current || Gitterstrom m. || courant m. de grille.

gridiron || Bratrost m.; Rost m. || gril m. / ~ (Shipb) || Kielbank f. || grillage m. de carénage. / ~ pendulum || Rostpendel m.; Kompensationspendel n. || pendule m. à gril.

grid-leak || Gitterableitung f. || résistance f. de fuite de grille; retour m. de grille. / ~

scheme of modulation || Gittergleichstrommodulation f. || modulation f. de la grille à courant continu.

grid-plate (Acc) || Gitterplatte f. || plaque f. à grilles *ou* à grillage. / accumulator ~ || Sammlergitterplatte f.; Akkumulatorgitterplatte f. || plaque f. à grillage d'accumulateur.

grid, shielded ~ pliotron (Tel) || Schirmgitterröhre f. || lampe f. à double grille. / ~ potential || Gitterspannung f.; Gitterpotential n. || potentiel m. de grille. / ~ rectification || Gittergleichrichtung f. || rectification f. par grille. / ~ shovel || Gitterschaufel f. || pelle f. à grille. / ~ support || Gitterträger m. || support m. en charpente métallique *ou* en treillis. / ~ tension *see* ~ voltage. / double ~ tube (Radio) || Doppelgitterröhre f. || lampe f. bigrille; bigrille f. / ~ type accumulator || Gittersammler m.; Gitterakkumulator m. || accumulateur m. à grillage.

grid-voltage || Gitterspannung f. | tension f. (de) grille. / negative priming ~ || negative Gittervorspannung f. || polarisation f. négative de grille. / priming ~ || Gittervorspannung f. || polarisation f. de la grille.

grievance || Klagegrund m. || grief m.

grill || Bratrost m. || gril m.

grillage (Build) || Gatter n.; Gitter n.; Gitterwerk n. || grille f.; treillis m. / ~ (Hydr arch) || Flechtwerk n.; Rost m. | clayonnage m.

grill inset || Rosteinsatz m. || grille f.

grind, to (To reduce to fine particles) || mahlen; vermahlen || moudre. / ~ (To sharpen) || schleifen; schärfen; wetzen || émoudre; aiguiser; affûter; meuler. / ~ (with emery) || abschmirgeln; einschleifen || roder (à l'émeri). / ~ (Curr) || abscheuern || quiosser. / ~ (Glassm) || klarschleifen; feinschleifen || doucir; adoucir. / ~ (Met) || schleifen; abschleifen || meuler; émoudre; rectifier à la meule. / ~ again the rolls pl. || die Walzen fpl. nachschleifen | finir les cylindres mpl. à la meule / ~ coarsely || grob zerkleinern || moudre grossièrement. / ~ the colour || die Farbe f. anreiben || broyer la couleur. / ~ to dust || zu Pulver n. zerreiben; pulverisieren || pulvériser. / ~ with emery || schmirgeln || émeriller; polir *ou* roder à l'émeri. / ~ the glass || Glas n. schleifen || tailler le verre. / ~-in || einschleifen || roder; tailler. / ~-in a cock || einen Hahn m. einschleifen || roder un robinet. / ~-in the piston || den Kolben m. einschleifen || roder le piston dans le cylindre. / ~ by the longitudinal traverse method || im Langschleifverfahren n. schleifen || rectifier par va-et-vient. / ~ the malt || das Malz schroten || moudre le malt. / ~-on the piece || ein Stück n. aufschleifen || roder la pièce sur son embase. / ~-out || ausschleifen || aléser à la meule. / ~-out into a spherical curvature || kugelförmig ausschleifen || rectifier sphériquement. / the piston grinds || der Kolben frißt || le piston grippe *ou* se mange. / ~ a plaster-floor || den Gipsestrich m. schleifen || dégrossir *ou* frotter une aire en plâtre. / ~ plate-glass || Spiegelglas schleifen || dresser les glaces. / ~ the rags || die Lumpen pl. zerreiben || broyer les chiffons mpl. / ~roughly || vorschleifen || dégrossir à la meule. / ~ with sand and water (Glassm) || matt schleifen || dépolir *ou* égriser le verre. / ~ the snuff || den

Schnupftabak m. mahlen ‖ râper *ou* pulvériser le tabac à priser. / ~-up ‖ verreiben ‖ malaxer; broyer. / ~ the valves ‖ die Ventile npl. einschleifen ‖ roder les soupapes fpl.

grinder (Grinding machine) ‖ Schleifmaschine f. ‖ machine f à rectifier *ou* à affûter *ou* à aiguiser. / ~ (Grindstone) ‖ Schleifstein m. ‖ pierre f. à aiguiser *ou* de rémouleur; meule f. / ~ (Person) ‖ Schleifer m.; Einschleifer m. ‖ meuleur m.; rodeur m. / file ~ ‖ Feilenschleifer m. ‖ émeuleur m. de limes. / foreman ~ ‖ Schleifmeister m. ‖ chef-meuleur m. / knife ~ ‖ Messerschleifer m. ‖ affûteur *ou* repasseur m. de couteaux. / knife ~ (Machine) ‖ Messerschleifmaschine f. ‖ machine f. à affûter les couteaux. / rough ~ ‖ Vorschleifer m. ‖ émouleur m. / sabre-blade ~ ‖ Schwertschleifer m. ‖ aiguiseur m. de lames de sabres. / scissors ~ ‖ Scherenschleifer m. ‖ affûteur m. *ou* repasseur m. de ciseaux; rémouleur m.; gagne-petit m. / scythe ~ ‖ Sensenschleifer m. ‖ aiguiseur m. de faux. / shears ~ *see* scissors ~. / shot ~ ‖ Kugelmühle f. ‖ broyeur m. à boulets. / tool ~ ‖ Werkzeugschleifer m. ‖ affûteur m. *ou* meuleur m. d'outils. / tool ~ (Machine) ‖ Werkzeugschleifmaschine f. ‖ machine f. à affûter les outils.

grinder conch, longitudinal ‖ Längsreiber m. ‖ broyeur m. longitudinal; conche f.

grinder spindle ‖ Schleifwelle f. ‖ arbre m. porte-meule.

grindery ‖ Schleiferei f. ‖ meulerie f.; aiguiserie f. / ~ for optical lenses ‖ optische Linsenschleiferei f. ‖ atelier m. pour la taille des lentilles optiques.

grinding ‖ Schleifen n.; Schliff m. ‖ affûtage m.; meulage m.; émoulage m. / ~ (Polishing) ‖ Schleifen n.; Polieren n.; Glanzschleifen n.; Schmirgeln n. ‖ polissage m.; polissure f. à l'émeri. / ~ (Colour) ‖ Zerreiben n. ‖ broiement m. / ~ (Met) ‖ Schleifriß m. ‖ strie f. sur la plaquette polie. / ~ (Mill) ‖ Vermahlen n.; Getreideschroterei f. ‖ mouture f. / ~ (Needl) ‖ Schauern n.; Scheuern n.; Polieren n. ‖ polissage m. / circular ~ *see* cylindrical ~. / coarse ~ ‖ grobkörnige Zerkleinerung f.; Grobzerkleinerung f. ‖ broyage m. grossier. / continuous ~ ‖ ununterbrochene Vermahlung ‖ broyage m. continu. / cylindrical ~ ‖ Rundschleifen n.; Wellenschliff m. ‖ rectification f. cylindrique *ou* circulaire *ou* d'arbres (cylindriques) *ou* de surfaces cylindriques. / cylindrical ~ between centres ‖ Wellenschliff m. zwischen Spitzen ‖ rectification f. d'arbres entre pointes. / dry ~ ‖ Trockenschleifen n. ‖ polissage m. à sec. / external cylindrical ~ ‖ Außenrundschliff m. ‖ rectification f. des surfaces cylindriques extérieures. / external taper ~ ‖ Außenschliff m. konischer Rundflächen ‖ rectification f. extérieure de surfaces circulaires coniques. / fine ~ ‖ Feinzerkleinerung f. ‖ broyage m. fin. / first ~ *see* rough ~. / to fit together by ~ one on the other ‖ aufeinander schleifen ‖ roder l'un sur l'autre. / ~ of the ink ‖ Farbverreibung f. ‖ broyage m. de couleurs. / intermittent ~ ‖ satzweise Vermahlung f. ‖ mouture f. périodique. / internal cylindrical ~ ‖ Innenrundschliff m. ‖ rectification f. des surfaces cylindriques intérieures. / plain cylindrical ~ ‖ Wellenschliff m. ‖ recti-

fication f. des arbres cylindriques. / ~ of rolls ‖ Schleifen n. von Walzen ‖ meulage m. de cylindres. / to give a final ~ to the rolls ‖ Walzen fpl. nachschleifen ‖ finir des cylindres mpl. à la meule. / rough ~ ‖ Vorschleifen n. ‖ meulage m. brut. / smooth ~ ‖ Blankschleifen n. ‖ meulage m. fin. / ~ from the solid ‖ Schleifen n. aus dem Vollen ‖ affûter dans la barre. / ~ of tools and milling cutters ‖ Schleifen n. von Schneidwerkzeugen und Fräsern ‖ meulage m. d'outils coupants et de fraises. / valve ~ ‖ Ventileinschleifen n. ‖ rodage m. de soupapes. / wet ~ ‖ Naßschleifen n. ‖ affûtage m. à l'eau; meulage m. en mouillant.

grinding apparatus ‖ Schleifapparat m. ‖ appareil m. à rectifier.

grinding attachment, cylindrical and internal ~ ‖ Vorrichtung f. zum Außen- und Innenrundschleifen ‖ dispositif m. à rectifier entre pointes et les intérieurs. / ~ for semi-circular form cutters ‖ Vorrichtung f. zum Schleifen von Pilzfräsern mpl. ‖ dispositif m. à affûter les fraises à champignon. / ~ for threading dies ‖ Gewindeschneidbackenschleifvorrichtung f. ‖ dispositif m. à affûter les coussinets de filières. / ~ for the throat on threading dies ‖ Vorrichtung f. zum Schleifen des Anschnittes an Gewindeschneidbacken ‖ dispositif m. à rectifier les entrées des coussinets de filières. / ~ for twist drills ‖ Vorrichtung f. zum Schleifen von Spiralbohrern m. ‖ dispositif m. à affûter les forets hélicoïdaux.

grinding bench (Glass) ‖ Schleifbank f. ‖ table f. à dresser les glaces. / ~ board ‖ Schleifbrett n. ‖ planchette f. à polir. / ~ box ‖ Mahlkranz m. ‖ anneau m. broyeur. / ~ charcoal ‖ Schleifkohle f. ‖ charbon m. pour adoucir. / ~ cloth ‖ Schleiftuch n. ‖ toile f. d'émeri. / corrugated ~ cone ‖ gerieffelter Mahlkonus m. ‖ cône m. broyeur cannelé. / ~ crystal ‖ Schleifkristall m. ‖ cristal m. à polir.

grinding cylinder ‖ Mahltrommel f. ‖ tambour m. broyeur. / ~ for erasers ‖ Schleiftrommel f. für Radiergummi ‖ tambour m. à polir la gomme à effacer.

grinding device ‖ Schleifvorrichtung f. ‖ dispositif m. à rectifier. / ~ diagram ‖ Mahldiagramm n. ‖ diagramme m. de mouture.

grinding disk ‖ Mahlscheibe f. ‖ meule f. / ~ of electro corundum and silicated carbide ‖ Schleifscheibe f. aus Elektrokorund und Siliziumkarbid ‖ disque m. à polir en électrocorindon et en carbure de silicium. / toothed ~ of chilled cast iron ‖ gezahnte Hartgußmahlscheibe f. ‖ disque m. broyeur denté de fonte en coquille.

grinding duty ‖ Mahlsteuer f. ‖ droits mpl. de mouture. / ~ glass ‖ Schleifglas n. ‖ verrerie f. pour taille; verre m. de gobeleterie.

grinding-in paste for cocks and valves ‖ Einschleifpaste f. für Hähne und Ventile ‖ pâte f. à roder pour robinets et soupapes.

grinding jig for sharpening the tools ‖ Halter m. zum Schleifen der Stähle ‖ montage m. pour l'affûtage des outils.

grinding lathe (Glass) *see* grinding bench.

grinding machine (Machine-tool) ‖ Schleifmaschine f. ‖ machine f. à rectifier. / ~ (Tool-sharpening) ‖ Schleifmaschine f.

‖ machine f. à affûter *ou* à aiguiser; affûteuse f. / ~ (Polishing machine) ‖ Schleifmaschine f.; Poliermaschine f. ‖ machine f. à polir. / ~ (Coarse grinding) ‖ Schleifmaschine f.; Grobschleifmaschine f. ‖ machine f. à meuler. / ~ (For castings) ‖ Schleifmaschine f.; Gußputzmaschine f. ‖ machine f. à ébarber. / ~ (Ore dress) ‖ Zerkleinerungsmaschine f. ‖ concasseur m.; broyeur m. / almond ~ ‖ Mandelreibmaschine f. ‖ machine f. à râper les amandes. / automatic ~ ‖ Schleifautomat m. ‖ machine f. à rectifier automatique(ment). / axle journal ~ ‖ Achsschenkelschleifmaschine f. ‖ machine f. à rectifier les fusées d'essieu. / ~ with belt drive ‖ Schleifmaschine f. mit Riemenantrieb ‖ machine f. à meuler commandée à courroie. / cam ~ ‖ Nockenschleifmaschine f. ‖ machine f. à rectifier les cames. / camshaft ~ ‖ Nockenwellenschleifmaschine f. ‖ machine f. à rectifier les arbres à cames. / centreless ~ ‖ spitzenlose Schleifmaschine f. ‖ machine f. à meuler sans pointes. / chasing tool ~ ‖ Gewindestrählerschleifmaschine f. ‖ machine f. à affûter les peignes à fileter. / circular ~ *see* cylindrical ~ / colour ~ ‖ Farbenreibmaschine f. ‖ machine f. à broyer les couleurs. / crankshaft ~ ‖ Kurbelwellenschleifmaschine f. ‖ machine f. à rectifier les arbres vilebrequins. / cutter ~ ‖ Messerschleifmaschine f. ‖ machine f. à affûter les couteaux. / cutter head ~ ‖ Messerkopfschleifmaschine f. ‖ machine f. à affûter les fraises à dents rapportées *ou* à affûter les têtes à couteaux / cylinder ~ ‖ Zylinderschleifmaschine f. ‖ machine f. à rectifier l'intérieur des cylindres. / cylindrical ~ ‖ Rundschleifmaschine f. ‖ machine f. à rectifier cylindriquement *ou* à rectifier les pièces cylindriques. / cylindrical ~ for locomotive axle journals ‖ Rundschleifmaschine f. für Lokomotivachsschenkel ‖ machine f. à rectifier cylindriquement les fusées déssieux de locomotives. / drill ~ ‖ Bohrerschleifmaschine f. ‖ machine f. à affûter les forets. / emery ~ ‖ Schmirgelschleifmaschine f. ‖ machine f. à polir à l'émeri. / ~ with flexible shaft ‖ Schleifmaschine f. mit biegsamer Welle ‖ machine f. à meuler avec arbre flexible. / gear ~ ‖ Zahnräderschleifmaschine f. ‖ machine f. à rectifier les engrenages. / ~ for gold leaf ‖ Anreibemaschine f. für Blattgold ‖ machine f. à broyer l'or en feuilles. / ~ with hand rest ‖ Schleifmaschine f. mit Handauflage ‖ machine f. à meuler avec support à main. / ink ~ *see* colour ~. / internal ~ ‖ Innenschleifmaschine f. ‖ machine f. à rectifier intérieurement. / internal and face ~ ‖ Innen- und Stirnflächenschleifmaschine f. ‖ machine f. à rectifier les alésages et les surfaces planes en bout. / journal ~ ‖ Zapfenschleifmaschine f. ‖ machine f. à rectifier les tourillons. / key shaft ~ ‖ Keilwellenschleifmaschine f. ‖ machine f. à rectifier les arbres à rainures. / knife ~ ‖ Messerschleifmaschine f. ‖ machine f. à aiguiser les couteaux. / lathe tool ~ ‖ Drehstahlschleifmaschine f. ‖ machine f. à affûter les outils de tours. / link and profile ~ ‖ Kulissen- und Kopierschleifmaschine f. ‖ machine f. à rectifier les coulisses et à copier. / locomotive crank

pin ~ ‖ Schleifwerk n. für Lokomotiv-kurbelzapfen ‖ machine f. à rectifier les tourillons de bielles motrices de locomotives. / milling cutter ~ ‖ Fräserschleifmaschine f. ‖ machine f. à affûter les fraises. / pendulum ~ ‖ Hängeschleifmaschine f. ‖ machine f. à meuler oscillante. / piston ring ~ ‖ Kolbenringschleifmaschine f. ‖ machine f. à rectifier les segments de piston. / piston ~ ‖ Kolbenschleifmaschine f. ‖ machine f. à rectifier les pistons. / piston rod ~ ‖ Kolbenstangenschleifmaschine f. ‖ machine f. à rectifier les tiges de piston. / plain ~ worked by a hand wheel ‖ einfache Handradschleifmaschine f. ‖ simple machine f. à meuler à volant à main. / planer type ~ ‖ Langflächenschleifmaschine f.; Doppelständerschleifmaschine ‖ machine f. à rectifier les surfaces planes; machine f. à rectifier à deux montants. / ~ for planishing operations ‖ Schleifmaschine f. zum Glätten ‖ machine f. à lisser. / ~ for plate-glass ‖ Schleifmaschine f. für Spiegelglas ‖ polisseuse f. à glaces. / portable ~ ‖ Handschleifmaschine f. ‖ machine f. à rectifier portative. / portable ~ for tracks ‖ fahrbare Schleifmaschine f. für Schienenstöße an Gleisen ‖ meule f. transportable à araser les abouts de rails. / roller ~ ‖ Walzenschleifmaschine f. ‖ machine f. à rectifier les cylindres de laminoirs. / sandstone ~ ‖ Sandsteinschleifmaschine ‖ meule f. de grès. / saw ~ ‖ Sägenschärfmaschine f. ‖ machine f. à affûter les scies. / / ~ for sharpening operations ‖ Schleifmaschine f. zum Schärfen ‖ machine f. à affûter. / single-housing ~ ‖ Einständerschleifmaschine f. ‖ machine f. à rectifier à un montant. / single-purpose ~ ‖ Schleifmaschine f. für Sonderzwecke ‖ machine f. à meuler spéciale. / surface ~ ‖ Flächenschleifmaschine f. ‖ machine f. à rectifier les surfaces planes. / surface ~ and polishing machine ‖ Flächenschleif- und -poliermaschine f. ‖ machine f. à mouler et à polir les surfaces. / threading die ~ ‖ Schneidbackenschleifmaschine f. ‖ machine f. à affûter les filières. / twist drill ~ ‖ Spiralbohrerschleifmaschine f. ‖ machine f. à affûter les forets hélicoïdaux. / universal ~ ‖ Universalschleifmaschine f. ‖ machine f. à rectifier universelle. / valve seat ~ ‖ Ventilsitzschleifmaschine f. ‖ machine f. à rectifier les sièges de soupapes. / vertical planetary spindle ~ ‖ Vertikalschleifmaschine f. mit Planetenspindel ‖ machine f. à rectifier à arbre vertical planétaire. / wheel and axle set ~ ‖ Achssatzschleifmaschine f. ‖ machine f. à rectifier. les trains de roues. / ~ for wood working ‖ Schleifmaschine f. für die Holzbearbeitung ‖ machine f. à meuler pour travailler les bois.

grinding and fluting machine with a column saddle ‖ Pilasterschleif- und Riffelmaschine f. ‖ polisseuse f. et canneleuse f. avec support pilastre.

grinding and polishing machine for all fabrics ‖ Schleif- und Poliermaschine f. für Gewebe aller Art ‖ machine f. à émeriser et polir tous les tissus.

grinding material ‖ Schleifmittel n. ‖ produit m. abrasif; matière f. à polir. / artificial ~ ‖ künstliches Schleifmittel n. ‖ produit m. artificiel pour le meulage.

grinding medium see grinding material. / ~ method (Mill) ‖ Mahlmethode f. ‖ méthode f. de mouture f.

grinding mill ‖ Mahlwalzwerk n. ‖ laminoir m. broyeur ou de mouture. / ~ (Glass) ‖ Reibkasten m.; Schleifkasten m. ‖ moëllon m. / ~ (Needl) ‖ Schleifmühle f.; Scheuermühle f. ‖ aiguiserie f. / ink ~ ‖ Farbmühle f. ‖ moulin m. à couleurs. / mineral manure ~ ‖ Düngermühle f. ‖ méthode f. de moulin m. à engrais.

grinding mortar ‖ Reibschale f. ‖ écuelle f. à broyer. / ~ motor ‖ Schleifmotor m. ‖ moteur-meule m.

grinding plant for the corn industry ‖ Mahlanlage f. für die Getreideindustrie ‖ installation f. de mouture pour l'industrie des céréales. / ~ for miller's industry ‖ Mahlanlage f. für die Müllereiindustrie ‖ installation f. de mouture pour l'industrie meunière. / ~ for minerals ‖ Mineralienmahlanlage f. ‖ broyeur m. pulverisateur de minéraux.

grinding plate ‖ Mahlplatte f. ‖ plaque f. de mouture ou de broyage; disque-broyeur m. / ~ for ball mills ‖ Mahlplatte f. für Kugelmühlen ‖ plaque f. triturante pour moulins à boulets. / ~ for ore samples ‖ Reibeplatte f. für Erzproben ‖ plaque f. de trituration pour essais de minerais.

grinding powder ‖ Schleifpulver n. ‖ poudre f. à rodage ou à meuler; poudre f. abrasive. / ~ residue ‖ Mahlrückstand m. ‖ résidu m. de la mouture. / ~ ring ‖ Mahlring m. ‖ anneau m. broyeur ou de mouture.

grindings pl. ‖ Schliff m.; Schleifsel n.; Abfälle mpl. beim Schleifen ‖ moulée f.

grinding sand ‖ Schleifsand m. ‖ sable m. à polir. / ~ stone see grindstone. / ~ support ‖ Schleifsupport m. ‖ support m. de polissage. / ~ surface (Mill) ‖ Mahlfläche f. ‖ feuillure f.; surface f. travaillante d'un broyeur. / ~ tool ‖ Schleifwerkzeug n. ‖ outil m. à affûter. / edge runner with stationary and rotary ~ track ‖ Kollergang m. mit feststehender und umlaufender Mahlbahn ‖ moulin m. à meules verticales avec plateau de roulement fixe et tournant. / ~ trough ‖ Schleifsteintrog m. ‖ auge f. de meule.

grinding wheel ‖ Schleifscheibe f.; Schleifrad n. ‖ meule f. / high production ~ ‖ Hochleistungsschleifrad n. ‖ meule m. à grand rendement. / shaped ~ ‖ Formschleifscheibe f. ‖ meule m. de forme.

grinding wheel diameter ‖ Schleifscheibendurchmesser m. ‖ diamètre m. de meule. / ~ guard ‖ Schutzhaube f. für Schleifscheiben ‖ couvre-meule m. / ~ head ‖ Schleifspindelstock m. ‖ poupée f. porte-meule. / ~ trough ‖ Schleifsteintrog m. ‖ auge f. de meule. / ~ trueing device ‖ Schleifsteineinrichtvorrichtung f. ‖ dispositif m. à dégrossir les meules.

grinding worm ‖ Brechschnecke f. ‖ vis f. broyeuse.

grindstone ‖ Schleifstein m. ‖ meule f.; pierre m. à aiguiser. / artificial ~ ‖ künstlicher Schleifstein m. ‖ meule f. artificielle. / hand ~ ‖ Handschleifstein m. ‖ pierre f. à affiler à la main. / ~ with trough ‖ Schleifstein m. mit Trog ‖ pierre f. de rémouleur avec auge.

grindstone turner ‖ Schleifsteindreher m. ‖ tourneur m. de meules.

grip ‖ Griff m. ‖ poignée f. / ~ (Cable) ‖ Ziehschlauch m. ‖ grip m. / ~ block ‖ Fangklotz m. ‖ taquet m. d'arrêt.

grip head ‖ Spannkopf m. ‖ tête f. à serrage. / rapid ~ ‖ Schnellspannkopf m. ‖ tête f. à serrage rapide.

gripper, auxiliary (Print) ‖ Vorgreifer m. ‖ dispositif m. de prise préalable.

gripper feed ‖ Greifervorschub m.; Zangenvorschub m. ‖ amenage m. par pinces.

gripping device (Dredger) ‖ Greifwerkzeug n. ‖ appareil m. de préhension. / ~ (Cage; Mine) ‖ Fangvorrichtung f. ‖ dispositif m. de parachute; arrête-cuffat m. / ~ (Elevator) ‖ Fangvorrichtung f. ‖ parachute m.; dispositif m. d'arrêt.

gripping jaw ‖ Spannbacke f. ‖ griffe f. ou mâchoire f. ou mordache f. de serrage.

grip type cock with bent outlet ‖ Griffhahn m. mit gebogenem Auslauf ‖ robinet m. à poignée à sortie courbe. / ~ with cap screw ‖ Griffhahn m. mit Überwurfmutter ‖ robinet m. à poignée avec écrou à chapeau. / ~ with sleeve ‖ Griffhahn m. mit Schlauchtülle ‖ robinet m. à poignée avec douille. / ~ with straight outlet ‖ Griffhahn m. mit geradem Auslauf ‖ robinet m. à poignée à sortie droite.

grist ‖ Mahlgut n. ‖ grains mpl. à moudre; produit m. broyé. / ~ (Yarn) ‖ Feinheit f. ‖ titre m. / fine ~ ‖ Feinschrot n. ‖ mouture f. fine.

grist fineness ‖ Mahlfeinheit f. ‖ degré m. de mouture.

gristle ‖ Knorpel m. ‖ cartilage m.

gristly ‖ knorpelig ‖ cartilagineux.

grist mill ‖ Getreidemühle f. ‖ moulin m. à blé.

grit (Sandstone) ‖ Sandstein m. ‖ grès m. / ~ ‖ Grieß m.; Gruß m.; Kies m.; grober Sand m. ‖ arène f.; gravier m.

grit layer (Road) ‖ Steinbewurf m. ‖ jetée f. / ~ masonry ‖ Sandsteinmauerwerk n. ‖ gresserie f. / ~ pavement ‖ Sandsteinpflaster n. ‖ pavé m. de grès. / ~ quarry ‖ Sandsteinbruch m. ‖ carrière f. de grès. / ~ rock see grit (Sandstone).

grits pl. ‖ Grütze f. ‖ gruau m. / ~ pl. of maize ‖ Maisgrütze f. ‖ grit m. de maïs.

grits mill ‖ Grießmühle f. ‖ moulin m. à semoule.

grit stone see grit.

gritty ‖ kiesig ‖ graveleux.

grizzly ‖ Grubenrost m. ‖ grille f. de mine.

groats pl. ‖ Hafergrütze f.; Grütze f. ‖ gruau m. d'avoine. / coarse wheaten ~ pl. ‖ grobes Weizenschrot n. ‖ froment m. égrugé. / to make ~ pl. ‖ Schrot machen ‖ rengrainer. / the ~ pl. swell up ‖ das Schrot oder die Grütze quillt ‖ la mouture gonfle.

groats mill ‖ Grützmühle f. ‖ semoulerie f.; moulin m. à gruau. / ~ miller ‖ Grützmüller m. ‖ semoulier m.

grocer ‖ Kolonialwarenhändler m. ‖ épicier m.

groceries pl. ‖ Kolonialwaren fpl. ‖ denrées fpl. coloniales. / mill for ~ ‖ Mühle f. für Kolonialwaren ‖ moulin m. à denrées coloniales.

grocers pl. ‖ Delikatessen fpl. ‖ comestibles mpl. fins.

grocer's scale ‖ Tafelwage f. ‖ balance f. de Roberval. / ~ shop see grocery.

grocery ‖ Kolonialwarenhandlung f.; Spezereienhandlung f. ‖ épicerie f. en gros.

groin ‖ Buhne f. ‖ épi m.; crèche f. / immerged ~ ‖ Tauchbühne f. ‖ épi m. plongeant. / ~ of a roof ‖ Grat m. eines Daches ‖ arête f. d'un toit.

groined vaulting ‖ Kreuzgewölbe n. ‖ voûte f. croisée.

groove, to ‖ kehlen; nuten; auskehlen ‖ creuser; rainer; canneler. / ~ a cask ‖ ein Faß n. gargeln *oder* gergeln ‖ jabler un tonneau. / ~ and tongue together (Join) ‖ verspunden ‖ assembler à rainure et languette.

groove ‖ Aushöhlung f.; Höhlung f. ‖ creux m.; cavité f. / ~ (Cable) ‖ Kaliberröhre f. ‖ cylindre m. de réception. / ~ (Join) ‖ Lippe f. ‖ entaille f. de jointure. / ~ (Mach) ‖ Rille f.; Hohlkehle f.; Nut(e) f.; Keilnut f.; Furche f.; Riefe f.; Kehle f. ‖ rainure f.; cannelure f.; creusure f.; gorge f.; mortaise f. / ~ (Roll mill) ‖ Riffelung f.; Kannelierung f. ‖ cannelure f. / angular ~ (Roll mill) ‖ Spitzbogenkaliber n. ‖ cannelure f. ogive; ogive f. / annular ~ ‖ Ringnut f. ‖ rainure f. annulaire. / circular ~ ‖ ringförmige Nut f. ‖ rainure f. annulaire. / female ~ (Roll mill) ‖ Mutterfurche f. ‖ cannelure f. femelle. / flange ~ ‖ Spurrille f.; Spurrinne f. ‖ ornière f. de passage *ou* de la voie. / half-round ~ ‖ halbrunde Rinne f. ‖ rainure f. semi-circulaire. / helicoidal ~ ‖ Schraubennut f. ‖ rainure f. hélicoïdale. / hollow ~ (Join) ‖ Hohlleiste f. ‖ cannelure f. / middle ~ (Roll mill) ‖ Mittelfurche f. ‖ cannelure f. médiane. / ~s pl. of a millstone ‖ Schärfen fpl. *oder* Rillen fpl. *oder* Häuschläge mpl. *oder* Furchen fpl. eines Mühlsteines ‖ rayons mpl. *ou* éveillures fpl. d'une meule. / oil ~ ‖ Ölnut f. ‖ patte f. d'araignée. / piston ring ~ ‖ Kolbenringnut f. ‖ alvéole m. de piston; rainure f. annulaire de piston. / pointed ~ *see* angular ~. / ~ of a sewing needle ‖ Kerbe f. einer Nähnadel ‖ cannelure f. d'une aiguille à coudre. / ~ of a sheave ‖ Scheibenrinne f. einer Blockscheibe ‖ gorge f. d'une poulie. / spiral ~ ‖ Spiralnut f. ‖ rainure f. hélicoïdale. / ~ of thread ‖ Gewinderille f. ‖ rainure f. de filetage. / to solder into a turned ~ ‖ in eine Ausdrehung f. einlöten ‖ souder dans une gorge tournée.

groove-and-tongue ‖ Spund m.; Nut f. und Feder f. ‖ rainure f. et languette f. / ~ joint ‖ Verspundung f. ‖ assemblage m. à rainure et languette.

groove- cutting chisel ‖ Nuteisen n. ‖ bec m. d'âne. / ~ cutting machine for cask grooves ‖ Faßkrösemaschine f. ‖ machine f. à jâbler les douves.

grooved ‖ gefurcht; geriffelt; gerillt ‖ cannelé. / ~ beam (Carp) ‖ ausgefalzter Balken m. ‖ poutre f. feuillée. / ~ drum ‖ Rillentrommel f. ‖ tambour m. à gorge. / ~ pulley ‖ Keilnutscheibe f. ‖ poulie f. à gorge. / ~ rail ‖ Rillenschiene f. ‖ rail m. à gorge. / ~ roll ‖ Kaliberwalze f.; kalibrierte Walze f.; Kammwalze f. ‖ cylindre m. calibré *ou* cannelé. / ~ roller gear ‖ Kammwalzengetriebe n. ‖ harnais m. d'engrenages de cylindres à cannelures. / ~ wheel ‖ Schnurscheibe f.; Kerbrolle f. ‖ poulie f. à gorge.

groove disk (Mach) ‖ Mittelscheibe f.; Leerscheibe f. ‖ fer m. à cheval; faussererondelle f.; faux-taillant m.; rondelle f. entre-deux. / ~ milling machine (Roll mill) ‖ Kaliberfräsmaschine f. ‖ fraiseuse-calibre f. / ~ planer (Join) ‖ Nutenhobler m. ‖ bouveteur m.

groover (Needl) ‖ Einfeiler m. ‖ évideur m.

groove ram (Needl) ‖ Kerbenstampfer m. ‖ estampeur m. des aiguilles à coudre.

grooving ‖ Riffeln n. ‖ cannelage m. / ~ (Join) ‖ Auskehlung f. ‖ évidement m. / ~ (Roll mill) ‖ Kalibrierung f. ‖ cannelure f. / ~ comb (Coop) ‖ Gargelkamm m. ‖ gravoir m.; jâbloir m. / ~ cutter ‖ Nutenfräser m. ‖ fraise f. à rainures. / ~ machine ‖ Nutmaschine f.; Nutenstoßmaschine f. ‖ machine f. à rainurer *ou* à mortaiser. / ~ plane ‖ Nuthobel m.; Spundhobel m.; Ausstoßhobel m. ‖ rabot m. à rainures; bouvet m. femelle *ou* à rainures. / ~ press ‖ Langfalzzudrückpresse f. ‖ presse f. à serrer les agrafes. / ~ saw (Coop) ‖ Gargelsäge f. ‖ scie f. à graver. / circular ~ saw ‖ Kreisfalzsäge f. ‖ scie f. circulaire à rainer. / ~ tool *see* ~ comb.

gross amount ‖ Rohbetrag m.; Bruttobetrag m. ‖ montant m. brut. / ~ cost ‖ Bruttopreis m.; Preis m. einschließlich Verpackung ‖ prix m. brut. / ~ freight ‖ Bruttofracht f.; Fracht f. einschließlich Verpackung ‖ fret m. brut. / ~ load ‖ Bruttobelastung f. ‖ charge f. brute.

grossly, to cut ~ ‖ roh bearbeiten ‖ ébaucher.

gross proceeds pl. ‖ Rohertrag m. ‖ rendement m. brut. / ~ profits pl. ‖ Rohgewinn m. ‖ bénéfices mpl. bruts. / ~ taxation ‖ Rohbesteuerung f. ‖ imposition f. brute (d'une contribution). / ~ tonnage ‖ Brutto-tonnengehalt m. ‖ tonnage m. brut.

grossular (Miner) ‖ Grossular m.; Kalkgranat m.; grüner Granat m. ‖ grenat m. calcarifère *ou* grossulaire; grossulaire m.

gross weight ‖ Rohgewicht n.; Bruttogewicht n. ‖ poids m. brut.

grotesque (gind of letters) ‖ Grotesk f. ‖ caractères mpl. grotesques.

grotto ‖ Grotte f. ‖ rocaille f.; grotte f. / ~ maker ‖ Grottenbauer m. ‖ constructeur m. de grottes.

ground, to (Dyer) ‖ grundieren ‖ piéter. / ~ (Electr) ‖ erden ‖ mettre à la terre. / ~ (Nav) ‖ auf Grund auflaufen ‖ toucher le fond; échouer; mouiller par la quille. / ~ (Paint) ‖ grundieren ‖ abreuver; apprêter; donner l'apprêt; imprimer. / ~-in (Textile print) ‖ eindrucken ‖ rentrer; imprimer.

ground (Glass) ‖ geschliffen ‖ taillé; poli. / ~ (Metal) ‖ geschliffen; scharf émoulu; aiguisé. / finely ~ (Chem) ‖ feinzerrieben ‖ finement broyé. / hollow ~ edge ‖ eingeschliffene Hohlkehle f. ‖ cannelure f. obtenue à la meule. / ~ and washed chalk ‖ Schlämmkreide f. ‖ blanc m. lavé; blanc de Meudon. / ~ coffee ‖ gemahlener Kaffee m. ‖ café m. moulu.

ground-down ‖ eingeschliffen ‖ rodé; dressé. / the surface was ~ with the utmost accuracy ‖ die Oberfläche wurde auf das sorgfältigste geschliffen ‖ la surface fut passée à la meule avec la précision la plus minutieuse.

ground-glass ‖ Milchglas n. ‖ verre m. dépoli. / ~ disk ‖ Mattscheibe f. ‖ verre m. dépoli; écran m. en verre dépoli. / ~ screen *see* ~ disk.

ground-in piston ‖ eingeschliffener Kolben m. ‖ piston m. rodé. / ~ stopper ‖ eingeschliffener Stöpsel m. ‖ bouchon m. rodé.

ground leather ‖ Ledermehl n. ‖ cuir m. en poudre. / ~ lens-surface ‖ geschliffene Glasfläche f. ‖ face f. de verre taillée. / ~ material ‖ Mahlgut n. ‖ mouture f.

ground-out vessel-shaped ‖ tonnenförmig ausgeschliffen ‖ évasé en forme de tonneau.

ground toilet soap ‖ pilierte Toiletteseife f. ‖ savon m. de toilette pilé.

ground ‖ Erdboden m.; Boden m.; Grund m. und Boden m. ‖ sol m.; terrain m. / ~ (Area) ‖ Grundfläche f.; Bodenfläche f. ‖ superficie f. du terrain. / ~ (Build) ‖ Baugrund m. ‖ sol m.; terrain m. / ~ (Dyer) ‖ Grund m. ‖ pied m. / ~ (Electr) ‖ Erdverbindung f.; Erdschluß m. ‖ connexion f. de terre; terre f. accidentelle. / ~ (Mine) ‖ Gebirge n.; Nebengestein n. ‖ terrain m. / ~ (Colour) ‖ Grundfarbe f. ‖ couleur f. de fond. / above ~ (Mine) ‖ über Tage ‖ au jour. / below ~ ‖ unter Tage ‖ au fond; sous terre; à l'intérieur. / to examine the ~ *see* to sound the ~. / to gain ~ (Mar) ‖ Grund m. gewinnen ‖ gagner du terrain. / to hack the ~ ‖ den Erdboden m. mit der Hacke loshauen ‖ attaquer le terrain au pic. / to level the ~ ‖ den Boden ebnen ‖ dresser la surface. / to lose ~ (Nav) ‖ Grund m. verlieren ‖ perdre du terrain. / no ~! (Mar) ‖ kein Grund! ‖ pas de fond! / to ram the ~ ‖ den Boden m. festrammen ‖ battre *ou* corroyer le sol m. / to sound the ~ (Mar) ‖ den Grund m. abloten *oder* auspeilen ‖ sonder le fond m. / to strike ~ (Mar) ‖ Grund m. werfen ‖ avoir *ou* trouver fond m. / to throw-out the ~ ‖ den Erdboden m. abwerfen ‖ jeter le terrain sur la berge. / to touch the ~ ‖ auf Grund m. sitzen ‖ échouer; faire côte; mouiller par la quille. / to wheel the ~ ‖ die Erde f. abkarren ‖ brouetter de la terre. / adjoining ~ ‖ angrenzendes Gelände n.; angrenzender Braugrund m. ‖ terrain m. voisin. / clayey ~ ‖ Lehmgrund m. ‖ fond m. d'argile. / bad-holding ~ ‖ schlechter Grund m. ‖ terrain m. de mauvaise tenue. / ~ for dismissal ‖ Kündigungsgrund m. ‖ motif m. de congédiement. / filled-up ~ ‖ aufgeschütteter Boden m. ‖ remblai m. / flinty ~ ‖ Kieselgrund m. ‖ fond m. de galets. / foul ~ ‖ ungesunder Boden m. ‖ fond m. dangereux *ou* malsain. / good holding ~ (Build) ‖ guter *oder* haltbarer Grund m. ‖ fond m. de bonne tenue. / hard ~ (Build) ‖ harter Grund m. ‖ fond m. dur. / hard stony ~ ‖ fester steiniger Boden m. ‖ terrain m. dure et pierreux. / heavy ~ ‖ fester Boden m. ‖ terrain m. résistant; sol m. ferme. / light ~ ‖ leichter Boden m. ‖ terrain m. mou ou léger. / light ~ partially interspersed with stones ‖ leichter teils mit Steinen durchsetzter Boden m. ‖ terrain m. léger partiellement parsemé de pierres. / loamy ~ ‖ Lehmboden m. ‖ terre limoneuse. / marshy ~ ‖ Sumpfboden m. ‖ terrain m. marécageux. / muddy ~ ‖ Schlammgrund m.; Schlickgrund m.; Modergrund m. ‖ fond m. mou; fond m. de vase; vase f. molle. / natural ~ ‖ gewachsener Boden m. ‖ terre f. naturelle. / open ondulating ~ ‖ flachwelliges Gelände n. ‖ terrain m. légèrement accidenté. / pebble ~ ‖ Kieselgrund m. ‖ fond m. de galets. / rocky ~ ‖ felsiger Grund m. ‖ fond m. de roche. / sandy ~ ‖ sandiger Boden m. ‖ terrain m. sablonneux. / ~ of the sea ‖ Meeresgrund m.; Meeresboden m. ‖ fond m. de la mer. / shifting ~ (Mar) ‖ Wellgrund m. aus

Triebsand ‖ fond m. mouvant. / slicky ~ see muddy ~. / slimy ~ see muddy ~. / soft ~ ‖ weicher Grund m. ‖ fond m. mou; terrain m. croulier. / solid ~ (Mine) ‖ festes Gebirge n. ‖ terrain m. ferme. / sticky ~ ‖ stickiger Grund m. ‖ fond m. tenace. / stony ~ ‖ Steingrund m. ‖ terrain m. rocheux; enrochement m.

ground acquisition ‖ Grunderwerb m. ‖ acquisition f. de terrains. / ~ auger ‖ Erdbohrer m. ‖ tarière f. / ~ bait (Fish) ‖ Grundköder m. ‖ appât m. de fond. / ~ beam (Build) ‖ Schwelle f.; Grundschwelle f.; Bodenschwelle f.; Grundbalken m. ‖ racinal m.; dormant m.; semelle f. / ~ clearance (of a vehicle) ‖ Bodenfreiheit f. ‖ espace m. libre au-dessus du sol; dégagement m. entre le véhicule et le sol. / ~ colour ‖ Grundfarbe f.; Grundierfarbe f. ‖ couleur f. d'apprêt; première couche f. / ~ connection ‖ Erdverbindung f. ‖ contact m. ou connexion f. de terre. / ~ detector (Electr) ‖ Erdschlußprüfer m.; Erdschlußanzeiger m. ‖ indicateur m. de (perte à la) terre; cherche-pertes m. de courant.

grounded simplex circuit (Tel) ‖ Einfachleitung f. mit Erde ‖ ligne f. simple avec terre. / ~ wire (against lightning) ‖ Erdseil n.; Blitzschutzseil n. ‖ fil m. de terre; fil m. paratonnerre.

ground excavation ‖ Bodenbaggerung f. ‖ excavation f. de terrain.

ground-floor ‖ Erdgeschoß n.; Untergeschoß n. ‖ rez-de-chaussée m.; bas-étage m.; basse-œuvre f. / raised ~ ‖ Hochparterre n. ‖ entresol m.

ground fog ‖ Bodennebel m. ‖ bruine f.; brouillard m. inférieur. / ~ frost ‖ Bodenfrost m. ‖ gelée f. du sol. / ~ haze ‖ Bodendunst m. ‖ brume f. de sol. / ~ humidity ‖ Bodenfeuchtigkeit f. ‖ humidité f. du sol. / ~ ice ‖ Grundeis n. ‖ glace f. du fond.

grounding (Colour) see ground colour. / ~ (Electr) ‖ Erdung f. ‖ mise f. à la terre. / ~ of working batteries ‖ Erdung f. von Betriebsbatterien ‖ mise f. à la terre des batteries de service. / ~ leakage resistance (Tel) ‖ Erdableitungswiderstand m. ‖ résistance f. d'écoulement à la terre. / ~ machine (Pap) ‖ Fonciermaschine f.; Grundiermaschine f. ‖ machine f. à foncer; fonceuse f. / ~ sleeve (Tel) ‖ Erdbuchse f. ‖ manchon m. de mise à la terre.

ground landing ‖ Bodenlandung f. ‖ atterissage m. sur le sol. / ~ leakage on two phases ‖ Doppelerdschluß m. ‖ mise f. à la terre sur deux phases. / ~ line ‖ Grundlinie f. ‖ base f. / ~ litter ‖ Waldbodenstreu f. ‖ litière f. en feuilles tombées. / ~ location (Tel) ‖ Erdschlußbestimmung f. ‖ détermination f. de la mise à la terre accidentelle. / ~ log (Mar) ‖ Grundlog n. ‖ loch m. du fond. / ~ mist ‖ Bodendunst m. ‖ brume f. de sol. / ~ moraine (Geol) ‖ Grundmoräne f. ‖ moraine f. profonde. / nature of the ~ (Agr) ‖ Bodenbeschaffenheit f. ‖ nature f. du sol. / nature of ~ (Build; Railw) ‖ Geländebeschaffenheit f.; Terrainbeschaffenheit f. ‖ nature f. du terrain. / ~ net (Radio) ‖ Erdnetz n. ‖ prise f. de terre. / ~ nut ‖ Erdnuß f. ‖ arachide f.; pistache f. de terre. / ~ peg (Wav) ‖ Ankerpfahl m. ‖ piquet m. de retenue. / ~ plan (Build) ‖ Lageplan m.; Grundriß m.; Grundplan m. ‖ plan m.; ichnographie f.

ground-plate ‖ Grundplatte f.; Bodenplatte f. ‖ plaque f. de fondation ou d'assise. / cast iron ~ ‖ gußeiserne Grundplatte f. ‖ soubassement m. en fonte de fer. / ~ of a frame-work ‖ Bundschwelle f. ‖ semelle f. d'assemblage; sablière f.

ground projector ‖ Bodenscheinwerfer m. ‖ aéro-phare m. / ~ register ‖ Grundbuch n. ‖ registre-terrien m.; cadastre m. / ~ rent (Agr) ‖ Grundpacht f. ‖ redevance f. emphytéotique. / ~ rent (Pol ec) ‖ Grundrente f. ‖ rente f. foncière. / ~ rope-traction ‖ Unterseilführung f. ‖ passage m. du câble par endessous. / ~ sea (Mar) ‖ Grundsee f. ‖ mer f. de fond. / ~ signal ‖ Weichensignal n. ‖ signal m. de branchement ou d'aiguille; indicateur m. de direction. / ~ sill of a framework see ground-plate of a framework. / ~ slide ‖ Erdriese f. ‖ glissoir m. / ~ speed (Aero) ‖ Bodengeschwindigkeit f. ‖ vitesse f. au sol. / subsidence of the ~ ‖ Bodensenkung f. ‖ tassement m. du sol. / ~ survey ‖ Geländeaufnahme f. ‖ levée f. (des plans) des terrains. / ~ swell (Mar) ‖ Grunddünung f. ‖ flot m. ou gain m. du fond. / ~ tackle (Mar) ‖ Grundgeschirr n. ‖ apparaux mpl. de mouillage. / ~ temperature ‖ Bodentemperatur f. ‖ température f. du sol. / ~ terminal (Electr) ‖ Erdklemme f. ‖ borne f. de terre. / ~ thread ‖ Grundfaden m. ‖ fil m. de fond. / ~ timber see ~ beam. / ~ value ‖ Bodenwert m. ‖ valeur f. locative du terrain. / ~ wall of a furnace ‖ Ofenstock m. ‖ mur m. de base d'un fourneau. / ~ water ‖ Grundwasser n. ‖ eau f. souterraine. / ~ water movement ‖ Grundwasserbewegung f. ‖ mouvement m. de l'eau souterraine. / ~ weather map ‖ Bodenwetterkarte f. ‖ carte f. du temps au sol. / ~ wind ‖ Bodenwind m. ‖ vent m. au sol. / ~ wire (Electr) ‖ Erdleitung f.; Rückleitung f. durch die Erde; Erddraht m. ‖ fil m. de masse ou à la terre; retour m. par la terre.

groundwork (Earthwork) ‖ Erdarbeit f. ‖ terrassement m. / ~ (Foundation) ‖ Grundmauerwerk n.; Gründung f. ‖ maçonnerie f. de fondement ou de soubassement.

group ‖ Gruppe f. ‖ groupe f. / ~ (Miner) ‖ Nest n. ‖ nid m. / ~ (of undertakers) ‖ Gruppe f.; Konsortium n. ‖ consortium m.; syndicat m.; groupe m. / by ~s pl. (Miner) ‖ nesterweise ‖ par nids mpl. / ~ of electrically driven refrigerating machines ‖ Batterie f. elektrisch angetriebener Kältemaschinen ‖ groupe m. de machines frigorifiques commandées par des électromoteurs. / in ~s pl. ‖ gruppenweise ‖ par groupes mpl. / ~ of power stations ‖ Kraftwerkegruppe f. ‖ groupe m. d'usines génératrices.

group calling key (Tel) ‖ Gruppeneinstelltaste f. ‖ clef f. d'appel commun. / ~ connection (Acc) ‖ Gruppenverbindung f. ‖ liaison f. ou réunion f. par groupes. / ~ drive ‖ gruppenweiser Antrieb m.; Gruppenantrieb m. ‖ commande f. par groupes. / power required for ~ drive ‖ Kraftbedarf m. bei gruppenweisem Antrieb ‖ force m. motrice à commande par groupes. / ~ driving see ~ drive.

grouped style (Print) ‖ Blocksatz m. ‖ composition f. en forme de carré; style m. carré.

group frequency (Radio) ‖ Gruppenfre-

quenz f. ‖ fréquence f. de groupe; fréquence f. de série.

grouping (Tel) ‖ Gruppierung f. ‖ armement m. / ~ the circuits on pole-lines ‖ Leitungsgruppierung f. ‖ armement m. de la ligne. / fresh ~ of the supply area ‖ Neugliederung f. des Stromversorgungsgebietes ‖ réorganisation f. du centre de consommation. / ~ increase (Aut tel) ‖ Gruppenzuschlag m. ‖ majoration f. du nombre des sélecteurs d'un groupe.

grout, to ~ with cement ‖ mit Zement m. ausgießen ‖ couler du ciment (dans les joints).

grout ‖ Zementbrei m. ‖ boullie f. ou lait m. de ciment.

grow, to ‖ wachsen; größer werden ‖ croître. / ~ (Plant) ‖ gedeihen; wachsen ‖ venir; croître. / ~ dull (Met) ‖ abblicken; (matt werden). ‖ cesser de faire l'éclair. / ~ fat (Soap) ‖ zäh werden ‖ s'engraisser. / ~-out (Corn) ‖ auswachsen ‖ germer. / the spring grows slack ‖ die Feder setzt sich ‖ le ressort se détend.

growing ‖ wachsend ‖ croissant. / slow ~ ‖ langsam wachsend ‖ à croissance f. lente. / ~ demand ‖ wachsende Anforderung f. ‖ demande f. croissante.

growing of the cable (Mar) ‖ Richtung f. der Ankerkette ‖ appel m. du câble. / ~ of lime ‖ Aufgehen n. oder Gedeihen n. oder Wachsen n. des Kalkes ‖ foisonnement m. de la chaux.

growing-out (Corn) ‖ Auswachsen n. ‖ germination f. du grain.

grown soil ‖ gewachsener Boden m. ‖ terrain m. naturel.

grown-out (Corn) ‖ ausgewachsen ‖ germé.

growth ‖ Wachsen n.; Wachstum m.; Wuchs m. ‖ croissance f. / ~ in length ‖ Längenwachstum n. ‖ croissance f. en longueur.

groyne (Hydr arch) ‖ Buhne f. ‖ clayonnage m.; crèche f.; éperon m.

grubber ‖ Grubber m.; Bodenmaschine f. ‖ machine f. à essarter. / tree stump ~ ‖ Baumstumpfrodemaschine f. ‖ machine f. à déraciner des troncs d'arbres.

grubbing axe ‖ Rodehacke f. ‖ houe f. à essarter.

grub-stone mortar ‖ Grundmörtel m.; Gußmörtel m.; Beton m. ‖ béton m.

grude stove ‖ Grudeofen m. ‖ fourneau m. Grude.

grummet ‖ Dichtungsring m. ‖ rondelle f. ou anneau m. de joint. / ~ thimble ‖ Seilkausche f. ‖ cosse f. pour cordages.

guadalcazarite ‖ Guadalcazarit m. ‖ guadalcazarite f.

guaiacol ‖ Guajakol n. ‖ gaïacol m. / ~ carbonate ‖ Guajakolkarbonat n. ‖ carbonate m. de gaïacol. / ~ derivatives pl. ‖ Guajakolderivate npl. ‖ dérivés mpl. de gaïacol.

guaiac resin ‖ Guajakharz n. ‖ résine f. de gaïac. / ~ wood ‖ Packholz n. ‖ bois m. de gaiac.

guag (Mine) ‖ verhauenes Feld n.; alter Mann m.; alte Baue f. ‖ vieux travaux mpl. ou ouvrages mpl.

guano ‖ Guano m. ‖ guano m. / artificial ~ of flesh, fishes or blood ‖ künstlicher Guano m. aus Fleisch, Fischen oder Blut ‖ guano m. artificiel de viande, de poisson ou de sang. / natural ~ ‖ natürlicher Guano m. ‖ guano m. naturel. / phosphatic ~ ‖ Phosphorguano m. ‖ guano m. phosphaté.

guaranine || Guaranin n. || caféine f.; guaranine f.; théine f.

guarantee, to || garantieren; sicherstellen; haften; avalieren || garantir; avaliser.

guarantee || Garantie f.; Bürgschaft f.; Sicherheit f.; Gewähr f. || garantie f.; caution f.; sûreté f. / ~ of payment || Delkredere n. || ducroire m. / ~ bill || Kautionswechsel m. || cautionnement m. par traite. / ~ company || Garantiegesellschaft f. || compagnie f. de garantie. / ~ fund || Garantiefonds m. || fonds m. de garantie. / ~ test || Garantieversuch m. || épreuve f. de garantie.

guaranty || Sicherheit f.; Delcredere n.; Garantieschein m.; Garantie f. || garantie f.; caution f.; sûreté f.; ducroire m.

guard || Wächter m. || garde m. / ~ (Auto) || Schutzblech n. || gardeboue m. / ~ (Mach) || Schutzvorrichtung f. || dispositif m. de protection; protecteur m.; garde f. / ~ (Print) || Falz m. || pli m.; onglet m. / ~ of the train || Schaffner m. des Zuges || conducteur m. de train. / ~ of wheel || Radkasten m. || couvre-roue m.

guard boat || Rondeboot n. || canot m. de ronde. / ~ channel for terminal poles || Schutzkanal m. am Abspanngestänge || gaîne f. de protection pour points de concentration aériens.

guarder (Railw) || Bahnwächter m. || garde-ligne m.

guarding of the line || Bewachen n. der Linie || surveillance f. de la ligne. / ~ effect || Abschirmwirkung f. || effet m. de parachute.

guard lock (Hydr arch) || Sicherheitstor n. oder Fluttor n. einer Schleuse || écluse f. de garde. / ~ plate || Schutzblech n. || plaque f. de garde. / ~ rail || Gegenschiene f.; Schutzschiene f.; Leitschiene f.; Sicherheitsschiene f. || contre-rail m. / ~ rail chair || Leitschienenstuhl m. || coussinet m. de contre-rail. / ~ razor || Rasierapparat m. || rasoir m. de sûreté ou mécanique de sûreté. / ~ ship || Wachtschiff n. || navire m. de garde. / ~ socket || Schutztülle f. || douille f. de protection. / ~ stone || Prellstein m. || borne f.; boute-roue f. / ~ wire || Fangdraht m. || fil m. de protection.

gudgeon || eingesetzter Zapfen m. || tourillon m. emmanché. / ~ of the rudder || Ruderöse fpl. || femelot m.; conassière f. ou femelle f. du gouvernail; étambot m.

guess rope || Schlepptau n. || câble m. de remorque. / ~ warp see ~ rope.

guidance, restricted ~ || zwangsläufige Führung f. || guidage m. forcé.

guide || Führung f. || guide m. / ~ (Cable) || Leitauge f. || œil-guide f. / ~ (Windmill) || Aufleiter m. || coursier m. / ~ for axle boxes || Achsbüchsenführung f. || guide m. de la boîte d'essieu. / band saw ~ || Bandsägeführung f. || guidage m. pour scies à ruban. / carriage ~ || Schlittenführung f. || guidage m. du chariot. / central ~ || Zentralführung f. || guidage m. central. / cross-head ~ || Kreuzkopfführung f. || guide m. ou guidage m. de la tête de crosse. / journal box ~ || Achslagerführung f. || guide m. de la boîte d'essieu. / mitre cutting ~ || Gehrungsanschlag m. || guide m. pour couper les onglets. / rear axle ~ || Hinterachsführung f. || guide f. de l'essieu arrière. / ~ of the reversing lever || Führungsbogen m. des Steuerungshebels || secteur m. du levier de changement de marche. / ~ of slide || Stößelführung f. || guidage m. du coulisseau. / adjustable ~ of slide || nachstellbare Stößelführung f. || guidage m. de coulisseau réglable. / ~ of the tabulator key levers || Tastenhebelführungsblech n. || guide m. des barres de touches du tabulateur. / upper ~ || Oberführung f. || guide m. supérieur.

guide angle || Führungswinkeleisen n. || cornière f. de guidage. / ~ bar || Führungsstange f.; Geradführungsstange f.; Leitstange f. || barre f. ou tige f. conductrice; barre directrice. / ~ blade (Turbine) || Leitschaufel f. || aube f. directrice. / Kaplan turbine with adjustable runner vanes and ~ blades || Kaplanturbine f. mit regelbarem Laufapparat und Leitapparat || turbine f. Kaplan avec appareil de marche réglable. / to cast in the ~ blades || die Leitschaufeln fpl. eingießen || couler les aubes fpl. directrices dans la roue. / ~ block || Führungsblock m.; Gleitblock m.; Führungsschlitten m. || bloc m. de guidage; coulisseau m. des glissières. / ~ bush || Führungsbuchse f. || douille f. ou boîte f. de guidage ou conductrice. / ~ comb || Kammführung f. || peigne m. de guidage. / ~ cord || Leitseil n. || laisse f.

guided || zwangläufig || conduit. / ~ within the link || in der Schwinge f. geführt || guidé dans la coulisse.

guide fork || Führungsbügel m.; Führungsgabel f. || étrier m. de guidage; fourchette-conductrice. / ~ head || Führungskopf m. || tête f. de guidage. / cylindrical ~ housing || zylindrischer Führungskörper m. || boîte f. de guidage cylindrique. / ~ iron || Führungseisen n. || fer-guide m. / ~ line of an alidade rule (Surv) || Richtlinie f. eines Dioptrinlineals || ligne f. de foi d'une alhidade. / ~ liner can be easily and cheaply replaced || Gleitfutter n. kann leicht und billig ausgewechselt werden || fourrure f. de glissement se laisse remplacer d'une manière facile et à des prix modérés. / ~ mark (Print) || Markzeichen n.; Rapportpunkt m. || point m. de repère. / ~ mechanism || Leitvorrichtung f. || dispositif m. de guidage. / ~ piece || Führungsstück n. || pièce f. de guidage. / ~ pin || Führungsstift m.; Rapportstift m. || cheville f. de guidage; repère m. / ~ post (Ram) || Läufer m.; Läuferrute f. || guide m.; montant m. / ~ pulley || Leitrolle f.; Führungsrolle f.; Umlenkrolle f. || poulie-guide f.; galet-guide m. / ~ rail (Mine) || Spurlatte f. || rail m. pour le guidage. / ~ rail (Railw) || Gegenschiene f.; Führungsschiene f.; Leitschiene f. || contre-rail m.; rail m. de sûreté; rail-guide m. / ~ ring || Führungsring m. || bague f. de guidage. / ~ rod (Loc) || Lenkhebel m. || levier m. de manœuvre. / ~ rod (Mach) || Führungsstange f.; Geradführungsstange f.; Leitstange f. || barre f. ou tige f. conductrice; barre f. directrice; règle f. ou tige des glissières. / ~ rod (Windmill) || Richtstange f. || mât m. de voilure. / ~ roll || Führungswalze f. || cylindre-guide m.; cylindre m. de guidage. / ~ roll for dredgers || Leiterrolle f. für Bagger || tambour m. de guidage pour dragues. / ~ roller || Umlenkrolle f. || galet m. ou poulie f. de renvoi. / ~ roller (Print) ||

Leitwalze f. || rouleau m. conducteur. / ~ roller bearing || Führungsrollenlager n. || palier m. de guidage à rouleau. / ~ rope || Schleppseil n.; Leitseil n. || guide-rope m.; laisse f.; amarre f. de touage.

guides pl. || Geradführung f.; Parallelleitung f. || glissière f.; guide m. / ~ pl. (Print) || Leitepunkte mpl. || points mpl. conducteurs; carrés mpl.

guide screw || Leitspindel f. || arbre m. fileté de tour. / ~ screw (Turn) || Schraubenpatrone f.; Patrone f. || pas m. de vis. / ~ shoe of crosshead || Gleitschuh m.; Gleitklotz m. || patin m. ou guide m. de crosse. / ~ spindle || Leitspindel f. || vis-mère f. / ~ surface || Führungsbahn f. || surface f. de guidage. / ~ tube for aerial || Antennenführungsrohr n. || tube m. guide d'antenne. / ~ vane || Leitschaufel f. || aube f. directrice.

guide wheel || Leitrad n. || satellite m. / ~ carrier || Leitradträger m. || support m. de la roue satellite. / ~ halves pl. (Turbine) || Leitrad n. in Hälften || roue f. directrice en moitiés. / ~ spindle || Leitradwelle f. || axe m. de la satellite.

guiding attachment for photographic work || Pointiereinrichtung f. für fotografische Aufnahmen || appareil m. de pointage pour la photographie. / ~ cable || Lenkseil n.; Schwenkseil n.; Leitseil n. || hauban m.; écharpe f. / ~ device || Leitvorrichtung f. || dispositif m. de guidage.

guiding eyepiece || Pointierungsokular n. || oculaire m. de pointage. / ~ head || Pointierungsokularkopf m. || porte-oculaire m. de pointage. / ~ head with revolving position collar || Pointierungsokularkopf m. mit Positionswinkeleinstellung || porte-oculaire m. de pointage avec cercle de calage pour l'angle de position.

guiding length || Führungslänge f. || longueur f. de guidage. / ~ light || Richtungsfeuer n. || feu m. de direction. / ~ microscope || Pointierungsmikroskop n. || microscope m. de pointage. / ~ piece for the radial axle box housing || Führung f. für Radialachslagergehäuse || glissière f. pour le support de boîte radiale d'essieu. / ~ telescope || Leitfernrohr n. || lunette-guide f. / ~ traverse || Führungstraverse f. || traverse f. de guidage.

guild || Innung f.; Gilde f. || corps m. de métier; corporation f.

guilloshing || Guillochieren n. || guillochage m.

guillotine || Guillotine f. || guillotine f. / ~ shear || Kurbelblechtafelschere f. || cisaille f. à guillotine.

guipure || Gipüre f. || guipure f.

guitar || Gitarre f.; Zupfgeige f.; Klampfe f. || guitare f.

gulch gold || Seifengold n. || or m. d'alluvion.

gulf || Golf m.; Meerbusen m. || golfe m.

Gulf Stream || Golfstrom m. || courant m. du golfe; golf-stream m.

gullet (Water channel) see gully. / ~ of a bridge || Brückenbogen m. || arc m. ou arche f. d'un pont. / ~ of a saw tooth || Einschweifung f. des Sägezahns || échancrure f. d'une dent de scie. / ~ tooth (Saw) || Wolfszahn m.; Hakenzahn m. || dent f. à crochet.

gully (Build) || Sinkloch n.; Gully m.; Kanaleinlauf m.; Absturzschacht m. || puisard m.; puits m. d'écoulement. / ~ (Hydr arch) || Wasserrinne f.; Wasser-

furche f.; Abzugskanal m. ‖ gouotlte f.;
rigole f.; ravine f. / ~ of a pulley ‖ Rille
f. einer Seilscheibe ‖ gorge f. d'une
poulie.

gully hole ‖ Schlammfang m.; Schlamm-
kasten m.; Ausflußöffnung f. eines Ab-
zugskanals ‖ bassin m. de dépôt; vasière f.

gum, to ‖ gummieren ‖ gommer.

gum ‖ Gummi n. ‖ gomme f. / ~ (Office) ‖
Büroleim m. ‖ colle f. de bureau. / ~
(Resin) ‖ Harz n. ‖ résine f. / acacine ~ ‖
Gummi n. arabikum ‖ gomme f. arabique.
/ artificial ~ ‖ Stärkegummi n. ‖ gomme f.
artificielle. / benzoïn ~ ‖ Benzoeharz n.
‖ gomme f. de benzoïne; benjoin m.

gum arabic ‖ Gummiarabikum n. ‖ gomme
f. arabique.

gumbelite ‖ Gumbelit m.; weißer Pyro-
phyllit m. ‖ gumbelite f.

gum drop ‖ Gummibonbon m. ‖ boule f. de
gomme. / ~ elastic ‖ Kautschuk n.;
Gummielastikum n. ‖ caoutchouc m.;
gomme f. élastique. / ~ gutta ‖ Gummi-
gutti n. ‖ gomme-gutte f. / ~ lac ‖ Gummilack n. ‖ gomme f. laque.

gummed cloth ‖ gummierter Stoff m. ‖
étoffe f. gommée. / ~ paper ‖ gummiertes
Papier n. ‖ papier m. gommé. / ~ tape ‖
gummiertes Band n. ‖ ruban m. gommé.

gummer (Bookb) ‖ Leimer m. ‖ encolleur m.

gumming ‖ Gummieren n. ‖ gommage m. /
~ machine ‖ Gummiermaschine f. ‖ machine f. à gommer. / ~ machine for boxes
‖ Klebemaschine f. für Schachteln ‖ colleuse f. pour boîtes. / ~ machine for envelopes ‖ Briefumschlagklebemaschine f.
‖ colleuse f. pour enveloppes à lettres.

gummous ‖ gummiartig ‖ gommeux.

gummy see gummous.

gum resin ‖ Gummiharz n. ‖ gomme-résine
f.; résine f. de caoutchouc. / ~ water for
marbling (Pap) ‖ Marmorierwasser n. ‖
gomme f. à marbrer.

gun ‖ Feuerwaffe f. ‖ arme f. à feu. / ~
(Cannon) ‖ Kanone f.; Geschütz n. ‖ canon m.; pièce f. (de canon); bouche f. à
feu. / ~ (Rifle) ‖ Gewehr n.; Flinte f. ‖
fusil m. / anti-tank ~ ‖ Kampfwagenabwehrgeschütz n. ‖ canon m. antichar. /
coast ~ ‖ Küstengeschütz n. ‖ pièce f. de
côte. / directing ~ ‖ Leitgeschütz n. ‖
pièce f. directrice; pièce-guide f. / double-
barrel infantry accompanying ~ ‖ dop-
pelrohriges Infanteriegeschütz n. ‖ canon
m. d'infanterie à double tube. / laying
the ~ for pointer fire ‖ Einzelrichten n.
des Geschützes ‖ pointage m. individuel
du canon. / military ~ ‖ Kriegsgewehr n. ‖
fusil m. de guerre. / pigeon shooting ~ ‖
Taubenflinte f. ‖ fusil m. de tir aux pigeons. / pivot ~ see revolving ~. / revolving ~ ‖ Drehgeschütz n. ‖ pièce f.
tournante ou (sur) pivot. / ~ on slide-
carriage ‖ Geschütz n. auf Rahmen-
lafette ‖ pièce f. sur affût à châssis. /
sporting ~ ‖ Jagdgewehr n. ‖ fusil m. de
chasse. / turning ~ see revolving ~. /
turret ~ ‖ Turmgeschütz n. ‖ pièce f. de
tourelle. / unlimbered ~ ‖ abgeprotztes
Geschütz n. ‖ pièce f. en batterie. / upper-
deck ~ ‖ Deckgeschütz n. ‖ pièce f. des
gaillards.

gun barrel (Cannon) ‖ Geschützrohr n. ‖
bouche f. à feu; canon m. / ~ (Rifle)
(Rifle) ‖ Gewehrlauf m.; Flintenlauf m.;
Lauf m. ‖ canon m. de fusil.

gunboat ‖ Kanonenboot n. ‖ canonnière f.;
chaloupe f. canonnière.

gun carriage ‖ Lafette f. ‖ affût m. / motor
caterpillar ~ ‖ Raupenkraftlafette f. ‖ affût m. automoteur à chenille. / universal
~ ‖ Rundumlafette f. ‖ affût m. universel.

gun carriage support ‖ Lafettenbefestigung
f. ‖ fixation f. de l'affût. / ~ carrier (Cannon) ‖ Artillerieträger m. ‖ camion m.
d'artillerie.

gun cartridge ‖ Kartusche f. ‖ gargousse f. /
shell of ~ ‖ Kartuschenhülse f. ‖ douille
f. de gargousse.

gun clip (Rifle) ‖ Gewehrträger m. ‖ porte-
fusil m.

gun cotton ‖ Schießbaumwolle f.; Pyroxylin n.; Schießwolle f. ‖ coton-poudre m.;
poudre-coton m.; fulmicoton m.; pyroxyline f.; nitrocellulose f.; coton m. explosif. / ~ boiler ‖ Schießwollekocher m. ‖
bouilleur m. de coton explosif. / ~ powder ‖ Schießwollepulver n. ‖ coton-poudre m.; poudre-coton m.

gun deck ‖ Batteriedeck m. ‖ faux-pont m.

gun fire ‖ Artilleriefeuer n. ‖ tir m. de l'artillerie. / ~ adjustment ‖ Artilleriefeuer-
regelung f. ‖ réglage m. du tir de l'artillerie.

gun fire control ‖ Feuerleitung f.; Ge-
schützrichten n. ‖ conduite f. du tir de
l'artillerie. / ~ equipment ‖ Feuerleitge-
rätausrüstung f. ‖ équipement m. des instruments pour la direction du tir de
l'artillerie. / ~ gyroscopic device for ~ ‖
Kreiselvorrichtung f. zum Geschützrich-
ten ‖ appareil m. gyroscopique de la conduite du tir de l'artillerie.

gun foundry ‖ Geschützgießerei f. ‖ fonderie f. de canons. / ~ furniture ‖ Gewehr-
teile mpl. ‖ pièces fpl. détachées pour fusils. / ~ layer ‖ Geschützrichter m. ‖ pointeur m. / ~ limber ‖ Geschützprotze f.;
Protze f. ‖ avant-train m. d'affût; caisson m. / ~ lock ‖ Schloß n.; Gewehrschloß n.; Flintenschloß n. ‖ platine f. /
maker see ~ smith. / ~ manufacturer ‖
Waffenfabrikant m. ‖ fabricant m. d'armes.

gun metal ‖ Kanonenmetall n.; Rotguß m.
‖ bronze m. rouge; bronze m. ou métal m.
à canons. / ~ case ‖ Rotgußgehäuse n. ‖
boîte f. en bronze rouge. / ~ wing ‖ Bron-
zeflügel m. ‖ aile m. en bronze.

gun mount see ~ carriage. / ~ mounting see
~ carriage.

gunnage ‖ Bestückung f. ‖ armement m. (de
canons).

gunnel see gunwale.

gunner ‖ Artillerist m.; Schütze m.; Kano-
nier m. ‖ artilleur m.; canonnier m. / machine ~ ‖ Maschinengewehrschütze m. ‖
mitrailleur m.

gunnet of a door ‖ Türband n. ‖ fiche f.
d'une porte.

gunny ‖ Sacktuch n.; Juteleinwand f.;
Sackleinwand f. ‖ grosse toile f.

gun port ‖ Stückpforte f. ‖ sabord m.

gunpowder ‖ Pulver n.; Schießpulver n. ‖
poudre f.; poudre f. noire ou à canon ou
de chasse (et de guerre). / ~ for the war,
shooting and mines ‖ Schießpulver n. für
Kriegs-, Jagd- und Bergwerkszwecke ‖
poudre f. de guerre, de chasse et de mine.
/ smokeless ~ ‖ rauchloses Schießpulver
n. ‖ poudre f. sans fumée.

gunpowder cake ‖ Pulverkuchen m. ‖ galette f. ou pâte f. ou lame f. de poudre.
/ ~ driver ‖ Pulverramme f. ‖ sonnette à
poudre. / ~ dust ‖ Pulverstaub m. ‖ poussier m. de poudre. / ~ factory ‖ Pulver-

fabrik f. ‖ moulin m. à poudre; poudrerie
f.; poudrière f. / ~ maker ‖ Pulvermüller
m. ‖ poudrier m. / ~ manufacture ‖ Pul-
verherstellung f.; Pulverfertigung f. ‖ fabrication f. de poudre. / ~ manufacturing
machine ‖ Schießpulverfabrikationsma-
schine f. ‖ machine f. pour la fabrication
de poudre à canon.

gun rod ‖ Ladestock m. ‖ baguette f. / ~
shot ‖ Schrot m.; n.; Flintenschrot m., n.
‖ plomb m. granulé ou de chasse; dragée
f.; grenaille f. de plomb. / ~ smith ‖ Büchsenmacher m. ‖ armurier m.

gun stock ‖ Gewehrschaft m. ‖ fût m. ou
bois m. de fusil. / ~ of walnut ‖ Gewehrschaft n. aus Walnußholz ‖ bois m. de
fusil en noyer.

gun support see ~ clip. / ~ truck ‖ Ge-
schützwagen m. ‖ truck m. au transport
des canons. / ~ turret ‖ Geschützturm m.
‖ tourelle f. des canons.

gunwale (Shipb) ‖ Schandeck n.; Schandeckel m. ‖ plat-bord m.

Gurjun balsam ‖ Kopaivabalsam m. ‖
baume m. de copahu.

gusher of natural gas ‖ Erdgasquelle f. ‖
source f. de gaz naturel.

gusset ‖ Eckblech n. ‖ gousset m. / ~ joint
connection ‖ Knotenblechverbindung f.
‖ assemblage m. par gousset. / ~ plate ‖
Knotenblech n.; Eckblech n. ‖ plaque f.
de jonction ou d'éclissage; gousset m. /
~ sheet see ~ plate. / ~ stay ‖ Eck-
verstrebung f. ‖ goussetage m.

gust ‖ Bö f.; Stoßwind m. ‖ rafale f. /
violent ~ ‖ schwere Bö f. ‖ violente
rafale f.; grain m. violent.

gust recorder ‖ Böenschreiber m. ‖ enregistreur m. de grains ou de rafales.

gut, to ~ and salt herrings pl. ‖ Heringe
mpl. kaken ‖ caquer les harengs mpl.

gut (Butch) ‖ Darm m. ‖ boyau m. / ~
(Gold-b) ‖ Goldschlägerhaut f. ‖ baudruche f. / ~ (Mar) ‖ enge Einfahrt f. ‖
goulet m. / animal ~ ‖ Tierdarm m. ‖
boyau m. d'animal.

gut assorting apparatus ‖ Darmsortier-
apparat m. ‖ appareil m. à trier les
boyaux. / ~ dressing ‖ Darmbereitung f.
‖ préparation f. des boyaux. / goods pl.
from ~s ‖ Waren fpl. aus Därmen ‖
ouvrages mpl. en boyaux. / ~ maker ‖
Darmschleimer m. ‖ boyaudier m. / ~
spinner ‖ Darmspinner m. ‖ fileur m. de
boyaux.

gut string ‖ Darmsaite f. ‖ corde f. en
boyau. / ~ making machine ‖ Darmsaitenherstellungsmaschine f. ‖ machine
f. à fabriquer des cordes en boyaux.

gutta-percha ‖ Guttapercha f. ‖ gutta-
percha f.; gutta f. / bleached ~ ‖ ge-
bleichte Guttapercha f. ‖ gutta-percha f.
blanchie. / dental ~ ‖ Zahnkautschuk m.
‖ gutta-percha f. dentaire. / ~ mixed
with other materials ‖ mit anderen Stoffen gemischte Guttapercha f. ‖ gutta-
percha f. mélangée à d'autres matières.
/ raw ~ ‖ rohe Guttapercha f. ‖ gutta-
percha f. brute.

gutta-percha articles pl. ‖ Guttapercha-
waren fpl. ‖ objects mpl. en gutta-percha.
~ bottle ‖ Guttaperchaflasche f. ‖ bouteille f. ou flacon m. en gutta-percha.

gutta-percha cable ‖ Guttaperchakabel n.
‖ câble m. sous gutta. / box for ~ ‖
Guttaperchakabelmuffe f. ‖ manchon m.
pour câble sous gutta.

gutta-percha cement || Guttaperchakitt m. || mastic m. à la gutta-percha. / ~ covering machine || Guttaperchaumpreß-maschine f.; Guttaperchapresse f. || machine f. pour la guttapercha; presse f. enveloppeuse à gutta. / electric cable covered with ~ || mit Guttapercha f. isoliertes Kabel || câble m. électrique isolé à la gutta. / ~ goods pl. *see* ~ articles. / ~ tooth cement || Guttaperchazahnkitt m. || gutta-percha f. pour les dents. / ~ trade machine || Guttaperchaindustriemaschine f. || machine f. pour l'industrie de la gutta-percha. / ~ worker || Guttaperchaarbeiter m. || ouvrier m. en gutta-percha.

gutter, to || auskehlen || canneler.

gutter || Furche f.; Rinne f.; Rille f. || rigole f.; goulotte f. / ~ (Build) || Dachrinne f.; Traufrinne f.; Gosse f. || gouttière f.; chéneau m. / ~ (Found) || Rinne f.; Gießrinne f. || échenau m.; canal m. de coulage. / ~ (Mach) || Hohlkehle f. || rainure f. cannelure f. / ~ (Road || Rinnstein m.; Rinne f.; Straßenrinne f.; Gosse f. || caniveau m.; rigole f. d'assèchement; ruisseau *ou* égout m. de rue. / long ~ to allow of simultaneously pouring out several crucibles || lange Rinne f. zum gleichzeitigen Ausgießen mehrerer Tiegel || rigole f. assez longue pour la coulée simultanée de plusieurs creusets. / long-stringed ~ (Build) || Strangfalzziegel m. || tuile f. à ourlet longitudinal. / wooden ~ || hölzerne Rinne f. || échenal m. *ou* enchenot m. *ou* rigole f. en bois.

gutter hook || Dachhaken m. || collier m. de gouttière. / ~ lead || Kehlblech n.; Bleiblech n. || noquet m. / ~ pipe || Dachrinne f.; Fallrohr n.; Dachröhre f. || tuyau m. de descente gouttière. / ~ ring || Dachrinnenkausche f. || anneau m. de gouttière. / ~ sticks pl. (Print) || Formatquadrate npl.; Hohlstege mpl. || bois mpl. de marge. / ~ stone *see* gutter (Road).

gutter tile || Falzziegel m.; Hohlziegel m.; Kehlziegel m. || tuile f. à onglet *ou* ondulée. / double ~ || doppelter Falzziegel m. || tuile f. à onglet double.

gut works pl. || Darmschleimerei f. || boyauderie f.

guy (Mar) || Backstag m. || haubans mpl. du beaupré. / ~ (Windmill) || Leitstrick m. || hauban m. / handling ~ (Aero) || Hochlaßtau n. || câble m. de transport du ballon flottant dans l'air. / stays pl. || Abspannmaterial n. || matériel m. d'étayage.

guy clamp (El line) || Ankerschelle f. || collier m. de l'hauban. / ~ wire hook (El line) || Ankerhaken m. || crochet m. de hauban.

gyle tun (Brew) || Anstellbottich m.; Gärbottich m. || cuve f. guilloire.

gymnastic apparatus || Turngerät n.; Gymnastikgerät n. || appareil m. de gymnastique. / ~ appliance *see* ~ apparatus. / ~ and fencing club || Turn- und Fechtklub m. || club m. de gymnastique et d'escrime. / ~ hall || Turnhalle f. || salle f. de gymnastique; gymnase m. couvert.

gymnastics pl., hygienic || Heilgymnastik f. || gymnastique f. hygiénique *ou* médicale. / apparatus for hygienic ~ || heilgymnastischer Apparat m. || appareil m. de gymnastique médicale.

gymnastic shoe || Turnschuh m. || chaussure f. de gymnastique *ou* pour gymnastes.

gymnast's shoe *see* gymnastic shoe.

gymnite || Gymnit m.; Deweylit m. || gymnite f.

gynecologic instrument || gynäkologisches Instrument n. || instrument m. gynécologique.

gynocardia oil || Gynokardiaöl n. || huile f. de gynocardia.

gypseous marl || Gipsmergel m. || marne f. gypseuse. / ~ spar || Gipsspat m.; Marienglas n.; Selenit m. || gypse m. cunéiforme *ou* en fer de lance *ou* spathique; sélénite f.

gypsum || Gips m. || plâtre m.; gypse m. / burnt ~ || gebrannter Gips m. || plâtre m. cuit. / compact ~ || Alabaster m.; körniger Gips m. || albâtre m.; gypse m. compact. / earthy ~ || Gipserde f. || gypse m. terreux. / fibrous ~ || Fasergips m. || gypse m. fibreux. / foliated granular ~ || schuppigkörniger Gips m. || gypse m. saccharoïde. / scaly foliated ~ || schaumiger Gips m.; Schneegips m. || chaux f. sulfatée niviforme. / snowy ~ *see* scaly foliated ~. / sparry ~ *see* gypseous spar. / specular ~ *see* gypseous spar. / unburnt ~ || ungebrannter Gips m. || plâtre m. cru.

gypsum articles pl. || Gipswaren fpl. || objets mpl. en plâtre. / ~ block || Gipsblock m. || bloc m. de plâtre. / ~ block culture || Gipsblockkultur f. || culture f. sur bloc de plâtre. / ~ burner || Gipsbrenner m. || cuiseur m. de plâtre. / ~ burning || Gipsbrennerei f. || cuite f. du plâtre. / ~ crusher || Gipsbrecher m. || concasseur m. à plâtre. / ~ deal || Gipsdiele f. || planche f. de gypse *ou* de plâtre. / ~ mill || Gipsmühle f. || moulin m. à plâtre. / ~ mould || Gipsmatrize f. || moule f. de plâtre. / ~ pit || Gipsgrube f.; Gipsbruch m. || plâtrière f.; carrière f. de plâtre. / ~ plate (Opt) || Gipsplättchen n. || lame f. de gypse. / ~ producing plant || Gipsbrennerei f. || installation f. de fabrication du plâtre. / ~ quarry *see* ~ pit. / ~ statue || Gipsfigur f.; Gipsstatue f. || figure f. *ou* statue f. en plâtre.

gyration || Wirbel m. || tournoiement m.; tourbillon m.; tourbillonnement m.

gyre || Kreis m.; Kreisbewegung f. || mouvement m. circulaire.

gyro centre || Kreiselschwerpunkt m. || centre m. de gravité du gyroscope.

gyro compass || Kreiselkompaß m. || compas m. *ou* boussole f. gyroscopique. / ~ for gunfire control || Kreiselkompaß m. für Feuerleitung || boussole f. gyroscopique pour la conduite de tir.

gyrolite || Gyrolith m. || gyrolite f.

gyrometer || Gyrometer n. || gyromètre m.

gyro movement || Kreiselbewegung f. || mouvement m. gyroscopique.

gyro oscillation || Kreiselschwingung f. || oscillation f. de gyroscope. / undamped ~ || ungedämpfte Kreiselschwingung f. || oscillation f. de gyroscope non amortie.

gyro pilot || Kreiselsteuergerät n. || appareil m. de pilotage gyroscopique.

gyroscope || Kreisel m.; Gyroskop n. || toupie f.; gyroscope m. / torpedo ~ || Torpedokreiselgeradlaufgerät n. || instrument m. gyroscopique pour torpilles. / centre of gravity of ~ the *see* gyro centre.

gyroscopic || gyroskopisch; Kreisel ... || gyroscopique. / ~ compass *see* gyro compass. / ~ direction indicator (Arm) || Kreiselfeuerleitanzeiger m. || indicateur m. gyroscopique. / ~ effect || Kreiselwirkung f. || effet m. de toupie. / ~ gunfire control || Kreiselvorrichtung f. zum Geschützrichten || appareil m. gyroscopique de la conduite du tir de l'artillerie. / ~ roll-and-pitch-recorder || Kreiselschlinger- und -stampfanzeiger m. || appareil m. enregistreur gyroscopique de roulis et de tangage.

gyrostatic *see* gyroscopic. / ~ compass *see* gyro compass.

H

H-iron ‖ Doppel-T-Eisen n.; I-Eisen n. ‖ fer m. double I ou T. / **~-pole** ‖ Doppelgestänge n. ‖ appui m. en H; ligne f. double. / **~-section armature** ‖ Doppel-T-Anker m. ‖ induit m. en I ou en double T. / **~-section iron** see H-iron. / **~-shackle** ‖ Steglasche f. in H-Form ‖ jumelles fpl. en une pièce.

HP (Horse power) ‖ Pferdestärke f.; Pferdekraft f. ‖ cheval-vapeur m.

haberdashery ‖ Kurzwaren fpl. ‖ mercerie f. / **~** from natural horsehair ‖ Bandwaren fpl. aus natürlichem Roßhaar ‖ merceries fpl. en crin naturel. / **~** of metal covered with thread ‖ Kurzwaren pl. aus mit Fäden übersponnenem Metall ‖ articles mpl. de mercerie en métal recouvert de fil.

habitable ‖ wohnlich ‖ comfortable; commode.

habitation building ‖ Wohngebäude n. ‖ bâtiment m. d'habitation.

hack, to ‖ aufgraben ‖ fouiller; découvrir. / **~** the ground ‖ den Erdboden m. loshauen ‖ creuser le terrain.

hack ‖ Hacke f. ‖ pioche f.; houe f. / **~** (Vehicle) ‖ Fiaker m. ‖ fiacre m.

hack file ‖ Messerfeile f. ‖ lime f. en couteau ou à dossière. / **~ iron** ‖ Schrotfäustel m. ‖ masse f. à tranche.

hackle, to ~ hemp ‖ Hanf m. hecheln ‖ sérancer ou peigner le chanvre.

hackle (Spinn) ‖ Hechel f. ‖ séran m.; peigne m.; sérançoir f.

hackler ‖ Hechler m. ‖ séranceur m.

hackling ‖ Hecheln n. ‖ sérançage m.; peignage m. / **~ machine** ‖ Hechelmaschine f. ‖ peigneuse f. mécanique; machine f. à peigner. / **~ room** ‖ Flachshechelei f. ‖ salle f. de peignage du lin.

hackney carriage ‖ Droschke f.; Fiaker m. ‖ voiture f. de place; fiacre m. / **~ driver** ‖ Droschkenkutscher m. ‖ cocher de fiacre m.

hacksaw ‖ Metallsäge f.; Handsäge f. ‖ scie f. à métaux ou à main. / power **~** ‖ Kaltsäge f.; Maschinensäge f. ‖ scie f. alternative; machine f. à scier alternative.

haddock smoker ‖ Schellfischräucherer m. ‖ fumeur m. de merluches.

hade, to (Mine) ‖ einfallen ‖ plonger; s'incliner; descendre.

hade (Mine) ‖ Einfallen n. ‖ pente f.; inclinaison f. / **~** flat **~** of a coal seam ‖ flaches Einfallen n. eines Kohlenflözes ‖ inclinaison f. ou pente f. faible d'une couche de houille.

haematite ‖ Hämatit m.; Roteisenerz n. ‖ oligiste m.; ématite f. / red **~** ‖ Roteisenstein m.; Eisenoxyd n. ‖ fer m. oligiste ou oxydé rouge; hématite m. rouge.

haft of the spade ‖ Spatenstiel m. ‖ manche m. de pelle.

hail, to ~ a ship ‖ ein Schiff preien oder anrufen ‖ héler ou appeler un vaisseau. / gun fired to hail a ship ‖ Preischuß m. ‖ coup m. de canon à héler un vaisseau.

hail ‖ Hagel m. ‖ grêle f. / **~** cloud ‖ Hagelwolke f. ‖ nuage m. de grêle. / damage done by **~** ‖ Hagelschlag m. ‖ coup m. de grêle; dommage m. causé par la grêle.

hailing, range of **~** ‖ Preidistanz f.; Rufweite f. ‖ portée f. de la voix.

hail-squall ‖ Hagelbö f. ‖ grain m. de vent à grêle ou accompagné de grêle.

hailstorm insurance ‖ Hagelschadenversicherung f. ‖ assurance f. contre la grêle.

hair (Animal) ‖ Haar n. ‖ crin m. / **~** (Person) ‖ Haar n. ‖ cheveu m. / **~** of the alpaca ‖ Pakohaar n. ‖ alpaga m.; poil m. d'alpaga. / **~** of animal origin ‖ Haar n. tierischen Ursprungs ‖ poil m. d'origine animale. / coarse **~** ‖ grobes Haar n. ‖ poil m. grossier. / coarse animal **~** ‖ grobes Tierhaar n. ‖ poil m. grossier d'animal. / cow **~** ‖ Kuhhaar n. ‖ poil m. de vache. / curled **~** ‖ Krollhaar n. ‖ poil m. ondulé. / **~** glued upon tissue ‖ auf Gewebe aufgeklebtes Haar n. ‖ poil m. collé sur tissus. / horse **~** ‖ Roßhaar n.; Pferdehaar n. ‖ crin m. (de cheval). / horse **~** stuff ‖ Roßhaarstoff m. ‖ étoffe f. ou tissu m. de crins. / human **~** ‖ Menschenhaar n. ‖ cheveux mpl. humains. / mane **~** ‖ Mähnenhaar n. ‖ poil m. de la crinière du cheval. / spun **~** ‖ Krollhaar n. ‖ crin m. frisé. / **~** for stuffing ‖ Füllhaar n.; Polsterhaar n. ‖ rembourrure f. / **~** of the tail (Horse) ‖ Schweifhaar n. ‖ crin m. de la queue. / woven **~** ‖ Haartuch n. ‖ étoffe f. ou tissu m. de crin. / woven **~** for furniture ‖ Möbelzeug n.; Roßhaarstoff m. zu Möbeln ‖ tissu m. ou étoffe f. de crin pour meubles.

hair, artist in **~** ‖ Haarkünstler m. ‖ artiste m. en cheveux. / **~** belting ‖ Haartreibriemen m. ‖ courroie f. en crin. / **~** bleacher ‖ Haarbleicher m. ‖ blanchisseur m. de cheveux. / **~** blowing ‖ Fellhaarbläserei f. ‖ soufflerie f. de poils. / **~** broom ‖ Haarbesen m. ‖ balai m. de crin. / **~** brush ‖ Haarbürste f. ‖ brosse f. à cheveux. / **~** brush (Painter) ‖ Haarpinsel m. ‖ pinceau m. en poils. / **~** carder (Spinn) ‖ Haarkrämpler m. ‖ cardeur m. de poils. / **~** clippers pl. ‖ Haarschneidemaschine f. ‖ tondeuse f.

hair cloth ‖ Haargewebe n.; Haartuch n. ‖ tissu m. de crin; tulle-cheveux m. / **~** for furniture ‖ Roßhaarmöbelstoff m. ‖ tissu m. de crin pour meubles. / **~** for sieves ‖ Haarsiebgeflecht m. ‖ toile f. en crin ou tissu m. de crin pour tamis.

hair, metal fittings pl. for **~** combs ‖ Kammbeschlag m. ‖ articles mpl. en métal pour peignes. / **~** compasses pl. ‖ Haarzirkel m. ‖ compas m. de précision. / **~** cordage ‖ Seilerarbeiten fpl. aus Haaren ‖ cordes fpl. en poils. / **~** cross (Opt) ‖ Fadenkreuz n. ‖ réticule m. / **~** cutter ‖ Haarschneider m. ‖ coupeur m. de cheveux. / **~** cutting ‖ Haarschneiderei f. ‖ couperie f. de poils. / designer in **~** ‖ Haarkunstflechter m. ‖ dessinateur m. en cheveux. / **~** divider ‖ Federzirkel m. ‖ compas m. à ressort.

hairdresser ‖ Friseur m.; Haarschneider m. ‖ coiffeur m. / **~'s appliance** ‖ Friseurapparat m. ‖ appareil m. à l'usage des coiffeurs. / **~'s chair** ‖ Frisierstuhl m. ‖ chaise f. pour coiffeurs. / **~'s scissors** pl. ‖ Frisörschere f. ‖ ciseau m. pour perruquiers.

hair dressing ‖ Haarbereitung f. ‖ préparation f. de cheveux. / **~** lamp ‖ Frisierlampe f. ‖ lampe f. à friser.

hair drier ‖ Haartrockner m. ‖ sèche-cheveux m. / **~** drying apparatus ‖ Haartrockengerät n. ‖ appareil m. à sécher les cheveux. / **~** dye ‖ Haarfarbe f.; Haarfärbemittel n. ‖ teinture f. pour les cheveux. / **~** dyeing tincture see **~** dye. / **~** dyer ‖ Haarfärber m. ‖ teinturier m. de cheveux. / **~** felt ‖ Haarfilz m. ‖ feutre m. de poil. / **~** hat ‖ Haarhut m. ‖ chapeau m. de poil. / **~** hygrometer ‖ Haarfeuchtigkeitsmesser m.; Haarhygrometer n. ‖ hygromètre m. à cheveux. / **~** jeweller in **~** ‖ Haarschmuckarbeiter m. ‖ bijoutier m. en cheveux. / **~** jewellery ‖ Fantasiewaren fpl. aus Haaren ‖ bijoux mpl. en cheveux. / **~** lace ‖ Haarnetz n. ‖ résille f. ou filet m. en cheveux; filet m. de front. / **~-like** structure ‖ haarartiges Gebilde n. ‖ structure f. capillaire. / **~** line (Print) ‖ Haarstrich m. ‖ délié m. / **~** net ‖ Haarnetz n. ‖ filet-coiffure m. / **~** oil ‖ Haaröl n. ‖ huile f. pour les cheveux. / **~** pin ‖ Haarnadel f. ‖ épingle f. à cheveux. / **~** plaiter ‖ Zopfflechter m. ‖ natteur m. de cheveux. / **~** pomade ‖ Haarpomade f. ‖ pommade f. pour les cheveux. / **~** pomatum see **~** pomade. / **~** puncheon ‖ Haarpunze f. ‖ matoir m. rayé. / **~** pyrites pl. ‖ Haarkies m.; Schwefelnickelkies m. ‖ pyrite f. capillaire; nickel m. sulfuré; millérite f. / **~** salt ‖ Haarsalz n. ‖ halotrichite f. / **~** scissors pl. ‖ Haarschere f. ‖ ciseaux mpl. à crins. / **~** side of the leather ‖ Haarseite f. oder Narbenseite f. des Leders ‖ côté m. poil du cuir. / **~** sieve ‖ Haarsieb n. ‖ tamis m. en crin. / **~** space (Print) ‖ Haarspatium n. ‖ espace m. fine. / **~** spinner ‖ Haarspinner m. ‖ fileur m. de poils. / **~** spinning ‖ Haarspinnerei f. ‖ filature f. de poils ou bourre ou crins. / **~** stroke see **~** line. / **~** tour ‖ Haartour f. ‖ tour m. de cheveux. / **~** tress ‖ Haarzopf m. ‖ tresse f. de cheveux. / **~** trigger lock (Gunsmith) ‖ Stechschloß m. ‖ platine f. à double détente. / **~** tuft ‖ Haarbusch m. ‖ crinière f. / **~** wash ‖ Haarwasser n. ‖ lotion f. pour les cheveux. / **~** washer ‖ Haarwäscher m. ‖ laveur m. de cheveux. / **~** work ‖ Haararbeit f. ‖ ouvrage m. en cheveux. / **~** worker ‖ Haararbeiter m. ‖ ouvrier m. en cheveux.

hairy cotton ‖ haarige Baumwolle f. ‖ coton m. velu. / **~** wool ‖ Haarwolle f. ‖ laine f. lisse.

hair yarn ‖ Haargarn n. ‖ fil m. de poils. / **~** spinning mill ‖ Haargarnspinnerei f. ‖ filature f. de fils de poils.

half ‖ halb ‖ demi. / ~ ‖ Hälfte f. ‖ moitié f. / ~ of axle ‖ Halbachse f.; Achsschenkel m. ‖ demi-essieu m. / ~ of the bearing ‖ Lagerhälfte f. ‖ moitié f. de palier. / ~ of bearing lining ‖ Lagerschalenhälfte f. ‖ moitié f. du coussinet. / ~ of the system (Electr) ‖ Netzhälfte f. ‖ demi-réseau m.

half-beam ‖ Halbbalken m. ‖ demi-bau m. / ~ binding ‖ Halbfranzband m.; Halblederband m. ‖ mi-reliure f. / ~ boiled ‖ halbgekocht ‖ mi-cuit.

half-boot ‖ Halbschuh m. ‖ soulier m. découvert. / ~ leg ‖ Halbstiefelschaft m. ‖ tige f. de bottine.

half-box (Found) ‖ Kastenhälfte f. ‖ demi-châssis m. / ~ breadth-plan (Shipb) ‖ Wasserlinienriß m. ‖ plan m. de flottaison; plan m. horizontal. / ~ calf see half-binding. / ~ chess (Bridge) ‖ Halbbrett n.; Halbpfosten m. ‖ demi-madrier m. / ~ cover ‖ Halbverdeck n. ‖ capote f.; demi-toit m.; demi-pavillon m. / ~ damask ‖ Halbseide f.; Seidendamast m.; Halbdamast m. ‖ damas-caffard m.; damassin m.; lampas m. / ~ deck (Shipb) ‖ Halbdeck n. ‖ demi-pont m.

half-double plate wheel ‖ Halbdoppelscheibenrad n. ‖ roue f. à disque semi-double.

half-finished goods pl. ‖ Halbzeug n. ‖ demi-produits mpl. / ~ products pl. see half-finished goods pl. / ~ steel-product ‖ A-Produkt n.; Halbzeugprodukt n. ‖ demi-produit m. en acier.

half-gang (Weav) ‖ halber Gang m. ‖ demi-portée f.

half-gross weight ‖ Halbrohgewicht n. ‖ poids m. demi-brut.

half-header (Mas) ‖ Quartierstein m.; Riemenstein m. ‖ demi-tuile f.; demi-brique f.

half-hitch (Mar) ‖ Halbstich m. ‖ demi-clef m. / ~ knot (Bridge) ‖ Schlüsselknoten m.; Halbstichschurz m. ‖ amarrage m. par demi-clefs.

half-hook shaped ‖ in Halbhakenform f. ‖ en forme f. du semi-crochet.

half-hose ‖ Locke f.; Halbstrumpf m. ‖ chaussette f.

half-linen ‖ Halbleinen n. ‖ mi-lin m.; demi-toile f. / ~ weaving manufacture ‖ Halbleinenweberei f. ‖ fabrication f. de tissus mi-toile.

half-mast high (Mar) ‖ halbmasts ‖ à mi-mât.

half-oval ‖ halboval ‖ demi-ovale.

half-painted phototypography ‖ Halbtonätzung f. ‖ phototypographie f. en demi-teinte.

half-pint ‖ Schoppen m. ‖ chopine f. / ~ place (Build) ‖ halber oder kurzer Podest m. ‖ demi-palier m. / ~ plank (Carp) ‖ Schalbrett n.; Tafelbrett n.; Dünnbrett n.; halbzölliges Brett n. ‖ planche f. ou ais m. d'un demi pouce d'épaisseur. / ~ porter see half-gang.

half-precious stone ‖ Halbedelstein m. ‖ pierre f. demi-précieuse.

half-principal see half-truss. / ~ roof see half-cover.

half-round ‖ halbrund ‖ demi-rond. / ~ iron ‖ halbrundes Eisen n. ‖ fer m. demi-rond. / ~ file ‖ Halbrundfeile f. ‖ lime f. demi-ronde. / ~ wood ‖ Halbholz n. ‖ bois m. mi-plat.

half-scoured see half-boiled.

half-silk ‖ Halbseide f. ‖ mi-soie f. / ~ weaving manufacture ‖ Halbseidenweberei f. ‖ fabrication f. de tissus mi-soie.

half-sized (Pap) ‖ halbgeleimt ‖ mi-collé.

half-staff high (Mar) ‖ halbstocks ‖ à mi-bâton.

half-story (Build) ‖ Halbgeschoß n. ‖ entre-sol m.; mezzanine f. / ~ stuff (Pap) ‖ Halbzeug n. ‖ pâte f. effilochée; demi-pâte f.; défilé m.

half-sunk traverser ‖ halbversenkte Drehscheibe. ‖ transbordeur m. à canivaux.

half-timer ‖ Halbtagsarbeiter m. ‖ ouvrier m. au demi-temps. / ~ tin ‖ Halbzinn n. ‖ étain m. bas.

half-tone (Print) ‖ Halbton m. ‖ demi-teinte f. / ~ engraving see ~ etching.

half-tone etching ‖ Autotypie f. ‖ autotypie f.; similigravure f. / ~ on copper ‖ Kupferätzung f. ‖ gravure f. sur cuivre. / printing of ~s ‖ Autotypiedruck m. ‖ impression f. de similigravure.

half-track tank ‖ Räderraupenkampfwagen m. ‖ char m. d'assaut à roues et à chenilles.

half-transparent ‖ halbdurchsichtig ‖ à demi-opaque; semi-opaque. / ~ truss (Carp) ‖ halbes Gebinde n.; Halbbinder m. ‖ demi-ferme f. de croupe. / ~ turret ‖ Halbturm m. ‖ demi-tourelle f.

half-twist bit ‖ Schneckenbohrer m. ‖ tarière f. hélicoïdale.

half-wet method (Met) ‖ Halbnaßverfahren n. ‖ voie f. ou méthode f. demi-sèche. / ~ treatment see half-wet method.

half-window ‖ Halbgeschoßfenster n. ‖ fenêtre f. mezzanine. / ~ wool ‖ Halbwolle f. ‖ mi-laine f.

half-woollen blanket ‖ Halbwolldecke f. ‖ couverture f. mi-laine.

half-wool weaving manufacture ‖ Halbwollweberei f. ‖ fabrication f. de tissus mi-laine.

halite ‖ Bergsalz n.; Steinsalz n. ‖ sel m. marin ou de gemme; halite f.

hall ‖ Saal m.; Halle f.; Raum m. ‖ salle f.; halle f.; hall m. / ~ to take dinners brought by the workers ‖ Saal m. zur Einnahme der mitgebrachten Mahlzeit ‖ réfectoire m. pour prendre les repas apportés. / glazed ~ ‖ verglaste Halle f. ‖ hall m. vitré. / iron ~ ‖ Eisenhalle f. ‖ halle f. en charpente métallique. / ~ with models ‖ Modellsaal m. ‖ salle f. de modèles. / ~ for railway station ‖ Bahnhofshalle f. ‖ hall m. pour gares ou pour voyageurs. / steel framed ~ construction ‖ Eisenhallenbau m. ‖ construction f. de halles métalliques.

halliard see halyard.

halloysite ‖ Halloysit m. ‖ halloysite f.

halo (Opt) ‖ Hof m. ‖ halo m.; couronne f.

halogen (Chem) ‖ Halogen n. ‖ halogène m. / ~ acid ‖ Halogensäure f. ‖ acide m. halogénique.

haloid salt ‖ Haloidsalz n. ‖ sel m. haloïde.

halometer ‖ Halometer n.; Salzwage f. ‖ halomètre m.

halometry ‖ Halometrie f.; Messung f. des Salzgehaltes ‖ halométrie f.

halotrichite (Miner) ‖ Eisenalaun m.; Bergbutter f. ‖ halotrichite f.; beurre m. de roche.

halt, to (Railw) ‖ halten ‖ stationner; arrêter.

halt (Railw) ‖ Haltestelle f.; Haltepunkt m.; Halbstation f. ‖ halte f.; point m. ou station f. d'arrêt.

halter ‖ Halfter f. ‖ licou m. / ~ chain ‖ Halfterkette f. ‖ chaîne f. de licou. / ~ rope see ~ strap. / ~ strap ‖ Halfter-

zügel m.; Halfterleine f.; Halfterriemen m.; Halfterstrick m. ‖ longe f. bouclée.

halting place see halt. / ~ station see halt.

halvan ore see halvans.

halvans pl. (Min) ‖ Pocherz n.; Pochgänge mpl. ‖ minerai m. (pauvre) à bocarder.; minerai m. de rebut.

halve, to ‖ halbieren ‖ partager en deux. / ~ (Wood) ‖ verblatten; abblatten ‖ assembler à mi-bois. / ~ the rails in the middle and to lap them together ‖ die Schienen überblatten ‖ assembler les rails à mi-fer.

halved belt ‖ gekreuzter Riemen m. ‖ courroie f. croisée ou demi-tordue. / ~ joint see halving.

halving ‖ Halbierung f. ‖ mi-partition f. / ~ (Wood) ‖ Überblattung f.; Anblattung f.; Verblattung f. ‖ assemblage m. ou entaille f. ou coupe f. à mi-bois. / ~ (Cast-iron) ‖ Überblattung f. ‖ assemblage m. à mi-fonte. / ~ (Wrought-iron) ‖ Überblattung f. ‖ assemblage m. à mi-fer.

halyard (Mar) ‖ Fall n. ‖ drisse. / studdingsail ~ ‖ Leesegelsfall n. ‖ drisse f. de bonnette.

ham ‖ Räucherschinken m.; Schinken m. ‖ jambon m. fumé. / ~ (of the knee) ‖ Kniekehle f. ‖ jarret m.

hame (Saddl) ‖ Kumtfeder f.; Kumtholz n.; Kumtthorn n. ‖ attelle f. (d'un collier). / horse collar ~ ‖ Kumt n.; Kummet n. ‖ collier m. de cheval. / iron ~ ‖ Kumteisen n. ‖ attelle f. en fer.

hame cover ‖ Kumtdecke f. ‖ couverture f. d'attelle. / loop of a ~ ‖ Kumtfederöse f.; Zugöse f. ‖ œil m. ou anneau m. d'attelle.

hames eye see hame, loop of a ~.

hame strap ‖ Kumt(feder)riemen m. ‖ courroie f. d'attelle. / ~ thong ‖ Kumtgurtriemen m. ‖ couplière f.

hammer, to ‖ hämmern; schmieden; treiben; schlagen ‖ marteler; faire au marteau; battre; forger; martiner. / ~ (To vibrate) ‖ schlagen; zittern; hämmern; stoßen ‖ tambouriner; marteler; cogner.

hammer, to cool-~ see to cold-~. / to cold-~ ‖ kaltschmieden; kalthämmern ‖ battre ou marteler à froid; écrouir. / to ~ even ‖ aushämmern; flachhämmern; platthämmern; strecken; recken; ausschmieden ‖ rabattre; aplatir. / to ~ even the iron ‖ das Eisen strecken oder recken ‖ étirer ou aplatir le fer sous le marteau. / to ~ the iron ‖ das Eisen schmieden oder hämmern ‖ forger ou battre le fer. / to ~ out ‖ ausschmieden; hämmern; strecken; richten ‖ forger ou étirer au marteau; marteler. / to ~ quickly ‖ kurz zuschlagen ‖ frapper un coup sec. / to ~ roughly ‖ grob abhämmern ‖ dégrossir au marteau.

hammer ‖ Hammer m. ‖ marteau m. / ~ (Larger) see lump ~. / ~ (Min) see miner's ~. / ~ (Pap) ‖ Stampfe m. ‖ maillet m.; pilon m. / ~ (Piano) ‖ Hammer m.; Klavierhammer m. ‖ marteau m. (de piano). / ~ (Typewr) ‖ Klöppel m. ‖ battant m. / about-sledge ~ ‖ Kreuzschlaghammer m. ‖ marteau m. à devant à panne en travers. / ball ~ ‖ Kugelhammer m. ‖ marteau m. rond. / ballast ~ ‖ Schotterschlegel m. ‖ marteau m. à concasser le ballast. / ball-face ~ see ball ~. / ball-pane ~ ‖ Hammer m. mit Kugelfinne ‖ marteau m. à panne sphérique. / bell ~ ‖ Klöppel m.; Glockenhammer m.; Schlegel m.; Schlaghebel m.; Schlag-

hammer m. ‖ marteau m. (de sonnerie); levier m. percuteur. / belt-driven ~ ‖ Transmissionshammer m.; durch Riemen angetriebener Hammer m. ‖ marteau m. à commande par courroie. / bench ~ ‖ Bankhammer m. ‖ marteau m. d'établi. / boiler scaling ~ see scaling ~. / bricklayer's ~ see mason's ~. / broad-face ~ ‖ Wulsthammer m. ‖ marteau m. de refoulement. / calking ~ see caulking ~. / cam ~ ‖ Daumenhammer m. ‖ marteau m. à cames. / carpenter's ~ ‖ Schreinerhammer m. ‖ marteau m. de charpentier. / casting cleansing ~ ‖ Gußputzhammer m. ‖ marteau m. à ébarber la fonte. / caulking ~ ‖ Meißelhammer m.; Kalfathammer m.; Stemmhammer m.; marteau m. à mâter; marteau m. matoir. / chasing ~ ‖ Treibhammer m.; Tellerhammer m. ‖ marteau m. à emboutir ou à chasser. / chipping ~ ‖ Schrothammer m. ‖ marteau m. à tranche. / claw ~ ‖ Klauenhammer m. ‖ Hammer m. mit Nagelziehklaue oder mit gespaltener Finne ‖ marteau m. fendu ou à panne fendue ou à dent. / clawed ~ see claw ~. / ~ of a clock bell see bell ~. / coal-pick ~ ‖ Abbauhammer m. ‖ marteau-pic m.; marteau m. minier. / ~with common pane ‖ Hammer m. mit gewöhnlicher Finne ‖ marteau m. à panne ordinaire. / compound spring ~ ‖ Verbundfederhammer m. ‖ marteau m. à ressort compound. / compressed air ~ see also pneumatic ~ ‖ Preßlufthammer m.; Drucklufthammer m.; Luftdruckhammer m. ‖ marteau m. ou mouton m. à air comprimé. / contact ~ ‖ Kontakthammer m. ‖ marteau m. de contact. / copper ~ ‖ Kupferhammer m. ‖ marteau m. en cuivre. / ~ operated by a crank ‖ Kurbelhammer m.; durch Kurbel angetriebener Hammer m. ‖ marteau m. à manivelle. / creasing ~ ‖ Sickenhammer m. ‖ marteau m. de suage ou à soyer. / cross-pane ~ ‖ Hammer m. mit Kreuzfinne ‖ marteau m. à panne de travers. / dead-stroke ~ see spring ~. / dowel socket ~ ‖ Dübellochhammer m. ‖ marteau m. pour trous à cheville. / drift ~ ‖ Dornhammer m.; Durchschlaghammer m.; Lochhammer m.; Durchtreiber m. ‖ chasse-coin m. / drilling ~ ‖ Bohrhammer m. ‖ marteau m. foreur; marteau m. perforateur. / drop ~ ‖ Fallhammer m.; Rahmenhammer m. ‖ marteau-pilon m. ou mouton m. (à friction); mouton m. à estamper ou marteau m. à chute libre. / electric-pneumatic ~ ‖ elektropneumatischer Hammer m. ‖ marteau m. électro-pneumatique. / engineer's ~ ‖ Maschinenbauerhammer m. ‖ marteau m. d'ajusteur. / enlarging ~ ‖ Streckhammer m. ‖ marteau m. à dégrossir. / file ~ ‖ Feilenhammer m. ‖ marteau m. à limes. / finishing ~ ‖ Polierhammer m. ‖ marteau m. finisseur / fitter's ~ see locksmith's ~. / flat (lump) ~ ‖ Streckhammer m.; Flachhammer m.; Flächhammer m. ‖ marteau m. plat ou à parer. / ~ with flat pane ‖ Hammer m. mit flacher Finne ‖ marteau m. à panne plate. / forge ~ ‖ Schmiedehammer m.; Hüttenhammer m. ‖ marteau-pilon m.; marteau m. de forge ou à forger; martinet m. / friction ~ ‖ Friktionshammer m. ‖ marteau m. à friction. / friction spring ~ ‖ Friktionsfederhammer m. ‖ marteau m. à ressort à friction. / front ~ ‖ Stirn-

hammer m. ‖ marteau m. frontal ou à chasse. / gas-driven ~ ‖ Gashammer m. ‖ marteau m. à gaz. / gong ~ see bell ~. / granulating ~ ‖ Stockhammer m.; Kraushammer m. ‖ marteau m. à granuler; boucharde f. / half-round ~ see ball-pane ~. / hand ~ ‖ Handhammer m.; Fausthammer m.; Fäustel n.; Schlegel m. ‖ marteau m. à main; martelet m. / heavy miner's ~ ‖ schweres Bohrfäustel n. ‖ massette f. lourde. / helving ~ ‖ Brusthammer m.; Aufwerfhammer m. ‖ marteau m. à devant ou à soulèvement. / helving ~ with a wooden shaft ‖ Aufwerfhammer m. mit hölzernem Stiel ‖ marteau m. à soulèvement à manche en bois. / hydraulic ~ ‖ Preßhammer m.; Wasserhammer m.; hydraulischer Hammer m. ‖ marteau m. à pression (hydraulique); marteau m. hydraulique. / laminated spring ~ ‖ Blattfederhammer m. ‖ marteau m. à ressort à lames. / large ~ of masons ‖ Boßhammer m.; Posseckel m. ‖ pic m. du maçon; gros marteau m. à briser. / lead ~ ‖ Bleihammer m. ‖ marteau m. ou masse f. en plomb. / light miner's ~ ‖ leichtes Bohrfäustel n. ‖ massette f. légère. / locksmith's ~ ‖ Schlosserhammer m.; Bankhammer m. ‖ marteau m. de serrurier ou d'établi. / lump ~ ‖ Schlegel m.; Schlägel m.; Fäustel n.; Knüppel m.; (eiserner) Hammer m. ‖ massette f.; masse f.; marteau m.; têtu m. / machine ~ see also power ~. ‖ Maschinenhammer m. ‖ marteau m. mécanique. / mash ~ see lump ~. / mason's ~ ‖ Maurerhammer m. ‖ marteau m. de maçon. / millstone ~ ‖ Mühlsteinpicke f. ‖ marteau m. pour moulins ou pour meules. / miner's ~ ‖ Fäustel n.; Handfäustel n.; Treibfäustel n.; Handschlegel m.; Bohrfäustel n. ‖ massette f.; masse f. à main. / pansing ~ ‖ Pinnhammer m. ‖ marteau m. à panne. / paring ~ ‖ Pritschhammer m .‖ marteau m. à ébarber. / paver's ~ ‖ Pflasterhammer m.; Steinsetzerhammer m. ‖ marteau m. de paveur ou d'assiette. / plane set ~ see square set ~. / planishing ~ ‖ Schlichthammer m.; Polierhammer m.; Planierhammer m. ‖ marteau m. à planer; flatoir m. / pneumatic ~ (Hand) ‖ Preßlufthammer m.; Drucklufthammer m. ‖ frappeur m. pneumatique ou à air comprimé. / pneumatic ~ (Mach) see also compressed-air ~ ‖ pneumatischer Hammer m.; Lufthammer m. ‖ marteau m. ou mouton m. pneumatique. / pneumatic coal pick ~ ‖ Preßluftabbauhammer m. ‖ marteau m. d'exploitation minière à air comprimé; marteau-pic m. pneumatique. / pneumatic scaling ~ ‖ Preßluftabklopfer m. ‖ détarteur m. pneumatique. / pneumatic spring ~ ‖ Druckluftfederhammer m. ‖ marteau m. à ressort pneumatique. / pointed ~ ‖ Spitzhammer m.; Keilhammer m. ‖ marteau m. à pointe ou à queue conique. / pointing ~ ‖ Anspitzhammer m. ‖ marteau m. à appointer. / power ~ ‖ Krafthammer m.; marteau-pilon m.; mouton m.; martinet m. / power ~ (Forge) see forge ~. / pressing ~ see hydraulic ~. / printing ~ (Typewr) ‖ Typenhammer m. ‖ touche f. à caractères. / punch ~ see drift ~. / rapid-action ~ ‖ Schnell(schlag)hammer m. ‖ marteau m. rapide; martinet m. / riveting ~ ‖ Niethammer m. ‖ rivoir m.;

marteau m. riveur ou rivoir ou à river. / rock-drill ~ (Min) ‖ Bohrhammer m.; Bohrfäustel m. ‖ marteau m. perforateur; massette f.; marteau m. de forage. / rounding ~ see round-set ~. / round-set ~ ‖ Ballhammer m.; Rundhammer m. ‖ marteau m. à balle; dégorgeoir m.; chasse f. ronde. / rust ~ see scaling ~. ‖ Schupphammer m.; Pickhammer m. ‖ marteau m. dérouilleur. / scale ~ see scaling ~. / scaling ~ ‖ Kesselsteinhammer m.; Pickhammer m.; Abklopfhammer m. ‖ marteau m. de piquage; marteau m. détrarteur; marteau-pic m.; picoche f. / scythe ~ ‖ Dengelhammer m. ‖ marteau m. à chapler les faux. / set ~ ‖ Setzhammer m.; Aufsatzhammer m. ‖ chasse f. / shaft ~ ‖ Stielhammer m. ‖ marteau m. à manche. / shingling ~ ‖ Zänghammer m. ‖ marteau m. cingleur; cinglard m. / sinking ~ ‖ Abteufhammer m. ‖ marteau m. perforateur de puits. / sledge ~ (Forg) ‖ Vorschlaghammer m.; Zuschlaghammer m.; Stielhammer m. ‖ marteau m. de frappeur; marteau m. à devant ou à frapper devant ou à manche. / sledge ~ (Railw) ‖ Schienenhammer m.; Schwellenhammer m. ‖ marteau m. à rails. / slow motion belly ~ ‖ Brusthammer m. mit langsamen Schlägen ‖ marteau m. frontal à coups lents. / small ~ (Min) see miner's ~. / smoothing ~ see planishing ~. / snap ~ ‖ Schellhammer m. ‖ bouterolle f. à œil ou à manche; chasse f. creuse. / spalling ~ ‖ Schrothammer m.; Zurichthammer m. ‖ marteau m. à dégrossir ou à tranche; dégrossisseur m. / spreading ~ see chasing ~. / spring ~ ‖ Federhammer m. ‖ marteau m. ou pilon m. à ressort. / square set ~ ‖ Setzhammer m. ‖ chasse f. carrée. / steam ~ ‖ Dampfhammer m. ‖ marteau-pilon m. à vapeur; marteau m. à vapeur. / steam ~ with a weight of tup of x kg ‖ Dampfhammer m. mit x kg Bärgewicht ‖ marteau-pilon m. à vapeur (poids du mouton x kg). / stone ~ ‖ Steinhammer m. ‖ casse-pierres m. / stone-mason's ~ ‖ Bossierhammer m.; Steinmetzhammer m. ‖ marteau m. de tailleur de pierres. / stretching ~ ‖ Spannhammer m. ‖ marteau m. à dresser. / tail ~ ‖ Schwanzhammer m. ‖ marteau m. à queue ou à levier. / tilt ~ see tail ~. / treadle ~ ‖ Fußhammer m.; Tritthammer m. ‖ marteau m. à pédale. / trip ~ see tail ~. / two edged ~ ‖ zweischneidiger Hammer m. ‖ marteau m. à deux biseaux. / two-handed ~ see sledge ~. / upholsterer's claw ~ ‖ Tapeziererhammer m. mit Nagelklaue ‖ marteau m. de tapissier avec tire-clou. / vertical ~ ‖ stehender oder senkrechter Hammer m. ‖ mouton m.; marteau m. vertical. / waller's ~ ‖ Abputzhammer m. ‖ marteau m. à repiquer. / wedge ~ ‖ Keilhammer m. ‖ chasse-coin m. / wedge-ended ~ see pointed ~. / wipple ~ ‖ Wipphammer m. ‖ martinet m. à ressort. / wooden ~ ‖ Holzhammer m. ‖ maillet m.; maillet m. en bois.

hammer axe ‖ Hammerbeil n. ‖ hachette f. à marteau. / carpenter's ~ ‖ Zimmermannshammerbeil. ‖ hacherau m. de charpentier.

hammer beam (Build) ‖ Stichbalken m. ‖ entrait m. retroussé.

hammer blow ‖ Hammerschlag m.; Schlag m. oder Stoß m. des Hammers ‖ coup m.

de marteau. / ~ (Railw) ‖ Schlag m.; Stoß m. ‖ choc m.

hammer body *see* ~ frame. / ~ bolt ‖ Hammerschraube f. ‖ boulon m. de fondation. / brake-control of ~ ‖ Regelung f. des Hammers durch Bremsen ‖ réglage m. du marteau par freinage. / ~ break *see* ~ interrupter. / ~ clip ‖ Eisenbeschlag m. am Hammer ‖ abras m. / ~ contact ‖ Hammerkontakt m. ‖ contact m. à marteau. / cushioning of a ~ ‖ Abfederung f. der Stöße des Hammers; Luftpufferung f. ‖ amortissement m. *ou* tamponnage des coups du marteau. / ~ cylinder ‖ Bärzylinder m. ‖ cylindre m. de mouton.

hammer-dress, to *see* to hammer.

hammer drill ‖ Bohrhammer m. ‖ marteau m. perforateur. / ~ driver (Forg) ‖ Hammerführer m. ‖ conducteur m. de marteau.

hammered ‖ gehämmert ‖ martelé; battu. / ~ glass ‖ gehämmertes Glas n. ‖ verre m. martelé. / ~ iron ‖ gehämmertes Eisen n. ‖ fer m. martelé. / ~ sheet iron ‖ gehämmertes Eisenblech n. ‖ tôle f. martelée.

hammer edge *see* hammer pane.

hammerer *see* hammerman.

hammer face (Hand hammer) ‖ Hammerbahn f.; Bahn f. des Hammers ‖ face f. *ou* aire f. du marteau. / ~ (Power hammer) ‖ Hammerbahn f. ‖ frappe f. du marteau. / interchangeable ~ ‖ auswechselbare Hammerbahn f. ‖ étampe f. *ou* frappe f. interchangeable du mouton.

hammer, flat side of a ~ *see* hammer face. / ~ frame ‖ Hammergerüst n.; Hammergestell n. ‖ chevalet m. *ou* bâti m. du pilon. / ~ furnace ‖ Schmiedeofen m. ‖ foyer m. à forger. / ~ guides pl. ‖ Hammerführung f.; Hammergleitbahn f.; Bärführung f. ‖ guidage m. de la masse; glissières fpl. *ou* coulisses fpl. du mouton. / ~ handle *see* ~ shaft.

hammer-harden, to ‖ hartschlagen; kalthämmern; kaltschmieden; federhart machen ‖ écrouir; battre à froid.

hammer-hardened iron ‖ kaltgeschmiedetes Eisen ‖ fer m. écroui.

hammer-hardening (Met) ‖ Kaltschmieden n.; Hartschlagen n. ‖ écrouissage m.; écrouissement m.

hammer hatchet *see* hammer axe.

hammer head (Hand hammer) ‖ Hammerkopf m. ‖ tête f. de marteau. / ~ (Power hammer) *see* ~ tup. / ~ chisel ‖ Schlageisen n. ‖ ciseau m. à froid. / ~ coverer (Piano) ‖ Hammerkopffilzer m. ‖ garnisseur m. de marteaux (de piano). / ~ crane ‖ Hammerkran m. ‖ grue-marteau f. / ~ eye ‖ Hammerauge n.; Hammeröhr n.; Hammerloch n. ‖ œil m. *ou* emmanchure f. d'un marteau.

hammer-headed dock crane ‖ Hammerwippkran m. ‖ grue-marteau f. à volée variable.

hammering ‖ Hämmern n. ‖ martelage m.; battage m. / water ~ ‖ Wasserschlag m. ‖ coup m. de bélier.

hammering test ‖ Ausbreit(e)probe f. ‖ essai m. d'élargissement.

hammer interrupter ‖ Hammerunterbrecher m. ‖ interrupteur m. à marteau. / magnetic ~ ‖ magnetischer Hammerunterbrecher m. ‖ interrupteur m. magnétique à marteau.

hammerman ‖ Hammerarbeiter m.; Hämmerer m. ‖ ouvrier m. de forge; marteleur m.; martineur m.

hammer mark (Sheet iron) ‖ Beule f. ‖ bosse f.

hammer mill (Forge) ‖ Hammerwerk n.; Hammerschmiede f.; Hammerhütte f.; Eisenhammer m.; Eisenhammerwerk n. ‖ forge f.; atelier m. de forge. / ~ (Ore dress) ‖ Schlagkreuzmühle f. ‖ broyeur m. à marteaux. / small ~ (Forge) ‖ Handhammerwerk n. ‖ petite forge f.

hammer, pane of the ~ ‖ Hammerfinne f.; Finne f. (des Hammers) ‖ panne f. *ou* queue f. du marteau. / ~ piston ‖ Hammerkolben m. ‖ piston m. du marteaupilon.

hammer-point riveting ‖ Spitzkopfnietung f. ‖ rivure f. conique *ou* à point de diamant.

hammer riveting ‖ Hammernietung f. ‖ rivetage m. au marteau. / ~ riveting machine ‖ Hammernietmaschine f. ‖ riveuse f. à marteau. / ~ scale ‖ Glühspan m.; Zunder m.; Hammerschlag m.; Eisenhammerschlag m. ‖ pailles fpl. (d'oxyde) de fer; battiture f. *ou* oxydes mpl. *ou* écailles fpl. de fer; mâchefer m. / ~ shaft ‖ Hammerstiel m. ‖ manche m. du marteau. / ~ shop *see* hammer mill. / ~ slag *see* ~ scale. / ~ smith ‖ Hammerschmied m.; Schmied m.; Grobschmied m.; Handschmied m. ‖ forgeron m.; forgeur m. (à la main); marteleur m. / ~ spindle (Electric bell) ‖ Hammerspindel f. ‖ axe m. du marteau. / ~ standard *see* ~ frame. / ~ striker *see* ~ face (Power hammer). / stroke of the ~ ‖ Hub m. *oder* Fallhöhe f. des Bären ‖ course f. de la frappe *ou* du mouton.

hammer tup ‖ Hammerbär m.; Bär m.; Hammerklotz m. ‖ mouton m.; masselotte f. / ~ guide ‖ Bärführung f. ‖ guidage m. de la masse tombante. / ~ stop ‖ Bäraufhaltevorrichtung f. ‖ arrêt m. du mouton.

hammer valve gear ‖ Hammersteuerung f. ‖ distribution f. du marteau-pilon. / ~ works pl. (Forge) *see* hammer mill (Forge).

hammock ‖ Hängematte f. ‖ hamac m.; branle f. / ~ with mattress ‖ Hängematte f. mit Matratze ‖ hamac m. matelassé *ou* à double fond.

hammock clew (Mar) ‖ Hängemattshahnpot f. ‖ araignée f. de hamac. / ~ girtline ‖ Hängemattsjolle f. ‖ ceinture f. de hamacs.

hamper ‖ Packkorb m.; Wagenkorb m. ‖ banne f. / bottle ~ ‖ Flaschenkorb m. ‖ panier m. à bouteilles. / in ~s pl. ‖ in Körben mpl. ‖ en paniers mpl.

ham smoker ‖ Schinkenräucherer m. ‖ fumeur m. de jambons.

hance *see* haunch.

hanch *see* haunch.

hand, to (Mar) ‖ bemannen ‖ mettre des hommes sur. / ~ over ‖ aushändigen ‖ remettre; délivrer. / ~ a sail ‖ ein Segel n. bergen *oder* einnehmen ‖ serrer une voile.

hand (Anatomy) ‖ Hand f. ‖ main f. / ~ (Clockm) ‖ Zeiger m. ‖ aiguille f. / ~ (Person) ‖ Arbeiter m.; Gehilfe m. ‖ ouvrier m. / adjusting by ~ ‖ Anstellung f. von Hand ‖ ajustage m. à la main. / clock ~ ‖ Uhrzeiger m. ‖ aiguille f. de montre *ou* de pendule. / charge ~ ‖ Vorarbeiter m. ‖ chef m. d'équipe. / large ~ ‖ große Schrift f. ‖ écriture f. grande; gros caractère m. / on ~ ‖ vorrätig ‖ en magasin m. / to put in ~ ‖ in Arbeit f.

geben ‖ mettre en mains fpl. / running ~ ‖ geschobene *oder* laufende *oder* liegende Schrift f. ‖ écriture f. coulée. / watch ~ *see* clock ~.

hand adjustment ‖ Handeinstellung f. ‖ réglage m. à la main. / ~ anode ‖ Handanode f. ‖ anode f. à main. / ~ anvil ‖ Handamboß m. ‖ enclumette f.; enclumeau m. / ~ axe ‖ Handaxt f. ‖ hache f. à main. / ~ bag ‖ Handtasche f.; Damentasche f. ‖ réticule m.; sac m. de dames *ou* à main. / ~ baggage *see* ~ luggage. / ~ ball thrust apparatus ‖ Handkugeldruckprüfapparat m. ‖ appareil m. d'essai à la bille à main. / ~ barrow (Frame) ‖ Trage f.; Tragbahre f. ‖ civière f. / ~ barrow (Handcart) ‖ Handkarren m.; Schubkarren m. ‖ cabrouet m.; galopin m.; charrette f. à bras. / ~ baster (Tail) ‖ Handhefter m. ‖ faufileur m. à la main. / ~ battery-switch ‖ Handzellenschalter m. ‖ réducteur m. à main. / ~ bell ‖ Handglocke f. ‖ clochette f.; sonnette f. / ~ bellows pl. ‖ Handblasebalg m. ‖ soufflet m. à main. / ~ blotter ‖ Löscher m. ‖ buvard m. à main. / ~ blow-pipe ‖ Handgebläse n. ‖ chalumeau m. à main. / ~ boring machine ‖ Handbohrmaschine f. ‖ perceuse f. à main. / ~ brace ‖ Brustleier f.; Brustbohrer m.; Faustleier f. ‖ vilebrequin m.; virebrequin m.; drille m. à arçon.

hand brake ‖ Handbremse f. ‖ frein m. à main. / ~ lever ‖ Handbremshebel m. ‖ levier m. de frein à main.

hand brick works ‖ Handziegelei f. ‖ briqueterie f. à la main. / ~ brush ‖ Handbürste f. ‖ brosse f. à main. / ~ capstan ‖ Gangspill n. ‖ cabestan m. à bras.

handcart ‖ Handkarre f.; Laufkarre f. ‖ charrette f. à bras; brouette f.

hand carver ‖ Handschnitzer m. ‖ sculpteur m. à la main. / ~ casting machine ‖ Handgießmaschine f. ‖ machine f. à fondre de mouvement à main. / ~ caulking ‖ Stemmen n. von Hand ‖ matage m. à la main. / ~ chain ‖ Handkette f. ‖ chaîne f. de manœuvre. / ~ circuit breaker ‖ Handausschalter m. ‖ interrupteur m. à main. / ~ clamping ‖ Spannen n. von Hand ‖ fixation f. à main. / ~ cloth printing ‖ Handdruckerei f. für Zeuge ‖ impression f. de tissus à la main. / ~ cold chisel ‖ Handmeißel m. ‖ ciseau m. à main. / ~ composition (Print) ‖ Handsatz m. ‖ composition f. à main. / ~ compositor (Print) ‖ Handsetzer m. ‖ compositeur m. à la main. / ~ cutting (Cloth) ‖ Handschnitt m. ‖ coupage m. à la main. / ~-cut peat ‖ Handtorf m. ‖ tourbe f. extraite à la main. / ~ drag ‖ Handbagger m. ‖ drague f. à main. / ~ dredger *see* ~ drag. / ~ drill ‖ Handbohrer m.; Drillbohrer m. ‖ perceuse f. *ou* perforatrice f. à main. / ~ drive ‖ Handbetrieb m. ‖ commande f. à la main. / ~ embroidering *see* ~ embroidery. / ~ embroidery ‖ Handstickerei f. ‖ broderie f. à la main.

hander-in (Weav) ‖ Fadenanreicher m. ‖ livreur m. de fils.

hand file ‖ Handfeile f. ‖ lime f. plate *ou* à main. / ~-forged ‖ von Hand f. geschmiedet ‖ forgé à la main. / the pieces pl. are partly ~-forged ‖ die Stücke npl. sind teilweise von Hand ge-

schmiedet || une partie des pièces est forgée à la main. / ~-forged screw || handgeschmiedete Schraube f. || vis f. forgée à la main. / ~-formed roofing-tile || handgeformter Dachziegel m. || tuile f. moulée à la main. / ~ forming || Handformerei f. || moulage m. à la main. / ~ glass || Handspiegel m. || glace f. à main. / ~ grenade || Handgranate f. || grenade f. à main. / ~ grinding-machine || Handschleifmaschine f. || machine f. à meuler à la main. / ~ grindstone || Handschleifstein m. || pierre f. à affiler à la main. / ~ grip || Handgriff m. || poignée f. / ~ guide || Handleiter m. || guide-main m. / ~ hammer || Handhammer m. || marteau m. à main. / ~ hold || Handgriff m. || poignée f. / ~ hole || Handloch n. || regard m. de visite. / ~ hole cover || Verschlußpfropfen m. des Handloches || tampon m. de fermeture du regard de visite. / ~ hook || Handhaken m. || croc m. à main.

handicap || Vorsprung m. || handicap m.

handicraft || Gewerbe n.; Handwerk n. || métier m.; profession f.

hand indexing device || Handteilapparat m. || appareil m. à diviser à la main. / ~ inking || Handfärbung f. encrage m. à main.

handiwork || Handarbeit f. || travail m. *ou* ouvrage m. manuel.

hand jack || Schraubenwinde f. || cric m. / ~ with claw || Klauenwinde f. || cric m. à crochet.

handkerchief || Taschentuch n. || mouchoir m. / ~ box || Taschentuchbehälter m. || étui m. à mouchoir de poche.

hand knitting || Häkelwaren fpl. || tricotage m. à la main. / ~ knitting machine || Handstrickmaschine f. || machine f. à tricoter à main. / ~ lace || Handspitze f. || dentelle f. à la main. / ~ ladle (Found) || Kelle f.; Gießkelle f. || poche f. de coulée; cuiller f. / ~ lamp || Handlampe f. || lampe f. baladeuse *ou* portative. / ~ lantern || Handlaterne f. || lanterne f. à main. / ~ lathe || Handdrehbank f. || tour m. à main.

handle, to || handhaben || manier; manipuler.

Handle with care! glass! ||Vorsicht! Glas! || attention! fragile!

handle (Door) || Drücker m.; Klinke f.; Türdrücker m. || loquet m.; clenche f. / ~ (Mach) || Haspel m.; Kurbel f. || manivelle f. / ~ (Tool) || Stiel m.; Heft n.; Griff m.; Handgriff m. || manche m.; poignée f. / ~ (Vessel) || Henkel m. || anse f. / ash wood ~ || Werkzeugheft n. aus Eschenholz || manche m. en frêne. / ball ~ || Kugelkurbel f. || manivelle f. équilibrée. / belt shifter ~ || Riemenrückergriff m. || poignée f. de passe-courroie. / bent ~ || gebogener Griff m. || poignée f. courbe. / with ~ coated with oil || mit geöltem Stiel m. || avec manche m. enduit de vernis. / copper ~ || Kupfergriff m. || poignée f. en cuivre. / cornelian wood ~ || Hartriegelwerkzeugheft n. || manche m. en cournouiller. / crank ~ || Handkurbel f. || manivelle f. / crutch ~ || Quergriffstiel m. || manche m. à béquille. / ~ of cycle || Fahrrad(hand)griff m. || poignée f. de bicyclette. / door ~ || Türgriff m. || poignée f. de porte. / fire door ~ || Heiztürgriff m. || poignée f. de porte de foyer. / flat ~ || flaches Werkzeugheft n. || manche m. plat. / ~ for

hammers || Hammerstiel m. || manche m. de *ou* à marteaux. / having a ~ || gestielt ||emmanché. / ~ for hoes || Hackenstiel m. || manche m. de *ou* à houes. / ~ of key || Tastenschlüssel m. || clé' f. de touche. / knife ~ || Messergriff m. || manche m. de couteau. / leather ~ || Ledergriff m. || manche m. en cuir. / lever ~ || Hebelgriff m. || manche m. de levier. / with loose ~ mit losem Stiel m. || avec manche m. rapporté. / manhole ~ || Mannlochgriff m. || poignée f. de couvercle de trou-d'homme. / ~ of an oar (Mar) || Handgriff m. eines Riemens; Rudergriff m. || main f. d'aviron. / octagone ~ ||achteckiges Werkzeugheft n. || manche m. à huit pans. / oval ~ || ovales Werkzeugheft n. || manche m. ovale. / removable ~ || Einsteckgriff m. || manche m. amovible. / ~ for removing shells (Arm) || Geschoßheber m. || tire-obus m.; crochet m. à poignée. / round ~ || rundes Werkzeugheft n. || manche m. rond. / screwed-on ~ || angeschraubter Handgriff m. || manche m. vissé. / shears ~ || Handgriff m. an der Tuchschere || manicle m.; manique f. / ~ for spades || Spatenstiel m. || manche m. de bêche. / star ~ || Kreuzgriff m. || croisillon m. / stick ~ || Stockgriff m. || poignée f. de canne. / twisted wire ~ || geflochtener Drahtgriff m. || poignée f. en fil tressé. / umbrella ~ || Schirmgriff m. || poignée f. de parapluie. / ~ of a window-sash || Fensterknopf m. || bouton m. *ou* tiroir m. *ou* olive f. de fenêtre. / window ~ (Wagon) || Fenstergriff || poignée f. de châssis. / wood ~ || Holzstiel m. || manche m. en bois. / wooden ~ of tool || Holzstiel m. eines Werkzeuges || manche m. d'outil en bois.

hand-lead || Handlot n. || sonde f. à main.

handle bar of cycle || Fahrradlenkstange f. || guidon m. de bicyclette. / ~ bender || Lenkstangenbieger m. || courbeur m. de guidons. / ~ stem || Lenkstangenrohr n.|| tube m. plongeur du guidon.

handled spade || Grabscheit n. mit Stiel || louchet m. emmanché.

handle latch *see* handle stop pin. / ~ maker (Pott) || Henkelformer m. || mouleur m. de poignées. / ~ machine for making ~s || Griffeherstellungsmaschine f. || machine f. à faire des poignées.

handler (Pott) || Henkelsetzer m. || poseur m. de poignées.

handle stop pin || Anschlagstift m. || arrêt m. de la poignée. / ~ switch || Kurbelumschalter m. || commutateur m. à manivelle.

hand lever || Handhebel m. || levier m. à main. / ~ brake || Handhebelbremse f. || frein m. à levier.

handling || Behandlung f. || traitement m.; maniement m. / coal ~ || Kohlenzufuhr f. || amenée f. du charbon. / ~ the job || Hantierung f. von Werkstücken || manutention f. des pièces. / ~ by one man || Einmannbedienung ~ || entretien m. par un seul homme. / ~ stone blocks || Transport m. von Steinblöcken || transport m. de blocs de pierre. / ~ of the work || Handhabung f. der Arbeitsstücke || manipulation f. des pièces.

handling crane || Verladekran m. || grue f. de manutention. / ~ cranes pl. for the unloading of the raw materials || Ausladebrücken fpl. für das Entladen der Rohstoffe || élévateur-transporteur

m. pour le déchargement des matières premières. / ~ crew (Aero) || Haltemannschaft f. || équipe f. de manœuvre. / hydraulic device for ~ the forgings || hydraulische Wendevorrichtung f. für Schmiedestücke || dispositif m. hydraulique à virer les pièces à forger. / ~ files pl. || Feilenbearbeitung f. || travail m. des limes. / wading and ~ plant for goods in bulks || Förderanlage f. für Massengüter || installation f. de manœuvre et de transport de marchandises en vrac. / ~ platform || Verladegerüst n.; Ausladebrücke f. || pont m. de chargement *ou* de déchargement.

hand loom || Handwebstuhl m. || métier m. à bras; métier à tisser à main. / ~ weaving || Handweben n. || tissage m. à main.

hand lubrication || Handschmierung f. || graissage m. à la main. / ~ lubricator || Handschmiervorrichtung f. || appareil m. de graissage à la main. / ~ luggage || Handgepäck n. || bagages mpl. à la main.

handly || handlich || facile à manier.

hand-made lace || handgeklöppelte Spitze f. || dentelle f. à fuseau à la main. / ~ paper || Büttenpapier n.; handgeschöpftes Papier n. || papier m. de Hollande *ou* à la main *ou* de cuve. / ~ picture || mit der Hand hergestelltes Bild n. || tableau m. fait à la main. / ~ rivet head || gehämmerter Nietkopf m. || tête f. de rivet martelée.

hand mast || Spiere f. || espar m.; mâtereau m. / ~ microscope || Handmikroskop n. || microscope m. à main. / ~ milling machine || Handfräsmaschine f. || fraiseuse f. avec chariots à main. / ~ mixing || Handmischung f. || mélange m. à la main. / ~ moulding || Handformerei f. || moulage m. *ou* façonnage m. à la main. / ~ moulding machine || Handformmaschine f. || machine f. à mouler à la main. / ~ net || Fanggarn n. || épuisette f. / ~ numbering machine || Handnummerungsmaschine f. / numéroteur m. à main. / ~ oiling || Handschmierung f. || graissage m. à la main.

hand-operated || von Hand f. angetrieben || commandé à la main. / ~ adjustment || Einstellung f. von Hand || ajustage m. à main. / ~ battery switch (Acc) || Handzellenschalter m. || réducteur-adjoncteur m. à la main. / ~ block tackle || Handflaschenzug m. || palan m. à main.

hand operation || Handbetrieb m. || marche f. à la main. / ~ or power driven || für Hand- oder Kraftbetrieb m. || à commande f. mécanique ou à main. / ~ painter (Pott) || Handmaler m. || peintre m. à la main. / ~ palm || Handfläche f. || paume f. de la main. / ~ perforator || Handlocher m. || perforateur m. à main.

handpick, to ~ ores pl. || Erze npl. von Hand scheiden || trier les minéraux à la main.

hand picking || Klaubarbeit f. || triage m. à la main. / ~ planer || Handabrichtmaschine f. || raboteuse f. à dégauchir *ou* à dresser à la main. / ~ power || Handbetrieb m. || commande f. à main. / ~ press || Handpresse f.; Stempelpresse f. || presse f. à bras. / ~ press for letterpress work || Buchdruckhandpresse f. || presse f. d'imprimerie à bras; presse f. à bras typographique. / ~-printed wall paper || Handdrucktapete f. || papier m.

peint à la main. / ~ printer ‖ Handdrucker m. ‖ imprimeur m. à la main. / ~ printing machine ‖ Handdruckmaschine f. ‖ imprimeuse f. à main. / ~ puddling ‖ Handpuddeln n. ‖ puddlage m. à la main. / ~ pump ‖ Handpumpe f. ‖ pompe f. à main. / ~ punch ‖ Durchschlag m. ‖ poinçon m. / ~ rail ‖ Laufstange f.; Handgriff m. eines Geländers ‖ main f. coulante ou courante; écuyer m. / ~ rail (Stairs) ‖ Treppengeländer n. ‖ balustrade f.; garde-fou m.; rampe f. d'ecsalier. / ~ railing see ~ rail. / ~ rail post ‖ Holzschaft m. am Geländer ‖ acrotère m. / ~ rake ‖ Handrechen m. ‖ râteau m. à main. / ~ rammer ‖ Handramme f. ‖ mouton m. à bras. / ~ reeler ‖ Handhaspeler m. ‖ dévideur m. à la main. / ~ riveter ‖ Handnieter m. ‖ riveur m. à la main. / ~ riveting ‖ Handnietung f. ‖ rivure f. à la main. / ~ ruff ‖ Manschette f. ‖ manchette f. / ~ ruffle see ~ ruff.

hands pl. **(Mar)** ‖ Bemannung f.; Mannschaft f. ‖ équipage m.; hommes mpl. / loading ~ ‖ Lademannschaft f. ‖ équipe f. de chargement.

hand safety guard ‖ Handschutzvorrichtung f. ‖ pare-main m. / ~ saw ‖ Fuchsschwanzsäge f.; Fuchsschwanz m. ‖ scie f. à main ou à manche. / ~ scale ‖ Handwage f. ‖ balance f. à main.

hand-screw ‖ Handwinde f. ‖ cric m. / ~ jack ‖ Fußwinde f. mit Schraube ‖ cric m. à vis. / ~ tap ‖ Handgewindebohrer m. ‖ taraud m. à main.

handsel ‖ Handgeld n.; Draufgeld n. ‖ denier m. à Dieu.

hand sewing-machine ‖ Handnähmaschine f. ‖ machine f. à coudre à manivelle. / ~ shears pl. ‖ Schere f. für Handbetrieb ‖ cisailles fpl. à main. / ~ sieve ‖ Handsieb n. ‖ crible m. / ~ sizer (Weav) ‖ Handschlichter m. ‖ encolleur m. à la main.

handsome ‖ ansehnlich ‖ considérable.

handspike ‖ Hebebaum m.; Brechstange f.; Handspake f. ‖ levier; anspect m. / ~ socket ‖ Brechstangenhülse f. ‖ emmanchement m. de levier de manœuvre. / ~ socket of the turn-table ‖ Drehbaumhülse f. der Drehscheibe ‖ douille f. de levier de la plaque tournante.

hand spinner ‖ Handspinner m. ‖ fileur m. à la main. / ~ spinning ‖ Handspinnerei f. ‖ filage m. à la main. / ~ splitter (Weav) ‖ Handaufwinder m. ‖ ensoupleur m. à la main. / ~ spooler ‖ Handspuler m. ‖ bobineur m. à la main. / ~-spun yarn ‖ Handgespinst n.; Handgarn n. ‖ fil m. de main. / ~ stamp ‖ Handstempel m. ‖ timbre m. à main. / ~ strap ‖ Handriemen m. ‖ courroie f. à main. / ~ tiller ‖ Handpinne f. ‖ barre f. franche. / ~ tippler ‖ Handwipper m. ‖ culbuteur m. à bras. / ~ tools pl. ‖ Handwerkszeug n. ‖ outillage m. / ~ track ‖ Handkarre f.; Laufkarre f. ‖ charrette f. à bras. / ~ traversing gear ‖ Handfahrwerk n. ‖ déplacement m. commandé à la main. / ~ travelling gear ‖ Handlaufkran m. ‖ pont m. roulant à commande à main. / ~ truck ‖ Rollwagen m. ‖ wagon-brouette f. / ~ turning gear ‖ Handdrehvorrichtung f. ‖ appareil m. de tournage à main. / ~ vacuum cleaner ‖ Staubsaugapparat m. ‖ aspirateur m. de poussière.

hand vice ‖ Feilkloben m.; Handkloben m. ‖ étau m. à main. / broad-chap ~ ‖ breitmäuliger Feilkloben m. ‖ étau m. à main à mâchoires larges. / dog-nosed ~ ‖ schmalmäuliger Feilkloben m. ‖ étau m. à main à mâchoires étroites. / ~ with handle ‖ Stielfeilkloben m. ‖ étau m. à main emmanché. / parallel ~ ‖ Parallelfeilkloben m. ‖ étau m. à main à mouvement parallèle. / square-nosed ~ see dog-nosed ~.

handwarm ‖ handwarm ‖ à la température à peine sensible à la main.

hand warper ‖ Handkettenscherer m. ‖ ourdisseur m. à la main. / ~ weaver ‖ Handweber m. ‖ tisseur m. à main.

hand wheel ‖ Handrad n. ‖ volant m. à main. / ~ with dished arms ‖ Handrad n. mit schrägen Armen ‖ volant m. à main avec croisillons obliques. / ~ with hollow rim ‖ Handrad n. mit hohlem Kranz ‖ volant m. à main avec volant creux.

hand whetstone ‖ Handwetzstein m. ‖ pierre f. à aiguiser à la main. / ~ winch ‖ Handwinde f. ‖ treuil m. à main; cric m. / ~ work ‖ Handarbeit f. ‖ travail m. ou ouvrage m. manuel.

hand-woven ware ‖ Handwebereierzeugnis n. ‖ produit m. de tissage à la main.

handwriting ‖ Handschrift f. ‖ écriture f. manuscrite. / ~ letters pl. ‖ Schrift f. ‖ écriture f.; caractères mpl. écrits.

handy ‖ handlich ‖ maniable; commode. / ~ design ‖ handliche Bauart f. ‖ construction f. aisément maniable.

handyman ‖ Handlanger m. ‖ manœuvre m.; aide-maçon m.

hang, to ‖ aufhängen ‖ étendre; suspendre. / ~ by gimbals ‖ kardanisch aufhängen ‖ suspendre à la cardan. / ~ the rudder ‖ das Ruder einhaken oder einhängen oder einsetzen ‖ monter le gouvernail.

hangar, locomotive ‖ Lokomotivschuppen m. ‖ remise f. de locomotives. / revolving-type ~ (Aero) ‖ drehbare Flugzeughalle f. ‖ hangar m. type orientable. / steel ~ ‖ Stahlflugzeughalle f. ‖ hangar m. d'acier.

hanger (Carp) ‖ Hängeeisen n. ‖ étrier m. / ~ (Hunting) ‖ Hirschfänger m. ‖ couteau m. de chasse. / ~ (Loc) ‖ Hängeeisen n. ‖ tige f. de suspension. / ~ (Mach) ‖ Hängestange f. ‖ tige f. ou barre f. de suspension. / ~ (Weav) ‖ Unterlitze f. ‖ maille f. d'en bas. / ~ (Wind mill) ‖ Hängedocke f. ‖ pilier m. suspendu. / pipe ~ ‖ Rohrschelle f. ‖ gâche f. étrier m. / spring ~ (Wagon) ‖ Federstütze f.; Federstuhl m. ‖ bride f. de ressort. / step ~ (Auto) ‖ Trittbretthalter m. ‖ patte f. de marche-pied.

hanger board ‖ Hängepaneel n. ‖ panneau m. suspendu. / ~ bracket ‖ Hängebock m. ‖ chaise f. suspendue.

hanging of charge (Met) ‖ Hängen n. der Gicht ‖ accrochage m. des charges. / ~ of tapestry ‖ Tapetenkleben n. ‖ tenture f. de tapisserie.

hanging bearing ‖ Hängelager n. ‖ palier m. pendant. ~ bridge ‖ Hängebrücke f. ‖ pont m. suspendu. / ~ clock ‖ Wanduhr f. ‖ pendule f. suspendue. / ~ floor ‖ Hängeboden m. ‖ soupente f. / ~ peel ‖ Aufhängekreuz n. ‖ ferlet m. / ~ pendulum clock ‖ Pendelwanduhr f. ‖ pendule f. murale. / ~ pipe carrier ‖ hängendes Gestängelager n. ‖ support m. de tringles suspendues. / ~ room ‖ Auf-

hängeboden; Trockenboden m. ‖ séchoir m.; étandage m.

hangings pl. ‖ Wandteppich m.; Wandbehang m. ‖ tapis m. de tenture; revêtement m. des murs. / embossed ~ ‖ gepreßte Tapete f. ‖ papier m. peint gaufré. / gilt ~ ‖ vergoldete Tapete f. ‖ papier m. peint doré. / japanned ~ ‖ gefirnißte Tapete f. ‖ papier m. peint verni. / leather ~ ‖ Ledertapete f. ‖ tapis m. en cuir. / paper ~ ‖ Tapete f.; Papiertapete f. ‖ papier m. peint. / satined ~ ‖ Glanztapete f. ‖ papier m. peint satiné. / silver ~ ‖ versilberte Tapete f. ‖ papier m. peint argenté.

hanging support see hanger (Wind mill). / ~ tie (Build) ‖ Hängeband n.; Zange f.; Hängeschiene f. ‖ moise f. pendante. / ~ tie (Mach) ‖ Hängeschiene f. ‖ rail m. suspendu. / ~ truss ‖ Hängewerk n. ‖ ferme f. à poinçons ou à clefs pendants. / ~ valve ‖ Scharnierventil n.; Klappenventil n. ‖ clapet m.; soupape f. à clapet ou à charnière. / ~ wall (Mine) ‖ Hangendes n. ‖ couche f. supérieure; toit m.

hank (Spinn) ‖ Nummer f. oder Zahl f. beim Haspeln des Garns ‖ échée f.; écheveau m. / little ~ ‖ Knäuel n. ‖ échevette f. / ~ knotter (Spinn) ‖ Dockenknüpferin f. ‖ torcheuse f. / ~ mercerizing machine ‖ Garnmerzerisiermaschine f. ‖ mercériseuse f. pour fils. / ~ polisher ‖ Garnglätter m. ‖ étriqueur m. / ~ winder ‖ Dockerin f. ‖ dévideuse f. d'écheveaux. / ~ wringer ‖ Garnausringer m. ‖ chevilleur m.

harbour ‖ Hafen m. ‖ port m. / commercial ~ ‖ Handelshafen m. ‖ port m. de commerce. / deep-sea ~ ‖ künstliche Reede f. ‖ rade f. endiguée. / inner ~ ‖ Binnenhafen m. ‖ port m. intérieur. / ~ of refuge ‖ Nothafen m. ‖ port m. de refuge. / ~ of refuge on a river ‖ Winterhafen m. eines Flusses ‖ port m. de refuge d'une rivière. / tidal ~ ‖ Fluthafen m. ‖ port m. à marée. / timber ~ ‖ Holzhafen m. ‖ port m. des bois.

harbour crane ‖ Hafenkran m. ‖ grue f. de port. / ~ depôt see ~ station. / ~ dues pl. ‖ Hafengeld n.; Ankergeld n. ‖ droit m. d'ancrage ou du port. / ~ fisherman ‖ Haffischer m. ‖ Wattenfischer m. ‖ pêcheur m. dans les bas-fonds ou dans les lagunes. / ~ gantry crane see ~ portal crane. / ~ labourer ‖ Hafenarbeiter m. ‖ ouvrier m. de port. / ~ lock ‖ Hafenschleuse f. ‖ Seeschleuse f. ‖ écluse f. à l'entrée d'un port. / ~ master ‖ Hafenmeister m. ‖ capitaine m. du port. / ~ master's man see ~ pilot. / ~ office ‖ Hafenamt n. ‖ bureau m. du port. / ~ pilot ‖ Hafenlotse m. ‖ pilote m. du port ou du mouillage ou lamaneur. / ~ portal crane ‖ Hafenportalkran m.; Uferportalkran m. ‖ grue f. à portique de port ou de quai. / ~ railway ‖ Hafenbahn f. ‖ chemin m. de fer de port. / ~ station ‖ Hafenbahnhof m. ‖ gare f. maritime ou d'un port. / ~ watch ‖ Hafenwache f. ‖ garde f. d'un port. / ~ works pl. ‖ Hafenarbeiten fpl. ‖ travaux mpl. de port.

hard ‖ hart ‖ dur; ferme; compact. / ~ (Brush) ‖ hart ‖ rude. / ~ (Mine) ‖ klemmig ‖ dur; compact; serré. / ~ (Mortar) ‖ abgebunden ‖ séché. / ~ as glass ‖ glashart ‖ dur comme le verre. / naturally ~ ‖ naturhart ‖ de dureté f. naturelle. / naturally ~ crucible steel ‖ naturharter

Tiegelstahl m. ‖ acier m. au creuset de dureté naturelle.

hard aluminium ‖ Duraluminium n. ‖ aluminium m. dur. / ~ brush ‖ Scheuerbürste f. ‖ brosse f. à frotter ou de frottement. / ~-burnt brick ‖ Hartbrandstein m. ‖ brique f. hollandaise. / ~-coal mine ‖ Anthrazitbergwerk n. ‖ mine f. d'anthrazite. / ~ copper ‖ Hartkupfer n. ‖ cuivre m. dur. / ~ crushing machine ‖ Hartzerkleinerungsmaschine f. ‖ broyeur m. à matière dure. / ~-drawn copper wire ‖ Hartkupferdraht m. ‖ fil m. de cuivre dur. / ~ drying brilliant oil ‖ Harttrockenglanzöl n. ‖ huile f. brillante de séchage dur. / ~ drying oil ‖ Harttrockenöl n. ‖ huile f. de séchage dur.

harden, to ‖ erhärten ‖ durcir; endurcir. / ~ (Mortar) ‖ abbinden; erhärten ‖ durcir; s'endurcir; prendre. / ~ (Steel) ‖ härten; abschrecken ‖ tremper.

hardened (Steel) ‖ gehärtet ‖ trempé. / steel capable of being ~ ‖ härtbarer Stahl m. ‖ acier m. pouvant être trempé ou prenant la trempe. / case-~ ‖ einsatzgehärtet; an der Oberfläche gehärtet ‖ trempé à la surface. / not ~ (Steel) ‖ nicht gehärtet ‖ non trempé. / ~ part went to about one third of the thickness ‖ die Härtung f. ging bis zu etwa einem Drittel der Plattendicke ‖ la dureté f. allait jusqu'au tiers environ de l'épaisseur de la plaque. / ~ resins pl. for varnishes ‖ Hartharze npl. für Lacke ‖ résines fpl. durcies pour vernis. / hard layer of ~ steel goes down fairly deep ‖ der Stahl m. erhält beim Härten eine ziemlich tiefgehende harte Schicht ‖ par la trempe l'acier se durcit à une assez grande profondeur.

hardener ‖ Härter m. ‖ trempeur m.

hardening (Chem) ‖ Härtung f. ‖ durcissement m. / ~ (Steel) ‖ Härtung f. ‖ trempe f.; durcissement m. / it is recommended to let down the steel in rape-seed oil after ~ ‖ nach der Härtung f. soll ein gelindes Anlassen n. des Stahles in Rüböl erfolgen ‖ il est à recommander d'adoucir l'acier m. à l'huile de colza après la trempe. / air ~ ‖ Lufthärtung f. ‖ trempe f. à l'air. / by means of ~ and subsequent annealing, higher resisting qualities may be obtained ‖ durch Härten n. und nachheriges Anlassen n. werden höhere Festigkeiten fpl. erzielt ‖ on obtient des résistances fpl. plus élevées quand on fait revenir l'acier m. après la trempe. / case ~ ‖ Einsatzhärtung f.; Oberflächenhärtung f. ‖ cémentation f. en paquet; trempe f. à la surface. / ~ of the concrete ‖ Erhärten n. des Betons ‖ durcissement m. du béton. / file ~ ‖ Feilenhärtung f. ‖ trempe f. de limes. / ~ of mortar ‖ Abbinden n. oder Erhärten n. des Mörtels ‖ prise f. ou durcissement m. du mortier. / oil ~ ‖ Ölhärtung f. ‖ trempe f. à l'huile. / water ~ ‖ Härtung f. in Wasser ‖ trempe f. à l'eau.

hardening agent ‖ Härtemittel n. ‖ moyen m. de trempe. / ~ boiler ‖ Härtekessel m. ‖ bassin m. de trempe. / ~ carbon ‖ Härtungskohle f. ‖ carbone m. de trempe. / ~ compound ‖ Härtemittel n. ‖ substance f. à tremper. / depth of ~ ‖ Härtetiefe f. ‖ épaisseur f. de trempe ou de couche durcie. / distortion due to ~ ‖ beim Härten auftretender Verzug m. ‖ déformation f. qui se produit lors de la trempe.

hardening furnace ‖ Härteofen m. ‖ four m. à tremper. / the heating is done in a ~ of convenient size ‖ die Erhitzung f. erfolgt in einem Härteofen von geeigneter Größe ‖ le chauffage a lieu dans un four à tremper d'un volume proportionné.

hardening installation ‖ Härteanlage f. ‖ installation f. de trempe. / ~-on the glazing (Porcel) ‖ Glattbrennen n. ‖ cuisson f. en couverte. / ~ preparation ‖ Härtepräparat n. ‖ préparation f. à tremper. / ~ process ‖ Härteverfahren n. ‖ procédé m. de trempe. / instructions pl. for the ~ process ‖ Härteanleitung f. ‖ instruction f. pour le procédé de trempe. / ~ shop ‖ Härtekammer f.; Härterei f. ‖ atelier m. de trempe. / ~ temperature ‖ Härtetemperatur f. ‖ température f. de trempe. / ~ trough ‖ Härtetrog m. ‖ cuve f. à tremper. / ~ water ‖ Härtewasser n. ‖ eau f. ou liquide m. de trempe.

hardest material ‖ dauerhaftester Werkstoff m. ‖ matière f. la plus résistante.

hard glass ‖ Hartglas n. ‖ verre m. dur. / heat-proof ~ ‖ hitzebeständiges Hartglas ‖ verre m. trempé insensible à la chaleur.

hard grade of steel ‖ harte Stahlmarke f. ‖ marque f. dure d'acier. / ~ iron ‖ Harteisen n. ‖ fonte f. dure.

hard lead ‖ Hartblei n. ‖ plomb m. dur ou durci. / ~ (Letter found) ‖ Hartblei n. ‖ matière f. forte. / ~ frame ‖ Hartbleirahmen m. ‖ cadre m. en plomb durci.

hard lime-stone ‖ harter Kalkstein m. ‖ pierre f. à chaux dure. / ~ metal ‖ Hartmetall n. ‖ métal m. dur.

hardness ‖ Härte f. ‖ dureté f. / the ~ of steel itself is not a brittle one ‖ die Härte des Stahls an sich ist keine spröde ‖ la dureté de l'acier en elle-même n'est pas fragile. / owing to the extraordinary ~, the wear and tear is very small ‖ infolge der außerordentlichen Härte f. ist der Verschleiß gering ‖ dureté f. très tenace, ce qui fait que l'usure f. des pièces est minime. / abrasive ~ ‖ Schneidhärte f. ‖ trempe f. active ou d'outil coupant. / Brinell ~ ‖ Brinellhärte f. ‖ dureté f. Brinell. / brittle ~ ‖ spröde Härte f. ‖ dureté f. cassante. / cutting ~ ‖ Schneidhärte f. ‖ trempe f. active ou d'outil coupant. / ~ going down fairly deep ‖ tiefgehende Härtung f. ‖ dureté f. d'une assez grande profondeur. / dynamic ~ ‖ dynamische Härte f. ‖ dureté f. dynamique. / to give to the castings external ~ ‖ den Gußstücken npl. Oberflächenhärtung geben ‖ donner aux moulages mpl. une dureté extérieure. / glass ~ ‖ Glashärte f. ‖ trempe f. glacée. / ~ of great cutting effect ‖ schneidhaltende Härte f. ‖ dureté f. tranchante ou d'outil coupant. / to indicate the ~ of steel ‖ den Härtegrad m. des Stahles angeben ‖ indiquer le degré de dureté de l'acier. / ~ of spring ‖ Federhärte f. ‖ dureté f. de ressort. / ~ on parts of the surface ‖ Oberflächenhärte f. ‖ dureté f. à certaines parties de la surface. / tough ~ ‖ zähe Härte f. ‖ dureté f. tenace. / uniform ~ ‖ gleichmäßige Härte f. ‖ dureté f. uniforme. / the ~ is uniform in the whole cross section ‖ die Härte ist im ganzen Querschnitt die gleiche ‖ la dureté est la même dans toute la section transversale. / uniformly penetrating its entire section ‖ den ganzen

Querschnitt durchdringende Härte f. ‖ dureté f. pénétrant toute la section. / ~ of water ‖ Härte f. des Wassers ‖ dureté f. de l'eau.

hardness, degree of ~ ‖ Härtegrad m. ‖ degré m. de dureté. / designation of the degree of ~ ‖ Härtegradbezeichnung f. ‖ désignation f. du degré de dureté. / ~ drop tester ‖ Fallhärteprüfer m. ‖ mouton m. à bille. / ~ test ‖ Härteprüfung f. ‖ examen m. de la dureté. / Brinell ~ test ‖ Brinellhärteprüfung f. ‖ essai m. de dureté d'après Brinell. / ~ test of inner surfaces ‖ Härteprüfung f. der Innenflächen ‖ essai m. de dureté des surfaces intérieures. / ~ tester for material ‖ Härteprüfung f. von Werkstoff ‖ essai m. de dureté de matériel. / ~ testing of material under drill ‖ Bohrleistungsversuch m. ‖ mesurage m. de charge à perçage. / ~ testing machine ‖ Härteprüfmaschine f. ‖ machine f. à vérifier la dureté.

hard nickel ‖ Hartnickel m. ‖ nickel m. dur. / ~ paper ‖ Hartpapier n. ‖ papier m. dur ou durci. / ~ pitch ‖ Hartpech n. ‖ brai m. / ~ porcelain ‖ Edelporzellan n.; Hartporzellan n. ‖ porcelaine f. dure. / ~ quality of steel ‖ harte Marke f. von Stahl ‖ marque f. dure d'acier.

hard rubber ‖ Hartgummi n. ‖ caoutchouc m. durci. / ~ cover ‖ Hartgummideckel m. ‖ couvercle m. en ébonite. / ~ disk ‖ Hartgummischeibe f. ‖ disque m. en caoutchouc durci. / ~ refilling pencil ‖ Hartgummifüllbleistift m. ‖ portemine m. en ébonite.

hards pl. ‖ Flachshede f.; Werg m. ‖ étoupe f. de lin; rebut m. de filasse.

hard silk winder ‖ Rohseidenwinder m. ‖ dévideur m. de soie grège.

hard-solder, to ‖ hartlöten ‖ braser.

hard solder ‖ Hartlot n.; Schlaglot n.; Messinglot n. ‖ brasure f.; soudure f. forte.

hard-soldered ‖ hartgelötet ‖ soudé fort.

hard spirit ‖ Hartspiritus m. ‖ alcool m. durci. / ~ steel ‖ Hartstahl m. ‖ acier m. dur. / ~ stone ‖ Hartstein m. ‖ pierre f. dure. / ~ subsoil ‖ harter Untergrund m. ‖ sous-sol m. compact ou dur.

hard-ware ‖ Kurzwaren fpl.; Stahlwaren fpl. ‖ quincaillerie f. / ~ for clothing industry ‖ Metallkurzwaren fpl. für die Bekleidungsindustrie ‖ quincaillerie f. en métal pour l'industrie des vêtements. / ~ works pl. ‖ Kleineisenzeugfabrik f. ‖ ferronnerie f.

hard water ‖ hartes Wasser n. ‖ eau f. dure ou crue.

hardwood ‖ Hartholz n. ‖ bois m. dur. / ~ plug-dowel ‖ Hartholzdübel m. ‖ cheville f. en bois dur. / ~ sleeper (Railw) ‖ Hartholzschwelle f. ‖ traverse f. en bois dur. / ~ tie see ~ sleeper. / ~ trenail (Railw) ‖ Hartholzschienenstuhlnagel m. ‖ cheville f. en bois dur pour coussinet.

hare's hair ‖ Hasenhaar n. ‖ poil m. de lapin.

harl ‖ Leinenbast m.; Pflanzenfasern fpl. ‖ filasse f. / flax ~ ‖ Flachsfaser f.; Flachsbast m. ‖ filasse f. du lin.

harmful ‖ schädlich ‖ nuisible.

harmless ‖ unschädlich ‖ inoffensif.

harmonic ‖ harmonisch ‖ harmonique. / ~ curve ‖ harmonische Kurve f. ‖ courbe f. harmonique. / ~ frequency ‖ harmonische Frequenz f. ‖ fréquence f. har-

monique. / ~ ringing ‖ abgestimmter Anruf m. ‖ appel m. harmonique.

harmonic ‖ Harmonische f.; Oberschwingung f. ‖ harmonique m.; oscillation f. harmonique. / triple ~ ‖ dreizahlige Harmonische f. ‖ harmonique f. de rang trois. / upper ~ ‖ Oberton m. ‖ son m. harmonique.

harmonica ‖ Harmonika f. ‖ harmonica m.

harmonist ‖ Orgelstimmer m. ‖ harmoniste m.

harmonium ‖ Harmonium n. ‖ harmonium m.

harmony ‖ Ebenmaß n. ‖ symmétrie f.; proportion f.; rhythme m. / ~ of colours ‖ Farbenharmonie f. ‖ harmonie f. de couleurs. / ~ whistle ‖ Zweiklangpfeife f. ‖ sifflet m. à deux notes.

harmotome ‖ Harmotom m.; Kreuzstein m.; Barytkreuzstein m. ‖ harmotome m.; pierre f. de croix.

harness (Saddl) ‖ Geschirr n. ‖ harnais m. / ~ (Weav) ‖ Webgeschirr n.; Zeug n.; Schaftwerk n. ‖ harnais m.; équipage m.; lisses fpl. / horse ~ ‖ Pferdegeschirr n. ‖ harnais m. des chevaux. / men's ~ for towing boats ‖ Kahnzugseil n. mit Halskoppel ‖ bretelle f. au halage des bâteau.

harness board (Weav) ‖ Harnischbrett n.; Löcherbrett n.; Schnürbrett n. ‖ planche f. d'arcades ou à trous. / ~ brusher (Weav) ‖ Harnischbürster m. ‖ brosseur m. de lames. / ~ cord ‖ Litzenheber m.; Arkade f. ‖ fil m. d'arcade; arcade f. / ~ fixer (Weav) ‖ Litzeneinrichter m.; Blatteinrichter m. ‖ lamier m.; monteur m. de lames; lamier-rotier m. / ~ fixing (Weav) ‖ Schnürung f. der Schäfte ‖ montage m. de lames. / ~ leather ‖ Geschirrleder n. ‖ cuir m. de harnachement ou de harnais. / ~ leveller (Weav) ‖ Weblitzenvorrichter m. ‖ appareilleur m. de maillons. / ~ loom weaver ‖ Harnischweber m. ‖ lissier m.; tisseur m. de lisses. / ~ maker ‖ Sattler m.; Geschirrmacher m.; Riemer m. ‖ harnacheur m. / ~ repairer (Weav) ‖ Weblitzenausbesserer m. ‖ remetteur m. de brins aux lisses. / ~ room ‖ Geschirrkammer f.; Sattelkammer f. ‖ sellerie f.

harp ‖ Harfe f. ‖ harpe f.

harpin (Shipb) ‖ Sente mit Schmiege ‖ lisse f. à double courbure. / bilge ~ (Shipb) ‖ Kimmsente f. ‖ lisse f. du fort.

harp maker ‖ Harfenbauer m. ‖ facteur m. de harpes.

harpoon, to ‖ harpunieren ‖ harponner.

harpoon ‖ Harpune f. ‖ harpon m.; foène f.; fouane f.

harpooner ‖ Harpunierer m. ‖ harponneur m.

harpoon gun ‖ Harpunenkanone f.; Harpunierkanone f. ‖ canon-harpon m.

harpsichord ‖ Spinett n. ‖ clavecin m.; piano m.

harrow, to ‖ eggen; Erde f. aufreißen ‖ herser.

harrow ‖ Egge f. ‖ herse f. / articulated ~ ‖ Gliederegge f.; Gelenkegge f. ‖ herse f. articulée. / bush ~ ‖ Buschegge f. ‖ herse f. en fascines ou en bois. / carriage of ~ ‖ Eggenrahmen m. ‖ traineau m. de herse. / chain ~ ‖ Kettenegge f. ‖ herse f. à chaîne; herse f. souple à chaînons. / disc ~ ‖ Scheibenegge f. ‖ herse f. à disques. / drill ~ ‖ Drillegge f.; Furchenegge f. ‖ herse f. semeuse; semoir m.

à herse. / fore-train of ~ see carriage of ~. / gang ~ see three-part ~. / levelling ~ ‖ Egge f. zum Ebnen des gepflügten Bodens ‖ herse f. aplanisseuse. / Norwegian ~ ‖ norwegische Egge f.; Rollenegge f. ‖ herse f. norvégienne ou roulante. / rotary ~ ‖ Rollenegge f. ‖ herse f. rotatoire. / ~ with spring teeth ‖ Federzahnegge f. ‖ herse f. à griffes de ressort. / three-part ~ ‖ dreiteilige Egge f. ‖ jeu m. de trois herses. / tooth of ~ ‖ Eggenzahn m.; Eggenzinke f. ‖ dent f. de herse. / wood ~ see bush ~.

harrowing ‖ Eggen n. ‖ hersage m.; hersement m.

harsh (Colour) ‖ hart; grell ‖ aigre. / ~ (Weav) ‖ hart; rauh ‖ rude; dur au toucher.

hartite ‖ Hartit m. ‖ hartite f.

hartshorn, artificial ~ ‖ künstliches Hirschhorn n. ‖ corne f. de cerf artificielle. / burnt ~ ‖ gebranntes Hirschhorn n. ‖ corne f. de cerf calcinée. / charred ~ see burnt ~. / oil of ~ ‖ Hirschhornöl n. ‖ huile f. de corne de cerf. / ~ salt ‖ Hirschhornsalz n. ‖ sel m. de corne de cerf; carbonate m. d'ammonium.

harvest ‖ Ernte f. ‖ récolte f. / the ~ of last year ‖ die letztjährige Ernte ‖ la récolte de la dernière année. / means pl. of conveying the ~ ‖ Erntetransportmittel n. ‖ dispositif m. de transport de récoltes.

harvester ‖ Mäher m. ‖ moissonneur m. / ~ (Machine) ‖ Erntemaschine f.; Mähmaschine f. ‖ moissonneuse f.; machine f. à récolter ou de récolte. / ~ for crops ‖ Getreideerntemaschine f. ‖ machine f. à récolter les céréales. / ~ for grass ‖ Graserntemaschine f. ‖ machine f. à récolter les herbes. / ~ for hay ‖ Heuerntemaschine f. ‖ machine f. à récolter le foin. / ~ and binding machine ‖ Bindemähmaschine f. ‖ moissonneuse-lieuse f. / ~ and reaper ‖ Mäh- und Garbenbindemaschine f. ‖ faucheuse-moissonneuse f. / ~ driver ‖ Getreidemähmaschinenführer m. ‖ moissonneur m. à la machine; conducteur m. de la moissonneuse.

harvesting seed crops ‖ Getreidemähen n. ‖ récolte f. des céréales.

harvesting machine see harvester.

harvest report ‖ Erntebericht m. ‖ bulletin m. de la récolte.

hashisch ‖ Haschisch n. ‖ haschisch m.

hasp ‖ Haspe f.; Klampe f.; Überwurf m. ‖ auberonnière f.; moraillon m.; happe f. / ~ and staple ‖ Kettel f. und Haspen m. ; Anwurf m. für ein Vorlegeschloß ‖ chaînette f. et picolet m. / ~ screw ‖ Klappschraube f. ‖ boulon m. à charnière.

hasten, to ‖ antreiben ‖ pousser; stimuler.

hat ‖ Hut m. ‖ chapeau m. / ~ (Printer's press) ‖ Krone f. ‖ chapeau m.; chapiteau m. / broad-brimmed ~ ‖ Hut m. mit breiter Krempe ‖ chapeau m. à large bord. / cloth ~ ‖ Stoffhut m. ‖ chapeau m. d'étoffe. / fancy ~ ‖ Fantasiehut m. ‖ chapeau m. fantaisie. / felt ~ ‖ Filzhut m. ‖ chapeau m. de feutre. / fur felt ~ ‖ Haarhut m. ‖ chapeau m. en feutre de poil. / ladies ~ ‖ Damenhut m. ‖ chapeau m. pour dames. / metal block for ~s ‖ Hutform f. aus Metall ‖ forme f. en métal pour la chapellerie. / mourning ~ ‖ Trauerhut m. ‖ chapeau m. de deuil. / narrow-brimmed ~ ‖ Hut m. mit schmaler Krempe ‖ chapeau m. à petit bord. / ~

for oxen ‖ Scheuklappe f. für Stiere ‖ tê-tière f. pour taureaux. / silken ~ ‖ Seidenhut m. ‖ chapeau m. de soie. / silk plush ~ ‖ Seidenhut m. ‖ chapeau m. en peluche de soie. / straw ~ ‖ Strohhut m. ‖ chapeau m. de paille. / ~ of uniform ‖ Uniformhut m. ‖ chapeau m. d'uniforme. / varnished ~ ‖ Lackhut m. ‖ chapeau m. vernis. / wood tissue ~ ‖ Hut m. aus Holzgewebe ‖ chapeau m. en tissu-bois.

hat band ‖ Hutband n. ‖ galon m. de chapellerie. / ~ block ‖ Hutform f.; Hutmacherform f. ‖ forme f. pour la chapellerie. / ~ blocker ‖ Hutformer m. ‖ dresseur m. de chapeaux. / ~ body ‖ Hutgerippe n. ‖ galette f. / ~ box see ~ case. / ~ brush ‖ Hutbürste f. ‖ brosse f. à chapeaux. / ~ buckle ‖ Hutschnalle f. ‖ boucle f. à chapeaux. / ~ case ‖ Hutschachtel f. ‖ étui m. ou boîte f. à chapeau.

hatch, to (Chicken) ‖ ausbrüten ‖ couver. / ~ (Drawing) ‖ schraffieren ‖ hacher.

hatch (Hydr arch) ‖ Schütz n.; Ziehschütz n.; Stauschütz n. ‖ vanne f.; pale f.; porte f. éclusière à coulisse. / ~ (Shipb) ‖ Luk n. ‖ écoutille f. / ~ to the hold (Shipb) ‖ Raumluk n. ‖ écoutille f. de la cale. / to lower the ~ ‖ das Schütz niederlassen ‖ baisser la pale. / ~ to the magazine (Shipb) ‖ Pulverluk n. ‖ écoutille f. de la soute aux poudres. / observation ~ ‖ Beobachtungsluke f. ‖ lucarne f. d'observation. / ship's ~ ‖ Schiffsluke f. ‖ écoutille f.

hatch bar (Shipb) ‖ Riegel m. eines Luks ‖ barre f. d'écoutille. / ~ crane ‖ Lukenkran m. ‖ grue f. d'écoutille.

hatched drawing ‖ schraffierte Zeichnung f. ‖ dessin m. haché. / the weld is ~ in the sketch ‖ die Schweißung ist in der Zeichnung schraffiert dargestellt ‖ la soudure est indiquée par des hachures sur le schéma.

hatchel, to ‖ hecheln ‖ peigner; sérancer.

hatchel ‖ Hechel f. ‖ séran m.; sérançoir m.; peigne-séran m.

hatcheling ‖ Hecheln n. ‖ sérançage m.; peignage m.

hatchet ‖ Axt f.; Beil n. ‖ hachette f.; cognée f.; hache f. / hammer ~ ‖ Hammerbeil n. ‖ hachette f. à marteau. / kitchen ~ ‖ Küchenbeil n. ‖ hachette f. de ménage. / ~ with spike behind ‖ Handbeil n. mit Spiker ‖ hache f. à pointe.

hatchet handle ‖ Axtstiel m. ‖ manche m. de hache.

hatchettine ‖ Hatchettin n. ‖ hatchetine f.; suif m. minéral ou de montagne.

hatching ‖ Schraffierung f. ‖ hachure f. / artificial ~ ‖ künstliches Ausbrüten n. ‖ couvée f. artificielle. / the tube walls are indicated by ~ ‖ die Rohrwände fpl. sind durch Schraffierung kenntlich gemacht ‖ les parois fpl. de tubes sont marquées en grisé sur ce schéma.

hatching apparatus ‖ Brutapparat m. ‖ couveuse f. / ~ house ‖ Brutanstalt f. ‖ couvoir m.; couveuse f.; incubateur m. / ~ machine ‖ Schraffiermaschine f. ‖ machine f. à hacher.

hatchway (Shipb) ‖ Luk n.; Niedergang m.; Niedergangsluk n. ‖ écoutille f. / ~ cover ‖ Lukendeckel m. ‖ couvercle m. d'écoutille. / ~ grating ‖ Lukengräting f. ‖ caillebottis m. d'écoutilles. / ~ lift ‖ Lukenaufzug m. ‖ monte-charge m. d'écoutille.

hat dyer ‖ Hutfärber m. ‖ teinturier m. en chapeaux. / ~ felter ‖ Hutwalker m. ‖ fouleur m. en chapeaux. / ~ hook ‖ Hut-haken m. ‖ porte-chapeau m. / electric ~ iron ‖ elektrisches Hutbügeleisen n. ‖ fer m. électrique à repasser les chapeaux. / ~ leather ‖ Hutleder n.; Schweißleder n. ‖ cuir m. à chapeaux. / ~ leather packing ‖ Winkelmanschette f. ‖ garniture f. coni-que. / ~ lining ‖ Hutfutter n. ‖ coiffe f. de chapeau. / ~ maker ‖ Hutmacher m. ‖ chapelier m.; ouvrier m. chapelier. / ~ making machine ‖ Hutherstellungsma-schine f. ‖ machine f. à fabriquer des chapeaux. / ~ manufactory ‖ Hutfabrik f. ‖ chapellerie f. / ~ money (Mar) ‖ Kapp-laken n. ‖ chapeau m. de maître.

hatpin ‖ Hutnadel f. ‖ épingle f. à cha-peau.

hat presser ‖ Hutpresser m. ‖ presseur m. de chapeau. / ~ rack ‖ Huthaken m. ‖ patère f. à chapeau. / ~ shape maker ‖ Hutformenmacherin f. ‖ formière f. de chapeaux de dame. / ~ stiffening ‖ Hut-steife f. ‖ empois m. de chapeau. / ~ stuff ‖ Hutstoff m. ‖ étoffe f. à chapeau. / ~ stump ‖ Hutstumpen m. ‖ forme f. de chapeau.

hatter ‖ Hutmacher m. ‖ chapelier m. / ~'s glue ‖ Hutmacherleim m. (die geringste Leimsorte) ‖ colle f. de chapeliers. / ~'s press form ‖ Hutumpreßform f. ‖ forme f. à presser les chapeaux.

hattery see hatting.

hatting ‖ Hutmacherei f. ‖ chapellerie f. / ~ machine ‖ Hutfabrikationsmaschine f. ‖ machine f. à faire des chapeaux.

hat trimmer ‖ Hutstepper m. ‖ garnisseur m. en chapeau. / ladies' ~ trimmer ‖ Damenhutgarniererin f. ‖ garnisseuse f. de chapeau pour dames. / ~ trimmings pl. ‖ Hutschmuck m.; Hutgarnitur f. ‖ garniture f. de chapeau.

hauerite ‖ Hauerit m. ‖ hauerite f.

haul, to (Flag) ‖ hissen ‖ haler; hisser. / ~ (Mar) ‖ den Kurs m. ändern ‖ changer de route. / ~ (Mine) ‖ fördern ‖ extraire. / ~ ashore ‖ ans Land n. ziehen ‖ haler à terre f. / ~ down ‖ herunterholen; nieder-holen ‖ haler bas. / ~ hand over hand ‖ Hand über Hand holen; palmen ‖ haler main sur main. / ~-in a rope ‖ ein Tau n. einholen ‖ recouvrer ou rentrer un cor-dage. / ~ a train by an electric loco-motive ‖ einen Zug m. mit einer elektri-schen Lokomotive ziehen ‖ remorquer un train m. par une locomotive électrique. / ~ the wind ‖ an den Wind m. gehen; anluven ‖ serrer le vent; loffer.

haulage ‖ Ziehen n.; Schleppen n. ‖ traînage m.; halage m.; touage m. / chain ~ (Mine) ‖ Kettenförderung f. ‖ extraction f. par chaîne. / chain ~ (River) ‖ Kettenförderung f. ‖ transport m. par chaîne flottante. / double ~ ‖ Doppelförderung f. ‖ extraction f. double. / electric canal ~ ‖ elektrischer Kanal-betrieb m.; elektrische Treidelei f. ‖ halage m. électrique sur les canaux. / ~ in exhaustions (Mine) ‖ Förderung f. in Abbauen ‖ extraction f. aux exploita-tions. / ~ of goods ‖ Güterbeförderung f.; Spedition f. ‖ transport m.; camionnage m. / ~ of trains ‖ Zugförderung f. ‖ re-morquage m. ou halage m. ou tirage m. de trains. / underground ~ ‖ Strecken-förderung f. ‖ roulage m. intérieur; traction f. dans les galeries.

haulage cable ‖ Förderseil n. ‖ câble m. d'extraction ou de transport. / ~ clip ‖ Schlepphaken m. ‖ tenaille f. d'attelage. / ~ rope see ~ cable. / ~ track ‖ Förder-strecke f. ‖ voie f. de roulage.

hauler (Mine) ‖ Schlepper m. ‖ rouleur m.; traîneur m.; galibot m. / steam ~ ‖ Dampfwinde f. ‖ treuil m. à vapeur.

hauling (Mine) ‖ Förderung f.; Schleppen n. ‖ extraction f.; transport m. / ~ through adit ‖ Stollenförderung f.; Strecken-förderung f. ‖ transport m. par les ga-leries.

hauling appliances pl. ‖ Fördergerät n. ‖ accessoires mpl. de transport. / ~ bridge ‖ Förderbrücke f. ‖ pont m. de transport / ~ cable ‖ Förderseil n. ‖ câble m. d'ex-traction ou de levage. / ~ costs pl. ‖ Förderkosten pl. ‖ frais mpl. d'exploita-tion. / ~-down cable (Aero) ‖ Haltetau n. ‖ câble m. de halage. / ~ drum ‖ Fördertrommel f. ‖ tambour m. de ma-chine d'extraction. / ~ engine ‖ Förder-maschine f. ‖ machine f. d'extraction. / ~ gear ‖ Fahrwinde f. ‖ treuil m. d'avancement. / ~ installation ‖ Förder-einrichtung f. ‖ installation f. de trans-port. / length of ~ ‖ Förderlänge f. ‖ distance f. de transport. / ~ load ‖ Förderlast f. ‖ charge f. à lever. / ~ ma-chine ‖ Fördermaschine f. ‖ machine f. d'extraction. / ~ output ‖ Förder-leistung f. ‖ capacité f. de transport ou d'extraction. / ~ plant ‖ Förderanlage f. ‖ installation f. de transport. / ~ rope ‖ Förderseil n. ‖ câble m. d'extraction. / ~ rope with wooden hand grip ‖ Zugstrick m. mit Querholz ‖ cordon m. de tirage avec menotte ou poignée en bois. / steel ~ rope ‖ stählernes Zugband n. ‖ ruban m. de tirage en acier. / ~ shaft (Mine) ‖ Förder-schacht m. ‖ fosse f. d'extraction. / ~ slip (Shipb) ‖ Aufschlepphelling f. ‖ cale f. à halage. / ~ stage ‖ Fördergerüst n. ‖ échafaudage m. d'extraction. / ~ steam-engine ‖ Dampffördermaschine f. ‖ machine f. d'extraction à vapeur. / ~ track ‖ Förderbahn f.; Grubenbahn f. ‖ chemin m. de fer d'exploitation ou de transport. / ~ truck ‖ Schleppwagen m.; Vorspannwagen m. ‖ remorqueur m. / ~ winch ‖ Förderwinde f. ‖ Förderhaspel f. ‖ treuil m. de transport ou d'extraction. / ~ windlass see ~ winch.

haulyard see halyard.

haunch (Build) ‖ Gewölbeschenkel m.; Bogenschenkel m. ‖ aisselle f. ou rein m. (de la voûte).

hausmannite ‖ Glanzbraunstein m. ‖ haus-mannite f.

haüyne ‖ Hauyn m. ‖ hauyne f.

hawk, to ‖ kolportieren ‖ colporter.

hawk (Mas) ‖ Scheibe f. (des Putzmaurers) ‖ oiseau m.

hawkbill ‖ Lötzange f. ‖ pincette f. ou tenaille f. à souder. / ~ pliers pl. see hawkbill.

hawker ‖ Kolporteur m. ‖ colporteur m.

hawling (Amplifier) ‖ Pfeifen n. ‖ chantage m.; sifflement m.

hawse ‖ Klüse f. ‖ écubier m. / clear ~ (Mar) ‖ klare Ketten fpl. ‖ chaînes fpl. dégagées. / foul ~ (Mar) ‖ unklare Ketten fpl. ‖ chaînes fpl. avec de tours.

hawse hole see hawse. / ~ pipe see hawse.

hawser ‖ Trosse f.; Kabeltau n.; Pferde-leine f. (zum Ziehen der Kähne); Treidel-leine f. ‖ haussière f.; aussière f.; grelin

m. de halage / ~ roller ‖ Trossenrolle f. ‖ tour m. pour amarres.

hawthorn ‖ Weißdornholz n.; Hagedorn m. ‖ aubépine f.

hay ‖ Heu n. ‖ foin m. / ~ barn ‖ Heu-scheuer f. ‖ fenil m. / ~ binder ‖ Heu-binder m. ‖ botteleur m. de foin. / ~ crop ‖ Mahd f. ‖ fauchaison f. ou fauchage m. de foin. / ~ fork ‖ Heugabel f. ‖ fourche à foin. / ~ harvest see ~ crop. / harvest-ing machine for ~ ‖ Heuerntemaschine f. ‖ machine f. de fenaison ou à récolter le foin. / ~kicker ‖ Heuwender m. ‖ faneuse f. / ~ loft ‖ Heuboden m. ‖ grenier m. à foin. / ~ maker ‖ Heu-wender m. ‖ tourne-foin m.; faneuse f. mécanique. / ~ press ‖ Heupresse f. ‖ presse f. à foin. / ~ rake ‖ Heurechen m. ‖ râteau m. à foin. / ~ tedder ‖ Heu-wendemaschine f. ‖ faneuse f.

haze ‖ Nebel m.; Höhenrauch m. ‖ brouil-lard m. sec; brume f.

hazel-nut ‖ Haselnuß f. ‖ noisette f. / ~ size ‖ Haselnußgröße f. ‖ grosseur f. de noi-sette. / ~ tree ‖ Haselnußbaum m. ‖ noisetier m.

hazel shrub ‖ Haselstrauchwerk n. ‖ noise-tier m.

hazel tree see hazel-nut tree.

hazy ‖ nebelig; diesig ‖ brumeux; nébuleux.

hazy picture ‖ unscharfes Bild n. ‖ image m. manquante de netteté.

head, to (Trees) ‖ kappen; köpfen ‖ éteter. / ~ a cask (Coop) ‖ ein Faß n. aus-böden ‖ foncer une futaille. / ~ the pin ‖ die Stecknadel anköpfen ‖ frapper la tête d'épingle. / ~-up wood in a wood-yard ‖ das Holz auf Lagerbäumen auf-stapeln ‖ enchanteler le bois.

head (Anatomy) ‖ Kopf m. ‖ tête m. / ~ (Beer) ‖ Haube f.; Spundhefe f. ‖ chapeau m. (de la bière entonnée). / ~ (Bridge) ‖ Kopfbalken m. ‖ chapeau m.; traverse f. / ~ (Build) ‖ Haube f. ‖ calotte f.; chapeau m.; bonnet m. / ~ (Dike) ‖ Kopf m. ‖ tête f. / ~ (Hydr arch) ‖ Gefälle n. ‖ hauteur f. de chute; chute f. / ~ (Letter) ‖ Kopf m. ‖ œil m. / ~ (Mach) ‖ Treib-stangenkopf m. ‖ tête f. de bielle. / ~ (Pump) ‖ Förderhöhe f. ‖ hauteur f. d'élévation. / ~ (Shipb) ‖ Topp m. ‖ tête f. / ~ (Of water) ‖ Fallhöhe f.; Druckhöhe f. ‖ chute f. (d'eau); hauteur f. (de chute). / ~ (Wind mill) ‖ Klappe f.; Haube f. ‖ toit m. du moulin. / auto-mobile ~ ‖ Kraftwagenverdeck n. ‖ dais m. ou capote f. pour automobiles. / boiler ~ ‖ Kesselboden m. ‖ fond m. de chaudière. / bolt ~ with feather ‖ Bolzenkopf m. mit Zunge ‖ tête f. de boulon à ergot. / buffer ~ ‖ Puffer-teller m.; Pufferscheibe f. ‖ disque m. ou plateau m. de tampon. / ~ of a carriage, which may be taken down ‖ Verdeck n. zum Abnehmen ‖ capot m. amovible d'une voiture. / ~ of a cask ‖ Faßboden m. ‖ fond m. de tonneau. / cheese ~ (Bolts and screws) ‖ Zylinder-kopf m.; runder Kopf m. ‖ tête f. cy-lindrique. / ~ of a comet ‖ Kopf m. eines Kometen ‖ tête f. d'une comète. / connect-ing-rod ~ ‖ Pleuelstangenkopf m. ‖ tête f. de bielle. / convex ~ (Typewr) ‖ ge-wölbter Kopf m. ‖ tête f. bombée. / counter-sunk ~ ‖ versenkter Schrauben-kopf m. ‖ tête f. de vis perdue ou fraisée ou noyée. / coupling ~ (Railw) ‖ Kuppel-kopf m. ‖ tête f. d'attelage. / cutter ~ ‖

Fräskopf m. ‖ tête f. porte-fraise. / cylinder ~ ‖ Zylinderkopf m.; Zylinderdeckel m. ‖ fond m. du cylindre. / dead ~ (Found) ‖ Gießkopf m.; verlorener Kopf m.; Anguß m. ‖ masselotte f.; tête f. perdue; jet m.; saumon m. / ~ of delivery ‖ Druckhöhe f. *oder* Förderhöhe f. einer Pumpe ‖ hauteur f. de refoulement. / ~ of a department ‖ Bürochef m. ‖ chef m. de bureau. / die ~ ‖ Schneidkluppe f.; Schneideisenhalter m. ‖ porte-coussinets m.; porte-filières m. / dividing ~ ‖ Teilkopf m.; Teilapparat m. ‖ poupée-diviseur f.; poupée f. à diviser. / feeding ~ *see* dead ~. / fillister ~ *see* cheese ~. / flat ~ (Typewr) ‖ flacher Kopf m. ‖ tête f. plate. / ~ of a gallery ‖ Feldort m. ‖ lieu m. de travail dans une galerie; fin f. d'une galerie. / geodetic ~ ‖ geodätisches Gefälle n. ‖ chute f. géodésique. / hexagonal ~ ‖ sechseckiger Kopf m. ‖ tête f. à six pans. / total hydraulic ~ ‖ Gesamtgefälle n.; gesamte Förderhöhe f. ‖ chute f. totale. / ~ of a level *see* ~ of a gallery. / lost ~ *see* dead ~. / ~ of a nail ‖ Nagelkopf m. ‖ tête f. d'un clou. / pin ~ ‖ Nadelkopf m. ‖ bouton m. *ou* tête f. d'épingle. / piston ~ ‖ Kolbenkörper m.; Kolbenboden m. ‖ corps m. *ou* fond m. de piston. / pit ~ ‖ Schachtöffnung f. ‖ gueule f. de mine. / poppet ~ (Turn) ‖ Reitstock m. ‖ poupée f. mobile *ou* coulissante. / ~ of the projectile ‖ Geschoßkopf m. ‖ tête f. du projectile. / ~ of the rail ‖ Schienenkopf m. ‖ champignon m. du rail. / width of ~ of rail ‖ Schienenkopfbreite f. ‖ largeur f. du champignon du rail. / ram ~ ‖ Stößelkopf m. ‖ tête f. *ou* porte-outil m. du coulisseau. / ~ of a river ‖ Quelle f. eines Flusses ‖ source f. d'un fleuve. / rivet ~ ‖ Nietkopf m. ‖ tête f. de rivet. / round ~ ‖ runder Kopf m.; Zylinderkopf m. ‖ tête f. ronde. / rounded ~ of rivet slightly cupped ‖ linsenförmiger Nietkopf m. ‖ tête f. à goutte de suif. / ~ of silk ‖ Bund m. Seide; Docke f. ‖ matteau m.; bouin m. de soie / ~ of a stone ‖ Kopfseite f. *oder* Stirnfläche f. *oder* Vorderseite f. eines Steins ‖ parement m. d'une pierre. / suction ~ of a pump ‖ Saughöhe f. einer Pumpe ‖ hauteur f. d'aspiration d'une pompe. / ~ of table ‖ Tabellenkopf m. ‖ tête f. de tableau. / vaulted ~ ‖ gewölbter Sturz m. ‖ larmier m. bombé et réglé. / ~ of water ‖ Druckhöhe f. *oder* Gefälle m. des Wassers ‖ hauteur f. de la chute; chute f. d'eau; charge f. d'eau. / ~ of a wedge ‖ Kopf m. *oder* Rücken m. eines Keiles ‖ tête f. d'un coin. / ~ of a window ‖ Fensterverschluß m.; Überdeckung f. eines Fensters ‖ fermeture f. d'une croisée.

headache ‖ Kopfschmerzen mpl. ‖ mal m. de tête.

head band ‖ Kopfband n.; Stirnreifen m. ‖ ruban m. serre-tête; bandeau m. frontal. / ~ bander ‖ Zigarrenbändchenumleger m. ‖ poseur m. de bagues. / ~ bay ‖ Oberhaupt n. einer Schleuse ‖ chambre f. *ou* tête f. d'amont *ou* d'écluse. / ~ beam (Hydr arch) ‖ Kopfbalken m.; Holm m. ‖ chapeau m.; sommier m.; traverse f. / ~ bearing ‖ Kopflager n. ‖ crapaudine f. supérieure. / releasable ~ bearing ‖ abschwenkbarer Lagerkopf m. ‖ palier m. basculant. / ~ brewer ‖ Braumeister m. ‖ maître brasseur m. / ~ brick ‖ Kopfziegel m. ‖ demi-boutisse en bri-

que. / ~ cellar-man ‖ Kellermeister m. ‖ chef m. de cave. / ~ clerk ‖ Geschäftsführer m.; Geschäftsleiter m. ‖ gérant m.; directeur m. / ~ collar ‖ Reithalfter f. ‖ licou m. / ~ collar rein ‖ Halfterzügel m. ‖ longe f. du licou *ou* bouclée; corde f. du licou. / ~ cord (Tel) ‖ Kopfhörerschnur f. ‖ cordon m. pour casque. / ~ covering ‖ Kopfbedeckung f. ‖ coiffure f. / ~ crown *see* ~ bay. / ~ cup ‖ Gegenhalter m. ‖ contre-bouterolle f.

head-dress of false hair ‖ Haaraufsatz m. ‖ postiche m.

header (Build) ‖ Binder m.; Tragstein m. ‖ boutisse f.; pierre f. parpaing. / ~ (Mas) ‖ Kopfstück n.; Scheinbinder m. ‖ fausse boutisse f. / ~ (Needl; Device) ‖ Stecknadelwippe f.; Knopfspindel f. ‖ têtoir m.; fuseau m. / ~ (Needl;Person) ‖ Kopfdreher m. ‖ enrouleur m.

head-flow, greatest ~ to the barrage ‖ höchster Zufluß m. zur Talsperre ‖ affluence f. maxima au barrage.

head gate (Sluice) ‖ Fluttor n.; Obertor n.; Vordertor n. ‖ porte f. d'amont *ou* de tête. / ~ gear (Horse) ‖ Zaumzeug n. ‖ garniture f. de tête de cheval. / ~ gear (Tel) ‖ Kopfbügel m. ‖ ressort m. de tête.

heading (Arch) ‖ Schluß m. ‖ fermeture f. / ~ (Mine) ‖ Richtstrecke f. ‖ galerie f. d'avancement. / ~ (Print) ‖ Kolumnentitel m. ‖ titre m. courant; ligne f. de tête. / bottom ~ ‖ Sohlstollen m. ‖ galerie f. de fond. / ~ of a cask ‖ Faßboden m. ‖ fond m. de tonneau. / ~ driven on the rise (Mine) ‖ schwebende Förderstrecke f. ‖ galerie f. de roulage montante. / under the ~ of ‖ unter der Firma f. ‖ sous la raison sociale de . . . .

heading course ‖ Binderschicht f.; Kopfschicht f. ‖ assise f. par boutisses. / cooper's ~ knife ‖ Küferschroppmesser m. ‖ plane f. à foncer pour tonneliers. / ~ lathe (Needl) ‖ Kopfrad n. ‖ tour m. à tête. / ~ machine (Needl) ‖ Knopfspindel f. ‖ fuseau m. / ~ stone (Build) ‖ Schlußstein m.; Bogenschluß m. ‖ clef m.; clausoir m.; mensole f. / ~ tool (Nailsm) ‖ Nageleisen n. ‖ cloutière f. / ~-up account ‖ Kontenüberschrift f. ‖ en-tête f. des comptes. / ~ weaving ‖ Zeugendenweberei f. ‖ tissage m. de chefs.

head lamp ‖ Scheinwerfer m. ‖ phare m. / to dim the ~s ‖ die Scheinwerfer mpl. abblenden ‖ baisser les phares mpl. / ~ support ‖ Scheinwerferstütze f. ‖ porte-lanterne f. / ~ support tie rod ‖ Verbindungsstange f. der Scheinwerfer ‖ tige f. d'accouplement des porte-lanternes.

head-leather, enamelled ~ for the heads of carriages ‖ lackiertes Verdeckleder n. ‖ cuir m. verni de capot.

head light (Med) ‖ Stirnlampe f. ‖ lampe f. frontale. / ~ (Railw) ‖ Lokomotivlaterne f. ‖ disque-lanterne f. / ~ (Signalling) ‖ Signallampe f. ‖ lampe f. de signaux. / acetylene gas ~ ‖ Azetylenscheinwerfer m. ‖ phare m. à acétylène. / ~ car ‖ Wagenlaterne f. ‖ phare m. de voiture. / ~ glare ‖ Blenden n. der Scheinwerfer ‖ éblouissement m. des phares. / ~ lens ‖ Scheinwerferlinse f. ‖ lentille f. de phare. / setting the ~s ‖ Einstellen n. der Scheinwerfer ‖ mise f. au foyer des lampes des phares.

headline (Print) ‖ Kopfzeile f. ‖ ligne f. en haut de page. / ~ (Print) ‖ Kolumnentitel m. ‖ titre m. courant; folio m.

head-man of the strikers ‖ Schmiedemeister m. ‖ maître-forgeron m.

head-miller ‖ Mühlenverwalter m. ‖ chef m. meunier.

head mirror ‖ Stirnspiegel m. ‖ miroir m. frontal. / ~ moulding (Build) ‖ Türverdachung f. ‖ corniche f. de porte. / ~ office (Tel) ‖ Hauptamt n. ‖ bureau m. central. / ~ offices pl. ‖ Hauptverwaltungsgebäude n. ‖ bâtiment m. *ou* bureaux mpl. de l'administration centrale.

headphone ‖ Kopfhörer m.; Kopffernhörer m. ‖ téléphone m. *ou* récepteur m. serretête; serre-tête m. (simple); écouteur m.; casque m. / ~ reproduction ‖ Kopfhörerwiedergabe f. ‖ audition f. à casque. / ~s pl. ‖ Doppelkopfhörer m. ‖ serre-tête m. double.

headpiece ‖ Helm m. ‖ casque m. / ~ (Mach) ‖ Kopfstück n. ‖ tête f. / ~ (Print) ‖ Kopfleiste f.; Zierleiste f. ‖ en-tête m.; remarque f. en vignette; fleuron m. / ~ (Radio; Tel) *see* headphone. / ~ (Textile) ‖ Kopfstock m. ‖ tétière f. / leather ~ (Aero; Auto) ‖ Sturzhelm m.; Lederhelm m. ‖ casque m. en cuir. / ~ telephone *see* headphone.

head pilot ‖ Oberlotse m. ‖ maître-pilote m. / position of the ~ ‖ Kopfhaltung f. ‖ position f. de la tête. / ~ rail (Balustrade) ‖ Brustriegel m.; Lehnriegel m. ‖ lisse f. de barrière; entretoise f. d'appui. / ~ rail (Carp) ‖ Türriegel m. im Fachwerk ‖ linteau m. en cloison. / ~ receiver ‖ Kopffernhörer m. ‖ téléphone m. *ou* récepteur m. serretête. / ~ resistance ‖ Stirnwiderstand m. ‖ résistance f. à l'avancement.

head-rest (Opt) ‖ Stirnbügel m. ‖ étrier m. frontal. / ~ (Phot) ‖ Stirnstütze f.; Kopfhalter m. ‖ appui-front m.; appui-tête m. / adjustable ~ (Opt) ‖ verstellbarer Kopfbügel m. ‖ arc m. réglable embrassant la tête.

head room for drawing the pistons ‖ zum Herausziehen der Kolben verfügbare Hubhöhe f. ‖ hauteur f. disponible à retirer les pistons.

heads pl. (Build) ‖ Traufschicht f.; Traufziegelreihe f. ‖ battellement m.

head sea ‖ Gegensee f. ‖ mer f. debout. / ~ set *see* ~phones. / ~ setter ‖ Titelsetzer m.; compositeur m. de titres. / ~ side (Print) ‖ Anlegesteg m. ‖ bois m. de marge d'un châssis. / ~ stall (Bridle-rein) ‖ Kopfgestell n.; Kopfstück m. ‖ tétière f. / ~ stamp (Nailsm) ‖ Kopfstempel m. ‖ estampe f. pour les têtes / ~ stiller ‖ Brennmeister m. ‖ maître-distillateur m. / ~ stock (Turn) ‖ Spindelkasten m.; Spindelstock m. ‖ poupée f. fixe. / ~ stone ‖ Kopfstein m. ‖ pavé m.; grès m. à pavés. / ~ straightening and jointing machine (Coop) ‖ Abrichthobel- und Fügemaschine f. für Bodenbretter ‖ machine f. à dresser et à jointer les fonds de fûts. / ~ sword ‖ Stollenwasser n. ‖ eau f. de galerie. / ~ telephone ‖ Kopffernhörer m. ‖ téléphone m. *ou* récepteur m. serre-tête. / ~ tie *see* ~ rail. / low- ~turbine ‖ Niederdruckturbine f. ‖ turbine f. pour chutes faibles.

headway, driving ~s pl. (Mine) ‖ Vorrichtungsarbeit f. beim Grubenbau ‖ travail m. préparatoire à la mine. / ~ free under crane ‖ lichte Höhe f. unter dem Kran ‖ dégagement m. sous grue.

head wheel (Mine) ‖ Seilscheibe f. ‖ molette f. / ~ wind ‖ Gegenwind m. ‖ vent m.

avant *ou* debout. / ~ works pl. ‖ Wasserschloß n.; Staubecken n. ‖ œuvres fpl. d'aménage. / ~ yeast ‖ Oberhefe f. ‖ levure f. superficielle.

heald ‖ Litze f.; Helfe f. ‖ lisse f.; maille f. / ~ cord ‖ Litzenschnur f. ‖ corde f. pour lisses. / ~ knitter (Weav) ‖ Schlingenaufleserin f. ‖ garnisseuse f. de maillons. / ~ knitting machine ‖ Litzenknüpfmaschine f.; Litzenstrickmaschine f. ‖ machine f. à tricoter les lisses; tricoteuse f. pour lisses.

health ‖ Gesundheit f. ‖ santé f. / public ~ ‖ öffentliche Gesundheitspflege f. ‖ hygiène f. publique. / ~ resort ‖ Luftkurort m. ‖ station f. climatérique.

heap, to ‖ anhäufen; aufstapeln ‖ entasser; amonceler; amasser. / ~ in a dry state ‖ trocken aufschütten ‖ charger à sec. / ~ up ‖ häufen ‖ entasser.

heap ‖ Haufen m. ‖ tas m.; amas m.; monceau m.; masse f. / ~ (Charc) ‖ Meiler m.; Kohlenmeiler m. ‖ meule f. de carbonisation; meule f. *ou* pile f. à charbon; charbonnière f. / ~ (Mine) ‖ Halde f. ‖ halde f.; halle f. / ~ of gravel ‖ Schottermasse f. ‖ masse f. de cailloux. / old ~ ‖ Althaufen m. ‖ vieille couche f. / ~ of ore ‖ Rösthaufen m. ‖ tas m. de minerai. / scrap ~ ‖ Haufen m. alten Eisens ‖ ferrailles fpl. amoncelées. / ~ of skins ‖ Schwitzhaufen m. ‖ échauffe f. de peaux. / waste ~ ‖ Halde f. ‖ halde f.; crassier m.

heap keeper ‖ Haldenarbeiter m. ‖ terrilleur m.

hearing ‖ Gehör n. ‖ ouïe f. / ~ (Tel) ‖ Hören n.; Deutlichkeit f. ‖ audition f. / ~ trumpet ‖ Hörrohr n. ‖ cornet m. acoustique.

hearse ‖ Leichenwagen m. ‖ corbillard m.

heart (Anatomy) ‖ Herz n. ‖ cœur m. / ~ (Carp) ‖ Kern m.; Kernholz n. ‖ cœur m. du bois. / ~ (Mach) ‖ Herzscheibe f.; Herznocken m. ‖ roue f. *ou* noix f. *ou* excentrique m. en cœur. / ~ (Mould) ‖ Kern m. ‖ noyau m. / ~ (Turn) ‖ Drehherz n. ‖ cœur m. de tour. / ~ (Wood) ‖ Mark n. ‖ moelle f. / ~ of the millstone ‖ Herzstück n. des Mühlsteines ‖ cœur m. de la meule. / ~ of a rope ‖ Herz n. *oder* Seele f. eines Taues ‖ mèche f. *ou* âme f. d'une corde. / ~ of a twisted column ‖ Kern m. einer gewundenen Säule ‖ noyau m. d'une colonne torse.

heart-beating ‖ Herzklopfen n. ‖ battements mpl. du cœur.

hearth (Blast-furnace) ‖ Herd m. (eines Hochofens) ‖ foyer m. (d'un haut-fourneau *ou* de fusion); sole f.; aire f. / ~ (Build) ‖ Ofenherd m. ‖ foyer m. / ~ (Forg) ‖ Schmiedefeuer n.; Schmiedeherd m. ‖ forge f.; feu m. *ou* âtre m. de forge; chaufferie f. / Catalan ~ ‖ katalonisches Luppenfrischfeuer n. ‖ foyer m. *ou* feu m. catalan. / ~ with charcoal-bed ‖ Löschfeuer n. ‖ feu m. brasque. / electric ~ ‖ Elektroesse f. ‖ four m. électrique. / ~ of a liquation-furnace ‖ Seigerherd m. ‖ four m. de liquation *ou* de ressuage. / furnace with movable ~ ‖ Ofen m. mit fahrbarem Herd ‖ four m. à sole mobile. / reciprocating ~ ‖ Schwingherd m. ‖ sole f. oscillante. / refining ~ ‖ Frischherd m. ‖ forge f. d'affinerie. / ~ of the refining-furnace ‖ Herd m. des Treibofens ‖ fond m. de coupelle. / ~ of the reverberatory furnace ‖ Herd m. des Flammofens ‖ foyer m. du four à réverbère.

hearth accretions pl. ‖ Herdansätze mpl. ‖ accroissements mpl. d'un four. / ~ ashes pl. ‖ Herd m. des Treibofens ‖ fond m. de coupelle. / ~ bricks pl. ‖ feuerfeste Steine mpl. ‖ briques fpl. réfractaires. / ~ casing ‖ Herdeinsatz m. ‖ caisse f. à foyer. / acid open ~ furnace ‖ saurer Martinofen m. ‖ four m. Martin acide. / basic open ~ furnace ‖ basischer Martinofen m. ‖ four m. Martin basique. / jacket of the ~ ‖ Herdmantel m. ‖ blindage m. du creuset. / ~ mould ‖ Herdform f. ‖ moule m. ouvert du foyer. / ~ plate ‖ Herdzacken m. ‖ plaque f. d'âtre. / ~ steel ‖ Rohstahl m. ‖ acier m. naturel *ou* brut. / ~ stone ‖ Herdstein m. eines Kamins ‖ pierre f. de cheminée.

heart rot ‖ Rotfäule f. ‖ pourriture f. rouge *ou* du cœur. / ~ shake (Wood) ‖ Kernriß m. ‖ fente f. au cœur. / ~-shaped ‖ herzförmig ‖ en forme de cœur m. / ~-shaped thimble ‖ Herzkausche f. ‖ cosse f. en forme de cœur. / ~ wheel ‖ Neoidenrad n.; Herzscheibe f. ‖ roue f. en cœur. / ~ wood ‖ Kernholz n. ‖ cœur m. du bois; bois m. de cœur.

heat, to (Forg) ‖ ausglühen; glühen ‖ rougir; recuire. / ~ (Oven) ‖ anheizen; heizen ‖ mettre le feu. / ~ (Met) ‖ erhitzen ‖ chauffer; rechauffer. / ~ over the bare flame ‖ in offener Flamme f. erhitzen ‖ chauffer à feu nu. / to begin ~ ‖ anheizen ‖ commencer à chauffer. / ~ a little ‖ anwärmen ‖ chauffer un peu. / ~ up the machine ‖ die Maschine anwärmen ‖ chauffer la machine. / the pieces pl. should be gradually and uniformly heated ‖ die Stücke npl. sind mäßig rasch und gleichmäßig zu erhitzen ‖ il faut chauffer les pièces fpl. lentement et régulièrement. / the bearing heats exceedingly ‖ das Lager wird übermäßig heiß ‖ le palier chauffe excessivement.

heat (Metal) ‖ Hitze f.; Glühhitze f.; Feuer n. ‖ chaude f. / ~ (Phys) ‖ Wärme f. ‖ chaleur f. / black ~ ‖ dunkle Hitze f. ‖ chaleur f. sombre. / ~ of a blast-furnace ‖ Feuer n. eines Hochofens ‖ feu m. d'un haut-fourneau. / bright-red ~ ‖ lebhafte Rotglühhitze f. ‖ rouge m. vif. / ~ carried away by excess of air ‖ mit der Luft abgeführte Wärme f. ‖ chaleur f. enlevée par l'excédent d'air. / to carry off the ~ ‖ die Wärme ableiten ‖ enlever la chaleur. / cherry-red ~ ‖ Kirschrotglut f. ‖ chaude f. rouge-cerise. / ~ of combination ‖ Bindungswärme f. ‖ chaleur f. de combinaison. / ~ of combustion ‖ Verbrennungswärme f. ‖ chaleur f. de combustion. / to conduct ~ ‖ Wärme f. leiten ‖ conduire la chaleur. / dark-red ~ ‖ Dunkelrotglut f. ‖ chaude f. sombre. / quantity of developed ~ ‖ Wärmetönung f. ‖ quantité f. de chaleur développée. / ~ of dissociation ‖ Dissoziationswärme f. ‖ chaleur f. de dissociation. / to eliminate ~ ‖ Wärme f. abführen ‖ enlever la chaleur. / emitted ~ ‖ abgegebene Wärme f. ‖ chaleur f. émise *ou* dégagée. / final mashing ~ ‖ Abmaischtemperatur f. ‖ température f. finale de saccharification. / finishing ~ (Malt) ‖ Abdarrtemperatur f. ‖ température f. finale de touraillage. / ~ of formation ‖ Bildungswärme f. ‖ chaleur f. de formation. / ~ of fuel ‖ Brennstoffwärme f. ‖ chaleur f. du combustible. / in full ~ ‖ in voller Hitze f. ‖ en pleine température f. / ~ of fusion

‖ Schmelzwärme f. ‖ chaleur f. de fusion. / to give ~ (Forg) ‖ das Eisen glühen; eine Hitze f. geben ‖ chauffer *ou* faire rougir le fer; donner une chaude au fer. / ~ of hot-house ‖ Treibhauswärme f. ‖ chaleur f. de serres chaudes. / ~ of hydration ‖ Hydrationswärme f. ‖ chaleur f. d'hydration. / the ingot was forged out under a press by means of x ~s ‖ der Block m. wurde in x Hitzen unter einer Presse ausgeschmiedet ‖ le lingot m. fut forgé sous une presse en x chauffes. / latent ~ ‖ gebundene Wärme f. ‖ chaleur f. latente. / lost ~ ‖ Abhitze f. ‖ chaleur f. perdue. / radiant ~ *see* radiating ~. / to radiate ~ ‖ Wärme f. ausstrahlen ‖ émettre la chaleur. / radiating ~ ‖ strahlende Wärme f. ‖ chaleur f. rayonnante. / ~ of reaction ‖ Reaktionswärme f. ‖ chaleur f. de réaction. / red ~ ‖ Glühhitze f. ‖ chaleur f. rouge; rouge m. / the rolling is done in one ~ ‖ das Auswalzen n. erfolgt in einer Hitze ‖ le laminage m. se fait dans une seule chaude. / rose-red ~ ‖ Hellrotglühhitze f. ‖ rouge m. claire. / sensible to ~ ‖ wärmeempfindlich ‖ sensible à la chaleur. / ~ of solidification ‖ Erstarrungswärme f. ‖ chaleur f. de solidification. / sparkling ~ *see* welding ~. / specific ~ ‖ spezifische Wärme f. ‖ chaleur f. spécifique. / terrestrial ~ ‖ Erdwärme f. ‖ chaleur f. terrestre. / total ~ ‖ Gesamtwärme f. ‖ chaleur f. totale. / to transmit ~ ‖ Wärme durchlassen ‖ laisser passer la chaleur. / uncombined ~ ‖ freie Wärme f. ‖ chaleur apparente. / upper ~ ‖ Oberfeuer n.; Oberhitze f. ‖ chaleur f. du gueulard. / ~ of vaporization ‖ Verdampfungswärme f. ‖ chaleur f. de vaporisation. / waste ~ ‖ abgehende Hitze f.: Abwärme f. ‖ chaleur f. perdue. / waste ~ recovering ‖ Abhitzerückgewinnung f. ‖ récupération f. de la chaleur perdue. / welding ~ ‖ Schweißhitze f. ‖ chaude f. suante *ou* soudante. / white ~ ‖ Weißglühhitze f. ‖ chaude f. blanche.

heat absorption ‖ Wärmeaufnahme f.; Wärmeeinstrahlung f. ‖ absorption f. de chaleur; chaleur f. absorbée. / ~ accumulator ‖ Wärmespeicher m. ‖ accumulateur m. de chaleur. / ~ capacity ‖ Wärmeaufnahmefähigkeit f.; Wärmekapazität f. ‖ capacité f. calorifique. / ~ coil (Tel) ‖ Feinsicherungspatrone f. ‖ bobine f. thermique. / ~ conductibility ‖ Wärmeleitung f.; Wärmeleitfähigkeit f. ‖ conductibilité f. calorifique. / ~-conducting alloy ‖ wärmeleitende Legierung f. ‖ alliage m. pyroforique. / ~-conducting power ‖ Wärmeleitungsvermögen n. ‖ pouvoir m. calorifique. / long-distance conveyance of ~ *see* long-distance ~ transmission. / deperdition of ~ ‖ Wärmeverlust m. ‖ déperdition f. de la chaleur. / ~ economizer ‖ Wärmeaustauschapparat m. ‖ échangeur m. de chaleur. / ~ economy ‖ Wärmewirtschaft f. ‖ économie f. calorique.

heated bearing ‖ warmgelaufenes Lager n. ‖ coussinet m. échauffé. / ~ circulating air ‖ Umlaufheizung f. ‖ chauffage m. à air de circulation. / boiler of ~ pipes pl. ‖ Heizrohrkessel m. ‖ chaudière f. de tuyaux chauds.

heat efficiency ‖ Heizeffekt m. ‖ rendement m. thermique *ou* en chaleur. / ~ emission ‖ Wärmeabgabe f. ‖ émission f. de

chaleur. / ~ equator ‖ Wärmeäquator m. ‖ équateur m. thermique.

heater (Device) ‖ Heizkörper m.; Vorwärmer m. ‖ calorifère m.; réchauffeur m. / ~ (Person) ‖ Heizer m. ‖ chauffeur m. / air ~ ‖ Lufterhitzer m. ‖ réchauffeur m. d'air. / counter-current feed ~ ‖ Gegenstromvorwärmer m. ‖ réchauffeur m. à contre-courant. / cream ~ ‖ Rahmvorwärmer m. ‖ réchauffeur m. de crème. / electric ~ ‖ elektrischer Heizkörper m. ‖ radiateur m. électrique. / feed water ~ ‖ Speisewasservorwärmer m. ‖ réchauffeur m. d'eau d'alimentation. / forge ~ ‖ Schmiedegeselle m. ‖ chauffeur m. de forge. / hot air ~ ‖ Warmluftheizungsanlage f. ‖ calorifère f. à air chaud. / hot water ~ ‖ Warmwasserheizungsanlage f. ‖ calorifère m. à eau chaude. / milk ~ ‖ Milchvorwärmer m. ‖ réchauffeur m. de lait. / steam ~ ‖ Dampfheizungsanlage f. ‖ calorifère f. à vapeur. / vehicle ~ ‖ Heizeinrichtung f. eines Wagens ‖ chaufferette f. de voiture. / mantle for ~s ‖ Heizkörperverkleidung f. ‖ revêtement m. de radiateurs.

heat exchange ‖ Wärmeaustausch m. ‖ échange m. de la chaleur. / ~ exchanger ‖ Wärmeaustauscher m. ‖ échangeur m. de chaleur ou de température. / ~ expansion ‖ Wärmeausdehnung f. ‖ dilatation f. thermique.

heath ‖ Heide f. ‖ bruyère f.; brande f.; lande f.

heath brush ‖ Heidebürste f. ‖ brosse f. en buisson ou de bruyère.

heather ‖ Heidekraut n. ‖ bruyère f. / ~ in bundles ‖ Heidekraut n. in Bündeln ‖ bruyères fpl. en paquets.

heath stone ‖ Heidenstein m. ‖ éricite f.

heating ‖ Beheizung f. ‖ chauffage m. / ~ (Chem; Phys) ‖ Erwärmung f. ‖ caléfaction f. / ~ (Forg) ‖ Hitze f.; Glühhitze f. ‖ chaude f. / ~ (Mach) ‖ Heißwerden n.; Heißlaufen n.; Erwärmung f. ‖ échauffement m. / ~ of a bearing ‖ Heißlaufen n. oder Warmlaufen n. eines Lagers ‖ échauffement m. d'un coussinet. / ~ with charcoal ‖ Holzkohlenfeuerung f. ‖ chauffage m. au charbon de bois. / electric ~ ‖ elektrische Beheizung f. ‖ chauffage m. électrique. / electric foundry machine ~ ‖ elektrische Gußmaschinenbeheizung f. ‖ chauffage m. électrique de machines à couler. / electrical ~ of engine pieces ‖ elektrische Heizung f. von Maschinenteilen ‖ chauffage m. électrique de pièces de machine. / ~ by electricity ‖ elektrische Heizung f. ‖ chauffage m. électrique. / excessive ~ ‖ starke Erhitzung f. ‖ échauffement m. exagéré. / ~ by gas ‖ Gasheizung f. ‖ chauffage m. au gaz. / the ~ is done in a hardening furnace of convenient size ‖ die Erhitzung f. erfolgt in einem Härteofen von geeigneter Größe ‖ le chauffage a lieu dans un four à tremper d'un volume proportionné. / ~ by hot air ‖ Warmluftheizung f. ‖ chauffage m. à air réchauffé. / ~ with hot-water pipes ‖ Warmwasserheizung f. ‖ chauffage m. à l'eau chaude. / local ~ ‖ örtliche Erwärmung f. ‖ échauffement m. local. / lower ~ ‖ Unterhitze f. ‖ chauffage m. par en bas. / ~ with peat see ~ with turf. / permanent ~ ‖ Dauerheizung f. ‖ chauffage m. continu. / ~ under pressure ‖ Druckerhitzung f. ‖ chauffage m. sous pression. / ~ by steam ‖ Dampfheizung f. ‖ chauffage m. à vapeur. / ~ with turf ‖ Torffeuerung f. ‖ chauffage m. à la tourbe. / upper ~ ‖ Oberhitze f. ‖ chauffage m. par en haut. / waste gas ~ ‖ Abgasheizung f. ‖ chauffage m. par les gaz perdus. / warm water ~ ‖ Warmwasserheizung f. ‖ chauffage m. à eau chaude.

heating apparatus ‖ Heizvorrichtung f. ‖ calorifère f.; appareil m. de chauffage. / ~ for eatables ‖ Speisenwärmer m. ‖ chauffe-plats m. / electric ~ ‖ elektrischer Heizapparat m. ‖ appareil m. de chauffage électrique. / hot-air ~ ‖ Luftheizung f. ‖ calorifère m. à air. / hot-water ~ ‖ Warmwasserheizung f. ‖ calorifère m. à eau chaude. / petrol ~ ‖ Benzinofen m. ‖ foyer m. à benzine.

heating arrangement ‖ Heizvorrichtung f. ‖ calorifère m. / ~ battery (Radio) ‖ Heizbatterie f. ‖ batterie f. de chauffage. / ~ body ‖ Heizkörper m. ‖ corps m. de chauffe. / ~ boiler ‖ Heizungskessel m. ‖ chaudière f. pour chauffage. / ~ boiler thermometer ‖ Heizungskesselthermometer n. ‖ thermomètre m. pour chaudières de chauffage. / ~ bowl ‖ Heizsonne f.; elektrische Sonne f. ‖ radiateur m. parabolique (de chaleur). / ~ capacity ‖ Heizfähigkeit f. ‖ capacité f. de chauffe. / ~ cartridge ‖ Heizpatrone f. ‖ cartouche f. de chauffage. / ~ chamber (Metal) ‖ Heizkammer f. ‖ chambre f. de chauffe. / ~ coil (Electr) ‖ Heizspule f. ‖ bobine f. de chauffage. / ~ coil (Mach) ‖ Heizschlange f. ‖ serpentin m. réchauffeur. / ~ coke (Found) ‖ Füllkoks m. ‖ coke m. d'allumage. / ~ coupling ‖ Heizkupplung f. ‖ raccordement m. pour circuits de chauffage.

heating-current circuit ‖ Heizstromkreis m. ‖ circuit m. de chauffage. / ~ intensity of repeater valves ‖ Heizstromstärke f. der Verstärkerröhren ‖ intensité f. du courant de chauffage des lampes amplificatrices. / ~ supervisory relay ‖ Heizstromkontrollrelais n. ‖ relais m. de contrôle du courant de chauffage. / ~ testing switch ‖ Heizstrommeßschalter m. ‖ interrupteur m. de mesure du courant de chauffage.

heating cushion see ~ pad. / ~ effect ‖ Heizeffekt m.; kalorimetrische Heizkraft f. ‖ effet m. calorimétrique ou calorifique. / ~ efficiency ‖ Anheizwirkungsgrad m. ‖ rendement m. pour la période d'échauffement.

heating element ‖ Heizelement n. ‖ élément m. de chauffage. / built-in ~ ‖ eingebautes Heizelement n. ‖ élément m. de chauffage monté. / mica ~ ‖ Glimmerheizelement n. ‖ élément m. de chauffage en mica.

heating filament ‖ Heizfaden m. ‖ filament m. de chauffage. / ~ flue ‖ Heizkanal m. ‖ carneau m. de chauffage. / ~ furnace ‖ Glühofen m. ‖ four m. à rechauffer ou à recuire. / ~ gas ‖ Heizgas n. ‖ gaz m. calorifère ou de chauffage. / excellent utilization of the ~ gases ‖ vorzügliche Ausnutzung f. der Heizgase ‖ utilisation f. parfaite des gaz de chauffage. / ~ grid see ~ grill. / ~ grill ‖ Heizgitter n. ‖ grille f. de chauffage. / ~ hour ‖ Heizstunde f. ‖ heure f. de chauffage. / ~ inset ‖ Heizeinsatz m. ‖ pièce f. de chauffage de rechange. / ~ installation ‖ Heizungsanlage f. ‖ installation f. de chauffage. / ~ jacket ‖ Heizmantel m. ‖ chemise f. de réchauffage. / ~ lamp ‖ Heizlampe f. ‖ lampe f. à chauffer. / ~ member ‖ Heizkörper m. ‖ corps m. de chauffage. / electric ~ machine ‖ elektrische Erwärmungsmaschine f. ‖ machine f. à réchauffer électriquement. / heavy ~ oil ‖ schweres Heizöl n. ‖ huile f. lourde pour chauffage. / ~ oven ‖ Heizofen m. ‖ calorifère m. / electric ~ pad ‖ elektrisches Heizkissen n. ‖ cataplasme m. électrique. / ~ pipe ‖ Heizrohr n. ‖ tuyau m. de chauffage ou de chaleur. / ribbed ~ pipe ‖ Rippenheizrohr n. ‖ tuyau m. de chauffage à ailettes. / ~ plant ‖ Heizungsanlage f. ‖ installation f. de chauffage. / ~ plate ‖ Heizplatte f. ‖ réchaud m.; dessous m. de plats; plaque f. chauffante. / electric ~ plate ‖ elektrische Heizplatte f. ‖ chaufferette f. électrique. / ~ power ‖ Heizkraft f. ‖ puissance f. calorifique ou de chauffe. / ~ register ‖ Dampfheizkörper m. ‖ radiateur m. / ~ resistance ‖ Heizwiderstand m. ‖ résistance f. de chauffage. / ~ section ‖ Heizstrang m. ‖ circuit m. de chauffage. / ~ spiral ‖ Heizspirale f. ‖ spirale f. de chauffage. / ~ station ‖ Heizwerk n. ‖ centrale f. de chauffage. / ~ stove ‖ Wärmofen m.; Trockenofen m. ‖ fourneau m. à sécher. / ~ surface ‖ Heizfläche f. ‖ surface f. de réchauffement ou de chauffe. / total ~ surface ‖ Gesamtheizfläche f. ‖ surface f. de chauffe totale. / ~ table ‖ Heiztisch m. ‖ table f. de chauffe. / ~ tape ‖ Heizband n. ‖ ruban m. de chauffe. / ~ technics pl. ‖ Heiztechnik f. ‖ technique f. du chauffage. / ~ test ‖ Glühprobe f. ‖ essai m. à la chaleur. / ~ tube ‖ Siederohr n.; Heizrohr n. ‖ bouilleur m.; tuyau m. de chauffage. / ribbed ~ unit ‖ Rippenheizelement n. ‖ élément m. de chauffage à ailettes. / electric ~ utensil ‖ elektrisches Heizgerät n. ‖ appareil m. de chauffage électrique. / ~ value ‖ Heizwert m. ‖ pouvoir m. calorifique. / determination of the ~ value ‖ Heizwertbestimmung f. ‖ détermination f. de la puissance calorifique. / ~ voltage ‖ Heizspannung f. ‖ tension f. de chauffage. / ~ wire ‖ Heizdraht m. ‖ fil m. de chauffage.

heat insulation composition ‖ Wärmeschutzmasse f. ‖ masse f. calorifuge. / ~ insulator ‖ Wärmeisoliermittel n. ‖ calorifuge m. thermique. / interchange of ~ ‖ Wärmeaustausch m. ‖ échange m. de chaleur. / ~ interchanging apparatus ‖ Wärmeaustauschapparat m. ‖ appareil m. échangeur de chaleur. / loss of ~ ‖ Wärmeverlust m. ‖ perte f. ou déperdition f. de chaleur. / ~ measuring device ‖ Wärmemeßeinrichtung f. ‖ installation f. calorimétrique. / mechanical equivalent of ~ ‖ mechanisches Wärmeäquivalent n. ‖ équivalent m. mécanique de la chaleur. / mechanical theory of ~ ‖ mechanische Wärmetheorie f. ‖ théorie f. dynamique de la chaleur. / ~ motor ‖ Wärmekraftmaschine f. ‖ moteur m. thermique. / ~-proof see ~-resisting. / ~ protective ‖ Wärmeschutzmittel n. ‖ moyen m. conservateur de chaleur. / ~ pump with evaporator ‖ Wärmepumpe f. mit Verdampfer ‖ pompe f. calorifique avec évaporateur. / ~ radiation ‖ Wärmestrahlung f. ‖ rayonnement m. thermique. / ~ resistance ‖ Heizwiderstand m. ‖ résistance f. de chauffage.

heat-resisting ‖ hitzebeständig ‖ résistant à la chaleur. / ~ castings pl. ‖ hitzebeständiger Guß m. ‖ moulages mpl. de résistance à hautes températures. / ~ retort ‖ hitzebeständige Retorte f. ‖ cornue f. d'une haute résistance à la temperature. / ~ steel ‖ hitzebeständiger Stahl m. ‖ acier m. résistant à la chaleur. / ~ tube ‖ hitzebeständiges Rohr n. ‖ tuyau m. allant au feu.

heat spectrum ‖ Wärmespektrum n. ‖ spectre m. de la chaleur. / ~ storage ‖ Wärmespeicherung f.; Wärmestauung f. ‖ accumulation f. ou emmagasinage m. de chaleur. / ~ storage boiler ‖ Wärmespeicherkessel m. ‖ chaudière f. accumulatrice d'eau chaude. / ~ storing stove ‖ Wärmespeicherofen m. ‖ poêle m. type accumulateur de chaleur. / ~ switch ‖ Heizschalter m. ‖ interrupteur m. de circuit de chauffage. / ~ transformation ‖ Wärmeumsatz m. ‖ échange m. de chaleur. / ~ transmission ‖ Wärmeübertragung f. ‖ transmission f. de chaleur. / long-distance ~ transmission ‖ Fernheizung f. ‖ transport m. ou transmission f. de chaleur à grande distance. / ~ treatment ‖ Wärmebehandlung f. ‖ traitement m. thermique. / ~ treatment furnace ‖ Wärmebehandlungsofen m. ‖ four m. pour le traitement thermique. / ~ unit ‖ Wärmeeinheit f.; Kalorie f. ‖ unité f. de chaleur; calorie f. / utilization of ~ ‖ Wärmeausnützung f. ‖ utilisation f. de la chaleur. / ~ value ‖ Heizwert m. ‖ pouvoir m. ou puissance f. calorifique.

heave, to ~ the cable ‖ das Ankertau einwinden ‖ virer le câble. / ~ in ‖ hieven; winden ‖ virer. / ~ the log ‖ loggen ‖ jeter le loch. / ~ up ‖ heben ‖ lever; élever; soulever.

heave (Mine) ‖ Verwerfung f. eines Ganges ‖ dérangement m. d'une couche. / to take a ~ (Mine) ‖ verworfen sein ‖ être dérangé.

heaver ‖ Hebebaum m.; Brechstange f.; Handspake f. ‖ levier m.; anspect. / locomotive ~ ‖ Lokomotivhebebock m. ‖ chèvre m. à locomotive. / ship's ~ ‖ Schiffshebebock m. ‖ chèvre f. à bateau.

heavier than air ‖ schwerer als Luft f. ‖ plus lourd que l'air m.

heaviness ‖ Schwere f. ‖ pesanteur f.; poids m.

Heaviside function (Tel) ‖ Stammfunktion f.; Stammgleichung f. ‖ fonction f. caractéristique de Heaviside. / ~ layer ‖ Heavisideschicht f. ‖ couche f. de Heaviside. / ~'s distorsionless line ‖ Heavisidesche verzerrungsfreie Leitung f. ‖ ligne f. sans distorsion de Heaviside. / ~'s expansion rule ‖ Heavisidesche Regel f. ‖ développement m. de Heaviside.

heavy ‖ schwer ‖ pesant; lourd. / ~ (Build) ‖ massiv; stark; massig ‖ massif. / ~ current ‖ Starkstrom m. ‖ courant m. fort. / ~ current laboratory ‖ Starkstromlaboratorium n. ‖ laboratoire m. pour courants forts. / ~ dragoon saddle ‖ deutscher Sattel m. ‖ selle f. allemande; selle de grosse cavallerie. / ~-duty engine ‖ Hochleistungsmotor m. ‖ moteur m. à haute capacité. / ~-duty machine ‖ Höchstleistungsmaschine f. ‖ machine f. à rendement maximum. / ~-duty upright drilling machine ‖ Ständerbohrmaschine f. für große Leistung ‖ machine f. à percer verticale de grande

puissance. / ~ hydrocarbon gas ‖ Äthylen n.; schweres Kohlenwasserstoffgas n. ‖ gaz m. oléfiant; gaz m. lourd d'hydrogène carburé. / ~ oil ‖ Schwerpetroleum n. ‖ pétrole m. lourd. / ~ oil motor ‖ Schwerölmotor m. ‖ moteur m. à pétrole lourd. / ~ petrol ‖ Schwerbenzin n. ‖ essence f. lourde. / ~ pressure ‖ starke Spannung f.; hoher Druck m. ‖ haute pression f. / to be of ~ sale ‖ schlechten oder schwierigen Absatz m. haben ‖ la marchandise est d'un placement difficile.

heavy spar ‖ Schwerspat m. ‖ spath m. lourd; baryte f.; baryte f. sulfatée. / fibrous ~ ‖ Faserbaryt m. ‖ baryte f. sulfatée fibreuse. / ~ mill ‖ Schwerspatmühle f. ‖ moulin m. à baryte.

heavy tank (War mat) ‖ schwerer Kampfwagen m.; Durchbruchtank m. ‖ char m. de combat lourd; char m. de rupture; tank m. écraseur.

heck (Weav) ‖ Leserost m.; Schergatter m. ‖ plot m. à grille. / ~ of a spinning wheel ‖ Gabel f. oder Flügel m. eines Spinnrades ‖ tréchoir m. ou ailette f. ou épinglier m. d'un rouet à filer. / ~ box of a warping mill (Weav) ‖ Führer m. oder Katze f. eines Scherrahmens ‖ plot m. ou giette f. d'un ourdissoir.

heckle, to see to hackle.

heckle see hackle.

hectare ‖ Hektar m. ‖ hectare m.

hectogram ‖ Hektogramm n. ‖ hectogramme m.

hectogramme see hectogram.

hectograph ‖ Hektograf m. ‖ hectographe m.

hectographic copy ‖ hektografischer Abzug m. ‖ copie f. hectographique.

hectographing paper ‖ Hektografenblatt n. ‖ papier m. hectographique.

hectograph mass ‖ Hektografenmasse f. ‖ composition f. hectographique.

hectoliter ‖ Hektoliter n. ‖ hectolitre m.

heddle ‖ Webelitze f.; Schaftlitze f.; Litze f.; Helfe f. ‖ lisse f. / ~ hook ‖ Einziehhaken m.; Einziehnadel f.; Reihhaken m. ‖ passette f. / ~ thread ‖ Litzenzwirn m.; Kammzwirn m. ‖ fil m. d'arcade.

hedenbergite ‖ Hedenbergit m. ‖ hédenbergite f.

hedge, to ~ with pales ‖ einpfählen ‖ palissader; entourer de palis m.

hedge ‖ (lebendiger) Zaun m.; Hecke f. ‖ haie f. / dead-wood ~ ‖ geflochtene Hecke f. ‖ haie f. de branches; haie f. sèche ou morte. / protective ~ ‖ Fangzaun m. ‖ haie f. ou palissade f. d'arrêt. / quick-set ~ ‖ lebende Hecke ‖ haie f. vive.

hedge cutter ‖ Heckenschneider m. ‖ tailleur m. de haies.

hedgehog transformer ‖ Igeltransformator m. ‖ transformateur m. hérisson.

hedging knife ‖ Hippe f. ‖ serpe f.; serpette f. / ~ shears pl. ‖ Heckenschere f. ‖ ciseaux mpl. de jardinier.

hedyphane ‖ Hedyphan m. ‖ hédiphane m.

heel, to ‖ überliegen ‖ s'incliner; donner de la bande. / ~ a vessel ‖ ein Schiff n. krängen oder überholen ‖ mettre un navire à la bande; incliner un navire.

heel (Mar) ‖ Krängung f. ‖ bande f. / ~ (Shoem) ‖ Absatz m. ‖ talon m. / ~ (Stocking) ‖ Ferse f.; Hacke f. ‖ talon m. / to give the ~ (Mar) ‖ rank unter Segel n. sein ‖ porter mal la voile. / ~ of the horse's foot ‖ Ferse f. am Pferdehufe ‖ talon m. du

cheval. / leather ~ (Shoem) ‖ Lederabsatz m. ‖ talon m. en cuir. / ~ of a mast ‖ Fuß m. eines Mastes ‖ pied m. d'un mât. / ~ of the rudder ‖ Ruderhacke f. ‖ talon m. de gouvernail. / ~ spliced ‖ verstärkte Ferse f. ‖ talon m. renforcé. / ~ of tongue ‖ Zungenwurzel f. ‖ talon m. d'aiguille. / wooden ~ (Shoe) ‖ Holzabsatz m. ‖ talon m. en bois.

heel chair of tongue (Switch) ‖ Drehstuhl m. ‖ coussinet m. de talon.

heeler (Shoem) ‖ Absatzaufnagler m. ‖ poseur m. de talons.

heeling (Mar) ‖ Krängung f. ‖ bande f. / ~ angle ‖ Krängungswinkel m. ‖ angle m. de bande. / ~ error (Mar) ‖ Krängungsfehler m. ‖ erreur f. de bande. / ~ magnet ‖ Krängungsmagnet m. ‖ aimant m. générateur du champ additionnel.

heel iron ‖ Stiefeleisen n. ‖ fer m. à talons. / ~ maker ‖ Absatzmacher m. ‖ monteur m. de talons. / ~ piece (Shoem) ‖ Absatzfleck m. ‖ talonnette f. / ~ piece stiffening ‖ Kappensteife f. ‖ empois m. de paton. / ~ post of a flood-gate ‖ Wendesäule f. eines Deichschleusentores ‖ poteau m. tourillon d'une porte d'écluse. / ~ tenon of a mast ‖ Fußzapfen m. eines Mastes ‖ tenon m. d'emplantation d'un mât. / ~ tip wire ‖ Stiefeleisendraht m. ‖ fil m. pour fers à bottes. / ~ trimmer (Shoem) ‖ Absatzfräser m. ‖ fraiseur m. de talons.

height ‖ Höhe f. ‖ hauteur f. / ~ (Hydr arch) ‖ nutzbares Gefäll n.; Druckhöhe f. ‖ hauteur f. de la chute; charge f. d'eau. / adjustable ~ ‖ verstellbare Höhe f. ‖ hauteur f. réglable. / ~ of barometer ‖ Luftdruckstand m.; Barometerstand m. ‖ hauteur f. barométrique. / ~ between decks ‖ Höhe f. zwischen den Decks ‖ hauteur f. d'entrepont. / ~ of a bowstring truss ‖ Pfeilhöhe f. eines Trägers mit gekrümmter Gurtung ‖ flèche f. d'une poutre en arc. / ~ of the centre of gravity ‖ Schwerpunktshöhe f. ‖ hauteur f. du centre de gravité. / ~ of the day ‖ lichte Höhe f.; Höhe f. im Lichten ‖ hauteur f. libre ou du jour. / ~ of fall ‖ Druckhöhe f.; Gefälle n.; Fallhöhe f. ‖ hauteur f. de la chute; chute f. ou charge f. d'eau. / ~ above floor ‖ Höhe f. über dem Boden ‖ hauteur f. au-dessus du sol. / ~ of focus ‖ Lichtpunkthöhe f. ‖ hauteur m. du point lumineux. / ~ of gib head ‖ Nasenhöhe f. ‖ hauteur f. du talon. / ~ of layer ‖ Schichthöhe f. ‖ hauteur f. de couche. / ~ of ledge (Railw) ‖ Spurhöhe f. ‖ hauteur f. du rebord. / ~ of a letter ‖ Buchstabenhöhe f. ‖ hauteur f. du caractère. / ~ of letters ‖ Schrifthöhe f.; Papierhöhe f. ‖ hauteur f. en papier. / ~ of level ‖ Bodenhöhe f. ‖ niveau m. du sol. / ~ of lift ‖ Förderhöhe f.; Hubhöhe f.; Hub m. ‖ course f.; course f. ou hauteur f. de levage. / ~ of lift of crane ‖ Hubhöhe f. beim Kran ‖ course f. de levage de grue. / ~ of the line ‖ Zeilenhöhe f. ‖ hauteur f. de la ligne. / metacentric ~ ‖ metazentrische Höhe f. ‖ hauteur f. métacentrique. / overall ~ ‖ Bauhöhe f. ‖ encombrement m. en hauteur. / ~ of projection ‖ Wurfhöhe f. ‖ hauteur f. du jet. / ~ of rail ‖ Schienenhöhe f. ‖ hauteur f. des rails. / ~ above sea level ‖ Höhe f. über dem Meeresspiegel; Seehöhe f. ‖ hauteur f.; altitude

f.; élévation f. au-dessus du niveau de la mer. / ~ of the shed (Weav) || Sprunghöhe f. || foule f. / ~ of stroke (Forg) || Hub m. || hauteur f. de chute. / suction ~ of a pump || Saughöhe f. einer Pumpe || hauteur f. d'aspiration d'une pompe. / ~ of swell || Stauhöhe f. || hauteur f. de barrage. / useful ~ || Nutzhöhe f. || hauteur f. utile. / ~ of a vault || Pfeilhöhe f. *oder* Stichhöhe f. einer Wölbung || flèche f. *ou* montée f. de voûte; voussure f. / ~ of velocity || Geschwindigkeitshöhe f. || hauteur f. de la vitesse. / ~ of waves || Wellenhöhe f. || hauteur f. des lames *ou* des ondes.

heighten, to ~ the yellow colour of brass in nitric acid (aqua fortis) || Messing n. abbrennen || décaper le laiton.

heightened || erhöht || relevé.

height finder || Höhenmesser m. || altimètre m. / ~ gauge || Höhenmaßstab m. || mesure m. de hauteur. / ~-to-paper (Print) || Schrifthöhe f. || hauteur f. en papier.

helical blower || Propellergebläse n. || soufflerie f. à hélice. / ~ curve || Schraubenlinie f. || hélice f. / ~ line || Spirale f.; Spirallinie f. || spirale f. / ~ pump || Schraubenpumpe f. || pompe f. hélicoïdale. / ~ spring || (zylindrische) Schraubenfeder f. || ressort m. à boudin *ou* en spirale *ou* hélicoïdal. / automatic machine for coiling ~ springs || Maschine f. zum selbsttätigen Wickeln von Schraubenfedern || machine f. automatique à enrouler les ressorts à boudin. / ~ tooth || Spiralzahn m. || dent f. hélicoïdale.

helicoid *see* helicoidal.

helicoidal || schraubenförmig || hélicé; hélicoïde; hélicoïdal. / ~ groove || Schraubennute f. || rainure f. hélicoïdale. / ~ parabola || parabolische Spirale f. || parabole f. hélicoïde; spirale f. parabolique. / ~ ventilator || Schraubenventilator m. || ventilateur m. hélicoïdal.

helicopter || Schraubenflieger m. || hélicoptère m. / ~ screw || Hubschraube f. || hélice f. sustentatrice.

heliocentric || heliozentrisch || héliocentrique.

heliochromogravure || Heliochromogravure f. || héliochromogravure f.

heliographic paper making machine || Lichtpauspapierherstellungsmaschine f. || machine f. à fabriquer des papiers photographiques.

heliographical printing || Lichtdruck m. || impression f. héliographique.

heliogravure || Heliogravüre f.; Fotogravüre f. || héliogravure f.; photogravure f.

heliometer || Heliometer n.; Sonnenmesser m. || héliomètre m.

helioscope || Helioskop n.; Sonnenglas n. || hélioscope m.

heliostate || Heliostat m. || héliostat m.; porte-lumière m.

heliotherapy || Höhensonnentherapie f. || héliothérapie f.

heliotrope || Heliotrop m. || héliotrope m.; agate f. ponctuée verte.

helium || Helium n. || hélium m. / ~ line || Heliumlinie f. || raie f. de hélium. / ~ valve || Heliumröhre f. || lampe f. hélium.

helix || Schraubenlinie f.; Spirale f. || hélice f.; spirale f.

hellebore root || Nießwurz f. || rhizome m. d'ellébore.

helm (Mar) || Ruder n.; Steuer n. || gouvernail m. / ~ (Tiller) || Ruderpinne f. || timon

m. *ou* barre f. du gouvernail. / to shift the ~ || Ruder n. legen || coucher la barre. / weather ~ || Luvruder n. || gouvernail m. arrivé.

helm angle || Ruderwinkel m. || angle m. de barre. / ~ coat || Ruderkragen m. || braie f. du gouvernail.

helmet || Helm m. || casque m. / colonial ~ || Kolonialhelm m.; Tropenhelm m. || casque m. colonial. / ~ for fire brigades || Feuerwehrhelm m. || casque m. de pompiers. / ~ for the tropics || Tropenhelm m. || casque m. pour pays chauds.

helmport || Ruderkoker m. || jaumière f.

helmsman || Rudergänger m.; Steuermann m. || timonier m. / ~ elevator || Höhensteuermann m. || timonier m. d'altitude.

help, to || raten; helfen || conseiller; porter remède; deviner.

helper || Gehilfe m. || aide m. / ~ (Build) || Handlanger m. || manœuvrier m.; journalier m.; journalier m. aide-maçon / ~ (Pap) || Holländergehilfe m. || sous-gouverneur m. / first ~ (Factory) || Betriebsassistent m. || premier aide m.

helve, to ~ an axe || ein Beil n. anschäften || monter une hache.

helve || Stiel m. (eines Werkzeuges) || manche m. / hammer ~ || Hammerstiel m. || manche m. d'un marteau.

helve hammer || Aufwurfhammer m. || marteau m. à creuser. / ~ with a wooden shaft || Aufwerfhammer m. mit hölzernem Stiel || marteau m. à creuser à manche en bois.

hem, to || säumen || ourler.

hematite || Hämatit m.; Blutstein m. || hématite f.; pierre f. sanguine; sanguine f. / ~ containing a small amount of manganese || manganarmes Hämatit n. || hématite f. à faible teneur en manganèse. / ~ iron || Hämatiteisen n. || fonte f. hématite. / ~ iron ore || roter Glaskopf m. || hématite f. rouge.

hemeralopia || Nachtblindheit f. || héméralopie f.

hemiheder || Halbflächner m. || hémièdre m.

hemihedral || halbflächig || hémiédrique. / ~ forms pl. || hemiedrische Formen fpl.; Hälftflächner m. || hémiédrie f.; cristaux mpl. hémièdres.

hemimorphism || Hemimorphie f. || hémimorphie f.

hemisphere || Halbkugel f.; Hemisphäre f. || hémisphère m. / southern ~ of the earth || südliche Halbkugel f. der Erde || hémisphère m. austral de la terre.

hemispherical || halbkugelförmig || hémisphérique. / ~ shape || Halbkugelgestalt f. || forme f. hémisphérique. / ~ vault || Kugelgewölbe n. || dôme m.; cul-defour m.; voûte f. sphérique.

hemitetrahexahedron || Pyritoeder n.; Pentagondodekaeder n. || hémihexatétraèdre m.; dodecaèdre m. pentagonal.

hemitrisoctahedron, trigonal || Pyramidentetraeder m.; Trigondodekaeder m. || tétraèdre m. pyramidé *ou* pyramidal; tétratrièdre m.; hémiicositétraèdre m.

hemmer (Person) || Säumerin f. || ourleuse f. / ~ (Sewing mach) *see* hemming rule.

hemming rule || Säummaschine f. || guide m. à ourler.

hemp || Hanf m. || chanvre m. / Bombay ~ || Bombayhanf m. || bombax m. / to break ~ || den Hanf m. brechen || broyer le chanvre. / clean ~ || Reinhanf m. || brin m.; chanvre m. net *ou* sérancé. / com-

mon ~ || Hanf m. || chanvre m. / cut ~ || Schnitthanf m. || chanvre m. coupé / dressed ~ || reinabgezogener Hanf m. || chanvre m. nettoyé. / ~ in fibres || Hanf m. in Fasern || chanvre m. en fibres. / hackled ~ || gekämmter Hanf m. || chanvre m. peigné. / half-clean ~ || Basthanf m. || chanvre m. cru. / Manila ~ || Manilahanf m. || chanvre m. de Manille. / raw ~ || roher Hanf m. || chanvre m. brut. / rough ~ || ungehechelter Hanf m. || chanvre m. brut. / retted ~ || gerösteter Hanf m. || chanvre m. roui. / scutched ~ || gepochter Hanf m. || chanvre m. teillé. / for shoemaker's thread || Schusterhanf m. || chanvre m. pour fil de cordonnerie. / Sisal ~ || Sisalhanf m. || sisal m. / spun ~ || Hanfgarn n. || fil m. de chanvre. / ~ in stems || Hanf m. in Stengeln || chanvre m. en tiges.

hemp bag || Hanftasche f. || sac m. en chanvre. / ~ cable || Hanfseil n. || corde f. *ou* cordage m. de chanvre. / ~ codille || Hanfhede f.; Hanfwerg n. || étoupe f. de chanvre. / ~ coiling || Hanfflechte f.; Hanfzopf m. || tresse f. de chanvre. / ~ comb || Hanfhechel f. || regayoir m. / coarsest ~ comb || Abzugshechel f.; Grobhechel f. || ébauchoir m.; grand peigne m., regayoir m.; séran m. / cord of ~ || Hanfseil n. || corde f. *ou* cordage m. de chanvre. / ~ core || Hanfseele f. || âme f. de chanvre. / ~ covering || Hanfumspinnung f. || guipage m. de chanvre. / ~ dresser || Hanfbereiter m. || chanvrier m.; préparateur m. de chanvre. / ~ dressing || Zubereitung f. von Hanf || préparation f. du chanvre. / ~ dressing machine || Hanfaufbereitungsmaschine f. || machine f. à préparer le chanvre.

hempen core *see* hemp core.

hemp fibre || Hanffaser f. || brin m. *ou* fibre f. de chanvre. / ~ hackle || Hanfhechel f. || regayoir m. / ~ hackling || Hanfhechelei f. || peignage m. de chanvre. / ~ hackling machine || Hanfhechelmaschine f. || machine f. à sérancer le chanvre. / ~ hards pl. *see* codille. / ~ hose || Hanfschlauch m. || tuyau m. en chanvre. / ~ linen || Hanfleinwand f. || toile f. de chanvre. / ~ linen weaving || Hanfleinwandweberei f. || tissage m. de toiles de chanvre. / ~ mill || Hanfreibemühle f. || moulin m. à broyer le chanvre. / ~ packing || Hanfliderung f. || garniture f. de chanvre. / ~ paper || Hanfpapier n. || papier m. de chanvre. / retting of ~ || Hanfröste f.; Rösten n. des Hanfes || rouissage m. du chanvre. / ~ rope || Hanfseil n. || corde f. *ou* câble m. en chanvre. / ~ rope grease || Hanfseilschmiere f. || graisse f. pour câbles de chanvre. / ~ seed || Hanfsame m. || graine f. de chanvre. / ~ seed oil || Hanföl n. || huile f. de chènevis *ou* de chanvre. / ~ spinner || Hanfspinner m. || fileur m. de chanvre. / ~ spinning || Hanfspinnerei f. || filature f. de chanvre. / ~ spinning machine || Hanfspinnereimaschine f. || machine f. à filer le chanvre. / ~ strap || Hanfgurt m. || sangle f. de chanvre. / ~ tow || Hanfhede f. || étoupe f. de chanvre. / ~ tow dressing || Zubereitung f. von Hanfbast || préparation f. de la filasse de chanvre. / ~ twister || Hanfzwirner m. || retordeur m. de chanvre. / ~ twist mill || Hanfzwirnerei f. || retorderie f. de chanvre. / ~ yarn || Hanfgarn n.;

Hanffaser f. || fil m. de chanvre. / ~ yarn weaver || Hanfgarnweber m. || tisseur m. de chanvre.

hemstitch || Hohlsaum m. || ourlet m. à jour. / ~ machine || Hohlsaummaschine f. || machine f. à ourler à jour.

henbane leaves pl. || Bilsenkraut n. || feuilles fpl. de jusquiame.

hen house || Hühnerstall m. || poulailler m.

hepar of sulphur || Schwefelleber f.; Kaliumsulfid n. || foie m. de soufre.

hepatic cinnabar || Korallenerz n. || mercure m. sulfuré bitumineux. / ~ water || Schwefelwasser n.; Schwefelleberwasser n. || eau f. sulfurée ou hépatique.

herald engraver || Wappenstecher m. || graveur m. héraldique.

heraldic chaser || Wappenziselör m. || ciseleur m. d'armoiries.

herbaceous plant || Staudenpflanze f. || plante f. herbacée. / ~ stalk || krautartiger Stengel m. || tige f. herbacée.

herb press || Kräuterpresse f. || presse f. à herbes.

herdsman || Hirt m. || berger m.; pâtre m.; bouvier m.; vacher m.; pasteur m.; gardien m. de troupeaux.

hermaphrodite brig || Briggschoner m. || brick-goélette f. / ~ tank (War mat) || Räderraupenkampfwagen m. || tank m. composite; char m. de combat composite.

hermetical || luftdicht || hermétique; fermé ou imperméable à l'air; fermé hermétiquement. / ~ fitter || Dosenschließer m. || sertisseur m.

hernial bandage (Med) || Bruchbandage f. || bandage m. herniaire.

herring, to gut and salt ~s || Heringe mpl. kaken || caquer des harengs mpl. / red ~ || Bückling m. || hareng m. saur ou fumé.

herring barreller || Heringspacker m. || encaqueur m. de harengs.

herringbone gear cutting machine || Pfeilräderfräsmaschine f. || machine f. à fraiser les roues à denture à chevrons. / ~ gear || Pfeilradgetriebe n. || engrenage m. hélicoïdal double. / ~ parquetry || Riemenparkett n.; Schiffsparkett n. || parquet m. à fougère. / ~ tooth || Winkelzahn m. || dent m. à chevron. / ~ wheel || Winkelzahnrad n. || roue f. à denture à chevrons.

herringer || Heringsfischer m. || pêcheur m. de harengs.

herring fisherman || Heringsfischer m. || pêcheur m. de harengs. / ~ fishing || Heringsfang m. || harengaison f. / ~ keg || Heringstonne f. || caque f. / ~ packer || Heringspacker m. || encaqueur m. / ~ salter || Heringssalzer m. || saleur m. de harengs. / ~ smoker || Heringsräucherer m. || préparateur m. de harengs fumés.

Hertzian doublet || Hertzscher Doppelpol m. || dipôle m. hertzien. / ~ oscillator || Hertzscher Sender m. || oscillateur m. hertzien.

Hertz' radiation integral || Hertzsche Funktion f. || fonction f. hertzienne.

hesitation, without || anstandslos || sans hésitation f. ou difficulté f.

heterodyne || Überlagerer m. || hétérodyne f. / ~ reception || Heterodynempfang m.; Überlagerungsempfang m. || réception f. hétérodyne.

heterogeneity || Ungleichförmigkeit f. || hétérogénéité f.

heterogeneous || heterogen || hétérogène.

heulandite || Heulandit m.; Stilbit m.; Blätterzeolit m. || heulandite f.; étilbite f.

hevea caoutchouc || Kautschukbaum m. || hévé m.; hévée f.

hew, to || hauen; schrämen || entailler. / ~ an ashlar with the pickhammer || einen Stein m. bespitzen || piquer une pierre avec la pointe d'une marteline. / ~ down the timber || Holz n. fällen || abattre le bois. / ~ a stone || einen Stein m. behauen || dresser une pierre. / ~ the trenches pl. || schrämen || entailler les couches fpl.

hew || Abhieb m. || épaufrure f.

hewer (Mine) || Hauer m.; Häuer; Kohlenhauer m.; Bergmann m. || piqueur m.; coupeur m.; mineur m.; hayeur m.; ouvrier m. de taille ou de fond; ouvrier m. abat(t)eur ou à veine.

hewing || Hauerarbeit f. || travail m. de havage. / building stone ~ || Bausteinbehauen n. || taille f. des pierres à bâtir.

hewing bench (Carp) || Haubank f. || calle f.

hewn timber || glattbehauenes Bauholz n. || bois m. dégrossi.

hexachlorethan || Hexachloräthan n. || hexachlorethane m.

hexagon || Sechseck n.; Sechsseit n. || hexagone m.

hexagonal || sechsseitig; sechseckig || à six pans mpl.; hexagonal; hexagone. / ~ bolt || Sechskantbolzen m. || boulon m. à six pans. / ~ head || Sechskantkopf m. || tête f. hexagonale. / ~ nut || Sechskantmutter f. || écrou m. hexagonal. / ~ perforation || Sechskantlochung f. || perforation f. hexagonale.

hexagon head || Sechskantkopf m. || tête f. à six pans ou hexagonale. / ~ spanner || Sechskantschlüssel m. || clef f. hexagonale.

hexahedron || Kubus m.; Würfel m. || cube m.; hexaèdre m.

hexakistetrahedron || Hexakistetraeder m. || hexakistétraèdre m.

hexamethylene-tetramine || Hexamethylentetramin n. || hexaméthylènetétramine m.

hexatomic || sechsatomig || hexatomique.

hib knob (Carp) || Helmstangenspitze f. || épi m. de faîte.

hiccough || Schlucken m. || hoquet m.

hickory wood || Hickory n.; Hickoryholz n. || hickory m.

hiddenite (Miner) || Hiddenit m. || hiddenite f.

hide || Fell n.; Haut f. || peau f. / fresh ~ || frische Haut f. || peau f. verte. / ~ too much fulled || zu stark gewalkte Haut f. || nerveuse f. / green ~ see fresh ~. / raw ~ see fresh ~. / ~ for tanning || Haut f. zum Gerben || peau f. pour le tannage. / untanned ~ || ungegerbte Haut f. || peau f. non tannée ou verte ou brute.

hide, articles pl. from ~s pl. || Waren fpl. aus Häuten || objets mpl. en peaux. / ~ parings pl. || Schabaas n.; Leimleder n.; Leimluder n. || écharnure f. de cuir. / ~ scrapings pl. see ~ parings. / ~ shreds pl. see ~ parings. / ~ sorter || Fellsortierer m. || lotisseur m. de peaux brutes.

Hienfong essence || Hienfongessenz f. || essence f. Hienfong.

high || hoch || haut. / ~ aerial || Hochantenne f. || antenne f. haute. / ~ building || Hochbau m. || construction f. en élévation. / ~ damping (Radio) || große Dämpfung f. || amortissement m. élevé. / ~ gearing ||

hohe Übersetzung f. || multiplication f. forte. / polish for ~ lustre || Hochglanzpolitur f. || poli m. à reflets. / ~ resistance (Electr) || großer oder hoher Widerstand m. || haute résistance f. / ~ reservoir || Hochbehälter m. || réservoir m. surélevé.

high (Meteor) || Hochdruckgebiet n. || zone f. de haute pression.

high-altitude climate || Höhenklima f. || climat m. des hautes altitudes. / ~ navigation of airships || Höhennavigation f. von Luftschiffen || repérage m. altimétrique de dirigeables.

high-breast water wheel || rückschlächtiges Wasserrad n. || roue f. hydraulique par derrière.

high-burst ranging || Einschießen n. mit hohen Sprengpunkten || réglage m. par coups fusants hauts.

high-capacity nibbling machine || Hochleistungsdekupiermaschine f. || machine f. à découper ou grignoteuse f. à grand rendement.

high-class || hochwertig || de haute valeur f.

high-duty boiler || Hochleistungskessel m. || chaudière f. à grand débit.

high efficiency milling machine || Hochleistungsfräsmaschine f. || machine f. à fraiser à grand rendement. / ~ smoothing-iron || Hochleistungsbügeleisen n. || fer m. à repasser de grand rendement.

higher || ober || supérieur. / next ~ || nächst-höher || immédiatement supérieur.

higher harmonics pl. || harmonische Oberschwingungen fpl. || ondes fpl. de l'harmonique supérieur. / ~ harmonic voltage || Oberspannung f. || tension f. de l'harmonique supérieur.

highest quality steel || höchstwertiger Stahl m. || acier m. de la plus haute qualité. / ~ tender || Höchstgebot n. || offre f. la plus haute.

high-forest || Hochwald m. || futaie f.

high-frequency || hochfrequent || à haute fréquence f.

high-frequency || Hochfrequenz f. || haute fréquence f.

high-frequency alternator || Hochfrequenzmaschine f.; Hochfrequenzerzeuger m. || alternateur m. ou générateur m. (à) haute fréquence. / ~ for wireless telegraphy || Hochfrequenzmaschine f. für drahtlose Telegrafie || machine f. ou alternateur m. haute fréquence pour télégraphie sans fil.

high-frequency amplifier || Hochfrequenzverstärker m. || amplificateur m. haute fréquence. / ~ amplifying || Hochfrequenzverstärkung f. || amplification f. haute fréquence. / ~ apparatus || Hochfrequenzapparat m. || appareil m. à haute fréquence. / ~ bridge || Hochfrequenzmeßbrücke f. || pont m. wheatstone à hautes fréquences. / ~ cable || Hochfrequenzlitze f.; Litzendraht m. || brin m. de ou à haute fréquence; litzendraht m. / ~ circuit || Hochfrequenzkreis m. || circuit m. haute fréquence. / ~ coil || Hochfrequenzspule f. || bobine f. à haute fréquence. / ~ current || Strom m. hoher Frequenz; Hochfrequenzstrom m. || courant m. à ou de haute fréquence. / ~ generator see ~ alternator. / ~ induction furnace || Hochfrequenzinduktionsofen m. || four m. à induction à haute fréquence. / ~ insulator || Hochfrequenzisolator m. || isolateur m. à haute fréquence.

/ ~ machine *see* ~ alternator. / ~ oscillation ‖ Hochfrequenzschwingung f. ‖ oscillation f. haute fréquence. / ~ resistance ‖ Hochfrequenzwiderstand m. ‖ résistance f. à haute fréquence. / ~ strengthener *see* high-frequency amplifier. / ~ transformer ‖ Hochfrequenztransformator m. ‖ transformateur m. à haute fréquence. / ~ tube ‖ Hochfrequenzröhre f. ‖ lampe f. haute fréquence. / ~ wire *see* ~ cable.

high-furnace *see* blast furnace.

high-grade ‖ erste Qualität f. ‖ de qualité f. supérieure. / production of ~ forgings and castings ‖ Herstellung f. hochwertiger Guß- und Schmiedestücke ‖ production f. de pièces forgées et coulées de qualité supérieure. / ~ iron ore ‖ hochhaltiges Eisenerz n. ‖ minerai m. de fer d'une teneur très haute.

high-level railway ‖ Hochbahn f. ‖ chemin m. de fer à voie aérienne.

highly carbonized surface ‖ kohlenstoffreiche Schicht f. ‖ couche f. carburée. / ~ concentrated ‖ hochkonzentriert ‖ à haut degré m. / ~ evacuated ‖ hochevakuiert ‖ à un degré m. de vide élevé. / ~ inductive shunt ‖ Shunt m. *oder* Nebenschluß m. mit hoher Selbstinduktion ‖ shunt m. à pouvoir inductif élevé. / ~ magnifying telescope ‖ stark vergrößerndes Fernrohr n. ‖ lunette f. à fort grossissement. / ~ polished ‖ hochglanzpoliert ‖ parfaitement poli *ou* brillant. / ~ volatile cooling agent ‖ hochflüchtiger Kälteträger m. ‖ agent m. frigorifique très volatil.

high-moor ‖ Hochmoor n. ‖ plateau m. marécageux.

high-pass filter (Radio) ‖ Kondensatorkette f.; Hochfrequenzsiebkette f. ‖ filtre m. haute-fréquence; filtre m. passe-haut.

high-potential *see* high-tension.

high-power gas engine ‖ Großgasmaschine f. ‖ moteur m. à gaz de grande puissance. / ~ station (Electr) ‖ Großkraftwerk n. ‖ station f. centrale de grande *ou* de haute puissance.

high-precision adjustment gear (Mach) ‖ Präzisionsschaltwerk n. ‖ engrenage m. de précision.

high-pressure (Electr) ‖ hochgespannt ‖ à haute tension f.

high-pressure (Mach; Meteor) ‖ Hochdruck m. ‖ haute pression f. / ~ (Electr) *see also* high-tension. ‖ Hochspannung f. ‖ haute tension f. / ~ air-compressor ‖ Hochdruckluftkompressor m. ‖ compresseur m. d'air à haute pression. / ~ air piping ‖ Hochdruckluftleitung f. ‖ tuyauterie f. d'air à haute pression. / ~ area (Meteor) ‖ Hochdruckgebiet n. ‖ zone f. de haute pression. / ~ current ‖ hochgespannter Strom m. ‖ courant m. sous haute tension. / ~ fittings pl. ‖ Hochdruckarmaturen fpl. ‖ garnitures fpl. *ou* armatures fpl. pour haute pression. / ~ gas ‖ Preßgas n. ‖ gaz m. surpressé. / ~ pipe cooler ‖ Hochdruckröhrenkühlapparat m. ‖ appareil m. frigorifique à tubes sous haute pression. / ~ piping ‖ Druckleitung f. ‖ tuyauterie f. à haute pression. / ~ pump ‖ Preßpumpe f. ‖ pompe f. à haute pression *ou* de pression hydraulique. / ~ pumping machine ‖ Hochdruckpumpmaschine f. ‖ pompe f. *ou* machine f. à pomper à haute pression.

high-pressure receiver ‖ Hochdruckbehälter m. ‖ réservoir m. à haute pression. /

seamless ~ ‖ nahtloser Hochdruckbehälter m. ‖ réservoir m. à haute pression sans soudure.

high-pressure steam ‖ hochgespannter Dampf m.; Hochdruckdampf m. ‖ vapeur f. à haute pression. / ~ steam boiler ‖ Hochdruckdampfkessel m. ‖ chaudière f. à vapeur à haute pression. / ~ steam engine ‖ Hochdruckdampfmaschine f. ‖ machine f. à vapeur à haute pression. / ~ steam turbine ‖ Hochdruckdampfturbine f. ‖ turbine f. à vapeur à haute pression. / ~ tube ‖ Hochdruckrohr n. ‖ tuyau m. à haute pression.

high-production grinding wheel ‖ Hochleistungsschleifrad n. ‖ meule f. à grand rendement.

high-quality casting ‖ hochwertiger Guß m. ‖ fonte f. de haute qualité.

highroad *see* highway.

high-sea fishery ‖ Hochseefischerei f. ‖ pêche f. en haute-mer. / ~ navigation ‖ Hochseeschiffahrt f. ‖ navigation f. hauturière. / ~ plane ‖ Überseeflugzeug n. ‖ avion m. *ou* hydravion m. de haute mer.

high-sensitive fuse ‖ Augenblickszünder m. ‖ fusée f. instantanée.

high-speed automatic transmitter ‖ Schnellsender m. für Kabelschrift ‖ transmetteur m. automatique à grande vitesse. / ~ chain loom ‖ Schnelläuferkettenstuhl m. ‖ métier m. chaîne à grande vitesse. / ~ computing machine ‖ Schnellrechenmaschine f. ‖ machine f. à calculer rapide. / ~ disc discharger (Radio) ‖ schnellumlaufende Scheibenfunkenstrecke f. ‖ éclateur m. à disque à grande vitesse. / ~ drilling machine ‖ Schnellbohrmaschine f. ‖ machine f. à percer rapide. / ~ folder (Office) ‖ Schnellfalzer m. ‖ plieuse f. à grande vitesse. / ~ gas engine ‖ schnellaufende Gasmaschine f. ‖ moteur m. à gaz à grande vitesse. / ~ lathe ‖ Schnellschnittdrehbank f. ‖ tour m. à grande vitesse. / ~ movement ‖ Eilbewegung f. ‖ mouvement m. rapide. / ~ planer ‖ Schnellhobler m. ‖ raboteur m. rapide. / ~ press (Print) ‖ Schnellpresse f. ‖ presse f. rapide. / ~ printing machine for letterpress work ‖ Schnellpresse f. für Buchdruck ‖ presse f. rapide pour imprimerie de livres. / ~ railway ‖ Schnellbahn f. ‖ ligne f. à service rapide. / receiver for Morse code ‖ Schnellmorseempfänger m. ‖ appareil m. récepteur Morse rapide. / ~ regulator (Electr) ‖ Schnellregler m. ‖ régulateur m. rapide. / ~ service ‖ Schnellbetrieb m. ‖ fonctionnement m. rapide. / ~ siphon recorder (Tel) ‖ Drehspulenschnellschreiber m. ‖ siphon-recorder m. rapide. / ~ steel ‖ Schnelldrehstahl m.; Schnellstahl m. ‖ acier m. rapide; acier m. à coupe rapide; acier m. pour tours à grande vitesse. / ~ telegraph ‖ Schnelltelegraf m. ‖ appareil m. télégraphique rapide. / ~ tool steel *see* ~ steel.

high-taper ‖ Königskerze f. ‖ bouillon m. blanc.

high-tension ‖ Hochspannung f. ‖ haute tension f. / ~ ammeter ‖ Hochspannungsstrommesser m. ‖ ampèremètre m. à haute tension. / ~ battery ‖ Hochspannungsbatterie f. ‖ batterie f. à haute tension. / ~ coil ‖ Hochspannungsspule f. ‖ bobine f. à haute tension. / ~ crossing ‖ Hochspannungskreuzung f. ‖ croisement

m. de haute tension. / ~ current ‖ Hochspannungsstrom m.; hochgespannter Strom m. ‖ courant m. sous haute tension. / dangerous interference with ~ current ‖ Gefährdung f. durch Hochspannungsstrom ‖ mise f. en danger par courant de haute tension. / ~ duct ‖ Hochspannungsdurchführung f. ‖ traversée f. à haute tension. / ~ engineering ‖ Hochspannungstechnik f. ‖ technique f. de la haute tension. / ~ generator ‖ Hochspannungserzeuger m. ‖ génératrice f. à haute tension. / ~ insulating material ‖ Hochspannungsisolierstoff m. ‖ matière f. isolante à haute tension. / ~ insulator ‖ Hochspannungsisolator m. ‖ isolateur m. à haute tension. / ~ insulator support ‖ Hochspannungsisolatorstütze f. ‖ support m. d'isolateur m. à haute tension. / ~ line ‖ Hochspannungsleitung f.; Starkstromleitung f. ‖ ligne f. à haute tension. / poles pl. for ~ lines ‖ Hochspannungsgestänge n. ‖ poteaux mpl. de lignes à haute tension. / ~ machine ‖ Hochspannungsmaschine f. ‖ machine f. à haute tension. / ~ magneto ‖ Hochspannungsmagnet m. ‖ magnéto m. à haute tension. / ~ network ‖ Hochspannungsnetz n. ‖ réseau m. de *ou* à haute tension. / ~ oil switch ‖ Hochspannungsölschalter m. ‖ interrupteur m. à huile à haute tension. / ~ plant ‖ Hochspannungsanlage f. ‖ installation f. à haute tension. / ~ power plant ‖ Überlandzentrale f. ‖ centrale f. interurbaine; centrale f. électrique régionale. / telephone line with protection against ~ ‖ hochspannungsgeschützte Fernsprechleitung f. ‖ ligne f. téléphonique protégée contre l'action de la haute tension. / ~ protective apparatus for ~ ‖ Hochspannungsschutzapparat m. ‖ appareil m. de protection de la haute tension. / ~ protector ‖ Hochspannungsschutz m. ‖ protection f. contre la haute tension. / ~ relay ‖ Hochspannungsrelais n. ‖ relais m. de haute tension. / ~ remote control switch ‖ Hochspannungsfernschalter m. ‖ téléinterrupteur m. de haute tension. / ~ source of supply (Electr) ‖ Hochspannungskraftquelle f. ‖ alimentation f. en courant de haute tension. / ~ switch ‖ Hochspannungsschalter m. ‖ interrupteur m. de haute tension. / ~ system *see* ~ network.

high-tension testing equipment ‖ Hochspannungsprüfeinrichtung f. ‖ installation f. pour essais sous haute tension. / ~ for generating surge pressures ‖ Hochspannungsprüfeinrichtung f. zur Erzeugung von Stoßspannungen ‖ dispositif m. d'essais de haute tension pour la production d'à-coups de tension.

high-tension tube fuse ‖ Hochspannungsröhrensicherung f. ‖ coupe-circuit m. à tube pour haute tension. / ~ winding ‖ Hochspannungswicklung f. ‖ enroulement m. *ou* bobinage m. à haute tension.

high tide *see* high-water (Mar).

high-vacuum ‖ Hochvakuum n. ‖ vide m. élevé. / ~ pump ‖ Hochvakuumpumpe f. ‖ pompe f. à vide élevé.

high-voltage *see* high-tension.

high-warp tapestry ‖ hochschäftige Tapete f. ‖ tapisserie f. de haute-lisse.

high-water (Mar) ‖ Flut f.; Flutzeit f.; Hochwasser n. ‖ marée f. montante; flot m. / ~ (of a river) ‖ Hochwasser n. (eines

Flusses) || grande crue f. (d'une rivière); haute-eau f. / ~ arch (Bridge) || Flutöffnung f. || débouché m. des hautes eaux. / ~ mark || Hochwassermarke f.; Hochwasserstandzeichen n. || marque f. des plus hautes eaux. / protection against ~ || Hochwasserschutz m. || protection f. contre les crues. / ~ turbine || Hochwasserrad n. || turbine f. pour hautes eaux.

highway || Landstraße f.; Chaussee f.; Kunststraße f.; Heerstraße f.; Hauptstraße f. || grand chemin m.; grande route f.; route f. nationale; chaussée f. / ~ bridge || Straßenbrücke f. || pont m. de chaussée. / ~ surveying office || Wegepolizei f. || police f. de la voirie.

high-wheeled || hochräderig || à grandes roues fpl.

hill || Anhöhe f.; montée f. / dangerous ~ || gefährliche Steigung f. || descente f. dangereuse.

hill-climbing || Bergfahrt f. || montée f.

hilly || hügelig || montueux. / ~ ground || bergiges Gelände n. || terrain m. montagneux.

hilt || Griff m.; Heft n. || manche m.; queue f.

hind-axle || Hinterachse f. || essieu m. (d')arrière; essieu m. de derrière. / ~ bolster || Hinterachsschale f. || hausse f.; sellette f. de derrière. / ~ carriage (Coachm) || Hintergestell n.; Hinterwagen m. || arrière-train m.; train m. de derrière. / ~ part of a vessel || Hinterschiff n. || arrière m. ou poupe f. d'un bateau. / ~ wheel || Hinterrad n. || roue f. de derrière ou d'arrière; arrière-roue f.

hinge (Joint) || Scharnier n.; Gelenk n. || charnière f.; jointure f. / ~ (Joint frame) || Scharnierband n. || fiche f. (simple). / ~ (Of a door) || Angel f.; Türangel f. || gond m. / ~ (Locksm) || Haspe f. || gond m. / ~ (Holdfast) || Fensterband n.; Türband n. || penture f. / bent ~ || gekröpftes Scharnierband n. || fiche f. coudée. / ~ with hook || Scharnierband n. mit Haken; Hakenband n.; Aufsetzband n.; Kegelband n. || fiche f. à gond ou à repos; penture f. à gond. / horizontal ~ || Horizontalgelenk n. || articulation f. horizontale. / ornamented ~ || Zierband n. || penture f. ornée. / ~ with pin || Scharnierband n. mit Vorstecker || fiche f. à double nœuds et à bouton.

hinge clip || Scharnierklemme f. || borne f. articulée.

hinged || klappbar; aufklappbar || relevable; à bascule; à charnières; à rabattement; articulé; rabattable; basculant. / ~ back || umlegbare Rücklehne f. || dossier m. réversible. / ~ bottom || Bodenklappe f.; Klappboden m. || trappe f. de fond; fond m. mobile ou à rabattement. / ~ bucket (Dredger) || Gelenkbecher m. || godet m. à charnières. / ~ bucket (Excavator) || Klappkübel m. || benne f. ouvrante. / ~ cover || Klappenverschluß m. || fermeture f. à clapet. / ~ door || Drehtür f. || porte f. pivotante. / ~ flap || um ein Gelenk drehbare Klappe f. || clapet m. à charnières. / ~ pushing-arm || umlegbare Ausdrückstange f. || tige f. de défournement articulée. / ~ table flap (Office) || herunterklappbarer Tischflügel m. || tablette f. réversible. / ~ tool holder || klappbarer Support m. || support m. porte-outil à charnières. / ~ tool post see ~ tool holder.

hinge device || Kippvorrichtung f. || dispositif m. basculant. / ~ joint see ~ (Joint). / ~ making machine || Scharnierherstellungsmaschine f. || machine f. à fabriquer les charnières. / ~ pin (Locksm) || Scharnierstift m. || cheville f. ou goujon m. de charnière. / ~ plate (Locksm) || Hakenband n.; Gehänge n.; Türband n. || paumelle f.; penture f. / ~ ring || Scharnierring m. || bague f. à charnière. / ~ saw || Gelenksäge f. || scie f. à charnière. / ~ stocks pl. (Locksm) || Scharnierkluppe f.; Scherkluppe f. || filière f. à charnière. / ~ wire see ~ pin.

hinny || Maulesel m. || mulet m.; bardot m.

hint || Andeutung f.; Anspielung f. || allusion f.

hinterland || Hinterland n. || hinterland m.

hip (Anatomy) || Hüfte f. || hanche f. / ~ (Bot) || Hagebutte f. || gratte-cul m.; fruit m. de l'églantier. / ~ (Build; Tiler) || Walm m. || croupe f. / conical ~ || Kegelwalm m. || croupe f. conique. / cylindrical ~ || Zylinderwalm m. || croupe f. cylindrique. / half ~ || Halbwalm m. || demi-croupe f. / partial ~ see half ~. / spherical ~ || Kugelwalm m. || croupe f. sphérique. / whole ~ || ganzer Walm m. || croupe f. pleine ou entière.

hip bead || Gratwulst m. || arêtière f. / ~ jack rafter || Gratschifter m. || empanon m. d'arête. / ~ lead see ~ sheet. / ~ peels pl. || Hagebuttenschalen fpl. || écorces fpl. de fruits de rosier sauvage.

hippuric acid || Hippursäure f. || acide m. hippurique.

hip rafter || Gratbalken m.; Gratsparren m.; Walmsparren m. || arêtier m.; coyer m.; chevron m. d'arête. / ~ ridge || Anfallpunkt m. || point m. d'appui; sommet m. / ~ roof || Walmdach n. || toit m. en croupe. / ~ sheet || Gratblech n. || bavette f. d'arête. / ~ slate || Kehlstein m. || ardoise f. cornière ou de noue. / ~ stone || Kehlstein m.; Kehlziegel m.; Walmziegel m. || tuile f. cornière ou de noue. / ~ strap (Saddl) || Schweberiemen m. eines Geschirrs || branche f. d'avaloire. / ~ tile || Walmziegel m.; Gratziegel m. || tuile f. de croupe.

hire, to || anwerben; dingen; in Lohn nehmen || engager. / ~ (Mar) || heuern || engager. / ~ (To rent) || mieten || louer.

hire || Lohn m.; Arbeitslohn m. || salaire m.; gage m.

hisingerite (Miner) || Hisingerit m.; Thraulit m. || hisingérite f.; thraulite f.

hissing of the arc || Zischen n. des Lichtbogens || sifflement m. de l'arc.

histogram (Statistics) || Staffelbild n. || histogramme m.

histological || histologisch || histologique.

historical || historisch || historique.

history, natural || Naturgeschichte f. || histoire f. naturelle.

hit, to || stoßen; schlagen || pousser; frapper.

hit (Shooting) || Treffer m. || coup m. touché.

hitch, to || befestigen; festmachen; anhängen || accrocher; fixer.

hitch (Mar) || Knoten m.; Stich m. || nœud m.; bouton m. / clove ~ || Schifferknoten m.; Webeleinstich m. || nœud m. de batelier ou d'artificier. / half ~ || Halbstich m. || demi-clef f. / running ~ see clove ~.

Hittorf tube (Electr) || Hittorfsche Röhre f. || tuyau m. ou tube f. d'Hittorf.

hive || Bienenstock m.; Bienenkorb m. || ruche f. / straw ~ || Bienenkorb m. aus Stroh || ruche f. en paille.

hive bee || Honigbiene f. || mouche f. à miel; abeille f. mellifique. / ~ maker || Bienenkorbmacher m. || rucheur m.

hiver || Imker m.; Bienenzüchter m. || apiculteur m.; éleveur m. d'abeilles.

hoar frost || Rauhreif m.; Reif m.; Rauhfrost m. || givre m.; gelée f. blanche; frimas m. / to become covered with ~ || bereifen || se couvrir de gelée blanche. / ~ line fault (El line) || Rauhreifstörung f. || dérangement m. dû au givre.

hoarseness || Heiserkeit f. || enrouement m.

hob, to ~ (Mach) || Gewinde n. schneiden || tarauder. / ~ (Shoem) see to hobnail.

hob (Horseshoer) see hobnail. / ~ (Shoem) see hobnail. / ~ (Master tap) see hob tap. / ~ (For screw chasers) || Kammfräser m. || fraise f. à peignes. / ~ (For worm wheels) || Schneckenfräser m.; Wurmfräser m. || vis-fraise f.; fraise f. hélicoïdale; fraise f. à vis-mère. / ~ ground relieved ~ || hinterschliffener Schneckenfräser m. || fraise f. à vis-mère détallonée. / ~ worm ~ see hob (For worm wheels).

hobnail, to ~ (Shoem) || (Schuhe mpl.) mit Nägeln beschlagen || ferrer.

hobnail (Horseshoer) || Hufnagel m. || clou m. à ferrer ou à cheval. / ~ (Shoem) || (eiserner) Schuhnagel m. || caboche f.; clou m. à souliers.

hob tap || Gewindebohrer m.; Schneidbohrer m. || taraud-mère m.

hod (Mas) || Mörteltrog m.; Tragmulde f. || auge f. à mortier. / ~ carrier (Mas) || Mörtelträger m.; Steinträger m. || bardeur m.

hod-man (Mas) see hod carrier. / ~ (Handy man) || Handlanger m. || manœuvre m.; aide-maçon m.

hoe, to (Agr) || hacken; häufeln || piocher; houer.

hoe (Agr) || Hacke f.; Karst m.; Gartenhaue f. || pioche f.; houe f.; hoyau m.; binette f. / ~ (Mine) || Hacke f.; Haue f.; Pickel m.; Picke f.; Erdhacke f.; Erdhaue f. || pic m.; pioche f.; houe f.; hoyau m. / forked ~ (Agr) || Gabelhaue f.; zweizinkige Hacke f. || bigot m.; bigorne f.; houe f. à deux fourchons. / gardener's ~ || Gartenhaue f. || binette f. / weeding ~ || Jäthacke f.; Jäthaue f. || serfouette f.; sarclet m.; sarclette f.; sarcloir m. / handle for ~s || Hackenstiel m. || manche m. de ou à houes.

hoeing machine || Hackmaschine f. || piocheuse f.

hoernesite (Miner) || Hörnesit m. (wasserhaltige arsensaure Magnesia) || hoernésite f.

Hoffmann's tincture || Hoffmannstropfen mpl. || liqueur m. d'Hoffmann.

hog, to ~ the bottom of a vessel || den Boden eines Schiffes schrubben || goreter la carène ou le fond d'un bâtiment.

hog || Schwein n. || porc m.; cochon m. / ~ (Mar) || Farke f.; Schrubber m.; Schiffsbesen m. || goret m.; balai-goret m.; fauber m.; faubert m. / ~ fat || Schweineschmalz n.; Schweinefett n. || graisse f. de porc; axonge f.; sain-doux m. / ~ pen see ~ sty.

hog's bristle || Schweinsborste f. || soie f. de porc.

hogshead (hog's head) || Oxhoft m. || demipièce f.; barrique f.

hog skin ‖ Schweinsleder n. ‖ peau f. *ou* cuir m. de porc.

hog's lard *see* hog fat.

hog sty ‖ Schweinestall m. ‖ étable f. à porcs; porcherie f.

hoist, to ‖ aufwinden; in die Höhe ziehen; hochheben; fördern ‖ hisser; (re)monter; (en)lever; extraire. / ~ (Mar) ‖ hissen; aufhissen; aufziehen; in die Höhe ziehen ‖ hisser; palanguer; haler. / ~ the flag ‖ die Flagge hissen ‖ hisser *ou* arborer le pavillon.

hoist ‖ Aufzug m.; Fahrstuhl m.; Lift m. ‖ ascenseur m.; élévateur m. / ~ (For Loads) ‖ Fahrstuhl m.; Lastenaufzug m.; Elevator m.; Hebewerk n.‖monte-charge m.; élévateur m. / ~ (Loading crane) ‖ Ladekran m.; Ladewinde f.; Lademast m.; Ladebaum m. ‖ grue f. (de chargement); guindal m.; guindoule f.; treuil m. d'embarquement. / ~ (Mar) ‖ Flaggenhöhe f.; Flaggentiefe f. ‖ guindant m. / ~ (Shipb) ‖ Heißmaschine f. ‖ machine f. de hissage. / air ~ ‖ pneumatischer Elevator m. ‖ monte-charges m. à air comprimé. / ash ~ ‖ Aschheißmaschine f. ‖ monte-escarbilles m. / blast furnace ~ ‖ Gichtaufzug m.; chargeur m. *ou* élévateur m. de haut-fourneau. / boat ~ ‖ Bootsheißmaschine f. ‖ monte-embarcations m. / cask ~ ‖ Faßwinde f. ‖ montefûts m.; treuil m. à fûts. / electric ~ ‖ Elektrozug m. ‖ palan m. électrique. / friction ~ ‖ Friktionsaufzug m. ‖ montecharge m. commandé par friction. / ice ~ ‖ Eisaufzug m. ‖ monte-glace m. / inclined ~ ‖ Schrägaufzug m. ‖ monte-charge m. incliné. / electrically driven inclined ~ ‖ elektrisch betriebener Schrägaufzug m. ‖ montecharge m. incliné à commande électrique. / loading ~ (Nav) *see* ~ (Loading crane). / mine ~ *see* hoisting engine. / motor ~ ‖ Motorwinde f. ‖ treuil m. à moteur. / travelling ~ ‖ Laufwinde f. ‖ treuil m. roulant. / vertical ~ ‖ senkrechter Aufzug m. ‖ élévateur-transporteur m. vertical. / warehouse ~ ‖ Speicherwinde f. ‖ treuil m. de magasin.

hoist engineer, blast furnace ‖ Gichtaufzugführer m. ‖ mécanicien m. du monte-charge de haut-fourneau.

hoist frame ‖ Fördergerüst n.; Förderturm m.; Fördergestell m. ‖ charpente f. du monte-charge.

hoisting ‖ Heben n. ‖ levage m. / ~ (Min) ‖ Schachtförderung f. ‖ extraction f.

hoisting apparatus ‖ Hebezeug n.; Hebewerk n.; Hebevorrichtung f.; Aufziehvorrichtung f.; Hebemaschine f. ‖ appareil m. *ou* engin m. de levage. / ~ for buildings ‖ Bauaufzug m. ‖ appareil m. à guinder; élévateur m. pour constructions. / electric ~ ‖ elektrische Hebevorrichtung f. ‖ engin m. de levage électrique. / universal ~ ‖ Universalhebezeug n. ‖ appareil m. de levage universel.

hoisting cage (Min) ‖ Förderkorb m. ‖ cage f. d'extraction. / ~ crab *see* ~ winch. / ~ device *see* ~ apparatus. / ~ engine (Min) ‖ Fördermaschine f. ‖ machine f. d'extraction. / ~ gear *see* ~ apparatus.

hoisting jack ‖ Fußwinde f.; Hebelade f.; Hebebock m. ‖ cric m. à crochet; vérin m. de levage. / hydraulic ~ ‖ hydraulische Hebelade f. ‖ vérin m. de levage hydraulique.

hoisting machine *see* hoisting apparatus. / ~ (Min) *see* hoisting engine. / ~ (Shipb) *see* hoist (Shipb)

hoisting magnet ‖ Hebemagnet m. ‖ électro m. de levage.

hoisting rope (Hoist) ‖ Aufzugseil n. ‖ câble m. de monte-charge *ou* d'ascenseur. / ~ (Min) ‖ Förderseil n. ‖ câble m. d'extraction *ou* de levage.

hoisting shaft (Min) ‖ Förderschacht m. ‖ puits m. d'extraction. / ~ tackle *see* ~ apparatus. / ~ velocity ‖ Hubgeschwindigkeit f. ‖ vitesse f. de levage. / ~ wheel ‖ Heberad n. ‖ roue f. élévatrice. / ~ winch ‖ Aufzugwinde f.; Hebewinde f. ‖ treuil m. de levage *ou* d'ascenseur.

hoist shaft ‖ Aufzugschacht m.; Fahrschacht m. ‖ puits m. d'élévateur; cage f. d'ascenseur.

hold, to ‖ halten; festhalten ‖ tenir. / ~ (Mach) ‖ feststellen ‖ arrêter; fixer. / ~ (To contain) ‖ enthalten; fassen ‖ contenir; renfermer; tenir. / ~ (To possess) ‖ besitzen ‖ posséder. / ~ up (Riveting) ‖ gegenhalten ‖ maintenir le rivet (au moyen de la contre-bouterolle).

hold (Shipb) ‖ Schiffsraum m.; Laderaum m. ‖ cale f.

hold-all ‖ Behälter m.; Necessaire n. ‖ nécessaire m.

hold beam (Shipb) ‖ Raumbalken m. ‖ bau m. de la cale.

holder (Clockm) ‖ Quadraturstift m. ‖ tenon m. / ~ (Cramp iron) ‖ Zwinge f. ‖ crampon m. / ~ (Glow lamp) ‖ Fassung f. ‖ douille f. / ~ (Mach tool) *see* holderdown. / ~ (Mar) ‖ Lastmann m.; Raumgast m. ‖ calier m. / ~ (Occupier) ‖ Inhaber m.; Besitzer m. ‖ possesseur m. / ~ (Reservoir) ‖ Behälter m.; Behältnis n.; Gefäß n. ‖ récipient m.; réservoir m. / ~ (Stand) ‖ Gestell n.; Halter m. ‖ support m. / ~ of a bill ‖ Wechselgläubiger m. ‖ créancier m. d'une lettre de change. / ink ~ ‖ Tintenbehälter m.; Tintenfaß n. ‖ encrier m. / ink ~ (Tel) ‖ Farbnapf m.; Farbschälchen n.; Farbgefäß n. ‖ godet m. à couleur. / mutton leg ~ ‖ Bratenhalter m. ‖ manche m. à gigot. / ~ for rolled paper (Office) ‖ Papierrollenhalter m. ‖ porte-rouleau m. pour rouleaux de papier. / sand ~ ‖ Sandtopf m. ‖ boîte f. à sable; sablier m. / the parts are firmly connected with the foundation through screwed ~s ‖ die Stücke npl. sind durch aufgeschraubte Halter mit dem Fundament fest verbunden ‖ les éléments mpl. sont reliés solidement aux fondations au moyen de supports boulonnés. / sheet-iron ~ ‖ Blechbüchse f.‖ boîte f. en tôle. / share ~ ‖ Aktionär m.; Aktieninhaber m.; Teilhaber m. einer Aktiengesellschaft ‖ actionnaire m. / ~ of stock *see* share ~. / three armed ~ for an air ship screw ‖ dreiarmiger Schraubenflügelhalter m. für ein Luftschiff ‖ cadre m. de pale à trois bras pour une hélice aérienne.

holder-down (Mach tool) *see also* holdingdown device ‖ Niederhalter m.; Abstreifer m. ‖ arracheur m.

holder-up (Riveting) ‖ Vorhalter m.; Gegenhalter m.; Nietstock m.; Nietkloben m. ‖ contre-bouterolle f; tas m. / pneumatic ~ ‖ Preßluft(niet)gegenhalter m. ‖ contre-bouterolle f. pneumatique. / ~ of riveting machine ‖ Anpreßstempel m.

oder Gegenhalter m. der Nietmaschine ‖ contre-bouterolle f. de la riveuse.

holdfast (Bench hook) ‖ Bankhaken m.; Bankeisen n.; Schließhaken m.; Bankhalter m. ‖ mentonnet m.; valet m.; greppe f. d'établi. / ~ (Cramp iron) ‖ Klammer f.; Zwinge f.; Klammerhaken m.; Klemmhaken m.; Kloben m.; Bankeisen n. ‖ crampon m.; patte f.; clameau m. / ~ (Flat-headed nail) ‖ flachköpfiger Nagel m. ‖ clou m. à patte. / ~ (Glue press) ‖ Schraub(en)zwinge f.; Leimzwinge f.; Schraubknecht m. ‖ serrejoint(s) m.; sergent m.; goberge f.; étreignoir m.; presse f. à main. / ~ (Hand vice) ‖ Klemmbock m.; Klemmeisen n.; Kloben m.; Feilkloben m.; Spannkluppe f. ‖ mordache f.; étau m. à main; détret m.; crampon m.

holding (Ware) ‖ Bestand m. ‖ état m.; effectif m.; inventaire m.

holding apparatus for elevator cages ‖ Aufsetzvorrichtung f. für Förderkörbe ‖ dispositif m. de chargement pour cages de montée. / ~ blotter (Office) ‖ Löschblatthalter m. ‖ porte-buvard m. / ~ bolt ‖ Riegelbolzen m.; Verbindungsbolzen ‖ boulon m. d'assemblage. / ~ company ‖ Holdinggesellschaft f.; Dachgesellschaft f. ‖ trust m. de valeurs.

holding device (Mach tool) ‖ Spannvorrichtung f. ‖ dispositif m. de fixation. / magnetic ~ ‖ Magnetspannvorrichtung f. ‖ dispositif m. de fixation magnétique.

holding-down arrangement *see* holdingdown device. / ~ bolt ‖ Verankerungsbolzen m.; Befestigungsbolzen m.; Ankerschraube f. ‖ boulon m. d'ancrage *ou* de fixation *ou* de fondation.

holding-down device (Mach tool) ‖ Niederhaltevorrichtung f.; Niederhalter m. ‖ dispositif m. arracheur; presse-fer m. / ~ for material ‖ Materialniederhalter m. ‖ presse-fer m. *ou* dispositif m. arracheur du matériel. / hydraulic ~ ‖ hydraulische Niederhaltevorrichtung f. ‖ dispositif m. répulseur hydraulique. / ~ for plates ‖ Plattendrücker m.; Plattenpresser m.; Blechniederhalter m.; Blechfesthaltung f. ‖ presse-tôle m.; serre-tôle m.

holding-down plate (Mach tool) ‖ Spannplatte f. ‖ Stützplatte f. ‖ plaque f. *ou* table f. de fixation; plateau m. de serrage cale f. / magnetic ~ ‖ Magnetspannplatte f. ‖ plateau m. de serrage magnétique.

holding plate *see* holding-down plate. / ~ ring ‖ Haltering m. ‖ bague f. de support. / ~ rope ‖ Halteseil n. ‖ câble m. de retenue. / ~ time (Tel) ‖ Belegungsdauer f. ‖ durée f. d'occupation.

holding-up cup ‖ Nietpfanne f. ‖ contre-bouterolle f. de turc. / ~ hammer *see* holder-up. / ~ lever ‖ Nietwippe f.; portetas m.; porte-contre-bouterolle m.; levier m. à bouterolle. / ~ tool *see* holder-up.

hold ladder (Shipb) ‖ Raumtreppe f. ‖ échelle f. de la cale.

holdover key (Tel) ‖ Haltetaste f. ‖ clef f. de garde *ou* d'arrêt. / ~ position (Tel) ‖ Haltestellung f. ‖ position f. d'arrêt.

hold pillar (Shipb) ‖ Raumstütze f. ‖ épontille f. de la cale.

hole, to ‖ aushöhlen; durchlöchern ‖ trouer. / ~ (Mine) ‖ schrämen; durchörtern ‖ entailler; haver; souchever; desserrer. / ~ a post (Mine) ‖ einen Kohlenpfeiler m. durchörtern ‖ desserrer un pilier. / ~ the

trenches pl. (Mine) || die Gänge mpl. unterschrämen || entailler les couches fpl.

hole || Loch n. || trou m. / ~ (Geol) || Mulde f.; Höhle f.; Grube f. || excavation f.; cavité f.; enfoncement m. / ~ (Mach) || Bohrung f. || forage m. / ~ (Of a reservoir) || Öffnung f.; Mündung f.; orifice m.; entrée f.; ouverture f.; bouche f. / ~ (Shipb) || Gatt n. || trou m.; mortaise f. / ~ for axle || Achsloch n. || alésage m. de l'essieu. / bore ~ || Bohrloch n. || trou m. de sonde; sondage m.; forage m. / circular ~ || rundes Loch n. || trou m. rond. / elongated ~ || längliches Loch n. || trou m. longitudinal. / to enlarge ~s pl. with a drift || Löcher npl. aufdornen || mandriner des trous mpl. / filling ~ || Einfüllöffnung f. || encoche f. d'entrée; trou m. de remplissage. / ~ in the fish plate for the bolt || Laschenloch n. || trou m. pour le boulon d'éclisse. / rectangular ~ || viereckiges oder rechteckiges Loch n. || trou m. rectangulaire. / rivet ~ || Nietloch n. || trou m. de rivet. / round ~ see circular ~. / shrinkage ~ || Lunker m. || retassure f. / taper ~ || konische Bohrung f. || forage m. conique.

hole board (Weav) || Harnischbrett n.; Löcherbrett n.; Schnürbrett n.; Scherbrett n.; Lesebrett n. || planche f. à trous ou d'arcades.

hole circle || Lochkreis m. || cercle m. de trou. / ~ diameter || Lochkreisdurchmesser m. || diamètre m. du cercle de trou.

holed || durchlöchert; löcherig; gelocht || troué; perforé. / ~ (Mine) || geschrämt || havé; souchevé; entaillé.

hole pitch (Rivetting) || Lochabstand m. || pas m. ou écartement m. des trous. / precise ~ || genaue Lochteilung f. || écartement m. de trous précis.

hole press || Lochpresse f. || presse f. à mandriner. / ~ punching machine || Lochstanze f.; Lochmaschine f.; Lochwerk n. || machine f. à poinçonner; poinçonneuse f.

holer (Mine) || Schrämhauer m.; Schramhauer m.; Schrämer m. || haveur m.; hayeur m.; soucheveur m.

holer's pick || Schramhaue f.; Keilhaue f. || pic m. à houille; haveresse .f; havricau m.; rivelaine f.

hole strip || Lochstreifen m. || bande f. perforée.

holey see holed.

holiday see also holyday || Feiertag m.; von Arbeit freier Tag; Ruhetag m. || jour m. férié; jour m. de repos. / to grant ~s to the workmen || Arbeitern mpl. Erholungsurlaub gewähren || accorder aux ouvriers mpl. des vacances.

holing || (Mine) || Durchhieb m.; Durchschlag m.; Vortrieb m.; Durchörtern m. || percement m.; traverse f. / ~ (Undercut) || Schram m. || entaille f.; havage m. / ~ (Undercutting) || Schrämen n.; Durchörtern n. || havage m.; souchevage m. / ~ pick see holer's pick.

holland || holländische oder ungebleichte Leinwand f. || gorgonelle f.; toile f. de Hollande; toile f. écrue. / brown ~ || halb oder wenig gebleichte Leinwand f. || blondines fpl.

Holland gin see hollands.

hollands (Distill) || Genever m.; feiner Wachholderbranntwein m. || genièvre m. (fin).

hollow, to || aushöhlen; vertiefen; hohl machen || creuser; évider; caver; excaver; gouger. / ~ out see to hollow. / ~ a precious stone || einen Edelstein m. ausschlägeln || creuser le dessous d'une pierre précieuse.

hollow || hohl || creux. / to forge ~ || hohlschmieden || forger à creux.

hollow adze || Hohldechsel f.; Breitbeil n. || aissette f. / ~ aerial line || Hohlseilfreileitung f. || ligne f. aérienne à câbles creux. / ~ axle || Hohlachse f. || essieu m. creux. / ~ ball || Hohlkugel f. || boule f. creuse. / ~ beam || Hohlbalken m. || poutre f. creuse. / ~ block || Hohlblock m. || bloc m. creux. / ~ brick || Hohlziegel m.; Hohlstein m.; Hohlziegelstein m. || brique f. ou pierre f. creuse. / ~ cast see ~ casting. / ~ casting || Hohlguß m.; Kernguß m. || fonte f. en creux; coulée f. à noyau; pièce f. coulée creuse ou de fonte creuse. / ~ column || Hohlständer m. || montant m. creux. / ~ concrete slab || Zementhohldiele f. || dalle f. creuse en ciment. / ~ concrete stone || Betonhohlstein m. || pierre f. creuse en béton. / ~ cutter || Außenfräser m. || fraise f. creuse ou extérieure. / ~ cylinder || Hohlzylinder m. || cylindre m. creux. / ~ drill || Hohlbohrer m. || mèche f. creuse. / ~ furnace-shovel || runde Kohlenschaufel f. || pelle f. à foyer ronde. / ~ glass || Hohlglas n. || verre m. creux.

hollow glass ware || Hohlglaswaren fpl.; gobeleterie f. / ~ for toilet service || Hohlglaswaren fpl. für den Toilettegebrauch || articles mpl. de gobeleterie pour le service de la toilette. / ~ works pl. || Hohlglaswarenfabrik f. || gobeleterie f.

hollow joint-wire (Clockm) || Scharnierröhrchen n. || tuyau m. à charnière. / ~ key || Hohlkeil m. || clavette f. creuse ou évidée. / ~ mill see ~ cutter. / ~ moulding (Arch) || Hohlleiste f. || cannelure f.; gorge f. / ~ part see ~ piece.

hollow piece || Hohlkörper m. || pièce f. creuse; corps m. creux. / drawing of irregularly-formed ~s || Ziehen n. unregelmäßig geformter Hohlkörper || estampage m. de corps creux de forme irrégulière. / seamless drawn ~ || nahtloser Hohlkörper m. || pièce f. creuse étirée sans soudure.

hollow piston || hohler Kolben m. || piston m. creux. / ~ place see hollow. / ~ plane see hollow-nosed plane. / ~ punch || Locheisen n. || emporte-pièce m.; découpoir m. / ~ road || Hohlweg m. || chemin m. creux. / ~ rod || hohle Stange f. || tige f. creuse. / ~ roofing-tile || Hohldachziegel m. || noue f. / ~ section || Hohlquerschnitt m. || profil m. creux. / ~ shaft || Hohlwelle f.; hohle Welle f. || arbre m. creux. / ~ sphere || Hohlkugel f. || sphère f. creuse. / ~ spoke || hohle Speiche f. || rayon m. creux.

hollow tile || Hohldachziegel m.; Hohlziegel m. || tuile f. creuse. / ~ (Gutter tile) || Falzziegel m. || tuile f. à onglet ou à coulisse.

hollow wall || Hohlmauer f. || mur m. creux. / ~ wheel set || Hohlradsatz m. || essieu m. monté creux.

hollow || Aushöhlung f.; Höhlung f.; Vertiefung f.; hohle Stelle f. || creux m.; cavité f.; enfoncure f.; excavation f. / ~ (Arch) || Hohlkehle f. || cavet m. / gorge f. / ~ (Geol) see hole (Geol). / ~ (Mach) || Hohlkehle f.; Rinne f.; Nut f.; Auskehlung f. || gorge f.; congé m.; cannelure f. / ~ of a rocket || Seele f. einer Rakete || âme f. ou creux m. d'une fusée.

hollow-drawn article || gezogener Hohlkörper m. || pièce f. emboutie creuse.

hollower (Turn) || Drechsler m. für Hohlkehlen || creuseur m. canneleur.

hollow-forged toothed rim || hohlgeschmiedeter Zahnkranz m. || couronne f. dentée forgée à creux.

hollow-grinding attachment || Hohlkehlenschleifgerät n. || dispositif m. pour affûtage concave.

hollow-ground || hohlgeschliffen || évidé; concave; creusé. / ~ edge || eingeschliffene Hohlkehle f. || cannelure f. obtenue à la meule. / ~ tool || Hohlkehlenstahl m. || outil m. à lèvre concave.

hollowing || Aushöhlen n. || creusement m.; évidement m.; évidage m. / ~ file || Hohlfeile f. || lime f. à évider. / ~ hammer || Klopfschlägel m. || enfonçoir m. / ~ knife (Coop) || Krummeisen n.; Krummesser m. || plane f. creuse ou à lame courbe.

hollowing machine || Aushöhlmaschine f. || machine f. à creuser ou à évider. / ~ boot last ~ || Schuhleistenaushöhlmaschine f. || machine f. à évider les embauchoirs à chaussures.

hollowing-out see hollowing.

hollow-nosed plane || Rundhobel m. || mouchette f.

hollow-spar || Hohlspat m. || hohlspath m.; mache f.; chiastolite f.

hollow-ware || Küchengeschirr n. || ustensiles mpl. de cuisine. / ~ presser (Pott) || Hohlformer m. || mouleur m. en faïence.

holly || Stechpalme f. || houx m. / ~ (Holm oak) || Steineiche f. || chêne m. rouvre ou vert; yeuse f.; rouvre m.

hollyhock || Stockrose f. || alcée f.; trémière f.; rose f. trémière.

holly leaf || Stechpalmenblatt n. || feuille f. de houx.

holly wood || Stechpalmenholz n. || houx m.; bois m. de houx.

holmite (Miner) || Holmit m. || holmite f.; crysophane f.

holmium (Chem) || Holmium n. || holmium m.

holm oak see holly (Holm oak).

holohedral (Miner; Math) || vollflächig; holoedrisch || holoédrique; holoèdre. / ~ crystals pl. || holoedrische Kristalle mpl. || cristaux mpl. holoédriques.

holohedric see holohedral.

holohedrism || Vollflächigkeit f.; holoedrische Beschaffenheit f.; Holoedrie f. || holoédrie f.

holohedron || Vollflächner m.; Holoeder m. || holoèdre m.

holohedry see holohedrism.

holosymmetric see holohedral.

holosymmetry see holohedrism.

holomorphic (Miner) || holomorphisch || holomorphique.

holster || Pistolenhalfter f. || fonte f. de selle; fourreau m. de pistolet. / ~ (Roll mill) || Ständer m.; Gestell n. || colonne f.

holy-day see also holiday || Feiertag m.; Festtag m. || jour m. de fête; fête f.

holystone, to ~ the deck || das Deck scheuern || briquer le pont.

holy-stone (Mar) || Scheuerstein m. || brique f. à main en pierre molle.

holy-water basin || Weihwasserkessel m.; bénitier m.

homatropine (Chem) ‖ Homatropin n. ‖ homatropine f.

home freight (Mar) ‖ Rückfracht f. ‖ fret m. de retour. / ~ **manufacture** ‖ Heimaterzeugnis n. ‖ produit m. indigène.

homeo ... see homœo ...

home produce ‖ Landeserzeugnis n. ‖ produit m. du pays ou indigène.

homer ‖ Brieftaube f. ‖ pigeon m. voyageur.

home signal (Railw) ‖ Einfahrsignal n. ‖ signal m. rapproché ou d'entrée.

homespun ‖ handgesponnen ‖ filé à la main ou à la maison. / ~ (Linen) see ~ linen. / ~ linen ‖ Hausleinwand f. ‖ toile f. de ménage.

home voyage (Nav) ‖ Rückreise f. ‖ voyage m. de retour.

homeward ‖ heimwärts ‖ vers la patrie. / ~ **bounder** (Nav) ‖ Heimatswimpel m. ‖ flamme f. longue (qu'on hisse quand on a reçu l'ordre de repatrier). / ~ **cargo** (Nav) ‖ Rückladung f. ‖ cargaison f. de retour. / ~ **freight** see home freight.

homewards see homeward.

home worker ‖ Heimarbeiter m. ‖ ouvrier m. à façon ou à domicile; façonnier m.

homilite (Miner) ‖ Homilit m. ‖ homilite f.

homodyne method (Radio) ‖ Homodynmethode f. ‖ méthode f. homodyne.

homœomorphic (Miner) ‖ homöomorph ‖ homéomorphe.

homœomorphism ‖ Homöomorphie f. ‖ homéomorphie f.

homœomorphous see homœomorphic.

homœopath (Med) ‖ see homœopathist.

homœopathic ‖ homöopathisch ‖ homéopathique. / ~ **glasses** pl. ‖ homöopathische Gläser pl. ‖ flacons mpl. homéopathiques.

homœopathist ‖ Homöopath m. ‖ homéopathe m.

homœopathy ‖ Homöopathie f. ‖ homéopathie f.

homœopolar composition (Chem) ‖ homöopolare Bindung f. ‖ combinaison f. homéopolaire.

homogeneity (Phys; Chem) ‖ Einheitlichkeit f.; Homogenität f.; Gleichförmigkeit ‖ homogénéité f.

homogeneous ‖ gleichförmig; gleichmäßig; homogen; einheitlich ‖ homogène; uniforme. / the ingot is absolutely ~ ‖ der Block m. ist völlig homogen ‖ le lingot est d'une homogénéité parfaite. / ~ **line** (Electr) ‖ homogene Leitung f. ‖ ligne f. homogène. / ~ **mixing** ‖ inniges Mischen n. ‖ mélange m. intime. / ~ **texture** ‖ gleichmäßiges Gefüge n. ‖ texture f. (parfaitement) homogène.

homogeneously lead-lined and tinned apparatus ‖ homogen verbleiter und verzinnter Apparat m. ‖ appareil m. plombé et étamé homogènement.

homogeneousness see homogeneity.

homohedral (Miner) see holohedral.

homological (Math) see homologous.

homologous ‖ gleichnamig; übereinstimmend; homolog ‖ correspondant; homonyme; homologue.

homology ‖ Homologie f. ‖ homologie f.

homomorphic (Biology) ‖ gleichartig; ähnlich; gleichförmig; gleichgestaltig ‖ homomorphe.

homomorphism ‖ Gleichartigkeit f. ‖ homomorphisme m.

homomorphous see homomorphic.

homomorphy see homomorphism.

homonymie ‖ gleichnamig; homonym ‖ homonyme.

homophonic ‖ gleichklingend; gleichlautend ‖ homophone.

homopolar (Electr) ‖ einpolig; unipolar ‖ unipolaire.

homotropine (Chem) ‖ Homotropin n. ‖ homotropine f.

homotypic (Biology) ‖ gleichartig im Aufbau m. ‖ homotype.

hone see honestone.

honest ‖ rechtschaffen; ehrenwert ‖ honnête.

honestone ‖ Schleifstein m.; Wetzstein m.; Schieferstein m. ‖ pierre f. à aiguiser; affiloire f. / ~ **quarry** ‖ Wetzsteinbruch m. ‖ carrière f. de pierres à aiguiser.

honey ‖ Honig m. ‖ miel m. / artificial ~ ‖ Kunsthonig m. ‖ miel m. artificiel. / natural ~ ‖ natürlicher Honig m.; Bienenhonig m. ‖ miel m. naturel. / purified ~ ‖ gereinigter Honig m. ‖ miel m. épuré. / strained ~ ‖ Schleuderhonig m. ‖ miel m. coulé.

honey aroma oil ‖ Honigaromaöl n. ‖ huile f. d'arome de miel.

honeycomb ‖ Wabe f.; Bienenwabe f.; Honigwabe f. ‖ rayon m. de miel; nid m. d'abeilles. / ~ **coil** ‖ Honigwabenspule f. ‖ bobine f. en nid d'abeilles.

honey-combed ‖ zellig; wabenförmig ‖ alvéolé.

honeycomb radiator ‖ Bienenkorbkühler m.; Wabenkühler m.; Zellenkühler m. ‖ radiateur m. à nid d'abeilles.

honey soap ‖ Honigseife f. ‖ savon m. au miel.

honeystone (Miner) ‖ Honigstein m.; Mellit m.; honigsaure Tonerde f. ‖ honigstein m.; pierre f. de miel; mellite f.

honey strainer ‖ Honigschleuder(maschine) f. ‖ machine f. centrifuge pour miel. / ~ **sugar** ‖ Honigzucker m. ‖ sucre m. de miel.

honeywort (Bot) ‖ Wachsblume f. ‖ cérinthe f.

honor see honour.

honoraria pl. see honorarium.

honoraries pl. see honorarium.

honorarium ‖ Honorar n. ‖ honoraires fpl.

honorary member ‖ Ehrenmitglied n. ‖ membre m. honoraire.

honour, to ~ a bill ‖ einen Wechsel m. einlösen ‖ honorer une lettre de change.

honour ‖ Rechtschaffenheit f. ‖ honnêteté f.

honourable ‖ reell ‖ réel; honnête; loyal.

hood (Techn) ‖ Haube f.; Kappe f.; Deckel m.; Dach m. ‖ capot m.; calotte f.; chapeau m.; bonnet m. / ~ (Auto) ‖ Haube f. ‖ capot m. / ~ (Chem) ‖ Abzug m.; Schutzdach n. ‖ hotte f. / ~ (Covering for the head) ‖ Kapuze f.; Haube f. ‖ capeline f.; bonnet m. / ~ (Wind mill) ‖ Klappe f.; Haube f. ‖ toit m. (du moulin). / ~ of a car ‖ Wagenplane f.; Wagenkappe f.; Wagendecke f.; Wagendach n. ‖ bâche f.; banne f. de voiture; capotage m. / motor ~ ‖ Motorhaube f. ‖ capot m. de moteur. / to put the ~ down ‖ das Verdeck öffnen ‖ baisser la capote. / to raise the ~ ‖ das Verdeck aufschlagen ‖ relever la capote. / ~ of a starling (Hydr arch) ‖ Haube f. eines Brückenpfeilerkopfes ‖ bonnet m. du bec d'un pilier de pont.

hood catch see ~ fastener. / ~ **fastener** (Mot) ‖ Haubenverschluß m.; Hauben-

schloß n. ‖ serrure f. ou attache f. de capot. / ~ **mould** see ~ moulding.

hood moulding (Build) ‖ Verdachung f.; (innere) Türverdachung f. ‖ entablement m. (du chambranle). / gabled ~ ‖ Verdachung f. mit Giebel ‖ entablement m. à pignon.

hoof ‖ Huf m.; Klaue f. ‖ sabot m.; corne f.; pied m.; ongle m.; griffe f.

hoof-bound ‖ hufzwängig ‖ encastelé. / to become ~ ‖ hufzwängig werden ‖ s'encasteler.

hoof knife ‖ Hufmesser n. ‖ couteau m. de maréchal. / ~ **meal** ‖ Klauenmehl n. ‖ poudre f. de griffes. / ~ **ointment** ‖ Huffett n.; Hufsalbe f. ‖ onguent m. pour pieds de cheval. / ~ **picker** ‖ Hufräumer m. ‖ curepied m.

hoof-shaped ‖ hufförmig; klauenförmig ‖ onguiforme.

hook, to ~ (To clasp) ‖ zuhaken; festhaken; einhaken ‖ agrafer. / ~ (To hitch) ‖ anhängen; aufhängen; festhaken; anhaken ‖ accrocher. / ~ **in** ‖ einhaken ‖ enclencher; accrocher. / ~ **up** (Textile) ‖ abketteln ‖ tourniller.

hook ‖ Haken m. ‖ crochet m.; croc m. / ~ (Cramp iron) ‖ Klammer m.; Haspe f.; Haken m.; Klammerhaken m. ‖ crochet m.; agrafe f.; crampon m.; griffe f. de serrage. / ~ (Door hinge) ‖ Türangel f.; Haspe f. ‖ gond m. / ~ (Tail) ‖ Haft m.; Kleiderhaft m.; Haken m. an Kleidern ‖ agrafe f.; crochet m. / ~ (Tile) ‖ Nase f. ‖ crochet m.; nez m. / ~ (Turn) see ~ tool. / to bring the ~ into a certain position ‖ den Haken m. einsteuern ‖ amener précisement le crochet. / coupling ~ (Railw) ‖ Kuppelhaken m. ‖ crochet m. d'attelage. / dog ~ see hook (Cramp iron) / eyeglass ~ ‖ Brillenhalter m.; Pincenezhaken m. ‖ porte-pince-nez m. / ~ with eyelet ‖ Haken m. mit Ose ‖ crochet m. à œillet. / front tow ~ (Trailer) ‖ Schlepphaken m. ‖ crochet m. d'avant. / ~ for gloves ‖ Haken m. für Handschule ‖ agrafe f. pour gants. / ~ of a gutter ‖ Rinneisen n.; Haken m. einer Dachrinne ‖ ferrement m. ou collier m. d'une gouttière. / iron ~ for drawing the charcoal ‖ Störhaken m. ‖ crochet m. pour tirer le charbon. / metallic ~ ‖ Metallhaken m. ‖ agrafe f. métallique. / use no ~s! ‖ nicht einhaken! ‖ manier sans crampon! / rave ~ (Mar) ‖ Nahthaken m.; Werghaken m. ‖ bec m. à corbin ou de corbeau. / rear tow ~ (Tractor) ‖ Zughaken m. ‖ crochet m. d'arrière. / sharp ~ ‖ scharfer Haken m. ‖ crochet m. tranchant. / ~ with socket ‖ Haken m. mit Hülse ‖ croc m. à douille. / tenter ~ ‖ Hakennagel m. ‖ clou m. à crochet. / ~ on the tumbler (Locksm) ‖ Zuhaltungshaken m. ‖ ergot m. d'arrêt. / weeding ~ ‖ Jäthacke f.; Jäthaue f. ‖ serfouette f.; sarclet m.; sarclette f.; sarcloir f.

hook and eye ‖ Haken m. und Ose f. ‖ agrafe f. et veillet m. / ~ **joint** ‖ Seilschloß n. ‖ attache f. de câbles; agrafe f. de jonction pour câbles. / ~ **nippers** pl. ‖ Zange f. für Haken und Ösen ‖ pince f. à agrafes et veillets mpl.

hook and hinge ‖ Angel f. und Band n.; Scharnier n. ‖ gond m. et penture f.

hook bolt ‖ Hakenschraube f. ‖ boulon m. à croc; crampon m. à vis.

hooked ‖ hakenförmig ‖ en forme f. de crochet. / ~ **fish plate** (Railw) ‖ Haken-

lasche f. ‖ éclisse f. à crampon. / ~ forceps pl. ‖ Hakenpinzette f. ‖ pince f. à griffes *ou* à crochets. / ~ key ‖ Hakenschlüssel m. ‖ clef f. à crochet. / ~ nail ‖ Hakennagel m. ‖ clou m. à crochet. / ~ retracting forceps ‖ Hakenzange f. ‖ pince f. à deux griffes. / ~ sole plate *see* ~ tie plate.

hooked tie plate (Railw) ‖ Krempenplatte f. ‖ plaque f. à rebord; selle f. à crochet. / ~ with tenon (Railw) ‖ Hakenzapfenplatte f. ‖ selle f. à crochet et crampon.

hooker (Metal) ‖ Aufhänger m.; Paketierofenmann m. ‖ attacheur m.; paqueteur m.

hook handle (Mine) ‖ Haspelhorn n. ‖ manivelle f. d'un treuil. / ~ head (Ropem) ‖ Hakenkrone f. ‖ croisille f. / ~ link chain ‖ Hakenkette f. ‖ chaîne f. à crochets. / ~ lock ‖ Hakenschloß n. ‖ verrou m. à crochet. / ~ making machine ‖ Hakenherstellungsmaschine f. ‖ machine f. à fabriquer les crochets. / ~ nail (Railw) ‖ Hakennagel m.; Krampnagel m.; Schienennagel m. ‖ crampon m. *ou* clou m. barbelé; tirefond m. / ~ plate track ‖ Hakenplattenoberbau m. ‖ superstructure f. à selles à crochet. / ~ screw ‖ Hakenschraube f. ‖ boulon m. à crochet. / ~ shuttle ‖ Hakenschütz n. ‖ crochet m. pour le tissage des tissus-crin. / ~ spanner ‖ Hakenschlüssel m. ‖ clef f. à crochet. / ~ spring ring ‖ Hakensprengring m. ‖ bague f. de sûreté à crochet. / ~ switch ‖ Hakenumschalter m. ‖ crochet m. commutateur. / ~ tool ‖ Hakenstahl m.; Drehhaken m. ‖ outil m. à crochet; crochet m. (du tourneur).

hoop, to ‖ Reifen mpl. aufschlagen ‖ cercler. / ~ a cask ‖ einen Faßreifen m. aufschlagen; ein Faß n. binden *oder* bereifen ‖ cercler *ou* pantalonner une futaille. / ~ a pile ‖ einen Pfahl beringen *oder* rinken ‖ fretter le pilot; garnir le pilot d'une frette.

hoop ‖ Öse f.; Ring m. ‖ anneau m.; boucle f.; œillet m. / ~ (Carp; Forg) ‖ Hirnring m.; Frette f.; Eisenband n.; Zwinge f. ‖ frette f.; virole f.; entrier m. / ~ (Cask) ‖ Band n.; Reif(en) m. ‖ cercle m.; cerceau m. / ~ (Found) ‖ Mantelring m. (einer Lehmform) ‖ cercle m. (du moule d'une bouche à feu). / ~ (Mach) ‖ Reif m.; Ring m. ‖ cercle m.; collier m.; virole f.; cerceau m. / ~ actuated by compressed air for overhead contact ‖ Druckluftbügelbetätigung f. für Stromabnehmer ‖ prise f. de courant à archet actionnée à l'air comprimé. / band iron ~ ‖ Bandeisenreif m. ‖ cercle m. en feuillard.

hoop of cask ‖ Faßreif m.; Faßband n. ‖ cercle m. de fût. / to drive the hoops pl. of casks ‖ Fässer npl. antreiben ‖ serrer les cercles mpl. de fûts. / to truss up the hoops pl. of casks ‖ Fässer npl. antreiben ‖ serrer les cercles mpl. de fûts.

hoop, iron ~ ‖ eiserner Reif m.; Eisenreif m. ‖ cerceau m. *ou* cercle m. de fer. / shrunkon ~ ‖ Schrumpfband n. ‖ frette f. / crank cheek with ~s shrunk on ‖ Kurbelarm m. mit Schrumpfbändern ‖ bras m. de manivelle muni de frettes. / ~ of spring ‖ Federbund m. ‖ bride f. de ressort. / ~ to strengthen any part ‖ Ring m., der zur Verstärkung einen anderen Teil umfaßt ‖ anneau m. entourant une pièce et la serrant fortement. / wooden ~ ‖

hölzerner Reif m. ‖ cerceau m. *ou* cercle m. de bois.

hoop cramp ‖ Reifzwinge f.; Schraubwinde f. ‖ traitoir m.; davier m. / ~ driver ‖ Fausttreiber m. ‖ chassoir m. / ~ driving machine ‖ Faßreifenantreibmaschine f.; Faßreifenauftreibmaschine f. ‖ machine f. à serrer les cercles de fûts; serreuse f. pour cercles de fûts. / ~ driving-on press ‖ Reifenaufziehpresse f. ‖ presse f. à monter les bandages. / ~ iron (Carp; Forg) ‖ Hirnring m.; Frette f.; Eisenband n.; Zwinge f. ‖ frette f.; virole f.; entrier m. / ~ machine ‖ Reifenaufziehmaschine f. ‖ machine f. à fretter.

hoop iron ‖ Bandeisen n. ‖ feuillard m.; fer m. feuillard. / ~ hoop ‖ Bandeisenreif m. ‖ cercle m. en feuillard. / ~ peg ‖ Bandeisendübel m. ‖ goujon m. de feuillard. / roughing device for ~ ‖ Schrappvorrichtung f. für Bandeisen ‖ dispositif m. à racler les bandes de fer. / ~ spool ‖ Bandeisenhaspel m., f. ‖ dévidoir m. pour fers plats. / tumbling device for ~ ‖ Schleuderapparat m. für Bandeisen ‖ appareil m. centrifuge pour nettoyer des bandes de fer.

hoop maker ‖ Reifenmacher m. ‖ cerclier m. / ~ net (Fish) ‖ Fischreuse f.; Reuse f. ‖ bire f.; nasse f.; panier m.; ablerette f. / ~ riveting machine ‖ Faßreifennietmaschine f. ‖ machine f. à river les cercles de fûts.

hoops pl. ‖ Bandeisen n. ‖ feuillard m.

hoop shearing and punching machine ‖ Bandeisenschere und -lochstanze f. ‖ machine f. à couper et à perforer les cercles. / ~ tongs pl. ‖ gekrümmte Schmiedezange f. ‖ tenailles fpl. crochues. / ~ wood ‖ Reifholz n. ‖ bois m. de cerclage; feuillard m.

hooter ‖ Signalhupe f. ‖ trompe f. à signaux.

hooting ‖ Hupensignal n. ‖ coup m. de sirène.

hop, to ~ the beer ‖ das Bier hopfen ‖ houblonner la bière.

hop ‖ Hopfen m. ‖ houblon m. / early ~ ‖ Augusthopfen m. ‖ houblon m. hâtif *ou* précoce.

hop back ‖ Hopfenseiher m. ‖ bac m. à houblon. / ~ bale ‖ Hopfenballen m. ‖ balle f. de houblon. / ~ cone ‖ Hopfenzapfen m. ‖ cône m. de houblon. / ~ cooling room ‖ Hopfenkühlraum m. ‖ chambre f. froide à houblon. / ~ drying ‖ Hopfendarren n. ‖ sécherie f. de houblon.

hopeful ‖ aussichtsreich ‖ plein de promesses fpl.

hopeite ‖ Hopeit m.; Zinkphyllit m. ‖ hopéite f.

hopeless ‖ aussichtslos ‖ sans chance f. de succès.

hop field ‖ Hopfenanpflanzung f.; Hopfenfeld n. ‖ houblonnière f. / ~ flour ‖ Hopfenmehl n. ‖ lupuline f. de houblon. / ~ garden *see* ~ field. / ~ grower ‖ Hopfenbauer m. ‖ planteur m. de houblon. / ~ jack ‖ Ausschlagbottich m. ‖ panier m. à houblon. / ~ kiln ‖ Hopfendarre f. ‖ touraille f. de séchage du houblon. / ~ oil ‖ Hopfenöl n. ‖ huile f. de houblon.

hopper (Mill) ‖ Trichter m.; Schütttrichter m.; Schüttrumpf m. ‖ trémie f.; trémie f. de déversement. / ~ with bell ‖ Glockenschütttrumpf m. ‖ trémie f. à cloche. / charging ~ ‖ Aufgabetrichter m.; Fülltrichter m.; Einfülltrichter m.; Aufschüttrumpf m. ‖ trémie f. de chargement *ou* de déversement. / discharging ~ ‖ Entladetrichter m. ‖ trémie f. de déchargement *ou* de déversement. / feed ~ *see* charging ~. / filling ~ *see* charging ~. / movable ~ ‖ fahrender Trichter m. ‖ trémie f. déplaçable. / receiving ~ ‖ Auffangtrichter m. ‖ entonnoir-récepteur m. / tipping ~ ‖ Kippmulde f. ‖ benne f. basculante.

hopper beam ‖ Rumpfbaum m.; Rumpfleiter f. ‖ trémion m.; trémillion m. / ~ boy (Mill) ‖ Mehlabkühler m.; Kühlrechen m. ‖ rateau m. refroidisseur. / ~ car ‖ Trichterwagen m. ‖ wagon m. à trémie. / ~ dredger ‖ Prahmbagger m. ‖ dragueur-porteur m. / ~ outlet ‖ Trichterauslauf m. ‖ bec m. de trémie. / ~-shaped crystals pl. ‖ Kastendrusen fpl. ‖ cristaux mpl. en forme d'escalier. / ~ wagon ‖ Trichterwagen m. ‖ wagon-trémie m.

hop sulphuring kiln ‖ Hopfenschwefeldarre f. ‖ touraille f. à soufrer le houblon. / ~ yard *see* hop field.

horary angle (Nav) ‖ Stundenwinkel m. ‖ angle m. horaire.

horizon (Astro) ‖ Gesichtskreis m.; Horizont m. ‖ horizon m. / ~ (Nav) ‖ Kimm m.; Horizont m. ‖ horizon m. / apparent ~ ‖ scheinbarer *oder* sichtbarer Horizont m. ‖ horizon m. physique *ou* sensible *ou* visible *ou* visuel. / artificial ~ ‖ künstlicher Horizont m. ‖ horizon m. fictif. / astronomical ~ ‖ astronomischer *oder* wahrer Horizont m. ‖ horizon m. astronomique *ou* rationnel *ou* vrai. / rational ~ *see* astronomical ~. / real ~ *see* astronomical ~. / sensible ~ *see* apparent ~. / visual ~ *see* apparent ~.

horizon glass (Sextant) ‖ Kimmspiegel m. ‖ miroir m. fixe.

horizontal ‖ wagerecht; horizontal ‖ horizontal. / ~ (Mach) ‖ liegend ‖ horizontal. / ~ (Min) ‖ söhlig ‖ horizontal. / ~ boring machine ‖ Wagerechtbohrmaschine f. ‖ machine f. horizontale à aléser. / ~ boring and milling machine ‖ Wagerechtbohr- und -fräsmaschine f. ‖ machine f. horizontale à aléser et à fraiser. / ~ brace ‖ Horizontalverband m. ‖ entretoisement m. horizontal. / ~ conveyance ‖ Horizontaltransport m. ‖ transport m. horizontal. / ~ deal frame ‖ Horizontalgatter n. ‖ scie f. horizontale alternative. / ~ displacement ‖ wagerechte Verschiebung f. ‖ déplacement m. horizontal. / ~ drilling machine ‖ Wagerechtbohrmaschine f. ‖ machine f. à percer horizontale. / ~ line ‖ Längslinie f. ‖ ligne f. horizontale. / ~ measure of a slope ‖ Ausladung f. *oder* Grundlinie f. einer Böschung ‖ reculement m. d'un talus. / ~ milling machine ‖ Wagerechtfräsmaschine f. ‖ machine f. horizontale à fraiser. / ~ plane ‖ Horizontalebene f. ‖ plan m. horizontal. / ~ projection ‖ Grundriß m. ‖ projection f. horizontale. / ~ rudder ‖ Höhensteuer n. ‖ gouvernail m. horizontal. / ~ shaft three-phase generator ‖ Drehstromgenerator m. mit horizontal liegender Welle ‖ génératrice f. triphasée à arbre horizontal.

/ ~ stiffening ‖ Horizontalversteifung f. ‖ renforcement m. horizontal. / ~ stripe ‖ Querstreifen m. ‖ rayure f. en laize.

horlogical school ‖ Uhrmacherschule f. ‖ école f. d'horlogerie.

horn ‖ Horn n. ‖ corne f. / ~ (Auto) ‖ Hupe f. ‖ cornet m.; trompe f. / ~ (Mus) ‖ Trompete f. ‖ corne f. / artificial ~ ‖ Kunsthorn n. ‖ corne f. artificielle. / artificial ~ producing machine ‖ Kunsthornherstellungsmaschine f. ‖ machine f. à fabriquer la corne artificielle. / buffalo ~ ‖ Büffelhorn n. ‖ corne f. de buffle. / ~ with flexible tube ‖ Schlauchhupe f. ‖ cornet m. à tube flexible. / hunting ~ ‖ Jagdhorn n. ‖ cor m. de chasse. / mechanical ~ ‖ mechanisch betätigte Hupe f. ‖ cornet m. mécanique. / ~ of plane ‖ Nase f. des Hobels ‖ poignée f. ou manche m. de rabot. / reversible ~ (Tool) ‖ drehbares Horn n. ‖ bigorne f. pivotante. / riveting ~ ‖ Niethorn n. ‖ bigorne f. à river.

hornbeam ‖ Weißbuche f. ‖ hêtre m. blanc; charme m.

hornblende ‖ Hornblende f.; Amphibol m. ‖ hornblende f.; amphibole f. / ~ granite ‖ Hornblendegranit m. ‖ granit m. amphibolique. / ~ slate ‖ Hornblendeschiefer m. ‖ amphibole f. schisteuse.

hornblock (Wagon) ‖ Bügelgleitbacken m.; Achsgabelbacken m.; Führungsbacken m. ‖ guide m. de boîte en forme d'arcade. / opening for ~ (Wagon) ‖ Achslagerausschnitt m. ‖ cage f. de boîte d'essieu. / ~ pedestal (Wagon) ‖ geschlossene Achslagerführung f. ‖ guide m. de boîte en forme d'arcade.

horn bulb ‖ Hupenball m. ‖ poire m. de trompe. / ~ button ‖ Hornknopf m. ‖ bouton m. en corne. / ~ chips pl. ‖ Hornspäne mpl. ‖ frisons mpl. de corne. / ~ comb ‖ Hornkamm m. ‖ peigne m. en corne. / ~ core ‖ Knochenzapfen m. ‖ cornillon m. / ~ cutter ‖ Hornschneider m. ‖ découpeur m. de corne. / ~ cutting ‖ Hornschneiden n. ‖ sciage m. de corne. / ~ dresser ‖ Hornarbeiter m. ‖ cornetier m. / ~ dressing ‖ Hornbearbeitung f. ‖ apprêt m. de corne. / ~ engraver ‖ Hornstecher m. ‖ graveur m. sur corne.

horner see horn dresser.

hornfels ‖ Hornfels m. ‖ cornéenne f.

horn goods pl. ‖ Hornwaren fpl. ‖ tabletterie f. de corne; objets mpl. en corne.

hornlike mass ‖ hornähnliche Masse f. ‖ masse f. cornée.

horn matter ‖ Hornsubstanz f. ‖ matière f. cornée. / ~ meal ‖ Hornmehl n. ‖ poudre f. de corne. / ~ mercury ‖ Quecksilberhornerz n.; Kalomel n. ‖ mercure m. corné ou muriaté. / ~ plate ‖ Hornplatte f. ‖ feuille f. de corne. / ~ polisher ‖ Hornschleifer m. ‖ polisseur m. de corne.

hornplate stay (Wagon) ‖ Strebe f.; Achsgabelsteg m. ‖ entretoise f. de la plaque de garde.

horn presser ‖ Hornpresser m. ‖ presseur m. de corne.

horns pl. ‖ Geweih n. ‖ bois m. de cerf. / ~ of the rack of a jack ‖ Gabel f. der Zahnstange einer Winde ‖ cornes fpl. de la crémaillère d'un cric.

horn-shaped ‖ hörnerartig ‖ en forme f. de corne.

horn silver ‖ Silberhornerz n. ‖ argent m. corné ou muriaté; chlorure m. d'argent.

horn spectacles pl. ‖ Hornbrille f. ‖ lunettes fpl. en corne ou en buffle. / ~ frame

Hornbrillengestell n. ‖ monture f. de lunettes en corne. / ~ rim ‖ Hornbrillengestell n. ‖ monture f. de lunettes en corne.

hornstone ‖ Hornstein m. ‖ silex m. corné. / ~ porphyry ‖ Hornsteinporphyr m. ‖ porhpyre m. kératique.

horntail ‖ Holzwespe f. ‖ sirèce f.

horn turner ‖ Horndrechsler m. ‖ tourneur m. sur corne. / ~ ware ‖ Hornwaren fpl. ‖ articles mpl. en corne. / ~ ware imitation ‖ Kunsthornware f. ‖ article m. en corne artificielle. / ~ waste ‖ Hornabfälle mpl. ‖ déchets mpl. de corne. / ~ worker ‖ Hornarbeiter m. ‖ cornassain m.; apprêteur m. de corne.

horse ‖ Pferd n. ‖ cheval m. / ~ (Build) ‖ Bock m.; Gerüst n.; Gestell n. ‖ âne m.; chevalet m.; tréteau m. / ~ (Hydr arch; Ferry) ‖ Galgen m.; Giermast m.; Portal n. einer fliegenden Brücke ‖ potence f. / ~ (Mar) ‖ Paard n.; Pferd n.; Rahepferd n. ‖ marchepied m. de vergue. / ~ (Metal) ‖ Schaleneisen n. ‖ carcas m. / ~ (Print) ‖ Anlegetisch m. ‖ table f. à papier. / wooden ~ (Curr) ‖ Streichbaum m.; Schabbaum m.; Gerbebaum m. ‖ chevalet m. des tanneurs.

horse appointments pl. see horse harness. / ~ bit ‖ Gebiß n. ‖ mors m. / ~ boat ‖ Fähre f.; Prahm m. ‖ passe-cheval m.; pont m. flottant. / ~ box (Wagon) ‖ Pferdetransportwagen m. ‖ wagon m. écurie ou à chevaux. / ~ breeder ‖ Pferdezüchter m. ‖ éleveur m. de chevaux. / ~ breeding ‖ Pferdezucht f. ‖ élevage m. de chevaux. / ~ butcher ‖ Roßschlächter m. ‖ boucher m. hippophagique. / ~ calk making machine ‖ Stollenherstellungsmaschine f. ‖ machine f. à fabriquer les crampons. / ~ capstan ‖ Göpel m.; Pferdegöpel m. ‖ manège m. / ~ chestnut tree ‖ Roßkastanienbaum m.; wilder Kastanienbaum m. ‖ marronnier m. d'Inde. / ~ clipper ‖ Pferdeschermaschine f. ‖ tondeuse f. à chevaux. / ~ cloth see ~ rug. / ~ collar (Saddl) ‖ Kumt n. ‖ collier m. de cheval. / ~ collar maker ‖ Kumtmacher m. ‖ bourrelier m. / ~ comb ‖ Striegel m. ‖ étrille f. / ~ draught ‖ Pferdebespannung f. ‖ attelage m. à chevaux. / ~ drive ‖ Pferdebetrieb m. ‖ traction f. à chevaux. / ~ dropping ‖ Pferdemist m. ‖ crottin m. de cheval. / ~ ferry see ~ boat.

horsefly oil ‖ Bremsenöl n. ‖ huile f. contre les taons.

horse gear ‖ Göpelbetrieb m. ‖ commande f. à manège. / ~ thresher ‖ Göpeldreschmaschine f. ‖ batteuse f. à manège.

horse gin see horse gear thresher. / ~ grease ‖ Kammfett n. ‖ graisse f. de cheval.

horsehair ‖ Roßhaar n. ‖ crin m. / artificial ~ ‖ künstliches Roßhaar n. ‖ crin m. artificiel. / curled ~ ‖ gekräuseltes Roßhaar n. ‖ crin m. frisé ou crépi. / drawn ~ ‖ glattes Roßhaar n. ‖ crin m. plat.

horsehair carpet ‖ Pferdehaarteppich m.; Roßhaarteppich m. ‖ tapis m. en crin. / ~ cloth ‖ Roßhaargewebe n.; Roßhaarstoff m. ‖ tissu-crin m.; drap m. de crin. / ~ comber ‖ Roßhaarkämmer m. ‖ peigneur m. de crins. / ~ pressed lid ‖ Roßhaarpreßdeckel m. ‖ couvercle m. de pression en crin. / ~ seating ‖ Möbelzeug n.; Roßhaarstoff m. für Möbel ‖ tissu m. de crin pour meubles. / ~ spinner ‖ Roßhaarspinner m. ‖ fileur m. de crin. / ~ spinning machine ‖ Roßhaarspinnerei-

maschine f. ‖ machine f. à filer le crin de cheval. / ~ tissue ‖ Gewebe n. aus Roßhaar ‖ tissu m. en crin.

horse harness ‖ Pferdegeschirr n. ‖ harnais m. de chevaux. / ~ haulage (Towing) ‖ Pferdezug m. ‖ halage m. à chevaux. / ~ hauling (Mine) ‖ Pferdebetrieb m. ‖ traction f. à chevaux. / ~ hide ‖ Pferdehaut f. ‖ peau m. de cheval. / ~ hoe ‖ Pferdehacke f. ‖ houe f. à cheval. / ~ keeper (Agr) ‖ Pferdeknecht m. ‖ garçon m. ou valet m. d'écurie. / ~ leather ‖ Roßleder n.; Pferdeleder n. ‖ cuir m. de cheval. / ~ letting-out ‖ Pferdevermietung f. ‖ louage m. de chevaux. / ~ load ‖ Saumtierladung f.; Saumtierlast f. ‖ somme f. / ~ mill ‖ Pferdemühle f. ‖ moulin m. à manège. / ~ nail ‖ Hufnagel m. ‖ clou m. à cheval. / ~ net ‖ Pferdenetz n. ‖ filet m. pour chevaux. / ~ omnibus ‖ Pferdeomnibus m. ‖ omnibus m. à chevaux. / ~ path ‖ Reitweg m. ‖ route f. muletière. / ~ picker ‖ Hufräumer m. ‖ cure-pied m. / ~ pond ‖ Pferdeschwemme f. ‖ abreuvoir m.

horse power (H. P.) ‖ Pferdestärke f.; PS. ‖ cheval-vapeur m.; C. V. / brake ~ ‖ Bremsleistung f. in PS; gebremste Pferdestärke f. ‖ cheval m. au frein. / effective ~ ‖ nutzbare oder effektive Pferdestärke f. ‖ cheval m. au frein. / indicated ~ ‖ indizierte Pferdestärke f. ‖ cheval m. indiqué. / nominal ~ ‖ nominelle Pferdestärke f. ‖ chevaux mpl. nominaux. / rating ~ ‖ Steuerpferdekraft f. ‖ puissance f. imposée en chevaux.

horse rake ‖ Pferdeharke f.; Pferderechen m. ‖ râteau m. à cheval. / ~ road see ~ path. / ~ rug ‖ Pferdedecke f. ‖ couverture f. de cheval; housse f.; caparaçon m.

horseshoe ‖ Hufeisen n. ‖ fer m. à cheval. / to bend a ~ ‖ das Hufeisen n. krümmen oder verengern ‖ voûter un fer à cheval. / ~ with calkings ‖ Hufeisen n. mit Stollen; Stolleneisen n. ‖ fer m. à cheval cramponné ou relevé.

horseshoe arch ‖ Hufeisenbogen m. ‖ arc m. outre-passé ou en fer à cheval. / pointed ~ arch ‖ Hufeisenspitzbogen m. ‖ ogive f. lancéolée. / ~ iron ‖ Hufstabeisen n. ‖ fer m. à maréchal. / ~ magnet ‖ Hufeisenmagnet m. ‖ aimant m. en fer à cheval. / ~ maker ‖ Hufschmied m. ‖ forgeur m. de fer à cheval; maréchal m. ferrant. / ~ making machine ‖ Hufeisenherstellungsmaschine f. ‖ machine f. à fabriquer les fers à cheval. / ~ manufacture ‖ Hufschmiede f. ‖ fabrique f. de fers à chevaux; maréchalerie f. / ~ nail ‖ Hufnagel m. ‖ clou m. à ferrer.

horseshoer ‖ Hufschmied m. ‖ maréchal m.

horseshoe roundhead ‖ Hufeisenrundbogen m. ‖ plein cintre m. outre-passé. / calk with screws for ~ ‖ Hufschraubstollen m. ‖ crampon m. à vis pour fers à cheval. / box for ~ smithing ‖ Hufbeschlagstand m. ‖ cabine f. de maréchalerie. / ~ spike ‖ Hufstollen m. ‖ crampon m. de fer à cheval. / ~ stand ‖ Hufeisenfuß m. ‖ pied m. en fer à cheval.

horse tail ‖ Schachtelhalm m. ‖ prêle f.; queue f. de cheval. / ~ trade ‖ Pferdehandel m. ‖ maquignonnage m.; commerce m. de chevaux. / ~ tramway ‖ Pferdebahn f. ‖ tramway m. à chevaux. / ~ trappings pl. see horse harness.

/ ~ way *see* ~ path. / ~ whims mechanism ‖ Pferdegöpel m. ‖ manège m. à chevaux.

horsewhip ‖ Reitpeitsche f. ‖ cravache f.

horticulture ‖ Gartenbau m.; Gärtnerei f. ‖ horticulture f. / ~ implement ‖ Gartengerät n. ‖ instrument m. pour horticulture. / machine for ~ ‖ Maschine f. für den Gartenbau ‖ machine f. pour l'horticulture.

hose ‖ Schlauch m. ‖ tuyau m.; manche f. / ~ (Stocking) (langer) Strumpf m.; Sportstrumpf m.; Stutzen mpl. ‖ bas m.; bas m. pour sport. / air supply ~ ‖ Luftansaugschlauch m. ‖ tuyau m. du supplément d'air. / cotton ~ ‖ Baumwollschlauch m. ‖ tuyau m. en coton. / ~ for fire engines ‖ Schlauch m. für Feuerspritzen ‖ tuyau m. pour pompes à incendie. / rubber ~ ‖ Gummischlauch m. ‖ tuyau m. en caoutchouc. / ~ of stuff ‖ Stoffschlauch m. ‖ tuyau m. en étoffe. / tarred canvas ~ ‖ Schlauch m. von geteertem Segeltuch ‖ manche f. de toile goudronnée. / ~ woven round with wire ‖ Panzerschlauch m. ‖ manche f. garnie de fil de fer.

hose bridge ‖ Schlauchrolle f.; Schlauchwagen m. ‖ chariot m. à tuyau. / ~ cart ‖ Schlauchwagen m. ‖ chariot m. à tambour pour tuyaux. / ~ chute ‖ Hosenrutsche f. ‖ goulotte f. *ou* manche f. de déversement; tubulure f. de chargement. / ~ cleaning apparatus ‖ Schlauchreinigungsapparat m. ‖ appareil m. à nettoyer les tuyaux flexibles. / ~ connector ‖ Schlauchbinder m. ‖ joint m. de tuyau. / ~ coupling ‖ Schlauchkupplung f. ‖ raccord m. de tuyau. / ~ drum ‖ Schlauchhaspel m. ‖ tambour m. à tuyaux. / ~ fittings pl. ‖ Schlaucharmatur f. ‖ armature f. pour tuyaux flexibles *ou* pour manches. / rubber ~ machine ‖ Gummischlauchmaschine f. ‖ machine f. à faire les tuyaux en caoutchouc. / ~ pipe ‖ Schlauchleitung f. ‖ conduite f. à tuyaux flexibles.

hosier ‖ Strumpfwirker m. ‖ chaussetier-bonnetier m.

hosiery ‖ Wirkwaren fpl.; Strumpfwaren fpl. ‖ articles mpl. de bonneterie; bonneterie f. / cotton ~ ‖ Baumwollstrickerei f. ‖ bonneterie f. de coton. / cut ~ ‖ zugeschnittene Wirkwaren fpl. ‖ bonneterie f. confectionnée. / linen ~ ‖ Leinenstrickerei f. ‖ bonneterie f. de lin. / oriental ~ ‖ orientalische Strumpfwaren fpl. ‖ bonneterie f. orientale. / seamless ~ ‖ nahtlose Strumpfwaren fpl. ‖ bonneterie f. sans couture. / silk ~ ‖ Seidenstrickerei f. ‖ bonneterie f. de soie. / ~ of silk ‖ Seidenstrumpfwaren fpl. ‖ bonneterie f. de soie. / wool ~ ‖ Wollstrickerei f. ‖ bonneterie f. de laine.

hosiery finishing ‖ Strickwarenappretur f. ‖ apprêt m. de bonneterie. / ~ fulling ‖ Wirkwarenwalke f. ‖ foulage m. de bonneterie. / ~ knitting machine ‖ Strumpfstrickmaschine f. ‖ machine f. à tricoter les bas. / ~ machine ‖ Wirkmaschine f. ‖ machine f. à bonneterie. / ~ maker ‖ Strumpfwarenwirker m. ‖ fabricant-bonnetier m. / manufacture of ~ ‖ Strumpffabrikation f. ‖ fabrication f. de bas. / / tubular ~ mill ‖ Rundwirkmaschine f. ‖ métier m. à tricotage circulaire. / ~ pressing ‖ Wirkwarenpresserei f. ‖ catissage m. de bonneterie. / ~ sewing ‖ Trikotnäherei f. ‖ couture f. de bonneterie.

hospital ‖ Hospital n.; Krankenhaus n. ‖ hôpital m.; hospice m.; hôtel-Dieu m.; maison-Dieu f. / barrack ~ ‖ Krankenbaracke f. ‖ hôpital-baraquement m. / ~ equipment ‖ Krankenhauseinrichtung f. ‖ installation f. pour hôpitaux. / ~ ship ‖ Hospitalschiff n.; Lazarettschiff n. ‖ navire m. hôpital. / ~ train ‖ Sanitätszug m. ‖ train m. d'ambulance.

host ‖ Hostie f. ‖ hostie f.

hostel, general ‖ Arbeiterheim n. ‖ maison-pension f.

hot ‖ heiß ‖ chaud. / the motor is getting ~ ‖ der Motor wird heiß ‖ le moteur s'échauffe. / too ~ ‖ zu heiß; übergar ‖ trop chaud. / white ~ ‖ weißglühend ‖ de chaleur blanche.

hot-air apparatus ‖ Heißluftapparat m. ‖ appareil m. à air chaud. / ~ conduit ‖ Warmluftkanal m. ‖ conduite f. d'air chaud. / ~ drying ‖ Warmlufttrocknung f. ‖ séchage m. à air réchauffée. / ~ engine ‖ Heißluftmotor m. ‖ moteur m. à air chaud. / ~ heater ‖ Warmluftheizungsanlage f. ‖ calorifère f. à air chaud. / ~ heating ‖ Warmluftheizung f. ‖ chauffage m. à air chaud. / ~ intake ‖ Warmlufteintritt m. ‖ entrée f. de l'air chaud. / ~ man ‖ Heißwindwärter m. ‖ régleur m. de l'air chaud. / ~ piping ‖ Heißwindrohrleitung f. ‖ tuyauterie f. à vent chaud. / ~ plant ‖ Heißluftheizung f. ‖ chauffage m. à air. / ~ sprayer ‖ Heißluftdusche f. ‖ douche f. à air chaud. / ~ stove ‖ Luftheizungsapparat m. ‖ calorifère m. à air. / ~ turbine ‖ Heißluftturbine f. ‖ turbine f. à air chaud.

hot bar shear ‖ Heißeisenschere f. ‖ cisailles fpl. à chaud pour le fer. / ~ bending test ‖ Warmbiegeprobe f. ‖ essai m. de ployage à chaud.

hot-blast apparatus ‖ Winderhitzer m. ‖ appareil m. à vent chaud. / Cowper ~ ‖ Cowperapparat m. für die Winderhitzung ‖ appareil m. Cowper pour le chauffage du vent.

hot-bulb ‖ Glühkopf m. ‖ bulbe m. chaud; calotte f. chaude. / cracking of the ~ ‖ Reißen n. des Glühkopfes ‖ déchirement m. du bulbe chaud *ou* de calotte chaude. / ~ ignition ‖ Glühkopfzündung f. ‖ allumage m. à calotte incandescent. / ~ motor ‖ Glühkopfmotor m. ‖ moteur m. à culasse incandescente. / lengthy pre-heating of ~ ‖ langwieriges Anheizen n. des Glühkopfes ‖ réchauffage m. préalable du bulbe d'allumage.

hot chisel ‖ Warmschrotmeißel m. ‖ ciseau m. à chaud; tranche f. à chaud- / ~ cutting of rolled iron ‖ Warmschneiden n. von Walzeisen ‖ coupe f. à chaud de fers laminés.

hotel bus *see* ~ omnibus. / ~ omnibus ‖ Hotelomnibus m. ‖ omnibus m. *ou* autobus m. d'hôtel.

hot-hammer, to ‖ warmschmieden ‖ battre à chaud.

hothouse ‖ Treibhaus n. ‖ serre f. chaude. / ~ gardener ‖ Treibhausgärtner m. ‖ jardinier m. de serre à chauffer. / heat of ~ ‖ Treibhauswärme f. ‖ chaleur f. de serres chaudes.

hot iron travelling saw ‖ Heißeisenschlittensäge f. ‖ scie f. à chaud à marche. / ~ metal saw ‖ Heißeisensäge f. ‖ scie f. à chaud pour le fer. / ~ nut press ‖ Warmmutternpresse f. ‖ presse f. à chaud pour écrous. / ~ plate ‖ Heizplatte

f. ‖ réchaud m.; dessous m. de plats; plaque f. chauffante. / ~ pressing machine ‖ Dekatiermaschine f. ‖ machine f. à décatir. / ~-riveted ‖ warmgenietet ‖ rivé à chaud.

hot-roll, to ‖ warmwalzen ‖ laminer à chaud.

hot saw ‖ Warmsäge f. ‖ scie f. à chaud. / ~-shearing ‖ Warmscheren n. ‖ cisailler à chaud. / ~-short ‖ rotbrüchig ‖ rouverin *ou* rouverain; cassant à chaud; métis. / ~-test ‖ Warmprobe f. ‖ essai m. à chaud. / ~-tube ignition ‖ Glührohrzündung f. ‖ allumage m. à tube incandescent. / ~ vulcanization ‖ heiße Vulkanisation f. ‖ vulcanisation f. à chaud.

hot water apparatus ‖ Heißwasserapparat m. ‖ appareil m. pour la préparation de l'eau chaude. / ~ bag ‖ Heißwasserbeutel m. ‖ sac m. à eau chaude. / ~ bottle ‖ Wärmflasche f. ‖ vessie f. à eau chaude; bouillotte f. / ~ can *see* ~ bottle. / ~ cistern ‖ Heißwassertank m. ‖ réservoir m. à eau chaude. / ~ frame *see* ~ spinning frame. / ~ heater ‖ Warmwasserheizungsanlage f. ‖ chauffage m. à eau chaude. / ~ heating ‖ Warmwasserheizung f. ‖ chauffage m. à l'eau chaude. / ~ plant ‖ Warmwasserbereitungsanlage f. ‖ installation f. de production d'eau chaude. / ~ pump ‖ Warmwasserpumpe f. ‖ pompe f. à eau chaude. / ~ radiator ‖ Warmwasserheizkörper m. ‖ radiateur m. à eau chaude. / ~ reservoir ‖ Heißwasserspeicher m. ‖ accumulateur m. *ou* réservoir m. d'eau chaude. / ~ spinning frame ‖ Warmwasserspinnmaschine f. ‖ métier m. à eau chaude. / ~ supplier ‖ Heißwassererzeuger m. ‖ chauffe m. d'eau très chaude. / ~ test ‖ Heißwasserprobe f. ‖ essai m. à l'eau chaude.

hot wire ‖ Hitzdraht m. ‖ fil m. thermique. / ~ ammeter ‖ Hitzdrahtstrommesser m. ‖ ampèremètre m. à fil chaud. / ~ instrument ‖ Hitzdrahtinstrument n. ‖ instrument m. à fil chaud. / ~ voltmeter ‖ Hitzdrahtspannungsmesser m. ‖ voltmètre m. à fil chaud.

hour ‖ Stunde f. ‖ heure f. / ~ of firing *see* ~ of heating. / ~ of heating ‖ Heizstunde f. ‖ heure f. de chauffe *ou* de chauffage. / ~ without heating ‖ heizfreie Stunde f. ‖ heure f. sans chauffage.

hour circle ‖ Stundenkreis m. ‖ cercle m. horaire. / ~ glass ‖ Sanduhr f. ‖ sablier m. / ~ hand ‖ Stundenzeiger m. ‖ aiguille f. des heures.

hourly capacity ‖ Stundenleistung f. ‖ rendement m. horaire. / ~ of a cooling tower ‖ Stundenleistung f. eines Kühlturmes ‖ puissance f. horaire d'un tour de réfrigération.

hourly output ‖ Stundenleistung f. ‖ rendement m. *ou* débit m. à l'heure.

housage ‖ Lagergeld n. ‖ magasinage m.; frais mpl. de magasinage.

house, to ‖ unterbringen; unter Dach und Fach bringen ‖ placer; loger; garer; mettre à l'abri m. / ~ a car ‖ einen Wagen m. unterstellen ‖ remiser *ou* garer une voiture.

house ‖ Haus n. ‖ maison f.; bâtiment m. / copper ~ ‖ Kupferhaus n. ‖ maison f. en cuivre. / ~ of the crew (Shipb) ‖ Roof m. ‖ rouf m. / ~ for one family only ‖ Einfamilienhaus n. ‖ maison f. de famille unique. / small ~ ‖ Kleinhaus n. ‖ maisonnette f. / steel ~ ‖ Stahlhaus n. ‖ maison

f. en acier. / transportable ~ || transportables Gebäude n. || maison f. transportable. / ~ for water plants || Wasserpflanzenhaus n. || maison f. de plantes aquatiques.

house agent || Hypothekenmakler m. || marchand m. d'hypothèque et d'immeubles.

house building enterprise || Baugeschäft n. || entreprise f. de bâtiments. / plan for ~ || Bebauungsplan m. || plan m. de construction.

house carpentry || Bautischlerei f. || menuiserie f. de bâtiment. / ~ connecting box || Hausanschlußkasten m. || boîte f. de raccordement pour maisons. / ~ decoration || Häuserschmuck m. || décoration f. de bâtiments. / ~ door || Haustür f. || porte f. de la maison *ou* de face. / ~ drainage || Hausleitung f. || canalisation f. de maison.

household || Haushalt m.; Haushaltung f. || ménage m.; usage m. domestique. / ~ articles pl. || Haushaltungsartikel mpl. || articles fpl. de ménage. / ~ coal || Hausbrandkohle f. || charbon m. de ménage. / ~ flat iron || Haushaltbügeleisen n. || fer m. à repasser de ménage. / ~ furnace || Haushaltfeuerung f. || four m. de ménage. / ~ furniture || Hausrat m. || ameublement m. / ~ grating machine || Reibmaschine f. für den Haushalt || machine f. de ménage à râper. / ~ implements pl. || Wirtschaftsgerät n. || ustensiles mpl. de ménage. / ~ linen || Hauswäsche f. || linge m. ordinaire *ou* de ménage. / ~ machine || Haushaltmaschine f.; hauswirtschaftliche Maschine f. || machine f. de ménage. / ~ requisite || Haushaltungsgegenstand m. || article m. de ménage. / ~ scales pl. || Hausstandswage f.; Küchenwage f. || balance f. de cuisine. / ~ soap || Haushaltseife f. || savon m. de ménage. / ~ utensil || Haushaltungsgerät n. || ustensile m. de ménage. / ~ utensils pl. of earthenware || Steingutgebrauchsgeschirr n. || service m. de ménage en faïence.

house, electrotechnical materials pl. for ~ installations || elektrotechnisches Hausinstallationsmaterial n. || matériaux mpl. électrotechniques pour installations de la maison. / ~ keeping school || Haushaltungsschule f. || école f. ménagère. / ~ keeping and cooking school || Koch- und Haushaltungsschule f. || école f. ménagère.

houseline *see* housing (Mar).

house mark || Hausmarke f. || marque f. de la maison. / ~ painter || Anstreicher m.; Tüncher m.; Dekorationsmaler m. || peintre m.; peintre-décorateur m. / ~ painting || Häuseranstrich m. || peinture f. en bâtiments. / ~ refuse dressing || Verarbeitung f. von Abfallstoffen || utilisation f. de vidanges. / ~ refuse incineration || Müllverbrennungsanstalt f. || usine f. d'incinération d'ordures ménagères. / ~ rent || Hauszins m. || loyer m. / ~ shoe || Hausschuh m. || chaussure f. d'appartement. / ~ telegraph || Haustelegraf m. || télégraphe m. domestique. / ~ telegraphy || Haustelegrafie f. || télégraphie f. domestique. / ~ telephone || Hausfernsprecher m.; Haustelefon n. || téléphone m. domestique. / ~ telephony || Haustelefonie f. || téléphonie f. domestique. / ~ transformer || Haustransformator m. || transformateur m. domestique. / ~ and kitchen utensils pl. || Haus- und Küchengeräte npl. ||

ustensiles mpl. de ménage et de cuisine. / ~ water pipe || Hauswasserleitung f. || tuyau m. de conduite domestique. / ~ water supply plant || Hauswasserversorgungsanlage f. || installation f. d'adduction d'eau pour maisons.

housing || Hülse f. || douille f. / ~ (Mar) || Hüsing f. || lusin m.; luzin m.; merlin m. à trois fils. / ~ (Roll mill) || Walzengerüst n. || cage f. de laminoir. / drive shaft ~ || Gelenkwellenrohr n. || jambe f. tubulaire de l'arbre à cardan. / ~ for a mill || Walzenständer m. || montant m. de laminoir. / rear axle ~ || Hinterachsgehäuse n. || carter m. du pont arrière. / ~ for a rolling mill || Walzenständer m. || montant m. de laminoir. / ~ for two rolls || Zweiwalzengerüst n. || cage f. à deux cylindres. / upper valve gear ~ || Zylinderkopfdeckel m. || couvercle m. supérieur des soupapes. / winch ~ || Windengehäuse n. || bâti m. du treuil.

housing bearers pl. *see* ~ frames. / ~ conditions pl. || Wohnungsverhältnisse npl. || conditions fpl. d'habitations. / ~ frames pl. (Roll mill) || Walzenständer mpl.; Ständergerüst n. || cages fpl.; fermes fpl.; montants mpl.; piliers mpl.; poupées fpl.; châssis m. / double two-~ mill with roughing train || Doppelduo n. mit Vorgerüst || double duo m. avec train ébaucheur. / ~ pillars pl. *see* ~ frames pl. / ~ posts pl. *see* ~ frames.

howel, to || abputzen || blanchir; raboter.

howel || Krummhaue f. || assette f.; essette f.; asseau m.; hachette f.; erminette f.

howelling || Abputzen n. || blanchiment m.; rabotage m.

howitzer || Haubitze f. || obus m. / breechblock of ~ || Haubitzkopf m. || tête f. d'obus.

howler || Summer m. || ronfleur m.; vibrateur m. / ~ for dialling tone || Summer m. für Amtszeichen || ronfleur m. bruit du réseau. / ~ connection (Tel) || Heuleranruf m. || appel m. par hurleur.

hub || Nabe f. || moyeu m. / ~ of armature (Electr) || Ankernabe f. || moyeu m. d'induit. / fan ~ || Windflügelnabe f. || moyeu m. de ventilateur. / rear wheel ~ || Hinterradnabe f. || moyeu m. de roue arrière. / serrated ~ || Kerbzahnnabe f. || moyeu m. à cannelures. / wheel ~ || Radnabe f. || moyeu m. de roue. / wooden ~ || Holznabe f. || moyeu m. en bois.

hub brake || Nabenbremse f. || frein m. du moyeu. / ~ mortising machine hand || Radnabenbohrer m. || mortaiseur m. de moyeux. / ~ turner || Radnabendrechsler m. || tourneur m. de moyeux.

hug, to ~ the land (Nav) || unter Land n. halten || serrer *ou* ranger la terre. / ~ the wind close (Nav) || dicht an den Wind m. halten || pincer ou serrer le vent.

Hughes telegraph apparatus || Hughes Telegraphenapparat m. || appareil m. Hughes télégraphique.

hulk || Hulk m. || fosse f.

hulking hammer || Abbauhammer m. || marteau m. d'excavation.

hull, to || aushülsen || écosser; écaler.

hull (Fruit) || Schale f. || coque f. / ~ (Insur) || Kasko m. || corps m. et quille. / ~ (Shipb) || Rumpf m. || coque f. / ~ of airship || Luftschiffkörper m. || carène f. du dirigeable. / iron ~ (Shipb) || eiserner Schiffsrumpf m. || coque f. de navires en fer. / ~ of a seagoing vessel

|| Seeschiffsrumpf m. || coque f. de bâtiment de mer. / ~ of a ship || Schiffskörper m.; Schiffsrumpf m. || corps m. *ou* coque d'un bateau. / the ~ of the ship is recessed forward and aft || der Schiffsrumpf ist vorn und achtern eingezogen || la coque du navire a des rentrants à l'avant et à l'arrière. / ~ of a vessel (Shipb) || Schiffsrumpf m. || coque f. d'un navire.

hulled grain || Graupe f. || grain m. mondé.

hully || Aalreuse f.; Aalkorb m. || nanse f.

hum || Summen n. || ronchonnement m.; ronflement m.

human || menschlich || humain.

humic acid || Humussäure f. || acide m. humique.

humidity || Wassergehalt m.; Feuchtigkeit f. || humidité f. / ~ of the air || Luftfeuchtigkeit f. || humidité f. de l'air. / separation of the ~ contained in the air || Abscheiden n. des in der atmosphärischen Luft enthaltenen Wasserdampfes || séparation f. des vapeurs d'eau contenues dans l'air. / ~ contents pl. of air || Luftfeuchtigkeitsgehalt m. || degré m. d'humidité de l'air. / degree of ~ || Feuchtigkeitsgehalt m. || degré m. d'humidité.

humming of gears || Heulen n. der Zahnräder || ronflement m. des engrenages. / ~ of the motor || Surren n. des Motors || bourdonnement m. du moteur. / ~ of telegraph wires || Summen n. der Telegrafendrähte || bourdonnement m. des lignes télégraphiques.

hump || Anschwellung f. || bosse f. / ~ (Railw; Mine) || Ablaufberg m. || dos m. d'âne.

humpy || höckerig || raboteux; bossu.

humus || Humus m.; Gartenerde f. || humus m.; terre f. végétale. / ~ deposit || Humusgestein n. || roche f. d'humus.

hundreds pl. || Hunderte mpl. || des centaines fpl.

hundredweight || Zentner m. || quintal m.

hung beef || Rauchfleisch n. || bœuf m. fumé; viande f. fumée.

hunt, to || jagen || chasser. / ~ (Radio) || abtasten || balayer.

hunt || Hetzjagd f. || chasse f. à courre.

hunter || Jäger m. || chasseur m.

hunting || Jagd f. || chasse f. / ~ coat || Jagdweste f. || gilet m. de chasse. / ~ device (Radio) || Abtastvorrichtung f. || dispositif m. de balayement. / ~ equipage || Jagdzeug n. || équipement m. de chasse f. / ~ glass || Jagdglas n. || jumelle f. de chasse. / ~ horn || Jagdhorn n. || cor m. de chasse. / ~ knife || Jagdmesser n. || couteau m. de chasse. / ~ movement (Aut tel) || freie Wahl f. || recherche f. libre. / ~ and shooting requisites pl. || Jagdrequisiten npl. || articles mpl. de chasse. / ~ spoon rolling mill || pendelnde Löffelwalzmaschine f. || laminoir m. à pendule à faire des cuillers *ou* cuillères. / ~ switch (Aut tel) || Freiwähler m. || sélecteur m. libre.

huntsman || Jäger m. || chasseur m. / ~'s dress || Jagdbekleidung f. || habillement m. de chasseurs.

hurdle || Horde f. || claie f. / drying ~ || Trockendarre f. || claie f. à sécher. / fruit ~ || Obsthorde f. || claie f. pour fruits. / lower finishing ~ (Malt) || Abdarrhorde f. || plateau m. de touraillage. / ~ for purifier || Horde f. für Reiniger || claie f. d'épurateurs. / ~ washer || Hordenwascher

m. || laveur m. de claie. / ~ work || Flechtzaun m. || clayon m.

hurdling || Einzäunung f.; Spalier n.; Staket n. || clôture f. de palis; espalier m.; palissade f.

hurricane || Orkan m. || ouragan m.

hurrier (Min) || Fördermann m.; Schlepper m.; Wagenstößer m. || rouleur m.; traîneur m.

hurst (Tool) || Hammerhülse f. || hulse f. *ou* hurasse f. *ou* hus m. du gros marteau de forge.

hurter (Bridge) || Landstoßbalken m.; Stoßschwelle f. || garde-sable m.

husbandman || Ackerbauer m. || cultivateur m.; laboureur m.

husbandry || Landwirtschaft f.; Ackerbau m. || agriculture f.; culture f. de la terre.

husk, to || enthülsen || décortiquer.

husk (Fruit) || Schale f.; Hülse f. || gousse f.; cosse f.; écale f. / cigarette ~ || Zigarettenhülse f. || enveloppe f. de cigarettes. / ~ of walnut || äußere grüne Walnußschale f. || brou m. de noix. / ~ determination || Hülsenbestimmung f. || détermination f. de la quantité des balles.

husky || dickhülsig; dickschalig || cossu; à écorce f. épaisse.

hustler (Sawm) || Blockwagenführer m. || binardeur m.

hut || Schuppen m.; Baracke f.; Bude f. || échoppe f.; hangar m.; hutte f.; loge f.

hutch (Bak) || Backtrog m. || pétrin m. / ~ (Box) || Koffer m. mit rundem Deckel || bahut m. / ~ (Mine) || Schachtfördergefäß n. || tonne f.; tine f. / ~ (Rabbit) || Kaninchenstall m. || clapier m. / coal ~ || Kohlenwagen m. || benne f. piocheuse *ou* à charbon.

hyacinth || Hyazinth m. || jacinthe f.; hyacinthe f.

hyalite || Gummistein m. || hyalithe f.

hyalography || Hyalografie f.; Hyalotypie f. || hyalographie f.

hyalophane || Barytfeldspat m.; Hyalophon m. || hyalophane m.; feldspath m. barytique.

hybrid coil (Tel) || Ringübertrager m.; Ausgleichübertrager m. || transformateur m. annulaire *ou* différentiel; bobine f. circulaire. / ~ terminal station of a four-wire circuit || Gabelpunkt m. einer Vierdrahtleitung || termineur m. d'un circuit à quatre fils.

hydrant || Hydrant m. || bouche f. à eau *ou* d'incendie.

hydrate || Hydrat n. || hydrate m. / ~ of lime || Kalkhydrat n. || hydrate m. de chaux; chaux f. hydratée. / ~ of potash || Ätzkali n. || potasse f. caustique.

hydrated || wasserhaltig || hydraté. / ~ amylene || Amylenhydrat n. || amylène m. hydraté. / ~ auric oxide || Goldsäure f. || acide m. aurique. / ~ oxide || Oxydhydrat n. || oxyde m. hydraté.

hydration || Hydration f. || hydration f. / ~ water || Hydratwasser n. || eau f. d'hydration.

hydratize, to || hydratisieren || hydrater.

hydraulic || hydraulisch || hydraulique. / ~ accumulator || hydraulischer Akkumulator m.; Druckwasserspeicher m. || accumulateur m. hydraulique; château m. d'eau de pression. / ~ admixture || hydraulischer Zuschlag m. || fondant m. hydraulique. / ~ brake || hydraulische

Bremse f. || frein m. hydraulique. / ~ central station || Elektrizitätswerk n. mit Wasserkraft || usine f. hydraulico-électrique. / ~ compressor || Wasserdruckkompressor m. || compresseur m. hydraulique. / ~ coupling || hydraulische Kupplung f. || accouplement m. hydraulique. / ~ crane || hydraulischer Kran m. || grue f. hydraulique. / ~ engine || Wasserkraftmaschine f. || moteur m. hydraulique. / ~ forging press || hydraulische Schmiedepresse f. || presse f. hydraulique à forger. / steam ~ forging press || hydraulische Dampfschmiedepresse f. || presse f. hydraulique à vapeur. / ~ frame press || hydraulische Rahmenpresse f. || presse f. hydraulique à emboutir des châssis. / ~ hammer || hydraulischer Hammer m. || marteau m. hydraulique. / ~ head || Wassersäulendruck m. || pression f. en colonne d'eau. / ~ ironwork || Eisenwasserbau m. || construction f. metallique pour ouvrage hydraulique. / ~ jack || hydraulischer Heber m. || vérin m. hydraulique. / ~ lifting device || hydraulisches Hebezeug n. || appareil m. de levage hydraulique. / ~ lime || hydraulischer Kalk m.; Wasserkalk m. || chaux f. hydraulique. / ~ machine || hydraulisch betriebene Maschine f. || machine f. hydraulique. / ~ motor || Wasserkraftmaschine f. || moteur m. hydraulique. / ~ press || hydraulische Presse f. || presse f. hydraulique. / quick acting ~ press || schnellaufende hydraulische Presse f. || presse f. hydraulique à marche rapide. / ~ pressure || Wasserdruck m. || pression f. hydraulique. / all parts pl. are obtained by ~ pressure || sämtliche Stücke npl. sind hydraulisch gepreßt || toutes les pièces fpl. sont obtenues par emboutissage hydraulique. / ~ pressure plant || Druckwasseranlage f. || installation f. hydraulique. / ~ ram || hydraulischer Widder m. || bélier m. hydraulique. / ~ riveting || hydraulische Nietung f. || rivure f. hydraulique. / ~ screw || Wasserschraube f. || vis f. d'Archimède. / ~ structure || Wasserbau m.; Wasserbauanlage f. || construction f. hydraulique. / ~ test || Wasserdruckprobe f. || épreuve f. hydraulique *ou* à pression hydraulique. / ~ working || Druckwasserbetrieb m. || fonctionnement m. hydraulique.

hydraulical *see* hydraulic.

hydraulics pl. || Hydraulik f. || hydraulique f.

hydride || Hydrid n. || hydrure m.

hydriodic acid || Jodwasserstoff m. || acide m. iodhydrique.

hydro (Hydropathic) *see* hydropathic establishment. / ~ (Hydroplane) *see* hydroplane.

hydrobromic acid || Bromwasserstoffsäure f.; Hydrobromsäure f. || acide m. bromhydrique *ou* hydrobromique.

hydrocarbide *see* hydrocarbon.

hydrocarbon || Kohlenwasserstoff m. || hydrocarbure f.; carbure m. d'hydrogène. / ~s pl. contained in the coal || in der Kohle f. enthaltene Kohlenwasserstoffe mpl. || hydrocarbures mpl. contenus dans le charbon. / light ~s pl. || leichte Kohlenwasserstoffe mpl. || hydrocarbures mpl. légères. / saturated ~s pl. || gesättigte Kohlenwasserstoffe mpl. || hydrocarbures mpl. saturés.

hydrocarbon gas || Kohlenwasserstoffgas n. || gaz m. d'hydrogène carburé.

hydrocarburet || Kohlenwasserstoff m. || hydrocarbure m.

hydrochloric acid || Chlorwasserstoff m.; Salzsäure f. || acide m. chlorhydrique. / ~ gas || Chlorwasserstoffgas n. || gaz m. acide chlorhydrique *ou* muriatique.

hydrocyanate of potassium || Zyankalium n. || prussiate m. de potasse; cyanure m. *ou* cyanide m. de potassium.

hydrocyanic acid || Zyanwasserstoffsäure f.; Blausäure f. || acide m. prussique *ou* cyanhydrique.

hydrodynamics pl. || Dynamik f. flüssiger Körper; Hydrodynamik f. || hydrodynamique f.

hydro-electric || hydroelektrisch || hydro-électrique. / ~ machine || hydroelektrische Maschine f. || machine f. hydro-électrique. / ~ power station || Wasserkraftwerk n. || usine f. hydro-électrique.

hydro-extracting machine *see* hydro-extractor.

hydro-extractor (Centrifugal) || Zentrifugaltrockenmaschine f.; Zentrifuge f. || essoreuse f.; essoureuse f. centrifuge. / acid ~ for cloth carbonizing || Säurezentrifuge f. für Tuchkarbonisation || essoreuse f. à succion d'acide pour la carbonisation des draps. / ~ minder || Zentrifugenführer m. || essoreur m.

hydroferrocyanic acid || Ferrozyanwasserstoffsäure f.; Eisenblausäure f. || acide m. ferrocyanique.

hydrofluoric acid || Fluorwasserstoffsäure f.; Flußsäure f. || acide m. fluorhydrique.

hydrofuge || feuchtigkeitvertreibend || hydrofuge.

hydrogas *see* water gas.

hydrogen || Wasserstoff m. || hydrogène m. / arseniureted ~ || Arsenwasserstoff m. || hydrogène m. arsénié. / mixture of ~ and oxygen || Knallgas n. || gaz m. tonnant *ou* oxyhydrique.

hydrogenation || Hydrierung f. || hydrogénation f.

hydrogen bottle || Wasserstofflasche f. || bouteille f. d'hydrogène. / formation of ~ || Wasserstoffentwicklung f. || formation f. d'hydrogène. / ~ gas || Wasserstoffgas n. || gaz m. hydrogène. / generation of ~ *see* production of ~.

hydrogenize, to || hydrieren; mit Wasserstoff m. anreichern *oder* sättigen || hydrogéner.

hydrogenizing process (Fat) || Härtungsverfahren n. || procédé m. d'hydrogénation.

hydrogen meter || Wasserstoffmesser m. || compteur m. de hydrogène. / ~ peroxide || Wasserstoffsuperoxyd n. || bioxyde m. d'hydrogène; eau f. oxygénée. / ~ phosphide || Phosphorwasserstoff m. | hydrogène m. phosphoré. / ~ producing plant || Wasserstoffgewinnungsanlage f. || installation f. pour la production du hydrogène. / production of ~ || Wasserstoffgaserzeugung f. || production f. de l'hydrogène. / electrolytic production of ~ || elektrolytische Herstellung f. von Wasserstoffgas || production f. électrolytique de l'hydrogène. / ~ soldering apparatus || Wasserstofflötapparat || chalumeau m. à hydrogène. / ~ sulphide || Schwefelwasserstoff m. || hydrogène m. sulfuré.

hydrographer (Mar) || Hydrograf m. || hydrographe m.

hydrographic map || Seekarte f. || carte f. marine.

hydrography ‖ Hydrografie f. ‖ hydrographie f.

hydrohematite ‖ Hydrohämatit m. ‖ hydrohématite f.

hydroiodic acid ‖ Jodwasserstoffsäure f. ‖ hydracide m. iodique.

hydrology ‖ Gewässerkunde f. ‖ hydrologie f.

hydrolysis ‖ Hydrolyse f. ‖ hydrolyse f.

hydrolytic ferment ‖ hydrolytisches Ferment n. ‖ ferment m. hydrolyte.

hydromagnesite ‖ Hydromagnesit m. ‖ hydrocarbonate m. de magnésie.

hydrometer ‖ Aräometer n.; Senkwage f. ‖ aréomètre m. / ~ with weight ‖ Gewichtsaräometer n. ‖ aréomètre m. à poids constant.

hydrometrical ‖ hydrometrisch ‖ hydrométrique. / ~ vane ‖ Strömungsmesser m. ‖ ailette f. hydrométrique.

hydrometry ‖ Wassermessung f. ‖ hydrométrie f.

hydro-oxygen light ‖ Drummond'sches Kalklicht n. ‖ lumière f. de Drummond.

hydropathic establishment ‖ Wasserheilanstalt f. ‖ établissement m. hydrothérapique.

hydrophite ‖ Eisengymnit m. ‖ hydrophite f.

hydrophone ‖ Unterwasserschallempfänger m. ‖ récepteur m. sous-marin du son.

hydroplane ‖ Wasserflugzeug n. ‖ hydravion m.

hydroquinone ‖ Hydrochinon n. ‖ hydroquinone f.

hydrorheostat ‖ Flüssigkeitsrheostat m. ‖ rhéostat m. à liquide.

hydrostatic ‖ hydrostatisch ‖ hydrostatique. / ~ level ‖ Senkwage f.; Aräometer n. ‖ aéromètre m.; pèse-liqueur m. / ~ pressure ‖ hydrostatischer Druck m. ‖ pression f. hydrostatique.

hydrostatics pl. ‖ Hydrostatik f. ‖ hydrostatique f.

hydrosulphite ‖ Hydrosulfit n. ‖ hydrosulfite m.

hydrosulphocyanic acid ‖ Rhodanwasserstoffsäure f.; Schwefelblausäure f.; Thiocyansäure f. ‖ acide m. hydrosulphocyanique ou sulfocyanhydrique.

hydrosulphuric acid ‖ Schwefelwasserstoff m.; Wasserstoffsulfid n. ‖ hydrogène m. sulfuré; acide m. hydrosulfurique; sulfure m. d'hydrogène.

hydrotherapeutic ‖ hydrotherapeutisch ‖ hydrothérapique. / ~ apparatus ‖ hydrotherapeutischer Apparat m. ‖ appareil m. d'hydrothérapie.

hydrous ‖ wässerig ‖ aqueux; aquifère. / ~ (Chem) ‖ wasserstoffhaltig ‖ hydrogéné. / ~ oxide of iron ‖ Rubinglimmer m. ‖ gœthite f. / ~ silicate of zinc ‖ Kieselgalmei m.; Kieselzinkerz n. ‖ hydrosilicate m. de zinc; zinc m. oxidé silicifère.

hydroxanthane ‖ Xanthanwasserstoff m. ‖ hydroxanthane m.

hydroxide (Chem) ‖ Hydroxyd n. ‖ hydrate m. / ~ of iron ‖ Eisenhydroxyd n. ‖ hydroxyde m. de fer.

hygienic ‖ hygienisch ‖ hygiénique. / ~ article ‖ hygienischer Artikel m. ‖ article m. hygiénique. / ~ gymnastics pl. ‖ Heilgymnastik f. ‖ gymnastique f. hygiénique. / apparatus for ~ gymnastics pl. ‖ heilgymnastischer Apparat m. ‖ appareil m. de gymnastique médicale. / industrial ~ plant ‖ gewerbehygienische Anlage f. ‖ installation f. hygiénique pour l'industrie. / from a ~ point ‖ in hygienischer Beziehung f. ‖ du point m. de vue hygiénique.

hygrometer ‖ Hygrometer n.; Feuchtigkeitsmesser m. ‖ hygromètre m. / hair ~ ‖ Haarhygrometer n. ‖ hygromètre m. à cheveux.

hygrometric ‖ hygrometrisch ‖ hygrométrique.

hygroscopic ‖ hygroskopisch ‖ hygroscopique.

hygroscopicity ‖ Hygroskopizität f. ‖ hygroscopicité f.

hyoscyamus ‖ Bilsenkraut n. ‖ jusquiame f.

hyperbola (Geom) ‖ Hyperbel f. ‖ hyperbole f. / equilateral ~ ‖ gleichseitige Hyperbel f. ‖ hyperbole f. équilatère.

hyperbole (Speech figure) ‖ Hyperbel f.; Übertreibung f. ‖ hyperbole f.

hyperbolic chart ‖ Tangenskarte f. ‖ abaque m. tangentiel. / ~ conoid ‖ Hyperboloid n. ‖ hyperboloïde m. / ~ line-angle ‖ Fortpflanzungsmaß n. ‖ affaiblissement m. complexe. / ~ wheel ‖ Hyperboloidenrad n. ‖ roue f. hyperbolique.

hyperboloid ‖ Hyperboloid n. ‖ hyperboloïde f. / ~ of revolution ‖ Rotationshyperboloid n. ‖ hyperboloïde m. de révolution.

hypermetropia ‖ Übersichtigkeit f. ‖ hypermétropie f.

hyperopic ‖ übersichtig ‖ hyperope.

hypersensible ‖ überempfindlich ‖ hypersensible.

hypersthene ‖ Hypersthen m.; Paulit m. ‖ hypersthène m.

hyphen ‖ Bindestrich m. ‖ trait m. d'union.

hypocaust ‖ Heizungsvorrichtung f. unter dem Fußboden ‖ hypocauste m.

hypochlorate ‖ unterchlorsaures Salz n. ‖ hypochlorate m.

hypochlorous acid ‖ unterchlorige Säure f. ‖ acide m. hypochloreux.

hypocycloid ‖ Hypozykloïde f. ‖ épicycloïde f. intérieure ou inférieure; hypocycloïde f.

hypodermic injection ‖ Einspritzung f. unter die Haut ‖ injection f. hypodermique. / ~ product ‖ Subkutanpräparat n. ‖ préparation f. souscutanée.

hypophosphite ‖ Hypophosphit n.; unterphosphorigsaures Salz n. ‖ hypophosphite m.

hyposulphite ‖ unterschwefligsaures Salz n. ‖ hyposulfite m.

hyposulphurous acid ‖ unterschweflige Säure f.; Thioschwefelsäure f. ‖ acide m. hyposulfureux.

hypotenuse ‖ Hypotenuse f. ‖ hypoténuse f.

hypothecary ‖ hypothekarisch ‖ hypothécaire.

hypothesis ‖ Hypothese f.; Annahme f. ‖ hypothèse f.

hypothetical ‖ hypothetisch ‖ hypothétique.

hypoxanthite ‖ Hypoxanthit m.; Terra di Siena f. ‖ hypoxanthite f.; terre f. de Sienne.

hyssop ‖ Ysop m. ‖ hysope f.; herbe d'hysope. / ~ leaves pl. ‖ Ysopkraut n. ‖ feuilles fpl. d'hysope.

hysteresis ‖ Hysteresis f. ‖ hystérésis f. / dielectric ~ ‖ Dielektrizitätshysteresis f. ‖ hystérésis f. diélectrique. / elastic ~ ‖ elastische Nachwirkung f. ‖ effet m. résiduel; résidu m. ou fatigue f. élastique.

hysteresis loop ‖ Hysteresisschleife f. ‖ boucle f. d'hystérésis. / ~ loss ‖ Hysteresisverlust m. ‖ perte f. par hystérésis. / loss of energy by ~ ‖ Energieverlust m. durch Hysteresis ‖ perte f. d'énergie par hystérésis. / phenomenon of ~ ‖ Nachwirkungserscheinung f. ‖ phénomène m. d'effet résiduel ou de fatigue.

I

ice ‖ Eis n. ‖ glace f. / artificial ~ ‖ künstliches Eis n.; Kunsteis n. ‖ glace f. artificielle. / production of artificial ~ ‖ Herstellung f. von Kunsteis ‖ fabrication f. de glace artificielle. / continental ~ ‖ Inlandeis n. ‖ glace f. continentale. / crystal ~ ‖ Kristalleis n.; Destillateis n. ‖ glace f. cristalline. / crystalline ~ see crystal ~. / ~ for domestic purposes ‖ Gebrauchseis n. ‖ glace f. de ménage. / inland~ see continental ~. / to manufacture ~ ‖ Eis n. herstellen ‖ produire de la glace. / the ~ melts ‖ das Eis taut auf; das Eis schmilzt ‖ la glace fond. / natural ~ ‖ Natureis n. ‖ glace f. naturelle. / opaque ~ ‖ Matteis n. ‖ glace f. opaque. / pack ~ ‖ Packeis n. ‖ banquise f. / plate ~ ‖ Platteneis n. ‖ glace f. en plaques. / transparent ~ ‖ Klareis n. ‖ glace f. transparente. / white ~ see opaque ~.

ice adze ‖ Eispickel m. ‖ erminette f. à glace; piolet m. / ~ age ‖ Eiszeit f. ‖ époque f. glaciaire. / ~ anchor ‖ Eisanker m. ‖ ancre f. pour la glace. / ~ axe see ~ adze. / ~ bag ‖ Eisbeutel m. ‖ sac m. ou vessie f. à glace.

iceberg ‖ Eisberg m. ‖ montagne m. de glace; glacier m.; iceberg m. / ~ service ‖ Eisnachrichtendienst m. ‖ service m. d'avertissement des glaces ou des icebergs; rapports mpl. sur la présence de glaciers.

ice block ‖ Eisblock m. ‖ bloc m. de glace. / ~ box ‖ Eiskasten m. ‖ glacière f.; caisse f. à glace. / ~ breaker ‖ Eisbrecher m. ‖ brise-glace m. / ~ calorimeter ‖ Eiskalorimeter n. ‖ calorimètre m. à glace. / ~ can ‖ Eiszelle f. ‖ moule m. ou mouleau m.; ou cellule f. à glace. / ~ cellar ‖ Eiskeller m. ‖ glacière f.; cave-glacière f.; soute f. à glaces. / ~ -cellar plant ‖ Eiskellereinrichtung f. ‖ installation f. de glacière. / ~ chest see ~ box. / ~ chute ‖ Eisrutsche f. ‖ table f. de démoulage de la glace; tréteau m. de décharge. / coat of ~ ‖ Eishülle f. ‖ couche f. de glace. / coating of ~ ‖ Eisbelag m. ‖ couche f. de glace.

ice-cold water ‖ eiskaltes Wasser n. ‖ eau f. glacée.

ice compressor ‖ Eiskompressor m.; Eisverdichter m. ‖ compresseur m. à glace.

ice-cooled ‖ eisgekühlt ‖ refroidi par de la glace.

ice and cooling installation ‖ Eis- und Kühlanlage f. ‖ installation f. glacière et frigorifique. / ~ **cream** ‖ Speiseeis n. ‖ crème f. glacée; glace f. / **equipment for production of ~ cream** ‖ Speiseeiserzeugungseinrichtung f. ‖ appareil m. à préparer la glace de confiserie. / ~ **cream factory** ‖ Rahmeisfabrik f. ‖ fabrique f. de crème glacée. / ~ **crushing machine** ‖ Eiszerkleinerungsmaschine f. ‖ machine f. à concasser la glace. / ~ **dispatch** ‖ Eisabgabe f. ‖ expédition f. de (la) glace. / **driving of ~** see floating of ~. / ~ **elevator** ‖ Eiselevator m. ‖ élévateur m. à glace. / ~ **fabrication** ‖ Eisfabrikation f. ‖ fabrication f. de glace. / ~ **factory** ‖ Eisfabrik f.; Eiswerk n. ‖ fabrique f. de glace. / ~ **fender** ‖ Eisfender m. ‖ pare-glace m. / ~ **float** ‖ Eisschwimmer m. ‖ nageur m. ou plongeur à glace. / **floating of ~** ‖ Eisgang m. ‖ débâcle f. / **floe** ‖ Eisscholle f. ‖ glaçon m.; glace f. flottante. / **formation of ~** ‖ Eisbildung f. ‖ formation f. de glace. / ~ **generator** ‖ Eisgenerator m. ‖ générateur m. à glace. / ~ **generator water** ‖ Eisgeneratorwasser n.; Salzsole f. ‖ eau f. de générateur à glace; saumure f. / ~ **goblet** ‖ Eisbecher m. ‖ gobelet m. en glace. / ~ **-guard** (Hydr arch) ‖ Eisbrecher m. ‖ brise-glace m. / ~ **harvest** ‖ Eisernte f. ‖ moisson f. de glace. / ~ **hoist** ‖ Eisaufzug m. ‖ monte-glace m. / ~ **-house** ‖ Eiskeller m.; Eishaus n. ‖ glacière f.; soute f. à glaces. / ~ **information service** ‖ Eisnachrichtendienst m. ‖ service m. d'avertissement des glaces ou des icebergs; rapports mpl. sur la présence de glaciers. / ~ **load** ‖ Eisbelastung f. ‖ charge f. de verglas. / **lump of ~** ‖ Eisklumpen m. ‖ glaçon m.

Iceland moss ‖ Isländisches Moos n. ‖ lichen m. d'Islande.

ice machine ‖ Eismaschine f. ‖ machine f. réfrigérante ou frigorifique ou à glace. / **ammonia ~** ‖ Ammoniakeismaschine f. ‖ machine f. à glace à ammoniaque. / **carbonic acid ~** ‖ Kohlensäureeismaschine f. ‖ machine f. à glace à acide carbonique. / **sulphurous acid ~** ‖ Schwefligsäureeismaschine f. ‖ machine f. à glace à acide sulfureux. / **vacuum ~** ‖ Vakuumeismaschine f. ‖ machine f. à froid à vide.

ice making ‖ Eiserzeugung f. ‖ fabrication f. de glace; glacerie f. / ~ **capacity** see ice production. / ~ **machine** ‖ Eismaschine f. ‖ machine f. à glace. / ~ **and refrigerating machine** ‖ Eiserzeugungs- und Kühlmaschine ‖ machine f. à glace et à froid.; machine f. de réfrigération. / ~ **plant** ‖ Eismaschinenanlage f. ‖ installation f. frigorifique. / ~ **tank** ‖ Eisgenerator m. ‖ générateur m. à glace.

ice manufacturing equipment ‖ Einrichtung f. für Eisfabriken ‖ installation f. de fabriques de glace. / **point of melting ~** ‖ Schmelzpunkt m.; Nullpunkt m. ‖ point m. de la glace fondante; point m. Zéro. / ~ **mould** ‖ Eiszelle f. ‖ mouleau m. à glace. / ~ **needle** ‖ Eisnadel f. ‖ aiguille f. de glace. / ~ **plant** ‖ Eiswerk n. ‖ usine f. à glace. / ~ **producer** ‖ Eisgenerator m. ‖ générateur m. de glace. / ~ **production** ‖ Eisgewinnung f. ‖ production f. de glace.

icer ‖ Fruchteiserzeuger m. ‖ pâtissier-glacier m.

ice and refrigerator plant ‖ Eis- und Kühlmaschine f. ‖ machine f. à glace et frigorifique. / ~ **reports** see ~ information service. / ~ **runway** see ~ chute. / ~ **safe** ‖ Eisschrank m.; Kühlschrank m. ‖ armoire f. glacière; glacière f. / ~ **saw** ‖ Eissäge f. ‖ scie f. à couper les glaces. / **sheet of ~** ‖ Eisscholle f. ‖ glaçon m.; nappe f. de glace. / ~ **shoot** see ~ chute. / ~ **shovel** ‖ Eisschaufel f. ‖ pelle f. à glace. / ~ **slide** see ~ chute. / ~ **spar** ‖ glasiger Feldspat m.; Eisspat m.; Sanidin m. ‖ feldspath m. vitreux; sanidine m. / ~ **store** see ~ cellar. / ~ **table** see ~ chute. / ~ **tank** see ~ box. / ~ **vinegar** ‖ Eisessig m. ‖ acide m. acétique glacé.

ice water ‖ Eiswasser n. ‖ eau f. glacée. / ~ **pump** ‖ Eiswasserpumpe f. ‖ pompe f. à eau glacée. / ~ **tank** ‖ Eiswasserkasten m.; Eiswasserreservoir n. ‖ réservoir m. à eau glacée.

ichnography ‖ Grundplan m.; Grundriß m. ‖ ichnographie f.; projection f. ichnographique.

ichthyocolla ‖ Fischleim m.; Hausenblase f. ‖ colle f. de poisson; ichthyocolle f.

ichthyolite ‖ Fischstein m. ‖ ichthyolithe f.

ichthyol soap ‖ Ichthyolseife f. ‖ savon m. à l'ichtyol.

icing-machine man ‖ Eiswerkarbeiter m. ‖ ouvrier m. à la machine à glace.

icosahedron ‖ Ikosaeder m., n.; Zwanzigflächner m. ‖ icosaèdre m.

idea ‖ Idee f.; Gedanke m. ‖ idée f.; notion f. / **interchange of ~s** ‖ Gedankenaustausch m. ‖ échange m. des opinions.

identical ‖ identisch ‖ identique.

identification light ‖ Fliegerlichtzeichen n. ‖ feu m. d'identification. / ~ **mark** ‖ Kennmarke f. ‖ plaque f. d'identité. / ~ **number of the required item** ‖ Bestellnummer f. des gewünschten Gegenstandes ‖ numéro m. de commande de l'article demandé.

identity ‖ Identität f. ‖ identité f. / **chromatic ~** ‖ Farbengleichheit f. ‖ identité f. chromatique. / **by proof of ~** ‖ gegen Legitimation f. ‖ contre ou sur légitimation f. / **to prove one's ~** ‖ sich ausweisen ‖ établir son identité f. / ~ **papers** pl. ‖ Legitimationspapiere npl. ‖ pièces fpl. d'identité. / **proof of ~** ‖ Ausweis m.; Legitimationspapier n. ‖ légitimation f.; pièce f. d'identité.

idiochromatic ‖ idiochromatisch ‖ idiochromatique.

idioelectric ‖ selbstelektrisch; idioelektrisch ‖ idio-électrique.

idle, to run ~ ‖ leerlaufen ‖ marcher à vide.

idle current ‖ wattloser Strom m. ‖ courant m. déwatté. / ~ **motion** ‖ Leergang m.; Leerlauf m. ‖ mouvement m. perdu; marchant m. ou marche f. à vide. / ~ **movement** ‖ toter Weg m. ‖ marche f. à vide. / ~ **period** (Electr) ‖ Leerlaufperiode f. ‖ période f. en circuit ouvert. / ~ **power** (Electr) ‖ Blindleistung f.; wattlose Leistung f. ‖ puissance f. réactive ou apparente ou déwattée.

idle running ‖ Leerlauf m. ‖ marche f. à vide. / ~ **current** ‖ Leerlaufstrom m. ‖ courant m. à vide. / ~ **pulley** ‖ leerlaufendes Rad n.; Losscheibe f. ‖ roue f. à vide; poulie f. folle. / ~ **wheel** ‖ leerlaufendes Rad n. ‖ roue f. marchant ou tournant à vide.

idle travel of ram ‖ Stößelüberlauf m. ‖ mouvement m. perdu du coulisseau.

igneous ‖ feurig flüssig ‖ igné.

ignite, to ‖ entzünden; Feuer n. fangen ‖ mettre en feu m.; prendre feu; s'enflammer; s'embraser. / ~ **(Mot)** ‖ zünden ‖ allumer. / ~ **irregularly** ‖ unregelmäßig zünden ‖ allumer ou amorcer irrégulièrement.

ignited ‖ glühend ‖ porté au rouge.

ignition ‖ Entzündung f.; Entflammen n. ‖ ignition f.; inflammation f. / ~ **(Mot)** ‖ Zündung f. ‖ allumage m. / **to adjust the ~** ‖ die Zündung einstellen ‖ régler l'allumage m. / **advanced ~** ‖ Vorzündung f.; Frühzündung f. ‖ avance f. à l'allumage; allumage m. avancé. / **arc ~** ‖ Lichtbogenzündung f. ‖ allumage m. par arc. / **automatic ~** ‖ automatische Zündeinrichtung f. ‖ allumage m. automatique. / **automobile ~** ‖ Automobilzündung f. ‖ allumage m. d'automobiles. / **battery ~** ‖ Batteriezündung f. ‖ allumage m. par accumulateur. / ~ **by compression** ‖ Kompressionszündung f. ‖ allumage m. par compression. / **delayed ~** ‖ Nachzündung f. ‖ allumage m. retardé. / **double-~** ‖ Doppelzündung f. ‖ allumage m. double. / **dual ~** see double ~. / **the ~ fails** ‖ die Zündungen fpl. bleiben aus ‖ il y a des ratés d'allumage. / ~ **of gas** ‖ Gasentzündung f. ‖ inflammation f. des gaz. / **hot-bulb ~** ‖ Glühkopfzündung f.; Glührohrzündung f. ‖ allumage m. à tube incandescent. / **hot-tube ~** see hot-bulb ~. / **the ~ intermits** ‖ die Zündung setzt aus ‖ l'allumage m. rate. / **the ~ is intermittant** ‖ die Zündung f. setzt zeitweilig aus ‖ l'allumage m. rate de temps en temps. / **low tension ~** ‖ Niederspannungszündung f. ‖ allumage m. à basse tension. / **magnetic plug ~** ‖ Magnetkerzenzündung f. ‖ allumage m. par bougie magnétique. / **magneto ~** ‖ Magnetzündung f. ‖ allumage m. par magnéto. / **make and break ~** ‖ Abreißzündung f. ‖ allumage m. à rupture. / **multipoint ~** ‖ Mehrfunkenzündung f. ‖ allumage m. à plusieurs étincelles. / **pre-~** ‖ Frühzündung f. ‖ allumage m. avancé ou prématuré. / **retarded ~** ‖ Nachzündung f.; Spätzündung f. ‖ retard m. à l'allumage; allumage m. retardé. / **self ~** ‖ Selbstzündung f. ‖ allumage m. spontané. / **single-spark ~** ‖ Einfunkenzündung f. ‖ allumage m. à simple étincelle. / ~ **by sparking plug** ‖ Kerzenzündung f. ‖ allumage m. par bougie. / **spontaneous ~** ‖ Selbstzündung f. ‖ allumage m. spontané; ignition f. spontanée. / **tube ~** see hot-bulb ~.

ignition, advance of the ~ ‖ Vorzündung f. ‖ avance f. à l'allumage. / ~ **apparatus** ‖ Zündvorrichtung f. ‖ appareil m. d'allumage. / ~ **apparatus for explosion motors** ‖ Zündapparat m. für Explosionsmotoren ‖ appareil m. d'allumage pour moteurs à explosion. / ~ **arrangement** ‖ Zündvorrichtung f. ‖ appareil m. d'allumage; allumeur m. / ~ **battery** ‖ Zündbatterie f. ‖ batterie f. d'allumage. / ~ **bolt** (Mot) ‖ Zündbolzen m. ‖ allumeur m. / ~ **cable** ‖ Zündkabel n. ‖ câble m. d'allumage. / ~ **cam** ‖ Zündnocken m. ‖ came f. d'allumage. / ~ **cartridge** ‖ Zündpatrone f. ‖ cartouche f. d'allumage. / ~ **chamber** ‖ Zündkammer f. ‖ chambre f. d'allumage ou de combustion. / ~ **coil** ‖ Zündspule f. ‖ bobine f. d'allumage.

/ ~ control-lever ‖ Zündungshebel m. ‖ manette f. d'allumage. / ~ current (Electr) ‖ Zündstrom m. ‖ courant m. d'allumage. / ~ device ‖ Zündvorrichtung f.; Zünder m. ‖ dispositif m. d'allumage; allumeur m. / auxiliary ~ device ‖ Hilfszündvorrichtung f. ‖ dispositif m. d'allumage auxiliaire. / ~ distributor ‖ Zündungsverteiler m.; Zündungsregler m. ‖ distributeur m. d'allumage. / ~ heat see ~ temperature. / ~ lever ‖ Zündhebel m. ‖ manette f. d'allumage. / ~ lighting generator ‖ Lichtmagnetzünder m. ‖ génératrice f. d'éclairage et d'allumage. / ~ loss ‖ Verlust m. bei der Zündung; Glühverlust m. ‖ perte f. à l'allumage. / ~ material for fire lighters ‖ Feuerzeuglunte f. ‖ mèche f. pour briquets. / ~ oil pipe ‖ Zündölleitung f. ‖ conduite f. d'huile d'allumage. / ~ piping ‖ Zündleitung f. ‖ conduite f. d'allumage. / measuring instrument for testing ~ pipings ‖ Meßinstrument n. zur Prüfung von Zündleitungen ‖ instrument m. de mesure à essayer les conduites d'allumage. / ~ plug ‖ Zündstöpsel m. ‖ bouchon m. allumeur. / point of ~ ‖ Entzündungspunkt m. ‖ point m. d'inflammation. / ~ point tester ‖ Zündpunktprüfer m. ‖ appareil m. pour le contrôle du point d'allumage. / ~ residue ‖ Glührückstand m. ‖ résidu m. de calcination. / ~ spark ‖ Zündfunke m. ‖ étincelle f. d'allumage. / ~ switch ‖ Magnetschalter m. ‖ commutateur m. de la magnéto. / ~ system ‖ Zündvorrichtung f. ‖ dispositif m. d'allumage. / ~ temperature ‖ Entzündungstemperatur f. ‖ température f. d'inflammabilité ou d'ignition. / ~ tension ‖ Zündspannung f. ‖ tension f. d'allumage. / ~ testing device ‖ Zündpunktprüfer m. ‖ essayeur m. du point d'inflammation. / ~ transformer ‖ Zündtransformator m. ‖ transformateur m. d'allumage. / ~ trouble ‖ Zündungsstörung f. ‖ dérangement m. dans l'allumage. / ~ tube ‖ Glührohr n. ‖ tuyau m. incandescent. / ~ turbine ‖ Gasturbine f. ‖ turbine f. d'ignition. / ~ wire ‖ Zündkerzenkabel n. ‖ câble m. d'allumage.

illegal ‖ rechtswidrig ‖ illégal.

illicit trading ‖ Schleichhandel m.; Schmugelei f. ‖ commerce m. de contrebande; contrebande f.

illipe oil ‖ Illipeöl n. ‖ huile f. d'illipé.

illness ‖ Krankheit f.; Erkrankung f. ‖ maladie f.

ill success ‖ Mißerfolg m. ‖ échec m.

illuminate, to ‖ beleuchten ‖ illuminer. / ~ (Opt) ‖ sich erhellen ‖ s'éclairer. / ~ (Paint) ‖ ausmalen; kolorieren ‖ enluminer. / ~ a cask ‖ ein Faß n. ausleuchten ‖ vister l'intérieur m. d'un fût à la lumière. / ~ indirectly ‖ mittelbar beleuchten ‖ éclairer indirectement. / ~ by vertical light ‖ durch Zenitlicht n. beleuchten ‖ éclairer par la lumière zénithale.

illuminated, to be ~ ‖ beleuchtet sein ‖ être illuminé ou éclairé.

illuminated advertising device ‖ Lichtreklamevorrichtung f. ‖ appareil m. lumineux de réclame. / ~ compass ‖ Leuchtkompaß m. ‖ compas m. lumineux. / ~ graticule ‖ beleuchtetes Fadenkreuz n. ‖ réticule m. éclairé. / ~ wind indicator ‖ beleuchteter Windanzeiger m. ‖ indicateur m. de vent éclairé.

illuminating ‖ leuchtend ‖ éclairant. / ~ apparatus ‖ Beleuchtungsapparat m. ‖ appareil m. d'éclairage ou d'illumination. / simplified ~ apparatus ‖ vereinfachter Beleuchtungsapparat m. ‖ appareil m. d'éclairage simplifié. / ~ arrangement on telescope ‖ Beleuchtungseinrichtung f. am Fernrohr ‖ appareil m. d'éclairage de la lunette. / ~ attachment ‖ Beleuchtungsvorrichtung f. ‖ appareil m. d'éclairage. / ~ body ‖ Beleuchtungskörper m. ‖ corps m. éclairant; appareil m. d'éclairage / ~ cone ‖ Beleuchtungskegel m. ‖ cône m. d'éclairage. / ~ device ‖ Beleuchtungseinrichtung f. ‖ appareil m. d'éclairage. / ~ effect ‖ Beleuchtungswirkung f. ‖ effet m. d'illumination. / ~ gas ‖ Leuchtgas n. ‖ gaz m. d'éclairage ou éclairant. / ~ lens ‖ Beleuchtungslinse f. ‖ lentille f. d'éclairage. / ~ mirror ‖ Beleuchtungsspiegel m. ‖ miroir m. d'éclairage. / ~ oil ‖ Leuchtöl n. ‖ huile f. d'éclairage. / ~ pencil of rays ‖ Beleuchtungsbüschel n.; Strahlenbüschel n. ‖ faisceau m. de rayons. / ~ power ‖ Leuchtvermögen n. ‖ pouvoir m. éclairant.

illumination ‖ Beleuchtung f. ‖ éclairage m. / ~ (Phot) ‖ Belichtung f. ‖ pose f.; éclairement m. / artificial ~ ‖ künstliche Beleuchtung f. ‖ éclairage m. artificiel. / bright field ~ ‖ Hellfeldbeleuchtung f. ‖ éclairage m. à fond clair. / ceiling ~ ‖ Deckenbeleuchtung f. ‖ éclairage m. de plafond; plafonnier m. / darkground ~ ‖ Dunkelfeldbeleuchtung f. ‖ éclairage m. à fond noir. / direct ~ ‖ direkte Beleuchtung f. ‖ éclairage m. central. / ~ of the entrance to the theatre ‖ Eingangsbeleuchtung f. des Theaters ‖ éclairage m. de l'entrée du théâtre. / ~ of fiducial lines (Opt) ‖ Fadenbeleuchtung f. ‖ éclairage m. des fils. / ~ by incident light ‖ Beleuchtung f. im auffallenden Licht ‖ éclairage m. par réflexion. / lateral ~ ‖ seitliche Beleuchtung f. ‖ éclairage m. latéral. / ~ number ‖ Nummernbeleuchtung f. ‖ éclairage m. de numéro. / oblique ~ ‖ schräg einfallende oder schiefe Beleuchtung f. ‖ éclairage m. oblique. / uniform ~ ‖ gleichmäßige Beleuchtung f. ‖ éclairage m. régulier.

illumination advertisements pl. ‖ Lichtreklame f. ‖ réclame f. lumineuse. / ~ article ‖ Illuminationsartikel m. ‖ article m. pour illumination / attachment for ~ of field of view ‖ Gesichtsfeldbeleuchtungseinrichtung f. ‖ dispositif m. pour l'éclairage du champ visuel. / ~ coupling ‖ Beleuchtungskupplung f. ‖ raccordement m. pour circuits d'éclairage. / device for the ~ of the mouth ‖ Mundbeleuchtungsapparat m. ‖ appareil m. pour éclairer la bouche.

illuminator (Opt) ‖ Beleuchtungsapparat m. ‖ appareil m. d'éclairage. / ~ (Person) ‖ Beleuchter m.; Illuminierer m. ‖ enlumineur m. / spectroscopic ~ ‖ Spektralbeleuchtungsapparat m. ‖ appareil m. pour l'éclairage monochromatique. / vertical ~ ‖ Vertikalilluminator m. ‖ illuminateur m. vertical.

illusory ‖ täuschend; trügerisch ‖ trompeur; illusoire.

illustrate, to ‖ illustrieren ‖ illustrer.

illustration ‖ Abbildung f.; Bebilderung f.; Illustration f. ‖ illustration f.; figure f. / ~ (Explanation) ‖ Erklärung f. ‖ explication f. / as the accompanying ~ shows ‖

wie nebenstehende Abbildung f. zeigt ‖ comme on le voit sur la figure ci-contre. / the opposite ~ ‖ nebenstehende Abbildung f. ‖ la figure ci-contre. / ~ printing press ‖ Illustrationsdruckpresse f.; Bilddruckpresse f. ‖ presse f. pour impression d'illustrations.

ilmenite ‖ Titaneisenerz n. ‖ ilménite f.

image ‖ Bild n.; Abbildung f. ‖ image f. / apparent ~ ‖ scheinbares Bild n. ‖ image f. virtuelle. / stopping down of the ~ of the crystal ‖ Abblendung f. des Kristallbildes ‖ diaphragme m. agissant sur l'image du cristal. / double ~ ‖ Doppelbild n. ‖ image f. double. / erect ~ ‖ aufrechtes Bild n. ‖ image f. droite. / ~ of the fundus of the eye ‖ Augenhintergrundbild n. ‖ image f. du fond de l'œil. / inverted ~ ‖ umgekehrtes Bild n. ‖ image f. renversée. / reduced ~ ‖ verkleinertes Bild n. ‖ image f. réduite. / ~ of the retina ‖ Netzhautbild n. ‖ image f. de la rétine. / ~ of the spark ‖ Funkenbild n. ‖ image f. de l'étincelle. / ~ of the stop ‖ Blendenbild n. ‖ image f. du diaphragme.

images pl. reflected together ‖ zusammengespiegelte Bilder npl. ‖ images fpl. concentrées par réflexion.

image carver ‖ Bildschnitzer m. ‖ sculpteur-imagier m. / ~ carving ‖ Bildschnitzerei f. ‖ sculpture f. en bois. / defect of the ~ ‖ Abbildungsfehler m. ‖ imperfection f. des images. / definition of the ~ ‖ Bildschärfe f. ‖ netteté f. de l'image. / ~ distance ‖ Bildweite f. ‖ distance f. de l'image.

image erecting ‖ Bildaufrichtung f. ‖ redressement m. de l'image. / ~ corneal microscope ‖ bildaufrichtendes Hornhautmikroskop n. ‖ microscope m. cornéen redresseur. / ~ microscope ‖ bildaufrichtendes Mikroskop n. ‖ microscope m. redresseur. / ~ tube ‖ bildaufrichtender Tubus m. ‖ tube m. redresseur.

image field ‖ Bildfeld n. ‖ champ m. de l'image. / ~ forming system ‖ abbildendes System n. ‖ système m. optique qui forme les images. / ~ reversion ‖ Bildaufrichtung f. ‖ redressement m. de l'image. / size of the ~ ‖ Bildgröße f. ‖ grandeur f. de l'image.

imaginary ‖ imaginär ‖ imaginaire. / ~ component of electric values ‖ Blindwert m. elektrischer Größen ‖ composante f. déwattée ou réactive des valeurs électriques.

imbed, to ‖ einlassen; einbetten ‖ enchâsser; noyer. / ~ a cable ‖ ein Kabel n. einbetten ‖ noyer un câble.

imbedded ‖ eingebettet ‖ noyé.

imbibe, to ‖ eintränken ‖ imbiber; plonger.

imbricated work ‖ Dachziegelverband m. ‖ imbrication f.

imbue, to ‖ leimtränken ‖ imbiber ou imprégner de colle forte.

imitate, to ‖ nachahmen; nachmachen; imitieren ‖ imiter.

imitated ‖ künstlich; nachgemacht ‖ artificiel; factice.

imitation ‖ Imitation f.; Nachahmung f. ‖ imitation f. / ~ bronze ‖ Kunstguß m. ‖ bronze m. d'art-imitation. / ~ diamond ‖ unechter Diamant m. ‖ faux diamant m. / ~ horn ware ‖ Kunsthornware f. ‖ article m. en corne artificielle. / ~ jewellery ‖ unechter Schmuck m. ‖ bijouterie f. fausse. / ~ leather ‖ Kunstleder n. ‖ simili-cuir m.; cuir m. artificiel.

/ ~ parchment ‖ Pergamin n. ‖ papier m. parcheminé. / ~ silk ‖ Kunstseide f. ‖ soie f. artificielle.

immalleable ‖ nicht hämmerbar ‖ non-malléable.

immediate payment ‖ sofortige Zahlung f. ‖ payement m. immédiat. / ~ reply ‖ umgehende Antwort f. ‖ réponse f. par retour du courrier.

immerged see also immersed ‖ versenkt liegend ‖ noyé.

immerse, to ‖ eintauchen ‖ immerger.

immersed body ‖ eingetauchter Körper m. ‖ corps m. immergé. / ~ compass ‖ Flüssigkeitskompaß m. ‖ compas m. à liquide.

immersion battery ‖ Tauchbatterie f. ‖ pile f. à immersion. / ~ boiling device ‖ Tauchsieder m. ‖ thermo-plongeur m. / ~ depth ‖ Eintauchtiefe f. ‖ profondeur f. d'immersion ou de submersion. / ~ fluid ‖ Tauchflüssigkeit f. ‖ Immersionsflüssigkeit f. ‖ liquide m. d'immersion. / ~ heater ‖ Tauchsieder m. ‖ thermo-plongeur m. / ~ objective ‖ Immersionsobjektiv n. ‖ objectif m. à immersion. / ~ pipe ‖ Eintauchröhre f. ‖ tuyau m. d'immersion; tube m. plongeur. / ~ refractometer ‖ Eintauchrefraktometer n. ‖ réfractomètre m. à immersion.

immiscible ‖ unmischbar ‖ non-miscible.

immobility ‖ Unbeweglichkeit f. ‖ immobilité f.

immovable ‖ unbeweglich; unverrückbar ‖ immobile; fixe.

immune ‖ immun ‖ immun.

immunity ‖ Immunität f. ‖ immunité f.

imp (Build) ‖ Gerüststange f. ‖ boulin m.; perche f.

impact ‖ Stoß m.; Zusammenstoß m. ‖ choc m. / direct ~ ‖ gerader Stoß m. ‖ choc m. direct. / oblique ~ ‖ schräger oder schiefer Stoß m. ‖ choc m. oblique. / ~ of the projectile ‖ Geschoßaufschlag m. ‖ chute f. du projectile.

impact bending test ‖ Schlagbiegeversuch m. ‖ essai m. au choc et à la flexion. / ~ folding test ‖ Schlagknickversuch m. ‖ essai m. au choc et au pliage. / ~ ionization (Radio) ‖ Stoßionisation f. ‖ ionisation f. par impact.

impact resistance ‖ Kerbschlagwiderstand m.; Kerbzähigkeit f. ‖ résistance f. au choc sur barreaux entaillés. / the ~ decreases as tensile strength increases ‖ die Kerbzähigkeit f. nimmt mit steigender Festigkeit ab ‖ la ténacité en barre aux entaillés diminue alors que la résistance à la rupture augmente. / ~ for each square centimeter section ‖ Kerbschlagwiderstand m. je Quadratzentimeter Querschnitt ‖ résistance f. au choc sur barreaux entaillés par centimètre carré de la section d'éprouvette.

impact speed of tup ‖ Auftreffgeschwindigkeit f. des Hammerbären ‖ vitesse f. du choc ou de frappe du mouton. / ~ surface ‖ Schlagfläche f. ‖ surface f. de frappe / ~ tensile test ‖ Schlagzugversuch m. ‖ essai m. au choc et à la traction.

impact test ‖ Schlagprobe f. ‖ essai m. de choc. / ~ on notched bars ‖ Kerbschlagprobe f. ‖ essai m. au choc sur barreaux entaillés.

impact testing machine of the pendulum type ‖ Pendelhammer m. ‖ marteau m. de pendule; mouton m. pendule.

impartial ‖ sachlich ‖ objectif.

impaste, to ~ (Paint) ‖ die Farbe dick auftragen ‖ empâter.

impedance ‖ Scheinwiderstand m.; Impedanz f. ‖ impédance f. / characteristic ~ ‖ Wellenwiderstand m. ‖ impédance f. caractéristique. / high ~ (Electr) ‖ große Drosselung f. ‖ haute ou forte impédance f.

impedance bond ‖ Drosselstoß m. ‖ connexion f. inductive; bond m. d'inductance. / ~ device ‖ Hemmvorrichtung f. ‖ dispositif m. de réactance. / ~ measuring bridge ‖ Scheinwiderstandsmeßbrücke f. ‖ pont m. Wheatstone pour la mesure d'impédance. / ~ relay ‖ Impedanzrelais n. ‖ relais m. à impédance.

impediment on the line (Railw) ‖ Hindernis n. auf freier Strecke ‖ obstacle m. sur la route.

impeller ‖ Windflügel m. ‖ ailette f.; aube f. / ~ of pump ‖ Pumpenflügelrad n. ‖ roue f. à ailettes.

impeller drive ‖ Windantrieb m. ‖ commande f. par moulinet à vent.

impenetrable ‖ undurchdringlich ‖ impénétrable.

imperfect ‖ mangelhaft ‖ défectueux; incomplet; insuffisant; imparfait. / very ~ ‖ höchst mangelhaft ‖ très défectueux.

imperfection ‖ Unvollkommenheit f. ‖ imperfection f.

Imperial barley ‖ Imperialgerste f. ‖ orge f. impériale.

imperishable ‖ unzerstörbar ‖ indestructible.

impermeable ‖ undurchlässig ‖ imperméable. / to make ~ ‖ dicht machen ‖ étancher; imperméabiliser.

impervious ‖ undurchdringlich ‖ imperméable. / ~ to light ‖ undurchdringlich für Licht; undurchsichtig ‖ impénétrable à la lumière; opaque. / ~ layer ‖ wasserundurchlässige Schicht ‖ couche f. imperméable. / ~ masonry ‖ wasserdichtes Mauerwerk n. ‖ maçonnerie f. hydraulique.

imperviousness of the ground ‖ Undurchlässigkeit f. des Bodens ‖ imperméabilité f. du sol.

impetus ‖ Antrieb m.; Stoß m.; Geschwindigkeitshöhe f. ‖ impulsion f.; choc m.; hauteur f. de la vitesse.

implement ‖ Werkzeug n. ‖ outil m. / ~ for superstructure (Railw) ‖ Oberbaugerät n. ‖ ustensile m. de superstructure.

implements pl. ‖ Gerät n.; Handwerkzeug n. ‖ ustensiles mpl.; outils mpl. / ~ for dismantling a nozzle ‖ Düsenausziehvorrichtung f. ‖ dispositif m. d'enlèvement du gicleur. / miner's ~ ‖ Gezähe n. ‖ outils mpl. de mineurs.

imponderous ‖ unwägbar ‖ impondérable.

import, to ‖ einführen ‖ importer.

import see importation.

importance ‖ Tragweite f. ‖ portée f.

importation ‖ Einfuhr f. ‖ importation f. / to prohibit the ~ ‖ die Einfuhr verbieten ‖ prohiber ou défendre l'importation f.

import duty ‖ Einfuhrzoll m. ‖ droits mpl. d'importation ou d'entrée.

imported coal ‖ eingeführte Kohle f. ‖ charbon m. d'importation.

importer ‖ Einführender m.; Einführer m.; Importör m. ‖ importateur m.

importing business ‖ Einfuhrgeschäft n. ‖ affaire f. d'importation. / ~ firm ‖ Importhaus n. ‖ maison f. d'importation.

import merchant ‖ Einfuhrhändler m. ‖ négociant m. importateur. / ~ permit ‖ Einfuhrerlaubnis f. ‖ permis m. d'importation. / ~ premium ‖ Einfuhrprämie f. ‖ prime f. d'importation. / ~ regulations pl. ‖ Einfuhrvorschriften fpl. ‖ règlements mpl. sur l'importation.

imports pl. ‖ Einfuhrwaren fpl. ‖ articles mpl. d'importation. / list of ~ ‖ Einfuhrliste f. ‖ liste f. des importations ou des objets à importer.

import trade ‖ Einfuhrhandel m. ‖ commerce m. d'importation.

impose, to (Print) ‖ ausschießen ‖ imposer. / ~ the pages ‖ die Seiten fpl. umbrechen ‖ remanier les pages fpl.

imposing stone (Print) ‖ Schließstein m. ‖ marbre m. de serrage.

imposition (Print) ‖ Formeinrichtung f. ‖ imposition f.

impossible ‖ unausführbar; unmöglich ‖ inexécutable; impossible (d'exécuter).

impost (Tax) ‖ Abgabe f. ‖ impôt m.; taxe f. / ~ (Arch) ‖ Bogenkämpfer m.; Kämpferschicht f. ‖ imposte f.; coussinet m. / bent ~ ‖ gekröpfter Kämpfer m. ‖ imposte f. recoupée. / curb ~ ‖ um einen Pfeiler herumgeführter Kämpfer m. ‖ imposte f. cintrée. / discontinuous ~ ‖ unterbrochener Kämpfer m. ‖ imposte f. coupée. / flat ~ ‖ eingezogener Kämpfer m. ‖ imposte f. mutilée. / mitred ~ see bent ~.

impost moulding ‖ Kämpfergesims n. ‖ imposte f. ornée.

impoverished (Metal) ‖ metallarm ‖ trop peu de métal; à très faible teneur en métal.

imp-pole ‖ Rüststamm m. ‖ baliveau m.; échasse f. d'échafaud.

impracticable see impossible.

imprecision of standard ‖ Ungenauigkeit f. der Eichung oder des Feingehalts ‖ imprécision f. d'étalonnage ou de titre.

impregnate, to ‖ imprägnieren; tränken ‖ imprégner. / ~ with creosote ‖ mit Kreosot n. tränken ‖ imprégner avec de la créosote. / ~ with soluble glass ‖ mit Wasserglas tränken; verkieseln ‖ imprégner de silicate de soude.

impregnated ‖ imprägniert ‖ imprégné. / ~ flower ‖ imprägnierte Blume f. ‖ fleur f. imprégnée.

impregnating of cables ‖ Kabeltränkung f. ‖ imprégnation f. des câbles. / ~ and fulling of the skins with fat ‖ Sämischgerben n.; Sämischgerberei f. ‖ chamoisage m.

impregnating apparatus ‖ Imprägnierapparat m. ‖ appareil m. à imprégner. / ~ boiler ‖ Imprägnierkessel m. ‖ chaudron m. d'imprégnation. / ~ compound ‖ Tränkmasse f. ‖ liquide m. d'imprégnation. / ~ compound for cables ‖ Tränkungsmasse f. für Kabel ‖ matériel m. d'imprégnation pour câbles. / ~ machine ‖ Imprägniermaschine f. ‖ machine f. à imprégner. / ~ oil ‖ Imprägnieröl n. ‖ huile f. à imprégner. / ~ pan ‖ Imprägnierpfanne f.; Imprägnierkessel m. ‖ cuve f. d'imprégnation. / ~ preparation ‖ Imprägniermittel n. ‖ préparation f. d'imprégnation. / ~ pressure cylinder for wood ‖ Holztränkkessel m. ‖ autoclave m. pour imprégner le bois. / ~ substance for wooden poles ‖ Tränkungsmittel n. für Holzstangen ‖ antiseptique m. pour l'imprégnation de poteaux en bois. / ~ tank ‖ Tränkungskessel m. ‖ chaudière f. d'imprégnation.

impregnation ‖ Durchtränkung f. ‖ imprégnation f. / ~ with chloride of zinc ‖ Imprägnierung f. mit Zinkchlorid ‖ imprégnation f. au chlorure de zinc. / ~ of sleepers (Railw) Tränkung f. der Schwellen; ‖ Schwellenimprägnierung f. ‖ injection f. de traverses. / ~ with sublimate ‖ Tränkung f. mit Sublimat ‖ imprégnation f. avec du sublimé. / ~ with sulphate of copper ‖ Imprägnierung f. *oder* Durchtränkung f. mit Kupfervitriol ‖ imprégnation f. au sulfate de cuivre. / ~ of telegraph poles ‖ Imprägnierung f. von Telegrafenstangen ‖ injection f. de poteaux télégraphiques. / ~ of wood with creosote ‖ Kreosotierung f. des Holzes; Tränkung f. des Holzes mit Kreosot ‖ injection f. des bois à la créosote.

impregnation, carbonic acid ~ apparatus ‖ Kohlensäureimprägnierapparat m. ‖ appareil m. pour l'imprégnation à acide carbonique. / method of ~ ‖ Tränkungsverfahren n. ‖ procédé m. d'imprégnation. / plant for ~ of timber ‖ Holzimprägnierungsanlage f. ‖ installation f. à imprégner le bois.

impress, to (Metal) ‖ einprägen; abdrücken ‖ imprimer; empreindre. / ~ (Print) ‖ abdrucken ‖ imprimer. / ~ a voltage ‖ eine Spannung f. aufdrücken ‖ imprimer une tension. / ~ on wax ‖ in Wachs n. abdrücken ‖ faire une empreinte dans la cire.

impressing on paper ‖ Druck m. auf Papier ‖ impression f. sur papier.

impression ‖ Eindruck m. ‖ impression f. / ~ (Print) ‖ Drucklegung f.; Abdruck m. ‖ impression f.; copie f. / coloured ~ ‖ Farbabdruck m. ‖ impression f. colorée. / ~ on the edge of a coin ‖ Rand m. einer Münze ‖ cordon m. d'une piéce de monnaie. / ~ of letters ‖ Buchstabenabdruck m. ‖ impression f. des caractères. / mat ~ ‖ Mattdruck m. ‖ impression f. mate. / new ~ ‖ Neudruck m. ‖ réimpression f. / rare and highly valued ~ ‖ seltener und hochwertiger Druck m. ‖ impression f. rare et de grande valeur. / soft ~ ‖ weicher Druck m. ‖ impression f. molle.

impression cylinder ‖ Druckzylinder m. ‖ cylindre m. presseur *ou* d'impression. / ~ die ‖ Petschaft n. ‖ cachet m. / ~ roller (Tel) ‖ Druckwalze f. ‖ cylindre m. imprimeur.

imprimatur ‖ Imprimatur n.; Druckerlaubnis f.; Druckreiferklärung f. ‖ patente f. d'imprimerie; brevet m.

imprint, to ‖ abdrucken; einprägen ‖ imprimer; empreindre.

improve, to ‖ verbessern ‖ améliorer.

improved model ‖ verbessertes Modell n. ‖ modèle m. amélioré.

improvement ‖ Verbesserung f.; Vervollkommnung f. ‖ amélioration f. / ~ of river ‖ Flußkorrektion f.; Flußregulierung f. ‖ régularisation f. de rivière. / ~ of the soil ‖ Bodenverbesserung f.; Melioration f. ‖ amélioration f. du sol.

improvement cutting ‖ Durchhieb m. ‖ coupe f. d'amélioration.

improvised compensation of compass ‖ behelfsmäßige Kompaßausgleichung f. ‖ compensation f. par des moyens de fortune du compas.

impulse ‖ Anregung f. ‖ impulsion f.; idée f.; proposition f. / ~ (Electr) ‖ Impuls m.; Schaltstoß m. ‖ impulsion f. / change of ~ ‖ Impulsveränderung f. ‖ variation f. de la quantité de mouvement.

impulse frequency tele-metering ‖ Impulsfrequenzfernmessung f. ‖ télémesure f. à fréquence d'impulsions. / ~ repeater (Aut tel) ‖ Stromstoßübertrager m. ‖ translateur m. d'impulsions. / ~ sending device (Aut tel) ‖ Stromstoßsendevorrichtung f. ‖ dispositif m. d'émission d'impulsion de courant. / ~ telegraphy ‖ Impulstelegrafie f. ‖ télégraphie f. par des impulsions de courant. / ~ turbine ‖ Aktionsturbine f. ‖ turbine f. à action.

impulsion ‖ Antrieb m. ‖ impulsion f.; choc m.; force f. impulsive.

impure ‖ unrein ‖ impure. / the steel was made ~ by small particles of ore, coal or slag ‖ der Stahl wurde durch kleine Teilchen von Erz, Kohle oder Schlacke verunreinigt ‖ l'acier m. était souillé de particules de minerai, de charbon ou de scorie.

impure gas ‖ unreines Gas n. ‖ gaz m. impur.

impurity ‖ Verunreinigung f.; Beimengung f. ‖ impureté f.; souillure f. / the steel is of a certain ~ ‖ der Stahl zeigt noch eine gewisse Unreinheit ‖ l'acier m. montre une certaine impureté.

impurities pl., free from mechanical ~ ‖ frei von mechanischen Verunreinigungen fpl. ‖ exempt d'impuretés fpl. mécaniques. / to keep water free from ~ ‖ die Gewässer npl. rein halten ‖ tenir les eaux fpl. propres.

imputrescible ‖ unverweslich ‖ imputrescible.

inaccessible ‖ unzugänglich ‖ inaccessible.

inaccurate ‖ ungenau ‖ inexact.

inactive (Market) ‖ lustlos; geschäftslos; matt ‖ lourd; sans affaires.

inappropriate ‖ unzweckmäßig ‖ impropre; inopportun.

inavailable ‖ nicht anwendbar; unanwendbar ‖ pas utilisable; inutilisable.

incalculable ‖ unberechenbar ‖ incalculable.

incandescence ‖ Entzündung f. ‖ allumage m.; inflammation f. / ~ (Electr) ‖ Glühen n. ‖ incandescence f. / ~ (Metal) ‖ Weißglühhitze f. ‖ chaleur f. blanche.

incandescent ‖ weißglühend ‖ incandescent; blanc au feu. / ~ article ‖ Glühstoff m. ‖ matériel m. incandescent. / ~ bulb ‖ Glühbirne f. ‖ ampoule f. incandescente. / ~ exploder ‖ Glühzünder m. ‖ exploseur m. à incandescence. / ~ gas burner ‖ Gasglühlichtbrenner m. ‖ bec m. à gaz à incandescence. / inverted ~ gas burner ‖ Invertgasglühlichtlampe f. ‖ lampe f. à incandescence par le gaz à bec renversé. / ~ gas lamp ‖ Gasglühlichtlampe f. ‖ lampe f. à incandescence par le gaz. / ~ gas light ‖ Gasglühlicht n. ‖ lumière f. à incandescence par le gaz.

incandescent lamp ‖ Glühlampe f. ‖ lampe f. à incandescence. / ~ for petroleum ‖ Petroleumglühlichtlampe f. ‖ lampe f. à incandescence au pétrole. / semi-~ ‖ Glimmlampe f. ‖ lampe f. semi-incandescente. / tubular ~ ‖ röhrenförmige Glühlampe f. ‖ lampe f. à incandescence tubulaire.

incandescent lamp, connection of ~s ‖ Glühlampenschaltung f. ‖ couplage m. *ou* montage m. *ou* connection f. des lampes à incandescence. / electrode for ~ manufacturing ‖ Elektrode f. für die Glühlampenfertigung ‖ électrode f. pour la fabrication des lampes à incandescence. / ~ making machine ‖ Glüh-lampenherstellungsmaschine f. ‖ machine f. à fabriquer les lampes à incandescence. / ~ tester ‖ Glühlampenprüfer m. ‖ instrument m. à essayer des lampes à incandescence.

incandescent light chemicals pl. ‖ Glühlichtchemikalien fpl. ‖ produits mpl. chimiques pour lumière à incandescence. / ~ mantle ‖ Glühstrumpf m. ‖ manchon m. de lampe à incandescence par le gaz. / ~ spirit-light ‖ Spiritusglühlicht n. ‖ lumière f. d'alcool à incandescence. / ~ vapour ‖ leuchtender Dampf m. ‖ vapeur f. incandescente.

incapable of further war-like service ‖ unbrauchbar für weitere Kriegszwecke mpl. ‖ incapable de servir dorénavant à la guerre.

incendiary bomb ‖ Brandbombe f. ‖ bombe f. incendiaire. / ~ material ‖ Brandstoff m. ‖ matériel m. incendiaire.

incense ‖ Weihrauch m. ‖ encens m.

inch ‖ Zoll m. ‖ pouce m. / ~ plank ‖ Mittelbrett n. ‖ planche f. d'un pouce d'épaisseur.

incidence of a ray ‖ Einfall m. eines Lichtstrahls ‖ incidence f. d'un rayon. / angle of ~ (Aero) ‖ Anstellwinkel m. ‖ angle m. d'incidence. / angle of ~ (Phys) ‖ Einfallwinkel m. ‖ angle m. d'incidence. / axis of ~ ‖ Einfallslot n. ‖ axe m. d'incidence; normale f. / ~ plane ‖ Einfallebene f. ‖ plan m. d'incidence.

incident (Phys) ‖ einfallend ‖ incident.

incident light ‖ einfallendes Licht n. ‖ lumière f. d'incidence; lumière f. incident. / ~ ray ‖ Einfallstrahl m. ‖ rayon m. incident.

incidental expenses pl. ‖ Nebenkosten pl. ‖ faux-frais mpl.; frais mpl. supplémentaires.

incinerate, to ‖ veraschen; einäschern ‖ incinérer; reduire en cendres.

incineration ‖ Einäscherung f. ‖ incinération f. / ~ dish ‖ Veraschungsschale f. ‖ capsule f. à *ou* d'incinération.

incipient fracture (Metal) ‖ Anbruch m. ‖ amorce f. de fissure.

incised, sharply ~ before bending ‖ vor dem Biegen n. scharf eingekerbt ‖ nettement entaillé avant la déformation *ou* avant le ployage.

incision ‖ Einschnitt m. ‖ incision f.; entaille f.; coupure f. / sharp ~ ‖ scharfe Einkerbung f. ‖ entaille f. en V.

inclinable ‖ neigbar; umlegbar; schrägstellbar ‖ inclinable; inclinant. / ~ adjustment ‖ Schrägeinstellbarkeit f. ‖ ajustage m. oblique. / ~ crank press ‖ neigbare Exzenterpresse f. ‖ presse f. à excentrique inclinable. / ~ (geared) eccentric press ‖ schrägstellbare Exzenterpresse f. ‖ presse f. excentrique à bâti inclinable.

inclination (Geogr) ‖ Neigung f.; Gefälle n. ‖ pente f.; inclinaison f. / ~ (Mine) ‖ Einfallen n. ‖ pente f.; inclinaison f. / ~ (Phys) ‖ Inklination f.; Neigung f. ‖ inclinaison f. / flat ~ of a coal bed ‖ flaches Einfallen n. eines Kohlenflözes ‖ pendage m. de plateur *ou* inclinaison f. faible d'une couche de houille. / earth's magnetic ~ ‖ erdmagnetische Inklination f. ‖ inclinaison f. magnétique terrestre. / ~ of jib ‖ Auslegerneigung f. ‖ inclinaison f. de la flèche. / ~ of letters ‖ Schiefstehen n. von Buchstaben ‖ inclinaison f. de caractères. / ~ of the rails ‖ Neigung f.

der Schienen ‖ écartement m. *ou* inclinaison f. des rails.

inclination, angle of ~ ‖ Neigungswinkel m. ‖ angle m. d'inclinaison. / **~ balance** ‖ Neigungswage f. ‖ balance f. à cadran. / **~ compass** ‖ Inklinationsbussole f. ‖ boussole f. d'inclinaison. / **measurement of ~ by means of the spirit level** ‖ Steigungsmessung f. mit der Wasserwage ‖ mesure f. de l'inclinaison au moyen du niveau à bulle d'air.

inclinatory needle ‖ Inklinationsnadel f. ‖ aiguille f. *ou* boussole f. d'inclinaison; inclinatoire m. magnétique.

incline, to ‖ neigen ‖ incliner. / **~ to heat** ‖ zum Heißlaufen n. neigen ‖ avoir une tendance f. à chauffer. / **~ a ship** ‖ ein Schiff n. krängen ‖ mettre un navire à la bande; incliner un navire.

incline ‖ Gefälle n.; Neigung f. ‖ inclinaison f.; pente f.; talus m. / **~ (Roof)** ‖ Gefälle n.; Neigung f. ‖ égout m. / **double ~ (Railw)** ‖ Ablaufberg m. ‖ dos m. d'âne. / **upwards ~** ‖ Aufwärtsseilung f. ‖ inclinaison f. pour l'élévation.

incline attendant (Mine) ‖ Bremsbergführer m. ‖ conducteur m. de plan incliné.

inclined ‖ abfallend; geneigt; schräg ‖ incliné. / **~ on the dip (Mine)** ‖ einfallend ‖ incliné en aval *ou* en descendant. / **~ arranged** ‖ schräg gelagert ‖ disposé *ou* couché obliquement. / **~ belt** ‖ Steigband n. ‖ bande f. de transport inclinée. / **~ bench** ‖ Schrägrampe f. ‖ rampe f. inclinée. / **~ bottom shaft** ‖ schrägliegende Untermesserwelle f. ‖ arbre m. à lames inférieur oblique. / **~ bucket-elevator** ‖ Schrägbecherwerk n. ‖ noria f. inclinée. / **~ coke bench** ‖ Koksrampe f. ‖ plancher m. de four à coke. / **~ grate** ‖ schräger Rost m. ‖ grille f. inclinée. / **~ lift for cupola furnaces** ‖ Schrägaufzug m. für Kuppelöfen ‖ monte-charge m. incliné pour cubilots.

inclined plane ‖ geneigte Fläche f.; schiefe Ebene f. ‖ plan m. incliné. / **~ (Mine)** ‖ Bremsberg m.; Schleppbahn f. ‖ plan m. incliné à traînage. / **~ on a canal** ‖ Schiffsaufzug m. ‖ ascenseur m. à bateau incliné *ou* à bateau funiculaire. / **~ with rope** ‖ geneigte Seilebene f. ‖ plan m. incliné à câble. / **accelerating device for self-acting ~s** ‖ Beschleunigungsvorrichtung f. für Ablaufberge ‖ dispositif m. d'accélération pour pentes de remisage. / **equipment for self-acting ~s** ‖ Bremsbergeinrichtung f. ‖ installation f. de pente à freinage. / **installation of ~** ‖ Bremsbergeinrichtung f. ‖ installation f. de plan incliné. / **~ switch (Railw)** ‖ Kletterweiche f.; Kletterkreuzung f. ‖ changement m. de voie à plans inclinés *ou* sans discontinuité de la voie principale.

inclined position ‖ Schiefstellung f. ‖ position f. inclinée. / **~ retort oven** ‖ Schrägretortenofen m. ‖ four m. à cornues inclinées. / **~ wall** ‖ schräge Wand f. ‖ mur m. incliné.

inclining experiment (Shipb) ‖ Krängungsversuch m. ‖ expérience f. de stabilité.

inclinometer ‖ Neigungsmesser m. ‖ inclinomètre m.

inclose, to ‖ einfriedigen ‖ enclore; enfermer.

inclosure ‖ Einfassung f. ‖ clôture f.; enceinte f.

included ‖ inbegriffen ‖ inclusivement; y inclus.

including ‖ einschließlich ‖ en plus; y compris.

inclusion ‖ Einschluß m. ‖ inclusion f.

inclusive of postage ‖ einschließlich Porto n. ‖ port m. compris.

inclusively ‖ inbegriffen ‖ inclusivement; inlcus.

incombustibility (Chem) ‖ Unverbrennbarkeit f. ‖ incombustibilité f.

incombustible (Chem) ‖ unverbrennbar ‖ incombustible. / **~ mineral** ‖ nicht brennbares Mineral n. ‖ minéral m. incombustible.

income ‖ Einkünfte fpl.; Einkommen n. ‖ revenus mpl. / **yearly ~** ‖ Jahreseinkommen n. ‖ revenu m. annuel.

income-tax ‖ Einkommensteuer f. ‖ impôt m. sur le revenu.

incoming current ‖ ankommender Strom m. ‖ courant m. d'arrivée. / **~ exchange (Tel)** ‖ Bestimmungsanstalt f. ‖ bureau m. d'arrivée. / **~ register (Tel)** ‖ Gesprächsbuch n. ‖ liste f. d'arrivée.

incompetency ‖ Unzuständigkeit f.; Inkompetenz f. ‖ incompetence f.

incomplete ‖ unvollständig; lückenhaft ‖ incomplet; défectueux.

incompressible ‖ unzusammendrückbar ‖ incompressible.

incongealable ‖ ungefrierbar ‖ incongealable.

inconsiderate ‖ rücksichtslos ‖ sans égards.

inconvenience ‖ Übelstand m. ‖ inconvénient m.

inconvenient ‖ unbequem; nachteilig ‖ gênant; inopportun.

incorrect ‖ unkorrekt ‖ incorrect. / **~ data** ‖ fehlerhafte Angabe f. ‖ donnée f. fausse *ou* incorrecte.

incorrectly measured length ‖ unrichtig gemessene Länge f. ‖ longueur f. mesurée inexactement.

increase, to ‖ vermehren ‖ augmenter. / **~ (Money)** ‖ auflaufen ‖ s'enfler; s'accumuler. / **~ the turnover** ‖ den Umsatz m. erhöhen ‖ augmenter le chiffre d'affaires. / **the bore had not increased** ‖ die Bohrung hatte sich nicht mehr vergrößert ‖ l'alésage m. ne s'était plus élargi.

increase ‖ Zunahme f.; Zuwachs m. ‖ augmentation f.; surcroît m.; accroissement m. / **~ (Phys)** ‖ Ausdehnung f. ‖ extension f.; étendue f. / **~ (Trade)** ‖ Aufschlag m.; Steigerung f. ‖ hausse f.; augmentation f.; renchérissement m. / **~ of breadth** ‖ Verbreiterung f. ‖ élargissement m. / **~ of efficiency by x to y %** ‖ Verbesserung f. des Wirkungsgrades um x bis y % ‖ augmentation f. du rendement de x à y %. / **~ of lift** ‖ Auftriebsvermehrung f. ‖ augmentation f. de force ascensionnelle *ou* de poussée. / **~ of load** ‖ Lastvermehrung f. ‖ augmentation f. de charge. / **~ of power** ‖ Leistungssteigerung f. ‖ augmentation f. de puissance. / **~ of pressure** ‖ Druckanstieg m. ‖ augmentation f. *ou* accroissement m. de pression. / **~ in price** ‖ Mehrpreis m. ‖ plus-value f. / **~ in resistance** ‖ Widerstandszunahme f. ‖ augmentation f. de résistance. / **~ of salary** ‖ Gehaltserhöhung f.; Gehaltszulage f. ‖ augmentation f. de traitement. / **~ in value** ‖ Werterhöhung f. ‖ élévation f. de valeur.

increasing of lime ‖ Aufgehen n. *oder* Wachsen n. des Kalkes (beim Löschen) ‖ foisonnement m. de la chaux.

increment (Math) ‖ Differential n. ‖ différentielle f.; quantité f. infiniment petite *ou* différentielle.

incrust, to ~ (Build) ‖ verkleiden; inkrustieren ‖ revêtir; incruster; encrouter.

incrusted ‖ überzogen; inkrustiert ‖ revêtu d'incrustations; incrusté.

incrustation ‖ Kruste f.; Überzug m. ‖ incrustation f. / **~ (Coal dress)** ‖ Sinterung f. ‖ solidification f. / **~ (Metal)** ‖ Schlackenbildung f. ‖ incrustation f. / **~ of boiler fur** ‖ Kesselsteinansatz m. ‖ dépôt m. de calcaire.

incrustation heat ‖ Sinterungshitze f. ‖ température f. de solidification.

incubator ‖ Brutapparat m.; Brutschrank m.; Brutofen m. ‖ couveuse f.; étuve f. bactériologique.

incunable ‖ Wiegendruck m.; Inkunabel f. ‖ incunable m.; impression f. primitive.

incur, to ‖ sich zuziehen; auf sich laden ‖ encourir; courir. / **~ a damage** ‖ eine Beschädigung f. verschulden ‖ être capable d'un dommage. / **~ a loss** ‖ Verlust m. erleiden ‖ subir des pertes fpl.

indanthrene ‖ Indanthren n. ‖ indanthrène m. / **~ colouring matter** ‖ Indanthrenfarbstoff m. ‖ matière f. colorante à l'indanthrène. / **~ dye** *see* **~ colouring matter.**

indefinite ‖ unbestimmt ‖ indéfini.

indelible ink ‖ unauslöschbare Farbe f. ‖ couleur f. indélébile.

indemnification ‖ Entschädigung f.; Schadloshaltung f. ‖ dédommagement m.; indemnité f.

indemnify, to ‖ entschädigen; schadlos halten ‖ dédommager; indemniser.

indemnity ‖ Abfindungssumme f. ‖ indemnité f.; payement m. par composition. / **~ for sickness** ‖ Krankenunterstützung f. ‖ secours m. en cas de maladie. / **~ for strikers** ‖ Streikunterstützung f. ‖ indemnité f. pour grévistes.

indemnity coal ‖ Reparationskohle f. ‖ charbon m. indemnitaire.

indent, to ‖ einzahnen ‖ adenter; endenter. / **~ a beam** ‖ einen Balken m. einzahnen *oder* verzahnen ‖ adenter *ou* entailler une poutre.

indent ‖ Kerbe f. ‖ coche f.; entaille f. / **~ (Carp)** ‖ Verzahnung f. ‖ adent m.; endent m.

indentation (Carp) ‖ Einschnitt m.; Zahnschnitt m. ‖ endenture f.; encoche f.; entaille f.; goujure f.

indented (Carp) ‖ gezahnt ‖ denté; endenté.

indenting mechanism (Typewr) ‖ Kolonnensteller m. ‖ sélecteur m. de colonnes.

independence ‖ Selbständigkeit f. ‖ indépendance f. / **~ of light in the inspection department** ‖ Unabhängigkeit f. von der Beleuchtung des Meßraumes ‖ indépendance f. à éclairage de l'atelier de contrôle. / **~ of subjective sources of mistakes** ‖ Unabhängigkeit f. von subjektiven Fehlerquellen ‖ indépendance f. à l'origines subjectives de fautes.

independent variable ‖ unabhängige Veränderliche f. ‖ variable f. indépendante.

indestructibility ‖ Unzerstörbarkeit f. ‖ indestructibilité f.

indestructible ‖ unverwüstlich; unzerbrechlich; unzerstörbar ‖ indestructible. / **prac-**

tically being ~ || fast unzerstörbar || presqu'indestructible.

index, to || registrieren || enregistrer.

index (Clockm) || Rücker m. || raquette f. / ~ (Math) || Potenzexponent m. || exposant m. / ~ (Pointer) || Zeiger m. || index m.: aiguille f. / ~ (Print) || Inhaltsverzeichnis n.; Sachregister n. || index m.; table f. des matières. / ~ (Surv) || Diopterlineal n. / alhidade f. / refractive ~ || Brechungsindex m.; Brechungsexponent m. || indice m. de réfraction. / subdivided alphabetical ~ || unterteiltes abeceliches Verzeichnis n. || registre m. alphabétique subdivisé.

index book || Briefordner m. || biblomane m.; classeur m. / ~ card || Karteikarte f. || carte f. ou fiche f. de classement.

indexing device (Toolm) || Teilapparat m. || appareil m. à diviser.

index notch || Indexkerbe f. || coche-index f. / ~ plate || Teilplatte f. || secteur m. gradué. / ~ plate with index pin || Teilscheibe f. mit Indexstift || disque m. diviseur avec verrou. / ~ strip || Registrierstreifen m. || bande f. de papier pour l'enregistrement. / ~ table (Mach tool) || Drehtisch m. || table f. tournante.

Indian cane || Bambus m.; Bambusrohr n. || bambou m. / ~ corn see also maize || Mais m. || mais m. / ~ hemp || indischer Hanf m. || chanvre m. des Indes. / ~ ink || Ausziehtusche f.; flüssige Tusche f. || encre f. de Chine; encre f. liquide pour dessin. / ~ red || Indischrot n. || rouge m. indien ou des Indes.

india-rubber; India rubber see also caoutchouc || Kautschuk m.; Gummi n. || caoutchouc m.; gomme f. élastique. / crude ~ || Rohgummi n. || caoutchouc m. brut. / electric wire covered with ~ || mit Kautschuk m. isolierter Draht || fil m. électrique isolé au caoutchouc. / vulcanized ~ || vulkanisiertes Gummi n. oder Kautschuk m. || caoutchouc m. vulcanisé.

india-rubber article || Gummiware f. || objet m. en caoutchouc; caoutchouc m. manufacturé. / moulded ~ article || geformtes Gummierzeugnis n. || article m. en caoutchouc moulé. / vulcanized ~ article || Hartgummiware f. || article m. en caoutchouc durci. / ~ ball || Gummiball m. || balle f. élastique ou en caoutchouc. / ~ ball machine || Gummiballmaschine f. || machine f. à faire des balles en caoutchouc. / ~ boot || Gummischuh m. || chaussure f. en caoutchouc. / ~ calender || Gummikalander m. || calandre f. à caoutchouc. / ~ coat || Gummimantel m.; Gummirock m. || imperméable m. / ~ collar || Gummikragen m. || faux-col m. en caoutchouc. / ~ comb || Gummikamm m. || peigne m. en caoutchouc. / ~ doll || Gummipuppe f. || poupée f. en caoutchouc. / ~ dress-shield || Gummischweißblatt n. || dessous m. de bras en caoutchouc. / ~ factory || Gummifabrik f. || fabrique f. de caoutchouc. / ~ glove || Gummihandschuh m. || gant m. en caoutchouc. / ~ goods pl. || Kautschukwaren fpl.; Gummiwaren fpl. || articles mpl. ou onvrages mpl. en caoutchouc. / ~ goods pl. for surgeons || chirurgische Gummiwaren fpl. || articles mpl. de chirurgie en caoutchouc. / ~ hose || Gummischlauch m. || tuyau m. en caoutchouc. / ~ linen || Gummiwäsche f. || linge m. en caoutchouc. / ~ mixer || Gummimischmaschinenarbeiter m. || mélangeur m. de caoutchouc. / mixing mill for ~ || Gummimischwalzwerk n. || laminoir m. mélangeur à caoutchouc. / ~ press || Gummipresse f. || presse f. à caoutchouc. / ~ refinery || Gummiraffinerie f. || raffinerie f. de caoutchouc. / ~ shoe || Gummigalosche f. || galoche f. ou chaussure f. en caoutchouc. / ~ solution || Gummilösung f. || solution f. de caoutchouc. / ~ stocking || Gummistrumpf m. || bas m. de gomme élastique. / ~ stopper || Gummipfropfen m. || bouchon m. en caoutchouc. / ~ thread || Gummifaden m. || fil m. de caoutchouc. / ~ toys pl. || Gummispielwaren fpl. || jouets mpl. en caoutchouc. / ~ trade machine || Gummiindustriemaschine f. || machine f. pour l'industrie du caoutchouc. / ~ worker || Gummiarbeiter m. || caoutchoutier m. / ~ wrap || Gummibinde f. || bande f. élastique.

indicate, to || anzeigen || indiquer.

indicating device || Anzeigevorrichtung f. || dispositif m. indicateur. / visible ~ || sichtbare Anzeigevorrichtung f. || indicateur m. visible.

indicating line || Meßstrich m.; Teilstrich m. || trait m. de mesurage ou d'indication ou de division. / ~ peg (Tel) || Hinweisstöpsel m. || fiche f. indicatrice.

indication of price || Preisangabe f. || indication f. de prix. / taking the ~ of thermometer || Thermometerablesung f. || lecture f. des degrés du thermomètre. / ~ magnet (Railw) || Rückmeldefeld n. || répétiteur m.

indicative horse-power || indizierte Pferdestärke f. || cheval m. indiqué.

indicator || Zähler m.; Registrierapparat m.; Gangzähler m. || compteur m. de tour; indicateur m.; totalisateur m.; enregistreur m. / ~ (Mach) || Indikator m. || indicateur m. / ~ showing back and front the kind and amount of transaction (Cash register) || Schauwerk n. mit Anzeige nach beiden Seiten || indicateurs mpl. visibles de deux côtés. / ~ on the top of a blast-furnace || Gichtanzeiger m. || indicateur m. des charges. / direction ~ || Fahrtrichtungsanzeiger m. || signalisateur m.; indicateur m. du sens de marche. / direction ~ for current || Stromrichtungszeiger m. || indicateur m. du sens de courant. / gyroscopic direction ~ (Arm) || Kreiselfeuerleitanzeiger m. || indicateur m. gyroscopique. / ~ for earthen electric contact || Erdschlußanzeiger m. || indicateur m. de contact électrique à la terre. / electric ~ || elektrischer Fernmelder m. || téléavertisseur m. électrique. / leakage ~ (Electr) see ~ for earthen electric contact. / ~ showing the length of stroke || Hublängenanzeiger m. || indicateur m. de longueur de course. / long-distance ~ || Fernzeiger m. || indicateur m. à distance. / speed ~ || Geschwindigkeitsmesser m. || compteur m. de vitesse. / stationary ~ (Clock) || feststehende Kontrolluhr f. || horloge f. de contrôle stationnaire. / water-level ~ || Wandstandsanzeiger m. || indicateur m. du niveau d'eau.

indicator balance || Zeigerwage f. || balance f. à cadran. / ~ case (Tel) || Anrufzeichenkästchen n. || tableau m. d'annonciateurs d'appel. / ~ cock || Indikatorhahn m. || robinet m. indicateur. / ~ cord || Indikatorschnur f. || cordon m. de l'indicateur. / detention gear of an ~ || Anhaltevorrichtung f. eines Indikators || dispositif m. d'arrêt d'un indicateur. / ~ diagram || Leistungsbild n. (einer Maschine); Indikatordiagramm n. || diagramme m. d'indicateur. / ~ frequency meter || Zeigerfrequenzmesser m. || fréquencemètre m. à aiguille. / ~ lag (A) || Anzeigeträgheit f. || inertie f. d'indication. / paper cylinder of the ~ || Indikatortrommel f. || cylindre m. porte-papier de l'indicateur. / ~ pencil || Indikatorstift m. || crayon m. de l'indicateur. / ~ point || Zeigernase f. || pointe f. de l'index. / ~ pulley || Indikatorrolle f. || poulie f. à gorge de l'indicateur. / ~ screw-plug || Indikatorschlußschraube f. || bouchon m. à vis d'indicateur. / ~ string see ~ cord.

indicial admittance || Kennleitwert m. || admittance f. indicatrice.

indict, to || anklagen || poursuivre en justice f.

indictment || Anklage f. || accusation f.; prévention f.; incrimination f. / bill of ~ || Anklageschrift f. || acte f. d'accusation.

indifferent || indifferent || indifférent; inerte; neutre. / ~ equilibrium || indifferentes Gleichgewicht n. || équilibre m. indifférent.

indigenous || einheimisch || indigène. / ~ population || ansässige oder einheimische Bevölkerung f. || population f. indigène.

indigo || Indigo m. || indigo m. / artificial ~ || künstlicher Indigo m. || indigo m. artificiel. / copper ~ || feuriger Indigo m. || indigo m. cuivré. / counterfeit ~ || unechter Indigo m. || indigo m. bâtard. / falsified ~ see counterfeit ~. / ~ in half squares || halbstückiger Indigo m. || indigo m. demi-pierré. / natural ~ || natürlicher Indigo m.|| indigo m. naturel. / precipitated ~ || Indigokarmin n. || indigo m. soluble; indigocarmine f. / pure ~ || Indigoblau n. || indigo m. bleu ou pur. / reduced ~ || Indigoweiß n. || indigo m. réduit ou blanc. / sandy ~ || sandiger Indigo m. || indigo m. sablé. / scorched ~ || verbrannter Indigo m. || indigo m. brûlé. / ~ in small fragments || kleinstückiger Indigo m. || indigo m. en grabeaux. / spotted ~ || punktierter Indigo m. || indigo m. piqueté. / striped ~ || gebänderter Indigo m. || indigo m. rubané. / sublimed ~ || abgezogener Indigo m. || indigo m. traité. / synthetic ~ || synthetischer oder künstlicher Indigo m. || indigo m. synthétique. / weathered ~ || ausgewitterter Indigo m. || indigo m. éventé.

indigo blue || Indigoblau n. || bleu m. d'indigo. / ~ brown || Indigobraun n. || brun m. d'indigo. / ~ carmine || Indigokarmin m.; blauer Karmin m. || indigo m. soluble. / ~ red ~ carmine || Purpurschwefelsäure f.; Phönicinschwefelsäure f.; Phönicin n. || phénicine f. / ~ composition || Indigotinktur f. || composition f. d'indigo. / ~ copper || Kupferindigo m.; Kovellin m. || covelline f. / ~ derivatives pl. || Indigopräparate npl. || dérivés mpl. de l'indigo. / ~ gluten || Indigoleim m. || matière f. glutineuse d'indigo.

indigoid || indigoartig || indigoïde.

indigo manufactory || Indigofabrik f.; Indigofertigungsstätte f. || indigoterie f. / ~ paper || Indigopapier n. || papier m.

d'indigo. / ~ paste || Indigokarmin m. || indigo m. soluble; indigocarmine f. / ~ purple see ~ carmine. / ~ red || Indigorot n. || rouge m. d'indigo. / ~ tincture || Indigotinktur f. || composition f. ou teinture f. d'indigo. / ~ vat || Indigoküpe f. || cuve f. d'Inde ou ~ white || Indigoweiß n. || indigo m. réduit ou blanc.

indirect arc furnace || indirekter elektrischer Lichtbogenofen m. || four m. à arc indirect.

indistinct || undeutlich || indistinct.

individual accounts pl. || Einzelkonten pl. || comptes mpl. individuels. / ~ drive || Einzelantrieb m. || commande f. individuelle. / ~ driving || s. ~ drive / ~ electric drive of machines || elektrischer Einzelantrieb m. von Maschinen || commande f. individuelle électrique de machines. / ~ measurement || Einzelmessung f. || mesure f. séparée.

indoor equipment || Innenausstattung f. || décoration f. intérieure. / ~ lighting for industrial places || Innenbeleuchtung f. für Industrieräume || diffuseur m. pour locaux industriels. / ~ lighting for offices || Bürobeleuchtung f. || diffuseur m. pour bureaux. / ~ lucette || Innenraumluzette f. || lucette f. d'intérieur.

indorse, to || indossieren || endosser.

indorsee || Indossat m.; Indossatar m.; Wechselübernehmer m. || endossé m.

indorsement || Indossament n.; Begebungsvermerk m.; Übertragungsvermerk m. || endossement m.

indorser || Indossant m.; Wechselbegebender m.; Wechselübertrager m. || endosseur m.

induce, to || induzieren || induire.

induced || induziert; sekundär || induit; secondaire. / ~ charge || induzierte Ladung f. || charge f. induite. / ~ current || Induktionsstrom m.; Sekundärstrom m. || courant m. induit. / ~ current at closing || Schließungsinduktionsstrom m. || courant m. induit de fermeture. / ~ draught || Saugzug m. || tirage m. induit ou par aspiration / ~ draught furnace || Saugzugfeuerung f. || foyer m. à tirage par aspiration. / ~ line || induzierte Leitung f. || ligne f. induite. / ~ winding || Induktorium n. || enroulement m. induit; inducteur m.

inducing current || Primärstrom m. || courant m. inducteur.

inductance || Induktivität f. || coefficient m. d'induction. / air-core ~ || Induktanz f. mit Luftkern || inductance f. sans fer. / distributed ~ || verteilte Induktanz f. || inductance f. répartie. / iron-core ~ || Induktanz f. mit Eisenkern || inductance f. avec noyau de fer. / mutual ~ || gegenseitige Induktanz f. || inductance f. mutuelle. / syntonizing ~ (Radio) || Abstimminduktanz f. || inductance f. de syntonisation.

inductance coil || Induktanzspule f. || bobine f. d'inductance.

induction || Induktion f. || induction f. / magnetic ~ || Magnetinduktion f. || induction f. magnétique. / ~ from power circuits || Induktion f. durch Starkstromanlagen || induction f. par des installations d'énergie électrique.

induction balance || Induktionswage f. || balance f. d'induction. / ~ coil || Induktionsspule f.; Induktor m. || bobine f. d'induction. / substitutional ~ coil || Nebenschlußdrosselspule f. || bobine f. de self de substitution. / ~ current || Induktionsstrom m. || courant m. d'induction ou induit. / ~ curve || Induktionskurve f. || courbe f. d'induction. / ~ flow see ~ flux. / ~ flux || Induktionsfluß m. || flux m. d'induction. / ~ furnace || Induktionsofen m. || four m. à induction. / low-frequency ~ furnace || Niederfrequenzinduktionsofen m. || four m. à induction à basse fréquence. / three-phase ~ generator || Asynchrondrehstromgenerator m. || génératrice f. triphasée asynchronique. / ~ interference || Störung f. durch Induktion || interférence f. par induction. / law of ~ || Induktionsgesetz n. || loi f. de l'induction. / ~ machine || Influenzmaschine f. || machine f. à influence. / ~ motor || Induktionsmotor m. || moteur m. d'induction. / ~ noise || Induktionsgeräusch n. || bruit m. d'induction. / ~ phenomena pl. of ~ || Induktionserscheinungen fpl. || phénomènes mpl. d'induction. / ~ process || Induktionsverfahren n. || procédé m. d'induction. / ~ relay || Induktionsrelais n. || relais m. à courants d'induction. / ~ spark of ~ || Induktionsfunke m. || étincelle f. d'induction.

inductive || induktiv || inductif. / ~ capacity || Induktionsvermögen n. || capacité f. inductrice. / ~ coupling (Radio) || induktive Kopplung f. || accouplement m. inductif. / ~ disturbance || Übersprechen n. || diaphonie f.; cross talk m.; interférence f. mutuelle. / ~ effects between high voltage lines and low voltage lines || gegenseitige Beeinflussung f. von Starkstromleitungen und Schwachstromleitungen || induction f. mutuelle de lignes d'énergie et de lignes à courant faible. / ~ load || induktive Belastung f. || charge f. inductive. / ~ power || Influenzkraft f. || pouvoir m. inducteur. / ~ reactance || induktiver Blindwiderstand m. || réactance f. inductive. / non-~ resistance || induktionsfreier Widerstand m. || résistance f. non-inductive. / ~ transmitter (Radio) || induktiv gekoppelter Sender m. || transmetteur m. à couplage inductif.

inductor, ringing ~ (Tel) || Läuteinduktor m. || inducteur m. magnéto-électrique. / static ~ || Induktionsapparat m. || appareil m. d'induction dynamique.

inductorium || Induktor m.; Induktionsapparat m. || inducteur m.

industrial || industriell || industriel.

industrial || Industrieller m.; Gewerbetreibender m. || industriel m.

industrial art || Kunstgewerbe n. || art m. industriel. / workshop for ~ || kunstgewerbliche Werkstätte f. || atelier m. d'art industriel. / ~ article || kunstgewerblicher Gegenstand m. || article m. d'art appliqué.

industrial branch || Arbeitszweig m.; branche f. industrielle. / ~ building || Industriebau m. || bâtiment m. industriel. / ~ exhibition || Gewerbeausstellung f. || exposition f. industrielle. / ~ film || Industriefilm m. || film m. d'industrie. / ~ furnace for gas || Industrieofen m. für Gasfeuerung || four m. industriel pour le chauffage au gaz. / ~ gas || Industriegas n. || gaz m. industriel. / ~ compression of ~ gases || Verdichtung f. von Industriegasen || compression f. des gaz de fabrique ou des gaz industriels. / ~ group || Industriegruppe f. || groupe m. d'industrie; groupement m. industriel. / ~ hygienic plant || gewerbehygienische Anlage f. || installation f. hygiénique pour l'industrie. / ~ mobilization || industrielle Mobilmachung f. || mobilisation f. industrielle. / ~ planning || Industrievereinheitlichung f. || unification f. industrielle. / ~ plant || industrielle Anlage f.; usine f.; établissement m. industriel. / ~ plant (Bot) || Gewerbepflanze f. || plante f. industrielle. / ~ power supply || Stromlieferung f. für Industrie || fourniture f. de courant à l'industrie. / ~ preparation || industrielle Zubereitung f. || préparation f. industrielle. / ~ preparedness for war || industrielle Kriegsbereitschaft f. || disposition f. industrielle en vue de la guerre. / ~ railway || Industriebahn f. || chemin m. de fer industriel; voie f. industrielle. / ~ school || Industrieschule f. || école f. industrielle. / ~ steel || handelsüblicher Stahl m. || acier m. de type commercial. / ~ utilization || gewerblicher Gebrauch m. || utilisation f. industrielle. / ~ war || Industriekrieg m.; Handelskrieg m. || guerre f. industrielle. / ~ wood || Werkholz n. || bois m. de travail.

industrially manufactured || industriell hergestellt || fabriqué industriellement.

industrious || erwerbstätig; arbeitsam; fleißig || industriel.

industry || Industrie f.; Fabrikation f.; Fertigung f. || industrie f.; fabrication f.; manufacture f.; industrie f. manufacturière. / ~ of artificial stones || Kunststeinindustrie f. || industrie f. des pierres artificielles. / ~ of building materials || Baustoffindustrie f. || industrie f. des matériaux de construction. / chemical ~ || chemische Industrie f. || industrie f. chimique. / domestic ~ || Hausindustrie f. || industrie f. à domicile. / ~ of furs || Rauchwarenindustrie f. || industrie f. de la fourrure. / ~ of jewels and ornament stones || Schmucksteinindustrie f. || industrie f. de pierres précieuses et de bijouterie. / ~ mining || Bergbauindustrie f. || Montanindustrie f. || industrie f. minière. / to promote the ~ || die Industrie f. fördern || encourager ou aider l'industrie; être favorable à l'industrie f. / ~ refrigerating ~ || Kälteindustrie f. || industrie f. frigorifique. / small ~ || Kleinindustrie f. || petite industrie f.

inegality (Math) || Ungleichung f. || inégalité f.

inequality || Ungleichheit f. || inégalité f.

inert || indifferent; gleichgültig; neutral || indifférent; inerte; neutre.

inertance || Inertanz f. || inertance f.

inertia || Trägheit f.; Beharrungsvermögen n.; Trägheitsvermögen n. || force f. d'inertie; inertie f. / ~ of needle || Anzeigeträgheit f. || inertie f. d'indication. / ~ of pointer see ~ of needle. / moment of ~ || Trägheitsmoment n. || moment m. d'inertie. / ~ starter || Schwungradanlasser m. || démarreur m. à volant.

inertness see inertia.

inexact || ungenau || inexact.

inexperienced people || Nichtfachleute pl. || personnel m. non-spécialisé. / in the hands of ~ || unter den Händen fpl. von Nichtfachleuten || entre les mains fpl. d'un personnel non-spécialisé.

inexploitable (Mine) || unbauwürdig || inexploitable.

inexplosive ‖ unexplodierbar ‖ inexplosible.

inextinguishable (Ink) ‖ unauslöschbar ‖ inextinguible.

infantry carrier ‖ Infanterielastwagen m. ‖ camion m. d'infanterie. / ~ howitzer ‖ Infanteriemörser m. ‖ mortier m. d'infanterie. / two-barrel ~ accompanying gun ‖ zweirohriges Infanteriegeschütz n. ‖ canon m. d'infanterie muni de deux tubes; canon m. d'infanterie à deux bouches.

infants' food ‖ Kindermehl n. ‖ farine f. lactée ou pour enfants. / ~ meal see ~ food. / ~ powder ‖ Kinderpuder m. ‖ poudre f. pour enfants. / ~ scale ‖ Säuglingswage f. ‖ balance f. pour nouveaux-nés. /~ school ‖ Kleinkinderschule f. ‖ école f. enfantine. / ~ sucking bottle ‖ Kindersaugflasche f. ‖ biberon m. pour enfants.

infected ‖ verseucht; angesteckt; infiziert ‖ infecté. / declared ~ ‖ für verseucht erklärt ‖ déclaré infecté.

infection ‖ Ansteckung f. ‖ contagion f.

infectious ‖ ansteckend ‖ contagieux.

inferior goods pl. ‖ Ramschware f.; minderwertige Waren fpl. ‖ marchandise f. de rebut. / ~ material ‖ minderwertiger Werkstoff m. ‖ matériel m. de qualité inférieure. / of ~ quality ‖ minderwertig ‖ d'une valeur ou d'une qualité inférieure. / ~ slack ‖ minderwertige Kohle f. ‖ charbon m. de rebut. / of ~ value see of ~ quality.

inferiority ‖ Minderwert m.; Minderwertigkeit f. ‖ moins-value f.

infernal stone ‖ Höllenstein m.; salpetersaures Silber n. ‖ pierre f. infernale; azotate m. d'argent.

infilling of vein ‖ Gangfüllung f. ‖ remplissage m. du filon.

infiltrate, to ‖ einsickern ‖ s'imbiber dans...; se perdre dans...; s'infiltrer.

infiltration ‖ Einsickerung f. ‖ infiltration f.

infinite ‖ unendlich ‖ infini.

infinity ‖ Unendlichkeit f. ‖ infinité f. / eye accommodated for ~ ‖ auf Unendlich akkommodiertes Auge n. ‖ œil m. accommodé à l'infini. / focussed for ~ ‖ auf Unendlich n. eingestellt ‖ réglé à l'infini ou sur les lointains. / to set to ~ ‖ auf Unendlich n. einstellen ‖ mettre au point sur l'infini. / set to ~ see focussed for ~.

infirmary ‖ Hospital n.; Krankenhaus n. ‖ hôpital m.; hospice m.; hôtel-Dieu m.; maison-Dieu f.

infirmity ‖ Gebrechen n.; Schwäche f. ‖ infirmité f.

inflame, to ‖ entflammen; entzünden ‖ enflammer; allumer. / ~ ‖ sich entflammen ‖ s'enflammer; prendre feu.

inflammability ‖ Entflammbarkeit f.; Entzündlichkeit f. ‖ inflammabilité f.

inflammable ‖ entflammbar; entzündlich; feuergefährlich ‖ inflammable; combustible. / easily ~ ‖ leicht entzündlich; feuergefährlich ‖ facilement inflammable. / spontaneously ~ ‖ selbstentzündlich ‖ inflammable spontanément.

inflammable gas ‖ brennbares Gas n. ‖ gaz m. combustible ou inflammable.

inflammable liquid ‖ feuergefährliche Flüssigkeit f. ‖ liquide m. inflammable. / storing of ~s ‖ Lagerung f. für feuergefährliche Flüssigkeiten ‖ magasin m. de liquides inflammables. / tapping for ~s ‖ Zapfstelle f. für feuergefährliche Flüssigkeiten ‖ poste m. de débit de liquides inflammables.

inflammable material ‖ entzündlicher Stoff m. ‖ matière f. inflammable.

inflammable metal ‖ Zündmetall n. ‖ métal m. d'allumage.

inflammables pl. ‖ Zündwaren fpl. ‖ (articles mpl. en) matières fpl. inflammables. / Bengal ~ ‖ bengalische Zündwaren fpl. ‖ feux mpl. ou inflammables mpl. de Bengale.

inflammation ‖ Entzündung f. ‖ allumage m.; inflammation f. / ~ of the eyes ‖ Augenentzündung f. ‖ inflammation f. des yeux. / point of ~ ‖ Entzündungspunkt m. ‖ point m. d'inflammation ou d'ignition. / ~ temperature ‖ Entzündungstemperatur f. ‖ température f. d'inflammation.

inflammatory affection ‖ entzündliche Erkrankung f. ‖ maladie f. infectueuse.

inflation (Airship) ‖ Füllung f. ‖ gonflement m. / ~ sleeve ‖ Füllansatz m. ‖ manche f. de gonflement ou de remplissage ou d'appendice.

inflexible ‖ unbiegsam ‖ inflexible.

inflection see inflexion.

inflexion of light ‖ Beugung f. des Lichtes ‖ diffraction f. de la lumière. / point of ~ (Math) ‖ Wendepunkt m. ‖ point m. d'inflexion.

influence ‖ Einwirkung f. ‖ influence f. / ~ which a company is able to exert on... ‖ Einfluß f. einer Gesellschaft auf... ‖ influence f. d'une société sur... / ~ of the earth surface on the propagation of the electromagnetic waves ‖ Bedeutung f. der Erdoberfläche für die Ausbreitung drahtloser Wellen ‖ influence f. de la surface de la terre sur la propagation des ondes électromagnétiques. / ~ of friction ‖ Reibungseinfluß m. ‖ influence f. de friction. / ~ of temperature ‖ Temperatureinfluß m. ‖ influence f. de la température. / under the ~ of ‖ unter dem Einfluß m. von ‖ sous l'influence f. de.

influence machine ‖ Influenzmaschine f. ‖ machine f. d'influence.

influences pl. deflecting the magnetic needle ‖ ablenkende Einflüsse mpl. auf die Magnetnadel ‖ influences fpl. déviatrices sur l'aiguille aimantée.

influx of workers ‖ Zuzug m. von Arbeitern ‖ affluence f. d'ouvriers.

inform, to ~ (Trade) ‖ benachrichtigen; avisieren ‖ informer; aviser; faire savoir; donner avis m.

information ‖ Auskunft f. ‖ renseignement m.; information f. / ~ (Law) ‖ Anzeige f. ‖ citation f.; signification f. / ~ (Trade) ‖ Benachrichtigung f.; Avis m. ‖ information f.; avis m. / exclusive ~ ‖ Privatauskunft f. ‖ renseignement m. personnel ou privé. / indefinite ~ ‖ unsichere Auskunft f. ‖ renseignement m. incertain. / to obtain ~ about ‖ sich erkundigen über ‖ prendre des renseignements mpl. sur.

information board (Tel) ‖ Auskunftstisch m.; Auskunftsstelle f.; Auskunftsplatz m. ‖ table f. ou place f. de renseignements. / ~ desk see ~ board. / one-way ~ message (Radio) ‖ Nachricht f. an Alle ‖ message m. d'information unilatéral. / ~ office ‖ Auskunftei f. ‖ agence f. d'information.

infra-red (Opt) ‖ infrarot ‖ infrarouge.

infringement ‖ Übergriff m. ‖ empiètement m.

infusible ‖ unschmelzbar ‖ infusible.

infusion ‖ Aufguß m.; Infusion f. ‖ infusion f. / ~ method (Brew) ‖ Aufgußverfahren n. ‖ méthode f. de brassage par infusion.

infusorial earth ‖ Kieselgur f.; Infusorienerde f. ‖ kieselguhr m.; farine f. fossile; terre f. d'infusoires. / ~ quarry ‖ Infusorienerdegewinnung f. ‖ carrière f. de farine fossile ou de kieselguhr.

ingate ‖ Gußtrichter m. ‖ jet m. de moulage.

ingot ‖ Gußblock m.; Barren m. ‖ lingot m. / absolutely close-grained ~ ‖ absolut dichter Gußblock m. ‖ lingot m. d'un grain parfaitement serré. / the ~ is absolutely homogeneous ‖ der Block m. ist völlig homogen ‖ le lingot est d'une homogénéité parfaite. / ~ with corners rounded off ‖ Block m. mit abgerundeten Kanten ‖ lingot m. à angles arrondis. / the ~ was forged out under an hydraulic press in x heats ‖ der Block m. wurde in x Hitzen unter einer Presse ausgeschmiedet ‖ le lingot m. fut forgé ou étiré sous une presse en x chauffes. / rough-rolled ~ ‖ vorgewalzter Block m. ‖ lingot m. dégrossi. / sectioned ~ ‖ zerteilter Gußblock m. ‖ lingot m. sectionné. / the ~s pl. are uniform throughout ‖ die Blöcke mpl. sind gleichförmig ‖ les lingots mpl. possèdent une régularité de texture.

ingot boring machine ‖ Blockausbohrmaschine f. ‖ foreuse f. pour blooms et lingots. / ~ breaker ‖ Blockbrecher m. ‖ machine f. à tronçonner les lingots; casse-lingot m. / ~ charging truck ‖ Blockwagen m. ‖ chariot m. à lingots. / ~ conveying device ‖ Blockvorrollvorrichtung f. ‖ dispositif m. à pousser les lingots dans le four à rechauffer; dispositif m. pour l'avance en roulant des lingots. / ~ cutting machine ‖ Blockschere f. ‖ machine f. à couper les lingots. / dead head of ~s ‖ Gußkopf m. ‖ masselotte f. / ~ drawing-out device ‖ Blockauszieher m. ‖ défourneuse f. de lingots. / pneumatic ~ dressing shop ‖ pneumatische Blockrüsterei f. ‖ atelier m. de nettoyage pneumatique des lingots. / ~ head heating device ‖ Blockkopfbeheizung f. ‖ réchauffage m. des têtes des lingots. / ~ iron ‖ Flußeisen n. ‖ fer m. en lingot. / ~ mould ‖ Kokille f.; Blockform f. ‖ coquille f.; lingotière f. / ~ parting and slicing machine ‖ Blockteil- und Trennmaschine f. ‖ machine f. à découper les lingots. / ~ pusher ‖ Blockausdrücker m. ‖ pousseuse f. des lingots. / ~ pushing device ‖ Blockdrücker m. ‖ dispositif m. pour l'enfournement des lingots. / putting on device for ~s ‖ Auflegevorrichtung f. für Blöcke ‖ dispositif m. de chargement de lingots. / ~ roughing lathe ‖ Blockschruppbank f. ‖ tour m. d'ébauchage pour lingots. / ~ shears pl. ‖ Blockschere f. ‖ cisaille f. à lingots. / size of an ~ ‖ Größe f. eines Blockes ‖ grosseur f. du lingot. / ~ slicing machine ‖ Blockteilmaschine f. ‖ machine f. à découper les lingots.

ingot steel ‖ Blockstahl m. ‖ acier m. en lingots. / production of ~ in an open hearth ‖ Flußstahlerzeugung f. auf offenem Herd ‖ production f. de l'acier en lingots dans un four à sole découvert. / sorts pl. of

~ ‖ Flußstahlarten fpl. ‖ qualités fpl. d'acier. / ~ tilter see ~ tipping device. / ~ tipping device ‖ Blockkipper m. ‖ culbuteur m. de lingots. / ~ turning machine ‖ Blockwendemaschine f. ‖ machine f. à retourner les lingots. / ~ withdrawing device ‖ Blockausziehvorrichtung f. ‖ défourneuse f. de lingots.

ingrain, to ‖ in der Wolle f. färben ‖ teindre en poil m.

ingredient ‖ Bestandteil m. ‖ partie f. constitutive ou intégrante; ingrédient m.

inhalation apparatus ‖ Inhalationsgerät n. ‖ appareil m. d'inhalation.

inhaler ‖ Inhalator m.; Inhalationsgerät n. ‖ inhalateur m.

inherent stability ‖ Eigenstabilität f. ‖ stabilité f. propre ou de la forme.

inheritance ‖ Erbschaft f. ‖ héritage m.

initial ‖ anfänglich; Anfangs-.... ‖ initial. / ~ acceptivity (Electr) ‖ Nennaufnahme f. ‖ puissance f. absorbée normale; réception f. nominale. / ~ adjustment ‖ Nulleinstellung f. ‖ mise f. au point zéro. / ~ amplification ‖ Anfangsverstärkung f. ‖ amplification f. initiale. / ~ capital ‖ Gründungskapital n. ‖ capital m. d'établissement. / ~ cost ‖ Anschaffungspreis m. ‖ prix m. d'acquisition. / ~ forging temperature ‖ Anfangsschmiedetemperatur f. ‖ température f. de début de forgeage. / ~ line ‖ Anfangszeile f. ‖ ligne f. de tête; première ligne f. / ~ letter ‖ Anfangsbuchstabe m. ‖ lettre f. initiale. / ~ permeability ‖ Anfangspermeabilität f. ‖ permeabilité f. initiale. / ~ position ‖ Ausgangsstellung f.; Ruhestellung f. ‖ position f. initiale ou de départ. / ~ pressure ‖ Anfangsdruck m. ‖ pression f. initiale. / ~ product ‖ Ausgangsprodukt n. ‖ produit m. de base. / ~ setting (Mach) ‖ Bereitschaftsstellung f. ‖ position f. d'attente. / ~ slope ‖ Anfangssteigung f. ‖ élévation f. ou montée f. initiale. / ~ step ‖ Anfangsstufe f. ‖ marche f. ou étage m. d'entrée. / ~ strain ‖ Anfangsbeanspruchung f. ‖ effort m. initial; charge f. initiale. / ~ temperature ‖ Anfangstemperatur f. ‖ température f. initiale. / ~ value ‖ Anfangswert m. ‖ valeur f. initiale. / ~ velocity ‖ Anfangsgeschwindigkeit f. ‖ vitesse f. initiale. / ~ voltage on charge ‖ Anfangsspannung f. der Ladung ‖ tension f. initiale de la charge. / ~ voltage on discharge ‖ Anfangsspannung f. der Entladung ‖ tension f. initiale de la décharge.

initial ‖ Initial n.; großer Anfangsbuchstabe m. ‖ initiale f. / ~ of assistant (Cash register) ‖ Verkäuferzeichen n. ‖ initiale f. du vendeur. / ~ key (Cash register) ‖ Buchstabentaste f. ‖ touche f. de lettre.

inject, to ‖ einspritzen; einblasen ‖ injecter; souffler. / ~ cement ‖ mit Zement m. torkretieren ‖ faire injection f. du ciment. / ~ fuel by means of a nozzle ‖ Brennstoff m. durch eine Düse einspritzen ‖ injecter le combustible à l'aide d'un gicleur.

injecting see injection.

injection ‖ Einspritzung f. ‖ injection f. / airless ~ ‖ luftlose Einspritzung f. ‖ injection f. mécanique.

injection air bottle ‖ Lufteinblaseflasche f. ‖ réservoir m. d'insufflation d'air. / ~ compressor ‖ Einblaseluftpumpe f. ‖ compresseur m. d'injection d'air. / ~ receiver

‖ Einblasegefäß n. ‖ réservoir m. d'insufflation.

injection cock ‖ Einspritzhahn m. ‖ robinet m. d'injection. / condensation by ~ ‖ Einspritzkondensation f. ‖ condensation f. par injection ou par mélange. / ~ condenser ‖ Einspritzkondensator m. ‖ condenseur m. par injection ou par mélange ou à jet. / ~ nozzle ‖ Einspritzdüse f. ‖ gicleur m. d'injection. / ~ outlet ‖ Einspritzöffnung f. ‖ orifice f. d'injection. / ~ valve ‖ Einspritzventil n. ‖ soupape f. d'injection. / ~ water ‖ Einspritzwasser n. ‖ eau f. d'injection.

injector ‖ Injektor m.; Strahlpumpe f. ‖ injecteur m.; pompe f. injectrice.

injure, to ‖ schädigen ‖ nuire; faire tort.

injured ‖ beschädigt; verletzt ‖ endommagé. / ~ (Mar) ‖ beschädigt ‖ avarié. / ~ person ‖ Unfallverletzter m. ‖ blessé m. par accident.

injurious ‖ schädlich ‖ nuisible; préjudiciable.

ink, to ~ (Drawing) ‖ austuschen; ausziehen ‖ encrer; passer à l'encre. / ~ (Print) ‖ Schwärze f. auftragen ‖ toucher la forme; encrer. / ~-in a drawing ‖ eine Zeichnung f. ausziehen ‖ passer un dessin à l'encre. / ~ the types ‖ die Farbe oder die Schwärze auf die Form auftragen ‖ encrer les lettres fpl.; toucher la forme.

ink ‖ Tinte f. ‖ encre f. / ~ (Print) ‖ Druckerschwärze f. ‖ encre f. d'imprimerie. / acidproof ~ ‖ säurefeste Tinte f. ‖ encre f. résistante aux acides. / autographic ~ ‖ autografische Tinte f. ‖ encre f. autographique. / black printing ~ ‖ Schwärze f. ‖ encre f. noire d'imprimerie. / China ~ ‖ Tusche f. ‖ encre f. de Chine. / drawing ~ ‖ Zeichentinte f. ‖ encre f. à dessiner. / duplicating ~ ‖ Vervielfältigungsfarbe f. ‖ encre f. à polycopier. / easy flowing ~ ‖ dünnflüssige Tinte f. ‖ encre f. fluide; encre f. coulant facilement. / fountain-pen ~ ‖ Füllfedertinte f. ‖ encre f. pour porte-plumes réservoir. / gallotannic ~ ‖ Gallustinte f. ‖ encre f. gallo-tannique. / ~ containing glycerine ‖ glyzerinhaltige Tinte f. ‖ encre f. à la glycérine. / indelible ~ (Print) ‖ unzerstörbare Farbe f. ‖ encre f. indélébile. / Indian ~ ‖ chinesische Tusche f. ‖ encre f. de Chine. / ~ made of lamp black ‖ Lampenrußfarbe f. ‖ couleur f. ou encre f. à base de fumée. / letter press ~ ‖ Schnellpressenfarbe f. ‖ encre f. pour presses rapides. / liquid drawing ~ ‖ Ausziehtusche f.; flüssige Tusche f. ‖ encre f. de Chine; encre f. liquide pour dessin. / lithographic ~ ‖ lithographische Tinte f. ‖ encre f. lithographique. / metallic ~ ‖ Metallfarbe f.; Metalltinte f. ‖ encre f. métallique. / mouldy ~ ‖ verschimmelte Tinte f. ‖ encre f. moisie. / permanent ~ ‖ beständige Tinte f. ‖ encre f. constante. / quick drying ~ ‖ schnell trocknende Tinte f. ‖ encre f. séchant rapidement. / sympathetic ~ ‖ sympathetische Tinte f. ‖ encre f. sympathique. / to take ~ ‖ die Druckerschwärze oder Druckfarbe auftragen ‖ encrer; toucher la forme. / viscous ~ ‖ dickflüssige Tinte f. ‖ encre f. visqueuse. / writing ~ ‖ Schreibtinte f. ‖ encre f. à écrire.

ink block (Print) ‖ Reiber m.; Farbläufer m.; Reibstein m. ‖ marbre f.; molette f.; broyon m. / ~ blotter ‖ Tintenlöscher m. ‖ tampon m. buvard. / ~ board

Farbenbrett n. ‖ palette f. à encre. / ~ bottle ‖ Tintenflasche f. ‖ bouteille f. pour encre. / ~ bottler ‖ Tintenflaschenabfüller m. ‖ embouteilleur m. d'encre. / ~-brush ‖ Tuschpinsel m. ‖ pinceau m. à laver. / ~ capacity of fountain-pen ‖ Tinteninhalt m. des Füllfederhalters ‖ encre f. contenue dans le stylographe. / ~ channel of fountain pen ‖ Tintenkanal m. des Füllfederhalters ‖ canal m. à encre du stylographe. / increased ~ distribution (Print) ‖ vermehrte Verreibung f. der Druckerschwärze ‖ distribution f. d'encre plus fine.

inker (Print) ‖ Auftragwalze f. ‖ rouleau m. toucheur.

ink eraser ‖ Tintengummi m. ‖ gomme f. à encre ou à effacer; gomme-grattoir m. / ~ filler (Fountain pen) ‖ Tintenfüller m. ‖ compte-gouttes m. / ~ fountain (Print) ‖ Farbenbehälter m. ‖ récipient m. à encre. / ~ glass ‖ Tintenglas n. ‖ verre m. pour encre. / ~ grinding mill ‖ Farbmühle f. ‖ broyeuse f. à couleurs.

inking ‖ Einfärbung f. ‖ encrage m.

inking apparatus (Print) ‖ Farbwerk n. ‖ encrage m.; mécanisme m. d'encrage. / removable ~ (Print) ‖ abfahrbares Farbwerk n. ‖ encrage m. mobile. / ~ of two rollers (Print) ‖ zweiwalziges Farbwerk n. ‖ encrage m. à deux rouleaux.

inking cylinder ‖ Farbzylinder m.; Farbrolle f. ‖ cylindre-encreur m.; rouleau-encreur m. / ~ device (Print) see inking apparatus. / ~ pad see ink-pad.

inking ribbon ‖ Farbband n. ‖ ruban-encreur m. / ~ crank ‖ Farbbandkurbel f. ‖ manivelle f. du ruban-encreur. / ~ length ‖ Farbbandlänge f. ‖ longueur f. du ruban-encreur.

inking roller ‖ Auftragwalze f. ‖ rouleau m. toucheur. / ~ stone ‖ Farbstein m. ‖ pierre f. à broyer la couleur.

ink jar ‖ Tintenkrug m. ‖ bouteille f. en grès à encre. / ~ manufactury ‖ Farbenfabrik f. ‖ fabrique f. d'encres ou de couleurs. / ~-pad ‖ Farbkissen n.; Stempelkissen n. ‖ coussin m. encreur; encreur m. pour timbres. / ~ pencil ‖ Tintenstift m. ‖ crayon-encre m.

inkpot ‖ Tintenfaß n. ‖ encrier m. / metal ~ ‖ Metalltintenfaß n. ‖ encrier m. en métal.

ink powder ‖ Tintenpulver n. ‖ encre f. en poudre; poudre f. d'encre. / ~ reservoir (Print) ‖ Farbbehälter m. ‖ réservoir m. à encre. / ~ roller (Print) ‖ Farbwalze f. ‖ rouleau m. encreur ou à encre. / ~ roller (Tel) ‖ Schwärzrolle f.; Farbrolle f. ‖ tampon m. / ~ slab ‖ Farbspachtel f. ‖ spatule f. à encre.

inkstand ‖ Tintenfaß n. ‖ encrier m. / artistic ~ ‖ Kunsttintenfaß n. ‖ encrier m. artistique. / self-closing ~ ‖ Tintenfaß n. mit Selbstverschluß ‖ encrier m. à fermeture automatique. / two-well ~ ‖ Doppeltintenfaß n. ‖ encrier m. à deux verres.

ink supply plant (Print) ‖ Farbversorgungsanlage f. ‖ installation f. pour la distribution de l'encre. / ~ tub see ~ reservoir.

inkwell ‖ Tintenbehälter m. ‖ réservoir m. à encre.

inkwriter (Tel) ‖ Farbschreiber m.; Schreibempfänger m. ‖ télégraphe m. écrivant; appareil m. Morse enregistreur.

inlaid cutter's pliers ‖ Vorschneider m. mit aufgesetzten Backen ‖ pince à couteaux rapportés. / ~ floor ‖ Parkettboden m. ‖

parquet m. de plancher; parqueterie f. /
~ linoleum || Einleglinoleum n. || linoléum
m. incrusté. / ~ press || Inlaidpresse f. ||
presse f. d'incrustation.

inlaid work || Einlegearbeit f.; Intarsie f.
|| marqueterie f.; placage m. / ~ of wood
|| Holzmosaik f. || marqueterie f. en bois.
/ wood for ~ || Holz n. für Einlegearbeiten
|| bois m. de marqueterie.

inland || landeinwärts || vers l'intérieur m.
/ ~ communication || Binnenverkehr m.
|| communication f. à l'intérieur. / at ~
end || landseitig || côté m. terre. / ~ fuel ||
einheimischer Brennstoff m. || combus-
tible m. indigène. / ~ lowlands pl. ||
Binnentiefland n. || pays m. bas *ou* plaine
f. à l'intérieur. / ~ navigation || Binnen-
schiffahrt f. || navigation f. intérieure. /
~ production || Inlandserzeugung f. || pro-
duction f. indigène. / ~ revenue || Akzise
f. || accise f.; octroi m.; contributions fpl.
indirectes. / ~ sea || Binnenmeer n. || mer
f. intérieure. / ~ trade || Binnenhandel m.
|| commerce m. intérieur. / ~ water ||
Binnengewässer n. || eaux fpl. continen-
tales.

inlay, to || furnieren || plaquer. / ~ with
black enamel || niellieren || nieller. / ~ a
floor || parkettieren || parqueter.

inlaying || Lambris f.; Getäfel n. || lambris
m.; boiserie f. / ~ of floors || Parkettie-
rung f.; Austäfelungsarbeit f. || parque-
tage m.

inlaying apparatus (Textile) || Durchschuß-
apparat m. || appareil m. à tramer. /
~ file || Intarsienfeile f. || lime f. à mar-
queterie. / ~ saw || kleine Schweifsäge f.;
Laubsäge f. || petite scie f. à contourner
ou à chantourner; scie à marqueterie *ou*
d'horloger.

inlet || Einführungsöffnung f. || ouverture f.
ou regard m. d'entrée. / brine ~ || Sole-
eintritt m. || entrée f. de la saumure. /
engine water ~ || Wassereinlaß m. || en-
trée f. d'eau.

inlet cam || Einlaßnocken m. || came f. d'ad-
mission. / ~ cam roller lever (Mot) ||
Einlaßnockenhebel m. || levier m. d'ad-
mission à came. / ~ chamber || Saugraum
m. || chambre f. d'aspiration. / ~ funnel ||
Einführungstrichter m. || pipe f. *ou* tube
f. d'entrée. / ~ hole || Luftsaugloch n. ||
trou m. d'aspiration. / ~ limit || Einlaß-
feld n. || champ m. d'entrée. / ~ manifold
|| Saugleitung f. || tuyauterie f. d'aspira-
tion. / ~ nipple || Durchführungstülle f. ||
douille f. de traversée. / ~ pipe || Saug-
rohr f. || tuyau m. d'aspiration. / ~ sluice
|| Einflußschleuse f. || écluse f. de chasse.
/ ~ stop cock || Eingußhahn m. || robinet
m. d'entrée. / ~ temperature || Eintritts-
temperatur f. || température f. d'admis-
sion. / ~ tube || Einführungsrohr n. || pipe
f. d'entrée.

inlet-valve || Saugventil n. || soupape f.
d'aspiration. / ammonia ~ || Ammoniak-
einlaßventil n. || vanne f. d'entrée d'am-
moniaque. / mechanically operated ~ ||
gesteuertes Saugventil n. || soupape f.
d'admission mécanique.

inn || Schenke f. || cabaret m.

inner charcoal of the heap (Charc) ||
Quandelkohle f. || charbon m. du côté
de la cheminée d'une meule. / ~ dam ||
Binnendeich m. || digue f. intérieure. /
removable ~ boiler || herausnehmbarer
Einsatzkessel m. || chaudière f. intérieure
mobile. / ~ covering || Futter n. || dou-

blure f. / ~ fish plate with round holes ||
Innenlasche f. mit runden Löchern ||
éclisse f. intérieure avec trous ronds. /
~ flame || Reduktionsflamme f. || flamme
f. de réduction. / ~ form (Print) || Wider-
druckform f. || forme f. d'impression au
verso. / ~ lining (Furnace) || Schacht-
futter n. || parois fpl.; chemise f. inté-
rieure.

inner pole armature || Innenpolanker m. ||
induit m. à pôles intérieurs. / ~ frame ||
Innenpolgehäuse n. || carcasse f. à pôles
intérieurs.

inner reflector || Innenreflektor m. || réflec-
teur m. intérieur. / ~ side || Innenseite f.
|| parement m. intérieur. / ~ sole (Shoem)
|| Brandsohle f. || semelle f. intérieure;
première f. / ~ span of a room || Innen-
weite f. eines Raumes || distance f. *ou* por-
tée f. de rez. / ~ width || lichte Weite
f. || largeur f. intérieure. / ~ wire || Innen-
leitung f. || fil m. intérieur.

inn-keeper's requisites || Gastwirtschafts-
artikel mpl. || articles mpl. d'hôtellerie.

innovation || Neuerung f. || innovation f.

inoculate, to || einimpfen || inoculer; en-
semencer. / ~ (Gard) || okulieren || écus-
sonner.

inodorous || geruchlos || inodore.

inoffensive, to render || ungefährlich
machen || rendre inoffensif.

inorganic || anorganisch || inorganique. /
~ acid || anorganische Säure f. || acide m.
inorganique. / ~ compound || Metallsalz
n. || sel m. inorganique.

inoxidable || nicht oxydierbar || inoxydable.

input || Kraftbedarf m. || énergie f. absor-
bée; puissance f. nécessaire. / ~ amplifier
|| Vorverstärker m. || amplificateur m.
d'entrée. / ~ circuit (Electr) || Eingangs-
stromkreis m. || circuit m. d'entrée. / ~
circuit (Radio) || Empfangsstromkreis m.
|| circuit m. de réception. / ~ circuit (Tel)
|| Aufnahmestromkreis m. || circuit m.
d'entrée. / ~ impedance of lines || Ein-
gangswiderstand m. von Leitungen || im-
pédance f. d'entrée de lignes. / ~ trans-
former (Tel) || Vorübertrager m. || trans-
formateur m. d'entrée.

inquartation (Met) || Quartation f. || in-
quart m.; inquartion f.; quartation f.

inquest || Untersuchung f. || enquête f. /
to make an ~ || eine Untersuchung f. an-
stellen || instituer une enquête.

inquire, to || anfragen; Umfrage f. halten ||
demander; s'informer; prendre des ren-
seignements. / ~ (Tel) || abfragen || prendre
la demande. / ~ into || nachforschen || re-
chercher; s'enquérir.

inquirer || Reflektant m. || amateur m.; in-
téressé m.

inquiry || Nachforschung f.; Umfrage f.;
Untersuchung f. || recherche f.; enquête f.
/ to make inquiries pl. || Umfrage f. hal-
ten || s'informer; prendre des renseigne-
ments.

inscription || Inschrift f. || inscription f.

inscriptions pl. (Coin) || Beschriftung f. || lé-
gende f. / marginal ~ (Coin) || Umschrift
f. || légende f.

insect flowers pl. || Insektenpulverblüten
fpl. || fleurs fpl. sauvages de pyrèthre.

insecticide see insect powder.

insect-powder || Insektenpulver n. || poudre
f. insecticide; insecticide m. / dry-spray-
ing of clothes with ~ || Einstäuben n. von
Kleidungsstücken mit Insektenpulver ||
dispersion f. de la poudre insecticide

pour la préservation de vêtements. / ~
blower || Insektenpulverspritze f. || souf-
flet m. de poudre insecticide.

insect wax || Insektenwachs n. || cire f. d'in-
sectes.

insensibility to awkward usage || Unemp-
findlichkeit f. gegen ungeschickte Be-
handlung || insensibilité f. à manipula-
tion malhabile.

insensible fermentation || Nachgärung f. ||
fermentation f. insensible; guillage m.

insensitive, to be ~ to dust and dirt || un-
empfindlich sein gegen Staub m. und
Schmutz m. || être à l'abri m. des pous-
sières.

insensitiveness to magnetism || Umemp-
findlichkeit f. gegen Magnetismus || in-
sensibilité f. contre le magnétisme.

insert, to || einbetten || encastrer; noyer. / ~
a drawing || einzeichnen || dessiner dans;
marquer.

inserted ceiling || Zwischendecke f. || faux-
plancher m. / ~ cover || Einsatzdeckel m.
|| couvercle m. inséré. / ~ crank-pin || ein-
gesetzter Kurbelzapfen m. || bouton m.
de manivelle emmanché.

inserted-teeth saw || Säge f. mit Einsatz-
zähnen || scie f. à dents rapportées.

inserter || Eindreher m. für Schrauben-
spundringe || emboutissoir m.

inserting, quick ~ and removing || schnel-
les Ein- und Ausspannen n. || serrage m.
et déserrage m. rapide.

insertion (Advertisement) || Anzeigenwesen
n. || annonces fpl. / ~ (Mach) || Einbau m.
|| montage m. / ~ of the coin || Einwurf
m. der Münze || introduction f. de la mon-
naie. / ~ of the paper || Einspannen n. des
Papiers || insertion f. du papier. / by ~ of
toothed gearings || unter Zwischenschal-
tung f. von Zahnradgetrieben || par
transmission f. d'engrenage.

insertion frame, small ~ || Einlegerähm-
chen n. || petit cadre m. *ou* petit tamis
m. à l'intérieur. / ~ funnel || Einsatz-
trichter m. || entonnoir m. d'insertion. /
~ step || Einsatzschale f. || coussinet m.
d'insertion.

inset diaphragm || Einhängeblende f. ||
diaphragme m. à suspendre. / ~ lock ||
Einstemmschloß n. || serrure f. enfoncée.

inside bearing || Innenlager n. || palier m.
intérieur. / ~ callipers pl. || Lochtaster
m. || compas m. d'intérieur. / ~ and out-
side callipers || Innen- und Außentaster
m. || maître m. de danse. / ~ chaser || In-
nensträhler m. || peigne m. femelle. / ~
compasses || Innentaster m.; Lochzirkel
m. || compas m. d'intérieur. / ~ crank ||
Innenkurbel f. || manivelle f. intérieure. /
~ cranks pl. which necessitate a cranked
axle || Innenkurbeln fpl., die Achskröp-
fung erfordern || manivelles fpl. inté-
rieures exigeant que l'essieu soit coudé. /
~ decorating article || Innendekora-
tionsartikel m. || article m. décoratif
d'intérieur. / ~ diameter || Innendurch-
messer m.; lichter Durchmesser m. ||
diamètre m. intérieur. / ~ door || Innen-
tür f. || porte f. intérieure. / ~ gearing ||
Innenantrieb m. || commande f. par
engrenages intérieurs. / ~ link || Innen-
glied n. || pièce f. à l'intérieur. / ~ plash-
ing of the fluid || Flüssigkeitsschlag m. ||
impact m. du liquide. / ~ running crab ||
innenlaufende Katze f. || chariot m. cir-
culant à l'intérieur. / ~ screw tool || In-
nengewindestahl m. || peigne m. femelle.

/ ~ thread || Innengewinde n. || filet m. intérieur.

insignificant || unansehnlich; unscheinbar || peu apparent; insignifiant.

insipid || geschmacklos || insipide; sans goût.

insole || Einlegesohle f. || semelle f. mobile *ou* intérieure. / ~ (of leather) || Brandsohle f. || première f.; semelle f. intérieure. / ~ out of old shoes (Shoem) || Altbrandsohle f. || vieille première f. / ~ liner (Shoem) || Brandsohlenfütterer m. || entoileur m. de premières.

insolubility || Unlöslichkeit f. || insolubilité f.

insoluble || unlösbar; unlöslich || insoluble. / rendering ~ || Unlöslichmachen n. || insolubilisation f.

insolvency || Insolvenz f.; Zahlungsunfähigkeit f.; Bankrott m.; Konkurs m. || faillite f.; banqueroute f.; insolvabilité f.

Insolvency Act || Konkursordnung f. || loi f. sur les faillites.

insolvent || insolvent; zahlungsunfähig; bankrott || banqueroutier; failli; insolvable. / to become ~ || fallieren || faillir; manquer.

inspect, to || besichtigen; prüfen || examiner; inspecter; visiter.

inspecting engineer || Abnahmebeamter m. || agent-réceptionnaire m.

inspection || Besichtigung f. || inspection f.; visite f.; examen m. / ~ (Mar) || Musterung f. || revue f.; appel m. / ~ for arms and gear || Musterung f. von Waffen und Gerät || inspection f. des armes et de l'outillage. / ~ of engines || Untersuchung f. von Maschinen || visite f. des machines. / facility of ~ of the power plant (Electr) || Übersichtlichkeit f. der Maschinenanlage || bonne disposition f. de l'installation des machines. / inside ~ of the boiler || innere Kesselrevision f. || visite f. intérieure de la chaudière. / ~ of internal surface || Innenmessung f. || mesurage m. d'intérieur. / rapid ~ || Schnellprüfung f.; Schnellrevision f. || contrôle m. rapide. / rapid ~ of pieces produced in series || Schnellprüfung f. von Massenteilen || contrôle m. rapide de pièces fabriquées en série.

inspection apparatus || Revisionswerkzeug n. || outil m. pour la révision. / ~ department || Revisionsabteilung f. || service m. de contrôle. / ~ gauge || Abnahmelehre f. || calibre m. *ou* jauge f. de réception. / ~ hole || Schauklappe f. || clapet m. de regard. / ~ pit || Arbeitsgrube f. || fosse f. de réparation. / ~ plate || Schaulochdeckel m. || couvercle m. de porte de visite. / ~ trolley (Railw) || Bahnmeisterwagen m. || wagonnet m. d'inspection.

inspector || Inspektor m. || inspecteur m. / ~ (Woman) || Aufseherin f. || visiteuse f. / ~ (Build) || Bauaufseher m. || surveillant m. des travaux. / ~ (Mach) || Abnahmebeamter m. || agent-réceptionnaire m. / ~ of mines || Revierbeamter m.; Bergwerksinspektor m. || inspecteur m. des mines. / ~ of permanent way || Bahnmeister m. || inspecteur m. de chemin de fer. / professional factory ~ || Betriebskontrollör m. || contrôleur m. de fabrication. / ~ of telegraphs || Telegrafeninspektor m. || inspecteur m. des télégraphes.

inspissate, to || eindicken || inspisser.

inspissation || Eindickung f. || épaississage m.

instability || Instabilität f. || instabilité f.

installation on board of a ship || Schiffsinstallation f. || installation f. à bord d'un navire. / ~ for bread factory || Brotfertigungsanlage f. || installation f. de panification. / ~ with by-product plant || Anlage f. mit Gewinnung von Nebenerzeugnissen || usine f. avec récupération de sous-produits. / ~ of hospitals || Krankenhauseinrichtung f. || installation f. pour hôpitaux. / easy ~ || leichter Einbau m. || montage m. facile. / electric ~ || elektrische Einrichtung f. *oder* Installation f. || installation f. électrique. / ~ for the protection of rooms || Raumschutzanlage f. || installation f. de protection pour des appartements. / ~ material || Installationswerkstoff m. || matériel m. d'installation.

installation, contracting for ~ || Installationsgeschäft n. || entreprise f. d'installations. / sanitary ~ implement || sanitärer Installationsgegenstand m. || appareil m. d'installations sanitaires. / ~ switch || Kleinschalter m. || interrupteur m. d'installation.

installed kilowatt || installiertes Kilowatt n. || kilowatt m. installé.

instalment || Teilzahlung f.; Anzahlung f.; Abschlagszahlung f.; Rate f. || acompte m.; payement m. partiel *ou* par acompte; terme m. / by ~s || ratenweise || par termes mpl. *ou* acomptes. / ~ of a delivery || Teilsendung f. || envoi m. partiel. / first ~ || Angeld n.; Anzahlung f. || arrhes fpl.; acompte m. / to make a first ~ || anzahlen || donner en acompte m.; donner à compte m. / to pay by ~s || in Raten fpl. bezahlen || payer par termes *ou* par acomptes. / payment by ~s || Ratenzahlung f. || payement m. par termes *ou* acomptes *ou* à tempérament. / business on the ~ system || Abzahlungsgeschäft n. || maison f. de vente à crédit *ou* à tempérament.

instance, for ~ || beispielsweise; zum Beispiel || par exemple.

instantaneous || momentan; augenblicklich || instantané. / ~ axis || Momentanachse f. || axe m. instantané. / ~ flight altitude || augenblickliche Flughöhe f. || altitude f. moyenne du vol. / ~ iris shutter || Irismomentverschluß m. || obturateur m. instantané à iris. / ~ load || Augenblicksbelastung f. || charge f. instantanée. / ~ plate (Phot) || Momentplatte f. || plaque f. pour instantanés.

instantaneous shutter (Phot) || Momentverschluß m. || obturateur m. pour instantané. / sector ~ || Sektormomentverschluß m. || obturateur m. instantané à secteurs. / ~ with variable sector spring || Momentverschluß m. mit verstellbarer Sektorweite || obturateur m. instantané à secteurs à ouverture variable. / release of ~ || Auslösung f. des Momentverschlusses || déclenchement m. de l'obturateur instantanée.

instantaneous stand-by || Momentanreserve f. || réserve f. momentanée. / ~ stereophotograph || Augenblicksstereoaufnahme f. || photographie f. stéréoscopique instantanée. / ~ stop-motion device || Momentausrücker m. || dispositif m. de débrayage instantané. / ~ view || Momentaufnahme f. || vue f. instantanée.

instep (of the foot) || Spann m. || cou-de-pied m.

institute || Anstalt f.; Institut n. | institut m. / anatomical ~ || Anatomie f. || institut m. d'anatomie. / scientific ~ || wissenschaftliche Anstalt f. || institut m. scientifique.

institution (Trade) || Gründung f. || fondation f.; établissement m.; création f. / technological ~ || Gewerbeschule f. || école f. professionnelle.

instruct, to ~ an apprentice || einen Lehrling m. anlernen || instruire un apprenti.

instructed, as ~ || auftragsgemäß || en conformité f. aux ordres.

instruction || Unterweisung f.; Anweisung f.; Anleitung f. || instruction f.; indication f.; ordre m.; enseignement m. / ~ for employment || Behandlungsvorschrift f. || mode m. d'emploi; instructions fpl. de service. / to give ~s || Anordnungen fpl. treffen || prendre des dispositions fpl. / speedy and distinct transmission of ~s || schnelle und eindeutige Übermittlung f. von Befehlen || transmission f. rapide et irréprochable des ordres. / ~ for use || Gebrauchsanweisung f. || mode m. d'emploi; instructions fpl. de service.

instructions pl. from the office of the Board of Works || baupolizeiliche Vorschriften f. || instructions f. de la police de construction.

instructional workshop || Lehrlingswerkstatt f. || atelier m. d'apprentissage.

instruction means || Lehrmittel npl. || moyens mpl. d'enseignement.

instructive film || Lehrfilm m. || film m. d'enseignement.

instrument || Gerät n.; Instrument n. || instrument m. / to set up an ~ || ein Instrument n. aufstellen || monter *ou* disposer *ou* mettre en place un instrument. / accessory ~ || Zusatzgerät n. || appareil m. complémentaire. / ~ for aerologial measuring || aerologisches Meßgerät n. || instrument m. pour des mesures aérologiques. / astronomical ~ || astronomisches Instrument n. || instrument m. astronomique *ou* d'astronomie. / ~ for calculating || Instrument n. zum Rechnen || instrument m. à calculer. / circular ~ || Kreismesser m. || instrument m. cyclométrique. / cosmographical ~ || kosmografisches Instrument n. || instrument m. de cosmographie. / dental ~ || zahnärztliches Instrument n. || instrument m. pour dentistes. / direct reading ~ || Instrument n. mit unmittelbarer Ablesung. || instrument m. à lecture directe. / drawing ~ || Zeicheninstrument n. || instrument m. de dessin. / flush-mounted edgewise pattern ~ || eingebautes Profilinstrument n. || instrument m. de profil encastré. / ~ of geodesy || Erdmeßinstrument n.; geodätisches Instrument n. || instrument m. géodétique *ou* de géodésie. / geodetical ~ *see* ~ of geodesy / hydrometric ~ || hydrometrisches Instrument n. || instrument m. hydrométrique. / ~ for investigating ore deposits || Instrument n. zur Untersuchung von Erzlagerstätten || instrument m. pour la recherche des gisements de minerais. / levelling ~ || Nivellierinstrument n. || instrument m. de nivellement. / ~ for technical levelling operations || Instrument n. für technische Einwägungen || instrument m. destiné aux nivellements techniques. / measuring ~ || Meßgerät n. || instrument m. de mesure *ou* de mesurage. / ~ for

measuring lengths ‖ Längenmeßgerät n. ‖ appareil m. pour la mesure des longueurs *ou* la mesure en long. / mechanical ~ of precision ‖ feinmechanisches Instrument n. ‖ instrument m. de précision. / meteorological ~ ‖ meteorologisches Instrument n. ‖ instrument m. météorologique. / nautical ~ ‖ nautisches Instrument n. ‖ instrument m. nautique. / ~ of observation ‖ Beobachtungsinstrument n. ‖ instrument m. d'observation. / optical ~ ‖ optisches Instrument n. ‖ instrument m. (d')optique. / photographic ~ ‖ fotografisches Instrument n. ‖ instrument m. de photographie. / ~ of polarization ‖ Polarisationsinstrument n. ‖ instrument m. de polarisation. / ~ of precision ‖ Präzisionsinstrument n. ‖ instrument m. de précision. / ~ ready for use ‖ Instrument n. in gebrauchsfertigem Zustande ‖ instrument m. tout monté et prêt à l'usage. / rustproof ~ ‖ nichtrostendes Instrument n. ‖ instrument m. inoxydable. / scientific ~ ‖ wissenschaftliches Instrument n. ‖ instrument m. scientifique. / short and handy ~ ‖ kurzes, handliches Instrument ‖ instrument m. d'une forme pratique et maniable./ stringed ~ ‖ Saiteninstrument n. ‖ instrument m. à cordes. / surveying ~ ‖ Feldmeßinstrument n. ‖ instrument m. d'arpentage. / suspended ~ ‖ Hängeinstrument n. ‖ instrument m. suspendu *ou* à suspension. / ~ for testing roundness and parallelism ‖ Gerät n. zum Prüfen der Zylindrizität f. ‖ appareil m. pour la vérification de l'ovalisation et de la cylindricité. / wind ~ (Mus) ‖ Blasinstrument n. ‖ instrument m. à vent.

instrumental error ‖ Instrumentfehler m.; Instrumentalfehler m. ‖ erreur f. d'instrument. / to correct the ~s ‖ die Instrumentalfehler mpl. berichtigen ‖ corriger les erreurs fpl. de l'instrument.

instrument board ‖ Instrumentenbrett n.; Schalttafel f. ‖ planche f. *ou* tablier m. de distribution *ou* d'instruments. / ~ body ‖ Instrumentenkörper m. ‖ corps m. de l'instrument. / ~ flying ‖ Blindfliegen n. ‖ vol m. aveugle. / ~ lamp ‖ Schalttafellaterne f. ‖ lampe f. de tablier. / ~ maker ‖ Feinmechaniker m. ‖ mécanicien m. de précision. / ~ pier ‖ Instrumentenpfeiler m. ‖ pilier m. pour les instruments.

instruments pl. ‖ Gerät n.; Gezähen.; Handwerkzeug n. ‖ ustensiles mpl.; outils mpl.

instrument table ‖ Instrumententisch m. ‖ table f. porte-instruments. / ~ tuner ‖ Instrumentenstimmer m. ‖ accordeur m. d'instruments de musique.

insubmersible ‖ unsinkbar ‖ insubmersible.

insufficient ‖ unbefriedigend; ungenügend ‖ peu satisfaisant; insuffisant.

insulate, to ‖ isolieren; absondern ‖ isoler.

insulated ‖ isoliert ‖ isolé.

insulating ‖ nichtleitend ‖ isolant; isolateur. / ~ accessories pl. ‖ Isolationszubehör n. ‖ matériel m. d'isolation. / ~ base cap ‖ Isoliersockel m. ‖ coulot m. isolant / ~ bottle ‖ Isolierflasche f. ‖ bouteille f. isolante./ ~ cement ‖ Isolierkitt m. ‖ ciment m. *ou* mastic m. isolant. / ~ chair ‖ Isolierschemel m. ‖ tabouret m. isolant. / heat ~ composition ‖ Wärmeschutzmasse f. ‖ composition f. calorifuge; calorifuge m. / ~ covering ‖ Isolationshülle f. ‖ revêtement m. isolant. / ~

filling paste ‖ Füllmasse f. für Isolierzwecke ‖ pâte f. de remplissage isolante. / ~ glove ‖ Gummihandschuh m. ‖ gant m. en caoutchouc. / ~ intermediate layer ‖ isolierende Zwischenlage f. ‖ couche f. isolante interposée. / ~ ledge ‖ Isolierleiste f. ‖ baguette f. d'isolement.

insulating material ‖ Isolierstoff m. ‖ matériel m. isolant. / artificial ~ ‖ künstlicher Isolationsstoff m. ‖ isolants mpl. synthétiques. / covering with ~ ‖ Abdecken n. mit Isolierstoff ‖ application f. d'un isolant (sur). / electric ~ ‖ elektrischer Isolierwerkstoff m. ‖ matière f. isolante électrique. / ~ for (electric) high-tension ‖ Hochspannungsisolierwerkstoff m. ‖ matière f. isolante pour courant à haute tension. / non-inflammable ~ ‖ flammensicherer Isolationsstoff m. ‖ isolant m. ignifuge. / plastic ~ ‖ plastischer Isolierstoff m. ‖ isolant m. plastique. / rubber ~ ‖ Gummiisolierung f.; Gummiisolationsstoff m. ‖ isolants mpl. à base de caoutchouc. / solid ~ ‖ fester Isolationsstoff m. ‖ isolant m. solide. / ~ making machine ‖ Maschine f. zum Herstellen von Isolierstoffen ‖ machine f. pour fabriquer des matières isolantes.

insulating muslin ‖ Isoliermull m. ‖ organdi m. d'isolation. / ~ pad ‖ Isolierpolster n. ‖ coussin m. isolant. / ~ paints pl. ‖ Isolieranstrich m. ‖ peinture f. isolante; enduit m. isolant. / ~ paper ‖ Isolierpapier n. ‖ papier m. isolateur. / ~ paper machine ‖ Isolierpapiermaschine f. ‖ machine f. à préparer le papier d'isolement. / ~ property ‖ Isoliervermögen n. ‖ pouvoir m. isolant. / ~ slab ‖ Isolierplatte f. ‖ plaque f. isolante. / ~ stool *see* ~ chair. / ~ support ‖ Isolierfuß m. ‖ pied m. isolant. / ~ switch ‖ Trennschalter m. ‖ sectionneur m.; isolateur m. / ~ tape ‖ Isolierband n. ‖ ruban m. isolant. / ~ tape producing machine ‖ Isolierbandherstellungsmaschine f. ‖ machine f. à fabriquer des rubans isolants. / ~ tube ‖ Isolierrohr n. ‖ tuyau m. *ou* tube m. isolant. / ~ tube coiling machine ‖ Isolierrohrwickelmaschine f. ‖ enrouleuse f. de tubes isolants. / ~ varnish ‖ Isolierfirnis m. ‖ vernis m. isolant.

insulation ‖ Isolierung f.; Isolation f. ‖ isolement m.; isolation f. / air-space paper ~ ‖ Papierluftraumisolierung f. ‖ isolement m. à air et au papier. / to burn off the ~ ‖ die Isolation abbrennen ‖ brûler l'isolant m. / dry ~ ‖ Trockenisolierung f. ‖ isolement m. sec. / ~ from earth ‖ Isolation f. gegen Erde ‖ isolation f. par rapport à la terre. / electric ~ ‖ elektrische Isolierung f. ‖ isolement m. électrique. / fibrous ~ ‖ Faserstoffisolierung f. ‖ isolement m. fibreux. / ~ for heat ‖ Wärmeschutzisolierung f. ‖ isolement m. de la chaleur. / india-rubber ~ ‖ Gummiisolierung f. ‖ isolement m. en caoutchouc. / ~ of masonry ‖ Mauerwerkisolierung f. ‖ isolation f. de maçonnerie. / rubber tape ~ ‖ Gummibandisolierung f. ‖ isolement m. en ruban de caoutchouc. / strengthened ~ ‖ verstärkte Isolation f. ‖ isolement m. renforcé.

insulation detector ‖ Isolationsprüfer m. ‖ détecteur m. d'isolement. / layer of ~ ‖ Isolierschicht f. ‖ couche f. d'isolation. / measurement of ~ ‖ Isolationsmessung f. ‖ mesure f. d'isolement. / electrical ~

parts pl. ‖ elektrische Isolationsteile mpl. ‖ pièces fpl. d'isolation électriques. / ~ resistance ‖ Isolationswiderstand m. ‖ résistance f. d'isolement. / very high ~ resistance ‖ sehr hoher Isolationswiderstand m. ‖ résistance f. d'isolement très élevée. / ~ test ‖ Isolationsprüfung f. ‖ essai m. d'isolement. / ~ tester ‖ Isolationsmesser m. ‖ essayeur m. d'isolement.

insulator (Apparatus) ‖ Nichtleiter m.; Isolator m. ‖ corps m. isolateur; diélectrique m.; isolateur m. / ~ (Material) ‖ Isolationsmittel n. ‖ isolant m. / bell-shaped ~ ‖ Glockenisolator m. ‖ isolateur m. en cloche; cloche f. isolante. / cleat ~ ‖ Klemmisolator m. ‖ isolateur m. à collier. / cold ~ ‖ Kälteisoliermittel n. ‖ calorifuge m. frigorifique. / ~ for cold ‖ Kälteschutzisolierung f. ‖ isolement m. du froid. / contact-wire ~ ‖ Fahrdrahtisolator m. ‖ isolateur m. de fil de contact. / cup ~ ‖ Glockenisolator m. ‖ cloche f. isolante. / double-cup ~ ‖ Doppelglockenisolator m. ‖ isolateur m. à double cloche. / double-shed ~ *see* double-cup ~. / flexible ~ ‖ biegsamer Isolator m. ‖ isolateur m. souple. / glass ~ ‖ Glasisolator m. ‖ isolateur m. en verre. / heat ~ ‖ Wärmeisoliermittel n.; Wärmeschutz m. ‖ calorifuge m. thermique; isolation f. calorifuge. / cork ~ for heat ‖ Wärmeschutzschale f. aus Kork ‖ enveloppe f. calorifuge en liège. / straw ~ for heat ‖ Wärmeschutz m. aus Stroh ‖ calorifuge f. en paille. / high-frequency ~ ‖ Hochfrequenzisolator m. ‖ isolateur m. à haute frequence. / high-tension ~ ‖ Hochspannungsisolator m. ‖ isolateur m. à haute tension. / petticoat ~ ‖ Glockenisolator m. ‖ isolateur m. à cloche. / pin ~ ‖ Stützisolator m. ‖ isolateur m. à ferrure. / porcelain ~ ‖ Porzellanisolator m. ‖ isolateur m. en porcelaine. / receiving ~ ‖ Isolator m. für Empfangsdraht ‖ isolateur m. de réception. / reel ~ ‖ Isolierrolle f. ‖ roulette f. isolante. / rod ~ ‖ Stabisolator m. ‖ isolateur m. à tige. / ~ for the fixing of room antenna ‖ Isolator m. für Befestigung von Zimmerantennen ‖ isolateur m. pour la pose des antennes de chambre. / screw cap ~ ‖ Isolator m. mit Schraubenkappe ‖ isolateur m. à couvercle fileté. / single-cup ~ ‖ Einfachglockenisolator m. ‖ isolateur m. à simple cloche. / soap-stone ~ ‖ Specksteinisolator m. ‖ isolateur m. en stéatite. / suspended ~ ‖ Pendelisolator m. ‖ isolateur m. suspendu *ou* à suspension. / swinging bell-shaped ~ ‖ Hängeglockenisolator m. ‖ isolateur f. à cloche à suspension. / transmitting ~ (Radio) ‖ Isolator m. für Sendeantenne ‖ isolateur m. de transmission.

insulator bracket ‖ Isolatorstütze f. ‖ support m. d'isolateur. / ~ chain ‖ Isolatorenkette f. ‖ chaîne f. d'isolateurs. / ~ pin *see* ~ spindle. / ~ spindle ‖ Isolatorstütze f. ‖ porte-isolateur m.; console f.; support m. / ~ string ‖ Isolatorenkette f. ‖ chaîne f. d'isolateurs.

insurable ‖ versicherbar ‖ assurable. / ~ value ‖ Versicherungswert m. ‖ valeur m. assurable.

insurance ‖ Versicherung f.; Assekuranz f. ‖ assurance f. / ~ against accidents ‖ Unfallversicherung f. ‖ assurance f.

contre les accidents. / additional ~ ‖ zusätzliche Versicherung f. ‖ assurance f. supplémentaire. / ~ of the aged and disabled workmen ‖ Invalidenversicherung f. ‖ assurance f. contre l'invalidité et la vieillesse. / ~ against breakage ‖ Versicherung f. gegen Bruch ‖ assurance f. contre la casse. / ~ of cargo ‖ Versicherung f. der Ladung ‖ assurance f. de la cargaison. / ~ upon the claims for wages (Mar) ‖ Versicherung der Heuerforderungen ‖ assurance f. des loyers des gens de mer. / covered by ~ ‖ durch Versicherung gedeckt ‖ couvert par assurance f. / fully covered by ~ ‖ durch Versicherung f. völlig gedeckt ‖ complètement couvert par assurance. / partially covered by ~ ‖ durch Versicherung f. teilweise gedeckt ‖ partiellement couvert par assurance. / ~ covers all risks ‖ die Versicherung f. deckt alle Risiken ‖ l'assurance f. couvre tous les risques. / ~ covers partial loss ‖ die Versicherung f. deckt teilweisen Verlust ‖ l'assurance f. couvre perte partielle. / ~ covers total loss only ‖ die Versicherung f. deckt nur gänzlichen Verlust ‖ l'assurance f. couvre perte totale seulement. / extra ~ ‖ Sonderversicherung f.; Nachversicherung f. ‖ assurance f. supplémentaire. / fire ~ ‖ Feuerversicherung f.; Brandassekuranz f. ‖ assurance f. contre l'incendie. / ~ on freight ‖ Frachtversicherung f. ‖ assurance f. sur fret. / ~ against damage by hail ‖ Versicherung f. gegen Hagelschaden ‖ assurance f. contre la grêle. / ~ on hull and appurtenances ‖ Kaskoversicherung f. ‖ assurance f. sur coque ou sur corps et quille. / ~ against leakage ‖ Versicherung f. gegen Leckage ‖ assurance f. contre fuite. / liability ~ ‖ Haftpflichtversicherung f. ‖ assurance f. contre les accidents causés à des ouvriers. / life ~ ‖ Lebensversicherung f. ‖ assurance f. vie. / marine ~ ‖ Seeversicherung f. ‖ assurance f. maritime. / old age pension ~ ‖ Altersversicherung f. ‖ assurance f. sur la vieillesse. / re-~ ‖ Rückversicherung f. ‖ réassurance f. / ~ against all risks ‖ Versicherung f. gegen alle Risiken ‖ assurance f. contre tous risques. / ~ against sickness ‖ Krankenversicherung f. ‖ assurance f. contre la maladie. / ~ against total loss ‖ Versicherung f. gegen gänzlichen Verlust ‖ assurance f. contre perte totale. / ~ against damage in transit ‖ Transportversicherung f. ‖ assurance f. contre les risques des transports. / ~ of value ‖ Wertversicherung f. ‖ assurance f. à la valeur. / ~ against war risks ‖ Versicherung f. gegen Kriegsgefahren ‖ assurance f. contre les risques de guerre.

insurance accounting ‖ Versicherungsbuchführung f. ‖ comptabilité f. des compagnies d'assurances. / ~ agent ‖ Assekuranzmakler m.; Versicherungsmakler m. ‖ courtier m. d'assurances. / amount of ~ ‖ Betrag m. der Versicherung ‖ montant m. de l'assurance. / ~ association ‖ Versicherungsverein m.; Lebensversicherungsgesellschaft f. ‖ société f. d'assurance. / ~ broker see ~ agent. / ~ charges pl. ‖ Versicherungsgebühr f. ‖ frais mpl. d'assurance. / ~ company ‖ Versicherungsgesellschaft f. ‖ compagnie f. d'assurance. / cost of ~ ‖ Versicherungskosten pl. ‖ coût m. de l'assurance. / extra rate of ~ ‖ Extraversicherungs-

prämiensatz m. ‖ taux m. supplémentaire d'assurance. / ~ policy ‖ Versicherungsschein m.; Police f. ‖ police f. d'assurance. / ~ premium ‖ Versicherungsprämie f. ‖ prime f. d'assurance. / ~ work ‖ Versicherungsarbeiten fpl. ‖ travaux mpl. d'assurances.

insure, to ‖ versichern ‖ assurer.

insured letter ‖ Wertbrief m. ‖ lettre f. chargée ou de valeur déclarée.

insurer ‖ Versicherer m. ‖ assureur m.

intact ‖ unbeschädigt; unversehrt ‖ intact.

intaglio ‖ Intaglio n.; Gemme f. ‖ intaille f.; gemme f. / ~ cylinder copper-coating plant (Print) ‖ Tiefdruckwalzenaufkupferungsanlage f. ‖ installation f. de cuivrage des cylindres d'impression en creux.

intake ‖ Einnehmen n. ‖ prise f.; recette f. / ~ ‖ Einmündung f. ‖ entrée f.; embouchure f. / hot air ~ ‖ Warmlufteintritt m. ‖ entrée f. de l'air chaud. / ~ chamber ‖ Wasserschloß n. ‖ bassin m. de charge; château m. d'eau.

integral (Math) ‖ Integral n. ‖ intégrale f. / ~ (Sign) see ~ sign. / definite ~ ‖ bestimmtes Integral n. ‖ intégrale f. définie. / indefinite ~ ‖ unbestimmtes Integral n. ‖ intégrale f. indéfinie.

integral calculus ‖ Integralrechnung f. ‖ calcul m. intégral. / ~ sign ‖ Integralzeichen n. ‖ signe m. intégral.

integrate, to ‖ integrieren ‖ intégrer.

integration ‖ Integration f. ‖ intégration f. / ~ by parts ‖ partielle oder teilweise Integration f. ‖ intégration f. par parties.

intelligence department ‖ Auskunftei f. ‖ bureau m. des renseignements.

intelligibility ‖ Verständlichkeit f.; Deutlichkeit f. ‖ intelligibilité f.; netteté f.; clarté f. / ~ of speech ‖ Verständlichkeit f. der Sprache ‖ articulation f.; clarté f. de la prononciation.

intend, to ~ (To project) ‖ projektieren; entwerfen ‖ projeter.

intending purchaser ‖ Reflektant m. ‖ amateur m.; intéressé m.

intense see intensive.

intensifier (Phot) ‖ Verstärker m. ‖ renforçateur m. / pressure ~ (Mach) ‖ Druckübersetzer m. ‖ multiplicateur m. de pression. / steam ~ ‖ Dampftreibapparat m. ‖ multiplicateur m. à vapeur. / steam pressure ~ ‖ Dampfdruckübersetzer m. ‖ multiplicateur m. de pression de vapeur.

intensify, to ~ (Phot) ‖ verstärken ‖ renforcer.

intensifying agent see intensifier. / ~ screen ‖ Verstärkungsschirm m. ‖ écran m. renforçateur.

intension see intensity.

intensity ‖ Intensität f.; Heftigkeit f.; Stärke f. ‖ intensité f. / ~ of the current ‖ Intensität f. des Stromes; Stromstärke f. ‖ intensité f. du courant. / field ~ ‖ Feldstärke f. ‖ intensité f. de champ. / horizontal ~ ‖ Horizontalintensität f. ‖ intensité f. horizontale. / intrinsic ~ of a lamp ‖ spezifische Helligkeit f. einer Lampe ‖ éclat m. spécifique d'une lampe. / ~ of irradiation ‖ Ausstrahlungsintensität f.; Ausstrahlungslebhaftigkeit f.; Ausstrahlungsstärke f. ‖ intensité f. lumineuse. / ~ of light ‖ Helligkeit f. ‖ clarté f.; éclat m. / photometric ~ ‖ fotometrische Lichtstärke f. ‖ intensité f. photométrique. / ~ of stress due to bending ‖ Biegespannung f. ‖ tension f. de flexion. / ~ of tone ‖ Tonstärke f. ‖ intensité f. de son. / total

~ of light ‖ Gesamthelligkeit f. ‖ clarté f. générale. / ~ of wind ‖ Windstärke f. ‖ intensité f. du vent.

intensity change (of light) ‖ Helligkeitsveränderung f. ‖ modification f. de la luminosité.

intensive ‖ intensiv; heftig; stark ‖ intensif; intense. / illumination by means of built-in ~ fittings ‖ Beleuchtung f. durch eingebaute Tiefstrahler oder Steilstrahler ‖ éclairage m. au moyen d'armatures à abat-jour profond masquées ou à rayonnement oblique. / ~ mixing of the fuel gas with the air of combustion ‖ innige Mischung f. des Heizgases mit der Verbrennungsluft ‖ mélange m. intime du gaz de chauffage avec l'air de combustion.

interaction ‖ Wechselwirkung f. ‖ action f. réciproque.

intercalation ‖ Zwischenschaltung f. ‖ intercalage m.

interception of malicious calls (Tel) ‖ Abfangen n. böswilliger Anrufe ‖ interception f. des communications malicieuses.

interchange, to ‖ auswechseln; austauschen; vertauschen ‖ échanger; remplacer.

interchange ‖ Auswechs(e)lung f.; Vertauschung f.; Austausch m. ‖ remplacement m.; échange m. / ~ of air ‖ Luftaustausch m. ‖ échange m. d'air. / ~ of current ‖ Stromaustausch m. ‖ échange m. de courant. / ~ of heat ‖ Wärmeaustausch m. ‖ échange m. de chaleur.

interchangeability ‖ Auswechselbarkeit f. ‖ interchangeabilité f.

interchangeable ‖ auswechselbar ‖ interchangeable; échangeable; remplaçable. / ~ between each other ‖ untereinander auswechselbar ‖ interchangeable mutuellement. / easily ~ ‖ leicht auswechselbar ‖ facilement interchangeable.

interchangeable coil set (Radio) ‖ auswechselbarer Spulensatz m. ‖ jeu m. de selfs interchangeables. / ~ fabrication ‖ Austauscharbeit f. ‖ fabrication f. interchangeable. / ~ parts pl. ‖ untereinander auswechselbare Teile mpl. ‖ pièces fpl. interchangeables.

interchanging of lenses pl. ‖ Auswechseln n. der Linsen ‖ échange m. des lentilles. / ~ of types ‖ Buchstabenauswechselung f. ‖ changement m. de caractères.

intercolumniation ‖ Säulenweite f. ‖ entrecolonne f.; entre-colonnement m.

intercommunication set ‖ Reihenanlage f. ‖ installation f. des postes en série.

interconnection (Tel) ‖ Verbindung f. ‖ interconnexion f.; communication f.

intercooler ‖ Zwischenkühler m. ‖ refroidisseur m. intermédiaire.

intercostal ‖ Interkostal n.; Zwischenblech n. ‖ tôle f. intercostale. / ~ plate see intercostal.

interest ‖ Verzinsung f. ‖ intérêt m. / accrued ~ see accumulated ~. / accumulated ~ ‖ aufgelaufene Zinsen mpl. ‖ intérêts mpl. accumulés. / to add the ~ to the capital ‖ die Zinsen mpl. zum Kapital schlagen ‖ joindre les intérêts au capital. / compound ~ ‖ Zinseszins m. ‖ intérêt m. composé. / electro-economic ~ ‖ elektrowirtschaftliche Beteiligung f. ‖ participation f. electro-économique. / to have an ~ ‖ teilhaben ‖ avoir part m.; participer. / ~ on money ‖ Geldzins m. ‖ intérêt m. ou rente f. en espèces. / ~ on mortgage ‖ Hypothekenzinsen mpl. ‖ intérêt m. sur

hypothèques. / particular ~ ‖ Sonderinteresse n. ‖ intérêt m. séparé ou particulier. / ~ payable on arrears ‖ Verzugs-
zinsen mpl. ‖ intérêts mpl. moratoires.
interest, balance of ~ ‖ Zinsensaldo m. ‖
solde m. des intérêts. / at bank rate of ~
‖ zum Bankzinsfuß m. ‖ intérêt m. au
taux de la banque. / ~ bearing security
‖ zinstragendes Papier n. ‖ valeur f. à
intérêts. / loss of ~ ‖ Zinsverlust m. ‖
perte f. d'intérêts. / main sphere of ~ /
Hauptinteressengebiet n. ‖ périmètre m.
d'influence.
interfere, to ~ (Phys) ‖ interferieren ‖ inter-
férer.
interference ‖ Interferenz f. ‖ interférence
f. / ~ (Between electric lines) ‖ Beeinflussung f. ‖ interférence f. mutuelle. / dangerous ~ with high-tension current ‖ Gefährdung f. durch Starkstrom ‖ mise f. en
danger par courant fort. / ~ from power
lines ‖ Störung f. durch Starkstrom ‖ perturbation f. par lignes à courant fort. / ~
in reception ‖ Interferenz f. im Empfang
‖ interférence f. dans la réception.
interference apparatus ‖ Interferenzgerät
n. ‖ appareil m. interférentiel. / ~ band ‖
Interferenzstreifen m. ‖ frange f. d'interférence. / ~ colour ‖ Interferenzfarbe f. ‖
couleur f. interférence. / ~ comparator
for absolute measurements ‖ Interferenz-
komparator m. für Absolutmessungen ‖
comparateur m. interférentiel pour mesures absolues. / ~ curve ‖ Interferenz-
kurve f. ‖ courbe f. d'interférence. / ~
factor ‖ Störfaktor m. ‖ facteur m. perturbateur. / ~ figure ‖ Interferenzerschei-
nung f. ‖ phénomène m. d'interférence. /
~ measuring apparatus ‖ Interferenzmeß-
gerät n. ‖ appareil m. de mesure interférentiel. / phenomenon of ~ ‖ Interferenz-
erscheinung f. ‖ phénomène m. d'interférence. / ~ prevention (Radio) ‖ Stö-
rungsverhinderung f. ‖ élimination f. des
interventions ou d'interférence. / ~ spectroscope ‖ Interferenzspektroskop n. ‖
spectroscope m. interférentiel. / ~ tube
‖ Interferenzröhre ‖ tuyau m. à l'interférence.
interferometer ‖ Interferometer n. ‖ interféromètre m. / combined gas and water ~
‖ vereinigtes Gas- und Wasserinterferometer n. ‖ interféromètre m. à eau et à
gaz. / ~ for gases ‖ Gasinterferometer n. ‖
interféromètre m. à gaz. / industrial ~ ‖
technisches Interferometer n. ‖ interféromètre m. technique. / laboratory ~ ‖ Laboratoriumsinterferometer n. ‖ interféromètre m. de laboratoire. / ~ for liquids
‖ Flüssigkeitsinterferometer n. ‖ interféromètre m. à liquides.
interim certificate ‖ Interimsschein m. ‖
certificat m. provisoire. / ~ dividend ‖
Interimsdividende f.; Zwischendividende
f. ‖ dividende m. intérimaire ou provisoire.
interior ‖ inner; inwendig; licht ‖ intérieur; interne. / ~ antenna ‖ Innenan-
tenne f. ‖ antenne f. intérieure. / ~ diameter ‖ lichter Durchmesser m. ‖ diamètre m. intérieur. / ~ flange ‖ Innen-
flansch m. ‖ bride f. intérieure. / ~ gilding
‖ Innenvergoldung f. ‖ dorure f. à l'intérieur. / ~ surface ‖ Innenfläche f. ‖ paroi
f. intérieure. / ~ view ‖ Innenansicht f. ‖
vue f. intérieure.
interior ‖ Inneres n. ‖ intérieur m.; dedans
m. / ~ of building ‖ Innenansicht f. eines

Hauses ‖ vue intérieure f. d'un édifice. /
~ of the earth ‖ Inneres n. der Erde ‖ intérieur m. de la terre.
interlace, to ‖ durchflechten ‖ enverger; entrelacer.
interleave, to ~ (Bookb) ‖ (mit Papier) durchschießen ‖ interfolier.
interleaved ‖ mit Papier durchschossen ‖
interfolié.
interline, to ~ (Print) ‖ die Schriftzeilen fpl.
durchschießen ‖ interligner les blancs
mpl.; espacer les lignes fpl.
interlinked current ‖ verketteter Strom m. ‖
courant m. enchaîné. / ~ piece ‖ Gleis-
zwischenstück n.; Verbindungsstück n. ‖
raccord m.; pièce f. intercalaire ou de
raccord. / the rails pl. are replaced by ~
steel-cast pieces ‖ die Schienen fpl. sind
durch Zwischenstücke aus Stahlguß er-
setzt ‖ les rails mpl. sont remplacés par
des pièces intercalaires en acier moulé. /
~ pressure ‖ verkettete Spannung f. ‖
tension f. entre phases reliées.
interlinking ‖ Verkettung f. ‖ interconn-
nexion f. / ~ of phases ‖ Phasenverkettung f. ‖ accouplement m. des phases.
interlinking point ‖ Verkettungspunkt m.
‖ point m. de jonction de phase.
interlock, to ‖ verriegeln; verblocken .‖
bloquer; verrouiller.
interlocking ‖ Verriegelung f.; Verblockung
f. ‖ verrouillage m. / ~ device ‖ Ver-
blockungssystem n. ‖ système m. de verrouillage. / ~ gear (Railw) ‖ Verschluß-
gitter n. ‖ gril m. d'enclenchement. / ~
points and signals pl. (Railw) ‖ Kontroll-
weichen fpl. und Signale npl. ‖ rails mpl.
mobiles et signaux mpl. de contrôle.
intermediary see also intermediate ‖
Zwischen-...; zwischenliegend ‖ intermédiaire. / ~ state ‖ Zwischenstufe f. ‖
stade m. intermédiaire. / ~ substance ‖
Zwischenprodukt n. ‖ produit m. intermédiaire.
intermediary (Trade) ‖ Zwischenhändler m.
‖ intermédiaire m.
intermediate ‖ zwischenliegend; Zwi-
schen-... ‖ intermédiaire. / ~ airway
beacon (Aero) ‖ Zwischenfeuer n. ‖ phare
m. intermédiaire. / ~ bearing ‖ Zwischenlager m. ‖ palier m. ou coussinet m. intermédiaire. / ~ chair (Railw) ‖ Zwischen-
stuhl m. ‖ chaise f. ou coussinet m. intermédiaire. / ~ circuit ‖ Zwischenkreis m. ‖
circuit m. intermédiaire. / ~ circuit condenser ‖ Kondensator m. im Zwischenkreis ‖ condensateur m. du circuit intermédiaire. / ~ cooling ‖ Zwischenkühlung
f. ‖ refroidissement m. intermédiaire. /
~ drawer (Spinn) ‖ Mittelstrecker m. ‖
étireur m. intermédiaire. / ~ flange ‖
Zwischenflansch m. ‖ bride f. intermédiaire. / ~ fly frame (Spinn) ‖ Mittelfleier
m. ‖ banc m. à broches intermédiaire. / ~
frequency transformer ‖ Zwischenfre-
quenztransformator m. ‖ transformateur
m. à moyenne fréquence. / ~ gear ‖ Zwischenschaltung f. ‖ transmission f. intermédiaire. / ~ gearing ‖ Vorgelege n. ‖ engrenage m. intermédiaire; renvoi m. (de
mouvement). / ~ grade ‖ Zwischenstufe
f. ‖ degré m. intermédiaire. / real ~ image
‖ reelles Zwischenbild n. ‖ image f. réelle
intermédiaire. / ~ implement for elevator
cages ‖ Zwischengeschirr n. für Förderkörbe ‖ pièces fpl. de commande pour
cages de montée. / ~ lever ‖ Zwischen-
hebel m. ‖ levier m. intermédiaire.

intermediate member ‖ Zwischenträger
m. ‖ entretoise f. / elastic ~ ‖ elastisches
Zwischenglied n. ‖ pièce f. intermédiaire élastique.
intermediate piece (Railway) ‖ Zwischenstück n. ‖ plot m. auxiliaire. / ~ between ahead and astern turbine ‖ Verbindungsstück n. zwischen Vor- und
Rückwärtsturbine ‖ pièce f. intercalaire
entre la turbine à marche avant et à
marche arrière.
intermediate position ‖ Mittelstellung f. ‖
position f. intermédiaire. / ~ reading ‖
Zwischenablesung f. ‖ lecture f. intermédiaire. / ~ resistance (Electr) ‖
Übergangswiderstand m. ‖ résistance f.
au passage.
intermediate shaft (Mach) ‖ Zwischenwelle f. ‖ arbre m. intermédiaire ou de
renvoi. / ~ (Mach tool) ‖ Vorgelegewelle f.
‖ arbre m. intermédiaire. / ~ (Shipb) ‖
Laufwelle f.; Tunnelwelle f.; Zwischenwelle f. ‖ arbre m. intermédiaire.
intermediate station ‖ Trennanstalt f. ‖
poste m. intermédiaire. / ~ story ‖
Halbgeschoß n. ‖ entresol m.; mezzanine f. / ~ straining ‖ Zwischenveranke-
rung f. ‖ ancrage m. intermédiaire. / ~
transmission ‖ Zwischentransmission f.
‖ transmission f. intermédiaire. / ~
wheel ‖ Zwischenrad n. ‖ roue f. intermédiaire. / ~ yield ‖ Zwischenergebnis
n. ‖ produit m. intermédiaire.
intermittent ‖ intermittierend; aussetzend
‖ intermittent. / ~ arc ‖ intermittierender
Lichtbogen m. ‖ arc m. intermittent. / ~
current ‖ intermittierender Strom m. ‖
courant m. intermittent. / ~ feed ‖
Sprungvorschub m. ‖ avance f. par intermittence. / ~ grinding ‖ satzweise Vermahlung f. ‖ mouture f. périodique. /
~ load ‖ intermittierende Belastung f. ‖
charge f. intermittente. / ~ rotation ‖
absatzweise Drehung f. ‖ rotation f. intermittente. / ~ service ‖ aussetzender Betrieb m. ‖ service m. intermittent. / ~
signal ‖ Blinksignal n. ‖ signal m. cligno-
tant. / ~ spring ‖ aussetzende Quelle f. ‖
source f. intermittente. / ~ working
mixer ‖ Mischer m. mit absatzweisem
Betrieb m. ‖ malaxeur m. à fonctionnement
discontinu.
intermitting see intermittent.
internal ‖ inner ‖ interne; intérieur. / ~
angle ‖ innerer Winkel m. ‖ angle m.
interne ou intérieur. / ~ capacity of a
coil ‖ Windungskapazität f. ‖ capacité f.
entre les enroulements d'une bobine. /
~ capacity of valves ‖ Röhrenkapazität
f. ‖ capacité f. intérieure des lampes. /
~ circular grinding ‖ Innenrundschliff m.
‖ rectification f. des surfaces cylindriques
intérieures. / ~ circular grinding machine ‖ Innenrundschleifmaschine f. ‖
machine f. à rectifier les alésages. /
~ combustion engine see ~ combustion
motor. / ~ combustion motor ‖ Verbrennungsmotor m.; Explosionsmotor m. ‖
moteur m. à combustion intérieure ou interne. / ~ dimensions pl. ‖ Innenabmes-
sungen pl. ‖ dimensions fpl. intérieures. /
~ explosion engine see ~ combustion motor.
/ ~ fluid motion ‖ Flüssigkeitsbewegung f.
im Innern ‖ mouvement m. interne du
liquide. / ~ furnace ‖ Innenfeuerung f. ‖
foyer m. intérieur. / ~ plunge-cut ~ grinder
‖ Einstechinnenschleifmaschine f. ‖ machine f. à rectifier les intérieurs.

internal grinding of straight bores ‖ Innenschliff m. zylindrischer Büchsen ‖ rectification f. intérieure de bagues cylindriques. / ~ of taper bores ‖ Innenschliff m. konischer Rundflächen ‖ rectification f. intérieure de surfaces circulaires coniques.

internal grinding machine ‖ Innenschleifmaschine f. ‖ machine f. à rectifier intérieurement. / ~ and face grinding machine ‖ Innen- und Stirnflächenschleifmaschine f. ‖ machine f. à rectifier les alésages et les surfaces planes en bout. / ~ spindle ‖ Innenschleifspindel f. ‖ arbre m. porte-meule à rectifier les intérieurs.

internal high-speed spindle ‖ Schnellaufinnenspindel f. ‖ broche f. intérieure à rotation rapide.

internal measuring instrument ‖ Innenmeßgerät n. ‖ appareil m. pour les mesures intérieures. / ~ with adjustable type rest ‖ Innenmeßgerät n. mit verstellbarem Anschlag ‖ appareil m. pour les mesures d'alésages à contact réglable. / ~ with fixed rest ‖ Innenmeßgerät n. mit festem Anschlag ‖ appareil m. pour les mesures d'alésages à contact fixe. / ~ for round rings ‖ Innenmeßgerät n. für sphärische Ringe ‖ appareil m. pour les mesures intérieures des anneaux sphériques. / ~ with spring type rest ‖ Innenmeßgerät n. mit federndem Anschlag ‖ appareil m. pour les mesures d'alésages à contact élastique. / ~ with three-point measuring head ‖ Innenmeßgerät n. mit Dreipunktmeßköpfen ‖ appareil m. pour mesures intérieures à têtes de contrôle à trois contacts.

internal micrometer ‖ Stichmaß n.; Innenmikrometer n. ‖ micromètre m. d'intérieur; jauge f. à coulisse. / ~ pinion with inside gearing ‖ Getriebe n. mit innerer Verzahnung ‖ engrenage m. à denture intérieure. / eyepiece free from ~ reflection ‖ reflexfreies Okular n. ‖ oculaire m. exempt de reflets. / ~ resistance ‖ innerer Widerstand m. ‖ résistance f. intérieure. / ~ shake ‖ Innenriß m. ‖ crevasse f. intérieure.

internal-tooth wheel ‖ Rad n. mit innerer Verzahnung ‖ roue f. à denture intérieure.

internally ‖ innen ‖ en dedans. / externally and ~ ‖ außen und innen ‖ au dehors et en dedans; à l'extérieur et à l'intérieur.

international competition ‖ internationaler Wettbewerb m. ‖ concours m. international. / ~ economic conference ‖ Weltwirtschaftskonferenz f. ‖ conférence f. économique internationale. / ~ electrotechnical commission ‖ internationale elektrotechnische Kommission f. ‖ commission f. électrotechnique internationale. / ~ meeting of telegraph and telephone engineers ‖ internationaler Kongreß m. der Fernmeldetechniker ‖ congrès m. international d'ingénieurs des télégraphes et des téléphones. / ~ telegraphic convention ‖ Welttelegrafenvertrag m. ‖ convention f. télégraphique internationale. / ~ telegraphic union ‖ Welttelegrafenverein m. ‖ union f. télégraphique internationale.

interocular distance ‖ Augenabstand m. ‖ distance f. entre les yeux.

interpoint, to ‖ interpunktieren ‖ mettre la ponctuation.

interpolate, to ‖ interpolieren ‖ interpoler.

interpolation ‖ Interpolation f. ‖ interpolation f. / ~ resistance ‖ Interpolationswiderstand m.; Eingrenzwiderstand m. ‖ résistance f. d'interpolation.

interpolator ‖ Interpolator m. ‖ interpolateur m.

interpole ‖ Wendepol m. ‖ pôle m. de commutation.

interpose, to ‖ zwischenlegen ‖ entreposer.

interposition of a lens ‖ Einschalten n. einer Linse ‖ interposition f. d'une lentille.

interpretation of an article of the law ‖ die Auslegung eines Gesetzesparagraphen ‖ interprétation f. d'un article de la loi.

interpupillary gauge ‖ Augenabstandmesser m. ‖ appareil m. pour mesurer l'écartement des yeux.

interrupt, to ‖ unterbrechen ‖ interrompre. / ~ a line ‖ eine Leitung f. unterbrechen ‖ couper ou interrompre un circuit.

interrupter (Electr) ‖ Unterbrecher m. ‖ interrupteur m. / ~ (Mot) ‖ Abreißhebel m. ‖ doigt m. de contact. / electrolytic ~ ‖ elektrolytischer Unterbrecher m. ‖ interrupteur m. électrolytique. / ~ with mercury ‖ Quecksilberausschalter m. ‖ disjoncteur m. au mercure. / rotary ~ ‖ Turbinenunterbrecher m. ‖ interrupteur m. tournant.

interrupting jack ‖ Unterbrecherklinke f. ‖ jack m. de rupture ou à rupture.

interruption ‖ Unterbrechung f. ‖ interruption f. / automatic ~ ‖ Selbstunterbrechung f. ‖ interruption f. automatique. / ~ of working ‖ Arbeitsunterbrechung f.; Betriebsunterbrechung f. ‖ interruption f. de travail ou de service.

interruption cable ‖ Notkabel n. ‖ câble m. provisoire.

intersect, to ~ (Math) ‖ sich schneiden ‖ se couper; s'entrecouper. / ~ (Mine) ‖ das Gebirge durchörtern ‖ percer le terrain.

intersected (Ground) ‖ durchschnitten ‖ coupé; difficile; fourré; accidenté. / an area is ~ by railway lines ‖ ein Gebiet n. wird von Eisenbahnlinien durchschnitten ‖ des lignes fpl. de chemin de fer traversent une région.

intersecting line ‖ Schnittlinie f. ‖ ligne f. d'intersection. / ~ point ‖ Schnittpunkt m. ‖ point m. d'intersection.

intersection ‖ Schnittpunkt m. ‖ intersection f. / ~ of the ground (Mine) ‖ Durchörterung f. des Gebirges ‖ percement m. souterrain.

intersectional line ‖ Schnittlinie f. ‖ ligne f. d'intersection. / ~ point ‖ Schnittpunkt m. ‖ point m. d'intersection.

intersection, point of ~ (Mine) ‖ Durchschlagspunkt m. ‖ point m. d'intersection. / ~ (Railw) ‖ Kreuzungspunkt m. ‖ point m. de croisement.

interstellar space ‖ Weltraum m.; Weltenraum m. ‖ espace m. de l'univers. / airvoid ~ ‖ luftleerer Weltenraum m. ‖ espace m. de l'univers vide d'air.

interstice ‖ Zwischenraum m. ‖ échappée f.; intervalle m.

interstices pl. of a millstone ‖ Schärfen fpl. oder Rillen fpl. oder Furchen fpl. eines Mühlsteins ‖ rayons mpl. de meule.

intertie (Carp) ‖ Riegel m.; Querband n. ‖ entretoise f.; lierne f. / ~ of a bay-work ‖ Riegel m. oder Querholz n. einer Fachwand; Wandriegel m. ‖ entretoise f. de cloison. / wooden ~ ‖ Querholz n.; Quer-

riegel m. ‖ traverse f. ou entretoise f. en bois.

intertwine, to ~ (Textile) ‖ verschlingen; verflechten ‖ entrelacer.

interurban telephony ‖ Ferntelefonie f. ‖ téléphonie f. interurbaine.

interval in the work ‖ Arbeitspause f. ‖ temps m. de repos; pause f.; récréation f. / ~ (Mus) ‖ Intervall n.; Pause f. ‖ intervalle m. / ~ between two scale lines of a graticule ‖ Abstand m. zwischen zwei Teilstrichen einer Strichplatte ‖ intervalle m. entre deux traits d'un micromètre. / at regular ~s pl. ‖ in regelmäßigen Abständen mpl. ‖ en intervalles régulières; périodiquement. / ~ of a shaft (Mine) ‖ Schachtfeld n.; Schachtverzug m. ‖ intervalle m. de puits.

intervene, to ‖ vermitteln; intervenieren ‖ intervenir.

intervening ‖ zwischenliegend ‖ intermédiaire. / ~ transformer ‖ Zwischentransformator m. ‖ transformateur m. intermédiaire.

interweave, to ‖ durchflechten ‖ enverger; entrelacer.

intimate ‖ innig ‖ intime.

intrados ‖ innere Wölbfläche f.; Bogenlaibung f. ‖ intrados m. / side against the ~ ‖ Laibungsseite f. eines Wölbesteins ‖ panneau m. de douelle.

intrinsic intensity of a lamp ‖ spezifische Helligkeit f. einer Lampe ‖ éclat m. spécifique d'une lampe.

introducing into existing machines ‖ Einbau m. an vorhandenen Maschinen ‖ montage m. aux machines existantes.

introduction (Steam) ‖ Einströmung f. ‖ introduction f.; admission f. / ~ (Trade) ‖ Einführungsschreiben n. ‖ lettre f. introduction; présentation f.

intrusion (Tel) ‖ Aufschaltung f. ‖ intrusion . f.

inturned ‖ gekröpft ‖ coudé.

inundate, to ‖ überfluten ‖ inonder.

inundation ‖ Überschwemmung f. ‖ inondation f. / ~ limit of river ‖ Hochwasserbett n. eines Flusses ‖ lit m. d'inondation d'une rivière.

invalid (Contract) ‖ nichtig; ungültig ‖ nul; vain.

invalid ‖ Kranker m. ‖ malade m. / ~ carriage ‖ Krankenwagen m. ‖ voiture f. de transport de malades.

invalidation ‖ Nichtigkeitserklärung f. ‖ annulation f.; cassation f.

invalidism ‖ Invalidität ‖ invalidité f.; infirmité f.

invalid wheel-chair ‖ Krankenfahrstuhl m. ‖ fauteuil m. roulant pour malades.

invent, to ‖ erfinden ‖ inventer.

invention ‖ Erfindung f. ‖ invention f.

inventor ‖ Erfinder m. ‖ inventeur m.

inventory ‖ Bestand m.; Ausrüstung f.; Inventar n. ‖ inventaire m. / ~ (Book) ‖ Bestandbuch n. ‖ état m. / continuous ~ ‖ laufende Inventur f. ‖ inventaire m. permanent. / to draw up an ~ (den) Bestand m. aufnehmen; inventarisieren ‖ inventorier.

invar ‖ Invar n. ‖ invar m.

invariant ‖ Invariante f.; Unveränderliche f. ‖ invariant m.

inverse current ‖ Gegenstrom m. ‖ contre-courant m.; courant m. inverse.

inversely proportional ‖ umgekehrt proportional ‖ inversement proportionnel.

inversion of strata ‖ Überkippung f. der Schichten ‖ renversement m. des couches. / ~ rule for lines ‖ Umkehrungssatz m. für Leitungen ‖ règle f. inverse pour lignes. / ~ telemeter ‖ Invertentfernungs. messer m. ‖ télémètre m. à renversement.

invert, to ‖ umkehren ‖ renverser; inverser. / ~ rails pl. ‖ Schienen fpl. umlegen *oder* umwenden ‖ renverser *ou* retourner des rails mpl.

inverted bow and chain girder ‖ Fischbauchträger m.; Linsenträger m. ‖poutre f. lenticulaire. / ~ comma ‖ Anführungszeichen n. ‖ guillemet m. / ~ gas-lamp ‖ Hängelichtgaslampe f. ‖ lampe f. à gaz à éclairage renversé / bec m. de gaz renversé. / ~ image ‖ umgekehrtes Bild n.; Kehrbild n. ‖ image f. renversée. / ~ incandescent gas-burner ‖ Invertgasglühlichtlampe f. ‖ lampe f. à gaz à bec renversé.

inverting ferment ‖ invertierendes Ferment n. ‖ ferment m. invertissant.

invert sugar ‖ Invertzucker m. ‖ sucre m. interverti.

invest, to ~ money ‖ Geld n. anlegen ‖ placer *ou* donner de l'argent m.

investigate, to ‖ erforschen ‖ étudier; investiger; rechercher; enquêter.

investigation ‖ Nachforschung f. ‖ recherche f.; enquête f. / magnetic ~ ‖ magnetische Untersuchung f. ‖ recherche f. magnétique.

investment of capital ‖ Kapitalanlage f. ‖ placement m. de fonds. / safe ~ ‖ mündelsichere Kapitalanlage f. ‖ placement m. de tout repos.

investment stock ‖ Anlagekapital n. ‖ mise f. de fonds.

inviolability of letters ‖ Briefgeheimnis n. ‖ secret m. des lettres.

invisible ‖ unsichtbar ‖ invisible.

invoice, to ‖ fakturieren ‖ facturer; donner facture f.

invoice ‖ Faktura f.; Rechnung f. ‖ calcul m.; compte m.; opération f.; facture f.; note f.; mémoire m. / ~ (Account purchased) ‖ Einkaufsrechnung f. ‖ compte m. d'achat. / to enclose ~ ‖ Verkaufsrechnung f. einschließen ‖ joindre facture f. / to legalize an ~ ‖ eine Faktura f. beglaubigen lassen ‖ faire légaliser une facture.

invoice amount ‖ Fakturbetrag m. ‖ montant m. de facture. / ~ book ‖ Fakturenbuch n. ‖ facturier m.; livre m. des factures. / copy of ~ ‖ Fakturabschrift f. ‖ copie f. de facture. / ~ typewriter ‖ Maschine f. zum Schreiben von Rechnungen ‖ machine f. à écrire les factures.

involve, to ~ (to envelop) ‖ umhüllen ‖ envelopper.

involute ‖ Evolvente f. ‖ développante f. / ~ of the cercle ‖ Kreisevolvente f. ‖ développante f. du cercle. / ~ toothing ‖ Evolventenverzahnung f. ‖ denture f. à développante.

inwards ‖ innen; innerhalb ‖ au dehors. / outwards and ~ ‖ außen und innen ‖ au dehors et en dedans; à l'intérieur et à l'extérieur.

iodargyrite ‖ Jodargyrit m.; Jodsilber n. ‖ iodure m. argentique *ou* d'argent; iodargyrite f.

iodide ‖ Jodid n.; Jodür n.; Jodverbindung f. ‖ iodide m.; iodure m. / argentic ~ *see* iodargyrite. / hydric ~ ‖ Jodwasserstoffsäure f. ‖ acide m. hydriodique *ou*

iodhydrique. / ~ of nitrogen ‖ Jodstickstoff m. ‖ azoture m. d'iode. / ~ of silver *see* iodargyrite.

iodides pl. ‖ Jodsalze npl. ‖ iodures mpl.

iodinate, to ‖ jodieren ‖ ioder.

iodine ‖ Jod n. ‖ iode m. / ~ test ‖ Jodprobe f. ‖ indice m. iodique; essai m. au iode. / ~ water ‖ Jodwasser n. ‖ eau f. hydriodatée.

iodize, to ‖ jodieren ‖ iodurer.

iodobromite ‖ Jodbromchlorsilber n. ‖ iodobromite m.

iodoform ‖ Jodoform n. ‖ iodoforme m. / ~ gauze ‖ Jodoformgaze f. ‖ gaze f. iodoformée.

iodyrite ‖ Jodsilber n. ‖ iodargure f.

ion ‖ Ion n. ‖ ion m. / migration of ~s ‖ Wanderung f. der Ionen ‖ migration f. *ou* transport m. des ions. / theory of ~s ‖ Ionentheorie f. ‖ théorie f. des ions.

ionizable ‖ ionisierbar ‖ ionisable.

ionization ‖ Ionisierung f.; Ionisation f. ‖ ionisation f. / ~ by collision ‖ Stoßionisation f. ‖ ionisation f. par les chocs. / ~ of a gas ‖ Ionisierung f. von Gasen ‖ ionisation f. d'un gaz.

ionization current ‖ Ionisationsstrom m. ‖ courant m. d'ionisation. / intensity of ~ ‖ Ionisationsstromstärke f. ‖ intensité f. du courant de ionisation.

ionize, to ‖ ionisieren ‖ ioniser.

ionizer ‖ Ionisierungsmittel n. ‖ ionisant m.

ionizing power ‖ Ionisierungskraft f. ‖ pouvoir m. ionisant.

ipecac root ‖ Ipekak n. ‖ racine f. d'ipécac.

ipecacuanha ‖ Ipekakuanha f.; Brechwurz f. ‖ ipécacuana m.

iridescent ‖ irisierend ‖ irisé.

iridic angle of the eye ‖ Kammerwinkel m. des Auges ‖ angle m. iridien de l'œil. / measuring the depth of the ~ chamber of the eye ‖ Tiefenmessung f. der Vorderkammer des Auges ‖ mesure f. de la profondeur de la chambre antérieure de l'œil.

iridium ‖ Iridium n. ‖ iridium m. / ~ point ‖ Iridiumspitze f. ‖ pointe f. en iridium.

iridosmine ‖ Osmiridium n.; Osmium-Iridium n. ‖ iridosmine f.

iris ‖ Regenbogenhaut f. ‖ iris m. / ~ diaphragm ‖ Irisblende f. ‖ diaphragme m. en forme d'iris. / ~ forceps pl. ‖ Irispinzette f. ‖ pince f. à iris. / ~ ground (Pap) ‖ Irisfond m. ‖ fond m. irisé *ou* ombré.

irisate, to ‖ irisieren ‖ iriser.

Irish moss ‖ Karaghenmoos n. ‖ lichen m. carragheen.

irising salt ‖ Irisiersalz n. ‖ sel m. à iriser.

iris oil ‖ Irisöl n. ‖ essence f. d'iris. / ~ root pea ‖ Fontanellkügelchen n. ‖ pois m. d'iris *ou* à cautères. / ~ scissors pl. ‖ Augenschere f. ‖ ciseaux mpl. à iridectomie. / instantaneous ~ shutter ‖ Irismomentverschluß m. ‖ obturateur m. instantané à iris.

iron, to ‖ aufbügeln; plätten ‖ repasser. / ~ the linen ‖ die Wäsche bügeln *oder* plätten ‖ repasser le linge.

iron (Made of iron) ‖ eisern; aus Eisen ‖ de fer; en fer.

iron ‖ Eisen n. ‖ fer m. / ~ (Join) ‖ Hobeleisen n. ‖ fer m. de rabot. / ~ (Tail) ‖ Plätteisen n. ‖ fer m. à repasser. / ~ for additions ‖ Zusatzeisen n. ‖ fonte f. crue d'addition. / bar ~ ‖ Stabeisen n. ‖ fer m. en barres. / basic ~ ‖ basisches Roheisen n. ‖ fonte f. basique. / brittle ~ ‖ brüchiges

Eisen n. ‖ fer m. cassant. / broken ~ ‖ Eisenabfälle mpl. ‖ferraille f.; débris mpl. de fer; riblons mpl. / carburized ~ ‖ gekohltes Eisen n. ‖ fer m. carburé; carbure m. de fer. / case-hardened ~ ‖ Kokillenguß n. ‖ fonte f. en coquilles; fer m. fondu en coquilles. / cast ~ ‖ Gußeisen n. ‖ fonte f. / cast ~ direct from the blastfurnace ‖ Hochofenguß m. ‖ fonte f. moulée *ou* moulée f. de première fusion. / charcoal ~ ‖ Holzkohleneisen n. ‖ fer m. au bois. / charcoal blast ~ ‖ Holzkohlenroheisen n. ‖ fonte f. au charbon de bois. / coarse-grained ~ ‖ grobkörniges Eisen n. ‖ fer m. à grain gros. / cold-blast cast ~ ‖ kalt erblasenes Gußeisen n. ‖ fonte f. à air froid; fer m. fondu à air froid. / cold-hammered ~ ‖ kaltgehämmertes *oder* hartgeschlagenes Eisen ‖ fer m. écroui. / corrugated ~ *see also* corrugated sheet ‖ Wellblech n. ‖ tôle f. ondulée. / crude ~ ‖ Gußeisen n.; Roheisen n. ‖ fonte f. crue *ou* en gueuses. / crystalline ~ ‖ Feinkorneisen n. ‖ fer m. aciéreux *ou* dur *ou* à texture grenue. / crystallized ~ ‖ kristallisches Eisen n. ‖ fer m. cristallisé. / double-T-~ ‖ Doppel-T-Eisen n. ‖ fer m. en T double. / ~ of a dovetail plane ‖ Grateisen n.; Grathobeleisen n. ‖ fer m. de bouvet mâle *ou* de feuilleret. / drossy ~ ‖ schlackenreiches Eisen n. ‖ fer m. gras. / ductile ~ ‖ schmiedbares Eisen n. ‖ fer m. ductile *ou* malléable. / fagotted ~ ‖ Alteisen n. ‖ fer m. de ferraille. / figured ~ ‖ Profileisen n.; Formeisen n. ‖ fer m. de profil *ou* profilé. / fine ~ ‖ Kleineisen n. ‖ petit fer m.; petits fers mpl. / fine-grained ~ *see* crystalline ~. / finished ~ ‖ Feineisen n. ‖ mazée f.; fonte f. mazée. / flat ~ ‖ Flacheisen n. ‖ larget m.; fer m. plat. / flue ~ ‖ Rohrblech n. ‖ tôle f. à tuyaux. / forged ~ ‖ Schmiedeeisen n. ‖ fer m. de forge. / ~ free from phosphorus and sulphur ‖ Eisen n. frei von Phosphor und Schwefel ‖ fonte f. exempte de phosphore et de soufre. / galvanized ~ ‖ verzinktes Eisen n. ‖ fer m. galvanisé. / grey cast ~ ‖ Grauguß m. ‖ fonte f. grise. / half round ~ ‖ halbrundes Eisen n. ‖ fer m. demi-rond. / hammered ~ ‖ gehämmertes Eisen n. ‖ fer m. martelé. / hammer-hardened ~ *see* cold-hammered ~. / hard ~ ‖ Harteisen n. ‖ fonte f. dure. / hoop ~ ‖ Bandeisen n. ‖ feuillard m. / horseshoe ~ ‖ Hufstabeisen n. ‖ fer m. à maréchal. / hot blast ~ ‖ heiß erblasenes Gußeisen n. ‖ fonte f. à air chaud; fer m. fondu à air chaud. / ingot ~ ‖ Flußeisen n. ‖ fer m. homogène; acier m. Siemens-Martin. / made of ~ eisern ‖ de fer m.; en fer m. / malleable ~ ‖ Schmiedeeisen n. ‖ fer m. forgé *ou* malléable. / manganese ~ ‖ Manganeisen n.; Eisenmangan n. ‖ ferromanganèse m. / merchant ~ ‖ Handelseisen n. ‖ fer m. marchand. / metallic ~ ‖ metallisches Eisen n. ‖ fer m. métallique. / meteoric ~ ‖ Meteoreisen n. ‖ fer m. météorique. / mixed gray and white ~ *see* mottled ~. / mottled ~ ‖ halbiertes Eisen n. ‖ fonte f. truitée *ou* mêlée *ou* masulée. / native ~ ‖ gediegenes Eisen n. ‖ fer m. natif. / pasty ~ ‖ teigiges Eisen n. ‖ fer m. pâteux. / pig ~ ‖ Roheisen n. ‖ fonte f. brute *ou* en gueuses. / poor in ~ ‖ eisenarm ‖ pauvre en fer m. / porous ~ ‖ luckiges *oder* poröses Eisen n. ‖ fer m. poreux. / puddled ~ ‖ Puddeleisen n. ‖ fer m. puddlé. / raw ~ ‖

Gußeisen n.; Roheisen n. ‖ fonte f. brute *ou* en gueuses. / red hot ~ ‖ (rot) glühendes Eisen n. ‖ fer m. chauffé au rouge. / refined ~ ‖ Frischeisen n. ‖ fer m. affiné. / right side round ~ ‖ Hohlkehleeisen n.; Anschlaghohlkehleeisen n. ‖ fer m. de congé. / rolled ~ ‖ Walzeisen n. ‖ fer m. laminé. / round ~ ‖ Rundeisen n. ‖ fer m. rond. / round nose ~ ‖ Hohlkehleeisen n. ‖ fer m. de rabot rond. / section ~ ‖ Formeisen n.; Profileisen n.; Fassoneisen n. ‖ fer m. profilé. / sharp ~ ‖ Scharfeisen n.; Schereisen n. ‖ fer m. taillant. / sheet ~ ‖ Eisenblech n. ‖ tôle f. en fer. / short ~ ‖ brüchiges Eisen n. ‖ fer m. cassant. / slaggy ~ ‖ schlackenreiches Eisen n. ‖ fer m. gras. / small ~ ‖ Kleineisen n. ‖ petit-fer m. / soft ~ ‖ Weicheisen n. ‖ fer m. doux *ou* mou *ou* soudable. / sow ~ ‖ Masselgrabeneisen n. ‖ gueuse f. mère; maître-calle f.; maître-câle m. / spathic ~ ‖ Spateisenstein m. ‖ fer m. oxydé carbonaté; sidérose f. / special ~ ‖ Profileisen n.; Formeisen n. ‖ fer m. de profil; fer m. profilé *ou* spécial. / spiegel ~ ‖ Spiegeleisen n. ‖ fonte f. spiegel. / square ~ ‖ Quadrateisen n. ‖ fer m. rectangulaire *ou* carré. / T-~ ‖ T-Eisen n. ‖ fer m. T. / to tap the ~ ‖ das Eisen abfangen ‖ faire couler la fonte. / thin sheet ~ ‖ dünnes Eisenblech n. ‖ tôle f. fine. / tilted ~ ‖ gehämmertes Stabeisen n. ‖ fer m. martelé *ou* forgé. / titanic ~ ‖ Titaneisen n. ‖ fer m. oxydulé titanifère. / weld ~ ‖ Schweißeisen n. ‖ fer m. soudable *ou* soudant. / wet ~ ‖ schlackenreiches Eisen n. ‖ fer m. gras. / white pig ~ ‖ Weißeisen n. ‖ fonte f. blanche ordinaire. / ~ for wire ‖ Drahteisen n. ‖ fer m. pour fil. / wrought ~ ‖ Schmiedeeisen n.; Schweißeisen n. ‖ fer m. forgé *ou* mou. / zinked ~ ‖ galvanisch verzinktes Eisen n. ‖ fer m. galvanisé *ou* zingué. / zores ~ ‖ Belageisen n.; Quadranteisen n. ‖ fer m. zorès.

iron acetate ‖ essigsaures Eisen n.; Eisenazetat n. ‖ acétate m. de fer. / ~ alum ‖ Eisenalaun m. ‖ alun m. de fer. / ~ analysis ‖ Eisenuntersuchung f. ‖ analyse f. du fer. / ~ angle ‖ Eisenwinkel m. ‖ équerre m. en fer. / nickel plated ~ anode ‖ vernickelte Eisenanode f. ‖ anode f. en fer nickelée. / ~ back of the hearth ‖ Feuerwand f. ‖ contre-cœur m. de foyer à fer. / ~ band ‖ Bandeisen n. ‖ feuillard m.; bande f. de fer. / ~ bar ‖ Eisenstange f. ‖ barre f. de fer. / ~ bars pl. rolled-up spirally ‖ spiralförmig gewickelte Eisenstäbe mpl. ‖ barres fpl. de fer roulées en spirale. / ~ barrel ‖ Eisenfaß n. ‖ tonneau m. en fer.

iron-bearing ‖ eisenführend ‖ ferrifère.
iron blacking ‖ Eisenschwärze f. ‖ gris m. de fer. / ~ block ‖ Sau f.; Eisenklumpen m. ‖ loup m.; bloc m. / ~ bloom ‖ Luppeneisen n. ‖ fer m. en loupes. / ~ borings pl. ‖ Eisenbohrspäne mpl. ‖ limaille f. de forage. / ~ box ‖ Eisenmuffe f. ‖ manchon m. en fer. / ~ bracket ‖ Eisenkonsole f.; Eisenleiste f. ‖ baguette f. *ou* console f. en fer. / ~ bridge ‖ Eisenbrücke f. ‖ pont m. en fer. / ~ buckle ‖ Eisenschnalle f. ‖ boucle f. en fer. / ~ building ‖ Eisenbauwerk n.; Eisenkonstruktion f.; Eisenbau m. ‖ charpente f. *ou* construction f. métallique. / ~ carbide ‖ Eisenkarbid n. ‖ carbure m. de fer. / ~ carbonate ‖ kohlensaures Eisen n. ‖ carbonate m. de fer.

iron castings ‖ Eisenguß m. ‖ fonte f. (grise). / chilled ~ ‖ Schalenguß m. ‖ fonte f. en coquille. / grey ~ ‖ Grauguß m. ‖ moulage m. en fonte grise.
iron cement ‖ Eisenzement m. ‖ ciment m. de fer. / ~ goods pl. ‖ Eisenzementwaren fpl. ‖ objets mpl. en ciment armé. / ~ work ‖ Eisenbetonarbeit f. ‖ travail m. en ferro-ciment m. *ou* en fer-béton m.
iron chest ‖ Kassette f. aus Eisenblech ‖ cassette f. en tôle. / ~ chimney ‖ Eisenkamin m. ‖ cheminée f. en tôle. / ~ chloride ‖ Eisenchlorid n. ‖ chlorure m. ferrique. / ~ chrysolite ‖ Fayalit m.; Eisenchrysolit m. ‖ fayalite f.; péridot m. ferrugineux. / ~ cinders pl. ‖ Eisenschlacke f. ‖ laitier m. *ou* scorie f. de fer. / ~ citrate ‖ zitronensaures Eisen n. ‖ citrate m. de fer.
iron-clad ‖ gepanzert ‖ cuirassé. / ~ cable ‖ Panzerkabel n. ‖ câble m. cuirassé. / ~ dynamo ‖ gekapselte Dynamomaschine f. ‖ dynamo f. cuirassée *ou* blindée. / ~ electric motor ‖ gekapselter Elektromotor m. ‖ moteur m. électrique blindé. / ~ motor ‖ gekapselter Motor m.; Panzermotor m. ‖ moteur m. cuirassé.
iron colour ‖ Eisenfarbe f. ‖ couleur f. de fer.
iron concrete ‖ Eisenbeton m. ‖ béton m. armé.
iron concrete goods pl. ‖ Eisenbetonwaren fpl. ‖ objets mpl. en béton armé. / ~ pole ‖ Eisenbetonpfahl m. ‖ pilier m. *ou* pieu m. en béton armé.
iron construction ‖ Eisenbau m. ‖ construction f. métallique *ou* en fer. / ~ core ‖ Eisenkern m. ‖ noyau m. en fer. / ~ core inductance ‖ Induktanz f. mit Eisenkern ‖ inductance f. avec noyau de fer. / ~ cross section ‖ Eisenquerschnitt m. ‖ section f. du fer. / ~ crucible ‖ Eisentiegel m. ‖ creuset m. en fer. / ~ cutter ‖ Eisenschere f. ‖ cisaille f. à fer. / ~ distributor ‖ Eisenhändler m. ‖ marchand m. de fer. / ~ door ‖ Eisentür f. ‖ porte f. en fer. / ~ dowel ‖ Eisendübel m. ‖ goujon m. de fer. / ~ driver (Coop) ‖ Eisenvollsetze f. ‖ chasse f. pleine en fer. / ~ dross ‖ Hammerschlacke f. ‖ crasse f. du marteau.
iron dust ‖ Eisenstaub m. ‖ limaille f. de fer. / ~ coil ‖ Eisenpulverkernspule f.; Staubkernspule f. ‖ bobine f. à noyau en limaille de fer comprimé.
ironed dray ladder ‖ eisenbeschlagene Schrotleiter f. ‖ poulain m. en bois ferré.
ironer ‖ Plätterin f.; Büglerin f. ‖ repasseuse f.
iron extract malate ‖ apfelsaurer Eisenextrakt m. ‖ extrait m. de malate de fer. / ~ filament ballast ‖ Eisenwiderstand m. ‖ résistance f. ballast. / ~ filings pl. ‖ Eisenfeilspäne mpl.; Eisenfeilicht n. ‖ limaille f. de fer.
iron fittings ‖ Eisenbeschläge mpl. ‖ ferrures fpl. / ~ for building purposes ‖ Baubeschläge mpl. ‖ ferrures fpl. de bâtiment. / ~ for doors ‖ eiserne Türbeschläge mpl. ‖ ferrures fpl. pour portes. / ~ for windows ‖ eiserne Fensterbeschläge mpl. ‖ ferrures fpl. pour fenêtres.
iron flattener ‖ Eisenquetscher m. ‖ aplatisseur m. de fer. / ~ flooring ‖ Eisenbelag m. ‖ recouvrement m. *ou* dallage m. *ou* platelage m. en fer. / ~ founder ‖ Eisengießer m. ‖ ouvrier m. de fonderie.

/ ~ foundry ‖ Eisengießerei f.; Eisenschmelzhütte f. ‖ fonderie f. de fer.
iron frame (Build) ‖ Eisengerippe n. ‖ pan m. en charpente métallique. / ~ (Shipb) ‖ eisernes Spant n. ‖ couple m. en fer. / built as ~ sheds ‖ in Eisenfachwerk n. ausgeführt ‖ construit en charpente f. métallique. / glazed ~ constructions pl. ‖ verglaste in Eisenkonstruktion gehaltene Hallen fpl. ‖ halls mpl. en charpente métallique vitrée.
iron framework ‖ Eisenfachwerk n. ‖ charpente f. métallique. / the shops pl. consist of ~ ‖ die Gebäude npl. sind aus Eisenfachwerk hergestellt ‖ les ateliers mpl. sont exécutés en charpente métallique.
iron framing ‖ Geripppe n. von Eisen ‖ charpente f. métallique. / ~ furnace shovel ‖ Ofenschaufel f. ‖ pelle f. de fourneau. / ~ furniture ‖ Eisenmöbel n. ‖ meuble m. en fer. / ~ furniture ‖ Eisenbeschläge mpl. ‖ garniture f. de fer; ferrure f.; armature f. de fer. / ~ galvanizing ‖ Eisenverzinkung f. ‖ galvanisation f. du fer. / ~ gauze ‖ Eisengaze f. ‖ gaze f. de fer; toile f. métallique (en fil de fer). / ~ girder ‖ Eisenträger m.; eiserner Träger m. ‖ poutre f. *ou* poutrelle f. *ou* support m. en fer. / ~ glaze ‖ Eisenglasur f. ‖ vernis m. pour fer. / ~ glycerine phosphate ‖ glyzerinphosphorsaures Eisen n. ‖ glyzérophosphate m. de fer. / ~ grate ‖ Eisengitter n. ‖ treillis m. en fer *ou* en charpente métallique. / ~ greenhouse ‖ eisernes Gewächshaus n. ‖ serre f. en fer. / ~ hall ‖ Eisenhalle f. ‖ halle f. en fer. / ~ hame ‖ Kumteisen n. ‖ attelle f. en fer. / ~ hook for drawing charcoal ‖ Störhaken m. ‖ crochet m. en fer pour tirer le charbon. / ~ hoop ‖ eisernes Band n. ‖ frette f. / ~ hull ‖ eiserner Schiffsrumpf m. ‖ coque f. de navires en tôle (de fer). / hydroxide of ~ ‖ Eisenhydroxid n. ‖ hydroxyde m. de fer. / ~ hypophosphite ‖ unterphosphorsaures Eisen n. ‖ hypophosphite m. de fer.
ironing board ‖ Plättbrett n. ‖ planche f. à repasser. / ~ machine ‖ Bügelmaschine f. ‖ machine f. à repasser. / ~ room ‖ Plättraum m. ‖ salle f. de repassage.
iron jacket ‖ Eisenmantel m. ‖ revêtement m. *ou* enveloppe f. en tôle. / enamelled ~ jacket ‖ emaillierter Eisenmantel m. ‖ revêtement m. en fer émaillé. / ~ lactate ‖ milchsaures Eisen n. ‖ lactate m. de fer. / ~ liquor ‖ Eisenbeize f. ‖ mordant m. de fer; bouillon m. noir. / ~ list ‖ Eisenliste f. ‖ liste f. du fer. / ~ loop ‖ Luppeneisen n. ‖ fer m. en loupes. / ~ loss ‖ Eisenverlust m. ‖ perte f. dans le fer. / magnetic oxide of ~ ‖ Eisenoxyduloxyd n. ‖ oxyde m. magnétique de fer.
iron manganate ‖ Eisenmanganat ‖ manganate m. ferrique. / peptonated ~ ‖ Eisenmanganatpeptonat n. ‖ manganate m. de fer peptoné. / saccharated ~ ‖ Eisenmanganatsaccharat n. ‖ manganate m. de fer saccharé.
iron manufacture ‖ Eisenerzeugung f.; fabrication f. du fer; métallurgie f. / ~ master ‖ Hüttenmeister m. ‖ maître m. des forges. / ~ material for telephone lines ‖ Eisenbauzeug n. für Telefonleitungen ‖ accessoires mpl. et charpente en fer pour lignes de téléphone. / ~ meal ‖ Eisenmehl n. ‖ farine f. de fer. / metallurgy of ~ ‖ Eisenhüttenkunde

f. || métallurgie f. du fer. / ~ mill || Hammerwerk m. || forge f. / ~ mine || Eisengrube f. || mine f. de fer. / ~ minium || Eisenmennige f. || minium m. de fer. / ~ monger || Eisenhändler m. || marchand m. de fer. / ~ mongery || Eisenwaren fpl. || quincaillerie f. / ~ mordant || Eisenbeize f. || mordant m. de fer. / ~ nail || Eisennagel m. || clou m. de fer. / ~ nickel accumulator || Eisennickelakkumulator m. || accumulateur m. au fer et au nickel. / ~ nipper || Kabelarschäkel m. || manille f. de tourne-vire. / ~ nitrate || salpetersaures Eisen n.; Eisennitrat n. || nitrate m. de fer. / ~ ochre, yellow || gelber Ocker m.; gelber Eisenocker m.; Berggelb n.; ockeriger Gelbeisenstein m. || ocre m. jaune; jaune m. de montagne; terre f. jaune; limonite m.

iron ore || Eisenerz n.; Eisenstein m. || minerai m. de fer. / brown ~ || Brauneisenerz n.; Brauneisenstein m. || hématite f. ou limonite f. brune; fer m. oxydé hydraté. / clay ~ || Nagelerz n.; Schindelerz n.; stengliger roter Toneisenstein m. || argile f. ferrugineuse en tiges. / compact brown ~ || Pecheisenerz n.; schlackiges Brauneisenerz n. || fer m. oxydé noir vitreux. / earthy red ~ || ockeriger Roteisenstein m.; Roteisenocker m. || fer m. oligiste rouge terreux. / fibrous red ~ || faseriger Roteisenstein m.; roter Glaskopf m.; Blutstein m.; roter Hämatit m. || fer m. oligiste concrétionné; hématite m. rouge; fer m. oligiste fibreux. / high grade ~ || hochhaltiges Eisenerz n. || minerai m. de fer d'une teneur très haute. / magnetic ~ || Magneteisenstein m. || fer m. magnétique. / manganiferous ~ || manganhaltiges Eisenerz n. || minerai m. de fer manganésé. / oligiste ~ || Glanzeisenstein m.; Eisenglanz m. || fer m. oligiste. / oolitic red ~ || oolithischer Roteisenstein m.; Minette f. || oligiste m. oolithique; minerai m. violet; minette f. / phosphatic ~ || phosphorhaltiges Eisenerz n. || minerai m. de fer phosphoreux. / red ~ || Roteisenstein m. || hématite f. rouge. / scaly red ~ || schuppiger Roteisenstein m.; Eisenschaum m.; Roteisenrahm || fer m. oxydé rouge brillant. / yellow ~ see iron ochre, yellow.

iron ore, allotment for ~ || Eisensteinfeld n. || concession f. pour minerai de fer. / ~ calciner || Eisensteinröster m. || calcinateur m. de minerai de fer. / deposit of ~ (Min) || Eisensteinlager n. || gisement m. de minerai de fer. / ~ mine || Eisensteingrube f.; Eisenerzgrube f. || mine f. de fer.

iron oxichloride || Eisenoxychlorid n. || oxychlorure m. de fer. / ~ oxide || Eisenoxyd n.; Tempererz n. || oxyde m. de fer; minerai m. pour fonte malléable. / ~ oxide black || Eisenoxydschwarz n. || noir m. d'oxyde fer. / ~ part of gas furnace || Gasofeneisenteil m. || pièce f. en fer pour fourneaux à gaz. / ~ paving || Eisenbelag m. || recouvrement m. ou dallage m. ou platelage m. en fer. / ~ peptonate || Eisenpeptonat n. || peptonate m. de fer. / ~ perchloride || Eisenperchlorid n. || perchlorure m. de fer. / ~ peroxide || Eisenperoxyd n. || ferrioxyde m. / ~ phosphate || Eisenphosphat n. || phosphate m. de fer. / ~ pig || Mulde f.; Eisenmassel f. || saumon m. de fer. / ~ pig cinder || Roheisenschlacke f. ||

laitier m. de fonte. / ~ pin || eiserner Stift m. || cheville f. en fer. / ~ pin of a rail chair || eiserner Bolzen m. eines Schienenstuhls || boulon m. ou cheville f. d'un coussinet de rail. / ~ pipe || Eisenröhre f. || tube m. en fer. / ~ pitch ore || Pecheisenerz n.; schlackiges Brauneisenerz n. || fer m. oxydé noir vitreux.

iron-plate || Eisenblechplatte f. || plaque f. de tôle. / black ~ || Schwarzblech n.; Sturzblech n. || tôle f. noir. / hammered ~ || gehämmertes Eisenblech n. || fer m. en lames forgé. / ~ of the liquation hearth || Seigerscharte f. || plaque f. de fer du four de liquation. / tinned ~ see white ~. / white ~ || Weißblech n. || ferblanc m.; feuille f. de ferblanc; tôle f. étamée.

iron plate box (Cable) || Eisenblechmuffe f. || manchon m. en tôle de fer.

iron plating || Ummantelung f.; Verkleidung f.; Panzerung f. || blindage m. / ~ pole || eiserne Telegrafenstange f. || poteau m. ou pylône m. de télégraphe en fer. / ~ Portland cement || Eisenportlandzement m. || ciment m. de laitier Portland. / ~ powder core coil || Eisenpulverkernspule f. || bobine f. à noyau en poudre de fer comprimé. / ~ preparation || Eisenpräparat n. || préparation f. ferrique. / ~ protochloride || Eisenchlorür n. || protochlorure m. de fer. / ~ protoxide || Eisenoxydul n. || protoxyde m. de fer.

iron pyrites pl. || Pyrit m.; Eisenkies m.; Schwefelkies m. || pyrite m.; fer m. sulfuré. / white ~ || Schreibkies m.; Wasserkies m.; Markasit m. || fer m. sulfuré blanc.

iron pyrites mine || Schwefelkiesgrube f. || mine f. de pyrite de fer.

iron pyrites recovery || Schwefelkiesgewinnung f. || extraction f. des pyrites de fer.

iron pyrolignite || holzessigsaures Eisen n. || pyrolignite m. de fer. / ~ rake || Eisenrechen m. || râteau m. en fer. / ~ rake (Brew) || Krücke f.; Maischharke f.; Rührharke f.; Malzkrücke f. || brassoir m. / ~ railing fitter || Eisengeländermacher m. || rampiste m. en fer. / ~ refuse || Eisenabfälle mpl. || ferraille f.; débris mpl. de fer; riblons mpl. / ~ resistance || Eisenwiderstand m. || résistance f. ballast ou en fer. / ~ rivet || Eisenniet n. || rivet m. en fer. / ~ rod || Hefteisen n. || pontil m. / ~ rolling mill || Eisenwalzwerk n. || laminoir m.; usine f. de laminage de fer. / ~ rust || Eisenrost m. || rouille f. de fer. / ~ rust cement || Eisenkitt m. || pouzzolane f. artificielle; mastic m. de fer. / ~ saccharate || Eisensaccharat n. || saccharate m. de fer. / carbonated ~ saccharate || kohlensaures Eisensaccharat n. || saccharate m. de fer carbonaté. / ~ salt || Eisensalz n. || sel m. de fer. / ~ scales pl. || Hammerschlag m.; Glühspan m.; Zunder m. || pailles fpl.; écailles fpl.; mâchefer m. / ~ scrap || Schmiedeeisenschrot m. || mitraille f. de fer. / ~ separator || Eisenseparator m. || séparateur m. électrique de ferraille. / ~ set hammer || Eisenvollsetze f. || chasse f. pleine en fer. / ~ shavings pl. || Eisenspäne mpl. || copeaux mpl. ou paille f. de fer. / ~ shears pl. || Eisenschere f. || cisaille f. à fer. / black ~ sheet || Schwarzblech n. || feuille f. de tôle. / ~ shell || Eisenblechmantel m. || revêtement m. ou enveloppe f. en tôle. / ~ shell of

furnace || Blechmantel m. des Ofens || enveloppe f. ou chemise f. en tôle de fourneau. / ~ ship || Eisenschiff n. || navire m. en fer. / ~ ship building || Eisenschiffbau m. || construction f. de navires en fer. / ~ shoe of the stamper || Pochschuh m.; Pocheisen n.; Pochstempelschuh m. || armure f. en fonte ou sabot m. en fonte du pilon d'un bocard. / cast ~ shot || Gußbruch m. || mitraille f. de fonte. / ~ sinter see ~ slag. / ~ slag || Eisenschlacke f.; Sinter m. || laitier m. ou scorie f. de fer. / ~ sleeper || Eisenschwelle f. || traverse f. métallique ou de fer.

ironsmith || Eisenschmied m. || forgeron m.; forgeur m. de fer.

iron spar || Spateisenstein m.; Eisenspat m. || fer m. spathique. / ~ sponge || Eisenschwamm m. || éponge f. de fer. / ~ staff (Forg) || Schweif m. || ringard m. / ~ stanchion || eiserne Stütze f. || chandelier m. de fer. / ~ stores pl. || Eisenlager n. || magasin m. ou dépôt m. de fer.

iron stone see also iron ore || Eisenerz n. || minerai m. de fer. / campact red ~ || dichter Roteisenstein n. || fer m. oligiste rouge compact. / fibrous red ~ || faseriger Roteisenstein m.; roter Glaskopf m.; Blutstein m.; roter Hämatit m. || fer m. oligiste concrétionné; hématite m. rouge; fer m. oligiste fibreux. / oolitic ~ || Linsenerz n. || fer m. oolithique. / oolitic calcareous ~ || Minette f. || minette f.; oolithe f. de fer calcifère. / red ~ || Roteisenstein m.; Roteisenerz n.; Eisenoxyd n. || fer m. oligiste ou oxydé rouge; hématite m. rouge. / sparry ~ || Spateisenstein m.; Eisenspat m. || fer m. spathique.

iron stone ware || Hartsteingut n. || faïence f. fine.

iron structure (Shipb) || Stellingsanlage f. || échafaudage m. métallique. / ~ builder || Eisenkonstruktör m. || constucteur m. en charpente métallique. / shop for ~s || Eisenkonstruktionswerkstatt f. || atelier m. de charpentes métalliques.

iron sulphate || Eisenvitriol n.; Eisensulfat n. || couperose f. verte; sulfate m. de fer. / ~ sulphide || Schwefeleisen n. || sulfure m. de fer. / ~ superstruction || Eisenhochbau m. || charpente f. métallique. / ~ tempering material || Eiseneinsatzmittel n. || produit m. à cementer l'acier ou à cementer le fer doux. / ~ tie || Maueranker m.; Stichanker m.; Zuganker m. || lien m. tirant ou tirant m. en fer. / ~ tool || Eisenwerkzeug n. || outil m. en fer. / ~ tower || Eisenturm m. || pylône m. métallique. / ~ trade || Eisenhandel m. || commerce m. de fer. / ~ trellis || eisernes Gitter n. || treillis m. de fer. / ~ tube || Eisenrohr n. || tuyau m. de fer. / ~ turner || Eisendreher m. || tourneur m. en fer. / ~ varnish || Eisenlack m. || vernis m. à fer; vernis m. noir. / ~ wall || Eisenwand f. || pan m. en charpente métallique.

iron-ware || Eisenwaren fpl. || articles mpl. en fer; quincailleries fpl. / small ~ industry || Kleineisenindustrie f. || industrie f. des objets de quincaillerie ou d'accessoires en fer. / ~ manufacture || Grobschmiede f. || ferronnerie f.

iron wedge || Eisenkeil m. || coin m. en fer. / ~ window || Eisenfenster n. || fenêtre f. en fer.

iron wire || Eisendraht m. || fil m. de fer. / drawn ~ || gezogener Eisendraht m. || fil

m. de fer tréfilé *ou* étiré. / **galvanized ~** ‖ galvanisierter *oder* verzinkter Eisendraht m. ‖ fil m. de fer galvanisé *ou* zingué.

iron wire armouring ‖ Eisendrahtarmierung f. ‖ armature f. en fil de fer. / **~ gauze** ‖ Eisendrahtnetz n. ‖ toile f. métallique.

ironwork ‖ Eisenbeschläge mpl. ‖ ferrures fpl. / **~ for carriages** ‖ Wagenbeschlagteile mpl. ‖ ferrures fpl. pour voitures. / **~ for doors and windows** ‖ Beschläge mpl. für Türen und Fenster ‖ quincaillerie f. de bâtiment./ **~ binders** pl. ‖ Wagenbeschläge mpl. ‖ ferrures fpl. de voiture. / **~ black** ‖ Eisenlack m. ‖ vernis m. noir.

ironworker ‖ Eisenarbeiter m.; Eisenkonstruktör m. ‖ ouvrier m. *ou* charpentier m. en fer.

ironworks pl. ‖ Hütte f.; Eisenhütte f. Hammerwerk n. ‖ usine f. métallurgie; fer; établissement m. sidérurgique.

irradiate, to ‖ belichten ‖ irradier. / **~ a patient** ‖ einen Kranken m. bestrahlen ‖ irradier un malade.

irradiation of the fundus of the eye ‖ Augenhintergrundbestrahlung f. ‖ bain m. de lumière du fond de l'œil.

irregular ‖ unpünktlich; unregelmäßig ‖ inexact; irrégulier. / **the piece proved too ~ in the core** ‖ das Stück erwies sich als zu ungleichmäßig im Kern ‖ la pièce f. manquait d'homogénéité au cœur. / **~ crystal surfaces** pl. ‖ unregelmäßige Kristallflächen fpl. ‖ surfaces fpl. cristallines irregulières. / **~ parting** ‖ ungleichmäßige Teilung f. ‖ division f. irrégulière *ou* non proportionnée.

irregularity ‖ Unregelmäßigkeit f. ‖ irrégularité f. / **degree of ~** ‖ Ungleichförmigkeitsgrad m. ‖ degré m. d'irrégularité.

irreversible ‖ nicht umkehrber; selbsthemmend ‖ irréversible.

irrigate, to ‖ berieseln; bewässern ‖ ruisseler; arroser. / **~ the meadows** ‖ die Wiesen berieseln ‖ irriger *ou* arroser les prés.

irrigating plant ‖ Bewässerungsanlage f. ‖ installation f. d'irrigation *ou* d'arrosage.

irrigation ‖ Berieselung f.; Bewässerung f. ‖ irrigation f. / **~ device** ‖ Berieselungseinrichtung f. ‖ installation f. d'arrosage. / **~ lock** ‖ Bewässerungsschleuse f. ‖ écluse f. d'irrigation. / **~ society** ‖ Berieselungsgesellschaft f.; Bewässerungsgesellschaft f. ‖ syndicat m. d'irrigation. / **~ squirt** ‖ Spülspritze f. ‖ seringue f. d'irrigation./ **~ works** pl. ‖ Bewässerungsanlage f. ‖ installation f. d'irrigation.

irrigator ‖ Irrigator m. ‖ irrigateur m.

isinglass ‖ Fischleim m.; Hausenblase ‖ colle f. de poisson; ichthyocolle. / **~ plaster** ‖ englisches Heftpflaster n. ‖ taf-

fetas m. anglais; emplâtre m. adhésif anglais.

isobarometric ‖ isobarometrisch ‖ isobarométrique.

isobornyl preparation ‖ Isobornylpräparat n. ‖ préparation f. d'isobornyle.

isobutyl preparation ‖ Isobutylpräparat n. ‖ préparation f. isobutylique.

isobutyric acid ‖ Isobuttersäure f. ‖ acide m. isobutyrique.

isochronal ‖ gleichdauernd; isochron ‖ isochrone.

isochronal line ‖ Zykloïde f. ‖ cycloïde f.

isochronism ‖ Isochronismus m.; Gleichdauer f.; Gleichzeitigkeit f.; Synchronismus m. ‖ isochronisme m.; synchronisme m.

isochronous see isochronal.

isochronous curve ‖ Tautochrone f.; Isochrone f. ‖ ligne f. *ou* courbe f. isochrone.

isodynamic ‖ isodynamisch ‖ isodynamique.

isogone ‖ Isogone f. ‖ ligne f. isogone.

isolate, to see to insulate.

isolated ‖ isoliert; vereinzelt ‖ isolé. / **~ bottle** ‖ Isolierflasche f. ‖ bouteille f. isolante. / **~ bottles manufacturing machine** ‖ Isolierflaschenherstellungsmaschine f. ‖ machine f. à fabriquer des bouteilles isolantes. / **~ dune** ‖ Einzeldüne f. ‖ dune f. isolée. / **~ pillar crane** ‖ freistehender Säulendrehkran m. ‖ grue f. pivotante isolée. / **~ wire** ‖ isolierter Draht m. ‖ fil m. isolé.

isolating see insulating.

isolation see insulation.

isolator see insulator.

isomorphous ‖ isomorph; gleichförmig; von gleicher Gestalt; homöomorph ‖ isomorphe.

isosceles ‖ gleichschenklig ‖ isocèle; isoscèle. / **~ triangle** ‖ gleichschenkliges Dreieck n. ‖ triangle m. isoscèle *ou* isocèle.

isothermal ‖ isotherm ‖ isotherme. / **~ air stratum** ‖ Luftschicht f. gleicher Wärme ‖ couche f. d'air isotherme. / **~ chart** ‖ Isothermenkarte f. ‖ carte f. des isothermes. / **~ compression** ‖ isothermische Verdichtung f. ‖ compression f. isotherme. / **~ curve** ‖ Wärmegleiche f.; Isotherme f. ‖ isotherme f. / **~ map** ‖ Isothermenkarte f. ‖ carte f. des isothermes. / **~ zone** ‖ Zone f. gleicher Wärme; isotherme Zone ‖ zone f. isotherme *ou* isothermique.

isotrop (Phys) ‖ isotrop ‖ isotrope.

isotropy ‖ Isotropie f. ‖ isotropie f.

issue, to ‖ ausfertigen ‖ dresser; rédiger. / **~ a bill** ‖ eine Tratte m. *oder* eine Rechnung f. ausstellen ‖ émettre une traite.

issue (Bank) ‖ Emission f. ‖ émission f. / **~** (Build) ‖ Abfluß m. ‖ épanchoir m.; déchargeoir m. / **~** (Print) ‖ Auflage f.;

Lieferung f. ‖ tirage m.; livraison f. / **~ of bank notes** ‖ Banknotenausgabe f. ‖ émission f. de billets de banque. / **~ of checks** ‖ Scheckausgabe f. ‖ émission f. de chèques *ou* de tickets. / **~ of a duplicate cheque** ‖ Ausstellen n. eines Duplikatschecks ‖ émission f. d'un duplicate de chèque. / **restricted ~ of shares** ‖ beschränkte Ausgabe f. von Aktien ‖ émission f. limitée d'actions. / **amount of ~** ‖ Emissionsbetrag m. ‖ montant m. de l'émission.

issuing bank ‖ Notenbank f. ‖ banque f. d'émission.

isthme ‖ Landenge f. ‖ isthmus m.; isthme m.

itacolumite ‖ Itakolumit m.; Gelenkquarz m. ‖ itacolumite f.; grès m. flexible.

Italian cloth ‖ Zanella n. ‖ drap m. italien. / **~ paste manufacture** ‖ Teigwarenfabrikation f. ‖ fabrication f. de pâtes alimentaires.

Italic (Print) ‖ Kursivschrift f.; Schrägschrift f. ‖ cursive m.; Italique f.

itamalic acid ‖ Itamalsäure f. ‖ acide m. itamalique.

item (Bookkeep) ‖ Posten m. in der Buchung ‖ écriture f.; entrée f.; article m.

iterative network ‖ Kettenleiter m. ‖ système m. itératif; filtre m.

itinerant map ‖ Reisekarte f. ‖ carte f. itinéraire.

ivorine ‖ Ivorin n. ‖ ivorine f.

ivory ‖ Elfenbein n. ‖ ivoire m. / **artificial ~** ‖ künstliches Elfenbein n. ‖ ivoire m. artificiel; éburine f. / **raw ~** ‖ rohes Elfenbein n. ‖ marfil m.; morfil m.

ivory articles pl. ‖ Elfenbeinwaren fpl. ‖ objets mpl. en ivoire. / **~-backed brushes** pl. ‖ Bürstenbinderwaren fpl. in Elfenbein ‖ brosserie f. montée sur ivoire. / **~ black** ‖ Elfenbeinschwarz n. ‖ noir m. d'ivoire. / **~ cardboard** ‖ Elfenbeinkarton m.; Elfenbeinpappe f. ‖ carton-ivoire m. / **~ carver** ‖ Elfenbeinschnitzer m. ‖ sculpteur m. en ivoire. / **~-coloured** ‖ elfenbeinfarbig ‖ ivoiré. / **~ comb** ‖ Elfenbeinkamm m. ‖ peigne m. en ivoire. / **~ cutter** ‖ Elfenbeinschnitzer m. ‖ découpeur m. d'ivoire. / **~ engraver** ‖ Elfenbeingravör m. ‖ graveur sur ivoire. / **~ goods** pl. ‖ Elfenbeinwaren fpl. ‖ articles mpl. *ou* objets mpl. en ivoire. / **~ maker** ‖ Elfenbeinarbeiter m. ‖ ivoirier m. / **~ nut** ‖ Steinnuß f. ‖ noix f. de corozo. / **~ painter** ‖ Elfenbeinmaler m. ‖ peintre m. sur ivoire. / **~ paper** ‖ Elfenbeinpapier n.; Bristolpapier n. ‖ carton m. ivoiré; papier m. Bristol / **~ polisher** ‖ Elfenbeinschleifer m. ‖ polisseur m. d'ivoire. / **~ turner** ‖ Elfenbeindrechsler m. ‖ tourneur m. sur ivoire; ivoirier m.

J

J-spindle ‖ J-Stütze f. ‖ support m. double en J. / ~ with terminal insulator ‖ Einschiebestütze f. ‖ support-tige m.

jaborandi leaves pl. ‖ Jaborandiblätter npl. ‖ feuilles fpl. de jaborandi.

jacaranda brown ‖ Jakarandabraun n. ‖ brun m. jacaranda. / ~ wood ‖ Jakarandaholz n.; Palisanderholz n. ‖ jacaranda m.; palissandre m.

jacconet (Cotton) ‖ Jakonett m. ‖ jaconas m.; jaconnat m.

jack, to (Join) ‖ schruppen; rauh behobeln ‖ corroyer; dégrossir l'ouvrage m. / ~ down the stuff (Join) ‖ Holz n. abhobeln ‖ dégrossir le bois.

jack ‖ Hebewinde f.; Handwinde f.; Schraubenwinde f.; Fußwinde f. ‖ cric m.; cric m. simple ou à crochet. / ~ (Build) ‖ Bock m.; Gerüst n.; Gestell n. ‖ chevalet m.; tréteau m. / ~ (Clockm) ‖ Hämmerchen n.; Anschläger m.; Heber m. einer Repetieruhr ‖ échappement m. de répétition. / ~ (Mar) ‖ Göschflagge f. ‖ pavillon m. de beaupré. / ~ (Tel) ‖ Klinke f. ‖ jack f.; poignée f. / ~ (Weav) ‖ Hakenstange f. ‖ tringle m. à crochet. / answering ~ (Tel) ‖ Abfrageklinke f. ‖ jack m. de réponse. / strip of answering ~ (Tel) ‖ Abfrageklinkenstreifen m. ‖ réglette f. de jacks de réponse. / barrel raising ~ ‖ Faßwinde f. ‖ treuil m. à fûts. / black ~ (Miner) ‖ Zinkblende f.; Schwefelzink n.; Blende f.; Sphalerit m. ‖ zinc m. sulfuré. / ~ for busy tone (Tel) ‖ Besetztklinke f. ‖ jack m. pour test signal. / chain ~ ‖ Kettenwinde f. ‖ cric m. à noix. / chain ~ with two hooks ‖ Doppelklauenwinde f. ‖ cric m. à noix avec deux pattes. / hand ~ with claw ‖ Handklauenwinde f. ‖ cric m. à crochet à main. / hop ~ ‖ Ausschlagbottich m. ‖ panier m. à houblon. / hydraulic ~ ‖ hydraulische Winde f. ‖ vérin m. hydraulique. / hydraulic hoisting ~ ‖ hydraulisches Hebewerk n. ‖ vérin m. de levage hydraulique. / lifting ~ ‖ Aufzugswinde f.; Hebewinde f. ‖ montecharge m. à main; vérin m. de levage. / rack and pinion ~ ‖ Zahnstangenwagenwinde f. ‖ cric m. à crémaillère. / screw ~ ‖ Wagenwinde f.; Schraubenwinde f. ‖ lève-roue m.; vérin m. / subscriber's ~ (Tel) ‖ Teilnehmerklinke f. ‖ jack m. d'abonné. / ~ of a warping mill ‖ Führer m. oder Katze f. eines Scherrahmens ‖ plot m. ou giette f. d'un ourdissoir.

jack chain ‖ Hemmkette f. ‖ chaîne f. d'enrayure. / ~ disc (Tel) ‖ Klinkenscheibe f. ‖ disque m. à cliquet.

jacket (Mach) ‖ Mantel m.; Wärmeschutz m.; Umhüllung f. (eines Maschinenteils) ‖ chemise f.; enveloppe f.; revêtement m. isolant. / ~ (Weav) ‖ Joppe f.; Wams n.; Jacke f. ‖ vareuse f.; pourpoint m. / blast furnace ~ ‖ Hochofenummantelung f. ‖ blindage m. pour hauts-fourneaux. / ~ of a boiler ‖ Mantel m. des Kessels ‖ enveloppe f. ou chemise f. d'une chaudière. / cooling ~ ‖ Kühlmantel m. ‖ chemise f. de réfrigérant. / heating ~ ‖ Heizmantel m. ‖ chemise f. de réchauffage. / shooting ~ ‖ Joppe f. ‖ vareuse f.; casaquin m. / water ~ ‖ Wassermantel m. ‖ chemise f. d'eau. / water-circulation ~ surrounding the electrodes ‖ Elektrodenwasserkühlmantel m. ‖ enveloppe f. à circulation d'eau entourant les électrodes. / water-heating ~ ‖ Wasserheizmantel m. ‖ chemise f. de réchauffage d'eau.

jacket body machine ‖ Jackenmaschine f.; Jackenstuhl m. ‖ machine f. ou métier m. pour camisoles. / ~ cooling ‖ Mantelkühlung f. ‖ refroidissement m. à chemise d'eau.

jacket sheet iron ‖ Mantelblech n. ‖ tôle f. d'enveloppe.

jack frame (Spinn) ‖ Spindelbank f.; Spulenmaschine f. ‖ banc m. à broches; boudinoir m. à bobines.

jack framing (Tel) ‖ Klinkenumrahmung f. ‖ ripolinage m. / ~ head (Mine) ‖ blinder Schacht m. ‖ puits m. intérieur. / numbering of ~s (Tel) ‖ Klinkennumerierung f. ‖ numérotage m. des jacks.

jack panel (Tel) ‖ Klinkenfeld n. ‖ panneau m. de jacks. / answering ~ (Tel) ‖ Abfragefeld n. ‖ champ m. de jacks locaux.

jack plane (Join) ‖ Rauhhobel m.; Schrupphobel m. ‖ rabot m. debout; riflard m. / ~ rafter (Carp) ‖ Halbsparren m.; Walmsparren m.; Schiftsparren m. ‖ empanon m. / ~ screw ‖ Schraubenwinde f.; Hebeschraube f. ‖ vérin m.; cric m. à vis. / ~ shaft ‖ Blindwelle f. ‖ faux-essieu m. / ~ sleeve (Tel) ‖ Klinkenbuchse f. ‖ douille f. de jack. / ~ spring (Tel) ‖ Klinkenfeder f. ‖ ressort m. de jack.

jack stay of a mast ‖ Stander m. ‖ draille f. ou mât m. de corde. / ~ of wood ‖ Kamm m. einer Rahe ‖ filière f. d'envergure en bois.

jacks' strip (Tel) ‖ Klinkenstreifen m. ‖ réglette f. de jacks. / band for ~s (Tel) ‖ Klinkengitter n. ‖ bande f. pour attacher des réglettes de jacks.

jack, testing equipment for ~s (Tel) ‖ Klinkenprüfeinrichtung f. ‖ dispositif m. d'essai des jacks. / ~ twine (Weav) ‖ Heber mpl.; Hebeschnur f. ‖ arcade f.; fil m. d'arcade.

jaconet see jacconet.

Jacquard ‖ Jacquardmaschine f. ‖ machine f. Jacquard. / tie up ~ ‖ Schaftmaschine f. ‖ métier m. à pédale genre Jacquard.

Jacquard attachment ‖ Jacquardvorrichtung f. ‖ dispositif m. à Jacquard. / ~ card lacer ‖ Jacquardkartenbinder m. ‖ enlaceur m. de cartons Jacquard. / ~ card puncher ‖ Jacquardkartenschläger m. ‖ perceur m. de cartons Jacquard. / ~ chain loom ‖ Jacquardkettenstuhl m. ‖ métier m. chaîne à Jacquard. / ~ circular knitting machine ‖ Jacquardrundstrickmaschine f. ‖ machine f. circulaire Jacquard à tricoter. / ~ fitter ‖ Jacquardmaschinenvorrichter m. ‖ jacquardier m. / ~ hand knitting machine ‖ Jacquardhandstrickmaschine f. ‖ machine f. Jacquard à tricoter à main. / ~ loom ‖ Jacquardstuhl m. ‖ métier m. Jacquard. / ~ weaver ‖ Jacquardweber m. ‖ tisserand m. au Jacquard. / ~ weaving ‖ Gebildweberei f. ‖ tissage m. au Jacquard; tissage m. des étoffes grand façonnées.

jade (Miner) ‖ Jadeit m. ‖ jade f.; jadéite f.

jag, to (Carp) ‖ einsetzen; einkerben ‖ poser; entailler; encocher.

jag ‖ Kerbe f.; Marke f.; Ausschnitt m. ‖ entaille f.; encoche f.; coche f.

jagging board (Ore dress) ‖ Setzherd m. ‖ table f. allemande ou à balais ou dormante.

jagging-out the sleepers ‖ Einschneiden n. der Schwellen ‖ entaillage m. des traverses.

jail, Pennsylvanian ~ ‖ Zellengefängnis n. ‖ prison f. à détention solitaire.

jake ‖ Abortanlage f. ‖ latrine f.

jalap ‖ Jalape f. ‖ jalap m. / ~ resin ‖ Jalappaharz m. ‖ résine f. de jalap. / ~ root ‖ Jalappawurzel f. ‖ racine f. de jalap.

jam, to (Mach) ‖ festfressen; klemmen ‖ gripper; serrer.

jam ‖ Marmelade f. ‖ marmelade f.

jamb (Build) ‖ Einfassung f. ‖ châssis m.; jambe f. / intermediate ~ ‖ Quaderpfeiler m.; Mittelpfeiler m. ‖ jambe f. étrière.

jamb lining of a door frame ‖ Türfutter n. ‖ fourrure f. d'huisserie. / ~ stone ‖ Quaderpfeiler m. ‖ jambe f. de maçonnage.

jam colours pl. ‖ Farben fpl. für Marmeladen ‖ couleurs fpl. pour marmelade.

jamesonite ‖ Jamesonit m.; Querspießglanz m. ‖ jamesonite f.

jam maker ‖ Konfitürenfabrikant m. ‖ confiturier m.

jamming ‖ Klemmung f. ‖ calage m. / ~ of the projectile in the bore ‖ Geschoßverkeilung f. ‖ obstruction f. de l'âme par le projectile. / ~ of upper against lower blade ‖ Aufreiten n. des Obermessers auf das Untermesser ‖ chevauchement m. de la lame supérieure sur la lame inférieure.

jam nut ‖ Gegenmutter f.; Stellmutter f. ‖ contre-écrou m.

japan, to ‖ mit japanischem Lack m. lackieren; lackieren ‖ vernir au vernis du Japon; vernir au four m.

japan ‖ Japanlack m. ‖ laque f. japonaise.

japanned goods pl. ‖ Lackwaren fpl. ‖ objets mpl. en laque. / ~ sheet iron articles pl. ‖ lackierte Blechwaren fpl. ‖ objets mpl. en tôle vernie.

japanner ‖ Lackierer m. ‖ laqueur m.

Japan wax ‖ Japanwachs n. ‖ cire f. du Japon.

jar ‖ Kruke f. ‖ pot m. de grès. / graduated ~ ‖ graduiertes oder kalibriertes Gefäß n. ‖ cloche f. graduée. / Leyden ~ ‖ Leydener Flasche f. ‖ bouteille f. électrique ou de Leyden. / lye ~ ‖ Laugentopf m. ‖ pot m. à lessive. / preserve ~ ‖ Einkochglas n. ‖ bocal m. à conserves. / stone ~ ‖ Kruke f. ‖ cruchon m. / wide-mouthed ~ ‖ weithalsige Flasche f. ‖ bouteille f. à col large.

jar diffusion ‖ Gefäßdiffusion f. ‖ diffusion f. en bocal.

jardinière ‖ Blumenschale f.; Blumentisch m.; Jardiniere f. ‖ jardinière f.

jargon (Miner) ‖ Jargon m.; heller Zirkon m. ‖ jargon m.; zircon m. de Ceylon.

jarring blow ‖ Prellschlag m. ‖ rebondissement m.

jaspe ‖ flammig meliert; jaspiert ‖ jaspé. / ~ paper ‖ gesprengtes Papier n.; Granitpapier ‖ papier m. jaspé.

jasper ‖ Jaspis m. ‖ jaspe m.; quartz-jaspe m. / black ~ ‖ schwarzer Jaspis m.; Kieselschiefer m.; lydischer Stein m. ‖ jaspe m. noir; pierre f. de Lydie ou de touche. / Egyptian ~ ‖ ägyptischer Kiesel m.; Nilkiesel m. ‖ jaspe m. égyptien. / ribbon ~ ‖ Bandjaspis m. ‖ jaspe m. rubané.

jasper agate ‖ Jaspisachat m. ‖ agate f. jaspée. / ~ opal ‖ Jaspisopal m.; Opaljaspis m. ‖ jaspe m. opale.

jaw (Anatomy) ‖ Kinnbacken m. ‖ mâchoire f. / ~ (Mach) ‖ Klemmbacke f. ‖ mordache f. de serrage. / chequered ~ ‖ geriffelte Brechbacke f. ‖ mâchoire f. cannelée ou à crans. / coupler ~ (Railw) ‖ Kuppelklaue f. ‖ griffe f. de la tête d'attelage. / crankshaft starting ~ ‖ Andrehklaue f. der Kurbelwelle ‖ griffe f. de mise en marche de la manivelle. / crushing ~ ‖ Brechbacke f. ‖ mâchoire f. broyeuse ou de concasseur. / movable ~ ‖ lose Brechbacke f. ‖ mâchoire f. mobile. / oscillating ~ ‖ bewegliche Brechbacke f. ‖ mâchoire f. mobile. / ~ of a rail-chair ‖ Backe f. eines Schienenstuhls ‖ saillie f. d'un coussinet de rails. / self-centering ~s pl. ‖ selbstzentrierte Klemmbacken fpl. ‖ mâchoires fpl. à centrage automatique. / slotted ~ ‖ Bremsgabel f. ‖ étrier m. du frein / starting crank ~ ‖ Klaue f. der Andrehkurbel ‖ griffe f. de la manivelle de mise en marche. / vice ~ ‖ Schraubstockspannbacke f. ‖ mâchoire f. d'étau.

jaw breaker ‖ Backenbrecher m. ‖ broyeur m. ou concasseur m. à mâchoires. / ~ chair (Railw) ‖ Drehstuhl m.; Gelenkstuhl m. ‖ coussinet m. de talon ou de rotation. / ~ chuck ‖ Backenfutter n. ‖ mandrin m. à griffes.

jaw coupling ‖ Klauenkupplung f. ‖ embrayage m. ou attelage m. à griffes. / ~ after American designs (Railw) ‖ Klauenkupplung f. nach amerikanischer Bauart ‖ attelage m. à griffes de construction américaine.

jaw crusher ‖ Backenbrecher m. ‖ concasseur m. à mâchoires. / depth of ~s pl. (Vice) ‖ Backenhöhe f. ‖ hauteur f. des mors. / the upper mouth of the ~s clasps the flanges of the rail ‖ das obere Klemmbackenmaul greift von beiden Seiten um den Schienenfuß ‖ le mors supérieur des mordaches embrasse de deux côtés le patin du rail. / width of ~s pl. (Vice) ‖ Backenbreite f. ‖ largeur f. des mors.

jeffersonite ‖ Jeffersonit m. ‖ jeffersonite f.

jellied ‖ gallertartig ‖ gélatineux.

jelly ‖ Gelee n.; Gallerte f. ‖ gélée f. / vegetable ~ ‖ Pektin n.; Pflanzengallert m. ‖ gelée f. végétale; pectine f. / ~-like ‖ gallertartig ‖ gélatineux.

jenny, mule ~ ‖ Mulemaschine f.; Mulespinnmaschine f. ‖ mull-jenny m. en fin; renvideur m.

jequirity ‖ Abrusbaumsame m. ‖ graine f. d'abrus.

jerk, by ~s pl. ‖ ruckweise ‖ en mouvement m. saccadé; par secousses fpl.; par à-coups mpl. / to start without ~s ‖ stoßfrei anlaufen ‖ démarrer sans à-coups mpl.

jerking action of the saw blade ‖ Werfen n. des Sägeblattes ‖ gondolement m. de la lame de scie. / ~ stop ‖ ruckweises Ausschalten n. ‖ mise f. hors circuit par retrait.

jersey ‖ Sweater m. ‖ sweater m.

jet, to ~ out (Arch) ‖ auskragen ‖ se jeter hors d'œuvre m.

jet ‖ Jett m.; jais m.; jayet m. / ~ (Mach) ‖ Düse f. ‖ gicleur m. / artificial ~ ‖ künstliches Jett m. ‖ jais m. artificiel. / auxiliary ~ ‖ Hilfsdüse f. ‖ gicleur m. auxiliaire. / compensating ~ ‖ Ausgleichdüse f. ‖ gicleur m. de compensation. / main ~ ‖ Hauptdüse f. ‖ gicleur m. principal. / pilot ~ ‖ Anlaßdüse f. ‖ gicleur m. de ralenti. / slow running ~ ‖ Leerlaufdüse f. ‖ gicleur m. de ralenti.

jet apparatus, compressed air ~ ‖ Druckluftstrahlapparat m. ‖ appareil m. à jet d'air comprimé.

jet condenser ‖ Einspritzkondensator m. ‖ condenseur m. par injection ou par mélange ou à jet. / ~ carburettor ‖ Düsenvergaser m. ‖ carburateur m. à gicleur. / ~ mine ‖ Jettgrube f. ‖ mine f. de jais. / ~ needle ‖ Düsennadel f. ‖ aiguille f. du gicleur. / ~ pipe ‖ Schlauchmundstück n. ‖ tubulure f. pour manches des pompes à incendie. / ~ pitching apparatus ‖ Pechspritzapparat m. ‖ injecto-poisseur m.; appareil m. pulvérisateur à poisser. / ~ propulsion ‖ Reaktionsantrieb m. ‖ propulsion f. par réaction. / ~ pump ‖ Strahlpumpe f. ‖ pompe f. à jet. / ~ sand blast ‖ Sandstrahlgebläse n. ‖ sableuse f. / ~ spraying ‖ Strahlzerstäubung f. ‖ pulvérisation f. par jet.

jetty (Hydr arch) ‖ Senkstückbau m. ‖ jetée f. / ~ (Nav) ‖ Landungsbrücke f.; Kai m.; Pier m. ‖ débarcadère f.; quai m. / ~ wall ‖ Kaimauer f. ‖ mur m. d'un quai.

jewel ‖ Edelstein m.; Schmuckstein m. ‖ pierre f. précieuse ou à bijou ou fine. / paste ~ ‖ künstlicher Edelstein m. ‖ imitation f. de pierre précieuse; pierre f. précieuse artificielle.

jewel case ‖ Schmuckkästchen n. ‖ écrin m. / ~ casket see ~ case. / ~ colourer ‖ Juwelenfärber m. ‖ mettre m. en couleur pour bijouterie. / ~ dealer ‖ Edelsteinhändler m. ‖ négociant m. de pierres précieuses. / ~ drilling machine (Clockm) ‖ Steinbohrmaschine f. ‖ perceuse f. pour joyaux ou pierres dures. / ~ factory ‖ Edelsteinschleiferei f. ‖ polissage m. de pierres précieuses.

jeweller ‖ Juwelier m. ‖ joaillier m.; bijoutier m.; orfèvre m. / ~ (Watchm) ‖ Steinmacher m. ‖ pierriste m. / ~'s red ‖ Englischrot n.; Polierrot n. ‖ rouge m. à polir. / ~'s refractometer ‖ Juwelierrefraktometer n. ‖ réfractomètre m. de bijoutier.

jewellery ‖ Juwelierwaren fpl.; Bijouteriewaren fpl. ‖ joaillerie f.; bijouterie f. / ~ of applied art ‖ kunstgewerblicher Juwelenschmuck m. ‖ bijouterie f. d'art. / gold ~ ‖ goldene Schmuckwaren fpl. ‖ bijouterie f. fine ou en or. / imitation ~ ‖ Fantasiesilberwaren fpl. ‖ bijouterie f. fausse. / mourning ~ ‖ Trauerschmuckwaren fpl. ‖ bijouterie f. de deuil. / plated ~ ‖ Dubleewaren fpl. ‖ bijouterie f. en doublé.

jewellery engraver ‖ Schmuckgravör m. ‖ graveur m. en bijoux. / ~ mounter ‖ Juwelenfasser m. ‖ monteur m. ou sertisseur m. en bijouterie.

jewelry see jewellery.

jewels pl. ‖ Juwelen npl. ‖ joyaux mpl.; pierres f. précieuses.

jewel setter ‖ Steinsetzer m.; Edelsteinsetzer m. ‖ sertisseur m. / ~ setting ‖ Steinsetzen n.; Edelsteinsetzen n. ‖ sertissage m.

Jewish slaughterer ‖ Schlächter m. ‖ sacrificateur m. juif.

Jews' pitch ‖ Judenpech n. ‖ bitume m. de Judée.

jib (Crane) ‖ Ausleger m. ‖ flèche f. / ~ (Mach) ‖ Schablone f. ‖ gabarit m. / ~ (Shipb) ‖ Klüver m. ‖ foc m. / flying ~ (Shipb) ‖ Außenklüver m. ‖ clinfoc m. / hinged ~ (Crane) ‖ aufklappbarer Ausleger m. ‖ flèche f. relevable. / on the top of the cranepost mounted on a turntable is a horizontal ~ which can run a circuit of x⁰ ‖ das obere Ende der Kransäule trägt auf einer Drehscheibe einen um x⁰ schwenkbaren Ausleger ‖ la tête du pilier porte sur une plaque tournante une flèche horizontale qui peut faire un tour de cercle de x⁰. / inner ~ (Shipb) ‖ Binnenklüver m. ‖ faux foc m. / lattice ~ ‖ Fachwerkausleger m. ‖ flèche f. en charpente. / luffing ~ ‖ Wippenausleger m. ‖ flèche f. relevable ou articulée. / storm ~ (Shipb) ‖ Sturmklüver m. ‖ foc m. de tourmente.

jib, angle of ~ ‖ Auslegerneigung f. ‖ inclinaison f. de la flèche. / ~ boom ‖ Klüverbaum m. ‖ bâton m. de foc; boute-hors m. du beaupré. / ~ crane ‖ Auslegerkran m. ‖ grue f. à flèche. / ~ halliard (Mar) ‖ Klüverfall n. ‖ drisse f. du grand foc. / ~ head sheave ‖ Schnabelrolle f. ‖ poulie f. de bec. / inclination of ~ see angle of ~. / length of ~ ‖ Auslegerlänge f. ‖ longueur f. de la flèche. / ~ man ‖ Klüvergast m. ‖ gabier m. du grand foc. / ~ sheet ‖ Klüverschote f. ‖ écoute f. du foc. / ~ stay (Crane) ‖ Auslegerstütze f. ‖ contrefiche f. / ~ stay (Shipb) ‖ Klüverleiter f.; Kluverstag m. ‖ draille f. ou étai m. du foc.

jig, to (Mar) ‖ stoßweise holen ‖ haler à secousses fpl. ou par à-coups.

jig (Ore dress) ‖ Setzmaschine f. ‖ crible m. hydraulique. / ~ (Mach) ‖ Bohrschablone f. ‖ calibre m.

jigged ore ‖ Setzgraupen fpl.; aufbereitetes oder gewaschenes Erz n. ‖ minerai m. lavé au crible.

jigger (Mould) ‖ Maschinenformer m. ‖ mouleur m. à la machine; calibreur m. / ~ (Ore dress) ‖ Setzmaschine f. ‖ crible m. / ~ (Pott) ‖ Drehscheibe f.; Töpferscheibe f. ‖ tour m. ou roue f. de potier. / ~ (Radio) ‖ Jigger m.; Kopplungstransformator m. ‖ jigger m.; transformateur m. d'oscillation. / balanced ~ (Radio) ‖ ausbalanzierter Jigger m. ‖ jigger compensé. / ~ with heaped windings (Radio) ‖ Jigger m. mit übereinandergestellten Wicklungen ‖ transformateur m. d'oscillations à enroulements superposés. / primary ~ (Radio) ‖ Primärjigger m. ‖ jigger m. primaire. / receiving ~ (Radio) ‖ Empfangsjigger m. ‖ jigger récepteur m. / secondary ~ (Radio)

|| Sekundärjigger m. || jigger m. secondaire. / transmitting ~ (Radio) || Sendejigger m. || jigger m. d'émission.

jigging conveyer || Schüttelrinne f. || convoyeur m. ou rigole f. à secousses. / ~ house (Ore dress) || Setzwäsche f. || lavage m. à la cuve ou au crible. / ~ machine (Ore dress) || Setzmaschine f. || jig m.; crible m. hydraulique; machine f. à cribler des minerais; sasseur m.

jig saw || Wippsäge f. || scie f. à pédale. / ~ screen || Schüttelsieb n. || crible m. oscillant. / ~ sieve || Setzsieb n. || tamis m. de lavage.

job || Arbeit f. || travail m.

jobber || Börsenmakler m.; Kursmakler m. || coulissier m. / ~ (Pap) || Ausgleicher m. || égaliseur m.

jobbing || Flickarbeit f. || travail m. à façon (à faire à la maison). / ~ machine (Print) || Akzidenzpresse f. || presse f. pour travaux de ville. / ~ man || Handlanger m. || manœuvre m.; aide-maçon m. / ~ office || Akzidenzdruckerei f. || imprimerie f. de travaux de ville.

job number || Arbeitsnummer f. || numéro m. du travail. / ~ printer || Akzidenzdrucker m. || imprimeur m. des travaux de ville. / ~ printing || Akzidenzdruck m. || impression f. des travaux de ville. / ~ time recorder || Akkordzeitregistrierapparat m. || appareil m. pour le contrôle du temps employé pour l'exécution des travaux à la tâche. / ~ turner || Fassondreher m. || tourneur m. à façon; décolleteur m. / ~ type (Print) || Akzidenzschrift f. || caractères mpl. de fantaisie. / ~ work (Mine) || Gedinge n. || ouvrage m. à forfait.

jog, to || einschneiden; einkerben || entailler. / ~ a rivet || ein Niet stauchen || aplatir un rivet.

jogger-up (Print) || Bogengeradeleger m. || appareil m. à égaliser les feuilles.

joggle, to || kröpfen; joggeln || épauler; couder. / ~ two beams || zwei Balken miteinander verschränken || réunir deux poutres en crémaillère.

joggled built beam || verdübelter Balken m. || poutre f. jointe par des clefs en bois.

joggle truss || Hängewerksbinder m. || ferme f. en arbalète ou à clefs pendants.

joggling || Kröpfung f.; Jogglung f. || épaulement m.; coude m. / ~ machine || Joggelmaschine f.; Kröpfmaschine f. || machine f. à épauler ou à joggliner. / plate ~ machine || Plattenkröpfmaschine f.; Plattenjoggelmaschine f. || machine f. à épauler les tôles. / ~ roll || Kröpfwalze f.; Joggelwalze f. || galet m. à épauler.

johannite || Johannit m.; Uranvitriol m. || johannite f.; urane m. sulfaté.

join, to || zusammenfügen || joindre; réunir. / ~ by cogging (Carp) || aufkämmen || assembler à entailles fpl. / ~ rafters together || anschiften || embrancher. / ~ by a scarf (Shipb) || verscherben || faire des écarts mpl. / ~ in series pl. (Electr) || hintereinanderschalten || monter en série f. / ~ by slit and tongue (Join) || zusammenscheren; zusammenzinken || enfourcher à rainure et languette. / ~ a thread || einen Faden m. anspinnen || joindre en filant. / ~ timbers by cogging || Holzstücke npl. verkämmen || assembler des bois mpl. de charpente à tenon et entaille. / ~ by a triangular notch (Carp) || aufklauen || empater ou fixer avec le grappin.

joined || fugendicht || jointif.

joiner || Tischler m.; Schreiner m. || menuisier m. / ~ building || Bautischler m. || menuisier m. de bâtiment. / master ~ || Schreinermeister m. || entrepreneur m. de menuiserie. / ship's ~ || Schiffszimmermann m. || charpentier m. de bateau. / theatrical ~ || Theaterschreiner m. || menuisier m. en décors de théâtre.

joiner's bench || Hobelbank f. || établi m. de menuisier. / ~ cement || Leimkitt m. || mastic m. à la colle. / ~ chisel || Tischlermeißel m. || ciseau m. de menuisier. / ~ glue || Tischlerleim m. || colle f. forte. / ~ putty || Holzkitt m. || mastic m. de colle; futée f. / ~ trade || Tischlerei f.; Schreinerei f. || menuiserie f. / ~ ware || Tischlerwaren fpl. || articles mpl. de menuisiers. / ~ work || Tischlerarbeit f. || ouvrage m. de menuiserie. / ~ work for buildings || Schreinerarbeiten fpl. für Gebäude || pièces fpl. de menuiserie pour bâtiments.

joinery || Schreinerkunst f.; Tischlerhandwerk n. || menuiserie f. / ~ for buildings || Bautischlerei f.; Bauschreinerei f. || menuiserie f. pour bâtiments. / ~ wood || Holz n. für Tischler || bois m. de menuiserie.

joining || Anschluß m. || jonction f.; assemblage m. / ~ (Build) || Verband m. im Holzwerk || assemblage m. / ~ (Carp) || Fuge f.; Verbindung f. || joint m.; assemblage m. / ~ in the circuit (Electr) || Einschaltung f. || mise f. en circuit. / ~ with key-piece (Carp) || Schurzwerk m. || assemblage m. à clef. / ~ of timbers || Holzverbindung f. || assemblage m. des bois.

joining line (Railw) || Anschlußgleis n. || voie f. de jonction. / ~ piece || Ansatzstück n. || pièce f. de jonction. / ~ pipe || Anschlußrohr n. || tuyau m. de raccordement. / ~ plate || Anschlußblech n. || lame f. d'assemblage. / ~ shackle || Verbindungsschäkel m. || manille f. d'assemblage. / ~ stone || Satzstein m. || pierre f. d'ajoute.

joint, to || fügen || jointer. / ~ (Mason) || ausfugen || jointoyer. / ~ by double rebating (Join) || mit Anschlag m. und Überschlag m. verbinden || refarder; assembler à recouvrement ou à mi-bois. / ~ timbers pl. || Holzstücke npl. verbinden || assembler des bois mpl. de construction.

joint (Cable) || Lötstelle f. || raccord m.; soudure f. / ~ (El line) || Verbindungsstelle f. || liaison f. des fils. / ~ (Geol) || Kluft f.; Ablösungsfläche f. || joint m. / ~ (Join) || Scharnier n. || charnière f. / ~ (Mach) || Gelenk n. || articulation f. / ~ (Mas; Join) || Fuge f. || joint m. / ~ (Rivet) || Nietfuge f. || joint m. / ~ (Tinm) || Naht f.; Saum m. || rapport m.; soudure f. / asbestos ~ || Asbestdichtung f. || garniture f. en amiante. / ball ~ || Kugelgelenk n. || joint m. à rotule. / ~ of bed (Build) || Lagerfuge f. eines Steins || Bettungsfuge f.; Ruhefuge f. || joint m. de lit ou d'assise. / bevel ~ || schräge Verbindung f. || assemblage m. en fausse coupe. / ~ with butt strap || Überlaschung f. || rivure f. à couvrejoint. / capped ~ || übereinandergelegte Fuge f. || joint m. superposé. / cardan ~ || Kreuzgelenk n. || joint m. de cardan. / cardan ~ of the ring type || ringförmiges Kardangelenk n. || joint m. de cardan sphérique. / chamfered ~ (Mas) || abgefaßte Fuge f. || joint m. démaigri ou flacheux ou chanfreiné. / covering

~ || überdeckende Fuge f.; Deckfuge f. || joint m. à recouvrement. / cross ~ || Kreuzverzapfung f. || joint m. en croix. / cross pin cardan ~ || Kardangelenk n. mit Kreuzzapfen || joint m. de cardan à croisillon. / diagonal ~ || Gehrfuge f.; Gehrstoß m. || assemblage m. à onglet; joint m. à onglet ou à mitre. / dry disk ~ || Trockengelenk n. || joint m. flexible sans graissage. / edge ~ || Eckverzapfung f. || assemblage m. d'angle. / expansion ~ || Ausdehnungsrohrverbindung f. || joint m. coulissant ou d'expansion. / fished ~ || verlaschter Schienenstoß m. || joint m. éclissé. / flanged edge ~ || Stirnstoß m. || joint m. à arête bridée. / foliated ~ see rebated ~. / fork ~ || Gabelgelenk n. || chape f. / frontal ~ (Mas) || Stirnfuge f. || joint m. de tête ou de face. / horizontal ~ see ~ of bed. / india rubber ~ || Gummidichtung f. || garniture f. en caoutchouc. / ~ with inside rivetted fish plates || Stoßverbindung f. mit innen angenieteten Stecklaschen || joint m. par éclisses Spalding, rivées intérieurement. / leaking ~ || undichte Fuge f. || joint m. défectif. / metal ~ || Metallgelenk n. || joint m. métallique. / mortise and tenon ~ || Verzapfung f. || assemblage m. à mortaise. / ~ of a pipe || Schlauchdichtung f. || joint m. de tuyau. / ~ of rails || Schienenstoß m. || joint m. des ornières ou des rails. / rebated ~ || überfalzte oder überplattete Fuge f. || joint m. feuillé à recouvrement. / ~ with reflection effects || Stoßstelle f. einer Leitung || jonction f. produisant des effets de réflexion. / ~ of rupture || Bruchfuge f. || joint m. de rupture. / scarf ~ (Railw) || schräger Stoß m. || joint m. biseauté. / square ~ || rechtwinklige Verbindung f. || assemblage m. carré. / ~ of a stone (Mas) || Fläche f. oder Fuge f. oder Seite f. eines Steines || panneau m. d'une pierre. / straight-glued ~ || stumpfe Fuge f.; Leimfuge f. || joint m. à plat-point. / supported ~ (Railw) || aufliegender Schienenstoß m. || joint m. des rails soutenu; joint m. sur appui. / suspended ~ (Railw) || schwebender Schienenstoß m. || joint m. en porte-à-faux. / temporary ~ (Tel) || Notbund m. || torsade f. provisoire. / ~ of tube || Röhrenstoß m. || jointure f. de tuyau. / universal ~ || Kreuzgelenk n.; Kardangelenk n. || joint m. de cardan; cardan m. / vertical ~ || senkrechte Fuge f.; Stoßfuge f. || joint m. montant ou vertical. / welded ~ || Schweißnaht f. || joint m. soudé; ligne f. de soudure; soudure f.

joint account || gemeinschaftliche Rechnung f. || compte m. à demi. / ~ bolt || Schließbolzen m.; Splintbolzen m. || boulon m. à clavette. / ~ box cover || Abzweigkastendeckel m. || couvercle m. de boîte de raccordement. / ~ coupling || Gelenkkupplung f. || accouplement m. à articulation. / ~ enterprise || gemeinschaftliches Unternehmen n. || entreprise f. en participation.

jointed, carefully || sorgfältig gefügt || bien assemblé. / ~ doll || Gliederpuppe f. || poupée f. articulée; marionnette f.

jointer (Coop) || Fügebank f. || colombe f. à joindre; varlope f. / ~ (Mas) || Fügeisen n.; Fugkelle f. || crochet m. à jointoyer. / ~'s tent (Tel) || Lötzelt n. || tente f. de soudeur.

joint file ‖ Scharnierfeile f. ‖ lime f. à charnière. / hollow edge ~ ‖ hohle Scharnierfeile f. ‖ lime f. à charnière concave. / round ~ ‖ runde Scharnierfeile f. ‖ lime f. à charnière ronde.

joint focussing arrangement to the eyepieces of a field glass ‖ Mitteltriebeinstellung f. der Okulare eines Feldstechers ‖ mise f. au point par molette centrale des oculaires d'une jumelle. / ~ glass ‖ Mitteltriebglas n. ‖ jumelle f. à molette centrale. / ~ wheel at the middle between the two bodies of a field glass ‖ Mitteltrieb m. eines Feldstechers ‖ molette f. centrale d'une jumelle.

joint fork ‖ Gelenkgabel f. ‖ fourche f. d'articulation.

joint-frame ‖ Scharnierband n.; Gelenkband n. ‖ fiche f. simple ou rivée ou à nœuds ou à lacet rivé; charnière f. / bent ~ ‖ gekröpftes Scharnierband n. ‖ fiche coudée. / ~ with pin ‖ Scharnierband n. mit Vorstecker ‖ fiche f. à double nœuds et à bouton.

joint, gap at ‖ Stoßfuge f. ‖ jeu m. du joint. / ~ hinge see ~ frame. / ~ hook (Carp; Join) ‖ Winkelmaß n. ‖ réglet m. du menuisier. / ~ hook-shaped axe (Carp) ‖ Bundaxt f. ‖ tire-boucler m.

jointing ‖ Dichtungsmittel n. ‖ joint m.; garniture f. / ~ (Action) ‖ Fügen n. ‖ jointage m. / ~ chamber (Tel) ‖ Kabelbrunnen m.; Lötbrunnen m. ‖ chambre f. de soudure ou de tirage. / ~ clamp ‖ Kluppe f.; Drahtkluppe f. ‖ pince f. / load in the ~s ‖ Knotenpunktbelastung f. ‖ charge f. aux nœuds.

jointing machine (Mach) ‖ Fügemaschine f. ‖ machine f. à rainer. / ~ (Wood) ‖ Fügemaschine f. ‖ machine f. à jointement ou à jointer. / head straightening and ~ (Coop) ‖ Abrichthobel- und Fügemaschine f. für Faßbodenbretter ‖ machine f. à dresser et à jointer les fonds de fûts. / stave ~ ‖ Faßdaubenfügemaschine f. ‖ machine f. à jointer les douves.

jointing mortar ‖ Fugenmörtel m. ‖ mortier m. à jointoyer. / ~ rule (Mas) ‖ Richtscheit n. zum Ausfugen ‖ règle f. de rejointoyeur. / ~ saw ‖ Fügesäge f. ‖ scie f. à jointer. / ~ sleeve (El line) ‖ Seilverbinder m.; Seilklemme f. ‖ manchon m. de jonction.

jointly and severally ‖ samt und sonders ‖ tous sans exception f.

joint-ownership ‖ Reedereigesellschaft f. ‖ association f. ou compagnie f. d'armateurs.

joint packing ‖ Flanschendichtung f. ‖ garniture f. de brides. / ~ part ‖ Anschlußstück n. ‖ pièce f. de raccord. / ~ pin ‖ Scharnierstift m. ‖ cheville f. ou goujon m. de charnière. / ~ pipe ‖ Verbindungsröhre f. ‖ tuyau m. de raccordement. / place of ~ ‖ Anschlußstelle f. ‖ point m. d'assemblage ou d'attache. / ~ plate ‖ Anschlußblech n. ‖ tôle f. ou plaque f. d'assemblage. / ~ plyer ‖ Scharnierzange f. ‖ béquettes fpl. à limer les charnières. / ~ proprietor ‖ Mitinhaber m. ‖ co-propriétaire m. / record of ~s ‖ Lötstellennachweis m. ‖ état m. des soudures. / ~ responsibility ‖ Solidarhaft f. ‖ responsabilité f. conjointe. / ~ ring ‖ Dichtungsring m. ‖ anneau m. ou rondelle f. de joint. / ~ rule ‖ Gelenkmaßstab m. ‖ mètre m. pliant. /

~ side (Shipb) ‖ Mallebene f. ‖ plan m. de gabariage ou du couple. / ~ sleeper (Railw) ‖ Stoßschwelle f. ‖ traverse f. de joint.

joint-stock ‖ Gesellschaftskapital n. ‖ capital m. ou fonds m. social. / ~ bank ‖ Aktienbank f. ‖ banque f. par actions. / ~ capital ‖ Aktienkapital n.; Gesellschaftsmittel npl. ‖ capital-actions m.; fonds m. social.

joint-stock company ‖ Aktiengesellschaft f. ‖ société f. anonyme ou par actions. / to assist in the amalgamation of enterprises in the form of a ~ ‖ an der Zusammenfassung f. von Unternehmungen in Form einer Aktiengesellschaft mitwirken ‖ aider à grouper des entreprises par la fondation d'une société anonyme. / ~ with limited liabilities ‖ Gesellschaft f. mit beschränkter Haftung ‖ société f. à responsabilité limitée. / to own the majority of the stock of a ~ ‖ die Majorität f. einer Aktiengesellschaft besitzen ‖ être titulaire m. de la majorité des parts d'une société anonyme.

joint tongue ‖ Feder f.; Federkeil m. ‖ languette f. à rainure. / ~ tool ‖ Scharniereisen n. ‖ outil m. à charnière. / ~ wire ‖ Scharnierröhrchen n. ‖ tuyau m. à charnière.

joist (Bridge) ‖ Querträger m. ‖ entretoise f. / ~ (Carp) ‖ Querbalken m. ‖ solive f. / ~ of a landing-place ‖ Podestbalken m. ‖ patin m. de palier d'un escalier.

joist rolling mill ‖ Trägerwalzwerk n. ‖ laminoir m. à poutrelles.

joists pl. ‖ Gebälk n. ‖ charpente f.

jolt, to (Forg) ‖ stauchen ‖ refouler.

jolter (Mould) ‖ Rüttelformmaschine f. ‖ machine f. à mouler à secousses.

jolting of building ‖ Gebäudeerschütterung f. ‖ vibration f. ou trépidation f. au bâtiment.

jolting machine ‖ Stauchmaschine f. ‖ machine f. à refouler.

jolt moulding machine see jolter.

jordanite ‖ Jordanit m. ‖ jordanite f.

joule ‖ Joule n. ‖ joule m.

journal (Mach) ‖ Zapfen m. ‖ tourillon m. / ~ (Paper) ‖ Zeitung f.; Tagebuch n. ‖ journal m. / axle ~ ‖ Achsschenkel m.; Achszapfen m.; Achshals m. ‖ fusée f. d'essieu. / ~ of axle bearing ‖ Achslagerschale f. ‖ coussinet m. de boîte d'essieu. / bedding of ~ into bearing by abradant ‖ Einschleifen n. des Zapfens in das Lager ‖ rodage m. du tourillon dans le coussinet à l'émeri. / inside ~ ‖ Achsschenkelinnenseite f. ‖ fusée f. d'essieu intérieure. / ~ king ‖ Königssäule f. ‖ pivot m. central. / technical ~ (Print) ‖ technische Fachzeitschrift f. ‖ revue f. technique.

journal box ‖ Achslager n. ‖ boîte f. d'essieu. / ~ guide ‖ Achslagerführung f. ‖ guide m. de la boîte d'essieu.

journal brass lining ‖ Lagerschalenausguß m. ‖ garniture f. antifriction du coussinet. / ~ folding machine (Print) ‖ Zeitungsfalzmaschine f. ‖ plieuse f. mécanique de journaux. / ~ grinding machine ‖ Zapfenschleifmaschine f. ‖ machine f. à rectifier les tourillons.

journalist ‖ Journalist m.; Zeitungsmann m. ‖ journaliste m.

journey ‖ Fahrt f. ‖ voyage m.; tournée f. / downward ~ (Aero) ‖ Niederfahrt f. ‖

descente f. / to set out on a ~ ‖ eine Reise f. antreten ‖ partir en voyage.

journey-man ‖ Geselle m.; Gehilfe m. ‖ compagnon m.; ouvrier-compagnon m. / ~ tailor ‖ Schneidergeselle m. ‖ ouvrier m. tailleur.

Juchten ‖ Juchten m.; Juchtenleder n. ‖ cuir m. de Russie; roussi m.; youftes mpl.

judging, way of ~ steel by the appearance of the fracture ‖ Stahlbeurteilung f. nach dem Bruchaussehen ‖ appréciations fpl. de l'acier basées sur l'aspect de la cassure.

judgement ‖ Gutachten n. ‖ avis m. / ~ (Judge) ‖ Urteilsspruch m.; Rechtsspruch m. ‖ arrêt m.; sentence f.; jugement m.

judicial ‖ gerichtlich ‖ judiciaire; juridique. / ~ inquiry ‖ gerichtliche Untersuchung f. ‖ enquête f. judiciaire.

judiciary see judicial.

judicious ‖ kritisch ‖ critique.

jug for milk ‖ Milchkrug m. ‖ cruchon m. à lait. / ~ for mineral water ‖ Mineralwasserkrug m. ‖ cruche f. à eaux minérales.

juice ‖ Saft m. ‖ jus m. / concrete ~ ‖ trockener Saft m. ‖ suc m. concret. / unfermented grape ~ ‖ ungegorener Traubensaft m. ‖ jus m. de raisin non fermenté. / viscous ~ ‖ dickflüssiger Saft m. ‖ suc m. visqueux.

juice pump ‖ Saftheber m. ‖ monte-jus m.

jumbling piece ‖ Zwischenstück n. ‖ pièce f. intermédiaire.

jump, to ~ into the groove ‖ in die Rast f. einspringen ‖ pénétrer dans l'encoche f. / ~ a pin ‖ einen Bolzen m. aufsitzen lassen ‖ refouler une cheville.

jump of temperature ‖ Temperatursprung m. ‖ saut m. de température.

jumper (Dress) ‖ Jumper m. ‖ jumper m. / ~ (Forge) ‖ Stauchhammer m. ‖ refouloir m. / ~ (Join) ‖ Rauhbank f. ‖ varlope f. / ~ (Mine) ‖ Bergwerksstange f. ‖ barre f. de mine. / ~ bit ‖ Stoßbohrer m. ‖ fleuret m.; perçoir m. à couronne. / ~ ring ‖ Schaltring m. ‖ anneau m. de fil jarretier. / ~ wire ‖ Schaltdraht m. ‖ fil m. volant.

jumping of the brushes ‖ Hüpfen n. der Bürsten ‖ soubresauts mpl. des balais. / ~ collars on wheel spokes ‖ Anstauchen n. der Bunde an Radspeichen ‖ refoulement m. des embases aux rayons de roues.

jump spark ‖ überspringender Funke m. ‖ étincelle f. disruptive.

junction (Railw) ‖ Anschluß m. ‖ raccordement m. / ~ (Tinm) ‖ Lötstelle f. ‖ soudure f. / ~ (Traffic) ‖ Knotenpunkt m. ‖ point m. de jonction. / ~ of lines ‖ Gleisanschluß m. ‖ jonction f. des voies. / ~ of lodes ‖ Zusammentreffen n. von Erzgängen ‖ rencontre f. des filons. / perfect ~ ‖ unlösbare Verbindung f. ‖ liaison f. indissoluble. / ring down ~ (Tel) ‖ Abfragebetrieb m. ‖ méthode f. du double appel. / ~ of rivets ‖ Nietverbindung f. ‖ assemblage m. par rivets.

junction box (Electr) ‖ Abzweigdose f.; Zwischenschaltdose f.; Abzweigkasten m.; Verbindungskasten m. ‖ boîte f. de raccordement ou de branchement ou de dérivation. / manhole ~ ‖ Schachtkabelkasten m. ‖ boîte f. de jonction à trou-d'homme. / ~ with replaceable fuse ‖ Anschlußdose f. mit auswechselbarer Sicherung ‖ boîte f. de jonction à coupe-circremcuit plaçable.

junction cable ‖ Verbindungskabel n. ‖ câble m. de jonction. / ~ canal ‖ Verbindungskanal m. ‖ canal m. de jonction. / ~ curve ‖ Verbindungskurve f. ‖ courbe f. de raccordement. / ~ jack (Tel) ‖ Verbindungsklinke f. ‖ jack m. de jonction.

junction line (Railw) ‖ Verbindungsbahn f.; Anschlußgleis n. ‖ ligne f. de jonction; voie f. de raccordement. / ~ (Tel) ‖ Verbindungsleitung f.; Amtsverbindungsleitung f. ‖ ligne f. de jonction ou de raccordement. / grouping of ~s (Tel) ‖ Leitungsbündel n. ‖ arrangement m. des jonctions. / ~ panel (Tel) ‖ Verbindungsleitungsfeld n. ‖ champ m. des lignes auxiliaires.

junction plate ‖ Verbindungsplatte f.; Knotenblech n. ‖ plaque f. de jonction ou d'éclissage; gousset m. d'assemblage. / point of ~ ‖ Knotenpunkt m. ‖ point m. de jonction. / ~ pole (Tel) ‖ Übergangsstange f. ‖ poteau m. de jonction. / ~ rail ‖ Anschlußschiene f.; Verbindungsschiene f. ‖ raccord m. de rail; rails mpl. convergents. / ~ rail end ‖ Anschlußende n. der Schiene; Anschlußschienenstück n. ‖ pièce f. de raccord de rail. / ~ railway ‖ Anschlußbahn f. ‖ chemin m. de fer de jonction. / ~ railway station ‖ Abzweigungsbahnhof m.; Anschlußbahnhof m. ‖ gare f. d'embranchement; station f. de jonction. / ~ service (Tel) ‖ Verbindungsverkehr m.; Verbindungsleitungsverkehr m. ‖ service m. de jonction; service m. des lignes auxiliaires. / ~ station see ~ railway station.

juniper (Tree) ‖ Wacholder m. ‖ genévrier m.; genièvre m. / ~ berry ‖ Wacholderbeere f. ‖ baie f. de genévrier. / ~ meat jelly ‖ Wacholdersulze f. ‖ gelée f. au ge-

nièvre. / ~ oil ‖ Wacholderöl n. ‖ essence f. de genévrier. / ~ wood ‖ Wacholderholz n. ‖ (bois m. de) genévrier m.

junk ‖ Schutt m.; Abfälle mpl.; Gerümpel n. ‖ décombres mpl.

junket (Fish) ‖ Aalreuse f.; Aalkorb m. ‖ nanse f.

junk ring ‖ Dichtungsring m.; Liderungsring m. ‖ couronne f.; rondelle f. ou anneau m. de joint. / cylinder ~ ‖ Zylinderdichtungsring m. ‖ segment m. de cylindre. / piston ~ ‖ Kolbendichtungsring m. ‖ garniture f. de segment de piston.

Jura-limestone ‖ Jurakalk m. ‖ calcaire m. jurassique.

Jurassic system (Geol) ‖ Jura m. ‖ jura m.

juristic ‖ juristisch ‖ juridique.

jury ‖ Jury f.; Preisgericht n.; Preisrichterkollegium n. ‖ jury m.

jury mast (Shipb) ‖ Notmast m. ‖ mât m. de fortune. / ~ rigging ‖ Nottakelage f. ‖ gréement m. de fortune. / ~ rudder ‖ Notruder n. ‖ gouvernail m. de fortune. / ~ scarf ‖ Notlasche f.; Notlaschung f.; Notscherbe f. ‖ écharpe f. de fortune.

justice ‖ Rechtspflege f. ‖ justice f. / Court of ~ ‖ Gericht n.; Gerichtshof m. ‖ tribunal m.; cour f. de justice.

justification (Juristics) ‖ Rechtfertigung f. ‖ justification f. / ~ (Print) ‖ Ausschließung f.; Justierung f.; Zeilenlänge f. ‖ justification f.

justifier (Print) ‖ Geviert n.; Quadrat n.; breites Spatium n. ‖ cadrat m.

justify, to ~ (Juristics) ‖ rechtfertigen ‖ justifier. / ~ (Print) ‖ ausschließen; justieren ‖ justifier.

justly ‖ mit Recht ‖ à bon droit; avec raison.

jut, to (Build) ‖ ausfluchten; hervorstehen ‖ désaffleurer; dépasser; avancer.

jute ‖ Jute f. ‖ jute m. / ~ in fibres ‖ Jute f. in Fasern ‖ jute m. en brins. / raw ~ ‖ rohe Jute f. ‖ jute m. brut. / scutched ~ ‖ gebrochene Jute f. ‖ jute m. teillé. / wormed ~ ‖ Jutetrense f. ‖ filet m. de jute; jute f. congrée.

jute bag ‖ Jutesack m. ‖ sac m. de jute. / ~ beater ‖ Juteschläger m. ‖ batteur m. de jute. / ~ bleaching ‖ Jutebleichung f. ‖ blanchiment m. de jute. / ~ carder ‖ Jutekardätscher m. ‖ cardeur m. de jute. / ~ covering ‖ Juteumspinnung f. ‖ guipage m. de jute. / ~ cutter ‖ Juteschneider m. ‖ coupeur m. de jute. / ~ goods pl. ‖ Jutewaren fpl.; Juteerzeugnisse npl. ‖ articles mpl. en jute. / ~ printing ‖ Jutedruck m. ‖ impression f. sur jute. / ~ roller ‖ Jutewalze f. ‖ cylindre m. à jute. / ~ scutcher ‖ Juteschläger m. ‖ batteur m. de jute. / ~ softener (Spinn) ‖ Juteeinweicher m. ‖ assouplisseur m. de jute. / ~ sorter ‖ Jutesortierer m. ‖ assortisseur m. de jute. / ~ spinner ‖ Jutespinner m. ‖ fileur m. de jute. / ~ spinning ‖ Jutespinnerei f. ‖ filature f. de jute. / ~ spinning machine ‖ Jutespinnereimaschine f. ‖ machine f. à filer la jute. / ~ stricker ‖ Jutekrempler m. ‖ briseur m. de jute. / ~ string see wormed ~. / ~ weaver ‖ Juteweber m. ‖ tisseur m. de jute. / ~ weaving ‖ Juteweberei f.; Weben n. von Jute ‖ tissage m. de jute. / ~ weaving (Tissue) ‖ Jutegewebe n. ‖ étoffe f. en jute. / ~ yarn ‖ Jutegarn n. ‖ fil m. de jute.

jutty ‖ Erker m. ‖ saillie f. / ~ room ‖ Erkerstube f. ‖ cul-de-lampe m.

juxtaposed to . . . ‖ angrenzend an . . . ‖ voisin de . . .; adjacent à . . .

juxtaposition (Geol) ‖ Anlagerung f. ‖ juxtaposition f.

K

kammererite (Miner) ‖ Kammererit m. ‖ kammérérite f.

kainite ‖ Kainit m. ‖ kaïnite f.

kaki (Fig) ‖ Kakifeige f. ‖ kaki m.

kakodylic acid ‖ Kakodylsäure f. ‖ acide m. kakodylique.

kakoxene (Miner) ‖ Kakoxen n. ‖ kakoxène m.

kaluszit ‖ Kaluszit m.; Syngenit m. ‖ kaluszite f.

kampylite ‖ Kampylit m. ‖ kampylite f.

kaneite (Miner) ‖ Arsenikmangan n. ‖ kanéite f.

Kankhura ‖ Kalluihanf m. ‖ caloée f.

kaolin ‖ Kaolin n.; Porzellanerde f.; Steinmark n. ‖ kaolin m.; terre f. porcelaine; lithomarge f. / ~ decanting machine ‖ Kaolinschlämmaschine f. ‖ machine f. à laver le kaolin.

kaolinite see kaolin.

kaolinization ‖ Kaolinisierung f. ‖ kaolinisation f.

Kaplan turbine ‖ Kaplanturbine f. ‖ turbine f. (de) Kaplan. / ~ with adjustable runner vanes and guide blades ‖ Kaplanturbine f. mit regelbarer Laufvorrichtung und Leitschaufeln ‖ turbine f. Kaplan à aubes et à appareil de marche réglable.

kapnite ‖ Kapnit m.; Eisenzinkspat m. ‖ kapnite f.

kapok (cotton) ‖ Kapok m. ‖ capoc m.

Karolus-cell ‖ Karolus-Zelle f. ‖ couple m. photoélectrique Karolus.

Kashmir (Weav) ‖ Kaschmir m. ‖ cachemire m. / ~ shawl (Weav) ‖ Kaschmirschal m. ‖ châle m. de cachemire. / ~ wool ‖ Kaschmirwolle f. ‖ laine f. de cachemire.

kathode see also cathode (Phys) ‖ Kathode f.; negativer Pol m. ‖ cathode m.; pôle m. négatif.

katigene dye ‖ Katigenfarbstoff m. ‖ couleur f. de catigène.

kedge (anchor) ‖ Wurfanker m.; Warpanker m. ‖ ancre f. à jet.

keel, to (Shipb) ‖ kielen ‖ effiler la quille.

keeled bottom of a boat ‖ gekielter Bootsboden m. ‖ fond m. du bateau en quille.

keel (Shipb) ‖ Kiel m. ‖ quille f. / to set the ~ ‖ den Kiel m. strecken ‖ poser la quille. / centre through-plate ~ ‖ durchlaufender Mittelkiel m. ‖ quille f. carlingue. / false ~ ‖ Loskiel m.; falscher Kiel m. ‖ fausse quille f. / upper false ~ ‖ Gegenkiel m. ‖ contre-quille f. intérieure.

keel plate bending press ‖ Kielplattenbiegemaschine f. ‖ machine f. pour cintrer les tôles de carène ou les plaques

de quille. / ~ scarf ‖ Kiellaschung f. ‖ écart m. de la quille.

keelson ‖ Kielschwein n. ‖ carlingue f.

keen (Tool) ‖ scharf; schneidend; tranchant; aigu; affilé. / ~ edge ‖ scharfe Kante f. ‖ bord m. à vive arête. / ~ edged ‖ scharfkantig; mit scharfen Kanten fpl. ‖ à vive arête f.; présentant d'angles vifs mpl. / ~ edge tool ‖ scharf schneidendes Werkzeug n. ‖ outil m. à tranchant affilé ou affûté.

keep, to ~ (To preserve) ‖ aufbewahren ‖ conserver. / ~ (To observe) ‖ innehalten; befolgen ‖ observer. / ~ (To be stable) ‖ sich halten ‖ se conserver. / ~ the course (Nav) ‖ Kurs m. halten; Kurs m. steuern ‖ aller ou porter à route f.; faire route. / the prices pl. keep firm ‖ die Kurse mpl. halten sich (fest); die Kurse mpl. bleiben fest ‖ les valeurs fpl. se maintiennent. / ~ the land aboard (Nav) ‖ unter Land halten ‖ serrer la terre; ranger la terre. / ~ the lead going (Nav) ‖ Land n. anloten ‖ s'approcher de la terre en sondant sans cesse. / ~ a mine in repair ‖ ein Bergwerk bauhaft halten ‖ entretenir une mine. / ~ the sea (Nav) ‖ die See gut halten ‖ tenir la mer. / ~ off the waters by timbering (Min) ‖ die

Wasser verdämmen; die Wasser durch Holzverdämmung abhalten ‖ cuveler les eaux. / ~ -up ‖ unterhalten ‖ entretenir. / ~ the weather-gage (Nav) ‖ luv halten; dicht an den Wind halten ‖ tenir le lof *ou* le vent. / ~ the wind *see* ~ the weather gage.

keep in a cool place! ‖ kühl aufbewahren! ‖ garder en lieu m. frais! / ~ dry! ‖ vor Nässe f. zu schützen!; trocken aufzubewahren! ‖ à préserver de l'humidité! / ~ upright! ‖ nicht stürzen! ‖ ne pas renverser!; ne pas laisser tomber!

keeper on the leather belt ‖ Schlaufe f. am Lederriemen ‖ passant m. *ou* coulant m. de la courroie en cuir.

keeping automatically the spectrograph with the diurnal motion of the sun ‖ automatische Nachführung f. des Spektrografen in der täglichen Bewegung der Sonne ‖ tournement m. du spectrographe pour suivre automatiquement le mouvement diurne du soleil. / ~ clean (Mill) ‖ Reinhaltung f. ‖ dégommage m. / ~ constant ‖ Konstanthaltung f. ‖ maintien m. de la constance. / ~ in good order ‖ Instandhaltung f. ‖ entretien m. / directions for ~ in good order ‖ Instandhaltungsanweisung f. ‖ instruction f. d'entretien. / ~ taut ‖ Straffhaltung f. ‖ conservation f. *ou* maintien m. de la tension. / ~ of telephone account ‖ Fernsprechrechnungsdienst m. ‖ service m. de la comptabilité téléphonique.

keepsake, travelling ‖ Reiseandenken n. ‖ souvenir m. de voyage.

keeve ‖ Bottich m. ‖ cuve f.

kefir ‖ Kefir m. ‖ kéfir m. / ~ fungus ‖ Kefirpilz m. ‖ fungus m. kéfir.

keg ‖ Banzen m.; Faß n. ‖ fût m.; tonneau m. / ~ beer ‖ Faßbier n. ‖ bière f. en fûts. / ~ beer store ‖ Faßbierlagerraum m. ‖ entrepôt m. pour de la bière en fûts. / ~ conveyor ‖ Faßförderer m.; Faßtransportör m. ‖ transporteur m. de fûts. / ~ filler ‖ Bierzapfer m. ‖ emplisseur m. de barils. / ~ funnel ‖ Faßtrichter m. ‖ entonnoir m. de fûts. / machine for removing pitch in ~s ‖ Faßentpichmaschine f. ‖ machine f. à dégoudronner les fûts. / ~ rinsing machine ‖ Faßspülmaschine f. ‖ machine f. à laver les fûts. / ~ sprinkler ‖ Faßausspritzer m. ‖ injecteur m. laveur de fûts. / ~ washer ‖ Faßwaschapparat m. ‖ appareil m. à laver les fûts.

kelloway (Geol) ‖ Kelloway m. ‖ étage m. callovien.

kelp (Chem) ‖ Kelp n.; rohe Soda f.; kalzinierte Asche f. von Blattang ‖ kelp m. / ~ cutter ‖ Seegrasfischer m. ‖ ramasseur m. de varech *ou* d'herbes marines.

kelson *s.* keelson.

Kelvin arrival curve (Tel) ‖ Thomsonkurve f. ‖ courbe f. de Thomson.

kennel of paving ‖ Gosse f.; Straßenrinne f.; Rinnstein m. ‖ rigole f. *ou* caniveau m. de pavé; ruisseau m. de rue.

kentledge (Shipb) ‖ Ballasteisen n. ‖ fer m. en saumon.

kepi ‖ Käppi n. ‖ képi m.

kerargyrite ‖ Hornerz n.; Hornsilber n.; Kerat n. ‖ argent m. corné *ou* muriaté; kérargyre f.

kerate *s.* kerargyrite.

keratometer ‖ Keratometer n. ‖ kératomètre m.

kerchief ‖ Kopftuch n. ‖ madras m.

kerf ‖ Kerbe f. ‖ encoche f.; entaille-coche f. / ~ of a saw ‖ Sägeschnitt m. ‖ chemin m. *ou* voie f. *ou* trait m. d'une scie.

kermes (Dyer) ‖ Kermes m. ‖ kermès m. / ~ (Miner) ‖ Antimonblende f.; Spießglanzblende f.; Rotspießglanz m.; Kermes m. ‖ antimoine m. rouge; antimoine m. oxydé sulfuré; kermès m. minéral.

kermesite *see* kermes (Miner).

kernel, to (Stone-c) ‖ kröneln; stocken ‖ bretteler; bretter.

kernel ‖ Kern m.; Obstkern m. ‖ pépin m. / ~ (Techn) ‖ Kern m. ‖ noyau m. / ~ (Met) ‖ Herz n. eines gerösteten Erzes ‖ partie f. intérieure de minerai demicalciné. / hard ~ ‖ harter Kern m. ‖ grain m. dur.

kernel oil ‖ Kernöl n. ‖ huile f. de noyau.

kerosene ‖ Kerosen n.; raffiniertes Leuchtpetroleum n. ‖ kérosène m.

kerve, to ~ (Coal) ‖ schlitzen; kerben; schrämen ‖ couper la coulaie; haver; sous-caver; souchever.

kerving (Coal) ‖ Schram m.; Schlitz m.; Kerb m. ‖ échancrure f. latérale; souscave f.

kettle ‖ Kochkessel m.; Wasserkessel m.; kleiner Kessel m. ‖ bouilloire f.; chaudron m.; petite chaudière f.; marmite f. / boiling ~ *see* kettle. / copper ~ ‖ kupferner Kochkessel m. ‖ bouilloire f. *ou* chaudron m. *ou* cuve f. en cuivre.

kettle hand (Dyer) ‖ Aussieder m. ‖ débouilleur m. / ~ maker's wire ‖ Kesselschmieddraht m. ‖ fil m. de chaudronnier. / ~ man (Brew) ‖ Biersieder m. ‖ cuiseur m. / ~ pan (Brew) ‖ Braupfanne f. ‖ brassin m. / ~ piece of a Bessemer converter ‖ Kesselstück n. einer Bessemerbirne ‖ creuset d'un convertisseur Bessemer. / ~ repairer ‖ Kesselflicker m. ‖ chaudronnier m. rhabilleur.

ketones pl. (Chem) ‖ Ketone npl. ‖ quétones pl.

key, to ‖ festkeilen ‖ claveter; caler; ajuster. / ~ (To fasten with a split pin) ‖ versplinten ‖ goupiller. / ~ (Print) ‖ füttern; unterlegen ‖ taquonner; rehausser. / ~ (Radio) ‖ eine Taste f. schließen ‖ manipuler *ou* fermer une clé. / ~-on the crank ‖ die Kurbel aufkeilen ‖ claveter la manivelle sur l'arbre. / ~ to the shaft ‖ auf der Welle f. festkeilen ‖ coincer *ou* caler sur l'arbre m. / ~ a transmitter (Radio) ‖ einen Sender tasten ‖ manier un transmetteur.

key (Locksm) ‖ Schlüssel m. ‖ clef f.; clé f. / ~ (Mach) ‖ Keil m.; Splint m. ‖ clavette f.; clef f.; coin m.; goupille f. / ~ (Tel) ‖ Taste f.; Manipulator m.; Telegrafentaste f.; Taster m. ‖ clef f.; manipulateur m. / ~ (Cotter) ‖ Schließkeil m.; Setzkeil m. ‖ clavette f. de serrage. / ~ (Piano; Typewriter) ‖ Taste f. ‖ touche f. / ~ (Spanner) ‖ Schraubenschlüssel m.; Schlüssel m. ‖ clef f. (à écrous). / ~ (Wooden plug) ‖ Dübel m. ‖ goupille f. en bois. / ~s pl. (Piano) ‖ Klaviatur f. ‖ clavier m. / ~ (Print) ‖ Unterlage f.; Fütterung f. ‖ tacon m.; taquon m. / to depress several ~s pl. at one time ‖ mehrere Tasten fpl. zu gleicher Zeit herunterdrücken ‖ abaisser plusieurs touches fpl. à la fois. / amount ~ (Cash register) ‖ Betragstaste f. ‖ touche f. de

montants. / barrel ~ *see* tubular ~. / bored ~ (Locksm) ‖ hohler Schlüssel m. ‖ clef f. forée *ou* creuse. / carriage release ~ (Typewr) ‖ Wagenauslösetaste f. ‖ touche f. de libération du chariot. / clock ~ ‖ Uhrschlüssel m. ‖ clef f. de pendule *ou* de montre. / dead ~ (Typewr) ‖ Leertaste f. ‖ touche f. morte. / ~ denoting kind of transaction (Cash register) ‖ Geschäftsarttaste f. ‖ touche f. de genre de l'opération. / ~s pl. disposed in x rows ‖ in x Reihen angeordnete Tasten fpl. ‖ touches fpl. disposées sur x rangées. / double ~ (Locksm) ‖ Dietrich m.; Nachschlüssel m. ‖ fausse clef f. / double ~ (Typewr) ‖ Doppeltaste f. ‖ touche f. double. / double-bored ~ ‖ doppelt gebohrter Schlüssel m. ‖ clef f. forée avec deux trous concentriques. / double-current ~ (Tel) ‖ Indotaste f. ‖ manipulateur m. à deux pôles. / duplex ~ (Tel) ‖ Gegensprechtaste f. ‖ manipulateur m. duplex. / easily-moved ~ (Typewr) ‖ leicht bewegliche Taste f. ‖ touche f. sensible à la frappe. / false ~ *see* double ~ (Locksm). / fixed ~ ‖ Paßfeder f. ‖ ressort m. d'ajustage; languette f. ajustée. / fixing ~ ‖ Verbindungskeil m. ‖ clavette f. d'assemblage. / flat ~ ‖ Flachkeil m. ‖ clavette f. conique sur méplat. / forced ~ ‖ verdrehter Schlüssel m. ‖ clef f. faussée. / forced-in ~ ‖ Treibkeil m. ‖ clavette f. chassée; clavette f. conique. / full ~ ‖ massiver Schlüssel m. ‖ clef f. pleine. / gib ~ *see* gib-headed ~. / gib-headed ~ (Mach) ‖ Nasenkeil m. ‖ clavette f. à talon. / gib-headed flat ~ ‖ Nasenflachkeil m. ‖ clavette f. à talon sur méplat. / gib-headed saddle ~ ‖ Nasenhohlkeil m. ‖ clavette f. creuse à talon. / group-calling ~ (Tel) ‖ Gruppeneinstelltaste f. ‖ clef f. d'appel commun. / holdover ~ (Tel) ‖ Haltetaste f. ‖ clef f. de garde *ou* d'arrêt. / hollow ~ *see* bored ~. / ~ for hydrants ‖ Hydrantenschlüssel m. ‖ clef f. de bouches à eaux. / initial ~ (Cash register) ‖ Buchstabentaste f. ‖ touche f. de lettre. / layed-in ~ ‖ Einlegekeil m. ‖ clavette f. encastrée. / magnetic ~ (Radio) ‖ magnetischer Taster m. ‖ manipulateur m. magnétique. / ~ for the manhole door ‖ Faßtürchenzieher m. ‖ clef f. pour tenir la portière de foudre. / open(ed) ~ ‖ offener Schraubenschlüssel m. ‖ clef f. ouverte. / piped ~ *see* bored ~. / projecting ~ (Typewr) ‖ hervorstehende Taste f. ‖ touche f. débordante. / ~ of the ram ‖ Schließeisen n.; Schlüssel m. des Kropfeisens ‖ louvette f. / registering ~ (Cash register) ‖ Registriertaste f. ‖ touche f. d'enregistrement. / rounded taper-sunk ~ *see* layed-in ~. / saddle ~ ‖ Hohlkeil m. ‖ clavette f. conique creuse. / sliding ~ ‖ Gleitfeder f. ‖ languette f. de guidage. / special ~ for resetting to zero ‖ Sonderschlüssel m. zum Einstellen auf Null ‖ clé f. spéciale de remise à zéro. / taper-sunk ~ ‖ Treibkeil m. ‖ clavette f. conique. / ~ for printing totals (Typewr) ‖ Summendrucktaste f. ‖ touche f. de totalisation. / tubular ~ ‖ Rohrschlüssel m. ‖ clef f. en bout; clef f. tubulaire. / unbored ~ (Locksm) ‖ Vollschlüssel m.; massiver Schlüssel m. ‖ clef f. pleine. / upright ~ (Typewr) ‖ aufrechtstehende Taste f. ‖ touche f. droite.

key arrangement (Typewr) ‖ Tastenanordnung f.; Anordnung f. der Tasten ‖ disposition f. des touches. / arrangement of the ~s according to the frequency of the letters ‖ Anordnung f. der Tasten nach der Häufigkeit der Buchstaben ‖ disposition f. des touches suivant la fréquence des caractères. / bank of ~s (Cash register) ‖ Tastenfeld n. ‖ touches fpl. / ~ bar ‖ Keilstahl m. ‖ barre f. ou acier m. à clavettes. / ~ bed see keyway. / ~ board ‖ Schlüsselleiste f.; Schlüsselbrett n. ‖ porte-clés m.; râtelier m. à clés.

keyboard (Typewr) ‖ Klaviatur f.; Griffbrett n.; Tastenbrett n.; Tastatur f. ‖ clavier m.; claviature f. / alphabetical ~ ‖ abeceliches oder alphabetisches Griffbrett n. ‖ clavier m. alphabétique. / complete ~ ‖ Volltastatur f. ‖ clavier m. complet. / extra ~ ‖ Zusatztastatur f. ‖ clavier m. supplémentaire.

keyboard composing machine (Print) ‖ Setzmaschine f. mit Tastatur ‖ machine f. à composer à clavier. / ~ maker ‖ Klaviaturmacher m. ‖ facteur m. de claviers. / ~ perforator (Tel) ‖ Tastenlocher m. ‖ perforateur m. à clavier. / ~ printing telegraph ‖ Tastenschnelltelegraf m. ‖ télégraph imprimeur m. à clavier transmetteur. / ~ typewriter ‖ Klaviaturschreibmaschine f. ‖ machine f. à écrire à clavier.

key borer ‖ Schlüsselbohrer m. ‖ foreur m. de clefs. / ~ button (Typewr) ‖ Tastenknopf m. ‖ bouton m. de touche. / ~ button lifter ‖ Tastenknopfheber m. ‖ tire-boutons m. pour touches. / ~ caster (Locksm) ‖ Schlüsselgießer m. ‖ fondeur m. de clefs. / ~ file (Locksm) ‖ Spaltfeile f.; Schlüsselfeile f. ‖ lime f. à clef. / ~ filer ‖ Schlüsselfeiler m. ‖ limeur m. de clefs.

key groove see also keyway ‖ Keilnut f.; Nut f. ‖ mortaise f. ou rainure f. de clavette. / ~ machine ‖ Nutenstoßmaschine f.; Keilnutenstoßmaschine f. ‖ machine f. à rainurer. / ~ milling machine ‖ Keilnutfräsmaschine f. ‖ machine f. à fraiser les rainures.

key handle ‖ Tastenschlüssel m. ‖ clé f. de touche.

key head (Mach) ‖ Keilnase f. ‖ talon m. de clavette. / ~ (Typewr) ‖ Tastenkopf m. ‖ tête f. de touche. / engraved ~ ‖ Tastenkopf m. mit gravierter Schrift ‖ tête f. de touche à caractères gravés. / plate covering the ~ ‖ Tastenknopf m. ‖ plaque f. recouvrant la touche. / ring of the ~ ‖ Tastenring m. ‖ bague f. ou virole f. de touche.

keyhole (Locksm) ‖ Schlüsselloch n. ‖ trou m. de serrure; entrée f. de la clef. / ~ (Dowel hole) ‖ Dübelloch n. ‖ trou m. de cheville. / ~ plate ‖ Schlüssellochdeckel m. ‖ cache-entrée m. / ~ saw ‖ Stichsäge f.; Spitzsäge f.; Lochsäge f. ‖ scie f. à guichet ou à main; passepartout m.

keying ‖ Festkeilen n.; Keilverbindung f. ‖ clavetage m. / ~ of the tie of a lockgate ‖ Keilvorrichtung f. der Zugstange am Schleusentor ‖ clavetage m. des tirants d'une porte d'écluse. / ~ the wheels on the axle ‖ Aufkeilen n. der Räder auf die Achse ‖ serrage m. à clavette ou clavetage m. des roues sur l'essieu.

---

keyless watch ‖ Remontoiruhr f. ‖ montre f. à remontoir.

key letter (Typewr) ‖ Tastenaufschrift f. ‖ caractère m. inscrit sur la touche. / ~ -lever ‖ Tastenhebel m.; Tastenarm m. ‖ barre f. de touche. / ~ lever adjustment pliers pl. ‖ Richtzange f. für Tastenhebel ‖ pince f. à ajuster ou à dresser les barres de caractères. / ~ metering (Tel) ‖ Tastenzählung f. ‖ comptage m. par clé. / ~ model (Shipb) ‖ Wasserlinienmodell m. ‖ modèle m. par tranches minces. / ~ pipe (Locksm) ‖ Schlüsselrohr n. ‖ canon m. d'une clef. / ~ pipe of a lock ‖ Schlüsselrohr n. eines Schlosses ‖ canon m. d'une serrure. / ~ plate see keyhead. / ~ punch (Office) ‖ Tastenlocher m. ‖ poinçonneuse f. à touches. / ~ ring ‖ Schlüsselring m. ‖ porte-clefs m. / ~ rod (Typewr) ‖ Tastenstange f. ‖ tige f. de touche. / ~ row (Keyboard) ‖ Tastenreihe f. ‖ rangée f. de touches. / ~ rows pl. arranged in steps ‖ stufenförmig angeordnete Tastenreihen fpl. ‖ rangées fpl. de touches superposées les unes au-dessous les autres. / ~ seat see keyway. / ~ seater ‖ Keilnutenziehmaschine f. ‖ machine f. à faire les rainures de clavettes. / ~ seating of parts ‖ Nuten n. von Werkstücken ‖ rainurage m. de pièces. / ~ seating machine ‖ Nutenziehmaschine f. ‖ machine f. à faire les rainures. / ~ set see keyboard. / ~ shaft grinding machine ‖ Keilwellenschleifmaschine f. ‖ machine f. à rectifier les arbres à rainures. / ~ shank see ~ pipe. / ~ socket (Electr) ‖ Fassung f. mit Hahn oder mit Schalter ‖ douille f. ou culot m. à clef. / ~ stone (Build) ‖ Gewölbestein m.; Schlußstein m. ‖ clef f. de voûte; clausoir m. / tapen of ~ ‖ Keilanzug m. ‖ serrage m. de la clavette. / thickness of ~ ‖ Keilhöhe f. ‖ épaisseur f. de clavette. / ~ watch ‖ Schlüsseluhr f. ‖ montre f. à clef.

keyway ‖ Keilnut f. ‖ rainure f. de clavette ou de clavetage. / ~ in the hub ‖ Nabennut f. ‖ rainure f. du moyeu. / ~ in the shaft ‖ Wellennut f. ‖ rainure f. de l'arbre. / tangent ~ for shock-like alternating thrust ‖ Tangentkeilnut f. für stoßartigen Wechseldruck ‖ rainure f. de clavette tangentielle pour pression alternative par à-coups.

keyway cutting ‖ Nutenfräsen n. ‖ fraisage m. des rainures. / ~ cutting machine see ~ planing machine. / depth of ~ ‖ Nuttiefe f. ‖ profondeur f. de la rainure. / ~ planing machine ‖ Keilnutenhobelmaschine f.; Keilnutenziehmaschine f. ‖ machine f. à raboter les rainures de clavettes; machine f. à rainurer. / width of ~ ‖ Nutbreite f. ‖ largeur f. de la rainure.

key, width of ‖ Keilbreite f. ‖ largeur f. de clavette. / ~ word ‖ Schlüsselwort n. ‖ mot m. significatif; code m. télégraphique; mot m. de code.

kibble ‖ Förderkübel m. ‖ barriquet m.; baquet m.; benne f.; tonne f.; seau m. d'extraction. / to fill and fasten the ~ (Min) ‖ den Förderkübel m. anschlagen ‖ charger et attacher les barriquets.

kibbling mill ‖ Schrotmühle f. ‖ moulin m. à égruger.

kick see back ~. / back ~ ‖ Rückschlag m.; Rückstoß m. ‖ contre-coup m.

---

kick starter ‖ Kickstarter m. ‖ démarreur m. à pied.

kiddle (Fish) ‖ Fischwehr n. ‖ écrille f.

kid leather ‖ Ziegenleder n. ‖ chevrotin m.

kidnapper (Mar) ‖ Seelenverkäufer f. ‖ périsoire m.

kidney ore ‖ faseriger Roteisenstein m.; roter Glaskopf m.; Blutstein m.; roter Hämatit m. ‖ fer m. oligiste concrétionné; hématite m. rouge; fer m. oligiste fibreux.

kid skin ‖ Zickelfell n. ‖ peau f. de chevreau.

kier (Wash) ‖ Bleichkessel m. ‖ chaudière f. à bouillir.

kieselguhr ‖ Kieselgur f.; Infusorienerde f. ‖ kieselguhr m.; farine f. fossile; terre f. d'infusoires. / ~ digging ‖ Kieselgurgewinnung f. ‖ extraction f. de kieselguhr.

kill, to ‖ schlachten ‖ abattre. / ~ lime ‖ Kalk m. (tot) löschen ‖ éteindre la chaux.

killing of wire (Tel) ‖ Geradelegen n. oder Ausrecken n. des Drahtes ‖ dressage m. ou traction f. du fil.

kiln, to ‖ ausdarren; darren ‖ touraillér. / ~ bricks ‖ Ziegel mpl. brennen ‖ cuire des briques. / ~ off (malt) ‖ abdarren ‖ maintenir la température finale.

kiln ‖ Ofen m. (zum Rösten und Brennen); Röstofen m.; Darrofen m.; Darre f. ‖ four m.; étuve f. séchoire; touraille f. / to discharge a ~ ‖ eine Darre abräumen ‖ décharger une touraille. / to load a ~ ‖ eine Darre beladen; auf eine Darre auftragen ‖ charger une touraille. / to unload a ~ see to discharge a ~. / alternate working ~ ‖ Ofen m. mit unterbrochenem Betriebe ‖ four m. discontinu. / annular ~ (Brick) ‖ Ringofen m. ‖ four m. circulaire ou annulaire. / continuous working ~ ‖ Ofen m. mit Dauerbetrieb ‖ four m. continu. / drying ~ ‖ Darre f. ‖ étuve f. séchoire; touraille f. / dumping ~ (Brew) ‖ Klapphorde f. ‖ touraille f. à jalousie. / field ~ (Brick) ‖ Feldbrandofen m. ‖ four m. de campagne. / one-floored ~ ‖ Einhordendarre f.; einhordige Darre f. ‖ touraille f. à un plateau. / ~ with overlying beds ‖ Etagenofen m. ‖ four m. à étages. / perpetual ~ see continuous working ~. / revolving tubular ~ see rotary tubular ~. / rotary ~ (Brick) see also annular ~ ‖ Ringofen m.; Drehofen m. ‖ four m. circulaire ou tournant. / rotary tubular ~ ‖ Drehrohrofen m. ‖ four m. tubulaire tournant. / running ~ see continuous working ~. / ~ combined with separated hurdles ‖ mit getrennten Horden kombinierte Darre f. ‖ touraille f. à plateaux séparés. / sweating ~ ‖ Darre f. zum Trocknen der Gerste; Gerstendarre f. ‖ touraille f. à sécher l'orge. / three-floored ~ ‖ Dreihordendarre f.; dreihordige Darre f. ‖ touraille f. à trois plateaux. / two-floored ~ ‖ Doppel(horden)darre f.; doppelhordige Darre f. ‖ touraille f. à deux plateaux. / yeast ~ ‖ Hefetrockner m. ‖ séchoir m. de levure.

kiln air ‖ Darrluft f. ‖ air m. chauffé d'une touraille. / ~ brick ‖ feuerfester Mauerstein m.; Ofenziegel m. ‖ brique f. réfractaire ou blanche. / ~ chamber ‖ Darraum m. ‖ chambre f. de touraille. / ~ controlling thermometer ‖ Darrkontrollthermometer n. ‖ thermomètre m. de contrôle d'une touraille. / discharging of a ~ ‖ Abräumen n. oder Abladen

n. einer Darre ‖ déchargement m. d'une touraille.

kiln-dried ‖ gedarrt ‖ séché au four *ou* à la touraille. / ~ malt ‖ Darrmalz n. ‖ malt m. touraillé; malt m. séché à la touraille; / ~ wood ‖ Darrholz n. ‖ bois m. séché au séchoir.

kiln-dry, to ‖ darren ‖ touraillert; sécher.

kiln floor ‖ Darrboden m.; Darrblech n. ‖ plateau m. de touraille. / ~ turner ‖ Darrwender m. ‖ retourneur m. d'une touraille.

kiln heating apparatus ‖ Darrheizvorrichtung f.; Darrheizapparat m. ‖ calorifère m. pour tourailles. / ~ heating pipe ‖ Darrheizrohr n. ‖ tuyau m. de chauffage d'une touraille. / ~ hurdle *see* ~ floor.

kilning (Brew) ‖ Ausdarren n.; Darrung f.; Darren n. ‖ touraillage m. / ~ drum ‖ Darrtrommel f. ‖ tambour m. à tourailler. / ~ temperature ‖ Darrtemperatur f. ‖ température f. de touraillage. / final ~ temperature ‖ Abdarrtemperatur f. ‖ température f. finale de touraillage. / ~ time ‖ Darrzeit f.; Darrdauer f. ‖ durée f. de touraillage.

kiln loading ‖ Auftragen n. auf die Darre ‖ chargement m. de la touraille.

kiln man (Agr) ‖ Korntrockner m. ‖ sécheur m. de grains. / ~ (Brew) ‖ Darrfax m. ‖ tourailleur m. / ~ (Metal) ‖ Tiegelbrenner m. ‖ cuiseur m. de creusets.

kiln rake (Found) ‖ Krücke f. ‖ crosse f.; écumoire f. / ~ recording thermometer ‖ Darrregistrierthermometer n. ‖ thermographe m. *ou* thermomètre-enregistreur m. d'une touraille. / ~ regulation ‖ Darrordnung f. ‖ règlement m. de touraillage. / ~ saw *see* ~ warming tub. / ~ system ‖ Darrsystem n. ‖ système m. de touraille. / combined lattice and trough ~ system ‖ mit Jalousien und Rinnen kombiniertes Darrsystem ‖ système m. de touraille combiné à jalousies et à rigoles. / ~ thermometer ‖ Darrthermometer n. ‖ thermomètre m. de touraille. / unloading of a ~ ‖ Abräumen n. *oder* Abladen n. einer Darre ‖ déchargement m. d'une touraille. / ~ ventilation ‖ Darrventilation f. ‖ ventilation f. de touraille. / ~ warming tub ‖ Darrsau f. ‖ cochon m. de touraille.

kilocalorie ‖ Kilokalorie f.; große Kalorie f. ‖ kilocalorie f.; grande calorie f.

kilogramme ‖ Kilogramm n. ‖ kilogramme m.

kilogrammeter ‖ Kilogrammeter m.; Meterkilogramm n. ‖ kilogrammètre m.; mètre m. kilogramme.

kilometre ‖ Kilometer n. ‖ kilomètre m.

kilovolt ‖ Kilovolt n. ‖ kilovolt m.

kilowatt ‖ Kilowatt n. ‖ kilowatt m. / installed ~ ‖ installiertes Kilowatt n. ‖ kilowatt m. installé. / curve of ~s ‖ Kilowattkurve f. ‖ courbe f. de puissance en kilowatts. / ~ hour ‖ Kilowattstunde f. ‖ kilowatt-heure m.

kind ‖ Art f.; Gattung f. ‖ genre m.; espèce f.; sorte f. / ~ of current (Electr) ‖ Stromgattung f. ‖ nature f. du courant. / ~ of goods ‖ Warengattung f. ‖ nature f. des marchandises. / ~ of service ‖ Betriebsart f. ‖ genre m. *ou* méthode m. de service. / ~ of timber ‖ Holzgattung f. ‖ espèce f. de bois.

kindergarten utensil ‖ Kindergartengerät n. ‖ ustensile m. pour jardins d'enfants.

kindle, to ‖ ~ a fire ‖ ein Feuer n. anmachen *oder* anzünden ‖ allumer les feux mpl.

kindness ‖ Gefälligkeit f. ‖ complaisance f.; obligeance f.

kinematics pl. ‖ Zwanglauflehre f.; Kinematik f. ‖ cinématique f.

kinematograph film ‖ Filmband n. ‖ film m. pour cinémas. / ~ projector ‖ Kinovorführungsgerät n. ‖ appareil m. de projection de cinéma.

kinetic (Phys) ‖ kinetisch ‖ cinétique. / ~ energy ‖ kinetische Energie f. ‖ énergie f. cinétique.

kinetics pl. ‖ Kinetik f. ‖ cinétique f.

king journal ‖ Königszapfen m.; Königssäule f.; Drehzapfen m. ‖ pivot m. central. / ~ pillar *see* ~ journal. / ~ pin *see* ~ journal.

kingpost ‖ Spannturm m. ‖ cabane f. *ou* pylône m. de haubannage. / ~ of a center (Build) ‖ Schlußpfosten m. eines Lehrgerüsts ‖ moise f. pendante de clef d'un cintre. / ~ truss ‖ Hängewerk n. mit einer Hängesäule ‖ armature f. *ou* ferme f. à un seul poinçon.

kink Cable ‖ Schleife f.; Kink f. ‖ tortillement m. / ~ (El line) ‖ Knick m. ‖ faux pli m.; brisure f.

kiosk ‖ Kiosk m.; Verkaufshäuschen m.; Verkaufsbude f. ‖ kiosque m.

kip ‖ rohes Kalbsfell n.; rohe Kalbshaut f. ‖ peau f. brute de veau.

Kirchhoff's laws pl. ‖ Kirchhoffsche Regeln fpl. ‖ lois fpl. de Kirchhoff.

kirsch ‖ Kirsch m.; Kirschwasser n.; Kirschbranntwein m. ‖ kirsch m.; eau f. de cerises.

kirschwasser *see* kirsch.

kirve, to ~ (Mine) *s.* to kerve.

kishy (Met) ‖ graphitisch ‖ limailleux; graphiteux; surcarburé. / ~ pig ‖ schwarzes Roheisen n. ‖ fonte f. limailleuse *ou* graphiteuse.

kit ‖ Handwerkzeug n.; Ausrüstung f.; Ausstattung f. ‖ trousseau m.; trousse f. / ~ bag ‖ Werkzeugtasche f. ‖ sac m. à outils.

kitchen ‖ Küche f. ‖ cuisine f. / small ~ ‖ Kleinküche f. ‖ petite cuisine f.

kitchen balance ‖ Haushaltswage f.; Wirtschaftswage f.; Küchenwage f. ‖ balance f. de cuisine. / ~ chimney ‖ Küchenschornstein m. ‖ cheminée f. de cuisine. / ~ chopper ‖ Hackbeil n.; Hackmesser n. ‖ couperet m. de cuisine. / ~ clock ‖ Küchenuhr f. ‖ horloge f. de cuisine. / ~ dresser ‖ Küchentisch m.; Küchenanrichte f. ‖ fourneau-crédence f. / ~ furniture ‖ Küchenmöbel npl. ‖ meubles mpl. de cuisine. / ~ garden ‖ Gemüsegarten m.; Küchengarten m. ‖ jardin m. potager; potager m. / ~ gardener ‖ Gemüsegärtner m. ‖ maraîcher m. / ~ gas stove ‖ Gaskochapparat m. ‖ fourneau m. de cuisine à gaz. / ~ hatchet ‖ Küchenbeil n. ‖ hachette f. de ménage. / ~ implement ‖ Küchengerät n. ‖ ustensile m. de cuisine. / ~ installation ‖ Kücheneinrichtung f. ‖ installation f. de cuisine. / ~ knife ‖ Küchenmesser n. ‖ couteau m. de cuisine. / ~ lift ‖ Speisenaufzug m. ‖ monte-plats m. / ~ linen ‖ Küchenwäsche f.; Küchenzeug n. ‖ linge m. de cuisine. / ~ machine ‖ Küchenmaschine f. ‖ machine f. de cuisine *ou* destinée à la cuisine; appareil m. de cuisine. / ~ mantle ‖ Rauchfang m. *oder* Schurz m. eines Küchenherdes ‖ hotte f. *ou* manteau m. de cheminée. / ~ plant ‖ Küchengewächs n. ‖ plante f. potagère.

/ ~ range *see also* ~ stove ‖ Kochherd m.; Küchenherd m. ‖ fourneau m. *ou* foyer m. de cuisine; cuisinière f. / ~ range de luxe ‖ Luxusküchenherd m. ‖ cuisinière f. de luxe. / ~ requisites pl. ‖ Küchengegenstände mpl. ‖ articles mpl. de cuisine. / ~ salt ‖ Siedesalz n. ‖ Kochsalz n. ‖ sel m. de saline; sel m. raffiné. / ~ salt (Miner) ‖ Kochsalz n. ‖ sel m. commun *ou* gemme; soude f. muriatée. / ~ scales pl. ‖ Haushaltwage f.; Küchenwage f. ‖ balance f. de cuisine. / ~ sets pl. of aluminium ‖ Aluminiumkochgeschirr n. ‖ batterie f. de cuisine *ou* bouillottes fpl. de cuisine en aluminium. / ~ stove ‖ Küchenofen m.; Kochherd m.; Sparherd m. ‖ fourneau m. *ou* four m. *ou* poêle f. de cuisine. / ~ towel ‖ Küchenhandtuch n. ‖ serviette f. de cuisine.

kitchen utensils pl. ‖ Kochgeschirr n.; Küchengeschirr n.; Küchengerät n. ‖ vaisselle f. de ménage; batterie f. de cuisine; ustensiles mpl. de cuisine. / ~ of aluminium ‖ Aluminiumkochgeschirr n. ‖ batterie f. de cuisine *ou* bouillottes fpl. de cuisine en aluminium. / enamelled ~ ‖ emailliertes Kochgeschirr n. ‖ ustensiles mpl. de cuisine émaillés. / house and ~ ‖ Haus- und Küchengeräte mpl. ‖ ustensiles mpl. de ménage et de cuisine.

kite ‖ Drachen m. ‖ cerf-volant m. / box ~ ‖ Kastendrachen m. ‖ cerf-volant m. cellulaire.

kite balloon ‖ Drachenballon m. ‖ ballon m. cerf-volant; drachenballon m.

knack ‖ Kunstgriff m.; Handwerkskniff m. ‖ coup m. de main; secret m.; truc m.

knacker ‖ Abdecker m. ‖ équarrisseur m.

knag in wood ‖ Knorren m. *oder* Knoten m. *oder* Knast m. im Holz ‖ nœud m. dans le bois.

knaggy (Wood) ‖ knorrig; knotig; knästig ‖ malandreux.

knapsack ‖ Rucksack m. ‖ bissac m.; sac m. de touriste. / ~ (Military) ‖ Tornister m. ‖ havre-sac m.; sac m. militaire. / ~ fittings pl. ‖ Tornisterbeschläge mpl. ‖ ferrures fpl. de sacs militaires. / ~ station (Radio) ‖ Tornisterstation f. ‖ poste m. de havre-sac; poste m. à dos d'homme.

knead, to ‖ kneten ‖ malaxer; pétrir. / ~ the dough ‖ den Teig m. kneten *oder* anmachen ‖ pétrir la pâte.

kneadable ‖ knetbar ‖ pétrissable.

kneader ‖ Kneter m. ‖ pétrisseur m.

kneading ‖ Kneten n. ‖ pétrissage m.; pétrissement m. / ~ arm ‖ Knetarm m. ‖ bras m. pétrisseur.

kneading machine (Bak) ‖ Knetmaschine f.; Knetwerk n. ‖ pétrin m. mécanique; machine f. à pétrir. / ~ (Build) ‖ Knetmaschine f. ‖ malaxeur m.; malaxeuse f. / ~ with brake device for the bowl at the bottom (Bak) ‖ Knetmaschine f. mit Trogbremse am Boden ‖ pétrin m. avec disposition de frein au fond de la cuve. / clay ~ ‖ Tonknetmaschine f. ‖ machine f. à pétrir la glaise.

kneading and mixing machine for concrete and mortar ‖ Knet- und Mischmaschine f. für Beton und Mörtel ‖ machine f. à pétrir et à mélanger le béton et le mortier.

kneading trough ‖ Backtrog m. ‖ pétrin m.

knee ‖ Knie n. ‖ genou m. / ~ (Mach) ‖ Knie n.; Kniestück n.; Kröpfung f. ‖

coude m.; épaulement m. / ~ (Mach tool) || Winkeltisch m. || console f. de table. / ~s pl. (Shipb) || Knieholz n. || courbes fpl. / to form a ~ || kröpfen || épauler; couder. / connecting ~ || Verbindungsknie n. || coude m. de jonction *ou* d'assemblage / iron ~ || eiserner Winkel m. || équerre f. en fer. / square ~ (Shipb) || Winkelknie n.; rechtwinkliges Knie n. || coude m. en équerre *ou* à angle droit.

kneecap (Anatomy) || Kniescheibe f. || rotule f. / ~ (Knee warmer) || Kniewärmer m. || genouillère f. / ~ (Shoem) || Knieriemen m.; Knieleder n. || genouillère f.; tire-pied m. / elastic ~ || Knieriemen m. aus elastischem Gewebe || genouillère f. en tissu élastique.

knee ham || Kniekehle f. || jarret m.

knee lever || Winkelhebel m. || levier m. coudé. / ~ embossing press for matrices with combined steam and electrical heating || Kniehebelmatrizenprägepresse f. mit dampfelektrischer Heizung || presse f. à empreindre à matrice et à genouillère avec chauffage combiné électrique et à vapeur.

kneestrap *see* kneecap (Shoem).

knettar (Mar) || Sackleine f. || aiguillette f. d'un sac de matelot.

knife || Messer n.; Schneide f. || couteau m. / ~ (Milling cutter) || Fräsmesser n. || lame f. de fraise / budding ~ || Pfropfmesser n.; Okuliermesser n. || greffoir m.; couteau m. à greffer. / butcher's ~ || Schlächtermesser n.; Schlachtmesser n. || couteau m. de boucher(s). / cabbage ~ || Krauthobel m. || couteau m. à choucroute. / carving ~ || Vorlegemesser n.; Fleischschneidemesser n.; Transchiermesser n. || couteau m. à découper. / ~ of the chaff cutter || Häckselklinge f.; Häckselmesser n. || couteau m. *ou* lame f. du hachepaille. / cheese ~ || Käsemesser n. || couteau m. à fromage. / double-handed cheese ~ || Doppelgriffkäsemesser n. || couteau m. à fromage à deux mains. / cooper's cleaning ~ || Küferputzmesser n. || plane f. curette pour tonneliers. / cooper's hollowing ~ || Küferkrummmesser n. || plane f. creuse pour tonneliers. / currier's ~ || Gerbermesser n. || couteau m. de corroyeur. / ~ for cutting mill-boards for book-covers (Bookb) || Schnitzer m. || couteau m. à rabaisser; pointe f. / ~ for cutting vines || Rebmesser n. || couteau m. à vendanges. / dessert ~ || Nachtischmesser n.; Dessertmesser n. || couteau m. à dessert. / drawing ~ (Wood working) || Abziehmesser n.; Ziehklinge f.; Schneidmesser n. mit zwei Handgriffen; Ziehmesser n.; Zugmesser n. || débordoir m.; couteau m. à deux manches; plane f. / ~ for dredging bucket || Baggereimermesser n. || bec m. de godet de drague. / fellow ~ || Kranzmesser n. || plane f. à déraser. / ~ with fixed blade *see* nonshutting ~. / grafting ~ || Okuliermesser n. || écussonnoir m.; greffoir m. / hunting ~ || Jagdmesser n. || couteau m. de chasse. / kitchen ~ || Küchenmesser n. || couteau m. de cuisine. / mincing ~ || Wiegemesser n. || hachoir m. / ~ of the mower || Mähmaschinenmesser n. || lame f. de la faucheuse. / ~ with myrtle-leaf blade || Lorbeerblattmesser n. || bistouri m. à double tranchant. / non-shutting ~ || nicht zusammenklappbares Messer n.; Messer n.

mit fester Klinge || couteau m. non fermant *ou* à lame fixe; poignard m. / ~ for opening tins of preserves || Konservendosenöffner m.; Dosenöffner m. || couteau m. pour boîtes de conserves; ouvreboîte m. / peeling ~ || Kartoffelschäler m.|| couteau m. à peler des pommes de terre. / ~ with pistol-handle || Messersäge f. mit Pistolenheft || scie f. à manche pistolet. / pocket ~ || Taschenmesser n. || couteau m. de poche. / probe-ended ~ || geknöpftes Messer n. || bistouri m. boutonné. / pruning ~ || Gartenmesser n. || serpe f. / radish ~ || Rettichschneider m. || coupe-radis m. / rag knives pl. of the rag engine || Messer npl. *oder* Schienen fpl. am Holländer || lames fpl. du moulin à cylindre de papeterie. / ~ of rustless steel || Messer n. aus nichtrostendem Stahl || couteaux m. en acier inoxydable. / saddler's and shoemaker's ~ || Sattler- und Schustermesser n.; Ledermesser n. || couteau m. à cuir. / sharp-pointed ~ (Med) || spitzes Messer n. || couteau m. à lame pointue. / shear ~ || Scherenmesser n. || lame f. de cisailles. / shoemaker's ~ || Schustermesser n. || tranchet m. / staves hollowing ~ *see* cooper's hollowing ~. / table ~ || Tafelmesser n.; Tischmesser n. || couteau m. de table. / ~ with two handles *see* drawing ~. / vine ~ || Winzermesser n. || serpette f. / ~ for workmen || Arbeitermesser n. || couteau m. pour ouvriers.

knife and fork || Besteck n. || couvert m. / ~ of alpaca metal || Alpakabesteck n. || couvert m. de table en métal alpaca. / ~ of Britannia metal || Britanniabesteck n. || couvert m. de table en métal Britannia. / ~ for fruit || Obstbesteck n. || couvert m. à fruits. / wooden ~ || Holzbesteck n. || couvert m. en bois.

knife, fork and spoon *see* knife and fork.

knife-blade || Messerklinge f.; Messerschneide f. || lame f. de couteau. / to finish upon a ~ || in einer dünnen Schneide f. endigen; messerschneidig endigen || finir en lame f. de couteau. / finishing of ~s || Feilen n. der Messerklingen || façonnage m. de lames de couteaux.

knife cleaning machine || Messerputzmaschine f. || machine f. à nettoyer les couteaux. / ~ crusher || Messerbrecher m. || concasseur m. à couteaux. / ~ disk || Messerscheibe f. || couteau m. circulaire. / shaft of the ~ disk || Messerscheibenwelle f. || arbre m. porte-couteau. / ~ driving shaft (Mowing mach) || Messerwelle f. || arbre m. de commande du couteau.

knife-edge || Schneide f. || tranchant m. de couteau. / ~ of a balance || Schneide f. am Wagebalken || couteau m. d'une balance. / ~ bearing || Schneidenlager n. || support m. à couteau.

knife file || Messerfeile f. || lime f. à couteau *ou* à dossière. / ~ for cross-cut saws || Schrotsägenschärffeile f. || lime f. à couteau pour passe-partout.

knife filer || Messerfeiler m. || limeur m. en coutellerie. / ~ forging || Messerschmiede f. || martelage m. *ou* forge f. de couteaux. / ~ graver (Engr) || Messerzeiger m. || onglette f.

knife grinder (Person) || Messerschleifer m. || affûteur m. *ou* repasseur m. de couteaux. / ~ (Apparatus) || Messerschleifvorrich-

tung f. || appareil m. *ou* dispositif m. à affûter les couteaux. / hand operated ~ || Messerschleifvorrichtung f. für Handbetrieb || dispositif m. à affûter les lames à la main.

knife grinding machine || Messerschleifmaschine f. || machine f. à aiguiser *ou* à affûter les couteaux.

knife handle || Messergriff m.; Messerheft n. || manche m. de couteau. / ~ cutter || Messerheftschneider m. || découpeur m. de manches de couteaux. / ~ filer || Messerheftfeiler m. || ébaucheur m. *ou* façonneur m. de manches de couteaux. / ~ shaper *see* ~ filer.

knife hardener || Messerhärter m. || trempeur m. de couteaux. / ~ indicator || Messerzeiger m. || aiguille f. en forme de couteau. / ~ polisher || Messerpolierer m. || polisseur m. de couteaux. / ~ polishing || Messerpolieren n. || polissage m. de couteaux. / ~ pruning saw || Messersäge f. || scie f. à couteau. / ~ regrinding || Messerschleifen n. || repassage m. de couteaux. / ~ sharpener (Tool) || Messerschärfer m. || affiloir m. / ~ smith || Messerschmied m. || forgeur m. de couteaux; coutelier m. / ~ steel || Messerstahl m. || acier m. à couteaux. / ~ switch (Electr) || Messerschalter m. || interrupteur m. à couteau. / double bladed ~ switch || Doppelmesserschalter m. || interrupteur m. à double couteau. / ~ tool *see* ~ graver.

knight-heads pl. (Shipb) || Ohrhölzer npl. || apôtres mpl.; hauts mpl. des apôtres.

knit, to || stricken || tricoter. / ~ by hand || mit der Hand stricken || tricoter à la main. / ~ on || anstricken || ajouter en tricotant. / ~ off || abstricken || dégager en tricotant.

knit-goods pl. || Strickwaren fpl.; Trikotwaren fpl. || bonneterie f. tricotée; articles mpl. en tricot; tricotages mpl. / ~ manufacture || Wirkwarenfertigung f. || fabrication f. de tricotages. / ~ manufacturer || Strickwarenfabrikant m. || fabricant m. d'articles tricotés.

knitted beret || Trikotmütze f. || béret m. tricoté. / fulled ~ cloth || gewalkter Trikotstoff m. || bonneterie f. drapée. / ~ dress || Strickkleid n. || robe f. tricotée. / ~ fabric || gewirkter Stoff m.; Strickware f. || tissu m. par mailles; tricotage m.; article m. tricoté. / ~ goods pl. *see* knit goods. / ~ shirt || Trikothemd n. || chemise f. en tricot. / ~ sporting clothes pl. || gestrickte Sport(be)kleidung f. || vêtement m. tricoté de sport.

knitter || Stricker m. || tricoteur m. / cuff ~ || Kunststricker m. || tricoteur m. de parements. / frame ~ || Maschinenstricker m. || tricoteur m. à la machine. / gobile rib ~ || Rippenstricker m. || tricoteur m. à la gobile. / plain ~ || Flachstricker m. || tricoteur m. en uni. / warp machine ~ || Trikotweber m. || tricoteur m. au métier à chaîne.

knitting || Trikotgewebe n. || tissu m. tricoté *ou* à mailles. / framework ~ *see* machine ~ / hand ~ || Häkelwaren fpl. || tricotage m. à la main. / machine ~ || Maschinenstrickerei f. || tricotage m. au métier.

knitting cotton || Strickbaumwolle f. || coton m. à tricoter.

knitting frame || Wirkmaschine f.; Wirkstuhl m. || métier m. rectiligne *ou* à tricot *ou* à tricoter. / self-acting ~ || mechanischer Wirkstuhl m. || métier m.

à tricoter mécanique. / ~ fitter ‖ Wirkstuhlvorrichter m. ‖ ajusteur m. *ou* mécanicien m. en métiers de bonneterie.
knitting industry ‖ Strickereiindustrie f. ‖ industrie f. de la maille. / ~ loom *see* ~ frame.
knitting machine ‖ Strickmaschine f. ‖ tricoteuse f. (mécanique); machine f. à tricoter. / ~ with double mechanism ‖ Doppelstrickmaschine f. ‖ machine f. à tricoter à double chute. / eightlock ~ ‖ Achtschloßstrickmaschine f. ‖ machine f. à tricoter à huit serrures. / frame ~ ‖ Kettelmaschine f. ‖ machine f. à mailler. / hand ~ ‖ Handstrickmaschine f. ‖ machine f. à tricoter à main. / hosiery ~ ‖ Strumpfstrickmaschine f. ‖ machine f. à tricoter les bas. / Jacquard ~ ‖ Jacquardstrickmaschine f. ‖ machine f. à tricoter Jacquard. / ~ for multicoloured Jacquard patterns ‖ Jacquard Buntmusterstrickmaschine f. ‖ machine f. Jacquard pour dessins multicolores. / purl ~ ‖ Noppenstrickmaschine f. ‖ machine f. à tricoter à nopes.
knitting machine chain ‖ Strickmaschinenkette f. ‖ chaîne f. de tricoteuse. / manufacture of ~s ‖ Strickmaschinenfertigung f. ‖ fabrication f. de tricoteuses. / ~ needle ‖ Strickmaschinennadel f. ‖ aiguille f. de machine à tricoter.
knitting machinery factory ‖ Strickmaschinenfabrik f. ‖ manufacture f. de machines à tricoter. / ~ needle ‖ Stricknadel f. ‖ aiguille f. à bas *ou* à tricoter. / ~ silk ‖ Strickseide f. ‖ soie f. à tricoter. / ~ wool ‖ Strickwolle f. ‖ laine f. à tricoter. / ~ yarn ‖ Strickgarn n. ‖ fil m. à tricoter.
knit tissue from silk ‖ Wirkstoff m. aus Seide ‖ tissu m. de bonneterie en soie.
knives pl. *see* knife.
knob ‖ Knopf m.; Knauf m. ‖ bouton m. / ~ (Tassel of a cap) ‖ Troddel f.; Quaste f. ‖ houppe f. / ~ (Bobbin insulator) ‖ Isolierrolle f. ‖ roulette f. isolante. / ~ of an arbor ‖ Hebedaumen m.; Wellendaumen m. ‖ mentonnet m. d'arbre. / door ~ ‖ Türgriff m.; Türknopf m. ‖ bouton m. de porte. / drawer ~ ‖ Schubladengriff m. ‖ coquille *ou* bouton m. de tiroir. / knurled ~ ‖ Zierknopf m. ‖ bouton m. molleté. / milled ~ ‖ gerändelter Knopf m. ‖ bouton m. molleté. / ~ of a tile ‖ Nase f. *oder* Haken m. eines Dachziegels ‖ crochet m. d'une tuile. / window ~ ‖ Fenstergriff m.; Fensterknopf m. ‖ bouton m. *ou* tiroir m. *ou* olive f. de fenêtre.
knob dowel ‖ Knopfdübel m. ‖ goujon m. à bouton. / handle ~ *see* Knob, door.
knobby (Wood) *see* knaggy.
knock, to ‖ pochen; klopfen; stoßen ‖ frapper. / ~ (Mot) ‖ klopfen ‖ cogner. / ~ off ‖ abhauen; abschlagen ‖ couper; détacher. / ~ off the boiler scales ‖ den Kesselstein m. abklopfen ‖ détartrer *ou* écailler *ou* piquer une chaudière. / ~ off the seams from the castings ‖ die Gußnaht abschlagen ‖ enlever la bavure; ébarber. / ~ off and to saw off runner and riser on the casting ‖ Einguß m.

und Steiger m. am Gußstück abschneiden ‖ couper l'évent m. et le jet m. des pièces coulées. / ~ over (Knitting) ‖ abschlagen ‖ abattre.
knock ‖ Schlag m.; Stoß m. ‖ coup m.; choc m. / piston ~ (Mot) ‖ Klopfen n. des Kolbens ‖ cognement m. du piston.
knocker (Door) ‖ Klöpfel m.; Türklopfer m. ‖ heurtoir m.
knocking ‖ klopfendes Geräusch n. ‖ bruit m. de frappe. / ~ (Motor) ‖ Klopfen n.; Hämmern n. ‖ tapage m.; cognement m. / ~ pl. (Mine) ‖ Gänge mpl.; geförderte Erzstücke npl. ‖ minerai m. brut *ou* en morceau. / ~ in the crank ‖ Kurbelschlag m. ‖ cogne m. *ou* coup m. dans la manivelle.
knocking-off (Boil) ‖ Abklopfen n. ‖ détartrage m.
knocking-over (Knitting) ‖ Abschlag m. ‖ abat(t)age m. / movable ~ ‖ beweglicher Abschlag m. ‖ abat(t)age m. mobile.
knocking-over bar ‖ Abschlagschiene f. ‖ barre f. d'abat(t)age. / ~ cam ‖ Abschlagexzenter m. ‖ came f. d'abat(t)age. / ~ comb ‖ Abschlagkamm m. ‖ peigne m. d'abat(t)age. / ~ device ‖ Abschlageinrichtung f. ‖ mécanique f. d'abat(t)age.
knot, to ‖ knoten ‖ faire un nœud; nouer. / ~ cords pl. ‖ Schnüre fpl. knüpfen ‖ nouer des lacets mpl.
knot ‖ Knoten m.; Schleife f.; Schlinge f. ‖ nœud m. / ~ (Mar) ‖ Knoten m. ‖ nœud; bouton m. / ~ (Weav) ‖ Docke f. ‖ torque m. / ~ (Wood) ‖ Ast m. ‖ nœud m. / to run *x* ~s pl. ‖ *x* Knoten mpl. laufen *oder* machen ‖ filer *x* nœuds de la ligne de loch. / ~s pl. in calcined ore ‖ Knoten mpl. beim Erzrösten; Krebse mpl.; ungeröstete Erzkerne mpl. ‖ nodules mpl. de minerais calcinés. / double-loop ~ ‖ Doppelknoten m. ‖ nœud m. à deux boucles. / false ~ ‖ falscher Knoten m.; Altweiberknoten m.; Waschweiberknoten m. ‖ faux nœud m. / ~s pl. in the glass ‖ Knoten mpl. im Glase ‖ nœuds mpl. *ou* nodules fpl. dans le verre brut. / ~ of the log-line ‖ Knoten m. der Logleine ‖ nœud m. de la ligne de loch. / overhand ~ ‖ einfacher Knoten m. ‖ nœud m. simple. / right ~ ‖ gerader *oder* rechter Knoten m.; Kreuzknoten m. ‖ nœud m. droit. / running ~ ‖ Schifferknoten m.; Seemannsknoten m. ‖ nœud m. de batelier *ou* d'artificier *ou* de tisserand. / sailor's ~ *see* running ~. / (little) ~ in silk yarn ‖ Seidengarnknötchen n. ‖ bouchon m. dans le fil de soie. / ~ in slate ‖ Knoten m. im Schiefer ‖ moelle f. d'ardoise. / ~ in wood ‖ Knorren m. *oder* Knoten m. *oder* Knast m. im Holz ‖ nœud m. dans le bois.
knot borer ‖ Astbohrer m. ‖ perceur m. de nœuds.
knotgrass herb ‖ russischer Knöterich m. ‖ herbe f. de persicaire.
knot hole (Wood) ‖ Astloch n. ‖ trou m. de nœud.
knot loosening device ‖ Knotenlösevorrichtung f. ‖ dispositif m. à défaire les nœuds.

knotted work ‖ Knüpfarbeit f.; Makrameearbeit f. ‖ ouvrage m. noué.
knotter (Grain binder) ‖ Knüpfer m. ‖ noueur m. / ~ of shawl fringes ‖ Schalfransenknüpferin f. ‖ noueuse f. de châles frangés.
knotter-hook (Grain binder) ‖ Knüpferhaken m. ‖ bec m. du noueur. / ~ cam ‖ Knüpferhakennocken m. ‖ came f. de commande du bec du noueur.
knotting (Paint) ‖ Grundanstrich m. ‖ première couche f. de peinture. / ~ machine (Weav) ‖ Knotmaschine f. ‖ machine f. à nouer.
knotty ‖ knorrig; knotig ‖ noueux; malandreux. / ~ wood ‖ ästiges Holz n. ‖ bois m. noueux.
knowledge of business ‖ Geschäftskenntnis f. ‖ connaissance f. des affaires. / expert ~ ‖ Sachkenntnis f.; Sachkunde f. ‖ connaissance f. des choses *ou* des faits. / ~ of a place ‖ Ortskenntnis f. ‖ connaissance f. d'un lieu. / special ~ *see* expert ~. / without our ~ ‖ ohne unser Vorwissen n. ‖ à notre insu.
knuckle, steering ~ ‖ Achsschenkel m.; Achszapfen m.; Achshals m. ‖ fusée f. d'essieu.
knuckle joint (Mach) ‖ Gelenk n. ‖ articulation f. / ~ pin ‖ Kuppelbolzen m. ‖ cheville f. d'attelage.
knurl, to ‖ rändeln; kordeln ‖ moleter.
knurled screw ‖ Rändelschraube f. ‖ vis f. moletée. / ~ with slot ‖ Rändelschraube f. mit Schlitz ‖ vis f. moletée fendue *ou* avec fente.
knurling tool ‖ Rändelwerkzeug n. ‖ outil m. à moleter.
kochelite ‖ Kochelit m. ‖ kochélite f.
koerzit ‖ Koerzit n. ‖ koerzit m.
koff ‖ Kuff f. ‖ koff m.
kola nut ‖ Kolanuß f. ‖ noix f. de cola; cola f. / ~ tablets pl. ‖ Kolatabletten fpl. ‖ tablettes fpl. de cola.
kollyrite ‖ Kollyrit m.; samische Erde f. ‖ collyrite f.; alumine f. hydratée silicifère.
koumiss ‖ Kumyß m.; gegorene Stutenmilch f. ‖ koumis m.
kousso flower ‖ Kussoblüte f. ‖ fleur f. de cousso.
Krarup cable ‖ Krarupkabel n. ‖ câble m. Krarup. / ~ line ‖ Krarupleitung f. ‖ ligne f. krarupisée. / ~ loading ‖ Verstärkung f. nach Krarup; Krarupisierung f. ‖ krarupisation f.
kräusen (Brew) ‖ Kräusen fpl. ‖ kræusen pl.
Kremnitz white ‖ Kremserweiß n. ‖ blanc m. d'argent *ou* de Krems.
kyanite ‖ Kyanit m.; Disthen m. ‖ kyanite f.; disthène f.
kyanize, to ‖ kyanisieren; mit Sublimat n. imprägnieren *oder* tränken ‖ kyaniser; cyaniser; imprégner au sublimé.
kyanizing of wood ‖ Tränkung f. *oder* Imprägnierung f. des Holzes mit Sublimat; Kyanisierung f. des Holzes ‖ imprégnation f. des bois au sublimé (corrosif).
kyrosite ‖ Kyrosit m. ‖ kyrosite f.

L

label, to ‖ auszeichnen; etikettieren ‖ étiqueter; marquer.

label ‖ Auszeichnung f.; Etikett n.; Beklebezettel m. ‖ marque f.; étiquette f. / attached ~ ‖ Anhänger m. ‖ étiquette f. de colis. / bone ~ ‖ Knochenschild n. ‖ étiquette f. en os. / ~ for bottles ‖ Flaschenschild n. ‖ étiquette f. à bouteilles. / gummed ~ ‖ gummiertes Etikett n. ‖ étiquette f. gommée. / numbered ~ for keys ‖ Schlüsselnummerschild n. ‖ plaque f. numerotée pour clefs. / woven ~ ‖ gewebtes Etikett n. ‖ étiquette f. tissée.

label boy ‖ Etikettenkleber m. ‖ étiqueteur m. / ~ and ticket fastener ‖ Kollianhänger m. ‖ étiquette f. pour colis. / ~ glue ‖ Etikettenleim m. ‖ colle f. pour étiquettes.

labelled ‖ etikettiert ‖ étiqueté.

labeller ‖ Etikettierer m.; Auszeichner m. ‖ étiqueteur m.

labelling machine ‖ Banderoliermaschine f.; Etikettiermaschine f. ‖ machine f. à banderoler *ou* à étiqueter.

label maker ‖ Etikettendrucker m. ‖ étiqueteur m.; imprimeur m. d'étiquettes. / ~ making machine ‖ Etikettenherstellungsmaschine f. ‖ machine f. à faire les étiquettes. / ~ varnish ‖ Etikettenlack m. ‖ vernis m. pour étiquettes.

labor (A) *see* labour.

laboratory ‖ Laboratorium n. ‖ laboratoire m. / ~ (Drug store) ‖ Offizin f. ‖ officine f. / chemical ~ ‖ chemisches Laboratorium n. ‖ laboratoire m. (d'analyses) chimique(s). / chemical ~ to control the current manufacture ‖ chemisches Laboratorium n. zur Überwachung der laufenden Fertigung ‖ laboratoire m. d'analyses chimiques pour contrôler la fabrication courante. / chemical and physical ~ ‖ chemisch-physikalisches Laboratorium n. ‖ laboratoire m. chimique et physique. / chemical and physical ~ for scientific and technical research ‖ chemisch-physikalische Versuchsanstalt f. ‖ laboratoire m. de recherches chimiques-physiques. / ~ for research ‖ Versuchsanstalt f. ‖ laboratoire m. des recherches. / steel works ~ ‖ Stahlwerkslaboratorium n. ‖ laboratoire m. d'aciérie.

laboratory apparatus ‖ Laboratoriumsgerät n.; Laboratoriumsapparat m. ‖ appareil m. de laboratoire. / ~ for sugar works ‖ Laboratoriumsapparat m. für Zuckerfabriken ‖ appareil m. pour laboratoires de sucreries.

laboratory assistant ‖ Laborant m.; Laboratoriumsassistent m. ‖ aide f. de laboratoire. / ~ crusher ‖ Laboratoriumsmühle f. ‖ broyeur m. *ou* triturateur m. de laboratoire. / ~ enlarging ‖ Laboratoriumserweiterung f. ‖ élargissement m. *ou* agrandissement m. du laboratoire. / ~ equipment ‖ Laboratoriumseinrichtung f. ‖ installation f. *ou* équipement m. de laboratoires. / ~ filter press ‖ Laboratoriumsfilterpresse f. ‖ filtrepresse

m. pour laboratoire. / ~ furnace ‖ Laboratoriumsofen m. ‖ fourneau m. pour laboratoires. / ~ furnishing *see* ~ equipment. / ~ glass ‖ Laboratoriumsglas n. ‖ verre m. de laboratoire. / ~ interferometer ‖ Laboratoriumsinterferometer n. ‖ interféromètre m. de laboratoire. / ~ man ‖ Laboratoriumsdiener m.; Laboratoriumsgehilfe m. ‖ garçon m. de laboratoire. / ~ microscope ‖ Laboratoriumsmikroskop n. ‖ microscope m. de laboratoire. / ~ outfit *see* ~ equipment. / ~ yield ‖ Laboratoriumsausbeute f. ‖ rendement m. au laboratoire.

labour ‖ Arbeit f. ‖ ouvrage m.; travail m.; main f. d'œuvre. / ~ is abundant ‖ Arbeitskräfte pl. sind reichlich vorhanden ‖ la main d'œuvre est abondante. / native ~ ‖ von eingeborenen *oder* einheimischen Arbeitern verrichtete Arbeit f. ‖ main-d'œuvre f. indigène. / to regulate ~ and to do away with all auxiliary details which could be spared ‖ den Arbeitsgang unter Beseitigung aller entbehrlichen Hilfs- und Nebenarbeiten leiten ‖ organiser le travail et supprimer tous les travaux auxiliaires et supplémentaires. / ~ is scarce ‖ Arbeitskräfte fpl. sind knapp ‖ la main d'œuvre est rare.

labourage ‖ Arbeitslohn m. ‖ salaire m.

labour bureau ‖ Arbeitsamt n. ‖ bureau m. de travail. / ~ charges pl. ‖ Arbeitskosten pl. ‖ frais mpl. de main-d'œuvre. / ~ efficiency ‖ Arbeitsleistung f. ‖ rendement m. effectué.

labourer ‖ Arbeiter m.; Tagelöhner m.; Hilfsarbeiter m. ‖ ouvrier m.; compagnon m.; journalier m.; manœuvre m. / manual ~ ‖ Handarbeiter m. ‖ manœuvre m. / skilled ~ ‖ gelernter Arbeiter m. ‖ artisan m.; homme de métier m. / unskilled ~ ‖ ungelernter Arbeiter m. ‖ simple ouvrier m.; manœuvre m.; manouvrier m.

labourer turnover ‖ Arbeiterwechsel m. ‖ changement m. d'ouvriers.

labouring ‖ werktätig ‖ actif; travailleur.

labour market ‖ Arbeitsmarkt m. ‖ bourse f. du travail. / ~ problem ‖ Arbeiterfrage f. ‖ question f. ouvrière. / ~ saving machinery ‖ arbeitsparende Maschine f. ‖ machine f. d'un rendement plus économique. / ~ scarcity of ~ ‖ Mangel m. an Arbeitskräften ‖ rareté f. de la main-d'œuvre.

Labrador feldspar ‖ Labradorfeldspat m.; Labradorit m. ‖ labradorite f.; feldspath m. chatoyant. / ~ stone ‖ Labradorstein m. ‖ pierre f. de Labrador; labrador m.

labyrinth box ‖ Labyrinthbüchse f. ‖ boîte f. à labyrinthe.

lac *see also* lacquer *and* varnish ‖ Lack m. ‖ laque f. / virginal ~ ‖ Jungfernmilch f.; Venusmilch f. ‖ lait m. virginal. / white ~ ‖ Weißlack m. ‖ laque f. blanche.

lace, to ‖ abbinden; zuschnüren ‖ ficeler; lacer. / ~ together ‖ schnüren; zusammen-

schnüren ‖ brêler; serrer; lier ensemble / ~ up *see* to lace.

lace ‖ Spitze f. ‖ dentelle f. / ~ (Shoem) ‖ Schnürriemen m.; Schnürsenkel m.; lacet m. (à bottines). / ~ (Textile) ‖ Tresse f. ‖ tresse f.; galon m. / bone ~ ‖ Klöppelspitze f. ‖ dentelle f. au(x) fuseau(x) *ou* au carreau. / bone ~ maker ‖ Spitzenklöpplerin f. ‖ dentellière f. aux fuseaux. / broad silk ~ ‖ Seidenspitze f. ‖ laize f. de soie. / Brussels ~ ‖ Brüsseler Spitze f. ‖ point m. de Bruxelles. / carriage and coach trimmings and ~s pl. ‖ Wagenposamenten pl. ‖ passementerie f. pour voitures. / filet ~ ‖ Filetspitze f. ‖ filet m.; dentelle-filet m. / hand ~ ‖ Handspitze f. ‖ dentelle f. à la main. / hand-made ~ ‖ handgeklöppelte Spitze f. ‖ dentelle f. au fuseau à la main. / ~ made from leather ‖ Schnürriemen m. aus Leder ‖ lacet m. en cuir. / machine-made ~ ‖ Tüllspitze f. ‖ dentelle f. imitation *ou* au métier. / Mechlin point ~ ‖ Brabanter Spitze f. ‖ point m. de Malines. / needle point ~ ‖ nadelgearbeitete *oder* Brüsseler *oder* Mechelner Spitze f. ‖ point m. à l'aiguille. / pasting ~ ‖ Kappnaht f. ‖ galon m. rabattu. / pillow ~ *see* bone ~. / to put ~s pl. on ‖ Spitzen fpl. annähen ‖ entoiler les dentelles. / round ~ (Saddl) ‖ Nahtschnur f.; Rundschnur f. ‖ rond-corde f.; galon m. (à coudre). / seaming ~ ‖ Abschlußborte f. ‖ galon m. couture. / Venetian ~ ‖ venetianische Spitze f. ‖ point m. de Venise.

lace attachment ‖ Petineteinrichtung f. ‖ dispositif m. à jour. / ~ blind ‖ Spitzenvorhang m. ‖ store m. en dentelles. / ~ boots pl. ‖ Schnürstiefel mpl. ‖ bottines fpl. à lacets; brodequins mpl. / ~ card puncher ‖ Spitzenkartenschläger m. ‖ perceur m. de cartes à dentelles. / ~ cleaner ‖ Spitzenreiniger m. ‖ nettoyeur m. de dentelles. / ~ cutterout ‖ Spitzenausschneiderin f. ‖ découpeuse f. en dentelle. / ~ designer ‖ Spitzenpatronenzeichner m. ‖ metteur m. en cartes pour dentelles. / ~ finishing ‖ Spitzenappretur f. ‖ apprêt m. de dentelles. / ~ glove ‖ Filethandschuh m. ‖ gant m. de filet. / ~ machine ‖ Spitzenstuhl m.; Häkelmaschine f.; Petinetmaschine f. ‖ métier m. à dentelles. / ~ and trimming machine *see* ~ trimming machine.

lace maker ‖ Spitzenwirkerin f.; Spitzengrundiererin f. ‖ réséleuse f. de dentelles; dentellière f. / ~ (Weav) ‖ Posamentier m.; Bortenwirker m.; Nestler m. ‖ passementier m.

lace manufacture ‖ Spitzenfabrikation f. ‖ fabrication f. de dentelles.

lace paper ‖ Papierspitze f. ‖ papier m. dentelle. / ~ maker ‖ Papierspitzenarbeiterin f. ‖ dentellière f. en papier.

lace purler ‖ Gardineneinfasserin f. ‖ entrepreneuse f. de picotage; picoteuse f. / ~ tagger ‖ Senkelbeschläger m. ‖ ferreur m. de lacets. / ~ trimming machine ‖ Posa-

mentenmaschine f. ‖ métier m. de passementerie; machine f. à passementerie. / ~ warp fabric ‖ durchbrochene Kettenware f. ‖ tricot m. chaîne à jour. / ~ washer ‖ Spitzenwäscher m. ‖ blanchisseur m. de dentelles. / ~ washing ‖ Spitzenwäscherei f. ‖ blanchissage m. de dentelles.

lacework ‖ Posamentierarbeit f. ‖ passement m.; passementerie f.

lace worker ‖ Spitzenmacherin f. ‖ dentellière f.

lace working ‖ Bortenwirkerei f. ‖ passementerie f.

lacework machinery ‖ Posamentiermaschine f. ‖ machine f. de passementerie. / ~ tool ‖ Posamentierwerkzeug n. ‖ outil m. de passementerie.

lachrymator ‖ Tränengas n. ‖ lacrymogène m.

lack of current ‖ Stromlosigkeit f. ‖ manque m. de courant. / ~ of provisions ‖ Nahrungsmangel m. ‖ disette f. d'aliments / ~ of uniformity ‖ Ungleichmäßigkeit f. ‖ inégalité f.

lacker, to see to lacquer.

lacker see lacquer.

lacmus see also litmus ‖ Lackmus n. ‖ tournesol m.

lacquer, to ‖ lackieren ‖ laquer; vernir.

lacquer see also lac and varnish ‖ Lackfarbe f. ‖ vernis m. de couleur; laque f. colorée. / colourless protecting ~ ‖ farbloser Schutzlack m. ‖ vernis m. protecteur incolore. / to protect by a coating of colourless ~ ‖ mit farblosem Schutzlack m. überziehen ‖ recouvrir d'un vernis protecteur incolore. / dull ~ ‖ Mattlack m. ‖ vernis m. mat.

lacquered ‖ lackiert ‖ laqué; verni / ~ goods pl. ‖ Lackwaren fpl. ‖ objets mpl. en laque ou en vernis.

lacquerer ‖ Lackierer m. ‖ laqueur m.; vernisseur m.

lacquering machine for wire ‖ Drahtlackiermaschine f. ‖ machine f. à laquer les fils.

lactate ‖ Laktat n.; milchsaures Salz n. ‖ lactate m. / antimony ~ ‖ milchsaures Antimon n. ‖ lactate m. d'antimoine.

lactic acid ‖ Milchsäure f. ‖ acide m. lactique.

lactine see lactose.

lactometer ‖ Milchmesser m.; Milchwage f.; Laktometer n. ‖ lactomètre m.; galactomètre m.; pèse-lait m.

lactoscope see lactometer.

lactose ‖ Laktose f.; Milchzucker m. ‖ lactose f.; sucre m. de lait.

lac varnish ‖ Lack m.; Lackfirnis m. ‖ laque f.; vernis m.

ladder ‖ Leiter f. ‖ échelle f. / ~ (Weav) ‖ Fallmasche f.; Laufmasche f. ‖ maille f. coulée. / dray ~ ‖ Schrotleiter f. ‖ poulain m. / fire ~ ‖ Feuerwehrleiter f. ‖ échelle f. des pompiers. / motor aerial ~ ‖ Motordrehleiter f. ‖ échelle f. automobile mécanique. / motor extension ~ see motor aerial ~. / observation ~ (Astron) ‖ Beobachtungsleiter f. ‖ échelle f. d'observation.

ladder beam ‖ Leiterbaum m. ‖ arbre m. d'échelle. / ~ maker ‖ Leitermacher m. ‖ échelleur m.

ladder rope (Build) ‖ Lenkseil n.; Leitseil n.; Schwenkseil n.; Schwungseil n. ‖ hauban m.; écharpe f. / ~ (Shipb) ‖ Fallreep n. ‖ coupée f.; échelle f. de coupée.

ladder step ‖ Leitersprosse f. ‖ échelon m. / ~ way (Mine) ‖ Fahrschacht m. ‖ puits m. de descente; fosse f. aux échelles.

lade, to ~ (Mar) see also to load ‖ laden; Ladung einnehmen ‖ charger; prendre chargement.

ladies' bag ‖ Damentasche f. ‖ sac m. à main de dame ou à dames; aumônière f. / ~ bow ‖ Damentaschenbügel m. ‖ fermoir m. pour sacs à dames.

ladies' band ‖ see ~ sanitary towels pl. / ~ boot ‖ Damenstiefel m. ‖ soulier m. de femme; bottine f.; / ~ bootmaker ‖ Damenstiefelmacher m. ‖ bottier m. pour dames. / ~ boots pl. and shoes pl. ‖ Damenschuhwaren fpl. ‖ chaussures fpl. pour dames. / ~ cap maker ‖ Haubennäherin f. ‖ chapeleuse f.

ladies' cloak ‖ Damenmantel m. ‖ manteau m. de dame. / cloth for ~ ‖ Damenmantelstoff m. ‖ étoffe f. pour manteaux de dames.

ladies' clothes pl. ‖ Damenkleidung f. ‖ vêtement m. pour dames.

ladies' dress ‖ Damenkleid n. ‖ robe f. de dame. / ~es pl. to order ‖ Damenkleider npl. nach Maß ‖ vêtements mpl. sur mesure pour dames.

ladies' handbag ‖ Damenhandtasche f. ‖ sacoche f. ou sac m. à main ou réticule m. pour dames.

ladies' handkerchief ‖ Damentaschentuch n. ‖ mouchoir m. de dame. / embroidered ~ ‖ gesticktes Damentaschentuch n. ‖ mouchoir m. brodé de dame.

ladies' hat ‖ Damenhut m. ‖ chapeau m. de dames.

ladies' hose see ladies' stocking.

ladies' knitted jacket ‖ Damenstrickjacke f. ‖ jaquette f. tricotée pour dames. / ~ linen ‖ Damenwäsche f. ‖ lingerie f. de dames. / ~ mantle see ~ cloak. / ~ outfit ‖ Damenkonfektion f. ‖ confection f. pour dames. / ~ ready-made dresses pl. ‖ Damenkonfektion f. ‖ confection f. pour dames. / ~ sanitary towels pl. ‖ Damenbinde f. ‖ serviette-éponge f. de dames. / ~ shirt ‖ Damenhemd n. ‖ chemise f. de dames.

ladies' stocking ‖ Damenstrumpf m. ‖ bas m. de dame ou de femme. / machine for making ~s ‖ Frauenstrumpfmaschine f. ‖ machine f. à bas de femme.

ladies' tailor ‖ Damenschneider m. ‖ tailleur m. pour dames. / requisites pl. for ~ tailoring ‖ Damenschneidereibedarf m. ‖ accessoires mpl. pour couturières. / ~ underwear ‖ Damenunterzeug n. ‖ dessous m. pour dames. / ~ work basket ‖ Nähkorb m. ‖ panier m. à ouvrage ou à coudre.

lading, bill of ~ (Mar) ‖ Verladeschein m.; Seefrachtbrief m.; Konnossement n. ‖ certificat m. de chargement; connaissement m.

ladle, to ~ out (Found) ‖ ausschöpfen ‖ puiser.

ladle ‖ Schöpflöffel m.; Schöpfkelle f.; Schöpfgefäß n.; Schöpfpfanne f. ‖ puisoir m.; puisette f.; cuiller f. ou poche f. à puiser. / ~ (Found) ‖ Gießlöffel m.; Gießkelle f.; Gießpfanne f. ‖ puisoir m.; puisette f.; creuset m.; cuiller f. de coulée ou à fondre; poche f. de coulée ou de fonderie. / ~ (Glassm) ‖ Einsetzlöffel m. ‖ estraquelle f. / casting ~ see ~ (Found). / foundry ~ see ~ (Found). / glass ~ ‖ Glaslöffel m. ‖ cuiller f. ou poche f. à verre

en fusion. / ~ with shank welded on ‖ Gießlöffel m. mit angeschweißtem Stielansatz ‖ poche f. avec douille raccordée par soudage. / slag ~ ‖ Schlackenpfanne f.; Schlackenkümpel m. ‖ poche f. à laitier; pot m. à scorie.

ladle barrel ‖ Schöpffäßchen n. ‖ puchoir m. / ~ brick ‖ Pfannenstein m. ‖ brique f. de poche de coulée. / ~ car see ~ carriage. / ~ carriage ‖ Gieß(pfannen)wagen m. ‖ chariot m. porte-poche de coulée; transporteur m. de la poche de coulée. / ~ cask ‖ Schöpffaß n. ‖ gomé m.; poche f. / ~ cleaner ‖ Gießlöffelreiniger m. ‖ nettoyeur m. de cuiller ou de poche de coulée. / ~ handling crane ‖ Gießkran m. ‖ grue f. pour la coulée.

ladleman ‖ Pfannenführer m. ‖ pocheur m.

ladle plug ‖ Gießpfannenpropfen m. ‖ bouchon m. de la poche de coulée. / ~ pourer ‖ Eingießer m. ‖ verseur m. de poches.

ladler ‖ Gießer m. ‖ pocheur m.

lady shorthand typist ‖ Stenotypistin f. ‖ sténo-dactylographe f.

lævulose ‖ Schleimzucker m.; Lävulose f. ‖ lévulose m.

lag, to ‖ sich verzögern ‖ être en retard; retarder. / ~ in phase ‖ in der Phase f. nacheilen ‖ être en retard de phase.

lag (Electr) ‖ Phasenverschiebung f. ‖ décalage m. / elastic ~ ‖ elastische Nachwirkung f. ‖ effet m. résiduel; résidu m. ou fatigue f. élastique. / ~ and lead of phases ‖ Phasenverschiebung f. ‖ décalage m. de phase.

lag, angle of ~ ‖ Verzögerungswinkel m. ‖ angle m. de retard. / correction for ~ ‖ Nachwirkungsberichtigung f. ‖ correction f. d'effet résiduel ou de fatigue.

lagging ‖ hemmend ‖ enrayant; refreinant; empêchant. / ~ current ‖ in der Phase nacheilender Strom m. ‖ courant m. décalé en arrière.

laggings pl. (Carp) ‖ Schalbretter npl.; Schallatten npl. ‖ couchis m.; madriers mpl. de boisage.

lagoon ‖ Haff n.; Lagune f. ‖ étang m.

lag phenomenon ‖ Nachwirkungserscheinung f. ‖ phénomène m. d'effet résiduel ou de fatigue.

laid paper ‖ geripptes Papier n. ‖ papier m. vergé.

lake ‖ See m.; Landsee m. ‖ lac m. / ~ (Paint) see lacquer. / ~ ore ‖ Limonit m.; Raseneisenstein m.; See-Erz n. ‖ fer m. des lacs; limonite f. / ~ reservoir ‖ See-reservoir n. ‖ réservoir m. naturel. / ~ sand ‖ Seesand m. ‖ sable m. de mer.

lama weaver ‖ Lamaweber m. ‖ tisseur m. de lama. / ~ yarn ‖ Lamagarn n. ‖ laine f. de lama.

lambskin ‖ Lammfell n. ‖ peau f. d'agneau.

lamb's wool ‖ Lammwolle f. ‖ laine f. agneline.

lamella ‖ Lamelle f. ‖ lamelle f.

lametta ‖ Lametta f. ‖ lamelles fpl.

laminable ‖ streckbar; walzbar ‖ extensible; laminable.

laminate, to ‖ lamellieren ‖ feuilleter. / ~ (Roll mill) ‖ walzen ‖ laminer.

laminated ‖ aus Platten fpl. geschichtet ‖ feuilleté. / ~ radiator arrangement for gradual heating ‖ Lamellenheizkörper m. für Stufenbeheizung ‖ calorifère m. à ailettes à chauffage réglable. / ~ ring (Aero) ‖ furnierter Holzring m. ‖ cercle m. en bois plaqué ou en bois de placage.

laminated spring ‖ Blattfeder f. ‖ ressort m. à lames. / ~ bending machine ‖ Blattfederbiegemaschine f. ‖ machine f. à cintrer les ressorts à lames. / ~ hammer ‖ Blattfederhammer m. ‖ marteau m. à ressort à lames.

lamination ‖ Lamellierung f. ‖ feuilletage m. / ~ (Roll mill) ‖ Strecken n. ‖ laminage m. / oblique ~ see transversal ~. / transversal ~ (Geol) ‖ falsche oder sekundäre oder transversale Schieferung f. ‖ clivage m.

lamp ‖ Lampe f.; Laterne f. ‖ lampe f. / ~ of acetate of amyle ‖ Amylazetatlampe f. ‖ lampe f. à l'acétate d'amyle. / alabaster ~ ‖ Alabasterlampe f. ‖ lampe f. en albâtre. / arc ~ ‖ Bogenlampe f. ‖ lampe f. à arc. / artistic ~ ‖ künstlerisch ausgeführte Lampe f. ‖ lampe f. artistique. / ~ of brass ‖ Lampe f. aus Bronze ‖ lampe f. en bronze. / carbon filament ~ ‖ Kohlefadenlampe f. ‖ lampe f. à filament de charbon. / clearing ~ (Tel) ‖ Schlußzeichenlampe f. ‖ lampe f. de clôture. / ~ of copper ‖ Lampe f. aus Kupfer ‖ lampe f. en cuivre. / dark room ~ ‖ Dunkelkammerlampe f. ‖ lampe f. à chambre noire. / gas filled ~ ‖ Glühlampe f. mit Gasfüllung ‖ lampe f. à atmosphère gazeuse. / glassblower's ~ ‖ Glasbläserlampe f.; Schmelzlampe f. ‖ lampe f. d'émailleur. / hand ~ ‖ Handlampe f. ‖ lampe f. portative ou baladeuse. / hanging ~ ‖ Hängelampe f. ‖ lampe f. suspendue ou à suspension; suspension f. / head ~ (Loc) ‖ Kopflaterne f.; Lokomotivlaterne f. ‖ falot m. ou lanterne f. de tête de locomotive. / ~ for illumination by transmitted light ‖ Durchleuchtungslampe f. ‖ lampe f. pour l'éclairage interne pénétrant. / incandescent ~ ‖ Glühlampe f. ‖ lampe f. à incandescence. / inspection ~ ‖ Ableuchtlampe f. ‖ lampe f. de visite. / instrument board ~ ‖ Schalttafellaterne f. ‖ lampe f. du tablier des instruments. / magnifier ~ ‖ Lupenlampe f. ‖ lampe f. à loupe. / metallic filament ~ ‖ Metallfadenlampe f. ‖ lampe f. à filament métallique. / miner's ~ ‖ Grubenlampe f. ‖ lampe f. de mineur. / oil ~ ‖ Öllampe f.; Petroleumlampe f. ‖ lanterne f. à huile; lampe f. à pétrole. / one-watt ~ ‖ einwattige Lampe f. ‖ lampe f. d'un watt. / ~ for operating position ‖ Apparatetischlampe f. ‖ lampe f. pour la place d'opération. / ~ for operations on the eye ‖ augenärztliche Operationslampe f. ‖ lampe f. pour les opérations ophtalmologiques. / pedestal ~ ‖ Ständerlampe f. ‖ lampe f. sur piédestal. / petroleum ~ see oil ~. / point light ~ ‖ Punktlichtlampe f. ‖ lampe f. ponctuelle. / portable ~ ‖ Handlampe f. ‖ lampe f. portative. / radiation ~ ‖ Bestrahlungslampe f. ‖ lampe f. radiateur ou pour bains de lumière. / safety ~ ‖ Sicherheitslampe f. ‖ lampe f. de sûreté. / side ~ ‖ Seitenlampe f. ‖ lampe f. de côté. / slit ~ ‖ Spaltlampe f. ‖ lampe f. à fente. / soldering ~ ‖ Lötlampe f.; Lötapparat m. ‖ appareil m. ou lampe f. à souder. / spirit ~ ‖ Spirituslampe f. ‖ lampe f. à alcool. / street ~ ‖ Straßenlaterne f. ‖ réverbère f. / suspended ~ see hanging ~. / swing(ing) ~ see hanging ~. / tail ~ ‖ Schlußlaterne f. ‖ lanterne f. arrière ou de queue. / test ~ ‖ Prüflampe f. ‖ lampe f. témoin; lampe f. d'essai. /

three filament ~ ‖ Dreifadenlampe f. ‖ lampe f. à trois filaments. / tungsten ~ ‖ Wolframlampe f. ‖ lampe f. au tungstène. / tuning ~ ‖ Abstimmlampe f. ‖ lampe f. de syntonisation. / ~ for wireless telegraphy ‖ Lampe f. oder Röhre f. für drahtlose Telegrafie; Radioröhre f. ‖ lampe f. pour télégraphie sans fil.

lampblack ‖ Kienruß m.; Ruß m.; Lampenruß m. ‖ noir m. de fumée. / fabrication of ~ ‖ Rußbrennerei f. ‖ fabrication f. de noir de fumée.

lamp bracket ‖ Laternenstütze f.; Lampenhalter m. ‖ porte-lanterne m.

lamp-burner case ‖ Lampenbrennerkapsel f. ‖ capsule f. de lampe.

lamp chimney ‖ Lampenzylinder m. ‖ verre m. de lampe. / ~ circuit ‖ Lampenstromkreis m. ‖ circuit m. des lampes. / ~ cleaner ‖ Laternenputzer m.; Laternenwäscher m. ‖ nettoyeur m. de lanternes. / ~ cover ‖ Lampenüberzug m. ‖ enveloppe m. de lanterne. / ~ detector ‖ Lampenprüfer m.; Prüflampe f. ‖ lampe f. témoin ou d'essai. / ~ examiner (Mine) ‖ Lampenmeister m. ‖ visiteur m. de lampes. / ~ fringe ‖ Lampenfranse f. ‖ frange f. de lampe. / ~ globe ‖ Lampenglocke f. ‖ ballon m. pour lampes.

lamp-holder (Electr) ‖ Lampenfassung f.; Fassung f. ‖ douille f. (de lampes). / ceiling ~ ‖ Deckenfassung f. ‖ douille f. de plafond. / watertight ~ ‖ wasserdichte Fassung f. ‖ douille f. étanche.

lamp hood ‖ Laternenkappe f. ‖ chapeau m. de lanterne.

lamplighter ‖ Laternenanzünder m.; Lampenanzünder m. ‖ allumeur m. de réverbères; allumeur m. / ~ (Apparatus) ‖ Lampenanzünder m. ‖ portefeu m.

lamp maker ‖ Lampenmacher m. ‖ fabricant m. de lampes; lampiste-ferblantier m. / ~-post ‖ Lampengestell n. ‖ lampadaire m. / ~ reflector ‖ Laternenspiegel m. ‖ réflecteur m. de lanterne.

lamprey ‖ Neunauge n. ‖ lamproie f. / ~ smoker ‖ Neunaugenröster m. ‖ fumeur m. de lamproies.

lamp shade ‖ Lampenschirm m. ‖ abatjour m. / ~ holder ‖ Schalenhalter m.; Lampenschirmhalter m. ‖ carcasse f. ou support m. d'abat-jour.

lamp shield ‖ Lampenkappe f. ‖ calotte f. de lampe. / ~ socket see ~ holder. / ~ stand ‖ Lampenfuß m. ‖ pied m. de lampe. / ~ strip ‖ Lampenstreifen m. ‖ réglette f. de lampes. / ~ switchboard (Tel) ‖ Glühlampenschrank m. ‖ tableau m. à signaux lumineux. / ~ wick ‖ Lampendocht m. ‖ mèche f. de lampe.

lanarkite (Miner) ‖ Lanarkit m. ‖ lanarkite f.

Lancashire boiler ‖ Zweiflammrohrkessel m. ‖ chaudière f. de Lancashire ou à deux tubes foyers.

lance ‖ Lanze f. ‖ lance f. / free ~ (Workman) ‖ freizügiger Arbeiter m. ‖ ouvrier m. ayant la liberté de résidence.

lancet ‖ Lanzette f. ‖ lancette f.; couteau m. lancéolaire. / curved ~ ‖ gebogene Lanzette f. ‖ couteau m. lancéolaire coudé.

lancet knife ‖ Lanzettmesser n. ‖ couteau m. lancéolé ou lancéolaire.

land, to ~ (Aero) ‖ landen ‖ atterrir. / ~ (Nav) ‖ landen ‖ aborder; accoster; atterrir. / ~ (To unload) ‖ löschen; ausschiffen; ausladen ‖ décharger; débarquer.

land ‖ Land n. ‖ terre f. / ~ (Landed property) ‖ Grundbesitz m.; Grundeigentum n.; Grundstück n.; Gut n.; Ländereien fpl. ‖ propriété f. foncière ou domaine ou territoriale; terre f.; bien m. / ~ (Soil) ‖ Grund m.; Boden m.; Gelände n.; Land n. ‖ terre f.; sol m.; champ m.; terrain m. / ~ (Mar) ‖ Küste f.; Ufer n. ‖ côte f.; littoral m.; rivage m.; terre f. / the extent of available ~ would not suffice ‖ die Größe des verfügbaren Geländes erschien nicht ausreichend ‖ l'étendue f. du terrain disponible ne paraissait pas suffisante. / by ~ and by water ‖ zu Wasser n. und zu Lande n. ‖ par mer f. et par terre f. / ~ liable to floods ‖ Überschwemmungsgebiet n. ‖ zône f. ou champ m. d'inondation. / high ~ (Mar) ‖ hohe Küste f.; hohes Land n. ‖ terre f. grosse ou haute. / irrigated ~ ‖ bewässertes Land n. ‖ champ m. arrosé.

land aeroplane ‖ Landflugzeug n. ‖ avion m. terrestre.

landau (Vehicle) ‖ Landauer m. ‖ landau m.

landaulet (Vehicle) ‖ Landaulet n. ‖ landaulet m.

land breeze ‖ Landwind m.; Landbrise f. ‖ vent m. ou brise f. de terre.

land carriage ‖ Frachtfuhre f.; Frachtwagen m. ‖ chariot m. ou voiture f. de roulage. / transportation by ~ ‖ Frachtfuhrwesen n. ‖ roulage m.

land fog ‖ Landnebel m. ‖ brouillard m. de campagne. / ~ holder ‖ Pächter m. ‖ amodiataire m.; amodiateur m.; fermier m.

landing (Aero) ‖ Landung f. ‖ atterrissage m. / ~ (Mar) ‖ Landen n.; Landung f. ‖ abord m.; atterrissage m. / ~ (Unloading) ‖ Löschen n.; Ausschiffung f.; Landung f. ‖ débarquement m. / ~ (Place) see landing place. / forced ~ (Aero) ‖ Notlandung f. ‖ atterrissage m. forcé. / ~ of a shaft (Mine) ‖ Schachthängebank f. ‖ pas m. de bure. / ~ of stairs ‖ Treppenabsatz m.; Podest m.; ‖ palier m.

landing base (Stairs) see ~ slab. / ~ charges pl. (Mar) ‖ Abladegebühr f.; Löschgebühr f. ‖ frais mpl. de déchargement. / ~ direction light (Aero) ‖ Landebahnfeuer n. ‖ feu m. d'atterrissage. / ~ field indicator ‖ Landungsfühler m. ‖ indicateur m. du terrain d'atterrissage. / ~ floodlight ‖ Landebahnleuchte f. ‖ projecteur m. ou batterie f. d'atterrissage.

landing gear ‖ Fahrgestell n. ‖ atterrisseur m. / retractable ~ ‖ einziehbares Fahrgestell n. ‖ atterrisseur m. escamotable ou orientable.

landing light (Aero) ‖ Landungslicht n. ‖ rampe f. ou phare m. d'atterrissage. / ~ place (Mar) ‖ Ausladeort m.; Landungsplatz m.; Löschplatz m. ‖ débarcadère f.; lieu m. de débarquement. / ~ projector (Aero) ‖ Landungsscheinwerfer m.; Landescheinwerfer m. ‖ projecteur m. ou phare m. d'atterrissage. / ~ shock ‖ Landungsstoß m. ‖ choc m. de ou à l'atterrissage. / ~ slab (Stairs) ‖ Treppenabsatz m.; Podestplatte f. ‖ plaque f. de palier. / ~ speed (Aero) ‖ Landegeschwindigkeit f. ‖ vitesse f. d'atterrissage. / ~ stage see ~ place. / ~ station (Mar) ‖ Landungsstation f. ‖ poste f. de débarquement. / ~ wheel axle (Aero) ‖ Laufradachse f. ‖ essieu m. de la roue d'atterrissage.

landmark (Nav) ‖ Bake f. ‖ balise f.; amer m. / ~ (Surv) ‖ Grenzstein m.; Markstein m. ‖ limite f.; borne f.

land measuring instrument ‖ Feldmeßgerät n. ‖ instrument m. d'arpentage. / ~ mortgage bank ‖ Bodenkreditbank f. ‖ crédit-foncier m. / ~-owner ‖ Landbesitzer m.; Landeigentümer m. ‖ propriétaire m. foncier. / ~-plane ‖ Landflugzeug n. ‖ avion m. terrestre. / ~ roller ‖ Ackerwalze f. ‖ émotteuse f.

landscape gardener ‖ Landschaftsgärtner m.; Gartenarchitekt m. ‖ jardinier m. ou architecte m. paysagiste.

landslip (Geol) ‖ Bergrutsch m.; Erdrutsch m. ‖ glissement m. ou chute f. de montagne.

land snail ‖ Landschnecke f. ‖ escargot m. (de terre); limaçon m. / ~ surveying ‖ Landmessung f.; Landaufnahme f. ‖ arpentage m.; levé m. du plan d'un terrain. / ~ surveyor ‖ Landmesser m.; Geodät m.; Geometer m. ‖ arpenteur m. / ~ tax ‖ Grundsteuer f. ‖ contribution f. foncière; impôt m. foncier. / ~ telescope ‖ terrestrisches Fernrohr n. ‖ lunette f. terrestre; longue-vue f. / ~-tenant ‖ Pächter m. ‖ amodiataire m.; amodiateur m.; fermier m. / ~ tie (Build) ‖ Erdanker m. ‖ tirant m. / ~ transport ‖ Landbeförderung f. ‖ transport m. par terre. / ~ turn (Meteor) ‖ Landbrise f. ‖ brose f. de terre.

land-type boiler ‖ Kessel m. für ortsfesten Betrieb ‖ chaudière f. fixe.

land wind ‖ Landwind m. ‖ vent m. de terre.

lane ‖ kleine Straße f.; Gasse f. ‖ ruelle f. / ~ (Forest) ‖ Durchhau m.; Schneise f. ‖ laie f.

language, national ~ ‖ Landessprache f. ‖ langue f. nationale. / open ~ ‖ offene Sprache f. ‖ langage m. clair.

lanoline ‖ Lanolin n. ‖ lanoline f. / ~ soap ‖ Lanolinseife f. ‖ savon m. à la lanoline.

lantern ‖ Laterne f. ‖ lanterne f. / ~ (Build) ‖ Dachaufsatz m.; Laterne f.; Turmaufsatz m. ‖ lanterne f.; tourelle f. ouverte; tour f. à cheval. / ~ of bicycle ‖ Fahrradlaterne f.; Fahrradlampe f. ‖ lanterne f. de bicyclette. / carriage ~ ‖ Wagenlaterne f. ‖ lanterne f. de voitures. / dark-room ~ ‖ Dunkelkammerlampe f. ‖ lampe f. de chambre noire. / ~ for illuminations ‖ Illuminationslaterne f. ‖ lanterne f. pour illuminations. / optical ~ ‖ Projektionslaterne f. ‖ lanterne f. de projection. / ship's ~ ‖ Schiffslaterne f.; Laterne f. ‖ fanal m. / stable ~ ‖ Stallaterne f. ‖ lanterne f. d'écurie. / storm ~ ‖ Sturmlaterne f. ‖ lanterne-tempête f. / street ~ ‖ Straßenlaterne f.; Laterne f. ‖ réverbère m. / street ~ for gas lighting ‖ Straßenlaterne f. für Gasbeleuchtung ‖ réverbère m. pour éclairage au gaz. / number and sign for street ~s pl. ‖ Straßenlaternennummer f. und -zeichen n. ‖ numéro m. et signe m. pour réverbères.

lantern bracket ‖ Laternenhalter m. ‖ porte-lanterne m. / ~ of cycle ‖ Fahrradlaternenhalter m. ‖ porte-lanterne m. pour bicyclette.

lantern gear (Mach) ‖ Triebstockkranz m. ‖ couronne f. de lanterne à fuseaux; lanterne f. ou roue f. à fuseaux.

lanthanite (Miner) ‖ Lanthanit m. ‖ lanthanite f.

lanthanium see lanthanum.

lanthanum (Chem) ‖ Lanthan n. ‖ lanthane m. / ~ salts pl. ‖ Lanthansalze npl. ‖ sels mpl. de lanthane.

lanthorn see lantern.

lanyard (Mar) ‖ Bindsel n.; Taljereep n.; Nähtau n. ‖ aiguillette f.; ride f.

lap, to ~ over ‖ übereinandergreifen ‖ chevaucher; recouvrir; assembler à recouvrement.

lap (Bookb) ‖ Falz m. ‖ pli m.; onglet m. / ~ (Spinn) ‖ Pack n.; Wickel m. ‖ nappe f.

lapidary ‖ Juwelier m.; Steinschneider m. ‖ joaillier m.; orfèvre m.; lapidaire m. / ~ of false and bad stones ‖ Steinschneider m. für unechte Steine ‖ lapidaire m. en faux. / ~ of precious stones ‖ Edelsteinschneider m. ‖ lapidaire m. en fin.

lapilli pl. ‖ Lapilli pl. ‖ lapilli npl.

lapis infernalis ‖ salpetersaures Silber n.; Höllenstein m. ‖ nitrate m. d'argent; pierre f. infernale.

lapis lazuli ‖ Lapislazuli m.; Lasurstein m. ‖ lapis-lazuli m.; pierre f. d'azur; lazulite m.

lapis-style ‖ Lapisdruck m. ‖ genre m. lapis.

lap joint ‖ Überlappungsverbindung f. ‖ joint m. à recouvrement.

lapp ‖ Fischband n.; Einsetzband n. ‖ fiche f. à vase.

lapped seam welding ‖ überlappte Nahtschweißung f. ‖ soudage m. de joints à recouvrement.

lapping engine (Textile) ‖ Bandleitungsmaschine f. ‖ réunisseuse f.

lapping-over ‖ Überschießen n.; Überlappen n. ‖ recouvrement m.; chevauchure f.

lap-welded ‖ überlappt geschweißt ‖ soudé à recouvrement. / ~ part ‖ überlappt geschweißtes Stück n. ‖ pièce f. soudée par recouvrement. / ~ tube ‖ überlappt geschweißtes Rohr n. ‖ tube m. soudé à recouvrement.

lap welding ‖ Überlappschweißung f. ‖ soudure f. à recouvrement.

larch ‖ Lärche f. ‖ mélèze m. / ~ wood ‖ Lärchenholz n. ‖ bois m. de mélèze.

lard ‖ Schmalz n. ‖ graisse f.; saindoux m. / artificial ~ ‖ Kunstschmalz n. ‖ saindoux m. artificiel. / natural ~ ‖ natürliches Schweineschmalz n. ‖ saindoux m. naturel.

lardery ‖ Fleischkammer f.; Speisekammer f. ‖ charnier m.

larding pin ‖ Spicknadel f. ‖ aigu lle f. pour coudre la volaille; aiguille f. à larder.

lard maker ‖ Speckräucherer m. ‖ saleur m. de lard. / ~ oil ‖ Specköl n. ‖ oléine f.

lardy ‖ speckig ‖ lardeux.

large ‖ groß; breit; geräumig ‖ grand. / ~ (Mar) ‖ raum ‖ largue. / ~ (Mine) ‖ mächtig ‖ épais; large. / ~ bomber ‖ Großbombenflugzeug n. ‖ gros avion m. de bombardement. / ~ coal ‖ Stückkohle f. ‖ gaillettes fpl. / ~ compressor ‖ Großkompressor m. ‖ compresseur m. de fortes dimensions.

large-berried coffee ‖ großbeeriger Kaffee m. ‖ café m. à gros grains.

large hammer (Forg) ‖ Poßhammer m. ‖ gros marteau m. de devant; pic m.

large kitchen furniture ‖ Großkücheneinrichtung f. ‖ installation f. de grande cuisine.

large locking angle ‖ große Wendefähigkeit f. ‖ grand angle m. de braquage.

large piece ‖ großes Stück n. ‖ gros morceau m.; grande pièce f. / in ~ pieces ‖ großstückig ‖ en gros morceaux mpl.

large power station ‖ Großkraftwerk n. ‖ usine f. d'énergie de forte puissance.

large-scale production ‖ Reihengroßfertigung f. ‖ fabrication f. en grandes séries.

large-surface plate ‖ Großoberflächenplatte f. ‖ plaque f. à grande surface.

larmier (Arch) ‖ Kranzleiste f. ‖ larmier m.

larva ‖ Larve f. ‖ larve. / to interrupt artificially the development of the ~ by cold storage (Silk) ‖ den Entwicklungsgang m. der Larve durch Kälteeinwirkung künstlich unterbrechen ‖ suspendre le développement de la larve par l'emploi du froid.

laryngeal mirror ‖ Kehlkopfspiegel m. ‖ miroir m. laryngien.

larynx ‖ Kehlkopf m. ‖ larynx m.

lascar (Mar) ‖ Laskar m.; indischer Matrose m. ‖ lascar m.

lash, to ‖ schnüren; zusammenschnüren ‖ brêler; serrer; ficeler. / ~ (Mar) ‖ seefest zurren ‖ amarrer à la mer.

lash ‖ Lasche f. ‖ entaille f. / ~ (Forest) ‖ Lasche f. ‖ entamure f.

lashing (Mar) ‖ Laschung f. ‖ amarrage m.; saisine f.; ligature f.; liure f. / ~ (Wind mill) ‖ Schwungleine f. ‖ hauban m. / ~ of a port ‖ Pfortenzurrung f. ‖ amarrage m. d'un sabord.

lasso ring ‖ Lassoring m. ‖ anneau m. de lasso.

last resource ‖ Notbehelf m. ‖ pis-aller m.; expédient m.

last runnings pl. (Brew) ‖ Glattwasser n. ‖ dernier bouillon m.; dernière trempe f. de lavage.

last (Ship) ‖ Last f. ‖ last m.; laste m. / ~ (Shoem) ‖ Leisten m. ‖ embauchoir m.

lasting ‖ nachhaltig; dauernd ‖ durable; efficace; persistant. / ~ dye ‖ haltbare oder echte Farbe f. ‖ colorant m. solide ou durable.

lasting (Weav) ‖ Lasting m.; Prunell m. ‖ prunelle f.; lasting m.

lastingness ‖ Dauerhaftigkeit f. ‖ durabilité f.; solidité f.; stabilité f.

last ironer ‖ Schuhleistenbeschläger m. ‖ ferreur m. de formes ou d'embauchoirs.

last line (Print) ‖ Ausgangszeile f. ‖ bout m. de ligne.

last maker ‖ Leistenschneider m. ‖ formier m. (d'embauchoirs).

latch ‖ Klinke f.; Drücker m.; Türdrücker m. ‖ loquet m.; clenche f. / falling ~ ‖ Fallklinke f.; Falle f. ‖ loqueteau m.

latch lever ‖ Sperrklinkenhebel m. ‖ manette f. du cliquet d'arrêt. / ~ member ‖ Verriegelungsglied n. ‖ pièce f. de verrouillage. / ~ needle machine ‖ Zungennadelmaschine f. ‖ machine f. à aiguilles à languettes.

late, at the ~st ‖ spätestens ‖ au plus tard. / to be too ~ ‖ sich verspäten ‖ s'attarder; être en retard.

lateen sail (Mar) ‖ Lateinsegel n. ‖ voile f. latine. / ~ yard (Mar) ‖ Lateinsegelrahe f. ‖ antenne f.

latent ‖ latent; gebunden ‖ latent.

lateral ‖ seitlich ‖ latéral. / ~ armour plate ‖ Seitenpanzerung f. ‖ blindage m. latéral. / ~ axis ‖ Querachse f. ‖ axe m. transversal. / ~ bearing surface ‖ seitliche Führungsfläche f. ‖ glissière f. latérale.

/ ~ drive ‖ seitlicher Antrieb m ‖ commande f. latérale. / ~ edge ‖ Seitenkante f.; Randkante f. ‖ arête f. latérale. / ~ face ‖ Randfläche f. ‖ face f. latérale. / ~ guide ‖ Seitenführung f. ‖ guidage m. latéral. / ~ plate ‖ Seitenplatte f. ‖ paroi f. latérale. / ~ play ‖ Beweglichkeit f. in axialer Richtung ‖ mobilité f. axiale; jeu m. latéral. / ~ position of the aeroplane ‖ Querlage f. des Flugzeuges ‖ position f. transversale de l'avion. / ~ pressure ‖ Seitendruck m. ‖ poussée f. latérale. / ~ sense ‖ Breitenrichtung. f. ‖ sens m. de la largeur. / ~ stability ‖ Querstabilität f. ‖ stabilité f. latérale. / ~ switch ‖ Seitenschalter m. ‖ commutateur m. latéral. / ~ trim ‖ Seitenlastigkeit f. ‖ centrage m. latéral; assiette f. latérale. / ~ wall of the brick-work in a gallery ‖ Scheibenmauer f. einer Streckenmauerung ‖ pied-droit m. d'un muraillement de galerie.

laterite ‖ Laterit m. ‖ latérite f.

latest pattern ‖ neueste Form f. ‖ dernière façon *ou* forme f.

lath, to ‖ belatten; mit Latten fpl. beschalen ‖ latter.

lath ‖ Latte f.; Leiste f. ‖ latte f. / ~ (Mine) ‖ Getriebepfahl m.; Pfahl m. ‖ palplanche f. / broad ~ (of schist) ‖ Schieferplatte f. ‖ latte f. volice *ou* de sciage.

lath cutter (Tiler) ‖ Schieferdeckerbeil n. ‖ hachotte f.

lathe (Mach tool) *see also* turning lathe ‖ Drehbank f. ‖ tour m. / ~ (Weav) ‖ Lade f.; Schlag m. ‖ battant m.; chasse f.; échasse f. / articles pl. made on ~s ‖ Drehereiwaren fpl. ‖ pièces fpl. faites au tour; articles mpl. tournés. / ~ automatic ~ ‖ Automat m. ‖ tour m. automatique *ou* à façonner automatique. / backing-off ~ ‖ Hinterdrehbank f. ‖ tour m. à façonner à profil invariable; tour m. à dépouiller *ou* à détalonner. / back knife gauge ~ ‖ Schablonendrehbank f. mit Hintermesser ‖ tour m. à gabarit avec couteau arrière. / to beat the ~ (Weav) ‖ die Lade f. anschlagen ‖ frapper le battant. / boring and reaming ~ ‖ Bohrdrehbank f. ‖ tour m. à percer et à aléser. / bulging ~ ‖ Drückbank f. ‖ tour m. à repousser les métaux. / centre ~ ‖ Spitzendrehbank f. ‖ tour m. à pointes. / centreless ~ ‖ spitzenlose Drehbank f. für Stangenmaterial ‖ tour m. sans pointe pour le travail des barres; tour m. sans pointe pour le décolletage. / copying ~ ‖ Kopierdrehbank f. ‖ tour m. à copier. / core-turning ~ ‖ Kerndrehbank f. ‖ tour m. à noyauter. / ~ with draw-in attachment ‖ Zangenspanndrehbank f. ‖ tour m. avec pinces de serrage. / <u>facing ~</u> ‖ Plandrehbank f. ‖ tour m. en l'air. / ~ for facing the axle ends ‖ Achsspiegeldrehbank f. ‖ tour m. à dresser les faces verticales aux extrémités des essieux. / ~ with hinge joints (Weav) ‖ Scharnierlade f. ‖ battant m. brisé. / ~ for hollow rods ‖ Gestängerohrdrehbank f. ‖ tour m. pour l'usinage des tiges de sondes tubulaires *ou* pour usinage des tubes de sondage. / hollow spindle ~ ‖ Hohlspindeldrehbank f. ‖ tour m. à arbre creux. / ~ with horizontal faceplate ‖ Karusselldrehbank f. ‖ tour m. vertical *ou* à plateau horizontal. / mechanic's ~ ‖ Mechanikerdrehbank f. ‖ tour m. pour mécaniciens. / multi-cut ~ ‖ Vielstahldrehbank f. ‖ tour m. à outils

multiples. / oval ~ ‖ Ovaldrehbank f. ‖ tour m. à tourner oval. / pipe slicing ~ ‖ Rohrabstechbank f. ‖ banc m. à tronçonner les tubes. / propeller ~ ‖ Propellerdrehbank f. ‖ tour m. à tourner les hélices. / railway wheel ~ ‖ Radsatzdrehbank f. ‖ tour m. pour essieux montés. / relieving ~ ‖ Hinterdrehbank f. ‖ tour m. à détalonner. / roll ~ ‖ Walzendrehbank f. ‖ tour m. à travailler des cylindres de laminoir. / roller ~ ‖ Trommeldrehbank f. ‖ tour m. à tambours. / slicing ~ ‖ Abstechbank f. ‖ tour m. à tronçonner *ou* à décolleter. / sliding and surfacing ~ ‖ Zugspindeldrehbank f. ‖ tour à charioter. / stud ~ ‖ Stehbolzendrehbank f. ‖ tour m. pour entretoises. / tube ~ ‖ Rohrdrehbank f. ‖ tour m. pour tubes. / turning ~ for rod jobs ‖ Drehbank f. für Stangenarbeiten ‖ tour m. pour travaux dans la barre; tour m. à décolleter. / turret ~ ‖ Revolverdrehbank f. ‖ tour m. revolver. / watchmaker's ~ ‖ Drehstuhl m. ‖ tour m. d'horloger. / wheel set ~ (Railw) ‖ Radsatzdrehbank f. ‖ tour m. pour trains de roues. / wheel tyre ~ ‖ Radreifendrehbank f. ‖ tour m. pour bandages de roues. / wood turning ~ ‖ Holzdrehbank f. ‖ tour m. à bois.

lathe chuck ‖ Drehbankfutter n.; Spannfutter n. ‖ chuck m.; mandrin m. de serrage. / ~ hand ‖ Dreher m.; Drehbankarbeiter m. ‖ tourneur m. / ~ testing tool machine ‖ Meßsupport m. ‖ support m. d'essais.

lathe tool ‖ Drehmeißel m.; Drehstahl m. ‖ outil m. à tourner; acier m. de tour. / ~ grinder ‖ Drehstahlschleifmaschine f. ‖ machine f. à affûter les outils de tours. / ~ and planer tool grinder ‖ Dreh- und Hobelstahlschleifmaschine f. ‖ machine f. à meuler et affûter les outils de tours et de raboteuses. / ~ holder ‖ Drehstahlhalter m. ‖ porte-outil m. de tour.

lathing ‖ Belattung f. ‖ lattis m.; lattie f.

lath-nail ‖ Lattennagel m. ‖ clou m. à lattes.

laths pl. (Mine) ‖ Pfähle mpl.; Verschalung f. ‖ garnissage m.; planches fpl.

lath sawyer ‖ Lattensäger m. ‖ scieur m. de lattes.

lathwork ‖ Lattengestell n. ‖ lattis m.

latitude ‖ geografische Breite f. ‖ latitude f. / ~ (Mar) ‖ Höhe f. eines Ortes ‖ hauteur f. / astronomical ~ of a star ‖ astronomische Breite f. eines Sternes ‖ latitude f. céleste d'une étoile. / in the ~ of ... (Mar) ‖ auf der Höhe f. von ... ‖ à la hauteur de ...; au large m. de ... / geographical ~ of a place ‖ geografische Breite f. eines Ortes ‖ latitude f. géographique d'un lieu. / ~ of the place of observation ‖ geografische Breite f. des Beobachtungsortes ‖ latitude f. du lieu d'observation.

latitude error ‖ Breitenfehler m. ‖ erreur f. de latitude.

latten *see also* brass ‖ Messing n. ‖ laiton m.; cuivre m. jaune; archal m. / ~ brass (Met) ‖ Messingblech n. ‖ laiton en feuilles *ou* en lames. / ~ clippings pl. ‖ Schrotmessing n. ‖ courtailles fpl.; laiton m. coupé en petites pièces; mitraille f. de cuivre jaune.

lattice, to ‖ vergittern ‖ treillisser; grillager; treillager. / ~ (Join) ‖ mit Holz n. vergittern ‖ grillager en bois; jalouser.

lattice ‖ Gitterwerk n.; Gitter n. ‖ treillis m. / iron wire ~ ‖ eisernes Drahtgitter n. ‖ grillage m. en fer. / wire ~ ‖ Drahtzaun m. ‖ grillage m. métallique. / wooden ~ ‖ Holzgitter n. ‖ grillage m. en bois.

lattice bridge ‖ Gitterbrücke f. ‖ pont m. à grille *ou* en treillis.

latticed bar ‖ Gitterstab m. ‖ barre f. de treillis *ou* de grille. / ~ brickwork ‖ gitterförmiges Mauerwerk n. ‖ maçonnerie f. grillagée.

lattice gate ‖ Gittertür f. ‖ porte f. à claire-voie. / ~ girder ‖ Fachwerkträger m.; Gitterträger m. ‖ poutre f. contre-fichée *ou* à treillis. / ~ jib ‖ Fachwerkausleger m. ‖ flèche f. en charpente. / ~ mast ‖ Gittermast m. ‖ pylône m. en charpente métallique; mât m. en treillis. / ~ partition ‖ Lattenwand f. ‖ cloison f. en claire-voie; cloison lattée. / ~ pole ‖ Gittermast m. ‖ poteau m. en treillis *ou* en charpente métallique. / ~ tower *see* ~ pole. / ~ truss ‖ Gitterbalken m. ‖ poutre f. en treillis *ou* en charpente métallique.

lattice-work ‖ Gitterwerk n.; Gitterkonstruktion ‖ treillis m.; charpente f. à treillis. / ~ (Partition) ‖ Lattenverschlag m. ‖ cloison f. lattée. / wooden ~ ‖ Holzstabgewebe n. ‖ treillis m. en bois. / wooden ~ making machine ‖ Holzstabgewebeherstellungsmaschine f. ‖ machine f. pour fabriquer des grillages en bois.

lattice-work mast ‖ Gittermast m. ‖ mât m. en treillis *ou* en charpente métallique.

laughing gas ‖ Lachgas n. ‖ gaz m. hilarant.

launch, to ‖ vom Stapel m. laufen lassen ‖ lancer; mettre à l'eau f.

launch (Mar) ‖ Barkasse f.; großes Boot n. ‖ barcasse f.; chaloupe f. / ~ (Shipb) ‖ Vorhelling f.; Seestapel m. ‖ avant-cale f. / steam ~ ‖ Dampfbarkasse f. ‖ chaloupe f. à vapeur.

launched ‖ vom Stapel gelassen ‖ lancé; mis à l'eau f. / to get ~ ‖ vom Stapel m. laufen ‖ être mis à l'eau f. / the ship was ~ down into the water ‖ das Schiff wurde vom Stapel gelassen ‖ le bâtiment a été mis à l'eau.

launching of a vessel ‖ Stapellauf m. ‖ lancement m. d'un navire. / ~ block (Shipb) ‖ Schmierkissen n. ‖ billot m. / ~ cradle ‖ Schlitten m.; Ablaufgerüst n. ‖ ber m.; berceau m. / ~ device (Aero) ‖ Startvorrichtung f. ‖ dispositif m. de lancement. / ~ tube ‖ Lanzierrohr n.; Ausstoßrohr n. ‖ tube m. de lancement. / ~ wedge (Shipb) ‖ Treibkeil m. (des Ablaufgerüstes) ‖ coin m. de ber.

laundering ‖ Wäscherei f. ‖ blanchissage m.

laundress ‖ Waschfrau f.; Wäscherin f. ‖ blanchisseuse f.; laveuse f. de linge. / fine ~ ‖ Feinwäscherin f. ‖ blanchisseuse f. de fin.

laundry ‖ Wäscherei f.; Waschanstalt f. ‖ blanchisserie f.; lavoir m. / floating ~ ‖ Waschboot n. ‖ bateau-lavoir m. / steam ~ ‖ Dampfwäscherei f. ‖ blanchisserie f. à vapeur.

laundry blue ‖ Waschblau n. ‖ papier m. à bleu d'outremer. / ~ machine ‖ Wäschereimaschine f. ‖ machine f. de blanchisserie. / ~ plant ‖ Wäschereianlage f. ‖ installation f. de blanchisserie.

laurel berry ‖ Lorbeerbeere f. ‖ baie f. de laurier. / ~ leaf ‖ Lorbeerblatt n. ‖ feuille f. de laurier. / ~ oil ‖ Lorbeeröl n. ‖ huile f. de laurier.

lava || Lava f. || lave f. / consolidated ~ || erstarrte Lava f. || lave f. solidifiée. / flowing ~ || fließende Lava f. || lave f. coulante. / liquid ~ || flüssige Lava f. || lave f. liquide. / scorified ~ || Schlacken- lava f. || lave f. scorifiée. / stony ~ || stein- artige *oder* steinige Lava f. || lave f. lithoïde. / vitreous ~ || Glaslava f.; Ob- sidian m. || laitier m. de volcan; lave f. vitreuse obsidienne; obsidienne f.

lava, current of ~ || Lavastrom m. || coulée f. de lave; courant m. de lave. / ~ paving stone || Lavapflasterstein m. || pavé m. en lave. / ~ quarry || Lavasteinbruch m. || carrière f. de lave. / stream of ~ || Lava- strom m. || coulée f. *ou* courant m. de lave.

lavatory || Abortanlage f. || latrine f.; lieu m. *ou* cabinet m. d'aisance. || basin || Stehabortbecken n. || cuvette f. d'urinoir.

lavender || Lavendel m. || lavande f. / ~ flower || Lavendelblüte f. || fleur f. de lavande. / ~ oil || Lavendelöl n. || essence f. de lavande.

law || Gesetz n.; Verordnung f. || loi f.; statut m. / contrary to ~ || gesetzwidrig; rechtswidrig || illégal; contraire à la loi. / to employ a workman contrary to the ~ || einen Arbeiter den gesetzlichen Vor- schriften entgegen beschäftigen || oc- cuper un ouvrier contrairement aux dis- positions légales. / to go to ~ || den Rechtsweg m. beschreiten || recourrir à la loi; recourrir à *ou* prendre la voie des tribunaux. / ~ relating to the in- stallation of antennas on the premises of others || Antennenrecht n. || loi f. con- cernant la pose d'antennes sur fond d'autrui; droit m. à l'antenne. / com- mon ~ || Landrecht n. || droit m. civil *ou* commun. / custom ~ || Zollgesetz n. || loi f. de douane. / ~ of exchange || Wechselrecht n. || droit m. de change. / ~ of gravitation || Gravitationsgesetz n. || loi f. de gravitation. / maritime ~ || Seerecht n. || droit m. maritime. / mari- time international ~ || internationales Seerecht n.; Völkerseerecht n. || droit m. maritime international. / ~ of nations || Völkerrecht n. || droit m. des gens; droit m. international. / ~ of refraction || Brechungsgesetz n. || loi f. de la réfraction.

laws pl. and regulations pl. relating to the service of telegraph and telephone || Fern- melderecht n. || lois fpl. et réglements mpl. concernant le service télégraphique et téléphonique. / ~ relating to the landing submarine cables || Kabellan- dungsrecht n. || lois fpl. réglant l'atter- rissage des câbles sous-marins.

law costs pl. || Gerichtskosten pl. || frais m. de justice.

lawful || rechtlich; rechtmäßig || juridique; honnête; loyal. / ~ holder || rechtmäßiger Inhaber m. || porteur m. *ou* détenteur m. légitime.

lawn (Garden) || Rasenplatz m. || gazon m.; boulingrin m. / ~ (Weav) || Schleiertuch n.; Linon m.; Batist m. || linon m.; voile m.; batiste f. de lin. / ~ mower || Rasen- mäher m. || tondeuse f. *ou* faucheuse f. de gazon.

lawns pl. and flower beds pl. || gärtnerische Anlagen fpl. || planches fpl. de fleurs et des glacis de gazon.

lawsuit || Rechtsfall m.; Prozeß m.; Rechtsstreit m. || procès m.; cas m. ligitieux; poursuite f. judiciaire. / ~ re- lating to bills of exchange || Wechsel-

klage f. || action f. fondée sur des lettres de change. / to carry on a ~ || prozes- sieren || plaider contre ...; être en procès avec ... / to conduct a ~ || einen Prozeß m. anstrengen *oder* einleiten || intenter un procès.

lawyer || Anwalt m. || avoué m.; avocat m.

laxative || Abführmittel n. || laxatif m.

lay, to || legen || jeter; mettre; placer; poser. / ~ at anchor || vor Anker m. liegen || être à l'ancre f.; être sur le fer. / ~ bare || frei- geben; aufdecken || découvrir. / ~ bare the scavenging port || den Eintrittsschlitz m. freigeben || découvrir la lumière d'ad- mission. / a bridge || eine Brücke bauen *oder* schlagen || construire *ou* établir *ou* jeter un pont. / ~ a cable || ein Kabel n. verlegen *oder* versenken || poser *ou* im- merger un câble. / ~ the covering || das Dach eindecken || poser la toiture. / ~ down the floor-boards || die Dielen fpl. legen || faire l'aire f.; poser les planches fpl. / ~ down as principle || als Grund- satz m. aufstellen || poser en principe m. / ~ down a road || eine Straße f. bauen || construire une route. / ~ down (A vessel) || auf Stapel m. legen || mettre sur cale f. / ~ flags pl. || die Fliesen fpl. legen || poser les carreaux mpl. / ~ flat || einebnen || niveler; égaliser; aplanir. / ~ the founda- tion of a building || ein Gebäude n. gründen; den Grund m. zu einem Ge- bäude legen || jeter les fondements mpl. d'un bâtiment. / ~ in good bond || ver- bandsmäßig *oder* in gutem Verband m. mauern || maçonner en liaison f.; poser en bonne liaison f.; liaisonner. / ~-in || sich eindecken mit ... || faire provision f. de...; ravitailler. / ~-in the supply of provisions (Mar) || Proviant m. einnehmen || faire des vivres mpl.; victuailler. / ~ a keel || den Kiel m. eines Schiffes legen *oder* strecken || mettre la quille en place *ou* sur le chan- tier. / ~ off a ship on the mould loft || ein Schiff n. auf dem Schnürboden abschla- gen || tracer un vaisseau à la salle. / ~-on gold-leaves || die Goldblätter npl. auf- legen || dorer dans la dorure en détrempe. / ~-on the pressing-board (Pap) || gaut- schen || coucher les feuilles fpl. de papier. / ~-out (Drawing) || aufreißen || faire l'épure f. / ~-out (Money) || auslegen || avancer; débourser. / ~-out a curve || eine Kurve f. abstecken || piqueter une courbe. / ~-out a garden || einen Garten m. anlegen || installer un jardin. / ~-out ridges pl. || rajolen || effondrer; rigoler la terre. / ~ pages pl. (Print) || Seiten fpl. einrichten || arranger les pages fpl. / ~ a pavement || pflastern || paver. / ~ a rope || ein Tau n. schlagen || commettre un cordage. / ~ a ship to || beidrehen; sich unter den Wind m. legen || se mettre à la cape. / ~ a ship in the road || ein Schiff n. auf die Reede legen || mettre un navire en rade. / ~ a ship on the stocks pl. || ein Schiff n. auf Stapel legen || mettre un navire sur le chantier. / ~ the string-pieces of a pile-work on the top of earth-piles || ein Pfahlwerk n. verhol- men || mettre les chapeaux mpl. par- dessus les pieux. / ~ the tiles || die Dach- ziegel mpl. einhängen || poser les tuiles fpl. / ~ the track || das Gleis verlegen || poser la voie.

lay beam (Weav) || Ladenklotz m. || masse f.; sommier m. / ~ day (Mar) || Liegetag m. || jour m. de planche.

layed-in key || Einlegekeil m. || clavette f. encastrée.

layer (Bot) || Steckling m.; Absenker m. || bouture f. / ~ (Geol; Mine) || Schicht f.; Flöz n. || couche f. / ~ (Print) || Einschie- ber m.; Leger m. || leveur m.; margeur m. / ~ of adhesive rubber || Klebgummi- schicht f. || couche f. de caoutchouc adhésive. / ~ of air || Luftschicht f. || couche f. *ou* pellicule f. d'air. / bottom ~ || untere Schicht f. || couche f. in- férieure. / ~ of broken stones (Road) || Kofferschicht f. || coffrage m.; couche f. de pierres cassées. / ~ of cement || Kitt- schicht f. || couche f. de ciment. / ~ of coal || Kohlenschicht f. || couche f. de charbon. / ~ of concrete || Betonschüt- tung f. || coulée f. de béton. / to drain a ~ || ein Flöz n. lösen || saigner *ou* démer- ger une veine. / ~ of earth || Erdschicht f. || couche f. de terre. / to find a ~ by bor- ing || eine Schicht f. anbohren || découvrir une couche par la sonde. / ~ of friable earth || mulmige Erdschicht f. || couche f. de terrain décomposé pulvérulant. / ~ of gravel (Road) || Lage f. Kies || couche f. de gravier. / ~ of grouting || Verguß- schicht f. || couche f. de remplissage. / a ~ of hard steel was cast on the plate || auf die Platte wurde eine Schicht harten Stahles aufgegossen || sur la plaque était coulée une couche d'acier dur. / in ~s || lagenweise || par couches fpl. / ~ of insula- tion || Isolierschicht f. || couche f. isolante. / intermediate ~ (Cable) || Beilauffaden m. || filin m. *ou* couche f. intermédiaire. / ~ of leaves || Federblatt n.; Federlage f. || lame f. *ou* feuille f. du ressort. / ~ of loam (Mould) || Lehmauftrag m.; Lehmaufschlag m. || couche f. d'argile *ou* de terre. / magnetic ~ || magné- tische Doppelschicht f. || feuille f. magné- tique. / ~ of oxide || Oxydschicht f. || cou- che f. d'oxyde. / ~ of paints || Farban- strich m. || couche f. de peinture. / ram- med ~ || Stampfschicht f. || couche f. damée. / sensitive ~ || empfindliche Schicht f. || couche f. sensible. / ~ of shale (Mine) || Bergemittel n. || laie f.; banc m. de schiste (entre deux veines). / sinking through a ~ || Durchteufen n. einer Schicht || creusement m. d'une cou- che. / ~ of sizing rubber *see* ~ of adhe- sive rubber. / ~ of snow || Schneedecke f. || couverture f. *ou* couche f. de neige. / ~ of straw-sheaves || Schaubenlage f. || ran- gée f. de javelles. / to uncover a ~ || ein Flöz n. erschürfen || découvrir une mine. / upper ~ || obere Schicht f. || couche f. supérieure. / ~ of warm air || warme Luft- schicht f. || couche f. d'air chaude. / ~ of weather-worn material || Verwitterungs- schicht f. || couche f. décomposée.

layer-in girl (Print) || Einlegerin f. || mar- geuse f.

layer-on (Print) || Einleger m.; Anleger m. || margeur m. / ~ girl (Print) || Anlegerin f. || margeuse f.

layer-out (Dyer) || Abdocker m. || déplieur m.

lay figure (Tail) || Puppe f. || mannequin m.; buste m.

laying || Anwurf m.; Bewurf m.; Putz m.; Spritzwurf m. || crépi m.; première cou- che f. d'enduit. / ~ of mains || Kabel- verlegung f. || pose f. des câbles. / ~ of an oil filled cable || Verlegung f. eines Öl- kabels || pose f. d'un câble à huile.

/ ~ pipes ‖ Rohrverlegung f. ‖ pose f. de tuyaux.

laying-day ‖ Liegetag m. ‖ jour m. de planche.

laying down (Cable) ‖ Verlegung f. ‖ pose f.

laying-in ‖ Einkellern n. ‖ encavage m.; mise f. sur chais.

laying-on of colours ‖ Farbenauftrag m.; Auftragen n. der Farbe ‖ application f. de la couleur.

laying-on apparatus for sheets ‖ Anlegeapparat m. für Papierbogen ‖ rectificateur m. de marge; appareil m. pour mettre les feuilles à la machine. / cement ~ machine ‖ Zementauftragemaschine f. ‖ machine f. à étaler le ciment. / ~ side (Print) ‖ Bedienungsseite f. ‖ côté m. de service.

laying-out apparatus for sheets (Print) ‖ Ablegeapparat m. für Papierbogen ‖ appareil m. pour mettre les feuilles de côté. / ~ length ‖ Baulänge f. ‖ longueur f. de pose.

laying-top (Ropem) ‖ Lehre f.; Hoofd n. ‖ gabieu m.; cabre f.

lazaretto ‖ Siechenhaus n.; Lazarett n. ‖ maladrerie f.; hôpital m.; infirmerie f.

lazar house *see* lazaretto.

lazulite ‖ Blauspat m.; Lazulith m. ‖ feldspath m. bleu; klaprothine f.; lazulite m. / ~ blue ‖ Lasurblau n. ‖ azur m.; (bleu m. d')outremer m.

lazy ‖ säumig; faul ‖ retardataire; paresseux.

lea of cotton-yarn ‖ Gebinde n. (von) Baumwollgarn ‖ échevette f. de fil de coton.

leaching plant ‖ Laugeeinrichtung f. ‖ installation f. de lessivage.

lead, to (Mach) ‖ voreilen ‖ avancer. / ~ (Tinm) ‖ verbleien ‖ plomber. / ~ and to caulk in ‖ mit Blei n. abdichten und verstemmen ‖ garnir de plomb m. et sertir. / ~ lights ‖ die Fenster npl. verbleien ‖ sceller les vitres fpl. au plomb. / ~ the lines (Print) ‖ die Schriftzeilen fpl. durchschießen ‖ interligner les blancs mpl.; espacer les lignes fpl. / ~ in phase ‖ in der Phase f. voreilen ‖ être en avance de phase.

lead ‖ Blei n. ‖ plomb m. / ~ (Electr) ‖ Leitung f. ‖ conducteur m.; ligne f. / ~ (Nav; Mas) ‖ Lot n.; Lotschnur f.; Bleilot n.; Senkblei n. ‖ plomb m.; fil m. à plomb; sonde f.; plomb m. à sonde. / ~ (Office) ‖ Bleimine f.; Grafitstift m. ‖ mine f. de plomb. / argentiferous ~ ‖ silberhaltiges Blei n. ‖ plomb m. argentifère *ou* d'œuvre. / black ~ ‖ Graphit m. ‖ graphite m.; plombagine f. / ~ of brushes ‖ Bürstenvorschub m. ‖ décalage m. en avant des balais. / ~ of the crank ‖ Voreilungswinkel m. der Kurbel ‖ avance f. angulaire de la manivelle. / crude ~ ‖ Rohblei n. ‖ plomb m. brut. / enriched ~ ‖ angereichertes Blei n. ‖ plomb m. affiné *ou* enrichi. / first ~ of a reverberatory furnace ‖ Jungfernblei n. ‖ plomb m. vierge. / furnace ~ ‖ Herdblei n. ‖ plomb m. de foyer. / hard ~ ‖ Hartblei n. ‖ plomb m. aigre *ou* durci. / hard ~ frame ‖ Hartbleirahmen m. ‖ cadre m. en plomb durci. / merchant ~ ‖ Weichblei n. ‖ plomb m. doux. / native ~ ‖ gediegenes Blei n. ‖ plomb m. natif. / oil ~ ‖ Ölrohr n. ‖ tuyau m. d'huile. / pattinsonized ~ ‖ pattinsoniertes Blei n. ‖ plomb m. affiné par cristallisation. / ~ in phase ‖ Phasenvoreilung f. ‖ avance

f. de phase. / pig ~ ‖ Blockblei n. ‖ plomb m. doux en saumon. / red ~ ‖ Bleimennige f. ‖ minium m. de plomb. / refined ~ ‖ Frischblei n.; raffiniertes Blei n. ‖ plomb m. raffiné. / rolled ~ ‖ Walzblei n. ‖ plomb m. laminé; sheet ~ ‖ Bleiblech n.; Bleipapier n. ‖ feuille f. de plomb; plomb m. laminé; plomb m. en feuilles. / skimmed ~ ‖ Abstrichblei n. ‖ plomb m. d'écumage. / slag ~ ‖ Antimonblei n.; Hartblei n. ‖ plomb m. aigre; spongy ~ ‖ Bleischwamm m. ‖ écume f. de plomb. / to take the ~ (Sport) ‖ die Führung übernehmen ‖ prendre l'avance *ou* la première place. / tinned ~ ‖ verzinntes Blei n. ‖ plomb m. étamé. / ~ for tracery of windows ‖ Bleiprofil n. für Glaserarbeit ‖ plombs mpl. de vitrerie. / under ~s pl. ‖ unter Bleiverschluß m. ‖ plombé. / white ~ ‖ Bleiweiß n. ‖ blanc m. de plomb; céruse f. / white ~ plant ‖ Bleiweißanlage f. ‖ installation f. de céruse. / ~ for windows ‖ Fensterblei n. ‖ plomb m. à vitres; bande f. de plomb pour fenêtres. / worked ~ ‖ Werkblei n. ‖ plomb m. usiné *ou* travaillé.

lead accumulator ‖ Bleiakkumulator m. ‖ accumulateur m. au plomb. / acetate of ~ ‖ Bleiazetat n. ‖ acétate m. de plomb. / alloy of ~ ‖ Bleilegierung f. ‖ alliage m. de plomb. / angle of ~ ‖ Voreilungswinkel m. ‖ angle m. d'avance. / ~ anode ‖ Bleianode f. ‖ anode f. de plomb. / ~ antimoniate ‖ antimonsaures Blei n. ‖ antimoniate m. de plomb. / ~ apparatus ‖ Bleiapparat m. ‖ appareil m. en plomb. / ~ armature ‖ Bleiarmatur f. ‖ armature f. en plomb. / ~ arseniate ‖ arsensaures Blei n. ‖ arséniate m. de plomb. / ~ ashes pl. ‖ Bleiasche f. ‖ cendre f. de plomb. / ~ ball making machine ‖ Bleikugelherstellungsmaschine f. ‖ machine f. pour fabriquer des balles de plomb. / basic acetate of ~ ‖ Bleiessig m. ‖ sous-acétate m. de plomb. / ~-bearing ‖ bleiführend ‖ plombifère. / ~ bismuth alloy ‖ Bleiwismutlegierung f. ‖ alliage m. plomb-bismuth. / ~ borate ‖ borsaures Blei n. ‖ borate m. de plomb. / ~ bottle ‖ Bleiflasche f. ‖ flacon m. *ou* bouteille f. en plomb. / ~ brush ‖ Bleibürste f. ‖ brosse f. à plomb. / ~ bullet ‖ Bleikugel f. ‖ balle f. de plomb. / ~ button ‖ Bleiknopf m. ‖ bouton m. en plomb. / ~ cable ‖ Bleikabel n. ‖ câble m. en plomb. / ~ cable press ‖ Bleikabelpresse f. ‖ presse f. pour câbles en plomb. / ~ calcining furnace ‖ Bleiofen m. ‖ four m. à plomb. / ~ cam ‖ Leitkurve f. ‖ came f. de chariotage. / screws pl. which are covered by soldered ~ caps ‖ Schrauben fpl., die durch aufgelötete Bleikappen verdeckt werden ‖ vis fpl. recouvertes de calottes en plomb soudées dessus. / carbonate of ~ ‖ Schwarzbleierz n.; Cerussit m. ‖ plomb m. carbonaté noir. / ~ chamber ‖ Bleikammer f. ‖ chambre f. de plomb. / ~ chromate ‖ chromsaures Blei n. ‖ chromate m. de plomb. / ~-coated sheet ‖ verbleites Blech n. ‖ tôle f. plombée. / ~ coffin ‖ Bleisarg m. ‖ cercueil m. en plomb. / ~ coil ‖ Bleischlange f. ‖ serpentin m. en plomb. / ~ colic ‖ Bleikolik f. ‖ colique f. saturnine *ou* de plomb. / ~ compass ‖ Bleistiftzirkel m. ‖ compas m. au crayon. / ~-containing ‖ bleihaltig ‖ plombifère. / ap-

paratus homogeneously covered with ~ ‖ Apparat m. mit homogener Verbleiung ‖ appareil m. à plombage homogène. / ~-covered cable ‖ Bleikabel n. ‖ câble m. sous plomb *ou* armé de plomb. / ~-covered rubber cable ‖ Gummibleikabel n. ‖ câble m. au caoutchouc sous plomb.

lead covering ‖ Eindecken n. mit Blei ‖ couverture f. en plomb. / ~ (Cable) ‖ Bleihülle f.; Bleimantel m. ‖ gaine f. de plomb. / ~ of roofs ‖ Bleibedachung f. ‖ toiture f. en plomb. / ~ plant ‖ Verbleiungsanlage f. ‖ installation f. de plombage.

lead cup ‖ Polschuh m. ‖ cosse f. polaire. / ~ cut-out ‖ Bleisicherung f. ‖ plomb m. fusible; coupe-circuit m. à plomb-fusible. / ~ cutter ‖ Bleischneider m. ‖ cisailles fpl. à plomb. / ~ deposit ‖ Bleischlamm m.; Bleischlammablagerung f. ‖ boue f. de plomb; dépôt m. de boue de plomb. / ~ desilverizing ‖ Bleientsilberung f. ‖ désargentation f. du plomb. / ~ dowel ‖ Bleidübel m. ‖ goujon m. de plomb. / ~ dust ‖ Bleistaub m. ‖ poussière f. de plomb. / end stirrup of ~ ‖ Endbleibügel m. ‖ ressort m. de calage en plomb.

leaden ‖ bleiern ‖ en plomb m. / ~ branch box ‖ Bleiverzweigungsmuffe f. ‖ manchon m. de branchement en plomb. / ~ case ‖ Bleimantel m. ‖ enveloppe f. *ou* germe f. en plomb. / ~ connecting box ‖ Bleiverbindungsmuffe f. ‖ manchon m. de jonction en plomb. / ~ dividing box ‖ Bleiaufteilungsmuffe f. ‖ manchon m. de répartition en plomb. / ~ end box ‖ Bleiabschlußmuffe f. ‖ manchon m. de fermeture en plomb. / ~ goods pl. ‖ Bleiwaren fpl. ‖ articles mpl. en plomb. / ~ pipe cable ‖ Bleirohrkabel n. ‖ câble m. sous gaine de plomb. / ~ seal ‖ Plombe f. ‖ plomb m.

leader (Newspaper) ‖ Leitartikel m. ‖ article m. de fond *ou* de tête; premier article m.

leaders pl. (Print) ‖ Leitepunkte mpl. ‖ points mpl. conducteurs; carrés mpl.

lead eraser ‖ Bleigummi m. ‖ gomme f. à crayon. / ~ extruding press for lead-covered cables ‖ Bleikabelpresse f. ‖ presse f. à filer les câbles à gaine de plomb. / ~ facing ‖ Bleiauskleidung f. ‖ revêtement m. en plomb. / ~ file ‖ Bleifeile f. ‖ écouenne f.; lime f. à plomb; écouane f. / ~ filings pl. ‖ Bleispäne mpl. ‖ copeaux mpl. *ou* limailles fpl. de plomb. / ~ fitting ‖ Bleieinfassung f. ‖ tenon m. / ~ foil ‖ Bleifolie f. ‖ feuille f. de plomb; plomb m. en feuilles. / ~ founder ‖ Bleigießer m. ‖ fondeur m. de plomb. / ~ foundry ‖ Bleigießerei f. ‖ fonderie f. de plomb. / ~ fuse ‖ Bleisicherung f. ‖ coupe-circuit m. à lame de plomb *ou* à plomb-fusible. / ~ glass ‖ Bleiglas n. ‖ anglésite f.; cristal m.; verre m. plombifère. / ~ glaze *see* ~ glazing. / ~ glazing ‖ Bleiglasur f. ‖ vernis m. plombifère *ou* de plomb. / ~ goods pl. ‖ Bleiwaren fpl. ‖ objets mpl. en plomb. / ~ gravel ‖ Bleigrieß m. ‖ gravier m. de plomb. / ~ gray ‖ Bleigrau n. ‖ gris m. de plomb. / ~ grid (Acc) ‖ Bleigitter n. ‖ grillage m. en plomb. / ~ hood for the cable end ‖ Bleikappe f. für das Kabelende ‖ calotte f. de plomb pour le bout du câble. / hyposulphite of ~ ‖ unterschwefligsaures Blei n. ‖ hyposulfite m. de plomb.

leading (Wind; Mar) ‖ raum ‖ largue.

leading ‖ Verbleien n.; Bleiverschluß m. ‖ plombage m. / ~ (Mine) ‖ Ader f.; Schnur f. ‖ veine f.; veinule f. / ~ of the exhaust gases ‖ Auspufführung f. ‖ canalisation f. ou conduite f. des gaz d'échappement.

leading article (Newspaper) ‖ Leitartikel m. ‖ article m. de fond ou de tête; premier article m. / ~ axle (Loc) ‖ Vorderachse f. ‖ axe m. ou essieu m. d'avant. / ~ bogie (Railw) ‖ Lenkachse f. ‖ essieu m. mobile. / ~ current ‖ in der Phase voreilender Strom m. ‖ courant m. décalé en avant. / ~ feature (Trade) ‖ Hauptmerkmal n. ‖ remarque f. principale. / ~ grate ‖ Leitrechen m.; Triftrechen m. ‖ allingre f.; gautier m.

leading-in of an overhead line ‖ oberirdische Leitungseinführung f. ‖ entrée f. d'une ligne aérienne.

leading-in cable ‖ Einführungskabel n. ‖ câble m. d'entrée. / ~ double bracket (El line) ‖ Einführungsdoppelstütze f. ‖ console f. double pour l'entrée de poste; console f. d'introduction. / ~ insulator ‖ Endisolator m. ‖ isolateur m. d'introduction. / ~ line ‖ Zufuhrgleis n. ‖ voie f. d'accès. / ~ manhole ‖ Einführungsbrunnen m. ‖ regard m. de visite du puits de canalisation. / ~ pole ‖ Einführungsgestänge n. ‖ appui m. tête de ligne. / ~ spindle see ~ double bracket.

leading light (Mar) ‖ Leitfeuer n. ‖ feu m. d'alignement. / ~ mark ‖ Seezeichen n. ‖ amer m.; balise f.; marque f. de guide. / ~ pulley ‖ Lenkrolle f. ‖ poulie f. de guidage. / ~ rod (Rifle) ‖ Schmirgelkolben m. ‖ polissoir m. / ~ screw lathe ‖ Leitspindeldrehbank f. ‖ tour f. à vis-mère. / ~ seaman ‖ Obermatrose m. ‖ gabier m. breveté. / ~ species pl. ‖ gangbarste oder gewöhnliche Sorten fpl. ‖ sortes fpl. ou marques fpl. les plus usitées ou les plus courantes. / ~ spindle (Turn) ‖ Leitspindel f. ‖ vis-mère f. / ~ tongs pl. ‖ Plombierzange f. ‖ pince f. à plomber. / ~ wheel ‖ Vorderrad n. ‖ roue f. avant. / ~ wind ‖ Rückwind m. ‖ vent m. arrière. / ~ worker ‖ Vorarbeiter m. ‖ chef m. d'équipe.

lead ion ‖ Bleiion n. ‖ ion m. de plomb. / ~ jacket ‖ Bleimantel m. ‖ chemise f. ou enveloppe f. ou revêtement m. en plomb. / ~ joint ‖ Bleidichtung f. ‖ joint m. au plomb. / ~ lignoleate ‖ holzölsaures Blei n. ‖ lignoléate m. de plomb. / ~ lining ‖ Bleiauskleidung f.; Bleieinlage f. ‖ revêtement m. ou garniture f. de plomb. / ~ linoleate ‖ leinölsaures Blei n. ‖ linoléate m. de plomb. / deep-sea ~ machine ‖ Tiefseelotmaschine f. ‖ machine f. pour sonder les grands fonds. / ~ magazine (Pencil) ‖ Minenkammer f. ‖ magasin m. à mines. / ~ manganese oleate ‖ ölsaures Bleimangan n. ‖ oléate m. de plomb et de manganèse. / ~ manganese resinate ‖ harzsaures Bleimangan n. ‖ résinate m. de plomb et de manganèse. / ~ manufacture ‖ Bleigewinnung f. ‖ métallurgie f. du plomb. / ~ meal ‖ Bleimehl n. ‖ farine f. de plomb. / ~ mine ‖ Bleigrube f.; Bleibergwerk n. ‖ mine f. de plomb. / ~ monoxide see ~ oxide. / ~ moulding matrice ‖ Bleimatrize f. ‖ flan m. en plomb. / ~ nail ‖ Bleinagel m. ‖ clou m. en plomb. / ~ nitrate ‖

salpetersaures Blei n. ‖ nitrate m. ou azotate m. de plomb.

lead ore ‖ Bleierz n. ‖ minerai m. de plomb. / antimonial sulphuretted ~ ‖ Spießglanzblei n. ‖ mine f. d'antimoine noire. / green ~ ‖ Pyromorphit m.; Grünbleierz n.; Braunbleierz n.; Phosphorblei n. ‖ pyromorphite f.; plomb m. phosphaté ou vert ou brun. / red ~ ‖ Rotbleierz n.; chromsaures Bleioxyd n.; Krokoit m. ‖ plomb m. chromaté.

lead ore oxide ‖ Bleiglätte f.; Bleioxyd n. ‖ litharge f.; massicot m.; protoxyde m. de plomb.

lead pencil ‖ Bleistift m. ‖ crayon m. / metallic ~ ‖ Bleigriffel m. ‖ crayon m. de plomb. / patent ~ ‖ Füllbleistift m. ‖ porte-mine m.

lead-pencil drawing ‖ Bleistiftzeichnung f. ‖ dessin m. au crayon.

lead peroxide ‖ Bleisuperoxyd n. ‖ peroxyde m. de plomb. / ~ pipe ‖ Bleirohr n. ‖ tuyau m. en plomb. / ~ pipe extruding press ‖ Bleirohrpresse f. ‖ presse f. pour tubes en plomb; presse f. à filer les tuyaux de plomb. / ~ plate ‖ Bleiplatte f. ‖ plaque f. en plomb. / ~ plating plant ‖ Verbleiungsanlage f. ‖ installation f. de plombage. / ~ plumbing ‖ Bleilöterei f. ‖ soudure f. au plomb. / ~ poisoning ‖ Bleivergiftung f. ‖ empoisonnement m. par le plomb; affection f. saturnine. / ~ powder ‖ Bleipulver n. ‖ poudre f. de plomb. / ~ press ‖ Bleipresse f. ‖ presse f. à plomb. / ~ press man ‖ Bleipressenarbeiter m. ‖ ouvrier m. de câbles sous plomb. / ~ pyrolignite ‖ holzessigsaures Blei n. ‖ pyrolignite m. de plomb. / ~ resinate ‖ harzsaures Blei n. ‖ résinate m. de plomb. / ~ rolling mill ‖ Bleiwalzwerk n. ‖ laminoir m. à plomb. / ~ salt ‖ Bleisalz n. ‖ sel m. de plomb. / English pitch ~ screw ‖ Zolleitspindel f. ‖ vis-mère f. au pas de pouce anglais.

lead seal ‖ Bleiplombe f. ‖ plomb m. à sceller; cachet m. en plomb. / ~ casting machine ‖ Plombengießmaschine f. ‖ machine f. à couler des plombes.

lead shears pl. ‖ Bleischere f. ‖ cisailles fpl. à plomb. / ~ sheath (Cable) ‖ Bleimantel m. ‖ gaine f. en plomb. / ~ shot ‖ Bleischrot n. ‖ plomb m. ou grenaille f. de chasse. / ~ slag ‖ Bleischlacke f. ‖ écume f. de plomb. / ~ smoke ‖ Bleirauch m. ‖ fumée f. de plomb. / ~ spring ‖ Bleibügel m. ‖ Bleifeder m. ‖ ressort m. en plomb. / ~ solder ‖ Lötblei n. ‖ plomb m. à souder. / ~ soldering ‖ Bleilöterei f. ‖ soudure f. au plomb. / ~ soldering shop ‖ Bleilöterei f. ‖ atelier m. à souder le plomb. / ~ soldier ‖ Bleisoldat m. ‖ soldat m. de plomb. / ~ stamping punch pliers ‖ Plombenzange f. ‖ pince f. à plomber. / sugar of ~ ‖ Bleizucker m. ‖ sucre de plomb m. / sulphate of ~ ‖ schwefelsaures Blei n.; Bleivitriol n. ‖ sulfate m. de plomb. / ~ sulphate residue ‖ Bleisulfatrückstand m. ‖ résidu m. de sulfate de plomb. / ~ toys pl. ‖ Bleispielwaren fpl. ‖ jouets mpl. en plomb. / ~ tree ‖ Bleibaum m. ‖ arbre m. de saturne.

lead tube ‖ Bleirohr n. ‖ tuyau m. ou tube f. de plomb. / ~ press ‖ Bleirohrpresse f. ‖ presse f. pour tubes en plomb.

lead vinegar ‖ Bleiessig m. ‖ vinaigre m. de plomb. / ~ washer ‖ Bleiunterlagscheibe f. ‖ rondelle f. de plomb. / ~ waste ‖ Bleiabfall m. ‖ déchet m. de plomb.

lead wire ‖ Bleidraht m. ‖ fil m. de plomb. / ~ press ‖ Bleidrahtpresse f. ‖ presse f. pour fils en plomb.

lead wool ‖ Bleiwolle f. ‖ laine f. de plomb. / ~ works pl. ‖ Bleihütte f. ‖ fonderie f. ou usine f. (de minerai) de plomb; plomberie f. / ~ yellow ‖ Bleigelb n. ‖ jaune m. de plomb.

leaf ‖ Blatt n. ‖ feuille f. / ~ (Leaves) see leaves. / ~ of a door ‖ Türflügel m. ‖ battant m. ou vantail m. d'une porte. / ~ of a pinion ‖ Triebstock m. eines Getriebes ‖ fuseau m. d'un pignon. / ~ of the spring ‖ Federblatt n. ‖ lame f. de ressort. / stripped ~ ‖ entripptes oder ausgeripptes Blatt n. ‖ feuille f. écôtée. / ~ of table ‖ Tischplatte f. ‖ planche f. de la table.

leaf aluminium ‖ Blattaluminium n. ‖ aluminium m. en feuilles. / ~ fibre ‖ Blattfaser f. ‖ fibre f. extraite de la feuille.

leaf gold ‖ Blattgold n. ‖ or m. en feuilles. / genuine ~ ‖ echtes Blattgold n. ‖ or m. véritable en feuilles.

leaf metal ‖ Blattmetall n. ‖ métal m. en feuilles. / ~ spring ‖ Blattfeder f. ‖ ressort m. à lames. / ~ valve ‖ Klappenventil n. ‖ clapet m.; soupape f. à clapet ou à charnière. / ~ white metal ‖ unechtes Blattsilber n. ‖ argent m. faux en feuilles. / ~ wood ‖ Laubholz n. ‖ bois m. feuillu.

league ‖ Meile f. ‖ lieue f.

League of Nations ‖ Völkerbund m. ‖ Société f. des Nations.

leak, to ‖ undicht sein; lecken; auslaufen ‖ laisser sortir ou laisser couler (l'eau); avoir une fuite. / ~ (Mar) ‖ leck sein; ein Leck n. haben; lecken ‖ faire (de l')eau f.; laisser entrer (l'eau). / ~ away see ~ out. / ~ off ‖ abfließen ‖ s'écouler. / ~ out (Gas) ‖ entweichen ‖ échapper. / ~ out (Liquids) ‖ auslaufen; tropfen; rinnen ‖ couler; fuir; s'écouler.

leak ‖ Undichtigkeit f. ‖ fuite f. / ~ (Mar) ‖ Leck n. ‖ voie f. d'eau. / slight ~ ‖ leichte Undichtigkeit f. ‖ petite fuite f.; non-étanchéité f. légère. / to spring a ~ (Mar) ‖ leck werden ‖ faire de l'eau; gagner de l'eau. / the ~ has stopped accidentally (Mar) ‖ das Leck n. hat sich zugezogen ‖ la voie f. d'eau a supé; la voie f. d'eau s'est bouchée. / to test for ~s pl. ‖ auf Dichtheit f. prüfen ‖ vérifier l'étanchéité.

leakage (of liquids) ‖ Auslaufen n.; écoulement m.; fuite f. / ~ (of reservoirs) ‖ Undichtheit f.; Leck n. ‖ fuite f. / ~ (Electr) ‖ Streuung; Streuungsverlust m. ‖ déperdition f. par dispersion. / ~ (Loss) ‖ Verlust m. ‖ perte f. / ~ (Mar) ‖ Leckage f. ‖ coulage m. / armature ~ ‖ Ankerstreuung f. ‖ dispersion f. de l'induit. / ~ of a canal ‖ Sickerverlust m. eines Kanals ‖ perte f. d'infiltration. / ~ of current ‖ Stromverlust m. ‖ perte f. ou fuite f. de courant. / ~ of electricity ‖ Elektrizitätsverlust m. ‖ perte f. d'électricité.

leakage conductance (Electr) see leakance. / ~ current ‖ entweichender Strom m.; Streustrom m. ‖ courant m. de fuite. / ~ flux ‖ Streufluß m. ‖ flux m. de dispersion. / ~ indicator ‖ Erdschlußprüfer m. ‖ indicateur m. de perte à la terre. / ~ pipe ‖ Leckleitung f. ‖ conduite f. à fuite. / ~ water ‖ Sickerwasser n. ‖ eau f. d'infiltration.

leakance (Electr) ‖ Ableitung f.; Verlust m. ‖ perditance f. / ~ constant (Electr) ‖ Verlustwinkel m. ‖ constante f. de perditance. / ~ meter ‖ Ableitungsmesser m. ‖ appareil m. de mesure de perditance.

leak detector ‖ Lecksucher m. ‖ appareil m. chercheur de fuite.

leaking (Mar) ‖ leckend ‖ faisant eau; fuyant; ayant une fuite.

leaking (Leak) see leakage.

leaky (Mar) ‖ undicht ‖ qui fait eau; non-étanche.

lean, to ~ over (Mar) ‖ krängen; überholen ‖ donner à la bande; pencher; se pencher.

lean ‖ mager ‖ maigre. / ~ (Shipb) ‖ scharf ‖ fin; maigre. / ~ bow (Shipb) ‖ scharfer Schiffsbug m. ‖ proue f. fine ou maigre. / ~ coke ‖ Magerkoks m. ‖ coke m. maigre. / ~ material ‖ Magerungsmittel n. ‖ matière f. d'atténuation.

lean-to (Arch) ‖ Abdach n.; Halbdach ‖ appentis m.

leap, to (Mine) ‖ sich verwerfen ‖ se déranger.

leaper (Ropem) ‖ Entwirrer m.; Folger m.; Nachhänger m. ‖ émerillon m.

leap year ‖ Schaltjahr n. ‖ année f. intercalaire ou bissextile.

learn, to ‖ lernen; erlernen ‖ apprendre.

learner ‖ Anfänger m.; Lehrling m. ‖ débutant m.; apprenti m.

learning, time of ‖ Lehrzeit f. ‖ temps m. d'apprentissage.

lease, to ~ (to let out on lease) ‖ verpachten; vermieten ‖ donner à ferme ou à loyer; affermer; louer. / ~ (to take on lease) ‖ pachten; mieten ‖ prendre à ferme ou à bail; affermer; louer; prendre à louage.

lease (Duration) ‖ Dauer f.; Frist f. ‖ durée f.; terme m. / ~ (Farm rent) ‖ Pacht f.; Pachtgeld n.; Miete f. ‖ ferme f.; fermage m.; bail m.; loyer m. / ~ (Leasehold deed) ‖ Pacht f.; Pachtvertrag m.; Miete f.; Mietvertrag m.; Mietkontrakt m. ‖ ferme f.; bail m.; contrat m. de ferme; louage m.; contrat m. de louage. / ~ (Letting on lease) ‖ Verpachtung f.; Pacht f.; Vermietung f.; Miete f. ‖ ferme f.; bail m. à ferme; arrentement m.; louage m. / ~ (Taking on lease) ‖ Pachtung f.; Pacht f.; Miete f.; Mieten n. ‖ prise f. à ferme; prise f. à bail; ferme f.; prise f. à louage. / ~ (Term of lease) ‖ Pachtzeit f.; Mietzeit f. ‖ temps m. du bail ou de louage. / ~ (Weav) ‖ Kreuz n.; Fadenkreuz n.; Gelese n.; Schrank m.; Rispe f. ‖ envergure f.; encroix m. / to let out on ~ see to lease. / to take on ~ see to lease.

leasehold see lease (Taking on lease). / ~ deed see lease (Leasehold deed).

leaseholder ‖ Pächter m.; Mieter m. ‖ fermier m.; preneur m.; locataire m.

lease pin (Weav) ‖ Kreuznagel m. ‖ cheville f. d'encroix. / ~ rod (Weav) ‖ Kreuzrute f.; Rute f.; Schiene f. ‖ baguette f. d'envergure. / term of ~ see ~ (Term of lease). / ~ tile (Build) ‖ Kreuzriegel m. ‖ tuile f. d'encroix.

leather, to ~ a piston ‖ einen Kolben m. beledern oder lidern ‖ garnir un piston de cuir.

leather ‖ Leder n. ‖ cuir m. / alumed ~ ‖ alaungares Leder n. ‖ cuir m. aluné ou mégissé. / artificial ~ ‖ Kunstleder n.; simili-cuir n.; cuir m. artificiel. / artistic ~ ‖ Leder n. für Kunstgegenstände ‖ cuir m. pour objets d'art. / ~ for auto-

mobiles ‖ Automobilleder n. ‖ cuir m. d'automobiles. / buff ~ ‖ Leder n. zum Polieren; Polierleder n. ‖ cuir m. à polir. / chamois ~ ‖ ölgares oder sämisches Leder n.; Ölleder n.; Waschleder n.; Wildleder n. ‖ cuir m. chamoisé; peau f. chamoisée ou à la chamoise. / chamois-dressed ~ ‖ sämischgares Leder n. ‖ peau f. chamoisée. / chrome ~ ‖ Chromleder n. ‖ cuir m. chromé. / chromed neat's ~ ‖ Chromrindleder n. ‖ cuir m. de vache chromé. / coloured ~ ‖ farbiges Leder n. ‖ cuir m. coloré ou de couleur. / corned ~ ‖ genarbtes Leder n. ‖ cuir m. grené. / dressed ~ ‖ zugerichtetes Leder n. ‖ cuir m. corroyé. / embossed ~ ‖ gepreßtes Leder n. ‖ cuir m. façonné. / fan ~ ‖ Fächerleder n. ‖ cuir m. à éventails. / furniture ~ ‖ Möbelleder n. ‖ cuir m. de meubles. / greased ~ ‖ geschmiertes Leder n. ‖ cuir m. gras. / ground oak-bark ~ see underground-tanned ~. / hat ~ ‖ Hutleder n. ‖ cuir m. à chapeaux. / Hungarian ~ ‖ ungarisches Leder n. ‖ cuir m. hongroyé. / imitation ~ ‖ Lederpappe f.; Kunstleder n. ‖ carton m. cuir; cuir m. artificiel. / ~ for inner soles ‖ Brandsohlenleder n. ‖ cuir m. pour demi-semelles. / kid ~ ‖ Ziegenleder n.; Bockleder n. ‖ chevreau m.; chevrotin m.; cabron m. / japanned ~ see patent ~. / lacquered ~ see patent ~. / made of ~ ‖ ledern ‖ de cuir m.; coriace. / morocco ~ ‖ Saffianleder n. ‖ maroquin m. / neat's ~ ‖ Rindsleder n. ‖ cuir m. de bœuf ou de vache; vache f. / of ~ see made of ~. / oil ~ see oiled ~. / oiled ~ ‖ ölgares Leder n. ‖ cuir m. travaillé à l'huile; cuir m. chamoisé. / oozed ~ ‖ lohgares Leder n. ‖ cuir m. jusé. / orthopedic ~ ‖ orthopädisches Leder n. ‖ cuir m. orthopédique. / parchment ~ ‖ Pergamentleder n. ‖ cuir m. parcheminé. / patent ~ ‖ Lackleder n.; Glanzleder n. ‖ cuir m. laqué ou verni; verni m. / patent ~ articles pl. ‖ Lacklederwaren fpl. ‖ objets mpl. en cuir verni. / perspiration ~ ‖ Schweißleder n. ‖ cuir m. à chapeau. / ~ prepared with oil see chamois ~. / russet upper ~ ‖ Fahlleder n. ‖ cuir m. tanné aux noix de galles; cuir m. à œuvre. / Russian ~ ‖ Juchtenleder n. ‖ cuir m. de Russie. / saddler's ~ ‖ Sattlerleder n. ‖ cuir m. de sellier. / scraped ~ ‖ gestrichenes Leder n. ‖ cuir m. étiré. / shaft ~ ‖ Schaftleder n. ‖ cuir m. mou pour tiges de bottes. / to shagreen the ~ ‖ das Leder mit Narben versehen ‖ chagriner le cuir. / shammy ~ see chamois ~. / shamoy ~ see chamois ~. / sleeked ~ ‖ Blankleder n. ‖ cuir m. avivé. / smooth ~ ‖ abgenarbtes Leder n. ‖ cuir m. lissé. / soft ~ ‖ Weichleder n. ‖ cuir m. mou. / sole ~ ‖ Sohlleder n. ‖ cuir m. à semelles. / strong ~ ‖ Starkleder n. ‖ cuir m. dur. / ~ for stuffbottomed furniture ‖ Polsterleder n. ‖ cuir m. pour meubles rembourrés. / to tallow ~ over a charcoal fire ‖ Leder n. abflammen ‖ donner le suif au cuir. / tallowed ~ ‖ Sattlerleder n. ‖ cuir m. en suif. / tanned ~ ‖ gegerbtes Leder n.; rohgares Leder n. ‖ cuir m. tanné ou en croûte. / ~ tanned in oozes ‖ in Lohbrühe gegerbtes Leder n. ‖ cuir m. jusé ou à la jusée. / tawed ~ ‖ Weißgarleder n.; weißgares Leder n. ‖ cuir m. mégissé ou blanc; peau f. passée en mégie. / tawed ~ for gloves ‖ weißgares Handschuhleder n.; Hand-

schuhleder n. ‖ cuir m. mégissé pour gants. / ~ for technical purposes ‖ technisches Leder n. ‖ cuir m. pour travaux techniques. / Turkey ~ ‖ Rauchleder n. ‖ cuir m. bronzé. / underground-tanned ~ ‖ lohgrubengegerbtes Leder n. ‖ cuir m. tanné dans la fosse. / upholstery ~ ‖ Möbelleder n. ‖ cuir m. d'ameublement. / upper ~ (Shoem) ‖ Oberleder n. ‖ avantpied m.; empeigne f. / varnished ~ ‖ Glanzleder n. ‖ cuir m. verni; verni m. / wash ~ ‖ Waschleder n. ‖ cuir m. rosette. / ~ for water fittings ‖ Lederscheibe f. für Wasserarmaturen ‖ rondelle f. en cuir pour robinets. / wax ~ ‖ Wichsleder n. ‖ cuir m. à cirer. / white ~ see tawed ~.

leather apron ‖ Lederschürze f.; Schurzfell n. ‖ tablier m. en cuir.

leather articles pl. with advertising matter ‖ Lederartikel mpl. mit Reklame ‖ articles mpl. de réclame en cuir. / ~ for dogs ‖ Lederartikel mpl. für Hunde ‖ articles mpl. en cuir pour chiens. / ~ for industrial purposes ‖ Lederwaren fpl. für gewerbliche Zwecke ‖ articles mpl. en cuir pour l'industrie.

leather bag ‖ Ledertasche f. ‖ sac m. ou pochette f. en cuir. / ~ maker ‖ Täschner m. ‖ ouvrier m. de sacs en cuir.

leather ball ‖ Lederball m. ‖ balle f. en cuir. / ~ beating machine ‖ Lederklopfmaschine f. ‖ machine f. à battre les cuirs.

leather belt ‖ Lederriemen m.; Ledergürtel m. ‖ courroie f.; ceinture f. en cuir. / endless ~ ‖ endloser Ledertreibriemen m. ‖ courroie f. en cuir sans fin.

leather blacker ‖ Lederschwarzfärber m. ‖ noircisseur m. de cuir. / ~ board ‖ Lederpappe f. ‖ carton m. cuir. / ~ boot lace ‖ Lederschnürsenkel m.; Schnürriemen m. aus Leder ‖ lacet m. en cuir. / ~-bottomed seat ‖ Ledersessel m. ‖ siège m. en cuir. / ~ box ‖ Lederkoffer m. ‖ malle f. ou valise f. en cuir. / ~ breeches ‖ Lederhose f. ‖ culotte f. de peau. / ~ bronzer ‖ Lederbronzör m. ‖ métalliseur m. sur cuir. / ~ bumper ‖ Lederpuffer m. ‖ amortisseur m. en cuir. / ~ button ‖ Lederknopf m. ‖ bouton m. en cuir. / ~ cam block ‖ Lederklotz m. ‖ sabot m. en cuir. / ~ cardboard ‖ Lederkarton m. ‖ carton-cuir m. / ~ case ‖ Lederfutteral n. ‖ étui m. en cuir. / ~ case for photographic apparatus ‖ Ledertasche f. für fotografische Apparate ‖ trousse f. en cuir pour appareils photographiques. / ~ casing see ~ case. / ~ cloth ‖ Ledertuch n. ‖ toile-cuir m. / ~ clothing ‖ Lederbekleidung f. ‖ vêtement m. en cuir. / ~ coal ‖ Lederkohle f. ‖ cuir m. torréfié; charbon m. de cuir. / ~ container ‖ Lederbehälter m. ‖ étui m. en cuir. / ~ corner ‖ Lederecke f. ‖ coin m. en cuir. / ~ currying ‖ Lederzurichtung f. ‖ corroyage m. de cuir. / ~ cushion ‖ Lederkissen n. ‖ coussin m. de cuir. / ~ cutter-out ‖ Lederstanzer m. ‖ découpeur m. de cuir. / agglomerated ~ cuttings ‖ gepreßte Lederabfälle mpl. ‖ déchets mpl. de cuirs agglomérés. / ~ doll ‖ Lederpuppe f. ‖ poupée f. en cuir.

leather drawing-knife ‖ Lederziehmesser n. ‖ plane f. à cuir. / double-channeled ~ ‖ zweiballiges Lederziehmesser n. ‖ plane f. à cuir à deux biseaux.

leather dresser ‖ Gerber m. ‖ chamoiseur m.; mégissier m.; tanneur m.

leather dressing cream ‖ Lederkrem m. ‖ crème f. pour cuir. / ~ grease ‖ Lederfett n. ‖ graisse f. pour cuir. / ~ oil ‖ Lederöl n. ‖ huile f. pour cuir.

leather driving-belt ‖ Ledertreibriemen m. ‖ courroie f. de transmission en cuir. / ~ drying plant ‖ Ledertrocknungsanlage f. ‖ installation f. de séchage de cuir. / ~ embossing ‖ Lederpressen n. ‖ gaufrage m. de cuir; coréoplastie f. / ~ enameller ‖ Lederemaillierer m. ‖ émailleur m. sur cuir. / ~ fat ‖ Lederfett n. ‖ graisse f. pour cuir. / ~ finishing ‖ Lederappretur f. ‖ apprêt m. de cuir. / ~ finishing ink *see* ~ finishing. / ~ fitting ‖ Lederbeschlag m. ‖ garniture f. de cuir. / ~ fold machine ‖ Lederfalzmaschine f. ‖ plieuse f. à cuir. / ~ frame ‖ Lederrahmen m. ‖ cadre m. en cuir. / ~ furniture ‖ Ledermöbel npl. ‖ meubles mpl. en cuir. / ~ gaiters pl. ‖ Ledergamaschen fpl. ‖ guêtres fpl. en cuir. / gilder on ~ (Bookb) ‖ Ledervergolder m. ‖ doreur m. sur cuir. / ~ gilding ‖ Ledervergoldung f. ‖ dorure f. sur cuir. / ~ girdle ‖ Ledergürtel m. ‖ ceinture f. de cuir. / ~ glove ‖ Lederhandschuh m. ‖ gant m. de peau *ou* de cuir. / ~ glue ‖ Lederleim m. ‖ colle f. de cuir.

leather goods pl. ‖ Lederwaren fpl. ‖ articles mpl. *ou* ouvrages mpl. en cuir. / ~ for technical objects ‖ technische Lederwaren fpl. ‖ articles mpl. techniques en cuir.

leather goods industry ‖ Lederwarenindustrie f. ‖ industrie f. travaillant le cuir. / ~ manufacturing ‖ Lederwarenfabrik f. ‖ fabrique f. d'objets en cuir.

leather grease ‖ Lederfett n. ‖ graisse f. pour le cuir. / ~ hammering ‖ Lederklopferei f. ‖ battage m. de cuirs. / ~ handle ‖ Ledergriff m. ‖ manche m. en cuir. / ~ hangings pl. ‖ Ledertapete f. ‖ tapis m. en cuir. / ~ head-gear ‖ Lederkopfbügel m. ‖ ressort m. de tête garni de cuir. / ~ heel ‖ Lederabsatz m. ‖ talon m. en cuir. / ~ helmet ‖ Lederhelm m. ‖ casque m. en cuir. / ~ hose ‖ Lederschlauch m. ‖ manche f. en cuir. / ~ imitation ‖ Lederimitation f. ‖ imitation f. de cuir.

leathering (Mach) ‖ Lederdichtung f.; Lederliderung f. ‖ garniture f. *ou* étoupage m. en cuir.

leather lac ‖ Lederlack m. ‖ vernis m. à cuir. / ~ leash for dogs ‖ Hundeleine f. aus Leder ‖ laisse f. en cuir pour chiens. / ~ lubricating material ‖ Lederschmiermittel n. ‖ matière f. pour la lubrification de cuir. / ~ measuring instrument ‖ Ledermeßinstrument n. ‖ instrument m. à mesurer le cuir.

leathern ‖ ledern ‖ de cuir m.

leatheroid ‖ Kunstleder n. ‖ simili-cuir m.; cuir m. artificiel.

leather packing ‖ Lederliderung f. ‖ garniture f. de cuir. / ~ packing collar ‖ Ledermanschette f.; Lederstulp m. ‖ garniture f. *ou* manchon m. *ou* joint m. embouti en cuir. / ~ pad ‖ Lederpuffer m. ‖ amortisseur m. en cuir. / ~ painter ‖ Ledermaler m. ‖ peintre m. sur cuir. / ~ paper ‖ Lederpapier n. ‖ papier-cuir m. / ~ pattern ‖ Lederschablone f. ‖ patron m. en cuir. / ~ pipe ‖ Lederschlauch m. ‖ tuyau m. en cuir. / ~ polishing material ‖ Lederputzmittel n. ‖ matière f. pour le nettoyage de cuir. / ~ pouch ‖ Lederbeutel m. ‖ pochette f. *ou* sac m. en cuir. / ~ preserving preparation ‖ Lederkonservierungsmittel n. ‖ produit m. conservateur de

cuir. / ~ pressing establishment ‖ Lederpreßwerk n. ‖ établissement m. à presser le cuir. / ~ pressman ‖ Lederpresser m. ‖ presseur m. de cuir. / ~ punching establishment ‖ Lederstanzwerk n. ‖ établissement m. à estamper et à emboutir le cuir. / ~ purse ‖ Lederbörse f. ‖ bourse f. en cuir. / refuse of ~ ‖ Lederabfall m. ‖ déchet m. de cuir. / ~ roller (Apparatus) ‖ Lederwalze f. ‖ rouleau m. en cuir. / ~ roller (Person) ‖ Lederwalzer m. ‖ lisseur m. de cuir. / ~ seat ‖ Ledersitz m. ‖ siège m. de cuir. / ~ sewing machine ‖ Ledernähmaschine f. ‖ machine f. à coudre les cuirs. / ~ shoe with wooden sole ‖ Lederüberschuh n. mit Holzsohle ‖ galoche f. en cuir avec semelle en bois. / ~ silverer ‖ Lederversilberer m. ‖ argenteur m. sur cuir. / ~ silvering ‖ Lederversilberung f. ‖ argenture f. sur cuir. / ~ skiving and splitting machine ‖ Lederschärf- und Spaltmaschine f. ‖ machine f. à parer et à refendre le cuir. / ~ slide-preventing device ‖ Ledergleitschutz m. ‖ antidérapant m. en cuir. / ~ slipper ‖ Lederhausschuh m. ‖ chausson m. en cuir. / ~ sole ‖ Ledersohle f. ‖ semelle f. en cuir. / sort of ~ ‖ Lederart f. ‖ espèce f. *ou* qualité f. de cuir. / ~ splitting machine ‖ Lederspaltmaschine f. ‖ machine f. à refendre le cuir. / ~ sport requisites pl. ‖ Ledersportartikel mpl. ‖ articles mpl. de sport en cuir. / ~ stamping establishment ‖ Lederprägewerk n. ‖ établissement m. à empreindre le cuir.

leather strap ‖ Lederriemen m.; Ledergürtel m. ‖ courroie f.; ceinture f. en cuir. / ~ for belts ‖ Ledernähriemen m. für Treibriemen ‖ lanière f. à coudre les courroies. / ~ for clogs ‖ Lederriemen m. für Holzschuhe ‖ bride f. de sabots.

leather strip cutting machine ‖ Lederstreifenschneidmaschine f. ‖ machine f. à couper les bandes de cuir. / ~ stuffing-box (Mach) ‖ Lederstopfbüchse f. ‖ manchon m. en cuir pour boîte de bourrage. / ~ surface measuring machine ‖ Ledermeßmaschine f. ‖ machine f. à mesurer les surfaces du cuir. / apparatus for measuring ~ thicknesses ‖ Lederdickenmeßvorrichtung f. ‖ appareil m. à mesurer l'épaisseur du cuir. / ~ thong for whips ‖ Peitschenriemen m. aus Leder ‖ courroie f. en cuir pour fouets. / ~ toys pl. ‖ Lederspielwaren fpl. ‖ jouets mpl. en cuir. / ~ treating establishment ‖ Lederzurichterei f. ‖ établissement m. à préparer le cuir *ou* à travailler le cuir.

leather trunk ‖ Lederkoffer m. ‖ malle f. *ou* valise f. en cuir. / ~ maker ‖ Lederkoffermacher m. ‖ coffretier m. en cuir.

leather valve ‖ Lederklappe f. ‖ clapet m. en cuir. / ~ varnish ‖ Lederlack m. ‖ laque f. pour cuir. / ~ varnisher ‖ Lederlackierer m. ‖ vernisseur m. sur cuir. / ~ washer ‖ Lederring m.; Lederscheibe f. ‖ rondelle f. de cuir. / ~ waste ‖ Lederabfall m. ‖ déchet m. de cuir. / ~ whitener ‖ Lederbleicher m. ‖ blanchisseur m. de cuir. / ~ whitening ‖ Lederbleiche f. ‖ blanchissage m. de cuirs. / ~ worker ‖ Lederarbeiter m. ‖ ouvrier m. en cuir. / ~ working machine ‖ Lederbearbeitungsmaschine f. ‖ machine f. à travailler le cuir. / ~ wrapper ‖ Ledermappe f. ‖ portefeuille m. en cuir.

leave, to ~ blank (Print) ‖ blank schlagen ‖ mettre du blanc. / ~ out ‖ auslassen ‖

omettre. / ~ a port ‖ in See f. gehen ‖ mettre *ou* prendre la mer. / ~ work without authority ‖ die Arbeit f. unbefugt verlassen ‖ quitter le travail sans permission.

leaven ‖ Sauerteig m.; Hefe f. ‖ levain m.

leaves pl. ‖ Flitter m. ‖ paillette f. / prepared ~ ‖ sterilisiertes Laub n. ‖ feuilles mpl. naturalisées.

leavings pl. ‖ Überbleibsel n.; Abfall m. ‖ reste m.; résidu m. / burnt ~ ‖ Röstrückstände mpl. ‖ résidus mpl. de grillage.

Leclanché cell ‖ Leclanché-Element n. ‖ élément m. de Leclanché.

lecture room ‖ Vortragsaal m. ‖ salle f. de conférences.

ledge of a plane ‖ Hobelanschlag m. ‖ joue f. du rabot.

ledge beam ‖ Firstbalken m. ‖ poutre f. de faîte.

ledger ‖ Hauptbuch n. ‖ grand livre m. / general ~ *see* ledger. / private ~ ‖ Bilanzbuch n. ‖ livre m. du bilan.

ledger blade ‖ Scherklinge f. ‖ lame f. tondeuse. / circular ~ ‖ runde Scherklinge f. ‖ lame f. tondeuse circulaire. / curved ~ ‖ gebogene Scherklinge f. ‖ lame f. tondeuse cintrée. / straight ~ ‖ gerade Scherklinge f. ‖ lame f. tondeuse droite.

lee ‖ Leeseite f. ‖ côté m. sous le vent. / a-~ ‖ leewärts; in Lee f. ‖ sous le vent.

leeboard (Shipb) ‖ Schwert n. ‖ aile f. de dérive; semelle f.

leech (Mar) ‖ Liek n. ‖ ralingue f. / ~ breeder ‖ Blutegelzüchter m. ‖ éleveur m. de sangsues. / ~ line (Mar) ‖ Gording f. ‖ cargue f. / ~ ropes pl. ‖ Leetaue npl.; Leetauwerk n. ‖ manœuvres mpl. de revers.

lees pl. (Brew) ‖ Glattwasser n. ‖ dernier bouillon m.; dernière trempe f. de lavage.

lee sheet ‖ Leeschot f. ‖ écoute f. de revers; écoute f. de sous le vent. / ~ shelter ‖ Windschatten m. ‖ ombre f. du vent; aire f. abritée du vent. / ~ shore ‖ Leeküste f.; ‖ côte f. *ou* terre f. sous le vent. / ~-side ‖ Leeseite f. ‖ côté m. sous le vent.

lees presser (Wine) ‖ Weinhefepresser m.; Treberpresser m. ‖ presseur m. de lies.

leeway ‖ Abtrift f.; Leeweg m. ‖ dérive f.

leeward ‖ leewärts; in Lee f. ‖ sous le vent.

left ‖ links ‖ à gauche. / to be ~ till called for ‖ postlagernd ‖ poste f. restante.

left-hand action ‖ Linksgang m. ‖ tournant m. à gauche. / ~ curve ‖ Linkskurve f. ‖ courbe f. à gauche. / ~ cutting ‖ linksschneidend ‖ coupant à gauche. / ~ deviation ‖ Linksabweichung f. ‖ déviation f. à gauche.

left-handed ‖ linksgängig ‖ à pas à gauche. / ~ arbor (Watchm) ‖ Linkser m. ‖ arbre m. à rebours. / ~ screw ‖ linksgängige Schraube f. ‖ vis f. à gauche. / spring with a ~ helice ‖ links gewundene Feder f. ‖ ressort m. enroulé à gauche.

left-hand milling cutter ‖ linksschneidender Fräser m. ‖ fraise f. coupant à gauche. / ~ motion ‖ Linkslauf m. ‖ marche f. à gauche. / ~ switch ‖ Linksweiche f. ‖ changement m. à gauche. / ~ thread ‖ Linksgewinde n. ‖ filet m. à gauche. / ~ threading attachment ‖ Linksgewindeschneideinrichtung f. ‖ dispositif m. de filetage à gauche. / ~ turn-out

(Railw) ‖ Linksweiche f. ‖ branchement m. *ou* aiguille f. à gauche.

left-side (Mach) ‖ Bedienungsseite f. ‖ côté m. au service.

leg ‖ Bein n.; Schenkel m. ‖ jambe f.; branche f. / ~ of an angle ‖ Schenkel m. eines Winkels ‖ côté m. d'un angle. / boot ~ ‖ Stiefelschaft m. ‖ tige f. de botte. / ~ of a tripod ‖ Gestellbein n. ‖ pied m. de support.

legacy ‖ Legat n. ‖ legs m.

legal ‖ rechtlich; rechtsgültig; gesetzmäßig ‖ juridique; honnête; loyal; légal; valide. / ~ action ‖ Rechtsverfahren n.; Rechtsfall m. ‖ procédure f.; cas m. ligitieuse. / to take ~ action ‖ den Rechtsweg m. beschreiten ‖ recourrir à la justice; prendre la voie tribunale. / ~ advice ‖ Rechtsbeistand m. ‖ conseil m. judiciaire; avocat m. / ~ charges pl. ‖ Gerichtskosten pl. ‖ frais mpl. judiciaires. / ~ claim ‖ Rechtsanspruch m. ‖ droit m.; titre m. / ~ dispute ‖ Rechtsstreit m. ‖ procès m.; litige m. / ~ expenses pl. ‖ Rechtskosten pl. ‖ frais mpl. de justice. / ~ force ‖ Rechtskraft f. ‖ force f. de loi. / ~ ground ‖ Rechtsgrund m. ‖ titre m.; cause f.; moyen m. / ~ obligation ‖ Rechtsverbindlichkeit f. ‖ obligation f. légale *ou* devant la loi. / ~ position ‖ Rechtslage f. ‖ situation f. légale. / ~ proceedings pl. ‖ Gerichtsverfahren n.; Rechtsverfahren n. ‖ procédure f. / ~ question ‖ Rechtsfrage f. ‖ question f. de droit. / place for delivery, payment and for ~ questions is N. N. ‖ Erfüllungsort m. für Lieferung und Zahlung sowie Gerichtsstand ist N. N. ‖ lieu m. pour livraison, payement et tribunal compétant est N. N. / ~ redress ‖ Rechtsmittel n. ‖ recours m. de droit. / ~ relations pl. between employer and employee ‖ Rechtsverhältnis n. zwischen Arbeitgeber und Arbeitnehmer ‖ convention f. légale entre employeur et employé. / ~ tender ‖ gesetzliches Zahlungsmittel n. ‖ valeur f. légale. / ~ time ‖ Einheitszeit f. ‖ temps m. fixé par la loi. / ~ title ‖ Rechtsanspruch m. ‖ droit m.; titre m.

legality ‖ Rechtsgültigkeit f. ‖ authenticité f.; validité f.; légalité f.

legalize, to ‖ legalisieren; beglaubigen ‖ légaliser; certifier.

legalized ‖ beglaubigt ‖ légalisé. / to have an invoice ~ ‖ eine Faktur f. beglaubigen lassen ‖ faire légaliser une facture.

legally binding ‖ rechtsgültig; rechtsverbindlich ‖ valide; légale et valide. / to become ~ ‖ rechtskräftig werden; Rechtskraft f. erlangen ‖ passer en force f. de loi.

legally valid ‖ rechtsverbindlich ‖ obligatoire; légal et valide. / ~ argument ‖ Rechtsgrund m. ‖ titre m.

legend (Coin) ‖ Rundschrift f. *oder* Umschrift f. einer Münze / légende f.

legging ‖ Gamasche f. ‖ guêtre f. / leather ~ ‖ Ledergamasche f. ‖ guêtre f. en cuir.

legging frame ‖ Längenstuhl m. ‖ cotière f.

legible ‖ leserlich ‖ lisible.

legislation ‖ Gesetzgebung f. ‖ législation f. / ~ for the benefit and protection of the working classes ‖ Arbeiterschutzgesetzgebung f. ‖ lois f. pour la protection de l'ouvrier.

leg manufacturing ‖ Schäftefabrikation f.; Schäftefertigung f. ‖ fabrication f. de tiges de chaussures.

legume ‖ Hülsenfrucht f. ‖ légume m. / dried ~ ‖ getrocknete Hülsenfrucht f. ‖ légume m. sec. / meal of ~s ‖ Mehl n. aus Hülsenfrüchten ‖ farine f. de légumes.

legumine ‖ Pflanzenkasein n. ‖ caseïne f. végétale; légumine f.

lemon ‖ Zitrone f. ‖ citron m. / ~ acid ‖ Zitronensäure f. ‖ acide m. citrique.

lemonade ‖ Limonade f. ‖ limonade f.

lemon-grass oil ‖ Lemongrasöl n. ‖ essence f. de lemongrass.

lemon juice ‖ Zitronensaft m. ‖ jus m. de citron. / ~ peel ‖ Zitronenschale f. ‖ écorce f. *ou* pelure f. de citron. / ~ wood ‖ Zitronenholz n. ‖ bois m. de citron.

lend, to ~ ‖ ausleihen ‖ prêter.

lender ‖ Ausleiher m. ‖ prêteur m.

lending business ‖ Leihgeschäft n. ‖ prêt m. sur gage *ou* sur nantissement. / ~ library ‖ Leihbücherei f. ‖ cabinet m. de lecture.

length ‖ Länge f. ‖ longueur f. / approximate ~ ‖ ungefähre Länge f. ‖ longueur f. approximative. / assorted ~s pl. ‖ assortierte Längen fpl. ‖ longueurs fpl. assorties. / average ~ ‖ durchschnittliche Länge f. ‖ longueur f. moyenne. / ~ of chain (Mar) ‖ Kettenlänge f. ‖ maillon m. *ou* longueur f. de chaîne. / ~ of cut ‖ Schnittlänge f. ‖ longueur f. de coupe. / to cut to ~ ‖ ablängen; auf Länge f. schneiden ‖ couper à longueur f. / to divide into commercial ~s ‖ auf handelsübliche Längen fpl. zerteilen ‖ débiter en longueurs fpl. commerciales. / exact ~ ‖ genaue Länge f. ‖ longueur f. exacte. / extra ~s pl. ‖ Extralängen fpl. ‖ longueurs fpl. supplémentaires. / ~ focal ~ ‖ Brennweite f. ‖ distance f. focale. / greatest ~ between centres ‖ größte Länge f. zwischen den Spitzen ‖ distance f. maximum *ou* longueur f. maximum entre pointes. / ground ~ ‖ Schleiflänge f. ‖ longueur f. à rectifier. / ~ of hauling ‖ Förderlänge f. ‖ distance f. de transport. / in ~s pl. of . . . ‖ in Längen fpl. von . . . ‖ en longueurs fpl. de . . . / ~ of jib ‖ Auslegerlänge f. ‖ longueur f. de la flèche. / ~ of the land chain (Surv) ‖ Kettenlänge f.; Kettenmaß n. ‖ chaînée f. / ~ measured incorrectly ‖ unrichtig gemessene Länge f. ‖ longueur f. mesurée inexactement. / ~ over all (Shipb) ‖ Länge f. über alles ‖ longueur f. hors tout. / ~ of a pace ‖ Schrittlänge f. ‖ longueur f. de pas. / ~ between perpendiculars ‖ Länge f. zwischen den Loten ‖ longueur f. entre perpendiculaires. / ~ of pointer ‖ Zeigerlänge f. ‖ longueur f. d'aiguille. / ~ of rail ‖ Länge f. der Schiene; Schienenlänge f. ‖ longueur f. du rail. / ~ of rivets depending on the joint thickness ‖ Nietlänge f. in Abhängigkeit von den Klemmlängen ‖ longueur f. des rivets dépendant de la longueur des bornes. / ~ of service ‖ Dienstalter n. ‖ ancienneté f. / ~ of strike (Mine) ‖ streichende Länge f. des Grubenfeldes ‖ longueur f. du champ d'exploitation. / ~ of stroke of piston ‖ Kolbenhub m. ‖ coup m. *ou* course f. du piston. / ~ of turret feed ‖ Revolverkopfdrehlänge f. ‖ course f. de chariotage de la tourelle. / usual ~ ‖ gewöhn-

liche Länge f. ‖ longueur f. habituelle. / ~ at water-line (Shipb) ‖ Länge f. in der Wasserlinie ‖ longueur f. à la flottaison. / working ~ ‖ Nutzlänge f. ‖ longueur f. utile.

lengthen, to ‖ verlängern; dehnen ‖ allonger; étendre. / ~ iron ‖ das Eisen strecken ‖ étirer le fer.

lengthened ‖ verlängert ‖ allongé; prolongé.

lengthening ‖ Verlängerung f. ‖ allongement m.; prolongement m. / ~ bar for compasses ‖ Zirkelverlängerung f. ‖ rallonge f. de compas. / ~ lever ‖ Verlängerungshebel m. ‖ levier m. de rallonge. / ~ rod ‖ Verlängerungsstange f. ‖ allonge f.

length gauge (Needl) ‖ Schachtmodell n. ‖ moule m. pour les brins. / ~ measuring instrument ‖ Längenmeßgerät n. ‖ appareil m. pour la mesure des longueurs *ou* en long. / ~ measuring machine ‖ Längenmeßmaschine f. ‖ machine f. à mesurer la longueur. / sense of ~ ‖ Längsrichtung f. ‖ sens m. de la longueur. / ~ shearing machine combined with nap lifting apparatus ‖ Langschermaschine f. mit vorgebautem Veloursrhebeapparat ‖ tondeuse f. longitudinale avec appareil à relever les poils du velours.

lengthwise ‖ längs ‖ en long; longitudinal.

lens ‖ Linse f. ‖ lentille f. / achromatic ~ ‖ achromatische Linse f. ‖ lentille f. achromatique. / achromatic quartz and fluorite ~ ‖ Quarzfluoritachromatlinse f. ‖ objectif m. achromatique quartzfluorine. / anastigmatic ~ ‖ anastigmatische Linse f. ‖ lentille f. anastigmatique. / annular ~ ‖ Ringlinse f. ‖ lentille f. annulaire. / back ~ ‖ Hinterlinse f. ‖ lentille f. arrière. / bevel edged ~es pl. ‖ facettierte Brillengläser npl. ‖ verres mpl. taillés à biseau. / bifocal ~ ‖ Bifokalglas n. ‖ verre m. à double foyer. / ~ for persons operated for cataract ‖ Glas n. für Staroperierte ‖ verre m. pour les opérés de la cataracte. / cemented ~es pl. ‖ verkittete Linsen fpl. ‖ lentilles fpl. collées; lentille f. constituée par verres accolés. / centring ~ ‖ Zentrierlinse f. ‖ lentille f. de centrage. / chromatic quartz ~ ‖ Quarzchromatlinse f. ‖ objectif m. chromatique en quartz. / coloured ~es pl. ‖ gefärbte Gläser npl. ‖ verres mpl. colorés. / colourless ~ transmitting the undiminished intensity of the light ‖ farbloses Glas n. von ungeschwächter Lichtdurchlässigkeit ‖ verre m. correcteur incolore de transparence parfaite. / compass ~ ‖ Kompaßglas n. ‖ verre m. pour boussoles. / concave ~ ‖ konkave Linse f; Hohllinse f. ‖ lentille f. concave. / concavo-concave ~ ‖ bikonkave Linse f. ‖ lentille f. biconcave *ou* concavo-concave. / concavo-convex ~ ‖ konkavkonvexe Linse f. ‖ lentille f. concavo-convexe; ménisque m. divergent. / condenser ~ ‖ Beleuchtungslinse f.; Kondensorlinse f. ‖ lentille f. de condensateur. / condensing ~ ‖ Sammellinse f. ‖ lentille f. convergente. / converging ~ *see* condensing ~. / convex ~ ‖ Sammellinse f.; Konvexlinse f. ‖ lentille f. convexe. / cylinder ~ ‖ Zylinderlinse f. ‖ lentille f. cylindrique. / dispersing ~ ‖ Zerstreuungslinse f. ‖ lentille f. divergente. / diverging ~ *see* dispersing ~.

/ double ~ ‖ Doppellinse f. ‖ lentille f. double. / double concave ~ / bikonkave Linse f. ‖ lentille f. biconcave *ou* concavo-concave. / dry ~es pl. ‖ Trockensystem n. ‖ objectifs mpl. à sec. / duplicate ~ ‖ Ersatzglas n. ‖ verre m. de rechange. / edged ~ ‖ facettiertes Glas n. ‖ verre m. taillé à biseau. / ~ for electric pocket lamps ‖ Linse f. für elektrische Taschenlampen ‖ lentille f. pour lampes électriques de poche. / to examine an object through a ~ ‖ einen Gegenstand m. durch die Lupe betrachten ‖ examiner un objet à la loupe. / eyepiece correcting ~ ‖ Okularaufsteckglas n. ‖ verre m. à emboîter sur l'oculaire. / field ~ ‖ Kollektivlinse f.; Vorderlinse f. ‖ verre m. de champ; lentille f. avant. / flat oval ~ ‖ flachovales Brillenglas n. ‖ verre m. à ovale aplati. / fluorite ~ ‖ Flußspatlinse f. ‖ lentille f. en spath fluor. / focussing ~ ‖ Einstellupe f. ‖ loupe f. de mise. / the ~es pl. are fogged ‖ die Gläser npl. laufen an ‖ les verres mpl. se couvrent de buée. / framed ~es pl. ‖ gefaßte Brillengläser npl. ‖ verres mpl. montés. / ~ entirely free from bubbles and streaks ‖ von Bläschen und Schlieren völlig freies Glas n. ‖ verre m. parfaitement exempt de bulles et de fils. / ~ entirely free from flaw in the glass ‖ von Glasfehlern völlig freies Glas n. ‖ verre m. parfaitement exempt de défauts (de verre). / ~ free from scratches ‖ von Kratzern freies Glas n. ‖ verre m. exempt d'éraflures. / ~ free from zones ‖ von Zonen freies Glas n. ‖ verre m. exempt de zones *ou* d'égratignures. / front ~ ‖ Vorsatzlinse f. ‖ lentille f. additionnelle. / front ~ operating as an objective ‖ Objektivvorsatzlinse f. ‖ lentille f. additionnelle se plaçant devant l'objectif. / globular ~ ‖ kugelige Linse f. ‖ lentille f. à surface sphérique. / ~ having in the upper portion a continuously deepening greyish brown colour ‖ Glas n. mit nach oben hin stetig zunehmender graubrauner Färbung ‖ verre m. de teinte gris-brunâtre de plus en plus foncée vers le haut. / ~ of headlight ‖ Glaslinse f. des Scheinwerfers ‖ lentille f. de phare. / illuminating ~ ‖ Beleuchtungslinse f. ‖ lentille f. d'éclairage. / ~ of great light transmitting capacity ‖ lichtstarke Beleuchtungslinse f. ‖ lentille f. d'éclairage très lumineuse. / magnifying ~ ‖ Lupe f. ‖ loupe f. / of small magnifying power ‖ Linse f. mit schwacher Vergrößerung ‖ lentille f. à grossissement faible. / meniscal ~ ‖ mondförmiges Glas n. ‖ (lentille f. à) ménisque m. / micoquille ~ ‖ Halbmuschelglas n. ‖ verre m. ménisque. / mounted ~ ‖ gefaßte Linse f. ‖ lentille f. sertie. / non-achromatic ~ ‖ nicht achromatische Linse f. ‖ lentille f. non achromatique. / nuclear ~ ‖ Kernlinse f. ‖ lentille f. nucléaire. / ophthalmoscope ~ ‖ Ophthalmoskoplinse f. ‖ lentille f. ophtalmoscopique. / optically defective ~ ‖ optisch mangelhaftes Glas n. ‖ verre m. optique défectueux. / ~ out of centre ‖ außermittige Linse f. ‖ lentille f. excentrique. / pantoscopic ~ ‖ pantoskopisches Glas n. ‖ verre m. pantoscopique. / plano-concave ~ ‖ plankonkave Linse f. ‖ lentille f. plan-concave *ou* concave-plane.

/ plano-convex ~ ‖ plankonvexe Linse f. ‖ lentille f. plan-convexe *ou* convexe-plane. / plano-spherical ~ ‖ Planglas n. ‖ verre m. ayant une face plane. / quartz fluorite ~es pl. ‖ Quarzfluoritobjektiv n. ‖ objectif m. en quartzfluorine. / rapid ~es pl. ‖ lichtstarkes Objektiv n. ‖ objectif m. lumineux. / ~ with ruled cross ‖ Strichkreuzglas n. ‖ verre m. à réticule. / semi-circular ~ ‖ Halbkugellinse f. ‖ lentille f. demi-boule. / sight correcting ~ for eyes ametropia ‖ Korrektionsglas n. für Fehlsichtige / verre m. correcteur pour amétropes. ‖ sight correcting ~ for spectacle wearers ‖ Korrektionsglas n. für Brillenträger ‖ verre m. correcteur pour porteurs de lunettes. / spectacle ~ ‖ Brillenglas n. ‖ verre m. de lunettes. / spherical ~ ‖ Kugellinse f. ‖ verre m. sphérique. / triple cemented ~ ‖ dreifach verkittete Linse f. ‖ lentille f. constituée par trois verres accolés. / umbral ~ with a uniform moderating effect at all viewing angles throughout every part of the disc ‖ für das ganze Blickfeld gleichmäßig gefärbtes Umbralglas n. ‖ verre m. umbral à coloration uniforme du champ total.

lens, absorbing capacity of a ~ ‖ Absorptionswert m. eines Glases ‖ capacité f. d'absorption d'un verre.

lens attachment ‖ Vorhänger m. ‖ face f. supplémentaire. / removable ~ ‖ abnehmbarer Vorhänger m. ‖ face f. supplémentaire amovible. / frame of the ~ ‖ Vorhängergestell n. ‖ monture f. des faces supplémentaires. / slipping off of the ~ ‖ Abnehmen n. des Vorhängers ‖ enlèvement m. de la face supplémentaire. / slipping on of the ~ ‖ Aufsetzen n. des Vorhängers ‖ mise f. en place de la face supplémentaire.

lens, axis of the ~ ‖ Linsenachse f. ‖ axe m. de la lentille. / ~ cap ‖ Objektivdeckel m. ‖ couvercle m. de l'objectif. / hinged ~ cap in dew mount ‖ aufklappbarer Objektivdeckel m. in der Taukappe ‖ couvercle m. à clapet de l'objectif dans le tube pare-rosée. / ~ combination ‖ Linsensystem n. ‖ système m. de lentilles. / ~ diameter ‖ Linsendurchmesser m. ‖ diamètre m. de la lentille. / fibrillation of the ~ ‖ Faserung f. der Linse ‖ fibres fpl. du cristallin. / ~ glass for lanterns ‖ Linsenglas n. für Laternen ‖ verre m. lenticulaire pour lanternes. / ~ head screw ‖ Linsenschraube f. ‖ vis f. à tête fraisée à goutte de suif. / ~ holder ‖ Linsenhalter m. ‖ portelentille m. / interposition of a ~ ‖ Einschalten n. einer Linse ‖ interposition f. d'une lunette. / ~ magnification ‖ Lupenvergrößerung f. ‖ grossissement m. de loupe. / ~ mount ‖ Linsenfassung f. ‖ monture f. de lentille. / pair of ~es ‖ Linsenpaar n. ‖ paire f. de lentilles. / ~-shaped ‖ linsenförmig ‖ lenticulaire. / ~ stand ‖ Lupenstativ n. ‖ statif m. de la loupe. / ground ~ surface ‖ geschliffene Glasfläche f. ‖ face f. de verre taillée. / polished ~ surface ‖ polierte Glasfläche f. ‖ face f. de verre polie. / system of ~es ‖ Linsensystem n. ‖ système m. de lentilles. / variable image erecting system of ~es ‖ verstellbares bildumkehrendes Linsensystem n. ‖ système m. de lentilles redresseur mobile.

lenticular ‖ linsenförmig ‖ lenticulaire.
lentiform *see* lenticular.
lentil ‖ Linse f. ‖ lentille f. / meal of ~s ‖ Linsenmehl n. ‖ farine f. de lentilles.
lentisk gum ‖ Mastixharz n. ‖ lentisque m.
leonine spun ‖ leonisches Gespinst n. ‖ tissu m. léonique.
lepidolite ‖ Lepidolith m.; Lithionglimmer m. ‖ lépidolithe f.
lessee ‖ Pächter m.; Mieter m. ‖ fermier m.
lessen, to ‖ schmälern; mindern ‖ rétrécir; diminuer; réduire.
lessor ‖ Verpächter m.; Vermieter m. ‖ bailleur m.; loueur m.
let, to ~ ‖ lassen; gestatten ‖ laisser. / ~ come down (Nav) ‖ sacken lassen ‖ mollir; larguer. / ~ down (Mine) ‖ hinablassen; hängen ‖ abaisser; descendre. / ~ down the fire ‖ den Ofen m. ausblasen ‖ mettre le fournau m. hors feu; arrêter *ou* refroidir le fourneau. / ~ down the steel ‖ den Stahl m. anlassen ‖ faire revenir *ou* recuire l'acier m. / it is recommended to ~ down the steel in rape-seed oil after hardening ‖ nach der Härtung f. soll ein gelindes Anlassen des Stahles in Rüböl erfolgen ‖ il est à recommander de faire revenir l'acier m. à l'huile de colza après la trempe. / ~ go ‖ loslassen ‖ larguer; lâcher; relâcher. / ~ in the spokes pl. ‖ die Speichen fpl. einzapfen ‖ empatter les rais mpl. / ~ off ‖ ablassen ‖ retirer.

letter ‖ Brief m. ‖ lettre f. / ~ (Print) ‖ Type f.; Buchstabe m. ‖ lettre f.; caractère m. d'imprimerie; type m. / to address a ~ to... ‖ einen Brief m. richten an ... ‖ adresser une lettre à ... / ~ for advertising purposes ‖ Buchstabe m. für Werbezwecke ‖ caractère m. pour réclame. / ~ of advice ‖ Avisbrief m. ‖ lettre f. d'avis. / ~ of application ‖ Bewerbungsschreiben n. ‖ demande f. de place. / by ~ ‖ brieflich ‖ par lettre f. / ~ of caoutchouc ‖ Kautschukbuchstabe m. ‖ lettre f. en caoutchouc. / capital ~ (Print) ‖ Majuskel f.; Titelbuchstabe m. ‖ lettre f. capitale *ou* majuscule; majuscule f. / circular ~ ‖ Rundschreiben n.; Zirkular n. ‖ lettre f. circulaire. / ~ of conveyance ‖ Frachtbrief m. ‖ lettre f. de voiture. / covering ~ ‖ Begleitbrief m. ‖ lettre f. d'envoi. / ~ of credit ‖ Kreditbrief m.; Akkreditiv n. ‖ lettre f. de crédit; accréditif m. / the ~ hangs (Print) ‖ die Schrift steht schief ‖ la lettre chevauche. / ~ of hypothecation ‖ Verpfändungsurkunde f. ‖ lettre f. hypothécaire. / ~ of indemnity ‖ Garantieschein m. ‖ cautionnement m.; promesse f. de garantie. / ~ of introduction ‖ Empfehlungsbrief m. ‖ lettre f. d'introduction. / kerned ~ (Print) ‖ unterschnittener Buchstabe m. ‖ caractère m. crêné. / lower case ~ ‖ Minuskel f. ‖ bas--de-casse m.; minuscule f. / metal ~ for signboards ‖ Firmenbuchstabe m. aus Metall ‖ lettre f. en métal à composer les plaques-adresses. / ~ containing money ‖ Geldbrief m. ‖ lettre f. chargée; chargement m. / ornamented ~ ‖ Zierbuchstabe m. ‖ lettre f. artistique à crochets. / pneumatic tube ~ ‖ Rohrpostbrief m. ‖ petit-bleu m. / ~ of recommendation ‖ Empfehlungsschreiben n. ‖ lettre f. de recommandation. / ~ of reference (Print) ‖ Nachweisebuchstabe m.; Hinweisbuchstabe m. ‖ lettre f. explica-

tive *ou* de renvoi. / registered ~ ‖ eingeschriebener Brief m. ‖ lettre f. recommandée. / ~ of renunciation ‖ Verzichterklärung f. ‖ lettre f. de renoncement *ou* de renonciation. / ~ requesting payment ‖ Mahnbrief m. ‖ réclamation f.; rappel m.; lettre f. de sommation *ou* de réclamation. / small ~ *see* lower case ~. / superior ~ ‖ Notenbuchstabe m.; Spaltenbuchstabe m.; Verweisungsbuchstabe m. ‖ lettrine f. supérieure f. / de notes supérieure f. / two-line ~ ‖ Majuskel f.; Titelbuchstabe m. ‖ lettre f. capitale *ou* majuscule; majuscule f. / ~ containing valuables ‖ Wertbrief m. ‖ lettre f. chargée *ou* de valeur déclarée.

letter bag ‖ Briefbeutel m. ‖ sac m. à lettres. / ~ balance ‖ Briefwage f. ‖ pèse-lettre m. / ~ basket ‖ Briefkorb m. ‖ corbeille f. *ou* panier m. à correspondance. / ~ blank ‖ Buchstabenweiß n. ‖ blanc m. de lettres. / ~ blank key (Typewr) ‖ Leertaste f. ‖ touche f. inopérante *ou* de lettre morte. / ~ book ‖ Briefbuch n. ‖ livre m. de correspondance. / ~ box ‖ Briefkasten m. ‖ boîte f. à lettres. / slot of a ~ box ‖ Briefeinwurf m. ‖ ouverture f. *ou* fente f. pour les lettres. / ~ bracket ‖ Briefklammer f.; Heftklammer ‖ attache-lettre f. crampon m. à lettres. / ~ brush (Print) ‖ Abziehbürste f. ‖ brosse f. d'imprimerie. / ~ card ‖ Kartenbrief m. ‖ carte-lettre f. / ~ carrier ‖ Briefträger m. ‖ facteur f. des lettres. / ~ case ‖ Brieftasche f. ‖ porte-cartes m.; portefeuille m. / ~ case (Print) ‖ Setzkasten m.; Schriftkasten m. ‖ casse f. / ~ clip ‖ Briefklammer f. ‖ attache-lettre f. / ~ closing machine ‖ Briefschließmaschine f. ‖ machine f. à fermer les lettres. / ~ copying book ‖ Kopierbuch n.; Briefkopierbuch n. ‖ livre m. de copies à lettres. / ~ cutter ‖ Schriftschneider m. ‖ graveur m. en caractères. / ~ engraver *see* ~ cutter. / ~ engraving ‖ Letternstich m. ‖ gravure n. en caractères d'imprimerie. / ~ file ‖ Briefordner m. ‖ classeur m. de lettres; biblorhapte m. / ~-head ‖ Briefkopf m. ‖ en-tête f. de lettre.

lettering brush ‖ Buchstabenpinsel m. ‖ pinceau m. à dessin. / ~ pen ‖ Zeichenfeder f. ‖ plume f. à dessin.

letter knife ‖ Brieföffner m. ‖ ouvre-lettre m.; couteau m. à lettres. / ~ opener *see* ~ knife. / ~ paper ‖ Briefpapier n. ‖ papier m. à lettres. / ~ post ‖ Briefpost f. ‖ poste f. aux lettres.

letter-press (Office) ‖ Briefkopierpresse f. ‖ presse f. à copier les lettres. / ~ ‖ Buchdruckerpresse f. ‖ presse f. typographique *ou* d'imprimerie.

letter presser ‖ Briefbeschwerer m. ‖ presse-papier m.

letter-press work, hand press for ~ ‖ Buchdruckhandpresse f. ‖ presse f. d'imprimerie à bras. / high speed printing machine for ~ ‖ Schnellpresse f. für Buchdruck ‖ presse f. rapide pour imprimerie de livres.

letter printer ‖ Briefdrucker m. ‖ imprime-lettres m.

letters pl. (Print) ‖ Schrift f. ‖ caractères mpl.; lettres fpl. / arrow-headed ~ *see* cuneiform ~. / the ~ are broken ‖ die Schrift fällt ab ‖ les lettres fpl. se couchent. / common ~ ‖ gewöhnliche Buchstabenschrift f. ‖ caractères mpl.

d'écriture ordinaire. / cuneiform ~ ‖ Keilschrift f.; Pfeilschrift f. ‖ caractères mpl. cunéiformes. / descending ~ ‖ geschwänzte Schrift f. ‖ lettres fpl. à queue. / double ~ ‖ Ligatur f. ‖ ligature f. / fat ~ ‖ fette Schrift f.; starker Schriftkegel m. ‖ caractère m. plein; corps m. gros. / Gothic ~ ‖ gotische Schrift f. ‖ caractère m. gothic *ou* allemand. / ~ of a large size ‖ grobe Schrift f. ‖ gros caractères mpl. / long ~ *see* descending ~. / ornamental ~ ‖ Leistenkasten m. ‖ caseau m.; bardeau m. / to pick up ~ ‖ Schrift f. aus dem Schriftkasten nehmen ‖ lever les lettres fpl. / ~ of reference ‖ Zeichenerklärung f. ‖ légende f. explicative. / to space the ~ ‖ die Schrift sperren ‖ espacer les lettres fpl. / to space the ~ closely ‖ die Schrift eng halten ‖ serrer les lettres fpl. / ~ of a small size ‖ kleine Schrift f. ‖ petits caractères mpl. / turned ~ ‖ Fliegenköpfe mpl.; Blockade f. ‖ lettres fpl. bloquées *ou* renversées; blocage m.

letters patent ‖ Patent n. ‖ brevet m. d'invention; patente f. nationale.

letter scales pl. ‖ Briefwage f. ‖ pèse-lettres m. / sender of a ~ ‖ Briefschreiber m.; Absender m. eines Briefes ‖ expéditeur m. *ou* auteur m. d'une lettre. / ~ spacing arrangement ‖ Sperrschriftvorrichtung f. ‖ dispositif m. pour l'écartement des lettres. / ~ stamp ‖ Buchstabenpunze f. ‖ poinçon m. à lettre. / ~ telegram ‖ Brieftelegramm n. ‖ lettre-télégramme f. / ~ weight ‖ Briefbeschwerer m. ‖ presse-papier m. / width of a ~ ‖ Buchstabenbreite f. ‖ largeur f. de caractère. / ~ writer *see* ~ sender.

letting-out in contract ‖ Verdingung f. einer Arbeit im ganzen ‖ (travail m. à) forfait m.

lettsomite ‖ Kupfersamterz n. ‖ cuivre m. velouté.

lettuce, wild ~ ‖ wilder Lattich m. ‖ scarole f.

leucite ‖ Leuzit m. ‖ leucite f.; amphigène m.

leucopyrite ‖ Arseneisen n. ‖ leucopyrite f.

leucosaphire ‖ Leukosaphir m.; weißer Saphir m. ‖ leucosaphire m.

Levanter ‖ heftiger Ostwind m. ‖ coup m. de vent d'Est.

Levantine soap root ‖ Levantiner Seifenwurzel f. ‖ racine f. de saponaire d'Orient.

level, to ‖ wagerecht machen; planieren; ebnen ‖ niveler; aplanir; égaliser. / ~ the ground ‖ den Boden m. ebnen ‖ dresser le sol *ou* le terrain *ou* la surface. / ~ the horizontal elevation of underground workings ‖ die Höhenlage von Grubenbauen nivellieren *oder* abwägen ‖ niveler *ou* dépendre pour tracer des plans de mines. / ~ a telescope ‖ ein Fernrohr n. richten ‖ braquer un télescope.

level ‖ eben ‖ uni; plan; à niveau m. / ~ (Mine) ‖ söhlig ‖ horizontal.

level ‖ Niveau n.; Horizontalebene f. ‖ niveau m. / ~ (Geom) ‖ wagerechte Linie f.; Horizontale f. ‖ ligne f. horizontale *ou* de niveau. / ~ (Hydr arch) ‖ Wassergang m.; Mühlgerinne n. ‖ bief m.; biez m. / ~ (Instrument) ‖ Wasserwage f.; Libelle f. ‖ niveau m. à bulle d'air *ou* à alcool. / ~ (Mas) ‖ Richtscheit n.; Setzlatte f. ‖ limande f.; règle f. / ~ (Mine; horizontal passage) ‖ Sohlenstrecke f.; Grundstrecke f. ‖ galerie f.;

voie f. ~ (Mine; horizontal plane) ‖ Sohle f.; Bausohle f.; Abbausohle f. ‖ niveau m.; étage m.; sol m. / ~ (Phys) ‖ Spiegel m. ‖ niveau m. / above the ~ ‖ über dem Niveau n. ‖ au-dessus du niveau m. / to adjust ~s ‖ auf gleiches Niveau n. einstellen ‖ égaliser les niveaux mpl. / apparent ~ ‖ scheinbare Horizontalebene f.; Scheinhorizont m. ‖ niveau m. apparent. / below the ~ ‖ unter dem Niveau n. ‖ en-dessous du niveau m. / box ~ *see* round spirit ~. / to bring to the ~ ‖ nivellieren; einebnen; planieren ‖ niveler; égaliser; aplanir. / control ~ ‖ Prüflibelle f. ‖ niveau m. de contrôle. / dead ~ ‖ ebene Fläche f. ‖ surface f. plane. / deck ~ ‖ Bordhöhe f. ‖ hauteur m. de l'appui. / to draw on the ~ ‖ auf ebener Strecke f. ziehen *oder* fördern ‖ remorquer en palier m. / float ~ ‖ Schwimmerniveau n. ‖ niveau m. du flotteur. / grate ~ ‖ Rostfläche f. ‖ grille f. / ~s pl. of the ground taken along a line ‖ Längennivellement n. ‖ nivellement m. composé; nivellement de l'axe d'un profil longitudinal. / higher ~ (Hydr arch) ‖ obere Haltung f. (einer Schleuse) ‖ bief m. supérieur; arrière-bief m. / ~ with horizontal circle ‖ Nivellierinstrument n. mit Horizontalkreis ‖ niveau m. à lunette muni d'un cercle horizontal. / ~ of ironworks ‖ Hüttensohle f. ‖ niveau m. d'usin métallurgique. / low ~ (Boiler) ‖ niedrigster Wasserstand m. ‖ bas niveau m. d'eau; niveau m. d'eau minimum. / lower ~ (Hydr arch) ‖ untere Haltung f. (einer Schleuse) ‖ bief m. inférieur; sous-bief m. / mason's ~ with plumb bob ‖ Maurersetzwage f.; Maurerrichtwage f. ‖ niveau m. de maçon à pendule *ou* à plomb. / miner's ~ *see* surveyor's ~. / normal ~ ‖ normaler Wasserstand m. ‖ niveau m. de l'eau normal. / oil ~ ‖ Ölstand m. ‖ niveau m. d'huile. / ~ of petrol ‖ Benzinstand m. ‖ niveau m. d'essence. / ~ of a pit ‖ Sohle f.; Abbausohle f. ‖ étage m.; horizon m.; sol m. / round spirit ~ ‖ Dosenlibelle f. ‖ niveau m. sphérique. / safety ~ ‖ Sicherheitswasserstand m. ‖ niveau m. d'eau de sûreté. / at sea-~ ‖ auf gleicher Höhe f. mit dem Meeresspiegel ‖ au niveau m. de la mer. / spirit ~ ‖ Wasserwage f. ‖ niveau m. à bulle d'air *ou* à alcool. / ~ of the street ‖ Straßenhöhe f.; Straßenniveau n. ‖ niveau m. de la rue *ou* de la route. / surveyor's ~ ‖ Markscheiderwage f.; Nivellierfernrohr n. ‖ niveau m. à plomb en demi-cercle *ou* à lunette; théodolite m. / tail water ~ ‖ Unterwasserspiegel m. ‖ niveau m. (de l'eau du bief) aval. / to take the ~ *see* to bring to the ~. / taking the ~s of points ‖ Einnivellieren n. von Punkten ‖ nivellement m. de points. / true ~ ‖ wahrer Horizont m.; rechte Horizontalebene f. ‖ niveau m. vrai. / ~ of underground water ‖ Grundwasserstand m. ‖ niveau m. de la nappe d'eau souterraine. / to determine the ~ of underground water ‖ den Grundwasserstand feststellen ‖ fixer le plan des eaux souterraines. / upper ~ *see* higher ~. / varying ~s of the line ‖ Höhenunterschiede mpl. der Bahn ‖ inégalité f. altimétrique de la ligne. / water ~ ‖ Wasserstand m. ‖ niveau m. d'eau.

level, change of ~ ‖ Wasserstandsänderung f. ‖ changement m. de niveau.

level crossing (Railw) ‖ Niveauübergang m.; Planübergang m. ‖ passage m. à niveau. / ~ in right angle ‖ rechtwinkeliger Niveauübergang m. ‖ passage m. à niveau à angle droit. / ~ on the skew ‖ schräger Niveauübergang m. ‖ passage m. à niveau oblique ou en biais.

level cutting ‖ Einschnitt m. in flachem Boden ‖ excavation f. ou tranchée f. de niveau. / ~ difference indicator ‖ Höhendifferenzmelder m. ‖ indicateur m. de la différence de niveau. / ~ height of ~ ‖ Bodenhöhe f. ‖ niveau m. du sol.

levelled ‖ planiert ‖ nivelé.

leveller ‖ Planierer m.; Nivellierer m. ‖ régaleur m.

levelling ‖ Nivellierung f. ‖ nivellement m. / barometrical ~ ‖ barometrische Höhenmessung f. ‖ nivellement m. barométrique. / ~ of the ground ‖ Planierung f. des Bodens ‖ rasement m. du terrain.

levelling arrangement ‖ Planiervorrichtung f. ‖ repaleuse f. / ~ bar ‖ Planierstange f. ‖ tige f. repaleuse. / ~ board ‖ Richtlatte f.; Richtscheit n. ‖ règle f. à niveler. / ~ device ‖ Horizontierung f. ‖ horizontalité f. / ~ instrument (Mas) ‖ Richtzeug n. ‖ instrument m. à dresser. / instrument (Surv) ‖ Nivellierinstrument n.; Fernglaslibelle f. ‖ instrument m. de nivellement; niveau m. à lunette. / ~ operation ‖ Flächenaufnahme f. ‖ levé m. de surface. / ~ pole ‖ Nivellierlatte f.; Meßstab m. ‖ mire f. à voyant; règle f.; verge f. / ~ rod ‖ Planierstange f. ‖ barre f. à planer. / ~ rule ‖ Nivelliermaßstab m. ‖ mire f. graduée. / ~ screw ‖ Stellschraube f. ‖ vis f. de calage.

levelling staff ‖ Nivellierlatte f. ‖ mire f. parlante ou graduée. / high precision ~ ‖ Feinnivellierlatte f. ‖ mire f. parlante de précision.

level luffing jib ‖ Wippausleger m. ‖ flèche f. à inclinaison variable. / ~ railway ‖ Niveaubahn f. ‖ chemin m. de fer au niveau du sol ou dans la plaine.

level section (Railw) ‖ wagerechte Strecke f. ‖ (chemin m. de fer ou ligne f. en) palier m. / ~ transportation ‖ Wagerechtförderung f. ‖ transport m. horizontal.

lever (Build) ‖ Hebebaum m.; Brechstange f.; Handspake ‖ levier m.; aspect m.; pince (f.) monseigneur; pince f. à pied de biche. / ~ (Mach) ‖ Hebel m. ‖ levier m. / ~ (Ropem) ‖ Knüppel m.; Drehstock m. ‖ gaton m. / ~ (Watch) ‖ Anker m. ‖ ancre m. / ~ for adjusting brushes ‖ Bürstenstellhebel m. ‖ levier m. de calage des balais. / ~ of balance ‖ Wagebalken m. ‖ fléau m. d'une balance. / bell crank ~ ‖ Winkelhebel m. ‖ levier m. coudé. / bent ~ see bell crank ~. / blocking ~ ‖ Arretierungshebel m.; Blockierungshebel m. ‖ levier m. de verrouillage ou de blocage ou d'enclenchement. / brake ~ (Auto) ‖ Bremshebel m. ‖ levier m. de frein. / brake hand ~ (Cycle) ‖ Handbremshebel m. ‖ manette f. de frein. / cam ~ ‖ Nockenhebel m.; Wälzhebel m. ‖ levier m. à came. / change speed ~ ‖ Getriebehebel m.; Getriebearm m. ‖ levier m. de commande. / clutch ~ ‖ Kupplungshebel m. ‖ levier m. de débrayage. / compensating ~ ‖ Ausgleichshebel m. ‖ levier m. compensateur. / ~ of contact ‖ Schalthebel m. ‖ levier m. interrupteur

ou de contact. / control ~ ‖ Schalthebel m. ‖ palonnier m. / counterbalance ~ ‖ Gegengewichtshebelarm m. ‖ (bras m. de) levier m. du contrepoids. / coupling ~ (Mach tool) ‖ Einrückhebel m. ‖ levier m. d'embrayage. / coupling ~ (Mot) ‖ Kupplungshebel m. ‖ levier m. de changement de vitesse. / disconnecting ~ ‖ Ausschaltungshebel m. ‖ levier m. de débrayage. / disengaging ~ ‖ Ausrückhebel m.; Ausrücker m. ‖ levier m. de débrayage. / double-armed ~ ‖ doppelarmiger Hebel m. ‖ levier m. à deux branches; levier m. double. / easing ~ ‖ Anlüfthebel m.; Anhebevorrichtung f. ‖ dispositif m. de soulèvement. / engaging ~ ‖ Kupplungshebel m.; Einschalthebel m.; Einrücker m. ‖ levier m. d'embrayage. / ~ with equal arms ‖ gleicharmiger Hebel m. ‖ levier m. à bras égaux. / feeler ~ ‖ Tasthebel m. ‖ levier m. à touche. / foot ~ ‖ Fußhebel m. ‖ pédale f. / forked ~ ‖ gegabelter Hebel m.; Gabelhebel m. ‖ levier m. fourchu. / gear shift ~ ‖ Schalthebel m. ‖ levier m. (d'embrayage) de vitesse. / governor ~ see regulating ~. / hand ~ ‖ Handhebel m. ‖ levier m. à main. / ignition ~ ‖ Zündhebel m. ‖ manette f. d'allumage. / interrupter ~ ‖ Unterbrecherhebel m. ‖ levier m. interrupteur. / knee ~ ‖ Winkelhebel m. ‖ levier m. coudé. / latch ~ ‖ Sperrklinkenhebel m. ‖ manette f. cliquet d'arrêt. / locking ~ ‖ Sperrhebel m.; Verschlußhebel m. ‖ levier m. d'enclenchement. / ~ for open exhaust ‖ Hebel m. für freien Auspuff ‖ levier m. pour échappement libre. / ~ with pawl ‖ Klinkhebel m. ‖ levier m. à cliquet. / pedal ~ ‖ Fußhebel m.; Tritthebel m. ‖ levier m. à pédale. / ratchet ~ ‖ Sperrhebel m. ‖ levier m. à cliquet ou à rochet. / reciprocating ~ ‖ Schwinghebel m. ‖ levier m. oscillant. / regulating ~ ‖ Regulatorhebel m.; Reglerhebel m. ‖ levier m. du régulateur. / releasing ~ see disengaging ~. / reversing ~ ‖ Steuerhebel m.; Umsteuerhebel m. ‖ levier m. de changement de marche. / rocking ~ ‖ Schwunghebel m. ‖ balancier m. / signal ~ ‖ Signalhebel m. ‖ levier m. de signal. / single-armed ~ ‖ einarmiger Hebel m. ‖ levier m. à bras unique; levier m. du troisième genre. / ~ for slabs ‖ Plattenhebel m. ‖ levier m. pour dalles. / spark hand ~ see ignition ~. / ~ with spherical end ‖ Kugelkurbel f. ‖ levier m. à boule. / standard ~ see regulating ~. / starting ~ ‖ Bedienungshebel m. ‖ levier m. de manœuvre. / steering ~ ‖ Lenkhebel m. ‖ levier m. de direction. / straight ~ ‖ gerader Hebel m. ‖ levier m. droit. / striking ~ ‖ Schlaghebel m. ‖ levier m. percuteur ou de percussion. / stripper ~ (Mach tool) ‖ Abstreiferarm m. ‖ levier m. d'arracheur. / strut ~ ‖ Stützhebel m.; Gegenhalter m. ‖ levier m. (de) soutien; levier m. contre-fiche. / ~ swinging the rack ‖ Rechenschwenkhebel m. ‖ levier m. commandant le mouvement des râteaux. / switch ~ (Railw) ‖ Umlegehebel m. für Weichen ‖ levier m. de manœuvre de l'aiguille. / testing ~ see easing ~. / throttle ~ ‖ Drosselhebel m.; Reglerhebel m. ‖ levier m. d'étranglement ou de régulateur. / throttle valve ~ ‖ Drosselklappenhebel m. ‖ levier m. du papillon. / toggle ~ ‖ Kniehebel m. ‖ levier m.

coudé ou à genouillère. / two-armed ~ ‖ Doppelhebel m.; zweiarmiger Hebel m. ‖ levier m. à deux bras; levier m. du premier genre. / tyre ~ ‖ Reifenauflegehebel m.; Reifenmontierhebel m. ‖ levier m. (de montage) de pneu. / uncoupling ~ ‖ Ausrückhebel m. ‖ levier m. déclencheur ou de dégagement ou de débrayage. / valve ~ ‖ Ventilhebel m. ‖ levier m. de soupape. / weight ~ ‖ Gewichtshebel m. ‖ levier m. à contrepoids. / working ~ ‖ Stellhebel m. ‖ levier m. de manœuvre.

leverage ‖ Hebelübersetzung f.; Hebelverhältnis n. ‖ transmission f. à levier; rapport m. des leviers. / ~ (Mach) ‖ Hebelkraft f. ‖ puissance f. du levier.

lever apparatus ‖ Hebelvorrichtung f. ‖ appareil m. à levier. / ~ arm ‖ Hebelarm m.; Schwengel m. ‖ bras m. de levier; bascule f. / ~ arrangement ‖ Hebelanordnung f. ‖ dispositif m. de levier. / ~ arresting device ‖ Hebelfangvorrichtung f. ‖ dispositif m. d'arrêt à levier. / ~ bearing ‖ Kurbellager n. ‖ palier m. de manivelle. / ~ brake ‖ Hebelbremse f. ‖ frein m. à levier. / ~ catch to close goods-vans for railways ‖ Daumenwelle f. zum Verschluß von Güterwagen ‖ arbre m. à cames pour la fermeture des wagons de marchandises. / ~ changeover switch ‖ Hebelumschalter m. ‖ commutateur m. à levier. / ~ dolly ‖ Nietwippe f. ‖ support m. à levier de contre-bouterolle. / ~ drawbridge ‖ Klappbrücke f. ‖ pont-levis m. à trappe. / ~ driving ‖ Hebelantrieb m. ‖ commande f. par levier. / ~ escapement ‖ (Watch) Ankerhemmung f. ‖ échappement m. à ancre. / ~ gauge ‖ Lehre f. mit Hebelübersetzung ‖ calibre m. ou jauge f. à levier. / ~ guide ‖ Hebelführung f. ‖ guide f. de ou à levier. / ~ hammer ‖ Hebelhammer m. ‖ marteau m. à levier. / ~ handle ‖ Hebelgriff m. ‖ manche m. ou poignée f. du levier. / ~ head (Foundation) ‖ Klemme f. ‖ clef f. de manœuvre du tube. / ~ key switch ‖ Kippschalter m. ‖ clé f. à bascule.

leverman, converter ~ ‖ Konverterkipper m. ‖ opérateur m. du convertisseur.

lever mechanism ‖ Hebelwerk n. ‖ mécanisme m. à levier. / ~ motion ‖ Hebelsteuerung f.; Hebelantrieb m. ‖ commande f. par levier. / ~ nail puller ‖ Hasenmaul n.; Nagelzieher m. ‖ arrachecious m. à levier. / ~ pin ‖ Hebelstift m. ‖ cheville f. de levier. / ~ point ‖ Klappweiche f. ‖ aiguille f. basculante. / ~ press ‖ Hebelpresse f. ‖ presse f. à levier. / ~ pump ‖ Hebelpumpe f. ‖ pompe f. à levier. / ~ punching machine ‖ Hebellochstanze f. ‖ poinçonneuse f. à levier. / ~ reversing gear ‖ Hebel(um)steuerung f. ‖ changement m. de marche à levier. / ~ rivetting machine ‖ Hebelnietmaschine f. ‖ riveuse f. à levier. / ~ rod ‖ Hebelstange f. ‖ tige f. de levier. / ~ spring loaded ~ safety valve ‖ Federsicherheitsventil n.; Sicherheitsventil n. mit Federbelastung ‖ soupape f. de sûreté à ressort. / ~ set keyboard ‖ Hebelstellwerk n. ‖ clavier m. à leviers. / ~ setting and jewelling (Watchm) ‖ Einsetzen n. der Steine in den Anker ‖ garnissage m. de l'ancre. / ~ shaft ‖ Kurbelwelle f. ‖ arbre m. à manivelle. / ~ shears pl. ‖ Hebelschere f. ‖

cisaille f. à levier. / ~ stand (Railw) ‖ Stellbock m. ‖ chevalet m. de manœuvre. / ~ stick ‖ Keilverschluß m. ‖ fermeture f. à coins. / ~ switch ‖ Kurbelausschalter m. ‖ interrupteur m. à manette. / ~ tension ‖ Hebelspannung f. ‖ tension f. par levier. / ~ watch ‖ Ankeruhr f. ‖ montre f. à ancre. / ~ weighted safety valve ‖ Sicherheitsventil n. mit Gewichtsbelastung ‖ soupape f. de sûreté à contre-poids. / ~ weir ‖ Klappenwehr n. ‖ batardeau m. à hausses mobiles.

levigate, to ‖ zu Pulver n. zerreiben; zerpulvern; pulverisieren ‖ léviger; pulvériser (finement).

levigation ‖ Pulverisieren n. ‖ lévigation f.

levulose ‖ Fruchtzucker m.; Lävulose f. ‖ levulose f.; sucre m. de fruit.

levy of a window-shutter ‖ Ladenflügel m.; Flügel eines Fensterladens ‖ battant m. de contrevent ou de volet.

levying, method of ~ duty ‖ Zollerhebungsverfahren n. ‖ méthode f. de perception des droits de douane.

lewis bolt ‖ Ankerschraube f.; Zwingkeil m. ‖ boulon m. d'ancrage.

Leyden jar ‖ Leydener Flasche f. ‖ bouteille f. électrique ou de Leyde. / battery of ~s ‖ Batterie f. Leydener Flaschen ‖ batterie f. de bouteilles de Leyde.

liabilities pl. (Trade) ‖ Passiva pl. ‖ passif m.; masse f. passive. / outstanding ~ ‖ Außenstände mpl. ‖ créances fpl.

liability ‖ Haftpflicht f. ‖ responsabilité f. / ~ of the administration ‖ Ersatzpflicht f. der Verwaltung ‖ responsabilité f. de l'administration. / ~ for damages caused by antennas ‖ Antennenhaftung f. ‖ responsabilité f. des dommages causés par des antennes. / ~ of the subscriber ‖ Ersatzpflicht f. des Fernsprechteilnehmers ‖ responsabilité f. de l'abonné.

liability insurance ‖ Haftpflichtversicherung f. ‖ assurance f. contre les accidents causés à des tiers; assurance f. à responsabilité civile.

liable ‖ haftbar; haftpflichtig ‖ responsable. / ~ for damages ‖ schadenersatzpflichtig ‖ responsable du dommage causé. / ~ to duty ‖ abgabenpflichtig ‖ soumis aux droits mpl. / ~ to postage ‖ portopflichtig ‖ soumis à la taxe. / ~ to recourse ‖ regreßpflichtig ‖ civilement responsable.

Lias ‖ Lias m. ‖ lias m.

liberty of action ‖ Handlungsfreiheit f. ‖ liberté f. d'action.

librarian ‖ Bücherwart m. ‖ bibliothécaire m.

library ‖ Bücherei f. ‖ bibliothèque f. / circulating ~ for the workmen and the staff ‖ Bücherei f. für Arbeiter und Beamte ‖ bibliothèque f. pour les ouvriers et employés. / technical ~ ‖ fachwissenschaftliche oder technische Bücherei f. ‖ bibliothèque f. technique.

licence see license.

license ‖ Lizenz f.; Erlaubnis f. ‖ licence f.; permis m. / ~ (Handicraft) ‖ Patent n.; Gewerbschein m. ‖ patente f.; licence f. / ~ (Print) ‖ Druckerlaubnis f.; Genehmigung f. ‖ patente f. d'imprimerie. / driving ~ ‖ Fahrschein m.; Führerschein m. ‖ permis m. de conduire. / ~ for the manufacture ‖ Recht n. zur Herstellung ‖ droit m. de fabrication. / to take a ~ ‖ ein Patent n. nehmen ‖ prendre une patente.

licensee ‖ Konzessionsinhaber m. ‖ concessionnaire m.

license support (Auto) ‖ Tragstütze f. für das Nummernschild ‖ support m. pour la plaque de license; support m. de numbros.

lichen, medicinal ~ ‖ Flechte f. zum Heilgebrauch ‖ lichen m. médicinal.

licker-in (Spinn) ‖ Vorreißer m. ‖ tambour m. briseur; briseur m.

licorice see also liquorice ‖ Süßholz n. ‖ réglisse f.

lid (Box) ‖ Deckel m. ‖ couvercle m. / ~ (Blast furnace) ‖ Kranz m. ‖ chapeau m. / ~ (Eye) ‖ Augenlid n.; Lid n. ‖ paupière f. / ~ of carburettor ‖ Vergaserdeckel m. ‖ couvercle m. du carburateur. / double-walled ~ ‖ doppelwandiger Dekkel m. ‖ couvercle m. à double paroi. / ~ of furnace ‖ Ofenschließstein m.; Ofendeckstein m. ‖ bouchoir m. d'un fourneau. / hinged ~ ‖ Klappdeckel m. ‖ couvercle m. à gonds ou à charnière. / ~ for pomatum glasses ‖ Deckel m. für Pomadengläser ‖ capsule f. pour verres à pommade. / protective ~ (Typewr) ‖ Schutzdeckel m. ‖ capot m. ou couvercle m. ou plaque f. de protection. / valve ~ ‖ Ventildeckel m. ‖ chapeau m. de soupape.

lid catch ‖ Deckelverriegelung f. ‖ verrouillage m. du couvercle. / ~ knob ‖ Deckelknauf m. ‖ bouton m. de couvercle. / ~ screw ‖ Deckelschraube f. ‖ vis f. de couvercle.

lie, to ~ the course (Nav) ‖ Kurs m. anliegen ‖ porter en route f. / ~ to (Nav) ‖ anlegen ‖ capéer; caposer.

lien ‖ Retentionsrecht n.; Pfandrecht n. ‖ droit m. de rétention ou de gage.

life ‖ Lebensdauer f.; Haltbarkeit f.; vie f.; durabilité f.; durée f. d'existence. / ~ of electrodes ‖ Lebensdauer f. von Elektroden ‖ durée f. des électrodes. / ~ of an element ‖ Lebensdauer f. eines Konstruktionsteiles ‖ durée f. de la pièce. / longer ~ of the blast-furnace ‖ größere Ofenreise f. ‖ campagne f. du fourneau plus grande.

life annuity ‖ Lebensrente f.; Leibrente f. ‖ rente f. viagère. / ~ assurance see ~ insurance. / ~ belt ‖ Rettungsgürtel m.; Rettungsring m. ‖ ceinture f. de sauvetage.

lifeboat ‖ Rettungsboot n. ‖ canot m. ou embarcation f. de sauvetage. / air-case of a ~ ‖ Luftkasten m. eines Rettungsbootes n. ‖ caisse f. à air d'un canot de sauvetage.

life buoy ‖ Rettungsboje f. ‖ bouée f. de sauvetage. / ~-guard of a locomotive ‖ Schienenräumer m. einer Lokomotive ‖ chasse-pierres m. ou garde m. d'une locomotive. / ~ insurance ‖ Lebensversicherung f. ‖ assurance f. pour la vie. / ~ insurance association ‖ Lebensversicherungsverein m.; Lebensversicherungsgesellschaft f. ‖ société f. d'assurance sur la vie.

lifeless ‖ leblos ‖ inanimé. / ~ (Market) ‖ matt; lustlos; geschäftslos ‖ lourd.

life line (Mar) ‖ Manntau n.; Laufstag n. ‖ garde-corps m.; garde-fou m. / ~ oil ‖ Lebensöl n. ‖ essence f. de vie. / ~ policy ‖ Lebensversicherungspolice f. ‖ police f. d'assurance sur la vie. / ~ preserver (Bludgeon) ‖ Totschläger m. ‖ casse-

tête m. / ~-saving jacket ‖ Schwimmweste f. ‖ gilet m. flotteur ou de sauvetage. / ~ train (Railw) ‖ Rettungszug m.; Hilfszug m. ‖ train m. de secours ou d'ambulance.

lift, to ~ ‖ heben ‖ lever; élever; soulever. / ~ (Mine) ‖ fördern ‖ remonter. / ~ the casting ‖ den Guß m. ausheben ‖ démouler la fonte. / ~ disks of copper ‖ Kupferscheiben fpl. reißen ‖ lever les rosettes fpl. de cuivre. / ~ off disks of pig-iron ‖ Roheisenscheiben fpl. reißen ‖ lever les plaques fpl. de fonte. / ~ out (Found) ‖ herausnehmen ‖ déchapper; démouler. / ~ out the form from the press (Print) ‖ die Form aus der Presse ausheben ‖ relever la forme de la presse. / ~ the pattern (Found) ‖ das Modell ausheben ‖ démouler le modèle. / ~ the sheets pl. (Pap) ‖ die Bogen mpl. legen ‖ lever les feuilles fpl. / ~ up ‖ aufheben ‖ ramasser; soulever. / ~ up (Mar) ‖ anlüften ‖ hisser.

lift (Aero) ‖ Auftrieb m. ‖ portance f. / ~ (Blast-furnace) ‖ Gichtaufzug m. ‖ monte-charge m. de gueulard. / ~ (Crane) ‖ Anhub m. ‖ soulèvement m.; hauteur f. de levage. / ~ (Goods elevator) ‖ Aufzug m. ‖ monte-charge m.; élévateur m. / ~ (Hydr arch) ‖ Schleuseneinsatz m.; Schleusenfall m. ‖ chute f.; sas m. / ~ (Mill) ‖ Panster n.; Panserzeug n. ‖ appareil m. à élever ou baisser une roue à aubes. / ~ (Mine) ‖ Pumpensatz m. ‖ jeu m. des pompes étagées; pompe f. installée dans une colonne élévatoire. / ~ (Passenger elevator) ‖ Fahrstuhl m. ‖ ascenseur m.; lift m. / ~ (Pump) ‖ Förderhöhe f. ‖ hauteur f. d'élévation. / ~ (Valve) ‖ Hub m.; Erhebung f. ‖ levée f.; soulèvement m.; course f. / auxiliary ~ of crane ‖ Hilfshubwerk n. eines Kranes ‖ petit levage m. d'une grue. / ~ of cam ‖ Nockenhub m. ‖ levée f. de la came. / ~ cellar ‖ Kelleraufzug m. ‖ monte-charges m. de cave. / dynamic ~ ‖ dynamischer Auftrieb m. ‖ poussée f. dynamique. / electric ~ ‖ elektrischer Aufzug m. ‖ ascenseur m. électrique. / freight ~ ‖ Lastenaufzug m.; Frachtaufzug m. ‖ monte-charge m. / hydraulic ~ ‖ hydraulischer Aufzug m. ‖ ascenseur m. hydraulique. / inclined ~ for cupola furnaces ‖ Schrägaufzug m. für Kuppelöfen ‖ monte-charge m. oblique pour cubilots. / ~ of key ‖ Tastenhub m. ‖ soulèvement m. de la touche. / ~ of a lifter ‖ Hub m. eines Pochstempels ‖ levée f. du pilon d'un bocard. / ~ of a lock ‖ Höhe f. des Schleusenabfalles ‖ chute f. de l'écluse. / main ~ (Crane) ‖ Hauptwindwerk n. / treuil m. principal; gros levage m. / maximum ~ ‖ größte Hubhöhe f. ‖ course f. maximum. / passenger ~ ‖ Personenaufzug m. ‖ ascenseur m. pour personnes. / pneumatic ~ ‖ pneumatischer Gichtaufzug m. ‖ monte-charge m. pneumatique. / ~ for railway wagons ‖ Aufzug m. für Eisenbahnfahrzeuge ‖ monte-charge m. pour voitures de chemins de fer. / ship ~ ‖ Schiffshebewerk n. ‖ monte-charge m. pour bateaux. / staircase ~ ‖ Treppenaufzug m. ‖ escalier-ascenseur m.; escalier m. roulant; tapis m. roulant. / static ~ ‖ statische Steigkraft f. ‖ force f. ascensionnelle statique. / total ~ (Aero) ‖ Gesamtauftrieb m. ‖ poussée f. totale.

/ total ~ (Crane) ‖ gesamte Hubhöhe f. ‖ hauteur f. totale de levée.

lift brake ‖ Fallbremse f. ‖ frein m. à chute. / ~ **bridge** ‖ Fahrstuhlbrücke f. ‖ plateforme f. de monte-charge. / ~ **car** ‖ Fahrbühne f. ‖ chariot m. transporteur / ~ **coefficient** (Aero) ‖ Auftriebszahl f. ‖ coefficient m. de portance. / ~ **control** ‖ Fahrstuhlsteuerung f. ‖ commande f. de l'ascenseur. / ~ **counter** ‖ Hubzähler m.; Umdrehungszähler m. ‖ compteur m. de tour.

lifter (Mach) ‖ Stößel m. ‖ taquet m. / ~ (Metal) ‖ Pochstempel m.; Stempel m. ‖ bocard m.; pilon m. / ~ (Print) ‖ Aufleger m.; Einschieber m.; Leger m. ‖ margeur m.; leveur m. / ~ **exhaust valve** ~ ‖ Auslaßventilheber m. ‖ toc m. de soupape d'échappement.

lift force formula ‖ Hubkraftformel f. ‖ formule f. de poussée. / ~ **frame** ‖ Fördergerüst n.; Förderturm m. ‖ charpente f. du monte-charges. / **height of** ~ ‖ Hubhöhe f. ‖ course f. ou hauteur f. de levage.

lifting ‖ Heben n.; Aufheben n. ‖ soulèvement m.; enlèvement m. / ~ **of the track** ‖ Freilegen n. der Gleise ‖ enlèvement m. des rails. / ~ **of the type bars** ‖ Lüften n. der Typenhebel ‖ soulèvement m. des tiges à caractères.

lifting apparatus ‖ Hebevorrichtung f. ‖ appareil m. de levage. / ~ **and lowering arrangement** ‖ Hebe- und Senkvorrichtung f. ‖ appareil m. de levage et de descente. / ~ **bar** ‖ Hebeschiene f. ‖ tige f. de levée. / ~ **bucket** ‖ Aufzugkasten m.; Aufzugkübel m. ‖ baquet m. d'ascenseur. / ~ **contrivance** ‖ Aushebevorrichtung f. ‖ dispositif m. de déchassage; extracteur m.; déchasseur m.

lifting device ‖ Hebevorrichtung f.; Abhebevorrichtung f. ‖ dispositif m. de relèvement ou de levage. / ~ **hydraulic** ~ ‖ hydraulisches Hebezeug n. ‖ appareil m. de levage hydraulique. / **pneumatic** ~ ‖ Preßlufthebezeug n. ‖ appareil m. de levage pneumatique. / **rolls** ~ ‖ Walzenaushebevorrichtung f. ‖ dispositif m. de soulèvement pour cylindres.

lifting force ‖ Auftriebskraft f.; Tragkraft f. ‖ force f. portante ou ascensionnelle ou de sustentation. / ~ **gas** ‖ Traggas n. ‖ gaz m. de gonflement ou de sustentation. / ~ **gear** ‖ Hubwerk n. ‖ mécanisme m. de levage. / ~ **gear of an ore loading bridge** ‖ Hubwerk n. einer Erzverladebrücke ‖ mécanisme m. de levage d'un pont de chargement de minerais. / ~ **jack** ‖ Hebebock m.; Bauwinde f. ‖ vérin m. de levage; cric m. / **rail** ~ **jack** ‖ Gleishebewinde f. ‖ treuil m. pour soulever les rails. / ~ **machine** ‖ Hebemaschine f. ‖ engin m. de levage. / ~ **magnet** ‖ Hebeelektromagnet m.; Lasthebemagnet m.; Magnethebezeug n. ‖ aimant m. de levage ou de suspension; électro m. de levage. / ~ **mechanism** ‖ Hebemechanismus m. ‖ mécanisme m. de soulèvement. / ~ **motor** ‖ Hubmotor m. ‖ moteur m. de levage. / ~ **movement** (Crane) ‖ Hubwindwerk n. ‖ mécanisme m. ou treuil m. de levage. / ~ **part** ‖ Huborgan n. ‖ organe m. de levage. / ~ **power** see ~ **force**. / ~ **pump** ‖ Hebepumpe f. ‖ pompe f. élévatoire. / ~ **rope** ‖ Rammtau n. ‖ câble m. de sonnette. / ~ **screw** ‖ Schraubenwinde f. ‖ vérin m.; cric m. à vis.

lifting table ‖ Hebetisch m. ‖ table f. de levage; tablier m. releveur ou à rouleaux. / ~ **with idle rollers** ‖ Hebetisch m. mit nicht angetriebenen Rollen ‖ releveur m. à rouleaux non commandés. / ~ **installation** ‖ Hebetischanlage f. ‖ installation f. de table de levage.

lifting tackle ‖ Hebevorrichtung f. ‖ appareil m. de levage.

lifting wheel for lifting beets ‖ Hubrad n. zum Heben von Rüben ‖ roue f. élévatoire pour betteraves. / ~ **for lifting potatoes** ‖ Hubrad n. zum Heben von Kartoffeln ‖ roue f. élévatoire pour pommes de terre.

lifting winch ‖ Hebewinde f.; Aufzugswinde f. ‖ treuil m. de levage. / ~ **for wagons and coaches** ‖ Wagenwinde f. ‖ engin m. de levage pour wagons.

lifting wire (Weav) ‖ Platine f.; Hebehaken m. ‖ crochet m. ou platine f. de levage.

lift-over ‖ Überhebvorrichtung f. ‖ dispositif m. de transbordement.

lift rope ‖ Aufzugseil n. ‖ câble m. d'ascenseur; câble m. de levage. / ~ **table** ‖ Rollentisch m. ‖ tablier m. à rouleaux. / **top limit of** ~ ‖ Hubbegrenzung f. ‖ limite f. supérieure de levage. / ~ **water wheel** (Mill) ‖ Pansterrad n. ‖ roue f. à aubes.

ligature (Print) ‖ Ligatur f. ‖ ligature f.

light, to ‖ anbrennen; anzünden ‖ allumer; mettre le feu à . . . / **to be lighted** ‖ beleuchtet sein ‖ être illuminé ou éclairé.

light ‖ hell; licht ‖ clair; éclatant. / ~ **blue** ‖ Hellblau n. ‖ bleu m. clair. / ~ **body** ‖ heller Körper m. ‖ corps m. clair ou de couleur claire. / ~ **green** ‖ Hellgrün n. ‖ vert m. clair. / ~ **grey** ‖ Hellgrau n. ‖ gris m. clair. / ~ **pink** ‖ Hellrosa n. ‖ rose m. clair. / ~ **red** ‖ Hellrot n. ‖ rouge m. clair. / ~ **shell** ‖ blondes Schildpatt n. ‖ écaille f. (de tortue) blonde.

light ‖ leicht ‖ léger; faible; petit. / ~ **aeroplane** ‖ Leichtflugzeug n. ‖ avion m. léger; moto-aviette f. / ~ **concrete** ‖ Leichtbeton m. ‖ béton m. léger. / ~ **with draught** ‖ leichtzügig; leichtlaufend ‖ à traction f. légère; de traction f. douce. / ~ **fabric** ‖ leichter oder dünner Stoff m. ‖ tissu m. léger. / ~ **fuel** ‖ Leichtbrennstoff m. ‖ combustible m. léger. / ~ **hydrocarbon gas** ‖ Grubengas n.; leichtes Kohlenwasserstoffgas n.; Sumpfgas n. ‖ gaz m. léger d'hydrogène carburé; gaz m. des marais; gaz m. méthane. / ~ **metal** ‖ Leichtmetall n. ‖ métal m. léger. / ~ **metal framework** ‖ Leichtmetallgerippe n. ‖ carcasse f. ou charpente f. en métal léger. / ~ **motor cycle** ‖ Leichtkraftrad n. ‖ motocyclette f. légère.

light oil ‖ Leichtöl n. ‖ huile f. légère; huile f. très fluide. / **raw** ~ ‖ rohes Leichtöl n. ‖ huile f. légère brute. / **refined** ~ ‖ raffiniertes Leichtöl n. ‖ huile f. légère raffinée.

light pattern ‖ leichte Bauart f. ‖ construction f. légère. / ~ **plane** ‖ Leichtflugzeug n. ‖ moto-aviette f.

light railway ‖ Feldbahn f. ‖ chemin m. de fer portatif. / ~ **wagon** ‖ Feldbahnwagen m. ‖ wagon m. ou wagonnet m. de chemin de fer portatif.

light soap ‖ schaumige Seife f. ‖ savon m. léger. / ~ **spar** ‖ Leichtspat m. ‖ blanc minéral m. léger. / ~ **weight Diesel**

rail car ‖ Leichtmetalldieseltriebwagen m. ‖ automotrice f. à moteur Diesel de construction en métaux légers. / ~ **yeast** ‖ Flughefe f. ‖ voltigeurs mpl.

light ‖ Licht n. ‖ lumière f. / ~ ‖ Laterne f. ‖ lanterne f. / ~ (Build) ‖ Öffnung f. ‖ jour m.; ouverture f. / ~ (Nav) ‖ Leuchtfeuer n. ‖ feu m. / **in the absence of** ~ ‖ unter Lichtabschluß m. ‖ à l'abri de la lumière. / **alternating** ~ ‖ Wechselfeuer n. ‖ feu m. alternatif. / **arc** ~ ‖ Bogenlampe f. ‖ lampe f. à arc. / **auxiliary** ~ ‖ Zusatzfeuer n. ‖ feu m. auxiliaire. / **back** ~ ‖ Rücklicht n. ‖ feu m. d'arrière. / **blue electric** ~ ‖ blaues elektrisches Licht n. ‖ lumière f. électrique bleue. / **boundary** ~ ‖ Umrandungsfeuer n. ‖ feu m. de délimitation de terrain. / **brush** ~ ‖ Büschellicht n. ‖ aigrette f. lumineuse. / **coastal** ~ ‖ Küstenfeuer n. ‖ feu m. de côte. / **cold** ~ ‖ kaltes Licht n. ‖ lumière f. froide. / **coloured** ~ ‖ farbiges Licht n. ‖ lumière f. colorée. / **Drummond's** ~ ‖ Drummondsches Licht n. ‖ lumière f. de Drummond. / **fixed** ~ (Nav) ‖ festes Feuer n. ‖ feu m. fixe. / **flashing** ~ ‖ Blinkfeuer n. ‖ phare m. à éclat. / **guiding** ~ (Mar) ‖ Richtungsfeuer n. ‖ feu m. de direction. / **head** ~ ‖ Stirnlampe f. ‖ lampe f. frontale. / **hidden** ~ ‖ verstecktes Fenster n.; Guckfenster n. ‖ (fenêtre f. ou châssis m. à) vue f. dérobée. / **high** ~ ‖ Oberlicht n.; hohes Seitenlicht n. ‖ jour m. d'en haut. / **incandescent** ~ ‖ Glühlicht n. ‖ éclairage m. à incandescence. / ~ **of petroleum** ‖ Petroleumglühlicht n. ‖ lumière f. de pétrole par incandescence. / **incident** ~ ‖ auffallendes Licht n. ‖ lumière f. incidente. / **landing direction** ~ ‖ Landebahnfeuer n. ‖ feu m. d'atterrissage. / **leading** ~ (Nav) ‖ Leitfeuer n. ‖ feu m. d'alignement. / **navigation** ~ ‖ Stellungslicht n. ‖ feu m. de position. / **obstruction** ~ ‖ Hindernisfeuer n. ‖ feu m. d'obstacle. / **occulting** ~ ‖ unterbrochenes Feuer n. ‖ phare m. à éclipse. / **polarized** ~ ‖ polarisiertes Licht n. ‖ lumière f. polarisée. / **revolving** ~ (Nav) ‖ Drehfeuer n. ‖ feu m. tournant. / ~ **generated by silent discharge** ‖ Glimmlicht n. ‖ lumière f. produite par l'effluve. / **sky-**~ ‖ Oberlicht n.; Dachfenster n.; lanterneau m.; tabatière f.; claire-voie f.; jour m. d'en haut. / **tidal** ~ ‖ Gezeitenfeuer n. ‖ feu m. de marée. / **transmitted** ~ ‖ durchfallendes Licht n. ‖ lumière f. transmise. / **ultraviolet** ~ ‖ ultraviolettes Licht n. ‖ lumière f. ultra-violette. / **white** ~ ‖ weißes Licht n. ‖ lumière f. blanche. / **white** ~ **which is also rich in blue rays** ‖ weißes an blauen Strahlen reiches Licht n. ‖ lumière f. blanche et riche en rayons bleus. / ~ **of a window** (Build) ‖ Lichtöffnung f. oder Lichte n. eines Fensters ‖ jour m. de fenêtre.

light antenna ‖ Lichtantenne f. ‖ antenne f. électrique. / **electro-therapeutic** ~ **apparatus** ‖ elektrischer Lichtheilapparat m. ‖ appareil m. électrophotothérapique. / ~ **ball** ‖ Leuchtkugel f. ‖ balle f. luisante ou à éclairer. / ~ **bath** ‖ Lichtbad n. ‖ bain m. de lumière. / ~ **beacon** ‖ Leuchtbake f. ‖ balise f. éclairée. / ~ **buoy** ‖ Leuchttonne f. ‖ bouée f. lumineuse ou éclairée ou portant un feu.

/ ~ cartridge carrier ‖ Leuchtpatronenträger m. ‖ porte-fusées éclairantes m. / ~ column ‖ Lichtsäule f. ‖ colonne f. lumineuse. / cone of ~ ‖ Lichtkegel m. ‖ cône m. de lumière. / distribution of ~ ‖ Lichtverteilung f. ‖ répartition f. *ou* distribution f. de la lumière. / ~ curve of distribution of ~ ‖ Lichtverteilungskurve f. ‖ courbe f. de répartition de la lumière. / ~ dynamo ‖ Lichtmaschine f. ‖ dynamo f. à éclairage. / ~ emission ‖ Lichtemission f. ‖ émission f. de lumière.

lighten, to ~ (Ship) ‖ leichtern ‖ alléger. / ~ down a piece of wood ‖ ein Holzstück n. verjüngen ‖ délarder une pièce de bois; démaigrir en biseau m.

lightening light ‖ Blinklicht n. ‖ feu m. à éclairs. / article for ~ purposes ‖ Beleuchtungsgegenstand m. ‖ article m. d'éclairage.

lighter than air ‖ leichter als Luft f. ‖ plus léger que l'air m. / ~ craft ‖ Aerostat m.; Gasluftfahrzeug n. ‖ aérostat m.; aéronat m.; aéronef m. plus léger que l'air.

lighter ‖ Anzünder m. ‖ allumoir m. / ~ (Pap) ‖ Hebellade f. ‖ levier m. / ~ (Shipb) ‖ Leichter m. ‖ allège m. / cigar-~ ‖ Zigarrenanzünder m. ‖ allumeur m. pour cigares. / ~ for distance ‖ Fernzünder m. ‖ allumeur m. à distance.

lighterage ‖ Löschgebühr f. ‖ gabarage m.

lightering wood (Mill) ‖ Tragbank f. ‖ poilier m.

lighterman ‖ Leichtermann m. ‖ gabarier m.

light filter, liquid ~ ‖ flüssiges Lichtfilter n. ‖ écran m. liquide coloré. / ~ trough (Photo) ‖ Vorsatzküvette f. ‖ cuve-écran m.

light flare ‖ Leuchtrakete f. ‖ fusée f. éclairante. / flux of ~ ‖ Lichtstrom m. ‖ rayonnement m. *ou* flux m. lumineux. / ~ fuse ‖ Lichtsicherung f. ‖ coupe-circuit m. d'éclairage.

lighthouse ‖ Leuchtturm m. ‖ phare m. / ~ apparatus ‖ Leuchtturmapparat m. ‖ appareil m. pour phares. / ~ keeper ‖ Leuchtturmwärter m. ‖ gardien m. de phare.

lighting ‖ Beleuchtung f. ‖ éclairage m. / airway ~ ‖ Flugstreckenbefeuerung f. ‖ balisage m. de ligne. / ~ for automobiles ‖ Automobilbeleuchtung f. ‖ éclairage m. pour automobiles. / emergency ~ ‖ Notbeleuchtung f. ‖ éclairage m. de secours. / indoor ~ for industrial places ‖ Innenbeleuchtung f. für Industrieräume ‖ diffuseur m. pour locaux industriels. / indoor ~ for offices ‖ Bürobeleuchtung f. ‖ diffuseur m. pour bureaux. / indoor ~ of shops ‖ Beleuchtung f. von Werkstätten ‖ éclairage m. des ateliers. / ~ for industrial places ‖ Fabrikbeleuchtung f. ‖ diffuseur m. pour locaux industriels. / ~ in the inspection department ‖ Beleuchtung f. des Prüfraumes *oder* Meßraumes ‖ éclairage m. de l'atelier du contrôle. / ~ of mines ‖ Grubenbeleuchtung f. ‖ éclairage m. de mine. / motor-car ~ ‖ Automobilbeleuchtung f. ‖ éclairage m. des automobiles. / railway carriage ~ ‖ Eisenbahnwagenbeleuchtung f. ‖ éclairage m. des trains. / ~ reflected of the ceiling ‖ Deckenbeleuchtung f.; Deckenlicht n.; Oberlicht n. ‖ éclairage m. par réflexion du plafond. / sewing machine ~ ‖ Nähmaschinenbeleuchtung f. ‖ lampes fpl. pour machines à coudre. / shop window ~ ‖ Schaufensterbeleuchtung f. ‖ éclairage m. des vitrines. / ~ of squares ‖ Platzbeleuchtung f. ‖ éclairage m. de places (publiques). / staircase ~ ‖ Treppenbeleuchtung f. ‖ éclairage m. des escaliers. / automatic device for ~ of staircases ‖ Treppenbeleuchtungsautomat m. ‖ appareil m. automatique pour l'éclairage des escaliers. / ~ of streets ‖ Straßenbeleuchtung f. ‖ éclairage m. dans les rues; éclairage m. public.

lighting article ‖ Beleuchtungsgegenstand m. ‖ article m. d'éclairage. / ~ cabin of the theatre ‖ Beleuchterloge f. des Theaters ‖ loge f. du préposé à l'éclairage du théâtre. / ~ cable ‖ Lichtkabel n. ‖ câble m. d'éclairage. / ~ car ‖ Beleuchtungswagen m. ‖ voiture f. d'éclairage. / ~ carbon ‖ Beleuchtungskohle f. ‖ charbon m. à lumière. / ~ circuit ‖ Lichtleitung f. ‖ circuit m. d'éclairage. / ~ connection ‖ Lichtkupplung f. ‖ raccordement m. *ou* connexion f. de la conduite d'éclairage. / ~ department ‖ Beleuchtungsabteilung f. ‖ section f. d'éclairage. / medical ~ device ‖ medizinischer Bestrahlungsapparat m. ‖ appareil m. d'éclairage médical. / ~ dynamo ‖ Lichtdynamo f. ‖ dynamo f. d'éclairage. / ~ engineering ‖ Beleuchtungstechnik f. ‖ technique f. d'éclairage. / ~ fitting for sewing machines ‖ Nähmaschinenbeleuchtung f. ‖ éclairage m. pour machines à coudre. / ~ fixture ‖ Beleuchtungskörper m. ‖ article m. d'éclairage. / ~ flame ‖ Leuchtflamme f. ‖ flamme f. éclairante. / ~ gas ‖ Leuchtgas n. ‖ gaz m. d'éclairage. / ~ gas motor ‖ Leuchtgasmotor m. ‖ moteur m. à gaz d'éclairage. / ~ glass ‖ Glas n. für Beleuchtungszwecke; Beleuchtungsglas n. ‖ verreries fpl. *ou* crystalleries fpl. d'éclairage. / ~ hour ‖ Beleuchtungsstunde f. ‖ heure f. d'éclairage. / ~ installation ‖ Beleuchtungsanlage f. ‖ installation f. d'éclairage. / ~ line ‖ Lichtnetz n. ‖ canalisation f. d'éclairage. / ~ line condenser (Radio) ‖ Lichtleitungskondensator m. ‖ condensateur m. fixe de réseau. / ~ mains pl. ‖ Lichtleitung f. ‖ canalisation f. d'éclairage. / small ~ outfits ‖ Kleinbeleuchtung f. ‖ petites installations fpl. d'éclairage. / period of ~ ‖ Brennzeit f. ‖ durée f. d'éclairage. / ~ plant ‖ Beleuchtungsanlage f. ‖ installation f. d'éclairage. / ~ point ‖ Lichtpunkt m. ‖ point m. lumineux. / ~ set ‖ Lichtaggregat n. ‖ groupe m. d'éclairage. / ~ switch ‖ Beleuchtungsschalter m. ‖ interrupteur m. d'éclairage. / ~ transformer ‖ Lichttransformator m. ‖ transformateur m. d'éclairage.

light intensity ‖ Lichtstärke f. ‖ pouvoir m. éclairant *ou* lumineux.

lightness (Brightness) ‖ Helligkeit f. ‖ degré m. de clarté. / ~ (Weight) ‖ Leichtigkeit f. ‖ légèreté f.; facilité f.

lightning ‖ Blitz m. ‖ éclair m. / destroyed by ~ ‖ durch Blitz m. zerstört ‖ détruit par la foudre. / destructive ~ ‖ Schadenblitz m. ‖ foudre f. *ou* éclair m. produisant des dégâts. / forked ~ ‖ gezackter Blitzstrahl m. ‖ foudre f. en zigzag *ou* en sillons. / globular ~ ‖ kugelförmiger Blitz m.; Kugelblitz m. ‖ éclair m. en boule. / ~ of silver ‖ Silberblick m. ‖ éclair m. de l'argent.

lightning apparatus for the stage ‖ Blitzerzeuger m. für die Bühne ‖ appareil m. à éclairs pour la scène.

lightning arrester *see also* lightning conductor *and* lightning protector ‖ Blitzableiter m.; Blitzschutzvorrichtung f. ‖ paratonnerre m.; parafoudre m. / carbon ~ ‖ Kohlenblitzableiter m. ‖ paratonnerre m. à charbon. / condenser ~ ‖ Kondensatorblitzschutzvorrichtung f. ‖ parafoudre m. à condensateur. / external ~ for communication circuits ‖ Außenblitzableiter m. für Fernmeldeleitungen ‖ parafoudre m. extérieur pour circuits télégraphiques et téléphoniques. / ~ with magnetic blow-out ‖ Blitzableiter m. mit magnetischer Funkenlöschung ‖ parafoudre m. à soufflage magnétique. / plate ~ ‖ Grobblitzableiter m. ‖ parafoudre m. à couteaux *ou* à plaques. / point ~ ‖ Spitzenblitzableiter m. ‖ parafoudre m. à pointes. / ~ for poles ‖ Stangenblitzableiter m. ‖ paratonnerre m. pour les poteaux. / rare gas ~ ‖ Edelgassicherung f. ‖ parafoudre m. à gaz rare. / wedge-shaped ~ ‖ Schneidenblitzableiter m. ‖ parafoudre m. à couteaux.

lightning arrester collar ‖ Blitzableiterschelle f. ‖ collier m. de paratonnerre. / ~ condenser ‖ Blitzschutzkondensator m. ‖ condensateur-parafoudre m. / ~ testing apparatus ‖ Blitzableiterprüfapparat m. ‖ pont m. de mesure pour parafoudres.

lightning call (Tel) ‖ Blitzgespräch n. ‖ conversation f. éclair.

lightning conductor *see also* lightning arrester ‖ Blitzableiter m.; Blitzableitererdleitung f. ‖ paratonnerre m.; mise f. à la terre *ou* conducteur m. de paratonnerre. / ~ cable ‖ Blitzableiterseil n. ‖ câble m. de parafoudre.

lightning discharge ‖ Blitzentladung f. ‖ décharge f. fulgurante; éclair m. / photography of ~ ‖ Blitzfotografie f. ‖ photographie f. d'éclair. / ~ point ‖ Blitzableiterspitze f. ‖ pointe f. de paratonnerre.

lightning protection ‖ Blitzschutz m. ‖ protection f. contre la foudre. / ~ fuse for open antenna ‖ Blitzschutzsicherung f. für Freiantennen ‖ fusible m. de parafoudre pour antennes extérieures.

lightning protector *see also* lightning arrester ‖ Blitzableiter m.; Blitzschutzvorrichtung f. ‖ parafoudre m.; paratonnerre m. / wedge-shaped ~ ‖ Schneidenblitzableiter m. ‖ paratonnerre m. à couteaux. / ~ point ‖ Blitzableiterspitze f. ‖ pointe f. de paratonnerre.

lightning rod ‖ Blitzableiterstange f. ‖ tige f. de paratonnerre. / ~ insulator ‖ Blitzableiterisolator m. ‖ isolateur m. de paratonnerre.

lightning stroke ‖ Blitzschlag m. ‖ coup m. de foudre. / ~ telegram ‖ Blitztelegramm n. ‖ télégramme m. éclair. / ~ tube ‖ Blitzröhre f. ‖ tube m. à éclairs.

light plant ‖ Lichtanlage f. ‖ installation f. d'éclairage.

light-proof casing ‖ lichtdichtes Gehäuse n. ‖ boîte f. étanche à la lumière. / ~ connecting sleeve ‖ Lichtmanschette f. ‖ manchon m. étanche à la lumière.

light-resisting ‖ lichtecht ‖ résistant à la lumière. / ~ room ‖ Lichtschacht m. ‖

évente f.; puits m. au jour. / ~ screen for steering-light (Mar) ‖ Blendschirm m. für Positionslichter ‖ abat-jour m. *ou* écran m. pour feux de route. / sensation of ~ ‖ Lichtreiz m. ‖ impression f. *ou* sensation f. de la lumière.

lightship (Nav) ‖ Feuerschiff n. ‖ bateau-feu m.

light signal ‖ Lichtzeichen n. ‖ signal m. lumineux; signe m. de feu. / ~ signalling installation ‖ Lichtsignalanlage f. ‖ installation f. de signalisation à lumière. / ~ socket ‖ Lichtanschlußdose f. ‖ boîte f. de jonction pour éclairage. / source of ~ ‖ Lichtquelle f. ‖ source f. lumineuse. / artificial source of ~ ‖ künstliche Lichtquelle f. ‖ source f. lumineuse artificielle. / natural source of ~ ‖ natürliche Lichtquelle f. ‖ source f. lumineuse naturelle. / stable to ~ ‖ lichtbeständig ‖ résistant à la lumière. / ~ subduing ‖ Lichtdämpfung f. ‖ atténuation f. de la lumière.

light-tight box ‖ lichtdichter Kasten m. ‖ boîte f. étanche à la lumière.

light, time flux of ~ ‖ Lichtmenge f. ‖ intensité f. de l'éclairage.

light tracing, half-cylinder for ~ ‖ Halbzylinder m. für Lichtpausapparate ‖ demi-cylindre m. pour appareils de calquage. / ~ establishment ‖ Lichtpausanstalt f. ‖ établissement m. de calquage.

light transmitting capacity of a field glass ‖ Lichtstärke f. eines Feldstechers ‖ intensité f. lumineuse *ou* luminosité f. d'une jumelle. / ~ capacity of the telescope ‖ Lichtstärke f. des Fernrohrs ‖ luminosité f. de la lunette. / ~ power of the telescope ‖ Lichtstärke f. des Fernrohrs ‖ luminosité f. de la lunette. / ~ quality of the glass ‖ Lichtdurchlässigkeit f. des Glases ‖ transparence f. du verre.

light treatment ‖ Lichtbehandlung f. ‖ traitement m. par les bains de lumière. / unit of ~ ‖ Lichteinheit f. ‖ unité f. de lumière.

light vessel *see* lightship.

light-wood oil (Chem) ‖ Kienholzöl n. ‖ essence f. de bois résineux.

ligne ‖ Linie f. ‖ ligne f.

ligneous ‖ holzartig ‖ ligneux.

lignification ‖ Verholzung f. ‖ lignification f.

lignified ‖ verholzt ‖ ligneux.

lignine works pl. ‖ Holzstoffabrik f. ‖ fabrique f. de pulpe de bois.

lignite ‖ Braunkohle f. ‖ lignite m. / bohemian ~ ‖ Schwarzkohle f.; Pechkohle f.; pechartige Braunkohle f. ‖ lignite m. bohémien; jais m. / ~ which is not briquettable in all ‖ lignitische nicht brikettierbare Braunkohle f. ‖ lignite m. que l'on ne peut pas comprimer en briquettes. / crude ~ *see* rough ~. / papyraceous ~ ‖ Papierkohle f.; Blätterkohle f. ‖ houille f. papyracée. / poor ~ of recent growth ‖ jüngere geringwertige Braunkohle f. ‖ lignite m. de formation récente et de peu de valeur. / rough ~ ‖ Rohbraunkohle f. ‖ lignite m. brut.

lignite beds pl. ‖ Braunkohlenvorkommen n. ‖ lignites mpl. / ~ briquette ‖ Braunkohlenbrikett n. ‖ lignite m. aggloméré; briquette f. en lignite. / carbonization of ~ ‖ Braunkohlenschwelung f. ‖ distillation f. lente du lignite. / ~ coke ‖ Grudekoks m. ‖ coke m. de lignite. /

~ dredger ‖ Braunkohlenbagger m. ‖ excavateur m. de lignite. / ~ dressing plant ‖ Braunkohlenaufbereitungsanlage f. ‖ installation f. de préparation du lignite. / ~ drying plant ‖ Braunkohlentrocknungsanlage f. ‖ installation f. de séchage du lignite. / ~ fields pl. ‖ Braunkohlenrevier n. ‖ district m. lignitifère. / ~ firing ‖ Braunkohlenfeuerung f. ‖ foyer m. à lignite. / ~ low-temperature carbonization power-station ‖ Braunkohlenschwelkraftwerk n. ‖ usine f. à distillation de lignite. / ~ mine ‖ Braunkohlengrube f. ‖ mine f. de lignite. / ~ mining machine ‖ Braunkohlengrubenmaschine f. ‖ machine f. pour mines de lignite. / ~ oil ‖ Braunkohlenöl n. ‖ huile f. de lignite. / ~ pit ‖ Braunkohlengrube f. ‖ mine f. de lignite. / ~ power plant ‖ Braunkohlenkraftwerk n. ‖ usine f. thermique à base de lignite. / extensive ~ resources owned by a company ‖ umfangreicher Braunkohlenbesitz m. einer Gesellschaft ‖ des réserves fpl. immenses de lignite dont dispose une société. / ~ tar ‖ Braunkohlenteer m. ‖ goudron m. de lignite. / ~ tar oil ‖ Braunkohlenteeröl n. ‖ huile f. de goudron de lignite. / ~ wax ‖ Montanwachs n. ‖ cire f. de lignite.

lignum vitae ‖ Pockholz n. ‖ bois m. de gaïac.

ligroine ‖ Ligroin n. ‖ ligroïne f.

like ‖ ähnlich ‖ du genre . . .; ressemblant.

likeness (Paint) ‖ Porträt n.; Bild n. ‖ portrait m.

lilac ‖ Fliederholz n. ‖ lilas m.

limb (Anatomy) ‖ Glied n. ‖ membre m. / ~ (Math) ‖ Gradbogen m.; Limbus m. ‖ arc m. gradué; limbe m. / artificial ~ ‖ künstliches Glied n. ‖ membre m. artificiel. / one ~ of the line ‖ Einzelleitung f. ‖ monoligne f. / ~ of syncline ‖ Muldenflügel m. ‖ flanc m. du synclinal.

lime, to ~ the grain ‖ das Getreide kalken ‖ chauler le blé.

lime (Bot) ‖ Lindenbaum m.; Linde f. ‖ tilleul m. / ~ (Miner) ‖ Kalk m. ‖ chaux f. / air-hardening ~ ‖ Luftkalk m. ‖ chaux f. durcissant à l'air. / anhydrous ~ *see* caustic ~. / artificial hydraulic ~ ‖ künstlicher hydraulischer Kalk m. ‖ chaux f. hydraulique artificielle. / to burn ~ ‖ Kalk m. brennen ‖ cuire la chaux. / burnt ~ ‖ Ätzkalk m.; gebrannter Kalk m. ‖ chaux f. calcinée *ou* vive; oxyde m. de calcium. / carbolic ~ ‖ Karbolkalk m. ‖ chaux f. carbolique. / caustic ~ ‖ Ätzkalk m.; gebrannter Kalk m. ‖ chaux f. vive *ou* caustique *ou* calcinée; oxyde m. de calcium. / dead ~ ‖ totgebrannter Kalk m. ‖ chaux f. morte ou surcuite. / dry-slaked ~ ‖ trocken gelöschter Kalk m. ‖ chaux f. étouffée. / fat ~ ‖ Fettkalk m.; Weißkalk m. ‖ chaux f. grasse. / fat white ~ ‖ fetter Weißkalk m. ‖ chaux f. grasse blanche. / hydraulic ~ ‖ hydraulischer Kalk m. ‖ chaux f. hydraulique. / to kill ~ ‖ Kalk m. verlöschen ‖ éteindre la chaux. / manuring ~ ‖ Düngekalk m. ‖ chaux f. pour engrais. / ~ of marble ‖ Marmorkalk m. ‖ chaux f. de marbre. / meagre ~ *see* poor ~. / nitrogenous ~ ‖ Kalkstickstoff m. ‖ chaux f. azotée. / ordinary ~ ‖ gewöhnlicher Kalk m. ‖ chaux f. ordinaire. / overburnt ~ *see* dead ~. / plashed ~ ‖ angemachter Kalk m. ‖ bain m. de chaux; chaux f. éteinte.

/ poor ~ ‖ magerer Kalk m. ‖ chaux f. maigre. / quick ~ *see* caustic ~. / quick slaking white ~ ‖ schnell löschender Weißkalk m. ‖ chaux f. blanche à extinction rapide. / shell ~ ‖ Muschelkalk m. ‖ chaux f. de coquilles. / to slake ~ ‖ Kalk m. löschen ‖ éteindre la chaux. / slaked ~ ‖ gelöschter Kalk m. ‖ chaux f. hydratée *ou* éteinte. / ~ slaked in the air ‖ abgestandener *oder* abgestorbener *oder* verwitterter Kalk m. ‖ chaux f. fusée. / the slaked ~ increases well ‖ der Kalk m. geht sehr auf *oder* gibt viel aus ‖ la chaux foisonne bien; la chaux se gonfle. / the slaked ~ swells *see* the slaked ~ increases well. / unslaked ~ *see* caustic ~. / Vienna polishing ~ ‖ Wiener Putzkalk m. ‖ chaux f. de Vienne à polir. / water ~ *see* hydraulic ~. / white ~ *see* fat ~.

lime acetate ‖ essigsaurer Kalk m. ‖ acétate m. de chaux. / ~ basin ‖ Kalkgrube f.; Kalkloch n. ‖ fosse f. *ou* bassin m. à chaux. / ~ bath ‖ Kalkbad n. ‖ bain m. de chaux. / ~ beater ‖ Rührkrücke f.; Kalkkrücke f.; Rudel n. ‖ mouve-chaux m.; rabot m. à chaux. / ~ bisulphite ‖ doppeltschwefligsaurer Kalk m. ‖ bisulfite m. de chaux f. / ~ blossom ‖ Lindenblüte f. ‖ fleur f. de tilleul. / ~ blue ‖ Kalkblau n. ‖ bleu m. à la chaux. / ~ burner ‖ Kalkbrenner m. ‖ chaufournier m. / ~ burning ‖ Kalkbrennerei f. ‖ chaufournerie f. / carbonate of ~ ‖ kohlensaurer Kalk m. ‖ carbonate m. de chaux. / ~ chest ‖ Kalkkasten m. ‖ caisse f. à chaux. / ~ citrate ‖ zitronensaurer Kalk m. ‖ citrate m. de chaux. / ~ colour ‖ Kalkfarbe f. ‖ couleur f. à la chaux. / ~ concrete ‖ Kalkbeton m. ‖ béton m. à la chaux. / ~ cream ‖ Kalkbrei m.; Schwödmasse f. ‖ enchaux m. / ~ crucible ‖ Kalktiegel m. ‖ creuset m. de chaux. / ~ cutting machine for chalky soil ‖ Kalkstechmaschine f. für Wiesenkalk ‖ machine f. d'extraction de chaux de prairies. / ~ glass ‖ Kalkglas n. ‖ verre m. à chaux. / ~ green ‖ Kalkgrün n. ‖ vert m. à la chaux. / ~ grinder ‖ Kalkmüller m. ‖ broyeur m. de chaux. / ~ grinding and bolting mill ‖ Kalkmühle f. ‖ broyage m. et bluterie f. de chaux. / ~ harmotome ‖ Kalkharmotom m.; Phillipsit m. ‖ harmotome m. calcaire; phillipsite f. / ~ hypophosphite ‖ unterphosphorigsaurer Kalk m. ‖ hypophosphite m. de chaux. / ~ juice ‖ Zitronensaft m. ‖ jus m. de citron. / ~ kiln ‖ Kalkofen m. ‖ four m. à chaux; chaufour m. / ~ kiln foreman ‖ Kalkbrennmeister m. ‖ maître-chaufournier m. / ~ light ‖ Kalklicht n. ‖ lumière f. de Drummond. / manuring with ~ ‖ Kalkdüngung f. ‖ chaulage m.; fumage m. à la chaux. / ~ marl ‖ Kalkmergel m. ‖ marne f. calcaire; chaux f. marneuse. / measuring device for ~ ‖ Abmeßapparat m. für Kalk ‖ appareil-doseur m. de chaux. / ~ milk ‖ Kalkmilch f. ‖ lait m. de chaux. / ~ mill ‖ Kalkmühle f. ‖ moulin m. à chaux. / ~ mortar ‖ Kalkmörtel m. ‖ mortier m. de chaux. / ~ paint ‖ Kalkfarbe f. ‖ peinture f. à la chaux. / ~ paste ‖ Kalkbrei m. ‖ pâte f. de chaux; chaux f. éteinte *ou* en pâte. / percentage of ~ ‖ Kalkgehalt m. ‖ teneur f. *ou* richesse f. en chaux. / ~ phenylate ‖ karbolsaurer Kalk m. ‖ phénate m. de

chaux. / ~ phosphate ‖ phosphorsaurer Kalk m. ‖ phosphate m. de chaux. / ~ pit see ~ basin. / ~ powder ‖ Kalkmehl n.; Staubkalk m. ‖ poudre f. de chaux. / ~ pyrolignite ‖ Graukalk m. ‖ pyrolignite m. de chaux.

limer ‖ Kalker m.; Tüncher m. ‖ chauleur m. / ~ (Curr) ‖ Kalker m. ‖ pelaineur m. / ~ (Mine) ‖ Kalksprenger m. ‖ chauleur m. de wagons.

lime rake ‖ Kalkschaufel f.; Kalkhaken m. ‖ mouve-chaux m. / ~-sand brick ‖ Kalksandziegel m. ‖ brique f. silico-calcaire; brique f. de sable et de chaux. / ~ shed ‖ Kalkschuppen m. ‖ hangar m. à chaud. / ~ slaker ‖ Kalklöscher m. ‖ extincteur m. de chaux. / ~ slaking box ‖ Kalklöschkasten m. ‖ caisse f. pour éteindre la chaux. / ~ slaking drum ‖ Kalklöschtrommel f. ‖ tambour m. d'extinction de chaux. / ~ soap ‖ Kalkseife f. ‖ savon m. à la chaux. / ~ soda feldspar ‖ Labradorit m.; Labradorfeldspat m.; Labradorstein m. ‖ labradorite f.; pierre f. de Labrador; feldspat m. Labrador ou opalin.

limestone ‖ Kalkstein m. ‖ pierre f. calcaire ou à chaux; calcaire m. / asphaltic ~ ‖ Asphaltkalkstein m. ‖ calcaire m. asphaltique. / bituminous ~ ‖ bituminöser Kalkstein m.; Stinkkalkstein m.; Stinkstein m. ‖ chaux f. carbonatée fétide; calcaire m. fétide. / carboniferous ~ ‖ Kohlenkalkstein m. ‖ chaux f. ou calcaire m. carbonifère. / common ~ ‖ dichter oder gemeiner Kalkstein m. ‖ calcaire m. compact. / cretaceous marly ~ ‖ Pläner m. ‖ calcaire m. marneux crétacé. / granular ~ ‖ Marmor m.; körniger Kalkstein m. ‖ chaux f. carbonatée saccharoïde; marbre m. / mountain ~ ‖ Kohlenkalkstein m. ‖ chaux f. ou calcaire m. carbonifère. / nummulitic ~ ‖ Nummulitenkalk m. ‖ chaux f. nummulitique. / siliceous ~ ‖ Kieselkalk m. ‖ calcaire m. siliceux. / tufaceous ~ ‖ Tuffkalk m. ‖ tuf m. calcaire.

limestone crushing plant ‖ Kalksteinzerkleinerungsanlage f. ‖ installation f. de broyage de calcaire. / ~ flux ‖ Kalkzuschlag m. ‖ castine f.; fondant m. calcaire. / ~ powder ‖ Kalksteinmehl n. ‖ poudre f. de pierres à chaux. / ~ quarry ‖ Kalksteinbruch m. ‖ carrière f. de pierre à chaux ou de castine ou de calcaire. / ~ roasting pan ‖ Steinhärtekessel m. ‖ chaudière f. pour le durcissement des pierres.

lime sulphite ‖ schwefligsaurer Kalk m. ‖ sulfite m. de chaux. / ~ tartrate ‖ weinsaurer Kalk m. ‖ tartrate m. de chaux. / ~ tungstate ‖ wolframsaurer Kalk m. ‖ tungstate m. de chaux. / ~ (covered) wagon ‖ Kalkdeckelwagen m. ‖ wagon m. couvert pour chaux. / ~ wash ‖ Kalkmilch f.; Tünche f. ‖ lait m. de chaux. / ~ water ‖ Ätzkalklösung f.; Kalkwasser n. ‖ eau f. de chaux. / ~ water colour ‖ Kalkfarbe f.; Wasserfarbe f. ‖ couleur f. détrempée avec le lait de chaux. / ~ weighing machine ‖ Kalkwage f. ‖ bascule f. à chaux. / ~ wood ‖ Lindenholz n. ‖ bois m. de tilleul. / ~-works equipment ‖ Kalkwerkeinrichtung f. ‖ installation f. d'usine à chaux. / ~-works plant ‖ Kalkwerksanlage f. ‖ installation f. de four à chaux.

liming ‖ Kalken n. ‖ chaulage m. / ~ (Curr) ‖ Kalken n. ‖ pelanage m. à la chaux; plainage m.; plamage m.

limit, to ‖ beschränken; abgrenzen ‖ limiter; confiner; borner.

limit (Math) ‖ Grenzwert m. ‖ limite f. / ~ (Mech) ‖ zulässiger Spielraum m. ‖ tolérance f.; limite f.; jeu m. / ~ (Trade) ‖ Mindestbetrag m. ‖ minimum m. / ~ (Trade) ‖ Höchstpreis m. ‖ prix m. maximum. / ~ of adhesion ‖ Adhäsionsgrenze f. ‖ limite f. par adhérence. / ~ in bending ‖ Biegegrenze f. ‖ limite f. de flexion. / ~ of current ‖ Höchststromstärke f. ‖ limite f. maximum du courant. / economical ~ ‖ Wirtschaftlichkeitsgrenze f. ‖ limite f. économique. / ~ of error ‖ Fehlergrenze f. ‖ limite f. d'erreur. / ~ of load ‖ Belastungsgrenze f. ‖ limite f. de charge. / lower ~ of measurement ‖ untere Meßgrenze f. ‖ limite f. inférieure de mesure. / upper ~ of measurement ‖ obere Meßgrenze f. ‖ limite f. supérieure de mesure. / ~ of proportionality ‖ Proportionalitätsgrenze f. ‖ limite f. de la proportionalité. / ~ of reading ‖ Ablesegrenze f. ‖ limite f. de lecture. / ~ of rotation ‖ Begrenzung f. der Drehung ‖ délimitation f. de rotation. / ~ of sensibility ‖ Empfindlichkeitsgrenze f. ‖ limite m. de sensibilité. / ~ of stability ‖ Stabilitätsgrenze f. ‖ limite f. de stabilité. / ~ of the trade-winds ‖ Passatgrenze f. ‖ limite f. de vents alizés. / ~ of weight ‖ Gewichtsgrenze f. ‖ limite f. de poids.

limitation ‖ Einengung f.; Begrenzung f. ‖ rétrécissement m.; limitation f.; limite f. / term of ~ ‖ Verjährungsfrist f. ‖ délai m. de prescription. / ~ case ‖ Grenzfall m. ‖ cas m. limite.

limited company ‖ Gesellschaft f. mit beschränkter Haftung (G. m. b. H.) ‖ société f. à responsabilité limitée. / ~ liability company s. ~ company. / ~ partnership ‖ Kommanditgesellschaft f. ‖ société f. en commandite. / ~ range of regulation ‖ begrenzter Regelbereich m. ‖ capacité f. de réglage limitée. / ~ submission ‖ beschränkte Ausschreibung ‖ appel m. limité à la concurrence. / ~ view ‖ beschränkte Fernsicht f. ‖ vue f. limitée.

limit gauge ‖ Toleranzlehre f.; Grenzlehre f. ‖ calibre m. de tolérance. / ~ mark ‖ Grenzstrich m. ‖ trait-limite m. / ~ stop (Tool mach) ‖ Anschlag m. ‖ arrêt m.; taquet m.; butoir m.; butée f.; ergot m.; équerre f. / ~ switch ‖ Endausschalter m. ‖ interrupteur m. de fin de course. / ~ value ‖ Grenzwert m. ‖ valeur f. limite.

limonite ‖ Berggelb n.; Bohnerz n.; Brauneisenstein m. ‖ limonite f.; hématite f. brune; fer m. oxydé hydraté; jaune m. de montagne; minerai m. de fer en grains. / ~ mine ‖ Raseneisensteingrube f. ‖ mine f. de fer limoneux.

limousine ‖ Limusine f. ‖ limousine f. D-front ~ ‖ Limousine f. mit runder Vorderscheibe ‖ limousine f. à parebrise arrondi.

limpid (liquid) ‖ klar ‖ claire; limpide.

limy ‖ kalkhaltig ‖ calcifère.

linaloe oil ‖ Linaloeöl n. ‖ essence f. de linaloés.

linch hoop (Coachm) ‖ Achsring m. ‖ anneau m. de bout d'essieu. / ~ nave hoop (Coachm) ‖ Nabenring m. ‖ frette f. au petit bout du moyeu. / ~ pin

(Coachm) ‖ Lünse f.; Achsnagel m. ‖ esse f.; asse f. / ~ pin with a linch pin cap (Coachm) ‖ Lünse f. mit Kotdeckel ‖ esse f. d'essieu à coiffe. / ~ washer (Coachm) ‖ Lünsscheibe f. ‖ rondelle f. du bout d'essieu.

lincrusta ‖ Linkrusta f. ‖ lincrusta m. / ~ factory ‖ Linkrustafabrik f. ‖ fabrique f. de lincrusta. / ~ goods pl. ‖ Linkrustawaren fpl. ‖ articles mpl. de lincrusta.

linden ‖ Linde f.; Lindenholz n. ‖ (bois m. de) tilleul m. / ~ bast ‖ Lindenbast m. ‖ écorce f. de tilleul. / ~ bast cordage ‖ Lindenbastseil n. ‖ corde f. d'écorce de tilleul.

line, to ~ (Bearing) ‖ ausgießen ‖ garnir. / ~ (Cloth) ‖ füttern ‖ doubler. / ~ (Mach) ‖ füttern ‖ revêtir; garnir. / ~ (Saddl) ‖ polstern ‖ matelasser; rembourrer. / ~ brasses pl. with white metal ‖ Lagerschalen fpl. mit Weißmetall ausgießen. ‖ antifrictionner les coussinets mpl. / ~ with brickwork ‖ mit Mauerwerk n. auskleiden ‖ garnir de maçonnerie f. / ~ with cloth ‖ mit Tuch n. ausschlagen ‖ garnir de drap m. / ~ out (Build) ‖ abfluchten; abvisieren ‖ dresser à ligne; aligner; jalonner. / ~ out straight lines upon a soil ‖ ein Gelände ausbaken ‖ aligner ou jalonner un terrain. / ~ out stuff (Carp) ‖ das Holz abschnüren oder schnüren ‖ aligner le bois. / ~ with paper ‖ mit Papier n. auslegen ‖ garnir de papier m. / ~ a round shaft with bricks ‖ einen runden Schacht m. mit Ziegeln ausmauern ‖ tuber un puits à l'aide d'un revêtement cylindrique en maçonnerie de briques. / ~ a shaft with wood-work ‖ einen Schacht m. auszimmern ‖ étrésillonner un puits de mine. / ~ a timber ‖ das Holz abschnüren ‖ cingler le bois; marquer le charpente au cordeau. / ~ up an engine ‖ eine Maschine f. aufstellen oder montieren oder adjustieren ‖ dresser ou ajuster une machine. / ~ with wicker-work ‖ beflechten ‖ enlacer des brins mpl. d'osier autour de . . .

line (Drawing) ‖ Linie f.; Strich m. ‖ trait m. / ~ (Electr) ‖ Leitung f. ‖ ligne f. / ~ (Mas) ‖ Mauerschnur f.; Schlagleine f. ‖ cordeau m.; fouet m.; ligne f. / ~ (Math) ‖ Linie f.; Reihe f. ‖ ligne f. / ~ (Print) ‖ weißer Raum m. ‖ blanc m. de la forme. / ~ (Railw) ‖ Strecke f.; Linie f.; Trasse f. ‖ voie f.; ligne f.; tracé m. / ~ (Road) ‖ Bauflucht f. ‖ alignement m. / ~ (Ropem) ‖ Leine f.; Strick m. ‖ ligne f.; corde f. / ~ (Surv) ‖ Meßband n.; Schnur f. zum Messen ‖ cordeau m. pour le mesurage (des lignes topographiques). / aerial ~ (Electr) ‖ Freileitung f. ‖ ligne f. aérienne. / ahead ~ (Mar) ‖ Kiellinie f. ‖ file f.; ligne f. de file. / artificial ~ ‖ künstliche Leitung f. ‖ ligne f. artificielle. / boundary ~ ‖ Grenzlinie f. ‖ ligne f. de séparation. / ~ on brackets (Electr) ‖ Leitung f. auf Konsolen ‖ ligne f. en consoles. / break ~ ‖ kurze Linie f. ‖ ligne f. abrégée. / ~ of business ‖ Geschäftskreis m. ‖ sphère f. d'activité. / city exchange ~ (Tel) ‖ Postleitung f. ‖ ligne f. au réseau. / to clear the ~ (Railw) ‖ freie Fahrt f. geben ‖ débloquer la ligne. / ~ in construction (Railw) ‖ im Bau befindliche Bahn f. ‖ ligne f. en construction. / continuously loaded ~ ‖ gleichmäßig belastete Lei-

tung f. ‖ ligne f. à charge continue. /
controlling ~ (Tel) ‖ Mithörleitung f. ‖
ligne f. d'observation. / ~ on cross arms
(Electr) ‖ Leitung f. auf Querträgern ‖
ligne f. disposé sur traverses. / to carry
current on the ~ ‖ die Leitung mit Strom
beschicken ‖ alimenter une ligne en
courant; porter du courant sur la ligne. /
curved ~ (Railw) ‖ krummer Strang m. ‖
voie f. courbe. / ~ of direction (Mech) ‖
Richtungslinie f. ‖ ligne f. de direction. /
~ of direction (Road) ‖ Bauflucht f.;
Baufluchtlinie f. ‖ alignement m. / ~
of direction of a force ‖ Kraftrichtung f.
‖ direction f. de la force. / dotted ~ ‖
gepunktete oder punktierte Linie f. ‖
ligne f. ponctuée ou pointillée. / double
~ ‖ Doppelleitung f. ‖ ligne f. double. /
~ of double curvature ‖ doppelt gekrümmte Linie f. ‖ courbe f. à double
courbure. / ~ drawn from an angle to
the middle of the opposite side ‖ Mittellinie f. eines Dreiecks ‖ médiane f. d'un
triangle / to drive the ~s (Print) ‖ die
Zeilen fpl. enger machen ‖ serrer les
lignes fpl. / ~ of equal barometrical
pressure ‖ Linie f. gleichen Luftdrucks ‖
ligne f. d'égale pression d'air. / fictive
~ ‖ imaginäre Linie f. ‖ emprunt m.; ligne
f. fictive. / ~ with fish-hooks ‖ Leine f.
mit Fischhaken ‖ ligne f. avec hameçons.
/ fishing ~ ‖ Fischerleine f. ‖ libouret
m.; ligne f. de pêche. / fixed ~ ‖ feste
Leitung f. ‖ ligne f. fixe. / ~ of force
‖ Kraftlinie f. ‖ ligne f. de force. / deflection of the ~s pl. of force ‖ Kraft
linienablenkung f. ‖ déviation f. des
lignes de force. / go and return ~ ‖
Hin- und Rückleitung f. ‖ ligne f. d'aller
et de retour. / ~ with heavy gradients
(Railw) ‖ Hügellandstrecke f. ‖ ligne f.
en pays accidenté. / ~ carrying very high
voltage ‖ Höchstspannungsleitung f. ‖
ligne f. à très haute tension. / horizontal
~ ‖ wagerechte Linie f.; Horizontale f.
‖ ligne f. horizontale. / ideal ~ (Electr) ‖
ideale Leitung f. ‖ ligne f. idéale. / induced ~ ‖ induzierte Leitung f. ‖ ligne f.
induite. / inducing ~ ‖ induzierende Leitung f. ‖ ligne f. inductrice. / ~ with
intermediate receiving-stations (Tel) ‖
Schleifleitung f.; Schleiflinie f. ‖ ligne f.
à embrechage. / to interrupt a ~ (Electr)
‖ eine Leitung f. unterbrechen ‖ interrompre un circuit. / ~ of intersection ‖
Schnittlinie f.; Durchschnittslinie f. ‖
ligne f. d'intersection. / intersectional ~
see line of intersection. / leading in ~ ‖
Zufuhrgleis n. ‖ voie f. d'accès. / mounted ~s pl. (Railw) ‖ zusammengesetzte
Gleise npl. ‖ voies fpl. montées ou com-
posées. / non-dissipative ~ see ideal, line. /
~ with numerous curves ‖ kurvenreiche
Linienführung f. ‖ ligne f. ayant de nombreuses courbes. / one limb of the ~ ‖ Einzelleitung f. ‖ monoligne f. / open wire ~s
‖ Freileitung f. ‖ ligne f. aérienne. / parallel ~ ‖ Parallele f.; parallele Gerade f. ‖
parallèle f. / perpendicular ~ ‖ senkrechte oder normale Linie; Lot n.;
Perpendikel n. ‖ ligne f. perpendiculaire; perpendiculaire f. / pipe ~ ‖ Rohrleitung f. ‖ tuyauterie f.; tuyautage m. /
portable ~ ‖ verlegbares Gleis n. ‖ voie f.
portative. / private ~ (Railw) ‖ In-
dustriegleis n.; Privatgleis n. ‖ voie f.
particulière; embranchement m. particulier ou d'usine. / ~ of rails ‖ Eisen-

bahngleis n. ‖ voie f.; rails mpl. / inner ~
of rails (Railw) ‖ innerer Schienenstrang
m. ‖ rangée f. de rails intérieure; voie f.
intérieure. / outer ~ of rails (Railw) ‖
äußerer Schienenstrang m. ‖ rangée f.
de rails extérieure; voie f. extérieure.
/ ranged ~ ‖ ausgefluchtete Linie f.
‖ ligne f. alignée. / right ~ ‖ gerade
Linie f.; Gerade f. ‖ ligne f. droite;
droite f. / ~ of rocks ‖ Riff n. ‖ écueil
m.; récif m. ‖ ligne f. de visée. / ~
Nebenbahn f.; Lokalbahn f.; Nebenlinie f.; Zweigbahn f. ‖ ligne f. secondaire ou d'intérêt local; chemin m. de
fer vicinal. / ~ of separation ‖ Scheide-
linie f. ‖ ligne f. de démarcation ou de
séparation ou de partage. / ~ of service
telephone ‖ Betriebsfernsprechleitung f.
‖ ligne f. du téléphone de service. /
~ of shafting (Mach) ‖ Wellenleitung f.
‖ ligne f. d'arbres. / ~ of sight (Arms)
‖ Visier linie f. ‖ ligne f. de visée. /
of sight (Opt) ‖ Blickrichtung f. ‖ direction f. de regard. / simple ~ ‖ Einfachleitung f. ‖ ligne f. simple. / smooth ~
‖ gleichmäßige Leitung f. ‖ ligne f.
homogène. / straight ~ ‖ gerade Linie f.;
Gerade f. ‖ ligne f. droite; droite f. /
submarine ~ ‖ unterseeische Leitung f.
‖ ligne f. sousmarine. / subterranean ~
‖ unterirdische Leitung f. ‖ ligne f.
souterraine; fil m. de terre; conduit m.
souterrain. / sunk ~ ‖ versenktes Gleis
n. ‖ voie f. noyée. / ~ of teeth on the
rack ‖ Rechenzackenreihe f. ‖ rangée f.
de dents du râteau. / telegraphic ~ ‖
telegraphische Leitung f. ‖ ligne f. télégraphique. / triple-rail ~ ‖ Dreischienengleis n. ‖ voie f. à trois rails. / underground ~ see subterranean ~. / uneven ~
‖ ungleichmäßige Leitung f. ‖ ligne f.
hétérogène. / up and down ~ ‖ Hin-
und Rückleitung f. ‖ ligne f. d'aller et
(de) retour. / vanishing ~ ‖ Fluchtlinie
f. ‖ ligne f. fuyante. / ~ of the vessel ‖
Linie f. des Schiffskörpers ‖ ligne f. ou
courbe f. du bateau. / ~ in the visible
spectrum ‖ Linie f. im sichtbaren Spektrum ‖ raie f. du spectre visible. / visual ~
‖ Gesichtslinie f. ‖ ligne f. visuelle. / ~
on wall brackets ‖ Leitung f. auf Wand-
konsolen ‖ ligne f. sur consoles murales.
/ whale ~ ‖ Harpunleine f. ‖ ligne f.
de harpon.
lines pl. (Shipb) ‖ Konstruktionszeichnung f. oder Riß m. (eines Schiffes) ‖
plan m. de construction (d'un bateau). /
~ convergent see ~ converging. / converging ~ ‖ konvergierende Linien fpl. ‖
lignes pl. convergentes. / cotidal ~ ‖
Isorachien fpl. ‖ lignes fpl. cotidales. / the
~ cut each other see the ~ intersect each
other. / divergent ~ see diverging ~.
/ diverging ~ ‖ divergierende Linien fpl.
‖ lignes fpl. divergentes. / the ~ intersect each other ‖ die Linien fpl. schneiden einander ‖ les lignes fpl. s'entre-
coupent. / isodynamic ~ ‖ Isodynamen
fpl. ‖ lignes fpl. isodynamiques. / iso-
gonic ~ ‖ Isogonen fpl. ‖ lignes fpl. iso-
gones. / isoclinic ~ ‖ Isoklinen fpl. ‖
lignes fpl. isoclines. / parallel ~ ‖ parallele
Linien fpl. ‖ lignes fpl. parallèles. / ~
of a vessel (Shipb) ‖ Linien fpl. eines
Schiffes ‖ lignes fpl. d'un bateau.
line adding key ‖ Zeilenaddiertaste f. ‖
touche f. pour l'addition en ligne.
linear ‖ linear ‖ linéaire. / ~ design see

drawing. / ~ drawing ‖ Linearzeichnung
f. ‖ dessin m. linéaire. / ~ element ‖
Linienelement n. ‖ élément m. de ligne.
/ ~ expansion ‖ Längenausdehnung f. ‖
dilatation f. linéaire. / ~ increase of
velocity of ascending currents ‖ lineare
Zunahme f. der Aufwindgeschwindig-
keit ‖ accroissement m. linéaire de la
vitesse du vent ascendant. / ~ measure
‖ Linearmaßstab m. ‖ mesure f. linéaire.
/ ~ perspective ‖ Linearperspektive f. ‖
perspective f. linéaire.
line arrester ‖ Leitungsblitzableiter m. ‖
parafoudre m. de ligne. / ~ bank ‖ Kon-
taktsatz m. ‖ banc m. des contacts. /
~ bar ‖ Leitungsschiene f. ‖ barre f.
d'amenée. / ~ battery (Fel) ‖ Linien-
batterie f. ‖ batterie f. de ligne. / straight
~ box ‖ einfache Verbindungsmuffe f.
‖ manchon m. de raccordement droit. /
~ calling current ‖ Linienrufstrom m. ‖
courant m. d'appel de ligne.
linecasting machine (Print) ‖ Zeilengieß-
maschine f. ‖ machine f. à fondre les
caractères en ligne.
line circuit for telegraphy ‖ Leitungs-
schaltung f. für Telegrafie ‖ circuit m.
télégraphique. / ~ commutator ‖ Linien-
umschalter m. ‖ commutateur m. de
ligne. / ~ conduit ‖ Leitung f. ‖ canalisation f. / ~ connection apparatus (Radio)
‖ Netzanschlußgerät n. ‖ boîte f. de
tension ou de raccordement. / ~ connector (Tel) ‖ Leitungswähler m. ‖ sélecteur m. final. / ~ consolidation ‖ Linien-
verstärkung f. ‖ consolidation f. de la
ligne. / ~ construction service ‖ Tele-
grafenbaudienst m. ‖ service m. de construction des lignes télégraphiques. / ~
counter ‖ Zeilenzähler m. ‖ compteur m.
de lignes. / crossing of ~s (Railw) ‖ Gleiskreuzung f. ‖ croisement m. de voie. /
portable ~ crossing (Railw) ‖ tragbarer
Wegeübergang m. ‖ croisement m. de voie
portative. / ~ current ‖ Linienstrom m.;
Netzstrom m. ‖ courant m. de ligne. /
curve of the ~ ‖ Gleiskrümmung f. ‖
courbe f. de la voie.
lined (Cloth) ‖ gefüttert ‖ doublé. / tin-~
case ‖ mit Blech ausgeschlagene Kiste f.
‖ caisse f. doublée de fer blanc. / ~ cloth ‖
Futtertuch n. ‖ doublure f. / ~ rinsing
pipe ‖ gefüttertes Spülversatzrohr n. ‖
tuyau m. de rinçage à revêtement.
line, direction of the ~ ‖ Linienführung f. ‖
tracé m. de la ligne. / ~ distributing
frame ‖ Leitungsverteiler m. ‖ distributeur m. de lignes. / ~ drawer ‖ Auf-
zeichner m. ‖ traceur m. de lignes / ~
drawing ‖ Strichzeichnung f.; Linearzeichnung f. ‖ dessin m. linéaire. / ~ drop
‖ Spannungsabfall m. ‖ chute f. de tension. / ~ electrode ‖ Netzelektrode ‖
électrode f. alvéolée. / ~ element ‖ Linienelement n. ‖ élément m. de ligne. / ~
equation ‖ Leitungsgleichung f. ‖ équation
f. de la ligne. / ~ equipment ‖ Strecken-
ausrüstung f. ‖ équipement m. de la ligne.
/ ~ fault ‖ Leitungsstörung f. ‖ dérangement m. dans la ligne ou de la ligne. /
~ faults pl. en masse (Tel) ‖ Massen-
störung f. ‖ dérangements mpl. en quantité. / ~ fault service ‖ Störungsdienst m.
‖ service m. des dérangements. / ~ finder
(Aut tel) ‖ Leitungssucher m.; Linien-
sucher m. ‖ chercheur m. de ligne. / ~
integral ‖ Linienintegral n. ‖ intégrale
f. linéaire.

linekeeper (Railw) ‖ Bahnwärter m. ‖ garde-ligne m.; garde-barrière m. / ~'s assistant (Railw) ‖ Hilfsbahnwärter m.; Hilfswärter m. ‖ garde m. auxiliaire. / ~'s box ‖ Bahnwärterhaus n. ‖ maison f. de gardes; maisonnette f. ou guérite f. de garde-barrière. / ~'s lodge see ~'s box. **line** loop ‖ Leitungsschleife f. ‖ bouche f.; lacet m. dans la ligne. / ~ loss ‖ Leitungsverlust m. ‖ perte f. de ligne ou à la ligne. **lineman** ‖ Telegrafenarbeiter m. ‖ ouvrier m. des lignes télégraphiques. **line,** map of ~ ‖ Streckenplan m. ‖ carte f. de la ligne. / network of ~s pl. ‖ Gleisnetz n. ‖ réseau m. de voies. **linen** ‖ Leinwand f. ‖ toile f. / ~ Wäsche f.; Weißwaren fpl. ‖ lingerie f.; linge m. / baby ~ ‖ Kinderwäsche f. ‖ lingerie f. d'enfants. / ~ for the bath ‖ Badewäsche f. ‖ linge m. de bain. / bed ~ ‖ Bettwäsche f. ‖ linge m. de lit. / body ~ ‖ Leibwäsche f. ‖ linge m. de corps. / brown ~ ‖ ungebleichte Leinwand f. ‖ toile f. écrue. / cotton warp ~ ‖ halbbaumwollene Leinwand f.; Halbleinen n. ‖ toile f. mixte; tissu m. mi-lin mi-coton. / covered ~ ‖ aufgelegte Leinwand f. ‖ toile f. superposée. / cream coloured ~ see half-bleached ~. / damask ~ ‖ Damastleinwand f. ‖ damassé m. / damask table-~ ‖ damastenes Tafelzeug n. ‖ linge m. damassé. / dirty ~ ‖ schmutzige Wäsche f. ‖ linge m. sale. / durable ~ ‖ Dauerwäsche f. ‖ linge m. durable ou économique ou permanent. / figured ~ ‖ Stangenleinwand ‖ toile f. ouvrée. / fine ~ ‖ feine Leinwand f. ‖ toile f. fine en lin. / flax ~ ‖ Flachsleinwand f. ‖ toile f. de lin. / ~ for gentlemen ‖ Herrenwäsche f. ‖ lingerie f. pour hommes. / glazed ~ ‖ Glanztuch n. ‖ toile f. gommée. / half-bleached ~ ‖ halbgebleichte Leinwand f. ‖ blondines fpl.; toile f. crémée. / ~ of hand-spun yarn ‖ Handgarnleinen n. ‖ toile f. de fil de main ou de fil à fuseau. / hemp ~ ‖ Hanfleinwand f. ‖ toile f. de chanvre. / home spun ~ ‖ Hausleinen n. ‖ toile f. tissue à domicile. / kitchen ~ ‖ Küchenwäsche f. ‖ linge f. de cuisine. / ladies' ~ ‖ Damenwäsche f. ‖ lingerie f. de dames. / multiple ~ ‖ mehrfache Leinwand f. ‖ toile f. multiple. / oiled ~ ‖ Ölleinen n. ‖ toile f. huilée. / packing ~ ‖ Packleinwand f. ‖ toile f. d'emballage. / plain ~ ‖ glattes Leinengewebe n. ‖ linge m. uni. / quilled ~ ‖ geköperte Leinwand f. ‖ toile f. croisée ou grainée; toile f. à grain d'orge ou ouvrée à petits dessins. / single ~ ‖ einfache Leinwand f. ‖ toile f. simple. / table ~ ‖ Tischwäsche f. ‖ lingerie f. de table. / ~ for toilet purposes ‖ Wäsche f. für Toilettezwecke ‖ lingerie f. de toilette. / tow ~ ‖ Wergleinwand f.; Hedeleinen n. ‖ toile f. d'étoupe. / tweeled ~ for trousers ‖ Hosendrell m. ‖ coutil m. pour pantalon. / unbleached ~ ‖ ungebleichte Leinwand f. ‖ toile f. écrue. / white China ~ ‖ chinesische Leinwand f. ‖ nunna m. / worker's ~ ‖ Arbeiterwäsche f. ‖ linge m. pour ouvriers. **linen** bag ‖ Wäschesack m. ‖ sac m. à linge. / ~ bandage ‖ Leinenbinde f. ‖ bande f. de toile. / ~ bed tick ‖ leinener Bettdrell m. ‖ coutil m. pour literie. / ~ bodice ‖ Korsettschoner m. ‖ cache-corset m. / ~ boiler ‖ Waschkessel m. ‖

chaudière f. pour le lessivage. / ~ border ‖ Einzugband n.; Durchzugband n. ‖ lacet m. / ~ button ‖ Wäscheknopf m. ‖ bouton m. pour lingerie. / ~ cap for screw bungs ‖ Dichtungsläppchen n. für Spunde ‖ rondelle f. graissée ou suifée pour bondes. / ~ cloth ‖ Leinwand f.; Linnen n.; Leinen n. ‖ toile f. de lin. / ~ cloth weaver ‖ Leinwandweber m. ‖ tisseur m. de toile de lin. / ~ cutter ‖ Wäscheschneider m. ‖ coupeur m. de lingerie. / ~ damask ‖ Leinendamast m. ‖ damas m.; linge m. damassé. / ~ disinfection plant ‖ Wäschedesinfektionsanlage f. ‖ installation f. pour la désinfection du linge. / ~ draper ‖ Schnittwarenhändler m. ‖ mercier m. / ~ drying room ‖ Wäschetrockenanstalt f. ‖ séchoir m. à linge. / ~ gauze ‖ Leinengaze f. ‖ gaze f. de fil. / ~ goods pl. ‖ Weißwaren fpl. ‖ toilerie f.; mercerie f.; lingerie f. / ~ ground weft ‖ Grundschuß m. aus Leinen ‖ trame f. de fond en lin. / ~ heald ‖ Litze f. aus Leinenzwirn ‖ lisse f. en câblé lin. / ~ hosiery ‖ Leinenstrikkerei f. ‖ bonneterie f. de lin. / ~ interlining ‖ Futterleinen n. ‖ toile f. doublure. / ~ ironing ‖ Plätten n. oder Bügeln n. der Wäsche ‖ repassage m. du linge. / ~ loom ‖ Leinwandstuhl m. ‖ métier m. à tisser la toile. / ~ mangling ‖ Wäschemangelanstalt f.; Rollstube f. ‖ calandrage m. de linge. / ~ manufacture ‖ Leinenindustrie f. ‖ industrie f. linière. / ~ marking ink ‖ Wäschezeichentinte f. ‖ encre f. à marquer le linge. / ~ mender ‖ Wäscheausbesserin f. ‖ raccommodeuse f. de linge. / ~ patch for screw bungs see ~ cap for screw bungs. / ~ piece of ~ ‖ Wäschestück n. ‖ pièce f. de lingerie. / ~ plush ‖ Leinenplüsch m. ‖ peluche f. de lin. / ~ rag paper ‖ Leinenpapier n. ‖ papier m. de chiffons de lin. / ~ seamstress ‖ Weißnäherin f. ‖ couseuse f. en lingerie. / ~ spinning ‖ Leinenspinnerei f. ‖ filature f. de lin. / ~ spooler ‖ Leingarnspuler m. ‖ bobineur m. de fil de lin. / ~ tape ‖ leinenes Band n.; Leinwandband n. ‖ ruban m. en lin. / ~ thread ‖ Leinenzwirn m. ‖ fil m. retors de lin. / ~ trade ‖ Flachshandel m. ‖ commerce m. du lin. / ~ twist tape ‖ Zwirnband n. ‖ ruban m. en lin retors. / ~ weaver ‖ Leinwandweber m. ‖ ouvrier m. toilier. / ~ weaving ‖ Leinenweberei f. ‖ tissage m. de toile ou du lin. / ~ yarn ‖ Leinengarn n. ‖ fil m. de lin. **line** packing machine ‖ Gleisstopfmaschine f. ‖ machine f. à damer sous les traverses de chemin de fer ou à bourrer les traverses. **line** parameter ‖ Leitungsparameter n. ‖ paramètre m. de ligne. **liner** (Bookb) ‖ Buchdeckelleimer m. ‖ couvreur m. / ~ (Mach) ‖ Futter n. ‖ garniture f. / ~ (Metal enamelling) ‖ Filetmaler m. ‖ fileteur m. / axle-box ~ ‖ Achslagergleitplatte f. ‖ fourrure f. de la boîte d'essieu. / cylinder ~ ‖ Zylinderbuchse f. ‖ fourrure f. du cylindre. / loose ~ (Gun) ‖ auswechselbares Futterrohr n. ‖ chemise f. de canon amovible. / removable ~ see loose ~. **line** relay (Fel) ‖ Linienrelais n. ‖ relais m. de ligne. **liner** locking screw ‖ Laufbuchsenhalteschraube f. ‖ boulon m. d'arrêt du fourreau.

line scale micrometer ‖ Strichmikrometer n. ‖ micromètre m. à tambour gradué. / ~ scheme ‖ Linienbild n. ‖ plan m. de la ligne. / portable ~ section (Railw) ‖ Gleisjoch n. ‖ châssis m. ou tronçon m. de voie. / ~ selector cable ‖ Linienwählerkabel n. ‖ câble m. pour appareils téléphoniques multiples. / ~ side (Electr) ‖ Leitungsseite f. ‖ côté f. de la ligne. / ~ signal (Tel) ‖ Anrufsignal n. ‖ signal m. d'appel. / ~ space gauge (Typewr) ‖ Zeilenabstandzeiger m. ‖ pointeau m. d'interligne. / ~ space lever (Typewr) ‖ Zeilenschalthebel m.; Zeileneinstellhebel m. ‖ levier m. d'interligne. / ~ spacing device (Typwr) ‖ Zeilenschaltung f. ‖ commande f. de l'interligne. / ~ spacing lever see ~ space lever. / ~ spacing mechanism (Typewr) ‖ Zeilenfortschaltmechanismus m. ‖ mécanisme m. d'interligne. / statement of ~s ‖ Liniennachweis m. ‖ carnet m. des lignes. / ~ telephony ‖ Drahttelefonie f. ‖ téléphonie f. par fil. / ~ throwing apparatus ‖ Leinenwurfapparat m. ‖ appareil m. à lancer des cordages. / ~ utilization ‖ Leitungsausnutzung f. ‖ utilisation f. des lignes. / ~ voltage ‖ Netzspannung f. ‖ tension f. du réseau. / ~ wire (Tel) ‖ Blockleitung f. ‖ conduite f. de l'appareil de bloc(age). **lingerie** ‖ Wäschefabrik f. ‖ (fabrication f. de) lingerie f. **lining** (Brake) ‖ Belag m. ‖ garniture f. / ~ (Brickwork) ‖ Ausmauerung f. ‖ maçonnage m. / ~ (Furnace) ‖ Ausfütterung f. ‖ revêtement m. / ~ (Mach) ‖ Futter n. ‖ fourrure f. / ~ (Mine) ‖ Markscheiderzug m.; Zug m. ‖ levé m.; tracé m. / ~ (Paper-hangings) ‖ Aufleimen n. ‖ marouflage m. / ~ (Tail) ‖ Futter n. ‖ doublure f. / acid ~ of the converter ‖ saures Konverterfutter n. ‖ revêtement m. acide du convertisseur. / ~ of the bearing ‖ Lagerausguß m. ‖ garniture f. du palier. / ~ for bedding ‖ Bettinlett n. ‖ coutil m. pour lit. / ~ of the brake-band ‖ Bremsbandbelag m. ‖ garniture f. du ruban de frein. / cotton ~ ‖ Baumwollfutterstoff m. ‖ doublure f. en coton. / furnace ~ of silica bricks ‖ saures Ofenfutter n. ‖ revêtement m. acide du four. / ~ of the hat ‖ Hutfutter n. ‖ coiffe f. du chapeau. / inner ~ (Metal) ‖ Kernschacht m.; Schachtfutter n. ‖ paroi f. intérieure ou de cuve; chemise f. intérieure. / leather ~ ‖ Belederung f. ‖ garniture f. en cuir. / moire ~ ‖ Moorleinen n. ‖ toile f. doublure moirée. / open ~ (Tail) ‖ Revers m.; Schoßumschlag m. ‖ retroussis m. / resistant basic ~ ‖ haltbares basisches Futter n. ‖ revêtement m. basique résistant. / ~ of a sail ‖ Saum m. oder Saumstreifen m. eines Segels ‖ gaine f. de voile. / ~ for shoes ‖ Schuhfutterstoff m. ‖ doublure f. pour chaussures. / steel ~ ‖ stählerner Einsatz m. ‖ chemise f. en acier. / stony ~ (Salt) ‖ Kesselstein m. ‖ Pfannenstein m. ‖ écailles fpl.; incrustation f.; tartre m. / ~ with tables ‖ Plattenverblendung f.; Plattenverkleidung f. ‖ revêtement m. en carreaux; tablement m. / unbleached ~ ‖ Franzleinen n. ‖ toile f. doublure écrue. / ~ of a well ‖ Brunneneinfassung f.; Brunnenrand m. ‖ margelle f. / wood ~ (Coupling) ‖ Holzschicht f. ‖ tablette f. de bois. / ~ with

woodwork ‖ hölzerner Schachtausbau m. ‖ cuvelage m. d'un puits. / woollen felt ~ ‖ Wollfilzeinlage f. ‖ garniture f. en feutre de laine.

lining and trimming ‖ Zutat f. ‖ fourniture f.; ingrédient m.

lining cloth ‖ Futterstoff m. ‖ (étoffe f. à) doublure f. / ~ goods ‖ Futterbarchent m. ‖ futaine f. pour doublure. / ~ linen ‖ Futterleinen n. ‖ toile f. doublure. / ~ maker (Cloth) ‖ Futternäher m. ‖ couseur m. de doublure. / shaft ~ material ‖ Schachtausbaustoff m. ‖ matériel m. de cuvelage du puits de mine. / ~ out ‖ Schnurschlag m. ‖ coup m. de cordeau; alignement m. / ~ plush ‖ Futterplüsch m. ‖ peluche f. à doublure. / ~ sawyer (Shoem) ‖ Futterzuschneider m. ‖ coupeur m. de doublure. / ~ taffety ‖ Futtertaft m. ‖ florence m. / wall of ~ (Blast furnace) ‖ Futtermauer f. ‖ paroi f. ou muraille f. de revêtement. / ~ wall (Build) ‖ Verkleidungsmauer f. ‖ mur m. de revêtement.

link, to (Weav) ‖ ketteln ‖ remmailler; entrelacer.

link (Chain) ‖ Kettenglied n.; Kettenschake f.; Kettenlasche f. ‖ chaînon m.; maille f.; maillon m.; anneau m. d'une chaîne. / ~ (Cuff) ‖ Manschettenknopf m. ‖ bouton m. de manchettes. / ~ (Mach) ‖ Kulisse f.; Schwinge f. ‖ coulisse f. / ~ (Ring) ‖ Öse f.; Ring m. ‖ anneau m.; boucle f. / bent ~ ‖ gekröpftes Glied n. ‖ membre m. coudé. / ~ of a chain ‖ Kettenglied n. ‖ maillon m. de la chaîne. / curved ~ ‖ gekrümmte Schwinge f. oder Kulisse f. ‖ coulisse f. cintrée ou coudée. / fixed ~ ‖ festgelagerte Schwinge f. oder Kulisse f. ‖ coulisse f. fixe. / forked ~ ‖ gabelförmige Schwinge f. oder Kulisse f. ‖ coulisse f. à fourche. / inside ~ ‖ Innenglied n. ‖ membre m. intérieur. / open ~ ‖ offene Schwinge f. oder Kulisse f. ‖ coulisse f. ouverte. / ~ of the pliers ‖ ovaler Ring m. zum Schließen der Schmiedezange‖chaînon m. de la tenaille à forger. / shackle ~ (Railw) ‖ Kupplungslasche f.; Kuppellasche f. ‖ bielle f. d'attelage. / slide ~ ‖ Kulisse f. ‖ coulisse f. / slotted ~ ‖ Schlitzkulisse f.; offene Kulisse f. ‖ coulisse f. fendue ou à fente. / ~ without stay-pin ‖ Kettenglied m. oder Schake f. ohne Steg ‖ maillon m. sans étai. / stem ~ ‖ Spindelgelenk n. ‖ articulation f. de la tige. / straight ~ ‖ gerades Glied n. ‖ membre m. droit. / suspension ~ ‖ Hängestange f. ‖ barre f. ou bielle f. de suspension. / unstudded ~ see ~ without stay-pin.

link bar ‖ Winkelgleithebel m. ‖ levier m. de coulisse coudé ou de glissement courbé. / ~ belt ‖ Gliederriemen m. ‖ courroie f. articulée ou à chaînons. / ~ block (Mach) ‖ Kulissenstein m.; Gleitklotz m. ‖ patin m. ou coulisseau m. de la crosse. / ~ box ‖ Gelenkmuffe f. ‖ boîte f. articulée. /~ bracket (Mach) ‖ Kulissenlager n.; Schwingenlager n. ‖ attache f. de coulisse. / ~ bush ‖ Schakenbuchse f. ‖ manchon m. de maillon. / centre position of the ~ ‖ Kulissentotlage f. ‖ point m. mort de la coulisse.

link chain ‖ Laschenkette f.; chaîne f. à maillons. / calibrated ~ ‖ kalibrierte Gliederkette f. ‖ chaîne f. calibrée à maillons.

link friction ‖ Kulissenreibung f. ‖ frottement m. de la coulisse. / ~ fuse with screw contacts ‖ Streifensicherung ɟ. mit Anschraubkontakten ‖ coupe-circuit m. à lamelle avec contacts vissés. / ~ and profile grinding machine ‖ Kulissen- und Kopierschleifmaschine f. ‖ machine f. à rectifier les coulisses et à copier. / ~ heald (Mar) ‖ Schlinghelfe f. ‖ maille f. à nœud simple; lisse f. coulante. / ~ holder (Electr) ‖ Streifenhalter m. ‖ porte-lamelle m.

linking (Chem) ‖ Bindung f. ‖ liaison f. / ~ (Weav) ‖ Ketteln n. ‖ remmaillage m. / double ~ (Chem) ‖ Doppelbindung f. ‖ liaison f. double. / ~ the warp yarn into chain ‖ Ketteln n. der gescherten Kette ‖ mise f. en mailles de la chaîne ourdie.

linking machine (Weav) ‖ Kettelmaschine f. ‖ remmailleuse f. / circular ~ ‖ Rundkettelmaschine f. ‖ remmailleuse f. circulaire.

linking point (Weav) ‖ Umschlingungspunkt m. ‖ point m. d'articulation ou d'attache.

linking-up of generating stations into a closed system ‖ Ausbau m. von Stromerzeugungsanlagen zu einem geschlossenen System ‖ conjugaison f. des usines génératrices pour former un seul système.

link lever ‖ Schlitzhebel m.; Kulissenhebel m. ‖ levier m. à coulisse. / lowering of the ~ ‖ Senkung f. der Kulisse ‖ abaissement m. ou descente f. de la coulisse. / ~ motion ‖ Kulissensteuerung f. ‖ distribution f. à coulisse.

link pin ‖ Lenkerbolzen m.; Gelenkzapfen m. ‖ tourillon m. d'articulation; boulon m. de guide. / ~ of roller chain ‖ Kettenbolzen m. ‖ goujon m. ou tourillon m. de chaîne.

link plate ‖ Kettenlasche f. ‖ maille f. de chaîne. / ~ polygon ‖ Seilpolygon n.; Seilzug m. ‖ polygone m. funiculaire. / reversing the ~ ‖ Umlegen n. der Kulisse ‖ renversement m. de la coulisse. / ~ tooth saw ‖ Kettensäge f. ‖ scie f. à dents articulées.

linnæite ‖ Linneit m.; Kobaltkies m. ‖ linnéite f.; cobalt m. sulfuré; koboldine f.

linnet hole (Glassm) ‖ Fuchs m. ‖ lunette f.

linography ‖ Öldruck m. ‖ linographie f.

linoleic acid ‖ Leinölsäure f. ‖ acide m. linoléique.

linoleum ‖ Linoleum n.; Korkteppich m. ‖ linoléum m. / ~ with imprinted or inlaid designs ‖ Linoleum n. mit aufgedruckten oder eingelegten Mustern ‖ linoléum m. à dessins imprimés ou à dessins incrustés. / unicolor ~ ‖ einfarbiges Linoleum n. ‖ linoléum m. unicolore.

linoleum carpet ‖ Linoleumteppich m. ‖ tapis m. de linoleum. / ~ engraving ‖ Linoleumschnitt m. ‖ gravure f. sur linoléum. / ~ making machine ‖ Linoleumherstellungsmaschine f. ‖ machine f. à fabriquer du linoléum. / ~ manufacturing ‖ Linoleumherstellung f. ‖ fabrication f. de linoléum. / ~ works pl. ‖ Linoleumfabrik f. ‖ fabrique d. de linoleum.

linon block ‖ Linonform f. ‖ forme f. en linon.

linophany ‖ Linophanie f. ‖ linophanie f.

linotype ‖ Linotype f. ‖ linotype f. / ~ composition ‖ Linotypesatz m. ‖ composition f. linotypique. / ~ machine ‖

Linotypesetzmaschine f. ‖ linotype m. / ~ man ‖ Linotypesetzer m. ‖ linotypiste m. / ~ operator see ~ man.

linseed ‖ Leinsame m. ‖ graine f. de lin. / ~ cake ‖ Leinkuchen m. ‖ tourteau m. de lin.

linseed oil ‖ Leinöl n. ‖ huile f. de lin. / boiled ~ ‖ gekochtes Leinöl n. ‖ huile f. de lin cuite. / thick ~ see linseed oil varnish.

linseed oil crusher ‖ Leinölschläger m. ‖ presseur m. d'huile de lin. / ~ substitute ‖ Leinölersatz m. ‖ huile f. de lin imitée. / ~ varnish ‖ Leinölfirnis m. ‖ vernis m. à l'huile de lin; vernis m. gras.

linseed sorter ‖ Leinsamenputzer m. ‖ trieur m. de graines de lin. / ~ varnish see linsed oil varnish.

lint ‖ Scharpie f. ‖ charpie f. / scraped ~ ‖ geschabte Scharpie f. ‖ charpie f. râpée.

lint doctor (Text print) ‖ Gegenrakel f.; Gegenschaber m. (einer Walzendruckmaschine) ‖ racle f. derrière; contreracle f.

lintel ‖ Fenstersturz m. ‖ linteau m. de fenêtre. / stone ~ ‖ Fenstersturz m. aus Stein ‖ linteau m. en pierre. / wooden ~ ‖ Fenstersturz m. aus Holz ‖ poitrail m.; linteau m. en bois.

lintel moulding ‖ Türsturzverzierung f. ‖ ornement m. de linteau.

linters pl. ‖ Baumwollsamenabfälle f.; Linters mpl. ‖ linters mpl.; bas-cotons mpl.

lip ‖ Lippe f. ‖ lèvre f. / ~ (Coachm) ‖ Kotlöffel m. ‖ cache-boue m.; couvre-moyeu. / ~ (Vessel) ‖ Ausguß m. ‖ bec m. / ~ of ladle ‖ Pfannenausguß m. ‖ bec m. de la poche. / ~ of tyre (Railw) ‖ Radreifenansatz m. ‖ talon m. du bandage.

liparit ‖ Liparit m.; Quarztrachyt m. ‖ liparite f.

lip glue ‖ Mundleim m. ‖ colle f. à bouche. / ~ salve ‖ Lippenpomade f. ‖ pommade f. pour les lèvres. / ~ valve ‖ Lippenventil n. ‖ soupape f. à lèvres.

liquate, to ‖ abseigern ‖ liquater; ressuer; achever de ressuer.

liquation ‖ Seigerprozeß m.; Seigerung f. ‖ liquation f.; ressuage m. / ~ cake see disk. / ~ disk ‖ Frischstück n.; Seigerstück n. ‖ disque m. de ressuage; pain m. de liquation. / ~ hearth ‖ Seigerherd m. ‖ four m. de liquation ou de ressuage. / ~ lead ‖ Seigerblei n. ‖ plomb m. de ressuage. / ~ pan ‖ Seigerpfanne f. ‖ chaudière f. de liquation ou de ressuage. / ~ thorn ‖ Seigerdorn m. ‖ épine f. de ressuage.

liquefaction ‖ Verflüssigung f. ‖ liquéfaction f. / ~ of air ‖ Luftverflüssigung f. / liquéfaction f. de l'air. / coal ~ ‖ Kohleverflüssigung f. ‖ liquéfaction f. du charbon. / plant for ~ of gases ‖ Anlage f. zur Verflüssigung von Gasen ‖ installation f. de liquéfaction des gaz. / ~ of gas mixtures ‖ Verflüssigung f. von Gasgemischen ‖ liquéfaction f. de mélanges de gaz. / partial ~ ‖ Teilverflüssigung f. ‖ liquéfaction f. partielle.

liquefaction column ‖ Trennungsapparat m. ‖ liquéfacteur-rectificateur m. / ~ process ‖ Verflüssigungsverfahren n. ‖ procédé m. de liquéfaction. / ~ temperature ‖ Verflüssigungstemperatur f. ‖ température f. de liquéfaction.

liquefied gas ‖ verflüssigtes Gas n. ‖ gaz m. liquéfié.

iquify, to ~ ‖ verflüssigen ‖ liquéfier. / ~ (Metal) ‖ schmelzen ‖ fondre; liquéfier; mettre en fusion f. / ~ by pressure ‖ durch Druck m. verflüssigen ‖ liquéfier par compression.

liquefying of industrial gases ‖ Verflüssigung f. von Industriegasen ‖ liquéfaction f. des gaz industriels. / ~ air ~ plant ‖ Luftverflüssigungsanlage f. ‖ installation f. de liquéfaction de l'air.

liqueur ‖ Likör m. ‖ liqueur f. / ~ distillery ‖ Likörfabrik f. ‖ distillerie f. de liqueur. / installation for ~ distilleries ‖ Likörfabrikeinrichtung f. ‖ équipement m. pour distilleries de liqueur. / ~ maker ‖ Likörfabrikant m. ‖ fabricant ou distillateur-liquoriste m. / ~ making ‖ Likörfabrikation f. ‖ distillerie f. de liqueurs. / ~ stand ‖ Likörschrank m. ‖ armoire f. à liqueurs. / ~ wine ‖ Likörwein m.; Dessertwein m. ‖ vin m. de liqueur ou de dessert.

liquid ‖ flüssig ‖ liquide. / ~ to a high temperature ‖ feuerflüssig ‖ liquide à haute température. / thinly ~ ‖ dünnflüssig ‖ très liquide.

liquid air ‖ flüssige Luft f. ‖ air m. liquide ou liquéfié. / installation for producing ~ ‖ Anlage f. zur Erzeugung von flüssiger Luft ‖ installation f. de production de l'air liquéfié.

liquidambar ‖ flüssiger Amber m. ‖ liquidambar m.

liquid ammonia ‖ flüssiges Ammoniak n. ‖ ammoniaque f. liquide. / ~ receiver ‖ Ammoniakbehälter m. ‖ réservoir m. à ammoniaque.

liquid caustic potash ‖ Kalilauge f. ‖ lessive f. de potasse caustique.

liquid fuels pl., production of ~ from coal ‖ Erzeugung f. von flüssigen Brennstoffen aus Kohle ‖ production f. ou obtention f. de combustibles liquides extraits du charbon. / recovery of ~ from coal ‖ Gewinnung f. flüssiger Brennstoffe aus Kohle ‖ récupération f. de combustibles liquides extraits du charbon. / underground tank plant for ~ ‖ unterirdische Tankanlage f. für flüssige Brennstoffe ‖ installation f. de réservoirs souterrains pour combustible liquides.

liquid glue ‖ flüssiger Leim m. ‖ colle f. liquide.

liquid manure barrel ‖ Jauchefaß n. ‖ tonneau m. ou baril m. à purin. / ~ cask see ~ barrel. / ~ implement ‖ Jauchegerät n. ‖ instrument m. à purin. / ~ pump ‖ Jauchepumpe f. ‖ pompe f. à purin. / ~ shovel ‖ Jaucheschöpfer m. / pelle f. à purin. / ~ spreader ‖ Jaucheverteiler m. ‖ distributeur m. de purin.

liquid polishing material for metal ‖ flüssiges Putzmittel n. für Metall ‖ brillant m. liquide pour métaux. / ~ product of condensation ‖ flüssiges Kondensationsergebnis n. ‖ produit m. liquide de condensation; liquide m. résultant. / ~ soap ‖ flüssige Seife f. ‖ savon m. liquide. / ~ state / flüssiger Zustand m. ‖ état m. liquide. / remaining in the ~ state ‖ flüssig bleiben ‖ rester liquide.

liquid ‖ Flüssigkeit f. ‖ liquide m. / alcoholic ~ ‖ weingeisthaltige Flüssigkeit f. ‖ liquide m. alcoolique. / to allow the ~ to settle ‖ die Flüssigkeit abstehen lassen ‖ laisser reposer ou laisser déposer le li-

quide. / cooling ~ ‖ Kühlflüssigkeit f. ‖ liquide m. de refroidissement. / ~ of crystallization ‖ Kristallflüssigkeit f. ‖ liquide m. de cristallisation. / fermentable ~ ‖ Gärflüssigkeit f. ‖ liquide m. fermentescible. / inflammable ~ ‖ feuergefährliche Flüssigkeit f. ‖ liquide m. inflammable. / noninflammable ~ ‖ unentzündbare Flüssigkeit f. ‖ liquide m. ininflammable. / to recover precious ~s pl. evaporating in the air ‖ in die Luft verdampfte wertvolle Flüssigkeiten fpl. wiedergewinnen / récupérer des liquides mpl. précieux évaporés à l'air.

liquid accumulator ‖ Flüssigkeitskraftspeicher m. ‖ accumulateur m. hydraulique d'énergie.

liquidate, to ‖ liquidieren ‖ liquider.

liquidation ‖ Liquidation f. ‖ liquidation f.

liquidation proceedings pl. ‖ Konkursverfahren n. ‖ procédure f. de faillite. / to commence ~ ‖ das Konkursverfahren n. eröffnen ‖ engager la procédure de faillite. / to stop ~ ‖ das Konkursverfahren n. einstellen ‖ suspendre la procédure de faillite.

liquidator ‖ Liquidator m. ‖ liquidateur m.

liquid condenser ‖ Flüssigkeitskondensator m. ‖ condensateur m. de liquide. / ~ cooler ‖ Flüssigkeitskühler m. ‖ refroidisseur m. de liquide. / ~ cycle ‖ Flüssigkeitskreislauf m. ‖ cycle m. du liquide. / ~ damping ‖ Flüssigkeitsdämpfung f. ‖ amortissement m. par liquide. / interferometer for ~s ‖ Flüssigkeitsinterferometer n. ‖ interféromètre m. à liquides.

liquidmeter ‖ Flüssigkeitsmesser m. ‖ pèse-liquide m.

liquid prism ‖ Flüssigkeitsprisma n. ‖ prisme m. à liquide. / ~ reversing and starting resistance ‖ Flüssigkeitsumkehranlaßwiderstand m. ‖ rhéostat m. de démarrage inverseur à liquide. / ~ siphon ‖ Flüssigkeitsheber m. ‖ syphon m. à liquides. / ~ slide-valve ‖ Flüssigkeitsschieber m. ‖ vanne f. à liquides. / ~ starting resistance ‖ Flüssigkeitsanlaßwiderstand m. ‖ rhéostat m. de démarrage à liquide. / tank for ~s ‖ Flüssigkeitsbehälter m. ‖ réservoir m. à liquides.

liquor see also liqueur ‖ Likör m. ‖ liqueur f. / boiling hot caustic ~ ‖ kochende Lauge f. ‖ lessive f. bouillante. / fine ~ (Sugar) ‖ Klärsel n. ‖ claire f.; clairée f. / ~ of a lixiviated matter ‖ Laugeflüssigkeit f. ‖ lessive f.; liquide m. / residuary ~ of spirits ‖ Schlempe f. ‖ vinasses fpl.

liquor ammonia ‖ wässeriges Ammoniak n.; Salmiakgeist m.; Salmiakspiritus m. ‖ esprit m. de sel ammoniaque. / ~ discharge ‖ abgeschleuderte Lauge f. ‖ liquide m. d'essorage.

liquorice ‖ Süßholz n. ‖ réglisse f. / ~ cake ‖ Lakritzenstengel m. ‖ bâton m. de réglisse. / ~ extract ‖ Lakritzensaftextrakt m. ‖ extrait m. de réglisse. / ~ juice ‖ Lakritzensaft m. ‖ jus m. de réglisse. / ~ factory ‖ Lakritzenfabrik f. ‖ réglisserie f. / ~ preparation ‖ Lakritzpräparate npl. ‖ préparation f. de réglisse. / ~ root ‖ Lakritzenwurzel f. ‖ racine f. de réglisse. / ~ sugar ‖ Süßholzzucker m. ‖ sucre m. de réglisse; glycyrrhizine f. / ~ wood ‖ Süßholz n. ‖ bois m. de réglisse.

liquor man (Dyer) ‖ Farbflottenbereiter m. ‖ préparateur m. de bains de teinture.

liroconite ‖ Linsenstein m.; Lirokonit m. ‖ liroconite f.

lisle goods pl. ‖ Florwaren fpl. ‖ articles mpl. en fil d'Écosse. / ~ stocking ‖ Florstrumpf m. ‖ bas m. en crêpe.

list, to ~ (Ship) ‖ überliegen ‖ s'incliner; donner de la bande.

list (Arch) ‖ Leiste f.; Saum m. ‖ filet m.; réglet m.; listel m. / ~ (Book) ‖ Verzeichnis n. ‖ liste f.; répertoire m.; index m. / ~ (Ropem) ‖ Spinnlappen m. ‖ paumelle f.; paumet m. / ~ (Ship) ‖ Krängung f. ‖ bande f. / ~ (Weav) ‖ Geweberand m.; Leiste f.; Sahlleiste f.; Sahlband n.; Kante f. ‖ lisière f. / ~ of contents ‖ Sachregister n. ‖ table f. des matières; répertoire m.; index m. / ~ of goods pl. ‖ Warenverzeichnis n. ‖ tableau m. des marchandises. / to make out a ~ ‖ eine Aufstellung f. machen ‖ rédiger une liste ou une spécification. / ~ of timber ‖ Holzliste f. ‖ borderau m. du bois.

listel (Arch) ‖ Leiste f.; Saum m. ‖ filet m.; réglet m.; listel m. / ~ (Join) ‖ Leistchen n.; Spitzleistchen n. ‖ feuillet m.

listen, to ~ in (Tel) ‖ mithören ‖ être à l'écoute.

listening (Tel) ‖ Abhörbarkeit f. ‖ réceptivité f. à l'écoute. / ~ button ‖ Mithörtaste f. ‖ clef f. d'écoute. / ~ key (Tel) ‖ Mithörschalter m. ‖ clef f. d'écoute. / ~ plug (Tel) ‖ Mithörstöpsel m. ‖ fiche f. d'écoute. / ~ point ‖ Lauschstelle f. ‖ poste m. d'écoute. / ~ relay (Tel) ‖ Mithörrelais n. ‖ relais m. d'écoute. / ~ station (Tel) ‖ Abhörstation f. ‖ station f. d'écoute. / ~ switch (Tel) ‖ Mithörschalter m. ‖ clef f. d'écoute. / ~ tube (Dictating machine) ‖ Hörschlauch m. ‖ tube m. acoustique.

lister ‖ Listendruckvorrichtung f. ‖ dispositif m. pour l'impression des listes.

listing machine ‖ selbstschreibende Rechenmaschine f. ‖ machine f. à additionner pourvue d'un mécanisme d'impression.

listoforme soap ‖ Listoformseife f. ‖ savon m. au listoforme.

list pot (Tinning) ‖ Saumtopf m.; Saumpfanne f. ‖ pot m. à lisière; chaudière f. à lisser. / ~ price ‖ Listenpreis m.; Katalogpreis m. ‖ prix m. du catalogue.

lists pl. (Balustrade) ‖ Lehnriegel m.; Brustriegel m. ‖ lisse f. de barrière; lice f. de barrière; entretoise f. d'appui.

lit (Build) ‖ Unterbettung f. ‖ couche f.; lit m.

liter ‖ Liter m. ‖ litre m.

literal ‖ wörtlich ‖ textuel; littéral; mot m. à mot.

literal calculus ‖ Buchstabenrechnung f. ‖ algèbre f.

literally ‖ buchstäblich ‖ littéralement.

literature ‖ Schrifttum n. ‖ littérature f.

litharge ‖ Bleiglätte f. ‖ litharge f. / commercial ~ ‖ Handelsbleiglätte f. ‖ litharge f. marchande ou en poudre. / gold ~ ‖ Goldglätte f. ‖ litharge f. d'or ou rouge. / red ~ see gold ~. / ~ for sale see commercial ~. / yellow-red ~ ‖ Goldglätte f. ‖ litharge f. jaune rouge.

litharge lead ‖ Frischblei n. ‖ plomb m. raffiné. / ~ plant ‖ Bleiglätteanlage f. ‖ installation f. de litharge. / ~ varnish ‖ Retuschierbutter f. ‖ vernis m. à litharge.

lithia ‖ Lithion n. ‖ lithine f. / ~ containing ‖ lithiumhaltig ‖ lithiné. / ~ feldspar

|| Lithionfeldspat m.; Petalit m. || feldspath m. à lithine; pétalite f. / ~ mica || Lepidolith m.; Lithionglimmer m. || mica m. violet; lithionite f. / ~ salt || Lithiumsalz n. || sel m. de lithium.

lithine *see* lithia.

lithionite || Lithionglimmer m.; Lepidolith m. || mica m. violet; lithionite f.

lithium || Lithium n. || lithium m.

lithochromic print || Ölbilderdruck m.; Ölfarbendruck m.; Chromolithographie f. || chromo-lithographie f.

lithochromics pl. || Farbensteindruck m.; Lithochromie f. || lithochromie f.

lithochromy || Lithochromie f.; Farbendruck m. || lithochromie f.; chromo m.

lithofracteur || Lithofraktör m. || lithofracteur m.

lithograph, to || lithografieren || lithographier.

lithographer || Lithograf m.; Steindrucker m.; Steinzeichner m. || lithographe m. / ~s stone || Lithografiestein m. || pierre f. lithographique.

lithographic || lithografisch || lithographique. / ~ artist || Lithograf m. || dessinateur m. lithographe. / ~ chalk || lithografische Kreide f. || crayon m. lithographique. / ~ cylinder || Lithografiewalze f. || cylindre m. à lithographier. / ~ material || Steindruckware f. || produit m. pour la lithographie. / ~ press || Steindruckpresse f. || presse f. lithographique. / ~ print || Lithografie f.; Steindruck m. || lithographie f. / ~ printer || Steindrucker m. || imprimeur m. lithographe. / ~ printing || Steindruckerei f. || lithographie f.; imprimerie f. lithographique. / ~ printing office *see* ~ printing works pl. / ~ printing works pl. || lithografische Anstalt f.; Steindruckerei f. || atelier m. d'impression lithographique; lithographie f. / ~ requirements pl. || lithografischer Bedarf m. || fournitures fpl. (pour les impressions) lithographiques.

lithographic stone || Lithografiestein m. || pierre f. lithographique. / ~ cementer || Steinkitter m. || doubleur m. de pierres lithographiques. / ~ dresser || Steinschleifer m. || préparateur m. de *ou* effaceur m. sur pierres lithographiques. / ~ pouncer || Steinpolierer m. || ponceur m. de pierres lithographiques. / ~ quarry || Lithografiesteinbruch m. || carrière f. de pierre lithographique. / ~ writer || Steinzeichner m. || écrivain m. lithographe.

lithography || Steindruck m. || lithographie f.; imprimerie f. lithographique. / auxiliary printing machine for ~ || Druckhilfsmaschine f. für Steindruck || machine f. auxiliaire d'impression pour lithographie.

lithology || Gesteinskunde f.; Lithologie f. || lithologie f.

lithophane || Lithofanie f.; Porzellanlichtbild n. || lithophanie f.

lithophany *see* lithophane.

lithopone || Lithopon n. || lithopone m. / ~ plant || Lithoponanlage f. || installation f. de lithopone. / ~ works pl. || Lithoponfabrik f. || fabrique f. de lithopone.

litho press || Steindruckpresse f. || presse f. pour lithographie. / ~ stone || Lithografiestein m. || pierre f. lithographique.

litigant || Prozeßpartei f. || plaideur m.

litmus || Lackmus n. || tournesol m. / ~ acid to ~ || sauer reagierend auf Lackmus n. || acide au tournesol m.

litmus paper || Lackmuspapier n. || papier m. de tournesol. / blue ~ || blaues Reagenspapier n. *oder* Lackmuspapier n. || papier m. de tournesol bleu. / red ~ (Chem) || rotes Lackmuspapier n. *oder* Reagenspapier n. || papier m. de tournesol rouge.

litre || Liter m. || litre m.

litter || Streu f. || litière f.

little by little || allmählich; nach und nach || peu à peu; successivement; graduellement.

littoral deposit || Küstenablagerung f. || sédiment (in) côtier. / ~ under water || Küstenschlick m. || limon m. côtier.

litzendraht (Electr) || Litzendraht m.; Hochfrequenzlitze f. || brin m. pour haute fréquence.

livelihood || Erwerb m.; Auskommen n.; Lebensunterhalt m. || travail m.; revenu m.

liveliness || Lebhaftigkeit f. || vivacité f.; intensité f.

live load || Verkehrslast f. || charge f. variable.

lively || lebhaft || vif; animé.

live oak || Steineiche f. || yeuse f.

liver || Leber f. || foie m. / ~ of sulphur || Schwefelleber f. || foie m. de soufre.

live roller || Rolle f. || rouleau m. / ~ bed || Rollbahn f. || rouleaux mpl. d'amenée. / ~ loading bed || Verladerollgang m. || train m. de rouleaux de chargement. / ~ train || Rollgang m. || transporteur m. à rouleaux.

liverstone || Hepatit m. || hépatite f.; pierre f. hépatique.

liverwort herb (Bot) || Leberkraut n. || herbe f. d'hépatique.

livery stables pl. || Mietstall m. || écurie f. de chevaux à louer.

live steam || Frischdampf m. || vapeur f. fraîche *ou* vive *ou* d'admission.

live steam turbine || Frischdampfturbine f. || turbine f. à vapeur fraîche *ou* à vapeur vive. / ~ with condensing plant only for power purposes || Frischdampfturbine f. mit Kondensation für reine Kraftzwecke || turbine f. à vapeur fraîche avec condensation destinée uniquement à la production d'énergie.

live-stock breeder || Viehzüchter m.; Tierhalter m. || éleveur m. de bétail. / ~ breeding || Viehzucht f. || élevage m. de bestiaux.

live wood || lebendes Holz n. || bois m. vivant.

living *see* livelihood.

living animal || lebendes Tier n. || animal m. vivant. / ~ room || Wohnraum m. || appartement m.; logement m.; chambre f.; habitacle m.

lixiviate, to (Met) || laugen; auslaugen || lessiver; lixivier. / ~ the wood || das Holz auslaugen || lessiver le bois.

lixiviated tan || ausgelaugte Gerberlohe f. || tan m. lessivé.

lixiviating arrangement || Auslaugvorrichtung f. || disposition f. de lavage *ou* de lessivage. / ~ liquor || Auslaugflüssigkeit f. || liquide m. laveur; lessive f.

lixiviation || Auslaugung f. || lavage m.; lessivage m. / ~ tank (Ore dress) || Laugebehälter m.; Laugebottich m.; Laugbecken n. || cuve f. de filtration; cuve f. à lessiver. / ~ vat *see* ~ tank.

lixivium || Laugeflüssigkeit f.; Lauge f. || lessive f.; liquide m.

lizard leather || Eidechsenleder n. || cuir m. de lézard.

lizard-stone (Miner) || Serpentinmarmor m. || marbre m. serpentin.

lizaric acid || Alizarin n.; Krappstoff m. || alizarine f.; pincoffine f.

llama's hair || Lamahaar n. || poil m. de lama.

Lloyd's rules and regulations pl. || Lloydvorschriften fpl. || règlement m. du Lloyd pour la construction des navires.

load, to ~ || laden; beladen || charger. / ~ (Blast-furnace) || füllen; laden; beschicken || charger; emplir. / ~ (Electr) || pupinisieren || pupiniser. / ~ (Goods) || einladen || charger. / ~ a battery || eine Batterie laden *oder* speisen || charger une batterie. / ~ the engine immediately after starting || den Motor m. sofort nach dem Anlassen belasten || charger le moteur dès sa mise en marche. / ~ a gun || ein Geschütz n. laden || charger un canon. / ~ the kiln (Brew) || auf die Darre auftragen || charger la touraille. / ~ a ship || ein Schiff befrachten *oder* laden || charger un bateau. / ~ a spring || eine Feder f. spannen || tendre *ou* serrer un ressort. / ~ a wagon || einen Wagen m. beladen oder laden || charger un chariot *ou* une voiture.

load || Last f.; Belastung f. || charge f. / ~ (Railw) || Ladegewicht n. || poids m. de charge. / ~ (Tel) || Leistung f. (im Betriebsdienst) || charge f. / ~ (Trade) || Fracht f. || charge f. / axle ~ || Achslast f. || charge f. par *ou* de l'essieu. / ~ on the building-ground || Belastung f. des Baugrundes || charge f. sur le terrain à bâtir. / to be in conformity with the ~ || der Belastung f. entsprechen || correspondre à la charge. / constant ~ || konstante Belastung f. || charge f. constante. / continuous ~ || Dauerlast f. || charge f. continue. / ~ per cm² cross-sectional area || Belastung f. je cm² Querschnitt || charge f. par cm² de section. / dead ~ || totes Gewicht n. || poids m. mort. / to distribute the ~ spl. || die Lasten fpl. verteilen || répartir la charge. / distributed ~ || verteilte Last f. || charge f. répartie; poids m. distribué. / evenly-distributed ~ || gleichmäßige Gewichtsverteilung f. || charge f. uniformément répartie. / full ~ || Vollbelastung f.; Vollast f. || pleine charge f. / at full ~ || voll belastet || à charge f. complète *ou* totale. / ~ on ground || Bodendruck m. || pression f. sur le sol. / half ~ || halbe Belastung f.; Halblast f. || demi-charge f. / the ~ increases slowly || die Last f. wächst allmählich an || la charge f. augmente *ou* accroît lentement *ou* continuellement. / instantaneous ~ || Augenblicksbelastung f. || charge f. instantanée. / intermittant ~ || intermittierende Belastung f. || charge f. intermittante. / ~ in the joints || Knotenpunktelast f. || charge f. aux nœuds d'assemblage *ou* aux points de jonction. / live ~ || bewegliche *oder* rollende Last f. || charge f. variable. / maximum ~ || Höchstbelastung f. || charge f. maximum. / no-~ current || Leerlaufstrom m. || courant m. à vide. / non-inductive ~ || induktionsfreie Belastung f. || charge f. non-inductive. / rising ~ || wachsende Belastung f. || charge f. (ac)croissante. / rolling ~ || bewegliche *oder* rollende Last

f. || charge f. variable. / ~ of a safety-valve || Belastungsgewicht n. eines Sicherheitsventils || poids m. de charge d'une soupape de sûreté. / ~ shown by counter || durch ein Meßinstrument angezeigte Last f. || charge f. indiquée par le compteur. / ~ per square meter || Belastung f. je Quadratmeter || charge f. par mètre carré. / to start under ~ || belastet anlaufen || démarrer en charge ou sous charge. / to start without ~ || leer anlaufen || démarrer sans charge f. / stipulated ~ || vorgeschriebene Belastung f. || charge f. prescrite. / test ~ || Probebelastung f. || charge f. d'épreuve. / total ~ || Gesamtbelastung f. || charge f. totale. / ultimate ~ || Bruchbelastung f. || charge f. de rupture. / ~ per unit of length || Belastung f. je Längeneinheit || charge f. par unité de longueur. / ~ per unit of surface || Flächenbelastung f. || charge f. par unité de surface ou au mètre carré. / useful ~ || Nutzlast f. || poids m. utile. / useful-performance ~ || Nutzleistungsbelastung f. || charge f. du rendement utile. / ~ per wheel || Belastung f. je Rad || charge f. par roue. / ~ on the wheel || Radbelastung f. || charge f. de la roue ou par la roue. / ~ due to wind-pressure || Windbelastung f. || charge f. due à la poussée du vent. / working ~ || zulässige Belastung f. || charge f. admissible.

load capacity || Ladefähigkeit f. || capacité f. de charge. / change of ~ || Belastungsschwankung f. || fluctuation f. de charge. / ~ characteristic || Belastungscharakteristik n. || caractéristique f. en charge. / ~ compensation || Belastungsausgleich m. || égalisation f. de charge. / ~ counting || Zählung f. der Lasten || comptage m. des charges. / ~ curve of feeders || Belastungskurve f. für Speiseleitungen || courbe f. de charge des lignes d'alimentation. / ~ diagram || Belastungsdiagramm n. || diagramme m. de charge. / discharging of ~ || Entlastung f. || déchargement m. / ~ dispatching plant || Lastverteileranlage f. || installation f. de répartition de charge / ~ distributing station for the whole network || Lastverteilungsstelle f. für das gesamte Netz || poste m. de distribution de charge pour le réseau entier. / ~ distribution || Lastverteilung f. || répartition f. de la charge. / division of ~ || Belastungsverteilung f. || répartition f. ou distribution f. de la charge.

loaded at . . . || verladen in . . . || chargé à . . . / when fully ~ || bei Vollast f. || en pleine charge; sous pleine charge.

loader, cage ~ (Metal) || Schrägaufzugarbeiter m. || encageur m. / wagon ~ (Mine) || Kohlenwagenfüller m. || chargeur m. de wagonnets ou de berlines.

load factor || Belastungsfaktor m. || facteur m. de charge. / ~ fluctuation || Belastungsschwankung f. || fluctuation f. de charge. / ~ hook || Lasthaken m. || crochet porte-charge m. / automatic ~ indication || selbsttätige Kraftmessung f. || mesurage m. de force automatique.

loading (Carriage) || Laden n.; Ladung f. || chargement m. / ~ (Chem) || Füllstoff m. || charge f. / ~ (Engine) || Belastung f. || charge f. / ~ (Mar) || Schiffsladung f. || cargaison f.; chargement m.; affrètement m. / ~ (Trade) || Fracht f. || charge

f. / ~ of conductors (Electr) || Belastung f. von Leitungen || charge f. des fils ou des conducteurs. / extra-light ~ || besonders geringe Belastung f. || charge f. particulièrement faible. / ~ of the kiln (Brew) || Auftragen n. auf die Darre; Beladen n. der Darre || chargement m. de la touraille. / medium-heavy ~ || mittelschwere Belastung f. || charge f. moyenne. / ~ per unit area || Flächeneinheitslast f. || charge f. par unité de surface ou par mètre carré.

loading band|| Verladeband n. || bande f. de chargement. / live roller ~ bed || Verladerollgang m. || train m. de rouleaux de chargement. / ~ berth || Ladestelle f. || embarcadère f. / ~ block || Ladeblock m. || bloc m. de chargement.

loading bridge || Verladebrücke f. || pont m. de chargement; pont-élévateur m. / coal ~ with x tons revolving crane having y feet radius || Verladebrücke f. für Kohle mit Drehkran von x t Tragfähigkeit und y m Ausladung || pont m. de chargement pour charbon avec grue pivotante de x t de puissance et y m de portée. / ~ with grab || Verladebrücke f. mit Greifer || pont m. de chargement avec benne preneuse.

loading bunker || Verladebunker m. || trémie f. de chargement. / ~ canal || Stichkanal m. || canal m. de garage. / ~ charges pl. || Ladegebühr f.; Ladegeld n. || droits mpl. ou frais mpl. de chargement. / ~ chute || Verladerinne f. || couloir m. de chargement.

loading coil || Ladungsspule f.; Pupinspule f. || bobine f. de charge; bobine f. Pupin. / ~ box || Pupinspulenkasten m. || boîte f. de bobines Pupin. / ~ section (Pupin) || Spulenfeld n. || section f. de charge. / ~ spacing || Spulenabstand m. || distance f. entre bobines de charge.

loading crane for smelting houses || Verladekran m. für Hüttenwerke || grue f. de chargement pour usines métallurgiques. / ~ device || Verladeeinrichtung f. || appareil m. de chargement. / ~ funnel || Hosenrutsche f. || goulotte f. ou tuyau m. de déversement; tubulure f. de chargement. / clearance ~ gauge (Railw) || Lademaß n. || gabari(t) m. de chargement. / ~ grab || Verladegreifer m. || benne f. de chargement. / ~ hatch || Ladeluke f. || écoutille f. de charge. / ~ hoist || Ladewinde f. || treuil m. d'embarquement. / ~ hopper || Sammeltasche f. || poche-accumulateur m. / ~ manner of ~ || Belastungsfall m. || cas m. de charge. / peak of the ~ || Spitzenbelastung f. || charge f. extrême ou maximum. / ~ place || Ladestelle f. || lieu m. de chargement.

loading plant || Verladeanlage f.; Verladevorrichtung f. || installation f. de chargement ou d'embarquement. / coke ~ || Koksverladeanlage f. || installation f. pour transporter le coke. / floating ~ || schwimmende Verladeanlage f. || installation f. flottante de chargement.

loading and handling plant for goods in bulk || Lade- und Förderanlage f. für Massengüter || installation f. de manœuvre et de transport de marchandises en vrac. / ~ platform (Harbour) || Ausladebrücke f. || appontement m. ou pont m. de déchargement. / ~ platform (Railw) || Ladebühne f.; Laderampe f. || rampe f.

ou plate-forme f. de chargement. / ~ and unloading platform || Aus- und Einladerampe f. || quai m. d'embarquement et de débarquement. / ~ point (Tel) || Spulenpunkt m. || point m. de bobines. / ~ ramp || Laderampe f. || rampe f. de chargement. / ~ resistance || Ballastwiderstand m. || résistance f. de charge. / ~ sidings pl. || Ladegleis n. || voie f. latérale ou de chargement. / ~ station || Verladehof m. || chantier m. d'expédition. / ~ test || Belastungsprobe f. || épreuve f. de chargement. / ~ track see ~ sidings pl. / ~ trough (Mine) || Einflußrinne f.; Lademulde f. || rigole f. ou pelle f. de chargement. / ~ unit || Spulensatz m. || jeu m. de self. / ~ wagon || Verladewagen m. || wagonnet-chargeur m. / ~ wharf || Laderampe f. || rampe f. de chargement. / ~ winch || Ladewinde f. || treuil m. de chargement.

load limit || Belastungsgrenze f. || charge f. limite ou maximum. / ~ line (Nav) || Tiefladelinie f. || ligne f. de flottaison en charge. / peak of ~ || Belastungsspitze f. || pointe f. de charge. / ~ porter || Lastträger m. || coltineur m.; portefaix m. / most unfavourable position of ~ || ungünstigster Belastungsfall m. || position f. la plus défavorable de la charge. / ~ pressure brake || Lastdruckbremse f. || frein m. actionné par le poids de la charge. / ~ rope || Lastseil n. || câble-porteur m.

loadstone || Magneteisenstein m.; natürlicher Magnet m. || aimant m. naturel; fer m. magnétique.

load test || Belastungsprobe f. || essai m. de charge. / ~ up to breaking || Belastungsprobe f. bis zum Bruch || essai m. de charge jusqu'à la rupture.

load train || Lastenzug m. || train m. routier. / ~ variation || Belastungsschwankung f. || variation f. ou fluctuation f. de charge. / considerable ~ variation || Belastungsstoß m. || variation f. brusque de charge. / ~ voltage || Belastungsspannung f. || tension f. ou voltage m. en charge. / ~ water-line || Ladewasserlinie f. || ligne f. de charge ou de flottaison en charge. / ~ weight (Aero) || Betriebsgewicht n. || poids m. en ordre de marche.

loaf of bread || Brot n.; Laib m. || pain m.

loaf bread || gesäuertes Brot n. || pain m. levé. / ~ sugar || Hutzucker m. || sucre m. en pains.

loam || Lehm m. || terre f. glaise ou à briques; argile f. / ~ beater (Found) || Lehmmesser n. || couteau m. pour battre les terres à mouler. / ~ castings pl. || Lehmguß m. || moulage m. en argile. / ~ coat (Mould) || Lehmauftrag m.; Lehmaufschlag m.; Lehmschicht f. || couche f. d'argile ou de terre. / ~ core || Lehmkern m. || noyau m. en terre. / ~ mortar || Lehmmörtel m. || mortier m. d'argile.

loam mould || Lehmform f. || moule m. en argile ou en terre. / to black-wash the ~ || die Lehmform anblaken oder schwärzen || flamber ou noircir le moule en argile. / to smoke the ~ see to black-wash the ~.

loam moulding || Lehmformerei f. || moulage m. en argile ou en terre. / ~ pit || Tongrube f.; Lehmgrube f. || carrière f. d'argile; glaisière f.

loamy ‖ lehmig; lettig ‖ glaiseux. / ~ ground ‖ Lehmboden m. ‖ terre f. glaiseuse *ou* limoneuse; sol m. argileux. / ~ sand ‖ Klebsand m. ‖ sable m. collant.

loan, to ~ ‖ ausleihen ‖ prêter. / ~ money on security ‖ auf Pfand leihen; lombardieren ‖ prêter sur gages.

loan ‖ Darlehen n.; Darlehn n. ‖ prêt m.; avance f. / as a ~ ‖ leihweise; als Darlehen ‖ à titre m. de prêt. / consolidated ~ ‖ Konsolidierungsanleihe f. ‖ emprunt m. de consolidation. / forced ~ ‖ Zwangsanleihe f. ‖ emprunt m. forcé. / to issue a ~ ‖ eine Anleihe f. ausgeben ‖ émettre un emprunt. / ~ of money ‖ Geldanleihe f. ‖ emprunt m. / mortgage ~ ‖ Hypothekendarlehen ‖ prêt m. hypothécaire. / to negotiate a ~ ‖ eine Anleihe f. abschließen ‖ contracter un emprunt. / public ~ ‖ öffentliche Anleihe f. ‖ emprunt m. publique. / unsecured ~ ‖ offener Kredit m. ‖ avance f. à découvert. / ~ from various directors of a company ‖ Arbeitgeberdarlehen n. ‖ prêt m. fait par la société.

loan bank ‖ Darlehnskasse f.; Darlehnsbank f. ‖ caisse f. de prêts. / ~ business ‖ Leihgeschäft n.; Lombardgeschäft n. ‖ prêt m. *ou* avance f. sur gages. / ~ certificate ‖ Darlehnsschein m. ‖ titre m. de prêt.

lobelia herb ‖ Lobeliakraut n. ‖ herbe f. de lobélie.

lobster ‖ Hummer m. ‖ homard m. / freshwater ~ ‖ Flußkrebs m. ‖ écrevisse f. de rivière. / spring ~ ‖ Languste f. ‖ langouste f.

lobster fisherman ‖ Hummerfischer m. ‖ pêcheur m. de homards.

local authority ‖ Ortsbehörde f. ‖ autorité f. locale. / ~ battery ‖ Ortsbatterie f. ‖ batterie f. locale. / ~ battery working ‖ Ortsbatteriebetrieb m. ‖ service m. à batteries locales. / ~ business ‖ Platzgeschäft n. ‖ commerce m. local. / ~-busy (Tel) ‖ ortsbesetzt ‖ occupé par une conversation urbaine. / ~ cable ‖ Ortskabel n. ‖ câble m. urbain. / ~ call (Tel) ‖ Ortsgespräch n. ‖ communication f. urbaine. / ~ circuit ‖ Ortsstromkreis m. ‖ circuit m. local. / ~ communication (Tel) ‖ Ortsverkehr m. ‖ service m. urbain. / ~ condition ‖ örtliche Bedingung f. ‖ condition f. locale. / ~ connecting cable ‖ Ortsverbindungskabel n. ‖ câble m. de jonction urbaine. / ~ connection (Tel) ‖ Ortsverbindung f. ‖ communication f. locale. / ~ custom ‖ Ortsgebrauch m. ‖ usage m. local. / ~ distiller ‖ Feldbrenner m. ‖ bouilleur m. de cru. / ~ exchange (Tel) ‖ Ortsamt n.; Ortsvermittlungsstelle f. ‖ bureau m. (central) urbain. / ~ group switch ‖ Ortsgruppenumschalter m. ‖ commutateur m. de groupe d'abonnés pour le service local. / ~ heating ‖ örtliche Erwärmung f. ‖ échauffement m. local. / ~ inspection ‖ Ortsbesichtigung f. ‖ inspection f. locale. / ~ junction line (Tel) ‖ Ortsverbindungsleitung f. ‖ ligne f. de raccordement urbaine. / ~ line (Tel) ‖ Ortsleitung f. ‖ ligne f. urbaine. / ~ load distributing station ‖ Lokallastverteilungsstelle f. ‖ poste m. pour la distribution de charge local. / ~ plant statement (Tel) ‖ Ortsnetzübersicht f. ‖ plan m. d'ensemble des lignes d'un réseau urbain. / ~ position (Tel) ‖ Ortsplatz m. ‖ position f. urbaine. / ~ rail-way ‖ Lokalbahn f. ‖ chemin m. de fer d'intérêt local. / electrified ~ railway ‖ elektrische Stadtbahn f. ‖ chemin m. de fer urbain électrifié; tramway m. électrique de ville. / ~ receiving ‖ Ortsempfang m. ‖ réception f. de l'émetteur local. / ~ requirements pl. ‖ örtlicher Bedarf m.; Platzbedarf m. ‖ besoins mpl. de la place. / ~ sender (Radio) ‖ Ortssender m. ‖ poste m. local. / ~ telegraph cable ‖ Ortstelegrafenkabel n. ‖ câble m. télégraphique urbain. / ~ telephonist (Tel) ‖ Ortsbeamtin f. ‖ opératrice f. urbaine. / ~ time ‖ Ortszeit f. ‖ heure f. du lieu. / ~ traffic (Railw) ‖ Lokalverkehr m. ‖ trafic m. local *ou* intérieur. / ~ traffic (Tel) ‖ Ortsverkehr m. ‖ trafic m. urbain. / ~ usance ‖ Platzgebrauch m. ‖ usance f. de la place. / ~ worker ‖ seßhafter Arbeiter m. ‖ ouvrier m. domicilié.

localization resistance ‖ Eingrenzwiderstand f. ‖ résistance f. d'interpolation. / ~ test ‖ Fehlerortsbestimmung f. ‖ localisation f. d'un dérangement.

localize, to ~ ‖ lokalisieren ‖ localiser. / ~ a fault ‖ einen Fehler m. eingrenzen ‖ localiser un défaut.

locating fence ‖ Auflageschiene f. ‖ barre f. d'appui. / ~ mark ‖ Strichmarke f. ‖ trait m. de repère.

location ‖ Lage f. ‖ position f. / ~ (Livelihood) ‖ Stellung f. ‖ emploi m. / ~ (Railw) ‖ Linienführung f.; Trassierung f. ‖ tracé m. / ~ beacon ‖ Ansteuerungsfeuer n. ‖ phare m. de terrain.

lock, to ~ ‖ verriegeln; abschließen ‖ bloquer; serrer; freiner. / ~ the carriage ‖ das Fuhrwerk lenken ‖ tourner *ou* braquer la voiture. / ~ in ‖ einschließen ‖ enfermer. / ~ against lateral shifting ‖ eine seitliche Verschiebung f. verhindern ‖ éviter un déplacement latéral. / ~ out ‖ aussperren ‖ renvoyer (en masse); congédier. / ~ out workmen ‖ Arbeiter mpl. aussperren ‖ exclure *ou* renvoyer des ouvriers mpl. / to come to a decision ~ out workmen ‖ eine Verabredung f. über Aussperrung von Arbeitern treffen ‖ concerter un accord pour l'exclusion des ouvriers. / ~ up capital ‖ Kapital n. festlegen ‖ immobiliser le capital. / ~ up the form (Print) ‖ die Form schließen ‖ serrer la forme.

lock ‖ Schloß n. ‖ serrure f. / ~ (Hydr arch) ‖ Schleuse f.; Kammerschleuse f. ‖ écluse f. à sas. / above the ~ ‖ oberhalb der Schleuse f. ‖ en amont de l'écluse f. / ~ box ‖ Einsteckschloß n. ‖ serrure f. auberonnière. / ~ in a canal ‖ Kanalschleuse f. ‖ écluse f. de canal. / ~ of carriage ‖ Lenkbarkeit f. eines Fuhrwerks m. ‖ tournant m. *ou* rayonnement m. *ou* braquage m. de la voiture. / case ~ ‖ Kastenschloß n. ‖ serrure f. à palastre. / cased ~ with sloping borders ‖ Kastenschloß n. mit schrägem Rand ‖ serrure f. à bosse. / ~ for cash register ‖ Kassensperrschloß n. ‖ serrure f. de caisse. / ~ with check-gates ‖ Drempelschleuse f.; Schlagschleuse f.; Schleuse f. mit Stemmtoren ‖ écluse f. busquée *ou* en éperon. / ~ with circular chamber ‖ Kesselschleuse f.; Trommelschleuse f. ‖ écluse f. à tambour; écluse f. ronde. / coverplate ~ ‖ Kistenschloß n. ‖ serrure f. de coffre. / dead ~ ‖ Riegelschloß n. *oder* Schubriegelschloß n. ohne Feder ‖ serrure f. à pêne sans ressort. / ~ for doors ‖ Türschloß n. ‖ serrure f. de porte. / double ~ (Hydr arch) ‖ Zwillingsschleuse f. ‖ écluses fpl. accolées. / double shutting ~ ‖ zweiseitig schließendes Türschloß n. ‖ serrure f. bénarde. / electrically operated ship ~ ‖ elektrisch betätigte Schiffsschleuse f. ‖ écluse f. de navigation à actionnement électrique *ou* à commande électrique. / ~ at the entrance of a dock ‖ Hafenschleuse f.; Seeschleuse f. ‖ écluse f. à l'entrée d'un port. / ~ with falling latch ‖ Fallschloß n. ‖ serrure f. à pêne dormant et à loquet. / false ~ ‖ blindes Schloß n. ‖ serrure f. fausse. / ~ fitted with ~ and key ‖ verschließbar ‖ fermant à clef. / French ~ ‖ Zuhaltungsschloß n. ‖ serrure f. à gâchette. / ~ for furniture ‖ Möbelschloß n. ‖ serrure f. pour meubles. / German spring-~ ‖ deutsches Schloß n.; Halbtourschloß n. ‖ serrure f. à demi-tour *ou* à pêne coulant *ou* à ressort; bec m. de canne; demi-tour m. / ~ of a gun ‖ Schloß n.; Gewehrschloß n.; Flintenschloß n. ‖ platine f. de fusil. / half-turning ~ *see* German spring ~. / hanging ~ ‖ Vorhängeschloß n. ‖ cadenas m. / nailed ~ ‖ angeschlagenes Schloß n. ‖ serrure f. posée *ou* à bosse. / once-turning ~ ‖ eintouriges Schloß n. ‖ serrure f. à un tour. / prime ~s (Spinn) ‖ Kernwolle f.; Oberwolle f.; Rückenwolle f. ‖ laine f. mère. / puzzle ~ ‖ Buchstabenschloß n.; Malschloß n.; Kombinationsschloß n. ‖ serrure f. secrète; cadenas m. secret *ou* à combinaison. / rabbeted ~ ‖ eingelassenes Schloß n. ‖ serrure f. affleurée. / ~ for safes ‖ Geldschrankschloß n. ‖ serrure f. pour coffre-forts. / safety ~ ‖ Sicherheitsschloß n. ‖ serrure f. de sûreté; antivol m. / spring ~ *see* German spring ~. / ~ steering (Auto) ‖ Einschlag m. (der Vorderräder) ‖ braquage m. des roues. / ~ with three bolts ‖ Schloß n. mit drei Riegeln ‖ serrure f. à trois pênes. / ~ for trunk ‖ Kofferschloß n. ‖ serrure f. à malles. / turning ~ ‖ Drehverschluß m. ‖ fermoir m. tournant. / ~ with turning-doors (Hydr arch) ‖ Drehtorschleuse f.; Kastenschleuse f. ‖ écluse f. à portes tournantes. / ventilation ~ ‖ Wetterschleuse f. ‖ écluse f. d'aération. / under ~ and key ‖ unter Verschluß m. ‖ sous clef f.; entreposé.

lock, angle of (Auto) ‖ Einschlagwinkel m. (der Lenkung) ‖ angle m. de braquage. / ~ bay ‖ Schleusenhaupt n. ‖ tête f. d'écluse. / ~ bolt ‖ Schloßriegel m. ‖ pêne m. / ~ box ‖ Schloßkasten m. ‖ palastre m.; palâtre m. / ~ building ‖ Schleusenbau m. ‖ construction f. d'écluses. / ~ canal *see* ~ channel. / ~ chain (Carriage) ‖ Hemmkette f. ‖ chaîne f. d'enrayure *ou* d'enrayage *ou* de freinage. / ~-chain hook ‖ Schließhaken m. *oder* Schneller m. einer Hemmkette ‖ clef f. de chaîne d'enrayage. / ~ chamber ‖ Schleusenkammer f. ‖ sas m.; neptune m.; bassin m. *ou* chambre f. d'écluse. / ~ channel ‖ Schleusenkanal m. ‖ canal m. de dérivation. / ~ corner cutter ‖ Sägeblatt n. zum Schneiden rechteckiger Nuten ‖ scie f. à défoncer les rainures droites. / ~ crown ‖ Schleusenhaupt n. ‖ tête f. d'écluse.

lockdoor, turning ~ of a sluice ‖ Drehtor n. einer Schleuse ‖ porte f. d'écluse tournante.

locked out ‖ ausgesperrt ‖ lock-outé; renvoyé.

locker ‖ Schrank m.; Kleiderschrank m. ‖ armoire f. / each man has a ~ for his clothes ‖ jeder Arbeiter m. hat einen Schrank für seine Kleider ‖ chaque homme a une armoire à lui pour ses habits. / steel ~ ‖ Stahlkleiderschrank m. ‖ garde-robe f. en tôle d'acier.

locker room ‖ Kleiderablage f. ‖ vestiaire m.

lockfiler's clamps pl. ‖ Reifkloben m. ‖ mordache f. ou tenaille f. à chanfrein.

lock fitter ‖ Schloßmontör m. ‖ monteur m. de serrures. / flight of ~s ‖ Kammerschleusentreppe f. ‖ écluses fpl. échelonnées.

lock gate ‖ Schleusentor n.; Stemmtor n. ‖ porte f. d'écluse ou busquée ou en éperon. / sliding ~ ‖ Schütz n.; Ziehschütz n.; Stauschütz n.; vanne f.; pale f.; porte f. éclusière à coulisse.

lock hatch of a pond ‖ Ablaß m. oder Schütz n. eines Teiches ‖ bande f. ou empellement m. ou vanne f. ou déversoir m. d'un étang. / ~ hook see ~ chain hook.

locking ‖ Verriegelung f. ‖ verrouillage m.; bloquage m. / automatic ~ ‖ selbsttätige Sperrung f. ‖ encliquetage m. automatique. / ~ of carriage ‖ Lenkbarkeit f. eines Fuhrwerks ‖ tournant m. ou rayonnement m. ou braquage m. de voiture.

locking angle, large ~ (Carriage) ‖ große Wendemöglichkeit f.; leichtes Umlenken n. ‖ grand angle m. de braquage.

locking apparatus ‖ Verriegelungsvorrichtung f. ‖ dispositif m. d'enclenchement. / ~ bar ‖ Verriegelungsstange f. ‖ barre f. de blocage. / ~ bush ‖ Haltebüchse f. ‖ douille f. d'arrêt. / ~ chain (Carriage) ‖ Hemmkette f. ‖ chaîne f. d'enrayure ou d'enrayage ou de freinage. / ~ circuit (Tel) ‖ Haltestromkreis m. ‖ circuit m. de collage.

locking device ‖ Sperrvorrichtung f.; Schließvorrichtung f.; Feststellvorrichtung f. ‖ dispositif m. d'enclenchement ou de fixation ou de fermeture. / ~ for boxes ‖ Kistenverschluß m. ‖ fermeture f. de caisses. / ~ for keys ‖ Tastenverriegelungsvorrichtung f. ‖ dispositif m. de verrouillage des touches. / oil-tight ~ ‖ öldichter Verschluß m. ‖ obturateur m. tanche à l'huile. / V-dump car with automatic ~ ‖ Muldenkipper m. mit selbsttätiger Muldenfeststellung ‖ basculeur m. à auge avec dispositif de fixation automatique de l'auge.

locking furniture ‖ Schließbeschlag m. ‖ fermeture f.; fermoir m. / ~ hook ‖ Verschlußhaken m. ‖ crochet m. de fermeture ou de verrou. / ~ mechanism ‖ Stellwerkapparat m. ‖ appareil m. de verrouillage ou d'enclenchement ou de blocage. / ~ nut see locknut. / ~ pin ‖ Sperrbolzen m.; Schnappstift m. ‖ goujon m. de blocage; cheville f. d'arrêt. / ~ plate ‖ Sicherungsblech n. ‖ tôle f. de sûreté. / ~ property (of carriage) see locking of carriage. / ~ relais (Electr) ‖ Halterelais n. ‖ relais m. de blocage. / ~ ring ‖ Verschlußring m.; Klemmring m. ‖ collier m. de serrage; bague f. de fermeture. / ~ ring of the handle-bar ‖ Lenkstangenklemmring m. ‖ collier m. de serrage du guidon. / ~ screw ‖ Verschlußschraube f. ‖ vis f. de fermeture. / ~ signal (Tel) ‖ Sperr-

zeichen n. ‖ signal m. de bloquage. / ~ switch ‖ Schalter m. mit Steckschlüsseleinrichtung ‖ interrupteur m. à clé amovible ou à clé à douille. / ~-up Festklemmen n. ‖ fixage m. / ~ washer ‖ federnde Unterlagscheibe f.; Sprengring m. ‖ rondelle f. élastique. / ~ wheel ‖ Festhaltezad n.; Arretierrad n. ‖ roue f. d'arrêt.

lock-jaw ‖ Starrkrampf m. ‖ tétanos m.

lock keeper ‖ Schleusenwärter m.; éclusier m. / ~ knob ‖ Verschlußknopf m. ‖ bouton m. de serrure. / ~ master ‖ Schleusenmeister m. ‖ maître-éclusier m.

locknut ‖ Gegenmutter f. ‖ contre-écrou m.; écrou m. de serrage ou de bloquage.

lock-out ‖ Aussperrung f. ‖ exclusion f.; lock-out m.; renvoi m.

lock plant ‖ Schleusenanlage f. ‖ installation f. d'écluse. / ~ plate of a gun-lock ‖ Schloßblech n. eines Gewehrschlosses ‖ corps m. de platine d'un fusil. / ~ receiver ‖ Schleusenempfänger m. ‖ poste-récepteur m. à écluse. / ~ safety device ‖ Schloßsicherung f. ‖ dispositif m. de sûreté pour serrures. / ~ saw ‖ Lochsäge f.; Stichsäge f.; Spitzsäge f. ‖ scie f. à guichet; scie f. égoïne. / ~ screw ‖ Schloßschraube f. ‖ boulon m. de fixation. / ~ sill ‖ Schlagschwelle f.; Schleusenschwelle f. ‖ seuil m. du busc; seuillet m. d'écluse.

locksmith ‖ Schlosser m. ‖ serrurier m. / ~ (Build) ‖ Bauschlosser m. ‖ serrurier m. en bâtiment. / artistic ~ ‖ Kunstschlosser m.; Kunstschmied m. ‖ serrurier m. d'art. / ~'s shop ‖ Schlosserwerkstatt f. ‖ atelier m. de serrurerie.

locksmith's work ‖ Schlosserarbeit f. ‖ (travail m. de) serrurerie f. / artistic ~ ‖ Kunstschlosserei f.; Kunstschmiede f. ‖ ferronnerie f. d'art. / ~ in building ‖ Bauschlosserei f. ‖ serrurerie f. de bâtiment.

lock spring ‖ Schloßfeder f. ‖ ressort m. de serrure. / ~ stitch ‖ Steppstich m. ‖ point m. de navette. / ~ switch see locking switch. / ~-up box ‖ verschließbarer Kasten m. ‖ caisse f. à serrure; caisse f. fermant à clé.

locomobile ‖ Lokomobile f. ‖ locomobile f.; machine f. locomobile. / gasolene ~ ‖ Benzinlokomobile f. ‖ locomobile f. à essence. / motor ~ ‖ Motorlokomobile f. ‖ locomobile f. à moteur.

locomotion ‖ Fortbewegung f. ‖ locomotion f.; mouvement m. progressif; marche f. / ~ engine ‖ Fortbewegungsmaschine f. ‖ engin m. de locomotion.

locomotive ‖ Lokomotive f. ‖ lovomotive f. / arranging ~ (Railw) ‖ Rangiermaschine f.; Verschiebemaschine f. ‖ machine f. à faire les manœuvres ou à composer les trains. / articulated ~ ‖ kurvenbewegliche Lokomotive f. ‖ locomotive f. à essieux couplés articulés. / articulated tank ~ ‖ Drehgestelltenderlokomotive f. ‖ locomotive-tender f. à essieux couplés articulés. / broad gauge ~ ‖ Breitspurlokomotive f. ‖ locomotive f. à voie large. / ~ with carrying axle ‖ Lokomotive f. mit Laufachse ‖ locomotive f. accouplée avec essieu porteur. / compressed-air ~ ‖ Druckluftlokomotive f. ‖ locomotive f. à air comprimé. / contractor's ~ ‖ Baulokomotive f. ‖ locomotive f. pour entrepreneurs. / crane ~ ‖ Kranlokomotive f. ‖ locomotive f. à grue. / ~ fitted up with crane gear ‖ Lokomotive f. mit Kran-

ausrüstung ‖ locomotive f. à grue. / Diesel~ ‖ Diesellokomotive f. ‖ locomotive f. Diesel ou à moteur Diesel. / eightwheel four-coupled passenger ~ ‖ ³/₄ gekuppelte Personenzuglokomotive f. ‖ locomotive f. à voyageurs à quatre essieux dont deux couplés. / eightwheel six-coupled ~ ‖ ³/₄ gekuppelte Lokomotive f. ‖ locomotive f. à quatre essieux dont trois couplés. / eight-wheel six-coupled goods tank ~ ‖ ³/₄ gekuppelte Güterzugtenderlokomotive f. ‖ locomotive-tender f. à marchandises à quatre essieux dont trois couplés. / electric ~ ‖ elektrische Lokomotive f. ‖ locomotive f. électrique. / express ~ ‖ Schnellzuglokomotive f. ‖ locomotive f. d'express. / fireless ~ ‖ feuerlose Lokomotive f. ‖ locomotive f. sans foyer. / freight ~ ‖ Güterzuglokomotive f. ‖ locomotive f. pour trains de marchandises. / goods ~ see freight ~. / ~ for gradients ‖ Bergbahnlokomotive f. ‖ locomotive f. pour fortes rampes. / heavy goods ~ ‖ schwere Güterzuglokomotive f. ‖ grosse locomotive f. pour trains de marchandises. / ~ with liquid combustible ‖ Lokomotive f. mit Ölfeuerung ‖ locomotive f. à combustibles liquides ou pour chauffage à huile. / mine ~ ‖ Grubenlokomotive f. ‖ locomotive f. pour mines. / motor ~ for light fuel ‖ Motorlokomotive f. für Leichtbrennstoffe ‖ locomotive f. à moteur pour combustibles légers. / ~ for oil fuel see ~ with liquid combustible. / passenger ~ ‖ Personenzuglokomotive f. ‖ locomotive f. pour trains de voyageurs. / to put an additional ~ ‖ eine Lokomotive f. vorspannen ‖ atteler une seconde machine. / rack ~ ‖ Zahnradlokomotive f. ‖ locomotive f. à roue dentée ou à crémaillère. / reciprocating ~ ‖ Kolbenlokomotive f. ‖ locomotive f. à piston. / road ~ ‖ Straßenlokomotive f. ‖ locomotive f. routière. / saturated steam tank ~ ‖ Naßdampftenderlokomotive f. ‖ locomotive-tender f. à vapeur saturée. / six-coupled compound ~ ‖ ³/₃ gekuppelte Verbundlokomotive f. ‖ locomotive f. compound à trois essieux couplés. / small-gauge ~ ‖ Schmalspurlokomotive f. ‖ locomotive f. pour voie étroite. / standard gauge ~ ‖ Normalspurlokomotive f. ‖ locomotive f. à voie normale. / steam ~ ‖ Dampflokomotive f. ‖ locomotive f. à vapeur. / ~ for steep gradients ‖ Gebirgslokomotive f. ‖ locomotive f. pour fortes rampes. / ~ with storage battery ‖ Akkumulatorenlokomotive f. ‖ locomotive f. à accumulateurs. / superheated express ~ ‖ Heißdampfschnellzuglokomotive f. ‖ locomotive f. d'express à vapeur surchauffée. / superheated steam tank ~ for passenger service ‖ Heißdampfpersonenzugtenderlokomotive f. ‖ locomotive-tender f. à vapeur surchauffée pour trains de voyageurs. / tender ~ ‖ Tenderlokomotive f. ‖ locomotive f. à tender. / turbine-driven ~ ‖ Turbinenlokomotive f. ‖ locomotive f. à turbine. / ~ in working order ‖ betriebsfähige Lokomotive f. ‖ locomotive f. prêt au service.

locomotive boiler ‖ Lokomotivkessel m. ‖ chaudière f. de locomotive. / building of ~s ‖ Lokomotivbau m. ‖ construction f. des locomotives. / ~ cleaner ‖ Lokomotivputzer m. ‖ nettoyeur m. de loco-

motives. / ~ coaling plant ‖ Lokomotiv-bekohlungsanlage f. ‖ installation f. de chargement de charbon pour locomotives. / special machine for the construction of ~s ‖ Lokomotivbausondermaschine f. ‖ machine f. spéciale pour la construction des locomotives. / ~ coupling device ‖ Lokomotivkuppelvorrichtung f. ‖ dispositif m. d'accouplement pour locomotives. / ~ crane ‖ Lokomotivkran m. ‖ grue f. sur locomotive; gruelocomotive f. / ~ crank ‖ Lokomotivkurbel f. ‖ manivelle f. de locomotive.

locomotive crank-axle ‖ Lokomotivkurbelachse f. ‖ essieu m. coudé de locomotive. / ~ with circular webs ‖ Lokomotivkurbelachse f. mit runden Kurbelblättern ‖ essieu m. de locomotive avec flasques circulaires. / ~ with oblique body ‖ Lokomotivkurbelachse f. mit schrägem Mittelstück ‖ essieu m. coudé de locomotive à corps oblique. / ~ with prismatic webs ‖ Lokomotivkurbelachse f. mit prismatischen Kurbelblättern ‖ essieu m. coudé de locomotive à flasques prismatiques. / single-throw ~ ‖ einfach gekröpfte Lokomotivkurbelachse f. ‖ essieu m. de locomotive à coude unique. / two-throw ~ ‖ doppelt gekröpfte Lokomotivkurbelachse f. ‖ essieu m. de locomotive à deux coudes.

locomotive crank pin grinding machine ‖ Schleifwerk n. für Lokomotivkurbelzapfen ‖ machine f. à rectifier les tourillons de bielles motrices de locomotives. / ~ driver ‖ Lokomotivführer m. ‖ mécanicien m. de locomotive. / ~ engine ‖ Lokomotive f. ‖ locomotive f.

locomotive firebox ‖ Lokomotivstehkessel m. ‖ boîte f. à feu ou foyer m. d'une locomotive. / back plate of ~ ‖ Rückwand f. eines Lokomotivstehkessels ‖ face f. ou plaque f. arrière de la boîte à feu d'une locomotive.

locomotive fireman ‖ Lokomotivheizer m. ‖ chauffeur m. de locomotive. / ~ fittings pl. ‖ Lokomotivarmatur f. ‖ armature f. pour locomotives. / ~ frame filling plate ‖ Lokomotivrahmenfutterstück n. ‖ fourrure f. de châssis de locomotive. / ~ furnace ‖ Planrostinnenfeuerung f. ‖ foyer m. intérieur à grille horizontale. / ~ hangar ‖ Lokomotivschuppen m. ‖ remise f. de locomotives. / ~ head light projector ‖ Lokomotivscheinwerfer m. ‖ projecteur m. de locomotive. / ~ heaver ‖ Lokomotivhebebock m. ‖ chèvre m. à locomotive. / ~ jack ‖ Lokomotivwinde f. ‖ vérin m. de locomotive. / ~ lifting crane ‖ Lokomotivhebekran m. ‖ pont m. roulant pour le transport des locomotives. / ~ man see ~ driver. / part of ~ ‖ Lokomotivteil m. ‖ pièce f. de locomotive. / ~ pre-heater joint grinding machine ‖ Dichtungsflächenschleifmaschine f. für Lokomotivvorwärmer ‖ machine f. à rectifier les surfaces de joints des réchauffeurs de locomotives. / ~ shed ‖ Lokomotivschuppen m. ‖ remise f. pour locomotives. / ~ shifting crane ‖ Lokomotivversatzkran m. ‖ grue f. à déplacer les locomotives. / steam driven ~ slewing crane ‖ Dampflokomotivdrehkran m. ‖ grue f. locomotive pivotante à vapeur. / ~ tender ‖ Lokomotivtender m. ‖ tender m. de locomotive. / ~'s testing stand ‖ Loko-

motivprüfstand m. ‖ plateforme f. d'essai pour locomotives. / ~ traction ‖ Lokomotivbespannung f. ‖ traction f. par locomotive.

locomotive traverser ‖ Lokomotivschiebebühne f. ‖ transbordeur m. de locomotive ou à locomotives. / ~ for electric drive ‖ Lokomotivschiebebühne f. mit elektrischem Antrieb ‖ transbordeur m. à locomotives actionné électriquement. / ~ for hand drive ‖ Lokomotivschiebebühne f. mit Handantrieb ‖ transbordeur m. à locomotives actionné à la main. / ~ running on a double rail ‖ Lokomotivschiebebühne f., auf zwei Schienensträngen laufend ‖ transbordeur m. à locomotives circulant sur deux rails.

locomotive washing plant ‖ Lokomotivauswaschanlage f. ‖ installation f. à nettoyer les locomotives. / ~ wheel set ‖ Lokomotivradsatz m. ‖ essieu m. monté pour locomotives. / ~ whistle ‖ Lokomotivpfeife f. ‖ sifflet m. d'alarme.

locus (Geom) ‖ geometrischer Ort m. ‖ lieu m.; lieu géométrique.

locust-tree ‖ unechte Akazie f.; Robinie f.; Schotendorn m. ‖ robinier m.

lode (Mine) ‖ Erzgang m. ‖ filon m. métallifère. / the ~ changes its course ‖ der Gang ändert das Streichen ‖ le filon change de direction. / ~ of copper ‖ Kupfergang m. ‖ veine f. de cuivre. / the ~ has crumpled ‖ der Erzgang m. hat sich verdrückt ‖ le filon m. a disparu. / dead ~ see poor ~. / one ~ is faulted by the other ‖ ein Gang m. verwirft einen anderen ‖ un filon m. est rejeté par un autre. / the ~ grows ‖ der Gang m. wird mächtiger ‖ le filon m. s'élargit ou prend du ventre ou gagne de la puissance. / ~ of medium dip ‖ flachfallender Gang m. ‖ filon m. à faible pendage. / ~ forming a network ‖ Netzgang m. ‖ filon m. en forme de réseau. / northern ~ (Mine) ‖ Mitternachtsgang m. ‖ filon m. vers le nord. / poor ~ ‖ tauber Gang m. ‖ faille f.; crain m. / ~ of steep dip ‖ tonnlägiger Gang m. ‖ filon m. à fort pendage. / superficial ~ ‖ Rasenläufer m. ‖ coureur m. de gazon. / vertical ~ ‖ seigerer Gang m. ‖ filon m. vertical.

lodesman ‖ Ganghäuer m. ‖ mineur m. d'un filon.

lodestone see loadstone.

lodge, to ~ a complaint ‖ Beschwerde f. oder Klage f. erheben ‖ porter plainte f.

lodge ‖ Hütte f.; Baracke f.; Bude f. ‖ hutte f.; baraque f.; loge f.; échoppe f.; cabane f.

lodging allowance ‖ Mietentschädigung f. ‖ indemnité f. de logement. / ~ installation ‖ Wohnungsinstallation f. ‖ installation f. de logement. / ~ house ‖ Schlafhaus n.; Hotel-garni n. ‖ maison f. à dortoir; hôtel m. garni.

loess ‖ Löß m.; Briz m. ‖ lœss m.

loft ‖ Dachstube f. ‖ grenier m.; mansarde f. / ~ (Trade) ‖ Speicher m. ‖ magasin m.; grenier m. / to lay off a ship on the mould ~ ‖ ein Schiff auf dem Schnürboden abschlagen ‖ tracer un bateau à la salle. / ~ door ‖ Bodentür f. ‖ porte f. de grenier.

loftsman (Pap) ‖ Aufhänger m. ‖ sécheur m.; étendeur m.

log (Nav) ‖ Log n.; Fahrtmesser m. ‖ loch m. / ~ (Wood) ‖ Kloben m.; Scheit n.; Klotz m. ‖ bûche f.; bloc m. / ground ~

(Nav) ‖ Grundlog n. ‖ loch m. de fond. / harpoon ~ see patent ~. / to heave the ~ ‖ das Log n. werfen; loggen ‖ jeter ou filer le loch. / ~ of mahogany ‖ Mahagoniholzblock m. ‖ bille f. d'acajou. / patent ~ ‖ Patentlog n. ‖ sillomètre m. / rail ~ (Nav) ‖ Relingslog n. ‖ loch m. de fortune. / round ~ ‖ Rundholz n. ‖ bois m. rond. / ~ of wood ‖ Blockholz n.; Holzscheit n. ‖ bille f. de bois; bûche f.

logarithm ‖ Logarithmus m. ‖ logarithme m. / Brigg's ~ see common ~. / common ~ ‖ gemeiner oder Brigg'scher Logarithmus m. ‖ logarithme m. ordinaire. / Napier's ~ see natural ~. / natural ~ ‖ natürlicher Logarithmus m. ‖ logarithme m. naturel ou népérien.

logarithmic ‖ logarithmisch ‖ logarithmique. / ~ decrement ‖ logarithmisches Dekrement n. ‖ décrément m. logarithmique.

logarithmically, to be graded ~ ‖ logarithmisch abgestuft ‖ échelonné logarithmiquement.

log board (Nav) ‖ Logbrett n.; Logtafel f. ‖ table f. de loch; casernet m. / ~ book ‖ Logbuch n.; Schiffstagebuch n. ‖ journal m. nautique ou de bord ou de navigation. / ~ carriage ‖ Blockwagen m.; chariot m. à grumes. / ~ cutter ‖ Holzhacker m. ‖ fendeur m. de bois. / ~ glass ‖ Logglas m. ‖ horloge f. du loch. / ~ hut ‖ zerlegbares Holzhaus n. ‖ maison f. de bois portatif ou démontable.

log line ‖ Logleine f. ‖ ligne f. de loch. / to pay out the ~ ‖ das Log n. werfen; loggen ‖ jeter ou filer le loch.

log maker ‖ Klotzarbeiter m. ‖ fendeur m. de bûches. / ~ roller (Sawmill) ‖ Aufbanker m. ‖ rouleur m. de troncs d'arbres.

logwood ‖ Blauholz n.; Kampescheholz n. ‖ bois m. de Campêche. / ~ extract ‖ Blauholzextrakt m. ‖ extrait m. de bois de Campêche.

löllingite ‖ Glanzarsenikkies m. ‖ löllingite f.

long ‖ lang ‖ long. / for a ~ time ‖ langfristig ‖ à longs jours; à longue échéance.

longboat ‖ Pinasse f. ‖ pinasse f.

long-burning lamp ‖ Dauerlampe f. ‖ lampe f. à longue durée.

long carriage (Weav) ‖ Langschlitten m. ‖ chariot m. en long.

long-cut wood ‖ Langholz n. ‖ bois m. de long.

long-dated (Bill of exchange) ‖ mit langem Ziel n. ‖ à longue échéance f.

long-distance calling signal ‖ Fernanrufzeichen n. ‖ signal m. d'appel interurbain. / ~ circuit (Tel) ‖ Fernamtsschaltung f. ‖ circuit m. interurbain. / ~ and local connector (Aut tel) ‖ Orts- und Fernleitungswähler m. ‖ connecteur m. local et interurbain. / ~ drop (Tel) ‖ Fernklappe f. ‖ annonciateur m. interurbain. / ~ flight ‖ Fernflug m. ‖ raid m. / ~ gun ‖ weitreichendes Geschütz n. ‖ canon m. à longue portée. / ~ indicator ‖ Fernzeiger m. ‖ indicateur m. à distance. / ~ jack lamp ‖ Fernklinkenlampe f. ‖ lampe f. de fin de conversation interurbaine. / ~ jack multiple ‖ Fernklinkenleitung f. ‖ ligne f. de jack multiple interurbaine. / ~ junction (Tel) ‖ Fernvermittlungsleitung f. ‖ ligne f. auxiliaire interurbaine ou intermédiaire. / ~ junction section ‖ Fernvermittlungsschrank m. ‖ tableau m. des lignes auxiliaires in-

terurbaines. / ~ line relay ‖ Fernanrufrelais n. ‖ relais m. d'appel interurbain. / ~ plug ‖ Fernstöpsel m. ‖ fiche f. interurbaine. / ~ position (Tel) ‖ Fernplatz m. ‖ position f. interurbaine. / ~ receiver (Radio) ‖ Fernempfänger m. ‖ récepteur m. à grandes distances. / ~ receiving (Radio) ‖ Fernempfang m. ‖ réception f. à grande distance. / ~ reconnaissance machine ‖ Fernaufklärungsflugzeug n. ‖ avion m. de reconnaissance à longue distance. / ~ section (Tel) ‖ Fernschrank m. ‖ tableau m. interurbain. / ~ station ‖ Großfunkstation f. ‖ poste f. de grandes distances. / ~ supply work ‖ Überlandzentrale f.; Fernkraftwerk n. ‖ centrale f. interurbaine. / ~ switch ‖ Fernnachwähler m. ‖ sélecteur m. interurbain. / ~ system for magneto-boards ‖ Fernleitungssystem n. für Klappenschränke ‖ système m. interurbain pour commutateurs à clapets. / ~ table ‖ Ferntisch m. ‖ table f. interurbaine. / ~ telephone cable ‖ Fernkabel n. ‖ câble m. téléphonique à grande distance. / ~ telephone cable net ‖ Fernkabelnetz n. ‖ réseau m. des câbles téléphoniques à grande distance.

long-duration test ‖ Dauerversuch m.; Dauerprüfung f. ‖ essai m. d'endurance.

long experience ‖ langjährige Erfahrung f. ‖ expérience f. de longues années. / ~ eye auger ‖ Stangenbohrer m. ‖ tarière f. torse. / ~ flaming coal ‖ langflammige Kohle f. ‖ charbon m. à longue flamme.

long-focus telescope ‖ langbrennweitiges Fernrohr n. ‖ lunette f. à long foyer.

longeron of the fuselage (Aero) ‖ Rumpfholm m. ‖ longeron m. de fuselage.

long grate ‖ Langrost m. ‖ grille f. longue.

longing rein ‖ Longe f.; Leine f. ‖ longe f.

longitude ‖ Länge f. ‖ longitude f. ‖ ~ by accounts ‖ gegißte Länge f. ‖ longitude f. estimée. / ~ by dead reckoning *see* by accounts. / east ~ ‖ östliche Länge f. ‖ longitude f. est *ou* orientale. / geocentric ~ ‖ geozentrische Länge f. ‖ longitude f. géocentrique. / heliocentric ~ ‖ heliozentrische Länge f. ‖ longitude f. héliocentrique. / terrestrial ~ ‖ geographische Länge f. ‖ longitude f. géographique *ou* géodésique. / west ~ ‖ westliche Länge f. ‖ longitude f. ouest *ou* occidentale.

longitude error ‖ Längenfehler m. ‖ erreur f. en longitude.

longitudinal ‖ die Länge betreffend; Längen- ‖ longitudinal. / ~ bench ‖ Längsbank f. ‖ banquette f. longitudinale. / ~ bulkhead ‖ Längsschott n. ‖ cloison f. longitudinale. / ~ crack ‖ Längsriß m. ‖ fissure f. longitudinale. / ~ feed ‖ Langvorschub m. ‖ avance f. de chariotage. / ~ friction device *see* ~ grinder conch. / ~ girder pressed in one piece ‖ aus einem Stück gepreßter Langträger m. ‖ longeron m. embouti d'une seule pièce. / ~ grinder conch (Bak) ‖ Längsreiber m. ‖ broyeur m. longitudinal; conche f. / ~ induction ‖ Längsinduktion f. ‖ induction f. longitudinale. / ~ machine (Cable) ‖ Longitudinalmaschine f.; Längsbedeckungsmaschine f. ‖ caoutchouteuse f. longitudinale. / ~ motion ‖ Längsverschiebung f.; Längsbewegung f. ‖ déplacement m. en longueur; course f. longitudinale. / ~ profiling apparatus ‖ Plankurvenfräsapparat m. ‖ appareil m. à fraiser les cames planes. ‖ ~ rivetting ‖

Längsnietung f. ‖ rivure f. longitudinale. / ~ runner ‖ Längsträger m. ‖ longeron m. du châssis m. / ~ seam ‖ Längsfalz m. ‖ couture f. longitudinale. / ~ section ‖ Längsschnitt m. ‖ coupe f. longitudinale. / ~ shearing-machine ‖ Longitudinalschere f. ‖ cisaille f. longitudinale. / ~ sheet cutter (Bookb) ‖ Längsschneider m. ‖ coupeur m. longitudinal de feuilles. / ~ sill ‖ Langschwelle f.; Holm f. ‖ longrine f. / ~ spring ‖ Längsfeder f. ‖ ressort m. longitudinal. / ~ stability ‖ Längsstabilität f. ‖ stabilité f. longitudinale. / ~ stiffening ‖ Längsversteifung f. ‖ renforcement m. *ou* raidissement m. longitudinal. / to resist rigidly~ stresses ‖ Längsbeanspruchungen fpl. unnachgiebig widerstehen ‖ résister rigidement aux efforts longitudinaux. / to grind by the ~ traverse method ‖ im Langschleifverfahren n. schleifen ‖ rectifier par va-et-vient. / ~ variation ‖ Längsschwingung f.; longitudinale Schwingung f. ‖ vibration f. longitudinale. / ~ view ‖ Längsansicht f. ‖ vue f. longitudinale. / ~ wave ‖ Längswelle f.; Longitudinalwelle f. ‖ onde f. longitudinale.

longitudinal (Railw) ‖ Längsschwelle f. ‖ longrine f. / ~ (Shipb) ‖ Längsspant n. ‖ membrure f. longitudinale; lisse f.

long lease ‖ Erbpacht f. ‖ bail m. héréditaire. / ~ loop apparatus ‖ Langmascher m. ‖ appareil m. pour rangée lâche. / ~ necked ‖ langhalsig ‖ à long col m.

long pane of a roof ‖ Langseite f. eines Daches ‖ long-pan m. de comble.

long pile shag (Weav) ‖ Felbel m. ‖ panne f.; peluche f.

long primer (Print) ‖ Korpus f.; Longprimer f. ‖ petit-romain m.; neuf m.

long-range gun ‖ weitreichendes Geschütz n. ‖ canon m. à longue portée.

long saw ‖ Dielensäge f.; Brettsäge f. ‖ scie f. de long.

long sight ‖ Weitsichtigkeit f. ‖ presbytie f.

long-sighted ‖ weitsichtig ‖ presbyte. / ~ eye ‖ übersichtiges Auge n. ‖ œil m. hypermétrope. / ~ slot burner ‖ Langlochbrenner m.; Schlitzbrenner m. ‖ brûleur m. à fente.

long-stapled cotton ‖ langstapelige Baumwolle f. ‖ coton m. longue-soie.

long-stroke steam engine ‖ langhubige Dampfmaschine f. ‖ machine f. à vapeur à longue course.

long-turning ‖ Langdrehen n. ‖ chariotage m.

long-wavy ray ‖ langwelliger Strahl m. ‖ rayon m. de grande longueur d'onde.

long wood ‖ Langholz n. ‖ bois m. de fil.

loof (Mar) ‖ Luvseite f. ‖ côté m. *ou* dessus m. du vent lof.

look, to ‖ blicken ‖ regarder; voir.

looking angle ‖ Sehwinkel m. ‖ angle m. visuel.

looking-glass ‖ Spiegel m. ‖ miroir m. / ~ cutter ‖ Spiegelglasschneider m. ‖ tailleur m. de glaces.

looking-hole (Furnace) ‖ Schauloch n. ‖ visière f.; regard m.

look out! ‖ Achtung! Obacht! ‖ gare! / ~ (Packing) ‖ vorgesehen! ‖ fragile!

lookout (Mar) ‖ Ausguck m. ‖ vigie f. / ~ car (Railw) ‖ Aussichtswagen m. ‖ wagon m. à guérite. / ~ man (Mas) ‖ Straßenwärter m. ‖ gardien m. de rue.

lookout telescope ‖ Aussichtsfernrohr n. ‖ lunette f. d'approche. / ~ binocular ‖ binokulares Aussichtsfernrohr n. ‖ lunette f. d'approche binoculaire. / ~ on stand for terrestrial observations ‖ Standaussichtsfernrohr n. für terrestrische Beobachtungen ‖ longue-vue f. d'approche pour des observations terrestres.

loom ‖ Webstuhl m. ‖ métier m. à tisser. / carpet weaver's ~ ‖ Teppichwebstuhl m. ‖ métier m. à tapis. / ~ of the clothier ‖ Tuchmacherstuhl m. ‖ métier m. du drapier. / hand ~ ‖ Handwebstuhl m. ‖ métier m. à bras *ou* à tisser à main. / Jacquard ~ ‖ Jacquardstuhl m. ‖ métier m. Jacquard. / ~ goods pl. manufactured on the ~ ‖ auf dem Webstuhl m. hergestellte Waren fpl. ‖ articles mpl. fabriqués sur le métier à tisser / metal gauze ~ ‖ Metalltuchwebstuhl m. ‖ métier m. à tisser les toiles métalliques. / ~ of oar ‖ Ruderschaft m.; Riemenschaft m. ‖ manche m. d'aviron. / power knitting ~ ‖ Strickmaschine f. ‖ tricoteuse f. mécanique. / ribbon ~ ‖ Bandwebstuhl m. ‖ métier m. à rubans. / round ~ ‖ Rundwebstuhl m. ‖ métier m. à tisser circulaire. / spindle ~ ‖ Klöppelmaschine f. ‖ métier m. à fuseaux. / to take from the ~ ‖ abbäumen ‖ dérouler. / weaving ~ ‖ Webstuhl m. ‖ métier m. à tisser. / ~ for weaving looped fabrics ‖ Webstuhl m. für Schlingenstoff ‖ métier m. pour tissus bouclés. / ~ for weaving plush ‖ Plüschwebstuhl m. ‖ métier m. pour peluche.

loom beam ‖ Weberbaum m. ‖ ensouple f.; déchargeoir m. / ~ fixer ‖ Webstuhlvorrichter m. ‖ régleur m. de métier m. / ~ fixing ‖ Webstuhlvorrichtung f. ‖ réglage m. de métiers à tisser. / ~ gaiter ‖ Webstuhlvorrichter m. ‖ metteur m. en main du métier.

looming ‖ Kimmung f.; Luftspiegelung f. ‖ mirage m.

loom motor ‖ Webstuhlmotor m. ‖ moteur m. de métier à tisser. / ~ mounter ‖ Webstuhlvorrichter m. ‖ ajusteur m. de métiers. / ~ switch ‖ Webstuhlschalter m. ‖ interrupteur m. de moteur de métier à tisser.

loop, to (Weav) ‖ ketteln ‖ remmailler; entrelacer. / ~ by hand (Weav) ‖ anschlagen ‖ faire l'ourlet m.

loop ‖ Öse f.; Schleife f. ‖ boucle f. / ~ (Metal) ‖ Luppe f.; Deul m. ‖ loupe f.; balle f. en fer. / ~ (Radio) ‖ Rahmenluftleiter m.; Rahmenantenne f. ‖ antenne f. à cadre. / ~ (Rope) ‖ Schleife f.; Schlinge f. ‖ lacet m. / ~ (Saddl) ‖ Öse f.; Kummetfeder f. ‖ anneau m.; œil m. d'attelle. / ~ (Tel) ‖ Doppelleitung f. ‖ boucle f. / ~ (Wire) ‖ Öse f.; Ring m. ‖ anneau m. / ~ of a door ‖ Türband n. ‖ penture f. / ear ~ ‖ Ohrspange f. ‖ boucle f. d'oreilles. / to form ~s ‖ Schlingen fpl. bilden ‖ former des serpentins mpl. / ~ of a hinge ‖ Lappen m. *oder* Ösenteil m. eines Türbandes ‖ platine f. de charnière. / line ~ (Electr) ‖ Leitungsschleife f. ‖ boucle f.; lacet m. / potential ~ ‖ Spannungsbauch m. ‖ ventre m. de tension. / to sink the ~s pl. (Textile) ‖ kulieren ‖ cueillir. / suspension ~ ‖ Aufhängebügel m. ‖ étrier m. de suspension. / ~ of a swingle-tree

clasp (Coachm) ‖ Ortscheitöse f. ‖ amorce f. de lamette de palonnier.

loop and hook ‖ Hakenband n.; Aufsetzband n.; Kegelband n. ‖ fiche f. à gond *ou* à repos; penture f. à gond. **loop antenna** ‖ Rahmenantenne f.; antenne f. à cadre. / breadth of the ~ (Tel) ‖ Schleifenbreite f. ‖ largeur f. de boucle. / ~ current (Tel) ‖ Schleifenstrom m. ‖ courant m. de boucle. / ~ decremeter (Tel) ‖ Schleifdämpfungsmesser m. ‖ appareil m. mesureur pour circuits bouclés. / ~ expansion pipe ‖ Rohrleitungsausgleichschleife f. ‖ boucle f. compensatrice de tuyaux. **looped fabric** ‖ Maschenware f. ‖ tissu m. à mailles. / ~ for toilet linen ‖ Schlingengewebe n. für Toilettewäsche ‖ tissu m. bouclé pour linge de toilette. **loop galvanometer** ‖ Schleifengalvanometer n. ‖ galvanomètre m. à boucle. **looping** (Weav) ‖ Maschenbildung f.; Ketteln n. ‖ rebouclement m.; remaillage m. / ~ machine ‖ Kettelmaschine f. ‖ remmailleuse f. / ~ mill ‖ Drahtstraße f. ‖ train m. à fil de fer. / ~ plush ‖ Schlingenplüsch m. ‖ peluche f. bouclé. / ~ wheel ‖ Maschenrad n. ‖ roue f. mailleuse. **loop maker** ‖ Posamentier m.; Bortenwirker m.; Nestler m. ‖ passementier m. / ~ plush ‖ Henkelplüsch m. ‖ peluche f. à brides. / ~ test (Electr) ‖ Schleifenmessung f. ‖ méthode f. de la boucle. / ~ wheel (Textile) ‖ Maschenleger m. ‖ mailleuse f. / ~ wheeler (Metal) ‖ Luppenfahrer m. ‖ porteur m. de loupes. / ~ wire (Electr) ‖ Schließungsdraht m. ‖ fil m. conjonctif.

loose ‖ lose; schlaff ‖ lâche; mou. / to get ~ ‖ locker werden ‖ se déserrer; se relâcher; se dresser. / to get ~ (Mach) ‖ Spiel m. *oder* Spielraum m. haben ‖ jouer; avoir du jeu. / getting ~ of the nut ‖ Lockern n. der Mutter im Betrieb ‖ desserrage m. de l'écrou par vibrations. / with ~ handle ‖ mit losem Stiel m. ‖ à manche m. rapporté. / to work ~ *see* to get ~. / ~ and woolly groats pl. ‖ lockeres und wolliges Schrot n. ‖ mouture f. légère et complète. **loose ashes** pl. ‖ lockere Asche f. ‖ cendre f. incohérente. / ~ bush ‖ Leerlaufbuchse f. ‖ douille f. de marche à vide / ~ countershaft pulley ‖ Losscheibe f. am Deckenvorgelege ‖ poulie f. folle du renvoi. / ~ coupling (Radio) ‖ lose *oder* schlaffe Kopplung f. ‖ couplage m. lâche; accouplement m. faible. / ~ flange (Tube) ‖ loser Bord m. ‖ bague f. *ou* bride f. folle. / ~ hogshead ‖ Gärfaß n. ‖ fût m. *ou* tonne f. de fermentation. / ~ leaf bookkeeping ‖ Buchführung f. auf losen Blättern ‖ comptabilité f. sur feuilles détachées. / ~ liner (Gun) ‖ auswechselbares Futterrohr n. ‖ chemise f. amovible. / ~ pin ‖ einsteckbarer Bolzen m. ‖ cheville f. démontable. / ~ road ‖ grundlose Straße f. ‖ route f. défoncée. / ~ roller bearing ‖ Losrollenlager n. ‖ palier m. à rouleaux libres. / ~ stuffing box ‖ Einsatzstopfbüchse f. ‖ tampon m. de boîte à bourrage. / ~ threads pl. of silk ‖ Zupfseide f. ‖ soie f. effilée. **loosely assembled quires** pl. (Pap) ‖ geholländerte Bogen mpl. ‖ feuilles fpl. détachées assemblées.

loosen, to ‖ lockern ‖ relâcher. / ~ (Spinn) ‖ auflockern ‖ désagréger. / ~ (Sugar) ‖ die Brote npl. schütteln ‖ locher. / ~ a bolt ‖ einen Bolzen m. lockern ‖ dégager *ou* déserrer un boulon. / ~ a cable ‖ ein Tau nachlassen *oder* schießen lassen ‖ larguer *ou* mollir un câble. / ~ the loops pl. (Weav) ‖ locker arbeiten ‖ desserrer les mailles fpl. / ~ the pages (Print) ‖ die Kolumnen fpl. auflösen ‖ délier les pages fpl. / ~ a sail ‖ ein Segel n. losmachen ‖ déferler *ou* larguer une voile. / ~ a screw ‖ eine Schraube f. lockern ‖ desserrer une vis / ~ the sheets of the courses (Mar) ‖ die Schoten fpl. aufstecken ‖ déborder les écoutes fpl. des basses voiles. / ~ the stock ‖ das Mahlgut lockern ‖ ramollir la marchandise. / ~ the wedge ‖ loskeilen ‖ déclaveter. / ~ wool by arsenic ‖ Wolle f. abgiften ‖ délainer à l'arsenic. **loosening** ‖ Auflockerung f. ‖ désagrégation f. / ~ (Sugar) ‖ Löschen n.; Schütteln n. ‖ lochage m. **lop, to** ~ (Wood) ‖ zopfen ‖ découper le petit bout. / ~ a tree ‖ einen Baum m. köpfen *oder* kappen *oder* lüften *oder* beschneiden ‖ écimer *ou* éhouper *ou* épointer *ou* étêter *ou* élaguer un arbre. **loper** (Ropem) ‖ Entwirrer m.; Nachlänger m. ‖ émerillon m. **lopped sleeper** (Railw) ‖ gekappte Schwelle f. ‖ traverse f. avec fermeture de tête. **lopping-shears** pl. ‖ Heckenschere f. ‖ ciseaux mpl. de jardinier. **Lord Mayor** ‖ Bürgermeister m. ‖ maire m. **lorgnette** pl. ‖ Lorgnette f. ‖ lorgnette f.; face-à-main m. **lorry** (Auto) ‖ Lastkraftwagen m. ‖ camion m. de transport. / ~ (Mine) ‖ Förderwagen m.; Hund m. ‖ berline f.; char m. à bennes. / ~ (Railw) ‖ offener Güterwagen m.; Lori f.; Lore f. ‖ truc m.; wagon m. en plate-forme. / break-down ~ ‖ Abschleppwagen m. ‖ dépanneuse f. / motor ~ ‖ Lastkraftwagen m. ‖ camion m. automobile. / platform ~ ‖ Pritschenwagen m. ‖ camion m. plat *ou* à plate-forme. **lose, to** ‖ einbüßen; verlieren ‖ perdre. / ~ (Chem) ‖ abgeben ‖ éliminer. / ~ one's bearings (Aero) ‖ sich verfliegen ‖ être désorienté; s'égarer. / ~ the course ‖ vom Fahrweg m. abkommen ‖ s'écarter de la route. / ~ ground (Mar) ‖ Grund m. verlieren ‖ perdre du terrain. **loss** ‖ Verlust m. ‖ perte f. / ~ (Damage) ‖ Nachteil m.; Schaden m. ‖ désavantage m.; préjudice m. / ~ (Deficit) ‖ Ausfall m.; Mindereinnahme f. ‖ perte f.; manque m. / ~ (Metal) ‖ Verlust m.; Abbrand m. ‖ perte f. au feu. **loss due to burning** ‖ Glühverlust m. ‖ perte f. au rouge. / ~ of capacity ‖ Kapazitätsschwund m. ‖ perte f. de capacité. / ~ of charge (Electr) ‖ Ladungsverlust m. ‖ perte f. de charge. / chimney ~ ‖ Schornsteinverlust m. ‖ pertes fpl. dans la cheminée. / ~ occasioned by a collision ‖ durch Kollision verursachter Schaden m. ‖ dommage m. causé par suite d'abordage. / ~ by cupellation ‖ Kapellenzug m.; Kapellenraub m. ‖ perte f. par la coupellation. / ~ eddy ‖ Wirbelverlust m. ‖ perte f. par tourbillons *ou* remous. / ~ of energy by hysteresis ‖ Energieverlust m. durch Hysteresis ‖ perte f.

d'énergie par hystérésis. / ~ of entrepôt ‖ Haldenverlust m. ‖ déchet m. de l'entrepôt *ou* de crasses. / estimated ~ ‖ veranschlagter Verlust m. ‖ perte f. estimée. / ~ of fall ‖ Gefällverlust m. ‖ perte f. de chute. / ~ by fire ‖ Brandschaden m. ‖ perte f. par incendie. / frictional ~ ‖ Reibungsverlust m. ‖ perte f. de friction; perte f. due à la friction / the plant stopped work after several years full of ~es ‖ der Betrieb wurde nach mehreren verlustreichen Jahren eingestellt ‖ après de plusieurs années riches en perte l'exploitation fut suspendue *ou* arrêtée. / to compensate for ~ of gas ‖ den Gasverlust m. ausgleichen ‖ compenser la perte *ou* la fuite de gaz. / ~ of heat ‖ Wärmeverlust m. ‖ perte f. de chaleur. / insulation against ~ of heat ‖ Wärmeschutz m. ‖ isolation f. calorifuge; isolement m. thermique. / to incur a ~ ‖ einen Verlust m. erleiden ‖ subir une perte f. / ~ of interest ‖ Zinsverlust m. ‖ perte f. d'intérêts. / ~ of lift ‖ Auftriebsverlust m. ‖ perte f. de force ascensionnelle *ou* de poussée. / ~ of material ‖ Stoffverlust m. ‖ perte f. de matière. / mechanical ~ ‖ mechanischer Verlust m. ‖ perte f. mécanique. / no-load ~ ‖ Leerlaufverlust m. ‖ perte f. à vide. / ~ by percussion ‖ Stoßverlust m. ‖ perte f. due aux secousses. / ~ of power ‖ Kraftverlust m. ‖ perte f. de force. / ~ of pressure ‖ Druckverlust m. ‖ perte f. de pression. / presumptive ~ (Mar) ‖ Verschollenheit f. ‖ présomption f. de perte. / to compensate for ~ of profit ‖ einen entgangenen Gewinn m. ersetzen ‖ compenser un manque à gagner. / ~ at a redheat ‖ Glühverlust m. ‖ perte f. au rouge. / ~ of sight ‖ Erblindung f. ‖ perte f. de la vue. / ~ of speed ‖ Geschwindigkeitsverlust m. ‖ perte f. de vitesse. / to suffer the ~ of ‖ einbüßen ‖ perdre. / to sustain a ~ ‖ einen Verlust m. erleiden *oder* tragen ‖ subir *ou* supporter une perte. / ~ of time ‖ Zeitverlust m. ‖ perte f. de temps. / ~ of voltage ‖ Spannungsverlust m. ‖ perte f. de tension électrique. / ~ of work ‖ Arbeitsverlust m. ‖ perte f. de travail.

löss ‖ Löß m. ‖ löss m.; glaise f.

lost ‖ verloren ‖ perdu. / ~ (Trade) ‖ ruiniert ‖ être ruiné *ou* flambé *ou* frit. / to be ~ ‖ verlorengehen ‖ se perdre; s'égarer. / to become ~ (Aero) ‖ sich verfliegen ‖ être désorienté; s'égarer.

lost motion (Mach) ‖ toter Gang m. ‖ perte f. de course; jeu m. / ~ of the screw ‖ toter Gang m. der Schraube ‖ jeu m. de la vis.

lost time (Tel) ‖ Verlustzeit f. ‖ temps m. perdu.

lot (Trade) ‖ Los n.; Posten m. ‖ lot m. / ~ of land ‖ Parzelle f. ‖ parcelle f.; lot m. de terrain. / the ~ is rejected ‖ das Los wurde verworfen *oder* ist zurückgewiesen ‖ le lot fut rebuté.

lottery ticket ‖ Lotterielos n. ‖ billet m. de loterie.

lotto ‖ Lottospiel n. ‖ jeu m. de loto.

lotus (Tree) ‖ Lotosbaum m. ‖ micoculier m. **loudspeaker** ‖ Lautsprecher m. ‖ haut-parleur m. / artillery ‖ Artillerielautsprecher m. ‖ haut-parleur m. de l'artillerie. / ~ with cellulose conus ‖ Lautsprecher m. mit Zellstoffkonus ‖ haut-

parleur m. à diaphragme conique en cellulose. / ~ with corrugated diaphragm || Riffellautsprecher m. || haut-parleur m. à membrane striée. / ~ without funnel || Lautsprecher m. ohne Trichter || haut-parleur m. sans pavillon. / giant ~ || Großlautsprecher m. || haut-parleur m. géant. / giant ~ of a wireless broadcasting station for music || Großlautsprecher m. einer Musikübertragungsanlage || haut-parleur m. géant d'une installation de transmission de musique. / ~ with paper conus || Lautsprecher m. mit Papierkonus || haut-parleur m. à diaphragme conique en papier. / plate ~ (Tel) || Blatthaller m. || haut-parleur m. avec plaque. / ribbon ~ (Tel) || Bandsprecher m. || haut-parleur m. avec bandelette. / ~ for the transmission of orders in power stations || Lautsprecher m. zur Übermittlung von Anordnungen in Kraftwerken || haut-parleur m. pour la transmission d'ordres dans les usines d'énergie.

loudspeaker, directional effect of ~s || Richtungseffekt m. bei Lautsprechern || effet m. directif en haut-parleurs. / ~ horn || Lautsprechertrichter m. || pavillon m. de haut-parleur; haut-parleur-pavillon m. / large area diaphragm for ~s || Großflächenmembrane f. für Lautsprecher || diaphragme m. à grande étendue pour haut-parleurs. / ~ reproduction || Lautsprecherwiedergabe f. || audition f. ou réception f. en haut-parleur. / ~ service || Lautsprecherbetrieb m. || service m. de haut-parleur. / ~ tube || Lautsprecherröhre f. || lampe f. de puissance pour hautparleur.

lounge || Ruhebett n.; Chaiselongue f. || lit m. de repos; chaise f. longue.

lounging chair || Lehnstuhl m.; Liegestuhl m.; Schaukelstuhl m. || chaise f. à dos; chaise f. longue; berceuse f.

louver (Auto) || Luftschlitz m. (in der Motorhaube) || persienne f. ou lumière f. (de capot). / ~ (Wind mill) || Jalousie f. || persienne f. / ~ roof || Laternendach n. || chapiteau m. de lanterne. / ~ turret || Dachreiter m. || lanterneau m.; tour f. à cheval. / ~ window || Schallfenster n. || baie f. de clocher; ouïe f.

lovage oil || Liebstöckelöl n. || essence f. de livèche. / ~ root || Liebstöckelwurzel f. || racine f. de livèche.

low || niedrig; tief || bas; peu élevé.

low (Meteor) || Niederdruckgebiet n. || zone f. de basse pression.

low aisle (Arch) || Nebenschiff n.; Seitenschiff n. || nef f. latérale ou base; petite nef f.; collatéral m.; bas-côté m.; contre-allée f.

low-alloy steel || niedrig legierter Stahl m. || acier m. à faible alliage.

low-boiling || niedrigsiedend || à bas point d'ébullition.

low-carbon pig || kohlenstoffarmes Roheisen n. || fonte f. brute à basse teneur de carbone.

low-compression engine || niedrigverdichtender Motor m.; Niederdruckmotor m. (Busch) || moteur m. à basse compression.

low construction || niedriger Bau m. || construction f. basse. / ~ content || niedriger Gehalt m. || faible teneur f. / ~ current protector || Feinspannungsschutz m. || coupe-circuit m. pour courant faible.

löweite || Löweit m. || löwéite f.

lower, to ~ (Build) || niedriger machen || abaisser. / ~ (Mach) || reduzieren || réduire. / ~ (Mine) || hinablassen; hängen || abaisser. / the cupola can be raised and lowered a little || die Kuppel ist um ein geringes Maß heb- und senkbar || dans une mesure restreinte la calotte peut être montée et descendue. / ~ down || herunterfieren || amener. / ~ down a sail || ein Segel n. streichen || amener une voile. / ~ the hatch || das Schütz niederlassen || baisser la pale. / ~ in price || verbilligen || baisser ou reduire le prix. / ~ a road || eine Straße f. tiefer legen || abaisser une route. / ~ the voltage || die Spannung erniedrigen || réduire ou diminuer la tension.

lower || niedriger; tiefer || inférieur; plus bas. / to be ~ (Surv) || niedriger gelegen sein || être en contre-bas d'un point. / next lowest || nächstniedrig || immédiatement inférieur.

lower and bending beam || Unterwange f. und Biegewange f. || tablier m. et coulisse f. inférieure.

lower bed of a stone (Mas) || Unterlager n. eines Steines || panneau m. de lit. / ~ boom (Bridge) || Untergurt m. || semelle f. inférieure. / ~ bush of the axle-box || Unterteil m. der Schmierbüchse || dessous m. d'une boîte à graisse. / ~ case (Print) || Unterkasten m. || bas-de-casse m. / ~ casing || Gehäuseunterteil m. || partie f. inférieure de l'enveloppe. / ~ chord || Untergurt m. || bride f. inférieure. / ~ cross girder || unterer Querträger m. || traverse f. inférieure.

lower deck || Zwischendeck n. || entrepont m.; faux-pont m. / ~ beam || Zwischendeckbalken m. || bau m. du faux-pont.

lower edge of a rail || Unterkante f. einer Schiene || dessous m. du rail.

lower flange (Bridge) || Untergurt m. || membrure f. inférieure. / ~ (Mach) || Unterflansch m. || semelle f. ou bride f. inférieure. / ~ of the rail || Fuß m. der Schiene; Schienenfuß m. || semelle f. ou patin m. du rail.

lower harmonic voltage (Electr) || Unterspannung f. || tension f. de l'harmonique fondamental. / ~-side (Electr) || unterspannungsseitig || côté m. de la basse tension.

lower part || unterer Teil m. || partie f. inférieure. / ~ port (Shipb) || Unterpforte f. || mantelet m. inférieur. / ~ roll || Unterwalze f. || cylindre m. inférieur. / ~ shaft of a blast-furnace || Unterschacht m. eines Hochofens || grand foyer m. d'un haut-fourneau. / ~ side of a dam || Binnenböschung f. || talus m. intérieur. / ~ spring || Unterfeder f. || ressort m. inférieur. / ~ story || Untergeschoß n. || rez-de-chaussée m.; bas étage m.; basse-œuvre f. / ~ transom (Carp) || Unterriegel m. (einer Fachwand) || premier épart m. / ~ window || Dachluke f. || lucarne f.; œillet m.; œillette f. / ~ wing (Aero) || Unterflügel m. || plan m. inférieur.

lowering || Senkung f. || abaissement m. / ~ of a constant (Phys) || Sinken n. einer Größe || abaissement m. d'une constante. / ~ of the piezometric level || Senkung f. des Grundwasserspiegels ||

abaissement m. du plan de l'eau souterraine. / ~ of price || Preisermäßigung f. || dégrèvement m.; rabais m.; baisse f.; réduction f. de prix. / ~ of the steam pressure || Sinken n. der Dampfspannung || réduction f. ou abaissement m. de la pression de la vapeur. / ~ of the voice || Senken n. der Stimme || abaissement m. de la voix.

lowering device || Niederlaßvorrichtung f.; Senkvorrichtung f. || dispositif m. d'abaissement ou de descente. / funeral ~ || Leichenversenkvorrichtung f. || dispositif m. pour descendre des cadavres.

lowering direction || Senkrichtung f. || sens m. de la descente. / ~ speed || Senkgeschwindigkeit f. || vitesse f. de descente. / ~ stage || Senkbühne f. || chargeur m. descendant.

lowest amount || Mindestbetrag m. || (montant m.) minimum m. / ~ temperature || tiefste Temperatur f. || température f. minimum; température f. extrême-minimum m. / ~ temperature || tiefste Temperatur f. || température f. extrêment basse. / ~ tender || Mindestgebot n. || offre f. la plus avantageuse. / ~ tenderer || Mindestfordernder m. || le moins offrant ou demandant.

low forest || Niederwald m. || taillis m. simple.

low-frequency || Niederfrequenz f. || basse fréquence f. / ~ amplifier || Niederfrequenzverstärker m. || amplificateur m. à basse fréquence. / ~ inductance || Induktanzspule f. niedriger Frequenz || bobine f. d'inductance du circuit à basse fréquence. / ~ strengthener || Niederfrequenzverstärker m. || amplificateur m. à basse fréquence. / ~ strengthening || Niederfrequenzverstärkung f. || amplification f. basse fréquence. / ~ transformer || Niederfrequenztransformator m. || transformateur m. à basse fréquence.

low-grade || geringhaltig || à faible teneur f. / ~ flour || Futtermehl n. || remoulage m.

low ground || Niederung f. || terrain m. bas.

low hearth, refining || Herdfrischen n. || affinage m. au bas-foyer.

löwigite || Löwigit m. || löwigite f.

lowland || Tiefland n. || bas pays m.; plaine f.

low pass filter (Radio) || Stromreiniger m.; Spulenleiter m. || filtre m. passe-bas. / ~ (Tel) || Drosselkette f. || chaîne f. de réactances.

low position of centre of gravity || Tieflage f. des Schwerpunktes || position f. en profondeur du centre de gravité.

low-potential source of supply || Niederspannungskraftquelle f. || source f. de force à basse tension.

low-power engine || Kleinkraftmaschine f. || machine f. de faible puissance. / ~ generator || Kleindynamomaschine f. || dynamo f. ou génératrice f. de faible puissance. / ~ motor || Kleinmotor m. || moteur m. de faible puissance.

low-pressure || Niederdruck m. || basse pression f. / ~ area || Tiefdruckgebiet n. || régime m. de dépression ou de basses pressions. / ~ boiler || Niederdruckkessel m. || chaudière f. à basse pression. / ~ heating || Niederdruckheizung f. || chauffage m. à basse pression. / ~ oil pump with electric drive || Niederdruckölpumpe f. mit elektrischem Antrieb ||

pompe f. à huile à basse pression à commande électrique.

low-pressure steam ‖ Niederdruckdampf m. ‖ vapeur f. à basse pression. / ~ boiler ‖ Niederdruckdampfkessel m. ‖ chaudière f. à vapeur à basse pression. / ~ engine ‖ Niederdruckdampfmaschine f. ‖ machine f. à vapeur à basse pression.

low-relief ‖ Flachrelief n. ‖ bas-relief m.

low resistance ‖ niedriger *oder* kleiner Widerstand m. ‖ basse *ou* faible résistance f.

lowry *see* lorry. / ~ tip wagon ‖ Kastenkippwagen m. ‖ basculeur m. à caisse.

low shoe ‖ Halbschuh m. ‖ soulier m. découvert.

low side *see* low aisle. / ~ speed arm ‖ Hebel m. der kleinen Geschwindigkeit ‖ levier m. de petite vitesse. / ~ stand ‖ Bodengestell n. ‖ chantier m.

low-temperature carbonizing furnace ‖ Schwelofen m. ‖ four m. à distillation lente. / ~ cooling plant ‖ Tiefkühlanlage f. ‖ installation f. frigorifique à basse température.

low-tension ‖ Niederspannung f. ‖ basse tension f. / ~ coil ‖ Niederspannungsspule f. ‖ bobine f. à basse tension. / ~ dynamo ‖ Niedrigspannungsdynamo f. ‖ dynamo f. à basse tension. / ~ generator ‖ Niederspannungserzeuger m. ‖ génératrice f. à basse tension. / ~ plant ‖ Niederspannungsanlage f. ‖ installation f. à basse tension. / ~ winding ‖ Niederspannungswindung f. ‖ enroulement m. à basse tension.

low-voltage switch ‖ Niederspannungsschalter m. ‖ interrupteur m. à basse tension.

low-warp tapestry ‖ tiefschäftige Tapete f. ‖ tapisserie f. de basse-lisse *ou* à basse lisse.

low-water ‖ Niederwasser n.; niedriger Wasserstand m. ‖ basse eau f.; étiage m. / ~ alarm (Boil) ‖ Speiserufer m.; Niedrigwasserstandrufer m. ‖ avertisseur m. d'alimentation à sifflet d'alarme. / ~ mark (Hydr arch) ‖ niedrigster Wasserstand m. ‖ étiage m.

low wine ‖ Blanquet m. ‖ blanquette f. / ~ wing monoplane ‖ Tiefdecker m. ‖ monoplan m. surbaissé *ou* à plans bas; avion m. à ailes surbaissées.

loxoclase ‖ Loxoklas m. ‖ loxoclase f.

loxodromics pl. (Nav) ‖ Kunst f. des Dwarskurses ‖ loxodromie f.

loxodromic spiral ‖ Dwarslinie f.; Loxodrome f.; Rhumblinie f. ‖ loxodromie f.

lozenge (Geom) ‖ Raute f.; Rhombus m. ‖ losange m.; rhombe m. / ~ (Med) ‖ Tablette f.; Pastille f. ‖ tablette f. comprimée; pastille f. / ~ cutter ‖ Karamellenschneider m. ‖ découpeur m. de caramels.

lozenged ‖ rautenförmig ‖ losangé. / ~ glass ‖ gerautetes Glas n. ‖ verre m. losangé.

lozenge maker ‖ Pastillenbereiter m. ‖ pastilleur m. / ~ moulding (Arch) ‖ Rautenstab m. ‖ moulure f. losangée.

lubricant ‖ schlüpfrig ‖ glissant; lubrifiant.

lubricant ‖ Schmiermittel n. ‖ matière f. lubrifiante; lubrifiant m.; graisse f. / to drain out the ~ ‖ das Schmieröl n. ablassen ‖ laisser couler l'huile f. / wire rope ~ ‖ Drahtseilschmiere f. ‖ graisse f. pour câbles métalliques.

lubricant distributer ‖ Ölverteiler m. ‖ distributeur m. de lubrifiant *ou* de graissage.

lubricate, to ‖ schmieren; ölen ‖ lubrifier; huiler.

lubricated water ‖ geschmiertes Wasser n. ‖ eau f. (rendue) lubrifiante.

lubricating, self-~ ‖ selbstschmierend ‖ à graissage automatique.

lubricating arrangement ‖ Schmiervorrichtung f. ‖ dispositif m. de graissage. / ~ axle-box extending from one axle journal to the other ‖ durchgehende Schmierbüchse f. ‖ graisseur m. enveloppant l'essieu tout entier. / ~ bearing ‖ Schmierlager n. ‖ palier-graisseur m. / ~ box ‖ Schmierlager n. ‖ boîte f. à graisse. / ~ box extending from one-axle journal to the other ‖ durchgehendes Schmierlager n. ‖ boîte f. à graisse enveloppant l'essieu tout entier. / ~ chain ‖ Schmierkette f. ‖ chaîne f. de graissage. / ~ device ‖ Schmiervorrichtung f. ‖ dispositif m. de graissage. / ~ gear ‖ Schmiervorrichtung f. ‖ appareil m. de graissage. / ~ grease ‖ Maschinenschmiere f. ‖ graisse f. à machines. / ~ liquid ‖ Schmierflüssigkeit f. ‖ lubrifiant m. liquide; liquide m. lubrifiant. / ~ oil ‖ Schmieröl n. ‖ huile f. lubrifiante *ou* de graissage. / ~ oil pump ‖ Schmierölpumpe f. ‖ pompe f. à huile. / ~ pump ‖ Schmierpumpe f. ‖ pompe f. de graissage. / ~ ring ‖ Schmierring m. ‖ bague f. de graissage. / ~ wick ‖ Schmierdocht m. ‖ mèche f. de graissage.

lubrication ‖ Ölung f.; Schmierung f. ‖ lubrification f.; graissage m. / central grease ~ ‖ zentrale Fettschmierung f. ‖ lubrification f. centrale à la graisse. / centralized ~ ‖ Zentralschmierung f. ‖ graissage m. centralisé. / centrifugal ~ ‖ Zentrifugalschmierung f. ‖ graissage m. centrifuge. / circulation system ~ ‖ Umlaufschmierung f. ‖ graissage m. par circulation. / continuous ~ ‖ Dauerschmierung f. ‖ lubrification f. continuelle; graissage m. continu. / ~ of crankshaft bearing ‖ Kurbellagerschmierung f. ‖ graissage m. du palier de vilebrequin. / drip-oil ~ ‖ Tropfölschmierung f. ‖ graissage m. à l'huile à compte-gouttes. / economical ~ ‖ sparsame Schmierung f. ‖ graissage m. économique. / fat ~ ‖ Fettschmierung f. ‖ lubrification f. à la graisse. / forced ~ ‖ Preßschmierung f.; Druckschmierung f. ‖ graissage m. sous pression. / fresh oil ~ ‖ Schmierung f. mit Frischöl ‖ graissage m. par huile fraîche. / graphite ~ ‖ Graphitschmierung f. ‖ lubrification f. *ou* graissage m. au graphite; graphitage m. / grease ~ ‖ Fettschmierung f. ‖ lubrification f. à la graisse. / hand ~ ‖ Handschmierung f. ‖ graissage m. à main. / internal ~ ‖ innere Schmierung f. ‖ graissage m. intérieur; lubrification f. intérieure. / oil ~ ‖ Ölschmierung f. ‖ graissage f. *ou* lubrification f. à l'huile; huilage m. / over ~ ‖ zu starke Ölung f. ‖ excès m. de graissage. / piping ~ ‖ Schmierleitung f. ‖ conduite f. de graissage. / piston ~ ‖ Kolbenschmierung f. ‖ graissage m. du piston. / ~ under pressure *see* forced ~. / pump type circulation ~ system ‖ Druckumlaufschmierung f. ‖ graissage m. par circulation forcée. / ring ~ ‖ Ringschmiervorrichtung f. ‖ graisseur m. à bague.

/ splash ~ ‖ Tauchbadschmierung f. ‖ graissage m. par barbotage.

lubrication cock ‖ Schmierhahn m. ‖ robinet m. graisseur. / ~ groove ‖ Schmiernute f. ‖ patte f. d'araignée. / ~ oil pump cam ‖ Schmierpumpennocken m. ‖ came f. de pompe à huile. / ~ ring ‖ Ölring m.; Schmierring m. ‖ bague f. de graissage.

lubricator ‖ Schmiervorrichtung f.; Schmierapparat m.; Öler m. ‖ dispositif m. de graissage; graisseur m.; lubrificateur m. / ~ (Railw) ‖ Schmierkapsel f. ‖ godet m. graisseur. / automatic ~ ‖ selbsttätige Schmiervorrichtung f.; Selbstöler m. ‖ graisseur m. automatique. / graphite ~ ‖ Graphitschmierapparat m. ‖ graisseur m. au graphite. / gravity feed ~ ‖ Tropfölapparat m. ‖ graisseur m. à comptegouttes. / ~ actuated by hand ‖ Handschmiervorrichtung f. ‖ appareil m. de graissage à la main. / motor car ~ ‖ Kraftwagenöler m. ‖ graisseur m. d'automobiles. / needle ~ ‖ Nadelschmiergefäß n. ‖ graisseur m. à aiguille. / ring ~ ‖ Ringschmiervorrichtung f. ‖ graisseur m. à bague. / rotating crank ~ ‖ umlaufendes Schmiergefäß n. ‖ graisseur m. rotatif. / selfacting ~ ‖ Selbstöler m. ‖ lubrificateur m. *ou* graisseur m. mécanique. / separate ~ ‖ Einzelöler m. ‖ graisseur m. séparé. / single ~ ‖ Einzelöler m. ‖ graisseur m. séparé. / ~ Stauffer ~ ‖ Staufferbüchse f. ‖ graisseur m. Stauffer. / steam-cylinder ~ ‖ Dampfzylinderschmiervorrichtung f. ‖ appareil m. pour le graissage des cylindres. / ~ for stophead ‖ Deckelschmiergefäß n. ‖ graisseur m. pour couvercle supérieur. / wick ~ ‖ Dochtschmierbüchse f. ‖ graisseur m. à mèche.

lubricator box ‖ Schmierbüchse f. ‖ boîte f. à graisse. / ~ cock ‖ Schmierhahn m. ‖ robinet m. de graissage; robinet-graisseur m. / ~ wheel ‖ Antriebsrad n. des Ölers ‖ roue f. de commande du graisseur.

lucifer match ‖ Schwefelhölzchen n. ‖ allumette f.

lucrative ‖ rentabel ‖ lucratif; d'un bon rapport m.

lucrativeness ‖ Rentabilität f. ‖ rendement m.

luff, to ~ (Mar) ‖ anluven ‖ loffer.

luff (Mar) ‖ Luvseite f. ‖ côté m. du vent. / a sharp ~ ‖ scharfes Anluven n. ‖ auloffée f.

luffing crane ‖ Wippkran m. ‖ grue f. à portée variable. / floating level ~ ‖ Schwimmwippkran m. ‖ grue f. flottante à volée variable.

luffing drum ‖ Einziehtrommel f. ‖ tambour m. à enrouler. / ~ gear (Hoisting engine) ‖ Einziehwerk n. ‖ dispositif m. de relevage. / level ~ jib ‖ Wippausleger m. ‖ flèche f. articulée.

lug ‖ Henkel m.; Öhr n.; Griff m. ‖ anse f.; oreille f. / ~ (Acc) ‖ Fahne f. ‖ nez m.; queue f. / ~ (Mach) ‖ Ansatz m. ‖ taquet m. / ~ (Measure of length) ‖ Rute f. (Maßstab) ‖ perche f. / boiler ~ ‖ Kesselpratze f. ‖ oreille f. de chaudière. / ~ of a collar ‖ Nase f. eines Ringes ‖ tenon m. d'une bague. / current carrying ~ (Acc) ‖ stromführende Fahne f. ‖ queue f. conductrice. / ~ of a ring *see* ~ of a collar. / ~ for supporting grate-bar bearers ‖ Rostbalkenträger m. ‖ galoche f. supportant le sommier de grille.

luggage ‖ Gepäck n. ‖ bagage m. / to book the ~ ‖ Gepäck n. aufgeben ‖ faire enregistrer les bagages mpl. / left ~ ‖ herrenloses Gepäck n. ‖ bagage m. abandonné. / missed ~ ‖ verschlepptes Gepäck n. ‖ bagage m. égaré. / to register the ~ see to book the ~. / small ~ ‖ Handgepäck n. ‖ bagages mpl. à main.

luggage car ‖ Gepäckwagen m. ‖ wagon-fourgon m. / ~ carrier ‖ Gepäckhalter m. ‖ porte-bagages m. / ~ compartment ‖ Gepäckraum m. ‖ compartiment m. à bagages. / ~ delivery office ‖ Gepäck-ausgabe f. ‖ bureau m. de délivrance des bagages. / ~ engine ‖ Güterzugmaschine f. ‖ locomotive f. à marchandises. / ~ examination ‖ zollamtliche Gepäck-revision f. ‖ vérification f. des bagages à la douane. / ~ freight ‖ Gepäckfracht f. ‖ frais mpl. de transport des bagages. / ~ grid ‖ Gepäckrast f.; Gepäckhalter m. ‖ porte-bagage m. / ~ grid at back ‖ Gepäckhalter m. am Wagenende ‖ porte-bagages m. arrière de l'automobile. / ~ insurance ‖ Gepäckversicherung f. ‖ assurance f. des bagages. / ~ lift ‖ Gepäck-aufzug m. / ~ monte-bagage m. / ~ office ‖ Gepäckannahme f.; Gepäckaufgabe f. ‖ guichet m. ou dépôt m. des bagages; enregistrement m. des bagages. / ~ rail on top of the cart ‖ Gepäckgalerie f. auf dem Wagendach ‖ galerie f. à bagages au toit de voiture. / ~ room (Ship) ‖ Gepäck-raum m. ‖ cale f. à bagages. / ~ ticket ‖ Gepäckschein m. ‖ bulletin m. de bagages. / ~ traffic ‖ Gepäckverkehr m. ‖ service m. des bagages. / ~ truck ‖ Gepäckkarren m. ‖ cabrouet m. ou diable m. à bagages. / ~ van ‖ Gepäck-wagen m.; Packwagen m. ‖ fourgon m.; wagon m. à bagages. / ~ wagon see ~ van.

lugger (Mar) ‖ Logger m. ‖ lougre m.

lug strap (Weav) ‖ Schlagriemen m. ‖ bride f. de chasse.

lukewarm ‖ lau; lauwarm ‖ tiède.

lull, to ~ ‖ sich beruhigen (lassen) ‖ (laisser) se calmer. / ~ (Wind) ‖ sich beruhigen; sich legen ‖ accalmir; tomber.

lull ‖ Windstille f. ‖ accalmie f.

lumachella marble ‖ Muschelmarmor m.; Lumachell m. ‖ lumachelle f.; marbre m. lumachelle ou coquillier.

lumber ‖ Bauholz n.; Nutzholz n. ‖ bois m. d'œuvre ou de construction ou de charpente. / ~ board (Weav) ‖ Harnisch-brett n.; Löcherbrett n.; Schnürbrett n. ‖ planche f. d'arcades. / ~ caliper ‖ Holz-meßgerät n. ‖ compas m. forestier. / ~ wagon ‖ Leiterwagen m. ‖ chariot m. à ridelles.

lumen ‖ Lumen n. ‖ lumen m. / ~ hour ‖ Lumenstunde f. ‖ lumen-heure m.

luminescence ‖ Lumineszenz f. ‖ lumi-nescence f.

luminescent substance ‖ lumineszierende Substanz f. ‖ substance f. luminescente.

luminosity ‖ Helligkeit f. ‖ luminosité f. / degree of ~ ‖ Helligkeitsstufe f. ‖ degré m. de clarté ou de luminosité. / varying of the ~ ‖ Helligkeitsänderung f. ‖ changement m. de la luminosité.

luminous advertising ‖ Lichtreklame f. ‖ réclame f. lumineuse. / ~ advertising board ‖ Leuchtschild n. ‖ enseigne f. lumineuse. / ~ buoy ‖ Leuchtboje f. ‖ bouée f. lumineuse. / ~ circuit diagram ‖ Leuchtschaltbild n. ‖ diagramme m. lumineux. / ~ colour ‖ Leuchtfarbe f. ‖

peinture f. lumineuse. / ~ compass ‖ Leuchtkompaß m. ‖ compas m. lumi-neux. / ~ current ‖ Glimmstrom m. ‖ courant m. de luminescence; effluves mpl. / ~ current discharge ‖ Glimm-stromentladung f. ‖ décharge f. d'efflu-ves. / ~ discharge current ‖ Glimmerent-ladungsstrom m. ‖ courant m. de décharge luminescente. / ~ flame ‖ leuchtende Flamme f. ‖ flamme f. éclairante. / ~ flower ‖ Leuchtblume f. ‖ fleur f. lumineuse. / ~ fountain ‖ be-leuchteter Springbrunnen m. ‖ fontaine f. lumineuse. / ~ indicator panel with remote control and telemetering devices ‖ Leuchtschaltbild n. mit Fernsteuerung und Fernmessung ‖ diagramme m. lu-mineux avec télécommande et mesure à distance.

luminous intensity, hemispherical ~ ‖ hemisphärische Lichtstärke f. ‖ inten-sité f. lumineuse hémisphérique. / spherical ~ ‖ sphärische Lichtstärke f. ‖ intensité f. lumineuse sphérique.

luminous phenomenon ‖ Leuchterschei-nung f. ‖ phénomène m. lumineux. / ~ point ‖ Leuchtpunkt m. ‖ point m. lumineux. / ~ spot ring condenser ‖ Leuchtbildkondensor m. ‖ condensateur m. à images lumineuses. / ~ stove ‖ Leuchtofen m. ‖ radiateur m. à feu visible.

lump (Glassm) ‖ Kölbchen n.; Ballen m.; Posten m. ‖ paraison f. / ~ (Metal) ‖ Klumpen m.; Deul m.; Luppe f. ‖ loupe f.; masse f. / ~ of fined steel (Metal) ‖ Schrei m.; Deul m. ‖ loupe f.; balle f.; masset m.; massé m. d'acier. / ~ of ice ‖ Eisklumpen m. ‖ glaçon m. / in ~s pl. ‖ stückig ‖ gailleteux; en morceaux. / ~ of lead ‖ Bleibrocken mpl. ‖ masse f. de plomb.

lump coal ‖ Würfelkohle f.; Füllkohle f. ‖ gailletin m.; grélat m.; gailettes fpl. / ~ lac ‖ Kuchenlack m. ‖ laque f. en plaques ou en gâteaux ou en masses. / ~-loaded open circuit ‖ Pupinfrei-leitung f. ‖ ligne f. aérienne pupinisée. / ~ maker ‖ Zigarrenwickelmacher m. ‖ poupière f. / ~ price ‖ Pauschalpreis m. ‖ prix m. global.

lump sugar ‖ Kochzucker m. ‖ lumps m.

lumpy semi-coke ‖ stückiger Halbkoks m. ‖ semi-coke m. en morceaux.

lunar caustic ‖ Höllenstein m. ‖ pierre f. infernale. / ~ crater ‖ Mondkrater m. ‖ cratère m. de la lune. / ~ distance ‖ Mondentfernung f. ‖ distance f. lunaire.

lunation ‖ Mondumlauf m. ‖ lunaison f.

lung ‖ Lunge f. ‖ poumon m.

lung protector ‖ Lungenschützer m. ‖ protège-poumons m.

lungwort herb ‖ Lungenkraut n. ‖ herbe f. de pulmonaire.

lupine ‖ Lupine f. ‖ lupin m. / ~ unbitter-ing apparatus ‖ Lupinenentbitterungs-apparat m. ‖ appareil m. à enlever l'amertume ou le principe amer du lupin.

lupulin ‖ Hopfenmehl n. ‖ lupuline f. ou glandules fpl. de houblon. / ~ gland of hop ‖ Hopfendrüse f. ‖ glande f. de houblon.

lurch, to ~ (Nav) ‖ ausscheren ‖ embarder.

lure ‖ Köder m. ‖ résure f.; appât m.; amorce f.

luring torch ‖ Lockfackel f. ‖ clairon m.

lustre ‖ Kronleuchter m. ‖ lustre m. / ~ (Miner) ‖ Glanz m. ‖ éclat m. / ~s pl. ‖ Leuchtergehänge n. ‖ pendeloque f. /

changeable ~ ‖ Schillerglanz m. ‖ éclat m. chatoyant ou changeant. / metallic ~ ‖ Metallglanz m. ‖ éclat m. ou brillant m. métallique. / natural ~ of the fur ‖ natürlicher Glanz m. des Pelzes ‖ lustre m. naturel de la fourrure. / resinous ~ ‖ Pechglanz m.; Harzglanz m.; Wachs-glanz m. ‖ éclat m. résineux ou cireux. / silky ~ ‖ Seidenglanz m. ‖ lustre m. ou éclat m. soyeux. / unctuous ~ ‖ fettiger Glanz m. ‖ éclat m. onctueux. / varying ~ see changeable ~.

lustre shrinking machine ‖ Preßglanz-dekatiermaschine f. ‖ machine f. à décatir et polir. / ~ with fixed hot press cylinders and corresponding wrapper cloth ‖ Preßglanzdekatiermaschine f. mit festgelagerten Dekatierzylindern und korrespondierendem Mitläufer ‖ machine f. à décatir et à polir à cylindres dé-catisseurs et à doublier correspondant.

lustring ‖ Lüstrine f. ‖ lustrine f. / fur ~ ‖ Lustrieren n. von Pelzen ‖ lustrage m. de fourrures.

lustrous ‖ glänzend ‖ lustré. / ~ coal ‖ Glanzkohle f. ‖ charbon m. luisant.

lutation ‖ Lutieren n.; Verkitten n. ‖ luta-tion f.; mastication f.

lute, to ‖ kitten; zusammenkitten; ver-kitten; lutieren ‖ luter; cimenter; mastiquer.

lute ‖ Kitt m.; Lutum n. ‖ ciment m.; lut m.; mastic m. / ~ (Mus) ‖ Laute f. ‖ luth m. / ~ ribbon ‖ Lautenband n. ‖ ruban m. de luths.

luting agent ‖ Kittstoff m.; Dichtungs-stoff m. ‖ mastic m.; ciment m.; étan-chéifiant m.

lux ‖ Lux n. ‖ lux m.

luxmeter ‖ Luxmesser m.; Beleuchtungs-messer m. ‖ luxmètre m.

luxury requirements pl. ‖ Luxusbedürfnis n. ‖ besoins mpl. de luxe.

lycopod ‖ Bärlapp m. ‖ lycopode m.

lycopodium powder ‖ Bärlappsamen mpl. ‖ soufre m. végétal.

Lydian stone ‖ Lydit m.; Kieselschiefer m. ‖ lydite f.; lydienne f.; pierre f. de Lydie.

lye (Chem) ‖ Lauge f. ‖ lessive f. / caustic ~ ‖ Seifensiederlauge f. ‖ lessive f. caustique. / caustic ~ of soda ‖ Ätznatronlauge f. ‖ lessive f. caustique de soude. / clear ~ ‖ klare Lauge f. ‖ lessive f. claire. / type-washing ~ ‖ Typenwaschlauge f. ‖ les-sive f. à laver les types.

lye boiler ‖ Laugekessel m. ‖ chaudière f. à lessive. / ~ cooler ‖ Laugenkühler m. ‖ réfrigérant m. à lessives. / ~ dissolving apparatus ‖ Laugeauflöseapparat m. ‖ appareil m. à dissoudre la lessive. / ~ dissolving tank ‖ Laugenauflöse-behälter m. ‖ réservoir m. pour dissoudre la soude caustique ou les lessives. / ~ graduating tank ‖ Laugenbereitungs-behälter m. ‖ réservoir m. à lessives. / ~ lever ‖ Laugenheber m. ‖ siphon m. de lessive. / ~ liquor ‖ Laugeflüssigkeit f. ‖ lessive f. liquide; liquide m. de lessive. / ~ fittings pl. for ~ pipe lines ‖ Laugen-leitungsarmatur f. ‖ accessoire m. pour conduites de lessive. / ~ pump ‖ Laugen-pumpe f. ‖ pompe f. à lessive.

lying-in hospital ‖ Wöchnerinnenheim n. ‖ maison f. d'accouchement; maternité f.

lying-window ‖ Querfenster n. ‖ fenêtre f. gisante.

lymph ‖ Lymphe f. ‖ lymphe f.; vaccin m.

lysol ‖ Lysol n. ‖ lysol m.

M

M-quadrat (Print) ‖ Quadrätchen n.; Geviert n.; M-Quadrat n. ‖ cadratin m. / / ~ teeth pl. (Of a saw) ‖ Stockzähne mpl. ‖ dents mpl. à M.

macadamize, to ‖ beschottern ‖ empierrer; macadamiser.

macadamized road ‖ Schotterstraße f. ‖ chaussée f. en empierrement; chaussée f. macadamisée.

macadamizing see macadam paving.

macadam paving ‖ Kiesbeschotterung f.; Makadamisierung f.; Kieselschlag m. ‖ macadam m.

macaronies pl. ‖ Makkaroni pl. ‖ macaronis mpl.

mace ‖ Muskatblüte f. ‖ macis m. / ~ butter ‖ Muskatbutter f. ‖ beurre m. de muscade.

macerate, to (Brew) ‖ einweichen; quellen ‖ tremper; mettre en trempe f.; mouiller; macérer. / ~ (Chem) ‖ einweichen; mazerieren ‖ macérer.

machinable ‖ bearbeitbar ‖ machinable; usinable. / soft ~ material ‖ bearbeitbarer weicher Werkstoff m. ‖ métal m. doux et susceptible d'être façonné ou usinable.

machinal ‖ maschinenmäßig ‖ mécanique; machinal.

machine, to ‖ bearbeiten ‖ travailler; usiner. / ~ (Milling cutter) ‖ fräsen ‖ fraiser. / ~ the edges ‖ die Kanten fpl. bearbeiten oder bestoßen ‖ dresser les arêtes fpl. / easily to be machined ‖ leicht zu bearbeiten ‖ facile à usiner. / the material may be easily machined after quenching ‖ der Werkstoff läßt sich nach dem Abschrecken noch ganz leicht bearbeiten ‖ l'acier m. se laisse encore facilement usiner après la trempe. / machined from the solid ‖ aus dem Vollen n. herausgearbeitet ‖ façonné dans du métal massif ou dans la masse.

machine ‖ Maschine f. ‖ machine f. / ~ for adjusting and keeping tools in good repair ‖ Maschine f. zur Erhaltung und Einstellung von Werkzeugen ‖ machine f. pour l'entretien et l'ajustage des outils. / agricultural ~ ‖ landwirtschaftliche Maschine f. ‖ machine f. agricole ou pour l'agriculture. / automatic ~ ‖ selbsttätige oder automatische Maschine f.; Automat m. ‖ machine f. automatique. / ~ for coiling helical springs ‖ Maschine f. zum selbsttätigen Wickeln von Schraubenfedern ‖ machine f. automatique à enrouler les ressorts à boudin. / auxiliary ~ (Electr) ‖ Zusatzmaschine f. ‖ machine f. supplémentaire. / ~ for bakehouses ‖ Maschine f. für Bäckereien ‖ machine f. pour boulangeries. / ~ for milling and marking the divisions at balance beams ‖ Wagebalkenfräs- und -teilmaschine f. ‖ machine f. à fraiser et à graduer les fléaux de balances. / belted ~ ‖ mit Riemen angetriebene Maschine f. ‖ machine f. commandée par courroie. / ~ for bending ‖ Biegemaschine f. ‖ machine f. à cintrer.

/ ~ for bending wood / Holzbiegemaschine f. ‖ machine f. à courber le bois. / bipolar ~ (Electr) ‖ zweipolige Maschine f. ‖ machine f. bipolaire. / ~ for casting moulds ‖ Formmaschine f. ‖ machine f. à mouler. / ~ for cast iron bending tests ‖ Gußeisenbiegemaschine f.; Maschine f. für Biegeproben aus Guß ‖ machine f. à essayer la fonte à la flexion. / chemigraphical ~ ‖ chemigrafische Maschine f. ‖ machine f. pour l'industrie chemigraphique. / closed ~ ‖ gekapselte Maschine f. ‖ machine f. étant fermée ou blindée. / colonial ~ ‖ Kolonialmaschine f. ‖ machine f. coloniale. / ~ for confectioners ‖ Konditoreimaschine f. ‖ machine f. pour confiseries. / ~ for cutting-out (Tail) ‖ Zuschneidemaschine f. ‖ machine f. à découper. / ~ running day and night ‖ (bei) Tag und (bei) Nacht laufende Maschine f.; Maschine f. in Tag- und Nachtbetrieb ‖ machine f. en marche de nuit et de jour. / directly operated ~ ‖ unmittelbar angetriebene Maschine f. ‖ machine f. à commande directe. / dynamo-electric ~ ‖ dynamoelektrische Maschine f. ‖ machine f. dynamoélectrique. / electric ~ ‖ elektrische Maschine f. ‖ machine f. électrique. / ~ for hand ‖ Maschine f. für Handbetrieb ‖ machine f. pour service manuel. / ~ for horticulture ‖ Maschine f. für den Gartenbau ‖ machine f. pour l'horticulture. / hydraulic ~ ‖ Wasserkraftmaschine f.; hydraulische Maschine f. ‖ machine f. hydraulique. / ~ for laundries ‖ Plättereimaschine f. ‖ machine f. à repasser. / ~ for mills ‖ Müllereimaschine f. ‖ machine f. de meunerie. / milling ~ ‖ Fräsmaschine f. ‖ machine f. à fraiser; fraiseuse f. / moulding ~ ‖ Formmaschine f. ‖ machine f. à mouler. / open-type ~ ‖ offene Maschine f.; Maschine f. offener Bauart ‖ machine f. de type non blindé. / ~ for the paper industry ‖ Maschine f. für die Papierindustrie ‖ machine f. pour l'industrie du papier. / ~ for power drive ‖ Maschine f. für Kraftbetrieb ‖ machine f. pour service à ou à commande par moteur. / ~ smoothing with pumice-stone (Hatt) ‖ Bimsmaschine f. ‖ ponceuse f. / ~ for repairing screw couplings ‖ Maschine f. zur Ausbesserung von Schraubenkupplungen ‖ machine f. pour réparer des accouplements à vis. / ~ for rope breaking tests ‖ Seilzerreißmaschine f. ‖ machine f. à soumettre les câbles à l'essai de rupture par la traction. / special ~ ‖ Sondermaschine f.; Spezialmaschine f. ‖ machine f. spéciale. / ~ for splitting willow and cane ‖ Spaltmaschine f. zum Spalten von Weiden und Rohr ‖ machine f. à fendre l'osier et le jonc. / ~ for testing building material ‖ Baustoffprüfmaschine f. ‖ machine f. pour l'essai des matériaux de construction. / ~ for testing cutlery ‖ Messerprüfmaschine f. ‖ machine f. à essayer les couteaux. / ~ for

testing durability arranged for alternate loads ‖ Dauerprüfmaschine f. für Wechselbelastung ‖ machine f. à essayer par charges alternatives en service continu. / ~ for testing tensile strength ‖ Zerreißfestigkeitsprüfmaschine f. ‖ machine f. pour essais de traction. / washing ~ ‖ Waschmaschine f. ‖ machine f. à laver.

machine attendant ‖ Maschinenwärter m. ‖ mécanicien m.; machiniste m.; conducteur m. de machines. / ~ bronze ‖ Maschinenbronze f. ‖ bronze m. pour machines. / ~ casting ‖ Maschinenguß m. ‖ fonte f. pour machines. / ~ composition (Print) ‖ Maschinensatz m. ‖ composition f. à la machine. / ~ compositer (Print) ‖ Maschinensetzer m. ‖ opérateur m.; compositeur m. à la machine. / construction of ~s ‖ Maschinenbau m. ‖ construction f. de machines. / ~ cutting (Cloth) ‖ Maschinenschnitt m. ‖ coupage m. à la machine.

machined ‖ bearbeitet ‖ usinée. / reughly ~ (Met) ‖ vorgearbeitet ‖ ébauché. / ~ from the solid ‖ aus dem Vollen n. gearbeitet ‖ façonné dans la masse. / ~ edge ‖ bearbeitete Kante f. ‖ arête f. usinée. / ~ hexagon nut ‖ blanke Sechskantmutter f. ‖ écrou m. hexagonal décolleté. / ~ stud ‖ blanke Stiftschraube f. ‖ vis f. de sûreté à goupille décolletée. / ~ washer (Mach) ‖ blanke Unterlagsscheibe f. ‖ rondelle f. décolletée.

machine data pl. ‖ Maschinendaten npl.; charakteristische Werte mpl. von Maschinen ‖ données fpl. caractéristiques de machines. / ~ detail ‖ Maschinenelement n. ‖ organe m. de machine. / ~ driller (Mine) ‖ Maschinenhauer m. ‖ haveur m. à la machine. / ~ drive ‖ Maschinenantrieb m. ‖ entraînement m. ou commande f. par moteur. / ~ elements pl. which even at high temperatures have still to be very resisting ‖ Konstruktionsteile mpl., die bei hoher Temperatur noch große Festigkeit besitzen müssen ‖ pièces fpl. qui doivent offrir une grande résistance même à très haute température. / ~ embroidery ‖ Maschinenstickerei f. ‖ broderie f. à la mécanique. / ~ engineer ‖ Maschineningeniör m. ‖ ingénieur m. mécanicien. / ~ equipment ‖ Maschinenpark m. ‖ parc m. à machines. / ~ factory ‖ Maschinenfabrik f. ‖ atelier m. de construction de machines. / ~ folder ‖ Maschinenfalzer m. ‖ plieur m. à la machine. / ~ grease ‖ Maschinenfett n. ‖ graisse f. pour machines.

machine gun ‖ Maschinengewehr n. ‖ mitrailleuse f. / air-cooled ~ ‖ luftgekühltes Maschinengewehr n. ‖ mitrailleuse f. à refroidissement d'air. / ~ in floor ‖ Bodenmaschinengewehr n. ‖ mitrailleuse f. tirant à travers le plancher. / water-cooled ~ ‖ wassergekühltes Maschinengewehr n. ‖ mitrailleuse f. à refroidissement d'eau.

machine gun bullet ‖ Maschinengewehrkugel f. ‖ balle f. de mitrailleuse.

/ to ~ withstand a rain of ~ bullets ‖ widerstandsfähig gegen Maschinengewehrgarben fpl. sein ‖ résister aux gerbes fpl. de balles de mitrailleuse. / ~ drive ‖ Maschinengewehrantrieb m. ‖ commande f. de mitrailleuse. / ~ nest ‖ Maschinengewehrnest n. ‖ nid m. de mitrailleuses. / ~ slit ‖ Maschinengewehrschlitz m. ‖ fente f. de mitrailleuse. / ~ stand ‖ Maschinengewehrstand m. ‖ poste m. de mitrailleuse. / ~ support ‖ Maschinengewehrträger m.; Maschinengewehrbock m. ‖ support m. de la mitrailleuse. / ~ tank ‖ Maschinengewehrtank m.; weiblicher Kampfwagen m. *oder* Tank m. ‖ char m. mitrailleuse; tank m. femelle. / ~ telescopic sight ‖ Zielfernrohr n. für Maschinengewehre ‖ lunette f. viseur pour mitrailleuse.

machine house ‖ Maschinenhaus n. ‖ bâtiment m. *ou* salle f. des machines. / ~ floor ‖ Maschinenhausfußboden m. ‖ sol m. *ou* planchet m. de la salle des machines.

machine inspector ‖ Maschinenaufseher m. ‖ surveillant m. *ou* inspecteur m. des machines. / ~ knife ‖ Maschinenmesser n. ‖ lame f. de machine. / ~ lace ‖ maschinengeklöppelte Spitze f. ‖ dentelle f. à fuseau mécanique.

machine-made ‖ maschinengearbeitet; mit der Maschine gefertigt ‖ travaillé mécaniquement *ou* à la machine. / ~ brick ‖ Maschinenziegel m. ‖ brique f. moulée à la machine. / ~ nail ‖ Maschinennagel m.; geschnittener Nagel m. ‖ clou m. découpé à froid de la tôle de fer. / ~ paper ‖ Maschinenpapier n.; maschinengefertigtes Papier n. ‖ papier m. (à la) mécanique *ou* fait à la machine. / ~ tooth ‖ maschinenarbeiteter Zahn m. ‖ dent f. formée à la machine.

machine maker ‖ Maschinenbauer m.; Mechaniker m. ‖ mécanicien m.; constructeur m.; constructeur-mecanicien m. / ~ making ‖ Maschinenbau m. ‖ construction f. de machines. / ~ manufactory ‖ Maschinenfabrik f. ‖ atelier m. de construction mécanique. / ~ minder ‖ Maschinenmeister m. ‖ conducteur m. / ~ mixer (Bak) ‖ Maschinenkneter m. ‖ pétrisseur m. à la machine. / ~ mixing (Concrete etc.) ‖ Maschinenmischung f. ‖ mélange f. à la machine. / ~ model ‖ Maschinenmodell n. ‖ modèle m. de machine. / ~ moulder ‖ Maschinenformer m. ‖ mouleur m. à la machine. / ~ moulding ‖ Maschinenformerei f. ‖ moulage m. à la machine. / ~ oil ‖ Maschinenöl n. ‖ huile f. pour machines. / ~ operator ‖ Maschinenarbeiter m. ‖ desservant m. de la machine *ou* (ouvrier-machine m.) / ~ pistol ‖ Maschinenpistole f. ‖ pistolet m. mitrailleur.

machine-printed wallpaper ‖ Maschinendrucktapete f. ‖ papier-peint m. mécanique *ou* à la machine.

machine pulley ‖ feste Riemenscheibe f. ‖ poulie f. fixe. / ~ repair workshop ‖ Maschinenreparaturwerkstatt f. ‖ atelier m. de réparation de machines. / ~ riveter ‖ Maschinennieter m. ‖ riveur m. à la machine. / ~ riveting ‖ Maschinennietung f. ‖ rivetage m. mécanique. / ~ room *see* ~ house.

machinery ‖ Maschinerie f. ‖ machinerie f.; mécanique f. / ~ (Gear) ‖ Getriebe n.; Triebwerk n. ‖ commande f. /

(Machine equipment) ‖ Maschinenpark m. ‖ parc m. à machines. / ~ for constructing the permanent way on railway ‖ Maschinenpark m. für den Eisenbahnoberbau ‖ machines fpl. pour la superstructure des chemins de fer.

machinery hall ‖ Maschinenhaus n. ‖ hall m. *ou* bâtiment m. *ou* pavillon m. des machines. / ~ maker *see* machine maker. / ~ oil *see* machine oil. / part of ~ ‖ Maschinenbestandteil m. ‖ pièce f. détachée de machine.

machine screw ‖ Maschinenschraube f. ‖ vis f. de machine. / ~ setter (Mach tool) ‖ Maschineneinrichter m.; Maschinensteller m. ‖ régleur m. de machines. / ~ shop ‖ mechanische Werkstatt f. ‖ hall m. d'usinage; atelier m. mécanique. / ~ sign plate ‖ Maschinenschild n. ‖ plaque f. pour machines. / ~-spinning ‖ Maschinenspinnerei f. ‖ filature f. mécanique *ou* à la machine.

machine-spun ‖ maschinengesponnen ‖ filé à la machine. / ~ yarn ‖ Maschinengarn n. ‖ fil m. mécanique *ou* fait à la machine.

machine surveyor (Railw) ‖ Maschineninspektor m. ‖ inspecteur m. de la traction. / ~ tenter ‖ Maschinenwärter m. ‖ mécanicien m.; machiniste m.; conducteur m. de machines. / ~ tester ‖ Maschinenprüfer m. ‖ ouvrier m. aux essais de machines. / ~ thermometer ‖ Maschinenthermometer n. ‖ thermomètre m. pour machines. / ~ thermometer with two branches ‖ Maschinenthermometer n. mit Doppelstutzen ‖ thermomètre m. pour machines avec tube double.

machine tool ‖ Werkzeugmaschine f.; Bearbeitungsmaschine f. ‖ machine-outil f. / labour-saving ~ ‖ arbeitsparende Werkzeugmaschine f. ‖ machine-outil f. économisant la main d'œuvre *ou* de type d'un rendement plus économique. / ~ for working metal ‖ Werkzeugmaschine f. zur Metallbearbeitung ‖ machine-outil f. à travailler les métaux. / single-purpose ~ ‖ Eintypwerkzeugmaschine f. ‖ machine-outil f. (pour modèle) unique. / special ~ ‖ Sonderwerkzeugmaschine f.; Spezialwerkzeugmaschine f. ‖ machine-outil f. spéciale. / special ~ for shipbuilding ‖ Schiffbausonderwerkzeugmaschine f. ‖ machine-outil f. speciale pour la construction des bateaux. / ~ for the working of metal, wood, stone or leather ‖ Werkzeugmaschine f. zur Metall-, Holz-, Stein- oder Lederbearbeitung ‖ machine-outil f. pour le travail des métaux, du bois, de la pierre ou du cuir.

machine-tool construction ‖ Werkzeugmaschinenbau m. ‖ construction f. des machines-outils. / ~ manufactory ‖ Werkzeugmaschinenfabrik f. ‖ atelier m. de construction de machines-outils.

machine work ‖ Maschinenarbeit f. ‖ ouvrage m. à la machine; travail m. mécanique. / ~ (Gear) ‖ Getriebe n.; Triebwerk n. ‖ commande f.

machine worker ‖ Maschinenarbeiter m. ‖ ouvrier m. à la machine m.

machine works pl. ‖ Maschinenfabrik f.; Maschinenbauanstalt f. ‖ atelier m. de construction de machines.

machining ‖ Bearbeitung f. ‖ façonnage m.; usinage m.; opération f. d'usinage. /

~ can only be done by grinding ‖ eine Bearbeitung kann nur durch Schleifen erfolgen ‖ le façonnage ne peut se faire que par émeulage. / clean and flawless surface ready for ~ ‖ reine porenfreie Bearbeitungsfläche f. ‖ surface f. de façonnage propre et sans pores. / ~ from the rough ‖ Vorarbeiten n. ‖ ébauchage m. / ~ (of) tubes ‖ Rohrbearbeitung f. ‖ travail m. des tubes.

machining allowance ‖ Bearbeitungszugabe f. ‖ surépaisseur f. pour l'usinage.

machinist ‖ Maschinist m.; Maschinenwärter m. ‖ machiniste m.

mackerel ‖ Makrele f. ‖ maquereau m. / ~ boat ‖ Makrelenfischer m. ‖ maquilleur m. / ~ sky ‖ Lämmerwolken fpl. ‖ nuage m. moutonneux.

mackintosh ‖ Gummimantel m. ‖ mackintosh m.; imperméable m.; manteau m. de caoutchouc. / ~ industry ‖ Gummimantelindustrie f. ‖ industrie f. des manteaux en caoutchouc.

mackle, to (Print) ‖ schmitzen; verschmieren ‖ maculer.

mackling (Print) ‖ Duplieren n.; Schmitzen n.; Verschmieren n. ‖ papillotage m.

maculature (Print) ‖ Ausschuß m.; Mißdruck m. ‖ maculation f.; impression f. incorrecte; rebut m. / ~ (Waste paper) ‖ Ausschußpapier n.; Makulatur f.; Schmutzpapier n. ‖ maculature f.

madapollam ‖ Schirting m.; Madapolam m. ‖ cretonne f.; madapolam m.; shirting m.

madder (Dyer) ‖ Krapp m.; Färberröte f.; Krapprot n. ‖ garance f. / artificial ~ ‖ Alizarin n. ‖ alizarine m.; garance f. artificielle.

madder bloom ‖ Krappblüte f. ‖ fleurs fpl. de garance. / ~ carmine ‖ Krappkarmin m.; roter Krapplack ‖ carmin m. de garance; laque f. de garance rouge. / ~-drying man ‖ Krappdarrer m. ‖ sécheur m. de garance. / ~ extract ‖ Krappauszug m. ‖ extrait m. de garance. / ~ flowers pl. *see* ~ bloom. / ~ lake ‖ Krapplack m. ‖ laque f. de garance. / red ~ lake ‖ roter Krapplack m.; Krappkarmin m. ‖ laque f. de garance rouge; carmin m. de garance. / ~ mill ‖ Krappmühle f. ‖ moulin m. à garance. / ~ purple ‖ Purpurin n.; Krappurpur m. ‖ purpurine f. / ~ red ‖ Krapp m.; Krapprot n. ‖ garance f. / ~ root ‖ Krappwurzel f. ‖ racine f. de garance. / ~ style (Textile print) ‖ Krappfarbendruck m. ‖ genre m. garance. / ~ yellow ‖ Krappgelb n. ‖ xanthine f.

made ‖ gemacht; angefertigt; hergestellt ‖ fait; fabriqué. / ~ in sections ‖ zerlegbar ‖ démontable. / ~ of iron ‖ eisern ‖ de fer m.; en fer m. / machine-~ ‖ maschinell hergestellt ‖ fait mécaniquement; fabriqué à la machine. / ~ to order ‖ auf Bestellung gemacht ‖ fait sur commande. / ~ of plate ‖ aus Blech n. ‖ de tôle f. / ~ of a single piece ‖ aus einem einzigen Stück n. gefertigt ‖ fait d'une seule pièce. / to be ~-up of ... (Chem) ‖ bestehen aus ... ‖ être constitué par ...

magazine ‖ Niederlage f.; Warenlager n.; Lagerhaus n. ‖ dépôt f.; entrepôt m.; magasin m. / ~ (Of a gun) ‖ Magazin n. ‖ magasin m. / ~ (Of a machine) ‖ Magazin n. ‖ magasin m. / ~ (Periodical publication) ‖ Zeitschrift f.; Magazin n. ‖ revue f.

magazine feed attachment ‖ Magazineinrichtung f. ‖ installation f. de magasin.

/ ~ rifle ‖ Magazingewehr n. ‖ fusil m. à magasin *ou* à répétition.

magenta red ‖ Anilinrot n. ‖ rouge m. d'aniline.

magic lantern ‖ Laterna magica f.; Zauberlaterne f. ‖ lanterne f. magique. / ~ slide ‖ Laterna-magica-Bild n. ‖ image f. pour lanternes magiques.

magistral (Met) ‖ Magistral n. ‖ magistral m.

magnalium ‖ Magnalium n. ‖ magnalium m.

magnecrystalline axis ‖ Magnetkristallachse f. ‖ axe m. magnécristallin. / ~ force ‖ Magnetkristallkraft f. ‖ force f. magnécristalline.

magnesia ‖ Bittererde f.; Talkerde f.; Magnesia f.; Magnesiumoxyd n. ‖ magnésie f. / ~ alum ‖ Magnesiaalaun m.; Talkerdealaun m. ‖ alun m. de magnésie. / carbonate of ~ ‖ Magnesiakarbonat n. ‖ carbonate m. de magnésie. / ~ cement ‖ Magnesiazement m. ‖ ciment m. de magnésie. / ~ citrate ‖ zitronensaure Magnesia f. ‖ citrate m. de magnésie. / ~ lime-stone ‖ Bitterkalk m.; Bitterspat m. ‖ chaux f. carbonatée magnésifère; dolomite f. / siliciferous carbonate of ~ ‖ quarziger Magnesit m.; Sepiolith m. ‖ magnésie f. carbonatée silicifère.

magnesian ‖ magnesiumhaltig ‖ magnésique. / ~ limestone ‖ Rauchkalk m. ‖ chaux f. carbonatée magnésifère; dolomie f. / ~ water ‖ Bitterwasser n. ‖ eau f. magnésifère.

magnesioferrite ‖ Magnesioferrit m.; Magnoferrit m. ‖ fer m. magnétique *ou* magnésifère; magnésioferrite f.

magnesite ‖ Magnesit m. ‖ magnésite f. / ~ brick ‖ Magnesitstein m. ‖ brique f. de magnésite. / ~ mill ‖ Magnesitmühle f. ‖ moulin m. à magnésite. / ~ works equipment ‖ Magnesitfabrikeinrichtung f. ‖ équipement m. de fabrique de magnésite.

magnesium ‖ Magnesium n. ‖ magnésium m. / ~ alloy ‖ Magnesiumlegierung f. ‖ magnésium-alliage m. / ~ chloride ‖ Chlormagnesium n. ‖ chlorure m. de magnésium. / ~ citrate ‖ zitronensaures Magnesium n. ‖ citrate m. de magnésium. / ~ compound ‖ Magnesiumverbindung f. ‖ combinaison f. de magnésie; magnésien m. / ~ fluoride ‖ Fluormagnesium n. ‖ fluorure m. de magnésium. / ~ light ‖ Magnesiumlicht n. ‖ lumière f. du magnésium. / ~ nitrate ‖ salpetersaures Magnesium n. ‖ nitrate m. de magnésium. / ~ oxide ‖ Magnesia f.; Bittererde f.; Magnesiumoxyd n.; Talkerde f. ‖ magnésie f.; oxyde m. de magnésium. / ~ perborate ‖ Magnesiumperborat n. ‖ perborate m. de magnésium. / ~ peroxide ‖ Magnesiumperoxyd n. ‖ peroxyde m. de magnésium. / ~ phosphate ‖ phosphorsaures Magnesium n. ‖ phosphate m. de magnésium. / ~ silicofluoride ‖ Kieselfluormagnesium m. ‖ silicofluorure m. de magnésium. / ~ stearate ‖ stearinsaures Magnesium n. ‖ stéarate m. de magnésium. / ~ sulphate ‖ Bittersalz n.; Magnesiumsulfat n. ‖ sulfate m. de magnésie; sel m. d'Epsom; sel m. amer.

magnet ‖ Magnet m. ‖ aimant m.; barreau m. magnétique. / artificial ~ ‖ künstlicher Magnet m. ‖ aimant m. artificiel. / band ~ ‖ Bandmagnet m. ‖ lame f. aimantée. / bell-shaped ~ ‖ Glockenmagnet m. ‖ aimant m. campanulé. / claw ~ ‖ Klauenmagnet m. ‖ aimant m.

à griffes. / closed ~ ‖ geschlossener Magnet m. ‖ aimant m. fermé. / compensating ~ ‖ Ausgleichmagnet m. ‖ aimant m. correcteur *ou* de compensation. / compound ~ ‖ zusammengesetzter Magnet ‖ aimant m. composé. / controlling ~ ‖ Richtmagnet m. ‖ aimant m. correcteur *ou* directeur. / crane ~ ‖ Lastmagnet m. ‖ aimant m. à monter la charge. / directing ~ see controlling ~. / fagot ~ see compound ~. / horse-shoe ~ ‖ hufeisenförmiger Magnet m.; Hufeisenmagnet m. ‖ aimant m. en fer-à-cheval. / lamellar ~ ‖ Lamellenmagnet m.; Blättermagnet m. ‖ aimant m. lamellaire *ou* feuilleté. / lifting ~ ‖ Lasthebemagnet m.; Magnethebezeug n. ‖ aimant m. de levage *ou* de suspension. / molecular ~ ‖ Molekularmagnet m. ‖ aimant m. moléculaire. / native ~ see natural ~. / natural ~ ‖ natürlicher Magnet m.; Magneteisenstein m.; Magnetit m. ‖ aimant m. naturel; fer m. magnétique; magnétite f. / permanent ~ ‖ bleibender *oder* permanenter Magnet m. ‖ aimant m. permanent. / plate ~ see band ~. / relay ~ ‖ Relaismagnet m.; Schützmagnet m. ‖ aimant m. du relais. / rotary ~ ‖ Drehmagnet m. ‖ électro m. de rotation.; aimant m. rotatif. / straight ~ ‖ Magnetstab m.; Stabmagnet m. ‖ barreau m. aimanté; aimant m. droit. / unclosed ~ ‖ nicht geschlossener Magnet m. ‖ aimant m. non fermé.

magnet core ‖ Magnetkern m. ‖ noyau m. aimanté *ou* d'un aimant. / ~ crane ‖ Magnetkran m. ‖ grue f. à aimant. / ~ frame ‖ Magnetgestell n. ‖ bâti m. magnétique. / ~ holder ‖ Magnethalter m. ‖ porte-aimant m.

magnetic ‖ magnetisch ‖ magnétique. / ~ circuit ‖ magnetischer Kreis m. ‖ circuit m. magnétique. / ~ compass ‖ Magnetkompaß m. ‖ compas m. magnétique. / ~ coupling (Mach) ‖ elektromagnetische Kupplung f. ‖ embrayage m. électromagnétique. / ~ detector (Radio) ‖ magnetischer Detektor m.; Magnetdetektor m. ‖ détecteur m. magnétique. / ~ drum separator ‖ Trommelmagnetscheider m. ‖ tambour m. séparateur magnétique. / ~ energy ‖ magnetische Energie f. ‖ énergie f. magnétique. / ~ equator ‖ magnetischer Äquator m.; Indifferenzzone f. ‖ équateur m. magnétique; zone f. neutre. / ~ field ‖ Magnetfeld n. ‖ champ m. magnétique. / ~ field flux ‖ Magnetkraftfluß m. ‖ flux m. de champ inducteur. / ~ field intensity ‖ magnetische Feldstärke f. ‖ intersité f. du champ magnétique. / ~ figure ‖ magnetisches Kraftlinienbild n. ‖ spectre m. magnétique. / ~ flux ‖ magnetischer Fluß m. ‖ flux m. magnétique. / ~ hammer break ‖ magnetischer Hammerunterbrecher m. ‖ interrupteur m. magnétique à marteau. / ~ holding device ‖ Magnetspannplatte f. ‖ plateau m. de serrage à aimant. / ~ induction ‖ Magnetinduktion f. ‖ induction f. magnétique. / ~ investigation ‖ magnetische Untersuchung f. ‖ recherche f. magnétique. / ~ iron see natural magnet. / ~ iron ore see natural magnet. / ~ iron-pyrites ‖ Magnetkies m.; Pyrrotin m.; Leberkies m. ‖ sulfure m. de fer magnétique; fer m. sulfuré magnétique; pyrite f. magnétique. /

~ iron-stone see natural magnet. / ~ meridian ‖ magnetischer Meridian m. ‖ méridien m. magnétique.

magnetic needle ‖ Magnetnadel f. ‖ aiguille f. aimantée *ou* de compas. / rhomboidal ~ ‖ rautenförmige Magnetnadel f. ‖ aiguille f. aimantée en forme de losange. / square-shaped ~ see rhomboidal ~. / action of the earth on a ~ ‖ kosmische Einwirkung f. auf die Magnetnadel ‖ influence f. cosmique sur l'aiguille aimantée. / influences pl. deflecting the ~ ‖ ablenkende Einflüsse mpl. auf die Magnetnadel ‖ influences fpl. déviatrices sur l'aiguille aimantée. / instrument employing ~ ‖ Magnetnadelinstrument n. ‖ instrument m. à aiguille aimantée.

magnetic permeability ‖ Magnetisierungskonstante f. ‖ perméabilité f. magnétique. / ~ polarity see ~ induction. / ~ pole ‖ Magnetpol m. ‖ pôle m. d'aimant. / ~ potential ‖ magnetisches Potential n. ‖ potentiel m. magnétique. / ~ roller ‖ Magnetwalze f. ‖ cylindre m. aimanté. / ~ screening device ‖ Magnetscheideanlage f. ‖ installation f. de triage magnétique. / ~ shunt (Tel) ‖ Nebenschluß m. mit erheblicher Induktivität ‖ dérivation f. d'induction. / ~ steel bar ‖ magnetischer Stahlstab m. ‖ barreau m. d'acier magnétique. / diurnal ~ variation ‖ tägliche magnetische Schwankung f. ‖ variation f. magnétique diurne.

magnetical see magnetic.

magnetisation see magnetization.

magnetise, to ~ see to magnetize.

magnetism ‖ Magnetismus m. ‖ magnétisme m. / to lose ~ ‖ den Magnetismus m. verlieren ‖ perdre le magnétisme. / to remove the ~ ‖ den Magnetismus m. entfernen ‖ enlever *ou* détruire le magnétisme. / to transfer the ~ ‖ den Magnetismus m. übertragen ‖ transmettre *ou* faire passer le magnétisme d'un corps sur un autre. / to transmit the ~ see to transfer the ~. / bound ~ ‖ gebundener Magnetismus m. ‖ magnétisme m. condensé. / ~ collected in the poles ‖ in den Polen angehäufter Magnetismus m. ‖ magnétisme m. accumulé dans les pôles. / free ~ ‖ freier Magnetismus m. ‖ magnétisme m. libre. / permanent ~ ‖ dauernder *oder* permanenter Magnetismus m. ‖ magnétisme m. permanent. / remanent ~ ‖ remanenter Magnetismus ‖ magnétisme m. rémanent *ou* résiduel. / residual ~ see remanent ~. / ~ of rocks ‖ Gesteinsmagnetismus m. ‖ magnétisme m. des roches. / ~ of rotation ‖ Rotationsmagnetismus m. ‖ magnétisme m. de rotation. / terrestrial ~ ‖ Erdmagnetismus m. ‖ magnétisme m. terrestre. / insensitiveness to ~ ‖ Unempfindlichkeit f. gegen Magnetismus ‖ insensibilité f. contre le magnétisme.

magnetite ‖ Magnetit m.; Magneteisenstein m. ‖ magnétite f.; fer m. magnétique (naturel). / ~ electrode ‖ Magnetitelektrode f. ‖ électrode f. en magnétite. / ~ mine ‖ Magneteisensteingrube f. ‖ mine f. de fer magnétique.

magnetizability ‖ Magnetisierbarkeit f. ‖ aimantabilité f.; susceptibilité f. de s'aimanter.

magnetizable ‖ magnetisierbar ‖ magnétisable.

magnetization ‖ Magnetisierung f. ‖ magnétisation f.; aimantation f. / new ~ ‖ Neumagnetisieren n. ‖ réaimantation f.

magnetization curve ‖ Magnetisierungskurve f. ‖ courbe f. d'aimantation. / intensity of ~ ‖ Magnetisierungsstärke f. ‖ intensité f. d'aimantation.

magnetize, to ‖ magnetisieren ‖ aimanter. / capable of being magnetized see magnetizable. / susceptibility of being magnetized see magnetizability.

magnetized bar ‖ Magnetstab m.; Stabmagnet m. ‖ barreau m. aimanté. / ~ needle see magnetic needle.

magnetizing current ‖ Magnetisierungsstrom m.; magnetisierender Strom m. ‖ courant m. d'aimantation; courant m. aimantisant.

magnet keeper ‖ Magnetanker m. ‖ armature f. d'un aimant.

magneto (Mot) ‖ Magnet m.; Magnetapparat m.; Magnetzünder m.; Zündapparat m.; Magneto m. ‖ magnéto f. / automatically timed ~ ‖ Magnet m. mit selbsttätiger Zündmomentverstellung ‖ magnéto f. à avance automatique. / chain-driven ~ ‖ Magnet m. mit Kettenantrieb ‖ magnéto f. à commande à chaîne. / ignition and lighting ~ ‖ Magnetapparat m. für Zündung und Beleuchtung ‖ magnéto f. à allumage et à éclairage. / starting ~ ‖ Anlaßmagnetapparat m.; Anlaßmagnet m. ‖ magnéto f. de mise en marche.

magneto board (Tel) ‖ Klappenschrank m. ‖ commutateur m. à clapets. / ~ bracket ‖ Magnetapparatstütze f. ‖ gousset m. de magnéto. / ~ carbon holder ‖ Kohlenhalter m. des Magnetapparats ‖ portecharbon m. de la magnéto. / ~ clutch ‖ Magnetmitnehmer m. ‖ griffe f. de magnéto. / ~ coupling ‖ Magnetkupplung f. ‖ accouplement m. de la magnéto. / ~ crank (Tel) ‖ Induktorkurbel f. ‖ manivelle f. de la magnéto d'appel. / ~ distributor ‖ Verteiler m. des Magnetapparates ‖ distributeur m. de la magnéto. / ~ drive ‖ Magnetantrieb m. ‖ commande f. de la magnéto.

magneto-electric ‖ magnetelektrische ‖ magnéto-électrique. / ~ gun ‖ magnetelektrisches Geschütz n. ‖ canon m. électro-magnétique.

magneto-electricity ‖ Magnetoelektrizität f.; Magnetelektrizität f. ‖ magnéto-électricité f.

magneto fire alarm system ‖ Induktorfeuermeldesystem n. ‖ système m. d'avertisseurs d'incendie à appel par magnéto. / ~ generator ‖ Kurbelinduktor m. ‖ magnéto f. d'appel.

magnetograph ‖ Magnetograf m. ‖ magnétographe m.

magneto ignition ‖ Magnetzündung f. ‖ allumage m. par magnéto.

magnetometer ‖ Magnetometer n. ‖ magnétomètre m. / bifilar ~ ‖ Bifilarmagnetometer n. ‖ magnétomètre m. bifilaire.

magnetomotive ‖ magnetomotorisch ‖ magnétomoteur. / ~ force ‖ magnetomotorische Kraft f. ‖ force f. magnétomotrice.

magneto spanner ‖ Magnetschlüssel m. ‖ clef f. à magnéto. / ~ strop ‖ Magnetbügel m. ‖ bride f. de la magnéto.

magnet regulator ‖ Magnetregler m. ‖ régulateur m. du champ (magnétique).

magnetron (Radio) ‖ Magnetron n. ‖ magnétron m.

magnet steel ‖ Magnetstahl m. ‖ acier m. à aimants. / ~ wheel ‖ Magnetrad n. ‖ roue f. magnétique. / ~ winding ‖ Magnetwicklung f. ‖ enroulement m. d'électro. / ~ yoke of ~ ‖ Magnetjoch n. ‖ culasse f. de l'aimant.

magnification (Opt) ‖ Vergrößerung f. ‖ grossissement m. / ~ (Tel) ‖ Verstärkung f. ‖ amplification f. / ~ due to an eyepiece ‖ Vergrößerung f. eines Okulares ‖ grossissement m. d'un oculaire. / effective ~ ‖ förderliche oder nutzbare Vergrößerung f. ‖ grossissement m. utile. / factorial ~ ‖ Lupenvergrößerung f. ‖ grossissement m. de ou à la loupe. / maximum ~ ‖ Maximalvergrößerung f. ‖ grossissement m. maximum. / minimum ~ ‖ Minimalvergrößerung f. ‖ grossissement m. minimum. / total ~ ‖ Gesamtvergrößerung f. ‖ grossissement m. total. / useful ~ see effective ~. / x-fold ~ ‖ x-fache Vergrößerung f. ‖ grossissement m. à x diamètres ou x fois. / coefficient of ~ ‖ Vergrößerungszahl f. ‖ coefficient m. de grossissement ou de grandissement. / limit of ~ ‖ Vergrößerungsgrenze f. ‖ limite f. de grossissement.

magnifier see also magnifying glass ‖ Vergrößerungsglas n.; Lupe f. ‖ microscope m. simple; loupe f. / adjustable ~ ‖ einstellbare Lupe f. ‖ loupe f. réglable; loupe f. permettant la ou de mise au point. / anastigmatic ~ ‖ anastigmatische Lupe f. ‖ loupe f. anastigmatique. / aplanatic ~ ‖ aplanatische Lupe f. ‖ loupe f. aplanétique. / botanical ~ ‖ botanische Lupe f. ‖ loupe f. de botanistes. / dissecting ~ ‖ Präparierlupe f. ‖ loupe f. à dissection. / double ~ ‖ Doppellupe f. ‖ loupe f. double. / ~ for the examination of the eyes ‖ Lupe f. für die Augenuntersuchung ‖ loupe f. pour l'examen des yeux. / field-glass ~ ‖ Feldstecherlupe f. ‖ jumelle-loupe f. / focussing ~ ‖ Einstellupe f. ‖ loupe f. de mise au point. / folding ~ ‖ Einschlaglupe f. ‖ loupe f. fermante ou pliante. / anastigmatic folding ~ ‖ anastigmatische Einschlaglupe f. ‖ loupe f. fermante anastigmatique. / aplanatic folding ~ ‖ aplanatische Einschlaglupe f. ‖ loupe f. aplanétique fermante. / ~ of high power ‖ starke Lupe f. ‖ loupe f. puissante. / measuring ~ ‖ Meßlupe f. ‖ loupe f. de mesure. / pocket ~ ‖ Taschenlupe f. ‖ loupe f. de poche. / ranging ~ with a ruled plate ‖ Visierlupe f. mit Strichplatte ‖ loupe f. de visée munie d'un réticule. / ~ with simple anastigmatic lens ‖ Lupe f. mit einfacher anastigmatischer Linse ‖ loupe f. constituée par une lentille simple anastigmatique. / spectacle ~ ‖ Brillenlupe f. ‖ loupe-lunettes f. / stereoscopic ~ ‖ stereoskopische Lupe f. ‖ loupe f. stéréoscopique. / telescopic ~ ‖ Fernrohrlupe f. ‖ téléloupe f. / tripod ~ ‖ Dreifußlupe f. ‖ loupe f. sur trépied.

magnifier lamp ‖ Lupenlampe f. ‖ lampe f. à loupes. / folding mount of a ~ ‖ Einschlagfassung f. einer Lupe ‖ monture f. fermante d'une loupe.

magnify, to (Opt etc.) ‖ vergrößern ‖ agrandir; grossir. / ~ (Electr) ‖ verstärken ‖ augmenter.

magnifying ‖ vergrößernd ‖ grossissant. /

highly ~ telescope ‖ stark vergrößerndes Fernrohr n. ‖ lunette f. à fort grossissement. / ~ x times ‖ x-fach vergrößernd ‖ grossissant x fois.

magnifying apparatus ‖ Vergrößerungsapparat m. ‖ appareil m. d'agrandissement. / ~ effect ‖ vergrößernde Wirkung f. ‖ action f. grossissante; effet m. grossissant. / ~ glass see also magnifier ‖ Lupe f.; Vergrößerungsglas n.; Leseglas n. ‖ loupe f.; verre m. grossissant.

magnifying lens ‖ Lupenlinse f. ‖ lentille f. de la loupe. / ~ for photographs ‖ Fotografielupe f. ‖ loupe f. pour examiner des photographies. / ~ for precious stones ‖ Steinlupe f.; Edelsteinlupe f. ‖ loupe f. pour lapidaires.

magnifying picture viewer ‖ Bildlupe f. ‖ loupe f. photoscopique.

magnitude (Math) ‖ Größe f. ‖ grandeur f.; quantité f.

magnoferrite ‖ Magnoferrit m. ‖ magnoferrite f.; magnésioferrite f.

mahaleb cherry ‖ Weichselkirsche f. ‖ griotte f.

mahogany ‖ Mahagoni n.; Mahagoniholz n. ‖ acajou m. / bastard ~ ‖ afrikanisches Mahagoni n.; Madeiramahagoni n. ‖ acajou m. bâtard. / mottled ~ ‖ moiriertes Mahagoniholz n. ‖ acajou m. moiré. / plain ~ ‖ glattes Mahagoniholz n. ‖ acajou m. uni. / plum pattern ~ ‖ geflammtes Mahagoniholz n. ‖ acajou m. flammé. / roe ~ ‖ gestreiftes Mahagoniholz n. ‖ acajou m. rubané. / speckled ~ see spotted ~. / spotted ~ ‖ geflecktes Mahagoniholz n. ‖ acajou m. moucheté. / ~ in trunks ‖ Mahagoniholz n. in Stämmen ‖ acajou m. en canons ou en billes. / veined ~ ‖ geädertes Mahagoniholz n. ‖ acajou m. veiné. / veneer pattern ~ ‖ Mahagonifurnierholz n. ‖ feuille f. de placage en acajou.

mahogany base ‖ Mahagonisockel m. ‖ base f. d'acajou. / ~ cabinet ‖ Mahagonischrank m. ‖ armoire f. en acajou. / ~ gum ‖ Mahagonibaumgummi m. ‖ gomme f. d'acajou. / ~ log ‖ Mahagoniholzblock m. ‖ bille f. d'acajou. / ~ plank ‖ Mahagonibohle f. ‖ madrier m. d'acajou.

maiden trip (Mar) ‖ erste Reise f. ‖ premier voyage m.

maid-servant ‖ Magd f. ‖ servante f.

mail ‖ Post f. ‖ poste f. / ~ (Knitting) ‖ Masche f. ‖ maille f. / ~ aeroplane ‖ Postflugzeug n. ‖ avion m. postal. / ~ bag ‖ Postbeutel m.; Postsack m.; Briefbeutel m. ‖ sac m. aux lettres. / ~ boat ‖ Paketboot n. ‖ paquebot m. / ~ car (Railw) ‖ Postwagen m.; Bahnpostwagen m.; Paketpostwagen m. ‖ malle-poste m.; wagon-poste m.; voiture f. postale. / ~ coach ‖ Postkutsche f. ‖ diligence f. / ~ day ‖ Posttag m. ‖ jour m. du courrier.

mailing machine ‖ Postversandmaschine f. ‖ machine f. pour l'expédition du courrier. / ~ tube ‖ Versandrolle f. ‖ tube m. de carton pour l'expédition postale.

mail plane see ~ aeroplane. / ~ service ‖ Postdienst m. ‖ service m. postale. / ~ station ‖ Postamt n. ‖ bureau m. de poste. / ~ steamer ‖ Postdampfer m. ‖ paquebot m. (à vapeur); vapeur m. postal; paquebot-poste m. / ~ train ‖ Postzug m. ‖ train-poste m. / ~ van (Railw) see ~ car.

main (Cable) ‖ Hauptkabel n. ‖ câble m. principal. / ~ (Pipe) ‖ Hauptleitungsrohr n. ‖ tuyau m. principal. / ~s pl. (Hydr arch) ‖ Stammleitung f. ‖ conduite f. principale. / electric ~s pl. ‖ elektrische Hauptleitung f. ‖ conduite f. principale. / exhaust steam ~s pl. ‖ Abdampfleitung f. ‖ conduite f. de vapeur d'échappement. / in the ~ ‖ zum größten Teil m. ‖ en majeure partie f.

main air-gate (Mine) ‖ Hauptwettertür f. ‖ porte f. d'aérage principale. / ~ beam (Carp) ‖ Hauptbalken m.; Binderbalken m. ‖ maîtresse f. poutre. / ~ bearing ‖ Hauptlager n.; Grundlager n. ‖ palier m. principal. / ~ bed (Mach) ‖ Hauptbett n. ‖ banc m. principal. / ~ boiler ‖ Hauptkessel m. ‖ chaudière f. principale. / ~ busbar ‖ Hauptsammelschiene f. ‖ barre f. omnibus principale.

main cable ‖ Hauptkabel n. ‖ câble m. principal. / ~ (Bridge) see main chain. / ~ (Tel) ‖ Fernsprechhauptkabel n. ‖ câble m. de transport. / ~ duct ‖ Hauptkabelkanal m. ‖ canalisation f. du câble principal.

main chain of a chain-bridge ‖ Tragkette f. oder Hängekette f. einer Kettenbrücke ‖ chaîne f. d'un pont suspendu. / ~ channel ‖ Sammelkanal m. ‖ égout m. principal; égout-collecteur m. / ~ connecting-rod (Steam eng) ‖ Haupttriebstange f. ‖ grande barre f.; grande verge f. de connexion. / ~ counter ‖ Hauptzähler m. ‖ compteur m. principal. / ~ couple (Carp) ‖ Dachbinder m. ‖ maîtresse f. ferme. / ~ current (Electr) ‖ Nutzstrom m. ‖ courant m. d'utilisation. / ~ deck (Shipb) ‖ Hauptdeck n. ‖ pont m. principal. / ~ dike ‖ Winterdeich m. ‖ digue f. insubmersible. / ~ distributing frame (Tel) ‖ Hauptverteiler m. ‖ répartiteur m. principal ou d'entrée. / ~ drain ‖ Hauptentwässerungsrohr n. ‖ égout m. collecteur. / ~ drum ‖ Hauptwalze f. ‖ cylindre m. principal. / ~ engine ‖ Hauptmaschine f. ‖ machine f. principale. / ~ engine of a motor freighter ‖ Hauptmaschine f. eines Motorfrachtschiffes ‖ moteur m. de propulsion d'un cargo. / ~ engine for propeller drive ‖ Hauptmaschine f. zum Antrieb von Schiffsschrauben ‖ machine f. principale pour la commande des hélices. / ~ equipment (Radio) ‖ Netzanschlußgerät n. ‖ bloc m. d'alimentation. / ~ feeder (Electr) ‖ Hauptspeiseleitung f. ‖ feeder m. principal. / ~ feeder (Hydr arch) ‖ Hauptspeisegraben m. ‖ canal m. (principal) d'alimentation. / ~ fermentation ‖ Hauptgärung f. ‖ fermentation f. principale. / ~ fire alarm ‖ Hauptfeuermelder m. ‖ avertisseur m. d'incendie principal. / ~ floor (Build) ‖ Hauptgeschoß n. ‖ bel-étage m.; étage m. premier.

main frame (Build) ‖ Bundwand f. ‖ cloison f. de charpente. / ~ (Mach) ‖ Hauptrahmen m. ‖ bâti m. ou cadre m. principal. / ~ (Shipb; Airpl) ‖ Hauptspant m. ‖ maître-couple m.

main front (Build) ‖ Hauptfront f. ‖ façade f. principale. / ~ fuse (Electr) ‖ Hauptsicherung f. ‖ coupe-circuit m. principal. / ~ fuse board ‖ Hauptsicherungstafel f. ‖ tableau m. principal des coupe-circuits. / ~ gate ‖ Haupttor n.; Portal n. ‖ porte f. principale ou majeure; portail m. / ~ girder ‖ Hauptbalken m.; Hauptträger

m.; Hauptlängsträger m. ‖ maîtresse f. poutre; poutre f. principale; longeron m. / ~ guard ‖ Hauptwache f. ‖ poste m. central. / ~ hoisting tackle ‖ Haupthubwerk n. ‖ treuil m. principal. / ~ hood beam (Windmill) ‖ Fugbalken m. ‖ poutre f.; entrait m. / ~ line (Railw) ‖ Hauptbahn f.; Hauptbahnlinie f. ‖ ligne f. principale; grande ligne. / ~ load distributing station for the whole network ‖ Hauptlastverteilungsstelle f. für das gesamte Netz ‖ poste m. principal de distribution de la charge pour le reseau entier. / ~ mast ‖ Großmast m. ‖ grand mât m. / ~ nozzle ‖ Hauptdüse f. ‖ gicleur m. principal. / ~ office (Tel) ‖ Vollamt n. ‖ bureau m. principal. / ~ part (Mach) ‖ Hauptteil m. ‖ élément m. principal ou essentiel.

main phase ‖ Hauptphase f. ‖ phase f. principale. / ~ winding ‖ Hauptphasenwicklung f. ‖ enroulement m. de phase principale.

main pipes pl. (Hydr arch) ‖ Stammleitung f. ‖ conduite f. principale. / ~ piston ‖ Arbeitskolben m. ‖ piston m. moteur. / ~ rail of a switch ‖ Hauptschiene f. oder feste Schiene f. einer Weiche ‖ rail m. fixe d'un changement de voie; rail m. contre-aiguille. / ~ regulator board ‖ Hauptreglertafel f. ‖ tableau m. du régulateur principal. / ~ repeater station ‖ Hauptverstärkeramt n. ‖ station f. de répéteurs principale. / ~ reservoir (Hydr arch) ‖ Hauptbehälter m. ‖ réservoir m. principal. / ~ river ‖ Hauptfluß m. ‖ rivière f. principale; fleuve m. principal. / ~ road ‖ Hauptstraße f.; Heerstraße f.; große Landstraße f. ‖ route f. nationale; grande route f.

main rod (Mach) ‖ Pleuelstange f.; Treibstange f. ‖ bielle f. motrice. / ~ (Mine) ‖ Hauptgestänge n.; Schachtgestänge n. ‖ maîtresse-tige f.; maître m. tirant. / ~ of a pump ‖ Hauptpumpenstange f. ‖ verge f. de pompe; maîtresse f. tige de pompe. / ~ head ‖ Treibstangenkopf m. ‖ tête f. de bielle.

main royal (Sail) ‖ Großroyal n. ‖ grand cacatois m.; grand perroquet m. volant. / ~ runner (Found) ‖ Hauptrinne f. ‖ chenal m. principal.

mains pl. ‖ Stammleitung f.; Hauptleitung f. ‖ conduite f. principale. / electric ~ ‖ elektrische Hauptleitung f. ‖ conduite f. électrique principale.

main sail ‖ Großsegel n. ‖ grande voile f. / ~ sewer see ~ channel. / ~ shaft ‖ Hauptwelle f.; Königswelle f. ‖ arbre m. principal (de commande); arbre m. de couche. / ~ shaft of a forging hammer ‖ Hammerwelle f. ‖ arbre m. moteur d'un marteau de forge. / ~ shafting ‖ Haupttransmission f. ‖ transmission f. principale. / ~ side frame (Loc) ‖ Rahmenwange f. ‖ longeron m. / ~ spar (Airpl) ‖ Hauptholm m. ‖ longeron m. principal. / ~ spring ‖ Haupt(antriebs)feder f. ‖ ressort m. principal (de commande). / ~ standard (Mach) ‖ Hauptständer m. ‖ montant m. principal. / ~ station (Tel) ‖ Hauptanschluß m. ‖ poste m. principal. / ~ support see ~ girder. / ~ switch (Electr) ‖ Netzumschalter m. ‖ commutateur m. principal. / ~ switchboard ‖ Hauptschalttafel f. ‖ tableau m. de distribution principal. / ~ switch station ‖ Hauptschaltwarte f. ‖ poste m. principal de manœuvre.

maintain, to ~ a battery ‖ eine Batterie speisen ‖ charger une pile. / ~ a preferential right ‖ ein Vorzugsrecht n. behaupten ‖ prétendre à un droit de priorité. / the prices pl. maintain their level ‖ die Kurse mpl. halten sich (fest) ‖ les valeurs fpl. se maintiennent.

maintaining constant ‖ Konstanthaltung f. ‖ maintien m. de la constance.

main tank see main reservoir.

maintenance ‖ Instandhaltung f.; Unterhaltung f. ‖ entretien m. / ~ of a connection (Tel) ‖ Aufrechterhaltung f. der Verbindung ‖ maintien m. de la communication. / ~ of the motor ‖ Instandhaltung f. des Motors ‖ entretien m. du moteur. / ~ of the permanent way (Railw) ‖ Unterhaltung f. des Oberbaues ‖ entretien m. de la voie. / ~ of the rails ‖ Instandhaltung f. der Gleise ‖ entretien m. de la voie.

maintenance charges pl. ‖ Unterhaltungskosten pl.; Instandhaltungskosten pl. ‖ frais mpl. d'entretien. / considerable saving of ~ charges ‖ erhebliche Ersparnisse fpl. an Unterhaltungskosten ‖ économie f. considérable de frais d'entretien. / ~ costs pl. see ~ charges. / ~ work ‖ Unterhaltungsarbeiten fpl. ‖ travaux mpl. d'entretien.

main thread ‖ Hauptfaden m. ‖ fil m. principal. / ~ timber (Carp) ‖ Hauptverbandstück n. ‖ maîtresse f. pièce d'une charpente. / ~ top (Mar) ‖ Großmars m. ‖ grande hune f. / ~ topgallant sail ‖ Großbramsegel n. ‖ grand perroquet m. / ~ topsail ‖ Großmarssegel n. ‖ grand hunier m. / ~ track (Railw) ‖ Hauptgleis n. ‖ voie f. principale. / ~ transmission ‖ Haupttransmission f. ‖ transmission f. principale. / ~ truss see ~ couple. / ~ tuning bottom (Radio) ‖ Hauptabstimmknopf m. ‖ bouton m. de réglage principal. / ~ turbine ‖ Hauptturbine f. ‖ turbine f. principale. / ~ valve ‖ Hauptventil n. ‖ soupape f. principale. / ~ vaulting (Build) ‖ Hauptgewölbe n. ‖ maîtresse-voûte f. / ~ voltage ‖ Hauptspannung f. ‖ tension f. principale. / ~ wall ‖ Hauptmauer f. ‖ maîtresse-muraille f. / ~ warp (Weav) ‖ Grundkette f. ‖ Bodenkette f. ‖ chaîne f. de fond. / ~ water ‖ Leitungswasser n. ‖ eau f. de conduite. / ~ wing rib (Airpl) ‖ Hauptflügelrippe f.; Hauptrippe f. des Flügels ‖ nervure f. principale de l'aile; nervure f. de compression de l'aile.

mainyard (mar) ‖ Großraa f. ‖ grande vergue f.

maize ‖ Mais m.; türkischer Weizen m. ‖ maïs m. / hulled ~ ‖ enthülster Mais m. ‖ maïs m. pelliculé.

maize flaking mill ‖ Maisflockenstuhl m. ‖ moulin m. à flocons de maïs. / ~ grits pl. ‖ Maisgrütze f. ‖ grit m. de maïs. / ~ meal ‖ Maismehl n. ‖ farine f. de maïs. / ~ mill ‖ Maismühle f. ‖ moulin m. à maïs. / ~ oil ‖ Maisöl n. ‖ huile f. de maïs. / ~ starch ‖ Maisstärke f. ‖ amidon m. de maïs. / ~ straw ‖ Maisstroh n. ‖ paille f. de maïs. / ~ working machine ‖ Maisbearbeitungsmaschine f. ‖ machine f. à travailler le maïs.

majolica ‖ Majolika f. ‖ majolique f. / ~ articles pl. ‖ Majolika f. ‖ objets mpl. en majolique. / ~ colours pl. ‖ Farben fpl. für Majolika ‖ couleurs fpl. pour majolica. / ~ painter ‖ Majolikamaler m. ‖

peintre m. en majolique. / ~ slab ‖ Majolikafliese f. ‖ carreau m. de majolique. / ~ ware ‖ Majolikawaren fpl. ‖ articles mpl. de majolique; majolica m.

major axis (Of an ellipse) ‖ Hauptachse f. ‖ grand-axe m.

majority ‖ Mehrheit f. ‖ pluralité f.; majorité f. / to own the ~ of the stock of a joint stock company ‖ die Mehrheit f. in einer Aktiengesellschaft besitzen ‖ être titulaire m. de la majorité des parts d'une société anonyme. / ~ of votes ‖ Stimmenmehrheit f. ‖ pluralité f. des voix; majorité f.

make, to ~ the dough ‖ den Teig m. kneten oder anmachen ‖ pétrir la pâte. / ~ even ‖ ebnen ‖ planer. / ~ fast (Nav) ‖ festmachen ‖ s'amarrer. / ~ fat ‖ mästen ‖ engraissern. / ~ flush with … ‖ bündig oder gleichmachen; in gleiche Flucht bringen mit … ‖ affleurer (deux surfaces). / ~ known ‖ verkündigen; bekanntmachen ‖ annoncer; publier; prononcer. / ~-out an account ‖ eine Rechnung f. ausschreiben oder ausfertigen oder ausziehen oder aufstellen ‖ dresser ou extraire ou établir un compte. / ~ an out (Print) ‖ ein Wort n. auslassen ‖ sauter un mot. / ~ ready (Print) ‖ zurichten ‖ mettre en train. / ~ sail ‖ Segel npl. setzen ‖ déployer les voiles fpl.; mettre les voiles fpl. au vent. / ~ for shore ‖ sich am Lande vertäuen ‖ s'amarrer à terre. / ~ a shot ‖ einen Schuß m. lösen ‖ tirer un coup. / ~ a traverse survey ‖ polygonisieren ‖ mesurer des polygones mpl. / ~ true (A machine) ‖ aufstellen ‖ montieren; adjustieren ‖ dresser; ajuster. / ~ up ‖ machen; zubereiten; fertigmachen; vollenden ‖ préparer; apprêter; finir. / ~up (Chem) ‖ ergänzen; vervollständigen ‖ compléter. / ~ up (Print) ‖ umbrechen; Umbruch m. machen ‖ mettre en pages. / ~ up the margin ‖ Format n. machen ‖ mettre la garniture. / ~ use of … ‖ ausnutzen ‖ tirer profit m.; exploiter. / ~ water (Mar) ‖ lecken ‖ faire eau; faire de l'eau.

make (Product) ‖ Erzeugnis n.; Produkt n.; Fabrikat n. ‖ produit m. / ~ (Production) ‖ Herstellung f.; Erzeugung f.; Anfertigung f. ‖ fabrication f.; production f.; confection f. / ~ (Structure) ‖ Ausführung f.; Bauart f. ‖ construction f. / best ~ ‖ bestes Fabrikat n. ‖ la meilleure marque. / inferior ~ ‖ minderwertiges Erzeugnis n. oder Fabrikat n. ‖ fabrication inférieure. / special ~ ‖ besondere Ausführung f. oder Bauart f. ‖ construction f. spéciale. / standard ~ ‖ gewöhnliche oder übliche oder normale Ausführung f. ‖ construction f. normale.

make and break (Electr) ‖ Abreißen n. ‖ rupture f. / mechanical ~ (Motor) ‖ Abreißvorrichtung f. ‖ dispositif m. de rupture mécanique.

make and break current ‖ zeitweilig unterbrochener Strom m. ‖ courant m. périodiquement interrompu. / ~ device ‖ Unterbrecherscheibe f. ‖ disrupteur m. / ~ ignition (Mot) ‖ Abreißzündung f. ‖ allumage m. à rupture ou par rupture.

make contact (Electr) ‖ Arbeitskontakt m. ‖ contact m. de travail.

make-ready (Print) ‖ Zurichtung f.; Zurichten n. ‖ mise f. en train; préparatifs mpl.

make-up (Print) ‖ Umbrechen n.; Umbruch m. ‖ mise f. en pages.

making ‖ Anfertigung f.; Fertigung f.; Herstellung f. ‖ fabrication f.; confection f.

making iron (Shipb) ‖ Dichteisen n.; Rabatteisen n. ‖ fer m. à calfat double ou cannelé.

making-up cost ‖ Aufmachungskosten pl. ‖ frais mpl. de confection.

Malabar tallow ‖ Pflanzentalg m. ‖ suif m. de Malabar.

malachite ‖ Malachit m.; Kupferspat m.; Atlaserz n.; Kupfergrün n. ‖ malachite f.; cuivre m. carbonaté vert. / earthy ~ ‖ Kupfergrün n.; erdiger Malachit m. ‖ cuivre m. carbonaté vert terreux. / fibrous ~ ‖ faseriger Malachit m. ‖ cuivre m. carbonaté vert fibreux.

malacolite ‖ grüner Augit m.; Malakolit m.; Sahlit m. ‖ diopside m. vert laminaire; malacolithe m.; sahlite f.

malate ‖ apfelsaures Salz n. ‖ malate m.

male hemp ‖ Kopfhanf m. ‖ chanvre m. mâle.

male screw ‖ Schraubenspindel f. ‖ vis f. / ~ of the press (Print) ‖ Preßspindel f. ‖ vis f. de la presse.

male tank ‖ männlicher Tank m. oder Kampfwagen m.; mit Kanonen bestückter Kampfwagen m. ‖ char m. canon.

malic acid ‖ Apfelsäure f. ‖ acide m. malique.

mall ‖ Schlegel m. ‖ massette f.

malleability ‖ Schmiedbarkeit f. ‖ malléabilité f.

malleable ‖ hämmerbar; schmiedbar; streckbar; dehnbar; geschmeidig ‖ malléable; ductile. / ~ castings pl. ‖ Temperguß m.; schmiedbarer Guß m. ‖ fonte f. malléable. / ~ cast iron see. ~ castings. / ~ iron ‖ Schmiedeeisen n.; schmiedbares Eisen n. ‖ fer m. forgé ou malléable ou forgeable. / cementation of ~ iron ‖ Glühfrischen n. ‖ cémentation f. de fer malléable.

malleableization see malleableizing.

malleableizing ‖ Glühfrischen n. ‖ cémentation f. de fer malléable.

mallet ‖ Holzhammer m. ‖ maillet m. / ~ (Hatt) ‖ Klopfer m.; Schlägel m.; Klöpfel m. ‖ battoir m.; maillet m. / ~ (Mine) ‖ Handfäustel m.; Fäustel m. ‖ massette f. / tanner's ~ ‖ Pumpkeule f. ‖ enfonçoir m. à tête.

mallow blossom ‖ Malvenblüte f. ‖ fleur f. de mauve. / ~ leaves pl. ‖ Malvenblätter npl. ‖ feuilles fpl. de mauve bleue.

malonic acid ‖ Malonsäure f. ‖ acide m. malonique.

malt, to (Brew) ‖ malzen; mälzen ‖ malter.

malt ‖ Malz n. ‖ malt m. / air-dried ~ ‖ Luftmalz n. ‖ malt m. séché à l'air. / ~ from barley ‖ Malz n. aus Gerste; Gerstenmalz n. ‖ malt m. d'orge. / black ~ ‖ Farbmalz n. ‖ farbmalz m.; malt m. grillé ou torréfié ou colorant. / black extract ‖ Farbmalzextrakt m. ‖ extrait m. de malt torréfié. / brown ~ see black ~. / bruised ~ ‖ Malzschrot m. ‖ drèche f.; drèche f.; drège f. / colour ~ see black ~. / flinty ~ ‖ Glasmalz n. ‖ malt m. vitreux. / glassy ~ see flinty ~. / green ~ ‖ Grünmalz n. ‖ malt m. vert. / Pilsen ~ ‖ Pilsnermalz n. ‖ malt m. de Pilsen. / roasted ~ see black ~. / steely ~ see flinty ~. /vitrified ~ see flinty ~. / withered ~ ‖ Schwelkmalz n. ‖ malt m.

(qu'on a assez) séché à l'air avant de le tourailler.

malt bruising mill ‖ Malzschrotmühle f. ‖ moulin m. à concasser le malt ou à broyer le malt. / ~ bruising plant ‖ Malzschroterei f. ‖ installation f. de broyage du malt. / ~ cleaning plant ‖ Malzputzereianlage f. ‖ installation f. de nettoyage du malt. / ~ coffee ‖ Malzkaffee m. ‖ café m. de malt. / ~ crusherman ‖ Malzquetscher m. ‖ ouvrier-broyeur m. de malt. / ~ crushing ‖ Malzquetschen n. ‖ broyage m. ou concassage m. de malt. / ~ crushing mill ‖ Malzquetsche f. ‖ concasseur m. à malt. / ~ crushing room ‖ Malzschroterei f.; Schroterei f. ‖ salle f. des concasseurs à malt. / ~ dressing plant ‖ Mälzereieinrichtung f. ‖ installation f. de maltage. / ~ drier see ~ kiln. / ~ drop ‖ Malzbonbon m. ‖ bonbon m. au malt. / ~ drying drum kiln ‖ Trommelmalzdarre f. ‖ touraille f. à tambour à sécher le malt. / ~ dye see black malt.

malted ‖ gemälzt ‖ malté. / ~ grain ‖ gemälztes Getreide n. ‖ grain m. malté. / ~ pepton ‖ Malzpepton n. ‖ malto-peptone f.

maltery installation ‖ Mälzereieinrichtung f. ‖ installation f. pour malteries.

malt extract ‖ Malzextrakt m.; Malzauszug m. ‖ extrait m. de malt. / ~ extract plant ‖ Malzextraktanlage f. ‖ installation f. à faire l'extrait de malt. / ~ floor ‖ Malztenne f. ‖ grenier m. à malt; germoir m. / cooling of ~ floors ‖ Kühlung f. der Malztennen ‖ refroidissement m. des germoirs.

maltha ‖ weiches Erdpech n. ‖ malthe f.

malt-house ‖ Mälzerei f. ‖ malterie f.

malt husks pl. ‖ Biertreber pl.; Malztreber pl.; Treber pl.; Schlempe f. ‖ drèche f.; drague f.

malting ‖ Malzen n.; Mälzen n. ‖ malterie f.; maltage m. / ~ plant ‖ Mälzereianlage f. ‖ installation f. de malterie.

malt kiln ‖ Malzdarre f.; Darre f. ‖ touraille f.

maltman ‖ Mälzer m. ‖ malteur m.

malt milk apparatus ‖ Malzmilchapparat m. ‖ appareil m. à lait de malt. / ~ mill ‖ Malzmühle f. ‖ moulin m. à malt. / ~ mill man ‖ Malzmüller m. ‖ ouvrier m. des moulins à malt. / ~ oar ‖ Malzwender m. ‖ appareil m. à retourner le malt; faneur m. à malt. / mechanical ~ oar ‖ mechanischer Malzwender m. ‖ faneur m. à malt mécanique.

maltose ‖ Maltose f.; Malzzucker m. ‖ maltose m. / ~ determination ‖ Maltosebestimmung f. ‖ dosage m. du maltose. / ~ sirup ‖ Maltosesirup m. ‖ sirop m. de maltose.

malt polishing plant ‖ Malzpolieranlage f. ‖ installation f. de polissage du malt. / ~ residuum see ~ husks. / ~ roaster ‖ Malzdarrer m. ‖ tourailleur m. / ~ shoveller ‖ Malzschaufler m. ‖ pelleteur m. de malt.

maltster ‖ Mälzer m. ‖ malteur m. / ~'s foreman ‖ Obermälzer m. ‖ chef m. malteur.

malt turner see ~ oar. / ~ vinegar ‖ Malzessig m. ‖ vinaigre m. de malt.

mamelonated ‖ warzenförmig ‖ mamelonné.

mammoth pump ‖ Mammutpumpe f. ‖ pompe f. système Mamouth.

man, to ~ a ship || ein Schiff n. bemannen || équiper un navire; mettre des hommes sur un navire.

man (Workman) || Arbeiter m. || ouvrier m. / ~ (Mar) || Gast m. || homme m.; marin m.; gabier m.; ouvrier m.

manage, to || verwalten || administrer; gérer. / ~ a firm || einer Firma f. vorstehen || diriger ou gérer ou administrer une maison. / well managed || gut geleitet || bien organisé ou administré.

manageable || (leicht) handlich || facile à manier; maniable; commode.

management (Manipulation) || Handhabung f. || manœuvre m.; maniement m. / ~ (Administration) || Leitung f.; Geschäftsführung f.; Bewirtschaftung f.; Verwaltung f. || gestion f.; administration f.; exploitation f.; aménagement m. / ~ (Board of managers) || Direktion f.; Verwaltung f.; Leitung f.; Vorstand m. || direction f.; directoire m.; gérance f.; administration f. / ~ of work || Bauführung f. || conduite f. des travaux. / theory of ~ || Betriebslehre f. || théorie f. d'aménagement.

manager || Geschäftsführer m. || gérant m. / ~ (Managing director) || Direktor m.; Betriebsdirektor m. || directeur m.; chef m. / ~ (Mine) || Obersteiger m. || maître-mineur m. / ~ of a business || Geschäftsführer m. || gérant m. / ~ of a department || Betriebsführer m. || chef m. de fabrication ou de service. / ~ of a firm || Geschäftsführer m.; Geschäftsleiter m. || gérant m.; directeur m. / general ~ || Generaldirektor m. || directeur m. général. / ~ entitled to sign for the firm || Prokurist m. || gérant m.: fondé m. de pouvoir. / works ~ || (technischer) Betriebsleiter m.; technischer Direktor m. || directeur m. techniques; chef m. de service des ateliers.

managing direction of the works || (technische) Leitung f. eines Werkes || direction f. technique d'une usine. / ~ director || geschäftsführender Direktor m. || directeur-gérant m. / ~ partner || geschäftsführender Teilhaber m. || associer-directeur m.

Manchester velvet || Manchester m. || velours m. de Manchester.

mandarin || Mandarine f. || mandarine f. / ~ oil || Mandarinenöl n. || essence f. de mandarines.

mandatary || Mandatar m.; Bevollmächtigter m. || mandataire f.

mandoline || Mandoline f.; kleine Laute f. || mandoline f. / ~ pick || Mandolinenplättchen n. || médiateur m. pour mandoline.

mandrel (Mach) || Dorn m.; Richtdorn m.; Steckdorn m. || mandrin m. / ~ (Mine) || Doppelkeilhaue f. || marteau m. à (deux) pointes; pic m. à deux branches. / expanding ~ || verstellbarer Dorn m. || mandrin m. à expansion. / grooved ~ || Riffelkloben m. || mandrin m. taillé. / rolling ~ || Walzdorn m. || mandrin m. de laminage ou d'étirage. / swing-out ~ / ausschwenkbarer Einlagedorn m. || mandrin m. supplémentaire articulé. / ~ which tapers at the rate of x mm to each y mm || kegeliger Dorn m., der sich auf je x mm Länge um y mm verjüngt || mandrin m. d'une conicité de y/x. / ~ for tube rolling mill || Ziehdorn m. in Röhrenwalzwerken || man-

drin m. d'étirage pour laminoir à tubes.

mandrel press || Dorneintreibpresse f.; Dornpresse f. || presse f. à mandrin; presse f. à enfoncer les mandrins. / ~ test || Dornprobe f. || essai m. de mandrinage. / ~ upsetting press || Dornstauchpresse f. || presse f. à refouler les mandrins.

mandril see mandrel.

manege || Reitbahn f.; Reitschule f. || manège m. / ~ saddle || Schulsattel m. || selle f. de manège ou à piquer.

maneuver, to ~ (Mar; Aero) || manövrieren (lassen) || (faire) manœuvrer.

maneuverability || Steuerfähigkeit f.; Wendigkeit f.; Manövrierfähigkeit f. || manœuvrabilité f.; maniabilité f.

maneuverible (Airpl) || wendig; manövrierfähig || maniable.

maneuvering capabilities pl. of a ship || Manövereigenschaften fpl. eines Schiffes || qualités fpl. evolutionnaires d'un navire. / ~ valve || Manövrierventil n. || soupape f. de manœuvre.

manganate || Manganat n.; mangansaures Salz n. || manganate m.

manganese || Mangan n. || manganèse m. / rich in ~ || manganhaltig || manganésifère. / black ~ || Schwarzmanganerz n.; Scharfmangan n. || oxyde m. de manganèse pyramidal. / cupreous ~ || Kupfermanganerz n. || manganate m. de cuivre; manganèse m. cuprifère. / red ~ || roter Braunstein m.; Manganspat m.; Diallogit m. || carbonate m. de manganèse; manganèse m. carbonaté; diallogite f. / siliceous ~ || Kieselmangan n.; Rhodonit m. || manganèse m. oxydé silicifère; rhodonite f.

manganese acetate || essigsaures Mangan n. || acétate m. de manganèse. / ~ black || Manganschwarz n. || noir m. de manganèse. / ~ borate || borsaures Mangan n. || borate m. de manganèse. / ~ bronze || Manganbronze f. || bronze m. manganésé ou au manganèse. / ~ brown || Manganbraun n. || brun m. de manganèse. / ~ carbonate || kohlensaures Mangan n. || carbonate m. de manganèse. / ~ chloride || Chlormangan n. || chlorure m. de manganèse. / ~ composition || Mangankitt m. || mastic m. au manganèse. / ~ copper || Mangankupfer n. || cuivre m. manganésé ou au manganèse. / ~ dioxide see ~ ore. / ~ garnet || granatförmiges Braunsteinerz n.; Mangangranat m. || spessartine f. / ~ iron || Eisenmangan n. || ferro-manganèse m. / ~ lignoleate || holzölsaures Mangan n. || lignoliate m. de manganèse. / ~ linoleate || leinölsaures Mangan n. || linoléate m. de manganèse. / ~ metal || Manganmetall n. || manganèse m. métallique. / ~ mine || Manganerzbergwerk n. || mine f. de manganèse.

manganese ore || Manganerz n.; Braunstein m.; Manganhyperoxyd n. || minerai m. de manganèse; pyrolusite f.; dioxyde m. ou peroxyde m. de manganèse. / compact and fibrous ~ || Psilomelan m.; Hartmanganerz n.; schwarzer Glaskopf m.; psilomélane m,; hydroxyde m. de manganèse barytifère ou de manganèse oxydé; manganèse m. barytique hydraté. / gray ~ || Braunmanganerz n.; Manganit m. || manganite f.

manganese ore anode || Mangandioxyd-anode f. || anode f. en peroxyde de man-

ganèse. / ~ cylinder || Braunsteinzylinder m. || cylindre m. de peroxide de manganèse.

manganese oxide || Manganoxyd n. || oxyde m. de manganèse. / ~ peroxide see ~ ore. / red carbonate of ~ || Manganspat m.; Rosenspat m. || manganèse m. carbonaté. / ~ resinate || harzsaures Mangan n. || résinate m. de manganèse. / ~ spar || Manganspat m. || carbonate m. de manganèse.

manganese steel || Manganstahl m. || acier m. (au) manganèse. / ~ is used both as castings and forgings for parts exposed to great wear and tear || Manganstahl m. findet sowohl in gegossenem als in geschmiedetem Zustande Verwendung für Teile, die starkem Verschleiß unterworfen sind || l'acier m. au manganèse s'emploie non seulement moulé mais aussi forgé pour (des) pièces exposées à forte usure. / ~ frog (Railw) || Manganstahlherzstück n. || cœur m. en acier manganèse. / ~ rail || Manganstahlschiene f. || rail m. en acier au manganèse.

manganese sulphate || schwefelsaures Mangan n. || sulfate m. de manganèse.

manganesiferous || manganhaltig || manganésifère. / ~ iron mine || Mangancisensteingrube f. || mine f. de fer manganésé.

manganesous see manganesiferous.

manganic oxide see manganese oxide.

manganine wire || Manganindraht m. || fil m. de manganine. / ~ resistance || Manganindrahtwiderstand m. || résistance f. en fil de manganine.

manganite see manganese ore.

manganocalcite || Manganokalzit m. || manganocalcite f.

manganous oxide || Manganoxydul n. || protoxyde m. de manganèse.

mangle, to ~ (Wash) || mangeln; rollen || calandrer; cylindrer.

mangle (Wash) || Mangel f.; Wäscherolle f. || calandre f. (à linge).

mangler || Mangler m. || calandreur m.; mangleur m.

manhole (Boil) || Fahrloch n.; Mannloch n.; Einsteigeloch n. || trou m. d'homme. / ~ (Cable) || Kabelbrunnen m. || chambre f. de soudure ou de tirage (de cables). / auxiliary ~ (Cable) || Hilfsbrunnen m. || chambre f. auxiliaire (de cables). / the boiler plates are supplied with ~ flanged inwards || die Kesselböden mpl. sind mit eingepreßtem Mannloch versehen || les fonds mpl. des chaudières sont fournis avec trou d'homme embouti. / ~ on pipe sewers || Einsteigeschacht m. einer Rohrleitung || puisard m. d'un tuyoutage ou d'une canalisation. / raised ~ || Fahrstutzen m. || trou m. d'homme surélévé. / ~ reinforced concrete ~ (Cable) || Kabelbrunnen m. aus Beton || chambre f. de tirage en béton armé. / ~ of a water conduit || Revisionsloch n. oder Wasserstube f. einer Wasserleitung || regard m. d'aqueduc.

manhole clothing || Mannlochverkleidung f. || recouvrement m. du trou d'homme. / ~ cover || Fahrdeckel m.; Mannlochdeckel m. || couvercle f. de trou d'homme. / complete ~ cover || vollständiger Mannlochverschluß m. || fermeture f. complète de trou d'homme. / cross bar of the ~ || Mannlochbügel m. || traverse f. de trou d'homme. / ~ dome || Fahrhut m. || dôme m. de trou d'homme. / ~ door || Ein-

steigetür f. || porte f. de trou d'homme. / ~ end plate || Mannlochboden m. || fond m. à trou d'homme. / ~ punch || Mannlochstanze f. || estampe f. à emboutir les trous d'hommes *ou* à faire des ouvertures de visite. / ~ punching machine || Mannlochstanze f. || poinçonneuse f. pour trous d'homme. / ~ ring || Mannlochversteifung f. || renforcement m. de trou d'homme. / ~ ventilation (Cable) || Lüften n. des Kabelbrunnens || ventilation f. de la chambre de soudure.

manicure article || Nagelpflegeartikel m.; Handpflegeartikel m. || article m. pour manicure *ou* pour manucure. / ~ case || Nagelpflegekasten m. || onglier m. / ~ instrument || Nagelpflegegerät n.; Nagelpflegeinstrument n. || instrument m. pour manucures. / ~ scissors pl. || Nagelschere f. || ciseaux mpl. à ongles.

manifest (of the cargo) || Manifest n. der Schiffsladung; Ladungsverzeichnis n. || déclaration f.; manifeste m.

manifold || mannigfach || varié; divers.

manifold (Graphics) || hektografischer Abzug m. || reproduction f. hectographique. / exhaust ~ || Auspuffleitung f. || tubulure f. d'échappement. / inlet ~ || Einlaßleitung f.; Saugleitung f. || tubulure f. d'admission.

manifolding machine || Vervielfältigungsmaschine f. || machine f. de reproduction. / ~ work || Vervielfältigungsarbeiten fpl. || travaux mpl. de polycopie *ou* de reproduction hectographique.

manifold pen || Durchschreibefeder f. || plume f. pour faire les copies au carbone.

manifold writer || Hektograf m.; grafischer Vervielfältigungsapparat m. || hectographe m.; appareil m. de reproduction graphique.

manilla *see* Manilla hemp.

Manilla hemp || Manilahanf m. || chanvre m. de Manille. / ~ rope || Manilatau n. || filin m. en abaca.

manioc root || Wurzel f. von Maniok || racine f. de manioc.

manipulation || Handhabung f. || maniement m.; manipulation f. / wrong ~ || falsche Handhabung f. || fausse manipulation f.

manna || Manna f. || manne f. / ~ sugar || Mannazucker m.; Mannit n. || mannite f.

mannequin || Probierdame f. || essayeuse f.; mannequin m.

manner of craftsmen || Handwerksbrauch m. || usage m. des gens du métier; usage m. du métier.

Mannesmann tube || Mannesmannrohr n. || tuyau m. Mannesmann. / ~ pole || Mannesmannrohrmast m. || poteau m. en tubes Mannesmann.

mannite *see* manna sugar.

manœuvre, to ~ (Mar) || manövrieren (lassen) || (faire) manœuvrer.

manœuvrability *see* maneuverability.

manœuvrible; manœuvring *see* maneuverible; maneuvering.

man-of-war || Kriegsschiff n. || bâtiment m. de guerre.

manometer || Druckmesser m.; Dampfdruckmesser m.; Manometer n.; Manomesser m. || manomètre m. / mercurial ~ || Quecksilbermanometer m.; Quecksilberdampfdruckmesser m. || manomètre m. à mercure. / metallic ~ || Metallmanometer n.; Metalldruckmesser m. || manomètre m. métallique. /

plate ~ || Plattendruckmesser m.; Plattenmanometer n. || manomètre m. à membrane métallique. / tube ~ || Röhrendruckmesser m.; Röhrenmanometer n. || manomètre m. à tube courbé.

manometric lift || manometrische Förderhöhe f. || hauteur f. manométrique de refoulement.

man power || Menschenkraft f. || force f. humaine *ou* d'homme.

man rope (Build) || Lenkseil n.; Schwungseil n.; Leitseil n.; Schwungseil n. || hauban m.; écharpe f. / ~ (Mar) || Manntau n. (am Laufsteg) || garde-corps m.; garde-fou m.; sauve-garde f.

mansard roof || Mansardendach n.; Gebrochenes Dach n. || toiture f. à la Mansard.

mantle || Mantel m. || manteau m. / ~ (of a chimney) || Rauchfang m.; Rauchmantel m.; Schurz m. || hotte f. *ou* manteau m. de cheminée. / ~ (Found) || Formmantel m.; Überform f. || surmoule m.; surtout m. / incandescent ~ || Glühstrumpf m. || capuchon m. *ou* manchon m. à incandescence. / ~ of a wall || Mauerhaupt n.; Mauermantel m.; Stirnseite f. einer Mauer || parement m. d'un mur.

mantle clock || Standuhr f. || pendule f. / ~ iron (Build) || Manteleisen n. || chemise f. en fer. / ~ piece (of a chimney) || Rauchfangholz n.; Schurzholz n. || poutre f. de hotte *ou* de manteau.

manual (Book) || Leitfaden m.; Handbuch n. || guide m.; manuel m.

manual fire engine || Handspritze f.; Handfeuerspritze f. || pompe f. à main. / ~ press (Print) || Handpresse f. || presse f. (typographique) à main. / ~ work || Handarbeit f. || main f. d'œuvre; travail m. manuel.

manufactory || Fabrik f. || fabrique f.; usine f.

manufacture, to || (fabrikmäßig) anfertigen *oder* verfertigen; fertigen || confectionner *ou* fabriquer (à l'atelier). / ~ into... || verarbeiten zu... || usiner; façonner. / manufactured article || Fertigerzeugnis n. || fabricat m.; produit m. fini *ou* usiné.

manufacture || Anfertigung f.; Fertigung f.; Herstellung f.; Fabrikation f. || fabrication f.; confection f. / ~s pl. || Erzeugnisse npl.; Fertigwaren fpl. || produits mpl. (fabriqués). / industrial ~ || fabrikmäßige Herstellung f. || fabrication f. industrielle. / ~ on a large scale || Großfabrikation f. || fabrication f. en grand. / present day ~ || neuzeitliche Fertigung f. || fabrication f. moderne. / ~ of parts || Fertigung f. *oder* Fabrikation f. von Einzelteilen || fabrication f. de pièces détachées. / ~ in series || Reihenfertigung f. || fabrication f. en série. / ulterior ~ || Weiterverarbeitung f. || fabrication f. ultérieure. / course of ~ || Arbeitsgang m. || cours m. de la fabrication. / process of ~ || Herstellungsverfahren n. || procédé m. de fabrication. /stage of ~ || Verarbeitungsstufe f. || état m. de fabrication.

manufacturer || Fabrikant m.; Hersteller m.; Erzeuger m. || fabricant m.; manufacturier m. producteur m.

manufacturing *see* manufacture. / ~ equipment || Fabrikeinrichtung f. || installation f. *ou* équipement m. de fabrique.

/ ~ expenses pl. || Erzeugungskosten pl.; Gestehungskosten pl. || prix m. de fabrication *ou* de revient. / ~ length (Cable) || Fabrikationslänge f. || longueur f. de fabrication. / ~ plant || industrielle Anlage f.; Fabrikanlage f. || établissements mpl. industriels; usine f. / progressive method of ~ || Fließarbeit f. || méthode f. progressive de fabrication. / ~ rights pl. || Herstellungsrecht n. || droit m. de fabrication. / ~ town || Fabrikstadt f. || ville f. industrielle. / ~ way || Herstellungsweise f. || mode f. de fabrication.

manure, to || düngen || engraisser.

manure || Dünger m.; Dung m.; Mist m. || engrais m.; fumier m. / ~ of animal origin || Tierdünger m.; Düngemittel n. tierischer Herkunft || engrais m. de provenance animale. / artificial ~ || Kunstdünger m.; künstliches Düngemittel n. || engrais m. artificiel *ou* chimique. / chemical ~ || chemisches Düngemittel n. || engrais m. chimique. / ~ from downs || Dünger m. aus Daunen || engrais m. de duvets. / ~ from feathers || Dünger m. aus Federn || engrais m. de plumes. / flower ~ || Blumendünger m. || engrais m. pour fleurs. / ~ from horsehair || Dünger m. aus Roßhaaren || engrais m. de crins. / liquid ~ || Jauche f. || purin m. / nitrogenous ~ || Stickstoffdünger m. || engrais m. azotique. / organic ~ || organischer Düngstoff m. || engrais m. organique. / ~ from slaughterhouse refuse || Dünger m. aus tierischen Abfällen || engrais m. de débris d'animaux. / vegetable ~ || pflanzlicher Dünger m.; Pflanzendünger m. || engrais m. (d'origine) végétal(e). / ~ from wastes || Dünger m. aus Abfällen || engrais m. de déchets. / ~ from wool wastes || Dünger m. aus Wollabfällen || engrais m. de déchets de laine.

manure distributor || Düngerstreuer m.; Düngerstreumaschine f. || sémoir m. *ou* distributeur m. d'engrais. / ~ drill machine || Reihendüngerstreuer m.; Düngerdrillmaschine f. || semoir m. en ligne à engrais. / ~ grinding mill *see* ~ mill. / ~ manufacturing || (Kunst-)Düngerherstellung f. || fabrication f. d'engrais (artificiel). / ~ manufacturing plant || Düngerherstellungsanlage f. || installation f. de fabrique d'engrais. / ~ mill || Düngermühle f. || moulin m. à engrais. / ~ spreader *see* ~ distributor / ~ works pl. || Düngerfabrik f. || fabrique f. d'engrais.

manurial value || Düngewert m. || valeur f. engraissante.

manuring with lime || Kalkdüngung f. || chaulage m.

manuring salt || Düngesalz n. || sel m. d'engrais; engrais m. salin.

manuscript || Manuskript n.; Handschrift f. || manuscrit m. / ~ (Print) || Manuskript n. || copie f.; manuscrit m. / curious ~ || seltenes Manuskript n. || manuscrit m. curieux.

manuscript copy || Manuskriptabschrift f. || copie f. à la main. / ~ cover || Handschriftenmappe f. || chemise f. pour manuscrits. / ~ holder (Print) || Manuskripthalter m. || visorium m.

many || mancherlei; vielerlei || différent; divers; beaucoup.

map || Landkarte f. || carte f. géographique. / aeroautical ~ || Flugkarte f. || carte f.

aéronautique. / ~ from air photographs ‖ Luftbildkarte f.; Luftbildplan m. ‖ carte f. *ou* plan m. de vues aériennes. / ~ of cable lay-out ‖ Kabellageplan m. ‖ plan m. ou croquis m. de pose du câble. / celestial ~ ‖ Himmelskarte f.; astronomische Karte f. ‖ carte f. céleste. / geographical ~ *see* map. / geological ~ ‖ Formationskarte f.; geologische Karte f. ‖ carte f. géologique; carte f. des couches géologiques. / hydrographical ~ ‖ Seekarte f.; Paßkarte f. ‖ carte f. marine *ou* nautique *ou* hydrographique. / military ~ ‖ Generalstabskarte f. ‖ carte f. d'Etat-major. / ordnance ~ *see* military ~. / orographical ~ ‖ Höhenkarte f. ‖ carte f. orographique. / schoolroom ~ ‖ Wandkarte f.; Schulwandkarte f. ‖ carte f. murale (pour écoles). / ~ showing seams ‖ Flözkarte f. ‖ carte f. stratigraphique. / ~ of the stars *see* celestial ~. / topographical ~ ‖ topografische Karte f. ‖ carte f. topographique. / ~ of the world ‖ Erdkarte f. ‖ carte f. géographique de la terre; mappe-monde f.

map case ‖ Kartentasche f. ‖ porte-carte m. / ~ colourer ‖ Landkartenmaler m. ‖ coloriste m. en cartes géographiques. / ~ holder ‖ Kartenhalter m. ‖ porte-cartes m.

maple ‖ Ahorn m. ‖ érable m. / ash-leaved ~ ‖ eschenähnlicher Ahorn m. ‖ érable m. négundo. / common ~ ‖ Feldahorn m. ‖ érable m. champêtre. / plane ~ ‖ Spitzahorn m. ‖ érable m. plan. / sugar ~ ‖ Zuckerahorn m. ‖ érable m. à sucre.

maple varnish ‖ Ahornlack m. ‖ vernis m. acéracé. / ~ wood ‖ Ahornholz n. ‖ bois m. d'érable. / ~ wood oil ‖ Ahornholzöl n. ‖ essence f. de bois d'érable.

map printer ‖ Kartendrucker m. ‖ imprimeur m. de cartes géographiques. / ~ printing ‖ Kartografie f.; Kartenzeichenkunst f. ‖ cartographie f. / ~ printing office ‖ kartografische Anstalt f.; Landkartendruckerei f. ‖ imprimerie f. de cartes géographiques. / ~ protector ‖ Kartenschoner m. ‖ porte-cartes f. / ~ sticker ‖ Landkartenaufzieher m. ‖ colleur m. de cartes géographiques. / ~ table ‖ Kartentisch m. ‖ table f. pour étaler la carte.

marble, to ‖ marmorieren; adern ‖ marbrer; veiner.

marble ‖ Marmor m.; körniger Kalkstein m. ‖ marbre m.; chaux f. carbonatée saccharoïde. / ~ (Toy) ‖ Murmel f.; Schusser m. ‖ marbre m. / artificial ~ ‖ Kunstmarmor m.; künstlicher Marmor m.; Gipsmarmor m. ‖ marbre m. artificiel *ou* factice. / compound ~ ‖ bunter Marmor m. ‖ marbre m. composé. / facing ~ ‖ Verblendmarmor m. ‖ marbre m. de parement. / granular ~ ‖ körniger Marmor m. ‖ marbre m. salignon. / red ~ ‖ roter Marmor m. ‖ marbre m. rouge. / ~ sawed in rough plates ‖ in rohe Platten zersägter Marmor m. ‖ marbre m. débité en plaques brutes. / statuary ~ ‖ Statuenmarmor m.; Büstenmarmor m. ‖ marbre m. statuaire *ou* de sculpteur.

marble articles pl. ‖ Marmorwaren fpl. ‖ objets mpl. en marbre. / ~ carving ‖ Marmorbildhauerei f. ‖ sculpture f. sur marbre. / ~ cement ‖ Marmorzement m.;

Alaungips m. ‖ ciment m. anglais; plâtre m. aluné. / ~ clock ‖ Marmoruhr f. ‖ pendule f. en marbre. / ~ cutter ‖ Marmorsäger m. ‖ marbrier m. / ~ cutting mill ‖ Marmorschneidemühle f. ‖ moulin m. à scier le marbre.

marbled ‖ marmoriert; gefleckt; gesprenkelt; madré; marbré; tacheté; tigré. / ~ paper ‖ marmoriertes Papier n. ‖ papier m. marbré.

marble dust ‖ Marmormehl n. ‖ poudre f. de marbre. / ~ engraver ‖ Marmorgravör m. ‖ graveur m. sur marbre. / ~ flag ‖ Marmorfliese f.; Marmorplatte f. ‖ dalle f. de marbre. / ~ goods pl. ‖ Marmorwaren fpl. ‖ objets mpl. en marbre. / ~ lime ‖ Marmorkalk m. ‖ chaux f. de marbre. / ~ meal ‖ Marmormehl n. ‖ marbre m. moulu. / ~ mill ‖ Marmormühle f. ‖ moulin m. à marbre. / ~ plate ‖ Marmorplatte f. ‖ dalle f. de marbre. / ~ polishing ‖ Marmorschleifen n. ‖ polissage m. de marbre. / ~ quarry ‖ Marmorbruch m. ‖ marbrière f.; carrière f. de marbre.

marbler ‖ Marmorierer m. ‖ marbreur m.

marble saw ‖ Marmorsäge f. ‖ scie f. à marbre. / ~ saw blade ‖ Marmorsägeblatt n. ‖ lame f. pour scies à marbre. / ~ sawing mill ‖ Marmorsägemühle f. ‖ scierie f. de marbre. / ~ sand for sawing ~ ‖ Marmorsägesand m. ‖ sable m. à scier le marbre. / ~ sawing yard *see* ~ sawing mill. / ~ sculpture ‖ Marmorplastik f. ‖ travail m. de sculpture en marbre. / ~ slab *see also* ~ flag ‖ Grabplatte f. ‖ pierre f. tombale. / ~ switchboard ‖ Marmorschalttafel f. ‖ tableau m. de distribution en marbre. / ~ tile ‖ Marmorfliese f. ‖ carreau m. de marbre. / ~ turner ‖ Marmordrechsler m. ‖ tourneur m. de marbre. / ~ working ‖ Marmorarbeit f. ‖ marbrerie f. / ~ workman ‖ Marmorarbeiter m. ‖ marbrier m. / ~ works pl. ‖ Marmorwerk n. ‖ marbrerie f.

marbling ‖ Marmorierung f. ‖ madrure f.; marbrure f.; madrage m.

marcasite ‖ Markasit m.; rhombischer Eisenkies n. ‖ marcassite f.

marchpane ‖ Marzipan m. ‖ massepain m. / ~ machine ‖ Marzipanmaschine f. ‖ machine f. à faire le massepain. / ~ manufacture ‖ Marzipanherstellung f. ‖ fabrication f. de massepain.

maregraph ‖ selbstregistrierendes Pegel n.; Maregraf m.; Flutzeiger m. ‖ échelle f. enrégistrante; marégraphe m.

marekanite ‖ Marekanit m.; edler kugelförmiger Obsidian m. ‖ marékanite f.

margaric acid ‖ Margarinsäure f. ‖ acide m. margarique.

margarine ‖ Margarine f.; Kunstbutter f. ‖ margarine f. / ~ cheese ‖ Margarinekäse m. ‖ fromage m. margariné. / ~ factory ‖ Margarinfabrik f.; Kunstbutterfabrik f. ‖ fabrique f. de margarine. / ~ maker ‖ Margarinefabrikarbeiter m. ‖ ouvrier-margarinier m. / ~ producing plant ‖ Margarinefabrikanlage f. ‖ établissement m. à fabriquer la margarine.

margarite (Miner) ‖ Margarit m.; Perlglimmer m. ‖ margarite m.; mica m. nacré.

margin (Print) ‖ Rand m. ‖ marge f. / ~ (Tel) ‖ Spielraum m. ‖ empiètement m. / to make-up the ~ (Print) ‖ Format

n. machen ‖ mettre la garniture. / ~ of manufacture (Mach tool) ‖ Einpaßzugabe f. ‖ surépaisseur f. / ~ of the paper sheet ‖ Rand m. des Papierbogens ‖ marge f. de la feuille de papier. / ~ for profit ‖ Gewinnmarge f.; Gewinnspanne f. ‖ marge f. de bénéfice.

marginal ‖ am Rande befindlich ‖ marginal. / ~ inscription of a coin ‖ Rundschrift f. *oder* Umschrift f. einer Münze ‖ légende f. d'une monnaie. / ~ note ‖ Randbemerkung f. ‖ note f. *ou* glosse f. marginale; manchette f. / ~ point ‖ Randpunkt m. ‖ point m. marginal. / ~ relay (Tel) ‖ träge wirkendes Relais n. ‖ relais m. à action différée. / ~ release (Typewr) ‖ Randauslösung f. ‖ déblocage m. des marges. / ~ stop (Typewr) ‖ Randhemmung f. ‖ butée f. marginale.

margin-sheet (Print) ‖ Unterlegbogen m. ‖ marge f.

Marie-Davy battery ‖ Quecksilberelement n. ‖ élément m. Marie Davy.

marinate, to ‖ marinieren; sauer einlegen ‖ mariner.

marine ‖ Marine f.; Seewesen n. ‖ marine f. / mercantile ~ ‖ Handelsmarine f. ‖ marine f. marchande *ou* du commerce.

marine auxiliary set ‖ Bordhilfsaggregat n. ‖ groupe m. auxiliaire de bord. / ~ blue ‖ Marineblau n. ‖ bleumarine m. / ~ board ‖ Seeamt n. ‖ bureau m. de la marine. / ~ boiler ‖ Schiffskessel m. ‖ chaudière f. marine. / ~ chronometer ‖ Schiffschronometer m.; Schiffsuhr f. ‖ chronomètre m. de bord *ou* de marine. / ~ compass ‖ Marinekompaß m.; Schiffskompaß m.; Bussole f. ‖ boussole f. marine; compas de navires. / ~ compass for watches ‖ Uhrmacherkompaß m. ‖ boussole f. d'horlogerie. / ~ Diesel engine ‖ Schiffsdieselmotor m. ‖ moteur m. Diesel marin. / ~ Diesel engine used for propulsion of crafts ‖ zum Antrieb von Fahrzeugen dienender Schiffsdieselmotor m. ‖ moteur m. Diesel marin appliqué à la propulsion des navires. / ~ engine ‖ Schiffsmaschine f. ‖ machine f. (de) marine. / ~ engine for superheated steam and jet condensing ‖ Schiffsmaschine f. für (Betrieb mit) Heißdampf und Einspritzkondensation ‖ machine f. marine avec surchauffe et condensation à injection. / ~ engineering ‖ Schiffsmaschinenbau m. ‖ construction f. de machines marines. / ~ glass ‖ Marineglas n. ‖ jumelles fpl. de marine. / ~ glue ‖ Schiffsleim m.; Marineleim m. ‖ glue f. *ou* colle f. marine. / ~ installation ‖ Schiffsanlage f. ‖ installation f. marine. / ~ insurance ‖ Seeversicherung f. ‖ assurance f. maritime. / ~ insurance policy ‖ Seeversicherungspolice f. ‖ police f. d'assurance maritime. / ~ motor ‖ Schiffsmotor m. ‖ moteur m. marine. / ~ oil motor ‖ Schiffsölmotor m. ‖ moteur m. à huile pour bateaux.

mariner ‖ Seemann m.; Matrose m. ‖ marin m. / ~'s compass *see* marine compass.

marine salt ‖ Seesalz n.; Siedesalz n. ‖ sel m. marin *ou* de mer; salmare f. / ~ shaft ‖ Schiffswelle f. ‖ arbre m. de couche. / ~ silk ‖ Seeseide f. ‖ soie f. marine. / ~ steam engine ‖ Schiffsdampfmaschine f. ‖ machine f. à vapeur pour bateaux; machine f. à vapeur marine. / ~ turbine ‖ Schiffsturbine f. ‖ turbine f. pour bateaux. / ~ twin boiler ‖ Schiffsdoppel-

kessel m. ‖ chaudière f. double marine. / ~-type boiler ‖ Schiffskessel m. ‖ chaudière f. marine.

maritime code ‖ Schiffsgesetzbuch n. ‖ code m. maritime.

marjoram ‖ Majoran m. ‖ marjolaine f. / Spanish ~ ‖ Origanumkraut n. ‖ herbe f. de marjolaine d'Espagne.

marjoram oil ‖ Majoranöl n. ‖ essence f. de marjolaine.

mark, to ‖ marken; auszeichnen; markieren; signieren ‖ marquer; signer. / ~ (Goldsm) ‖ stempeln ‖ timbrer; estampiller. / ~ (Print) ‖ mit Markzeichen n. versehen ‖ repérer. / ~ (Techn) ‖ vorreißen ‖ troussequiner. / ~ by beacons (Mar) ‖ das Fahrwasser n. abbacken ‖ baliser; mettre des balises. / ~ with the centre punch ‖ ankörnen ‖ amorcer au pointeau (un trou). / ~ with ciphers ‖ beziffern ‖ chiffrer. / ~ with a number ‖ benummern; numerieren ‖ numéroter. / ~-out (Join; Carp) ‖ anreißen ‖ marquer; tracer. / ~-out straightlines upon soil ‖ ein Gelände ausbaken ‖ aligner ou jalonner un terrain. / ~ the pages of a book ‖ ein Buch n. foliieren oder paginieren ‖ numéroter les feuilles fpl. d'un livre. / ~ by pickets or by stakes ‖ abpfählen ‖ piqueter. / ~ products pl. ‖ Fabrikate npl. stempeln ‖ marquer des produits mpl.

mark ‖ Abzeichen n.; Marke f.; Merkzeichen n.; Stempel m. ‖ marque f.; insigne m.; repère m. / ~ (Characteristic) ‖ Merkmal n. ‖ marque f.; caractère m. / ~ (Signature) ‖ Signatur f. ‖ marque f.; étiquette f. / ~ (Trade mark) ‖ Fabrikmarke f.; Handelsmarke f. ‖ marque f. de fabrique. / ~ (Mar) ‖ Marke f.; Seezeichen n.; Landmarke f.‖ amer m.; marque f.; remarque f. / ~ of correction (Print) ‖ Korrekturzeichen n. ‖ marque f. des correcteurs ou de la correction. / counter ~ ‖ Gegenzeichen n. ‖ contremarque f. / ~ of distinction see distinguishing ~. / distinguishing ~ ‖ Unterscheidungsmerkmal n.; Wahrzeichen n.; Kennzeichen n. ‖ marque f. distinctive; signe m. caractéristique. / flood ~ ‖ Hochwassermarke f. ‖ marque f. de la pleine mer ou des plus hautes eaux. / high-water ~ see flood ~. / land ~ (Surv) ‖ Grenzstein m. ‖ borne f.; borne f. limitrophe. / ~ of suspension (Print) ‖ Gedankenstriche m. ‖ moins m.; tiret m.; trait m. de plume. / trade ~ ‖ Handelsmarke f. ‖ marque f. de fabrique.

marker (Hydr arch) ‖ Pegel n. ‖ échelle f. d'eau ou fluviale; marque m. d'eau. / ~ (Met) ‖ Meßmeister m. ‖ marqueur m. / ~ (Needl) ‖ Nadellocher m. ‖ marqueur m. / ~ (Weav) ‖ Markierer m. ‖ marqueur m.

market ‖ Markt m.; Absatzgebiet n. ‖ marché m.; débouché m. / ~ (Place) ‖ Markt m.; Marktplatz m. ‖ foire f.; marché m. / to be in the ~ ‖ reflektieren auf . . . ‖ être acheteur de . . . / to bring on the ~ / auf den Markt m. bringen ‖ lancer sur le marché. / covered ~ ‖ Markthalle f. ‖ halle f. de marché. / the ~ is depressed ‖ der Markt ist gedrückt ‖ le marché est lourd. / to open new ~s pl. ‖ neue Handelsbeziehungen fpl. anbahnen ‖ entamer de nouvelles relations fpl. commerciales. / principal ~ ‖ Hauptabsatzgebiet n. ‖ débouché m. principal.

/ to find a ready ~ ‖ guten Absatz m. finden ‖ être d'une bonne défaite; être de bonne vente. / high quality goods, that even with a smaller ~ turn out high profits ‖ hochwertige Erzeugnisse npl., die auch bei geringem Umsatz lohnenden Nutzen bringen ‖ produits mpl. de haute qualité qui, malgré leur faible vente, n'en donnaient pas moins un profit rémunérateur.

marketable ‖ verkäuflich; marktgängig ‖ vendable; courant.

market bag ‖ Markttasche f. ‖ sac m. à provisions. / ~ fluctuations pl. ‖ Marktschwankungen fpl. ‖ fluctuations fpl. du marché. / ~ gardening ‖ Gartenbaubetrieb m.; Handelsgärtnerei f. ‖ (établissement m. d')horticulture f. / ~ hall ‖ Markthalle f. ‖ halle f. de marché; marché m. couvert. / ~ house ‖ Kaufhaus n. ‖ grand magasin m. / ~ place ‖ Markt m.; Marktplatz m. ‖ foire f.; marché m.

market price ‖ Marktpreis m. ‖ prix m. courant ou du marché. / ~ (Bourse) ‖ Börsenkurs m. ‖ cours m. de la bourse. / below ~ ‖ unter Marktpreis m. ‖ en dessous du cours du marché. / ~ of discount ‖ Privatdiskont m. ‖ escompte m. hors banque.

market rate see market price.

market report ‖ Marktbericht m. ‖ bulletin m. du marché. / ~ (Bourse) ‖ Kursbericht m.; Börsenbericht m. ‖ bulletin m. de la bourse.

market stall see ~ standing. / ~ standing ‖ Marktbude f.; Marktstand m. ‖ baraque f.; échoppe f.; boutique f.; loge f. de foire. / ~ standing builder ‖ Marktbudenbauer m. ‖ constructeur m. de baraques pour marchés. / state of ~ ‖ Geschäftslage f. ‖ situation f. des affaires ou du marché.

marking of projectile ‖ Eindrücke mpl. der Züge auf einem Geschoß m. ‖ empreinte f. sur un projectile. / ~ of the cloth ‖ Markieren n. des Tuches ‖ marquage m. de drap. / ~ of tools ‖ Zeichnen n. von Werkzeugen ‖ marquage m. des outils.

marking awl ‖ Reißahle f.; Reißspitze f. ‖ pointe f. à tracer; traceret m. / ~ device ‖ Zeichenvorrichtung f. ‖ dispositif m. à marquer. / ~ gauge ‖ Reißmaß n.; Reißstock m.; Reißmodel m. ‖ tracequin m.; trusquin m.

marking-ink ‖ Zeichentinte f. ‖ encre f. à marquer. / ~ for linen ‖ Wäschezeichentinte f. ‖ encre f. à marquer le linge.

marking machine ‖ Markiermaschine f. ‖ machine f. à marquer.

marking pincers pl. for sheep ‖ Markierzange f. für Schafe ‖ pince f. à marquer les moutons. / ~ pole (Surv) ‖ Piket n.; Absteckpfählchen n. ‖ piquet m.; taquet m. / ~ strip (Tel) ‖ Bezeichnungsstreifen m. ‖ réglette f. de porte-étiquettes. / ~ tool ‖ Anreißwerkzeug n.; Anreißnadel f. ‖ outil m. à marquer; pointe f. à tracer.

mark pile (Surv) ‖ Markierpfahl m. ‖ fiche f.; piquet m. / ~ pole (Railw) ‖ Haltepfahl m.; Markierpfahl m. ‖ poteau m. d'arrêt. / ~ scraper ‖ Reißnadel f. ‖ pointe f. à tracer.

markstone (Mine) ‖ Lochstein m. ‖ pierre f. de borne.

marl, to (Agr) ‖ mergeln; mit Mergel m. düngen ‖ marner. / ~ down (Mar) ‖ marlen; (Taue) umwickeln ‖ merliner; transfiler.

marl ‖ Mergel m. ‖ marne f. / argillaceous ~ ‖ Mergelton m. ‖ marne f. argileuse; argile f. marne. / bituminous ~ ‖ bituminöser Mergel m.; Stückmergel m. ‖ marne f. bitumineuse. / calcareous ~ ‖ Kalkmergel m. ‖ marne f. calcaire. / clayey ~ see argillaceous. / compact ~ ‖ Mergelstein m.; verhärteter Mergel m. ‖ marne f. compacte. / earthy ~ ‖ Mergelerde f. ‖ marne f. cendrée. / red ~ ‖ Keupermergel m. ‖ marne f. irisée. / slaty ~ ‖ Mergelschiefer m.; Schiefermergel m. ‖ schiste m. marneux; marne f. schisteuse.

marline (Mar) ‖ Marling f.; Marlleine f.; merlin m. / ~ for sacks ‖ Sackband n. ‖ cordon m. à sac; ficelle f. pour fermer les sacs.

marl pellets pl. ‖ Tongalle f. ‖ taches fpl. d'argile. / ~ pit ‖ Mergelgrube f. ‖ marnière f.; carrière de marne f. / ~ slate see slaty marl. / spheroidal concretions of ~ ‖ Kalksteinnieren fpl.; Mergelnieren fpl. ‖ dés mpl. de van Helmont; marne f. sphéroïdale cloisonnée.

marlstone see marl, compact ~.

marly clay ‖ Mergelton m. ‖ argile f. marneuse. / ~ sandstone ‖ Mergelsandstein m. ‖ grès m. marneuse.

marmalade ‖ Marmelade f. ‖ marmelade f. / ~ boiling vessel ‖ Marmeladenkochkessel m. ‖ chaudron m. pour marmelade. / ~ maker ‖ Marmeladekocher m.; Muskocher m. ‖ cuiseur m. de marmelade.

marmolite ‖ Marmolith m. ‖ marmolite f.

marmorite ‖ Marmorit m.; Milchglas n. ‖ marmorite f.

maroon ‖ Kastanienbraun n.; Dunkelrotbraun n. ‖ couleur f. rouge foncé.

marquee stuff ‖ Markisenstoff m. ‖ étoffe f. pour persiennes.

marquetry ‖ Holzeinlegearbeit f. ‖ marqueterie f. / ~ cutting ‖ Holzausschnittarbeit f. ‖ découpage m. de marqueterie. / ~ inlayer ‖ Holzeinleger m. ‖ incrusteur m. sur bois.

married worker ‖ verheirateter Arbeiter m. ‖ ouvrier m. marié.

marrow ‖ Mark n. ‖ moëlle f.

marseille (Weav) ‖ Pikee m. ‖ piqué m.

marsh ‖ Sumpf m. ‖ marais m. / drained ~es pl. ‖ Polder m.; eingedeichte Landfläche f.; polder m.; vaste plaine f. protégée par des digues.

marshalling yard ‖ Verschiebebahnhof m.; Rangierbahnhof m. ‖ gare f. de triage.

marsh gas ‖ Sumpfgas n.; Erdgas n. ‖ gas m. (des) marais. / ~ motor ‖ Erdgasmotor m. ‖ moteur m. à gaz naturel.

marshland ‖ Sumpfgebiet n. ‖ terrain m. marécageux.

marshmallow ‖ Eibischzucker m. ‖ pâte f. de guimauve. / ~ flower ‖ Eibischblüte f. ‖ fleur f. de guimauve. / ~ root ‖ Eibischwurzel f. ‖ racine f. de guimauve.

marshman ‖ Marschbauer m.; Moorbauer m. ‖ habitant m. d'un pays marécageux.

marshy ‖ sumpfig; moorig ‖ marécageux. / ~ ground ‖ Sumpfboden m.; Moorboden m.; mooriger Boden m.; Marschboden m. ‖ terrain m. ou sol m. marécageux. / cultivation of ~ ground ‖ Moorkultur f. ‖ culture f. des marais. / ~ land see ~ ground. / ~ soil see ~ ground.

martensite (Met) ‖ Martensit m. ‖ martensite f.

martite ‖ Martit m. ‖ martite f.

marzipan *see* marchpane.

mash, to ~ (in) (Brew) ‖ einmaischen; maischen; einteigen ‖ empâter; démêler; brasser; encuver. / **~ off** ‖ abmaischen ‖ réunir les matières en cuve; transvaser.

mash (Brew) ‖ Maische f. ‖ moût m. / **~ apparatus** ‖ Maischapparat m. ‖ appareil m. à moût. / **~ charger** ‖ Maischbehälter m. ‖ réservoir m. à moût. / **~ goods** pl. ‖ Einmaischquantum n. ‖ versement m.

mashing (Brew) ‖ Einmaischen n. ‖ empâtage m. / **final ~ heat** ‖ Abmaischtemperatur f. ‖ température f. finale de saccharification. / **second ~** ‖ Aufmaischen n. ‖ brassage m.

mashing-in ‖ Einteigen n.; Einmaischen n. ‖ empâtage m.

mashing-off ‖ Abmaischen n. ‖ réunion f. des matières en cuve; transvasement m.

mashing tun ‖ Maischbottich m. ‖ bac m. à moût; brassin m.; cuve f. matière.

mash liquor (Brew) ‖ Guß m. ‖ eau f. de trempe.

mash machine ‖ Maischmaschine f. ‖ machine f. à moût *ou* à vaguer. / **~ with movable rakes** ‖ Maischmaschine f. mit beweglichen Schaufeln ‖ machine f. à vaguer à palettes mobiles. / **~ with solid rakes** ‖ Maischmaschine f. mit festen Schaufeln ‖ machine f. à vaguer à palettes fixes.

mashman ‖ Maischer m. ‖ mélangeur m.

mash tub ‖ Maischepfanne f. ‖ chaudière f. à trempe. / **~ tun** *see* mashing tun. / **~ water** *see* mash liquor.

mask ‖ Maske f. ‖ masque m. / **fencing ~** ‖ Fechtmaske f. ‖ masque m. d'escrime. / **gas ~** ‖ Gasmaske f. ‖ masque m. à gaz. / **gas ~ with chemical filter** ‖ Gasmaske f. mit Chemikalfilter ‖ masque m. à gaz à filtres chimiques. / **slaughtering ~** ‖ Schlachtmaske f. ‖ masque m. d'abattoir.

maslin (Agr) ‖ Mischfrucht f.; Mengkorn n. ‖ Mischkorn n. ‖ méteil m.; blé m. mêlé. / **~ flour** ‖ Mehl n. von Mengkorn ‖ farine f. de méteil.

mason ‖ Maurer m. ‖ maçon m. / **~'s broom** ‖ Netzpinsel m.; Annässer m.; Annetzer; Quast m. ‖ balai m. / **~'s brush** *see* **~'s broom.** / **yellow ~'s colour** ‖ Mauergelb m. ‖ badigeon m.; badigeon m. jaune. / **~'s labourer** ‖ Maurergehilfe m.; Handlanger m. ‖ aide-maçon m.

masonry ‖ Mauerwerk n.; Gemäuer n.; Mauerung f. ‖ maçonnerie f.; muraillement m. / **~** (Mason's work) ‖ Maurerarbeit f. ‖ travail m. du maçon. / **~ above the crown of formation** (Railw) ‖ Hochbau m. ‖ construction f. au-dessus du sol. / **artificial-stone ~** ‖ Mauerwerk n. aus künstlichen Steinen ‖ maçonnerie f. en pierres artificielles. / **bound ~** ‖ in Verband aufgeführtes Mauerwerk n. ‖ maçonnerie f. en liaison. / **cemented ~** ‖ Mauerung f. in Mörtel ‖ liaison f. de joint. / **dry ~** ‖ trockene Mauerung f. ‖ liaison f. à sec. / **fire-proof ~** ‖ feuerfeste Ausmauerung f. ‖ maçonnerie f. réfractaire. / **impervious ~** ‖ wasserdichtes Mauerwerk n. ‖ maçonnerie f. im-

perméable à l'eau. / **~ without mortar** *see* dry **~**. / **natural-stone ~** ‖ Mauerwerk n. aus natürlichen Steinen ‖ maçonnerie f. en moëllons. / **quarry-stone ~** ‖ Bruchsteinmauerwerk n. ‖ maçonnerie f. à moëllons. / **rough ~** (Blast furnace) ‖ Rauhgemäuer n. ‖ massif m. en maçonnerie.

masonry dam (Hydr arch) ‖ Staumauer f. ‖ barrage m. en maçonnerie. / **insulation of ~** ‖ Mauerwerkisolierung f. ‖ isolation f. de maçonnerie.

mass ‖ Masse f.; Menge f.; Volumen n. ‖ masse f. / **~ (of ore)** ‖ Mittel n. ‖ milieu m. / **~ of the air** ‖ Luftmasse f. ‖ masse f. d'air. / **~ of cold air** ‖ Kaltluftmasse f. ‖ masse f. d'air froid. / **gravitational ~** ‖ Schweremasse f.; Gravitationsmasse f. ‖ masse f. gravitationnelle. / **isolated ~ (of ore)** ‖ schwebendes Mittel n. ‖ milieu m. isolé. / **sterile ~ (of ore)** ‖ Gestein n. *oder* taubes Mittel n. ‖ milieu m. stérile. / **unfixed ~es** pl. (Mine) ‖ rollige Massen fpl. ‖ terrains mpl. mouvants.

mass action (Phys) ‖ Massenwirkung f. ‖ action f. des masses. / **law of ~** ‖ Massenwirkungsgesetz n. ‖ loi m. d'action des masses.

massage ‖ Massage f.; Knetkur f. ‖ massage m. / **~ apparatus** ‖ Massagegerät n. ‖ appareil m. de massage.

massicot ‖ Massikot n.; Neugelb n.; Bleigelb n.; Königsgelb n. ‖ massicot m.

massive ‖ massig; massiv; dicht; fest ‖ massif.

mass production ‖ Massenerzeugung f.; Massenfertigung f.; Massenherstellung f. ‖ fabrication f. en masse. / **~ of cheap ware** ‖ billige Massenerzeugung f. ‖ fabrication f. à bas prix et par grandes quantités. / **~ of installation materials** ‖ Massenfertigung f. von Installationsmaterial ‖ fabrication f. en grandes quantités de matériel d'installation.

mass velocity (Phys) ‖ Massengeschwindigkeit f. ‖ vitesse f. de la masse.

massy *see* massive.

mast, to ~ a ship ‖ ein Schiff bemasten ‖ mâter un navire.

mast ‖ lange Stange f.; Mast m. ‖ poteau m.; perche f.; mât m. / **~ (Shipb)** ‖ Mast m. ‖ mât m. / **~s** pl. (Shipb) ‖ Rundholz n.; espars mpl. ‖ spars mpl. ‖ Bemastung f.; Mastwerk n. ‖ mâture f. / **to cut away a ~** ‖ einen Mast kappen ‖ couper un mât. / **to secure a ~** ‖ einen Mast sichern ‖ assujettir un mât. / **~ of iron-concrete** ‖ Eisenbetonmast m. ‖ poteau m. en béton armé. / **jury ~** ‖ Notmast m. ‖ mât m. de fortune. / **lattice-work ~** ‖ Gittermast m. ‖ mât m. en treillis. / **portable ~** ‖ versetzbarer oder tragbarer Mast m. ‖ mât m. portatif. / **spring ~** ‖ gesprungener Mast m. ‖ mât m. craqué. / **steel sectional ~** ‖ Stahlmast m. in Teilen ‖ mât m. d'acier à sections.

mast crane ‖ Mastenkran m. ‖ grue f. à poteau *ou* à soulever les mâts.

master, to ~ a gradient ‖ eine Steigung überwinden ‖ franchir une rampe.

master (of an apprentice) ‖ Lehrherr m.; Lehrmeister m. ‖ patron m.; maître m. / **~ of the pilot station** ‖ Lotsenkommandör m. ‖ commandeur m. de la station de pilotes. / **~ of a vessel** ‖ Schiffsführer m.; Seeschiffer m.; Kapitän m. ‖ capitaine m. de navire; patron m.

master builder ‖ Architekt m.; Baumeister m. ‖ constructeur m.; architecte m. / **~'s certificate** (Mar) ‖ Kapitänszertifikat n. ‖ brevet m. de capitaine. / **~ clock** ‖ Zentraluhr f.; Normaluhr f.; Hauptuhr f.; Mutteruhr f. ‖ horloge f. centrale *ou* régulatrice *ou* principale; horloge f. mère. / **~ feed chuck with jaws** (Mach tool) ‖ Vorschubpatronenkörper m. mit Einsätzen ‖ corps m. de pinces d'avance avec mors. / **~ furrow** (Mill) ‖ Hauptfurche f. ‖ maître-rayon m. / **~ key** (Locksm) ‖ Hauptschlüssel m. ‖ passepartout m. / **~ mariner** *see* **~ of a vessel.** / **~ mason** ‖ Maurermeister m. ‖ maître-maçon m. / **~'s office** ‖ Meisterzimmer n. ‖ bureau m. du chef d'atelier. / **~ painter** ‖ Malermeister m. ‖ maître-peintre m. / **~ pattern** (Found) ‖ Originalmodell n. ‖ modèle m. original. / **~ piece** ‖ Musterstück n. ‖ pièce f. originale.

masterpiece ‖ Meisterstück n. ‖ chef m. d'œuvre.

master smelter (Found) ‖ Gießermeister m. ‖ maître-fondeur m. / **~ spring chuck with jaws** ‖ Spannpatronenkörper m. mit Einsätzen ‖ corps m. de pinces de serrage avec mors. / **~ tap** (Tool) ‖ Gewindebohrer m. ‖ taraud-mère m. / **~ worker** ‖ Meister m.; Werkführer m.; Werkmeister m. ‖ contremaître m.

masterwort root ‖ Meisterwurz f. ‖ racine f. d'impératoire.

mastery of the sea ‖ Seeherrschaft f. ‖ maîtrise f. des mers.

masthead ‖ Topp m. ‖ tête f. du mât.

mast heel ‖ Mastfuß m. ‖ pied m. d'un mât.

mastic *see also* cement *and* putty ‖ Kitt m.; Mastix m. ‖ mastic m.; lut m.; ciment m. / **acid proof ~** ‖ säurefester Kitt m. ‖ mastic m. résistant aux acides.

mastic asphaltum ‖ Gußasphalt m. ‖ asphalte m. mastic.

masticate, to ~ ‖ kauen; (zerkleinern) ‖ triturer; mâcher. / **~ (To knead)** ‖ kneten ‖ pétrisser.

masticator (For masticating meat) ‖ Fleischhackmaschine f. ‖ masticateur m.; hâchon à viande. / **~ (For kneading caoutchouc etc.)** ‖ Knetwerk n.; Knetmaschine f. ‖ machine f. à pétrir; pétrisseuse f.

mastic cement ‖ Zementmastix m.; Steinkitt m.; Mastixzement m. ‖ ciment m.; mastic m. / **~ manufacturing machinery** ‖ Kittherstellungsmaschine f. ‖ machine f. à fabriquer le mastic. / **~ varnish** ‖ Mastixfirnis m. ‖ vernis m. à mastiquer.

masting sheers pl. ‖ Mastenkran m. ‖ grue f. à soulever les mâts.

mast insulator ‖ Mastisolator m. ‖ isolateur m. de mât. / **~ maker** ‖ Mastenbauer m.; Mastmacher m. ‖ mâtier m.; mâteur m. / **~ mooring gear** ‖ Mastfesselgeschirr n. ‖ gréement m. *ou* agrès m. *ou* équipage m. *ou* dispositif m. d'amarrage au mât. / **~ plate bending machine** ‖ Mastplattenbiegemaschine f. ‖ machine f. à cintrer les plaques pour mâts de navires. / **~ powder for cattle** ‖ Viehmastpulver n. ‖ engrais m. en poudre pour les bestiaux.

masut ‖ Masut m. ‖ mazout m.

mat, to ~ (To interlace) ‖ flechten; verflechten ‖ natter; tresser; entrelacer. / **~ (To dull)** ‖ matt schleifen; mattieren ‖ dépolir; mater; matir.

mat ‖ Matte f. ‖ natte f.; paillasson m.; paillet m. / ~ of alfa ‖ Spartgrasmatte f. ‖ natte f. en alfa. / ~ of felt ‖ Fußmatte f. aus Filz ‖ paillasson m. en feutre. / ~ of rush ‖ Rohrmatte f.; Binsenmatte f. ‖ natte f. en jonc. / screen ~ (Hydr arch) ‖ Sinkmatte f. ‖ claie f. en forme de natte. / sewed ~ ‖ gewebte Matte f. ‖ tapis m. tissu. / step ~ (Auto; Coachm) ‖ Trittbrettmatte f. ‖ tapis-décrottoir m. / straw ~ ‖ Strohmatte f. ‖ paillasson m. / thrummed ~ ‖ gespickte Matte f. ‖ paillet m. lardé ou cordé. / wrought ~ ‖ englische Matte f. ‖ paillet m. tortu.

match, to ~ (To fumigate with sulphur) ‖ schwefeln ‖ soufrer; mécher. / ~ (Join) ‖ passend verbinden; anpassen; falzen ‖ bouveter; adapter; assembler.

match ‖ Zündholz n.; Streichholz n. ‖ allumette f. / ~ (Arm) ‖ Lunte f.; Zündschnur f. ‖ mèche f. (à canon); corde f. à feu; cordeau m. de mise en feu. / ~ (Coop) ‖ Schwefelgarn n. ‖ mèche f. soufrée. / Bengal ~es pl. ‖ bengalische Zündhölzer npl. ‖ allumettes fpl. du Bengale. / quick ~ ‖ Schwefelfaden m. ‖ mèche f. soufrée ou de soufre; fil m. soufré. / slow ~ see ~ (Arm). / stearin ~ ‖ Stearinzündholz n. ‖ allumette f. en stéarine. / sulphurated ~ see quick ~. / Swedish ~ ‖ schwedisches Zündholz n. ‖ allumette f. suédoise. / wax ~ ‖ Wachszündholz n. ‖ allumette f. en cire. / ~ of wood ‖ Zündholz n. (aus Holz) ‖ allumette f. en bois.

match box ‖ Zündholzschachtel f. ‖ boîte f. à allumettes. / table ~ ‖ Tischzündholzbehälter m. ‖ porte-allumettes m. de table. / ~ filling machine ‖ Zündholzschachtelfüllmaschine f. ‖ machine f. à remplir les boîtes à allumettes. / ~ making machine ‖ Zündholzschachtelfertigungsmaschine f. ‖ machine f. à fabriquer les boîtes à allumettes / ~ sanding-machine ‖ Zündholzschachtelbesandungsmaschine f. ‖ machine f. à sabler les boîtes d'allumettes.

match cord ‖ Feuerzeuglunte f. ‖ mèche-amorce f. de briquet. / ~ dipper ‖ Zündholztunker m. ‖ trempeur m. d'allumettes. / ~ dipping ‖ Eintauchen n. der Zündhölzer in die Zündmasse ‖ trempage m. d'allumettes.

matching of casks ‖ Schwefeln n. der Fässer ‖ soufrage m. des fûts.

match maker ‖ Zündholzarbeiter m. ‖ allumettier m. / ~ making machine ‖ Zündholzherstellungsmaschine f.; Zündholzherstellungsmaschine f. ‖ machine f. à fabriquer les allumettes. / ~ manufacturing ‖ Zündholzherstellung f. ‖ fabrication f. d'allumettes. / ~ paper ‖ Zündholzpapier n. ‖ papier m. pour allumettes. / ~ plane ‖ Spundhobel m.; Falzhobel m. ‖ bouvet m. (à joindre). / ~ splint ‖ Holzdraht m. ‖ fil m. en bois. / cutter of splints for ~es ‖ Zündholzschneider m. ‖ découpeur m. de bois d'allumettes. / ~ wood ‖ Zündhölzchenholz n.; Holz n. für Zündhölzer ‖ bois m. pour allumettes. / wood prepared for ~ ‖ für Zündhölzer zugerichtetes Holz n. ‖ bois m. préparé pour allumettes.

mate (Fellow workman) ‖ Geselle m.; Gehilfe m. ‖ compagnon m.; ouvrier-compagnon m.; camarade m.; aide m. / ~ (Mar) ‖ Matrose m.; Steuermann m.; Maat m. ‖ marinier m.; pilote m.; acide m.

maté ‖ Mate f.; Paraguaytee m. ‖ maté m.

material ‖ stofflich; materiell; körperlich ‖ matériel.

material ‖ Werkstoff m.; Material n.; Rohstoff m. ‖ matériel m.; matière f.; matériau m. / ~s pl. (Build) ‖ Baustoffe mpl. ‖ matériaux mpl. de construction. / bituminous ~ ‖ bituminöser Werkstoff m. ‖ matière f. bitumineuse. / ~s pl. to be filled up ‖ Spülgut n. ‖ matières fpl. à remblayer. / ~ to fill up the empty galleries (Mine) ‖ Versatz m.; Versatzstoffe mpl. ‖ remblais mpl. / filtering ~ (Railw) ‖ durchlässiges Material n. ‖ matériel m. perméable. / fossil ~ ‖ fossiler Rohstoff m. ‖ matière f. fossile. / ~ for mine equipment ‖ Grubeneinrichtungsgegenstände mpl. ‖ matériel m. d'installation de mines. / mineral ~ ‖ mineralischer Stoff m. ‖ matière f. minérale. / moderately hard ~ ‖ mittelharter Werkstoff m. ‖ matière f. mi-dure. / permeable ~ ‖ filterfähiger Werkstoff m. ‖ matière f. filtrable. / plastic ~ ‖ plastischer Stoff m. ‖ matière f. plastique. / ~ for rail permanent way ‖ Eisenbahnoberbaustoffe mpl. ‖ matériaux mpl. de superstructure de chemin de fer. / raw ~ ‖ Rohstoff m.; Werkstoff m. ‖ matière f. première. / starting ~ see raw ~. / used ~ ‖ Altmaterial n. ‖ (matériel m. usagé); vieux matériaux mpl. / ~s pl. to be washed ‖ Waschgut n. ‖ matières fpl. à laver.

material consumption ‖ Materialverbrauch m. ‖ consommation f. de matériaux. / ~ man ‖ Materialverwalter m. ‖ magazinier m. / ~ requisition ‖ Werkstoffbestellung f. ‖ commande f. de matériel. / selection of ~ ‖ Materialauswahl f. ‖ choix m. des matériaux. / ~ store ‖ Werkstofflagerplatz m ; Materialienlagerplatz m. ‖ dépôt m. de matériaux. / ~ stress ‖ Werkstoffbeanspruchung f.; Materialbeanspruchung f. ‖ effort m. sur les matériaux. / ~ test ‖ Werkstoffprüfung f. ‖ épreuve f. de matériaux. / ~ testing apparatus ‖ Werkstoffprüfvorrichtung f.; Materialprüfer m.; Werkstoffprüfer m. ‖ appareil m. pour l'essai des matériaux. / ~ testing laboratory ‖ Werkstoffprüfungsanstalt f. ‖ laboratoire m d'essai des matériaux. / ~ testing machine ‖ Werkstoffprüfmaschine f.; Materialprüfmaschine f. ‖ machine f. à essayer (la résistance) des matériaux.

maternity hospital ‖ Wöchnerinnenheim n. ‖ maternité f.; maison f. d'accouchement.

mathematical ‖ mathematisch ‖ mathématique. / ~ formula ‖ mathematische Formel f. ‖ formule f. mathématique. / ~ geography ‖ mathematische Erdkunde f. ‖ géographie f. mathématique. / ~ instrument ‖ mathematisches Instrument n. ‖ instrument m. de mathématique. / ~ instruments pl. ‖ Reißzeug n. ‖ compas mpl.; étui m. de mathématique. / box of ~ instruments see ~ instruments pl. / ~ signs pl. (Print) ‖ mathematische Zeichen npl. ‖ signes mpl. mathématiques. / ~ work ‖ mathematische Arbeit f. ‖ travail m. mathématique.

mathematics pl. ‖ Mathematik f. ‖ mathématiques fpl. ‖ mathématique f.

matico leaves pl. ‖ Matikoblätter npl. ‖ feuilles fpl. de matico.

mat maker ‖ Mattenflechter m. ‖ nattier m.

matrass (Chem) ‖ Glaskolben m.; Destillierkolben m. aus Glas ‖ matras m.; ballon m. en verre.

matrice see matrix.

matrix (Print) ‖ Form f.; Matrize f.; Schriftmutter f.; Gießmutter f. ‖ matrice f.; moule f. / ~ (Forge; Coin) ‖ Matrize f.; Unterlage f.; Unterstanze f.; Unterstempel m. ‖ matrice f. (à découper). / ~ (of the ore) ‖ Ganggestein n.; Gangart f. ‖ gangue f.; matrice f.; roche f. mère. / ~ (Roll mill) ‖ Mutterkaliber n. ‖ matrice f. / stamping ~ (Forg) ‖ Gesenk n. ‖ matrice f. à estamper.

matrix band ‖ Matrizenband n. ‖ métal laminé en feuilles à découper en flans. / ~ dryer ‖ Matrizentrockner m. ‖ sécheur m. de flans. / ~ drying drum with fan ‖ Matrizentrockner m. mit Ventilator ‖ séchoir m. à ventilateur pour flans. / ~ striking press ‖ Matrizenprägepresse f.; Schlagpresse f. ‖ presse f. à empreindre ou à estamper.

mat-surface paper for art printing ‖ Mattkunstdruckpapier n. ‖ papier m. mat pour impression artistique.

matter (Med) ‖ Eiter m. ‖ pus m. / ~ (Phys) ‖ Stoff m.; Materie f. ‖ matière f.; substance f. / ~ (Print) ‖ Satz m.; Schriftsatz m. ‖ composition f. / close ~ (Print) ‖ enger Satz m. ‖ composition f. pleine. / dead ~ ‖ Satz m. zum Ablegen; Ablegesatz m. ‖ types mpl. à distribuer. / elementary ~ of a rock ‖ Grundmasse f. eines Gesteins ‖ pâte f. (première) d'un minéral composé. / leaded ~ (Print) ‖ durchschossener Satz m. ‖ composition f. interlignée. / live ~ ‖ ungedruckter Satz f. ‖ composition f. vierge. / narrow ~ ‖ enger oder enggehaltener Satz m. ‖ composition f. serrée ou rapprochée. / open ~ ‖ Satz m. mit viel Durchschuß; stark durchschossener Satz m. ‖ composition espacée. / parallel ~ ‖ Satz nebeneinander ‖ composition f. opposée. / running-on ~ ‖ glatter Satz m. ‖ composition f. simple. / solid ~ see close ~. / standing ~ ‖ stehender Schriftsatz m.; Stehsatz m. ‖ composition f. permanente.

matting ‖ Mattenstoff m. ‖ matière f. à nattes / esparto grass ~ ‖ Sparterie f.; Spartgrasgeflecht m. ‖ sparterie f.

matting salt ‖ Mattsalz n. ‖ sel m. de mattage. / ~ tool (Engr) ‖ Mattpunze f. ‖ matoir m.

mattled iron ‖ halbiertes Eisen n. ‖ fonte f. truitée.

mattock ‖ Breithacke f.; Breithaue f. ‖ houe f.; pic m. d'avaleur; piochehache f.

mattress ‖ Matratze f. ‖ matelas m.; sommier m. / ~ (Hydr arch) ‖ Senkstück n.; Packwerk n. ‖ clayonnage m. claie f. / hair ~ ‖ Roßhaarmatratze f. matelas m. de crin. / horsehair ~ see hair ~. / spring ~ ‖ Federmatratze f.; Sprungfedermatratze f. ‖ sommier m. métallique ou élastique. / steel wire ~ ‖ Stahldrahtmatratze f. ‖ sommier m. (métallique) en fil d'acier.

mattress maker ‖ Matratzenmacher m. ‖ matelassier m.

maturing of magnets ‖ Altern n. von Magneten ‖ maturation f. des aimants.

maturity (Bill of exchange) ‖ Verfallzeit f.; Fälligkeit f. ‖ échéance f.; exigibilité f. / ~ (Forest) ‖ Haubarkeitsalter n. ‖

âge m. d'exploitabilité. / date of ~ ‖ Fälligkeitstag m.; Verfalltag m. ‖ échéance f.; terme m. de l'échéance.

mat work ‖ Netzwerk n.; Flechtwerk n. ‖ nattes fpl.; treillis m.; clayonnage m.

maul ‖ schwerer Hammer m.; Muskeule f. ‖ mailloche f.

maundril (Mine) ‖ Doppelkeilhaue f. ‖ pic m. à deux pointes.

maxim (Trade) ‖ Wahlspruch m. ‖ devise f.

maximal ‖ höchst; maximal; im höchsten Grade m. ‖ maximal.

maximum ‖ Maximum n.; höchster Grad m.; größte Menge f.; äußerste Grenze f.; Höhepunkt m. ‖ maximum m. / ~ (amount) ‖ Höchstbetrag m. ‖ la plus forte somme f.; maximum m. / ~ amount of work ‖ Höchstleistung f. ‖ maximum m. de rendement. / ~ amplitude of the oscillation ‖ Höchstausschlag m. der Schwingung ‖ déviation f. maximum de l'oscillation. / ~ and minimum-thermometer ‖ Maximum- und Minimum-thermometer n. ‖ thermomètre m. à maxima et à minima. / ~ area ‖ Maximalfeld n. ‖ champ m. d'exploitation maximum. / ~ automatic device (Electr) ‖ Maximalautomat m. ‖ disjoncteur m. à maxima. / ~ deflection of the oscillation see ~ amplitude of the oscillation. / ~ demand meter (Electr) ‖ Höchstverbrauchsmesser m. ‖ compteur m. à maximum. / ~ output ‖ Höchstleistung f. ‖ maximum m. de rendement. / ~ price ‖ Höchstpreis m. ‖ prix m. maximum. / ~ temperature ‖ Höchsttemperatur f.; Höchstwärmegrad m. ‖ température f. maximum. / ~ thermometer ‖ Höchstwärmegradmesser m.; Maximalthermometer n. ‖ thermomètre m. à maximum. / ~ value ‖ Höchstwert m.; Maximalwert m. ‖ valeur f. maxima. / ~ working capacity ‖ nutzbarer Inhalt m. ‖ capacité f. utile.

may flowers pl. ‖ Maiblumen fpl. ‖ muguets mpl.

mayonnaise ‖ Mayonnaise f. ‖ mayonnaise f. / ~ stirring machine ‖ Mayonnaiserührmaschine f. ‖ batteuse f. pour mayonnaise.

mazout see also masut ‖ Masut n. ‖ mazout m.

mead ‖ Met m. ‖ hydromel m. (vineux).

meadow ‖ Wiese f. ‖ pré m. / irrigated ~ ‖ Rieselwiese f.; Kunstwiese f.; künstlich bewässerte Wiese f. ‖ pré m. irrigué. / prepared ~ see irrigated ~ ‖ Kunstwiese f. ‖ pré m. irrigué.

meadow clover ‖ Wiesenklee m.; Rotklee m.; Futterklee m. ‖ trèfle m. commun (des prés). / ~ ground ‖ Wiesenboden m.; Wiesenland n. ‖ sol m. de prairie. / ~ iron ore ‖ Raseneisenerz n.; Wiesenerz n.; Sumpferz n. ‖ limonite f.; fer m. limoneux. / ~ land see ~ ground. / ~ ore see ~ iron ore.

meagre ‖ mager ‖ maigre.

meagreness (of supplies) ‖ Knappheit f. ‖ manque f.; modicité f. ‖ pénurie f.

meal ‖ Mehl n. ‖ farine f. / aromatized ~ ‖ aromatisiertes Mehl n. ‖ farine f. aromatisée. / coarse ~ ‖ Schrotmehl n. ‖ grosse farine f.; gruau m. / fermenting ~ ‖ Triebmehl n. ‖ farine f. fermentante. / infant's ~ ‖ Kindermehl n. ‖ farine f. lactée.

meal bench (Mill) ‖ Mehlbank f. ‖ plancher m. à farine. / ~ dust ‖ Staub-

mehl n. ‖ farine f. folle. / ~ groats pl. ‖ Grießmehl n. ‖ farine f. de semoule; grésillon m. / ~ powder composition (Pyrot) ‖ rascher Satz m. ‖ composition f. vive. / ~ times pl. ‖ Arbeitspause f. ‖ temps m. de repos; pause f.; récréation f. / ~ tub (Mill) ‖ Mehlkasten m.; Mehltonne f. ‖ huche f. à mouture; arche f.; récipient m. extérieur ou à boulange.

mean ‖ mittler(e); mittelmäßig; durchschnittsmäßig ‖ moyen. / ~ annual temperature ‖ mittlere Jahreswärme f. ‖ température f. moyenne de l'année. / ~ deviation ‖ durchschnittliche Abweichung f. ‖ écart m. moyen. / ~ error ‖ mittlerer Fehler m. ‖ erreur f. moyenne. / ~ height ‖ mittlere Höhe f. ‖ hauteur f. moyenne. / ~ level of the sea see ~ sea level. / ~ number ‖ Durchschnittszahl f.; Mittelwert m. ‖ moyenne f. / ~ paper ‖ Mittelpapier n.; Papier n. mittlerer Güte ‖ papier m. de qualité moyenne. / ~ place (Astron) ‖ mittlerer Ort m. ‖ lieu m. moyen. / ~ power ‖ mittlere Leistung f. ‖ puissance f. ou productivité f. moyenne. / ~ pressure see mean-pressure. / ~ proportional ‖ mittlere Proportionale f.; geometrisches Mittel n. ‖ moyen m. proportionnel; moyenne f. proportionnelle. / ~ sea leves ‖ (mittlerer) Meeresspiegel m.; Normalnull f. ‖ niveau m. moyen de la mer. / ~ value ‖ Mittelwert m. ‖ valeur f. moyenne. / ~ yield ‖ Durchschnittsertrag m. ‖ rendement m. moyen.

mean ‖ Durchschnitt m.; Mittel n.; Mittelwert m. ‖ moyen m.; moyenne f. / ~ (Math) ‖ Mittel n. ‖ moyen m.; moyenne f. / arithmetical ~ ‖ arithmetisches Mittel n. ‖ moyenne f. arithmétique. / geometrical ~ ‖ geometrisches Mittel n.; mittlere Proportionale f. ‖ moyenne f. proportionnelle ou géométrique. / harmonical ~ ‖ harmonisches Mittel n. ‖ moyenne f. harmonique.

mean-pressure boiler ‖ Mitteldruckkessel m. ‖ chaudière f. à moyenne pression. / ~ burner unit with **premixing of gas and air** ‖ Mitteldruckbrenner m. mit Gas-Luft-Vermischung ‖ brûleur m. à moyenne pression à prémélange d'air et de gaz. / ~ engine ‖ Mitteldruckmaschine f. ‖ machine f. à moyenne pression.

means pl. ‖ Hilfsmittel npl.; Mittel npl.; Werkzeug n. ‖ moyens mpl. / ~ (Money) ‖ Geldmittel npl.; ressources fpl. / without ~ ‖ mittellos ‖ dépourvu de ressources. / ~ of conveyance see ~ of transport. / ~ of protection ‖ Abdeckmasse f. ‖ matière f. protectrice. / ~ of transport ‖ Beförderungsmittel n.; Transportmittel n. ‖ moyen m. de transport.

measled (Meat) ‖ finnig ‖ ladre. / slightly ~ ‖ schwachfinnig ‖ quelque peu ladre.

measurable ‖ meßbar ‖ mesurable.

measure, to ‖ messen; vermessen ‖ mesurer; métrer. / ~ an angle (Surv) ‖ einen Winkel m. messen oder aufnehmen ‖ mesurer ou relever ou observer un angle. / ~ with the chain ‖ mit der Kette messen ‖ chaîner. / ~ clothes pl. ‖ Stoffe mpl. messen ‖ auner ou métrer ou mesurer les étoffes fpl. / ~ the depth of a shaft with the plumb line ‖ einen Schacht m. abseigern ‖ aplomber un puits m. / ~ the depth of water ‖ die Tiefe f. des Wassers messen ‖ mesurer ou sonder la

profondeur de l'eau. / ~ a ground ‖ ein Feld n. ausmessen ‖ arpenter un terrain. / ~ the length and width ‖ der Länge und Breite nach ausmessen ‖ mesurer en long et en large. / ~ out a liquid ‖ eine Flüssigkeit abmessen ‖ mesurer un liquide. / ~ with precision ‖ mit Genauigkeit messen ‖ mesurer ou vérifier avec précision. / ~ a ship ‖ ein Schiff ausmessen oder vermessen ‖ jauger un navire. / ~ the timber ‖ das Holz vermessen ‖ cuber le bois. / ~ the tonnage of a ship see ~ a ship.

measure ‖ Maß n. ‖ mesure f. / ~ (Rule) ‖ Maßstab m. ‖ règle f. divisée. / ~ (Print) ‖ Kolumnenbreite f. ‖ largeur f. de colonne. / ~ of altitude (Build) ‖ Höhenmaß n. ‖ élévation f.; mesure f. de la hauteur. / ~ of capacity ‖ Hohlmaß n. ‖ mesure f. pour les liquides; mesure f. de capacité. / coal ~s pl. (Mine) ‖ Kohlengebirge n. ‖ terrain m. houiller. / ~ of contraction see ~ of shrinkage. / cubic ~ ‖ Kubikmaß n.; Raummaß m.; Festmaß n. ‖ mesure f. pour les solides et les bois. / ~s pl. of the day (Build) ‖ Lichtenmaß n. ‖ échappée f. du jour. / dry ~ see ~ of capacity. / ~ of elevation see ~ of altitude. / ~ of length see linear ~. / linear ~ ‖ Längenmaß n. ‖ mesure f. de longueur. / long ~ see linear ~. / made to ~ ‖ maßhaltig ‖ porté au juste. / to put down all the ~s in the sketch ‖ sämtliche Maße npl. in die Skizze eintragen ‖ inscrire toutes les mesures fpl. sur le croquis. / ~ of shrinkage ‖ Schwindmaß n. ‖ mesure f. de retraite. / solid ~ see cubic ~. / square ~ see superficial ~. / standard ~ ‖ Normalmaßstab m. ‖ mesure f. normale. / superficial ~ ‖ Flächenmaß n. ‖ mesure f. de superficie; mesure f. carrée.

measured ‖ gemessen ‖ mesuré. / ~ in the clear (Build) ‖ im Lichten gemessen ‖ pris dans l'œuvre; mesuré ou pris dans le jour. / ~ within the walls see ~ in the clear. / ~ angle ‖ gemessener Winkel m. ‖ angle m. mesuré.

measurement see also measuring ‖ Messung f.; Vermessung f.; Ausmessung f.; Abmessung f. ‖ mesure f.; mesurage m. / mensuration f. / ~ (Mine) ‖ Markscheidung f. ‖ arpentage m. souterraine; mesurage m. d'une mine. / ~ (Drawing) ‖ Vermessung f.; Aufnahme f. ‖ mesurage m.; levé m. / ~ with cadmium ‖ Kadmiummessung f. ‖ mesure f. au cadmium. / calorimetric ~ ‖ kalorimetrische Messung f. ‖ mesure f. au calorimètre. / double ~ of charge ‖ Doppelkraftmessung f. ‖ mesurage m. double de charge. / comparative ~ ‖ vergleichende Messung f. ‖ mesure f. comparative. / ~ of displacements of spectrum lines ‖ Messung f. von Linienverschiebungen im Spektrum ‖ mesurement m. des déplacements des raies spectroscopiques. / exact ~ ‖ genaue Messung f. ‖ mesure f. exacte. / ~ of inclination by means of the spirit level ‖ Steigungsmessung f. mit der Wasserwage ‖ mesure f. de l'inclinaison au moyen du niveau à bulle d'air. / ~ of resistance ‖ Widerstandsmessung f. ‖ détermination f. de la résistance. / ~ of a ship ‖ Schiffsvermessung f. ‖ jaugeage m. d'un navire. / ~ of submarine cable ‖ (elektrische)

Seekabelmessung f. || mesure f. (électrique) des câbles sous-marins. / superficial ~ || Flächenvermessung f. || mesure f. ou mesurage m. de surfaces planes ou de la superficie d'un terrain. / synchronous ~ || gleichzeitige Messung f. || mesure f. synchrone. / ~ of thickness || Dickenmessung f. || mesure f. de l'épaisseur. / ~ of wind near the ground || Bodenwindmessung f. || mesure f. du vent au sol; mesure f. au sol de l'intensité du vent. / ~ of wind at high altitude || Höhenwindmessung f. || mesure f. du vent en altitude ou à une certaine altitude.

measurement, accuracy of ~ || Genauigkeit f. der Messung; Meßgenauigkeit f. || précision f. de la mesure. / high accuracy of ~ || große Meßgenauigkeit f. || haut degré m. de précision de la mesure. / central office of ~ || Meßzentrale f. || centrale f. de mesure. / importance of a ~ || Wichtigkeit n. einer Messung || importance f. d'une mesure. / operation of ~ || Meßtätigkeit f. || opération f. de mesurage. / rapidity of ~ || Meßgeschwindigkeit f. || rapidité f. de mesure. / ~ || ton (Shipb) || Raumtonne f. || tonneau m. d'encombrement.

measurer, automatic || selbsttätiger Meßapparat m. || appareil-mésureur m. automatique. / ~ of cloth || Stoffmesser m. || mesureur m. des étoffes. / ~ of mines || Markscheider m. || arpenteur m. des mines; géomètre m. souterrain. / sand and lime ~ || Meßapparat m. für Sand und Kalk || doseur m. pour sable et chaux.

measure, system of ~s || Maßsystem n. || système m. de mesure. / ~ tape || Bandmaß n. || mètre m. en ruban.

measuring see also measurement || Messung f.; Messen n.; Vermessung f. || mesure f.; mesurage m. / for ~ || zum Messen n. || pour mesurer. / ~ of altitudes || Höhenmessung f. || hypsométrie f. / ~ of base || Bodenstärkenmessung f. || mesurage m. de l'épaisseur du fond. / ~ between two points || Messung f. zwischen zwei Punkten || contrôle m. entre deux points. / ~ the depth of the iridic chamber of the eye || Tiefenmessung f. der Vorderkammer des Auges || mesure f. de la profondeur de la chambre antérieure de l'œil. / ~ of distances pl. || Abstandsmessung f. || mesure f. de distances. / ~ the inductive effects between high voltage lines and low voltage lines || Messen n. der gegenseitigen Beeinflussung von Starkstromleitungen und Schwachstromleitungen || mesure f. de l'induction mutuelle de lignes d'énergie et de lignes à courant faible. / ~ of the outside diameter || Außenrundmessung f. || mesurage m. à diamètre extérieur. / ~ by sight || Augenmaß n. || estimation f. à vue d'œil; coup m. d'œil. / ~ of spectrograms || Ausmessung f. von Spektrogrammen || mesurage m. des photographies spectrales. / ~ in V block || Messung f. im Schwenkprisma || contrôle m. avec vé à bascule.

measuring apparatus || Meßgerät n.; Meßvorrichtung f. || appareil m. de mesure. / ~ for astro-photography || Meßvorrichtung f. für astrofotografische Aufnahmen || appareil m. pour la mensuration des clichés astrophotographiques. / ~ for leather thicknesses || Leder-

dickenmeßvorrichtung f. || appareil m. à mesurer l'épaisseur de cuir.

measuring appliance || Vermessungsgerät n. || instrument m. d'arpentage. / ~ bridge (Electr) || Meßbrücke f. || pont m. de mesure. / ~ bridge with lever switches || Kurbelmeßbrücke f. || pont m. de mesure à manettes. / ~ cable || Prüfkabel n. || câble m. de mesure. / ~ can see measure of capacity. / ~ chain || Meßkette f.; Lachterkette f. || chaîne f. à mesurer ou d'arpenteur(s). / ~ circuit (Electr) || Meßstromkreis m. || circuit m. de mesure. / ~ clock || Meßuhr f. || compteur m. / ~ coil || Meßspule f. || bobine f. de mesure. / ~ cord || Meßschnur f. || corde f. à mesurer. / ~ cylinder || Meßzylinder m. || éprouvette f. en verre; éprouvette f. graduée; cylindre m. gradué.

measuring device || Meßvorrichtung f.; Meßgerät n.; Meßapparat m. || appareil m. de mesure. / electric ~ || elektrische Meßvorrichtung f. || appareil m. de mesure électrique. / paving stone ~ || Pflastersteinmeßmaschine f. || machine f. à mesurer les pavés. / ~ for sand || Abmeßapparat m. für Sand || trémie f. à doser ou doseur m. ou appareil-mésureur m. de sable.

measuring equipment || Meßausrüstung f. || garniture f. de mesurage. / ~ fault || Meßfehler m. || erreur f. de mesure. / ~ flask (Chem) || Meßkloben m. || ballon m. gradué ou de mesure. / ~ hopper || Meßtrichter m. || trémie f. de dosage.

measuring instrument || Meßinstrument n.; Meßgerät n.; Meßwerkzeug n. || instrument m. de mesure ou de mesurage ou de mensuration. / ~ (Surv) || Vermessungsgerät n. || instrument m. d'arpentage. || accurate ~ see precision ~. / cylinder ~ || Zylindermeßgerät n. || appareil m. à mésurer les cylindres. / geodetic ~ || geodätisches Meßwerkzeug n. || instrument m. de mesure géodésique. / internal ~ || Innenmeßgerät n. || appareil m. pour les mesures intérieures. / leather ~ || Ledermeßgerät n. || instrument m. à mesurer le cuir. / optical ~ || optisches Meßwerkzeug n. || appareil m. optique de mensuration. / precision ~ || Feinmeßgerät n.; Genauigkeitsmeßwerkzeug n.; Präzisionsmeßwerkzeug n. || instrument m. à mesurer de précision; instrument m. de précision à mesurer. / recording ~ || registrierendes Meßinstrument n. || instrument m. de mesure enregistreur.

measuring line || Meßstrich m. || trait m. de mesurage ou d'indication. / ~ (Electr) || Meßleitung f. || circuit m. de mesure. / variable artificial ~ (Electr) || variable künstliche Meßleitung || circuit m. artificiel variable de mesure.

measuring machine || Meßmaschine f. || machine f. de mesure. / ~ (Weav) || Meßmaschine f. || machine f. à auner ou de métrage. / leather surface ~ || Ledermeßmaschine f. || machine f. à mesurer les surfaces de cuir. / length ~ || Längenmeßmaschine f. || machine f. à mesurer la longueur. / ~ for ribbons || Meßmaschine f. für Bänder || machine f. servant au mesurage de rubans. / ~ for thread pitches || Meßmaschine f. für Gewindesteigungen || machine f. à mesurer les pas des filetages.

measuring microscope || Meßmikroskop n. || microscope m. de mesure. / ~ pin || Meßschnabel m.; Tastbolzen m. || touche f. de mesure ou à méplat. / ~ pin with ball || Tastbolzen m. mit Kugel || touche f. à bille. / ~ plate || Meßplatte f. || plaque f. de mesure. / ~ point || Tastpunkt m. || point m. de contact. / ~ range || Meßbereich m. || portée f. de mesure. / wide ~ range || weiter Meßbereich m. || portée f. ample de mesure. / ~ rod || Maßstab m. || règle f. / ~ rod for the descent of the charges (Met) || Gichtmaß n.; Gichtmesser m. || sonde f.; bécasse f. / ~ rule || Maßstock m. || jauge f. / ~ screw || Meßschraube f. || vis f. de mesure. / ~ staff || Meßlatte f. || règle f. divisée ou d'arpenteur; latte f. de mesure.

measuring system || Meßverfahren n. || système m. de mesure. / three-point ~ || Dreipunktmeßverfahren n. || système m. de mesure avec trois points de contact.

measuring tape || Meßband n.; Meßschnur f.; Bandmaß n. || ruban-mesure m.; mètre m. à ruban. / steel ~ || Stahlmeßband n.; Stahlbandmaß n. || ruban m. d'arpentage en acier mètre m. en ruban d'acier.

measuring tin for benzine || Benzinmeßkanne f. || bidon m. à mesurer l'essence. / ~ tool see also measuring instrument || Meßwerkzeug n. || outil m. de mesure. / ~ transformer (Tel) || Meßtransformator m.; Meßübertrager m. || transformateur m. de mesure ou de symétrie. / ~ vessel (Chem) || Meßgefäß n. || burette f. / ~ voltage || Meßspannung f. || tension f. de mesure. / ~ wheel || Meßrad n. || podomètre m.; rouet m. d'arpenteur. / electro-dynamic ~ work || elektrodynamisches Meßwerk n. || engregistreur m. avec mécanisme électrodynamique.

meat || Fleisch n.; Schlachtfleisch n. || viande f. (de boucherie). / canned ~ || Büchsenfleisch n.; Fleischkonserve f. || conserve f. de viande. / cooked ~ || gekochtes Fleisch n. || viande f. cuite. / fresh ~ || frisches Fleisch n. || viande f. fraîche. / fresh cooled ~ || frisch gekühltes Fleisch n. || viande f. fraîche congelée. / fresh-killed ~ || frisch geschlachtetes Fleisch n. || viande f. fraîchement abattue. / ~ in fresh state preserved by a refrigerating process || in frischem Zustand durch Kühlverfahren haltbargemachtes Fleisch n. || viande f. conservée à l'état frais par un procédé de congélation. / frozen ~ || Gefrierfleisch n. || viande f. congelée ou réfrigérée ou frigorifiée. / preserved ~ || Fleischkonserve f.; haltbar gemachtes Fleisch n. || conserve f. de viande; viande f. conservée. / preserved ~ in tins || eingemachtes Fleisch n. in Büchsen; Büchsenfleisch n. || conserves fpl. de viande en boîtes. / salted ~ || gesalzenes Fleisch n.; Salzfleisch n. || viande f. salée. / smoked ~ || geräuchertes Fleisch n.; Räucherfleisch n. || viande f. fumée. / tinned ~ see canned ~.

meat broth || Fleischbrühe f.; Fleischsuppe f. || bouillon m. / ~ (Extract of meat) || Fleischsaft m. || extrait m. ou jus m. de viande. / ~ with peptone gelatine || Fleischsaftpeptongelatine f. || extrait m. de viande peptonisé et gélatinisé. / ~ gelatine || Fleischsaftgelatine f. || extrait m. de viande gélatinisé.

meat canning *see* canned meat. / ~ carrying ‖ Fleischbeförderung f.; Fleischtransport m. ‖ transport de viande f. / ~ chamber ‖ Fleischzelle f. ‖ chambre f. à viande. / ~ chill room *see* ~ cooling room.

meat chopper (Person) ‖ Fleischhacker m. ‖ hacheur m. de viande. / ~ (Machine) ‖ Fleischhackmaschine f. ‖ hache-viande m.; machine f. à hacher la viande.

meat cooling room ‖ Fleischkühlraum m.; Fleischkühlhalle f. ‖ salle f. frigorifique à viande; chambre f. froide à viande. / ~ cutter ‖ Fleischer m.; Fleischhauer m. ‖ dépeceur m. de viande. / ~ decoction *see* meat broth. / ~ extract ‖ Fleischextrakt m. ‖ extrait m. de viande. / ~ flour ‖ Fleischmehl n. ‖ farine f. de viande. / ~ flour factory ‖ Fleischmehlfabrik f. ‖ fabrique f. de farine de viande. / ~ freezing room ‖ Fleischgefrierraum m. ‖ chambre f. de congélation (pour) de (la) viande. / ~ gelatine ‖ Fleischgelatine f. ‖ gélatine f. de viande. / ~, ham and sausage slicing machine ‖ Aufschnittschneidemaschine f. ‖ machine f. à couper la charcuterie. / ~ hook ‖ Fleischhaken m. ‖ crochet m. pour la viande. / ~ inspection ‖ Trichinenschau f.; Fleischbeschauung f. ‖ inspection f. des viandes. / ~ jelly ‖ Sülze f.; Sulze f.; Fleischsülze f. ‖ viande f. à la gelée; gelée f. / ~ jelly manufacturer ‖ Sulzer m. ‖ fabricant m. de viande en gelée. / ~ peptone ‖ Fleischpepton n. ‖ peptone f. de viande. / preparation of ~ ‖ Zubereitung f. von Fleisch ‖ préparation f. de viande. / ~ press ‖ Fleischpresse f. ‖ presse-viande f. / ~ safe ‖ Speiseschrank; Vorratschrank; Aufbewahrungsschrank m.; Fliegenschrank m. ‖ garde-manger m.; armoire f. à viande. / ~ salter ‖ Fleischpökler m. ‖ salaisonnier m. / ~ screen *see* ~ safe. / ~ ham and sausage slicing machine ‖ Aufschnittschneidemaschine f. ‖ machine f. à couper la charcuterie. / ~ smoker ‖ Fleischräucherer m. ‖ fumeur m. de viande. / ~ store ‖ Fleischproviant m. ‖ provision f. en viande. / ~ van ‖ Fleischwagen m.; Fleischtransportwagen m. ‖ voiture f. à transporter la viande. / ~ waste ‖ Fleischabfälle mpl. ‖ déchets mpl. de viande. / ~ waste and carcass utilization plant ‖ Fleischabfälle- und Kadaververwertungsanlage f. ‖ installation f. pour l'utilisation des déchets de viande et des cadavres d'animaux. / ~ water ‖ Fleischwasser n. ‖ bouillon m. de viande. / ~ water gelatine ‖ Fleischwassergelatine f. ‖ bouillon m. gélatinisé.

mechanic ‖ Mechaniker m. ‖ mécanicien m. / ~ (Artisan) ‖ Handwerker m. ‖ homme m. de métier; ouvrier m.; artisan m. / ~'s lathe ‖ Mechanikerdrehbank f. ‖ tour m. pour mécaniciens.

mechanical ‖ mechanisch ‖ mécanique. / ~ (By routine) ‖ handwerksmäßig ‖ technique; machinalement. / ~ adjustment ‖ mechanische Verstellung f. ‖ réglage m. mécanique. / ~ advertising figure for show-windows ‖ mechanische Schaufensterreklamefigur f. ‖ figurine f. mécanique de réclame pour devantures. / ~ apparatus of precision ‖ feinmechanischer Apparat m. ‖ appareil m. de précision. / ~ chair ‖ verstellbarer Lehn-

stuhl m. ‖ fauteuil m. mécanique. / ~ device ‖ mechanische Vorrichtung f. ‖ dispositif m. mécanique. / ~ drive ‖ mechanischer Antrieb m. ‖ commande f. mécanique. / ~ energy ‖ mechanische Energie f. ‖ énergie f. mécanique. / ~ engineering ‖ Maschinenbau m. ‖ construction f. mécanique. / ~ furnace ‖ mechanische Feuerung f. ‖ foyer m. mécanique. / ~ gas-seller ‖ Gasautomat m. ‖ distributeur m. automatique de gaz. / ~ instrument of precision ‖ feinmechanisches Instrument n. ‖ instrument m. de précision. / ~ method ‖ mechanische Methode f. ‖ méthode f. mécanique. / ~ moulding ‖ Maschinenformerei f. ‖ moulage m. à la machine. / ~ part ‖ maschineller Teil m. ‖ partie f. mécanique. / ~ plant ‖ maschinelle Einrichtung f. ‖ installation f. mécanique. / ~ power ‖ mechanische Leistung f. ‖ rendement m. utile. / ~ press (Print) ‖ Schnellpresse f. ‖ presse f. mécanique. / ~ principle ‖ mechanisches Prinzip n. ‖ principe m. de mécanique. / ~ toy ‖ mechanisches Spielzeug n. ‖ jouet m. mécanique. / ~ workshop ‖ mechanische Werkstatt f. ‖ atelier m. mécanique; usine f.

mechanically driven conveyor trough ‖ mechanisch bewegte Förderrinne f. ‖ rigole f. de transport à commande mécanique. / ~ operated tilting support ‖ maschinell verstellbare Kippvorrichtung f. ‖ inclinaison f. mécanique du bâti; dispositif-basculeur m. automatique.

mechanician *see* mechanic.

mechanics pl. ‖ Mechanik f. ‖ mécanique f. / ~ of elastic fluids ‖ Mechanik f. gasförmiger Körper; Aeromechanik f. ‖ mécanique f. des fluides aériformes. / ~ of fluids ‖ Mechanik f. flüssiger Körper; Hydromechanik f. ‖ mécanique f. des fluides; hydraulique f. / ~ of precision ‖ Feinmechanik f. ‖ mécanique f. de précision. / ~ of rigid bodies ‖ Mechanik f. fester Körper ‖ mécanique f. des corps solides.

mechanism ‖ Mechanismus m.; Vorrichtung f.; Einrichtung f. ‖ mécanisme m. / motive ~ ‖ Bewegungsmechanismus m. ‖ appareil m. de mouvement. / recording ~ ‖ Registriervorrichtung f. ‖ mécanisme m. enregistreur.

mechanist *see* mechanic.

mechanization ‖ Mechanisierung f.; Verkraftung f. ‖ mécanisation f. / army ~ and motorization ‖ Heeresverkraftung f. ‖ mécanisation f. et motorisation f. de l'armée.

mechanize, to ‖ mechanisieren; verkraften ‖ mécaniser.

mechanized army ‖ verkraftetes Heer n. ‖ armée f. mécanique. / ~ fighting brigade ‖ mechanisierte Kampfbrigade f. ‖ brigade f. de combat mécanisée. / ~ forces pl. (Arm) ‖ mechanisierte Streitkräfte fpl. ‖ forces fpl. motorisées.

medal ‖ Medaille f.; Denkmünze f.; Schaumünze f. ‖ médaille f. / ~ engraver ‖ Medaillengravör m. ‖ graveur m. en médailles. / ~ engraving ‖ Medaillengravierung f. ‖ gravure f. en médailles.

medallion ‖ Medaillon n. ‖ médaillon m.

medal manufacturing ‖ Medaillenprägeanstalt f. ‖ fabrique f. de médailles.

median (Statistics) ‖ Zentralwert m.; Mittelwert m.; Mediane f. ‖ médiane f. / ~ line ‖ Mittellinie f. ‖ ligne f. médiane.

mediation ‖ Zwischenschaltung f. ‖ intermédiaire m.; interpolation f.

medical ‖ ärztlich ‖ médical. / ~ apparatus pl. and instruments ‖ medizinische Apparate mpl. und Instrumente npl. ‖ appareils mpl. et instruments mpl. de médecine. / ~ capsule ‖ Arzneikapsel f.: pharmazeutische Kapsel f. ‖ capsule f. médicamenteuse. / ~ confectionery ‖ medizinische Zuckerwaren fpl. ‖ sucreries fpl. *ou* dragées fpl. médicamenteuses. / ~ examination ‖ ärztliche Untersuchung f. ‖ visite f. médicale; perquisition f. / ~ instrument ‖ medizinisches Instrument n. *oder* Gerät n. ‖ instrument m. de médecine. / ~ investigation ‖ medizinische Untersuchung f. ‖ recherche f. médicale. / ~ lozenges pl. *see* ~ confectionery. / ~ plant ‖ Heilpflanze f. ‖ plante f. médicinale. / ~ plant gatherer ‖ Heilpflanzensammler m. ‖ ramasseur m. *ou* herborisateur m. de plantes médicinales. / ~ plaster ‖ Pflaster n. ‖ emplâtre m. / ~ quartz lamp ‖ Quarzlampe f. für medizinische Zwecke ‖ lampe f. de quartz à l'usage médicale. / ~ soap ‖ Arzneiseife f.; Medizinalseife f.; medizinische Seife f. ‖ savon m. médicamenteux *ou* médicinal. / ~ wine ‖ Arzneiwein m.; Medizinalwein m. ‖ vin m. médical *ou* médicinal. / ~ wood ‖ Arzneiholz n. ‖ bois m. médicinal.

medicated soap *see* medical soap. / ~ wine *see* medical wine.

medicinal *see also* medical ‖ medizinisch ‖ médicamenteux; médical.

medicine ‖ Arznei f.; Heilmittel n.; Medizin f. ‖ médecine f.; médicament m. / homœopathic ~ ‖ homöopathisches Arzneimittel n. ‖ médicament m. homéopathique.

medicine chest ‖ Medizinkiste f.; Medizinkasten m.; Hausapotheke f.; Arzneikasten m. ‖ coffre m. de médicaments; pharmacie f. domestique. / pocket ~ ‖ Taschenapotheke f. ‖ pharmacie f. de poche. / portable ~ ‖ Reiseapotheke f. ‖ pharmacie f. de voyage. / ~ for ships ‖ Schiffsapotheke f. ‖ coffre m. de médicaments *ou* pharmacie f. pour bateaux. / ~ for the tropics ‖ Tropenapotheke f. ‖ pharmacie f. pour pays chauds.

medicine dropper ‖ Tropfglas n. ‖ compte-goutte m. / ~ glass ‖ Arzneiglas n.; Medizinglas n. ‖ verre m. à médecine. / ~ glassware ‖ medizinische Glasware f. ‖ verrerie f. pour usages médicaux. / ~ spoon ‖ Arzneilöffel m.; Medizinlöffel m. ‖ cuiller f. à médecine. / ~ stopper ‖ Medizinkorken m. ‖ bouchon m. pour verres à médicaments.

medico-optical instrument ‖ medizinisch-optisches Instrument n. ‖ instrument m. médico-optique.

medlar ‖ Mispel f. ‖ nèfle f.

medium (Math) *see* mean (Math). / ~ (Phys) ‖ Medium n.; Mittel n. ‖ milieu m. / dense ~ ‖ dichtes Mittel n. *oder* Medium n. ‖ milieu m. dense. / rare ~ ‖ dünnes Medium n. *oder* Mittel n. ‖ milieu m. rare. / refractive ~ ‖ brechendes Mittel n. ‖ milieu m. réfringent. / resistent ~ ‖ widerstehendes Mittel n. ‖ milieu m. résistant. / subtile ~ *see* rare ~.

medium contact (Electr) ‖ mittlerer Kontakt m. ‖ contact m. médian.

medium-faced (Print) ‖ halbfett ‖ demigras.

medium iron rolling mill ‖ Mitteleisenwalzwerk n.; Mitteleisenstraße f. ‖ laminoir m. à pièces de dimensions moyennes *ou* à fers moyens. / ~ iron train *see* ~ iron rolling mill. / ~ paper ‖ Medianpapier n. ‖ grand papier m. / ~ pitch cutter bar (Harvester) ‖ Schneidbalken m. für Mittelschnitt ‖ barre f. coupeuse pour coupe moyenne. / ~ size ‖ Mittelgröße f. ‖ grosseur f. moyenne.

medium-soft ‖ mittelweich ‖ mi-tendre.

medium tank (War mat) ‖ mittelschwerer Kampfwagen m. ‖ char m. médium; médium tank m.

medullary ray (Wood) ‖ Markstrahl m. ‖ rayon m. médullaire.

meerschaum ‖ Meerschaum m. ‖ écume f. de mer. / false ~ ‖ unechter Meerschaum m. ‖ écume f. de mer fausse. / regenerated ~ ‖ regenerierter Meerschaum m. ‖ écume f. de mer recomposée *ou* récuperée.

meerschaum articles pl. ‖ Meerschaumwaren fpl. ‖ objets mpl. en écume de mer. / ~ cutter ‖ Meerschaumschnitzer m. ‖ sculpteur m. sur écume de mer. / ~ engraver ‖ Meerschaumgravör m. ‖ graveur m. sur écume de mer.

meerschaum pipe ‖ Meerschaumpfeife f. ‖ pipe f. en écume de mer. / ~ turner ‖ Meerschaumpfeifendrechsler m. ‖ tourneur m. de pipes d'écume de mer.

meet, to ~ (Vein of mines) ‖ sich scharen ‖ se réunir.

meeting ‖ Versammlung f. ‖ assemblée f. / ~ of shareholders ‖ Generalversammlung f. ‖ assemblée f. d'actionnaires. / statutory ~ ‖ satzungsmäßige Versammlung f. ‖ assemblée f. statutaire.

megabromite ‖ Megabromit m. ‖ mégabromite f.

megadyne ‖ Megadyne f. ‖ mégadyne f.

megaphone ‖ Megafon n.; Sprachrohr n. ‖ mégaphone m.

megger *see* megohmmeter.

megohmmeter ‖ Megohmmesser m. ‖ mégohmmètre m.

megrim crayon ‖ Migränestift m. ‖ crayon m. antimigraine.

Meidinger cell (Electr) ‖ Meidinger-Element n. ‖ élément m. de Meidinger.

Meissen china ‖ Meißener Porzellan n. ‖ porcelaine f. de Meissen.

melalite insulator ‖ Melalithisolator m. ‖ isolateur m. en mélalithe.

melanite (Miner) ‖ Melanit m. ‖ mélanite f.

melissa blossom ‖ Melissenblüte f. ‖ fleur f. de mélisse. / ~ oil ‖ Melissenöl n. ‖ essence f. de mélisse.

mellite (Miner) ‖ Honigstein m.; Mellit m. ‖ mellite f.

mellowness (Brew) ‖ Auflösung f.; Gare f. ‖ désagrégation f.; dissolution f.; friabilité f.

melodious ‖ wohlklingend ‖ harmonieux; sonore; euphonique.

melon ‖ Melone f. ‖ melon m. / ~ peel ‖ Melonenschale f. ‖ écorce f. de melon. / ~seed oil ‖ Melonenöl n. ‖ huile f. de petit béraff *ou* de melon.

melt, to ~ *see also* to smelt ‖ schmelzen ‖ (se) fondre; liquéfier; mettre en fusion. / ~ (Fuse) ‖ durchschmelzen ‖ fondre. / ~ down ‖ einschmelzen ‖ refondre. / parts pl. as are supposed to be melted into

glass ‖ Konstruktionsteile mpl., die in Glas eingeschmolzen werden sollen ‖ pièces fpl. destinées à être moulées dans le verre. / the ice melts ‖ das Eis taut auf ‖ la glace fond. / ~ off ‖ abschmelzen ‖ séparer par fusion. / ~ together (Chem) ‖ einschmelzen; zusammenschmelzen ‖ fondre ensemble. / melted cement ‖ Schmelzzement m. ‖ ciment m. fondu.

melt ‖ Schmelze f.; geschmolzene Masse f. ‖ masse f. fondue.

melter *see also* smelter (Glass) ‖ Glasschmelzer m. ‖ surveillant m. du four de verrerie. / ~ (Met) ‖ Tiegelgießer m. ‖ fondeur m.; verseur m. / ~ pin (Found) ‖ Haltestift m. ‖ épingle f. du fondeur.

melting *see also* smelting ‖ schmelzend ‖ fondant. / high ~ ‖ hochschmelzend ‖ à point de fusion élevé.

melting *see also* smelting ‖ Schmelzen n. ‖ fonte f. / ~ of snow ‖ Schneeschmelze f. ‖ fonte f. de la neige. / ~ of sugar ‖ Zuckerschmelzen n. ‖ fonte f. du sucre.

melting charge ‖ Schmelzgut n.; Schmelzstoff m.; Einsatz m.; Beschickung f. ‖ matière f. à fondre *ou* de fusion; charge f. / ~ cone ‖ Brennkegel m.; Schmelzkegel m.; Segerkegel m. ‖ cône m. pyrométrique. / ~ crucible ‖ Schmelzkessel m. ‖ marmite f. *ou* chaudière f. à fusion. / ~ furnace ‖ Schmelzofen m.; Gießofen m. ‖ four m. à fondre; fourneau m. de fusion. / ~ heat ‖ Schmelzwärme f. ‖ chaleur f. de fusion. / ~ house ‖ Gießerei f. ‖ fonderie f. / ~ kettle for glass smelting ‖ Glasschmelzhafen m. ‖ creuset m. à fondre le verre. / ~ ladle ‖ Schmelzlöffel m. ‖ cuiller m. à fondre. / ~ material ‖ Schmelzgut n. ‖ matériel m. de fusion. / ~ pan ‖ Schmelzpfanne f. ‖ poche f. de coulée.

melting point ‖ Schmelzpunkt m.; Fließpunkt m. ‖ point m. de fusion. / apparatus for determination of the ~ ‖ Schmelzpunktbestimmungsapparat m. ‖ appareil m. pour la détermination du point de fusion. / to lower the ~ ‖ den Schmelzpunkt m. erniedrigen ‖ abaisser le point de fusion.

melting pot (Met) ‖ Schmelztiegel m.; Gießpfanne f. ‖ creuset m. / ~ (Letter f) ‖ Schmelzkessel m. ‖ fondoir m.

melting process ‖ Schmelzvorgang m. ‖ procédé m. de fusion. / process of ~ out ‖ Ausschmelzverfahren ‖ procédé m. d'extraction par fusion. / ~ trough ‖ Schmelztrog m. ‖ cuve f. du cubilot. / ~ water ‖ Schmelzwasser n. ‖ eau f. de fonte.

member ‖ Glied n. ‖ membre m. / artificial ~ ‖ künstliches Glied n. ‖ membre m. artificiel. / cross ~ (Mach) ‖ Querträger m. ‖ traverse f. / ~ of the directorate ‖ Mitglied n. des Direktoriums ‖ membre m. du comité de direction. / ~ of an equation ‖ Seite f. einer Gleichung ‖ membre m. d'une équation. / intermediate ~ (Mach) ‖ Zwischenträger m. ‖ entretoise f. / ~ of the machine ‖ Maschinenteil m. ‖ organe m. de machine. / ~s pl. of a machine which have plain regular shapes ‖ einfach geformte Maschinenteile mpl. ‖ organes mpl. de machine de forme simple et régulière. / ~ of office staff ‖ Bürobeamter m. ‖ employé m. de bureau. / side ~ (Mach) ‖ Längsträger m. ‖ longeron m.

membrane ‖ Membran f. ‖ membrane f. / ~ of the cell (Wood) ‖ Zellwand f. ‖ membrane f. de la cellule. / choroid ~ (Opt) ‖ Aderhaut f. ‖ membrane f. choroïde. / transparent ~ ‖ durchsichtige Membrane f. ‖ membrane f. transparente.

membrane pump ‖ Membranpumpe f. ‖ pompe f. à membrane.

memo-book *see* memorandum book.

memorandum (Trade) ‖ Memorandum n.; Nota f. ‖ bordereau m.

memorandum book ‖ Notizbuch n.; Merkbuch n. ‖ carnet m.; calepin m.; livre m. de notes. / ~ with rings ‖ Loseblätter-Notizbuch n. mit Heftringen ‖ calepin m. à feuilles détachées et à anneaux.

men pl. (Mar) ‖ Leute pl.; Mannschaft f. ‖ monde m. / ~ (Mine) ‖ Belegschaft f. ‖ équipe f. (des mineurs).

mend, to ‖ ausbessern ‖ réparer; raccommoder. / ~ (To retouch) ‖ nacharbeiten; nachbessern; retuschieren ‖ retoucher; réparer; rejuster. / ~ cast-iron ‖ Gußeisen n. schweißen ‖ souder la fonte.

mender ‖ Ausbesserer m. ‖ réparateur m.; rhabilleur m.; raccommodeur m. / linen ~ ‖ Wäscheausbesserin f. ‖ raccommodeuse f. de linge.

mending (Cloth) ‖ Stopfen n.; Ausbessern n. ‖ réparation f.; raccommodage m. / garment ~ ‖ Kleiderausbesserung f. ‖ raccommodage m. de vêtements.

mending tailor ‖ Flickschneider m. ‖ tailleur m. raccommodeur.

meniscal ‖ halbmondförmig; mondförmig; meniskenförmig ‖ en forme de ménisque.

meniscus ‖ Meniskus m.; Flüssigkeitskuppe f. ‖ ménisque m. / ~ of mercury ‖ Quecksilberkuppe f. ‖ ménisque m. de mercure.

mensuration *see* measurement *and* measuring.

mental calculations pl. ‖ Kopfrechnen n. ‖ calcul m. mental. / ~ computation *see* ~ calculations.

menthol ‖ Menthol n. ‖ menthol m. / ~ pencil ‖ Mentholstift m.; Migränestift m. ‖ crayon m. de menthol *ou* à migraine. / ~ valerianate ‖ baldriansaures Menthol n. ‖ valérianate m. de menthol.

mephitic gas (Mine) ‖ böse Wetter npl.; Nachschwaden mpl. ‖ gaz mpl. méphitiques *ou* délétérés.

mercantile discount ‖ Nachlaß m.; Rabatt m.; rabais m.; remise f. / ~ fleet ‖ Handelsflotte f. ‖ flotte f. marchande. / ~ marine ‖ Handelsmarine f. ‖ marine f. marchande *ou* du commerce. / ~ system ‖ Merkantilsystem n. ‖ système m. mercantile.

mercerization *see* mercerizing.

mercerize, to ‖ merzerisieren ‖ merceriser.

mercerized cotton ‖ merzerisierte Baumwolle f. ‖ coton m. mercerisé.

mercerizing ‖ Merzerisierung f.; Merzerisierung f. ‖ mercerisage m.; similisage m. / ~ machine ‖ Merzerisiermaschine f. ‖ machine f. à merceriser. / ~ press ‖ Merzerisierungspresse f. ‖ presse f. de mercérisage.

merchandise *see also* goods *and* ware ‖ Ware f.; Waren fpl. ‖ marchandise f.

merchant ‖ Handeltreibender m.; Kaufmann m. ‖ commerçant m.; marchand m.; négociant m.

merchant bar ‖ Stabeisen n.; Eisenstange f. ‖ fer m. marchand. / ~ bar mill ‖ Stabeisenstraße f. ‖ train m. pour fers marchants. / ~ flag (Mar) ‖ Handelsflagge f. ‖ pavillon m. marchand.

/ ~ fleet ‖ Handelsflotte f. ‖ flotte f. marchande. / ~ iron ‖ Grobeisen n.; Handelseisen n. ‖ fer m. marchand. / ~ lead ‖ Weichblei n. ‖ plomb m. doux.

merchantman ‖ Handelsschiff n.; Frachtschiff n.; Kauffahrer m.; Kauffahrteischiff n. ‖ bateau-marchand m.; navire m. de commerce; bâtiment m. de transport.

merchant marine ‖ Handelsmarine f. ‖ marine f. marchande. / ~ mill ‖ Kunstmühle f. ‖ moulin m. de commerce. / ~ navy see ~ marine. / ~ rolls pl. ‖ Grobeisenwalzwerk n. ‖ train m. marchand. / ~ service ‖ Seehandel m. ‖ commerce m. maritime. / ~ ship see merchantman. / ~ steamer ‖ Handelsdampfer m. ‖ vapeur m. marchand. / ~ tailor ‖ Schneidermeister m. ‖ marchand-tailleur m. / ~ vessel see merchantman.

mercurammonium chloride ‖ Merkuriammoniumchlorid n. ‖ précipité m. blanc infusible.

mercurial (Chem) ‖ quecksilberhaltig ‖ mercuriel. / ~ air pump ‖ Quecksilberluftpumpe f. ‖ pompe f. pneumatique à mercure. / ~ gauge (Steam eng) ‖ Quecksilbermanometer n.; Quecksilberdampfdruckmesser m. ‖ manomètre m. à mercure. / ~ level (Surv) ‖ Quecksilberwage f. ‖ niveau m. à mercure. / ~ metal (Chem) ‖ Quecksilbermetall n. ‖ mercure m. métallique. / ~ ointment ‖ Quecksilbersalbe f.; graue Salbe f. ‖ onguent m. mercuriel. / ~ ore ‖ Quecksilbererz n. ‖ minerai m. de mercure. / ~ pendulum ‖ Quecksilberpendel n. ‖ pendule m. de mercure. / pneumatic ~ trough ‖ pneumatische Quecksilberwanne f. ‖ cuve f. hydrargyro-pneumatique. / ~ trough of porcelain ‖ Quecksilberwanne f. aus Porzellan ‖ cuve f. en porcelaine pour mercure.

mercuric chloride ‖ Quecksilberchlorid n. ‖ chlorure m. mercurique; sublimé m. corrosif; protochlorure m. de mercure. / ~ fulminate ‖ Knallquecksilber n. ‖ mercure m. fulminant; fulminate m. de mercure. / ~ jodide ‖ Quecksilberjodid n. ‖ protoiodure m. de mercure. / ~ nitrate ‖ Quecksilbernitrat n.; Merkurinitrat n. ‖ azotate m. de protoxyde de mercure; deutonitrate m. de mercure. / ~ oxide ‖ Quecksilberoxyd n. ‖ protoxyde m. de mercure; oxyde m. (rouge) de mercure. / ~ salt ‖ Quecksilbersalz n.; Quecksilberpräparat n. ‖ sel m. de mercure. / black ~ sulphide ‖ Quecksilbermohr m. ‖ éthiope m. minéral. / red ~ sulphide ‖ künstlicher Zinnober m.; Quecksilbersulfid n. ‖ cinabre m. artificiel; sulfure m. de mercure.

mercurous chloride ‖ Quecksilberchlorür n. ‖ chlorure m. mercureux. / ~ oxide ‖ Quecksilberoxydul n. ‖ oxyde m. mercureux; oxydule m. de mercure.

mercury ‖ Quecksilber n. ‖ mercure m.; vif-argent m. / native ~ ‖ gediegenes Quecksilber n. ‖ mercure m. natif. / y millimeters of ~ ‖ y Millimeter Quecksilbersäule f. ‖ y millimètres de hauteur barométrique.

mercury air pump ‖ Quecksilberluftpumpe f. ‖ pompe f. (pneumatique) à mercure. / ~ arc rectifier (Radio) ‖ Quecksilberdampfgleichrichter m. ‖ redresseur m. à vapeur de mercure. / ~

breaker (Electr) ‖ Quecksilberwippe f.; Quecksilberschalter m. ‖ interrupteur m. ou commutateur m. à mercure. / ~ chloride see mercuric chloride. / ~ cleaning apparatus ‖ Quecksilberreinigungsvorrichtung f. ‖ appareil m. à purifier le mercure. / ~ columm ‖ Quecksilbersäule f. ‖ colonne f. de mercure ou barométrique. / ~ commutator ‖ Quecksilberkommutator m. ‖ commutateur m. à mercure. / ~ contact ‖ Quecksilberkontakt m. ‖ contact m. à mercure. / ~ converter ‖ Quecksilbergleichrichter m. ‖ convertisseur m. à mercure. / ~ cyanide ‖ Quecksilberzyanid n. ‖ cyanure m. de mercure. / ~ dichloride see mercuric chloride. / ~ fulminate see mercuric fulminate. / ~ interrupter ‖ Quecksilberunterbrecher m. ‖ interrupteur m. à mercure. / ~ jet interrupter ‖ Quecksilberstrahlunterbrecher m. ‖ interrupteur m. à jet de mercure. / ~ lamp (Electr) ‖ Quecksilberlampe f. ‖ lampe f. à mercure. / ~ manometer (Steam eng) ‖ Quecksilbermanometer n. ‖ Quecksilberdampfdruckmesser m. ‖ manomètre m. à mercure. / ~ mine ‖ Quecksilberbergwerk n. ‖ mine f. de mercure. / ~ muriate ‖ Quecksilberchlorür n. ‖ chlorure m. mercureux. / ~ ore ‖ Quecksilbererz n.; Queckerz n. ‖ minerai m. de mercure. / ~ plating ‖ Feuerversilberung f. ‖ argenture f. au feu. / ~ process ‖ Quecksilberverfahren n. ‖ procédé m. à l'anode de mercure. / ~ protocyanide ‖ Quecksilberzyanür n. ‖ protocyanure m. de mercure. / ~ pump see ~ air pump. / ~ spark gap ‖ Quecksilberfunkenstrecke f. ‖ éclateur m. à mercure. / ~ tilting tube ‖ Quecksilberkippröhre f. ‖ tube m. à mercure basculant. / ~ tipping tube see ~ tilting tube.

mercury vapour ‖ Quecksilberdampf m. ‖ vapeur f. de mercure. / ~ arc rectifier ‖ Quecksilberdampfgleichrichter m. ‖ redresseur m. à (vapeur de) mercure. / ~ lamp ‖ Quecksilberdampflampe f. ‖ lampe f. à vapeur de mercure. / ~ quartz lamp ‖ Quarzquecksilberlampe f. ‖ lampe f. de quartz aux vapeurs de mercure.

meridian ‖ Meridian m. ‖ méridien m. / ~ (Surv) ‖ Mittagslinie f. ‖ méridienne f. / ~ line ‖ Mittagslinie f. ‖ méridienne f. / ~ radius ‖ Meridianradius m. ‖ rayon m. méridien. / ~ section ‖ Meridianschnitt m. ‖ section f. méridienne.

merino (Sheep) ‖ Merinoschaf n. ‖ mérinos m. / ~ (Wool) ‖ Merino m.; Merinowolle f. ‖ mérinos m.; laine f. mérine.

meroxene ‖ Meroxen m. ‖ méroxène m.

merry-go-round ‖ Karussell n. ‖ carrousel m. forain.

mesaconic acid ‖ Mesakonsäure f. ‖ acide m. mésaconique.

mesh, to (Mach) ‖ ineinandergreifen; im Eingriff m. ‖ engrener.

mesh ‖ Masche f.; Maille f. / ~es pl. (Fish) ‖ Maschen fpl. (eines Netzes); Netzwerk n. ‖ gueules fpl. de raie. / iron wire ~ ‖ Eisendrahtgeflecht n. ‖ treillis m. en fil de fer. / non-elastic ~ ‖ nicht elastische Masche f. ‖ maille f. non élastique.

mesh connection (Electr) ‖ Deltaschaltung f. ‖ couplage m. en triangle ou en delta.

meshed ‖ maschig; netzartig ‖ à mailles fpl.; maillé. / close-~ ‖ engmaschig ‖ à mailles fpl. serrées.

meshing of the teeth ‖ Ineinandergreifen n. der Zähne ‖ engrènement m. des dents.

mesh, row of ~es ‖ Maschenreihe f. ‖ rangée f. de mailles. / ~ tissue ‖ Maschengewebe n. ‖ tissu m. à mailles. / width of ~ ‖ Maschenweite f. ‖ largeur f. des mailles.

meshy see meshed.

mesitine-spar (Miner) ‖ Mesitinspat m. ‖ mésitine f.

meslin (Agr) ‖ Mengkorn n.; Mischfrucht f. ‖ méteil m.; mouture f.; blé m. mêlé.

mesolite ‖ Mesolit m.; Skolezit m. ‖ mésolite f.; scolésite f.

mesotype (Miner) ‖ Natrolit m.; Mesotyp m. ‖ natrolite f.; mesotype f.

Mesozoic group (Geol) ‖ mesozoische Gruppe f. ‖ groupe m. mésozoïque.

message ‖ Bericht m.; Mitteilung f. ‖ rapport m.; bulletin m. / wireless ~ ‖ Funkspruch m.; Funkmeldung f.; Radiogramm n. ‖ radio m.; sans fil m.

message rate (Tel) ‖ Einzelgesprächsgebühr f. ‖ taxe f. de conversation. / ~ register (Tel) ‖ Gesprächszähler m. ‖ compteur m. de conversations.

messenger ‖ Laufbursche m.; Bote m. ‖ garçon m. de course ou de magasin; galopin m. / ~ strand see ~ wire. / ~ wire for aerial cables ‖ Tragseil n. für Luftkabel ‖ corde f. de suspension pour câbles aériens. / ~ wire clamp ‖ Tragseilschelle f. ‖ plaque f. de serrage ou bride f. de la corde de suspension.

mess room (Mar) ‖ Messe f. ‖ carré m.: poste m.

metacentre ‖ Metazentrum n. ‖ métacentre m. / latitudinal ~ ‖ Quermetazentrum n. ‖ métacentre m. latitudinal. / longitudinal ~ ‖ Längenmetazentrum n. ‖ métacentre m. longitudinal.

metachromotype ‖ Abziehbild n. ‖ décalcomanie f. / ~ process ‖ Abziehbilderverfahren n. ‖ procédé m. de décalcomanie.

metacinnabarite ‖ Metazinnabarit m. ‖ métacinnabarite f.

metal, to ~ (Road) ‖ beschottern ‖ empierrer; ballaster.

metal ‖ Metall n. ‖ métal m. / ~ (Alloy of copper) ‖ Bronze f.; Metallegierung f. ‖ bronze m.); métal m. / ~ (Glassm) ‖ geläuterte Glasmasse f. ‖ métal m.; verre m. affiné. / ~ (Road) ‖ Beschotterung f. ‖ matériaux mpl. d'empierrement. / alkaline ~ ‖ Alkalimetall n. ‖ métal m. alcalin. / base ~ ‖ unedles Metall n. ‖ métal m. commun ou ignoble. / copper ~ ‖ Kupferstein m. ‖ matte f. de cuivre; régule m. de cuivre. / Dutch ~ ‖ Flittergold n.; Rauschgold n. ‖ clinquant m.; oripeau m. / expanded ~ ‖ Streckmetall n. ‖ métal m. déployé. / fine ~ ‖ Feineisen n. ‖ fer m. fin ou raffiné, fin métal m. / granulated ~ ‖ Granalien fpl. grenailles fpl. / gun ~ ‖ Rotguß m. ‖ bronze m. rouge. / hammered ~ ‖ gehämmertes Blech n. ‖ tôles fpl. martelées. / heavy ~ ‖ Schwermetall n. ‖ métal m. lourd ou pésant. / light ~ ‖ Leichtmetall n. ‖ métal m. léger. / rich in lines ‖ linienreiches Metall n. ‖ métal m. riche en raies. / monumental ~ ‖ Erz n.; Statuenmetall n. ‖ métal m. à statues. / native ~ ‖ gediegenes Metall n.; Jungfernmetall n. ‖ métal m. natif ou vierge. / noble ~ see precious ~. / non-

~ ‖ Nichtmetall n. ‖ métalloïde m. / non-corrosive ~ ‖ nichtoxydbares Metall n. ‖ métal m. inoxydable. / non-ferrous ~ ‖ Nichteisenmetall n. ‖ métal m. non ferrugineux ou non ferrique. / ~ of the family of platinum ‖ Metall n. aus der Platingruppe ‖ métal m. de la famille de platine. / precious ~ ‖ Edelmetall n. ‖ métal m. précieux ou noble. / Rose's ~ ‖ Rose's Metall n. ‖ métal m. de Rose. / specular ~ ‖ Spiegelmetall n. ‖ métal m. spéculaire; métal m. à miroirs. / virgin ~ see native ~. / waste ~ ‖ Gekrätz n.; Krätze f.; Metallabfälle mpl. ‖ déchet m. / white ~ (Antifriction metal) ‖ (weißes) Lagermetall n.; Weißmetall n. ‖ métal m. blanc; métal m. antifriction; antifriction f. / white ~ (Argentan) ‖ Neusilber n.; Weißguß m. ‖ metal m. blanc; maillechort m.; argentan m.

metal alloy ‖ Metallegierung f. ‖ alliage m. métallique. / ~ analysis ‖ Metalluntersuchung f. ‖ analyse f. des métaux. / ~ ashes pl. ‖ Metallasche f. ‖ cendre f. métallique. / ~ bagging twine ‖ Metallsackbinde f. ‖ ligature f. métallique pour sacs. / ~ bar ‖ Metallstange f. ‖ barre f. métallique. / ~ bar extrusion press ‖ Metallstangenpresse f. ‖ presse f. à profiler les barres métalliques. / ~ bath ‖ Metallbad n. ‖ bain m. métallique ou de métal. / ~ bedstead ‖ Metallbettstelle f. ‖ lit m. metallique ou en métal. / ~ bindings pl. ‖ Metallbeschlag m. ‖ monture f. métallique.

metal board ‖ Metallschild n.; Schild n. aus Metall ‖ plaque f. en métal. / chemically engraved ~ ‖ (chemisch) geätztes Metallschild n. ‖ plaque f. en métal gravée chimiquement ou à l'eau forte.

metal box ‖ Metallbüchse f.; Metallschachtel f. ‖ boîte f. en métal ou en étain. / ~ free from iron ‖ eisenfreie Metallbüchse f. ‖ boîte f. métallique exempte de fer.

metal broker ‖ Metallmakler m. ‖ courtier m. en métaux. / ~ bronzing ‖ Metallbronzieren n. ‖ bronzage m. sur métaux. / ~ button ‖ Metallknopf m. ‖ bouton m. en métal. / ~ carbide ‖ Metallkarbid n. ‖ carbure m. des métaux. / ~ case (Watch) ‖ Metallgehäuse n. ‖ boîte f. en métal. / ~ casting ‖ Metallguß m. ‖ fonte f. métallique (non ferreuse). / ~ cement ‖ Metallzement m. ‖ ciment m. métallique. / ~ chaser ‖ Metallziselierer m. ‖ ciseleur m. sur métaux. / ~ chasing ‖ Metallpreßarbeit f. ‖ repoussage m. de métaux. / ~ circular saw ‖ Metallkreissäge f. ‖ scie f. circulaire à métaux. / ~ cleaning material ‖ Metallscheuermittel n. ‖ matière f. à nettoyer les métaux. / ~ cleaning and polishing material ‖ Metallputzmittel n. ‖ matière f. à nettoyer et à polir les métaux. / ~ cloth ‖ Metalltuch n. ‖ toile f. ou tissu m. métallique. / ~ colouring ‖ Metallfärbung f. ‖ coloration f. des métaux. / ~ construction (Build) ‖ Metallbau m.; Metallkonstruktion f. ‖ construction f. métallique. / ~ core ‖ Metallseele f. ‖ âme f. en métal.

metal covering ‖ Metallbekleidung f.; revêtement m. métallique. / ~ (Tiler) ‖ Metallbedachung f.; Metalleindeckung f. ‖ couverture f. ou toiture f. en métal.

metal cup for tea glasses ‖ Teeglashalter m. aus Metall ‖ support m. métallique pour verres à thé. / ~ cutter-out ‖ Metall-

ausschneider m. ‖ découpeur m. de métaux. / ~ cutting-out ‖ Metallblechstanzen n. ‖ découpage m. de métaux en feuilles. / ~ deck cover (Airpl) ‖ Deckelblech n. ‖ tôle f. de revêtement du pont. / ~ diaphragm ‖ Metallmembran f. ‖ diaphragme m. métallique.

metal disk ‖ Metallscheibe f.; Metallplatte f. ‖ disque m. en métal. / perforated ~ ‖ durchlochte Metallplatte f. ‖ disque m. en métal perforé.

metal dowsing rod ‖ Wünschelrute f. aus Metall ‖ baguette f. divinatoire en métal. / ~ drain (Found) ‖ Einguß m.; Gußgerinne n. ‖ chenal m.; échenal m. / ~ drawing ‖ Metallziehen n. ‖ étirage m. de métaux. / ~ drill ‖ Metallbohrer m. ‖ foret m. à métaux. / ~ embosser ‖ Metallpräger m. ‖ ciseleur-repousseur m. / ~ embroidery ‖ Perlstickerei f. ‖ broderie f. métallique. / ~ enameller ‖ Metallemaillierer m. ‖ émailleur m. sur métaux. / ~ enamelling manufacture ‖ Emaillierwerk n. ‖ émaillerie f. industrielle. / ~ engraver ‖ Metallgravör m. ‖ graveur m. sur métal. / ~ envelope ‖ Metallhülle f. ‖ enveloppe f. métallique. / ~ etching ‖ Metallätzung f.; Ätzen m auf Metall ‖ gravure f. sur métal à l'eau forte. / ~ exchange ‖ Metallbörse f. ‖ bourse f. des métaux. / ~ fancy goods pl. ‖ Metallgalanteriewaren pl. ‖ articles mpl. de fantaisie en métal. / ~ fastening (Mach) ‖ Metallverbolzung f. ‖ chevillage m. en métal. / fatigue of ~ ‖ Ermüdung f. des Metalls ‖ fatigue f. du métal.

metal filament ‖ Metallfaden m. ‖ fil m. métallique. / ~ lamp ‖ Metallfadenlampe f. ‖ lampe f. à filament métallique.

metal fitter ‖ Maschinenschlosser m. ‖ ajusteur-mécanicien m. / ~ fittings pl. ‖ Metallbeschläge mpl. ‖ garnitures fpl. ou armatures fpl. en métal. / ~ foil ‖ Metallfolie f. ‖ feuille f. en métal. / ~ founder ‖ Metallgießer m. ‖ fondeur m. en métaux. / ~ foundry ‖ Metallgießerei f. ‖ fonderie f. de métaux. / ~ fuse ‖ Metallzünder m. ‖ fusée f. métallique. / fusion of ~s ‖ metallische Verbindung f. ‖ alliage m. de métaux. / ~ gauge ‖ Blechlehre f.; Metallschablone f. ‖ jauge f. ou patron m. en métal.

metal gauze ‖ Metalltuch n.; Drahtgaze f. ‖ toile f. métallique. / ~ loom ‖ Metalltuchwebstuhl m. ‖ métier m. à tisser les toiles métalliques. / ~ stretching machine ‖ Metalltuchstreckmaschine f. ‖ machine f. à dresser les toiles métalliques.

metal gilder ‖ Metallvergolder m. ‖ doreur m. sur métaux. / ~ gilding ‖ Metallvergoldung f. ‖ dorure f. sur métaux.

metal goods pl. ‖ Metallwaren fpl. ‖ quincaillerie f.; articles mpl. ou ouvrages mpl. en métal. / silverplated ~ ‖ versilberte Metallwaren fpl. ‖ objets mpl. en métal argentés.

metal hose ‖ Metallschlauch m. ‖ tube m. métallique flexible. / ~ ignition lighter ‖ Feuerzeug n. aus Metall ‖ briquet m. d'allumage en métal. / ~ joint ‖ Metalldichtung f. ‖ garniture f. métallique. / ~ label ‖ Metallanhängeschild n.; Metalletikett n. ‖ étiquette f. en métal. / ~ lacquer (Chem) ‖ Metallack m. ‖ vernis m. pour métaux. / ~ leaves pl. ‖ Blattmetall n. ‖ métal m. en feuilles.

metalled ‖ beschottert ‖ empierré; caillouté. / ~ road ‖ Chaussee f.; Land-

straße f.; Schotterstraße f. ‖ chaussée f.; route f. empierrée.

metallic ‖ metallisch; metallartig ‖ métallique. / ~ ashes pl. see metal ashes. / ~ box see metal box. / ~ brush ‖ Metallbürste f. ‖ balais m. métallique. / ~ cartridge ‖ Metallpatrone f. ‖ cartouche f. métallique. / ~ circuit (Tel) ‖ Doppelleitung f. ‖ circuit m. bifilaire ou à double fil. / ~ circuit operation ‖ Doppelleitungsbetrieb m. ‖ transmission f. à fil double. / ~ cover ‖ Metallüberzug m. ‖ depôt m. galvanique. / ~ element (Chem) ‖ metallisches Element n. ‖ élément m. métallique. / ~ filament lamp see metal filament lamp. / ~ frame ‖ Metallrahmen m. ‖ cadre m. métallique. / ~ induction protection in cables ‖ Metallinduktionsschutz m. in Kabeln ‖ protection f. d'induction métallique des câbles. / ~ ink ‖ Metallfarbe f. ‖ encre f. métallique. / ~ lustre ‖ Metallglanz m. ‖ brillant m. métallique. / ~ oxide ‖ Metalloxyd n. ‖ oxyde m. métallique ou de métal. / ~ packing (of the piston) ‖ Metalliderung f.; Metallpackung f. ‖ garniture f. ou joint m. métallique. / ~ paper ‖ Metallpapier n. ‖ papier m. métallique. / ~ part ‖ Metallteil m. ‖ pièce f. métallique. / ~ poison ‖ Metallgift n. ‖ poison m. métallique. / ~ product ‖ Metallwerkerzeugnis n. ‖ produit m. métallurgique. / ~ residues pl. ‖ Metallrückstände mpl. ‖ résidus mpl. métalliques. / ~ salt ‖ Metallsalz n. ‖ sel m. inorganique.

metallic sounding ‖ metallisch klingend ‖ rendant un son métallique.

metalliferous ‖ metallhaltig; metallführend ‖ métallifère. / ~ residue ‖ metallhaltiger Rückstand m. ‖ résidu m. métallifère. / ~ vein ‖ Erzgang m. ‖ filon m. métallifère.

metalliform see metallic and metalline.

metalline ‖ metallisch; metallartig ‖ métallique.

metalling ‖ Beschotterung f. ‖ empierrement m. / stones pl. for ~ (of) roads ‖ Steine mpl. zur Beschotterung von Straßen ‖ pierres fpl. pour l'empierrement des routes.

metallist ‖ Metallarbeiter m. ‖ ouvrier m. en métaux.

metallization ‖ Metallisierung f. ‖ métallisation f.

metallize, to ‖ metallisieren ‖ métalliser.

metallized ‖ metallisiert ‖ métallisé. / ~ cigarette paper ‖ metallisiertes Zigarettenpapier n. ‖ papier m. à cigarettes métallisé. / ~ paper ‖ metallisiertes Papier n. ‖ papier m. métallisé.

metallochromy ‖ galvanische Metallfärbung f. ‖ métallochromie f.

metallographist ‖ Metallograf m. ‖ métallographe m.

metallography ‖ Metallografie f. ‖ métallographie f. / ~ of iron ‖ Metallografie f. des Eisens ‖ métallographie f. du fer.

metalloid ‖ Nichtmetall n.; Metalloid n. ‖ non-métal m.; métalloïde m.

metallometric balance ‖ metallometrische Wage f. ‖ balance f. métallométrique.

metallophone ‖ Metallofon n. ‖ métallophone m.

metallurgic(al) ‖ metallurgisch; hüttenmännisch ‖ métallurgique. / ~ machine ‖ Hüttenmaschine f. ‖ machine f. métallurgique. / ~ plant ‖ hüttenmännischer

oder metallurgischer Betrieb m. ‖ usine f. métallurgique.

metallurgical works pl. ‖ Metallhütte f.; Hüttenwerk n. ‖ usine f. métallurgique. / auxiliary machine for ~ ‖ Hüttenwerkhilfsmaschine f. ‖ machine f. auxiliaire pour usines métallurgiques. / ~ equipment ‖ Hüttenwerkseinrichtung f. ‖ installation f. d'usines métallurgiques. / ~ plant ‖ Hüttenwerksanlage f. ‖ usine f. métallurgique.

metallurgist ‖ Hüttenmann m.; Metallurg m. ‖ métallurgiste m.

metallurgy ‖ Hüttenkunde f.; Metallurgie f.; Hüttenwesen n. ‖ métallurgie f. / ~ of iron ‖ Eisenhüttenkunde f. ‖ métallurgie f. du fer.

metal maker (Letter-f.) ‖ Schmelzer m. fondeur m. / ~ mount (Letter-f) ‖ Bleifuß m. ‖ pied m. de plomb. / ~ object cast in a mould ‖ gegossener Rohling m. ‖ pièce f. brute coulée *ou* moulée; pièce f. venue de fonte *ou* de fonderie; objet m. coulé. / ~ oxide ‖ Metalloxid n. ‖ oxyde m. métallique. / ~ packing (of a piston) ‖ Metalliderung f.; Metallpackung f. ‖ garniture f. métallique. / ~ packing ring ‖ Metalldichtungsring m. ‖ anneau m. métallique de joint. / ~ painter ‖ Metallmaler m. ‖ peintre m. sur métaux. / ~ painting ‖ Metallmalerei f. ‖ peinture f. sur métaux. / ~ paper ‖ Metallpapier n. ‖ papier m. métallique; papier-métal m. / ~ pattern ‖ Metallschablone f. ‖ patron m. métallique. / ~ perchloride ‖ Metallchlorid n. ‖ perchlorure m. des métaux. / ~ pipe ‖ Metallröhre f.; Metallrohr n. ‖ tube m. métallique. / ~ plating ‖ Plattierung f. ‖ placage m. sur métaux. / ~ polish ‖ Metallputzmittel n. ‖ brillant m. *ou* pâte f. à nettoyer les métaux. / ~ polisher ‖ Metallpolierer m. ‖ polisseur m. sur métaux. / ~ polishing ‖ Polieren n. von Metallwaren ‖ polissage m. sur métaux. / ~ polishing material ‖ Metallputzmittel n.; Metallpoliermittel n. ‖ matière f. *ou* moyen m. à polir les métaux. / ~ porcelain ‖ Metallporzellan n. ‖ porcelaine f. montée. / ~ press for making round bars ‖ Stangenpresse f. ‖ presse f. à profiler à chaud les barres rondes. / ~ printing machine ‖ Metalldruckpresse f. ‖ machine f. pour la métallographie. / ~ printing works pl. ‖ Metalldruckerei f. ‖ imprimerie f. sur métaux. / ~ punching ‖ Metallochung f. ‖ perforation f. de métaux. / ~ punch pliers pl. ‖ Metallochzange f. ‖ pince f. emportepièce pour métaux. / ~ purse ‖ Metallbörse f. ‖ bourse f. en tissu métallique. / ~ refinement ‖ Metallveredelung f. ‖ raffinage m. des métaux. / ~ reflector ‖ Metallspiegel m. ‖ réflecteur m. métallique. / protective ~ ribbon ‖ Metallschutzband n. ‖ ruban m. de métal protecteur. / ~ rolling mill ‖ Metallwalzwerk n. ‖ laminoir m. pour métaux. / ~ rule ‖ Metallineal n. ‖ règle f. métallique. / ~ scouring cloth ‖ Metallputztuch n. ‖ tissu m. métallique servant à nettoyer les métaux. / ~ screen ‖ Metallschirm m. ‖ écran m. en métal. / ~ screw ‖ Metallschraube f. ‖ vis f. à métaux. / ~ shearing ‖ Metallschneiden n. ‖ découpage m. de métaux. / ~ sheathing (Shipb) ‖ Metallbeschlag m. ‖ doublage m. en métal.

metal sheet ‖ Metallplatte f. ‖ plaque f. *ou* feuille f. *ou* lame f. *ou* planche f. de métal. / ~ poster ‖ Blechplakat n. ‖ affiche n. en tôle.

metal sieve ‖ Metallsieb n.; Drahtsieb n. ‖ crible m. en métal. / ~ silverer ‖ Metallversilberer m. ‖ argenteur m. sur métaux. / ~ silver plating ‖ Metallversilberung f. ‖ argenture f. sur métaux. / ~ slag ‖ Metallschlacke f. ‖ scorie f. / ~ slate ‖ Metallschiefer m. ‖ ardoise f. métallique. / ~ sleeper (Railw) ‖ Eisenschwelle f. ‖ traverse f. métallique. / ~ compound ~ slide ‖ Metallkreuzschlitten m. ‖ chariot m. à deux mouvements rectangulaires. / ~ smelting works pl. ‖ Metallschmelzwerk n. ‖ fonderie f. de métaux. / ~ stamping press ‖ Metallprägepresse f. ‖ presse f. à **estamper** des métaux. / ~ stamping workshop ‖ Metallprägeanstalt f. ‖ atelier m. d'estampage de métaux. / ~ stone (Mine) ‖ sandiger Schieferton m. des Kohlengebirges ‖ schiste m. houiller quartzeux. / ~ strip mill ‖ Metallwalzwerk n. ‖ laminoir m. à métal. / ~ stud ‖ Metallsteg m. ‖ entretoise f. métallique. / ~ sweep (of the mower) ‖ Anhaublech n. ‖ tôle f. à moissonner en andain. / ~ texture ‖ Metallgewebe n. ‖ tissu m. métallique. / thickness of ~ (Found) ‖ Wanddicke f. des Gußstücks ‖ épaisseur f. de métal. / ~ tile ‖ Metallziegel m. ‖ tuile f. métallique. / ~ track ‖ Rollbahn f. ‖ voie f. de roulement. / ~ tubbing ‖ Tübbingausbau m.; Verrohrung f. eines Ausbaues ‖ tubage m. en fonte. / ~ tube ‖ Metallrohr n.; Metallröhre f. ‖ tuyau m. *ou* tube m. métallique. / flexible ~ tube *see* metal hose. / ~ turner ‖ Metalldreher m. ‖ tourneur m. sur métaux. / ~ turning ‖ Metalldreherei f. ‖ tournage m. sur métaux. / ~ type (Print) ‖ Metalltype f. ‖ caractère m. métallique. / ~ ungreasing apparatus ‖ Metallentfettungsvorrichtung f. ‖ appareil m. de dégraissage de métaux. / ~ varnishing ‖ Lackieren n. *oder* Lackierung f. von Metallen ‖ vernissage m. sur métaux. / ~ wing (Airpl) ‖ Metallflügel m. ‖ aile f. en métal.

metal wire ‖ Metalldraht m. ‖ fil m. métallique. / ~ drawing ‖ Metalldrahtzieherei f. ‖ tréfilage m. de fils métalliques. / ~ pot cleaner ‖ Metallfadentopfreiniger m. ‖ nettoie-casserole m. en fil métallique.

metal wool ‖ Metallwolle f. ‖ laine f. métallique. / ~ work of applied art ‖ kunstgewerbliche Metallarbeit f. ‖ travail m. d'art en métal. / ~ worker (Mach) ‖ Metallarbeiter m. ‖ ouvrier m. en métaux. / ~ worker (Metallurgical works) ‖ Hüttenarbeiter m. ‖ ouvrier m. d'usine metallurgique. / ~ working ‖ Metallbearbeitung f. ‖ travaillage n. des métaux. / ~ working machine ‖ Metallbearbeitungsmaschine f. ‖ machine f. à travailler les métaux. / ~ works pl. ‖ Metallhütte f. ‖ usine f. métallurgique.

metamorphic (Geol) ‖ metamorphisch ‖ métamorphosé.

metamorphism (Geol) ‖ Metamorphose f.; Metamorphismus m. ‖ métamorphisme m.; métamorphose f.

metaphosphate ‖ metaphosphorsaures Salz n. ‖ métaphosphate m.

meteor *see* meteoric stone.

meteoric iron ‖ Meteoreisen n. ‖ fer m. météorique. / ~ stone (Astron) ‖ Meteor-stein m. ‖ aérolithe m.; météorolithe; pierre f. météorique; météorite f.

meteorolite *see* meteoric stone.

meteorological information ‖ Wetterberatung f. ‖ information f. météorologique. / ~ instrument ‖ meteorologisches Gerät n. *oder* Instrument n. ‖ instrument m. météorologique *ou* de météorologie.

meteorological service ‖ Wetterdienst m. ‖ service m. météorologique. / ~ concerning the atmosphere at high altitudes ‖ Höhenwetterdienst m. ‖ service m. météorologique pour les couches supérieures de l'atmosphère. / ~ for aviation ‖ Flugwetterdienst m. ‖ service m. météorologique de l'aviation. / ~ for ships ‖ Schiffswetterdienst m. ‖ service m. météorologique pour la navigation *ou* la marine.

meteorological station ‖ Wetterdienststelle f. ‖ station f. de prévision du temps *ou* météorologique.

meteorology ‖ Meteorologie f.; Wetterkunde f. ‖ météorologie f.

meter ‖ Meter n. ‖ mètre m. / ~ (Counting apparatus) ‖ Zähler m. ‖ compteur m. / ~ (Measuring apparatus) ‖ Messer m.; Meßapparat m. ‖ appareil m. mesureur *ou* de mesure *ou* à mesurer. / ~ calibrated in . . . ‖ Messer m. mit Einteilung für . . . ‖ instrument m. de mesure avec graduation pour . . . / ~ (Counter) with click action ‖ Zählwerk n. mit springenden Ziffern ‖ minuterie f. à chiffres mobiles. / ~ for duration of conversation (Tel) ‖ Gesprächszeitmesser m. ‖ compteur m. du temps de conversation. / electric ~ ‖ elektrischer Zähler m. ‖ compteur m. électrique. / running ~ ‖ laufendes Meter n. ‖ mètre m. courant. / per running ~ ‖ je laufendes Meter n. ‖ par mètre m. courant. / per ~ run *see* per running ~. / theoretic weight per ~ run ‖ rechnungsmäßiges Metergewicht n. ‖ poids m. théorique par mètre courant. / agreement by ~ ‖ Akkord m. nach Maß ‖ marché m. au mètre.

meter armature ‖ Zähleranker m. ‖ induit m. de compteur. / ~ board ‖ Zählertafel f. ‖ panneau m. de compteurs. / ~ box ‖ Zählergehäuse n. ‖ boîte f. pour compteurs. / ~ calibrating equipment ‖ Zählerprüfeinrichtung f. ‖ dispositif m. d'essai de compteurs. / ~ case ‖ Zählergehäuse n. ‖ boîte f. du compteur. / ~ constant ‖ Zählerkonstante f. ‖ constante f. du compteur.

metering ‖ Messen n.; Messung f. ‖ mesurage m. / ~ of the feed water consumption ‖ Messen n. des Speisewasserverbrauches. ‖ mesure f. de la consommation de l'eau d'alimentation. / repeated ~ (Tel) ‖ Mehrfachzählung f. ‖ comptage m. répété.

metering diaphragm ‖ Meßmembran f. ‖ membrane f. d'organe de mesure.

meter key (Tel) ‖ Zähltaste f. ‖ clé f. de comptage. / ~ lamp (Tel) ‖ Zählerüberwachungslampe f. ‖ lampe f. de comptage. / ~ relay (Tel) ‖ Zählerrelais n.; Zählerschütz n. ‖ relais m. de compteur.

meter rule ‖ Metermaß n. ‖ mètre m. / ~ hinged ~ ‖ zusammenlegbares Metermaß n. ‖ mètre m. pliant. / ~ in wood ‖ Metermaß n. aus Holz ‖ mètre m. en bois.

meter ton ‖ Metertonne f. ‖ tonne-mètre f.

meter works pl. ‖ Zählerfabrik f. ‖ fabrique f. de compteurs.

methanol ‖ Methanol n. ‖ méthanol m.

method ‖ Verfahren n.; Methode f.; Prozeß n. ‖ méthode f.; procédé m. / analytical ~ ‖ analytisches Verfahren n. ‖ méthode f. analytique. / ~ of approximation ‖ Annäherungsverfahren n.; Annäherungsmethode f. ‖ méthode f. d'approximation. / ~ of calculation ‖ Berechnungsverfahren n. ‖ procédé m. de calcul. / ~ of comparison ‖ Vergleichungsverfahren n.; Vergleichungsmethode f. ‖ méthode f. de comparaison. / ~ of compensation ‖ Kompensationsverfahren n. ‖ méthode f. de compensation. / centrobaric ~ ‖ Guldin'sche Regel f. ‖ méthode f. centrobarique; théorèmes mpl. de Guldin. / ~ of drying ‖ Eintrocknungsverfahren n. ‖ méthode f. de dessication. / improved ~ ‖ verbessertes Verfahren n. ‖ méthode f. améliorée. / inferior ~ ‖ minderwertiges Verfahren n. ‖ méthode f. inférieure. / ~ of least squares ‖ Verfahren n. der kleinsten Quadrate ‖ méthode f. des plus petits carrés. / ~ of manufacturing ‖ Herstellungsverfahren n. ‖ mode f. de fabrication. / not a good ~ ‖ kein gutes Verfahren n. ‖ pas une bonne méthode f. / ~ of obtaining ‖ Darstellungsverfahren n. ‖ méthode f. de préparation. / ~ of operation (Mach) ‖ Arbeitsweise f. ‖ mode f. de fonctionnement. / ~ of sparging ‖ Härteverfahren n. ‖ conduite f. de la trempe. / synthetical ~ ‖ synthetisches Verfahren n. ‖ méthode f. synthétique. / ~ of undetermined coefficients ‖ Verfahren n. der unbestimmten Koeffizienten ‖ méthode f. des coefficients indéterminés. / ~ of variable area (Sound film) ‖ Schwarzweißverfahren n. ‖ procédé m. noir-blanc ou tranversal. / ~ of variable density (Sound film) ‖ Schattierungsverfahren n. ‖ procédé m. d'intensité. / ~ of working ‖ Arbeitsweise f. ‖ méthode f. ou mode f. de travail. / ~ of working of telegraph circuits ‖ Betriebsweise f. der Telegrafie ‖ manière f. d'opérer les circuits télégraphiques.

methodical ‖ planmäßig; methodisch ‖ méthodique.

methyl ‖ Methyl n. ‖ méthyle m.

methyl acetate ‖ Methylazetat n.; essigsaures Methyl n. ‖ acétate m. de méthyle.

methylacetic ether ‖ Essigsäuremethylester m. ‖ éther m. méthylacétique.

methyl alcohol see also methylated spirit ‖ Methylalkohol m. ‖ alcool m. méthylique.

methylamidophenol ‖ Methylamidophenol n. ‖ méthylamidophénol m.

methyl aniline ‖ Methylanilin n. ‖ méthylaniline f.

methylate, to ‖ vergällen ‖ dénaturer.

methylated alcohol see ~ spirit. / ~ spirit ‖ Holzgeist m.; Methylalkohol m. ‖ alcool m. méthylique; esprit m. de bois.

methyl benzoate ‖ benzoasures Methyl n. ‖ benzoate m. de méthyle. / ~ bromide ‖ Brommethyl n. ‖ bromure m. de méthyle. / ~ chloride ‖ Chlormethyl n. ‖ chlorure m. de méthyle.

methylene ‖ Methylen n. ‖ méthylène m. / ~ bromide ‖ Brommethylen n. ‖ bromure m. de méthylène. / ~ chloride ‖ Chlormethylen n. ‖ chlorure m. de

méthylène. / ~ iodide ‖ Jodmethylen n. ‖ iodure m. de méthylène.

methylic alcohol see methylated spirit. / ~ ether ‖ Holzäther m. ‖ éther m. méthylique.

methyl iodide ‖ Jodmethyl n. ‖ iodure m. de méthyle. / ~ salicylate ‖ salizylsaures Methyl n. ‖ salicylate m. de méthyle.

metol ‖ Metol n. ‖ méthol m.

metre see meter.

metric ‖ metrisch ‖ métrique. / ~ fine thread ‖ metrisches Feingewinde n. ‖ filetage m. métrique fin. / ~ pitch lead screw ‖ metrische Leitspindel f. ‖ vis-mère f. métrique. / ~ system ‖ metrisches System n. ‖ système m. métrique. / ~ thread ‖ metrisches Gewinde n. ‖ filetage m. métrique.

metronome (Phys) ‖ Metronom n.; Taktmesser m. ‖ batteur m. de mesure; métronome m.

metropolis ‖ Weltstadt f. ‖ ville f. cosmopolite; métropole f.

metropolitan railway ‖ Stadtbahn f. ‖ chemin m. de fer urbain, métropolitain m.; chemin m. de fer circulaire.

Mexican grass ‖ Aloehanf m. ‖ chanvre m. d'aloès.

mew oil ‖ Bärwurzöl n. ‖ essence f. de méum.

mezzanine (Build) ‖ Halbgeschoß n.; Zwischengeschoß n. ‖ entre-sol m.; mezzanine f.

mezzotint ‖ Kupferstich m. in Schabmanier; Schwarzkunst f. ‖ gravure f. au noir. / ~ scraper ‖ Mezzotintoschaber m. ‖ grattoir m. ou ébarboir m. employé dans la gravure au noir.

miargyrite ‖ Rubinblende f.; Miargyrit m. ‖ miargyrite f.

mica (Glimmer) ‖ Glimmer m.; Katzensilber n.; Frauenglas n. ‖ mica m. / ~ (Specular gypsum) ‖ Marienglas n. ‖ mica m. / wound on ~ ‖ auf Glimmer m. aufgewickelt ‖ sur mica m. / argentine ~ ‖ Silberglimmer m.; Kaliglimmer m. ‖ mica m. argentin. / biaxial ~ see rhombic ~. / black ~ see magnesia ~. / lithia ~ ‖ Lepidolith m.; Lithionglimmer m. ‖ lépidolithe f. / magnesia ~ ‖ Magnesiaglimmer m.; einachsiger Glimmer m. ‖ mica m. magnésia; biotite f. / perl ~ ‖ Margarit m.; Perlglimmer m. ‖ margarite m.; mica m. nacré. / potash ~ ‖ Kaliglimmer m.; russisches Frauenglas m. ‖ muscovite f. / pressed ~ ‖ Preßglimmer m. ‖ mica m. comprimé. / rhombic ~ ‖ zweiachsiger Glimmer m. ‖ mica m. rhombique. / sheet ~ ‖ Plattenglimmer m. ‖ mica m. en feuilles. / uniaxial ~ see magnesia ~.

micaceous copper ‖ Kupferglimmer m. ‖ cuivre m. micacé ou mica.

mica collar ‖ Glimmerring m. ‖ bague f. de mica. / ~ condenser ‖ Glimmerkondensator m. ‖ condensateur m. à mica. / ~ cover ‖ Glimmerdeckel m. ‖ couvercle m. en mica. / ~ dielectricum ‖ Glimmerdielektrikum n. ‖ diélectrique m. au mica. / ~ filler ‖ Glimmerfüllung f. ‖ remplissage m. de mica. / ~ film ‖ Glimmerplättchen m. ‖ lame f. de mica; paillette f. / ~ foil machine ‖ Mikafoliummaschine f. ‖ machine f. à préparer des feuilles de mica. / ~ goggles pl. ‖ Schutzbrille f. mit Glimmerscheiben ‖ lunettes pl. de protection en mica. / ~ goods pl. ‖ Glimmerwaren fpl. ‖ articles mpl. en

mica. / ~ heating element ‖ Glimmerheizelement n. ‖ élément m. de chauffage en mica. / ~ hornfels ‖ Glimmerhornfels m. ‖ reche f. de corne micacée. / ~ mine ‖ Glimmerbruch m. ‖ carrière f. de mica.

micanite ‖ Mikanit n. ‖ micanite f. / ~ linen ‖ Mikanitleinwand f. ‖ toile f. de mica. / ~ paper ‖ Mikanitpapier n. ‖ papier m. micanite. / ~ producing machine ‖ Mikanitherstellungsmaschine f. ‖ machine f. à fabriquer de micanite.

mica schist ‖ Glimmerschiefer m. ‖ micaschiste m.; schiste m. micacé. / argillaceous ~ ‖ Tonglimmerschiefer m.; Phyllit m. ‖ micaschiste m. argilleux. / calcareous ~ ‖ Kalkglimmerschiefer m. ‖ schiste m. micacé calcaire.

mica slate see mica schist. / ~ spangle ‖ Glimmerflitter m. ‖ paillette f. en mica. / ~ spark-plug ‖ Glimmerkerze f. ‖ bougie f. en mica. / ~ spectacles pl. ‖ Glimmerbrille f. ‖ lunettes fpl. en mica. / ~ strip ‖ Glimmerstreifen m. ‖ bande f. de mica. / ~ top see ~ cover. / ~ utilizing machine ‖ Glimmerverwertungsmaschine f. ‖ machine f. pour l'utilisation du mica.

micro arc-lamp ‖ Mikrobogenlampe f. ‖ microlampe f. à arc.

microbe ‖ Mikrobe f. ‖ microbe m.

micro-burner ‖ Mikrobrenner m. ‖ micro-brûleur m.

microchemical analysis ‖ mikrochemische Analyse f. ‖ analyse f. microchimique.

micro-chronometer ‖ Kurzzeitmesser m. ‖ appareil m. pour mesurer les petits intervalles de temps.

microcline ‖ Mikroklin m. ‖ microcline f.

micro-colorimeter ‖ Mikrokolorimeter n. ‖ micro-colorimètre m.

microcosmic salt ‖ Phosphorsalz n. ‖ sel m. de phosphore.

microcrystalline ‖ mikrokristallinisch ‖ microcristallin.

micrographic apparatus ‖ mikrografisches Gerät n.; mikrografischer Apparat m. ‖ appareil m. de micrographie.

micrography ‖ Mikrografie f. ‖ micrographie f.

microlite ‖ Mikrolith m. ‖ microlite f.

micro-manipulator ‖ Mikromanipulator m. ‖ micromanipulateur m.

micrometer (Opt) ‖ Mikrometer n. ‖ micromètre m. / ~ (Gauge) ‖ Schraubenlehre f. ‖ palmer m. / circular ~ ‖ Kreismikrometer n. ‖ micromètre m. circulaire. / contrast ~ ‖ Kontrastmikrometer n. ‖ micromètre m. à contraste. / cross bar ~ ‖ Kreuzstabmikrometer n. ‖ micromètre m. à barres croisées. / crossline ~ ‖ Netzmikrometer n. ‖ micromètre m. à réseau. / eyepiece cross line ~ ‖ Okularnetzmikrometer n. ‖ micromètre-oculaire m. à réseau. / eyepiece screw ~ ‖ Okularschraubenmikrometer n. ‖ oculaire-micromètre m. à vis. / linear ~ / lineares Mikrometer n. ‖ micromètre m. linéaire. / line scale ~ ‖ Strichmikrometer n. ‖ micromètre m. à traits. / ring ~ ‖ Ringmikrometer n. ‖ micromètre m. annulaire. / spark ~ ‖ Funkenmikrometer n. ‖ micromètre m. à étincelles. / stage ~ ‖ Objektmikrometer n. ‖ micromètre-objectif m.

micrometer adjustment ‖ Mikrometereinstellung f. ‖ réglage m. par vis micrométrique. / ~ casing ‖ Mikrometergehäuse n. ‖ boîte f. de micromètre.

micrometer drum ‖ Meßtrommel f. ‖ tambour m. de mesure. / ~ with type inking recording attachment ‖ Meßtrommel f. mit Typendruckregistriereinrichtung ‖ tambour m. de mesure avec dispositif enregistreur par caractères d'impression. / ~ reading ‖ Meßtrommelablesung f. ‖ lecture f. du tambour de mesure.

micrometer eyepiece ‖ Meßokular n.; Mikrometerokular n. ‖ oculaire m. de micromètre. / ~ with focussing collar graduated in terms of diopters ‖ Mikrometerokular n. mit Dioptrieneinstellung ‖ oculaire m. pour micromètres muni d'une graduation en dioptries.

micrometer gauge see micrometer (Gauge).

micrometer screw ‖ Mikrometerschraube f. ‖ vis f. micrométrique ou à micromètre.

micrometric(al) calipers pl. ‖ Mikrometerzirkel m. ‖ micromètre m. / ~ spark discharger ‖ Mikrometerfunkenstrecke f. ‖ éclateur m. à intervalle micrométrique. / ~ spark gap see ~ spark discharger.

micrometrically movable ‖ mikrometrisch verstellbar ‖ à mouvement micrométrique. / ~ slide ‖ mikrometrisch verstellbarer Blendschieber m. ‖ volet m. à mouvement micrométrique.

micron ‖ Mikromillimeter m. ‖ micron m.

micro-organism ‖ Mikroorganismus m. ‖ microorganisme m.

microphone ‖ Mikrofon n. ‖ microphone m. / electromagnetic ~ ‖ elektromagnetisches Mikrofon n. ‖ microphone m. électromagnétique. / electrostatic ~ ‖ elektrostatisches Mikrofon n. ‖ microphone m. électrostatique.

microphone amplifier ‖ Mikrofonverstärker m. ‖ amplificateur m. microphonique. / ~ apparatus ‖ Mikrofonapparat m. ‖ appareil m. microphone. / ~ buzzer ‖ Mikrofonsummer m. ‖ ronfleur m. / ~ cap ‖ Sprechtrichter m. ‖ calotte f. du microphone. / ~ cell ‖ Mikrofonelement n. ‖ élément m. de microphone. / ~ theory of ~ contacts ‖ Mikrofontakttheorie f. ‖ théorie f. des contacts du microphone.

microphotographic apparatus for visual and ultraviolet light ‖ mikrofotografischer Apparat m. für sichtbares und ultraviolettes Licht ‖ appareil m. de microphotographie pour les radiations visibles et ultraviolettes. / ~ installation ‖ mikrofotografische Einrichtung f. ‖ installation f. pour la microphotographie.

microscope ‖ Mikroskop n.; Kleinseher m. ‖ microscope m. / auxiliary ~ ‖ Hilfsmikroskop n. ‖ microscope m. auxiliaire. / ball stage ~ ‖ Kugelmikroskop n. ‖ microscope m. à articulations. / class ~ ‖ Kursmikroskop n. ‖ microscope m. de travaux pratiques. / comparison ~ for spectra ‖ Vergleichsmikroskop n. für Spektren ‖ microscope m. pour la comparaison des spectres. / compound ~ ‖ zusammengesetztes Mikroskop n. ‖ microscope m. composé. / coordinate reading ~ for measuring watch parts ‖ Koordinatenmeßmikroskop n. für Taschenuhrbestandteile ‖ microscope m. pour la mesure des coordonnées destiné à l'horlogerie. / corneal ~ ‖ Hornhautmikroskop n. ‖ microscope m. cornéen. / corneal ~ for image erecting ‖ bildaufrichtendes Hornhautmikroskop n. ‖ microscope m. cornéen redresseur. /

demonstration ~ ‖ Vorführungsmikroskop n.; Demonstrationsmikroskop n. ‖ microscope m. de démonstration. / ~ for diagnostic purposes ‖ Mikroskop n. für diagnostische Zwecke ‖ microscope m. pour la diagnostie. / dipping ~ ‖ Eintauchmikroskop n. ‖ réfractomètre m. à immersion. / ~ for evaluating negatives ‖ Negativmeßmikroskop n. ‖ microscope m. de mesure pour négatifs; microscope m. pour la mensuration des négatifs. / eye ~ ‖ Augenmikroskop n. ‖ microscope m. ophtalmologique. / eyepiece ~ ‖ Okularmikroskop n. ‖ microscope m. oculaire. / focussing ~ ‖ Einstellmikroskop n. ‖ microscope m. de mise au point. / hand ~ ‖ Handmikroskop n. ‖ microscope m. à main. / ~ with a heatable plate ‖ Heizmikroskop n. ‖ microscope m. à platine chauffante. / ~ with hinged body ‖ schwenkbares Mikroskop n. ‖ microscope m. à genouillère. / image-erecting ~ ‖ bildaufrichtendes Mikroskop n. ‖ microscope m. redresseur. / meat inspector's ~ ‖ Fleischbeschaumikroskop n. ‖ microscope m. à trichines. / metallurgical ~ ‖ Metallmikroskop n. ‖ microscope m. pour l'examen des métaux. / petrological class and laboratory ~ ‖ mineralogisches Kurs- und Arbeitsmikroskop n. ‖ microscope m. minéralogique pour travaux pratiques et de laboratoire. / pharmacological ~ ‖ Mikroskop n. für Apotheker ‖ microscope m. pour pharmaciens. / plate culture ~ ‖ Plattenkulturmikroskop n. ‖ microscope m. pour plaques de culture. / polarizing ~ ‖ Polarisationsmikroskop n. ‖ microscope m. polarisant. / prism ~ ‖ Prismenmikroskop n. ‖ microscope m. à prismes. / reading ~ ‖ Ablesemikroskop n. ‖ microscope m. de lecture. / reflecting ~ ‖ Spiegelmikroskop n. ‖ microscope m. à réflecteur. / research ~ ‖ Forschungsmikroskop n. ‖ microscope m. pour les recherches scientifiques. / school ~ ‖ Schulmikroskop n. ‖ microscope m. pour écoles. / seed ~ ‖ Getreidelupe f. ‖ loupe f. pour l'examen des grains. / single or simple ~ ‖ Lupe f.; Vergrößerungsglas n. ‖ microscope m. simple; loupe f. / skin ~ ‖ Hautmikroskop n. ‖ microscope m. dermatologique. / solar ~ ‖ Sonnenmikroskop n. ‖ microscope m. solaire. / stereoscopic dissecting ~ ‖ stereoskopisches Präpariermikroskop n. ‖ microscope m. à dissection stéréoscopique. / table ~ ‖ Tischmikroskop n. ‖ microscope m. de table. / telescopic ~ ‖ Fernrohrmikroskop n. ‖ télémicroscope m. / travelling ~ ‖ Reisemikroskop n. ‖ microscope m. de voyage. / ultra ~ ‖ Ultramikroskop n. ‖ ultramicroscope m. / ~ with wide tube ‖ Mikroskop n. mit weitem Tubus ‖ microscope m. à large tube.

microscope arc lamp ‖ Mikroskopierbogenlampe f. ‖ microlampe f. à arc. / axis of the ~ ‖ Mikroskopachse f. ‖ axe m. du microscope. / design of the ~ ‖ Mikroskopbau m. ‖ construction f. d'un microscope. / exit-pupil of a ~ ‖ Austrittspupille f. eines Mikroskopes ‖ pupille f. de sortie d'un microscope. / illuminating apparatus of a ~ ‖ Beleuchtungsapparat m. eines Mikroskopes ‖ appareil m. d'illumination d'un microscope. / ~ lamp ‖ Mikroskopierlampe f. ‖ lampe f. pour

la microscopie. / incandescent ~ lamp ‖ Mikroskopierglühlampe f. ‖ microlampe f. à incandescence. / incandescent gas ~ lamp ‖ Mikroskopiergasglühlampe f. ‖ microlampe f. à gaz à manchon incandescent. / ~ objective ‖ Mikroskopobjektiv n. ‖ objectif m. de microscope. / ~ point-light lamp ‖ Mikroskopierpunktlichtlampe f. ‖ microlampe f. ponctuelle. / ~ stage ‖ Tisch m. des Mikroskopes ‖ platine f. du microscope. / ~ tube ‖ Mikroskoptubus m. ‖ tube m. du microscope. / type of ~ ‖ Mikroskoptyp m. ‖ type m. de microscope.

microscopic ‖ mikroskopisch ‖ microscopique. / ~ image ‖ Mikroskopbild n. ‖ image f. formée par le microscope. / ~ optics pl. see microscopy. / ~ preparation ‖ mikroskopisches Präparat n. ‖ préparation f. microscopique. / ~ utensils pl. ‖ mikroskopisches Zubehör n.; mikroskopische Geräte npl. ‖ ustensiles fpl. microscopiques.

microscopy ‖ Mikroskopie f. ‖ microscopic f. / slit lamp ~ ‖ Spaltlampenmikroskopie f. ‖ étude f. microscopique au moyen de la lampe à fente. / polarizing ~ ‖ Polarisationsmikroskopie f. ‖ microscopie f. en lumière polarisée.

microspectroscope ‖ Mikrospektroskop n. ‖ microspectroscope m.

microspectroscopic eyepiece ‖ Mikrospektralokular n. ‖ oculaire m. microspectroscopique.

micro-tasimeter (Electr) ‖ Mikrotasimeter n. ‖ microtasimètre m.

mid (File) ‖ halbschlicht ‖ demi-doux.

middle ‖ mittel, mittlere(r) ‖ intermédiaire. / ~ aisle (Of a church) ‖ Hauptschiff n.; Mittelschiff n. ‖ grande nef f.; haute nef f.; nef f. centrale ou principale. / ~ bay of a shop ‖ Mittelhalle f. einer Werkstatt ‖ nef f. centrale d'un atelier. / ~ body of the telescope ‖ Fernrohrmittelstück n. ‖ partie f. médiane de la lunette. / ~ line (Mar) ‖ Kiellinie f. ‖ file f.; ligne f. de file.

middleman (Trade) ‖ Zwischenhändler m. ‖ intermédiaire m.

middle ‖ Mittelteil m.; mittlerer Teil m.; Mittelstück n. ‖ partie f. centrale. / ~ piece see ~ part. / ~ plain ‖ Mittelebene f. ‖ plan m. médian. / ~ plank ‖ Kernbrett n. ‖ planche f. de cœur. / ~ pressure see mean-pressure. / ~ price (Exchange) ‖ Mittelkurs m. ‖ cours m. moyen. / ~ roll (Rolling mill) ‖ Mittelwalze f. ‖ cylindre m. médial n.

middleshot water wheel ‖ mittelschlächtiges Wasserrad n.; Kropfrad n. ‖ roue f. hydraulique de côté.

middle wing (Airpl) ‖ Mittelflügel m. ‖ plan m. médian; aile f. médiane. / ~ traverse (El line) ‖ Mittelriegel m. ‖ entretoise f.

midship frame ‖ Hauptspant n.; Mittelspant n.; Nullspant n. ‖ maître-couple m.

midships ‖ mittschiffs ‖ au milieu du navire.

migraine pencil ‖ Migränestift m. ‖ crayon m. antimigraine.

migration of ions ‖ Wanderung f. der Ionen ‖ transport m. des ions.

mikrolin ‖ Mikrolin m. ‖ microline f.

mild ‖ mild; sanft; weich ‖ doux. / ~ quality of steel ‖ weiche Stahlmarke f.; weicher Stahl m. ‖ acier m. doux.

mild steel ‖ Flußstahl m.; Flußeisen n. ‖ acier m. doux *ou* fondu *ou* homogène. / ~ for automatics ‖ Automatenweichstahl m. ‖ acier m. doux de décolletage. / ~ in ingots ‖ Flußstahl m. in Blöcken ‖ acier m. doux en lingots. / ~ capable of being hardened ‖ härtbarer Flußstahl m. ‖ acier m. homogène prenant la trempe. / weldable ~ ‖ schweißbarer Flußstahl m. ‖ acier m. homogène soudable.

mild steel plate ‖ Flußstahlblech n. ‖ tôle f. en acier doux. / ~ flanged for fireboxes ‖ gebördeltes Feuerblech n. aus Flußstahl ‖ tôle f. à brides en acier doux pour foyer.

mildew ‖ Schimmel m.; Meltau m. ‖ mildiou m.

mildewed ‖ mit Meltau m. behaftet ‖ rouillé.

mile ‖ Meile f. ‖ mille m.; lieue f. / nautical ~ ‖ Knoten m.; Seemeile f. ‖ mille m. marin; nœud m. / statute ~ ‖ englische Meile f. ‖ mille m. d'Angleterre.

mileage indicator *see* mileage recorder. / extra ~ rate (Tel) ‖ Leitungszuschlag m. ‖ redevance f. d'entretien de la ligne.

mileage recorder ‖ Meilenzähler m.; Kilometerzähler m. ‖ odomètre m.; compteur m. kilométrique. / ~ for bicycles ‖ Fahrradkilometerzähler m. ‖ compteur m. kilométrique pour bicyclettes.

mile mark *see* milestone.

mileometer *see* mileage recorder.

mile post *see* milestone.

milestone ‖ Meilenstein m. ‖ (borne f.) milliaire m.; pierre f. milliaire. / ~ ‖ Kilometerstein m. ‖ borne f. kilométrique.

milfoil flowers pl. ‖ Schafgarbenblüten f. ‖ fleurs fpl. de millefeuille.

military airplane ‖ Militärflugzeug n. ‖ aéroplane m. militaire. / ~ braid ‖ Militärposamenten pl. ‖ passementerie f. militaire. / ~ bridge ‖ Kriegsbrücke f. ‖ pont m. militaire. / ~ button ‖ Uniformknopf m. ‖ bouton m. d'uniforme. / ~ car ‖ Militärfahrzeug n. ‖ voiture f. militaire. / ~ direction of telegraphs ‖ Etappentelegrafendirektion f. ‖ direction f. des télégraphes d'étape. / ~ equipment ‖ Heeresausrüstung f.; Militärausrüstung f.; Militäreffekten pl. ‖ équipement m. militaire; effets mpl. militaires. / ~ gun ‖ Militärgewehr n. ‖ fusil m. de guerre. / ~ rifle *see* ~ gun. / portable ~ station (Radio) ‖ tragbare Militärstation f. ‖ poste f. militaire transportable. / ~ telegraph section ‖ Feldtelegrafenabteilung f. ‖ section f. télégraphique de campagne. / ~ telegraphy ‖ Militärtelegrafie f. ‖ télégraphie f. militaire. / ~ trappings pl. *see* ~ equipment.

milk, to ‖ melken ‖ traire. / ~ the battery (Acc) ‖ die Batterie zum Kochen bringen ‖ survolter la batterie.

milk ‖ Milch f. ‖ lait m. / ~ of almonds ‖ Mandelmilch f. ‖ lait m. d'amandes. / coagulated ~ ‖ geronnene Milch f. ‖ lait m. caillé. / condensed ~ ‖ Dauermilch f.; kondensierte Milch f. ‖ lait m. condensé. / curdled ~ *see* coagulated ~ ‖ geronnene Milch f. ‖ lait m. caillé. / fresh ~ ‖ frische Milch f.; Frischmilch f. ‖ lait m. frais. / ~ of lime ‖ Kalkmilch f. ‖ lait m. de chaud; échaudage m. / malt ~ ‖ Malzmilch f. ‖ lait m. de malt. / peptonized ~ ‖ peptonisierte Milch f. ‖ lait m. peptonisé

/ powdered ~ ‖ Trockenmilch f. ‖ lait m. en poudre. / preserved ~ ‖ konservierte Milch f. ‖ lait m. conservé. / skimmed ~ ‖ entrahmte Milch f. ‖ lait m. écrémé. / sterilized ~ ‖ sterilisierte Milch f. ‖ lait m. stérilisé. / ~ of sulphur ‖ Schwefelmilch f. ‖ lait m. de soufre. / whole ~ ‖ Vollmilch f. ‖ lait m. complet.

milk boiler ‖ Milchsieder m. ‖ chauffe-lait m. / ~ bucket ‖ Milcheimer m. ‖ seau m. *ou* baquet m. à lait. / ~ chocolate ‖ Milchschokolade f. ‖ chocolat m. au lait. / ~ cooling plant ‖ Milchkühlanlage f. ‖ installation f. frigorifique de laiterie. / ~ cooling plant for low temperature ‖ Milchtiefkühlanlage f. ‖ installation f pour la réfrigération intense du lait. / ~ cooling room ‖ Milchkühlraum m. ‖ chambre f. froide à lait. / ~ extract ‖ Milchauszug m.; Milchextrakt m. ‖ extrait m. de lait. / ~ farm ‖ Meierei f. ‖ laiterie f. / ~ fat refractometer ‖ Milchfettrefraktometer n. ‖ refractomètre m. à lait.

milk glass ‖ Milchglas n. ‖ verre m. opale. / ~ bottle ‖ Milchglasflasche f. ‖ bouteille f. en verre opale. / ~ jar ‖ Milchglasdose f. ‖ boîte f. en verre opale.

milk heater ‖ Milchvorwärmer m. ‖ réchauffeur m. pour lait.

milkiness (Chem) ‖ Trübung f. ‖ opacité f.; trouble m.

milking house ‖ Milchwirtschaft f., Molkerei f. ‖ laiterie f. / ~ machine ‖ Melkmaschine f. ‖ machine f. à traire.

milk jug ‖ Milchkrug m. ‖ cruchon m. à lait.

milkmaid ‖ Milchmädchen n. ‖ laitière f.

milkman ‖ Milchmann m. ‖ laitier m.

milk poise ‖ Milchwage f. ‖ Milchmesser m. ‖ lactomètre m.; pèse-lait m.; galactometer m. / ~ pot ‖ Milchkanne f. ‖ pot m. à lait. / ~ powder ‖ Milchpulver n. ‖ lait m. en poudre. / ~ preserve ‖ Milchkonserve f. ‖ lait m. conservé. / ~ sieve ‖ Milchsieb n. ‖ passoir m. à lait. / ~ sugar (Chem) ‖ Milchzucker m. ‖ sucre m. de lait. / ~ tester ‖ Milchprüfer m. ‖ pèse-lait m. / ~ transporting can ‖ Milchtransportkanne f. ‖ bidon m. pour le transport de lait.

milky ‖ milchig ‖ laiteux. / ~ quartz ‖ Milchquarz m. ‖ quartz m. laiteux; quartz m. blanc de lait.

mill, to ‖ mahlen ‖ moudre. / ~ (Pap) ‖ mahlen; walzen ‖ broyer; cylindrer. / ~ (To fraise) ‖ fräsen ‖ fraiser. / ~ (Coins) ‖ rändeln ‖ cordonner. / ~ off ‖ abfräsen ‖ enlever à la fraise. / ~ teeth of gear wheels *or* of spur wheels ‖ Zahnräder npl. fräsen ‖ fraiser des engrenages mpl.; tailler des engrenages à la fraise.

mill ‖ Mühle f. ‖ moulin m. / ~ (Coin) ‖ Spindelwerk n.; Stoßwerk n. ‖ balancier m. à pièces. / ~ (Glassm) ‖ Reibkasten m.; Schleifkasten m. ‖ moëllon m. / ~ (Manufactory) ‖ Fabrik f.; Werk n. ‖ fabrique f.; usine f. / ~ (Milling cutter) ‖ Fräsmaschine f.; Fräse f. ‖ fraise f. / ~ (Rolling mill) ‖ Walzwerk n. ‖ laminoir m. / ~ to stop the ~ ‖ die Mühle f. zum Stillstand bringen ‖ arrêter le moulin. / alabaster ~ ‖ Alabastermühle f. ‖ moulin m. à albâtre. / ~ of armour-plates ‖ Panzerplattenwalzwerk n. ‖ train m. à blindages. / ball ~ ‖ Kugelmühle f. ‖ broyeur m. *ou* moulin m. à boulets. / ~ for bars ‖ Stabeisenstraße f. ‖ train m.

pour fers marchands. / billet ~ ‖ Knüppelwalzwerk n. ‖ laminoir m. à billettes. / blooming ~ ‖ Blockwalzwerk n.; Blockstraße f.; Grobwalzwerk n. ‖ laminoir m. à blooms; train m. dégrossisseur. / bone ~ ‖ Knochenmühle f. ‖ moulin m. à os. / breaking down ~ (Met) ‖ Blockbrecher m. ‖ casseur m. de lingots. / break roller ~ ‖ Schrotwalzenstuhl m. ‖ broyeur m. / bruising ~ ‖ Quetschmühle f. ‖ moulin m. écraseur. / case ~ (Windmill) ‖ Kochermühle f. ‖ moulin m. à colonne creuse *ou* de support. / cement ~ ‖ Zementmühle f. ‖ moulin m. pour ciment. / centrifugal ~ ‖ Schleudermühle f. ‖ broyeur m. centrifuge; désintégrateur. m. / chalk ~ ‖ Kreidemühle f. ‖ moulin m. à craie. / ~ for chemicals ‖ Chemikalienmühle f. ‖ moulin m. pour produits chimiques. / coal-dust ~ ‖ Kohlenstaubmühle f. ‖ pulvérisateur m. à charbon. / cogging ~ *see* blooming ~. / cold rolling ~ ‖ Kaltwalzwerk n. ‖ laminoir m. à froid. / ~ for common use ‖ Gemeindemühle f. ‖ moulin m. banal. / compound ~ ‖ Verbundmühle f. ‖ moulin m. combiné. / cone ~ ‖ Konusmühle f.; Glockenmühle f. ‖ moulin m. à cône. / continuous ~ ‖ kontinuierliches Walzwerk n. ‖ train m. continu. / cork ~ ‖ Korkmühle f.; Korkmüllerei f. ‖ moulin m. à liège. / corn ~ ‖ Getreidemühle f. ‖ moulin m. (à blé). / cross beater ~ ‖ Schlagkreuzmühle f. ‖ moulin m. à croisillon percuteur. / disk ~ ‖ Scheibenmühle f. ‖ moulin m. à disque. / dish ~ ‖ Tellermühle f. ‖ moulin m. à plateaux. / ~ for drugs ‖ Drogenmühle f. ‖ moulin m. à drogues. / edge runner ~ ‖ Kollergang m.; Kollermühle f. ‖ moulin m. à meules verticales. / enamel ~ ‖ Emaillemühle f. ‖ moulin m. à émail. / finishing ~ ‖ Fertigstraße f.; Nachwalzwerk n. ‖ laminoir m. *ou* train m. finisseur. / ~ for fluor spar ‖ Flußspatmühle f. ‖ moulin m. à spath fluorine. / ~ with fluted rolls ‖ Riffelwalzenstuhl m. ‖ moulin m. à cylindres cannelés. / four-roll ~ (Roll mill) ‖ Vierwalzenstraße f. ‖ laminoir m. à quatre cylindres. / four-roller ~ (Ore dress) ‖ Vierwalzenmühle f. ‖ broyeur m. à quatre cylindres. / fruit ~ ‖ Obstfruchtmühle f.; Obstpresse f. ‖ moulin m. à fruits (charnus); presse f. à fruits. / girder ~ ‖ Trägerwalzwerk n. ‖ laminoir m. à poutrelles. / girder and section ~ ‖ Formeisenwalzwerk n. ‖ laminoir m. à fers profilés et cornières. / grape ~ ‖ Traubenmühle f. ‖ moulin m. à raisins. / graphite ~ ‖ Graphitmühle f. ‖ moulin m. à graphite. / grinding ~ ‖ Mahlwerk n. ‖ laminoir m. broyeur. / grits ~ ‖ Grießmühle f. ‖ moulin m. à semoule. / ~ for groceries ‖ Kolonialwarenmühle f. ‖ moulin m. à denrées coloniales. / gypsum ~ ‖ Gipsmühle f. ‖ moulin m. à plâtre. / hammer ~ ‖ Hammermühle f. ‖ moulin m. à marteaux. / ~ worked by hand ‖ Handmühle f. ‖ moulin m. à bras. / heavy spar ~ ‖ Schwerspatmühle f. ‖ moulin m. à baryte. / ~ for industrial and agricultural purposes ‖ Mühle f. für Industrie und Landwirtschaft ‖ moulin m. pour l'industrie et l'agriculture. / kibbling ~ ‖ Schrotmühle f. ‖ moulin m. pour égruger. / large ~ ‖ Großmühle f. ‖ gros moulin m. / lime ~ ‖ Kalkmühle f. ‖ moulin m. à chaux.

/ magnesite ~ || Magnesitmühle f. || moulin m. à magnésite. / maize ~ || Maismühle f. || moulin m. à maïs. / maize flaking ~ || Maisflockenstuhl m. || moulin m. à flocons de maïs. / malt ~ || Malzmühle f. || moulin m. à malt. / manure ~ || Düngermühle f. || moulin m. à engrais. / marble ~ || Marmormühle f. || moulin m. à marbre. / ~ with millstones || Mühle f. mit Mahlgängen || moulin m. à meules. / mint ~ || Münzwalzwerk n. || laminoir m. monnayeur. / mixing ~ || Mischwalzwerk n. || laminoir m. mélangeur. / mortar ~ || Mörtelmühle f. || moulin m. à mortier. / oil ~ || Ölmühle f. || moulin m. à huile. / peat ~ || Torfmühle f. || moulin m. à tourbe. / pendulum ~ || Pendelmühle f. || moulin m. à pendule. / ~ for pickle barley || Graupenmühle f. || moulin m. à mouler et perler l'orge. / pin beater ~ || Schlagstiftmühle f. || moulin m. à battoirs à chevilles. / pipe ~ || Röhrenwalzwerk n. || laminoir m. à tubes. / plate ~ || Grobblechstraße f. || train m. à grosses tôles. / plating ~ || Plattierwalzwerk n. || laminoir m. pour tôles plaquées. / pop-corn and rice ~ || Röstmais- und Reismühle f. || moulin m. à riz et à maïs. / poppy ~ || Mohnmühle f. || moulin m. à pavot. / preliminary ~ || Vorwalzwerk n. || laminoir m. dégrossisseur. / ~ for the rolling of rails || Schienenwalzwerk n. || train m. de laminoirs pour rails. / reversing ~ || Umkehrstraße f. || train m. réversible. / rice ~ || Reismühle f. || moulin m. à riz. / roller ~ || Walzenmühle f. || moulin m. à cylindres. / rolling ~ || Walzwerk n.; Straße f. || laminoir m.; train m. / ~ for rough-grinding corn || Schrotmühle f. || moulin m. à égruger. / roughing ~ || Vorstraße f. || train m. dégrossisseur. / rye flour ~ || Roggenmühle f. || moulin m. à seigle. / screw ~ || Schraubenmühle f. || moulin m. à vis. / sheet ~ || Feinblechstraße f. || train m. à fines tôles. / sheet bar ~ || Platinenwalzwerk n. || laminoir m. à largets. / sheet billet ~ see sheet bar ~. / sizing ~ || Maßwalzwerk n. || laminoir m. de précision. / slabbing ~ || Luppenwalzwerk n. || laminoir m. à blooms. / small iron ~ || Feineisenstraße f.; Feineisenwalzwerk n. || laminoir m. ou train m. à petits fers. / small section ~ see small iron ~ || Feineisenstraße f. || train m. à petits fers. / smooth roller ~ || Glattwalzenstuhl m. || convertisseur m.; moulin m. à cylindres lisses. / ~ with smooth rolls see smooth roller ~. / special ~ || Sonderwalzwerk n. || laminoirs mpl. spéciaux. / ~ for spices || Gewürzmühle f. || moulin m. à épices. / spoon-rolling ~ || Löffelwalze f. || laminoir m. à cuillers. / steam ~ || Dampfmühle f. || moulin m. à vapeur. / three-high ~ || Dreiwalzenstraße f. || train m. trio. / tire ~ || Radreifenwalzwerk n. || laminoir m. à bandages. / trass ~ || Traßmühle f. || moulin m. à trass. / trunk ~ (Windmill) ~ || Kochermühle f. || moulin m. à colonne creuse ou de support. / tube ~ || Rohrmühle f. || moulin m. à tube; tube m. broyeur. / tube reducing ~ || Rohrreduzierwalzwerk n. || laminoir m. pour la réduction des tuyaux. / two-high ~ || Zweiwalzenstraße f. || train m. duo. / two-roller ~ (Cornmill) || Zweiwalzenstuhl m. || mou-

lin m. à deux cylindres. / tyre ~ see tire ~. / vertical ~ || Kollergang m.; Kollermühle f. || meule f. verticale; tordoir m. / water ~ || Wassermühle f. || moulin m. à eau. / wind ~ || Windmühle f. || moulin m. à vent. / wire rolling ~ || Drahtstraße f. || train m. à fils.

mill bar (Met) || Platine f.; Rohschiene f.; fer m. ébauché (plat). / ~ of rough steel || Rohstahlschiene f. || acier m. ébauché.

millboard || Glanzkarton m. || carton m. glacé ou lissé.

mill cake || Pulverkuchen m. || galette f. ou pâte f. ou lame f. de poudre. / ~ carpenter || Mühlzimmermann m. || charpentier m. moulageur. / ~ casing || zweiteiliges Mahlgehäuse n. || enveloppe f. de moulin en deux pièces. / ~ chips pl. || Walzsplitter mpl. || paille f. de laminage. / ~ cinder || Walzsinter m. || battitures fpl. des laminoirs; croûte f. de laminage.

milled border || gefräster Rand m. || bord m. fraisé. / ~ edge || gerieffter Rand m. || bord m. moleté. / ~ edge of a coin || Kräuselung f. oder Rand m. einer Münze || cordon m. d'une monnaie. / ~ head for rotating circle || Scheibenknopf m. für Kreisdrehung || bouton m. actionnant le mouvement circulaire. / ~ head screw || Kordelschraube f. || vis f. moletée. / ~ nut || gerändelte Mutter f. || écrou m. moleté.

mill dust || Mühlenstaub m.; Mahlstaub m. || farine f. folle.

millefiori || Glasmosaik f. || verre m. mosaïque; verre m. mille-fleurs.

mill engine man (Roll mill) || Walzwerkmaschinist m. || mécanicien m. de laminoir.

miller || Müller m. || meunier m. / ~ (Mach) see milling machine. / ~'s charges pl. || Mahlgeld n. || (prix m. de) mouture f. / ~'s industry || Müllereiindustrie f. || industrie f. meunière. / ~'s machine || Müllereimaschine f. || machine f. de meunerie. / ~'s product || Müllereierzeugnis n. || produit m. de la minoterie. / ~'s trade || Müllereigewerbe n. || meunerie f.

millerite || Schwefelnickel m.; Haarkies m. || millérite f.; nickel m. sulfuré; pyrite f. capillaire.

millery || Müllerei f. || meunerie f. / machine for ~ see miller's machine / ~ machine || Müllereimaschine f. || machine f. de minoterie.

millet || Hirse f. || millet m. / ~ mill || Hirsemühle f. || moulin m. à millet. / ~ straw || Hirsestroh n. || paille f. de millet.

mill foreman (Roll mill) || Walzwerkmeister m. || chef m. lamineur. / ~ frame (Of the windmill) || Stuhl m. oder Bock m. der Mühle || chaise f. ou pylone m. du moulin. / ~ furnace (Met) || Schweißofen m. || four m. ou fourneau m. à réchauffer; four à souder le fer. / ~ hand || Mühlenarbeiter m. || garçon m. meunier. / ~ hopper || Einlauftrichter m.; Rumpf m. || entonnoir m. d'entrée; trémie f.

milliammeter || Milliamperemesser m. || milliampéremètre m.

milliampere || Milliampere n. || milliampère m.

milligram || Milligramm n. || milligramme m. / tenth of ~ || Dezimilligramm n. || dixième f. de milligramme.

millimeter || Millimeter m. || millimètre m. / ~ paper || Millimeterpapier n. || papier m. quadrillé.

millimetre see millimeter.

millinery || Putzwaren fpl.; Modewaren fpl. || modes mpl.; nouveautés fpl. / ~ and fashion articles pl. || Putz- und Modewaren fpl. || objets mpl. de toilette et de la mode. / ~ business || Putzgeschäft n. || magasin m. de modes.

milling || Mahlen n.; Müllerei f. || mouture f. / ~ (Mach) || Fräsen n.; Abfräsen n. || fraisage m. / ~ carriage || Frässchlitten m.; Frässpindelstock m. || chariot m. porte-fraise; tête f. porte-fraise. / ~ course || Mahlgang m. || mouvement m. de la meule.

milling cutter || Fräser m. || fraise f. / ~ backed for clearance || hinterdrehter Fräser m. || fraise f. dégagée. / conical type ~ || kegeliger Fräser m. || fraise f. conique. / ~ end-on type || Planfräser m. || fraise f. en bout à surfacer. / ~ for grooving || Schlitzfräser m. || fraise f. à canneler. / plain ~ || Walzenfräser m. || fraise f. cylindrique. / rotating ~ || rotierender oder umlaufender Fräser m. || fraise f. rotative. / screw cutting ~ || Gewindefräser m. || fraise f. à tailler les filets. / ~ of shaped type || Profilfräser m. || fraise f. à profiler. / slot ~ || Nutenfräser m. || fraise f. pour rainures. / ~ for toothed-gear cutting || Fräser m. zum Zähneschneiden || fraise f. module pour tailler les engrenages. / twin ~ || zweiteiliger Fräser m. || fraises fpl. accouplées. / ~ for worms || Schneckenfräser m. || fraise f. à vis-mère.

milling cutter driving gear || Fräserantriebsrad n. || roue f. de commande de la fraise. / ~ grinding machine || Fräserschleifmaschine f. || machine f. à affûter les fraises. / ~ spindle of a portal milling machine || Frässpindel f. eines Portalfräswerkes || broche f. de fraisage d'une fraiseuse à montant.

milling depth || Frästiefe f. || profondeur f. à fraiser. / ~ device || Fräsvorrichtung f. || dispositif m. de fraisage.

milling machine || Fräsmaschine f. || fraiseuse f.; machine f. à fraiser. / ~ (Coin) || Kräuselwerk n., Rändelwerk n. || machine f. à cordonner ou à tranche. / ~ (Fulling machine) || Walke f. || foulon m. / automatic ~ || Fräsautomat m.; selbsttätige Fräsmaschine f. || machine f. à fraiser automatique. / ~ with beaters for treating carbonized cloth || Klopfwalke f. zum Reinigen karbonisierter Tuche || foulon m. servant au nettoyage des draps carbonisés. / circular ~ || Rundfräsmaschine f. || fraiseuse f. circulaire; machine f. à fraiser circulaire. / column ~ || Ständerfräsmaschine f. || fraiseuse f. à montant. / copy ~ || Kopierfräsmaschine f. || fraiseuse f. à copier ou à reproduire. / ~ for dies and punches || Fräsmaschine f. für Gesenke und Schnitte || fraiseuse f. pour étampes et matrices. / double ~ || Doppelfräsmaschine f. || fraiseuse f. à deux broches horizontales. / ~ groove (Roll mill) || Kaliberfräsmaschine f. || fraiseuse-calibre f. / high-efficiency ~ || Hochleistungsfräsmaschine f. || machine f. à fraiser à grand rendement. / horizontal ~ || Wagerechtfräsmaschine f. || machine f. horizontale à fraiser. / key-groove ~ || Keilnutfräsmaschine f. à fraiser les rainures de clavettes. / automatic keyway ~ || selbsttätige Keilnutenfräsmaschine f. || machine f. automatique à fraiser

les rainures de clavettes. / plain horizontal ~ || Einfachwagerechtfräsmaschine f. || fraiseuse f. horizontale simple. / plano ~ || Langtischfräsmaschine f. || fraiseuse f. raboteuse. / profile ~ || Kopierfräsmaschine f. || fraiseuse f. à copier. / rail ~ || Schienenfräser m. || fraiseuse f. pour rails. / rotary ~ || Rundlauffräsmaschine f. || fraiseuse f. à plateau horizontal rotatif. / slot ~ || Nutenfräsmaschine f. || machine f. à fraiser les rainures. / ~ for stone working || Fräsmaschine f. für die Steinbearbeitung || fraiseuse f. pour travailler la pierre. / ~ for test rods || Probenfräsmaschine f. || fraiseuse f. pour éprouvettes. / universal ~ || Universalfräsmaschine f. || fraiseuse f. universelle. / various-purpose ~ || Mehrzweckfräsmaschine f. || fraiseuse f. pour fabrications variées. / vertical ~ || Senkrechtfräsmaschine f. || fraiseuse f. verticale.

milling process || Mahlverfahren n. || procédé m. de mouture. / to speed-up the ~ || die Vermahlung f. beschleunigen || accélérer la mouture.

milling shop || Fräserei f. || atelier m. de fraisage. / ~ tool (Turn) || Rändelgabel f.; Rändeleisen n. || porte-molette m. / ~ wheel || Rändelrad n.; Rändelscheibe f. || molette f.

millivolt || Millivolt n. || millivolt m.

mill laboratory || Mühlenlaboratorium n. || laboratoire m. du meunier. / ~ machine || Müllereimaschine f. || machine f. de meunerie. / ~ machinery maker || Mühlenbauer m. || constructeur m. de moulins. / ~ man (Curr) || Fellwalker m. || fouleur m. de peaux. / ~ manager || Fabrikdirektor m. || directeur de fabrique f. ou de manufacture f. ou d'usine f. / ~ owner || Fabrikbesitzer m. || industriel m.; manufacturier m.; usinier m. / ~ pond || Mühlteich m. || étang m. de moulin. / ~ post of the ~ || Stuhl n. oder Bock m. der Mühle || chaise f. ou pylone m. du moulin. / ~ race (Of a corn mill) || Mühlengerinne n. || coursier m.; auge f.; bief m. / ~ scale || Walzsinter m. || battitures fpl. des laminoirs; croûte f. de laminage. / ~ splinters pl. || Walzsplitter mpl. || paille f. de laminage.

mill-spun yarn || Maschinengarn n. || fil m. mécanique.

millstone || Mühlstein m.; Mahlstein m.; Mahlscheibe f.; Mahlgang m. || meule f.; pierre f. meulière. / to dress the ~ || den Mühlstein m. schärfen || rhabiller la meule. / to recut the grooves of the ~ || die Einlauffurchen fpl. im Mühlstein nachhauen || faire strier la meule à nouveau. / bottom ~ || Bodenstein m. || meule f. dormante. / upper ~ (Wind mill) || Läufer m. || meule f. courante.

millstone box || Mühlsteinbuchse f. || boîtard m. / ~ cutter || Mühlsteinarbeiter m. || rhabilleur m. de meules. / ~ dresser || Mühlsteinschleifer m. || piqueur m. de meules. / ~ dressing || Mühlsteinschärfen n. || rhabillage m. de meules de moulins. / ~ eye || Mühlsteinloch n. || œillet m. de meule. / ~ furrow || Mühlsteinfurche f. || silon m. d'une meule. / ~ grit (Miner) || Kohlensandstein m.; Mühlsandstein m. || meulière de grès; grès m. houiller ou à meules. / ~ hammer || Mühlsteinpicke f. || marteau m. pour moulins. / heart of ~ || Herzstück n. des

Mühlsteines || cœur m. de la meule. / ~ piercer || Mühlsteinbohrer m.; Steinbohrer m. || perce-meule m.; bonnet m. de prêtre. / ~ quarry || Mühlsteingrube f. || carrière f. de pierre meulière. / ~ quartz || Mühlsteinquarz m. || caillouasse f. / ~ rock || Mühlkalkstein m. || meulière f. / running balance of ~ || laufendes Gleichgewicht n. des Mühlsteins || équilibre m. de la meule en marche. / skirt of a ~ || Mahlbahn f. || feuillure f. / ~ spindle || Mühlsteinspindel f. || fer m. de meule. / standing balance of ~ || ruhendes Gleichgewicht n. des Mühlsteins || équilibre m. de la meule en repos.

mill train || Walzenstraße f. || train m. de laminoir.

millwright || Modelltischler m. || modeleur m. / ~ (Builder of cornmills) || Mühlenbauer m. || constructeur m. de moulins. / ~ industry || Mühlenbauindustrie f. || industrie f. de construction des moulins.

mimetesite || Mimetesit m.; Arsenbleierz n.; Grünbleierz n. || plomb m. arséniaté; mimétésite f.

mimetite || Flockenerz n.; Mimetit m.; Arsenbleispat m. || mimétite f.

mimosa bark || Rinde f. vom Mimosenbaum || écorce f. du mimosa.

mincer || Fleischhackmaschine f. || hache-viande m.

mincing knife || Wiegemesser n.; Hackmesser n. || hachoir m. / two-bladed ~ || Doppelschneidewiegemesser n. || Zweischneidenwiegemesser n. || hachoir m. à deux lames.

mincing machine || Fleischhackmaschine f. || hache-viande m.; machine f. à hacher la viande.

minder || Wärter m. || soigneur m. / ~ (Mine) || Füllortarbeiter m. || avanceur m. ou ravanceur m. de berlines; homme m. de chambre. / gill box ~ (Wool) || Rohstrecker m. || gill-boxeur m. / finishing box ~ (Wool) || Feinstrecker m. || finisseur m. de laine.

mindfor || Mindfor n. || mindfor m.

mine, to || untergraben; unterhöhlen; unterminieren || miner; creuser.

mine || Bergwerk n.; Zeche f.; Grube f. || mine f. / ~ (Mar) || Mine f. || mine f. / amalgamated ~ || konsolidiertes Bergwerk n. || mine f. fusionnée ou consolidée. / to begin a ~ || ein Bohrloch niederbringen || commencer à faire une fouille. / to blast a ~ || einen Sprengschuß abfeuern || faire sauter une mine. / coal ~ || Kohlenzeche f.; Kohlenbergwerk n. || mine f. de charbon ou de houille; houillère f.; charbonnage m. / contact ~ || Kontaktmine f. || torpille f. automatique. / defensive ~ || Gegenmine f. || contre-mine f.; mine f. défensive. / to discover a ~ || fündig werden; ein Bergwerk fündig machen || découvrir une mine. / exhausted ~ || erschöpfte Zeche f. || mine f. épuisée. / the ~ is fiery || das Bergwerk hat schlagende Wetter || la mine est grisouteuse. / iron ore ~ || Eisensteingrube f. || mine f. de fer. / lignite ~ || Braunkohlengrube f. || mine f. de lignite. / old ~ || verlassenes Bergwerk n. || mine f. abandonnée. / overcharged ~ || überladene Mine f. || fourneau m. surchargé; globe m. de compression. / ~ of precious stones || Edelmetallbergwerk n. || mine f. à métaux précieux. / to keep a ~ in repair || ein Bergwerk bauhaft halten ||

entretenir une mine. / single ~ (Mar) || einfache Mine f. || fourneau m. isolé. / surcharged ~ see overcharged ~. / to survey a ~ || eine Grube f. aufnehmen || lever le plan d'une mine. / undercharged ~ || Quetschmine f. || fourneau m. souschargé; camouflet m. / to work a ~ || eine Grube f. abbauen; eine Lagerstätte f. ausbeuten || exploiter une mine. / to work a ~ on behalf of the state || ein Bergwerk n. für Rechnung des Staates betreiben || exploiter une mine pour le compte de l'Etat.

mine action || Bergwerksaktie f.; Bergwerksanteil m. || action f. de mine. / administration of a ~ || Bergwerksverwaltung f. || administration f. d'une mine. / ~ air || Wetter npl. || air m. de mine. / air in ~s || Grubenluft f. || air m. des galeries de mine. / ~ car || Grubenwagen m. || wagonnet m. de mine. / ~ chamber (Military mining) || Minenkammer f. || fourneau m. ou chambre f. de mine. / charge of a ~ || Minenladung f. || charge f. de mine. / ~ damage || Bergschaden m. || dommage m. causé par l'exploitation minière. / downcast of a ~ || einziehender Schacht m. eines Bergwerkes || puits m. descendant d'une mine. / ~ drainage || Wasserhaltung f. || épuisement m. / ~ dues pl. || Bergwerksabgabe f. || contribution f. minière. / economics pl. of ~ || Grubenhaushalt m. || économie f. de la mine. / ~ equipment material || Grubeneinrichtungsgegenstände mpl. || matériel m. d'installation de mines. / ~ exploder || Minenzünder m. || exploseur m. de mine. / ~ fan || Wettermaschine f.; Grubenventilator m. || ventilateur m. de mine. / ~ field (Mar) || Minenfeld n.; Minensperre f. || barrage m. en mines. / ~ fire || Grubenbrand m. || feux mpl. dans une mine; incendie f. dans une mine. / ~ foreman || Obersteiger m. || maître-mineur m. / ~ foreman's assistant || Steiger m. || maître-mineur m.; porion m. / ~ fuse || Minenzünder m. || détonateur m. de mine. / ~ fuse with electric ignition || Minenentzünder m. mit elektrischer Zündung || détonateur m. de mine à allumage électrique. / ~ gas || Grubengas n.; schlagende Wetter npl. || grisou m. / ~ hoist || Fördermaschine f. || machine f. d'extraction. / ~ kibble || Abteufkübel m.; Grubenkübel m. || seau m. de mine. / ~ levelling staff || Grubennivellierlatte f. || règle f. de nivellement pour mines. / ~ locomotive || Grubenlokomotive f. || locomotive f. de mine ou pour mines. / ~ manager || Bergwerksdirektor m. || directeur m. de mine. / ~ office || Bergamt n. || conseil m. des mines. / ~ officer || Bergbeamter m. || ingénieur m. des mines. / ~ owner || Bergwerksbesitzer m. || exploitant m. ou propriétaire m. de mine. / ~ plan || Grubenbild n. || plan m. de la mine. / ~ product || Bergwerkserzeugnis n. || produit m. de la mine. / ~ projector (War mat) || Minenwerfer m. || lance-mine m. / ~ prop || Grubenstempel m. || étançon m. pour mines. / ~ pumping || Wasserhaltung f. || épuisement m. de la mine.

miner || Bergarbeiter m.; Bergmann m.; Grubenarbeiter m. || ouvrier m. du fond; mineur m. / ~ in the strip mining || Tagebauarbeiter m. || mineur m. à ciel ouvert. / ~s pl. || Belegschaft f. || équipe f. des

mineurs. / ~'s association ‖ Knappschaft f. ‖ corps m. des mineurs. / ~'s bar ‖ Bohrstange f. ‖ barre f. de mine. / ~'s benefit fund ‖ Knappschaftskasse f. ‖ caisse f. de secours des mineurs. / ~'s cap ‖ Schachthut m. ‖ bonnet m. de mineur. / ~'s car see ~'s truck. / ~'s code of law ‖ Bergrecht n. ‖ code m. de mineur. / ~'s compass ‖ Grubenkompaß m.; Hängekompaß m. ‖ boussole f. ou poche f. / ~'s hammer ‖ Handfäustel m.; Fäustel m. ‖ masse f. à main; massette f. / ~'s hat see ~'s cap. / ~'s implements pl. ‖ Gezähe n. ‖ outils mpl. de mineurs. / ~'s lamp ‖ Grubenlampe f. ‖ lampe f. de mineur ou de mines. / ~'s passenger car ‖ Personenwagen m. für Grubenbetriebe ‖ wagonnet m. pour le personnel dans les mines. / ~'s pike ‖ Haue f. ‖ pic m. / ~'s powder ‖ Sprengpulver n.; Grubenpulver n. ‖ poudre f. de mine. / ~'s society ‖ Knappschaft f. ‖ corps m. des mineurs. / ~'s spade ‖ Minenspaten m.; Stecheisen n. ‖ écoupe f.; escope f. / ~'s tools pl. ‖ Gezähe n. (für Bergleute) ‖ outils mpl. de mineurs. / ~'s (trade-)union ‖ Bergarbeiterverband m. ‖ union f. des mineurs. / ~'s truck ‖ Förderwagen m.; Hund m.; Förderhund m.; Grubenwagen m. ‖ berline f.; wagonnet m. / ~'s wedge ‖ Pfändekeil m. ‖ coin m. du coffrage. / ~'s work (Military mining) ‖ Minenarbeit f. ‖ fouille f.

mine rail ‖ Grubenschiene f. ‖ rail m. de mine. / ~ railway ‖ Grubenbahn f. ‖ chemin m. de fer minier ou de mine; voie f. de mine.

mineral ‖ mineralisch ‖ minéral. / ~ acid ‖ mineralische Säure f.; Mineralsäure f. ‖ acide m. minéral. / ~ black ‖ Mineralschwarz n. ‖ noir m. minéral. / ~ caoutchouc ‖ elastisches Erdpech n. ‖ caoutchouc m. minéral. / ~ coal ‖ Steinkohle f. ‖ houille f.

mineral colour ‖ Erdfarbe f.; Mineralfarbe f. ‖ terre f. colorante; couleur f. minérale. / calcined ~ ‖ gebrannte Erdfarbe f. ‖ couleur f. minérale calcinée. / ground ~ ‖ gemahlene Erdfarbe f. ‖ couleur f. minérale pulvérisée. / washed ~ ‖ geschlämmte Erdfarbe f. ‖ couleur f. minérale lavée.

mineral colouring matter ‖ Mineralfarbstoff m. ‖ matière f. colorante d'origine minérale.

mineral deposit ‖ Lagerstätte f. ‖ gisement m. / flat lying ~ ‖ flache Lagerstätte f. ‖ gisement m. horizontal. / vertical ~ ‖ stehende Lagerstätte f. ‖ gisement m. vertical ou en dressant.

mineral detector (Radio) ‖ Mineraldetektor m. ‖ détecteur m. à minéral. / ~ dye ‖ Mineralfarbe f. ‖ couleur f. minérale. / ~ ethiops ‖ Quecksilbermohr m. ‖ éthiope m. minéral. / ~ fuel ‖ mineralischer Brennstoff m. ‖ combustible m. minéral. / ~ green ‖ Mineralgrün n. ‖ vert m. minéral. / ~ kingdom ‖ Mineralreich n. ‖ règne m. minéral.

mineral manure ‖ Dünger m. aus mineralischen Stoffen; Steinmehl n. ‖ engrais m. minéral. / ~ grinding ‖ Steinmehlherstellung f. ‖ broyage m. d'engrais minéraux. / ~ grinding mill ‖ Düngermühle f. ‖ moulin m. à engrais minéraux.

mineral material ‖ mineralischer Stoff m. ‖ matière f. minérale. / articles pl. of ~

materials ‖ Waren fpl. aus mineralischen Stoffen ‖ objets mpl. en matières minérales. / ~ mill ‖ Mineralmühle f. ‖ moulin m. pour minéraux.

mineral oil ‖ Mineralöl n.; Erdöl n. ‖ huile f. minérale; bitume m. liquide ; naphte m. / crude ~ ‖ Roherdöl n. ‖ naphte f. brute.

mineral oil producing plant ‖ Mineralölgewinnungsanlage f. ‖ installation f. pour l'extraction d'huile minérale. / ~ spring ‖ Mineralölquelle f.; Erdölquelle f. ‖ source f. ou gisement m. d'huile minérale. / ~ varnish ‖ Mineralölfirnis m. ‖ vernis m. à l'huile minérale. / ~ well see ~ spring.

mineral origin ‖ mineralischer Ursprung m. ‖ origine f. minérale. / ~ peat ‖ Mineraltorf m. ‖ tourbe f. minérale. / ~ pigment quarry ‖ Farberdegewinnung f. ‖ extraction f. de terre à couleur.

mineral pitch ‖ Asphaltstein m.; Erdpech n.; Asphalt m. ‖ goudron m. minéral; asphalte m. / ~ extraction ‖ Erdpechgewinnung f. ‖ extraction f. de goudron minéral.

mineral resources pl. ‖ Bodenschätze mpl. ‖ richesses fpl. mineralogiques ou souterraines. / ~ royalty ‖ Bergregal n. ‖ droit m. régalien sur les mines. / ~ salt ‖ Mineralsalz n.; Steinsalz n.; Bergsalz n. ‖ sel m. minéral (ou marin ou [de] gemme); halite f. / ~ spring ‖ Mineralquelle f. Mineralbrunnen m. ‖ source f. minérale ou thermale. / ~ superphosphate ‖ Superphosphat n. ‖ superphosphate m. de chaux.

mineral tar ‖ Bergteer m.; Mineralteer m. ‖ goudron m. minéral; malthé f.; pétrole f. tenace. / ~ oil ‖ Bergteeröl n. ‖ huile f. de goudron minéral.

mineral vein ‖ Mineralgang m. ‖ filon m. minéral.

mineral water ‖ Mineralwasser n. ‖ eau f. minérale ou minéralisée. / artificial ~ ‖ künstliches Mineralwasser n. ‖ eau f. minérale artificielle. / producing of artificial ~ ‖ Herstellung f. von künstlichem Mineralwasser ‖ fabrication f. d'eau minérale artificielle. / natural ~ ‖ natürliches Mineralwasser n. ‖ eau f. minérale naturelle. / enriching of natural ~ ‖ Anreicherung f. von natürlichem Mineralwasser ‖ enrichissement m. d'eau minérale naturelle.

mineral water apparatus ‖ Mineralwasserapparat m. ‖ appareil m. à fabriquer les eaux minérales. / ~ producing plant ‖ Mineralwasserherstellungsanlage f. ‖ installation f. à fabriquer des eaux minérales. / ~ salt (for hygienic and healing purposes) ‖ Mineralwassersalz n. (für hygienische und Heilzwecke) ‖ sel m. d'eau minérale (pour usages hygiéniques et curatifs).

mineral wax ‖ Ozokerit m.; Erdwachs n.; Bergwachs n. ‖ ozokérite f.; cire f. fossile ou minérale. / ~ wealth ‖ Bodenschätze mpl. ‖ produits mpl. du soussol; richesses fpl. souterraines. / ~ wool ‖ Mineralwolle f. ‖ laine f. minérale. / ~ yellow ‖ Kaisergelb n. ‖ jaune m. minéral.

mineral ‖ Mineral n. ‖ minéral m. / ~s pl. ‖ Berggut n. ‖ minéraux mpl. / to bore a ~ ‖ ein Mineral n. erbohren ‖ découvrir un minéral par sondage. / ~ excluded from the right of disposal of the surface owner ‖ vom Verfügungsrecht des

Grundbesitzers ausgeschlossenes Mineral ‖ minéral m. exclu du droit de disposer du propriétaire du terrain. / to get ~s pl. without being authorised thereto ‖ unbefugt Mineralien npl. gewinnen ‖ extraire des minéraux sans autorisation. / incombustible ~ ‖ nicht brennbares Mineral n. ‖ minéral m. incombustible. / to prove the existence of a ~ in respect of its quantity and constitution ‖ ein Mineral n. in Menge und Beschaffenheit nachweisen ‖ constater l'existense d'un minéral en qualité et quantité. / reserved ~ ‖ vorbehaltenes Mineral n. ‖ minéral m. réservé. / ~s pl. lying stratified one above the other ‖ schichtenweise übereinandergelagerte Mineralien ‖ minéraux mpl. en couches superposées. / to withdraw from the owner the right of disposal of ~s ‖ Mineralien npl. der Verfügung des Eigentümers entziehen ‖ soustraire au propriétaire du terrain de la disposition sur les minéraux.

mineral extraction ‖ Gewinnung f. von Mineralien ‖ extraction f. de minéraux. / to reserve the right of ~ ‖ sich die Gewinnung von Mineralien npl. vorbehalten ‖ se réserver l'extraction des minéraux.

mineral grinding plant ‖ Mineralienmahlanlage f. ‖ broyeur m. pulvérisateur de minéraux.

mineralize, to ‖ versteinern ‖ minéraliser.

mineralizer (Chem) ‖ Mineralbildner m. ‖ agent m. minéralisateur.

mineralogical ‖ mineralogisch ‖ minéralogique. / ~ collection ‖ mineralogische Sammlung f. ‖ collection f. de minéralogie. / ~ investigation ‖ mineralogische Untersuchung f. ‖ recherche f. minéralogique.

mineralogist ‖ Mineraloge m. ‖ minéralogiste m. / ~'s hammer ‖ Gesteinhammer m. ‖ marteau m. de géologue. / ~'s hammer with pic ‖ Pickelhammer m. ‖ marteau m. de géologue à pointes.

mineralogy ‖ Mineralogie f.; Gesteinkunde f.; Gesteinlehre f. ‖ minéralogie f.

miners pl. ‖ Belegschaft f. (im Bergwerk) ‖ équipe f. des mineurs.

miner's . . . see under miner.

mine rubbish ‖ Berge mpl. ‖ éboulis m.; remblai m. / tipping device for ~ rubbish ‖ Bergekipper m. ‖ tombereau m. à éboulis. / ~ safety device ‖ Grubensicherungseinrichtung f. ‖ installation f. de sûreté pour mines. / school of ~s ‖ Bergakademie f.; Bergschule f. ‖ école f. des mines. / set of ~s pl. ‖ Grubenfeld n.; verliehenes Feld n. ‖ lot m.; terrain m. alloué. / ~ shaft ‖ Förderschacht m. ‖ fosse f. d'extraction. / equivalent opening of the ~ shaft ‖ äquivalente Grubenweite f. ‖ orifice m. équivalent de mine; section f. équivalente du puits. / ~ share ‖ Bergwerksanteil n. ‖ part m. de mine. / costbook ~ share ‖ Kux m.; Kuxe f. ‖ part m. de mine; valeur m. minière. / ~ signalling gear ‖ Grubensignalvorrichtung f. ‖ appareil m. de signaux de mine. / ~ survey ‖ Grubenmessung f. ‖ levé m. minier. / ~ surveying appliances pl. ‖ Markscheiderzeug n. ‖ outillage m. d'arpenteur de mine. / ~ surveying instrument ‖ Markscheideinstrument n. ‖ instrument m. pour arpenteurs de mines. / ~ surveying operations pl. ‖ markscheiderische Ar-

beiten fpl. ‖ travaux mpl. d'arpentage de mines. / ~ surveyor ‖ Markscheider m. ‖ arpenteur m. des mines; géomètre m. souterrain. / ~ timber ‖ Grubenholz n. ‖ bois m. (de charpente) pour les mines. / ~ tools pl. ‖ Grubengezähe n. ‖ outils mpl. de mineur.

minette ‖ Minette f. ‖ minette f.; oolithe f. de fer calcifère.

mine tubbing ‖ Grubenverschalung f.; Grubenvertonnung f. ‖ cuvelage m. de mine. / ~ ventilation ‖ Grubenbewetterung f.; Wetterführung f.; Grubenlüftung f. ‖ aérage m. de mine. / ~ ventilator ‖ Wettermaschine f.; Grubenventilator m. ‖ ventilateur m. de mines. / ~ water ‖ Grubenwasser n.; Bergwasser n. ‖ eaux fpl. de mine. / to let the ~ water drain off ‖ das Grubenwasser n. versickern lassen ‖ laisser les eaux fpl. se perdre par infiltration. / ~ wood see ~ timber. / ~ worker see miner. / ~ workers' rights pl. ‖ Bergarbeiterrecht n. ‖ code m. ou statut m. ou droit m. de l'ouvrier mineur. / working of ~s ‖ Bergbau m.; Grubenbetrieb m. ‖ exploitation f. des mines.

miniature field glass ‖ Taschenfeldstecher m.; Miniaturfeldstecher m. ‖ jumelle f. de poche. / ~ painting ‖ Kleinmalerei f.; Miniaturmalerei f. ‖ (peinture f. en) miniature f. / ~ toys pl. ‖ Miniaturspielwaren fpl. ‖ jouets-miniatures mpl.

minimeter ‖ Minimeter n. ‖ minimètre m.

minimeter instrument for testing the distance between the pin holes and the piston head ‖ Minimetergerät n. zum Messen der Kompressionshöhe ‖ appareil m. à minimètre pour la vérification de la hauteur de compression. / ~ to check the pitch of gears ‖ Minimetergerät n. zum Prüfen der Teilung von Zahnrädern ‖ appareil m. à minimètre pour la vérification du pas des roues dentées. / ~ to check the pitch of threads ‖ Minimetergerät n. zum Prüfen der Steigung an Gewinden ‖ appareil m. à minimètre pour la vérification du pas des filetages. / ~ for testing the squareness of the piston pin holes with the axis of the piston ‖ Minimetergerät n. zum Prüfen der Lage der Kolbenbolzenbohrung zur Kolbenachse ‖ appareil m. à minimètre pour la vérification de la position de l'alésage des tourillons de pistons par rapport à l'axe du piston.

minimum ‖ Mindestmaß n.; Minimum n. ‖ minimum m. / ~ (Sum) ‖ Mindestbetrag m. ‖ minimum m. / ~ audibility (Tel) ‖ Minimalhörbarkeit f. ‖ audibilité f. minimum. / ~ characteristics pl. ‖ Mindestwert m. ‖ chiffre m. minimum qualitatif. / ~ compressivestrain ‖ Mindestdruckfestigkeit f. ‖ résistance f. minimum à la compression. / ~ current (Electr) ‖ Minimalstrom m. ‖ minimum m. de courant. / ~ current cut-out ‖ Minimalstromausschalter m. ‖ disjoncteur m. à minimum de courant. / ~ cut-out ‖ Minimalausschalter m. ‖ interrupteur m. à minimum. / ~ fee ‖ Mindestgebühr f. ‖ taxe f. minimum. / ~ limit of a curve (Railw) ‖ Mindestgrenze f. einer Kurve ‖ minimum m. de limite d'une courbe. / ~ magnification ‖ Mindestvergrößerung f.; Minimalvergrößerung f. ‖ grossissement m. minimum. /

~ rate (Tel) ‖ Mindestgesprächgebühr f. ‖ minimum m. de conversations. / ~ starting voltage ‖ Mindestanlaßspannung f. ‖ tension f. initiale de démarrage. / ~ strength per square millimeter of the primitive section ‖ Mindestfestigkeit f. je Quadratmillimeter des ursprünglichen Querschnitts ‖ résistance f. minimum par millimètre carré de la section primitive. / ~ stroke ‖ Mindesthub m. ‖ course f. minimum. / ~ temperature ‖ niedrigster Wärmegrad m.; ‖ température f. minimum. / ~ thermometer ‖ Niedrigstwärmegradmesser m.; Minimumthermometer m. ‖ thermomètre m. à minimum. / ~ value ‖ Mindestwert m.; Minimalwert m. ‖ valeur f. minimum. / ~ wages pl. ‖ Mindestlohn m. ‖ salaire m. minimum. / ~ weight ‖ Mindestgewicht n. ‖ poids m. minimum.

mining ‖ Bergbau m.; Grubenbetrieb m.; Bergwerksbetrieb m. ‖ exploitation f. des mines. / to be free for ~ ‖ bergfrei sein ‖ être libre pour l'exploitation minière. / ~ in open cuts ‖ Tagebau m. ‖ ouvrage m. ou exploitation f. à ciel ouvert. / ~ by the surface owner ‖ Eigentümerbergbau m. ‖ exploitation f. par le propiétaire du fonds. / strip ~ see ~ in open cuts. / underground ~ ‖ Bergbau m. unter Tage ‖ exploitation f. des mines souterraines.

mining administration ‖ Grubenverwaltung f. ‖ administration f. des mines.

mining area ‖ Grubenfeld n. ‖ champ m. minier. / boundary between adjoining ~s pl. ‖ Grubenfeldgrenze f. ‖ limite f. d'une mine. / to determine the boundary of a ~ ‖ die Grenze f. eines Grubenfeldes festlegen ‖ fixer la limite d'un champ minier. / to exploit ~s pl. for private economic purposes ‖ Felder npl. privatwirtschaftlich ausbeuten ‖ exploiter des terrains mpl. miniers à titre d'entreprise privée.

mining association ‖ Kohlengrubenvereinigung f. ‖ comité m. des houillères. / ~ cable ‖ Grubenkabel n. ‖ câble m. minier. / ~ captain ‖ Obersteiger m. ‖ maître-mineur m. / ~ car ‖ Grubenförderwagen m. ‖ wagonet m.; benne f. / patent of ~ claims pl. ‖ Bergwerksverleihung f. ‖ concession f. d'exploitation. / ~ company ‖ Grubengesellschaft f.; Gewerkschaft f. ‖ compagnie f. minière; société f. des exploitants ou d'exploitation. / ~ department ‖ Bergbauverwaltung f.; Bergverwaltung f. ‖ administration f. des mines. / ~ district ‖ Bergrevier n.; Revier n. ‖ district m. de mines. / ~ engineer ‖ Bergingeniör m. ‖ ingénieur m. des mines. / ~ expert ‖ Sachverständiger m. im Bergbau; Bergbausachverständiger m. ‖ expert m. en affaires minières. / ~ industry ‖ Bergwerksindustrie f.; Bergbauindustrie f.; Montanindustrie f. ‖ industrie f. minière. / ~ installation ‖ Bergwerkseinrichtung f. ‖ installation f. de mine. / ~ lamp ‖ Grubenlampe f. ‖ lampe f. pour mineurs. / ~ law ‖ Berggesetz n.; Bergordnung f. ‖ loi f. minière. / ~ legislation ‖ Berggesetzgebung f. ‖ législation f. minière. / ~ locomotive ‖ Grubenlokomotive f. ‖ locomotive f. de mine.

mining machine ‖ Bergwerksmaschine f. ‖ machine f. minière ou de mine. / ~ for lignite mines ‖ Braunkohlengruben-

maschine f. ‖ machine f. pour mines de lignite.

mining plant ‖ Bergwerksanlage f. ‖ établissement m. de mine. / ~ product ‖ Bergbauerzeugnis n.; Grubenerzeugnis n. ‖ produit m. minier ou de mine. / ~ property ‖ Grubenbesitz m. ‖ propriété f. minière. / ~ prospecting enterprise ‖ Mutungsgesellschaft f. ‖ entreprise f. de recherches minières. / ~ pump ‖ Bergwerkspumpe f. ‖ pompe f. pour mines. / ~ railway ‖ Grubenbahn f.; Bergwerksbahn f. ‖ chemin m. de fer minier. / ~ report ‖ Grubenbericht m. ‖ rapport m. minier.

mining right ‖ Bergbaugerechtigkeit f.; Bergbaufreiheit f.; Bergbaurecht n. ‖ privilège m. d'exploitation minière; droit m. d'exploitation. / to grant the ~ ‖ das Bergwerkseigentum verleihen ‖ concéder la propriété minière. / to grant ~s pl. for a mineral ‖ ein Mineral n. verleihen ‖ concéder le droit à l'extraction d'un minéral.

mining, science of ~ ‖ Bergbaukunde f. ‖ science f. des mines. / ~ share ‖ Kux m.; Bergwerksanteil m. ‖ part f. de mine; valeur f. minière. / ~ statistics pl. ‖ Bergbaustatistik f.; Montanstatistik f. ‖ statistique f. minière. / ~ timber see mine timber. / ~ truck ‖ Grubenwagen m.; Förderwagen m. ‖ wagonnet m. de mine. / waste in ~ ‖ Abbauverlust m. ‖ déchet m. d'exploitation. / ~ wax ‖ Montanwachs n. ‖ cire f. minérale ou de lignite. / ~ work ‖ Bergbauarbeit f. ‖ travail m. de mine.

minion (Size of type) ‖ Kolonel f. ‖ mignonne f.

minister ‖ Minister m. ‖ ministre m. / Minister of Finances ‖ Finanzminister m. ‖ ministre m. des finances.

Ministry of Labour ‖ Arbeitsministerium n. ‖ ministère m. du travail.

minium ‖ Mennige f.; rotes Bleioxyd n. ‖ minium m.; plomb m. oxydé rouge. / substitute for ~ ‖ Mennigeersatz m. ‖ minium m. factice.

miniver ‖ Feh n.; sibirisches Eichhorn n. ‖ petit-gris m.

minks pl. ‖ Nerzfell n. ‖ vison m.

minor axis (Of an ellipse) ‖ Nebenachse f. ‖ petit axe m.

minority ‖ Minderheit f.; Minderzahl f. ‖ nombre m. inférieur; minorité f.

mint, to ‖ ausprägen; ausmünzen; münzen ‖ monnayer; frapper des monnaies.

mint (Botany) ‖ Minze f. ‖ menthe f.

Mint (Minthouse) ‖ Münze f. ‖ Monnaie f.; hôtel m. des monnaies.

mintage ‖ Prägeschatz m.; Schlagschatz m. ‖ rendage m.; brassage m.

mint alcohol ‖ Pfefferminztropfen mpl. ‖ alcool m. de menthe. / ~ bronze ‖ Münzbronze f. ‖ bronze m. de monnaie. / ~ drop ‖ Pfefferminzplätzchen n. ‖ pastille f. de menthe. / ~ gold ‖ Münzgold n. ‖ or m. de monnaie ou de vaisselle.

minting ‖ Münzen n.; Münzkunst f. ‖ monnayage m.; art m. de monnayer.

minting mill ‖ Spindelwerk n.; Stoßwerk n.; Prägewerk n. ‖ balancier m. à frapper la monnaie.

mint mill ‖ Münzwalzwerk n.; Münzstreckwerk n. ‖ laminoir m. monnayeur ou monétaire. / ~ warden ‖ Münzwardein m. ‖ essayeur m.

minuscula ‖ Minuskel f.; kleiner Anfangs-buchstabe f. ‖ lettre f. minuscule; minuscule f.

minus sign ‖ Minuszeichen n. ‖ signe m. moins.

minute, to ‖ protokollieren ‖ verbaliser; enregistrer.

minute ‖ fein; sehr klein; zierlich ‖ ténu; minime; extrêmement petit.

minute ‖ Minute f. ‖ minute f. / ~ of arc ‖ Bogenminute f. ‖ minute f. d'arc. / reading to one ~ ‖ auf eine Minute f. ablesbar ‖ donnant la minute. / ~ of time ‖ Zeitminute f. ‖ minute f. de temps. / ~s pl. (Protocol) ‖ Protokoll n. ‖ procès-verbal m.; protocole m.

minute current ‖ Schwachstrom m. ‖ courant m. faible. / ~ hand (Clockm) ‖ Minutenzeiger m. ‖ aiguille f. à minutes. / ~ wheel (Clockm) ‖ Minutenrad n.; großes Bodenrad n. ‖ roue f. de minute ou de chaussée. / ~ wheel work ‖ Minuten-werk n. ‖ minuterie f. / ~ works pl. ‖ Zeigerwerk n. ‖ minuterie f.

miocene (system) ‖ Miozän n. ‖ système m. miocène.

mirage ‖ Kimmung f.; Luftspiegelung f. ‖ mirage m.

mirbane essence ‖ Mirbanöl n. ‖ essence f. de mirbane.

mire ‖ Straßenkot m.; Kot m. ‖ crotte f.; boue f. / ~ (Geol) ‖ Schlamm m. ‖ bourbe f., limon m.; vase f.

mirror ‖ Spiegel m. ‖ miroir m.; glace f. / ~ (Reflector) ‖ Rückstrahler m.; Reflek-tor m. ‖ réflecteur m. / automobile ~ ‖ Kraftwagenspiegel m. ‖ avertisseur m. ou espion m. pour automobiles. / auxi-liary ~ ‖ Hilfsspiegel m. ‖ miroir m. auxiliaire. / collimator ~ ‖ Kollimator-spiegel m. ‖ miroir-collimateur m. / concave ~ ‖ Hohlspiegel m.; Brenn-spiegel m.; Konkavspiegel m. ‖ miroir m. concave. / condenser ~ ‖ Kondensor-spiegel m. ‖ miroir-condensateur m. / convex ~ ‖ Konvexspiegel m. ‖ miroir m. convexe. / framed ~ ‖ gerahmter Spie-gel m. ‖ miroir m. encadré. / gilt-framed ~ ‖ goldgerahmter Spiegel m.; Gold-rahmenspiegel m. ‖ miroir m. à cadre doré. / gold ~ ‖ Goldspiegel m. ‖ miroir m. en or. / head ~ ‖ Stirnspiegel m. ‖ miroir m. frontal. / ~ of magnet ‖ Mag-netspiegel m. ‖ miroir m. d'aimant. / metallic ~ ‖ Metallspiegel m. ‖ miroir m. métallique ou de métal. / observation ~ ‖ Beobachtungsspiegel m. ‖ miroir m. d'observation. / observation ~ (Auto) see rear-vision ~. / parabolic ~ ‖ Parabol-spiegel m. ‖ miroir m. ou réflecteur m. parabolique. / to set the ~s pl. parallel ‖ die Spiegel mpl. parallel stellen ‖ placer les miroirs mpl. parallèlement. / parallel-ground ~ ‖ parallel geschliffener Spie-gel m. ‖ miroir m. poli parallèlement. / plane ~ ‖ Planspiegel m.; ebener Spie-gel m. ‖ miroir m. plan. / plane-ground or -polished ~ ‖ eben geschliffener Spiegel m. ‖ miroir m. poli planement. / rear reflecting ~ ‖ Spion m. ‖ espion m.; avert-isseur m. / rear-vision ~ ‖ Rückspiegel m.; Rückblickspiegel m. ‖ rétroviseur m. / shaving ~ ‖ Rasierspiegel m. ‖ miroir m. à raser. / ~ of short focal length ‖ kurzbrennweitiger Spiegel m. ‖ miroir m. de court foyer. / silvered ~ ‖ Silberspiegel m. ‖ miroir m. argenté. / spherical concave ~ ‖ sphärischer Hohl-

spiegel m. ‖ miroir m. concave sphérique. / wall ~ ‖ Wandspiegel m. ‖ miroir m. mural.

mirror comparator ‖ Spiegellehre f. ‖ com-parateur m. à miroir. / ~ extensometer (Assaying) ‖ Spiegelapparat m. ‖ appa-reil m. à miroir. / ~ foil ‖ Spiegelbelag m. ‖ étain m. en feuilles pour miroirs. / ~ galvanometer ‖ Spiegelgalvanomes-ser m. ‖ galvanomètre m. à miroir.

mirror glass ‖ Spiegelglas n. ‖ verre m. à glaces; glace f. / bevelled ~ ‖ an den Kan-ten geschliffenes Spiegelglas n. ‖ glace f. biseautée. / cambered ~ ‖ gewölbtes Spie-gelglas n. ‖ glace f. bombée. / curved ~ ‖ ge-bogenes Spiegelglas n. ‖ glace f. courbée. / engraved ~ ‖ graviertes Spiegelglas n. ‖ glace f. gravée. / foiled ~ ‖ belegtes Spiegelglas n. ‖ glace f. étamée. / gilt ~ ‖ vergoldetes Spiegelglas n. ‖ glace f. dorée. / plated ~ ‖ versilbertes Spiegel-glas n. ‖ glace f. argentée. / platinized ~ ‖ platiniertes Spiegelglas n. ‖ glace f. platinée. / reinforced ~ ‖ verstärktes Spiegelglas n. ‖ glace f. armée.

mirror handle ‖ Spiegelgriff m. ‖ manche m. pour miroir. / ~ iron ‖ Spiegeleisen n. ‖ fonte f. miroitante. / ~ magnet ‖ Spiegelmagnet m. ‖ aimant m. à miroir. / ~ method ‖ Spiegelmethode f.; Spiegel-verfahren n. ‖ méthode f. à miroir. / ~ observation ‖ Spiegelbeobachtung f. ‖ observation f. par miroir. / ~ polisher ‖ Spiegelglasschleifer m. ‖ polisseur m. de glaces. / ~ polishing ‖ Spiegelglas-schleifen n. ‖ polissage m. de glaces ou des miroirs. / ~ position finder ‖ Spiegel-ortungsgerät n. ‖ chercheur m. de posi-tion à miroir. / ~ quicksilvering ‖ Spiegel-belegung f. ‖ étamage m. des glaces. / ~ reading ‖ Spiegelablesung f. ‖ lecture f. au miroir. / ~ reflex camera ‖ Spiegel-reflexkammer f. ‖ chambre f. à miroir de visée. / ~ scale ‖ Spiegelskale f. ‖ échelle f. à miroir ou sur miroir.

mirrorscope ‖ Rückspiegel m. ‖ miroir m. d'observation.

mirror, setting of the ~ ‖ Spiegelstellung f. ‖ position f. du miroir. / to correct the setting of the ~ ‖ die Spiegelstellung f. berichtigen ‖ rectifier la position du miroir. / ~ speaker (Tel) ‖ Sprechgalva-nomesser m. ‖ galvanomètre m. récep-teur. / ~ symmetry ‖ Spiegelgleichheit f. ‖ symétrie f. par rapport à un miroir. / extra-axial ~ zone ‖ außeraxialer Spie-gelabschnitt m. ‖ portion f. marginale de miroir.

miry sand ‖ Schlammsand m.; Modder-sand m. ‖ sable m. vasard.

miscalculation ‖ Rechenfehler m.; Kalku-lationsfehler m. ‖ erreur m. en calcul.

miscibility ‖ Mischbarkeit f. ‖ miscibilité f.

miscible ‖ mischbar ‖ miscible.

misdirect, to ‖ falsch adressieren ‖ mal adresser.

misenite ‖ Misenit m. ‖ misenite f.

misfire, to ~ (Gun) ‖ versagen ‖ rater. / ~ (Mot) ‖ aussetzen ‖ manquer.

misfire (Gun; Motor) ‖ Aussetzer m.; Ver-sager m. ‖ raté m. / ~ ignition ‖ Fehl-zündung f. ‖ allumage m. défectueux.

mishap ‖ Unfall m. ‖ accident m.; contre-temps m.

mislead, to ‖ irreführen ‖ égarer; four-voyer.

misoperation (Of the machine) ‖ falsche Handhabung f. ‖ fausse manipulation f.

mispickel (Miner) ‖ Mispickel m.; Arsen-kies m. ‖ arséno-sulfure m. de fer; arsénopyrite f.

misprint, to ‖ verdrucken; fehldrucken ‖ transposer; imprimer à contre-sens.

misprint ‖ Druckfehler m. ‖ faute f. d'im-pression.

miss, to ~ the connexion (Railw) ‖ den Anschluß m. versäumen oder verpassen oder verfehlen ‖ manquer la correspon-dance.

missing (Bad shot) ‖ Fehlschuß m. ‖ coup m. bleu ou manqué.

missy (Miner) ‖ Gelbeisenerz n.; Gelb-eisenstein m. ‖ fer m. oxydé jaune.

mist ‖ Nebel m.; Bodennebel m. ‖ brunine f.; brume f.; brouillard m. de terre. / ~ due to water vapour ‖ Dunstnebel m. ‖ brouillard m. de fumées ou de vapeurs légères.

mistake ‖ Irrtum m. ‖ erreur f.; méprise f. / to make a ~ ‖ einen Fehler m. machen; sich versehen ‖ se tromper; faire une bévue. / ~ in calculation ‖ Rechenfehler m.; Kalkulationsfehler m. ‖ erreur f. en calcul. / ~ in the drawing ‖ Konstruktionsfehler m. ‖ vice f. de con-struction; erreur f. dans le dessin.

mistaken ‖ irrtümlich ‖ erroné; faux.

mistletoe ‖ Mistelkraut n.; Mistel f. ‖ gui m.

mist streaks pl. see ~ swath. / ~ swath ‖ Nebelschwaden m. ‖ brouillard m. flottant.

misty ‖ nebelig ‖ nébuleux; brument.

misunderstanding ‖ Irrtum m.; Mißver-ständnis n. ‖ erreur m.; méprise f.

misuse, to ‖ mißbrauchen ‖ abuser. / ~ confidence ‖ das Vertrauen mißbrauchen ‖ abuser de la confiance.

miter see mitre.

mitre, to ~ (Join) ‖ auf Gehrung f. ver-binden ‖ faire un assemblage d'onglet.

mitre (Join) ‖ Gehrung f. ‖ onglet m.; biais m. / ~ (Conical valve) ‖ Kegelventil n. ‖ mitre f.; obturateur m. conique. / ~ block ‖ Gehrlade f.; Gehrungsstoßlade f. ‖ boîte f. de mitre ou à onglet. / ~ box see ~ block. / ~ cut ‖ Gehrschnitt m. ‖ coupe f. à onglet.

mitre cutting guide ‖ Gehrungsanschlag m. ‖ guide m. pour couper les onglets. / tilting ~ ‖ abklappbarer Gehrungs-anschlag m. ‖ butée f. d'onglets bascu-lante.

mitre cutting machine ‖ Gehrungsstanz-maschine f. ‖ machine f. à découper les onglets. / ~ cutting shears pl. ‖ Gehrungs-schere f. ‖ cisaille f. à onglets. / ~ dove-tail ‖ Gehrungszinke f. ‖ queue f. perdue. / ~ gate ‖ Stemmtor n. ‖ porte f. bus-quée ou en éperon. / ~ gear ‖ Winkel-getriebe n. ‖ engrenage m. d'angle.

mitreing machine see mitre block.

mitre joint ‖ Gehrfuge f.; Gehrstoß m.; Stoß m. auf Gehrung ‖ assemblage m. à onglet; joint m. à onglet ou à mitre. / ~ by slit and tongue ‖ Zusammen-fügen n. auf Gehrung ‖ assemblage m. à bois de fil.

mitre line ‖ Gehrungslinie f. ‖ ligne f. de mitre ou d'onglet. / ~ plane ‖ Gehrungs-hobel m. ‖ rabot m. à mitre ou à on-glet; guillaume m. à onglet. / ~ quoin see ~ joint. / ~ quoin cut see ~ cut. / ~ rule ‖ Gehrmaß n.; Schmiegwinkel m.; Schmiege f.; Schrägmaß n. ‖ béveau m.; équerre f. à mitre. / ~ square ‖ festes

Gehrdreieck n.; Winkellineal n. von 45 Grad ‖ équerre f. onglet *ou* mitre. / ~ sill (Of a sluice) ‖ Schlagschwelle f.; Schleusenschwelle f. ‖ seuil m. du busc; seuillet m. d'écluse. / ~ valve ‖ Kegelventil n. ‖ mitre f.; obturateur m. conique.

mitre wheel ‖ Kegelrad n. ‖ roue f. conique; roue f. d'angle. / ~ gearing ‖ Winkelgetriebe n.; Kegelradgetriebe n. ‖ engrenage m. d'équerre *ou* d'angle.

mitten ‖ Fausthandschuh m.; fingerloser Handschuh m. ‖ mitaine f.; moufle f.

mix, to ‖ vermengen; mischen; vermischen; durchmischen ‖ mêler; mélanger. / ~ colours pl. ‖ Farben fpl. mischen ‖ mélanger les couleurs fpl. / ~ ores pl. and fluxes pl. ‖ möllern ‖ mélanger les minerais avec les fondants. / ~ thoroughly ‖ durchmischen ‖ mélanger complètement *ou* uniformement. / ~ together *see* to mix. / ~ with ... ‖ sich mischen mit ... ‖ se mélanger à ...

mixed ‖ gemischt ‖ mélangé; mixte. / ~ colours pl. ‖ Mischfarben fpl. ‖ couleurs fpl. binaires. / ~ construction ‖ Gemischtbau m.; Mischbau m. ‖ construction f. mixte. / ~ grain ‖ Mengfutter n.; Mangfutter n. ‖ dragées fpl. / ~ manure ‖ Kompost m.; Mischdünger m. ‖ engrais m. composé. / ~ process ‖ halbtrocknes Verfahren n. ‖ procédé m. mixte. / ~ stand (Forest) ‖ Mischbestand m. ‖ peuplement m. mélangé. / ~ system (Acc) ‖ gemischter Betrieb m. ‖ système m. mixte.

mixer ‖ Mischer m.; Mischwerk n.; Mischmaschine f. ‖ mélangeur m. / ~ drum ‖ Trommelmischmaschine f. ‖ mélangeur m. à tambour. / gravity ~ ‖ Freifallmischer m. ‖ malaxeur m. à chute libre. / intermittent-working ~ ‖ Mischer m. mit stoßweisem Betrieb ‖ malaxeur m. à fonctionnement discontinu *ou* à fonctionnement intermittent. / ~ with orifice for the entraining medium ‖ Strahldüsenmischer m. ‖ mélangeur m. à buse à jet. / pig-iron ~ ‖ Roheisenmischer m. ‖ mélangeur m. de fonte. / tipping trough ~ ‖ Kipptrogmischmaschine f. ‖ malaxeur m. à auge basculante. / ~ with uninterrupted drive ‖ Mischer m. mit ununterbrochenem Betrieb ‖ mélangeur m. à fonctionnement continu.

mixer plant with two pig iron mixers ‖ Mischeranlage f. mit zwei Roheisenmischern ‖ installation f. de mélangeurs avec deux mélangeurs de fonte.

mixer shaft ‖ Mischerwelle f.; Mischungswelle f. ‖ arbre m. du mélangeur.

mixing ‖ Mischung f.; Mischen n. ‖ mélange m. / ~ of coal ‖ Kohlenmischung f. ‖ mélange m. de charbon. / ~ of composition (Pyrot) ‖ Ansetzen n. des Satzes ‖ mélange m. des matières. / homogeneous ~ ‖ inniges Mischen n. ‖ mélange m. intime. / machine-~ ‖ Maschinenmischung f.; Mischen n. mit der Maschine ‖ mélange m. à la machine. / ~ of the ores ‖ Mischung f. der Erze ‖ mélange m. des minerais. / ~ of the ores and fluxes ‖ Möllerung f.; Mischung f. der Erze mit Zuschlag ‖ mélange n. des minerais avec les fondants; / **mixing** battery (Heat) ‖ Mischbatterie f. mitigeur m. / ~ blade ‖ Mischflügel m.; Mischerschaufel f. ‖ palette f. mélangeuse; palette f. de malaxage. /

chamber (Mot) ‖ Mischkammer f.; Mischraum m. ‖ chambre f. de mélange. / ~ cock ‖ Mischhahn m. ‖ robinet m. mélangeur. / ~ drum ‖ Mischtrommel f. ‖ tambour m. mélangeur. / ~ hopper ‖ Spültrichter m. ‖ entonnoir m. de mélange. / ~ house (Met) ‖ Möllerboden m. ‖ halle f. aux mélanges.

mixing machine ‖ Mischmaschine f. ‖ mélangeuse f.; machine f. à mélanger. / clay ~ ‖ Tonmischmaschine f. ‖ machine f. à mélanger l'argile. / ~ for concrete and mortar ‖ Mischmaschine f. für Beton und Mörtel ‖ mélangeuse f. pour béton et mortier. / kneading and ~ for concrete and mortar ‖ Knetmaschine f. und Mischmaschine f. für Beton und Mörtel ‖ machine f. à malaxer et à mélanger le béton et le mortier. / conglomerate ~ (Glassm) ‖ Gemengemischmaschine f. ‖ machine f. à mélanger la composition du verre. / dough ~ ‖ Teigmischmaschine f. ‖ machine f. à mêler la pâte. / ~ with shaking sifter ‖ Mischer m. mit Schüttelsieb ‖ mélangeur m. avec tamis à sesousses. / tip trough ~ ‖ Kipptrogmischmaschine f. ‖ mélangeur m. à auge basculante.

mixing mill ‖ Mischwalzwerk n. ‖ laminoir m. mélangeur; broyeur m. malaxeur *ou* mélangeur. / ~ for india rubber ‖ Gummimischwalzwerk n. ‖ laminoir m. mélangeur pour caoutchouc.

mixing, order of ~ ‖ Mischungsvorgang m.; Mischungsreihenfolge f. ‖ marche f. de mélange. / ~ pipette ‖ Mischpipette f. ‖ pipette-mélangeuse f.

mixing plant ‖ Mischeranlage f. ‖ installation f. des mélangeurs. / coal ~ ‖ Kohlenmischanlage f. ‖ installation f. à mélanger le charbon. / coal ~ (Mixing tower) ‖ Kohlenmischturm m. ‖ tour f. de mélangeurs à houille. / pneumatic ~ ‖ pneumatische Mischanlage f. ‖ installation f. de mélangeur pneumatique.

mixing runner ‖ Mischkollergang m. ‖ mélangeur m. à meules verticales. / ~ shaft ‖ Rührwelle f. ‖ arbre m. de mélangeur. / ~ sieve ‖ Trommelsieb n. ‖ tambour m. cribleur. / ~ table (Pyrot) ‖ Reibetafel f. ‖ table f. à égruger; table f. à rebord. / ~ tower ‖ Mischturm m. ‖ tour f. de mélangeurs. / ~ trough (Acc) ‖ Mischbottich m. ‖ cuve f. à mélange. / ~ vessel ‖ Mischgefäß n.; Mischkessel m. ‖ récipient m. *ou* chaudron m. à mélanger. / ~ water (Mas) ‖ Anmachewasser n. ‖ eau f. pour le malaxage. / ~ worm ‖ Mischschnecke f. ‖ vis f. mélangeuse .

mixture ‖ Mischung f.; Gemenge n.; Gemisch n. ‖ mélange m. / ~ (of gases) ‖ Gemisch n.; Gasgemisch n. ‖ mélange m. de gaz. / ~ (Met) ‖ Möller m.; Möllerung f. ‖ lit m. de fusion; mélange f. / anti-freezing ~ ‖ kältebeständiges Gemisch n. ‖ mélange m. incongelable. / dry ~ ‖ Trockenmischung f. ‖ mélange m. sec. / ~ of equal parts ‖ Gemisch m. von gleichen Teilen ‖ mélange m. à parties égales. / explosive ~ ‖ explosibles *oder* explosionfähiges *oder* explodierbares Gemisch n. ‖ mélange m. tonnant *ou* explosif.

mixture of gases ‖ Gasgemisch n. ‖ mélange m. de gaz. / optically sensitive ~ ‖ optisch empfindliches Gasgemisch n. ‖ mélange m. de gaz sensible au point de vue optique. / optical analysis of a ~ ‖

optische Analyse f. eines Gasgemisches ‖ analyse f. optique d'un mélange de gaz.

mixture of gas oil and crude oil ‖ Mischung f. von Gasöl mit Rohöl ‖ mélange m. de gasoil avec de l'huile brute. / ~ of gas oil and residual fuels ‖ Mischung f. von Gasöl und Erdölrückständen ‖ mélange m. de gasoil et de résidus de distillation. / intensive ~ of the fuel gas with the combustion air ‖ innige Mischung f. des Heizgases mit der Verbrennungsluft ‖ mélange m. intime du gas de chauffage avec l'air de combustion. / ~ of iron ‖ Eisenmischung f. ‖ espèce f. de fonte. / ~ of metals ‖ Legierung f. ‖ alliage m. / overrich ~ (Mot) ‖ gasreiches Gemisch n. ‖ mélange m. trop riche. / ~ of sand and lime ‖ Kalksandgemisch n. ‖ mélange m. de sable et de chaux. / saturated ~ (Mot) ‖ gesättigtes Gemisch n. ‖ mélange m. saturé. / uniform ~ ‖ gleichmäßige Mischung f. ‖ mélange m. uniforme. / weak ~ (Mot) ‖ gasarmes Gemisch n. ‖ gaz m. trop pauvre.

mixture, proportions pl. of ~ ‖ Mischungsverhältnis n. ‖ proportion f. d'alliage *ou* de mélange.

mizen ‖ Besan m.; Besansegel m. ‖ voile f. d'artimon. / ~ mast ‖ Kreuzmast m.; Besanmast m. ‖ mât m. d'artimon; artimon m. / ~ sail *see* mizen. / ~ stay ‖ Kreuzstag n. ‖ étai m. du mât d'artimon. / ~ top ‖ Kreuzmars m.; Besanmars m. ‖ hune f. d'artimon.

mizzen *see* mizen.

mobile ‖ beweglich ‖ mobile. / ~ warfare ‖ Bewegungskrieg m. ‖ guerre f. de mouvement.

mobilization ‖ Mobilmachung f. ‖ mobilisation f. / industrial ~ ‖ industrielle Mobilmachung f. ‖ mobilisation f. industrielle.

mobility ‖ Beweglichkeit f. ‖ mobilité f. / ~ of labour ‖ Freizügigkeit f. ‖ libre circulation f.; liberté f. de domicile.

mocha ‖ Mokka m.; Mokkakaffee m. ‖ café m. de Moka; du moka m. / ~ coffee *see* mocha.

Mocha stone ‖ Mokkastein m.; Baumachat m. ‖ agate f. moka; pierre f. de moka.

mock ‖ falsch; nachgeahmt; unecht ‖ faux; imité. / ~ lead ‖ Zinkblende f.; Blende f. ‖ fausse galène f.; blende f. / ~ seam ‖ falsche Naht f. ‖ fausse couture f. / ~ window ‖ blindes Fenster n. ‖ fenêtre f. feinte *ou* borne *ou* aveugle. / ~ worsted ‖ Halbgarn n. ‖ cardé-peigné m.; peignécardé m. ‖ peigné-mixte m.

model, to ‖ formen; gestalten; modellieren ‖ façonner; former; modeler; bosseler. / ~ (Engrav) ‖ modellieren ‖ lanter; lamter. / ~ in wax ‖ in Wachs modellieren ‖ modeler en cire.

model ‖ Schablone f.; Muster n.; Modell n. ‖ échantillon m.; calibre m.; modèle m.; façon f. / ~ (Sculpt) ‖ Model m. ‖ maquette f. / ~ for anatomical courses of instruction ‖ anatomisches Lehrmodell n. ‖ modèle m. anatomique pour l'enseignement de l'anatomie. / experimental ~ ‖ Versuchsmodell n. ‖ modèle m. d'essai. / improved ~ ‖ verbessertes Modell n. ‖ modèle m. amélioré. / ~ de luxe ‖ Luxusmodell n.; Luxusausführung f. ‖ modèle m. de luxe. / original ~ ‖ erstes Modell n. ‖ modèle m. primitif. / regular ~ ‖ gewöhnliches Modell n. ‖ modèle m. régulier.

/ ~ of a ship ‖ Schiffsmodell n. ‖ modèle m. de navire. / ~ of an historical ship ‖ historisches Schiffsmodell n. ‖ modèle m. de navire historique.

model brick (Building stone) ‖ Probestein m. ‖ échantillon m. / ~ drawing ‖ Zeichnen n. nach Vorlagen *oder* Modellen ‖ dessin m. d'après modèle *ou* sur object.

modeller (Join) ‖ Modelltischler m. ‖ modeleur m. / ~ (Found) ‖ Modellör m.; Modellierer m. ‖ modeleur m. / ~'s clay ‖ Modellierton m. ‖ argile f. de modeleur.

modelling board for the core (Found) ‖ Kernbrett n. *oder* Schablone f. zur Bildung des Kernes ‖ échantillon m. de noyau. / ~ paste ‖ Modelliermasse f. ‖ pâte f. à modeler. / ~ wax ‖ Modellierwachs n. ‖ cire f. à modeler.

model maker *see* modeller (Join). / ~ stone *see* model brick. / ~ varnish ‖ Modelllack m. ‖ vernis m. pour modèles.

moderate, to ‖ mäßigen ‖ modérer. / ~ (A movement) ‖ mäßigen; langsamer machen; verlangsamen ‖ ralentir.

moderate ‖ mäßig; gemäßigt ‖ modéré. / ~ in price ‖ preiswert ‖ économique; bon marché. / ~ly hard material ‖ mittelharter Werkstoff m. ‖ matière f. mi-dure.

moderating glass ‖ Blendglas n. ‖ verre m. foncé. / ~ of a dense neutral tint glass ‖ Blendglas n. aus dunklem Neutralglas ‖ verre m. neutre sombre.

moderator (Mach) ‖ Moderator m. ‖ modérateur m.

modern ‖ neuzeitlich; zeitgemäß; modern ‖ moderne; à la mode. / ~ working method ‖ neuzeitliches Bearbeitungsverfahren n. ‖ usinage m. avec un outillage perfectionné. / ~ workshop ‖ neuzeitlich eingerichtete Werkstätte f. ‖ atelier m. moderne.

modernization ‖ Modernisierung f. ‖ modernisation f. / a complete ~, alteration, and enlargement was begun ‖ ein vollständiger Umbau m. der Anlage wurde begonnen ‖ il fut procédé à une reconstruction et à la modernisation de l'usine.

modernize, to ‖ modernisieren ‖ moderniser.

modification ‖ Abänderung f.; Umänderung f.; Veränderung f. ‖ modification f. / without any ~ ‖ ohne jede Abänderung ‖ sans modification.

modul *see* module.

modulate, to ‖ abstimmen; modulieren ‖ accorder; moduler.

modulated (Radio) ‖ abgestimmt ‖ accordé. / ~ antenna ‖ abgestimmte Antenne f. ‖ abgestimmter Luftleiter m. ‖ antenne f. accordée.

modulation (Radio) ‖ Abstimmung f.; Modulation f. ‖ accord m.; modulation f. / to improve the clearness of ~ ‖ die Abstimmschärfe f. erhöhen ‖ augmenter la sélectivité. / perfect ~ ‖ vollkommene Modulation f. ‖ modulation f. parfaite.

modulator ‖ Modulator m. ‖ modulateur m. / ~ tube (Radio) ‖ Steuerröhre f. ‖ lampe-modulatrice f.

module ‖ Modulus m.; Modul m.; Einheitsmaß n. ‖ module m. / ~ of direct elasticity ‖ Zugelastizitätsmodul n. ‖ module m. d'élasticité à la traction. / ~ of transverse elasticity ‖ Gleitmodul m. ‖ module m. d'élasticité au glissement. / ~ of a gear ‖ Modul m. eines Zahnrades ‖ module m. d'un engrenage.

modulus *see* module.

mofette (Geol) ‖ Mofette f.; Kohlensäureexhalation f. ‖ mofette f.

mohair ‖ Mohärziegenhaar n.; Kamelhaar n.; Angorahaar n. ‖ poil m. de chèvre mohair.

moiree (Weav) ‖ Moiré m. ‖ moiré m.; moirée f.; étoffe f. moirée. / metallic ~ ‖ Metallmohr m. ‖ moiré m. métallique.

moist ‖ naß; feucht ‖ humide; mouillé. / ~ chamber ‖ Feuchtkammer f. ‖ chambre f. humide. / ~ concrete ‖ erdfeuchter Beton m. ‖ béton m. à consistance de terre humide.

moisten, to ‖ anfeuchten; befeuchten; naß machen; besprengen ‖ humecter; humidifier; mouiller. / ~ (To impregnate) ‖ tränken ‖ imprégner; imbiber.

moistener ‖ Anfeuchter m. ‖ humecteur m. / finger ~ (Office) ‖ Fingeranfeuchter m. ‖ humecteur m. digital.

moistening of the soil ‖ Durchfeuchtung f. des Bodens ‖ pénétration f. du sol par l'humidité. / ~ apparatus ‖ Befeuchtungsvorrichtung f. ‖ humidificateur m.; humecteur m. / air ~ chamber ‖ Luftbefeuchtungskammer f. ‖ chambre f. d'humidification. / ~ machine ‖ Feuchtmaschine f. ‖ machine f. à humecter. / method of ~ ‖ Anfeuchtungsverfahren n. ‖ méthode f. d'humectage. / ~ plant ‖ Befeuchtungsanlage f. ‖ installation f. d'humectation.

moistness *see* moisture.

moisture ‖ Feuchtigkeit f.; Nässe f. ‖ humidité f. / free of ~ ‖ frei von Feuchtigkeit ‖ sans humidité. / to become covered with ~ (at the surface) ‖ (sich) mit Feuchtigkeit f. beschlagen ‖ se couvrir d'une couche de rosée. / ~ of air ‖ Luftfeuchtigkeit f. ‖ humidité f. de l'air. / ~ of coals ‖ Naßgehalt m. der Kohlen ‖ humidité f. des charbons.

moisture content ‖ Feuchtigkeitsgehalt m. ‖ degré m. *ou* quantité f. d'humidité; teneur f. en humidité. / excess of ~ ‖ Feuchtigkeitsüberschuß m. ‖ excès m. d'humidité.

molar motion (Phys) ‖ Massenbewegung f. ‖ mouvement m. des masses.

molasse (Geol) ‖ Molasse f.; lockerer Sandstein m. ‖ molasse f.

molassed fodder ‖ Melassefutter n. ‖ fourrage m. mélassé.

molasses pl. (Sug) ‖ Melasse f.; Zuckersirup m.; Zuckerhonig m. ‖ mélasse f. / fodder with ~ ‖ Futter n. mit Melassezusatz ‖ fourrage m. mélassé.

molasses desugarization ‖ Melasseentzuckerung f. ‖ désucratage m. / rapid and thorough extraction of the sugar from the ~ ‖ schnelle und gründliche Entzuckerung f. der Melasse ‖ désucration f. rapide et radicale de la mélasse; désucratage f. rapide et radicale. / ~ factory ‖ Melassefabrik f. ‖ fabrique f. de mélasse. / ~ refiner ‖ Melasseraffinierer m. ‖ raffineur m. de mélasse. / ~ refinery ‖ Melasseraffinerie f. ‖ raffinerie f. de mélasse.

mold *see* mould.

mole (Gramme-molecule) ‖ Mol n.; Grammolekül n. ‖ molécule-gramme f. / ~ (Hydr arch) ‖ Mole f.; Hafendamm m. ‖ môle m. / ~ (Zoology) ‖ Maulwurf m. ‖ taupe f. / berth ~ ‖ Hellingmole f. ‖ môle m. de cale.

mole catcher ‖ Maulwurffänger m. ‖ chasseur m. de taupes; taupier m.

molecular ‖ molekular ‖ moléculaire. / ~ aggregate (Chem) ‖ molekulare Assoziation f. ‖ association f. moléculaire. / ~ air pump ‖ Molekularluftpumpe f. ‖ pompe f. à air moléculaire. / ~ force ‖ Molekularkraft f. ‖ force f. moléculaire. / ~ transformation ‖ Molekülumlagerung f. ‖ déplacement m. moléculaire. / ~ weight ‖ Molekulargewicht n. ‖ Molekülgewicht n. ‖ poids m. moléculaire.

molecule ‖ Molekel f.; Molekül n. ‖ molécule f.

mole head ‖ Molenkopf m. ‖ tête f. de la jetée.

moleskin ‖ Maulwurfsfell n. ‖ peau f. de taupe. / ~ (Cotton fabric) ‖ englisches Leder n.; Moleskin n. ‖ peau f. de taupe; molesquine f.; cuir m. anglais.

molette of turbine shaft ‖ Wellenstern m. der Turbine ‖ molette f. d'arbre de turbine.

mollusc ‖ Weichtier n.; Molluske f. ‖ mollusque m.

molten *see also* to melt ‖ geschmolzen ‖ fondu. / ~ mass (Chem) ‖ Schmelze f. ‖ masse f. fondue.

molybdate of ammonium ‖ Ammoniummolybdat n. ‖ molybdate m. d'ammoniaque. / ~ of lead ‖ Gelbbleierz n.; Molybdänblei n. ‖ molybdate m. de plomb; plomb m. jaune *ou* molybdaté.

molybdenite ‖ Molybdänglanz m.; Schwefelmolybdän n.; Wasserblei n. ‖ molybdénite f.

molybdenum ‖ Molybdän n. ‖ molybdène m. / ~ ore ‖ Molybdänerz n. ‖ minerai m. de molybdène.

molybdic acid ‖ Molybdänsäure f. ‖ acide m. molybdique. / ~ ochre ‖ Molybdänocker m.; Wasserbleiocker m. ‖ molybdène m. oxydé; oxyde m. de molybdène.

molybdine *see* molybdic ochre.

moment ‖ Zeitpunkt m.; Augenblick m.; Moment m. ‖ moment m. / ~ (Mech) ‖ Moment n. ‖ moment m. / ~ of a couple ‖ Moment n. eines Kräftepaares ‖ moment m. d'un couple. / ~ of flexion ‖ Biegungsmoment n. ‖ moment m. de flexion. / ~ of a force ‖ Moment n. einer Kraft; statisches Moment n. ‖ moment m. d'une force. / ~ of inertia ‖ Trägheitsmoment n. ‖ moment m. d'inertie. / magnetic ~ ‖ magnetisches Moment n. ‖ moment m. magnétique. / pitching ~ (Aero) ‖ Kippmoment n. ‖ moment m. de tangage. / ~ of resistance ‖ Widerstandsmoment n. ‖ moment m. de résistance. / ~ of sail ‖ Segelmoment n. ‖ moment m. de voilure. / ~ of stability ‖ Stabilitätsmoment n. ‖ moment m. de stabilité. / virtual ~ ‖ virtuelles Moment n. ‖ moment m. virtuel.

momentary value ‖ Augenblickswert m. ‖ valeur f. instantanée.

momentum *see also* moment ‖ Triebkraft f. ‖ force f. motrice. / ~ change ‖ Impulsänderung f. ‖ variation f. de la quantité de mouvement.

monatomic *see* monoatomic.

monazite sand ‖ Monazitsand m. ‖ sable m. de monazit.

mondamin(e) ‖ Mondamin n.; feines Maismehl n. ‖ mondamine m.

monetary standard ‖ Münzfuß m. ‖ titre m. des monnaies.

money || Geld n.; Vermögen n.; Geldmittel npl. || argent m.; fonds mpl. / ~ (Coin) || Münze f.; Geldstück n. || monnaie f. / ~ paid in advance || Geldvorschuß m. || avance f. de fonds. / ~ at call || Tagesgeld n. || dépôt m. à vue. / earnest ~ || Angeld n. || arrhes fpl. / to invest ~ || Geld n. anlegen oder hineinstecken || placer ou donner de l'argent m.; s'intéresser avec du capital. / to invest ~ in a business || Geld n. in ein Geschäft stecken || mettre de l'argent dans une entreprise. / to put ~ in a business see to invest ~ in a business. / to raise ~ || Geld n. aufbringen || se procurer de l'argent m. / ready ~ || bares Geld n. || argent m. comptant.

money bag || Geldtasche f.; Geldtäschchen n.; Geldbeutel m. || porte-monnaie m.; bourse f. / ~ balance see ~ scales. / ~ box || Sparbüchse f. || tire-lire m. / ~ broker || Geldmakler m. || courtier m.; agent m. de change. / ~ case see ~ bag. / ~ changer || Wechsler m.; Geldwechsler m. || changeur m.; banquier m. / ~ changer's office || Wechselstube f. || boutique f. de changeur. / ~ changing || Geldwechseln n. || change m.; banque f. / ~ counter || Geldzähler m. || compteur m. (de monnaie). / ~ counting machine || Geldzählmaschine f. || machine f. à compter la monnaie. / ~ counting and rolling machine || Geldzähl- und Packmaschine f. || machine f. à compter et à empaqueter la monnaie. / ~ drawer (Of a cash register) || Geldschublade f. || tiroir m. de monnaie. / ~ due || Geldforderung f. || demande f. d'argent; créance f. / ~ loan || Geldanleihe f. || emprunt m. / ~ market || Geldmarkt m. || marché m. monétaire. / ~ market (Stock exchange) || Kapitalmarkt m. || marché m. des capitaux; marché m. financier. / ~ matters pl. || Geldangelegenheiten fpl. || affaires fpl. d'argent. / ~ order || Geldanweisung f. || mandat m.; chèque m. / ~ order (Mail) || Postanweisung f. || mandat m. de poste. / ~ rolling machine || Geldrollmaschine f. || machine f. à empaqueter la monnaie. / ~ scales pl. || Münzwage f.; Geldwage f. || balance f. pour monnaie; trébuchet m. / ~ transaction || Geldgeschäft n. || affaire f. d'argent ou de banque. / ~ transport car || Geldtransportwagen m. || voiture f. pour le transport d'argent. / ~ want of ~ || Geldmangel m. || disette f. d'argent.

Monier arch || Moniergewölbe n. || voûte f. système Monier. / ~ building || Monierbau m. || construction f. système Monier. / ~ construction see ~ building. / ~ reinforced concrete ceiling || Monierdecke f. aus Eisenbeton || plancher m. Monier en béton armé. / ~ slab || Monierplatte f. || dalle f. Monier. / ~ vault || Moniergewölbe n. || voûte f. système Monier. / ~ wall || Monierwand f. || paroi f. système Monier.

monitor desk (Tel) || Auskunfttisch m. || table f. des renseignement.

monitorial device (Tel) || Mitleser m. || transcripteur m. à contrôle.

monitoring board (Tel) || Mithörschrank m. || tableau m. d'écoute. / ~ coil || Mithörspule f. || bobine f. d'écoute. / ~ device || Mithöreinrichtung f. || arrangement m. d'écoute. / ~ jack || Mithörklinke f. || jack m. d'écoute.

monk (Fuse of a mine) || Mönch m.; Minenzünder m. || moine m. / ~ (Print) || verschmierte Stelle f. (im Druck) || feinte f.; moine m.

monkey (Pile driver) || Rammbär m.; Fallblock m.; Rammblock m.; Hoyer m. || mouton m.; billot m. de batte. / ~ spanner see ~ wrench. / ~ wrench || Universalschlüssel m.; Engländer m.; Franzose m. || clef f. anglaise ou universelle.

monoacid || einsäurig || monoacide.

monoatomic || einatomig || monoatomique.

monoatomicity || Einatomigkeit f. || monoatomicité f.

monobasic || einbasisch || monobasique.

monocentric eyepiece || monozentrisches Okular n. || oculaire m. monocentrique.

monochloracetic acid || Monochloressigsäure f. || acide m. monochloracétique.

monochromatic || einfarbig; monochromatisch || monochromatique. / ~ source of light || monochromatische Lichtquelle f. || source f. lumineuse monochromatique.

monochromator || Monochromater m.; Monochromator m. || monochromateur m. / ~ for visible and ultra-violet light || Monochromater m. für sichtbares und ultraviolettes Licht || monochromateur m. pour la lumière visible et ultraviolette.

monocle || Einglas n.; Monokel n. || monocle m. / ~ case || Einglasfutteral n. || étui m. pour monocles.

monocular || monokular; einäugig || monoculaire. / ~ field glass || monokularer Feldstecher m. || longue vue f. monoculaire. / ~ field glass for use with one eye || monokularer Feldstecher m. für einäugigen Gebrauch || longue-vue f. monoculaire pour un seul œil. / ~ lookout telescope || monokulares Aussichtsfernrohr n. || lunette f. d'approche monoculaire. / ~ telescopic magnifier || monokulare Fernrohrlupe f. || téléloupe f. monoculaire. / ~ tube (Opt) || monokularer Tubus m. || tube m. monoculaire.

monogram embroiderer || Monogrammstickerin f. || chiffreuse-brodeuse f.

mono-gyro compass || Einkreiselkompaß m. || compas m. monogyroscopique ou à un seul gyroscope.

monohydric (Chem) || einwertig || monovalent. / ~ alcohol || einwertiger Alkohol m. || alcool m. monovolant.

monoline (Print) || Monoline f. || monoline f.

mononuclear || einkernig || à un (seul) noyau.

monophase alternomotor || Einphasenwechselstrommotor m. || moteur m. à courant alternatif monophasé. / ~ current || Einphasenstrom m. || courant m. monophasé. / ~ network || Einphasennetz n. || réseau m. monophasé.

monoplane (Airpl) || Eindecker m. || monoplan m.; appareil m. monoplan. / cantilever ~ || freitragender Eindecker m. || monoplan m. en porte à faux. / highwing ~ || Hochdecker m. || monoplan m. à ailes surélevées. / ~ with low set wing || Tiefdecker m. || monoplan m. surbaissé ou à plans bas; avion m. à ailes surbaissées. / low-wing ~ see ~ with low set wing. / semi-high wing ~ || Mitteldecker m. || avion m. à ailes à demisurélevées.

monoplane aeroplane see monoplane.

monopolist || Alleinvertreter m.; Monopolist m. || représentant m. exclusif.

monopolize, to || monopolisieren || monopoliser.

monopoly || Monopol n.; Alleinhandel m. || monopole m. / to exercize a ~ || ein Monopol n. ausüben || exercer un monopole. / Government ~ || Regierungsmonopol n. || monopole m. du Gouvernement. / ~ of the State relating to telegraphy || Telegraphenhoheitsrecht n. || monopole m. télégraphique d'Etat.

monorail (Railway) || Einschienenbahn f. || chemin de fer m. monorail; monorail m. / ~ bucket crab (running on an overhead track) || Einschienengreiferkatze f. (auf Hochbahn) || chariot m. monorail à grappin (sur voie surélevée). / ~ crab || Einschienenlaufkatze f. || chariot m. de (grue) monorail. / ~ railway see monorail (Railway) || Einschienenbahn f. || chemin de fer m. à un rail ou monorail. / ~ track || Einschienenlaufbahn f. || voie f. de roulement monorail. / ~ travelling crab || Einschienenhängekatze f. || chariot m. monorail suspendu.

monosilicate || Singulosilikat n. || monosilicate m.

monotype (composing machine) || Monotypesetzmaschine f.; Monotype f. || monotype m. / ~ casting machine || Monotypegießmaschine f. || monotypefondeuse f. / ~ operator || Monotypesetzer m. || monotypiste m.

monotyper see monotype operator.

monovalent (Chem) see monohydric.

montebrasite || Montebrasit m. || montebrasite f.

montejus || Saftheber m.; Montejus m. || monte-jus m.

month || Monat m. || mois m. / by the ~ || monatlich; allmonatlich; monatweise || par mois m.

monthly || monatlich || mensuel. / ~ balance || Monatsbilanz f.; Monatsausweis m.; Monatsabschluß m. || balance f. mensuelle. / ~ report || Monatsbericht m.; monatlicher Bericht m. || rapport m. mensuel ou du mois. / ~ statement || Monatsauszug m. || relevé m. mensuel; relevé m. de fin de mois.

monument || Denkmal n.; Monument n. || monument m. / architectural ~ || Baudenkmal n. || monument m. architectonique.

monumental metal || Statuenmetall n.; Erz n. || métal m. à statues. / ~ stone || Denkmalsstein m. || pierre f. de monument.

moon || Mond m. || lune f. / crescent ~ see waxing ~. / waning ~ || abnehmender Mond m. || lune f. décroissante. / waxing ~ || zunehmender Mond m. || lune f. croissante.

moon camera || Mondkamera f. || chambre f. noire pour la lune. / ~ knife (Curr) || Schlichtmond m. || lunette f.

moonless night || mondlose Nacht f. || nuit f. sans lune.

moon steel || Mondstahl m. || acier m. au croissant. / ~ stone || Mondstein m.; Adular m. || feldspath m. nacré; pierre f. de lune.

moor, to ~ a vessel || ein Schiff vertäuen || amarrer un bâtiment. / ~ alongside a quay || sich längsseits einer Mole vertäuen || s'amarrer le long d'un quai.

moor ‖ Moor n.; Sumpf m.; Morast m. ‖ marais m. / ~ canal ‖ Entwässerungskanal m. eines Torfmoores ‖ canal m. de marais; wateringue m. / ~ coal ‖ Moorkohle f. ‖ houille f. limoneuse; lignite m. terne en partie.

mooring(s pl.) (Mar) ‖ Hafenanker m.; Moorings fpl. ‖ amarres fpl. / ~ block ‖ Mooringsblock m. ‖ aurai m. / ~ cable (Aero) ‖ Verankerungskabel n. ‖ câble m. d'amarrage. / ~ ring ‖ Festmachering m. ‖ anneau m. d'amarrage. / ~ rope ‖ Halteleine f. ‖ câble m. de retenue ou d'amarrage; amarre f. / ~ swivel ‖ Mooringsschäkel m. ‖ émerillon m. d'affourche.

moory ‖ moorig ‖ marécageux.

mop ‖ Putzlappen m.; Scheuerlappen m. ‖ torchon m.; chiffon m. à nettoyer. / ~ (Mar) ‖ Quast m.; Schmierquast m. ‖ pinceau m.; guipon m.; broche f.

moraine (Geol) ‖ Moräne f. ‖ moraine f. / lateral ~ ‖ Seitenmoräne f. ‖ moraine f. latérale ou marginale.

morality (In the legal terminology) ‖ gute Sitten fpl. ‖ bonnes mœurs fpl. / action contrary to ~ ‖ gegen die guten Sitten fpl. verstoßende Handlung f. ‖ action f. contraire aux bonnes mœurs.

morbid ‖ krankhaft ‖ maladif.

mordant, to ‖ beizen ‖ mordancer.

mordant ‖ beißend; ätzend; scharf ‖ mordant.

mordant ‖ Beize f.; Ätzmittel n.; Beizmittel n. ‖ mordant m.; morsure f. / aluminous ~ ‖ Alaunbeize f. ‖ mordant m. d'alun. / iron ~ ‖ Eisenbeize f. ‖ mordant m. de fer. / spirit ~ ‖ Spiritusbeize f. ‖ mordant m. à l'alcool. / wood ~ ‖ Holzbeize f. ‖ mordant m. pour bois.

mordant dye ‖ Beizenfarbstoff m. ‖ colorant m. sur mordant.

mordanter ‖ Beizer m. ‖ mordanceur m. / wool ~ ‖ Wollbeizer m. ‖ mordanceur m. de laine.

mordanting ‖ Beizen n.; Beizung f. ‖ mordançage m.

mordant maker see mordanter. / ~ yellow ‖ Beizengelb n. ‖ jaune m. sur mordant.

more than . . . ‖ mehr als . . . ‖ plus de . . .; plus que . . .

morgue ‖ Leichenschauhaus n.; Leichenhalle f. ‖ morgue f. / cooled inspection cell in a ~ ‖ gekühlte Schauzelle f. in einem Leichenschauhaus ‖ cellule f. refroidie dans une morgue.

morion (Min) ‖ Morion m.; dunkler Rauchtopas m. ‖ morion m.

morling (Wool) ‖ Sterblingswolle f. ‖ laines fpl. mortes.

morning gown ‖ Morgenrock m. ‖ matinée f. / ~ print ‖ Morgenblatt n.; Morgenzeitung f. ‖ feuille f. ou journal m. de matin. / ~ shift ‖ Frühschicht f. ‖ équipe f. du matin.

morocco see ~ leather. / ~ articles pl. ‖ Saffianwaren fpl. ‖ objets mpl. en maroquinerie. / ~ leather ‖ Marokkoleder n.; Saffian m. ‖ cuir m. maroquiné; maroquin m. / ~ paper ‖ Saffianpapier n.; Marokinpapier n. ‖ papier m. maroquiné. / ~ tanner ‖ Saffiangerber m. ‖ maroquineur m. / ~ tanning ‖ Saffiangerben n. ‖ maroquinerie f.

morphia ‖ Morphium n. ‖ morphium m.

morphine (Chem) ‖ Morphin n.; Morphium n. ‖ morphine f. / ~ acetate ‖ essigsaures Morphin n.; Morphinazetat n. ‖ acétate m. de morphine. / ~ hydrochlorate ‖ salzsaures Morphium n. oder Morphin n. ‖ chlorhydrate m. de morphine.

morphinic acid ‖ Morphinsäure f. ‖ acide m. morphinique.

morphotropy ‖ Morphotropie f. ‖ morphotropie f.

Morse apparatus ‖ Morsetelegraf m.; Morsegerät n. ‖ appareil m. Morse. / ~ inker ‖ Schreibempfänger m. ‖ appareil m. Morse enregistreur. / ~ inkwriter see ~ inker. / ~ reception ‖ Morseempfang m. ‖ réception f. Morse. / ~ wireless high speed ~ reception ‖ drahtloser Schnellmorseempfang m. ‖ réception f. rapide de signes Morse radiodiffusés. / ~ safety circuit in fire alarm installations ‖ Morsesicherheitsschaltung f. in Feuermeldeanlagen ‖ montage m. Morse de garantie dans les installations d'avertisseurs d'incendie. / ~ signal ‖ Morsezeichen n. ‖ signal m. Morse; caractère m. Morse. / contracted ~ signal ‖ abgekürztes Morsezeichen n. ‖ signal m. Morse abrégé. / ~ writing ‖ Morseschrift f. ‖ écriture f. morse.

mortality ‖ Sterblichkeit f.; Sterblichkeitsziffer f. ‖ mortalité f. / ~ per 1000 persons employed underground ‖ Sterblichkeit f. unter 1000 Arbeitern unter Tage ‖ mortalité f. par 1000 ouvriers du fond.

mortar (Mixture of lime, sand, and water ‖ Mörtel m. ‖ mortier m. / ~ (Vessel for crushing with a pestle) ‖ Mörser m.; Reibschale f. ‖ mortier m. / ~ (Ore dress) ‖ Pochgerinne n.; Pochtrog m.; Stampftrog m. ‖ huche f. / ~ (Shell gun) ‖ Mörser m. ‖ mortier m. / ~ air ~ (Mas) ‖ Luftmörtel m. ‖ mortier m. à la chaux ou durcissant à l'air. / asphalt ~ ‖ Asphaltmörtel m. ‖ mortier m. d'asphalte. / bad ~ ‖ Halbmörtel m. ‖ mortier m. bâtard. / cement ~ ‖ Zementmörtel m. ‖ mortier m. au ciment. / finishing ~ ‖ Putzmörtel m. ‖ mortier m. d'achèvement. / fire-proof ~ ‖ feuerfester Mörtel m. ‖ mortier m. réfractaire. / frozen ~ ‖ gefrorener Mörtel m. ‖ mortier m. congelé. / hydraulic ~ ‖ hydraulischer Mörtel m. ‖ mortier m. hydraulique; ciment m. romain. / jointing ~ ‖ Fugenmörtel m. ‖ mortier m. à jointoyer. / ~ (made) of lime and brick dust ‖ Ziegelmehlmörtel m. ‖ mortier m. à la chaux et à la poussière de briques. / ~ of lime and sand ‖ Kalksandmörtel m. ‖ mortier m. à la chaux et au sable. / meagre ~ ‖ magerer Mörtel m. ‖ mortier m. maigre. / ~ of over-burnt lime-grains ‖ Krumpmörtel m. ‖ repous m. / ~ of plaster ‖ Gipsmörtel m. ‖ plâtre-ciment m. / quickly-hardening ~ ‖ schnell bindender Mörtel m. ‖ mortier m. à prise rapide / rich ~ ‖ fetter Mörtel m. ‖ mortier m. gras. / slowly-hardening ~ ‖ langsam bindender Mörtel m. ‖ mortier m. à prise lente. / strong hydraulic ~ ‖ Zementmörtel m. ‖ mortier m. de ciment. / ~ of wax ‖ Wachskitt m. ‖ mortier m. de cire.

mortar admixture ‖ Mörtelzusatz m. ‖ addition f. pour mortier. / ~ bath ‖ Kalkbad n.; Mörtelbad n. ‖ bain m. de chaux. / ~ beater ‖ Kalkschaufel f.; Rührkrücke f.; Kalkkrücke f. ‖ mouvechaux m.; croc m. ou rabot m. à chaux. / ~ funnel ‖ Kalkrutsche f. ‖ trémie f. à

mortier. / ~ joint ‖ Mörtelfuge f. ‖ joint m. au mortier. / ~ kneading machine ‖ Mörtelknetmaschine f. ‖ machine f. à pétrir le mortier. / layer of ~ ‖ Mörtelschicht f. ‖ couche f. de mortier. / making of the ~ ‖ Mörtelaufbereitung f. ‖ préparation f. du mortier. / ~ mill ‖ Mörtelmühle f. ‖ moulin m. à mortier. / ~ mixer ‖ Mörtelmischer m. ‖ mélangeur m. à mortier. / ~ mixing machine ‖ Mörtelmischmaschine f. ‖ malaxeur m. à mortier. / ~ pestle ‖ Mörserkeule f.; Stößel m.; Stampfe f. ‖ pilon m. (du mortier). / ~ pillar ‖ Mörtelständer m. ‖ pilier m. à mortier. / ~ trough ‖ Mörteltrog m. ‖ auge f. de mortier. / ~ tub ‖ Kalkkübel m.; Mörtelkübel m. ‖ cuve f. à mortier. / ~ vase ‖ Mörtelbehälter m.; Mörtelgefäß n. ‖ vase m. à mortier. / ~ worker (Pav) ‖ Fugenausstreicher m. ‖ ficheur m.

mortgage, to ‖ verpfänden; zum Pfand verschreiben; mit Hypotheken fpl. belasten ‖ hypothéquer; donner en nantissement ou en hypothèque. / ~ property ‖ eine Hypothek f. verbriefen ‖ hypothéquer des immeubles mpl.

mortgage ‖ Hypothek f.; Pfandverschreibung f.; Grundpfand n. ‖ hypothèque f. / first ~ ‖ erste Hypothek f. ‖ première hypothèque f. / on ~ ‖ hypothekarisch ‖ hypothécaire. / to raise a ~ ‖ eine Hypothek f. aufnehmen ‖ lever une hypothèque. / security ~ ‖ Sicherungshypothek f. ‖ hypothèque f. de garantie.

mortgage authority ‖ Grundbuchbehörde f. ‖ administration f. des hypothèques. / ~ bank ‖ Hypothekenbank f. ‖ banque f. hypothécaire. / ~ bond ‖ Pfandbrief m.; Pfandverschreibung f. ‖ lettre f. de gage; titre m. ou cédule f. hypothécaire.

mortgage book ‖ Grundbuch n.; Hypothekenregister n. ‖ registre m. des hypothèques. / to make a note in the ~ ‖ einen Vermerk m. im Grundbuch machen ‖ faire une inscription dans le registre des hypothèques.

mortgaged ‖ hypothekarisch verpfändet oder belastet ‖ hypothéqué.

mortgage deed see mortgage bond.

mortgagee ‖ Hypothekengläubiger m.; Hypothekar m. ‖ créancier m. hypothécaire.

mortgage loan ‖ Hypothekendarlehen n. ‖ prêt m. hypothécaire. / paying off a ~ ‖ Amortisierung f. einer Hypothek ‖ purge f. d'hypothèque. / ~ registrar ‖ Hypothekenverwalter m. ‖ conservateur m. des hypothèques.

mortgager ‖ Hypothekenschuldner m.; Pfandschuldner m. ‖ débiteur m. hypothécaire.

mortice see mortise.

mortise, to ‖ einzapfen; verzapfen; einlassen ‖ mortaiser; emmortaiser; faire une mortaise; enchâsser. / ~ letters pl. (Print) ‖ Buchstaben mpl. ausklinken ‖ entailler des caractères mpl. / ~ a tenon ‖ einen Zapfen m. einlochen ‖ mortaiser le trou d'un tenon.

mortise ‖ Keilnut f.; Nut f.; Zapfenloch n. ‖ mortaise f.; rainure f. / to cut ~s pl. ‖ Zapfenlöcher npl. stemmen ‖ mortaiser.

mortise axe ‖ Stichaxt f.; Stoßaxt f.; bisaiguë f.; piochon m. / ~ bolt ‖ Zapfennagel m. ‖ dent-de-loup f.

mortise chisel ‖ Lochbeitel m.; Stechbeitel m. ‖ bédane m.; ciseau m. fort; bec m. d'âne. / chairmaker's ~ ‖ Stuhlmacherlochbeitel m. ‖ bec m. d'âne pour chaisiers. / firmer ~ ‖ Lochbeitelchen n. ‖ bec m. d'âne. / with oval polster ‖ Lochbeitel m. mit Eierband ‖ bec m. d'âne à embase ovale.

mortised ‖ verzapft ‖ assemblé à mortaise et tenon; mortaisé.

mortise gauge ‖ Zapfenstreichmaß n. ‖ trusquin m. d'assemblage. / ~ joint ‖ Verzapfung f. ‖ assemblage m. à mortaise. / ~ lock ‖ Einsteckschloß n. ‖ serrure f. à fourreau ou à mortaise; serrure f. cachée ou encloisonnée ou enfaillée. / ~ sounding tool ‖ Zapfensonde f. ‖ quilboquet m. / ~ wheel ‖ Zahnrad n. mit Winkelzähnen ‖ roue f. à chevron.

mortising ‖ Verzapfung f. ‖ assemblage m. à tenon et mortaise. / ~ machine ‖ Stemmaschine f.; Zapfenlochmaschine f. ‖ machine f. à mortaiser; mortaiseuse f.

mortuary see morgue.

mosaic ‖ Mosaik f. ‖ mosaïque f. / ~ of glass ‖ Glasmosaik f. ‖ mosaïque f. de verre.

mosaic flag ‖ Mosaikfliese f.; Mosaikplatte f ‖ carreau m. (en) mosaïque. / ~ floor ‖ Mosaikfußboden m. ‖ parquet m. mosaïque. / ~ gold ‖ Musivgold n. ‖ or m. musif ou de Judée ou mosaïque. / ~ silver ‖ Musivsilber n. ‖ argent m. musif. / ~ tile see ~ flag. / ~ work ‖ Mosaikarbeit f.; Musivarbeit f. ‖ mosaïque f.; ouvrage m. à la mosaïque. / ~ worker ‖ Mosaikarbeiter m. ‖ mosaïste m.

moss ‖ Moos n. ‖ mousse f. / ~ (Peat) ‖ Torf m. ‖ tourbe f. / ~ agate (Miner) ‖ Moosachat m. ‖ agate f. mousseuse. / ~ dresser (Basketry) ‖ Moosarbeiter m. ‖ apprêteur m. de mousse. / ~ dyeing ‖ Moosfärben n.; Färben n. von Moos ‖ teinture f. de mousse. / ~ gatherer ‖ Moossammler m. ‖ moussier m. / ~ green ‖ Moosgrün n. ‖ vert m. de mousse. / ~ litter ‖ Torfstreu f. ‖ tourbe f. pour litière.

moth ‖ Motte f. ‖ taigne f.

mother, to ‖ schimmeln; vermodern ‖ (se) moisir; (se) chancir.

mother (Chem) ‖ Essigmutter f.; Essiggärungshefe f. ‖ mère f. (de vinaigre).

mother-gate (Mine) ‖ Förderstrecke f.; Hauptförderstrecke f. ‖ galerie f. principale ou de roulage; voie f. de roulage. / inclined ~ ‖ Flaches n.; flach einfallende Förderstrecke f. ‖ descenderie f.; vallée f.; toret f.

mother-lie ‖ Mutterlauge f. ‖ eau f. mère; eaux fpl. mères; lessive-mère f. / argentiferous ~ ‖ reiche silberhaltige Mutterlauge f. ‖ eau f. mère argentifère. / cupriferous ~ ‖ kupferhaltige Mutterlauge f. ‖ eau m. mère de cuivre. / ferriferous ~ ‖ eisenhaltige Mutterlauge f. ‖ eau f. mère ferrifère.

mother-liquor see mother-lie.

mother-lye see mother-lie.

mother-of-pearl ‖ Perlmutter f. ‖ nacre f. (de perles). / ~ button ‖ Perlmutterknopf m. ‖ bouton m. en nacre. / ~ engraver ‖ Perlmuttergravör m. ‖ graveur m. sur nacre. / ~ goods pl. ‖ Perlmutterwaren fpl. ‖ articles mpl. en nacre; tabletterie f. en nacre. / ~ polisher ‖ Perlmutterschleifer m. ‖ polisseur m. de nacre.

mother-of-vinegar ‖ Essigmutter f. ‖ mère f. (de vinaigre).

mother-substance (Chem) ‖ Muttersubstanz f. ‖ substance mère f.; substance f. génératrice.

mother wool ‖ Rückenwolle f. vom Schaf ‖ laine f. mère ou prime, mère f. laine.

moth preventative ‖ Mottenschutzmittel n.; Mottenvertilgungsmittel n. ‖ préservatif m. contre les taignes.

motion ‖ Gang m. ‖ marche f.; fonction f. / ~ (Gear) ‖ Mechanismus m.; Triebwerk n. ‖ mécanisme m. / ~ (Proposal) ‖ Antrag m. ‖ motion f.; propos m. / to put a ~ ‖ einen Antrag m. stellen ‖ présenter une motion; proposer. / to put in ~ ‖ in Bewegung setzen ‖ mettre en marche ou en train. / to set into ~ see to put in ~. / accelerated ~ ‖ beschleunigte Bewegung f. ‖ mouvement m. accéléré. / angular ~ ‖ Winkelbewegung f. ‖ mouvement m. angulaire. / apparent ~ ‖ scheinbare Bewegung f. ‖ mouvement m. apparent. / backward and forward ~ ‖ hin- und hergehende Bewegung f. ‖ mouvement m. de va-et-vient. / circular ~ ‖ Kreisbewegung f. ‖ mouvement m. circulaire. / composed ~ see compound ~. / compound ~ ‖ zusammengesetzte Bewegung f. ‖ mouvement m. composé. / decreasing ~ ‖ verzögerte Bewegung f. ‖ mouvement m. retardé. / eccentric ~ ‖ exzentrische Bewegung f. ‖ mouvement m. excentrique. / electric ~ ‖ Bewegung f. durch elektrischen Antrieb ‖ mouvement m. à force électrique. / free ~ ‖ Spiel n.; Spielraum m. ‖ jeu m.; liberté f. / guided ~ ‖ Zwangsläufigkeit f.; zwangsläufige Bewegung f. ‖ mouvement m. forcé. / heart ~ ‖ herzförmige Bewegung f. ‖ mouvement m. en courbe de Vaucanson. / idle ~ ‖ Leergang m. ‖ mouvement m. à vide. / intermittent ~ ‖ Wechselbewegung f. ‖ mouvement m. intermittent. / longitudinal ~ ‖ Längsverschiebung f. ‖ déplacement m. en longueur. / lost ~ ‖ toter Gang m. ‖ perte f. de course. / ~ by motor ‖ Motorbetrieb m. ‖ marche f. au moteur. / oscillatory ~ ‖ schwingende Bewegung f. ‖ mouvement m. oscillatoire. / parallel ~ ‖ Parallelbewegung f. ‖ mouvement m. parallèle. / pendulous ~ ‖ pendelnde Bewegung f. ‖ mouvement m. oscillant. / periodical uniform ~ ‖ gleichförmig wiederkehrende Bewegung f. ‖ mouvement m. périodiquement uniforme. / pitching ~ (Ship) ‖ Stampfbewegung f. ‖ mouvement m. de tangage. / planetary ~ ‖ Planetenbewegung f. ‖ mouvement m. planétaire. / progressive ~ ‖ Fortbewegung f. ‖ locomotion f.; mouvement m. progressif; marche f. / ~ of projection ‖ Wurfbewegung f. ‖ mouvement m. d'un projectile. / retrograde ~ ‖ Rückgang m.; Rückwärtsbewegung f. ‖ marche f. en arrière. / reversing ~ ‖ Kehrbewegung f. ‖ mouvement m. alternatif. / rolling ~ (Ship) ‖ Rollbewegung f.; Rollen n. ‖ mouvement m. de roulis. / rotating ~ ‖ drehende Bewegung f.; Drehbewegung f. ‖ mouvement m. de rotation. / self-acting ~ ‖ selbsttätige Bewegung f. ‖ mouvement m. automatique. / simple ~ ‖ einfache Bewegung f. ‖ mouvement m. simple. / slackening ~ ‖ abnehmende Bewegung f. ‖ mouvement m. de relâche. / slow ~ ‖

feine Verstellung f.; Feinbewegung f. ‖ mouvement m. lent. / ~ to and fro ‖ Hin- und Herbewegung f. ‖ mouvement m. de va-et-vient. / ~ of translation ‖ fortschreitende Bewegung f. ‖ mouvement m. de translation. / uniform ~ ‖ gleichförmige Bewegung f. ‖ mouvement m. uniforme. / uniformly accelerated ~ ‖ gleichförmig beschleunigte Bewegung f. ‖ mouvement m. uniformément accéléré. / uniformly decreasing ~ ‖ gleichförmig verzögerte Bewegung f. ‖ mouvement m. uniformément retardé. / uniformly variable ~ ‖ gleichförmig veränderliche Bewegung f. ‖ mouvement m. uniformément varié. / up and down ~ ‖ Auf- und Abbewegung f. ‖ montée f. et descente f. / variable ~ ‖ veränderliche Bewegung f. ‖ mouvement m. varié. / variable increasing ~ ‖ veränderlich beschleunigte Bewegung f. ‖ mouvement m. variable accéléré. / vertical ~ of bodies ‖ freier Fall m. der Körper ‖ chute f. libre des corps. / wave ~ ‖ Wellenbewegung f. ‖ mouvement m. ondulatoire. / ~ of the wind ‖ Windbewegung f. ‖ mouvement m. du vent; déplacement m. du vent.

motionless ‖ unbeweglich; bewegungslos; still ‖ tranquille; inactif; silencieux; immobile.

motion picture(s pl.) ‖ Film m.; Filmstreifen m. ‖ film m. / ~ camera ‖ Filmaufnahmekammer f. ‖ chambre f. de prise de vues. / ~ projector ‖ Filmvorführungsgerät n. ‖ appareil m. de projection de cinéma.

motion, quantity of ‖ Bewegungsgröße f. ‖ quantité f. de mouvement. / transmission of ~ ‖ Bewegungsübertragung f. ‖ transmission f. de mouvement.

motive ‖ bewegend ‖ moteur. / ~ mechanism (Mech) ‖ Bewegungsmechanismus m. ‖ appareil m. de mouvement. / ~ power ‖ Triebkraft f.; bewegende Kraft f.; Betriebskraft f. ‖ force f. motrice. / ~ water (Water wheel) ‖ Aufschlagwasser n. ‖ eau f. motrice.

motive ‖ Bewegrund m.; Anlaß m. ‖ motif m. / without any ~ ‖ ohne jeden Anlaß m. ‖ sans aucun motif m.; sans aucune raison.

motor ‖ Motor m.; Kraftmaschine f. ‖ moteur m. / the ~ is getting hot ‖ der Motor wird heiß ‖ le moteur s'échauffe. / the ~ comes out of step ‖ der Motor fällt außer Tritt m. ‖ le moteur ne tourne plus synchroniquement. / to let the ~ run itself in ‖ den Motor m. einlaufen lassen ‖ laisser tourner le moteur pour qu'il se fasse. / to utilize the ~ fully ‖ den Motor m. voll ausnutzen ‖ utiliser la pleine puissance du moteur. / the ~ runs without load ‖ der Motor läuft leer ‖ le moteur marche à vide. / the ~ works quietly and without shocks ‖ der Motor arbeitet ruhig und stoßfrei ‖ le moteur travaille tranquillement et sans choc. / air-cooled ~ ‖ luftgekühlter Motor m. ‖ moteur m. à refroidissement d'air ou à refroidissement par air. / alternating current ~ ‖ Wechselstrommotor m. ‖ moteur m. à courant alternatif; alternomoteur m. / asynchronous starting ~ ‖ Asynchronanwurfmotor m. ‖ moteur m. de démarrage asynchrone. / balanced ~ ‖ ausgewuchteter Motor m. ‖ moteur m. équilibré. / ~ with centrifu-

gal pump for filling tank and for watering ‖ Motor m. mit Schleuderpumpe, zum Kesselfüllen und Sprengen verwendbar ‖ groupe moteur-pompe m. centrifuge servant au remplissage du réservoir et à l'arrosage. / ~ with commutating poles ‖ Motor m. mit Wendepolen ‖ moteur m. avec pôles de commutation. / ~ of compact design ‖ Motor m. von gedrängter Bauart ‖ moteur m. de construction compacte. / compoundwound ~ ‖ Doppelschlußmotor m. ‖ moteur m. compound. / constant speed ~ ‖ Motor m. mit unveränderlicher Umlaufzahl ‖ moteur m. à vitesse constante. / continuous current ~ ‖ Gleichstrommotor m. ‖ électromoteur m. à courant continu. / crude oil ~ ‖ Rohölmotor m. ‖ moteur m. à huile brute. / ~ for cycles ‖ Fahrradeinbaumotor m. ‖ moteur m. pour bicyclettes. / ~ with cylinders cast in pairs ‖ Motor m. mit paarweise (zusammen)gegossenen Zylindern ‖ moteur m. à cylindres fondus en paires. / direct current ~ ‖ Gleichstrommotor m. ‖ moteur m. à courant continu. / domestic ~ ‖ Haushaltmotor m. ‖ moteur m. de ménage. ‖ double-commutator ~ ‖ Doppelkollektormotor m. ‖ moteur m. à deux collecteurs. ‖ double-diameter piston-type ~ ‖ Differentialkolbenmotor m. ‖ moteur m. à piston différentiel. / doublepiston~‖Motor m. mit gegenläufigen Kolben ‖ moteur m. à deux pistons opposés. / electric ~ ‖ Elektromotor m. ‖ moteur m. électrique; électromoteur m. / explosion ~ ‖ Explosionsmotor m. ‖ moteur m. à explosion. / extra small ~ ‖ Kleinstmotor. ‖ moteur m. de très petite puissance. / flange ~ ‖ Flanschmotor m. ‖ moteur m. à bride. / forced induction ~ ‖ Motor m. mit zwangsläufiger Gaszuführung ‖ moteur m. à alimentation forcée. / frame-supported ~ ‖ Motor m. mit Rahmenaufhängung ‖ moteur m. suspendu au chassis. / four-cylinder ~ ‖ Vierzylindermotor m. ‖ moteur m. à quatre cylindres. / four-stroke ~ ‖ Viertaktmotor m. ‖ moteur m. à quatre temps. / gasoline ~ ‖ Benzinmotor m. ‖ moteur m. à essence. / heavy oil ~‖ Schwerölmotor m. ‖ moteur m. à huile lourde. / heavy traffic ~ ‖ Straßenlokomotive f. ‖ tracteur m. / high-compression ~ ‖ Motor m. mit hoher Kompression ‖ moteur m. à haute compression. / horizontal ~ ‖ Motor m. mit liegenden Zylindern ‖ moteur m. horizontal. / hot-air ~ ‖ Heißluftmotor m. ‖ moteur m. à air chaud. / hot-bulb ~ ‖ Glühkopfmotor m. ‖ moteur m. à culasse incandescente. / hydraulic ~ ‖ Wasserkraftmaschine f.; hydraulischer Motor m. ‖ moteur m. hydraulique. / I-head ~ ‖ Motor m. mit hängenden Ventilen ‖ moteur m. à soupapes (commandées) par le haut. / ~ for independent electric drive ‖ elektrischer Einbaumotor m. ‖ moteur m. électrique pour montage dans les bâtis de machines. / induction ~ ‖ Induktionsmotor m. ‖ moteur m. d'induction. / internal combustion ~ ‖ Verbrennungsmotor m. ‖ moteur m. à combustion. / iron clad ~ ‖ eingekapselter Motor m. ‖ moteur m. blindé. / kitchen ~ ‖ Küchenmotor m. ‖ moteur m. pour cuisine. / lifting ~ ‖ Hubmotor m. ‖ moteur m. de levage. / long-stroke ~ ‖ Motor m. mit langem

Hub ‖ moteur m. à longue course. / low-compression ~ ‖ Motor m. mit niedriger Kompression ‖ moteur m. à basse compression. / low-speed ~ ‖ langsam laufender Motor m. ‖ moteur m. à faible vitesse. / manual ~ (Radio) ‖ Handdrehmaschine f. ‖ moteur m. à commande manuelle. / marine ~ ‖ Schiffsmotor m. ‖ moteur m. marine pour bateaux. / multi-cylinder ~ ‖ Mehrzylindermotor m. ‖ moteur m. polycylindrique ou à plusieurs cylindres. / over-cooled ~ ‖ überkühlter Motor m. ‖ moteur m. surrefroidi. / portable ~ ‖ fahrbarer Motor m. ‖ moteur m. sur roues. / radial-type ~ ‖ Sternmotor m. ‖ moteur m. en étoile. / reversible ~ ‖ umsteuerbarer Motor m. ‖ moteur m. réversible. / ~ with reversing poles ‖ Motor m. mit Wendepolen ‖ moteur m. à pôles auxiliaires de commutation. / revolving cylinder ~ ‖ Motor m. mit umlaufenden Zylindern; Umlaufmotor m. ‖ moteur m. à cylindres rotatifs; moteur m. rotatif. / ~ with separate cylinders ‖ Motor m. mit einzeln stehenden Zylindern ‖ moteur m. à cylindres séparés. / series ~ ‖ Reihenschlußmotor m. ‖ moteur m. série. / ~ for shaking trough ‖ Schüttelrutschenmotor m. ‖ moteur m. pour couloir incliné à secousses. / ~ with short-circuit rotor ‖ Motor m. mit Kurzschlußanker m. ‖ moteur m. avec induit en court-circuit. / short-stroke ~ ‖ Motor m. mit kurzem Hub ‖ moteur m. à faible course. / shunt-wound ~ ‖ Nebenschlußmotor m. ‖ moteur m. shunt ou en dérivation. / single-cylinder ~ ‖ Einzylindermotor m. ‖ moteur m. monocylindrique. / six-cylinder ~ ‖ Sechszylindermotor m. ‖ moteur m. à six cylindres. / slow-speed ~ ‖ langsam laufender Motor m. ‖ moteur m. à faible vitesse. / small ~ ‖ Kleinmotor m. ‖ moteur m. de petite puissance ou de faible puissance. / smoothly running ~ ‖ ruhig gehender Motor m. ‖ moteur m. marchant silencieusement. / stationary ~ ‖ ortsfester oder stationärer Motor m. ‖ moteur m. fixe. / ~ with subdivided magnet winding ‖ Motor m. mit unterteilter Magnetwicklung ‖ moteur m. à enroulement subdivisé d'électroaiment. / synchronous ~ ‖ Synchronmotor m. ‖ moteur m. synchronique. / two-cycle ~ see two-stroke ~. / two-cylinder ~ ‖ Zweizylindermotor m. ‖ moteur m. à deux cylindres. / two-stroke ~ ‖ Zweitaktmotor m. ‖ moteur m. à deux temps. / two-stroke reversible ~ ‖ umsteuerbarer Zweitaktmotor m. ‖ moteur m. reversible à deux temps. / V-type ~ ‖ Motor m. mit V-förmig angeordneten Zylindern ‖ moteur m. à cylindres en V. / valveless ~ ‖ ventilloser Motor m. ‖ moteur m. sans soupapes. / variable-speed ~ ‖ Regelmotor m. ‖ moteur m. à vitesse réglable. / ventilated ~ ‖ ventilierter Motor m. ‖ moteur m. ventilé. / vertical ~ ‖ stehender Motor m. ‖ moteur m. vertical. / water-cooled ~ ‖ wassergekühlter Motor m. ‖ moteur m. à refroidissement d'eau ou à refroidissement par eau. / wind ~ ‖ Windmotor m. ‖ moteur m. à vent. / x-cylinder ~ ‖ x Zylindermotor m. ‖ moteur m. à x cylindres. / ~ with x

cylinders in line ‖ Motor m. mit x Zylindern hintereinander ‖ moteur m. à x cylindres en ligne.

motor aerial ladder ‖ Motordrehleiter f. ‖ échelle f. automobile mécanique. / ~ alternator disk set ‖ Wechselstromgenerator m. mit rotierender Funkenstrecke ‖ groupe m. moteur alternateur avec éclateur à disque. / ~ arc-lamp ‖ Motorbogenlampe f. ‖ lampe f. à arc à moteur. / ~ bicycle see ~ cycle. / ~ blower ‖ Motorgebläse n. ‖ soufflerie f. à moteur.

motor boat ‖ Motorboot n. ‖ canot m. automobile; bateau m. moteur ou automobile. / ~ headlight ‖ Motorbootscheinwerfer m. ‖ phare m. pour canots automobiles. / ~ industry ‖ Motorbootindustrie f. ‖ industrie f. des bateaux automobiles.

motor bracket ‖ Motorgrundplatte f. ‖ plaque f. de fondation du moteur. / ~ breaker ‖ Motorbrecher m. ‖ concasseur-moteur m. / ~ bus ‖ Kraftomnibus m.; Autobus m. ‖ autobus m.; omnibus-automobile m. / ~ bus driver ‖ Kraftomnibusführer m. ‖ chauffeur d'autobus m. / ~ caisson ‖ Munitionskraftwagen m. ‖ voiture-caisson f. automotrice.

motor car ‖ Kraftwagen m.; Automobil n.; Auto n.; Motorwagen m. ‖ automobile f.; m.; auto f. / ~ (Railw) Triebwagen m. ‖ voiture f. motrice. / armoured ~ ‖ gepanzertes Kampffahrzeug n.: Panzerkraftwagen m.; Panzerauto n. ‖ voiture f. de combat blindée. / delivery ~ ‖ Lieferauto n.; Liefer(kraft)wagen m. ‖ Geschäftswagen m. ‖ automobile f. ou voiture f. de livraison. / electric ~ ‖ elektrischer Kraftwagen m.; Elektrowagen m. ‖ voiture f. électrique. / ~ with front-wheel drive ‖ Auto n. mit Vorderradantrieb ‖ automobile f. à commande de l'essieu d'avant. / ~ of high efficiency ‖ Kraftwagen m. für große Leistungen ‖ automobile f. de rendement élevé. / petrol-electric ~ (Railw) ‖ petrolelektrischer Triebwagen m. ‖ automotrice f. pétroléoélectrique. / small ~ ‖ Kleinauto m.; Kleinkraftwagen m. ‖ auto f. de faible puissance; voiturette f.

motor car accessories pl. ‖ Kraftwagenzubehörteile mpl. ‖ accessoires mpl. d'automobile. / ~ body ‖ Karosserie f.; Autokarosserie f.; Kraftwagenaufbau m.; Kraftwagenkasten m. ‖ carosserie f. d'automobile. / ~ building see ~ construction. / ~ can ‖ Kraftwagenkanister m. ‖ bidon m. d'automobiles. / ~ chain ‖ Kraftwagenkette f. ‖ chaîne f. d'automobiles. / ~ construction ‖ Kraftwagenbau m.; Automobilbau m. ‖ construction f. des automobiles. / ~ construction machine ‖ Kraftwagenbaumaschine f. ‖ machine f. pour la construction d'automobiles. / ~ driver ‖ Fahrer m.; Führer m.; Schoffer m. ‖ chauffeur m. d'automobile. / ~ exhibition ‖ Kraftwagenausstellung f.; Automobilausstellung f. ‖ exposition f. automobile ou d'automobiles. / ~ flag ‖ Kraftwagenflagge f. ‖ pavillon m. d'automobiles. / ~ heating ‖ Kraftwagenheizung f. ‖ chauffage m. d'automobile. / ~ horn ‖ Kraftwagenhupe f.; Autohupe f. ‖ cornet m. d'automobile. / ~ industry ‖ Kraftwageninudstrie f.; Auto-

mobilindustrie f. ‖ industrie f. de construction d'automobiles. / ~ lamp ‖ Kraftwagenlampe f.; Autolampe f.; Kraftwagenlaterne f. ‖ lanterne f. d'automobiles. / ~ leather ‖ Automobilleder n. ‖ cuir m. pour automobiles. / ~ letting-out ‖ Kraftwagenvermietung f. ‖ location f. d'automobiles. / ~ lighting ‖ Kraftwagenbeleuchtung f. ‖ éclairage m. des automobiles. / ~ lubricator ‖ Kraftwagenöler m. ‖ graisseur m. d'automobiles. / ~ part ‖ Kraftwagenersatzteil m.; Autoersatzteil m. ‖ pièce f. détachée d'automobiles. / ~ parts pl. ‖ Automobilteile mpl.; Kraftwagenzubehörteile mpl. ‖ accessories mpl. d'automobiles. / pneumatic tyre for ~s ‖ Kraftwagenluftreifen m.; Automobilluftreifen m. ‖ pneu m. ou pneumatique m. d'automobile. / ~ polish ‖ Kraftwagenpolitur f. ‖ produits mpl. à polir les carosseries d'automobiles. / ~ radiator ‖ Kraftwagenkühler m.; Automobilkühler m. ‖ radiateur m. d'automobile. / ~ repair shop ‖ Kraftwagenausbesserungswerkstatt f.; Automobilreparaturwerkstatt f. ‖ atelier m. de réparation d'automobiles.

motor carriage ‖ Motorwagen m.; Motorfahrzeug n. ‖ voiture f. automobile.

motor car slide preservers pl. ‖ Gleitschutz m. für Kraftfahrzeuge ‖ autodérapants mpl. d'automobiles. / ~ speaking trumpet ‖ Kraftwagensprachrohr n. ‖ tuyau m. acoustique d'automobiles. / ~ storage room ‖ Kraftwagenschuppen m.; Automobilgarage f. ‖ garage m. d'automobiles. / ~ stuffing material ‖ Kraftwagenpolsterwerkstoff m. ‖ matériaux mpl. de rembourrage pour automobiles. / tipping device for ~s ‖ Kippvorrichtung f. für Kraftwagen ‖ dispositif m. basculeur pour automobiles. / ~ tyre ‖ Kraftwagenreifen m.; Automobilreifen m.; Autoreifen m. ‖ bandage m. (en caoutchouc) d'automobile. / ~ tyres pl. ‖ Kraftwagenbereifung f.; Autobereifung f. ‖ bandages mpl. (en caoutchouc) ou pneumatiques mpl. pour automobiles. / vulcanisator for ~ tyres ‖ Vulkanisator m. für Autoreifen ‖ vulcanisateur m. pour bandages d'automobiles. / ~ trunk ‖ Automobilkoffer m. ‖ coffre m. d'automobile. / ~ wheel ‖ Kraftwagenrad m. ‖ roue f. pour automobile.

motor caterpillar gun mount ‖ Raupenkraftlafette f. ‖ affût m. automoteur à chenille.

motor coach (Railw) ‖ Triebwagen m. ‖ voiture f. motrice. / ~ of the suburban railway ‖ Triebwagen m. der Vorortbahn ‖ voiture f. motrice du chemin de fer de banlieue. / ~ train of the circular railway ‖ Triebwagenzug m. der Ringbahn ‖ train m. à voitures motrices du chemin de fer de ceintures.

motor converter ‖ Umformeraggregat n. ‖ moteur-générateur m. / ~ crane truck ‖ Motorkranwagen m. ‖ grue f. automobile. / ~ cultivator with rotary knives ‖ Bodenfräse f. ‖ motoculteur m. à fraises rotatives.

motor cycle ‖ Kraftrad m.; Motorrad n.; motocyclette f.; motocycle m. / ~ with back-wheel drive ‖ Motorrad n. mit Hinterradantrieb ‖ motocyclette f. à transmission sur l'arrière. / delivery ~ ‖ Liefermotorrad n.; Lieferkraftrad n. ‖

motocyclette f. de livraison. / ~ with front-wheel drive ‖ Motorrad n. mit Vorderradantrieb ‖ motocyclette f. à transmission sur l'avant. / ~ with side-car ‖ Motorrad n. mit Beiwagen ‖ motocyclette f. avec side-car ou avec voiture latérale.

motor cycle belt ‖ Motorradriemen m. ‖ courroie f. de motocyclettes. / ~ dynamo ‖ Motorraddynamo f. ‖ dynamo f. de motocyclette.

motor cyclist ‖ Kraftradfahrer m.; Motorradfahrer m. ‖ motocyclist m.

motor drive ‖ Motorantrieb m. ‖ commande f. par moteur. / individual ~ ‖ Motoreinzelantrieb m. ‖ commande f. individuelle par moteur.

motor-driven ‖ mit Motorantrieb ‖ actionné par moteur.

motor dust cart ‖ Müllkraftwagen m. ‖ chariot m. automobile pour le transport des ordures.

motor engine ‖ Kraftmaschine f. ‖ machine f. motrice. / ~ construction ‖ Kraftmaschinenbau m. ‖ construction f. de machines motrices.

motor extension ladder ‖ Motordrehleiter f. ‖ échelle f. automobile mécanique. / ~ fire-brigade vehicle ‖ Feuerwehrmotorfahrzeug n. ‖ pompe f. à incendie automobile. / ~ fire engine ‖ Motorfeuerspritze f.; Motorspritze f. ‖ pompe f. à incendie à moteur; pompe f. automobile. / ~ fish cutter ‖ Motorfischkutter m. ‖ cutter m. de pêche à moteur. / ~ fishing smack ‖ Motorfischerboot n. ‖ bâteau m. de pêche automobile. / ~ flight ‖ Motorflug m. ‖ vol m. avec moteur. ~ force see ~ power. / ~ frame ‖ Motorgehäuse n. ‖ bâti m. de moteur. / ~ freight car see ~ lorry. / ~ freighter ‖ Motorfrachtschiff n. ‖ cargo m. à moteur. / main engine of a ~ freighter ‖ Hauptmaschine f. eines Motorfrachtschiffes ‖ moteur m. de propulsion d'un cargo. / ~ gas ‖ Kraftgas n. ‖ gaz m. pauvre. / ~ generator ‖ Motorgenerator m.; Motorumformer m. ‖ moteur-générateur m. ~ generator set ‖ Motordynamo f.; Motorgeneratoraggregat n.; Motorgenerator m. ‖ groupe m. motogénérateur; station f. à moteur générateur. / ~ goggles pl. ‖ Kraftfahrerschutzbrille f. ‖ lunettes fpl. (protectrices) d'automobilistes. / ~ grass mower ‖ Motorgrasmäher m. ‖ faucheuse f. à moteur. / ~ hoist ‖ Motorwinde f. ‖ treuil m. à moteur. / ~ hood (Auto) ‖ Motorhaube f. ‖ capot m. de moteur. / ~ hood heater ‖ Kühlerwärmer m. ‖ chauffe-radiateur m. à mettre en dessous du capot.

motorization ‖ Verkraftung f.; Motorisierung f. ‖ motorisation f. / army mechanization and ~ ‖ Heeresverkraftung f. ‖ mécanisation f. et motorisation f. de l'armée.

motorized battery (War mat) ‖ motorisierte Batterie f. ‖ batterie f. motorisée.

motor launch ‖ Motorbarkasse f.; Motorboot n. ‖ chaland m. automobile; canot m. automobile. / ~ locomobile ‖ Motorlokomobile f. ‖ locomobile f. motrice. / ~ locomotive ‖ Motorlokomotive f. ‖ locomotive f. à moteur. / ~ locomotive for light fuel ‖ Motorlokomotive f. für Leichtbrennstoffe ‖ locomotive f. à moteur pour combustibles légers.

motor lorry (Electr) ‖ Kraftkarren m. ‖ chariot m. à moteur. / ~ (Auto) ‖ Lastkraftwagen m.; Lastauto n. ‖ camion m. automobile. / special ~ for transporting long cut timber and long sections ‖ Sonderkraftwagen m. zur Beförderung von Langhölzern und langen Formeisen ‖ camion m. spécial pour le transport de long bois et de longs profilés.

motor lorry train ‖ Kraftwagenlastzug m. ‖ convoi m. de camions-automobiles. / ~ wheel ‖ Lastkraftwagenrad n. ‖ roue f. de camion-automobile.

motor maintenance ‖ Instandhaltung f. des Motors ‖ entretien m. du moteur. / ~ man (Electr railway) ‖ Wagenführer m. ‖ conducteur m. / ~ noise ‖ Motorenlärm m. ‖ bruit m. de moteur. / ~ number ‖ Motornummer f. ‖ numéro m. du moteur. / ~ oil ‖ Motor(en)öl n.; Treiböl n. ‖ huile f. à moteurs. / light ~ oil ‖ leichtes Treiböl n. ‖ huile f. légère pour moteurs. / ~ omnibus see ~ bus. / ~ plant ‖ Motoranlage f.; Kraftmaschinenanlage f. ‖ installation f. motrice; établissement m. à force motrice. / ~ plough ‖ Motorpflug m. ‖ charrue f. à moteur. / ~ power ‖ treibende oder motorische Kraft f.; force f. motrice. / ~ protection switch ‖ Motorschutzschalter m. ‖ interrupteur m. de protection pour moteurs. / ~ sailing vessel ‖ Motorsegler m.; Segelschiff n. mit Hilfsmotor ‖ voilier m. à moteur auxiliaire. / ~ return carriage (Typewr) ‖ Wagen m. mit motorischem Umkehrantrieb ‖ chariot m. à retour automatique. / ~ sawing device ‖ Motorkraftsäge f. ‖ scie f. à moteur. / ~ scooter ‖ Motorroller m. ‖ moto-scooter m. / ~ set ‖ Motorensatz m. ‖ groupe m. des moteurs. / ~ shaft ‖ Motorwelle f. ‖ arbre m. (de) moteur. / ~ ship ‖ Motorschiff n. ‖ bateau m. à moteur. / ~ sled (Motordriven) ‖ Motorschlitten m. ‖ traîneau m. automobile. / ~ sled (Motor conveying) ‖ Motorschleife f. ‖ traîneau m. porte-moteur. / ~ sleigh see ~ sled. / ~ spirit ‖ Benzin n. ‖ essence f. / ~ sprinkler ‖ Motorsprengwagen m. ‖ arroseuse f. automobile; automobile f. d'arrosage. / ~ sprinkler with brush and washing roll ‖ Motorsprengmaschine f. mit Kehr- und Waschwalze ‖ automobile f. d'arrosage avec cylindre à brosse et cylindre-laveur. / ~ starter ‖ Motoranlasser m. ‖ démarreur m. à moteur. / ~ street brush see ~ street cleanser. / ~ street cleanser ‖ selbstfahrende Straßenkehrmaschine f.; Motorstraßenkehrmaschine f. ‖ autobalayeuse f.; balayeuse f. automobile. / ~ stretcher ‖ Motortrage f. ‖ brancard m. du moteur. / ~ supply meter ‖ Verbrauchszähler m. ‖ compteur m. de consommation. / ~ sweeper see ~ street cleanser. / ~ syren ‖ Motorsirene f. ‖ sirène f. à moteur. / ~ syringe ‖ Motorspritze f. ‖ pompe f. à moteur. / ~ tank truck ‖ Kesselkraftwagen m.; Behälterkraftwagen m.; Tankkraftwagen m. ‖ tonneau m. d'eau automobile; camion m. automobile à citerne. / ~ tank truck of x cubic meters capacity ‖ Kesselkraftwagen m. für x cbm Inhalt ‖ camion m. automobile à citerne d'une capacité de x m³. / ~ tender ‖ Motortender; Motorbarkasse f. ‖ chaland m. automobile. / ~ terminal board (Electr) ‖ Motorklemmbrett n. ‖ planchette f. à

bornes pour moteurs. / ~ thrasher see ~ threshing machine. / ~ threshing machine || Motordreschmaschine f. || moto-batteuse f.; batteuse f. à moteur. / ~ tillage cutter ||. Ackerfräser m. mit Motorantrieb; Motorackerfräser m. || fraiseuse f. de labour à moteur. / ~ touring trunk || Automobilkoffer m.; Automobilreisekoffer m. || coffre m. d'automobile. / ~ traction || Kraftzug m. || traction f. mécanique. / ~ tractor || Motorzugmaschine f.; Kraftschlepper m.; Krafttrecker m.; Motortrecker m. || tracteur m. à moteur. / treatment of the ~ || Wartung f. des Motors || soins mpl. à donner au moteur. / ~ tricycle for luggage || Kraftdreirad n. für Gepäckbeförderung || tricar m.; tricycle m. de bagages. / ~ truck (Auto) see ~ lorry. / ~ truck (Electr) || Motorkarren m. || chariot m. porte-moteur. / ~ tug || Motorschlepper m.; Motorschleppschiff n. || remorqueur m. automobile. / ~ type || Motortype f. || type m. du moteur. / ~ vehicle || Kraftfahrzeug n.; Motorfahrzeug n. || véhicule m. à moteur; automobile f. / ~ water car see ~ sprinkler. / ~ winch || Motorwinde f. || treuil m. à moteur. / ~ works pl. || Motorenfabrik f. || ateliers mpl. de construction de moteurs. / ~ yacht || Motorjacht f. || yacht m. automobile.

mottled || geadert; marmoriert || madré; marbré. / ~ (Met) || halbiert || truité. / ~ iron || halbiertes Eisen n. || fonte f. truitée. / ~ mahogany || moiriertes Mahagoniholz n. || acajou m. moiré.

mottling || Marmorierung f. || madrure f.; marbrure f.; madrage m.

mould, to || formen; abformen; modellieren || mouler; modeler; façonner; former. / ~ (To get mouldy) || verschimmeln; schimmeln; schimmelig werden || (se) moisir; (se) chancir; se couvrir de moisissure. / ~ (Shipb) || abmallen, mallen || gabarier. / ~ the brick-clay || Ton m. streichen || mouler l'argile. / ~ bricks || Ziegel mpl. streichen || mouler des briques fpl. / ~ from a casting || über das Modell formen || surmouler. / ~ in wax || in Wachs modellieren || modeler en cire.

mould (Botany) || Schimmelpilz m.; Schimmel m. || moisissure f. / ~ (For casting) || Form f.; Gießform f. || moule m. / ~ (Forg) || Gesenk n.; Nonne f. || estampe f.; étampe f.; matrice f. / ~ (Candle) || gegossenes Licht n. || chandelle f. moulée. / ~ (Garden) || Gartenerde f. || terre f. de jardin. / ~ (Pap) || Papierform f. || forme f.; moule m. / ~ (Needl) || Knopfspindel f. || fuseau m. / ~ (Shipb) || Mall n. || gabarit m. / to black the ~ (Found) || die Form f. mit Kohlenstaub bestreuen || saupoudrer le moule. / covered with ~ || schimmelig || chanci; couvert de moisissure; moisi. / to face the ~ see to black the ~. / ~ baking || Backform f. || forme f. pour boulangerie. / ~ for bottles and hollow glass || Form f. für Flaschen und Hohlglaswaren || moule m. pour flaconneries et gobeleteries. / cast iron ~ || Kokille f.; eiserne Gußschale f. || coquille f.; moule m. en fonte. / chill ~ see cast iron ~. / ~ without core || Stürzform f. || moule m. à renverser. / dead ~ || verlorene Gießform f. || moule m. perdu. / dry-sand ~ || Massenform f. || moule m.

en sable gras. / exterior ~ || Mantel m. (einer Lehmform) || chape f.; manteau m.; surmoule m. / finishing ~ (Gold-b) || zweite Häuteform f.; Dünnschlag m. || second ou dernier chaudret m. / first and second ~ (Gold-b) || Pergamentform f.; Quetschform f. || caucher m. / ~ for flower pots || Form f. für Blumentöpfe || moule m. pour pots à fleurs. / folding ~ || Klappform f. || moule m. à bascule. / ~ of gut (Gold-b) || Häuteform f. || chaudret m. / ~ for insulators || Preßform f. für Isolatoren || matrice f. d'isolateurs. / iron ~ || eiserne Form f. || moule m. en fer. / loam ~ || Lehmform f. || moule m. en argile. / ~ for pastry || Konditorform f. || moule m. pour confiserie. / ~ for making pipes (Tinm) || Röhrenform f.; Rohrwalze f. || tondin m. / ~ for plate glass || Form f. für Spiegelglas || moule m. à glaces. / ~ for rubber || Gummiform f. || moule m. pour caoutchouc. / splitting ~ (Fission fungus) || Spaltpilz m.; Bakterium n. || schizomycète m.; champignon m. dédoublant. / stamping ~ || Prägeform f. || forme f. à estamper. / tin plate ~ || Blechform f. || moule m. en fer-blanc. / vegetable ~ || Pflanzenerde f. || terreau m. / ~ of vellum (Gold-b) || Pergamentform f.; Quetschform f. || caucher m.

mould board of a plough || Riester m. oder Röster m. oder Streichbrett n. eines Pfluges || oreille f. ou versoir m. d'une charrue. / hollow-shouldered ~ || geschweiftes Streichbrett n. || versoir m. à épaules creuses. / square-shouldered ~ || gerades Streichbrett n. || versoir m. à épaules carrées.

mould candle || gegossenes Licht n. || chandelle f. moulée ou au moule. / ~ casting || Formguß m. || moulage m. en châssis. / ~ cistern (Sugar) || Formtrog m.; Formback n. || bac m. à formes. / ~ core || Formkern m.; Kern m. || marron m.; noyau m. du moule. / ~ drying stove || Trockenofen m. || étuve f.

moulded depth (Shipb) || Seitenhöhe f. || creux m. sur quille. / ~ piece || Formstück n. || moulage m. en acier.

moulder, to (Wood) || verfaulen || vermouler.

moulder || Former m. || mouleur m. / ~ (Print) || Abformer m. || mouleur m. de clichés. / floor ~ || Dammgrubenformer m. || mouleur m. en fosse. / machine-~ || Maschinenformer m. || mouleur m. à la machine. / plate ~ || Plattenformer m. || mouleur m. de plaques. / wood ~ || Leistenmacher m. || moulurier m.

moulder's blacking || Gießereischwärze f. || noir m. pour fonderies. / ~ punch || Formenmesser m.; Formermesser n. || poinçon m. / ~ tool || Formerwerkzeug n. || outil m. de mouleur.

mouldering (Mine) || gebrech; bröcklig || cassant, fragile.

mouldery || Formerei f.; Formen n. || moulage m.

mould frame (Pap) || Formrahmen m. || affût m. ou fût m. de la forme. / ~ (Ice machine) || Eiszellenrahmen m. || cadre m. de mouleaux.

mould holder (Glassm) || Glasformenschieber m. || teneur m. de moules.

mould-hoop of a loam mould || Mantelring m. einer Lehmform || frette f. d'un moule en argile.

moulding (For casting) || Formerei f.; Formen n.; Gußformerei f. || moulage m. / ~ in boxes || Kastenformerei f. || moulage m. en châssis ou avec châssis. / ~ with clay-sheets (Porcel) || Schwartenformerei f. || moulage m. à la croûte. / dry-sand ~ || Massenformerei f. || moulage m. en sable gras. / hand ~ || Handformerei f. || façonnage m. à la main. / mechanical ~ || Maschinenformerei f. || moulage m. à la machine. / in metal moulds || Formen n. oder Formerei f. in Schalen || moulage m. en coquilles. / open sand ~ || Herdformerei f. || moulage m. à découvert. / ~ with sheets (Pott) || Schwartenformerei f. || moulage m. à la croûte. / steel ~ || Stahlgußformerei f. || moulage m. d'acier.

moulding (Join) || Zierleiste f.; Kehlung f.; Nutleiste f. || moulure f.; planche f. à rainure. / concave ~ || Hohlkehle f.; Hohlleiste f. || gorge f.; cavet f.; membre m. creux. / ~ for frames || Rahmenleiste f. || baguette f. d'encadrement; tringle f. pour cadres. / furniture ~ || Möbelleiste f. || moulure f. pour meubles. / gilt ~ || Goldleiste f. || baguette f. dorée. / polished ~ || Politurleiste f. || tringle f. polie.

moulding bench || Formbank f. || banc m. de moulage. / ~ board || Modellplatte f.; Formbrett n.; Modellbrett n. || planche f. de fond. / ~ box || Formkasten m.; Gießkasten m.; Gußkasten m.; Gießlade f. || châssis m. (de moulage). / ~ clay || Formlehm m. || argile f. à mouler. / ~ cutter || Kehlmesser n. || fraise à moulurer. / ~ edge (Shipb) || Mallkante f. || gabariage m. / ~ hall (Found) || Formhalle f. || hall m. de moulage. / ~ house || Formerei f. || moulerie f.; atelier m. de moulage. / ~ loam || Formlehm m. || argile f. à mouler.

moulding machine (Found) || Formmaschine f. || machine f. à mouler. / ~ (Join) || Kehlmaschine f.; Kehlhobelmaschine f.; Gesimsmaschine f. || raineuse f.; machine f. à moulurer ou à moulures. / ~ (Rubber) || Spritzmaschine f. || machine f. d'aspersion. / dough ~ || Teigwirkmaschine f. || machine f. à pétrir la pâte; rouleur m. de pâte. / ~ for foundries || Gießereiformmaschine f. || machine f. à mouler. / gear wheel ~ || Zahnräderformmaschine f. || machine f. à former les moules pour roues dentées.

moulding material || Formerstoff m.; Formsand m. || sable m. à mouler. / preparing machine for ~ || Aufbereitungsmaschine f. für Formsand || machine f. à préparer le sable de moulage.

moulding pin || Formerstift m. || pointe f. de mouleur.

moulding plane (Join) || Leistenhobel m. || bouvet m. à doucine; tarabiscot m.

moulding press || Formpresse f. || presse f. à mouler. / hydraulic ~ (Found) || hydraulische Formpresse f. || presse f. hydraulique à mouler.

moulding rammer || Stampfer m. || damoir m. / ~ room || Gießraum m.; Formerei f. || moulerie f.; atelier m. de moulage.

moulding sand || Formsand m.; Gießereisand m.; Modellsand m. || sable m. de moulage ou de fonderie ou des fondeurs. / to prepare the ~ || den Formsand m. zurichten || préparer le sable de moulage.

/ ~ for iron ‖ Gießereiformsand m. ‖ sable m. de moulage pour la fonte.

moulding sand dressing plant ‖ Formsand-aufbereitungsanlage f. ‖ installation f. de préparation de sable de moulage. / ~ tester ‖ Formsandprüfer m. ‖ appareil m. à essayer le sable de moulage.

moulding side (Shipb) ‖ Mallebene f. ‖ plan m. de gabariage *ou* du couple. / ~ stake *see* moulding pin.

mould loft floor (Shipb) ‖ Mallboden m.; Schnürboden m. ‖ salle f. des gabarits *ou* à tracer. / ~ maker (Found) *see* moulder. / ~ maker (Pott) ‖ Former ‖ ouvrier-mouleur m. / ~ press ‖ Form-presse f. ‖ presse f. à mouler. / ~ sander ‖ Sandstreuer m. ‖ sableur m. de moules.

mouldy ‖ schimm(e)lig ‖ moisi; chanci; couvert de moisissure. / ~ (Smelling fusty) ‖ muffig ‖ qui sent le moisi. / to get ~ ‖ verschimmeln; schimmeln ‖ (se) moisir; (se) chancir; se couvrir de moisissure.

mound (of earth) ‖ Wall m.; Erdwall m. ‖ remblai m. / ~ (of slag) ‖ Halde f.; Schlackenhalde f. ‖ crassier m. / ~ of boulder-stones (Geol) ‖ Schuttkegel m.; Schutthalde f. ‖ cône m. de déjection.

mound area (Met) ‖ Röststadel m.; Rost-stadel m. ‖ aire f. murée *ou* de grillage.

mount, to ~ (To ascend) ‖ steigen; auf-steigen; hinaufsteigen ‖ monter. / ~ (To rise) ‖ aufsteigen; sich erheben; steigen; in die Höhe f. steigen ‖ mon-ter; s'élever. / ~ (To fit up) ‖ aufstellen; montieren ‖ monter. / ~ (To rise in amount) ‖ sich belaufen auf...; be-tragen ‖ se monter à...; faire; s'élever à... / ~ a block ‖ ein Klischee n. *oder* ein Galvano n. *oder* einen Abklatsch m. aufklotzen ‖ clouer un cliché.

mount ‖ Fassung f. ‖ monture f. / ~ of a block (Print) ‖ Klischeefuß m. ‖ pied m. d'un cliché. / folding ~ ‖ Einschlag-fassung f. ‖ monture f. pliante.

mountain ‖ Berg m. ‖ mont m.; montagne f. / ~ ash ‖ Eberesche f. ‖ sorbier m. / ~ artillery ‖ Gebirgsartillerie f. ‖ ar-tillerie f. de montagne. / ~ blue ‖ Berg-blau n. ‖ cendre f. bleue. / ~ breeze ‖ Bergwind m. ‖ brise f. *ou* vent m. de montagne. / ~ cloud ‖ Bergwolke f. ‖ nuage m. montagneux. / ~ cork ‖ Berg-kork m. ‖ amiante m. / ~ green ‖ Berg-grün n. ‖ vert m. de montagne; chryso-cole m. / ~ locomotive ‖ Gebirgslokomo-tive f. ‖ locomotive f. de rampes. / pres-sure of ~ mass ‖ Gebirgsdruck m. ‖ pres-sion f. des terrains. / ~ oil ‖ Bergöl n. ‖ huile f. de montagne.

mountainous ‖ gebirgig ‖ montagneux.

mountain railway ‖ Gebirgsbahn f. ‖ chemin m. de fer alpin; chemin m. de fer de montagne.

mountains pl. ‖ Gebirge n. ‖ (chaîne f. de) montagnes fpl. / ~ produced by erosion ‖ Abtragungsgebirge n. ‖ roches fpl. de destruction. / height of the ~ ‖ Gebirgs-höhe f. ‖ hauteur f. des montagnes.

mountain soap ‖ Bergseife f. ‖ savon m. de montagne. / ~ sun ‖ Höhensonne f.; Bergsonne f. ‖ soleil m. d'altitude *ou* de montagne. / artificial ~ sun ‖ künst-liche Höhensonne f. *oder* Bergsonne f. ‖ soleil m. de montagne artificiel. / ~ wax (Chem) ‖ Montanwachs n. ‖ cire f. de lignite. / ~ wind ‖ Bergwind m. ‖ brise

f. *ou* vent m. de montagne. / ~ yellow ‖ Berggelb n.; gelber Ocker m. ‖ jaune m. de montagne; limonite f.

mounted ‖ gefaßt; eingefaßt ‖ monté; serti. / ~ lens ‖ gefaßte Linse f. ‖ lentille f. sertie. / ~ on... ‖ gelagert auf...; *oder* in... ‖ supporté par...; appuyé sur...; logé dans...; monté sur... / ~ on wooden spools ‖ auf Holzspulen fpl. gewickelt ‖ enroulé sur des bobines fpl. de bois.

mounting (Fitting-up) ‖ Aufstellung f.; Montierung f.; Montage f. ‖ montage m. / ~ (Rising) ‖ Aufsteigen n.; Aufstieg m. ‖ montée f.; élévation f.; ascension f. / ~ (Of spectacles etc. *or* of precious stones) ‖ Fassung f.; Einfassung f. ‖ monture f. / ~ (Weav) ‖ Zeug n.; Geschirr n.; Webgeschirr n. ‖ équipage m.; harnais m.; lisses fpl.; remisse m. / ~ of fusee chains ‖ Schneckenkettenzusammennie-tung f. ‖ montage m. de chaînes à fusées. / metal ~ ‖ Metallfassung f. ‖ monture f. métallique. / ~ in nested cells ‖ Füllfassung f. ‖ monture f. fourrée. / nickel-plated ~ ‖ vernickelte Fassung f. ‖ monture f. nickelée. / ~ of wheels (Railw) ‖ Auflaufen n. des Rades ‖ montée f. de la roue sur les rails.

mounting apparatus for blocks (Print) ‖ Klischeenagelvorrichtung f.; Vorrich-tung f. zum Aufklotzen von Klischees ‖ appareil m. à clouer les clichés. / ~ bracket ‖ Befestigungsschelle f. ‖ collier m. de fixation. / ~ hoop (Coop) ‖ Auf-setzreif m. ‖ bâtissoir m. / ~ plate ‖ Be-festigungsplatte f. ‖ plaque f. de fixa-tion.

mountings pl. ‖ Beschläge mpl. ‖ arma-tures fpl.; garnitures fpl.; monture f. / boiler ~ ‖ Kesselarmaturen fpl. ‖ garni-tures fpl. de chaudières. /. ~ for car-riages ‖ Wagenbeschläge fpl. ‖ ferrures fpl. pour carrosseries. / ~ of casement ‖ Fensterbeschläge mpl. ‖ armatures fpl. de vitrage; ferrures fpl. de croisée. / ~ of a door ‖ Türbeschläge mpl. ‖ ferrures fpl. de portes. / engine ~ ‖ Armaturen pl. ‖ garnitures fpl. pour machines et chaudières. / iron ~ ‖ Eisenbeschläge mpl. ‖ ferrures fpl.

mounting shop ‖ Montagewerkstatt f.; Montagehalle f. ‖ atelier m. de montage *ou* d'ajustage.

mounts pl. *see* mountings pl.

mourning band ‖ Trauerbinde f. ‖ crêpe m. de deuil. / ~ bead ‖ Trauerperle f. ‖ perle f. de deuil. / ~ crêpe ‖ Trauerflor m. ‖ crêpe m. de deuil. / ~ dress ‖ Trauer-kleid n.; Trauerkleidung f. ‖ habit m. de deuil. / ~ hat ‖ Trauerhut m. ‖ chapeau m. de deuil. / ~ jewelry ‖ Schwarz-schmuckwaren fpl. ‖ bijouterie f. de deuil. / ~ jewellery ‖ Trauergalanterie-waren fpl. ‖ bijouterie f. de deuil. / ~ paper ‖ Trauerpapier n. ‖ papier m. de deuil. / ~ paper articles pl. ‖ Trauer-papierwaren fpl. ‖ papeteries fpl. pour deuil. / ~ pin ‖ Trauernadel f. ‖ épingle f. de deuil. / ~ warehouse ‖ Trauer-magazin n. ‖ maison f. de deuil.

mouse destroyer ‖ Mäusevertilgungsmittel m. ‖ produit-destructeur m. de souris. / ~ trap ‖ Mausefalle f. ‖ souricière f.

mousseline de laine ‖ Wollmusselin m. ‖ mousseline f. de laine.

mousse rubber ‖ Kautschukschaum m. ‖ cautschouc-mousse m.

moustache pincers pl. ‖ Bartklemme f. ‖ pince f. à moustaches. / ~ pomade ‖ Bartwichse f. ‖ pommade f. à moustache. / ~ trainer ‖ Bartbinde f. ‖ relève-moustache m.

mouth ‖ Mund m. ‖ bouche f. / ~ of a canal ‖ Kanalmündung f. ‖ embouchure f. d'un canal. / ~ of a blast furnace ‖ Gicht-öffnung f. eines Hochofens ‖ bure f. d'un haut fourneau. / ~ of the jaws ‖ Klemm-backenmaul n. ‖ bouche f. de mordache. / the ~ of the jaws is rolled at an angle corresponding to the incline of the rail foot ‖ das Klemmbackenmaul ist der Neigung des Schienenfußes entsprechend schräg ausgewalzt ‖ la bouche de mor-dache est laminée obliquement con-formément à l'inclinaison du patin. / ~ of a pit (Mine) ‖ Mundloch n. des Stol-lens ‖ ouverture f. *ou* embouchure f. d'une mine. / ~ of a plane (Join) ‖ Keil-loch n. eines Hobels ‖ lumière f. d'un rabot. / ~ of a river ‖ Flußmündung f. ‖ embouchure f. d'une rivière. / ~ of a tunnel ‖ Tunnelmundloch n. ‖ entrée f. d'un tunnel. / ~ of the twyer ‖ Öff-nung f. im Rüssel eines Hochofen-gebläses ‖ bouche f. *ou* œil m. de la tuyère. / ~ of a vessel ‖ Öffnung f. *oder* Mündung f. eines Gefäßes ‖ ouverture f. *ou* orifice m. *ou* entrée f. d'un vase. / ~ of a wind instrument ‖ Mundstück n. eines Blasinstrumentes ‖ embouchure f. *ou* embouchoir m. d'un instrument à vent.

mouth-and-foot disease ‖ Maul- und Klauenseuche f. ‖ piétin m. et muguet m.

mouth glue ‖ Mundleim m. ‖ colle f. à bouche. / ~ mirror ‖ Mundspiegel m. ‖ miroir m. buccal. / ~ organ ‖ Mund-harmonika f. ‖ guimbarde f.; harmonica f. à bouche.

mouthpiece ‖ Mundstück n. ‖ embouchure f.; embouchoir m.; bout m. / ~ (Tel) ‖ Schalltrichter m. ‖ cornet m.; porte-voix m.; pavillon m. / clarionet ~ ‖ Klarinettenmundstück n. ‖ bec m. de clarinette. / ~ of a pipe ‖ Pfeifenmund-stück n. ‖ bout m. de pipe. / ~ of the speaking tube (Dictating machine) ‖ Mündstück n. des Sprechschlauchs ‖ embouchure f. du tube acoustique.

mouth screen (Blast furnace) ‖ Gicht-mantel m.; Gichtmauer f. ‖ mur m. de bataille; batailles fpl. / ~ sponge ‖ Mundschwamm m. ‖ éponge f. pour la bouche. / ~ wash ‖ Mundwasser n. ‖ collu-toire m.; eau f. pour rincer la bouche; eau f. dentrifice.

movability ‖ Beweglichkeit f. ‖ mobilité f.

movable ‖ beweglich; lose; verstellbar; verschiebbar ‖ mobile; coulant; avan-çable. / vertically ~ ‖ der Höhe f. nach verstellbar ‖ réglable en hauteur f. / ~ in a very wide range ‖ in sehr weiten Grenzen fpl. verstellbar ‖ déplaçable *ou* mobile dans de très larges limites fpl.

movable beam (Railw) ‖ Wiegeträger m.; Wiegebalken m. ‖ traverse f. mobile. / pressed ~ beam ‖ gepreßter Wiegebalken m. ‖ traverse f. mobile emboutie. / ~ buffer (Railw) ‖ beweglicher Prellbock m. ‖ heurtoir m. mobile. / ~ crane ‖ Fahr-kran m. ‖ grue f. sur rails. / ~ disk discharger ‖ Abreißfunkenstrecke f. ‖ éclateur m. à électrodes tournantes. / ~ driving ‖ verstellbarer Antrieb m. ‖ commande f. réglable. / ~ erecting system

of lenses ‖ verstellbares bildumkehrendes Linsensystem n. ‖ système m. de lentilles redresseur réglable *ou* mobile. / ~ grate ‖ beweglicher Rost m. ‖ grille f. mobile. / ~ load ‖ bewegliche Belastung f. ‖ charge f. mobile. / ~ wheel set ‖ Einstellradsatz m. ‖ essieu m. monté mobile.

move, to ‖ bewegen ‖ mouvoir. / ~ forward and backward ‖ vor- und zurückbewegen ‖ mouvoir en avant et en retour. / ~ to and fro ‖ hin und her bewegen ‖ déplacer dans un sens et dans l'autre. / the pieces must be moved to and fro in the water ‖ die Stücke npl. müssen im Wasser hin und her bewegt werden ‖ il faut bien agiter les pièces fpl. dans l'eau.

move ‖ Bewegung f. ‖ mouvement m. / room to ~ ‖ Bewegungsfreiheit f. ‖ place f. à mouvements.

moveable *see* movable.

moved air ‖ bewegte Luft f. ‖ air m. agité *ou* en mouvement.

movement ‖ Bewegung f. ‖ mouvement m. / to be forced to follow a ~ ‖ eine Bewegung f. zwangläufig mitmachen ‖ accomplir obligatoirement un mouvement. / ~ in blank (Watchm) ‖ Rohuhrwerk n. ‖ mouvement m. en blanc. / ~ of boulders ‖ Geschiebeführung f. ‖ transport m. de débris minéraux. / ~ of the centre of pressure ‖ Druckpunktwanderung f. ‖ déplacement m. du centre de pression. / circular ~ ‖ Kreisbewegung f. ‖ mouvement m. circulaire. / ~ in a curved line ‖ krummlinige Bewegung f. ‖ mouvement m. curviligne. / ~ of divining rod ‖ Wünschelrutenausschlag m. ‖ mouvement m. *ou* réaction f. de la baguette divinatoire. / ~ free from play ‖ spielfreier Gang m. ‖ marche f. sans jeu. / reciprocating ~ ‖ Hin- und Herbewegung f. ‖ mouvement m. de va-et-vient. / ~ of the ribbon (Typewr) ‖ Farbbandtransport m. ‖ avance f. du ruban encré. / self-acting ~ ‖ selbsttätige Bewegung f. ‖ mouvement m. automatique. / shaking ~ ‖ Rüttelbewegung f.; Schüttelbewegung f. ‖ mouvement m. oscillant. / ~ in a straight line ‖ geradlinige Bewegung f. ‖ mouvement m rectiligne. / turbulent ~ ‖ Wirbelbewegung f. ‖ mouvement m. turbulent.

movement control mechanism ‖ Steuermechanismus m.; Steuervorrichtung f. ‖ mécanisme m. de manœuvre. / freedom of ~ ‖ Freizügigkeit f. ‖ liberté f. de résidence. / periodicity of ~ ‖ Bewegung f. in gleichen Zeiträumen ‖ periodicité f. des mouvements. / return of ~ ‖ Bewegungsumkehr f. ‖ renversement m. du mouvement.

mover ‖ Motor m. ‖ moteur m. / prime ~ (Mech) ‖ Hauptkraft f.; Urkraft f.; Hauptantrieb m. ‖ force f. primordiale *ou* primitive; grande force f. / prime ~ (Mach) ‖ Antriebsmaschine f. ‖ machine f. motrice.

moving *see* movement. / ~ apparatus *see* ~ device. / ~ car ‖ Möbelwagen m. ‖ voiture f. de déménagement; tapissière f.

moving coil (Electr) ‖ Drehspule f.; bewegliche Spule f. ‖ bobine f. (à cadre) mobile. / ~ ammeter ‖ Drehspulamperemesser m.; Drehspulstrommesser m.; d'Arsonvalscher Amperemesser m.; ammètre m. à bobine mobile. / ~ galvanometer ‖ Drehspulgalvanomesser m. ‖ galvanomètre m. à cadre mobile.

moving device ‖ Verschiebevorrichtung f. ‖ dispositif m. de déplacement. / hydraulic ~ ‖ hydraulische Verschiebevorrichtung f. ‖ dispositif m. hydraulique de déplacement. / tilting and ~ for rolling mills ‖ Kant- und Verschiebevorrichtung f. für Walzwerke ‖ culbuteur m. et dispositif m. de déplacement pour laminoirs.

moving force ‖ Treibkraft f.; Triebkraft f. ‖ force f. impulsive *ou* motrice.

moving gear ‖ Fahrvorrichtung f. ‖ mécanisme m. d'avancement.

moving iron ammeter ‖ Dreheisenamperemesser m.; Dreheisenstromzeiger m. ‖ ampèremètre m. à fer doux. / ~ voltmeter ‖ Dreheisenvoltmesser m.; Dreheisenspannungsanzeiger m. ‖ voltmètre m. à fer doux.

moving magnet galvanometer ‖ Nadelgalvanomesser m.; Drehmagnetgalvanomesser m. ‖ galvanomètre m. à aimant mobile *ou* à aiguille. / ~ part ‖ Triebwerkteil m. ‖ organe m. de commande. / ~ picture *see* motion picture. / ~ power ‖ bewegende *oder* treibende Kraft ‖ force f. motrice. / ~ spring ‖ Antriebsfeder f. ‖ ressort m. de commande. / ~ water (Mach) ‖ Aufschlagwasser n. ‖ eau f. motrice.

mow, to ‖ mähen; abmähen ‖ faucher.

mower (Machine) ‖ Mähmaschine f.; Erntemaschine f.; Grasmäher m. ‖ faucheuse f.; machine f. à faucher; moissonneuse f. / ~ (Person) ‖ Mäher m.; Schnitter m. ‖ faucheur m.; moissonneur m. / corn ~ ‖ Mähmaschine f. für Getreide; Getreidemähmaschine f. ‖ moissonneuse-javeleuse f.; moissonneuse f. / grass ~ ‖ Grasmäher m.; Grasmähmaschine f. ‖ faucheuse f. / ~ motor ‖ Motormähmaschine f. ‖ faucheuse f. à moteur.

mower driver ‖ Mähmaschinenführer m. ‖ moissonneur m. à la machine; conducteur m. de la moissonneuse *ou* de la faucheuse. / ~ knife ‖ Mähmaschinenmesser n. ‖ lame f. de faucheuse.

mowing (Grass after cutting) ‖ Mahd f. ‖ fauchaison f.; fauchage f.; foin m.

mowing machine *see also* mower (Machine) ‖ Mähmaschine f. ‖ faucheuse f. mécanique; moissonneuse f.

mt ‖ Metertonne f.; mt ‖ tonne-mètre f.; tm.

mucic acid ‖ Milchsäure f.; Schleimsäure f. ‖ acide m. mucique; acide saccholactique.

mucilage ‖ Pflanzenschleim m.; flüssiger Leim m. ‖ mucilage m.; colle f. liquide. / ~ for offices ‖ Büroleim m.; Klebstoff m. ‖ colle f. de bureau.

mucilage bottle with sponge top ‖ Leimflasche f. mit Schwammverschluß ‖ bouteille f. à colle à bouchon-éponge. / ~ holder ‖ Klebstoffbehälter m. ‖ récipient m. à colle.

mucilaginous ‖ schleimig ‖ mucilagineux.

muck (Dung in moist state) ‖ (feuchter) Mist m.; Dung m. ‖ engrais m.; fumier m. / ~ (Mine) ‖ erdige Kohle f.; taubes Kohl n. ‖ terre f. houille; crave f.

mucous ‖ schleimig ‖ mucilagineux. / ~ membrane ‖ Schleimhaut f. ‖ muqueuse f.

mud ‖ Schlamm m.; Schlick m.; bourbe f.; limon m.; vase f. / ~ (in the road) ‖ Straßenkot m.; Kehricht m. ‖ crotte f.; boue f. / ~ of the lead chamber ‖ Bleikammerschlamm m. ‖ boue f. de la chambre de plomb. / to stick in the ~ ‖

im Schlick m. festsitzen ‖ s'envaser; s'embourber.

mud box ‖ Schlammtopf m. ‖ réservoir m. à boue. / ~ coal ‖ Schlammkohle f. ‖ charbon m. limoneux. / ~ cock ‖ Schlammhahn m. ‖ robinet m. d'ébouage. / ~ collector ‖ Schlammsammler m. ‖ collecteur m. de boue. / ~ discharging device ‖ Schlammablaßvorrichtung f. ‖ appareil m. à évacuer les boues. / ~ drying ‖ Schlammtrocknung f. ‖ séchage m. des schlamms.

muddy ‖ schlammig ‖ boueux. / ~ sand ‖ Moddersand m.; Schlammsand m. ‖ sable m. vasard. / ~ water ‖ schlammiges *oder* schlammhaltiges Wasser n. ‖ eau f. boueuse.

mud exhauster ‖ Schlammsauger m. ‖ aspirateur m. de boue. / ~ for cleaning sewers ‖ Kanalschlammsauger m. ‖ aspirateur m. de boue pour le nettoyage des égouts.

mud guard (Coachm.; Auto) ‖ Kotblech n.; Kotflügel m.; Schutzblech n.; Schmutzfänger m. ‖ garde-boue m.; aile f. / ~ for cycles ‖ Schutzblech n. für Fahrräder; Fahrradschutzblech n. ‖ garde-boue m. pour bicyclettes. / wooden ~ ‖ Kotfänger m. aus Holz ‖ garde-boue m. en bois.

mud guard bracket ‖ Kotflügelstütze f. ‖ support m. de l'aile.

mudhole (Boiler) ‖ Schlammablaß m.; Schlammloch n. ‖ trou m. de décharge des boues; trou m. d'ébouage; orifice m. des impuretés. / ~ cover ‖ Schlammlochdeckel m. ‖ couvercle m. de l'orifice des impuretés *ou* de la décharge des boues.

mud lighter (Hydr arch) ‖ Baggerprahm m.; Modderprahm m. ‖ gabare f. à vase. / soft ~ ooze (Mar) ‖ Schlammgrund m.; Schlickgrund m.; Moddergrund m. ‖ fond m. de vase; vase f. molle. / ~ plug (Loc) ‖ Reinigungspfropfen m.; Reinigungsstöpsel m.; Schraub(en)pfropf m. ‖ bouchon m. à vis de nettoyage. / ~ pocket *see* mudhole. / ~ protector *see* mud guard. / ~ pump ‖ Schlammpumpe f.; Schmutzwasserpumpe f. ‖ pompe f. à boue *ou* à vase suceuse. / ~ shovel ‖ Schlammschaufel f. ‖ pelle f. à vase. / ~ volcano (Geol) ‖ Schlammvulkan m.; Salse f. ‖ salse f.

mud-wall, to ‖ mit Lehm m. und Stroh n. mauern; wellern ‖ torcher; bousiller.

mud wall ‖ Lehmstrohwand f.; Lehmwand f.; Wellerwand f. ‖ Mauerwerk n. in Strohlehm ‖ mur m. de bousillage; maçonnerie f. en torchis *ou* en bousillage.

mud walling *see also* mud wall. ‖ Lehmbau m.; Wellerwerk n. ‖ bousillage m.

muff (Cloth) ‖ Muff m. ‖ manchon m. / ~ (Techn) *see also* socket ‖ Muffe f. ‖ manchon m. / fixing ~ ‖ Verbindungsmuffe f. ‖ manchon m. d'assemblage.

muff coupling ‖ Muffenkupplung f. ‖ accouplement m. par manchon.

muffle (Chem; Met) ‖ Muffel f. ‖ moufle m. / to pass through the ~ (Enamel) ‖ einbrennen ‖ passer au feu.

muffle furnace ‖ Muffelofen m.; Probierofen m. ‖ fourneau m. à coupelle; four m. à moufles. / electric ~ ‖ elektrischer Muffelofen m. ‖ fourneau m. à moufles électrique. / gas-fired ~ ‖ gasgeheizter Muffelofen m. ‖ fourneau m. à moufle au gaz.

muffler (Auto) ‖ Auspufftopf m.; Schalldämpfer m. ‖ pot m. silencieux. / ~ explosion ‖ Knallen n. im Auspufftopf ‖ pétarade f. au pot d'échappement.

muff rolling mill ‖ Muffenwalzwerk n. ‖ laminoir m. à manchons.

mug ‖ Becher m.; Krug m. ‖ timbale f.

mugwort leaves pl. ‖ Beifußkraut n. ‖ feuilles fpl. d'armoise.

mule ‖ Maultier n. ‖ mulet m. / ~ (Spinn) ‖ Mulemaschine f.; Jennymaschine f. ‖ mull-jenny m. / self-acting ~ ‖ Selfaktor m.; selbstspinnende Mulemaschine f. ‖ mull-jenny m. renvideur; renvideur m mull-jenny selfacting; mull-jenny m. automate.

mule doubler (Spinn) ‖ Mulezwirnmaschine f. ‖ mull-jenny m. à retordre. / ~ fitter ‖ Webereimechaniker m. ‖ mécanicien m. pour tissages. / ~ spinner ‖ Mulespinner m. ‖ fileur m. au mule-jenny. / ~ twist ‖ Muletwist n. ‖ fil m. du métier mull-jenny. / double ~ twist ‖ geschleiftes Garn n. ‖ fil m. de coton légèrement retors.

muley saw ‖ Blockbandsäge f. ‖ scie f. alternative. / ~ sawyer ‖ Blockbandsäger m. ‖ conducteur m. de scie alternative.

mull (Textile) ‖ Mull m.; feiner Musselin m. ‖ mousseline f.; mulle f.

mulled ale ‖ Warmbier n. ‖ bière f. chaude. / ~ wine ‖ Glühwein m. ‖ vin m. chaud.

mullein (Botany) ‖ Königskerze f. ‖ bouillon m. blanc.

muller (Mirror making) ‖ Reibkasten m.; Schleifkasten m. ‖ moëllon m.

mullion (Of a window) ‖ Mittelpfosten m. ‖ meneau m.; montant m. de milieu.

mull madder (Dyer) ‖ Mullkrapp m. ‖ garance-mulle f.; billon m.; garance f. courte.

mull muslin ‖ Mull n.; (feiner) Musselin m. ‖ organdi m.; mulle f.; mousseline f.

multi-chambered ‖ mehrstufig ‖ polyétagé; à étages m. multiples.

multi-circuit switch (Electr) ‖ Serienschalter m. ‖ commutateur m. multiple ou à combinaison.

multi-coloured writing (Typewr) ‖ mehrfarbige Schrift f. ‖ écriture f. polychrome.

multi-coulor press (Print) ‖ Vielfarbendruckpresse f. ‖ presse f. multicolore. / ~ rotary printing machine ‖ Mehrfarbenrotationsdruckmaschine f. ‖ machine f. rotative pour illustrations en plusieurs couleurs.

multi-cut lathe ‖ Vielstahldrehbank f. ‖ tour m. à outils multiples.

multi-cylinder motor ‖ Mehrzylindermotor m. ‖ moteur m. polycylindrique.

multi-engine(d) ‖ mehrmotorig ‖ multimoteur.

multi-engine plane ‖ Mehrmotorenflugzeug n. ‖ avion m. multimoteur.

multi-motor drive ‖ Mehrmotorenantrieb m.; mehrmotoriger Antrieb m. ‖ commande f. multiple ou par moteurs multiples. / electric ~ of a paper making machine ‖ elektrischer Mehrmotorenantrieb m. einer Papiermaschine ‖ commande f. électrique multiple d'une machine à papier.

multinomial (Math) ‖ Aggregat n.; Polynom n. ‖ polynôme m.

multiphase see also polyphase ‖ mehrphasig; Mehrphasen . . . ‖ polyphasé. / ~

generator ‖ Mehrphasengenerator m. ‖ génératrice f. polyphasée.

multiplane (Aeroplane) ‖ Mehrdecker m.; Vieldecker m. ‖ multiplan m.; appareil m. multiplan.

multiple ‖ Vielfaches n.; Mehrfaches n. ‖ multiple m. / in ~ (Electr) ‖ parallel ‖ en parallèle.

multiple see also multiplex ‖ vielfach; mehrfach ‖ multiple. / ~-blade saw frame ‖ Vollgatter n.; Bundgatter m. ‖ scie f. verticale alternative ou à cadre à plusieurs lames. / ~ connection (Tel) ‖ Vielfachschaltung f. ‖ connection f. multiple. / ~ connection of cable wires ‖ Multiplexverteilung f. von Kabeladern ‖ connection f. multiple des fils de câble. / ~ cord ‖ Mehrfachschnur f. ‖ cordon m. multiple. / ~ die ‖ Mehrfachgesenk n. ‖ étampe f. multiple. / ~-disk clutch ‖ Lamellenkupplung f. ‖ embrayage m. à disques. / ~ distribution box ‖ Vielfachdose f. ‖ boîte f. multiple de distribution. / ~ drilling machine ‖ Vielspindelbohrmaschine f. ‖ perceuse f. à broches multiples. / ~ field (Tel) ‖ Vielfachfeld n. ‖ champ m. avec multiplage. / ~ jack (Tel) ‖ Vielfachklinke f. ‖ jack m. multiple. / ~-lever cash register ‖ Mehrzählerhebelkontrollkasse f. ‖ totalisatrice f. multiple à leviers. / ~ operation (Tel) ‖ Vielfachbetrieb m. ‖ exploitation f. avec multiplage. / ~ reception (Radio) ‖ Vielfachempfang m. ‖ réception f. multiple. / ~ request line apparatus ‖ Mehrfachanschlußapparat m. ‖ poste m. à appel multiple. / ~ spark gap ‖ unterteilte Funkenstrecke f. ‖ éclateur m. en série; éclateur m. fractionné.

multiple-spindle . . . ‖ mehrspindlig ‖ à plusieurs broches. / ~ centering machine ‖ mehrspindlige Zentriermaschine f. ‖ machine f. à centrer à plusieurs broches. / ~ drill ‖ mehrspindlige Bohrmaschine f. ‖ machine f. à percer à broches multiples. / ~ drill head ‖ Mehrspindelbohrkopf m. ‖ plateau m. d'alésage à forets multiples. / ~ full automatic machine ‖ Mehrspindelvollautomat m. ‖ machine f. automatique à broches multiples. / ~ semi-automatic machine ‖ Mehrspindelhalbautomat m. ‖ machine f. semi-automatique à broches multiples.

multiple spooling machine (Spinn) ‖ Fachmaschine f. ‖ machine f. à bobinage multiple.

multiple-stage boiler ‖ Etagenkessel m. ‖ chaudière f. à étages. / ~ compression ‖ mehrstufige Verdichtung f. ‖ compression f. à plusieurs étages.

multiple suspension (El line) ‖ Vielfachaufhängung f. ‖ suspension f. multiple. / ~ switchboard (Tel) ‖ Vielfachumschalter m. ‖ table f. de commutateur multiple. / ~ telegram ‖ Mehrfachtelegramm n. ‖ télégramme m. multiple. / ~ telephony with high frequent carrier currents ‖ Mehrfachsprechen m. mit hochfrequenten Trägerströmen ‖ téléphonie f. multiple à l'aide des courants porteurs à haute fréquence. / ~-tone transmitter (Radio) ‖ Vieltonsender m. ‖ transmetteur m. à fréquences multiples. / ~ tool holder ‖ Mehrstahlhalter m. ‖ porte-outils pour outils multiples. / ~ transmission (Tel) ‖ Vielfachübertragung f. ‖ transmission f. multiple. / ~ transmission on lines ‖ Mehrfachbetrieb m. auf Leitungen ‖

transmission f. multiple sur lignes. / ~ tube (Radio) ‖ Mehrfachröhre f. ‖ lampe f. multiple. / ~ tuner (Radio) ‖ Vielfachabstimmvorrichtung f. ‖ syntonisateur m. multiple.

multiple-type typewriter ‖ Vieltypenschreibmaschine f. ‖ machine f. à écrire à caractères multiples.

multiple-unit control ‖ Vielfachsteuerung f. ‖ commande f. multiple.

multiple V-gear ‖ Keilrädergetriebe n. ‖ engrenage m. à friction; transmission f. à friction par engrenages coniques.

multiple-way switch ‖ Mehrwegumschalter m. ‖ commutateur m. à plusieurs directions.

multiplex see also multiple ‖ vielfach; mehrfach ‖ multiple; multiplex. / ~ apparatus (Tel) ‖ Mehrfachtelegraf m. ‖ appareil m. multiple. / ~ printing telegraph ‖ schreibender Vielfachtelegraf m. ‖ télégraphe m. imprimeur multiple.

multiplex telegraphy ‖ Mehrfachtelegrafie f. ‖ télégraphie f. multiplex. / selective ~ ‖ wechselseitige Mehrfachtelegrafie f. ‖ télégraphie f. multiplex alternative. / simultaneous ~ ‖ gleichzeitige Mehrfachtelegrafie f. ‖ télégraphie f. multiplex simultanée.

multiplex telephony ‖ Mehrfachtelefonie f. ‖ téléphonie f. multiplex. / ~ (Tel) ‖ Multiplexbetrieb m. ‖ transmission f. multiplex.

multiplication (Math) ‖ Multiplikation f.; Vervielfachung f. ‖ multiplication f. / ~ energy of the yeast ‖ Vervielfältigungsenergie f. der Hefe ‖ énergie f. de multiplication de la levure. / ~ sign ‖ Multiplikationszeichen n. ‖ signe m. de multiplication.

multiplicator (Math) ‖ Multiplikator m. ‖ multiplicateur m.

multiplier see multiplicator.

multiply, to ~ (Math) ‖ multiplizieren; vervielfachen ‖ multiplier. / ~ (Tel) ‖ vielfach schalten ‖ multiplier.

multiplying arrangement ‖ Multipliziervorrichtung f. ‖ dispositif m. multiplicatif. / ~ machine ‖ Multipliziermaschine f. ‖ machine f. à multiplier.

multipolar (Electr) ‖ mehrpolig ‖ multipolaire.

multishape ‖ vielgestaltig ‖ multiforme.

multistage ‖ mehrstufig ‖ à plusieurs étages. / ~ pump ‖ Mehrstufenpumpe f. ‖ pompe f. à plusieurs étages.

multitubular boiler ‖ Heizrohrkessel m.; Feuerrohrkessel m.; vielrohriger Kessel m. ‖ chaudière f. multitubulaire (à tubes de fumée). / ~ with removable fire box and smoke tubes ‖ ausziehbarer Flammrohrsiederöhrenkessel m. ‖ chaudière f. avec boîte à feu et faisceau tubulaire amovible.

multure (Mill) ‖ Mahlgeld n. ‖ mouture f.

mummy (Grafting wax) ‖ Baumwachs n. ‖ cire f. à greffer.

mungo wool ‖ Mungo m.; Kunstwolle f.; Schoddy m. ‖ laine f. Renaissance. / ~ spinning mill ‖ Kunstwollspinnerei f.; Schoddyspinnerei f. ‖ filature f. de déchets de laine ou de laine Renaissance.

municipal ‖ städtisch ‖ municipal. / ~ airport ‖ städtischer Flughafen n.; Stadtflughafen m. ‖ aérodrome m. de ville. / ~ area ‖ Weichbild n. einer Stadt; Stadtgebiet n. ‖ banlieue f. / ~ loan ‖ Stadtanleihe f. ‖ emprunt m. de ville.

/ ~ theatre ‖ Stadttheater n.; städtisches Schauspielhaus n. *oder* Theater n. ‖ théâtre m. municipal.

munition *see also* ammunition ‖ Munition f.; Schießbedarf m. ‖ munition f.; munitions fpl.

mural decoration ‖ Wanddekoration f.; Wandzierat m. ‖ décoration f. murale. / ~ painting ‖ Wandmalerei f. ‖ peinture f. murale. / ~ quadrant (Astron) ‖ Mauerquadrant m. ‖ quart m. de cercle mural.

murexide (Chem) ‖ Murexid n. ‖ murexide f.

muriate of ammonia ‖ Salmiak m. ‖ ammoniaque f. muriatée; chlorure m. d'ammonium; sel m. ammoniac; salmiac m. / ~ of mercury ‖ Quecksilberhornerz n.; Kalomel n. ‖ mercure m. corné *ou* muriaté.

muriatic acid ‖ Chlorwasserstoff m.; Salzsäure f. ‖ acide m. chlorhydrique *ou* muriatique; chlorure m. d'hydrogène. / ~ condensing plant ‖ Salzsäurekondensationsanlage f. ‖ installation f. de condensation d'acide muriatique. / ~ gas ‖ Chlorwasserstoffgas n. ‖ gaz m. d'acide chlorhydrique. / ~ manufacturing plant ‖ Salzsäuregewinnungsanlage f. ‖ installation f. pour la fabrication d'acide muriatique.

muring ‖ Mauerwerk n.; Gemäuer n.; Mauerung f. ‖ maçonnerie f.; ouvrage m. de maçonnerie; maçonnage m.; murage m.; muraillement m.

murkiness ‖ Dunkelheit f.; Finsternis f. ‖ obscurité f.; ténèbres fpl. / ~ of the air ‖ Lufttrübung f. ‖ obscurissement m. *ou* assombrissement m. atmosphérique.

murrey ‖ dunkelrot; braunrot ‖ rouge brun; mordoré; rouge foncé.

muscle ‖ Muskel m. ‖ muscle m.

muscovado ‖ Farinzucker m. ‖ cassonade f.

muscovite (Miner) ‖ Muscovit m.; Kaliglimmer m.; weißer Glimmer m. ‖ muscovite f.

muscular current ‖ Muskelstrom m. ‖ courant m. musculaire. / ~ power ‖ Muskelkraft f. ‖ force f. musculaire.

museum ‖ Museum n. ‖ musée m.

mushroom ‖ (eßbarer) Pilz m.; Champignon m. ‖ champignon m. comestible. / dried ~ ‖ getrockneter Pilz m. ‖ champignon m. séché. / preserved ~ ‖ konservierter Pilz m. ‖ champignon m. conservé.

mushroom anchor (Mar) ‖ Schildanker m.; Pilzanker m. ‖ crapaud m. d'amarrage. / ~ cultivation ‖ Pilzzucht f. ‖ culture f. de champignons de couche. / ~ farmer ‖ Pilzzüchter m.; Champignonzüchter m. ‖ champignonniste m. / ~ gatherer ‖ Pilzsammler m. ‖ ramasseur m. de champignons. / ~ valve (Mach) ‖ Kegelventil n. ‖ soupape f. à siège conique; soupape f. conique.

music ‖ Musik f. ‖ musique f. / ~ (Note) ‖ Note f. ‖ note f. / accompaning ~ ‖ Begleitmusik f. ‖ musique f. d'accompagnement.

musical box ‖ Spieldose f. ‖ boîte f. à musique.

musical instrument ‖ Musikinstrument n. ‖ instrument m. de musique. / ~ with direct percussion ‖ Musikinstrument n. mit direktem Schlag ‖ instrument m. de musique à percussion directe. / mechanical ~ ‖ mechanisches *oder* auto-

matisches Musikwerk n.; Orchestrion n. ‖ orchestrion m.; instrument m. de musique mécanique. / ~ playing mechanically by means of rolls ‖ mittels Walzen mechanisch spielendes Musikinstrument n. ‖ instrument n. de musique jouant mécaniquement à l'aide de cylindres. / toy ~ ‖ Musikinstrument n. für Kinder ‖ instrument m. de musique pour enfants.

musical instrument case ‖ Kasten m. für Musikinstrumente ‖ étui m. pour instruments de musique. / ~ maker ‖ Instrumentenmacher m. ‖ facteur m. d'instruments de musique; luthier m. / ~ polisher ‖ Instrumentenlackierer m. ‖ polisseur m. en lutherie.

musical spark tester (Radio) ‖ Tonprüfer m. ‖ essayeur m. de son. / ~ transmitter (Radio) ‖ Tonsender m. ‖ poste m. transmetteur à étincelles musicales.

musical string ‖ Musiksaite f. ‖ corde f. d'instrument à musique. / ~ toys pl. ‖ Musikspielwaren fpl. ‖ jouets mpl. à musique.

music automaton ‖ Musikautomat m. ‖ automate m. de *ou* à musique. / ~ book ‖ Notenheft n.; Notenbuch n. ‖ cahier m. *ou* livre m. de musique. / ~ case ‖ Musiknotenmappe f. ‖ porte-musique m. / ~ engraver ‖ Notenstecher m. ‖ graveur m. de musique. / ~ engraving ‖ Notenstechen n. ‖ gravure f. de musique. / ~ film ‖ Musikfilm m. ‖ film m. musical. / ~ paper ‖ Notenpapier n. ‖ papier m. à musique. / ~ printer ‖ Notendrucker m.; Musikaliendrucker m. ‖ imprimeur m. de musique. / ~ printing ‖ Notendruck m. ‖ impression f. musicale *ou* de musique. / ~ printing office ‖ Notendruckerei f.; Musikaliendruckerei f. ‖ imprimerie f. de musique. / ~ stand ‖ Notenständer m. ‖ pupitre m. à musique. / ~ string *see* musical string. / ~ typewriter ‖ Notenschreibmaschine f. ‖ machine f. à écrire les notes à musique. / ~ wire ‖ Drahtsaite f.; metallene Saite f.; Saitendraht m. ‖ corde f. métallique; fil m. à instrument.

musk ‖ Moschus m.; Bisam m. ‖ musc m.

musket ‖ Muskete f.; Flinte f. ‖ mousqueton m. / ~ shot ‖ Flintenschuß m. ‖ coup m. de fusil.

musk seed ‖ Bisamkörner n. ‖ semences fpl. d'ambrette.

muslin ‖ Musselin m. ‖ mousseline f. / embroidered ~ ‖ gestickter *oder* lanzierter Musselin m.; Mullstickerei f. ‖ broderie f. sur mousseline; mousseline f. brodée *ou* lancée. / figured ~ ‖ broschierter Musselin m. ‖ mousseline f. brochée.

muslin bandage ‖ Mullbinde f. ‖ bande f. de mousseline. / ~ glass ‖ Musselinglas n. ‖ verre m. mousseline. / ~ weaver ‖ Musselinweber m. ‖ mousselinier m. / ~ weaving ‖ Musselinweberei f. ‖ fabrication f. de tissus de mousseline.

mussel ‖ Miesmuschel f.; (zweischalige) Muschel f. ‖ moule f.; coquille f. / full ~ ‖ volle Muschel f. ‖ coquillage m. plein.

mussel culture ‖ Muschelzucht f. ‖ mytiliculture f. / ~ culturist ‖ Muschelzüchter m. ‖ cultivateur m. *ou* éleveur m. de moules; bouchoteur m. / ~ gatherer ‖ Muschelfischer m. ‖ pêcheur m. de coquilles *ou* de coquillages *ou* de moules.

must, to ‖ schimmeln; schimmelig werden ‖ (se) moisir; (se) chancir.

must (of wine) ‖ Weinmost m.; Most m. ‖ moût m. (de vin).

mustache *see* moustache.

mustard ‖ Mostrich m.; Senf m. ‖ moutarde f. / ~ flour ‖ Senfmehl n. ‖ farine f. de moutarde. / ~ gas ‖ Senfgas n. ‖ gaz m. moutarde; Ypérite m.; sulfure m. d'éthyle dichloré. / ~ maker ‖ Senfmacher m. ‖ moutardier m. / ~ meal *see* ~ flour. / ~ mill ‖ Senfmühle f. ‖ moulin m. à moutarde. / ~ oil ‖ Senföl n. ‖ essence f. de moutarde. / ~ paper ‖ Senfpapier n.; Senfpflaster n. ‖ papier m. moutarde; papier m. sinapisé; sinapisme m.; emplâtre m. de moutarde. / ~ seed ‖ Senfsamen mpl.; Senfsaat f. ‖ graines fpl. de moutarde.

muster-book (Mar) ‖ Musterrolle f. ‖ rôle m. d'équipage.

mustering (Mar) ‖ Musterung f. ‖ revue f.; appel m.

musting implement ‖ Mostereigerätschaft f. ‖ ustensile m. de cidrerie.

musty ‖ muffig; schimmelig; dumpfig ‖ moisi; chanci; humide. / to turn ~ ‖ stocken; schimmeln ‖ se moisir; se chancir.

mutation of energy ‖ Energieumformung f. ‖ mutation f. de l'énergie.

mutilated telegram ‖ verstümmeltes Telegramm n.; verstümmelte Drahtnachricht f. ‖ dépêche f. mutilée; télégramme m. mutilé.

mutilation of a telegram ‖ Verstümmelung f. eines Telegramms ‖ mutilation f. d'une dépêche.

mutton, leg of ~ ‖ Hammelkeule f. ‖ gigot m.

mutual ‖ wechselseitig; gegenseitig ‖ réciproque; mutuel. / ~ benefit society *see* mutual society. / ~ capacity (Electr) ‖ Betriebskapazität f. ‖ capacité f. mutuelle. / ~ debt ‖ Gegenschuld f. ‖ dette f. passive. / ~ inductance ‖ gegenseitige Induktanz f. ‖ inductance f. mutuelle. / ~ obligations pl. ‖ gegenseitige Leistungen fpl. ‖ obligations fpl. réciproques. / ~ reaction ‖ Wechselwirkung f. ‖ action f. réciproque. / ~ relation(ship) ‖ Wechselbeziehung f. ‖ rapport m. réciproque. / ~ (benefit) society ‖ Gesellschaft f. auf Gegenseitigkeit; Gegenseitigkeitsgesellschaft f. ‖ société f. mutuelle.

muzzle ‖ Maulkorb m. ‖ muselière f. / cattle ~ ‖ Viehmaulkorb m. ‖ muselière f. pour bestiaux.

muzzle brake (Gun) ‖ Mündungsbremse f. ‖ frein m. de bouche. / ~ compensator (Gun) ‖ Mündungsausgleicher m. ‖ compensateur m. de bouche. / ~ loader (Gun) ‖ Vorderlader m. ‖ arme f. à feu se chargeant par la bouche.

mycelium ‖ Hausschwamm m. ‖ mycose f.

myriare ‖ Quadratkilometer n. ‖ myriare m.

myristic acid ‖ Myristinsäure f. ‖ acide m. myristique.

myrobalan ‖ Myrobalane f. ‖ myrobalan m. / ~ extract ‖ Myrobalanenauszug m. ‖ extrait m. de myrobalans.

myrrh ‖ Myrrhe f. ‖ myrrhe f.

myrtle ‖ Myrte f. ‖ myrica m.

myrtle-leaf-bladed knife (Med) ‖ Lorbeerblattmesser n. ‖ bistouri m. à double tranchant.

myrtle oil ‖ Myrtenöl n. ‖ essence f. de myrthe.

myrtle wax ‖ Myrikawachs n.; Myrtenwachs n. ‖ cire f. de Myrica.

N

N-quadrat n. (Print) ‖ Halbgeviert n. ‖ demi-cadratin m.

nacelle ‖ Gondel f. ‖ nacelle f. / ~ of the balloon ‖ Ballonkorb m. ‖ nacelle f. de ballon. / engine ~ bracing ‖ Motorgondelverstrebung f. ‖ mâture f. de bâti-moteur.

nacre ‖ Perlmutter f. ‖ nacre f. / ~ goods pl. ‖ Perlmutterwaren fpl. ‖ objets mpl. en nacre.

nacreous ‖ perlmutterartig ‖ nacré. / ~ paper ‖ Perlmutterpapier n. ‖ papier m. nacré.

nactalopia ‖ Nachtblindheit f. ‖ héméralopie f.

nadir ‖ Nadir m. ‖ nadir m.

nagyagite ‖ Nagyagit m.; Blättererz n. ‖ nagyagite f.; argent m. vierge en lames; tellure m. natif auro-plombifère.

nail, to ‖ nageln ‖ clouer. / ~ (Shipb) ‖ spiekern ‖ clouer. / ~ a lock ‖ ein Schloß n. anschlagen ‖ clouer une serrure. / ~ a plank on the ribs (Shipb) ‖ eine Planke auf die Spanten festspiekern ‖ coudre un bordage sur les couples.

nail ‖ Nagel m.; Stift m. ‖ clou m.; cheville f.; pointe f. / ~ (Anatomy) ‖ Nagel m. ‖ ongle m. / ~ (Mar) ‖ Nagel m.; Spieker m. ‖ clou m. / ~ (Mine) ‖ Bohrnadel f.; Räumnadel f.; Schießnadel f. ‖ épinglette f. *ou* barre f. à mine. / barbed ~ ‖ Nagel m. mit Widerhaken ‖ cheville f. barbelée. / bellied ~ ‖ ausgebauchter Nagel m. ‖ cheville f. renflée. / brass ~ ‖ Messingstift m. ‖ clou m. en laiton. / cast ~ ‖ gegossener Nagel m. ‖ clou m. fondu. / cast iron ~ ‖ gußeiserner Nagel m. ‖ clou m. (fondu) en fonte (de fer). / copper ~ ‖ Kupfernagel m. ‖ clou m. de cuivre. / cut ~ ‖ geschnittener Nagel m.; Polsternagel m. ‖ clou m. coupé *ou* découpé. / fancy ~ ‖ Dekorationsnagel m.; Ziernagel m. ‖ clou m. de fantaisie. / forged ~ ‖ geschmiedeter Nagel m.; Schmiedenagel m. ‖ clou m. forgé. / ~ for furniture ‖ Polsternagel m. ‖ clou m. pour meubles rembourrées. / hand-made ~ ‖ geschmiedeter Nagel m. ‖ clou m. forgé. / horse ~ ‖ Hufnagel m. ‖ clou m. à cheval. / machine-made ~ ‖ geschnittener Nagel m. ‖ clou m. découpé. / ~ for plaster of Paris flooring slabs ‖ Gipsdielenstift m. ‖ clou m. à dalles en plâtres. / prismatical ~ ‖ prismatischer Nagel m. ‖ cheville f. prismatique. / ~ for reed ‖ Rohrstift m. ‖ clou m. à roseau. / ship ~ ‖ Schiffsnagel m. ‖ clou m. de navires. / shoe ~ ‖ Schuhnagel m. ‖ clou m. pour chaussures. / ~ for stone board ‖ Steinpappstift m. ‖ clou m. à cartonpierre. / the ~ takes ‖ der Nagel zieht an ‖ le clou prend. / upholstering ~ ‖ Polsternagel m. ‖ clou m. de tapisserie. / wire ~ ‖ Drahtstift m. ‖ pointe f. de Paris. / wooden ~ (Carp) ‖ Holznagel m.; Pflock m. ‖ cheville f. de bois; goujon m. / wooden ~ (Shipb) ‖ Holznagel m. ‖ gournable f. / wrought ~ ‖ schmiedeeiserner Nagel m. ‖ clou m. forgé.

nailable ‖ nagelbar ‖ pénétrable aux clous.

nail bore ‖ Nageldocke f.; Nageleisen n.; Nagelform f. ‖ clouère f.; clouière f.; cloutière f.; clouvière f.; étampe f. de cloutier. / ~ brush ‖ Nagelbürste f. ‖ brosse f. à ongles, / ~ catcher ‖ Nagelfänger m. ‖ arrache-clou m. / ~ claw ‖ Nagelzieher m.; Nagelheber m.; Nagelauszieher m. ‖ arrache-clou m.; tire-clou m. / ~ cleaner ‖ Nagelreiniger m. ‖ lime f. à ongle. / ~ cutter ‖ Nagelschneider m. ‖ coupe-ongle m. / ~ drawer ‖ Nagelzieher m.; Nagelheber m.; Nagelauszieher m. ‖ arrache-clou m.; tire-clou m. / ~ drawer (Mar) ‖ Kuhfuß m. ‖ pied m. de chèvre; loup m.

nailer ‖ Nagelschmied m. ‖ cloutier m. / scissors ~ ‖ Schraubeneinsetzer m. ‖ monteur m. de vis en coutellerie.

nail extractor ‖ Nagelzieher m. ‖ arrache-clou m. / ~ file ‖ Nagelfeile f. ‖ lime f. à ongles. / ~ head ‖ Nagelkopf m. ‖ tête f. de clou. / ~ hole ‖ Nagelloch n. ‖ trou m. de clou.

nailing ‖ Nagelung f. ‖ clouture f. / ~ (Shipb) ‖ Spiekerung f. ‖ clouage m.

nail iron ‖ Nageleisen n. ‖ fer m. en barres pour clous. / ~ maker ‖ Nagelschmied m. ‖ cloutier m. / machine for making ~s ‖ Nagelherstellungsmaschine f. ‖ machine f. à faire les clous. / ~ mandrel ‖ Nagledocke f.; Nageleisen n.; Nagelform f. ‖ clouère f.; clouière f.; cloutière f.; clouvière f.; étampe f. de cloutier. / ~ manufacture ‖ Nagelschmiede f. ‖ clouerie f. / ~ mould *see* ~ mandrel. / ~ nippers pl. ‖ Nagelzange f. ‖ arrache-clou m. / ~ puller ‖ Nagelzieher m. ‖ tire-clou m.; arrache-clou m. / ~ rod *see* ~ iron. / ~ scissors pl. ‖ Nagelschere f. ‖ ciseaux mpl. à ongles; onglier m. / ~-shaped ‖ nagelförmig ‖ cloudiforme. / ~ smith ‖ Nagelschmied m. ‖ cloutier m. / ~ smith's chisel ‖ Blockmeißel m.; Stockmeißel m. ‖ tranchet m. / ~ tool ‖ Nageldocke f.; Nageleisen n.; Nagelform f. ‖ clouère f.; clouière f.; cloutière f.; étampe f. de cloutier. / ~ works pl. ‖ Nagelfabrik f.; Nagelschmiede f. ‖ clouterie f. / ~ wrench with flat shank ‖ Kistenöffner m. mit Vierkantstiel ‖ ciseau m. à déballer à manche plate. / ~ wrench with round shank ‖ Kistenöffner m. mit rundem Stiel ‖ ciseau m. à déballer à manche ronde.

nakrit ‖ Nakrit m. ‖ nakrite f.

name, to ‖ ~ a ship ‖ ein Schiff n. taufen ‖ baptiser un navire.

name, by the ~ of ‖ unter der Firma f. ‖ sous la raison sociale. / of the same ~ ‖ gleichnamig ‖ de même nom. / ~ of the street ‖ Straßenname m. ‖ nom m. de la rue.

name band ‖ Namenband n. ‖ ruban m. à inscription. / ~ block ‖ Firmenklischee n. ‖ galvano m. de la maison *ou* à firme.

named ‖ namhaft ‖ dénommé; notable.

name plate ‖ Firmenschild n. ‖ plaque f. de la maison; enseigne f. / enamelled ~ ‖ Emailschild n. ‖ plaque f. émaillée. / ~ for machinery ‖ Firmenschild n. für Maschinen ‖ plaque f. (à inscriptions) pour machine. / metal ~ ‖ Metallschild n. ‖ plaquette f. en métal. / ~ on plaited wirework ‖ Drahtluftschild n,; Namenschild n. auf Drahtgewebe ‖ plaque-adresse f. sur treillis en fil métallique.

nankeen ‖ Nanking m. ‖ nanquin m.; nankin. / prime ~ ‖ feiner Nanking m. ‖ nankinet m.

nankeen cotton ‖ Nankingbaumwolle f. ‖ coton m. nankin.

nankin *see* nankeen.

nap (Weav) ‖ Noppe f.; Samtmasche f.; Flordecke f. ‖ boucle f.; poil m.

naphtha ‖ Naphtha n.; f.; Erdöl n.; Steinöl n. ‖ naphte m.; bitume m. liquide; huile f. minérale. / crude ~ ‖ Rohöl n. ‖ pétrole m. brut. / high-flash ~ ‖ Naphtha f. mit hohem Flammpunkt ‖ naphte m. à point d'inflammation élevé.

naphthacene ‖ Naphthacen n. ‖ naphtacène m.

naphtha, distillation product of ~ ‖ Erdöldestillat m. ‖ produit m. de la distillation du pétrole. / ~ distiller ‖ Petroleumraffinierer m. ‖ raffineur m. de pétrole.

naphthagil ‖ Ozokerit m.; Erdwachs n.; Bergwachs n. ‖ ozokérite f.

naphthalene *see* naphthaline.

naphthaline ‖ Naphthalin n. ‖ naphtaline f. / ~ engine ‖ Naphthalinmotor m. ‖ moteur m. à naphtaline. / segregation of ~ ‖ Naphthalinausscheidung f. ‖ séparation f. de la naphtaline. / ~ washer ‖ Naphthalinwascher m. ‖ laveur m. de naphtaline. / ~ washing cylinder ‖ Naphthalinwascher m. ‖ laveur m. de naphtaline. / ~ washing oil ‖ Naphthalinwaschöl n. ‖ huile f. à laver la naphtaline. / ~ works pl. ‖ Naphthalinfabrik f. ‖ fabrique f. de naphtaline.

naphtha pitch ‖ Naphthapech n. ‖ poix f. de naphte. / ~ production ‖ Naphthagewinnung f. ‖ production f. de naphte.

naphthenic acid ‖ Naphthensäure f. ‖ acide m. naphténique.

naphthol ‖ Naphthol n. ‖ naphtol m.

Napier's bone ‖ Rechenstab m. ‖ baguette f. à calculer.

napkin ‖ Mundtuch n. ‖ serviette f. / paper ~ ‖ Papiermundtuch n. ‖ serviette f. en papier.

Naples yellow ‖ Neapelgelb n. ‖ jaune m. de Naples.

nap lifting apparatus ‖ Velourhebeapparat m. ‖ appareil m. à relever les velours. / ~ lifting machine with revolving straightedge table ‖ Velourhebemaschine f. mit drehbarem Linealtisch ‖ machine f. à velouter avec table à règle tournante. / ~ pattern ‖ Noppenmuster n. ‖ dessin m. nopes.

napping machine ‖ Rauhmaschine f. ‖ garnisseuse f.; lainerie f.; laineuse f.; machine f. à lainer.

nap warp ‖ Polkette f.; Samtkette f. ‖ chaîne f. de poil; poil m.

narcosis ‖ Narkose f.; künstliche Betäubung f. ‖ narcose f.; anesthésie f.

narcotic ‖ Betäubungsmittel n.; Narkotikum n. ‖ narcotique m.

narcotine ‖ Narkotin n. ‖ narcotine m.

narrow ‖ eng ‖ étroit.

narrowed ‖ schmal; verengt ‖ resserré. / ~ goods pl. ‖ geminderte Ware f. ‖ tricot m. diminué.

narrow gauge ‖ Schmalspur f. ‖ voie f. étroite. / ~ carriage ‖ Schmalspurwagen m. ‖ wagon m. à voie étroite. / ~ engine ‖ Schmalspurlokomotive f. ‖ machine f. pour voie étroite. / ~ line ‖ Schmalspurbahn f. ‖ chemin m. de fer à voie étroite / ~ locomotive ‖ Kleinbahnlokomotive f.; Schmalspurmaschine f. ‖ locomotive f. à voie étroite. / ~ railway ‖ Schmalspurbahn f.; Kleinbahn f. ‖ chemin m. de fer à voie étroite; petit chemin m. de fer. / ~ railway wagon ‖ Kleinbahnwagen m.; Schmalspurbahnwagen m. ‖ wagon m. à voie étroite. / ~ wagon *see* ~ railway wagon.

narrow-heeled, to become ~ ‖ hufzwängig werden ‖ s'encasteler.

narrow-heeledness ‖ Hufzwang m. ‖ encastelure f.

narrowing ‖ Verengung f.; Minderung f. ‖ rétrécissement m.; diminution f. / ~ chain (Textile) ‖ Deckkette f. ‖ chaîne f. à diminuer. / ~ machine ‖ Mindermaschine f. ‖ diminueuse f.

narrow-neck bottle ‖ enghalsige Flasche f. ‖ bouteille f. à col étroit. / ~ pass ‖ Hohlweg m. ‖ chemin m. creux. / ~-pitch cutter bar ‖ Schneidbalken m. für Tiefschnitt ‖ barre f. coupeuse pour coupe profonde. / ~ sunken road ‖ Hohlweg m. ‖ chemin m. creux.

nasal douche ‖ Nasendusche f. ‖ douche f. nasale.

nascent ‖ aus einer Verbindung f. freiwerdend; naszierend ‖ naissant. / ~ state ‖ Entstehungszustand m. ‖ état m. naissant.

nation ‖ Nation f. ‖ nation f. / League of Nations ‖ Völkerbund m. ‖ Société f. des Nations.

national census ‖ Volkszählung f. ‖ recensement m. (de la population). / National Defence Act ‖ nationales Verteidigungsgesetz n. ‖ loi f. de la défense nationale. / ~ economy ‖ Volkswirtschaft f.; Volkswirtschaftslehre f. ‖ économie f. politique. / ~ flag ‖ Nationalflagge f. ‖ pavillon m. national.

nationalization ‖ Verstaatlichung f. ‖ étatisation f.

national language ‖ Landessprache f. ‖ langue f. nationale. / ~ prosperity ‖ Volkswohlstand m.; Volksvermögen n. ‖ fortune f. publique. / ~ wealth *see* ~ prosperity.

native ‖ natürlich ‖ natif. / ~ (Geogr) ‖ inländisch ‖ indigène; intérieur. / ~ (Miner) ‖ gediegen ‖ natif; vierge; pur. / ~ alum extraction ‖ Alaunerzbergwerk n. ‖ carrière f. d'alun natif. / ~ iron ‖ gediegenes Eisen n. ‖ fer m. natif. / ~ lead ‖ gediegenes Blei n. ‖ plomb m. natif. / ~ metal ‖ gediegenes Metall n. ‖ métal m. natif. / ~ silver ‖ gediegenes Silber n. ‖ argent m. natif. / ~ substance ‖ Naturprodukt n. ‖ principe m. immédiat; produit m. naturel.

natrolite ‖ Natrolith m. ‖ natrolithe f.

natural ‖ natürlich ‖ naturel. / ~ cement ‖ Naturzement m. ‖ ciment m. naturel. / ~ cork plate ‖ Naturkorkplatte f. ‖ plaque f. en liège véritable. / ~ draught ‖ natürlicher Zug m. ‖ tirage m. naturel. / ~ flowers pl. ‖ Naturblumen fpl. ‖ fleurs fpl. naturelles. / ~ frequency ‖ Eigenschwingungszahl f. ‖ nombre ‖ m. d'oscillations propres. / ~ gas ‖ Erdgas n. ‖ gaz m. naturel. / ~ ground ‖ gewachsener Boden m. ‖ terre f. naturelle. / ~ hard crucible steel ‖ naturharter Tiegelstahl m. ‖ acier m. au creuset de dureté naturelle. / ~ hard steel ‖ naturharter Stahl m. ‖ acier m. de dureté naturelle. / ~ history ‖ Naturgeschichte f. ‖ histoire f. naturelle. / ~ history colourer ‖ Kolorist m. naturwissenschaftlicher Darstellungen ‖ coloriste m. d'histoire naturelle. / ~ ice ‖ Natureis n. ‖ glace f. naturelle. / ~ oscillation ‖ Eigenschwingung f. ‖ oscillation f. propre. / ~ philosopher ‖ Naturforscher m. ‖ naturaliste m. / ~ produce ‖ Naturerzeugnis n. ‖ produit m. naturel. / ~ sciences pl. ‖ Naturwissenschaften fpl. ‖ sciences fpl. naturelles. / of ~ size ‖ in natürlicher Größe f. ‖ de grandeur f. réelle; en grandeur naturelle. / ~ slope ‖ Böschungswinkel m. ‖ inclinaison f. d'un talus. / ~ stone ‖ Naturstein m. ‖ pierre f. naturelle. / ~ vibration ‖ Eigenschwingung f. ‖ oscillation f. propre. / ~ vibration of aerials ‖ Eigenschwingung f. von Antennen ‖ oscillation f. propre d'antennes. / ~ wave ‖ Eigenwelle f. ‖ onde f. naturelle.

naturalist ‖ Tierkonservator m. ‖ naturaliste m.

nature of the ground ‖ Bodenbeschaffenheit f. ‖ nature f. du terrain. / ~ of a locality ‖ Ortsbeschaffenheit f. ‖ nature f. du lieu. / ~ of rock ‖ Ausbildung f. des Gesteins ‖ formation f. de la roche. / ~ of the soil ‖ Bodenart f. ‖ nature f. *ou* qualité f. du sol *ou* du terrain.

naught ‖ Null f. ‖ zéro.

naumannite ‖ Naumannit m.; Selensilber n.; Selensilberblei n. ‖ argent m. séléniuré; séléniure m. d'argent; naumannite.

nausea ‖ Brechneigung f. ‖ envie f. de vomir.

nautical ‖ nautisch ‖ nautique. / ~ compass ‖ Schiffskompaß m. ‖ boussole f. marine; compas m. de navires. / ~ instrument ‖ nautisches Instrument n. ‖ instrument m. nautique. / ~ mile ‖ Seemeile f. ‖ mille m. marin. / ~ school ‖ Navigationsschule f. ‖ école f. d'hydrographie. / ~ signal ‖ Schiffahrtszeichen n. ‖ signal m. de navigation.

nautics pl. ‖ Nautik f.; Schiffskunde f. ‖ navigation f.

naval armament, agreement on ~ limitation ‖ Flottenbeschränkungsvertrag m. ‖ traité m. de la limitation des armements navals; accord m. sur la limitation des armements navals.

naval brass ‖ Marinemetall n. ‖ métal m. marin. / ~ construction ‖ Schiffbau m. ‖ construction f. navale. / ~ constructions pl. ‖ Schiffsbauten mpl. ‖ constructions fpl. navales. / ~ constructor ‖ Schiffsbauingeniör m. ‖ ingénieur m. de la marine. / ~ disarmament conference ‖ Flottenabrüstungskonferenz f. ‖ conférence f. pour le désarmement naval. / ~ dockyard ‖ Marinewerft f.; Seearsenal n. ‖ chantiers mpl. de la marine; arsenal m. maritime. / ~ fire control device ‖ Marinefeuerleitgerät n. ‖ appareil m. de conduite de tir naval. / ~ limitation treaty ‖ Flottenbeschränkungsvertrag m. ‖ traité m. de la limitation des armements navals; accord m. sur la limitation des armements navals. / ~ matters pl. ‖ Seewesen n. ‖ art m. naval. / ~ port ‖ Kriegshafen m. ‖ port m. ne guerre; port m. militaire. / ~ power ‖ Seemacht f. ‖ puissance f. maritime; force f. *ou* puissance f. navale. / ~ supremacy ‖ Seeherrschaft f. ‖ maîtrise f. des mers. / ~ tactics pl. ‖ Seetaktik f. ‖ tactique f. navale.

nave ‖ Nabe f. ‖ moyeu m. / ~ (Arch) ‖ Schiff n. ‖ nef f. / ~ of a fly wheel ‖ Nabe f. eines Schwungrades ‖ noyau m. *ou* moyeu m. d'un volant. / main ~ (Build) ‖ Hauptschiff n. ‖ travée f. principale. / ~ of the shop ‖ Werkstattschiff n. ‖ travée f. d'un hall. / supporting ~ ‖ Tragnabe f. ‖ moyeu m. de support m. / the ~s pl. of wheels have bushes cast in ‖ die Radnaben fpl. erhalten eingegossene Büchsen ‖ les moyeux mpl. sont garnis de boîtes rapportées par coulées.

nave borer ‖ Nabenbohrer m. ‖ quillier m.; foret m. de moyeu. / ~ box ‖ Nabenbuchse f. ‖ boîte f. de moyeu. / ~ disc ‖ Nabenscheibe f. ‖ disque m. de moyeu. / ~ hole ‖ Nabenloch n. ‖ emboîture f. du moyeu.

navel bandage ‖ Nabelbinde f. ‖ ceinture f. *ou* cordon m. ombilicale.

nave paddle ‖ Nabenschaufel f. ‖ palette f. de moyeu. / ~ turner ‖ Radnabendrechsler m. ‖ tourneur m. de moyeux.

navigable ‖ schiffbar ‖ navigable. / ~ for rafts ‖ flößbar ‖ flottable.

navigate, to ‖ navigieren ‖ naviguer.

navigating ‖ Lotsen n. ‖ pilotage m.; la manage m. / ~ of a ship ‖ Navigierung f. eines Schiffes ‖ l'art m. de piloter le navire. / ~ equipment ‖ Navigationsausrüstung f. ‖ installation f. *ou* équipement m. de navigation.

navigation ‖ Nautik f.; Schiffahrtskunde f.; Schiffahrt f. ‖ navigation f. / highaltitude ~ ‖ Höhennavigation f. ‖ repérage m. altimétrique. / inland ~ ‖ Binnenschiffahrt f. ‖ navigation f. intérieure. / ~ on rivers ‖ Flußschiffahrt f. ‖ navigation f. fluviale. / sea ~ ‖ Seeschiffahrt f. ‖ navigation f. de mer *ou* maritime.

navigation company ‖ Schiffahrtsgesellschaft f. ‖ compagnie de navigation f. / ~ equipment ‖ Navigationsausrüstung f. ‖ installation f. *ou* équipement m. de navigation. / ~ light ‖ Positionslampe f.; Stellungslicht n. ‖ feu m. de position. / ~ lock ‖ Schiffahrtsschleuse f. ‖ écluse f. pour la navigation. / ~ police ‖ Schiffspolizei f. ‖ police f. de la navigation. / ~ service ‖ Hafendienst m. ‖ service des ports m. / ~ signal *see* nautical signal.

navigator ‖ Franz m.; Flugzeugbeobachter ‖ observateur m.

navigraph ‖ Kursschreiber m. ‖ navigraphe m.; enregistreur m. de route.

navvy (Railw) ‖ Eisenbahnarbeiter m. ‖ terrassier m. de la voie. / steam ~ ‖ Trockenbagger m.; Ausschachtmaschine f.; Exkavator m.; Löffelbagger m. ‖ drague f. sèche; excavateur m.; drague f. à cuiller; grue-terrassier f.

navy ‖ Marine f. ‖ marine f. / ~ (Ships) ‖ Flotte f. ‖ flotte f.; armée f. navale. / ~ of the League of Nations ‖ Völkerbundflotte f. ‖ flotte f. de la Société des Nations. / merchant ~ ‖ Handelsmarine f. ‖ marine f. marchande *ou* de commerce.

near ‖ nahe ‖ près. / ~ (Mar) ‖ hart ‖ dur; fort; près.

nearer, the rolls pl. are brought ~ together after each pass ‖ die Walzen fpl. werden nach jedem Stich nähergestellt ‖ les cylindres mpl. se rapprochent l'un à l'autre à chaque passage.

near point ‖ Nahepunkt m. ‖ punctum m. proximum.

neat ‖ sauber ‖ net. / ~ casting ‖ sauberes Gußstück n. ‖ moulage m. d'exécution soigneuse.

neat ‖ Rindvieh n.; Hornvieh n. ‖ bœuf m.

neat's foot oil ‖ Klauenfett n. ‖ huile f. de pied de bœuf. / ~ leather ‖ Vacheleder n. ‖ cuir m. de vache.

nebula ‖ Nebelfleck m. ‖ nébuleuse f.

nebule ‖ Nebel m. ‖ brouillard m.; brume f.; non-vue f.

necessary ‖ notwendig; nötig; erforderlich ‖ nécessaire. / ~ power ‖ Kraftbeanspruchung f. ‖ force f. nécessaire. / ~ room ‖ Raumbedarf m. ‖ place f. nécessaire; encombrement m.

necessary ‖ Latrine f. ‖ lieux pl. m. d'aisance; latrine f. / ~ pit ‖ Latrine f. ‖ lieux pl. m. d'aisance; latrine f.

neck ‖ Hals m. ‖ cou m.; col m.; gorge f.; collet m. / ~ (Tyre) ‖ Füllansatz m. ‖ manche f. de gonflement *ou* de remplissage *ou* appendice. / ~ (Still) ‖ Röhrenverbindung f. ‖ tubulure f. / ~ of an axle (Coachm) ‖ Achsschenkel m. ‖ fusée f. d'essieu; (tourillon m.) / ~ of a Bessemer convertor ‖ Hals m. einer Bessemerbirne ‖ bec m. d'un convertisseur Bessemer. / ~ of the broach ‖ Räumnadelhals m. ‖ tige f. de la broche. / ~ of a column ‖ Säulenhals m. ‖ col m. *ou* gorge f. de colonne. / ~ of a hatchet ‖ Kehle f. einer Axt ‖ néron m. d'une hache. / the stresses pl. in the ~ of the pin ‖ die Beanspruchung f. in der Kehle des Zapfens ‖ les efforts mpl. au congé du tourillon. / ~ of a retort (Chem) ‖ Retortenhals m. ‖ col m. *ou* cou m. d'une cornue. / ~ of a shaft ‖ Hals m. einer Welle ‖ gorge f. *ou* congé m. d'un arbre. / sharp ~ between journal and crank cheek ‖ scharfer Übergang m. zwischen Zapfen und Hohlkehle ‖ congé m. fortement marqué entre le tourillon et le flasque de manivelle. / ~ of a vessel ‖ Halsverengung f. eines Gefäßes ‖ écolète f. *ou* écolette f. d'un vase.

neck cord ‖ Platinenschnur f. ‖ collet m.

necked, wide-~ bottle ‖ weithalsige Flasche f. ‖ bouteille f. à goulot large.

neckerchief ‖ Halstuch n. ‖ fichu m.

neck groove of insulator ‖ Drahtlager m. am Isolator ‖ gorge f. de l'isolateur. / ~ groove binding ‖ Halsbindung f. ‖ ligature f. sur la gorge de l'isolateur. / ~ journal ‖ Halszapfen m. ‖ tourillon m. à collets.

necklace ‖ Halskette f. ‖ collier m.

necklet ‖ Halsband n.; Halskette f. ‖ collier m.

neck ring ‖ Grundring m. ‖ bague f. du fond.

necktie ‖ Schlips m.; Krawatte f.; Binder m. ‖ cravate f. / ~ cloth ‖ Krawattenstoff m. ‖ étoffe f. à cravates. / ~ machine ‖ Krawattenmaschine f. ‖ machine f. à cravates. / ~ material ‖ Krawattenstoff m. ‖ étoffe f. pour cravates.

neck twines pl. ‖ Heber mpl. ‖ arcades fpl. / ~ wire ‖ Drahthalsband n. ‖ collier m. en fil de fer.

nectarine ‖ Nektarine f. ‖ brugnon m.

needle ‖ Nadel f. ‖ aiguille f. / ~ (Mine) ‖ Schießnadel f.; Räumnadel f. ‖ épinglette f. / baiting ~ ‖ Ködernadel f. ‖ aiguille f. à amorcer. / ~ of a balance ‖ Zunge f. einer Wage ‖ languette f. / barbed ~ ‖ Bartnadel f. ‖ aiguille f. à barbe. / bolt rope ~ ‖ Lieknadel f. ‖ aiguille f. à ralinguer. / ~ for circular frame ‖ Rundstuhlnadel f. ‖ aiguille f. de métier rond. / compass ~ ‖ Magnetnadel f. ‖ aiguille f. aimantée *ou* de compas. / darning ~ ‖ Stopfnadel f ‖ aiguille f. à repriser *ou* à ravauder. / ~ with drilled eye ‖ Nähnadel f. mit gebohrtem Ohr ‖ aiguille f. à œil percé au foret. / embroidering machine ~ ‖ Stickmaschinennadel f. ‖ aiguille f. pour métiers à broder. / embroidery ~ ‖ Sticknadel f. ‖ aiguille f. à broder. / eye-pointed ~ ‖ Lochnadel f. ‖ aiguille f. percée. / fancy ~ ‖ Schmucknadel f. ‖ aiguille f. de fantaisie. / ~ for gramophone ‖ Grammofonnadel f. ‖ aiguille f. pour gramophone. / knitting ~ ‖ Stricknadel f. ‖ aiguille f. à tricoter. / knitting machine ~ ‖ Strickmaschinennadel f. ‖ aiguille f. de tricoteuse. / loading ~ (Mine) ‖ Bohrnadel f.; Raumnadel f. ‖ Räumnadel f.; Schießnadel f. ‖ épinglette f. / long narrow channelled ~ ‖ Ritznadel f.; Rute f.; Samtnadel f. ‖ fer m. *ou* épingle f. d'un métier à velours; baguette f. / ~ for mechanical work ‖ Maschinenarbeitsnadel f. ‖ aiguille f. *ou* épingle f. pour les ouvrages de tricoteuse. / ~ for needle work ‖ Handarbeitsnadel f. ‖ aiguille f. *ou* épingle f. pour les ouvrages manuels ou de dames. / packing ~ ‖ Packnadel f. ‖ aiguille f. à empaqueter. / roping ~ ‖ Lieknadel f. ‖ aiguille f. à ralinguer. / round-eye ~ ‖ Nadel f. mit rundem Öhr ‖ aiguille f. à chas rond. / sewing ~ ‖ Nähnadel f. ‖ aiguille f. à coudre. / ~ for sewing machines ‖ Nähmaschinennadel ‖ aiguille f. pour machines à coudre. / silver-eyed ~ ‖ Nadel f. mit silbernem Öhr ‖ aiguille f. à tête d'argent. / stocking frame ~ ‖ Strickmaschinennadel f. ‖ aiguille f. pour métiers à bas. / ~ for talking machines ‖ Nadel f. für Sprechmaschinen ‖ aiguille f. pour machines parlantes. / undamped ~ ‖ ungedämpft schwingende Nadel f. ‖ aiguille f. folle. / Y-~ ‖ Y-Nadel f. ‖ aiguille f. à l'Y. / the ~ yaws ‖ die Magnetnadel schwankt hin und her ‖ l'aiguille f. vacille *ou* danse *ou* va-et-vient.

needle beard ‖ Nadelbart m. ‖ barbe f. d'aiguille. / ~ bed ‖ Nadelbett n. ‖ lit m. d'aiguilles. / ~ bending nippers pl. ‖ Nadelbiegezange f. ‖ coudoir m. d'aiguilles. / ~ board ‖ Nadelbrett n. ‖ planchette f. des aiguilles. / ~ bronzer ‖ Nadelbronzierer m. ‖ bronzeur m. d'aiguilles. / ~ bush (Mot) ‖ Nadelhülse f. ‖ douille f. d'aiguille. / ~ case ‖ Nadelbüchse f. ‖ étui m. à aiguilles. / ~ caster ‖ Nadelsetzer m. ‖ monteur m. d'aiguilles. / ~ cylinder ‖ Nadelzylinder m. ‖ cylindre

m. à aiguilles. / ~ distance ‖ Nadelentfernung f. ‖ écartement m. des aiguilles. / ~ dressing plyers pl. ‖ Nadelrichtzange f. ‖ pince f. à dresser les aiguilles. / ~ driller ‖ Nadelaugendriller m. ‖ drilleur m. d'aiguilles. / ~ embroiderer ‖ Plattstichstickerin f. ‖ brodeuse f. en plumetis. / ~ embroidering ‖ Nadelstickerei f. ‖ broderie f. à l'aiguille. / ~ eye ‖ Nadelöhr n. ‖ chas m. d'aiguille. / ~ file ‖ Nadelfeile f. ‖ lime f. à aiguille. / ~ filer ‖ Nadelfeiler m. ‖ limeur m. d'aiguilles. / ~ fracture ‖ nadelförmiger Bruch ‖ fracture f. aciculaire *ou* aiguillé *ou* hérissé. / ~ full ‖ eingefädelte Nadel f. ‖ aiguillée f. / ~ gauge ‖ Nadelmaß n. ‖ jauge f. à piquer. / ~ holder ‖ Nadelhalter m. ‖ porte-aiguille m. / machine for the ~ industry ‖ Nadelindustriemaschine f. ‖ machine f. pour l'industrie des aiguilles. / ~ instrument ‖ Nadelapparat m. ‖ appareil m. à aiguilles. / ~ iron ore ‖ Nadeleisenerz n.; Pyrrhosiderit m. ‖ fer m. hydro-oxydé *ou* oxydé hydraté. / ~ lubricator ‖ Nadelöler m. ‖ graisseur m. à aiguilles.

needle machine, latch ~ ‖ Zungennadelmaschine f. ‖ machine f. à aiguilles à languettes. / spring ~ ‖ Spitzennadelmaschine f. ‖ machine f. à aiguilles à bec.

needle maker ‖ Nadelmacher m.; Nadler m. ‖ aiguillier m. / ~'s articles pl. ‖ Nadlerwaren fpl. ‖ articles mpl. de l'aiguillier.

needle ore ‖ Nadelerz n.; Belonit m. ‖ bélonite f.; aciculite f.; aikinite f. / ~ paperer ‖ Nadeleinpacker m. ‖ empaqueteur m. d'aiguilles. / interchangeable ~ point of the compass ‖ auswechselbare Nadelspitze f. des Kompasses ‖ pointe f. sèche interchangeable du compas. / ~ polisher ‖ Nadelpolierer m. ‖ polisseur m. d'aiguilles. / ~ row ‖ Nadelreihe f. ‖ rangée f. d'aiguilles. / ~ shank ‖ Nadelschaft m. ‖ tige f. d'aiguille. / ~-shaped ‖ nadelförmig ‖ aciculaire. / ~ sorter ‖ Nadelsortierer m. ‖ trieur m. d'aiguilles. / ~ splitter ‖ Nadelaufreiher m. ‖ enfileur m. d'aiguilles. / ~ spring ‖ Nadelfeder f. ‖ ressort m. à *ou* d'aiguille.

needless, to avoid all ~ transport to and fro ‖ zur Vermeidung f. aller entbehrlichen Hin- und Hertransporte ‖ à éviter tout transport de va-et-vient inutile.

needle stamper ‖ Kerbenstampfer m. ‖ estampeur m. des aiguilles à coudre. / ~ tapestry ‖ Nadelstickerei f. ‖ tapisserie f. à l'aiguille. / ~ telegraph ‖ Zeigertelegraf m. ‖ télégraphe m. à aiguille *ou* à cadran. / threading of the ~ ‖ Einfädeln n. der Nadel ‖ enfilage m. de l'aiguille. / ~ thread take-up ‖ Nadelfadenzug m. ‖ crochet m. tendeur de file. / ~ valve ‖ Nadelventil n. ‖ soupape f. à pointeau m. / ~ valve seat ‖ Nadelventilsitz m. ‖ siège m. du pointeau. / ~ weir ‖ Nadelwehr n. ‖ barrage m. à aiguilles. / ~ wire ‖ Nadeldraht m. ‖ fil m. pour aiguilles. / ~ work ‖ Näharbeit f.; Handarbeit ‖ ouvrage m. de couture; travail m. à l'aiguille. / ~ works pl. ‖ Nadelfabrik f. ‖ aiguillerie f.

needy ‖ notdürftig ‖ à peine suffisant; indigent.

negative ‖ negativ ‖ négatif. / ~ acceleration ‖ Verzögerung f. ‖ retardation f. / ~ paper ‖ Negativpapier n. ‖ papier m. sensible négatif. / ~ photograph ‖ Negativ n.;

negative Fotografie f. ‖ image f. *ou* épreuve f. négative. / ~ picture ‖ negative Fotografie f. ‖ image f. *ou* épreuve f. négative. / ~ plate ‖ Minusplatte f.; negative Platte ‖ plaque f. négative. / ~ varnish ‖ Negativlack m. ‖ vernis m. négatif.

negative ‖ Negativ n. ‖ film m. *ou* épreuve f. négatif; image f. négative. / ~ reading microscope ‖ Negativmeßmikroskop n. ‖ microscope m. de mesure pour négatifs; microscope m. pour la mensuration des négatifs. / ~ viewing apparatus ‖ Betrachtungsapparat m. für Negative ‖ appareil m. pour examiner les négatifs.

negatron ‖ Negatron n. ‖ négatron m.

neglect, to ‖ vernachlässigen ‖ négliger.

negligent ‖ säumig ‖ retardataire.

negotiability ‖ Übertragbarkeit f. ‖ transmissibilité f.

negotiate, to ‖ verhandeln ‖ négocier. / ~ a loan ‖ eine Anleihe f. abschließen ‖ contracter un emprunt.

negotiation, to abandon ~s ‖ Verhandlungen fpl. aufgeben ‖ abandonner les négociations fpl. / to enter into ~s ‖ in Unterhandlung f. *oder* in Verbindung f. treten ‖ entrer en relations fpl.

neighbourhood traffic ‖ Nachbarortverkehr m. ‖ trafic m. avec les réseaux voisins.

neighbouring position ‖ Nachbarstellung f. ‖ position f. voisine.

Neocomian ‖ Neokom n. ‖ néocomien m.

neolite ‖ Neolith m. ‖ néolite f.

neon ‖ Neon n. ‖ néon m.

neotype ‖ Neotyp m. ‖ néotype m.

Neozoic group ‖ neozoische Gruppe f. ‖ groupe f. néozoïque.

nepheline ‖ Nepheline m.; Eläolith m.; Davyn m.; Fettstein m. ‖ néphéline f.; elæolite f.; pierre f. grasse.

nephelinite ‖ Nephelinit m. ‖ néphélinite f.

nephelite ‖ Fettstein m. ‖ néphéline f.

nephrite ‖ Nephrit m.; Beilstein m. ‖ néphrite f.; jade m.

Nernst lamp ‖ Nernstlampe f. ‖ lampe f. Nernst.

nerve ‖ Nerv m. ‖ nerf m. / ~ (Arch) ‖ Gewölbrippe f.; Rippe f. ‖ côte f.; lierne f.; nerf m. *ou* nervure f. de voûte. / corneal ~ ‖ Hornhautnerv m. ‖ nerf m. de la cornée.

nerve broach ‖ Nervnadel f. ‖ tire-nerfs m. / ~ fibre ‖ Nervenfaser f. ‖ fibre f. de nerfs. / ~ tissue ‖ Nervensubstanz f. ‖ substance f. des nerfs.

nervine (Preparation) ‖ Nerven(beruhigungs)mittel n. ‖ nervin m.; calmant m.

nervose (Bot) ‖ gerippt ‖ nervé; nerveux.

nervous ‖ nervös ‖ nerveux. / ~ current ‖ Nervenstrom m. ‖ courant m. des nerfs. / ~ system ‖ Nervensystem n. ‖ système m. nerveux. / central ~ system ‖ Zentralnervensystem n. ‖ système m. nerveux central.

nervure (Arch) ‖ Gewölberippe f.; Rippe f. ‖ nervure f. de voûte; côte f.; lierne f. / ~ (Bot) ‖ Blattrippe f.; Blattnerv m. ‖ côte f. d'une feuille.

nest ‖ Nest n. ‖ nid m. / ~ of boiler tubes ‖ Kesselrohrbündel n. ‖ faisceau m. de tubes à chaudières. / ~ of gear wheels ‖ ein Satz m. Zahnräder ‖ équipage m. d'engrenages. / ~ of ore ‖ Erznest n. ‖ nid m. de minerai.

net amount ‖ Nettobetrag m. ‖ montant m. *ou* produit m. net. / ~ cost Nettopreis

m. ‖ prix m. net. / ~ earnings pl. ‖ Reingewinn m. ‖ produit m. *ou* bénéfice m. net. / ~ equivalent (Tel) ‖ Restdämpfung f. ‖ équivalent m. net; atténuation f. résiduelle. / ~ freight ‖ Nettofracht f. ‖ fret m. net. / ~ gain ‖ Reinertrag m. ‖ produit m. net. / ~ load ‖ Nettolast f. ‖ poids m. net. / ~ price ‖ Nettopreis m. ‖ prix m. net. / ~ proceeds pl. ‖ Reinertrag m.; Nettoertrag m. ‖ produit m. net. / ~ produce ‖ Reinertrag m.; Reingewinn m. ‖ bénéfice m. *ou* produit m. net. / ~ profit ‖ Reinertrag m.; Reingewinn m. ‖ bénéfice m. *ou* produit m. net. / ~ receipt ‖ Nettoeinnahme f. ‖ recette f. nette. / ~ sale ‖ Nettoverkauf ‖ vente pl. nette. / ~ tonnage ‖ Nettotonnengehalt m. ‖ tonnage m. net. / ~ weight ‖ Nettogewicht n.; Reingewicht n. ‖ poids m. net. / ~ yield ‖ Nettoerträgnis f. ‖ rendement m. net.

net ‖ Netz n. ‖ filet m. / anchored ~ ‖ Standnetz n.; Stellnetz n. ‖ filet m. *ou* rets m. sédentaire. / ball ~ ‖ Ballnetz n. ‖ filet m. à balles. / broché ~ ‖ broschierter Tüll m. ‖ tulle m. broché. / close-meshed ~ ‖ engmaschiges Netz n. ‖ palliole f. / enclosing ~ ‖ Einschlußnetz n. ‖ rets m. d'enceinte. / figured ~ ‖ Mustertüll m. ‖ tulle m. façonné. / ~ for the fishery ‖ Fischereinetz n. ‖ filet m. de pêche. / fixed ~ ‖ Standnetz n.; Stellnetz n. ‖ filet m. *ou* rets m. sédentaire. / game bag ~ ‖ Jagdtaschennetz n. ‖ filet m. de carnassière. / ~ for luggage ‖ Gepäcknetz n. ‖ filet m. à bagages. / plain ~ ‖ glatter Tüll m. ‖ tulle m. uni. / railway ~ ‖ Gleisnetz n.; Eisenbahnnetz n. ‖ réseau m. de voies ferrées. / surrounding ~ ‖ Einschlußnetz n. ‖ rets m. d'enceinte.

net embroidering ‖ Tüllstickerei f. ‖ broderie m. au crochet sur tulle *ou* tissu à jour; broderie f. de tulle. / ~ fisherman ‖ Netzfischer m. ‖ pêcheur m. au filet. / ~ folder ‖ Spitzenwicklerin f. ‖ plieuse f. de dentelles. / ~ knitter ‖ Netzstricker m. ‖ tricoteur m. de filets. / ~ knotter ‖ Netzknoter m. ‖ noueur m. de filets. / ~ machine ‖ Tüllwebstuhl m. ‖ métier m. à tulle. / ~ maker ‖ Netzmacher m.; Netzstricker m. ‖ filetier m.; laceur m. / bobbin ~ maker ‖ Tüllweber m. ‖ tulliste m. / ~ masonry ‖ Netzverband m. ‖ maçonnerie f. maillée *ou* en échiquier *ou* réticulée; ouvrage m. réticulé. / ~ mender (Fish) ‖ Netzflicker m. ‖ réparateur m. de filets. / ~ mender (Textile) ‖ Spitzenausbesserin f. ‖ raccommodeuse f. à l'écru de dentelles. / ~ mending ‖ Netzausbesserung f. ‖ raccommodage m. de filets. / ~ netting machine ‖ Netzknüpfmaschine f. ‖ machine f. de nouage de filets.

netting ‖ Geflecht n. ‖ grillage m. / sieve ~ ‖ Siebgewebe n. ‖ toile f. de tamisage. / wire ~ ‖ Drahtgeflecht n. ‖ treillis m. *ou* toile f. métallique; (réseau m. en fil).

netting machine ‖ Netzknüpfmaschine f. ‖ machine f. à nouer les filets. / wire ~ ‖ Drahtflechtmaschine f. ‖ machine f. à faire les grillages.

netting needle ‖ Filetnadel f. ‖ aiguille f. à filet; navette f. à fileter. / ~ washing ‖ Netzwäscherei f. ‖ lavage m. de filets. / ~ wire ‖ Webedraht m. ‖ fil m. à filet *ou* à toile métallique.

net tissue, tulle like ~ ‖ tüllartiges Netzgewebe n. ‖ tissu m. façon tulle.

nettle ‖ Nessel f. ‖ ortie f. / dead ~ ‖ Taubnessel f. ‖ lamier m. / ~ fibre ‖ Nesselfaser f. ‖ fibre f. d'ortie. / ~ fibre weaver ‖ Nesselweber m. ‖ tisseur m. de fil d'ortie. / ~ herb ‖ Brennesselkraut n. ‖ herbe f. d'orties. / ~ ribbon ‖ Nesselband n. ‖ ruban m. en fil d'ortie.

net washer ‖ Netzwäscher m. ‖ laveur m. de filets. / ~ weaving ‖ Tüllweberei f. ‖ tissage m. de tulle.

network (Arch) ‖ Netzwerk n. ‖ nattes fpl.; treillis m. / closed ~ (Electr) ‖ geschlossenes Netz n. ‖ réseau m. fermé. / direct current ~ ‖ Gleichstromnetz n. ‖ réseau m. à courant continu. / distribution ~ ‖ Verteilungsnetz n. ‖ réseau m. de distribution. / extensive ~ of standard and narrow gauge lines of railroads ‖ weitverzweigtes Netz n. normal- und schmalspuriger Schienenstränge ‖ réseau m. étendu de voies normales et étroites. / five-wire ~ ‖ Fünfleiternetz n. ‖ réseau m. à cinq fils. / to develop a ~ of high-capacity transmission and connection lines ‖ ein Netz n. leistungsfähiger Übertragungsleitungen und Kupp(e)lungsleitungen ausbauen ‖ développer un réseau m. puissant de lignes de transmission et d'interconnections. / high-tension ~ ‖ Hochspannungsnetz n. ‖ réseau m. à haute tension. / to improve existing ~s pl. ‖ bestehende Leitungsnetze npl. verbessern ‖ améliorer les réseaux mpl. existants. / ~ of lines ‖ Gleisnetz n. ‖ réseau m. de voies. / lode forming a ~ (Mine) ‖ Netzgang m. ‖ filon m. en forme de réseau. / monophase ~ ‖ Einphasennetz n. ‖ réseau m. monophasé. / ~ of railways ‖ Eisenbahnnetz n.; Gleisnetz n. ‖ réseau m. de chemins de fer *ou* de voies ferrées. / ~ of streams and torrents ‖ Flußnetz n.; Flußsystem n. ‖ réseau m. fluvial *ou* de cours d'eau. / ~ fusion of the local supply ~s ‖ Zusammenfassung f. der lokalen *oder* örtlichen Versorgungsnetze ‖ groupement m. des réseaux de distribution locaux. / telemetering ~ ‖ Fernmeßnetz n. ‖ système m. de télémesure. / three-phase ~ with four wires ‖ Dreiphasenvierleiternetz n. ‖ réseau m. triphasé à quatre fils. / three-wire ~ ‖ Dreileiternetz n. ‖ réseau m. à trois fils. / traction ~ ‖ Straßenbahnnetz n. ‖ réseau m. de tramway. / two-wire ~ ‖ Zweileiternetz n. ‖ réseau m. à deux fils.

network calling machine ‖ Netzrufmaschine f. ‖ machine f. d'appel à réseau. / ~ map ‖ Netzplan m. ‖ plan m. du réseau.

neutral carbonate of potassium ‖ Pottasche f.; kohlensaures Kalium n. ‖ potasse f.; carbonate m. de potasse. / of ~ equilibrium ‖ statisch indifferent ‖ d'équilibre neutre. / ~ fat ‖ Neutralfett n. ‖ corps m. gras neutre. / ~ fibre ‖ neutrale Faser f. ‖ fibre f. neutre. / absorbing filter of ~ glass ‖ neutrales Blendglas n. ‖ verre m. sombre à teinte neutre.

neutral line ‖ Neutrale f.; neutrale Linie f.; Nullinie f. ‖ ligne f. neutre; axe m. neutre. / ~ of a magnet ‖ Indifferenzzone f. eines Magnets ‖ zone f. neutre d'un aimant.

neutral position ‖ Ruhestellung f.; Nullstellung f. ‖ position f. neutre *ou* de repos. / ~ tint ‖ Neutralfärbung f. ‖ teinte neutre. / ~ wire system ‖ Mittelleitersystem n. ‖ système m. à trois fils. / ~ zone ‖ Indifferenzzone f. ‖ zone f. neutre.

29*

neutrality ‖ Neutralität f. ‖ neutralité f.

neutralization ‖ Neutralisieren n. ‖ neutralisation f. / ~ plant ‖ Neutralisationsanlage f. ‖ installation f. de neutralisation.

neutralize, to ‖ neutralisieren ‖ neutraliser.

neutralizing condenser ‖ Neutralisierungskondensator m. ‖ condensateur m. de neutralisation.

neutro apparatus (Radio) ‖ Neutrogerät n. ‖ récepteur m. neutrodyne.

neutrodyne ‖ Neutrodyn n. ‖ neutrodyne m. / ~ receiver ‖ Neutrodynempfänger m. ‖ récepteur m. neutrodyne.

new building ‖ Neubau m. ‖ nouvelle construction f. / ~ construction ‖ Neubau m. ‖ nouvelle construction f. / ~ formation ‖ Neubildung f. ‖ formation f. nouvelle. / ~ milk butter ‖ Frischmilchbutter f. ‖ beurre m. de premier lait. / ~ moon ‖ Neumond m. ‖ nouvelle lune f. / ~ snow ‖ Neuschnee m. ‖ neige f. tombée récemment.

Newfoundland fisherman ‖ Neufundlandfischer m. ‖ pêcheur m. de Terre-Neuve.

Newfoundlandman ‖ Neufundlandfahrer m. ‖ Terre-Neuvier m.

newness ‖ Neuheit f. ‖ nouveauté f.

newshouse ‖ Zeitungsdruckerei f. ‖ imprimerie f. d'un journal.

newspaper ‖ Zeitung f. ‖ journal m. / eight-page ~ ‖ achtseitige Zeitung f. ‖ journal m. de huit pages. / old ~s pl. ‖ alte Zeitungen fpl. ‖ vieux journaux mpl.

newspaper article ‖ Zeitungsartikel m. ‖ article m. de journal. / ~ folder ‖ Zeitungsfalzerin f. ‖ plieuse f. de journaux. / ~ holder ‖ Zeitungshalter m. ‖ porte-journaux m. / ~ printing ‖ Zeitungsdruck m. ‖ impression f. de journaux. / ~ printing establishment ‖ Zeitungsdruckerei f. ‖ imprimerie f. de journaux. / ~ rotary machine ‖ Zeitungsrotationsdruckmaschine f. ‖ machine f. rotative à imprimer des journaux. / ~ wrapper ‖ Kreuzband n. ‖ sous-bande f. / ~ work ‖ Zeitungsdruck m. ‖ impression f. de journaux.

New Zealand flax ‖ Neuseelandflachs m. ‖ lin m. de la nouvelle Zélande.

nib (Bird) ‖ Schnabel m. ‖ bec m. / ~ (Coffee) ‖ Kaffeebohne f. ‖ grain m. de café. / nickel-plated ~ ‖ vernickelte Spitze f. ‖ bec m. nickelé. / ~ of a writing pen ‖ Spitze f. einer Schreibfeder ‖ pointe f. d'une plume métallique.

nibbling machine ‖ Dekupiermaschine f.; Aushaumaschine ‖ machine f. à découper; grignoteuse f. / high-capacity ~ ‖ Hochleistungsdekupiermaschine f. ‖ machine f. à découper ou grignoteuse f. à grand rendement.

niccolite ‖ Rotnickelkies m.; Kupfernickel m. ‖ nickéline f.

niche ‖ Nische f.; Halbkuppel f. ‖ niche f. / ~ vaulting ‖ Chorgewölbe n. ‖ voûte f. en niche.

nichrotherm steel ‖ Nichrothermstahl m. ‖ acier m. nichrotherm.

nick, to ~ out ‖ ausfurchen; furchen ‖ sillonner.

nick ‖ Kerbe f. ‖ encoche f.; entaille-coche f.

nicked, bar sharply ~ with a chisel ‖ mit dem Meißel scharf eingekerbter Stab m. ‖ barreau m. nettement entaillé au ciseau ou au burin.

nickel, to ‖ vernickeln ‖ nickeler.

nickel ‖ Nickel n. ‖ nickel m. / arsenical ~ ‖ Rotnickelkies m.; Arseniknickel n. ‖

nickel m. arsenical. / hard solid ~ ‖ Hartnickel m. ‖ nickel m. pur trempé. / to plate with ~ see to nickel. / pure ~ en apparatus ‖ Reinnickelapparat m. ‖ appareil m. en nickel pur. / spongy ~ ‖ Nickelschwamm m. ‖ nickel m. spongieux.

nickel alloy ‖ Nickellegierung f. ‖ alliage m. de nickel. / ~ anode ‖ Nickelanode f. ‖ anode f. de nickel. / ~ apparatus ‖ Nickelapparat m. ‖ appareil m. en nickel. ~ arseniate ‖ Nickelblüte f.; arsensaures Nickel n. ‖ nickel m. arséniaté. / ~ bath ‖ Vernickelungsbad n. ‖ bain de nickelage. / ~ bolt ‖ Nickelbolzen m. ‖ boulon m. de nickel. / ~ bronze ‖ Nickelbronze f. ‖ bronze m. au nickel. / ~ copper ‖ Nickelkupfer n. ‖ alliage m. de nickel et de cuivre.

nickeled ‖ vernickelt ‖ nickelé. / ~ steel spectacles pl. ‖ vernickelte Stahlbrille f. ‖ lunettes fpl. en acier nickelé.

nickel electrode ‖ Nickelelektrode f. ‖ électrode f. de nickel. / ~ foil ‖ Nickelpapier n. ‖ papier m. nickelé. / ~ galvanoplastic plant ‖ Nickelgalvanoplastikanlage f. ‖ installation f. de galvanoplastic au nickel. / ~ glance ‖ Nickelglanz m.; Arsennickelglanz m. ‖ nickel m. arseniosulfuré. / ~ goods pl. ‖ Nickelwaren fpl. ‖ articles mpl. en nickel.

nickelgreen ‖ Nickelgrün n. ‖ vert m. de nickel.

nickel gymnite ‖ Nickelgymnit m. ‖ nickelgymnite f.

nickelic oxide ‖ Nickeloxyd n. ‖ sesquioxyde m. de nickel.

nickelin ‖ Nickelin n. ‖ nickeline m.

nickeling ‖ Vernickelung f. ‖ nickelage m. / rapid ~ ‖ Schnellvernicklung f. ‖ nickelage m. rapide. / ~ salt ‖ Vernickelungssalz n. ‖ sel m. de nickelage. / ~ workshop ‖ Vernickelungsanstalt f. ‖ atelier m. de nickelage.

nickelize, to ‖ vernickeln ‖ nickeler.

nickelizer ‖ Vernickler m. ‖ nickeleur m.

nickel iron ‖ Nickeleisen n. ‖ ferronickel m. / ~ matrice ‖ Nickelmatrize f. ‖ matrice f. en nickel. / ~ metal ‖ Nickelstein m. ‖ matte f. de nickel. / ~ mine ‖ Nickelerzbergwerk n. ‖ mine f. de nickel m. / ~ ochre ‖ Nickelocker m. ‖ nickel-ocre m.; nickel m. arséniaté. / ~ oxide ‖ Nickeloxyd n. ‖ oxyde m. de nickel. / ~ plate ‖ Nickelblech n. ‖ tôle f. de nickel.

nickel-plated ‖ vernickelt ‖ nickelé. / ~ bolt ‖ vernickelter Bolzen m. ‖ boulon m. nickelé. / ~ mounting ‖ vernickelte Fassung f. ‖ monture f. nickelée.

nickel plating ‖ Vernickelung f. ‖ nickelage m. / heavy ~ ‖ Starkvernickelung f. ‖ nickelage m. extra fort. / ~ plant ‖ Vernickelungsanlage f. ‖ installation f. de nickelage.

nickel powder ‖ Nickelpulver n. ‖ nickel m. en poudre. / ~ production ‖ Nickelgewinnung f. ‖ métallurgie f. du nickel. / ~ pyrites pl. ‖ Nickelkies m.; Millerit m. ‖ nickel m. sulfuré; millerite f. / ~ rivet ‖ Nickelniet n. ‖ rivet m. de nickel. / ~ rod ‖ Nickelstange f. ‖ barre f. en nickel. / ~ rolling mill ‖ Nickelwalzwerk n. ‖ laminoir m. à nickel. / ~ salts pl. ‖ Nickelsalze npl. ‖ sels mpl. de nickel. / ~ sheet ‖ Nickelblech n. ‖ feuille f. en nickel. / ~ speiss ‖ Nickelspeise f. ‖ speiss m. de nickel.

nickel steel ‖ Nickelstahl m. ‖ acier m. au nickel. / ~ high in Ni ‖ hochprozentiger

Nickelstahl m. ‖ acier m. (au nickel) à haute teneur en Ni. / ~ pendulum ‖ Nickelstahlpendel m. ‖ pendule m. en acier au nickel.

nickel strip ‖ Nickelstreifen m. ‖ ruban m. de nickel. / sulphide of ~ ‖ Nickelkies m.; Millerit m. ‖ nickel m. sulfuré; millerite f. / ~ vessel ‖ Nickelgefäß n. ‖ récipient m. de nickel. / ~ ware ‖ Nickelwaren fpl. ‖ articles mpl. ou objets mpl. en nickel. / ~ wire ‖ Nickeldraht m. ‖ fil m. de nickel. / ~ works pl. ‖ Nickelhütte f. ‖ fonderie f. de minerai de nickel.

nicotine ‖ Nikotin n. ‖ nicotine f. / free from ~ ‖ nikotinfrei ‖ sans nicotine f. / cigarette poor in ~ ‖ nikotinarme Zigarette f. ‖ cigarette f. à basse teneur en nicotine.

niello (Engr) ‖ Schwarzschmelz m. ‖ nielle f. / to work in ~ ‖ niellieren ‖ nieller.

niello engraving ‖ Niello n.; schwarz ausgefüllte Metallgravierung f. ‖ niellure f. / ~ making (Watch) ‖ Niellierung f. ‖ niellage m.

Niger seed ‖ Nigersame m. ‖ graine f. de niger.

nigging chisel, broad ~ (Stone cutter) ‖ Scharriereisen n. ‖ ciseau m. à charruer.

night airplane ‖ Nachtflugzeug n. ‖ avion m. nocturne. / ~-blind eye ‖ nachtblindes Auge n. ‖ œil m. héméralope. / ~ blue ‖ Nachtblau n. ‖ bleu m. de nuit. / ~ bomber ‖ Nachtbombenflugzeug n. ‖ avion m. de bombardement de nuit. / ~ bombing aeroplane see ~ bomber. / ~ call ‖ Nachtruf m. ‖ appel m. de nuit. / ~ candle ‖ Nachtlicht n. ‖ veilleuse f. / ~ circuit ‖ Nachtschaltung f. ‖ couplage m. pour l'éclairage de nuit. / ~ connection (Tel) ‖ Nachverbindung f. ‖ liaison f. de nuit. / ~ current ‖ Nachtstrom m. ‖ courant m. de nuit. / ~ flying ‖ Nachtflug m. ‖ vol m. de nuit. / ~ glass ‖ Nachtglas n. ‖ binocle m. ou lunette ou jumelle f. de nuit. / ~ gown ‖ Nachtkleid n.; Nachtrock m. ‖ robe f. de chambre. / ~ lamp ‖ Nachtlampe f. ‖ lampe f. de nuit. / ~ light ‖ Nachtlicht n. ‖ veilleuse f.

nightman (Sewage removal) ‖ Senkgrubenreiniger m. ‖ vidangeur m. des fosses.

night photograph ‖ Nachtaufnahme f. ‖ photographie f. nocturne ou de nuit. / ~ position (Tel) ‖ Nachtplatz n. ‖ position f. de nuit. / ~ service ‖ Nachtdienst m. ‖ service m. de nuit.

nightshade berry ‖ Nachtschattenbeere f. ‖ baie f. de morelle.

night shift ‖ Nachtschicht f. ‖ travail m. de nuit. / ~ shift (Persons) ‖ Nachtschicht f. ‖ équipe f. de nuit. / ~ shirt ‖ Nachthemd n. ‖ chemise f. de nuit. / ~ signal ‖ Nachtsignal n. ‖ signal m. de nuit. / ~ signalling plant ‖ Nachtsignaleinrichtung f. ‖ dispositif m. de signal de nuit.

nightsoil man see nightman.

night tariff ‖ Nachttarif m. ‖ tarif m. de nuit. / ~ thunderstorm ‖ Nachtgewitter n. ‖ orage m. de nuit. / ~ trunk position (Tel) ‖ Nachtfernschrank m. ‖ groupe m. interurbain de nuit. / ~ watch ‖ Nachtwächter m. ‖ garde m. de nuit; veilleur m. / ~ wind ‖ Nachtwind m. ‖ vent m. de nuit. / ~ work ‖ Nachtschicht f.; Nachtarbeit f. ‖ poste m. ou travail m. de nuit.

ninepin ‖ Kegel m. ‖ quille f.

niobite ‖ Niobit n. ‖ niobite f.

nipper ‖ Hebedaumen m.; Nocken m. ‖ came f.; camme f. / ~ gauge (Print) ‖ Längenlinie f. ‖ réglette m. de longueur.

nippers pl. ‖ Kneifzange f. ‖ pince f.; tenailles fpl. / champagne ~ ‖ Champagnerzange f. ‖ pince f. à champagne. ‖ cutting ~ ‖ Beißzange f.; pince f. coupante. / flat-nosed and cutting ~ ‖ Flachzange f. mit Seitenschneider ‖ pince f. plate et coupante. / grating ~ ‖ Käfigbauerzange f. ‖ pince f. de cagiste. / nail ~ ‖ Nagelzange f.; Nagelzieher m. ‖ arrache-clou m. / ~ for needle bending ‖ Nadelbiegezange f. ‖ coudoir m. d'aiguilles. / side-cutting ~ ‖ Seitenschneider m. ‖ pince f. coupante de côté. / ticket ~ ‖ Kartenlochzange f.; Kartenlocher m.; Knipszange f. ‖ pince f. de contrôle. / weaver's ~ ‖ Noppzange f.; Weberzange f.; Klüppchen n. ‖ pincette f. du tisserand.

nipple ‖ Nippel m. ‖ raccord m. (fileté). / double ~ ‖ Doppelnippel m. ‖ double raccord m. fileté. / machine for making ~s ‖ Nippelmaschine f. ‖ machine f. à faire des raccords. / reduction ~ ‖ Reduktionsnippel m. ‖ raccord m. intermédiaire à pas de vis différent. / spoke ~ ‖ Speichennippel m. ‖ écrou m. de rayon. / wall ~ ‖ Wandnippel m. ‖ raccord m. mural.

nipple threading machine ‖ Nippelgewindeschneidmaschine f. ‖ machine f. à fileter les raccords *ou* nipples.

Nitra lamp ‖ Nitralampe f. ‖ lampe f. Nitra.

nitrate, to ‖ nitrieren ‖ nitrer.

nitrate ‖ Nitrat n.; salpetersaures Salz n. ‖ nitrate m.; azotate m. / ~ of lime ‖ Kalksalpeter m.; Nitrokalzit m. ‖ azotate m. de chaux; chaux f. nitratée; nitrocalcite m. / ~ of potash ‖ Kalisalpeter m. ‖ nitre m.; salpêtre m.; azotate m. de potasse. / ~ of silver ‖ Silbernitrat n. ‖ nitrate m. d'argent. / ~ of soda ‖ Natronsalpeter m. ‖ nitrate m. de soude. / ~ refinery ‖ Salpeterraffinerie f. ‖ raffinerie f. de salpêtre.

nitrating apparatus ‖ Nitrierapparat m. ‖ appareil m. de nitrification. / ~ cotton ‖ Nitrierbaumwolle f. ‖ coton m. à nitrer.

nitration ‖ Nitrierung f. ‖ nitration f.

nitre ‖ Kalisalpeter m. ‖ nitre m.; salpêtre m.; azotate m. de potasse. / ~ in bars ‖ Stangensalpeter m. ‖ nitre m. en baguettes. / ~ refined ‖ gereinigter *oder* raffinierter Salpeter m. ‖ salpêtre m. raffiné. / ~ paper ‖ Salpeterpapier n. ‖ papier m. nitré *ou* azotaté. / ~ plantation ‖ Salpeterplantage f. ‖ nitrière f.; salpêtrière f.

nitriary *see* nitre plantation.

nitric ‖ salpetersauer ‖ azotique.

nitric acid ‖ Salpetersäure f. ‖ acide m. azotique *ou* nitrique. / fuming ~ ‖ rauchende Salpetersäure f. ‖ acide m. azotique fumant. / synthetical ~ ‖ Luftsalpetersäure f. ‖ acide m. azotique synthétique.

nitric acid plant ‖ Salpetersäureherstellungsanlage f. ‖ installation f. à fabriquer l'acide azotique.

nitric ether ‖ Salpeteräther m. ‖ éther m. azotique. / ~ sulphuric acid ‖ Nitriersäure f. ‖ mélange f. sulfonitrique.

nitride, to ‖ nitrieren ‖ nitrurer.

nitrided steel ‖ Nitrierstahl m. ‖ acier m. nitruré.

nitride hardening ‖ Nitrierhärtung f. ‖ trempe f. par nitruration.

nitriding process ‖ Nitrierverfahren n. ‖ nitruration.

nitrite ‖ Nitrit n.; salpetrigsaures Salz n. ‖ nitrite m.

nitro-barite ‖ Barytsalpeter m. ‖ nitrobarite f. / ~-benzol ‖ Nitrobenzol n. ‖ nitrobenzol m.; nitrobenzène f. / ~-calcite ‖ Kalksalpeter m. ‖ azotate m. de chaux. / ~-cellulose ‖ Nitrozellulose f. ‖ nitrocellulose f. / ~-dye ‖ Nitrofarbstoff m. ‖ colorant m. nitré. / ~-gelatine ‖ Sprenggallerte f.; Sprenggelatine f. ‖ dynamite-gomme f.

nitrogen ‖ Stickstoff m. ‖ azote m.; nitrogène m. / aerial ~ ‖ Luftstickstoff m. ‖ azote m. atmosphérique. / ~ of lime ‖ Kalkstickstoff m.; Kalziumcyanamid n. ‖ cyanamide m. de chaux; cyamide m. calcique.

nitrogen, containing ~ ‖ stickstoffhaltig ‖ azoté. / ~ determination ‖ Stickstoffbestimmung f. ‖ dosage m. de l'azote; détermination f. du nitrogène. / ~ determining apparatus ‖ Stickstoffbestimmungsapparat m. ‖ appareil m. de dosage de l'azote. / ~ iodide ‖ Jodstickstoff m. ‖ iodure m. d'azote.

nitrogenous ‖ stickstoffhaltig ‖ azoté. / ~ lime works equipment ‖ Kalkstickstoff-fabrikeinrichtung f. ‖ installation f. de fabrique du cyanamide. / ~ manure ‖ Stickstoffdünger m. ‖ engrais m. azoté.

nitrogen producing plant ‖ Stickstoffgewinnungsanlage f. ‖ installation f. de la production d'azote. / production of ~ ‖ Stickstoffgewinnung f. ‖ production f. d'azote.

nitro-glycerine ‖ Nitroglyzerin n. ‖ nitroglycérine f. / ~-magnesite ‖ Nitromagnesit m.; Magnesiasalpeter m. ‖ nitromagnésite f. / ~-muriatic acid ‖ Königswasser n. ‖ acide m. nitro-chlorhydrique; eau f. régale. / ~ powder ‖ rauchschwaches Pulver n. ‖ coton-poudre m. / ~-saccharose ‖ Knallzucker m. ‖ nitrosaccharose m. / ~ silk ‖ Kollodiumseide f. ‖ soie f. de collodion.

nitrotoluene ‖ Nitrotoluol n. ‖ nitrotoluène m.

nitrous acid ‖ salpetrige Säure f. ‖ acide m. azoteux. / ~ ether ‖ Salpetersäureäthylester m. ‖ éther m. azoteux. / ~ fume ‖ nitroser Dampf m. ‖ vapeur f. azoteuse. / ~ gas ‖ nitroses Gas n. ‖ gas m. azoteux. / ~ oxide ‖ Lachgas n. ‖ gaz m. hilarant.

nob in wood ‖ Knorren m. *oder* Knoten m. *oder* Ast m. im Holz ‖ nœud m. dans le bois.

no-backlash ‖ frei von totem Gang m. ‖ exempt de tout jeu.

noble ‖ edel ‖ noble; précieux. / ~ earths pl. ‖ Edelerden fpl. ‖ terres fpl. nobles *ou* précieuses.

nodal point ‖ Knotenpunkt m. ‖ point m. nodal. / ~ point of plant ‖ Netzknotenpunkt m. ‖ centre m. d'alimentation.

node ‖ Knoten m. ‖ nœud m. / potential ~ ‖ Spannungsknoten m. ‖ nœud m. de tension.

no-delay service (Tel) ‖ Schnellverkehr m.; wartezeitloser Fernsprechverkehr m. ‖ trafic m. sans délai.

nodule (Geol) ‖ Niere f.; Knolle f. ‖ rognon m.; globule m. oblong; nodule m.

nog, to ~ the bay-work with bricks ‖ das Fachwerk mit Ziegeln ausmauern ‖ hourder les pans mpl. de bois.

noil (Spinn) ‖ Kämmling m. ‖ blousse f. / ~s pl. of carded waste-silk ‖ Seidenwerg n.; Stumpen mpl. ‖ déchet m. de cardette. / ~ dyeing ‖ Kämmlingswollefärberei f. ‖ teinture f. de blousses.

noise ‖ Geräusch n. ‖ bruit m. / ~ of bearings ‖ Brummen n. der Lager ‖ ronflement m. *ou* grincement m. des paliers. / ~ of click (Tel) ‖ Knackgeräusch n. ‖ bruit m. de clic. / ~ of exhaust ‖ Auspuffgeräusch n. ‖ bruit m. d'échappement. / extraneous ~ ‖ Nebengeräusch n. ‖ bruit m. étranger.

noise elimination ‖ Störungsbefreiung f. ‖ élimination f. de bruits.

noiseless ‖ geräuschlos ‖ silencieux. / ~ gliding ‖ geräuschloses Gleiten n. ‖ glissage m. silencieux. / ~ running ‖ geräuschloser *oder* ruhiger Gang m. ‖ allure f. *ou* marche f. silencieuse. / ~ working ‖ geräuschloses Arbeiten n. ‖ travail m. silencieux. / ~ working of the motor ‖ geräuschloses Arbeiten n. des Motors ‖ marche f. silencieuse du moteur.

noise measurement set ‖ Geräuschmesser m. ‖ appareil m. de mesure des bruits. / suppression of the ~ ‖ Abdämpfung f. der Geräusche ‖ amortissement m. du bruit.

noise voltage ‖ Geräuschspannung f. ‖ tension f. perturbatrice. / ~ tester ‖ Geräuschspannungsmesser m. ‖ appareil m. de mesure de la tension perturbatrice.

noisy ‖ geräuschvoll ‖ bruyant. / ~ running (Mach) ‖ geräuschvoller *oder* unruhiger Gang m. ‖ marche f. bruyante *ou* par à coups.

no-load ‖ Leerlauf m. ‖ marche f. à vide. / ~ bushing ‖ Leerlaufbuchse f. ‖ douille f. de marche à vide. / ~ characteristic ‖ Leerlaufcharakteristik f. ‖ caractéristique f. à circuit ouvert. / ~ current ‖ Leerlaufstrom m. ‖ courant m. à vide. / ~ cut-out ‖ Nullausschalter m.; Leerlaufausschalter m. ‖ disjoncteur m. *ou* interrupteur m. à zéro. / ~ friction ‖ Leergangsreibung f. ‖ frottement m. à vide. / ~ impedance ‖ Leerlaufwiderstand m. ‖ impédance f. en circuit ouvert. / ~ loss ‖ Leerlaufverlust m. ‖ perte f. à vide. / ~ nozzle ‖ Leerlaufdüse f. ‖ tuyère f. à vide. / ~ switch ‖ Nullausschalter m. ‖ interrupteur m. à zéro. / ~ test ‖ Leerlaufversuch m. ‖ essai m. à vide. / ~ time ‖ Leerzeit f. ‖ temps m. à vide. / ~ work ‖ Leergangsbewegung f. ‖ marche f. à vide.

nominal acceptivity ‖ Nennaufnahme f. ‖ réception f. nominale. / ~ current ‖ Nennstrom m. ‖ courant m. nominal. / ~ current strength ‖ Nennstromstärke f. ‖ nombre m. d'ampères / ~ diameter of case ‖ Gehäusenenndurchmesser m. ‖ diamètre m. nominal de la boîte. / ~ output ‖ Nennleistung f. ‖ débit m. nominal. / ~ tension ‖ Nennspannung f. ‖ tension f. nominale. / ~ torsional moment ‖ Nenndrehmoment n. ‖ moment m. de torsion nominal. / ~ value ‖ Nennwert m. ‖ valeur f. nominale. / ~ voltage ‖ Nennspannung f. ‖ tension f. nominale.

nomogram ‖ Fluchtlinientafel f. ‖ abaque m.; nomogramme m.

nomography ‖ Nomografie f. ‖ nomographie f.

non-acceptance ‖ Annahmeverweigerung f.; Nichtannahme f. ‖ refus m. d'acceptation; non-acceptation f.

nonagon ‖ Neuneck n. ‖ ennéagone m.

non-baking coal ‖ Magerkohle f. ‖ houille f. maigre.

non-balanced ‖ unausgeglichen ‖ non compensé.

non-conducting composition ‖ Isoliermittel n. ‖ isolant m.

non-conductor (Electr) ‖ Nichtleiter m. ‖ corps m. isolateur; diélectrique m.

non-coupled axle ‖ Laufachse f. ‖ essieu m. porteur.

non-delivery ‖ Nichtlieferung f. ‖ non-accomplissement m.; non-livraison f.

non-drying oil ‖ nichttrocknendes Öl n. ‖ huile f. non-sécheuse.

non-exploding reservoir ‖ explosionssicheres Gefäß n. ‖ vase m. à l'épreuve d'explosion; récipient m. inexplosible. / ~ vessel see ~ reservoir.

non-ferrous metal ‖ Nichteisenmetall n. ‖ métal m. non ferrique.

non-fulfilment ‖ Nichterfüllung f. ‖ inexécution f.

non-hygroscopic paper ‖ nichtwasseranziehendes Papier n. ‖ papier m. non hygroscopique.

non-inductive ‖ induktionsfrei ‖ sans induction f. / ~ load ‖ induktionsfreie Belastung f. ‖ charge f. non-inductive. / ~ shunt ‖ induktionsfreier Nebenschluß m. ‖ shunt m. non inductif.

non-inflammable gas ‖ unbrennbares Gas n. ‖ gaz m. non inflammable.

non-interchangeable ‖ unauswechselbar ‖ non-interchangeable.

nonius ‖ Nonius m. ‖ nonius m.; vernier m. / ~ graduation ‖ Noniusteilung ‖ trait m. de division de vernier. / ~ zero ‖ Noniusnullpunkt m. ‖ zéro m. du vernier.

non-listing calculating machine ‖ nichtschreibende Rechenmaschine f. ‖ machine f. à additioner ne comportant pas de mécanisme d'impression.

non-magnetic steel ‖ unmagnetischer Stahl m. ‖ acier m. non-magnétique.

non-malleable cast-iron ‖ nicht schmiedbarer Guß m. ‖ fonte f. non malléable.

non-metal ‖ Nichtmetall n. ‖ non-métal m.; métalloïde m.

non-oxidizing annealing and hardening furnace ‖ Blankhärte- und Glühofen m. ‖ four m. à tremper et à cémenter. / generator for producing ~ safety gases ‖ Erzeugungsmaschine f. für nicht oxydierende Schutzgase ‖ générateur m. à produire les gaz protecteurs non oxydants.

nonpareil ‖ Nonpareille f. ‖ nonpareille f.; corps m. six.

non-payment ‖ Nichtzahlung f. ‖ non-paiement m.

non-poisonous colour ‖ giftfreie Farbe f. ‖ couleur f. anodine. / ~ for toys ‖ giftfreie Farbe f. für Spielwaren ‖ couleur f. sans poison pour jouets. / ~ for victuals ‖ giftfreie Farbe f. für Genußmittel ‖ couleur f. inoffensive pour les aliments.

non-return flap ‖ Rückschlagklappe f. ‖ clapet m. de retenue. / ~ valve ‖ Rückschlagventil n. ‖ soupape f. de retenue.

non-rigidity of the connection ‖ Unstarrheit f. der Verbindung ‖ non-rigidité f. de la liaison ou du couplage.

non-skid (Auto) ‖ Gleitschutz m. ‖ antidérapant m. / ~ band ‖ Gleitschutz-

reifen m. ‖ antidérapant m. / ~ chain ‖ Schneekette f. ‖ chaîne f. antidérapante. / ~ cover with rivets ‖ Gleitschutzdecke f. mit Nieten ‖ protecteur m. antidérapant à rivets. / ~ device ‖ Gleitschutzvorrichtung f. ‖ antidérapant m. / ~ tyre ‖ Gleitschutzreifen m. ‖ pneu m. antidérapant.

non-spherical ‖ asphärisch ‖ asphérique.

non-stop flight ‖ Dauerflug m.; Flug m. ohne Halt ‖ vol m. de durée ou sans escale. / ~ train ‖ durchgehender Zug m. ‖ train m. direct.

non-striker ‖ Streikbrecher m. ‖ non-gréviste m.

non-synchronous gap (Radio) ‖ asynchrone Funkenstrecke f. ‖ éclateur m. non synchrone.

nontronite ‖ Nontronit m. ‖ nontronite f.

non-uniform combustion ‖ ungleichmäßige Verbrennung f. ‖ combustion f. non uniforme. / ~ movement ‖ ungleichförmige Bewegung f. ‖ mouvement m. non uniforme.

non-volatile ‖ nicht flüchtig ‖ non-volatil.

non-vortical ‖ wirbellos ‖ sans remous mpl.; sans tourbillons mpl.

non-working time ‖ Ruhezeit f. ‖ temps m. de repos.

noodle machine ‖ Nudelmaschine f. ‖ machine f. à faire des nouilles.

noontide ‖ Mittagszeit f.; Mittag m. ‖ heure f. de midi ou du dîner.

noose ‖ Schleife f.; Schlinge f. ‖ lacet m.

no parking ‖ Parkverbot n. ‖ parc m. interdit; interdit de parquer.

noria ‖ Noria f.; Paternosterwerk n. ‖ noria f.; patenôtre f.

normal ‖ normal ‖ normal.

normal (Geom) ‖ Normale f. ‖ normale f. / to return to ~ ‖ in die Ruhelage f. zurückkehren ‖ retourner en position f. normale.

normal acceleration ‖ Normalbeschleunigung f. ‖ accélération f. normale. / ~ admission ‖ normale Füllung f. ‖ admission f. normale. / ~ bend (Press) ‖ Normalbug m. ‖ pli m. normal. / ~ breadth of a river ‖ Normalbreite f. eines Flußes ‖ largeur f. du profil amélioré d'une rivière. / ~ curve ‖ Normalkurve f. ‖ courbe f. normale. / ~ cut-off see ~ admission. / ~ drive ‖ Normalantrieb m. ‖ commande f. normale. / ~ electrode ‖ Normalelektrode f. ‖ électrode f. normale. / ~ field of a mine where coal was recently disclosed by borings ‖ neuaufgeschlossenes Normalfeld n. einer Grubenanlage ‖ concession f. normale d'une installation minière récemment reconnue par sondages. / ~ force ‖ Senkrechtkraft f. ‖ force f. normale. / ~ freight ‖ Normalfracht f. ‖ fret m. normal. / ~ head turbine ‖ Normalgefällerad n. ‖ turbine f. pour eaux normales.

normalization ‖ Normung f. ‖ normalisation f.

normal level ‖ normaler Wasserstand m. ‖ niveau m. normal de l'eau. / ~ load (Tel) ‖ Regelleistung f. ‖ charge f. normale. / ~ output ‖ Normalleistung f. ‖ débit m. normal. / ~ power see ~ output. / ~ scheme of erection ‖ Normalaufstellung f. ‖ réglage m. normal. / ~ solution ‖ Normallösung f. ‖ solution f. normale. / ~ speed ‖ Normalgeschwindigkeit f. ‖ vitesse f. normale. / ~ state ‖

Normalzustand m. ‖ état m. normal. / ~ steam ‖ Normaldampf m. ‖ vapeur f. normale. / ~ value ‖ Normalwert m. ‖ valeur f. normale. / ~ wind pressure ‖ Normaldruck m. des Windes ‖ pression f. normale du vent. / ~ working ‖ normaler Betrieb m. ‖ marche f. normale.

north ‖ Nord m.; Norden m. ‖ nord m. / to the ~ see northern.

northern ‖ nördlich ‖ septentrional; boréal; au ou du nord; arctique. / ~ light ‖ Nordlicht n. ‖ aurore f. boréale.

north latitude ‖ Nordbreite f. ‖ latitude f. nord. / ~-magnetic ‖ nordmagnetisch ‖ magnétique nord. / ~ pole ‖ Nordpol m. ‖ pole m. nord. / North Sea ‖ Nordsee f. ‖ mer f. d'Allemagne ou du nord.

Norway maple ‖ Spitzahorn m.; Leinbaum m. ‖ érable m. platane ou plane.

nose ‖ Nase f. ‖ nez m. / ~ (Shipb) ‖ Bug m. ‖ proue f. / ~ (Tile) ‖ Nase f. ‖ crochet m. / ~ of a Bessemer convertor ‖ Hals m. einer Bessemerbirne ‖ bec m. d'un convertisseur Bessemer./~ of a chisel ‖ Meißelschneide f. ‖ bec m. ou nez m. de burin. / to make the ~ (Met) ‖ die Nase bilden ‖ former le nez. / to smelt with a ~ ‖ mit Nase f. schmelzen ‖ fondre à nez m. / ~ of the twyer (Met) ‖ Formrüssel m. ‖ museau m. de la tuyère.

nose bag maker ‖ Futterbeutelschneider m. ‖ pochiste m. / ~ band (Saddl) ‖ Nasenriemen m.; Nasenband n. ‖ muserolle f. / ~ bit ‖ Löffelbohrer m.; Hohlbohrer m. mit Zahn ‖ mèche-cuiller f.; tarière f. à cuiller.

nose-bleeding ‖ Nasenbluten n. ‖ saignement m. du nez.

nose dip (Aero) ‖ Sturzflug m. ‖ vol m. piqué. / ~ dive see ~ dip. / ~-heaviness ‖ Kopflastigkeit f. ‖ lourdeur f. de nez ou de l'avant. / ~-heavy ‖ buglastig ‖ chargé à la proue; lourd du nez m.; à nez m. lourd. / ~ key ‖ Hakenkeil m. ‖ contre-clavette f. / ~ piece (Saddl) see ~ band. / ~ piece of engine ‖ Motorvorderteil m. ‖ nez m. du moteur. / ~ radiator ‖ Bugkühler m. ‖ radiateur m. caudal. / ~ shape of the ~ ‖ Nasenform f. ‖ forme f. du nez.

nostril ‖ Nasenloch n. ‖ narine f.

notable ‖ namhaft ‖ dénommé; notable.

notary (public) ‖ Notar m. ‖ notaire m. / drawn up by a ~ ‖ notariell ‖ notarié; passé devant notaire m.

notary's fees pl. ‖ Notariatsgebühren fpl. ‖ droits mpl. notariaux. / ~ office ‖ Notariat n. ‖ notariat m.

notation ‖ Bezeichnung f. ‖ détermination f.; désignation f. / unit of ~ ‖ Maßeinheit f. ‖ unité f. de mesure.

notch, to ‖ einschneiden; einkerben ‖ entailler; encocher. / ~ (Carp) ‖ einsetzen ‖ insérer; poser. / ~ a cask ‖ ein Faß n. gargeln oder gergeln ‖ jabler un tonneau.

notch ‖ Einschnitt m.; Kerbe f. ‖ encoche f.; entaille f.; goujure f. / ~ (Arm) ‖ Kimme f.; Visiereinschnitt m. ‖ encoche f.; entaille f. / ~ (Blade) ‖ Scharte f.; brèche f. / ~ (Join) ‖ Schere f. ‖ entaille f. / full of ~es ‖ schartig ‖ ébréché. / round ~ ‖ Rundkerb m. ‖ entaille f. à fond arrondi. / V-shaped ~ ‖ Scharfkerb m. ‖ entaille f. en V.

notch bending test ‖ Kerbbiegeprobe f. ‖ essai m. au choc sur éprouvette entaillée.

notched to the centre ‖ bis zur Mitte f. eingekerbt ‖ entaillé jusqu'au milieu m.

/ the fish plate was ~ (Railw) ∥ die Stoßlasche wurde ausgeklinkt ∥ les éclisses fpl. des joints recurent des encoches. / ~ galloon ∥ Polstergurt m. ∥ lézarde f. / piece sharply ~ before bending ∥ vor dem Biegen scharf eingekerbtes Stück n. ∥ éprouvette f. nettement entaillée avant la déformation ou le ployage.

notcher (Coop) ∥ Kimmhobel m.; Gergelmesser n. ∥ jabloire f.

notching ∥ Einkerben n. ∥ entaillage m. / ~ of the fish plate ∥ Laschenausklinkung f. ∥ encochement m. ou entaille f. de l'éclisse. / ~ the flange of the rail ∥ Einklinkung f. des Schienenfußes ∥ encochement m. ou entaille f. du patin de rail.

notching attachment, swivelling ~ ∥ drehbare Ausklinkvorrichtung f. ∥ dispositif m. pivotant pour gruger.

notching knife ∥ Kerbmesser n. ∥ cochoir m. / ~ machine ∥ Nutenstanzmaschine f.; Ausklinkmaschine f. ∥ machine f. à encocher ou à gruger; grugeoir m. / ~ and mitre-cutting machine ∥ Ausklink- und Gehrungsstanzmaschine f. ∥ machine f. à gruger et découper les onglets. / ~ saw ∥ Kerbsäge f. ∥ scie f. à entailler.

notch shock test ∥ Kerbschlagbiegeprobe f. ∥ essai m. de flexion au choc sur éprouvette entaillée / ~ wheel (Clockm) ∥ Zählrad n. ∥ roue-compteur f.

notchy ∥ schartig ∥ ébréché.

note, to ∥ vermerken ∥ noter; prendre note f. de.

note ∥ Vermerk m.; Anmerkung f. ∥ remarque f. / ~ (Bill) ∥ Rechnung f. ∥ calcul m.; compte m.; facture f.; note f. / ~ (Mus) ∥ Note f.; Tonzeichen n. ∥ note f. / cut-in ~s pl. (Print) ∥ eingelaufene Marginalnoten fpl. ∥ notes fpl. marginales dans le texte. / engraved ~ ∥ gestochene Musiknote f. ∥ note f. gravée. / ~ of hand ∥ Schuldbrief m.; Schuldschein m.; Schuldverschreibung f. ∥ reconnaissance f. / to make a ~ of ∥ vormerken ∥ prendre note f. de. / marginal ~ ∥ Marginalie f.; Randbemerkung f.; Randglosse f.; Seitenanmerkung f. ∥ glosse f. ou note f. marginale; manchette f. / printed ~ ∥ gedruckte Musiknote f. ∥ note f. imprimée. / pure ~ ∥ reiner Ton m. ∥ note f. pure. / side ~ see marginal ~. / to take ~ of see to note.

notebook ∥ Notizbuch n. ∥ livre m. de notes; carnet m.

note-paper ∥ Briefpapier n. ∥ papier m. à lettres. / ~ box ∥ Briefpapierschachtel f. ∥ boîte f. pour papier à lettres. / ~ monogram stamper ∥ Briefpapiermonogrammpräger m. ∥ chiffreur m. à monogramme de papier à lettres.

notes pl. ∥ Noten fpl. ∥ musique f.

note stand ∥ Notenpult n. ∥ pupitre m. à musique. / ~ and wave tuning ∥ Abstimmung f. von Tonhöhe und Welle ∥ syntonisation f. de la note et de l'onde.

notice ∥ Ankündigung f.; Nachricht f.; Notiz f. ∥ notice f. / ~ (Contract) ∥ Kündigung f. ∥ congé m. / to give ~ ∥ aufkündigen ∥ donner congé m. / to give ~ of appeal ∥ Berufung f. einlegen ∥ interjeter appel m. / to agree upon a period of ~ of dismissal ∥ eine Kündigungsfrist f. vereinbaren. ∥ fixer un délai m. de congédiement. / to observe the length of a ~ of dismissal ∥ die Kündi-

gungsfrist f. einhalten ∥ tenir le délai m. de congédiement. / x-days ~ of dismissal ∥ x-tägige Kündigungsfrist f. ∥ délai m. de congédiement de x jours. / ~ of sale by auction ∥ Versteigerungsbekanntmachung f. ∥ publication f. de vente aux enchères.

noticeable ∥ wahrnehmbar ∥ perceptible.

notice board ∥ Warnungstafel f. ∥ tableau m. d'avis.

notices pl. to mariners (Radio) ∥ Gefahrmeldedienst m. für die Schiffahrt ∥ avis m. aux navigateurs.

notification ∥ Meldung f.; Ansage f.; Bekanntmachung f. ∥ annonce f.; rapport m.; avis; publication f. / ~ within a given period ∥ befristete Anzeige f. ∥ avis m. dans un certain délai.

notify, to ∥ ankündigen; benachrichtigen ∥ notifier; informer; aviser; faire savoir.

notion ∥ Idee f. ∥ idée f.; notion f.

nougat ∥ Nougat n. ∥ nougat m. / machine for the manufacture of ~ masses ∥ Maschine f. zur Herstellung von Nougatmasse ∥ machine f. pour la fabrication de nougat.

nought ∥ Null f. ∥ nulle f.; zéro m.

nourishing ∥ nahrhaft ∥ nourrissant; nutritif. / ~ preparation ∥ Nährmittel n. ∥ produit m. alimentaire.

novelty ∥ Neuheit f. ∥ nouveauté f.

no-volt release ∥ Nullspannungsauslösung f. ∥ déclenchement m. à tension nul.

noxious animal catching ∥ Vertilgung f. schädlicher Tiere ∥ destruction f. d'animaux nuisibles.

nozzle ∥ Düse f. ∥ gicleur m.; jet m.; ajoutage m.; buse f.; tuyère f. / acceleration ~ ∥ Beschleunigungsdüse f. ∥ tuyère f. d'accélération. / adjustable ~ ∥ verstellbare Düse f. ∥ tuyère f. à vapeur ajustable. / air ~ ∥ Luftdüse f. ∥ buse f. d'air. / ~ of a blowpipe ∥ Lötrohrspitze f. ∥ buse f. du chalumeau. / burner ~ ∥ Brennerdüse f. ∥ tuyère f. de brûleur. / to choke a ~ ∥ eine Düse f. verstopfen ∥ boucher un gicleur. / circular ~ ∥ runde Düse f. ∥ tuyère f. ronde ou circulaire. / combining ~ ∥ Mischdüse f. ∥ tuyère f. de mélange. / ~ on dome ∥ Domstutzen m. ∥ tubulure f. de dôme. / expanding ~ ∥ Expansionsdüse f. ∥ tuyère f. évasée ou divergente. / feed ~ (Boiler) ∥ Speisestutzen m. ∥ tubulure f. d'alimentation. / fuel ~ ∥ Brennstoffdüse f. ∥ ajutage m. ou buse f. pour combustible liquide. / injector ~ ∥ Einspritzdüse f. ∥ tuyère f. d'injection. / main ~ ∥ Hauptdüse f. ∥ gicleur m. principal. / no-load ~ ∥ Leerlaufdüse f. ∥ tuyère f. à vide. / reversing ~ ∥ Umlenkungsdüse f. ∥ Umkehrkanal m. ∥ tuyère f. d'inversion. / slow-running ~ ∥ Nebendüse f. für Langsamgang ∥ gicleur m. de ralenti. / spray ~ ∥ Streudüse f. ∥ tuyère f. / square ~ ∥ quadratische Düse f. ∥ tuyère f. carrée. / steam ~ ∥ Dampfdüse f. ∥ tuyère f. à vapeur. / ventilation ~ (Mine) ∥ Bewetterungsdüse f. ∥ tuyère f. d'aération.

nozzle angle ∥ Anstellwinkel m. der Düse ∥ angle m. de la tuyère. / atomization by the ~ ∥ Zerstäubung f. durch die Düse ∥ pulvérisation f. par le gicleur. / ~ atomizer ∥ Düsenzerstäuber m. ∥ pulvérisateur m. à tuyère. / ~ block ∥ Düsenblock m. ∥ bloc m. de tuyère. / ~ body ∥ Düsenkörper m. ∥ corps m. de tuyère. / ~ channel ∥ Düsenkanal m. ∥ canal m. ou

conduit m. de tuyère. / cleaning tool for ~ ∥ Düsenreiniger m. ∥ nettoyeur m. du gicleur. / ~ contraction ∥ Düseneinschnürung f. ∥ étranglement m. de la tuyère. / ~ exhaust box ∥ Düsenauspufftopf m. ∥ pot m. d'échappement. / ~ fittings pl. ∥ Düsengarnitur f. ∥ garniture f. de tuyère. / ~ flap ∥ Düsenzunge f. ∥ languette f. ou langue f. de tuyère. / ~ inclination ∥ Düsenneigung f. ∥ inclinaison f. de la tuyère. / ~ loss ∥ Düsenverlust m. ∥ perte f. dans les tuyères. / ~ mixing burner ∥ Kreuzstrombrenner m. ∥ brûleur m. à jets croisés. / ~ mixing gas burner ∥ Kreuzstromgasbrenner m. ∥ brûleur m. à gaz à jets croisés. / exchanging of a ~ part ∥ Auswechseln n. eines Düsenteils ∥ changement m. d'une pièce du gicleur. / ~ passage see ~ channel. / ~ pressure ∥ Düsenspannung f. ∥ pression f. à la tuyère. / ~ ring ∥ Düsenring m. ∥ bague f. ou anneau m. de tuyère. / testing of the ~ ∥ Düsenprüfung f. ∥ essai m. du gicleur.

nuclear charge ∥ Kernladung f. ∥ charge f. du noyau. / ~ lens ∥ Kernlinse f. ∥ lentille f. nucléaire.

nucleus ∥ Kern m. ∥ noyau m.; nucléus m. / crystal ~ ∥ Keim m. eines Kristalles ∥ germe m. cristallin.

nugget of gold (Gold) ∥ Klumpen m.; Goldklumpen m. ∥ masse f. d'or. / ~ of lime ∥ Kalknest m. ∥ poche f. de chaux.

null and void ∥ null und nichtig ∥ nul et non avenu; de nul effet m.

nullity of a patent ∥ Ungültigkeit f. eines Patentes ∥ nullité f. d'un brevet.

null method ∥ Nullverfahren n. ∥ méthode f. de zéro.

number, to ∥ numerieren; beziffern ∥ numéroter; chiffrer. / ~ (To count) ∥ zählen ∥ compter. / ~ continuously ∥ durchlaufend benummern ∥ numéroter d'un bout à l'autre.

number ∥ Zahl f.; Anzahl f. ∥ nombre m.; rang m.; numéro m. / ~ (Booksell) ∥ Lieferung f.; Heft n. ∥ livraison f. / ~ of blows per minute ∥ Schlagzahl f. in der Minute ∥ nombre m. de coups par minute. / ~ in brackets ∥ eingeklammerte Zahl f. ∥ chiffre m. placé entre parenthèses. / ~ of coils of a spring see ~ of turns of a spring. / complex ~ ∥ komplexe Zahl f. ∥ expression f. complexe. / consecutive ~ see current ~. / current ~ ∥ laufende Nummer f. ∥ numéro m. d'ordre ou de série. / ~ of cycles ∥ Periodenzahl f. ∥ nombre m. de périodes. / even ~ ∥ gerade Zahl f. ∥ nombre m. pair. / ~ of the frog ∥ Herzwinkel m. ∥ angle m. du croisement. / in ~s pl. (Booksell) ∥ heftweise; lieferungsweise ∥ par livraisons fpl. / irrational ~ ∥ irrationale Zahl f. ∥ nombre m. irrational ou sourd. / ~ of kilowatt hours ∥ Kilowattstundenzahl f. ∥ nombre m. de kilowatt-heures. / ~ of kind ∥ Gattungsnummer f. ∥ numéro m. de la série. / ~ of the machine ∥ Maschinennummer f. ∥ matricule f. de la machine. / ~ of men per day ∥ tägliche Belegschaft f. ∥ personnel m. journalier. / mixed ~ (Math) ∥ gemischter Bruch m. ∥ nombre m. fractionnaire. / odd ~ ∥ ungerade Zahl f. ∥ nombre m. impair. / ~ of periodicity ∥ Periodenzahl f. ∥ nombre m. des périodes. / ~ of persons employed ∥ Zahl f. der beschäftigten Personen ∥ nombre m. des personnes employées.

/ polygonal ~ || Polygonalzahl f. || nombre m. polygone. / rational ~ || rationale Zahl f. || nombre m. rationnel. / ~ of revolutions || Drehzahl f.; Umdrehungszahl f. || nombre m. de tours. / to express in round ~s || eine Zahl f. abrunden || arrondir un nombre. / running ~ || laufende Nummer f. || numéro m. d'ordre *ou* de série. / smaller ~ || Minderzahl f.; Minderheit f. || nombre m. inférieur; minorité f. / ~ and sign for street lanterns || Straßenlaternennummer f. und -zeichen n. || numéro m. et marque m. pour réverbères. / ~ of teeth || Zähnezahl f. || nombre m. des dents. / ~s pl. of teeth of the change wheels || Zähnezahlen fpl. der Wechselräder || nombres mpl. de dents des pignons de la boîte des vitesses. / ~ of turns of a spring || Windungszahl f. einer Feder || nombre m. d'enroulements d'un ressort. / ~ of vibrations || Schwingungszahl f. || nombre m. de vibrations.

number apparatus || Nummernapparat m. || appareil m. de transmission des numéros *ou* de numérotage.

numbered key label || Schlüsselnummernschild n. || plaque f. numérotée pour clefs.

numberer (Office) || Nummernwerk n. || numéroteur m. / ~ (Print) || Seitenzahlensetzer m. || numéroteur m.; folioteur m.; paginateur m.

number, illumination of ~ || Nummernbeleuchtung f. || éclairage m. du numéro.

numbering || Numerierung f. || numération f. / ~ apparatus || Nummerngeber m. || numéroteur m. / ~ machine || Nummerungsmaschine f. || machine f. à numéroter. / ~ machine operator || Maschinenseitenzahlsetzer m. || numéroteur m. à la machine. / ~ nail || Bezeichnungsnagel m. || clou m. estampillé; marque f. distinctive. / workshop for ~ || Werkstatt f. für Nummerngebung || atelier m. de numérotage.

number key || Nummerntaste f. || clef f. numerotée. / ~ luminous indicator || Nummernlichttableau n. || tableau m. lumineux des numéros. / ~ peg (Surv) || Nummerpfahl m. || pieu m. numéroté. / ~ perforating machine || Zahlenlochmaschine f. || machine f. à perforer les chiffres. / ~ plate || Nummernschild n. || plaque f. à numéro *ou* de contrôle. / ~ printer || Zahlendruckvorrichtung f. || dispositif m. imprimant les chiffres. / ~ pronunciation || Zahlenaussprache f. || prononciation f. des nombres. / ~ stone || Nummerstein m. || pierre f. numerotée. / third power of a ~ || Kubikzahl f.; Kubus m. || cube m. d'un nombre; nombre m. cube. / ~ type bar || Zifferntypenhebel m. || barre f. à chiffres.

numeral || numerisch || numérique.

numerator || Zähler m. eines Bruches || numéroteur m.

numerical aperture || numerische Apertur f. || ouverture f. numérique. / ~ com-

parison || zahlenmäßiger Vergleich m. || comparaison f. numérique.

numismatic || numismatisch || numismatique.

nurling tool || Rändelgabel f. || portemolette m.

nursery garden || Baumschule f. || pépinière f. / ~ gardener || Baumschulgärtner m. || pépiniériste m. / plant of ~ || Baumschulenpflanze f. || plante f. de pépinière.

nursing requisites pl. || Krankenpflegeartikel mpl. || articles mpl. pour le soin des malades. / ~ and sickroom utensils pl. || Krankenpflegegegenstände mpl. || articles mpl. pour les soins aux malades.

nut (Bot) || Nuß f.; Haselnuß f. || noix f. / ~ (Mach) || Mutter f. || écrou m. / adjusting ~ || Stellmutter f.; Lochmutter f. || écrou m. à trous *ou* de fixage. / bolt with head and ~ || Bolzen m. mit Kopf und Mutter; Mutterschraube f. || boulon m. à tête et écrou. / butterfly ~ *see* winged ~ || Flügelmutter f. || écrou m. à ailettes *ou* à oreilles. / cap ~ || Überwurfmutter f. || écrou m. à chape. / capped ~ || Kapselmutter f. || chapeau m. d'écrou fileté. / ~ for carving || Nuß f. zum Schnitzen || noix f. à tailler. / castel ~ || Kronenmutter f. || écrou m. crénelé *ou* cannelé. / castellated ~ *see* castel ~. / check ~ *see* lock ~. / circular ~ *see* adjusting ~. / clasp ~ || Mutterschloß n. || écrou m. (embrayable sur la) de vis-mère. / collar ~ || Bundmutter f. || écrou m. à collet. / counter-sunk-head and ~ || versenkte Schraube f. || boulon m. noyé. / coupling ~ (Railw) || Kupplungsmutter f. || écrou m. de tendeur. / to fix a ~ with a splitpin || eine Mutter f. versplinten || goupiller un écrou. / flange ~ *see* collar ~. / fly ~ *see* winged ~. / guiding ~ || Führungsmutter f. || écrou m. guide-tiroir. / hexagonal ~ || Sechskantmutter f. || écrou m. à six pans. / jam ~ *see* lock ~. / knurled ~ || gerändelte Mutter f. || écrou m. molleté. / lock ~ || Gegenmutter f. || contreécrou m. / milled ~ *see* knurled ~. / piston ~ || Kolbenschraube f. || vis f. du piston. / raw ~ || rohe Mutter f. || écrou m. brut. / screw ~ || Schraubenmutter f. || écrou m. fileté. / six-paned ~ *see* hexagonal ~. / sleeve ~ (Railw) || Stangenschloß n.; Schraubenschloß n. || écrou m. de réglage de la tringle de frein. / split ~ || Schlitzmutter f. || écrou m. fendu. / square ~ || Vierkantmutter f. || écrou m. carré. / steering ~ || Lenkmutter f. || écrou m. de direction. / thumb ~ *see* winged ~. / ~ for tightening springs || Spannmutter f. || écrou m. de tension des ressorts. / ~ with two holes || Zweilochmutter f. || écrou m. à deux trous. / winged ~ || Flügelmutter f. || écrou m. à oreilles.

nutation || Nutation f. || nutation f.

nut bevelling machine || Mutterabkantmaschine f. || machine f. à ébarber les écrous. / ~ burr removing machine ||

Mutternabgratmaschine f. || machine f. à ébarber les écrous. / ~ coal || Nußkohle f. || gailletteries fpl.; noisettes fpl. / cold press for forming and upsetting ~s || Kaltpresse f. zum Formen und Stauchen von Muttern || presse f. à froid pour former et fouler des écrous. / ~ cracker || Nußknacker m. || cassenoisettes m. / ~ forging machine || Mutternschmiedemaschine f. || machine f. à forger les écrous.

nutgall || Gallapfel m. || noix f. de galle.

nut lock || Schraubensicherung f. || blocage m. d'écrou. / ~ maker || Mutternmacher m. || fabricant m. d'écrous. / making ~s || Mutternfertigung f. || fabrication f. d'écrous. / machine for making ~s || Mutternherstellungsmaschine f. || machine f. à fabriquer les écrous. / ~ making press || Mutternpresse f. || presse f. à faire les écrous.

nutmeg || Muskatnuß f. || noix f. de muscade. / ~ butter || Muskatbutter f. || beurre m. de muscade. / ~ oil || Muskatnußöl n. || essence m. de muscade.

nut milling automaton || Mutternfräsautomat m. || fraiseuse f. automatique à écrous. / ~ mordant || Nußbeize f. || mordant m. noyer. / ~ oil || Nußöl n. || huile f. de noix. / ~ picker (Mine) || Kohlenausklauber m. || ramasseur m. de gailleteries. / ~ press || Mutternpresse f. || machine f. à estamper des écrous. / cold ~ press || Kaltmutternpresse f. || presse f. à froid à estamper les écrous.

nutriment colouring || Nahrungsmittelfarbe f. || couleur f. pour denrées alimentaires. / preparation of ~s || Nährmittelzubereitung f. || préparation f. de produits alimentaires.

nutritive salt || Nährsalz n. || sel m. nutritif.

nutritious || nahrhaft || nourrissant; nutritif.

nuts pl. (Clockm) || Spindellappen mpl. || palettes fpl. / ~ pl. (Coal) || Nüsse fpl.; Nußkohle f. || braisettes fpl.; noisettes fpl.; gailleteries. / ~ medicinal ~ pl. || Nüsse fpl. für medizinische Zwecke || noix fpl. médicamenteuses.

nut shaping machine || Mutterfräsmaschine f. || machine f. à dresser *ou* à tailler les écrous. / ~ shaver || Mutternfräser m. || fraiseur m. d'écrous. / ~ tapper || Mutterngewindeschneider m. || taraudeur m. d'écrous. / ~ tapping machine || Mutterschneidmaschine f. || machine f. à tarauder les écrous. / ~ tree leaf || Nußbaumblatt n. || feuille f. de noyer. / ~ washer || Mutternscheibe f. || rondelle f. d'écrou. / ~ wood || Nußbaumholz n. || noyer m. / ~ wrench || Mutter(n)schlüssel m.; Schraubenschlüssel m. || clef f.

nux vomica || Brechnuß f. || noix f. vomique.

nyctalopia || Tagblindheit f.; Nachtsichtigkeit f. || nyctalopie f.

nymphæa || Seerose f.; Wasserrose f. || nénuphar m.

O

oak ‖ Eiche f. ‖ chêne m. / ~ (Wood) ‖ Eichenholz n. ‖ bois m. de chêne. / cork ~ ‖ Korkeiche f. ‖ chêne-liège m. / of ~ ‖ eichen ‖ de chêne m.; en bois m. de chêne.

oak apple ‖ Eichengallapfel m. ‖ noix f. de galle.

oak bark ‖ Eichenrinde f. ‖ écorce f. de chêne. / ~ (Curr) ‖ Eichenlohe f.; Gerberlohe f. ‖ tan m. / ~ tannin ‖ Eichenrindentannin n. ‖ tannin m. d'écorce de chêne.

oak base ‖ Eichensockel m. ‖ base f. en chêne. / ~ coppice wood ‖ Eichenschälwald m. ‖ tallis m. de chêne à écorce. / ~ cover ‖ Eichenholzdeckel n. ‖ couvercle m. en bois de chêne.

oaken ‖ eichen ‖ de chêne m.; en bois m. de chêne. / thick ~ board see ~ plank. / ~ joist see ~ plank. / ~ plank ‖ Eichenbohle f. ‖ madrier m. en chêne.

oak extract ‖ Eichenauszug m.; Eichenholzextrakt m. ‖ extrait m. de chêne. / ~ forest m. ‖ Eichenwald m. ‖ forêt f. de chênes. / ~ gall ‖ Gallapfel m.; Galle f. ‖ galle f.; noix f. de galle. / ~ root ‖ Eichenwurzel f. ‖ racine f. de chêne. / ~ stave ‖ Eichenholzdaube f. ‖ douve f. en chêne. / ~ timber ‖ Eichenholz n. ‖ bois m. de chêne.

oak tree ‖ Eiche f. ‖ chêne m. / ~ (Wood) ‖ Eichenholz n. ‖ bois m. de chêne. / ~ oil see oak wood oil.

oakum ‖ Werg n. ‖ étoupe f. / tarred ~ ‖ geteertes Werg n. ‖ étoupe f. goudronnée.

oakum carder ‖ Wergkämmer m. ‖ cardeur m. d'étoupes. / ~ maker ‖ Wergmacher m. ‖ étoupier m. / ~ picker ‖ Wergzupfer m. ‖ éplucheur m. d'étoupes. / ~ sorter ‖ Wergsortierer m. ‖ trieur m. d'étoupes. / ~ waste hand ‖ Wergarbeiter m. ‖ ouvrier-enleveur m. d'étoupes.

oak varnish ‖ Eichenholzlack m. ‖ vernis m. de chêne.

oak wood ‖ Eichenholz n. ‖ bois m. de chêne; chêne m. / ~ for cask making ‖ Eichenfaßholz n. ‖ bois m. de chêne pour fûts. / ~ oil ‖ Eichenholzöl n. ‖ huile f. de bois de chêne.

oar, to ‖ rudern; riemen; rojen ‖ nager; ramer.

oar ‖ Riemen m.; Ruder n. ‖ aviron m.; rame f. / ~ (Brew) ‖ Rührharke f.; Krücke f.; Maischharke f. ‖ fourquet m.; vague f.; brassoir m. / double-banked ~ ‖ kurzer Riemen m. ‖ aviron m. à couple. / green malt ~ ‖ Grünmalzwender m. ‖ appareil m. à retourner le malt vert. / malt ~ ‖ Malzwender m. ‖ appareil m. pour le retournage du malt. / to ship the ~s ‖ die Riemen auslegen ‖ border les avirons. / single-banked ~ ‖ langer Riemen m. ‖ aviron m. à pointe.

oar maker ‖ Rudermacher m. ‖ avironnier m.

oast ‖ Darre f. (für Malz oder Hopfen) ‖ touraille f.

oat cleansing mill ‖ Haferflockenmühle f. ‖ moulin m. à décortiquer l'avoine. / ~ crusher ‖ Haferquetsche f. ‖ concasseur m. ou aplatisseur m. d'avoine. / ~ grit see oatmeal. / ~ groats pl. see oatmeal.

oath, to take an ~ ‖ schwören ‖ jurer; prêter serment.

oatmeal ‖ Hafergrütze f. ‖ farine f. ou gruau m. d'avoine.

oat mill ‖ Hafermühle f. ‖ moulin m. à avoine.

oats pl. ‖ Hafer m. ‖ avoine f. / crushed ~ ‖ Haferflocken fpl. ‖ fleurs fpl. d'avoine. / wild ~ ‖ Taubhafer m. ‖ folle avoine f.

oat sieve ‖ Hafersieb n. ‖ crible m. à avoine. / ~ straw ‖ Haferstroh n. ‖ paille f. d'avoine.

object, to ‖ reklamieren ‖ réclamer.

object ‖ Objekt n. ‖ objet m. / ~ of art ‖ Kunstgegenstand m. ‖ objet m. d'art. / ~ of art made of glass ‖ Glaskunstgegenstand m. ‖ objet m. d'art en verre. / ~ of collection ‖ Sammlungsgegenstand m. ‖ objet m. de collection.

object distance ‖ Objektabstand m. ‖ distance f. frontale. / ~ drawing ‖ Zeichnen n. nach Vorlagen oder Modellen ‖ dessin m. d'après modèle ou sur objet.

object glass see also objective ‖ Objektiv n. ‖ objectif m. / ~ without mount ‖ Objektivlinse f. ohne Fassung ‖ objectif m. non monté. / chromatic aberration of the ~ ‖ chromatische Abweichung f. des Objektivs ‖ aberration f. chromatique de l'objectif. / type of ~ ‖ Objektivtyp n. ‖ type m. d'objectif.

objection ‖ Reklamation f.; Mahnung f. ‖ réclamation f.

objectionable ‖ tadelhaft; tadelnswert ‖ blâmable; répréhensible.

objective ‖ Objektiv n. ‖ objectif m. / achromatic ~ ‖ Achromat n. ‖ objectif m. achromatique. / astigmatically corrected ~ ‖ astigmatisches korrigiertes Objektiv n. ‖ objectif m. astigmate corrigé. / astro-photographic ~ ‖ astrofotografisches Objektiv n. ‖ objectif m. astrophotographique. / ~ composed of several lenses ‖ mehrlinsiges Objektiv n. ‖ objectif m. composé de plusieurs lentilles. / ~ corrected for the photochemically active blue violet region of the spectrum ‖ für den fotografisch wirksamen blauvioletten Teil des Spektrums korrigiertes Objektiv n. ‖ objectif m. corrigé pour la région bleu-violette du spectre dont l'action sur la plaque photographique est la plus intense. / ~ of short focal length ‖ kurzbrennweitiges Objektiv n. ‖ objectif m. à court foyer. / immersion ~ ‖ Immersionsobjektiv n. ‖ objectif m. à immersion. / photographic ~ ‖ fotografisches Objektiv n. ‖ objectif m. photographique. / ~ of an enhanced resolving power ‖ Objektiv n. mit gesteigertem Auflösungsvermögen ‖ objectif m. de pouvoir resolvant plus grand. / telescope ~ ‖ Fernrohrobjektiv n. ‖ objectif m. de lunette. / ~ of great transmitting capacity ‖ lichtstarkes Objektiv n. ‖ objectif m. lumineux. / two-lens ~ ‖ zweiteiliges Objektiv n. ‖ ob-

jectif m. à deux lentilles. / two-lens apochromatic telescope ~ of glasse without secondary spectrum ‖ zweiteiliges apochromatisches Fernrohrobjektiv n. ‖ objectif m. de lunette apochromatique à deux lentilles en verre exempt de spectre secondaire.

objective, revolving ~ changer ‖ Objektivrevolver m. ‖ revolver m. à objectifs. / ~ changing device ‖ Objektivwechselvorrichtung f. ‖ dispositif m. pour le changement des objectifs. / distance between the ~s ‖ Objektivabstand m. ‖ écartement m. des objectifs. ‖ ~ prism ‖ Objektivprisma n. ‖ prisme-objectif m. / ~ prism spectrograph ‖ Objektivprismaspektrograf m. ‖ spectrographe m. à prisme-objectif. / ~ protector glass ‖ Objektivschutzglas n. ‖ verre m. protégeant l'objectif. / ~ sun-shade ‖ Objektivsonnenblende f. ‖ diaphragme m. pour le soleil s'adaptant sur l'objectif; diaphragme-objectif m. protégeant des rayons solaires.

object marker ‖ Objektivmarkierapparat m. ‖ marqueur m. pour objets. / plane of the ~ ‖ Objektebene f. ‖ plan m. de la préparation ou plan m. objet. / ~ point ‖ Objektpunkt m. ‖ point-objet m. / ~ slide ‖ Objektträger m. ‖ lame f. porte-objet. / ~ stage ‖ Objekttisch m. ‖ platine f.

oblate spheroid ‖ abgeplattetes Rotationsellipsoid n. ‖ ellipsoïde m. de révolution aplati; gloïde m.

obligation ‖ Schuldverschreibung f. ‖ obligation f. / ~ to give notice ‖ Anzeigepflicht f. ‖ obligation f. de notifier. / ~ to make good any loss by breakage ‖ Ersatzverbindlichkeit f. für Bruch ‖ garantie f. pour le bris.

obligatory ‖ obligatorisch ‖ obligatoire.

obligingness ‖ Gefälligkeit f. ‖ complaisance f.; obligeance f.

oblique ‖ schiefwinklig ‖ à angle oblique. / ~ crank axle ‖ schrägschenklige Kurbelachswelle f. ‖ essieu m. coudé à flasque oblique. / ~ ray ‖ flach auffallender Strahl m. ‖ rayon m. oblique. / ~ suspension rod (Bridge) ‖ Zugdiagonale f. ‖ tige f. inclinée.

oblique-angled ‖ schiefwinklig ‖ à angle m. oblique.

obliquity of the ecliptic ‖ Schiefe f. der Ekliptik ‖ obliquité f. de l'écliptique.

oblong ‖ länglich ‖ oblong; longitudinal. / ~ perforation ‖ Schlitzlochung f. ‖ perforation f. allongée. / ~ size ‖ Langformat n. ‖ format m. oblong.

oboe ‖ Hoboe f.; Oboe f.; Hochflöte f. ‖ hautbois m.

observation ‖ Beobachtung f. ‖ observation f. / aerial ~ ‖ Lufterkundung f.; Luftbeobachtung f. ‖ observation f. aérienne. / astronomical ~ ‖ astronomische Beobachtung f. ‖ observation f. astronomique. / continued ~ ‖ längeres Beobachten n. ‖ observation f. prolongée. / to eliminate errors from ~s ‖

die Messungen fpl. von Fehlern befreien ‖ éliminer les erreurs de mesures. / for ~s pl. ‖ für Beobachtungen fpl. ‖ pour des observations fpl. / terrestrial ~ ‖ Geländebeobachtung f. ‖ observation f. terrestre. / ~ for time (Nav) ‖ Zeitbestimmung f. (durch Stundenwinkel) ‖ observation f. de longitude.

observation basket ‖ Fesselballonkorb m. ‖ nacelle f. d'observation. / ~ car see ~ basket. / ~ chair ‖ Beobachtungsstuhl m. ‖ chaise f. d'observation. / ~ desk (Tel) ‖ Überwachungsstelle f. ‖ poste m. de surveillance. / ~ distance ‖ Beobachtungsabstand m. ‖ distance f. d'observation. / ~ hatch ‖ Beobachtungsluke f. ‖ lucarne f. d'observation. / instrument of ~ ‖ Beobachtungsinstrument n. ‖ instrument m. d'observation. / ~ ladder ‖ Beobachtungsleiter f. ‖ échelle f. d'observation. / ~ line ‖ Beobachtungsleitung f. ‖ ligne f. d'observation. / ~ mirror ‖ Beobachtungsspiegel m. ‖ miroir m. d'observation. / ~ mirror (Auto) ‖ Rückblickspiegel m. ‖ rétroviseur m.; avertisseur m. / ~ revolving platform ‖ drehbare Beobachtungsbühne f. ‖ plateforme f. d'observation tournante. / rising ~ platform ‖ in der Höhe verstellbare Beobachtungsbühne f. ‖ plancher m. d'observation mobile en hauteur. / ~ post ‖ Beobachtungsstelle f. ‖ poste m. d'observation. / ~ tube ‖ Beobachtungsrohr n. ‖ tube m. ou lunette f. d'observation.

observatory ‖ Sternwarte f. ‖ observatoire m. / ~ caisson ‖ Beobachtungsmunitionswagen ‖ caisson-observatoire m.

observatory dome ‖ Sternwartenkuppel f. ‖ coupole f. de l'observatoire. / rotating ~ from seat of telescope ‖ Drehen m. der Sternwartenkuppel vom Sitz des Fernrohres aus ‖ faire tourner la coupole de l'observatoire à partir du siège de la lunette. / wooden ~ ‖ Holzkuppel f. einer Sternwarte ‖ coupole f. en bois d'un observatoire.

observatory dome ring ‖ Sternwartenkuppelkranz m. ‖ couronne f. de la coupole de l'observatoire. / sheathing of an ~ ‖ Außenverkleidung f. einer Sternwartenkuppel ‖ revêtement m. extérieure d'une coupole d'observatoire.

observatory time service ‖ Sternwartezeitdienst m. ‖ service m. horaire d'observatoire.

observe, to ‖ beobachten ‖ observer. / ~ an angle ‖ einen Winkel m. messen oder aufnehmen ‖ mesurer ou relever ou observer un angle. / ~ with the naked eye ‖ mit unbewaffnetem Auge n. beobachten ‖ observer à l'œil m. nu. / ~ the time of delivery ‖ die Lieferzeit f. innehalten ‖ respecter le délai de livraison.

observer ‖ Beobachter m. ‖ observateur m. / aerial ~ ‖ Flugzeugbeobachter m. ‖ observateur m. d'avion. / ~ chronometer ‖ Beobachtungsuhr f. ‖ montre f. ou horloge f. pour les observations. / ~'s horizon ‖ Vermessungshorizont m. ‖ horizon m. de mesure.

obsidian ‖ Feuerkiesel m.; Obsidian m. ‖ pierre f. obsidienne; jaspe m. volcanique.

obsolete ‖ veraltet ‖ suranné. / ~ war ship ‖ veraltetes Kriegsschiff n. ‖ bâtiment m. de guerre vieilli ou suranné.

obstruction ‖ Verstopfung f. ‖ obstruction f. / ~ light ‖ Hindernisfeuer n. ‖ feu m. d'obstacle.

obtaining the coal in the open workings ‖ Kohlengewinnung f. im Tagebau ‖ extraction f. du charbon à ciel ouvert.

obtuse (Geom) ‖ stumpf ‖ obtus. / ~ angle ‖ stumpfer Winkel m. ‖ angle m. obtus. / ~-angled ‖ stumpfwinklig ‖ obtusangle. / ~-angled triangle ‖ stumpfwinkliges Dreieck n. ‖ triangle m. à angle obtus ou obtusangle ou amblygone.

obverse (Coin) ‖ Vorderseite f. ‖ page f. belle ou impaire; recto m.

obviate, to ‖ verhüten; schützen vor . . . ‖ empêcher; préserver de. . . .

obvious reason ‖ erkennbarer Grund m. ‖ cause f. apparente.

occlude, to ‖ okkludieren; einschließen; verstopfen ‖ occluder.

occlusion ‖ Okklusion f.; Einschließung f. ‖ occlusion f.

occultation (Astron) ‖ Verfinsterung f. ‖ occultation f.

occulting light ‖ unterbrochenes Feuer n. ‖ phare m. à éclipse.

occupation ‖ Beruf m. ‖ profession f. / nature of ~ ‖ Art f. der Beschäftigung ‖ genre m. d'occupation.

occupier ‖ Besitzer m. ‖ possesseur m.

occurrence ‖ Vorkommnis n. ‖ cas m.; événement m.; occurrence f.

ocean ‖ Weltmeer n. ‖ océan m. / ~ cable ‖ Ozeankabel n. ‖ câble m. transocéanique. / ~ climate ‖ Seeklima n. ‖ climat m. maritime ou marin. / ~-going steam tug ‖ Hochseeschleppdampfer m. ‖ bateau m. remorqueur à vapeur de haute mer.

oceanic current ‖ Meeresströmung f. ‖ courant m. marin.

ochre ‖ Ockerfarbe f.; Ocker m. ‖ ocre f.; terre f. d'ombre. / antimonial ~ ‖ Antimonocker m.; Spießglanzocker m. ‖ antimoine m. oxydé terreux. / red ~ ‖ ockeriger Roteisenstein m.; Roteisenocker m. ‖ fer m. oligiste rouge terreux. / yellow ~ ‖ gelber Ocker m.; gelber Eisenocker m. ‖ ocre m. jaune; terre f. jaune; limonite m.

ochre maker ‖ Ockerbrenner m. ‖ briocheur m. / ~ washing ‖ Ockerschlämmerei f. ‖ laverie f. d'ocre.

octaedron see octahedron.

octagon ‖ Achteck n. ‖ octogone m.

octagonal ‖ achteckig ‖ octogonal. / ~ spanner ‖ Achtkantschlüssel m. ‖ clef f. à huit pans.

octagon bar ‖ Achtkantstab m. ‖ barre f. à huit pans. / ~ handle ‖ achteckiges Werkzeugheft n. ‖ manche m. à huit pans.

octahedron ‖ Oktaeder n. ‖ octaèdre m. / pyramidal ~ ‖ Pyramidenoktaeder n.; Triakisoktaeder n. ‖ octaèdre m. pyramidé; octatrièdre m.

octant ‖ Oktant m. ‖ octant m.

octave ‖ Oktave f. ‖ octave f.

octavo ‖ Oktavformat n. ‖ in-octavo m. / large ~ ‖ Großoktav n. ‖ grand octavo m.

oculist ‖ Augenarzt m. ‖ oculiste m.; médecin m. oculiste.

odd ‖ überzählig ‖ excédant; en trop. / ~ (Math) ‖ ungerade ‖ impair. / ~ number ‖ ungerade Zahl f. ‖ nombre m. impair.

odometer ‖ Kilometerzähler m. ‖ compteur m. kilométrique.

odoriferous matter ‖ Riechstoff m. ‖ matière f. odorante.

odorous flower ‖ wohlriechende Blume f. ‖ fleur f. odoriférante.

œnometer ‖ Weingärungsmesser m. ‖ œnomètre m.

off (Elektr) ‖ aus(geschaltet) ‖ ouvert. / ~ its feet (Print) ‖ krummer Satz m. ‖ caractères mpl. tombantes.

offal ‖ Abfall m. ‖ issues fpl.

offer, to ‖ anbieten ‖ offrir. / ~ for sale ‖ ausbieten; zum Verkauf m. anbieten ‖ mettre en vente f.

offer ‖ Offerte f. ‖ offre f. / exceptional ~ ‖ Sonderangebot n. ‖ offre f. de faveur. / fair ~ ‖ billiges Angebot n. ‖ bonne offre f. / highest ~ ‖ Höchstgebot n. ‖ offre f. la plus élevée; enchère f. la plus forte. / increased ~ ‖ vermehrtes Angebot n. ‖ augmentation f. d'offres.

offering most ‖ meistbietend ‖ au plus offrant.

office ‖ Büro n. ‖ bureau m.; office m. / ~ of the Board of Works ‖ Baupolizei f. ‖ police f. de construction. / building ~ ‖ Baubüro n. ‖ bureau m. de l'ingénieur. / ~ of fault localization (Tel) ‖ Eingrenzungsamt n. ‖ bureau m. de localisation des dérangements. / registered ~ ‖ Hauptniederlassung f.; Sitz m. ‖ siège m. social.

office appliances pl. ‖ Kanzleibedarf m. ‖ articles mpl. de bureau. / ~ book ‖ Bürobuch n. ‖ livre m. de bureau. / ~ boy ‖ Bürodiener m. ‖ garçon m. de bureau. / head ~ building ‖ Hauptverwaltungsgebäude n. ‖ bâtiment m. d'administration centrale. / ~ chair ‖ Bürostuhl m. ‖ chaise f. de bureau. / ~ clip ‖ Büronadel f. ‖ épingle f. de bureau. / ~ desk ‖ Schreibtisch m. ‖ bureau m. / ~ fastener ‖ Büronadel f. ‖ épingle f. de bureau. / ~ furnishing ‖ Bürobedarf m. ‖ fourniture f. de bureau. / ~ furniture ‖ Büromöbel npl. ‖ meubles mpl. de bureau. / ~ hours pl. ‖ Bürozeit f.; Bürostunden fpl.; Geschäftsstunden fpl. ‖ heures fpl. de bureau. / ~ lighting ‖ Bürobeleuchtung f. ‖ éclairage m. de bureau. / semi-indirect ~ lighting ‖ halbindirekte Bürobeleuchtung f. ‖ éclairage m. de bureau semi-indirect. / ~ machine ‖ Büromaschine f. ‖ machine f. de bureau. / ~ manifolding machine ‖ Bürovervielfältigungsmaschine f. ‖ machine f. de bureau pour la réproduction; multiplicateur m. de bureau. / ~ organization ‖ Büroorganisation f. ‖ organisation f. pour bureaux. / ~ paste ‖ Büroleim m. ‖ colle f. de bureau. / ~ prefix (Tel) ‖ Amtsname m. ‖ nom m. du bureau. / ~ printing machine ‖ Bürodruckmaschine f. ‖ **machine f. d'impression pour bureau.**

officer ‖ Beamter m. ‖ fonctionnaire m.; employé m. / ~ in command (Mar) ‖ Kommandant m. ‖ commandant m.; capitaine m.

office requisite ‖ Bürogegenstand m. ‖ objet m. de bureau. / member of ~ staff ‖ Bürobeamter m. ‖ employé m. de bureau. / ~ stand ‖ Bürotisch m. ‖ table f. de bureau. / ~ supplies pl. ‖ Bürobedarfswaren fpl. ‖ fournitures fpl. de bureau. / leather ~ supplies pl. ‖ Bürogegenstände mpl. aus Leder ‖ articles mpl. de bureau en cuir. / ~ supplies pl. in wood ‖ Bürogegenstände mpl. aus Holz ‖ articles mpl. de bureau en bois. / ~ typewriter ‖ Büro-

schreibmaschine f. ‖ machine f. à écrire de bureau. / ~ utensil ‖ Büroartikel m. ‖ article m. de bureau. / ~ wire stitcher ‖ Bürohefter m. ‖ appareil m. à relier pour bureaux.

official ‖ amtlich ‖ officiel. / to be responsible for the execution of ~ instructions ‖ für die Befolgung gesetzlicher Vorschriften verantwortlich sein ‖ être responsable de l'observation des prescriptions légales. / ~ receiver ‖ Syndikus m. ‖ syndic m.

official ‖ Beamter m. ‖ fonctionnaire m.; employé m. / government ~ ‖ Staatsbeamter m. ‖ fonctionnaire m. de l'Etat.

offing ‖ offene See f. ‖ large m.; mer f. ouverte. / to get the ~ ‖ die offene See gewinnen ‖ gagner le large.

off-peak hour ‖ Sperrstunde f.; Sperrzeit f. ‖ heures fpl. pendant lesquelles on tend à diminuer la consommation. / ~ time see ~ hour.

off-position ‖ Ausschaltstellung f. ‖ position f. d'arrêt.

offset ‖ versetzt ‖ désaxé.

offset ‖ Gegenforderung f. ‖ créance f. en compensation; demande f. reconventionnelle. / ~ (Build) ‖ vorspringende Grundschicht f. ‖ assise f. saillante; empattement m. d'une fondation.

offset machine ‖ Gummidruckmaschine f.; Offsetpresse f. ‖ machine f. d'impression par caoutchouc; machine f. offset. / sheet fed ~ ‖ Bogenoffsetpresse f. ‖ machine-offset f. pour la marge de feuilles.

offset press see offset machine.

offset printing ‖ Gummidruck m.; Offsetdruck m. ‖ impression f. offset ou sur caoutchouc. / small ~ ‖ Kleinoffsetdruck ‖ petite impression f. offset.

offset printing press ‖ Offsetdruckpresse f. ‖ presse f. typographique offset.

offset roller ‖ Offsetwalze f. ‖ rouleau m. offset. / ~ rotary printing machine ‖ Offsetrotationsdruckmaschine f. ‖ machine-offset f. rotative.

off-side (Machine) ‖ Antriebsseite f. ‖ côté f. de la commande.

offward (Mar) ‖ seewärts ‖ du côté de la mer; vers la mer ou le large.

ogee ‖ Kehlstoß m.; Kehlleiste f. ‖ talon m.; gueule f. renversée; cymaise f. lesbienne. / ~ arch ‖ Spitzbogen m. ‖ arc m. gothique. / common ~ iron with quirk bead ‖ Karnieseisen m. mit Rundstab ‖ fer m. de talon à baguette. / common ~ iron with square ‖ Karnieseisen n. mit Platte ‖ fer m. de boudin à baguette. / quirk ~ iron ‖ Karnieseisen n.; Karnieshobeleisen ‖ fer m. doucine. / quirk ~ iron with square and bead ‖ Karnieseisen n. mit Platte und Rundstab ‖ fer m. de doucine à baguette. / ~ plane ‖ Karnieshobel m. ‖ grain m. d'orge; rabot m. à doucine; mouchette f.; doucine f. / ~ plane with bead ‖ Karnieshobel m. mit Stäbchen ‖ doucine f. guimpée.

ogive, false ~ type of shell ‖ Haubengeschoß n.; Geschoß n. mit falscher Bogenspitze ‖ projectile m. à fausse ogive.

ohm ‖ Ohm n. ‖ ohm m.

ohmmeter ‖ Ohmmeter n. ‖ ohmmètre m.

Ohm's law ‖ Ohmsches Gesetz n. ‖ loi f. d'Ohm.

oil, to ‖ ölen; schmieren ‖ huiler; lubrifier; graisser.

oil ‖ Öl n. ‖ huile f. / acacia ~ ‖ Akazienöl n. ‖ huile f. ou essence f. d'acacia. / acorn ~ ‖ Eichenkernöl n. ‖ huile f. de

gland. / animal ~ ‖ tierisches Öl n. ‖ huile f. (d'origine) animale. / antique ~ ‖ antikes Öl n. ‖ huile f. antique. / apricot kernel ~ ‖ Aprikosenkernöl n. ‖ huile f. de noyaux d'abricots. / ~ for (fire) arms ‖ Waffenöl n. ‖ huile f. pour armes. / aromatic ~ ‖ aromatisches Öl n. ‖ huile f. aromatique. / artificial ~ of bitter almonds ‖ künstliches Bittermandelöl n. ‖ essence f. d'amandes amères artificielle. / ~ for automobiles ‖ Autoöl n. ‖ huile f. pour automobiles. / baobab ~ ‖ Affenbrotbaumöl n. ‖ huile f. de baobab. / ~ for bicycles ‖ Fahrradöl n. ‖ huile f. pour bicyclettes. / blackened ~ ‖ geschwärztes Öl n. ‖ huile f. noircie. / bleached ~ ‖ gebleichtes Öl n. ‖ huile f. blanchie. / boiled ~ ‖ gekochtes Öl n. ‖ huile f. cuite ou épaisse. / boiled linseed ~ ‖ gekochtes Leinöl n. ‖ huile de lin cuite. / Brazil nut ~ ‖ Paranußöl n. ‖ huile f. de noix du Brézil. / calamus ~ ‖ Kalmusöl n. ‖ essence f. d'acore. / castor ~ ‖ Rizinusöl n. ‖ huile f. de ricin. / circuit breaker ~ ‖ Schalteröl n. ‖ huile f. pour interrupteurs. / coal tar ~ ‖ Steinkohlenteeröl n. ‖ huile f. de goudron. / with handle coated with ~ ‖ mit geöltem Stiel m. ‖ avec manche m. enduit de vernis. / Diesel ~ ‖ Dieselöl n. ‖ huile f. Diesel; Dieseloil f. / dry ~ ‖ Trockenöl n. ‖ siccatif m. / dust ~ ‖ staubbindendes Öl n. ‖ huile f. antipoussière. / easily fusible ~ ‖ dünnflüssiges Öl n. ‖ huile f. très fluide ou minée. / essential ~ ‖ ätherisches Öl n. ‖ huile f. éthérée. / factory of essential ~s ‖ Fabrik f. ätherischer Öle ‖ fabrique f. d'huiles éthérées. / fat ~ ‖ fettes Öl n. ‖ huile f. grasse. / fixed ~ ‖ gehärtetes Öl n. ‖ huile f. hydrogénée. / to free the ~ from all septic matter ‖ das Öl von allen fäulniserregenden Stoffen befreien ‖ débarrasser l'huile f. de toutes les matières septiques. / ~ free from acid ‖ säurefreies Öl n. ‖ huile f. neutre. / ~ free from resin ‖ harzfreies Öl n. ‖ huile f. exempte de résine; huile f. non-résinifère. / fuel ~ ‖ Brennöl n. ‖ huile f. à brûler. / full of ~ ‖ mit Öl n. gefüllt ‖ rempli d'huile f. / garlic ~ ‖ Knoblauchöl n. ‖ essence f. d'ail. / gas ~ ‖ Gasöl n. ‖ gasoil f. / gas engine cylinder ~ ‖ Gasmotorenzylinderöl n. ‖ huile f. pour cylindre de moteurs à gaz. / green ~ ‖ Grünöl n. ‖ huile f. verte. / half-drying ~ ‖ halbtrocknendes Öl n. ‖ huile f. demi-sécheuse. / ~ of hartshorn ‖ Hirschhornöl n. ‖ huile f. de corne de cerf. / heavy ~ ‖ Schweröl n. ‖ huile f. lourde. / ~ of high boiling point ‖ hochsiedendes Öl n. ‖ huile f. à point d'ébullition élevé. / hydrogenated ~ ‖ gehärtetes Öl n. ‖ huile f. hydrogénée. / illuminating ~ ‖ Leuchtöl n. ‖ huile f. d'éclairage. / ~ from inland fuel material ‖ Öl n. aus einheimischem Brennstoff ‖ huile f. (extraite de combustible indigène. / lamp ~ ‖ Leuchtpetroleum n. ‖ pétrole m. ordinaire. / light ~ ‖ Leichtöl n. ‖ huile f. légère. / lignite tar ~ ‖ Braunkohlenteeröl n. ‖ huile f. de goudron de lignite. / ~ of low boiling point ‖ Öl n. mit niedrigem Siedepunkt ‖ huile f. à point d'ébullition bas. / ~ for lowering printing ink ‖ Öl n. zum Verdünnen der Druckerschwärze ‖ huile f. pour étendre l'encre

d'imprimerie. / lubricating ~ ‖ Schmieröl n. ‖ huile f. de graissage ou à graisser. / maple tree ~ ‖ Ahornholzöl n. ‖ essence f. de bois d'érable. / medium ~ ‖ Mittelöl n. ‖ huile f. moyenne. / mineral ~ ‖ Mineralöl n.; Erdöl n. ‖ huile f. minérale. / heavy mineral ~ ‖ schweres Mineralöl ‖ huile f. minérale lourde. / motor ~ ‖ Motor(en)öl n.; Treiböl n. ‖ huile f. pour moteurs ou pour force motrice. / neat's foot ~ ‖ Klauenfett n. ‖ huile f. de pied de bœuf. / non-drying ~ ‖ nichttrocknendes Öl n. ‖ huile f. non-sécheuse. / nut ~ ‖ Nußöl n. ‖ huile f. de noix. / oxidized ~ ‖ oxydiertes Öl n. ‖ huile f. oxydée. / quick drying ~ ‖ Trockenöl n. ‖ huile f. siccative. / paraffin ~ ‖ Paraffinöl n. ‖ huile f. de paraffine. / to remove the ~ ‖ entölen ‖ déshuiler. / to remove all solid and liquid impurities from the ~ ‖ dem Öl n. alle festen und flüssigen Fremdkörper entziehen ‖ débarrasser l'huile f. de toutes les impuretés solides et liquides. / resin ~ ‖ Kienöl n. ‖ huile f. de pin. / ~ of roses ‖ Rosenöl n. ‖ huile f. de roses. / salad ~ ‖ Speiseöl n. ‖ huile f. alimentaire ou de table. / scented ~ ‖ parfümiertes Öl n. ‖ huile f. parfumée. / ~ for sewing machines ‖ Nähmaschinenöl n. ‖ huile f. pour machines à coudre. / shale ~ ‖ Schieferöl n. ‖ huile f. de schiste. / soluble ~ ‖ wasserlösliches Öl n. ‖ huile f. soluble. / to supply with ~ ‖ mit Öl n. beliefern ‖ fournir l'huile f. / ~ for technical purposes ‖ technisches Öl n. ‖ huile f. à l'usage technique. / terpeneless essential ~ ‖ terpenfreies ätherisches Öl n. ‖ essence f. déterpénée. / textile ~ ‖ Textilöl n. ‖ huile f. textile. / thick ~ ‖ dickflüssiges Öl n. ‖ huile f. peu fluide; huile épaisse. / thick fuel ~ ‖ dickflüssiges Heizöl n. ‖ huile f. épaisse de combustion. / ~ thrown about ‖ fortgeschleudertes Öl n. ‖ graissage m. projeté. / Turkey red ~ ‖ Türkischrotöl n. ‖ huile f. rouge d'Andrinople. / ~ of turpentine ‖ Terpentinöl n. ‖ huile f. térébenthine. / ~ for typewriters ‖ Schreibmaschinenöl n. ‖ huile f. pour machine à écrire. / vegetable ~ ‖ Pflanzenöl n. ‖ huile f. végétale. / viscous ~ see thick ~. / ~ of vitriol ‖ Schwefelsäure f. ‖ acide m. sulfurique. / factory of ~ of vitriol ‖ Vitriolhütte f. ‖ fabrique f. d'huile de vitriol; vitriolerie f. / volatile ~ ‖ ätherisches oder flüchtiges Öl n. ‖ huile f. essentielle ou volatile. / volatile ~ of mustard ‖ Senföl n. ‖ huile f. de moutarde. / walnut ~ ‖ Walnußöl n. ‖ huile f. de noix. / water-soluble ~ ‖ wasserlösliches Öl n. ‖ huile f. soluble à l'eau. / apparatus for water-soluble ~s pl. ‖ Anlage f. für wasserlösliche Öle ‖ installation f. pour huiles solubles dans l'eau. / ~ for wool oiling ‖ Öl n. für Wolleinfettung ‖ huile f. d'ensimage.

oil atomizer ‖ Öldüse f. ‖ gicleur m. d'huile. / ~ bath ‖ Ölbad n. ‖ bain m. d'huile. / ~ black ‖ Ölschwarz n. ‖ noir m. d'huile. / ~ bleaching plant ‖ Ölbleichanlage f. ‖ installation f. de blanchissement d'huiles. / ~ bottle ‖ Ölflasche f. ‖ bouteille f. d'huile. / ~ box ‖ Schmierbüchse f. ‖ boîte f. à graisse. / ~ brush ‖ Ölpinsel m. ‖ pinceau m. à huile. / ~ burner ‖ Ölbrenner m. ‖ brûleur m. à huile. / ~ burning ‖ Ölfeuerung f. ‖ chauffage m. à l'huile. / ~ cable head ‖

Ölendverschluß m. ‖ tête f. de câble avec isolation à huile.

oil-cake ‖ Ölkuchen m. ‖ tourteau m.; pain m. d'olives. / ~ breaker ‖ Ölkuchenbrecher m. ‖ broyeur m. pour tourteaux *ou* gâteaux d'huile. / ~ crusher *see* ~ ~ breaker. / ~ crushing ‖ Ölkuchenbrechen n. ‖ concassage m. de tourteaux. / ~ crushing for manure ‖ Vermahlen n. von Ölkuchen für Bodendüngung ‖ broyage m. de tourteaux pour engrais. / ~ mill ‖ Ölkuchenmühle f. ‖ broyeur m. de tourteaux. / ~ mill plant ‖ Ölkuchenmahlanlage f. ‖ moulin m. de tourteaux.

oil can ‖ Ölkanne f. ‖ burette f. à huile. / ~ cap ‖ Helmöler m. ‖ graisseur m. à casque. / ~ cardboard ‖ Ölkarton m. ‖ carton m. huilé. / ~ catcher ‖ Ölfänger m. ‖ collecteur m. d'huile. / ~ centrifuge ‖ Ölzentrifuge f. ‖ essoreuse f. d'huile. / ~ circulating lubrication ‖ Ölumlaufschmierung f. ‖ graissage m. à circulation d'huile. / ~ cleaning apparatus ‖ Ölreiniger m. ‖ purificateur m. d'huile. / ~ cleaning centrifuge ‖ Ölreinigungszentrifuge f. ‖ épurateur m. d'huile à force centrifuge; essoreuse f. d'huile.

oil-cloth ‖ Wachstuch n. ‖ toile f. cirée. / ~ (Suit) ‖ Ölanzug m. ‖ vêtement m. huilé. / case of ~ ‖ Wachstuchfutteral n. ‖ étui m. en toile cirée. / ~ factory ‖ Wachstuchfabrik f. ‖ fabrique f. de toiles cirées. / ~ making machine ‖ Wachstuchherstellungsmaschine f. ‖ machine f. à fabriquer les toiles cirées.

oil coke ‖ Petroleumkoks m. ‖ coke m. de pétrole. / ~ collector ‖ Ölfänger m. ‖ collecteur m. d'huile. / ~ colour ‖ Ölfarbe f. ‖ couleur f. à l'huile. / ~ consumption ‖ Ölverbrauch m. ‖ consommation f. d'huile. / ~ conveying plant ‖ Ölfördereinrichtung f. ‖ installation f. à transporter l'huile. / ~ cooler ‖ Ölkühler m. ‖ réfrigérant m. *ou* refroidisseur m. d'huile. / ~ cover ‖ Ölüberzug m. ‖ enduit m. à base d'huile. / ~ crust ‖ Ölkruste f. ‖ croûte f. d'huile. / ~ cup ‖ Schmierbüchse f.; Ölgefäß n. ‖ boîte f. à graisse. / flap-covered ~ cup ‖ Klappenöler m. ‖ graisseur m. à clapet. / dehydration of ~ ‖ Öltrocknen n. ‖ séchage *ou* essorage m. de l'huile. / deposit burnt on ‖ festgebrannter Ölrückstand m. ‖ résidu m. d'huile (adhérent après avoir été) brûlé. / dregs pl. of ~ ‖ Ölhefe f. ‖ lie f. d'huile. / pipe for ~ drippings pl. ‖ Leckölleitung f. ‖ conduite f. d'huile de fuite. / ~ duct ‖ Ölleitung f. ‖ canalisation f. d'huile. / ~ engine ‖ Ölmaschine f. ‖ moteur m. à huile.

oiler ‖ Öler m. ‖ graisseur m.; godet m. graisseur; huileur m. / automatic ~ ‖ automatischer Öler m.; Selbstöler m. ‖ godet m. à l'huile automatique. / drip ~ ‖ Tropföler m. ‖ graisseur m. à compte-gouttes. / ring ~ ‖ Ringöler m. ‖ graisseur m. à bague f. / self ~ ‖ Selbstöler m. ‖ graisseur m. automatique. / sight-feed ~ ‖ Tropföler m. ‖ graisseur m. compte-gouttes. / spring ~ ‖ Federschmierbüchse f. ‖ graisseur m. de ressort.

oil-expelling worm press ‖ Ölschneckenpresse f. ‖ presse f. à huile à vis sans fin. / ~ extracting plant ‖ Ölgewinnungsanlage f. ‖ installation f. d'extraction des huiles. / ~ factory ‖ Ölfabrik f. ‖ huilerie f. /

~ feed ‖ Ölzuführung f. ‖ canalisation f. d'huile. / ~ self-feeder ‖ Selbstöler m. ‖ lubrifacteur m. mécanique; graisseur m. mécanique. / ~ felt pad ‖ Ölfilz m. ‖ feutre m. huilé. / ~ field ‖ Naphtafeld n.; Erdölfeld n. ‖ terrain m. pétrolifère. / ~ filler cap ‖ Ölfüllstutzen m. ‖ bouchon m. de remplissage d'huile. / ~ filler lid ‖ Deckel m. des Öleinfüllstutzens ‖ couvercle m. de l'orifice de remplissage d'huile. / ~ filter ‖ Ölfilter n. ‖ filtre m. à huile. / ~ fired furnace ‖ Ölfeuerung f. ‖ foyer m. *ou* fourneau m. à huile.

oil firing apparatus ‖ Ölfeuerungsapparat m. ‖ brûleur m. à l'huile lourde. / ~ installation ‖ Ölfeuerungsanlage f. ‖ installation f. de chauffage à l'huile. / ~ plant *see* ~ installation.

oil fuel ‖ Brennöl n. ‖ huile f. à brûler. / ~ funnel ‖ Öltrichter m. ‖ entonnoir m. à huile. / ~ furnace *see* ~ fired furnace.

oil gas ‖ Ölgas n. ‖ gaz m. d'huile. / ~ burner ‖ Ölgasbrenner m. ‖ brûleur m. à gaz d'huile. / ~ tar ‖ Ölgasteer m. ‖ goudron m. de gaz d'huile.

oil gauge glass ‖ Ölstandsglas n. ‖ (tube m. de) verre m. du niveau d'huile. / ~ gilding ‖ Ölvergoldung f. ‖ dorure f. à l'huile. / ~ graves pl. ‖ Öltrester mpl.; Olivenölkuchen m. ‖ grignons npl. / ~ groove ‖ Ölnut f.; Schmiernut f. ‖ patte f. d'araignée. / ~ groove cutting machine ‖ Schmiernutenziehmaschine f. ‖ machine f. à faire les pattes d'araignée. / ~ gun ‖ Ölspritze f. ‖ seringue f. d'huile. / ~ hand pump ‖ Handölpumpe f. ‖ pompe f. de graissage à main. / ~ hardening ‖ Ölhärtung f. ‖ trempe f. à l'huile. / ~ hardening plant ‖ Ölhärtungsanlage f. ‖ installation f. d'hydrogénisation d'huile. / ~ hole ‖ Schmierloch n.; Ölloch n. ‖ canal m. *ou* trou de graissage. / ~ hole screw ‖ Schmierschraube f. ‖ bouchon m. de graisseur à vis.

oiling ‖ Schmieren n.; Ölen n. ‖ graissage m.; lubrification f. / ~ under pressure ‖ Preßölschmierung f. ‖ graissage m. à l'huile sous pression. / separate ~ ‖ Einzelschmierung f. ‖ graissage m. séparé. / ~ of wool ‖ Einfettung f. der Wolle ‖ ensimage m. de laine.

oiling apparatus ‖ Schmierapparat m. ‖ graisseur m.; appareil m. *ou* dispositif m. de graissage. / ~ economical ~ ‖ Ölsparapparat m. ‖ graisseur m. économique.

oiling engine ‖ Schmierapparat m. ‖ appareil m. de graissage. / ~ point ‖ Schmierstelle f. ‖ point m. de graissage. / ~ willow ‖ Ölwolf m. ‖ ensimeuse f.

oil insulator ‖ Ölisolator m. ‖ isolateur m. à huile. / ~ joint ‖ Ölabdichtung f. ‖ joint m. à huile.

oil lamp ‖ Petroleumlampe f. ‖ lampe f. à pétrole. / ~ burner ‖ Petroleumlampenbrenner m. ‖ bec m. pour lampes à pétrole.

oil leather ‖ Waschleder n.; Sämischleder n. ‖ cuir m. chamoisé. / ~ manufactory ‖ Sämischgerberei f. ‖ chamoiserie f. / ~ manufacturer ‖ Sämischgerber m.; Weißgerber m. ‖ chamoiseur m.

oillet ‖ Schlitzfenster n. ‖ œillet m.; lézarde f.

oil level ‖ Ölstand m. ‖ niveau m. d'huile. / ~ level gauge ‖ Ölstandanzeiger m. ‖ indicateur m. d'huile. / ~ linen ‖ Ölleinwand f.; Ölleinen n. ‖ toile f. huilée. / ~ linen making machine ‖ Ölleinenher-

stellungsmaschine f. ‖ machine f. à fabriquer des toiles huilées. / ~ maker ‖ Ölmüller m. ‖ huilier m.; presseur m. d'huile. / ~ manufacturing plant ‖ Ölfabrik f. ‖ huilerie f.

oil mill ‖ Ölmühle f.; Ölschlägerei f. ‖ huilerie f.; moulin m. à huile; tordoir m. d'huile. / ~ colonial ~ ‖ Kolonialölmühle f. ‖ moulin m. à huile pour les colonies. / ~ producing vegetable oils for edible purposes ‖ Speiseölfabrik f. ‖ fabrique f. d'huiles comestibles; huilerie f.

oil milling service ‖ Ölmühlenbetrieb m. ‖ service m. de l'huilerie.

oil outlet ‖ Ölablaß m. ‖ vidange f. d'huile. / ~ overflow ‖ Ölüberlauf m. ‖ tuyau m. de trop-plein d'huile. / ~ pad ‖ Schmierkissen m. ‖ tampon m. de graissage. / ~ paint ‖ Ölfarbe f. ‖ peinture f. à l'huile.

oil painting ‖ Ölanstrich m. ‖ peinture f. à l'huile. / first coat in ~ ‖ erster Ölanstrich m. ‖ première couche de peinture f. à l'huile.

oil pan ‖ Ölsumpf m. ‖ fond m. du carter formant réservoir. / ~ drain cock ‖ Ölablaßhahn m. ‖ robinet m. de vidange du carter inférieur. / ~ drain plug ‖ Ölablaßstopfen m. ‖ bouchon m. de vidange du carter inférieur.

oil paper ‖ Ölpapier n. ‖ papier m. huilé *ou* transparent. / ~ making plant ‖ Ölpapierherstellungsanlage f. ‖ installation f. de fabrication du papier huilé.

oil pipe ‖ Ölrohr n. ‖ tube m. d'huile. / ~ pole shoe ‖ Ölpolschuh m. ‖ pièce f. polaire à huile. / ~ press ‖ Ölpresse f. ‖ presse f. à huile.

oil pressing plant on the cage system ‖ Ölpreßanlage f. nach dem Seihersystem ‖ presse à huile du système à cages. / ~ on the plate press system ‖ Ölpreßanlage f. nach dem Plattenpressensystem ‖ presse f. à huile du système à plateaux.

oil pressure gauge ‖ Ölmanometer m. ‖ manomètre m. à huile. / ~ priming ‖ Ölgrund m. ‖ impression f. à l'huile. / ~ print ‖ Öldruck m. ‖ impression f. à l'huile. / ~ mineral ~ producing plant ‖ Mineralölgewinnungsanlage f. ‖ installation f. d'extraction d'huile minérale.

oil pump ‖ Ölpumpe f.; Schmierpumpe f. ‖ pompe f. à huile *ou* de graissage. / ~ safety valve ‖ Ölpumpensicherheitsventil m. ‖ soupape f. de sûreté de la pompe à huile.

oil purifier ‖ Ölreiniger m. ‖ épurateur m. d'huile. / ~ centrifugal ~ ‖ Ölreinigungszentrifuge f. ‖ épurateur m. d'huile à force centrifuge.

oil refiner ‖ Ölsieder m. ‖ raffineur m. d'huiles; épurateur-huilier m. / ~ refinery ‖ Ölraffinerie f. ‖ raffinerie f. d'huile. / ~ refining plant *see* ~ refinery. / ~ refrigerator ‖ Ölkühler m. ‖ réfrigérant m. à huile. / ~ residuum of ~ ‖ Ölbodensatz m. ‖ résidus mpl. d'huile. / ~ retaining ring (Crankshaft) ‖ Ölspritzring m. ‖ égoutteur m. / ~ return pipe ‖ Ölrücklaufrohr n. ‖ tube m. de retour d'huile. / ~ ring ‖ Schmierring m. ‖ bague f. *ou* anneau m. de graissage. / ~ saving bearing ‖ Ringschmierlager n. ‖ palier m. graisseur à bague. / ~ seal (Crankcase) ‖ Ölfangring m. ‖ anneau m. (du carter) de retenue d'huile. / ~ seed ‖ ölhaltiger Samen m. ‖ graine f. oléagineuse. / ~ self-feeder ‖ Selbstöler m. ‖ graisseur m. mécanique. / ~ separating centrifuge ‖

Entölungszentrifuge f. ‖ centrifuge f. de déshuilage. / ~ separating installation ‖ Entölungsanlage f. ‖ installation f. de déshuilage.

oil separator ‖ Entöler m.; Ölabscheider m. ‖ déshuileur m.; séparateur m. d'huile. / ~ for exhaust steam ‖ Abdampfentöler m. ‖ déshuileur m. de la vapeur d'échappement. / ~ for steam ‖ Dampfentöler m. ‖ déshuileur m. de vapeur; séparateur m. d'huile de la vapeur.

oil seringue see oil gun.

oilshale ‖ Ölschiefergestein n. ‖ schiste m. ardoisier huileux.

oil share ‖ Erdölaktie f. ‖ pétrolifère f. / ~ silk ‖ Ölseide f. ‖ soie f. huilée. / dipping plant for ~ silk ‖ Tauchanlage f. für Ölseide ‖ installation f. d'immersion de soie huilée. / ~ skin cloth ‖ Ölanzug m. ‖ vêtements mpl. huilés. / ~ soap ‖ Ölseife f. ‖ savon m. à l'huile. / ~ splash guard ‖ Ölschutzblech n. ‖ plaque f. d'huile; pare-huile m. / ~ switch ‖ Ölschalter m. ‖ interrupteur m. à huile. / ~ steamer ‖ Öldampfer m. ‖ navire m. pétrolier.

oilstone ‖ Ölstein m. ‖ pierre f. à huile.

oil strainer ‖ Ölsieb n.; Ölfilter n. ‖ filtre m. d'huile. / to replenish the ~ supply ‖ das Öl n. erneuern ‖ renouveler l'huile f.

oil tank ‖ Ölbehälter m. ‖ réservoir m. à huile. / ~ for transformers ‖ Transformatorenölbehälter m. ‖ réservoir m. à l'huile de transformateurs.

oil tanker ‖ Tankdampfer m. für Petroleum ‖ bateau m. pétrolier; navire-tank m. à pétrole. / ~ tester ‖ Ölprüfer m. ‖ essayeur m. d'huile.

oil testing apparatus ‖ Öluntersuchungsapparat m. ‖ appareil m. d'examination d'huiles. / ~ device ‖ Ölprüfeinrichtung f. ‖ dispositif m. d'essai d'huile. / ~ machine ‖ Ölprüfmaschine f. ‖ machine f. à essayer les huiles.

oil thrower ‖ Spritzring m. ‖ bague f. de projection.

oil-tight ‖ öldicht ‖ étanche à huile f. / with ~ locking device ‖ mit öldichtem Verschluß m. ‖ à obturateur m. étanche à l'huile.

oil transformer ‖ Öltransformator m. ‖ transformateur m. à huile. / small ~ tube ‖ Ölröhrchen n. ‖ petit tube m. d'huile. / ~ varnish ‖ Ölfirnis m.; Leinölfirnis m.; Öllack m. ‖ vernis m. gras ou à l'huile. / ~ way ‖ Ölkanal m.; Ölrinne f. ‖ gouttière f. de graissage. / ~ well ‖ Ölbohrloch n. ‖ puits m. à pétrole. / ~ white ‖ Ölweiß n. ‖ blanc m. d'huile. / ~ wiper ‖ Ölabstreifer m. ‖ racloir m. d'huile.

oily ‖ ölig ‖ huileux. / ~ (Wine) ‖ schleimig ‖ gras. / ~ ink ‖ Öltinte f. ‖ encre f. oléique.

ointment ‖ Salbe f. ‖ onguent m.; pommade f. / boric ~ ‖ Borsalbe f. ‖ onguent m. borique; vaseline f. boriquée. / burns ~ ‖ Brandsalbe f. ‖ onguent m. contre brûlures. / machine for grating ~s to the utmost fineness ‖ Maschine f. zum Feinzerreiben von Salben ‖ machine f. à broyer ou à moudre les onguents. / perfumed ~ ‖ parfümierte Salbe f. ‖ onguent m. parfumé.

ointment box ‖ Salbenbüchse f. ‖ boîte f. à onguent. / ~ braying and mixing machine ‖ Salbenreib- und -mischmaschine f. ‖ machine f. à broyer et à mélanger des onguents. / ~ grinder ‖ Salbenreib-

maschine f. ‖ broyeuse f. à onguents. / ~ jar ‖ Salbentopf m. ‖ pot m. à pommade ou à onguent. / ~ jug see ~ jar. / ~ spatula ‖ Salbenspatel m. ‖ spatule f. à onguent.

okonite ‖ Okonit n. ‖ okonite m.

old age pension insurance ‖ Altersversicherung f. ‖ assurance f. sur la vieillesse. / ~ brass ‖ Altmessing n. ‖ laiton m. vieux. / ~ English (Print) ‖ gotische Schrift f. ‖ caractère m. gothique ou allemand. / ~ floor see ~ heap. / ~ heap ‖ Althaufen m. ‖ vieille couche f. / ~ material ‖ Altmaterial n. ‖ vieux matériaux mpl.; matériel m. usagé. / ~ metal ‖ Altmetall n. ‖ vieux métal m. / ~ river bed ‖ Altwasser n.; altes Flußbett n. ‖ ancien lit m. ou bras m. d'une rivière. / plant for the utilization of ~ rubber ‖ Betriebsanlage f. zur Verarbeitung von Altgummi ‖ installation f. pour l'utilisation du vieux caoutchouc. / ~ steel scrap ‖ Alteisenstoffe mpl. ‖ ferraille f.; vieux fers mpl. (hors d'usage). / ~ timber / Altholz n. ‖ vieux bois m. / ~ tin ‖ Altzinn n. ‖ étain m. à verges. / ~ waste paper ‖ Altpapier n. ‖ papier m. à la cuve; vieux papiers mpl. / ~ workings pl. (Mine) ‖ verhauenes Feld n.; alter Mann m. ‖ vieux travaux mpl. ou ouvrages mpl.

oleate ‖ Oleat n. ‖ oléate m.

olefiant gas ‖ Äthylen n. ‖ éthylène m.; gaz m. oléifiant.

oleic acid ‖ Ölsäure f. ‖ acide m. oléique.

oleine ‖ Oleïn n. ‖ oléine f. / ~ soap ‖ Oleïnseife f. ‖ savon m. à l'oléine.

oleographic paper ‖ Öldruckpapier n. ‖ papier m. à peindre ou à chromolithographie ou à l'oléographie.

oleography ‖ Öldruck m. ‖ chromolithographie ou oléographie f.

oleomargarine ‖ Oleomargarine f. ‖ oléomargarine f.

oleo-pneumatic shock absorber ‖ Ölstoßdämpfer m. ‖ amortisseur m. oléopneumatique.

olibanum ‖ Oliban n. ‖ oliban m.

oligiste iron ore ‖ Glanzeisenstein m.; Eisenglanz m. ‖ fer m. oligiste. / ~ mine ‖ Glanzeisenerzgrube f. ‖ mine f. de fer oligiste ou spéculaire.

oligocene ‖ Oligocän n. ‖ syssème m. oligocène.

oligoclas ‖ Oligoklas m. ‖ oligoclase m.

oligon spar ‖ Oligonspat m. ‖ oligonite f.; sieérose m. manganésifère.

olive ‖ Olive f. ‖ olive f. / ~ black ‖ olivschwarz ‖ noir olive. / ~ brown ‖ olivenbraun ‖ brun olive. / ~ crushing mill ‖ Ölquetsche f. ‖ détritoir m. d'olives. / ~ green ‖ olivgrün ‖ vert olive. / ~ gum ‖ Olivenbaumgummi m. ‖ gomme f. d'olivier. / ~ husks pl. ‖ Öltrester npl.; Olivenölkuchen m. ‖ grignons mpl.

olivenite ‖ Olivenit m.; Olivenerz n. ‖ olivénite f. / fibrous ~ ‖ Holzkupfererz n. ‖ olivénite f. fibreuse.

olive oil ‖ Olivenöl n. ‖ huile f. d'olive. / ~ fatty acid ‖ Olivenölfettsäure f. ‖ acide m. gras d'huile d'olives. / ~ soap ‖ Olivenölseife f. ‖ savon m. à l'huile d'olives.

olive pipe (Clockm) ‖ ovale Hülse f. für die Zeigerstellung ‖ canon-olive m. / ~ salter ‖ Oliveneinsalzer m. ‖ saleur m. d'olives. / ~ wood ‖ Olivenholz n. ‖ bois m. d'olivier.

olivine ‖ Olivin m.; Chrysolit m. ‖ péridot m.; olivine f.; chrysolithe f.

ombro madder ‖ unberaubter Krapp m. ‖ garance f. non-robée.

omission ‖ Auslassung f. ‖ omission f.

omit, to ‖ auslassen ‖ omettre.

omnibus ‖ Omnibus m. ‖ omnibus m. / double-deck ~ ‖ Decksitzomnibus m. ‖ autobus m. à impériale.

on (Electr) ‖ ein(geschaltet) ‖ fermé. / ~ board ‖ an Bord ‖ à bord. / ~ its feet (Print) ‖ aufrechtstehender Satz m. ‖ composition f. droite.

once, at ~ ‖ sofort; auf einmal ‖ immédiatement; en une seule fois.

one and a half times ‖ anderthalbfach ‖ une fois f. et demie.

one-floored ‖ einhordig ‖ à un plateau. / ~ kiln ‖ Einhordendarre f. ‖ touraille f. à un plateau.

one-hole range ‖ Einlochherd m. ‖ cuisinière f. à un trou.

one-horse agricultural machine ‖ Einspännerlandmaschine f. ‖ machine f. agricole à un seul cheval. / ~ carriage ‖ Einspänner m. ‖ charrette f.

one-man bus ‖ Einmannomnibus m. ‖ omnibus m. desservi par le conducteur.

one-meter drill ‖ Einmeterbohrer m. ‖ foret m. d'un mètre de long.

one-objective binocular microscope ‖ monoobjektivbinokulares Mikroskop n. ‖ microscope m. binoculaire à un seul objectif.

one-piece wheel ‖ Vollrad n. ‖ roue f. pleine ou d'une seule pièce.

one-rail crab ‖ Einschienenlaufkatze f. ‖ chariot m. de grue monorail.

one-sided wearing ‖ einseitige Abnutzung f. ‖ usure f. unilatérale ou par un seul côté.

one-side rope ‖ eintrumiges Seil n. ‖ câble m. à un brin.

one-spindled ‖ einspindlig ‖ à broche f. unique. / ~ centering machine ‖ einspindlige Zentriermaschine f. ‖ machine f. à centrer à une seule broche.

onion ‖ Zwiebel f. ‖ oignon m. / ~ burning ‖ Zwiebelrösterei f. ‖ fabrique f. d'oignons grillés. / ~ peeler ‖ Zwiebelschäler m. ‖ peleur m. d'oignons. / ~ roaster ‖ Zwiebelröster m. ‖ grilleur m. d'oignons.

onsetter (Mine) ‖ Anschläger m. ‖ moulineur m.; accrocheur m.; metteur m. en cayat.

onyx ‖ Onyx m. ‖ onyx m.

oolite ‖ Rogenstein m.; Oolith m. ‖ calcaire m. oolithique. / upper ~ ‖ weißer Jura m.; Malm m. ‖ jurassique m. supérieur. / ~ quarry ‖ Rogensteinbruch m. ‖ carrière f. d'oolithe.

oolitic iron mine ‖ Rogeneisensteingrube f. ‖ mine f. de fer oolithique. / ~ ore ‖ Minette f. ‖ minette f.

ooze, to ‖ durchsickern ‖ échapper; suinter; filtrer. / ~ the hides ‖ Leder n. lohen ‖ juser le cuir.

ooze (Curr) ‖ Lohbrühe f.; saure Beize f. ‖ jus m. de tannée; passement m. / ~ (Geol) ‖ Schlick m. ‖ vase f. molle.

oozed leather ‖ lohgares Leder n. ‖ cuir m. jusé.

oozing see ooze.

opacity ‖ Undurchsichtigkeit f. ‖ opacité f. / ~ (Glass) ‖ Glanzlosigkeit f. ‖ dépoli m.

opal ‖ Opal m. ‖ opale f. / common ~ ‖ gemeiner Opal m. ‖ opale f. commune. / ferruginous ~ ‖ Jaspisopal m.; Opal-

jaspis m.; Eisenopal m. || opale f. ferrugineuse. / noble ~ || edler Opal m.; Edelopal m. || opale f. (irisée) noble. / precious ~ *see* noble ~. / white ~ || Milchopal m. || opale f. blanc de lait.

opalescence || Opalisieren n. || opalescence f.

opal glass || Opalglas n.; Milchglas n. || verre m. opale. / ~ globe || Opalglasglocke f. || globe m. en verre opale. / ~ plate || Milchglasplatte f. || lame f. de verre opalin.

opaline || Opalin m.; Milchglas n. || opaline f.; marmorite m.

opalize, to || opalisieren || opaliser.

opal-jasper || Jaspisopal m.; Opaljaspis m.; Eisenopal m. || jaspe-opale f.

opaque || undurchsichtig || opaque. / to make ~ by grinding || matt schleifen || dépolir; égriser. / ~ colour || Deckfarbe f. || couleur f. non transparente; couleur opaque. / ~-ground || matt geschliffen || dépoli. / ~ ice || Matteis n. || glace f. opaque.

opaqueness (Colour) || Deckkraft f. || opacité f. / ~ pigment *see* colour.

open, to || öffnen || ouvrir. / ~ a chamber (Mine) || eine Abbaustrecke f. aufhauen || ouvrir une taille. / ~ a circuit || einen Stromkreis m. öffnen *oder* unterbrechen || ouvrir un circuit. / ~ the kiln (Glass) || den Glasofen m. durchstoßen || nettoyer le fourneau. / ~ a line for traffic || dem Verkehr m. übergeben || livrer à la circulation. / ~ the mould || die Gußform losbrechen || dévêtir un objet fondu. / ~ out (Locksm) || aufreiben || étamper. / ~ up (Mine) || aufschließen || ouvrir.

open || offen || ouvert. / ~ (Electr) || ausgeschaltet || déclenché; mis hors circuit m. / ~ (Vessel) || ungedeckt || ouvert.

open-air manometer || offenes Luftmanometer n. || manomètre m. à air libre. / ~ transformer plant || Freilufttransformatorenanlage f. || groupe m. de transformateurs à l'air libre.

open antenna || offene Antenne f. || collecteur m. ouvert. / ~ arc || nackter Lichtbogen m. || arc m. en air libre. / ~ boat || offenes Boot n. || bâtiment m. non-ponté. / ~ cast (Mine) || Tagebau m. || travail m. à ciel ouvert. / ~ cell (Electr) || offene Zelle f. || élément m. ouvert.

open-circuit voltage || Leerlaufspannung f. || tension f. de marche à vide. / ~ working (Tel) || Arbeitsstrombetrieb m. || télégraphie f. à courant de travail.

open-coil armature || Anker m. mit offener Wicklung || induit m. à circuit ouvert. / ~ credit || Blankokredit m. || crédit m. à découvert. / ~ digging (Mine) || Tagebau m. || exploitation f. à ciel ouvert. /~face construction || offene Bauart f. || construction f. ouverte.

open-front eccentric press || Einständerexzenterpresse f. || presse f. excentrique à col de cygne. / ~for direct fly-wheel drive || Einständerexzenterpresse f. für direkten Schwungradantrieb || presse f. excentrique à col de cygne à commande directe sur le volant.

open-gap plate and scrap metal shearing machine || Ausladungsblech- und Schrotschere f. || cisaille f. col de cygne à tôles et à mitraille.

open goods wagon || offener Güterwagen m. || wagon m. découvert.

open-hearth furnace || Siemens-Martin-Ofen m. || four m. Siemens-Martin. /

acid ~ furnace || saurer Martinofen m. || four m. Martin acide. / basic ~ furnace || basischer Martinofen m. || four m. Martin basique ./ ~ pig || Siemens-Martin-Stahl m. || acier m. Siemens-Martin. / ~ plant || Martinwerk n. || aciérie f. Martin. / ~ process || Martinprozeß m. || procédé m. Martin. / ~ steel || Martinstahl m. || acier m. Martin.

Open here! || hier öffnen! || ouvrir par ici!

open line || freie Strecke f. || pleine voie f. / ~ balancing network || Freileitungsnachbildung f. || équilibreur m. d'un circuit aérien.

open link chain || Gliederkette f.; Schakenkette f. || chaîne f. à maillons. / ~ undulating ground || flachwelliges Gelände n. || terrain m. légèrement ondulé. / ~ pit || Tagbau m. || exploitation f. à ciel ouvert. / ~ sand casting || Herdguß m. || moulage m. à découvert; fonte f. à découvert. / ~ sand moulding || Herdformerei f. || moulage m. à découvert. / ~ sheet delivery (Print) || Planoausleger m. || sortie f. de feuilles à plat. / ~ side planer || Einständerhobelmaschine f. || raboteuse f. à montant unique. / ~-type machine || Maschine f. offener Bauart f. || machine f. ouverte. / ~ wagon || Kastenwagen m. || wagon m. à caisse. / ~-worked (Arch) || durchbrochen || percé à-jour. / ~ work hemmer || Hohlsaumnäherin f. || ourleuse f. à jour.

open working (Mine) || Tagebau m. || ouvrage m. *ou* exploitation f. à ciel ouvert. / obtaining the coal in the ~s || Kohlengewinnung f. im Tagebau m. || extraction f. du charbon à ciel ouvert.

open, in the ~ (Mine) || über Tage m. || au jour m.; à ciel ouvert.

opened, the mines were ~ by an extensive pit plant || Erschließung f. der Gruben durch umfangreiche Schachtanlagen || ouverture f. des gisements par de nombreux puits. / recently ~ strip mine || neuaufgeschlossener Tagebau m. || mine f. à ciel ouvert mise en exploitation récemment.

opener (Textile) || Öffner m. || ouvreuse f.; machine f. à ouvrir la laine *ou* le coton. / ~ for spinning mills || Öffner m. für Spinnereien || ouvreuse f. pour filatures. / ~ for tin boxes || Büchsenöffner m. || ouvre boîte m.

opening || Öffnung f. || ouverture f. / ~ (Bridge) || Spannweite f. || ouverture f. / ~ (Business) || Eröffnung f. || ouverture f. / ~ (Forest) || Lichtung f. || clairière f. / ~ (Mine) || Schurf m.; Schürfung f. || fouille f. de recherche. / ~ for axle box guide || Achslagerausschnitt m. || cage f. de la boîte d'essieu. / ~ of a business || Geschäftseröffnung f. || ouverture f. d'une maison. / ~ provided in the crank disc || ausgespartes Kurbelblatt n. || plateau m. évidé; évidement m. des flasques de manivelle. / ~ of the dies || Klemmbackenöffnung f. || ouverture f. des mâchoires. / ~ equivalent ~ of the mine shaft || äquivalente Grubenweite f. || orifice m. équivalent de mine; section f. équivalente du puits. / ~ for hornblock || Achslagerausschnitt m. || cage f. de la boîte d'essieu. /~ in the ocular plate || Okularöffnung f. || œilleton m. / ~ for the sale || Absatzmöglichkeit f. || possibilité f. de vente. / ~ of the screw

spanner || Maulweite f. des Schraubenschlüssels || ouverture f. de la clef à écrou. / ~ in wheel centre || Aussparung f. im Radkörper || évidement m. du centre de roue. / ~ through the woods || Durchschlag m. im Walde; Schneise f. || percée f.; trouée f. / ~ and fore-winning (Mine) || Aus- und Vorrichtungsarbeiten fpl. || travaux mpl. de premier établissement.

opening bit || Reibahle f.; Räumahle f. || alésoir m.; équarrissoir m.; broche f. / ~ capital || Grundkapital n. || capital m. d'apport. / ~ machine || Öffnungsmaschine f. || ouvreuse f. mécanique. / ~ mill (Spinn) || Reißwolf m. || diable m.; loup m. / ~-out a hole (Locksm) || Auftreiben m. eines Loches || opération f. d'étamper un trou. / ~ price || Eröffnungskurs m.; Anfangskurs m. || premier cours m.

opera glass || Opernglas n. || jumelles fpl.; lorgnette f. / ~ hat || Klapphut m. || chapeau m. claque; gibus m.

operate, to || betätigen; handhaben || opérer; actionner; manœuvrer.

operated, electrically ~ || elektrisch angetrieben || actionné à l'électricité. / ~ by means of a treadle || Fußbetrieb m. || marche f. à pédale.

operating || Arbeitsvorgang m. || procédé m. de travail. / ~ of signals and switches by distance || Fernbedienung f. von Signalen und Weichen || télé-commande f. des aiguilles et signaux.

operating circuit || Arbeitsstromkreis m. || circuit m. de travail. / minimal ~ costs pl. || geringe Betriebskosten pl. || des frais mpl. d'exploitation très avantageux. / ~ equipment || Bedienungsvorrichtung f. || équipement m. pour l'opération de la machine. / ~ handle || Bedienungshebel m. || levier m. de manœuvre. / ~ observation (Tel) || Betriebsüberwachung f. || surveillance f. de l'exploitation. / ~ position || Arbeitsstellung f. || position f. de travail. / ~ room (Cinema) || Vorführraum m. || cabine f. de projection. / ~ room (Tel) || Bedienungszimmer n. || salle f. de manipulation et réception. / ~ stand || Steuerstand m. || plate-forme f. de manœuvre. / ~ switch || Betriebsschalter m. || interrupteur m. de service. / ~ table (Med) || Operationstisch m. || table f. d'opération. / ~ table (Radio) || Apparatetisch m. || table f. de manipulation. / ~ time of the relay || Anzugszeit f. des Relais || temps m. d'actionnement du relais et du contracteur. / ~ tool || Bedienungswerkzeug n. || outil m. de commande *ou* de service *ou* de manœuvre.

operation || Betätigung f. || opération f.; manœuvre f. / ~ by compressed air || Preßluftbetätigung f. || commande f. à air comprimé. / convenient ~ || bequeme Bedienung f. || manipulation f. simple. / ~ of a current transmission plant || Betrieb m. einer Stromübertragungsanlage f. || exploitation f. d'une station de distribution de courant. / effortless ~ || leichte Handhabung f. || manipulation f. facile. / ~ of an engine || Betrieb m. eines Motors || fonctionnement m. d'un moteur. / ~ of measurement || Meßtätigkeit f. || opération f. de mesurage. / out of ~ || außer Betrieb m. || hors service m. / sensitive ~ || feinfühlige Führung f. || commande f. précise. / separate

~ (Chem) ‖ Einzelbestimmung f. ‖ opération f. séparée. / ~ of signals ‖ Bedienung f. von Signalen ‖ manœuvre f. des signaux. / simple ~ ‖ einfache Bedienung f. ‖ service m. simple. / in a single ~ ‖ in einem Arbeitsgang m. ‖ en une seule opération f. / uniting the localities of successive ~s räumliche Zusammenlegung f. ineinandergreifender Arbeitsvorgänge ‖ réunion f. dans un même local des opérations intéressant un même produit. / ~ of switches ‖ Bedienung f. von Weichen ‖ manœuvre f. des aiguilles. / vibrationless ~ ‖ erschütterungsfreier Gang m. ‖ marche f. exempte de vibrations.

operation furniture (Med) ‖ Operationsmobiliar n. ‖ ameublement m. pour salles d'opérations. / method of ~ ‖ Arbeitsweise f. ‖ mode f. de fonctionnement. / mode of ~ of machines ‖ Arbeitsweise f. der Maschinen ‖ mode m. de fonctionnement des machines.

operator (Radio) ‖ Funker m. ‖ opérateur m. / answering ~ ‖ Abfragebeamtin f. ‖ opératrice f.

operator's circuit ‖ Abfragestromkreis m. ‖ circuit m. de demande. / ~ jack ‖ Anschalteklinke f.; Anschlußklinke f.; Abfrageklinke f. ‖ jack m. d'opératrice. / ~ platform ‖ Bedienungspodest n. ‖ poste m. de commande. / ~ plug ‖ Anschaltestöpsel m.; Abfragestöpsel m. ‖ fiche f. jumelle ou de demande ou de réponse. / ~ set ‖ Abfragesystem n. ‖ poste m. micro-téléphonique. / ~ telephone set ‖ Abfrageeinrichtung f.; Abfragegehäuse n.; Abfrageapparat m. ‖ poste m. d'opératrice; appareil m. de réponse. / ~ transmitter ‖ Brustmikrophon n. ‖ microphone m. plastron.

ophthalmic hospital ‖ Augenklinik f. ‖ clinique f. ophtalmologique. / ~ of the university ‖ Universitätsaugenklinik f. ‖ clinique f. ophtalmologique de l'université.

ophthalmological instrument ‖ ophthalmologisches Instrument n. ‖ instrument m. d'ophtalmologie.

ophthalmology ‖ Ophthalmologie f.; Augenheilkunde f.; Augenlehre f. ‖ ophtalmologie f.

ophthalmometer ‖ Augenmesser m. ‖ optomètre m.

ophthalmoscope ‖ Ophthalmoskop n.; Augenspiegel m. ‖ ophtalmoscope m. / concave ~ ‖ hohler Augenspiegel m. ‖ ophtalmoscope m. concave. / demonstration ~ ‖ Demonstrationsaugenspiegel m. ‖ ophtalmoscope m. de démonstration. / hand ~ ‖ Handophthalmoskop n. ‖ ophtalmoscope m. à main. / plane ~ with protection rim ‖ ebener Augenspiegel m. mit Schutzrand ‖ ophtalmoscope m. plan avec rebord de protection. / simplified large ~ ‖ vereinfachtes großes Ophthalmoskop n. ‖ grand ophtalmoscope m. simplifié. / stand ~ ‖ Standophthalmoskop n. ‖ ophtalmoscope m. sur pied.

ophthalmoscopy, application of the ~ ‖ Verwendung f. von Augenspiegeln ‖ ophtalmoscopie f. / ~ lens ‖ Ophthalmoskoplinse f. ‖ lentille f. ophtalmoscopique. / binocular telescopic ~ magnifier ‖ binokulare Ophthalmoskopfernrohrlupe f. ‖ téléloupe f. binoculaire de l'ophtalmoscope. / central ~ method ‖

zentrische Ophthalmoskopie f. ‖ ophtalmoscopie f. centrée. / eccentric ~ method ‖ azentrische Ophthalmoskopie f. ‖ ophtalmoscopie f. excentrique.

opinion ‖ Anschauung f. ‖ vue f.; opinion f. / ~ (Expert) ‖ Gutachten n. ‖ avis m. / to agree with an ~ ‖ sich einer Ansicht f. anschließen ‖ se rallier à l'opinion f. / to express an ~ ‖ ein Urteil n. abgeben ‖ manifester une opinion. / to give an ~ ‖ begutachten ‖ expertiser.

opium ‖ Opium n. ‖ opium m. / containing ~ ‖ opiumhaltig ‖ opiacé.

opopanax ‖ Opopanax m. ‖ opopanax m. / ~ oil ‖ Opopanaxöl n. ‖ essence f. d'opopanax.

opponent ‖ Gegenpartei f. ‖ parti m. opposé; adversaire m.; opposition f.

opportune ‖ zweckmäßig ‖ convenable; conforme au but. / ~ (Up to date) ‖ zeitgemäß ‖ opportun; de circonstance; de notre époque.

oppose, to ~ the claim of a concessionaire by means of another concession ‖ dem Anspruch m. des Muters eine andere Mutung entgegenhalten ‖ opposer une autre demande à la prétention du requérant.

opposed ‖ entgegengesetzt ‖ opposé. / ~-piston Diesel engine ‖ Doppelkolbendieselmotor m. ‖ moteur m. Diesel à double piston.

opposite ‖ gegenteilig ‖ contraire. / ~ (Electr) ‖ ungleichnamig ‖ de signes mpl. contraires. / ~ cranks pl. ‖ gegenläufige Kurbeln fpl. ‖ manivelles fpl. contraires. / ~ force ‖ entgegengesetzt gerichtete Kraft f. ‖ force f. opposée. / ~ illustration ‖ nebenstehende Abbildung f. ‖ la figure ci-contre. / ~ party ‖ Gegenpartei f. ‖ parti m. opposé; adversaire m.; opposition f. / field of ~ polarity ‖ ungleichnamiges Feld n. ‖ champ m. de polarité contraire.

opposite (Contrary) ‖ Gegensatz m.; Gegenteil n. ‖ opposition f.; contraire m.; contraste m.

oppositely, acting ~ ‖ entgegengesetzt wirkend ‖ agissant en sens contraire.

oppressive air (Mine) ‖ matte Wetter npl. ‖ air m. lourd ou méphitique.

optical ‖ optisch ‖ optique. / ~ analysis of a mixture of gases ‖ optische Analyse f. eines Gasgemisches ‖ analyse f. optique d'un mélange de gaz. / ~ apparatus ‖ optischer Apparat m. ‖ appareil m. optique. / ~ axis ‖ optische Achse ‖ axe m. optique. / ~ bench ‖ optische Bank f. ‖ banc m. d'optique. / ~ glass ‖ optisches Glas n. ‖ verre m. optique. / ~ correction of the defective sight ‖ optische Berichtigung f. des fehlsichtigen Auges ‖ correction f. optique de l'amétropie. / ~ goods pl. ‖ optische Waren fpl. ‖ articles m. d'optique. / ~ instrument ‖ optisches Instrument n. ‖ instrument m. d'optique. / ~ lens grindery ‖ optische Linsenschleiferei f. ‖ atelier m. pour la taille des lentilles optiques. / ~ lantern ‖ Projektionslaterne f. ‖ lanterne f. de projection.

optical measuring instrument ‖ optisches Meßinstrument n. ‖ appareil m. optique de mensuration. / ~ for mineralogical investigations ‖ optisches Meßinstrument n. für mineralogische Untersuchungen ‖ instrument m. optique de mensuration pour les recherches minéralogiques.

optical particulars of a field glass ‖ optische Eigenschaften fpl. eines Feldstechers ‖ constantes fpl. optiques d'une jumelle. / ~ phenomenon ‖ Lichterscheinung f. ‖ phénomène m. optique. / ~ plummet ‖ optisches Lot n. ‖ plomb m. optique. / ~ sensibility to variation ‖ optische Unterschiedsempfindlichkeit f. ‖ sensibilité f. optique differentielle. / ~ square ‖ Winkelspiegel m. ‖ goniomètre m. à réflecteur. / ~ structure of the eye ‖ optischer Bau m. des Auges ‖ système m. optique de l'œil. / ~ telegraph ‖ optischer Telegraf m. ‖ télégraphe m. optique. / ~ transference ‖ optische Übersetzung f. ‖ transmission f. ou transfert m. optique. / ~ transmission see ~ transference.

optically active ‖ optisch aktiv ‖ optiquement actif. / ~ defective lens ‖ optisch mangelhaftes Glas n. ‖ verre m. optique défectueux. / ~ sensitive mixture of gases ‖ optisch empfindliches Gasgemisch n. ‖ mélange m. de gaz sensible au point de vue optique.

optician ‖ Optiker m. ‖ opticien m. / component ~ ‖ Fachoptiker m. ‖ opticien m. spécialiste.

opticians pl. ‖ optische Artikel mpl. ‖ articles mpl. d'optique; opticiens mpl.

optics pl. ‖ Optik f. ‖ optique f. / microscopic ~ ‖ Mikroskopie f. ‖ microscopie f.

option ‖ Option f. ‖ option f. / ~ with ~ of refusal ‖ mit Vorkaufsrecht n. ‖ avec option f. de refuser. / extension of ~ ‖ Optionsverlängerung f. ‖ prolongation f. d'option. / ~ money ‖ Optionsgeld n. ‖ dépôt m. de fonds garantissant l'option. / renewal of ~ ‖ Optionserneuerung f. ‖ renouvellement m. de l'option.

optometer ‖ Optometer n. ‖ optomètre m.

orange ‖ Apfelsine f. ‖ orange f. / blood ~ ‖ Blutapfelsine f. ‖ orange f. rouge.

orange blossom water ‖ Orangenblütenwasser n. ‖ eau f. de fleur d'oranger. / ~ house ‖ Orangerie f. ‖ orangerie f. / ~ leave ‖ Orangenblatt n. ‖ feuille f. d'orange. / ~ minium ‖ Orangemennig m.; englischer Mennig m. ‖ minium m. orange; minium m. anglais. / ~ oil ‖ Pomeranzenöl n. ‖ essence f. d'orange. / sweet ~ oil ‖ Orangenschalenöl n. ‖ essence f. de Portugal. / ~ peel ‖ Apfelsinenschale f. ‖ écorce f. ou pelure f. d'orange. / ~ peeler ‖ Orangenschälerin f. ‖ éplucheuse f. d'oranges.

oratory ‖ Bethaus n. ‖ oratoire m.

orbit ‖ Augenhöhle f. ‖ orbite f. / ~ of the earth (Astron) ‖ Erdbahn f. ‖ orbite f. terrestre.

orchard ‖ Obstgarten m. ‖ verger m.

orchardist ‖ Obstbaumzüchter m. ‖ pomiculteur m.

orchard watchman ‖ Obstbaumwärter m. ‖ gardien m. de verger m.

orchella ‖ Orseille f. ‖ orseille f.

orchestrion ‖ Orchestrion n. ‖ orchestrion m.

orchid grower ‖ Orchideenzüchter m. ‖ cultivateur m. d'orchidées.

orchil ‖ Orseille f. ‖ orseille f.

orchilla see orchil.

orchil liquor ‖ Orseilleextrakt m. ‖ extrait m. d'orseille.

order, to ~ beforehand ‖ vorausbestellen ‖ commander d'avance. / ordered by . . .

bestellt durch ... || commandé par ...
/ when ordering please state ... || bei Be-
stellung f. wird um die Angabe gebeten ||
on est prié d'indiquer dans la commande.
order (Business) || Auftrag; Anweisung f. ||
commission f.; ordre m. / ~ (Law) ||
Mandat n. || mandat m.; plein pou-
voir m. / ~ (Math) || Ordnung f. || ordre
m. / ~ (Military) || Orden m. || décoration
f.; ordre m. / alphabetic ~ || alphabe-
tische Reihenfolge f. || ordre m. alpha-
bétique. / by ~ || im Auftrage m. ||
d'ordre m.; de la part. / ~ for delivery ||
Bezugsschein m. || bulletin m. de
livraison; bon m. / in due ~ || ordnungs-
mäßig || réglementaire; régulier. / to
execute an ~ || einen Auftrag m. aus-
führen || exécuter une commande. / firm
~ || fester Auftrag m. || ordre m. ferme.
/ it should be expressly stated in the ~ ||
es ist bei Bestellung f. ausdrücklich zu
bemerken || il faut avoir soin m. de le
spécifier expressément dans la com-
mande. / to give an ~ || bestellen ||
donner des ordres mpl. / to give ~s pl.
|| Anordnungen fpl. treffen || prendre des
dispositions fpl. / ~ of letters || Buch-
stabenfolge f. || suite f. des caractères.
/ out of ~ || nicht in Ordnung f. || pas en
règle f. / when placing an ~ || bei Auf-
tragerteilung f. || en cas m. de commande.
/ to put in ~ || ordnen || assortir; mettre
en ordre m.; régler. / to put in ~ (Dis-
putes) || schlichten || arranger; aplanir.
/ receiving ~ against debtor in bank-
ruptcy proceedings || Veräußerungs-
verbot n. || ordonnance f. de mise sous
sequestre. / reverse ~ || umgekehrte
Reihenfolge f. || ordre m. ou suite f.
inversé(e). / on special ~ || auf besondere
Bestellung f. || sur commande f. spéciale.
/ supplementary ~ || Nachbestellung f. ||
commande f. ultérieure. / ~ of traffic ||
Verkehrsordnung f. || règlement m. du
trafic.
order form || Bestellformular n.; Bestell-
vordruck m. || formulaire m. pour com-
mandes. / ~ form (Machine) || Laufzettel
m. || fiche f. de travail. / ~ number || Be-
stellnummer f. || numéro m. de commande.
orders pl. || Verhaltungsmaßregel f. || in-
struction f.
order slip || Auftragzettel m. || fiche f. de
travail. / transmission of ~s in power
stations || Übermittlung f. von Anord-
nungen in Kraftwerken || transmission f.
des ordres dans les usines d'énergie.
order wire || Dienstleitung f. || ligne f. de
service. / long distance ~ || Ferndienst-
leitung f. || ligne f. de service interurbain.
order wire button || Dienstleitungstaste f. ||
touche f. de ligne d'ordres. / ~ button
strip || Diensttastenstreifen m. || réglette
f. des touches des lignes d'ordres. /
~ jack || Dienstabfrageklinke f. || jack m.
de ligne d'ordres. / ~ key || Dienst-
leitungstaste f. || bouton m. pour la
ligne d'ordres. / ~ lamp || Dienst(anruf)-
lampe f. || lampe f. des lignes d'ordres. /
~ operation || Dienstleitungsbetrieb m. ||
exploitation f. par ligne de service. / ~
panel || Dienstleitungsfeld n. || champ m.
de jacks de service. / ~ relay || Dienst-
anrufrelais n. || relais m. des lignes
d'ordres. / ~ signal || Dienstanrufzeichen
n. || annonciateur m. de lignes d'ordres.
/ ~ switch || Dienstleitungsumschalter m.
|| clef f. de ligne d'ordres.

ordinary || gewöhnlich || ordinaire. / ~
share || Stammaktie f. || action f. d'ori-
gine. / ~ ceramic || Grobkeramik f. || céra-
mique f. industrielle. / ~ light contain-
ing red rays || gewöhnliches rothaltiges
Licht n. || lumière f. ordinaire com-
prenant le rouge.
ordinate || Ordinate f. || ordonnée f.
ordnance, automotive ~ || Kraftartillerie f.
/ artillerie f. mécanique. / ~ committee ||
Geschützkomitee n. || comité m. de
l'ordonnance.
ore || Erz n. || minerai m. / aluminous ~ ||
Alaunerz n. || minerai m. d'alun. /
to assay an ~ || eine Erzprobe f. nehmen ||
essayer un minerai. / bog ~ || Rasen-
eisenstein m. || limonite f. / brown iron
~ || Brauneisenstein m. || hématite f.
brune. / to cob the ~ || Erz n. scheiden ||
trier le minerai. / to crush ~ || Erz n.
pochen || broyer le minerai. / diluvial ~ ||
Seifenerz n.; Wascherz n. || minerai m.
d'alluvion ou de lavage. / to draw ~s pl. ||
Erz npl. fördern || extraire les minerais
mpl. / earthy ~ || Mulm m. || mine f.
pulvérulante. / ~ in grains || Graupen-
erz n. || mine f. en grains. / high-grade ~ ||
hochhaltiges Erz n. || minerai m. de
qualité supérieure. / high-phosphorus ~ ||
hochphosphorhaltiges Erz n. || minerai
m. à haute teneur en phosphore. / to
jig ~ || Erz n. waschen || laver le minerai.
/ iron ~ || Eisenerz n.; Eisenstein m. ||
minerai m. de fer. / kidney ~ || Nierenerz
n. || minerai m. en rognons. / lead ~ || Blei-
erz n. || minerai m. de plomb. / low-grade
~ || minderwertiges Erz n. || minerai m.
de qualité inférieure. / magnetic iron ~ ||
Magneteisenstein m. || fer m. magnéti-
que. / manganese ~ || Braunstein m. ||
pyrolusite f. / nodular ~ see kidney ~.
/ oligiste iron ~ || Glanzeisenstein m.;
Eisenglanz m. || fer m. oligiste. / pile of
~ || Erzhaufen m. || lot m. de minerai. /
to put aside the roasted ~ || das geröstete
Erz wegräumen || porter le minerai grillé
à côté du fourneau. / raw roasted ~ ||
roher Rost m. || minerai m. grillé brut. /
red iron ~ || Roteisenstein m. || hématite
f. rouge. / refractory ~ || strengflüssiges
Erz n. || minerai m. réfractaire. / roasted
~ || Garerz n. || mine f. grillée. / ~ in
sight || anstehendes Erz n. || minerai m.
en vue. / to stamp ~ || Erz n. pochen ||
bocarder ou briser ou écraser le minerai.
/ stamped ~ || Pochmehl n. || poussière f.
ou sable m. de bocard. / sufficiently
roasted ~ || garer Rost m. || minerai m.
assez grillé. / titaniferous ~ || titan-
haltiges Erz n. || minerai m. titanifère. /
white copper ~ || Weißkupfererz n. ||
cubane m.
ore analysis || Erzuntersuchung f.; Erz-
analyse f. || analyse f. des minerais. / ~
assaying || Erzprobenehmen n. || essai
m. des minerais. / average sample of ~ ||
Durchschnittsprobe n. des Erzes ||
échantillon m. moyen du minerai. /
average value of ~ || Durchschnittswert
m. des Erzes || valeur f. moyenne du
minerai. / ~ bearing || erzhaltig || métalli-
fère. / ~ breaker || Erzbrecher m. || con-
casseur m. de minerais. / ~ briquetting
plant || Erzbrikettierungsanlage f. || in-
stallation f. à agglomérés le minerai. /
bunch of ~ || Gangstock m. || filon m.
en forme d'amas. / ~ bunker || Erztasche
f. || poche f. à minerais. / ~ burner || Erz-

röster m. || calcinateur m. ou grilleur m. de
minerai. / ~ calciner || Erzröster m. || calci-
nateur m. ou grilleur m. de minerai. /
classification of ~ || Erzbezeichnung f. ||
désignation f. du minerai. / ~ crusher ||
Erzpocher m. || bocardeur m. or broyeur
m. or pileur m. de minerai. / ~ crush-
ing mill || Erzerkleinerungsanlage f. ||
broyeur m. de minerai. / ~ crushing
shoe || Pochschuh m. || sabot m. de
bocard. / ~ deposit || Erzfeld n.; Erz-
lagerstätte f. || champ m. d'exploita-
tion; gîte m. métallifère dépôt m. de
minerai. / ~ dresser || Aufbereitungs-
arbeiter m. || ouvrier m. de préparation
de minerai. / ~ dressing plant || Erzauf-
bereitungsanlage f. || installation f. pour
le traitement de minerai. / ~ dressing
works pl. || Aufbereitungsanstalt f. für
Erze || usine f. de préparation de minerai.
/ ~ dust || Mulm m. || minerai m. pulvé-
rulent. / ~ elevator || Erzbecherwerk n. ||
noria f. à minerai. / ~ filler || Erzlader m.
|| chargeur m. de minerai. / ~ funnel ||
Erzaufgabetrichter m. || trémie f. pour
minerai. / ~ house || Möllerboden m. ||
halle f. aux mélanges.
ore loading bridge || Erzverladebrücke f. ||
pont de chargement de minerai. /
lifting gear of an ~ || Hubwerk n. einer
Erzverladebrücke || mécanisme m. de
levage d'un pont de chargement de
minerai.
ore mill || Erzmühle f. || moulin m. à
minerai. / ~ picker || Erzklauber m. ||
trieur m. de minerai; scheideur m. / ~
pocket || Erznest n. || nid m. de minerai.
/ production of ~ || Erzproduktion f. ||
production f. de minerai. / ~ roaster ||
Erzröster m. || calcinateur m. ou grilleur
m. de minerai. / installation of ~
roasting || Erzröstungsanlage f. || instal-
lation f. de grillage de minerai.
ore sample || Erzprobe f. || essai m. de
minerai. / grinding plate for ~s || Reibe-
platte f. für Erzproben f. || plaque f. de
trituration pour essais de minerai.
ore separator || Erzscheider m. || sépara-
teur m. de minerai. / electromagnetic ~
|| elektromagnetischer Erzscheider m. ||
séparateur m. électromagnétique de
minerai.
ore slag || Rohschlacke f. || scorie f. pauvre.
/ small ~ slime || Schlich m. || schlich m.;
schlick m. / ~ smelting || Rohschmelzen
n. || fonte f. crue. / ~ stamper || Erz-
pocher m. || broyeur m. de minerai. / ~
treatment || Erzverhüttung f. || traite-
ment m. du minerai. / ~ wagon || Erz-
wagon m. || wagonnet m. à minerai. / ~
washer || Erzwäscher m. || laveur m. ou
débourbeur m. de minerai. / ~ washery ||
Erzwäsche f. || lavoir m. du minerai. / ~
working || Erzverarbeitung f. || travail m.
du minerai. / ~ yard || Erzlagerplatz m. ||
dépôt m. de minerais.
organ || Orgel f. || orgue m. / church ~ ||
Kirchenorgel f. || orgue m. d'église. /
mechanical ~ || mechanische Orgel f. ||
orgue m. mécanique. / pipe ~ || Pfeifen-
orgel f. || orgue m. à tuyaux. / portable ~
|| tragbare Orgel f. || orgue f. porta-
tive.
organ bellow || Orgelblasebalg m. || soufflet
m. d'orgue. / ~ builder || Orgelbauer m. ||
facteur m. d'orgues. / ~ case || Orgel-
gehäuse n. || buffet m. ou cabinet m.
d'orgue. / ~ factory || Orgelbauanstalt f.

|| fabrique f. d'orgues. / ~ finisher || Orgelausbauer m. || monteur m. d'orgue.
organic || organisch || organique. / ~ acid || organische Säure f. || acide m. organique. / ~ chemistry || organische Chemie f. || chimie f. organique. / ~ compound || organisches Präparat n. || composé m. organique.
organism of fermentation || Gärungsorganismus m. || organisme m. de fermentation.
organization || Organisation f. || organisation f. / ~ of workers || Zusammenschluß m. der Arbeitnehmer || coalition f. ou organisation f. des ouvriers.
organize, to || organisieren || organiser. / compulsion to organize || Organisationszwang m. || contrainte f. au syndicat. / freedom to organize || Koalitionsfreiheit f. || liberté f. à l'organisation.
organized worker || organisierter Arbeiter m. || ouvrier m. syndiqué.
organizer || Organisator m. || organisateur m.
organizing ability || Organisationstalent n. / talent m. organisateur.
organ loft || Orgelbühne f.; Orgelchor n. || tribune f. d'orgue.
organogeneous sedimentary rock || organogenes Sedimentgestein n. || roche f. sédimentaire organogénique.
organotherapeutical preparations pl. || organotherapeutische Präparate npl. || préparations fpl. organothérapeutiques.
organ pipe || Orgelpfeife f. || tuyau m. d'orgue. / ~ maker || Orgelpfeifenarbeiter m. || ouvrier m. en tuyaux d'orgue.
organ reed || Orgelzunge f. || anche f. pour orgue. / ~ stop || Orgelzug m. || registre m.; jeu m. d'orgue.
organzine || Organsinseide || organsin m.
orient || Ost m.; Osten m.; Ostpunkt m. || est m.
oriental || östlich || oriental.
orientation || Orientierung f. || orientation f.
orifice || Mündung f.; Öffnung f. || orifice m.; ouverture f.; entrée f. / ~ of conduit || Kanalmundstück n. || orifice m. de conduite. / ~ of exit || Ausflußöffnung f. || orifice m. de sortie.
origanum oil || Dostenöl n.; Origanumöl n. || essence f. d'origan.
origin || Ursprung m.; Herkunft f. || origine f. / ~ of coordinates || Koordinatennullpunkt m. || origine des coordonnés. / ~ of force || Angriffspunkt m. einer Kraft || point d'application d'une force. / ~ of a river || Quelle f. eines Flusses || source f. d'un fleuve.
original copy || Urschrift f. || original m. / ~ extract || ursprünglicher Extraktgehalt m. || teneur f. originale en extrait. / ~ length || ursprüngliche Länge f. || longueur f. primitive. / ~ packing || Originalverpackung f. || emballage m. d'origine. / after removing the load the leaf of spring resumed its ~ shape || nach der Entlastung f. ging das Federblatt in die ursprüngliche Lage zurück || après déchargement la lame de ressort reprit sa position primitive. / ~ works pl. of artists || Originalwerke npl. von Künstlern || œuvres fpl. originales d'artistes.
originating exchange || Anmeldeanstalt f. || bureau m. de départ.
origin, certificate of ~ || Ursprungszeugnis n. || certificat m. d'origine. / country of ~ || Ursprungsland n. || pays m. d'origine.

orle (Arch) || Saum m. || orle m.; orlet m. / ~ of a metal plate || Rundkante f. oder Saum m. oder Rundfalz m. einer Metallplatte || membran m. ou ourlet m. d'une plaque de métal.
Orlean || Orlean m. || rocou m.; roucou m.
orlop deck || Plattformdeck n.; Raumdeck n. || plate-forme f. de la cale.
ormolu || Malergold n.; Muschelgold n. || or m. d'applique.
ornament, to || verzieren; ausschmücken; schmücken || orner; ornementer; décorer.
ornament || Verzierung f.; Ornament n. || ornement m. / Christmas tree ~s pl. || Christbaumschmuck m. || ornements mpl. ou verreries fpl. pour arbres de Noël. / ~ for hats || Hutschmuck m. || ornement m. pour chapeaux. / ~ for trellis fences || Gitterornament n. || ornement m. de grille.
ornamental fitting for furniture || Möbelverzierung f. || ornement m. pour meubles. / ~ glass || Ornamentglas n. || verre m. ornemental. / ~ iron || Ziereisen n. || fer m. à dessin. / ~ object || Ziergegenstand m. || objet m. d'ornement. / ~ painter || Dekorationsmaler m. || peintre-décorateur m. / ~ shrub || Zierstrauch m. || arbuste m. d'ornement. / ~ thread || Verzierungsfaden m. || fil m. d'ornement. / ~ tree || Zierbaum m. || arbre m. d'ornement. / ~ types pl. || Zierschrift f. || caractères mpl. de fantaisie.
ornamentation || Ausschmückung f.; Verzierung f.; Verzieren n. || ornementation f. / china ~ || Auftragen n. von Reliefs auf Porzellan || pastillage m.
ornamenter || Verzierer m. || décorateur m. / china ~ || Porzellanformer m.; Modellierer m. || pastillageur m.
ornament maker || Gipsbildhauer m. || ornemaniste m. / machine for the manufacture of ~s for Christmas trees || Christbaumschmuckherstellungsmaschine f. || machine f. à fabriquer les verreries pour arbres de Noël.
orpiment || Auripigment n.; Rauschgelb n.; Operment n. || orpiment m.; arsenic m. sulphuré jaune. / red ~ || Realgar n.; m.; rote Arsenblende f.; Rotrauschgelb n. || arsenic m. sulphuré rouge; réalgar m.
orreries pl. || Planetarium n. || planétaire m.
orris root || Veilchenwurzel f. || racine f. d'iris de Florence.
orseille || Orseille f.; Orseilleextrakt m. || orseille f. / violet ~ || violette Orseille f. || orseille f. violette.
orseille weeds pl. || Färberflechte f.; Lackmusflechte f. || orseille f.
orthite (Miner) || Orthit m.; Allanit m.; Zerin n. || orthite f.; allanite f.; cérine f.
orthoclase (Miner) || Orthoklas m.; rechtwinkeliger Feldspat m. || orthoclase f.; orthose f.
orthogonal || rechtwinklig || orthogonal; rectangulaire ; rectangle.
orthopædic see orthopedic.
orthopedia see orthopedy.
orthopedic apparatus || orthopädischer Apparat m. || appareil m. d'orthopédie. / ~ boot maker || orthopädischer Schuhmacher m. || bottier m. orthopédiste. / ~ instrument || orthopädisches Instrument n. || appareil m. d'orthopédie. / ~ leather || orthopädisches Leder n. || cuir m. orthopédique.

orthopedics pl. see orthopedy.
orthopedist || Orthopäde m. || orthopédiste m.
orthopedy || Orthopädie f. || orthopédie f.
orthophosphoric acid || Phosphorsäure f. || acide m. orthophosphorique.
orthoptic (Diopter) || Diopter m. || dioptre m.
orthoscopic || orthoskopisch || orthoscopique.
orthoscopic eyepiece || orthoskopisches Okular n. || oculaire m. orthoscopique. / ~ with focussing collar graduated in terms of diopters || orthoskopisches Okular n. mit Dioptrieneinstellung || oculaire m. orthoscopique muni d'une graduation en dioptries.
orthoscopic micrometer eyepiece || orthoskopisches Mikrometerokular n. || oculaire m. orthoscopique pour micromètres.
oscillate, to || schwingen; oszillieren; vibrieren || osciller; vibrer. / ~ about the steady position || um die Ruhelage f. schwingen || osciller autour de l'état d'équilibre. / to arrange to oscillate || pendelnd lagern || monter oscillant.
oscillating || oszillierend; schwingend || oscillant. / ~ apparatus || Schaukelapparat m. || appareil m. basculeur. / ~ chute || Schwingrutsche f.; Schüttelrutsche f. || glissière f. oscillante.
oscillating circuit (Electr) || Schwingungskreis m. || circuit m. oscillant. / closed ~ || geschlossener Schwingungskreis m. || circuit m. oscillant fermé. / modulated ~ || abgestimmter Schwingungskreis m. || circuit m. oscillant accordé. / open ~ || offener Erregerkreis m. || circuit m. oscillant ouvert.
oscillating conveying channel || Schwingförderrinne f. || gouttière f. de transport oscillante. / ~ conveyor || Schwingrinne f.; Schüttelrutsche f. || gouttière f. oscillante. / ~ cylinder || schwingender Zylinder m. || cylindre m. oscillant. / ~ feeding apparatus || Schüttelspeiseapparat m. || trémie f. d'alimentation à secousses. / ~ jaw || bewegliche Brechbacke f. || mâchoire f. mobile.
oscillating motion (Railw; Engine) || Schwanken n. || roulis m. / irregular ~ of the engine (Railw) || Schlingern n. der Lokomotive || lacet m. de la locomotive.
oscillating rod || Schaukelstange f. | tige f. basculante. / ~ roll || schwingende Walze f. || cylindre m. oscillant. / ~ shaft || pendelnde Welle f. || arbre m. pendulaire. / ~ table || Schüttelherd m. || table f. oscillante. / ~ transmitting circuit || schwingender Sendungskreis m.; Sendererregerkreis m. || circuit m. oscillant d'émission. / ~ twisting machine || Torsionsschwingungsmaschine f. || machine f. oscillante à torsion.
oscillation || Schwingung f.; Oszillation f. || oscillation f.; va-et-vient m. / continuous ~ || ungedämpfte Schwingung f. || oscillation f. entretenue. / coupled ~s pl. || gekoppelte Schwingungen fpl. || oscillations fpl. liées. / ~ in coupled circuits || Kopplungsschwingung f. || oscillation f. de circuits accouplés. / damped ~ || gedämpfte Schwingung f. || oscillation f. amortie. / electric ~ || elektrische Schwingung f. || oscillation f. électrique. / electromagnetic ~ ||

elektromagnetische Schwingung f. ||
oscillation f. électromagnétique. / forced
~ || erzwungene Schwingung f. || oscilla-
tion f. contrainte. / free ~ || freie Schwin-
gung f. || oscillation f. libre. / fundamen-
tal ~ || Grundschwingung f. || oscillation
f. fondamentale. / high-frequency ~ ||
Hochfrequenzschwingung f. || oscilla-
tion f. haute fréquence. / isochronal ~ ||
isochrone oder synchrone Schwingung
f. || oscillation f. isochrone ou synchrone.
/ natural ~ || Eigenschwingung f. ||
oscillation f. propre. / ~ of the pendulum
|| Pendelschwingung f. || oscillation f. du
pendule. / ~ of the plump bob || Lot-
schwankung f. || oscillation f. du plomb.
/ ~s pl. of a ship || Schiffsschwingungen
fpl. || oscillations fpl. d'un navire. /
undamped ~ || ungedämpfte Schwin-
gung f. || oscillation f. non-amortie.
oscillation, amplitude of ~ || Schwingungs-
weite f. || amplitude f. d'oscillation. /
maximum amplitude of the ~ || Höchst-
ausschlag m. der Schwingung || déviation
f. maxima de l'oscillation. / ~ analyzer ||
Schwingungsprüfer m. || analyseur m.
d'oscillations. / arc of ~ || Schwingungs-
bogen m. || arc m. des oscillations. /
asunder of ~s see ~ analyzer. / axis
of ~ || Schwingungsachse f. || axe m.
d'oscillation. / centre of ~ || Schwin-
gungsmittelpunkt m. || centre m.
d'oscillation. / ~ circuit see oscillating
circuit. / ~ coil || Schwingspule f. || self f.
oscillatrice. / ~ damper || Schwingungs-
dämpfer m. || amortisseur m. d'oscilla-
tions. / damping of the ~ || Abklingen n.
der Schwingung || décroissement m. ou
amortissement m. de l'oscillation. / ~
generator || Schwingungserzeuger m. ||
oscillateur m. / ~ pair || Schwingungs-
paar n. || ensemble m. de deux oscilla-
tions. / period of ~ || Schwingungsdauer
f. || durée f. d'oscillation. / phase of ~ ||
Schwingungsphase f. || phase f. d'oscilla-
tion. / time of ~ see period of ~. / ~
transmitting circuit || Sendererregerkreis
m. || circuit m. oscillant d'émission. /
~ valve detector || Ventilröhrendetek-
tor m. || détecteur m. à valve d'oscillation
oscillator (Phys) || Oszillator m.; Schwin-
gungserzeuger m. || oscillateur m. / ~ (Tel)
|| Summer m. || ronfleur m.; oscillateur m.
/ electromagnetic ~ || Magnetsummer m.
|| oscillateur m. électromagnétique. / ~
for measuring and signalling purposes ||
Wechselstromerzeuger m. für Meß- und
Rufzwecke || oscillateur m. pour les
mesures et la signalisation. / open ~ ||
offener Oszillator m. || oscillateur m.
ouvert.
oscillatory || schwingend || oscillant;
oscillatoire. / ~ circuit || Schwingungs-
kreis m. || circuit m. oscillatoire. / ~
coil || Schwingungsspule f. || bobine f.
oscillatrice. / ~ discharge || oszillierende
Entladung f. || décharge f. oscillatoire.
/ ~ motion || schwingende Bewegung f. ||
mouvement m. oscillatoire. / ~ trans-
former || Schwingungsumformer m. ||
transformateur m. oscillant.
oscillograph || Oszillograf m. || oscillographe
m. / mirror ~ || Spiegeloszillograf m. ||
oscillographe m. à miroir.
osculate, to (Geom) || oskulieren || osculer.
osculating circle (Geom) || Oskulations-
kreis m.; Berührungskreis m. || cercle m.
osculateur.

osculation (Geom) || Oskulation f.; Be-
rührung f. höherer Ordnung || oscula-
tion f.
osculatory (Geom) || oskulierend; be-
rührend || osculateur.
osier || Weide f.; Bandweide f.; Korb-
weide f. || osier m.; saule m. / peeled ~ ||
geschälte Weide f. || osier m. écorcé ou
pelé.
osier bottle || Korbflasche f. || bouteille f.
clissée. / ~ ground || Weidenpflanzung f. ||
oseraie f. / ~ holt see ~ ground. / ~ plaiter
|| Weidenflechter m. || tresseur m. d'osier.
/ ~ switch see ~ twig. / ~ twig || Weiden-
rute f.; Weidenzweig m. || verge f.
d'osier ou de saule. / ~ work || Korb-
macherware f. || vannerie f.
osmide || Osmiumverbindung f. || osmide
m.
osmiridium (Miner) || Osmiridium n.;
Iriodosmin n.; Osmiumiridium n. ||
osmiridium m.; osmiure m. d'iridium;
iridosmine f.; iridosmium m.
osmium || Osmium n. || osmium m. / ~
alloy || Osmiumlegierung f. || alliage m.
d'osmium. / ~ iridium see osmiridium.
osmose || Osmose f. || osmose f. / electric ~
|| elektrische Osmose f. oder Endosmose f.
|| osmose f. ou endosmose f. électrique.
osmosis see osmose.
osmotic || osmotisch || osmotique. / ~ pres-
sure || osmotischer Druck m. || pression
f. osmotique.
Osram lamp || Osramlampe f. || lampe f.
osram.
ossein(e) || Osseïn n.; Knochensubstanz f. ||
osséine. f. / ~ plant || Osseïnanlage f. ||
installation f. d'osséine.
osselet || Sepia f. || seiche f.; sépia f.
osseous || knochig; knöchern || osseux.
ossification || Knochenbildung f. || ossifi-
cation f.
ossuary || Beinhaus n. || ossuaire m.
osteal see osseous.
ostensory || Monstranz f. || ostensoir(e) m.;
soleil m.
ostentation implement in precious metal
for churches || Prunkgerät n. aus Edel-
metall für kirchliche Zwecke || ustensile
m. somptueux en métal précieux ou pur
pour églises.
osteocolla || Knochenleim m. || ostéocolle
f.; gélatine f. d'os. / ~ (Miner) see osteolite.
osteolite (Miner) || Knochenstein m. ||
ostéolithe f.
ostracean (Oyster) || Auster f. || huître f.
ostrich feather || Straußfeder f. || plume f.
d'autruche.
otalgia; otalgy || Ohrenschmerz m. ||
otalgie f.
other || anders; auf andere Weise || autre-
ment.
otological instrument || Ohreninstrument
n. || instrument m. otique.
otology || Ohrenkunde f. || otologie f.
otto || Blütenessenz f.; Blumenessenz f.;
Blütenöl n. || essence f. de fleurs.
ottoman || Ottomane f.; Liegestuhl m. ||
ottomane f.
ounce || Unze f. || once f. / by the ~ || unzen-
weise || à l'once.
out || aus; zu Ende; leer || fini; vide.
out of || außer; ohne || sans. / ~ a job see ~
work. / ~ work || arbeitslos || sans travail.
out (Print) || Leiche f.; Auslassung f. (von
Wörtern) || bourdon m.; omission f. /
to make an ~ (Print) || ein Wort aus-
lassen || sauter un mot.

outbid, to ~ see to overbid.
outcrop (Mine) || zu Tage liegend; über
Tage || (étant) situe au jour; superficiel.
outcrop (Mine) || Tagebau m. || exploita-
tion f. ou travail m. à ciel ouvert. / ~ of
a seam || Ausgehen n. eines Flözes || sope
f. d'une couche.
outdoor illumination || Außenbeleuchtung
f. || éclairage m. extérieur. / ~ tempéra-
ture || Außentemperatur f. || tempéra-
ture f. extérieure. / ~ transformer sta-
tion || Freiluftumspannwerk n. || poste
m. de transformation extérieur.
outer bearing || Außenlager n. || palier m.
extérieur.
outer casing of a blast-furnace || Rauh-
gemäuer n. eines Hochofens || enveloppe
f. ou manteau m. ou muraillement m. d'un
haut-fourneau. / ~ of a furnace || Ofen-
stock m. || manteau m. d'un fourneau. /
~ for travellers' bags || Reisetaschen-
bügel m. || monture f. de valises de
voyage.
outer conductor see outer wire. / ~ dock ||
Außenhafen m. || avant-port m. / ~
down || Vordüne f. || dune f. de garde. /
~ edge || Außenrand m. || bord m. ex-
térieur. / ~ fish plate with elongated
holes || Außenlasche f. mit Langlöchern ||
éclisse f. extérieure avec trous oblongs.
/ ~ flame || Oxydationsflamme f. || feu
m. d'oxydation. / ~ globe || Außenglocke
f. || globe m. extérieur. / ~ harbour ||
Vorhafen m. || avant-port m. / ~ layer ||
Außenschicht f. || couche f. extérieure.
/ ~ main see outer wire. / ~ plank || Sei-
tenbrett n. || planche f. extérieure. /
portion || äußerer Teil m. || partie f.
extérieure. / ~ radius (Railway curve) ||
äußerer Radius m. oder Halbmesser m. ||
grand rayon m.
outer stack (Furnace) see outer casing. /
~ of a wall || Mauerhaupt n.; Mauer-
mantel m.; Stirnseite f. der Mauer ||
parement m. d'un mur.
outer support (Bearing) || Gegenlager n. ||
contre-support m. / ~ (Riveting) ||
Gegenhalterstütze f. || bretelle f. de
bouterolle.
outer wall || Umfassungsmauer f. || mur m.
de pourtour. / ~ of a volcano || äußerer
Ringwall m. eines Vulkans || rempart m.
circulaire d'un volcan.
outer wire || Außenleiter m. || conducteur
m. extérieur. / ~ voltage || Außenleiter-
spannung f. || voltage m. du conducteur
extérieur.
outfit || Ausrüstung f. || équipement m.;
équipage m.; outillage m. / works ~ ||
Fabrikeinrichtung f. || installation f. de
fabriques.
outgoing air || Abluft f. || air m. sortant. /
~ current || abgehender Strom m. ||
courant m. de départ. / ~ secondary
switch (Aut tel) || Mischwähler m. ||
sélecteur m. de mélange. / ~ stream ||
Ebbstrom m. || courant m. de jusant;
courant m. de sortie.
outlast, to || überdauern || survivre à ...
outlay, modern machines pl. which yielded
twice the work with less ~ || neuzeitliche
Maschinen fpl., die bei geringeren Be-
triebskosten das Doppelte leisteten ||
machines fpl. de construction moderne
qui avec des dépenses courantes moin-
dres avaient un rendement double.
outlay expenses pl. || Anlagekosten pl. ||
frais mpl. d'installation.

outlet ‖ Abfluß m.; Abzug m. ‖ sortie f.; échappement m.; écoulement m.; décharge f. / ~ (Exit) ‖ Ausgang m. ‖ sortie f. / ~ (Hydr arch) ‖ Auslaßschleuse f. ‖ colateur m. / ~ (Kitchen) ‖ Ausguß m. ‖ décharge f.; évier m. / ~ (Trade) ‖ Absatzgebiet n. ‖ débouché m.; marché m. / brine ~ ‖ Soleaustritt m. ‖ sortie f. de la saumure. / to find a good ~ ‖ guten Absatz m. finden ‖ être d'une bonne (défaite *ou*) de bonne vente. / to open new ~s pl. ‖ neue Handelsbeziehungen fpl. anbahnen ‖ entamer de relations fpl. commerciales. / rain ~ (Hydr arch) ‖ Notauslaß m. einer Kanalisation ‖ bonde f. de sortie d'eau de pluie. / water ~ ‖ Wasserauslaß m. ‖ sortie f. d'eau.

outlet channel ‖ Freigerinne n.; Freilauf m. ‖ lancière f.; déchargeoir m. / ~ cock ‖ Ablaufhahn m. ‖ robinet m. d'écoulement. / ~ fitting ‖ Auslaufarmatur f. ‖ déversoir m. / ~ gutter ‖ Ablaufrinne f. ‖ rigole f. d'écoulement. / ~ junction ‖ Auslaufstutzen m. ‖ tubulure f. de sortie. / ~ pipe ‖ Auslaßrohr n.; Abzugrohr n. ‖ tuyau m. d'échappement *ou* d'écoulement. / ~ slide ‖ Auslaßschieber m. ‖ tiroir m. d'évacuation.

outlet sluice ‖ Fluchtschleuse f.; Abzugsschleuse f. ‖ écluse f. de fuite *ou* de chasse. / adjustable ~ ‖ Regelablaßschütz n. ‖ vanne f. régulatrice.

outlet tube ‖ Ablaßrohr n. ‖ tube m. de décharge.

outlet valve ‖ Auslaßventil n. ‖ soupape f. de décharge. / ammonia ~ ‖ Ammoniakauslaßventil n. ‖ vanne f. de sortie d'ammoniaque. / ~ on steeping tanks (Brew) ‖ Ausweichventil n. ‖ soupape f. de décharge de décuvage.

outline ‖ Schattenriß m.; Schattenbild n. ‖ silhouette f. / sharp ~ ‖ scharfe Ausbildung f. ‖ exactitude f. des plus infimes détails.

outline drawing ‖ Umrißzeichnung f.; dessin m. en contours.

outlook ‖ Aussicht f.; Ausblick m. ‖ vue f.; perspective f. / ~ envelope ‖ Fensterbriefumschlag m. ‖ enveloppe f. à fenêtre.

outlying department ‖ Außenverwaltung f. ‖ service m. extérieur. / ~ lands pl. ‖ Vorgelände n. ‖ avant-plaine f.; avant-pays m.

out of ‖ außer; ohne ‖ sans.

out-of-date ‖ abgelaufen; verfallen ‖ arrêté; échu.

out-of-focus image ‖ unscharfes Bild n. ‖ image f. floue *ou* manquant de netteté.

out-of-order tone (Tel) ‖ Gestörtzeichen n. ‖ signal m. de dérangement.

out-of-work ‖ arbeitslos ‖ sans travail m. / to be ~ ‖ arbeitslos sein ‖ être sans travail; chômer.

out-of-work ‖ Arbeitslose(r) m. ‖ sans-travail m. / ~ benefit *see* ~ pay. / ~ pay ‖ Arbeitslosenunterstützung f. ‖ allocation f. de chômage.

outport ‖ Außenhafen m. ‖ avant-port m.

output ‖ Arbeitsleistung f.; Produktion f. ‖ production f. / ~ (Beer) ‖ Ausstoß m. ‖ débit m.; vente f. / ~ (Mach) ‖ Leistung f. ‖ effet m.; rendement m. / ~ (Mine) ‖ Förderung f. ‖ extraction f.; production f. / ~ (Radio) ‖ Endleistung f. ‖ puissance f. terminale. / annual ~ ‖ Jahresleistung (sfähigkeit) f. ‖ production f. annuelle. / average ~ ‖ Durchschnittsleistung f.;

durchschnittliche Produktion f. ‖ rendement m. moyen. / for cheaper ~ ‖ zwecks billigerer Erzeugung ‖ en vue f. d'une fabrication à meilleur marché. / continuous ~ ‖ Dauerleistung f. ‖ débit m. continu. / daily ~ ‖ Tagesleistung f. ‖ rendement m. par jour; débit m. journalier. / daily ~ of a machine ‖ Tagesleistung f. einer Maschine ‖ rendement m. journalier d'une machine. / ~ per day *see* daily ~. / decreased ~ ‖ verminderte Produktion f. ‖ rendement m. diminué. / driving ~ (Ship) ‖ Antriebsleistung f. ‖ puissance f. de propulsion. / ~ per hour ‖ stündliche Leistung ‖ rendement m. *ou* débit m. par heure. / hourly ~ *see* ~ per hour. / incoming ~ ‖ aufgenommene Leistung f. ‖ puissance f. fournie. / increased ~ ‖ vermehrte Produktion f. ‖ rendement m. augmenté. / increased ~ (Mach) ‖ erhöhte Leistung f. ‖ débit m. plus élevé. / increased ~ of raw lignite ‖ verstärkte Förderung f. von Rohbraunkohle ‖ extraction f. augmentée de lignite. / the ~ increased rapidly ‖ die Ausbeute f. stieg schnell ‖ l'extraction f. augmenta rapidement. / ~ to keep-on the simmer ‖ Fortkochzahl f. ‖ chiffre f. de maintien de l'ébullition. / maximum ~ ‖ Höchstleistung f. ‖ puissance f. maximum. / maximum ~ of a motor ‖ Höchstleistung f. eines Motors ‖ puissance f. maxima d'un moteur. / ~ of a motor ‖ Motor(en)leistung f.; Leistung f. eines Motors ‖ puissance f. d'un moteur. / normal ~ ‖ normale Leistung f. ‖ débit m. de régime. / outgoing ~ (Mach) ‖ abgegebene Leistung f. ‖ puissance f. disponible. / rated ~ ‖ Normalleistung f. ‖ puissance f. *ou* productivité f. normale. / reduced ~ ‖ Minderleistung f. ‖ débit m. réduit. / temporary maximum ~ ‖ vorübergehende Spitzenleistung f. ‖ puissance f. maximum transitoire. / total ~ ‖ Gesamtleistung f. ‖ débit m. total. / total ~ of the motors ‖ Gesamtleistung f. der Motoren ‖ puissance f. totale des moteurs. / ~ per week ‖ wöchentliche Leistung f. ‖ rendement m. par semaine. / ~ per year ‖ Jahresproduktion f. ‖ rendement m. par an.

output amplifier ‖ Leistungsverstärker m. ‖ amplificateur m. de puissance. / ~ meter ‖ Leistungsmesser m. ‖ compteur m. de rendement. / ~ plate ‖ Leistungsschild n. ‖ écusson m. / power of ~ ‖ Leistungsvermögen n. ‖ capacité f. de production. / ~ recorder ‖ Leistungsschreiber m. ‖ enregistreur m. de rendement. / ~ test ‖ Leistungsversuch m. ‖ essai m. de puissance. / ~ transformer (Tel) ‖ Nachübertrager m.; Ausgangstransformator m. ‖ transformateur m. de sortie *ou* au bout.

outreach of a crane ‖ Ausladung f. eines Krans ‖ portée f. d'une grue.

outrigger (Post) ‖ Mastausleger m. ‖ bras m. de poteau. / ~ (Shipb) ‖ Luvbaum m. ‖ boute-hors m.

outshot of porcelain ‖ Ausschußgeschirr n. ‖ porcelaine f. de rebut.

outshot hemp ‖ Ausschußhanf m. ‖ chanvre m. outshot.

outside of a wall ‖ Mauerfront f.; Außenseite f. einer Mauer ‖ parement m. *ou* côté m. de devant d'un mur.

outside box ‖ Außenlager n. ‖ palier m. extérieur. / ~ chaser ‖ Außenstrehler m. ‖

peigne m. mâle. / ~ crank ‖ Außenkurbel f. ‖ manivelle f. extérieure. / ~ diameter ‖ Außendurchmesser m. ‖ diamètre m. extérieur. / ~ link ‖ Außenglied n. ‖ membre m. extérieur. / ~ plate (Shipb) ‖ Außenhautplatte f. ‖ plaque f. d'enveloppe. / ~ planking (Shipb) ‖ Außenbeplankung f. ‖ bordé m. extérieur (en bois). / ~ temperature ‖ Außentemperatur f. ‖ température f. extérieure. / ~ wall without plastering ‖ unverputzte Außenmauer f. ‖ mur m. extérieur sans enduit.

outstanding ‖ rückständig ‖ en retard m.; redevable. / ~ debts pl. ‖ Außenstände pl. ‖ actifs mpl.; créances fpl. / ~ liabilities pl. *see* ~ debts.

outward grid (Radio) ‖ Außengitter n. ‖ grille f. extérieure.

outwards and inwards ‖ außen und innen ‖ au dehors et en dedans.

oval ‖ oval; eiförmig ‖ ovale; oviforme; ovoïde. / ~ chuck ‖ Ovalwerk n.; Oval(spann)futter n. ‖ poupée f. de tour ovale. / ~ file ‖ Ovalfeile f. ‖ lime f. ovale. / ~ frame ‖ Ovalrahmen m. ‖ cadre m. oval. / ~ handle ‖ Werkzeugheft n. ovaler Form ‖ manche m. ovale.

oval-head countersunk screw ‖ Linsensenkschraube f. ‖ vis f. à tête fraisée bombée. / ~ countersunk wood screw ‖ Linsen(senk)holzschraube f. ‖ vis f. à bois à tête fraisée bombée *ou* à goutte de suif.

oval hole ‖ Langloch n. ‖ trou m. oblong. / ~ lathe ‖ Ovaldrehbank f. ‖ tour m. à (tourner en) ovale. / ~ line ‖ Ovale f. ‖ ovale m. / ~ shape ‖ ovale Form f. ‖ forme f. ovale. / ~ turning ‖ Passigdrehen n.; Ovaldrehen n. ‖ tourner en ovale. / ~-turning lathe *see* oval lathe.

ovalization ‖ Unrundwerden n. ‖ ovalisation f.

ovary ‖ Fruchtknoten m. ‖ ovaire m.

oven ‖ Ofen m. ‖ four m.; fourneau m. / baker's ~ *see* baking ~. / baking ~ ‖ Backofen m. ‖ four m. (à cuire *ou* de boulangerie). / confectionery ~ ‖ Konditorherd m. ‖ four m. pâtissier. / gas ~ ‖ Gasofen m. ‖ four m. à gaz.

oven building ‖ Ofenbau m. ‖ fumisterie f. / ~ coke ‖ Ofenkoks m. ‖ coke m. de four à coke. / ~ drawer (Coke oven) ‖ Koksofenentleerer m. ‖ défourneur m.; déchargeur m. de fours. / ~ filler (Coke oven) ‖ Koksofenfüller m. ‖ chargeur m. de fours. / ~ man ‖ Ofeneinsetzer m. enfourneur m. / ~ plant ‖ Ofenanlage f. ‖ installation f. de fours. / range of ~s ‖ Ofenbatterie f. ‖ batterie f. de fours. / ~ tenter (Bak) ‖ Backofenheizer m. ‖ chauffeur m. de four de boulanger.

overall ‖ Arbeitsanzug m.; Schmutzkittel m. ‖ blouse f. / ~ height (Crane) ‖ Bauhöhe f. ‖ hauteur f. totale.

overall length ‖ Gesamtlänge f.; Ausmaß n. ‖ longueur f. totale. / ~ (Shipb) ‖ Länge f. über alles ‖ longueur f. horstout.

overall loss (Tel) ‖ Betriebsdämpfung f. ‖ affaiblissement m. effectif; pertes fpl. de transmission. / ~ transmission loss (Tel) ‖ Restdämpfung f. ‖ affaiblissement m. effectif. / ~ width ‖ Baubreite f. ‖ largeur f. de l'installation.

overarch, to ‖ einwölben ‖ envoûter; voûter.

overbid, to ‖ überbieten ‖ renchérir sur; surenchérir.

overbridge ‖ Überführung f. ‖ viaduc m.; passage m. par dessus.

overburden, to ‖ überbürden ‖ surcharger. / ~ a valve ‖ ein Ventil überlasten ‖ surcharger une soupape.

overburden removal of ~ of a coal seam ‖ Abräumen n. eines Kohlenflözes ‖ déblaiement m. d'une couche de charbon.

overcast (Tail) ‖ überwendliche Naht f. ‖ rentraiture f.

overcharge, to ~ the accumulator ‖ den Akkumulator m. überladen ‖ surcharger l'accumulateur. / ~ the ink (Print) ‖ Schwärze f. zu dick auftragen ‖ charger trop d'encre.

overcharge ‖ Überlastung f. ‖ surcharge f. / ~ (Electr) see overcharging.

overcharging ‖ Überladung f. ‖ surcharge f.; surchargement m.

overcoat ‖ Überzieher m. ‖ paletot m.

over-compounded ‖ überkompoundiert ‖ hypercompoundé.

over-compounding ‖ Überkompoundierung f. ‖ hypercompound m.

overcool, to ‖ überkühlen ‖ surrefroidir.

overcorrection ‖ Überkorrektion f. ‖ surcorrection f.

over-done (Flax or hemp) ‖ überrottet ‖ détérioré par un rouissage trop prolongé.

overdraft ‖ Überdisposition f. ‖ découvert m.

overdrawn ‖ übertrieben ‖ outré; exorbitant.

overdue ‖ überfällig ‖ en souffrance f.

overestimate, to ‖ zu hoch schätzen ‖ surtaxer.

overexert, to ~ see to overwork.

overexposure ‖ Überbelichtung f. ‖ surexposition f.

overfall (Hydr arch) see overfall weir. / incomplete ~ ‖ Grundwehr n. ‖ déversoir m. incomplet.

overfall dyke ‖ Überlaufdeich m. ‖ digue f. à déversoir.

overfalls pl. (Mar) ‖ überbrechende Seen fpl. ‖ lames fpl. déferlantes.

overfall weir ‖ Überfallwehr n.; Stauwehr n. ‖ batardeau m.

overflow, to (Aut tel) ‖ durchdrehen ‖ outre-passer. / ~ (River) ‖ übertreten; austreten ‖ déborder.

overflow ‖ Überlauf m. ‖ trop-plein m. / ~ (Aut tel) ‖ Überlauf m. ‖ congestion f. / ~ of a dam ‖ Überlauf m. einer Sperrmauer ‖ déversoir m. d'un mur de barrage. / stormwater ~ (Hydr arch) ‖ Notauslaß m. ‖ bonde f. de pluie.

overflow channel ‖ Überlaufrinne f. ‖ couloir m. de trop-plein. / ~ cone ‖ Überlaufkonus m. ‖ embout m. / ~ contact (Aut tel) ‖ Überlaufkontakt m. ‖ contact m. à signal de congestion.

overflowed ‖ überflutet ‖ débordé.

overflow flask ‖ Auslaufflasche f. ‖ flacon m. de trop-plein.

overflow pipe ‖ Überlaufrohr n. ‖ tuyau m. de trop plein. / ~ (Hydr arch) ‖ Überschleusleitung f. ‖ canalisation f. ou conduite f. de dégorgement.

overfolded rocks pl. ‖ Überfaltungsgebirge n. ‖ montagne f. formée par des plis de recouvrement.

overfolding (Geol) ‖ Überfaltung f. ‖ recouvrement m.

overfreight ‖ Überfracht f. ‖ surcharge f.

overground line see overhead line. / ~ workings pl., undertaking of ~ ‖ Hoch-

bauunternehmung f. ‖ entreprise f. de constructions au-dessus du sol.

overgrow, to ~ with wood ‖ bewalden ‖ boiser.

overhand knot (Mar) ‖ Sackstich m. ‖ nœud m. à plein poing. / ~ stope (Mine) ‖ Firstenstoß m. ‖ gradin m. renversé.

overhang ‖ Überhang m. ‖ porte f. à faux. / ~ eccentric press ‖ ausladende Exzenterpresse f. ‖ presse f. à excentrique avec bâti à col de cygne ou en porte-à-faux.

overhanging ‖ überhängend ‖ en porte à faux. / ~ (Surv) ‖ aus dem Lote n. ‖ surplombant. / ~ arm (Mach) ‖ Ausleger m. ‖ bras m. radial. / ~ side (Mine) ‖ überhängender Stoß m. ‖ paroi f. surplombante.

overhaul, to ‖ nachsehen; überholen ‖ examiner. / ~ an engine at regular intervals ‖ eine Maschine f. in regelmäßigen Abständen nachsehen ‖ vérifier une machine périodiquement. / ~ the rigging ‖ Takelage f. nachsehen ‖ parcourir le gréement.

overhaul ‖ Überholung f. ‖ révision f. général.

overhauled ‖ instandgesetzt; überholt ‖ examiné.

overhauling see overhaul. / ~ work ‖ Überholungsarbeit f. ‖ travail m. d'entretien.

overhead ‖ oberirdisch ‖ aérien.

overhead cable ‖ Freileitung f. ‖ câble m. aérien. / self-supporting ~ ‖ freitragendes Luftkabel n. ‖ câble m. aérien à auto-support. / sleeve of a self-supporting ~ ‖ Muffe f. für ein freitragendes Luftkabel ‖ manchon m. pour câble aérien à auto-support. / straining of a self-supporting ~ ‖ Abspannung f. eines freitragenden Luftkabels ‖ arrêt m. d'un câble aérien à auto-support.

overhead contact, hoop actuated by compressed air for ~ ‖ Druckluftbügelbetätigung f. für Stromabnehmer ‖ prise f. de courant à archet actionnée à l'air comprimé.

overhead distribution point ‖ Freileitungsverteilungspunkt m. ‖ point m. de distribution aérienne. / ~ frog ‖ Fahrdrahtweiche f. ‖ changement m. de voie du fil de contact.

overhead line (Electr) ‖ Freiluftleitung f.; Luftleitung f.; oberirdische Leitung f. ‖ ligne f. aérienne. / ~ (Railw) see overhead railway. / ~ fittings pl. ‖ Freileitungsarmatur f. ‖ accessoire m. pour ligne aérienne.

overhead line material ‖ Oberleitungsmaterial n. ‖ matériel m. de ligne aérienne. / ~ for tramways ‖ Straßenbahnoberleitungsmaterial n. ‖ matériel m. pour lignes de contact de tramway.

overhead railway ‖ Hochbahn f. ‖ chemin m. de fer aérien. / electric ~ ‖ elektrische Hochbahn f. ‖ chemin m. de fer aérien électrique.

overhead rope railway ‖ Seil(hänge)bahn f.; Drahtseilbahn f. ‖ transporteur m. aérien à câble. / electro ~ ‖ Elektrohängebahn f. ‖ transporteur m. aérien électrique.

overhead single-rail line ‖ Einschienenhochbahn f.; Schwebebahn f. ‖ voie f. monorail suspendue. / ~ tank ‖ Hochbehälter m. ‖ réservoir m. surélevé. / ~ track ‖ Schwebebahngeleise n. ‖ voie f. surélevée ou de chemin de fer aérien.

overhearing (Tel) ‖ Mitsprechen n. ‖ cross talk entre combinant et combiné. / to insert for ~ ‖ sich zum Mithören n. einschalten ‖ se mettre sur l'écoute.

overheat, to ‖ überhitzen ‖ surchauffer. / ~ (Mach) ‖ heißlaufen ‖ surchauffer.

overheater (Steam) ‖ Dampfüberhitzer m. ‖ surchauffeur m. / ammonia ~ ‖ Überhitzer m. für Ammoniak ‖ surchauffeur m. d'ammoniaque.

overheating (Steam) ‖ Überhitzung f. ‖ surchauffe f. / ~ (Tel) ‖ Überheizung f. ‖ surchauffement m. / sensitive to ~ (Tel) ‖ gegen Überheizung f. empfindlich ‖ sensible au surchauffage m.

overhung flywheel ‖ fliegendes Schwungrad n. ‖ volant m. monté en porte à faux. / ~ girder ‖ Kragträger m. ‖ poutre f. en encorbellement. / ~ shears pl. ‖ Schere f. mit fliegender Lagerung ‖ cisailles fpl. en parte à faux.

overland cable ‖ oberirdisches Kabel n. ‖ câble m. aérien. / ~ route ‖ Landstraße f. ‖ grand chemin m.; grande route f. / ~ service ‖ Überlandverkehr m. ‖ communication f. inter-urbaine.

overlap (Geol) ‖ Überschiebung f. ‖ recouvrement m. / ~ (Mach) ‖ Überlappung f. ‖ recouvrement m. / ~ (Railw) ‖ Schutzstrecke f. ‖ distance f. de protection.

overlapping of oscillations ‖ Überlagerung f. der Schwingungen ‖ recouvrement m. des oscillations.

overlaying of tortoise shell ‖ Schildpattauflage f. ‖ applique f. écaille.

overload ‖ Überlast f.; Überlastung f. ‖ surcharge f. / permissible ~ ‖ zulässige Überbelastung f. ‖ surcharge f. admissible. / to stand 100 % ~ ‖ um 100 % überlastbar sein ‖ avoir une capacité f. de surcharge de 100 %. / breaking by ~ ‖ Bruch m. bei Überlastung ‖ rupture f. en cas de surcharge.

overload capacity ‖ Überlastungsfähigkeit f.; Überlastbarkeit f. ‖ capacité f. de surcharge. / ~ circuit breaker ‖ Maximalausschalter m. ‖ interrupteur m. à maximum.

overloaded ‖ überlastet ‖ surchargé.

overloading ‖ Überlastung f. ‖ surcharge f.

overload spring (Railw) ‖ Pufferfeder f. ‖ ressort m. supplémentaire. / ~ switch ‖ Höchststromschalter m. ‖ interrupteur m. à maximum.

overlooker ‖ Aufseher m. ‖ surveillant m.

overman ‖ Aufseher m.; Vorarbeiter m. / ~ (Mine) ‖ Steiger m. ‖ maître-mineur m.; porion m.

overpass ‖ Straßenüberführung f. ‖ passerelle f.

overpressure ‖ Überdruck m. ‖ surpression f.

overproduction ‖ Überproduktion f. ‖ surproduction f.

overretted (Spinn) ‖ verrottet ‖ détérioré par un rouissage trop prolongé.

overretting (Spinn) ‖ Überrotte f. ‖ excès m. de rouissage.

overrigged (Shipb) ‖ übertakelt ‖ de trop gros agrès.

oversea ‖ transatlantisch ‖ transatlantique. / ~ plane ‖ Überseeflugzeug n. ‖ avion m. ou hydravion m. de haute mer. / ~ trade ‖ Überseehandel m. ‖ commerce m. d'outremer.

overseer ‖ Aufseher m. ‖ surveillant m. / ~ (Build) ‖ Bauaufseher m.; Polier m. ‖

inspecteur m. (des travaux de construction); maître m. ouvrier; contremaître m. / ~ (Print) ‖ Druckereifaktor m.; Faktor m. ‖ prote m. / ~ of department ‖ Abteilungsvorsteher m. ‖ chef m. de rayon. / ~ of the line ‖ Bahnaufseher m.; Oberbahnwärter m. ‖ garde-chef m.; garde-ligne m. / ~ of a mine ‖ Steiger m. ‖ maître-mineur m.

overset, to (Mar) ‖ kentern ‖ chavirer; sombrer; faire capot; capoter.

overshot water wheel ‖ oberschlächtiges Wasserrad n. ‖ roue f. hydraulique en dessus.

oversize ‖ Übergröße f. ‖ dimension f. renforcée. / ~ pieces pl. ‖ Überkorn n. ‖ grains mpl. trop grands. / ~ tyre ‖ Luftreifen m. in Übergröße ‖ pneu m. surprofilé.

oversleeve ‖ Überärmel m. ‖ fausse manche f.; garde-manche f.

overspeed test ‖ Schleuderversuch m. ‖ essai m. d'emballement.

oversprinkler (Brew) ‖ Anschwänzer m.; Anschwänzkreuz n. ‖ croix f. écossaise.

overstress, to ~ the fabric ‖ den Stoff n. übermäßig beanspruchen ‖ faire travailler le tissu d'une manière excessive.

overstressing ‖ Überbeanspruchung f. ‖ surtension f.

overtask, to see to overwork.

overthrust (Geol) ‖ Überschiebung f. ‖ recouvrement m. / ~ transverse to the strike (Geol) ‖ Querüberschiebung f. ‖ faille f. inverse transversale. / ~ mass (Geol) ‖ Überschiebungsmasse f. ‖ masse f. inversée ou de glissement.

overtime ‖ Überstunden fpl. ‖ heures fpl. supplémentaires. / to work ~ ‖ Überstunden fpl. machen ‖ faire des heures fpl. supplémentaires de travail.

overturn, to ‖ umkippen ‖ culbuter; renverser. / ~ (Carp) ‖ kanten; umkanten ‖ rouler sur la carne; cabaner; renverser; faire l'abatage m. / Do not overturn! (Package) ‖ nicht stürzen! ‖ pas renverser!

overturning by means of a lever ‖ Umkanten n. mittels Hebel ‖ abatage m. par levier.

overturning moment ‖ Kippmoment n. ‖ moment m. de renversement.

overweight ‖ Übergewicht n.; Mehrgewicht n. ‖ excédent m. de poids; surpoids m.

overwhelmed with work ‖ mit Arbeit f. überhäuft ‖ surchargé de travail m.

overwork, to ‖ überanstrengen ‖ surmener.

overwound ‖ überdreht ‖ foiré. / the nut has been ~ ‖ die Mutter ist überdreht worden ‖ l'écrou m. a été foiré.

overzinc, to ‖ verzinken ‖ zinguer.

owl-light ‖ Abenddämmerung f. ‖ crépuscule m. du soir.

own, to ‖ besitzen ‖ posséder. / ~ shares pl. in a company ‖ Anteile mpl. einer Gesellschaft besitzen ‖ posséder des parts dans une société.

owner ‖ Besitzer m. ‖ possesseur m. / ~ of the firm ‖ Firmeninhaber m. ‖ propriétaire m. de la maison. / part ~ ‖ Miteigentümer m. ‖ co-propriétaire m. / ~ of a ship ‖ Reeder m. eines Schiffes ‖ armateur m.; armateur m. propriétaire d'un navire. / sole ~ ‖ Alleineigentümer m. ‖ propriétaire m. unique.

owner-farmed land ‖ Landwirtschaftsbetrieb m. mit Eigenland ‖ propriété f. agricole.

ownership, assignment of ~ ‖ Besitzeinweisung f. ‖ envoi m. en possession. / right of ~ ‖ Eigentumsrecht n. ‖ droit m. de propriété.

ox ‖ Rind n.; Ochs(e) m. ‖ bœuf m. / grunting ~ ‖ Yak m. ‖ yak m.; yack m.

oxalate ‖ Oxalat n. ‖ oxalate m. / ~ of ammonium ‖ Ammoniumoxalat n. ‖ oxalate m. d'ammoniaque. / acid ~ of potassium ‖ Kleesalz n.; Sauerkleesalz n.; oxalsaures Kalium n. ‖ bioxalate de potasse; sel m. d'oseille.

oxalic acid ‖ Oxalsäure f. ‖ acide m. oxalique. / ~ works pl. ‖ Oxalsäurefabrik f. ‖ fabrique f. d'acide oxalique.

oxalite ‖ Oxalit m. ‖ fer m. oxalaté.

oxen breeder ‖ Ochsenzüchter m. ‖ éleveur m. de bœufs. / ~ driver ‖ Ochsentreiber m. ‖ conducteur de bœufs m.

ox-eye ‖ Ochsenauge n. ‖ œil m. de bœuf.

Oxford clay ‖ Oxfordton m. ‖ masse f. argileuse oxfordienne.

ox gall ‖ Ochsengalle f. ‖ fiel m. de bœuf.

oxidable ‖ oxydierbar ‖ oxydable.

oxidate, to ‖ oxydieren ‖ oxyder; oxygéner.

oxidation ‖ Oxydation f. ‖ oxydation f.; oxygénation. / ~ fermentation ‖ Oxydationsgärung f. ‖ fermentation f. d'oxydation. / ~ flame ‖ Oxydationsflamme f. ‖ flamme f. oxydante. / ~ step ‖ Oxydationsstufe f. ‖ degré m. d'oxydation.

oxide ‖ Oxyd n. ‖ oxyde m. / ~ of aluminium ‖ Aluminiumoxyd n. ‖ oxyde m. d'aluminium. / ~ of barium ‖ Bariumoxyd n. ‖ (prot)oxyde m. de barium. / ~ of calcium ‖ Ätzkalk m.; gebrannter Kalk m. ‖ chaux f.; oxyde m. de calcium; chaux f. vive ou caustique ou calcinée. / ferric ~ ‖ Eisenoxyd n. ‖ oxyde m. ferrique. / ferrous ~ ‖ Eisenoxydul n. ‖ fer m. oxydulé. / ~ of iron see ferric ~. / lead ~ ‖ Bleiglätte f.; litharge f. ‖ manganic ~ ‖ Manganoxyd n. ‖ oxyde m. manganique; sesquioxyde m. de manganèse. / manganous ~ ‖ Manganoxydul n.; Manganooxyd n. ‖ oxyde m. manganeux; protoxyde m. de manganèse. / ~ of mercury ‖ Quecksilberoxyd m. oxyde m. de mercure. / metallic ~ ‖ Metalloxyd n. ‖ oxyde m. métallique. / mixed ~ ‖ Oxyduloxyd n. ‖ oxyde m. salin. / red ~ of lead ‖ Mennige f.; Bleimennige f. ‖ minium n.; oxyde m. rouge de plomb. / red ~ of mercury ‖ rotes Quecksilberoxyd n. ‖ protoxyde m. de mercure; oxyde m. rouge de mercure. / to remove the film of ~ ‖ die Oxydschicht f. entfernen ‖ enlever la couche f. d'oxyde. / siliciferous ~ of manganese ‖ Kieselmangan m.; Rhodonit m. ‖ manganèse m. silicieux; rhodonite f. / ~ of zinc ‖ Zinkoxyd n. ‖ oxyde m. de zinc.

oxide cathode ‖ Oxydkathode f. ‖ cathode f. à oxydes. / ~-coated filament ‖ Oxyddraht m. ‖ filament m. recouvert d'oxyde. / ~-coated filament valve ‖ Oxydröhre f. ‖ lampe f. d'oxyde. / film of ~ ‖ Oxydschicht f. ‖ couche f. d'oxyde. / layer of ~ see film of ~.

oxidizability ‖ Oxydierbarkeit f. ‖ oxydabilité f.

oxidizable ‖ oxydierbar ‖ oxydable.

oxidize, to ‖ oxydieren ‖ oxyder; oxygéner. / ~ off ‖ herausoxydieren; fortoxydieren ‖ éliminer par oxydation f. / the material oxidizes only very little ‖ der

Werkstoff rostet nur wenig ‖ le métal ne se rouille que très peu.

oxidized brass mounting ‖ oxydierte Messingfassung f. ‖ monture f. cuivre oxydé. / ~ water ‖ Wasserstoffsuperoxyd n. ‖ eau f. oxygénée; bioxyde m. d'hydrogène.

oxidizer see oxidizing agent.

oxidizing ‖ oxydierend ‖ oxydant. / ~ agent ‖ Oxydationsmittel n. ‖ oxydant m. / ~ blast smelting ‖ oxydierendes Schmelzen n. ‖ fusion f. oxydante. / ~ flame see oxidation flame. / ~ plant ‖ Oxydationseinrichtung f. ‖ installation f. d'oxydation.

ox-muzzle salad ‖ Ochsenmaulsalat m. ‖ salade f. de museau de bœuf.

ox-tongue ‖ Rinderzunge f.; Ochsenzunge f. ‖ langue f. de bœuf. / ~ boiler ‖ Ochsenzungekocher m. ‖ cuiseur m. de langue de bœuf.

oxychloride ‖ Oxychlorid n. ‖ oxychlorure m.

oxygen ‖ Sauerstoff m. ‖ oxygène m. / active ~ ‖ aktiver Sauerstoff m.; Ozon n. ‖ oxygène m. actif; ozone m. / to obtain from the atmosphere ~ of highest purity ‖ reinsten Sauerstoff m. aus atmosphärischer Luft gewinnen ‖ obtenir l'oxygène m. de la plus haute pureté de l'air atmosphérique. / ozonized ~ ‖ Ozonsauerstoff m. ‖ oxygène m. ozonisé.

oxygen apparatus ‖ Sauerstoffapparat m. ‖ appareil m. à oxygène. / ~ appliance ‖ Sauerstoffgerät n. ‖ appareil m. à oxygène.

oxygenate, to see to oxidize.

oxygenated water ‖ Wasserstoffsuperoxyd n. ‖ bioxyde m. d'hydrogène; eau f. oxygénée.

oxygenation see oxidation.

oxygen compressor ‖ Sauerstoffkompressor m. ‖ compresseur m. d'oxygène. / ~ enzyme ‖ Sauerstoffenzym n. ‖ oxydase f. / ~ flask ‖ Sauerstoffflasche f. ‖ bouteille f. à oxygène. / ~ inhaling apparatus ‖ Sauerstoffrettungsapparat m. ‖ appareil m. de sauvetage à l'oxygène.

oxygenize, to see to oxidize.

oxygen lance ‖ Schneidbrenner m. ‖ brûleur m. à découper. / ~ meter ‖ Sauerstoffmesser m. ‖ compteur m. d'oxygène. / ~ plant ‖ Sauerstoffanlage f. ‖ installation f. pour la fabrication d'oxygène. / ~ producing plant ‖ Sauerstoffgewinnungsanlage f. ‖ installation f. à produire l'oxygène. / ~ producing plant with multiple rectification ‖ Sauerstoffgaserzeugungsanlage f. mit mehrfacher Luftzerlegung ‖ installation f. pour la production d'oxygène à rectification multiple. / ~ production of ~ ‖ Sauerstoffgewinnung f. ‖ production f. d'oxygène.

oxygon ‖ spitzwinkliges Dreieck n. ‖ triangle m. acutangle.

oxygonal ‖ spitzwinklig ‖ acutangle.

oxyhydrogen blow-pipe ‖ Knallgasgebläse n. ‖ chalumeau m. oxhydrique.

oxyhydrogen gas ‖ Knallgas n. ‖ gaz m. fulminant ou explosif ou oxyhydrique. / formation of ~ ‖ Knallgasentwicklung f. ‖ formation f. de gaz détonnant. / ~ voltameter ~ ‖ Knallgasvoltameter n. ‖ voltamètre m. à gaz oxyhydrique.

oylet ‖ Schlitzfenster n. ‖ œillet m.; lézarde f.

oyster ‖ Auster f. ‖ huître f.

oyster bed ‖ Austernbank f. ‖ huîtrière f.; parc m. à huîtres; banc m. d'huîtres; claire f.; crassat m. / ~ labourer ‖ Austernbankarbeiter m. ‖ ouvrier m. du parc m. à huîtres. / lessee of ~ ‖ Austernbankpächter m. ‖ fermier m. d'huîtrière ou du parc m. à huîtres.

oyster catcher ‖ Austernfischer m. ‖ pêcheur m. ou dragueur m. d'huîtres. / ~ culture ‖ Austernzucht f. ‖ ostréiculture f. / ~ culturist ‖ Austernzüchter m. ‖ ostréiculteur m.; parceur m. ou éleveur m. d'huîtres. / ~ dredger see ~ catcher. / ~ ground see oyster bed. / ~ knife ‖ Austernmesser n. ‖ couteau m. à ouvrir les huîtres. / ~ shell ‖ Austernschale f. ‖ écaille f. d'huître.

ozocerite ‖ Ozokerit m.; Bergwachs n.; Erdwachs n. ‖ ozokérite f.; cire f. fossile. / purified ~ ‖ gereinigtes Ozokerit n. ‖ ozokérite f. purifiée. / refined ~ ‖ Zeresin n. ‖ cire f. fossile raffinée.

ozokerite see ozocerite.

ozone ‖ Ozon n. ‖ ozone m. / rich in ~ ‖ ozonreich ‖ riche en ozone m. / ~ bleaching apparatus ‖ Ozonbleichapparat m. ‖ appareil m. de blanchiment à l'ozone. / ~ paper ‖ Ozon(reagens)papier n. ‖ papier m. à ozone. / percentage of ~ in the air ‖ Ozongehalt m. der Luft ‖ teneur m. ou richesse f. de l'air en ozone; quantité f. d'ozone dans l'air. / ~ water see ozonized water.

ozoniferous ‖ ozonhaltig ‖ ozonifère; ozonisé.

ozoning plant ‖ Ozonanlage f. ‖ installation f. de préparation d'ozone.

ozonize, to ‖ ozonisieren ‖ ozoniser.

ozonized air ‖ ozonhaltige Luft f. ‖ air m. ozonisé. / ~ oxygen ‖ Ozonsauerstoff m. ‖ oxygène m. ozonisé. / ~ water ‖ Ozonwasser n. ‖ eau f. ozonisée.

ozonizer ‖ Ozonisator m. ‖ ozoniseur m.

ozonizing, water ~ ‖ Wasserozonisierung f. ‖ ozonisage m. de l'eau. / ~ tube ‖ Ozonröhre f. ‖ tube m. ozoniseur.

ozonograph ‖ (selbst)schreibendes Ozonoskop n. ‖ ozonographe m.

ozonometer ‖ Ozonmesser m.; Ozonometer n. ‖ ozonomètre m.

ozonoscope ‖ Ozonoskop n. ‖ ozonoscope m.

P

pace, to ~ a distance ‖ eine Strecke f. abschreiten ‖ mesurer une distance au pas.

paceometer ‖ Schrittzähler m. ‖ pedomètre m.

pack, to ‖ verpacken; einpacken; packen; emballer. / ~ (To caulk) ‖ abdichten ‖ calfater; calfeutrer. / ~ (Mach) ‖ lidern ‖ étouper. / ~ (Railw) ‖ unterstopfen ‖ bourrer. / ~-over again ‖ umpacken ‖ changer l'emballage; réemballer. / ~ the sleepers ‖ die Schwellen fpl. unterstopfen ‖ bourrer les traverses fpl.

package ‖ Packung f.; Verpackung f.; emballage m. / ~ (Railw) ‖ Frachtstück n. ‖ paquet m.; colis m. / trade ~ (Barrel) ‖ Transportfaß n. ‖ fût m. d'expédition; tonneau m. de transport; barrique f.

package sealer ‖ Paketschließer m. ‖ appareil m. à fermer les paquets.

pack cloth ‖ Sackleinwand f.; Packleinen n. ‖ toile f. à sacs ou d'emballage; serpillière f. / ~ weaving ‖ Packleinenweberei f. ‖ tissage m. de toile d'emballage.

packed in boxes ‖ in Schachteln fpl. verpackt ‖ emballé dans des boîtes fpl. / to be carefully ~ ‖ sorgfältig zu verpacken ‖ à emballer soigneusement. / frames pl. carefully ~ for transport ‖ für den Transport sorgfältig verpackte Rahmen mpl. ‖ châssis mpl. bien emballés pour le transport. / ~ in bundles ‖ in Bündeln npl. verpackt ‖ enveloppé en faisceaux mpl. / ~ in cases ‖ in Kisten fpl. verpackt ‖ emballé en caisses fpl. / ~ compactly ‖ lückenlos verpackt ‖ emballé bien serré. / ~ in crates ‖ in Lattenverschlägen mpl. verpackt ‖ emballé en caisses fpl. à claire voie. / ~ in mats ‖ in Matten fpl. verpackt ‖ enveloppé de mattes fpl.

packer ‖ Verpacker m. ‖ emballeur m. / ~ (Mine) ‖ Versatzarbeiter m. ‖ remblayeur m.; restapleur m.; releveur m. de terres.

packet, to ‖ paketieren ‖ paqueter.

packet ‖ Paket n.; Bündel n. ‖ paquet m. ~ (Print) ‖ Schriftstück n.; stückweiser Satz m. ‖ paquet m. / ~ boat ‖ Paketboot n.; Postdampfer m. ‖ paquebot m. / ~ closing disk ‖ Paketverschlußscheibe f. ‖ disque m. à paquets. / ~ filling machine ‖ Paketfüllmaschine f. ‖ machine f. à remplir les paquets.

packeting machine ‖ Verpackungsmaschine f. ‖ machine f. d'emballage.

packfilm ‖ Packfilm m. ‖ filmpack m.

pack ice ‖ Packeis n. ‖ banquise f.

packing ‖ Verpackung f. ‖ emballage m. / ~ (Mach) ‖ Packung f.; Dichtung f.; Liderung f. ‖ garniture f.; bourrage m.; étoupage m.; joint m. / ~ (Mine) ‖ Bergeversatz m. ‖ remblai m.; remblayage m. / ~ of asbestos ‖ Asbestdichtung f. ‖ garniture f. d'asbeste. / ~ the ballast below the sleepers ‖ Nachstopfen n. der Gleise ‖ bourrage m. de la voie. / ~ in cases ‖ Verpackung f. in Kisten ‖ emballage m. en caisses / cord ~ ‖ Schnurpackung f. ‖ garniture f. de cordon. / ~ to preserve against dust ‖ Staubdichtung f. ‖ garniture f. pour empêcher l'entrée de poussière. / felt ~ ‖ Filzdichtung f. ‖ joint m. de feutre. / fibre ~ ‖ Fiberdichtung f. ‖ joint m. de fibre. / flexible metallic ~ ‖ biegsame Metallpackung f. ‖ bourrage m. métallique flexible. / hemp ~ ‖ Hanfdichtung f.; Hanfliderung f. ‖ étoupage m. ou garniture f. de chanvre. / hydraulic ~ ‖ Hydraulikpackung f. ‖ garniture f. hydraulique. / hydraulic ~ of the wastes (Mine) ‖ Spülversatzverfahren n. ‖ remblayage m. hydraulique. / including ~ ‖ einschließlich Verpackung f. ‖ emballage m. compris. / india-rubber ~ ‖ Kautschukpackung f.; Kautschukdichtung f. ‖ garniture f. de caoutchouc. / leather ~ ‖ Lederliderung f. ‖ garniture f. de cuir. / machine for ~ the line ‖ Gleisstopfmaschine f. ‖ machine f. à bourrer les traverses de chemin de fer. / metallic ~ ‖ Metalldichtung f.; Metallliderung f. ‖ garniture f. métallique. / not including ~ ‖ Verpackung f. nicht einbegriffen ‖ emballage m. non compris. / paper ~ ‖ Papierverpackung f. ‖ emballage m. en papier. / piston ~ ‖ Kolbenliderung f.; Kolbenpackung f. ‖ garniture f. de piston. / pneumatic ~ (Mine) ‖ Blasversatz m. ‖ remblayage m. pneumatique. / seaproof ~ ‖ seemäßige Verpackung f. ‖ emballage m. (pour transport) maritime. / ~ of sheet metal ‖ Blechverpackung f. ‖ emballage m. en fer-blanc. / ~ for stuffing box ‖ Stopfbüchsenpackung f. ‖ garniture f. de presse-étoupe. / ~ and ranging the track ‖ Heben n. und Richten n. der Gleise ‖ relèvement m. et redressement n. de la voie.

packing balk ‖ Rödelbalken m. ‖ poutrelle f. de guindage. / ~ barrel ‖ Packfaß n. ‖ tonneau m. d'emballage. / ~ bolt (Mach) ‖ Packungsbolzen m. ‖ boulon m. de serrage.

packing box ‖ Packschachtel f. ‖ boîte f. d'emballage. / ~ branding machine ‖ Packkistenbrennmaschine f. ‖ machine f. à marquer les caisses d'emballage au fer rouge.

packing bush ‖ Stopfbuchse f. ‖ presseétoupe f.; douille f. de garniture. / ~ canvas ‖ Packleinwand f.; Packleinen n. ‖ toile f. d'emballage.

packing case ‖ Verpackungskiste f.; Packkiste f. ‖ caisse f. d'emballage. / dismounted ~ ‖ zerlegte Packkiste f. ‖ caisse f. d'emballage démontée. / mounted ~ ‖ zusammengesetzte Packkiste f. ‖ caisse f. d'emballage montée.

packing cloth see ~ canvas. / ~ cord ‖ Dichtungsschnur f. ‖ corde f. de bourrage. / ~ cost ‖ Verpackungskosten pl. ‖ frais mpl. d'emballage. / ~ device ‖ Verpackungsapparat m. ‖ appareil m. à empaqueter. / ~ and labelling device ‖ Etikettierungs- und Verpackungsapparat m. ‖ appareil m. à empaqueter et à étiqueter. / ~ disc ‖ Dichtungsscheibe f. ‖ rondelle f. de joint. / ~ glass ‖ Verpackungsglas n. ‖ verre m. d'emballage. / ~ grease ‖ Fettpräparat n. für Packungen ‖ graisse f. pour étoupages. / ~ groove ‖ Dichtungsrille f. ‖ rainure f. de garniture. / ~ linen ‖ Packleinen n. ‖ toile f. d'emballage.

packing machine ‖ Verpackungsmaschine f. ‖ machine f. d'emballage. / barrel ~ ‖ Faßpackungsmaschine f. ‖ machine f. à empaqueter en tonneaux. / ~ for dry products ‖ Packmaschine f. für

Trockenerzeugnisse ‖ machine f. à emballer les produits secs. / railway line ~ ‖ Gleisstopfmaschine f. ‖ machine f. à bourrer les voies de chemin de fer. / ~ for small packages ‖ Beutelpackmaschine f. ‖ machine f. à remplir les sachets.

packing material ‖ Verpackungsmaterial n. ‖ matériel m. d'emballage. / ~ (Mach) ‖ Dichtungsmaterial n. ‖ matériel m. d'étoupage.

packing needle ‖ Packnadel f. ‖ aiguille f. à emballer.

packing paper ‖ Packpapier n. ‖ papier m. d'emballage ou à emballer. / common ~ ‖ gewöhnliches Packpapier n. ‖ papier m. d'emballage ordinaire. / tar-covered ~ ‖ geteertes Packpapier n. ‖ papier m. d'emballage goudronné. / varnished ~ ‖ gefirnißtes Packpapier n. ‖ papier m. d'emballage verni. / waxed ~ ‖ gewachstes Packpapier n. ‖ papier m. d'emballage ciré.

packing press ‖ Packpresse f. ‖ presse f. à emballer.

packing ring ‖ Dichtungsring m. ‖ anneau m. de garniture. / cylinder ~ ‖ Zylinderdichtungsring m. ‖ anneau m. ou bague f. de garniture de cylindre. / ~ press ‖ Dichtungsringandrückmaschine f. ‖ machine f. à poser les joints en caoutchouc. / ~ press and edge turning-up machine ‖ Dichtungsringandrück- und Randvorrollmaschine f. ‖ machine f. combinée à poser les joints et à rouler le bord.

packing, rubber for ~s pl. ‖ Dichtungsgummi n. ‖ caoutchouc m. pour garnitures. / ~ sleeve ‖ Rillenbuchse f. ‖ fourrure f. à rainures. / cleaning apparatus for the ~ space of boiler tube closures ‖ Dichtungsringreiniger m. für Siederohrverschlüsse ‖ nettoyeur m. de surfaces de joints aux tubes bouilleurs. / ~ stick ‖ Packstock m. ‖ cheville f. à tourniquet. / ~ wadding ‖ Packwatte f. ‖ ouate f. d'emballage. / ~ washer ‖ Packungsscheibe f. ‖ rondelle f. de garniture. / weight of ~ ‖ Gewicht n. der Verpackung; Umschließungsgewicht n. ‖ poids m. de l'emballage. / ~ worm ‖ Packungszieher m. ‖ tire-bourre m.; tire-étoupe m.

pack rags pl. ‖ Packhadern mpl.; Packlumpen mpl. ‖ chiffons-emballages mpl.

pack-saddle ‖ Packsattel m.; Saumsattel m. ‖ bât m. / ~ maker ‖ Sattler m. für Packsättel oder Saumsättel ‖ bâtier m.

packthread ‖ Bindfaden m.; Kordel f.; Packschnur f.; Spagat m. ‖ ficelle f.; fil m. d'emballage.

paco hair ‖ Pakoshaar n. ‖ alpaga m.; poil m. d'Alpaga.

pad, to (Text print) ‖ aufklotzen ‖ plaquer. / ~ (Upholstery) ‖ polstern ‖ matelasser; rembourrer.

pad (Chem) ‖ Flausch m. ‖ tampon m. / ~ (Saddl) ‖ Sattelkissen n.; Packkissen n. ‖ panneau m.; coussinet m. de charge. / ~ (Upholstery) ‖ Polster n. ‖ rembourrage m. / ~ (Tel) ‖ Verlängerungsleitung f. ‖ ligne f. artificielle de complément. / foot pedal ~ ‖ Pedalauflage f. ‖ patin m. à pédale. / insulating ~ ‖ Isolierpolster n. ‖ rembourrement m. isolant. / ornamental ~ of a harness ‖ Kammkissen n.; Zierkissen n. ‖ mantelet m. / ~ of straw ‖ Strohsack m. ‖ paillasse f.

padded ‖ gepolstert ‖ rembourré.

padding ‖ Wattierung f. ‖ ouatage m. / ~ (Saddl) ‖ Füllhaar n. ‖ bourre f. / ~ (Text print) ‖ Aufklotzen n. ‖ placage m.

paddle ‖ Schaufel f.; Radschaufel f. ‖ aile f.; aube f.; palette f.; aileron m.; alichon m. / ~ (Metal) ‖ Kratze f.; Rührstange f. ‖ perche f. à brasser; râble m.; brassoir m. / feathering ~ ‖ bewegliche Schaufel f. ‖ aube f. articulée ou mobile. / nave ~ ‖ Nabenschaufel f. ‖ palette f. de moyeu.

paddle beam ‖ Radbalken m. ‖ grand-bau m. porte-roue.

paddle board ‖ Radschaufel f. ‖ aube f. de la roue; palette f. / cycloidal ~ ‖ zykloïdische Schaufel f. ‖ palette f. cycloïdale.

paddle bolt ‖ Schaufelbolzen m. ‖ crochet m. d'aube; boulon m. à croc. / ~ hole (Sluice) ‖ Freiarche f. ‖ auge f. ou conduit m. d'une écluse. / ~ ring ‖ Schaufelradring m. ‖ cercle m. entourant une roue à palettes. / ~ shaft ‖ Radachse f.; Radwelle f. ‖ arbre m. de roue. / ~ steamer ‖ Raddampfer m. ‖ bateau m. à roues. / ~ valve (Hydr arch) ‖ Schleusenschütze f. ‖ vanne f.; vantelle f.

paddle wheel ‖ Schaufelrad n. ‖ roue f. à aubes ou à palettes. / common ~ ‖ Rad n. mit festen Schaufeln ‖ roue f. à aubes fixes. / feathering ~ ‖ Rad n. mit beweglichen Schaufeln ‖ roue f. à aubes articulées ou à palettes mobiles. / radial ~ ‖ Rad n. mit festen Schaufeln ‖ roue f. à aubes fixes.

paddle wheel boss ‖ Radnabe f. eines Schaufelrades ‖ tourteau m. d'une roue à aubes. / ~ ring ‖ Radring m. eines Schaufelrades ‖ cercle m. entourant une roue à palettes. / ~ steamer see paddle steamer.

pad grimper ‖ Lederpresse f. ‖ presse f. à cuir.

padlock, to ‖ anschließen ‖ attacher; cadenasser.

padlock ‖ Hängeschloß n.; Vorhängeschloß n.; Vorlegeschloß n. ‖ cadenas m. / automatic ~ ‖ Vorhängeschloß n. mit Rücksprungbügel ‖ cadenas m. automatique. / combination-letter ~ ‖ Kombinationsschloß n.; Vexierschloß n. ‖ cadenas m. à combinaisons à lettres. / round-shackle ~ ‖ Vorhängeschloß n. mit rundem Bügel ‖ cadenas m. à anse ronde. / square-shackle ~ ‖ Vorhängeschloß n. mit Vierkantbügel ‖ cadenas m. à anse carrée. / turning-shackle ~ ‖ Vorhängeschloß n. mit Drehbügel ‖ cadenas m. à anse tournante.

padlock smith ‖ Vorhängeschloßmacher m. ‖ cadenassier m.

pad saw ‖ Fuchsschwanzsäge f.; Fuchsschwanz m. ‖ scie f. à main ou à manche. / backed ~ ‖ deutscher Fuchsschwanz m. mit Rücken ‖ scie f. à dossière ou à dos. / English ~ ‖ englischer Fuchsschwanz m.; Blattsäge f.; Biberschwanz m. ‖ scie f. à manche d'égoïne; sciotte f.

padstone ‖ Auflagerstein m. ‖ pierre f. d'appui.

page, to ~ **a book** ‖ ein Buch n. foliieren oder paginieren ‖ numéroter les feuilles fpl. d'un livre.

page ‖ Kolumne f.; Schriftseite f.; Seite f. ‖ page f. / blank ~ ‖ Blankseite f. oder erste Seite f. eines Buches ‖ fausse-page f.; page f. blanche. / even ~ ‖ Kehrseite f.; Rückseite f.; gerade Seite f. ‖ Kolumne f. ‖ page f. paire; verso m. /

first ~ ‖ erste Seite f. ‖ folio m. recto. / foul ~ ‖ fehlerhafte Druckseite f. ‖ page f. mal venue. / odd ~ ‖ Vorderseite f.; page f. belle ou impaire; recto m. / slur ~ ‖ Schmutzseite f. ‖ fausse-page f.

page cord (Print) ‖ Kolumnenschnur f. ‖ ficelle f.

pager ‖ Seitenzahlensetzer m. ‖ numéroteur m.; folioteur m.; paginateur m.

page setter ‖ Seiteneinrichter m. ‖ metteur m. en pages.

paging ‖ Paginierung f.; Seitenzahlensetzen n. ‖ pagination f. / ~ machine ‖ Paginiermaschine f. ‖ machine f. à paginer. / workshop for ~ ‖ Werkstatt f. für den Druck der Seitenzahlen ‖ atelier m. de pagination.

paid (Post) ‖ franko; frei ‖ franco de port m.; franco. / ~ time (Tel) ‖ bezahlte Sprechzeit f. ‖ temps m. payé.

paid-out (Cash register) ‖ Ausgabe f. ‖ dépenses fpl. / ~ capital ‖ ausgezahltes Kapital n. ‖ capital m. remboursé.

pail ‖ Eimer m.; seau m. / ~ for jams ‖ Marmeladeneimer m. ‖ seau m. à marmelades. / wooden ~ ‖ Holzeimer m. ‖ seau m. en bois.

pail handle ‖ Eimerbügel m. ‖ anse m. de seau.

pain ‖ Schmerz m. ‖ douleur f.

paint, to ‖ bemalen; malen; anstreichen ‖ peindre; peinturer. / ~ several coats ‖ mehrmals anstreichen ‖ donner plusieurs couches de couleur. / ~ thin ‖ die Farbe mager auftragen ‖ donner une faible couche de couleur.

paint ‖ Anstrich m.; Farbe f.; Anstrichfarbe f. ‖ peinture f.; couleur de peinture. / ~ (Cosmetic) ‖ Schminke f. ‖ fard m. / acid-proof ~ ‖ säurefeste Farbe f.; säurebeständiger Anstrich m. ‖ peinture f. inattaquable aux acides. / antifouling ~ ‖ fäulnissichere Farbe f. ‖ peinture f. anti-viciée. / ~ fire-proof ‖ feuersichere Farbe f. ‖ peinture f. ignifuge. / ready mixed ~ ‖ anstrichfertige Farbe f. ‖ couleur f. préparée pour la peinture. / water-proof ~ ‖ wasserfester Anstrich m. ‖ peinture f. hydrofuge; couleur f. lucidonique. / weather-proof ~ ‖ wetterfeste Farbe f. ‖ peinture f. à l'épreuve des intempéries. / ~ for wood ‖ Holzanstrich m. ‖ peinture f. à bois.

paint box ‖ Tuschkasten m.; Farb(en)kasten m. ‖ boîte f. ou carte-échantillon f. de couleurs.

paint brush ‖ Pinsel m.; Malerpinsel m. ‖ pinceau m. à peindre. / ~ hair ‖ Pinselhaar n. ‖ crin m. pour pinceau à peindre. / ~ maker ‖ Pinselmacher m. ‖ pinceautier m.

paint drier ‖ Sikkativ n. ‖ siccatif m.

painted wooden art work ‖ Holzmalerei f. ‖ travail m. d'art en bois peint.

painter ‖ Maler m. ‖ peintre m. / ~ (Bridge) ‖ Spanntau n. ‖ traversière f.; amarre f.; écharpe f. / ~ (Mar) ‖ Fangleine f. ‖ bosse f. / ~ on fabrics ‖ Stoffmaler m. ‖ peintre m. sur étoffes. / ~ of glass ‖ Glasmaler m. ‖ peintre m. sur verre.

painter's canvas ‖ Malleinwand f. ‖ toile f. à peindre. / ~ colour ‖ Malerfarbe f. ‖ couleur f. pour peintres. / ~ glue ‖ Malerleim m. ‖ colle f. pour peintres. / ~ lye ‖ Malerlauge f. ‖ lessive f. de peintres.

paint grinder ‖ Farbenreibmaschine f.; Farbenmühle f. ‖ broyeur m. à couleurs.

painting ‖ Anstrich m.; Farbe f. ‖ peinture f. / ~ (Picture) ‖ Gemälde n. ‖ peinture f. ‖ tableau m. / article for ~ on cloth ‖ Stoffmalereiartikel m. ‖ article m. pour la peinture sur étoffe. / coach ~ ‖ Anstreichen n. von Wagen ‖ peinture f. de voitures. / distemper ~ ‖ Aquarellmalerei f.; Wasserfarbenmalerei f. ‖ peinture f. à gouache ou à l'aquarelle ou en détrempe. / ~ of glass ‖ Glasmalerei f. ‖ peinture f. sur verre.

painting in oil ‖ Ölmalerei f. ‖ peinture f. à l'huile. / ~ (Build) ‖ Ölanstrich m.; Ölfarbenanstrich m. ‖ peinturage m. à l'huile. / ~ strewed over with sand ‖ besandeter Ölanstrich m. ‖ peinturage m. au sable.

painting on porcelain ‖ Porzellanmalerei f. ‖ peinture f. sur porcelaine. / ~ in size colours ‖ Leimfarbenmalerei f. ‖ peinture f. à la colle. / ~ with transparent colours ‖ Malerei f. mit Lasurfarben ‖ peinture f. à couleurs transparentes. / water colour ~ ‖ Gemälde n. mit Wasserfarben; Aquarell(gemälde) n. ‖ aquarelle f.

painting apparatus, compressed air ‖ Farbenspritzpistole f. mit Preßluft ‖ appareillage m. à peindre par air comprimé. / ~ book for children ‖ Kindermalbuch n. ‖ cahier m. à illustrer pour enfants. / ~ colour ‖ Anstrichfarbe f. ‖ couleur f. de peinture. / ~ machine ‖ Anstreichmaschine f. ‖ badigeonneuse f.

paint mill ‖ Farbenmühle f. ‖ broyeur m. à couleurs. / ~ remover ‖ Farbenabbeizmittel n. ‖ décapant m. de peinture.

paint spraying pistol ‖ Spritzpistole f. zum Anstreichen ‖ pistolet m. à peinturer. / ~ system ‖ Farbenspritzverfahren n. ‖ peinture f. pneumatique.

pair, in ~s pl. ‖ paarweise ‖ par paires fpl.; par couples fpl. / ~ of bevel wheels ‖ Kegelräderpaar n. ‖ paire f. de roues coniques. / ~ of cylinders cast in ~s ‖ paarweis (zusammen)gegossene Zylinder mpl. ‖ cylindres mpl. fondus par deux. / ~ of compasses ‖ Zirkel m. ‖ compas m. / ~ of cords for long distance positions (Tel) ‖ Fernschnurpaar n. ‖ dicorde f. interurbaine. / ~ of lenses ‖ Linsenpaar n. ‖ paire f. de lentilles. / ~ of needles ‖ Nadelpaar n. ‖ couple f. d'aiguilles. / ~ of scales ‖ Balkenwage f. ‖ balance f. ordinaire. / ~ of shafts ‖ Gabeldeichsel f. ‖ limonière f. / ~ of sights ‖ Diopter n. ‖ dioptre f.; viseur m.; pinnule f. / ~ of wheels ‖ Räderpaar n. ‖ paire f. de roues.

pakfong see paktong.

paktong ‖ Neusilber n. ‖ argentan m.; argent m. d'Allemagne; packfong m.; maillechort m.

palace ‖ Schloß n. ‖ château m.; palais m.

palæobotany ‖ vorgeschichtliche Pflanzenkunde f.; Paläobotanik f. ‖ paléophytologie f.

palæontologic(al) ‖ paläontologisch ‖ paléontologique.

palæontology ‖ Paläontologie f.; Versteinerungskunde f. ‖ paléontologie f.

palagonite ‖ Palagonit m. ‖ palagonite m.

palatable ‖ schmackhaft ‖ savoureux.

palate ‖ Gaumen m. ‖ palais m.

pale, to (Hydr arch) ‖ verpfählen; Pfähle mpl. schlagen ‖ palifier; enfoncer des pieux.

pale ‖ blaß ‖ pâle. / ~ coffee ‖ blasser Kaffee m. ‖ café m. pâle.

pale ‖ Pfahl m. ‖ pal m.; pieu m.; pilot m. / ~ for fences ‖ Zaunpfahl m. ‖ palis m.; petit pieu m. pointu.

paling ‖ Pfahlzaun m. ‖ clôture f. en échalas. / ~ (Hydr arch) ‖ Verpfählung f. ‖ palification f. / wooden ~ ‖ Pfahlzaun m.; Staket n. ‖ échalier m.; estacade f.

palisade, to ‖ einpfählen ‖ palissader; entourer de palis m.

palisade ‖ Einpfählung f.; Einzäunung f.; Pfahlhecke f.; Pfahlzaun f.; Spalier n.; Staket n. ‖ clôture f. de palis; espalier m.; palissade f.

pall (Arch) ‖ Gabelkreuz n. ‖ pairle m.

palladium ‖ Palladium n. ‖ palladium m. / ~ bath ‖ Palladiumbad n. ‖ bain m. de palladiumage. / ~ gold ‖ Palladiumgold n.; Porpezit m. ‖ palladium m. aurifère.

pallet (Paint) ‖ Palette f.; Farbenbrett n. ‖ palette f. / ~ (Pott) ‖ Drehscheibe f.; Töpferscheibe f. ‖ tour m. ou roue f. de potier. / ~ of the escapement (Clockm) ‖ Hemmungslappen m.; Spindellappen m. ‖ aile f. de l'échappement; paillette f.

palm, to ~ **in** (Mar) ‖ Hand über Hand hissen; palmen ‖ hisser main sur main.

palm ‖ Palme f.; Palmbaum m. ‖ palmier m. / ~ of the hand ‖ Handfläche f. ‖ paume f. de la main.

palmarosa oil ‖ Palmarosaöl n. ‖ essence f. de palme rose.

palm butter ‖ Palmöl n.; Palmbutter f. ‖ huile f. de palme. / ~ fibre ‖ Palmfaser f. ‖ fibre f. de palmier. / ~ fruit dressing machine ‖ Palmfruchtaufbereitungsmaschine f. ‖ machine f. à travailler les fruits de palmiers.

palmitic acid ‖ Palmitinsäure f. ‖ acide m. palmitique.

palmitin ‖ Palmitin n. ‖ palmitine f. / ~ candle ‖ Palmitinkerze f. ‖ bougie f. palmitique.

palm kernel oil ‖ Palmkernöl n. ‖ huile f. de noix d'Inde.

palm nut ‖ Palmnuß f. ‖ noix f. de palmier. / ~ oil ‖ Palmkernöl n. ‖ huile f. de noix de palmier.

palm oil ‖ Palmöl n. ‖ huile f. de palme. / ~ tree ‖ Palme f.; Palmbaum m. ‖ palmier m.

pamphlet ‖ Flugblatt n.; Flugschrift f.; Broschüre f. ‖ feuille f. volante; pamphlet m.; brochure f. / ~ stitching machine ‖ Broschürenheftmaschine f. ‖ machine f. à coudre les brochures.

pan (Kitchen) ‖ Pfanne f. ‖ casserole f. / ~ (Print) ‖ Setzschiff n. ‖ galée f. / ~ (Techn) ‖ Kessel m.; Pfanne f. ‖ chaudière f. / amalgamating ~ ‖ Amalgamierpfanne f. ‖ cuve f. d'amalgamation. / ~ of balance ‖ Wagschale f. ‖ plateau m. ou bassin m. de balance. / crystallizing ~ ‖ Kristallisationsgefäß f. ‖ Kristallisationskessel m. ‖ cristallisoir m. / electric ~ ‖ elektrischer Kochtopf m. ‖ marmite f. chauffée électriquement. / evaporating ~ ‖ Abdampfgefäß n.; Abdampfschale f. ‖ capsule f. évaporatoire ou à évaporation. / galvanization ~ ‖ Galvanisierpfanne f. ‖ cuve f. de galvanisation. / ~ for carrying glowing coke ‖ Transportgefäß n. für glühenden Koks ‖ récipient m. de transport du coke ardent. / ~ of the pivot ‖ Angelring m.; Türangelpfanne f. ‖ crapaudine f.; piton m. / ~ for roasting limestone ‖ Steinhärte-

kessel m. ‖ chaudière f. pour le durcissement des pierres. / stirring ~ ‖ Rührpfanne f. ‖ marmite f. à agitateur. / the ~ can be tilted at an angle of x° ‖ die Pfanne kann um x° geschwenkt werden ‖ l'auge f. peut être basculée à l'angle de x°. / tin plate warming ~ ‖ Wärmepfanne f. oder Wärmekessel m. aus Weißblech ‖ chauffe-plat m. en ferblanc. / washing ~ ‖ Waschpfanne f. ‖ cuve f. de lavage.

Panama bark ‖ Panamarinde f. ‖ bois m. de Panama. / ~ hat ‖ Panamahut m. ‖ panama m.

panary fermentation ‖ Brotgärung f. ‖ fermentation f. panaire.

pancake, to (Aero) ‖ durchsacken ‖ descendre à plat.

pancake ‖ Pfannkuchen m.; flacher Eierkuchen m. ‖ omelette f. / ~ (Leather cardboard) ‖ Lederpappe f. ‖ carton m. cuir.

pane, to (Mar) ‖ pinnen, finnen ‖ panner.

pane ‖ Platte f.; Tafel f. ‖ dalle f.; lame f.; plaque f.; table f. / ~ (Of a door etc.) ‖ Füllung f. ‖ panneau m.; pan m. / ~ (Window) ‖ Fensterscheibe f. ‖ vitre f.

pane of glass ‖ Fensterscheibe f. aus Glas; Glasscheibe f. ‖ carreau m. de vitre; vitre f.; panneau m. de vitre ou de verre. / bulged ~ ‖ bauchige Fensterscheibe f. ‖ vitre f. bombée.

pane of a hammer ‖ Hammerfinne f.; Pinne f. ‖ panne f. du marteau. / rhombic ~ (Glassm) ‖ Rautenglas n. ‖ rhombe m. de vitre; vitre f. rhomboïde. / ~ of a roof ‖ Dachfläche f. ‖ plan m. de comble. / ~ of a stone ‖ Fläche f. oder Fuge f. oder Seite f. eines Steines ‖ panneau m. d'une pierre. / ~ of a wall ‖ Mauerstrecke f.; Mauerfeld n. ‖ pan m. de mur.

panel (Join) ‖ Füllung f. ‖ panneau m. / ~ (Saddl) ‖ Sattelkissen n. ‖ panneau m. / ~ of a trussed beam ‖ Feld n. eines Fachwerkträgers ‖ cloison f. d'une poutre américaine. / cased ~ ‖ eingestemmte Füllung f. ‖ panneau m. de menuiserie. / door ~ ‖ Türverkleidung f. ‖ panneau m. de porte. / false ~ ‖ blinde Füllung f. ‖ faux panneau m. / flush ~ (Join) ‖ bündige Füllung f. ‖ panneau m. plein. / luminous indicator ~ with remote control and telemetering devices ‖ Leuchtschaltbild n. mit Fernsteuerung und Fernmessung ‖ diagramme f. lumineux avec télécommande et mesure à distance.

panel board ‖ Füllbrett n.; Parkettafel f. ‖ carreau m. de parquet. / ~ form ‖ Paneelform f. ‖ forme f. de panneau.

panelled ‖ getäfelt; in Füllungen geteilt ‖ pannelé.

panelling ‖ Getäfel n.; Täfelwerk n.; Täfelung f. ‖ boiserie f.; lambris m. / wood ~ ‖ Holztäfelung f. ‖ revêtement m. en bois.

panel stake ‖ Schalholz n.; Stakholz n.; Wellerholz n. ‖ palançon m.; palençon m.; polisson m. / ~ switch ‖ Stangenwähler m. ‖ sélecteur m. à panneau. / ~ telephone system ‖ Stangenwählersystem n. ‖ système m. de sélecteurs à panneau. / ~ work (Mine) ‖ Pfeilerabbau m. in Bauabteilungen ‖ travail m. par compartiment ou par chambres isolées.

pan greaser (Bak) ‖ Backformeneinfetter m. ‖ graisseur m. de moules.

panic on the exchange || Börsenkrach m. || débâcle f.; krach m.

panne || Panne f. || panne f.

pannel see panel.

panorama || Rundblick m. || vue f. panoramique.

panoramic exposure (Phot) || Rundblickaufnahme f. || prise f. de vue panoramique. / ~ sight || Rundblickvisiervorrichtung f.; Rundblickaufsatz m. || hausse f. panoramique.

pan riveting || Flachkopfnietung f. || rivure f. à tête plate.

pan scales pl. (Salt) || Pfannenstein m.; Satzschuppen fpl. || écailles fpl. / to knock off the ~ || den Kesselstein m. abklopfen || détartrer ou piquer une chaudière.

pansy herb || Stiefmütterchenkraut n. || herbe f. de pensée.

pan tile || Dachziegel m. || tuile f.; panne f.; tuile f. flamande. / much recurved ~ || Fittichziegel m. || tuile f. flamande très recourbée.

pan tiling || Eindeckung f. mit Dachpfannen; Pfannendach n. || couverture f. en tuiles flamandes.

pantograph || Storchschnabel m.; Pantograf m. || pantographe m. / ~ designer (Weav) || Pantografenzeichner m. || pantographiste-dessinateur m.

pantoscope || Pantoskop n. || pantoscope m.

pantoscopic lens || pantoskopisches Glas n. || verre m. pantoscopique.

pantry (Mar) || Pantry f.; Anrichteraum m. || office f.

pap (Bookb) || Mehlkleister m.; Kleister m.; Pappe f. || colle f. de farine; pâte f.; bouillie f.

paper, to || austapezieren || tapisser. / ~ the pins || die Stecknadeln fpl. einbriefen || encarter les épingles fpl.

paper || Papier n. || papier m. / albuminized ~ || Albuminpapier n. || papier m. albuminé. / art ~ || Kreidepapier n. || papier m. crayonneux. / ~ for making artificial flowers || Blumenpapier n. || papier m. à fleurs. / ~ for art printing || Kunstdruckpapier n. || papier m. pour impression artistique. / ~ for banknotes and stamps || Banknoten- und Wertzeichenpapier n. || papier m. pour billets de banque et pour valeurs. / bichrome-metallized ~ || zweifarbig metallisiertes Papier n. || papier m. métallisé bicolore. / bituminized ~ || bitumiertes Papier n.; Asphaltpapier n. || papier m. bitumé. / black-edged ~ || Trauerpapier n. || papier-deuil m. / blotting ~ || Löschpapier n. || papier m. buvard. / sheet of blotting ~ || Löschblatt n. || feuille f. de papier buvard. / ~ for book-printing || Buchdruckpapier n. || papier m. d'impression de livres. / Brazil wood ~ || Braunholzpapier n. || papier m. fait de bois brun. / cambric ~ || Seidenpapier n. || papier m. de soie. / carbon ~ || Kohlepapier n. || papier m. au charbon; papier-carbone m. / charcoal ~ || Kohlezeichenpapier n. || papier m. pour le dessin au fusain. / charged ~ || beschwertes Papier n. || papier m. chargé. / chemically prepared ~ || chemisch präpariertes Papier n. || papier m. chimique. / cloth-mounted ~ || Leinenpapier n. || papier m. entoilé. / coated ~ || gestrichenes Papier n. || papier m. couché. / coloured ~ || Buntpapier n. || papier m. color(i)é ou de couleur ou de fantaisie. / copper plate ~ || Kupferdruckpapier n. || papier m. pour taille douce. / copying ~ || Kopierpapier n.; Durchschreibepapier n.; Durchschlagpapier n. || papier m. autocopiste ou à copier. / corrugated ~ || Wellpapier n. || papier m. ondulé. / ~ for covers || Umschlagpapier n. || papier m. de couverture. / ~ for decorating shop windows || Schaufensterdekorationspapier n. || papier m. décoratif pour étalages. / dialytic ~ || dialytisches Papier n. || papier m. dialytique. / drawing ~ || Zeichenpapier n. || papier m. à dessin ou à dessiner. / drawing ~ in rolls || Zeichenpapier n. in Rollen || papier m. à dessin en rouleaux. / duplicating ~ see copying ~. / electric ~ || elektrisches Papier n. || papier m. electrique. / enamelled ~ || emailliertes Papier n. || papier m. émaillé. / endless ~ || Rollenpapier n.; endloses Papier n. || papier m. continu ou sans fin. / ~ for envelopes || Briefumschlagpapier n. || papier m. à enveloppes. / fancy ~ || Buntpapier n.; Vorsatzpapier n. || papier m. color(i)é ou de fantaisie. / filigreed ~ || Papier n. mit Wasserzeichen || papier m. filigrané. / filtering ~ || Filtrierpapier n. || papier m. à filtrer. / fine ~ || Luxuspapier n. || papier m. de luxe. / fleecy ~ || Wolkenpapier n. || papier m. pelucheux. / fools cap ~ || Kanzleipapier n. || papier m. de chancellerie. / fumigating ~ || Räucherpapier n. || papier m. d'arménie. / glazed ~ || Glanzpapier n.; glasiertes oder satiniertes Papier n.; Taftpapier n. || papier m. glacé ou lustré ou satiné. / gilt ~ see gold ~. / gold ~ || Goldpapier n. || papier m. doré. / grained ~ || gekörntes Papier n. || papier m. grainé. / grey ~ || Fließpapier n. || papier m. gris. / layer of gray ~ || Fließpapierunterlage f. || double m. de papier gris. / gummed ~ || gummiertes Papier n. || papier m. gommé. / half-sized ~ || halbgeleimtes Papier n. || papier m. mi-collé. / handmade ~ || Büttenpapier n.; Handpapier n.; geschöpftes Papier n. || papier m. puisé ou à la cuve. / hard ~ || Hartpapier n. || papier m. durci. / impregnated ~ || imprägniertes Papier n. || papier m. imprégné. / India ~ || Japanpapier n. || papier m. du Japon. / irized ~ || Irispapier n. || papier m. irisé. / ~ for journals || Zeitungspapier n. || papier m. pour ou à journaux. / lace ~ || Papierspitze f. || papier-dentelle m. / laid ~ || geripptes Papier n. || papier m. vergé. / leather ~ || Lederpapier n. || papier-cuir m. / lighting ~ || leuchtendes Papier n. || papier m. lumineux. / linen ~ || Leinenpapier n. || papier m. de chiffons de lin. / lining ~ || Makulatur f. beim Tapezieren || dessous m.; papier m. gris dans le collage des papiers peints sur les murs. / marbled ~ || marmoriertes Papier n. || papier m. marbré. / marbled ~ made by tipping || getupftes Papier n. || papier m. guilloché. / mat art ~ || Mattkunstdruckpapier n. || papier m. mat pour impression artistique. / metal ~ || Metallpapier n. || papier m. métallique. / mill-finished ~ || maschinenglattes Papier n. || papier m. glacé à la machine. / music ~ || Notendruckpapier n. || papier m. à musique. / mustard ~ || Senfpapier n. || papier m. sinapisme. / ~ for newspapers see ~ for journals. / non-hygroscopic ~ || nicht wasseranziehendes Papier n. || papier m. non-hygroscopique. / oiled ~ || geöltes Papier n.; Ölpapier n.; Firnispapier n. || papier m. huilé ou verni. / old waste ~ || Altpapier n. || vieux papiers mpl.; papier m. à la cuve. / ornamented ~ || gepreßtes Papier n. || papier m. gaufré. / packing ~ || Packpapier n. || papier m. d'emballage. / photographic ~ || fotografisches Papier n. || papier m. photographique. / plated ~ || versilbertes Papier n. || papier m. argenté. / ~ for posters || Affichenpapier n. || papier m. à affiches. / pressed ~ || gepreßtes Papier n. || papier m. pressé. / ~ for printing || Druckpapier n. || papier m. d'impression. / ~ in reels || Rollenpapier n.; endloses Papier n. || papier m. en bobines ou continu ou sans fin. / ~ in rolls see ~ in reels. / ruled ~ || liniiertes Papier n. || papier m. réglé ou ligné. / scratch ~ || Konzeptpapier n. || papier m. brouillon. / sensitive ~ see sensitized ~. / sensitized ~ || lichtempfindliches Papier n. || papier m. sensible à la lumière. / silk ~ || Seidenpapier n. || papier m. de soie. / sized ~ || geleimtes Papier n. || papier m. collé. / ~ for slate pencils || Griffelpapier n. || papier m. pour crayons en ardoise. / soap ~ || Seifenpapier n. || papier m. savonné. / soft-sized ~ || halbgeleimtes Papier n. || papier m. demi-collé. / special ~ || Spezialpapier n.; Sonderpapier n. || papier m. spécial. / machine for special ~ manufacturing || Maschine f. zur Herstellung von Sonderpapier || machine f. pour la fabrication de papiers spéciaux. / stained ~ || Buntpapier n. || papier m. de couleur. / stamped ~ || Stempelpapier n. || papier m. timbré ou marqué. / standard ~ || Normalpapier n. || papier m. normal. / straw ~ || Strohpapier n. || papier m. de paille. / striped ~ || gestreiftes Papier n. || papier m. rayé. / stuff-coloured ~ || Naturpapier n. || papier m. coloré à la pâte. / sulphurized ~ || mit Schwefelsäure behandeltes Papier n. || papier m. sulfurisé. / supercalendered ~ || satiniertes Papier n. || papier m. satiné. / ~ for technical purposes || technisches Papier n. || papier m. pour usage technique. / thick blotting ~ || Löschkarton m.; Fließpapier n. || carton m. buvard. / thin ~ || dünnes Papier n. || papier m. mince. / thin typewriting ~ || Durchschlagpapier n. || papier m. pelure. / tissue ~ || Seidenpapier n. || papier m. de soie. / toilet ~ || Klosettpapier n. || papier m. hygiénique. / tracing ~ || Pauspapier n. || papier m. à calquer. / transparent ~ || durchsichtiges Papier n. || papier m. transparent. / two-coloured ~ || zweifarbiges Papier n. || papier m. bicolore. / ~ for typewriters || Schreibmaschinenpapier n. || papier m. à écrire à la machine. / unsized ~ || ungeleimtes Papier n. || papier m. non-collé. / used ~ || Altpapier n. || vieux papiers mpl. / ~ varnished with oil-varnish || Wachstuchpapier n. || papier-toile f. cirée. / vegetable-sized ~ || harzgeleimtes Papier n. || papier m. à encollage végétal. / veined ~ || geädertes Papier n. || papier m. veiné. / vellum ~ || Pergamentpapier n. || papier m. parchemin ou parcheminé.

/ velveted ~ ‖ samtartiges Papier n. ‖ papier m. velouté. / wall ~ ‖ Tapete f.; Tapetenpapier n. ‖ papier m. peint. / wall ~ hanger ‖ Tapezierer m. ‖ étendeur m. de papiers peints. / waste ~ ‖ Ausschußpapier n.; Makulatur f.; Papierabfall m.; Altpapier n. ‖ maculature f.; papier m. vieux *ou* de rebut. / watered ~ ‖ Metallpapier n. ‖ papier m. moiré métallique. / wood-pulp ~ ‖ Holzpapier n. ‖ papier m. de pâte de bois. / mechanical wood-pulp ~ ‖ Holzschliffpapier n. ‖ papier m. de pâte mécanique. ~ / without wood-pulp cellulose ‖ holzfreies Papier n. ‖ papier m. exempt de pâte au bois. / wrapping ~ ‖ Einschlagpapier n.; Einwickelpapier n. ‖ papier m. d'emballage *ou* à emballer. / writing ~ ‖ Schreibpapier n. ‖ papier m. colle *ou* à écrire. / ~ covered with zinc powder ‖ galvanisches Papier n. ‖ papier m. galvanisé.

paper articles pl. ‖ Papierwaren fpl. ‖ articles mpl. divers en papier. / ~ of attire *see* paper linen. / ~ for industrial purposes ‖ Papierwaren fpl. für gewerbliche Zwecke ‖ articles mpl. industriels en papier.

paper bag ‖ Papierbeutel m.; Papiertüte f. ‖ cornet m. *ou* sac m. en papier. / to glue ~s pl. ‖ Tüten fpl. kleben ‖ coller des sacs mpl. en papier.

paper bag factory ‖ Tütenfabrik f. ‖ fabrique f. de sacs en papier. / ~ machine *see* ~ manufacturing machine. / ~ maker ‖ Tütenmacher m. ‖ ouvrier m. de sacs en papier. / ~ manufacture ‖ Tütenherstellung f. ‖ fabrication f. de sacs en papier. / ~ manufacturing machine ‖ Papierbeutelmaschine f.; Tütenmaschine f. ‖ machine f. à faire les sacs en papier. / ~ paster ‖ Tütenkleber m. ‖ colleur m. de sacs en papier.

paper balloon ‖ Papierballon m. ‖ ballon m. en papier. / ~ barrel ‖ Papierfaß n. ‖ fût m. *ou* tonneau m. en papier. / ~ bobbin ‖ Papierspule f. ‖ bobine f. en carton. / ~ box ‖ Papierschachtel f. ‖ boîte f. en carton. / ~ button ‖ Papierknopf m. ‖ bouton m. en papier. / ~ cable *see* ~-covered cable. / ~ capsule ‖ Papierkapsel f. ‖ capsule f. en papier. / ~ cell for silk worms ‖ Papierzelle f. für Seidenraupen ‖ cellule f. en papier pour vers à soie. / ~ clip ‖ Büroklammer f.; Heftklammer f.; Büronadel f.; Briefklammer f. ‖ pince f. de bureau; attache-feuilles m. / ~ coal ‖ Papierkohle f.; Blätterkohle f. ‖ houille f. papyracée. / ~ collar ‖ Papierkragen m. ‖ col m. en papier. / ~ condenser ‖ Papierkondensator m. ‖ condensateur m. en papier. / ~ container ‖ Papierbehälter m. ‖ récipient m. à papier. / ~ cornet *see* paper bag. / ~-covered cable ‖ Papierkabel n.; Kabel n. mit Papierumhüllung ‖ câble m. sous papier. / ~ cotton-covered cable ‖ Papierbaumwollkabel n. ‖ câble m. sous papier et coton. / ~ credit ‖ Wechselkredit m. ‖ crédit m. en banque. / ~ cuff ‖ Papiermanschette f. ‖ manchette f. en papier.

paper currency ‖ Papierwährung f. ‖ cours m. *ou* valeur f. du papier-monnaie. / ~ (Money) ‖ Papiergeld n. ‖ papier-monnaie m.

paper cutter ‖ Papierschneidevorrichtung f. ‖ coupe-papier m. / ~ cutter-out ‖

Papierausschneider m.; Papierausschläger m. ‖ découpeur m. de papier.

paper cutting machine ‖ Papierschneidemaschine f. ‖ coupe-papier m. mécanique; coupeuse f. de papier. / cross ~ ‖ Querpapierschneidemaschine f. ‖ coupeuse f. de papier en travers. / longitudinal ~ ‖ Langpapierschneidmaschine f. ‖ coupeuse f. de papier en long.

paper cutting-out ‖ Stanzen n. von Papier ‖ découpage m. de papier. / ~ cylinder of the goffering machine ‖ Gegenwalze f. der Gaufriermaschine ‖ contre-partie f. *ou* contre-épreuve f. d'une machine à gaufrer. / ~ destroying machine ‖ Papiervernichtungsmaschine f. ‖ machine f. pour la destruction du papier. / ~ embroidery ‖ Papierstickerei f. ‖ papierbroderie f. / ~ envelope ‖ Briefumschlag m. ‖ enveloppe f. de lettre.

paperer *see* paper hanger.

paper fastener ‖ Papierhefter m. ‖ pince f. pour le brochage de papier. / clipless ~ ‖ klammerloser Papierhefter m. ‖ pince f. perforante pour le brochage de papier sans attache.

paper feed ‖ Papiernachschub m. ‖ déplacement m. du papier. / ~ finger ‖ Papierfinger m. ‖ guide-papier m. / ~ finisher ‖ Papierzubereiter m. ‖ apprêteur m. de papier. / ~ flag ‖ Papierfahne f. ‖ drapeau m. en papier.

paper flower ‖ Papierblume f. ‖ fleur f. en papier. / ~ punching ‖ Ausschneiden n. von Papierblumen ‖ découpage m. de fleurs en papier.

paper folder ‖ Falzbein n. ‖ plioir m. / ~ (Person) ‖ Papierfalzerin f. ‖ plieuse f. de papier.

paper garland ‖ Papiergirlande f. ‖ guirlande f. en papier. / ~ glazer ‖ Papierglätter f.; Glätter m.; Satinierer m. ‖ glaceur m. *ou* satineur m. de papier. / ~ glazing works pl. ‖ Papiersatinieranstalt f. ‖ atelier m. de satinage de papier. / ~ goods pl. ‖ Papierwaren fpl. ‖ papeteries fpl.; articles mpl. en papier. / ~ guide (Typewr) ‖ Papierführung f. ‖ guide m. de papier. / ~ guide roll (Typewr) ‖ Papierführungsrolle f. ‖ rouleau m. de guidage du papier. / ~ gummer ‖ Papiergummierer m. ‖ gommeur m. de papier.

paper hanger ‖ Tapezierer m.; Tapezier m. ‖ tenturier m.; tapissier m. / ~'s work ‖ Tapeziererarbeit f. ‖ travail m. de tapissier.

paper hanging ‖ Tapetenkleben n. ‖ collage m. *ou* tapisserie f. de papiers peints.

paper hangings pl. ‖ Tapete f.; Papiertapete f. ‖ papier m. peint; papier-tenture m.; tapisserie f. / irized ~ ‖ Iristapete f. ‖ papier m. irisé. / scarlet ~ ‖ Scharlachtapete f. ‖ tapis m. teint en écarlate.

paper hangings pl., single breadth of ~ ‖ einfach breite Tapetenbahn f. ‖ pan m. de papier peint de simple largeur. / ~ designer ‖ Tapetenmusterzeichner m. ‖ dessinateur m. en papiers peints. / ~ enameller *see* ~ varnisher. / ~ varnisher ‖ Tapetenfirnisser m. ‖ vernisseur m. de papiers peints.

paper holder ‖ Bogenhalter m. ‖ porte-papier m. / ~ industry ‖ Papierindustrie f. ‖ industrie f. papetière. / ~ insulating tube ‖ Papierisolierrohr n. ‖ tube m. isolant en papier. / ~ insulation ‖ Papier-

isolierung f. ‖ isolement m. de papier. / ~ kind of ~ ‖ Papiersorte f. ‖ sorte f. de papier. / ~ knife ‖ Papiermesser n. ‖ couteau m. à papier. / ~ lantern ‖ Papierlaterne f.; Lampion m. ‖ lanterne f. vénitienne; lampion m. / ~ lead-covered cable *see* paper-covered cable. / ~ letter ‖ Papierbuchstabe m. ‖ lettre f. en papier.

paper linen ‖ Papierwäsche f. ‖ linge m. en papier. / ~ with addition of spinning materials ‖ Papierwäsche f. in Verbindung mit Spinnstoffen ‖ linge m. en papier avec addition de matières textiles.

paper machine ‖ Papiermaschine f. ‖ machine f. à papier. / ~ tenter ‖ Papiermaschinenarbeiter ‖ conducteur m. de machine à papier.

paper maker ‖ Papierarbeiter m. ‖ ouvrier-papetier m.

paper making machine ‖ Papierherstellungsmaschine f. ‖ machine f. à fabriquer le papier.

paper manufactory ‖ Papierfabrik f. ‖ papeterie f.

paper manufacture ‖ Papierfertigung f.; Papierherstellung f. ‖ fabrication f. du papier. / machine for ~ ‖ Maschine f. zur Papierherstellung ‖ machine f. pour la fabrication du papier. / materials pl. for ~ ‖ Stoffe mpl. zur Papierbereitung ‖ matières fpl. servant à la fabrication du papier.

paper manufacturer ‖ Papierfabrikant m. ‖ fabricant-papetier m. / ~ manufacturing *see* paper manufacture. / ~ merchant ‖ Papierhändler m. ‖ marchand m. papetier. / ~ mill ‖ Papiermühle f.; Papierfabrik f. ‖ papeterie f.; moulin m. à papier. / ~ money ‖ Papiergeld n. ‖ papier-monnaie m. / ~ moving system (Tel) ‖ Papierführung f. ‖ système m. d'entraînement du papier. / ~ mulberrytree ‖ Papiermaulbeerbaum m. ‖ broussonétie f. / ~ napkin ‖ Papierserviette f.; Papiermundtuch n. ‖ serviette f. en papier. / ~ outfit ‖ Papierausstattung f. ‖ papeterie f. / ~ parchment ‖ Pergamentpapier n. ‖ papier m. parcheminé. / ~ pattern ‖ Schnittmuster n. ‖ patron m. / ~ piercer ‖ Locher n. ‖ perçoir n.

paper pulp ‖ Papiermasse f. ‖ pâte f. à papier. / chemical ~ ‖ chemische Papiermasse f. ‖ pâte f. à papier chimique. / ~ in dry state ‖ Papiermasse f. in trockenem Zustand ‖ pâte f. à papier à l'état sec. / ~ from esparto ‖ Papiermasse f. aus Spartgras ‖ pâte f. à papier d'alfa. / humid ~ ‖ feuchte Papiermasse f. ‖ pâte f. à papier humide. / mechanical ~ ‖ mechanische Papiermasse f. ‖ pâte f. à papier mécanique. / ~ from rags ‖ Papiermasse f. aus Lumpen ‖ pâte f. à papier de chiffons. / semi-chemical ~ ‖ halbchemische Papiermasse f. ‖ pâte f. à papier mi-chimique. / ~ from straw ‖ Papiermasse f. aus Stroh ‖ pâte f. à papier de paille. / ~ from wood ‖ Papiermasse f. aus Holz ‖ pâte f. à papier de bois.

paper pulp bleacher ‖ Papiermassebleicher m. ‖ blanchisseur m. de pâte à papier.

paper punch ‖ Papierlocher m. ‖ perforateur m. de papier.

paper rags pl. ‖ Lumpen mpl.; Hadern mpl.; Stratzen mpl. ‖ chiffons mpl.; drilles fpl.; pilot m. / coarse ~ ‖ grobe

graue Ausschußlumpen mpl. ‖ boulongeon m.

paper ribbon ‖ Papierband n. ‖ ruban m. *ou* bande f. de papier. / coiled-up ~ ‖ Papierschlange f. ‖ serpentin m. de papier.

paper rod (Typewr) ‖ Papierandrückstange f. ‖ barre f. presse-papier. / ~ roller ‖ Papierwalze f. ‖ cylindre m. à papier.

paper roller-blind ‖ Papierrolladen m. ‖ store m. en papier. / ~ stainer ‖ Papierrolladendrucker m. ‖ imprimeur m. de stores en papier.

paper roof ‖ Pappdach n. ‖ toit m. en carton. / tarred ~ ‖ geteertes Pappdach n.; mit Dachpappe belegtes Dach n. ‖ toit m. en carton bitumé. / waterproof ~ ‖ wasserdichtes Pappdach n. ‖ toiture f. en carton bitumé.

paper ruling ‖ Papierliniieren n. ‖ réglage m. de papier.

paper ruling machine ‖ Liniiermaschine f. ‖ régleuse f. / ~ tenter ‖ Liniiermaschinenarbeiter m. ‖ régleur m. à la machine.

paper sack ‖ Papiersack m. ‖ sac m. en papier. / ~ manufacturing machine ‖ Papiersackherstellungsmaschine f. ‖ machine f. pour la fabrication de sacs en papier.

paper scale ‖ Papierwage f. ‖ balance f. à papier. / ~ scissors pl. ‖ Papierschere f. ‖ ciseaux mpl. à papier. / scraps pl. of ~ ‖ Papierschnitzel npl. ‖ rognures fpl. de papier. / ~ seal ‖ Papiersiegel n. ‖ cachet m. en papier.

paper sheet ‖ Papierbogen m. ‖ feuille f. de papier. / margin of the ~ ‖ Rand m. des Papierbogens ‖ marge f. de la feuille de papier.

paper, size of ~ ‖ Papierformat n. ‖ format m. du papier. / ~ sizing ‖ Papierleimung f. ‖ collage m. de papier. / ~ sleeve ‖ Papierröhrchen n. ‖ manchon m. *ou* tube m. de papier. / ~ snake ‖ Papierluftschlange f. ‖ serpentin m. de carnaval. / ~ sole ‖ Papiersohle f. ‖ semelle f. en papier. / sort of ~ ‖ Papierart f. ‖ sorte f. de papier. / ~ spinning factory ‖ Papierspinnerei f. ‖ filature f. de papier. / ~ spool for spinning shops ‖ Papierspule f. für Spinnereien ‖ busette f. en papier pour filatures. / ~ staining machine ‖ Fonciermaschine f.; Grundiermaschine f. ‖ machine f. à foncer; fonceuse f. / ~ stamp ‖ Papierstampfe f. ‖ pilon m. à papier. / ~ stamper ‖ Papierpräger m. ‖ estampeur m. sur papier. / ~ strip ‖ Papierstreifen m. ‖ bande f. de papier. / ~ support (Typewr) ‖ Papierstütze f. ‖ support m. pour le papier. / ~ table (Typewr) ‖ Papierauflegeblech n. ‖ table f. porte-papier. / ~ table cloth ‖ Papiertischtuch n. ‖ nappe f. en papier. / ~ tape ‖ Papierband n. bande f. *ou* ruban m. de papier. / ~ tearing device ‖ Papierabreißvorrichtung f. ‖ coupe-papier m. / ~ terminal cable ‖ Papierabschlußkabel n. ‖ câble m. de fermeture sous papier. / ~ testing ‖ Papierprüfung f. ‖ essai m. mécanique du papier. / ~ testing apparatus ‖ Papierprüfungsapparat m. ‖ appareil m. à essayer le papier. / ~ texture ‖ Papiergewebe n. ‖ texture f. du papier. / ~ trade ‖ Papierhandel m. ‖ papeterie f.; commerce de papeterie f. / automatic ~ transport (Typewr) ‖ selbsttätiger Papiertransport m. ‖ entraînement m.

automatique du papier. / ~ tube ‖ Papierhülse f.; Papierröhre f. ‖ tube m. en papier *ou* en carton. / ~ twine ‖ Papierbindfaden m.; Papiergarn n. ‖ ficelle f. en papier. / ~ waste ‖ Papierabfall m. ‖ déchet m. de papier. / ~ weaving factory ‖ Papierweberei f. ‖ tissanderie f. de papier. / ~ weight ‖ Papierbeschwerer m. ‖ presse-papier m. / wetting machine for ~ ‖ Anfeuchtmaschine f. für Papier ‖ machine f. à humecter le papier. / ~ wholesale trade ‖ Papiergroßhandel m. ‖ commerce m. de papier en gros. / ~ wood ‖ Papierholz n. ‖ bois m. à papier. / ~ working industry ‖ papierverarbeitende Industrie f. ‖ industrie f. du papier. / ~ working machine ‖ Papierverarbeitungsmaschine f. ‖ machine f. pour papeteries. / ~ works pl. ‖ Papierfabrik f. ‖ papeterie f.

paper yarn ‖ Papiergarn n. ‖ fil m. de papier. / ~ plate spinning machine ‖ Papiergarntellerspinnmaschine f. ‖ machine f. à plateau à filer le fil en papier. / ~ spinning mill ‖ Papiergarnspinnerei f. ‖ filature f. de fils de papier.

papier-mâché ‖ Papiermaché n.; Papierstoff m. ‖ papier-mâché m.; cartonpâte m. / lacquered ~ ‖ lackiertes Papiermaché ‖ carton m. laqué.

papier-mâché box ‖ Papiermachédose f. ‖ boîte f. en carton-pâte. / ~ button ‖ Papierstoffknopf m. ‖ bouton m. de papier-mâché. / ~ goods pl. ‖ Papiermachéwaren fpl. ‖ articles mpl. en carton verni *ou* en papier-mâché.

papier-mâché ware ‖ Papiermachéwaren fpl. ‖ objets mpl. en carton-pâte. / lacquered ~ ‖ Lackpapiermachéwaren fpl. ‖ objets mpl. en carton laqué.

papilla ‖ Ausstülpung f.; Auswuchs m. ‖ excroissance f.; protubérance f.

pappy ‖ breiig ‖ pâteux.

par, at ~ ‖ paritätisch ‖ égalitaire. / ~ of exchange ‖ Wechselpari n. ‖ pair m. du change.

parabola ‖ Parabel f. ‖ parabole f. / ~ of a higher order ‖ Parabel f. höherer Ordnung ‖ parabole f. d'ordre *ou* de degré supérieur.

parabolic ‖ parabolisch ‖ parabolique. / ~ concave mirror ‖ parabolischer Hohlspiegel m. ‖ miroir m. parabolique concave. / ~ crossing ‖ Parabelweiche f. ‖ croisement m. parabolique.

parabolic glass reflector ‖ Glasparabolspiegel m. ‖ réflecteur m. parabolique en verre. / silvered ~ ‖ parabolischer Glassilberspiegel m. ‖ miroir m. en verre parabolique argenté. / with silvered front surface ‖ Glasparabolspiegel m. mit versilberter Vorderfläche ‖ réflecteur m. parabolique en verre à surface avant argentée.

parabolic mirror ‖ Parabolspiegel m. ‖ miroir m. *ou* réflecteur m. parabolique. / ~ reflector *see* ~ mirror. / ~ shape ‖ Parabelgestalt f. ‖ forme f. parabolique. / ~ symbol ‖ parabolische Spirale f. ‖ parabole f. hélicoïde; spirale f. parabolique.

paraboloid ‖ Paraboloid n. ‖ paraboloïde m. / revolution ~ ‖ Rotationsparaboloid n. ‖ paraboloide m. de révolution.

paraboloid condenser ‖ Paraboloidkondensator m. ‖ condensateur m. parabolique.

paracentric ‖ parenzentrisch ‖ paracentrique.

parachute (Aero) ‖ Fallschirm m. ‖ parachute m. / ~ (Lift) ‖ Fangvorrichtung f. ‖ parachute m. de la cage d'ascenseur. / back-type ~ ‖ Rückenkissenfallschirm m. ‖ parachute m. sac dorsal. / seat pack ~ ‖ Sitzkissenfallschirm m. ‖ parachute m. sac-coussin *ou* sac-siège. / seat-type ~ *see* seat pack ~. / ~ flare ‖ Fallschirmrakete f. ‖ fusée f. à parachute.

parade (Build) ‖ Hochplan m.; Freiheit f. ‖ esplanade f.

paraffin ‖ Paraffin n. ‖ paraffine f. / soft ~ ‖ Ozokerit m.; Bergwachs n.; Erdwachs n. ‖ ozocérite f.; cire f. fossile.

paraffin candle ‖ Paraffinkerze f. ‖ bougie f. *ou* chandelle f. de paraffine.

paraffined ‖ paraffiniert ‖ paraffiné.

paraffin lacquer ‖ Paraffinlack m. ‖ laque f. de paraffine. / ~ oil ‖ Paraffinöl n. ‖ huile m. de paraffine. / ~ ointment ‖ Paraffinsalbe f. ‖ onguent m. de paraffine. / ~ paper ‖ Paraffinpapier n. ‖ papier m. paraffiné. / ~ paper manufacturing machine ‖ Paraffinpapierherstellungsmaschine f. ‖ machine f. à fabriquer le papier paraffiné. / ~ plant *see* ~ works. / ~ works pl. ‖ Paraffinfabrik f. ‖ fabrique f. de paraffine.

paragon ‖ Paragon f. ‖ petit paragon m. / small ~ ‖ kleine Paragon f. ‖ dix-huit m.

paragonite ‖ Paragonit m.; Natronglimmer m. ‖ paragonite f.

paragraph (Print) ‖ Absatz m.; Paragraf m. ‖ alinéa m. / to make a new ~ (Print) ‖ einen Absatz m. machen ‖ faire un paragraphe.

paraldehyde ‖ Paraldehyd n. ‖ paraldéhyde.

parallactic axis ‖ parallaktische Achse f. ‖ axe m. parallactique.

parallax ‖ Parallaxe f. ‖ parallaxe f. / ~ of altitude ‖ Höhenparallaxe f. ‖ parallaxe f. de hauteur. / annual ~ ‖ jährliche Parallaxe f.; heliozentrische Parallaxe f. ‖ parallaxe f. annuelle *ou* héliocentrique. / to correct the ~ ‖ die Parallaxe f. beseitigen ‖ écarter la parallaxe. / diurnal ~ ‖ tägliche *oder* geozentrische Parallaxe f. ‖ parallaxe f. diurne *ou* géocentrique. / free from ~ ‖ parallaxenfrei ‖ exempt de parallaxe. / horizontal ~ ‖ Horizontalparallaxe f. ‖ parallaxe f. horizontale. / to see the bubble without ~ trough prisms ‖ die Libelle parallaxenfrei durch Prismen beobachten ‖ observer la nivelle sans parallaxe par l'intermédiaire de prismes.

parallax refractometer ‖ Parallaxenrefraktometer n. ‖ réfractomètre m. à parallaxe.

parallel ‖ parallel ‖ parallèle. / with ~ faces ‖ planparallel ‖ à faces fpl. parallèles. / to set ~ ‖ parallel stellen ‖ placer parallèlement. / ~ with ‖ parallel zu ‖ parallèle à.

parallel bedding ‖ Parallelstruktur f. ‖ structure f. parallèle. / ~ circle ‖ Parallelkreis m. ‖ parallèle m. / connection in ~ ‖ Parallelschaltung f. ‖ couplage m. en parallèle. / ~ motion ‖ Parallelbewegung f. ‖ mouvement m. parallèle. / ~ ohm method ‖ Parallelohmmethode f. ‖ réception f. à résistance parallèle. / ~ perspective ‖ Parallelperspektive f. ‖ vue f. de face. / ~ precision gauge block ‖ Parallelendmaß n. ‖ jauge f. étalon. / ~ projection ‖ Parallelprojektion f. ‖ projection f. parallèle. / ~ rail ‖ Doppel-T-

Schiene f. ‖ rail m. à champignon symétrique *ou* parallèle. / ~ shears pl. ‖ Parallelschere f. ‖ cisaille f. à guillotine. / ~ shift of systems of coordinates ‖ Parallelverschiebung f. der Koordinatensysteme ‖ déplacement m. parallélique des systèmes de coordonnées. / ~ vice ‖ Parallelschraubstock m. ‖ étau m. parallèle. / ~ vice with straight or angular tightening ‖ Maschinenparallelschraubstock m. für gerade und Winkelspannung ‖ étau m. parallèle à serrage droit ou angulaire. / ~ working (Electr) ‖ Parallelbetrieb m. ‖ marche f. en parallèle.

parallel ‖ Parallele f. ‖ parallèle f. / to connect in ~ ‖ in den Nebenschluß m. legen ‖ monter en dérivation f. / in ~ (Electr) ‖ nebeneinander ‖ en parallèle f.

parallelism ‖ Parallelität f. ‖ parallélisme m.

parallelogram ‖ Parallelogramm n. ‖ parallélogramme m. / ~ of forces ‖ Kräfteparallelogramm n. ‖ parallélogramme m. des forces. / right-angled ~ ‖ Rechteck n. ‖ rectangle m. / ~ of velocities ‖ Geschwindigkeitsparallelogramm n.; Parallelogramm n. der Geschwindigkeiten ‖ parallélogramme m. des vitesses.

parallelepipedon ‖ Parallelepipedon n.; Quader n. ‖ parallélopipède m.

paralysis ‖ Lähmung f. ‖ paralysie f.

paramagnetic ‖ paramagnetisch ‖ paramagnétique.

paramagnetism ‖ Paramagnetismus m. ‖ paramagnétisme m.

parament ‖ Parament n. ‖ parement m.

parameter ‖ Parameter m. ‖ paramètre m.

paramorphose ‖ Paramorphose f. ‖ paramorphose f.

paramount ‖ überwiegend ‖ prépondérant.

parangon ‖ Parangon f. ‖ petit-parangon m.

parapet ‖ Brüstung f.; Balustrade f. ‖ appui m.; parapet m.; balustrade f. / ~ of a bridge ‖ Brückengeländer n. ‖ parapet m. *ou* garde-fou m. d'un pont. / ~ of a lining wall ‖ Brustwehr f. einer Futtermauer ‖ parapet m. d'un mur de revêtement. / ~ of a window ‖ Fensterbrüstung f. ‖ appui m. d'une fenêtre.

Para rubber ‖ Paragummi n. ‖ Para m.; caoutschouc m. de Para.

paraselene ‖ Nebenmond m. ‖ parasélène f.

parasol ‖ Sonnenschirm m. ‖ parasol m.

parcel, to ‖ parzellieren ‖ parceller; morceler; lotir.

parcel ‖ Kollo n.; Stückgut n. ‖ coli(s) m. / ~ sent by book post ‖ Kreuzbandsendung f. ‖ envoi m. sous bande. / ~ of land ‖ Parzelle f. ‖ parcelle f. / registered ~ containing valuables ‖ Wertpaket n. ‖ paquet m. avec valeur déclarée. / ~ of shares ‖ Aktienpaket n. ‖ paquet m. d'actions.

parcel balance ‖ Paketwage f. ‖ balance f. à paquets. / ~ closing stamp ‖ Paketverschlußmarke f. ‖ marque f. de fermeture pour paquets.

parcels delivery ‖ Paketbeförderung f.; factage m. / ~ (Railw) ‖ Gepäckausgabe f. ‖ distribution f. des bagages. / ~ company ‖ Paketfahrtgesellschaft f. ‖ entreprise f. de distribution des colis postaux.

parchment ‖ Pergament n. ‖ parchemin m. / animal ~ ‖ tierisches Pergament n. ‖ parchemin m. animal. / drum ~ ‖ Trom-

melleder n. ‖ peau f. pour tambours. / vegetable ~ ‖ pflanzliches Pergament n. ‖ parchemin m. végétal.

parchment glue ‖ Pergamentleim m. ‖ colle f. au parchemin. / to give a coating of ~ glue ‖ mit Pergamentleim m. bestreichen ‖ enduire de colle au parchemin. / ~ leaf ‖ Pergamentblatt n. ‖ feuille f. de parchemin. / ~ leather ‖ Pergamentleder n. ‖ cuir m. parcheminé. / ~-like ‖ pargamentartig ‖ parcheminé. / ~ maker ‖ Pergamentmacher m. ‖ parcheminier m.

parchment paper ‖ Pergamentpapier n. ‖ papier m. parcheminé. / ~ manufacturing machine ‖ Pergamentpapierherstellungsmaschine f. ‖ machine f. à fabriquer le papier parcheminé.

parchment sieve ‖ Schrotsieb n. ‖ guillaume m. / ~ size ‖ Pergamentleim m. ‖ colle f. au baquet. / ~ substitute ‖ Pergamentersatz m. ‖ parchemin m. d'imitation.

pare, to (Join) ‖ abschachteln ‖ rogner. / ~ (Parchment ‖ schaben ‖ raturer. / ~ quarry stones ‖ Bruchsteine mpl. abschalen ‖ ébouziner les moëllons mpl.

parent company ‖ Stammfirma f. ‖ société-mère f.

parentheses pl. (Punctuation) ‖ runde Klammern fpl. ‖ parenthèses fpl. / to put in ~ ‖ einklammern; in (runde) Klammern fpl. setzen ‖ mettre en parenthèse *ou* entre (deux) parenthèses.

parenthesis ‖ Parenthese f.; Einschaltung f. ‖ parenthèse f.

parget ‖ Gipsstuck m.; Stuck m. ‖ enduit m. en plâtre; plâtre m.; stuc m. / ~ stone ‖ Gips m. ‖ gypse m.; plâtre m.

Parian ‖ Parian n. ‖ parian m.

paring axe ‖ schwere Schälaxt f. ‖ cognée f. à blanchir. / ~ hammer ‖ Pritschhammer m. ‖ marteau m. à ébarber. / ~ knife (Bookb) ‖ Schärfemesser n. ‖ paroir m.; couteau m. à parer. / ~ machine ‖ Nutstoßmaschine f.; Stoßmaschine f. ‖ machine f. à mortaiser; raboteuse f. verticale. / ~ man (Bookb) ‖ Lederschärfer m. ‖ pareur m. / ~ plough ‖ Schälpflug m. ‖ dégazonnoir m.; charrure f. à peler.

parings pl. (Curr) ‖ Aas n.; Abschabsel n. ‖ écharnure f. / ~ (Join) ‖ Hobelspäne mpl. ‖ copeaux mpl.; planure f.; raboture f. / ~ of leaf gold ‖ Goldabfall m.; Krätze ‖ bractéole f.; rognure f. de feuilles d'or.

paring tool ‖ Krummeißel m. ‖ ciseau m. à bride.

Paris blue ‖ Pariser Blau n. ‖ bleu m. de Paris. / ~ green ‖ Schweinfurter Grün n. ‖ vert m. de Paris.

parity ‖ Parität f. ‖ parité f.

park ‖ Park m. ‖ parc m.

parking, no ‖ Parkverbot n.! ‖ interdit de parquer!

park walk ‖ Parkweg m. ‖ chemin m. de parc.

parlour ‖ Empfangszimmer n.; Salon m. ‖ parloir m. / ~ car ‖ Salonwagen m. ‖ voiture-salon f. / ~ game ‖ Gesellschaftsspiel n. ‖ jeu m. de société.

parquet ‖ Parkett n.; Parkettfußboden m. ‖ parquet m. / hard wood ~ ‖ Parkett n. aus Hartholz ‖ parquet m. en bois dur. / soft wood ~ ‖ Parkett n. aus Weichholz ‖ parquet n. en bois tendre.

parquet inlayer ‖ Mosaikfußbodenleger m. ‖ mosaïste m. en parquets. / ~ planing

machine ‖ Hobelmaschine f. für Parkett ‖ raboteuse f. pour parquets.

parquetry ‖ Parkett n.; Täfelung f. ‖ parquet m. / ~ machine ‖ Parkettschleifmaschine f. ‖ machine f. à poncer les parquets.

par rate of exchange ‖ Parikurs m. ‖ pair m. du change.

parsley ‖ Petersilie f. ‖ persil m. / ~ herb ‖ Petersilienkraut n. ‖ herbe f. de persil. / ~ oil ‖ Petersilienöl n. ‖ huile f. de persil.

part, to (Ore dress) ‖ scheiden ‖ séparer.

part ‖ Teil m. ‖ partie f.; portion f. / ~ (Math) ‖ Periode f. ‖ période f. / ~ of boots ‖ Schuhbestandteil m. ‖ partie f. de chaussures. / ~ shaped like a box ‖ trogförmiges Stück n. ‖ pièce f. en forme d'une auge. / ~ carrying current ‖ stromführender Teil m. ‖ partie f. sous tension; pièce f. porte-courant. / ~ of cycle ‖ Fahrradteil m. ‖ pièce f. de bicyclette. / essential ~ ‖ Hauptbestandteil m.; wesentlicher Bestandteil m. ‖ partie f. principale *ou* intégrante; élément m. *ou* ingrédient m. *ou* composant m. *ou* organ m. principal. / detached ~s pl. ‖ Einzelteile mpl.; einzelne Teile mpl. ‖ pièces fpl. détachées. / in ~s pl. ‖ teilweise ‖ en partie f.; partiellement. / in ~s pl. (Booksell) ‖ heftweise; lieferungsweise ‖ par livraisons fpl. / ~ of locomotive ‖ Lokomotivteil m. ‖ pièce f. détachée de locomotive. / lower ~ ‖ unterer Teil m. ‖ partie f. inférieure. / ~ of machinery ‖ Maschinenbestandteil m. ‖ piece f. détachée de machine. / main ~ *see* essential ~. / ~ by measure ‖ Maßteil m. ‖ partie f. en volume; volume m. / mechanical ~ ‖ maschineller Teil m. ‖ partie f. mécanique. / middle ~ ‖ mittlerer Teil m. ‖ partie f. centrale. / for our ~ ‖ unsererseits ‖ de notre côté m. *ou* part f. / ~ of projectile ‖ Geschoßteil m. ‖ partie f. d'un projectile. / rotating ~ ‖ rotierender Teil m. ‖ élément m. rotatif. / separate machine ~s pl. ‖ einzelne Teile mpl. einer Maschine ‖ pièces fpl. détachées d'une machine. / spare ~ ‖ Ersatzteil m. ‖ pièce f. de rechange *ou* de réserve. / to take ~ ‖ teilnehmen ‖ prendre part m.; participer. / upper ~ ‖ Oberteil n. ‖ partie f. supérieure. / ~ by weight ‖ Gewichtsteil m. ‖ partie f. en poids.

partial ‖ partiell; teilweise ‖ partiellement; en partie f. / ~ alienation ‖ Teilenteignung f. ‖ expropriation f. partielle. / ~ discharge ‖ teilweise Entladung f. ‖ décharge f. partielle. / ~ disturbance ‖ Teilstörung f. ‖ perturbation f. partielle; composante f. de perturbation. / ~ fraction ‖ Teilbruch m. ‖ fraction f. partielle. / ~ impulse ‖ Reststromstoß m. ‖ impulsion f. incomplète. / ~ pressure ‖ Teilspannung f. ‖ tension f. partielle. / ~ vortex ‖ Teilwirbel m. ‖ tourbillon m. composant.

partially reflected ray ‖ partiellreflektierter Strahl m. ‖ rayon m. partiellement réfléchi.

participate, to ‖ teilhaben; teilnehmen; sich beteiligen ‖ avoir *ou* prendre m. part m.; participer à.

participation in profits ‖ Teilnahme f. am Gewinn ‖ participation f. aux bénéfices.

particle ‖ Partikel f.; Teilchen n. ‖ particule f. / ~ of water ‖ Stromfaden m. ‖ filet m. d'eau.

particular interest ‖ Sonderinteresse n. ‖ intérêt m. séparé *ou* particulier.

parting ‖ Trennung f. ‖ séparation f. / dry ~ (Ore dress) ‖ Scheidung f. auf trockenem Wege ‖ départ m. par la voie sèche. / ~ of gold and of silver ‖ Scheidung f. von Gold und Silber ‖ départ m. de l'or et de l'argent m.

parting gold ‖ Scheidegold n. ‖ or m. de départ. / ~ sand ‖ Streusand m.; Formsand m. ‖ sable m. sec *ou* en poudre *ou* de moulage.

parting tool ‖ Geißfuß m. ‖ carrelet m.; burin m. triangulaire. / bent ~ ‖ gebogener Geißfuß m. ‖ burin m. triangulaire coudé. / print cutter's bent ~ ‖ gebogener Graviergeißfuß m. ‖ burin m. triangulaire coudé pour graveur. / spoon bit ~ ‖ Löffelgeißfuß m. ‖ burin m. triangulaire à cuiller. / straight ~ ‖ gerader Geißfuß m. ‖ burin m. triangulaire droit.

parting work ‖ Scheideanstalt f. ‖ atelier m. de départ.

partition ‖ Scheidewand f. ‖ paroi f.; cloison f. / ~ of baywork ‖ Scheidewand f. in Fachwerk ‖ cloison f. mitoyenne *ou* en charpente. / ~ of boards ‖ Bretterverschlag m. ‖ cloison f. en planches. / latticed ~ ‖ Lattenverschlag m. ‖ cloison f. en lattes. / ~ of planks *see* ~ of boards. / staked ~ ‖ Stackwand f. ‖ cloison f. clayonnée. / wooden ~ ‖ Verschlag m. ‖ cloison m. de charpente.

partition rock (Mine) ‖ Nebengestein n. ‖ roche f. des parois. / ~ wall ‖ Scheidemauer f. ‖ mur m. de refend *ou* de séparation; cloison f. massive.

partly finished articles pl. ‖ Zwischenerzeugnisse npl. ‖ demi-produits mpl.

partner ‖ Geschäftsteilhaber m. ‖ associé m. / to admit as ~ ‖ als Teilhaber m. aufnehmen ‖ associer. / ~ in a firm ‖ Mitinhaber m. ‖ codétenteur m.; copropriétaire m.; compagnon m. / to become ~ of a firm ‖ als Teilhaber m. eintreten ‖ entrer comme associé. / sleeping ~ ‖ stiller Teilhaber m. *oder* Gesellschafter m. ‖ associé m. commanditaire.

partnership, to dissolve ~ ‖ eine Genossenschaft f. aufheben ‖ dissoudre une association coopérative. / to enter into ~ ‖ sich assoziieren ‖ s'associer. / to form a ~ ‖ eine Handelsgesellschaft f. errichten ‖ constituer *ou* former une société f. de commerce. / limited ~ ‖ Kommanditgesellschaft f. ‖ société f. en commandite.

part-owner (Mar) ‖ Schiffspartner m. ‖ co-bourgeois m.

part-payment ‖ Teilzahlung f. ‖ payement m. partiel. / as ~ ‖ abschläglich ‖ à compte m.; en déduction f.

party line ‖ Gesellschaftsleitung f.; Gemeinschaftsanschluß m. ‖ ligne f. pour groupe d'abonnés. / ~ central office equipment ‖ Leitungswählerschaltung f. für Zweiganschlüsse ‖ connecteur m. pour party lines.

party wall ‖ Brandmauer f. ‖ mur m. massif *ou* mitoyen.

par value ‖ Pariwert m. ‖ valeur f. au pair.

pass, to ‖ überholen ‖ dépasser. / ~ to the debit ‖ anrechnen ‖ compter; mettre *ou* passer en compte m. / ~ in lee ‖ leewärts vorbeisegeln ‖ doubler *ou* passer par sous le vent. / ~-over in silence ‖

mit Stillschweigen n. übergehen ‖ passer sous silence f. / ~ a vessel through a lock / ein Schiff n. durchschleusen ‖ écluser un bateau. / ~ through the muffle ‖ (Enamel) ‖ einbrennen ‖ passer au feu. / ~ through a station ‖ einen Bahnhof m. durchfahren ‖ brûler une station.

pass ‖ Passierschein m.; Paß m. ‖ laissez-passer m.; coupe-file m.; permis m. / ~ (Topogr) ‖ Sattel m.; Joch n. ‖ croupe f.; crête f.; ligne f. de faîte. / low ~ filter (Tel) ‖ Niederfrequenzsiebkette f. ‖ filtre m. basse fréquence. / the rolls pl. are brought nearer together after each ~ ‖ die Walzen fpl. wurden nach jedem Stich nähergestellt ‖ les cylindres mpl. furent rapprochés l'un à l'autre à chaque passage.

passaba fibre ‖ Piassavafaser f. ‖ fibre f. de piassava.

passage ‖ Durchgang m.; Gang m.; Durchlaß m. ‖ passage m. / ~ (Mach) ‖ Dampfkanal m. ‖ canal m. / ~ (Mar) ‖ Durchfahrt f. ‖ passage m.; traversée f. / ~ (Radio) ‖ Durchgriff m. ‖ rapport m. / difficult ~ ‖ schwierige Durchfahrt f. ‖ passe f. difficile. / inlet and outlet ~ (Sluice) ‖ Umlauf m. ‖ aqueduc m. / ~ for vessels ‖ Schiffsdurchlaß m. ‖ passe f. navigable.

passage door ‖ Durchgangstür f. ‖ porte f. de passage. / ~ section ‖ Durchgangsprofil n. ‖ profil m. de passage.

pass-bye (Railw) ‖ Kletterweiche f. ‖ changement m. de voie provisoire *ou* à plans inclinés.

passementerie ‖ Posamentierwaren fpl. ‖ passementerie f. / ~ from natural horsehair ‖ Posamentierwaren fpl. aus natürlichem Roßhaar ‖ passementeries fpl. en crin naturel.

passenger ‖ Reisender m. ‖ voyageur m. / ~ (Mar) ‖ Passagier m. ‖ passager m. ‖ ~ cabin (Aero) ‖ Fluggastraum m. ‖ cabine f. des passagers. / ~ cabin (Mar) ‖ Passagierraum m. ‖ salon m. *ou* cabine f. des passagers.

passenger car (Aero) ‖ Fluggastgondel f. ‖ nacelle f. pour voyageurs. / ~ (Auto) ‖ Verkehrsfahrzeug n. ‖ véhicule m. de communication. / ~ (Railw) ‖ Personenwagen m. ‖ voiture f. à voyageurs. / open ~ ‖ offener Personenwagen m. ‖ voiture f. ouverte à voyageurs. / ~ with x seats crosswise ‖ Personenwagen m. mit x Quersitzen ‖ wagon m. avec x places en travers.

passenger carriage *see* passenger car. / ~ communication apparatus ‖ Zugsignalvorrichtung f. ‖ appareil m. d'intercommunication. / ~ hall ‖ Bahnhofshalle f. ‖ hall m. de gare *ou* pour voyageurs. / ~ launch ‖ Personenbeförderungsboot n. ‖ vedette f. à passagers. / ~ lift ‖ Personenaufzug m. ‖ ascenseur m. pour personnes. / ~ motor car ‖ Personenkraftfahrzeug n. ‖ automobile m. à voyageurs. / ~ service ‖ Personenbeförderung f. ‖ transport m. de voyageurs. / ~ ship ‖ Passagierschiff n. ‖ navire m. à passagers. / ~ steamer ‖ Fahrgastdampfer m. ‖ vapeur m. à passagers. / ~ traffic ‖ Personenverkehr m. ‖ trafic-voyageur m. / ~ train ‖ Personenzug m. ‖ train m. de voyageurs. / ~ transport ‖ Personenbeförderung f. ‖ transport m. de voyageurs. / ~ transshipment ‖ Überschiffung f. der Passagiere ‖ transborde-

ment des passagers m. / ~ vessel *see* ~ ship.

passer (Roll mill) ‖ Schnapper m. ‖ passeur m.

passing the spectrum in review ‖ Durchmusterung f. des Spektrums ‖ exploration f. du spectre. / ~ threads pl. into combs ‖ Blattstechen n. ‖ empeignage m.

passing axle ‖ durchgehende Achse f. ‖ arbre m. traversant. / ~ colour ‖ Übergangsfarbe f. ‖ teinte f. de passage. / ~ bolt ‖ durchgehender Bolzen m. ‖ boulon m. traversant. / ~ machine ‖ Passiermaschine f. ‖ passoire f.

passive ‖ passiv ‖ passif.

passivity ‖ Passivität f. ‖ passivité f. / ~ of the iron ‖ Passivität f. des Eisens ‖ passivité f. du fer. / ~ of metals ‖ Passivität f. von Metallen ‖ passivité f. des métaux.

pass-key ‖ Hauptschlüssel m. ‖ passe-partout m.

passport for abroad ‖ Auslandspaß m. ‖ passeport m. pour l'étranger. / compulsory ~ system ‖ Paßzwang m. ‖ obligation f. de se munir d'un passeport.

paste, to ‖ kleben; kleistern; pappen; bekleben; aufkleben ‖ coller; empâter; cartonner. / ~ (Acc) ‖ pastieren ‖ tartiner. / ~ on ‖ aufkleben ‖ coller. / ~ plates (Acc) ‖ Platten fpl. schmieren ‖ tartiner des plaques.

paste ‖ Paste f.; Klebstoff m.; Kleister m. ‖ pâte f.; colle f. de pâte *ou* d'amidon. / ~ (Bak) ‖ Paste f.; Teig m. ‖ pâte f. / ~ (Glassm) ‖ Glasfluß m.; Glasmasse f. ‖ fluor m. / ~ (Pott) ‖ Masse f.; Tonmasse f. ‖ pâte f.; pâte f. céramique. / artificial ~ (Glassm) ‖ Glasstein m.; künstlicher Edelstein m.; Simili m. ‖ pierre f. précieuse artificielle. / ~ for conservation of boots and shoes ‖ Paste f. zur Erhaltung von Schuhwerk ‖ pâte f. pour l'entretien des chaussures. / extra glossy ~ ‖ Hochglanzpaste f. ‖ composition f. à ravivage. / ~ of lime ‖ Kalkbrei m. ‖ pâte f. de chaux; chaux f. gâchée *ou* en pâte. / modelling ~ ‖ Modelliermasse f. ‖ pâte f. à modeler. / ~ of plaster ‖ Gipsbrei m. ‖ pâte f. de plâtre; plâtre m. gâché. / scented ~ ‖ parfümierte Paste f. ‖ pâte f. parfumée. / starch ~ ‖ Stärkekleister m. ‖ colle f. d'amidon. / to work in ~ (Bookb) ‖ kleben; kleistern; pappen; bekleben ‖ coller; empâter; cartonner.

pasteboard ‖ Pappe f. ‖ carton m. / ~ for bootmaking ‖ Pappe f. für die Schuhmacherei ‖ carton m. pour la cordonnerie. / compressed ~ ‖ gepreßte Pappe f. ‖ carton m. comprimé. / couched ~ ‖ gekautschte Pappe f. ‖ carton m. couché. / elastic ~ ‖ elastische Pappe f. ‖ carton m. élastique. / glazed ~ ‖ Glanzpappe f. ‖ carton m. glacé. / ~ from paper waste ‖ Pappe f. aus Abfallpapier ‖ carton m. de deuxième moulage. / perforated ~ ‖ durchlochte Pappe f. ‖ carton m. perforé. / ~ for shoe manufacture ‖ Schuhkarton m. ‖ carton m. pour chaussures. / white paper covered ~ ‖ mit weißem Papier überzogene Pappe f. ‖ carton m. blanchi.

pasteboard article maker ‖ Kartonagenarbeiter m. ‖ cartonnier m. / ~ box ‖ Pappschachtel f. ‖ boîte f. en carton. / ~ card ‖ Pappkarte f. ‖ carton m. / ~ casting machine ‖ Pappengußmaschine

f. ‖ machine f. à couler le carton. / ~ factory ‖ Pappenfabrik f. ‖ fabrique f. de carton. / ~ goods pl. ‖ Pappwaren fpl. ‖ cartonnages mpl. / ~ maker ‖ Papparbeiter m. ‖ cartonnier m. / ~ making machine ‖ Pappeherstellungsmaschine f. ‖ machine f. à fabriquer le carton. / machine for ~ manufacture ‖ Maschine f. für Pappefertigung ‖ machine f. à fabriquer le carton. / ~ moulder ‖ Pappeformer m. ‖ mouleur m. en cartonnages. / ~ press ‖ Pappepresse f. ‖ presse f. à carton. / ~ roll ‖ Papprolle f. ‖ roulette f. en carton. / ~ and paper toys pl. ‖ Pappe- und Papierspielwaren fpl. ‖ jouets mpl. en carton et en papier. / ~ tube ‖ Papprohr n. ‖ tube m. cylindrique en carton. / work in ~ ‖ Papparbeit f. ‖ cartonnage m.; cartonnerie f.; ouvrage m. de carton.

paste bowl (Book) ‖ Kleistertopf m. ‖ pot m. à colle. / ~ brush ‖ Kleisterpinsel m. ‖ brosse f. à colle. / ~ colours pl. ‖ Teigfarben fpl. ‖ couleurs fpl. en pâte.

pasted plate (Acc) ‖ Masseplatte f.; pastierte Platte f. ‖ plaque f. empâtée.

paste, falling out of the ~ (Acc) ‖ Ausfallen n. der Masse ‖ chute f. de la masse. / ~ goods pl. ‖ Teigwaren fpl. ‖ pâtes fpl. alimentaires.

pastel ‖ Pastell n. ‖ pastel m. / ~ (Crayon) ‖ Farbstift m.; Buntstift m.; Pastellstift m. ‖ crayon m. de couleur; ou à pastel m. / ~ chalk ‖ Pastellkreide f. ‖ craie f. pour peintures au pastel. / ~ colour ‖ Pastellfarbe f. ‖ couleur f. à pastel m. / ~ crayon ‖ Pastellstift m.; Farbstift m. ‖ crayon m. à pastel. / ~ painting ‖ Pastellmalerei f.; Trockenmalerei f. ‖ peinture f. à pastel. / ~ paper ‖ Pastellpapier n. ‖ papier m. pumicif. / ~ picture ‖ Pastellgemälde n. ‖ pastel m.; tableau m. à pastel.

paste jewel ‖ künstlicher Edelstein m. ‖ imitation f. de pierre précieuse; pierre f. précieuse artificielle. / ~ knife ‖ Pappmesser n.; Schnitzer m. ‖ pointe f. ou couteau m. à rabaisser. / ~ manufacturing machine ‖ Teigwarenmaschine f. ‖ machine f. à faire les pâtes alimentaires. / ~ plate see pasted plate. / ~ pot ‖ Kleistertiegel m. ‖ jatte f. à pâte.

paster ‖ Leimer m.; Aufkleber m. ‖ colleur m.; encolleur m.

paste roller hand ‖ Teigmangler m. ‖ étendeur m. de pâte au rouleau. / ~ stone ‖ Similistein m. ‖ pierre f. de strass.

pasteurizer ‖ Pasteurisierer m. ‖ pasteurisateur m.

pasteurizing apparatus ‖ Pasteurisierapparat m. ‖ appareil m. de pasteurisation. / bottle ~ ‖ Pasteurisierapparat m. für Flaschen ‖ appareil m. à pasteuriser les bouteilles; pasteurisateur m. de bouteilles.

paste ware see ~ goods. / ~ work ‖ Papparbeit f. ‖ cartonnage m.; cartonnerie f.; ouvrage m. de carton.

pastille ‖ Pastille f. ‖ pastille f.

pasting of paper hangings on the walls ‖ Aufkleben n. der Tapeten ‖ collage m. des papiers peints sur les murs. / ~ of silks ‖ Verkleistern n. von Seide ‖ empâtement m. de soies.

pasting apparatus ‖ Klebeapparat m. ‖ appareil m. à coller. / ~ lace ‖ Kappnaht f. ‖ galon m. rabattu.

pasting machine (Acc) ‖ Aufstreichmaschine f. ‖ machine f. à tartiner. / ~ (Bookb) ‖ Klebemaschine f. ‖ machine f. à coller. / ~ for cardboard and pasteboard ‖ Beklebemaschine f. für Karton und Pappe ‖ machine f. à coller du papier sur carton et cartonnages. / plate ~ for dynamo plates ‖ Blechbeklebemaschine f. für Dynamobleche ‖ machine f. à coller du papier sur les tôles de dynamo.

pastry ‖ Backwerk n. ‖ pâtisserie f. / fine ~ ‖ feine Backware f. ‖ pâtisserie f. fine.

pastry baker ‖ Kuchenbäcker m. ‖ pâtissier-cuisinier m. / ~ delivery apparatus ‖ Speiseautomat m. ‖ distributeur m. de pâtisseries. / ~ mould ‖ Backform f. ‖ moule m. de confiserie.

pasturage ‖ Trift f.; Weide f. ‖ pâturage m.; pâture f.

pasture ‖ Viehfutter n. ‖ fourrage m.; pâture f. / ~ (Land) ‖ Weide f. ‖ pâture f. / green ~ ‖ Grünfutter n. ‖ fourrage m. vert.

pasture land see pasturage.

pasty ‖ breiig; pappig ‖ pâteux. / ~ (Iron) ‖ teigig ‖ pâteux. / the wrought iron remained partly ~ ‖ das Schmiedeeisen blieb zum Teil teigig ‖ le fer forgé restait en partie pâteux.

pasty condition ‖ teigartiger Zustand m. ‖ état m. pâteux. / ~ iron ‖ teigiges Eisen n. ‖ fer m. pâteux. / ~ sediment of oil ‖ Ölschlamm m. ‖ crasse f. d'huile. / ~ soap ‖ Schmierseife f. ‖ savon m. en pâte ou vert.

patch (Pap) ‖ Musche f. ‖ écaille f. / adhesive ~ ‖ Klebpflaster n. ‖ pièce f. collante; taffetas m. collant; emplâtre f. / blow-out ~ ‖ Mantelmanschette f. ‖ corset m. de pneu. / sticking ~ see adhesive ~.

patched ‖ geflickt ‖ rapiécé.

patchouli herb ‖ Patschulikraut n. ‖ herbe f. de patchouly. / ~ oil ‖ Patschuliöl n. ‖ essence f. de patchouly.

patent, to ‖ patentieren ‖ breveter.

patent ‖ Patent n. ‖ brevet m. / additional ~ ‖ Zusatzpatent n. ‖ brevet m. additionnel. / ~ applied for ‖ Patent n. angemeldet ‖ brevet m. a été demandé. / the claim of the ~ is clearly expressed in the following points ‖ der Patentanspruch ist in folgenden Hauptpunkten klar ausgedrückt ‖ les points mpl. principaux qui sont à protéger par le brevet résultent clairement de la demande. / to grant a ~ ‖ ein Patent n. erteilen ‖ accorder un brevet. / ~ has been granted ‖ Patent n. ist erteilt worden ‖ le brevet m. a été accordé. / ~ of invention ‖ Erfindungspatent n. ‖ brevet m. d'invention. / to protect the invention an application for a ~ was made ‖ zum Schutze der Erfindung ist ein Patent n. angemeldet ‖ voulant protéger l'invention on avait fait enregistrer une demande de brevet m. / provisional ~ ‖ vorläufiges Patent n. ‖ brevet m. provisoire. / ~ has been rejected ‖ Patent n. ist verweigert worden ‖ brevet m. a été refusé. / to secure a ~ for ‖ Patent n. erwirken für ‖ obtenir un brevet m. de . . . / to take-out a ~ ‖ ein Patent n. nehmen ‖ prendre un brevet d'invention. / utility ~ ‖ Gebrauchsmuster n. ‖ brevet m. d'application.

patentable ‖ patentierbar ‖ brevetable.

patent agent ‖ Patentanwalt m. ‖ agent m. de ou en brevets. / ~ axle (Coachm) ‖ Patentachse f. ‖ essieu m. à l'huile; essieu m. breveté. / ~ bobbin (Spinn) ‖ Glanzzwirn m. ‖ coton-cordonnet m.; fil m. glacé. / copy of ~ specification ‖ Exemplar n. der Patentschrift ‖ copie f. du brevet. / cost of ~ ‖ Kosten pl. des Patentes ‖ coût m. du brevet. / ~ coupling ‖ Schraubenkupplung f. ‖ Patentkupplung f.; attelage m. ou tendeur m. à vis. / demand for a ~ ‖ Patentgesuch n. ‖ demande f. de brevet. / drawing of ~ ‖ Patentzeichnung f. ‖ dessin m. du brevet.

patented ‖ patentiert ‖ breveté. / not ~ ‖ nicht patentiert ‖ pas breveté. / ~ construction ‖ patentierte Konstruktion f. ‖ construction f. brevetée. / ~ machine ‖ patentierte Maschine f. ‖ machine f. brevetée.

patentee ‖ Patentinhaber m. ‖ titulaire m. ou possesseur m. d'un brevet.

patent fastener ‖ Druckknopf m. ‖ bouton m. à pression. / ~ fees pl. ‖ Patentgebühren fpl. ‖ taxes fpl. des brevets.

patent fuel ‖ Brikett n. ‖ briquette f. / ~ making machinery ‖ Brikettmaschine f. ‖ machine f. à briquettes.

patent, infringement of ~ ‖ Patentverletzung f. ‖ contrefaçon f. du brevet. / ~ law ‖ Patentgesetz n.; Patentrecht n. ‖ loi f. sur les brevets d'invention. / ~ lawyer see patent agent. / ~ lead ‖ Patentlot n. ‖ sonde f. spéciale. / ~ leather ‖ Lackleder n.; Glanzleder n. ‖ cuir m. verni ou laqué. / ~ leather boot ‖ Lackstiefel m. ‖ botte f. de cuir verni. / model of ~ ‖ Patentmodell n. ‖ modèle m. du brevet. / ~ office ‖ Patentamt n. ‖ bureau m. des brevets. / ~ pencil ‖ Patentstift m. ‖ crayon m. breveté. / ~ right ‖ Patentrecht n. ‖ droit m. des brevets. / royalties pl. for the use of ~s ‖ Lizenzgebühren fpl. ‖ redevance f. payée pour l'usage d'un brevet d'invention.

patent specification ‖ Patentschrift f.; Patentbeschreibung f. ‖ lettre f. patente; description f. de brevet; spécification f. / several other combinations are possible all of which are described in the ~ ‖ in der Patentschrift sind verschiedene andere Kombinationen fpl. beschrieben ‖ les cas mpl. qui pourraient se présenter sont étudiés dans la description de brevet. / ~ supplementary ~ ‖ Nachtrag m. zu einer Patentschrift ‖ supplément m. à une lettre patente.

patent strong yarn ‖ Eisengarn n. ‖ fil m. à coudre extra-fort. / ~ survey ‖ Patentuntersuchung f. ‖ examen m. du brevet. / validity of ~ ‖ Gültigkeit f. des Patents ‖ validité f. du brevet.

paternoster (Arch) ‖ Paternoster n.; Rosenkranz m.; beperlter Rundstab m. ‖ chapelet m.; collier m.; fusarolle f.; patenôtre f.; perles fpl. / ~ elevator (Dredger) ‖ Becherwerk n. ‖ élévateur m. à godets. / ~ elevator (Lift) ‖ Paternosteraufzug m. ‖ monte-charge m. à chaîne sans fin. / ~ lift see ~ elevator. / ~ work ‖ Paternosterwerk n.; Noria f. ‖ patenôtre f.; élévateur m. à augets; noria f.

path ‖ Fußweg m.; Pfad m. ‖ sentier m. / ~ (Wind motor) ‖ Laufbahn f. ‖ chemin m. de roulement. / ~ of the current ‖ Stromweg m. ‖ parcours m. du courant.

/ ~ of rays in spectroscope || Strahlengang m. im Spektroskop || marche f. des rayons dans le spectroscope. / ~ of spark || Funkenbahn f. || parcours m. de l'étincelle.

pathogenic || krankheitserregend || pathogène.

pathological changes pl. || pathologische Veränderungen fpl. || modifications fpl. pathologiques.

patina || Patina f. || patine f.

patrol instrument || Patrouillenapparat m. || appareil m. pour patrouilles.

patron || Lehrherr m. || maître m.

patten || Pantine f.; Holzschuh m. || sabot m. / ~ (Build) || Latsche f.; Anlage f.; Mauerrecht n. || empattement m.; fondation f. en saillie.

pattern || Modell n.; Muster n.; Probestück n. || modèle m.; échantillon m. / ~ (Cloth) || Schnittmuster n. || patron m. / ~ (Found) || Gußmodell n. || modèle m. / ~ (Tool) || Schablone f.; Lehre f. || échantillon m.; calibre m. / ~ (Weav) || Muster n.; Dessin n. || dessin m.; figure f. / according to ~ || nach Muster n. || selon modèle m. / current ~ || fortlaufendes Muster n. || dessin m. continu. / to cut to ~ || nach einer Schablone f. ausschneiden || profiler sur modèle m. / foundry ~ || Gußmodell n. || modèle m. de fonderie. / horizontal stripe ~ || Ringelmuster n. || dessin m. à rayures horizontales en travers. / latest ~ || neueste Form f. || dernière forme f. / leather ~ || Lederschablone f. || patron m. en cuir. / light ~ || leichte Bauart f. || construction f. légère. / ~ for the mould || Gußmodell n. || modèle m. de fonderie. / plain ~ || glattes Muster n. || dessin m. uni. / pricked ~ used in pouncing (Paint) || Schablone f.; Karton m. || poncis m.; poncif m.; calque m. / registered ~ || eingetragene Schutzmarke f. || marque f. déposée. / repeated ~ || wiederkehrendes Muster n. || dessin m. diapré ou gaufré. / reserved ~ || Spitzmuster n. || dessin m. à regard ou à retour. / rose engine ~ || Guillochierung f. || guillochis m. / show room ~ || Ausstellungsmodell n. || modèle m. pour expositions. / vertical stripe ~ || Langstreifenmuster n. || dessin m. à rayures verticales. / wood ~ maker || Modellschreiner m. || menuisier-modeleur m.; menuisier-mécanicien m.

pattern board || Modellbrett n. || planche f. à modèles. / ~ card || Musterkarte f. || carte f. d'échantillons. / ~ card maker || Musterkartenarbeiter m. || perforeur m. de cartes d'échantillons. / ~ designer || Musterzeichner m. || dessinateur m. de modèles. / ~ drawer see ~ designer. / ~ drawing || Modellzeichnung f. || dessin m. de modelage. / ~ folder || Schnittmusterfalzerin f. || plieuse f. de patrons. / ~ half || Modellhälfte f. || moitié f. de modèle. / ~ lifter || Modellheber m. || poignée f. pour enlever les modèles.

pattern maker || Modelltischler m.; Modellschreiner m. || modeleur m. / brass ~ || Bronzemodellör m. || modeleur m. pour le bronze. / foundry ~ || Gießereimodelltischler m. || modeleur m. de fonderie. / ~'s shop || Modellschreinerei f. || atelier m. de modelage.

pattern mechanism || Musterapparat m. || appareil m. à échantillons. / ~ paper primed with oil || Patronenpapier n.; Schab-

lonenpapier || imprimure f.; papier m. à l'huile. / ~ perforating machine || Musterstechmaschine f. || machine f. à perforer les échantillons. / ~ plate see ~ board. / set of ~s || Modellsatz m. || jeu m. de modèles. / ~ shop || Modelltischlerei f. || atelier m. de modelage. / ~ storage || Modellboden m. || magasin m. de modèles. / ~ turner || Modelldrechsler m. || modeleur-tourneur m. / ~ varnish || Modellack m. || laque m. à modèle. / ~ weaver || Musterweber m. || échantillonneur m.; tisseur-échantillonneur m.

paucity of money || Geldknappheit f. || pénurie f. d'argent.

paumelle (Curr) || Krispelholz n. || paumelle f.; pommelle f.; marguerite f. à la main; grènetoir m.

paunch (Mar) || gewebte Matte f.; Stoßmatte f. || sangle f. / ~ of a bell || Kranz m. oder Anschlag m. einer Glocke || bord m. ou panse f. d'une cloche.

pause (Print) || Gedankenstrich m. || moins m.; trait m. suspensif ou de suspension.

pave, to || pflastern || paver. / ~ a floor with tiles || mit Fliesen fpl. belegen || carreler.

pavement || Pflaster n. || pavement m.; pavé m. / artificial stone ~ || Kunststeinpflaster n. || pavé m. aggloméré. / asphalt ~ || Asphaltpflaster n. || pavé m. d'asphalte. / bricked ~ || Ziegelsteinpflaster n. || pavage m. en briques. / diamond ~ || Pflaster n. im Schlagverband oder Rautenverband || pavé m. de carreaux rangés en losange. / fancy ~ || gemustertes Pflaster n. || pavé m. panneau ou de fantaisie. / ~ of paving tiles || Plattenbelag m. || carrelage m.; dallage m. / rubble ~ || unregelmäßiges Pflaster n. || pavé m. en blocages. / ~ of a seam || Liegendes n. eines Flözes || mur m. ou lit m. d'une couche. / square-dressed ~ || bossiertes Pflaster n. || pavage m. rangé. / wood ~ || Holzpflaster n. || pavé m. de bois.

pavement, bed of the ~ || Sandbett n. || lit m. du pavé; couche f. de sable sous le pavé. / ~ rammer || Pflasterramme f.; Stampfe f. || dame f. de paveur.

paver (Build) || Fliesenleger m.; Plattenleger m. || carreleur m. / ~ (Road) || Steinsetzer m. || paveur m.

paver's dressing hammer || Zurichtehammer m. || épinçoir m.; marteau m. du paveur. / ~ trowel || Pflasterkelle f. || décentoir m. / ~ work || Pflasterarbeit f.; Pflasterung f. || ouvrage m. du paveur; pavage m.

pavilion || Zelt n. || tente f. / ~ (Brilliant) || Krone f.; Hauptfacette f. (eines Edelsteins) || pavillon m. / ~ roof || Zeltdach n. || toit m. en pavillon.

paving || Pflasterarbeit f.; Pflasterung f. || pavage m. / concrete ~ || Betonstraßendecke f. || pavage m. en béton. / ~ of the railway stations || Abpflasterung f. der Bahnhöfe || pavage m. des gares.

paving beetle see ~ rammer. / ~ brick || Pflasterziegel m. || brique f. ou carreau m. à paver. / ~ clinker || Pflasterklinker m. || brique f. dure à paver. / ~ flag || Pflasterplatte f. || carreau m. de pavage. / ~ material || Pflastermaterial n.; Pflasterwerkstoffe m.; Pflasterbaustoffe m. || matériaux mpl. de pavage. / ~ rammer || Pflasterramme f.; Stampfe f. || dame f. de paveur.

paving stone || Pflasterstein m.; Kopfstein m. || pavé m. de pierre; grès m. à paver. / dressed ~ || fertig bearbeiteter Pflasterstein m. || pavé m. piqué. / ~ for the gutter || Gossenstein m.; Rinnenstein m. || jumelle f. de pavement. / ~ of half-size || Pflasterstein m. von halber Dicke || pavé m. refendu ou de deux. / ~ from natural stone || Pflasterstein m. aus natürlichem Stein || pavé m. en pierre naturelle. / rough ~ || roher Pflasterstein m. || pavé m. brut. / squared ~ || bossierter Pflasterstein m. || pierre f. échantillonnée. / standard ~ || Normalpflasterstein m. || pavé m. d'échantillon normal.

paving stone measuring device || Pflastersteinmeßmaschine f. || machine f. à mesurer les pavés. / ~ quarry || Pflastersteinbruch m. || carrière f. de pavés.

paving tile || Fliese f.; Kachel f. || carreau m. de fayence ou glacé. / coloured ~ || farbige Tonfliese f.; farbige Kachel f. || carreau m. coloré en mosaïque.

paving tile maker || Fliesenarbeiter m. || carrelier m.

paviour see paver.

pawl || Klinke f.; Sperrklinke f. || cliquet m. / disengaging ~ || ausrückbare Sperrklinke f. || cliquet m. débrayable. / rotary ~ || Drehklinke f. || jack m. rotatif.

pawl coupling || Klinkenapparat m. || appareil m. à cliquet. / ~ guide || Sperrklinkenführung f. || guide f. de cliquet d'arrêt. / ~ lever || Sperrklinkenhebel m. || manette f. du cliquet d'arrêt. / ~ rim (Shipb) || Pallscheibe f. || cercle m. des linguets.

pay, to || zahlen || payer; solder. / ~ || bezahlen || honorer; rémunérer. / able to pay || solvent || solvable. / ~ on account || anzahlen || donner en acompte m.; donner à compte m. / ~ the costs pl. || die Kosten pl. tragen || supporter les frais mpl. / ~ down || auszahlen || payer; verser. / ~ for || frankieren; freimachen || affranchir; payer le port. / ~-in || einzahlen || payer; verser. / ~ in full || ausbezahlen || achever de payer; liquider. / ~-off (Bank) || tilgen || amortir, / ~-off (Wages) || ablohnen || payer et congédier. / ~-off the crew (Mar) || die Mannschaft abbezahlen oder abmustern || licencier l'équipage m. / ~-out || auszahlen || payer; verser. / ~-out a cable || ein Kabel n. abrollen oder auslegen || dérouler un câble. / ~-out a sum || einen Betrag m. auszahlen || verser un montant. / ~ wages || löhnen || payer le salaire ou la solde. / it pays to work it even in small units || die Verarbeitung f. lohnt sich selbst in kleinen Einheiten || l'industrie f. rend même si elle est entreprise sur une petite échelle.

pay || Löhnung f. || payement m.; paye f. / ~ (Mar) || Heuer f. || loyer m.; salaire m.; paye f.; solde f. / day's ~ || Tagelohn m. || salaire m. journalier.

payable || zahlbar || payable. / ~ on delivery || zahlbar bei Lieferung f. || payable à la livraison. / ~ deposit (Mine) || abbauwürdige Mächtigkeit f. || puissance f. d'un gisement justifiant l'exploitation.

pay bill || Lohnzettel m. || bordereau m. de paye. / ~ cheque || Lohnscheck m. || chèque m. de paye. / ~ day || Lohntag m.; Zahltag m. || jour m. de paye ou de payement.

payee ‖ Zahlungsempfänger m. ‖ personne f. recevant paiement; porteur m. / ~ (Check) ‖ Remittent m. ‖ remetteur m.

pay envelope ‖ Lohntüte f. ‖ enveloppe f. de paye.

payer ‖ Bezahler m. ‖ payeur m. / bad ~ ‖ säumiger Zahler m. ‖ client m. lent à payer; mauvais payeur m.

paying ladle ‖ Pechlöffel m. ‖ cuiller f. à brai. / ~ mechanism with slot ‖ Geldautomat m. ‖ distributeur m. automatique.

paying-out drum ‖ Auslegetrommel f. ‖ tambour m. dérouleur ou de déroulement. / ~ machine (Cable) ‖ Auslegemaschine f.; Kabelmaschine f. ‖ machine f. à poser les câbles; machine f. de pose. / ~ and picking-up gear ‖ Seekabelwinde f. ‖ machine f. pour immerger et lever les câbles sous-marins.

pay load ‖ zahlende Nutzlast f. ‖ charge f. utile payante.

payment ‖ Auszahlung f. ‖ payement m.; paiement m.; versement m. / ~ (Bank) ‖ Inkasso n. ‖ encaissement m.; recouvrement m. / ~ (Mail) ‖ Porto n. ‖ port m. / ~ (Wages) ‖ Löhnung f. ‖ payement m.; paye f. / ~ on account ‖ Abschlagszahlung f.; Ratenzahlung f. ‖ acompte m.; payement m. par termes ou à compte; versement m. partiel. / ~ to account of ‖ Zahlung f. für Rechnung ‖ versement m. à valoir. / additional ‖ ~ Nachzahlung f. ‖ payement m. ultérieur. / ~ in advance ‖ Vorschußzahlung f. ‖ payement m. d'avance. / ~ of balance ‖ Restzahlung f. ‖ payement m. pour solde. / ~ after a certain date ‖ Nachtragszahlung f. ‖ payement m. à effectuer après un certain délai. / ~ of dividend ‖ Dividendenausschüttung f. ‖ distribution f. du dividende. / immediate ~ ‖ sofortige Zahlung f. ‖ payement m. immédiat. / ~ by instalments ‖ Ratenzahlung f.; Teilzahlung f. ‖ payement m. par termes ou par acomptes. / to keep up one's ~ ‖ eine Zahlung f. einhalten ‖ être exact à payer; observer le délai. / part ~ ‖ Teilzahlung f. ‖ paiement m. partiel. / ~ of postage ‖ Frankierung f. ‖ affranchissement m. / to press for ~ ‖ zur Zahlung f. auffordern ‖ inviter à payer. / to provide for ~ ‖ für Deckung f. sorgen ‖ pourvoir à la couverture f. / to suspend ~ ‖ die Zahlungen fpl. einstellen ‖ suspendre les paiements mpl. / ~ in terms see ~ by instalment. upon ~ ‖ bei Bezahlung f. ‖ sur payement m. / ~ of wages ‖ Lohnzahlung f. ‖ payement m. des salaires. / your ~ ‖ Ihre Zahlung f. ‖ votre payement.

payment, condition of ~ ‖ Zahlungsbedingung f. ‖ condition f. de payement. / stopping of ~ ‖ Zahlungssperre f. ‖ arrêt m. dans les versements.

pay ore ‖ abbauwürdiges Erz n. ‖ minerai m. profitable ou payant. / ~ roll ‖ Lohnliste f. ‖ fiche f. ou feuille f. ou liste f. de paye.

pea ‖ Erbse f. ‖ pois m. / ~ coal ‖ Perlkohle f. ‖ charbon m. menu.

peacock coal ‖ Glanzkohle f. ‖ charbon m. luisant.

peace, terms pl. of ~ ‖ Friedensbedingungen fpl. ‖ conditions fpl. de paix.

peace treaty ‖ Friedensvertrag m. ‖ traité m. de paix.

peach ‖ Pfirsich m. ‖ pêche f. / ~ stone ‖ Pfirsichkern m. ‖ noyau m. de pêche.

pea husks removing machine ‖ Erbsenausschälmaschine f. ‖ machine f. à écosser les pois.

peak ‖ Spitze f. ‖ pointe f.; comble m. / ~ of a cap ‖ Mützenschirm m. ‖ visière f. de casquette. / ~ of the curve ‖ Kurvenscheitelpunkt m. ‖ apogée m. de la courbe. / ~ of load ‖ Belastungsspitze f. ‖ pointe f. de charge. / ~ of the mountain ‖ Berggipfel m. ‖ sommet m. ou cime f. de la montagne.

peak consumption, to cope with the ~ ‖ den Spitzenbedarf m. decken ‖ couvrir la pointe.

peak current compensation ‖ Spitzendeckung f. ‖ compensation f. de pointe de courant. / ~ generating station see peak load power station. / provision of an adequate supply of ~ at reasonable prices ‖ ausreichende Bereitstellung f. von Spitzenstrom zu befriedigenden Preisen ‖ mise f. à disposition de courant de pointe suffisant à des prix satisfaisants.

peak electrode (Radio) ‖ Bauchelektrode f. ‖ électrode f. ventrale.

peak load ‖ Spitzenbelastung f. ‖ charge f. de pointe. / ~ plant see ~ power station. / ~ power station ‖ Spitzenkraftwerk n. ‖ usine f. de pointe.

peak value ‖ Höchstwert m.; Spitzenwert m. ‖ valeur f. maximum.

peanut ‖ Erdnuß f. ‖ arachide f. / ~ oil ‖ Erdnußöl n. ‖ huile f. d'arachides.

pear ‖ Birne f. ‖ poire f.

pearl ‖ Perle f. ‖ perle f. / ~ (Nacre) ‖ Perlmutter n. ‖ nacre f. / ~ (Print) ‖ Perlschrift f. ‖ perle f. / artificial ~ ‖ Glasperle f.; künstliche Perle f. ‖ perle f. artificielle. / false ~ ‖ unechte Perle f. ‖ perle f. fausse. / fine ~ ‖ echte Perle f. ‖ perle f. fine. / real ~ see fine ~.

pearlash ‖ Pottasche f. ‖ carbonate m. de potasse.

pearl barley ‖ Perlgraupen fpl. ‖ orge m. perlé. / ~ maker ‖ Graupenmüller m. ‖ fabricant m. d'orge perlé. / ~ manufacture ‖ Graupenschälerei f. ‖ fabrication f. d'orge perlé.

pearl borer ‖ Perlenbohrer m. ‖ perceur m. de perles. / ~ boring ‖ Perlenbohren n. ‖ perçage m. de perles. / ~ button ‖ Perlmutterknopf m. ‖ bouton m. de nacre. / ~ button cutter ‖ Perlmutterknopfdrechsler m. ‖ tourneur m. de boutons de nacre. / ~ coal ‖ Perlkohle f. ‖ charbon m. menu. / ~ diver ‖ Perlenfischer m. ‖ pêcheur m. d'huîtres perlières ou de perles. / ~ goods pl. ‖ Perlenerzeugnisse npl. ‖ articles mpl. en perles.

pearling mill ‖ Graupenmühle f. ‖ moulin m. à mouler et à perler l'orge. / ~ punch ‖ Perlpunze f. ‖ perloir m.

pearl mica ‖ Margarit m.; Perlglimmer m. ‖ margarite m.; mica m. nacré. / ~ oats pl. see pearl barley. / ~ oyster dredger see ~ diver. / ~ shell ‖ Perlmutter f. ‖ nacre f. / ~ spar ‖ Perlspat m. ‖ spath m. perlé.

pearlstone ‖ Perlit m.; Perlstein m. ‖ perlite f.

pearl strings pl. ‖ Perlposamenten pl. ‖ passementerie f. en perles.

pearly ‖ perlartig ‖ perlaire. / ~ (Met) ‖ gekörnt ‖ grenu. / ~ lustre ‖ Perlmutterglanz m. ‖ éclat m. nacré.

pear-shaped ‖ birnenförmig ‖ en forme f. de poire. / ~ converter ‖ birnenförmiger Konverter m. ‖ convertisseur m. en forme de poire.

pear switch ‖ Birnenausschalter m. ‖ interrupteur m. à poire.

pear tree ‖ Birnbaum m. ‖ poirier m. / wild ~ ‖ Holzbirnbaum m. ‖ poirier m. sauvage.

pear tree wood ‖ Birnbaumholz n. ‖ bois m. de poirier.

peasant ‖ Bauer m. ‖ paysan m.

pea sausage ‖ Erbswurst f. ‖ saucisse f. aux pois.

pease pl. see pea.

pease-meal ‖ Erbsmehl n. ‖ farine f. de pois.

pea sheller ‖ Schotenaushülserin f. ‖ écosseuse f. de petits pois. / ~-sized ‖ erbsengroß ‖ de la grosseur d'un pois. / ~ splitter ‖ Erbsenschrotmüller m. ‖ fabricant m. de pois concassés.

peastone ‖ Pisolit m.; Erbsenstein m. ‖ pisolite f.; pierre f. de pois.

peat ‖ Torf m. ‖ tourbe f. / bituminous ~ ‖ Specktorf m. ‖ tourbe f. grasse. / black ~ ‖ Pechtorf m. ‖ tourbe f. limoneuse. / to dig ~ ‖ Torf m. stechen ‖ extraire la tourbe. / dredged ~ ‖ Baggertorf m. ‖ tourbe f. draguée. / fibrous ~ ‖ Fasertorf m. ‖ tourbe f. fibreuse. / hand-cut ~ ‖ Handtorf m. ‖ tourbe f. à la main. / lamellated ~ ‖ Blättertorf m. ‖ tourbe f. feuilletée. / moulded ~ ‖ Streichtorf m. ‖ tourbe f. moulée. / powdered ~ ‖ Torfmehl n. ‖ poudre f. de tourbe. / pressed ~ ‖ Preßtorf n. ‖ tourbe f. comprimée. / transformation into ~ ‖ Vermoorung f. ‖ transformation f. en tourbière.

peat board ‖ Torfpappe f. ‖ carton m. de tourbe. / ~ bog master ‖ Torfmeister m. ‖ maître m. de tourbière. / ~ briquette ‖ Torfbrikett n. ‖ briquette f. de tourbe. / ~ briquetting ‖ Torfbrikettierung f. ‖ briquettage m. de tourbe. / ~ carrier ‖ Torfträger m. ‖ porteur m. de tourbe. / ~ coal ‖ Torfkohle f. ‖ charbon m. de tourbe. / ~ conveyor ‖ Torffördermittel n. ‖ appareil m. de manutention de la tourbe. / ~ cutter ‖ Torfstecher m. ‖ tourbier m. / ~ digger see ~ cutter. / ~ digging ‖ Torfgewinnung f. ‖ extraction f. de tourbe. / ~ drag ‖ Torfbagger m. ‖ puchette f. / ~ firing ‖ Torffeuerung f. ‖ chauffage m. à la tourbe. / furnace for ~ ‖ Feuerung f. für Torf ‖ foyer m. à tourbe. / ~ litter factory ‖ Torfstreufabrik f. ‖ fabrique f. de litières de tourbe. / ~ mill ‖ Torfmühle f. ‖ moulin m. à tourbe. / ~ moor ‖ Torfmoor n. ‖ marais m. tourbeux. / ~ moss ‖ Torfstreu f. ‖ tourbe f. pour litière. / ~ power plant ‖ Torfkraftwerk n. ‖ usine f. thermique à base de tourbe. / ~ press ‖ Torfpresse f. ‖ presse f. à mottes de tourbe. / ~ producing machine ‖ Torfgewinnungsmaschine f. ‖ machine f. pour l'extraction de la tourbe. / ~ press for producing ~ ‖ Presse f. für die Torfgewinnung ‖ presse f. pour l'extraction de la tourbe. / ~ tar ‖ Torfteer m. ‖ goudron m. de tourbe. / ~ wool ‖ Torfwolle f. ‖ tourbe f. ligneuse; laine f. de tourbe.

peaty soil ‖ Torfboden m. ‖ sol m. tourbeux.

pebble || Kiesel m. || caillou m. / Egyptian ~ || ägyptischer Jaspis m.; Kugeljaspis m. || jaspe m. égyptien. / ~ ground || Kiesgrund m.; Keigrund m.; Kegelgrund m. || fond m. de cailloutage. / ~ manganese || Mangansuperoxyd n.; Braunstein m. || peroxyde m. de manganèse. / ~ mill || Kugelmühle f. || broyeur m. à boulets. / ~ powder || Kieselpulver n. || poudre-pebble f. / ~ stone for grinding mills || Flintstein m. für Kugelmühlen || silex m. rond ou galet m. sphérique pour broyeurs. / washer for ~ works pl. || Wäsche f. für Kieswerke || laveur m. pour usines de gravier.

pebbling machine (Curr) || Krispelmaschine f. || machine f. à rebrousser; marguerite f. mécanique.

pechblende || Pechblende f. || pechblende f,

peck (Measure) || Metze f. || minot m.

pectic acid || Pektinsäure f. || acide m. pectique.

pectine || Pektin n.; Pflanzengallerte f. || gelée f. végétale; pectine f.

pectolite || Pektolit m. || pectolite f.

pectoral cure || Hustenmittel n. || médicament m. antitousseux.

pecuniary, to demand ~ compensation || eine Geldentschädigung f. verlangen || exiger une indemnité en argent. / ~ difficulties pl. || Geldverlegenheit f. || embarras m. financier ou d'argent; gêne f. / ~ loss || Geldverlust m. || perte f. pécuniaire.

pedal || Fußhebel m.; Pedal n. || pédale f. / adjustable ~ || nachstellbares Pedal n. || pédale f. ajustable. / brake ~ || Bremspedal n. || pédale m. du frein. / clutch ~ || Kupplungsfußhebel m. || pédale f. d'embrayage. / reverse ~ || Pedal n. für Rückwärtsgang || pédale f. de marche arrière.

pedal cone || Pedalkonus m. || cône m. de pédale. / ~ plate || Pedalplatte f. || patin m. de pédale. / ~ rubber || Pedalgummi n. || caoutchouc m. pour pédales. / ~ shaft || Fußhebelwelle f. || axe m. de pédale. / ~ switch || Tretschalter m. || commutateur m. à pédale.

pedestal (Arch) || Fußgestell n. einer Säule; Piedestal n.; Säulenständer m. || piédestal m. / ~ (Mach) || Untersatz m. || socle m. / ~ (Railw) || Bügelgleitbacke f.; geschlossene Achslagerführung f. || guide m. de boîte en forme d'arcade. / ~ (Shipb) || Lagerbock m. || chaise f. de palier. / ~ of a column || Fuß m. oder Basis f. einer Säule || base f. d'une colonne. / ~ of a couple of millstones || Gestell n. eines Mahlganges || piédestal m. d'une paire de meules. / wooden ~ || Holzkonsole f. || socle m. en bois.

pedestal bearing || Rumpflager n. || palier m. ordinaire. / ~ bearing type || Bocklagertype f. || type f. de palier à chaise. / ~ binder (Railw) || Strebe f.; Achsgabelsteg m. || entretoise f. de plaque de garde. / ~ lamp || Stehlampe f.; Ständerlampe f. || lampe f. à pied ou de parquet. / supporting roller in ~ box-form || kastenartiger Stützrollenträger m. || support m. des rouleaux d'appui en forme de carter. / ~ tie bar || Achsgabelsteg m. || entretoise f. de plaque de garde.

pedicure, article for ~ || Fußpflegeartikel m. || article m. pour pédicures.

pediment || Giebel m.; Ziergiebel m. || fronton m.

pedometer || Schrittzähler m. || pédomètre m.

pecker || Abfühlnadel f. || goujon m.; aiguille f.

peel, to || schälen || écorcer peler. / ~ fruits || Früchte fpl. abschälen || peler des fruits mpl. / ~-off (Plastering) || abbröckeln; abblättern || s'écailler; s'écaler. / ~-off the bark || abrinden || écorcer; écrouter. / ~-off the bast of the hemp || den Hanf m. schälen || teiller ou tiller le chanvre. / ~-off the grain (Curr) || das Fell abnarben || effleurer la peau.

peel || Schale f.; Hülse f. || gousse f.; cosse f.; écale f. / ~ (Pap) || Aufhängekreuz n. || ferlet m.

peeled || abgeschält || écorcé; pelé.

peeler || Abschäler m. || écorceur m. / ~ (Peas) || Ausleserin f. || éplucheuse f.

peeling (Silkworm) || Häutung f. || mue f. / wood ~ || Holzabrinden n. || écorçage m. du bois en forêt.

peeling device || Schälvorrichtung f. || dispositif m. diviseur. / ~ drum || Schältrommel f. || dépulpeur m. / ~ knife || Schälmesser n. || couteau m. diviseur ou à peler.

peeling machine, bark ~ || Rindenschälmaschine f. || machine f. à décortiquer. / almond ~ || Mandelschälmaschine f. || râpeuse f. mécanique pour amandes. / ~ for splints || Schälmaschine f. für Holzdrahterzeugung || machine f. à dérouler le bois pour la fabrication de copeaux à tiges. / ~ for wood || Schälmaschine f. für Holz || machine f. à décortiquer le bois.

peel steam baking oven heated by wood || Einschießdampfbackofen m. für Holzheizung || four m. de boulangerie à vapeur pour enfourner et pour chauffage au bois.

peeping window || verstecktes Fenster n.; Guckfenster n. || fenêtre f. à vue dérobée.

peg, to || anstiften; verdübeln || goujonner.

peg (Bolt) || Bolzen m. || boulon m. / ~ (Coop) || Dübel m. || goujon m. / ~ (Carp) || Holznagel m.; Pflock m. || cheville f. de bois; goujon m. / ~ (Locksm) || Keil m.; Schlüssel m.; Splint m. || clavette f.; clef f. / ~ (Mach) || Knagge f.; Daumen m.; Steuerdaumen m.; Nase f. || came f.; taquet m. / ~ (Peg ladder) || Sprosse f. || ranche f. / clothes ~ || Wäscheklammer f. || épingle f. à linge en bois. / hoop iron ~ || Bandeisendübel m. || goujon m. de feuillard. / lease ~ (Weav) || Kreuznagel m. || cheville f. d'encroix. / ~ for musical instruments || Wirbel m. für Musikinstrumente || cheville f. d'instruments de musique. / vent ~ on the wine cask || Zwicker m. am Weinfaß || fausset m. au tonneau de vin. / ~ of a violin || Geigenwirbel m. || cheville f. de violon. / wooden ~ || Holzdübel m.; Holzstift m. || cheville f. en bois. / wooden ~ making machine || Holzstiftmaschine f. || machine f. à faire des goupilles en bois.

pegamoid || Pegamoid n. || pégamoïd m.

peganite || Peganit m. || péganite f.

peg beam (Wheelbarrow) || Riegelholz n. || échelier m.

pegging || Verdübelung f. || assemblage m. à goujons. / ~ rammer || Stampfer m. des Formers || batte f. du mouleur.

peg hole || Dübelloch n. || trou m. de goujon. / ~ boring machine || Dübellochbohrmaschine f. || machine f. à forer les trous de goujons.

peg ladder || Stangenleiter f.; Stockleiter f. || casse-cou m.; échelier m.; rancher m. / ~ of a crane || Kranleiter f. || rancher m. ou échelier m. de grue.

pegmatite || Pegmatit m.; Schriftgranit m. || pegmatite f.; granite m. graphique.

pegmatolite || Pegmatolit m. || pegmatolite f.

peg turner || Wirbeldrechsler m. || tourneur m. de chevilles. / ~ wood || Pflockholz n. || fenton m.

pelerine || Umhang m. || pèlerine f.

pellicle of cocoons || Kokonhäutchen n. || pellicule f. de cocons.

pellitory root || Bertramwurzel f. || racine f. de pyrèthre.

pelting rain || Platzregen m. || pluie f. battante.

Pelton water wheel || Peltonrad n. || roue f. hydraulique de Pelton.

peltry || Pelzwerk n. || pelleterie f. / ~ imbued with an antiseptic paste to prevent rotting || Pelzwerk n. mit einer antiseptischen Masse zur Verhinderung der Fäulnis überzogen || pelleterie f. enduite d'une pâte antiseptique pour éviter la détérioration. / ~ dressed with ashes || mit Asche behandeltes Pelzwerk n. || pelleterie f. saupoudrée de cendres. / ~ for clothing || Pelzwerk n. für Kleidung || pelleterie f. pour l'habillement. / dressed ~ || verarbeitetes Pelzwerk n. || pelleterie f. ouvrée. / dried ~ || getrocknetes Pelzwerk n. || pelleterie f. séchée. / ~ for finery || Pelzwerk n. zum Schmuck || pelleterie f. pour la parure. / ~ for furniture || Pelzwerk n. zur Möbelausstattung || pelleterie f. d'ameublement. / purified ~ || gereinigtes Pelzwerk n. || pelleterie f. purifiée. / raw ~ || rohes Pelzwerk n. || pelleterie f. brute. / ready-made ~ || konfektioniertes Pelzwerk n. || pelleterie f. confectionnée. / ~ in skins or skin-parts sewed together || Pelzwerk n. in zusammengenähten Fellen oder Fellteilen || pelleterie f. en peaux ou en parties de peaux cousues ensemble. / softened ~ || geschmeidig gemachtes Pelzwerk n. || pelleterie f. assouplie. / tawed ~ || weißgegerbtes Pelzwerk n. || pelleterie f. passée en mégie. / ~ in form of triangles or squares || Pelzwerk n. in Spitz- oder Quadratform || pelleterie f. en forme de touloupes ou carrés. / varnished ~ || glänzend gemachtes Pelzwerk n. || pelleterie f. lustrée.

pen, to ~ the water || das Wasser anstauen || hausser ou élever les eaux fpl.

pen || Schreibfeder f. || plume f. à écrire. / ball-pointed ~ || Kugelspitzfeder f. || plume f. à pointe sphérique. / double-pointed ~ || Parallelfeder f. || plume f. à deux becs. / drawing ~ || Zeichenfeder f. || plume f. à dessin. / fountain ~ || Füllfederhalter m. || stylographe m. / gold ~ || Goldfeder f. || plume f. en or. / gold fountain ~ || Goldfüllfederhalter m. || plume f. à réservoir en or. / gold-plated ~ || Golddoubléschreibfeder f. || plume f. en or doublé. / ~ with a long nib || Schreibfeder f. mit langer Spitze || plume f. à long bec. / manifold ~ || Durchschreib-

feder f. ‖ plume f. pour faire les copies au carbone. / mapping ~ ‖ Landkartenschreibfeder f. ‖ plume f. pour cartes géographiques. / ~ for writing music ‖ Notenfeder f. ‖ plume f. à écrire la musique. / ~ with one slit ‖ einspaltige Schreibfeder f. ‖ plume f. à une entaille. / quick writing ~ ‖ Schnellschreibfeder f. ‖ plume f. à écrire rapide. / roundpoint ~ ‖ Rundspitzfeder f. ‖ plume f. à pointe ronde. / round hand ~ ‖ Rundschriftfeder f. ‖ plume f. à ronde. / singlepointed ~ ‖ Schreibfeder f. mit einer Spitze ‖ plume f. à un seul bec. / ~ for sketching ‖ Skizzierfeder f. ‖ plume f. à croquis. / steel ~ ‖ Stahlfeder f. ‖ plume f. d'acier. / ~ for technical purposes ‖ Feder f. für technische Zwecke ‖ plume f. pour usage technique. / ~ with three slits ‖ dreigespaltene Schreibfeder f. ‖ plume f. à trois entailles. / turn-up point ~ ‖ Löffelschreibfeder f. ‖ plume f. à pointe relevée. / writing ~ ‖ Schreibfeder f. ‖ plume f. à écrire.

penalty for delayed delivery ‖ Verzugsstrafe f. ‖ pénalité f. pour retard. / ~ for non-fulfilment of a contract ‖ Konventionalstrafe f. ‖ amende f. contractuelle. / ~ for non-performance of a contract see ~ for non-fulfilment of a contract.

pen and ink drawing ‖ Federzeichnung f. ‖ dessin m. à la plume.

pen block ‖ Federhalterablage f. ‖ poseplume m.

penbox, wooden ~ ‖ Federkästchen n. aus Holz ‖ boîte f. à plumes en bois.

pen boxer ‖ Schreibfedernpacker m. ‖ emboîteur m. de plumes.

pencil ‖ Schreibstift m.; Bleistift m. ‖ crayon m. / ~ (Geom; Opt) ‖ Büschel n. ‖ faisceau m. / ~ (Paint) ‖ Pinsel m. ‖ pinceau m. / artist's ~ ‖ Malerstift m. ‖ crayon m. pour artistes. / blue ~ ‖ Blaustift m. ‖ crayon m. bleu. / ~ for making carbon copies ‖ Durchschreibstift m. ‖ crayon m. pour faire les copies. / coloured ~ ‖ Farbstift m. ‖ crayon m. de couleur. / coloured copying ~ ‖ farbiger Kopierstift m. ‖ crayon m. de couleur à copier. / ~ with eraser ‖ Bleistift m. mit Radiergummi ‖ crayon m. avec gomme à effacer. / fitchet ~ ‖ Iltispinsel m. ‖ pinceau m. en poils de putois. / half red and half blue ~ ‖ halb rot und halb blauer Bleistift m. ‖ crayon m. mi-rouge et mi-bleu. / hard ~ ‖ harter Bleistift m. ‖ crayon m. dur. / hard rubber refilling ‖ Hartgummifüllbleistift m. ‖ porte-mine m. en ébonite. / lead ~ ‖ Bleistift m. ‖ crayon m. (noir). / ~ of light ‖ Lichtstrahlenbüschel n. ‖ faisceau m. de rayons. / ~ for painting ‖ Pinsel m. zum Malen ‖ pinceau m. ou brosse f. à peindre. / patent lead ~ ‖ Patentbleistift m. ‖ crayon m. breveté. / propelling ~ ‖ Drehbleistift m. ‖ porte-mine m. à vis. / ~ of rays ‖ Strahlenbüschel n. ‖ faisceau m. de lignes. / ~ of rays (Light) ‖ Lichtbüschel n. ‖ faisceau n. lumineux. / red ~ ‖ Rotstift m. ‖ crayon m. rouge. / ~ with red lead ‖ Bleistift m. mit roter Farbmine ‖ crayon m. à mine rouge. / slate ~ ‖ Schieferstift m.; Griffel m. ‖ crayon m. ardoise. / ~ of artificial slate ‖ Schreibstift m. aus künstlichem Schiefer ‖ crayon m. d'ardoise factice. / ~ of natural slate ‖ Schreibstift m. aus natür-

lichem Schiefer ‖ crayon m. d'ardoise naturelle. / soft ~ ‖ weicher Bleistift m. ‖ crayon m. mou.

pencil blue ‖ Kastenblau n.; Schilderblau n. ‖ bleu m. de pinceau ou d'application. / ~ drawing ‖ Bleistiftzeichnung f. ‖ dessin m. au crayon. / ~ factory ‖ Bleistiftfabrik f. ‖ manufacture f. de crayons. / ~ holder ‖ Bleistifthalter m. ‖ portecrayon m. / ~ industry ‖ Bleistiftindustrie f. ‖ industrie f. du crayon. / ~ machine ‖ Bleistiftmaschine f. ‖ machine f. à faire les crayons. / ~ maker ‖ Bleistiftarbeiter m. ‖ crayonnier m. / ~ point ‖ Einlage f. ‖ mine f. / ~ point for pair of compasses ‖ Bleieinsatz m. für Zirkel ‖ mine f. de crayon m. de compas. / ~ pointer see ~ sharpener. / ~ polisher ‖ Bleistiftpolierer m. ‖ polisseur m. de crayons. / ~ protector ‖ Bleistiftschoner m. ‖ protège-pointe m. pour crayon. / ~ sharpener ‖ Bleistiftanspitzer m. ‖ taille-crayon m. / blade for ~ sharpener ‖ Klinge f. für Bleistiftspitzer ‖ lame f. de taille-crayons. / ~ sharpening machine ‖ Bleistiftanspitzmaschine f. ‖ taille-crayons m. mécanique. / ~ stick ‖ Pinselstiel m. ‖ hampe f. de pinceau. / ~ trade machine ‖ Bleistiftindustriemaschine f. ‖ machine f. pour l'industrie des crayons.

pen cleaner ‖ Federreiniger m. ‖ essuie-plumes. / ~ cutter ‖ Schreibfederstanzer m. ‖ découpeur m. de plumes.

pendant (Lamp) ‖ Hängelampe f. ‖ lampe f. à suspension; suspension f. / ~ (Watch) ‖ Gehäuseknopf m. ‖ pendant m. / broad ~ (Mar) ‖ Stander m. ‖ guidon m.

pendant cord ‖ Pendelschnur f. ‖ cordon m. de suspension. / ~ winding watch ‖ Remontoiruhr f. ‖ montre f. à remontoir.

pendulous ‖ pendelartig ‖ pendulaire.

pendulum ‖ Pendel n. ‖ pendule m. / ~ (Clock) ‖ Pendel n.; Perpendikel n. ‖ pendule m. / to arrange after the fashion of a ~ ‖ pendelnd lagern ‖ articuler comme un pendule. / ballistic ~ ‖ ballistisches Pendel n. ‖ pendule m. balistique. / circular ~ ‖ konisches Pendel n. ‖ pendule m. conique. / compensated ~ ‖ Kompensationspendel n. ‖ pendule m. compensé. / compound ~ ‖ zusammengesetztes Pendel n. ‖ pendule m. composé. / conical ~ ‖ konisches Pendel n. ‖ pendule m. conique. / electric ~ ‖ elektrisches Pendel n. ‖ pendule m. électrique. / hydrometrical ~ ‖ hydrometrisches Pendel n.; Stromquadrant m. ‖ pendule m. hydrométrique. / simple ~ ‖ einfaches oder mathematisches Pendel n. ‖ pendule m. simple. / wooden bar ~ ‖ Holzstangenpendel m. ‖ pendule m. à tige en bois.

pendulum appliance for electric lamps ‖ Pendelvorrichtung f. für elektrische Lampen ‖ appareil m. à pendule pour lampes électriques. / ~ ball (Clockm) ‖ Pendellinse f. ‖ lentille f. / ~ bearing ‖ Pendellager n. ‖ palier m. de balanciers. / ~ bob see ~ ball.

pendulum clock ‖ Penduhr f. ‖ pendule f. / ~ hanging ~ ‖ Pendelwanduhr f. ‖ pendule f. murale. / standing ~ ‖ Pendelstanduhr f. ‖ pendule f. de cheminée.

pendulum, coupling box of a ~ ‖ Muffe f. eines Pendels ‖ manchon m. fixe d'un pendule. / ~ disturbance ‖ Pendel-

störung f. ‖ dérangement m. du pendule. / ~ grinding machine ‖ Hängeschleifmaschine f. ‖ machine f. à meuler oscillante. / ~ hydro-extractor ‖ Pendelzentrifuge f. ‖ essoreuse f. oscillante. / ~ level ‖ Pendelwage f. ‖ niveau m. à pendule. / ~ meter ‖ Pendelzähler m. ‖ compteur m. à pendule. / ~ mill ‖ Pendelmühle f. ‖ broyeur m. pendule. / oscillation of the ~ ‖ Pendelschwingung f. ‖ oscillation f. du pendule. / ~ ram impact testing machine ‖ Pendelschlagwerk n. ‖ mouton m. pendule. / ~ rod ‖ Pendelstange f. ‖ tige f. du pendule. / ~ saw ‖ Pendelsäge f. ‖ scie-pendule f. / swing of the ~ ‖ Pendelschwingung f. ‖ oscillation f. du pendule. / ~ wire ‖ geplätteter Stahldraht m. ‖ fil m. plat.

penetrate, to ‖ durchdringen; eindringen ‖ pénétrer.

penetrating power ‖ Durchdringungsvermögen n. ‖ force f. de pénétration. / ~ radiation ‖ durchdringende Strahlung f. ‖ rayonnement m. pénétrant. / ~ therapy ‖ Tiefentherapie f. ‖ thérapie f. pénétrante.

penetration ‖ Durchdringung f. ‖ pénétration f. / depth of ~ ‖ Eindringungstiefe f. ‖ profondeur f. de pénétration.

pen grinder ‖ Schreibfederschleifer m. ‖ aiguiseur m. de plumes. / ~ hardener ‖ Schreibfederhärter m. ‖ trempeur m. de plumes.

penholder ‖ Federhalter m. ‖ porte-plume m. / ~ with cork-tip ‖ Federhalter m. mit Korkgriff ‖ porte-plume m. à bout de liège. / hard rubber ~ ‖ Hartgummifederhalter m. ‖ porte-plume m. en ébonite. / ~ with rubber tip ‖ Federhalter m. mit Gummigriff ‖ porte-plume m. à bout de caoutchouc. / plug of the ~ ‖ Schreibfederträger m. ‖ galet m. du porte-plume.

peninsula ‖ Halbinsel f. ‖ presqu'île f.; péninsule f.

penknife ‖ Federmesser n.; Radiermesser n. ‖ canif m.; grattoir m.

Pennine ‖ Pennin m. ‖ pennine f.

penny-in-the slot ‖ Verkaufsautomat m. ‖ automate m.; distributeur m. automatique. / ~ electricity meter ‖ Elektrizitätsselbstverkäufer m. ‖ compteur m. électrique à prépaiement.

pennyroyal oil ‖ Poleiöl n. ‖ essence f. de pouliot.

pen point slitter ‖ Schreibfederspalter m. ‖ fendeur m. de plumes. / ~ polisher ‖ Schreibfederpolierer m. ‖ polisseur m. de plumes.

pension ‖ Pension f. ‖ pension f.

pension fund for workmen, their widows and orphans ‖ Arbeiter-Pensions-, Witwen- und Waisenkasse f. ‖ caisse f. de pensions de retraite pour ouvriers, leurs veuves et orphelins.

pen steel roller ‖ Schreibfederstahlwalzer m. ‖ lamineur m. d'acier pour plumes.

pentachlorethan ‖ Pentachloräthan n. ‖ pentachlorætane m.

pentagon ‖ Fünfeck n. ‖ pentagone m.

pentagonal ‖ fünfeckig; fünfkantig ‖ pentagonal.

pentatron ‖ Pentatron n. ‖ pentatron m. / ~ tube ‖ Fünfelektrodenröhre f. ‖ lampe f. pentatron.

penthouse ‖ Schauer n.; Schutzdach n. ‖ hangar m.; angar m.; échoppe f. ouverte.

pentlandite ‖ Eisennickelkies m. ‖ pentlandite f.

pen tray ‖ Federhalterablage f. ‖ poseplume m.

pent roof ‖ Halbdach n.; Pultdach n. ‖ comble m. à potence; toit m. en appentis *ou* à un seul égout.

penumbra ‖ Halbschatten m. ‖ pénombre f.

pen wiper ‖ Tintenwischer m. ‖ essuie-plumes m.

people, inexperienced ~ ‖ Nichtfachmann m. ‖ personnel m. non-specialisé. / ~'s **bank** ‖ Genossenschaftsbank f. ‖ banque f. coopérative.

peperine ‖ Peperin n. ‖ pépérine f.; pépérin m.

pepper ‖ Pfeffer m. ‖ poivre m. / **black ~ oil** ‖ Pfefferöl n. ‖ essence f. de poivre. / ~ **in corns** ‖ Pfeffer m. in Körnern ‖ poivre m. en grains. / **ground ~** ‖ gemahlener Pfeffer m. ‖ poivre m. moulu.

pepper box ‖ Pfefferbüchse f. ‖ poivrière f. / ~ **(grinding) mill** ‖ Pfeffermühle f. ‖ moulin m. à poivre.

peppermint ‖ Pfefferminz n. ‖ menthe f. poivrée. / ~ **drops** pl. ‖ Pfefferminzplätzchen npl. ‖ pastilles fpl. de menthe. / ~ **herb** ‖ Pfefferminze f. ‖ herbe f. de menthe poivrée. / ~ **liqueur** ‖ Pfefferminzschnaps m. ‖ liqueur f. de menthe poivrée. / ~ **oil** ‖ Pfefferminzöl n. ‖ essence f. de menthe poivrée.

pepsin ‖ Pepsin n. ‖ pepsine f.

peptone ‖ Pepton n. ‖ peptone f. / ~ **of meat** ‖ Fleischpepton n. ‖ peptone f. de viande.

peptonization ‖ Stärkeabbau m. ‖ dégradation f. de l'amidon. / ~ **temperature** ‖ Abbautemperatur f. ‖ température f. de dégradation *ou* de peptonisation.

pequin ‖ Pequin m. ‖ péquin m.; étoffe f. de Chine.

perambulator ‖ Kinderwagen m. ‖ voiture f. d'enfants; poussette f. / **doll's ~** ‖ Puppenwagen m. ‖ voiture f. de poupées.

perborate ‖ Perborat n. ‖ perborate m.

percale ‖ Perkal m. ‖ percale f.

percaline imbued with paste ‖ mit Kleister überzogenes Perkalin n. ‖ percaline f. enduite de colle.

perceive, to ‖ wahrnehmen ‖ apercevoir.

percent of grade ‖ Steigungsverhältnis n. in Vomhundert ‖ taux m. de la rampe en %. / **yield ~** ‖ Ertrag m. vom Hundert ‖ rendement m. pour cent.

percentage (Bank) ‖ Prozentsatz m.; Zinsfuß m. ‖ pourcentage m.; taux m. / ~ **(Chem)** ‖ Prozentgehalt m.; Gehalt m. ‖ teneur m. pour cent. / ~ **of acid** ‖ Säuregehalt m. ‖ pourcentage m. en acide. / ~ **of ashes** ‖ Aschengehalt m. ‖ teneur m. en cendres. / ~ **of carbon** ‖ Kohlenstoffgehalt m. ‖ teneur f. en carbone. / ~ **of coupling (Radio)** ‖ prozentuale Kopplung f. ‖ expression f. procentuelle du couplage. / **alloy of low ~** ‖ niedrigprozentige Legierung f. ‖ alliage m. à faible teneur. / ~ **of moisture** ‖ Feuchtigkeitsgehalt m. ‖ degré m. d'humidité. / ~ **of ozone in the air** ‖ Ozongehalt m. der Luft ‖ teneur m. *ou* richesse f. de l'air en ozone; quantité f. d'ozone dans l'air.

perceptible ‖ wahrnehmbar ‖ perceptible. / **scarcely ~ temperature** ‖ kaum wahrnehmbare Wärme f. ‖ température f. à peine perceptible.

perch (Carriage) ‖ Langbaum m.; Lenkbaum m. ‖ flèche f.; logne f. / ~ **(Surv)** ‖ Meßrute f.; Meßstange f. ‖ perche f.; règle f.; verge f.

percher (Weav) ‖ Warendurchseher m. ‖ percheur m.

perchlorate ‖ Perchlorat n. ‖ perchlorate m. / ~ **of ethylene** ‖ Perchloräthylen n. ‖ perchloréthylène m.

perchlorethylenum *see* perchlorate of ethylene.

perchloric acid ‖ Perchlorsäure f.; Überchlorsäure f. ‖ acide m. perchlorique.

perchloride of carbon ‖ Tetrachlorkohlenstoff m. ‖ tétrachloride m. de charbon. / ~ **of mercury** ‖ Quecksilberchlorid n. ‖ sublimé m. corrosif; perchlorure m. de mercure.

perch pin (Carriage) ‖ Langbaumvorstecker m. ‖ esse f. de flèche. / ~ **plate** ‖ Langbaumblech n.; Streichschiene f. ‖ plaque f. de flèche.

percolation basin ‖ Versickerungsbecken n. ‖ bassin m. d'infiltration.

percolator ‖ Seihetrichter m. ‖ entonnoir m. à filtre; filtre m.

percussion borer ‖ Stoßbohrer m. ‖ fleuret m. à percussion; barre f. à mine. / ~ **cap** ‖ Zündhütchen n.; Sprengkapsel f.; Zündkapsel f. ‖ amorce f. *ou* capsule f. fulminante. / **shell for ~ caps** ‖ Sprengkapselhülse f. ‖ capsule d'amorce. / ~ **drill** *see* ~ borer. / ~ **drilling machine** ‖ Stoßbohrmaschine f. ‖ perforatrice f. à percussion. / ~ **feeder** ‖ Stoßaufgabevorrichtung f. ‖ appareil m. chargeur à choc. / ~ **fuse** ‖ Perkussionszünder m. ‖ fusée f. à percussion. / ~ **instrument** ‖ Schlaginstrument n. ‖ instrument m. de frappe. / ~ **lever** ‖ Stoßhebel m. ‖ levier m. de frappe. / ~ **mill** ‖ Schlagkreuzmühle f. ‖ broyeur m. de percussion à crosse. / ~ **mortar** ‖ Diamantmörser m. ‖ mortier m. d'Abiche. / ~ **power** ‖ Stoßkraft f. ‖ force m. percutante *ou* de poussée.

perfect ‖ vollkommen; tadelfrei; tadellos ‖ parfait; irréprochable; sans défauts mpl. / ~ **combustion** ‖ vollkommene Verbrennung f. ‖ combustion f. complète. / ~ **modulation** ‖ vollkommene Modulation f. ‖ modulation f. parfaite.

perfect, to ‖ vervollkommnen ‖ perfectionner. / ~ **(Print)** ‖ widerdrucken ‖ imprimer au verso.

perfected ‖ vervollkommnet ‖ perfectionné.

perfecting (Print) ‖ Widerdruck m. ‖ retiration f.; réimpression f.; impression f. au verso. / ~ **cylinder (Print)** ‖ Widerdruckzylinder m. ‖ cylindre m. de deux *ou* de seconde. / ~ **form (Print)** ‖ Widerdruckform f. ‖ forme f. d'impression *ou* verso.

perfection ‖ Vollkommenheit f. ‖ perfection f.

perforate, to ‖ lochen; perforieren ‖ perforer.

perforated ‖ durchlöchert; gelocht; perforiert ‖ troué; perforé. / ~ **bottom** ‖ Siebboden m. ‖ fond m. à tamis. / ~ **girder** ‖ gelochter Träger m. ‖ poutre f. perforée. / ~ **iron plate** ‖ durchlochtes Eisenblech n. ‖ tôle f. perforée. / ~ **plate** ‖ gelochtes Blech n. ‖ tôle f. perforée. / ~ **sheet iron** ‖ Gitterblech n. ‖ tôle f. perforée. / ~ **strip recorder** ‖ Lochstreifenempfänger m. ‖ récepteur-perforateur m.

perforating device for postage stamps ‖ Briefmarkendurchlocheinrichtung f. ‖ dispositif m. de perforation des timbres. / ~ **effect** ‖ Durchschlagskraft f. ‖ force f. de pénétration. / ~ **machine** ‖ Perforiermaschine f. ‖ machine f. à perforer. / ~ **press** ‖ Perforierpresse f. ‖ presse f. à perforer. / ~ **rule** ‖ Perforierlinie f. ‖ filet m. à perforer.

perforation ‖ Durchlochung f. ‖ perforation f. / **alternating ~** ‖ Zickzacklochung f. ‖ perforation f. en quinconce. / ~ **placed in diagonals** ‖ diagonal stehende Lochung f. ‖ perforation f. en diagonale. / ~ **of the paper** ‖ Papierlochung f. ‖ perforation f. du papier.

perforator ‖ Locher m. ‖ perforateur m. / **hand ~** ‖ Handlocher m. ‖ perforateur m. à main.

perform, to ‖ leisten; ausführen ‖ faire; effectuer; accomplir; produire. / ~ **quarantine** ‖ Quarantäne f. halten ‖ être en quarantaine f. *ou* en seraine f.; faire la quarantaine.

performance, optical ~ ‖ optische Leistung f. ‖ action f. optique.

perfume ‖ Riechstoff m.; Duftstoff m.; Wohlgeruch m. ‖ parfum m. / **artificial ~** ‖ künstlicher Riechstoff m. *oder* Duftstoff m. ‖ parfum m. artificiel. / **to extract the ~** ‖ den Riechstoff m. ausziehen ‖ extraire le parfum. / **natural ~** ‖ natürlicher Riechstoff m. *oder* Wohlgeruch m. *oder* Duftstoff m. ‖ parfum m. naturel. / **synthetic ~** ‖ synthetischer Riechstoff m. ‖ parfum m. synthétique.

perfume, box of ~ ‖ Räucherbüchse f. ‖ boîte f. à parfums. / ~ **colour** ‖ Riechstofffarbe f. ‖ couleur f. pour la parfumerie.

perfumed benzene ‖ wohlriechendes Benzin n. ‖ benzine f. parfumée. / ~ **cigarette** ‖ parfümierte Zigarette f. ‖ cigarette f. parfumée. / ~ **fat** ‖ wohlriechendes Fett n. ‖ graisse f. parfumée. ‖ ~ **ointment** ‖ parfümierte Salbe f. ‖ onguent m. parfumé. / ~ **pomade** ‖ parfümierte Pomade f. ‖ pommade f. parfumée.

perfume distiller ‖ Riechstoffdestillierer m. ‖ distillateur-parfumeur m. / **factory of ~s** ‖ Parfümfabrik f. ‖ fabrique f. de parfumeries. / ~ **flask** ‖ Parfümflakon m. ‖ flacon m. à parfum. / ~ **label** ‖ Parfümetikett n. ‖ étiquette f. pour parfumeries. / ~ **powder crusher** ‖ Gewürzmühle f. ‖ broyeur m. d'aromates.

perfumer ‖ Parfümör m. ‖ parfumeur m.

perfumery ‖ Parfümerien fpl.; Duftstoffe mpl. ‖ parfumeries fpl. / **label for ~** ‖ Parfümerieetikett n. ‖ étiquette f. pour la parfumerie. / ~ **manufacture** ‖ Parfümerieherstellung f. ‖ fabrication f. de parfumerie. / ~ **plant** *see* factory of perfumes. / ~ **soap** ‖ Parfümerieseife f. ‖ savon m. parfumée.

perfumes pl. *see* perfumery.

perfume sprayer ‖ Parfümzerstäuber m. ‖ pulvérisateur m. de parfum.

pergamyn ‖ Pergamyn n. ‖ pergamyne m.

periclas ‖ Periclas m.; Periklas m. ‖ périclase f.

pericline ‖ Periklin m.; Albit m. ‖ péricline f.; péricline f.

peridot ‖ Peridot m. ‖ péridot m.

perihelion ‖ Perihelium n. ‖ périhélie m.

perihelium *see* perihelion.

perilla oil ‖ Perillaöl n. ‖ huile f. de pérille.

perimeter || Umfang m. || périmètre m. / ~ of a concession (Mine) || Schlagkreis m. || cercle m. *ou* la périmètre m. de la concession.

perimorphose || Perimorphose f. || périmorphose f.

perineal electrode (Med) || Dammelektrode f. || électrode f. périnéale.

period || Periode f.; Zeitabschnitt m. || période f. / ~ (Phys) || Schwingungsdauer f. || période f. || période f. de admission || Einströmperiode f. || période f. de pleine admission. / ~ of contact (Auto) || Eingriffsdauer f. || durée f. d'engrènement. / ~ of cooling || Abkühlungszeit f. || période f. de refroidissement. / idle ~ (Electr) || Leerlaufperiode f. || période f. en circuit ouvert. / ~ of lighting || Brennzeit f. || durée f. d'éclairage. / ~ of oscillation || Schwingungsdauer f. || durée f. d'oscillation. / ~ of presumptive loss (Mar) || Verschollenheitsfrist f. || délai m. de présomption de perte. / ~ of the rolling motion (Mar) || Rollperiode f. || période f. de roulis. / ~ of service || Benutzungsdauer f. || durée f. d'utilisation. / ~ of service on trial || Probedienstzeit f. || temps m. de service à l'essai.

periodical || periodisch || périodique. / ~ current || periodischer Strom m. || courant m. périodique. / ~ publication || periodische Veröffentlichung f. || publication f. périodique. / ~ supplementary charging (Acc) || Ladung f. mit Ruhepausen || chargement m. suivi de repos. / ~ uniform motion || gleichförmig wiederkehrende Bewegung f. || mouvement m. périodiquement uniforme.

periodical || Zeitschrift f. || revue f.; périodique f.

periodicity || Periodizität f. || fréquence f.; périodicité f. / ~ of movement || Bewegung f. in gleichen Zeiträumen; Periodizität f. der Bewegung || périodicité f. des mouvements.

periphery || Peripherie f.; Umfang m. || circonférence f.; périphérie f. / ~ of a circle || Kreisumfang m.; Kreisperipherie f. || circonférence f. du cercle.

periscope || Periskop n. || périscope m. / ~ for submarines || Sehrohr n. für Unterseeboote || périscope m. de sous-marin.

periscopic || periskopisch || périscopique. / ~ concave || periskopisch konkav || périscopique-concave. / ~ convex || periskopisch konvex || périscopique-convexe.

perishable || verderblich || périssable. / ~ goods pl. || leicht verderbliche Ware f. *oder* Lebensmittel npl. || marchandise f. sujette à se gâter; comestibles mpl. périssables.

peristyle || Peristyl m.; Säulengang m. || péristyle m.

periwig || Perücke f. || perruque f.

permanency of the ink || Beständigkeit f. der Tinte || durabilité f. de l'encre.

permanent || fest; dauernd; beständig; nachhaltig; bleibend || permanent; fixe; solide; durable; persistant. / ~ call || Dauerbelegung f.; Dauerruf m. || appel m. permanent. / ~ charging (Acc) || Dauerladung f. || charge f. de durée. / ~ deformation || dauernde Formänderung f. || déformation f. permanente. / ~ fire brigade || Berufsfeuerwehr f. || corps m. de sapeurs-pompiers. / ~ load || ständige Belastung f. || charge f. permanente.

/ ~ magnet || permanenter Magnet m.; Dauermagnet m. || aimant m. permanent. / ~ position || Lebensstellung f. || position f. pour la vie. / ~ service || Dauerbetrieb m. || service m. continu. / ~ set || Dehnungsrest m. || allongement m. permanent.

permanent way (Railw) || Bahnoberbau m.; Oberbau m. || superstructure f.; voie f. permanente. / machinery for constructing the ~ || Eisenbahnoberbaumaschinen fpl. || machines fpl. pour la superstructure de chemin de fer. / ~ material || Oberbaumaterial n. || matériel m. de voie. / dressing machine for ~ materials || Aufbereitungsmaschine f. für Eisenbahnoberbaustoffe || machine f. à préparer les matériaux de superstructure de chemin de fer.

permanent white || Barytweiß n.; Permanentweiß n. || blanc m. fixe *ou* permanent.

permanganate || Permanganat n. || permanganate m. / ~ of potassium || Kaliumpermanganat n. || permanganate m. de potassium.

permanganic acid || Übermangansäure f. || acide m. permanganique.

permeability || Permeabilität f.; Durchlässigkeit f. || perméabilité f. / initial ~ || Anfangspermeabilität f. || perméabilité f. initiale. / great magnetic ~ || große magnetische Permeabilität f. || haute perméabilité f. magnétique. / the ~ under low magnetizing force is remarkably high || die Permeabilität f. bei niedriger Magnetisierung ist bemerkenswert hoch || la perméabilité à faible aimantation est remarquablement grande. / ~ of a medium || Permeabilität f. eines Mediums || perméabilité f. d'un milieu. / reversible ~ || reversible Permeabilität f. || perméabilité f. réversible.

permeability curve || Permeabilitätskurve f. || courbe f. de perméabilité.

permeable || durchlässig || perméable. / ~ layer || wasserdurchlässige Schicht f. || couche f. perméable à l'eau. / ~ material || filterfähiger Werkstoff m. || matière f. filtrable.

permeameter || Durchlässigkeitsmesser m.; Permeameter n. || perméamètre m.

permission || Erlaubnis f. || permission f. / to sue for the ~ of working a mine || die Erlaubnis für den Abbau einer Grube beantragen || demander la concession d'une mine.

permit || Passierschein m.; Erlaubnisschein m. || laissez-passer m.; coupefile m.; permis m. / ~ of exportation || Ausfuhrerlaubnis f. || permis m. d'exportation. / government ~ || staatlicher Erlaubnisschein m. || permis m. de gouvernement. / official ~ || amtlicher Erlaubnisschein m. || permis m. officiel.

permutation || Permutation f. || permutation f. / ~ coupling (Railw) || Umsteckkupplung f. || attelage m. à griffes et à vis permutable.

Pernambuco wood || Fernambukholz n. || bois m. de Brésil *ou* de Fernambouc; brésillet m.

peroxide || Superoxyd n. || peroxyde m. / ~ of mercury || Quecksilberoxyd n.; Merkurioxyd n. || protoxyde m. *ou* bioxyde m. de mercure.

peroxol oxygen salt || Peroxolsauerstoffsalz n. || sel m. d'oxygène-peroxol.

perpender (Build) || Vollbinder m. || parpaing m.; pierre f. parpaigne.

perpendicular || senkrecht; lotrecht || à plomb m.; vertical; perpendiculaire. / ~ (Mine) || seiger || d'aplomb; perpendiculaire. / to draw a ~ line || ein Lot n. fällen || abaisser une perpendiculaire.

perpendicular || Lot n. || perpendiculaire f. / ~ (Phys) || Einfallslot n. || axe m. d'incidence; normale f. / ~ (Shipb; Clockm) || Perpendikel n. || perpendiculaire f. / bottom of a ~ || Fußpunkt m. eines Lotes || pied m. d'une perpendiculaire. / length between ~s || Länge f. zwischen den Loten || longueur f. entre perpendiculaires.

perpend stone (Build) || Vollbinder m. || parpaing m.; pierre f. parpaigne. / ~ wall || Mauer f. aus lauter Binderschichten || mur m. de parpaing.

perpetual snow || ewiger Schnee m. || neige f. éternelle.

perpetuum mobile || Perpetuum n. mobile || mouvement m. perpétuel.

perron || Freitreppe f. || perron m.

perrotine (Text print) || Perrotine f. || perrotine f.

perry || Birnenwein m. || poiré m.

Persian berry || Gelbbeere f. || graine fpl. de Perse. / ~ red || persischrot || rouge de Perse.

persist, to ~ in a refusal to carry-out duties || sich weigern, den obliegenden Verpflichtungen fpl. nachzukommen || se refuser de remplir ses obligations fpl.

person in arrears || Restant m. || reliquataire m. / ~ entitled to the use of water || Wasserberechtigter m. || ayant droits mpl. aux eaux.

personal delivery || eigenhändig || en mains fpl. propres. / ~ error || Beobachtungsfehler m. || erreur f. personnelle.

perspective || perspektivisch || perspectif. / ~ drawing || perspektivische Zeichnung f. || dessin m. perspectif. / ~ glass || Fernglas n.; Fernrohr n. || lunette f. d'approche. / ~ representation || perspektivische Darstellung f. || représentation f. perspective.

perspective || Perspektive f. || perspective f. / aerial ~ || Luftperspektive f. || perspective f. aérienne. / isometrical ~ || isometrische *oder* isoperimetrische Perspektive f. || espèce f. de vue d'angle. / linear ~ || Linearperspektive f. || perspective f. linéaire. / oblique ~ || Akzidentalperspektive f. || perspective f. en vue accidentelle. / parallel ~ || Parallelperspektive f. || perspective f. en vue de face. / principal ~ || Hauptperspektive f. || perspective f. principale.

perspiration leather || Schweißleder n. || cuir m. à chapeaux.

persulphuric acid || Überschwefelsäure f. || acide m. persulfurique.

perthite || Perthit m. || perthite f.

pertinax || Pertinax m. || pertinax m.

Peru balsam *see* Peruvian balsam.

Peruvian balsam || Perubalsam m. || baume m. du Pérou *ou* péruvien. / ~ oil || Perubalsamöl n. || huile f. de baume du Pérou.

Peruvian bark || Chinarinde f. || quinquina m.; quina m.

pervious || durchlässig || perméable. / ~ ground || durchlässiger Boden m. || sol m. perméable.

perviousness ‖ Durchlässigkeit f. ‖ perméabilité f.

pesage ‖ Wägegeld n. ‖ droits mpl. de pesage.

pessimist ‖ Schwarzseher m. ‖ pessimiste m.

pestilence ‖ Seuche f.; epidemische Krankheit; Pest f. ‖ peste f. maladie f. épidémique.

pestle ‖ Stößer m.; Stössel m. ‖ pilon m. de mortier. / ~ for smoothing hides ‖ Taucherstange f. ‖ enfonçoir m. / ~ stamp ‖ Stampfe f.; Stempel m. ‖ maillet m.; pilon m.

petal ‖ Blumenkronenblatt n. ‖ pétale m.

petalite ‖ Lithionfeldspat m.; Petalit m. ‖ feldspath m. à lithine; pétalite f.

pet cock for pumps ‖ Pumpenprobierhahn ‖ robinet m. d'essai de pompes.

petitgrain oil ‖ Petitgrainöl n. ‖ essence f. de petitgrain.

petrification ‖ Versteinerung f. ‖ pétrification f.

petrificative ‖ versteinernd ‖ pétrifiant.

petrified ‖ versteinert ‖ pétrifié.

petrify, to ‖ versteinern ‖ pétrifier.

petrifying ‖ versteinernd ‖ pétrifiant.

petrography ‖ Petrografie f.; Gesteinskunde f. ‖ pétrographie f.

petrol ‖ Benzin n. ‖ benzine f.; essence f. / heavy ~ ‖ Schwerbenzin n. ‖ essence f. lourde. / ~ cleaning centrifuge ‖ Benzinreinigungszentrifuge f. ‖ épurateur f. de benzine centrifuge. / ~ coke ‖ Petrolkoks m. ‖ coke m. de pétrole. / ~ consumption ‖ Benzinverbrauch m. ‖ consommation f. d'essence. / ~ cooking apparatus ‖ Benzinkocher m. ‖ appareil m. à cuire à essence. / ~ -electric motor car ‖ benzinelektrischer Triebwagen m. ‖ automotrice f. pétroléo-électrique. / ~ engine ‖ Benzinmotor m. ‖ moteur m. à essence.

petroleum ‖ Erdöl n.; Petroleum n. ‖ pétrole m.; naphte m.; bitume m. liquid. / the tanks containing the ~ are placed outside the hull of the boat ‖ das Petroleum n. ist in außenbords liegenden Behältern untergebracht ‖ le pétrole est chargé dans des réservoirs disposés à l'extérieur de la coque. / lamp ~ ‖ Leuchtpetroleum n. ‖ pétrole m. lampant. / raw ~ ‖ Roherdöl n.; rohes Petroleum n. ‖ pétrole m. brut; naphte m.; bitume m. liquide. / refined ~ ‖ gereinigtes Petroleum n. ‖ pétrole m. épuré ou raffiné.

petroleum burner ‖ Petroleumbrenner m. ‖ brûleur m. à pétrole. / ~ can ‖ Petroleumkanne f. ‖ bidon m. à pétrole. / ~ cooker ‖ Petroleumkocher m. ‖ fourneau m. à pétrole. / ~ distillation ‖ Petroleumdestillation f. ‖ distillation f. du pétrole. / ~ engine see ~ motor. / ~ ether ‖ Petroläther m. ‖ éther m. de pétrole. / ~ heating apparatus ‖ Petroleumofen m. ‖ fourneau m. à pétrole. / ~ incandescent lamp ‖ Petroleumglühlichtlampe f. ‖ lampe f. à incandescence au pétrole. / ~ incandescent light see ~ incandescent lamp. / ~ jelly ‖ Vaseline f. ‖ vaseline f. / ~ locomobile ‖ Petroleumlokomobile f. ‖ locomobile f. à pétrole. / ~ motor ‖ Petroleummotor m. ‖ moteur m. à pétrole. / ~ obtaining plant ‖ Petroleumindustrieanlage f. ‖ installation f. pour l'industrie du pétrole. / ~ pitch ‖ Petrolpech n. ‖ poix f. de pétrole. / ~ refinery ‖ Petroleumraffinerie f. ‖ raffinerie f. de pétrole. / ~ refining plant see ~ refinery. / ~

set-off device ‖ Petrolör m. ‖ pétroleur m. / ~ spirit ‖ Petroläther m. ‖ éther m. de pétrole. / ~ tank ‖ Petroleumbehälter m. ‖ soute f. ou cuve f. à pétrole. / ~ tank steamer ‖ Petroleumtankdampfer m. ‖ navire-pétrolier m.; navire-tank m. à pétrole; pétrolier m. / ~ tar ‖ Petroleumteer m. ‖ goudron m. de pétrole. / ~ tar distillation ‖ Petroleumteerdestillation f. ‖ distillation f. des goudrons de pétrole. / ~ testing apparatus ‖ Petroleumuntersuchungsapparat m. ‖ appareil m. d'examination de pétrole. / ~ well ‖ Petroleumbohrloch n.; Ölbohrung f. ‖ puits m. à pétrole.

petrol filter ‖ Benzinfilter m.; Benzinreiniger m. ‖ filtre m. à essence. / ~ funnel with filter ‖ Benzintrichter m. mit Sieb ‖ entonnoir m. à essence avec tamis. / ~ gauge ‖ Benzinstandanzeiger m. ‖ indicateur m. de niveau d'essence. / ~ heating apparatus ‖ Benzinofen m. ‖ fourneau m. à benzine. / ~ lamp ‖ Benzinlampe f. ‖ lampe f. à essence. / ~ level ‖ Benzinstand m. ‖ niveau m. d'essence. / ~ meter ‖ Kraftstoffmesser m.; Brennstoffuhr m. ‖ jauge f. d'essence. / ~ motor ‖ Benzinmotor m. ‖ moteur m. à essence. / ~ obtaining plant ‖ Benzinindustrieanlage f. ‖ installation f. pour l'industrie de la benzine.

petrological class and laboratory microscope ‖ mineralogisches Kurs- und Arbeitsmikroskop n. ‖ microscope m. minéralogique pour travaux pratiques et de laboratoire.

petrology / Gestein(s)kunde f.; Felskunde f.; Petrografie f. ‖ pétrographie f.; minéralogie f.; pétrologie f.

petrol pressure indicator ‖ Benzindruckmesser m. ‖ manomètre m. à essence. / ~ tank ‖ Benzinbehälter m. ‖ réservoir m. à essence. / ~ tank support ‖ Benzinbehälterstütze f. ‖ support m. du réservoir à essence. / ~ vault ‖ Benzinkeller m. ‖ cave f. à essence.

petrosilex ‖ Felsit m. ‖ pétrosilex m.

petticoat ‖ Unterrock m. ‖ jupon m.; cotillon m.

pewter (Dishes) ‖ Zinngeschirr n. ‖ vaisselle f. d'étain. / ~ (Metal) ‖ Hartzinn m.; Weißmetall n. ‖ potin m.; métal m. blanc. / grey ~ ‖ Graumetall n. ‖ potin m. gris.

pewter dishes pl. ‖ Zinngeschirr n. ‖ poterie f. d'étain.

phaeton ‖ Phaeton n. ‖ phaéton m.

phantom circuit ‖ Phantomleitung f.; Viererleitung f. ‖ ligne f. combinée. / ~ coil ‖ Viererspule f. ‖ bobine f. à circuits fantômes. / ~ coil plate ‖ Viererspulenplatte f. ‖ panneau m. de bobines fantômes. / ~ connection ‖ Viererschaltung f. ‖ connexion f. fantôme. / ~ operation ‖ Viererbetrieb m. ‖ fonctionnement m. en duplex. / ~ repeating coil ‖ Doppelsprechübertrager m. ‖ translateur m. de combinaison. / ~ telephone connection ‖ Doppelsprechschaltung f. ‖ connexion f. téléphonique en duplex. / ~ telephony ‖ Doppelsprechen n. ‖ double conversation f. / ~ transposition ‖ Platzwechsel m. ‖ croisement m. des circuits; rotation f.

pharmaceutical ‖ arzneilich; pharmazeutisch ‖ pharmaceutique. / ~ extract ‖ pharmazeutischer Extrakt m. ‖ extrait m. pharmaceutique. / cosmetic ~ industry ‖ kosmetisch pharmazeutische Industrie f. ‖ industrie f. cosmétique-pharmaceutique. / ~ machine ‖ pharmazeutische Maschine f. ‖ machine f. pharmaceutique. / ~ paper ‖ pharmazeutisches Papier n. ‖ papier m. pharmaceutique. / ~ preparation ‖ pharmazeutisches Präparat n. ‖ préparation f. pharmaceutique. / ~ product ‖ pharmazeutisches Produkt n. ‖ produit m. pharmaceutique. / synthetic ~ product ‖ synthetisches pharmazeutisches Erzeugnis n. ‖ produit m. médicamenteux synthétique. / ~ speciality ‖ pharmazeutische Spezialität f. ‖ spécialité f. pharmaceutique.

pharmacist ‖ Pharmazeut m.; Apotheker m. ‖ pharmacien m.; apothicaire m.

pharmacolite (Miner) ‖ Pharmakolith m. ‖ pharmacolithe f.

pharmacological microscope ‖ Mikroskop n. für Apotheker ‖ microscope m. pour pharmaciens.

pharmacy ‖ Pharmazie f.; Apotheke f. ‖ pharmacie f.

pharynx and œsophagus instrument ‖ Schlund- und Speiseröhreninstrument n. ‖ instrument m. pour le gosier et l'œsophage.

phase ‖ Phase f. ‖ phase f. / to be in ~ ‖ in gleicher Phase f. sein ‖ être en phase. / to be out of ~ ‖ in der Phase f. verschoben sein ‖ être déphasé. / ~ of the current ‖ Stromphase f. ‖ phase f. du courant. / to differ in ~ ‖ ungleiche Phase f. haben ‖ différer en phase f. / to lag in ~ ‖ in der Phase f. nacheilen ‖ être en retard de phase f. / to lead in ~ ‖ in der Phase f. voreilen ‖ être en avance de phase. / any load of ~s ‖ beliebig belastete Phasen fpl. ‖ phases fpl. non uniformément chargées. / equally loaded ~s ‖ gleichbelastete Phasen pl. ‖ phases fpl. uniformément chargées. / ~ of oscillation ‖ Schwingungsphase f. ‖ phase f. d'oscillation.

phase angle ‖ Phasenwinkel m. ‖ angle m. de phase. / angle of lead in ~ ‖ Phasenvoreilungswinkel m. ‖ angle m. d'avance de phase. / change of ~ (El line) ‖ Phasenmaß n. ‖ mesure f. angulaire. / ~ coincidence ‖ Phasengleichheit f. ‖ concordance f. de phase. / ~ constant ‖ Phasenkonstante f. ‖ mesure f. angulaire linéique. / ~ equality ‖ Phasengleichheit f. ‖ concordance f. de phases. / ~ equalizer ‖ Phasenausgleicher m. ‖ égalisator m. de phases. / ~ difference ‖ Phasenunterschied m. ‖ différence f. de phases. / angle of ~ difference ‖ Phasenverschiebungswinkel m. ‖ angle m. de déphasage. / ~ displacement ‖ Phasenverschiebung f. ‖ déphasage m. / leading ~ displacement, due to over-excitation ‖ voreilende Phasenverschiebung f. durch Übererregung ‖ déphasage m. en avant par surexcitation. / ~ displacement angle ‖ Phasenverschiebungswinkel m. ‖ angle m. de déphasage. / one ~ ground ‖ Phasenerdschluß m. ‖ terre f. accidentelle sur un des conducteurs. / interlinking of ~s ‖ Phasenverkettung f. ‖ accouplement m. des phases. / ~ lag ‖ Phasennacheilung f. ‖ retard m. de phases. / ~ lamp ‖ Phasenlampe f. ‖ lampe-témoin f. / lead in ~ ‖ Phasenvoreilung f. ‖ avance f. de phase. / leakage on one ~ see one ~ ground.

phasemeter ‖ Phasenmesser m. ‖ phasemètre m.

phase, velocity of ~ propagation ‖ Phasengeschwindigkeit f. ‖ vitesse f. de phase. / **~ recorder** ‖ Phasenschreiber m. ‖ phasemètre m. enregistreur. / **~ rule** ‖ Phasenregel f. ‖ règle f. des phases. / **~ selector** ‖ Phasenwähler m. ‖ commutateur m. de phases. / **~ shifter** ‖ Phasenverschieber m. ‖ décaleur m. de phases. / **~ transformer** ‖ Phasentransformator m. ‖ transformateur m. de phase. / **time of ~ transmission** ‖ Phasenlaufzeit f. ‖ temps m. de la progression de phase. / **~ voltage** ‖ Phasenspannung f. ‖ voltage m. ou tension f. de phase. / **~ voltmeter** ‖ Phasenvoltmesser m. ‖ phasenvoltmètre m.

pheasantry ‖ Fasanerie f. ‖ faisanderie. f.

pheasant shooter ‖ Fasanenjäger m. ‖ chasseur m. de faisans.

phelloplastics pl. ‖ Phelloplastik f.; Korbmodellierkunst f. ‖ phelloplastique f.

phenakite ‖ Phenakit m. ‖ phénakite f.

phenetidine ‖ Phenetidin n. ‖ phénétidine f. / **~ citrate** ‖ zitronensaures Phenetidin n. ‖ citrate m. de phénétidine.

phenic acid ‖ Phenol n. ‖ acide m. phénique.

phenicine ‖ Phönizinschwefelsäure f.; Purpurschwefelsäure f.; Indigopurpur m. ‖ phénicine f.

phenol ‖ Phenol n. ‖ phénol m. / **~ dye** ‖ Phenolfarbstoff m. ‖ colorant m. au phénol. / **~ ether** ‖ Phenoläther n. ‖ éther m. phénolique. / **~ soap solution** ‖ Phenolseifenlösung f. ‖ savon m. liquide phénolique.

phenomenon ‖ Erscheinung f.; Vorgang m. ‖ phénomène m. / **~ of discharge** ‖ Entladungserscheinung f. ‖ phénomène m. de décharge. / **~ of hysteresis** ‖ Nachwirkungserscheinung f. ‖ phénomène m. d'effet résiduel ou de fatigue ou d'hystérésis. / **~ of lag** see **~ of hysteresis.**

phenylacetic acid ‖ Phenylessigsäure f. ‖ acide m. phénylacétique.

phenylcinchonincarbonic acid ‖ Phenylcinchoninkarbonsäure f. ‖ acide m. phénylcinchonincarbonique.

phenylethylic alcohol ‖ Phenylaethylalkohol m. ‖ alcool m. phényléthylique.

phial ‖ Phiole f.; Ampulle f.; Fläschchen n. ‖ petite bouteille f.; flacon m.; fiole f.; ampoule f.

phillipsite ‖ Phillipsit m.; Kalkharmotom m. ‖ christianite f.; harmotome m. calcaire; phillipsite f.

philosophy, natural ~ ‖ Naturwissenschaft f. ‖ sciences fpl. naturelles.

phlegm ‖ Phlegma n. ‖ flegme m.

phlogopite ‖ Phlogopit m. ‖ phlogopite f.

phloroglucine ‖ Phlorogluzin n. ‖ phloroglucine f.

phœnicine ‖ Purpurschwefelsäure f.; Phönicinschwefelsäure f.; Phönicin n. ‖ phénicine f.

pholerite ‖ Pholerit m. ‖ pholérite f.

phone, to ‖ fernsprechen; telefonieren ‖ téléphoner. / **~ up** ‖ anrufen; antelefonieren ‖ appeler; téléphoner.

phone ‖ Fernsprechapparat m.; Telefon n. ‖ appareil m. téléphonique; téléphone m.

phonic wheel ‖ Tonrad n. ‖ roue f. phonique.

phonograph ‖ Fonograf m.; Grammofon n.; Sprechmaschine f. ‖ phonographe m.

phonolite ‖ Phonolith m.; Klingstein m. ‖ phonolithe m. / **~ tuff** ‖ Phonolithtuff m. ‖ phonolithe-tuf m.

phonoscope ‖ Phonoskop n. ‖ phonoscope m.

phorometer ‖ Phorometer n. ‖ phoromètre m.

phosgene ‖ Phosgen n.; Karbonylchlorid n. ‖ phosgène m.; oxychlorure m. de carbon. / **~ gas** ‖ Phosgengas n. ‖ gaz m. phosgène.

phosgenite ‖ Chlorbleispat m. ‖ phosgénite f.

phosphate ‖ Phosphat n. ‖ phosphate m. / **~ of ammonium** ‖ Ammoniumphosphat n. ‖ phosphate m. d'ammoniaque. / **~ of calcium** ‖ Kalziumphosphat n. ‖ phosphate m. calcique. / **~ of copper** ‖ Pseudomalachit m.; Phosphorkupfer n.; Phosphorkupfererz n. ‖ cuivre m. phosphaté. / **~ of iron** ‖ Eisenphosphat n. ‖ phosphate m. de fer. / **~ of lead** ‖ Pyromorphit m.; Grünbleierz n.; Braunbleierz n.; Phosphorblei n. ‖ pyromorphite f.; plomb m. phosphaté ou brun ou vert. / **~ of lime** ‖ phosphorsaurer Kalk m.; Phosphorit m. ‖ chaux f. phosphatée. / **~ of lime quarry** ‖ Phosphoritgrube f. ‖ carrière f. de phosphate de chaux. / **~ of soda** ‖ Natriumphosphat n. ‖ phosphate m. de soude. / **~ of yttrium** ‖ Ytterspat m. ‖ cénotime m.; phosphate m. d'yttrium.

phosphate-bearing ‖ phosphathaltig ‖ phosphaté.

phosphated earth, preparation of ~ ‖ Phosphaterdezubereitung f. ‖ préparation f. de phosphates bruts.

phosphate mill ‖ Phosphatmühle f. ‖ moulin m. à phosphates bruts; broyeur m. de phosphates. / **~ miner** ‖ Phosphoritgräber m. ‖ phosphatier m.

phosphatic slag ‖ phosphorhaltige Schlacke f. ‖ scorie f. phosphatée.

phosphide ‖ Phosphid n. ‖ phosphure m. / **hydrogen ~** ‖ Phosphorwasserstoff m. ‖ hydrogène m. phosphoré.

phosphite ‖ phosphorigsaures Salz n. ‖ phosphite m.

phosphorbronze ‖ Phosphorbronze f. ‖ bronze m. phosphoreux.

phosphoresce, to ‖ phosphoreszieren ‖ entrer en phosphorescence f.

phosphorescence ‖ Phosphoreszenz f. ‖ phosphorescence f. / **~ of the sea** ‖ Meeresleuchten n. ‖ phosphorescence f. de la mer.

phosphorescent ‖ phosphoreszierend ‖ phosphorescent.

phosphoric ‖ phosphorhaltig ‖ phosphoré.

phosphoric acid ‖ Phosphorsäure f. ‖ acide m. phosphorique. / **soluble ~** ‖ lösliche Phosphorsäure f. ‖ acide m. phosphorique soluble.

phosphoric acid determination ‖ Phosphorsäurebestimmung f. ‖ dosage m. de l'acide phosphorique.

phosphoric anhydride ‖ Phosphorsäureanhydrid n. ‖ anhydride m. phosphorique. / **~ pig** ‖ Phosphorroheisen n. ‖ fonte f. phosphoreuse. / **~ salt** ‖ Phosphat n.; phosphorsaures Salz n. ‖ phosphate m.

phosphorite ‖ Phosphorit m.; Apatit m. ‖ chaux f. phosphatée; apatite m. / **~ washing works** pl. ‖ Phosphoritwäsche f. ‖ laverie f. d'apatite.

phosphorized ‖ phosphorhaltig ‖ phosphoré.

phosphorous ‖ phosphorig ‖ phosphoreux. / **~ acid** ‖ phosphorige Säure f. ‖ acide m. phosphoreux. / **~ composition attendant** ‖ Phosphorteigbereiter m. ‖ chimiqueur m. / **~ ore** ‖ phosphorhaltiges Erz ‖ minerai m. phosphoreux. / **high-~ ore** ‖ hochphosphorhaltiges Erz n. ‖ minerai m. à haute teneur en phosphore.

phosphorus ‖ Phosphor m. ‖ phosphore m. / **amorphous ~** ‖ amorpher oder roter Phosphor m. ‖ phosphore m. rouge ou amorphe. / **free from ~** ‖ phosphorfrei ‖ exempt de phosphore m. / **metallic ~** ‖ metallischer Phosphor m. ‖ phosphore m. métallique. / **red ~** see amorphous **~.** / **white ~** ‖ weißer Phosphor m. ‖ phosphore m. blanc.

phosphorus chloride ‖ Chlorphosphor m. ‖ chlorure m. de phosphore. / **~ match** ‖ Phosphorzündhölzchen n. ‖ allumette f. phosphorique. / **~ metal** ‖ Phosphormetall n. ‖ métal m. phosphoreux. / **~ oxichloride** ‖ Phosphoroxychlorid n. ‖ oxychlorure m. de phosphore. / **~ pentasulphide** ‖ Phosphorpentasulfid n. ‖ pentasulfure m. de phosphore. / **~ removal of the ~** ‖ Phosphorabscheidung f. ‖ élimination f. du phosphore. / **sesquisulfide of ~** ‖ Phosphorsesquisulfid n. ‖ sesquisulfure m. de phosphore. / **~ trichloride** ‖ Phosphortrichlorid n. ‖ trichlorure m. de phosphore. / **~ trisulphide** ‖ Phosphortrisulfid n. ‖ trisulfure m. de phosphore.

phosphotungstic acid ‖ Phosphorwolframsäure f. ‖ acide m. phosphotungstique.

photo, to see to photograph.

photo see photograph.

photo album ‖ Lichtbilderalbum n. ‖ album m. pour photos.

photochemical ‖ fotochemisch ‖ photochimique.

photochemically active blue violet region of the spectrum ‖ fotografisch wirksamer blauvioletter Teil m. des Spektrums ‖ région f. bleu-violette du spectre dont l'action sur la plaque photographique est la plus intense.

photochemistry ‖ Fotochemie f. ‖ photochimie f.

photochromatic ‖ fotochromatisch ‖ photochromatique.

photochrome ‖ Fotochrom n. ‖ photochromie f.

photoceramic establishment ‖ fotokeramische Anstalt f. ‖ atelier m. photocéramique.

photo copying apparatus ‖ fotografischer Kopierapparat m.; Lichtpausapparat m. ‖ appareil m. photographique à copier; appareil m. à tirer les bleus.

photo dryer ‖ Fototrockner m. ‖ appareil sécheur m. de photos.

photo-electric ‖ lichtelektrisch ‖ photoélectrique. / **~ cell** ‖ fotoelektrische Zelle f. ‖ cellule f. ou élément m. photoélectrique. / **~ cell amplifier** ‖ Fotozellenverstärker m. ‖ amplificateur m. de cellule photoélectrique.

phot oengraving ‖ Lichtdruck m.; Fotogravüre f. ‖ photogravure f.

photo enlarger ‖ Lichtbildvergrößerer m. ‖ agrandisseur m. photographe. / **studio for ~ enlarging** ‖ Lichtbildvergrößerungsanstalt f. ‖ atelier m. d'agrandissements photographiques.

photogalvanography ‖ Fotogalvanografie f. ‖ photogalvanographie f.

photogen ‖ Fotogen n. ‖ photogène m.
photogrammetric working-out apparatus ‖ fotogrammetrisches Auswertungsgerät n. ‖ ustensile m. d'évaluation photogrammétrique.
photogrammetry ‖ Lichtbildauswertung f. photogrammétrie f.
photograph, to ‖ aufnehmen; fotografieren ‖ photographier.
photograph ‖ Lichtbild n.; Fotografie f. ‖ photographie f. / restitution of aerial ~s ‖ Entzerrung f. von Flugaufnahmen ‖ restitution f. des photographies aériennes. / daylight ~ ‖ Tagesaufnahme f. ‖ photographie f. diurne ou pendant le jour. / night ~ ‖ Nachtaufnahme f. ‖ photographie f. noctale ou pendant la nuit. / propaganda ~ ‖ Werbeaufnahme f.; Reklamefotografie f. ‖ photographie f. de réclame.
photographer ‖ Fotograf m. ‖ photographe m. / ~'s studio ‖ fotografisches Atelier n. ‖ atelier m. de photographie.
photograph frame ‖ Lichtbildrahmen m. ‖ cadre m. de photographie. / ~ glazing ‖ Lichtbildsatinieren n. ‖ satinage m. de photographies. / ~ glazer ‖ Lichtbildsatinierer m.; Fotografiesatinierer m. ‖ satineur m. de photographies.
photographic ‖ fotografisch ‖ photographique. / ~ aeroplane ‖ Luftbildflugzeug n. ‖ avion m. de photographie. / ~ apparatus ‖ fotografischer Apparat m. ‖ appareil m. photographique. / ~ article ‖ fotografischer Artikel m. ‖ article m. pour la photographie. / ~ artist ‖ Kunstfotograf m. ‖ artiste m. photographe. / ~ assistant ‖ Kopierer m. ‖ tireur m. / ~ basin of cardboard ‖ Fotografieschale f. aus Pappmaché ‖ bassine f. de carton-pâte pour photographes. / ~ cardboard ‖ fotografischer Karton m. ‖ carton m. pour la photographie. / ~ chemicals pl. ‖ fotografische Chemikalien fpl. ‖ produits mpl. chimiques pour la photographie. / ~ dry plate ‖ fotografische Trockenplatte f. ‖ plaque f. photographique sèche. / ~ enamel ‖ fotografisches Schmelzbild n.; Emaillephotographie f. ‖ émail m. photographique. / ~ eyepiece ‖ fotografisches Okular n. ‖ oculaire m. photographique. / ~ glass scale ‖ Glasdiapositiv n. ‖ diapositive f. en verre. / ~ implement ‖ fotografischer Bedarfsartikel m. ‖ accessoire m. photographique. / ~ impression ‖ fotografischer Aufdruck m. ‖ impression f. photographique. / ~ industry ‖ fotografische Industrie f. ‖ industrie f. photographique. / ~ instrument ‖ fotografisches Instrument n. ‖ instrument m. de photographie. / ~ magnifying apparatus ‖ fotografischer Vergrößerungsapparat m. ‖ appareil m. d'agrandissement pour la photographie. / ~ object glass ‖ fotografisches Objektiv n. ‖ objectif m. photographique. / ~ objective see ~ object glass. / ~ paper ‖ fotografisches Papier n. ‖ papier m. photographique. / ~ paper covered with a reproducing paste ‖ fotografisches Papier n. mit einer Reproduktionsmasse überzogen ‖ papier m. photographique recouvert d'une pâte à reproduire. / ~ plate ‖ fotografische Platte f. ‖ plaque f. photographique. / ~ plate of glass ‖ fotografische Platte f. aus Glas ‖ plaque f. en verre pour la photographie. / ~ plate photometer ‖

Schwärzungsfotometer n. ‖ photomètre m. à noircissement. / ~ printing ‖ Lichtpausverfahren n. ‖ procédé m. photographique à copier. / ~ puzzle ‖ Vexierfotografie f. ‖ photographie f. à surprise. / ~ recording ‖ fotografische Aufschreibung f. ‖ enregistrement m. photographique. / ~ refractor ‖ fotografischer Refraktor m. ‖ réfracteur m. photographique. / ~ surveying ‖ Fotogrammetrie f.; Lichtbildmeßkunst f.; Lichtbildvermessung f. ‖ photogrammétrie f. / ~ telescope with ultraviolet triplets ‖ fotografisches Fernrohr n. mit Ultraviolettriplets ‖ lunette f. photographique munie de triplets ultraviolets. / ~ trade machine ‖ Fotoindustriemaschine f. ‖ machine f. pour l'industrie photographique.
photography ‖ Fotografie f.; Aufnahme f. (Verfahren) ‖ photographie f. / aerial ~ ‖ Fliegerfotografie f. ‖ photographie f. aérienne. / astro ~ ‖ Astrofotografie f. ‖ astrophotographie f. / industrial ~ ‖ Industriefotografie f. ‖ photographie f. industrielle. / ~ of lightning ‖ Blitzfotografie f. ‖ photographie f. d'éclair. / ~ of the sun's corona ‖ Sonnenkoronaaufnahme f. ‖ photographie f. de la couronne solaire. / ~ with ultraviolet light ‖ Fotografie f. mit ultraviolettem Licht ‖ photographie f. en lumière ultra-violette.
photogravure ‖ Fotogravüre f. ‖ photogravure f. / universal one-reel ~ rotary machine ‖ Universaleinrollentiefdruckrotationsmaschine f. ‖ machine f. rotative en creux universelle à une bobine.
photolithograph, to ‖ fotolithografieren ‖ photolithographier.
photolithograph ‖ Fotolithograf m. ‖ photolithographe m.
photolithography ‖ Fotolithografie f.; Lichtsteindruck m. ‖ photolithographie f.; phototypie f.
photolitho paper ‖ Lichtdruckpapier n. ‖ papier m. pour phototypie.
photomechanical apparatus ‖ fotomechanischer Apparat m. ‖ appareil m. photomécanique. / ~ establishment for ~ process ‖ Reproduktionsanstalt f. ‖ atelier m. de reproduction photomécanique. / ~ process ‖ fotomechanisches Verfahren n.; Reproduktionsverfahren n. ‖ procédé m. photomécanique ou de reproduction.
photometer, to ‖ fotometrieren ‖ photométrer.
photometer ‖ Fotometer n.; Lichtmesser m. ‖ photomètre m. / flicker ~ ‖ Flimmerfotometer n. ‖ photomètre m. à éclats. / glow lamp ~ ‖ Glühlampenfotometer m. ‖ photomètre m. pour lampes à incandescence. / gradation ~ ‖ Stufenfotometer n. ‖ photomètre m. graduel. / grease spot ~ ‖ Fettfleckfotometer m. ‖ photomètre m. à tache d'huile. / photographic plate ~ ‖ Schwärzungsfotometer n. ‖ photomètre m. à noircissement. / recording photo-electrical ~ ‖ lichtelektrisches Registrierfotometer n. ‖ photomètre m. enregistreur photo-électrique. / retinal ~ see visual ~. / visual ~ ‖ Lichtsinnprüfer m. ‖ appareil m. pour examiner la sensibilité pour la lumière.
photometric ‖ fotometrisch ‖ photométrique. / ~ bench ‖ Fotometerbank f. ‖ banc m. photométrique. / ~ measure-

ment ‖ Fotometrierung f. ‖ photométrisation f.
photometry ‖ Fotometrie f.; Lichtmessung f. ‖ photométrie f. / spectro ~ ‖ Spektralfotometrie f. ‖ photométrie f. spectrale.
photomicrograph ‖ Mikrofotogramm n. ‖ microphotogramme f.
photomicrographic camera ‖ mikrofotografischer Apparat m. ‖ appareil m. de microphotographie. / ~ mechanical stage ‖ mikrofotografischer Kreuztisch m. ‖ platine f. à chariot de microphotographie.
photomicrography ‖ Mikrofotografie f. ‖ microphotographie f.
photo paper see photographic paper.
photophone ‖ Fotofon n. ‖ photophone m.
photoprinting on china ‖ Porzellanfotografie f. ‖ photocéramique f.
phototelegraph ‖ Fototelegraf m. ‖ phototélégraphe m.
phototheodolite ‖ Fototheodolit m. ‖ photothéodolite m.
phototype high speed press ‖ Lichtdruckschnellpresse f. ‖ presse f. rapide pour phototypie. / ~ printing establishment ‖ Lichtdruckanstalt f. ‖ atelier m. d'impression phototypique.
phototypography ‖ Fototypografie f. ‖ phototypographie f. / half-painted ~ ‖ Halbtonätzung f. ‖ phototypographie f. en demi-teinte.
photozincography ‖ Fotozinkografie f. ‖ photozincographie f.
phthalic acid ‖ Phthalsäure f. ‖ acide m. phtalique.
phthisis of lead smelters ‖ Bleikolik f. ‖ colique f. de plomb.
physical ‖ körperlich ‖ corporel. / ~ (Phys) ‖ physikalisch ‖ physique. / ~ apparatus ‖ physikalischer Apparat m. ‖ appareil m. de physique. / ~ condition ‖ Aggregatzustand m. ‖ état m. physique. / ~ exertion ‖ körperliche Anstrengung ‖ fatigue f. corporelle. / ~ and chemical formula ‖ physikalisch-chemische Formel f. ‖ formule f. physique et chimique. / ~ instrument ‖ physikalisches Instrument n. ‖ instrument m. de physique. / chemical and ~ laboratory for scientific and technical research ‖ chemisch-physikalische Versuchsanstalt f. ‖ laboratoire m. de recherches chimiques-physiques. / ~ quality ‖ physikalische Eigenschaft f. ‖ propriété f. physique. / institute of ~ science ‖ physikalisches Institut n. ‖ institut m. de physique.
physician ‖ Arzt m. ‖ médecin m.
physicist ‖ Physiker m. ‖ physicien m.
physicochemical ‖ physikalisch-chemisch ‖ physico-chimique.
physics pl. ‖ Physik f. ‖ physique f.
physiological ‖ physiologisch ‖ physiologique. / ~ analysis ‖ physiologische Analyse f. ‖ analyse f. physiologique. / ~ product ‖ physiologisches Präparat n. ‖ produit m. physiologique.
physiologist for colour investigation ‖ Farbenphysiologe m. ‖ physiologiste m. de la théorie des couleurs.
phytochemistry ‖ Pflanzenchemie f. ‖ chimie f. végétale; phytochimie f.
piano ‖ Klavier n. ‖ piano m. / automatic ~ ‖ selbstspielendes Klavier n. ‖ piano m. automatique. / mechanical ~ see automatic ~. / player ~ see automatic ~.

/ toy ~ ‖ Kinderklavier n. ‖ piano m. pour enfants.

piano accessories pl. ‖ Klavierzubehör f. ‖ fournitures fpl. pour pianos. / ~ case ‖ Klaviergehäuse n. ‖ caisse f. du piano. / ~ case maker ‖ Klavierschreiner m. ‖ caissier m. en piano. / ~ chair ‖ Klaviersessel m. ‖ tabouret m. de pianos. / ~ cloth ‖ Klaviertuch n. ‖ drap m. à piano. / ~ cord ‖ Klaviersaite f. ‖ corde f. à piano. / ~ fittings pl. ‖ Klavierbeschläge mpl. ‖ garnitures fpl. de piano. / ~ fittings fixer ‖ Klavierbeschlaganschläger m. ‖ ferreur m. de piano.

pianoforte see piano.

piano hammer ‖ Klavierhammer m. ‖ marteau m. de piano. / ~ hammer leather ‖ Klavierhammerleder n. ‖ cuir m. pour les marteaux de piano. / ~ parts pl. and accessories pl. for the ~ industry ‖ Klavierbaubestandteile mpl. ‖ pièces fpl. détachées de pianos. / ~ lamp ‖ Klavierlampe f. ‖ lampe f. de piano. / ~ maker ‖ Klavierbauer m. ‖ facteur m. de pianos. / ~ maker's bit ‖ Klavierbauerbohrer m. ‖ mèche f. pour facteur de pianos. / ~ maker's nippers pl. ‖ Klavierbauerzange f.; Klaviersaitenzange f. ‖ pince f. de facteur de piano. / ~ mechanism ‖ Klaviermechanik f. ‖ mécanique f. pour pianos. / ~ moving ‖ Klaviertransport m. ‖ transport de pianos m. / ~ plate ‖ Pianoplatte f. ‖ planche f. pour pianos. / ~ stool see ~ chair. / stop pin for ~s ‖ Klavierstift m. ‖ goujon m. pour pianos. / ~ tuner ‖ Klavierstimmer m. ‖ accordeur m. de pianos. / ~ wire ‖ Klaviersaitendraht m. ‖ corde f. à piano. / ~ worker ‖ Klavierarbeiter m. ‖ ouvrier m. en pianos.

pianzite ‖ Pianzit m. ‖ pianzite f.

piassava ‖ Pikaba f.; Piassava f. ‖ picaba f.; piazzava f. / ~ broom ‖ Piassavabesen m. ‖ balai m. de piazzava. / ~ brush ‖ Piassavabürste f. ‖ brosse f. en piazzava. / ~ fibre ‖ Piassavafaser f. ‖ fibre f. de piazzava.

pic ‖ Spitzhacke f. ‖ pioche f.

pica (Print) ‖ Cicero m. ‖ cicéro m. / double ~ (Print) ‖ Sekunda f.; Textschrift f. ‖ gros-paragon m.; texte m. / eight-lines ~ (Print) ‖ Real f. ‖ double-canon m. / ~ mine-lines (Print) ‖ Imperial f. ‖ gros double canon m.; triple canon m.; soixante-douze m. / small ~ (Print) ‖ Garmond f. ‖ corps-dix m. / two-lines double ~ (Print) ‖ Mittelkanon m.; Saban f.; Trismegist m. ‖ trismégiste m.

piccolo (Mus) ‖ Pikkolo m.; Pickelflöte f. ‖ petite flûte f.

pick, to (Cloth) ‖ noppen; belesen ‖ épincer; épinceler; épinceter. / ~ (Mine) ‖ ausklauben; klauben ‖ épierrer; trier à la main. / ~ the hides ‖ die Felle n. verlesen ‖ éplucher les peaux f. / ~ a lock ‖ ein Schloß n. mit dem Dietrich öffnen ‖ crocheter une serrure. / ~ ores pl. ‖ Erz n. klauben oder scheiden ‖ trier le minerai m. à la main. / ~-out (Ore dress) ‖ ausklauben; auslesen ‖ trier. / ~-out the letters ‖ Buchstaben mpl. herausheben ‖ piquer les lettres fpl. / ~-out slate (Coal) ‖ Steine mpl. auslesen ‖ épierer le charbon. / ~ to pieces ‖ zerpflücken ‖ effeuiller. / ~-up aufheben ‖ rammasser; lever. / ~-up (Mach) ‖ auf Touren fpl. kommen ‖ remonter. / ~-up the cable ‖ das Telegrafenkabel auf-

holen ‖ retirer le câble. / ~-up the sound ‖ den Ton m. aufnehmen ‖ recueillir le son. / ~ the wool ‖ die Wolle pflücken oder verlesen oder auszupfen ‖ épluser ou pluser ou trier ou égraterroner la laine.

pick ‖ Hacke f.; Spitzhacke f. ‖ pioche f.; pic m. / ~ (Mine) ‖ Haue f.; Keilhaue f. ‖ pioche f. / ~ (Print) ‖ Spieß m.; Schmutzbuchstabe f. ‖ espace f. haute. / ~ (Railw) ‖ Stopfhacke f. ‖ pioche f. à bourrer. / ~ (Weav) ‖ Schützenschlag m.; Schuß m.; Schlag m. ‖ passée f. / beater ~ (Railw) ‖ Stopfhacke f. ‖ pioche f. à bourrer. / double ~ (Mine) ‖ Doppelspitzhacke f. ‖ pic m. à deux pointes. / double ~ (Railw) ‖ Doppelstopfhacke f. ‖ pioche f. à bourrer à deux pointes. / double-pointed ~ (Stone cutter) ‖ Steinpickel m. mit zwei Spitzen ‖ grosse pointe f. / flat ~ ‖ Stopf- und Spitzhacke f. ‖ pioche f. et pic m. / mandolin ~ ‖ Mandolinenplättchen n. ‖ médiateur m. pour mandoline. / pneumatic ~ ‖ Preßluftspitzhacke f. ‖ pic m. à air comprimé. / single ~ ‖ einfache Spitzhacke f.; Flachhacke f. ‖ pic m. ou pioche f. simple. / tampering ~ ‖ Stopfhacke f. ‖ pic m. à bourrer. / universal ~ ‖ Doppelkeilhacke f. ‖ marteau m. à (deux) pointes; pic m. à deux branches.

pickaxe see pick.

picked almonds pl. ‖ ausgesuchte Mandeln fpl. ‖ amandes fpl. triées à la main. / ~ man ‖ ausgesuchter Arbeiter m. ‖ ouvrier m. consommé.

picker (Ore dress) ‖ Klaubhammer m. ‖ marteau m. de triage. / ~ (Weav) ‖ Webvogel m.; Treibvogel m. ‖ taquet m. / ~ of a ribbon loom ‖ Rechen m. oder Treiber m. eines Bandwebstuhles ‖ chasse-navettes m. d'un métier à rubans. / thread ~ ‖ Fadenklauber m. ‖ éplucheuse f. de filaments.

pickeringite ‖ Magnesiaalaun m.; Talkerdealaun m. ‖ alun m. de magnésie.

picket, to ‖ abpfählen ‖ piqueter.

picket (Surv) ‖ Absteckpfahl m.; Pflock m. ‖ piquet m.; jalon m. / ~ (Vineyard) ‖ Pfahl m.; Weinbergspfahl m.; Winzerspfahl m. ‖ pieu m.; échalas m.

pick hammer ‖ Brechhammer m. ‖ têtu m.; décintroir m. / ~ (Ore dress) ‖ Klaubhammer m. ‖ marteau m. de triage. / ~ (Stone cutter) ‖ Spitzhaue f. mit einer Spitze ‖ marteline f.

pick helve ‖ Hackenstiel m. ‖ manche m. de pioche.

picking of wool ‖ Entkletten n. der Wolle ‖ épaillage m. ou échardonnage m. ou égratteronnage m. de laine.

picking band ‖ Sortierband n.; Leseband n. ‖ bande f. de triage. / ~ and loading belt ‖ Lese- und Verladeband n. ‖ bande f. de triage et de chargement. / ~ drum ‖ Läutertrommel f. ‖ tambour m. de triage. / ~-out slate (Coal) ‖ Steinausklaubung f. ‖ épierrage m. / ~ plant ‖ Sortieranlage f. ‖ atelier m. de triage. / ~ table (Ore dress) ‖ Lesetisch m.; Klaubetisch m. ‖ table f. de triage. / ~-up of motor ‖ Auftourenkommen n. des Motors m. ‖ reprise f. du moteur. / ~-up of a submarine cable ‖ Hebung f. eines Kabels ‖ relèvement m. d'un câble.

pickle, to ‖ pökeln ‖ mariner. / ~ (Metal) ‖ abbeizen; beizen ‖ décaper; dérocher. / ~ brass ‖ Messing n. abbrennen ‖ relever la couleur du laiton.

pickle ‖ Salzlake f. ‖ saumure f.

pickled cucumber ‖ Salzgurke f. ‖ concombre m. au sel.

pickler ‖ Essigkonservenfabrikant m. ‖ fabricant m. de pickles. / ~ (Metal) ‖ Beizer m. ‖ dérocheur m.; décapeur m. / wire ~ ‖ Drahtbeizer m. ‖ décapeur m. des fils de fer.

pickling ‖ Abbeizen n. ‖ décapage m.; dérochage m. / ~ basket ‖ Beizkorb m. ‖ panier m. de décapage. / electrolytical ~ bath ‖ elektrolytisches Beizbad n. ‖ bain m. de dérochage électrolytique. / ~ compound ‖ Beizmittel n. ‖ produit m. décapant. / ~ crate see ~ basket. / ~ machine ‖ Beizmaschine f. ‖ machine f. à décaper. / ~ medium (Clothm) ‖ Dekapiermittel n. ‖ produit m. à décaper. / ~ plant ‖ Beizanlage f. ‖ installation f. de décapage. / plate ~ plant ‖ Plattenbeizanlage f. ‖ installation f. à décaper les tôles. / ~ tub ‖ Pökelkufe f. ‖ saloir m. / ~ tub (Metal) ‖ Beizbottich m. ‖ bac m. de décapage.

picklock ‖ Dietrich m.; Nachschlüssel m. ‖ fausse clef f.; rossignol m.; crochet m.

pick-off gear type ‖ Wechselradtyp m. ‖ modèle m. de roues de rechange.

pick pincers pl. (Clothm) ‖ Noppeisen n. ‖ épincette f.; brucelles fpl.

picks pl. (Print) ‖ Putzen mpl.; Zwiebelfische mpl. ‖ pâtés mpl. / ~ (Weav) ‖ Schüsse mpl. ‖ passées fpl.

pick-up ‖ Tonabnehmer m. ‖ pick-up m. / ~ device ‖ Tonabnahmevorrichtung f. ‖ dispositif m. de captage du son. / ~ system ‖ Seilpost f. ‖ transporteur m. à corde.

picot edge (Textile) ‖ Zackenrad n. ‖ revers m. à picot.

picotite ‖ Chromspinell m.; Picotit m. ‖ picotite f.

picrate ‖ Pikrat n. ‖ picrate m.

picric acid ‖ Pikrinsäure f. ‖ acide m. picrique. / ~ powder ‖ Pikratpulver n. ‖ poudre f. picrique.

picrolite ‖ Pikrolith m.; Serpentin m. ‖ picrolite f.; serpentin m.

picrosmine ‖ Pikrosmin m. ‖ picrosmine f.

pictorial postcard see picture postcard.

pictorial ‖ illustrierte Zeitung f. ‖ journal m. illustré.

picture ‖ Bild n.; Gemälde n. ‖ image f.; peinture f.; tableau m. / ~ (Film) ‖ Film m.; Filmbild n.; Filmstreifen m. ‖ film m. / ~s pl. (Film) ‖ Film m.; Filmvorstellung f.; Filmspiel n. ‖ film m. / handmade ~ ‖ handgemaltes Bild n. ‖ tableau m. fait à la main. / motion or moving ~ see ~s (Film). / negative ~ ‖ Negativ n. ‖ épreuve f. négative. / photographic ~ ‖ Lichtbild n.; Fotografie f. ‖ épreuve f. photographique. / ~ on porcelain ‖ Porzellanbild n. ‖ peinture f. sur porcelaine. / positive ~ ‖ Positiv n. ‖ épreuve f. positive. / silent ~s pl. ‖ stummer Film m. ‖ film m. muet. / talking ~s pl. ‖ Sprechfilm m.; Tonfilm m.; tönender Film m. ‖ film m. parlant.

picture book ‖ Bilderbuch n. ‖ livre m. d'images. / ~ eye ‖ Bilderöse f. ‖ œillet m. d'images. / ~ frame ‖ Bilderrahmen m. ‖ cadre m. pour tableaux. / ~ framer ‖ Bildereinrahmer m. ‖ encadreur m. de tableaux. / ~ gallery ‖ Gemäldegalerie f.; Bildergalerie f. ‖ galerie f. de tableaux / ~ house ‖ Lichtspieltheater n.; Kino n.; Lichtspiele npl. ‖ cinématographe m.;

cinéma m. / ~ palace see ~ house. / ~ play || Film m.; Filmspiel n. || film m. / ~ post card || Ansichtspokarte f. || carte f. postale illustrée. / ~ print || Bilderbogen m. || feuille f. d'image. / ~ projector || Bildprojektor m.; Projektionsapparat m. || appareil-projecteur m. d'image. / motion ~ projector || Kinovorführungsapparat m. || appareil m. de projection de cinéma. / ~ radio telegraphy || Bild(rund)funk m. || phototélégraphie f. sans fil; transmission f. d'images à distance par télégraphie sans fil. / rectification of the ~ || Entzerrung f. der Bildaufnahme || correction f. de la distorsion de la photographie. / restitution of the ~ see rectification of the ~. / ~ restorer || Gemälde-restaurateur m. || retaurateur m. de tableaux. / ~ restoring || Gemälderestauration f. || restauration f. de tableaux. / ~ room see ~ gallery.

pictures pl. see under picture.
picture sheet || Bilderbogen m. || feuille f. d'images. / apparatus for moving ~ shows pl. || Kinoapparat m. || appareil m. cinématographique. / ~ telegraph || Bildübertragungsgerät n. || appareil m. téléphotographique. / receiver station for a ~ telegraph installation || Empfangseinrichtung f. einer Bildtelegrafenanlage || équipement m. de réception d'une installation de téléphotographie. / transmitter station for a ~ telegraph installation || Gebeeinrichtung f. einer Bildtelegrafenanlage || équipement m. d'émission d'une installation de téléphotographie. / ~ telegraphy || Bildtelegrafie f. || phototélégraphie f.; téléphotographie f. / ~ theatre see ~ house. / ~ varnish || Gemäldelack m. || vernis m. pour tableaux. / magnifying ~ viewer || Bildlupe f. || loupe f. photoscopique.
picturize, to || verfilmen || cinématographier.
piece || Stück n. || morceau m.; pièce f. / ~ (Glassm) || Külbchen n.; Ballen m.; Posten m.; paraison f. / arched (Carp) ~ || Krummholz n. || bois m. courbé ou bombé; courbe f. / connecting ~ || Zwischenstück n. || pièce f. intermédiaire. / ~ of construction in dismounted state || Konstruktionsstück n. in zerlegtem Zustand || pièce f. de construction à l'état démonté. / field ~ (Arm) || Feldgeschütz n. || pièce f. de campagne. / ~ of forged work || geschmiedeter Rohling m. || pièce f. brute forgée ou venue de forge. / fugitive ~ || Pamphlet n.; Flugblatt n.; Flugschrift f. || pamphlet m.; volante f. / to go to ~s pl. || auseinanderfallen || tomber en morceaux mpl. / in ~s pl. || in Stücken npl. || en morceaux mpl. / in large ~s pl. || großstückig || en gros morceaux mpl. / to be machined see ~ of work. / in one ~ || einstückig; aus einem Stück n. || d'une pièce f. / made of one ~ see piece, in one. / pressed-out of one ~ || aus einem Stück gepreßt || embouti en une seule pièce. / ~ produced in series || Massenteil m. || pièce f. fabriquée en série. / to take to ~s pl. || in einzelne Teile mpl. zerlegen || démonter (en pièces fpl. détachées). / ~ to be tested || Prüfling m. || pièce f. à vérifier. / ~ of thin sections || dünnwandiges Gußstück n. || pièce f. à parois de faibles épaisseurs. / ~ without hard skin of casting || Gußstück n. ohne harte Kruste || pièce f. coulée sans croûte dure de coulée. / ~

of work || Werkstück n. || pièce f. à usiner.
piece cutter || Stückenschneidemaschine f. || coupeuse f. en morceaux.
pieced leads pl. (Print) || Durchschuß m. || interlignes mpl. coupés.
piece dyeing || Stückfärberei f. || teinture f. d'étoffes. / ~ goods pl. || Stückgut n. || marchandises fpl. en colis ou en cueillette. / ~ goods crane || Stückgutkran m. || grue f. pour colis. / ~ goods traffic || Stückgutverkehr m. || trafic m. par colis séparés. / ~ hand (Print) || Stücklohnsetzer m. || compositeur m. aux pièces.
piecer (Weav) || Anknüpfer m. || rattacheur m. / ~ and doffer (Weav) || Anspinner m. || rattacheur m.
piece rate || Stückzeit f. || durée f. de l'usinage d'une pièce. / ~ roller (Weav) || Roller m. || enrouleur m. d'étoffes. / ~ wage || Akkordlohn m.; Stücklohn m. || prix m. à la pièce ou à la tâche.
piecework || Stückarbeit f.; Akkordarbeit f. || travail m. à la pièce ou à forfait ou à la tâche.
pieceworker || Akkordarbeiter m. || tâcheron m. / ~ (Mine) || Gedingearbeiter m. || ouvrier m. à la tâche; tâcheron m.
piece, working by the ~ || im Akkord m. arbeiten || travailler à la pièce ou à la tâche.
piecework price || Stücklohn m. || salaire m. aux pièces ou à la tâche. / the labour is paid for at ~ prices || die Arbeit wird im Stücklohn ausgeführt || le travail est exécuté à la tâche. / ~ rates pl. see ~ price.
pier || Landungsplatz m.; Pier m.; Schiffslände f.; Ladedamm m.; Löschplatz m. || quai m. de déchargement; débarcadère f. / ~ (Bridge) || Pfeiler m. || culée f.; massif m.; pied-droit m.; pilier m. / ~ (Build) || Hauptpfeiler m. || jambage m.; maître m. pilier. / ~ of a bridge || Brückenpfeiler m. || pied-droit m. ou pilier m. ou massif m. d'un pont. / floating ~ || schwimmender Pfeiler m. || support m. flottant. / ~ of a harbour || Hafendamm m.; Mole f.; Wellenbrecher m. || môle m. de port; jetée f. ou digue f. d'un port. / intermediate ~ || Zwischenpfeiler m. || pied-droit m. intermédiaire. / projecting ~ || Gurtpfeiler m.; Stützpfeiler m. || dosseret m. de voûte. / reinforcing ~ see projecting ~. / ~ in the river || Strompfeiler m. || pile f. placée dans le courant.
pierce, to || lochen || percer; perforer.
pierced, a hole was ~ through the disk under a hammer || die Scheibe f. wurde unter einem Hammer gelocht || le disque m. fut percé sous un marteau.
piercer || Durchschlag m.; Locher m. || poinçon m. / ~ (Arm) || Räumnadel f. || épinglette f. / ~ (Drill) || Nagelbohrer m. || vrille f.
piercing a tunnel || Tunneldurchstechung f. || percement m. d'un tunnel. / ~ bushing || Führungsbuchse f. || douille f. de guidage. / ~ saw || Stichsäge f.; Spitzsäge f. || scie f. à guichet.
pier head || Molenkopf m. || musoir m.; tête f. de jetée. / ~ table || Pfeilertisch m.; Spiegeltisch m. || trumeau m. à table.
pies pl. (Print) || Putzen mpl.; Zwiebelfische mpl. || pâtés mpl.

piezo-electric || piezoelektrisch || piézoélectrique.
piezo-electricity || Piezoelektrizität f. || piézoélectricité f.
piezometer || Piezometer n. || piézomètre m.
piezometric level, lowering of the ~ || Senkung f. des Grundwasserspiegels || abaissement m. du niveau d'eau souterrain.
pig || Schwein n. || porc m. / ~ (Blast furnace; Met) see also pigs || Massel f. || gueuse f.; barre f. / ~ (Metal) see also pig iron || Roheisen n. || fonte f.; fonte f. crue. / basic ~ see phosphoric ~. / to break up the ~ || die Massel brechen || casser la gueuse. / charcoal ~ || Holzkohlenroheisen n. || fonte f. crue en charbon de bois. / cold-blast ~ || kalt erblasenes Roheisen n. || fonte f. au vent froid. / electric ~ || Elektroroheisen n. || fonte f. crue au four électrique. / forge ~ || Puddelroheisen n. || fonte f. crue de puddlage; fer m. puddlé. / foundry ~ || Gießereiroheisen n. || fonte f. crue de fonderie ou de moulage. / ~ of iron || Eisenmassel f. || sauman m. de fer. / low-carbon ~ || kohlenstoffarmes Roheisen n. || fonte f. brute à basse teneur de carbone. / mottled ~ || halbiertes Roheisen n. || fonte f. truitée ou mêlée ou masulée. / phosphoric ~ || Phosphorroheisen n. || fonte f. phosphoreuse. / refined ~ || Feineisen n. || mazée f.; fonte f. mazée. / steel ~ || Stahleisen n.; Rohstahleisen n. || fonte f. aciéreuse. / Swedish ~ || schwedisches Eisen n. || fonte f. suédoise. / Thomas ~ || Thomasroheisen n. || fonte f. crue Thomas.
pig bed || Gießbett n. || lit m. de coulée. / ~ bed dressing machine || Gießbettaufbereitungsmaschine f. || machine f. à préparer les lits de coulée. / ~ breaker || Masselbrecher m. || casseur m. de gueuse. / to break up the pig-iron by means of a ~ breaker || das Roheisen mittels eines Masselbrechers zerkleinern || casser la fonte par une casse-gueuses. / ~ breeder || Schweinezüchter m. || éleveur m. de porcs.
pigeon || Taube f. || pigeon m. / ~ breeder || Taubenzüchter m. || éleveur m. de pigeons. / ~ breeding || Taubenzucht f. || élevage m. ou éducation f. de pigeons. / ~ house || Taubenhaus n. || colombier m. / ~ shooting || Taubenschießen n. || tir m. aux pigeons. / live ~ trap || Ablaßkäfig m. für Taubenschießen || boîte f. à pigeons.
piggery || Schweinestall m. || porcherie f.
pig iron see also pig (Met) || Roheisen n. || fonte f.; fonte f. crue. / black ~ || schwarzes Roheisen n. || fonte f. limailleuse ou graphiteuse ou surcarburée. / ~ for castings see foundrq ~. / foundry ~ || Gießereiroheisen n. || fonte f. de moulage. / granular ~ || körniges Roheisen n. || fonte f. grenue. / grey ~ || graues Roheisen n. || fonte f. grise. / light grey ~ || hellgraues Roheisen n. || fonte f. grise claire. / malleable ~ || schmiedbares Gußeisen n.; Temperguß m. || fonte f. malléable. / ~ rich in manganese and silicium || an Mangan und Silizium reiches Roheisen n. || fonte f. riche en manganèse et en silicium. / ordinary ~ || gewöhnliches Roheisen n. || fonte f. brute ordinaire. / ~ with the lowest percentage of phosphorus || phosphorfreies Roheisen n. || fonte f. renfer-

mant un minimum de phosphore. / ~ for refining ‖ Frischereiroheisen n.; Roheisen n. zum Verfrischen ‖ fonte f. d'affinage. / to remelt ~ ‖ Roheisen n. umschmelzen ‖ refondre le fer cru. / specular ~ ‖ Spiegeleisen n.; Hartfloß n. ‖ fonte f. miroitante *ou* spéculaire. / white ~ ‖ weißes Roheisen n. ‖ fonte f. blanche; fer m. cru blanc.

pig iron castings pl. ‖ Eisenguß m. ‖ fonte f. grise. / ~ mixer ‖ Roheisenmischer m. ‖ mélangeur m. de fonte. / automatic control valve for ~ mixers ‖ selbsttätige Mischersteuerung f. für Roheisenmischer ‖ distributeur m. automatique pour mélangeurs de fonte. / refining of ~ ‖ Roheisenfrischung f. ‖ affinage m. de la fonte.

pig killer ‖ Schweinestecher m. ‖ abatteur m. *ou* tueur m. de porcs. / ~ lead ‖ Blockblei n.; Muldenblei n. ‖ plomb m. en saumons. / ~ machine ‖ Masselgießmaschine f. ‖ machine f. à couler les gueuses.

pigment ‖ Pigment n.; Farbstoff m. ‖ pigment m.; matière f. colorante. / mineral~ ‖ mineralische Pigmentfarbe f. ‖ pigment m. minéral. / vitrifiable ~ ‖ Schmelzfarbe f. ‖ couleur f. vitrifiable.

pigment gilding ‖ Beizvergoldung f. ‖ dorure f. au mordant.

pig moulding machine ‖ Masselformmaschine f. ‖ machine f. à mouler les gueuses. / ~ nut *see* peanut.

pigs pl. for casting ‖ Masseln fpl. zum Gießen ‖ gueuses fpl. de moulage. / ~ for fining ‖ Masseln fpl. zum Verfrischen ‖ gueuses fpl. de affinage. / ~ for the production of steel ‖ Masseln fpl. zur Stahlbereitung ‖ gueuses fpl. pour la production de l'acier.

pigskin ‖ Schweinsleder n. ‖ cuir m. de porc.

pigsty ‖ Schweinestall m. ‖ porcherie f.

pig yard crane ‖ Eisenmasselkran m. ‖ grue f. d'étalage de gueuses.

pike (Mine) ‖ Haue f.; Keilhaue f. ‖ pioche f. / turn ~ ‖ Zollschranke f. ‖ barrière f. de péage.

pikeman (Mine) ‖ Hauer m. ‖ abatteur m.; haveur m.; piqueur m.

pilaster ‖ Pilaster m.; Wandpfeiler m. ‖ pilastre m. / isolated ~ ‖ freistehender Pilaster m. ‖ pilastre m. isolé.

pilchard ‖ Sardine f. ‖ sardine f. / ~ fishing ‖ Sardinenfischerei f. ‖ pêche f. de la sardine.

pile, to (Hydr arch) ‖ auspfählen; verpfählen ‖ piloter. / ~-metal ‖ paketieren ‖ mettre en paquets m. / ~-up ‖ stapeln; aufhäufen ‖ empiler. / ~-up wood in a wood yard ‖ das Holz auf Lagerbäumen aufstapeln *oder* auf Lagerbäume legen ‖ enchanteler le bois. / the coal is piled up ‖ die Kohle f. wird aufgehäuft ‖ la houille est entassée.

pile ‖ Pfahl m. ‖ pilot·m.; pieu m.; pilotis m. / ~ (Charc) ‖ Holzkohlenmeiler m.‖ meule f. / ~ (Metal) ‖ Paket n. ‖ paquet m. / ~ (Print) ‖ Kreuz n.; Aufhängekreuz n. ‖ étendoir m. / ~ (Vineyard) ‖ Pfahl n.; Winzerpfahl m.; Weinbergpfahl m. ‖ échalas m. / ~ (Weav) ‖ Flor m.; Pole f. ‖ poil m. / the earth that covers the ~ (Charc) ‖ Meilerdecke f. ‖ couverture f. d'une meule. / false ~ *see* pile block. / ferruled ~ ‖ beschuhter Pfahl m. ‖ pieu m. ferré. / filling ~ of

a cofferdam ‖ Binnenpfahl m. *oder* Füllpfahl m. *oder* innerer Pfahl m. eines Fangdammes ‖ pilotis m. de remplage *ou* de retenue d'un bâtardeau. / gauged ~ of a cofferdam ‖ Bordpfahl m. *oder* äußerer Pfahl m. eines Fangdammes ‖ pilotis m. de bordage d'un bâtardeau. / grooved ~ ‖ Nutenpfahl m. einer Spundwand; Spundpfahl m. ‖ palplanche f. / the ~ sits on the ground ‖ der Pfahl m. sitzt auf ‖ le mouton m. refuse; le pieu m. refuse le mouton. / horizontal ~ (Charc) ‖ liegender *oder* langer Meiler m. ‖ meule f. couchée. / ~ of ore ‖ Los n.; Erzhaufen m. ‖ lot m. de minerai. / reinforced concrete ~ ‖ Eisenbetonpfahl m. ‖ pieu m. en béton armé. / ~ of saggers (Porcel) ‖ Kapselstoß m. ‖ pile f. de cassettes. / ~ of sea salt ‖ Haufen m. von Seesalz ‖ camelle f.; vache f.; pilot m. de sel marin. / standard ~ of a cofferdam ‖ Bordpfahl m. *oder* äußerer Pfahl m. eines Fangdammes ‖ pilotis m. de bordage d'un bâtardeau. / ~ of steel (Met) ‖ Garbe f.; Zange f. ‖ trousse f. *ou* fagot m. *ou* paquet m. d'acier. / ~ of substructure ‖ Grundpfahl m. ‖ pieu m. *ou* pilot m. de fondation. / tubular ~ ‖ Rohrpfahl m. ‖ pieu m. creux. / ~ of wood ‖ Holzstapel m. ‖ tas m. de bois.

pile block (Hydr arch) ‖ Aufsetzer m.; Rammknecht m.; Rammjungfer f. ‖ faux-pieux m.; faux-pilot m. / ~ bridge ‖ Pfahlbrücke f. ‖ pont m. de pilots. / ~ brushing device ‖ Strichbürstvorrichtung f. ‖ brosse f. pour l'effet de poil. ‖ ~ carpet ‖ Plüschteppich m.; Samtteppich m. ‖ tapis m. velouté. / ~ cut carpet ‖ geschnittener Samtteppich m. ‖ moquette f. veloutée coupée. / ~ deliverer ‖ Stapelausleger m. ‖ metteur m. de feuilles sur pile. / ~ drawing engine *see* ~ withdrawing engine.

pile driver ‖ Pfahlramme f. ‖ marteaupilon m.; sonnette f. / hand ~ ‖ Handramme f. ‖ mouton m. à bras. / ~ with tripping device ‖ Fallwerk n. *oder* Ramme f. mit Ausklinkvorrichtung ‖ sonnette f. à déclic.

pile driving ‖ Rammarbeit f.; Pfahlschlagen n.; Pfählen n. ‖ battage m. de pieux. / ~ machine *see* pile driver.

pile engine *see* ~ driver. / ~ ferrule ‖ Pfahlschuh m. ‖ armature f. *ou* chaussure f. d'un pieu; lardoire f. *ou* sabot m. de pilot. / ~ foundation *see* piling foundation. / ~ framing ‖ Pfahlrost m. ‖ pilotis m. / ~ lifting apparatus (Weav) ‖ Velourhebeapparat m. ‖ appareil m. à relever le poil de velours. / lifting machine with revolving straight-edge table ‖ Velourhebemaschine f. mit drehbarem Linealtisch ‖ machine f. à velouter avec table à règle tournante. / ~ planking ‖ Spundwand f. ‖ cloison f. avec palée; file f. de palplanches. / point of the ~ ‖ Pfahlspitze f. ‖ pointe f. du pieu.

piler (Metal) ‖ Paketierer m.; Paketbinder m. ‖ paquetier m.

pile reinforcement ‖ Pfahlbewehrung f. ‖ armature f. de pieu.

piles pl. ‖ Vorratshaufen m. ‖ tas m. sur parc; emmagasinage m. en tas.

pile sinking with water jet ‖ Eintreiben n. von Pfählen mit Wasserspülung ‖ enfoncement m. de pieux avec injection d'eau. / ~ warp (Weav) ‖ Oberkette f.; Polkette f.; Sammetkette f. ‖ chaîne f.

de poil. / ~ weaving ‖ Samtweberei f. ‖ tissage m. de velours. / ~ withdrawer ‖ Hebel m. zum Ausziehen eingerammter Pfähle ‖ levier m. arrache-pieux. / ~ withdrawing engine ‖ Ausziehmaschine f. für Pfähle; Pfahlausheber m. ‖ machine f. à arracher les pilotis. / ~ withdrawing lever *see* ~ withdrawer. / ~ wood ‖ Pfahlholz n. ‖ bois m. de pilotage. / ~ work ‖ Pfahlbau m.; Bohlwand f.; Pfahlreihe f.; Pfahlwerk n. ‖ pilotage m.; palée f. / ~ work of substructure ‖ Grundpfählung f. ‖ basse-palée f.; palée-basse f. / ~ working ‖ Verpfählung f. ‖ palification f.

pilgrim step rolling mill ‖ Pilgerschrittwalzwerk n. ‖ laminoir m. à pas de pèlerin.

piling ‖ Pfahlrammen n. ‖ battage m. de pieux. / canal ~ ‖ Kanalspundwand f. ‖ palplanche f. de canal. / ~ of hides ‖ Schwitzen n. der Häute ‖ travail m. à la jusée; travail m. à l'échauffe. / ~ and welding of iron ‖ gerben ‖ corroyer. / ~ through quicksand ‖ Abtreibearbeit f. *oder* Getriebearbeit f. im Schwimmsand ‖ méthode f. par palplanches dans les sables mouvants.

piling foundation ‖ Pfahlfundament n. ‖ fondation f. sur pieux. / scrap ~ machine ‖ Schrotpaketiermaschine f. ‖ machine f. à paqueter les déchets de fer. / ~ steel ‖ Spundwandeisen n. ‖ fer m. à palplanches. / ~ stores pl. ‖ Vorratshaufen m. ‖ tas m. sur parc; emmagasinage m. en tas.

pill ‖ Pille f. ‖ globule m.; pilule f.

pillar ‖ Säule f. ‖ colonne f.; pilier m.; montant m.; poteau m. / ~ (Bridge) ‖ Stütze f.; Pfeiler m. ‖ pilier m. / ~ (Mine) ‖ Pfeiler m. ‖ pilier m.; massif m. / adjoining ~ (Arch) ‖ Nebenpfeiler m. ‖ pilier m. latéral. / ~ of a blast-furnace ‖ Pfeiler m. *oder* Vierpaß m. eines Hochofens ‖ pilier m. de cœur d'un hautfourneau. / cast-iron ~ ‖ gußeiserne Säule f. ‖ colonne f. en fonte. / corner ~ of body (Auto) ‖ Ecksäule f. der Karosserie ‖ montant m. de bout de caisse. / foundation ~ ‖ Grundpfeiler m. ‖ pilier m. de fondation. / imbedded ~ ‖ flacher Pilaster m. ‖ pilastre m. engagé. / inserted ~ *see* imbedded ~. / king ~ ‖ Königssäule f.; Königszapfen m. ‖ pivot m. central. / round-shafted ~ ‖ Rundpfeiler m. ‖ pilier m. rond. / ~ for sighting-on ‖ Signalstein m. ‖ borne f. de signalisation. / smooth round-shafted ~ ‖ glattrunder Pfeiler m.; Säule f. ‖ colonne f. monocylindrique; pilier-monocylindre m. / ~ of a telescope ‖ Stativsäule f. eines Fernrohres ‖ colonne f. du pied d'une lunette. / wooden ~ ‖ Holzträger m. ‖ pilier m. en bois.

pillar bolt ‖ Stehbolzen m. ‖ entretoise f. / ~ crane ‖ Säulendrehkran m. ‖ grue f. à colonne. / ~ file ‖ Dickflachfeile f. ‖ lime f. à pilier. / ~ rolling mill ‖ Pilarwalzwerk n. ‖ laminoir m. à colonnes. / ~ stand ‖ Säulenstativ n. ‖ support m. à pilier.

pill box ‖ Pillenschachtel f. ‖ boîte f. à pilules.

pillion ‖ Soziussitz m. ‖ pillion m.

pill making machine ‖ Pillenmaschine f. ‖ machine f. à faire les pilules.

pillow ‖ Kopfkissen n. ‖ oreiller m. / ~ case ‖ Kissenüberzug m. ‖ taie f. / ~ cod

(Coachm) ‖ Ohrkissen n. ‖ custode f. / ~ fustian ‖ glatter Barchent m. ‖ futaine f. à deux envers. / ~ lace ‖ Klöppelspitze f. ‖ dentelle f. au fuseau ou au carreau.

pill vial ‖ Pillenglas n. ‖ tube m. en verre pour pilules.

pilot, to ~ a ship out to sea ‖ ein Schiff aus einem Hafen auslotsen ‖ piloter un navire hors d'un port.

pilot (Aero) ‖ Führer m. ‖ pilote m. / ~ (Nav) ‖ Lotse m. / gyro ~ ‖ Kreiselsteuergerät n. ‖ appareil m. de pilotage gyroscopique. / outward ~ ‖ Seelotse m. ‖ hauturier m.; pilot m. hauturier.

pilotage ‖ Lotsendienst m. ‖ pilotage m. / ~ (Fee) ‖ Lotsengeld n. ‖ droit m. de pilotage. / compulsory ~ ‖ Lotsenzwang m. ‖ pilotage m. obligatoire.

pilotage authorities pl. ‖ Lotsenbehörde f. ‖ autorité f. de pilotage. / certificate of ~ ‖ Lotsenpatent n. ‖ certificat m. de pilotage.

pilot boat ‖ Lotsenboot n. ‖ bateau m. pilote. / ~ burner ‖ Sparbrenner m.; Sparflämmchen n. ‖ brûleur m. à veilleuse. / ~ cloth ‖ Fries m.; Flausch m. ‖ frise f. / ~'s cockpit ‖ Führerraum m. ‖ cabine f. ou poste m. du pilote. / ~ flag ‖ Lotsenflagge f. ‖ pavillon m. de pilote.

piloting ‖ Lotsen n. ‖ pilotage m.; lamanage m.

pilot jet ‖ Leerlaufdüse f. ‖ gicleur m. secondaire. / ~ lamp (Tel) ‖ Kontrolllampe f.; Platzlampe f. ‖ lampe f. de contrôle. / ~'s license (Aero) ‖ Flugzeugführerschein m. ‖ brevet m. de pilote. / ~ master (Nav) ‖ Oberlotse m. ‖ maître m. pilote. / ~ relay ‖ Kontrollrelais n. ‖ relais m. des lampes pilotes ou de controle. / ~'s seat (Aero) ‖ Führersitz m. ‖ siège m. du pilote. / ~ signal ‖ Lotsensignal n. ‖ signal m. de pilote. / ~ station ‖ Lotsenstation f. ‖ station f. des pilotes. / ~ steamer ‖ Lotsendampfer m. ‖ bateau m. pilote à vapeur. / ~ tap ‖ Gewindebohrer m. ‖ taraud m. / ~ valve (Hydr press) ‖ Vorfüllventil n. ‖ soupape f. d'admission préalable. / ~ vessel see ~ boat. / ~ wire ‖ Meßdraht m. ‖ témoin m.

pimelite ‖ Pimelith m. ‖ pimélite f.

pimento ‖ Piment m. ‖ piment m.

pimpernel root ‖ Pimpinellwurzel f. ‖ racine f. de boucage.

pimple metal ‖ blasiger Kupferstein m.; Pimpelstein m. ‖ matte f. vésiculeuse.

pin, to ‖ anstiften; verdübeln ‖ goujonner.

pin (Carp) ‖ Vorstecker m.; Pflock m.; Stift m.; Holznagel m.‖ goupille f.; goujon m. / ~ (Carp) ‖ Dübel m. ‖ tenon m.; goujon m.; cheville f. / ~ (Door) ‖ Angel f. ‖ pivot m. / ~ (Insulator) ‖ Stütze f. ‖ porte-isolateur m.; console f.; support m. / ~ (Nailsm) ‖ Stift m.; Stiftnagel m. ‖ pointe f. / ~ (Needle) ‖ Stecknadel f. ‖ épingle f. / ~ (Rivet) ‖ Niet n. ‖ rivet m. / brass ~ ‖ Messingstecknadel f. ‖ épingle f. en laiton. / cross head ~ ‖ Gabelzapfen m. ‖ tourillon m. de fourche. / dowel ~ ‖ Prisonstift m. ‖ cheville f. de répérage. / drawing ~ ‖ Reißnagel m.; Reißstift m. ‖ punaise f. / ~ with enamelled head ‖ Stecknadel f. mit Emaillekopf ‖ épingle f. à tête émaillé. / fulcrum ~ ‖ Drehzapfen m. ‖ tourillon m. / ~ with glass head ‖ Stecknadel f. mit Glaskopf ‖ épingle f. à tête en verre. / hair ~ ‖ Haarnadel f. ‖ épingle f. à cheveux. / hat ~ ‖ Hutnadel f. ‖ épingle f. à chapeau. / to head the ~ ‖ die Stecknadel anköpfen ‖ frapper la tête d'épingle. / inserted crank ~ ‖ eingesetzter Kurbelzapfen m. ‖ bouton m. de manivelle emmanché. / iron ~ ‖ Dorn m. ‖ chevillette f. / ~ of a joint ‖ Bolzen m. oder Stift m. eines Scharnieres ‖ clavette f. d'une charnière. / knuckle ~ ‖ Kuppelbolzen m. ‖ cheville f. d'attelage. / loose ~ ‖ einsteckbarer Bolzen m. ‖ cheville f. démontable. / measuring ~ ‖ Meßschnabel m. ‖ touche f. de mesure. / ~ with ornaments ‖ Nadel f. mit Verzierungen ‖ épingle f. avec ornements. / ~ with round eye (Door) ‖ Angelring m.; Türangelpfanne f. ‖ crapaudine f.; piton m. / safety ~ ‖ Sicherheitsnadel f. ‖ épingle f. de sûreté ou de nourrice. / steel ~ ‖ Stahlstecknadel f. ‖ épingle f. en acier. / stop ~ ‖ einsteckbarer Bolzen m. ‖ loose or stop pin. / terminal double ~ (Tel) ‖ Abspanndoppelstütze f. ‖ console f. double d'arrêt. / threaded ~ ‖ Gewindestift m. ‖ goupille f. avec filetage / truck centre ~ ‖ Drehzapfen m. ‖ pivot m. du bogie. / wooden ~ (Carp) ‖ Holznagel m. ‖ cheville f. de bois; gournable f.

pin and skittle ‖ Kegelspiel n. ‖ jeu m. de quilles. / ~ beater centrifugal mill ‖ Schlagstiftmaschine f. ‖ moulin m. à goujons percuteurs. / ~ bolt ‖ Federbolzen m. ‖ boulon m. de ressort; goujon m.

pince-nez ‖ Kneifer m. ‖ binocle m.; pince-nez m.

pincers pl. ‖ Zange f.; Kneifzange f.; Beißzange f. ‖ tenailles fpl.; pince f. / ~ (Med) ‖ Pinzette f. ‖ pincette f. / ~ (Print) ‖ Punkturzange f.; Korrekturzange f. ‖ pinc(ett)e f. / ~ for correction (Print) ‖ Korrigierzange f. ‖ pince. f à correction. / farrier's ~ ‖ Hufzange f. ‖ tricoises fpl. / goldsmith's ~ ‖ Goldschmiedzange f. ‖ molette f. / ~ of a pile engine ‖ Schere f. oder Scherenhaken m. einer Ramme ‖ pince f. à déclic. / shoeing ~ see farrier's ~.

pinch, to ‖ abzwicken ‖ couper à la pince.

pinch bar ‖ Brecheisen n.; Brechstange f. ‖ barre f. de levier; pied-de-biche m.

pinchbeck ‖ Tombak m. ‖ tombac m.

pinch cock (Chem) ‖ Quetschhahn m. ‖ pince f. pressante.

pincher see pinch bar.

pinching-out of a stratum ‖ Auskeilen n. einer Schicht ‖ amincissement m. en coin d'une couche.

pinch stopper ‖ Griffstöpsel m. ‖ bouchon m. à poignée.

pin clutch ‖ Drehkeilkupplung f. ‖ embrayage m. à clavette tournante.

pincushion ‖ Nadelkissen n. ‖ pelote f.

pin drill ‖ Zapfenbohrer m. ‖ tarière f.

pine (Tree) ‖ Kiefer f.; Föhre f. ‖ pin m. / ~ (Wood) ‖ Kiefernholz n. ‖ bois m. de pin; pin m.

pineapple ‖ Ananas f. ‖ ananas m. / ~ fibre ‖ Ananasfaser f. ‖ fibre m. d'ananas.

pine bark extract ‖ Fichtenrindenextrakt m. ‖ extrait m. d'écorce de pin. / ~ bung ‖ Fichtenholzspund m. ‖ bonde f. en bois de sapin. / ~ kernel ‖ Pinienkern m. ‖ pignon m.

pine needle gatherer ‖ Fichtennadelnsammler m. ‖ ramasseur m. d'aiguilles de pins. / ~ oil ‖ Kiefernadelöl n.; Fichtennadelöl n. ‖ essence f. d'aiguilles de pin. / ~ preparation ‖ Fichtennadelpräparat n. ‖ préparation f. de bourgeons de pin. / ~ wool ‖ Waldwolle f. ‖ laine f. de pin sylvestre.

pine oil ‖ Terpentinöl n. ‖ huile f. de térébenthine. / ~ resin ‖ Fichtenharz n. ‖ résine f. de pin. / ~ soot ‖ Kienruß m. ‖ noir m. de fumée. / ~ sprouts pl. ‖ Fichtensprossen fpl. ‖ bourgeons mpl. de sapin. / ~ tar ‖ Kienteer m. ‖ goudron m. de pin. / ~ tar pitch ‖ Kienteerpech m. ‖ poix f. de goudron de pin. / ~ torch ‖ Kienspan m. ‖ perluau m.

pine wood ‖ Kiefernholz n.; Kienholz n.; Fichtenholz n. ‖ bois m. de pinastre ou pin ou de sapin rouge. / white ~ ‖ Weißtannenholz n. ‖ bois m. de sapin.

pine wool ‖ Fichtennadelwolle f. ‖ coton m. de pin.

pinguite ‖ Pinguit m. ‖ pinguite f.

pin head ‖ Stecknadelkopf m. ‖ tête f. d'épingle. / ~ hinge ‖ Aufsatzband n. ‖ fiche f. à vase; paumelle f. / ~ hole ‖ Nagelloch n. ‖ enlaçure f.

pinion (Mach) ‖ Ritzel n. ‖ pignon m. / ~ (Roll mill) ‖ Kammwalze f. ‖ cylindre de laminoir avec pignon m. à chevron. / axle drive bevel ~ ‖ kleines Differentialantriebskegelrad n. ‖ pignon m. de commande du différentiel. / differential spider ~ ‖ Ausgleichkegelrad n. ‖ satellite m. du différentiel. / gear ~ ‖ Getriebezahnrad n. ‖ pignon m. de la boîte à vitesse. / ~ with inside gearing ‖ Zahnrad n. mit Innenverzahnung ‖ pignon m. à denture intérieure. / ~ of report (Watchm) ‖ Schneckenzapfen m.; Zeigerstange f. ‖ pinion m. de renvoi. / six-toothed ~ (Watchm) ‖ Sechser m. ‖ pignon m. à six dents. / ~ of slewing gear ‖ Drehwerkritzel n. ‖ pignon m. du dispositif de rotation. / staggered driving ~s ‖ gegeneinander versetzte Zahnräder fpl. ‖ pinions mpl. à dents placés alternativement.

pinion box ‖ Triebkasten m. ‖ boîte f. à pignon. / ~ end core head ‖ Ankerflansch m. der Triebseite ‖ flasque m. de l'induit du côté de la commande. / ~ file (Watchm) ‖ Triebfeile f. ‖ lime f. à pignon. / ~ head for focussing eyepiece upon cross lines ‖ Okularfokussierung f. auf dem Fadenkreuz ‖ mise f. au point de l'oculaire sur le réticule. / ~ hobbing machine ‖ Ritzelwalzenfräsmaschine f. ‖ machine f. à tailler les pignons par fraisemère. / ~ shaft ‖ Ritzelwelle f. ‖ arbre m. de pignon. / ~ wire ‖ Riffeldraht m.; Triebstahldraht m. ‖ fil m. à pignons.

pinite ‖ Pinit m. ‖ pinite f.

pinitride ‖ Pinitrid m. ‖ pinitride f.

pink, to (Join) ‖ ausbohren ‖ évider.

pink ‖ rosa ‖ rose.

pinking punch ‖ Zackenausstecher m. ‖ emporte-pièce m. à denteler.

pink salt ‖ Pinksalz n.; Ammoniumzinnchlorid n. ‖ bichlorure m. d'étain.

pin length ‖ Nadellänge f.; Nadelschaft m. ‖ hanse f. / ~ lock ‖ Schloß n. mit Rohrschlüssel ‖ serrure f. à broche. / ~ maker (Nailsm) ‖ Stiftmacher m. ‖ fabricant m. de pointes. / ~ maker (Needle) ‖ Stecknadelmacher m. ‖ épinglier m.

pinnace ‖ Pinasse f. ‖ péniche f.; pinasse f.

pinned ‖ zusammengenagelt ‖ cloué.

pin point (Text print) ‖ Rapportstift m. ‖ repère m. / ~ pointer ‖ Strecknadel-spitzer m. ‖ empointeur m. d'épingles.
pins pl. ‖ Nadlerwaren fpl. ‖ épinglerie f.
pin setter (Spinn) ‖ Kardensetzer m. ‖ bouteur m. de cardes.
pintle ‖ Fingerling m. ‖ doigtier m.
pin way of the grain ‖ Hirnseite f. des Holzes ‖ côté m. de la moëlle; coupe f. transversale. / ~ weir ‖ Nadelwehr n. ‖ barrage m. à aiguilles. / ~ wheel gear ‖ Triebstockgetriebe n. ‖ denture f. en fuseau pour l'engrenage de la cheville. / ~ wire ‖ Nadeldraht m. ‖ fil m. à moule.
pip ‖ Obstkern m. ‖ pépin m.
pipe, to ‖ pfeifen ‖ siffler.
pipe *see also* tube ‖ Röhre f.; Rohr n. ‖ tuyau m.; tube m.; conduit m. / ~ (Glassm) ‖ Glasmacherpfeife f.; Pfeife f. ‖ canne f. de verrier; felle f. / ~ (Metal) ‖ Lunker m. ‖ retassement m. / ~ (Organ) ‖ Pfeife f. ‖ tuyau m. / ~ (Road) ‖ Röhrendurchlaß m. ‖ aquéduc m. / ~ (Tobacco) ‖ Pfeife f. ‖ pipe f. / additional ~ ‖ Ansatzrohr n. ‖ tuyau m. additionel. / articulated ~s pl. ‖ Gliederröhren fpl. ‖ tubes mpl. articulés. / bent ~ ‖ gebogenes Rohr n. ‖ Knierohr n. ‖ tuyau m. courbé. / bifurcated ~ *see* forked ~. / blast ~ ‖ Ausblasrohr n. ‖ tuyère f. d'échappement. / blow ~ ‖ Lötrohr n. ‖ chalumeau m. / branch ~ ‖ Abzweigrohr n. ‖ embranchement m. en tube. / butt-welded ~ ‖ stumpfgeschweiß-tes Rohr n. ‖ tuyau m. soudé à rapproche-ment; tube m. soudé par contact. / carburettor hot air ~ ‖ Luftwärmer m. ‖ prise f. d'air chaud. / cast iron ~ ‖ guß-eisernes Rohr n. ‖ tuyau m. en fonte. / cement ~ ‖ Zementrohr n. ‖ tuyau m. en ciment. / connecting ~ ‖ Anschluß-rohr n. ‖ conduite f. *ou* tuyauterie f. de jonction. / copper ~ ‖ Kupferrohr n. ‖ tuyau m. en cuivre. / copper blow ~ ‖ kupfernes Lötrohr n. ‖ chalumeau m. à souder en cuivre. / discharging ~ *see* blast ~. / earthen ~ ‖ Tonrohr n. ‖ tuyau m. en poterie *ou* en argile. / exhaust ~ ‖ Auspuffrohr n. ‖ tuyau m. d'échappement. / expansion ~ ‖ Deh-nungsrohr n. ‖ tube m. de compensation. / feed ~ ‖ Speiserohr n. ‖ tube m. d'ali-mentation. / flexible ~ ‖ Schlauch m. ‖ tuyau m. flexible. / forked ~ ‖ Gabel-rohr n. ‖ tuyau m. bifurqué. / gas ~ ‖ Gasrohr n. ‖ tuyau m. à gaz. / hammered ~ ‖ gehämmertes Rohr n. ‖ tuyau m. martelé. / hawse ~ (Mar) ‖ Ankerklüse f. ‖ écubier m. / inlet ~ ‖ Saugrohr n.; Einlaßrohr n. ‖ tuyau m. d'aspiration *ou* d'admission. / iron ~ ‖ Eisenrohr n. ‖ tube m. en fer. / ~ of the key ‖ Schlüssel-rohr n. ‖ canon m. d'une clef. / lead ~ ‖ Bleirohr n. ‖ tuyau m. en plomb. / leather ~ ‖ Lederschlauch m. ‖ tuyau m. en cuir. / meerschaum ~ ‖ Meerschaum-pfeife f. ‖ pipe f. en écume de mer. / imitation meerschaum ~ ‖ Tabakpfeife f. aus Meerschaumnachahmung ‖ pipe f. en imitation d'écume de mer. / metal ~ ‖ Metallrohr n. ‖ tube m. métallique. / ~ for oil drippings ‖ Leckölleitung f. ‖ con-duite f. pour huile de fuite. / outlet ~ ‖ Abzugrohr n. ‖ tuyau m. d'écoulement. / ~ of a rag engine ‖ Rinne f. im Aufsatz eines Holländers zum Wasserabfluß ‖ dalot m. du chapiteau d'un moulin à cylindre. / ribbed ~ ‖ Rippenrohr n. ‖

tuyau m. à ailettes. / lined rinsing ~ (Mine) ‖ gefüttertes Spülversatzrohr n. ‖ tuyau m. revêti à rinçure. / rising ~ ‖ Steigrohr n. ‖ tuyau m. de refoule-ment. / riveted ~ ‖ genietetes Rohr n. ‖ tuyau m. rivé. / seamless ~ ‖ nahtloses Rohr n.; Mannesmannrohr n. ‖ tube m. sans soudure. / socket ~ ‖ Muffenrohr n. ‖ tuyau m. à manchon. / steel ~ ‖ Stahl-rohr n. ‖ tuyau m. en acier. / ~ of supply *see* feed ~. / waste ~ ‖ Abflußrohr n. ‖ tuyau m. de décharge *ou* de dégorge-ment. / water ~ ‖ Wasserrohr n. ‖ tuyau m. d'eau. / welded ~ ‖ geschweißtes Rohr n. ‖ tube m. soudé. / wooden ~ ‖ Holzrohr n. ‖ conduite f. en bois. / wooden ~ (Tobacco) ‖ Holzpfeife f. ‖ pipe f. en bois. / wooden ~ turner (To-bacco) ‖ Pfeifendrechsler m. ‖ tourneur m. de pipes en bois. / wrought iron ~ ‖ schmiedeeisernes Rohr n. ‖ tube m. en tôle. / zinc ~ ‖ Zinkrohr n. ‖ tuyau m. en zinc.
pipe bending press ‖ Rohrbiegepresse f. ‖ presse f. à courber les tubes. / ~ bowl ‖ Pfeifenkopf m. ‖ tête f. de pipe. / ~ bowl carver ‖ Pfeifenknopfschnitzer m. ‖ sculpteur m. de têtes de pipes. / ~ box (Wheel) ‖ Nabenbuchse f. ‖ boîte f. de moyeu. / ~ branching ‖ Rohrverzweigung f. ‖ branchement m. de tuyaux. / ~ burst ‖ Rohrbruch m. ‖ rupture f. d'un tuyau. / ~ calibrating press ‖ Rohrkalibrierpresse f. ‖ presse f. à calibrer les tubes. / ~ clamp ‖ Rohrschelle f. ‖ étrier m.; collier m. / ~ clay ‖ Pfeifenton m. ‖ terre f. à pipes. / ~ cleaner ‖ Pfeifenreiniger ‖ cure-pipe m. / ~ clip *see* ~ clamp. / ~ closer ‖ Rohr-verschluß m. ‖ obturateur m. de tuyau. / single ~ conduit (Cable) ‖ Vollrohrkanal m. ‖ conduite f. unitaire. / ~ connection ‖ Rohrverbindung f. ‖ raccord m. à tuyaux. / cross section of ~ ‖ Rohrquer-schnitt m. ‖ section f. du tube. / ~ cutter ‖ Rohrschneider m. ‖ tenailles fpl. pour couper les tuyaux; coupe-tubes m. / diameter of a ~ ‖ Rohrdurchmesser m.; Rohrweite f. ‖ diamètre m. d'un tuyau. / ~ drain ‖ Röhrendrän m. ‖ drain m. / ~ elbow bending machine ‖ Knierohr-biegemaschine f. ‖ machine f. à faire les coudes des tuyaux de poêle. / ~ expand-ing press ‖ Rohraufweitpresse f. ‖ presse f. à mandriner les tuyaux. / ~ fit-ter *see* ~ layer. / ~ flange ‖ Rohr-flansch m. ‖ bride f. de tuyaux. / ~ fracture *see* ~ burst. / ~ frame ‖ Rohr-rahmen m. ‖ châssis m. en tuyaux. / ~ hole (Found) ‖ Lunkerung f. ‖ retasse-ment m.; retassure f. / ~ hook ‖ Rohr-haken m. ‖ crampon m. pour tubes. / ~ isolating valve ‖ Rohrbruchventil n. ‖ soupape f. à rupture de conduite. / screwed ~ joint ‖ Rohrverschraubung f. ‖ assemblage m. de tuyaux à vis. / ~ jumping press ‖ Röhrenstauchpresse f. ‖ presse f. à refouler les tuyaux. / ~ key ‖ Rohrschlüssel m. ‖ clef f. à tube. / ~ layer ‖ Rohrleger m. ‖ poseur m. de conduites. / laying of ~s ‖ Rohrverlegung f. ‖ pose f. de tuyaux.
pipe line ‖ Rohrleitung f. ‖ tuyauterie f. / ~ for acids and lyes ‖ Rohrleitung f. für Säuren und Laugen ‖ tuyauterie f. pour acides et lessives. / ascending and des-cending ~s for heating water ‖ Steig- und Falleitungen fpl. für Heizwasser ‖ tuyauterie fpl. de monté et de descente

de l'eau de chauffage. / turbine ~ ‖ Turbinenrohrleitung f. ‖ conduite f. de turbines. / connection piece for ~s ‖ Ver-bindungsstück n. für Rohrleitungen ‖ raccord m. pour tuyauteries.
pipe maker ‖ Pfeifenmacher m. ‖ fabricant m. de pipes; pipier m. / ~ mill ‖ Röhren-walzwerk n. ‖ laminoir m. à tubes. / ~ organ ‖ Pfeifenorgel f. ‖ orgue m. à tuyaux. / ~ shearing machine ‖ Rohr-abstechmaschine f. ‖ coupe-tuyau m. / ~ slicing lathe ‖ Rohrabstechbank f. ‖ banc m. à tronçonner les tubes. / ~ straightening press ‖ Rohrrichtpresse f. ‖ presse f. à dresser les tubes. / ~ system with shut-off device ‖ absperrbarer Lei-tungsstrang m. ‖ tronçon m. de la canali-sation à fermeture. / ~ testing machine ‖ Rohrprüfmaschine f. ‖ machine f. à es-sayer les tubes. / ~ testing press ‖ Rohr-prüfpresse f. ‖ presse f. à éprouver les tubes. / ~ tongs pl. ‖ Rohrzange f. ‖ tenailles fpl. à gaz. / ~ trench ‖ Rohrgraben m. ‖ tranchée f. de tubes.
pipette ‖ Pipette f.; Stechheber m. ‖ pi-pette f. / filtering ~ ‖ Filterpipet f. ‖ pipette f. à filtration. / graduate ~ ‖ Meßpipette f. ‖ pipette f. graduée. / mixing ~ ‖ Mischpipette f. ‖ pipette-mélangeur f.
pipe vice ‖ Rohrschraubstock m. ‖ étau m. pour tubes. / ~ weaver ‖ Schlauchweber m. ‖ tisseur m. de tuyaux. / ~ welding ‖ Rohrschweißung f. ‖ soudage m. de tubes. / rough-shaper of ~ wood ‖ Pfeifen-holzzubereiter m. ‖ ébaucheur m. de pipes. / ~ wood sawyer ‖ Pfeifenholz-säger m. ‖ débiteur m. de bois pour pipes. / ~ wrapping ‖ Rohrbandagierung f. ‖ enveloppement m. de tubes. / ~ wrapping in asphalt ‖ Asphaltieren n. von Röhren ‖ goudronnage m. de tubes.
pip fruit ‖ Kernobst n. ‖ fruits mpl. à pépins.
piping *see also* pipe line ‖ Rohrleitung f. ‖ conduite f.; tuyauterie f. / ~ inside the buildings ‖ Leitung f. in Gebäuden ‖ con-duite f. dans les bâtiments. / ~ for mines ‖ Rohrleitung f. für den Bergbau ‖ tu-yauterie f. pour mines. / ~ in steel ‖ Lun-kerbildung f. ‖ retassure f. / underground ~ ‖ Erdleitung f. ‖ conduite f. souter-raine.
piqué ‖ Pikee m. ‖ piqué m. / silk ~ ‖ Seidenpikee m. ‖ tissu m. piqué soie.
piqué saddle ‖ Schulsattel m. ‖ selle f. de manège *ou* à piquer.
pirn (Weav) ‖ Einschußspule f. ‖ cannette f.; sépoule f. ‖ spoule f.; époulle f.; volue f. / immovable ~ (Weav) ‖ Schleifspule f. ‖ cannette f. à défiler.
pirn shuttle (Weav) ‖ Schütz n. mit Lauf-spule ‖ navette f. à dérouler.
pirogue (Mar) ‖ Piroge f. ‖ pirogue f.
Pisang wax ‖ Bananenwachs n. ‖ cire f. de Pisang.
pisanite ‖ Kupfereisenvitriol n. ‖ pisanite f.
piscicultor ‖ Fischzüchter m. ‖ piscicul-teur m.
pisciculture ‖ Fischzucht f. ‖ pisciculture f.; élevage f. de poissons.
pisciculturist *see* piscicultor.
pisé work ‖ Lehmstampfbau m.; Hasten-werk n. ‖ manière f. de bâtir en pisé; œuvre f. pisée; coffre m. en pisé; con-struction f. en pisé.
pisolite ‖ Pisolit m.; Erbsenstein m. ‖ piso-lite f.; pierre f. de pois.

pissasphalt ‖ Bergteer m. ‖ goudron m. minéral; malthé f.; pétrole f. tenace.

pistachio ‖ Pistazie f. ‖ pistache f. / ~ nut dryer ‖ Pistaziendörrer m. ‖ pistachier m.

pistil (Botanic) ‖ Stempel m. ‖ pistil m.

pistol ‖ Pistole f. ‖ pistolet m. / air ~ ‖ Luftpistole f. ‖ pistolet m. souffleur. / automatic ~ ‖ Selbstladepistole f. ‖ pistolet m. automatique. / blow ~ ‖ Ablaspistole f. ‖ pistolet m. souffleur. / flare ~ ‖ Leuchtpistole f. ‖ pistolet m. pour fusées d'éclairage. / machine ~ ‖ Maschinenpistole f. ‖ pistolet m. mitrailleur. / oil ~ ‖ Ölspritze f. ‖ seringue f. d'huile. / spraying ~ ‖ Spritzpistole f. ‖ pistolet m. pulvérisateur. / spraying ~ for painting ‖ Spritzpistole f. zum Anstreichen ‖ pistolet m. à peinturer. / welding ~ ‖ Schweißpistole f. ‖ pistolet m. à souder.

pistol handle ‖ Pistolenheft n. ‖ manche f. de pistolet. / ~ pipe ‖ Pistolenrohr n. ‖ tuyau m. de pistolet.

piston (Mach) ‖ Kolben m. ‖ piston m. / die-forged ~ ‖ im Gesenk geschmiedeter Kolben m. ‖ piston m. matricé. / differential ~ ‖ Stufenkolben m.; Differentialkolben m. ‖ piston m. différentiel. / double diameter ~ see differential ~. / ejecting ~ ‖ Ausstoßkolben m. ‖ piston-éjecteur m. / full ~ ‖ massiver oder voller Kolben m. ‖ piston m. plein. / to grind the ~ in ‖ den Kolben m. einschleifen ‖ roder le piston dans le cylindre. / the ~ grinds ‖ der Kolben frißt ‖ le piston se grippe. / ground-in ~ ‖ eingeschliffener Kolben m. ‖ piston m. rodé. / hemp-packed ~ ‖ Kolben m. mit Hanfliderung ‖ piston m. à garniture de chanvre. / hollow ~ ‖ durchbrochener Kolben m. ‖ piston m. percé ou perforé ou foré. / to leather a ~ ‖ einen Kolben m. beledern ‖ garnir un piston de cuir. / ~ with leather packing ‖ Kolben m. mit Lederliderung ‖ piston m. à garniture de cuir. / ~ with metallic packing ‖ Kolben m. mit Metalliderung ‖ piston m. à garniture métallique. / ~ working in opposite direction ‖ gegenläufiger Kolben m. ‖ piston m. convergent. / perforated ~ see hollow ~. / to put-in the ~ ‖ den Kolben m. einbringen ‖ mettre le piston en place. / the ~ can easily be reground ‖ der Kolben kann leicht nachgeschliffen werden ‖ le piston m. peut se roder facilement. / to remove the ~ ‖ den Kolben m. herausziehen ‖ retirer le piston. / the ~ seizes see the ~ grinds. / self-distributing ~ ‖ selbststeuernder Kolben m. ‖ piston m. à distribution automatique / solid ~ see full ~. / submerged ~ ‖ versenkt liegender Kolben m. ‖ piston m. noyé.

piston barrel see ~ body. / ~ bearing ‖ Kolbenlager n. ‖ coussinet m. de piston. / ~ blower ‖ Kolbengebläse n. ‖ soufflerie f. à piston. / ~ blowing apparatus see ~ blower. / ~ blowing engine ‖ Kolbengebläse n. ‖ machine f. soufflante à piston; soufflet m. cylindrique; soufflerie f. à piston cylindrique. / ~ body ‖ Kolbenkörper m. ‖ corps m. de piston. / ~ bolt removing device ‖ Kolbenbolzenentferner m. ‖ appareil m. à retirer les boulons de piston. / ~ boss bushing ‖ Kolbenbuchse f. ‖ coussinet m. de bossage du piston. / ~ caulking ring ‖ Kolbendichtungsring m.

‖ couronne f. de garniture de piston. / ~ clearance ‖ Kolbenspiel n. ‖ jeu m. du piston. / ~ cover ‖ Kolbendeckel m. ‖ couvercle m. ou plateau m. du piston. / ~ displacement ‖ Kolbenverdrängung f. ‖ déplacement m. de piston. / ~ feeder ‖ Kolbenaufgabevorrichtung f. ‖ feeder m. à piston. / ~ grinding machine ‖ Kolbenschleifmaschine f. ‖ machine f. à rectifier les pistons. / ~ head ‖ Kolbenboden m. ‖ fond m. du piston. / ~ head plate ‖ Kolbenkopfplatte f. ‖ plaque f. de tête du piston. / ~ heating ‖ Kolbenheizung f. ‖ échauffement m. du piston. / ~ joint spring ‖ Dichtungskolbenpaßfeder f. ‖ ressort m. de la bague d'antifriction. / ~ knock ‖ Klopfen n. des Kolbens ‖ cognement m. du piston. / ~ lubricating connection ‖ Kolbenschmierstutzen m. ‖ embouchure f. de graissage de piston. / ~ lubrication ‖ Kolbenschmierung f. ‖ graissage m. du piston. / ~ packing ‖ Kolbenliderung f.; Kolbenpackung f. ‖ garniture f. de piston. / ~ packing leather ‖ lederne Kolbenmanschette f. ‖ cuir m. embouti du piston. / ~ packing ring see ~ ring. / ~ pin ‖ Kolbenbolzen m. ‖ axe m. de piston. / ~ pin bearing ‖ Kolbenbolzenlager n. ‖ coussinet m. de pied de bielle. / ~ pin bushing ‖ Kolbenbolzenbuchse f. ‖ coussinet m. de pied de bielle. / ~ pin lock screw ‖ Kolbenbolzensicherungsschraube f. ‖ vis f. de fixation de l'axe de piston. / ~ play ‖ Kolbenspiel n. ‖ jeu m. du piston. / ~ pump ‖ Kolbenpumpe f. ‖ pompe f. à piston. / double ~ pump ‖ Doppelkolbenpumpe f. ‖ pompe f. à double piston.

piston ring ‖ Kolbenring m. ‖ bague f. ou segment m. de piston. / diagonally cut ~ ‖ Kolbenring m. mit schräger Stoßfuge ‖ segment m. coupé en sifflet. / lap-ended ~ ‖ Kolbenring m. mit überlapptem Stoß ‖ segment m. à coupure en recouvrement. / the ~ sticks ‖ der Kolbenring m. sitzt fest ‖ le segment m. de piston est calé.

piston ring grinding machine ‖ Kolbenringschleifmaschine f. ‖ machine f. à rectifier les segments de piston. / ~ groove ‖ Kolbennute f. ‖ rainure f. annulaire de piston. / ~ pin ‖ Kolbenringstift m. ‖ goupille f. du segment de piston. / ~ tightener ‖ Kolbenringspanner m. ‖ appareil m. tendeur de segment de piston.

piston rod ‖ Kolbenstange f. ‖ tige f. de piston. / break-down of the ~ ‖ Kolbenstangenbruch m. ‖ rupture f. de la tige de piston. / ~ grinding machine ‖ Kolbenstangenschleifmaschine f. ‖ machine f. à rectifier les tiges de pistons. / ~ nut ‖ Kolbenstangenmutter f. ‖ écrou m. de la tige de piston.

piston slide valve ‖ Kolbenschieber m. ‖ vanne f. de distribution à tiroir-piston; piston-tiroir m. / ~ speed ‖ Kolbengeschwindigkeit f. ‖ vitesse f. de piston. / ~ speed curve ‖ Kolbengeschwindigkeitskurve f. ‖ courbe f. de vitesse du piston. / ~ spring ‖ Kolbenfeder f. ‖ ressort m. de piston. / ~ steam engine ‖ Kolbendampfmaschine f. ‖ machine f. à piston. / stroke of ~ ‖ Kolbenhub m. ‖ course f. du piston. / ~ surface ‖ Kolbenfläche f. ‖ surface f. de piston. / ~ valve see ~ slide valve. / ~ wrench ‖ Kolbenschlüssel m. ‖ clef f. du piston.

pit, to ‖ anfressen ‖ gripper.

pit (Found) ‖ Gießgrube f. ‖ fosse f. de coulée. / ~ (Mine) ‖ Bergwerk n.; Zeche f.; Grube f. ‖ mine f. / ~ (Shaft) ‖ Schacht m. ‖ puits m. / ~ (Spinn) ‖ Rottegrube f. ‖ routoir m. / ~ (Theater) ‖ Parterre loge f. ‖ baignoire f. / ~ cock ~ ‖ Pilotensitz m. ‖ poste m. du pilote. / ~ of examination see inspection ~. / foundry ~ ‖ Dammrube f. ‖ fosse f. de coulée. / inspection ~ (Auto) ‖ Arbeitsgrube f. ‖ fosse f. de réparation. / ~ for preparing the clay (Pott) ‖ Sumpf m. ‖ marcheux m. / soaking ~ ‖ Durchweichungsgrube f. ‖ pit m. chauffé. / winding ~ see working ~. / ~ being worked out ‖ erschöpfte Zeche ‖ mine f. épuisée. / working ~ (Min) ‖ Förderschacht m. ‖ puits m. d'extraction. / working ~ for pumping and ventilation ‖ Schacht m. für Wasserhaltung und Wetterführung ‖ puits m. d'épuisement des eaux et de la ventilation.

pita ‖ Pitehanf m. ‖ pitte f; aloès pitte m.; chanvre m. des Indes.

pit bottom ‖ Füllort m. ‖ chambre f. de chargement. / ~ box (Theater) ‖ Parterre loge f. ‖ baignoire f. / ~ burning ‖ Grubenverkohlung f. ‖ carbonisation f. dans des fosses ou en fosses.

pitch, to ‖ pechen; verpechen; verteeren ‖ goudronner. / ~ (Brass) ‖ anstellen ‖ mettre en levain. / ~ (Mar) ‖ verpichen ‖ brayer. / ~ (Road) ‖ beschottern ‖ perreyer. / ~ piles ‖ Pfähle mpl. eintreiben ‖ enfoncer les pieux mpl.

pitch ‖ Pech n. ‖ poix f.; brai m. / ~ (Inclination) ‖ Neigung f.; Abdachung f.; Steigung f.; Abfall m. ‖ pente f. / ~ (Lathe) ‖ Spitzenhöhe f. ‖ hauteur f. des pointes. / ~ (Mach) ‖ Zahnteilung f. ‖ pas m. à engrenage. / ~ (Mountain) ‖ Gipfel m.; Spitze f.; Höhe f. ‖ hauteur f. / ~ (Mus) ‖ Tonhöhe f. ‖ hauteur f. d'un son. / ~ (Screw) ‖ Steigung f.; Steighöhe f.; Ganghöhe f. ‖ pas m. (d'une vis). / beech tar ~ ‖ Buchenholzpech n. ‖ poix f. de goudron de hêtre. / bituminous ~ ‖ Asphaltpech n. ‖ poix f. d'asphalte. / circular ~ ‖ Teilung f. ‖ pas m. circulaire ou circonférentiel. / common black ~ ‖ Schusterpech n.; gemeines schwarzes Pech n. ‖ brai m. gras; poix f. noire. / compressed ~ ‖ Stampfasphalt m. ‖ asphalte m. comprimé. / dry ~ ‖ Hartpech n. ‖ brai m. sec. / minimeter instrument to check the ~ of gears ‖ Minimetergerät n. zum Prüfen der Teilung von Zahnrädern ‖ appareil m. minimètre pour la vérification du pas des roues dentées. / liquid ~ ‖ Harzpech n. ‖ résine f. commune; brai m. sec; poix-résine f.; poix f. grasse. / mean ~ ‖ mittlere Steigung f. ‖ pas m. moyen. / mineral ~ ‖ Asphalt m.; Erdpech n.; Judenpech n. ‖ asphalte m.; bitume m. solide; goudron m. minéral. / moist ~ ‖ Weichpech n. ‖ brai m. gras. / navy ~ ‖ Schiffspech n. ‖ goudron m. à calfater. / precise hole-~ (Techn) ‖ genaue Lochteilung f. ‖ écartement m. de trous précis. / ~ of rivets ‖ Nietenfernung f.; Nietteilung f. ‖ écartement m. ou espacement m. des rivets. / ~ of a screw ‖ Ganghöhe f. einer Schraube ‖ pas m. d'une vis. / to determine the ~ of a screw with the utmost accuracy ‖ die Ganghöhe einer Schraube mit möglichster Genauigkeit auswerten ‖ déterminer le pas d'une vis avec la plus grande précision possible. / ~ of

a short cord winding ‖ verkürzter Wicklungsschritt m. ‖ pas m. raccourci. / ~ of thread ‖ Gewindesteigung f. ‖ pas m. de vis. / ~ of a toothed wheel ‖ Zahnteilung f. eines Zahnrades ‖ pas m. d'engrenage. / uniform ~ ‖ konstante Steigung f. ‖ pas m. constant. / ~ of the vanes (Wind mill) ‖ Flügelabstand m. ‖ écartement m. des palettes. / variable ~ ‖ veränderliche Steigung f. ‖ pas m. variable. / ~ of a vault ‖ Pfeilhöhe f. oder Stichhöhe f. einer Wölbung ‖ flèche f. ou montée f. de voûte; voussure f.

pitch and tar ‖ Schiffsteer m. ‖ brai m. gras.

pitch, angle of ‖ Steigungswinkel m. ‖ angle m. de tangage ou d'inclinaison.

pitchblende ‖ Pechblende f. ‖ urane m. oxydulé; péchurane f.

pitch boat ‖ Pechboot n. ‖ pigoulière f.; pégoulière f. / ~ breaker ‖ Pechbrecher m. ‖ concasseur m. de poix. / ~ burner ‖ Pechsieder m. ‖ fondeur m. de poix. / ~ chain ‖ Vaucanson'sche Kette f.; Bandkette f. ‖ chaîne f. à la Vaucanson. / ~ chain of a dredger ‖ Eimerkette f. eines Baggers ‖ chaîne f. sans fin d'un cure-môle. / ~ circle ‖ Teilkreis m. ‖ cercle m. primitif. / ~ circle of teeth ‖ Zahnkreis m. ‖ cercle m. primitif de la denture. / ~ coal ‖ Pechkohle f. ‖ jais m.; houille f. piciforme. / ~ coke ‖ Pechkoks m. ‖ coke m. de poix. / ~ diameter ‖ Teilkreisdurchmesser m. ‖ diamètre m. du cercle primitif.

pitched roof ‖ Dach n. mit hoher Steigung; steiles Dach n. ‖ toit m. à forte pente.

pitcher ‖ Krug m. ‖ cruche f. / ~ maker ‖ Kannenmacher m. ‖ broquier m.

pitch-fir ‖ Fichte f.; Rottanne f. ‖ sapin m. rouge; épicéa m.

pitchfork ‖ Heugabel f.; Forke f. ‖ fourche f.

pitch grinding mill ‖ Pechmühle f. ‖ moulin m. à brai. / ~ house ‖ Pechhütte f.; Pechküche f. ‖ pigoulière f.; pégoulière f.

pitching ‖ Bepichen n. ‖ goudronnage m. / ~ (Brew) ‖ Anstellen n. ‖ mise en levain. / ~ (Earthw) ‖ Pflaster n. von Böschungen ‖ perré m. / ~ (Ship) ‖ Stampfen n. ‖ tangage m. / dry stone ~ (Build) ‖ Steinpackung f.; Steinsatz m. ‖ perré f. en pierre sèche.

pitching apparatus ‖ Verpichapparat m.; Pichapparat m. ‖ appareil m. à goudronner. / cask ~ apparatus ‖ Faßpichapparat m. ‖ appareil m. à goudronner les fûts. / borer ‖ Meißelbohrer m. ‖ pistolet m. / ~ indicator (Aero) ‖ Kippzeiger m. ‖ indicateur m. de tangage.

pitching machine ‖ Bepichmaschine f. ‖ machine f. à goudronner. / automatic ~ ‖ Pichautomat m. ‖ goudronneur m. automatique. / capacity of the automatic ~ up to x kegs per hour ‖ Leistung f. des Pichautomaten bis x Faß je Stunde ‖ rendement m. du goudronneur automatique jusqu'à x fûts par heure. / cask ~ ‖ Faßpichmaschine f. ‖ machine f. à goudronner les fûts. / steam ~ ‖ Dampfpichmaschine f. ‖ goudronneur m. à vapeur.

pitching moment (Aero) ‖ Kippmoment n. ‖ moment m. de tangage. / ~ motion (Aero; Nav) ‖ Stampfbewegung f.; Stoßbewegung f.; Querschwingung f. ‖ mouvement m. de tangage; tangage m. / period of ~ motion (Ship) ‖ Stampfperiode f. ‖ période f. de tangage. /

~ temperature ‖ Anstelltemperatur f. ‖ température f. de mise en levain. / ~ tool (Clockm) ‖ Geradhängmaschine f.; Plantiermaschine f. ‖ machine f. ou outil m. à planter. / ~ tub ‖ Anstellbütte f. ‖ cuve f. guilloire. / ~ wort ‖ Anstellwürze f. ‖ moût m. / ~ yeast ‖ Anstellhefe f. ‖ levain m.

pitch kettle ‖ Pechkessel m. ‖ chaudière f. à poix. / ~ ladle ‖ Pechlöffel m.; Pechkelle f.; Pechpfanne f. ‖ cuiller f. à brai; pucheure m. / metric ~ lead screw ‖ metrische Leitspindel f. ‖ vis-mère f. métrique. / ~ mark (Text print) ‖ Rapportstift m. ‖ repère m. / ~ mop (Shipb) ‖ Pechquast m. ‖ penne f. à brai; guipon m. à calfat. / ~ note (Mus) ‖ Grundton m. ‖ son m. fondamental; tonique f. / ~ ore ‖ Uranpechblende f. ‖ urane m. oxydulé. / ~ pan see ~ laddle. / ~ paper ‖ Wachspapier n. ‖ papier m. ciré.

pitch pine ‖ Pechpinie f. ‖ pitchpin m. / ~ wood ‖ Pitchpineholz n. ‖ bois m. de pitchpin.

pitch pipe ‖ Stimmpfeife f. ‖ diapason m. à bouche. / ~ polishing method of glass ‖ Pechpolitur f. des Glases ‖ polissage m. de verre à la poix. / gyroscopic roll and ~ recorder ‖ Kreiselschlinger- und -Stampfanzeiger m. ‖ appareil m. enregistreur gyroscopique de roulis et de tangage. / machine for removing ~ from kegs ‖ Faßentpichmaschine f. ‖ machine f. à dégoudronner les fûts. / ~ shed ‖ Pechhütte f.; Pechküche f. ‖ pigoulière f.; pégoulière f.

pitch spraying apparatus ‖ Pechspritzapparat m. ‖ injecto-poisseur m.; appareil m. à poisser. / ~ combined with rolling machine ‖ mit einer Rollmaschine kombinierter Pechspritzapparat m. ‖ appareil m. à poisser combiné avec une machine à rouler.

pitch spraying station ‖ Bepichstation f. ‖ station f. à poisser.

pitchstone ‖ Pechstein m. ‖ rétinite f.

pitch thread ‖ Pechdraht m. ‖ fil m. poissé. / ~ torch ‖ Pechfackel f. ‖ torche f. de poix; torchère f.

pitchy fir oil ‖ Edeltannenöl n. ‖ essence f. de sapin résineux.

pit coal ‖ Steinkohle f. ‖ charbon m. de terre; houille f. / contents pl. of bitumen in ~ ‖ Bitumengehalt m. der Steinkohle ‖ teneur f. en produits bitumineux de la houille. / to burn ~ ‖ Steinkohle f. verfeuern ‖ employer ou brûler de la houille.

pit coal ashes pl. ‖ Steinkohlenasche f. ‖ cendre f. de charbon. / ~ dressing ‖ Steinkohlenaufbereitung f. ‖ préparation f. de la houille. / ~ drying plant ‖ Steinkohlentrocknungsanlage f. ‖ installation f. de séchage de la houille. / ~ dust ‖ Steinkohlenstaub m. ‖ fraisil m. / ~ industry ‖ Steinkohlenindustrie f. ‖ industrie f. de la houille. / ~ power station ‖ Steinkohlenkraftwerk n. ‖ usine f. thermique à base de houille.

pit eye see pit head. / ~ furnace ‖ Schachtofen m. ‖ fourneau m. à cuve. / ~ gravel ‖ Grubenkies m. ‖ gravier m. de carrière.

pith (Wood) ‖ Mark n. ‖ moëlle f. / ~ ball electroscope ‖ Holundermarkelektroskop n. ‖ électroscope m. à balles de sureau.

pit head ‖ Füllort m. ‖ chambre f. de chargement. / ~ baths pl. ‖ Waschkaue f. ‖ lavoir m. à douches. / ~ building ‖

Schachtgebäude n. ‖ bâtiment m. de la mine. / ~ frame ‖ Fördergerüst n. ‖ charpente f. de monte-charges ou de puits. / ~ price ‖ Preis m. frei Grube ‖ prix m. sur le carreau de la mine. / ~ winch ‖ Förderhaspel m. ‖ treuil m. de mines.

pit lamp ‖ Grubenlampe f. ‖ lampe f. de mine. / ~ life saving station ‖ Grubenrettungsstation f. ‖ station f. de sauvetage des mines.

pitman ‖ Knappe m.; Bergmann m.; Kohlenhauer m. ‖ mineur m.; piqueur m. / ~ (Metal) ‖ Tiefofenarbeiter m. ‖ ouvrier m. des pits.

pit mouth ‖ Bergwerkseingang m. ‖ ouverture f. ou entrée f. d'une mine. / ~ (Shaft) ‖ Hängebank f. ‖ margelle f. ou recette f. d'un puits; palier m. de déchargement; pas m. de bure.

Pitot tube ‖ Staudruckdüse f. ‖ tube m. de Venturi.

pit planing machine ‖ Grubenhobelmaschine f. ‖ raboteuse f. à fosse. / ~ props pl. ‖ Stempelholz n.; Grubenholz n. ‖ bois m. de mine. / ~ ring ‖ Schachtring m. ‖ anneau m. de puits. / ~ rope ‖ Förderseil n. ‖ câble m. d'extraction. / ~ sand ‖ Grubensand m. ‖ sable m. de fouille.

pit saw (Carp) ‖ Brettsäge f.; Schrotsäge f. ‖ passe-partout m.; scie f. de long. / ~ (Mine) ‖ Grubensäge f. ‖ scie f. de mine. / quarryman's ~ ‖ Steinhauerkransäge f. ‖ scie f. de carrier.

pitticite ‖ Pittizit m.; Eisensinter m. ‖ pittizite f.; fer m. oxydé résinite.

pitting (Metal) ‖ Körnung f. ‖ piqué m.

pit water ‖ Grubenwässer npl. ‖ eau f. de mine. / ~ work ‖ Grubenarbeit f. ‖ travail m. à la mine.

pivot (Crane) ‖ Königszapfen m. ‖ pivot m. / ~ (Door) ‖ Angelzapfen m. ‖ pivot m. / ~ (Mach) ‖ Spurzapfen m.; Zapfen m. ‖ pivot m. / the ~ together with casing is cast in one piece ‖ der Zapfen ist mit dem Trog in einem Stück gegossen ‖ le pivot et la cuve sont meulés d'une seule pièce. / central ~ ‖ Königssäule f.; Königszapfen m. ‖ pivot m. central. / central ~ of a turntable ‖ Drehzapfen m. der Drehscheibe ‖ pivot m. de la plaque tournante. / front-axle ~ ‖ Vorderachszapfen m. ‖ pivot m. de l'essieu avant. / ~ of a horizontal shaft ‖ Tragzapfen m. ‖ tourillon m. / ~ of the tongue (Railw) ‖ Drehpunkt m. bei Weichen ‖ centre m. de l'aiguille. / to turn upon a ~ ‖ sich auf einem Zapfen m. drehen ‖ pivoter. / watch ~ ‖ Uhrzapfen m. ‖ pivot m. de montre.

pivot bearing ‖ Drehlager n. ‖ coussinet m. de pivot. / ~ broach (Clockm) ‖ Zapfenreibahle f. ‖ alésoir m. à pivots.

pivoted ‖ drehbar eingesetzt ‖ pivotant.

pivoter ‖ Zapfendreher m. ‖ pivoteur m.

pivot file (Watchm) ‖ Zapfenfeile f. ‖ lime f. à pivots.

pivoting bearing ‖ Kipplager n. ‖ palier m. pivotant ou à bascule.

pivot lathe ‖ Spitzendrehbank f. ‖ tour m. à pointes. / ~ pin ‖ Scharnierzapfen m. ‖ goujon m. de charnière. / ~ tooth ‖ Stiftzahn m. ‖ dent f. de pivot.

placard ‖ Anschlagzettel m.; Plakat n. ‖ affiche f.; placard m. / ~ printing establishment ‖ Plakatdruckerei f. ‖ imprimerie f. d'affiches. / ~ sticker ‖ Zettel-

ankleber m. ‖ afficheur m.; colleur m. d'affiches.

place, to ‖ unterbringen ‖ loger; placer; caser; poser. / ~ the ceiling (Shipb) ‖ wegern ‖ vaigrer. / ~ in layers ‖ in Schichten fpl. einbringen ‖ introduire par couches fpl. / ~ the paving stones in position ‖ das Pflaster setzen ‖ poser le pavé. / ~ under water ‖ unter Wasser n. lagern ‖ mettre sous l'eau f.

place ‖ Ort m. ‖ lieu m.; place f. / apparent ~ of an object ‖ scheinbarer Ort m. eines Gegenstandes ‖ lieu m. apparent d'un objet. / ~ for delivery, payment and legal questions is N. N. ‖ Erfüllungsort m. für Lieferung und Zahlung sowie Gerichtsstand ist N. N. ‖ lieu m. pour livraison, payement et tribunal compétant est N. N. / ~ of exchange ‖ Wechselplatz m. ‖ place f. cambiste. / to give a ~ ‖ anstellen ‖ assigner un poste; engager. / ~ of joint ‖ Anschlußstelle f. ‖ point m. d'assemblage *ou* d'attache. / ~ where the material is used ‖ Verwendungsstelle f. ‖ lieu m. d'utilisation. / ~ of quarantine ‖ Quarantänehafen m.; Quarantäneplatz m. ‖ port m. de quarantaine. / ~ for the sale ‖ Ausgabestelle f. ‖ débit m.; bureau m. d'émission.

place brick ‖ Weichbrand m. ‖ brique f. de rebut.

placed side by side ‖ nebeneinander angeordnet ‖ installé l'un à côté de l'autre.

placer miner ‖ Goldwäscher m. ‖ orpailleur m.

placing (Employment) ‖ Stellung f. ‖ emploi m.; place f.

plagioclase ‖ Plagioklas m. ‖ plagioklase f.

plague ‖ Seuche f. ‖ peste f.

plaid ‖ Plaid n.; Hülle f. (aus Tuch) ‖ plaid m.

plain ‖ eben; glatt; flach ‖ plain; uni; plat. / to make ~ ‖ klar *oder* deutlich machen; verdeutlichen ‖ rendre clair; élucider.

plain aerial ‖ einfacher Luftleiter m. ‖ antenne f. simple. / ~ colour sizer (Pap) ‖ Grundierer m. ‖ fonceur m. / ~ cylindrical grinding ‖ Wellenschliff m. ‖ rectification f. des arbres cylindriques. / ~ horizontal milling machine ‖ Einfachwagerechtfräsmaschine f. ‖ fraiseuse f. horizontale simple. / ~ joint ‖ stumpfe Fuge f.; Leimfuge f. ‖ joint m. à plat-point. / ~ machine ‖ Maschine f. zur Herstellung glatter Ware ‖ machine f. pour maille unie. / ~ mahogany ‖ glattes Mahagoniholz n. ‖ acajou m. uni. / ~ moulding of a window ‖ Fenstergesims n. ‖ corniche f. *ou* dessus m. d'une fenêtre. / ~ pattern ‖ glattes Muster n. ‖ dessin m. uni. / ~ printing ‖ Flachdruck m. ‖ impression f. sur une surface plate. / ~ ribbon ‖ glattes Band n. ‖ ruban m. uni. / ~ striped goods pl. (Textil) ‖ glatte Ringelware f. ‖ tricot m. rayé en maille unie. / ~ super-refined steel ‖ unlegierter Edelstahl m. ‖ acier m. spécial non allié. / ~ tile ‖ Biberschwanz m.; Flachziegel m. ‖ tuile f. plate *ou* à crochet. / ~ tool holder ‖ einfacher Support m. ‖ support m. porte-outil simple. / ~ tool post *see* ~ tool holder.

plain ‖ Flachland n. ‖ plaine f. / elevated ~ ‖ Hochebene f. ‖ plaine f. élevée; plateau m. / ~ of a wall ‖ Mauerfläche f.; Mauerflucht f. ‖ nu m. d'un mur.

plaint ‖ Klageschrift f. ‖ plainte f.

plaintiff ‖ Ankläger m. ‖ accusateur m.

plait, to ‖ flechten ‖ natter; tresser.

plait ‖ Zopf m. ‖ natte f.; tresse f.

plaited cord ‖ geflochtene Leine f. ‖ corde f. tressée; drisse f. / ~ heel ‖ plattierte Ferse f. ‖ talon m. en maille vanisée. / ~ line *see* ~ cord. / ~ shoe ‖ Strohpantoffel m. ‖ chausson m. de nattes. / ~ tress (Hair) ‖ Haarflechte f. ‖ natte f. de cheveux. / ~ tubular goods pl. ‖ plattierte Schlauchware f. ‖ tricot m. tubulaire en maille vanisée. / ~ wirework for enclosures ‖ Drahtgeflecht n. zur Einfriedigung ‖ treillis m. en fil métallique pour enceintes.

plaiter ‖ Plisseepresser m. ‖ plisseur m. / osier ~ ‖ Weidenflechter m. ‖ tresseur m. d'osier. / reed ~ ‖ Rohrflechter m. ‖ tresseur m. de roseaux.

plaiting fabrics ‖ Plissieren n. von Stoffen ‖ plissage m. d'étoffes. / ~ machine ‖ Flechtmaschine f. ‖ machine f. à tresser. / ~ material ‖ Flechtstoff m. ‖ matière f. à tresser. / ~ tackle ‖ Plattiervorrichtung f. ‖ dispositif m. à vaniser. / ~ thread ‖ Plattierfaden m. ‖ fil m. de vanisage.

plan, to ‖ entwerfen; einen Plan m. machen ‖ projeter.

plan ‖ Grundriß m.; Plan m.; Riß m. ‖ plan m.; tracé m. / according to ~ ‖ plangemäß ‖ selon plan m. / ~ back ‖ Hinteransicht f. ‖ élévation f. de derrière. / ~ for a building ‖ Bauriß m. ‖ plan m. d'un bâtiment. / ~ of coal seams ‖ Flözkarte f. ‖ carte f. stratigraphique. / ~ of the concession (Mine) ‖ Situationsriß m. der Mutung ‖ plan m. de la demande de concession. / ~ construction ‖ Konstruktionszeichnung f. ‖ plan m. de construction. / to draw up a ~ for the current supply of the whole area ‖ einen Stromversorgungsplan m. für das Gesamtgebiet aufstellen ‖ établir un plan m. de distribution de courant pour toute la région. / ~ of the deck ‖ Decksplan m. ‖ plan m. du pont. / ~ of the diagonals (Shipb) ‖ Sentenriß m. ‖ plan m. des lisses plans. / geometrical ~ *see* ground ~. / ground ~ ‖ geometrischer Grundriß m. ‖ plan m. géométral *ou* objectif. / ~ for housebuilding ‖ Bebauungsplan m. ‖ plan m. de construction. / ~ of management ‖ Wirtschaftsplan m. ‖ plan m. d'exploitation. / ~ of a mine ‖ Grubenriß m.; Grubenbild n.; Grubenplan m. ‖ plan m. de mine. / perspective ~ ‖ perspektivischer Grundriß m. ‖ plan m. perspectif. / pipework *see* ~ of tubing. / principal ~ ‖ Generalplan m.; Übersichtsplan m. ‖ plan m. principal. / ~ of site ‖ Situationsplan m. ‖ tracé m. général. / to take a ~ ‖ einen Grundriß m. aufnehmen ‖ lever un plan m. / ~ of tubing ‖ Rohrplan m.; Rohrleitungsplan m. ‖ plan m. de tuyautage. / vertical ~ ‖ Aufriß m.; Profil n.; Vertikalprojektion f. ‖ élevation f.; projection f. verticale.

planar ‖ Planar n. ‖ planar m.

planchet ‖ Münzplatte f.; Scheibe f. für Münzen ‖ flan m. / ~ file ‖ Schrotfeile f. ‖ lime f. à ébarber.

plan drawing ‖ Planzeichnung f. ‖ dessin m. des plans *ou* des cartes topographiques.

plane ‖ eben; flach ‖ plan; plain; plat.

plane, to ‖ ebnen; planieren; abgleichen ‖ aplanir; planer; coller. / ~ (Carp)

hobeln; abhobeln ‖ raboter; aplanir avec le rabot. / ~ (Types) ‖ bestoßen ‖ couper. / ~ down the form (Print) ‖ die Form klopfen ‖ taquer la forme. / ~-off ‖ abhobeln ‖ raboter. / -off timber ‖ Holz n. abhobeln *oder* schruppen ‖ dégrossir le bois. / ~ over ‖ überschlichten ‖ replanir. / ~ roughly (Join) ‖ schruppen ‖ corroyer; dégrossir l'ouvrage. / ~ a switch tongue ‖ eine Weichenzunge abhobeln ‖ raboter une aiguille.

plane ‖ Ebene f.; Fläche f. ‖ plan m. / ~ (Aero) ‖ Tragfläche f.; Flügel m. ‖ plan m.; surface f. portante *ou* sustentatrice. / ~ (Airplane) ‖ Flugzeug n. ‖ aéroplane m.; avion m. / ~ (Bot) *see* plane tree. / ~ (Curr) ‖ Ziehmesser n. ‖ couteau m. à deux mains; plane f. droite. / ~ (Join) ‖ Hobel m. ‖ rabot m. / ~ (Mine) ‖ Förderstrecke f. ‖ voie f. de roulage. / balancing ~ (Aero) ‖ Stabilisierungsfläche f. ‖ aile f. stabilisatrice; stabilisateur m. / bevelling ~ ‖ Gehrungshobel m. ‖ guillaume m. d'onglet. / ~ of cleavage (Miner) ‖ Spaltungsfläche f. ‖ face f. du clivage. / cornice ~ ‖ Kehlhobel m.; Simshobel m.; Karnieshobel m. ‖ grain m. d'orge; mouchette f.; rabot m. à moulures. / ~ of a crystal ‖ Kristallfläche f. ‖ face f. *ou* facette f. *ou* plan m. d'un cristal. / dovetail ~ ‖ Grathobel m. ‖ bouvet m. mâle; feuilleret m.; rabot m. à languette. / ~ of flotation ‖ Schwimmebene f. ‖ plan m. de flottaison. / focal ~ ‖ Brennebene f. ‖ plan m. focal. / grooving ~ *see* plough ~. / ~ with handle ‖ Hobel m. mit Nase ‖ varlope f. / hollowing ~ ‖ Kehlhobel m. ‖ gorget m.; gorgefouille f.; varlope f. onglée. / hollownosed ~ ‖ Rundhobel m. ‖ mouchette f. / horizontal ~ ‖ Horizontebene f. ‖ plan m. horizontal. / inclined ~ (Build) ‖ Rampe f.; Steigung f.; Auffahrt f. ‖ rampe f.; chemin m. taluté. / inclined ~ (Geom) ‖ geneigte Fläche f.; schiefe Ebene f. ‖ plan m. incliné. / inclined ~ (Mine) ‖ Bremsberg m.; Schleppbahn f. ‖ plan m. incliné de trainage. / inclined ~ with rope ‖ Seilebene f. ‖ plan m. à câble. / to give iron to the ~ *see* plane iron, to drive-on the. / lateral ~ of a crystal ‖ Kristallseitenfläche f. ‖ face f. latérale d'un cristal. / light ~ ‖ Leichtflugzeug n. ‖ moto-aviette f. / long ~ ‖ Fügehobel m. ‖ varlope f.; galère f. / moulding ~ *see* cornice ~. / ogee ~ *see* cornice ~. / ~ of osculation ‖ Oskulationsebene f. ‖ plan m. osculateur. / overhanging ~ (Aero) ‖ überhängende Tragfläche f. ‖ plan m. débordant. / ~ that takes to pieces ‖ zerlegbarer Hobel m. ‖ bouvet m. brisé *ou* démontable. / plough ~ ‖ Nuthobel m. ‖ bouvet m. à approfondir *ou* à rainure *ou* femelle; rabot m. à rainure. / plough ~ with movable stop ‖ Nuthobel m. mit verstellbarem Auflauf ‖ bouvet m. à approfondir *ou* à dégorger. / plough and tongue ~ ‖ Nuthobel m. und Federhobel m.; voller Spundhobel m. ‖ bouvet m. à fourchement. / ~ of polarization ‖ Polarisationsebene f. ‖ plan m. de polarisation. / principal ~ ‖ Hauptebene f. ‖ plan m. principal. / ~ of projection ‖ Projektionsebene f. ‖ plan m. de projection. / pursuit ~ ‖ Jagdflugzeug n. ‖ avion m. de chasse. / rail ~ ‖ Schienenhobel m. ‖ rabot m. pour rails. / reed ~ ‖ Hobel m. für nebeneinanderliegende Rundstäbe ∖ rabot m.

à chantourner. / ~ of reference ‖ Einstellebene f. ‖ plan m. de mise au point. / ~ of reflection ‖ Reflexionsebene f. ‖ plan m. de réflexion. / ~ of refraction ‖ Brechungsebene f. ‖ plan m. de réfraction. / ~ propelled by rockets ‖ Raketenflugzeug n. ‖ avion m. à réaction; avion-fusée m. / round ~ ‖ Rundstabhobel m. ‖ rabot m. à boudin. / round-nosed ~ ‖ Rauhhobel m.; Schrupphobel m. ‖ rabot m. debout; riflard m. / round-sole ~ *see* hollowing ~. / router ~ ‖ Grundhobel m. ‖ guimbarde f. / screw plough ~ ‖ Nuthobel m. mit Stellung ‖ bouvet m. brisé *ou* de deux pièces. / spout ~ ‖ runder Hobel m. ‖ rabot m. rond. / supporting ~ (Aero) ‖ Tragfläche f. ‖ plan m. (sustentateur); surface f. sustentatrice. / sustaining ~ *see* supporting ~. / tail ~ (Aero) ‖ Schwanzfläche f. ‖ plan m. de queue. / top and bottom ~ (Aero) ‖ obere und untere Tragfläche f. ‖ plan m. supérieur et inférieur. / advanced training ~ ‖ Übungsflugzeug n. ‖ avion m. d'entraînement. / primary training ~ ‖ Schulflugzeug n. ‖ avion m. école. / two and a half ~ ‖ Zweieinhalbdecker m. ‖ triplan m. ayant un de ses plans plus petits que les autres. / vegetable ~ ‖ Gemüsehobel m. ‖ rabot m. à légumes. / vertical ~ (Geom) ‖ vertikale *oder* senkrechte Ebene f. ‖ plan m. vertical.

plane aerial ‖ Flächenantenne f. ‖ antenne f. en nappe. / ~ angle ‖ Flächenwinkel m. ‖ angle m. plan *ou* rectiligne.

planed board ‖ gehobeltes Brett n. ‖ planche f. rabotée. / ~ soap ‖ gehobelte Seife f.; Seifenspäne mpl. ‖ savon m. râpé.

plane element ‖ Flächenelement n. ‖ élément m. de surface. / face of ~ *see* sole of ~. / ~ glass mirror silvered in front ‖ Planglasspiegel m. mit Vorderversilberung ‖ miroir m. plan en verre argenté sur la face avant. / ~ grate ‖ Planrost m. ‖ grille f. horizontale. / ~ grate bar ‖ Planroststab m. ‖ barreau m. de grille horizontale. / ~ ground mirror ‖ eben geschliffener Spiegel m. ‖ miroir m. poli planement.

plane iron ‖ Hobeleisen n. ‖ fer m. de rabot. / double ~ ‖ doppeltes Hobeleisen n. ‖ fer m. double de rabot. / to drive on the ~ ‖ das Hobeleisen n. vortreiben ‖ donner du fer au rabot. / moulding ~ ‖ Kehleisen n.; Kehlhobeleisen n. ‖ fer m. à moulure *ou* de grain d'orge. / round-nosed ~ ‖ Schrupphobeleisen n. ‖ fer m. du rabot *ou* du riflard. / to set-in the ~ ‖ das Hobeleisen n. einsetzen ‖ mettre en fût le fer de rabot. / skew and bevelled ~ ‖ Plattbankeisen n. ‖ fer m. de rabot plate-bande.

plane iron grinder ‖ Hobeleisenschleifmaschine f. ‖ machine f. à affûter les lames de raboteuses.

plane maple ‖ Spitzahorn m. ‖ érable m. plane.

plane mirror ‖ Planspiegel m. ‖ miroir m. plan. / ~ of speculum metal ‖ Planspiegel m. aus Spiegelmetall ‖ miroir m. plan en métal spéculaire.

plane parallel glass ‖ Planparallelglas n. ‖ verre m. à faces planes et parallèles. / ~ plate ‖ planparallele Glasplatte f. ‖ lame f. de verre planparallèle.

plane, apparatus for testing the ~ parallelism of glass plates ‖ Gerät n. zur Prüfung der Planparallelität f. von Glasplatten ‖ appareil m. pour le contrôle des lames planparallèles. / ~ polished mirror *see* ~-ground mirror.

planer (Build) ‖ Planierer m. ‖ régaleur m. / ~ (Mach tool) ‖ Hobelmaschine f. ‖ raboteuse f. / ~ (Mach tool worker) ‖ Hobler m. ‖ raboteur m. / ~ (Print) ‖ Klopfholz n. ‖ taquoir m. / high-speed ~ ‖ Schnellhobler m. ‖ raboteuse f. rapide. / open side ~ ‖ Einständerhobelmaschine f. ‖ raboteuse f. à montant unique. / plate edge ~ ‖ Blechkantenhobelmaschine f. ‖ machine f. à chanfreiner les tôles. / rail ~ ‖ Schienenhobel m. ‖ rabot-rails m.; limeuse f. pour rails. / single-purpose ~ ‖ Hobelmaschine f. für Sonderzwecke ‖ raboteuse f. spéciale.

planer knife ‖ Hobelmesser n. ‖ lame f. de raboteuse. / ~ proof (Print) ‖ Klopfholzabzug m. ‖ épreuve f. du taquoir.

planer-type grinding machine ‖ Doppelständerschleifmaschine f. ‖ machine f. à deux montants à rectifier. / ~ horizontal spindle grinding machine ‖ Langflächenschleifmaschine f. mit Horizontalschleifkopf ‖ machine f. à rectifier les surfaces planes avec porte-meule horizontal. / ~ vertical spindle grinding machine ‖ Langflächenschleifmaschine f. mit Vertikalschleifkopf ‖ machine f. à rectifier les surfaces planes avec porte-meule vertical.

plane set-hammer ‖ Setzhammer m. ‖ chasse f. carrée. / sole of ~ ‖ Hobelbahn f.; semelle f. de fût de raboteuse. / ~ stock ‖ Hobelkasten m.; Kasten n. ‖ fût m. de rabot.

planet ‖ Planet m. ‖ planète f.

plane table ‖ Meßtisch m. ‖ planchette f. / ~ survey ‖ Aufnahme f. nach dem Meßtisch ‖ levé m. à la planchette. / ~ survey sheet ‖ Meßtischblatt n. ‖ feuille f. topographique levée à la planchette.

planet camera ‖ Planetenkamera f. ‖ chambre f. noire pour les planètes. / ~ gear ‖ Planetengetriebe n. ‖ engrenage m. planétaire. / ~ stirring mechanism ‖ Planetenrührwerk n. ‖ agitateur m. planétaire. / ~ wheel ‖ Planetenrad n. ‖ roue f. planétaire. / ~ wheel motion ‖ Planetenbewegung f. ‖ mouvement m. planétaire.

plane tree ‖ Platane f. ‖ platane m.

plane wood ‖ Platanenholz n. ‖ bois m. de platane.

planimeter, to ~ a diagram ‖ ein Diagramm n. planimetrieren ‖ planimétrer un diagramme.

planimeter ‖ Planimeter n. ‖ planimètre m. / polar ~ ‖ Polarplanimeter n. ‖ planimètre m. polaire.

planimetry ‖ Planimetrie f. ‖ planimétrie f.

planing the edges of plates ‖ Behobeln n. der Blechkanten ‖ rabotage m. des arêtes de tôles. / ~ of sleepers ‖ Hobeln n. der Schwellen ‖ entaillage m. des traverses.

planing bench ‖ Hobelbank f. ‖ établi m. de menuisier. / ~ device for reamers ‖ Reibahlenabziehvorrichtung f. ‖ dispositif m. d'aiguisage pour alésoirs. / ~ hammer ‖ Pritschhammer m.; Abrichthammer m. ‖ marteau m. de parage. / ~ knife ‖ Hobelmesser n. ‖ lame f.

de rabot. / ~ knife (Curr) ‖ Schabeisen n. ‖ lame f. à planer.

planing machine ‖ Hobelmaschine f. ‖ raboteuse f. / ~ for bevel wheels ‖ Kegelradhobelmaschine f. ‖ raboteuse f. à tailler des pignons coniques. / cask ~ ‖ Faßabhobelmaschine f. ‖ machine f. à blanchir *ou* à raboter les fûts. / key way ~ ‖ Nutenziehmaschine f. ‖ machine f. à rainurer. / parallel ~ ‖ Tangentialhobelmaschine f.; Langhobelmaschine f. ‖ raboteuse f. longitudinale. / plate edge ~ ‖ Kantenhobelmaschine f. ‖ machine f. à raboter *ou* chanfreiner les tôles. / switch blade ~ ‖ Weichenzungenhobelmaschine f. ‖ raboteuse f. d'aiguilles de changement de voie. / wood ~ ‖ Holzhobelmaschine f. ‖ machine f. à raboter le bois.

planing machine hand ‖ Maschinenhobler m. ‖ raboteur m. à la machine. / ~ table ‖ Hobelmaschinenschlitten m. ‖ chariot m. de machine à raboter. / ~ tenter *see* ~ hand.

planing machinist *see* planing machine hand. / ~ mill ‖ Hobelwerk n. ‖ atelier m. de rabotage. / ~ and thicknessing machine ‖ Dicktenhobelmaschine f. ‖ machine f. à raboter tirant des bois d'épaisseur. / ~ tool ‖ Hobelmeißel m. ‖ outil m. raboteur; acier m. de raboteuse. / ~ work ‖ Hobelarbeit f. ‖ travail m. de rabotage.

planish, to ‖ schlichten ‖ planer.

planished sheet-iron ‖ poliertes Eisenblech n. ‖ tôle f. lustrée.

planisher ‖ Planierer m. ‖ planeur m.

planishing hammer ‖ Schlichthammer m. ‖ marteau m. à planer. / ~ knife (Coop) ‖ Geradeisen n. ‖ plane f. à lame droite. / ~ mill (Wiredr) ‖ Plättmühle f. ‖ moulin m. à écacher *ou* à laminer le fil. / ~ stake ‖ Polierstock m. ‖ bas m. à planer.

planisphere ‖ Planisphärium n.; Planiglob n. ‖ planisphère m.

plank, to (Carp) ‖ verschalen; dielen ‖ planchéier. / ~ (Hatt) ‖ (ver)filzen feutrer. / ~ (Mine) ‖ verzimmern ‖ coffrer / ~ a ship ‖ ein Schiff n. beplanken ‖ border un navire.

plank (Carp) ‖ Planke f.; Diele f.; Brett n.; Bohle f. ‖ planche f.; ais m.; madrier m. / ~ (Coin) ‖ Münzplatte f.; Scheibe f.; flan m. / ~ (Mine) ‖ Schwarte f.; dosse f. ~ (Print) ‖ Laufbrett n.; Schienenbrett n. ‖ table f. d'une presse; berceau m. / to boil a ~ ‖ eine Planke f. garmachen ‖ chauffer un bordage. / exterior ~ ‖ Außenplanke f. ‖ bordage m. du revêtement extérieur. / fir ~ ‖ tannene Bohle f. ‖ tavillon m. / half-inch ~ ‖ Schalbrett n.; halbzölliges Brett n. ‖ planche f. *ou* ais m. d'un mi-pouce d'épaisseur. / interior ~ (Shipb) ‖ Garnierplanke f.; Wegerungsplanke f. ‖ bordage m. du revêtement intérieur; vaigre f. / middle ~ ‖ Kernbrett n. ‖ planche f. de cœur. / oaken ~ ‖ eichene Bohle f. ‖ madrier m. en chêne. / outer ~ ‖ Seitenbrett n. ‖ planche f. extérieure. / ~ of a ship ‖ Schiffsplanke f. ‖ bordage m. d'un navire. / squared ~ ‖ vollkantiges Brett n. ‖ madrier m. équarri. / thick ~ ‖ Bohle f. ‖ madrier m. / ~ two inches thick ‖ zwei Zoll dickes Brett n. ‖ planche f. de deux pouces d'épaisseur. / wooden ~ ‖ Holzdiele f.;

Holzbohle f.; Holzplanke f. ‖ planche f. en bois.

plank bed ‖ Ladepritsche f. ‖ plateau m. de chargement. / ~ bit with square shank ‖ Holzbohrer m. mit Vierkant-kopf ‖ mèche f. de menuisier à tête carrée. / ~ butt (Shipb) ‖ Plankenstoß m. ‖ about m. de bordages.

planked ceiling ‖ Schaldecke f. ‖ plafond m. cloisonné.

plank end ‖ Plankenende n. ‖ bout m. *ou* tête f. de bordage. / ~ flattener (Coin) ‖ Plätthammer m.; Schrötlingshammer m. ‖ flatoir m.

planking ‖ Verschalung f. ‖ cuvelage m. / ~ of a bridge ‖ Brückenbelag m. ‖ plancher m. *ou* platelage m. d'un pont. / ~ of a lock gate ‖ Schleusentorbeklei-dung f. ‖ doublage m. d'une porte d'écluse. / covered with timber ~ ‖ durch Holzbelag m. abgedeckt ‖ garni d'une couche de madriers. / the turntable is covered with a timber ~ ‖ die Dreh-scheibe durch einen Bohlenbelag ab-gedeckt ‖ la plaque tournante est garnie d'une couverture en madriers.

plank log ‖ Sägeblock m. ‖ bloc m. de sciage; membrure f.; doubleau m. / ~ nail ‖ Brettnagel m.; Dielennagel m. ‖ clou m. à planche. / ~ nail (Shipb) ‖ Plankenspieker m.; Spieker m. ‖ clou m. à madrier *ou* à planche. / ~ revetment ‖ Bretterbekleidung f. ‖ revêtement m. en charpente. / ~ sawing ‖ Langholzsägen n. ‖ sciage m. de long. / ~ sawyer ‖ Brett-schneider m. ‖ scieur m. de long. / ~ seam (Shipb) ‖ Plankennaht f. ‖ couture f. entre les virures de bordage. / ~ sheer (Shipb) ‖ Schandeck n.; Schandeckel m. ‖ plat-bord m. / ~ sheer rail (Shipb) ‖ Schandeckelleiste f. ‖ liston m. du plat-bord. / ~ timber ‖ Klotz n.; Sägeblock m. ‖ bloc m. en bois. / ~ tubbing ‖ wasser-dichter Schachtausbau m. durch Holz-ringe ‖ cuvelage m. circulaire en bois; tubage m. en bois.

plankways-cut wood ‖ Langholz n. ‖ bois m. de fil.

planning, industrial ~ ‖ Industrieverein-heitlichung f. ‖ unification f. industrielle.

plano-concave ‖ plankonkav ‖ plancon-cave.

plano-convex ‖ plankonvex ‖ planconvexe.

plano-cylindrical ‖ planzylindrisch ‖ plan-cylindrique.

planometer ‖ Richtplatte f.; Planometer n. ‖ planomètre m.

plano-milling machine ‖ Langtischfräs-maschine f. ‖ fraiseuse f. raboteuse.

plan sailing ‖ Plansegeln n. ‖ navigation f. à l'aide d'un carte plane.

plansifter ‖ Plansichter m. ‖ plansichter m.; blutoir m. horizontal. / double-box ~ ‖ Zweikastenplansichter m. ‖ plansichter m. à deux caisses. / free-swinging ~ ‖ freischwingender Plansichter m. ‖ plan-sichter m. autobalanceur. / two-case ~ *see* double box ~.

plansifting machine *see* plansifter.

plant, to ‖ pflanzen ‖ planter. / ~ struts pl. (Mine) ‖ absteifen ‖ étrésilloner.

plant (Bot) ‖ Pflanze f. ‖ plante f. / ~ (Mach) ‖ Anlage f.; Betrieb m. ‖ établis-sement m.; installation f.; fabrique f.; usine f. / ~ (Mine) ‖ Grubenanlage f. ‖ établissement m. de mines. / ~ (Tool) ‖ Werksgerät n. ‖ outillage m. / ~ for broken stone manufacturing ‖ Schotter-

werkanlage f. ‖ installation f. de con-cassage de pierres. / ~ with the recovery of by-products ‖ Anlage f. mit Gewinnung f. von Nebenerzeugnissen ‖ usine f. avec récupération des sous-produits. / ~ for producing carbonic acid ‖ Kohlensäure-gewinnungsanlage f. ‖ installation f. de production d'acide carbonique. / ~ for chemical manure manufacturing ‖ Ge-winnungsanlage f. für künstliche Dünge-mittel ‖ installation f. pour la fabrication d'engrais chimique. / complete ~ ‖ voll-ständige Anlage f. ‖ installation f. com-plète. / ~ worked by a contractor ‖ von einem Unternehmer betriebenes Werk n. ‖ l'usine f. dont l'exploitation est confiée à un entrepreneur. / ~ for decoration (Bot) ‖ Schmuckgras n. ‖ arbuste m. pour décoration. / ~ without distribution boxes (Tel) ‖ starres Kabeladernetz n. ‖ réseau m. souterrain sans boîtes de coupure. / ~ electric ~ ‖ elektrische Be-triebsanlage f. ‖ installation f. électrique. / ~ for electric light ‖ elektrische Licht-anlage f. ‖ installation f. d'éclairage électrique. / ~ of the forges ‖ Hammer-werk n. ‖ forge f. / herbaceous ~ (Bot) ‖ Staudenpflanze f. ‖ plante f. herbacée. / hydraulic ~ ‖ hydraulische Anlage f. ‖ installation f. hydraulique. / industrial ~ ‖ industrielle Anlage f. ‖ établissement m. industriel. / industrial ~ (Bot) ‖ Ge-werbepflanze f. ‖ plante f. industrielle. / ~s pl. for use in industry and medicine (Bot) ‖ Pflanzen fpl. für Gewerbe- und Heilgebrauch ‖ plantes fpl. industrielles et médicinales. / living ~ (Bot) ‖ lebende Pflanze f. ‖ plante f. vivante. / manu-facturing ~ *see* industrial ~. / mechani-cal ~ ‖ maschinelle Anlage f. ‖ installa-tion f. mécanique. / medicinal ~ (Bot) ‖ Pflanze f. zum Heilgebrauch ‖ plante f. employée en médecine. / ~ for preparing moulding sand ‖ Formsandbereitungs-anlage f. ‖ préparation f. du sable de moulage. / ~ of nursery (Bot) ‖ Baum-schulenpflanze f. ‖ plante f. de pépinière. / ~ for use in perfumery (Bot) ‖ Pflanze f. zur Verwendung in der Parfümerie ‖ plante f. utilisée en parfumerie. / pre-pared ~ (Bot) ‖ sterilisierte Pflanze f. ‖ plante f. naturalisée. / ~ at rest ‖ Anlage f. außer Betrieb ‖ mine f. inoccupée *ou* hors service. / the ~ stopped work after several years of heavy losses ‖ der Betrieb wurde nach mehreren verlust-reichen Jahren eingestellt ‖ après plu-sieurs années riches en pertes l'exploita-tion f. fut suspendue.

plantation ‖ Anpflanzung f.; Pflanzung f.; plantation f. / implements pl. for ~s ‖ Plantagengerät n. ‖ ustensiles fpl. pour plantations.

plant, capacity of ~ ‖ Leistungsfähigkeit f. der Betriebseinrichtung ‖ capacité f. de l'installation.

planter (Agr machine) ‖ Pflanzmaschine f. ‖ planteuse f. / ~ (Agr person) ‖ Pflanzer m. ‖ planteur m.

plant, expenditure on ~ ‖ Ausgaben pl. für die Betriebseinrichtung ‖ dépenses fpl. en travaux d'installation.

planting machine, turnip ~ ‖ Rübenpflanz-maschine f. ‖ planteur m. mécanique de betteraves.

plant oil ‖ Pflanzenöl n. ‖ huile f. végétale. / ~ protective ~ ‖ Pflanzenschutzmittel n. ‖ produit m. préservatif pour plantes. /

~ switchboard ‖ Stationsschalttafel f. ‖ tableau m. de distribution de la centrale.

plan verification ‖ Planung f. ‖ établisse-ment m. d'un projet.

plashed lime ‖ angemachter Kalk m. ‖ bain m. de chaux.

plasma ‖ Plasma m.; grüner Chalcedon m. ‖ plasma m.

plaster, to ‖ putzen; verputzen ‖ plâtrer; enduire. / ~ the mortar ‖ den Mörtel m. auftragen ‖ épigeonner le mortier. / ~ the sound proof floor ‖ den Fehlboden m. verfüllen ‖ entrevoûter la solive. / ~ a wall ‖ eine Mauer verputzen ‖ en-duire un mur.

plaster (Arch) ‖ Verputz m.; Putz m. ‖ enduit m.; crépi m. / ~ (Med) ‖ Pflaster n. ‖ emplâtre m. / ~ (Miner) ‖ Gips m. ‖ plâtre m. / ~ (Mould) ‖ Abdruckmasse f.; ‖ plâtre m. / conductor in the ~ ‖ im Putz verlegte Leitung f. ‖ conducteur m. enrobé dans le plâtrage. / court ~ ‖ Englisches Pflaster n. ‖ taffetas m. gommé. / ~ for facings ‖ Fassadenputz m. ‖ enduit m. de parement. / ~ of Paris ‖ gebrannter Gips m. ‖ plâtre m. cuit *ou* de Paris. / sculptor's ~ ‖ Bildhauergips m. ‖ plâtre m. de moulage. / finely sifted ~ ‖ feingesiebter Gips m. ‖ plâtre m. au sas *ou* fin. / unsifted ~ ‖ grober *oder* un-gesiebter Gips m. ‖ plâtre m. gros; gros plâtre m. / well-burnt ~ ‖ fetter Gips m. ‖ plâtre m. gras.

plaster bandage ‖ Gipsbinde f. ‖ bandage m. en plâtre. / ~ block ‖ Gipsplatte f. ‖ car-reau m. de plâtre. / ~ burning ‖ Gips-brennerei f. ‖ cuisson f. du plâtre. / ~ cast ‖ Gipsabguß m. ‖ plâtre m.; moule m. en plâtre. / ~ casting factory ‖ Gipsformerei f. ‖ atelier m. de moulage en plâtre. / ~ ceiling ‖ Stuckdecke f. ‖ plafond m. de plâtre. / ~ ceiling enter-prise ‖ Gipsdeckenwerk n. ‖ entreprise f. de plafonnage.

plasterer ‖ Stukkatör m. ‖ plafonneur m. / ~'s trowel ‖ Kelle f. zum Bewerfen mit Putz ‖ grande aplanissoire f.

plaster floor ‖ Gipsestrich m. ‖ aire f. en plâtre. / to pour, to beat and to rub the ~ floor ‖ den Gipsestrich m. gießen, schlagen und schleifen ‖ étendre, battre et frotter l'aire f. / hardening of the ~ ‖ Gipserhärtung f. ‖ durcissage m. du plâtre. / ~ hatchet ‖ Gipshaue f.; Tünchhacke f. ‖ hacherau m. / ~ image ‖ Gipsfigur f. ‖ figure f. en plâtre.

plastering ‖ Putz m.; Verputz m. ‖ enduit m.; crépissage m. / ~ on the ceiling ‖ Gips-putz m. an der Decke ‖ crépi m. *ou* en-duit m. de plâtre au plafond. / rough ~ ‖ Rauhputz m. ‖ enduit m. fouetté.

plastering contractor ‖ Stukkatörgeschäft n. ‖ entrepreneur m. de plafonnage. / ~ material ‖ Verkittungsmaterial n. ‖ matière f. à mastiquer.

plaster kiln ‖ Gipsofen m. ‖ four m. à plâtre. / ~ mortar ‖ Stuckmörtel m. ‖ mortier-stuc. / ~ mould ‖ Gipsform f. ‖ moule m. *ou* coquille f. en plâtre. / ~ mould (Letter-f) ‖ Gipsmatrize f. ‖ moule m. de plâtre. / ~ moulding ‖ Gips-formen n. ‖ moulage m. en plâtre. / ~ partition ‖ Gipswand f. ‖ cloison f. en plâtre. / ~ paste ‖ Gipsbrei m. ‖ bouillie f. de plâtre. / ~ slab ‖ Gipsdiele f. ‖ planche f. de plâtre. / ~ stucco ‖ Gipsputz m.; Gipsstuck m. ‖ enduit m. en plâtre. / ~ ware ‖ Gipswaren fpl. ‖ objets mpl. en

plâtre. / ~ works pl. ‖ Gipswerk n. ‖ usine f. à plâtre.

plastic ‖ bildsam; plastisch ‖ plastique. / ~ art ‖ Bildhauerei f. ‖ art m. plastique. / ~ article of wax ‖ Formerarbeit f. aus Wachs ‖ article m. plastique en cire. / ~ clay ‖ Modellierton m. ‖ terre f. à modeler. / ~ insulating material ‖ plastisches Isoliermaterial n. ‖ isolant m. plastique. / ~ material ‖ plastischer Stoff m. ‖ matière f. *ou* substance f. plastique. / ~ substance *see* ~ material. / ~ wax for dental surgery ‖ plastisches Wachs n. für zahnärztliche Zwecke f. ‖ cire f. plastique pour chirurgie dentaire.

plasticity ‖ Bildsamkeit f. ‖ plasticité f.

plastics pl. ‖ Plastik f. ‖ plastique f.

platband ‖ Streifen m.; Borte f. ‖ plate-bande f.; voûte f. plate. / ~ (Gard) ‖ Rabatte f. ‖ plate-bande f.; bordure f.

plate, to ‖ plattieren ‖ plaquer. / ~ with nickel ‖ vernickeln ‖ nickeler. / ~ with silver ‖ versilbern ‖ argenter.

plate (Arch) ‖ Platte f.; Tafel f. ‖ dalle f.; lame f.; plaque f.; table f. / ~ (Dishes) ‖ Teller m. ‖ assiette f. / ~ (Engr) ‖ Kupferstich m.; Kupferstichplatte f. ‖ estampe f. *ou* gravure f. en taille-douce; gravure f. en cuivre. / ~ (Geol) ‖ Flöz n. ‖ nappe f. / ~ (Iron) ‖ Blech n. ‖ tôle f. / ~ (Mach) ‖ Unterlagsplatte f. ‖ plaque f.; platine f.; selle f. / ~ (Phot) ‖ Platte f. ‖ plaque f. / ~ (Radio) ‖ Anode f. ‖ anode f.; plaque f. / ~ for accessories ‖ Zubehörblech n. ‖ tôle f. pour accessoires. / ~ of anvil ‖ Amboßbahn f. ‖ face f. *ou* table f. de l'enclume. / armour ~ ‖ Panzerplatte f. ‖ plaque f. de blindage *ou* cuirassée. / artificial stone ~ ‖ Kunststeinplatte f. ‖ plaque f. en pierre factice. / ~ covering the balance (Watchm) ‖ Unruhedeckplatte f. ‖ coq m. / bed ~ *see* foundation ~. / bent ~ ‖ gebogene Platte f. ‖ plaque f. cintrée. / boiler shell ~ ‖ Kesselwand f. ‖ paroi f. de chaudière. / bottom ~ ‖ Bodenblech n. ‖ tôle f. de fond. / brass ~ ‖ Messingblech n. ‖ tôle f. de laiton. / buckled ~ ‖ Buckelblech n. ‖ tôle f. bombée *ou* emboutie à panneau. / bullet-proof ~ ‖ kugelsicheres Blech n. ‖ tôle f. à l'épreuve des balles d'armes de feu. / centre ~ ‖ Drehzapfenlager n.; Drehgestellzapfenlager n.; ‖ crapaudine f. de pivot du bogie. / ~ of centre piece ‖ Tischaufsatzplatte f. ‖ plateau m. de table. / channelled ~ *see* checquered ~. / ~ for chasing work ‖ Blech n. für Zieharbeiten ‖ tôle f. pour emboutissage. / checquered ~ ‖ Riffelblech n. ‖ tôle f. gaufrée. / collecting ~ ‖ Sammelplatte f. ‖ plaque f. collectrice. / corrugated ~ ‖ Wellblech n. ‖ tôle f. gaufrée *ou* ondulée. / ~ cut to size ‖ nach Maß n. geschnittene Platte f. ‖ plaque f. coupée à dimension. / deep ~ ‖ tiefer Teller m. ‖ assiette f. creuse. / to dish the ~ ‖ Blech n. kümpeln ‖ (cintrer *ou*) emboutir la tôle. / dished ~ *see* buckled ~. / dry ~ (Phot) ‖ Trockenplatte f. ‖ plaque f. sensible. / earthen ware stone ~s pl. for furniture ‖ Möbeleinlageplatten fpl. ‖ carreaux mpl. en faïence pour meubles. / electro-silver ~ ‖ galvanisch versilbertes Geschirr n. ‖ vaisselle f. plaquée d'argent *ou* argentée. / ~ of epergne *see* ~ of centre piece. / etched ~ (Engr) ‖ radierte Platte f. ‖ estampe f. gravée à l'eau-forte. / ~ of firebox shell ‖ Feuer-

buchsenwand f. ‖ tôle f. de boîte à feu. / flanged ~ ‖ Kümpelblech n. ‖ tôle f. à bord tombé; tôle f. emboutie. / foundation ~ ‖ Fundamentplatte f.; Grundplatte f. ‖ plaque f. de fondation. / glass ~ ‖ Glasplatte f. ‖ plaque f. en verre. / goffered ~ ‖ Waffelblech n. ‖ tôle f. gaufrée. / hammered ~ ‖ gehämmertes Blech n. ‖ tôle f. martelée. / hooked tie ~ ‖ Hakenplatte f. ‖ selle f. à crochet. / junction ~ ‖ Knotenblech n. ‖ gousset m. d'assemblage. / ~ with large surface (Acc) ‖ Großoberflächenplatte f. ‖ plaque f. à grande surface. / lead ~ ‖ Bleiplatte f. ‖ plaque f. en plomb. / lobe ~ *see* foundation ~. / made of ~ ‖ blechern ‖ de tôle f. / main ~ of a lock ‖ Schloßblech n. ‖ palâtre m. d'une serrure. / measuring ~ ‖ Meßplatte f. ‖ plaque f. de mesure. / ~ measuring the space between the rail ends ‖ Zwischensteckblech n. zum Ausmessen der Schienenstöße ‖ cale f. pour mesurer la largeur des joints de rails. / medium ~ ‖ Mittelblech n. ‖ tôle f. moyenne. / ~ of metal ‖ Metallplatte f. ‖ plaque f. *ou* table f. de métal. / negative ~ (Acc) ‖ Minusplatte f.; negative Platte f. ‖ plaque f. négative. / ~ opposed to the twyer plate of a German refining forge ‖ Gichtzacken m. *oder* Raststein m. eines Feinfeuers ‖ contre-vent m. du creuset d'un feu d'affinerie allemand. / ~ outside (Shipb) ‖ Außenhautplatte f. ‖ plaque f. d'enveloppe. / perforated ~ ‖ gelochtes Blech n. ‖ tôle f. perforée. / photographic ~ ‖ fotografische Platte f. ‖ plaque f. photographique. / positive ~ (Acc) ‖ Plusplatte f. ‖ plaque f. positive. / pressed ~ ‖ gepreßtes Blech n. ‖ tôle f. emboutie. / punched ~ *see* perforated ~. / the ~s also fit to rails of similar bases ‖ die Platten fpl. passen auch zu Schienen mit ähnlicher Fußbreite ‖ les selles fpl. conviennent aussi aux rails de largeur d'appui semblable. / ~ of refined copper ‖ Garscheibe f. ‖ rosette f. *ou* lame f. de cuivre affiné. / rolled ~ ‖ Walzblech n. ‖ tôle f. laminée. / ~ for safes ‖ Geldschrankblech n. ‖ tôle f. pour coffres-forts. / sensitive ~ (Phot) ‖ lichtempfindliche Platte f. ‖ plaque f. sensible. / sensitized ~ ‖ lichtempfindlich gemachte Platte f. ‖ plaque f. sensibilisée. / ~ for set of teeth ‖ Gebißplatte f. ‖ plaque f. de dentier. / shallow ~ (Dishes) ‖ flacher Teller m. ‖ assiette f. plate. / slices pl. cut-off from ~s ‖ Abschnitte mpl. von Platten ‖ rognures fpl. des plaques. / sole ~ *see* foundation ~. / ~ for spring band ‖ Federbundplatte f. ‖ plaque f. pour bride de ressort. / steel ~ ‖ Stahlblech n. ‖ tôle f. d'acier. / stiffening ~ ‖ Verstärkungsblech n. ‖ tôle f. *ou* plaque f. de renfort. / supporting ~ ‖ Stützscheibe f. ‖ verre m. support. / thick ~ ‖ Grobblech n. ‖ grosse tôle f. / thin ~ ‖ Feinblech n. ‖ tôle f. mince *ou* fine. / tie ~ ‖ Unterlagsplatte f. ‖ selle f. / toggle ~ ‖ Druckplatte f. ‖ plaque f. de compression. / ~ of unglazed earthenware ‖ Tonteller m. ‖ assiette f. poreuse. / U-shaped ~ ‖ Kanalblech n. ‖ tôle f. en U. / ~ of the valve ‖ Ventilplatte f. ‖ plaque f. de la soupape. / wedge-shaped ~ ‖ keilförmige Platte f. ‖ plaque f. de forme conique. / ~ with wedge-shaped groove ‖ Keil-

nutplatte f. ‖ plaque f. à rainure conique.

plateau (Geol) ‖ Hochebene f. ‖ plaine f. élevée; plateau m.

plate band winder ‖ Bandeisenwickler m. ‖ enrouleur m. de feuillard. / ~ basket ‖ Tellerkorb m. ‖ ménagère f. / ~ battery ‖ Anodenbatterie f. ‖ batterie f. de plaques. / ~ bearing ‖ Schildlager n. ‖ palier m. sur bouclier. / ~ bending machine ‖ Blechbiegemaschine f. ‖ machine f. à cintrer les tôles. / ~ bending press ‖ Plattenbiegepresse f. ‖ presse f. à cintrer les tôles. / ~ bending test ‖ Blechbiegeprobe f. ‖ essai m. de cintrage de la tôle. / ~ brass ‖ Messingblech n. ‖ laiton m. en feuilles *ou* en lames. / ~ buffer ‖ Plattenpuffer m. ‖ tampon m. à plaques. / ~ cleaner (Acc) ‖ Plattenputzer m. ‖ nettoyeur m. de plaques. / ~ closer ‖ Blechschließer m. ‖ serre-tôles m. / ~ clutch ‖ Scheibenkupplung f. ‖ embrayage m. par plateaux. / ~ compressor brake ‖ Lamellenbremse f. ‖ frein m. à lames centrales. / ~ condenser ‖ Plattenkondensator m. ‖ condensateur m. à lames. / ~ conveyor ‖ Plattenband n. ‖ transporteur m. à tablier sans fin. / ~ cooler ‖ Blechtafelkühler m. ‖ refroidisseur m. de tôles. / ~ covering ‖ Blechbelag m. ‖ platelage m. en tôle. / ~ cupboard ‖ Tellerschrank m. ‖ chauffeassiettes m. / ~ current measuring switch ‖ Anodenstrommeßschalter m. ‖ clé f. de mesure du courant de plaque. / cutter of ~s pl. (Phot) ‖ Plattenschneider m. ‖ coupeur m. de plaques. / ~ cutting shears pl. ‖ Blechschere f. ‖ cisaille f. de ferblantier.

plated ‖ plattiert ‖ doublé; plaqué. / copper-~ ‖ mit Kupfer n. plattiert oder beschlagen; verkupfert ‖ plaqué de cuivre; cuivré. / electro-~ ‖ auf galvanischem Wege metallisiert *oder* mit einer Metallschicht überzogen ‖ couvré d'une couche déposée électriquement. / gold-~ ‖ vergoldet ‖ doré. / metal-~ *see* plated. / nickel-~ ‖ vernickelt ‖ nickelé. / silver-~ ‖ (galvanisch) versilbert ‖ argenté; galvanisé (à l'argent). / tin-~ ‖ verzinnt ‖ étamé. / zinc-~ ‖ (galvanisch) verzinkt ‖ galvanisé (au zinc); zingué.

plated copper ‖ versilbertes Kupfer n. ‖ cuivre m. argenté. / ~ goods pl. ‖ plattierte Ware f. ‖ articles mpl. plaqués; plaqués mpl. / ~ jewellery ‖ Dubleewaren fpl. ‖ bijouterie f. en doublé. / ~ seat (Join) ‖ Furniersitz m. ‖ siège m. en plaqué.

plate detacher ‖ Tellerdetaschör m. ‖ détacheur m. à plateaux. / ~ disc wheel ‖ Blechscheibenrad n. ‖ roue f. à disque en tôle. / ~ doubling machine ‖ Blechdoppler m. ‖ plieuse f. de tôles. / ~ drying rack (Phot) ‖ Plattentrockengestell n. ‖ châssis m. pour sécher les plaques. / ~ edge planer ‖ Blechkantenhobelmaschine f. ‖ machine f. à chanfreiner les tôles. / ~ planing the edges of ~s ‖ Behobeln n. der Blechkanten ‖ rabotage m. des arêtes de tôles. / ~ edge planing machine *see* ~ edge planer. / ~ electrometer ‖ Plattenelektrometer n. ‖ électromètre m. à feuilles. / ~ cupped ~ end (Dishes) ‖ vertiefter Tellerboden m. ‖ fond m. bombé. / ~ filament circuit ‖ Anodenkreis m. ‖ circuit m. de plaque. / ~ flanging machine ‖ Blechbördelma-

schine f. ‖ machine f. à border les tôles. / ~ form (Blast furnace) ‖ Gichtbühne f. ‖ plate-forme f. du haut-fourneau. / ~ frame ‖ Blechrahmen m. ‖ châssis m. en tôle. / ~ gauger ‖ Blechlehre f. ‖ calibre m. à tôles. / ~ girder ‖ Blechträger m. ‖ poutre f. en tôles et cornières. / pressed ~ girder ‖ gepreßter Blechträger m. ‖ longeron m. en tôle emboutie.

plate glass ‖ Spiegelglas n. ‖ glace f.; verre m. à glaces. / bent ~ ‖ gebogenes Spiegelglas n. ‖ glace f. cintrée. / to grind ~ ‖ Spiegelglas n. schleifen ‖ dresser les glaces f. / polished ~ ‖ geschliffenes Spiegelglas n. ‖ glace f. poli.

plate glass cutter ‖ Spiegelglasschneider m. ‖ découpeur m. de glaces. / ~ disc ‖ Spiegelglasplatte f. ‖ plaque f. de glace. / ~ grinding machine ‖ Schleifmaschine f. für Spiegelglas ‖ polisseuse f. de glaces. / ~ mould ‖ Form f. für Spiegelglas ‖ moule m. à glaces. / transport of ~ ‖ Spiegelglastransport m. ‖ transport m. de glaces.

plate glazer (Pap) ‖ Papiersatinierer m. ‖ satineur m. de papier. / ~ heating furnace ‖ Blechglühofen m. ‖ four m. à recuire la tôle. / ~ ice ‖ Platteneis n. ‖ glace f. en plaques. / ~ iron ‖ Walzblech n. ‖ tôle f. laminée. / ~ joggling machine ‖ Plattenkröpfmaschine f.; Plattenjoggelmaschine f. ‖ machine f. à épauler les tôles. / ~ keel ‖ Plattenkiel m. ‖ quille f. en tôle. / ~ layer ‖ Schienenleger m. ‖ poseur m. de la voie. / ~ lifting device ‖ Blechhebevorrichtung f. ‖ dispositif m. à soulever les tôles. / ~ lightning arrester ‖ Plattenblitzableiter m. ‖ parafoudre m. à plaques. / ~ lightning conductor see ~ lightning arrester. / rapid ~ locking-up device ‖ Plattenschnellspannvorrichtung f. ‖ dispositif m. de serrage rapide de clichés. / ~ lug (Acc) ‖ Plattenfahne f. ‖ queue f. de plaque. / ~ magnet ‖ Bandmagnet m. ‖ aimant m. plat en lames. / ~ maker (Arm) ‖ Systemmacher m. ‖ systémeur m. / ~ maker (Metal) ‖ Blechschmied m. ‖ tôlier m. / ~ mill ‖ Grobblechstraße f.; Blechwalzwerk n. ‖ train m. à grosses tôles; laminoir m. à tôle. / ~ moulder ‖ Plattenformer m. ‖ mouleur m. de plaques.

platen machine ‖ Flachdruckmaschine f. ‖ machine f. pour l'impression à forme plate. / ~ minder ‖ Tiegeldrucker m. ‖ minerviste m.

platen press ‖ Tiegeldruckpresse f. ‖ presse f. d'impression à platine. / ~ printing press see ~ press.

plate painter ‖ Blechstreicher m. ‖ marqueur m. de tôles. / ~ paper ‖ Kupferdruckpapier n. ‖ papier m. pour taille-douce. / ~ pasting machine for dynamo plates ‖ Blechbeklebemaschine f. für Dynamobleche ‖ machine f. à coller du papier sur les tôles de dynamo. / ~ pewter ‖ Tellerzinn n.; Hartmetall n. ‖ potin m. d'assiettes ou à vaisselle. / ~ pickling plant ‖ Plattenbeizanlage f. ‖ installation f. de décapage des tôles. / ~ planing machine ‖ Blechhobelmaschine f. ‖ machine f. à raboter les tôles. / ~ protection ‖ Plattenschutz m. ‖ protection f. par plaques. / ~ punching machine ‖ Blechlochmaschine f. ‖ machine f. à poinçonner les tôles; découpoir m. à tôle. / ~ puzzle lock ‖ Scheibenschloß n. ‖ serrure f. à rondelles. / ~ racks pl. ‖

Plattenlager n. ‖ magasin m. de tôles. / ~ rail ‖ flache Schiene f.; Plattschiene f. ‖ bande f. plate; rail m. plat. / ~ roll ‖ Blechwalze f.; Glattwalze f. ‖ cylindre m. à tôle. / ~ roller ‖ Blechwalzer m. ‖ lamineur m. ou doubleur m. de tôles. / ~ rolling-mill ‖ Eisenblechwalzwerk n. ‖ laminoir m. à tôles. / ~ shearer ‖ Blechzuschneider m. ‖ découpeur m. de tôle. / ~ shearing machine ‖ Plattenschere f. ‖ cisaille f. à tôles. / ~ shears pl. ‖ Blechschere f. ‖ cisaille f. à tôles. / double standard ~ shears pl. ‖ Zweiständerblechschere f. ‖ cisaille f. à tôles à double montant. / ~ slot ‖ Schlitzplatte f. ‖ plaque f. à fente. / paper yarn ~ spinning machine ‖ Papiergarntellerspinnmaschine f. ‖ machine f. à plateau pour le filetage du fil en papier. / splitting of ~s ‖ Schneiden n. von Blechtafeln ‖ coupe f. de feuilles de tôles. / ~ splitting machine ‖ Blechschere f. ‖ cisaille f. à tôles. / ~ splitting, section shearing, and mitre cutting machine ‖ Blech-, Profileisen- und Gehrungsschere f. ‖ cisaille f. à tôles combinée avec cisaille à profiles et à onglets. / ~ spring ‖ Blattfeder f. ‖ ressort m. à lames. / boring of ~ spring ‖ Bohrung f. der Scheibenfeder ‖ trou m. de rondelle Belleville (ou de ressort à lames). / ~ straightener ‖ Blechrichter m. ‖ planeur m. de tôles. / ~ straightening machine ‖ Richtmaschine f. für Bleche ‖ machine f. à planer les tôles. / strip shears for dividing strips and ~s ‖ Streifenschere f. zum Teilen von Streifen und Platten ‖ cisailles fpl. à bandes pour découper les bandes et les plaques. / strip shears for trimming strips and ~s ‖ Streifenschere f. zum Besäumen von Streifen und Platten ‖ cisailles fpl. à bandes pour replier les bandes et les plaques. / ~ supporting table ‖ Blechauflagetisch m. ‖ table f. pour placer la tôle. / ~ sweeper ‖ Blechputzer m. ‖ nettoyeur m. de tôles. / thickness of ~ ‖ Blechdicke f.; Blechstärke f. ‖ épaisseur f. de tôle. / ~ trimming shears pl. ‖ Saumschere f. ‖ cisaille f. à rogner les tôles. / ~ voltage ‖ Anoden(gleich)spannung f. ‖ tension f. de plaque. / ~ warmer ‖ Tellerwärmer m. ‖ réchaud m.; chauffe-assiettes m. / weight of ~ ‖ Plattengewicht n. ‖ poids m. de plaque. / ~ wheel mill ‖ Radscheibenwalzwerk n. ‖ atelier m. de laminoir de disques de roues. / ~ work ‖ Blecharbeiten fpl. ‖ tôlerie f. / ~ working machine ‖ Blechbearbeitungsmaschine f. ‖ machine f. à travailler les tôles.

platform ‖ Plattform f. ‖ plate-forme f. / ~ (Mach) ‖ Laufbühne f. ‖ plate-forme f. / ~ (Mar) ‖ Laufbrücke f. ‖ pont m. roulant. / ~ (Railw) ‖ Ladeplatz m. ‖ quai m. de chargement. / ~ (Reaper) ‖ Gatter n. ‖ tablier m. / ~ (Speaker) ‖ Tribüne f.; Rednerbühne f. ‖ tribune f. / charging ~ ‖ Beschickungsbühne f. ‖ plateforme f. de chargement. / ~ with chequered plate covering (Railw) ‖ Wagenplattform f. mit Riffelblech ‖ plate-forme f. couverte de tôle striée. / closed ~ ‖ geschlossene Plattform f. ‖ plateforme f. fermée. / delivery ~ ‖ Abzughängebank f. ‖ palier m. d'exploitation. / ~ of flat iron bars (Railw) ‖ Flacheisenplattform f. ‖ plate-forme f. en fers plats. / fore ~ ‖

Vorderplattform f. ‖ plateforme f. extrême de devant. / handling ~ ‖ Ausladebrücke f. ‖ pont m. de déchargement. / ~ fitted with hand rails ‖ Bühne f. mit Schutzgeländer ‖ plate-forme f. avec rampe. / loading ~ ‖ Laderampe f.; Laderampe f. de chargement. / ~ for loading goods see loading ~. / ~ for the operator ‖ Bedienungsbühne f. ‖ plate-forme f. de service ou de commande. / removable wooden ~ ‖ abnehmbarer Holzboden m. ‖ fond m. en bois démontable. / service ~ see ~ for the operator. / ~ of a station ‖ Bahnsteig m. ‖ trottoir m. du quai des voyageurs. / swinging ~ ‖ Schwenkbühne f. ‖ pont-levis m. / travelling ~ ‖ Schiebebühne f. ‖ plate-forme f. roulante.

platform balance ‖ Brückenwage f.; Straßenwage f. ‖ pont m. à bascule; bascule f. / ~ car (Railw) ‖ Lore f. ‖ fardier m.; truck m.; wagon m. à plateforme découverte. / ~ clock ‖ Bahnsteiguhr f. ‖ horloge f. de quai. / ~ deck ‖ Plattformdeck n.; Raumdeck n. ‖ plateforme f. de la cale. / ~ roof (Railw) ‖ Bahnsteigbedachung f. ‖ couverture f. ou toiture f. du quai. / ~ truck ‖ Rollwagen m. ‖ truck m. roulant. / ~ wagon ‖ Plattformwagen m. ‖ wagon plateforme m. / ~ wagon of pressed plates ‖ Plattformwagen m. aus gepreßten Blechen ‖ wagon m. à plate-forme en tôle emboutie.

platina see platinum.

platinate, to ‖ platinieren ‖ platiner.

platinating ‖ Platinierung f. ‖ platinage m.

plating see metal ~. / ~ cadmium ~ ‖ Kadmiumüberzug m. ‖ doublé m. de cadmium. / ~ of the hats ‖ Plattieren n. oder Überziehen n. der Hüte ‖ dorage m. des chapeaux. / iron ~ ‖ Panzerung f. ‖ blindage m. / ~ on iron ‖ Plattierung f. auf Eisen ‖ plaqué m. sur fer. / metal ~ ‖ Plattierung f. ‖ placage m. sur métaux.

plating mill ‖ Plattierwalzwerk n. ‖ laminoir m. à plaquer. / ~ thread ‖ Deckfaden m. ‖ fil m. à vaniser.

platinic acid ‖ Platinsäure f. ‖ acide m. platinique. / ~ chloride see platinum chloride. / ~ oxide see platinum oxide.

platiniferous ‖ platinhaltig ‖ platinifère.

platinit ‖ Platinit n. ‖ platinite m.

platinocyanide ‖ Platincyanür n. ‖ platinocyanure m.

platinotype ‖ Platindruck m. ‖ platinotypie f.

platinous chloride ‖ Platinchlorür n. ‖ chlorure m. de platine.

platinum ‖ Platin n. ‖ platine m. / burnish ~ ‖ Glanzplatin n. ‖ platine m. imité. / native ~ ‖ Platin n.; Polyxen n. ‖ platine m. natif. / spongy ~ ‖ Platinschwamm n. ‖ mousse f. ou éponge f. de platine.

platinum anode ‖ Platinanode f. ‖ anode f. de platine. / ~ apparatus ‖ Platinapparat m. ‖ appareil m. en platine. / ~ articles pl. ‖ Platinwaren fpl. ‖ objets mpl. en platine. / ~ basin see ~ dish. / ~ bath ‖ Platinbad n. ‖ bain m. de platinage. / ~-bearing see platiniferous. / ~ black ‖ Platinschwarz n. ‖ noir m. de platine. / ~ chloride ‖ Platinchlorid n. ‖ bichlorure m. de platine. / ~ contact ‖ Platinkontakt m. ‖ contact m. en platine. / ~ crucible ‖ Platintiegel m. ‖ creuset m. en platine. / ~ dish ‖ Platinschale f. ‖

coupelle f. de platine. / ~ foil ‖ Platinblech n. ‖ tôle f. de platine. / ~ incineration dish ‖ Platinveraschungsschale f. ‖ cuvette f. en platine pour incinération. / ~ iridium ‖ Platiniridium n. ‖ platine iridium m. / ~ iridium wire netting ‖ Platiniridiumdrahtnetz n. ‖ tamis m. de platine iridé. / ~ metal ‖ Platinmetall n. ‖ métal m. de platine. / ~ mine ‖ Platingrube f. ‖ mine f. de platine. / ~ oxide ‖ Platinoxyd n. ‖ oxyde m. platinique ou de platine. / ~ paper ‖ Platinpapier n. ‖ papier m. platiné. / ~ pin ‖ Platinnadel f. ‖ aiguille f. en platine. / ~ pin tooth ‖ Platin(stift)zahn m. ‖ dent f. à pointe en platine. / ~ plated ‖ platinplattiert ‖ doublé ou plaqué de platine. / ~ plating ‖ Platinierung f. ‖ platinage m. / ~ plating plant ‖ Platinierungsanlage f. ‖ installation f. de platinage. / ~ refiner ‖ Platinscheider m. ‖ affineur m. de platine. / ~ refining ‖ Platinscheidung f. ‖ affinage m. de platine. / ~ rolling mill ‖ Platinwalzwerk n. ‖ laminage m. de platine. / ~ salt ‖ Platinsalz n. ‖ sel m. de platine. / ~ sheet ‖ Platinblech n. ‖ tôle f. de platine; platine m. laminé. / ~ substitute anode ‖ Platinersatzanode f. ‖ anode f. en faux platine. / ~ ware ‖ Platingerät n. ‖ ustensile m. en platine. / ~ wire ‖ Platindraht m. ‖ fil m. de platine. / ~ wire spiral ‖ Platindrahtspirale f. ‖ spirale f. en fil de platine.

platymeter ‖ Platymeter n. ‖ platymètre m.

play, to ‖ spielen ‖ jouer. / ~ from one colour into another ‖ schillern ‖ chatoyer; miroter; jeter des reflets variés.

play ‖ Spiel n. ‖ jeu m. / ~ (Mach) ‖ Spielraum m.; Spiel n. ‖ jeu m. / to have ~ ‖ Spiel n. oder Spielraum m. haben ‖ avoir du jeu. / lateral ~ ‖ seitliches Spiel n. ‖ mouvement m. latéral. / ~ of the piston ‖ Kolbenspiel n. ‖ jeu m. du piston. / radial ~ ‖ Beweglichkeit f. in radialer Richtung ‖ jeu m. radial. / ~ of the type bars ‖ Spiel n. der Typenhebel ‖ jeu m. des tiges à caractères.

play bill ‖ Theaterzettel m. ‖ programme m. / ~ costume ‖ Theaterkostüm n. ‖ costume f. de théâtre.

playing ball made of india-rubber ‖ Gummispielball m. ‖ balle f. à jouer en caoutchouc.

playing card ‖ Spielkarte f. ‖ carte f. à jouer. / ~ calenderer ‖ Spielkartenglätter m. ‖ lisseur m. de cartes à jouer. / ~ checker ‖ Spielkartenprüferin f. ‖ recouleuse f. de cartes à jouer. / ~ cutter ‖ Spielkartenschneider m. ‖ découpeur m. de cartes à jouer. / gilder of ~ edges ‖ Spielkartenrandvergolder m. ‖ doreur m. de tranches des cartes à jouer. / examiner see ~ checker. / ~ glazer ‖ Spielkartenglätter m. ‖ lisseur m. de cartes à jouer. / ~ maker ‖ Spielkartenfabrikant m. ‖ fabricant m. de cartes à jouer. / ~ paster ‖ Spielkartenkleber m.; Spielkartenleimer m. ‖ colleur m. de cartes à jouer. / ~ rounder-off ‖ Spielkarteneckenabrunder m. ‖ corneur m. ou arrondisseur m. de cartes à jouer. / ~ shuffler ‖ Spielkarten(zusammen)legerin f. ‖ mêleuse f. de cartes à jouer. / ~ sorter ‖ Spielkartensortiererin f. ‖ tableuse f. de cartes à jouer.

playing counter ‖ Spielmarke f. ‖ jeton m. de jeu. / ~ die ‖ Spielwürfel m. ‖ dé m. à jouer.

plaything ‖ Spielzeug n. ‖ jouet m.

plea ‖ Rechtseinwand m.; Rechtfertigungsgrund m. ‖ raison m. justificative; dénégation f. / on the ~ ‖ unter dem Vorwande m. ‖ sous prétexte m. de. / ~ of nullity ‖ Nichtigkeitsklage f. ‖ demande f. en nullité.

pleaded, contentiously ~ ‖ streitig verhandelt ‖ débattu contradictoirement.

pleasant ‖ ansprechend; angenehm ‖ agréable.

pleasure yacht ‖ Vergnügungsjacht f. ‖ yacht m. de plaisance.

pleating machine see plissé machine.

pleistocene era ‖ Pleistozän n. ‖ ère f. pleïstocène.

plentiful ‖ reichlich; reichhaltig ‖ ample; suffisant; abondant.

pleochroism ‖ Pleochroismus m. ‖ pléochroïsme m.

pleonaste ‖ schwarzer Spinell m.; Pleonast m.; Ceylanit m. ‖ ceylanite f.; pléonaste m.

pliable ‖ biegbar; biegsam ‖ pliable; flexible.

pliant rule ‖ Kurvenlineal n. ‖ règle f. courbe ou pliante ou montée; pistolet m.

pliers pl. ‖ Zange f. ‖ tenaille f.; pince f. / bending ~ ‖ Biegezange f. ‖ pince f. à cintrer. / comb ~ ‖ Hechlerzange f. ‖ pince f. de peigneur. / convex ~ (Goldsm) ‖ Schienenzange f. ‖ pince f. convexe. / crucible ~ with olive nose ‖ Tiegelzange f. ‖ pince f. à creuset à olive. / eyelet punch ~ ‖ Ösenzange f. ‖ pince f. à œillet. / flat ~ ‖ Flachzange f. ‖ pince f. plate. / flat clockmaker's ~ ‖ Flachzange f. für Uhrmacher ‖ pince f. plate d'horloger. / flat-nosed ~ ‖ Flachzange f.; Drahtzange f. mit flachen Backen ‖ pince f. plate. / gas ~ ‖ Gasrohrzange f. ‖ pince f. à gaz. / gasburner ~ ‖ Gasbrennerzange f.; Brennerzange f. ‖ pince f. à brûleur. / insulated ~ ‖ isolierte Zange f. ‖ pince f. isolante. / ~ for inserting key buttons (Typewr) ‖ Zange f. für Tastenknöpfe ‖ pince f. pour appliquer les touches. / lead stamping punch ~ ‖ Plombenzange f. ‖ pince f. à plomber. / metal punch ~ ‖ Metalllochzange f. ‖ pince f. emporte-pièce pour métaux. / pointed ~ ‖ Spitzzange f. ‖ pince f. pointue. / punch ~ ‖ Lochzange f. ‖ pince f. emporte-pièce. / punch ~ with bent handles ‖ Lochzange f. mit geschweiften Schenkeln ‖ pince f. emporte-pièce à branches coudées. / revolving four-hole punch ~ ‖ Revolverkopflochzange f. mit vier Lochhülsen ‖ pince f. emporte-pièce avec étoile à quatre tubes. / round ~ ‖ Rundzange f. ‖ pince f. ronde. / round-nosed ~ (Electr) ‖ Flügelklemme f. ‖ pince f. à oreilles. / round-nosed ~ (Mach) ‖ Drahtzange f. mit runden Backen ‖ pince f. ronde. / round-pointed rosary ~ ‖ Rosenkranzzange f. ‖ pince f. à chapelet bec effilé. / saddler's ~ ‖ Sattlerzange f. ‖ pince f. pour sellier. / staple ~ ‖ Drahtreiterklemme f. ‖ pince f. à cavalier. / stockinger's ~ ‖ Wirkerzange f. ‖ pince f. de bonnetier. / turnscrew gas ~ ‖ Gasrohrzange f. mit Schraubenzieher ‖ pince f. à gaz à tournevis.

plinth ‖ Bühne f.; Podium n. ‖ plateforme f.

pliocene ‖ Pliozän n. ‖ système m. pliocène.

pliotron ‖ Pl(e)iotron n. ‖ pleiotron m.

plissé machine ‖ Plissiermaschine f. ‖ machine f. à plisser; plisseuse f. mécanique.

plot ‖ Grundriß m.; Plan m.; Riß m. ‖ délinéation f.; plan m.; tracé m. / ~ of land ‖ Parzelle f. ‖ parcelle f.; lot m. de terrain. / ~ of a mine ‖ Markscheiderriß m. ‖ plan m. de mine.

plotting by polar coordinates ‖ Aufnahme f. nach dem Polverfahren ‖ levé m. de plan par la méthode polaire. / ~ by rectangular coordinates ‖ Aufnahme f. nach dem Koordinatenverfahren ‖ levé m. de plan par la méthode des coordonnées rectangulaires. / spotting and ~ method ‖ Feuerleitverfahren n. ‖ système m. ou méthode f. de conduite du tir.

plough, to ‖ ackern; pflügen ‖ labourer; remuer avec la charrue. / ~ the sea lange auf derselben Stelle in See kreuzen ‖ battre la mer. / ~ and tongue together (Coop) ‖ verspunden ‖ bouveter.

plough ‖ Pflug m. ‖ charrure f. / ~ (Bookb) ‖ Hobel m.; Beschneidhobel m. ‖ rognoir m.; couteau m. à rogner. / ~ enters too deeply ‖ der Pflug greift zu tief ein ‖ la charrue est sur son nez. / draining ~ see plough, trench. / motor ~ ‖ Motorpflug m. ‖ charrue f. automobile. / paring ~ ‖ Rasenpflug m. ‖ charrue f. à peler; dégazonnoir m. / ridge ~ ‖ Häufelpflug m. ‖ charrue-buttoir f. / ~ with several shares ‖ mehrschariger Pflug m. ‖ polysoc m. / sliding ~ (Elektr) ‖ Schleifkontakt m.; Schlittenkontakt m. ‖ collecteur m. courseur; contact m. frotteur. / steam ~ ‖ Dampfpflug m. ‖ charrue f. à vapeur. / trench ~ ‖ Grabenpflug m.; Rigolpflug m. ‖ charrue f. à effondrer. / turn wrest ~ ‖ Kehrpflug m. ‖ charrue f. tourne-oreille.

plough beam ‖ Grendel m. oder Grindel m. eines Pfluges; Pflugbaum m. ‖ arbre m. ou flèche f. ou haie f. d'une charrue. / ~ bit (Join) ‖ Nuteisen n.; Nuthobeleisen n. ‖ fer m. du bouvet à approfondir ou à rainures.

ploughing ‖ Pflügen n. ‖ labourage m. / ~ implement ‖ Ackergerät n. ‖ instrument m. aratoire.

plough knife (Bookb) ‖ Hobel m.; Beschneidhobel m. ‖ rognoir m.; couteau m. à rogner.

ploughland ‖ Ackerboden m. ‖ sol m.; terroir m.; terre f. arable ou labourable ou végétable.

ploughman ‖ Landarbeiter m. ‖ laboureur m.

plough plane ‖ Spundhobel m. ‖ bouvet m. / ~ with movable stop ‖ Nuthobel m. mit verstellbarem Auflauf ‖ bouvet m. à approfondir ou à dégorger.

ploughshare ‖ Pflugschar f. ‖ soc m.

ploughtail ‖ Pflugsterze f. ‖ manche m.; ou mancheron m. de la charrue.

plow, to see to plough.

plow see plough.

pluck, to ‖ rupfen ‖ arracher; plumer.

plucker (Spinn) ‖ Reißwolf m. ‖ diable m. loup m.

plucking instrument ‖ Zupfinstrument n. ‖ instrument m. à tirer. / ~ machine for turnip tops ‖ Rübenblätterzerreißmaschine f. ‖ machine f. à déchiqueter

les feuilles de betteraves. / ~ woman ‖ Ausrupferin f. ‖ ébouqueuse f.

plug, to ‖ verspunden ‖ tamponner; boucher. / ~ a hollow tooth ‖ einen Zahn m. plombieren ‖ plomber une dent. / ~ the tap hole (Found) ‖ das Stichloch verschließen ‖ boucher le trou de coulée.

plug ‖ Pflock m.; Dübel m. ‖ cheville f.; fiche f.; tenon m.; goujon m. / ~ (Coop) ‖ Spund m.; Faßspund m. ‖ bondon m.; tape f. / ~ (Electr) ‖ Stöpsel m. ‖ fiche f. / ~ (Found) ‖ Gießpfropfen m. ‖ boucher m.; bouchon m. / charging ~ (Electr) ‖ Ladestöpsel m. ‖ bouchon m. de charge. / ~ of a cock ‖ Hahnkegel m.; Hahnschlüssel m. ‖ noix f. ou clef f. d'un robinet. / connecting ~ (Electr) ‖ Kontaktstöpsel m. ‖ cheville f. de contact. / ~ for making crucibles ‖ Kern m. zur Tiegelfertigung ‖ noyau m. pour la fabrication de creusets. / four-part ~ (Tel) ‖ vierteiliger Stöpsel m. ‖ fiche f. quadruple. / ~ for hand lamp ‖ Steckdose f. zur Handlampe ‖ contact m. à fiches de lampe baladeuse. / the inner ~s are soft ‖ die inneren Schmelzpfropfen mpl. sind weich ‖ les bouchons mpl. fusibles intérieurs se sont ramollis. / magnetic spark ~ ‖ Magnetkerze f.; Kerze f. ‖ bougie f. magnétique. / the outermost ~ begins melting ‖ der äußerste Schmelzpfropfen beginnt zu schmelzen ‖ le bouchon m. fusible extérieur commence à couler / ~ of the penholder ‖ Schreibfederträger m. ‖ galet m. du porte-plume. / ~ with protection collar (Electr) ‖ Stecker m. mit Schutzkragen ‖ boîte f. de contact à col protecteur. / radiator draw-off ~ ‖ Kühlerablaßschraube f. ‖ bouchon m. de vidange du radiateur. / the ~ ran out ‖ der Schmelzpfropfen ist ausgeflossen ‖ le bouchon fusible s'est écoulé. / screwed ~ ‖ verschraubbarer Verschlußstopfen m. ‖ bouchon m. à vis. / short-circuit ~ ‖ Kurzschlußstöpsel m. ‖ plot m. de mise en court-circuit. / spark ~ ‖ Zündkerze f. ‖ bougie f. / ~ of three-way cock ‖ Dreiweghahnküken n. ‖ noix f. de robinet à trois voies. / ~ of two-way cock ‖ Durchgangshahnküken n. ‖ noix f. de robinet droit. / wooden ~ ‖ Holzdübel m. ‖ tampon m. en bois.

plug box (Electr) ‖ Steckdose f.; Steckkontakt m. ‖ prise f. de courant; boîte f. de contact à fiche. / ~ bush (Radio) ‖ Steckbuchse f. ‖ douille f. de prise. / ~ cartridge (Tel) ‖ Steckpatrone f. ‖ cartouche f. à insertion. / ~ commutator (Tel) ‖ Stöpselwähler m. ‖ commutateur m. à fiche. / ~ contact ‖ Stöpselkontakt m. ‖ contact m. à cheville. / ~ device (Electr) ‖ Steckvorrichtung f. ‖ dispositif m. de prise de courant. / ~ fuse ‖ Stöpselsicherung f. ‖ coupe-circuit m. à cartouche. / ~ gauge ‖ Kaliberbolzen m. ‖ calibre m. tampon.

plugged ‖ zugestopft ‖ bouché.

plug maker (Cork) ‖ Korkschneider m.; Pfropfenschneider m. ‖ bouchonnier m. / ~ ramming machine ‖ Bodenstampfmaschine f. ‖ machine f. à damer les fonds. / ~ ramming machine for converters ‖ Bodenstampfmaschine f. für Konverter ‖ machine f. à damer les fonds de convertisseurs. / ~ restored indicator (Tel) ‖ Rückstellklappe f. ‖ volet m. à relèvement automatique. / ~ seat (Tel) ‖ Stöpselsitzplatte f. ‖ plaque f. de repos

à fiches. / ~ seat switch ‖ Stöpselsitzschalter m. ‖ commutateur m. à fiche de repos. / ~ socket ‖ Steckdose f. ‖ prise f. de courant à fiches. / ~ switch ‖ Stöpselschalter m. ‖ commutateur m. à cheville. / ~-type cut-out ‖ Patronensicherung f. ‖ coupe-circuit m. à bouchon. / ~ valve ‖ Kegelventil n. ‖ soupape f. à siège conique.

plum ‖ Pflaume f. ‖ prune f.

plumassier ‖ Federschmuckmacher m. ‖ plumassier m.

plumb, to ‖ loten; bleien; einloten; abloten ‖ plomber; prendre l'aplomb. / ~ (Nav) ‖ peilen ‖ sonder.

plumb ‖ Lot n. ‖ plomb m.; fil m. à plomb; sonde f.

plumbago ‖ Grafit m. ‖ graphite m.; plombagine f. / ~ crucible ‖ Grafittiegel m. ‖ creuset m. de plombagine ou en graphite.

plumb bob see plummet. / ~ chamber man ‖ Bleikammerarbeiter m. ‖ surveillant m. des chambres de plomb.

plumber ‖ Klempner m. ‖ plombier m. / ~'s pliers ‖ Klempnerzange f. ‖ pince f. de plombier. / ~'s solder ‖ Lötmörtel m. ‖ soudure f. au plomb.

plumbery ‖ Klempnerei f. ‖ plomberie f.

plumbic acetate ‖ Bleizucker m. ‖ sucre m. de Saturne; acétate m. de plomb. / ~ gum ‖ Bleigummi n. ‖ hydroaluminate m. de plomb.

plumbiferous ‖ bleihaltig ‖ plombifère.

plumbing in a shaft ‖ Schachtlotung f. ‖ prise f. de l'aplomb dans un puits. / ~ over a station mark in the floor ‖ Einlotung f. über der Sohle ‖ réglage m. par le fil à plomb au dessus du sol. / ~ telescope ‖ Fernrohrlot n. ‖ instrument m. de prise d'aplomb à lunette.

plumb line see plummet.

plumbo-calcite ‖ Plumbokalzit m. ‖ plumbocalcite f. / ~-resinite see plumbic gum.

plumb rule ‖ Schrotwage f. ‖ chas m.; niveau m.

plum distiller ‖ Pflaumensieder m. ‖ distillateur m. de prunes.

plume ‖ Schmuckfeder f. ‖ plume f. pour parures.

plum gum ‖ Pflaumenbaumgummi m. ‖ gomme f. de prunier.

plummer block ‖ Rumpflager n.; Stehlager n. ‖ palier m. ordinaire.

plummet ‖ Lot n.; Bleilot n.; Senklot n.; Senkblei n.; Richtlot n. ‖ plomb m.; fil m. à plomb. / ~ freely suspended ~ ‖ frei hängendes Lot n. ‖ plomb m. à suspension libre. / to measure the depth of a shaft with a ~ ‖ einen Schacht m. abseigern ‖ aplomber un puits. / optical ~ ‖ optisches Lot n. ‖ plomb m. optique. / pear-shaped ~ ‖ birnenförmiges Lot n. ‖ plomb m. en forme de poire. / to set the pole perpendicular by the ~ ‖ den Meßstab m. nach dem Lot senkrecht aufstellen ‖ planter le jalon verticalement au moyen du fil à plomb. / portable ~ ‖ Taschenlot n. ‖ plomb m. de poche. / ~ with reel ‖ Lot n. mit Abrollung ‖ plomb m. à câble sur poulie. / to steady the ~ ‖ das Lot n. feststellen ‖ fixer le plomb. / to try with the ~ see to plumb. / position of rest of the ~ ‖ Ruhelage f. des Lotes ‖ position f. de repos du plomb. / to draw-up the ~ wire ‖ das Lotseil aufholen ‖ relever le câble à plomb. / ~ wire centering disc ‖ Lotteller m. ‖ plateau m. du plomb.

plumming line ‖ Lotriß m.: lotrechte oder vertikale Linie f. ‖ ligne f. à plomb.

plum pattern mahogany ‖ geflammtes Mahagoni n. ‖ acajou m. flammé. / ~ stone ‖ Pflaumenkern m. ‖ noyau m. de prune. / ~ tree wood ‖ Zwetschenbaumholz n. ‖ bois m. de prunier.

plumula ‖ Federchen n. ‖ plumule f.

plunge, to ‖ eintauchen ‖ tremper.

plunge battery ‖ Tauchbatterie f. ‖ batterie f. à immersion. / ~-cut internal grinder ‖ Einstechinnenschleifmaschine f. ‖ machine f. à rectifier les intérieurs.

plunger ‖ Plunger m.; Tauchkolben m. ‖ (piston-)plongeur m. / ~ running in metal guides ‖ Plunger m. in Metallführung ‖ piston-plongeur m. guidé par des coulisses métalliques.

plunger bucket see plunger. / ~ jig ‖ hydraulische Setzmaschine f. ‖ crible m. hydraulique. / ~ piston see plunger. / ~ platen release ‖ Stechwalze f. ‖ cylindre m. à tige plongeante. / ~ pump ‖ Tauchkolbenpumpe f.; Plungerpumpe f. ‖ pompe f. à piston-plongeur. / ~-type line switch ‖ Keith-Vorwähler m. ‖ Keithprésélecteur m.

plunging siphon ‖ Stechheber m. ‖ tâtevin m.

plus ‖ zuzüglich ‖ en plus.

plush ‖ Plüsch m. ‖ peluche f. / cut ~ ‖ Schneidplüsch m.; gerissener Plüsch m. ‖ peluche f. coupée. / loop ~ ‖ Henkelplüsch m. ‖ peluche f. à brides. / silk ~ ‖ Seidenplüsch m. ‖ peluche f. soie. / uncut ~ ‖ ungerissener Plüsch m. ‖ peluche f. bouclée. / warp ~ ‖ Kettplüsch m. ‖ peluche f. par chaîne. / weft ~ ‖ Schußplüsch m. ‖ peluche f. par trame.

plush carpet ‖ Plüschteppich m.; Samtteppich m. ‖ tapis m. velouté. / ~ carpet weaver ‖ Plüschteppichweber m. ‖ moquettier m. / ~ cover ‖ Plüschdecke f. ‖ couverture f. de peluche. / ~ cutting machine ‖ Plüschschneidemaschine f. ‖ machine f. à couper la peluche. / ~ friezer ‖ Plüschkräusler m. ‖ ratineur m. de peluche. / ~ weaver ‖ Plüschweber m. ‖ tisseur m. de peluche. / ~ weaving mill ‖ Plüschweberei f. ‖ tissage m. de peluche.

plus and minus limits pl. ‖ Plus- und Minusgrenze f. ‖ tolérances mpl. maxima et minima. / ~ sign ‖ Pluszeichen n. ‖ signe m. plus.

Plutonic ‖ plutonisch ‖ plutonique.

pluviometer ‖ Regenmesser m. ‖ pluviomètre m.; udomètre m.

pluviometry ‖ Regenmessung f. ‖ pluviométrie f.; ombrométrie f.

ply, to ‖ duplieren ‖ doubler. / ~ between ‖ den Verkehr m. vermitteln zwischen ‖ faire le service entre.

plyer of a draw-bridge ‖ Schwengel m. oder Wippe f. einer Zugbrücke ‖ bascule f. à fléau; flèche f. d'un pont à bascule.

plyers pl. see pliers.

plywood ‖ Sperrholz n. ‖ bois m. contreplaqué. / ~ framework ‖ Sperrholzgerippe n. ‖ carcasse f. ou charpente f. en contreplaqué. / ~ planking ‖ Sperrholzbeplankung f. ‖ revêtement m. en contreplaqué. / ~ ring ‖ furnierter Holzring m. ‖ anneau m. en bois plaqué ou en bois de placage.

pneumatic ‖ pneumatisch ‖ pneumatique. / ~ accumulator ‖ Luftdruckakkumulator m. ‖ accumulateur m. à air comprimé. / ~ air plant ‖ Saugluftanlage f. ‖ instal-

lation f. pneumatique. / ~ balancing for weighing machines || Luftdruckentlastung f. für Wagen || décharge f. pneumatique pour bascules. / ~ brake || pneumatische Bremse f.; Druckluftbremse f. || frein m. pneumatique. / ~ car || preßluftgetriebener Wagen m. || voiture f. *ou* wagon m. commandé(e) à l'air comprimé. / ~ chipping hammer || Druckluftmeißel m. || marteau m. pneumatique à buriner. / ~ chisel hammer *see* ~ chipping hammer. / ~ coal pick hammer || Preßluftabbauhammer m. || marteau m. d'exploitation minière à air comprimé; marteaupic m. pneumatique. / ~ control valve || Druckluftkontrollventil n. || soupape f. de contrôle pour air comprimé. / ~ dispatch *see* ~ post. / ~ door closing device || Drucklufttürschließer m. || ferme-porte f. à air comprimé. / ~ drill || Preßluftbohrhammer m. || marteau-foreur m. à air comprimé. / ~ drilling hammer *see* ~ drill. / ~ equilibrator || Druckluftausgleicher m. || équilibreur m. à air comprimé. / ~ feeding arrangement (Print) || pneumatische Bogenführung f. || transport m. pneumatique de feuilles. / ~ foot switch || pneumatischer Fußschalter n. || commutateur m. pneumatique à pédale. / ~ hammer || Drucklufthammer m.; Preßlufthammer m. || marteau m. pneumatique à air comprimé *ou*. / ~ lifting device || Preßlufthebezeug n. || appareil m. de levage pneumatique. / ~ machine || Preßluftmaschine f. || machine f. pneumatique. / ~ operation of flaps for self-tippings || Druckluftklappenbetätigung f. für Selbstentlader || clapet m. pour déchargeurs automatiques commandé à l'air comprimé. / ~ packing (Mine) || Blasversatz m. || remblayage m. pneumatique. / ~ pickaxe || Preßluftspitzhacke f. || pic m. à air comprimé. / ~ post *see also* pneumatic tube || Rohrpost f. || poste f. pneumatique *ou* tubulaire. / ~ post plant || Rohrpostanlage f. || installation f. de poste pneumatique / ~ pump || Luftpumpe f. || pompe f. pneumatique *ou* à air. / ~ rammer || Druckluftstampfer m. || dame f. *ou* pilou m. *ou* fouloir m. à air comprimé. / ~ ramming hammer || Preßluftrammhammer m. || dame f. pneumatique pour enfoncer. / ~ riveting hammer || Preßluftniethammer m. || marteau m. à river à air comprimé. / ~ riveting machinery || pneumatische Nietanlage f. || riveuses fpl. pneumatiques. / the plant is equipped with ~ riveting and drilling machinery || die Werkstatt ist mit pneumatischen Niet- und Bohranlagen ausgerüstet || l'atelier m. est équippé de riveuses et foreuses pneumatiques. / ~ rock drill || Preßluftgesteinsbohrmaschine f. || perceuse f. de roches à air comprimé. / ~ sand strewing device || Druckluftsandstreuer m. || sablerie f. à air comprimé. / ~ scaling hammer || Preßluftabklopfer m. || détartreur m. pneumatique. / ~ spring hammer || Druckluftfederhammer m. || marteau m. à ressort pneumatique. / ~ telegraph *see* ~ post. / ~ ticket carrier || Zettelrohrpost f. || poste f. pneumatique pour fiches. / ~ tool || Preßluftwerkzeug n.; Druckluftwerkzeug n. || outil m. pneumatique *ou* à air comprimé. / ~ trough (Phys) || pneumatische Wanne f. || cuve f. pneumatique.

pneumatic tube || Rohrpoströhre f.; Röhre f. einer Rohrpostanlage || tube m. *ou* tuyau m. pneumatique. / ~ blower || Rohrpostgebläse n. || soufflerie f. pour postes pneumatiques. / ~ card || Rohrpostkarte f. || carte f. pneumatique. / ~ collecting station || Rohrpostsammelstelle f. || station f. collectrice de postes pneumatiques. / ~ distributing station || Rohrpostverteilerstelle f. || station f. de distribution de postes pneumatiques. / ~ letter || Rohrpostbrief m. || petit bleu m. / ~ post conveyer || Rohrpostanlage f. || installation f. de poste tubulaire *ou* pneumatique.

pneumatic tubes pl. *see* pneumatic post plant.

pneumatic tyre || Pneumatik m.; Luftreifen m. || (bandage m.) pneumatique m.; pneu m. / ~ for motor cars || Automobilluftreifen m. || pneu m. *ou* pneumatique m. d'automobile. / ~ rim || Luftreifenfelge f. || jante f. pour pneu.

pneumatic winch || Lufthaspel m.; Preßlufthaspel m. || treuil m. à air comprimé.

pneumatic (Subst) *see* pneumatic tyre.

pneumonia || Lungenentzündung f. || pneumonie f.

poacher || Wilderer m.; Wilddieb m. || braconnier m.

pocket || Tasche f. || poche f. / ~ for cards || Kartentasche f. || poche f. pour cartes. / ~ of ore || Erznest n.; Erztasche f. || poche f. de minerai. / small ~ || Täschchen n. || pochette f.

pocket accumulator || Taschenakkumulator m. || accumulateur m. de poche. / ~ almanac || Taschenkalender m. || calendrier m. de poche. / ~ ammeter || Taschenstrommesser m.; Taschenamperemeter n. || ampèremètre m. de poche. / ~ book || Brieftasche f. || portefeuille m. / ~ calculator || Taschenrechenmaschine f. || appareil m. à calculer de poche. / ~ calendar for advertizing purposes || Taschenkalender m. für Reklamezwecke || calendrier-pochette m. de réclame. / ~ chronometer || Taschenuhr n. || chronomètre m. de poche. / ~ compass || Taschenbussole f. || boussole f. de poche. / ~ counter || Taschenzähler m. || compteur m. de poche. / ~ cutter || Taschenzuschneider m. || coupeur m. de poches. / ~ dictionary || Taschenwörterbuch n. || dictionnaire m. portatif *ou* de poche. / ~ edition || Taschenausgabe f. || édition f. de poche. / ~ handkerchief || Taschentuch n. || mouchoir m. / ~ knife || Taschenmesser n. || canif m.; couteau m. de poche.

pocket lamp || Taschenlampe f. || lampe f. de poche. / ~ battery || Taschenlampenbatterie f. || batterie f. de lampe de poche. / ~ bulb || Taschenlampenglühbirne f. || ampoule f. de lampe de poche. / ~ case || Taschenlampenhülse f. || corps m. pour lampes de poche.

pocket lighter || Taschenfeuerzeug m. || briquet m. de poche. / ~ magnifier || Taschenlupe f. || loupe f. de poche. / ~ magnifier for use in the hand || Taschenlupe f. für den freihändigen Gebrauch || loupe f. de poche employée à la main. / ~ maker || Taschenmacher m. || confectionneur m. de poches. / ~ medicine chest || Taschenapotheke f. || pharmacie f. de poche. / ~ metal ignition lighter || Taschenfeuerzeug n. für Metallzündung

briquet m. de poche pour allumage au métal. / ~ mirror || Taschenspiegel m. || miroir m. *ou* glace f. de poche. / ~ refractometer || Taschenrefraktometer n. || réfractomètre m. de poche. / ~ size || Taschenformat n. || format m. portatif. / ~ telescope || Fernglas n.; Fernrohr n. || lunette f. d'approche. / ~ typewriter || Reiseschreibmaschine f. || machine f. à écrire de voyage.

pockwood || Pockholz n. || bois m. de gaïac.

pod || Schote f. || cosse f.

poil, long ~ || Pelzsammet m.; Felper m. || panne f.; peluche f.

point, to || anspitzen || aiguiser; appointer; empointer. / ~ flat the joints (Mas) || die Fugen f. glattstreichen || jointoyer. / ~ a pile (Carp) || einen Pfahl m. spitzen || tailler en pointe un pieu. / ~ the pins || die Stecknadeln fpl. zuspitzen || empointer les épingles fpl. / ~ with steel || stählen || acérer; armer d'acier m. / ~ a telescope || ein Fernrohr n. richten || braquer un télescope. / ~ the tiles || die Ziegel fpl. verstreichen || sceller les tuiles fpl. en mortier.

point || Stift m.; Spitze f. || pointe f. / ~ (Nail) || Nagelspitze f. || pointe f. d'un clou. / ~ (Print) || Punkturspitze f.; Punktur f. || pointure f.; pointe f. du tympan. / ~ (Punctuation) || Punkt m. || point m. / ~s pl. (Railw) *see also* points || Weiche f. || changement m. (de voie). / accidental ~ (Opt) || Fluchtpunkt m. || point m. de fuite *ou* fuyant *ou* de concours *ou* accidentel *ou* évanouissant. / ~ of application || Angriffspunkt m. || point m. d'application d'une force. / automatic ~s pl. || selbsttätige Weiche f. || changement m. de voie automatique. / black ~ (Glass) || Glaskoralle f. || grain m. de verre noir. / ~ of blade de-forged from the rolled section || aus dem Walzenprofil im Gesenk ausgeschmiedetes Zungenende || pointe f. d'aiguille obtenue de la section laminée par forgeage en matrice. / ~ of blade welded-on || angeschweißtes Zungenende n. || pointe f. d'aiguille raccordée au rail par soudage. / broad ~ of a pen || breite Spitze f. || pointe f. large. / ~ of compasses || Zirkelspitze f. || pointe f. d'un compas. / conjugate ~ || konjugierter Punkt m. || point m. conjugué. / ~ of contact || Berührungspunkt m. || point m. de contact. / ~ fixed by its co-ordinates || durch seine Koordinaten festgelegter Punkt m. || point m. donné par ses coordonnées. / ~ of a crossing || Herzspitze f. || pointe f. de cœur. / dead ~ || toter Punkt m. || point m. mort. / ~ where the declivity changes || Gefällswechsel m. || passage m. d'une déclivité à une autre. / ~ of departure (Surv) || Ausgangspunkt m. || point m. de départ. / to determine the position of a ~ || die Lage f. eines Punktes eindeutig festlegen || indiquer nettement la position d'un point. / essential ~ || Anhaltspunkt m. || point m. d'appui. / ~ of the feeler || Tasterspitze f. || pointe f. du calibre. / fixed ~ || Fixpunkt m.; Festpunkt m. || point m. fixé *ou* de repère. / flexible ~ (Pen) || biegsame Spitze f. || pointe f. flexible. / freezing ~ || Gefrierpunkt m.; Nullpunkt m. || point m. de congélation. / ~ of ignition || Flammpunkt m. || point m. d'ignition. / ~ of impact || Stoß(mittel)-

punkt m. ‖ centre m. de percussion. / ~ of inflammation ‖ Entzündungspunkt m. ‖ point m. d'inflammation. / ~ of interruption (Electr) ‖ Unterbrechungsstelle f. ‖ point m. d'interruption. / ~ of intersection ‖ Schnittpunkt m. ‖ point m. d'intersection. / ~ of intersection (Mine) ‖ Durchschlagspunkt m. ‖ point m. d'intersection. / isolated ~ ‖ isolierter Punkt m. ‖ point m. isolé. / ~ of junction ‖ Knotenpunkt m. ‖ point m. de jonction. / material ~ ‖ materieller Punkt m. ‖ point m. matériel. / ~ of measuring ‖ Meßpunkt m. ‖ point m. de mesurage. / melting ~ of ice ‖ Eispunkt m.; Nullpunkt m. ‖ point m. de fusion de la glace. / multiple ~ ‖ vielfacher Punkt m. ‖ point m. multiple. / neutral ~ (Electr) ‖ Sternpunkt m. ‖ point m. neutre. / nickel-plated ~ ‖ vernickelte Spitze f. ‖ pointe f. nickelée. / nodal ~ (Phys) ‖ Knotenpunkt m. ‖ nœud m. / ~ of observation ‖ Beobachtungspunkt m. ‖ poste m. d'observation. / oval ~ (Pen) ‖ Flachspitze f. ‖ pointe f. ovale. / ~ of regression (Geom) ‖ Haltepunkt m. ‖ point m. d'arrêt. / ~ of revolution ‖ Drehpunkt m. ‖ point m. de rotation. / ~ of saturation ‖ Sättigungspunkt m. ‖ point m. de saturation. / ~ of sight ‖ Augenpunkt m. ‖ point m. de vue. / singular ~s pl. ‖ ausgezeichnete Punkte mpl. ‖ points mpl. singuliers. / slip ~s pl. (Railw) ‖ Kreuzungsweiche f. ‖ traversée f. à aiguilles. / ~ of solidification ‖ Erstarrungspunkt m. ‖ point m. de solidification. / stationary ~ (Geom) ‖ Umkehrpunkt m. oder Rückkehrpunkt m. einer Kurve ‖ point m. de rebroussement. / ~ of support ‖ Stützpunkt m. ‖ point m. d'appui. / ~ of support of a bridge ‖ Auflager n. einer Brücke ‖ appui m. d'un pont. / ~ of suspension ‖ Unterstützungspunkt m. ‖ point m. de suspension. / ~ of time ‖ Zeitpunkt m. ‖ moment m. / ~s pl. for tramways ‖ Straßenbahnweiche f. ‖ changement m. de voie pour tramways. / ~ of transfer ‖ Übergabestelle f. ‖ point m. de transfert. / vanishing ~ see visual ~. / ~ vertically beneath a given point ‖ Lotpunkt m. ‖ point m. de prise d'aplomb. / economic ~ of view ‖ wirtschaftlicher Gesichtspunkt m. ‖ point m. de vue économique. / visual ~ (Opt) ‖ Fluchtpunkt m. ‖ point m. de fuite ou fuyant ou de concours ou accidentel ou évanouissant.

pointal (Glassm) ‖ Hefteisen n. ‖ pontil m.

point, determination of a ~ ‖ Punktbestimmung f. ‖ fixation f. d'un point.

pointed ‖ zugespitzt, spitz ‖ pointu; aigu. / ~ arch ‖ Spitzbogen m. ‖ ogive f.; arc m. gothique. / ~ bolt ‖ Scharfbolzen m. ‖ boulon m. à pointe affilée.

pointed chisel ‖ Spitzmeißel m. ‖ aiguille f. / ~ with hexagon shank ‖ Spitzmeißel m. mit sechskantigem Einsteckende ‖ aiguille f. à queue hexagonale. / ~ with round shank ‖ Spitzmeißel m. mit rundem Einsteckende ‖ aiguille f. à queue cylindrique.

pointed drill ‖ Spitzbohrer m. ‖ foret m. à langue d'aspic. / ~ pin ‖ Tastspitze f. ‖ pointe f. de contact. / ~ radiator ‖ Spitzkühler m. ‖ radiateur m. coupe-vent.

pointer (Clock) ‖ Zeiger m. ‖ aiguille f. / ~ (Mas) ‖ Fassadenmaurer m. ‖ jointoyeur m. / ~ (Needl) ‖ Nadelspitzer m. ‖ em-

pointeur m. d'aiguilles. / ~ (Print) ‖ Punktierer m. ‖ pointeur m. / ~ (Shipb) ‖ Kettenbrecher m.; Kettengabel f. des Gangspills ‖ désengreneur m. / adjustable ~ showing the cutting-off direction ‖ verstellbarer Zeiger m. auf Schnittrichtung ‖ index m. réglable pour la direction de coupe. / to bring the ~ into full action ‖ den Zeiger m. zum vollen Ausschlag bringen ‖ faire osciller entièrement l'aiguille f. / the ~ moves ‖ der Zeiger m. schlägt aus ‖ l'aiguille f. oscille. / the ~ of the instrument begins to move across the scale ‖ der Zeiger m. des Instruments beginnt über der Skale zu spielen ‖ l'aiguille f. de l'instrument commence à osciller ou à jouer devant le cadran.

pointer base ‖ Zeigerlagerung f. ‖ portée f. de l'index ou de l'aiguille. / ~ eyepiece ‖ Zeigerokular n. ‖ oculaire m. indicateur. / laying the gun for ~ fire ‖ Einzelrichten n. des Geschützes ‖ pointage m. individuel du canon. / ~ instrument ‖ Zeigerinstrument n. ‖ instrument m. à aiguille. / ~ reading ‖ Zeigerablesung f. ‖ lecture f. des indications de l'aiguille. / ~ support ‖ Zeigerlagerung f. ‖ portée f. de l'index ou de l'aiguille. / ~ telegraph ‖ Zeigertelegraf m. ‖ télégraphe m. à cadran. / ~ typewriter ‖ Zeigerschreibmaschine f. ‖ machine f. à écrire à tige indicatrice.

point focal spectacle lens ‖ punktuell abbildendes Brillenglas n. ‖ verre m. correcteur à image ponctuelle. / ~ hole ‖ Punkturloch n. ‖ trou m. de pointure. / ~ image ‖ Bildpunkt m. ‖ point-image m. / ~ indicator (Railw) ‖ Weichensignal n. ‖ signal m. d'aiguille; indicateur m. de direction.

pointing of pins ‖ Zuspitzen n. der Stecknadeln ‖ empointage m. des épingles. / ~ engine ‖ Spitzmaschine f.; Anspitzmaschine f. ‖ dégrossisseuse f. / ~ hammer ‖ Anspitzhammer m. ‖ marteau m. à appointer.

point lace ‖ genähte Spitze f. ‖ dentelle f. à point. / ~ lever (Railw) ‖ Weichenhebel m. ‖ levier m. d'aiguillage. / ~ light lamp ‖ Punktlichtlampe f. ‖ lampe f. ponctuelle. / ~ light lamp casing ‖ Punktlichtlampengehäuse n. ‖ cage f. de la lampe ponctuelle. / ~ locking (Railw) ‖ Weichensicherung f. ‖ blocage m. des aiguilles. / ~ motor (Railw) ‖ Weichenantriebsmotor m. ‖ moteur m. de commande d'aiguille. / ~-nose shovel with long socket ‖ Schaufel f. mit langer Spitztülle ‖ pelle f. à longue douille pointue. / ~ opening (Print) ‖ Punkturöffnung f. ‖ crenure f. / ~ paper (Weav) ‖ Partonenpapier n.; Tupfpapier n.; Musterpapier n. ‖ papier m. à patron ou quadrillé ou rayé. / ~ planing machine (Railw) ‖ Weichenzungenhobelmaschine f. ‖ raboteuse f. de pointe d'aiguilles de changement de voie; raboteuse f. pour pointer d'aiguilles. / ~ plate (Print) ‖ Punkturschere f. ‖ ciseaux mpl. de pointures. / ~ rail (Railw switch) ‖ Leitzunge f. ‖ rail m. mobile.

points pl. (Railw) ‖ Weiche f. ‖ changement m. (de voie). / automatic ~ ‖ selbsttätige oder automatische Weiche f. ‖ changement m. de voie automatique. / slip ~ ‖ Kreuzungsweiche f. ‖ traversée f. à ai-

guilles. / ~ for tramways ‖ Straßenbahnweiche f. ‖ changement m. de voie pour tramways.

pointsman (Railw) ‖ Weichensteller m. ‖ aiguilleur m.

points operating device ‖ Weichenstellvorrichtung f. ‖ dispositif m. d'aiguillage; appareil m. de manœuvre d'aiguilles. / set of ~ ‖ Weichenstraße f. ‖ groupe m. de changements de voie.

point screw ‖ Punkturschraube f. ‖ vis f. de pointures. / ~ seamer ‖ Stepper m. ‖ piqueur m. / ~ spur (Print) ‖ Punkturspitze f.; Punktur f. ‖ pointure f.; pointe f. du tympan. / ~ switch ‖ Zungenweiche f. ‖ changement m. de voie à aiguille(s). / three-throw ~ switch (Railw) ‖ verschränkte Doppelweiche f. ‖ changement m. à aiguille double. / ~ tool (Turn) ‖ Spitzstahl m. ‖ grain m. d'orge. / ~ welding ‖ Punktschweißung f. ‖ soudage m. par points.

poison, to ‖ vergiften ‖ empoisonner.

poison ‖ Gift n. ‖ poison m. / metallic ~ ‖ Metallgift n. ‖ poison m. métallique.

poisoned corn ‖ Giftgetreide n. ‖ blé m. empoisonné.

poison gas warfare ‖ Gaskrieg m.; Giftgaskrieg m. ‖ guerre f. aux gaz.

poisoning ‖ Vergiftung f. ‖ empoisonnement m.

poisonous, non- ~ colour ‖ giftfreie Farbe f. ‖ couleur f. exempte de poison.

poison tower (Metal) ‖ Giftfang m.; Giftturm m. ‖ cheminée f. pour l'arsenic.

poke, to ‖ schüren ‖ attiser. / ~ the fire ‖ Feuer n. anschüren ‖ activer ou aviver le feu.

poker ‖ Schüreisen n.; Stocheisen n.; Feuerhaken m. ‖ ringard m.; tisonnier m. / fire ~ ‖ Feuerhaken m. ‖ attisoir m. / mechanical ~ ‖ Stochapparat m. ‖ appareil m. d'attisage. / small ~ ‖ kleiner Schürhaken m. ‖ râblot m.

polar ‖ polar ‖ polaire. / ~ adjustment ‖ Polhöhenverstellung f. ‖ réglage m. de la latitude. / ~ angle ‖ Polarwinkel m. ‖ angle m. décrit par le rayon polaire. / ~ axis ‖ Polarachse f. ‖ axe m. polaire. / ~ flattening ‖ Abplattung f. an den Polen ‖ aplatissement m. polaire. / ~ circle ‖ Polarkreis m. ‖ cercle m. polaire. / ~ co-ordinates pl. ‖ Polarkoordinaten fpl. ‖ coordonnées fpl. polaires. / ~ curve ‖ Polarkurve f. ‖ courbe f. polaire. / ~ distance ‖ Polardistanz f.; Poldistanz f. ‖ distance f. au pol. / ~ equation ‖ Polargleichung f. ‖ équation f. polaire. / ~ line ‖ Polstrahl m. ‖ rayon m. polaire. / ~ moment of inertia ‖ polares Trägheitsmoment n. ‖ moment m. d'inertie polaire. / ~ planimeter ‖ Polarplanimeter m. ‖ planimètre m. polaire. / ~ subnormal ‖ Polarsubnormale f. ‖ sousnormale f. polaire. / ~ subtangent ‖ Polarsubtangente f. ‖ soustangente f. polaire.

polar (Geom) ‖ Polare f. ‖ polaire f. / ~ reciprocal ‖ reziproke Polare f. ‖ polaire f. réciproque.

polarimeter ‖ Polarimeter m. ‖ polarimètre m.

polarity ‖ Polarität f. ‖ polarité f. / ~ indicator ‖ Stromrichtungsanzeiger m. ‖ indicateur m. de sens du courant.

polarizable ‖ polarisierbar ‖ susceptible de se polariser. / ~ state ‖ Polarisierbarkeit f. ‖ susceptibilité f. de se polariser.

polarization ‖ Polarisation f. ‖ polarisation f. / electrolytic ~ ‖ elektrolytische Polarisation f. ‖ polarisation f. électrolytique. / ~ of light ‖ Lichtpolarisation f. ‖ polarisation f. de la lumière. / rotating ~ ‖ Rotationspolarisation f. ‖ polarisation f. rotatoire. / ~ of a wave (Radio) ‖ Wellenpolarisation f. ‖ polarisation f. d'une onde.

polarization battery ‖ Polarisationsbatterie f. ‖ pile f. de polarisation. / ~ cell ‖ Polarisationszelle f. ‖ pile f. polarisable. / ~ current ‖ Polarisationsstrom m. ‖ courant m. de polarisation. / ~ tube ‖ Polarisationsröhre f. ‖ tuyau m. à polarisation.

polarize, to ‖ polarisieren ‖ polariser.

polarized light ‖ polarisiertes Licht n. ‖ lumière f. polarisée.

polarizer ‖ Polarisator m. ‖ polariseur m.

polarizing ‖ polarisierend ‖ polariseur. / ~ apparatus ‖ Polarisationsapparat m. ‖ appareil m. de polarisation. / ~ attachment ‖ Polarisationsvorrichtung f. ‖ appareil m. de polarisation. / ~ microscope ‖ Polarisationsmikroskop n. ‖ microscope m. polarisant. / studying with the ~ microscope ‖ polarisationsmikroskopische Untersuchung f. ‖ examen m. en lumière polarisée. / ~ microscopy ‖ Polarisationsmikroskopie f. ‖ microscopie f. en lumière polarisée. / ~ solar prism see ~ sun prism. / ~ sun prism ‖ Polarisationssonnenprisma n. ‖ hélioscope m. de polarisation.

pole, to ~ copper ‖ Kupfer n. polen ‖ travailler le cuivre à la perche.

pole (Astron; Elektr) ‖ Pol m. ‖ pôle m. / ~ (Build) ‖ Pfahl m.; Pfosten m. ‖ poteau m.; pieu m.; pilot m. / ~ (Coachm) ‖ Deichsel f. ‖ timon m. / ~ (El line) ‖ Stange f.; Telegrafenstange f.; Mast m. ‖ poteau m. de télégraphe; perche f. / ~ (Geom) ‖ Pol m. ‖ pôle m. / ~ (Surv) ‖ Meßstange f. ‖ perche f.; règle f.; verge f. / ~ (Vine) ‖ Rebstecken m.; Rebpfahl m. ‖ échalas m. / A-shaped ~ ‖ gekuppelte Stange f.; A-Mast m. ‖ poteau m. couplé ou en A. / boundary ~ ‖ Grenzpfahl m. ‖ poteau m. frontière. / ~ set in concrete ‖ einbetonierter Mast m. ‖ poteau m. encastré en béton. / consequent ~s pl. ‖ Folgepole mpl. ‖ pôles mpl. conséquents. / coupled ~ ‖ Kuppelstange f. ‖ poteau m. moisé. / distributing ~ ‖ Überführungssäule f. ‖ tour m. de dispersion. / double ~ ‖ Doppelstütze f. ‖ appui m. double. / earthed ~ ‖ geerdeter Pol m. ‖ pôle m. à la terre. / ~ of iron concrete ‖ Eisenbetonpfahl m. ‖ pilier m. ou pieu m. en béton armé. / ~ like ~ ‖ gleichnamiger Pol m. ‖ pôle m. de même nom. / ~ with line stays ‖ Linienfestpunkt m. ‖ appui m. d'arrêt. / N-~ ‖ N-förmige Stütze f. ‖ appui m. en N. / negative ~ ‖ Kathode f.; negativer Pol m. ‖ cathode m.; pôle m. négatif. / to set the ~ perpendicular by the plummet ‖ den Meßstab m. nach dem Lot senkrecht aufstellen ‖ planter le jalon verticalement au moyen du fil à plomb. / positive ~ ‖ Anode f.; positiver Pol m. ‖ anode m.; pôle m. positif. / pyramidal ~ ‖ pyramidenförmige Stütze f. ‖ appui m. en pyramide. / to range in a ~ ‖ einen Stab m. einfluchten ‖ aligner un jalon. / reinforced concrete ~

(Electr line) ‖ Eisenbetonmast m. ‖ mât m. en béton armé. / reversing ~ ‖ Wendepol m. ‖ pôle m. de commutation. / ~ for sighting-on ‖ Signalstange f. ‖ perche f. de signalisation. / single ~ ‖ Einzelstütze f. ‖ appui m. simple. / ~ with socle ‖ angeschuhte Stange f. ‖ poteau m. allongé. / stayed ~ ‖ verankerte Stütze f. ‖ appui m. haubanné. / strutted ~ ‖ Stange f. mit Streben ‖ poteau m. à contrefiches. / telegraph ~ ‖ Telegrafenstange f. ‖ poteau m. de télégraphe. / terminal ~ ‖ Abspanngestänge n.; Abspannstütze f. ‖ appui m. d'arrêt. / training ~ see terminal ~. / tubular ~ ‖ Rohrmast m. ‖ poteau m. tubulaire. / unlike ~ ‖ ungleichnamiger Pol m. ‖ pôle m. de nom contraire. / wooden ~ ‖ Telegrafenstange f. aus Holz ‖ poteau m. de télégraphe en bois.

pole and flag (Surv) ‖ Richtstange f.; Signalstange f. ‖ jalon m. / armature with salient ~es ‖ Polanker m. ‖ armature f. à pôles saillants. / base of a ~ ‖ Mastsockel m. ‖ socle m. de poteau. / ~ bolt (Coachm) ‖ Scherbolzen m. ‖ boulon m. de timon ou d'assemblage. / ~ bridge ‖ Rundholzbrücke f.; Stangenbrücke f. ‖ pont m. de rondins. / ~ cap ‖ Pfahlkappe f. ‖ chapeau m. métallique; chapiteau m. d'un poteau. / ~ casing for dynamos ‖ Polgehäuse n. für Dynamomaschinen ‖ revêtement m. polaire de dynamo. / ~ changer ‖ Polwechsler m. ‖ inverseur m. de pôles; vibrateur m. générateur. / to change the ~ connections pl. ‖ die Pole mpl. umschalten ‖ regrouper les pôles mpl. / ~ diagram (El line) ‖ Stangenbild m ‖ carnet m. d'armement. / ~ diagram in local telephone plants ‖ Gruppierungsbild n. für Ortsfernsprechnetze ‖ schéma m. des groupements de lignes d'abonnés / ~ distance (El line) ‖ Stützpunktabstand m. ‖ distance f. entre les appuis. / ~ drill ‖ Stützenbohrer m. ‖ tarière f. / erection of the ~s ‖ Aufstellung f. der Stützen ‖ plantation f. des appuis. / ~-finding paper ‖ Polsuchpapier n. ‖ papier m. pôle. / ~ fittings pl. ‖ Gestängeausrüstung f. ‖ armement m. de l'appui. / ~ holder (Coachm) ‖ Scherholz n. ‖ armon m. / ~ hole ‖ Stangenloch n. ‖ fouille f.; fosse f. / ~ hook (Coachm) ‖ Zughaken m. ‖ crochet m. de timon ou de bout de timon. / ~ lathe ‖ Drechslerwippe f. ‖ tour m. à perche. / ~ line ‖ Gestängeleitung f. ‖ ligne f. de poteaux. / ~ maker (Coachm) ‖ Deichselmacher m. ‖ ouvrier m. en timons. / ~ mast ‖ Pfahlmast m. ‖ mât m. à pible. / ~ painting ‖ Stangenanstrich m. ‖ peinture f. des poteaux. / ~ pin see ~ bolt. / ~ plate (Carp) ‖ Sparrensohle f.; Dachschwelle f.; Fußrahmen m. ‖ plate-forme f. de comble. / establishment for ~ preservation ‖ Tränkungsanstalt f. ‖ établissement m. pour l'imprégnation des bois. / preparation of ~s ‖ Stangenzubereitungf. ‖ préparation f. des poteaux. / ~ reinforcement ‖ Verstärken von Telegrafenstangen ‖ consolidation f. des poteaux. / ~ ring in halves ‖ Polring m. in zwei Teilen ‖ anneau m. de pôles en deux parties. / ~ shoe ‖ Polschuh m. ‖ pièce f. polaire; cosse f. / ~ smith (Coachm) ‖ Deichselbeschläger m. ‖ ferreur m. de brancards.

pole-star ‖ Nordstern m.; Polarstern m. ‖ étoile f. polaire.

pole statistics pl. ‖ Stangenstatistik f. ‖ statistique f. des poteaux. / ~ step (El line) ‖ Steigeisen n. ‖ échelons mpl. / ~ store ‖ Stangenlagerplatz m. ‖ dépôt m. de poteaux. / strength of a ~ ‖ Polstärke f. ‖ intensité f. de pôle. / adjustable ~ support ‖ Deichselausgleichvorrichtung f. ‖ équilibreur m. de timon. / ~ switch ‖ Mastschalter m. ‖ interrupteur m. de poteau. / ~ terminal ‖ Polklemme f. ‖ borne f. d'élément. / ~ test box ‖ Untersuchungsstange f. ‖ poteau m. de coupure. / ~ timber ‖ Stangenholz n. ‖ rondin m. / ~ transformer ‖ Masttransformator m. ‖ transformateur m. à mât. / ~ truck (Coachm) ‖ Deichseltragrolle f. ‖ roue f. porteuse de timon.

polianite ‖ Polianit m. ‖ polianite f.; manganèse m. oxydé hydraté.

police ‖ Polizei f. ‖ police f. / ~ box ‖ Polizeimelder m. ‖ avertisseur m. de police. / ~ launch ‖ Polizeiboot n. ‖ vedette f. de la police. / ~ radio service ‖ Polizeifunkdienst ‖ service m. radioélectrique de la police. / ~ signal system ‖ Polizeirufanlage f. ‖ installation f. d'appel de la police. / switchboard for ~ signal systems ‖ Umschalteschrank m. für Polizeirufanlagen ‖ commutateur m. téléphonique pour installations d'appel de la police. / ~ vessel ‖ Hafenwachschiff n. ‖ bâtiment m. garde-port; navire m. de surveillance.

policy ‖ Police f.; Versicherungsschein m. ‖ police f. d'assurance. / floating ~ ‖ laufende Police f. ‖ police f. flottante. / insurance ~ ‖ Police f.; Versicherungsschein m. ‖ police f. d'assurance. / surrender value of an insurance ~ ‖ Rückkaufswert m. einer Versicherungspolice ‖ valeur f. de rachat d'une police d'assurance. / to take-out a ~ ‖ eine Police f. nehmen ‖ prendre une police.

polish, to ‖ polieren ‖ polir. / ~ (Leather) ‖ wichsen ‖ cirer. / ~ with emery ‖ abschmirgeln ‖ polir à l'émeri. / ~ gun shot ‖ Flintenschrot m. polieren ‖ lustrer ou roder la dragée. / ~ with pumice stone ‖ mit Bimsstein m. schleifen; bimsen ‖ poncer; polir à la ponce. / the surface was polished with extreme accuracy ‖ die Oberfläche wurde mit äußerster Genauigkeit poliert ‖ la surface fut polie avec la précision la plus minutieuse.

polish ‖ Politur f. ‖ poli m. / aluminium ~ ‖ Aluminiumputzmittel n. ‖ produit m. à nettoyer l'aluminium. / steel of a capacity for taking a high ~ ‖ Stahl m. von hoher Polierfähigkeit ‖ acier m. susceptible de prendre un poli remarquable. / ~ for high lustre ‖ Hochglanzpolitur f. ‖ poli m. à reflets. / motor car ~ ‖ Kraftwagenpolitur f.; Autolack m. ‖ vernis m. d'automobile.

polishable ‖ polierfähig ‖ polissable.

polished ‖ poliert; glänzend ‖ poli. / ~ with emery ‖ abgeschmirgelt ‖ poli à l'émeri m. / highly ~ ‖ Hochglanz m. poliert ‖ poli à reflets mpl. / ~ lacquer ware ‖ Schleiflackwaren fpl. ‖ articles mpl. vernis et polis. / ~ lens surface ‖ polierte Glasfläche f. ‖ face f. de verre polie. / ~ material ‖ blankgezogenes Material n. ‖ matériel m. étiré poli blanc. / brightly ~ steel ‖ hochpolierter Stahl m. ‖ acier m. avec un polissage intense. / ~ washer

‖ blanke Scheibe f. ‖ rondelle f. décolletée.

polisher ‖ Polierer m. ‖ polisseur m. / ~ (Tool) ‖ Polierscheibe f. ‖ polissoire f. / brush ~ ‖ Bürstenpolierer m. ‖ polisseur m. de bois de brosses. / ~ of concave glass ‖ Hohlglasschleifer m. ‖ polisseur de verres mpl. concaves. / ~ of convex glass ‖ Rundglasschleifer m. ‖ polisseur de verres mpl. convexes. / felted ~ (Glassm) ‖ befilztes Polierholz n. ‖ estrique f.; lustroir m.; moëllette f. / floor ~ ‖ Bohnermaschine f. ‖ cireuse f. / knife ~ ‖ Messerpolierer m. ‖ polisseur m. de couteaux. / scissors ~ ‖ Scherenpolierer m. ‖ polisseur m. de ciseaux.

polisher's frame ‖ Polierrahmen m. ‖ châssis m. pour tenir les objets à polir.

polishing ‖ Polieren n. ‖ polissage m. / ~ (Needl) ‖ Schauern n.; Scheuern n.; Polieren n. ‖ polissage m. / knife ~ ‖ Messerpolieren n. ‖ polissage m. de couteaux. / metal ~ ‖ Polieren n. von Metallwaren ‖ polissage m. de métaux. / scissors ~ ‖ Scherenpolieren n. ‖ polissage m. de ciseaux. / ~ of silk goods ‖ Glätten n. von Seidenzeug ‖ polissage m. de tissus de soie.

polishing agent see polish. / ~ barrel ‖ Polierfaß n.; Poliertrommel f. ‖ tambour m. à polir. / ~ bit ‖ Kaliberbohrer m.; Schlichtbohrer m. ‖ alésoir m.; polissoir m. / ~ brick ‖ Putzstein m. ‖ brique f. à polir. / ~ brush ‖ Polierbürste f.; Putzbürste f. ‖ brosse f. à polir. / ~ composition ‖ Poliermasse f. ‖ composition f. à polir. / ~ device for spring wires ‖ Federdrahtpoliereinrichtung f. ‖ installation f. à polir le fil à ressorts. / ~ disk see ~ wheel. / ~ drum ‖ Poliertrommel f. ‖ tambour m. de polissage. / ~ file ‖ Polierfeile f. ‖ brunissoir f.; carrelette f. / ~ hammer ‖ Polierhammer m. ‖ marteau m. à polir. / ~ head ‖ Polierkopf m. ‖ nez m. de polissage. / cold ~ ink ‖ Kaltpolierfarbe f. ‖ vernis m. à polir au froid. / ~ iron ‖ Polierstahl m.; brunissoir m. / ~ iron (Cable) ‖ Glätteisen n. ‖ fer m. à polir. / ~ leather ‖ Schleifleder n. ‖ cuir m. à polir. / ~ machine ‖ Poliermaschine f. ‖ machine f. à polir; polisseuse f. / cylinder ~ machine ‖ Zylinderpoliermaschine f. ‖ machine f. à polir l'intérieur des cylindres. / ~ machine for knives ‖ Messerputzmaschine f. ‖ machine f. à nettoyer les couteaux. / raw, granulated, washed or ground ~ material ‖ rohes, gekörntes, geschlämmtes oder gemahlenes Poliermittel n. ‖ matière f. à polir brute, granulée, lavée ou moulue. / ~ mill (Needl) ‖ Scheuermühle f. ‖ table f. à polissage. / ~ motor ‖ Poliermotor m. ‖ moteur m. à polir ou pour appareils de polissage. / ~ paste (Leather) ‖ Wichse f.; cirage m. / ~ paste (Metal) ‖ Putzpomade f. ‖ pâte f. à nettoyer. / malt ~ plant ‖ Malzpolieranlage f. ‖ installation f. pour le lissage du malt. / ~ powder ‖ Polierpulver n. ‖ poudre f. à polir. / ~ powder mill ‖ Glättemühle f. ‖ moulin m. pour la fabrication de poudre à polir. / ~ red ‖ Polierrot n. ‖ rouge m. à polir. / ~ roll ‖ Polierwalze f. ‖ cylindre m. de laminoir à polir. / ~ slate ‖ Polierschiefer m. ‖ schiste m. tripoléen ou à polir. / ~ stick

(Shoem) ‖ Glättholz n.; Glättschiene f. ‖ buis m. / ~ stone ‖ Polierstein m. ‖ meule f. polissoire. / ~ substance ‖ Poliermittel n. ‖ produit m. à polir. / ~ wax ‖ Bohnerwachs n. ‖ cirage m. / ~ wheel ‖ Polierscheibe f. ‖ disque m. à polir. / ~ wheel (Needl) ‖ Schleifmühle f. ‖ aiguiserie f. / flannel ~ wheel ‖ Flanellpolierscheibe f. ‖ roue f. de polissage en flanelle. / ~ wool ‖ Putzwolle f. ‖ déchets mpl. de coton.

political economy ‖ Volkswirtschaft f.; Volkswirtschaftslehre f. ‖ économie f. politique.

poll, to ~ a tree ‖ einen Baum m. köpfen oder abästen ‖ ébrancher ou épointer un arbre.

pollard ‖ Kopfholz n. ‖ têtard m.

poll tax ‖ Kopfsteuer f. ‖ capitation f.

pollution (River) ‖ Verunreinigung f. ‖ contamination f.

Pollux ‖ Pollux m. ‖ Pollux m.

polybasite ‖ Polybasit m.; Schwarzgültigerz m. ‖ polybasite f.

polychromy ‖ Mehrfarbigkeit f. ‖ polychromie f. / ~ (Print) ‖ Vielfarbendruck m. ‖ impression f. en plusieurs couleurs ou en polychromie.

polycrase ‖ Polykras m. ‖ polycrase f.

polydymite ‖ Polydymit m. ‖ polydymite f.

polygon ‖ Polygon n.; Vieleck n. ‖ polygone m. / ~ circumscribed about a circle ‖ einem Kreise umschriebenes Polygon n.; Tangentenvieleck n. ‖ polygone m. circonscrit au cercle. / ~ of forces ‖ Kräftepolygon n. ‖ polygone m. des forces. / funicular ~ ‖ Seilpolygon n. ‖ polygone m. funiculaire. / ~ inscribed in a circle ‖ einem Kreise eingeschriebenes Polygon n.; Sehnenvieleck n. ‖ polygone m. inscrit au cercle.

polygonal bar ‖ Mehrkantstab m. ‖ barre f. polygonale. / ~ number ‖ Polygonalzahl f. ‖ nombre m. polygone. / ~ suspension ‖ Polygonaufhängung f. ‖ suspension f. polygonale.

polygon connection (Electr) ‖ Vieleckschaltung f. ‖ montage m. en polygone.

polygraph ‖ Polygraf m. ‖ polygraphe m.

polygraphy ‖ polygrafisches Gewerbe n. ‖ polygraphie f.

polyhalite ‖ Polyhalit m. ‖ polyhalithe f.

polyhedron ‖ Vielflächner m.; Polyeder n. ‖ polyèdre m.

polyphase ‖ mehrphasig ‖ polyphasé. / ~ alternator ‖ Mehrphasenwechselstromgenerator m. ‖ alternateur m. polyphasé. / ~ current ‖ Mehrphasenstrom m. ‖ courant m. polyphasé. / ~ rectifier ‖ Gleichrichter m. für Mehrphasenstrom ‖ redresseur m. pour courant polyphasé.

polytechnical society ‖ Gewerbeverein m. ‖ société f. industrielle.

polyvalent (Chem) ‖ mehrwertig ‖ polyvalent.

pomade ‖ Pomade f. ‖ pommade f. / perfumed ~ ‖ parfümierte Pomade f. ‖ pomade f. perfumée. / scented ~ see perfumed ~.

pomade box ‖ Pomadedose f.; Salbenbüchse f. ‖ boîte f. à pommade.

pomatum see pomade.

pomegranate ‖ Granatapfel m. ‖ grenade f.

pommel ‖ Knauf m.; Knopf m. ‖ pommeau m. / ~ (Curr) ‖ Krispelholz n. ‖ pommelle f.; marguerite f. à la main; grènetoir m. / ~ (Saddl) ‖ Sattelknopf m. ‖ pommeau m.

pomplemous ‖ Pompelmuse f. ‖ pamplemousse f.

Poncelet's water wheel ‖ Ponceletrad n. ‖ roue f. Poncelet.

poncho ‖ Poncho m. ‖ poncho m.

pond ‖ Teich m.; Weiher m.; Tümpel m. ‖ étang m.; pièce f. d'eau; mare f.; flaque f. d'eau. / cooling ~ ‖ Kühlbecken n. ‖ bassin m. de refroidissement. / ~ for fish ‖ Fischteich m. ‖ vivier m.

ponderable ‖ wägbar ‖ pondérable.

ponderal analysis ‖ Gewichtsanalyse f. ‖ analyse f. pondérale ou gravimétrique.

ponderous ‖ schwer ‖ pesant; lourd.

pond fishing ‖ Teichfischerei f. ‖ pêche f. en étang. / ~ grate ‖ Teichrechen m. ‖ grille f. d'étang. / ~ lily ‖ Wasserrose f.; Seerose f.; Wasserlilie f. ‖ nénuphar m. / ~ plug (Hydr arch) ‖ Ablaßschütze f.; Grundablaß m. ‖ bonde f. ou décharge f. de fond.

poniard ‖ Dolch m. ‖ poignard m.

pontage ‖ Brückenzoll m.; Brückengeld n. ‖ pontonage m.

pontlevis ‖ Zugbrücke f. ‖ pont-levis m.

pontoon ‖ Ponton m.; Brückenboot n. ‖ Brückenschiff n. ‖ ponton m. / ~ (Shipb) ‖ Prahm m. ‖ prame f.; ponton m. / wooden ~ ‖ hölzerner Ponton m. ‖ ponton m. en bois.

pontoon bottom ‖ Pontonboden m. ‖ fond m. de ponton. / ~ bridge ‖ Pontonbrücke f.; Schiffbrücke f. ‖ pont m. de pontons ou de bateaux. / ~ carriage ‖ Pontonwagen m. ‖ haquet m. à bâteau ou à ponton. / ~ dredger ‖ Pontonschwimmbagger m. ‖ excavateur m. flottant monté sur ponton. / ~ wagon see ~ carriage.

pony car ‖ Ponywagen m. ‖ voiture f. à poneys. / ~ engine (Railw) ‖ Rangiermaschine f.; Rangierlokomotive f.; Verschiebemaschine f. ‖ coucou m.; machine f. à faire les manœuvres.

pool (Pond) see pond. / ~ (Race) ‖ Totalisator m. ‖ totalisateur m. / ~ (Trade) ‖ Interessengemeinschaft f. ‖ convention f.; syndicat m.; communauté f. d'intérêts. / cess ~ ‖ Senkgrube f. ‖ puisard m.; fosse f. de nettoyage.

pool counter ‖ Spielmarke f. ‖ jeton m.; fiche f.

pooled to buy ‖ zwecks Kaufes zu einem Syndikat n. vereinigt ‖ syndiqués pour l'achat. / ~ to sell ‖ zwecks Verkaufes zu einem Syndikat n. vereinigt ‖ syndiqués pour la vente.

poonahlite (Miner) ‖ Poonalit m. ‖ poonalite f.

poop (Shipb) ‖ Kampanje f. ‖ dunette f. / ~ of a roof ‖ Gebinde n. oder Bundgesperre n. eines Dachstuhls ‖ ferme f. de comble d'un toit.

poop deck ‖ Schanzdeck n. ‖ plat-bord m. / ~ lantern ‖ Achterlaterne f.; Hecklaterne f. ‖ fanal m. de poop.

poor in iron ‖ eisenarm ‖ pauvre en fer. / ~ in seams pl. ‖ flözarm ‖ pauvre en couches fpl.

poor coal ‖ minderwertige Kohle f. ‖ charbon m. de peu de valeur. / ~ gas ‖ Generatorgas m. ‖ gaz m. pauvre. / ~ slag ‖ Rohschlacke f. ‖ scorie f. pauvre ou crue; laitier m. pauvre.

pop corn ‖ Röstmais m. ‖ maïs m. à griller. / ~ and rice mill ‖ Mais- und Reismühle f. ‖ moulin m. à riz et à maïs.

pop gun ‖ Knallbüchse f. ‖ canonnière f.

poplar ‖ Pappel f.; Pappelholz n. ‖ peuplier m.; grisard m. / black ~ ‖ Schwarzpappel f. ‖ liardier m.; peuplier m. noir. / Canadian ~ ‖ kanadische Pappel f. ‖ peuplier m. du Canada. / water ~ see black ~. / white ~ ‖ Weißpappel f. ‖ peuplier m. blanc.

poplar bark ‖ Pappelrinde f. ‖ écorce f. de peuplier. / ~ bud ‖ Pappelknospe f. ‖ bourgeon m. de peuplier. / ~ tree ‖ Pappelbaum m.; Pappel f. ‖ peuplier m.

poplar wood ‖ Pappelholz n. ‖ bois m. de peuplier. / ~ disc ‖ Pappelholzscheibe f. ‖ disque m. en peuplier.

poplin ‖ Popelin n.; Poplin n. ‖ popeline f. de laine.

poppet (Shipb) ‖ Schlittenständer m. ‖ colombier m.

poppet head (Lathe) ‖ Docke f. ‖ poupée f. / ~ (Mach) ‖ Rollkloben m. ‖ poulie. / ~ (Min) ‖ Seilscheibengerüst n.; Abstützung f. ‖ belle-fleur f.; chevalement m.; chevalet m.; châssis m. à molettes.

poppet leg (Min) ‖ Strebe f. oder Pfosten m. oder Schenkel m. des Seilscheibengerüstes ‖ montant m. de chevalement.

poppet valve ‖ Schnarchventil n.; Schnüffelventil n.; Rohrventil n. ‖ soupape f. tubulaire ou à manchon ou à déclic. / ~ gear ‖ Schnarchventilsteuerung f. ‖ distribution f. à soupape à déclic.

popping in the carburettor ‖ Knallen n. im Vergaser ‖ pétarade f. au carburateur.

poppy ‖ Mohn m. ‖ pavot m.; coquelicot m. / ~ mill ‖ Mohnmühle f. ‖ moulin m. à pavot. / ~ oil ‖ Mohnöl n. ‖ huile f. de pavot; huile f. d'œillette. / ~ pounder ‖ Mohnstampfer m. ‖ broyeur m. de pavot.

poppy seed ‖ Mohnsamen m. ‖ graine f. d'œillette. / ~ oil see poppy oil.

popularly ‖ gemeinverständlich ‖ populairement.

population ‖ Bevölkerung f. ‖ population f. / indigenous ~ ‖ ansässige Bevölkerung f. ‖ population f. domiciliée ou indigène.

populous ‖ volkreich ‖ populeux.

porcelain ‖ Porzellan n. ‖ porcelaine f. / artistic ~ ‖ Kunstporzellan n. ‖ porcelaine f. artistique. / biscuit baked ~ ‖ verglühtes Porzellan n. ‖ dégourdi f. / electrotechnical ~ ‖ elektrotechnisches Porzellan n.; Elektroporzellan n. ‖ porcelaine f. électro-technique. / hard ~ ‖ Steinporzellan n.; Feldspatporzellan; Edelporzellan n. ‖ porcelaine f. dure. / ~ for hotels ‖ Hotelporzellan n. ‖ porcelaine f. à l'usage des hôtels. / household ~ ‖ Gebrauchsporzellan n. ‖ porcelaine f. pour ménage. / soft ~ ‖ weiches Porzellan n. ‖ porcelaine f. tendre. / tender ~ see soft ~.

porcelain borer ‖ Porzellanbohrer m. ‖ perceur m. de porcelaine f. / ~ button ‖ Porzellanknopf m. ‖ bouton m. en porcelaine. / ~ calcining furnace ‖ Porzellanbrennofen m. ‖ four m. à cuire la porcelaine. / ~ caster ‖ Porzellangießer m. ‖ mouleur m. en porcelaine. / ~ cell ‖ Porzellanküvette f. ‖ cuve f. en porcelaine. / ~ clay ‖ Porzellanerde f.; Kaolin n. ‖ terre f. à porcelaine; kaolin m. / ~ colour ‖ Porzellanfarbe f. ‖ couleur f. pour porcelaine. / ~ crucible ‖ Porzellantiegel m. ‖ creuset m. en porcelaine. / ~ cup ‖ Isolierglocke f. ‖ cloche f. en porcelaine. / ~ decoration ‖ Porzellandekorieren n. ‖ décoration f. de la porce-

laine. / ~ earth see ~ clay. / ~ evaporating basin ‖ Porzellaneindampfschale f. ‖ cuvette f. d'évaporation en porcelaine. / ~ fancy goods ‖ Porzellanwaren fpl. ‖ objets mpl. en porcelaine. / ~ filter plate ‖ Siebplatte f. aus Porzellan ‖ plaque f. à filtrer en porcelaine. / ~ funnel ‖ Nutsche f.; Saugfilter m. ‖ entonnoir-filtre m. / ~ gilder ‖ Porzellangoldmaler m. ‖ doreur m. sur porcelaine.

porcelain goods pl. ‖ Porzellanwaren fpl. ‖ articles mpl. en porcelaine. / electro-technical ~ ‖ Porzellanwaren fpl. für die Elektrotechnik ‖ articles mpl. en porcelaine électrotechniques.

porcelain industry ‖ Porzellanindustrie f. ‖ industrie f. de la porcelaine. / ~ insulator ‖ Porzellanisolator m. ‖ isolateur m. en porcelaine. / ~ jasper ‖ Porzellanjaspis m.; jaspe m. porcelaine; porcelanite f. / ~ maker ‖ Porzellanarbeiter m. ‖ ouvrier m. porcelainier. / ~ manufacturing plant ‖ Porzellanfabrik f. ‖ fabrique f. de porcelaines. / ~ plate ‖ Porzellanschild n. ‖ plaque f. en porcelaine. / ~ polisher ‖ Porzellanschleifer m. ‖ useur m. sur porcelaine. / ~ thread guide ‖ Porzellanfadenführer m. ‖ guide-fil m. en porcelaine. / ~ thrower ‖ Porzellandreher m. ‖ tourneur m. de porcelaine. / ~ tooth ‖ Porzellanzahn m. ‖ dent f. en porcelaine. / ~ tube ‖ Porzellanrohr n. ‖ tube m. en porcelaine / ~ utensils pl. ‖ Porzellangerät n. ‖ ustensiles fpl. en porcelaine. / ~ varnish ‖ Porzellanlack m. ‖ vernis m. porcelaine. / ~ ware see porcelain goods.

porch ‖ Portal n. ‖ portail m.

porcupine (Textile) ‖ Nadelwalze f. ‖ peigne m. circulaire.

pore ‖ Pore f. ‖ pore m.

pork ‖ Schweinefleisch n. ‖ viande f. de porc; du porc m. / ~ butcher ‖ Schweinemetzger m. ‖ charcutier m. / ~ butchery ‖ Schweinemetzgerei f. ‖ charcuterie f. / ~ fat ‖ Schweinefett n. ‖ saindoux m.; graisse f. de porc. / ~ pie ‖ Schweinefleischpastete f. ‖ pâte m. de porc.

porose see porous.

porosity ‖ Porosität f. ‖ porosité f.

porous ‖ porös; porig ‖ poreux. / ~ (Met) luckig; blasig ‖ poreux; lacuneux; avec soufflure. / ~ iron ‖ luckiges Eisen n. ‖ fer m. caverneux ou poreux. / ~ plate ‖ Tonteller m. ‖ assiette f. poreuse en terre cuite. / ~ pot (Chem) ‖ Tonzelle f. ‖ vase m. poreux.

porousness see porosity.

porpezite ‖ Porpezit m. ‖ porpézite f.

porphyrite ‖ Porphyrit m.; quarzfreier Porphyr m. ‖ porphyrite f.

porphyry ‖ Porphyr m. ‖ porphyre m. / red ~ ‖ roter Porphyr m. ‖ porphyre m. rouge.

porphyry quarry ‖ Porphyrbruch m. ‖ carrière f. de porphyre. / ~ roller ‖ Porphyrwalze f. ‖ rouleau m. de porphyre.

porringer ‖ Schale f ; Napf m.; Suppenschüssel f. ‖ écuelle f.

port ‖ Hafen m.; Seehafen m. ‖ port m. / ~ (Build) ‖ Pforte f. ‖ porte f. / ~ (Larboard) ‖ Backbord n. ‖ bâbord m. / ~ (Mach) see steam ~. / ~ (Portwine) see portwine. / ~ (Shipb) ‖ Stückpforte f.; Ladepforte f. ‖ sabord m. / ~ of arrival ‖ Ankunftshafen m. ‖ port m. d'arrivée. / ~ of call ‖ Anlaufhafen m. ‖ port m. d'escale. / crown ~ ‖ Oberhafen m. ‖ port

m. supérieur. / ~ of departure ‖ Abgangshafen m. ‖ port m. de départ. / ~ of discharge ‖ Löschungshafen m. ‖ port m. de déchargement. / ~ of distress ‖ Nothafen m. ‖ port m. de refuge, bon port m. de relâche forcée. / exhaust ~ (Auto) ‖ Auspufföffnung f. ‖ lumière f. d'échappement. / free ~ ‖ Freihafen m. ‖ port m. franc. / inlet ~ (Auto) ‖ Einlaßkanal m. ‖ lumière f. d'admission. / intermediate ~ ‖ Zwischenhafen m. ‖ port m. intermédiaire. / land-locked ~ ‖ eingeschlossener Hafen m. ‖ port m. entouré de terres. / natural ~ ‖ natürlicher Hafen m. ‖ havre m. ou port m. naturel. / ~ of quarantine ‖ Quarantänehafen m. ‖ port m. de quarantaine. / ~ of registry ‖ Heimatshafen m. ‖ port m. d'en registrement. / ~ of sailing ‖ Abfahrtshafen m. ‖ port m. de départ. / steam ~ ‖ Dampfkanal m. ‖ lumière f. d'admission; orifice f. de vapeur.

portability ‖ Fahrbarkeit f.; Tragbarkeit f. ‖ mobilité f. / easy ~ ‖ leichte Tragbarkeit f. oder Fahrbarkeit f. ‖ facilité f. de transport.

portable ‖ fahrbar; tragbar; versetzbar ‖ portatif; transportable. / ~ boiler ‖ Lokomobilkessel m. ‖ chaudière f. locomobile. / ~ breaker ‖ fahrbarer Brecher m. ‖ concasseur m. sur chariot. / ~ forge ‖ Feldschmiede f. ‖ forge f. volante. / ~ lamp ‖ Handlampe f. ‖ lampe f. portative. / ~ line ‖ verlegbares Gleis n. ‖ voie f. portative. / ~ line section ‖ Gleisjoch n. ‖ châssis m. ou tronçon m. de voie. / ~ mast ‖ versetzbarer Mast m. ‖ mât m. portatif. / ~ medicine chest ‖ Reiseapotheke f. ‖ pharmacie f. de voyage. / ~ military station (Tel) ‖ tragbare Militärstation f. ‖ poste f. militaire transportable. / ~ motor ‖ Fahrmotor m. ‖ moteur m. de locomotion. / ~ organ ‖ tragbare Orgel f. ‖ orgue f. portative. / ~ railway ‖ Feldbahn f. ‖ chemin m. de fer portatif. / ~ riveting machine ‖ ortsbewegliche Nietmaschine f. ‖ riveuse f. portative. / ~ section of track see ~ line section. / ~ stage lamp ‖ Versatzständer m. für Bühnenbeleuchtung ‖ lampe f. mobile de coulisse pour l'éclairage de la scène. / ~ station for wireless telegraphy ‖ fahrbare Station f. für drahtlose Telegrafie ‖ poste m. radiotélégraphique transportable. / ~ steam engine ‖ Dampflokomobile f.; Lokomobile f. ‖ locomobile f. (à vapeur). / ~ telephone set for troublemen ‖ Streckenfernsprecher m. ‖ station f. téléphonique portative pour le surveillant des lignes. / ~ track for a frequent and quick displacement ‖ leicht bewegliches Gleis n. für häufiges und schnelles Verlegen ‖ voie f. facilement mobile pour poser et changer souvent et rapidement. / ~ type ‖ fahrbare Ausführung f. ‖ monté sur roues fpl.

portage ‖ Hafengeld n. ‖ portage m.

portal ‖ Portal n. ‖ portail m. / ~ crane ‖ Portalkran m. ‖ grue f. à portique. / ~ gate see portal. / ~ milling machine ‖ Portalfräswerk n. ‖ fraiseuse f. à portique. / ~ revolving crane ‖ Portaldrehkran m. ‖ grue f. pivotante à portique.

portative force (Magnet) ‖ Tragkraft f. ‖ force f. portante.

port captain ‖ Hafenkapitän m.; Hafenmeister m. ‖ capitaine m. de port.

/ ~ charges pl. ‖ Hafengeld n.; Ankergeld n. ‖ droit m. d'ancrage *ou* de port. / ~ closing (Shipb) ‖ Pfortenverschluß m. ‖ fermeture f. des mantelets de sabord. / ~ dues pl. *see* ~ charges. / ~ engine nacelle ‖ Backbordmotorgondel f. ‖ nacelle f. du moteur bâbord.

porter ‖ Pförtner m.; Portier m.; Torhüter m. ‖ concierge m.; portier m. / ~ (Brew) ‖ Porter m.; Porterbier n. ‖ porter m. / ~ (Forg) ‖ Kehrstange f.; Hebeisen n. ‖ ringard m. / ~ (Linen weav) ‖ Gang m. ‖ compte m.; portée f. / ~ (Luggage) ‖ Gepäckträger m. ‖ commissionnaire m.; facteur m.; porteur de bagage.

porterage ‖ Transport m.; Befördern n. ‖ transport m. / ~ (Post) ‖ Bestellgebühr f.; Bestellgeld n. ‖ factage m. / ~ (Railw) ‖ Rollgebühr f.; Rollgeld n. ‖ tarif m. de camionnage.

porter's lodge ‖ Pförtnerraum m.; Pförtnerloge f. ‖ loge f. du concierge *ou* du pipelet.

portfolio ‖ Aktentasche f.; Ledermappe f.; Briefmappe f. ‖ serviette f.; portefeuille f. / ~ leather ‖ Aktenmappenleder n. ‖ cuir m. à portefeuille. / ~ manufacturing ‖ Aktenmappenherstellung f. ‖ manufacture f. de portefeuilles.

port-hole light (Nav) ‖ Bullauge n.; Ochsenauge n. ‖ œil m. de bœuf.

portière ‖ Zimmerportiere f. ‖ portière f. d'appartement.

portion ‖ Teil m.; Anteil m. ‖ partie f.; portion f.; part f. / ~ (Trade) ‖ Quote f. ‖ quote-part f.; cote f. / outer ~ ‖ äußerer Teil m. ‖ portion f. extérieure.

Portland cement ‖ Portlandzement m. ‖ ciment m. de Portland. / iron ~ ‖ Eisenportlandzement m. ‖ ciment m. de laitier Portland.

Portland clinker ‖ Portlandklinker m. ‖ brique f. de ciment Portland. / ~ limestone ‖ Portlandkalk m. ‖ calcaire m. portlandien. / ~ rock *see* ~ limestone.

port lid (Shipb) ‖ Pfortendeckel m.; Pfortenluke f.; Stückpfortenklappe f. ‖ mantelet m. de sabord.

portmanteau ‖ Mantelsack m. ‖ valise f.

port pendant (Mar) ‖ Pfortenhanger m. ‖ itague f. des mantelets.

portrait ‖ Porträt n. ‖ portrait m. / ~ film ‖ Porträtfilm m. ‖ portrait-film m. / ~ photographer ‖ Porträtfotograf m. ‖ photographe-portraitiste m.

port sanitary service ‖ Hafensanitätsdienst m. ‖ service m. sanitaire de port. / ~ tackle (Mar) ‖ Pfortenaufholer m.; Pfortentalje f. ‖ palan m. *ou* palanquin m. de sabord.

portwine ‖ Portwein m. ‖ porto m.; vin m. de Porto.

position ‖ Lage f.; Stand m.; Stellung f.; Position f. ‖ position f. / ~ (Existence) ‖ Existenz f. ‖ existence f. / ~ by bearing ‖ gepeiltes Besteck n. ‖ position f. de relèvement. / closed ~ ‖ Einschaltstellung f. ‖ position f. fermée. / ~ of constraint ‖ Zwanglage f. ‖ état m. de contrainte. / ~ of crank ‖ Kurbelstellung f. ‖ position f. de la manivelle. / ~ of equilibrium ‖ Gleichgewichtslage f. ‖ position f. d'équilibre. / to find a ~ (Aero) ‖ orten ‖ s'orienter. / to find one's ~ (Aero) ‖ franzen ‖ repérer. / ~ of a firm ‖ Geschäftslage f. einer Firma ‖ situation f. des affaires d'une

maison. / to fix a ~ *see* to find a ~. / ~ of flight ‖ Fluglage f. ‖ position f. *ou* régime m. en vol. / inclined ~ ‖ Schiefstellung f. ‖ position f. inclinée. / in an inclined ~ ‖ in geneigter Lage f. ‖ dans une position inclinée. / ~ of load ‖ Lage f. der Last ‖ position f. de la charge. / open ~ ‖ Ausschaltstellung f. ‖ position f. ouverte. / to determine the ~ of a place ‖ die Lage f. eines Ortes bestimmen ‖ fixer la position d'un lieu. / relative ~s pl. ‖ gegenseitige Lage f. ‖ position f. réciproque. / ~ of rest of the plummet ‖ Ruhelage f. des Lotes ‖ position f. de repos du plomb. / ~ at right angles ‖ rechtwinklige Lage f. ‖ position f. orthogonale. / ~ in space ‖ Lage f. im Raum ‖ position f. dans l'espace. / ~ of the sun ‖ Sonnenstand m. ‖ position f. du soleil. / ~ of the wheels ‖ Radstand m. ‖ écartement m. des essieux.

position, determination of ~ by one base line ‖ Ortsbestimmung f. mit einer Standlinie ‖ détermination f. de la position au moyen d'une ligne de base. / determination of ~ by measurement of star's altitude ‖ Ortsbestimmung f. durch Gestirnhöhenmessung ‖ détermination f. astronomique du point.

position circle divided into degrees ‖ Positionskreis m. mit Gradteilung ‖ cercle m. de position divisé en degrés. / ~ divided into minutes of arc ‖ Positionskreis m. mit Bogenminutenteilung ‖ cercle m. de position divisé en minutes d'arc.

position filar micrometer ‖ Positionsfadenmikrometer n. ‖ micromètre m. de position.

position finder ‖ Ortungsgerät n. ‖ chercheur m. de position. / mirror ~ ‖ Spiegelortungsgerät n. ‖ chercheur m. de position à miroir.

position finding ‖ Ortung f. ‖ orientation f. / ~ joint ‖ zwangsläufige Kupplung f. ‖ accouplement m. à guidage forcée. / ~ lantern ‖ Positionslaterne f. ‖ fanal m. de route; fanal m. de position. / ~ meter (Tel) ‖ Platzzähler m. ‖ compteur m. de contrôle. / ~ warfare ‖ Stellungskrieg m. ‖ guerre f. de position. / ~ wiring ‖ Platzbelegung f. im Fernsprechbetrieb ‖ garnir la position avec des lignes.

positive ‖ positiv ‖ positif. / ~ plate (Acc) ‖ Plusplatte f. ‖ plaque f. positive. / ~ printing ‖ Positivdruck m. ‖ impression f. positive.

positive ‖ Positiv n. ‖ image f. positive; épreuve f. positive.

possess, to ‖ besitzen ‖ posséder.

possession ‖ Besitz m. ‖ possession f. / to stay entrance into ~ of property by taking legal action ‖ die Besitznahme des Grundstückes durch Beschreitung des Rechtsweges aufhalten ‖ retarder la prise de possession du terrain par recours aux voies légales. / now in ~ ‖ gegenwärtig im Besitz m. ‖ maintenant en possession f. / taking ~ ‖ Besitznahme f. ‖ prise f. de possession.

possessor ‖ Besitzer m. ‖ possesseur m. / ~ of the usufruct of a property ‖ Nutzungsberechtigter m. eines Grundstückes ‖ usufruitier m. d'un terrain.

posset ‖ Molke f. ‖ petit lait m.

possibility of overloading ‖ Überlastungsmöglichkeit f. ‖ possibilité f. de surcharger. / ~ of use ‖ Verwendungsfähigkeit f. ‖ possibilité f. d'utilisation.

possible ‖ tunlich; möglich ‖ faisable; practicable; possible. / as far as ~ ‖ nach Möglichkeit f. ‖ dans la mesure du possible.

post ‖ Pfahl m.; Pfosten m.; Ständer m.; Balken m. ‖ poteau m.; pieu m. / ~ (Electr line) ‖ Stange f.; Telegrafenstange f. ‖ poteau m. télégraphique. / ~ (Employment) ‖ Stellung f. ‖ emploi m.; place f. / ~ (Mail) ‖ Post f. ‖ poste f. / ~ (Pap) ‖ Pauscht m.; Bauscht m. ‖ porse f. / ~ (Vineyard) ‖ Weinbergspfahl m.; Winzerspfahl m. ‖ échalas m. / to be at one's ~ ‖ auf seinem Posten m. sein ‖ être à son poste. / ~ of a crane ‖ Säule f. eines Kranes ‖ arbre m. d'une grue; poinçon m. d'une grue; pivot m. central. / direction ~ ‖ Wegweiser m. ‖ poteauguide m. / ~ of a hammer stand ‖ Hammergerüstsäule f. ‖ montant m. du bâti du marteau. / ~ of the mill ‖ Stuhl m. *oder* Bock m. der Mühle ‖ chaise f. *ou* pylone m. du moulin. / by parcel ~ ‖ als Postpaket n. ‖ par colis m. postal. / pneumatic ~ plant ‖ Rohrpostanlage f. ‖ installation f. de poste pneumatique. / by return ~ ‖ postwendend ‖ par retour du courrier. / ~ royal (Mill) ‖ Königssäule f.; Königszapfen m. ‖ pivot m. central. / rudder ~ (Shipb) ‖ Hintersteven m. ‖ étambot m. / ~ as a traveller ‖ Stellung f. als Reisender ‖ place f. de voyageur de commerce. / tubular ~ ‖ Rohrpost f. ‖ poste f. pneumatique *ou* tubulaire. / white ~ (Pap) ‖ weißer Pauscht m. ‖ porse f. blanche.

postage ‖ Briefporto n.; Porto n. ‖ port m.; affranchissement m. / additional ~ ‖ Strafporto n. ‖ surtaxe f. postale. / ~ free ‖ portofrei ‖ franc de port. / including ~ ‖ einschließlich Porto n. ‖ port m. compris. / payment of ~ ‖ Frankierung f. ‖ affranchissement m.

postage stamp ‖ Briefmarke f. ‖ timbreposte m. / box for ~s ‖ Briefmarkenschachtel f. ‖ boîte f. pour timbres-poste. / ~ delivery apparatus ‖ Briefmarkenautomat m. ‖ distributeur m. de timbresposte.

postal charges pl. ‖ Portokosten pl. ‖ frais mpl. de port. / ~ cheque ‖ Postscheck m. ‖ chèque m. postal. / ~ permit printing attachment ‖ Freimachungsmaschine f. ‖ dispositif m. pour l'affranchissement du courrier. / ~ scale ‖ Briefwage f. ‖ pèse-lettre m.

Postal Union ‖ Weltpostverein m. ‖ union f. postale universelle.

post card ‖ Postkarte f. ‖ carte f. postale. / illustrated ~ *see* picture ~. / picture ~ ‖ Ansichtspostkarte f. ‖ carte f. postale illustrée. / Postal Union ~ ‖ Weltpostkarte f. ‖ carte f. de l'union postale universelle. / ~ with voices ‖ Postkarte f. mit Stimme ‖ carte f. postale à voix. / view ~ *see* picture ~.

post card, fine cardboard for ~s ‖ Postkartenkarton m. ‖ carton m. pour cartes postales. / ~ colourer ‖ Ansichtskartenmaler m. ‖ coloriste m. en cartes postales illustrées. / ~ holder ‖ Postkartenhalter m. ‖ porte-cartes m. / ~ size ‖ Postkartenformat n. ‖ format m. de carte postale. / ~ stand ‖ Postkartenständer m. ‖ étalage m. pour cartes postales.

post car(riage) *see* post office carriage. / ~ crane ‖ Säulenkran m. ‖ grue f. pivotante; grue f. à pivot.

post-dated cheque ‖ vorausdatierter Scheck m. ‖ chèque m. post-daté.

poster ‖ Plakat n.; Anschlagzettel m. ‖ placard m.; affiche f. / metal sheet ~ ‖ Blechplakat n. ‖ affiche f. en tôle. / border for ~s ‖ Plakateinfassung f. ‖ encadrement m. d'affiches.

poste restante ‖ postlagernd ‖ poste restante.

poster type ‖ Plakatschrift f. ‖ caractères mpl. pour affiches.

post-free *see* postage free.

posting ‖ Aufgabe f. ‖ remise f.

post light (Mar) ‖ Backbordlicht n. ‖ feu m. de bâbord.

postman ‖ Briefbote m.; Postbote m. ‖ facteur distributeur m.

post mark ‖ Briefstempel m.; Poststempel m. ‖ timbre m. de la poste. / ~ master ‖ Postdirektor m. ‖ directeur m. des postes. / ~ mill *see* ~ wind mill. / ~ motor car ‖ Postauto n. ‖ autobus m. *ou* automobile f. du service de la poste. / ~ octavo (Print) ‖ englisches Romanformat n. ‖ format m. de roman anglais.

post office ‖ Postamt n. ‖ bureau m. de poste; hôtel des postes. / ~ authorities pl. ‖ Postdirektion f. ‖ direction f. des postes. / ~ box ‖‖ Postschließfach n. ‖ boîte f. postale. / ~ carriage (Railw) ‖ Postwagen m. ‖ wagon-poste m.; voiture f. postale. / ~ cheque account ‖ Postscheckkonto n. ‖ compte m. de chèque postale. / ~ clerk ‖ Postbeamter m. ‖ employé m. des postes. / ~ order ‖ Postanweisung f. ‖ mandat m. de poste. / ~ van service ‖ Postkraftwagenverkehr m. ‖ service m. des automobiles postaux.

post paid ‖ portofrei; frankiert ‖ port payé.

postpone, to ‖ verschleppen; aufschieben ‖ retarder.

postponement ‖ Aufschub m. ‖ remise f.

postscript ‖ Nachschrift f. ‖ post-scriptum m.

post stone (Geol) ‖ Kohlensandstein m. ‖ grès m. houiller; querelle f.

posture (Position) *see* position. / ~ (Fine arts) ‖ Akt m. ‖ pose f.

post wagon *see* post office carriage. / ~ windmill ‖ deutsche Mühle f.; Bockmühle f.; Ständermühle f. ‖ moulin m. allemand *ou* à pile *ou* à pylone.

pot ‖ Topf m. ‖ pot m. / ~ (Glassm) ‖ Glashafen m.; Hafen m.; Schmelzhafen m. ‖ creuset m.; padelin m.; pot m. / ~ for adding carbonate of sodium ‖ Sodazusatztopf m. ‖ pot m. pour ajouter le carbonate de sodium. / crystallizing ~ ‖ Kristallisationsgefäß n.; Kristallisationskessel m. ‖ cristallisoir m. / earthen ~ ‖ irdener Topf m. ‖ pot m. en terre cuite. / melting ~ (Sulphur) ‖ Läuterungskessel m. ‖ chaudière f. de raffinage. / spherical ~ ‖ Rundkessel m. ‖ chaudière f. ronde. / pots pl. and pans pl. ‖ Küchengeschirr n. ‖ batterie f. de cuisine.

potash (Potassium carbonate) ‖ Pottasche f.; Kaliumkarbonat n. ‖ carbonate m. de potasse; potasse f. / ~ (Potassium oxide) ‖ Kali n.; Kaliumoxyd ‖ potasse f. / anhydrous ~ ‖ Kali n.; Kaliumoxyd n. ‖ potasse f. / artificial ~ ‖ künstliche Pottasche f. ‖ potasse f. artificielle. / ~ from beet root molasses ‖ Rübenschlempepottasche f. ‖ potasse f. de vinasses de betteraves. / caustic ~ *see* potassium

hydroxide. / crude ~ ‖ rohe *oder* schwarze Pottasche f.; Pottaschenfluß m.; / salin m. / stick ~ ‖ Stangenkali n. ‖ potasse f. en crayons. / ~ from suint ‖ Wollschweißpottasche f. ‖ potasse f. de suint.

potash ... *see also* potassium ... / ~ alum ‖ Kalialaun n. ‖ alun m. de potasse. / ~ bulbs pl. ‖ Kaliapparat m. ‖ appareil m. *ou* tube m. à potasse. / ~ glass ‖ Kaliglas n. ‖ verre m. à base de potasse. / ~ lye *see* potassium hydroxide. / ~ maker ‖ Pottaschesieder m. ‖ ouvrier m. de potasse. / ~ manure ‖ Kalidüngesalz n. ‖ engrais m. potassique. / ~ mica ‖ Muscovit m.; Kaliglimmer m. ‖ muscovite f. / ~ mine ‖ Kalibergwerk n. ‖ mine f. de sel de potasse. / ~ production ‖ Kalierzeugung f. ‖ production f. de l'industrie potassière. / red prussiate of ~ *see* potassium ferricyanide. / ~ salt for manuring ‖ Kalidüngesalz n. ‖ sel m. potassique d'engrais. / ~ soap ‖ Kaliseife f. ‖ savon m. à base de potasse. / ~ vat (Dyer) ‖ Pottaschküpe f. ‖ cuve f. à la potasse. / ~ water ‖ Sodawasser n. ‖ eau f. de soude *ou* de Seltz; soda m. / ~ works pl. ‖ Kaliwerke npl. ‖ fabrique f. de sel de potasse. / yellow prussiate of ~ *see* potassium ferrocyanide.

potass *see* potash.

potassa *see* potash.

potassic ‖ kaliumhaltig ‖ potassique. / raw natural ~ salt ‖ rohes natürliches Kalisalz n. ‖ sel m. de potasse brut naturel.

potassium ‖ Kalium n. ‖ potassium m. / ~ ... *see also* potash ... / ~ acetate ‖ essigsaures Kalium n.; Kaliumacetat n. ‖ acétate m. de potassium. / ~ alum *see* potash alum. / ammonia persulphide ‖ Kaliumammoniumpersulfat n. ‖ persulfate m. d'ammonium et de potassium. / ~ arseniate ‖ arsensaures Kalium n. ‖ arséniate m. de potasse. / ~ bicarbonate ‖ doppeltkohlensaures Kalium n.; Kaliumbikarbonat n. ‖ bicarbonate m. de potasse. / ~ bichromate ‖ Kaliumbichromat n. ‖ bichromate m. de potasse. / ~ bioxalate ‖ Kleesalz n. ‖ bioxalate m. de potasse; sel m. d'oseille. / ~ bitartrate ‖ Kaliumbitartrat n.; weinsteinsaures Kalium n. ‖ tartrate m. acide de potassium; bitartrate m. de potasse. / ~ bromate ‖ bromsaures Kalium n. ‖ bromate m. de potasse. / ~ bromide ‖ Bromkalium n.; Kaliumbromid n. ‖ bromure m. potassique. / ~ carbonate *see* potash. / ~ chlorate ‖ chlorsaures Kali n.; Kaliumchlorat n. ‖ chlorate m. de potasse. / ~ chloride ‖ Kaliumchlorid n.; Chlorkalium n. ‖ chlorure m. de potasse. / ~ chloride plant ‖ Chlorkaliumanlage f. ‖ installation f. de chlorure de potasse. / ~ chromate ‖ chromsaures Kalium n. ‖ chromate m. de potasse. / ~ citrate ‖ zitronensaures Kalium n. ‖ citrate m. de potasse. / ~ cyanide ‖ Cyankalium n.; blausaures Kalium n. ‖ prussiate m. de potasse; cyanure m. de potassium. / ~ ferricyanide ‖ Ferricyankalium n.; rotes Blutlaugensalz n. ‖ ferricyanure m. de potasse. / ~ ferrocyanide ‖ Ferrocyankalium n.; gelbes Blutlaugensalz n. ‖ ferrocyanure m. de potassium. / ~ fluoride ‖ Fluorkalium n. ‖ fluorure m. de potassium. / ~ hydroxide ‖ Ätzkali n.; Kaliumhydrat n.; Kalilauge f. ‖

hydrate m. *ou* caustique de potasse: potasse f. à la chaux. / ~ hypophosphite ‖ unterphosphorigsaures Kalium n. ‖ hypophosphite m. de potasse. / ~ hyposulphite ‖ unterschwefligsaures Kalium n. ‖ hyposulfite m. de potasse. / ~ iodide ‖ Jodkalium n. ‖ iodure m. de potassium. / ~ magnesium sulphate ‖ schwefelsaure Kalimagnesia f.; Kaliummagnesiumsulfat n. ‖ sulfate m. de potasse et de magnésie. / ~ manganate ‖ mangansaures Kalium n.; Kaliummanganat n. ‖ manganate m. potassique *ou* de potasse. / ~ monochromate ‖ neutrales *oder* gelbes chromsaures Kalium n. ‖ chromate m. de potasse; chromate neutre de potasse. / ~ nitrate ‖ Kalisalpeter m. ‖ nitrate m. de potasse. / ~ oxalate ‖ oxalsaures Kalium n. ‖ oxalate m. potassique *ou* de potasse. / ~ perchlorate ‖ überchlorsaures Kalium n.; Kaliumperchlorat n. ‖ perchlorate m. de potasse. / ~ permanganate ‖ übermangansaures Kalium n.; Kaliumpermanganat n. ‖ permanganate m. de potasse. / ~ persulphate ‖ überschwefelsaures Kalium n. ‖ persulfate m. de potasse. / ~ phosphate ‖ phosphorsaures Kalium n.; Kaliumphosphat n. ‖ phosphate m. de potasse. / protoxide of ~ ‖ Kaliumoxyd n. ‖ protoxyde m. de potassium. / ~ silicate ‖ Kaliwasserglas n.; Kaliumsilikat n. ‖ silicate m. de potasse. / ~ sulphate ‖ Kaliumsulfat n. ‖ sulfate m. de potasse. / ~ sulphite ‖ schwefligsaures Kalium n. ‖ sulfite m. de potasse. / ~ tartrate ‖ weinsaures Kalium n. ‖ tartrate m. de potasse. / ~ xanthogenate ‖ xanthogensaures Kalium n. ‖ xanthogénate m. de potasse.

potato ‖ Kartoffel f. ‖ pomme f. de terre. / ~ crusher ‖ Kartoffelquetsche f. ‖ écraseur m. de pommes de terre. / cultivator of ~es *see* ~ grower. / ~ cutting machine ‖ Kartoffelschneider m. ‖ machine f. à hacher les pommes de terre. / ~ digging machine ‖ Kartoffelerntemaschine f. ‖ arracheur m. de pommes de terre. / ~ flour ‖ Kartoffelmehl n. ‖ fécule f. de pommes de terre. / ~ gathering machine ‖ Kartoffelerntemaschine f. ‖ machine f. à récolter les pommes de terre. / ~ grower ‖ Kartoffelbauer m. ‖ planteur m. de pommes de terre. / ~ peeling machine ‖ Kartoffelschälmaschine f. ‖ machine f. à éplucher les pommes de terre. / ~ planting implement ‖ Kartoffelkulturgerät n. ‖ ustensils mpl. pour la culture des pommes de terre. / ~ planting machine ‖ Kartoffellegemaschine f. ‖ planteur m. mécanique de pommes de terre. / ~ sorting drum ‖ Kartoffelauslesemaschine f. ‖ appareil m. *ou* tambour m. à assortir les pommes de terre. / ~ steamer ‖ Kartoffeldämpfer m. ‖ étuve f. pour pommes de terre. / ~ starch ‖ Kartoffelstärke f. ‖ amidon m. de fécule. / ~ starch manufactury ‖ Kartoffelmehlfabrik f. ‖ féculerie f. / ~ washer ‖ Kartoffelwäsche f. ‖ laveur m. de pommes de terre. / ~ working machine ‖ Kartoffelverarbeitungsmaschine f. ‖ machine f. pour la préparation industrielle des pommes de terre.

pot cleaner ‖ Topfreiniger m. ‖ cure-casserole m. / metal wire ~ ‖ Metallfadentopfreiniger m. ‖ nettoie-casserole m. en fil métallique.

potential ∥ potentiell ∥ potentiel. / ~ energy ∥ potentielle Energie f. ∥ énergie f. potentielle. / ~ function ∥ Potentialfunktion f. ∥ fonction f. potentielle.

potential ∥ Potential n. ∥ potentiel m. / disruptive ~ ∥ Funkenpotential n. ∥ potentiel m. disruptif. / earth ~ ∥ Erdpotential n. ∥ potentiel m. terrestre. / electric ~ ∥ elektrisches Potential n. ∥ potentiel m. électrique. / magnetic ~ ∥ magnetisches Potential n. ∥ potentiel m. magnétique. / velocity ~ ∥ Geschwindigkeitspotential n. ∥ potentiel m. de vitesse.

potential, antinode of ~ ∥ Spannungsgegenknoten m. ∥ antinœud m. de tension. / ~ difference ∥ Spannungsdifferenz f.; Potentialdifferenz f. ∥ différence f. de potentiel. / difference of ~ at the terminals ∥ Potentialdifferenz f. an den Klemmen ∥ différence f. de potentiel aux bornes. / ~ drop ∥ Spannungsabfall m. ∥ chute f. de tension ou de potentiel. / ~ energy ∥ potentielle Energie f. ∥ énergie f. potentielle. / fall of ~ see ~ drop. / ~ function ∥ Potentialfunktion f. ∥ fonction f. potentielle. / ~ governor see ~ regulator. / ~ gradient ∥ Potentialgefälle n. ∥ gradient m. du potentiel. / ~ loop ∥ Spannungsbauch m. ∥ ventre m. de tension. / ~ node ∥ Spannungsknoten m. ∥ nœud m. de tension. / ~ regulator ∥ Spannungsregler m.; Spannungsteilertransformator m. ∥ régulateur m. de tension; autotransformateur-diviseur m. / ~ transformer ∥ Spannungswandler m. ∥ transformateur m. de tension.

potentiometer ∥ Potentiometer n. ∥ potentiomètre m. / ~ of telephone repeaters ∥ Schwächungswiderstand m. für Telefonverstärker ∥ potentiomètre m. pour amplificateurs téléphoniques.

pot filler (Metal) ∥ Tiegelfüller m. ∥ remplisseur m. de creusets. / ~ furnace (Glassm) ∥ Hafenofen m. ∥ four m. à pots ou à creusets.

potherb ∥ Küchengewächs n. ∥ plante f. potagère.

potmaker ∥ Tiegelmacher m. ∥ creusetier m.; fabricant m. de creusets.

potsherd ∥ Scherbe f.; Topfscherbe f. ∥ tesson m. de poterie.

potter ∥ Töpfer m. ∥ potier m.

potter's beetle ∥ Tonschlägel m. ∥ batte f. / ~ clay extraction ∥ Töpfertongewinnung f. ∥ extraction f. de terre à poterie. / ~ earth ∥ Ton m. ∥ argile f. / ~ lathe ∥ Töpferscheibe f. ∥ roue f. de potier. / ~ ore ∥ Glasurerz n.; Bleiglanz m. ∥ alquifoux m. / ~ ware see pottery. / ~ wheel see ~ lathe.

pottery ∥ Töpferware f.; Tonware f. ∥ poterie f. / artistic ~ ∥ Kunststeingut n. ∥ poterie f. artistique. / coarse ~ ∥ irdenes Geschirr n.; gemeine Tonware f. ∥ poterie f. commune. / glazed ~ ∥ glasierte Töpferware f. ∥ plommure f. / hygienic ~ ∥ Gesundheitsgeschirr n. ∥ poterie f. de santé; hydiocérame m. / potteries pl. for technical purposes ∥ technische Tonwaren fpl. ∥ articles mpl. de poterie à l'usage technique. / scraps pl.

pottery decorator ∥ Tonwarenmaler m. ∥ peintre m. sur poteries. / raw materials pl. of ~ ∥ keramische Rohstoffe mpl. ∥ matières fpl. premières de l'industrie céramique et de la poterie. / scraps pl.

of ~ ∥ Bruchstücke npl. von Tonwaren ∥ débris m. de poterie.

pouch ∥ Beutel m.; Täschchen n. ∥ sachet m.; bourse f.; petit sac m.; pochette f. / ~ for shipping ∥ Versandtasche f. ∥ sac m. à échantillon. / tobacco ~ ∥ Tabakdose f. ∥ blague f. à tabac. / ~ heel apparatus ∥ Keilfersenvorrichtung f. ∥ dispositif m. à faire les talons américains. / ~ varnish ∥ Lederlack m. ∥ vernis m. à cuir.

poultry ∥ Geflügel n. ∥ volaille f. / ~ breeding ∥ Geflügelzucht f. ∥ aviculture f. / ~ cooling room ∥ Geflügelkühlraum m. ∥ chambre f. froide à volaille. / ~ coop ∥ Geflügelhaus n.; Hühnerstall m. ∥ poulailler m. / ~ farmer ∥ Geflügelzüchter m. ∥ aviculteur m.; basse-courrier m. / ~ farm supplies pl. ∥ Geflügelfarmbedarf m. ∥ articles mpl. pour l'élevage de volaille. / ~ fattener ∥ Geflügelmäster m. ∥ engraisseur m. de volaille. / ~ fattening ∥ Geflügelmästerei f. ∥ engraissement m. ou engraissage de volailles. / ~ keeper see ~ farmer. / ~ keeping ∥ Geflügelzucht f. ∥ élevage m. de volaille; aviculture f. / ~ plucker ∥ Geflügelrupferin f. ∥ plumeuse f. de volailles. / ~ plucking ∥ Geflügelrupfen n. ∥ plumage m. de volailles. / ~ shears pl. ∥ Geflügelschere f. ∥ ciseaux mpl. à découper la volaille. / ~ shop ∥ Geflügelhandlung f. ∥ magasin m. de volailles. / ~ yard ∥ Hühnerhof m.; Geflügelhof m. ∥ basse-cour f. / ~ yard equipment ∥ Geflügelzuchteinrichtung f. ∥ installation f. de basse-cour.

pounce, to (Drawing) ∥ durchpausen ∥ calquer; poncer.

pound, to ~ an ore ∥ Erz n. pochen oder stampfen ∥ bocarder un minerai.

pound ∥ Pfund n. ∥ livre f. / foot ~ ∥ Fußpfund n. ∥ livre-pied m.

pounded ∥ gestoßen ∥ pilé.

pounder (Mortar) ∥ Stößel m. ∥ pilon m.

pounding ∥ Zerstampfen n. ∥ broiement m.; pilage m.; bocardage m. / ~ machine ∥ Pochhammer m. ∥ bocard m. / ~ trough ∥ Pochtrog m. ∥ huche f. de bocard.

pour, to ∥ gießen ∥ verser; couler; jeter. / ~ from the bottom ∥ steigend gießen ∥ couler en source. / ~ on the end ∥ stehend gießen ∥ couler debout. / ~ in ∥ eingießen ∥ verser dans. / ~ off ∥ abgießen ∥ couler; verser. / ~ out ∥ ausgießen ∥ déverser; couler. / ~ from the top ∥ fallend gießen ∥ couler en plan incliné. / ~ water again ∥ Wasser n. nachgießen ∥ remettre de l'eau f.

pourer (Found) ∥ Gießer m. ∥ verseur m.; fondeur m. / ladle ~ see steel ~. / steel ~ ∥ Stahlgießer m. ∥ verseur m. d'acier ou de poches.

pouret (Chem) ∥ Bürette f. ∥ burette f.

pouring ∥ Guß m. ∥ coulée f.; coulage m.; jet m. / ~ bed ∥ Gießbett n. ∥ lit m. de coulée. / ~ conveyor ∥ Gießband n. ∥ convoyeur m. de coulée. / ~ crane ∥ Gießkran m. ∥ grue f. de coulée. / ~ machine ∥ Gießmaschine f. ∥ machine f. à couler ou de fonderie. / ~ method ∥ Gießtechnik f. ∥ méthode f. de coulage. / ~-out several crucibles simultaneously ∥ gleichzeitiges Ausgießen n. mehrerer Tiegel ∥ coulée f. simultanée de plusieurs creusets.

powder, to ∥ pulverisieren ∥ pulvériser.

powder ∥ Pulver n. ∥ poudre f. / ~ (Cosmetic) ∥ Puder m. ∥ poudre f. / angular ~ ∥ eckiges Pulver n. ∥ poudre f. angulaire ou anguleuse. / aromatic ~ ∥ wohlriechender Puder m. ∥ poudre f. de senteur. / ~ of Cassius ∥ Goldpurpur m.; Cassiuspurpur m. ∥ pourpre m. de Cassius. / cementation ~ see cementing ~. / cementing ~ ∥ Einsatzpulver n.; Härtepulver n. ∥ poudre f. à cémenter. / coarse-grained ~ ∥ grobkörniges Pulver n. ∥ poudre f. à gros grains. / emery ~ ∥ Schmirgelpulver n.; Schmirgel m. ∥ poudre f. d'émeri. / face ~ ∥ Puder m. ∥ poudre f. de riz. / to fall in ~ (Chem) ∥ zerfallen ∥ se désagréger. / fine-grained ~ ∥ feinkörniges Pulver n. ∥ poudre f. fine. / fulminating ~ ∥ Knallpulver n. ∥ poudre f. fulminante. / glass ~ ∥ pulverisiertes Glas n. ∥ verre m. pilé; poudre f. de verre. / glazed ~ ∥ geglättetes oder poliertes Pulver n. ∥ poudre f. polie. / granulated ~ ∥ Kornpulver n. ∥ poudre f. grenée. / granulated and sieved ~ ∥ gekörntes und gesiebtes Pulver n. ∥ égalisures fpl. / gun ~ ∥ Schießpulver n.; Schwarzpulver n. ∥ poudre f. à canon. / gun cotton ~ ∥ Schießbaumwolle f. ∥ coton-poudre m. / in ~ ∥ in Pulverform f. ∥ en poudre f. / large-grained ~ ∥ grobkörniges Pulver n. ∥ poudre f. à gros grains. / offensive ~ ∥ brisantes Pulver n. ∥ poudre f. brisante ou offensive. / prismatic ~ ∥ Würfelpulver n. ∥ poudre f. prismatique ou en dés. / slow-burning ~ ∥ langsam verbrennendes Pulver n. ∥ poudre f. à combustion lente. / small-grained ~ see fine-grained ~. / smokeless ~ ∥ rauchloses Pulver n. ∥ poudre f. sans fumée. / ~ of zinc ∥ Zinkstaub m. ∥ poussière f. de zinc.

powder barrel ∥ Pulverfaß n.; Pulvertonne f. ∥ baril m. à poudre. / ~ box ∥ Puderdose f. ∥ boîte f. à poudre. / ~ composition ∥ Pulvermasse f. ∥ pâte f.

powdered, coarsely ~ ∥ grobgepulvert ∥ grossièrement pulvérisé. / finely ~ ∥ feingepulvert ∥ finement pulvérisé.

powdered coal ∥ Kohlenstaub m. ∥ poussière f. de charbon. / ~ liquorice ∥ Süßholz pulver n. ∥ coco m. en poudre. / ~ milk ∥ Trockenmilch f. ∥ lait m. en poudre. / ~ turf making ∥ Torfmullherstellung f. ∥ fabrication f. de poussier de tourbe.

powder dust ∥ Pulverstaub m. ∥ poussier m. de poudre. / ~ factory ∥ Pulverfabrik f. ∥ poudrerie f. / ~-impulse catapult ∥ pulvergetriebener Katapult m. ∥ catapulte f. de lancement à poudre. / ingredients for ~ ∥ Pulversatz m. ∥ composition f. de la poudre à canon. / ~ ink ∥ Puderfarbe f. ∥ couleur f. à poudrer. / ~ machine ∥ Pudermaschine f. ∥ machine f. à poudrer. / ~ magazine attendant ∥ Pulververwalter m. ∥ garde m. magasin des explosifs. / ~ mill ∥ Pulvermühle f. ∥ moulin m. à poudre; poudrerie f. / ~ paper ∥ Puderpapier n. ∥ papier m. poudré. / ~ puff ∥ Puderquaste f. ∥ houppe f. à poudrer. / ~ storage vessel for the industry of explosives ∥ Pulverfaß n. für die Sprengstoffindustrie ∥ barrique f. à poudre pour l'industrie des explosifs. / ~ sugar ∥ gestoßener Zucker m. ∥ sucre m. en poudre. / weighing machine ∥ Pulverdosiermaschine f. ∥ machine f. à doser la poudre.

powdery ∥ pulverartig; pulverisiert ∥ pulvérulant. / ~ snow ∥ Pulverschnee m. ∥ neige f. en poudre.

power ∥ Kraft f. ∥ force f.; puissance f.; pouvoir m.; énergie f. / ~ (Math) ∥ Po-

tenz f. ‖ puissance f. / ~ (Mine) ‖ Mächtigkeit f. ‖ puissance f. / adhesive ~ ‖ Adhäsionskraft f. ‖ adhérence f.; adhésion f. / animal ~ ‖ tierische Kraft f. ‖ force f. animale. / apparent ~ ‖ scheinbare Leistung f. ‖ débit m. apparent; puissance f. apparente.

power of attorney ‖ Vollmacht f. ‖ plein-pouvoir m.; mandat m. général. / to exercise ~ ‖ eine Vollmacht f. ausüben ‖ exercer un mandat général. / unlimited ~ ‖ Blankovollmacht f. ‖ blanc-seing m.

power of a crane ‖ Tragvermögen n. eines Hebekrans ‖ puissance f. d'une grue. / cutting ~ ‖ Schnittkraft f. ‖ force f. de coupe. / driving ~ ‖ Antriebskraft f. ‖ force f. motrice. / effective ~ ‖ Nutzkraft f. ‖ force f. effective. / fourth ~ of a number ‖ vierte Potenz einer Zahl ‖ quatième puissance d'un nombre. / full ~ (Of attorney) ‖ Vollmacht f. ‖ plein-pouvoir m. / to give full ~s to ... ‖ bevollmächtigen ‖ autoriser; donner plein-pouvoir m. / with full ~ ‖ mit voller Kraft f. ‖ à toute volée f.; à toute puissance f. / installed ~ ‖ installierte Leistung f. ‖ puissance f. installée. / lifting ~ (Aeroplane) ‖ Steigvermögen n.; Tragvermögen n. ‖ force f. ascensionnelle ou sustentatrice. / lifting ~ (Crane) ‖ Tragkraft f. ‖ puissance f. de levée ou d'élévation. / lifting ~ (Pump) ‖ Liefermenge f. ‖ débit m. de refoulement. / moving ~ ‖ treibende Kraft f. ‖ force f. motrice. / ~ of a number ‖ Potenz f. einer Zahl ‖ puissance f. d'un nombre. / ~ of output ‖ Leistungsvermögen n. ‖ capacité f. de production. / pressing ~ ‖ Druckkraft f. der Presse ‖ force f. de pression. / pulling ~ ‖ Durchzugskraft f. ‖ puissance f. d'entraînement. / purchasing ~ ‖ Kaufkraft f. ‖ pouvoir m. d'achat. / radiated ~ ‖ Strahlungsenergie f. ‖ énergie f. de rayonnement. / to raise to a ~ (Math) ‖ auf eine Potenz f. erheben ‖ élever à une puissance. / real ~ ‖ wirkliche Leistung f. ‖ puissance f. effective ou réelle ou active. / refractive ~ ‖ Brechungsvermögen n. ‖ pouvoir m. réfringent. / ~ of repulsion ‖ Rückstoßkraft f. ‖ force f. de répulsion. / required ~ ‖ Kraftbedarf m. ‖ puissance f. ou force f. nécessaire. / ~ required under cut ‖ Kraftverbrauch m. unter Schnitt ‖ force f. absorbée en travail. / ~ required for group drive ‖ Kraftbedarf m. bei gruppenweisem Antrieb ‖ force f. motrice avec commande par groupes. / ~ required by the machine ‖ Kraftbedarf m. der Maschine ‖ force f. nécessaire de la machine. / ~ required by a pump ‖ Leistungsbedarf m. der Pumpe ‖ force f. demandée par une pompe. / second ~ of a number ‖ zweite Potenz f. oder Quadrat n. einer Zahl ‖ deuxième puissance ou carré m. d'un nombre. / ~ of seeing ‖ Sehschärfe f.; Sehvermögen n. ‖ acuité f. visuelle. / stray ~ ‖ Leerlaufarbeit f. ‖ dépense f. à vide. / sustaining ~ ‖ Widerstandskraft f. ‖ force f. résistante; résistance f. / third ~ of a number ‖ dritte Potenz f. oder Kubus m. einer Zahl ‖ cube m. ou troisième puissance d'un nombre. / useful ~ ‖ Nutzleistung f. ‖ débit m. utile. / visual ~ ‖ Sehkraft f. ‖ vue f.

power basin with banked-up water level for a power plant ‖ Betriebsbecken n. mit

Stau für eine Kraftanlage ‖ réservoir m. de service avec haussement de niveau pour une usine hydraulique. / ~ bus-bar ‖ Kraftschiene f. ‖ barre f. (omnibus) de force motrice. / ~ cable ‖ Kraftkabel n. ‖ câble m. de transport de force. / ~ car battery ‖ Triebwagenbatterie f. ‖ batterie f. pour chariots à force-motrice. / ~ circuit ‖ Kraftstromkreis m. ‖ circuit m. de force motrice. / ~ circuit interference with communication line ‖ Beeinflussung f. der Fernmeldeleitung durch Starkstromanlagen ‖ influence f. perturbatrice engendrée dans la ligne de communication par les installations d'énergie électrique. / ~ consumption ‖ Kraftverbrauch m. ‖ consommation f. d'énergie ou de force. / low ~ consumption ‖ geringer Kraftbedarf m. ‖ force f. absorbée minime. / with a minimum of ~ consumption ‖ bei geringstem Kraftbedarf m. ‖ consommant un minimum de puissance.

power current cable ‖ Starkstromkabel n. ‖ câble m. pour courant fort. / ~ entering communication lines ‖ Stromübergang m. von Starkstromanlagen auf Fernmeldeleitungen ‖ passage m. de courant fort aux lignes de communication. / ~ fuse ‖ Starkstromsicherung f. ‖ coupe-circuit m. pour courants forts.

power, device for controlling the distribution of ~ ‖ Einrichtung f. zur Überwachung der Energieverteilung ‖ installation f. pour le contrôle de la répartition de l'énergie. / ~ drive ‖ Maschine f. für Kraftbetrieb ‖ machine f. à commande par moteur. / hand or ~ driven ‖ für Hand- oder Kraftbetrieb m. ‖ à commande f. mécanique ou à main. / ~ drop see power loss. / ~ engine ‖ Kraftmaschine f. ‖ machine f. motrice.

power factor ‖ Leistungsfaktor m. ‖ facteur m. de puissance. / ~ indicator ‖ Leistungsfaktoranzeiger m. ‖ indicateur m. du facteur de puissance. / ~ meter ‖ Leistungsfaktormesser m. ‖ instrument m. à mesurer le facteur de puissance.

powerful ‖ stark; kräftig ‖ puissant; fort.

power feed ‖ Kraftfutter n. ‖ fourrage m. vigoureux. / ~ gas ‖ Kraftgas n. ‖ gaz m. industriel ou pour moteurs / the gas is used for ~-generating purposes ‖ das Gas wird zur Krafterzeugung ausgenutzt ‖ le gaz est utilisé pour la production d'énergie. / apparatus for controlling the generation of ~ ‖ Apparat m. zur Überwachung der Energieerzeugung ‖ appareil m. pour le contrôle de la production d'énergie. / ~ hammer ‖ Krafthammer m. ‖ marteau-pilon m.

power house ‖ Kraftanlage f.; Kraftwerk n. ‖ centrale f. de force motrice. / electric ~ ‖ elektrische Zentrale f. ‖ centrale f. d'électricité.

power, increase of ~ ‖ Leistungssteigerung f. ‖ augmentation f. de puissance. / kind of ~ ‖ Antriebsart ‖ genre m. de commande. / ~ level ‖ Leistungspegel n. ‖ niveau m. de puissance.

power loom ‖ mechanischer Webstuhl m.; Maschinenwebstuhl m. ‖ métier m. mécanique. / ~ weaver ‖ Maschinenweber m. ‖ tisseur m. ou tisserand m. à la mécanique. / ~ weaving ‖ mechanische oder maschinelle Weberei f. ‖ tissage m. mécanique.

power loss ‖ Kraftverlust m. ‖ perte f. de puissance. / ~ operating of signals and switches ‖ Kraftstellung f. von Signalen und Weichen ‖ commande f. mécanique des aiguilles et signaux. / ~ plant see ~ station. / ~ producer ‖ Kraftquelle f. ‖ source f. d'énergie. / ~ quick cross traverse ‖ Eilgang m. für Querfahren ‖ déplacement m. ou mouvement m. rapide dans le sens transversal. / ~ quick elevating motion of drilling spindle ‖ Schnellhub m. der Bohrspindel ‖ dispositif m. de relevage rapide de la broche de perçage. / ~ quick longitudinal traverse ‖ Eilgang m. für Längsfahren ‖ déplacement m. ou mouvement m. rapide longitudinal par boîtes d'avances. / ~ quick traverse of the table ‖ Eilgang m. des Tischschlittens ‖ mouvement m. rapide de la table. / ~ rail ‖ Arbeitsschiene f. ‖ rail m. de prise de courant. / ~ reserve ‖ Kraftreserve f. ‖ réserve f. de puissance. / ~ room ‖ Maschinenraum m. ‖ salle f. des machines. / source of ~ ‖ Kraftquelle f.; Energiequelle f. ‖ source f. d'énergie.

power station ‖ Kraftanlage f.; Kraftwerk n. ‖ usine f. ou centrale f. de force motrice. / electric ~ ‖ Elektrizitätswerk n. ‖ station f. électrique. / fully automatic asynchronous ~ ‖ vollautomatisches Asynchronkraftwerk n. ‖ centrale f. électrique asynchrone entièrement automatique. / high-tension ~ ‖ Überlandzentrale f. ‖ centrale f. électrique régionale. / large ~ ‖ Großkraftwerk n. ‖ usine f. d'énergie de forte puissance. / ~ for medium utilization ‖ Kraftwerk n. für mittlere Belastung ‖ usine f. de charge moyenne. / pump fed ~ ‖ Pumpspeicherwerk n. ‖ usine f. d'accumulation par pompage. / to take over a ~ ‖ ein Kraftwerk n. übernehmen ‖ prendre à sa charge une station de force.

power supply (Radio) ‖ Netzanschluß m. ‖ alimentation f. par le secteur. / ~ for the industry ‖ Stromlieferung f. für Industrie ‖ fourniture f. de courant industriel.

power transformer ‖ Leistungstransformator m. ‖ transformateur m. à grande puissance. / ~ transmission ‖ Kraftübertragung f. ‖ transmission f. de puissance ou de force motrice. / ~ transmission engineering ‖ Kraftübertragungstechnik f. ‖ manipulation f. de l'énergie; technique f. de la transmission d'énergie. / ~ unit ‖ Krafteinheit f. ‖ unité f. de puissance.

pox ‖ Pocken fpl. ‖ variole f.; petite vérole f.

pozzolana ‖ Puzzolanerde f. ‖ pouzzolane f. / artificial ~ ‖ künstliche Puzzolanerde f.; Eisenkitt m. ‖ pouzzolane f. artificielle.

pozzuolana see pozzolana.

practicability ‖ Durchführbarkeit f. ‖ possibilité f. d'exécution.

practicable ‖ ausführbar ‖ practicable; exécutable.

practical experience ‖ praktische Erfahrung f. ‖ expérience f. pratique. / ~ geology ‖ angewandte Geologie f. ‖ géologie f. pratique. / ~ unit ‖ praktische Einheit f. ‖ unité f. pratique. / to put to ~ use ‖ in die Praxis f. umsetzen ‖ mettre en pratique. / ~ yield ‖ praktische Ausbeute f. ‖ rendement m. pratique.

practice, to see to practise.

practice ‖ Praxis f. ‖ pratique f.; usage m.; habitude f. / successful in ~ ‖ in der Praxis f. bewährt ‖ éprouvé par la pratique.

practice buzzer ‖ Übungssummer m. ‖ vibrateur m. d'apprentissage.

practise, to ~ an art ‖ eine Kunst f. ausüben ‖ professer un art.

praise, to ‖ anpreisen ‖ vanter; prôner.

praise ‖ Anpreisung f. ‖ réclame f.; boniment m.

praline ‖ gebrannte Mandel f. ‖ praline f. / ~ maker ‖ Hersteller m. von gebrannten Mandeln ‖ pralineur m.

prase-opal ‖ Prasopal m.; lauchgrüner Opal m. ‖ prasopale f.

pratique (Mar) ‖ Praktika f.; Verkehrserlaubnis f. ‖ pratique f. / to admit to ~ ‖ Praktika npl. erteilen ‖ accorder le permis de travail; admettre à la libre pratique.

precaution ‖ Vorsicht f.; Vorsichtsmaßregel f. ‖ précaution f. / not to be used without special ~s pl. ‖ nicht ohne besondere Vorsichtsmaßregeln fpl. benutzen ‖ ne pas s'en servir sans précautions spéciales. / to take ~s pl. ‖ Maßregeln fpl. oder Vorkehrungen fpl. treffen ‖ prendre des mesures m.

precedent ‖ Präzedenzfall m. ‖ précédent m.

precincts pl. ‖ Weichbild n. ‖ banlieue f.

precious ‖ wertvoll ‖ précieux.

precious metal ‖ Edelmetall n. ‖ métal m. précieux. / ~ assaying ‖ Edelmetallprobe f. ‖ essai m. de métaux précieux. / ~ refiner ‖ Edelmetallscheider m. ‖ affineur m. de métaux précieux.

precious stone ‖ Edelstein m. ‖ pierre f. précieuse. / cut ~ ‖ geschliffener Edelstein m. ‖ pierre f. précieuse taillée. / facetted ~ ‖ facettierter Edelstein m. ‖ pierre f. précieuse à facettes. / not set ~ ‖ ungefaßter Edelstein m. ‖ pierre f. précieuse non-montée. / polished ~ ‖ polierter Edelstein m. ‖ pierre f. précieuse polie. / rounded ~ ‖ abgerundeter Edelstein m. ‖ pierre f. précieuse arrondie. / worked ~ ‖ bearbeiteter Edelstein m. ‖ pierre f. précieuse travaillée.

precious stone borer ‖ Edelsteinbohrer m. ‖ perceur m. de pierres précieuses. / ~ engraving ‖ Edelsteingravierung f. ‖ gravure f. en pierres fines.

precipitability (Chem) ‖ Fällbarkeit f. ‖ précipitabilité f.

precipitable (Chem) ‖ abscheidbar; ausfällbar ‖ précipitable.

precipitant (Chem) ‖ Niederschlagsmittel n.; Fällungsmittel n.; Fällbad n. ‖ précipitant m.; réactif m. de précipitation.

precipitate, to ‖ fällen; ausfällen ‖ précipiter.

precipitate (Chem) ‖ Niederschlag m. ‖ précipité m. / white ~ of mercury ‖ weißer Quecksilberniederschlag ‖ précipité m. blanc de mercure.

precipitated, freshly ~ (Chem) ‖ frisch gefällt ‖ fraîchement précipité.

precipitating the dust from waste gases of a rotary kiln ‖ Entstaubung f. von Drehofenabgasen ‖ dépoussiérage m. de gaz d'échappement d'un four tournant. / ~ substance see precipitant.

precipitation ‖ Niederschlag m.; Ausfällung f. ‖ précipité m.; précipitation f. / annual ~ ‖ jährliche Niederschlagsmenge f. ‖

quantité f. annuelle de précipitation. / ~ of an aqueous solution ‖ Ausscheidung f. aus wässeriger Lösung ‖ séparation f. d'une solution aqueuse. / curdy ~ ‖ käsiger Niederschlag m. ‖ précipité m. caillé. / electrical ~ of tar mists ‖ elektrische Niederschlagung f. der Teernebel ‖ précipitation f. électrique des brouillards de goudron. / electrolytic ~ ‖ elektrolytischer Niederschlag m. ‖ précipité m. électrolytique. / the ~ peals off ‖ der Niederschlag m. blättert ab ‖ le dépôt m. s'écaille. / pulverulent ~ ‖ pulverförmiger Niederschlag m. ‖ précipité m. pulvérulent. / mechanical conveyor for the ~s of sand ‖ mechanische Fördereinrichtung für Bergeversatz ‖ engin m. mécanique de transport pour éboulis de remblayage.

precipitation, distribution of ~ ‖ Niederschlagsverteilung f. ‖ distribution f. de la pluie. / electricity of ~s ‖ Niederschlagselektrizität f. ‖ électricité f. des précipitations. / ~ electrode ‖ Niederschlagselektrode f. ‖ électrode f. de précipitation. / ~ tank ‖ Niederschlagsgefäß n. ‖ bassin m. à précipitation.

precise ‖ genau; bestimmt ‖ précis; juste. / ~ calculation ‖ genaue Berechnung f. ‖ calcul m. précis ou juste. / ~ mechanical work ‖ feinmechanische Arbeit f. ‖ travail m. de fine mécanique. / ~ traversing ‖ Polygonierung f. ‖ tachéométrie f.

precision ‖ Genauigkeit f.; Präzision f. ‖ exactitude f.; précision f.; justesse f. / ~ of measurements ‖ Genauigkeit f. bei Messungen ‖ précision f. de mesures.

precision adjustment ‖ Präzisionsschaltwerk n. ‖ rouage m. de précision. / ~ ammeter ‖ Präzisionsstrommesser m.; Präzisionsampèremeter n. ‖ ampèremètre m. de précision. / ~ balance ‖ Präzisionswaage f. ‖ balance f. de précision. / ~ drawing ‖ Präzisionszieherei f. ‖ étirage m. de précision. / ~ gear ‖ Genauigkeitsverzahnung f.; Präzisionsverzahnung f. ‖ denture f. de précision. / ~ high-speed drilling machine ‖ Präzisionsschnellbohrmaschine f. ‖ perceuse f. rapide de précision. / ~ hoisting gear ‖ Feinhubwerk n. ‖ dispositif m. de levage de précision. / ~ instrument ‖ Präzisionsinstrument n.; Genauigkeitsinstrument n. ‖ instrument m. de précision. / mechanical apparatus of ~ ‖ feinmechanischer Apparat m. ‖ appareil m. mécanique de précision. / mechanical instrument of ~ ‖ feinmechanisches Instrument n. ‖ instrument m. de précision. / ~ measuring instrument ‖ Präzisionsmeßinstrument n. ‖ instrument m. de mesure de précision. / ~ mechanics pl. ‖ Feinmechanik f. ‖ mécanique f. de précision. / ~ mechanician ‖ Präzisionsmechaniker m.; Feinmechaniker m. ‖ mécanicien m. de précision. / astronomical ~ pendulum clock ‖ astronomische Präzisionspendeluhr f. ‖ horloge f. astronomique de précision à pendule. / ~ regulator with seconds stroke ‖ Präzisionsregler m. mit Sekundenschlag ‖ régulateur m. de précision battant la seconde. / ~ scales pl. see ~ balance. / ~ screw ‖ Präzisionsschraube f. ‖ vis f. de précision. / ~ slit ‖ Präzisionsspalt m. ‖ fente f. de précision.

precision tool ‖ Präzisionswerkzeug n. ‖ outil m. ou instrument m. de précision. / ~ for gauging ‖ Präzisionsmeßwerkzeug n. ‖ instrument m. de précision à me-

surer. / ~ maker ‖ Präzisionswerkzeugfabrikant m. ‖ fabricant m. d'instruments de précision.

precision watch ‖ Präzisionstaschenuhr f. ‖ montre f. de précision. / ~ weight ‖ Präzisionsgewicht n. ‖ poids m. de précision. / ~ work ‖ Präzisionsarbeit f. ‖ travail m. de précision.

preclassification ‖ Vorklassierung f. ‖ classement m. préalable.

precluding, tenacious alloy ~ the possibility of bringing cutting tools into operation upon them ‖ zähe, durch Werkzeuge unangreifbare Legierung f. ‖ alliage m. très tenace et inattaquable par l'outil.

precombustion chamber ‖ Vorverbrennungskammer f. ‖ chambre f. de précombustion.

precool, to ~ to extremely low temperatures ‖ zu sehr tiefen Temperaturen vorkühlen ‖ préréfrigérer à des températures fpl. très basses.

precooling ‖ Vorkühlung f. ‖ refroidissement m. préalable; préréfrigération f. / ~ of coke oven gas ‖ Vorkühlen n. des Koksofengases ‖ préréfrigération f. du gaz des fours à coke.

precooling plant ‖ Vorkühlanlage f. ‖ installation f. de préréfrigération.

preeminence ‖ Vorrang m. ‖ préséance f.; priorité f.

prefer, to ~ ‖ vorziehen ‖ préférer.

preference ‖ Priorität f. ‖ priorité f.; obligation f. privilégiée. / ~ share ‖ Prioritätsaktie f. ‖ action f. privilégiée.

preferential duty ‖ Begünstigungszoll m. ‖ droit m. différentiel. / ~ rate ‖ Vorzugspreis m. ‖ prix m. exceptionnel ou de faveur. / to maintain a ~ right ‖ ein Vorzugsrecht n. behaupten ‖ prétendre à un droit de priorité. / ~ treatment ‖ Vorzugsbehandlung f. ‖ bénéfice m.; traitement m. privilégié.

preferred stock see preference share.

preheat, to ‖ vorwärmen; anheizen ‖ chauffer préalablement ou à l'avance; avant-chauffer.

preheated air ‖ vorgewärmte Luft f. ‖ air m. chauffé préalablement.

preheater ‖ Vorwärmer m. ‖ réchauffeur m. / air ~ ‖ Winderhitzer m. ‖ appareil m. à air chaud. / exhaust steam ~ ‖ Abdampfvorwärmer m. ‖ réchauffeur m. à vapeur d'échappement. / feed water ~ ‖ Speisewasservorwärmer m. ‖ réchauffeur m. d'eau d'alimentation. / flue gas ~ ‖ Rauchgasvorwärmer m. ‖ réchauffeur m. à gaz de fumée.

preheating ‖ Vorwärmung f. ‖ réchauffage m.; avant-chauffage m. / ~ of feed water ‖ Speisewasservorwärmung f. ‖ réchauffage m. de l'eau alimentaire. / ~ of hot-bulb ‖ Anheizen n. des Glühkopfes ‖ réchauffage m. du bulbe d'allumage.

prehensible ‖ ausziehbar; Teleskop … ‖ s'allongeant à volonté; extensible.

preignition ‖ Frühzündung f. ‖ allumage m. prématuré.

prejudice ‖ Schädigung f. ‖ tort m.; préjudice m.

prejudicial ‖ schädlich ‖ nuisible; préjudiciable.

preliminary ‖ vorläufig; Vor … ‖ préliminaire; préalable; préparatoire. / ~ breaker ‖ Vorbrecher m. ‖ avant-concasseur m.; avant-broyeur m. / ~ breaking ‖ Vorbrechen n. ‖ avant-broyage m. / ~ cleaning ‖ Vorreinigung f. ‖ épuration f.

préalable. / ~ drying ‖ Vortrocknung f. ‖ préséchage m. / ~ feed valve for hydraulic presses ‖ Vorfüllventil n. für hydraulische Pressen ‖ soupape f. d'admission préalable pour presses hydrauliques. / ~ geological examination ‖ geologische Vorarbeiten fpl. ‖ travaux mpl. géologiques préparatoires. / ~ inspection (El line) ‖ Vorrevision f. ‖ révision f. préalable. / ~ mill ‖ Vorwalzwerk n. ‖ laminoir m. dégrossisseur. / ~ project ‖ Vorentwurf m. ‖ avant-projet m. / ~ washing ‖ Vorwäsche f. ‖ lavage m. préparatoire.

premature ‖ vorzeitig ‖ prématuré. / ~ wear ‖ vorzeitige Abnutzung f. ‖ usure f. prématurée.

premises pl. ‖ Grundstück n. ‖ fonds m. de terre.

premium ‖ Prämie f. ‖ prime f. / ~ (Trade) ‖ Agio n.; Aufgeld n. ‖ agio m. / annual ~ ‖ Jahresprämie f. ‖ prime f. annuelle. / ~ on gold ‖ Goldagio n.; Goldaufgeld n. ‖ agio m. sur l'or. / import ~ ‖ Einfuhrprämie f. ‖ prime f. d'importation. / ~ of insurance ‖ Versicherungsprämie f. ‖ prime f. d'assurance. / producing ~ ‖ Produktionsprämie f. ‖ prime f. à la production.

premium bonus system ‖ Prämienlohnsystem n. ‖ système m. de salaire à prime.

premixing ‖ Vormischung f. ‖ pré-mélange m.

prepaid ‖ vorausbezahlt ‖ payé d'avance. / ~ reply ‖ Antwort f. bezahlt ‖ réponse f. payée.

preparation ‖ Vorbereitung f. ‖ préparation f.; prédisposition f. / ~ (Dressing; Making) ‖ Zubereitung f.; Bereitung f. ‖ préparation f.; apprêt m. / ~ (Chem) ‖ Präparat n. ‖ préparation f. / ~ (Wood) ‖ Imprägnierung f.; Präparieren n. ‖ imprégnation f. / anatomical ~ ‖ anatomisches Präparat n. ‖ préparation f. anatomique. / ~ for asthma ‖ Asthmapräparat n. ‖ médicament m. contre l'asthma. / to be in ~ ‖ in Vorbereitung f. sein ‖ être en préparation f. / ~ of concrete ‖ Betonbereitung f. ‖ préparation f. du béton. / dosed ~ ‖ dosiertes Präparat n. ‖ préparation f. dosée. / to make ~s pl. ‖ Anstalten fpl. machen ‖ se préparer à... / ~ of meat ‖ Zubereitung f. von Fleisch ‖ préparation f. de viande. / microscopic ~ ‖ mikroskopisches Präparat n. ‖ préparation f. microscopique. / ~ of nutriments ‖ Nährmittelzubereitung f. ‖ préparation f. alimentaire. / ~ of ores ‖ Erzaufbereitung f. ‖ traitement m. des minerais. / ~ in the pure state ‖ Reindarstellung f. ‖ préparation f. à l'état de pureté. / ~ of the raw material ‖ Aufbereitung f. der Rohstoffe ‖ préparation f. des matières premières. / ~ of the weft ‖ Vorbereitung f. des Einschlags ‖ tramage m.; préparation f. du fil de trame. / ~ of wood ‖ Imprägnierung f. oder Tränkung f. von Holz ‖ injection. f. ou imprégnation f. ou imbibition f. de bois. / ~ of wood with sulphate of copper ‖ Imprägnierung f. des Holzes mit Kupfersulfat ‖ préparation f. des bois au sulfate de cuivre. / zoological ~ ‖ zoologisches Präparat n. ‖ préparation f. zoologique.

preparation, means pl. of ~ ‖ Gewinnungsweise f. ‖ mode m. de préparation. / ~ utensils pl. ‖ Präparierutensilien npl. ‖ trousse f. à dissection.

preparator ‖ Präparator m. ‖ empailleur m.

preparatory machine for weaving ‖ Webereivorbereitungsmaschine f. ‖ machine f. préparatoire aux tissages. / ~ work of a railway ‖ Vorarbeiten fpl. zu einem Eisenbahnbau ‖ études fpl. d'un projet de chemin de fer.

prepare, to ‖ vorbereiten; präparieren ‖ préparer. / ~ (Chem) ‖ darstellen ‖ préparer. / ~ (Mortar) ‖ anmachen ‖ préparer. / ~ ores pl. ‖ Erz n. aufbereiten ‖ traiter un minerai. / ~ skins by the application of oil ‖ sämisch gerben ‖ chamoiser; passer en chamois. / ~ the vat (Dyer) ‖ die Küpe zurichten ‖ poser la cuve. / ~ the way for... ‖ anbahnen ‖ frayer le chemin; ouvrir; préparer.

preparedness for war ‖ Kriegsbereitschaft f. ‖ préparation f. à la guerre. / industrial ~ ‖ industrielle Kriegsbereitschaft f. ‖ préparation f. industrielle en vue de la guerre.

preparer (Spinn) ‖ Vorspinner m. ‖ prépareur m.

preparing of composition (Pyrot) ‖ Ansetzen n. des Satzes ‖ mélange m. des matières. / ~ with fat (Curr) ‖ Sämischgerben n.; Sämischgerberei f. ‖ chamoisage m.

preparing machine ‖ Vorbereitungsmaschine f.; Aufbereitungsmaschine f. ‖ machine f. de préparation. / ~ machine for moulding material ‖ Aufbereitungsmaschine f. für Formsand ‖ machine f. à préparer les matières de moulage. / ~ plant for fibre materials ‖ Aufbereitungsanlage f. für Faserstoffe ‖ atelier m. de préparation de fibrines ou de matières de fibre. / ~ vat (Pott) ‖ Anmachebottich m. ‖ gâchoir m.

prepay, to ‖ vorausbezahlen ‖ payer d'avance. / ~ (Mail) ‖ frankieren; freimachen ‖ affranchir; payer le port.

prepayment ‖ Vorauszahlung f. ‖ paiement m. en avance. / ~ electricity meter ‖ Elektrizitätsselbstverkäufer m.; Automatenzähler m. ‖ compteur m. électrique à prépaiement.

preponderance ‖ Überlegenheit f. ‖ supériorité f. / ~ (Weight) ‖ Übergewicht n. ‖ poids m. supplémentaire.

preponderate, to ‖ überwiegen ‖ prépondérer.

preponderating ‖ überwiegend ‖ prépondérant.

presbyopia ‖ Weitsichtigkeit f. ‖ presbytie f. / increasing ~ ‖ zunehmende Weitsichtigkeit f. ‖ presbytie f. augmentante.

presbyopic ‖ weitsichtig ‖ presbyte.

prescribed dimensions pl. ‖ vorgeschriebene Abmessungen fpl. ‖ dimensions fpl. préscrites.

prescription ‖ Verordnung f. ‖ ordonnance f. / ~ (Law) ‖ Verjährung f. ‖ prescription f.; surannation f. / ~ of spectacles ‖ Brillenverordnung f. ‖ ordonnance f. des lunettes. / ~ instrument for sight testing and giving the ~ for spectacles ‖ Instrument n. für Sehprüfung und Brillenverordnung ‖ instrument m. pour l'examen de la vue et l'ordonnance de lunettes. / ~ form ‖ Rezeptvordruck m. ‖ bulletin m. d'ordonnance.

preselection (Aut tel) ‖ Vorwahl f. ‖ présélection f.

preselector (Aut tel) ‖ Vorwähler m. ‖ présélecteur m. / ~ system (Aut tel) ‖ Vorwählersystem n. ‖ système m. à présélecteurs.

presence ‖ Anwesenheit f. ‖ présence f.

present ‖ anwesend ‖ présent. / ~ day manufacture ‖ neuzeitliche Fertigung f. ‖ fabrication f. moderne.

present, to ~ a bill of exchange ‖ einen Wechsel m. präsentieren oder vorlegen ‖ présenter une lettre f. de change.

present ‖ Geschenkartikel m. ‖ article m. de cadeau.

presentation ‖ Zustellung f. ‖ remise f.; notification f.

preservation ‖ Konservierung f.; Haltbarmachung f. ‖ conservation f. / to attain a better ~ of the fibre ‖ bessere Erhaltung f. der Faser erzielen ‖ conserver mieux la fibre. / prepared for ~ ‖ auf Haltbarkeit f. zubereitet ‖ préparé pour assurer la conservation. / ~ of wood see preserving of timber.

preservative ‖ Konservierungsmittel n. ‖ préservatif m. / ~ (Preventive) ‖ Präservativ n.; Schutzmittel n. ‖ préservatif m.

preserve, to ‖ haltbar machen ‖ conserver. / ~ fruits pl. ‖ Früchte fpl. einmachen ‖ confire des fruits mpl.

preserve (Jam) ‖ Kompott n. ‖ compote f.

preserved in fresh state by a refrigerating method ‖ in frischem Zustand durch ein Kälteverfahren n. haltbar gemacht ‖ conservé à l'état frais par un procédé frigorifique. / ~ fish ‖ Fischkonserven fpl. ‖ conserves fpl. de poissons. / ~ fruits pl. ‖ Obstkonserven fpl. ‖ conserves fpl. de fruits. / ~ meat ‖ Fleischkonserven fpl. ‖ conserves fpl. de viande. / ~ meat in tins ‖ eingemachtes Fleisch n. in Blechbüchsen ‖ conserves fpl. de viande en boîtes. / ~ milk manufactory ‖ Milchsterilisieranstalt f. ‖ fabrique f. de lait stérilisé. / ~ provisions pl. ‖ eingemachte Eßwaren fpl. ‖ conserves fpl. alimentaires. / ~ soup ‖ Suppenkonserven fpl. ‖ soupe f. conservée. / ~ vegetables ‖ Gemüsekonserven fpl. ‖ conserves fpl. de légumes.

preserve, factory of ~s ‖ Konservenfabrik f. ‖ fabrique f. de conserves alimentaires. / ~ glass see ~ jar. / ~ jar ‖ Einkochglas n. ‖ verre m. à conserves. / closing for ~ jars ‖ Einkochglasverschluß m. ‖ fermeture f. pour bocaux à conserves. / ~ making implement ‖ Einkochgerät n. ‖ ustensile m. pour l'ébullition des conserves. / ~ manufactory see factory of preserves.

preserves pl. ‖ Konserven fpl. ‖ conserves fpl.

preserve tin ‖ Konservendose f. ‖ boîte f. à conserves. / ~ trade machine ‖ Konservenindustriemaschine f. ‖ machine f. pour l'industrie des conserves.

preserving of timber ‖ Holzkonservierung f. ‖ conservation f. des bois. / ~ the yeast ‖ Konservieren n. der Hefe ‖ conservation f. de la levure.

preserving apparatus ‖ Einkochapparat m. ‖ appareil m. d'ébullition de conserves. / ~ glass ‖ Konservenglas n. ‖ verre m. à conserves. / ~ preparation ‖ Konservierungspräparat m. ‖ conserver. / floor ~ preparation ‖ Fußbodenpflegemittel n. ‖ enduit m. pour l'entre-

tien des planchers. / ~ tube ‖ Schutzhülse f. ‖ tube f. de protection.
president of the company ‖ Vorsitzender m. der Gesellschaft ‖ président m. de la société.
presignal ‖ Vorsignal n. ‖ signal m. avancé.
press, to ‖ pressen; einpressen ‖ presser. / ~ cloth ‖ das Tuch pressen ‖ catir le drap. / ~ cold ‖ kalt pressen ‖ presser à froid. / ~ the cover against the glued back of the booklet ‖ den Umschlag m. gegen den geleimten Rücken der Broschüre drücken ‖ presser la couverture contre le dos de la brochure enduit de colle. / ~ glass ‖ das Glas pressen ‖ mouler le verre. / ~ out ‖ auspressen ‖ pressurer. / ~ for payment ‖ zur Zahlung f. auffordern ‖ inviter à payer. / ~ the rough tiles pl. ‖ die Ziegel mpl. nachpressen ‖ rebattre les tuiles fpl. / ~ tyres pl. (Bicycle) ‖ Radreifen mpl. aufpressen ‖ monter les bandages mpl. à la presse.
press ‖ Presse f. ‖ presse f. / ~ (Join) ‖ Schrank m.; Schrein m. ‖ armoire f. / ~ (Print) ‖ Feuchtstein m. ‖ serre-feuilles m. / amalgam ~ ‖ Amalgampresse f. ‖ presse f. pour amalgame. / ~ for building armature and transformer cores ‖ Presse f. für den Anker- und Transformatorenbau ‖ presse f. à serrer les tôles d'induits et de transformateurs. / armour bending ~ ‖ Panzerplattenbiegepresse f. ‖ presse f. à cintrer les blindages. / ~ for asbestos and cement plates ‖ Asbestzementplattenpresse f. ‖ presse f. à plaques en ciment d'amiante. / automatic ~ ‖ automatische Presse f. ‖ presse f. automatique. / baling ~ ‖ Ballenpresse f. ‖ presse f. pour balles. / ball ~ ‖ Kugelpresse f. ‖ presse f. à billes. / beam bending ~ ‖ Balkenbiegepresse f. ‖ presse f. à cintrer les poutrelles. / bending ~ ‖ Biegepresse f. ‖ presse f. à courber. / horizontal bending and forming ~ ‖ horizontale Biege- und Formpresse f. ‖ bulldozer m. pour cintrer et former. / ~ for the trade of beverages ‖ Presse f. für die Getränkeindustrie ‖ presse f. pour l'industrie des boissons. / bolt ~ ‖ Schraubenbolzenpresse f. ‖ presse f. à faire les boulons. / book perfecting ~ ‖ Presse f. für gleichzeitigen Druck auf beiden Seiten ‖ presse f. à double impression. / brick ~ ‖ Ziegelpresse f. ‖ presse f. à briques. / ~ for placing the buckle on springs ‖ Federbundaufziehpresse f. ‖ presse f. à monter les brides de ressorts. / ~ for building materials ‖ Presse f. für Baustoffe ‖ presse f. pour matériaux de construction. / ~ for taking off the burr ‖ Abgratpresse f. ‖ presse f. d'ébarbage. / calender roller ~ ‖ Kalanderwalzenpresse f. ‖ presse f. à rouleaux. / carbon moulding ~ ‖ Kohlenstiftpresse f. ‖ presse f. à filer le charbon. / ~ for cardboard works ‖ Presse f. für Pappfabriken ‖ presse f. pour cartonneries. / celluloid ~ ‖ Zelluloidpresse f. ‖ presse f. pour celluloïd. / centering ~ ‖ Zentrierpresse f. ‖ presse f. à centrer. / cider ~ ‖ Apfelkelter f. ‖ presse f. à cidre ou pressoir m. / cold ~ for forming and upsetting rivets ‖ Kaltpresse f. zum Formen und Stauchen von Nieten ‖ presse f. à froid pour former et fouler des rivets. / column ~ ‖ Säulenpresse f. ‖ presse f. à colonnes. / concrete testing ~ for compression, buckling and bending tests ‖ Betonprüfpresse f. für

Druck-, Knick- und Biegeversuche ‖ presse f. à essayer le béton à la compression, à la rupture et à la flexion. / copper plate ~ ‖ Kupferdruckpresse f. ‖ presse f. en taille-douce. / ~ for copper plate printing ‖ Tiefdruckpresse f. ‖ presse f. à impression en taille-douce. / copying ~ ‖ Kopierpresse f. ‖ presse f. à copier. / ~ for corrugated iron sheets ‖ Wellblechpresse f. ‖ presse f. à tôle ondulée. / cotton baling ~ ‖ Baumwollballenpresse f. ‖ presse f. à mettre le coton en balles. / crank ~ ‖ Kurbelpresse f. ‖ presse f. à manivelle. / crank-operated ~ ‖ Kurbelpresse f. ‖ presse f. à manivelle. / double-sided eccentric ~ ‖ Doppelständerexzenterpresse f. ‖ presse f. à excentrique à double montant. / drawing ~ ‖ Ziehpresse f. ‖ presse f. à étirer ou à emboutir. / eccentric ~ ‖ Exzenterpresse f. ‖ presse f. à excentrique. / eccentric gitting ~ ‖ Exzenterabkneifpresse f. ‖ presse f. à excentrique pour entailler; coupe-jet m. / edge finishing ~ see ~ for taking off the burr. / electrode ~ ‖ Elektrodenpresse f. ‖ presse f. à électrodes. / employed ~ ‖ gehende Presse f. ‖ presse f. roulante. / extruding ~ see drawing ~. / filter ~ ‖ Filterpresse f. ‖ filtre-presse f. / filtering ~ see filter ~. / flanging ~ ‖ Kümpelpresse f. ‖ presse f. à emboutir. / foot ~ ‖ Fußpresse f. ‖ presse f. à pédale. / frame bending ~ ‖ Rahmenbiegepresse f. ‖ presse f. à cintrer les châssis. / friction ~ ‖ Friktionspresse f. ‖ presse f. à friction. / ~ for galvanoplastic ‖ Presse f. für Galvanoplastik ‖ presse f. pour la galvanoplastie. / ~ for glued boards ‖ Leimzwinge f.; Schraubzwinge f. ‖ sergent m.; serre-joint m.; presse f. à serrer. / hand ~ ‖ Handpresse f. ‖ presse f. à bras. / hand ~ for letterpress work ‖ Buchdruckhandpresse f. ‖ presse f. d'imprimerie à bras. / hay ~ ‖ Heupresse f. ‖ presse f. à foin. / housekeeping ~ ‖ Haushaltpresse f. ‖ presse f. de ménage. / hydraulic ~ ‖ hydraulische Presse f. ‖ presse f. hydraulique. / hydraulic bending ~ of x tons pressing power ‖ hydraulische Biegepresse f. mit x ts Druckkraft ‖ presse f. hydraulique à cintrer d'une force de x tonnes. / hydraulic forging ~ ‖ hydraulische Schmiedepresse f. ‖ presse f. hydraulique à forger. / hydraulic stamping ~ ‖ hydraulische Gesenkpresse f. ‖ presse f. hydraulique à estamper. / hydraulic tube drawing ~ ‖ hydraulische Rohrziehpresse f. ‖ presse f. hydraulique à étirer les tuyaux. / ~ for illustration printing ‖ Bilddruckpresse f. ‖ presse f. pour impression d'illustrations. / inclinable eccentric ~ ‖ schrägstellbare Exzenterpresse f. ‖ presse f. à excentrique à bâti inclinable. / inclinable geared eccentric ~ see inclinable eccentric ~. / ~ of a joiner's bench ‖ Zange f. einer Hobelbank ‖ presse f. d'un établi de menuisier. / keel plate bending ~ ‖ Kielplattenbiegemaschine f. ‖ machine f. à cintrer les tôles de carène. / ~ for dismounting bands from laminated springs ‖ Federbundabziehpresse f. ‖ presse f. pour le démontage des brides de ressorts. / lead cable ~ ‖ Bleikabelpresse f. ‖ presse f. à câbles sous plomb. / lead pipe ~ ‖ Bleirohrpresse f. ‖ presse f. à filer les tuyaux de plomb. / lead wire ~ ‖ Bleidrahtpresse f. ‖ presse f.

pour fils en plomb. / letter ~ see printing ~. / litho ~ see lithographic ~. / lithographic ~ ‖ Steindruckpresse f. ‖ presse f. lithographique. / mandrel ~ ‖ Dorneintreibepresse f.; Dornpresse f. ‖ presse f. à mandrin ou à enfoncer les mandrins. / mandrel upsetting ~ ‖ Dornstauchpresse f. ‖ presse f. à fouler les mandrins. / metal ~ for making round bars ‖ Stangenpresse f. ‖ presse f. à filer les barres rondes. / metal pipe ~ ‖ Metallrohrpresse f. ‖ presse f. à filer les tuyaux métalliques. / metal rod ~ see profile ~ for metal rods. / ~ for stamping monograms ‖ Monogrammprägepresse f. ‖ presse f. à monogrammes. / multicolour ~ ‖ Vielfarbendruckpresse f. ‖ presse f. multicolore. / nut making ~ ‖ Mutternpresse f. ‖ presse f. à faire les écrous. / oil ~ ‖ Ölpresse f. ‖ presse f. à huile. / open-front eccentric ~ ‖ Einständerexzenterpresse f. ‖ presse f. à excentrique à col de cygne. / open-front eccentric ~ for direct flywheel drive ‖ Einständerexzenterpresse f. für direkten Schwungradantrieb ‖ presse f. à excentrique à col de cygne à commande directe sur le volant. / ~ for ore working ‖ Presse f. für die Erzverarbeitung ‖ presse f. à travailler le minerai. / packing ~ ‖ Packpresse f.; Paketierpresse f. ‖ presse f. à paqueter. / ~ for paper works ‖ Presse f. für Papierfabriken ‖ presse f. pour papeteries. / peat ~ ‖ Torfpresse f. ‖ presse f. à mottes de tourbe. / perforating ~ ‖ Perforierpresse f. ‖ presse f. à perforer. / ~ for pharmaceutic articles ‖ Presse f. für pharmazeutische Artikel ‖ presse f. à comprimés pharmaceutiques. / pipe expanding ~ ‖ Rohraufweitpresse f. ‖ presse f. à élargir les tuyaux. / pipe jumping ~ ‖ Röhrenstauchpresse f. ‖ presse f. à refouler les tuyaux. / pipe testing ~ ‖ Röhrenprobierpresse f. ‖ presse f. à essayer les tuyaux. / ~ for the pit coal industry ‖ Presse f. für die Steinkohlenindustrie ‖ presse f. pour l'industrie de la houille. / plate bending ~ ‖ Plattenbiegepresse f. ‖ presse f. à cintrer les tôles. / the ~ is adapted for a pressure of about x tons ‖ die Presse f. ist für einen Druck von x tons gebaut ‖ la presse est construite pour une pression de x tons. / printing ~ ‖ Druckerpresse f. ‖ presse f. à imprimer ou typographique. / profile ~ for metal rods ‖ Metallstrangpresse f. ‖ presse f. à profiler à chaud les barres métalliques. / proof ~ (Print) ‖ Abziehapparat m. ‖ appareil m. à tirer les épreuves. / ~ for the provision industry ‖ Presse f. für die Nahrungsmittelindustrie ‖ presse f. pour l'industrie des produits alimentaires. / pulp ~ ‖ Pülpepresse f. ‖ presse f. à pulpe. / quick-acting hydraulic ~ ‖ schnelllaufende hydraulische Presse f. ‖ presse f. hydraulique à marche rapide. / rapid-action screw ~ ‖ Schnellspindelpresse f. ‖ presse f. à vis rapide. / reaming ~ ‖ Aufweitpresse f. ‖ presse f. à mandriner. / ~ for relief print ‖ Presse f. für Prägedruck ‖ presse f. pour imprimer en relief. / reprinting ~ ‖ Umdruckpresse f. ‖ presse f. de réimpression. / riveting ~ ‖ Nietpresse f. ‖ presse f. à river. / automatic rivet making ~ ‖ automatische oder selbsttätige Nietenpresse f. ‖ presse f. automatique à faire les rivets. / roller ~

‖ Walzenaufziehpresse f. ‖ presse f. à caler les cylindres. / ~ for putting off roller shells ‖ Walzenabziehpresse f. ‖ presse f. à décaler les chemises de cylindres. / roofing tile ~ ‖ Dachziegelpresse f. ‖ presse f. à tuiles. / rotary ~ ‖ Rotationspresse f. ‖ presse f. rotative. / ~ for sauerkraut ‖ Sauerkrautpresse f. ‖ presse f. à choucroute. / scrap baling ~ ‖ Schrottpaketierpresse f. ‖ presse f. à paqueter les ferrailles. / scrap baling ~ (Gun) ‖ Schrottpaketierpresse f. ‖ presse f. à paqueter la mitraille. / screw ~ ‖ Spindelpresse f. ‖ presse f. à vis. / ~ for manufacturing slabs of asphalt ‖ Presse f. zur Herstellung von Asphaltplatten ‖ presse f. pour la fabrication de plaques en asphalte. / sleeper ~ ‖ Schwellenpresse f. ‖ presse f. à traverses. / ~ for printing slips ‖ Fahnenpresse f. ‖ presse f. pour placards. / ~ for mounting solid tyres ‖ Presse f. zum Aufmontieren der Vollreifen ‖ presse f. pour le montage de pneus pleins. / spring band dismantling ~ ‖ Federbundabziehpresse f. ‖ presse f. à démonter les brides de ressort. / spring band mounting ~ ‖ Federbundaufziehpresse f. ‖ presse f. à monter les brides de ressorts. / stamping ~ ‖ Stanzpresse f. ‖ presse f. à étamper; étampeuse f. / ~ with steam-heated plates ‖ Heizplattenpresse f. ‖ presse f. à plaques chauffées. / steam-hydraulic ~ ‖ dampfhydraulische Presse f. ‖ presse f. hydraulique à vapeur. / steam-hydraulic forging ~ ‖ hydraulische Dampfschmiedepresse f. ‖ presse f. de forge hydraulique à vapeur. / straightening ~ ‖ Richtpresse f. ‖ presse f. à dresser. / straw ~ ‖ Strohpresse f. ‖ presse f. à paille. / the stroke of the ~ can be adjusted to any required height ‖ der Hub m. der Presse ist in den weitesten Grenzen verstellbar ‖ la course du piston-comprimeur est réglable dans des limites très larges. / to take out of the ~ ‖ aus der Presse f. nehmen ‖ retirer de la presse. / tire ~ ‖ Reifenpresse f. ‖ presse f. à bandage. / toggle lever ~ ‖ Kniehebelpresse f. ‖ presse f. à genouillère. / toggle lever ~ with automatic regulation of pressure ‖ Kniehebelpresse f. mit automatischer Druckregulierung ‖ presse f. à levier coudé ou à genouillère avec réglage automatique de la pression. / trace ~ ‖ Strangpresse f. ‖ boudineusepeloteuse f. / trimming ~ ‖ Abgratpresse f. ‖ presse f. à ébarber; ébarbeuse f. / ~ for making tubes ‖ Rohrpreßmaschine f. ‖ presse f. à tuyaux. / tyre ~ see tire ~. / veneering ~ ‖ Furnierpresse f.; Sperrholzpresse f. ‖ presse f. à plaquer. / wheel ~ ‖ Räderaufziehpresse f. ‖ presse f. à caler les roues. / ~ for putting-off wheels ‖ Räderabziehpresse f. ‖ presse f. à décaler les roues. / wine ~ ‖ Weinpresse f. ‖ presse f. à vin. / wooden ~ ‖ Holzpresse f. ‖ presse f. en bois. / ~ for zigzag feed ‖ Zickzackpresse f. ‖ presse f. pour avance en zigzag.

press bar ‖ Preßstange f. ‖ levier m. de presse. / ~ bottom (Wine press) ‖ Kelterboden m. ‖ maie f.

press button ‖ Druckknopf m. ‖ bouton m. de pression. / ~ control ‖ Druckknopfsteuerung f. ‖ dispositif m. de manœuvre à boutons-poussoirs. / ~-controlled variable-speed motor ‖ druckknopfgesteuerter Regelmotor m. ‖ moteur m.

réglable à commande par bouton-poussoir.

press clamp ‖ Druckklemme f. ‖ borne f. à pression. / ~ cloth ‖ Preßtuch n. ‖ tissu m. de pressure. / ~ conversation by subscription ‖ Presseabonnementsgespräch n. ‖ conversation f. de presse par abonnement. / ~ cylinder ‖ Preßzylinder m. ‖ cylindre m. compresseur ou de presse.

pressed, cold ‖ kaltgepreßt ‖ pressé à froid. / ~ out from one piece with the flange ‖ aus einem Stück n. mit Flansch gepreßt ‖ embouti en une seule pièce avec les flasques. / ~ dust core coil ‖ Massekernspule f. ‖ bobine f. à noyau en poudre de fer comprimé.

pressed girder ‖ Preßträger m. ‖ longeron m. embouti. / U-shaped ~ ‖ U-förmiger Preßträger m. ‖ longeron m. embouti en U. / ~ underframe ‖ Preßträgeruntergestell n. ‖ châssis m. à longerons emboutis.

pressed glass ‖ Preßglas n. ‖ verre m. comprimé. / ~ glass base ‖ Sockel m. aus Preßglas ‖ socle m. en verre comprimé. / ~ hard glass for ship windows ‖ Preßhartglas n. für Schiffsfenster ‖ verre m. moulu trempé pour hublots. / ~ part ‖ Preßteil m.; Preßstück n. ‖ pièce f. en tôle emboutie; pièce estampée. / ~ piece see ~ part. / ~ plate girder ‖ gepreßter Blechträger m. ‖ longeron m. en tôle emboutie. / ~ rubber article with cloth enclosed ‖ Gummipreßformartikel m. mit Stoffeinlage ‖ article m. matricé en caoutchouc avec intérieur en étoffe. / ~ screw ‖ gepreßte Schraube f. ‖ vis f. pressée. / ~ steel crucible ‖ Tiegel m. aus Preßstahl ‖ creuset m. acier comprimé.

presser (Bookb) ‖ Bandschläger m. ‖ batteur m.

presserman (Pap) ‖ Presser m. ‖ pressier m.

presser plate (Weav) ‖ Preßblech n. ‖ tôle f. à encoches. / ~ rollers pl. (Pap) ‖ Satinierwerk n. ‖ laminoir m. en lissoir; laminoir m.; glaceur m.; satineuse f.

press filter ‖ Druckfilter n. ‖ filtre m. à pression. / ~ gilding ‖ Preßvergoldung f. ‖ dorage m. à pression. / ~ head ‖ Preßkopf m. ‖ chapeau m. de presse.

pressing ‖ Preßarbeit f.; Pressen n. ‖ travail m. de pressage; pressage m.; cartissage m. / ~ (Product) ‖ Preßling m.; Preßstück n. ‖ pièce f. estampée ou pressée. / by ~ a button ‖ durch Druck m. auf einen Knopf ‖ en appuyant sur un bouton. / cloth ~ ‖ Pressen n. von Geweben ‖ pressage m. d'étoffes. / cold ~ ‖ Kaltpressen n. ‖ écatissage m. / ~ nuts pl. ‖ Mutternpressen n. ‖ estampage m. des écrous. / ~-on the wheel on the axle ‖ Aufpressen n. des Rades auf die Achse ‖ calage m. à la presse de la roue sur l'essieu. / ~ the raw glass discs into their mould ‖ Pressen n. der Rohglasscheiben ‖ moulage m. des disques de verre bruts.

pressing bag of horsehair-cloth ‖ Preßbeutel m. oder Preßsack m. aus Roßhaartuch ‖ étreinte f.; étrindelle f.; étendelle f.; sac m. de crin; cabas m. / ~ bar ‖ Preßstange f. ‖ barre f. de presse. / ~ board ‖ Preßspan m. ‖ carton m. glacé;

presspan m. / ~ boards pl. (Bookb) ‖ Preßbretter npl. ‖ membrures fpl.; ais mpl. à presser. / ~ brush (Weav) ‖ Andrückbürste f. ‖ brosse f. conductrice. / ~ cylinder ‖ Preßcylinder m. ‖ cylindre m. de presse. / ~ hammer ‖ Preßhammer m. ‖ marteau m. hydraulique. / ~ lever ‖ Andrückhebel m. ‖ levier m. de pression. / ~ machine (Pap) ‖ Abpreßmaschine f. ‖ presse f. à reliure. / ~ machine minder ‖ Maschinenpresser m. ‖ presseur m. à la presse hydraulique. / ~ plant ‖ Preßanlage f.; Preßwerk n.; Preßwerksanlage f. ‖ installation f. de presses. / ~ plant for cotton ‖ Baumwollpreßanlage f. ‖ installation f. de presses à coton. / iron ~ plates pl. (Oil press) ‖ Preßbleche npl. ‖ semelles fpl. / ~ power ‖ Druckkraft f. ‖ force f. de pression. / hydraulic bending press of x ~ power ‖ hydraulische Biegepresse f. mit x ts Druckkraft ‖ presse f. hydraulique à cintrer d'une force de x tonnes. / ~ rod ‖ Druckstange f. ‖ barre f. de pression. / ~ roll ‖ Preßwalze f.; Druckwalze f. ‖ rouleau m. compresseur; cylindre m. comprimeur. / ~ roller paste ‖ Buchdruckerwalzenmasse f. ‖ pâte f. à rouleaux d'imprimerie. / ~ screw ‖ Druckschraube f. ‖ vis f. de pression.

pression ‖ Druck m. ‖ pression f.; poussée f. / lateral ~ of a vault ‖ Seitenschub m. eines Gewölbes ‖ effort m. latéral ou poussée f. d'une voûte. / ~ on the support ‖ Auflagerdruck m. ‖ pression f. sur l'appui. / ~ bottle ‖ Druckflasche f. ‖ bouteille f. ou flacon m. de pression.

press jack (Print) ‖ Preßbengel m. ‖ barre f. de la presse. / ~ key (Tel) ‖ Drucktaste f. ‖ contact m. à pression.

pressman (Glass) ‖ Glaspresser m. ‖ mouleur m. de verre à la presse. / ~ (Print) ‖ Preßmeister m.; Zubereiter m. ‖ pressier m. / ~ (Weav) ‖ Presser m. ‖ presseur m.

press mantle (Wine) ‖ Keltermantel m. ‖ manteau m. / ~ mould ‖ Preßfutter n.; Preßform f. ‖ moule m. de presse. / ~ piston ‖ Preßkolben m. ‖ piston m. de presse. / ~ plate ‖ Preßplatte f.; Preßtisch m. ‖ rouleau-compresseur m.; plateau m. de presse. / ~ plunger see ram. / ~ power ‖ Preßdruck m. ‖ puissance f. de pression. / ~ proof revisor (Print) ‖ Schlußrevisor m. ‖ reviseur m. de tierces. / ~ pump ‖ Preßpumpe f. ‖ pompe f. de presse. / ~ ram ‖ Preßstempel m. ‖ poinçon m. / ~ reviser (Print) ‖ Revisor m. ‖ tierceur m. / ~ rod ‖ Druckstelze f. ‖ tige f. de pression. / ~ screw ‖ Preßschraube f. ‖ vis f. de pression. / ~ setter (Clothm) ‖ Preßkartoneinleger m. ‖ cartonneur m. / ~ shop ‖ Pressenhaus n. ‖ salle f. des presses. / ~ spahn see pressing board. / ~ stick (Print) ‖ Preßbengel m. ‖ barreau m. de la presse. / ~ stone (Print) ‖ Setzstein m. ‖ marbre m. / ~ stud see button. / ~ switch (Electr) ‖ Druckschalter m. ‖ interrupteur m. à pression.

press-telegram ‖ Pressetelegramm n. ‖ télégramme m. de presse.

pressure ‖ Druck m. ‖ pression f.; poussée f. / ~ (Steam; Electr) ‖ Spannung f. ‖ tension f.; pression f. / absolute ~ ‖ absoluter Druck m. ‖ pression f. absolue. / ~ of the air see atmospheric ~. / ~ from all sides ‖ allseitiger Druck m. ‖

pression f. de tous côtés. / atmospheric ~ ‖ atmosphärischer Druck m.; Atmosphärendruck m. ‖ pression f. de l'atmosphère. / ~ of axle ‖ Achs(en)druck m. ‖ charge f. d'essieu. / back ~ ‖ Gegendruck m. ‖ contre-pression f. / boiler ~ ‖ Kesseldruck m. ‖ pression f. dans la chaudière. / the ~ of compressed air is too low ‖ der Luftdruck m. ist zu niedrig ‖ là pression f. d'air est trop faible. / constant ~ ‖ gleichbleibender Druck m. ‖ pression f. constante. / dangerous ~ ‖ gefährliche Spannung f. ‖ tension f. dangereuse. / the ~ decreases ‖ der Druck nimmt ab ‖ la pression baisse. / under diminished ~ ‖ im luftverdünnten Raum m. ‖ sous pression f. réduite. / the ~ drops ‖ der Druck sinkt ‖ la pression f. tombe. / dynamic ~ ‖ Staudruck m. ‖ pression f. dynamique. / ~ of earth (Build) ‖ Erddruck m. ‖ poussée f. des terres. / effective ~ as shown by the manometer ‖ vom Manometer angezeigte Spannung f. ‖ pression f. effective indiquée par le manomètre. / ~ reached at end of compression ‖ Kompressionsenddruck m. ‖ degré m. de compression final. / exhaust gas ~ ‖ Auspuffgasdruck m. ‖ pression f. des gaz d'échappement. / ~ of the finger ‖ Fingerdruck m. ‖ pression f. du doigt. / ~ of the ground see ~ of the soil. / high ~ ‖ Hochdruck m. ‖ pression f. haute. / higher ~ ‖ höherer Druck m. ‖ pression f. élevée. / to hold a ~ ‖ den Druck m. halten ‖ tenir la pression. / the ~ is applied to the hub of the wheels ‖ der Druck m. erfolgte auf die Nabe der Räder. ‖ la pression est portée sur le moyeu des roues. / hydraulic ~ ‖ Wasserdruck m. ‖ pression f. hydraulique. / hydrostatic ~ ‖ hydrostatischer Druck m. ‖ pression f. hydrostatique. / impressed ~ ‖ aufgedrückte Spannung f. ‖ pression f. appliquée. / interlinked ~ ‖ verkettete Spannung f. ‖ tension f. entre phases reliées. / ~ applied to the key (Typewr) ‖ auf die Taste ausgeübter Druck m. ‖ pression f. exercée sur la touche. / lateral ~ of the ground ‖ Seitendruck m. des Gebirgs ‖ poussées fpl. latérales du terrain. / low ~ ‖ Niederdruck m. ‖ basse pression f. / low ~ (Electr) ‖ Niederspannung f. ‖ basse tension f. / mean ~ ‖ mittlerer Druck m. ‖ pression f. moyenne. / mean effective ~ ‖ mittlerer Arbeitsdruck m. ‖ pression f. moyenne de travail. / ~ for measuring ‖ Meßdruck m. ‖ pression f. de mesurage. / middle ~ see mean ~. / osmotic ~ ‖ osmotischer Druck m. ‖ pression f. osmotique. / primary ~ ‖ Primärspannung f. ‖ tension f. primaire. / ~ of saturation ‖ Sättigungsdruck m. ‖ pression f. de saturation. / secondary ~ (Electr) ‖ Sekundärspannung f. ‖ tension f. secondaire. / ~ of the soil (Geol) ‖ Gebirgsdruck m. ‖ pression f. du terrain. / static ~ ‖ statischer Druck m. ‖ pression f. statique. / steam ~ ‖ Dampfdruck m.; Dampfspannung f. ‖ pression f. de la vapeur. / superincumbent ~ of the ground ‖ Firstendruck m. des Gebirges ‖ poussées fpl. du faîte. / ~ of the system ‖ Netzspannung f. ‖ tension f. du réseau. / terminal ~ ‖ Auspuffdruck m. ‖ pression f. finale. / to be under ~ ‖ unter Druck m. stehen ‖ être sous pression f. /

working ~ ‖ Arbeitsdruck m. ‖ pression f. de travail ou effective.
pressure airship ‖ Pralluftschiff n. ‖ dirigeable m. souple. / ~ balancing valve ‖ Druckausgleichventil n. ‖ soupape f. de compensation de pression. / ~ boiler ‖ Druckkessel m. ‖ chaudière f. à pression. / centre of ~ ‖ Druckmittelpunkt m. ‖ centre m. de pression. / ~ compensating device ‖ Druckausgleicheinrichtung f. ‖ dispositif m. de compensation de pression. / decrease of ~ ‖ Druckabfall m. ‖ chute f. ou diminution f. de pression. / diminution of the ~ see decrease of ~. / ~ disc ‖ Druckscheibe f. ‖ rondelle f. de pression. / distribution of ~ ‖ Druckverteilung f. ‖ distribution f. de pression. / fall of ~ see decrease of ~. / fluctuation of ~ ‖ Druckschwankung f. ‖ fluctuation f. de pression.
pressure gauge ‖ Manometer n.; Druckmesser m. ‖ manomètre m. / diaphragm ~ ‖ Plattenfedermanometer n. ‖ manomètre m. à diaphragme ondulé.
pressure gauge connection ‖ Druckmesseranschluß m.; Manometeranschluß m. ‖ taraudage m. pour manomètre.
pressure gear ‖ Druckübersetzer m. ‖ transmetteur m. de pression. / ~ governor ‖ Druckregler m. ‖ régulateur m. de pression. / increase of ~ ‖ Druckanstieg m. ‖ augmentation f. ou accroissement m. de pression. / ~ petrol ~ indicator ‖ Benzindruckmesser m. ‖ manomètre m. à essence. / ~ intensifier ‖ Druckverstärker m. ‖ multiplicateur m. de pression. / ~ lever ‖ Druckhebel m. ‖ levier m. de pression. / loss of ~ ‖ Druckverlust m. ‖ perte f. de pression. / ~ lubrication ‖ Druckschmierung f. ‖ graissage m. sous pression. / ~ oiler ‖ Drucköler m. ‖ servo-graisseur m.; graissage m. centrale sous pression.
pressure pipe ‖ Druckrohr n. ‖ conduit m. de refoulement ou de pression. / ~ connection ‖ Druckrohranschluß m. ‖ connection f. du conduit de pression. / ~ line ‖ Druckrohrleitung f. ‖ conduite f. de tuyaux de refoulement.
pressure plate ‖ Druckplatte f. ‖ plaque f. de pression. / ~ pump ‖ Druckpumpe f. ‖ pompe f. de pression. / ~ recorder ‖ Druckschreiber m. ‖ appareil m. enregistreur de la pression. / ~ reducer see ~ reducing apparatus. / ~ reducing apparatus ‖ Druckminderapparat m.; Druckreduzierapparat m. ‖ appareil m. à réduire la pression; réducteur m. de pression. / ~ reducing valve ‖ Druckminderventil n.; Reduzierventil n. ‖ soupape f. réductrice.
pressure regulator ‖ Druckregler m. ‖ régulateur m. de pression. / beer ~ ‖ Bierdruckregler m. ‖ régulateur m. de pression de bière. / gas ~ ‖ Gasdruckregler m. ‖ régulateur m. de pression de gaz.
pressure release ‖ Druckauslösung f. ‖ décharge f. de pression. / ~ reservoir ‖ Druckspeicher m. ‖ réservoir m. de pression. / ~ ring (Aero) ‖ Druckring m. ‖ anneau m. ou bague f. ou rondelle f. de pression. / ~ screw ‖ Druckschraube f. ‖ vis f. de pression. / ~ speed indicator ‖ Staudruckfahrtmesser m. ‖ indicateur m. de vitesse à pression dynamique. / ~ spring ‖ Druckfeder f. ‖ ressort m. de pression. / ~ stage ‖ Druckstufe f. ‖ degré m. de pression. / ~ stamp ‖ Druck-

stempel m. ‖ poinçon m. de compression. / ~ stirrup ‖ Druckbügel m. ‖ arc m. de pression. / ~ syringe ‖ Handdruckspritze f. ‖ seringue f. à main. / ~ tank ‖ Druckbehälter m. ‖ réservoir m. sous pression. / ~ test ‖ Druckprobe f. ‖ épreuve f. de pression. / ~ test pump ‖ Probedruckpumpe f.; Preßpumpe f. ‖ pompe f. (foulante) d'épreuve. / ~ transformation in loudspeakers ‖ Drucktransformation f. bei Lautsprechern ‖ transformation f. de pression en haut-parleurs. / ~ transformer ‖ Spannungstransformator m. ‖ transformateur m. de tension. / speed of ~ transmission ‖ Druckübertragungsgeschwindigkeit f. ‖ vitesse f. de transmission de la pression. / proportion of ~ transmittance ‖ Druckübersetzungsverhältnis n. ‖ rapport m. de multiplication de la pression. / ~ tube joint ‖ Druckstutzen m. ‖ culotte f. / ~ vacuum gauge ‖ Vakuummanometer n. ‖ manovacuomètre m. / ~ vessel ‖ Druckgefäß n. ‖ vase m. de pression. / ~ water plant ‖ Druckwasseranlage f. ‖ installation f. d'eau sous pression. / ~ wave ‖ Druckwelle f. / onde f. de pression. / ~ welding ‖ Preßschweißung f. ‖ soudage m. à pression.
press water conduit ‖ Druckwasserleitung f. ‖ conduite f. d'eau sous pression. / ~ work (Print) ‖ Druckarbeit f. ‖ ouvrage m. fait à la presse. / ~ working (Porcel) ‖ Preßformerei f. ‖ moulage m. à la presse. / ~ yeast ‖ Preßhefe f. ‖ levure f. sèche. ‖ ~ yeast works pl. ‖ Preßhefefabrik f. ‖ fabrique f. de levure comprimée.
presumed assets pl. ‖ Sollbestand m. ‖ effectif m. théorique.
presumptive loss (Mar) ‖ Verschollenheit f. ‖ présomption f. de perte. / period of ~ (Mar) ‖ Verschollenheitsfrist f. ‖ délai m. de présomption de perte.
pretence ‖ Ausrede f. ‖ excuse f.; fauxfuyant m. / on false ~s pl. ‖ unter Vorspiegelung f. falscher Tatsachen ‖ débiter des mensonges m.; sous faux-fuyant m. / under the ~ ‖ unter dem Vorwande m. ‖ sous prétexte m.
pretentious ‖ anspruchsvoll ‖ exigeant; prétentieux.
pretext see pretence.
prevailing ‖ üblich ‖ usuel; d'usage; usité. / ~ wind ‖ vorherrschender Wind m. ‖ vent m. dominant. / ~ wind direction ‖ vorherrschende Windrichtung f. ‖ direction f. dominante du vent.
prevalent ‖ weitverbreitet ‖ très répandu dans le public. / ~ wind see prevailing wind.
prevent, to ‖ verhüten; verhindern ‖ empêcher; préserver de. / ~ the air from entering ‖ das Eindringen n. von Luft verhindern ‖ empêcher l'air m. de pénétrer. / ~ fermentation ‖ den Gärungsprozeß m. vermeiden ‖ empêcher la fermentation.
prevention of accidents ‖ Unfallverhütung f. ‖ mesures fpl. préventives contre les accidents. / ~ of freezing in of manhole covers ‖ Schutz m. gegen Einfrieren von Brunnendeckeln ‖ protection f. contre la gelée des couvercles de puits.
preventive ‖ Schutzmittel n. ‖ préservatif m. / ~ against accidents ‖ Unfallverhütungsvorrichtung f. ‖ dispositif-protecteur m. contre les accidents. / ~ measure ‖

Schutzmaßnahme f. ‖ mesure f. de protection.

previous raising device (Weav) ‖ Vorrauhapparat m. ‖ dispositif m. d'avant-lainage. / ~ testing ‖ Vorprüfung f. ‖ essai m. préalable.

price ‖ Preis m. ‖ prix m. / ~ (Exchange) ‖ Kurs m. ‖ cours m.; change m. /~s pl. are advancing ‖ die Preise mpl. steigen ‖ les prix mpl. sont à la hausse. / average ~ ‖ Durchschnittspreis m. ‖ prix m. moyen. / buying ~ ‖ Kaufpreis m. ‖ prix m. d'achat. / cash ~ ‖ Preis m. bei Barzahlung ‖ prix m. au comptant. / current ~ ‖ üblicher Preis m. ‖ prix m. courant. / to cut ~s pl. ‖ schleudern ‖ vendre à vil prix; gâcher la marchandise. / ~s pl. are declining ‖ die Preise mpl. fallen ‖ les prix mpl. sont en baisse. / ~s pl. will be quoted on demand ‖ Preise mpl. werden auf Anfrage mitgeteilt ‖ nous envoyons des devis sur demande mpl. / ~ is subject to discount of ‖ der Preis m. unterliegt einem Rabatt von ‖ le prix m. est sujet à un escompte de. / fair ~ ‖ annehmbarer Preis m. ‖ prix m. raisonnable. / the ~s pl. keep firm (Exchange) ‖ die Kurse mpl. halten sich ‖ les valeurs fpl. se maintiennent. / fixed ~ ‖ fester Preis m. ‖ prix m. fixe. / former ~ ‖ früherer Preis m. ‖ prix m. antérieur. / to force ~s pl. ‖ die Preise mpl. in die Höhe treiben ‖ faire hausser les prix mpl. / the ~ depends on the goods ‖ der Preis richtet sich nach der Ware ‖ le prix se règle sur la marchandise. / the ~ is high ‖ der Preis stellt sich hoch ‖ le prix est élevé. / the ~ rules high see the ~ is high. / highest possible ~ ‖ höchstmöglicher Preis m. ‖ prix m. le plus élevé possible. / ~ must include ‖ der Preis m. muß einschließen ‖ le prix m. doit comprendre. / invoice ~ ‖ Fakturapreis m. ‖ prix m. de facture. / list ~ ‖ Katalogpreis m.; Listenpreis m. ‖ prix m. du catalogue. / the lowest ~ ‖ der äußerste Preis m. ‖ le dernier prix. / lump ~ ‖ Pauschalpreis m. ‖ prix m. global. / the ~s pl. maintain their level see the ~s pl. keep firm. / making-up ~ ‖ Liquidationspreis m. ‖ prix m. de liquidation. / market ~ ‖ Marktpreis m. ‖ prix m. du marché. / at a maximum ~ of ‖ zu einem Maximalpreis m. von ‖ à un prix maximum de. / middle ~ ‖ Mittelkurs m. ‖ cours m. moyen. / at a minimum ~ of ‖ zu einem Minimalpreis m. von ‖ à un prix minimum de. / net ~ ‖ Nettopreis m. ‖ prix m. net. / ~ is strictly net ‖ der Preis ist absolut netto ‖ le prix m. est strictement net. / ~ leaves no profit ‖ der Preis läßt keinen Nutzen ‖ le prix ne laisse pas de bénéfice. / ~ must not exceed ‖ der Preis darf nicht übersteigen ‖ le prix ne doit pas excéder. / ~ does not include ‖ der Preis schließt nicht ein ‖ le prix ne comprend pas. / opening ~ ‖ Eröffnungskurs m.; Anfangskurs m. ‖ premier cours m. / to reach a ~ ‖ einen Preis m. erzielen ‖ atteindre un prix. / reasonable ~ ‖ annehmbarer Preis m. ‖ prix m. raisonnable. / ~ has risen to ‖ der Preis ist gestiegen auf ‖ le prix s'est élevé à. / to send up ~s pl. ‖ die Preise mpl. in die Höhe treiben ‖ faire hausser les prix mpl. / selling ~ ‖ Verkaufspreis m. ‖ prix m. de vente. / the ~ varied widely ‖ der Preis

schwankte auf und nieder ‖ le prix avait des soubresauts. / ~ ex warehouse ‖ Preis m. ab Lager ‖ prix m. ex magasin.

price, advance in ~ of ... ‖ Preiserhöhung f. von ... ‖ hausse f. de prix de ... / change in ~ ‖ Preisänderung f. ‖ changement m. de prix. / ~ board ‖ Preisschild n. ‖ écriteau m. de prix. / ~ board for shop windows ‖ Schaufensterpreisschild n. ‖ indicateur m. de prix pour étalages. / ~ current on change ‖ Börsenkurs m. ‖ cours m. de la bourse. / ~ limit ‖ Preisgrenze f. ‖ limites fpl. de prix. / ~ list ‖ Preisliste f.; Preisverzeichnis n. ‖ prix-courant m. / latest ~ list ‖ neueste Preisliste f. ‖ dernier prix-courant m. / ~ reduction ‖ Preisermäßigung f. ‖ réduction f. de prix. / ~ reduction of ~s ‖ Preisabbau m. ‖ réduction f. de prix. / rise in ~s (Exchange) ‖ Hausse f. ‖ hausse f. / upswing of ~s ‖ Haussebewegung f. ‖ mouvement m. de hausse.

prick, to ‖ durchstechen ‖ priquer. / ~ the chart (Nav) ‖ die Karte pricken ‖ pointer la carte. / ~ the coal ‖ das Feuer anschüren ‖ activer ou aviver le feu.

pricked by worms ‖ wurmstichig ‖ vermoulu. / to get ~ ‖ wurmstichig werden ‖ se vermouler.

pricker (Build) ‖ Sondiereisen n. ‖ tige f. de sondage. / ~ (Mine) ‖ Schießnadel f.; Räumnadel f. ‖ épinglette f. / ~ (Saddl) ‖ Ahle f.; Pfriem m. ‖ alène f.

pricking machine (Weav) ‖ Schablonenstechmaschine f.; Tüpfelmaschine f. ‖ machine f. à piquer les cartons.

prima (Print) ‖ Prima f. ‖ premier folio m.

primary ‖ primär ‖ primaire. / ~ alternating current ‖ Primärwechselstrom m. ‖ courant m. alternatif primaire. / ~ amplifier (Radio) ‖ Grundverstärker m. ‖ amplificateur m. d'entrée. / ~ battery ‖ Primärbatterie f. ‖ batterie f. primaire. / ~ cell (Electr) ‖ Primärelement n. ‖ élément m. primaire; pile f. électrique. / ~ circuit ‖ Primärstromkreis m. ‖ circuit m. primaire. / ~ coil ‖ Primärspule f. ‖ bobine f. ou enroulement m. primaire. / ~ current ‖ Primärstrom m. ‖ courant m. primaire. / ~ inductance ‖ Primärinduktanz f. ‖ inductance f. primaire. / ~ jigger ‖ Primärjigger m. ‖ primaire m. de transformateur d'oscillations; jigger m. primaire. / ~ machine ‖ Hauptmaschine f. ‖ machine f. principale. / ~ pressure (Electr) ‖ Primärspannung f. ‖ tension f. primaire. / ~ share ‖ Prioritätsaktie f. ‖ action f. de priorité. / ~ still ‖ Vordestiller m. ‖ appareil m. de première distillation ou de distillation préalable; appareil m. prédistillateur. / ~ switching ‖ Primärschaltung f. ‖ circuit m. primaire. / ~ winding ‖ Primärwicklung f. ‖ enroulement m. primaire.

prima sheet (Print) ‖ Primablatt n. ‖ feuille f. première.

prime, to (Paint) ‖ grundieren ‖ abreuver; apprêter; donner l'apprêt; imprimer.

prime colour ‖ Grundfarbe f. ‖ couleur f. de fond. / ~ cost ‖ Selbstkostenpreis m. ‖ prix m. de revient ou coûtant. / ~ number ‖ Primzahl f. ‖ nombre m. premier.

primer, carburettor ~ ‖ Tipper m. des Vergasers ‖ poussoir m. de carburateur. / great ~ (Print) ‖ Tertia f. ‖ gros romain m. / two lines great ~ (Print) ‖ Kanonschrift f. ‖ deux points de gros-romain m.

priming (Paint) ‖ Grundierung f. ‖ imprimure f.; première couche f. / ~ (Pyrot) ‖ Zündsatz m.; Zündmasse f. ‖ composition f. fulminante; matière f. fulminante. / colour ~ ‖ Grundfarbe f. ‖ première couche f. de couleur. / ~ in oil ‖ erster Ölanstrich m. ‖ imbu m. / ~ of a shell ‖ Geschoßzündung f. ‖ amorce f. d'un projectile.

priming cartridge ‖ Zündpatrone f. ‖ cartouche f. d'amorce. / ~ cock ‖ Kompressionshahn m.; Zischhahn m. ‖ robinet m. de compression. / ~ pump ‖ Einspritzpumpe f. ‖ pompe f. d'injection. / ~ valve ‖ Sicherheitsventil n. ‖ soupape f. de sûreté.

principal ‖ hauptsächlich ‖ principal; essentiel. / ~ airway beacon ‖ Hauptfeuer n. ‖ phare m. principal. / ~ axis ‖ Hauptachse f. ‖ axe m. principal. / ~ beam (Carp) ‖ Hauptbalken m.; Binderbalken m. ‖ maîtresse f. poutre. / ~ circuit ‖ Grundschaltung f. ‖ schéma m. de principe. / ~ eduction canal ‖ Hauptabflußgraben m. ‖ canal m. principal de colature. / ~ front ‖ Hauptfront f. ‖ façade f. principale. / ~ perspective ‖ Hauptperspektive f. ‖ perspective f. principale. / ~ rafter ‖ Hauptsparren m. ‖ maître-chevron m. / ~ railway ‖ Hauptbahn f. ‖ ligne f. principale. / ~ ray ‖ Hauptstrahl m. ‖ rayon m. principal. / ~ section ‖ Hauptschnitt m. ‖ section f. principale. / ~ warehouse ‖ Hauptniederlage f. ‖ dépôt m. central. / ~ wind ‖ Hauptwind m. ‖ vent m. dominant. / ~ workshop ‖ Hauptwerkstatt f.; Zentralwerkstatt f. ‖ atelier m. principal.

principle ‖ Prinzip n. ‖ principe m. / ~ of action and reaction ‖ Prinzip n. der Wirkung und Gegenwirkung ‖ principe m. de l'action et de la réaction. / ~ of construction ‖ Konstruktionsprinzip n. ‖ principe m. de construction. / to lay down as ~ ‖ als Grundsatz m. aufstellen ‖ poser en principe m. / mechanical ~ ‖ mechanisches Prinzip n. ‖ principe m. de mécanique. / ~ of virtual velocities ‖ Prinzip n. der virtuellen Geschwindigkeiten ‖ principe m. des vitesses virtuelles.

print, to ‖ drucken; abdrucken ‖ imprimer; tirer. / ~ waste ‖ makulieren; fehldrucken ‖ faire de la maculature.

print ‖ Abdruck m.; Druck m. ‖ impression f.; empreinte f. / ~ (Engr) ‖ Druckplatte f. ‖ estampe f.; gravure f. / ~ (Forg) ‖ Gesenk n. ‖ estampe f.; étampe f. / ~ (Found) ‖ Kernlager n.; Kernauge n. ‖ logement m. du noyau. / blue ~ ‖ Blaupause f. ‖ bleu m.; photocopie f. bleue. / ~ in brimstone ‖ Schwefelabdruck m.; Schwefelpaste f. ‖ empreinte f. ou ectype f. en soufre. / ~ with concealed letters ‖ Abdruck m. mit überlegter Schrift ‖ cache-lettre m.; épreuve f. avec la lettre supprimée. / engraving-like ~ ‖ stichähnlicher Druck m. ‖ simili-gravure f. / ~ with erased letters ‖ Abdruck m. mit ausradierter Schrift ‖ épreuve f. avec la lettre grattée. / lithographic ~ ‖ Stein(ab)druck m. ‖ gravure f. lithographique. / photographic ~ ‖ fotografischer Abdruck m. oder Abzug m. ‖ planche f. phototypique. / white ~ ‖ Weißpause f. ‖ blanc m.

print cutter ‖ Formschneider m. ‖ metteur m. sur bois.

printed glass ‖ Druckglas n. ‖ verre m. imprimé. / ~ matter ‖ Drucksache f. ‖ imprimé m. / we beg leave to send you some ~ matter herewith ‖ wie übersenden anbei einige Drucksachen fpl. ‖ permettez-nous de joindre à la présente quelques imprimés mpl. / ~ sheet ‖ Druckbogen m. ‖ feuille f. d'impression.

printer ‖ Drucker m.; Buchdrucker m. ‖ imprimeur m. / ~ (Printing device) ‖ Druckvorrichtung f. ‖ mécanisme m. d'impression. / copper-plate ~ ‖ Kupferdrucker m. ‖ imprimeur m. en taille-douce. / journeyman ~ ‖ Buchdruckergehilfe m. ‖ compagnon m. imprimeur. / piratical ~ ‖ Nachdrucker m. ‖ contrefacteur m. / two-colour ~ ‖ Zweifarbendrucker m. ‖ imprimeur m. bicolore. / ~ and typesetter ‖ Druck- und Setzmaschine f. ‖ machine f. à imprimer combinée avec machine à composer.

printer magnet (Tel) ‖ Druckmagnet m. ‖ électro m. imprimeur.

printer's flower ‖ Randverzierung f. ‖ bordure f.; vignette f. / ~ machine ‖ Buchgewerbemaschine f. ‖ machine f. pour l'art graphique. / ~ peel ‖ Aufhängekreuz n. ‖ étendoir m. / ~ supply ‖ Buchdruckgerät n. ‖ ustensiles mpl. d'imprimerie. / ~ type ‖ Drucktype f. ‖ caractère m. d'imprimerie. / ~ varnish ‖ Buchdruckfirnis m. ‖ vernis m. typographique f. ou d'imprimerie.

printing ‖ Druck m.; Drucken n. ‖ impression f. / ~ from aluminium plates ‖ Aluminiumdruck m. ‖ aluminographie f. / ~ on back ‖ Widerdruck m. ‖ impression f. au verso; réimpression f.; retiration f. / ~ in black ‖ Schwarzdruck m. ‖ impression f. en noir. / ~ booklets pl. ‖ Broschürendruck m. ‖ impression f. de brochures. / chromatic ~ see ~ in colours. / cloth ~ ‖ Zeugdruck m. ‖ impression f. sur étoffes. / machine ~ on cloths ‖ mechanischer oder maschineller Zeugdruck m. ‖ impression f. mécanique sur étoffes. / ~ in colours ‖ Buntdruck m.; Farbendruck ‖ chromotypie f.; chromotypographie f.; impression f. en (plusieurs) couleurs. / copper-plate ~ ‖ Kupferdruck m.; Tiefdruck m. ‖ impression f. en taille-douce. / ~ cut work ‖ Illustrationsdruck m. ‖ impression f. d'illustrations. / cylinder ~ ‖ Walzendruck m. ‖ impression f. au rouleau. / embossed ~ for blinds ‖ Blindendruck m. ‖ impression f. en relief pour aveugles. / ~ on enamelled metal ‖ keramischer Druck m. ‖ impression f. sur céramique. / ~ from an etched surface ‖ Ätzdruck m. ‖ première épreuve f. d'une étampe. / flat ~ ‖ Flachdruck m. ‖ impression f. à plat ou à forme plate. / ~ with four colours ‖ Vierfarbendruck m. ‖ impression f. en quatre couleurs. / ~ of half-tone engravings ‖ Autotypiedruck m. ‖ impression f. de similigravure. / hand ~ ‖ Handdruck m. ‖ impression f. à la main; genre m. à la main. / heliographic ~ ‖ Lichtfarbendruck m. ‖ impression m. héliographique. / label ~ ‖ Etikettdruck m. ‖ impression f. des étiquettes. / letter press ~ ‖ Typendruck m. ‖ impression f. typographique. ‖ ~ by machine ‖ Maschinendruck m. ‖ impression f. à la machine. / multicolour ~ ‖ Mehrfarbendruck m. ‖ impression f. en plusieurs couleurs. / device for ~ numbers ‖ Zahlendruckeinrichtung f. ‖ dis-

positif m. pour l'impression numérique. / page ~ ‖ Abdruck m. auf Blätter ‖ impression f. sur feuilles. / ~ of periodicals pl. ‖ Zeitschriftendruck m. ‖ impression f. de périodiques. / plain ~ see flat ~. / plate ~ see also copper-plate ~ ‖ Tiefdruck m.; Kupferdruck m. ‖ impression f. en creux. / ~ receipt on sale slips ‖ Quittungsdruck m. ‖ impression f. sur fiche. / in relief ‖ Reliefdruck m. ‖ impression f. en relief. / seven-colour ~ ‖ Siebenfarbendruck m. ‖ impression f. en sept couleurs. / stereotype ~ ‖ Plattendruck m. ‖ stéréotypage m.; stéréotypie f. / three-colour ~ ‖ Dreifarbendruck m. ‖ trichromie f. / rotation machine for ~ tickets (Railw) ‖ Fahrkartenrotationsdruckmaschine f. ‖ machine f. rotative à imprimer de billets. / tin plate ~ high speed machine ‖ Blechdruckschnellpresse f. ‖ presse f. rapide pour l'impression sur tôles. / transfer ~ ‖ Abziehbilderdruck m. ‖ décalcomanie f. / two-colour ~ ‖ Zweifarbendruck m. ‖ impression f. en deux couleurs. / ~ of the warp (Text print) ‖ Kettendruck m. ‖ impression f. des chaînes ou sur chaîne. / ~ of wool fabrics ‖ Wolldruck m. ‖ impression f. des tissus de laine. / zink ~ high speed machine ‖ Zinkdruckschnellpresse f. ‖ presse f. rapide pour zincographie.

printing accessories pl. ‖ Buchdruckereizubehör n. ‖ accessoires mpl. d'imprimerie. / ~ apparatus ‖ Druckvorrichtung f. ‖ appareil m. à imprimer. / art of ~ ‖ Buchkunst f. ‖ art m. du livre. / ~ block ‖ Buchdruckklischee n. ‖ cliché m. d'impression. / ~ box for children ‖ Kinderdruckerei f. ‖ boîte f. d'imprimerie pour enfants. / ~ cam ‖ Druckdaumen m. ‖ came f. d'imprimerie. / ~ character ‖ Druckereiletter f. ‖ caractère m. d'imprimerie. / ~ cloth ‖ Farbtuch n. ‖ toile f. d'encrage. / factory for ~ colours ‖ Druckfarbenfabrik f. ‖ fabrique f. de couleurs d'impression. / ~ cylinder ‖ Auftragwalze f. ‖ cylindre m. à imprimer. / ~ device for printing numbers ‖ Zahlendruckeinrichtung f. ‖ dispositif m. pour l'impression numérique. / ~ disk (Tel) ‖ Schreibrädchen n.; Schreibscheibe f. ‖ molette f. / ~ expenses pl. ‖ Druckkosten pl. ‖ frais mpl. d'impression. / ~ fly press ‖ Buchdruckschnellpresse f. ‖ presse f. d'imprimerie rapide. / ~ hammer (Typewr) ‖ Typenhammer m. ‖ marteau m. à caractères. / ~ house ‖ Buchdruckerei f. ‖ imprimerie f. / ~ implements pl. ‖ Buchdruckgerät n. ‖ ustensiles mpl. d'imprimerie. / ~ ink ‖ Druckerschwärze f.; Druckfarbe f. ‖ encre f. d'imprimerie ou typographique. / ~ and lithographic inks pl. ‖ Farben fpl. für Buch- und Steindruck ‖ encres fpl. d'imprimerie et lithographiques. / ~ letter ‖ Buchdruckletter f. ‖ caractère m. d'imprimerie. / ~ letter of lead ‖ Buchdruckletter f. aus Blei ‖ caractère m. d'imprimerie en plomb.

printing machine ‖ Schnellpresse f.; Druckmaschine f.; Buchdruckermaschine ‖ machine f. à imprimer ou d'imprimerie ou d'impression. / blue ~ ‖ Blaupausemaschine f. ‖ machine f. à tirer les bleus. / auxiliary ~ for chemitypy ‖ Druckhilfsmaschine f. für Chemigraphie ‖ machine f. auxiliaire d'impression pour

chimigraphie. / block ~ (Text print) ‖ Modelldruckmaschine f. ‖ machine f. à planche. / colour ~ ‖ Farbendruckmaschine f. ‖ machine f. pour l'impression en couleurs. / copper-plate ~ ‖ Kupferdruckmaschine f.; Kupferdruckschnellpresse f. ‖ presse f. d'imprimeur en taille-douce. / cylinder ~ ‖ Walzendruckmaschine f. ‖ machine f. à imprimer au rouleau. / flat-bed ~ for job work ‖ Akzidenzschnellpresse f. ‖ machine f. à imprimer en blanc. / hand ~ ‖ Handdruckmaschine f. ‖ imprimeuse f. à main. / high-speed ~ see printing machine. / ~ for letterpress work ‖ Schnellpresse f. für Buchdruck ‖ presse f. rapide pour imprimerie des livres. / auxiliary ~ for lithography ‖ Druckhilfsmaschine f. für Steindruck ‖ machine f. auxiliaire d'impression pour lithographie. / metal ~ ‖ Metalldruckmaschine f. ‖ machine f. pour la métallographie. / multi-colour rotary ~ ‖ Mehrfarbenrotationsdruckmaschine f. ‖ machine f. rotative pour illustrations en plusieurs couleurs. / office ~ ‖ Bürodruckmaschine f. ‖ machine f. de bureau pour l'impression. / offset rotary ~ ‖ Offsetrotationsdruckmaschine f. ‖ machine-offset f. rotative. / ~ for pasteboard tickets ‖ Fahrkartendruckmaschine f. für steife Karten ‖ machine f. à imprimer les billets en carton. / plate ~ ‖ Tiefdruckmaschine f. ‖ machine f. à imprimer en creux. / platen ~ ‖ Flachdruckmaschine f.; Tiegeldruckmaschine f.; Tiegeldruckpresse f. ‖ machine f. d'impression) à platine ou à forme plate. / rotary ~ ‖ Rotations(druck)maschine f. ‖ machine f. (à imprimer) rotative. / rotary newspaper ~ ‖ Zeitungsrotationsmaschine f. ‖ machine f. à rotation pour imprimer les journaux. / auxiliary ~ for stereotyping ‖ Druckhilfsmaschine f. für Stereotypie ‖ machine f. auxiliaire d'impression pour stéréotypie. / ~ for tin goods ‖ Blechdruckmaschine f. ‖ machine f. à imprimer articles en tôle. / types pl. of metal for ~s ‖ Metalltypen fpl. für Druckmaschinen ‖ lettres fpl. en métal pour machines à imprimer.

printing machine minder ‖ Maschinendrucker m. ‖ conducteur m. typographe.

printing office ‖ Druckerei f. ‖ imprimerie f. / art ~ ‖ Kunstdruckerei f. ‖ imprimerie f. d'art. / lithographic ~ ‖ Steindruckerei f. ‖ atelier m. de lithographie; imprimerie f. lithographique. / music ~ ‖ Notendruckerei f. ‖ imprimerie f. de musique.

printing paper ‖ Druckpapier n. ‖ papier m. à imprimer ou d'impression ou sans colle. / ~ (Phot) ‖ Lichtpauspapier n. ‖ papier m. héliographique ou photocalque. / copper-plate ~ ‖ Kupferdruckpapier n. ‖ papier m. pour impression en taille-douce. / thin ~ ‖ Dünndruckpapier n. ‖ papier m. pour impression délicate. / roll of ~ ‖ Rolle f. Druckpapier ‖ rouleau m. de papier d'impression.

printing press ‖ Druckerpresse f. ‖ presse f. typographique ou à imprimer. / copperplate ~ ‖ Kupferdruckpresse f. ‖ presse f. pour taille-douce. / high-speed ~ see also printing machine ‖ Schnellpresse f.; Druckmaschine f. ‖ presse f. (à la) mécanique; presse f. (à imprimer) rapide; machine f. à imprimer ou d'impression ou d'imprimerie. / offset ~ ‖ Offsetdruckpresse f. ‖ presse f. typographique offset.

/ platen ~ ‖ Tiegeldruckpresse f. ‖ presse f. d'impression à platine. / rotary ~ ‖ Rotations(druck)presse f. ‖ presse f. (à imprimer) rotative.

printing process ‖ Druckverfahren n. ‖ procédé m. d'impression. / ~ roller ‖ Farbenwalze f.; Auftragewalze f. ‖ rouleau m.; toucheur m. / ~ roller's composition ‖ Buchdruckwalzenmasse f. ‖ pâte f. à rouleaux. / ~ stamp for wood printing ‖ Druckstempel m. für Holzdruck ‖ timbre m. xylographique ou à imprimer sur bois. / ~ table ‖ Drucktisch m. ‖ table f. de presse. / ~ telegraph ‖ Drucktelegraf m. ‖ télégraphe m. imprimeur; typotélégraphe m. / ~ trade ‖ Druckgewerbe n. ‖ typographie f. / ~ trade worker ‖ Druckereiarbeiter m. ‖ ouvrier m. typographique. / ~ type ‖ Letter f. ‖ lettre f. ou caractère d'imprimerie. / ~ type which imitates written letters ‖ Schreibschrift f. ‖ caractère m. d'écriture. / ~ varnish ‖ Buchdruckfirnis m. ‖ vernis m. typographique. / ~ works pl. ‖ typographische Anstalt f. ‖ atelier m. de typographie.

priority ‖ Priorität f.; Vorrang m. ‖ priorité f.; obligation f.; préséance f. / ~ document ‖ Prioritätsbeleg m. ‖ document m. de priorité.

prior right, to grant a ~ ‖ ein Vorzugsrecht n. einräumen ‖ accorder un droit de préférence. / ~ of purchase ‖ Vorkaufsrecht n. ‖ droit m. de préemption.

prism ‖ Prisma n. ‖ prisme m. / comparison ~ ‖ Vergleichsprisma n. ‖ prisme m. de comparaison. / direct-vision ~ ‖ geradsichtiges Prisma n. ‖ prisme m. à vision directe. / double ~ ‖ Doppelprisma n. ‖ double prisme m. / double ~ for angles of 90⁰ and 180⁰ ‖ Prismenkreuz n. ‖ équerre f. à prismes. / drawing ~ ‖ Zeichenprisma n. ‖ prisme m. à dessiner. / erecting ~ ‖ bildumkehrendes Prisma n. ‖ prisme m. redresseur. / set of erecting ~s ‖ Prismenumkehrsatz m. ‖ système m. de prismes redresseurs. / set of erecting ~s with reflecting surfaces ‖ Prismenumkehrsatz m. mit Reflektionsflächen ‖ système m. de prismes redresseurs à surfaces réfléchissants. / eyepiece ~ ‖ Okularprisma n. ‖ prisme m. oculaire. / hexagonal ~ ‖ sechsseitiges Prisma n. ‖ prisme m. hexagonal. / ~ of high dispersion ‖ Prisma n. von hoher Dispersion ‖ prisme m. de grande dispersion. / ~ of light flint ‖ Leichtflintprisma n. ‖ prisme m. en flint léger. / objective ~ ‖ Objektivprisma n. ‖ prisme-objectif m. / oblique ~ ‖ schiefes Prisma n. ‖ prisme m. oblique. / pentagonal ~ ‖ fünfseitiges Prisma n. ‖ prisme m. pentagonal. / rectangular ~ see prismatic square. / reflecting ~ ‖ Spiegelprisma n. ‖ prisme m. réflecteur. / rhombic ~ ‖ rhombisches Prisma n. ‖ prisme m. rhombique. / right ~ ‖ gerades Prisma n. ‖ prisme m. droit. / square ~ ‖ vierseitiges Prisma n. ‖ prisme m. quadrangulaire. / triangular ~ ‖ dreiseitiges Prisma n. ‖ prisme m. triangulaire. / triple ~ ‖ dreiteiliges Prisma n. ‖ prisme m. à trois verres.

prism apparatus ‖ Prismengerät n. ‖ appareil m. à prismes.

prismatic ‖ prismatisch ‖ prismatique. / ~ arseniate of copper ‖ Pharmakochalzit m.; Olivenit m. ‖ cuivre m. arséniaté en octaèdres aigus. / ~ deviation ‖ prisma-

tische Ablenkung f. ‖ déviation f. prismatique. / ~ glass ‖ Prismenglas n. ‖ verre m. prismatique. / ~ objective spectrograph ‖ Objektivprismaspektrograf m. ‖ spectrographe m. à prisme-objectif. / ~ shape ‖ Prismengestalt f. ‖ forme f. prismatique. / ~ spectacles pl. ‖ prismatische Brille f. ‖ lunettes fpl. à verres prismatiques. / ~ square ‖ Winkelprisma n. ‖ équerre f. à prisme.

prismatic-cylindrical ‖ prismatisch zylindrisch ‖ prismatique-cylindrique.

prism binocular ‖ Prismenfeldstecher m. ‖ jumelle f. à prismes. / ~ bridge (Acc) ‖ Prismensteg m. ‖ sommet m. de prisme. / ~ casing ‖ Prismenfassung f. ‖ monture f. du prisme. / exceptionally clear type ~ crown glass ‖ besonders klares Prismenkronglas n. ‖ verre m. Crown à prismes parfaitement transparent. / ~ microscope ‖ Prismenmikroskop n. ‖ microscope m. à prismes. / ~ rotator ‖ Prismenrotator m. ‖ rotateur m. à prismes. / ~ spectroscope ‖ Prismenspektroskop n. ‖ spectroscope m. à prismes. / ~ stage ‖ Prismentisch m. ‖ tablette f. porte-prisme. / ~ telescope ‖ Prismenfernrohr n. ‖ longue-vue f. à prismes.

prisoner ‖ Angeklagter m.; Gefangener ‖ incriminé m.; prisonnier m.

pritchel ‖ Hufeisendorn m.; Lochdorn m. ‖ poinçon m.

private automatic branch exchange ‖ private Selbstanschlußanlage f. ‖ installation f. automatique privée. / ~ automatic exchange ‖ Selbstanschlußnebenstellenanlage f. ‖ installation f. supplémentaire automatique. / ~ bank ‖ Privatbank f. ‖ banque f. privée. / ~ branch exchange ‖ Fernsprechnebenstellenanlage f. ‖ installation f. téléphonique supplémentaire. / ~ call ‖ Privatgespräch n. ‖ communication f. privée. / ~ company ‖ Privatgesellschaft f. ‖ société f. privée. / to exploit something for ~ economic purposes ‖ etwas privatwirtschaftlich ausbeuten ‖ exploiter quelque chose au titre d'entreprise privée. / ~ ledger ‖ Bilanzbuch n. ‖ livre m. du bilan. / ~ property ‖ Privatvermögen n. ‖ fortune f. privée. / ~ service ‖ Privatbetrieb m. ‖ exploitation f. par entreprise privée. / ~ telegraph plant ‖ Privattelegrafenanlage f. ‖ installation f. télégraphique privée. / ~ telephone branch exchange ‖ Privatfernsprechanlage f. ‖ installation f. téléphonique privée. / ~ telephone plant ‖ Privatfernsprechnetz n. ‖ réseau m. téléphonique privé. / ~ wire (Tel) ‖ c-Ader f. ‖ fil m. de rest.

privilege ‖ Vorrecht n.; Privileg n.; Sonderrecht n. ‖ privilège m. / ~ of the author ‖ Urheberrecht n. ‖ droit m. d'auteur.

privileged work (Tel) ‖ bevorrechtigte Anlage f. ‖ installation f. privilégiée.

privy tub ‖ Latrinentonne f. ‖ tonneau m. pour fosses d'aisance.

prize ‖ Preis m.; Prämie f. ‖ prix m. / ~ (Mar) ‖ Prise f. ‖ prise f. / ~ court (Mar) ‖ Prisengericht n. ‖ cour f. ou tribunal m. des prises. / ~ essay ‖ Preisschrift f. ‖ mémoire m. couronné. / ~ money (Mar) ‖ Prisengeld n. ‖ part m. de prise.

probability ‖ Wahrscheinlichkeit f. ‖ probabilité f. / theory of ~ ‖ Wahrschein-

lichkeitsrechnung f. ‖ calcul m. des probabilités.

probable ‖ wahrscheinlich; mutmaßlich ‖ probable. / ~ error in measuring ‖ wahrscheinlicher Fehler m. der Messung ‖ erreur f. probable de mesure.

probation ‖ Probe f.; Prüfung f. ‖ épreuve f. / time of ~ ‖ Probezeit f. ‖ stage m.

probe (Med) ‖ Sonde f. ‖ sonde f.

probe-ended knife (Med) ‖ geknöpftes Messer n. ‖ bistouri m. boutonné.

problem ‖ Aufgabe f. ‖ tâche f.

procedure ‖ Arbeitsweise f. ‖ mode m. opératoire; méthode f. de travail. / ~ (Law) ‖ Rechtsweg m. ‖ procédure f. civile.

proceeding (Chem) see process. / ~ (Law) ‖ Verfahren n.; Prozeß m. ‖ démarche f.; procès m. / to abandon the ~ ‖ das Verfahren einstellen ‖ abandonner la démarche. / to take legal ~ ‖ einen Prozeß m. anstrengen ‖ intenter un procès.

proceeds pl. ‖ Erlös m. ‖ produit m. / entire ~ ‖ Gesamtertrag m. ‖ produit m. ou revenu m. total. / gross ~ ‖ Bruttoertrag m. ‖ produit m. brut. / net ~ ‖ Nettoertrag m. ‖ produit m. net.

process (Chem; Met) ‖ Prozeß m.; Methode f.; Verfahren n. ‖ méthode f.; opération f.; procédé m. / ~ (Transaction) ‖ Vorgang m. ‖ marche f.; processus m.; procédé m.; phénomène m.; réaction f. / acid ~ (Met) ‖ saures Verfahren m. ‖ procédé m. acide. / basic ~ (Met) ‖ basisches Verfahren n. ‖ procédé m. basique. / ~ of coking ‖ Verkokungsvorgang m. ‖ marche f. de la carbonisation. / ~ of combustion ‖ Verbrennungsvorgang m. ‖ marche f. de la combustion. / ~ of conversion ‖ Umwandlungsvorgang m. ‖ processus m. de conversion; phénomène m. de transformation. / ~ of extraction ‖ Gewinnungsverfahren n. ‖ procédé m. d'extraction. / ~ of fermentation ‖ Gärverfahren n. ‖ procédé m. de fermentation. / ~ of fuel conversion ‖ Brennstoffumwandlungsverfahren n. ‖ procédé m. de transformation de combustibles. / induction ~ ‖ Induktionsverfahren n. ‖ procédé m. d'induction. / legal ~ ‖ Prozeß m. ‖ procès m. / ~ for the production of liquid fuels from coal ‖ Verfahren n. zur Gewinnung von flüssigen Brennstoffen aus Kohle ‖ méthode f. pour extraire de combustibles liquides du charbon. / ~ of manufacture ‖ Herstellungsverfahren n. ‖ procédé m. de fabrication. / ~ of printing ‖ Druckverfahren n. ‖ procédé m. d'impression. / ~ of production see ~ of manufacture. / ~ of separation ‖ Trennungsvorgang m. ‖ marche f. de séparation. / ~ of setting ‖ Erhärtungsvorgang m. ‖ processus m. de la prise. / wet ~ ‖ nasses Verfahren n. ‖ procédé m. par voie humide.

proclaim, to ‖ verkündigen ‖ annoncer; publier; prononcer.

proctor ‖ Prokurator m.; Faktor m.; Geschäftsverwalter m.; Sach(ver)walter m. ‖ procureur m.; gérant m.

procuration ‖ Prokura f. ‖ procuration f. / ~ (Fee) see ~ money. / joint ~ ‖ Gesamtprokura f. ‖ procuration f. collective. / per ~ ‖ in Stellvertretung f. ‖ par procuration f. / to sign per ~ ‖ in Stellvertretung f. oder per Prokura f. zeichnen ‖ signer ou souscrire par procuration f.

procuration fee see ~ money. / ~ money ‖ Provision f.; Maklergebühr f. ‖ provision f.

procurator *see* proctor.

procure, to ‖ anschaffen ‖ acheter; faire l'acquisition f. de.

produce, to ‖ erzeugen ‖ produire. / ~ (Mine) ‖ gewinnen ‖ extraire. / ~ with great advantage ‖ vorteilhaft herstellen ‖ fabriquer avantageusement.

produce *see also* product ‖ Produkt n.; Erzeugnis n. ‖ produit m. / ~ (Bank) ‖ Belauf m. ‖ montant m. / ~ (Yield) ‖ Ertrag m.; Gewinn m. ‖ rapport m. bénéfice m. / agricultural ~ ‖ landwirtschaftliches Erzeugnis n. ‖ produit m. agricole; denrée f. / animal ~ ‖ tierisches Erzeugnis n. ‖ produit m. animal. / home ~ (Country) ‖ Landesprodukt n. ‖ produit m. indigène *ou* du pays. / inland ~ *see* home ~. / natural ~ ‖ Naturerzeugnis n. ‖ produit m. naturel. / vegetable ~ ‖ pflanzliches Erzeugnis n. ‖ produit m. végétal.

produce broker ‖ Produktenmakler m. ‖ courtier m. en produits.

produced in series ‖ in Reihen fpl. hergestellt ‖ construit en série f.

produce market ‖ Produktenmarkt m. ‖ marché m. des produits *ou* des denrées. / ~ merchant ‖ Produktenhändler m. ‖ négociant m. en produits.

producer ‖ Erzeuger m. ‖ producteur m. / ~ (Phot) ‖ Entwickler m. ‖ générateur m. / to remove the slag from the ~ ‖ den Generator m. entschlacken ‖ dégrasser le foyer.

producer gas ‖ Generatorgas n. ‖ gaz m. de gazogènes.

producing ‖ Produktion f.; Erzeugung f. ‖ production f. / ~ premium ‖ Produktionsprämie f. ‖ prime f. à la production.

product ‖ Erzeugnis n.; Produkt n. ‖ produit m. / ~ (Math) ‖ Produkt n. ‖ produit m. / by-~ ‖ Nebenprodukt n. ‖ produit m. accessoire; dérivé m.; sous-produit m. / chemical ~ ‖ chemisches Erzeugnis n. ‖ produit m. chimique. / dairy ~ ‖ Molkereierzeugnis n. ‖ produit m. de la laiterie. / ~ of different composition ‖ Erzeugnis n. verschiedener Zusammensetzung ‖ produit m. de composition diverse. / ~s pl. to be dressed ‖ Aufbereitungsgut n. ‖ matières fpl. à préparer *ou* à traiter. / finished ~ ‖ Fertigerzeugnis n. ‖ produit m. fini. / finished and semi-finished ~s pl. ‖ ganz und halb fertige Waren fpl. ‖ marchandises fpl. finies et mi-produites. / first ~ ‖ Anfangserzeugnis n.; Anfangsergebnis n. ‖ produit m. initial. / first-runnings ~ *see* first ~. / ~ coming from grinding ‖ vom Vermahlen herrührendes Erzeugnis n. ‖ produit m. provenant de la mouture. / half-finished ~ ‖ Halbzeug n. ‖ demi-produit m. / intermediate ~ ‖ Zwischenerzeugnis n. ‖ produit m. intermédiaire. / pharmaceutical ~ ‖ pharmazeutisches Erzeugnis n. ‖ produit m. pharmaceutique. / principal ~ ‖ Haupterzeugnis n. ‖ produit m. principal. / raw ~ ‖ rohes Erzeugnis n.; Rohprodukt n. ‖ produit m. brut. / ~ of solidification ‖ Erstarrungsgebilde n. ‖ produit m. de la solidification. / synthetic ~ for photography ‖ synthetisches Erzeugnis n. für photographische Zwecke ‖ produit m. synthétique pour la photographie. / washed ~ (Mine) ‖ Waschprodukt n. ‖ produit m. lavé.

production (Producing) ‖ Erzeugung f.; Produktion f. ‖ production f. / ~ (Product) *see* product. / ~ of coke ‖ Koks-

erzeugung f. ‖ fabrication f. du coke. / ~ estimated ~ ‖ schätzungsweise Erzeugung f. *oder* Produktion f. ‖ production f. évalué. / falling-off ~ ‖ abnehmende Erzeugung f. *oder* Produktion f. ‖ production f. diminuante. / increasing ~ ‖ zunehmende Erzeugung f. *oder* Produktion f. ‖ production f. augmentante. / ~ of liquid fuels from coal ‖ Erzeugung f. von flüssigen Brennstoffen aus Kohle ‖ production f. de combustibles liquides extraits du charbon. / ~ in masses ‖ Massenfertigung f. ‖ production f. en grandes séries. / large enterprise for the ~ in masses ‖ für die Massenerzeugung bestimmte Anlage f. ‖ usine f. destinée à une production industrielle d'objets en masse. / ~ of a mine ‖ Förderleistung f. einer Grubenanlage ‖ production f. d'une installation minière. / ~ of ozone ‖ Ozonerzeugung f. ‖ production f. d'ozone. / ~ of steel by the basic Thomas process ‖ Stahlerzeugung f. nach dem basischen Thomasverfahren ‖ fabrication f. de l'acier d'après le procédé basique Thomas.

production, branch of ‖ Produktionszweig m. ‖ branche f. de production. / cost of ~ ‖ Herstellungskosten pl. ‖ frais mpl. de fabrication. / process of ~ ‖ Werdegang m.; Produktionsvorgang m. ‖ procédé m. de fabrication. / rate of ~ ‖ Produktionssatz m. ‖ taux m. de production. / ~ reports pl. ‖ Produktionsbericht m. ‖ rapport m. sur la production. / ~ time ‖ Fertigungszeit f. ‖ temps m. d'usinage.

productive bed ‖ abbauwürdige Lagerstätte f. ‖ couche f. productive. / ~ capacity ‖ Ertragsfähigkeit f. ‖ capacité f. de production. / ~ coal formation ‖ abbauwürdige Steinkohlenlagerung ‖ formation f. houillère productive.

productiveness ‖ Ergiebigkeit f. ‖ débit m.; rendement m.

profession ‖ Beruf m.; Handwerk n. ‖ profession f.; métier m.

professional ‖ gewerbsmäßig; berufsmäßig ‖ par profession. / ~ clothing ‖ Berufskleidung f. ‖ vêtement m. d'ouvriers *ou* de travail. / ~ factory inspector ‖ Betriebskontrollör m. ‖ contrôleur m. d'usine. / ~ school ‖ Fachschule f. ‖ école f. professionnelle.

proficiency, certificate of ‖ Befähigungszeugnis n. ‖ certificat m. de capacité.

profilated brick ‖ Formziegel m. ‖ brique f. profilée.

profile, to (Join) ‖ auskehlen ‖ profiler.

profile ‖ Profil n.; Schnitt m.; Querschnitt m.; Seitenansicht f. ‖ profil m.; coupe f. / ~ of surface ‖ Bodenprofil n. ‖ relief m. *ou* profil m. du sol.

profiled iron rolling mill ‖ Profileisenwalzwerk n. ‖ laminoir m. pour fers profilés.

profiled wire ‖ Fassondraht m. ‖ fil m. façonné.

profile drawing ‖ Profilzeichnung f. ‖ tracé m. du profil. / ~ form (Railw) ‖ Profillehre f.; gabarit m. / ~ milling machine ‖ Kopierfräsmaschine f. ‖ fraiseuse f. à copier. / ~ press for metal rods ‖ Metallstrangpresse f. ‖ presse f. pour profiler à chaud les métaux en barres. / ~ tool steel ‖ Formstahl m. ‖ outil m. profilé.

profile-turned piece ‖ Formdrehteil m. ‖ pièce f. façonnée au tour.

profile turning work ‖ Fassondreherei f. ‖ atelier m. de profilage en tour. / ~ wire ‖ Profildraht m. ‖ fil m. profilé.

profiling apparatus, circular ‖ Zylinderkurvenfräsapparat m. ‖ appareil m. à fraiser les cames cylindriques. / longitudinal ~ ‖ Plankurvenfräsapparat m. ‖ appareil m. à fraiser les cames planes.

profiling cutter ‖ Profilfräser m. ‖ fraise f. profilée.

profit ‖ Ertrag m.; Gewinn m. ‖ profit m.; bénéfice m. / ~ brought forward from ... ‖ Gewinnvortrag m. aus ... ‖ report m. de bénéfices de ... / estimated ~s pl. ‖ veranschlagter Gewinn m. ‖ évaluation f. des bénéfices. / gross ~ ‖ Bruttogewinn m.; Rohgewinn m. ‖ bénéfice m. brut. / to bring-in large ~s pl. ‖ großen Verdienst m. einbringen ‖ faire de gros bénéfices. / to make a ~ ‖ profitieren ‖ profiter. / net ~s pl. ‖ Reinertrag m.; Gewinn m. ‖ bénéfices mpl. nets. / to reckon upon reasonable return ~ on the capital invested ‖ mit einer guten Ausnützung f. der Kapitalsanlage rechnen ‖ compter sur un rendement rationnel du capital investé. / to share in ~s pl. ‖ am Gewinn m. beteiligt sein ‖ participer aux bénéfices mpl. / year's ~ ‖ Jahresgewinn m. ‖ bénéfice m. de l'année.

profitable ‖ vorteilbringend; vorteilhaft; rentabel ‖ avantageux; profitable; lucratif.

profitableness ‖ Rentabilität f. ‖ rendement m.

profit and loss account ‖ Gewinn- und Verlustrechnung f. ‖ compte m. de gain et de perte.

profit, balance of ‖ Gewinnsaldo m. ‖ solde m. de bénéfice. / margin of ~ ‖ Gewinnmarge f.; Gewinnspanne f. ‖ écart m. de bénéfice. / share of the ~ ‖ Anteil m. am Gewinn ‖ part m. dans les bénéfices. / ~ sharing ‖ Gewinnbeteiligung f. ‖ participation f. aux bénéfices. / ~ taking ‖ Gewinnrealisierung f. ‖ réalisation f. des bénéfices.

proforma invoice ‖ Proformarechnung f. ‖ compte m. simulé. / ~ transaction ‖ Scheingeschäft n. ‖ opération f. fictive.

program *see* programme.

programme ‖ Programm n. ‖ programme m. / rolling ~ ‖ Walzprogramm n. ‖ programme m. de laminage.

progress ‖ Fortschritt m.; Wachstum n. ‖ progression f.; avance f. / rapid ~ of modern life ‖ rascher Fortschritt m. unserer Zeit ‖ progrès m. rapide des temps modernes.

progression ‖ Fortbewegung f. ‖ locomotion f.; mouvement m. progressif; marche f. / ~ (Math) ‖ Progression f. ‖ progression f. / geometrical ~ ‖ geometrische Progression f. *oder* Reihe f. ‖ progression f. géométrique.

prohibit, to ~ the import ‖ die Einfuhr verbieten ‖ prohiber l'importation f.

prohibition (Trade) ‖ Sperre f. ‖ suspension f.; blocus m. / ~ (A) ‖ Prohibition f.; Alkoholverbot n. ‖ prohibition f. / ~ to combine ‖ Koalitionsverbot n. ‖ interdiction f. de coalition. / ~ of exportation ‖ Ausfuhrverbot n. ‖ défense f. d'exporter. / to remove a ~ of exportation ‖ ein Ausfuhrverbot n. aufheben ‖ abroger une défense d'exporter. / ~ to import goods ‖ Einfuhrverbot n. ‖ défense f. d'importer. / ~ to prospect ‖ Schürfverbot n. ‖ interdiction f. de recherches. / ~ of trade ‖ Handelsverbot

n.; Handelssperre f. ‖ interdiction f. du commerce.

prohibitory circuit (Tel) ‖ Verhinderungsschaltung f. ‖ circuit m. prohibitif. / ~ duty ‖ Sperrzoll m. ‖ droit m. prohibitif.

project, to ‖ schleudern; werfen ‖ lancer; jeter. / ~ (A design, a building etc.) ‖ entwerfen; projektieren; ersinnen; planen ‖ projeter. / ~ (Build) ‖ auskragen ‖ faire saillir. / ~ (Geom) ‖ projizieren ‖ projeter.

project ‖ Entwurf m. ‖ projet m. / ~ of a telephone plant ‖ Entwurf m. einer Fernsprechanlage ‖ projet m. d'une installation téléphonique.

projected scale ‖ projizierte Skale f. ‖ échelle f. de projection.

projectile ‖ Geschoß n. ‖ projectile m. / bursting ~ ‖ Explosivgeschoß n. ‖ projectile m. éclatant. / ~ with coloured cloud of smoke ‖ Geschoß n. mit farbiger Rauchwolke ‖ projectile m. à nuage d'explosion coloré. / gas ~ ‖ Gasgeschoß n. ‖ projectile m. à gaz. / ~ with guide rings ‖ Geschoß n. mit Pressionsführung ‖ projectile m. forcé. / rifled ~ ‖ gezogenes Geschoß n. ‖ obus m. rayé.

projectile curve ‖ Flugbahn f. ‖ trajectoire f. / ~ graze ‖ Geschoßaufschlag m. ‖ point m. de chute d'un projectile. / part of ~ ‖ Geschoßteil m. ‖ partie f. d'un projectile.

projecting see project. / ~ apparatus ‖ Projektionsapparat m. ‖ appareil m. de projection. / ~ beam ‖ überstehender Balken m. ‖ poutre f. saillante.

projecting edge (Met) ‖ Grat m. ‖ barbe f.; barbure f.; bavure f. / ~ (Railw) ‖ vorspringender Wulst m. ‖ saillie f.

projecting screen ‖ Projektionswand f. ‖ écran m. pour la projection.

projection ‖ Projektion f. ‖ projection f. / ~ (Arch) ‖ Auslandung f. ‖ projecture f. / bracket ~ ‖ Auskragung f. ‖ saillie f.; encorbellement m. / ~ of a crane ‖ Ausladung f. eines Krans ‖ portée f. d'une grue. / forked ~ ‖ gegabelte Nase f. ‖ appendice m. en forme de fourche. / gnomonical ~ ‖ gnomonische Projektion f. ‖ projection f. gnomonique. / orthographic ~ ‖ orthografische Projektion f. ‖ projection f. orthographique. / polar ~ ‖ Polarprojection f. ‖ projection f. polaire. / stereographic ~ ‖ stereografische Projektion f. ‖ projection f. stéréographique. / upright ~ ‖ Aufriß m.; Vertikalprojektion f. ‖ projection f. verticale.

projection apparatus ‖ Projektionsapparat m. ‖ appareil m. de projection. / ~ for advertising purposes ‖ Projektionsapparat m. für Reklamezwecke ‖ appareil m. de projection de réclame. / ~ for macroprojection ‖ Projektionsgerät n. für Makroprojektion ‖ appareil m. de macroprojection. / ~ for microprojection ‖ Projektionsgerät n. für Mikroprojektion ‖ appareil m. de microprojection.

projection, plane of ~ ‖ Projektionsebene f. ‖ plan m. de projection.

projection screen ‖ Projektionsschirm m. ‖ écran m. de projection. / solar ~ see sun ~. / sun ~ ‖ Sonnenprojektionsschirm m. ‖ écran m. pour la projection du soleil; écran m. de projection pour le soleil.

projector ‖ Scheinwerfer m. ‖ phare m.; projecteur m. / ~ (Cinema) ‖ Projek-

tionsapparat m. ‖ appareil m. de projection. / cinema ~ see motion picture ~. / ~ for an even film passage ‖ Projektor m. für gleichmäßigen Filmablauf ‖ projecteur m. pour un déroulement continu du film. / ground ~ ‖ Bodenscheinwerfer m. ‖ aéro-phare m. / kinematograph ~ see motion picture ~. / motion picture ~ ‖ Kinovorführungsapparat m. ‖ appareil m. de projection de cinéma. / picture ~ ‖ Bildprojektor m. ‖ projecteur m. pour l'image. / sound ~ ‖ Tonprojektor m. ‖ projecteur m. pour l'émission sonore. / sound picture ~ ‖ Tonbildprojektor m. ‖ projecteur m. pour la reproduction sonore.

projector arc ‖ Scheinwerferbogenlampe f. ‖ arc m. du projecteur. / electric ~ lamp ‖ elektrische Projektionslampe f. ‖ lampe f. électrique de projection.

prolong, to ‖ verlängern ‖ prolonger.

prolongation ‖ Verlängerung f. ‖ prolongation f. / ~ (Trade) ‖ Fristverlängerung; Fristgewährung f. ‖ délai m.; sursis m.; prorogation f. / ~ of bills of exchange ‖ Wechselverlängerung f. ‖ prolongation f. d'une lettre de change.

promenade deck ‖ Promenadendeck n. ‖ pont-promenade m. / ~ gardener ‖ Stadtgärtner m. ‖ jardinier m. de promenades.

prominence spectroscope ‖ Protuberanzenspektroskop n. ‖ spectroscope m. à protubérances.

promise ‖ Zusage f.; Versprechen n. ‖ assentiment m.; promesse f.

promising ‖ aussichtsreich ‖ plein de promesses fpl. / most ~ ‖ vielversprechend ‖ prometteur.

promissory note ‖ Schuldbrief m.; Schuldschein m.; Schuldverschreibung f. ‖ reconnaissance f. / ~ (Bank) ‖ Solawechsel m. ‖ seule f. de change.

promontory ‖ Vorgebirge n. ‖ promontoire m.; cap m.

promote, to ‖ vorwärts bringen ‖ faire avancer.

promoter (Company) ‖ Gründer m. ‖ promoteur m.

promotion, expenses pl. of ~ ‖ Gründungskosten pl. ‖ frais mpl. de promotion. / ~ money see promotion, expenses of ~.

prompt ‖ künktlich ‖ ponctuel; exact; prompt.

promptness ‖ Raschheit f.; Pünktlichkeit f. ‖ promptitude f.; célérité f.

promptnote ‖ Mahnzettel m. ‖ avertissement m.; semonce f.

prong ‖ Kralle f. ‖ griffe f. / ~ (Fork) ‖ Zinke f. ‖ dent m.; fourchon m.

pronged ‖ gezinkt ‖ adenté.

prong hoe (Agr) ‖ Karst m. ‖ houe f.

Prony's brake ‖ Bremszaum m.; Pronyscher Zaum m. ‖ frein m. de Prony. / ~ dynamometer ‖ Bremsdynamometer n. ‖ frein m. dynamométrique; dynamomètre m. à frein de Prony.

proof ‖ beständig ‖ inattaquable à; à l'épreuve de; résistant à. / rain ~ ‖ regendicht ‖ imperméable à la pluie.

proof ‖ Beweis m. ‖ preuve f.; argument m. / ~ (Math) ‖ Probe f. ‖ preuve f. / ~ (Print) ‖ Probeabzug m.; Probedruck m. ‖ épreuve f. / artist's ~ (Print) ‖ Abdruck m. vor aller Schrift ‖ épreuve f. d'artiste ou de remarque; épreuve f. ou estampe f. ou gravure f. avant toutes lettres. / the ~ looks blotted

‖ die Schrift schmutzt ‖ les caractères mpl. bavochent. / clean ~ (Print) ‖ Reindruck m. ‖ seconde f.; seconde épreuve f. / ~ of the discovery (Mine) ‖ Nachweis m. der Fündigkeit ‖ preuve f. de la découverte. / documentary ~ ‖ Beweisstück n. ‖ pièce f. justificative. / ~ by evidence ‖ Zeugenbeweis m. ‖ preuve f. testimoniale. / first ~ (Print) ‖ erste Korrektur f.; Korrekturbogen m.; Fahnenabzug m. ‖ première f.; première épreuve f.; première typographique. / forced ~ ‖ Gewaltversuch m. ‖ essai m. à outrance. / ~ of identity ‖ Ausweis m.; Legitimation f. ‖ légitimation f. / ~ of the necessary educational training ‖ Nachweis m. der erforderlichen Vorbildung f. ‖ preuve f. de la préparation nécessaire. / to pull-off a ~ ‖ einen Bürstenabzug m. machen ‖ faire ou tirer une épreuve. / to read a ~ ‖ Korrektur f. lesen ‖ lire des épreuves fpl. / to strike-off a ~ with the beating brush ‖ einen Bürstenabzug m. machen ‖ tirer à la brosse. / to take a ~ see to pull-off a ~ (Print) / ~ to the utmost see violent ~. / violent ~ ‖ Gewaltprobe f. ‖ épreuve f. à outrance.

proof bank (Arm) ‖ Geschoßfang m.; Kugelfang m. ‖ butte f. du polygone. / ~ impression ‖ Probeabzug m.; Probedruck m.; Korrekturabzug m. ‖ épreuve f.; impression f. d'essai.

proofing machine (Print) ‖ Andruckmaschine f. ‖ machine f. à tirer les épreuves.

proof lead ‖ Probierblei n. ‖ plomb m. d'essai. / ~ paper ‖ Abziehpapier n. ‖ papier m. à épreuve. / ~ pile (Print) ‖ Probepfahl m. ‖ pieu m. d'essai. / ~ press (Print) ‖ Abziehapparat m.; Abziehpresse f. ‖ appareil m. ou presse f. à tirer les épreuves. / ~ print see ~ impression. / ~ puller (Print) ‖ Fahnenabzieher m. ‖ tireur m. d'épreuves. / ~ reader (Print) ‖ Korrekturleser m. ‖ correcteur m.

proof-sheet ‖ Probebogen m.; Fahne f.; Korrekturabzug m. ‖ épreuve f. / to beat off a ~ ‖ einen Korrekturbogen m. abziehen ‖ tirer une épreuve à la brosse. / to strike off a ~ see to beat off a ~.

prop, to ‖ abstützen; abfangen ‖ étançonner. / ~ with a buttress ‖ mit einem Strebepfeiler m. stützen ‖ buter; étançonner par un pieu. / ~ a ship on the stocks ‖ ein Schiff auf Stapel n. abstützen ‖ accorer un vaisseau sur le chantier. / ~ up ‖ abspreizen ‖ étayer.

prop ‖ Pfahl m.; Baumpfahl m. ‖ tuteur m. / ~ (Vineyard) ‖ Pfahl m.; Rebpfahl m. ‖ échalas m. / ~ of a cart ‖ Bergstütze f.; Schleppstock m.; Hemmstütze f. ‖ servante f. d'un chariot. / headed ~ (Build) ‖ Kopfspreize f. ‖ étrésillon m. coudé; étrier m. en bois. / mine ~ ‖ Grubenstempel m. ‖ étançon m. de mines. / ~ of a wall ‖ Stütze f. einer Mauer f. ‖ appui m. d'un mur.

propaganda ‖ Propaganda f.; Werbung f.; Reklame f. ‖ propagande f.; réclame f. / to make ~ ‖ werben ‖ faire de la propagande.

propaganda drawing ‖ Werbezeichnung f. ‖ dessin m. de réclame. / ~ photo ‖ Werbeaufnahme f. ‖ photographie f. de réclame.

propagate, to ‖ sich fortpflanzen ‖ se propager.

propagation ‖ Fortpflanzung f. ‖ propagation f. / ~ of sound in the free atmosphere ‖ Schallfortpflanzung f. in der freien Luft ‖ propagation f. du son dans l'air libre.

propagation constant ‖ Fortpflanzungskonstante f. ‖ constante f. de propagation. / time of ~ (Tel) ‖ Laufzeit f. ‖ temps m. de propagation. / ~ velocity ‖ Fortpflanzungsgeschwindigkeit f. ‖ vitesse f. de propagation.

propeller (Aero) ‖ Propeller m.; Luftschraube f. ‖ hélice f. d'avion. / ~ (Shipb) ‖ Schiffsschraube f. ‖ hélice f. (de bateau). / aluminium ~ (Aero) ‖ Luftschraube f. aus Aluminium ‖ hélice f. d'avion en aluminium. / hydraulic ~ ‖ Reaktionsschiffsschraube f. ‖ propulseur m. hydraulique. / reversible ~ ‖ Drehflügelschraube f. ‖ hélice f. réversible. / tractor ~ ‖ Zugschraube f. ‖ hélice f. tractive. / wooden ~ (Aero) ‖ Holzluftschraube f. ‖ hélice f. d'avion en bois.

propeller blade ‖ Schraubenflügel m. ‖ lame f. d'hélice. / ~ bolt ‖ Schraubenbolzen m. ‖ boulon m. d'hélice. / ~ boss ‖ Schraubennabe f. ‖ moyeu m. d'hélice. / main engine for ~ drive ‖ Hauptmaschine f. zum Antrieb von Schiffsschrauben ‖ machine f. principale pour la commande des hélices. / ~ efficiency ‖ Schraubenwirkungsgrad m. ‖ rendement m. de l'hélice. / ~ fan ‖ Schraubenventilator m. ‖ ventilateur m. hélicoïdal. / ~ hub see ~ boss. / ~ lathe (Aero) ‖ Propellerdrehbank f. ‖ tour m. à hélices. / ~ performance ‖ Schraubenleistung f. ‖ rendement m. de l'hélice. / ~ pitch ‖ Schraubensteigung f. ‖ pas m. de l'hélice. / ~ pump ‖ Schraubenschaufler m.; Kreispumpe f. ‖ pompe f. à hélice.

propeller shaft ‖ Schraubenwelle f.; Triebwelle f. ‖ arbre m. de l'hélice ou d'entraînement. / ~ (Auto) ‖ Kardanwelle f. mit zwei Gelenken ‖ arbre m. à deux cardans. / distance between centre line and centre of ~ ‖ Entfernung f. von Mitte bis Mitte der Schraubenwelle ‖ écartement m. de l'axe en axe d'arbre porte-hélice.

propeller thrust ‖ Schraubendruck m. ‖ poussée f. de l'hélice.

propelling machine (Shipb) ‖ Antriebsmaschine f. ‖ machine f. de propulsion. / ~ pencil ‖ Drehbleistift m. ‖ porte-mine m. à vis.

property ‖ Besitztum n.; Besitzung f. ‖ propriété f.; possession f. / to assign the right to use a ~ ‖ das Benutzungsrecht des Grundstückes zusprechen ‖ adjuger le droit d'usage d'un terrain. / to be the ~ of the surface owner ‖ im Eigentum n. des Grundbesitzers stehen ‖ être pars fundi. / to be bound to acquire a ~ ‖ zum Erwerb m. eines Grundstückes verpflichtet werden ‖ être obligé à l'acquisition f. d'un terrain. / to break up a ~ by the cession of portions thereof ‖ ein Grundstück n. durch Abtretungen parzellieren ‖ morceler un terrain par la cession de parcelles. / to cede the ownership of a ~ ‖ das Eigentum am Grundstück abtreten ‖ céder la propriété du terrain. / to compulsorily cede a ~ ‖ ein Grundstück n. zwangsweise abtreten ‖ céder un terrain par voie d'expropriation. / to decrease the value of a ~ by its use ‖ durch Benutzung f. des Grundstückes seinen Wert mindern ‖ diminuer la valeur d'un terrain par l'usage. / to dispose freely of a ~ ‖ frei über ein Grundstück n. verfügen ‖ disposer librement d'un terrain. / ~ divisible among the creditors ‖ Teilungsmasse f. ‖ actif m. / exclusive ~ ‖ Alleinbesitz m. ‖ propriété f. / freehold ~ ‖ Freigrundbesitz m. ‖ pleine propriété f. / leasehold ~ ‖ Pachtgrundbesitz m. ‖ propriété f. à bail. / to require the surface owner to cede a ~ ‖ den Grundbesitzer m. zur Abtretung eines Grundstückes anhalten ‖ obliger le propriétaire à la cession du terrain. / to round-off one's ~ ‖ seinen Besitz m. abrunden ‖ arrondir son domaine.

property, capitalized value of ‖ Ertragswert m. ‖ valeur f. de rendement ou de la capitalisation d'une rente. / ~ tax ‖ Vermögenssteuer f. ‖ impôt m. sur le capital.

proportion ‖ Proportion f.; Verhältnis n. ‖ proportion f. / alloy containing copper in the ~ of x to y per cent ‖ das Metall enthält Kupfer im Verhältnis von x bis y Prozent ‖ alliage m. contenant du cuivre en proportion de x à y pour cent. / continual ~ ‖ stetige Proportion f. ‖ proportion f. continue. / in diminished ~s pl. ‖ in verjüngtem Maßstabe m. ‖ à échelle f. réduite. / ~ of gross to net load (Railw; Shipb) ‖ Ladeziffer f. ‖ rapport m. du poids mort au poids utile. / harmonical ~ ‖ harmonische Proportion f. ‖ proportion f. harmonique. / in ~ to ‖ nach Maßgabe f. von ‖ conformément à; selon; en raison de. / ~ of ingredients for gunpowder ‖ Mischungsverhältnis n. des Pulversatzes ‖ dosage m. de la poudre à canon. / small ~ of phosphorus ‖ schwache Beimengung f. von Phosphor ‖ phosphore m. en faible proportion. / in small ~s pl. ‖ in bescheidenem Maße n. ‖ dans des proportions fpl. restreintes.

proportional ‖ proportional ‖ proportionnel. / directly ~ ‖ unmittelbar oder direkt proportional ‖ directement proportionnel. / inversely ~ ‖ umgekehrt proportional ‖ inversement proportionnel.

proportionate ‖ ebenmäßig ‖ symétrique; proportionné.

proposal ‖ Antrag m. ‖ proposition f. / ~ for settlement ‖ Vermittlungsvorschlag m. ‖ proposition f. de consiliation ou d'arrangement.

propose, to ‖ anregen; vorschlagen ‖ animer; stimuler; proposer. / ~ as candidate ‖ als Kandidat m. aufstellen ‖ porter ou présenter comme candidat m.

proposition see proposal. / ~ (Math) ‖ Lehrsatz m. ‖ proposition f.; théorème m.

propping (Carp) ‖ Stützwerk n. ‖ soutènement m. / ~ of a beam ‖ Absteifung f. eines Balkens ‖ étayement m. d'une poutre.

proprietor ‖ Besitzer m. ‖ possesseur m. / joint ~ ‖ Mitinhaber m. ‖ co-propriétaire m.

proprietorship ‖ Eigentumsrecht n. ‖ droit m. de propriété. / ~ (Print) ‖ Verlagsrecht n. ‖ droit m. d'impression ou de publication. / reservation of ~ ‖ Eigentumsvorbehalt m. ‖ réserve f. de propriété.

propriety ‖ Schicklichkeit f. ‖ convenance f.

props pl., pit ‖ Grubenholz n. ‖ bois m. de mine. / ~ and supports pl. of a roof ‖ Dachstuhl m. ‖ ferme f.

propulsion ‖ Vortrieb m. ‖ impulsion f.; translation f.; avancement m.; poussée f. en avant.

propwood ‖ Schlagholz n. ‖ bois m. d'étais ou de soutènement.

propylalcohol ‖ Propylalkohol m. ‖ alcool m. propylique.

propylite ‖ Propylit m. ‖ propylite f.

proscenium ‖ Proszenium n. ‖ avant-scène f.

prosecute, to ~ a firm ‖ eine Firma f. gerichtlich verklagen ‖ porter plainte f. contre une maison.

prosopite ‖ Prosopit n.; Steinmark n. ‖ prosopite f.

prospect, to ‖ schürfen; graben ‖ fouiller; faire des fouilles fpl. / ~ for common account ‖ auf gemeinschaftliche Rechnung f. schürfen ‖ faire des recherches fpl. à compte commun. / authority to ~ ‖ Schürfbefugnis f. ‖ autorisation f. de faire des fouilles. / freedom to ~ ‖ Schürffreiheit f. ‖ liberté f. des recherches. / to request a permit to ~ ‖ die Genehmigung zum Schürfen einholen ‖ solliciter la permission de faire des recherches. / prohibition to ~ ‖ Schürfverbot n. ‖ interdiction f. de recherches.

prospect ‖ Ausblick m.; Aussichten fpl. ‖ perspective f. / ~ glass ‖ Fernrohr n.; Fernglas n. ‖ lunette f. / ~ hole ‖ Schürfloch n. ‖ trou m. de sondage.

prospecting licence ‖ Schürfschein m. ‖ permis m. de prospecter.

prospecting operations pl. ‖ Schürfbetrieb m. ‖ travaux mpl. de recherches. / to cease ~ ‖ die Schürfarbeiten fpl. einstellen ‖ arrêter les fouilles fpl.

prospecting work ‖ Schürfarbeit f. ‖ travail m. de recherche.

prospective customer ‖ Reflektant m.; Interessent m. ‖ amateur m.; intéressé.

prospect shaft ‖ Schürfschacht m. ‖ puits m. de recherche.

prospectus ‖ Prospekt m. ‖ prospectus m.

prosperity, national ‖ Volkswohlstand m. ‖ fortune f. publique.

prosperous ‖ schwunghaft; gedeihlich ‖ florissant.

prosthesis ‖ Prothese f.; künstliches Glied n. ‖ prothèse f. / dental ~ ‖ künstliches Gebiß n.; Zahnprothese f. ‖ dentier m.; prothèse f. dentaire. / baseplate for dental ~ ‖ Gebißplatte f. ‖ plaque f. base pour dentiers.

protect, to ‖ schützen; beschützen ‖ protéger. / ~ by a coating of colourless varnish ‖ mit farblosem Schutzlack m. überziehen ‖ recouvrir d'un vernis protecteur incolore. / ~ against draught ‖ vor Zugluft f. schützen ‖ protéger contre le courant d'air. / ~ a river bank ‖ das Ufer befestigen ‖ défendre la rive. / ~ against the sun ‖ vor Sonnenschein m. schützen ‖ abriter contre le soleil.

protected ‖ geschützt ‖ protégé. / ~ from dust ‖ vor Staub m. geschützt ‖ imperméable à la poussière. / ~ by law ‖ gesetzlich geschützt ‖ protégé par la loi.

protected area (Forest) ‖ Schutzbezirk m. ‖ triage m.; district m. ou périmètre m. de protection. / ~ cruiser ‖ geschützter Kreuzer m. ‖ croiseur m. protégé.

protecting cover ‖ Schutzdeckel m. ‖ couvercle-protecteur m. / ~ for tripod head ‖ Schutzhülle f. für den Stativkopf ‖ enveloppe f. protectrice de la tête du trépied.

protecting device ‖ Schutzvorrichtung f. ‖ appareil m. de protection. / ~ edge ‖ Schutzrand m. ‖ bordure f. de protection. / ~ glass which moderates the visible glaring rays ‖ die sichtbaren blendenden Strahlen abschwächendes Schutzglas n. ‖ verre m. protecteur affaiblissant les radiations visibles éblouissantes. / ~ network ‖ Schutznetz n. ‖ filet m. protecteur. / ~ sheating (Cable) ‖ Schutzhülle f. ‖ enveloppe f. protectrice. / ~ sheet roller ‖ Schutzblechwalze f. ‖ cylindre m. à écran de tôle. / ~ sluice ‖ Sperrschleuse f. ‖ écluse f. de garde. / ~ varnish ‖ Deckfirnis m. ‖ vernis m. protecteur. / ~ wire ‖ Schutzdraht m. ‖ fil m. protecteur.

protection ‖ Schutz m. ‖ protection f. / ~ for dunes by planting ‖ Dünenschutz m. durch Bepflanzung ‖ protection f. des dunes par plantation. / ~ for dunes by tree planting ‖ Dünenschutz m. durch Aufforstung f. ‖ protection f. des dunes par plantation sylvestre. / ~ against gas ‖ Gasschutz m. ‖ protection f. contre le gaz. / ~ against high water ‖ Hochwasserschutz m. ‖ protection f. contre les crues. / ~ from radiation ‖ Schutz m. gegen Strahlung ‖ protection f. contre les radiations. / ~ against rust ‖ Rostschutz m. ‖ protection f. contre la rouille. / ~ of the stock (El line) ‖ Fußschutz m. ‖ préservation f. du pied des poteaux. / ~ against strong current ‖ Starkstromschutz m. ‖ protection f. contre les courants forts. / ~ of trade marks ‖ Markenschutz m. ‖ protection f. des marques de fabrique.

protectional layer ‖ Schutzschicht f. ‖ couche f. protectrice.

protection, area of ‖ Schutzkreis m. ‖ rayon m. préservé ou de protection. / ~ cage ‖ Schutzkorb m. ‖ cage f. de protection.

protectionism ‖ Schutzzollsystem n. ‖ protectionnisme m.

protectionist ‖ Anhänger m. des Schutzzollsystems ‖ protectionniste m.

protection strip ‖ Schutzleiste f. ‖ bande f. protectrice. / ~ suit ‖ Grubenanzug m.; Schachtanzug m.; Schutzanzug m. ‖ vêtement m. de protection. / ~ switch with resistance ‖ Schutzschalter m.; Vorstufenschalter m. ‖ interrupteur m. de sécurité pourvu de résistance.

protective apparatus for high tension ‖ Hochspannungsschutzapparat m. ‖ appareil m. de protection pour haute tension. / ~ appliance see protective device. / ~ belt of trees ‖ Waldschutzstreifen m. ‖ rideau m. protecteur d'arbres. / ~ cap ‖ Schutzkapsel f. ‖ capsule f. protective. / ~ casing ‖ Schutzhülle f. ‖ enveloppe f. protectrice. / ~ coating to counteract the effects of the weather ‖ Schutzanstrich m. gegen Witterungseinflüsse ‖ couche f. de peinture pour protéger contre les influences atmosphériques.

protective clothing ‖ Schutzkleidung f. ‖ vêtement m. protecteur. / ~ for bee keepers ‖ Imkerschutzbekleidung f. ‖ vêtement m. de protection pour éleveurs d'abeilles.

protective device ‖ Schutzvorrichtung f. ‖ dispositif m. de protection. / ~ against accidental contact between communication lines and power lines ‖ Berührungsschutz m. zwischen Fernmeldeleitungen und Starkstromleitungen ‖ mesure f. de protection contre le contact accidentel entre les lignes de communication et celles d'énergie électrique. / ~ with oxygen supply ‖ Sauerstoffschutzgerät n. ‖ appareil m. protecteur à oxygène.

protective duty ‖ Schutzzoll m. ‖ droit m. protecteur. / ~ facing for toothed wheels ‖ Schutzverkleidung f. für Zahnräder ‖ appareil m. protecteur pour engrenages. / ~ fence ‖ Schutzgeländer m. ‖ garde-corps m. / ~ gas ‖ Schutzgas n. ‖ gaz m. protectif ou de protection. / ~ glove ‖ Schutzhandschuh m. ‖ gant m. protecteur. / ~ grating ‖ Schutzgitter n. ‖ treillis m. de protection; grilles fpl. protectrices. / ~ ground (Electr) ‖ Schutzerdung f. ‖ protection f. par mise à la terre.

protective hood ‖ Schutzhaube f. ‖ capot m. de protection; capuchon m. / wooden ~ ‖ Holzschutzkasten m. ‖ couvercle m. de protection en bois.

protective lid (Typewr) ‖ Schutzdeckel m. ‖ plaque f. de protection. / ~ metallic coating ‖ metallischer Schutzüberzug m. ‖ revêtement m. protecteur métallique. / ~ shield ‖ Schutzschild m. ‖ protecteur m.; plaque f. protectrice; bouclier m. / ~ spectacles pl. ‖ Schutzbrille f. ‖ lunettes fpl. protectrices. / ~ system see protectionism.

protective tube ‖ Schutzrohr n. ‖ tube m. protecteur. / ~ for pyrometers ‖ Pyrometerschutzrohr n.; Hitzemesserschutzrohr n. ‖ tube m. protecteur de pyromètre.

protector ‖ Schutzvorrichtung f. ‖ protecteur m. / ~ tube see protective tube.

proteid ‖ Proteid n.; Eiweißstoff m. ‖ albumine f.; protéide m. / ~ turbidity ‖ Eiweißtrübung f. ‖ trouble m. d'albumine ou des matières azotées.

protein ‖ Protein n.; Eiweiß n. ‖ protéine f. / ~ body ‖ Eiweißkörper m. ‖ protéine f. / ~ determination ‖ Eiweißbestimmung f. ‖ dosage m. des matières albuminoïdes. / ~ flour ‖ Proteinmehl n. ‖ farine f. de protéine.

proteolysis ‖ Eiweißabbau m. ‖ protéolyse f.

proteolytical diastase ‖ proteolytische Diastase f. ‖ diastase f. protéolytique.

protest, to ‖ protestieren ‖ protester.

protest ‖ Protest m.; Einspruch m. ‖ protestation f.; protêt m. / extended ~ see ship's ~. / to note ~ ‖ Protest m. notieren; Einspruch m. erheben ‖ noter procès m. verbal ou protêt m. / ship's ~ ‖ Seeprotest m.; Verklarung f. ‖ rapport m. de mer.

protoacetate of iron ‖ essigsaures Eisenoxydul n. ‖ protoacétate m. de fer.

protochloride ‖ Chlorür n. ‖ protochlorure m. / ~ of mercury ‖ Quecksilberchlorür n.; Kalomel n.; Merkurochlorid n. ‖ calomel m.; sous-chlorure m. de mercure; mercure m. doux. / ~ of tin ‖ Zinnchlorür n. ‖ protochlorure m. d'étain ‖ chlorure m. stanneux.

protocol ‖ Protokoll n. ‖ procès-verbal m.; protocole m.

protocyanide of gold ‖ Goldcyanür n. ‖ protocyanure m. d'or. / ~ of iron ‖ Eisenzyanür n. ‖ ferrocyanure m. .

protogyne ‖ Protogyn m.; Talkgranit m. ‖ protogyne f.; granit m. talqueux.

protoplasm ‖ Protoplasma n. ‖ protoplasma m.; protoplasme m.

protosulphide, native ~ of nickel ‖ Schwefelnickel m.; Haarkies m. ‖ nickel m. sulfuré; pyrite f. capillaire.

protoxide ‖ Oxydul n. ‖ protoxyde m. / ~ of carbon ‖ Kohlenoxyd n. ‖ oxyde m. de carbone. / ~ of cobalt ‖ Kobaltoxydul n. ‖ protoxyde m. de cobalt. / ~ of iron ‖ Eisenoxydul n. ‖ oxyde m. ferreux; protoxyde m. de fer. / ~ of lead ‖ Bleiglätte f.; Bleioxyd n. ‖ litharge f.; massicot m.; protoxyde m. de plomb. / ~ of manganese ‖ Manganoxydul n. ‖ protoxyde m. de manganèse. / ~ of nitrogen ‖ Stickstoffoxydul(gas) n.; Lachgas n. ‖ protoxyde m. d'azote. / ~ of uranium ‖ Pechblende f. ‖ urane m. oxydulé.

protractor ‖ Anlegegoniometer n. ‖ goniomètre m. d'application. / ~ (Semicircular) ‖ Gradbogen m.; Transporteur m. ‖ demi-cercle m. gradué; rapporteur m.

protrusive motion ‖ Stoßbewegung f. ‖ mouvement m. saccadé de propulsion.

protuberance ‖ Ausstülpung f.; Protuberanz f. ‖ excroissance f.; protubérance f.

proustite ‖ lichtes Rotgültigerz; Arsensilberblende f.; Proustit m. ‖ argent m. arséniaté sulfuré.

prove, to see also to test ‖ erproben; prüfen; feststellen; beweisen ‖ éprouver; mettre à l'épreuve f. / ~ (Math) ‖ die Probe machen ‖ faire la preuve. / ~ one's identity ‖ sich ausweisen ‖ établir son identité.

provender ‖ Trockenfutter n. ‖ fourrage m. sec.; sec m.

prover (Met) ‖ Probierlöffel m. ‖ éprouvette f.

provide, to ‖ anschaffen ‖ acheter; faire l'acquisition f. de. / ~ with ‖ beliefern; versehen mit ‖ livrer; fournir.

provided with ‖ versehen mit; ausgestattet mit ‖ muni de.

provident benefit ‖ Arbeitslosenunterstützung f. ‖ dédommagement m. ou secours m. de chômage.

proving ground ‖ Schießplatz m. ‖ polygone m. de tir. / ~ machine ‖ Festigkeitsprüfmaschine f. ‖ machine f. à essayer la résistance.

provision ‖ Vorkehrung f.; Schutzmaßnahme f. ‖ mesure f.; préparatifs mpl. / ~s pl. concerning the protection of submarine cables ‖ Kabelschutzrecht n. ‖ règlements mpl. concernant la protection des câbles sous-marins.

provisional receipt ‖ Interimsquittung f. ‖ reçu m. provisoire.

provision basket ‖ Eßkorb m. ‖ panier m. à provisions. / ~ industry ‖ Nahrungsmittelindustrie f.; Lebensmittelindustrie f. ‖ industrie f. de produits alimentaires.

provisions pl. ‖ Nahrungsmittel pl.; Lebensmittel pl. ‖ produits mpl. alimentaires. / ~ (Mar) ‖ Proviant m. ‖ vivres mpl.; provisions fpl. / daily ~ ‖ täglicher Proviant m. ‖ vivres mpl. de consommation. / dainty ~ ‖ Feinkostware f. ‖ comestibles mpl. fins. / to furnish with ~

see to supply with ~. / preserved ~ ‖ eingemachte Eßwaren fpl. ‖ conserves fpl. alimentaires. / to supply with ~ ‖ mit Proviant m. versehen *oder* versorgen ‖ fournir en vivres mpl. ravitailler.

provision trade ‖ Nahrungsmittelindustrie f. ‖ industrie f. des produits alimentaires. / ~ works equipment ‖ Nahrungsmittelfabrikeinrichtung f. ‖ installation f. à fabriquer des produits alimentaires.

proxy ‖ Bevollmächtigter m.; Stellvertreter m. ‖ mandataire m.; plénipotentiaire m. / by ~ ‖ in Stellvertretung f. ‖ par procuration f.

prune, to ~ a tree ‖ einen Baum m. stutzen ‖ étêter *ou* égayer un arbre.

prune distillery ‖ Zwetschenwasserbrennerei f. ‖ distillerie f. d'eau-de-vie de prunes. / ~ dryer ‖ Zwetschendörrer m. ‖ sécheur m. de pruneaux.

prunelle salt ‖ Prunellensalz n.; geschmolzener Salpeter m. ‖ sel m. de prunelle; cristal m. minéral.

pruning ‖ Aufpfropfung f. ‖ greffage m. / ~ knife ‖ Gartenmesser n. ‖ serpe f. / ~ saw ‖ Gärtnersäge f.; Baumsäge f.; Astsäge f. ‖ scie f. de jardinier *ou* à greffer. / ~ shears pl. ‖ Baumschere f.; Gartenschere f. ‖ sécateur m.; ciseaux mpl. du jardinier.

Prussian blue ‖ Berlinerblau n.; Preußischblau n. ‖ bleu m. de Prusse. / ~ mixed with mineral matters ‖ Berliner Blau n. mit mineralischen Stoffen verschnitten ‖ bleu m. de Prusse additionné de matières minérales.

Prussian cap (Build) ‖ preußische Kappe f. ‖ voussette f. entre nervures.

prussiate ‖ Zyansalz n. ‖ cyanure m.; prussiate m. / ~ of potash ‖ Zyankalium n. ‖ prussiate m. de potasse; cyanure m. *ou* cyanide m. de potassium. / red ~ of potash ‖ Ferrizyankalium n.; rotes Blutlaugensalz n. ‖ ferricyanure m. potassique.

prussic acid ‖ Blausäure f.; Zyanwasserstoffsäure f. ‖ acide m. prussique *ou* cyanhydrique.

psilomelane ‖ schwarzer Glaskopf m.; Psilomelan m. ‖ psilomélane m.; oxyde m. manganique. / ~ containing potash ‖ Kalipsilomelan m. ‖ peroxyde m. de manganèse potassé.

psychrometer ‖ Psychrometer n.; Feuchtigkeitsmesser m. ‖ psychromètre m. / tele-~ ‖ Fernpsychrometer n. ‖ télépsychromètre m.

public ‖ öffentlich ‖ public. / ~ abattoir ‖ städtischer Schlachthof m. ‖ abattoir m. municipal. / ~ auction ‖ öffentliche Versteigerung f. ‖ encan m. public. / ~ carriages pl. ‖ öffentliches Fuhrwesen n. ‖ voitures fpl. publiques.

public electricity supply, influence of a company on the ~ ‖ Einfluß m. einer Gesellschaft auf die Elektrizitätswirtschaft ‖ influence f. d'une société sur l'économie électrique. / rational amalgamation of the existing ~ system ‖ rationelle Zusammenfassung f. der bestehenden elektrowirtschaftlichen Anlagen ‖ exploitation f. rationnelle d'installations électriques existantes.

public funds pl. ‖ Staatspapiere npl. ‖ effets mpl. publics; valeurs fpl. / ~ house ‖ Schankwirtschaft f. ‖ débit m. de boissons. / force of ~ opinion of the civilized world ‖ Macht f. der öffentlichen Meinung der zivilisierten Welt ‖ force f. d'opinion publique des nations civilisées. / ~ plant (Tel) ‖ öffentliches Netz n. ‖ réseau m. public. / ~ regulations pl. ‖ behördliche Vorschriften fpl. ‖ instructions fpl. de la police. / ~ road ‖ Verkehrsweg m. ‖ chemin m. public. / ~ sale ‖ Auktion f. ‖ vente f. publique. / ~ soup kitchen ‖ Volksküche f. ‖ cuisine f. publique. / ~ station (Tel) ‖ öffentliche Sprechstelle f. ‖ poste m. public. / ~ supply (Tel) ‖ Netzanschluß m. ‖ branchement m. du réseau. / ~ telephone plant ‖ öffentliches Fernsprechnetz n. ‖ réseau m. téléphonique public. / ~ thoroughfare ‖ offene Straße f. ‖ pleine rue f. / ~ waters pl. ‖ öffentliches Gewässer n. ‖ eaux fpl. du domaine public. / ~ works contracting ‖ Baugeschäft n.; Baugesellschaft f. ‖ entreprise f. de construction.

publication ‖ Bekanntmachung f.; Veröffentlichung f. ‖ avis m.; publication f. / ~ (of a book) ‖ Verlag m. (eines Buches) ‖ impression f.; publication f. et vente f. (d'un livre). / annual ~ ‖ Jahrgang m. ‖ année f. / periodical ~ ‖ periodische Veröffentlichung f. ‖ publication f. périodique.

publicity ‖ Öffentlichkeit f. ‖ publicité f. / ~ letter ‖ Werbebrief m. ‖ lettre f. de propagande.

publish, to ‖ herausgeben; veröffentlichen ‖ publier; éditer; faire paraître. / ~ at one's own expense ‖ im Selbstverlag m. herausgeben ‖ s'éditer.

publisher ‖ Verleger m.; Herausgeber m. ‖ éditeur m. / ~ (Book sell) ‖ Verlagsbuchhändler m. ‖ libraire-éditeur m. / art ~ ‖ Kunstverlag m. ‖ éditeur m. d'art. / ~ in foreign languages ‖ Verlag m. für Fremdsprachen ‖ imprimerie f. de langues étrangères. / ~'s business *see* publishing business. / firm of ~s ‖ Verlagsbuchhandlung f.; Verlagsgeschäft n.; Verlagsanstalt f.; Verlag m. ‖ librairie f. de fonds *ou* d'éditeur.

publishing business ‖ Verlagsbuchhandel m. ‖ commerce m. d'éditeur. / ~ expenses pl. ‖ Verlagskosten pl. ‖ frais mpl. de publication. / ~ firm *see* firm of publishers. / ~ trade *see* ~ business.

pudding ‖ Pudding m. ‖ pouding m. / ~ powder ‖ Puddingpulver n. ‖ poudre f. à pouding.

puddingstone ‖ Nagelfluh f. ‖ poudingue m. / ~ quarry ‖ Nagelfluhsteingrube f. ‖ carrière f. de poudingues.

puddle, to ‖ puddeln ‖ puddler; pudler.

puddle (Build) ‖ Lettenschlag m.; Lehmschlag m.; Tonschlag m. ‖ couche f. battue; corroi m. de terre glaise. / ~ bar ‖ Rohschiene f.; Luppe f. ‖ fer m. ébauché. / to roll ~ bars ‖ Rohschienen fpl. *oder* Luppen fpl. walzen ‖ ébaucher le fer au laminoir. / ~ bar roller ‖ Luppenwalzer m. ‖ ébaucheur m.

puddled iron ‖ Puddeleisen n. ‖ fer m. puddlé. / shop for the manufacture of ~ *see* puddling works.

puddled steel ‖ Puddelstahl m. ‖ acier m. puddlé.

puddler ‖ Puddler m. ‖ puddleur m. / underhand ~ ‖ Hilfspuddler m. ‖ aidepuddleur m.

puddle rolling mill ‖ Puddelwalzwerk n. ‖ train m. ébaucheur *ou* de puddlage; laminoir m. ébaucheur. / ~ rolls pl. ‖ Puddelwalzen fpl. ‖ cylindres mpl. ébaucheurs *ou* dégrossisseurs. / ~ steel *see* puddled steel. / ~ wall (Hydr arch) ‖ Lehmschlagwand f. ‖ mur m. battu de terre glaise; massif m. en argile corroyée.

puddling ‖ Puddelbetrieb m.; Puddeln n. ‖ puddlage m. / ~ of fibrous iron ‖ Sehneneisenpuddeln n. ‖ puddlage m. de fer fibreux. / ~ of fine-grained iron ‖ Feinkornpuddeln n. ‖ puddlage m. de fer à grain fin. / ~ of foundation walls ‖ Verfüllung f. der Grundmauern ‖ remblai m. des fondations. / ~ of steel ‖ Stahlpuddeln n. ‖ puddlage m. d'acier.

puddling bar *see* puddle bar. / ~ furnace ‖ Puddelofen m. ‖ four m. à puddler. / revolving ~ furnace ‖ rotierender Puddelofen m. ‖ four-puddleur m. rotatif. / ~ machine ‖ Puddelmaschine f. ‖ puddleur m. mécanique. / ~ process ‖ Puddelverfahren n. ‖ procédé m. de puddlage. / ~ rolls pl. *see* puddle rolls. / ~ slag ‖ Puddelschlacke f. ‖ scorie f. de puddlage. / ~ works pl. ‖ Puddelwerk n. ‖ atelier m. de puddlage.

puff ‖ Puderquaste f. ‖ houppe f. / ~ box ‖ Puderquastenbüchse f. ‖ boîte f. à houppe.

pugging (Pott) ‖ Kneten n. ‖ pétrissage m.

pug mill ‖ Mörtelmischer m. ‖ machine f. à malaxer le mortier.

pull, to ‖ ziehen ‖ tirer. / ~ (Mar) ‖ rudern ‖ nager; ramer. / ~ along ‖ schleifen; schleppen ‖ traîner. / ~-down a building ‖ ein Gebäude n. abbrechen *oder* niederreißen ‖ démolir *ou* raser un bâtiment. / ~ flax ‖ den Flachs ziehen ‖ arracher du lin. / ~-off a proof ‖ einen Probedruck m. abziehen ‖ faire une épreuve. / ~-off the bast of the hemp ‖ den Hanf m. schälen *oder* schleißen ‖ teiller *ou* tiller le chanvre. / ~-out the hemp with the hand ‖ den Hanf pellen *oder* schälen *oder* schleißen ‖ teiller *ou* tiller le chanvre. / ~ the form in pieces (Print) ‖ die Form f. ausschlachten ‖ désosser. / ~ in slips (Print) ‖ in Fahnen fpl. abziehen ‖ placarder. / ~-up ‖ hissen; heißen; aufziehen ‖ hisser; palanquer; haler. / the tank had to be pulled down ‖ der Behälter mußte abgebrochen werden ‖ il a fallu démolir le réservoir.

pull ‖ Zug m.; Stoß m. ‖ traction f.; tirage m. / ~ (Print) ‖ Satz m. ‖ coup m. / ~ of belt ‖ Riemenzug m. ‖ traction f. sur la courroie. / ~ of centrifugal forces ‖ Beanspruchung f. durch Fliehkraft ‖ effort m. de la force centrifuge. / first ~ (Print) ‖ erster Satz m. ‖ premier coup m. / second ~ (Print) ‖ zweiter Satz m. ‖ second coup m. / side ~ ‖ seitlich auftretender Zug m. ‖ force f. latérale.

pull back cable ‖ Rückzugsseil n. ‖ câble m. tracteur en arrière. / ~ contact ‖ Zugkontakt m. ‖ contact m. à tirage.

puller out, crucible ~ ‖ Tiegelzieher m. ‖ tireur m. de creusets.

pulley (Belt pulley) ‖ Riemenscheibe f.; Riemscheibe f. ‖ poulie f. / ~ (Rope pulley) ‖ Seilrolle f.; Seilscheibe f.; Rolle f.; Scheibe f. ‖ poulie f. (à corde). / ~ (Tackle) *see* ~ block. / back-shaft ~ (Spinn) ‖ Mandausenscheibe f. ‖ main f. douce. / belt ~ ‖ Riemenscheibe f. ‖ poulie f. / brake ~ ‖ Bremsscheibe f. ‖ poulie f. de frein. / cast iron ~ ‖ gußeiserne Riemenscheibe f. ‖ poulie f. en

fonte. / chain ~ || Kettenrolle f. || poulie f. à chaîne. / cone ~ || Kugelriemenscheibe f. || poulie f. à cône. / constant speed ~ || Einscheibe f. || monopoulie f. / double-groove ~ || zweirillige Seilscheibe f. || poulie f. à deux gorges. / driving cone ~ || Antriebstufenscheibe f. || poulie f. de commande à gradins. / expanding ~ || Ausdehnungsriemenscheibe f. || poulie f. extensible. / fast ~ || Festscheibe f. || poulie f. fixe. / flywheel ~ || Schwungradscheibe f. || poulie-volant f. / grooved ~ || Seilrolle f.; Nutrolle f. || poulie f. à gorge. / guide ~ || Leitrolle f. || galet-guide m. / fan driving ~ || Ventilatorriemenscheibe f. || poulie f. de ventilateur. / fast ~ || Festscheibe f. || poulie f. fixe. / loose ~ || Losscheibe f. || poulie f. folle. / metallic ~ || . Metallscheibe f. || poulie f. en métal. / movable ~ see loose ~. / one-groove ~ || einrillige Seilscheibe f. || poulie f. à une gorge. / overhung ~ || fliegende Scheibe f. || poulie f. en porte-à-faux. / quadruple-groove ~ || vierrillige Seilscheibe f. || poulie f. à quatre gorges. / rope ~ || Seilrolle f.; Seilscheibe f. || poulie f. à corde ou à câble. / single ~ || Einscheibe f. || monopoulie f. / split ~ || geteilte Riemscheibe f. || poulie f. divisée ou bipartite. / strap driving ~ || Antriebsriemenscheibe f. || poulie f. de commande à courroie. / stretching ~ || Spannrolle f. || poulie f. de tension; tendeur m. de courroie. / tension ~ see stretching ~. / tightening ~ see stretching ~. / triple-groove ~ || dreirillige Seilscheibe f. || poulie f. à trois gorges. / wire ~ (Railway signal gear) || Drahtseilrolle f. || poulie-guide f.; galet-guide m. / wooden ~ || Holzriemenscheibe f. || poulie f. en bois.

pulley, axle of the ~ || Bocknagel m. || axe m. de la poulie. / ~ block || Flaschenzug m.; Rollenzug m.; Talje f. || moufle f.; palan m. / ~ bracket || Rollenbock m. || support m. de galet.

pulley drive, constant speed ~ see pulley drive, single. / single ~ || Einscheibenantrieb m. || commande f. par monopoulie.

pulley frame (Mach) || Gehäuse n. eines Flaschenzuges || caisse f. ou chape f. ou corps m. d'une moufle. / (Mine) || Seilscheibengerüst n. || charpente f. des molettes.

pulley shaft (Mine) || Förderschacht m. || puits m. d'extraction.

pulley sheave see pulley block.

pull fastener || Reißverschluß m. || curseurs-bloqueurs mpl.

pulling, wool ~ of skins || Entwollen n. der Felle || délainage m. de peaux.

pulling boat || Ruderboot n. || bateau m. à rames.

pulling-down (Build) . || Abbruch m.; Abbruchsarbeit f. || démolition f. / ~ enterprise || Abbruchsunternehmen n. || entreprise f. de démolition.

pulling-in (Cable) || Einziehen n. || tirage m.; halage m.

pulling ladder || Schrotleiter f. || poulain m. / ~ power || Durchzugskraft f. || puissance f. d'entraînement.

pullock (Carp) || Netzriegel m.; Schußriegel m. || boulin m. ou traverse f. d'échafaudage.

pull, brake ~ rod || Bremslasche f. || bielle f. de frein. / ~ rope || Zugseil n. || câble-

tracteur m. / ~ switch || Zugschalter m. || interrupteur m. à tirette.

pulp (Pap) || Papierbrei m.; Pülpe f. || pâte f.; pulpe f. / ~ catcher (Pap) || Pülpefänger m. || puiseur m. de pâte. / ~ drying plant || Pülpetrocknungsanlage f. || installation f. de séchage de pulpes. / ~ engine knife || Holländermesser n. || lame f. de moulin à cylindre.

pulper || Rohrmühle f. || épulpeur m.

pulp grinder see pulp mill.

pulpit (Arch) || Kanzel f. || chaire f.; tribune f. sacrée. / ~ (Metal) || Kanzel f.; Manövrierbühne f. || banc m. de manœuvre.

pulp machine see ~ mill. / ~ meter (Pap) || Zeugregulator m.; Stoffregler m. || régulateur m. ou mésureur m. ou distributeur m. d'étoffe / ~ mill || Ganzholländer m. || pile f. raffineuse; cylindre m. broyeur. / ~ preparing machine || Papierstoffbereitungsmaschine f. || machine f. pour la préparation de la pâte à papier. / ~ press || Pülpepresse f.; Papierstoffpresse f. || presse f. à pulpe. / ~ strainer (Pap) || Knotenfänger m.; Zeugsichter m. || machine f. à boutons; épurateur m. de pâte à papier.

pulpy || breiig; breiartig || pâteux; comme de la bouillie.

pulsate, to || pulsieren; schlagen || vibrer; battre.

pulsating current || pulsierender Strom m. || courant m. pulsatoire. / ~ stress || stoßweise Beanspruchung f. || effort m. intermittent ou pulsatoire.

pulsation || Schwingen m.; Pulsieren n.; kurzer Stoß m. || pulsation f.

pulsatory || pulsierend; stoßweise || pulsatoire; pulsatif. / ~ current see pulsating current.

pulse || Puls m. || pouls m.

pulsometer || Pulsometer n. || pulsomètre m.

pulverizable || pulverisierbar || pulvérisable.

pulverization || Pulverisieren n. || pulvérisation f.

pulverizator see pulverizer.

pulverize, to || pulverisieren; pulvern || pulvériser; atomiser.

pulverized, finely ~ || feingepulvert || finement pulvérisé.

pulverized coal || Kohlenstaub m. || poussière f. de charbon. / ~ furnace || Kohlenstaubfeuerung f. || foyer m. à charbon pulvérisé.

pulverized fuel || pulverförmiger oder staubförmiger Brennstoff m.; Ölstaub m. || combustible m. pulvérisé. / ~ firing || Staubkohlenfeuerung f. || chauffage m. au poussier de charbon.

pulverizer || Pulverisierapparat m.; Zerstäuber m. || pulvérisateur m.; vaporisateur m.

pulverizing appliance see pulverizer. / ~ cylinder || Trommelmühle f. || moulin m. à tambour. / ~ mill || Kleinmühle f.; Pulvermühle f. || moulin m. de pulvérisation.

pulverulent || pulverförmig; feinpulvrig || pulvérulent. / ~ precipitation || pulverförmiger Niederschlag m. || précipité m. pulvérulent.

pumicate, to || abbimsen || poncer; polir à la ponce.

pumice || Bimsstein m. || pierre-ponce f. / artificial ~ || künstlicher Bimsstein m. || pierre-ponce f. artificielle. / natural ~ ||

natürlicher Bimsstein m. || pierre-ponce f. naturelle. / to polish with ~ || mit Bimsstein m. schleifen; bimsen || poncer; polir à la ponce. / to rub with ~ || mit Bimsstein m. abreiben || enlever à la pierre-ponce.

pumice, articles pl. of ~ || Bimssteinwaren fpl. || articles mpl. en pierreponce. / ~ cloth || Bimssteintuch n. || toile f. à pierre-ponce. / ~ concrete || Bimsbeton m. || béton m. de pierre-ponce. / ~ concrete ceiling || Bimsbetondecke f. || plafond m. en béton de pierre-ponce. / ~ concrete roof || Bimsbetondach n. || toiture f. en béton de pierre-ponce. / ~ paper || Bimssteinpapier n. || papierponce m. / ~ powder || Bimssteinpulver n. || poudre f. de pierre-ponce. / ~ product || Bimssteinerzeugnis n. || produit m. de pierre-ponce. / ~ quarry || Bimssteingrube f. || carrière de pierre-ponce. / ~ sand || Bimssand m. || sable m. de pierre-ponce. / ~ soap || Bimssteinseife f. || savon-ponce m.

pumice stone see pumice.

pump, to || pumpen || pomper. / ~-out || auspumpen || vider à la pompe. / ~-up a tyre || einen Reifen m. aufpumpen || gonfler un pneu. / ~ a vessel dry || ein Schiff m. auspumpen || agréner un navire.

pump || Pumpe f. || pompe f. / acid ~ || Säurepumpe f. || pompe f. à acides. / air ~ || Luftpumpe f. || pompe f. à air. / air ~ (Laboratory) || Entlüftungspumpe f. || trombe f. pneumatique. / ball ~ || Kugelpumpe f. || pompe f. à boulet. / bilge ~ || Bilgepumpe f.; Lenzpumpe f. || pompe f. de cale. / the ~ blows || die Pumpe zieht Luft || la pompe prend de l'air. / boiler feed ~ || Kesselspeisepumpe f. || pompe f. d'alimentation de chaudière. / bore hole ~ || Abteufpumpe f. || pompe f. de sondage. / building ~ || Baupumpe f. || pompe f. de construction ou à épuisement. / ~ for canalization || Pumpmaschine f. für Kanalisation || pompe f. pour canalisation. / ~ with capacity of . . . || Pumpe f. mit einer Leistung von . . . || pompe f. d'un débit de . . . / centrifugal ~ || Kreiselpumpe f.; Zentrifugalpumpe f. || pompe f. centrifuge. / chain ~ || Kettenpumpe f.; Paternosterpumpwerk n. || pompe f. à chapelet; chapelet m. / chlorine water ~ || Chlorwasserpumpe f. || pompe f. pour eau de chlore. / choked ~ || unklare Pumpe f. || pompe f. engorgée. / circulating ~ || Umlaufpumpe f.; Zirkulationspumpe f. || pompe f. de circulation. / circulating ~ for cooling water || Kühlwasserumlaufpumpe f. || pompe f. de circulation pour l'eau de refroidissement. / ~ with claper valve || Pumpe f. mit Schlagventil || pompe f. à soupape m. d'arrêt. / cooling water ~ || Kühlwasserpumpe f. || pompe f. à eau de refroidissement. / diaphragm ~ || Membranpumpe f.; Diaphragmapumpe f. || pompe f. à diaphragme. / diffusion ~ || Diffusorpumpe f. || pompe f. à diffusion. / donkey ~ || Hilfspumpe f. || pompe f. à vapeur auxiliaire. / double-acting ~ || doppeltwirkende Pumpe f. || pompe f. à double effet. / double three-throw pressure ~ || Doppeldrillingspreßpumpe f. || pompe f. double à compression à trois plongeurs. / the ~ draws || die Pumpe faßt oder zieht || la pompe marche

ou est en prise *ou* est chargée. / duplex ~ ‖ Duplexpumpe f. ‖ pompe f. duplex. / efficient ~ ‖ leistungsfähige Pumpe f. ‖ pompe f. à grand débit. / electrically driven ~ ‖ elektrisch betriebene Pumpe f. ‖ électropompe f.; pompe f. à commande par électromoteur. / engine-driven ~ ‖ Motorpumpe f. ‖ pompe f. à moteur. / feed~ ‖ Speisepumpe f. ‖ pompe f. d'alimentation. / the ~ does not fetch ‖ die Pumpe faßt nicht ‖ la pompe ne prend pas. / filter press ~ ‖ Filterpreßpumpe f. ‖ pompe f. de filtre-presse. / fire ~ ‖ Feuerlöschpumpe f. ‖ pompe f. à incendie *ou* à feu. / foot ~ ‖ Fußpumpe f. ‖ pompe f. à pied. / forcing ~ ‖ Druckpumpe f. ‖ pompe f. foulante; refouleur m. / foul – *see* choked ~. / to free the ~ from obstacles ‖ Pumpenverstopfungen fpl. beseitigen ‖ déboucher *ou* dégorger la pompe. / to free the ~ from water ‖ die Pumpe entleeren ‖ franchir *ou* épuiser la pompe. / fresh water ~ ‖ Frischwasserpumpe f. ‖ pompe f. à eau fraîche. / gear-driven ~ ‖ Zahnradpumpe f. ‖ pompe f. rotative à engrenage. / hand ~ ‖ Handpumpe f. ‖ pompe f. à main. / hard rubber ~ ‖ Hartgummipumpe f. ‖ pompe f. en caoutchouc durci. / house-type ~ ‖ Hauswasserpumpe f.; Hofpumpe f. ‖ pompe f. ménagère. / helical ~ ‖ Schraubenpumpe f. ‖ pompe f. hélicoïdale. / high-pressure ~ ‖ Hochdruckpumpe f.; Preßpumpwerk n. ‖ pompe f. à haute pression. / horizontal three-throw pressure ~ ‖ liegende Drillingspreßpumpe f. ‖ pompe f. horizontale à compression à trois plongeurs. / juice ~ ‖ Saftheber m. ‖ monte-jus m. / ~ for liquid manure ‖ Jauchepumpe f. ‖ pompe f. à purin. / lubricating oil ~ ‖ Schmierölpumpe f. ‖ pompe f. à huile. / mechanically driven ~ ‖ maschinell angetriebene Pumpe f. ‖ pompe f. à commande mécanique. / mercury air ~ ‖ Quecksilberluftpumpe f. ‖ trombe f. à mercure. / motor-driven ~ ‖ Motorpumpe f. ‖ pompe f. à moteur. / multistage ~ ‖ Mehrstufenpumpe f. ‖ pompe f. à plusieurs étages. / oil ~ ‖ Ölpumpe f. ‖ pompe f. à huile. / oil hand ~ ‖ Handölpumpe f. ‖ pompe f. de graissage à main. / piston ~ ‖ Kolbenpumpe f. ‖ pompe f. à piston. / plunger ~ ‖ Tauchkolbenpumpe f. ‖ pompe f. à piston plongeur. / power-type ~ ‖ Motorluftpumpe f. ‖ gonfleur m. mécanique. / press ~ ‖ Preßpumpe f. ‖ pompe f. de pression. / pressure ~ ‖ Druckpumpe f. ‖ pompe f. de pression. / pressure test ~ ‖ Probedruckpumpe f. ‖ pompe f. (foulante) d'épreuve. / propeller ~ ‖ Schraubenschaufler m. ‖ pompe f. à hélice. / ~ for the removal of fæces ‖ Pumpe f. für die Fäkalienabfuhr ‖ pompe f. pour l'enlèvement des matières fécales. / rinsing ~ ‖ Spülpumpe f. ‖ pompe f. à laver *ou* de balayage. / road-side gasoline ~ ‖ Zapfsäule f. für Benzin; Benzin(zapf)pumpe f. ‖ distributeur m. d'essence. / rotary ~ ‖ Rotationspumpe f.; Rundlaufkolbenpumpe f. ‖ pompe f. rotative *ou* rotatoire. / scavenging ~ (Gas engine) ‖ Spülpumpe f. ‖ pompe f. de balayage. / semi-rotative ~ ‖ Flügelpumpe f. ‖ pompe f. mi-rotative. / simplex ~ ‖ Simplexpumpe f. ‖ pompe f. simplex. / sinking ~ ‖ Abteufpumpe f. ‖ pompe f.

de sondage. / steam feed ~ ‖ Dampfspeisepumpe f. ‖ pompe f. d'alimentation à vapeur. / ~ of steam pressure ‖ mit gespanntem Dampf arbeitende Pumpe f. ‖ pompe f. à pression de vapeur. / syrup ~ ‖ Siruppumpe f. ‖ pompe f. à mélasses. / turbine ~ ‖ Turbinenpumpe f.; Turbopumpe f. ‖ turbopompe f. / twin ~s pl. ‖ Zwillingspumpe f. ‖ pompes fpl. jumelles. / two-way ~ ‖ Zweiwegpumpe f. ‖ pompe f. à deux voies. / tyre ~ ‖ Reifenpumpe f. ‖ pompe f. à pneumatique; gonfle-pneus m.; gonfleur m. de pneus. / vacuum ~ ‖ Vakuumpumpe f. ‖ pompe f. à vide. / valveless ~ ‖ ventillose Pumpe f. ‖ pompe f. sans clapet. / vane-type ~ ‖ Flügelpumpe f. ‖ pompe f. à palette. / vertical three-throw pressure ~ ‖ stehende Drillingspreßpumpe f. ‖ pompe f. verticale à compression à trois plongeurs. / water ~ ‖ Wasserpumpe f. ‖ pompe f. à eau. / water circulating ~ ‖ Kühlwasserpumpe f. ‖ pompe f. d'eau de circulation. / wing ~ ‖ Flügelpumpe f. ‖ pompe f. à ailettes. / wooden ~ ‖ Holzpumpe f. ‖ pompe f. en bois.

pump bit ‖ Löffelbohrer m.; Röhrenbohrer m. ‖ cuiller f. à pompes; rouanne f. de pompe; perçoir m. à curette. / ~ body ‖ Pumpengehäuse n. ‖ corps m. de pompe. / ~ bolt ‖ Pumpenbolzen m. ‖ cheville f. de pompe. / ~ borer *see* ~ bit. / ~ box ‖ Pumpenkasten m. ‖ caisse f. de pompe. / ~ brake *see* ~ handle. / ~ case ‖ Pumpengehäuse n. ‖ chapelle f. *ou* caisse f. *ou* enveloppe f. de pompe. / delivery of the ~ per hour ‖ Pumpenleistung f. in der Stunde; stündliche Fördermenge f. *oder* Liefermenge f. ‖ refoulement m. *ou* débit m. par heure. / ~ drill ‖ Drillbohrer m.; Rennspindel f. ‖ drille f.; trépan m. / ~ drive ‖ Pumpenantrieb m. ‖ commande f. de pompe. / ~ driver ‖ Pumpenführer m. ‖ conducteur m. de pompes.

pumped out ‖ abgepumpt; ausgepumpt ‖ épuisé à la pompe.

pump fed basin ‖ Rückpumpbecken n. ‖ bassin m. de refoulement. / ~ power station ‖ Pumpspeicherwerk n. ‖ usine f. d'énergie automatisée.

pump fittings pl. ‖ Pumpenarmatur f. ‖ armature f. pour pompes. / ~ handle ‖ Pumpenschwengel m. ‖ brimbale f. / ~ hook ‖ Pumpenhaken m. ‖ croc m. de pompe. / ~ hose ‖ Pumpenschlauch m. ‖ manche à pompe. / ~ housing ‖ Pumpengehäuse n. ‖ corps m. de pompe. / impeller of ~ ‖ Pumpenflügelrad n. ‖ roue f. à ailettes.

pumping ‖ Wasserhaltung f.; Flüssigkeitsförderung f. ‖ pompage m.; élévation f. des eaux.

pumping engine ‖ Pumpmaschine f. ‖ machine-pompe f. / ~ delivering x cubic meters of water per minute ‖ Pumpmaschine f. mit x m³ minutlicher Leistung ‖ machine f. à pomper d'un débit x m³ par minute. / high-pressure ~ ‖ Hochdruckpumpmaschine f. ‖ pompe f. *ou* machine f. à pomper à haute pression. / ~ underground ~ (Mine) ‖ unterirdische Wasserhaltungsmaschine f. ‖ machine f. d'épuisement souterraine. / ~ for water works ‖ Wasserhaltungsmaschine f. ‖ machine f. pour l'adduction d'eau.

pumping plant ‖ Pumpwerk n. ‖ installation f. de pompes. / ~ station of a

regional water work ‖ Förderstation f. einer Landeswasserversorgung ‖ poste m. de mise en charge des pompes d'une distribution d'eau d'une région.

pump injection (Diesel) ‖ Druckeinspritzung f. ‖ injection f. sous pression. / ~ kettle ‖ Saugkorb m. einer Pumpe ‖ crépine f. d'une pompe.

pumpkin ‖ Kürbis m. ‖ courge f. / ~ pip ‖ Kürbiskern m. ‖ pépin m. de citrouille *ou* de courge.

pump leather ‖ Ledermanschette f. einer Pumpe ‖ cuir m. embouti de pompe. / ~ maker ‖ Brunnenmacher m.; Pumpenfabrikant m. ‖ fontenier m.; fontainier m.; fabricant m. de pompes.

pumpman ‖ Pumper m. ‖ pompier m.

pump nail ‖ Pumpenspiker m. ‖ clou m. de pompe. / ~ piston ‖ Pumpenkolben m. ‖ piston m. de pompe. / ~ piston chamber ‖ Pumpenzylinder m. ‖ corps m. *ou* cylindre m. de pompe. / ~ plunger ‖ Plunger m.; Tauchkolben m. ‖ piston m. plongeur. / ~ rod ‖ Pumpenstange f. ‖ tige f. *ou* verge f. *ou* lance f. de pompe. / ~ rods pl. ‖ Pumpengestänge n. ‖ tiges fpl. de pompes.

pumps pl. for gymnastics ‖ Turnschuhe mpl. ‖ chaussures fpl. de gymnastes.

pump spear *see* ~ rod. / ~ spindle ‖ Pumpenwelle f. ‖ arbre m. de pompe. / ~ storage plant *see* ~ storage station.

pump storage station ‖ Pumpspeicherwerk n. ‖ station f. d'accumulation d'eau par pompage. / basin for a ~ ‖ Becken n. für ein Pumpspeicherwerk ‖ réservoir m. d'une usine d'accumulation par pompage. / equalizing basin for a ~ ‖ Ausgleichbecken n. für ein Pumpspeicherwerk ‖ bassin m. compensateur d'une usine d'accumulation par pompage.

pump sump ‖ Pumpensumpf m. ‖ fosse f. de pompe. / ~ tack *see* ~ nail. / ~ telecontrol ‖ Pumpenfernsteuerung f. ‖ télécommande f. des moteurs de pompes. / total amount of delivery of ~ ‖ Liefermenge f. einer Pumpe ‖ débit m. d'une pompe. / ~ tube ‖ Pumpenrohr n. ‖ tuyau m. de pompe. / ~ valve ‖ Pumpenventil n. ‖ soupape f. de pompe. / ~ wagon ‖ Pumpenwagen m. ‖ voiture f. à pompe. / ~ water ‖ Brunnenwasser n. ‖ eau f. de puits. / ~ well ‖ Pumpensood m. ‖ archipompe f.

punch, to ‖ durchlochen; lochen; stanzen ‖ percer; déboucher; poinçonner. / ~ deep ‖ tief lochen ‖ estamper maigre. / ~ flat ‖ dicht lochen ‖ estamper gras. / ~ out ‖ ausstanzen ‖ poinçonner; découper.

punch ‖ Durchschlag m.; Locher m.; Lochstempel m.; Stempel m. ‖ chasse-pointes m.; découpoir m.; poinçon m. / ~ (Letter-f) ‖ Patrize f.; Stempel m.; Schriftstempel m. ‖ poinçon m. / ~ (Liquor) ‖ Punsch m. ‖ punch m. / bringing down the ~ to centre mark ‖ Niederstellen n. des Stempels auf Körnermarke ‖ amenage m. du poinçon sur le coup de pointeau. / centre ~ ‖ Körner m. ‖ pointeau m. / automatic centre ~ ‖ Federdruckkörner m. ‖ pointeau m. automatique. / to cut ~es pl. ‖ Stempel mpl. schneiden ‖ graver des poinçons mpl. / drift ~ ‖ Dorn m.; Durchschlag m. ‖ chassoir m.; chasse-clavette m. / drift ~ (Forge) ‖ Durchtreiber m. ‖ mandrin m. de forge *ou* à chaud. / hollow ~ ‖ Locheisen n. ‖ emporte-pièce m. / lever ~

|| Hebelstanze f. || poinçonneuse f. à levier. / manhole ~ || Mannlochstanze f. || estampe f. à faire des regards *ou* des ouvertures de visite. / prick ~ *see* centre ~. / ~ for ramming in a pile || Aufsatz m. zum Einrammen eines Pfahles || faux-pilot m. pour enfoncer un pieu. / round ~ || Kreisbohrer m.; Kreisausheber m. || coupe-cercle m. / ~ for stamping rosettes || Rosettenstempel m. || rosetier m.

Punch and Judy show || Kasperletheater n. || théâtre m. de marionnettes.

punch cutter || Stempelschneider m. || graveur m. de poinçons. / ~ cutting machine || Stempelschneidemaschine f. || machine f. à graver les caractères.

punched || gestanzt || estampé. / ~ articles pl. in lots || Stanzmassenartikel mpl. || articles mpl. découpés en séries. / ~-out round || rund ausgestanzt || découpé rond. / ~ plate || durchlochtes Blech n. || tôle f. perforée.

puncheon (Brew) || Gärfaß n. || fût m. *ou* tonne f. de fermentation. / ~ (Needl) || Lüfter m. || boutereau m. / round ~ || Ringdorn m. || mandrin m. rond.

puncher || Stanzer m. || poinçonneur m. / ~ (Aut tel) || Schriftlocher m. || compositeur-perforateur m. / ~ (Bookb) || Stanze f. || perçoir m. / ~ (Mach) *see* punching machine. / sheet iron ~ || Blechstanzer m. || poinçonneur m. de tôle. / tin plate ~ || Weißblechstanzer m. || perceur m. de fer-blanc.

puncher-chisel || Locheisen n.; Räumeisen n. || équarrissoir m.

punch hammer || Döppel m.; Döpper m. || marteau m. à cuvette. / swing out ~ holder || ausschwenkbarer Stempelhalter m. || porte-poinçon m. oscillant.

punching, metal ~ || Metallstanzen n. || perçage m. de métaux.

punching apparatus || Stanzapparat m. || dispositif m. de poinçonnage. / ~ capacity || Stanzleistung f. || rendement m. de poinçonnage. / ~ device || Stanzeinrichtung f. || dispositif m. pour poinçonner. / ~ die || Lochkaliber n. || calibre m. de perçage. / ~ effect || Durchschlagswirkung f. || effet m. de poinçonnage. / ~ instrument || Stanzwerkzeug n. || poinçon m. instrument m. à poinçonner. / ~ iron (Join) || Rändeleisen n. || molettes fpl. / ~ knife for cardboard || Stanzmesser n. für Pappe || couteau m. d'estampage pour carton. / ~ magnet || Stanzmagnet m. || aimant m. performateur *ou* poinçonneur.

punching machine || Stanzmaschine f.; Lochmaschine f.; Lochstanze f. || machine f. à découper *ou* à poinçonner; poinçonneuse f. / ~ (Weav) || Vorstechmaschine f. || machine f. à piquer. / ~ with bevel shears || Lochstanze f. mit Gehrungsschere || poinçonneuse f. avec cisailles pour coupes obliques. / fish plate ~ || Laschenlochmaschine f. || machine f. à percer les éclisses. / lever ~ || Hebellochstanze f. || poinçonneuse f. à levier. / ~ for manholes *see* punch, manhole ~. / multiple ~ || Mehrstempellochmaschine f. || poinçonneuse f. à poinçonnage multiple. / ~ with multiple punching attachment || Stanze f. mit Mehrstempellochapparat || poinçonneuse f. multiple. / multiple rectilinear ~ || Vielstempelreihenlochmaschine f. || poin-

çonneuse f. multiple rectiligne. / ~ requiring only one operator || Einmannlochmaschine f. || poinçonneuse f. conduite par un seul homme. / plate ~ || Blechlochmaschine f. || découpoir m. à tôle. / ~ with profile iron shears || Lochstanze f. mit Gehrungsschere || poinçonneuse f. avec cisailles pour coupes obliques. / rectilinear ~ || Reihenlochmaschine f. || poinçonneuse f. rectiligne. / universal ~ || Universallochmaschine f. || poinçonneuse f. universelle. / universal horizontal ~ || horizontale Universallochmaschine f. || poinçonneuse f. universelle horizontale. / universal horizontal ~ combined with bending and straightening press || horizontale Universallochmaschine f. mit Biege- und Richtpresse || poinçonneuse f. universelle horizontale combinée avec presse à cintrer et à dresser. / hoop shearing and ~ || Bandeisenschere und Lochstanze f. || machine f. à couper et à perforer les feuillards.

punching and notching machine, universal double-ended ~ || doppelte Universalloch- und Ausklinkmaschine f. || poinçonneuse f. double universelle-grugeoir.

punching press || Lochpresse f. || poinçonneuse f.

punching recorder for fire and police alarm systems || Lochschreiber m. für Feuermelde- und Polizeirufanlagen || appareil m. enregistreur par perforation pour installations d'incendie et d'appel de police.

punchings pl. || Stanzabfälle mpl. || débouchures fpl.

punching, fixed table for ~ || festgegossener Stanztisch m. || table f. de poinçonnage fixe. / ~ tools pl. || Lochwerkzeug n.; Schnittwerkzeug n. || outils mpl. à poinçonner *ou* à découper. / ~ work || Stanzarbeit f. || travail m. de poinçonnage.

punch manufacturer (Liquor) || Punschfabrikant m. || fabricant m. de punch. / centre ~ mark || Körnerschlag m. || coup m. de pointeau.

punch pliers pl. || Lochzange f. || pince f. emporte-pièce. / ~ with bent handles || Lochzange f. mit geschweiften Schenkeln || pince f. emporte-pièce avec branches coudées.

punch pliers tube || Lochhülse f. für Lochzange || tube m. pour pince à emporte-pièce.

punch press man || Lochmaschinenarbeiter m. || poinçonneur m. à la machine.

punctual || pünktlich || ponctuel; exact.

punctuate, to || interpunktieren || mettre la ponctuation.

punctuation || Interpunktion f. || ponctuation f.

puncture, to || durchlochen; durchstoßen || ponctionner; percer; perforer; trouer.

puncture of tyre || Reifendefekt m. || crevaison f. du pneumatique.

punctured (Cable) || durchgeschlagen || brûlé. / ~ tyre || zerplatzter Luftreifen m. || pneu m. piqué *ou* percé *ou* crevé.

puncture-proof (Tyre) || nagelsicher || à l'épreuve de (toute) perforation par un clou.

pungent (Chem) || stechend || piquant.

punt (Glassm) || Hefteisen n. || pontil m.

pupil || Schüler m. || élève m. / ~ (Eye) || Pupille f. || pupille f. / exit ~ || Austritts-

pupille f. || pupille f. de sortie; cercle m. *ou* anneau m. oculaire. / distance between the ~s || Pupillenabstand m. || écartement m. pupillaire.

pupilloscope || Pupilloskop n. || pupilloscope m.

Pupin coil for open lines || Pupinfreileitungsapparat m. || appareil m. pour la pupinisation des lignes aériennes.

pupinization of cables || Pupinisierung f. der Kabel || pupinisation f. des câbles. / plan of ~ || Pupinisierungsplan m. || plan m. de pupinisation. / ~ point || Pupinspulenpunkt m. || point m. de pupinisation.

pupinize, to || pupinisieren; Pupinspulen fpl. einschalten || pupiniser.

puppet || Gliederpuppe f.; Puppe f. || poupée f. articulée; marionnette f.; poupée f. / ~ (Turn) || Docke f. || poupée f.

purchase, to || einkaufen; erwerben || acheter. / to intend to purchase || reflektieren auf . . . || avoir des vues fpl sur . . .

purchase (Mach) || Hebelarm m.; Hebevorrichtung f. || bras m. de levier; pouvoir m. de multiplication. / ~ (Trade) || Aufkauf m.; Anschaffung f.; Kauf m. || achat m. / to cancel a ~ || einen Kauf m. rückgängig machen || révoquer un achat. / ~ of current || Strombezug m.; Fremdbezug m. von Strom || achat m. de courant. / ~ of land || Grundankauf m. || achat m. de terrain. / ~ cost || Ankaufspreis m.; Einkaufspreis m. || prix m. d'achat. / ~ price *see* ~ cost.

purchaser || Käufer m. || client m.; client-acheteur m.

purchasing department || Bestellungsabteilung f.; Einkaufsabteilung f. || service m. des commandes. / ~ power || Kaufkraft f. || pouvoir m. d'achat.

pure || rein || pur. / ~ (Miner) || gediegen || vierge; pur; natif. / absolutely ~ || absolut rein || absolument pure. / ~ cast iron || geläutertes Roheisen n. || fonte f. épurée. / ~ culture || Reinkultur f. || culture f. pure. / ~ steel || reiner Stahl m. || acier m. particulièrement pur.

pureness || Reinheit f. || pureté f.; netteté f.

purge cock || Entleerungshahn m. || robinet m. de décharge *ou* de purge.

purificating drum || Läutertrommel f. || tambour m. de purification.

purification || Reinigung f.; Klärung f. || purification f.; décantation f. / ~ of caoutchouc || Reinigen n. des Kautschuks || déchiquetage m. du caoutchouc. / ~ of gas || Reinigung f. des Gases || épuration f. du gaz. / ~ of gas mixtures || Reinigung f. von Gasgemischen || purification f. de mélanges de gaz. / ~ of juice || Raffinieren n. des Saftes || épuration f. *ou* raffination f. du jus. / ~ of sewage || Abwasserreinigung f. || épuration f. des eaux d'égout. / plant for the ~ of waste-water *see* purifying plant, waste water ~.

purification machine || Reinigungsmaschine f. || machine f. à nettoyer.

purification plant || Reinigungsanlage f. || installation f. d'épuration. / sewage ~ || Kläranlage f. || installation f. de clarification des eaux d'égout.

purifier || Reinigungsapparat m.; Waschapparat m. || épurateur m.; lavoir m.; purificateur m.; sasseur m. / ~ (Oil) || Ölabscheider m. || dégraisseur m. / air ~ || Luftreinigungsgerät n. || appareil m.

pour la purification de l'air. / bolter ~ see sieve ~. / feed-water ~ || Speisewasserreiniger m. || épurateur m. d'eau d'alimentation. / semolina ~ || Grießputzmaschine f. || épurateur m. de semoules. / sieve ~ || Siebreiniger m. || sasseur m.; épurateur m. sasseur.

purify, to || reinigen; läutern; klären || purifier; épurer. / ~ discharge water || die Abwässer npl. reinigen || épurer les eaux fpl. d'écoulement. / ~ the oil || das Öl reinigen || épurer ou filtrer l'huile f.

purifying plant for blast furnace gas || Hochofengasreinigungsanlage f. || installation f. d'épuration des gaz de hautsfourneaux. / waste water ~ || Abwasserreinigungsanlage f. || installation f. d'épuration des eaux ménagères.

purity || Reinheit f. || pureté f.; netteté f. / chromatic ~ || Farbenreinheit f. || pureté des teintes. / the steel is remarkable by its ~ || der Stahl m. zeichnet sich durch größte Reinheit aus || l'acier m. est remarquable par son extrême pureté.

purl (Wiredr) || Kantille f.; Drahtspirale f. || cannetille f.

purlin (Carp) || Pfette f. || panne f. / ~ roof || Pfettendach n.; Pfettendachgebinde n. || comble m. à pannes. / ~ knitting machine || Noppenstrickmaschine f. || machine f. à tricoter à nopes.

purple || Purpur m.; Purpurfarbe f. || pourpre f; m); couleur f. de pourpre. / gold ~ || Goldpurpur m. || pourpre m. de Cassius.

purple ore || Kiesabbrand m. || pyrite f. traitée; résidus mpl. de pyrites grillées. / ~ oxide || Oxydrot n. || rouge m. oxyde. / ~ ribbon (Typewr) || Violettfarbband n. || ruban-encreur m. violet. / ~ wood || Purpurholz n. || bois m. d'amaranthe violet.

purpose of employment || Verwendungszweck m. || but m. d'emploi.

purpuric acid || Purpursäure f. || acide m. purpurique.

purse || Geldbörse f.; Geldtäschchen n. || bourse f.; porte-monnaie m. / leather ~ || Lederbörse f. || bourse f. en cuir. / metal ~ || Metallbörse f. || bourse f. en mailles métalliques.

purse maker || Geldbörsenmacher m. || boursier m. / ~ net || Beutelnetz n. || havenau m.; havenet m. boudeux. / ~ sewer || Geldtaschennäherin f. || couseuse f. en porte-monnaie.

pursuit airplane || Jagdflugzeug n. || aéroplane m. de poursuite.

purveyance works pl. || Lieferwerk n. || fabrique f. de livraison.

purveyor || Lieferant m. || pourvoyeur m.; fournisseur m.

push, to || drücken || pousser. / ~-on by setters || staken || gaffer.

push || Schub m.; Druck m. || poussée f. / ~ of a vault || Seitenschub m. eines Gewölbes || effort m. ou poussée f. d'une voûte.

pushbar || Stoßstange f. || poussoir m.

push button || Druckknopf m. || bouton m. poussoir. / ~ control || Druckknopfsteuerung f. || dispositif m. de manœuvre à boutons poussoirs; commande f. par bouton de pression. / ~ fire alarm installation || Druckknopffeuermeldeanlage f. || avertisseur m. d'incendie à boutons poussoirs. / ~ switch || Druck-

knopfschalter m. || interrupteur m. à bouton poussoir. / ~ switchboard || Druckknopfschrank m. || tableau m. à boutons poussoirs.

push car (Railw) || Rollwagen m. zum Handbetrieb || wagon-brouette f.; wagon m. de terrassement; lory m. / ~ cart || Steinkarren m. || diable m.

pusher (Met) || Ausdrücker m.; Ausdrückmaschine f. || défourneur m. mécanique; défourneuse f.; machine f. défourneuse. / ingot ~ || Blockausdrücker m. || défourneuse f. de lingots. / tram ~ (Mine) || Förderwagenaufschieber m. || appareil m. à remonter les wagonnets de transport.

pusher airscrew see ~ propeller. / ~ propeller || Druckpropeller m.; Druckschraube f.; Treibschraube f. || hélice f. propulsive. / ~ screw see ~ propeller.

push feeder || Schubaufgabevorrichtung f. || feeder m. à poussée. / ~ heating furnace || Stoßofen m. || four m. à secousses ou à poussée.

pushing forward of the rails in a longitudinal direction || Verschieben n. des Gleises in der Längsrichtung || déplacement m. des rails en sens longitudinal.

pushing blade || Mitnehmerschaufel f. || palette f. d'entraînement.

pushing device, stationary ~ for cooling furnaces || feste Stoßvorrichtung f. für Kühlöfen || dispositif m. de choc fixe pour fours de refroidissement. / transportable ~ for annealing furnaces || fahrbare Stoßvorrichtung f. für Glühöfen || dispositif m. mobile de choc pour fours à recuire. / wagon ~ || Wagenstoßvorrichtung f. || dispositif m. de choc pour wagons.

pushing machine || Ausdrückmaschine f.; Ausstoßmaschine f. || machine-défourneuse f. / ~ ram || Ausdrückstange f.; tige-défourneuse f. / ~ rod see ~ ram. / ~ trough || Schubförderrinne f. || auge f. transporteuse.

push key (Typewr) || Schubtaste f. || touche f. à glissière. / ~ pole for wagons || Stoßbaum m. für Wagen || levier m. de poussée pour wagons. / ~ pull connection (Tel) || Gegentaktschaltung f. || couplage m. de push pull.

push rod || Schubstange f. || bielle f. de poussée. / valve ~ || Ventilanhubstange f. || tige f. de galet de soupape.

push rods pl. || Gestänge n. || tringles fpl.

push-up (Met) || Stauchung f. || refoulement m.

put, to || legen || mettre; placer; poser. / ~ a beam || einen Balken m. einziehen || traverser une poutre. / ~ in boards || in Pappe f. binden || cartonner. / ~ the bread into the oven || das Brot in den Backofen einschieben || enfourner le pain. / ~ the bricks in the kiln || die Ziegel mpl. in den Ofen einsetzen; den Brand m. einfahren || enfourner les briques fpl. / ~ a buoy || eine Boje f. auslegen || placer une bouée. / ~ by || reservieren || réserver; tenir en réserve. / ~ in circuit || einschalten || mettre en circuit m. / ~ into circulation (Bank) || girieren || endosser; virer. / ~-down the flooring || die Dielen fpl. legen || poser les planches fpl. / ~-down the helm || das Ruder in Lee legen || mettre la barre dessous. / ~ the joists || die Balken mpl. einbringen || mettre en place les solives fpl.; mettre dedans les

solives fpl. / ~ hops in the beer || das Bier hopfen || houblonner la bière. / ~ on the last (Shoem) || auf den Leisten m. schlagen || mettre ou monter sur forme f. / ~-off || vertrösten || faire prendre patience f. / ~-on || ansetzen || mettre; appliquer; ajuster. / ~-on armour plates (Shipb) || panzern || cuirasser. / ~-on the block (Hatt) || den Filz m. setzen oder anstoßen || dresser le feutre. / ~-on the brakes pl. || die Bremsen fpl. anziehen || freiner. / ~ out of gear || loskuppeln; auskuppeln || désembrayer. / ~ on laces || Spitzen fpl. annähen || entoiler les dentelles fpl. / ~ out of circuit || ausschalten || mettre hors de circuit m. / ~ on paste board || auf Pappe f. stecken || encartonner. / ~ to practical use || in die Praxis f. umsetzen || mettre en pratique. / ~ to sea || in See f. gehen || mettre ou prendre la mer. / ~ a sheet into the hand-press || einen Bogen m. auf die Handpresse auflegen || poser une feuille dans la presse à bras. / ~ a ship on the stocks || ein Schiff n. auf Stapel legen || mettre un navire sur le chantier. / ~ the steam on (Steam engine) || die Dampfmaschine anlassen || faire entrer la vapeur f. / ~ into swaths || in Schwaden fpl. legen || enjaveler; javeler. / ~ in tune || stimmen || mettre d'accord m. / ~ a wall upon vaults || eine Mauer f. auf Bogen satteln || monter un mur sur des arcs. / ~-up as candidate || als Kandidat m. aufstellen || porter ou présenter comme candidat m.

putlog (Build) || Gerüststange f. || boulin m.; perche f. / ~ (Carp) || Schußriegel m.; Netzriegel m. || boulin m. ou traverse f. d'échafaudage. / ~ hole || Gerüststangenloch n. || trou m. de boulin ou de traverse.

putrefaction || Verwesung f.; Fäulnis f. || putréfaction f. / ~ process || Fäulnisprozeß m. || putréfaction f.

putrefactive bacterium || Fäulnisbakterium n. || bactérie f. putride.

putridity || Fäulnis f. || putréfaction f.

putrify, to || verwesen; faulen || se putréfier.

puttee || Übergamasche f. || molletière f.

putter (Mine) || Schlepper m. || traîneur m.; rouleur m.; esclauneur m.; hiercheur m.

putting into service || Inbetriebsetzung f. || mise f. en marche. / ~ out of circuit || Ausschaltung f. || mise f. hors circuit. / ~ together || Zusammenstellung f. || rapprochement m.; assemblage m.; combinaison f.; association f.

putting-down machine for transporting barren rock || Absetzmaschine f. für den Abraumbetrieb || machine f. d'enlèvement pour travaux de déblaiement.

putting-on device for ingots || Auflegevorrichtung f. für Blöcke || dispositif m. de chargement pour blocs.

putty, to || spachteln || mastiquer.

putty || Kitt m. || mastic m. / asbestos ~ || Asbestkitt m. || mastic m. d'amiante. / glazier's ~ || Fensterkitt m.; Glaserkitt m. || mastic m. à vitres; lut m. de vitrier. / joiner's ~ || Holzkitt m. || mastic m. de colle; futée f.

putty oil || Kittöl n. || huile f. pour mastics. / ~ paste || Spachtelkitt m. || masse f. ou mastic m. de rebouchage.

puzzle article || Scherzartikel m. || article m. de surprise. / ~ lock || Kombinationsschloß n. || serrure f. secrète ou à surprise.

pycnite ‖ Pyknit m.; stengeliger Topas m.; Stangenstein m. ‖ pycnite f.

pyramid ‖ Pyramide f. ‖ pyramide f. / square ~ ‖ vierseitige Pyramide f. ‖ pyramide f. quadrangulaire. / triangular ~ ‖ dreiseitige Pyramide f.; Tetraeder n. ‖ pyramide f. triangulaire.

pyramidal poplar ‖ Pyramidenpappel f. ‖ peuplier m. pyramidal.

pyramid crane ‖ Pyramidenkran m. ‖ gruepyramide f. tournante. / ~ stand ‖ Pyramidenstativ n. ‖ pied-pyramide m. / ~ support ‖ Pyramidenstütze f. ‖ pylône m. en pyramide. / ~ switchboard (Tel) ‖ Pyramidenschrank m. ‖ tableau m. en pyramide.

pyrargyrite ‖ dunkles Rotgültigerz n.; Antimonsilberblende f.; Pyrargyrit m. ‖ argent m. antimonié sulfuré; pyrargyrite f.

pyrites ‖ Pyrit m.; Schwefelkies m. ‖ pyrite f. / arsenical ~ ‖ Arsenkies m. ‖ fer m. arsenical. / capillary ~ ‖ Schwefelnickel m.; Haarkies m. ‖ nickel m. sulfuré; pyrite f. capillaire. / copper ~ ‖ Kupferkies m. ‖ pyrite f. cuivreuse. / iron ~ ‖ Eisenkies m.; Schwefelkies m. ‖ pyrite f.; fer m. sulfuré. / magnetic ~ ‖ Magnetkies m.; Leberkies m. ‖ fer m. sulfuré magnétique; pyrite f. magnétique. / roasted ~ ‖ Kiesabbrand m. ‖ pyrites fpl. grillées.

pyrites detector ‖ Pyritdetektor m. ‖ détecteur m. à pyrite. / ~ furnace ‖ Pyritofen m. ‖ four m. pour pyrites.

pyroacetic spirit ‖ Essiggeist m.; Azeton n. ‖ esprit m. pyro-acétique; acétone f.

pyroacid ‖ brenzliche Säure f.; Brenzsäure f. ‖ acide m. pyrogéné.

pyrocatechine ‖ Brenzkatechin n. ‖ pyrocatéchine f.

pyroelectric ‖ pyroelektrisch ‖ ˉpyroélectrique.

pyroelectricity ‖ Pyroelektrizität f. ‖ pyroélectricité f.

pyrogallic acid ‖ Pyrogallussäure f. ‖ acide m. pyrogallique.

pyrogallol see pyrogallic acid.

pyrognostical analysis ‖ pyrognostische Analyse f. ‖ analyse f. pyrognostique.

pyrography ‖ Brandmalerei f.; Holzbrandmalerei f. ‖ pyrogravure f.

pyroligneous acid ‖ Holzessig m. ‖ acide m. pyroligneux.

pyrolignite ‖ holzessigsaures Salz n. ‖ pyrolignite m.

pyrolusite ‖ Pyrolusit m. ‖ pyrolusite f.

pyrometer ‖ Pyrometer n. ‖ pyromètre m. / distant reading ~ ‖ Fernpyrometer n. ‖ télépyromètre m.; pyromètre m. (pour indication) à distance. / flue ~ ‖ Fuchspyrometer n. ‖ pyromètre m. de carneau. / optical ~ ‖ optisches Pyrometer m. ‖ pyromètre m. optique. / radiant heat ~ ‖ Strahlungspyrometer n. ‖ pyromètre m. à rayonnement. / recording ~ ‖ Registrierpyrometer n. ‖ pyromètre m. enregistreur.

pyrometer protection tube ‖ Pyrometerschutzrohr n. ‖ gaine f. de pyromètre; tuyau m. protecteur de pyromètre.

pyrometric cone ‖ Brennkegel m.; Schmelzkegel m.; Segerkegel m. ‖ cône m. pyrométrique. / ~ effect ‖ Heizwert m. ‖ effet m. pyrométrique; pouvoir m. calorifique.

pyromorphite ‖ Pyromorphit m.; Phosphorblei n. ‖ pyromorphite f.; plomb m. phosphaté.

pyromucic acid ‖ Pyroschleimsäure f. ‖ acide m. pyromucique ou pyrosaccholactique.

pyrope ‖ Pyrop m.; böhmischer Granat m. ‖ pyrope m.; grenat m. pyrope ou magnésien.

pyrophone f. ‖ Pyrophon n. ‖ pyrophone m.

pyrophoric ‖ luftentzündlich ‖ pyrophorique.

pyrophosphate ‖ pyrophosphorsaures Salz n.; Pyrophosphat n. ‖ pyrophosphate m.

pyrophorous alloy ‖ pyrophore Legierung f. ‖ alliage m. pyrophorique.

pyrophyllite ‖ Pyrophyllith m.; strahliger Talk m. ‖ pyrophyllite f.

pyrophysalite ‖ Pyrophysalith m. ‖ pyrohpysalite f.

pyropissite ‖ Pyropissit m.; Wachskohle f. ‖ pyropissite f.

pyrosaccholactic acid see pyromucic acid.

pyroscope ‖ Pyroskop n. ‖ pyroscope m.

pyrosulphuric acid ‖ rauchende Schwefelsäure f. ‖ acide m. sulfurique fument.

pyrotechnical ‖ feuerungstechnisch ‖ pyrotechnique. / ~ laboratory ‖ pyrotechnisches Laboratorium n. ‖ laboratoire m. de pyrotechnie. / ~ pistol ‖ Leuchtpistole f. ‖ pistolet m. éclairant.

pyrotechnics pl. ‖ Pyrotechnik f.; Feuerwerkerei f. ‖ pyrotechnie f.

pyrotechnist ‖ Pyrotechniker m.; Kunstfeuerwerker m. ‖ pyrotechnicien m.

pyroxene ‖ Augit m.; Pyroxen m. ‖ pyroxène m.

pyroxyline ‖ Pyroxylin n.; Schießbaumwolle f. ‖ coton-poudre m.; poudrecoton m.; pyroxyle m. / ~ lacquer ‖ Nitrozelluloselack ‖ vernis m. à la nitrocellulose.

pyrrhosiderite ‖ Nadeleisenerz n.; Pyrrhosiderit m. ‖ fer m. hydro-oxyde ou oxydé hydraté.

pyrrhotine see pyrites, magnetic ~.

Pythagorean theorem ‖ pythagoreischer Lehrsatz m.; Pythagoras m. ‖ théorème m. du carré de l'hypoténuse; proposition f. de Pythagore.

Q

quad ‖ Vierfache n. ‖ quadruple m. / ~ (Geom) see quadrangle. / ~ (Print) see also quadrat ‖ Quadrat n.; Ausschluß m.; Ausschließung f. ‖ cadrat m. / ~ (Tel) ‖ Adervierer m. ‖ quadrette f. / ~ case ‖ Quadratenkasten m. ‖ casseau m. à cadrats.

quadrangle (Geom) ‖ Viereck n.; Vierseit n. ‖ quadrangle m. / ~ (Build) ‖ Häuserblock m. ‖ îlot m. de maisons. / ~ (Yard) ‖ viereckiger (von Gebäuden umschlossener) Hof m. ‖ préau m.

quadrangular ‖ viereckig; vierkantig; vierseitig ‖ quadrangulaire.

quadrant (Geom) ‖ Quadrant m.; Kreisquadrant m.; Viertelkreis m. ‖ quart m. ou arc m. de cercle. / ~ (Mar; Astro) ‖ Quadrant m.; octant m.; cadran m.; quartier m. / ~ (Spinn) ‖ Quadrant m. ‖ secteur m. / Hadley's ~ (Astro) ‖ Spiegelquadrant m. ‖ cadran m. ou quartier m. de Hadley. / mural ~ (Astro) ‖ Mauerquadrant m. ‖ quart m. de cercle mural. / pendulum ~ ‖ Pendelquadrant m. ‖ quart m. de cercle à pendule. / sinical ~ (Astro) ‖ Reduktionsquadrant m. ‖ quartier m. de réduction. / spirit-level

~ ‖ Libellenquadrant m. ‖ quart m. de cercle à niveau à bulle d'air. / toothed ~ (Mach) ‖ Zahnbogen m. ‖ secteur m. denté.

quadrantal triangle ‖ Quadrantendreieck n. ‖ triangle m. sphérique dont l'un des côtés est égal à un quart de cercle.

quadrant electrometer ‖ Quadrant(en)-elektrometer n. ‖ électromètre m. à quadrant.

quadrat (Print) ‖ Quadrat n.; Ausschluß m.; Ausschließung f. ‖ cadrat m. / em-~ see m-~. / en-~ see n-~. / m-~ ‖ Geviert n. ‖ cadratin m. / n-~ ‖ Halbgeviert n. ‖ demi-cadratin m.

quadrate (Geom) ‖ quadratisch ‖ carré; quadratique.

quadrate (Geom) ‖ Quadrat n. ‖ carré m. / ~ (Math) ‖ Quadrat n.; Quadratzahl f. ‖ nombre m. carré. / ~ (Astro) see quadrature (Astro).

quadratic (Math) ‖ quadratisch; zweiten Grades ‖ quadratique; du second degré.

quadratically, to multiply ‖ quadrieren; zur zweiten Potenz erheben ‖ élever au carré ou à la seconde puissance.

quadrature (Geom; Build) ‖ Quadratur f.; Ausvierung f.; Vierung f. ‖ quadrature f. / ~ (Astro) ‖ Quadratur f.; Geviert n.; Geviertschein m. ‖ quadrature f.

quadric (Math) see quadratic.

quadricycle ‖ Vierrad n.; vierräderiges Fahrrad n. ‖ cyclecar m.; quadricycle m.

quadrilateral see quadrangle and quadrangular.

quadruple, to ‖ vervierfachen ‖ quadrupler.

quadruple ‖ vierfach ‖ quadruple. / ~ prism quartz spectrograph ‖ Vierprismenquarzspektrograf m. ‖ spectrographe m. à quatre prismes en quartz. / ~-screw liner ‖ Vierschraubendampfer m. ‖ vapeur m. à quatre hélices.

quadruple ‖ Vierfache n. ‖ quadruple m.

quadruplet ‖ Viersitzer m.; viersitziges Fahrrad n. ‖ quadruplette f.

quadruplex system ‖ Vierfachbetrieb m.; Quadruplexbetrieb m. ‖ installation f. quadruplex. / ~ telegraphy ‖ Vierfachtelegrafie f. ‖ télégraphie f. quadruplex. / ~ working (Tel) ‖ Doppelgegensprechbetrieb n.; Quadruplexbetrieb m. ‖ transmission f. quadruplex.

quadruplicate, to see to quadruple.

quads pl. (Print) ‖ großer Ausschluß m.; Quadraten npl. ‖ cadrats mpl.

quagmire ‖ Moorboden m.; Sumpfboden m.; Sumpfland n.; Marschland n. ‖ sol m. marécageux.

quail ‖ Wachtel f.; caille f. / ~ call see ~ pipe. / ~ pipe ‖ Wachtelpfeife f.; Wachtellocke f. ‖ courcaillet m.; cailler m. / ~ shot ‖ Flintenschrot n. ‖ plombs m. de chasse.

quaker colour ‖ Mausgrau n.; mausgraue Farbe f. ‖ gris-de-souris m.; couleur f. de souris.

qualification ‖ Befähigung f.; Qualifikation f. ‖ qualification f.; capacité f. / to lose ~ ‖ die Befähigung einbüßen ‖ perdre l'aptitude f.

qualify, to ‖ sich eignen ‖ convenir à; se prêter à.

qualitative ‖ qualitativ ‖ qualitatif. / ~ analysis ‖ qualitative Analyse f. ‖ analyse f. qualitative.

quality ‖ Qualität f.; Güte f.; Beschaffenheit f.; Eigenschaft f. ‖ qualité f. / average ~ ‖ durchschnittliche Qualität f. ‖ qualité f. moyenne. / fair average ~ ‖ gute Durchschnittsqualität f. ‖ bonne qualité f. moyenne. / ~ guaranteed ‖ Qualität f. garantiert ‖ qualité f. garantie. / inferior ~ ‖ minderwertige Qualität f. ‖ qualité f. inférieure. / medium ~ ‖ Mittelqualität f. ‖ qualité f. moyenne. / equal to sample ‖ mustergleiche Qualität f. ‖ qualité f. conforme à l'échantillon. / standard ~ ‖ mustergleiche Qualität f. ‖ qualité f. type. / ~ of a tone ‖ Klangfarbe f. ‖ timbre m. (du son). / usual commercial ~ ‖ handelsübliche Güte f. oder Qualität f. ‖ qualité f. courante ou de commerce ou commerciale.

quality coefficient ‖ Gütekoeffizient m.; Güteziffer f.; Gütefaktor m. ‖ cœfficient m. qualitatif ou de qualité; chiffre m. qualitatif. / ~ factor see quality coefficient. / specification for ~ ‖ Gütevorschrift f.; Qualitätsvorschrift f. ‖ prescription f. pour la qualité.

quant ‖ Spazierstock m. ‖ badine f.; canne f.

quantitative ‖ mengenmäßig; quantitativ ‖ quantitatif. / ~ analysis ‖ quantitative Analyse f. ‖ analyse f. quantitative. / ~ spectrum analysis ‖ quantitative Spektralanalyse f. ‖ analyse f. spectrale quantitative.

quantity ‖ Quantität f.; Menge f.; Masse f.; Größe f. ‖ quantité f. / ~ (Math) ‖ Größe f. ‖ grandeur f.; quantité f. / affirmative ~ ‖ positive Größe f. ‖ quantité f. positive. / ~ of air required for combustion ‖ zur Verbrennung notwendige Luftmenge f. ‖ quantité f. d'air nécessaire à la combustion. / ~ of air supplied ‖ zugeführte Luftmenge f. ‖ quantité f. d'air admise. / commensurable ~ ‖ kommensurable Größe f. ‖ quantité f. commensurable. / constant ~ ‖ konstante Größe f.; Konstante f. ‖ constante f.; quantité f. constante. / ~ of cooling water required ‖ Kühlwasserbedarf m. ‖ quantité f. d'eau nécessaire au refroidissement. / ~ of current ‖ Strommenge f. ‖ quantité f. du courant. / differential ~ ‖ unendlich kleine Größe f.; Differential n. ‖ quantité f. différentielle ou infiniment petite. / electric ~ ‖ Elektrizitätsmenge f. ‖ quantité f. d'électricité. / imaginary ~ ‖ imaginäre Größe f. ‖

quantité f. imaginaire. / incommensurable ~ ‖ inkommensurable Größe f. ‖ quantité f. incommensurable. / infinitesimal ~ see differential ~. / irrational ~ ‖ irrationale Größe f. ‖ quantité f. irrationelle. / ~ of motion ‖ Größe f. der Bewegung ‖ quantité f. du mouvement. / negative ~ ‖ negative Größe f. ‖ quantité f. négative. / positive ~ ‖ positive Größe f. ‖ quantité f. positive. / ~ raised (Pump) ‖ Fördermenge f. ‖ quantité f. débitée. / rational ~ ‖ rationale Größe f. ‖ expression f. ou quantité f. rationelle. / with regard to ~ (Trade) ‖ quantitativ ‖ par rapport à la quantité. / ~ to be transported ‖ Fördermenge f. ‖ quantité f. à transporter. / unlimited ~ ‖ beliebig große Menge f. ‖ quantité f. illimitée. / variable ~ ‖ variable oder veränderliche Größe f. ‖ quantité f. variable. / ~ of water ‖ Wassermenge f. ‖ volume m. d'eau.

quantity detector of a battery ‖ Batterieprüfer m. ‖ vérificateur m. de pile. / ~ production ‖ Massenfertigung f. ‖ fabrication f. en grande série.

quantum ‖ Quantum n. ‖ quantum m. / ~ theory ‖ Quantentheorie f. ‖ théorie f. des quanta.

quarantine ‖ Quarantäne f. ‖ quarantaine f.; seraine f. / ~ (Station) ‖ Quarantänestation f. ‖ station f. de quarantaine; santé f. / to pass ~ ‖ Quarantäne f. halten ‖ être en quarantaine f. ou en seraine f.; faire la quarantaine f. / to perform ~ see to pass ~.

quarantine boat ‖ Quarantäneboot n. ‖ bateau m. de santé. / ~ flag ‖ Quarantäneflagge f. ‖ signal m. de quarantaine. / ~ officer ‖ Quarantänebeamte m. ‖ officier m. de santé. / ~ service ‖ Quarantänedienst m. ‖ service m. de quarantaine f. / ~ station ‖ Quarantänehaus n.; Quarantänestation f. ‖ station f. de quarantaine; santé f.

quarrel ‖ Streit m. ‖ querelle f. / ~ (Geom) ‖ Raute f.; losange m. / ~ (Glaz) ‖ Glaserdiamant m. ‖ diamant m. (pour couper le verre et les glaces); diamant m. de vitrier.

quarrier see quarryman.

quarry, to ‖ Steine mpl. brechen ‖ tirer des carrières fpl.; travailler à la carrière.

quarry ‖ Steinbruch m. ‖ carrière f. / limestone ~ ‖ Kalksteinbruch m. ‖ carrière f. de castine ou de pierre calcaire ou de pierre à chaux. / ~ of slate ‖ Schieferbergwerk n.; Schieferbruch m.; Schiefersteinbruch m. ‖ ardoisière f.; carrière f. d'ardoise.

quarry blacksmith ‖ Steinbruchschmied m. ‖ forgeron m. carrier.

quarryman ‖ Steinbrucharbeiter m.; Steinbrecher m. ‖ carrier m.; carreyeur m. / ~ in a quarry of sandstone ‖ Sandsteinbrucharbeiter m. ‖ grésier m.

quarryman's hammer ‖ Steinhammer m. ‖ casse-pierre(s) m.

quarry master ‖ Steinbruchmeister m. ‖ maître-carrier m. / ~ miner ‖ Steinsprenger m. ‖ carrier m. mineur. / owner of a ~ ‖ Steinbruchbesitzer m. ‖ propriétaire m. de carrière. / ~ sap ‖ Grubenwasser n. ‖ eau f. de carrière.

quarry stone ‖ Bruchstein m.; Werkstein m. ‖ moellon m.; pierre f. de taille.

quarry stone masonry ‖ Bruchsteinmauerwerk n. ‖ maçonnerie f. en moellons. /

dam of ~ ‖ Sperrmauer f. in Bruchsteinmauerung ‖ mur m. du barrage en pierre de carrière.

quartan ‖ Viertelmaß n.; Quart n.; Viertel n. ‖ pinte f.

quartation (Met) ‖ Scheidung f. durch die Quart; Quartation f. ‖ inquart m.; inquartation f.; quartation f.

quart bottle ‖ Quartflasche f. ‖ bouteille f. d'une pinte.

quarter ‖ Viertel n. ‖ quart m. / ~ (Lodgings) ‖ Quartier n.; Wohnung f.; Aufenthalt m. ‖ quartier m.; logement m. / ~ (Year) ‖ Quartal n.; Jahresviertel n.; Vierteljahr n. ‖ trimestre m. / ~ of a brick ‖ Viertelziegel m. ‖ nicoteux m. / ~ of the moon ‖ Mondviertel n. ‖ quartier m. de lune. / ~ of a ship ‖ Windvierung f. eines Schiffes ‖ hanche f. / winding ~ see quarter pace. / ~ of the year see ~ (Year).

quarterage ‖ Vierteljahrsgehalt n. ‖ (salaire m. du) trimestre m.

quarter bend (Pipe) ‖ Krümmer m. ‖ coude m. en équerre. / ~ binding ‖ Band m. mit Lederrücken; Halblederband m.; Halbfranzband m. ‖ demi-reliure f. / ~ boat ‖ Seitenboot n. ‖ embarcation f. de côté. / ~ deck ‖ Quarterdeck n.; Achterdeck n.; Hinterdeck n. ‖ demipont m.; château m. de derrière. / ship with raised ~ deck ‖ Schiff n. mit erhöhtem Achterdeck ‖ navire m. à pont surélevé. / ~ gallery (Mar) ‖ obere Seitengalerie f.; Heckgalerie f. ‖ clavecins mpl. de la galerie.

quartering ‖ Vierteilen n. ‖ écartèlement m. / ~ (Carp) ‖ Sparrenholz n. ‖ bois m. de chevrons; moises fpl. / ~ (Lodgings) see quarter (Lodgings). / ~ gauge ‖ Nachprüfvorrichtung f. ‖ dispositif m. de contrôle.

quartermaster ‖ Quartiermeister m. ‖ quartier-maître m.

quartern see quartan.

quarter pace (Stairs) ‖ Viertelwend(el)ung f.; Viertelspodest n. ‖ quartier m. tournant. / ~ partition ‖ Bundwand m.; Fach(werk)wand f.; Riegelwand f. ‖ cloison f. en charpente. / ~ port (Mar) ‖ Heckpforte f. ‖ sabord m. d'arcasse ou de retraite. / ~ stuff ‖ vierzölliges Brett n. ‖ planche f. de quatre pouces d'épaisseur.

quartile see quadrature.

quarto ‖ Quart(format) n.; Quarto n. ‖ inquarto m. / ~ book ‖ Quartband m. ‖ -volume m.) inquarto m. / sheet of paper in ~ ‖ Quartblatt n. ‖ feuille f. inquarto.

quart pot see quartan.

quartz ‖ Quarz m. ‖ quartz m. / arenaceous ~ see quartz sand. / blue ~ ‖ Saphirquarz m.; blauer Quarz m.; Liderit m. ‖ sidérite m. / flexible ~ ‖ Itakolumit m.; Gelenkquarz m.; elastischer Sandstein ‖ itacolumite m.; grès m. élastique ou flexible. / fused ~ see melted ~. / gold(-bearing) ~ ‖ goldhaltiger Quarz m.; Goldkies m. ‖ quartz m. aurifère. / granular ~ ‖ körniger Quarz m. ‖ quartz m. grenu. / melted ~ ‖ geschmolzener Quarz m. ‖ quartz m. fondu. / milky ~ ‖ Milchquarz m. ‖ quartz m. blanc de lait. / pulverized ~ ‖ feingemahlener Quarz m. ‖ quartz m. finement moulu. / resinous ~ ‖ Fettquarz m. ‖ quartz m. résinite. / smoky ~ ‖ Rauchquarz m.; Rauchtopas m. ‖ fausse-topaze f.; topaze f. enfumée ou

occidentale. / spongiform ~ || Schwimm-kiesel m. || quartz m. nectique.

quartz cell || Quarzkammer f. || chambre f. en quartz. / ~ condenser || Quarzkonden-sor m. || condensateur m. en quartz. / ~ crusher || Quarzmühle f. || broyeur m. de quartz. / ~ eyepiece || Quarzokular n. || oculaire m. en quartz.

quartz fluorite || Quarzfluorit m.; Quarz-fluß m. || quartz-fluorine m.; fluor m. quartzeux. / ~ lens || Quarzfluoritlinse f. || lentille f. en quartz-fluorine.

quartz glass || Quarzglas n. || verre m. quartzeux. / ~ thermometer || Quarz-glasthermometer n. || thermomètre m. en verre quartzeux. / ~ vessel || Quarz-glasgefäß n. || vase m. en verre-quartz.

quartziferous || quarzhaltig || quartzifère.

quartzite || Quarzit m.; Quarzfels m.; Quarzgestein n.; Quarzstein m. || quart-zite m.; roche f. quartzeuse. / ~ lining || Quarzitausfütterung f. || revêtement m. en quartzite. / ~ quarry || Quarzitbruch m. || carrière f. de quarzite.

quartz lamp || Quarz(glas)lampe f. || lampe f. de quartz. / medical ~ || Quarzlampe f. für medizinische Zwecke || lampe f. de quartz à l'usage médicale. / mercury vapour ~ || Quarzquecksilberlampe f. || lampe f. aux vapeurs de mercure.

quartz lens || Quarzlinse f. || lentille f. en quartz. / ~ mica rock || Quarzglimmer-fels m. || roche f. de quartz micacé.

quartzose see quartziferous.

quartzous see quartziferous.

quartz plate || Quarzplatte f. || plaque f. ou lame f. en quartz. / ~ porphyry || quarzführender Porphyr m.; Felsit m. || porphyre m. quartzifère. / ~ powder || Quarzpulver n. || quartz m. en poudre. / ~ prism || Quarzprisma n. || prisme m. en quartz. / ~ quarry || Quarzgrube f. || carrière f. de quartz. / ~ rock || Quarz-gestein n.; Quarzstein m. || roche f. quartzeuse. / ~ rock-salt || Quarzstein-salz n. || quartz-sel-gemme m.

quartz rock-salt lens || Quarzsteinsalz-linse f. || lentille f. en quartz-sel-gemme. / achromatic ~ || Quarzsteinsalzachro-matlinse f. || lentille f. en quartz-sel-gemme achromatique. / four-lens achromatic ~ || vierteilige Quarzstein-salzachromatlinse f. || lentille f. achro-matique à quatre lentilles en quartz-sel-gemme.

quartz sand || Quarzsand m. || sable m. quartzeux. / ~ sinter || Quarzsinter m. || quartz-agate m. thermogène. / ~ spectro-graph || Quarzspektrograf m. || spectro-graphe m. à optique en quartz. / ~ thread || Quarzfaden m. || filament m. de quartz. / ~ wave-meter || Quarz-wellenmesser m. || ondemètre m. à quartz. / ~ window || Quarzfenster n. || fenêtre f. en quartz.

quartzy || quarzig; quarzhaltig || quart-zeux; quartzifère. / ~ sandstone || Quarz-sandstein m. || grès m. quartzeux.

quassia || Quassie f.; Quassia f.; Bitter-holzbaum m. || quassia m.; quassie f. / ~ wood || Quassiaholz m.; Bitterholz n. || bois m. de quassia amara ou de quassie.

quassin(e) || Quassin n. || quassine f.

quassite see quassin(e).

quaternary || quaternär || quaternaire.

quaver, to || vibrieren; zittern || vibrer; trembler.

quaver || Achtelnote f. || croche f.

quay (Hydr arch) || Mole f.; Wellen-brecher m.; Kai m. || môle m. de port; digue f. ou jetée f. d'un port; quai m. / ~ (Mar) || Kai m.; Hafendamm m.; Uferdamm m.; Landungsplatz m.; An-furt f.; Schiffslände f.; Ladedamm m. || embarcadère m. / unloading ~ || Lösch-platz m. || débarcadère f.; lieu m. de débarquement; quai m. de décharge-ment.

quayage (Duty) || Landungszoll m. || droits mpl. de débarquement. / ~ (Quay plant) || Kaianlage f. || quais mpl.

quay crane || Uferkran m.; Hafenkran m. || grue f. de quai. / ~ labourer || Kaiarbei-ter m. || ouvrier m. de quai. / ~ wall || Kaimauer f. || mur m. d'un quai.

quebracho extract || Quebrachoauszug m. || extrait m. de québracho. / ~ wood || Quebrachoholz n. || bois m. de qué-bracho.

queen post || Sprengewerkstütze f. || poin-çon m. latéral; clef f. pendante latérale. / ~ roof || Hängewerk n. mit zwei Säu-len || double arbalète f. / ~ truss || Hänge-werk n. mit zwei Hängesäulen || arma-ture f. à clefs pendantes ou à deux poinçons.

quench, to (Metal) || abschrecken || re-froidir brusquement; tremper. / ~ the ashes || die Aschen fpl. ablöschen || éteindre les cendres fpl. / ~ the coke || den Koks löschen oder ablöschen || éteindre le coke. / ~ the steel in water || Stahl m. ablöschen oder härten || tremper l'acier m. dans l'eau.

quenched in oil || in Öl n. abgeschreckt oder gehärtet || refroidi à l'huile f. / ~ in water || in Wasser n. abgeschreckt oder gehärtet || refroidi à l'eau f.

quenched charcoal || Löschkohle f. || char-bon m. de braise.

quenched spark || Löschfunke m. || étin-celle f. étouffée. / ~ gap || Löschfunken-strecke f. || éclateur m. pour étincelle étouffée. / ~ system || Löschfunken-system n. || système f. à étincelle étouf-fée.

quenching || Ablöschung f. || extinction f. / the ~ in water raises the tensile strength of steel || das Abschrecken des Stahles in Wasser erhöht die Festigkeit || la trempe à l'eau de l'acier en augmente la résistance.

quenching bath || Härtebad n. || bain m. à tremper. / ~ bench || Löschrampe f. || rampe f. d'extinction. / ~ car || Lösch-wagen m. || chariot m. d'extinction. / ~ coil || Löschdrossel f. || extinction f. à réactance. / ~ drum || Löschtrommel f. || tambour m. d'extinction. / ~ effect || Löschwirkung f. || effet m. extincteur. / ~ pit || Löschgrube f. || fosse f. d'ex-tinction. / ~ press || Härtepresse f. || presse f. à tremper. / ~ tank || Lösch-gefäß n. || réservoir m. d'extinction. / ~ tower || Löschturm m. || tour f. d'ex-tinction. / ~ tub || Löscheimer m. || seau m. / coke ~ wagon || Kokslösch-wagen m. || wagon m. pour l'extinction du coke. / ~ water || Löschwasser m. || eau f. d'extinction.

quercitin(e) || Querzetin n. || quercétine f.

quercitron || Querzitron n. || quercitrone f. / ~ bark || Querzitronrinde f.; Färber-rinde f. || écorce f. de quercitrone.

query || Beanstandung f. || réclamation f.

question sheet || Fragebogen m. || question-naire m.

quick || schnell; rasch || rapide; instan-tané. / ~ break || plötzliches Ausschalten n. || rupture f. brusque. / ~ breaker see ~ break switch. / ~ break switch || Mo-mentschalter m. || interrupteur m. à rupture brusque. / ~ filer || Schnell-hefter m. || relieur m. rapide. / ~ in-spection || schnelle Messung f. || Schnell-messung f. || mesurage m. à grande vitesse. / ~ returns pl. || schneller Um-satz m. || vente f. ou débit m. rapide. / ~ stopping || schnelles Anhalten n. || arrêt m. rapide.

quick-acting brake equipment || Schnell-bremsausrüstung f. || garniture f. d'ap-pareils de frein rapide.

quick ash || Flugasche f. || cendre f. mou-vante ou volante. / ~ collector || Flug-aschenfänger m. || collecteur m. de cen-dre volante.

quick-casting machine || Schnellgieß-maschine f. || machine f. de fonderie à grande vitesse.

quick-drying ink (Print) || Schnelltrocken-farbe f. || encre f. siccative.

quicken, to ~ a building || ein Gebäude n. in seiner Umgebung freimachen || égayer ou dégager un bâtiment.

quick-growing || schnell wachsend; von schnellem Wuchs m. || à croissance f. rapide.

quicklime || ungelöschter Kalk m. || chaux f. vive.

quick match || Zündschnur f. || mèche f.; cordeau m. porte-feu; cordon Bickford. / ~ for miner's lamps || Zündband n. für Wetterlampen || mèche f. pour lampes de mineurs.

quicksand || Schwimmsand m.; Flugsand m. || mouvants mpl.; sables mpl. mouvants ou coulants. / freezing shaft sinking in ~ || Schachtabteufen n. im Schwimmsand mit Hilfe des Gefrierverfahrens || fon-çage m. des puits par congélation à tra-vers de couches de terres mouvantes.

quickset enclosure || Einzäunung f. mit lebenden Hecken || clôture f. vive.

quicksilver || Quecksilber n. || mercure m. / native ~ || gediegenes Quecksilber n. || mercure m. natif.

quicksilver gilding || Quecksilbervergol-dung f. || dorure f. au feu. / ~ mine || Quecksilberbergwerk n. || mine f. de mercure. / ~ works pl. || Quecksilber-hütte f. || usine f. à mercure.

quiescent source of sound || ruhende Schallquelle f. || source f. sonore immo-bile.

quiet, to (Chem) || abstehen lassen || lais-ser se calmer ou se reposer ou se déposer.

quiet || geräuschlos || silencieux. / ~ cut || ruhiger Schnitt m. || coupe f. sans broutage. / ~ running (Mach) || ruhiger Gang m. || marche f. silencieuse.

quill || Federkiel m. || tuyau m. de plume. / ~ (Mach) || Hülse f. (für Kupplungen) || gaine f. tubulaire. / ~ for silks || Seiden-garnspule f. || cannettes f. à trame de soie.

quillai bark || Quillayarinde f. || écorce f. de Panama. / ~ extract || Quillaya-extrakt m. || extrait m. d'écorce de Panama.

quill, articles from ~s || Waren fpl. aus Federkielen || articles mpl. en tuyaux de plumes. / ~ bit (Carp; Join) || Hohl-

bohrer m. ‖ évidoir m.; mèche-cuiller f. / ~ brush ‖ Federkielbürste f. ‖ brosse f. en tuyaux de plumes.

quiller (Lace-m) ‖ Rüschenmacher m. ‖ rucheur m.

quill feather ‖ Kielfeder f. ‖ plume f. à tige.

quilling (Lace-m) ‖ Rüsche f. ‖ ruche f.

quill pen see quill.

quilt, to ‖ steppen ‖ piquer.

quilt ‖ Steppdecke f. ‖ couverture f. ouatée ou piquée; courte-pointe f. / cotton ~ ‖ baumwollene Bettdecke f. ‖ couvre-lit m. coton.

quilted coat ‖ Stepprock m. ‖ vêtement m. piqué. / ~ cover see quilt.

quilter ‖ Stepper m. ‖ piqueur m.

quilting ‖ Pikee m. ‖ piqué m. / ~ needle ‖ Steppnadel f. ‖ aiguille f. à piquer. / ~ seam ‖ Steppnaht f. ‖ couture f. piquée. / ~ weaver ‖ Pikeeweber m. ‖ tisseur m. de piqué.

quilt work ‖ Polsterung f. ‖ rembourrage m.

quince ‖ Quitte f. ‖ coing m. / ~ pip ‖ Quittenkern m. ‖ pépin m. de coing.

quincunx (Build) ‖ Rautenstellung f.; Gefünfte n. ‖ quinconce m.; échiquier m. diagonal.

quinine ‖ Chinin n. ‖ quinine f.; quina f.; quinquina f.

quinotannic acid ‖ Gerbstoff m.; Tannin n. ‖ acide m. quinotannique.

quire (Bookb) ‖ Bogensatz m.; Lage f. (eines ungebundenen Buches) ‖ ensemble m. de feuilles. / loosely assembled ~s pl. ‖ geholländerte Bogen mpl. ‖ feuilles fpl. attachées assemblées. / served ~s pl. (Bookb) ‖ fadengehefteter Bogensatz m. ‖ ensemble m. de feuilles cousu au fil.

quirk head (Arch) ‖ Halskehle f.; stehende Hohlkehle f. ‖ Einziehung f. ‖ gorge f.

quite modern ‖ hochmodern ‖ du dernier genre m.; de la dernière mode.

quitrent ‖ Erbpachtgeld n. ‖ redevance f.

quittance ‖ Quittung f.; Empfangsbescheinigung f. ‖ acquit m.; quittance f.

quod ‖ Gefängnishof m. ‖ préau m.

quoin, to ‖ keilen; verkeilen ‖ coincer; caler. / ~ (Print) ‖ (die Form) einkeilen ‖ arrêter ou assujettir avec de coins.

quoin ‖ Keil m. ‖ coin m.; cale f. / ~ (Build) ‖ Eckstein m.; Kropfstein m.; Keilstein m.; Winkelstein m. ‖ pierre f. d'encoignure. / ~ (Corner) ‖ ausspringende oder vorspringende Ecke f. ‖ coin m.;

angle m. / ~ (Print) ‖ Schließkeil m.; Steg m. ‖ coin m. de serrage. / hollow ~ ‖ einspringende Ecke f. ‖ encoignure f. / inclined ~ (Print) ‖ Keilsteg m. ‖ bois m. de corps. / iron ~ ‖ eiserner Keil m. ‖ coin m. de fer. / mitre ~ (Join) ‖ Gehrung f.; Gierung f.; Gehre f. ‖ onglet m.; anglet m.; biais m.; biaisement m. / mitred ~ see mitre ~. / wooden ~ ‖ hölzerner Keil m. ‖ coin m. de bois; cale f. en bois.

quoin chase (Print) ‖ Keilrahmen m. ‖ châssis m. à coins.

quoit ‖ Diskus m.; eiserne oder hölzerne Wurfscheibe f. ‖ disque m.

quota ‖ Quote f. ‖ quote-part f.; cote f.

quotation ‖ Preisangabe f.; Angebot n. ‖ mise f. à prix; offre f. / special ~ ‖ Sonderangebot n. ‖ offre f. spéciale ou de faveur.

quotation marks pl. ‖ Anführungszeichen npl.; Gänsefüßchen pl. ‖ guillemets mpl.

quote, to ‖ anführen; zitieren ‖ citer; rapporter. / ~ (Trade) ‖ notieren; im Kurse stehen ‖ côter. / ~ a price ‖ ein Preisangebot n. machen ‖ faire un prix; côter.

quotient ‖ Quotient m. ‖ quotient m.

quotum ‖ Anteil m.; Anteilsverhältnis n. ‖ part f.; portion f.

R

rabbet, to ‖ ausfalzen; mit einem Falz m. versehen ‖ feuiller; plier.

rabbet (Carp) ‖ Falz m.; Fuge f.; Anschlag m. ‖ feuillure f. / ~ (Hammer) ‖ Reitel m.; Stoßreitel m. ‖ buttoir m.; rabbat m.; ressort m. / ~ of the keel ‖ Kielsponung f. ‖ rablure f. de la quille.

rabbet beam ‖ Reitelsattel m. ‖ rabat m. supérieur. / ~ iron ‖ Simseisen n. ‖ fer m. de guillaume. / ~ plane ‖ Falzhobel m. ‖ feuilleret m.; rabot-feuilleret m.; guillaume m. / ~ stand ‖ hintere Gerüstsäule f.; Reitelsäule f. ‖ colonne f. de rabat. / ~ wall ‖ Fensterlaibung f.; Anschlagmauer f. ‖ embrasure f.

rabbit ‖ Kaninchen n. ‖ lapin m. / tame ~ ‖ Hauskaninchen n. ‖ lapin m. domestique. / ~ breeder ‖ Kaninchenzüchter m. ‖ léporiculteur m. / ~ breeding ‖ Kaninchenzucht f. ‖ léporiculture f. / ~ glue ‖ Hasenleim m. ‖ colle f. de lapin. / ~ hair ‖ Kaninchenhaar n. ‖ poil m. de lapin. / ~ skin ‖ Kaninchenfell n. ‖ peau f. de lapin.

rabble, to (Metal) ‖ umrühren ‖ brasser.

rabble (Metal) ‖ Kratze f.; Rührstange f.; Krücke f. ‖ perche f. à brasser; râble m.; brassoir m.; ringard m.

rabbling ‖ Umrühren n. ‖ brassage m.

Rabitz texture ‖ Rabitzgewebe n. ‖ toile f. métallique Rabitz. / ~ wall ‖ Rabitzwand f. ‖ mur m. Rabitz. / ~ wire-netting ‖ Rabitzgeflecht n. ‖ treillis m. métallique Rabitz.

racahout ‖ Racahout n. ‖ racahout m.

race, to (Engine) ‖ durchgehen ‖ s'emballer.

race (Hydr arch) ‖ Abflußrinne f.; Entwässerungsgraben m. ‖ canal m.; rigole f. / ~ (Mill) ‖ Fluder m.; Mühlengerinne n. ‖ bief m.; biez m. / ~ (Sport) ‖ Rennen n. ‖ course f. / ~ (Trade) ‖ Wettbewerb m. ‖

concours m. / ~ (Weav) ‖ Schützenbahn f. ‖ lit m. ou voie f. ou chemin m. de la navette. / ball ~ (Mach) ‖ Kugelkorb m.; Kugelring m.; Führungsring m. ‖ bague f. ou rondelle f. à billes; anneau m. de roulement à billes. / circular ~ (Swing bridge) ‖ Laufkranz m. ‖ couronne f. de roulement / exchangeable ~ ‖ auswechselbarer Laufring m. ‖ anneau m. à billes interchangeable. / ~ of external ring of ball bearing ‖ Laufrille f. des Kugellageraußenringes ‖ gorge f. de roulement de la bague extérieure du roulement à billes. / the outside ~ is adjustable ‖ der äußere Laufring ist einstellbar ‖ l'anneau m. de roulement extérieure est ajustable.

race board see race (Weav). / ~ course (Mill) ‖ Mühlengerinne n.; Fluder m. ‖ auge f.; bief m.; biez m.; coursier m.; coursière f. / ~ glass ‖ Rennglas n. ‖ jumelles fpl. de courses. / ~ ground see ~ track.

racer (Aero) ‖ Rennflugzeug n. ‖ avion m. de concours. ~ (Auto) ‖ Rennwagen m. ‖ voiture f. de course. / ~ (Person) ‖ Rennfahrer m. ‖ coureur m.

race track ‖ Rennbahn f.; Rennstrecke f. ‖ champ m. de course.

rachis (Bot) ‖ Fruchtspindel f.; Hauptstiel m. ‖ rachis m.

racing of the engine ‖ Durchgehen n. des Motors ‖ emballement m. du moteur.

racing aeroplane ‖ Rennflugzeug n. ‖ avion m. de concours ou de course. / ~ car ‖ Rennwagen m. ‖ voiture f. de course. / ~ knife ‖ Ritzeisen n.; Krabber m. ‖ rouanne f. à marquer. / ~ plane see ~ aeroplane. / ~ yacht ‖ Rennjacht f. ‖ yacht m. de course.

rack, to (Brew) ‖ abfüllen ‖ mettre en fûts mpl.; soutirer. / ~ (Clothm) ‖ recken ‖

‖ étendre; étirer. / ~ (Curr) ‖ ausrecken ‖ retaler. / ~ clear (Beer) ‖ lauter fassen ‖ traverser la bière tombée. / ~ green (Beer) ‖ grün fassen ‖ traverser la bière verte.

rack ‖ Gestell n.; Gerüst n.; Stativ n. ‖ chevalet m.; tréteau m.; pied m. / ~ (Agr) ‖ Futterraufe f. ‖ râtelier m. / ~ (Gymnastics) ‖ Reck n.; Turnreck n. ‖ barre f. fixe. / ~ (Househ) ‖ Kleiderständer m. ‖ porte-habit m. / ~ (Mach) ‖ Zahnstange f. ‖ crémaillère f. / ~ (Mine) ‖ Kehrrad n. ‖ roue f. à deux directions. / ~ (Spinn) ‖ Rocken m.; Spinnrocken m. ‖ quenouille f. / ~ for automatic telephone exchanges ‖ Gestell n. für selbsttätige Fernsprechzentralen ‖ support m. pour centraux téléphoniques interurbains. / circular ~ ‖ Rundraufe f. ‖ râtelier m. circulaire. / cooling ~ ‖ Kühlbett n. ‖ lit m. refroidisseur. / feeding ~ ‖ Raufe f.; Futterraufe f. ‖ râtelier m.; doublière f. / ~ of ship's screw lifting gear ‖ Führung f. des Schiffsschraubenheberahmens ‖ guide m. ou coulisse f. de l'appareil pour hisser l'hélice d'un navire.

rack and pinion ‖ Zahnstangengetriebe n. ‖ engrenage m. à crémaillère. / ~ mechanism ‖ Zahn- und Triebbewegung f. ‖ mouvement m. crémaillère et pignon.

rack bar of a jack ‖ Windenstock m. ‖ crémaillère f. du cric.

rack cutting machine ‖ Zahnstangenfräsmaschine f. ‖ machine f. à fraiser les crémaillères ou à tailler les crémaillères.

racker (Beer) ‖ Abzieher m.; Bierzapfer m. ‖ soutireur m.; emplisseur m. de barils. / ~ (Brew) ‖ Abfüllapparat m. ‖ soutireuse f. / ~ (Wine) ‖ Weinzieher m.; Weinabzieher m. ‖ tireur m. de vins.

racket ‖ Racket n.; Tennisschläger m. ‖ raquette f.; battoir m. de paume. / ~ (Ski) ‖ Schneeschuh m.; Ski m.; Schi m. ‖ raquette f.; ski m.; patin m. / ~ cord ‖ Tennisschlägersaite f. ‖ corde f. pour raquettes. / ~ maker ‖ Rakettmacher m. ‖ raquetier m. / ~ press ‖ Racketpresse f. ‖ presse f. pour raquettes.

rack gear ‖ Zahnstangenantrieb m. ‖ engrenage m. à crémaillère.

racking (Brew) ‖ Abziehen n.; Abfüllen n.; Abzapfen n. ‖ mise f. en bouteilles *ou* en fûts; soutirage m. / ~ (Clothm.) ‖ Recken n. ‖ étendage m.; étirage m. / ~ (Wind mill) ‖ Schwungleine f. ‖ hauban m.

racking apparatus ‖ Abfüllapparat m.; Abziehvorrichtung f. ‖ soutireuse f. / back-pressure ~ ‖ Abfüllapparat m. mit Gegendruck ‖ soutireuse f. à contre-pression. / ~ for casks ‖ Faßfüllapparat m. ‖ appareil m. de soutirage *ou* soutireuse f. pour fûts. / isobarometric ~ ‖ isobarometrische Faßfüllvorrichtung f. ‖ soutireuse f. isobarométrique pour fûts.

racking balk ‖ Rödelbalken m. ‖ poutrelle f. de guidage. / ~ bench (Brew) ‖ Abfüllständer m.; Abfüllbock m. ‖ chevalet m. de soutirage. / ~ bock *see* ~ bench. / ~ cellar ‖ Abziehkeller m. ‖ cave f. de soutirage. / ~ cock ‖ Abziehhahn m. ‖ robinet m. de soutirage. / ~ faucet ‖ Faßzwickel m. ‖ fausset m. / ~ gut ‖ Abfülldarmschlauch m. ‖ boyau m. de soutirage. / ~ hose *see* ~ pipe. / ~ machine ‖ Abfüllmaschine f. ‖ machine f. de soutirage *ou* à soutirer; machine f. à remplir les bouteilles. / ~ pipe ‖ Abfüllschlauch m. ‖ tuyau m. de soutirage. / ~ plant ‖ Abfüllanlage f. ‖ installation f. de soutirage. / ~ room ‖ Abziehraum m. ‖ salle f. de soutirage. / ~ square ‖ Abfüllbütte f. ‖ cuve f. de soutirage.

rack jack ‖ Zahnstangen(wagen)winde f. ‖ cric m. à crémaillère. / ~ ladder of a crane-tripod ‖ Leiterbalken m. eines Krandreifußes ‖ rancher m. d'une grue à trois pieds.

rack locomotive ‖ Zahnradlokomotive f. ‖ locomotive f. à crémaillère.

rack-mounted ‖ auf Gestellen npl. montiert ‖ monté sur bâtis mpl.

rack planer tool attachment ‖ Kammstahleinrichtung f. ‖ dispositif m. de taillage par peigne.

rack rail ‖ Zahnstange f. ‖ crémaillère f. / ~ locomotive *see* rack locomotive.

rack railway ‖ Zahnradbahn f. ‖ chemin m. de fer à crémaillère.

rack rope ‖ Reitelleine f.; Rödelleine f. ‖ commande f. de guidage. / ~ stick ‖ Rödelholz n.; Reitelholz n. ‖ garrot m. *ou* billot m. de guidage. / ~ tool ‖ Kammstahl m. ‖ outil-peigne m. / ~ wagon ‖ Leiterwagen m. ‖ chariot m. à ridelles. / ~ wagon for children ‖ Kinderleiterwagen m. ‖ chariot m. à ridelles pour enfants. / ~ wheel ‖ Sperrad n. ‖ roue f. à rochet. / ~ work ‖ Zahngetriebe n. ‖ pignon m. denté.

radial ‖ radial ‖ radial. / ~ acceleration ‖ Fliehkraftbeschleunigung f. ‖ accélération f. centrifuge. / ~ admission ‖ radiale Beaufschlagung f. ‖ admission f. radiale. / ~ arrangement of the type bars ‖ sternartig gelagerte Typenhebel mpl. ‖ tiges fpl. à caractères disposées en étoile. / ~ axle-box housing ‖ Radialachslager-

gehäuse n. ‖ support m. de boîte radiale d'essieu. / ~ bearing ‖ Querlager n. ‖ roulement m. transversal. / ~ brick for chimneys ‖ Radialziegelstein m. für Schornsteine ‖ brique f. radiale pour cheminées. / ~ drill *see* ~ drilling machine. / ~ drilling machine. ‖ Radialbohrmaschine f. ‖ machine f. à percer radiale; perceuse f. radiale. / ~ drilling and countersinking machine ‖ Radialbohr- und Versenkmaschine f. ‖ machine f. radiale pour perçage et fraisage de trous en ligne. / ~ end axle (Railw) ‖ kurvenbewegliche Endachse f. ‖ essieu m. arrière à orientation radiale par engrenages.

radial engine (Electr) ‖ Sternmotor m. ‖ moteur m. en étoile. / fixed ~ ‖ feststehender Sternmotor m. ‖ moteur m. fixe en étoile.

radial expansion (Gun) ‖ radiale Ausdehnung f.; Kaltreckung f. ‖ expansion f. radiale; autofrettage m. / ~ motor *see* radial engine. / ~ play ‖ Beweglichkeit f. in radialer Richtung ‖ mobilité f. radiale.

radial position of the axles forced by coupling bar of the motor car ‖ Radialstellung f. der Achsen zwangsläufig durch Kupplung vom Motorwagen aus ‖ orientation f. des essieux commandée par l'attelage de la voiture motrice. / ~ of wheel sets ‖ radiale Einstellung der Radsätze mpl. ‖ orientation f. radiale des essieux montés.

radial stone ‖ Schachtstein m. ‖ pierre f. radiale. / to resist flexible ~ stresses pl. ‖ Radialbeanspruchungen fpl. nachgiebig widerstehen ‖ résister élastiquement aux efforts radiaux. / ~ -type engine *see* ~ type motor. / ~ -type motor ‖ Sternmotor m. ‖ moteur m. en étoile. / ~ wheel set ‖ radial einstellbarer Lenkradsatz m. ‖ essieu m. radial monté.

radially, to adjust ~ ‖ radial einstellen ‖ mettre en position f. radiale. / ~ arranged around the motor crank-shaft ‖ sternförmig um die Motorwelle angeordnet ‖ disposé en étoile f. autour de l'arbre moteur.

radiant energy ‖ strahlende Energie f. ‖ énergie f. rayonnante. / instrument for measuring ~ ‖ Apparat m. zur Strahlenmessung ‖ appareil m. pour la mesure des radiations.

radiant field stop ‖ Leuchtfeldblende f. ‖ diaphragme m. limitant le champ lumineux. / ~ filament ‖ Leuchtdraht m. ‖ filament m. lumineux.

radiate, to ‖ bestrahlen; strahlen ‖ irradier; rayonner. / ~ heat ‖ Wärme f. abgeben ‖ rayonner *ou* irradier de la chaleur.

radiated ‖ strahlig ‖ radié. / ~ power (Radio) ‖ Strahlungsleistung f. einer Antenne ‖ puissance f. de rayonnement.

radiating ‖ strahlend ‖ rayonnant. / open ~ circuit (Radio) ‖ offener Strahlungskreis m. ‖ circuit m. radiant ouvert. / ~ source ‖ Strahlungsquelle f. ‖ source f. de rayonnement.

radiation ‖ Strahlung f.; Bestrahlung f.; radiation f.; irradiation f.; rayonnement m. / absorbed ~ ‖ absorbierende Strahlung f. ‖ radiation f. absorbée. / to decompose the ~ into the elements of its spectrum ‖ die Lichtstrahlung f. in ihr Spektrum zerlegen ‖ décomposer la radiation f. lumineuse dans son spectre. / dif-

fuse ~ ‖ zerstreute Strahlung f. ‖ rayonnement m. diffus. / to diminish the ~ ‖ die Strahlung f. verringern ‖ atténuer la radiation. / ~ of heat ‖ Wärmestrahlung f. ‖ rayonnement m. de (la) chaleur. / scattered ~ ‖ Streustrahlung f. ‖ rayonnement m. dispersé. / ultra-violet ~ ‖ ultraviolette Strahlung f. ‖ radiation f. ultraviolette. / treatment by ultra-violet ~ ‖ Ultraviolettbestrahlung f. ‖ traitement m. par les radiations ultraviolettes. / visible ~ ‖ sichtbare Strahlung f. ‖ radiation f. visible. / ~ of waves ‖ Ausstrahlung f. der Wellen ‖ radiation f. des ondes.

radiation apparatus ‖ Bestrahlungsapparat m. ‖ appareil m. radiateur. / eye ~ ‖ Augenbestrahlungsapparat m. ‖ appareil m. pour l'irradiation de l'œil.

radiation, atmospheric influence upon the ~ ‖ Einfluß m. der Atmosphäre auf die Wellenausbreitung ‖ influence f. de l'atmosphère sur la propagation des ondes. / characteristic of ~ ‖ Fernwirkungscharakteristik f. ‖ caractéristique f. de l'effet de radiation. / decrement of ~ ‖ Strahlungsdekrement n. ‖ décrément m. de rayonnement. / effect of ~ ‖ Strahlenwirkung f. ‖ effet m. de radiation. / efficiency of ~ (Radio) ‖ Strahlungswirkungsgrad m. ‖ rendement m. du rayonnement. / ~ energy ‖ Strahlungsenergie f. ‖ énergie f. radiée. / energy of ~ in sunlight ‖ Strahlungsenergie f. des Sonnenlichtes ‖ radiation f. solaire.

radiation lamp ‖ Bestrahlungslampe f. ‖ lampe f. radiateur *ou* pour bains de lumière. / ~ for therapeutical purposes ‖ Bestrahlungslampe f. für therapeutische Zwecke ‖ lampe f. d'irradiation à l'usage thérapeutique.

radiation pyrometer ‖ Strahlungspyrometer n. ‖ pyromètre m. optique. / ~ receiver ‖ Strahlungsempfänger m. ‖ radio-récepteur m. / ~ resistance ‖ Strahlungswiderstand m. ‖ résistance f. de rayonnement. / ~ stove ‖ Strahlungsofen m. ‖ radiateur m. / ~ thermometer ‖ Strahlungsthermometer n. ‖ thermomètre m. à radiation *ou* pour la mesure des radiations. / yellow-green region of ~ ‖ gelb-grüner Strahlenbereich m. ‖ région f. vert-jaune des radiations.

radiator (Auto) ‖ Kühler m. ‖ radiateur m. / ~ (Build) ‖ Heizkörper m.; Radiator m. ‖ radiateur m.; calorifère m. / booster ~ ‖ Hilfskühler m. ‖ radiateur m. auxiliaire. / cast iron ribbed ~ ‖ gußeiserner Rippenheizkörper m. ‖ radiateur m. à ailettes en fonte. / cellular ~ ‖ Zellenkühler m.; Bienenkorbkühler m. ‖ radiateur m. à nid d'abeilles. / draw-out ~ ‖ einziehbarer Kühler m. ‖ radiateur m. amovible *ou* escamotable. / electric ~ ‖ elektrischer Wärmestrahler m. ‖ radiateur m. électrique. / extractable ~ *see* draw-out ~. / front-mounted ~ (Auto) ‖ Kühler m. vorn am Wagen ‖ radiateur m. à l'avant. / gilled tube ~ ‖ Rippenrohrkühler m. ‖ radiateur m. à tuyaux garnis d'ailettes. / honeycomb ~ ‖ Wabenkühler m.; Bienenkorbkühler m. ‖ radiateur m. à nid d'abeilles. / lateral ~ ‖ Ohrenkühler m. ‖ radiateur m. auriculaire. / motor-car ~ ‖ Automobilkühler m. ‖ radiateur m. d'automobile. / pointed ~ ‖ Spitzkühler m. ‖ radiateur m. coupe-vent. / retractable ~ *see* draw-out ~. / ribbed ~

(Auto) ‖ Lamellenkühler m. ‖ radiateur m. à ailettes de tôle. / ~ for stove pipes ‖ Ofenrohrradiator m. ‖ radiateur m. pour tuyaux de poêle. / tubular ~ ‖ Rohrkühler m. ‖ radiateur m. à tuyaux. / underslung ~ ‖ Hängekühler m. ‖ radiateur m. à éclipse. / V-fronted ~ ‖ Spitzkühler m. ‖ radiateur m. en coupe-vent.

radiator, laminated ~ arrangement for gradual heating ‖ Lamellenkalorifer m. für Stufenbeheizung ‖ calorifère m. à ailettes à chauffage réglable. / ~ boring and milling machine ‖ Radiatorenfräs- und -bohrmaschine f. ‖ machine f. à fraiser et à percer les éléments de radiateur. / ~ bracket ‖ Kühlerfuß m. ‖ support m. de radiateur. / ~ bumper rod (Auto) ‖ Kühlerschutzbügel m. ‖ parechocs m. de radiateur. / ~ coil ‖ Schlangenkühler m. ‖ radiateur m. à serpentins. / ~ core ‖ Kühlerblock m. ‖ faisceau m. de radiateur. / ~ coupling ‖ Radiatorverschraubung f. ‖ boulonnage m. du radiateur. / ~ cover ‖ Kühlerhaube f.; Kühlerabdeckung f. ‖ couvre-radiateur m.; capot m. du radiateur. / ~ damper see radiator shutter. / ~ drain cock ‖ Kühlerablaßhahn m. ‖ robinet m. de vidange du radiateur. / ~ draw-off plug ‖ Kühlerablaßschraube f. ‖ bouchon m. de vidange du radiateur. / ~ fan ‖ Kühlerventilator m. ‖ ventilateur m. du radiateur. / ~ filler cap ‖ Kühlereinfüllstutzen m. ‖ orifice m. de remplissage du radiateur. / ~ fin ‖ Kühlrippe f. ‖ ailette f. de refroidissement. / ~ guard ring ‖ Kühlerschutzring m. ‖ bague f. de protection du radiateur. / moulding machine for ~s ‖ Radiatorenformmaschine f. ‖ machine f. à mouler des radiateurs. / ~ pipe ‖ Heizrohr n. ‖ tuyau m. de chauffage. / sectional core of ~ ‖ Teilblockkühler m. ‖ nid m. sectionné du radiateur. / ~ section drilling, milling and tapping machine ‖ Radiatorenbohr-, -fräs- und -gewindeschneidemaschine f. ‖ machine f. à percer, fraiser et tarauder les éléments de radiateurs. / ~ shell ‖ Kühlerrahmen m. ‖ cadre m. du radiateur.

radiator shutter ‖ Kühlerklappe f.; Kühlluftregler m. ‖ volet m. du radiateur. / lever for operating the ~ ‖ Kühlerklappenhebel m. ‖ levier m. de commande des volets de radiateur.

radiator stay ‖ Kühlerstrebe f. ‖ tirant m. de radiateur. / ~ tank ‖ Wasserkasten m. des Kühlers ‖ réservoir m. du radiateur. / ~ tapping machine ‖ Radiatorengewindeschneidmaschine f. ‖ machine f. à tarauder les éléments de radiateurs. / ~ thermometer ‖ Kühlwasserthermometer n. ‖ thermomètre m. de l'eau de refroidissement. / ~ tube ‖ Kühlrohr n. ‖ tube m. de refroidissement. / water inlet of ~ ‖ Kühlereinlaßstutzen m. ‖ entrée f. de l'eau du radiateur. / water outlet of ~ ‖ Kühlerausflußstutzen m. ‖ sortie f. d'eau du radiateur.

radical ‖ Radikal n. ‖ radical m. / ~ sign ‖ Wurzelzeichen n. ‖ signe m. radical. / ~ vinegar ‖ Eisessig m. ‖ vinaigre m. radical.

radii pl. of a mill-stone ‖ Rillen fpl. oder Hauschläge mpl. oder Furchen fpl. eines Mühlsteins ‖ rayons mpl. ou éveillures fpl. d'une meule.

radio ‖ Radio n. ‖ radio m. / aircraft ~ ‖ Fliegerfunkerei f. ‖ radiotélégraphie f.

d'aéroplane. / beam ~ ‖ Richtungstelegrafie f. ‖ radiotélégraphie f. dirigée.

radio accessories pl. ‖ Radiozubehör n. ‖ accessoires mpl. pour radio; pièces fpl. détachées pour appareils radiophoniques.

radioactive ‖ radioaktiv ‖ radioactif. / ~ constant ‖ Radioaktivitätskonstante f. ‖ constante f. radioactive. / ~ matter ‖ radioaktiver Stoff m. ‖ matière f. rayonnante. / ~ mineral ‖ radioaktives Gestein n. ‖ minéral m. radio-actif. / ~ substance ‖ radioaktive Substanz f. ‖ substance f. radioactive.

radioactivity ‖ Radioaktivität f. ‖ radioactivité f.

radio amateur ‖ Funkfreund m.; Funkliebhaber m.; Radioamateur m. ‖ radioamateur m. / ~ apparatus ‖ Radioapparat m. ‖ appareil m. à radio.

radio beacon ‖ Funkfeuer n. ‖ radiophare m. / circular ~ ‖ Kreisfunkfeuer n. ‖ radiophare m. circulaire. / directional ~ ‖ Richtfunkfeuer m.; Richtfunkbaken m. ‖ radiophare m. de direction.

radio bearing ‖ drahtlose Peilung f.; Funkpeilung f. ‖ radio-goniométrie f.; relèvement m. par télégraphie sans fil. / ~ station ‖ Funkpeilstation f. ‖ poste f. de relèvement par télégrauhie sans fil.

radiocommunication by bifurcation ‖ Funkgabelverkehr m. ‖ communication f. radio-électrique à bifurcation. / ~ station ‖ Funkstelle f. ‖ station f. de radio-communication.

radio compass ‖ Radiokompaß m.; Richtungsfinder m. ‖ radio-boussole f.; boussole f. sans fil.

radio component part ‖ Radioeinzelteil m. ‖ accessoire m. pour postes de télégraphie sans fil.

radio control of moving objects ‖ Fernlenkung f. ‖ commande f. à grande distance des objets mouvants par télégraphie sans fil. / ~ of vessels and airships ‖ drahtlose Fernlenkung f. von Wasserfahrzeugen und Luftfahrzeugen ‖ commande f. à distance par télégraphie sans fil des embarcations navales et aériennes.

radio detector ‖ Radiodetektor m. ‖ détecteur m. (de radiophonie). / ~ direction finder ‖ Radioortungsgerät n.; Funkpeiler m. ‖ radio-chercheur m. de position; radiogoniomètre m. à cadre mobile. / ~ direction finding service ‖ Funkpeildienst m. ‖ service m. radiogoniométrique.

radio duplex-service ‖ drahtloser Duplexverkehr m. ‖ service m. duplex radiotélégraphique. / ~ engineering ‖ Rundfunktechnik f. ‖ technique f. de la radioélectricité.

radiogoniometer ‖ Radiogoniometer n. ‖ radiogoniomètre m.

radiogoniometric ‖ radiogoniometrisch ‖ radiogoniométrique.

radiography ‖ Radiografie f.; Röntgenaufnahme f. ‖ radiographie f.

radio ground communication ‖ Funkverbindung f. ‖ liaison f. radio-terrestre. / ~ headphone ‖ Radiokopfhörer m. ‖ casque m. de radiophonie. / ~ letter telegram ‖ Funkbrief m. ‖ radio lettre-télégramme m.

radiology ‖ Röntgenlehre f.; Röntgenologie f. ‖ radiologie f.; science f. des rayons de Röntgen.

radio loud-speaker ‖ Lautsprecher m. für Radio; Radiolautsprecher m. ‖ hautparleur m. pour radiophonie. / ~ mast ‖ Antennenmast m. ‖ mât m. d'antenne. / ~ officer ‖ Funkoffizier m. ‖ officier m. d'antenne.

radiophony (Tel) ‖ Radiofonie f. ‖ radiophonie f.

radio receiver ‖ Funkempfänger m.; Radioempfänger m. ‖ radio-récepteur m. / ~ with crystal detector ‖ Radioempfänger m. mit Kristalldetektor ‖ radiorecepteur m. à detection sur galène. / ~ battery ‖ Funkempfangsbatterie f. ‖ batterie f. pour radio-récepteur.

radioscope ‖ Radioskop n.; Durchleuchtungsapparat m. ‖ radioscope m. / ~ box ‖ Durchleuchtungskasten m. ‖ boîte f. radioscopique. / ~ chair ‖ Durchleuchtungsstuhl m. ‖ chaise f. radioscopique. / ~ table ‖ Durchleuchtungstisch m. ‖ table f. radioscopique.

radioscopy ‖ Durchleuchtung f. ‖ radioscopie f.

radio service ‖ Funkdienst m. ‖ service m. de radiocommunication. / ~ set ‖ funkentelegrafische Gruppe f. ‖ groupe m. radio.

radio station ‖ Funkstation f.; Rundfunkstation f.; Sender m. ‖ radio-station f.; station f. de télégraphie sans fil; poste m. radiotélégraphique. / coast ~ ‖ Küstenfunkstelle f. ‖ radio-station f. côtière. / long-distance ~ ‖ Großfunkstelle f. ‖ station f. de télégraphie sans fil à grande distance. / private experimental ~ ‖ private Versuchsfunkstelle f. ‖ radio-station f. expérimentale privée.

radio synoptic weather message ‖ Wettersammelfunkspruch m. ‖ radiogramme m. météorologique collectif.

radiotelegram ‖ Funktelegramm n. ‖ radiogramme m.

radiotelegraph cart ‖ Funkenkarren m. ‖ voiture f. radiotélégraphique.

radiotelegrahic long distance station / radiotelegrafische Großstation f. ‖ poste f. radiotélégraphique de grande distance. / ~ operating table ‖ radiotelegrafischer Bedienungstisch m. ‖ table f. de manipulation radiotélégraphique. / ~ plant ‖ radiotelegrafische Anlage f. ‖ installation f. radiotélégraphique. / ~ station ‖ Funkstelle f. ‖ poste f. radiotélégraphique.

radiotelegraphy ‖ Funktelegrafenbetrieb m.; Funkverkehr m. ‖ télégraphie f. sans fil. / military ~ ‖ Feldfunktelegrafie f. ‖ télégraphie f. sans fil militaire. / laws pl. and regulations pl. relating to ~ ‖ Funkrecht n. ‖ lois fpl. et règlements mpl. concernant la télégraphie sans fil.

radio telephonic transmitter ‖ Telefoniesender m. ‖ émetteur m. radiotéléphonique. / ~ telephony ‖ drahtlose Telefonie f. ‖ radiotéléphonie f.

radiotherapeutics pl. ‖ Röntgentherapie f. / radiothérapie f.

radio tower ‖ Funkturm m.; Antennengroßmast m. ‖ tour f. d'antenne; radiophare m. / ~ transmission service ‖ Funkübertragungsdienst m. ‖ service m. de transmission radioélectrique. / telegraphic ~ transmitter ‖ Telegrafiesender m. ‖ émetteur m. radiotélégraphique. / control of ~ transmitters ‖ Tasteinrichtung f. für Hochfrequenzsender ‖ mani-

pulation f. des émetteurs. / ~ tube ‖ Radioröhre f. ‖ tube m. pour radio; lampe f. pour radiophonie. / ~ weather service ‖ funkentelegrafischer Wetterdienst m. ‖ service m. radiotélégraphique de prévision du temps.

radish ‖ Rettich m. ‖ radis m.; raifort m. / ~ knife ‖ Rettichschneider m. ‖ couperadis m.

radium ‖ Radium m. ‖ radium m. / ~ luminous article ‖ Radiumleuchtartikel m. ‖ article m. lumineux à base de radium.

radius (Crane) ‖ Ausladung f.; Reichweite f. ‖ portée f. / ~ (Geom) ‖ Radius m.; Halbmesser m. ‖ rayon m.; demidiamètre m. / ~ of action ‖ Reichweite f.; Aktionsradius m. ‖ rayon m. d'action. / ~ of arm of the drilling machine ‖ Schwinghalbmesser m. der Bohrmaschine ‖ rayon m. du bras radial pivotant de la perceuse. / ~ of a crane ‖ Ausladung f. eines Krans ‖ portée f. d'une grue. / ~ of crane arm ‖ Arbeitsbereich m. des Kranes ‖ rayon m. de portée de la grue. / ~ of curvature of the concave mirror ‖ Krümmungshalbmesser m. des Hohlspiegels ‖ rayon m. de courbure du miroir concave. / ~ of curve ‖ Krümmungsradius m. ‖ rayon m. de courbure. / ~ of dishing ‖ Wölbungsradius m. ‖ rayon m. des fonds bombés. / ~ of inertia ‖ Trägheitshalbmesser m. ‖ rayon m. d'inertie. / inside ~ of dished ends of boiler ‖ Radius m. der inneren Wölbung des Kesselbodens ‖ rayon m. intérieur des fonds bombés de chaudière. / meridian ~ ‖ Meridianradius m. ‖ rayon m. méridien. / prime ~ (Geom) ‖ Polarachse f. ‖ axe m. polaire. / ~ of a slewing crane ‖ Drehkreis m. eines Drehkranes ‖ cercle m. balayé par une grue tournante fixe. / small flanging ~ ‖ kleiner Eckradius m. ‖ petit rayon m. de bride. / the crane has a useful ~ of x meters ‖ der Kran hat eine nutzbare Ausladung von x m ‖ la grue a une portée utile de x m. / effective ~ of the wireless tower ‖ Reichweite f. des Funkturmes ‖ rayon m. utile de la tour de radiodiffusion. / within a ~ of ‖ im Umkreis m. von ‖ dans un rayon de.

radius bar ‖ Gegenlenker m. ‖ bras m. de rappel. / ~ gauge ‖ Radiuslehre f. ‖ gabarit m. pour les différents rayons. / ~ rod ‖ Hinterachsschubstange f. ‖ bielle f. de poussée. / ~ vector ‖ Polstrahl m.; Radiusvektor m. ‖ rayon m. vecteur.

raffia bast ‖ Raphiabast m. ‖ écorce f. de raphia.

raft, to ‖ flößen ‖ flotter. / ~ wood ‖ Holz n. flößen ‖ flotter du bois.

raft ‖ Floß n. ‖ radeau m.; train m. de bois. / bridge ~ ‖ Brückenfloß n. ‖ support m. flottant. / ~ of casks ‖ Faßfloß n.; Tonnenfloß n. ‖ radeau m. de tonneaux ou de barriques. / narrow ~ ‖ Schleusenfloß n. ‖ éclusée f. / temporary ~ ‖ Notfloß n. ‖ radeau m. circonstances.

raft bridge ‖ Floßbrücke f. ‖ pont m. de radeaux.

rafter ‖ Dachsparren m. ‖ chevron m. / empty ~ see intermediate ~. / intermediate ~ ‖ Leersparren m. ‖ chevron m. intermédiaire. / principal ~ (Carp) ‖ Hauptsparren m. ‖ maîtrechevron m. / principal ~ of a purlinroof ‖ Pfettenträger m.; Bindesparren

des Pfettendachs ‖ arbalétrier m. d'un comble à pannes.

rafter nail ‖ Sparrennagel m. ‖ dent m. de loup.

rafters pl. ‖ Dachgesparre n. ‖ chevron m. ou charpente f. du toit. / ~ of a trussframe ‖ Streben fpl. eines Hängewerkes ‖ arbalétriers mpl. d'une armature ou d'une ferme à poinçons.

raft-port (Shipb) ‖ Ladepforte f. ‖ sabord m. de charge.

raftsman ‖ Flößer m.; Floßführer m. ‖ flotteur m.

raft-wood ‖ Floßholz n. ‖ bois m. flotté ou volant ou de flottage ou de train.

rag ‖ Lumpen m.; Lappen m. ‖ torchon m.; chiffon m.; haillon m.; drille f. / ~ assorter see rag sorter. / ~ ball ‖ Filzbällchen n. ‖ tampon m. ou bouchon m. de feutre. / ~ boiler ‖ Hadernkocher m.; Lumpenkocher m. ‖ chaudière f. à haillons; lessiveur m. de chiffons. / ~ bolt ‖ Steinschraube f. ‖ boulon m. de scellement; cheville f. à barbe. / ~ bucket (Pap) ‖ Lumpenbütte f.; Zuber m. ‖ gerlon m. / ~ cleaning machine ‖ Lumpenreinigungsmaschine f. ‖ machine f. à nettoyer les chiffons. / ~ cutter ‖ Lumpenschneider m. ‖ coupeur m. de chiffons. / ~ cutting ‖ Lumpenreißerei f. ‖ effilochage m. de chiffons. / ~ cutting machine (Pap) ‖ Lumpenschneider m.; Hadernschneider m. ‖ dérompoir m.; délisseuse f. mécanique; coupe-chiffons m.; machine f. à couper les chiffons. / ~ cylinder driver (Pap) ‖ Halbholländerführer m. ‖ défileur m. / ~ devil ‖ Lumpenwolf m. ‖ machine f. à défiler ou à rompre les chiffons. / ~ dressing plant (Pap) ‖ Holländer m. ‖ installation f. pour la préparation des chiffons. / ~ engine (Pap) ‖ Holländer m.; Stoffmühle f. ‖ pile f. ou moulin m. à cylindre; cylindre m. / ~ gatherer ‖ Lumpensammler m. ‖ chiffonier m.; drillier m.

ragged ‖ geschuppt; gezackt ‖ barbu; grillé.

ragman ‖ Lumpensammler m. ‖ chiffonnier m.; drillier m.

rag picker see ragman. / ~ pulp ‖ Lumpenstoff m. ‖ pâte f. de chiffons. / ~ pump ‖ Kettenpumpe f. ‖ pompe f. à chapelet.

rags pl. ‖ Lumpen mpl.; Hadern mpl.; drilles fpl.; chiffons mpl. / boiled ~ ‖ gekochte Lumpen mpl. ‖ bouillie f. / ~ for cleaning ‖ Putzlappen m.; Putzwerg n.; Putzwolle f. ‖ étoupe f. ou chiffons mpl. ou coton m. à nettoyer. / devilled ~ ‖ gewolfte Lumpen mpl. ‖ cobre m. / ~ impregnated with gold-solution ‖ Goldzunder m.; mit Goldauflösung durchtränkte Leinenlumpen ‖ or m. en chiffons ou en drapeaux.

rag sorter ‖ Lumpensortierer m. ‖ trieur m. de chiffons; chiffonnier m.; délisseur m.

ragstone ‖ Pläner m. ‖ moëllon m. gisant ou feuilleté ou lamineuse ou marneux ou schisteux. / ~ laid contrary to the cleaving-grain ‖ hochkantiger Pläner m. ‖ moëllon m. posé en délit. / ~ laid in vaulting ‖ Wölbpläner m. ‖ moëllon m. en coupe. / ~ work ‖ Plänermauerwerk n. ‖ maçonnerie f. en pierre marneuse.

rag-tearing engine see rag devil. / ~ woman ‖ Lumpenzupferin f. ‖ effilocheuse f.

rag tub see ~ bucket. / ~ washer ‖ Lumpenwäscher m. ‖ laveur m. de chiffons. / ~ washerman see ~ washer. / ~ wheel ‖ Schwabbelrad n. ‖ disque m. à multiple

épaisseurs de chiffons. / ~ work ‖ rauhes Mauerwerk n. ‖ maçonnerie f. en moëllons bruts. / ~ working machine ‖ Lumpenverarbeitungsmaschine f. ‖ machine f. pour la préparation des chiffons.

raider, commerce ~ (Mar) ‖ Handelszerstörer m. ‖ croiseur m. corsaire.

rail, to ‖ gittern; einfriedigen ‖ mailler. / ~ (Railw) ‖ Schienen fpl. legen ‖ mettre les rails mpl.

rail (Carp) ‖ Geländerriegel m. ‖ lisse f. d'appui. / ~ (Railw) ‖ Schiene f.; Schienenstrang m. ‖ rail m. / ~ (Shipb) ‖ Reling f. ‖ lisse f. de bastingage. / ~ of a baywork ‖ Riegel m. oder Querholz n. einer Fachwand; Wandriegel m. ‖ entretoise f. de cloison. / to bend ~s pl. ‖ Schienen fpl. abrichten ‖ courber les rails mpl. / ~ with bottom discharge tubs ‖ Standbahn f. mit Bodenentladern ‖ voie f. fixe à déchargement par le fond. / broadfooted ~ ‖ breitbasige Schiene f.; Vignolesschiene f. ‖ rail m. Vignole ou à patin. / bull-headed ~ ‖ Doppelkopfschiene f. ‖ rail m. à double champignon. / by ~ ‖ mit Eisenbahn f. ‖ par chemin m. de fer. / check ~ ‖ Leitschiene f.; Zwangschiene f. ‖ contre-rail. / ~ embedded in concrete ‖ in Beton eingebettete Schiene f. ‖ rail m. encastré en béton. / crane ~ ‖ Kranschiene f. ‖ rail m. pour grues. / curve ~ ‖ Kurvenschiene f. ‖ rail m. pour courbes. / curved ~ ‖ gebogene Schiene f. ‖ rail m. cintré. / double-headed ~ ‖ Doppelkopfschiene f. ‖ rail m. à double champignon. / drilled ~ ‖ gebohrte Schiene f. ‖ rail m. percé. / to exchange the damaged ~s ‖ die beschädigten Schienen fpl. auswechseln ‖ remplacer les rails mpl. avariés. / fastening the ~s to the sleepers with bolts and spikes ‖ Nagelung f. der Schienen ‖ clouage m. des rails. / fastening the ~s to timber sleepers ‖ Befestigung f. der Schienen auf hölzernen Schwellen ‖ fixation f. des rails aux traverses en bois. / filled section ~ ‖ Blockschiene f.; Vollschiene f. ‖ rail m. de section pleine ou en U renversé. / fished ~s pl. ‖ verlaschte Schienen fpl. ‖ rails mpl. éclissés. / flat-headed ~ ‖ Schiene f. mit flachem Kopf ‖ rail m. à surface plane. / grooved ~ ‖ Rillenschiene f. ‖ rail m. à ornière ou à gorge. / guide ~ ‖ Spurlatte f. ‖ rail m. de guidage. / rolled section for guide and check ~s ‖ Zwangsschienenwinkel m. ‖ cornières fpl. pour contrerails. / to insert a ~ ‖ eine Schiene f. einlegen ‖ mettre en place f. un rail. / junction ~ ‖ Anschlußschiene f. ‖ raccord m. de rail. / junction ~ end ‖ Anschlußende n. der Schiene; Anschlußschienenstück n. ‖ pièce f. de raccord de rail. / manganese steel ~ ‖ Manganstahlschiene f. ‖ rail m. en acier au manganèse. / ~ for mines ‖ Grubenschiene f. ‖ rail m. pour mines. / parallel ~ ‖ Parallelschiene f. ‖ rail m. parallèle ou à champignon symétrique. / plate ~ ‖ Flachschiene f. Plattschiene f. ‖ bande f. plate; rail m. plat. / ~ for portable railways ‖ Feldbahnschiene f. ‖ rail m. pour voies portatives. / relay ~ ‖ Relaisschiene f. ‖ rail m. de relais. / ~ resting upon chairs ‖ Stuhlschiene f. ‖ rail m. à champignon. / rigid ~ of the switch (Railw) ‖ Anschlagschiene f.; Stockschiene f.; Backenschiene f.; unbewegliche Schiene f. des Wechsels ‖ rail m. d'applique; rail m.

fixe d'excentrique. / to run off the ~s ‖ entgleisen ‖ sauter hors des rails; sortir des rails; dérailler. / to set and straighten ~s ‖ Schienen ausrichten ‖ dresser les rails mpl. / single-headed ~ see single-T ~. / single T-~ ‖ T-Schiene f. ‖ rail m. à simple champignon. / standard-gauge ~ ‖ Vollbahnschiene f. ‖ rail m. pour voie normale. / stock ~ see rigid ~. / three-~-track ‖ Dreischienengleis n. ‖ voie f. à trois rails. / tramway ~ ‖ Straßenbahnschiene f. ‖ rail m. de tramway; rail m. à ornière. / transverse ~ ‖ Querschiene f. ‖ rail m. transversal. / the ~s pl. of the line are uninterrupted throughout ‖ die Schienen f. sind ununterbrochen durchgeführt ‖ les rails mpl. sont posés sans interruption de la ligne. / Vignoles ~ ‖ breitbasige Schiene f.; Vignolesschiene f. ‖ rail m. Vignole ou à patin. / the ~ warps ‖ die Schiene f. wirft sich ‖ le rail se déjète. / wearing ~ ‖ Abnutzungsschiene f. ‖ rail m. d'usure. / wing ~ see rigid ~. / wooden ~ ‖ Querholz n.; Querriegel m. ‖ traverse f. ou entretoise f. en bois.

rail anchor ‖ Schienenklemme f. ‖ dispositif m. d'ancrage des rails. / screw ~ ‖ Schraubenklemme f. ‖ dispositif m. d'ancrage à vis. / universal ~ (Railw) ‖ Einheitsklemme f. ‖ dispositif m. d'ancrage universel. / wedge ~ ‖ Keilklemme f. ‖ dispositif m. d'ancrage à coin.

rail, barrow tongs pl. for ~s ‖ Schienentragzange f. ‖ porte-tenaille m. pour rails.

rail base ‖ Schienenfuß m. ‖ patin m. de rail. / flange of ~ ‖ Flansch m. des Schienenfußes ‖ aile f. du patin de rail. / width of ~ ‖ Schienenfußbreite f. ‖ largeur f. du patin de rail.

rail bender ‖ Schienenbiegemaschine f. ‖ machine f. à cintrer les rails; cintreuse f. pour rails. / ~ bond ‖ Schienenverbinder m. ‖ éclisse f. ou joint m. de rails. / ~ bond tester ‖ Schienenstoßprüfer m. ‖ essayeur m. d'éclissages ou de joints de rails. / ~ boring machine ‖ Schienenbohrmaschine f. ‖ machine f. à percer les rails. / ~ brake ‖ Schienenbremse f. ‖ frein m. de rail; sabot-frein m. / ~ breakage ‖ Schienenbruch m. ‖ rupture f. de rail. / ~ bridge ‖ Gleisbrücke f. ‖ pont m. de voie. / ~ car ‖ Schienenfahrzeug n.; Triebwagen m. ‖ automotrice f. ou voiture f. à rail. / ~ carrier (Mine) ‖ Schienenträger m. ‖ meneur m. de rails. / ~ chair ‖ Schienenstuhl m. ‖ coussinet m. de rail. / ~ cramp ‖ Schienenklammer f. ‖ crampon m. des rails. / ~ creeping ‖ Wandern n. der Schienen ‖ marche f. des rails. / crossing of ~s ‖ Gleiskreuzung f. ‖ croisement m. de voies. / ~ cycle ‖ Eisenbahnfahrrad n.; Draisine f. ‖ draisine f. / depth of ~ ‖ Schienenhöhe f. ‖ hauteur f. du rail. / end of ~ (Railw) ‖ Schienenende n. ‖ bout m. ou extrémité f. du rail. / ~ fastening ‖ Schienenbefestigung f. ‖ montage m. des rails. / ~ fish plate ‖ Schienenlasche f. ‖ éclisse f. de rails. / foot of ~ ‖ Schienenfuß m. ‖ patin m. du rail. / bent-up ~ foot ‖ aufgebogener Schienenfuß m. ‖ patin m. du rail recourbé. / ~ gauge ‖ Spurweite f. des Bahngleises ‖ écartement m. des rails ou de la voie. / getting off the ~s ‖ Entgleisung f. ‖ déraillement m. / ~ guard of a locomotive ‖ Schienenräumer m. der Lokomotive ‖ chasse-pierres m.

ou chasse-corps m. ou garde m. d'une locomotive.

rail head ‖ Schienenkopf m. ‖ champignon m. du rail. / width of the ~ ‖ Kopfbreite f. der Schiene ‖ largeur f. du champignon du rail.

rail, height of ~ ‖ Schienenhöhe f. ‖ hauteur f. des rails. / ~ hoisting engine see ~ lifting jack.

railing (Build) ‖ Geländer n. ‖ garde-fou m.; garde-corps m.; balustrade f. / ~ (Mill) ‖ Griesholm m. ‖ traverse f. de poteaux. / ~ (Shipb) ‖ Reling f.; Geländer n. ‖ lice f. ou lisse f. d'appui; barrière f.; garde-corps m. / ~ of a bridge ‖ Brückengeländer n. ‖ parapet m. ou garde-fou m. d'un pont. / ~ for stairs ‖ Treppengeländer n. ‖ rampe f. ou balustrade f. d'escalier. / ~ of a well ‖ Brunnengeländer n. ‖ margelle f.

railing bar (Railw) ‖ Schrankenstange f. ‖ lisse f.; barrière f. / ~ iron ‖ Geländerstange f. ‖ lisse f. de garde-corps. / ~ joint ‖ Schienenverbindung f. ‖ assemblage m. ou éclissage m. des rails. / ~ layer (Mine) ‖ Schienenleger m. ‖ déferailleur m.; metteur m. de rails. / ~ layer (Railw) ‖ Oberbauarbeiter m. ‖ ouvrier m. ou poseur m. de la voie.

railless trolley car ‖ schienenloser Motorwagen m.; Drahtautobus m. ‖ trolleybus m.

rail lifter ‖ Gleisheber m. ‖ pince f. à rails. / ~ lifting jack ‖ Gleishebewinde f. ‖ levier m. pour rails; treuil m. pour soulever les rails. / lower edge of a ~ ‖ Unterkante f. einer Schiene ‖ dessous m. du rail. / length of ~ ‖ Schienenlänge f. ‖ longueur f. du rail.

rail line, double ~ ‖ Doppelschienengleis n. ‖ voie f. à rails doubles. / triple ~ ‖ Dreischienengleis n. ‖ voie f. à trois rails.

rail mill ‖ Schienenwalzwerk n. ‖ laminoir m. à rails.

rail milling machine ‖ Schienenfräser m. ‖ fraiseuse f. pour rails. / ~ motor car ‖ Triebwagen m. ‖ automotrice f. à rail.

rail permanent way ‖ Eisenbahnoberbau m. ‖ superstructure f. des chemins de fer. / material for ~ ‖ Eisenbahnoberbaumaterial n. ‖ matériaux mpl. de superstructure de chemin de fer. / ~ wagon ‖ Bahndienstwagen m. ‖ wagonet m. de service de la voie.

rail piece of a door-frame ‖ Querfries m. eines Türrahmens ‖ traverse f. d'un cadre de porte. / ~ pile (Metal) ‖ Schienenpaket n. ‖ paquet m. ou trousse f. pour rail. / ~ plane ‖ Schienenhobel m. ‖ rabot-lime m. pour rails. / ~ post ‖ Geländerpfosten m.; Geländerstütze f. ‖ barreau m. de garde-corps; montant m. / press ‖ Schienenrichtpresse f. ‖ presse f. à dresser les rails. / ~ return ‖ Schienenrückleitung f. ‖ retour m. par les rails.

railroad ‖ Bahn f.; Eisenbahn f. ‖ chemin m. de fer. / ~ artillery ‖ Eisenbahnartillerie f. ‖ artillerie f. sur voie ferrée. / ~ ballast ‖ Eisenbahnschotter m. ‖ ballast m. pour voies ferrées. / ~ net with a very lively traffic ‖ Bahnnetz n. mit lebhaftem Eisenbahnbetrieb ‖ réseau m. de chemins de fer d'un trafic très vif / ~ permanent way ‖ Eisenbahnoberbau m. ‖ superstructure f. de chemin de fer. / ~ system of ~s ‖ Bahnnetz n. ‖ réseau m. de chemins de fer. / ~ tracks pl. see ~ permanent way. / ~ vehicle ‖

Schienenfahrzeug n. ‖ véhicule m. ou voiture f. sur rails.

rail rolling mill (Roll) ‖ Schienenwalzwerk n. ‖ laminoir m. ou train m. à rails.

rail saw ‖ Schienensäge f. ‖ scie f. à rails. / ~ seat ‖ Schienenlagerung f. ‖ assise f. des rails. / to form ~ seats in the wooden sleepers ‖ die Holzschwellen fpl. dechseln ‖ saboter les traverses fpl. / ~ section ‖ Schienenprofil n. ‖ section f. ou profil m. du rail. / ~ shifting machine ‖ Gleisrückmaschine f. ‖ machine f. à avancer les rails. / ~ slewer ‖ Schienenrücker m. ‖ appareil m. à déplacer les rails longitudinalement. / ~ spike ‖ Schienennagel m.; Hakennagel m. ‖ crampon m. de rail. / ~ straightener ‖ Richtmaschine f. für Schienen ‖ appareil m. à rectifier ou à dresser des rails. / ~ straightening machine see ~ straightener. / ~ and girder straightening press ‖ Schienen- und Trägerrichtpresse f. ‖ presse f. à dresser les rails et poutrelles. / ~ templet ‖ Schienenprofilschablone f. ‖ gabarit m. de profil pour rails. / ~ tongs pl. ‖ Schienenzange f. ‖ tenaille f. à rails. / top of ~ ‖ Schienenoberkante f. ‖ dessus m. du rail. / ~ track ‖ Oberbau m. ‖ superstructure f. / tread of ~ ‖ Schienenbahn f. ‖ chemin m. à rails. / upper edge of a ~ ‖ Oberkante f. einer Schiene ‖ dessus m. du rail. / wagon running on ~s ‖ Gleisfahrzeug n. ‖ véhicule m. sur rails.

railway ‖ Bahn f.; Eisenbahn f. ‖ chemin m. de fer. / adhesion ~ ‖ Adhäsionseisenbahn f. ‖ chemin m. de fer à adhérence. / aerial ~ ‖ Hängebahn f. ‖ chemin de fer m. aérien. / auxiliary ~ ‖ Hilfsbahn f. ‖ chemin m. de fer provisoire; voie f. auxiliaire. / cable ~ ‖ Kabelbahn f. ‖ funiculaire f. / chain ~ built on the ground ‖ bodenständige Kettenbahn f. ‖ funiculaire m. fixe à chaîne. / circular ~ ‖ Ringbahn f. ‖ chemin m. de fer de ceinture. / colonial ~ ‖ Kolonialbahn f. ‖ ligne f. coloniale. / contractor's ~ ‖ Feldbahn f. ‖ chemin m. de fer portatif; Decauville m. / electric ~ ‖ elektrische Bahn f. oder Eisenbahn f. ‖ tramway m. ou chemin m. de fer électrique. / electrified ~ ‖ elektrische Vollbahn f. ‖ chemin m. de fer électrifié. / elevated ~ ‖ Hochbahn f. ‖ chemin m. de fer levé. / encircling ~ see circular ~. / field ~ see portable ~. / forest ~ ‖ Forstbahn f. ‖ chemin m. de fer forestier. / harbour ~ ‖ Hafenbahn f. ‖ chemin m. de fer de port. / high-speed ~ ‖ Schnellbahn f. ‖ ligne f. à service rapide; chemin m. de fer à grande vitesse. / industrial ~ ‖ Industriebahn f. ‖ chemin m. de fer industriel. / light ~ ‖ Sekundärbahn f. ‖ chemin m. de fer vicinal. / local ~ ‖ Lokalbahn f. ‖ chemin m. de fer local. / mine ~ ‖ Grubenbahn f. ‖ chemin m. de fer minier. / narrow-gauge ~ ‖ Schmalspurbahn f.; Kleinbahn f. ‖ chemin m. de fer à voie étroite; petit chemin m. de fer. / ~ in operation ‖ Bahn f. im Betriebe ‖ ligne f. en exploitation. / overhead ~ ‖ Hängebahn f. ‖ transporteur m. aérien à câble. / portable ~ ‖ Feldbahn f. ‖ chemin m. de fer portatif; voie f. portative. / principal ~ ‖ Hauptbahn f. ‖ ligne f. principale. / rope ~ built on the ground ‖ bodenständige

Seilbahn f. ‖ funiculaire f. fixe à câble. / secondary ~ ‖ Nebenbahn f.; Sekundärbahn f. ‖ chemin m. de fer secondaire. / single-phase ~ ‖ Einphasenbahn f. ‖ chemin m. de fer à courant monophasé. / standard ~ ‖ Normalbahn f.; Vollspurbahn f. ‖ chemin m. de fer normal. / suburban ~ ‖ Vorortbahn f. ‖ chemin m. de fer de banlieue. / suspended ~ see suspension ~. / suspension ~ ‖ Hängebahn f. ‖ chemin m. de fer aérien ou suspendu. / suspension ~ for persons ‖ Personenschwebebahn f. ‖ transporteur m. suspendu à câbles pour personnes. / ~ with rails of timber ‖ Holzschienenbahn f. ‖ route f. à ornières en bois. / underground ~ ‖ Untergrundbahn f. ‖ chemin m. de fer souterrain. / wooden ~ ‖ Holzbahn f. ‖ voie f. en bois.
railway accident ‖ Eisenbahnunglück n. ‖ accident m. de chemin de fer. / ~ alarm ‖ Eisenbahnläutewerk n. ‖ sonnerie f. de chemin de fer. / ~ artillery ‖ Eisenbahnartillerie f. ‖ artillerie f. sur voie ferrée. / ~ board ‖ Eisenbahndirektion f. ‖ direction f. des chemins de fer. / body of ~ ‖ Bahnkörper m. ‖ superstructure f. de voie. / ~ brake ‖ Eisenbahnbremse f. ‖ frein m. de chemin de fer. / ~ braker ‖ Eisenbahnbremser m. ‖ garde-frein m. de chemin de fer. / ~ bridge ‖ Eisenbahnbrücke f. ‖ pont m. ou viaduc m. de chemin de fer. / ~ buffer ‖ Eisenbahnpuffer m. ‖ tampon m. de choc pour wagons de chemin de fer. / ~ building enterprise ‖ Eisenbahnbauunternehmung f. ‖ entreprise f. de construction de chemin de fer. / ~ car ‖ Eisenbahnwagen m. ‖ voiture f. de chemin de fer. / ~ car brake ‖ Eisenbahnwagenbremse f. ‖ frein m. pour wagons de chemin de fer. / ~ carriage ‖ Eisenbahnwagen m. ‖ wagon m. ou voiture f. de chemin de fer. / fittings pl. for ~ carriages ‖ Eisenbahnwagenarmatur f. ‖ armature f. pour wagons de chemin de fer. / ~ carriage lighting ‖ Eisenbahnwagenbeleuchtung f. ‖ éclairage m. des trains. / ~ carriage wheel ‖ Eisenbahnwagenrad n. ‖ roue f. pour wagons de chemin de fer. / circle of ~ ‖ Ringgleis n. ‖ réseau m. circulaire de chemin de fer. / ~ clerk ‖ Eisenbahnbeamter m. ‖ employé m. de chemin de fer. / ~ collision ‖ Eisenbahnzusammenstoß m. ‖ collision f. de chemin de fer. / ~ company ‖ Eisenbahngesellschaft f. ‖ compagnie f. de chemin de fer. / ~ connection ‖ Eisenbahnverbindung f. ‖ communication f. par chemin de fer. / ~ construction ‖ Eisenbahnbau m.; Bahnbau m. ‖ construction f. de chemins de fer ou de la voie. / ~ construction and operation rules ‖ Eisenbahnbau- und -betriebsordnung f. ‖ règlement m. de construction et d'exploitation des voies ferrées. / ~ construction works pl. ‖ Eisenbahnbauunternehmung f. ‖ entreprise f. de construction de chemin de fer. / ~ contractor ‖ Eisenbahnbauunternehmer m. ‖ entrepreneur m. de construction de chemin de fer. / ~ conveyance ‖ Eisenbahntransport m. ‖ transport m. par chemin de fer. / automatic ~ coupling ‖ selbsttätige Eisenbahnwagenkupplung f. ‖ attelage m. automatique de wagons de chemin de fer. / ~ crossing ‖ Gleiskreuzung f.; Eisenbahnkreuzung f. ‖ croisement m. de chemin de fer; traversée f. de voie. /

~ earth-works pl. ‖ Eisenbahnunterbau m. ‖ substructure f. ou infrastructure f. de chemin de fer. / ~ enterprise ‖ Eisenbahnunternehmen n. ‖ entreprise f. de chemin de fer. / ~ equipment ‖ Ausrüstungsteile mpl. für Bahnen ‖ matériel m. pour chemins de fer. / ~ ferry ‖ Eisenbahnfähre f.; Trajektschiff n. ‖ bac m. porte-train; ferry-boat m.; bateau m. transbordeur. / ~ gate ‖ Eisenbahnschranke f. ‖ barrière f. de voie. / ~ gauge ‖ Spurweite f. ‖ largeur f. de voie f. / ~ generator ‖ Bahndynamomaschine f. ‖ dynamo f. de traction. / ~ goods pl. ‖ Frachtgut n. ‖ marchandise f. en petite vitesse. / ~ granting ‖ Eisenbahnverleihung f. ‖ concession f. de chemin de fer. / ~ ground ‖ Eisenbahngelände n. ‖ terrains mpl. appartenant au chemin de fer. / ~ heating member ‖ Bahnwagenheizkörper m. ‖ appareil m. de chauffage pour wagons de chemin de fer. / ~ implements pl. ‖ Streckenwerkzeug n. ‖ outillage m. de la voie. / ~ improvement shop ‖ Eisenbahnausbesserungswerk n. ‖ atelier m. de réparation des chemins de fer. / ~ lantern ‖ Eisenbahnlaterne f. ‖ lanterne f. de chemin de fer. / ~ law ‖ Eisenbahnrecht n. ‖ législation f. des chemins de fer. / ~ lighting ‖ Eisenbahnwagenbeleuchtung f. ‖ éclairage m. de wagons de chemin de fer.
railway line ‖ Bahnlinie f. ‖ ligne f. ou tracé m. de chemin de fer. / an area is intersected by ~s ‖ ein Gebiet n. wird von Eisenbahnlinien durchschnitten ‖ des lignes fpl. de chemin de fer traversent une région. / main ~ ‖ Haupteisenbahnlinie f. ‖ grande ligne f. de chemin de fer.
railway line packing machine ‖ Gleisstopfmaschine f. ‖ machine f. à bourrer les voies de chemin de fer.
railway loan ‖ Eisenbahnanleihe f. ‖ emprunt m. de chemin de fer. / ~ management ‖ Eisenbahnverwaltung f. ‖ administration f. des chemins de fer. / ~ material ‖ Eisenbahnbaustoff m. ‖ matériel m. de chemins de fer. / ~ motor ‖ Bahnmotor m. ‖ moteur m. de traction. / ~ net see ~ network. / ~ network ‖ Gleisnetz n. ‖ réseau m. de voies. / ~ overbridge ‖ Eisenbahnüberführung f. ‖ passage m. supérieur de chemin de fer. / ~ plant ‖ Bahnanlage f. ‖ installation f. du chemin de fer. / ~ radio service ‖ Eisenbahnfunkdienst m. ‖ service m. radioélectrique des chemins de fer. / ~ rate ‖ Eisenbahntarif m. ‖ tarif m. de chemin de fer. / ~ Red-Cross wagon ‖ Eisenbahnsanitätswagen m. ‖ wagon m. de croix rouge. / ~ regulation ‖ Eisenbahnverordnung f. ‖ règlement m. de chemin de fer. / ~ requirement ‖ Eisenbahnbedarf m. ‖ accessoires mpl. de chemin de fer. / ~ rolling stock ‖ Eisenbahnmaterial n. ‖ matériel m. de chemin de fer. / ~ safeguarding plant ‖ Eisenbahnsicherungsanlage f. ‖ installation f. de sécurité pour le chemin de fer. / installation of ~ safety appliances see ~ safeguarding plant. / ~ section ‖ Eisenbahnstrecke f. ‖ section f. de chemin de fer. / ~ service ‖ Eisenbahnbetrieb m. ‖ exploitation f. ou entreprise f. de chemin de fer. / ~ share ‖ Eisenbahnaktie f. ‖ action f. de chemin de fer. / ~ siding ‖

Anschlußgleis n. ‖ voie f. de jonction. / ~ signal ‖ Eisenbahnsignal n. ‖ signal m. de chemin de fer. / ~ signal lamp ‖ Eisenbahnsignallampe f. ‖ lanterne f. de signaux pour chemins de fer. / ~ sleeper ‖ Eisenbahnschwelle f. ‖ traverse f. de chemin de fer. / ~ station ‖ Eisenbahnstation f.; Bahnhof m. ‖ station f. de chemin de fer; gare f. / ~ statistics pl. ‖ Eisenbahnstatistik f. ‖ statistique f. des chemins de fer. / ~ substructure for ~s ‖ Eisenbahnunterbau m. ‖ infrastructure f. pour chemins de fer. / ~ superstructure for ~s ‖ Eisenbahnoberbau m. ‖ superstructure f. de chemins de fer. / ~ supervision of ~s pl. ‖ Eisenbahnoberaufsicht f. ‖ surveillance f. de chemins de fer. / ~ system ‖ Eisenbahnnetz n. ‖ réseau m. de chemins de fer ou de voies ferrées. / ~ tank wagon ‖ Eisenbahnkesselwagen m. ‖ wagon m. réservoir de chemin de fer. / ~ tariff ‖ Eisenbahntarif m. ‖ tarif m. de chemin de fer. / ~ technics pl. ‖ Eisenbahntechnik f. ‖ technique f. de chemin de fer. / ~ telegraph ‖ Eisenbahntelegraf m. ‖ télégraphe m. du chemin de fer. / ~ telegraph office ‖ Eisenbahntelegrafenanstalt f. ‖ bureau m. télégraphique du chemin de fer. / ~ telegraph station ‖ Eisenbahntelegrafenstation f. ‖ station f. télégraphique de chemins de fer. / ~ terminus ‖ Endbahnhof m. ‖ gare f. terminale. / ~ territory ‖ Bahngebiet n. ‖ terrain m. appartenant au chemin de fer. / ~ ticket ‖ Eisenbahnfahrkarte f. ‖ billet m. ou ticket m. de chemin de fer. / ~ ticket delivery apparatus ‖ Fahrkartenselbstgeber m.; Fahrkartenautomat m. ‖ distributeur m. de tickets de chemin de fer. / ~ tire of iron bars rolled up spirally and welded together ‖ Eisenbahnreifen m. aus spiralförmig gewickelten und geschweißten Eisenstäben ‖ bandage m. de chemin de fer fait de barres en fer roulées en spirale et puis soudées. / ~ tools pl. ‖ Streckenwerkzeug n. ‖ outillage m. de la voie. / ~ track ‖ Gleisstrang m. ‖ file f. de rails. / ~ traffic ‖ Eisenbahnverkehr m. ‖ trafic m. des chemins de fer. / ~ train ‖ Eisenbahnzug m. ‖ train m.; convoi m. / transport by ~ ‖ Eisenbahntransport m. ‖ transport m. par chemin de fer. / ~ tyre see ~ tire. / ~ wagon ‖ Eisenbahnwagen m. ‖ wagon m. de chemin de fer. / narrow-gauge ~ wagon ‖ Kleinbahnwagen m. ‖ wagon m. à voie étroite. / wagon for suspension ~ ‖ Hängebahnwagen m. ‖ wagon m. suspendu. / ~ watchman ‖ Bahnwärter m. ‖ garde-voie m.; garde-barrière m. / ~ wheel ‖ Eisenbahnrad n. ‖ roue f. pour wagons de chemin de fer. / ~ wheelset lathe ‖ Radsatzdrehbank f. ‖ tour m. pour essieux montés. / ~ work ‖ Eisenbahnarbeit f. ‖ travaux mpl. à la voie ferrée. / ~ worker ‖ Bahnarbeiter m. ‖ ouvrier m. de chemin de fer. / ~ workshop ‖ Eisenbahnwerkstatt f. ‖ atelier m. du chemin de fer.
rail, web of ~ ‖ Schienensteg m. ‖ âme f. d'un rail. / ~ welder ‖ Schienenschweißer m. ‖ soudeur m. de rails. / ~ welding ‖ Schienenschweißung f. ‖ soudure f. de rails. / ~ working machine ‖ Schienenbearbeitungsmaschine f. ‖ machine f. à travailler les rails.
rain ‖ Regen m. ‖ pluie f. / owing to absence of ~ ‖ wegen Regenmangel m. ‖

en raison f. du manque de pluie. /
drizzling ~ ‖ Staubregen m. ‖ bruin m. /
general ~ ‖ Landregen m. ‖ pluie f.
générale. / lasting ~ see general ~. /
to withstand a ~ of machine-gun
bullets ‖ widerstandsfähig gegen Maschinengewehrgarben fpl. sein ‖ résister aux gerbes fpl. des balles de
mitrailleuses. / universal ~ see general ~.

rain-awning ‖ Regensegel n.; Regenzelt n.
‖ taud m.

rainbow ‖ Regenbogen m. ‖ arc-en-ciel m.
/ ~ ground (Text print) ‖ Irisfond m. ‖
fond m. irisé ou ombré.

rain chamber (Metal) ‖ Regenkammer f. ‖
chambre f. à pluie. / ~ cloud ‖ Regen-
wolke f. ‖ nuage m. de pluie. / ~ drop ‖
Regentropfen m. ‖ goutte f. de pluie.

rainfall ‖ Regenfall m. ‖ pluie f.; chute f.
de pluie. / annual ~ ‖ jährlicher Niederschlag m.; jährliche Regenmenge f. ‖
quantité f. de pluie annuelle. / amount
of ~ ‖ Niederschlagsmenge f. ‖ quantité
f. d'eau tombée. / depth of ~ ‖ Regen-
höhe f. ‖ quantité f. d'eau tombée. /
distribution of ~ ‖ Niederschlagsverteilung f. ‖ distribution f. de la pluie.

rainfull ‖ niederschlagsreich ‖ riche de
condensations fpl. ou en pluies fpl.;
à fortes condensations fpl.

rain gauge ‖ Regenmesser m. ‖ pluviomètre m. / self-registering ~ ‖ selbst-
aufzeichnender Regenmesser m. ‖ pluvio-
graphe. m.

rain measurement ‖ Regenmessung f. ‖
pluviométrie f.; ombrométrie f. / ~
outlet ‖ Notauslaß n. (einer Kanalisation) ‖ bonde f. de pluie. / ~ pipe ‖ Abfallrohr n.; Dachrohr n. ‖ tuyau m. de
descente ou de gouttières. / ~ plant ‖
Beregnungsanlage f. ‖ installation f.
pluviale ou d'arrosage.

rain-proof ‖ regendicht ‖ protégé à la
pluie. / ~ finish ‖ regensichere Ausführung f. ‖ exécution f. étanche à
la pluie.

rains pl., abundant ~ ‖ reichlicher Regenfall m. ‖ pluies fpl. abondantes. / excessive ~ ‖ übermäßiger Regenfall m. ‖
pluies fpl. excessives.

rain sheet ‖ Regenwolkenbank f. ‖ couche
f. de nuages pluvieuses. / shortage of ~ ‖
Regenmangel m. ‖ manque m. de pluie.
/ ~ shower ‖ Regenschauer m. ‖ ondée
f.; averse f.; giboulée f. / ~ squall ‖
Regenbö f. ‖ grain m. de pluie. / ~ time
see ~ rainy season. / ~ water ‖ Regen-
wasser n. ‖ eau f. pluviale ou de
pluie. / ~ water head ‖ Rinnenkasten
m. ‖ trémie f. de gouttière.

rainy ‖ regnerisch ‖ pluvieux. / ~ season ‖
Regenzeit f. ‖ saison f. des pluies; hivernage m.

raise, to ‖ errichten ‖ ériger. / ~ (Curr) ‖
schwellen; treiben ‖ gonfler; travailler
à l'orge. / ~ (Mine) ‖ fördern ‖ extraire.
/ ~ (Phys) ‖ steigern ‖ augmenter. ‖ ~
the cloth ‖ das Tuch rauhen ‖ garnir ou
lainer ou tirer le drap à la perche. / ~
to the cube ‖ kubieren ‖ cuber; élever
au cube m. / ~ the frames (Shipb) ‖ die
Spanten npl. aufrichten ‖ lever les
couples mpl. / ~ the grain (Curr) ‖
krispeln ‖ rebrousser; crêpir. / ~ the
ground ‖ Boden m. aufschütten ‖ remblayer. / ~ the hatch ‖ das Schütz aufziehen ‖ lever la pale. / ~ money ‖ Geld n.

aufbringen ‖ se procurer de l'argent m.
/ ~ the nap of cloth ‖ den Stoff m. aufkratzen oder aufrauhen oder kardätschen
‖ égratigner ou lainer ou garnir l'étoffe f.
/ ~ to a power ‖ potenzieren ‖ élever à
une puissance. / ~ a number to the
second power ‖ eine Zahl f. quadrieren ‖
former le carré d'un nombre; élever un
nombre au carré. / ~ a timberwork ‖ das
Zimmerwerk richten oder aufbringen ‖
monter un toit; lever la charpente. / ~
up (Build) ‖ in die Höhe richten ‖ lever;
ériger. / ~ up (Trade) ‖ aufbringen ‖ trouver; réussir. / ~ a wall ‖ eine Mauer f.
aufhöhen ‖ donner de l'exhaussement m.
à un mur; rehausser un mur. / ~ with
a windlass ‖ aufhaspeln ‖ guinder.

raise of salary ‖ Gehaltserhöhung f. ‖
augmentation f. des appointements ou
de traitement. / ~ in wages ‖ Lohnerhöhung f. ‖ augmentation f. des salaires.

raised ‖ hochköpfig ‖ saillant. / the cupola
can be ~ and lowered a little ‖ die Kuppel ist um ein geringes Maß heb- und
senkbar ‖ dans une mesure restreinte la
calotte peut se monter et descendre. / ~
manhole ‖ Fahrstutzen m. ‖ trou m.
d'homme surélevé. / ~ platform ‖ Laufbühne f. ‖ plancher m. à claire voie. / ~
wing on the corner of a house ‖ Eck-
flügel m. eines Hauses ‖ pavillon m.
angulaire d'une maison.

raisin ‖ Rosine f. ‖ raisin m. sec.

raising ‖ Erhöhung f.; Steigerung f. ‖ (re)-
haussement m.; élévation f. / ~ (Mine) ‖
Schachtförderung f. ‖ extraction f. / ~
(Print) ‖ Herauslangen n.; Greifen n. ‖
levée f. / cloth ~ ‖ Rauhen n. des Tuches
‖ lainage m. ou garnissage m. de drap. /
~ a dam higher ‖ Aufkadung f. eines
Deiches ‖ exhaussement d'une digue. /
~ of salary ‖ Gehaltserhöhung f.; Ge-
haltsaufbesserung f. ‖ augmentation f.
des appointements ou de traitement ou
des émoluments. / ~ of the timberwork
‖ Heben n. oder Richten n. des Zimmer-
werks ‖ levage m. de la charpente. / ~
the trusses of a bridge ‖ eine Brücke f.
einbringen ‖ monter les poutres fpl. d'un
pont; mettre en place un pont. / ~ the
voice ‖ Heben n. der Stimme ‖ élévation
f. de la voix. / ~ of wages ‖ Lohnerhöhung
f. ‖ augmentation f. des salaires; amélioration f. des émoluments. / ~ of water ‖
Wasserförderung f. ‖ élévation f. d'eau.

raising apparatus (Coop) ‖ Aufsatzform f.
‖ forme f. d'assemblage. / ~ apparatus
(Weav) ‖ Rauhapparat m. ‖ appareil m.
à lainer. / ~ beam ‖ Rauhbaum m. ‖
perche f. / ~ card ‖ Tuchkarde f. ‖ car-
dinal m.

raising device before the machine (Weav) ‖
vorgebauter Vorrauhapparat m.; Postier-
apparat m. ‖ dispositif m. d'avant-
lainage à l'entrée de la machine. / ~
for both sides of the fabric ‖ Vorrichtung f. zum gleichzeitigen Rauhen n.
beider Seiten des Gewebes ‖ dispositif
m. pour le lainage simultané des deux
côtés de tissu. / previous ~ (Weav) ‖
Vorrauhapparat m. ‖ dispositif m.
d'avant-lainage.

raising force formula ‖ Hubkraftformel f. ‖
formule f. de poussée. / ~ and lowering
gear for arcs ‖ Bogenlampenaufzug m. ‖
moulinet m. de hissage pour arcs.

raising machine ‖ Rauhmaschine f. ‖
laineuse f. / teasel ~ (Weav) ‖ Roll-

kardenrauhmaschine f. ‖ laineuse f. à
tambours à chardons roulants.

raising-up ‖ Rohaufbrechen n. ‖ soulèvement m.; désornage m.

raisin wine ‖ Rosinenwein m. ‖ vin m. de
raisins secs.

rak ‖ Arrak m. ‖ arack m.; rack m.

rake, to ~ up the ground ‖ den Boden m.
aufgraben ‖ fouiller ou creuser le terrain.

rake (Agr) ‖ Harke f.; Rechen m. ‖
râteau m. / ~ (Brew) ‖ Rührharke f.;
Krücke f.; Maischharke f. ‖ fourquet m.;
vague g.; brassoir m. / ~ (Metal) ‖ Ofenkrücke f.; Rührstange f. ‖ croc m. à feu;
râble m. / ~ (Railw) ‖ Schneerechen m. ‖
râteau m. à neige. / ~ (Saltpetre) ‖ Scharre
f. ‖ ratissoire f. / hay ~ ‖ Heurechen m. ‖
râteau m. à foin. / horse ~ ‖ Pferderechen m. ‖ râteau m. à cheval. / self-
dumping horse ~ ‖ Pferderechen m. mit
selbsttätiger Entleerung ‖ râteau m. à
cheval à décharge automatique. / iron
~ ‖ Eisenrechen m. ‖ râteau m. en fer.
/ ~ of the masts ‖ Fall m. der Masten ‖
inclinaison f. des mâts. / side-delivery ~
‖ Schwadenrechen m. ‖ râteau m. à
andain.

rake basket ‖ Rechenkorb m. ‖ râteau m.
/ ~ cleaning machine ‖ Rechenreini-
gungsmaschine f. ‖ machine f. à nettoyer
les écrilles. / ~ conveyor ‖ Rechen-
förderer m. ‖ transporteur m. à râteaux.

raker (Mas) ‖ Kratzeisen n. ‖ racloir m.;
grattoir m. / ~ (Mine) ‖ Krätzer m.;
Löffelräumer m. ‖ curette f. de mineur;
tire-sable m.

rake stand ‖ Rechenständer m. ‖ colonne f.
de râteaux. / ~ tooth ‖ Rechenzinke f. ‖
dent m. de râteau.

raking plant ‖ Rechenanlage f. ‖ installation f. à râteaux.

rakings pl. (Mine) ‖ Nußkohle f.; Würfelkohle f. ‖ gailletteries fpl.; petites gaillettes fpl.; grélassons mpl.

ram, to ‖ stampfen; rammen ‖ damer;
fouler. / ~-down piles ‖ Pfähle mpl. eintreiben ‖ mettre en fiche les pieux mpl.
/ ~ the earth ‖ die Erde rammen oder
feststampfen ‖ battre ou damer la terre.
/ ~ the ground ‖ den Boden m. festrammen ‖ battre ou damer l'aire f. /
~-in a pile ‖ einen Pfahl m. einrammen ‖
enfoncer ou battre un pieu. / ~ the
pavement ‖ das Pflaster rammen ‖
battre le pavé.

ram ‖ Ramme f.; Fallblock m.; Bär m.;
Rammbär m. ‖ mouton m.; marteau m.
pilon. / ~ (Mine) ‖ Tauchkolben m.;
Plunger m. ‖ piston-plongeur m.; plongeur m. / ~ (Shipb) ‖ Ramme f.; Sporn
m. ‖ éperon m. / common ~ ‖ Zug-
ramme f. ‖ sonnette f. à tiraude. / hydraulic ~ ‖ hydraulischer Widder m. ‖
bélier m. hydraulique. / steam ~ ‖ Dampf-
ramme f. ‖ mouton m. à vapeur. / water
~ ‖ hydraulischer Widder m. ‖ bélier m.
hydraulique.

ram block ‖ Rammbär m. ‖ mouton m. /
~ bow (Shipb) ‖ Rammbug m. ‖ avant
m. en éperon. / ~ breeder ‖ Schafbock-
halter m. ‖ propriétaire m. d'un bélier.
/ ~ crane ‖ Fallwerkskran m. ‖ grue f.
à bélier. / ~ engine see ram.

ramie ‖ Ramie f.; Chinagras n. ‖ ramie f.;
china-grass m. / ~ bleaching ‖ Ramie-
bleichung f. ‖ blanchiment m. de la
ramie. / ~ fibre ‖ Ramie f. ‖ ramie f. / ~
spinning ‖ Ramiespinnerei f. ‖ filature f.

de ramie. / ~ yarn ‖ Ramiegarn n. ‖ fil m. de ramie.

rammed concrete ‖ gestampfter Beton m. ‖ béton m. damé. / ~ layer ‖ Stampfschicht f. ‖ couche f. damée. / ~ stuff ‖ gestampfte Masse f. ‖ masse f. damée.

rammer ‖ Rammbär m.; Ramme f. ‖ mouton m.; billot m. de batte; bélier m. à pilotage. / ~ (Hand) ‖ Stampfer m.; Handramme f. ‖ batte f.; cogneux m.; mouton m. à bras; dame f.; pilon m. / ~ (Shipb) ‖ Ramme f. (für Kielklötze) blin m. / earth ~ ‖ Handramme f. ‖ dame f.; demoiselle f. / moulding ~ ‖ Stampfer m. ‖ damoir m. / paving ~ see earth ~. / pneumatic ~ ‖ Druckluftstampfer m. ‖ dame f. à air comprimé.

rammer foot ‖ Stampffuß m. ‖ pilette f. / ~ log ‖ Rammbär m.; Rammklotz m.; Bär m. ‖ mouton m.; bélier m.; bilot m. de batte.

ramming ‖ Stampfen n. ‖ damage m. / ~ depth of a pile ‖ Rammtiefe f. eines Pfahles ‖ fichée f. d'un pieu. / pneumatic ~ hammer ‖ Preßlufttrammhammer m. ‖ marteau m. pneumatique pour enfoncer. / plug ~ machine for converters ‖ Bodenstampfmaschine f. für Konverter ‖ machine f. à damer les fonds de convertisseurs.

ramp ‖ Rampe f.; rampe f. / ~ for climbing ‖ Auflaufzunge f. ‖ aiguille f. en rampe. / ~ a pair of ~ points pl. ‖ ein Paar n. Rampenspitzen fpl. ‖ paire f. de rampes d'accès. / ~ stroke ‖ Stößelhub m. ‖ course f. du coulisseau.

rampart (Fortification) ‖ Wall m. ‖ rempart m.

rancher ‖ Viehhüter m. ‖ herberger m.

rancid ‖ ranzig ‖ rance. / to grow ~ ‖ ranzig werden ‖ rancir. / ~ butter ‖ ranzige Butter f. ‖ beurre m. rance.

rancidity ‖ Ranzigkeit f. ‖ rancidité f.

random length ‖ Fabrikationslänge f. ‖ longueur f. tout-venant. / ~ stone foundation ‖ Gründung f. auf Steinschüttung ‖ fondation f. à pierres perdues ou par enrochements.

range, to ~ the coast ‖ längs der Küste f. fahren ‖ côtoyer une terre; descendre une côte. / ~ from x to y ‖ zwischen x und y schwanken ‖ s'échelonner entre x et y. / ~-out a curve ‖ eine Kurve f. abstecken ‖ piqueter une courbe. / ~ straight lines pl. with a telescope ‖ mit dem Fernrohr n. die Geraden abstecken ‖ tracer des lignes fpl. droites au moyen de la lunette.

range ‖ Reihe f. ‖ rangée f. / ~ (Ballistics) ‖ Wurfweite f.; Schußweite f. ‖ portée f. / ~ (Build) ‖ Herd m. ‖ poêle f. / ~ (Mar) ‖ Geschirr n. (eines Mastes) ‖ agrès mpl.; apparaux mpl.; armement m. / ~ (Pott) ‖ Brennofen m. / poêle f. / ~ (Road) ‖ Bauflucht f.; Baufluchtlinie f. ‖ alignement m. / ~ (Tel) ‖ Reichweite f. ‖ portée f. / ~ of balance ‖ Meßbereich m. der Wage ‖ portée f. ou amplitude f. de mesure de la balance. / double ~ ‖ Doppelherd m. ‖ cuisinière f. à double foyers. / ~ of frequencies ‖ Frequenzbereich m. ‖ bande f. de fréquences. / ~ of hills ‖ Hügelkette f. ‖ chaînon m.; chaîne f. de collines. / ~ of investigation ‖ Untersuchungsbereich m. ‖ région f. examinée. / large kitchen ~ ‖ Großküchenherd m. ‖ grande cuisinière f. / approximate ~ of measure-

ment ‖ ungefährer Meßbereich m. ‖ étendue f. approximative des mesures. / ~ of nominal tension ‖ Nennspannungsbereich m. ‖ étendue f. de la tension nominale. / ~ of ovens ‖ Ofenbatterie f. ‖ batterie f. de fours. / public institute ~ ‖ Anstaltsherd m. ‖ cuisinière f. pour établissements. / limited ~ of regulation ‖ begrenzter Regelbereich m. ‖ capacité f. de réglage limitée. / ~ of revolutions ‖ Drehzahlbereich m. ‖ limites fpl des nombres de tours. / set-up ~ ‖ Aufsatzherd m. ‖ cuisinière f. à galerie. / ~ of sight ‖ Sehweite f. ‖ portée f. de vue. / ~ up to ‖ ausreichend ‖ suffisant. / vertical ~ (Gun) ‖ Steighöhe f.; senkrechte Reichweite f. ‖ portée f. verticale. / ~ of vision ‖ Augenweite f. ‖ écartement m. des yeux. / visual ~ ‖ Sehweite f.; Weite f. des deutlichen Sehens ‖ distance f. de la vue distincte; portée f. visuelle. / augmentation of ~ (Radio) ‖ Verbesserung f. der Reichweite ‖ augmentation f. de la portée.

ranged line ‖ ausgefluchtete Linie f. ‖ ligne f. affleurée.

range finder ‖ Entfernungsmesser m. ‖ télémètre m.; compteur m. de distance. / stereoscopic ~ ‖ Stereotelemeter n. ‖ télémètre m. stéréoscopique.

ranger ‖ Klassiererin f. ‖ rangeuse f.

range system with conical inlet ‖ Etagensystem n. mit konischem Einlaufsfeld ‖ étage m. avec chemin d'introduction conique. / graphic ~ table ‖ grafische Schußtafel f. ‖ table f. graphique de tir.

ranging (Print) ‖ Ordnen n. ‖ alignement m.; rangette f. / flash ~ ‖ Lichtmessen n. ‖ repérage m. par la lueur. / ~-out curves ‖ Kurvenabsteckung f. ‖ tracé m. de courbes.

ranging magnifier with ruled plate ‖ Visierlupe f. mit Strichplatte ‖ loupe f. de visée munie d'un réticule. / ~ pole (Surv ‖ Fluchtstab m.; Visierstab m. ‖ jalon m. ou piquet m. de mire. / travelling ~ pole ‖ Reisefluchtstab m.; tragbarer Fluchtstab m. ‖ jalon m. portatif. / ~ rod see ~ pole. / ~ section ‖ Meßtrupp m. ‖ section f. de repérage.

rank of piles ‖ Pfahlreihe f.; Pfahlwand f. ‖ file f. de pieux; rang m. de pilots.

rap, to ~ the sand of the moulding box ‖ den Sand m. im Formkasten rütteln ‖ décotter le sable du châssis de moulage.

rap of cotton-yarn ‖ Gebinde n. Baumwollgarn ‖ échevette f. de fil de coton.

rape (Agr) ‖ Rübsen m.; Raps m. ‖ navette f.; colza m. / ~ (Bot) ‖ Traubenkamm m.; Weinkamm m. ‖ raffe f.; rafle f.; râpe f. / ~ cake ‖ Rapskuchen m. ‖ tourteau m. de colza. / ~ mill ‖ Rapsmühle f. ‖ moulin m. à colza. / ~ oil ‖ Rüböl n. ‖ huile f. de colza.

rapeseed ‖ Rapssamen m.; Rübsamen m. ‖ graine f. de ravison ou de navette. / ~ oil ‖ Rüböl n. ‖ huile f. de navette. / ~ oil substitute ‖ Rübölersatz m. ‖ huile f. de colza artificielle.

raphia ‖ Raphia f. ‖ raphia m.

rapid ‖ schnell; rasch ‖ rapide; brusque. / ~-action screw press ‖ Schnellspindelpresse f. ‖ presse f. à vis rapide. / ~ boiling device ‖ Schnellkocher m. ‖ bouilloire f. rapide. / ~ change of temperature ‖ rasche Temperaturänderung f. ‖ variation f. rapide de la température. / ~ combustion ‖ lebhafte Ver-

brennung f. ‖ combustion f. vive. / ~ composing apparatus (Print) ‖ Schnellsetzer m. ‖ appareil m. à composer rapide. / ~ discharger (Railw) ‖ Schnellentladewagen m.; Schnellentlader m. ‖ déchargeur m. rapide. / ~ discharger leaving no remainder ‖ restlos entleerender Schnellentladewagen m. ‖ déchargeur m. rapide à déchargement complet. / ~ dissolver ‖ Schnellauflöser m. ‖ appareil m. destiné à la dissolution rapide. / ~ inspection ‖ Schnellprüfung f.; Schnellrevision f.; schnelle Messung f. ‖ contrôle m. rapide; mesurage m. rapide. / ~ inspection of pieces produced in series ‖ Schnellprüfung f. von Massenteilen ‖ contrôle m. rapide de pièces fabriquées en série. / ~ lens ‖ lichtstarkes Objektiv n. ‖ objectif m. lumineux. / ~ method ‖ Schnellverfahren n. ‖ méthode f. rapide. / ~ reading (Weav) ‖ Schnellstechen n. ‖ lisage m. accéléré. / ~ sand filter ‖ Sandschnellfilter n. ‖ filtre m. rapide à sable. / ~-tightening vice ‖ Schnellspannschraubstock m. ‖ étau m. à serrage instantané. / ~ typewriter ‖ Schnellschreibmaschine f. ‖ machine f. à écrire rapide.

rapidity of measurement ‖ Meßgeschwindigkeit f. ‖ rapidité f. de mesurage. / ~ of telegraphic signalling ‖ Telegrafiergeschwindigkeit f. ‖ rapidité f. de succession des signaux télégraphiques. / ~ of vaporization ‖ Verdampfungsgeschwindigkeit f. ‖ rapidité f. de la vaporisation. / ~ of writing ‖ Schreibschnelligkeit f. ‖ vitesse f. de la machine à écrire.

rapier ‖ Rapier n.; Stoßdegen m. ‖ rapière f.

rapper ‖ Türklopfer m. ‖ heurtoir m.

rapping-in (Moulding box) ‖ Rütteln n. ‖ décottage m.

rare ‖ selten ‖ rare.

rare gas ‖ Edelgas n. ‖ gaz m. rare. / ~ lightning arrester ‖ Edelgassicherung f. ‖ parafoudre m. à gaz rare. / ~ tube ‖ Edelgasröhre f. ‖ lampe f. à gaz spécial.

rarefaction ‖ Verdünnung f. ‖ raréfaction f.

rarefied air ‖ verdünnte Luft f. ‖ air m. raréfié.

rarefy, to ‖ verdünnen ‖ raréfier.

rareness ‖ Seltenheit f.; rareté f.

rarity see rareness.

rashness ‖ Übereilung f. ‖ précipitation f.

rasp, to ‖ raspeln ‖ râper.

rasp ‖ Raspel f. ‖ râpe f. / crooked ~ ‖ Krummraspel f. ‖ égohine f.; égoïne f. / fine ~ see rasp, wood. / mechanical ~ (Dyer) ‖ Farbholzraspelmaschine f. ‖ machine f. à râper les bois de teinture. / ~ for widening grooves ‖ Nutenraspel f.; Innenreißer m. ‖ ramasse f. / wood ~ ‖ Holzraspel f. ‖ écouane f. ou écouenne f.; ou râpe f. à bois.

raspberry ‖ Himbeere f. ‖ framboise f. / ~ juice ‖ Himbeersaft m. ‖ sirop m. de framboises.

rasping ‖ Raspeln n. ‖ râpage. / ~ (Bak) ‖ geriebene Brotkruste f. ‖ chapelure f.

rasping file ‖ Raspel f. ‖ râpe f. / ~ machine ‖ Raspelmaschine f. ‖ machine f. à râper. / ~ machine for dyeing wood ‖ Raspelmaschine f. für Farbhölzer ‖ machine f. à râper pour bois colorants. / ~ machine for tanning wood ‖ Raspelmaschine f. für Gerbhölzer ‖ machine f. à râper pour bois tannants.

/ machine for ~ colours and varnishes to the highest degree of fineness ‖ Maschine f. zum Feinzerreiben von Farben und Lacken ‖ machine f. à broyer les couleurs et vernis à la plus grande finesse.

rat, to ‖ unter Tarif m. arbeiten ‖ travailler au-dessous du tarif m.

ratan *see* rattan.

rat-catcher ‖ Rattenfänger m. ‖ chasseur m. *ou* preneur m. de rats.

ratch ‖ gezahnte Sperrstange f. ‖ crémaillère f.

ratchet ‖ Sperradvorrichtung f.; Sperrwerk n. ‖ encliquetage m. / ~ brace ‖ Bohrknarre f.; Bohrratsche f. ‖ vilebrequin m. à cliquet; racagnac m. / ~ drill *see* ~ brace. / ~ gear ‖ Sperrwerk n. ‖ mécanisme m. d'encliquetage. / ~ lever ‖ Ratschenhebel m. ‖ levier m. à rochet *ou* à cliquet. / ~ motion ‖ Sperrklinkenschaltwerk n. ‖ encliquetage m.; rochet m. / ~ release ‖ Sperradauslösung f. ‖ relâchement m. par cliquet. / ~ tooth ‖ Sperrzahn m. ‖ dent f. d'arrêt. / ~ wheel ‖ Zahnscheibe f.; Klinkrad n.; Sperrad n. ‖ roue f. à rochet *ou* à cliquet.

rat destroyer ‖ Rattenvertilgungsmittel n. ‖ mort f. aux rats.

rate, to ‖ abschätzen ‖ estimer. / ~ (Auto) ‖ in Klassen fpl. einteilen; klassifizieren ‖ classer; classifier. / ~ (Flax) ‖ rösten ‖ naiser; rouir.

rate (Exchange) ‖ Kurs m. ‖ cours m.; change m. / ~ (Sport) ‖ Grad m. der Geschwindigkeit ‖ intensité f. / ~ (Ship) ‖ Lauf m.; Fahrgeschwindigkeit f. ‖ erre f. / ~ (Telephone) ‖ Gebühr f. ‖ taux m.; tarif m.; prix m. / ~ (Trade) ‖ Rate f. ‖ prorata m.; quote-part f. / at the ~ of ‖ im Verhältnis n. von ‖ au taux m. de; à raison f. de. / ~ per cent ‖ Hundertsatz m.; Prozentsatz m.; Zinsfuß m. ‖ taux m. pour cent. / ~ of the chronometer ‖ Gang m. des Chronometers ‖ marche f. du chronomètre. / ~ of climb (Aero) ‖ Steigzeit f. ‖ temps m. de montée. / ~ of discount ‖ Diskontsatz m. ‖ taux m. de l'escompte. / ~ of exchange ‖ Wechselkurs m. ‖ cours m. de change. / current ~ of exchange ‖ Tageskurs m. ‖ change m. du jour. / ~ of expansion ‖ Füllungsgrad m. ‖ degré m. de détente *ou* d'expansion. / ~ of flow ‖ Durchflußgeschwindigkeit f. ‖ vitesse f. de passage. / ~ of freight ‖ Frachtsatz m. ‖ taux m. du fret. / ~ of interest ‖ Zinsfuß m. ‖ taux m. de l'intérêt. / ~ of postage ‖ Portosatz m. ‖ tarif m. *ou* taxe f. des ports. / ~ of a ship ‖ Klasse f. eines Schiffes ‖ classement m. d'un navire. / ~ of wage ‖ Lohnsatz m.; Lohntarif m. ‖ tarif m. (de salaire).

rate notification (Tel) ‖ Gebührenausgabe f. ‖ notification f. de la taxe. / ~ payer ‖ Steuerzahler m. ‖ contribuable m. / remission of ~s ‖ Gebührenermäßigung f. ‖ réduction f. des frais.

rates pl. of pay for certain classes of work (Mine) ‖ Gedingelohn m. ‖ salair m. à la tâche *ou* à la journée. / piecework ~ ‖ Akkordlohn m. ‖ salaire m. aux pièces *ou* à la tâche. / ~ of pilotage ‖ Lotsentarif m. ‖ droit m. des pilotes; taxe f. *ou* tarif m. de pilotage.

rate tariff, measured (Tel) ‖ Gesprächsgebührentarif m. ‖ tarif·m. à conversation taxée.

ratification of treaty ‖ Vertragsbestätigung f. ‖ ratification f. du traité.

ratify, to ‖ ratifizieren ‖ ratifier.

rating (Ship) ‖ Klasse f. ‖ division f. / ~ (Flax) ‖ Rösten n.; Röten n. ‖ naisage m.; rouissage m. / ~ formula ‖ Steuerformel f. ‖ formule f. de taxation. / ~ horse power ‖ Steuerleistung f. ‖ puissance f. fiscale.

ratio (Math) ‖ Verhältnis n. ‖ raison f. / ~ of the aperture of the lens and the focal length ‖ Verhältnis n. der Objektivöffnung zur Brennweite ‖ rapport m. entre l'ouverture de l'objectif et la distance focale. / average ~ of stroke to bore ‖ mittleres Hubverhältnis n. ‖ rapport m. moyen de course à alésage. / ~ of compression ‖ Kompressionsverhältnis n.; Verdichtungsverhältnis n. ‖ taux m. de compression. / ~ of cylinder volumes ‖ Zylinderinhaltverhältnis n. ‖ rapport m. des cylindrées. / ~ of gear ‖ Übersetzungsverhältnis n. ‖ rapport m. d'engrenage *ou* de démultiplication. / irrational ~ ‖ irrationales Verhältnis n. ‖ raison f. irrationelle. / rational ~ ‖ rationales Verhältnis n. ‖ raison f. rationelle. / ~ of stroke to diameter ‖ Verhältnis n. des Hubes zum Durchmesser ‖ rapport m. entre la course et le diamètre. / transformer ~ (Electr) ‖ Umwandlungsverhältnis n.; Übersetzungsverhältnis (eines Transformators) ‖ rapport m. *ou* coefficient m. de transformation.

ration ‖ Ration f. ‖ ration f.

rational ‖ vernunftgemäß; rationell ‖ conforme à la raison; rationnel. / ~ (Math) ‖ rational ‖ rationnel.

rationalization ‖ Rationalisierung f. ‖ rationalisation f.; organisation f. rationnelle.

rationalize, to ‖ rationalisieren ‖ rationaliser.

ration bread ‖ Kommißbrot n. ‖ pain m. de munition.

ratsbane ‖ Rattengift n. ‖ mort f. aux rats.

rat-tail ‖ Rattenschwanzfeile f. ‖ queue f. de rat.

rattan ‖ spanisches Rohr n.; Rotang m. ‖ rotin m.; rotang m. / pith of ~ ‖ Mark n. vom spanischen Rohr; Peddig n. ‖ moelle f. de rotin.

ratteen, to ‖ ratinieren ‖ friser; ratiner.

ratteening machine ‖ Ratiniermaschine f. ‖ ratineuse f.

rattling machine tenter ‖ Maschinengußputzer m. ‖ nettoyeur m. de fonte à la machine.

rat trap ‖ Rattenfalle f. ‖ nasse f. ratière.

ravelin ‖ Wallschild m.; Ravelin n.; halber Mond m. ‖ demi-lune f.; ravelin m.

ravelled cane ‖ gefasertes Rohr n. ‖ canne f. filée.

ravelling machine ‖ Lumpenwolf m. ‖ effilocheuse f.

ravine ‖ Klamm f.; Schlucht f. ‖ ravin m.; gorge f.

raw ‖ roh ‖ cru; brut; écru. / ~ brick ‖ Rohziegel m. ‖ brique f. crue. / ~ brine ‖ Rohsole f. ‖ eaux fpl. vierges. / ~ cotton ‖ Rohbaumwolle f. ‖ coton m. brut. / ~ film ‖ Rohfilm m. ‖ film m. brut. / ~ flax ‖ roher Flachs m. ‖ lin m. brut. / ~ fur ‖ Rohpelz m. ‖ fourrure f. brute.

raw glass ‖ Rohglas n. ‖ verre m. brut. / ~ in balls ‖ Rohglas n. in Kugeln ‖ verre m. brut en boules. / ~ in segments ‖ Rohglas n. in Segmenten ‖ verre m. brut

en segments. / pressing the ~ discs into their mould ‖ Pressen n. der Rohglasscheiben ‖ moulage m. des disques de verre bruts.

raw helium ‖ Rohhelium n. ‖ hélium m. brut. / ~hide ‖ Rohhaut f. ‖ cuir m. vert. / ~hide pinion (Mach) ‖ Rohhautritzel m. ‖ pignon m. en cuir vert. / ~ ingot ‖ Rohblock m. ‖ lingot m. brut. / ~ iron ‖ Masseleisen n.; Roheisen n. ‖ fonte f. crue; fer m. cru. / increased output of ~ lignite ‖ verstärkte Förderung f. von Rohbraunkohle ‖ production f. augmentée de lignites bruts.

raw material ‖ Rohstoff m.; Rohmaterial n.; Ausgangsstoff m.; Ausgangsmaterial n. ‖ matière f. première *ou* brute. / ~ (Met) ‖ Beschickungsmaterial n. ‖ matières fpl. brutes. / good ~s pl. of uniform quality ‖ gute und gleichmäßige Rohstoffe mpl. ‖ des matières fpl. brutes de bonne et invariable qualité. / industry subsisting on native ~s pl. ‖ auf einheimische Rohstoffe gestützte Industrie f. ‖ industrie f. alimentée par de matières premières du pays. / ~s pl. of pottery ‖ keramische Rohstoffe mpl. ‖ matières fpl. premières de l'industrie céramique et de la poterie.

raw material bin ‖ Erztasche f. ‖ poche f. à minerais. / provision of the ~s pl. *see* supply of ~s. / supply of ~s pl. ‖ Rohstoffversorgung f. ‖ approvisionnement m. en matières premières.

raw matte (Met) ‖ Rohstein m.; Kupferrohstein m. ‖ matte f. brute de cuivre; métal m. brut de cuivre. / ~ metal *see* raw matte. / ~ ore ‖ Roherz n. ‖ minerai m. cru. / ~ produce ‖ rohes Erzeugnis n. ‖ produit m. brut. / ~ silk ‖ Rohseide f. ‖ soie f. grège. / ~ silk winder ‖ Rohseidenwinder m. ‖ dévideur m. de soie grège. / ~ skin wheel ‖ Rohhautrad n. ‖ roue f. de peau brute. / ~ slag ‖ Rohschlacke f. ‖ scorie f. pauvre *ou* crue; laitier m. pauvre. / ~ steel ‖ Rohstahl m. ‖ acier m. brut *ou* naturel. / ~ steel in ingots ‖ Rohstahl m. in Blöcken ‖ acier m. brut en lingots. / ~ sugar ‖ Rohzucker m. ‖ sucre m. brut. / ~ tobacco ‖ Rohtabak m. ‖ tabac m. brut. / ~ wood-spirit ‖ Rohholzgeist m. ‖ alcool m. méthylique brut.

ray ‖ Strahl m. ‖ rayon m. / actinic ~s pl. ‖ aktinische Strahlen mpl. ‖ rayons mpl. actiniques. / broken ~ ‖ gebrochener Lichtstrahl m. ‖ rayon m. réfracté *ou* rompu. / to change the direction of the ~ ‖ die Strahlenrichtung f. ändern ‖ changer la direction du rayon. / ~ of any given colour ‖ einfarbiger Lichtstrahl m. ‖ rayon m. monochromatique. / ~ deflected by diffraction ‖ durch Beugung abgelenkter Strahl m. ‖ rayon m. dévié par diffraction. / ~ of heat ‖ Wärmestrahl m. ‖ rayon m. de chaleur. / ~ incident at small anle ‖ flach auffallender Strahl n. ‖ rayon m. tombant dans une direction voisine de la surface. / ~ of light ‖ Lichtstrahl m. ‖ rayon m. delumière. / ~ of long wave length ‖ langwelliger Strahl m. ‖ rayon m. de grande longueur d'onde. / marginal ~ ‖ Randstrahl m. ‖ rayon m. marginal. / medullary ~ ‖ Markstrahl m.; Spiegelfaser f. ‖ rayon m. médullaire. / to modify the direction of the ~ *see* to change the direction of the ~. / polarized ~ ‖ polarisierter Strahl m. ‖ rayon m. polarisé. / principal ~ ‖ Haupt-

strahl m. ‖ rayon m. principal. / reflected ~ ‖ reflektierter *oder* zurückgeworfener Strahl m. ‖ rayon m. réfléchi. / refracted ~ ‖ gebrochener Strahl m. ‖ rayon m. réfracté *ou* rompu. / scattered ~s pl. ‖ Streustrahlen mpl. ‖ rayons mpl. dispersés. / ~ of short wave length ‖ kurzwelliger Strahl m. ‖ rayon m. de faible longueur d'onde. / totally reflected ~ ‖ total reflektierter Strahl m. ‖ rayon m. totalement réfléchi.

ray burner ‖ Strahlenbrenner m. ‖ bec m. à rayons. / course of the ~s ‖ Verlauf m. der Strahlen ‖ marche f. des rayons.

rayon ‖ Kunstseide f. ‖ soie f. artificielle.

ray, path of ~s in spectroscope ‖ Strahlengang m. im Spektroskop ‖ marche f. des rayons dans le spectroscope. / trace of the ~s *see* path of ~s.

raze, to (Build) ‖ abtragen; abreißen ‖ démanteler; démolir; raser. / ~ (Fortress) ‖ schleifen; rasieren ‖ démolir; démanteler. / ~ a ship ‖ ein Schiff n. abwracken ‖ déborder un navire.

razor ‖ Rasiermesser n. ‖ rasoir m. / hollow-ground ~ ‖ hohlgeschliffenes Rasiermesser n. ‖ rasoir m. évidé. / safety ~ ‖ Sicherheitsrasierapparat m. ‖ rasoir m. mécanique *ou* de sûreté.

razor blade ‖ Rasierklinge f. ‖ lame f. de rasoir. / ~ paste ‖ Streichriemenpaste f. ‖ pâte f. à cuir de rasoir. / ~ strap *see* ~ strop ‖ Streichriemen m. ‖ cuir m. à rasoir.

reach, to ‖ reichen; sich erstrecken ‖ atteindre; arriver à. / ~ layers pl. by digging ‖ Lagerstätten fpl. erschürfen ‖ découvrir des mines fpl.

reach (Canal) ‖ Haltung f. ‖ bief m.; biez m. / ~ (Gun) ‖ Reichweite f. ‖ portée f. / ~ of call (Tel) ‖ Rufweite f. ‖ portée f. de l'appel. / ~ of grab (Dredger) ‖ Grabweite f. ‖ portée f. de fouille. / ~ of talk ‖ Sprechweite f. ‖ portée f. de la voix.

reach-over edge ‖ vorspringende Kante f. ‖ saillie f.; rebord m. en saillie.

react, to ‖ reagieren; eine Reaktion f. eingehen ‖ réagir; produire une réaction f.

reactance ‖ Blindwiderstand m.; Reaktanz f. ‖ réactance f. / ~ coil ‖ Drosselspule f.; Selbstinduktionsspule f. ‖ bobine f. de réactance *ou* de self-induction. / ~ joint ‖ Drosselstoß m. ‖ joint m. de réactance.

reacting ‖ rückwirkend ‖ rétroactif.

reaction ‖ Rückwirkung f.; Gegenwirkung f.; Reaktion f. ‖ réaction f. / ~ (Blow) ‖ Rückschlag m. ‖ contre-coup m. / acid ~ ‖ saure Reaktion f. ‖ réaction f. acide. / armature ~ ‖ Ankerrückwirkung f. ‖ réaction f. de l'induit. / basic ~ ‖ basische Reaktion f. ‖ réaction f. basique. / chemical ~ ‖ chemischer Vorgang m. ‖ réaction f. chimique. / chief ~ ‖ Hauptreaktion f.; Hauptrückwirkung f. ‖ réaction f. principale. / elastic ~ ‖ elastischer Widerstand m. ‖ réaction f. élastique. / neutral ~ ‖ neutrale Reaktion f. ‖ réaction f. neutre. / principle of action and ~ ‖ Prinzip n. der Wirkung und Gegenwirkung ‖ principe m. de l'action et de la réaction.

reaction coil ‖ Rückkopplungsspule f. ‖ bobine f. de réaction. / ~ coupling ‖ Rückkopplung f. ‖ couplage m. de réaction. / ~ mixture ‖ Reaktionsgemisch n. ‖ mélange m. réactif. / ~ propulsion ‖ Reaktionsantrieb m. ‖ pro-

pulsion f. par réaction. / ~ tower ‖ Reaktionsturm m. ‖ tour f. de réaction. / ~ turbine ‖ Überdruckturbine f.; Aktionsturbine f. ‖ turbine f. à action. / ~ wheel ‖ Reaktionsrad n. ‖ roue f. à réaction.

reactive current ‖ Blindstrom m. ‖ courant m. réactif *ou* déwatté. / to supply ~ to the network ‖ Blindstrom m. in das Netz abgeben ‖ débiter du courant m. déwatté *ou* réactif dans le réseau. / to take ~ from the network ‖ Blindstrom m. dem Netz entnehmen ‖ absorber du courant m. déwatté *ou* réactif du réseau.

reactive paper ‖ Reaktionspapier n. ‖ papier m. réactif. / recorder for registering ~ volt-amperes ‖ Schreibgerät n. zur Aufzeichnung der Blindleistung ‖ enregistreur m. de puissance réactive.

reactivity ‖ Reaktivität f. ‖ réactivité f. / to lose the ~ ‖ die Reaktionsfähigkeit f. verlieren ‖ perdre le pouvoir *ou* la capacité de réaction.

reactor (Radio) ‖ strombegrenzende Reaktanz f.; Drossel f. ‖ réacteur m.; tamponneur m.

read, to ‖ lesen ‖ lire. / ~ off ‖ ablesen ‖ faire une lecture f. / ~ a proof ‖ Korrektur f. lesen ‖ lire des épreuves mpl.

readable ‖ lesbar ‖ lisible.

reader (Print) ‖ Korrektor m. ‖ correcteur m.

reader-in (Weav) ‖ Einleser m. ‖ ouvrier m. du lisage.

readiness, immediate ~ for service ‖ sofortige Betriebsbereitschaft f. ‖ mise f. en route instantanée; entrée f. en fonctionnement immédiate.

reading ‖ Ablesen n.; Ablesung f. ‖ lecture f. / ~ (Print) ‖ Lesen n. ‖ lecture f. / ~ at 180° to each other ‖ diametrale Ablesung f. ‖ lecture f. diamétrale. / ~ the adding devices by simple pressure of key ‖ Ablesen n. der Addierwerke durch einfachen Druck auf die Taste ‖ lecture f. des totalisateurs par simple pression sur la touche. / ~ of the angle ‖ Ablesung f. des Winkels ‖ lecture f. de l'angle. / ~ of the barometer ‖ Ablesen n. des Barometerstandes ‖ lecture f. de l'hauteur du baromètre. / ~ the circles to a double minute ‖ Ablesung f. der Kreise auf eine Doppelminute ‖ lecture f. des cercles sur la minute double. / direct ~ ‖ direkte *oder* unmittelbare Ablesung f. ‖ lecture f. directe. / to take the mean of two ~s ‖ aus zwei Ablesungen das Mittel nehmen ‖ prendre la moyenne de deux lectures. / ~ to one minute ‖ auf eine Minute f. ablesbar ‖ donnant la minute. / mirror ~ ‖ Spiegelablesung f. ‖ lecture f. au miroir. / ready ~ ‖ schnelles Ablesen n. ‖ lecture f. vive. / rough ~ ‖ Rohablesung f. ‖ lecture f. brute. / ~ a telegram by sound ‖ Aufnahme f. eines Telegramms nach dem Gehör ‖ lecture f. d'une dépêche au son.

reading desk ‖ Lesepult n. ‖ pupitre m. / ~ device ‖ Ableseeinrichtung f. ‖ dispositif m. de lecture. / ~ distance ‖ Leseabstand m. ‖ distance f. de lecture. / ~ error ‖ Ablesefehler m. ‖ erreur f. de lecture. / ~ eye glass *see* ~ pince-nez. / ~-in (Weav) ‖ Einlesen n. ‖ lisage m. de dessins pour tissus. / ~ lamp ‖ Ableselampe f. ‖ lampe f. de lecture. / ~ microscope ‖ Ablesemikroskop n. ‖ microscope m. de lecture. / ~ microscope, reading to one minute of arc ‖ auf eine Bogen-

minute ablesbares Ablesemikroskop n. ‖ miscroscope m. de lecture donnant la minute d'arc. / ~ pince-nez ‖ Nahklemmer m.; Nahkneifer m. ‖ pince-nez m. pour voir de près. / ~ room ‖ Lesehalle f.; Lesesaal m. ‖ salle f. de lecture. / ~ spectacles pl. ‖ Lesebrille f.; Nahbrille f. ‖ lunettes fpl. pour lecture *ou* pour voir de près. / ~ telescope ‖ Ablesefernrohr n. ‖ lunette f. de lecture.

ready ‖ fertig; bereit ‖ prêt; apprêté; préparé. / always ~ for instant use ‖ immer gebrauchsfertig ‖ toujours prêt à l'usage m. / ~ for conveyance ‖ transportfähig ‖ transportable. / ~ for driving ‖ fahrbereit ‖ prêt à conduire. / ~ for immediate use ‖ sofort gebrauchsfertig ‖ prêt immédiatement à l'usage m. / to make the form ~ (Print) ‖ die Form zurichten ‖ poser la forme; faire le registre. / ever ~ for operation ‖ stets betriebsbereit ‖ toujours prêt à fonctionner. / ~ for press ‖ druckfertig ‖ bon à tirer; prêt pour l'impression. / ~ to sail ‖ segelklar ‖ prêt à prendre la mer. / ~ packed for shipment ‖ seemäßig verpackt ‖ avec emballage m. maritime *ou* pour l'exportation par mer. / ~ for use ‖ gebrauchsfertig; in gebrauchsfertigem Zustand m. ‖ prêt à l'usage m. *ou* à entrer en service. / ~ for working ‖ betriebsfertig; betriebsbereit ‖ prêt à marcher *ou* à fonctionner.

ready-made ‖ vorrätig; zum Gebrauch m. fertig ‖ prêt à servir; confectionné; livré prêt à l'emploi m. / ~ clothes pl. ‖ Konfektion f. ‖ confection f. / ~ clothes pl. for gentlemen ‖ Herrenkonfektion f. ‖ confection f. pour hommes. / ~ clothing maker ‖ Konfektionsschneider m. ‖ tailleur-confectionneur m. / ~ fur ‖ Konfektionspelz m. ‖ fourrure f. confectionnée.

ready-mixed paint ‖ anstrichfertige Farbe f. ‖ couleur f. préparée pour la peinture. / ~ money ‖ bares Geld n. ‖ argent m. comptant. / for ~ money ‖ gegen bar ‖ au comptant m. / ~ reckoner ‖ Rechenknecht m. ‖ barème m.; barrême m.

reagent ‖ Reagens n. ‖ réactif m.

real ‖ reell; tatsächlich; wirklich ‖ réel; effectif. / ~ (Math) ‖ reell ‖ réel. / ~ component (Electr) ‖ Wirkwert m. (elektrischer Größen) ‖ composante f. effective *ou* réelle.

real estate ‖ Grundbesitz m.; Grundeigentum m. ‖ propriété f. foncière. / ~ credit association ‖ Bodenkreditbank f. ‖ crédit-foncier m.

real extract ‖ wirklicher Extraktgehalt m. ‖ teneur f. réelle en extrait. / ~ facts pl. of case ‖ Sachverhalt m.; Sachverhältnis n. ‖ état m. des choses. / ~ gravity *see* real extract. / ~ image ‖ reelles Bild n. ‖ image f. réelle. / ~ intermediate image ‖ reelles Zwischenbild n. ‖ image f. réelle intermédiaire. / ~ measure ‖ Sollmaß n. ‖ dimension f. théorique. / ~ power ‖ wirkliche Kraft f. ‖ puissance f. effective *ou* réelle. / ~ power (Electr) ‖ Wirkleistung f.; Wattleistung f. ‖ puissance réelle. / ~ property *see* real estate. / ~ servitude ‖ Reallast f. ‖ charge f. immobilière. / ~ size ‖ Sollmaß n. ‖ dimension f. théorique. / ~ weight ‖ Nutzlast f. ‖ poids m. utile.

realgar ‖ Realgar n.; roter Arsenik m. ‖ réalgar m.; sulfure m. rouge d'arsenic.

realizable ‖ realisierbar ‖ réalisable.

realization ‖ Verwertung f.; Realisierung f. ‖ réalisation f.; mise f. en valeur. / ~ of slag ‖ Schlackenverwertung f. ‖ utilisation f. de scorie.

realty ‖ Grundbesitz m. ‖ propriété f. territoriale *ou* foncière. / charge upon a ~ ‖ dingliche Belastung f. ‖ charge f. réelle.

ream, to ‖ erweitern; räumen ‖ aléser; équarrir; élargir. / ~ the hole ‖ die Bohrung aufreiben ‖ calibrer l'alésage m. / ~ rivet holes ‖ Nietlöcher npl. aufreiben ‖ aléser des trous mpl. de rivets.

ream (Pap) ‖ Ries n. ‖ rame f.

reamed bolt ‖ Paßschraube f. ‖ boulon m. ajusté.

reamer ‖ Reibahle f. ‖ alésoir m. / conical ~ ‖ konische Reibahle f. ‖ alésoir m. conique. / cylindrical ~ ‖ walzenförmige oder zylindrische Reibahle f. ‖ alésoir m. cylindrique. / hand ~ ‖ Handreibahle f. ‖ alésoir m. à main. / machine ~ ‖ Maschinenreibahle f. ‖ alésoir m. à commande mécanique.

reamer bit ‖ Reibahle f. ‖ louche f.; alésoir m.; élargisseur m. / ~ holder ‖ Reibahlenhalter m. ‖ porte-alésoir m. / planing device for ~s ‖ Reibahlenabziehvorrichtung f. ‖ dispositif m. d'aiguisage pour alésoirs.

reaming ‖ Aufreiben n. ‖ alésage m. / ~ press ‖ Aufweitpresse f. ‖ presse f. à mandriner. / rolling mill for ~ tubes ‖ Rohraufweitewalzwerk n. ‖ laminoir m. à élargir les tubes.

reap, to ‖ ernten; mähen ‖ moissonner; récolter; faucher.

reaper ‖ Getreidemäher m.; Getreidemähmaschine f. ‖ moissonneuse f. / ~ (Person) ‖ Schnitter m. ‖ faucheur m.; moissonneur m. / ~ and binder ‖ Getreidemäher m. und Garbenbinder m.; Bindemäher m. ‖ moissonneuse-lieuse f. / ~ for hay ‖ Grasmähmaschine f. ‖ faucheuse f.

reap hook *see* reaping hook.

reaping hook ‖ Sichel f. ‖ faucille f.

reaping machine ‖ Getreidemäher m.; Getreidemähmaschine f. ‖ moissonneuse-javeleuse d.; moissonneuse f. / ~ driver ‖ Getreidemähmaschinenführer m. ‖ moissonneur m. à la machine; conducteur m. de la moissonneuse.

rear, to ‖ sich erheben ‖ se redresser. / ~ silk worms pl. ‖ Seidenwürmer mpl. ziehen ‖ élever des vers mpl. à soie.

rear ‖ Rückseite f. ‖ arrière m.

rear-axle (Auto) ‖ Hinterachsbrücke f. ‖ pont m. arrière. / ~ (Railw) ‖ Hinterachse f. ‖ essieu m. d'arrière. / ~ drive ‖ Hinterradantrieb m. ‖ commande m. des roues arrières. / ~ guide ‖ Hinterachsführung f. ‖ guide f. pour l'essieu arrière. / ~ housing ‖ Hinterachsgehäuse n. ‖ carter m. du pont arrière. / ~ tie-bar ‖ Hinterachsunterzug m. ‖ Hinterachsstrebe f. ‖ tendeur m. *ou* contre-fiche f. du pont arrière. / ~ tie-rod *see* ~ tie-bar. / ~ tube (Auto) ‖ Hinterachsrohr n. ‖ tube m. du pont arrière. / ~ tube (Railw) ‖ Hinterachsrohr n. ‖ tube m. de l'essieu arrière.

rear cycle fork ‖ Hinterradgabel f. ‖ fourche f. de roue arrière. / ~ drive ‖ rückwärts liegender Antrieb m. ‖ commande f. à l'arrière. / ~ guide ‖ hintere Führung f. ‖ guidage ‖ m. arrière.

rearing bees pl. ‖ Bienenzucht f. ‖ apiculture f. / ~ form (Coop) ‖ Aufsatzform f. ‖ forme f. d'assemblage.

rear lamp bracket ‖ Halter m. für Schlußlaterne ‖ porte-lanterne m. arrière. / ~ side rod (Loc) ‖ hintere Kuppelstange f. ‖ bielle f. d'accouplement d'arrière. / ~ spar ‖ Stirnholm m. ‖ arêtier m. / ~ tow hook ‖ Zughaken m. ‖ crochet m. d'arrière. / ~ view ‖ Rückansicht f. ‖ vue f. arrière. / ~ vision mirror ‖ Rückspiegel m. ‖ retroviseur m. / ~ wheel ‖ Hinterrad n. ‖ roue f. arrière. / ~ wheel set ‖ Hinterradsatz m. ‖ train m. de roues arrière.

reason ‖ Vernunft f.; vernünftiger Grund m. ‖ raison f.; motif m. / obvious ~ ‖ erkennbarer Grund m. ‖ cause f. apparente. / with ~ ‖ mit Recht n. ‖ à bon droit m.; avec raison f.

reasonable ‖ verständig; vernunftgemäß ‖ conforme à la raison. / ~ (Trade) ‖ mäßig ‖ modéré; sobre. / ~ price ‖ annehmbarer Preis m. ‖ prix m. raisonnable.

re-assemble, to ‖ wieder zusammen bauen; wieder versammeln ‖ remonter; rassembler.

rebag, to ‖ umsacken ‖ changer de sac m.

rebate, to (Join) ‖ ausfalzen ‖ feuiller.

rebate ‖ Rabatt m. ‖ rabais m.; remise f. / ~ (Door) ‖ Falz m.; Anschlag m. ‖ feuillure f. / to allow ~ ‖ Rabatt m. bewilligen ‖ accorder un rabais. / ~ cannot be allowed ‖ Rabatt m. kann nicht gewährt werden ‖ on ne peut accorder de rabais m. / special ~ ‖ Sonderrabatt m. ‖ rabais m. de faveur. / ~ subject to a ~ of x % ‖ einem Rabatt m. von x % unterliegend ‖ sujet à une diminution f. de x %. / with a ~ of x % ‖ mit einem Rabatt m. von x % ‖ avec une diminution f. de x %. / without any ~ ‖ ohne jeden Rabatt m. ‖ sans diminution f.

rebatement ‖ Ermäßigung f. ‖ remise f.

rebate plane ‖ Falzhobel m. ‖ feuilleret m.; rabot-feuilleret m.; guillaume m. / side ~ ‖ Wandhobel m. ‖ guillaume m. de côté.

rebed, to ~ bearings pl. ‖ Lager npl. nachpassen ‖ ajuster des coussinets mpl.

rebending of the spring (Locksm; Clockm) ‖ Nachspannen n. der Feder ‖ reserrage m. du ressort.

reboil, to ‖ abkochen ‖ bouillir; rebouillir.

rebore, to ‖ nachbohren ‖ aléser.

rebroadcasting ‖ Ballsenden n. ‖ retransmission f. d'une radiodiffusion.

rebuild, to ~ the foundation ‖ ein Gebäude n. unterfahren *oder* mit neuen Grundmauern versehen ‖ reprendre les fondements mpl. d'un édifice.

rebuilding ‖ Umbau m. ‖ reconstruction f. / cost of ~ ‖ Neubaukosten pl. ‖ frais mpl. de reconstruction.

rebuilt ‖ neuaufgebaut ‖ reconstruit. / the plant was entirely ~ ‖ das Werk wurde vollständig umgebaut ‖ l'atelier m. fut reconstruit totalement.

rebushing the bearings ‖ Ausbuchsen n. der Lager ‖ emboîtage m. des coussinets.

re-calculation ‖ Neuberechnung f. ‖ calcul m. fait à nouveau.

recall, to (Contract) ‖ aufkündigen ‖ donner congé m. / ~ (Ware) ‖ abrufen ‖ rappeler. / ~ capital ‖ Kapital n. kündigen ‖ révoquer du capital. / ~ from circulation ‖ außer Kurs m. setzen ‖ mettre hors de circulation f.

recall (Contract) ‖ Aufkündigung f. ‖ congé m.; révocation f. / ~ (Tel) ‖ Rückruf m. ‖ rappel m. / ~ (Ware) ‖ Abruf m. ‖ rappel m.

recast, to ‖ umgießen; einschmelzen ‖ refondre.

recasting ‖ Wiedereinschmelzen n. ‖ refonte f.

receipt, to ‖ quittieren ‖ acquitter.

receipt ‖ Empfang m. ‖ réception f. / ~ (Trade) ‖ Quittung f.; Quittungsformular n.; Quittungsvordruck m. ‖ acquit m.; quittance f. / ~ (Ware) ‖ Ablieferungsschein m.; Empfangsschein m. ‖ bulletin m. *ou* bon m. *ou* ordre m. de livraison; reçu m. / to acknowledge the ~ ‖ den Empfang m. anzeigen ‖ accuser la réception. / ~ of delivery ‖ Aufgabeschein m. ‖ récépissé m.; reçu m. / ~ in duplicate ‖ Doppelquittung f. ‖ quittance f. double. / ~ in full ‖ Generalquittung f. ‖ quittance f. finale. / ~ of money ‖ Geldeinnahme f. ‖ recette f. / on ~ ‖ gegen Quittung f. ‖ sur quittance f. / no printing ~ on slips (Cash register) ‖ kein Quittungsdruck m. ‖ pas d'impression f. sur fiche f.

receipt, acknowledgment of ~ ‖ Empfangsbestätigung f. ‖ accusé m. de réception.

receipts pl. ‖ Einnahme f. ‖ recette f.; revenu m. / additional ~ ‖ Mehreinnahme f. ‖ excédent m. de recette. / book for entering ~ ‖ Einnahmebuch n. ‖ livre m. des recettes. / estimated ~ ‖ Solleinnahme f. ‖ recette f. brute. / supposed ~ *see* estimated ~. / total ~ ‖ Gesamteinnahme f. ‖ recette f. totale *ou* brute.

receive, to ‖ empfangen ‖ recevoir. / ~ (Money) ‖ einnehmen ‖ recevoir; toucher.

received on account (Cash register) ‖ Rechnungseinzahlung f. ‖ versement m.

receiver (Chem) ‖ Sammelgefäß n.; Rezipient m. ‖ récipient m. / ~ (Print) ‖ Ausleger m. ‖ receveur m. / ~ (Sugar ‖ Füllbecken n. ‖ Saftbehälter m. ‖ bassin m. à suc. / ~ (Tel) ‖ Hörer m. ‖ récepteur m. / ~ (Trade) ‖ Empfänger m.; Ladungsempfänger m. ‖ destinataire m.; consignataire m. / ~ air ‖ Windkessel m. ‖ réservoir m. d'air. / balanced ~ (Radio) ‖ ausbalancierter Empfänger m. ‖ récepteur m. compensé. / continuous wave ~ (Radio) ‖ Empfänger m. für ungedämpfte Wellen ‖ récepteur m. pour ondes non amorties. / directional ~ ‖ Richtempfänger m. ‖ récepteur m. dirigé. / head ~ (Radio) ‖ Kopfhörer m. ‖ casque m. / ~ with high-frequency strengthening ‖ Empfänger m. mit Hochfrequenzverstärkung ‖ récepteur m. à haute fréquence. / high-pressure ~ ‖ Hochdruckbehälter m. ‖ réservoir m. à haute pression. / local ~ ‖ Ortsempfänger m. ‖ appareil m. à recevoir le poste local. / ~ with low-frequency strengthening ‖ Empfänger m. mit Niederfrequenzverstärkung ‖ récepteur m. à amplification à basse-fréquence. / moving coil ~ (Tel) ‖ Spulentelefon n. ‖ récepteur m. à bobines. / ~ for multiple tubes ‖ Mehrfachröhrenempfänger m. ‖ récepteur m. pour lampes universelles. / official ~ ‖ amtlicher Konkursverwalter m. ‖ syndic m. officiel de la faillite. / radio ~ ‖ Radioempfänger m. ‖ radiorécepteur m. / seamless high-pressure ~ ‖ nahtloser

Hochdruckbehälter m. ‖ réservoir m. à haute pression sans soudure. / simple ~ ‖ Primärfunkempfänger m. ‖récepteur m. avec un seul circuit. / small ~ ‖ Kleinfernhörer m. ‖ petit récepteur m. / ~ between high- and low-pressure steam ‖ Dampfaufnehmer m. ‖ réservoir m. intermédiaire de vapeur. / ~ of steel ‖ Stahlbehälter m. ‖ recipient m. en acier. / ~ with x stepped low-frequency strengthening (Radio) ‖ Empfänger m. mit x-Stufen Niederfrequenzverstärkung ‖ récepteur m. à x étages d'amplification basse fréquence. / stoneware ~ ‖ Kondensationsgefäß n.; große irdene Flasche f. ‖ bonbonne f. en grès.

receiver arrangement (Tel) ‖ Empfangsvorrichtung f. ‖ dispositif m. de réception. / ~ case (Tel) ‖ Fernhörerkapsel f. ‖ capsule f. ou boîtier m. du récepteur. / ~ casing fitted with a closing cover in two parts ‖ Empfängergehäuse n. mit zweiteiligem Abschlußdeckel ‖ cage f. du récepteur munie d'une porte à deux battants. / hook of the ~ (Tel) ‖ Hörergabel f. ‖ fourchette f. de téléphone.

receivership ‖ Konkursverwaltung f. ‖ syndicat m. de la faillite.

receiver station for a picture telegraph installation ‖ Empfangseinrichtung f. einer Bildtelegrafenanlage ‖ équipement m. de réception d'une installation de téléphotographie.

receiving, directed ~ ‖ gerichteter Empfang m. ‖ réception f. dirigée. / ~ without distortion ‖ unverzerrter Empfang m. ‖ réception f. sans déformation. / long-distance ~ ‖ Fernempfang m. ‖ réception f. à grande distance. / short-distance ~ ‖ Ortsempfang m. ‖ réception f. régionale.

receiving aerial (Radio) ‖ Empfangsdraht m.; Empfangsantenne f. ‖ antenne f. de réception. / ~ amplifier ‖ Empfangsverstärker m. ‖ amplificateur m. de réception. / ~ apparatus ‖ Empfangsapparat m. ‖ appareil m. récepteur. / ~ apparatus for water level tele-indicators ‖ Empfänger m. für Wasserstandsfernmelder ‖ récepteur m. pour télé-indicateurs de niveau d'eau. / ~ box ‖ Ablegekasten m. ‖ caisse f. de réception. / change of connection for ~ (Radio) ‖ Umschaltung f. für Empfang ‖ commutation f. pour la réception. / ~ distributor ‖ Empfangsteiler m. ‖ distributeur m. de réception. / ~ end ‖ Empfangsende n. ‖ extrémité f. réceptrice. / ~ floor (Build) ‖ Hauptgeschoß n. ‖ bel étage m. / ~ hopper ‖ Auffangtrichter m. ‖ entonnoir-récepteur m. / ~ insulator ‖ Isolator m. für Empfangsdraht ‖ isolateur m. de réception. / ~ jigger ‖ Empfangsjigger m. ‖ jigger récepteur m. / ~ office ‖ Annahmestelle f. ‖ bureau m. de consignation. / ~ perforator ‖ Empfangslocher m. ‖ perforateur m. d'arrivée. / ~ plant for long-distance receiving ‖ Empfangsanlage f. für Fernempfang ‖ réception f. pour recevoir les postes lointains. / ~ relay ‖ Empfangsrelais n. ‖ relais m. de réception. / ~ and milling separator for grains ‖ Getreideaspirationsreinigungsmaschine f. ‖ nettoyeur aspirateur m. du blé. / ~ set ‖ Empfänger m. ‖ appareil m. récepteur. / double-circuit ~ set (Radio) ‖ Sekundärfunkempfänger m. ‖ récepteur m. avec deux

circuits. / ~ ship ‖ Kasernenschiff n. ‖ caserne f. flottante. / ~ station ‖ Empfangsstation f. ‖ poste m. récepteur. / ~ stone ‖ Aufnahmestein m. ‖ pierre f. pour la prise de la dictée. / ~ system ‖ Empfangsanlage f. ‖ installation f. de réception. / ~ trial ‖ Empfangsversuch m. ‖ essai m. de réception. / ~ tube ‖ Empfängerröhre f. ‖ lampe f. pour récepteur.

recently opened cast working ‖ neuaufgeschlossener Tagebau m. ‖ mine f. à ciel ouvert récemment installée.

receptacle for beer working out of the bung hole ‖ Gärbecken n.; Gärhaube f. ‖ cuvelle f. / ~ for office use ‖ Bürokasten m. ‖ casier m. pour bureau. / ~ for perfume ‖ Parfümbehälter m. ‖ porte-parfum m. / ~ for warehouse ‖ Lagerkasten m. ‖ casier m. pour magasin. / ~ for waste ‖ Müllkasten m. ‖ caisse f. pour les résidus.

reception ‖ Aufnahme f.; Empfang m. ‖ réception f.; admission f. / intermediate circuit ~ ‖ Zwischenkreisempfang m. ‖ réception f. par circuit intermédiaire. / multiple ~ ‖ Vielfachempfang m. ‖ réception f. multiple. / ~ by sound ‖ Hörempfang m. ‖ réception f. auditive. / ~ by tape ‖ Schreibempfang m. ‖ réception f. sur bande. / ~ with vibrating relay ‖ Vibrationsempfang m. ‖ réception f. à relais vibrateur.

reception, interference in ~ ‖ Interferenz f. im Empfang ‖ interférence f. dans la réception. / ~ test set ‖ Stationsprüfer m. ‖ circuit m. vérificateur de réception.

recess, to ‖ vertiefen; einsenken ‖ défoncer. / the hull of the ship is recessed forward and aft ‖ der Schiffsrumpf ist vorn und achtern eingezogen ‖ la coque du navire a des rentrants à l'avant et à l'arrière.

recess (Build) ‖ Rücksprung m. ‖ renfoncement m.; retrait m. / ~ (Hydr arch) ‖ Nische f. ‖ niche f. / ~ (Mach) ‖ Auskehlung f.; Vertiefung f. ‖ gorge f.; creux m. / binding ~ (Electr) ‖ Bandagenute f. ‖ creux m. de frette ou de bandage. / ~ of a ring ‖ Aussparung f. eines Ringes ‖ évidement m. d'une bague.

recess head (Join) ‖ Hohlkehle f. ‖ gorge f.

recessing tool holder ‖ Einstechstahlhalter m. ‖ porte-outil m. à saigner.

recharge, to ‖ wiederaufladen ‖ recharger.

recharge (Acc) ‖ Nachladen n. ‖ charge ment m. supplémentaire.

recharging ‖ Nachladung f. ‖ rechargement m.

recipient (Found) ‖ Sammelherd m. ‖ récipient m.

reciprocal ‖ reziprok; wechselseitig ‖ réciproque; inverse. / ~ debt ‖ Gegenschuld f. ‖ dette f. passive.

reciprocal ‖ reziproker Wert m.; Gegenstück n. ‖ réciproque f.; inverse m. / ~ of the voltage amplification factor ‖ Durchgriff m. ‖ inverse m. du facteur d'amplification de potentiel.

reciprocate, to ‖ hin- und hergehen; umwechseln ‖ réciproquer; alterner.

reciprocating engine ‖ Maschine f. mit hin- und hergehender Bewegung; Kolbenmaschine f. ‖ machine f. à mouvement alternatif. / ~ hearth ‖ Schwingherd m. ‖ sole f. oscillante. / ~ locomotive ‖ Kolbenlokomotive f. ‖ lovomotive f. à piston. / ~ movement ‖ Hin- und

Herbewegung f. ‖ mouvement m. alternatif ou de va-et-vient. / ~ saw ‖ Gattersäge f. ‖ scie f. alternative.

reciprocity ‖ Gegenseitigkeit f. ‖ réciprocité f.; mutualité f.

reckon, to ‖ rechnen; zählen ‖ calculer; compter.

reckoner (Spinn) ‖ Zählerin f. ‖ compteuse f.

reckoning ‖ Berechnung f. ‖ calcul m.; compte m. / dead ~ ‖ Gissung f.; Koppelkurs m. ‖ gisement m.; route f. estimée. / to work the ~ ‖ das Besteck berechnen ‖ calculer le point.

reclamation ‖ Zurückforderung f.; Reklamation f. ‖ réclamation f.

recognizable only in a thin section ‖ nur im Schliff m. wahrnehmbar ‖ reconnaissable seulement dans la lame mince.

recognize, to ‖ erkennen; anerkennen ‖ reconnaître.

recoil, to ‖ zurückstoßen; zurückschlagen ‖ reculer; repousser.

recoil ‖ Rückstoß m. ‖ contre-coup m.; recul m.; repoussement m. / ~ rod (Typewr) ‖ Prellanschlag m. ‖ tige f. de butée. / ~ stop ‖ Stoßreitel m.; Rückstoßhemmung f.; Begrenzungspuffer m. ‖ butée f. de recul.

recoin, to ‖ umprägen ‖ refrapper.

recommend, to ‖ empfehlen; befürworten ‖ recommander. / ~ (Ware) ‖ anpreisen ‖ vanter; prôner.

recommendable ‖ empfehlenswert ‖ recommandable.

recommendation ‖ Empfehlung f. ‖ recommandation f.

recomposition (Print) ‖ neuer Satz m.; Wiedergesetztes n. ‖ récomposition f.

reconducting bar ‖ Rückführungsstange f. ‖ tige f. de rappel. / ~ belt ‖ Rückführungsband n. ‖ bande f. de reconstitution. / ~ shoot ‖ Rückführungsrutsche f. ‖ glissière f. de reconduite.

reconduction band ‖ Zusammenführungsband n. ‖ bande f. de reconduite.

reconnaissance airplane ‖ Aufklärungsflugzeug n. ‖ avion m. de reconnaissance ou d'éclairage. / long-distance ~ machine ‖ Fernaufklärungsflugzeug n. ‖ avion m. de reconnaissance à longue distance. / ~ tank ‖ Erkundungskampfwagen m. ‖ char m. de reconnaissance.

re-cooler ‖ Nachkühler m. ‖ condenseur m. complémentaire.

re-cooling plant ‖ Rückkühlanlage f. ‖ réfrigérant m. de retour.

reconstruction ‖ Wiederaufbau m. ‖ reconstruction f. / ~ of the plant ‖ Umbau m. einer Anlage ‖ reconstruction f. des anciens ateliers et chantiers.

record, to ‖ registrieren; aufschreiben; aufzeichnen ‖ enregistrer; noter; consigner.

record ‖ Urkunde f.; Dokument n.; Aufzeichnung f. ‖ document m.; acte m.; pièce f.; dessin m. / ~ (Mus) ‖ Grammofonplatte f.; Schallplatte f. ‖ disque m. de gramophone. / synchronous ~ ‖ zeitgleich ablaufende Schallplatte f. ‖ disque m. tournant en synchronisme. / ~ for talking machines ‖ Schallplatte f. ‖ disque m. pour machines parlantes ou de gramophone.

recorder (Mach) ‖ Zähler m.; Gangzähler m.; Registrierapparat m. ‖ compteur m.; indicateur m.; totalisateur m.; enregistreur m. / ~ (Tel) see record opera-

tor. / ~ for registering active volt-amperes ‖ Schreibgerät n. zur Aufzeichnung der Wirkleistung ‖ enregistreur m. de la puissance active. / distant ~ ‖ Fernzähler m. ‖ compteur m. à distance. / ~ for gas leakages ‖ Kontrollapparat m. für Gasverlust ‖ appareil m. contrôleur des fuites de gaz. / gyroscopic roll and pitch ~ ‖ Kreiselschlinger und Stampfanzeiger m. ‖ appareil m. enregistreur gyroscopique de roulis et de tangage. / perforated strip ~ ‖ Lochstreifenempfänger m. ‖ récepteur-perforateur m. / ~ for water level tele-indicators ‖ Wasserstandsregistrierapparat m. ‖ enregistreur m. pour télé-indicateurs de niveau d'eau.

recorder signalling key for cable-code ‖ Doppeltaste f. für den Kabelbetrieb ‖ manipulateur m. pour câble-code.

recording an account (Cash register) ‖ Zahlung f. ‖ versements mpl. / ~ by diagram ‖ Diagrammaufzeichnung f. ‖ traçage m. par diagramme. / photographic ~ ‖ fotografische Aufschreibung f. ‖ enregistrement m. photographique.

recording altimeter ‖ Höhenschreiber m. ‖ altimètre m. enregistreur. / ~ ammeter ‖ Registrierstrommesser m. ‖ ampèremètre m. enregistreur. / ~ apparatus ‖ Registrierapparat m. ‖ appareil m. enregistreur; enregistreur m. / ~ attachment ‖ Registriereinrichtung f. ‖ dispositif m. enregistreur. / ~ counter ‖ Registrierkasse f. ‖ caisse f. (d'argent) enregistreuse. / ~ desk ‖ Meldetisch m. ‖ table f. annotatrice. / ~ detector ‖ registrierender Detektor m. ‖ détecteur enregistreur m. / ~ device ‖ Registrierapparat m.; Selbstschreiber m. ‖ enregistreur m. / ~ device for observatory time ‖ Sternwartezeitregistriereinrichtung f. ‖ dispositif m. d'enregistrement pour le service chronométrique d'observatoire. / ~ disk ‖ Zählscheibe f. ‖ disque m. compteur. / ~ drum ‖ Schreibtrommel f. ‖ tambour-enregistreur m. / ~ instrument ‖ registrierendes Instrument n. ‖ instrument m. enregistreur. / ~ measuring instrument ‖ schreibendes Meßgerät n. ‖ appareil m. de mesure écrivant. / ~ mechanism ‖ Registriervorrichtung f. ‖ mécanisme m. enregistreur. / ~ photo-electrical photometer ‖ lichtelektrisches Registrierfotometer n. ‖ photomètre m. enregistreur photo-électrique. / ~ section (Tel) ‖ Meldeschrank m. ‖ groupe m. d'annotatrices. / ~ signal indicator ‖ registrierender Signalzeiger m. ‖ transmetteur-enregistreur m. / ~ strip rotary printing machine ‖ Registrierstreifenrotationsdruckmaschine f. ‖ machine f. rotative à imprimer des bandes enregistreuses. / ~ telegraph (Tel) ‖ Registriertelegraf m. ‖ télégraphe m. enregistreur. / ~ trunk for line fault service ‖ Meldeleitung für den Störungsdienst ‖ fil m. avertisseur du service des dérangements. / ~ voltmeter ‖ registrierender Spannungsmesser m. ‖ voltmètre m. enregistreur. / ~ wax for talking machine records ‖ Schallplattenaufnahmewachs n. ‖ cire f. servant à faire les disques de gramophone. / ~ wire ‖ Meldeleitung f. ‖ ligne f. d'ordre.

record line ‖ Meldeleitung f. ‖ ligne f. annotatrice. / ~ office ‖ Meldeamt n. ‖

bureau m. d'enregistrement des demandes. / ~ office plant ‖ Archiveinrichtung f. ‖ ameublement m. d'archives. / ~ operator ‖ Meldebeamtin f. ‖ annotatrice f. / ~ position ‖ Meldeplatz m. ‖ position f. d'annotatrice.

records pl. ‖ Archiv n. ‖ archives fpl.
record section with selectors ‖ Melde-verteileramt n. ‖ bureau m. d'enregistrement avec sélecteurs. / ~ station ‖ Meldestelle f. ‖ table f. d'annotatrice. / ~ telephonist (Tel) ‖ Meldebeamtin f. ‖ annotatrice f. / ~ traffic ‖ Meldeverkehr m. ‖ trafic m. d'enregistrement.

recoupling effect ‖ Rückkopplungserscheinung f. ‖ phénomène m. de recouplage.
recourse ‖ Regreßrecht n. ‖ recours m. / to reserve right of ~ ‖ sich den Regreßanspruch m. vorbehalten ‖ se réserver le recours.
recover, to ‖ wiedergewinnen ‖ récupérer. / ~ a metal by annealing ‖ ein Metall n. nachglühen ‖ faire revenir un métal recuit. / ~ average ‖ Ersatz m. für Havarie erhalten ‖ recouvrer des avaries fpl. / ~ a metal by tempering *see* ~ a metal by annealing.
recovering ‖ Wiedergewinnung f.; Rückgewinnung f. Abfangen n. ‖ récupération f.; recouvrement m. / ~ of sludge ‖ Schlammgewinnung f. ‖ récupération f. des schlamms. / waste heat ~ ‖ Abhitzerückgewinnung f. ‖ récupération f. de la chaleur perdue.
recovering plant ‖ Wiedergewinnungsanlage f. ‖ installation f. de récupération.
recovery ‖ Wiedergewinnung f.; Rückgewinnung f. ‖ récupération f. / ~ of the airship ‖ Einholen n. des Luftballons ‖ halage m. du ballon. / ~ of by-products ‖ Gewinnung f. von Nebenerzeugnissen ‖ récupération f. de sous-produits. / ~ of liquid fuels from coal ‖ Gewinnung f. flüssiger Brennstoffe aus Kohle ‖ récupération f. de combustibles liquides extraits du charbon. / ~ of money due ‖ Inkasso n. ‖ encaissement m.; recouvrement m. / ~ of potash ‖ Pottaschegewinnung f. ‖ récupération f. du carbonate de potasse.
recovery plant for by-products ‖ Rückgewinnungsanlage f. für Nebenerzeugnisse ‖ installation f. de récupération des sous-produits.
recreation ‖ Erholung f. ‖ récréation f. / ~ park ‖ Erholungsgarten m. ‖ parc m. de récréation.
recrystallization ‖ Umkristallisation f. ‖ recristallisation f.
recrystallize, ‖ umkristallisieren ‖ recristalliser.
rectal syringe ‖ Klistierspritze f. ‖ seringue f. à lavement.
rectangle ‖ Rechteck n. ‖ rectangle m.
rectangled *see* rectangular.
rectangular ‖ rechtwinklig ‖ rectangulaire; rectangle. / ~ prism ‖ Winkelprisma n. ‖ équerre f. à prismes. / basin of nearly ~ section made of concrete ‖ Betonbassin n. von annähernd rechteckigem Querschnitt ‖ bassin m. *ou* auge f. en béton de section approximativement rectangulaire. / ~ stove tile ‖ rechteckige Ofenkachel f. ‖ carreau m. à poêles glacé rectangulaire.
rectification ‖ Rektifikation f.; Richtigstellung f. ‖ rectification f. / ~ (Chem) ‖ Rektifikation f. ‖ rectification f. / ~

(Electr) ‖ Gleichrichtung f. ‖ redressement m. / ~ of a curve ‖ Rektifikation f. einer Kurve ‖ rectification f. d'une courbe. / ~ of the picture ‖ Entzerrung f. der Bildaufnahme ‖ correction f. de la distorsion de la prise de vue.
rectified (Electr) ‖ gleichgerichtet ‖ redressé. / ~ current (Electr) ‖ gleichgerichteter Strom m. ‖ courant m. redressé.
rectifier (Chem) ‖ Rektifizierapparat m. ‖ appareil m. à rectifier. / ~ (Electr) ‖ Gleichrichter m. ‖ redresseur m. / aluminium cell ~ ‖ Aluminiumzellengleichrichter m. ‖ redresseur m. avec anode en aluminium. / alternating current ~ ‖ Wechselstromgleichrichter m. ‖ redresseur m. de courant alternatif. / alternating current relay ~ ‖ Wechselstromrelaisgleichrichter m. ‖ redresseur m. avec relais à courant alternatif. / copper plate ~ ‖ Plattengleichrichter m. ‖ redresseur m. à plaques de cuivre. / current ~ ‖ Gleichrichter m. ‖ rectificateur m. *ou* redresseur m. de courant. / electrolytic ~ ‖ elektrolytischer Gleichrichter m. ‖ redresseur m. électrolytique. / glow cathode ~ ‖ Glühkathodengleichrichter m. ‖ lampe f. redresseuse à cathode incandescente. / glow discharge ~ ‖ Glimmlichtgleichrichter m. ‖ redresseur m. à effluves; lampe f. redresseuse à fluorescence. / mechanic ~ ‖ mechanischer Gleichrichter m. ‖ redresseur m. mécanique. / mercury vapour arc ~ ‖ Quecksilberdampfgleichrichter m. ‖ redresseur m. à (vapeur de) mercure. / synchronously rotating ~ ‖ Kommutatorgleichrichter m. ‖ permutatrice f.; redresseur m. synchrone.
rectifier cell ‖ Gleichrichterzelle f. ‖ élément m. redresseur. / four-cell switch for ~s ‖ Vierzellenschaltung f. für Gleichrichter ‖ réducteur m. à quatre éléments. / ~ plant ‖ Gleichrichteranlage f. ‖ installation f. de redressement. / ~ plate ‖ Gleichrichterplatte f. ‖ panneau m. de redresseur. / ~ power station ‖ Gleichrichterwerk n. ‖ poste m. de redressement. / ~ railway ‖ Gleichrichterbahn f. ‖ chemin m. de fer à courant continu redressé. / ~ strengthener ‖ Richtverstärker m. ‖ amplificateur m. dirigeable.
rectifier tube ‖ Gleichrichterröhre f. ‖ tube m. redresseur. / ~ for line connection instruments ‖ Gleichrichterröhre f. für Netzanschlußgeräte ‖ valve f. redresseuse pour boîtes d'alimentation de tension anodique.
rectify, to (Chem) ‖ rektifizieren ‖ rectifier. / ~ (Electr) ‖ gleichrichten ‖ redresser. / ~ a curve ‖ eine Kurve f. rektifizieren ‖ rectifier une courbe.
rectifying (Electr) ‖ Gleichrichtung f. ‖ détection f. / ~ a short-circuit ‖ Kurzschlußentfernung f. ‖ suppression f. du court-circuit.
rectifying amplifier ‖ Richtverstärker m. ‖ amplificateur-redresseur m. / ~ apparatus ‖ Rektifizierapparat m. ‖ appareil m. de rectification. / ~ cell ‖ Gleichrichterzelle f. ‖ appareil m. redresseur. / ~ strengthener ‖ Richtverstärker m. ‖ amplificatrice-détectrice f. / ~ tube ‖ Ventilröhre f. ‖ tube m. rectificatif. / ~ telegram ‖ Berichtigungstelegramm n. ‖ télégramme m. rectificatif.
rectilineal *see* rectilinear.

rectilinear ‖ geradlinig ‖ rectiligne.

rectometer (Weav) ‖ Rektometer n. ‖ rectomètre m.

recuperating plant ‖ Rückgewinnungsanlage f. ‖ installation f. de récupération.

recuperative furnace ‖ Rekuperativfeuerung f. ‖ foyer m. à récupération de chaleur.

recurrent network ‖ Kettenleiter f. ‖ chaîne f. de circuits identiques.

recut, to ~ files pl ‖ Feilen fpl. aufhauen ‖ retailler les limes fpl.

recutting files ‖ Feilenhauen n. ‖ retaillage m. de limes.

red ‖ rot ‖ rouge. / ~ beech ‖ Rotbuchenholz n. ‖ hêtre m. rouge. / ~ brass see ~ bronze. / ~ bronze see also ~ copper ‖ Rotguß m.; Tombak m. ‖ bronze m. rouge; cuivre m. ou laiton m. rouge. / ~ bronze fittings pl. ‖ Rotgußarmatur f. ‖ garniture f. en cuivre rouge. / ~ caster ‖ Rotgießer m. ‖ fondeur m. en cuivre rouge. / ~ chalk ‖ Rötel m.; Rotstein m. ‖ craie f. ou crayon m. rouge. / ~ chalk pencil ‖ Rotstift m.; Rötel m. ‖ rubrique f.; crayon m. rouge; arcanne f. / ~ copper foundry ‖ Rotgießerei f. ‖ fonderie f. de cuivre rouge. / ~ copper ore ‖ Rotkupfererz n.; Kupferrot n. ‖ cuivre m. rouge ou oxydulé. / ~ copper smith ‖ Rotschmied m. ‖ chaudronnier m. en cuivre rouge. / ~ crayon see ~ chalk pencil. / ~ flax ‖ Rösteflachs m.; Rotteflachs m. ‖ lin m. naisé ou de rouissage.

red heat ‖ Glühhitze f.; Rotglut f. ‖ chaleur f. rouge ; rouge m. / beginning ~ ‖ anfangende Rotglut f. ‖ rouge m. naissant. / bright ~ ‖ lebhafte Rotglut f. ‖ rouge m. vif. / to bring to a ~ ‖ auf Glühhitze f. bringen ‖ porter au rouge m. / the steel may be permanently kept at ~ ‖ der Stahl kann dauernd auf Rotglut gehalten werden ‖ l'acier m. peut être porté constamment à la chaleur de rouge. / to work at ~ ‖ rotwarm bearbeiten ‖ travailler les pièces fpl. à la chaleur rouge.

red hematite mine ‖ Roteisensteingrube f. ‖ mine f. d'hématite rouge. / ~ herring ‖ Bückling m. ‖ hareng m. saur ou fumé. / ~ iron ochre ‖ Eisenmennige f. ‖ minium m. de fer. / ~ iron ore ‖ Roteisenstein m. ‖ hématite f. rouge.

red lead ‖ Bleimennige f. ‖ minium m. de plomb. / ~ pencil see red chalk pencil. / ~ plant ‖ Mennigeanlage f. ‖ installation f. de minium. / ~ putty ‖ Mennigkitt m. ‖ mastic m. au minium. / ~ substitute ‖ Bleimennigeersatz m. ‖ minium m. de plomb imité.

red light ‖ rotes Licht n. ‖ feu m. rouge. / ~ liquor ‖ Rotbeize f.; essigsaure Tonbeize f. ‖ mordant m. de rouge. / ~ metal see ~ bronze. / ~ mordant ‖ Rotbeize f.; essigsaure Tonbeize f. ‖ mordant m. de rouge. / ~ ochre ‖ rotocker ‖ ocre rouge. / ~ oxide ‖ Englischrot n. ‖ rouge m. anglais. / ~ oxide of copper ‖ Rotkupfererz n.; Kupferrot n. ‖ cuivre m. rouge ou oxydulé. / ~ pencil ‖ Rotstift m. ‖ crayon m. rouge. / ~ pepper ‖ Paprika m. ‖ piment m.; poivre m. rouge; paprika m. / ~ pigment ‖ Rot n. ‖ rouge m. / ~ print ‖ Rotpause f. ‖ calque m. en rouge. / ~ prussiate of potash ‖ rotes Blutlaugensalz n. ‖ cyanoferride m. de potassium; cyanure m. rouge de potassium et de fer; ferricyanure m. de

potassium. / ~ rot ‖ Rotfäule f.; Naßfäule f. ‖ pourriture f. rouge ou de cœur. / ~-rotted ‖ rotfaul ‖ pourri au cœur. / ~ sandstone ‖ Rotsandstein m. ‖ grès m. rouge. / ~ sandstone quarry ‖ Rotsandsteinbruch m. ‖ carrière f. de grès rouge. / ~ silver ‖ Rotgültigerz n.; Rotgulden n.; Silberblende f. ‖ argent m. rouge. / ~ stripiness ‖ Rotstreifigkeit f. ‖ bandes fpl. de pourriture rouges.

red ‖ Rot n. ‖ rouge m. / ~ (Metal) ‖ Rotglut f.; Rotglühhitze f. ‖ chaude f. rouge; rouge m. / ~ brown ‖ Englischrot n.; Braunrot n. ‖ rouge m. d'Angleterre. / dark ~ ‖ dunkle Rotglut f. ‖ rouge m. sombre. / English ~ see dark ~. / iron ~ ‖ Eisenrot n. ‖ rouge m. de fer. / jeweller's ~ ‖ Pariser Rot n.; Polierrot n. ‖ rouge m. à polir. / Paris ~ see jeweller's ~. / Persian ~ ‖ Persischrot n. ‖ rouge m. de Perse. / Pompeian ~ ‖ Pompejanischrot n. ‖ rouge pompéien. / Venetian ~ ‖ Venezianischrot n. ‖ rouge m. de Venise.

red-abstracting filter ‖ rotarmes Filter n. ‖ écran m. pauvre en rouge.

reddle see red chalk.

reddsman (Mine) ‖ Versatzarbeiter m. ‖ remblayeur m.; restapleur m.; releveur m. de terres.

redeem, to ‖ amortisieren ‖ amortir.

redemption (Capital) ‖ Rückzahlung f. ‖ remboursement m. / ~ (Loan) ‖ Amortisation f. ‖ amortissement m.; rachat m. / ~ (Trade) ‖ Auslösung f. ‖ dégagement m.; acquittement m. / ~ of shares ‖ Aktieneinziehung f. ‖ rachat m. ou amortissement m. d'actions. / ~ fund ‖ Schuldentilgungskasse f. ‖ caisse f. de remboursement. / right of ~ ‖ Rückkaufsrecht n. ‖ droit m. de réméré; réméré m. / term of ~ ‖ Tilgungsfrist f. ‖ terme m. d'amortissement. / terms pl. of ~ ‖ Tilgungsplan m. ‖ plan m. d'amortissement.

red-free electric light ‖ rotfreies elektrisches Licht n. ‖ lumière f. électrique exempte de rouge. / ~ filter ‖ Rotfreifilter n. ‖ écran m. exempt de rouge. / ~ lamp ‖ Rotfreilampe f. ‖ lampe f. exempte de rouge. / ~ light ‖ Rotfreilicht n. ‖ lumière f. exempte de rouge. / abundance of rays of short waves in the ~ light ‖ Reichhaltigkeit f. des Rotfreilichtes an kurzwelligen Strahlen ‖ richesse f. de la lumière exempte de rouge en rayons de petite longueur d'onde.

red-heat see under red.

red-hot (Metal) ‖ rotglühend ‖ rouge; chauffé au rouge. / to become ~ ‖ rotglühend werden ‖ être porté au rouge m. / to remain ~ ‖ rotglühend bleiben ‖ rester rouge-cerise.

redistil, to ‖ rektifizieren ‖ rectifier.

redistillation ‖ Läuterung f.; Rektifikation f. ‖ rectification f.

red-purple ‖ rotviolett ‖ rouge violet.

redress, to (Millstone) ‖ schärfen; picken; pillen ‖ repiquer; rhabiller.

redressed current ‖ gleichgerichteter Strom m. ‖ courant m. redressé.

red-rotted (Wood) ‖ rotfaul ‖ pourri au cœur.

redruthite ‖ Kupferglanz m.; Kupferglas f.; Schwefelkupfer n. ‖ cuivre m. sulfuré ou vitreux.

red-short ‖ rotbrüchig ‖ rouverin; rouverain; cassant à chaud; métis.

reduce, to ‖ schmälern; verkleinern; reduzieren ‖ rétrécir; diminuer; réduire. / ~ (Math) ‖ reduzieren; vereinfachen ‖ réduire. / ~ (Phot) ‖ abschwächen ‖ affaiblir. / ~ (Trade) ‖ umrechnen ‖ réduire; calculer. / ~ a drawing (Draw) ‖ eine Zeichnung verkleinern ‖ réduire un dessin. / ~ lead ‖ das Blei frischen ‖ réduire ou refondre le plomb. / ~ to powder ‖ pulvern; zu Pulver reiben ‖ pulvériser. / ~ in price ‖ verbilligen ‖ baisser le prix. / ~ the strength (Chem) ‖ verarmen ‖ appauvrir.

reduced ‖ vermindert ‖ réduit. / ~ image ‖ verkleinertes Bild n. ‖ image f. réduite. / ~ output ‖ Minderleistung f. ‖ débit m. réduit. / ~ scale ‖ verjüngter Maßstab m. ‖ échelle f. réduite. / ~ square hole ‖ verjüngtes Vierkantloch n. ‖ trou m. carré à épanouissement.

reducer (Phot) ‖ Abschwächer m. ‖ affaiblisseur m.

reducibility ‖ Reduzierbarkeit f. ‖ réductibilité f.

reducible ‖ reduzierbar ‖ réductible.

reducing drawings ‖ Verkleinerung f. von Zeichnungen ‖ réduction f. des dessins. / ~ first costs ‖ Verbilligung f. der Anschaffungskosten ‖ réduction f. des frais d'achat. / process of ~ iron ores in a revolving furnace with producer gas firing ‖ Reduktion f. von Eisenerzen in einem rotierenden Ofen mit Generatorgasfeuerung ‖ procédé m. (que consistait) à réduire le minerai de fer dans un four rotatif pourvu d'un générateur à gaz.

reducing agent ‖ Reduktionsmittel n. ‖ agent m. de réduction; réducteur m. / ~ coupling box ‖ Absatzmuffe f. ‖ manchon m. de réduction. / ~ flame ‖ Reduktionsflamme f. ‖ flamme f. réductrice ou de réduction. / ~ gas ‖ reduzierendes Gas n. ‖ gaz m. réducteur ou de réduction. / ~ piece ‖ Reduzierstück n. ‖ pièce f. de réduction. / ~ pipe ‖ Übergangsrohr n. ‖ raccord m. avec réduction de diamètre. / ~ power ‖ Reduzierfähigkeit f. ‖ pouvoir m. réducteur. / ~ resistance ‖ Abschwächungswiderstand ‖ résistance f. modératrice.

reducing rolling mill ‖ Flachmahlstuhl m. ‖ moulin m. pour monture basse. / tube ~ ‖ Reduzierwalzwerk n. für Rohre ‖ laminoir m. à réduire les tubes.

reducing scale ‖ Verjüngungsmaßstab m. ‖ échelle f. de réduction. / ~ socket ‖ Reduziereinsatz m. ‖ pièce f. de réduction. / ~ transformer ‖ Abwärtstransformator m. ‖ dévolteur m.; transformateur-réducteur m. / ~ valve ‖ Reduzierventil n.; Druckminderventil n. ‖ soupape f. de réduction.

reduction ‖ Reduktion f. ‖ réduction f. / ~ (Chem) ‖ Abbau m.; Reduktion f. ‖ dissolution f.; désagrégation f. / ~ (Math) ‖ Reduktion f.; Vereinfachung f. ‖ réduction f. / ~ (Phot) ‖ Verkleinerung f, ‖ réduction f. / ~ of area (Tensile test) ‖ Einschnürung f.; Querschnittverminderung f. ‖ contraction f. / ascending ~ (Math) ‖ Reduktion f. oder Zurückführung f. auf eine höhere Benennung ‖ réduction f. ascendante ou à la dénomination supérieure. / ~ in charges ‖ Gebührenermäßigung f. ‖ réduction f. des frais. / descending ~ ‖ Reduktion f. oder Zurückführung f. auf eine niedrigere Benennung ‖ réduction f. descendante

ou à une dénomination inférieure. /
~ by eliquation ‖ Seigerprozeß m.; Seigerung f. ‖ liquation f.; ressuage m.;
ségrégation f. / ~ into gas ‖ Vergasung f.
‖ gazéification f. / to make a ~ ‖ nachlassen ‖ rabattre. / ~ of output ‖ Pro-
duktionsverminderung f. ‖ réduction f.
de rendement. / ~ of prices ‖ Preisabbau m. ‖ réduction f. des prix. / ~ of
salary ‖ Gehaltsabzug m. ‖ réduction f.
sur le traitement. / ~ of strength (Chem)
‖ Verarmung f. ‖ appauvrissement f. /
~ of weight ‖ Gewichtsverminderung f.
‖ réduction f. de poids.
reduction compasses pl. ‖ Reduzierzirkel
m. ‖ compas m. à coulisse *ou* de réduction. / ~ cone ‖ Einsatzfutter n. ‖ cône
m. de réduction. / ~ factor ‖ Reduktions-
faktor m. ‖ facteur m. de réduction. /
~ gear ‖ Reduktionsgetriebe n.; Unter-
setzungsgetriebe n. ‖ engrenage m. réducteur; démultiplicateur m. / ~ gear-
(ing) (Mach) ‖ Vorgelege n.; Zahnradvorgelege n. ‖ engrenage m. *ou* transmission
f. intermédiaire; renvoi m. (de mouvement); contre-arbre m. / ~ gear ratio ‖
Untersetzungsverhältnis n. ‖ rapport m.
de réduction. / ~ plant ‖ Reduktionsan-
lage f. ‖ installation f. de réduction. / ~
product (Chem) ‖ Abbauprodukt n.; Auf-
löseprodukt n. ‖ produit m. de dissolution *ou* de dégradation. / ~ ratio ‖ Übersetzungsverhältnis n. ‖ rapport m. de
transmission. / ~ tacheometer ‖ Reduktionstachometer n. ‖ tachéomètre m.
auto-réducteur. / zone of ~ ‖ Reduktions-
zone f. ‖ zone f. *ou* région f. de réduction.
reductive (Chem; Met) ‖ Reduktionsmittel
n. ‖ agent m. de réduction.
redwood extract ‖ Rotholzextrakt m. ‖
extrait m. de bois rouge.
re-dyeing ‖ Umfärben n. ‖ bisage m.
reed ‖ Ried n.; Schilf n.; Schilfrohr n. ‖
ros m.; roseau m.; canne f. / ~ (Mine) ‖
Gangspalte f.; Spalte f. ‖ fente f.; fissure f. / ~ (Mus) ‖ Rohrzunge f. ‖ anche f.
en roseau. / ~ (Weav) ‖ Rietkamm m.;
Webeblatt n. ‖ peigne m.; ros m. /
metallic ~ (Weav) ‖ Metallblatt n. ‖
peigne m. métallique. / organ ~ ‖ Orgel-
zunge f. ‖ anche f. pour orgue. / to provide a wall with ~s ‖ eine Wand f. aus-
rohren ‖ revêtir une muraille de roseaux.
/ ~ for warping frame ‖ Kamm m. für
Scherrahmen ‖ peigne m. d'ourdissoirs. /
~ for weaving looms ‖ Weberkamm m.;
Rietkamm m. ‖ rot m. pour métiers à
tisser.
reed articles pl. ‖ Rohrwaren fpl. ‖ articles
mpl. en roseau. / ~ blade (Weav) ‖ Riet-
stab m. ‖ lame f. à tisser. / bundle of ~s
for thatching ‖ Dachschaube f.; Strohschaube f. ‖ javelle f. / ~ maker (Weav)
‖ Webeblattsetzer m. ‖ peigneron m.
/ ~ mat ‖ Rohrmatte f. ‖ natte f. de
roseau. / ~ mounting (Weav) ‖ Blatt-
einsetzen n. ‖ montage m. de peignes à
tisser. / ~ pipe (Mus) ‖ Rohrflöte f.;
Rohrpfeife f.; Zungenpfeife f. ‖ tuyau m.
à anche. / ~ plaiter ‖ Rohrflechter m. ‖
tresseur m. de roseaux.
reed plane ‖ Stabhobel m. für nebeneinander liegende Rundstäbe ‖ rabot m. à
chantourner des baguettes accouplées. /
~ iron with square and cove ‖ Doppel-
karnieseisen n. ‖ fer m. de doucine à
couvrejoints.

reed roll ‖ Rietwalze f. ‖ cylindre m. à
laminer les ros. / ~ tissue ‖ Rohrgewebe
n. ‖ tissu m. de roseaux.
reef (Geol) ‖ Riff n.; Sandbank f.; Untiefe
f. ‖ récif m.; bas-fond m.; banc m. de
sable. / ~ (Sail) ‖ Reef n.; Reff n. ‖ ris m.
/ fringing ~ ‖ Küstenriff n. ‖ récif m.
côtier *ou* en bordure.
reef band ‖ Reffband n.; Reefband n. ‖
raban m. *ou* bande f. de ris. / ~ knot ‖
Weberknoten m.; Kreuzknoten m. ‖
nœud m. plat *ou* de tisserand. / ~ line ‖
Reffleine f. ‖ filière f. de ris. / ~ tackle ‖
Refftalje f. ‖ palan m. de ris.
reel, to (Spinn) ‖ abhaspeln; haspeln;
weifen ‖ dévider. / ~-off ‖ abhaspeln ‖
dévider. / ~-off from the bobbin ‖ von
der Spule f. abhaspeln ‖ dérouler de la
bobine.
reel (Mill) ‖ Beutelmaschine f.; Mehl-
maschine f. ‖ bluterie f. / ~ (Spinn) ‖
Haspel m.; Weife f. ‖ asplet m.; dévidoir m. / ~ of copying ribbon ‖ Farbbandspule f. ‖ bobine f. de ruban encré. /
~ of log ‖ Logrolle f. ‖ rouleau m. *ou* tour
m. du loch. / ~ of the rope-maker ‖
Haspel m. des Reepschlägers ‖ touret m.
du cordier. / wind-up and wind-off ~ ‖
Auf- und Abwickelhaspel m. ‖ enrouleur
m. et dévidoir m.
reel carriage ‖ Drahtrollenwagen m. ‖
chariot m. à bobines. / ~ drive ‖ Rollen-
antrieb m. ‖ commande f. de bobines.
reeler, hand ~ (Spinn) ‖ Handhaspeler m.
‖ dévideur m. à la main.
reel frame (Tedder) ‖ Trommelrahmen m.
‖ châssis m. porte-tambour.
reeling (Spinn) ‖ Haspeln n. ‖ dévidage m.
reeling machine ‖ Abhaspelmaschine f. ‖
machine f. à dévider. / ~ for paper ‖
Rollmaschine f. für Papier ‖ machine f.
enrouleuse pour papier. / reverse ~ for
stockinets ‖ Umkehraufrollmaschine f.
für Trikotstoffe ‖ machine f. à changement de marche pour enrouler les
tricotages.
reel insulator ‖ Isolierrolle f. ‖ roulette f.
isolante. / ~ jack ‖ Kabeltrommelwinde
f. ‖ vérin m. à vis de câble. / ~ lifting
device ‖ Rolleneinhebevorrichtung f. ‖
dispositif m. pour la mise en place des
bobines. / ~ star (Print) ‖ Rollenstern m.
‖ étoile f. à bobines. / ~ stump (Print) ‖
Rollenrest m. ‖ reste m. de bobine. / ~
tedder ‖ Trommelheuwender m. ‖
râteau-faneur m. à tambour.
reem, to ~ the seam (Shipb) ‖ die Naht f.
aufschlagen ‖ ouvrir la couture.
re-enter, to (Text print) ‖ eindrucken ‖
rentrer.
reep see reap.
reeve, to ~ a rope in a block ‖ ein Tau in
einen Block einscheren ‖ passer un cordage dans une poulie.
re-exportation ‖ Wiederausfuhr f. ‖ ré-
exportation f.
refer, to ~ to ‖ sich beziehen auf ‖ se rapporter à / ~ points pl. to a system of
coordinates ‖ Punkte mpl. auf ein Koordinatensystem beziehen ‖ rapporter
des points mpl. à un système de coordonnées.
referee ‖ Schiedsmann m. ‖ arbitre m.
reference ‖ Referenz n. ‖ référence f. / ~
(Print) ‖ Hinweis m. ‖ renvoi m. / ~
mark ‖ Anmerkungszeichen n. ‖ renvoi

m. de notes. / plane of ~ ‖ Einstellebene
f. ‖ plan m. de mise au point. / ~ title
of the catalogue ‖ Katalogbezeichnung
f. ‖ désignation f. du catalogue.
refill, to ‖ nachfüllen ‖ rajouter; remplir.
refilling ‖ Neubekleiden n. von Walzen ‖
regarnissage m. de cylindres.
refilling composition ‖ Füllmasse f. ‖ composition f. de remplissage.
refine, to (Distill; Glassm) ‖ läutern; rektifizieren ‖ épurer. / ~ (Metal) ‖ raffinieren; verfeinern; veredeln ‖ raffiner.
/ ~ oil ‖ Öl n. läutern *oder* raffinieren ‖
épurer l'huile f. / ~ saltpetre ‖ Salpeter m. läutern ‖ raffiner le salpêtre.
/ ~ steel ‖ Stahl m. garmachen *oder*
gärben *oder* raffinieren ‖ corroyer *ou*
raffiner l'acier m. / ~ sugar ‖ Zucker m.
raffinieren *oder* läutern *oder* sieden ‖
raffiner le sucre. / ~ tin ‖ Zinn gattern
‖ corroyer l'étain m.
refined ‖ raffiniert; geläutert; gar ‖ raffi-
niné; pur. / ~ lead ‖ raffiniertes Blei
n. ‖ plomb m. raffiné. / ~ pig ‖ raffiniertes Roheisen n. ‖ mazée f.; fonte f.
mazée. / ~ steel ‖ raffinierter Stahl m.;
Edelstahl m. ‖ acier m. fin *ou* raffiné. /
~ sugar ‖ raffinierter Zucker m. ‖ sucre
m. raffiné. / ~ zone (Met) ‖ Übergangszone f. ‖ zone f. raffinée.
refinement of metal surface *see also* refining
‖ Behandlung f. *oder* Verfeinerung f. der
Metalloberfläche ‖ traitement m. de la
surface des métaux.
refiner with rollers (Bak) ‖ Walzenreib-
maschine f. ‖ broyeuse f. *ou* broyeur m.
à rouleaux. / ~ master (Metal) ‖ Frisch-
meister m. ‖ maître m. affineur.
refinery ‖ Raffinerie f. ‖ raffinerie f. / ~
(Metal) ‖ Raffinierfeuer n.; Feineisen-
feuer n. ‖ finerie f.; fourneau m. de
finerie *ou* de raffinerie. / sugar ~ ‖ Zuckerraffinerie f. ‖ raffinerie f. de sucre. / distillation and ~ of tar ‖ Teerzerlegung f.
‖ traitement m. du goudron.
refinery plant ‖ Raffinieranlage f. ‖ installation f. de raffinerie.
refining (Chem) ‖ Scheidung f.; Raffi-
nierung f. ‖ affinage m.; raffinage m.
/ ~ (Glassm) ‖ Läutern n. ‖ affinage m.;
raffinage m. / ~ (Metal) ‖ Raffinieren n.;
Läuterung f. ‖ mazéage m.; raffinage m.;
affinage m. / ~ (Steel) ‖ Gärben n. ‖
corroyage m. / ~ of copper containing
lead ‖ Raffination f. bleihaltigen Kupfers
‖ raffinage m. du cuivre contenant du
plomb. / ~ in low hearth ‖ Herdfrischen
m. ‖ affinage m. au bas-foyer. / ~ of
metals ‖ Läutern n. *oder* Reinigen n. der
Metalle ‖ affinage m. des métaux. / ~
of pig-iron ‖ Läutern n. des Roheisens ‖
raffinage m. de la fonte. / ~ in the reverberatory furnace ‖ Flammofenfrischen
n. ‖ affinage m. de la fonte au four à
réverbère. / ~ of saltpetre ‖ Reinigung f.
des Salpeters ‖ raffinage m. du salpêtre.
/ ~ of silver ‖ Silberscheidung f. ‖ affinage m. de l'argent; coupellation f. / ~
of steel ‖ Raffinieren n. *oder* Gärben n.
des Stahls ‖ affinage m. de l'acier. / ~ by
the wet way ‖ Scheidung f. auf nassem
Wege ‖ affinage m. par la voie humide.
refining assay (Metal) ‖ Garprobe f. ‖
essai m. de raffinage. / ~ boiler (Salt-
peter) ‖ Läuterungskessel m. ‖ chaudière f. de raffinage. / ~ charges pl. ‖
Raffinierungskosten pl. ‖ frais mpl. de
raffinage. / ~ cinders pl. ‖ Frischschlacke

f.; Schmiedesinter m. ‖ scorie f. des feux d'affinerie; laitier m. de forge. / ~ furnace ‖ Raffinierofen m. ‖ four m. d'affinage. / ~ hearth ‖ Frischherd m.; Garherd m. ‖ forge f. d'affinerie. / machinery for ~ steel strip ‖ Bandstahlverfeinerungsmaschine f. ‖ affineuse f. mécanique pour feuillards d'acier. / ~ man ‖ Eisenfrischer m. ‖ affineur m. / ~ plant ‖ Raffinationsanlage f. ‖ installation f. de raffinerie. / ~ plant for vegetable oils ‖ Raffinationsanlage f. für vegetabilische Öle ‖ installation f. de raffinage d'huiles végétales. / ~ process ‖ Veredelungsvorgang m.; Veredelungsverfahren n. ‖ opération f. ou procédé m. de raffinage.

refit, to ‖ wiederinstandsetzen ‖ réparer.
refitment ‖ Wiederinstandsetzung f. ‖ réparation f.
refitted ‖ wiederinstandgesetzt ‖ réparé.
reflatten, to ~ wire ‖ den Draht m. nachplätten ‖ repasser le fil.
reflect, to ‖ zurückwerfen ‖ réfléchir.
reflected ray ‖ reflektierter oder zurückgeworfener Strahl m. ‖ rayon m. réfléchi.
reflecting ‖ reflektierend ‖ réfléchissant. / ~ circle ‖ Reflexionszirkel m.; Reflexionskreis m. ‖ cercle m. de réflexion. / ~ exophthalmometer ‖ Spiegelexophthalmometer n. ‖ exophtalmomètre m. à miroirs. / ~ galvanometer ‖ Spiegelgalvanometer n. ‖ galvanomètre m. à miroir. / ~ grating ‖ Reflexionsgitter n. ‖ réseau m. réflecteur. / ~ instrument ‖ Spiegelinstrument n. ‖ instrument m. à réflecteur. / ~ level ‖ Spiegelwage f. ‖ niveau m. réflecteur. / ~ power ‖ Reflexionsvermögen n. ‖ pouvoir m. réflecteur ou réfléchissant.
reflecting prism ‖ Spiegelprisma n. ‖ prisme m. réflecteur ou à réflexion. / ~ in swingout mount ‖ wegklappbares Reflexionsprisma n. ‖ prisme m. réflecteur amovible.
reflecting screen ‖ reflektierender Schirm m. ‖ écran m. réflecteur. / ~ spectrometer for heat rays ‖ Spiegelspektroskop n. für Wärmestrahlen ‖ spectromètre m. à miroirs pour les radiations calorifiques. / ~ spectroscope ‖ Spiegelspektroskop n. ‖ spectroscope m. à miroirs. / ~ surface ‖ Reflexionsfläche f. ‖ surface f. réfléchissante. / ~ telescope ‖ Spiegelteleskop n. ‖ télescope m. à réflexion. / ~ wall ‖ spiegelnde Wand f. ‖ paroi f. miroitante.
reflection ‖ Zurückwerfung f. ‖ réflexion f. / absence of ~ in the image ‖ Reflexlosigkeit f. des Bildes ‖ absence f. de reflets dans l'image. / total ~ ‖ Totalreflexion f. ‖ réflexion f. totale.
reflection factor ‖ Reflexionsfaktor m. ‖ facteur m. de réflexion. / ~ loss ‖ Reflexionsverlust m. ‖ perte f. par réflexion.
reflector ‖ Reflektor m.; Scheinwerfer m.; Rückstrahler m.; Reflexionsspiegel m. ‖ réflecteur m.; projecteur m.; miroir m. réflecteur. / ceiling ~ ‖ Deckenreflektor m.; Deckenlichtspiegler m. ‖ réflecteur m. de plafond. / elliptical ~ ‖ elliptischer Reflexionsspiegel m. ‖ miroir m. plan elliptique. / glow-lamp ~ ‖ Glühlampenreflektor m.; Glühlampenlichtspiegler m. ‖ réflecteur m. de lampe à incandescence. / ~ of headlamp ‖ Reflektor m. des Scheinwerfers ‖ miroir-projecteur m. de phare. / parabolic ~ ‖ Parabolspiegel m. ‖ miroir m. ou réflecteur m. parabolique.

reflector tube (Telescope) ‖ Spiegelrohr n. ‖ tube m. du télescope. / ~ with silvered glass reflector ‖ Spiegelrohr n. mit Glassilberspiegel ‖ tube m. du télescope muni d'un miroir en verre argenté. / ~ clip ‖ Spiegelrohrhalter m. ‖ armature f. portant le tube du réflecteur.
reflex ‖ Reflex m. ‖ réflexe m. / undisturbed by ~es pl. ‖ reflexfrei ‖ sans réflexe m.
reflexion see reflection.
reflex receiver ‖ Reflexempfänger m. ‖ récepteur m. réflexe.
reflux ‖ Rückfluß m. ‖ reflux m. / with ~ ‖ rückfließend ‖ à reflux.
reform, to ‖ umgestalten ‖ transformer.
refound, to ‖ umgießen ‖ refondre.
refract, to (Ray) ‖ brechen; ablenken ‖ réfracter.
refracted ray ‖ gebrochener Strahl m. ‖ rayon m. réfracté.
refracting ‖ brechend ‖ réfringent. / double ~ ‖ doppelbrechend ‖ biréfringent.
refracting angle ‖ brechender Winkel m. ‖ angle m. réfringent. / ~ power ‖ Brechungsvermögen n. ‖ pouvoir m. réfringent.
refraction ‖ Brechung f. ‖ réfraction f. / ~ (Power) ‖ Brechungskraft f.; Ablenkungskraft f. ‖ force f. de réfraction. / double ~ ‖ Doppelbrechung f. ‖ double réfraction f.; biréfringence f. / ~ of light ‖ Lichtbrechung f.; Strahlenbrechung f. ‖ réfraction f. de la lumière.
refraction index ‖ Brechungsverhältnis n. ‖ indice m. de réfraction. / ~ law ‖ Brechungsgesetz n. ‖ loi f. de la réfraction.
refractive ‖ optisch dicht; brechend ‖ réfringent.
refractive index ‖ Brechungsindex m.; Brechungsexponent m. ‖ indice m. de réfraction.
refractive power ‖ Brechungsvermögen n.; Brechkraft f. ‖ pouvoir m. réfringent; réfringence f. / ~ of different types of glass ‖ Brechungsvermögen n. der verschiedenen Glassorten ‖ réfringence f. des diverses sortes de verre. / ~ of a gem ‖ Lichtbrechungsvermögen n. eines Edelsteins ‖ indice m. de réfraction d'une pierre précieuse.
refractometer ‖ Refraktometer n. ‖ réfractomètre m. / crystal ~ ‖ Kristallrefraktometer n. ‖ réfractomètre m. à cristaux. / dipping ~ see immersion ~. / hemisphere crystal ~ ‖ Halbkugelkristallrefraktometer n. ‖ réfractomètre m. demi-boule à cristaux. / immersion ~ ‖ Eintauchrefraktometer n. ‖ réfractomètre m. à immersion. / industrial ~ ‖ Betriebsrefraktometer n. ‖ réfractomètre m. industriel. / jeweller's ~ ‖ Juwelierrefraktometer n. ‖ réfractomètre m. de bijoutier. / milk fat ~ ‖ Milchfettrefraktometer n. ‖ réfractomètre m. à crème de lait. / parallax ~ ‖ Parallaxenrefraktometer n. ‖ réfractomètre m. à parallaxe. / pocket ~ ‖ Taschenrefraktometer n. ‖ réfractomètre m. de poche. / sugar ~ ‖ Zuckerrefraktometer m. ‖ réfractomètre m. à sucre. / ~ with variable refracting angle ‖ Refraktometer n. mit veränderlichem brechendem Winkel ‖ réfractomètre m. à angle réfringent variable.
refractor ‖ Refraktor m. ‖ réfracteur m. / ~ for testing the optical quality of astronomical instruments ‖ Refraktor m. zur Prüfung astronomischer Optik ‖ ré-

fracteur m. pour l'examen de l'optique astronomique. / photographic ~ ‖ fotografischer Refraktor m. ‖ réfracteur m. photographique. / double photographic ~ ‖ fotografischer Doppelrefraktor m. ‖ réfracteur m. double photographique. / ~ of special mechanical construction ‖ Refraktor m. von besonderer mechanischer Konstruktion ‖ réfracteur m. d'une construction mécanique spéciale. / ~ with stress relieving system to telescopes and to equatorial axes ‖ Refraktor m. mit Entlastungssystem der Fernrohre und Achsen ‖ réfracteur m. avec dispositif pour décharger les lunettes et les axes.
refractoriness ‖ Feuerfestigkeit f.; schwere Schmelzbarkeit f. ‖ qualité f. ou nature f. réfractaire.
refractor standard with equatorial head ‖ Refraktorsäule f. mit parallaktischem Achsensystem ‖ colonne f. de réfracteur portant la monture équatoriale.
refractory ‖ feuerbeständig ‖ réfractaire; apyre. / ~ (Met) ‖ schwer schmelzbar; strengflüssig ‖ réfractaire. / ~ brick ‖ feuerfester Stein m.; Schamottstein m. ‖ brique f. réfractaire. / ~ clay ‖ Schamotte f. ‖ argile m. réfractaire. / ~ cooking utensil ‖ feuerfestes Kochgeschirr n. ‖ batterie f. de cuisine en poterie réfractaire. / ~ sand ‖ feuerfester Sand m. ‖ sable m. réfractaire. / ~ stone ‖ feuerfester Stein m. ‖ brique f. ou pierre f. réfractaire.
refractory ‖ feuerfestes Erzeugnis n. ‖ produit m. réfractaire.
refrangibility ‖ Brechungsvermögen n.; Brechbarkeit f. ‖ réfrangibilité f.; indice m. de réfraction.
refreshing drink ‖ erfrischendes Getränk n. ‖ boisson f. rafraîchissante.
refrigerate, to ‖ abkühlen ‖ refroidir; réfrigérer.
refrigerated boat ‖ Kühlschiff n. ‖ bateau m. frigorifique.
refrigerating ‖ Kühlung f. ‖ réfrigération f. / ~ agent ‖ Kältemedium n. ‖ agent m. réfrigérant. / ~ capacity ‖ Kälteleistung f. ‖ puissance f. frigorifique. / ~ industry ‖ Kälteindustrie f. ‖ industrie f. frigorifique.
refrigerating machine ‖ Kältemaschine f.; Eismaschine f. ‖ machine f. réfrigérante ou frigorifique ou à froid. / ammonia ~ ‖ Ammoniakkältemaschine f. ‖ machine f. frigorifique à ammoniaque. / ether ~ ‖ Ätherkühlmaschine f. ‖ machine f. frigorifique à éther. / ~ for manufacturing ice cream ‖ Eismaschine f. für Speiseeisbereitung ‖ machine f. frigorifique pour la fabrication de glace alimentaire. / large ~ ‖ Großkältemaschine f. ‖ machine f. frigorifique de grande puissance. / small ~ ‖ Kleinkältemaschine f. ‖ machine f. frigorifique de faible puissance.
refrigerating machinery ‖ Kältemaschinenanlage f. ‖ installation f. de machines frigorifiques.
refrigerating medium ‖ Kältemittel n. ‖ agent m. frigorifique. / ~ mixture of salt and crushed ice ‖ Kältemischung f. aus Salz und zerkleinertem Eis ‖ mélange m. frigorifique de sel et de glace concassée. / ~ plant ‖ Gefrieranlage f.; Kühlwerk n. ‖ installation f. de congélation; usine f. frigorifique. / sinking to the ~ process ‖ Schachtabteufung f. nach dem Gefrierverfahren ‖ creusement m.

de puits de mine par congélation. / ~ unit ‖ Kälteeinheit f. ‖ frigorie f.

refrigeration ‖ Kälteerzeugung f.; Kühlung f. ‖ production f. du froid; réfrigération f. / mechanic ~ ‖ maschinelle Kühlung f. ‖ réfrigération f. mécanique. / producing mechanic ~ at reasonable costs ‖ wohlfeile Erzeugung f. künstlicher Kälte ‖ production f. rationnelle du froid artificiel.

refrigerator ‖ Abkühlapparat m. ‖ réfrigérant m.; réfroidisseur m. / automatic ~ ‖ Kälteautomat m. ‖ automate m. frigorifique. / electric ~ ‖ elektrischer Kühlschrank m. ‖ armoire f. réfrigérante électrique.

refrigerator capacity ‖ Kühlvermögen n. ‖ capacité f. de l'appareil frigorifique. / ~ car ‖ Kühlwagen m. ‖ wagon m. frigorifique.

refrigeratory *see* refrigerator.

refuel, to ‖ tanken; neuen Brennstoff m. einnehmen ‖ ravitailler (en combustible).

refuelling ‖ Tanken n.; Einnehmen n. von neuem Brennstoff ‖ ravitaillement m.

refund, to ‖ rückvergüten ‖ bonifier.

refund of taxes ‖ Rückvergütung f. von Steuern ‖ bonification f. des contributions.

refusal ‖ abschlägige Antwort f.; Absage f. ‖ réponse f. défavorable; refus m. / ~ of acceptance ‖ Annahmeverweigerung f. ‖ refus m. d'acceptation.

refuse, to ‖ ausschlagen; abschlägig bescheiden ‖ refuser; écarter une demande. / ~ the service ‖ den Dienst m. verweigern ‖ refuser le service.

refuse (Mach tool) ‖ Abfall m.; Ausschuß m. ‖ déchet m.; rognures fpl. / ~ (Metal) ‖ Schlacke f. ‖ scorie f. / ~ (Pap) ‖ Altpapier n. ‖ papier m. à la cuve; vieux papiers mpl. / ~ (Print) ‖ Papierausschuß m.; Makulatur f. ‖ bardot m. / ~ (Public hygiene) ‖ Abfallstoff m. ‖ excrément m. / ~ (Street) ‖ Müll n. ‖ gadoue f.; balayures fpl. / ~ (Sweepings) ‖ Hausmüll n.; Kehricht m. ‖ ordures fpl.; balayures fpl. / animal ~ ‖ tierischer Abfall m. ‖ déchet m. d'animaux. / ~ of leather ‖ Lederabfall m. ‖ déchet m. de cuir.

refuse burning furnace ‖ Müllverbrennungsofen m. ‖ four m. à brûler les ordures. / ~ destructor ‖ Müllverbrennung f.; Müllfeuerung f. ‖ incinérateur m. d'ordures; foyer m. à ordures. / ~ exhauster ‖ Müllsauger m. ‖ aspirateur m. d'ordures. / ~ olive oil ‖ Olivennachöl. n. ‖ huile f. d'olive de ressence. / removal of ~ ‖ Müllabfuhr f. ‖ enlèvement m. des ordures ménagères. / ~ removing cart ‖ Straßenkehrichtabfuhrwagen m.; Schlammabfuhrwagen m. ‖ chariot m. pour l'enlèvement des ordures. / ~ transportation ‖ Müllabfuhr f. ‖ transport m. de gadoues. / ~ wood ‖ Wildholz n. ‖ bois m. de rebut. / ~ wool ‖ Zackelwolle f. ‖ laine f. de rebut.

regard, with ~ to ‖ bezogen auf; rücksichtlich ‖ relativement à; à l'égard m. de.

regarding ‖ bezüglich ‖ concernant; touchant.

regatta ‖ Regatta f.; Bootwettfahrt f. ‖ régate f.

regenerable cell (Electr) ‖ aufladbares Element m. ‖ pile f. régénérable.

regenerate, to (Chem) ‖ regenerieren; auffrischen ‖ régénérer; récupérer. / ~ oil by the gravity settling process after it has been used for lubricating purposes ‖ gebrauchtes Schmieröl n. durch das Absitzverfahren reinigen ‖ régénérer par décantation l'huile ayant déjà servi au graissage.

regenerating *see* regeneration.

regeneration (Chem) ‖ Regenerierung f.; Auffrischung f. ‖ régénération. f. / ~ of current ‖ Stromrückgewinnung f. ‖ récupération f. de courant.

regenerative circuit (Radio) ‖ Rückkopplung f. ‖ couplage m. en arrière; rétroaction f. / ~ firing ‖ Regenerativfeuerung f. ‖ chauffage m. à régénération. / ~ gas firing ‖ Regenerativgasfeuerung f. ‖ chauffage m. à régénération à gaz. / ~ gas furnace ‖ Regenerativgasofen m. ‖ four m. régénérateur à gaz. / ~ heating for the crucible melting process ‖ Regenerativfeuerung f. für den Tiegelschmelzprozeß ‖ utilisation f. des régénérateurs à gaz pour le procédé de la fusion au creuset.

regenerator (Metal) ‖ Regenerator m.; Regeneratorofen m.; Regenerativofen m. ‖ four m. régénérateur au gaz.

regime of a river ‖ Wasserstandsverhältnisse fpl. eines Flusses ‖ régime m. d'un fleuve.

region of downward currents ‖ Abwindfeld n. ‖ champ m. de vent descendant. / ~ of invisible radiation ‖ unsichtbarer Strahlenbereich m. ‖ région f. des radiations invisibles. / ~ of the spectrum ‖ Spektralbereich m. ‖ région f. du spectre.

register, to ‖ registrieren ‖ enregistrer. / ~ the luggage ‖ Gepäck n. aufgeben ‖ faire enregistrer les bagages mpl.

register (Boil) ‖ Luftschieber m.; Zugschieber m.; Rauchklappe f. ‖ registre m. de tirage. / ~ (Book) ‖ Registerbuch n. ‖ registre m. / ~ (Organ) ‖ Register n.; Orgelzug m. ‖ registre m.; jeu m. / ~ (Print) ‖ Register n. ‖ registre m. / cash ~ ‖ Registrierkasse f. ‖ caisse f. enregistreuse. / ~ closed (Cash register) ‖ Kasse f. abgeschlossen ‖ caisse f. fermée. / ~ of contents ‖ Inhaltsverzeichnis n.; Register n. ‖ index m.; table f. des matières. / ~ issues duplicate checks ‖ die Kasse gibt den Scheck mit Doppelaufdruck aus ‖ la caisse émet un ticket double. / Lloyd's ~ ‖ Lloydregister n. ‖ registre m. de Lloyd.

register controller (Aut tel) ‖ Steuerschalter m. ‖ commutateur m. aiguilleur.

registered address ‖ Telegrammkurzanschrift f. ‖ adresse f. conventionnelle. / ~ capital ‖ eingetragenes Kapital n. ‖ capital social *ou* nominal. / ~ depth (Shipb) ‖ Vermessungstiefe f. ‖ creux m. du registre. / ~ design ‖ Gebrauchsmuster n. ‖ brevet m. pour modèles d'utilité; modèle m. déposé. / ~ letter ‖ eingeschriebener Brief m.; Einschreibbrief m. ‖ lettre f. recommandée. / ~ name ‖ gesetzlich geschützter Name m. ‖ nom m. déposé. / ~ office ‖ Hauptniederlassung f.; Sitz m.; siège m. social. / ~ pattern ‖ Musterschutz m. ‖ modèle m. d'utilité. / ~ share ‖ auf den Namen lautendes Papier n. ‖ titre m. nominatif. / ~ trade mark ‖ eingetragene Schutzmarke f. ‖ marque f. de fabrique enregistrée.

registering ‖ registrierend ‖ enregistreur.

registering of luggage ‖ Gepäckannahme f. ‖ engregistrement m. des bagages.

registering apparatus ‖ Registrierapparat m. ‖ enregistreur m. / apparatus for ~ rapidly passing occurrences ‖ Registriergerät n. für schnell verlaufende Vorgänge ‖ appareil m. enregistreur pour événements rapides. / ~ balloon ‖ Registrierballon m. ‖ ballon m. enregistreur *ou* explorateur. / ~ instrument ‖ Registrierinstrument n. ‖ appareil m. enregistreur. / ~ key ‖ Registriertaste f. ‖ touche f. d'enregistrement.

register sheet (Print) ‖ Registerbogen m.; Zurichtebogen m. ‖ feuille f. de registre. / ~ system (Aut tel) ‖ Registersystem n. ‖ système m. muni d'enregistreurs. / ~ ton ‖ Registertonne f. ‖ tonneau m. de registre.

registrar of mortgages ‖ Hypothekenverwalter m. ‖ conservateur m. des hypothèques.

registration ‖ Eintragung f.; Registrierung f. ‖ enregistrement m. / ~ of the felling ‖ Holzaufnahme f. ‖ inventaire m. de la coupe de bois.

registration fee ‖ Einschreibegebühr f. ‖ taxe f. de recommandation. / ~ form ‖ Meldezettel m. ‖ lettre f. d'avis. / ~ office ‖ Meldeamt n. ‖ bureau m. de police *ou* du recensement.

registry, certificate of ~ ‖ Beilbrief m. / certificat m. de construction. / ship's certificate of ~ ‖ Schiffszertifikat n. ‖ acte m. de nationalité.

reglet (Arch) ‖ Leistchen n.; Riemchen n. ‖ feuillet m. / ~ (Print) ‖ Holzreglette f. ‖ réglette f.

regrind, to ‖ nachschleifen ‖ rebroyer; remoudre. / ~ a valve ‖ ein Ventil n. nachschleifen ‖ roder une soupape.

regrinding ‖ Nachschleifen n. ; rodage m. / knife ~ ‖ Messerschleifen n. ‖ repassage m. de couteaux. / scissors ~ ‖ Scherenschleifen n. ‖ repassage m. de ciseaux.

regular ‖ regulär; regelmäßig ‖ régulier. / ~ alignment of the letters ‖ Zeilengeradheit f. ‖ rectitude f. des lignes. / ~ business connection ‖ regelmäßige Geschäftsverbindung f. ‖ relations fpl. d'affaires suivies. / ~ delivery ‖ regelmäßige Lieferung f. ‖ fourniture f. régulière. / ~ parting ‖ gleichmäßige Teilung f. ‖ division f. régulière *ou* proportionnelle.

regularity ‖ Stetigkeit f. ‖ régularité f.; continuité f.; constance f.

regulate, to ‖ regeln ‖ régler.

regulating *see also* regulation ‖ Regulierung f. ‖ réglage m./~the clockwork ‖ Regelung f. des Uhrwerks ‖ réglage m. du mouvement d'horlogerie. / ~ of the steam distribution ‖ Regulierung f. der Dampfverteilung ‖ réglage m. de la distribution de vapeur.

regulating ‖ regelnd ‖ régulateur. / ~ apparatus ‖ Regelapparat m. ‖ appareil m. régulateur. / apparatus ~ the circulation of tissue ‖ Stofflaufregelapparat m. ‖ dispositif m. pour le réglage de la marche du tissu introducteur m. / ~ cell ‖ Zellenschalterelement n. ‖ élement de réduction *ou* de réglage. / ~ cock ‖ Regelhahn m. ‖ robinet m. modérateur *ou* régulateur. / ~ cock for radiators ‖ Regelhahn m. für Heizkörper ‖ robinet m. de réglage pour radiateurs. / ~ device for the fabric ‖

Stofflaufregelapparat m. ‖ dispositif m. pour le réglage de la marche du tissu; introducteur m. / ~ feed valve ‖ Schieberdruckregler m. ‖ soupape f. d'alimentation automatique. / ~ plane ‖ Lageplan m; Bauebene f. ‖ plan m. de site. / ~ resistance ‖ Regelwiderstand m. ‖ résistance f. réglable. / ~ rod (Wind mill) ‖ Streichstange f. ‖ perche f. / ~ sector ‖ Regelsegment n. ‖ secteur m. de réglage. / ~ shaft ‖ Reglerwelle f. ‖ arbre m. de réglage. / ~ siphon ‖ Regelsiphon m.; Ablaßdücker m. ‖ épanchoir m. à siphon. / ~ transformer ‖ Regeltransformator m. ‖ transformateur m. de régulation. / ~ valve ‖ Reduzierventil n. ‖ détendeur m.

regulation ‖ Regelung f. ‖ réglage m. / ~ (Order) ‖ Verordnung f. ‖ règlement m. / according to ~s pl. ‖ vorschriftsmäßig ‖ selon les règlements mpl. / air ~ ‖ Luftregelung f. ‖ réglage m. d'air. / contrary to ~s pl. ‖ vorschriftswidrig ‖ contraire aux prescriptions fpl. *ou* aux règlements mpl. / for ~ ‖ einstellbar ‖ réglable. / ~ of pressure ‖ Druckregelung f. ‖ réglage m. de la pression. / automatic ~ of the wind turbine ‖ Selbstregelung f. der Windturbine ‖ régulation f. automatique de la turbine aérienne. / ~s pl. relating to exchange ‖ Wechselordnung f. ‖ règlements mpl. *ou* loi f. sur les lettres de change.

regulation step ‖ Regelstufe f. ‖ degré m. de réglage.

regulator ‖ Regler m. ‖ régulateur m.; régleur m. / ~ (Clockm) ‖ Rücker m. ‖ raquette f. / automatic ~ ‖ selbsttätiger Regler m. ‖ régulateur m. automatique. / draught ~ ‖ Zugregler m. ‖ régulateur m. de tirage. / electric ~ ‖ elektrischer Regler m. ‖ régulateur m. électrique. / exhaust-steam pressure ~ ‖ Abdampfdruckregler m. ‖ régulateur m. de pression de la vapeur d'échappement. / feed water ~ ‖ Speisewasserregler m. ‖ régulateur m. d'alimentation. / gas ~ ‖ Gasregler m. ‖ régulateur m. de gaz. / ~ at the head of a channel ‖ Einlaßschleuse f. eines Bewässerungskanals ‖ martellière f. / pressure ~ ‖ Druckregler m. ‖ régulateur m. de pression. / ~ for stage-lighting ‖ Bühnenlichtregler m. ‖ régulateur m. d'éclairage de scène. / temperature ~ ‖ Temperaturregler m. ‖ régulateur m. de température. / turbine ~ ‖ Turbinenregler m. ‖ régulateur m. de turbine.

regulator, main ~ board ‖ Hauptreglertafel f. ‖ tableau m. du régulateur principal. / ~ slide ‖ Reglerschieber m. ‖ tiroir m. de régulateur. / ~ valve rod ‖ Zugstange f. des Reglers ‖ tringle f. de manœuvre du régulateur.

reguline (Chem) ‖ regulinisch ‖ régulin.

regulus (Chem) ‖ Metallkorn n. ‖ culot m. / ~ (Metal) ‖ Speise f. ‖ speiss m. / ~ of copper ‖ Kupferstein m. ‖ matte f. de cuivre.

rehabilitate, to ‖ rehabilitieren ‖ réhabiliter.

rehardened ‖ von neuem gehärtet ‖ trempé à nouveau.

reheat, to ‖ wiedererhitzen ‖ réchauffer. / ~ (Steel) ‖ anlassen; nachlassen; tempern ‖ adoucir; ramollir; recuire.

reheating ‖ nochmalige Erwärmung f. ‖ réchauffage m. / ~ fire ‖ Schweißfeuer n. ‖ foyer m. *ou* feu m. à réchauffer.

reheating furnace ‖ Wärmeofen m. ‖ four m. à réchauffer. / ~ for iron wires ‖ Glühofen m. für Eisendrähte ‖ four m. à recuire les fils métalliques.

reheating furnace slag *see* ~ scoria. / ~ hearth ‖ Glühofen m. ‖ four m. réchauffeur. / ~ scoria ‖ Schweißofenschlacke f. ‖ scorie f. de réchauffage.

reimburse, to ‖ entschädigen ‖ dédommager; indemniser.

reimbursement ‖ Rückvergütung f.; Rückzahlung f. ‖ remboursement m.; indemnisation f. / ~ (Mail) ‖ Nachnahme f. ‖ remboursement m. / annual ~ ‖ Jahresabrechnung f. ‖ liquidation f. de la fin d'année. / ~ of expenses ‖ Spesennachnahme f. ‖ remboursement m. des frais.

reimportation ‖ Wiedereinfuhr f. ‖ réimportation f.

reimpression (Print) ‖ Neuauflage f. ‖ réimpression f.

rein ‖ Zaum m.; Zügel m. ‖ bride f.; rêne f. / ~ of a blacksmith's tongs ‖ Schmiedezangengriff m. ‖ branche f. d'une tenaille de forgeron.

reinforce, to ‖ verstärken; bewehren ‖ renforcer; armer. / ~ at the bearings ‖ an den Auflagern npl. verstärken ‖ renforcer aux appuis mpl.

reinforced ‖ verstärkt ‖ renforcé. / doubly ~ ‖ doppelt bewehrt ‖ à double armature f.

reinforced concrete ‖ Eisenbeton m. ‖ béton m. armé. / ~ bending machine ‖ Betonbiegemaschine f. ‖ machine f. servant à cintrer le béton. / ~ building-company ‖ Eisenbetonbaugesellschaft f. ‖ société f. de construction en béton armé. / ~ construction ‖ Eisenbetonbau m. ‖ construction f. en béton armé. / ~ floor ‖ Eisenbetondecke f. ‖ plancher m. en béton armé. / ~ pile ‖ Eisenbetonpfahl m. ‖ pilier m. *ou* pieu m. en béton armé. / ~ ship ‖ Eisenbetonschiff n. ‖ navire m. en béton armé. / skeleton of ~ ‖ Eisenbetongerippe n. ‖ ossature f. en béton armé. / tube of ~ ‖ Eisenbetonrohr n. ‖ tuyau m. en béton armé.

reinforced seam ‖ Wulstnaht f. ‖ joint m. renforcé *ou* surépaissé.

reinforcement ‖ Verstärkung f. ‖ renforcement m. / ~ (Met) ‖ Wulst m. ‖ bourrelet m.; renflement m. / diagonal ~ ‖ Diagonalbewehrung f. ‖ armature f. dans le sens diagonal.

reinforcement plate ‖ Verstärkungsplatte f. ‖ plaque f. de renfort.

reinforcing insole for boots and shoes ‖ Schuhgelenkeinlage f. ‖ garniture f. pour jointures de chaussures. / ~ iron ‖ Betoneisen n. ‖ fer m. à béton. / ~ metal ‖ Armierungseisen n. ‖ barre f. d'armature. / ~ strengthener (Med) ‖ Verstärkungseinlage f. ‖ renfort m. pour dentiers.

reinstate, to ‖ rehabilitieren ‖ réhabiliter.

reinsurance ‖ Rückversicherung f. ‖ réassurance f. / amount of ~ ‖ Rückversicherungsbetrag m. ‖ montant m. de réassurance. / ~ policy ‖ Rückversicherungspolice f. ‖ police f. de réassurance. / ~ premium ‖ Rückversicherungsprämie f. ‖ prime f. de réassurance.

reiteration (Print) ‖ Widerdruck m. ‖ retiration f.; réimpression f.; impression f. au verso.

reject, to ‖ verwerfen ‖ rejeter; rebuter. / ~ a cheque ‖ einen Scheck m. zurückweisen ‖ refuser un chèque.

rejections pl. ‖ Ausschußware f. ‖ marchandises fpl. de rebut.

rejector circuit ‖ Drosselkreis m. ‖ circuit m. de réactance.

rejoinder ‖ Gegenschrift f. ‖ réplique f.; réfutation f.

relapse ‖ Rückschlag m. ‖ contre-coup m.; revirement m.

relate, to ~ to ‖ sich beziehen auf ‖ se rapporter à.

relating to business ‖ geschäftlich ‖ concernant les affaires fpl.; commercial.

relation (Math) ‖ Relation f.; Verhältnis n. ‖ relation f. / in ~ to ‖ bezogen auf ‖ relativement à.

relationship, mutual ~ ‖ Wechselbeziehung f. ‖ rapport m. réciproque.

relative ‖ relativ ‖ relatif. / ~ positions pl. ‖ gegenseitige Lage f. ‖ position f. réciproque.

relativity, theory of ~ ‖ Relativitätstheorie f. ‖ théorie f. de la relativité.

relax, to ‖ erschlaffen ‖ relâcher.

relaxed state of accommodation ‖ Akkommodationsruhe f. ‖ relâchement m. de l'accommodation.

relay, to (Build) ‖ umlegen ‖ remanier. / ~ completely ‖ vollständig umlegen ‖ remanier à bout. / ~ the paving of a street ‖ eine Straße f. umpflastern ‖ remanier le pavé d'une rue. / ~ a roof ‖ ein Dach n. umdecken ‖ reposer une toiture.

relay (Mach) ‖ Hilfstriebwerk n. ‖ servomoteur m.; relais-moteur m. / ~ (Tel) ‖ Relais n.; Schütz n. ‖ relais m. / absence-of-current ~ ‖ Stromlosigkeitsrelais n. ‖ relais m. de manque de courant. / alarm ~ for plate potential ‖ Anodenalarmrelais n. ‖ relais m. de contrôle du potentiel de plaque. / ~ without armature ‖ ankerloses Relais n. ‖ relais m. sans armature. / change-over ~ ‖ Umschalterelais n. ‖ relais m. commutateur. / clearing ~ ‖ Schlußrelais n. ‖ relais m. de fin de conversation. / ~ of clearing section ‖ Abrückrelais n. ‖ relais m. débloqueur. / cut-off ~ (Tel) ‖ Trennrelais n. ‖ relais m. de coupure. / deferring ~ ‖ Zeitrelais n. ‖ relais m. à action différée. / discriminating ~ ‖ Selektivschutz m. ‖ protection f. par relais discriminateur. / electrostatical ~ ‖ elektrostatisches Relais n. ‖ relais m. électrostatique. / fuse supervisory ~ ‖ Sicherungskontrollrelais n. ‖ relais m. pilote avertisseur. / ~ for heavy current ‖ Fernschalter m. ‖ relais m. à courant de grande intensité. / high-tension ~ ‖ Hochspannungsrelais n. ‖ relais m. pour haute tension. / ~ with horizontal coils ‖ liegendes Relais n. ‖ relais m. à bobines horizontales. / hot-wire ~ ‖ Hitzdrahtrelais n. ‖ relais m. thermique. / impedance ~ ‖ Impedanzrelais n. ‖ relais m. à impédance. / induction-current ~ ‖ Induktionsrelais n. ‖ relais m. à courant d'induction. / meter ~ ‖ Zählerrelais n. ‖ relais m. de compteur. / neutral ~ ‖ neutrales Relais n. ‖ relais m. neutre. / pilot ~ ‖ Steuerrelais n. ‖ relais m. de contrôle. / polarized ~ ‖ polarisiertes Relais n. ‖ relais m. polarisé. / the ~ releases ‖ das Relais n. fällt ab ‖ le relais m. laisse retomber son armature. / repeating ~ ‖ Übertragungsrelais n. ‖ relais m. translateur; répétiteur m. / ringing ~ ‖ Läuterelais n. ‖ relais m. d'appel. / slow-release ~ ‖ Verzögerungsrelais n. ‖ relais m. à fonc-

tionnement retardé. / sounder ~ || Klopferrelais n. || répéteur m. frappeur. / supervisory ~ || Kontrollrelais n. || relais m. pilote. / time-delay ~ || Zeitrelais n. || relais m. à temps. / ~ with vertical coils || stehendes Relais n. || relais m. à bobines verticales.

relay action || Relaiswirkung f. || action f. de relais. / ~ bell || Relaisglocke f. || sonnerie f. à relais. / ~ clock || Relaisuhr f. || horloge f. à relais. / ~ clock central station || Relaisuhrenzentrale f. || centrale f. d'horloges à relais. / ~ combination || Relaissystem n. || système m. automatique tout à relais. / ~ connection || Relaisschaltung f. || couplage m. de relais. / ~ construction || Relaisbau m. || construction f. de relais. / ~ frame || Relaisschrank m. || armoire f. de relais. / ~ key || Relaistaste f. || manipulateur m. à relais. / ~ magnet || Relaismagnet m. || aimant m. du relais. / ~ master clock || Relaishauptuhr f. || horloge f. principale à relais. / ~ plate (Tel) || Relaisplatte f. || panneau m. de relais. / ~ rail || Relaisschiene f. || rail m. de relais. / ~ releasing time || Relaisabfallzeit f. || temps m. de déplacement des relais. / ~ repeater for ringing currents in repeater work || Relaisrufübertragung f. für den Verstärkerbetrieb || translation f. d'appel par relais en service des répéteurs. / ~ sender || Relaiszahlengeber m. || émetteur m. des impulsions par relais. / ~ set || Relaissatz m. || dispositif m. des relais. / ~ working diagram || Erregungsdiagramm n. || tableau m. des relais.

release, to (Electr) || abfallen || décoller; relâcher; retomber. / ~ (Gear) || auslösen || débrayer; désembrayer; déclencher; déclancher. / ~ (Mach) || lockern || desserrer. / ~ the brake || die Bremse lösen || desserrer ou débloquer le frein. / ~ capital || Kapital n. flüssig machen || mobiliser du capital. / ~ the hook of the groove || den Haken m. aus der Rast f. auslösen || débrayer le crochet de l'encoche f. / ~ a spring || eine Feder f. entspannen || détendre un ressort.

release (Electr) || Abfallen n. || décollage m.; relâchement m. / ~ (Mach) || Lösung f.; Ausklinkung f. || déclanchement m.; échappement m. / ~ (Tel) || Auslösung f. || libération f. / ~ back ~ (Aut tel) || Rückauslösung f. || libération f. inverse. / ~ of brake device || Auslösung f. der Hemmvorrichtung || déclenchement m. du dispositif de freinage. / compression ~ || Kompressionsnocken m. || came f. de décompression. / ~ of goods || Freigabe f. von Waren || libération f. de marchandises. / hand ~ (Arms) || Handabzug m. || gâchette f. à la main. / instantaneous engaging and ~ || augenblickliches Einrücken n. und Ausrücken n. || embrayage m. et débrayage m. instantané. / ~ to instantaneous shutter || Auslösung f. des Momentverschlusses || déclenchement m. de l'obturateur instantané. / no-volt ~ || Nullspannungsauslösung f. || déclenchement m. à tension nul. / ~ of a vessel || Freigabe f. eines Schiffes || restitution f. d'un bâtiment.

release arrangement || Auslösevorrichtung f. || dispositif m. de déclenchement. / ~ catch || Auslösknagge f. || ergot m. de déclenchement. / ~ cord || Reißleine f. || corde f. de déchirure. / ~ current || Auslösestrom m. || courant m. de déclenchement. / ~ current strength || Auslösestromstärke f. || ampérage m. de déclenchement. / ~ gear || Auslösevorrichtung f. || dispositif m. de déclenchement ou de désamorçage. / ~ gear (Mach) || Ausklinksteuerung f. || distribution f. à déclic. / ~ handle || Auslösegriff m. || poignée f. de déclenchement. / ~ key || Freigabetaste f. || touche f. de libération. / ~ lever || Auslösehebel m. || levier m. de déclenchement. / ~ magnet || Auslöseelektromagnet m. || électro m. de déconnexion. / ~ mechanism (Typewr) || Schaltwerk n. || mécanisme m. de déclenchement.

releaser || Auslöser m. || débrayeur m.

release retardation || Auslöseverzögerung f. || retardation f. de déclenchement. / ~ spring || Rückzugfeder f. || ressort m. de rappel. / ~ trigger || ausrückbare Sperrklinke f. || cliquet m. de dégagement. / ~ valve || Auslöseventil n. || valve f. de purge ou de dégagement.

releasing || Auslösung f. || déconnexion f.; déclenchement m. / ~ circuit || Auslösestromkreis m. || circuit m. de déconnexion. / ~ device (Typewr) || Abstellvorrichtung f. || dispositif m. de débrayage. / ~ gear || Ausklinkvorrichtung f. || déclic m. / ~ key || Auslösetaste f. || touche f. de déclenchement. / ~ lever || Ausrückhebel m.; Ausrücker m. || levier m. de débrayage. / ~ mechanism || Auslösemechanismus m. || déclic m.; mécanisme m. de déclic; déclenchement m. / ~ relay || Entkupplungsrelais n. || relais m. de désembrayage.

relevant || sachdienlich; sachgemäß || convenable; pratique.

reliability || Zuverlässigkeit f. || certitude f.; authenticité f.; sûreté f.; solidité f. / ~ (Trade) || Kreditfähigkeit f. || solvabilité f. / ~ in service || Betriebssicherheit f.; Zuverlässigkeit f. || sécurité f. au service. / working ~ || Betriebssicherheit f. || sécurité f. du fonctionnement.

reliable || zuverlässig || éprouvé; certain. / ~ (Trade) || kreditwürdig || digne de crédit m. / ~ working || zuverlässige Arbeitsweise f. || fonctionnement m. sûr.

relief || Hochbild n.; Relief n. || relief m. / low ~ (Arch; Sculpt) || Flachrelief n. || bas-relief n. / ~ of the surface || Oberflächenrelief n. || relief m. superficiel.

relief engraving || Reliefgravierkunst f. || gravure f. en relief.

relief-grind, to || hinterschleifen || dépouiller à la meule.

relief grinding machine || Hinterschleifmaschine f. || machine f. à dépouiller à la meule.

relief, hour of ~ (Mine) || Lösestunde f.; Wechselstunde f. || heure f. de relais. / ~ map || Hochbildkarte f.; Reliefkarte f. || carte f. en relief.

relief-mill, to || hinterfräsen || dépouiller à la fraise.

relief milling machine || Hinterfräsmaschine f. || machine f. à dépouiller à la fraise.

relief print || Prägedruck m. || gaufrure f. / ~ printing machine || Reliefdruckmaschine f. || machine f. à l'impression en relief. / ~ society || Unterstützungsverein m. || société f. de bienfaisance. / ~ surface || Bodenprofil n. || relief m. ou profil m. du sol. / ~ valve || Entlastungsventil n.; Überdruckventil n. || soupape f. de soulagement. / ~ work || Notstandsarbeit f. || travail m. d'urgence.

relieve, to || hinterdrehen || dépouiller: dégager.

relieving apparatus || Hinterdrehapparat m. || appareil m. à dépouiller ou à détalonner. / ~ arch || Stützbogen m. || arc m. de soutènement. / ~ attachment || Hinterdrehapparat m. || appareil m. à détalonner. / ~ cam || Hinterdrehkurve f. || courbe f. de détalonnage. / ~ gear || Feststellvorrichtung f. || appareil m. de calage. / ~ lathe || Hinterdrehbank f. || tour m. à détalonner.

relievo see relief.

religious picture || religiöses Bild n. || image f. religieuse.

reloading charges pl. || Umladegebühr f. || frais mpl. de transbordement.

reluctance || magnetischer Widerstand m.; Reluktanz f. || réluctance f.

rely, to ~ on || sich stützen auf || appuyer sur.

remain, to || bleiben; verbleiben || rester; demeurer. / ~ at anchor || vor Anker m. liegen bleiben || demeurer sur le fer.

remains pl. (Geol) || Rückstände mpl.; irdische Überreste mpl. || restes mpl.; vestiges mpl. / ~ (Metal) || Ofenansätze mpl.; Rückstände mpl. || débris mpl.; résidus mpl. / fossil ~ || Abdruck m. einer Versteinerung || empreinte f. d'un fossil; vestiges mpl. fossiles. / ~ of plants || Pflanzenreste m. || restes mpl. ou débris mpl. de plantes.

remainder || Rest m.; Überbleibsel n. || reste m. / ~ (Bank) || Saldo m. || solde m.; reliquat m.

remanence || Remanenz f. || aimantation f. rémanente.

remanent || remanent || rémanent. / ~ magnetism || remanenter Magnetismus m. || magnétisme m. rémanent.

remark || Vermerk m. || remarque f.

remarkable || bemerkenswert; sehenswert || digne d'être vu; remarquable.

remblai || Anschüttung f. || remblai m.

remedied || abgeholfen || remédié. / will be ~ || wird abgeholfen werden || sera remédié.

remedy, to || abhelfen || remédier; mettre ordre m. à . . .

remedy || Abhilfe f. || remède m. / ~ || Hilfsmittel n. || remède m.; ressource f. / ~ (Coin) || Remedium n.; Toleranz f. || remède m.; tolérance f. / (Med)) Arzneimittel n.; Heilmittel n. || médicament m. / compressed ~ || komprimiertes Arzneimittel n. || médicament m. comprimé. / prepared ~ || zubereitetes Arzneimittel n. || médicament m. préparé.

remelt, to || umschmelzen || refondre. / ~ the pig-iron || das Roheisen umschmelzen || refondre la fonte crue.

remelted steel || umgeschmolzener Stahl m. || acier m. refondu.

remelting || Wiedereinschmelzen n. || refonte f.

remetal, to || die Lager npl. wiederausgießen || réantifrictionner.

reminder (Trade) || Mahnung f.; Erinnerung f. || avertissement m.; sommation f.

remission of fees || Gebührenermäßigung f. || réduction f. de frais.

remit, to || remittieren; anweisen || assigner; affecter; faire des remises fpl.

remittance || Rimesse f. || remise f.; envoi m. / counter ~ || Gegenrimesse f. || retour m.

remitter ‖ Remittent m. ‖ remetteur m.

remnants pl. ‖ Rest m.; Überbleibsel npl. ‖ coupon m. / ~ of hides (Curr) ‖ Fellspäne mpl. ‖ effleurures fpl.

remote ‖ entfernt; fern; entlegen ‖ écarté; éloigné. / ~ adjusting device for pendulum clocks ‖ Ferneinstellvorrichtung f. für ˙Pendeluhren ‖ dispositif m. de réglage à distance des horloges à pendule.

remote control ‖ Fernsteuerung f.; Fernbedienung f.; Fernkontrolle f. ‖ télécommande f.; contrôle m. à distance. / ~ (Radio) ‖ Tastung f. ‖ manipulation f. à distance. / ~ device ‖ Fernsteuereinrichtung f. ‖ appareillage m. de commande à distance. / ~ plant ‖ Fernsteueranlage f. ‖ installation f. de télécommande.

remote group switch ‖ Ferngruppenumschalter m. ‖ commutateur m. de groupe d'abonnés pour le service interurbain. / ~ reading water level indicator ‖ Anzeigegerät n. der Wasserstandfernmeldeanlage ‖ appareil m. indicateur de l'installation de téléindication du niveau d'eau.

remote signalling ‖ Fernmeldung f. ‖ télécommunication f. / ~ plant ‖ Fernsignalanlage f.; Fernmeldeanlage f. ‖ installation f. de télétransmission des signaux ou de signalisation sur grandes distances.

remote switch ‖ Fernschaltapparat m.; Fernschalter m. ‖ téléinterrupteur m.; interrupteur m. à commande à distance. / ~ with pilot lamp ‖ Fernschaltung f. mit Kontrolllampe ‖ commutateur m. à distance avec lampe de contrôle.

remould, to ‖ neuformen; umgestalten ‖ remouler.

remount, to (Mach) ‖ wieder aufstellen; wieder versehen mit ‖ remonter.

removable ‖ fortschaffbar; beweglich ‖ mobile; amovible; démontable. / ~ by washing ‖ auswaschbar ‖ lavable. / ~ cutter ‖ verschiebbare Schneidzunge f. ‖ languette f. déplaçable. / ~ handle ‖ Einsteckgriff m. ‖ manche m. amovible. / ~ lens attachment ‖ abnehmbarer Vorhänger m. ‖ face f. supplémentaire amovible. / ~ liner ‖ auswechselbares Futterrohr n. ‖ chemise f. amovible. / ~ side ‖ abnehmbare Seitenwand f. ‖ paroi f. de côté démontable. / ~ table (Mach tool) ‖ wegnehmbarer Tisch m. ‖ table f. démontable.

removal (El line) ‖ Verlegung f. ‖ transfert m. / ~ of ashes ‖ Aschenabfuhr f. ‖ enlèvement m. des cendres. / ~ of feces ‖ Fäkalienabfuhr f. ‖ enlèvement m. des matières fécales. / institution of ~ of house garbage ‖ Müllabfuhranstalt f. ‖ entreprise f. d'enlèvement des ordures ménagères. / ~ of overburden of a mine ‖ Abraum m. einer Grubenanlage ‖ décapage m. d'une installation minière. / ~ of the phosphorus ‖ Phosphorabscheidung f. ‖ élimination f. du phosphore. / ~ of the substances possessing smell ‖ Entfernung f. von Geruchstoffen ‖ élimination f. des matières odorantes. / ~ by suction ‖ Absaugung f. ‖ essorage m.

removal expenses pl. ‖ Umzugskosten pl. ‖ frais m. de déménagement.

remove, to ‖ abnehmen; entfernen; aufräumen ‖ enlever; écarter; retirer; déporter. / ~ (Dwelling) ‖ umziehen ‖ changer de domicile; déménager. / ~

the bitter substances ‖ entbittern ‖ enlever les principes mpl. amers. / ~ the borax by dilute sulphuric acid after soldering ‖ den Borax nach dem Löten abbeizen ‖ dérocher le borax après le soudage. / ~ a bridge ‖ eine Brücke abbrechen ‖ enlever ou replier un pont. / ~ the carbon ‖ frischen ‖ décarburer. / ~ dust ‖ entstauben ‖ dépoussiérer; enlever la poussière. / ~ grease ‖ entfetten ‖ dégraisser. / ~ all solid and liquid impurities from oil ‖ dem Öl n. alle festen und flüssigen Fremdkörper entziehen ‖ débarrasser l'huile f. de toutes les impuretés solides et liquides. / ~ into new premises ‖ sein Geschäft n. verlegen ‖ transférer sa maison. / ~ the ore ‖ das Erz abbauen ‖ exploiter le minerai. / ~ parallax ‖ die Parallaxe f. beseitigen ‖ écarter la parallaxe. / ~ a prohibition of exportation ‖ ein Ausfuhrverbot n. aufheben ‖ abroger une défense d'exporter. / ~ the receiver (Tel) ‖ den Hörer m. abheben oder abnehmen ‖ enlever ou décrocher le récepteur. / ~ the resin from a tree ‖ einen Baum abharzen ‖ soutirer la résine d'un arbre. / ~ the tyres ‖ die Reifen mpl. abmontieren ‖ enlever ou démonter les bandages mpl.

removing grits from brass castings ‖ Abkneifen n. von Messinggußtrichtern ‖ entaille f. des masselottes en laiton. / ~ old paint ‖ alte Farbe f. entfernen ‖ enlever des anciennes couches fpl. de couleur. / quick inserting and ~ ‖ schnelles Ein- und Ausspannen n. ‖ serrage m. et déserrage m. rapide. / ~ soil in thin layers ‖ Abtragen n. von Erdmassen geringer Mächtigkeit ‖ enlèvement m. des couches de terre peu épaisses. / grains reraking and ~ machine ‖ Aufhack- und Austrebermaschine f. ‖ piocheur-dédrcheur m.

remunerate, to (Damages) ‖ entschädigen ‖ dédommager; indemniser. / ~ (Wages) ‖ besolden ‖ rémunérer.

remuneration ‖ Entschädigung f.; Honorar n. ‖ dédommagement m.; indemnité f.; rémunération f.; honoraires mpl.

render, to (Build) ‖ berappen; abputzen ‖ crépir. / ~ unclean ‖ verunreinigen ‖ salir.

rendering, cement ~ ‖ Zementverputz m. ‖ enduit m. de ciment.

rendering-concrete ‖ Spritzbeton m. ‖ crépi m. en béton.

renew, to ‖ erneuern; auswechseln ‖ renouveler; remplacer. / ~ a bill ‖ einen Wechsel m. prolongieren ‖ prolonger une lettre de change.

renewal ‖ Erneuerung f.; Auswechseln n. ‖ renouvellement m. / ~ of a bill of exchange ‖ Prolongation f. eines Wechsels ‖ renouvellement m. d'une lettre de change / where excessive wear and tear requires frequent ~s of the material ‖ da, wo ein überaus starker Verschleiß m. eine häufige Auswechslung f. der Maschinenteile erfordert ‖ là où une usure trop rapide nécessite un remplacement fréquent du matériel. / ~ of the rolling stock (Railw) ‖ Erneuerung f. der Betriebsmittel ‖ renouvellement m. du matériel roulant. / ~ of sleepers (Railw) ‖ Auswechseln n. der Schwellen ‖ renouvellement m. des traverses.

rennet ‖ Lab n. ‖ présure f. / ~ in natural state ‖ Lab n. in natürlichem Zustand ‖

présure f. à l'état naturel. / ~ maker ‖ Labzubereiter m. ‖ présurier m. / ~ preparation ‖ Labpräparat n. ‖ préparation f. de présure.

renounce, to ‖ verzichten auf... ‖ renoncer à... / ~ a right ‖ auf ein Recht n. verzichten ‖ renoncer à un droit.

renovate, to ‖ erneuern; wiederherstellen ‖ renouveler.

renovation ‖ Erneuerung f. ‖ renouvellement m. / cost of ~ ‖ Wiederherstellungskosten pl. ‖ frais m. de réparation.

renowned ‖ namhaft; berühmt ‖ renommé; notable.

rent, to ‖ pachten ‖ prendre à ferme f. ou à bail m.

rent (Annuity) ‖ Annuität f.; Rente f. ‖ annuité f.; rente f. / ~ (Geol) ‖ Riß m.; Spalte f. ‖ déchirure f.; fendille f.; fissure f.; gerçure f.; fêlure f. / ~ (Hire) ‖ Miete f. ‖ louage m.; loyer m. / ground ~ ‖ Grundrente f. ‖ rente f. foncière. / subject to ~ ‖ zinspflichtig ‖ tributaire.

rental ‖ Mietgeld n.; Pacht f. ‖ fermage m. / monthly ~ ‖ monatliche Pacht f. ‖ ferme f. mensuelle.

rental charge ‖ Pachtbetrag m. ‖ taux m. de la ferme.

renterer ‖ Tuchstopfer m. ‖ rentrayeur.

rentering ‖ Stoßnaht f. ‖ rentraiture f. / ~ of cloth ‖ Stopfen n. des Tuches ‖ stoppage m. de drap.

rent tax ‖ Hauszinssteuer f. ‖ impôt m. sur le loyer. / money obtained from ~es ‖ Kapitalien npl. aus der Hauszinssteuer ‖ capitaux mpl. provenant de l'impôt sur le loyer.

renunciation ‖ Verzicht m.; Verzichtleistung f. ‖ renonciation f.; désistement m. / letter of ~ ‖ Verzichterklärung f. ‖ lettre f. de renoncement.

reorganization ‖ Sanierung f. ‖ réorganisation f.

reorganize, to ‖ neugestalten; reorganisieren; sanieren ‖ réorganiser.

repack, to ‖ umpacken ‖ changer d'emballage m.; réemballer.

repair, to ‖ ausbessern ‖ réparer. / ~ shoes pl. ‖ Schuhe mpl. flicken ‖ raccommoder des chaussures fpl. / ~ the tympan (Print) ‖ den Deckel m. ausbessern ‖ faire une reluire.

repair ‖ Reparatur f.; Ausbesserung f. ‖ réparation f. / emergency ~ ‖ Notausbesserung f. ‖ réparation f. improvisée. / in good ~ ‖ in gutem Ausbesserungszustand m. ‖ en bon état m. de réparation. / to keep a mine in repair ‖ eine Grube f. bauhaft halten ‖ entretenir une mine. / the engine suffered from a constant need of ~s ‖ die Maschine krankte an oftmaligen Reparaturen ‖ la machine dut souvent subir des réparations. / in need of ~ ‖ baufällig ‖ délabré. / roadside ~ (Auto) ‖ behelfsmäßige Ausbesserung f. ‖ réparation f. de fortune. / temporary ~ ‖ einstweilige Ausbesserung f. ‖ réparation f. temporaire.

repaired ‖ ausgebessert ‖ réparé.

repairer ‖ Ausbesserer m. ‖ réparateur m.; rhabilleur m.; raccommodeur m. / ~ (Mine) ‖ Reparaturhauer m. ‖ répareur m.

repair gang (Tel) ‖ Sondertrupp m. ‖ épuipe f. spéciale de contrôle des lignes.

repairing of the armature ‖ Ankerausbesserung f. ‖ réparation f. de l'induit.

/ ~ the boiler ‖ Kesselausbesserung f.; Kesselreparatur f. ‖ réparation f. *ou* raccommodage m. des chaudières. / not worth ~ ‖ nicht der Ausbesserung f. wert ‖ ne vaut pas la peine d'être réparé. / ~ of watch cases ‖ Uhrgehäuseausbesserung f. ‖ rhabillage m. de boîtes de montres.

repairing automobile ‖ Ausbesserungskraftfahrzeug n.; Reparaturkraftfahrzeug ‖ automobile f. de réparation. / ~ material for cycle hoses ‖ Fahrradflickzeug n. ‖ matériel m. à réparer les chambres de bicyclettes. / ~ shop ‖ Reparaturwerkstätte f. ‖ atelier m. de réparation.

repair outfit ‖ Flickkasten m. ‖ nécessaire m. pour réparation. / ~ part ‖ Ersatzteil m. ‖ pièce f. de rechange *ou* de réserve.

repairs pl., contract of ~ ‖ Ausbesserungsvertrag m. ‖ contrat m. de réparation. / cost of ~ ‖ Ausbesserungskosten pl. ‖ frais mpl. des réparations. / estimate for ~ ‖ Ausbesserungskostenanschlag m. ‖ évaluation f. des réparations.

repair shop ‖ Ausbesserungswerkstätte f.; Reparaturwerkstatt f. ‖ atelier m. de réparations. / ~ truck ‖ Werkstattwagen m. ‖ camion m. atelier. / ~ work ‖ Ausbesserungsarbeit f.; Instandsetzungsarbeit ‖ travail m. de réparation. / machine ~ workshop ‖ Maschinenreparaturwerkstatt f. ‖ atelier m. de réparation de machines.

reparation ‖ Ersatzleistung f. ‖ réparation f. / ~ coal ‖ Reparationskohle f. ‖ charbon m. indemnitaire. / ~ account ‖ Reparationskonto n. ‖ compte f. des réparations.

repaste, to ‖ repastieren ‖ retartiner.

repayable ‖ rückzahlbar ‖ remboursable.

repayment ‖ Rückzahlung f.; Rückvergütung f. ‖ remboursement m. / ~ (Loan) ‖ Tilgung f. ‖ amortissement m.

repeat, to ‖ wiederholen; repetieren ‖ répéter; reproduire. / ~ an order ‖ nachbestellen ‖ faire une commande supplémentaire.

repeated ‖ oftmalig; wiederholt ‖ fréquent. / ~ stresses pl. ‖ Schwingungsbeanspruchung f.; dynamische Beanspruchung f. ‖ effort m. d'oscillation.

repeater (Roll mill) ‖ Umführung f. ‖ dispositif m. de retour. / ~ (Tel) ‖ Verstärker m. ‖ amplificateur m. / ~ (Watchm) ‖ Repetieruhr f. ‖ montre f. à répétition. / cord circuit ~ (Tel) ‖ Schnurverstärker m. ‖ amplificateur m. à cordons. / double-bridge two-way ~ (Tel) ‖ Doppelbrückenverstärker m. ‖ répéteur m. à deux fils en double pont. / four-wire ~ (Tel) ‖ Vierdrahtverstärker m. ‖ répéteur m. à quatre fils. / four-wire intermediate ~ ‖ Vierdrahtzwischenverstärker m. ‖ amplificateur m. intermédiaire à quatre fils. / high-frequency ~ ‖ Hochfrequenzverstärker m. ‖ amplificateur m. (à) haute fréquence. / intermediate ~ ‖ Zwischenverstärker m. ‖ répéteur m. intermédiaire. / low-frequency ~ ‖ Niederfrequenzverstärker m. ‖ amplificateur m. (à) basse fréquence. / multi-stage ~ ‖ Kaskadenverstärker m.; mehrstufiger Verstärker m. ‖ amplificateur m. à (plusieurs) étages. / one-valve two-way ~ ‖ Einröhrenzwischenverstärker m. ‖ répéteur m. embroché *ou* réversible à une lampe. / resistance ~ ‖ Widerstandsverstärker m. ‖ amplifica-

teur m. à résistance. / terminal ~ ‖ Endverstärker m. ‖ amplificateur m. d'extrémité. / three-valve ~ ‖ Dreiröhrenverstärker m. ‖ amplificateur m. à trois lampes. / two-valve two-wire ~ (Tel) ‖ Doppelrohrzwischenverstärker m. ‖ amplificateur m. à deux lampes et à deux fils. / two-wire two-way ~ ‖ Zweidrahtzwischenverstärker m. ‖ répéteur m. réversible à deux fils.

repeater, two-wire ~ board ‖ Zweiröhrenzwischenverstärkerschrank m. ‖ commutateur m. de répéteurs à deux fils. / ~ circuit ‖ Verstärkerschaltung f. ‖ système m. de répéteurs. / ~ connection ‖ Verstärkerschaltung f. ‖ dispositif m. d'amplificateurs. / voice-operated ~ device ‖ Umsteuerschaltung f. für Verstärker ‖ système m. de répéteurs dirigeables. / switching by voice-operated ~ devices ‖ Umschalteverfahren n. im Verstärkerbetrieb ‖ procédé m. de commutation par des courants téléphoniques en service des répéteurs.

repeatered circuit (Tel) ‖ mit Verstärkern versehene Leitung f. ‖ circuit m. muni d'amplificateurs.

repeater gain ‖ Verstärkungsgrad m. ‖ gain m. d'amplification. / ~ measuring ‖ Verstärkungsmessung f. ‖ mesure f. d'amplification; kerdométrie f. / ~ position ‖ Schnurverstärkerplatz m. ‖ position f. des amplificateurs sur cordon. / ~ rack ‖ Verstärkergestell n. ‖ bâti m. de répéteurs. / ~ room (Tel) ‖ Verstärkersaal m. ‖ salle f. de répéteurs. / ~ section ‖ Verstärkerabschnitt m.; Verstärkerfeld n. ‖ tronçon m. d'amplification; section f. entre deux amplificateurs. / ~ set ‖ Verstärkersatz m. ‖ système m. d'appareils répéteurs.

repeater station ‖ Verstärkeramt n.; Verstärkerstation f. ‖ station f. amplificatrice; poste m. amplificateur. / secondary ~ ‖ Nebenverstärkeramt n. ‖ station f. de répéteurs secondaire. / building-up of ~ ‖ Aufbau m. des Verstärkeramtes ‖ ensemble m. d'une station de répéteurs. / ~ test desk ‖ Meßschrank m. für Verstärkerämter ‖ table f. de mesure pour répéteurs.

repeater system ‖ Verstärkerschaltung f. ‖ système m. de répéteurs. / two-wire ~ ‖ Zweiröhrenschaltung f. für Verstärker ‖ système m. de répéteurs à deux fils.

repeater valve ‖ Verstärkerröhre f. ‖ lampe f. amplificatrice.

repeating (Weav) ‖ Rapport m.; Musterwiederkehr f. ‖ rapport m. du dessin. / ~ a bell signal (Railw) ‖ Rückmeldung f. ‖ réplique f. / ~ axis ‖ Repetitionsachse f. ‖ axe m. de répétition. / ~ circle ‖ Repetitionskreis m. ‖ cercle m. répétiteur. / ~ coil ‖ Abzweigspule f.; Übertrager m.; Übertragerspule f. ‖ transformateur m. (différentiel); bobine f. d'induction. / ~ method ‖ Repetitionsverfahren n. ‖ méthode f. de répétition. / ~ relay ‖ Übertragungsrelais n. ‖ relais m. de transmission. / ~ signal (Railw) ‖ Rückmeldesignal n. ‖ signal m. de réplique. / ~ theodolite with altitude circle ‖ Repetitionstheodolit m. mit Höhenkreis ‖ théodolite m. répétiteur avec limbe vertical. / ~ watch ‖ Repetieruhr f. ‖ montre f. à répétition.

repeat key (Typewr) ‖ Wiederholtaste f. ‖ touche f. de répétition. / ~ order ‖ Nach-

bestellung f. ‖ commande f. *ou* ordre m. supplémentaire.

repel, to (Phys) ‖ abstoßen ‖ repousser. / ~ mutually ‖ sich gegenseitig abstoßen ‖ se repousser mutuellement.

repelling power ‖ Repulsionskraft f. ‖ force f. répulsive.

repercussion ‖ Rückwirkung f. ‖ réaction f.

repetend (Math) ‖ Periode f. ‖ période f.

replace, to ‖ ersetzen ‖ remplacer; substituer.

replaceable ‖ ersetzbar ‖ remplaçable.

replacement ‖ Zurückstellung f.; Ersetzen n. ‖ remplacement m. / electrical ~ ‖ elektrische Abstellung f. ‖ disparition f. électrique. / mechanical ~ ‖ mechanische Abstellung f. ‖ disparition f. mécanique.

replace part ‖ Ersatzteil m.; Reserveteil m. ‖ pièce f. de rechange *ou* de réserve.

replacing ‖ Austausch m. ‖ remplacement m. / ~ the rolls ‖ Auswechseln n. der Walzen ‖ changement m. des cylindres. / ~ (a wagon) upon the track ‖ Aufgleisung f. ‖ remise f. sur rails. / ~ calendar block ‖ Kalenderersatzblock m. ‖ bloc m. de remplacement de calendrier.

replenish, to ‖ auffüllen ‖ faire le plein. / ~ oil supply ‖ das Öl n. erneuern ‖ renouveler l'huile f.

replenishing cup ‖ Einfülltopf m. ‖ pot m. de remplissage.

reply, to ~ to a letter ‖ auf einen Brief m. antworten ‖ répondre à une lettre.

reply Antwort f.; Rückäußerung f. ‖ réponse f.; réplique f. / a ~ is requested ‖ um Antwort f. wird gebeten ‖ réponse f., s'il vous plaît. / ~ by return ‖ umgehende Antwort f. ‖ réponse f. par retour du courrier. / ~ by wire ‖ Drahtantwort f. ‖ réponse f. télégraphique *ou* par fil. / written ~ ‖ Gegenschrift f. ‖ réplique f.; réfutation f.

reply bell (Railw) ‖ Rückmeldeläutewerk n. ‖ sonnerie f. de réplique *ou* de répétition. / ~ coupon ‖ Antwortschein m. ‖ coupon-réponse m.

report, to ‖ berichten; Bericht m. erstatten ‖ référer; rapporter; faire un rapport.

report ‖ Bericht m.; Referat n. ‖ rapport m. / annual ~ ‖ Jahresbericht m.; Geschäftsbericht m. ‖ rapport m. annuel *ou* de gestion. / ~ of approbation ‖ Abnahmebericht m. ‖ relation f. de réception. / ~ of a bank ‖ Bankbericht m. ‖ situation f. d'une banque. / ~ concerning ~ ‖ Bericht m. über… ‖ rapport m. concernant… / daily ~ ‖ täglicher Bericht m. ‖ rapport m. quotidien. / expert's ~ ‖ Gutachten n. eines Sachverständigen ‖ rapport m. d'expert. / favourable ~ ‖ günstige Auskunft f. ‖ bon renseignement m. / ~ on firms ‖ Erkundigung f. über Geschäftsverhältnisse ‖ demande f. de renseignements. / geological ~ ‖ geologisches Gutachten n. ‖ expertise f. géologique. / market ~ ‖ Marktbericht m. ‖ rapport m. du marché. / official ~ ‖ amtlicher Bericht m. ‖ rapport m. officiel. / ~ of test ‖ Prüfungsbericht m. ‖ rapport m. d'épreuve. / ~ is unreliable ‖ Bericht m. ist unzuverlässig ‖ rapport m. est peu digne de foi.

reporter ‖ Referent m. ‖ rapporteur m.

represent, to ~ a firm ‖ einer Firma f. vorstehen ‖ conduire *ou* gouverner une maison.

represented, to be ~ on the board of a company ‖ im Präsidium n. einer Ge-

sellschaft vertreten sein ‖ faire partie f. de la présidence d'une société.

representation ‖ Darstellung f. ‖ représentation f. / ~ (Trade) ‖ Agentur f. ‖ agence f. succursale. / ~ of a concern ‖ Vertretung f. eines Konzerns ‖ représentation f. d'un groupe. / ~ of employees ‖ Betriebsvertretung f. ‖ syndicat m. des ouvriers; représentation f. des employés. / perspective ~ ‖ perspektivische Darstellung f. ‖ représentation f. perspective.

representative ‖ Abgeordneter m. ‖ envoyé m.; mandataire m.; représentant m. ‖ works' ~ ‖ Werkvertreter m. ‖ représentant m. de l'entreprise.

reprimand, to ‖ rügen ‖ blâmer.

reprimand ‖ Rüge f.; Tadel m. ‖ blâme m.

reprint, to ‖ nachdrucken ‖ réimprimer.

reprint ‖ neuer Abdruck m.; Wiederabdruck m. ‖ réimpression f.

reprints pl., either partly or in full, forbidden ‖ Nachdruck m. auch auszugsweise verboten ‖ reproduction f., même par extraits, interdite.

reprinting press ‖ Umdruckpresse f. ‖ presse f. de réimpression.

reprisals pl. ‖ Repressalien fpl. ‖ représailles fpl.

reproduce, to ‖ reproduzieren; nachbilden ‖ reproduire.

reproducing method ‖ Reproduktionstechnik f. ‖ procédé m. de reproduction. / ~ stone ‖ Wiedergabestein m. ‖ pierre f. de réception.

reproduction ‖ Wiedererzeugung f.; Nachbildung f. ‖ reproduction f. / ~ of illustrations is subject to the consent of ... ‖ Wiedergabe f. von Abbildungen ist ohne Zustimmung nicht gestattet ‖ il est interdit de reproduire des illustrations fpl. sans consentement. / ~ of the types ‖ Abdruck m. der Typen ‖ reproduction f. des caractères. / process of ~ ‖ Reproduktionsverfahren n. ‖ procédé m. de reproduction.

reps ‖ Rips m. ‖ reps m.

reptile ‖ kriechend ‖ reptile.

reptile ‖ Reptil n.; Kriechtier n. ‖ reptile m. / ~ catcher ‖ Reptilienfänger m. ‖ chasseur m. de reptiles. / genuine ~ leather ‖ echtes Reptilleder n. ‖ cuir m. de reptile véritable. / ~ skin ‖ Kriechtierhaut f. ‖ peau f. de reptile.

republish, to ~ a book ‖ ein Buch n. neu auflegen ‖ réimprimer un livre.

repulp, to (Pap) ‖ einstampfen ‖ mettre au pilon.

repulse ‖ Rückstoß m. ‖ recul m.; rebuffade f.

repulsing ‖ zurückschlagend ‖ repoussant.

repulsion (Phys) ‖ Abstoßung f. ‖ répulsion f. / electrical ~ ‖ elektrische Abstoßung f. ‖ répulsion f. électrique. / force of ~ ‖ Abstoßungskraft f. ‖ force f. répulsive. / magnetic ~ ‖ magnetische Abstoßung f. ‖ répulsion f. magnétique.

repulsive force ‖ Abstoßungskraft f. ‖ force f. répulsive.

repulsiveness see repulsive force.

repurchase ‖ Rückkauf m. ‖ rachat m. / right of ~ ‖ Rückkaufsrecht n. ‖ droit m. de réméré; réméré m.

repurchased ‖ zurückgekauft ‖ racheté.

reputation ‖ Renommee n.; Ruf m. ‖ réputation f.; renom m. ‖ renommée f.

repute see reputation.

request, to ‖ bitten; ersuchen ‖ demander.

request ‖ Bitte f.; Anliegen n. ‖ demande f. / ~ for communication ‖ Gesprächsanmeldung f. ‖ demande f. de communication. / ~ for a respite ‖ Stundungsantrag m.; Stundungsgesuch n. ‖ demande f. de délai.

request apparatus (Tel) ‖ Rückfrageapparat m. ‖ poste m. à double appel.

requested, as ~ ‖ wunschgemäß ‖ selon votre désir m.

request equipment (Tel) ‖ Rückfrageeinrichtung f. ‖ dispositif m. à double appel. / ~ time (Tel) ‖ Anmeldezeit f. ‖ temps m. d'enregistrement.

require, to ‖ fordern; verlangen ‖ exiger; demander. / no longer require it ‖ nicht mehr darauf reflektieren ‖ renoncer à.

required ‖ gesucht; verlangt; benötigt ‖ à souhait m.; à besoin m.; demandé. / if ~ ‖ auf Wunsch m. ‖ par ailleurs; sur demande f.

requirement ‖ Nachfrage f.; Erfordernis n.; Verwendungszweck m. ‖ exigence f.; besoin m.; nécessité f.; demande f.

requirements pl., chemigraphical ~ ‖ chemigrafische Bedarfsartikel mpl. ‖ articles mpl. pour l'industrie chémigraphique. / ~ for the graphical trade ‖ grafische Fachartikel mpl. ‖ articles mpl. concernant l'art graphique. / local ~ ‖ Platzbedarf m. ‖ besoins mpl. de la place. / to meet the ~ ‖ den Ansprüchen mpl. genügen ‖ répondre aux exigences fpl. / ~ for mines ‖ Grubenbedarf m. ‖ ustensiles mpl. de mines. / railway ~ ‖ Eisenbahnbedarf m. ‖ articles mpl. de chemin de fer. / ~ for smelting-works ‖ Hüttenbedarf m. ‖ ustensiles mpl. de fonderies. / in case of special ~ ‖ im Falle besonderer Anforderungen fpl. ‖ dans le cas de besoins mpl. spéciaux. / to supply the whole ~ from indigenous materials ‖ den Gesamtbedarf m. aus einheimischen Stoffen decken ‖ alimenter ou fournir la consommation totale en matières indigènes. / technical ~ for machines ‖ technische Bedarfsartikel m. ‖ accessoires fpl. techniques pour machines. / ~ for theatrical stages ‖ Bühnenbedarf m. ‖ ustensiles mpl. de théâtre. / ~ for weaving ‖ Webereibedarfsartikel mpl. ‖ accessoires mpl. tissages.

requisite ‖ Bedarfsgegenstand m. ‖ matière f. accessoire. / ~s pl. for the textile industry ‖ Bedarfsartikel mpl. für die Textilindustrie ‖ accessoires mpl. de l'industrie textile.

requisite power ‖ Kraftbedarf m. ‖ force f. nécessaire.

rerailing (a wagon) (Railw) ‖ Aufgleisung f. ‖ remise f. sur rails. / ~ plate ‖ Aufgleisungsschuh m. ‖ plaque f. d'accès. / ~ ramp ‖ Aufgleisungsschuh m. ‖ plaque f. d'accès. / double ~ ramp ‖ zweiseitige Aufgleisungsplatte f. ‖ plaque f. d'accès double.

rerake, to ~ the draff ‖ Treber pl. aufhacken ‖ piocher ou piquer les drêches fpl.

reraking (Draff) ‖ Aufhacken n. ‖ piochage m.; piquage m. / grains ~ and removing machine ‖ Aufhack- und Austrebermaschine f. ‖ piocheur-dédrêcheur m.

resack, to ‖ umsacken ‖ changer de sac m.

rescue, to ‖ befreien; retten ‖ sauver; secourir.

rescue apparatus ‖ Rettungsapparat m. ‖ appareil m. ou engine m. de sauvetage. / ~ party ‖ Rettungsmannschaft f. ‖ équipe f. de sauvetage. / ~ regulation ‖ Rettungseinrichtung f. ‖ installation f. de sauvetage. / ~ station ‖ Rettungsstation f. ‖ poste m. de sauvetage ou de secours.

rescuer ‖ Retter m.; Befreier m. ‖ sauveteur m.

research ‖ (wissenschaftliche) Untersuchung f.; Forschung f. ‖ recherche f. / ~ of materials ‖ Werkstoffuntersuchung f. ‖ épreuve f. ou analyse f. de matériaux. / scientific ~ ‖ wissenschaftliche Untersuchung f. ‖ recherche f. scientifique.

research committee ‖ Komitee n. für Erfindungen ‖ comité m. des inventions. / ~ laboratory ‖ Forschungslaboratorium n. ‖ laboratoire m. de recherches. / ~ microscope ‖ Forschungsmikroskop n. ‖ microscope m. pour les recherches scientifiques. / ~ purpose ‖ Forschungszweck m. ‖ but m. d'investigation. / ~ work ‖ Forschungsarbeit f. ‖ travail m. de recherche ou d'exploration; enquête f.

reservation of proprietorship ‖ Eigentumsvorbehalt m. ‖ réserve f. de propriété.

reserve, to ‖ reservieren; aufsparen; offenhalten ‖ réserver; tenir en réserve f.

reserve ‖ Reserve f. ‖ réserve f. / ~ (Bank) ‖ Rücklage f. ‖ fonds m. de réserve. / ~ (Print) ‖ Kehrseite f.; Rückseite f. ‖ page f.; verso m. / ~ (Text print) ‖ Deckpapp m.; Schutzpapp m. ‖ réserve f. / coloured ~ ‖ farbige Reserve f. ‖ réserve f. à couleur. / fat ~ ‖ fette Reserve f. ‖ réserve f. épaisse. / hidden ~ ‖ stille Reserve f. ‖ réserve f. latente. / ~ prescribed by law ‖ gesetzliche Rücklage f. ‖ réserve f. légale. / special ~ ‖ Sonderrücklage f. ‖ réserves fpl. spéciales. / white ~ ‖ weiße Reserve f. ‖ réserve f. blanche.

reserve account ‖ Reservekonto n. ‖ compte m. de réserve. / ~ battery ‖ Reservebatterie f. ‖ batterie f. de réserve. / ~ capacity ‖ Reserveleistung f. ‖ réserve f. de pouvoir. / ~ capital ‖ Reservekapital n. ‖ capital m. de réserve. / ~ machine ‖ Reservemaschine f. ‖ machine f. de réserve. / ~ needle ‖ Ersatznadel f. ‖ aiguille f. de rechange. / ~ part ‖ Ersatzstück n. ‖ pièce f. de réserve; élément m. de rechange. / ~ piece ‖ Reservestück n. ‖ pièce f. de réserve. / ~ power ‖ Kraftüberschuß m.; Energiereserve f. ‖ réserve f. de puissance. / ~ sieve ‖ Reservesieb n. ‖ tamis m. de rechange. / ~ style (Text print) ‖ Reservagedruck m. ‖ impression f. avec réserves. / ~ water tank ‖ Aushilfswasserbehälter m. ‖ récipient m. de réserve d'eau.

reserved, all rights ~ ‖ alle Rechte npl. vorbehalten ‖ tous droits mpl. réservés. / ~ mineral ‖ vorbehaltenes Mineral n. ‖ minéral m. réservé.

reservedly ‖ unter Vorbehalt m. ‖ sauf bonne fin f.

reserve fund ‖ Reservefonds m. ‖ fonds m. de réserve.

reservoir ‖ Behälter m.; Reservoir n.; Sammelgefäß n. ‖ réservoir m.; bassin m. / volume of the ~ of a barrage ‖ Stauinhalt m. einer Talsperre ‖ capacité f. de retenu d'un barrage d'une vallée. / collecting ~ ‖ Auffangschale f. ‖ bac m. collecteur. / compressed air ~ ‖ Druckluftbehälter m. ‖ réservoir m. à air com-

primé. / distributing ~ (Hydr arch) ||
Hochbehälter m. || réservoir m. à distri-
bution d'eau. / gas ~ || Gasbehälter m. ||
réservoir m. à gaz. / high-level service ~
|| Hochbehälter m. || réservoir m. surélevé
à distribution d'eau. / low-level service
~ || Rücklaufreservoir n. || réservoir m. à
retour d'eau. / non-exploding ~ || explo-
sionssicheres Gefäß n. || vase m. à l'é-
preuve d'explosion; récipient m. in-
explosible. / ~ of sheet || Sammelbehälter
m. aus Blech || réservoir m. en tôle.

reservoir embankment || Staudamm m. ||
barrage m. en terre. / ~ keeper || Wasser-
turmwärter m.; Sammelbeckenwärter m.
|| garde-réservoirs m.

resetting to zero (Cash register) || Stellen n.
auf Null || mise f. à zéro.

resetting counter || Nullstellenzähler m. ||
compteur m. de remise à zéro. / ~ key ||
Rückstelltaste f. || clef f. de disparition.

resharpening of files || Feilenhauen n. ||
retaillage m. de limes.

reship, to || als Rückfracht f. versenden ||
réexpédier.

residence station (Tel) || Wohnungsan-
schluß m. || poste m. dans l'habitation
de l'abonné.

resident || ansässig; seßhaft || établi; do-
micilié; sédentaire. / ~ worker || seßhafter
Arbeiter m. || ouvrier m. domicilié.

residual || zurückbleibend; übrig || rési-
duel; résiduaire. / ~ magnetism || rema-
nenter Magnetismus m. || magnétisme m.
résiduel.

residuary product || Abfallerzeugnis n. ||
produit m. résiduel.

residue || Rest m.; Rückstand m. || reste
m.; résidu m. / burnable ~ || brennbarer
Rückstand m. || résidu m. combustible. /
coke-like ~ of distillation || koksähn-
licher Destillationsrückstand m. || résidu
m. de distillation similaire au coke.
/ ~ of distilled sulphur || Schwefel-
schlacke f. || crasse f. de soufre. / ~
of extraction || Auslaugrückstand m. ||
résidu m. d'extraction. / metallic or
metalliferous ~ from the treatment of
ores || metallischer oder metallhaltiger
Rückstand m. von der Verhüttung von
Erzen || résidu m. métallique ou mé-
tallifère provenant du traitement des
minerais. / metalliferous ~ || metall-
haltiger Rückstand m. || résidu m. mé-
tallifère. / roasting ~ || Abbrand m. ||
résidu m. de grillage. / purifying steel of
its ~ of slag by remelting it || den
Stahl m. durch Umschmelzen von den
Schlackenresten reinigen || purifier l'acier
des restes de scorie par fusion. /
solid ~ of evaporation || Abdampfrück-
stand m. || résidu m. solide d'évapora-
tion.

residue beer in chip cask || Abseihbier n. ||
fonds mpl. de cuves ou de foudres.

residuum see also residue || Bodensatz m. ||
résidu m. / ~ of oil || Ölbodensatz m. ||
dépôt m. d'huile.

resignation || Verzicht m.; Verzichtleistung
f. || renonciation f.; désistement m.

resin || Harz n.; Kolophonium n. || résine
f.; colophane f. / artificial ~ || Kunstharz
n. || résine f. artificielle. / common ~ ||
gelbes Fichtenharz n. || résine f. jaune;
poix résine f. / fiddler's ~ || Geigenharz
n.; Kolophonium n. || arcanson m.;
colophane f. / ~ of fir || Föhrenharz
n. || résine f. du pin. / exploitation of pine

|| Fichtenharzgewinnung f. || exploita-
tion f. de résine de pins. / to remove the ~
from a tree || einen Baum m. abharzen ||
soutirer la résine d'un arbre. / white ~ ||
Galipot n.; weißes Fichtenharz n. || gali-
pot m. / yellow ~ || gelbes Fichtenharz n.
|| résine f. jaune; poix-résine f.

resin acid || Harzsäure f. || acide m. ré-
sineux. / ~ burner || Harzsieder m. || fon-
deur m. de résine. / ~ cake || Harzkuchen
m. || disque m. de résine. / ~ collector ||
Fichtenharzeinsammler m. || résinier m.;
gemmier m. / ~ distillation || Harzdestill-
lation f. || distillation f. de résine. / ~ duct
|| Harzkanal m. || canal m. résinifère. /
~ gas || Harzgas n. || gaz m. de résine.

resiniferous fir tree || Harztanne f. || sapin
m. résineux.

resinification || Verharzung f. || résinifica-
tion f. / to avoid ~ || die Verharzung f.
vermeiden || empêcher la résinification.

resinify, to || verharzen || se résinifier.

resin maker || Fichtenharzeinsammler m.
|| résinier m.; gemmier m. / ~ odour ||
Harzgeruch m. || odeur f. de résine.

resin oil || Harzöl n.; Kienöl n. || huile f.
de résine ou de pin. / ~ distiller || Kien-
ölbrenner m. || distillateur m. d'huile de
pin.

resinous || harzig || résineux. / highly ~ ||
harzreich || riche en résine f.

resinous cement || Harzkitt m.; Harz-
zement m. || mastic m. résineux; ciment m.
à la résine. / ~ compound || Harzmasse
f. || masse f. résineuse. / ~ forest || Nadel-
wald m. || forêt f. résineuse. / ~ lustre ||
Pechglanz m. || éclat m. résinéux. / ~
putty || Harzkitt m. || mastic m. résineux.

resin oven || Harzofen m. || four m. à
résine. / ~ produce || Harzprodukt n. ||
produit m. résineux. / ~ rod || Harz-
stab m. || bâton m. de résine. / ~ soap ||
Harzseife f. || savon m. à la résine. / ~
tapping || Harzgewinnung f. || résinage
m. / ~ torch || Pechfackel f. || torche f.
de résine.

resist, to || widerstehen || résister. / ~ ri-
gidly longitudinal stresses || Längsbean-
spruchungen fpl. unnachgiebig wider-
stehen || résister rigidement aux efforts
mpl. longitudinaux.

resist (Text print) || Reservage f.; Reserve
f.; Schutzbeize f. || réservage m.; réserve
f.

resistance (Electr) || Widerstand m. || ré-
sistance f. / ~ (Mech) || Festigkeit f.;
Widerstand m. || résistance f. / ~ to
abrasive wear || Verschleißfestigkeit f. ||
résistance f. à l'usure. / adjustable~|| Rheo-
stat m.; Regelwiderstand m. || rhéostat
m.; résistance f. réglable. / ~ of air || Luft-
widerstand m. || résistance f. de l'air. /
~ to breaking || Bruchfestigkeit f. || ré-
sistance f. à la rupture. / bridge ~ ||
Brückenwiderstand m. || résistance f.
à pont. / ~ to compressive strain ||
Druckfestigkeit f. || résistance f. à la
compression. / critical ~ || kritischer
Widerstand m. || résistance f. critique.
/ ~ to crushing by cross-breaking ||
Knickfestigkeit f. || résistance f. au
flambage. / ~ in curves (Railw) ||
Kurvenwiderstand m. || résistance f.
que les courbes opposent aux trains.
/ to cut-out ~ || den Widerstand m. aus-
schalten || mettre de la résistance hors
circuit. / ~ to cutting || Schnittwider-
stand m. || résistance f. à la coupe. /

~ of earth plate || Ausbreitungswider-
stand m. || résistance f. des plaques à la
terre. / electric ~ || elektrischer Wider-
stand m. || résistance f. électrique. /
exterior ~ || äußerer Widerstand m. ||
résistance f. extérieure. / fixed ~ || Ball-
astwiderstand m. || résistance f. fixe. / ~
of graphite || Grafitwiderstand m. || ré-
sistance f. de graphite. / ~ to heat ||
Hitzebeständigkeit f. || résistance f.
aux températures élevées. / ~ to the
effects of heat || Wärmebeständigkeit
f. || résistance f. contre les effets de la
chaleur. / high ~ || hoher oder großer
Widerstand m. || haute résistance f. /
high-frequency ~ || Hochfrequenzwider-
stand m. || résistance f. de haute fré-
quence. / high-ohmic ~ || Hochohm-
widerstand m. || résistance f. aux valeurs
ohmiques très élevées. / impact ~ ||
Kerbzähigkeit f. || ténacité f. sur barreaux
entaillés. / impact ~ expressed in kilo-
gram-meters for each square centimeter
section || Schlagwiderstand m. in Meter-
kilogramm für 1 qcm Querschnitt der
Probe || résistance au choc en kilo-
grammètres par cm² de section des
éprouvettes. / intermediate ~ || Über-
gangswiderstand m. || résistance f. au
passage. / internal ~ || innerer Wider-
stand m. || résistance f. intérieure. /
liquid ~ || flüssiger Widerstand m. || ré-
sistance f. liquide. / low ~ || niedriger oder
kleiner Widerstand m. || faible résistance
f. / ~ of the material || Werkstoffestig-
keit f. || résistance f. du matériel. /
negative ~ || negativer Widerstand m. ||
résistance f. négative. / non-inductive ~ ||
induktionsfreier Widerstand m. || ré-
sistance f. non-inductive. / reducing ~ ||
Abschwächungswiderstand m. || ré-
sistance f. modératrice. / ~ for reversing
starter || Umkehranlaßwiderstand m. ||
rhéostat m. inverseur de démarrage. /
rolling ~ (Railw) || Schienenreibung f. ||
friction f. des rails. / ~ to rolling (Aero) ||
Fahrwiderstand m. || résistance f. de roule-
ment. / ~ to rupture || Bruchfestigkeit f. ||
résistance f. à la rupture. / series ~ ||
Vorschaltewiderstand m. || résistance f.
additionnelle. / ~ to shearing strain ||
Scherfestigkeit f. || résistance f. au cisaille-
ment. / ~ against shocks || Stoßfestig-
keit f. || résistance f. aux chocs. / to
short-circuit the ~ || den Widerstand m.
kurzschließen || mettre la résistance en
court-circuit. / ~ with sliding contact ||
Schiebewiderstand m. || résistance f. à
contact glissant. / solid ~ || fester Wider-
stand m. || résistance f. solide. / steadying
~ || Vorschaltwiderstand m. || rhéostat
m. amortisseur. / stepped ~ || abgestuf-
ter Widerstand m. || résistance f. éta-
gée. / to switch-in ~ || Widerstand m.
einschalten || mettre de la résistance en
circuit. / ~ to tearing || Reißfestigkeit f. ||
résistance f. à la déchirure. / ~ to tensile
strain || Zugfestigkeit f. || résistance f.
à la traction. / three-decade ~ || Drei-
dekadenwiderstand m. || résistance f. de
trois décades. / ~ to torsional strain
|| Dreh(ungs)festigkeit f. || résistance
f. à la torsion. / total ~ || Gesamt-
widerstand m. || résistance f. totale. /
useful ~ || Nutzdämpfung f || amortisse-
ment m utile / ~ to wear || Verschleiß-
festigkeit f ; Verschleißwiderstand m. ||
résistance f à l'usure. / ~ of steel against

wear and tear ‖ Verschleißhärte f. des Stahles ‖ dureté f. de résistance à l'usure de l'acier. / ~ of a whole circuit ‖ Gesamtwiderstand m. ‖ résistance f. de tout un circuit.

resistance alloy ‖ Widerstandslegierung f. ‖ alliage m. pour rhéostats *ou* pour résistance. / amount of ~ ‖ Widerstandswert m. ‖ valeur f. de la résistance. / ~ amplifier ‖ Widerstandsverstärker m. ‖ amplificateur m à résistance / ~ box ‖ Widerstandskasten m. ‖ boîte f. de résistance. / decimal ~ box ‖ Dekadenwiderstandskasten m. ‖ boîte f. de résistance en décades. / ~ coil ‖ Widerstandsspirale f. ‖ spire f. de résistance. / standard ~ coil ‖ Normalwiderstand m. / étalon m. d'Ohm. / ~ coupling ‖ Widerstandskopplung f. ‖ accouplement m. par impédance *ou* à résistance. / ~ furnace ‖ Widerstandsofen m. ‖ four m. à résistance. / ~ head ‖ Widerstandshöhe f. ‖ hauteur f. de résistance. / ~ holder ‖ Widerstandshalter m. ‖ support m. de résistance. / measurement of ~ ‖ Widerstandsmessung f. ‖ détermination f. *ou* mesure f. de la résistance. / set for ~ measurements ‖ Widerstandsmesser m. ‖ appareil m. pour la mesure de résistances. / ~ metal for high temperatures ‖ Widerstandsmetall n. für hohe Temperaturen ‖ métal m. de résistance pour hautes températures. / moment of ~ ‖ Widerstandsmoment n. ‖ moment m. résistant. / ~ repeater *see* ~ strengthener. / ~ step ‖ Widerstandsstufe f. ‖ étage m. *ou* degré m. de résistance. / ~ strengthener ‖ Widerstandsverstärker m. ‖ amplificateur m. à résistance. / ~ tape ‖ Widerstandsband n. ‖ ruban m. métallique pour résistance. / ~ welding machine ‖ Widerstandsschweißapparat m. ‖ appareil m. à souder par résistance. / ~ wire ‖ Heizdraht m.; Widerstandsdraht m. ‖ fil m. métallique pour résistance. / work of ~ ‖ Widerstandsarbeit f. ‖ travail m. résistant.

resistant ‖ widerstandsfähig ‖ résistant. / the steel is extremely ~ to the action of boiling hot caustic solutions and dilute acids ‖ der Stahl ist außerordentlich widerstandsfähig gegen den Angriff von kochenden Laugen und verdünnten Säuren ‖ l'acier m. est très résistant contre l'influence de lessives bouillantes et d'acides étendus. / fire-~ ‖ feuersicher ‖ ignifugé. / ~ basic lining ‖ haltbares basisches Futter n. ‖ revêtement m. basique durable.

resisting (Text print) ‖ Reservieren n. ‖ réservage m. / non-~ ‖ unecht ‖ non-résistant.

resistivity ‖ Leitungswiderstand m. ‖ résistivité f. / ~ against fire ‖ Feuerbeständigkeit f. ‖ résistance f. au feu.

resist past (Text print) ‖ Schutzpapp m.; Deckpapp m.; Reservagepapp m. ‖ réserve f.

resolution ‖ Beschluß m. ‖ résolution f.; conclusion f. / ~ (Chem) ‖ Spaltung f. ‖ dédoublement m. / ~ of forces ‖ Zerlegung f. von Kräften ‖ décomposition f. de forces.

resolve, to ‖ beschließen ‖ résoudre; décider. / ~ (Chem) ‖ spalten ‖ dédoubler (graisses). / ~ (Math) ‖ lösen ‖ résoudre.

/ ~ the force ‖ die Kraft f. zerlegen ‖ décomposer la force f.

resolving very close binaries ‖ Auflösung f. engster Doppelsterne ‖ séparation f. des étoiles doubles très approchées. / ~ power ‖ Auflösungsvermögen n. / ~ (Chem) ‖ dissolvant *ou* résolvant; solubilité f.

resonance ‖ Resonanz f. ‖ résonnance f. / ~ curve ‖ Resonanzkurve f. ‖ courbe f. de résonnance. / ~ peak ‖ Resonanzspitze f. ‖ pointe f. à résonnance. / ~ transformer ‖ Resonanzinduktor m. ‖ transformateur m. à résonnance.

resonator ‖ Resonator m. ‖ résonateur m.

resorcin ‖ Resorzin n. ‖ résorcine f.

resorcinate explosive cartridge ‖ Resorzinatsprengkapsel f. ‖ capsule f. explosive à la résorcine.

resource, as a last ~ ‖ in letzter Instanz f. ‖ en dernier ressort m.

resources pl. ‖ Erwerbsmittel npl. ‖ ressource f.

respectability ‖ Reellität f. ‖ solidité f.; honnêteté f.

respectable firm ‖ reelles Geschäft n. ‖ affaire f. sérieuse; maison f. respectable.

respecting ‖ bezüglich ‖ concernant; touchant.

respirator ‖ Respirator m.; Atmungsapparat m. ‖ appareil m. respiratoire *ou* de respiration; respirateur m.

respite ‖ Verlängerung f. einer Frist; Nachfrist f. ‖ prolongation f.; prorogation f.; délai m.; sursis m. / to accord a ~ ‖ Frist f. gewähren ‖ accorder un répit. / to apply for a term of ~ ‖ Stundung f. verlangen ‖ demander un répit m. / to grant ~ for payment ‖ stunden ‖ accorder un délai. / grant of ~ ‖ Fristverlängerung f.; Fristgewährung f. ‖ délai m.; sursis m.; prorogation f.

respondent ‖ Berufungsbeklagter m. ‖ intimé m.

responsibility ‖ Haftpflicht f. ‖ responsabilité f. / joint ~ ‖ Gesamtverbindlichkeit f. ‖ responsabilité f. solidaire. / without any ~ ‖ ohne Verbindlichkeit f. ‖ sans responsabilité f.

responsible ‖ haftbar ‖ responsable.

rest ‖ Ruhe f. ‖ repos m. / ~ (Spectacles) ‖ Nasensteg m. ‖ pont m.; nez m. / at ~ ‖ in Ruhe f. ‖ au repos m.

restaurant ‖ Bierhalle f.; Restaurant n. ‖ restaurant m.; débit m. de bière. / ~ car ‖ Speisewagen m. ‖ wagon-restaurant m.

rest cask ‖ Restfaß n. ‖ foudre m. aux restes. / ~ contact ‖ Ruhekontakt m. ‖ contact m. de repos.

rested on ... ‖ gelagert auf *oder* in ... ‖ supporté par; appuyé sur; logé dans; monté sur ...

rest fermentation ‖ Rastgärung f. ‖ arrêt m. *ou* interruption f. de la fermentation.

resting place (Build) ‖ Podest m.; Treppenabsatz m. ‖ palier m. *ou* repos m. d'escalier. / ~ (Mine) ‖ Ruhebühne f. ‖ repos m. / trussed ~ ‖ gesprengter Podest m. ‖ claveau m. d'escalier.

resting spore ‖ Dauerspore f. ‖ spore f. durable.

restitution of aerial photographs ‖ Entzerrung f. von Flugaufnahmen ‖ restitution f. des photographies aériennes. / ~ of the picture ‖ Entzerrung f. der Bildaufnahme ‖ correction f. de la distorsion de la prise de vue. / ~ of a vessel ‖ Freigabe f. eines Schiffes ‖ restitution f. d'un bâtiment.

restocking ‖ Aufforstung f. ‖ boisement m.

restore, to (Arch) ‖ wiederherstellen ‖ restaurer. / ~ (Millstone) ‖ schärfen; picken; pillen ‖ repiquer; rhabiller. / ~ (Relay) ‖ abfallen ‖ décoller; relâcher; retomber.

restoring the previous condition ‖ Wiederherstellung f. des früheren Zustandes ‖ rétablissement m. en l'état antérieur.

rest, position of ~ ‖ Ruhelage f. ‖ position f. de repos.

restrainer ‖ Verzögerer m. ‖ retardateur m.

restrict, to ‖ einschränken ‖ restreinder; limiter.

restricted guidance ‖ zwangsläufige Führung f. ‖ guidage m. forcé. / ~ line ‖ nicht berechtigte Leitung f. ‖ ligne f. non-autorisée. / ~ space ‖ Platzmangel m. ‖ manque m. de place.

restriction, armament ~ ‖ Rüstungsbeschränkung f. ‖ limitation f. des armements.

result ‖ Ergebnis n.; Resultat n. ‖ résultat m.; produit m. / final ~ of the calculation ‖ Endergebnis n. der Rechnung ‖ résultat m. final du calcul. / ~ of the operation ‖ Ergebnis n. der Rechnung; Rechnungsergebnis n. ‖ résultat m. de l'opération. / ~ of test ‖ Versuchsergebnis n.; Prüfungsergebnis n. ‖ résultat m. de l'examen *ou* d'épreuve.

resultant ‖ resultierende Kraft f.; Resultante f. ‖ force f. résultante; résultante f.

resulting ‖ resultierend ‖ résultant.

resumption of work ‖ Wiederaufnahme f. der Arbeit ‖ reprise f. du travail.

ret, to (Flax) ‖ rösten; rötten ‖ naiser; rouir.

retail ‖ Kleinhandel m.; Kleinverkauf m. ‖ détail m.; vente f. en détail. / ~ business ‖ Kleinhandel m.; Einzelverkauf m. ‖ vente f. en détail; débit m. / ~ dealer ‖ Detailhändler m. ‖ détaillant m. / ~ goods pl. ‖ Schnittwaren fpl. ‖ mercerie f.

retailer *see* retail dealer.

retailing ‖ Detailverkauf m. ‖ vente f. au détail. / ~ apparatus (Restaurant) ‖ Ausschankapparat m. ‖ appareil m. de débit.

retail price ‖ Einzelpreis m. ‖ prix m. de détail. / sale by ~ ‖ Einzelverkauf m. ‖ vente f. en détail; débit m. / ~ selling *see* sale by ~.

retail trade ‖ Detailhandel m. ‖ commerce m. de détail. / dressed for the ~ ‖ in Aufmachung f. für den Kleinverkauf ‖ conditionné pour la vente en détail.

retain, to ‖ halten ‖ rappeler; retenir. / ~ the water ‖ das Wasser anstauen ‖ hausser *ou* élever les eaux.

retaining clip ‖ Haltebügel m. ‖ étrier m. de retenue. / ~ dam ‖ Stauwehr n. ‖ digue-réservoir m. / ~ hoop ‖ Arbeitsreif m. ‖ cercle m. de travail.

retaining ring ‖ Sprengring m. ‖ bague f. de retenue; agrafe f. circulaire *ou* annulaire. / screw-in ~ of interchanging lenses ‖ einschraubbarer Haltering m. zum Auswechseln der Linsen ‖ bague f. filetée permettant d'interchanger les lentilles. / ~ rolling machine ‖ Sprengringeinwalzmaschine f. ‖ machine f. à caler les bagues de retenue.

retaining spring ‖ Haltefeder f. ‖ ressort m. de retenue. / ~ valve ‖ Rückschlagventil n.; Absperrventil n. ‖ clapet m. *ou* soupape f. de retenue. / ~ wall ‖ Schutzmauer f. ‖ épaulement m.

/ ~ works pl. ‖ Stauwerk n.; Talsperre f. ‖ travaux mpl. de défense; ouvrages mpl. d'endiguement.

retanner ‖ Nachgerber m. ‖ retanneur m.

retard, to ‖ verzögern; verlangsamen ‖ retarder; ralentir. / ~ the current ‖ den Strom m. drosseln ‖ donner du retard m. au courant.

retardation ‖ Verzögerung f. ‖ retardation f. / ~ of the speed ‖ Abnahme f. der Geschwindigkeit ‖ diminution f. de la vitesse.

retarded ‖ verzögert ‖ retardé. / ~ ignition ‖ Spätzündung f. ‖ retard m. à l'allumage.

reticle (Opt) ‖ Fadenkreuz n. ‖ réticule m.

reticulated ‖ gestrickt ‖ tricoté. / ~ bond (Build) ‖ Netzverband m. ‖ maçonnerie f. maillée ou en échiquier ou réticulée; ouvrage m. réticulé. / ~ glass ‖ Filigranglas n. ‖ verre m. filigrané. / ~ work see ~ bond.

re-tighten, to ~ the screw ‖ die Schraube wieder anziehen ‖ reserrer la vis.

retile, to ~ a roof ‖ ein Ziegeldach umdecken ‖ remanier une toiture en tuiles.

retina ‖ Netzhaut f. ‖ rétine f. / detachment of the ~ ‖ Netzhautablösung f. ‖ décollement m. de la rétine. / image of the ~ ‖ Netzhautbild n. ‖ image f. de la rétine.

retinal camera ‖ Netzhautkammer f. ‖ chambre f. rétinienne.

retinasphalt ‖ Retinasphalt m.; Retinit m. ‖ rétinasphalte m.; rétinite f.

retinite see retinasphalt.

retire, to ~ from business. ‖ das Geschäft aufgeben ‖ cesser le commerce.

retirement ‖ Austritt m. ‖ retraite f.

retiring from business ‖ Aufgabe f. des Geschäftes ‖ cessation f. du commerce.

retort, to ‖ mittels einer Retorte f. trennen ‖ distiller.

retort ‖ Retorte f. ‖ cornue f.; retorte f. / clay ~ ‖ Tonretorte f. ‖ cornue f. en terre. / earthen ~ ‖ irdene Retorte f. ‖ cornue f. de grès. / ~ of fire-clay ‖ Schamotteretorte f. ‖ retorte f. en terre réfractaire. / pivoting ~ ‖ Kippretorte f. ‖ cornue f. pivotante. / rotatory ~ ‖ rotierende Retorte f. ‖ retorte f. à fond mobile. / tubulated ~ ‖ Röhrenretorte f. ‖ cornue f. ou retorte f. tubulée. / welded ~ ‖ geschweißte Retorte f. ‖ cornue f. soudée.

retort brick ‖ Retorte stein m. ‖ brique f. à cornue. / ~ carbon ‖ Elementkohle f.; Retortenkohle f. ‖ charbon m. de cornue. / ~ cement ‖ Retortenkitt m. ‖ mastic m. à cornues. / charging machine for ~s ‖ Lademaschine f. für Retorten ‖ machine f. à charger les cornues. / ~ coke ‖ Gaskoks m. ‖ coke m. de gaz; coke m. d'usine à gaz. / ~ contact method ‖ Retortenkontaktverfahren n. ‖ procédé m. de contact dans des cornues. / ~ contact process see ~ contact method. / ~ emptier ‖ Retortenleerer m. ‖ déchargeur m. de cornues. / ~ filler ‖ Retortenfüller m. ‖ chargeur m. de cornues. / ~ furnace ‖ Retortenofen m. ‖ four m. à cornue. / ~ graphite ‖ Retortengraphit m. ‖ graphite m. de cornue. / ~ house ‖ Retortenhaus n. ‖ hall m. des cornues. / inclined ~ oven ‖ Schrägretortenofen m. ‖ four m. à cornues inclinées.

retouch, to ‖ retuschieren ‖ retoucher; enjoliver. / ~ a photographic negative ‖ ein fotografisches Negativ retuschieren ‖ retoucher un cliché photographique.

retoucher ‖ Retuschierer m. ‖ retoucheur m.

retouching ‖ Retusche f. ‖ retouche f. / ~ medium ‖ Retuschierfarbe f. ‖ encre f. pour retouche.

retractable landing gear (Aero) ‖ einziehbares Fahrgestell n. ‖ atterrisseur m. escamotable. / ~ radiator ‖ einziehbarer Kühler m. ‖ radiateur m. amovible ou escamotable.

retractile spring ‖ Abreißfeder f. ‖ ressort m. antagoniste.

retractor (Med) ‖ Wundhaken m. ‖ écarteur m.

retransmission of a message to follow ‖ Nachsendung f. einer Depesche ‖ réexpédition f. d'un télégramme à faire suivre.

retransmit, to ~ a message ‖ eine Depesche f. nachsenden ‖ faire suivre ou réexpédier une dépêche.

retreading (Auto) ‖ Neubelegen n. der Laufdecke ‖ rechapage m.

retreating part (Build) ‖ Hinterflügel m. ‖ arrière-corps m.

retrograde ‖ rückläufig ‖ rétrograde. / ~ motion ‖ Rückwärtsbewegung f. ‖ marche f. en arrière.

retrospective force ‖ rückwirkende Kraft f.; Rückwirkung f. ‖ effet m. rétroactif; réaction f.

retting of flax ‖ Flachsröste f.; Rösten n. des Flachses ‖ rouissage m. du lin. / after- ~ of flax ‖ Nachrotte f. des Flachses ‖ curage m. de lin. / ~ of hemp ‖ Hanfröste f.; Rösten n. des Hanfes ‖ rouissage m. du chanvre. / mixed ~ ‖ gemischte Rotte f. ‖ rouissage m. à la méthode mixte. / warm-water ~ ‖ Warmwasserrotte f. ‖ rouissage m. à l'eau chaude.

return, to ‖ zurückkehren ‖ retourner. / ~ to normal ‖ in die Ruhelage f. zurückkehren ‖ retourner en position f. normale. / ~ a salute ‖ Salut m. erwidern ‖ rendre un salut. / when empty return to . . . ‖ leer zurück nach . . . ‖ vide retour à . . . .

return ‖ Rückgabe f. ‖ restitution f. / ~ (Arch) ‖ Einsprung m. ‖ saillie f. / ~ (Build) ‖ Seitenflügel m. ‖ aile f. latérale. / ~ (Mach) ‖ Rücklauf m. ‖ course-arrière f. / ~ of a bank ‖ Bankbericht m. ‖ situation f. d'une banque. / ~ of charges ‖ Erstattung f. von Gebühren ‖ remboursement m. de taxes. / ~ by frame ‖ Rückleitung f. durch die Masse ‖ retour m. par la masse. / hand ~ ‖ Handrückzug m. ‖ retour m. à main. / ~ of insurance-premium ‖ Ristorno n. ‖ ristorne f. / ~ of the key (Typewr) ‖ Zurückschlagen n. der Taste ‖ retour m. de la touche. / ~ of the lever ‖ Rückgang m. des Hebels ‖ recul m. du levier. / by ~ of mail ‖ postwendend ‖ par retour m. du courrier. / ~ of movement ‖ Bewegungsumkehr f. ‖ renversement m. du mouvement. / rail ~ ‖ Rückleitung f. durch die Schienen ‖ retour m. par les rails. / ~ of the type bars ‖ Zurückfallen n. der Typenhebel ‖ retour m. des tiges à caractères.

return cable ‖ Rückleitungskabel n. ‖ câble m. de retour. / ~ cargo ‖ Rückfracht f. ‖ port m. ou charge f. de retour. / ~ conductor ‖ Rückleiter m. ‖ conducteur m. de retour. / ~ crank ‖ Gegenkurbel f. ‖ contre-manivelle f.

return current (Electr) ‖ Rückstrom m. ‖ courant m. de retour. / ~ cut-out ‖ Rückstromausschalter m. ‖ disjoncteur m. à retour de courant. / ~ relay ‖ Rückstromschütz n. ‖ relais m. d'inversion de courant.

return earth ‖ See-Erde f. (im Kabelbetrieb) ‖ terre f. au large. / ~ feeder ‖ Schienenspeisekabel n. ‖ câble m. de retour. / ~ flow ‖ Rückströmung f. ‖ courant m. ou écoulement m. de retour. / ~ freight ‖ Rückfracht f. ‖ port m. ou charge f. de retour. / ~ gear ‖ Rücklaufgetriebe n. ‖ harnais m. de renversement de marche. / ~ motion see ~ movement. / ~ movement ‖ Rücklauf m. ‖ mouvement m. de retour; course f. arrière. / ~ piping ‖ Rückleitung f. ‖ conduite f. de retour. / ~ pulley ‖ Umleitungsrolle f. ‖ galet m. de retour. / ~ pump ‖ Rückförderpumpe f. ‖ pompe f. de refoulement. / ~ remittance ‖ Gegenrimesse f. ‖ retour m.

returns pl. (Bank) ‖ Geldumsatz m.; Geldverkehr m. ‖ mouvement m. ou maniement m. de fonds; roulement m.; mouvement m. ou circulation f. monétaire. / ~ (Trade) ‖ Erwerb m. ‖ gain m.; profit m. / estimated ~ ‖ veranschlagter Ertrag m. ‖ évaluation f. des rendements. / malt ~ ‖ Schlempe f.; Malztreber pl. ‖ drêche f. / quick ~ ‖ schneller Umsatz m. ‖ vente f. ou débit m. rapide.

return service ‖ Gegenleistung f. ‖ équivalent m.; revanche f. / ~ shock (Tel) ‖ Rückschlag m. ‖ choc m. en retour. / ~ speed ‖ Rücklaufgeschwindigkeit f. ‖ vitesse f. de retour. / ~ stroke (Mach tool) ‖ Leerlaufhub m. ‖ course f. de retour. / ~ ticket ‖ Rückfahrkarte f.; Rückfahrschein m. ‖ billet m. d'aller et retour; aller et retour m. / ~ valve (Steam) ‖ Rückschlagklappe f.; Rückschlagventil n. ‖ clapet m. ou soupape f. de retenue. / ~ wire (Electr) ‖ Rückleiter m. ‖ fil m. de retour.

revalorization ‖ Aufwertung f. ‖ revalorisation f.

reveal (Build) ‖ äußere Fensterleibung f. ‖ tableau m.

revenue ‖ Einkommen n. ‖ revenu m. / approximate ~ ‖ ungefähres Einkommen n. ‖ revenu m. approximatif. / inland ~ ‖ Akzise f. ‖ accise f.; octroi m.; contributions fpl. indirectes.

revenue cutter ‖ Zollkutter m. ‖ cotre m. de la douane. / ~ law ‖ Zollgesetz n. ‖ loi f. douanière. / ~ office ‖ Rentamt n. ‖ bureau m. des finances; recette f.

reverberatory furnace ‖ Flammofen m. ‖ four m. à reverbère. / refining in the ~ ‖ Flammofenfrischen n. ‖ affinage m. de la fonte au four à réverbère.

reverberatory kiln see reverberatory furnace.

reversal of the magnetic needle ‖ Umschlagen n. der Magnetnadel ‖ retournement m. de l'aiguille aimantée.

reverse, to (Auto) ‖ rückwärts fahren ‖ renverser la marche. / ~ (Electr) ‖ umpolen ‖ inverser les pôles. / ~ (Mach) ‖ umsteuern ‖ renverser. / ~ an engine ‖ eine Maschine f. umsteuern ‖ changer la marche d'une machine. / ~ by x° ‖ um x° versetzen ‖ faire tourner de x°.

reverse ‖ Gegenteil n. ‖ contraire m. / ~ (Coin) ‖ Rückseite f. ‖ verso m. / ~ (Print) ‖ Kehrseite f.; Rückseite f.; gerade Seite f.; Kolumne f. ‖ page f. paire; verso m. / ~ (Trade) ‖ Rückschlag m. ‖ contre-coup m.; revirement.

reverse current ‖ Gegenstrom m. ‖ courant m. contraire. / ~ ‖ Rückstrom m. ‖ courant m. de retour. / ~ switch ‖ Rückstromausschalter m. ‖ interrupteur m. à retour de courant.

reverse cylinder ‖ Umkehrwalze f. ‖ cylindre-inverseur m. / ~ gear (Auto) ‖ Rückwärtsgang m. ‖ marche f. arrière. / ~ motion (Mach) ‖ Rückwärtsgang m. ‖ marche f. (en) arrière. / ~ movement ‖ Rückwärtsfahrt f. ‖ marche f. arrière. / ~ order ‖ umgekehrte Reihenfolge f. ‖ ordre m. inverse.

reverser (Electr) ‖ Wender m. ‖ inverseur m.

reverse reeling machine for stockinets ‖ Umkehraufrollmaschine f. für Trikotstoffe ‖ machine f. enrouler les tricotages. / ~ transformation ‖ Rückverwandlung f. ‖ transformation f. inverse.

reversibility ‖ Umkehrbarkeit f. ‖ réversibilité f.

reversible (Chem) ‖ umkehrbar ‖ réversible. / ~ (Mach) ‖ umsteuerbar ‖ réversible. / ~ cell ‖ umkehrbares Element n. ‖ pile f. réversible. / ~ crossing ‖ umwendbare Kreuzung f. ‖ croisement m. à retournement. / ~ delivery (Print) ‖ umsteuerbarer Ausleger m. ‖ sortie f. de feuilles réversible. / each mill is driven by a ~ engine ‖ jede Straße f. wird durch einen umsteuerbaren Motor angetrieben ‖ chaque train m. est actionné par une machine avec mouvement réversible. / ~ lamp ‖ Kipplampe f. ‖ lampe f. à bascule. / ~ marine Diesel engine ‖ umsteuerbarer Schiffsdieselmotor m. ‖ moteur m. marin Diesel directement réversible. / ~ motor ‖ umsteuerbarer Motor m. ‖ moteur m. réversible. / ~ pilot coupler ‖ Hornschwenkkopf m. ‖ attelage m. à griffe pivotante. / ~ propeller ‖ Drehflügelschraube f.; Wendeschraube f. ‖ hélice f. réversible. / ~ screw arrangement for not reversible motors ‖ Wendeschraubenanlage f. für nicht umsteuerbare Motoren ‖ mécanisme m. de l'hélice réversible pour moteurs non réversibles. / ~ shears pl. ‖ Umkehrschere f. ‖ cisaille f. réversible. / ~ spirit level ‖ Wendelibelle f. ‖ libelle f. ou nivelle f. réversible. / ~ switch ‖ Kippschalter m. ‖ interrupteur m. à bascule.

reversing of the movements ‖ Umkehren n. der Bewegungen ‖ renversement m. des mouvements. / ~ bar ‖ Umsetzschubstange f. ‖ tige f. de renversement. / ~ device ‖ Umsteuervorrichtung f. ‖ dispositif m. de changement de marche. / ~ engine ‖ umsteuerbare Maschine f. ‖ machine f. réversible.

reversing gear ‖ Umsteuerung f.; Umsteuergetriebe n. ‖ mécanisme m. de changement de marche ou de renversement. / friction ~ ‖ Friktionswendegetriebe n. ‖ dispositif m. de renversement de marche à friction.

reversing handle ‖ Umsteuerhebel m. ‖ levier m. de renversement ou de changement de marche. / ~ key (Tel) ‖ Doppeltaste f. für den Kabelmeßdienst ‖ manipulateur m. double. / lever ~ ‖ Umstell-

hebel m. ‖ levier m. de renversement. / ~ lever (Loc) ‖ Steuerstange f.; Steuerungshebel m. ‖ levier m. de relevage. / ~ mill ‖ Umkehrstraße f. ‖ train m. réversible. / ~ pole ‖ Wendepol m. ‖ pôle m. de commutation. / ~ relay for clock installations ‖ Stromwenderelais n. für Uhrenanlagen ‖ relais m. d'inversion de courant pour installations d'horloges électriques. / ~ and starting resistance ‖ Umkehranlaßwiderstand m. ‖ rhéostat m. de démarrage inverseur. / ~ rolling mill ‖ Reversierwalzwerk n. ‖ laminoir m. réversible. / ~ shaft ‖ Umsteuerwelle f. ‖ arbre m. de changement de marche. / ~ sheave ‖ Umlenkscheibe f. ‖ poulie f. de renvoi. / ~ slide ‖ Wendesupport m. ‖ support m. à renversement. / ~ strip mill ‖ Umkehrbandwalzwerk n. ‖ laminoir m. réversible à feuillards.

reversing switch ‖ Polwender m.; Stromwender m. ‖ inverseur m. de polarité. / ~ drum ‖ Umschaltwalze f. ‖ cylindre m. de renversement de marche.

reversing valve ‖ Umsteuerschieber m. ‖ tiroir m. de distribution. / ~ (Mot) ‖ Umsteuerventil n. ‖ soupape f. de renversement de marche.

reversion ‖ Anwartschaft f. ‖ expectative f.; droit m. / ~ (Electr) ‖ Umpolung f. ‖ inversion f. des pôles. / ~ of the image ‖ Bildaufrichtung f. ‖ redressement m. de l'image.

revertive control (Aut tel) ‖ Rückkontrolle f. ‖ contrôle m. reversible. / ~ selection ‖ Impulsrücksendung f. ‖ renvoi m. d'impulsion.

review ‖ Zeitschrift f. ‖ revue f.

revise, to ‖ korrigieren; Korrektur lesen; revidieren ‖ corriger; lire ou relire une épreuve.

revise ‖ Revisionsbogen f. ‖ seconde f.; seconde épreuve f. / second ~ ‖ zweiter Revisionsbogen m. ‖ tierce f.

revise proof ‖ Korrekturabzug m. ‖ épreuve f. au rouleau ou à la brosse.

reviser ‖ Korrektor m. ‖ correcteur m.

revision ‖ Revision f. ‖ révision f.

revive, to ‖ wieder beleben ‖ raviver. / ~ copper ‖ das Kupfer frischen ‖ rafraîchir le cuivre. / ~ the fire ‖ Feuer n. anschüren ‖ activer ou aviver le feu. / ~ litharge ‖ die Bleiglätte frischen ‖ raffiner la litharge.

revocation ‖ Aufkündigung f. ‖ congé m.; révocation f.

revolution ‖ Umwälzung f.; Umschwung f.; Revolution f. ‖ révolution f.; revirement m. / ~ (Mach) ‖ Umdrehung f.; Umlauf m. ‖ révolution f.; tour m. / complete ~ ‖ volle Umdrehung f. ‖ tour m. complet. / ~s pl. per minute ‖ Drehzahl f. in der Minute ‖ tours-minute m. / axis of ~ ‖ Umdrehungsachse f. ‖ axe m. de rotation. / ~ counter ‖ Umdrehungszähler m. ‖ Tachometer n. ‖ compte-tours m.; tachymètre m. / ~ ellipsoid ‖ Rotationsellipsoid n. ‖ ellipsoïde m. de révolution. / ~ indicator ‖ Drehzahlmesser m. ‖ compte-tours m. / decrease in the number of ~s ‖ Abfallen n. der Drehzahl ‖ diminution f. du nombre de tours. / rise in the number of ~s (Mach) ‖ Steigen n. der Drehzahl ‖ accroissement m. du nombre de tours. / ~ paraboloid ‖ Drehungsparaboloid n. ‖ paraboloïde m. de révolution. / range of ~s ‖

Drehzahlbereich m. ‖ limites fpl. des nombres de tours. / time of ~ ‖ Umlaufzeit f. ‖ durée f. d'une révolution.

revolve, to ‖ kreisen; drehen; rotieren ‖ tourner; circuler; tourillonner; pivoter.

revolver ‖ Revolver m. ‖ revolver m. / double-~ (Telescope) ‖ zweifacher Revolver m. ‖ revolver m. double.

revolver box (Weav) ‖ Revolverlade f.; Drehwebstuhl m. ‖ battant m. revolver; métier m. à revolver. / ~ holster ‖ Revolvertasche f. ‖ fonte f. de revolver. / ~ loom ‖ Drehwebstuhl m.; Revolverstuhl m. ‖ métier m. revolver.

revolving bath ‖ Karussellbad n. ‖ bain m. tournant. / ~ breech ‖ Revolvergehäuse n. ‖ rempart m. de carcasse. / ~ chair ‖ Drehstuhl m.; Drehsessel m. ‖ taburet m.; chaise f. à vis. / ~ colour disc ‖ Farbglasrevolver m. ‖ revolver m. à verres colorés. / ~ crane ‖ Drehkran m. ‖ grue f. pivotante. / hand-worked ~ crane ‖ von Hand getätigter Drehkran m. ‖ grue f. à bras ou pivotante à main. / ~ cylinders pl. ‖ umlaufende Zylinder mpl. ‖ cylindres mpl. rotatifs. / ~ eyepiece changer furnished with three eyepieces ‖ Okularrevolver m. mit drei Okularen ‖ revolver m. à trois oculaires. / ~ eyepiece head ‖ Okularrevolver m. ‖ revolver-oculaire m. / ~ four-hole punch pliers ‖ Revolverkopflochzange f. mit vier Lochhülsen ‖ pince f. emporte-pièce avec étoile à quatre tubes. / ~ frame ‖ Getriebequerhaupt n. ‖ traverse f. tournante. / ~ grate ‖ Drehrost m. ‖ grille f. tournante. / ~ grate gas producer ‖ Drehrostgaserzeuger m. ‖ gazogène m. à grille tournante. / ~ head for lathe ‖ Revolverkopf m. für Drehbank ‖ tourelle f. révolver pour la tour. / ~ illuminated advertising device ‖ drehbare Lichtreklamevorrichtung f. ‖ appareil m. tournant pour l'illumination de réclames. / ~ iron instrument ‖ Dreheiseninstrument n. ‖ instrument m. de mesure à fer tournant. / ~ light ‖ Drehfeuer n. ‖ feu m. tournant. / ~ objective changer ‖ Objektivrevolver m. ‖ revolver m. à objectifs. / ~ one-ledged gantry-crane ‖ schwenkbarer Halbportalkran m. ‖ grue f. tournante à demi-portique. / ~ pistol ‖ Revolver m. ‖ pistolet m. revolver ou à cylindre tournant; revolver m. / ~ press ‖ Revolverpresse f. ‖ presse f. revolver. / ~ screen ‖ Drehsieb n. ‖ trommel-classeur m. / ~ seat ‖ Drehsitz m. ‖ siège m. tournant. / ~ shackle suspension ‖ Drehbügelaufhängung f. ‖ suspension f. à étrier tournant. / ~ shutter ‖ Rolladen m. ‖ volet m. à rideaux. / ~ slide-rest (Turn) ‖ Kugelsupport m. ‖ chariot m. circulaire ou tournant ou pivotant. / ~ stirrup ‖ drehbarer Bügel m. ‖ étrier m. pivotant. / ~ table press ‖ Drehtischpresse f. ‖ presse-revolver f. / ~ tubular kiln ‖ Drehrohrofen m. ‖ four m. tubulaire tournant. / ~ wagon crane ‖ Wagendrehkran m. ‖ grue f. pivotante sur wagon.

rewash, to ‖ nachwaschen ‖ relaver.

rewashing ‖ Nachwäsche f. ‖ relavage m. / ~ machine ‖ Nachwaschsetzmaschine f. ‖ machine f. de relavage; releveur m.

rewind, to ‖ umspulen ‖ rebobiner.

rewinding ‖ Neuwicklung f. ‖ réfaction f. des enroulements. / ~ the ribbon

(Typewr) ‖ Umspulen n. des Farbbandes ‖ rembobinage m. du ruban.

rewinding machine (Pap) ‖ Umrollmaschine f. ‖ machine f. à enrouler; enrouleur m. / ~ mechanism ‖ Spulmechanismus m. ‖ mécanisme m. d'enroulement *ou* de bobinage.

rhabdite ‖ Nadeleisen n.; Rhabdit m. ‖ rhabdite f.

Rhætian system ‖ rhätisches System n. ‖ système m. rhétien.

rhea ‖ Rhea f.; indische *oder* grüne Nessel f. ‖ ramie f. verte; ortie f. des Indes.

rheostat ‖ Rheostat m.; Regelwiderstand m. ‖ rhéostat m.; résistance f. réglable. / decimal ~ ‖ Dekadenrheostat m. ‖ rhéostat m. en décades. / series ~ ‖ Serienwiderstand m. ‖ rhéostat m. en série. / ~ for starting ‖ Anlaßwiderstand m. ‖ rhéostat m. de mise en marche; résistance f. de démarrage.

Rheum ‖ Rhabarber m. ‖ rhubarbe f.

rhind (Mill) ‖ Haue f.; Mühlhaue f. ‖ anille f.

rhinestone ‖ Similistein m. ‖ pierre f. de strass.

rhinological instrument ‖ Naseninstrument n. ‖ instrument m. rhinologique.

rhodicite ‖ Kalkborazit m.; Rhodizit m. ‖ rhodicite f.

rhodium ‖ Rhodium n. ‖ rhodium m. / ~ gold ‖ Rhodiumgold n. ‖ rhodium m. aurifère; rhodite f. / ~ oil ‖ Rosenholzöl n. ‖ essence f. de bois de rose.

rhodochrome (Miner) ‖ Rhodochrom n. ‖ rhodochrome m.

rhodochrosite ‖ Manganspat m.; Rhodochrosit m. ‖ manganèse m. carbonaté.

rhodonide ‖ Rhodanmetall n.; Rhodanverbindung f.; Schwefelzyanverbindung f. ‖ sulfocyanide m.

rhodonite ‖ Manganstein m.; Mangankiesel m. ‖ rhodonite f.

rhomb ‖ Raute f.; Rhombus m. ‖ losange m.

rhombic ‖ rhombisch; rautenförmig ‖ rhombique. / ~ dodecahedron ‖ Rhombendodekaeder m. ‖ dodécaèdre m. rhomboïdal. / ~ mica ‖ Phlogopit m. ‖ phlogopite f.

rhombohedral ‖ rhomboedrisch ‖ rhomboédrique.

rhombohedron ‖ Rhomboeder n. ‖ rhomboèdre m.

rhomboid ‖ Rhomboid n. ‖ rhomboïde m.

rhomboidal ‖ rautenförmig; rhombisch ‖ rhomboïdal. / ~ magnetic needle ‖ rautenförmige Magnetnadel f. ‖ aiguille f. aimantée en forme de losange.

rhomb spar ‖ Magnesitspat m.; Rautenspat m. ‖ magnésite f. cristalline.

rhombus see rhomb.

rhubarb ‖ Rhabarber m. ‖ rhubarbe f. / ~ root ‖ Rhabarberwurzel f. ‖ racine f. de rhubarbe.

rhumb ‖ Kompaßstrich m.; Rhumb m. ‖ aire f. *ou* rumb m. de vent. / ~ line ‖ Dwarslinie f.; Loxodrome f.; Rhumblinie f. ‖ loxodromie f. / ~ point see rhumb.

rhynd see rhind.

rib, to ‖ riffeln; mit Rippen fpl. versehen ‖ rainurer; canneler.

rib (Aero) ‖ Spiere f. ‖ nervure f. / ~ (Anatomy) ‖ Rippe f. ‖ côte f. / ~ (Arch) ‖ Gewölbrippe f.; Rippe f. ‖ côte f.; lierne f.; nerf m. *ou* nervure f. de voûte. / ~ (Mach) ‖ Rippe f. ‖ nervure f. /

(Mill) ‖ Leiste f. ‖ liteau m. / ~ (Mine) ‖ Sicherheitspfeiler m. ‖ pilier m. de sûreté. / ~ (Shipb) ‖ Spant n. ‖ couple m. / ~ on the back of a book ‖ Rippe f. auf dem Buchrücken ‖ nerf m. sur le dos d'un livre. / cast-on ~ ‖ angegossene Rippe f. ‖ ailette f. venue de fonte. / ~ of a centering ‖ Gerüstrippe f. ‖ ferme f. de cintre. / compression ~ of the wing *see* rib, principal. / cooling ~ ‖ Kühlrippe f. ‖ ailette f. de refroidissement. / diagonal ~ ‖ Diagonalrippe f. ‖ nervure f. diagonale. / ~ of a dome ‖ Kuppelrippe f. ‖ côte f. de dôme. / principal ~ of the wing ‖ Hauptrippe f. des Flügels ‖ nervure f. principale d'aile; nervure f. de compression d'aile. / stiffening ~ ‖ Verstärkungsrippe f. ‖ nervure f. de renforcement.

riband (Shipb) ‖ Sente f. ‖ lisse f. (de construction et d'exécution). / intermediate ~ (Shipb) ‖ Zwischensente f. ‖ lisse f. intermédiaire.

ribband *see* riband.

ribbed ‖ gerippt; geriffelt ‖ nervuré; ridé; gaufré; cannelé. / transversally ~ ‖ quer gerippt ‖ à nervures fpl. transversales.

ribbed back-plate (Dent) ‖ gerippte Gaumenplatte f. ‖ plaque f. de dentier à nervure. / ~ glass ‖ geripptes Glas n.; Rippenglas n. ‖ verre m. rayé *ou* à nervures. / ~ heating member *see* radiator. / ~ hose ‖ Ränderstrumpf m. ‖ bas m. à côte. / ~ pipe ‖ Rippenrohr n. ‖ tuyau m. à ailettes. / ~ plate ‖ Rippenplatte f. ‖ plaque f. à nervures. / ~ radiator ‖ Rippenheizkörper m. ‖ radiateur m. à ailettes. / ~ spring leaf ‖ Rippenfeder f. ‖ ressort m. à nervures. / ~ spring steel ‖ gerippter Federstahl m. ‖ acier m. à ressorts cannelé. / ~ stocking ‖ gerippter Strumpf m. ‖ bas m. à côtes.

ribbing machine (Stockinger) ‖ Rändermaschine f. ‖ machine f. à bords-côtes.

ribbon ‖ Band n. ‖ ruban m. / ~ (Typewr) ‖ Farbband n. ‖ ruban(-encreur) m. / ~ for caps ‖ Mützenband n. ‖ ruban m. pour casquettes. / ~ for cigarette machines ‖ Zigarettenmaschinenband n. ‖ ruban m. pour machines à cigarettes. / elastic ~ ‖ Gummiband n. ‖ ruban m. élastique; élastique m. / endless ~ ‖ endloses Band n. ‖ ruban m. sans fin. / ~ for gentlemen's hats ‖ Herrenhutband n. ‖ ruban m. pour chapeaux d'hommes. / inking ~ (Typewr) ‖ Farbband n. ‖ ruban-encreur m. / inking ~ crank (Typewr) ‖ Farbbandkurbel f. ‖ manivelle f. du ruban-encreur. / non-drying ~ (Typewr) ‖ nicht eintrocknendes Farbband n. ‖ ruban m. ne desséchant pas. / non-fading ~ (Typewr) ‖ nicht verblassendes Farbband n. ‖ ruban m. indélébile. / plain ~ ‖ glattes Band n. ‖ ruban m. uni. / purple ~ (Typewr) ‖ Violettfarbband n. ‖ ruban-encreur m. violet. / red ~ (Typewr) ‖ rotes Farbband n. ‖ ruban-encreur m. rouge. / steel in the shape of ~ ‖ Stahl m. in Bandform ‖ acier m. sous forme de ruban. / silk ~ (Typewr) ‖ seidenes Farbband n. ‖ ruban-encreur m. en soie. / stationary ~ ‖ feststehendes Band n. ‖ ruban m. fixe. / student's ~ ‖ Studentenband n. ‖ ruban m. pour étudiants. / three-coloured ~ (Typewr) ‖ Dreifarbband n. ‖ ruban m. tricolore. / torn ~

(Typewr) ‖ abgenutztes Farbband n. ‖ ruban m. déchiré. / ~ for typewriters ‖ Band n. für Schreibmaschinen ‖ ruban m. pour machines à écrire. / unwound ~ (Typewr) ‖ abgelaufenes Farbband n. ‖ ruban m. arrivé à fin de course. / ~ of wadding for packing ‖ Watteband n. zum Dichten ‖ ruban m. d'ouate pour calfeutrer. / ~ of wadding for upholstering ‖ Watteband n. zum Polstern ‖ ruban m. d'ouate pour rembourrer. / ~ of wax ‖ Wachsband n. ‖ ruban m. de cire. / the ~ is wound up (Typewr) ‖ das Farbband ist aufgelaufen ‖ le ruban est enroulé.

ribbon arrangement (Typewr) ‖ Farbbandanordnung f. ‖ disposition f. du ruban. / ~ conveyor ‖ Gurtförderer m. ‖ bande f. transporteuse. / ~ cutting machine ‖ Bandschneidemaschine f. ‖ machine f. à refendre les rubans.

ribboner ‖ Bortennäher m. ‖ poseur m. de rubans.

ribbon folder ‖ Bandwickler m. ‖ enrouleur m. de rubans. / ~ guide ‖ Bandführung f. ‖ guidage m. du ruban. / ~ lightning ‖ Bandblitz m. ‖ éclair m. en forme de ruban.

ribbon loom ‖ Bandwebstuhl m. ‖ métier m. à ruban. / ~ gaiter ‖ Bandstuhleinrichter m. ‖ appareilleur m. de métiers à rubans.

ribbon mechanism (Typewr) ‖ Bandmechanismus m. ‖ mécanisme m. du ruban. / movement of the ~ (Typewr) ‖ Farbbandtransport m. ‖ avance f. du ruban. / direction of the movement of the ~ (Typewr) ‖ Laufrichtung f. des Farbbandes ‖ direction f. de la course du ruban. / ~ plate ‖ schmales Schloßblech n. ‖ tôle f. à palâtre. / ~ reverse lever ‖ Farbbandumstellhebel m. ‖ levier m. de renversement de marche du ruban. / ~ road ‖ Straße f. in Windungen ‖ route f. en lacets. / ~ spool (Pap) ‖ Spule f. für Bänder ‖ tambour m. à enrouler les rubans. / ~ spool (Typewr) ‖ Farbbandspule f. ‖ bobine f. de ruban-encreur. / ~ support (Typewr) ‖ Farbbandträger m. ‖ support m. du ruban-encreur. / ~ transmitter (Tel) ‖ Bandmikrophon n. ‖ microphone m. avec plaque. / ~ weaver ‖ Bandweber m. ‖ rubanier m. / ~ weaving ‖ Bandweberei f. ‖ rubanerie f. / ~ weaving loom ‖ Bandwebstuhl m. ‖ métier m. à rubans.

rib flange ‖ Deckleiste f. ‖ lisse f. de recouvrement; couvre-joint m.

rib plate ‖ Rippen(unterlags)platte f. ‖ platine f. à nervures. / ~ milling machine ‖ Rippenplattenfräsmaschine f. ‖ fraiseuse f. pour selles à nervures. / ~ track ‖ Rippenplattenoberbau m. ‖ superstructure f. à selles à nervures. / ~ track with vertical rail ‖ Rippenplattenoberbau m. mit senkrecht stehender Schiene ‖ superstructure f. à selles à nervures avec rail vertical.

rib saw ‖ Schweifsäge f.; Stellsäge f. ‖ scie f. à tourner *ou* à chantourner. / ~ system of a vault ‖ Rippengestell n. eines Gewölbes ‖ croisée f. de voûte. / ~ top (Stockinger) ‖ Strumpfrand m. ‖ bordcôte m. / ~ top frame (Stockinger) ‖ Ränderstuhl m. ‖ métier m. pour bordscôtes. / ~ web ‖ Rippensteg m. ‖ âme f. de nervure.

rice ‖ Reis m. ‖ riz m. / broken ~ ‖ Bruchreis m. ‖ riz m. concassé. / ground ~ ‖ Reisgrieß m. ‖ semoule f. de riz. / ~ in the

husk ‖ Reis m. im Stroh ‖ riz m. en paille. / peeled ~ ‖ enthülster Reis m. ‖ riz m. pelé. / polished ~ ‖ polierter Reis m. ‖ riz m. glacé. / unhusked ~ ‖ nicht enthülster Reis m. ‖ riz m. non pelé.

rice flour ‖ Reismehl n. ‖ farine f. de riz. / ~ mill ‖ Reismühle f. ‖ moulin m. à riz; rizerie f. / ~ miller ‖ Reismüller m. ‖ décortiqueur m. de riz; rizeur m. / ~ powder ‖ Reispuder m. ‖ poudre f. de riz. / ~ starch ‖ Reisstärke f. ‖ amidon m. de riz. / ~ straw ‖ Reisstroh n. ‖ paille f. de riz.

rich ‖ reichhaltig ‖ abondant; riche; bien assorti. / ~ in ozone ‖ ozonreich ‖ riche en ozone m. / ~ coal ‖ Fettkohle f. ‖ charbon m. gras; houille f. grasse. / ~ slag ‖ Garschlacke f. ‖ scorie f. douce ou riche; laitier m. riche.

riddle, to ‖ sieben; aussieben; rättern ‖ cribler; tamiser.

riddle ‖ Durchschlag m.; Rätter m. ‖ crible m.; tamis m.

ride (Drive) ‖ Fahrt f. ‖ voyage m.; course f. / ~ (Riding ground) ‖ Reitbahn f. ‖ carrière f.; lice f. / ~ (Window) ‖ Fensterband n. ‖ penture f.

rideau ‖ Bodenerhebung f. ‖ rideau m.

rider (Coachm) ‖ Lenkschemel m.; Wendeschemel ‖ lisoir m. / ~ (Mach) ‖ Reiter m. ‖ cavalier m. / ~ (Mine) ‖ Salband n. eines Ganges ‖ éponte f. ou paroi f. ou salbande f. d'un filon.

rider of the balance ‖ Reiterchen n. der Wage ‖ cavalier m. d'une balance. / ~ cask ‖ Sattelfaß n. ‖ foudre m. gerbé.

ridge, to ~ anew ‖ neu befirsten ‖ renfaîter. / ~ a house ‖ ein Dach n. verfirsten ‖ enfaîter une maison.

ridge (Build) ‖ First m.; Dachfirst m. ‖ faîte m. / ~ (Found) ‖ Hauptrinne f. ‖ chenal m. principal. / ~ (Geol) ‖ Gebirgskamm m. ‖ crête f.; arête f. / ~ (Mar) ‖ Riff n.; Sandbank f.; Untiefe f. ‖ récif m.; bas-fond m.; banc m. de sable. / to lay a ~ ‖ befirsten ‖ enfaîter. / ~ of an open working (Mine) ‖ Rand m. eines Tagebaues ‖ crête f. d'une exploitation à ciel ouvert.

ridge beam ‖ Kappe f.; Kopfbalken m. ‖ chapeau m.; traverse f. / ~ circuit (Lightning arrester) ‖ Firstleitung f. ‖ circuit m. des faîtes. / ~ covering ‖ Verfirstung f. ‖ enfaîtage m.

ridged roof ‖ Satteldach n.; zweihängiges Dach n. ‖ toit m. en battière; comble m. à deux pentes.

ridge drill ‖ Drillmaschine f. ‖ semoir m. en lignes. / laying of a ~ ‖ Befirstung f. ‖ enfaîtement m. / ~ lead ‖ Bleifirstplatte f. ‖ enfaîtement m. en plomb.

ridgeless ‖ stufenlos ‖ sans échelon m. / ~ seam ‖ wulstlose Naht f. ‖ joint m. sans surépaisseur. / ~ seam welding ‖ wulstlose Nahtschweißung f. ‖ soudage m. de joints sans surépaisseur.

ridge piece (Build) ‖ Firstbalken m. ‖ poutre f. de faîte. / ~ piece (Hydr arch) ‖ Kopfbalken m. ‖ chapeau m.; traverse f. / ~ plate ‖ Firstblech n. ‖ plomb m. ou tôle f. de faîtage. / ~ plough ‖ Häufelpflug m. ‖ charrue f. à butter; charruebuttoir f. / ~ purlin ‖ Firstpfette f. ‖ panne f. faitière. / ~ rib ‖ Scheitelrippe f. ‖ nervure f. de sommet; grande lierne f. / ~ tile (Build) ‖ Firstziegel m. ‖ tuile f. faîtière. / ~ turret ‖ Dachreiter m. ‖ tour f. à cheval.

ridging ‖ Firsteindeckung f.; Verfirstung f. ‖ faîtage m.

ridiculously low price ‖ Schleuderpreis m. ‖ vil prix m.; prix m. de déballage.

riding (Print) ‖ Krummstehen n. ‖ chevauchement m. / ~ cut through a forest ‖ Durchhau m. oder Schneise f. im Walde ‖ laie f.

riding bed (Coachm) ‖ Lenkschemel m.; Wendeschemel m. ‖ lisoir m. / ~ bolster see riding bed. / ~ boot ‖ Reitstiefel m. ‖ botte f. à l'écuyère. / ~ costume ‖ Reitanzug m. ‖ costume m. de cavalier. / ~ habit ‖ Reitkleid n. ‖ habit m. d'amazone; amazone f. / ~ house see ~ school. / ~ school ‖ Reitbahn f. ‖ carrière f.; lice f.; manège m. / ~ school snaffle ‖ Schultrense f. ‖ filet m. de manège. / ~ stick ‖ Reitstock m. ‖ cravache f. / ~ whip ‖ Reitpeitsche f. ‖ cravache f.

riffler ‖ Raumfeile f.; Riffelfeile f. ‖ lime f. à archet; riflard m.; rifloir m.

rifle ‖ Gewehr n.; Büchse f. ‖ fusil m.; carabine f. / automatic ~ ‖ Selbstladegewehr n. ‖ fusil m. automatique. / military ~ ‖ Kriegsgewehr n. ‖ fusil m. de guerre. / repeating ~ ‖ Repetiergewehr n. ‖ fusil m. à répétition. / semiautomatic ~ ‖ Halbselbstladegewehr n. ‖ fusil m. mitrailleur.

rifle, back-sight of ~ ‖ Visier n. des Gewehres ‖ hausse f. de fusil. / ~ barrel ‖ gezogener Flintenlauf m. ‖ canon m. rayé ou rainé ou carabiné. / bolt of ~ ‖ Gewehrbolzen m.; Nadelbolzen m. ‖ Verrou m. de fermeture d'un fusil. / bore of ~ ‖ Gewehrkaliber n. ‖ calibre m. du canon. / chamber of ~ ‖ Patronenkammer f. ‖ chambre f. du fusil. / ~ cleaner ‖ Gewehrputzstock m. ‖ baguette f. à nettoyer les fusils.

rifled barrel see rifle barrel. / ~ projectile ‖ gezogenes Geschoß n. ‖ obus m. rayé.

rifle, extractor of ~ ‖ Patronenzieher m. eines Gewehres ‖ extracteur m. de cartouche d'un fusil. / ~ firing pin ‖ Schlagbolzen m. oder Schlagstift m. eines Gewehres ‖ percuteur m. d'un fusil. / foresight of ~ ‖ Korn n. des Gewehres ‖ guidon m. supérieur d'un fusil. / ~ grenade ‖ Gewehrgranate f. ‖ grenade f. à fusil. / lock of a ~ ‖ Gewehrschloß n. ‖ platine f. / magazine of ~ ‖ Gewehrmagazin n. ‖ magasin m. d'un fusil. / ~ maker ‖ Büchsenmacher m. ‖ armurier m. / muzzle of ~ ‖ Gewehrmündung f. ‖ bouche f. de canon du fusil. / notch on sight of ~ ‖ Visiereinschnitt m. eines Gewehres ‖ cran m. de mire d'un fusil. / ~ rocket ‖ Rotationsrakete f. ‖ fusée f. à rotation. / sight of a ~ ‖ Visier n. eines Gewehres ‖ hausse f. d'un fusil. / ~ trigger ‖ Gewehrabzughahn m. ‖ détente f. d'un fusil.

rifling of projectile ‖ Riefelung f. des Geschosses ‖ rayage m. du projectile.

rifling machine ‖ Gewehrlaufziehmaschine f. ‖ rayeuse f. à canons de fusils. / pitch of the ~ ‖ Drallsteigung f. ‖ pas m. des rayures. / ~ rod ‖ Ziehstange f. ‖ tringle f.

rift, to ‖ aufreißen; reißen; Risse mpl. bekommen ‖ se crévasser; se fendre.

rift valley (Geol) ‖ Graben m. ‖ renforçage m.

rig, to ~ out a vessel ‖ ein Schiff n. ausrüsten ‖ équiper un navire. / ~ the yards pl. ‖ die Rahen zutakeln ‖ garnir ou gréer les vergues fpl.

rigged ‖ betakelt ‖ gréé. / high-~ ‖ hoch getakelt ‖ à mâture f. haute. / low-~ ‖ niedrig getakelt ‖ à mâture f. basse.

rigger ‖ Takelmeister m. ‖ gréeur m. / ~'s shop ‖ Takelrei f. ‖ atelier m. des gréeurs.

rigging (Airpl) ‖ Verspannung f. ‖ haubanage m. / ~ (Mar) ‖ Schiffsausrüstungsgegenstände mpl. ‖ articles mpl. de marine; agrès mpl. de navires. / ~ (Sail ship) ‖ Zutakelung f. ‖ garniture f. / jury ~ ‖ Nottakelage f. ‖ gréement m. de fortune. / to put the ~ on the masthead ‖ Wanten npl. auflegen ‖ capeler les haubans mpl. / running ~ ‖ laufendes Gut n. ‖ manœuvres fpl. courantes. / ship's ~ ‖ Schiffstakelung f. ‖ montage m. d'agrès ou de manœuvres de navire. / standing ~ ‖ stehendes Gut n. ‖ manœuvres fpl. dormantes.

rigging, measuring and rolling machine ‖ Dublier-, Meß- und Wickelmaschine f. ‖ machine f. à dosser, mesurer et enrouler. / ~ screw ‖ Spannschloß n. ‖ tendeur m. / ~ works pl. ‖ Takelagearbeiten fpl. ‖ travaux mpl. de matelotage.

right, to ~ the helm ‖ das Ruder mittschiffs legen ‖ dresser la barre. / ~ the machine (Airpl) ‖ die Maschine wieder aufrichten ‖ redresser l'appareil m. / ~ a ship ‖ ein Schiff wieder aufrichten ‖ dresser ou redresser un navire.

right ‖ gerade ‖ droit. / ~ by the plummet ‖ senkrecht; lotrecht ‖ à plomb; vertical; perpendiculaire. / ~ before the wind ‖ platt vor dem Winde m. ‖ droit de l'arrière m.

right angle ‖ rechter Winkel m. ‖ angle m. droit. / at ~ ‖ im rechten Winkel m. ‖ à angle m. droit. / at ~ to the direction of flight ‖ quer zur Flugrichtung f. ‖ dans un sens m. perpendiculaire à la direction du vol. / bent at a ~ ‖ rechtwinklig gebogen ‖ courbé à angle m. droit.

right-angle bend (Pipe) ‖ Kniestück n. ‖ coude m. d'équerre. / ~ bevel gearing ‖ Winkelgetriebe n. ‖ engrenage m. d'équerre.

right-angled ‖ rechtwinklig ‖ rectangulaire; rectangle. / ~ isosceles triangle ‖ gleichschenklig-rechtwinkliges Dreieck n. ‖ triangle m. rectangle isoscèle. / ~ triangle ‖ rechtwinkliges Dreieck n. ‖ triangle m. rectangle ou rectangulaire.

right-angle position ‖ rechtwinklige Lage f. ‖ position f. orthogonale.

right ascension circle reading to one minute ‖ auf eine Minute ablesbarer Stundenkreis m. ‖ cercle m. horaire permettant de lire la minute. / ~ with vernier ‖ Stundenkreis m. mit Noniusablesung ‖ cercle m. horaire à vernier.

right crank, the central crank is set at x⁰ with the ~ and y⁰ with the left ‖ die Mittelkurbel ist gegen die Rechtskurbel um x⁰ und gegen die Linkskurbel um y⁰ versetzt ‖ la manivelle centrale est déplacée de x⁰ par rapport à la manivelle droite et de y⁰ par rapport à la manivelle gauche.

right-cutting ‖ rechtsschneidend ‖ coupant à droite.

right-hand ‖ rechtshändig; rechtsgewunden ‖ à droite f. / ~ action ‖ Rechtsgang m. ‖ tournant m. à droite. / ~ curve ‖ Rechtskurve f. ‖ courbe f. à droite. / ~ motion ‖ Rechtslauf m. ‖ marche f. à droite. / ~ switch (Railw) ‖ Rechts-

weiche f. ‖ branchement m. *ou* croisement m. à droite. / ~ turn-off (Railw) ‖ Rechtsweiche f. ‖ branchement m. à droite.

right-handed, spring with ~ helices ‖ rechtsgewundene Spiralfeder f. ‖ ressort m. à boudin en hélice à droite. / ~ screw ‖ rechtsgängiges Gewinde n. ‖ filet m. à droite. / ~ screw thread ‖ Rechtsgewinde n.; rechtsgängiges Gewinde n. ‖ filet m. à droite.

right-lined ‖ geradlinig ‖ rectiligne.

right shape ‖ Schick m. ‖ habileté f.; chic m. / ~ side (Mach) ‖ Antriebsseite f. ‖ côté m. de commande.

right ‖ Anrecht n.; Gerechtsame f. ‖ droit m.; titre m.; franchise f.; légitimité f. / to acquire the ~ to construct a dam ‖ das Staurecht n. erwerben‖acquérir le droit de barrage. / all ~s pl. reserved ‖ alle Rechte npl. vorbehalten ‖ touts droits mpl. réservés. / ~ of appeal ‖ Einspruchsrecht n. ‖ droit m. d'opposition. / by ~s pl. ‖ von Rechts wegen ‖ de plein droit m. / ~ of the discoverer ‖ Finderrecht n. ‖ droit m. de découverte. / ~ of dismissal ‖ Kündigungsrecht n. ‖ droit m. de congédiement. / ~ of disposal by the owner of the soil ‖ Verfügungsrecht n. des Grundeigentümers ‖ droit m. pour le propriétaire de disposer du fonds. / to enjoy a ~ ‖ ein Recht n. genießen ‖ jouir d'un droit. / to exercise a ~ ‖ ein Recht n. ausüben ‖ exercer un droit. / ~ of foreclosure by creditors ‖ Beschlagnahmerecht n. der Gläubiger ‖ droit m. de saisie des créanciers. / ~ of joint use in a channel for the discharge of water ‖ Anspruch m. auf Mitbenutzung einer Abwässerleitung ‖ droit m. à l'usage en commun d'une conduite d'eau d'écoulement. / to lose a ~ ‖ eines Rechtes n. verlustig gehen ‖ perdre un droit. / ~ of a mortgage ‖ Pfandrecht n. ‖ droit m. de saisie attaché à l'obligation foncière. / ~ of ownership ‖ Eigentumsrecht n. ‖ droit m. de propriété. / ~ of pasturage ‖ Weiderecht n. ‖ droit m. de pâturage. / ~ of preemption ‖ Vorkaufsrecht n. ‖ droit m. de préemption. / ~ to enter premises (Tel) ‖ Tretrecht n. ‖ droit m. d'accès. / ~ of prevention ‖ Präventionsrecht n. ‖ droit m. de prévention. / prior ~ of purchase *see* ~ of pre-emption. / ~ to prospect ‖ Schürfrecht n. ‖ droit m. de recherche. / to renounce a ~ ‖ auf ein Recht n. verzichten ‖ renoncer à un droit. / ~ of retention ‖ Retentionsrecht n. ‖ droit m. de rétention. / subscription ~ ‖ Bezugsrecht n. ‖ droit m. d'achat *ou* de prise. / to transfer a ~ ‖ ein Recht n. übertragen ‖ passer un droit. / transfer of a ~ subject to compensation ‖ Übertragung f. eines Rechtes gegen Entgelt ‖ transfert m. d'un droit contre redevance. / transfer of a ~ for a prescribed period ‖ Übertragung f. eines Rechtes auf Zeit ‖ transfert m. d'un droit à temps. / ~ of translation ‖ Übersetzungsrecht n. ‖ droit m. de traduction. / ~ of transport ‖ Durchfahrtsrecht n. ‖ droit m. de passage. / ~ of using ‖ Nutzungsrecht n. ‖ droit m. d'usage. / ~ of use in a water course ‖ Benutzungsrecht n. am Wasserlauf ‖ droit m. d'usage sur un cours d'eau.

right-angle *see under* right.

rightful owner ‖ rechtmäßiger Inhaber m. ‖ porteur m. *ou* détenteur m. légitime.

right-hand *see under* right.

rightness ‖ Richtigkeit f. ‖ exactitude f.; régularité f.

rights pl. and privileges pl. ‖ Rechte npl. und Pflichten fpl. ‖ droits mpl. et devoirs mpl.

rigid ‖ starr ‖ rigide; raide; engourdi. / ~ (Aero) ‖ formhaltend ‖ rigide. / ~ (Mach) ‖ standfest ‖ stable. / ~ (Mech) ‖ starr ‖ non déformable. / ~ airship ‖ starres Luftschiff n.; Starrluftschiff n. ‖ dirigeable m. rigide; rigide m. / ~ fastening ‖ feste Verbindung f. ‖ assemblage m. rigide. / ~ frame construction ‖ starrer Rahmen m. ‖ bâti m. rigide. / formula relating to ~ system ‖ Rahmenformel f. ‖ formule f. relative aux systèmes rigides.

rigidity (Mech) ‖ Starrheit f. ‖ rigidité f. / ~ of the connection ‖ Starrheit f. der Verbindung ‖ rigidité f. de la liaison *ou* du couplage. / ~ of the frame (Airpl) ‖ Festigkeit f. *oder* Steifheit f. des Gestells ‖ rigidité f. du châssis *ou* des membrures.

rills pl. of a millstone ‖ Rillen fpl. *oder* Hauschläge mpl. eines Mühlsteins ‖ rayons mpl. *ou* éveillures fpl. d'une meule.

rill sowing ‖ Rillensaat f. ‖ semis m. en rigoles.

rim, to ~ a wheel ‖ ein Rad n. befelgen ‖ janter une roue.

rim (Edge) ‖ Rand m.; Kante f. ‖ tranche f.; bord f. / ~ (Wheel) ‖ Felge f. ‖ jante f. / ~ balance ~ (Watchm) ‖ Ring m. der Unruh einer Uhr ‖ anneau m. du balancier d'une montre. / ~ of a chain wheel ‖ Kettenradkranz m. ‖ couronne f. dentée. / clincher ~ ‖ Wulstfelge f. ‖ jante f. à talon. / demountable ~ *see* detachable ~. / detachable ~ ‖ abnehmbare Feile f. ‖ jante f. amovible. / divided ~ ‖ geteilte Felge f. ‖ jante f. sectionnée *ou* bipartie *ou* fendue. / to drive-on the ~ ‖ auf der Felge f. fahren ‖ rouler dégonflé. / drop-base ~ ‖ Tiefbettfelge f. ‖ jante f. base creuse. / expanding ~ ‖ dehnbare Felge f. ‖ jante f. extensible. / extra ~ ‖ Reservefelge f. ‖ jante f. de réserve. / to fit the ~ on ‖ die Felge aufziehen ‖ monter la jante. / flanged ~ ‖ umgekrempelter Rand m. ‖ bord m. bridé. / flywheel ~ ‖ Schwungradkranz m.; Schwungring m. eines Schwungrades ‖ couronne f. ou anneau m. d'un volant; limbe m. / ~ of guide blading ‖ Leitkranz m. ‖ couronne f. directrice. / hollow-forged toothed ~ ‖ hohlgeschmiedeter Zahnkranz m. ‖ couronne f. dentée forgée à creux. / one-piece ~ ‖ feste Felge f. ‖ jante f. normale. / pressed-on ~ ‖ aufgepreßte Felge f. ‖ jante f. calée *ou* embattue à la presse. / ~ for solid tyre ‖ Felge f. für Vollgummireifen ‖ jante f. pour plein. / spectacle ~ ‖ Brillenrand m. ‖ cercle m. des lunettes. / split ~ *see* divided ~. / straight-side ~ ‖ Geradseitfelge f. ‖ jante f. straight-side. / ~ of a toothed wheel ‖ Zahnradkranz m. ‖ jante f. de roue d'engrenage. / ~ of wheel ‖ Radfelge f. ‖ jante f. de roue; sous-bandage m. / width of wheel ~ ‖ Radkranzbreite f. ‖ largeur f. de jante de roue. / wood ~ ‖ Holzfelge f. ‖ jante f. en bois.

rim, bead of ~ ‖ Felgenrand m. ‖ rebord m. de jante. / ~ brake ‖ Felgenbremse f. ‖ frein m. sur jante.

rime ‖ Reif m.; Rauhfrost m. ‖ gelée f. blanche; givre m. / to cover with ~ ‖ bereifen ‖ couvrir de gelée f. blanche.

rimer ‖ Reibahle f. ‖ alésoir m.

rim fermentation ‖ Randgärung f. ‖ fermentation f. périphérique.

rimless folder ‖ randloser Klemmer m. ‖ pince-nez m. à verres nus. / ~ holder (Spectacles) ‖ randloses Gestell n. ‖ monture f. sans cercle. / ~ spectacles pl. ‖ randlose Brille f. ‖ lunettes fpl. à verres nus.

rim lock ‖ Kastenschloß n. ‖ serrure f. encloisonnée.

rimmed folder ‖ Klemmer m. mit Randfassung ‖ pince-nez m. à cercle.

rim, thickness of the ~ ‖ Radfelgenstärke f. ‖ épaisseur f. de la jante. / ~ type ‖ Felgenart f. ‖ type m. de jante.

rincing machine ‖ Nachbohrmaschine f. ‖ machine f. à calibrer les trous percés.

rind ‖ Rinde f. ‖ écorce f. / ~ of a tree ‖ Borke f. eines Baumes; Baumrinde f. ‖ écorce f. d'un arbre. / ~ gall ‖ Holzkropf m. ‖ excroissance f. du tronc.

ring, to (Bell) ‖ läuten ‖ sonner. / ~ (Tree) ‖ beringen ‖ cerner. / ~ off (Tel) ‖ abklingeln; abläuten ‖ annoncer la fin de la conversation. / ~ up (Tel) ‖ anrufen; anklingeln; antelefonieren ‖ appeler; téléphoner; avertir.

ring ‖ Ring m. ‖ anneau m. / ~ (Chem) ‖ geschlossene Kette f.; Ring m. ‖ chaîne f. fermée. / ~ (Mach) ‖ Reif m. ‖ cercle m.; collier m.; virole f.; cerceau m. / ~ (Mill) ‖ Kranz m. ‖ couronne f. / ~ (Moon) ‖ Hof m. ‖ halo m.; couronne f. / annual ~s pl. of timber ‖ Jahresringe mpl. im Holz ‖ cercles mpl. annuels *ou* couches fpl. de bois; couches ligneuses annuelles. / ~ of bars of various sections ‖ Ring m. aus gewalzten Profilstäben ‖ anneau m. fabriqué de barres profilées. / ~ of the case (Watchm) ‖ Gehäusebügel m. ‖ anneau m. de la boîte. / to cast ~s pl. in vertical moulds with a very big runner ‖ Ringe mpl. in stehender Form mit hohem Kopf gießen ‖ couler des anneaux mpl. dans un moule vertical avec une haute masselotte. / ~ cast into a slot ‖ eingegossener Ring m. ‖ anneau m. coulé dans une enrayure. / check ~ ‖ Anschlagring m. ‖ collier m. de butée. / ~ set with a diamond ‖ Brillantring m. ‖ bague f. de diamant. / distance ~ ‖ Zwischenring m.; Distanzring m. ‖ bague f. d'écartement. / ~ for dredging bolt ‖ Baggerbolzenring m. ‖ anneau m. pour boulons de dragues. / ~ for dredging bolts forged of manganese steel ‖ aus Hartstahl geschmiedeter Ring m. für Baggerbolzen ‖ anneau m. pour boulons de dragues, forgé en acier au manganèse. / eccentric ~ ‖ Exzenterring m. ‖ collier m. *ou* bride f. d'excentrique. / edge runner ~ ‖ Kollerring m. ‖ anneau m. de meules verticales. / felt ~ ‖ Filzring m. ‖ rondelle f. en feutre. / ~ with flaps ‖ Lappenring m.; Flanschring m. ‖ anneau m. à pattes. / flywheel ~ ‖ Schwungring m. eines Schwungrades ‖ couronne f. d'un volant. / to furnish with ~s ‖ ringeln ‖ anneler; boucler. / ~ with graduation in degrees ‖ Ring m. mit Gradteilung ‖ anneau m. avec graduation en degrés. / guard ~

(Electr) ‖ Packungsring m.; Dichtungsring m. ‖ anneau m. de joint. / intermediate ~ ‖ Zwischenring m. ‖ bague f. de raccord. / junk ~ ‖ Packungsring m.; Dichtungsring m. ‖ garniture f. *ou* joint m. du piston. / metal ~ ‖ Metallring m. ‖ bague f. métallique. / nave ~ ‖ Nabenring m. ‖ frette f. de moyeu. / nozzle ~ ‖ Düsenkranz m. ‖ couronne f. de tuyères. / oil retaining ~ ‖ Ölspritzring m. ‖ anneau-égoutteur m. / packing ~ (Mach) ‖ Dichtungsring m.; Packungsring m.; Dichtungsscheibe f. ‖ segment m. de piston; anneau m. de serrage *ou* de joint; garniture f. *ou* joint m. de piston; rondelle f. de garniture. / ~ supporting the pen ‖ Federhalterring m. ‖ bague f. du porte-plume. / ~ of piston ‖ Kolbenring m. ‖ segment m. *ou* bague f. de piston. / ~ on pliers ‖ Zangenring m. ‖ coulant m. d'une tenaille à boucle. / retaining ~ ‖ Sprengring m. ‖ bague f. de retenue. / ~ of Saturn ‖ Saturnring m. ‖ anneau m. de Saturne. / scraper ~ ‖ Abstreifring m. ‖ segment m. racleur. / shrunk-on ~ ‖ warm aufgepreßter Ring m. ‖ frette f. calée à chaud. / slip ~ ‖ Schleifring m. ‖ bague f. frottante *ou* collectrice. / spring ~ ‖ Sprengring m. ‖ bague f. de retenue; agrafe f. circulaire *ou* annulaire. / steel ~ ‖ Stahlöse f. ‖ anneau m. en acier. / stiffening ~ ‖ Versteifungsring m. ‖ anneau m. de renforcement *ou* de consolidation. / taper shell ~ ‖ kegelförmiger Kesselschuß m. ‖ virole f. conique de chaudière. / wiper ~ ‖ Nockenring m. ‖ collier m. racloir *ou* lécheur. / ~ of wire ‖ Drahtring m. ‖ botte f. *ou* torche f. de fil.

ring binder ‖ Ringordner m. ‖ classeur m. à anneaux. / ~ bolt ‖ Augbolzen m.; Ringbolzen m. ‖ cheville f. à boucle; piton m. à anneau. / ~ burner ‖ Ringbrenner m. ‖ couronne f. de gaz. / ~ clamp ‖ Ringklemme f. ‖ machoire f. de serrage circulaire. / ~ clip ‖ Rundklammer f. ‖ attache f. ronde. / ~ closing ‖ Ringbildung f. ‖ cyclisation f. / ~ doubling frame ‖ Ringzwirnmaschine f. ‖ métier m. à retordre à anneau. / ~-down junction (Tel) ‖ Abfragebetrieb m. ‖ exploitation f. du double appel. / ~-down operation (Tel) ‖ Induktoranruf m. ‖ appel m. magnétique.

ringed, coarsely ~ (Wood) ‖ grobjährig ‖ aux cernes mpl. larges. / cast-steel ~ roll ‖ Ringwalze f. aus Gußstahl ‖ cylindre m. annulaire en acier fondu.

ring fastener ‖ Befestigungsring m. ‖ bague f. de fixation. / ~ flower ‖ Ringelblume f. ‖ fleur f. de souci. / ~ gauge ‖ Kaliberring m. ‖ bague f. étalon. / ~ gearing ‖ Ringführung f. ‖ guide f. de crosse circulaire. / ~ hook ‖ Ringhaken m. ‖ piton m.

ringing current ‖ Rufstrom m. ‖ courant m. d'appel. / ~ connection ‖ Rufstromschaltung f. ‖ circuit m. du courant d'appel. / ~ indicator ‖ Rufstromanzeiger m. ‖ signal m. du courant d'appel. / signal of missing ~ ‖ Signal n. für fehlenden Rufstrom ‖ signal m. de manque du courant d'appel. / ~ source ‖ Rufstromquelle f. ‖ source f. du courant d'appel. / ~ transformer ‖ Rufstromübertrager m. ‖ transformateur m. de courant d'appel. / ~ transmission ‖ Rufstromübertragung f. ‖ translation f. de courant d'appel.

ringing device ‖ Läutwerk n.; Hammerglocke f. ‖ sonnerie f.; timbre m. / ~ junction working (Tel) ‖ markierter Dienstleitungsbetrieb m. ‖ méthode f. de l'appel direct sur les lignes auxiliaires. / ~ key (Tel) ‖ Rufschalter m.; Ruftaste f. ‖ clé f. d'appel. / ~ line (Tel) ‖ Weckerleitung f. ‖ ligne f. de sonnerie. / ~ and signalling machine (Aut tel) ‖ Ruf- und Signalmaschine f. ‖ machine f. d'appel et de signaux. / ~ pilot lamp ‖ Rufstromkontrollampe f. ‖ lampe f. de contrôle d'appel. / ~ relay ‖ Läuterelais n. ‖ relais m. d'appel. / ~ signal audible at calling (Aut tel) ‖ beim Anrufen hörbares Rufsignal n. ‖ signal m. indicateur d'appel. / ~ signal alarm relay ‖ Rufstromalarmrelais n. ‖ relais m. avertisseur pour coupe-circuit de courant d'appel. / ~ tone ‖ Freizeichen n. ‖ signal m. de ligne libre.

ringlet ‖ Löckchen n. ‖ mèche f.

ring lock ‖ Ringschloß n. ‖ cadenas m. à rouleaux. / ~ lubricating bearing ‖ Ringschmierlager n. ‖ palier m. de graissage annulaire; palier-graisseur m. à bague. / ~ lubrication ‖ Ringschmierung f. ‖ lubréfaction f. *ou* graissage m. à bague. / ~ lubricator ‖ Ringschmiervorrichtung f. ‖ graisseur m. à bague. / ~ main ‖ Ringleitung f. ‖ canalisation f. circulaire. / ~ maker ‖ Ringmacher m. ‖ baguiste m. / ~ making machine ‖ Ringherstellungsmaschine f. ‖ machine f. à fabriquer les anneaux. / ~ micrometer ‖ Ringmikrometer n. ‖ micromètre m. annulaire. / ~ mill ‖ Ringmühle f. ‖ broyeur m. à anneau. / ~-off system (Tel) ‖ Abklingelsystem n. ‖ système m. de fin de communication par sonnerie. / ~ oiler ‖ Ringöler m. ‖ graisseur m. à bagues. / ~ oiler bearing ‖ Ringschmierlager n. ‖ palier m. à graissage par bagues. / ~ oven ‖ Ringofen m. ‖ four m. circulaire. / ~ pivot ‖ Ringzapfen m. ‖ pivot m. annulaire. / ~ roller mill ‖ Ringwalzenmühle f. ‖ moulin m. à cylindres annulaires. / ~ rope ‖ Kettenstopper m. ‖ bosse f. de câble. / ~ screw mandrel (Turn) ‖ aufgesteckte Schraubenpatrone f. ‖ manchon m.

ring-shaped disk ‖ ringförmige Scheibe f. ‖ disque m. annulaire. / ~ iron core ‖ ringförmiger Eisenkern m. ‖ noyau m. circulair en fer.

ring sight ‖ Kreiskimme f. ‖ hausse f. annulaire. / ~ spinner ‖ Ringspinner m. ‖ fileur m. au continu à anneaux. / ~ spinning frame ‖ Ringspinnmaschine f. ‖ métier m. continu à anneaux; continu m. à filer. / ~ stopper ‖ Ringstopper m. ‖ barbarasse f. / ~ throstle *see* ~ spinning frame ‖ Ringspinnmaschine f. ‖ continu m. à filer.

ring transformer ‖ Ringübertrager m. ‖ transformateur m. annulaire. / ~ box ‖ Übertragerkästchen n. ‖ petite boîte f. pour translations.

ring twisting frame ‖ Ringzwirnmaschine f. ‖ métier m. à retordre à anneau. / ~ wall ‖ Kernschacht m.; Schachtfutter n. ‖ paroi f. *ou* chemise f. intérieure. / ~ warp ‖ Ringkette f. ‖ chaîne f. *ou* chaîne f. pliée en boucles. / ~ welding ‖ Ringschweißung f. ‖ soudage m. d'anneaux. / ~-wound armature ‖ Ringanker m. ‖ induit m. en anneau.

rinse, to ‖ spülen; abwaschen ‖ laver;

rincer; dégorger. / ~ casks pl. ‖ faßschwanken ‖ rincer les fûts mpl.

rinser, cask ~ ‖ Faßspülgerät n. ‖ rinceuse f. de fûts.

rinsing ‖ Spülen n. ‖ rinçage m. / ~ casks pl. ‖ Faßschwanken n. ‖ rinçage m. des fûts. / conical ~ ‖ Konusspülung f. ‖ rinçage m. conique. / return ~ ‖ Rückspülung f. ‖ rinçage m. de retour.

rinsing apparatus for w. c. ‖ Abortspülapparat m. ‖ appareil m. à chasse d'eau pour cabinets. / ~ box for water closets ‖ Abortspülkasten m. ‖ caisse f. de chasse d'eau de W.C. / ~ buddle ‖ Rührfaß n.; Schlämmfaß n. ‖ cuve f. à rincer. / ~ machine (Dyer) ‖ Spülmaschine f. ‖ machine f. à décrasser *ou* à rincer; rinceuse f. / cask ~ machine ‖ Faßspülmaschine f. ‖ machine f. à laver les fûts; rinceuse f. de fûts. / ~ pump ‖ Spülpumpe f. ‖ pompe f. à laver *ou* de balayage. / boiler ~ pump ‖ Kesselspülpumpe f. ‖ pompe f. à laver les chaudières. / ~ screen ‖ Abbraussieb n. ‖ tamis m. d'arrosage. / ~ table ‖ Spültisch m. ‖ table f. à rincer la vaisselle. / ~ water ‖ Spülwasser n. ‖ eau f. d'arrosage.

rip, to ‖ reißen; zerreißen ‖ déchirer; crever. / ~ the anchor ‖ den Anker m. losreißen ‖ faire déraper l'ancre f. / ~ off a strake of planks (Shipb) ‖ einen Plankengang m. abbrechen ‖ dépecer *ou* découdre une file de bordage.

ripener, cream ~ ‖ Rahmreifer m. ‖ appareil m. à faire fermenter la crème.

ripping chisel ‖ Stechbeitel m. ‖ ciseau m. fort; entailloir m. / ~ panel (Balloon) ‖ Reißbahn f. ‖ panneau m. de déchirure. / ~ saw ‖ Spaltsäge f.; Brettsäge f. ‖ scie f. de long *ou* à refendre.

ripple, to ‖ riffeln ‖ canneler. / ~ flax ‖ Flachs reffeln *oder* riffeln ‖ dréger *ou* égruger la lin.

ripple ‖ Welligkeit f.; Kräuselung f. ‖ ondulation f. / ~ frequency ‖ Wellenfrequenz f. ‖ fréquence f. d'ondes. / ~ mark ‖ Wellenfurche f.; leichte Wellenspur f. ‖ ripple-mark f.; sillon m. ondulée. / ~ sea ‖ Kabbelsee f.; kabbelige See f. ‖ clapotage m.; clapotis m.

riprap foundation ‖ Gründung f. auf Steinschüttung ‖ fondation f. à pierres perdues *ou* par enrochements.

rip saw ‖ Schrotsäge f. ‖ scie f. à buches.

rise, to (Aero) ‖ in die Höhe steigen ‖ monter; s'élever en altitude f. / ~ (Astro) ‖ aufgehen ‖ se lever. / ~ (Chem) ‖ gären; aufgehen ‖ fermenter; entrer en fermentation. / ~ (River) ‖ anschwellen ‖ (s')enfler; croître. / ~ in the back (Mine) ‖ aufhauen ‖ travailler au-dessous de la tête. / ~ up ‖ hissen; heißen; aufziehen ‖ hisser; palanquer; haler.

rise (Aero) ‖ Anstieg m. ‖ montée f.; ascension f. / ~ (Exchange) ‖ Hausse f. ‖ hausse f. / ~ (River) ‖ Anschwellung f. ‖ crue f. / ~ (Trade) ‖ Aufschlag m. ‖ augmentation f.; renchérissement m. / ~ of an arch ‖ Stich m. *oder* Wölbung f. eines Bogens ‖ flèche f. d'un arc. / ~ of a bowstring truss ‖ Pfeilhöhe f. eines Trägers mit gekrümmter Gurtung ‖ flèche f. d'une poutre en arc. / ~ of current ‖ Stromzunahme f. ‖ accroissement m. du courant. / ~ of salary ‖ Gehaltserhöhung f. ‖ augmentation f. des appointements *ou* de traitement *ou* des émoluments. / ~ of temperature ‖ Tem-

peraturzunahme f.; Temperaturerhöhung f. ‖ augmentation f. *ou* hausse f. de température. / ~ of a vault ‖ Bogenhöhe f. eines Gewölbes ‖ montée f. *ou* flèche f. d'une voûte. / ~ in wages ‖ Lohnerhöhung f. ‖ augmentation f. des salaires; amélioration f. des émoluments.

riser (Found) ‖ Steiger m. ‖ évent m. / glass frame ~ ‖ Fensterriemen m. ‖ tirant m. de glace. / to knock off and saw off runner and ~ on castings ‖ Einguß m. und Steiger m. am Gußstück abschneiden ‖ couper l'évent m. et le jet m. de coulée. / ~ of a vein (Min) ‖ Verwerfung f. eines Ganges ‖ dérangement m. d'une couche.

riser board of a staircase ‖ Setzbrett n. einer Holztreppe; Futterbrett n. ‖ ais m. de contre-marche.

rising ‖ Aufsteigen n.; Emporsteigen n. ‖ élévation f.; ascension f. / ~ (Rebellion) ‖ Aufruhr m.; Aufstand m.; Empörung f.; Revolte f. ‖ rébellion f.; révolte f.; insurrection f.; sédition f. / ~ (Astro) ‖ Aufgang m.; Aufgehen n.; Aufsteigen n. ‖ ascension f.; lever m. / ~ proportionately ‖ im Verhältnis n. steigend ‖ montant progressivement. / ~ simultaneous ~ of the punch and lowering of the table ‖ gleichzeitiges Hochgehen n. des Stempels und Niedergehen n. des Preßtisches ‖ mouvements mpl. simultanés de montée du porte-poinçon et de descente de la table. / ~ of a vault ‖ Stichhöhe f. eines Gewölbes ‖ montée f. de voûte. / ~ of the water ‖ Schwellen n. *oder* Wachsen n. des Wassers ‖ crue f. des eaux.

rising floor for astronomical instruments ‖ Hebebühne f. für astronomische Instrumente ‖ plancher m. mobile pour les instruments astronomiques. / dynamic ~ force ‖ dynamische Steigkraft f. ‖ force f. ascensionnelle dynamique. / static ~ force ‖ statische Steigkraft f. ‖ force f. ascensionnelle statique. / ~ ground (Build) ‖ Rampe f.; Auffahrt f. ‖ rampe f.; chemin m. taluté. / ~ ground (Geol) ‖ Bodenerhebung f. ‖ rideau m. / ~ main (Electr) ‖ Steigleitung f. ‖ colonne f. montante. / ~ pipe ‖ Steigrohr n. ‖ tuyau m. de refoulement. / ~ scaffold bridge ‖ Fahrbrücke f. ‖ Laufbrücke f. ‖ pont m. d'échafaudage. / casting in ~ stream ‖ Gießen n. in steigendem Strom ‖ coulée f. en source. / ~ timber in a boat ‖ Krummholz n. *oder* Kniestück n. eines Flußfahrzeuges ‖ courbe f. de bateau.

risk, to ‖ wagen; aufs Spiel n. setzen; riskieren ‖ risquer.

risk ‖ Risiko n. ‖ risque m. / at company's ~ ‖ auf Gefahr f. der Gesellschaft ‖ aux risques mpl. de la compagnie. / at consignee's ~ ‖ auf Gefahr f. des Empfängers ‖ aux risques mpl. du consignataire. / at owner's ~ ‖ auf Gefahr f. des Eigentümers ‖ aux risques mpl. du propriétaire. / at sender's ~ ‖ auf Gefahr f. des Absenders ‖ aux risques mpl. de l'expéditeur. / at your ~ ‖ auf Ihre Gefahr f. ‖ à vos risques mpl. / ~ and peril of the seas ‖ Seegefahr f. ‖ fortune f. de mer; péril m. de la mer.

risked ‖ gewagt ‖ risqué.

ristorno ‖ Ristorno n. ‖ ristorne f.

rival ‖ Konkurrent m. ‖ concurrent m.; compétiteur m.

rivalry ‖ Konkurrenz f. ‖ concurrence f.

rive, to ‖ spalten ‖ fendre; refendre.

riven stave ‖ gespaltene Daube f. ‖ douve f. fendue.

river ‖ Fluß m.; Strom m. ‖ rivière f.; fleuve m. / navigable ~ ‖ schiffbarer Fluß m. ‖ rivière f. navigable. / tributary ~ ‖ Nebenfluß m. ‖ affluent m.

river bed ‖ Flußbett n. ‖ lit m. de fleuve. / ~ bend ‖ Flußkrümmung f. ‖ coude m. de rivière. / ~ boat ‖ Flußschiff n. ‖ bateau m. fluvial *ou* de fleuve; bâtiment m. de rivière. / ~ cable ‖ Flußkabel n. ‖ câble m. sousfluvial. / ~ cable box ‖ Flußkabelmuffe f. ‖ manchon m. pour câble fluvial. / ~ clay ‖ Flußton m. ‖ argile f. fluviale. / ~ cleaner ‖ Binsenarbeiter m. im Fluß ‖ faucardeur m. / ~ crossing ‖ Flußübergang m. ‖ traversée f. de rivière. / ~ course ‖ Flußlauf m. ‖ cours m. d'un fleuve. / ~ craft ‖ Fahrzeug n. der Binnenschiffahrt ‖ navire m. de navigation intérieure. / ~ dredge ‖ Flußbagger m. ‖ drague f. fluviale. / ~ fisherman ‖ Flußfischer m. ‖ pêcheur m. en eau douce *ou* en rivière. / ~ fishing ‖ Flußfischerei f. ‖ pêche f. en rivière *ou* fluviale. / ~ flats pl. ‖ Flußniederung f. ‖ bas-fond m. de fleuve *ou* fluvial. / ~ gold ‖ Waschgold n. ‖ or m. de lavage. / ~ gravel ‖ Flußkies m. ‖ gravier m. de rivière. / ~ head ‖ Flußquelle f. ‖ source f. de fleuve. / ~ mouth ‖ Flußmündung f. ‖ bouche f. *ou* embouchure f. de fleuve. / ~ navigation ‖ Flußschiffahrt f. ‖ navigation f. de rivière. / ~ pilot ‖ Binnenlotse m.; Flußlotse m.; Revierlotse m. ‖ pilote m. lamaneur. / ~ rands pl. *see* ~ flats. / ~ retting ‖ Rotten n. in fließendem Wasser ‖ rouissage m. à eau courante. / ~ sand ‖ Flußsand m. ‖ sable m. de rivière. / ~ station ‖ Flußkraftwerk n.; Laufwasserkraftwerk n. ‖ usine f. fluviale *ou* hydraulique. / ~ steamer ‖ Flußdampfer m. ‖ vapeur m. de rivière. / ~ valley *see* ~ flats. / ~ water ‖ Flußwasser n. ‖ eau f. de rivière.

rivet, to ‖ nieten; vernieten ‖ river; riveter. / ~-on the flange ‖ den Flansch m. annieten ‖ appliquer la bride par des rivets. / ~ the plates ‖ die Platten fpl. vernieten ‖ river les tôles fpl.

rivet ‖ Niet n. ‖ rivet m. / ~ (Locksmith) ‖ Dorn m. ‖ chevillette f. / copper ~ ‖ Kupferniet n. ‖ rivet m. en cuivre. / countersunk ~ ‖ versenktes Niet n. ‖ rivet m. noyé. / to draw ~s pl. ‖ abnieten; die Niete npl. herausschlagen ‖ dériver; dériveter. / iron ~ ‖ Eisenniet n. ‖ rivet m. en fer. / metal ~ ‖ Metallniete f. ‖ rivet m. de métal. / cold press for forming and upsetting ~s ‖ Kaltpresse f. zum Formen und Stauchen von Nieten ‖ presse f. à froid à former et fouler des rivets.

rivet control apparatus ‖ Nietkontrollapparat m. ‖ appareil m. à contrôler le rivetage. / ~ controlling device ‖ Nietkontrollvorrichtung f. ‖ dispositif m. de contrôle pour rivures.

riveted ‖ genietet ‖ riveté. / hot-~ ‖ warm genietet ‖ rivé à chaud.

riveted joint ‖ Nietung f.; Nietnaht f. ‖ rivure f. / ~ of difficult accessibility ‖ schwer zugängliche Nietverbindung f. ‖ rivure f. d'accès difficile.

riveted seam, double-~ ‖ doppelte Nietnaht f. ‖ rivure f. double *ou* à deux rangs; couture f. à double rivure. / single-~ ‖ einfache Nietnaht f. ‖ rivure f. à un rang; couture f. à simple rivure.

riveter (Mach tool) ‖ Nietmaschine f. ‖ riveuse f.; riveteuse f.; machine f. à river. / ~ (Person) ‖ Nieter m. ‖ riveur m. / hand ~ ‖ Handnieter m. ‖ riveur m. à la main. / machine ~ ‖ Maschinennieter m. ‖ riveur m. à la machine. / steam ~ ‖ Dampfnietmaschine f. ‖ riveuse f. à vapeur. / set of ~s ‖ Nietkolonne f. ‖ équipe f. de riveurs.

rivet forge ‖ Nietwärmofen m.; Nietglühofen m.; Nietfeuer n. ‖ four m. à rivets / ~ forging machine ‖ Nietenschmiedemaschine f. ‖ machine f. à forger les rivets. / ~ furnace *see* ~ forge. / ~ head ‖ Nietkopf m. ‖ tête f. de rivet. / ~ hearth *see* ~ forge. / ~ heater ‖ Nietenwärmer m. ‖ chauffeur m. de rivets. / electric ~ heater ‖ elektrischer Nietwärmer m. *oder* Nietwärmofen m. ‖ machine f. électrique à chauffer les rivets. / ~ hole ‖ Nietloch n. ‖ trou m. de rivet. / ~ hole driller ‖ Nietenlochbohrer m. ‖ perceur m. de trous de rivets.

riveting ‖ Nietung f.; Nietnaht f.; Nieten n. ‖ rivure f.; rivetage m. / chain ~ ‖ Kettennietung f. ‖ rivetage m. en chaine *ou* à rivets parallèles. / cold ~ ‖ kalte Nietung f. ‖ rivure f. à froid. / conical ~ ‖ Spitzkopfnietung f. ‖ rivure f. conique *ou* à point de diamant. / conical point ~ ‖ Spitzkopfnietung f. ‖ rivure f. conique *ou* à point de diamant. / double ~ ‖ doppelte *oder* zweireihige Nietung f. ‖ rivure f. double *ou* à deux rangs. / electric ~ ‖ elektrische Punktschweißung f. ‖ rivetage m. électrique. / hammer point ~ *see* conical point ~. / hydraulic ~ ‖ hydraulische Nietung f. ‖ rivure f. hydraulique. / machine ~ ‖ Maschinennietung f. ‖ rivetage m. mécanique. / single ~ ‖ einfache Nietung f. ‖ rivure f. simple *ou* à un rang. / triple ~ ‖ dreifache *oder* dreireihige Nietung f. ‖ rivure f. triple *ou* à trois rangs. / water-tight ~ ‖ wasserdichte Nietung f. ‖ rivure f. étanche. / zigzag ~ ‖ Zickzacknietung f. ‖ rivure f. en quinconce.

riveting apparatus ‖ Nietapparat m. ‖ dispositif m. *ou* appareil m. à river. / crownplate ~ ‖ Deckennietapparat m. ‖ appareil m. à river le ciel de foyer. / ~ for locomotive-tubes ‖ Kesselnietzwinge f. ‖ serre-tubes m.

riveting clamp ‖ Nietkluppe f. ‖ mordache f. à river; presse f. à river; pince f. à rivets. / ~ die ‖ Döpper m. ‖ bouterolle f.; chasse-rivet m.

riveting hammer ‖ Niethammer m. ‖ marteau-riveur m.; marteau m. à river; rivoir m. / pneumatic ~ ‖ Preßluftniethammer m. ‖ marteau m. à river à air comprimé.

riveting horn ‖ Niethorn n. ‖ bigorne f. à river. / ~ knob ‖ Vorhalter m. beim Nieten ‖ contre-bouterolle f.

riveting machine ‖ Nietmaschine f. ‖ riveuse f.; machine f. à river. / electrohydraulic ~ ‖ elektrohydraulische Nietmaschine f. ‖ riveuse f. électro-hydraulique. / hydraulic ~ ‖ hydraulische Nietmaschine f. ‖ riveuse f. hydraulique. / portable ~ ‖ tragbare Nietmaschine f. ‖ riveuse f. mobile *ou* portative. / special

~ ‖ Spezialnietmaschine f. ‖ riveuse f. spéciale. / stationary ~ ‖ feststehende oder ortsfeste Nietmaschine f. ‖ riveuse f. fixe. / toggle-joint ~ ‖ Kniehebelnietmaschine f. ‖ riveuse f. à col de cygne. / tongs ~ ‖ Zangennietmaschine f. ‖ riveuse f. à pince.

riveting machinery ‖ Nietanlage f. ‖ installation f. de riveuses. / pneumatic ~ ‖ pneumatische Nietanlage f. ‖ riveuses fpl. pneumatiques.

riveting press ‖ Nietpresse f. ‖ presse f. à river. / ~ punch ‖ Nietpunze f.; Nietmeißel m. ‖ poinçon m. à river. / ~ set ‖ Döpper m.; Nietstempel m.; Nietzieher m. ‖ chasse-rivet m.; bouterolle f. à rivet. / ~ snap tool ‖ Nietdöpper m. ‖ bouterolle f. / ~ stock ‖ Nietstöckchen n.; Nietbank f.; Nietplatte f. ‖ banc m. à river. / ~ tongs pl. ‖ Nietzange f. ‖ pince f. à rivets. / ~ tool ‖ Nietwerkzeug n. ‖ outil m. à river.

rivet iron ‖ Nieteisen n. ‖ fer m. à rivets. / junction of ~s ‖ Nietverbindung f. ‖ assemblage m. par rivets. / ~ length depending on the joint thickness ‖ Nietlänge f. in Abhängigkeit von den Klemmlängen ‖ longueur f. des rivets dépendant de la longueur des bornes. / making ~s pl. ‖ Nietenherstellung f. ‖ fabrication f. derivets. / ~ making machine ‖ Nietenherstellungsmaschine f. ‖ machine f. à fabriquer les rivets. / ~ making press ‖ Nietenpresse f. ‖ presse f. à faire les rivets. ‖ ~ point ‖ Schließkopf m. ‖ tête f. fermante ou serrante. / row of ~s ‖ Nietreihe f. ‖ file f. ou ligne f. ou rang m. de rivets. / staggered row of ~s ‖ versetzte Nietreihe f. ‖ rang m. de rivets alternés. / ~ setter ‖ Nietenzieher m. ‖ chasse-rivets m. / ~ shaft ‖ Nietschaft m. ‖ tige f. ou corps m. de rivet. / ~ shank see ~ shaft. / ~ shearing machine ‖ Nietenschere f. ‖ cisaille f. à rivets. / ~ stamp ‖ Nietstempel m. ‖ chasse-rivet m.; bouterolle f. à rivet.

rivetted see riveted.

rivetter see riveter.

rivetting see riveting.

rivet tongs pl. ‖ Nietzange f. ‖ tenaille f. à rivets. / ~ wire ‖ Nietendraht m. ‖ fil m. (métallique) pour rivets.

rivina berry ‖ Rivinabeere f. ‖ baie f. de rivina.

riving knife ‖ Spaltklinge f. ‖ coutre m.

rivulet ‖ Bach m. ‖ ruisseau m.; ruisselet m.

road (Highway) ‖ Straße f.; Weg m. ‖ route f.; chemin m.; voie f. / ~ (Mar) ‖ Reede f. ‖ rade f. / ~ (Mine) ‖ Förderstrecke f. ‖ galerie f.; voie f. / ~ (Railw) ‖ Strecke f.; Gleis n. ‖ voie f. / barrelled ~ ‖ gewölbte Straße f. ‖ route f. bombée. / to block-up a ~ ‖ eine Straße f. sperren ‖ barrer une route. / brick ~ ‖ Klinkerstraße f. ‖ chaussée f. de briques. / broken stone ~ ‖ Schotterstraße f. ‖ chaussée f. en empierrement ou à la Mac-Adam ou macadamisée. / ~ broken-up by traffic ‖ ausgefahrene Straße f. ‖ route f. défoncée par le roulage. / cambered ~ ‖ gewölbte Straße f. ‖ chaussée f. en dos d'âne. / concave ~ ‖ hohle Straße f. ‖ route f. creuse. / concrete ~ ‖ Betonstraße f. ‖ route f. bétonnée. / surface layer of the concrete ~ surface ‖ Oberschicht f. der Betonstraßendecke ‖ couche f. supérieure du revêtement de la route en béton. /

~ crossing the railway over a bridge ‖ Eisenbahnüberbrückung f. ‖ passage m. par dessus le chemin de fer. / cross-over ~ ‖ Verbindungsgleis n. ‖ voie f. de jonction. / to be driven from a ~ (Mar) ‖ von einer Reede f. treiben ‖ dérader. / Government ~ ‖ Landstraße f. ‖ route f. gouvernementale. / gravelled ~ ‖ Kiesstraße f.; Schotterstraße f. ‖ chemin m. ferré. / heavy ~ ‖ ausgefahrene Straße f. ‖ route f. fatiguée. / to lay down a ~ ‖ eine Straße f. bauen oder anlegen ‖ établir ou construire une route. / to lower a ~ ‖ eine Straße f. tiefer legen ‖ abaisser une route. / macadamized ~ ‖ makadamisierter Weg m. ‖ macadam m. / national ~ ‖ Staatsstraße f. ‖ route f. nationale. / open ~ (Mar) ‖ offene Reede f. ‖ rade f. foraine ou ouverte. / paved ~ ‖ gepflasterte Straße f. ‖ route f. pavée. / private ~ ‖ Privatweg m. ‖ chemin m. rural ou particulier. / provincial ~ ‖ Kreisstraße f. ‖ route f. départementale. / sheltered ~ (Mar) ‖ geschützte Reede f. ‖ rade f. close ou fermée. / tarred ~ ‖ geteerter Weg m.; Asphaltstraße f. ‖ chaussée f. goudronnée. / ~ through a wood ‖ Holzweg m. ‖ chemin m. qui sert au charriage du bois d'une forêt.

road ballast ‖ Straßenschotter m. ‖ ballast m. de route. / ~ bed (Railw) ‖ Bahnkörper m. ‖ assiette f. de la voie; terre-plein m. / ~ breaker ‖ Straßenaufreißer m. ‖ piccheur m. de rues. / ~ building ‖ Straßenbau m.; Wegebau m. ‖ construction f. de routes. / ~ covering of a bridge ‖ Brückenbelag m. ‖ plancher m. ou platelage m. d'un pont. / ~ crossing (El line) ‖ Wegekreuzung f.; Straßenüberführung f. ‖ traversée f. de route. / ~ curve ‖ Wegkurve f. ‖ virage m. / ~ dung removal ‖ Straßenreinigung f. ‖ nettoyage m. des rues. / ~ engine ‖ Straßenmaschine f. ‖ machine f. routière. / ~ locomotive ‖ Straßenlokomotive f. ‖ locomotive f. routière.

road making ‖ Straßenbau m. ‖ construction f. de routes. / ~ machine ‖ machine f. pour la construction des routes. / ~ material ‖ Straßenbaumaterial n. ‖ matériaux mpl. de construction pour routes. / ~ plough ‖ Straßenpflug m.; Straßenaufreißer m. ‖ charrue f. pour chemins; piocheuse f. scarificateur.

roadman ‖ Straßenarbeiter m. ‖ cantonnier m.

road map ‖ Wegekarte f. ‖ carte f. routière.

road metal ‖ Straßenbeschotterung f.; Kleinschlag m. ‖ couche f. de pierres concassées; pierres fpl. concassées. / ~ plant ‖ Schottermühle f. ‖ casse-pierre m. / ~ preparing machine ‖ Schottermaschine f. ‖ casse-pierres m. à balast.

road roller ‖ Straßenwalze f. ‖ écraseur m.; rouleau m. compresseur; cylindre m. pour chaussée. / steam ~ ‖ Dampfstraßenwalze f. ‖ rouleau m. compresseur à vapeur.

road roller construction ‖ Straßenwalzenbau m. ‖ construction f. de rouleaux compresseurs.

road scarifier see road making plough.

roadside station ‖ Haltepunkt m. der Eisenbahn ‖ point m. d'arrêt du chemin de fer.

roadstead ‖ Reede f. ‖ rade f.

roadster ‖ offener Zweisitzer m. ‖ roadster m.; routière f. à deux places.

road surface, clearance above ‖ Bodenfreiheit f. ‖ dégagement m. au-dessus du sol. / ~ tarring machine ‖ Straßenteermaschine f. ‖ machine f. à goudronner les routes. / ~ traction engine ‖ Straßenbahnlokomotive f. ‖ locomotive f. routière. / ~ turn ‖ Wendung f. ‖ virage m.

roadway ‖ Fahrweg m. ‖ chaussée f. / macadamized ~ ‖ Makadamstraße f. ‖ macadam m.; route f. macadamisée.

road wheel ‖ Wagenrad n. ‖ roue f. de voiture.

roast, to ‖ ausglühen; abschwelen; rösten ‖ griller; calciner; torréfier. / ~ thoroughly ‖ abrösten ‖ griller complètement.

roasted ore ‖ geröstetes Erz n. ‖ minerai m. grillé.

roaster, coffee ~ ‖ Kaffeeröstmaschine f. ‖ brûloir m. pour café. / ~ slag ‖ Schwarzkupferschlacke f. ‖ scorie f. du cuivre brut; scorie f. grillée.

roasting (Ore dress) ‖ Rösten n.; Zubrennen n.; Brennen n. ‖ rôtissage m.; grillage m.; calcination f.; rouissage m. / ~ in heaps ‖ Rösten n. in Haufen ‖ grillage m. ou rôtissage m. en tas. / ~ of ores ‖ Rösten n. der Erze ‖ grillage m. des minerais. / oxidation ~ ‖ oxydierendes Rösten n. ‖ grillage m. d'oxydation.

roasting bed ‖ Röstbett n. ‖ lit m. de grillage. / ~ charge ‖ Röstposten m. ‖ charge f. de rôtissage ou de grillage. / ~ dish ‖ Glühschale f. ‖ test m. à rôtir. / ~ furnace ‖ Röstofen m. ‖ four m. de grillage; fourneau m. de calcinage. / ~ implement ‖ Röstgerät n. ‖ ustensile m. de torréfaction. / ~ oven ‖ Bratrohr n. ‖ rôtissoire f.; four m. à rôtir. / ~ pan ‖ Bratpfanne f. ‖ poêle f. à frire; sauteuse f. / ~ place ‖ Röststätte f. ‖ grilloir m. / ~ plant (Ore dress) ‖ Rösteinrichtung f. ‖ installation f. de grillage. / ~ process ‖ Röstarbeit f. ‖ méthode f. de grillage. / ~ and reduction process ‖ Röstreduktionsprozeß m. ‖ méthode f. de grillage et de réduction. / ~ residue ‖ Abbrand m. ‖ résidu m. de grillage. / ~ spot see ~ place. / ~ test ‖ Röstprobe f. ‖ essai m. de grillage. / ~ test (Dish) see ~ dish.

Roberval scales pl. ‖ oberschalige Tafelwage f. ‖ balance f. de Roberval.

robust construction ‖ unverwüstliche Ausführung f. ‖ exécution f. robuste.

rock (Geol) ‖ Gestein n.; Felsen m. ‖ rocher m.; roche f. / ~ (Nav) ‖ Klippe f. ‖ écueil m; rocher m.; roche f. / ~ (Spinn) ‖ Spinnrocken m. ‖ quenouille f. / barren ~ ‖ Abraum m. ‖ découverte f.; mortterrain m.; déblai m.; lit m. de décombres. / bedded ~ ‖ geschichtetes Gestein n. ‖ roche f. stratifiée. / carboniferous ~ ‖ Steinkohlengebirge n. ‖ roche f. carbonifère. / ~ crossing a lode ‖ Quergestein n. ‖ roche f. transversale à un filon. / eruptive ~ ‖ Eruptivgestein n. ‖ roche f. éruptive. / fast ~ ‖ festes Gebirge n. oder Gestein n. ‖ roche f. compacte ou dure ou solide. / folded ~s pl. ‖ Faltengebirge n. ‖ montagne f. formée par plissement. / hard ~ ‖ Knauer m.; Hartklamm m. ‖ roche f. très dure. / ~ of lime forming algæ ‖ Algenbank f. ‖ roche f. d'algues calcaires. / organo-

geneous sedimentary ~ ‖ organogenes Sedimentgestein n. ‖ roche f. sédimentaire organogénique. / partition ~ ‖ Nebengestein n. ‖ roche f. des parois. / primary ~ ‖ Urgebirge n.; Urfels m. ‖ roche f. primitive. / primitive ~ ‖ Grundgestein n. ‖ roche f. originaire. / secondary ~ ‖ Flözgebirge n. ‖ roche f. stratiforme *ou* secondaire. / sedimentary ~ ‖ sedimentäres Gestein n.; Sedimentgestein n. ‖ roche f. sédimentaire. / stratified ~ *see* bedded ~.

rock asphalt ‖ Erdasphalt m. ‖ bitume m. de roche. / ~ borer ‖ Gesteinbohrer m. ‖ foreuse f. de roches. / ~ boring machine ‖ Gesteinsbohrmaschine f. ‖ machine f. à percer les roches. / ~ cellar ‖ Felsenkeller m. ‖ cave f. creusée dans le roc. / ~ chisel ‖ Kuttermeißel m.; Felsmeißel m. ‖ burin m. brise-rocs. / ~ cork ‖ Korkstein m. ‖ liège m. durci. / ~ crystal ‖ Bergkristall m. ‖ cristal m. de roche.

rock cutting machine ‖ Schrämmaschine f. ‖ machine f. haveuse. / ~ driver ‖ Maschinenhauer m. ‖ haveur m. à la machine.

rock dredger ‖ Steinbagger m. ‖ dérocheuse f.

rock drill ‖ Gesteinsbohrer m.; Gestein(s)bohrmaschine f. ‖ perforateur m.; perforatrice f.; machine f. à percer la roche. / electro-pneumatic ~ ‖ elektropneumatische Gesteinbohrmaschine f. ‖ machine f. électro-pneumatique à forer les roches. / ~ worked by hand ‖ Handbohrmaschine f. ‖ perforatrice f. à main. / pneumatic ~ ‖ Preßluftgesteinbohrmaschine f. ‖ perceuse f. de roches à air comprimé.

rock drill hammer ‖ Bohrhammer m. ‖ marteau m. perforateur m.

rock drilling ‖ Gesteinsbohrung f. ‖ forage m. de roches. / ~ hammer ‖ Gesteinbohrhammer m. ‖ marteau-foreur m. pour roches.

rock drilling machine ‖ Gesteinbohrmaschine f. ‖ foreuse f. pour roches. / compressed air ~ ‖ Preßluftgesteinbohrmaschine f. ‖ perceuse m. de roches à air comprimé. / electro-pneumatic ~ ‖ elektropneumatische Gesteinbohrmaschine f. ‖ machine f. électro-pneumatique à forer les roches.

rock drilling plant ‖ Gesteinbohranlage f. ‖ installation f. de perforateurs.

rock drill upsetting machine ‖ Gesteinsbohrerstauchmaschine f. ‖ machine f. à refouler les forets à roches. / ~ sharpening machine ‖ Gesteinsbohrerschärfmaschine f. ‖ machine f. à aiguiser les forets à roches.

rocker ‖ Schwinghebel m.; Kipphebel m. ‖ bascule f. / valve ~ ‖ Ventilschwinghebel m. ‖ balancier m. de soupape.

rocker arm ‖ Unterbrecherhebel m. ‖ levier m. interrupteur. / ~ shaft ‖ Schwinghebelwelle f. ‖ arbre m. de levier oscillant. / ~ support ‖ Schwinghebelbock m. ‖ support m. du levier oscillant.

rocket ‖ Rakete f. ‖ fusée f. / rotatory ~ ‖ Rotationsrakete f. ‖ fusée f. à rotation. / ~ for touring and flying ‖ Rakete f. für Fahrt und Flug ‖ fusée f. volante pour la route et le vol.

rocket apparatus for life saving ‖ Raketenapparat m. für Rettungswesen ‖ canon m. porte-amarre. / ~ carriage ‖ Raketenwagen m. ‖ caisson m. à fusées.

rocket case ‖ Raketenhülse f. ‖ cartouche f. *ou* douille f. de fusée volante; gobelet

m. / sheet-iron ~ ‖ eiserne Raketenhülse f. ‖ cartouche f. en tôle de la fusée de guerre.

rocket case paper ‖ Raketenhülsenpapier n. ‖ papier m. à cartouches de fusées.

rocket composition ‖ Raketensatz m. ‖ composition f. fusante. / ~ lightball ‖ Leuchtrakete f. ‖ fusée f. lumineuse *ou* à éclairer *ou* d'éclairage. / ~ mould ‖ Raketenform f.; Raketenstock m. ‖ moule m. de fusée. / ~ paper ‖ Raketenhülsenpapier n. ‖ papier m. à cartouches de fusées. / ~ problem ‖ Raketenproblem n. ‖ problème m. des fusées volantes.

rocket-propelled (air) plane ‖ Raketenflugzeug n. ‖ avion m. à réaction; avion-fusée m. / ~ space ship ‖ Raketenraumschiff n. ‖ avion m. à réaction; avion-fusée m.

rock fall ‖ Felssturz m. ‖ éboulement m. de roches. / ~ flint ‖ Quarz m. ‖ quartz m. / formation of ~ ‖ Gesteinsbildung f. ‖ formation f. de roche.

rocking ‖ Schwankung f. ‖ tangage m.; balancement m. / ~ commutator ‖ Wippe f. ‖ basculeur m. / ~ cylinder ‖ schwingender Zylinder m. ‖ cylindre m. oscillant. / ~ device ‖ Schaukelapparat m. ‖ basculeur m. / ~ horse ‖ Schaukelpferd n. ‖ cheval m. à bascule. / ~ lever ‖ Schwinghebel m. ‖ balancier m. / ~ motion ‖ Schaukelbewegung f. ‖ mouvement m. de bascule. / ~ screen ‖ Schüttelsieb n. ‖ tamis m. à secousses. / ~ tree (Weav) ‖ Ladenstock m.; Ladenprügel m. ‖ bâton m.; porte-battant m.

rock island ‖ Klippeninsel f. ‖ rocher m. isolé. / ~ lime ‖ Bergkreide f. ‖ craie f. de roche. / magnetism of ~s ‖ Gesteinsmagnetismus m. ‖ magnétisme m. des roches.

rockman ‖ Steinhauer m. ‖ rocteur m.

rock moisture ‖ Bergfeuchtigkeit f. ‖ humidité f. des roches. / ~ oil ‖ Naphta f. ‖ Erdöl n.; Steinöl n. ‖ naphte m.; bitume m. liquide; huile f. minérale.

rock salt ‖ Bergsalz n.; Steinsalz n. ‖ sel m. minéral *ou* gemme; halite f. / ~ deposits pl. ‖ Abraumsalz n. ‖ sel m. de déblai. / ~ mine ‖ Salzbergwerk n. ‖ mine f. de sel; saline f. / ~ miner ‖ Steinsalzhauer m. ‖ piqueur de sel m. / plate of ~ with laquer film ‖ Steinsalzplatte f. mit Schutzlack ‖ lame f. de sel gemme vernie.

rockwood ‖ Holzasbest m. ‖ asbeste m. ligniforme.

rocky ‖ felsig ‖ rocheux.

rod ‖ Stab m.; Stange f. ‖ barre f.; barreau m.; bâton m.; tige f. / ~ (Carp) ‖ Deckleiste f. ‖ couvre-joint m. / ~ (Mach) ‖ Lenker m.; Gegenlenker m. ‖ guide m. / ~s pl. (Mine) ‖ Gestänge n. ‖ tiges fpl. / ~ (Surv) ‖ Meßrute f. ‖ Meßstange f.; Absteckpfahl m. ‖ perche f.; règle f.; piquet m.; jalon m. / adjusting ~ ‖ Justierstange f.; Einstellstange f. ‖ gabarit m. de réglage. / brake ~ ‖ Bremsstange f.; Bremsgestänge n. ‖ tige f. de frein. / brak-pull ~ (Railw) ‖ Bremslasche f. ‖ bielle f. du frein. / clearing ~ (Fire arms) ‖ Patronenzieher m. ‖ baguette f. à décharger. / connecting ~ ‖ Treibstange f.; Pleuelstange f. ‖ bielle f. / coupling ~ ‖ Kuppelstange f. ‖ bielle f. d'accouplement. / deep-boring ~s pl. ‖ Bohrgestänge n.; Gestänge n. für Tiefbohrungen ‖ tiges fpl. des sondage. / ebonite ~ ‖ Hartgummistange f. ‖ ba-

guette f. en ébonite. / eccentric ~ ‖ Exzenterstange f. ‖ tige f. d'excentrique. / float ~ ‖ Schwimmerspindel f. ‖ tige f. du flotteur. / front coupling ~ ‖ vordere Kuppelstange f. ‖ bielle f. d'avant. / guide ~ ‖ Führungsstange f.; Lenkhebel m. ‖ barre f. de guidage; levier m. de manœuvre. / ignition ~ ‖ Zündgestänge n. ‖ tige f. du rupteur. / ~ covering a joint ‖ Deckleiste f.; Fugenleiste f. ‖ couvre-joint m. / lightning ~ ‖ Blitzableiter m. ‖ tige f. de paratonnerre. / piston ~ ‖ Kolbenstange f. ‖ tige f. de piston. / pump ~s pl. ‖ Pumpengestänge n. ‖ tiges fpl. de pompes. / rack ~ ‖ Zahnstange f. ‖ crémaillère f. / radius ~ ‖ Hinterachsschubstange f. ‖ bielle f. de poussée. / regulating ~ (Windmill) ‖ Streichstange f. ‖ perche f. / screwed ~ ‖ Stange f. mit Gewinde ‖ tige f. filetée. / slide valve ~ ‖ Schieberstange f. ‖ tige f. du tiroir. / spring ~ ‖ elastischer Stab m. ‖ lame f. / steering gear connecting ~ (Auto) ‖ Lenkstange f. ‖ barre f. de direction. / steering knuckle tie ~ ‖ Spurstange f. ‖ barre f. d'accouplement. / ~ for stuffing collars ‖ Polsterstange f. ‖ verge f. à enverger. / suspending ~ ‖ Hängestange f. ‖ tige f. de suspension. / valve ~ ‖ Ventilstange f. ‖ tige f. de soupape.

rod arm, steering knuckle gear ~ ‖ Lenkhebel m. ‖ levier m. de manœuvre des roues.

rod barrier ‖ Einlegeschranke f. ‖ barrière f. à lisse suspendue. / ~ boring ‖ Gestängebohren n. ‖ sondage m. à tige rigide. / ~ chisel ‖ Stielmeißel m.; Schrotmeißel m. ‖ tranche f. à manche; ciseau m. à chaud. / ~ copper ‖ Stangenkupfer n. ‖ cuivre m. en barres. / ~ coupling ‖ Stangenkupplung f. ‖ accouplement m. de tiges.

rodding ‖ Gestänge n. ‖ tiges fpl.

rod head grinding machine ‖ Stangenkopfschleifmaschine f. ‖ machine f. à rectifier les têtes de bielles. / ~ insulator ‖ Stabisolator m. ‖ isolateur m. à tige. / ~ iron ‖ Krauseisen n.; Zaineisen n. ‖ fer m. en barres dentelé; verge f. crénelée; fercarillon m.; barre f. de fer crêpée. / turning lathe for ~ jobs pl. ‖ Drehbank f. für Stangenarbeiten ‖ tour m. pour travaux dans la barre. / ~ mill ‖ Drahtstraße f. ‖ train m. à fil.

rods pl. (Met) ‖ Stabeisen n.; Stangeneisen n. ‖ fer m. en barres. / ~ (Mine) ‖ Gestänge n. ‖ tiges fpl. / deep-boring ~ ‖ Bohrgestänge n.; Gestänge n. für Tiefbohrungen ‖ tiges fpl. pour les sondages (profonds). / pump ~ ‖ Pumpengestänge n. ‖ tiges fpl. de pompes.

rod shaft (Mine) ‖ Pumpenschacht m.; Wasserhaltungsschacht m. ‖ puits m. d'épuisement *ou* d'exhaure; bure f. aux pompes. / ~ steel ‖ Stabstahl m.; Stangenstahl m. ‖ acier m. en barres. / ~ turning machine ‖ Rundstabmaschine f. ‖ machine f. à faire les bâtons ronds.

roe of cod ‖ Rogen m. vom Kabeljau ‖ rogue m. de morue.

roe mahogany ‖ scheckiges *oder* gesprenkeltes *oder* gestreiftes Mahagoniholz n. ‖ acajou m. rubané.

roll, to ‖ rollen; wälzen ‖ rouler. / ~ (Aero) ‖ wälzen ‖ rouler; voler en tonneau; avoir du roulis. / ~ (Mar) ‖ schlingern ‖ rouler. / ~ (Roll mill) ‖ walzen ‖ laminer.

/ ~ cold ‖ kaltwalzen ‖ laminer à froid. / ~ the copper billet to bars ‖ den Kupferbarren m. zu Stangen auswalzen ‖ laminer le lingot de cuivre en barres. / ~ the dough ‖ den Teig m. wälgern ‖ étendre la pâte avec le rouleau. / ~ the form (Print) ‖ die Druckerschwärze *oder* Druckfarbe auftragen ‖ encrer; toucher la forme. / ~ hot ‖ warmwalzen ‖ laminer à chaud. / ~ out ‖ auswalzen ‖ laminer. / ~ up ‖ aufrollen ‖ enrouler. / ~ up a rope ‖ ein Tau n. in einer Scheibe aufrollen ‖ glèner; faire une glène de filin. / ~ wire ‖ Draht m. walzen ‖ laminer le fil métall'que.

roll *see also* roller (Mach) ‖ Rolle f.; Walze f. ‖ rouleau m.; cylindre m./~ (Mar) ‖ Scheibe f. von Tauwerk ‖ glène f.; glène de filin. / ~ (Spinn) ‖ Locke f. ‖ loquette f.; ploque f. / ~ (Window air draught) ‖ Wulst m. ‖ bourrelet m.; renflement m. / brush ~ ‖ Bürstenwalze f. ‖ brosse f. cylindrique. / chilled iron ~ ‖ Hartgußwalze f. ‖ cylindre m. en fonte coulée en coquille. / ~ of clay ‖ Lettennudel f. ‖ tige f. en terre glaise. / coarse-crushing ~s ‖ Grobwalzwerk n. ‖ cylindres mpl. dégrossisseurs. / cold ~ ‖ Kaltwalze f. ‖ cylindre m. de (laminoir pour) laminage à froid. / two ~s pl. which can be displaced toward each other ‖ zwei gegeneinander verschiebbare Walzen fpl. ‖ deux cylindres mpl. déplaçables l'un en regard de l'autre. / engraved ~ ‖ gravierte Walze f. ‖ cylindre m. gravé. / felt ~ ‖ Wickelwalze f. ‖ cylindre m. en feutre. / female ~ ‖ Mutterfurche f. ‖ cannelure f. femelle. / to give a final grinding to ~s ‖ Walzen fpl. nachschleifen ‖ finir des cylindres mpl. à la meule. / finishing ~s pl. ‖ Stabwalzwerk n. ‖ cylindres mpl. finisseurs. / first ~s pl. ‖ Luppenstraße f. ‖ train m. ébaucheur. / ~ producing the flange by lateral pressure ‖ Walze f., die von der Seite her den Spurkranz anstaucht ‖ rouleau m. refoulant (le métal pour faire) le boudin. / fluted ~ ‖ geriffelte Walze f. ‖ cylindre m. cannelé. / forged ~ ‖ geschmiedete Walze f. ‖ cylindre m. de laminoir forgé. / gelatine-coated ~ ‖ mit Gelatine überzogene Walze f. ‖ rouleau m. recouvert de gélatine. / grooved ~ ‖ Kaliberwalze f.; kalibrierte Walze f. ‖ cylindre m. calibré *ou* cannelé. / middle ~ ‖ Mittelwalze f. ‖ cylindre m. médian. / music ~ for automatic pianos ‖ Notenrolle f. für Klavierspielapparate ‖ rouleau m. de musique pour les appareils à jouer le piano. / the ~s pl. are brought nearer together after each pass ‖ die Walzen fpl. werden nach jedem Stich nähergestellt ‖ les cylindres mpl. s'approchent l'un à l'autre à chaque passage. / oscillating ~ ‖ schwingende Walze f. ‖ cylindre m. oscillant. / polishing ~ ‖ Polierwalze f. ‖ cylindre m. poliseur. / ~s pl. for rails ‖ Schienenwalzen fpl. ‖ cylindres mpl. pour rails. / rough ~s pl. *see* first ~. / ~ for filling with sand ‖ Walzenrad n. m't Sandfüllung ‖ rouleau m. à être rempli avec du sable. / ~ carrying sensitized paper ‖ Walze f. mit lichtempfindlichem Papier ‖ cylindre m. avec papier sensible à la lumière. / sheet ~ ‖ Blechwalze f. ‖ cylindre m. de laminoir à tôle. / smooth ~ ‖ glatte Walze f. ‖ cylindre m.

lisse. / with ~s pl. arranged in staggered order ‖ mit versetzter Walzenanordnung f. ‖ avec rouleaux mpl. placés en position alterne. / ~ of tobacco ‖ Tabakrolle f. ‖ carotte f.; boudin m. *ou* rouleau m. de tabac. / ~ for weaver's reeds ‖ Rietwalze f. ‖ cylindre m. pour laminer des rots.

roll blotter ‖ Tintenlöscher m. ‖ portepapier buvard m. / ~ brimstone ‖ Stangenschwefel m. ‖ soufre m. en canon. / ~ butter ‖ gepfundete Butter f. ‖ beurre m. en livre. / ~ caliber ‖ Walzenkaliber n. ‖ calibre m. de cylindre de laminoir. / ~ coverer (Spinn) ‖ Pergamentaufzieherin f. ‖ colleuse f. de parchemin. / ~ designing ‖ Walzenkalibrierung f. ‖ calibrage m. des cylindres de laminoir. / ~ dough dividing machine ‖ Brötchenteigteilmaschine f. ‖ machine f. à diviser la pâte pour petits pains. / ~ dough moulding machine ‖ Brötchenteigwirkmaschine f. ‖ machine f. à rouler la pâte pour petits pains.

rolled ‖ gewalzt ‖ cylindré; laminé. / cold-~ ‖ kaltgewalzt ‖ laminé à froid. / ~ bar ‖ gewalzter Stab m. ‖ barre f. laminée. / ~ iron ‖ Walzeisen n. ‖ fer m. laminé. / ~ lead ‖ Walzblei m. ‖ plomb m. laminé. / ~ lead box ‖ Walzbleimuffe f. ‖ manchon m. de plomb laminé. / tree protector of ~ metal ‖ Baumschützer m. aus Streckmetall ‖ grillage m. de protection en métal étiré pour arbres. / ~ section for guide and check rails ‖ Zwangsschienenwinkel m. ‖ cornières fpl. pour contrerails. / ~ tube ‖ gewalzte Röhre f. ‖ tube m. laminé. / ~ wire ‖ gewalzter Draht m.; Walzdraht m. ‖ fil m. laminé.

roll engraver ‖ Druckwalzengravör m. ‖ graveur m. sur cylindres.

roller *see also* roll (Mach) ‖ Führungsrolle f.; Leitrolle f. ‖ galet m.; roulette f.; poulie f. / ~ (Mar) ‖ Rollbank f. ‖ chevalet m. à rouleau. / ~ (Mine) ‖ Fördermann m.; Schlepper m.; Wagenstößer m. ‖ escaneur m.; rouleur m.; traîneur m. / ~ (Porcel) ‖ Mangelholz n.; Rollholz n. ‖ billette f.; rouleau m. / ~ (Print) ‖ Druckwalze f. ‖ rouleau m. / ~ (Roll mill) ‖ Walze f. ‖ rouleau m.; cylindre m. / ~ (Sea) ‖ Brandungswelle f. ‖ vague f. de ressac. / ~ (Spinn) ‖ Walzendreher m. ‖ couvreur m. de cylindres. / adjusting ~ ‖ Einstellwalze f. ‖ tambour m. de réglage. / ~ for bitumen pavements ‖ Walze f. für Bitumendecken ‖ rouleau m. cylindreur pour revêtements bitumineux. / case-hardened ~ ‖ Hartgußwalze f. ‖ cylindre m. en fonte durci à la surface. / delivering ~ (Spinn) ‖ Zugwalze f. ‖ cylindre-étireur m. / draw-in ~ ‖ Einzugswalze f. ‖ rouleau m. inférieur. / flatting ~ ‖ Plättwalze f.; Streckwalze f. ‖ aplatissoire f. / fluted ~ ‖ Riffelwalze f. ‖ cylindre m. cannelé. / inking ~ (Print) ‖ Auftragwalze f. ‖ rouleau m. toucheur. / ~ with interchangeable segments ‖ Walze f. mit auswechselbaren Segmenten ‖ cylindre m. muni de segments interchangeables. / leather ~ (Lithography) ‖ Lederwalze f. ‖ rouleau m. en cuir. / magnetic ~ ‖ Magnetwalze f. ‖ cylindre m. aimanté. / puddled bar ~ ‖ Luppenwalzer m. ‖ ébaucheur m. / road ~ ‖ Straßenwalze f. ‖ rouleau-écraseur m.; rouleau m. compresseur. / ~ for roller bearing ‖

Rolle f. für Rollenlager ‖ rouleau m. pour boîte à rouleaux. / smooth ~ ‖ Ackerwalze f. ‖ rouleau m. de labourage. / straightening ~ ‖ Richtwalze f. ‖ cylindre m. planeur. / submersion ~ (Hydr arch) ‖ Versenkwalze f. ‖ rouleau m. submersible. / superheated steam three-wheel ~ ‖ Heißdampfdreiradwalze f. ‖ rouleau m. compresseur à trois rouleaux à vapeur surchauffée. / supporting ~ ‖ Stützwalze f. ‖ rouleau m. d'appui. / tension ~ ‖ Spannrolle f. ‖ poulie f. de tension; tendeur m. de courroie. / withdrawal ~ ‖ Ausgangswalze f. ‖ rouleau m. d'appui.

roller beam (Weav) ‖ Kettenbaum m. ‖ ensouple f.

roller bearing ‖ Rollenlager n. ‖ palier m. *ou* roulement m. à rouleaux. / conical ~ ‖ konisches Rollenlager n. ‖ roulement m. à rouleaux coniques. / roll-pin ~ ‖ Walzenzapfenrollenlager n. ‖ palier m. à rouleaux pour tourillons des cylindres. / ~ with shoulder ‖ Schulterrollenlager n. ‖ roulement m. à rouleaux avec épaulement.

roller bearing grease box ‖ Rollenschmierbüchse f. ‖ graisseur m. à rouleaux.

roller blind ‖ Rollvorhang m.; Rouleau n. ‖ store m. en étoffe. / iron ~ ‖ eiserner Rolladen m.; eiserne Rolljalousie f. ‖ jalousie f. en fer. / sheet iron ~ ‖ Rolladen m. *oder* Rolljalousie f. aus Eisenblech ‖ jalousie f. en tôle.

roller blind shutter (Phot) ‖ Schlitzmomentverschluß m. ‖ obturateur m. instantané à rideau.

roller bracket ‖ Rollenbock m. ‖ support m. de galet. / ~ bridge ‖ Rollbrücke f. ‖ pont m. roulant. / ~ buckle (Saddl) ‖ Walzenschnalle f. ‖ boucle f. roulante; boucle f. à rouleau. / ~ card (Spinn) ‖ Walzenkrempel f. ‖ carde f. à hérissons. / ~ carriage (Print) ‖ Walzenwagen m. ‖ chariot m. porte-rouleaux toucheurs. / ~ chain ‖ Rollenkette f. ‖ chaîne f. à rouleaux. / chair on ~s for sick persons ‖ Krankenfahrstuhl m. (zum Schieben) ‖ chaise f. roulante pour malades. / ~ coating ‖ Walzenüberzug m. ‖ enduit m. de cylindre.

roller composition (Print) ‖ Walzenmasse f. ‖ pâte f. à rouleaux. / mincing machine for ~ ‖ Walzenmassezerkleinerungsmaschine f. ‖ machine f. à morceler la pâte à rouleaux.

roller contact ‖ Rollenkontakt m. ‖ contact m. à rouleaux *ou* à roulette. / ~ conveyor ‖ Rollenförderer m. ‖ transporteuse f. à rouleaux. / ~ cooling bed ‖ Rollenkühlbett n. ‖ refroidissoir m. à rouleaux. / ~ crusher ‖ Quetschwalzwerk n. ‖ concasseur m. à cylindres. / ~ delivery bed ‖ Ablaufrollgang m. ‖ train m. de rouleaux d'évacuation. / ~ dryer ‖ Walzentrockner m.; Rollentrockner m. ‖ sécheur m. à cylindres. / ~ electrode ‖ Rollenelektrode f. ‖ électrode f. à rouleau. / ~ feed bed ‖ Auflaufrollgang m. ‖ train m. de rouleaux d'accès. / ~ feeder ‖ Walzenaufgabevorrichtung f. ‖ feeder m. à cylindres. / ~ frame ‖ Walzenstuhl m. ‖ râtelier m. à rouleaux; cage m. de laminoir.

roller gear ‖ Rollenlaufwerk n. ‖ chariot m. de roulement à cylindres. / grooved ~ ‖ Kammwalzengetriebe n. ‖ harnais m. d'engrenages à chevron.

roller gear bed || Rollgang m. || rouleau m. d'amenée. / ~ for cases || Rollbahn f. für Kisten || tapis m. roulant pour caisses.
roller gear table || Rolltisch m. || table f. à rouleaux.
roller gin (Spinn) || Walzenegreniermaschine f. || roller-gin m. / ~ grate || Walzenrost m. || grille f. à cylindre. / ~ grinder (Chem) || Walzenreibmaschine f. || broyeur m. à cylindre. / ~ grinding machine || Walzenschleifmaschine f. || machine f. à rectifier ou à rôder les cylindres de laminoirs. / ~ guide || Rollenführung f. || guidage m. sur galets. / ~ lathe || Trommeldrehbank f. || tour m. à tambours. / ~ leather || Walzenleder n. || cuir m. à rouleaux. / ~ leather cutting machine || Walzenlederstreifenschneidmaschine f. || machine f. à couper les bandes de cuir pour cylindres. / ~ machine (Pap) || Rollmaschine f. || machine f. enrouleuse. / material for ~s || Walzenmasse f. || pâte f. à cylindres.
roller mill || Walzenmühle f.; Holländermühle f. || moulin m. à cylindres. / reducing ~ || Flachmahlstuhl m. || moulin m. pour monture basse. / ring ~ || Ringwalzenmühle f. || moulin m. à cylindres annulaires. / ~ with rolls lying side by side || Walzwerk n. mit nebeneinander liegenden Walzen || laminoir m. à cylindres juxtaposés.
roller path (Mill) || Flur m.; Flurbalken m. || chemin m. ou poutre f. de roulement. / ~ press || Walzenaufziehpresse f. || presse f. à caler les cylindres. / ~ rack (Print) || Walzenständer m. || raquette f. porte-rouleau. / ~ ring || Walzenring m. || anneau m. à cylindre. / ~ screen || Rollenklassiersieb n. || crible-classeur m. à rouleaux. / ~ shell || Walzenmantel m. || enveloppe f. à cylindre. / ~ skate || Rollschuh m. || patin m. à roulettes; patinette f.
roller sluice || Rollschütz n. || vanne f. à rouleaux. / ~ gate which can be raised and lowered || versenkbares Rollschütz n. || vannette f. à rouleaux submersible. / ~ weir || Walzenschützenwehr n. || barrage m. à cylindres.
roller table || Walztisch m. || table f. de laminoir. / ~ transporter || Transportrollgang m. || rouleaux mpl. d'amenage. / turntable on ~s || Rollendrehscheibe f. || plaque f. tournante sur galets. / ~ vat || Rollfaß n. || tonneau m. à rouleaux. / ~ washing machine (Print) || Walzenwaschmaschine f. || machine f. à laver les rouleaux. / ~ way || Rollenbahn f. || voie f. à galets. / ~ weir || Walzenwehr n. || barrage m. à cylindre.
rolley (Mine) || Förderwagen m.; Hund m. || chariot m.; berline f.; galliot m. / ~ way (Mine) || Förderstrecke f. || galerie f. ou voie f. de roulage. / ~ way man (Mine) || Schienenleger m.; Schienenwärter m. || raccomodeur m.; défrailleur m.; metteur m. de rails.
roll film || Rollfilm m. || roll-film m.; pellicule f. en bobines. / ~ container || Rollfilmkassette f. || châssis m. à bobine muni d'un rouleau de film.
roll gaiter || Wickelgamasche f. || bande f. molletière. / ~ grinder || Walzenriffeler m. || tailleur m. de cylindres.
roll grinding machine || Walzenschleifmaschine f. || machine f. à rectifier les

cylindres. / ~ (Bak) || Semmelmühle f. || machine f. à chapelure.
roll housing || Walzenständer m. || cage f. à cylindres de laminoir. / ~ of pilgrim mill || Pilgergerüst n. || cage f. de laminoir à pas de pèlerin.
rolling (Metal) || Walzen n.; Strecken n. || laminage m. / ~ (Ship) || Schlingerbewegung f. || roulis m. / ~ in smooth water || Rollen n. in ruhigem Wasser || roulis m. en eau calme. / ~ metal with tar onto road surface || Einwalzen n. von Splitt bei Oberflächenteerung || aplanissement m. ou cylindrage m. du cailloutis sur la surface de route goudronnée. / the ~ is done in one heating only || das Auswalzen n. erfolgt in einer einzigen Hitze || le laminage m. se fait dans une seule chaude. / ~ of steel ingots || Walzen n. von Stahlblöcken || laminage m. de lingots d'acier.
rolling crusher || Walzenbrecher m. || concasseur m. giratoire; broyeur m. à cylindres. / across the direction of ~ || quer zur Walzrichtung f. || transversalement au laminage. / along the direction of ~ || längs zur Walzrichtung f. || dans le sens du laminage. / ~ door || Rolltür f. || porte f. roulante. / ~ engine for clay and gravel || Walzwerk n. für Ton und Kies || laminoir m. à argile et gravier. / ~ firing barrage || Feuerwalze f. || barrage m. de tir roulant. / ~ friction || Schienenreibung f. || friction f. des rails. / ~ hoop for barrels || Rollreifen m. für Fässer || cercle m. de roulement pour tonneaux.
rolling machine (Pap) || Satiniermaschine f.; Kalander m. || laminoir m. en lissoir; laminoir m.; glaceur n.; satineuse f. / ~ (Roll mill) || Walzmaschine f. || laminoir m. / ~ for fabrics || Quetschmaschine f. für Stoffe || machine f. à exprimer les étoffes. / retaining ring ~ || Sprengringeinwalzmaschine f. || machine f. à caler les bagues de retenue. / thread ~ || Gewindewalzmaschine f. || machine f. à cylindrer le filet. / tube ~ || Rohrwalzmaschine f. || laminoir m. à tubes.
rolling machine operator || Biegemaschinenarbeiter m. || cintreur m. à la machine.
rolling mandril || Walzdorn m. || mandrin m.
rolling mill || Walzwerk n. || laminoir m. / ~ (Machine) || Walzmaschine f.; Walz(en)straße f.; Walzwerk n. || laminoir m.; machine f. ou train m. de laminage; train m. (de laminoir); machine f. à laminer ou à cylindrer. / armour plate ~ || Panzerplattenwalzwerk n. || laminoir m. à plaques de blindage. / brass ~ || Messingwalzwerk n. || laminoir m. à laiton. / cold ~ || Kaltwalzwerk n. || laminoir m. à froid. / continuous ~ || kontinuierliches Walzwerk n. || laminoir m. à mouvement continu. / copper ~ || Kupferwalzwerk n. || laminoir m. à cuivre; usine f. de laminage de cuivre. / ~ for corrugated tubes || Wellrohrwalzwerk n. || laminoir m. à tubes ondulés. / dessinier ~ || Dessinierwalzwerk n. || laminoir m. à dessins. / ~ for disc wheels || Scheibenräderwalzwerk n. || laminoir m. à disques de roues. / equalizing ~ || Egalisierwalzmaschine f. || machine f. à cylindres égaliseurs. / fine-iron ~ || Feineisenwalzwerk n. || laminoir m. ou train m. à petits fers. / finishing ~ ||

Justierwalzwerk n. || laminoir m. à calibrer. / German silver ~ || Argentanwalzwerk n. || laminoir m. de maillechort. / gold ~ || Goldwalzwerk n. || laminoir m. à or. / ~ of great speed || Schnellwalzwerk n. || train m. à grande vitesse. / hunting spoon ~ || pendelnde Löffelwalzmaschine f. || laminoir m. à pendule à faire des cuillers. / iron ~ || Eisenwalzwerk n. || usine f. de laminage de fer; laminoir m. à fer. / joist ~ || Trägerwalzwerk n. || laminoir m. à poutrelles. / medium iron ~ || Mitteleisenwalzwerk n. || laminoir m. à fer de dimension moyenne. / metal ~ || Metallwalzwerk n. || laminoir m. à métaux. / ~ for the Mint || Münzwalzwerk n. || laminoir m. monnayeur ou monétaire. / muffs ~ || Muffenwalzwerk n. || laminoir m. à manchons. / nickel ~ || Nickelwalzwerk n. || laminoir m. à nickel. / pilgrim step ~ || Pilgerschrittwalzwerk n. || laminoir m. à pas de pèlerin. / ~ for pipes || Röhrenwalzwerk n. || laminoir m. à tuyaux. / plate ~ || Eisenblechwalzwerk n. || laminoir m. à tôles. / profiled iron ~ || Profileisenwalzwerk n. || laminoir m. pour fers profilés. / rail ~ || Schienenwalzwerk n. || laminoir m. à rails. / ~ for reaming tubes || Rohraufweitewalzwerk n. || laminoir m. à mandriner les tubes. / reversing ~ || Umkehrwalzwerk n.; Reversierwalzwerk n. || laminoir m. réversible. / section ~ || Formeisenwalzwerk n. || laminoir m. pour profilés. / ~ for sharpening wire || Anspitzwalzwerk n. für Draht || laminoir m. à appointer des fils. / ~ for sheet billets || Platinenwalzwerk n. || laminoir m. à tôlesplatines. / slant ~ || Schrägwalzwerk n. || laminoir m. oblique. / slitting ~ || Schneideisenwalzwerk n. || laminoir m. à refendre. / small iron bar ~ || Feineisenwalzwerk n. || laminoir m. à petits fers. / ~ for smoothing tubes || Röhrenglättwalzwerk n. || laminoir m. à polir les tubes. / spoon ~ || Löffelwalze f. || laminoir m. à cuillers. / steel ~ || Stahlwalzwerk n. || usine f. de laminage d'acier. / thin sheet ~ || Feinblechwalzwerk n. || laminoir m. à tôles minces. / tube ~ || Rohrwalzwerk n. || laminoir m. à tubes. / equipment for tube ~s || Rohrwalzwerkeinrichtung f. || installation f. pour laminoirs à tubes. / machine for tube ~s || Rohrwalzwerkmaschine f. || machine f. pour laminoirs à tubes. / tube reducing ~ || Reduzierwalzwerk n. für Rohre || laminoir m. à réduire les tubes. / turbine blade ~ || Turbinenschaufelwalzwerk n. || laminoir m. à aubes de turbine. / universal ~ || Universalwalzwerk n. || laminoir m. universel. / weld tube ~ || Schweißrohrwalzwerk n. || laminoir m. à souder les tubes. / wire ~ || Drahtstraße f. || train m. pour fils. / zinc ~ || Zinkwalzwerk n. || laminoir m. à zinc.
rolling mill, auxiliary machine for ~s || Walzwerkhilfsmaschine f. || machine f. auxiliaire pour laminoirs.
rolling mill engine || Walzenzugmaschine f. || moteur m. ou machine f. motrice de laminoirs. / gas ~ || Gaswalzenzugmaschine f. || moteur m. à gaz pour laminoirs. / tandem reversing ~ || Tandemreversierwalzenzugmaschine f. || machine f. tandem réversible de laminoir.
rolling mill equipment || Walzwerkeinrichtung f. || installation f. de laminoir.

/ ~ furnace ‖ Walzwerkofen m. ‖ four m. de laminoirs. / twin armature ~ motor ‖ Doppelankerwalzmotor m. ‖ moteur m. de laminoir à double induit. / ~ path drive ‖ Walzenstraßenantrieb m. ‖ commande f. de train de laminoir. / ~ plant ‖ Walzwerksanlage f. ‖ installation f. laminoir. / ~ scale ‖ Glühspan m. ‖ pailles fpl. d'oxyde de fer. / ~ shops pl. ‖ Walzwerkshallen fpl. ‖ halls mpl. de laminoirs.

rolling moment (Aero; Nav) ‖ Quermoment n. ‖ moment m. de roulis.

rolling motion ‖ Rollbewegung f. ‖ mouvement m. de roulis. / easy ~ ‖ sanfte Rollbewegung f. ‖ mouvement m. doux de roulis. / ~ of a ship ‖ Rollbewegung f. eines Schiffes ‖ mouvement m. de roulis *ou* roulis m. d'un navire. / uneasy ~ ‖ harte *oder* heftige Rollbewegung f. ‖ mouvement m. dur de roulis.

rolling-and-pitching-motion ‖ Schlinger- und Stampfbewegung f. ‖ mouvement m. de roulis et de tangage.

rolling-out ‖ Auswalzen n. ‖ laminage m. / ~ of the rings into tires ‖ Auswalzen n. der Ringe zu Radreifen ‖ laminage m. des anneaux mpl. en bandages de roue.

rolling pirn ‖ Laufspule f. ‖ cannette f. à dérouler. / ~ plant ‖ Walzwerk n. ‖ laminoirs mpl. / ~ programme ‖ Walzprogramm n. ‖ programme m. de laminage. / ~ stick ‖ Mangelholz n.; Rollholz n. ‖ billette f.; rouleau m.

rolling stock (Railw) ‖ rollendes Material n.; Betriebsmittel npl. ‖ matériel m. roulant. / railway ~ ‖ rollendes Eisenbahnmaterial n. ‖ matériel m. roulant de chemin de fer.

rolling stock clearance gauge ‖ Umgrenzungslinie f. für Eisenbahnfahrzeuge ‖ gabarit m. des wagons de chemin de fer.

rolling surface (Wheel) ‖ Lauffläche f. ‖ surface f. de roulement. / ~ tape measure ‖ Bandmaß n.; Rollmaß n. ‖ mesure f. en ruban. / ~ thread cutter ‖ Gewindewalze f. ‖ cylindre m. à laminer les filets. / ~ track bedding ‖ Gleisbettung f. ‖ ballastage m. de la voie. / ~ train ‖ Walzenstraße f.; Walzenstrecke f. ‖ laminoir m. *ou* train m. à cylindres; train m. de laminoir.

rolling-up contrivance ‖ Aufrollvorrichtung f. ‖ dispositif m. d'enroulement.

rolling works pl. for handling precious metals ‖ Walzwerk n. für Edelmetalle ‖ laminoir m. pour métaux précieux.

roll lathe ‖ Walzendrehbank f. ‖ tour m. à travailler les cylindres; tour m. pour cylindres de laminoir. / ~ lifting device ‖ Walzenaushebevorrichtung f. ‖ dispositif m. de soulèvement pour cylindres. / machine for making ~s ‖ Rollenherstellungsmaschine f. ‖ machine f. pour la fabrication des rouleaux. / ~ motor ‖ Rollmotor m. ‖ moteur m. roulant. / ~ pin roller bearing ‖ Walzenzapfenrollenlager n. ‖ palier m. à rouleaux pour tourillons des cylindres. / gyroscopic ~ and pitch recorder ‖ Kreiselschlinger- und Stampfanzeiger m. ‖ appareil m. enregistreur gyroscopique de roulis et de tangage. / ~ repairer ‖ Walzendreher m. ‖ tourneur m. de cylindres. / ~ shaft *see* ~ spindle. / ~ sheet-iron ‖ Rollenblech n. ‖ tôle f. en rouleaux. / ~ shell ‖ Walzenring m. ‖ anneau m. de cylindre. / ~ shutter ‖ Rolladen m. ‖ jalousie f.; rouleau m.

/ ~ shutter fitting ‖ Rollvorhangbeschlag m. ‖ garniture f. de rouleau. / ~ skate ‖ Rollschuh m. ‖ patin m. à roulette. / ~ spindle ‖ Walzenspindel f. ‖ arbre m. de cylindres *ou* à réglage. / ~ sulphur ‖ Stangenschwefel m. ‖ soufre m. en canons. / ~ tape measure ‖ Rollbandmaß n. ‖ mètre m. à ruban. / ~ tenter ‖ Walzwerkarbeiter m. ‖ lamineur m. / ~ tobacco ‖ Rollentabak m. ‖ tabac m. en andouilles *ou* en rouleaux *ou* roulé. / ~ top desk ‖ Schreibtisch m. mit Rollverschluß ‖ bureau m. à rideaux. / ~ turner ‖ Walzendreher m. ‖ tourneur m. de cylindres de laminoir. / ~ turning shop ‖ Walzendreherei f. ‖ atelier m. de tournage de cylindres. / ~ wobbler milling machine ‖ Walzenzapfenfräsmaschine f. ‖ machine f. à fraiser les tourillons des cylindres de laminoir. / ~ working machine ‖ Walzenbearbeitungsmaschine f. ‖ machine f. à travailler les cylindres de laminoir.

Roman balance with hook ‖ Schnellwage f. mit Haken ‖ romaine f. à crochet. / ~ cement ‖ Romanzement m. ‖ ciment m. romain. / ~ numerals pl. ‖ römische Zahlen fpl. ‖ chiffres mpl. romains.

Roman type ‖ Antiqua f. ‖ caractère m. romain. / bold ~ (Print) ‖ fette Antiqua f. ‖ romain m. gras.

Romanesque (Arch) ‖ romanisch ‖ roman.

ronde (Print) ‖ Ronde f. ‖ italique f.; lettres fpl. italiques.

Röntgen apparatus ‖ Röntgenapparat m. ‖ appareil m. Röntgen. / ~ photo ‖ Durchleuchtungsbild n. ‖ image f. radioscopique. / ~ rays pl. ‖ Röntgenstrahlen mpl. ‖ rayons mpl. de Röntgen. / ~ ray sickness ‖ Röntgenkater m. ‖ mal m. des irradiations pénétrantes. / ~ ray technics pl. ‖ Röntgentechnik f. ‖ technique f. des rayons Röntgen. / ~ tube ‖ Röntgenröhre f. ‖ tube m. (de) Röntgen. / watercooled ~ tube ‖ Röntgenröhre f. mit Wasserkühlung ‖ tube m. Röntgen avec refroidissement à l'eau.

roof, to ‖ bedachen ‖ couvrir.

roof (Build) ‖ Dach n. ‖ toit m.; comble m. / ~ (Mine) ‖ Hangendes n. ‖ couche f. supérieure; toit m. de la veine *ou* d'une couche. / ~ arched ‖ Tonnendach n.; tonnenförmiges Dach n. ‖ toit m. en berceau. / barrel ~ see arched ~. / to bring down the ~ (Mine) ‖ zu Bruche m. bauen ‖ faire ébouler les débris mpl. / corrugated sheet ~ ‖ Wellblechdach n. ‖ toiture f. en tôle ondulée. / curb ~ see mansard ~. / flat ~ ‖ flaches Dach n. ‖ toiture f. plate; toit m. *ou* comble m. plat; terrasse f. / ~ for furnaces ‖ Gewölbe n. für Feuerungen ‖ voûte f. de foyers. / gable ~ ‖ Giebeldach n. ‖ toit m. à pignon. / ~ of a gallery ‖ Firste f. eines Stollens ‖ faîte m. d'une galerie. / groined ~ ‖ gerippte Decke f.; Rippendecke f. ‖ plafond m. à nervures. / hip ~ ‖ Walmdach n. ‖ toit m. en croupe. / ~ of manhole (Tel) ‖ Brunnenabdeckung f. ‖ plafond m. de chambre de tirage. / mansard ~ ‖ gebrochenes Dach n.; Mansardendach n. ‖ comble m. brisé *ou* coupé *ou* en mansarde. / to cause the ~ of a mine to fall in ‖ den Bruch niedergehen lassen *oder* einstürzen ‖ ébouler *ou* faire ébouler les débris mpl. / omnibus ~ ‖ Omnibusverdeck n. ‖ impériale f. / ~ of the outside

firebox ‖ Stehkesseldecke f. ‖ ciel m. de boîte à feu. / ~ of planks ‖ Bretterdach n. ‖ toit m. de planches. / ~ over a pulpit ‖ Kanzeldach n. ‖ abat-voix m. / pumice-concrete ~ ‖ Bimsbetondach n. ‖ toiture f. en béton de pierre-ponce. / ridged ~ ‖ Satteldach n. ‖ toit m. en batière *ou* à deux égouts. / saddle ~ ‖ Satteldach n. ‖ toit m. en batière *ou* à deux égouts. / ~ of a seam ‖ Hangendes n. ‖ toit m. d'une couche; couche f. supérieure. / slated ~ ‖ Schieferdach n. ‖ toit m. couvert en ardoise. / suspended ~ ‖ Hängedecke f. ‖ ciel m. suspendu. / thatched ~ ‖ Strohdach n. ‖ toit m. en chaume. / umbrella-shaped ~ ‖ schirmförmige Dachspitze f. ‖ toit m. *ou* toiture f. conique. / water-proof paper ~ ‖ Pappdach n. ‖ toiture f. en carton bitumé. / ~ of wood-cement ‖ Holzzementdach n. ‖ toit m. en ciment à pâte de bois. / wooden ~ ‖ Holzdach n. ‖ toit m. en bois. / zinc-covered ~ ‖ Zinkdach n. ‖ toit m. couvert en zinc.

roof bar ‖ Deckenträger m. ‖ tirant m. du ciel de foyer. / ~ batten ‖ Dachlatte ‖ latte f. de toit. / ~ beam ‖ Dachbalken m. ‖ poutre f. du toit. / ~ and floor beam ‖ Querträger m. ‖ entretoise f. / ~ boarding ‖ Dachschalung f. ‖ planchéiage m. de toit. / ~ bracket ‖ Dachgesimsstütze f. ‖ ferrure f. *ou* support m. de corniche. / ~ construction ‖ Dachkonstruktion f. ‖ construction f. de toiture. / ~ covering ‖ Dachdeckung f. ‖ couverture f.

roofed, room ~ with glass ‖ glasüberdeckte Halle f. ‖ hall m. vitré.

roofer ‖ Dachdecker m. ‖ couvreur m.

roof garden ‖ Dachgarten m. ‖ jardin m. suspendu. / ~ glazing ‖ Glaseindeckung f. ‖ couverture f. en verre. / ~ gutter ‖ Dachrinne f. ‖ gouttière f.

roofing ‖ Bedachung f. ‖ couverture f. de toit; toiture f. / asbestos-protected metal ~ ‖ Dachdeckung f. aus asbestgeschütztem Blech ‖ couverture f. de toit en tôle protégée d'asbeste. / fireproof ~ ‖ feuersicheres Eindecken n. ‖ couverture f. incombustible.

roofing cardboard ‖ Dachpappe f. ‖ carton m. pour toitures. / ~ felt ‖ Dachfilz m. ‖ feutre m. pour toitures. / tarred ~ felt ‖ geteerte Dachpappe f. ‖ carton m. bitumé. / ~ glass ‖ Dachglas n. ‖ verre m. de toiture. / ~ nail ‖ Dachnagel m. ‖ clou m. de toit. / ~ paper ‖ Dachpappe f. ‖ carton m. de toiture. / ~ sheet ‖ Dachblech n. ‖ tôle f. de toiture. / ~ slate ‖ Dachschiefer m. ‖ ardoise f. tégulaire *ou* pour toitures; schiste m. tabulaire. / ~ supplies pl. ‖ Dachdeckereibedarf m. ‖ articles mpl. pour couvreurs.

roofing tile ‖ Dachziegel m. ‖ tuile f. / cement ~ ‖ Zementdachziegel m. ‖ tuile f. de ciment. / grooved ~ ‖ Dachfalzziegel m. ‖ tuile f. en onglet pour toitures. / hand-formed ~ ‖ handgeformter Dachziegel m. ‖ tuile f. moulée à la main. / hollow ~ ‖ Hohldachziegel m. ‖ tuile f. creuse.

roofing tile machine ‖ Dachsteinmaschine f. ‖ machine f. à tuiles. / ~ press ‖ Dachziegelpresse f. ‖ presse f. à tuiles.

roof-like ‖ dachartig ‖ en forme f. de toit.

roof, pane of a ~ ‖ Dachfläche f. ‖ plan m. de comble. / ~ principal ‖ Dachrahmen m. ‖ panne f. de comble. / ~ protecting shoe ‖

Dachschuh m. ‖ savate f. / ~ rock ‖ ‖ Dachgebirge n. ‖ toit m. de la veine *ou* d'une couche. / ~ seat ‖ Decksitz m. ‖ impériale f. / ~ sheet ‖ Stehkesseldecke f. ‖ ciel m. de boîte à feu. / ~ sill ‖ Dachunterzug m. ‖ sablière f. du toit. / ~ slate ‖ Dachschiefer m. ‖ ardoise f.; schiste m. tégulaire. / ~ slating ‖ Dachdeckung f. ‖ couverture f. / ~ slope of ~ ‖ Dachneigung f. ‖ pente f. du toit. / ~ standard ‖ Dachgestänge n. ‖ montant m. sur toiture. / ~ timber ‖ Dachgesparre n. ‖ chevron m. *ou* charpente f. du toit. / ~ trap door ‖ Aussteigeluke f. ‖ lucarne f. / ~ tree ‖ Firstbalken m. ‖ poutre f. de faîte. / ~ truss ‖ Dachbinder m. ‖ ferme f.

room (Build) ‖ Raum m.; Gelaß n. ‖ salle f.; pièce f. / ~ (Phys) ‖ Raum m. ‖ espace m. / boiler ~ ‖ Kesselraum m. ‖ salle f. de chauffe *ou* des chaudières. / ~ dry ‖ trockener Raum m. ‖ local m. sec. / engine ~ ‖ Maschinenraum m.; Maschinensaal m. ‖ salle f. des machines. / large ~ ‖ Saal m. ‖ salle f.; salon m. / ~ for live models ‖ Aktsaal m. ‖ salle f. des modèles vivants. / ~ for machines ‖ Maschinensaal m. ‖ salle f. des machines. / ~ to move ‖ Bewegungsfreiheit f. ‖ place f. à mouvements. / necessary ~ ‖ Raumbedarf m. ‖ encombrement m. / small ~ ‖ Kammer f. ‖ chambre f. / ~ for the spectators ‖ Theatersaal m. ‖ salle f. d'audition. / ~ taken up ‖ Raumbedarf m. ‖ encombrement m.

room antenna ‖ Zimmerantenne f. ‖ antenne f. de chambre. / ~ heating ‖ Raumheizung f. ‖ chauffage m. de chambres. / ~ heating stove ‖ Raumheizofen m. ‖ poêle m. à grande surface de chauffe. / ~ noise ‖ Raumgeräusch n. ‖ bruit m. local. / ~ painter ‖ Zimmermaler m. ‖ peintre m. d'appartements. / ~ stove ‖ Zimmerofen m. ‖ poêle m. / ~ temperature ‖ Zimmertemperatur f. ‖ température f. intérieure *ou* du laboratoire. / ~ toilet ‖ Zimmerabort m. ‖ chaise f. percée d'appartement.

root, to ~ out ‖ ausroden ‖ essarter.

root (Hydr arch) ‖ Widerlager n. ‖ culée f. / ~ (Math; Bot) ‖ Wurzel f. ‖ racine f. / ~ of the blade ‖ Schaufelfuß m. ‖ racine f. de l'aube. / cube ~ ‖ Kubikwurzel f. ‖ racine f. cubique. / to cut the ~s ‖ die Wurzeln fpl. abhauen ‖ couper les racines fpl. / edible ~ ‖ eßbare Wurzel f. ‖ racine f. alimentaire *ou* comestible. / to extract a ~ (Math) ‖ eine Wurzel f. ziehen ‖ extraire une racine. / ~ for forage ‖ Wurzel f. für Futterzwecke ‖ racine f. fourragère. / square ~ ‖ Quadratwurzel f. ‖ racine f. carrée.

root bruiser *see* ~ cutter. / ~ cutter ‖ Wurzelschneidemaschine f. ‖ couperacines m. / ~ cutter (Person) ‖ Wurzelschneider m. ‖ coupeur m. de racines. / ~ cutting machine *see* ~ cutter. / ~ decay ‖ Wurzelfäule f. ‖ pourriture f. des racines. / drawing-out of a ~ (Math) ‖ Wurzelziehen n. ‖ extraction f. d'une racine.

rooter, turnip ~ ‖ Rübenroder m. ‖ essarteuse f. de betteraves.

root pin ‖ Wurzelstift m. ‖ cheville f. de racine. / ~ pin for a dental prosthesis ‖ Wurzelstift m. für ein Zahnersatzstück ‖ cheville f. de racine d'une prothèse dentaire. / ~ sign (Math) ‖ Wurzelzeichen n. ‖ radical m. de racines. / ~ stamp mill ‖

Stampfwerk n. für Wurzeln ‖ moulin m. à pilons pour racines. / ~ stock (Bot) ‖ Wurzelstock m. ‖ rhizome m. / ~ stud *see* root pin.

rope, to ~ a sail ‖ ein Segel n. einlieken ‖ ralinguer une voile.

rope ‖ Seil n.; Strick m.; Tau n.; Reep n. ‖ corde f.; cordage m.; câble m. / carrying ~ ‖ Tragseil n. ‖ câble m. porteur. / to coil-up a ~ ‖ ein Seil n. aufwickeln *oder* aufschießen ‖ lover un cordage. / coiled ~ ‖ aufgeschossenes Tau n. ‖ glène f. / closing ~ ‖ Schließseil n. ‖ câble m. de fermeture. / continuous ~ ‖ durchlaufendes Seil n. ‖ câble m. continu. / damaged ~ ‖ abgenutztes Tau n. ‖ cordage m. étrivé. / drawing ~ (Spinn) ‖ Manndausenseil n. ‖ corde f. de main douce. / driving ~ ‖ Treibseil n. ‖ câble f. à transmission. / endless ~ ‖ Seil n. ohne Ende ‖ corde f. sans fin. / grappling ~ ‖ Fangleine f. ‖ amarre f. / hauling ~ ‖ Förderseil n. ‖ câble m. d'extraction. / hawserlaid ~ ‖ troßweise geschlagenes Tau n. ‖ cordage m. commis en aussière. / hempen ~ ‖ Hanftau n. ‖ cordage m. de chanvre. / holding ~ ‖ Halteseil n. ‖ câble m. de retenue. / immerged ~ (Hydr arch) ‖ Tauereiseil n. ‖ câble m. immergé de touage. / laid ~ ‖ kabelweise geschlagenes Tau n. ‖ cordage m. commis en câble. / left-hand ~ ‖ linksgeschlagenes Tau n. ‖ cordage m. commis de droite à gauche. / mast ~ ‖ H(e)ißtau n. ‖ drisse f.; guinderesse f. / one-side ~ ‖ eintrümiges Seil n. ‖ câble m. à un brin. / pointed ~ ‖ gespitztes Tau n. ‖ cordage m. en queue de rat. / right-hand ~ ‖ rechtsgeschlagenes Tau n. ‖ cordage m. commis de gauche à droite. / running ~ ‖ laufendes Tau n. ‖ manœuvre f. courante. / shroud-laid ~ ‖ vierschäftiges Tau n. ‖ cordage m. en quatre. / steel-wire ~ ‖ Stahldrahtseil n. ‖ câble m. en fil d'acier. / straw ~ ‖ Strohseil n. ‖ tresse f. en paille; éclisse f. en paille. / tarred ~ ‖ geteertes Tau n. ‖ cordage m. goudronné. / three-stranded ~ ‖ dreischäftiges Tau n. ‖ cordage m. commis en trois. / tow ~ ‖ Schlepptau n. ‖ câble m. de remorque. / ~ twisted the wrong way ‖ verkehrt gedrehtes Tau n. ‖ garochoir m. / used ~ ‖ halbgeschlissenes Tau n. ‖ filin m. usé. / wire ~ ‖ Drahtseil n. ‖ câble m. métallique. / wood wool ~ ‖ Holzwollseil n. ‖ tresse f. en laine de bois. / yard ~ *see* mast ~.

rope acidifying plant (Weav) ‖ Strangsäureeinrichtung f. ‖ installation f. à aciduler en boyaux. / ~ balancing sheave ‖ Seilausgleichrolle f. ‖ galet m. compensateur de tension du câble. / ~ band (Shipb) ‖ Anschlagbändsel n.; Rahebändsel n. ‖ raban m. d'envergure *ou* de faix *ou* de têtière. / ~ block ‖ Taukloben m. ‖ moufle m. à corde. / ~ brake ‖ Seilbremse f. ‖ frein m. à corde. / ~ breaking ‖ Seilbruch m. ‖ rupture f. du câble. / ~ bridge ‖ Seilbrücke f. ‖ pont m. suspendu à câbles. / ~ catch ‖ Seilklemme f. ‖ pince f. à cordages. / ~ clamp *see* ~ clip. / ~ clip ‖ Seilklemme f. ‖ borne f. de cordage; étrier m. / ~ coupling ‖ Seilschloß n. ‖ attache f. de câbles; agrafe f. de jonction pour câbles. / ~ driving ‖ Seiltrieb m. ‖ mise f. en marche par corde; commande f. *ou* traction f. par câble. / ~ drums pl. identical in shape

and size ‖ vollkommen gleiche Seiltrommeln fpl. ‖ tambours mpl. de câbles identiques. / ~ end ‖ Tauende n. ‖ bout m. d'une manœuvre. / ~ fender ‖ Seilfender m.; Taufender m. ‖ défense f. en cordage. / ~ grab ‖ Seilgreifer m. ‖ benne f. à câble. / ~ grease ‖ Seilfett n. ‖ graisse f. pour câbles. / ~ guide ‖ Seilführung f. ‖ guide-câble m. / ~ handle ‖ Strickgriff m. ‖ poignée f. en corde. / ~ haulage ‖ Seilförderung f. ‖ transport m. par câble. / ~-hauled grab ‖ Seilzugkatze f. ‖ chariot m. mu par câbles. / ~ hauling ‖ Seilbetrieb m. ‖ transport m. par câble. / ~ ladder ‖ Strickleiter f. ‖ échelle f. de cordes *ou* en cordage. / ~ line ‖ Seilstrang m. ‖ brin m. de câble. / ~ loop ‖ Schlinge f. ‖ boucle f. de câble.

ropemaker ‖ Seiler m.; Reepschläger m. ‖ cordier m. / foreman ~ ‖ Seilermeister m. ‖ maître cordier m.; maître m. de roue.

ropemaker's articles pl. ‖ Seilerwaren fpl. ‖ articles mpl. de corderie. / ~ assistant ‖ Radjunge m. ‖ aidefileur m.; tourneur m. de roue. / ~ goods pl. ‖ Seilerwaren fpl. ‖ corderies fpl.

rope-making ‖ Seilerei f.; Tauschlagen n. ‖ corderie f. / ~ by hand ‖ Handseilerei f. ‖ fabrication f. des cordages à la main. / mechanical ~ ‖ Maschinenseilerei f. ‖ corderie f. mécanique.

rope-making machine ‖ Seilermaschine f. ‖ machine f. pour corderie; engin m. à commettre les cordages.

rope post plant ‖ Seilpostanlage f. ‖ installation f. de poste funiculaire; transport m. de lettres par câble. / ~ pulley ‖ Seilscheibe f. ‖ poulie f. à câble. / ~ pulley block ‖ Seilflaschenzug m. ‖ moufle m. à câble.

roper ‖ Reepschläger m.; Seiler m. ‖ cordier m.

rope railway ‖ Seilbahn f. ‖ chemin m. de fer funiculaire. / ~ built on the ground ‖ bodenständige Seilbahn f. ‖ funiculaire m. fixe à câble.

rope relief ‖ Seilentlastung f. ‖ déchargement m. du câble. / ~ roller ‖ Seilrolle f. ‖ poulie f. à corde.

ropery ‖ Seilerbahn f.; Reeperbahn f.; Tauschlägerei f.; Seilerei f. ‖ corderie f.; atelier m. de cordier. / ~ machine ‖ Seilermaschine f. ‖ machine f. pour corderie.

ropes pl. ‖ Tauwerk n. ‖ cordage m.

rope scouring machine ‖ Strangwaschmaschine f. ‖ machine f. à laver en boyaux. / ~ with solid cast iron framework for x meter washing width ‖ Strangwaschmaschine f. mit massiven gußeisernen Gestellwänden für x-Meter Waschbreite ‖ machine f. à laver en boyaux avec bâtis massifs en fonte d'une largeur à laver de x mètres.

rope setter ‖ Riemenaufleger m. ‖ metteur m. de courroies. / ~ sheave ‖ Zugseiltragrolle f. ‖ rouleau-porteur m. de câble. / slackening of the ~ ‖ Schlappwerden n. des Seiles ‖ mou m. du câble. / ~ splice ‖ Seileinband m. ‖ enveloppe f. de cordage. / ~ splicer ‖ Seilspleißer m. ‖ épisseur m. de cordes. / ~ squeezing apparatus (Weav) ‖ Strangausquetschapparat m. ‖ appareil m. à exprimer en boyaux. / ~ stopper (Mar) ‖ Taustopper m. ‖ bosse f. à bouton. / ~ strop ‖ Tautropp m. ‖ estrope f. en cordage. / ~ suspension bridge ‖ Seilbrücke f. ‖ pont m. suspendu

à câbles. / ~ testing machine ‖ Seilprüfmaschine f. ‖ machine f. à essayer les câbles. / upper ~ traction ‖ Oberseilführung f. ‖ passage m. du câble par dessus. / ~ walk see ropery. / ~ ware ‖ Seilerwaren fpl. ‖ article m. de corderie; cordage m. / ~ washing machine see ~ scouring machine.

ropeway ‖ Seilbahn f. ‖ transporteur m. à câble. / aerial ~ ‖ Drahtseilbahn f. ‖ funiculaire m.; chemin m. de fer funiculaire. / ~ car ‖ Seilbahnwagen m. ‖ wagonnet m. de chemin de fer aérien. / ~ crane ‖ Seilbahnkran m. ‖ grue f. à câble aérien.

rope winch ‖ Seilwinde f. ‖ treuil m. à câble. / ~ window ‖ Seilfenster n. ‖ fenêtre f. du passage du câble. / ~ wire ‖ Seildraht m. ‖ fil m. pour câbles. / ~ yard see ropery.

rope yarn ‖ Kabelgarn n. ‖ fil m. de caret. / ~ for tying bags ‖ Sackband n. ‖ cordon m. de sac; ficelle f. pour fermer les sacs. / white ~ ‖ weißes, ungeteertes Garn n. ‖ fil m. blanc à cordage.

rope-yarn spinner ‖ Seilspinner m. ‖ fileur m. de fil de caret.

ropiness (Wine) ‖ Fettsein n.; Dickflüssigkeit f. ‖ graisse f.; viscosité f.

roping needle ‖ Lieknadel f. ‖ aiguille f. à ralingue. / ~ twine ‖ Liekgarn n. ‖ fil m. à ralingue. / tarred ~ twine ‖ geteertes Takelgarn n. ‖ fil m. à voile goudronné.

ropy (Wine) ‖ schleimig; fadenziehend; klebrig; fett; ölig ‖ gras; filant. / to get ~ (Wine) ‖ fett oder schleimig werden ‖ tourner au gras ou à la graisse.

ropy fermentation ‖ Schleimgärung f. ‖ fermentation f. visqueuse.

rosace ‖ Rosette f. ‖ rosace f.; roson m.

rosary ‖ Rosenkranz m. ‖ rosaire m.; chapelet m. / ~ trompe ‖ Kettengebläse n.; Paternostergebläse n. ‖ chapelet m.

rose (Bot) ‖ Rose f. ‖ rose f. / ~ (Watering can) ‖ Brause f.; crépine f. / attar of ~s ‖ Rosenöl n. ‖ essence f. de rose. / ~ colour ‖ Rosa f. ‖ rose m.

rose copper ‖ Garkupfer n.; Scheibenkupfer n.; Rosettenkupfer n. ‖ cuivre m. rosette; rosette f. de cuivre affiné. / cakes pl. of ~ ‖ Rosettenkupfer n. ‖ gâteau m. de rosette; plaque f. de cuivre; rosette f. / ~ disk ‖ Garscheibe f. ‖ rosette f. ou lame f. de cuivre affiné.

rose diamond ‖ Rosendiamant m.; Rosette f. ‖ diamant m. en rose; rose f.

rose engine ‖ Guillochiermaschine f.; Patronendrehbank f. ‖ machine f. à guillocher; tour m. à guillocher ou à rosettes. / ~ pattern ‖ Guillochierung f. ‖ guillochis m.

rose garden ‖ Rosenschule f. ‖ roseraie f. / ~ grower ‖ Rosengärtner m. ‖ rosiériste m. / ~ honey ‖ Rosenhonig m. ‖ miel m. mercurial ou rosat. / ~ leave ‖ Rosenblatt n. ‖ feuille f. de rose.

rosemary ‖ Rosmarin m. ‖ romarin m. / ~ leave ‖ Rosmarinblatt n. ‖ feuille f. de romarin. / ~ oil ‖ Rosmarinöl n. ‖ essence f. de romarin.

rose oil ‖ Rosenöl n. ‖ huile f. de roses. / ~ quartz ‖ Rosenquarz m. ‖ quartz m. rose.

rose-red heat ‖ Hellrotglühhitze f. ‖ rouge m. clair.

rose spar ‖ Manganspat m.; Rosenspat m. ‖ manganèse m. carbonaté. / ~ stamp ‖ Rosettenstempel m. ‖ rosetier m.

roset ‖ Patrone f. ‖ rosette f.; couronne f. / ~ (Jewel) see rose diamond.

rosette ‖ Rosette f.; Einsetzrose f. ‖ rosace f.; roson m. / ~ copper ‖ Rosettenkupfer n. ‖ cuivre m. rosette. / to produce ~ copper ‖ rosettieren; Scheiben reißen oder schleißen ‖ faire des rosettes fpl.

rosewood ‖ Palisanderholz n.; Jakarandaholz n. ‖ palissandre m. / ~ oil ‖ Rosenholzöl n. ‖ essence f. de bois rhodien.

rosin see also resin ‖ Harz n.; Kolophonium n. ‖ résine f.; colophane f. / artificial ~ ‖ Kunstharz n. ‖ résine f. artificielle. / common ~ ‖ Harzpech n. ‖ résine f. commune; brai m. sec; poix-résine f.; poix f. grasse. / white ~ ‖ weißes Pech n. ‖ poix f. blanche.

rosin distilling plant ‖ Harzdestillationsanlage f. ‖ installation f. de distillation de résine.

rosing (Dyer) ‖ Rosieren n.; Schönen n. ‖ rosage m.

rosin mastic ‖ Harzkitt m. ‖ mastic m. résineux. / ~ oil ‖ Harzöl n. ‖ huile f. de résine. / ~ product ‖ Harzprodukt n. ‖ produit m. résineux. / ~ size ‖ Harzleim m. ‖ colle f. de résine. / ~ soap ‖ Harzseife f. ‖ savon m. résineux ou à la résine.

rot, to ‖ verfaulen ‖ pourrir.

rot ‖ Fäulnis f.; Fäule f. ‖ pourriture f. / ~ at branches ‖ Astfäule f. ‖ pourriture f. des branches. / dry ~ ‖ trockene Fäulnis f.; Trockenfäule f. ‖ pourriture f. sèche. / ~ of poles ‖ Fäulnis f. der Holzstangen ‖ pourriture f. des poteaux. / red ~ ‖ Rotfäule f.; Naßfäule f. ‖ pourriture f. rouge. / wet ~ ‖ nasse Fäulnis f. ‖ pourriture f. humide.

rotary ‖ sich drehend; rotierend ‖ rotatif; tournant; rotatoire. / ~ addresser ‖ Rotationsanschriftenmaschine f. ‖ machine f. rotative à écrire les adresses. / triple-beam ~ beacon ‖ dreistrahliges Drehfeuer n. ‖ feu m. tournant à trois faisceaux.

rotary blower ‖ Umlaufgebläse n. ‖ soufflerie f. rotative. / high-pressure ~ ‖ Hochdruckkapselgebläse n. ‖ soufflerie f. rotative à haute pression.

rotary breaker ‖ Kreiselbrecher m. ‖ concasseur m. rotatif. / ~ coal drilling machine ‖ Kohlendrehbohrmaschine f. ‖ perceuse f. tournante à houille. / ~ compressor ‖ Rotationskompressor m. ‖ compresseur m. rotatif. / ~ converter ‖ Einankerumformer m.; Drehumformer m. ‖ commutatrice f.; convertisseur m. / ~ copying press ‖ Kopierpresse f. mit Trommel ‖ presse f. rotative à copier. / ~ crane ‖ Drehkran m. ‖ grue f. tournante ou pivotante ou à pivot. / ~ crusher ‖ Rundbrecher m. ‖ concasseur m. à cône oscillant.

rotary current ‖ Drehstrom m. ‖ courant m. triphasé.

rotary-current-continuous-current converter ‖ Drehstromgleichstromeinankerumformer m. ‖ commutatrice f. à courant triphasé en courant continu. / ~ dynamotor ‖ Drehstromgleichstromumformer m. ‖ dynamoteur m. à courant triphasé en courant continu.

rotary-current measuring-instrument ‖ Drehstrommeßinstrument n. ‖ instrument m. de mesure à courant triphasé. / ~ transformer ‖ Drehstromtransformator m. ‖ transformateur m. triphasé.

rotary disc ‖ Laufplatte f. ‖ plateau m. tournant. / ~ drum ‖ drehbare Trommel f. ‖ tambour m. rotatif. / ~ engine ‖ Umlaufmotor m. ‖ moteur m. rotatif. / ~ field ‖ Drehfeld n. ‖ champ m. tournant. / ~ furnace ‖ Drehofen m. ‖ four m. tournant. / ~ grate ‖ drehbarer Rost m. ‖ grille f. tournante. / ~ kiln ‖ Ringofen m. ‖ four circulaire. / ~ knitting frame hand ‖ Rundstuhlwirker m. ‖ tricoteur m. au métier circulaire. / ~ magnet ‖ Drehelektromagnet m. ‖ électro m. de rotation.

rotary machine ‖ Rotationsmaschine f. ‖ machine f. rotative. / ~ for producing illustrated work ‖ Rotationsmaschine f. für Illustrationsdruck ‖ machine f. rotative pour illustrations. / tapeless ~ ‖ bänderlose Rotationsmaschine f. ‖ machine f. rotative sans cordon.

rotary machine hand (Print) ‖ Rotationsmaschinenarbeiter m. ‖ conducteur m. de rotative. / ~ ink ‖ Rotationsdruckfarbe f. ‖ encre f. pour presses rotatives. / ~ printing ‖ Rotationsdruck m. ‖ impression f. par machine rotative.

rotary milling machine ‖ Rundlauffräsmaschine f. ‖ fraiseuse f. à plateau horizontal rotatif. / ~ motion ‖ Achsendrehung f. ‖ rotation f. de l'axe. / ~ newspaper printing machine ‖ Zeitungsrotationsmaschine f. ‖ machine f. rotative à imprimer les journaux. / ~ pawl ‖ Drehklinke f. ‖ jack m. rotatif. / ~ planer ‖ Drehbank f. ‖ tour m. à dresser. / ~ power ‖ Drehungsvermögen n. ‖ pouvoir m. rotatoire.

rotary press (Mach tool) ‖ Revolverpresse f. ‖ presse f. révolver. / ~ (Print) ‖ Rotationspresse f. ‖ presse f. rotative. / high-speed ~ with electric multi-motor drive ‖ Rotationspresse f. mit elektrischem Mehrmotorenantrieb ‖ presse f. rotative rapide à commande électrique multiple.

rotary printing machine ‖ Rotationsdruckmaschine f. ‖ presse f. rotative à imprimer; machine f. typographique rotative. / ~ for cash blocks ‖ Kassenblockrotationsdruckmaschine f. ‖ machine f. rotative à imprimer les blocs de caisse. / fifteen-reel unit ~ in line arrangement ‖ Fünfzehnrollenrotationsdruckmaschine f. in Reihenanordnung ‖ machine f. rotative à 15 unités d'impression en agrégat. / single-width ~ ‖ einfachbreite Rotationsdruckmaschine f. ‖ machine f. typographique rotative de simple largeur.

rotary pump ‖ Umlaufpumpe f. ‖ pompe f. rotative. / ~ shears pl. ‖ Kreisschere f. ‖ cisaille f. circulaire ou cylindrique. / ~ slide valve ‖ Drehschieber m. ‖ tiroir m. rotatif. / ~ snow plough ‖ Schneeschleuder f. ‖ chasse-neige m. rotatif. / ~ squeezer ‖ Luppenmühle f. ‖ moulin m. à cingler; cingleur m. rotatif. / ~ switch ‖ Drehwähler m. ‖ sélecteur m. rotatif. / ~ system (Aut tel) ‖ Drehwählermaschinensystem n. ‖ système m. rotatif. / ~ table surface grinding machine ‖ Rundflächenschleifmaschine f. ‖ machine f. à rectifier les surfaces circulaires planes.

rotary tipper ‖ Kreiselwipper m. ‖ culbuteur m. rotatif. / ~ for trams ‖ Kreiselwipper m. für Grubenwagen n. ‖ culbuteur m. rotatif pour wagonnets de mine. / ~ on wheels ‖ fahrbarer Kreiselwipper m. ‖ culbuteur m. basculant sur roues.

rotary transformer ‖ Motorgenerator m. ‖ transformateur m. rotatif. / ~type printer ‖ Rotationstypendrucker m. ‖ machine f. rotative à caractères. / ~ wagon tipper ‖ Kreiselkipper m. ‖ basculeur m. circulaire.

rotate, to ‖ kreisen; rotieren ‖ tourner; circuler.

rotating on a vertical axis ‖ um eine vertikale Achse f. drehbar ‖ tournant autour d'un axe vertical. / ~ anemometer ‖ Drehungswindmesser m. ‖ anémomètre m. à rotation. / ~ button ‖ drehbarer Knopf m. ‖ bouton m. rotatif. / ~ drum ‖ umlaufende Trommel f. ‖ tambour m. tournant. / ~ field ‖ Drehfeld n.; rotierendes Feld n. ‖ champ m. tournant. / ~ gap ‖ rotierende Funkenstrecke f. ‖ éclateur m. rotatif. / ~ motion ‖ drehende Bewegung f. ‖ mouvement m. de rotation. / ~ moulding table ‖ schwenkbare Formplatte f. ‖ table f. de moulage orientable. / ~ part ‖ rotierender Teil m. ‖ élément m. rotatif. / ~ shear ‖ rotierende Schere f. ‖ cisaille rotative. / ~ spoon rolling mill ‖ rotierende Löffelwalzmaschine f. ‖ laminoir m. tournant pour faire des cuillers. / ~ tool ‖ umlaufendes Werkzeug n. ‖ outil m. tournant. / ~ voltage cut-out ‖ umlaufende Spannungssicherung f. ‖ coupecircuit m. de tension rotatif. / ~ wheel vane anemograph ‖ Windradanemograf m. ‖ anémographe m. à moulinet.

rotation ‖ Drehung f. ‖ rotation f. / clockwise ~ ‖ Rechtsdrehung f.; Rechtslauf m. ‖ rotation f. à droite. / counter-clockwise ~ ‖ Linkslauf m.; Linksdrehung f. ‖ rotation f. à gauche. / intermittent ~ ‖ absatzweise Drehung f. ‖ rotation f. intermittente. / ~ in the liquor tanks (Washing machine) ‖ Flottenzirkulation f. ‖ circulation f. d'eau dans les réservoirs de flottation. / ~ about the position axis ‖ Positionswinkeldrehung f. ‖ rotation f. dans l'angle de position. / positive ~ ‖ positiver Drehungssinn m. ‖ sens m. positif de rotation. / to set a body in ~ ‖ einen Körper m. in Drehung versetzen ‖ mettre un corps m. en rotation; faire tourner un corps.

rotation, axis of ~ ‖ Umdrehungsachse f. ‖ axe m. de rotation. / centre of ~ ‖ Drehpunkt m. ‖ centre m. de rotation. / direction of ~ ‖ Drehrichtung f.; Drehsinn m. ‖ direction f. ou sens m. de rotation. / changeable direction of ~ ‖ wechselnde Drehrichtung f. ‖ renversement m. de marche. / direction of ~ of spindle ‖ Spindeldrehsinn m. ‖ sens m. de rotation de la broche. / wrong direction of ~ ‖ falsche Drehrichtung f. ‖ sens m. inversé. / ~ machine for printing tickets ‖ Fahrkartenrotationsdruckmaschine f. ‖ machine f. rotative à imprimer de billets. / ~ plane of type wheel ‖ Bewegungsebene f. des Typenrades ‖ plan m. de rotation de la roue à caractères. / ~ roller ‖ Rollenbahn f. ‖ transporteur m. à rouleaux. / ~ roller path ‖ Rollkranz m. ‖ couronne f. de roulement des galets. / speed of ~ ‖ Umdrehungsgeschwindigkeit f. ‖ vitesse f. de rotation. / velocity of ~ ‖ Drehgeschwindigkeit f. ‖ vitesse f. de rotation.

rotative see rotary.

rotator (Metal) ‖ rotierender Puddelofen m. ‖ four m. à puddler rotatif.

rotatory see rotary.

rotor (Electr) ‖ Läufer m.; Anker m.; Rotor m. ‖ rotor m.; induit m. / ~ body ‖ Ankerkörper m. ‖ corps m. de l'induit. / ~ wheel (Turbine) ‖ Laufrad n. ‖ roue f. mobile; rotor m. / ~ winding ‖ Rotorwicklung f. ‖ enroulement m. du rotor. / ~ wing ‖ Rotorflügel m. ‖ aile f. de rotor.

rot protection ‖ Fäulnisschutz m. ‖ protection f. contre la pourriture.

rotten ‖ verfault ‖ pourri. / ~ (Wood) ‖ stockig ‖ échauffé. / ~ earth ‖ Modererde f. ‖ terre f. pourrie. / ~ stone ‖ englische Erde f.; englischer Tripel m. ‖ terre f. pourrie. / ~ wood ‖ Schrammholz n. ‖ bois m. spongieux ou fongueux.

rottenness ‖ Fäulnis f. ‖ pourriture f. / ~ of poles see rot of poles. / ~ of the top end of poles ‖ Zopffäule f. ‖ pourriture f. du sommet des poteaux.

rotting ‖ Verwesung f. ‖ putréfaction f. / ~ of poles see rot of poles. / ~ of the wood ‖ Fäule f. des Holzes ‖ pourriture f. du bois.

rotunda ‖ Rotunde f.; Rundbau m. ‖ rotonde f.

rouge (Mach) ‖ Pariser Rot n.; Polierrot n. ‖ rouge m. à polir. / ~ (Paint) ‖ Schminkrot n.; Schminke f. ‖ rouge m. en feuilles; fard m.

rough, to (Glassm) ‖ rauhschleifen ‖ dégrossir; dépolir; égriser. / ~-down (Roll mill) ‖ vorwalzen ‖ ébaucher; dégrossir

rough ‖ rauh; roh; unbearbeitet ‖ rugueux; brut. / ~ average ‖ annähernder Durchschnitt m. ‖ moyenne f. approximative. / ~ brickwork see ~ masonry. / ~ coal ‖ rohe Förderkohle f.; Rohkohle f. ‖ houille f. crue; tout-venant m. / ~ cut (File) ‖ grober Hieb m. ‖ grosse taille f. / ~ cut (Turn) ‖ Schruppschnitt m. ‖ coupe f. à dégrossir. / ~ handling ‖ rauhe Behandlung f. ‖ traitement m. brutal. / ~ masonry ‖ Feldsteinmauerwerk n.; Rauhgemäuer m.; rohes Mauerwerk n. ‖ massif m. brut en maçonnerie; hourdage m.; maçonnerie f. en galets ou en moellons. / ~ reading ‖ Rohablesung f. ‖ lecture f. brute. / ~ roll ‖ rauhe Walze f. ‖ cylindre m. à surface rugueuse. / ~ smalls pl. ‖ Fördergrus m. ‖ petit tout-venant m. / ~ steel ‖ Rohstahl m. ‖ acier m. naturel. / ~ walling see ~ masonry. / ~ weather ‖ Unwetter n. ‖ gros temps m. / ~ woollen cloth ‖ Loden m. ‖ drap m. brut; loden m.

rough vorgearbeitetes Stück n. ‖ ébauche f. / the parts pl. are supplied in the ~ ‖ die Teile mpl. werden roh geliefert ‖ les pièces fpl. sont fournies brutes. / to furnish a tube bored in the ~ ‖ ein Rohr n. vorgebohrt liefern ‖ livrer un tube avec un forage d'ébauchage.

rough-cast, to ‖ berappen ‖ gobeter.

rough-casting ‖ Berappung f.; Bewurf m. ‖ gobetage m.; crépi m.; crepissure f.

rough-cut, to ~ the wood ‖ das Holz (aus dem Gröbsten) zurichten ‖ ébaucher le bois.

rough-down, to (Roll mill) ‖ vorwalzen ‖ ébaucher; dégrossir.

roughen, to ‖ aufrauhen ‖ gratter.

rough-finished nut ‖ rohe oder schwarze Mutter f. ‖ écrou m. brut d'étampage.

rough-floor (Brew) ‖ Greifhaufen m. ‖ couche f. prenant.

rough-grind, to (Metal) ‖ aus dem Groben

schleifen ‖ dégrossir à la meule; émoudre. / ~ (Mill) ‖ schroten ‖ égruger.

rough-grinder ‖ Vorschleifer m. ‖ émouleur m. / ~ grinding ‖ Vorschleifen n. ‖ émoulage m.

rough-hew, to ~ the timber with the axe ‖ das Holz mit der Axt f. grob behauen ‖ dégrossir ou dégauchir le bois à la hache.

roughing (Glassm) ‖ Schruppschliff m.; Rauhschleifen n. ‖ dégrossissage m. / ~ and smoothing of glass ‖ Glasveredlung f.; Glasverfeinerung f. ‖ affinage m. de verre.

roughing cylinder (Roll mill) ‖ Präparierwalze f.; Luppenwalze f.; Puddelwalze f. ‖ cylindre m. à cingler; cylindre m. cingleur ou dégrossisseur ou ébaucheur. / ~ device for hoop iron ‖ Schrappvorrichtung f. für Bandeisen ‖ dispositif m. à racler le feuillard. / ~-down ‖ Vorarbeiten n. ‖ ébauchage m. / ~-down of piles ‖ Vorhämmern n. der Pakete ‖ serrage m. des paquets. / ~-down roll ‖ Rohschienenvorwalze f. ‖ cylindre m. dégrossisseur. / ~ hole (Metal) ‖ Schlakkensumpf m. ‖ trou m. de laitier. / ~ machine for bar work ‖ Schruppmaschine f. für Stangenarbeit ‖ dégrossisseuse f. ou machine f. d'ébauche pour le travail de la barre. / ~ mill (Agr) ‖ Schrotmühle f. ‖ moulin m. à égruger. / ~ mill (Roll mill) ‖ Vorstraße f. ‖ train m. dégrossisseur. / ~ roll see ~ cylinder. / ~ steel (Turn) ‖ Schruppmeißel m. ‖ outil m. dégrossisseur; acier m. à dégrossir. / cranked ~ tool ‖ gekröpfter Schruppstahl m. ‖ outil m. d'ébauche renvoyé. / ~ tool box ‖ Schruppstichelhaus n. ‖ porte-outil m. (à charioter) à ébaucher.

roughly, the ingots pl. were ~ forged octangularly under steam hammers ‖ die Blöcke mpl. wurden unter dem Dampfhammer achtkantig vorgereckt ‖ les lingots mpl. furent forgés préalablement par des marteaux-pilons sur un profil octogonal. / ~ forged axle ‖ roh geschmiedete Achse f. ‖ essieu m. brut de forge. / ~ worked object ‖ (grob) vorgearbeiteter Gegenstand m. ‖ objet m. ébauché; pièce f. ébauchée.

rough-machine, to ‖ vorarbeiten ‖ ébaucher.

rough-machined shaft ‖ vorgearbeitete Welle f. ‖ arbre m. ébauché.

roughness ‖ Rauheit f. ‖ aspérité f.; rugosité f. / ~ (Glassm) ‖ Glanzlosigkeit f. ‖ dépoli m.

rough-plane, to ‖ vorhobeln ‖ dégrossir à la raboteuse.

rough-planer ‖ Abrichtmaschinenarbeiter m. ‖ dégauchisseur m.

rough-plaster, to ‖ rauh putzen ‖ hourder.

rough-plastering ‖ Rauhputz m. ‖ enduit m. fouetté.

rough-pressed wheel ‖ vorgepreßtes Rad n. ‖ roue f. ébauchée à la presse.

rough-pressing of discs from blocks ‖ Vorpressen n. der Blöcke zu Scheiben ‖ ébauchage m. des blocs à disques à la presse.

rough-rolled ingot ‖ vorgewalzter Block m. ‖ lingot m. ébauché au laminoir.

rough-sketch ‖ Handskizze f. ‖ minute f.

rough-towel ‖ Frottierhandtuch n. ‖ essuiemains m. floconneux à frotter.

rough-turn, to ‖ abdrehen ‖ dégrossir. / ~ the axle ‖ die Achse vordrehen ‖ dégrossir l'essieu m.

rough-work, to || vorarbeiten || dégrossir; ébaucher.

roulette || Roulett n. || roulette f. / ~ (Geom) || Rollkurve f. || roulette f.

rounce (Print) || Kurbel f. || manivelle f.

round, to (Math) || abrunden || arrondir. / ~-off || runden; abrunden || arrondir.

round || rund || rond; arrondi; circulaire. / ~ bend || Rundbug m. || cintre m.; pli m. arrondi. / ~ coal scoop || gewölbte Kohlenschaufel f. || pelle f. à charbon ronde. / ~ corner || rundkantig; à coin m. arrondi. / ~ corner bevelling machine || Eckenrundstoßmaschine f. || machine f. à arrondir les coins. / in ~ figures || in runden Ziffern fpl. || en chiffres mpl. ronds. / ~ file || Rundfeile f. || lime f. ronde. / ~ handle || rundes Werkzeugheft n. || manche m. rond. / ~ iron || Rundeisen n. || fer m. rond. / ~ iron bending apparatus || Betoneisenbieger m. || appareil m. à plier le fer rond. / ~ iron cutting apparatus || Betoneisenschneider m. || appareil m. à découper le fer rond. / ~ iron shearing machine || Rundeisenschere f. || cisaille f. à fers ronds. / ~-leg compasses pl. || Zirkel m. mit runden Schenkeln || compas m. à branches rondes. / ~-link chain || Rundgliederkette f. || chaîne m. à maillons ronds. / ~ log || Rundholz n. || bois m. rond. / ~ notch || Rundkerb m. || entaille f. à fond arrondi. / ~ opening || Rundlochung f. || perforation f. ronde. / ~-plate buffer || bogenförmiger Puffer m. || tampon m. à plaque à surface courbe. / ~ pliers pl. || Rundzange f. || pince f. ronde. / ~-shackle padlock || Vorhängeschloß n. mit rundem Bügel || cadenas m. à anse ronde. / ~ slide || Ringschieber m. || tiroir m. rond. / ~ spoke || Rundspeiche f. || rayon m. rond. / ~ timber || Rundholz m. || bois m. rond. / grab for ~ timber || Greifer m. für Rundholz || grappin m. pour rondin. / ~ wire || Runddraht m. || fil m. à section circulaire. / ~ wood see ~ timber.

round || Rundstab m. || baguette f. ronde. / ~ of beam || Decksbalkenbucht f. || bouge m. de barrot.

roundabout || rundherum || à la ronde.

round-counter || Munitionszähler m. || compteur m. de munitions.

rounded || abgerundet || arrondi. / ~ cheek || abgerundete Kurbelwange f. || flasque m. arrondi. / ~-off || rundlich || arrondi.

round-elastic || Gummifaden m. || fil m. en caoutchouc.

rounder || Rundpresser m.; Rundschläger m. || arrondisseur m.

roundhand || Rundschrift f. || ronde f.

roundhead, horseshoe ~ || Hufeisenrundbogen m. || plein cintre m. outre-passé. / ~ wood screw || Halbrundholzschraube f. || vis f. à bois à tête goutte de suif.

rounding || abrundend || arrondissant.

rounding || Krümmung f.; Bogen m.; Bucht f. || courbure f.; cintre m.; tonture f. / ~ machine || Abrundmaschine f. || machine f. à arrondir. / universal folding, ~ and box-forming machine || Universalabkant-, Rund- und Kastenbiegemaschine f. || machine f. universelle à plier, rouler et former des boîtes.

rounding-off (Drawing) || Abrunden n. || congé m. / ~ (Mech) || Wälzung f. || arrondissage m. / ~ the thread || Gewindeabrundung f. || arrondi m. du filet.

rounding-off machine || Abkantmaschine f. || machine f. à émousser.

rounding tool (Forg) || Rundgesenk n. || estampe f. ou étampe f. ronde.

rounding-up machine || Wälzmaschine f. || machine f. à arrondir.

roundness and parallelism, to check the ~ || auf Zylindrizität f. prüfen || vérifier l'ovalisation f. et la cylindricité f. / ~ testing instrument || Gerät n. zum Prüfen der Zylindrizität || appareil m. à vérifier l'ovalisation et la cylindricité.

round-nosed || rundnasig; rund || à angle m. arrondi. / ~ pliers pl. || Drahtzange f. mit runden Backen || pince f. ronde. / ~ pliers pl. (Electr) || Flügelklemme f. || pince f. à oreilles.

round-off file || Wälzfeile f. || lime f. cabinette ou à arrondir.

round-plane || Rundstabhobel m. || rabot m. à boudin. / ~ rod planing machine || Rundstabhobelmaschine f. || raboteuse f. mécanique pour bois rond. / ~ rolling machine (Pap) || Umrollmaschine f. || machine f. à enrouler; enrouleur m.

rounds pl. || Stabeisen n. || fer m. en barres.

round-sawing machine || Rundsägemaschine f. || scie f. à chantourner mécanique. / ~ shave || Rundschaber m. || plane f. à genoux. / ~ shot || Kupfergranalien fpl. || cuivre m. en graines ou en dragées. / ~ steel pen || Rundschriftfeder f. || plume f. à écriture ronde.

rouse, to ~ the fire || Feuer n. anschüren || activer ou tisonner le feu. / ~ in the vat || im Gärbottich m. aufziehen || aérer en cuve f.

route (Tel) || Leitweg m. || route f. / en ~ || unterwegs || en route f.

router gage (Join) || Adernkratzer m.; Nutenreißer m. || trusquin m. à filet. ~ iron || Grundeisen n. (am Grundhobel) || fer m. de guimbarde.

routine || Geschäftsgang m. || marche f. des affaires; opération f. / by ~ || handwerksmäßig || technique; machinal. / ~ of duties on board || Routine f. oder Dienstbetrieb m. an Bord || distribution f. du service journalier à bord. / ~ works control || laufende Betriebskontrolle f. || contrôle m. courant de l'exploitation.

routing machine || Nut(en)stoßmaschine f. || machine f. à mortaiser ou à faire les rainures.

routining of the construction works and of the lines (Tel) || Überwachung f. der Bauarbeiten und des Linienzustandes || surveillance f. des travaux de construction et des lignes.

rove, to || vorspinnen || filer en gros.

rover || Vorgarnspinner m. || boudineur m.

roving || Vorgespinst n. || mèche f. / fine ~ || Vorgespinstgarn n. || fil m. doux.

roving bobbin with presser || Preßspule f. || bobine f. comprimée. / ~ carrier || Spulenträger m. || porteur m. de bobines. / ~ frame || Vorspinnmaschine f. || métier m. en gros; méchoir m.

row, to || rudern || nager; ramer. / ~ (Clothm) || rauhen || garnir; lainer; tirer à la perche.

row || Reihe f. || rang m.; file f.; rangée f. / ~ (Build) || Flucht f.; Bauflucht f.; Baufluchtlinie f. || alignement m. / ~ of beads || Perlenschnur f.; Rosenkranz m. || collier m.; chapelet m.; rosaire m.; perles fpl. / ~ of beaters (Breaker) || Stabreihe f. || rangée f. de barres. / ~ of bricks || Ziegel-

schicht f. || assise f. ou couche f. de briques. / ~ of carriages || Wagenzug m. || train m. de wagonnets. / ~ of fascines || Faschinenschicht f. || rang m. de fascines. / ~ of holes || Lochreihe f. || rangée f. ou ligne f. de trous. / ~ of keys || Tastenreihe f. || rangée f. de touches. / ~ of meshes || Maschenreihe f. || rangée f. de mailles. / ~ of naps || Samtmaschenreihe f. || bouclé m. / ~ of piles || Pfeilerreihe f.; Pfahlreihe f.; Pfahlwand f. || file f. de pieux; rang m. de pilots. / ~ of rivets || Nietnaht f.; Nietreihe f. || file f. ou ligne f. de rivets.

rower || Ruderer m.; Bootsgast m. || rameur m.; canotier m.

rowing || Rudersport m. || canotage m. / ~ (Clothm) || Rauhen n. || garnissage m.; lainage m.

rowing boat || Ruderboot n. || bateau m. à rames. / ~ with outboard(slung)motor || Ruderboot n. mit Außenbordmotor || bateau m. à moto-godille.

rowing club || Ruderklub m. || société f. de canotage. / ~ sport boat || Rudersportboot n. || canot m. de sport à avirons.

rowlock see row fork.

royal || Obersegel n. || cacatois m. / ~ mast || Royalstänge f. || mât m. du cacatois. / ~ post || Königssäule f.; Königszapfen m. || pivot m. central.

royalties pl. for the use of patents || Lizenzgebühren fpl. || redevance f. payée pour l'usage d'un brevet d'invention.

rub, to || reiben || frotter. / ~ (Metal) || polieren; schleifen || fourbir; frotter; polir. / ~ with emery || schmirgeln || émeriller; polir ou roder à l'émeri. / ~ to rub in (Med) || zum Einreiben n. || pour frictionner. / ~ with oil || einölen || huiler. / ~ the pitch out of the screw bung || Fässer npl. ausreiben || gratter la poix des bondes à vis. / ~ a plaster floor || den Gipsestrich m. schleifen || dégrossir ou frotter une aire en plâtre. / ~ with pumice || abbimsen || poncer; polir à la ponce. / ~ the stone || den Stein m. schleifen || frotter la pierre. / ~ the wire || den Draht m. richten oder gerade richten || dresser ou redresser ou tirer le fil métallique.

rubber (Aero) || Kleinluftschiff n. || dirigeable m. de petites dimensions. / ~ (Auto) || Gummireifen m. || pneu m.; pneumatique m.; bandage m. en caoutchouc. / ~ (Bookb) || Reibeisen n.; Rückeneisen n. || frottoir m. / ~ (Electr) || Reibkissen n. || coussin m.; frottoir m. / ~ (File) || Armfeile f. || carreau m.; lime f. à bras. / ~ (Gold-b) || Reibelappen m. || frottoir m. / ~ (Grinding) || Abziehstein m.; Schleifstein m.; Wetzstein m. || pierre f. à adoucir ou à aiguiser ou à repasser; repasseur m. / ~ (Gum) || Gummi n. || caoutchouc m.; gomme f. / ~ (Join) || Raspel f. || râpe f.; lime f. mordante. / ~ (Mar) || Rubber m.; Glätteisen n. || frottoir m. / ~ (Match) || Reibfläche f. || surface f. de frottement. / ~ (Mechanic) || Polierer m.; Schleifer m. || polisseur m.; repasseur m.; brunisseur m. / ~ (Mill) || Beutelmaschine f.; Mehlmaschine f. || bluterie f. / ~ (Needl) || Streicheisen n.; Richt-

eisen n. ∥ râpe f.; règle f. à jour. / ~ (Office) ∥ Radiergummi m. ∥ gomme f. à effacer; gomme-grattoir m. / ~ (Pyrot) ∥ Mengholz n. ∥ écrémoire f. / ~ (Rag for cleaning) ∥ Wischtuch n.; Scheuerlappen n.; Putzlappen m. ∥ torchon m.; frottoir m.; chiffon m. à nettoyer. / ~ (Rough bath towel) ∥ Frottierhandtuch n. ∥ serviette f. floconneuse à frotter; frottoir m. / coarse ~ ∥ rauher Schleifstein m. ∥ pierre f. rugueuse. / crêpe ~ ∥ Kreppgummi n. ∥ caoutchouc m. crêpé. / hard ~ ∥ ébonit m.; caoutchouc m. durci. / India ~ ∥ Kautschuk m. ∥ caoutchouc m. / India ~ factory ∥ Gummifabrik f. ∥ fabrique f. de caoutchouc. / ~ for packings ∥ Dichtungsgummi n. ∥ caoutchouc m. pour garnitures. / Para ~ ∥ Paragummi n. ∥ Para m. / raw ~ ∥ Rohgummi n. ∥ caoutchouc m. brut. / reclaimed ~ see regenerated ~. / regenerated ~ ∥ Regeneratgummi n.; regeneriertes Gummi n. ∥ caoutchouc m. régénéré. / scrap ~ ∥ Gummiabfälle mpl. ∥ déchets mpl. de caoutchouc. / seamless ~ ∥ nahtloses Gummi n. ∥ caoutchouc m. sans couture. / semi-soft ~ ∥ halblinder Schleifstein m. ∥ pierre f. demi-douce ou demi-rugueuse. / sheet ~ ∥ Plattengummi n. ∥ caoutchouc m. en plaques; feuilles fpl. de caoutchouc. / smoked sheet ~ ∥ geräuchertes Gummi n. in Blättern ∥ feuilles fpl. de caoutchouc fumé. / soft ~ (Grinding) ∥ linder Schleifstein m. ∥ pierre f. douce. / soft ~ (Gum) ∥ Weichgummi n.; weiches Gummi n. ∥ caoutchouc m. mou ou tendre. / transparent ~ ∥ durchsichtiges Gummi n. ∥ caoutchouc m. transparent. / unsmoked sheet ~ ∥ ungeräuchertes Gummi n. in Blättern ∥ feuilles fpl. de caoutchouc non fumé. / used ~ ∥ Altgummi n. ∥ vieux caoutchouc m. / vulcanized ~ ∥ vulkanisiertes Gummi n. ∥ gomme f. vulcanisée. / ~ from waste see regenerated ~.

rubber articles pl., surgical ∥ Gummiwaren fpl. für chirurgische Zwecke ∥ articles mpl. de chirurgie en caoutchouc. / ~ band ∥ Gummiband n. ∥ ruban m. élastique. / ~ bandage ∥ Gummibinde f. ∥ bande f. élastique ou de caoutchouc. / ~ bathing cap ∥ Gummibademütze f. ∥ bonnet m. de bain en caoutchouc. / ~ belt (Mach) ∥ Gummitreibriemen m. ∥ courroie f. en caoutchouc. / ~ belt (Med) ∥ Gummigürtel m. ∥ ceinture f. élastique. / ~ blanket ∥ Gummituch n. ∥ drap m. de caoutchouc. / ~ buffer ∥ Gummipuffer m. ∥ buttoir m. ou tampon m. en caoutchouc. / ~ buffers pl. for typewriters ∥ Schreibmaschinenschalldämpfer m. ∥ pieds mpl. amortisseurs pour machine à écrire. / ~ cable ∥ Gummikabel n. ∥ câble m. sous caoutchouc. / ~ cape ∥ Gummiregenrock m. ∥ imperméable m.; caoutchouc m. / ~ cement ∥ Kautschukkitt m. ∥ mastic m. au caoutchouc. / ~ cloth ∥ Gummistoff m. ∥ étoffe f. caoutchoutée. / ~ connecting tube ∥ Schlauchverbindung f. ∥ raccord m. de caoutchouc. / ~ cord ∥ Gummiseil n. ∥ câble m. de caoutchouc. / ~ cord springing ∥ Gummilitzenfederung f. ∥ amortissement m. par torons de caoutchouc. / ~ core ∥ Gummieinlage f. ∥ noyau m. de caoutchouc. / ~ cork ∥ Kautschukstopfen m. ∥ bouchon m. de caoutchouc. / ~ cover ∥ Gummiüberzug m. ∥ couverture f. de caoutchouc. / ~ cover for cylinders ∥ Walzenbezug m. ∥ recouvrement m. pour cylindres. / ~-covered ∥ mit Gummi m. überzogen ∥ à revêtement m. en caoutchouc. / ~-covered cable ∥ Gummikabel n. ∥ câble m. sous caoutchouc. / ~ covering ∥ Gummibelag m.; Gummihülle f. ∥ revêtement m. ou garniture f. de caoutchouc. / ~ cylinder for typewriters ∥ Gummiwalze f. für Schreibmaschinen ∥ rouleau m. en ébonite pour machines à écrire. / ~ denture ∥ Kautschukgebiß n.; Kautschukgebißstück n. ∥ dentier m. en caoutchouc. / ~ disk ∥ Gummischeibe f. ∥ disque m. caoutchouc. / ~ fabric ∥ Gummistoff m. ∥ tissu m. caoutchouté. / ~ factory ∥ Gummifabrik f. ∥ fabrique f. de caoutchouc. / ~ feed roller ∥ Gummifarbwalze f. ∥ rouleau m. d'encrage en caoutchouc. / gas-retaining ~ film ∥ gasundurchlässige Gummischicht f. ∥ couche f. de caoutchouc imperméable au gaz. / ~ finger ∥ Gummifingerling m. ∥ doigtier m. en caoutchouc. / ~ glove ∥ Gummihandschuh m. ∥ gant m. de caoutchouc. / ~ goods pl. ∥ Gummiwaren fpl. ∥ articles mpl. en caoutchouc. / ~ heel ∥ Gummiabsatz m. ∥ talon m. en caoutchouc. / ~ hose ∥ Gummischlauch m. ∥ tube m. en caoutchouc. / ~ hose for artificial flowers ∥ Gummiröhrchen n. für Kunstblumen ∥ tube m. en caoutchouc pour fleurs artificielles. / ~ hose machine ∥ Gummischlauchmaschine f. ∥ machine f. à tuyaux en caoutchouc. / ~ insulating material ∥ Gummiisolationsstoff m. ∥ isolants mpl. à base de caoutchouc. / ~ insulating tape ∥ Gummiisolierband n. ∥ ruban m. isolant de caoutchouc. / ~ joint ∥ Gummidichtung f.; Gummipackung f. ∥ garniture f. ou joint m. en caoutchouc. / ~ manufacture of ∥ Gummifabrikat n. ∥ article m. de caoutchouc. / ~ manufacturing ∥ Gummiherstellung f. ∥ fabrication f. de caoutchouc. / ~ mat ∥ Gummiteppich m. ∥ tapis m. de caoutchouc. / grooved ~ matting ∥ gerippter Gummiteppich m. ∥ tapis m. de caoutchouc à côtes.

rubberoid millboard ∥ Ruberoidpappe f. ∥ carton m. rubéroïde. / ~ sheating ∥ Ruberoidbedeckung f. ∥ recouvrement m. en rubéroïde.

rubber packing ∥ Gummipackung f. ∥ garniture f. ou bourrage m. ou joint m. en caoutchouc. / ~ packing ring ∥ Gummidichtungsring m. ∥ anneau m. de joint en caoutchouc. / ~ painter on ~ ∥ Gummimaler m. ∥ peintre m. sur caoutchouc. / ~ painting on ~ ∥ Gummimalerei f. ∥ peinture f. sur caoutchouc. / ~ plate ∥ Gummiplatte f. ∥ plaque f. de caoutchouc. / ~ playing ball ∥ Gummispielball m. ∥ balle f. à jouer en caoutchouc. / ~ punching works pl. ∥ Gummistanzwerk n. ∥ établissement. m. à estamper le caoutchouc. / ~ resin ∥ Gummiharz n. ∥ gomme-résine f.; résine f. de caoutchouc.

rubber ring ∥ Gummiring m. ∥ bague f. de caoutchouc. / ~ for preserve glasses ∥ Konservenglasgummiring m. ∥ rondelle f. en caoutchouc pour verres à conserves. / ~ packing ∥ Gummiringdichtung f. ∥ garniture f. ou joint m. en caoutchouc.

rubber roll ∥ Gummiwalze f. ∥ rouleau m. en caoutchouc. / ~ roller ∥ Gummimöbelrolle f. ∥ roulette f. en caoutchouc pour meubles. / ~ rolling machine ∥ Kautschukwalzwerk n. ∥ laminoir m. à caoutchouc.

rubbers pl. ∥ Gummischuhe mpl. ∥ chaussures fpl. de caoutchouc; caoutchoucs mpl.

rubber shock absorber ∥ Gummistoßdämpfer m.; Gummipuffer m. ∥ amortisseur m. de choc; tampon m. en caoutchouc. / ~ shoe (Typewr) ∥ Gummifuß m. ∥ pied m. en caoutchouc. / ~ sole ∥ Gummisohle f. ∥ semelle f. en caoutchouc. / ~ solution ∥ Gummilösung f. ∥ dissolution f. de caoutchouc. / ~ sponge ∥ Gummischwamm m. ∥ éponge f. de caoutchouc.

rubber stamp ∥ Kautschukstempel m.; Gummistempel m. ∥ timbre m. en caoutchouc. / ~ box ∥ Stempelkissenschachtel f. ∥ boîte f. à tampons. / ~ ink ∥ Stempeltinte f. ∥ encre f. pour timbres.

rubber standard ∥ Gumminorm f. ∥ norme f. ou règle f. pour le caoutchouc. / ~ stopper ∥ Kautschukstopfen m.; Gummipfropfen m. ∥ bouchon m. en caoutchouc. / ~ surgical plaster ∥ Kautschukpflaster n. ∥ emplâtre m. au caoutchouc. / ~ surrogate ∥ Gummiersatzstoff m. ∥ succédané m. de caoutchouc. / ~ syringe ∥ Gummispritze f. ∥ seringue f. en caoutchouc. / ~ tape ∥ Gummiband n. ∥ ruban m. en caoutchouc; élastique m. / ~ tape insulation ∥ Gummibandisolierung f. ∥ isolement m. en ruban de caoutchouc. / ~ teat ∥ Gummisauger m. ∥ tétine f. en caoutchouc. / ~ tire see rubber tyre. / ~ track of the tank ∥ Gummikette f. des Kampfwagens ∥ chenille f. souple ou au caoutchouc ou Kegrésse du char de combat. / ~ tree ∥ Gummibaum m. ∥ arbre m. à caoutchouc.

rubber tube ∥ Gummischlauch m. ∥ tube m. en caoutchouc. / ~ of air-valve ∥ Ventilschlauch m. ∥ tube m. en caoutchouc pour soupape. / ~ with metal sleeve ∥ Gummischlauch m. mit Metallhülse ∥ tube m. en caoutchouc à douille métallique. / ~ wiring ∥ Gummischlauchleitung f. ∥ cordon m. à conducteur isolé par tuyau de caoutchouc.

rubber tubing ∥ Gummischlauch m. ∥ tuyau m. de caoutchouc. / ~ type ∥ Gummibuchstabe m. ∥ caractère m. en caoutchouc.

rubber tyre ∥ Gummireifen m. ∥ bandage m. en caoutchouc; pneu m. / ~ for cycles ∥ Fahrradgummireifen m. ∥ bandage m. de roues en caoutchouc pour cycles. / ~ for motor cars ∥ Gummireifen m. für Kraftfahrzeuge ∥ bandage m. en caoutchouc pour automobiles. / solid ~ ∥ Vollgummireifen m. ∥ bandage m. plein en caoutchouc.

rubber union ∥ Schlauchkupplung f. ∥ raccord m. en tuyau de caoutchouc. / ~ varnish ∥ Gummilack m. ∥ gomme-laque f.; vernis m. au caoutchouc. / ~ vessel for fire service ∥ Feuereimer m. aus Kautschuk ∥ seau m. à incendie en caoutchouc. / ~ warper ∥ Gummifadenscherer m. ∥ ourdisseur m. en caoutchouc. / ~ washer ∥ Gummiunterlage f. ∥ rondelle f. en caoutchouc. / ~ weaving goods pl. with metal inserts. ∥ Gummiwebwaren fpl. mit Metalleinlagen ∥ articles mpl. tissus en caoutchouc avec doublure métallique. / ~ working ∥ Gummibearbeitung f. ∥ préparation f. du caoutchouc.

/ ~ woven goods pl. ‖ Gummiweb- und -wirkwaren fpl. ‖ tissus mpl. de caoutchouc.

rubbing ‖ Friktion f.; Reibung f. ‖ frotte- ment m. / ~ (Metal) ‖ Polieren n.; Schlei- fen n.; Schmirgeln n. ‖ adoucissage m.; opération f. de doucir *ou* de frotter *ou* de polir *ou* de roder. / ~ brush (Househ) ‖ Wichsbürste f. ‖ brosse f. à cirer. / ~ brush (Mach) ‖ Drahtbürste f. ‖ gratte- brosse f.; brosse f. métallique *ou* déra- peuse. / ~ cloth ‖ Reibelappen m. ‖ frot- toir m. / ~ contact ‖ Streichkontakt m. ‖ contact m. à frottement. / ~ plate of the kid shoe (Aero) ‖ Schleifplatte f. des Spornschuhs ‖ patin m. du sabot. / ~ strake (Shipb) ‖ Scheuerleiste f. ‖ cordon m.; boudin m.; bourrelet m. / ~ surface (Millstone) ‖ Mahlfläche f. ‖ face f. travaillante. / ~ ware ‖ Frottier- artikel m. ‖ article m. pour frotter. / ~ wax ‖ Bohnerwachs n.; Bohnwachs n.; Polierwachs n. ‖ encaustique m.

rubbish ‖ Abfall m. ‖ déchet m.; rognure. / ~ (Build) ‖ Schutt m.; Gerölle n. ‖ éboulis m.; décombres mpl. / ~ (Geol) ‖ Gerölle n.; Geschiebe n.; grober Kies m.; Rollstein n. ‖ cailloux mpl. roulés; galet m. / ~ (Househ) ‖ Hausmüll n. ‖ ordures fpl. ménagères. / ~ (Mine) ‖ taubes Gestein n.; Gangmasse f.; gangue f.; matière f. *ou* roche f. stérile. / tipping device for mine~ ‖ Bergekipper m. ‖ tombereau m. à éboulis. / ~ of old plaster (Mas) ‖ Kalk- schutt m.; Mulm m. ‖ gravat m.; gravoir m. de plâtre. / ~ of walling ‖ verfallenes Gemäuer n. ‖ masure f.

rubbish carrying ‖ Schuttabfuhr f. ‖ trans- port de gravats m. / ~ depository ‖ Schutt- abladeplatz m. ‖ décharge m. publique.

rubble (Build) ‖ Bruchstein m. ‖ moellon m. / ~ (Geol) ‖ Gerölle n.; Geschiebe n.; grober Kies m.; Rollstein n. ‖ cailloux mpl. roulés; galet m. / ~ (Mas) ‖ Kalk- schutt m.; Mulm m. ‖ gravat m.; gravois m. de plâtre. / ~ (Mine) ‖ Abraum m. ‖ couche f.; lit m. de terre *ou* de dé- combres. / ~ (Road) ‖ Steinschutt m. ‖ pierraille f. / ~ filter ‖ Kiesfilter m. ‖ filtre m. à gravier. / ~ stone ‖ Füllstein m. ‖ blocaille f. / ~ walling ‖ Feldsteinmauer- werk n.; rauhes Mauerwerk n. ‖ maçon- nerie f. en galets; hourdage m.; maçon- nage m. en moellons. / ~ work ‖ Füll- mauerwerk n. ‖ remplage m.; muraille f. de remplage. / ~ work (Road) ‖ Stein- packung f. ‖ empierrement m.

rubellite ‖ Rubellit m.; roter Turmalin m. ‖ rubellite f.

rubidium ‖ Rubidium n. ‖ rubidium m. / ~ alum ‖ Rubidiumalaun m. ‖ alun m. de rubidium.

rubin fluor ‖ Rubinglas n. ‖ faux-rubis m.

rubric ‖ Rubrik f. ‖ rubrique f.

ruby (Miner) ‖ edler Korund m.; Rubin m. ‖ corindon m. hyalin; rubis m. / ~ (Print) ‖ Parisienneschrift f. ‖ parisienne f.; sédanoise f. / ~ blende ‖ Rotgültig n.; Rubinblende f. ‖ argent m. antimonié sulfuré; blende f. rouge. / ~ silver *see* ~ blende. / ~ sulphur ‖ rotes Schwefel- arsen n.; Realgar n. ‖ arsenic m. sulfuré rouge; réalgar m.

ruche ‖ Rüsche f. ‖ ruche f. / ~ machine ‖ Rüschenmaschine f. ‖ machine f. à rucher.

rudder (Shipb; Aero) ‖ Ruder n.; Steuer- ruder n.; Steuer n. ‖ gouvernail m.;

---

timon m. / balanced ~ ‖ Balanzeruder n.; entlastetes Ruder n. ‖ gouvernail m. équilibré. / bow ~ ‖ Bugruder n. ‖ gou- vernail m. de devant. / depth ~ ‖ Tiefen- ruder n.; Horizontalruder n. ‖ gouver- nail m. de profondeur. / Flettner ~ ‖ Flettnerruder n. ‖ gouvernail m. Flettner. / jury ~ ‖ Notruder n. ‖ gouvernail m. de fortune. / lift ~ ‖ Steigruder n. ‖ gouver- nail m. de montée. / ~ of a ship ‖ Schiffs- steuer n. ‖ gouvernail m. du navire. / temporary ~ *see* jury ~. / the ~ turns on x pintles ‖ das Ruder n. dreht sich auf x Fingerlingen ‖ le gouvernail tourne sur x aiguillots. / twin ~ ‖ Zwillingsruder n. ‖ gouvernails mpl. jumeaux. / verti- cal ~ ‖ Seitensteuer n. ‖ gouvernail m. vertical.

rudder brace ‖ Ruderöse f. ‖ conassière f. du gouvernail. / ~ bracing ‖ Ruderver- spannung f. ‖ haubanage m. de l'em- pennage. / ~ case ‖ Ruderkoker m. ‖ tube m. de jaumière. / ~ coat ‖ Ruder- kragen m. ‖ braie f. du gouvernail. / ~ connection ‖ Ruderanschluß m. ‖ bâti m. d'union des gouvernails. / ~ crank ‖ Ruderhebel m. ‖ guignole f. du gouver- nail. / ~ frame (Shipb) ‖ Ruderrahmen m. ‖ châssis m. *ou* cadre m. de gouver- nail. / ~ head ‖ Ruderkopf m. ‖ tête f. du gouvernail. / ~ heel ‖ Ruderhacken m.; Rudersohle f. ‖ talon m. de gouver- nail. / ~ helmsman (Aero) ‖ Seitensteuer- mann m. ‖ timonier m. de direction. / ~ in- dicator ‖ Ruderlageanzeiger m. ‖ indica- teur m. de position du gouvernail. / ~ mould ‖ Rudermall n. ‖ gabarit m. du gouvernail. / ~ pintle ‖ Ruderfingerling m. ‖ aiguillot m. du gouvernail. / ~ planes pl. ‖ Steuerflächen fpl. ‖ surfaces f. de gouvernail. / plate of the ~ ‖ Ruderblatt n. ‖ tôle f. du gouvernail. / ~ plating ‖ Beplattung f. des Ruders ‖ bordé m. du gouvernail.

rudder post ‖ Hintersteven m. ‖ étambot m.; porte-gouvernail m. / ~ bracket ‖ Ruderbock m. ‖ palier m. extérieur. / lower part of the ~ ‖ unterer Ruder- schaft m. ‖ mèche f. inférieure. / up- per part of the ~ ‖ oberer Ruderschaft m. ‖ mèche f. supérieure.

rudders pl. ‖ Rudergeschirr n. ‖ gouver- nails mpl.

rudder spindle ‖ Ruderschaft m. ‖ mèche f. du gouvernail. / ~ stops pl. ‖ Hemm- klampen fpl.; Stoppklampen fpl. ‖ arrêts mpl. du gouvernail. / ~ telegraph ‖ Rudertelegraf m. ‖ télégraphe m. du gouvernail. / ~ trunk ‖ Ruderkoker m. ‖ jaumière f.

ruddle ‖ Rötel m. ‖ craie f. rouge.

rudge teeth pl. ‖ Kerbverzahnung f. ‖ den- ture f. rudge.

rue (Bot) ‖ Raute f. ‖ rue f. / ~ herb ‖ Raute f.; herbe f. de rue. / ~ oil ‖ Rau- tenöl n. ‖ essence f. de rue.

ruff, to ~ the mirror glasses pl. ‖ die Spie- gelgläser npl. rauhschleifen ‖ dégrossir des glaces fpl.

ruff ‖ Halskrause f. ‖ collarette f.

ruffing (Glassm) ‖ Rauhschleifen n. ‖ dégrossi m.; dégrossissage m.

ruffle ‖ Manschette f.; Krause f. ‖ man- chette f.; ruche f.

rug ‖ Wolldecke f.; Decke f.; Schlafdecke f. ‖ couverture f. en laine. / carriage ~ ‖ Wagendecke f.; Plane f. ‖ bâche f. / floor ~ ‖ Fußteppich m. ‖ tapis m. de

---

pied. / fur ~ (Carpet) ‖ Pelzteppich m. ‖ tapis m. fourré. / fur ~ (Covering) ‖ Pelzdecke f. ‖ couverture f. en fourrure. / horse ~ ‖ Pferdedecke f. ‖ couverture f. pour chevaux. / skin ~ ‖ Fellteppich m. ‖ tapis m. en fourrure. / sleeping ~ ‖ Schlafdecke f. ‖ couverture f. de nuit. / travelling ~ ‖ Reisedecke f. ‖ couver- ture f. de voyage.

rug-carpet ‖ Plüschteppich m.; Samt- teppich m. ‖ tapis m. velouté *ou* de peluche.

rugged ‖ uneben ‖ inégal.

ruille ‖ Mauerkehle f.; Kalkleiste f. ‖ ruellée f.; ruillée f.

ruin, to ‖ ruinieren ‖ ruiner; gâter; abîmer.

ruin ‖ Verfall m. ‖ décadence f.; ruine f.

ruined ‖ ruiniert ‖ ruiné; flambé; frit.

ruin marble ‖ Ruinenmarmor m. ‖ marbre m. ruiniforme.

ruinous by water ‖ unterwaschen; unter- spült ‖ déchaussé.

ruins pl. ‖ Schutt m.; Gerölle n. ‖ éboulis m.; décombres mpl.

rule, to ‖ liniieren ‖ régler.

rule (Drawing) ‖ Lineal n. ‖ règle f. / ~ (Mas) ‖ Richtscheit n. ‖ règle f. / ~ (Math) ‖ Formel f. ‖ formule f. / ~ (Mea- sure) ‖ Maßstab m. ‖ règle f. divisée; mètre m. / ~ (Print) ‖ Kolumnenmaß n. ‖ longueur f.; mesure f. d'une réglette. / ~ (Standard) ‖ Norm f.; Regel f. ‖ norme f.; règle f. / against the ~s pl. ‖ ordnungs- widrig ‖ contraire à l'ordre m.; irrégu- lier. / as a ~ ‖ in der Regel f. ‖ ordinaire- ment. / bevel ~ ‖ Schmiege f. ‖ sauterelle f.; équerre f. mobile. / conjoined ~ of three (Math) ‖ Kettenregel f. ‖ règle f. conjointe. / contrary to ~ ‖ regelwidrig ‖ contraire à la règle. / fixed ~ ‖ Satzung f. ‖ statut m.; dogme m. / folding pocket ~ ‖ Zollstock m.; zusammenlegbarer Maß- stab m. ‖ réglet m.; mètre m. pliant. / gol- den ~ *see* ~ of proportion. / iron ~ (Text print) ‖ Liniierzirkel m. ‖ compas m. à tracer dans l'impression des cravates. / joint ~ *see* pliant ~. / long narrow-chan- nelled ~ (Weav) ‖ Rute f.; Samtnadel f. ‖ épingle f.; fer m. d'un métier à velours. / metal ~ ‖ Metallineal n. ‖ règle f. métallique. / mitre ~ *see* bevel ~. / x-part ~ ‖ x-teiliges Metermaß n. ‖ mètre m. à x branches. / parallel ~ ‖ Parallellineal n. ‖ règle f. à parallèles. / pliant ~ ‖ Klappmaßstab m.; Gelenk- maßstab m. ‖ mètre m. pliant. / pocket ~ *see* folding pocket ~. / ~ of propor- tion ‖ Regeldetri f. ‖ règle f. de trois. / slide ~ ‖ Rechenschieber m. ‖ règle f. à calcul. / standard ~ ‖ Normallineal n. ‖ règle f. normale. / ~ of three *see* ~ of proportion. / ~ of thumb (Sugar) ‖ Fingerprobe f. ‖ preuve f. du filet. / ~ of thumb (Electr) ‖ Daumenregel f. ‖ règle f. du pouce. / two-scale ~ ‖ Doppel- maßstab m. ‖ double règle f. graduée. / waved ~ ‖ Schlangenlinie f. ‖ filet m. ondulé.

rule bender (Print) ‖ Linienbiegeapparat m. ‖ appareil m. à cintrer les filets. / ~ cutting machine ‖ Linienschneidma- schine f. ‖ taille-filets m.

ruled line ‖ vorgedruckte Linie f. ‖ ligne f. imprimée au préalable. / ~ paper ‖ lini- iertes Papier n. ‖ papier m. ligné *ou* réglé.

rule drawing frame ‖ Linienziehbank f. ‖ établi m. à tirer les filets. / machine

for making ~s ‖ Maßstabmaschine f. ‖ machine f. pour la fabrication des mètres pliants. / ~ paper (Weav) ‖ Patronenpapier n.; Tupfpapier n.; Musterpapier n. ‖ papier m. à patron *ou* quadrillé *ou* rayé; carte f.

ruler (Drawing) ‖ Lineal n. ‖ règle f. / ~ (Print) ‖ Linienzieher m.; Linienreißer m. ‖ tire-ligne m. / ~ (Surv) ‖ Setzlatte f. ‖ règle f. de nivellement. / ~ for the elevation of rails ‖ Richtscheit n. für die Schienenerhöhung ‖ règle f. de surhaussement des rails. / graduated ~ ‖ Lineal n. mit Maßeinteilung ‖ règle f. graduée. / long narrow-channelled ~ (Weav) ‖ Ritznadel f.; Samtnadel f. ‖ épingle f. d'un métier à velours; baguette f. / parallel ~ ‖ Parallellineal n. ‖ règle f. à parallèles.

rules pl. (Instructions) ‖ Vorschriften fpl.; Verordnungen fpl. ‖ règles fpl.; instructions fpl. de service; prescriptions fpl.; règlements mpl. / ~ for boiler house ‖ Kesselhausvorschriften fpl. ‖ instructions fpl. de service pour la chaufferie. / ~ for the building of vessels ‖ Bauvorschriften fpl. der Schiffsklassifikationsgesellschaften ‖ règlement m. pour la construction des navires.

rule shoot board (Print) ‖ Linienhobel m. ‖ biseautoir m. pour filets.

ruling arrangement ‖ Liniiervorrichtung f. ‖ dispositif m. de réglage. / ~ device (Typewr) ‖ Linienzieher m. ‖ dispositif m. pour tracer les lignes. / ~ ink ‖ Liniierfarbe f. ‖ encre f. à régler. / ~ machine ‖ Liniiermaschine f. ‖ machine f. à régler *ou* rayer; régleuse f.

ruling pen ‖ Reißfeder f. ‖ tire-ligne m. / double-line ~ ‖ Doppelziehfeder f. ‖ tire-ligne m. double.

rullock ‖ Rudergabel f.; Dolle f. ‖ toletière f.; dame f.

rum ‖ Rum m. ‖ rhum m.

rumble ‖ Bauschutt m. ‖ décombres mpl.

rum distillery ‖ Rumfabrik f. ‖ distillerie f. de rhum; rhumerie f. / ~ essence ‖ Rumessenz f. ‖ essence f. de rhum.

rumpel with beaters for treating carbonized cloth ‖ Klopfwalke f. zum Reinigen karbonisierter Tuche ‖ foulon m. servant au nettoyage des draps carbonisés.

run, to ‖ laufen; fahren; verkehren ‖ courir. / ~ (Geol) ‖ streichen ‖ aller. / ~ (Mach) ‖ laufen; umlaufen; arbeiten; marchen; fonctionner; travailler; tourner. / ~ (Mine) ‖ zu Bruche m. bauen; einstürzen ‖ faire ébouler les débris mpl.; ébouler. / ~ an adit (Mine) ‖ einen Stollen m. treiben ‖ percer une galerie. / ~ aground ‖ auf Grund m. auflaufen ‖ s'échouer; mouiller par la quille. / ~ down (Clockm) ‖ ablaufen ‖ détendre. / ~ down (Water) ‖ fallen ‖ baisser; descendre / ~ the battery down ‖ die Batterie zu stark entladen ‖ décharger la batterie à l'excès. / ~ down the coast ‖ längs der Küste f. fahren ‖ côtoyer une terre; descendre une côte. / ~ engine slowly up to full load ‖ die Maschine f. langsam auf Vollast gehen lassen ‖ mettre lentement le moteur en pleine charge. / ~ foul of a vessel ‖ ein Schiff n. ansegeln ‖ tomber sur un navire. / ~ idle ‖ leerlaufen ‖ marcher à vide.

run, to ~ **in** (Chem) ‖ einströmen ‖ couler. / ~ (Mach) ‖ einlaufen ‖ rôder. / ~ the

clamps with lead ‖ die Klammern fpl. mit Blei vergießen ‖ sceller les crampons mpl. au plomb. / ~ the metal at the bottom (Found) ‖ mit dem Steigrohr n. gießen ‖ couler à cale *ou* en siphon. / to let the motor run itself in ‖ den Motor m. einlaufen lassen ‖ laisser tourner le moteur pour qu'il se fasse. / ~ a rope ‖ ein Tau n. längs Deck holen ‖ traîner *ou* tirailler *ou* filer une corde.

run, to ~ **light** ‖ unbelastet laufen ‖ marcher à vide. / ~ the motor up to full speed ‖ den Motor m. auf volle Umdrehungszahl bringen ‖ porter le moteur à sa vitesse de régime.

run, to ~ **off** (Chem) ‖ abläutern ‖ filtrer. / ~ (Found) ‖ gießen ‖ couler. / ~ the cinder ‖ die Schlacke abstechen ‖ faire écouler le laitier. / ~ the pig-iron ‖ den Hochofen m. abstechen ‖ percer le hautfourneau; faire la percée du hautfourneau; faire couler la fonte. / ~ the rails ‖ entgleisen ‖ sauter hors *ou* sortir des rails; dérailler. / ~ the scoria ‖ die Schlacken fpl. abwerfen ‖ haler le laitier *ou* les scories fpl.

run, to ~ **out** ‖ abfließen; ablaufen; auslaufen; schmelzen ‖ s'écouler. / ~ a cable ‖ ein Kabel m. ausrollen ‖ dérouler un câble.

run, to ~ **past** the signal ‖ ein Haltesignal n. überfahren ‖ forcer *ou* brûler un signal d'arrêt. / ~ in plaster ‖ unter Putz m. verlegen ‖ enrober dans le plâtre. / ~ the roof (Mine) ‖ den Bruch m. niedergehen lassen ‖ faire ébouler les débris mpl. / ~ roughly ‖ abschruppen ‖ dégrossir au tour. / ~ short ‖ ausgehen ‖ manquer; s'épuiser. / ~ synchronous ‖ synchron laufen ‖ marcher synchrone. / ~ through a sharp curve ‖ eine scharfe Kurve f. durchfahren ‖ franchir une courbe à faible rayon. / ~ through a station ‖ einen Bahnhof m. durchfahren ‖ brûler une station. / ~ true ‖ kreisförmig *oder* zentrisch laufen ‖ tourner rond. / ~ untrue ‖ unrund laufen; schlagen ‖ être excentré *ou* désaxé; ne pas tourner rond. / ~ up (Money) ‖ auflaufen ‖ s'enfler; s'accumuler. / ~ up a wall ‖ eine Mauer f. aufhöhen ‖ exhausser un mur. / ~ before the wind ‖ vor dem Winde m. segeln ‖ courir *ou* faire vent arrière.

run ‖ Ansturm m. ‖ assaut m. / ~ (Hydr arch) ‖ Rinnsal n. ‖ chenal m.; cours m. d'eau; lit m. / continuous ~ of a curve ‖ stetiger Verlauf m. einer Kurve ‖ continuité f. d'une courbe. / ~ of mine ‖ Rohkohle f. ‖ houille f. crue. / ~ of a vessel ‖ Linien fpl. *oder* Form f. eines Schiffes ‖ lignes fpl. d'un bâtiment.

runabout ‖ leichter offener Wagen m.; Fabrikkarren m. ‖ camionnette f.

run-down ‖ abgelaufen ‖ au bas. / ~ battery ‖ erschöpfte Batterie f. ‖ batterie f. déchargée.

rundle ‖ Sprosse f. einer Raufe ‖ roulon m.

rung ‖ Steigsprosse f. ‖ cheville f.; ranche f.

runner (Boring implement) ‖ Hebeklaue f.; Gehänge f. ‖ clef f. de revelée; grappin m. aux sondages. / ~ (Errand boy) ‖ Laufbursche m. ‖ garçon m. de magasin. / ~ (Found) ‖ Gießkopf m.; Einguß m.; Ansatz am Gußstück ‖ jet m.; masselotte f. / ~ (Mach) ‖ Laufrolle f.; galet m. de roulement. / ~ (Mar) ‖ Drehreep m. ‖ itague f. d'un palan.

/ ~ (Mill) ‖ Läufer m. ‖ meule f. courante; surmeule f. / ~ (Ore dress) ‖ Läufer m. ‖ meules fpl. de broyeur. / ~ (Print) ‖ Zeilenzähler m. ‖ nombre m. de lignes. / ~ (Spinn) ‖ Läuferwalze f.; Schnellwalze f. ‖ volant m. / longitudinal ~ of carriage body ‖ Längsträger m. der Karosserie ‖ brancard m. longitudinal de carrosserie. / to cast rings pl. in vertical moulds with a very big ~ ‖ Ringe mpl. in stehender Form mit hohem Kopf gießen ‖ des anneaux mpl. furent coulés dans un moule vertical avec une haute masselotte. / chilled iron ~ ‖ Hartgußläufer m. ‖ meule f. de fonte coulée en coquille. / double-conical ~ ‖ doppelkonischer Läufer m. ‖ meule f. biconique. / edge ~ (Mill) ‖ Kollergang m. ‖ moulin m. à meules verticales. / edge ~ with stationary and rotary grinding track ‖ Kollergang m. mit feststehender und umlaufender Mahlbahn ‖ broyeur m. à meules verticales avec plateau de roulement fixe et tournant. / edge ~ for wet grinding ‖ Naßkollergang m. ‖ broyeur m. à meules verticales par voie humide. / to knock off ~ and riser on castings ‖ Einguß m. und Steiger m. am Gußstück abschneiden ‖ couper l'évent m. et le jet des pièces coulées. / mixing ~ ‖ Mischkollergang m. ‖ mélangeur m. à meules verticales. / stone ~ ‖ Steinläufer m. ‖ meule f. en pierre.

runner frame ‖ Läuferzarge f. ‖ cuve f. de broyeur à meules. / ~ head ‖ Gußtrichter m. ‖ jet m. *ou* entonnoir m. de coulée.

runner-on of wagons (Mine) ‖ Schlepper m. von Kohlenwagen ‖ rouleur m. de berlines; wagoneur m.

runner rail ‖ Laufschiene f. ‖ chemin m. de roulement. / ~ ring ‖ Läuferring m. ‖ bandage m. de meule.

runners pl. and risers pl. ‖ Eingüsse mpl. und Steiger mpl. ‖ jets mpl. de coulée.

runner shaft ‖ Rotorwelle f. ‖ arbre m. de rotor. / ~ stick ‖ Gußlochzapfen m.; Gußzapfen m. ‖ modèle m. du jet. / ~ stone ‖ Laufstein m.; Läufer m. ‖ surmeule f. / Kaplan turbine with adjustable ~ vanes pl. and guide blades pl. ‖ Kaplanturbine f. mit regelbarem Laufapparat und Leitapparat ‖ turbine f. Kaplan avec appareil de marche réglable.

runnet ‖ Lab n. ‖ présure f.

running (Mach) ‖ Gang m. ‖ mouvement m; marche f. / backward ~ ‖ Rückwärtsgang m.; Rücklauf m. ‖ marche f. arrière; recul m. / ~ down of battery ‖ Selbstentladung f. *oder* Erschöpfung f. der Batterie ‖ déchargement m. *ou* épuisement m. de la batterie. / when ~ empty ‖ bei Leerlauf m. ‖ à vide m. / first ~ (Distill) ‖ Vorlauf m. ‖ avant-coulant m. / forward ~ ‖ Vorwärtsgang m. ‖ marche f. avant. / full-load continuous ~ ‖ dauerndes Laufen n. unter Vollast ‖ marche f. continue à pleine charge. / ~ hot ‖ Heißlaufen m. ‖ échauffement m. / intermittent ~ ‖ intermittierender Betrieb m. ‖ service m. discontinu *ou* intermittent. / last ~ (Distill) ‖ Nachlauf m. ‖ après-coulant m.; eau-de-vie f. dernière; repasses fpl. / ~ light (Mach) ‖ Leergang m. ‖ marche f. à vide. / ~ of a machine ‖ Gang m. einer Maschine ‖ marche f. d'une machine. / noiseless ~ ‖ geräuschloser Gang m. ‖ allure f. *ou* marche f. silencieuse. / noisy

~ ‖ geräuschvoller Gang m. ‖ allure f.
ou marche f. bruyante. / in ~ order ‖
betriebsfertig ‖ prêt au service; en
ordre m. de marche. / ~-out of the saw
blade ‖ Verlaufen n. der Säge ‖ déviation
f. de la lame de scie. / parallel ~ ‖
Parallellaufen n.; Synchronlaufen n. ‖
fonctionnement m. en parallèle. / ~ of
the selector (Aut tel) ‖ Anlaufen n. des
Wählers ‖ mouvement m. du sélecteur.
/ silent ~ (Mach) ‖ ruhiger Gang m. ‖
marche f. silencieuse. / smooth ~ of the
motor ‖ leichter Gang m. des Motors ‖
marche f. silencieuse du moteur. / smooth
~ without any rocking (Vehicle) ‖ ruhiger
schaukelfreier Lauf m. ‖ marche f.
silencieuse et stable. / ~ smoothly ‖
stoßfrei laufend ‖ marchant aisément. /
~ the soap into the frames ‖ Ausgießen
n. der Seife in die Formkasten ‖ coulage
m. du savon dans les moules. / soft ~
(Mach) ‖ leichter Gang m. ‖ marche f.
douce *ou* légère. / steady ~ in parallel ‖
sicherer Parallellauf m. ‖ mise f. en
parallèle aisée. / synchronous ~ *see*
parallel ~. / uniform ~ (Mach) ‖ regelmäßiger Gang m. ‖ marche f. normale
ou régulière. / unquiet ~ (Mach) ‖ unruhiger Gang m. ‖ marche f. par à-coups.
/ ~ without load (Mach) ‖ Leerlauf m.
‖ marche f. à vide.

running balance of millstone ‖ laufendes
Gleichgewicht n. des Mühlsteins ‖ équilibre m. de la meule en marche. / ~ board
‖ Trittbrett n.; Laufbrett n. ‖ marchepied m. / ~ bridge ‖ Laufbrücke f. ‖
pont m. roulant. / ~ costs pl. ‖ Betriebskosten pl. ‖ frais mpl. de fonctionnement *ou* d'exploitation. / ~ day ‖ Liegetag m. ‖ jour m. de planche. / direction
of ~ ‖ Drehrichtung f. ‖ direction f. *ou*
sens m. de rotation. / ~ edge ‖ Fahr-
kante f. ‖ côté m. du roulement. / ~
expenses pl. ‖ Betriebskosten pl. ‖ frais
mpl. d'exploitation. / ~ gear ‖ Laufwerk
n. ‖ mécanisme m. de roulement. / ~
knot (Mar) ‖ Laufknoten m.; Schiebe-
knoten m.; Schifferknoten m. ‖ nœud
m. coulant. / ~ meter ‖ laufendes Meter
n. ‖ mètre m. courant. / per ~ meter
‖ je laufendes Meter n. ‖ par mètre m.
courant. / ~ number ‖ laufende Nummer
f. ‖ numéro m. d'ordre *ou* de série. / in
~ order ‖ betriebsfertig ‖ prêt au service.
/ ~ part of the after-body of a ship ‖
Schärfe f. des Hinterschiffes ‖ aissade f.
de la poupe d'un navire. / ~ position
‖ Fahrtstellung f. ‖ position f. de marche.
/ ~ sheave ‖ Führungsrolle f.; Leitrolle
f. ‖ galet m.; roulette f. / ~ speed (Mach)
‖ Umlaufgeschwindigkeit f.; Umdrehungszahl f. ‖ vitesse f. de rotation. /
~ speed (Vehicles) ‖ Fahrgeschwindig-

keit f. ‖ vitesse f. de marche. / clearance
limit of ~-through ‖ Durchfahrtprofil
n. ‖ gabarit m. de libre passage. / ~
time of an installation ‖ Betriebszeit f.
einer Anlage ‖ durée f. de service d'une
installation. / ~ title ‖ Kolumnentitel m.
‖ titre m. courant; ligne f. de tête. / ~
water ‖ fließendes Wasser n. ‖ eau f.
courante.

runway (Aero) ‖ Startbahn f. ‖ voie f. de
départ. / ~ (Bridge) ‖ Laufsteg m. ‖
passerelle f. / ~ (Mine) ‖ Schurre f. ‖
glissoir m. / ~ (Railw) ‖ Einschienen-
bahn f.; Schwebebahn f. ‖ voie f. suspendue; monorail aérien.

Rupert's drop ‖ Glasträne f.; Glastropfen
m. ‖ goutte f. de verre; larme f. batavique.

rupture ‖ Bruch m. ‖ rupture f. / ~ (Mine)
‖ Zubruchegehen n. ‖ éboulement m. /
~ of rails ‖ Schienenbruch m. ‖ rupture
f. des rails. / elongation at ~ ‖ Bruchdehnung f. ‖ allongement m. de rupture.

rural ‖ ländlich ‖ rural. / ~ line (Tel) ‖
Farmerleitung f. ‖ ligne f. rurale. / ~
mansion ‖Landhaus n. ‖ manoir m. / ~
plant (Electr) ‖ Landzentrale f. ‖ bureau
m. rural. / ~ postman ‖ Landbriefträger
m. ‖ facteur m. rural. / ~ telephone
plant ‖ Landfernsprechnetz n. ‖ téléphone m. rural; centrale f. rurale.

rush (Bot) ‖ Rohr n.; Binse f. ‖ jonc m. /
~ (Haste) ‖ Eile f.; Eilarbeit f. ‖ hâte
f.; affaire f. à la hâte. / ~ (Trade) ‖ Zudrang m.; Zulauf m. ‖ affluence f.; concours m. / ~ of current ‖ Stromstoß m.
‖ à-coup m. de courant. / varnished ~ ‖
Lackrohr n. ‖ jonc m. verni.

rush bridge ‖ Schilfbrücke f. ‖ pont m. de
joncs. / ~ cutter ‖ Binsenschneider m. ‖
coupeur m. de joncs. / ~ mat ‖ Binsen-
decke f. ‖ natte f. de jonc. / ~ plait ‖
Rohrgeflecht n. ‖ tresse f. de jonc. /
~ plate trunk ‖ Rohrplattenkoffer m. ‖
malle f. en plaques de jonc. / ~ reed
goods pl. ‖ Schilfwaren fpl. ‖ articles
mpl. en roseau. / ~ rope ‖ Binsenstrick
m. ‖ cordage m. de jonc. / ~ splitter ‖
Binsenspalter m. ‖ fendeur m. de jonc.

rusk ‖ Zwieback m. ‖ biscuit m. / ~ cutting
machine ‖ Zwiebackschneidmaschine f. ‖
machine f. à couper les biscottes.

russet upper leather ‖ rotgares Leder n.;
Fahlleder n. ‖ cuir m. à œuvre.

Russia-leather ‖ Juchtenleder n. ‖ cuir m.
de Russie.

rust, to ‖ rosten; verrosten; einrosten ‖ se
rouiller; s'enrouiller. / ~ into ‖ fest-
rosten ‖ se rouiller.

rust ‖ Rost m. ‖ rouille f. / to clean from ~
‖ rostfrei machen ‖ dérouiller. / fixed ~ ‖
fest anhaftender Rost m. ‖ rouille f. fixe.
/ iron ~ ‖ Eisenrost m. ‖ rouille f. de fer.

/ loose ~ ‖ loser Rost m. ‖ rouille f.
détachée. / to remove ~ ‖ Rost m. entfernen ‖ enlever la rouille f.; dérouiller.

rust cement ‖ Eisenkitt m.; Eisenzement
m. ‖ potée f. de fer *ou* de rouille.

rusted ‖ verrostet; rostig ‖ rouillé.

rust, resistance to the effects of the ~ ‖
widerstandsfähig gegen Rostbildung f. ‖
résistant aux effets de la rouille; résistance f. à la rouille. / to examine for ~
formation ‖ auf Rostansatz m. prüfen ‖
vérifier s'il n'y a pas de dépôts mpl. de
rouille.

rusting ‖ Rosten n.; Verrostung f. ‖ rouille
f. / ~ of the parts ‖ Verrosten n. *oder*
Verrostung f. der Teile ‖ rouille f. des
éléments.

rustless steel ‖ nichtrostender Stahl m. ‖
acier m. inoxydable.

rustlessness ‖ Rostsicherheit f. ‖ inoxyda-
bilité f.

rust preventative ‖ Rostschutzmittel n. ‖
moyen m. *ou* enduit m. antirouille;
moyen de protection contre la rouille. /
~ preventing ‖ rostschützend ‖ protégeant contre la rouille. / ~ preventive
see ~ preventative.

rustproof ‖ nicht rostend; widerstandsfähig gegen Rostbildung ‖ antirouille;
résistant aux effets de la rouille; résistance f. à la rouille. / ~ instrument ‖
nichtrostendes Instrument n. ‖ instrument m. inoxydable. / a steel alloy ~
even in damp air ‖ eine auch in feuchter
Luft nicht rostende Stahllegierung f. ‖
alliage m. d'acier inoxydable même à
l'air humide.

rust protection ‖ Rostschutz m. ‖ protection f. contre la rouille. / ~ removing
material ‖ Rostentfernungsmittel n. ‖
produit m. à dérouiller *ou* à enlever la
rouille. / ~ removing plant ‖ Ent-
rostungsanlage f. ‖ installation f. de
dérouillement. / ~ -resisting tube ‖ nichtrostendes Rohr n. ‖ tuyau m. inoxydable.

rusty ‖ rostig ‖ rouillé; enrouillé.

rut ‖ Gleis n.; Radspur f.; Furche f. ‖
ornière f.

Ruths steam storage plant with vertical
Ruths accumulators ‖ Ruthsdampfspeicheranlage f. mit stehenden Ruthsspeichern ‖ installation f. d'accumulateurs de chaleur Ruths comportant des
accumulateurs Ruths verticaux.

rye ‖ Roggen m. ‖ seigle m. / ~ flour ‖
Roggenmehl n. ‖ farine f. de seigle. /
~ flour mill ‖ Roggenmühle f. ‖ moulin
m. à seigle. / ~ straw ‖ Roggenstroh n. ‖
paille f. de seigle.

rynd ‖ Haue f.; Mühleisen n. ‖ anille f. /
balance ~ ‖ schwebende Haue f. ‖ anille f.
à balance. / stiff ~ ‖ feste Haue f. ‖ anille
f. fixe.

S

S-hook ‖ S-Haken ‖ crochet m. en S.
S-iron ‖ S-Eisen n. ‖ fer m. en S.
S-piece ‖ S-Stück n. ‖ pièce f. en S; tuyau m. en S; raccord m. en S.
S-rounding (Shipb) ‖ S-Bucht f. ‖ courbure f. *ou* courbe f. en S.
S-shaped ‖ S-förmig ‖ en forme d'S.
S-tube ‖ S-Stück n.; S-Rohr n. ‖ pièce f. en S; tuyau m. en S.
sabadilla seed ‖ Sabadillsamen m. ‖ sémence f. de sabadille.
sabine herb ‖ Sadebaumblatt n. ‖ herbe f. de sabine.
sable dyer ‖ Zobelfärber m. ‖ teinturier m. de zibeline.
sabot ‖ Holzschuh m. ‖ sabot m.
sabre ‖ Säbel m. ‖ sabre m. / ~ blade ‖ Säbelklinge f. ‖ lame f. de sabre. / ~ blade grinder ‖ Säbelschleifer m. ‖ aiguiseur m. de lames de sabres.
saccharate ‖ Saccharat n. ‖ saccharate m.
saccharic acid ‖ Zuckersäure f. ‖ acide m. saccharique.
sacchariferous ‖ zuckerhaltig ‖ saccharifère.
saccharify, to ‖ verzuckern; süßen ‖ saccharifier.
saccharimeter ‖ Zuckermesser m.; Saccharimeter m. ‖ saccharimètre m.
saccharin ‖ Saccharin n. ‖ saccharine f. / ~ food ‖ zuckerhaltiges Nährmittel n. ‖ préparation f. alimentaire au sucre.
saccharometer *see* saccharimeter.
saccholactic acid ‖ Milchsäure f.; Schleimsäure f. ‖ acide m. saccholactique.
sacerdotal ornament ‖ Priesterschmuck m. ‖ ornement m. sacerdotal.
sack ‖ Sack m. ‖ sac m. / to put into ~s ‖ sacken; in Säcke mpl. füllen ‖ mettre en sac m.; ensacher. / small ~ ‖ Beutel m. ‖ sachet m.; bourse f.; petit sac. / standard gauge ~ ‖ Normalsack m. ‖ sac m. normal. / ~ from tissue for the packing of goods ‖ Sack m. aus Gewebe zur Verpackung von Waren ‖ sac m. de tissu pour l'emballage de marchandises.
sack beating plant ‖ Sackausklopfanlage f. ‖ installation f. à battre les sacs. / ~ closing ‖ Sackverschluß m. ‖ fermeture f. à sacs. / ~ cloth ‖ Sackleinen n. ‖ toile f. à sacs. / ~ cloth weaving ‖ Sackleinenweberei f. ‖ tissage m. de toile à sacs. / ~ dusting machine ‖ Sackausklopfmaschine f. ‖ machine f. à épousseter les sacs.
sacked ‖ eingesackt ‖ ensaché.
sack filler, balance with ~ and sack lifter ‖ Schütt- und Absackwage f. ‖ balance f. à remplissage et à pesage.
sack filling apparatus ‖ Sackfüllgerät n. ‖ ensacheur m. / attachment for filling into ~s ‖ Absackvorrichtung f. ‖ dispositif m. d'ensachage *ou* à décharger aux sacs. / ~ filling screw (Mill) ‖ Absackschnecke f. ‖ hélice f. d'ensachage. / ~ gross weigher ‖ Bruttoabsackwage f. ‖ balance f. d'ensachage à poids brut.
sacking *see* sack cloth. / ~ mechanism ‖ Absackvorrichtung f. ‖ dispositif m.

d'ensachage. / ~ scale ‖ Sackwage f. ‖ bascule f. d'ensachage.
sack mender ‖ Sackflickerin f. ‖ raccommodeuse f. de sacs. / ~ net weigher ‖ Nettoabsackwage f. ‖ balance f. d'ensachage à poids net. / ~ packing machine ‖ Sackpackmaschine f. ‖ machine f. à emballer en sacs. / ~ posser (Mill) ‖ Säcker m. ‖ ensacheur m.; ensachoir m. / ~ shaking plant ‖ Sackausschüttelanlage f. ‖ installation f. d'époussetage des sacs. / ~ shutter of wire ‖ Sackverschluß m. aus Draht ‖ fermeture f. de sacs en fils de fer. / ~ tie ‖ Sackband n. ‖ cordon m. de sac; ficelle f. pour fermer les sacs. / transporting ~ trolley ‖ Sackkarren m. ‖ chariot m. à transporter les sacs. / ~ weigher ‖ Absackwage f. ‖ balance f. d'ensachage; ensacheuse f.
saddening (Textile) ‖ Irisdruck m. ‖ irisé m.; ombré m.
saddle, to ‖ satteln ‖ seller.
saddle ‖ Sattel m. ‖ selle f. / ~ of a bicycle ‖ Fahrradsattel m. ‖ selle f. de bicyclette. / English ~ without back board ‖ englischer Sattel m. ohne Rückenbogen ‖ selle f. anglaise à nez coupé. / ~ of heavy cavalry ‖ deutscher Sattel m. ‖ selle f. allemande *ou* de grosse cavallerie. / luggage ~ ‖ Handsattel m. ‖ sellette f. / ~ of a seam (Geol) ‖ Sattel m. eines Flözes ‖ selle f. d'une couche. / shaded ~ ‖ schattierter Sattel m. ‖ selle f. ombrifère.
saddle bag ‖ Satteltasche f. ‖ sacoche f. / ~ beams pl. ‖ Sattelhölzer npl. ‖ empanons mpl. / ~ blanket ‖ Woilach f. ‖ couverture f. en laine. / ~ bottom (Railw) ‖ Sattelboden m. ‖ fond m. en dos d'âne. / ~-bottomed self-emptying truck unloading to both sides ‖ Sattelbodenselbstentlader m. für Entladung nach beiden Seiten ‖ auto-déchargeur m. à fond en dos d'âne se vidant sur les deux côtés. / ~ cloth ‖ Satteldecke f. ‖ housse f. de main; couverture f. de selle. / ~ flap ‖ Satteltasche f. ‖ quartier m. de selle; sacoche f. / ~ middle ~ girth ‖ Mittelgurt m. ‖ seconde sangle f. / ~ grate (Metal) ‖ Sattelrost m.; Schweinerücken m. ‖ grille f. à deux plaines. / ~ key ‖ Hohlkeil m. ‖ clavette f. conique creuse. / ~ maker *see* saddler. / ~ nail ‖ Sattelnagel m.; Sattelzwecke f. ‖ clou m. de sellier. / ~ pin socket ‖ Sattelmuffe f. ‖ manchon m. de support de la selle. / ~ plate ‖ Sattelscheibe f. ‖ rondelle f. à ados.
saddler ‖ Sattler m. ‖ sellier m. / carriage ~ ‖ Wagensattler m. ‖ sellier m. en voitures.
saddle roof ‖ Satteldach n. ‖ toit m. en batière; toit à deux égouts. / ~ room ‖ Geschirrkammer f.; Sattelkammer f. ‖ sellerie f.
saddler's awl ‖ Ahle f.; Pfriem m. ‖ alêne f. / ~ goods pl. *see* saddlery. / ~ hammer ‖ Sattlerhammer m. ‖ marteau m. de sellier. / ~ and shoemaker's knife ‖ Sattler- und Schustermesser n. ‖ couteau m. à

cuir. / ~ leather ‖ Sattlerleder n. ‖ cuir m. de sellier. / ~ machine ‖ Sattlereimaschine f. ‖ machine f. de sellerie. / ~ needle ‖ Sattlernadel f. ‖ aiguille f. du sellier. / ~ pliers pl. ‖ Sattlerzange f. ‖ pince f. pour sellier. / ~ sewing machine ‖ Sattlernähmaschine f. ‖ machine f. à coudre pour selliers. / ~ tack ‖ Sattlernagel m.; Sattlerzwecke f. ‖ clou m. de sellier. / ~ tool ‖ Sattlerwerkzeug n. ‖ outil m. pour selliers.
saddlery ‖ Sattlerwaren fpl. ‖ articles mpl. de sellerie; sellerie f.
saddles, harnesses and riding equipment ‖ Sattel- und Reitzeug n. ‖ équipement m. d'équitation.
saddle tree ‖ Sattelbaum m. ‖ arçon m.
sadiron ‖ Bügeleisen n. ‖ fer m. à repasser. / ~ for spirit ‖ Spiritusbügeleisen n. ‖ fer m. à repasser à l'alcool.
safe (Trade) ‖ kreditwürdig ‖ digne de crédit m. / ~ from vibrations ‖ erschütterungsfrei ‖ exempte de vibrations fpl.
safe ‖ Geldschrank m.; Safe m. ‖ coffre-fort m. / concrete-made ~ ‖ Geldschrank m. aus Beton ‖ coffre-fort m. en béton. / fireproof ~ ‖ feuerfester Geldschrank m. ‖ coffre-fort m. incombustible.
safe deposit ‖ Wertgelaß n.; Tresor m. ‖ chambre f. forte. / ~ armour plate ‖ Tresorplatte f. ‖ plaque f. pour caveau de sûreté. / fittings pl. for ~s ‖ Geldschrankbeschläge mpl. ‖ ferrures fpl. de coffre-forts.
safeguard, as a ~ against ‖ zur Sicherheit f. gegen ‖ en garantie f. de. / ~ of a locomotive ‖ Schienenräumer m. ‖ chasse-pierres m.; garde m. d'une locomotive.
safeguarding ‖ Sicherung f. ‖ sûreté f.; assurance f.; garantie f. / ~ by shearing pin ‖ Abscherstiftsicherung f. ‖ dispositif m. de sécurité par goupille. / railway ~ plant ‖ Eisenbahnsicherungsanlage f. ‖ installation f. de sécurité de chemin de fer.
safe investment ‖ mündelsichere Kapitalanlage f. ‖ placement m. de tout repos *ou* de père de famille. / ~ key ‖ Geldschrankschlüssel m. ‖ clef f. de coffre-fort. / ~ lock ‖ Geldschrankschloß n. ‖ serrure f. pour coffre-fort. / ~ plate for ~s ‖ Geldschrankblech n. ‖ tôle f. pour coffres-forts. / protective equipment for ~s ‖ Tresorsicherung f. ‖ contact m. de sûreté contre l'effraction. / without exceeding the ~ stress in the material ‖ ohne die Sicherheitsbeanspruchung f. des Werkstoffes zu überschreiten ‖ sans excéder l'effort m. de sécurité des matériaux.
safety ‖ Sicherheit f. ‖ sûreté f.; sécurité f. / service for the ~ of aircraft ‖ Flugsicherungsdienst m. ‖ service m. de sécurité aérienne. / ~ against explosion accidents ‖ Explosionssicherheit f. ‖ sûreté f. contre les explosions. / ~ of working ‖ Betriebssicherheit f. ‖ sécurité f. de service.
safety apparatus ‖ Sicherheitsvorrichtung f. ‖ appareil m. de sûreté. / ~ for cages

(Mine) || Fangvorrichtung f. für Förderkörbe || parachute m. pour cages d'extraction. / ~ for winding engines || Sicherheitsapparat m. für Fördermaschinen || appareil m. de sûreté pour machines d'extraction.

safety appliance of a lift || Fangvorrichtung f. eines Fahrstuhles || parachute m. d'un ascenseur. / ~ arrangement for fire alarm systems || Sicherheitseinrichtung f. in der Feuermeldeanlage || disposition f. de sécurité dans les installations d'avertisseurs d'incendie. / ~ belt (Aero) || Anschnallgurt m. || ceinture f. de sangle. / ~ belt (Mar) || Rettungsgürtel m. || ceinture f. de sauvetage. / ~ board || Sicherheitsbrett n. || planchette f. de sûreté. / ~ boat || Rettungsboot n. || embarcation f. de sauvetage. / ~ bolt || Sperriegel m. || verrou m. de sûreté. / ~ bottle || Sicherheitsflasche f. || flacon m. de garde. / ~ bow (El line) || Fangbügel m. || étrier m. de garde. / ~ brake (Crane) || Sicherheitsbremse f. || frein m. de sûreté. / ~ brake (Railw) || Notbremse f. || frein m. de sûreté. / spring ~ brake || Federsicherheitsbremse f. || frein m. de sûreté à ressort. / ~ buoy || Rettungsboje f. || bouée f. de sauvetage. / ~ catch || Sicherheitsriegel m. || cliquet m. de sûreté. / ~ chain || Sicherheitskette f. || chaîne f. de sûreté. / ~ charge (Acc) || Sicherheitsladung f. || charge f. de sûreté. / ~ code || Sicherheitsvorschriften fpl. || réglements mpl. de prévoyance contre les accidents. / ~ coefficient || Sicherheitskoeffizient m. || coefficient m. de sûreté. / ~ coil || Schutzspule f. || bobine f. de protection. / ~ connector || Sicherheitseinrücker m. || éclencheur m. de sûreté. / ~ contact for safe || Geldschranksicherung f. || contact m. de sûreté pour coffre-fort. / ~ coupling (Electr) || Überlastungskupplung f. || accouplement m. de surcharge. / ~ coupling (Railw) || Sicherheitskupplung f. || attelage m. de sûreté. / ~ crook || Sicherheitshaken m. || crochet m. de sûreté. / ~ cutout see fuse. / ~ detonator || Sicherheitszünder m. || détonateur m. de sûreté.

safty device || Sicherungseinrichtung f.; Schutzvorrichtung f. || appareil m. de sûreté ou de protection. / ~ for door locks || Türschloßsicherung f. || dispositif m. de sûreté pour serrures de portes. / gas ~ || Gassicherheitsapparat m. || appareil m. de sûreté contre les gaz. / mine ~ || Grubensicherungseinrichtung f. || installation f. de sûreté pour mines.

safety disconnector || Notausrücker m. || débrayeur m. supplémentaire. / ~ doorlocking device || Sicherheitstürverschluß m. || dispositif m. de sécurité de fermeture pour portes. / ~ earthing || Sicherheitserdung f. || prise f. de terre de sécurité. / ~ embankment || Schutzdamm m. || digue f. de garantie. / ~ explosive || Sicherheitssprengstoff m. || explosif m. de sûreté. / factor of ~ || Sicherheitsfaktor m. || coefficient m. de sécurité. / ~ flange || Sicherungsflansch m. || bride f. de sécurité. / ~ type fountain pen || Sicherheitsfüllfederhalter m. || stylographe m. ou porte-plume m. réservoir de sûreté. / ~ fuse (Electr) || Sicherung f.; Abschmelzsicherung f. || fusible m. de sûreté; coupe-circuit à fusible. / ~ fuse (Mine) || Sicherheitszündschnur f. ||

mèche f. de sûreté. / machine for producing non-oxidizing ~ gases || Erzeugungsmaschine f. für nicht oxydierende Schutzgase || machine f. à produire les gaz protecteurs non oxydants. / ~ gripper (Crane) || Sicherheitsbügel m. || étrier m. de sûreté. / ~ guard || Schutzvorrichtung f. || dispositif m. de protection. / ~ hook (Railw, coupling) || Kupplungssicherheitshaken m. || crochet m. d'attelage de sûreté. / ~ lamp || Sicherheitslampe f. || lampe f. de sûreté. / cabin for the examination of the ~ lamps (Mine) || Lampenstube f. || lampisterie f. / ~ lock || Sicherheitsschloß n. || serrure f. de sûreté; anti-vol m. / ~ mask || Gesichtsmaske f. || masque m. de sûreté. / ~ match || Sicherheitszündholz n. || allumette f. de sûreté. / ~ mirror || Schutzspiegel m. || miroir m. protecteur. / ~ pillar (Mine) || Sicherheitspfeiler m. || pilier m. de sûreté. / ~ pin || Sicherheitsnadel f. || épingle f. de nourrice. / ~ pipe || Sicherheitsröhre f. || tuyau m. de sûreté. / ~ plug || Bleipfropf m. || plomb m. de sûreté. / ~ powder see ~ explosive. / ~ rail || Schutzschiene f.; Leitschiene f.; Sicherheitsschiene f. || contre-rail m. / ~ razor || Rasierapparat m. || rasoir m. mécanique ou de sûreté. / ~ regulations pl. || Sicherheitsvorschriften fpl. || prescriptions fpl. relatives à la sécurité. / ~ rope (Railw) || Notleine f.; Zugleine f. || corde f. de sûreté. / ~ rules pl. see ~ regulations. / ~ screw || Sicherungsschraube f. || boulon m. de retenue. / ~ sliding bolt || Sicherheitsschubriegel m. || targette f. pêne de sûreté. / ~ spark gap || Sicherheitsfunkenstrecke f. || parcours m. d'étincelle de sûreté. / ~ specifications pl. (Trade) || Sicherheitsvorschriften fpl. || instructions fpl. de sécurité. / ~ spectacles || Schutzbrille f. || lunettes fpl. de protection. / ~ spring for locomotive valves || Lokomotivsicherheitsventilfeder f. || ressort m. de sûreté pour soupapes de locomotives. / ~ switch || Entgleisungsweiche f.; Sicherungsweiche f. || dérailleur m.; changement m. de voie de sûreté. / ~ tap (Steam) || Sicherheitshahn m. || robinet m. de sûreté. / ~ train || Rettungszug m.; Hilfszug m. || train m. de secours.

safety valve || Sicherheitsventil n. || soupape f. de sûreté. / external ~ || äußeres Sicherheitsventil n. || soupape f. de sûreté extérieur. / lever ~ || Sicherheitsventil n. mit Hebel || soupape f. de sûreté à levier. / spring ~ || Sicherheitsventil n. mit Feder || soupape f. de sûreté à ressort. / ~ of a steam boiler || Dampfkesselsicherheitsventil n. || soupape f. de sûreté de chaudière à vapeur.

safety valve weight || Sicherheitsventilbelastung f. || contre-poids m. de soupape de sûreté.

safety whistle || Alarmpfeife f. || sifflet m. d'alarme. / ~ winch || Sicherheitswinde f. || treuil m. de sûreté.

safe working, absolutely ~ || unbedingte Betriebssicherheit f. || sécurité f. absolue du fonctionnement.

safflorite || Safflorit m.; faseriger Speiskobalt m.; Eisenkobaltkies m. || safflorite f.

safflower || Safflorpflanze f.; Färberdistel f.; falscher Safran m. || carthame m.; safran m. bâtard. / Indian ~ || Gelb-

wurzel f.; Curcuma f. || safran m. ou souchet m. des Indes; curcuma m.

saffron || Safran m. || safran m. / ~ of antimony || Spießglanzsafran m. || crocus m. d'antimoine. / ~ mill || Safranmühle f. || moulin m. à safran.

sag, to || sich senken || fléchir; abaisser; descendre. / ~ to leeward || nach Lee sacken || aller en dérive f.

sag (Electr) || Leitungshang m. || flèche f. / ~ of cable || Durchhang m. des Seiles || flèche f. du câble.

sage || Salbei f. || sauge f. / ~ leaf || Salbeiblatt n. || feuille f. de sauge.

sagenite || Rutil m.; Sagenit m. || sagénite f.

sage oil || Salbeiöl n. || essence f. de sauge.

sagged || gesenkt || fléchi; abaissé; descendu.

sagger || Kapsel f. || cassette f.

sagging || sich senkend || fléchissant; descendant.

sago || Palmenstärke f.; Sago m. || sagou m.; tapioca m. / equipment for ~ works pl. || Sagofabrikeinrichtung f. || équipement m. de fabrique de tapioca.

sag, table of ~s pl. (Mech) || Durchhangtabelle f. || table f. des flèches.

sahlite || Sahlit m.; Malakolith m. || malacolithe m.; pyroxène m.

sail, to || segeln || aller à la voile; cingler; faire route; naviguer. / ~ (Depart) || auslaufen || sortir. / ~ on a bowline || beim Winde m. segeln || aller ou courir à la bouline; courir vers le vent. / ~ in || einsegeln || entrer à la voile. / ~ out || aussegeln || sortir à la voile. / ~ before the wind || vor dem Winde m. laufen oder segeln || courir vent m. arrière. / ~ by the wind see ~ on a bowline.

sail || Segel n. || voile f. / ~ (Windmill) || Flügel m. || volant m.; aile f.; palette f. / after ~ || Achtersegel n. || voile f. arrière. / boat's ~ || Bootssegel n. || voile f. d'embarcation. / curved ~ (Windmill) || gewölbter Flügel m. || palette f. cintrée. / to douse a ~ see to strike a ~. / drawing ~ || tragendes Segel n. || voile f. qui porte. / drift ~ || Schleppsegel n.; Treibsegel n. || ancre f. flottante. / fixed ~ (Windmill) || unbeweglicher oder fester Flügel m. || aube f. ou palette f. fixe. / fore ~ || Focksegel n. || misaine f. / furled ~ || festgemachtes Segel n. || voile f. serrée. / to haul down a ~ || ein Segel n. niederholen || carguer une voile. / to hand a ~ see to take a ~ in. / loose ~ || losgemachtes Segel n. || voile f. larguée. / lower ~ || Untersegel n. || basse voile f. / to lower down a ~ see to strike a ~. / main ~ || Großsegel n. || grande voile f. / to make ~ see to set ~. / ready to ~ || segelklar || prêt à prendre la mer. / to set ~ || Segel npl. setzen || déployer les voiles fpl.; mettre les voiles fpl. au vent. / a set ~ || beigesetztes Segel n. || une voile f. établie. / all ~s pl. set || unter vollen Segeln npl. || à pleines voiles fpl.; toutes voiles fpl. dehors. / ~s of a ship || Segelwerk n. eines Schiffes || voilure f. d'un navire. / to shorten ~s pl. || Segel npl. kürzen || diminuer de voiles fpl. / smooth ~ (Windmill) || glatter Flügel m. || palette f. plane. / spare ~ || Reservesegel n. || voile f. de rechange. / square ~ || Quersegel n. || voile f. carrée. / upper corner of a square ~ || Nock f. eines Quersegels || empointure f. ou point m. d'envergure d'une voile carrée. / stay ~ ||

Stagsegel n. ‖ voile f. d'étai. / steel ~
(Windmill) ‖ Stahlflügel m. ‖ palette f.
d'acier. / stowed ~ ‖ festgemachtes Segel
n. ‖ voile f. serrée. / to strike a ~ ‖ ein
Segel streichen ‖ amener une voile. /
studding ~ ‖ Leesegel n. ‖ bonnette f. /
to take a ~ in ‖ ein Segel n. bergen *oder*
einnehmen ‖ serrer une voile. / top ~ ‖
Obersegel n. ‖ voile f. haute. / topgallant
~ ‖ Bramsegel n. ‖ perroquet m. / under
~ ‖ unter Segel n. ‖ sous voiles. / to get
under ~ ‖ unter Segel n. gehen ‖ mettre
à la voile *ou* sous voiles. / to put under ~
see to get under ~. / unfurled ~ *see*
loose ~. / wooden ~ (Windmill) ‖
Holzflügel m. ‖ palette f. en bois.

sail, area of ~s ‖ Segelfläche f. ‖ surface f.
de voilure. / bunt of a ~ ‖ Buk m. eines
Segels ‖ fonds mpl. d'une voile. / varnished ~ canvas ‖ gefirnißtes Segeltuch
n. ‖ toile f. à voile enduite de vernis.

sail-cloth ‖ Segeltuch n. ‖ voile f.; toile f.
à voile. / ~ ware ‖ Segeltuchware f. ‖
articles mpl. de toile à voile. / ~ weaving
‖ Weben n. des Segeltuches ‖ tissage m.
de toile à voile.

sailer *see* sailing vessel.

sail, framework of the ~ (Windmill) ‖
Flügelgerippe n. ‖ ossature f. de l'aile.

sailing (Departure) ‖ Auslauf m. ‖ départ
m. / to get ready for ~ ‖ sich segelfertig
oder segelklar machen ‖ être en appareillage m.

sailing boat ‖ Segelboot n. ‖ bateau m. à
voiles. / ~ flight ‖ Segelflug m. ‖ vol m.
à voile. / ~ ship *see* sail'ng vessel.

sailing vessel ‖ Segelschiff n. ‖ voilier m.;
navire m. à voiles. / ~ with auxiliary
steam power ‖ Schiff n. mit Hilfsdampf-
maschine ‖ navire m. mixte. / Diesel
engined ~ ‖ Dieselmotorsegelschiff n. ‖
voilier m. à moteur auxiliaire Diesel.

sailing yacht ‖ Segeljacht f. ‖ yacht m.
à voile.

sailmaker ‖ Segelmacher m. ‖ voilier m.

sailmaker's bodkin ‖ Segelmacherpfriem
m. ‖ marprime f.; poinçon m. de voilier.
/ ~ crew ‖ Segelmachersgasten mpl. ‖
ouvriers mpl. voiliers. / ~ palm ‖ Hand-
platte f.; Segelhandschuh m. ‖ paumelle
f. *ou* paumet m. des voiliers.

sail making ‖ Segelmacherhandwerk n. ‖
voilerie f. / ~ mast (Windmill) ‖ Mast m.
oder Rute f. der Flügel ‖ bras m. des
ailes. / ~ needle ‖ Segelnadel f. ‖ aiguille f.
à voiles. / large ~ needle ‖ Gatnadel f. ‖
aiguille f. à faire des œillets.

sailor ‖ Matrose m. ‖ matelot m.

sailor's bag ‖ Kleidersack m. ‖ sac m. de
matelot. / ~ knot ‖ Kreuzknoten m. ‖
nœud m. marin. / ~ outfit ‖ Seemanns-
ausrüstung f. ‖ costumes mpl. et accessoires mpl. pour marins.

sail, plan of ~s ‖ Segelplan m.; Segelriß m.
‖ plan m. de voilure. / ~ room ‖ Segel-
koje f. ‖ chambre f. *ou* soute f. aux
voiles. / ~ tackle ‖ Segeltakel m. ‖ cartahu m. pour enverguer les huniers. /
~ twine ‖ Segelgarn n. ‖ fil m. à voile. /
working moment of the ~s ‖ Arbeits-
moment n. der Flügel ‖ moment m.
moteur des ailes.

sal acetosellæ ‖ Kleesalz n.; Sauerkleesalz
n.; oxalsaures Kalium n. ‖ sel m.
d'oseille; bioxalate m. de potasse.

salad ‖ Salat m. ‖ salade f. / ~ oil ‖ Speiseöl
n. ‖ huile f. de table *ou* à salade. / ~ oil
from Provence ‖ Provenceröl n. ‖ huile f.
de Provence. / ~ servers pl. ‖ Salatbesteck n. ‖ couvert m. à salade.

salamander (Metal) ‖ Ofensau f. ‖ loup m.

sal ammoniac ‖ Salmiak m. ‖ sel m.
ammoniac. / ~ pastilles pl. ‖ Salmiak-
pastillen pl. ‖ pastilles fpl. au sel ammoniac. / ~ plant ‖ Salmiakanlage f. ‖
installation f. de sel ammoniac. / ~ slag ‖
Salmiakschlacke f. ‖ scories fpl. du sel
ammoniac.

salary ‖ Gehalt n. ‖ salaire m.; appointements mpl. / to earn a ~ ‖ Gehalt n. beziehen ‖ toucher un salaire. / to be entit-
ed to ~ ‖ Anspruch m. auf Gehalt haben ‖ avoir droit m. à un traitement.

salary, increase of ~ ‖ Gehaltszulage f. ‖
augmentation f. de salaire. / reduction
of ~ ‖ Gehaltsabzug m. ‖ réduction f. sur
le traitement.

sale ‖ Verkauf m.; Vertrieb m.; Absatz m.
‖ vente f.; débit m. / ~ by auction ‖ Verkauf m. an den Meistbietenden ‖ vente f.
à l'enchère. / ~ in blank ‖ Blankogeschäft
n. ‖ marché m. à découvert. / brisk ~ ‖
schneller Absatz m. ‖ bon débit m. /
commanding a ready ~ ‖ leicht verkäuflich ‖ d'un bon débit m. / compulsory ~ by
auction ‖ Zwangsversteigerung f. ‖ vente
f. forcée. / exclusive ~ ‖ Alleinverkauf m.
‖ vente f. exclusive. / to expose goods pl.
for ~ ‖ Waren fpl. zum Verkauf aufstellen ‖ étaler des marchandises fpl.
pour vente. / for ~ ‖ verkäuflich ‖ à
vendre. / forced ~ ‖ Zwangsverkauf m. ‖
vente f. forcée. / to offer for ~ ‖ ausbieten ‖ mettre en vente. / to find
prompt ~ ‖ guten Absatz m. finden ‖ être
d'une bonne défaite *ou* de bonne vente. /
public ~ ‖ Auktion f. ‖ vente f. publique;
enchère f. / rapid ~ ‖ schneller Absatz
m. ‖ bon débit m. / ready ~ ‖ schneller
Umsatz m. ‖ vente f. *ou* débit m. rapide.
/ ~ by retail ‖ Einzelverkauf m. ‖ vente f.
en détail; débit m. / ~ by royalty ‖ Verkauf m. nach Taxe ‖ vente f. au tarif. /
to be of slow ~ ‖ schlechten *oder* schwierigen Absatz m. haben ‖ la marchandise
est d'un placement difficile. / to meet
with a slow ~ *see* to be of slow ~. /
~ of standing trees ‖ Verkauf m. auf dem
Stamm ‖ vente f. sur pied. / total ~ ‖
Gesamtabsatz m. ‖ total m. de vente. /
~ of wood ‖ Holzverkauf m. ‖ vente f.
de bois.

saleable ‖ marktgängig; verkäuflich ‖
de vente courante *ou* facile.

sale, conditions pl. of ~ ‖ Verkaufsbedingungen pl. ‖ conditions fpl. de vente. /
deed of ~ ‖ Verkaufsurkunde f. ‖ acte
m. de vente.

salep ‖ Salep m. ‖ salep m.

sales pl. ‖ Absatz m. ‖ débit m.; vente f.
/ account of ~ ‖ Verkaufsabrechnung
f. ‖ compte m. de vente. / ~ analysis
(Comm) ‖ Verkaufsanalyse f. ‖ analyse f. des ventes. / ~ force ‖ Verkaufs-
personal n. ‖ personal m. occupé à la
vente.

salesman ‖ Verkäufer m. ‖ vendeur m.

sales manager ‖ kaufmännischer Direktor
m. ‖ chef m. de l'exploitation. / ~ note ‖
Schlußschein m. ‖ bordereau m. de
vente.

sales organization ‖ Verkaufsorganisation
f. ‖ organisation f. de la vente. / ~ of a
concern ‖ Verkaufsorganisation f. eines
Konzerns ‖ organisation f. de vente d'un
groupe.

sales sheet ‖ Verkaufszettel m. ‖ fiche f.
de vente. / ~ value ‖ Verkaufswert m. ‖
valeur f. venale.

salicine ‖ Salizin n. ‖ salicine f.

salicyl ‖ Salizyl n. ‖ salicyle f.

salicylate ‖ Salizylpräparat n. ‖ salicylate
m.

salicylic acid ‖ Salizylsäure f. ‖ acide m.
salicylique.

saliferous ‖ salzhaltig ‖ salifère.

salifiable base ‖ salzfähige Base f. ‖ base
f. salifiable.

salina *see* saline (Subst).

saline ‖ salzig; salzhaltig ‖ salin; salzu-
gineux. / ~ solution ‖ Salzlösung f. ‖ solution f. saline.

saline (Subst) ‖ Saline f.; Salzwerk n.
‖ saline f.; saunerie f.

salinometer ‖ Salzwage f.; Solewage f. ‖
pèse-sel m.; salinomètre m.

salinous ‖ salzig; salzhaltig ‖ salin; salsugi-
neux.

salmon ‖ Lachs m. ‖ saumon m. / ~ fisherman ‖ Lachsfischer m. ‖ pêcheur m. de
saumons. / ~ trout ‖ Lachsforelle f. ‖
truite f. saumonnée.

saloon ‖ Salon m. ‖ salon m. / ~ carriage
‖ Salonwagen m. ‖ wagon-salon m. / ~
deck ‖ Salondeck n. ‖ pont-salon m.

sal prunella ‖ Prunellensalz n. ‖ sel m.
prunelle; cristal m. minéral.

salt, to ‖ salzen; einsalzen ‖ saler.

salt ‖ gesalzen; salzig ‖ salé. / ~ butter ‖
gesalzene Butter f. ‖ beurre m. salé.

salt ‖ Salz n. ‖ sel m. / acid ~ ‖ saures
Salz n. ‖ sel m. acide. / alkaline ~ ‖
alkalisches Salz n. ‖ sel m. alkalin. /
basic ~ ‖ basisches Salz n. ‖ sel m.
basique. / bath ~ ‖ Badesalz m. ‖ sel
m. pour bains. / black ~ ‖ Ochras m.;
Pottaschenfluß m.; rohe Pottasche f. ‖
salin m.; cassoudes fpl.; casottes fpl. /
artificial Carlsbad ~ ‖ künstliches Karls-
bader Salz n. ‖ sel m. de Carlsbad artificiel. / caustic ~ ‖ Beizsalz n. ‖ sel m.
caustique. / common ~ ‖ Kochsalz n. ‖
sel m. commun. / double ~ ‖ Doppelsalz
n. ‖ sel m. double. / dried ~ ‖ getrocknetes Salz n. ‖ sel m. séché. / feed ~ ‖
Nährsalz n. ‖ sel m. nutritif. / iron ~ ‖
Eisensalz n. ‖ sel m. de fer. / neutral ~
‖ neutrales Salz n. ‖ sel m. neutre. /
preserving ~ ‖ Konservierungssalz n. ‖
sel m. à conserver. / ~ of rare earths ‖
Salz n. von seltenen Erden ‖ sel m. de
terres rares. / rock ~ ‖ Steinsalz n. ‖ sel
m. gemme *ou* de roche. / ~ of silver
oxide ‖ Silbersalz n. ‖ sel m. d'argent. /
~ of sorrel ‖ Kleesalz n. ‖ sel m. d'oseille.

salt and pepper ‖ Doppelstreuer m. ‖ salière
f. et poivrière f. combinée.

salt bath furnace ‖ Salzbadofen m. ‖ four
m. au bain de sel. / ~ beef ‖ Pökelfleisch
n. ‖ bœuf m. salé. / ~ breaker *see* ~ crusher.
/ ~ cake ‖ Salzkuchen m. ‖ salignon m.
/ ~ case (Mar) ‖ Salzdose f.; Salzlade f.
‖ boîte f. à sel. / ~ caster ‖ Salzstreuer
m. ‖ poudrière f. à sel. / ~ content of
sea breeze ‖ Salzgehalt m. des See-
windes ‖ teneur f. en sel de la brise de
mer *ou* du vent de mer. / ~ crusher ‖
Salzbrecher m. ‖ broyeur m. *ou* concasseur m. de sel. / ~ dissolver ‖ Salz-
löser m. ‖ dissolveur m. de sel. / ~ dome
‖ Salzhorst m. ‖ redressement m. salin.

salted ‖ gesalzen ‖ salé. / ~ for preservation
during conveyance ‖ zur Haltbar-
machung während der Beförderung ge-

salzen ‖ salé pour assurer la conservation pendant le transport.

salter, skin ~ ‖ Häutesalzer m. ‖ saleur m. de peaux.

saltern ‖ Saline f.; Salzsiedewerk n.; Salzwerk n. ‖ salin m.; saline f.; saunerie f.

salt gage see salinometer. / ~ **garden** ‖ Salzgarten m.; Seesalzwerk n. ‖ marais m. salant. / ~ **gauge** see salinometer. / ~ **glace** (Pott) ‖ Salzlasur f. ‖ vernis m. en sel commun. / ~ **grained sludge** ‖ salzbreiartige Masse f. ‖ boue f. saline. / ~ **grained sludge with admixture of clay** ‖ salzbreiartige Masse f. mit Tonbeimengung ‖ boue f. saline additionée d'argile. / ~ **heap** ‖ Salzhaufen m. ‖ pilot m. de sel.

salting see salt pond.

salt lake ‖ Salzsee m. ‖ lac m. salant ou salifère. / ~ **load** (Geol) ‖ Salzflöz m., n. ‖ couche f. de sel. / ~ **maker** ‖ Salzsieder m. ‖ salinier m.; saunier m. / ~ **marsh** see **garden**. / ~ **meat** ‖ Pökelfleisch n. ‖ salaison f.; viande f. salée. / **cooling cell for** ~ **meat** ‖ Kühlzelle f. für Pökelfleisch ‖ chambre f. froide pour des salaisons. / ~ **mine** ‖ Salzbergwerk n. ‖ mine f. de sel. / ~ **mining** ‖ Salzbergbau m. ‖ exploitation f. des mines de sel. / ~ **mill** ‖ Salzmühle f. ‖ moulin m. à sel.

saltpetre ‖ Kalisalpeter m. ‖ potasse f. nitratée; salpêtre m. / ~ **of the first boiling** ‖ Salpeter m. vom ersten Sud ‖ salpêtre m. brut ou de première cuite. / **native** ~ ‖ Gayerde f.; Gaysalpeter m.; Kehrsalpeter m. ‖ salpêtre m. de houssage. / **swept** ~ see **native** ~.

saltpetre lye ‖ Salpeterlauge f. ‖ lessive f. de salpêtre. / ~ **plantation** ‖ Salpeterplantage f. ‖ nitrière f.; salpêtrière f. / ~ **refiner** ‖ Salpetersieder m. ‖ raffineur m. de salpêtre.

salt pond ‖ Salzsumpf m.'‖ marais m. salant.

saltpork ‖ Pökelschweinefleisch n.; Salzschweinefleisch n. ‖ porc m. salé.

salt preparation ‖ Salzaufbereitung f. ‖ préparation f. du sel. / ~ **producing plant** ‖ Salzgewinnungsanlage f. ‖ installation f. de saunerie. / ~ **provisions** pl. ‖ (Ein-)Gesalzenes n. ‖ vivres mpl. de campagne; salaison m. / ~ **refinery** ‖ Salzraffinerie f. ‖ raffinerie f. de chlorure de sodium. / ~ **shed** ‖ Salzspeicher m. ‖ magasin m. à sel. / **concentrated** ~ **solution** ‖ gesättigte Kochsalzlösung f. ‖ solution f. de sel de cuisine saturée. / **spirit of** ~ ‖ Salzsäure f. ‖ esprit m. de sel ou acide chlorhydrique. / ~ **spoon** ‖ Salzlöffel m. ‖ cuiller m. à sel. / ~ **spring** ‖ Salzquelle f. ‖ source f. salée. / ~ **stirrer** ‖ Salzkrücke f. ‖ râble m. du saunier. / **stock-shaped body of** ~ see ~ **dome**. / ~ **stove** ‖ Salzsiedeofen m. ‖ poêle f. à sel. / ~ **stratum** see ~ **load**. / ~ **water** ‖ Salzwasser n. ‖ eau f. salée. / ~ **workman** ‖ Salinenarbeiter m. ‖ ouvrier m. de saline ou saunier.

saltworks pl. ‖ Saline f. ‖ saline f.; saunerie f. / ~ **owner** ‖ Salinenbesitzer m.; Salzwerksbetreiber m. ‖ propriétaire m. saunier.

salutary lighting apparatus ‖ Lichtheilapparat m. ‖ appareil m. d'éclairage thérapeutique.

salute ‖ Salut m. ‖ salut m.

salvage (Mar) ‖ Bergung f. ‖ sauvetage m. / ~ **of vessels** ‖ Schiffsbergung f. ‖ sauvetage m. de navires.

salvage apparatus ‖ Rettungsapparat m. ‖ appareil m. de sauvetage. / **charges** pl. **for** ~ see ~ **money**. / ~ **company** ‖ Bergungsgesellschaft f. ‖ société f. de sauvetage. / ~ **crane ship** ‖ Bergungskranschiff n. ‖ bateau m. de sauvetage à grue. / ~ **money** ‖ Bergegeld n. ‖ droit m. de sauvetage. / ~ **steamer** ‖ Bergungsdampfer m. ‖ vapeur m. de sauvetage.

salvarsane ‖ Salvarsan n. ‖ salvarsane m.

salve like ‖ salbenartig ‖ analogue à une pommade; pateux.

salver ‖ Auftragbrett n. ‖ plateau m. de service.

sal volatile ‖ Riechsalz n. ‖ sel m. volatile.

samarskite ‖ Samarskit m.; Uranotantal m.; Yttroilmenit m. ‖ samarskite f.

sample, to ‖ ausproben ‖ essayer; éprouver. / ~ (Metal) ‖ Probe f. nehmen ‖ échantillonner.

sample ‖ Muster n.; Probe f. ‖ échantillon m. / **average** ~ ‖ Durchschnittsprobe f. ‖ échantillon m. moyen. / ~ **of bending work** ‖ Biegeprobe f. ‖ éprouvette f. de pliage. / **counter** ~ ‖ Gegenmuster n. ‖ échantillon m. de contrôle. / **equal to** ~ ‖ dem Muster n. entsprechend ‖ conforme à l'échantillon. / ~ **of fabric** (Weav) ‖ Arbeitsmuster n. ‖ échantillon m. de travail. / ~ **taken at hazard** ‖ Stichprobe f.; échantillon m. pris au hazard. / **ladled-out** ~ (Metal) ‖ Schöpfprobe f. ‖ échantillon m. puisé. / **not according to** ~ ‖ nicht dem Muster n. entsprechend ‖ non conforme à l'échantillon m. / ~ **of no value** ‖ Muster n. ohne Wert ‖ échantillon m. sans valeur. / ~ **of ores** ‖ Erzprobe f. ‖ échantillon m. ou prise f. d'essai de minerais. / ~ **of certain quality** ‖ Qualitätsmuster n. ‖ échantillon m. de qualité. / ~ **of the soil** ‖ Bodenprobe f. ‖ échantillon m. du sol. / **to take** ~s pl. see **to sample**.

sample bag ‖ Musterbeutel m. ‖ sachet m. à échantillons. / ~ **book** ‖ Musterbuch n. ‖ livre m. d'échantillons. / ~ **book with interchangeable samples** ‖ Musterbuch n. mit auswechselbaren Proben ‖ livre m. d'échantillons avec échantillons interchangeables. / ~ **card** ‖ Musterkarte f. ‖ carte f. d'échantillons. / **collection of** ~s ‖ Musterkollektion f.; Musterzusammenstellung f. ‖ collection f. d'échantillons. / ~ **envelope** ‖ Musterumschlag m. ‖ enveloppe f. à échantillons. / ~ **fair** ‖ Mustermesse f. ‖ foire f. d'échantillons. / ~ **perforation machine** ‖ Musterstechmaschine f.; Perforiermaschine f. ‖ machine f. à perforer les échantillons et accessoires. / ~ **pouch** see **sample bag**.

sampler (Met) ‖ Probenehmer m. ‖ échantillonneur m.

sample room ‖ Musterlager n. ‖ bureau m. d'échantillons. / ~ **sheet** ‖ Papiermusterbogen m. ‖ feuille f. de modèle. / ~ **trunk** ‖ Musterkoffer m. ‖ coffre m. à échantillons. / ~ **tube** ‖ Probeglas n.; Reagensglas n. ‖ verre m. à d'essayer ou à d'essai.

sampling (Met) ‖ Probenahme f. ‖ prise f. d'échantillon; échantillonnage f.

sanatorium ‖ Kurhaus n.; Heilstätte f.; Sanatorium n. ‖ sanatorium m.; sanatoire m.

sand, to ‖ sanden; mit Sandüberzug versehen ‖ sabler. / ~ **the stone** (Stone-c) ‖ den Stein m. schleifen ‖ frotter ou polir la pierre ou le grès.

sand ‖ Sand m. ‖ sable m. / **alluvial** ~ ‖ Schwemmsand m.; Flußsand m. ‖ sable m. de rivière ou de ravine. / **bleached** ~ ‖ Bleichsand m. ‖ sable m. blanchi. / ~ **for casting purposes** ‖ Sand m. für Gußzwecke ‖ sable m. pour fonderies. / **coarse** ~ ‖ Kies m.; Grus m.; Grieß m.; grober Sand m. ‖ gravier m. / **downs** ~ ‖ Dünensand m. ‖ sable m. des dunes. / **drift** ~ ‖ Flugsand m. ‖ sable m. emporté par le vent. / **dry** ~ (Found) ‖ Formmasse f. ‖ masse f. / **dry sifted** ~ ‖ gesiebter trockener Sand m. ‖ sable m. sec criblé. / **dug** ~ see **pit** ~. / ~ **for enamel manufactures** ‖ Sand m. für Schmelzglaswerke ‖ sable m. pour émailleries. / **facing** ~ (Found) ‖ feingesiebter Sand m. ‖ sable m. fin de moulage. / **fine** ~ (Ore dress) ‖ Schlamm m.; Schlämme f. ‖ matières fpl. fines; (mouces fpl. fines); schlamm m. / **very fine** ~ ‖ Staubsand m. ‖ sablon m. / ~ **for founders** ‖ Gießsand m. ‖ sable m. des fondeurs; sablon m. / **glass** ~ see **vitreous** ~. / **granitic** ~ ‖ Granitsand m. ‖ sable f. de granit. / **green** ~ (Found) ‖ grüner oder magerer oder nasser Formsand m. ‖ sable m. vert ou maigre. / **green** ~ (Geol) ‖ Grünsand m.; Grünsandstein m. ‖ grès m. vert. / **loamy** ~ (Found) ‖ fetter Formsand m. ‖ sable m. gras. / **miry** ~ see **muddy** ~. / **moulding** ~ ‖ Formsand m. ‖ sable m. de moulage. / **moulding** ~ **for steel** ‖ Stahlformsand m. ‖ sable m. de moulage pour acier. / **plant for preparing moulding** ~ ‖ Formsandbereitungsanlage f. ‖ préparation f. du sable de moulage. / **muddy** ~ ‖ Moddersand m.; Schlammsand m. ‖ sable m. vasard. / **parting** ~ (Found) ‖ trockener Sand m.; Streusand m. ‖ sable m. sec. / **pit** ~ ‖ Grubensand m.; gegrabener Sand ‖ sable m. de fouille ou fouillé. / **quick** ~ ‖ Triebsand m.; Schwimmsand m. ‖ sable m. mouvant. / **refractory** ~ ‖ feuerfester Sand m. ‖ sable m. réfractaire. / **scouring** ~ ‖ Tünchsand m.; Scheuersand m. ‖ sablon m. ou sable m. très menu. / **to screen the** ~ see **to sift the** ~. / **screened dry** ~ ‖ gesiebter trockener Sand m. ‖ sable m. sec tamisé. / **sharp** ~ ‖ scharfer Sand m. ‖ sable m. cru. / **to sift the** ~ ‖ den Sand m. sieben ‖ passer le sable au crible ou par le crible. / **slimy** ~ see **muddy** ~. / **sticky** ~ ‖ klebriger Sand m. ‖ sable m. (de nature) collant. / **vitreous** ~ ‖ Glassand m. ‖ sable m. de verrerie. / **volcanic** ~ ‖ vulkanischer Sand m. ‖ sable m. volcanique.

sand agglutinant ‖ Sandbindemittel n. ‖ agglutinant m. de sable.

sandal ‖ Sandale f. ‖ sandale f.

sandalwood ‖ Sandelholz n. ‖ bois m. de santal; santal m. / ~ **oil** ‖ Sandelholzöl n. ‖ essence f. de santal.

sandarac ‖ Sandarak m. ‖ sandaraque f.

sand-bag ‖ Sandsack m. ‖ sac m. à sable. / ~ **revetment** ‖ Sandsackbekleidung f. ‖ revêtement m. en sac à sable.

sand-bank ‖ Sandbank f. ‖ banc m. de sable; ensablement m.; barre f.; hautfond m.

sand bath ‖ Sandbad n. ‖ bain m. de sable. / **furnace of the** ~ ‖ Sandbadofen m. ‖ fourneau m. à bain de sable.

sand bed ‖ Sandbett n. ‖ lit m. de sable.

sand-blast ‖ Sandstrahl m. ‖ jet m. de sablage. / **steam jet** ~ ‖ Dampfsandstrahlgebläse n. ‖ sableuse f. à vapeur.

sand-blast apparatus ‖ Sandstrahlgebläse n. ‖ appareil m. à jet de sable; sableuse f. / ~ machine see sand-blast apparatus.

sand box ‖ Sandkasten m. ‖ sablière f. / ~ (Loc) ‖ Sandstreuapparat m. ‖ boîte f. à sable. / ~ cover ‖ Sandkastenverschluß m. ‖ fermeture f. de la sablière.

sand cart ‖ Sandwagen m.; Sandkarren m. ‖ culbuteur m. ou tomberau m. à sable. / ~-cast ‖ in Sand m. gegossen ‖ moulé au sable. / casting in ~ see ~ casting. / ~ casting ‖ Sandguß m. ‖ moulage m. en sable. / ~ sintering ~ coal ‖ sinternde Sandkohle f. ‖ houille f. maigre sableuse. / crunching of ~ ‖ Knirschen n. des Sandes ‖ crissement m. du sable. / ~ crusher ‖ Sandstampfer m. ‖ broyeur m. de sable. / ~ crushing ‖ Streusandzubereitung f. ‖ broyage m. de sable. / ~ digger ‖ Sandgräber m. ‖ sablier m. / ~ distributor ‖ Sandverteiler m. ‖ doseur m. de sable. / ~ down see ~ dune. / ~ dragger see ~ digger. / ~ dressing plant ‖ Sandaufbereitungsanlage f. ‖ installation f. de préparation du sable. / ~ drying oven ‖ Sandtrockenofen m. ‖ four m. à sécher le sable. / ~ drying stove see ~ drying oven. / ~ dune ‖ Sanddüne f. ‖ dune f. de sable.

sander see sandpapering machine. / mould ~ (Found) ‖ Sandstreuer m. ‖ sableur m. de moules.

sand filter ‖ Sandfilter n. ‖ filtre m. à sable. / ~ getting plant ‖ Sandgewinnungsanlage f. ‖ installation f. d'extraction de sable. / ~ glass ‖ Sanduhr f. ‖ sablier m. / ~ grain ‖ Sandkorn n. ‖ grain m. de sable. / ~ grass ‖ Sandgras m. ‖ herbe f. poussante sur le sable. / ~ hole ‖ Sandloch n. ‖ camelot m.; moye f.

sanding ‖ Sandstreuen n. ‖ ensablement m. / ~ device ‖ Sandstreuvorrichtung f. ‖ sablière f. / ~ machine see sandpapering machine. / ~ pipe ‖ Sandstreurohr n. ‖ tuyau m. à sable.

sandiver ‖ Glasgalle f. ‖ fiel m. ou sel m. de verre; suin m.

sand jet blower see sand blast apparatus. / layer of ~ (Road) ‖ Sanddecke f. ‖ couche f. de sable; ensablement m.

sand. lime brick factory ‖ Kalksandsteinfabrik f. ‖ fabrique f. de briques silicocalcaires. / ~ making machine ‖ Kalksandsteinmaschine f. ‖ machine f. pour la fabrication de briques silico-calcaires. / manufacture of ~s ‖ Kalksandsteinfabrikation f. ‖ fabrication f. de briques silico-calcaires.

sand, measuring device for ~ ‖ Abmeßapparat m. für Sand ‖ appareil-doseur m. de sable. / ~ mill ‖ Sandmühle f. ‖ moulin m. à sable. / ~ mixer ‖ Sandmischer m. ‖ mélangeur m. de sable. / ~ mixing machine ‖ Sandmischmaschine f. ‖ machine f. à mélanger le sable.

sand-mould ‖ Sandform f. ‖ moule m. en sable maigre ou sec. / open ~ ‖ offene Form f.; Herdform f. ‖ moule m. découvert.

sand moulder ‖ Sandformer m. ‖ sableur m. / ~ moulding ‖ Sandformerei f. ‖ moulage m. en sable. / ~ packer ‖ Sandsacker m. ‖ ensacheur m. de sable.

sandpaper ‖ Sandpapier n. ‖ papier m. sablé. / ~ grinding machine see sandpapering machine.

sandpapering machine ‖ Sandpapierschleifmaschine f.; Sandschleifmaschine f. ‖ machine f. à poncer au papier de verre. / ~ and polishing machine ‖ Sandschleifund Poliermaschine f. ‖ machine f. à poncer et à polir au papier de verre ou sablé.

sand pin ‖ Formerstift m. ‖ pointe f. de mouleur. / ~ pipe (Geol) ‖ geologische Orgel f. ‖ puits m. naturel; orgue f. géologique. / ~ pit ‖ Sandgrube f. ‖ carrière f. de sable m.; sablière f. / ~ polishing machine see sandpapering machine. / ~ preparing ‖ Sandaufbereitung f. ‖ préparation f. du sable. / ~ pump ‖ Sandpumpe f. ‖ pompe f. à sable. / ~ pump dredger ‖ Pumpenbagger m. ‖ dragueur m. à pompes. / ~ screener ‖ Sandabsieber m. ‖ cribleur m. de sable. / ~ shovel ‖ Sandschaufel f. ‖ pelle f. à col de cygne. / ~ sieve ‖ Sandsieb n. ‖ tamiseur m.; tamis m. à sable. / rotary ~ sieve ‖ Sandsiebmaschine f. ‖ machine f. à tamiser le sable. / ~ soap ‖ Sandseife f. ‖ savon m. sablé.

sandstone ‖ Sandstein m. ‖ grès m. / argillaceous ~ ‖ toniger Sandstein m.; Tonsandstein m. ‖ grès m. argilleux. / artificial ~ ‖ Kunstsandstein m. ‖ grès m. artificiel. / calcareous ~ ‖ kalkiger Sandstein m. ‖ grès m. calcaire. / fireproof ~ ‖ feuerfester Sandstein m. ‖ grès m. réfractère. / flexible ~ ‖ elastischer Sandstein m.; Gelenkquarz m. ‖ grès m. flexible ou élastique. / iron ~ ‖ eisenhaltiger Sandstein m. ‖ grès m. ferreux. / marly ~ ‖ Mergelsandstein m. ‖ grès m. marneux. / new red ~ see variegated ~. / quartzy ~ ‖ Quarzsandstein m.; Kieselsandstein m. ‖ grès m. quartzeux. / shelly ~ ‖ Muschelsandstein m. ‖ grès m. coquiller. / upper cretaceous ~ (Geol) ‖ Quadersandstein m. ‖ grès m. crétacé supérieur. / upper new red ~ (Geol) ‖ Röt n. ‖ grès m. bizarré supérieur. / variegated ~ (Geol) ‖ Buntsandstein m. ‖ grès m. bigarré; nouveau grès m. rouge.

sandstone apparatus ‖ Sandsteinapparat m. ‖ appareil m. en grès. / ~ grinding machine ‖ Sandsteinschleifmaschine f. ‖ meule f. de grès. / ~ masonry ‖ Sandsteinmauerwerk n. ‖ maçonnerie f. de grès. / ~ pavement ‖ Sandsteinpflaster n. ‖ pavé m. de grès. / ~ quarry ‖ Sandsteinbruch m. ‖ carrière f. de grès. / quarryman for ~ ‖ Sandsteinbehauer m. ‖ grésonnier m.

sand strewing device, pneumatic ‖ Druckluftsandstreuer m. ‖ sablerie f. à air comprimé. / ~ trap (Pap) ‖ Sandfang m. ‖ sablier m. / ~ tube (Loc) ‖ Sandröhre f. ‖ tuyau m. à sable. / ~ washer ‖ Sandwaschmaschine f. ‖ laveur m. de sable. / ~ washery ‖ Sandwäsche f. ‖ laverie f. de sable.

sandwich paper ‖ Butterbrotpapier n. ‖ papier m. pour tartines (au beurre).

sandy ‖ sandig ‖ sablonneux. / ~ ground ‖ Sandboden m. ‖ terrain m. sablonneux. / ~ marl ‖ Sandmergel m. ‖ marne f. sablonneuse.

sanidine ‖ Sanidin m.; glasiger Feldspat m.; Eisspat m. ‖ sanidine f.; feldspath m. vitreux.

sanitary ‖ sanitär ‖ sanitaire. / ~ apparatus ‖ gesundheitstechnischer Apparat m. ‖ appareil m. sanitaire. / ~ articles pl. ‖ hygienische Artikel mpl. ‖ marchandises fpl. sanitaires. / ~ device ‖ sanitäre Vorrichtung f. ‖ appareil m. sanitaire. / ~ equipment ‖ sanitäre Ausrüstung f. ‖ équipment m. sanitaire. / ~ installation see ~ plant. / ~ plant ‖ hygienische oder sanitäre Anlage f. ‖ installation f. d'hygiène ou sanitaire. / fittings pl. for ~ plants ‖ Armatur f. für sanitäre Anlagen ‖ garniture f. pour installations sanitaires. / ~ towel ‖ Damenbinde f. ‖ serviette f. hygiénique.

santalol ‖ Santalol n. ‖ santalol m.

santonine preparation ‖ Santoninpräparat n. ‖ préparation f. à la santonine.

sap, to ~ the foundation ‖ den Grund m. untergraben ‖ saper les fondations.

sap ‖ Nährsaft m.; Holzsaft m. ‖ sève f.

sapanwood ‖ Sapanholz n. ‖ bois m. de sapan ou Japon; sapan m.

sap brown ‖ Saftbraun n. ‖ mordant m. noyer de sève. / ~ colour ‖ Saftfarbe f. ‖ couleur f. de sève. / ~ green ‖ Blasengrün n.; Saftgrün n. ‖ vert m. de vessie ou de sève ou d'iris.

saponifiability ‖ Verseifbarkeit f. ‖ tendance f. à la saponification; saponifiabilité f.

saponifiable ‖ verseifbar ‖ saponifiable.

saponification ‖ Seifenbildung f.; Verseifung f. ‖ saponification f. / ~ by lime ‖ Verseifung f. mit Kalk ‖ saponification f. à la chaux. / sulphuric ~ ‖ Verseifung f. mit Schwefelsäure ‖ saponification f. à l'acide sulphurique.

saponification factor ‖ Verseifungsgrad m. ‖ degré m. de saponification.

saponify, to ‖ verseifen ‖ saponifier.

saponite ‖ Seifenstein m. ‖ saponite f.

sapphire ‖ Saphir m.; blauer Korund m. ‖ saphir m.; corindon m.

sappiness (Tree) ‖ Saftgehalt m. ‖ teneur f. en sève.

sappy wood ‖ saftreiches Holz n. ‖ bois m. riche en sève.

sap rot (Wood) ‖ Blaufäule f. ‖ pourriture f. bleue ou de l'aubier.

sapwood ‖ Splintholz n. ‖ bois m. d'aubier; aubier m. / ~ rot ‖ Splintfäule f. ‖ pourriture f. de l'aubier.

sarcolithe ‖ Sarkolith m. ‖ sarcolithe f.; analcime f.

sardine fishing ‖ Sardinenfischerei f. ‖ pêche f. de la sardine. / ~ preserver ‖ Sardinenzubereiter m. ‖ sardinier m.

sardines pl. in oil packing ‖ Ölsardinen fpl. ‖ sardines fpl. à l'huile.

sardine works pl. ‖ Sardinenfabrik f. ‖ confiserie f. de sardines; sardinerie f.

sarsaparilla ‖ Sassaparilla f. ‖ salsepareille f.

sartorite ‖ Bleiarsenglanz m. ‖ sartorite f.

sash ‖ Gurt m.; Schärpe f. ‖ sangle f.; écharpe f. / ~ (Build) ‖ Fensterrahmen m. ‖ châssis m. de fenêtre; croisée f.; cadre m. de croisée. / dead ~ ‖ stehender Fensterflügel m.; toter Flügel m. ‖ châssis m. dormant ou mort. / sliding ~ ‖ Schiebeflügel m. ‖ châssis m. coulant ou à coulisse ou à guillotine. / turning ~ ‖ Drehflügel m. ‖ battant m. ou vantail m. de fenêtre.

sash bolt ‖ Fensterriegel m. ‖ targette f. de fenêtre. / sliding ~ door ‖ Rolltür f. ‖ porte f. à coulisse. / ~ fastener ‖ Fensterreiber m.; Vorreiber m. ‖ happe f. de fenêtre. / ~ frame see sash. / ~ gate (Sluice) ‖ Schütz n.; Schutzfalle f. ‖ vanne f. de barrage. / ~ lock (Hydr arch) ‖ Schützenschleuse f. ‖ écluse f. à vanne. / ~ mortise chisel ‖ Glaserbeitel m. ‖ ciseau m. de vitrier.

/ ~ saw ‖ Schlitzsäge f. ‖ scie f. à tenon. / ~ square ‖ Fensterfach n.; Fensterfeld n. ‖ panneau m. à verre *ou* à vitre *ou* de fenêtre. / sliding ~ window ‖ Schiebefenster n. ‖ fenêtre f. à coulisse.

sassafras oil ‖ Sassafrasöl n. ‖ essence f. de sassafras. / ~ root ‖ Sassafrasholz n.; Fenchelholz n. ‖ bois m. de sassafras.

satchel ‖ Schulmappe f. ‖ sac m. d'écolier.

sateen ‖ englisches Leder n. ‖ peau f. de taupe; cuir m. anglais.

satellite of Saturn ‖ Saturnmond m. ‖ satellite m. de Saturne.

satine, to (Pap) ‖ satinieren ‖ glacer; satiner.

satin ‖ Satin m.; Atlas m. ‖ satin m. / woollen ~ ‖ Wollatlas m. ‖ satin m. de laine; satin m. zéphyr.

satined ‖ atlasartig; satiniert ‖ satiné.

satinette ‖ Satinette f. ‖ satinette f.

satining apparatus (Pap) ‖ Glättmaschine f. ‖ lissoir m.; machine f. à lisser *ou* à satiner *ou* à glacer *ou* à lustrer.

satin paper for cards ‖ geglättetes Kartenpapier n. ‖ carton m. glacé *ou* satiné. / ~ spar ‖ Faserkalk m. ‖ chaux f. carbonatée fibreuse. / ~ tweel *see* satin. / ~ weaver ‖ Atlasweber m. ‖ satinaire m. / ~ wood ‖ Satinholz n.; Satinetholz n. ‖ bois m. satiné.

satisfactory arrangement ‖ befriedigende Abmachung f. ‖ arrangement m. satisfaisant.

saturate, to ‖ sättigen ‖ saturer.

saturated (Chem) ‖ gesättigt ‖ saturé. / ~ air ‖ gesättigte Luft f. ‖ air m. saturé. / ~ mixture ‖ gesättigtes Gemisch n. ‖ mélange m. saturé.

saturated steam ‖ gesättigter Dampf m. ‖ vapeur f. saturée. / ~ tank locomotive ‖ Naßdampftenderlokomotive f. ‖ locomotive-tender f. à vapeur saturée.

saturating device ‖ Sättiger m. ‖ saturateur m.

saturation ‖ Sättigung f. ‖ saturation f. / to dissolve to ~ (Chem) ‖ bis zur Sättigung f. auflösen ‖ dissoudre jusqu'à saturation f. / magnetic ~ ‖ magnetische Sättigung ‖ saturation f. magnétique. / ~ of the market ‖ Übersättigung f. des Marktes ‖ sursaturation f. du marché.

saturation, bath of ~ ‖ Sättigungsbad n. ‖ bain m. de saturation. / ~ current ‖ Sättigungsstrom m. ‖ courant m. de saturation. / ~ current in repeater valves ‖ Sättigungsstrom m. der Verstärkerröhren ‖ courant m. de saturation en tubes-à-vide. / ~ curve ‖ Sättigungskurve f. ‖ caractéristique f. de saturation. / ~ deficit ‖ Sättigungsunterschuß m. ‖ déficit m. de saturation. / degree of ~ of air ‖ Luftsättigungsgrad m. ‖ degré m. de saturation de l'air. / ~ plant ‖ Saturationsanlage f. ‖ installation f. de saturation. / ~ value ‖ Sättigungswert m. ‖ valeur f. de saturation.

sauce (Tobacco) ‖ Sauce f.; Beize f. ‖ sauce f. / ~ extract ‖ Tunkenauszug m. ‖ extrait m. de sauce.

saucepan, traveller's electric ~ ‖ elektrischer Reisekocher m. ‖ bouillotte f. de voyage électrique.

saucer (Porcel; Pott) ‖ Untertasse f. ‖ soucoupe f.

saucing machine, grain ~ ‖ Getreidebeizmaschine f. ‖ machine f. à corroder le blé.

sauerkraut ‖ Sauerkraut n. ‖ choucroute f. / ~ manufacturer ‖ Sauerkrautfabrikant m. ‖ choucroutier m.

saunderswood ‖ Santelholz n. ‖ santal m. rouge; bois m. de santal.

sausage ‖ Wurst f. ‖ saucisse f. / ~ boiler ‖ Wurstkessel m. ‖ marmite f. à saucisses. / ~ factory ‖ Wurstfabrik f. ‖ fabrique f. de saucisses. / ~ frying apparatus ‖ Würstchenbratapparat m. ‖ gril m. à saucisses. / ~ machine ‖ Wurstfüllmaschine f. ‖ machine f. à (remplir les) saucisses. / ~ maker ‖ Wurstmacher m. ‖ fabricant m. de saucisses *ou* de saucissons. / ~ skin maker ‖ Wurstdarmbereiter m. ‖ fabricant m. de peaux *ou* de boyaux à saucisses.

saussurite ‖ Bitterstein m. ‖ saussurite f.

save, to ~ goods ‖ Güter npl. bergen ‖ sauver des marchandises fpl.

saving, labour ~ machinery ‖ arbeitsparende Maschine f. ‖ machine-outil f. (d'un rendement plus économique *ou*) économisant la main-d'œuvre.

saving ‖ Ersparnis f. ‖ économie f. / considerable ~ ‖ erhebliche Ersparnis f. ‖ économie f. considérable. / ~ of fuel ‖ Brennstoffersparnis f. ‖ économie f. de combustible. / ~ of maintenance costs ‖ Ersparnis f. an Betriebskosten ‖ économie f. de frais d'exploitation. / ~ of material ‖ Werkstoffersparnis f. ‖ économie f. de matière. / ~ of power ‖ Kraftersparnis f. ‖ gain m. de puissance. / ~ of space ‖ Platzersparnis f. ‖ économie f. de place. / ~ of time ‖ Zeitersparnis f. ‖ économie f. de temps. / ~ of time and expenses ‖ Ersparnis f. an Zeit und Kosten ‖ économie f. de temps et d'argent. / ~ of time and wages ‖ Ersparnis f. an Zeit und Arbeitslohn ‖ économie f. de temps et de salaire. / ~ of wages ‖ Lohnersparnis f. ‖ économie f. de salaires. / ~ in water ‖ Wasserersparnis f. ‖ économie f. d'eau. / ~ in working costs ‖ Betriebsunkosteneinsparung f. ‖ économie f. dans les frais d'exploitation.

saving apparatus ‖ Rettungsapparat m. ‖ appareil m. de sauvetage.

savings pl. ‖ Ersparnisse fpl. ‖ épargne f. / ~ bank ‖ Sparkasse f. ‖ caisse f. d'épargne.

savory herb ‖ Pfefferkraut n. ‖ herbe f. de sarriette. / ~ oil ‖ Bohnenkrautöl n. ‖ essence f. de sarriette.

savoury ‖ schmackhaft ‖ savoureux.

saw, to ‖ sägen ‖ scier. / ~ off ‖ absägen ‖ scier. / ~ out ‖ aussägen ‖ découper à la scie. / ~ square a plank ‖ ein Brett n. säumen ‖ équarrir une planche. / ~ timber ‖ Holz n. schneiden *oder* zuschneiden ‖ couper *ou* débiter *ou* découper le bois.

saw ‖ Säge f. ‖ scie f. / arm ~ *see* hand ~. / band ~ ‖ Bandsäge f. ‖ scie f. à ruban. / bow ~ ‖ Bügelsäge f. ‖ scie f. à étrier *ou* à archet. / broken space ~ ‖ Lochsäge f. ‖ scie f. à guichet; égohine f. / circular ~ ‖ Kreissäge f. ‖ scie f. circulaire. / circular ~ bench for cutting metal ‖ Tischkreissäge f. für Metall ‖ scie f. circulaire à métaux à table. / circular ~ for mines ‖ Grubenkreissäge f. ‖ scie f. circulaire pour mines. / circular ~ with several saw blades ‖ Kreissäge f. mit mehreren Sägeblättern ‖ scie f. circulaire à coupe multiple *ou* à plusieurs lames. / cold ~ ‖ Kaltsäge f. ‖ scie f. à froid. / cold circular ~ ‖ Kaltkreissäge f. ‖ scie f. circulaire à froid. / compass ~ ‖ Laubsäge f. ‖ scie f. à chantourner. / cross-cut ~ ‖ Zugsäge f.; Schrotsäge f. ‖ passe-partout m. / curvelinear ~ ‖ Frettsäge f. ‖ scie f. à échancrer *ou* à évider *ou* à tournefond. / frame ~ ‖ Gestellsäge f.; Spannsäge f.; Gattersäge f. ‖ scie f. à châssis *ou* montée *ou* alternative. / framed whip ~ ‖ Örtersäge f. ‖ scie f. à débiter; scie f. montée à débiter. / fret ~ ‖ Laubsäge f. ‖ scie f. à chantourner. / grooving ~ ‖ Nutsäge f. ‖ scie f. à mortaiser. / hand ~ ‖ Handsäge f. ‖ scie f. à main. / hinge ~ ‖ Gelenksäge f. ‖ scie f. à charnière. / hot ~ *see* warm ~. / joiner's ~ ‖ Schreinersäge f. ‖ scie f. de menuisier. / knife pruning ~ ‖ Messersäge f. ‖ scie f. à couteau. / metal circular ~ ‖ Metallkreissäge f. ‖ scie f. circulaire à métaux. / mounted ~ *see* frame ~. / notched ~ ‖ schartige Säge f. ‖ scie f. édentée. / notching ~ ‖ Kerbsäge f. ‖ scie f. à entailler. / ~ for parallel sawing ‖ Besäumsäge f. ‖ scie f. à dresser et à équarrir. / pendulum ~ ‖ Pendelsäge f. ‖ scie-pendule f. / pruning ~ ‖ Baumsäge f. ‖ scie f. de jardinier *ou* à greffer. / revolving cross cut ~ ‖ rotierende Ablängsäge f. ‖ scie f. tournante à tronçonner. / sliding ~ ‖ Schlittensäge f. ‖ scie f. à chariot. / snicked ~ *see* notched ~. / stave shortening ~ ‖ Daubenabkürzsäge f. ‖ scie f. à raccourcir les douves. / stone ~ ‖ Steinsäge f. ‖ scie f. à pierre. / surgeon's ~ ‖ Wundarztsäge f. ‖ scie f. de chirurgien. / sweep ~ ‖ Schweifsäge f. ‖ scie f. à échancrer. / tenon ~ ‖ Zapfensäge f. ‖ scie f. à arraser *ou* à couper les tenons. / toothless ~ ‖ zahnlose Säge f. ‖ scie f. sans dents. / warm ~ ‖ Warmsäge f. ‖ scie f. à chaud.

saw bench ‖ Sägebank f. ‖ chevalet m. de scieur.

saw blade ‖ Sägeblatt n. ‖ lame f. de scie. / endless ~ ‖ Bandsägeblatt n. ‖ lame f. continue *ou* de scie à ruban. / ~ for fret saws ‖ Sägeblatt n. für Laubsägen ‖ lame f. pour scies à chantourner. / ~ with inserted steel teeth ‖ Sägeblatt n. mit eingesetzten Stahlzähnen ‖ lame f. de scie à dents en acier rapportées. / ~ palet ‖ Sägenblech n. ‖ tôle f. pour scies.

saw block *see* saw log. / ~ bow ‖ Sägenbügel m. ‖ archet m. de scie. / ~ cut ‖ Sägenschnitt m. ‖ trait m. de scie.

saw dust ‖ Sägemehl n. ‖ sciure f. de bois. / ~ desiccating machine ‖ Sägemehlentwässerungsmaschine f. ‖ séchoir m. de sciure de bois.

sawed wood ‖ gesägtes Holz n. ‖ bois m. scié.

saw file ‖ Sägenfeile f. ‖ lime f. à scies. / ~ filer ‖ Sägenschärfer m. ‖ affûteur m. de scies. / ~ frame ‖ Rahmen m. für Sägegatter ‖ châssis m. de scie alternative. / ~ gate *see* ~ frame. / ~ grinding machine ‖ Sägenschleifmaschine f. ‖ machine f. à affûter les scies. / ~ handle ‖ Sägengriff m. ‖ manche m. de scie.

sawing ‖ Sägen n. ‖ sciage m. / ~ of fire wood ‖ Sägen n. von Brennholz ‖ sciage m. de bois de chauffage. / ~ plank ‖ Langholzsägen ‖ sciage m. de long.

sawing machine ‖ Sägemaschine f. ‖ machine f. à scier; scie f. mécanique. / ~ mill *see* sawmill.

saw log ‖ Sägeblock m. ‖ billot m. *ou* grume f. *ou* bloc m. de sciage.

sawmill ‖ Sägerei f.; Sägemühle f.; Schneidemühle f.; Sägegatter n. ‖ scierie f.; moulin m. à scierie. / horizontal ~ ‖ Horizontalgatter n. ‖ scierie f. horizontale; scie f. à grume. / ~ to measure and price ‖ Lohnsägerei f.; Lohnsägemühle f. ‖ scierie f. à façon. / mechanical ~ ‖ mechanische Sägemühle f. ‖ scierie f. mécanique. / with one saw ‖ Mittelgatter n.; Blockgatter n. ‖ scierie f. verticale à une lame au milieu. / ~ with rollers ‖ Sägemühle f. mit Walzenvorschub ‖ scierie f. à amenée par cylindres. / ~ with several blades ‖ Vollgatter n.; Bundgatter n. ‖ scierie f. à plusieurs lames. / steam ~ ‖ Dampfsägemühle f. ‖ scierie f. à vapeur. / ~ with two blades ‖ Doppelgatter n. ‖ scierie f. à deux lames. / wind-driven ~ ‖ Schneidemühle f. mit Windbetrieb ‖ moulin m. à vent pour scierie; scierie f. mécanique à vent.

sawmill equipment ‖ Sägewerkeinrichtung f. ‖ équipement m. de scierie. / ~ machine ‖ Sägewerksmaschine f. ‖ machine f. pour scierie.

sawn ‖ gesägt ‖ scié. / ~ stave ‖ gesägte Faßdaube f. ‖ douve f. sciée. / ~ timber ‖ Schnittholz n. ‖ bois m. de sciage.

saw notch ‖ Sägeschnitt m. ‖ trait m. de scie. / ~ pad ‖ Sägengriff m. ‖ manche m. de scie. / ~ pit ‖ Sägegrube f. ‖ fosse f. des scieurs de long. / ~ pit frame ‖ Schneiderost m. ‖ chevalet m. des scieurs de long; tréteau m. / ~ sash *see* ~ frame. / ~ set ‖ Schränkeisen n. ‖ fer m. à contourner *ou* à donner la voie à la scie. / ~ set pliers pl. ‖ Schränkzange f. ‖ tenaille f. à contourner *ou* à donner la voie à la scie. / automatic ~ sharpener ‖ Sägenselbstschärfer m. ‖ machine f. automatique à affûter les scies.

saw sharpening ‖ Sägenschärfen n. ‖ affûtage m. de scies. / ~ machine ‖ Sägenschärfmaschine f. ‖ machine f. à affûter les scies. / ~ wheel ‖ Sägenschärfscheibe f. ‖ meule f. pour l'affûtage des scies.

saw stave jointer ‖ Daubenfügesäge f. ‖ scie f. à jointer les douves. / ~ table ‖ Sägetisch m. ‖ table f. à scier.

saw teeth pl., to set the ~ ‖ das Sägeblatt schränken ‖ donner de la voie à la lame de scie. / ~ setting machine ‖ Schränkmaschine f. für Sägeblätter ‖ machine f. à donner la voie aux dents (de lames) de scies.

saw tooth ‖ Sägezahn m. ‖ dent f. de scie. / ~ roof (Build) ‖ Scheddach n. ‖ toit m. en shed *ou* en forme de dent de scie. / ~ roof in iron construction ‖ Scheddach n. in Eisenkonstruktion ‖ toit m. en forme de dent de scie en charpente métallique.

sawyer ‖ Holzsäger m. ‖ scieur m. de bois. / ~ (Bookb) ‖ Einschneider m. ‖ grecqueur m. / band ~ ‖ Bandsäger m. ‖ scieur m. à la scie à ruban. / gang ~ ‖ Gattersäger m. ‖ conducteur m. de scie alternative *ou* à plusieurs lames.

sawyer's jack *see* saw log.

Saxon tin (Met) ‖ Tafelzinn n. ‖ étain m. en briques.

saxophone ‖ Saxophon n. ‖ saxophone m.

scab ‖ Räude f. ‖ rogne f.; gale f. / ~ remedy ‖ Räudemittel n. ‖ médicament m. de gale.

scabbard of a knife ‖ Scheide f. eines Messers ‖ gaîne f. *ou* fourreau m. *ou* étui m. d'un couteau.

scaffold, to ‖ rüsten; berüsten; mit einem Gerüst versehen ‖ échafauder.

scaffold (Build) ‖ Gerüst n.; Baugerüst n.; Arbeitsgerüst n. ‖ échafaud m.; échafaudage m. / flying ~ *see* hanging ~. / hanging ~ ‖ fliegendes Gerüst n. ‖ Hängegerüst n. ‖ échafaud m. *ou* pont m. volant. / movable ~ *see* hanging ~. / sawyer's ~ ‖ Schrotrost m. des Brettsägers ‖ baudet m. *ou* hout m. *ou* tréteau m. du scieur de long.

scaffold beam ‖ Gerüstbalken m. ‖ poutre f. d'échafaudage. / rising ~ bridge ‖ Laufbahn f.; Auflauf m. des Gerüstes ‖ pont m. d'échafaudage.

scaffolder ‖ Gerüstbauer m. ‖ échafaudeur m.

scaffolding *see also* scaffold ‖ Rüstung f.; Aufrichtung f. eines Gerüstes ‖ échafaudage m. / ~ a blast-furnace ‖ Rostschlagen n. beim Hochofen ‖ grillage m. des hauts-fourneaux. / erecting ~ ‖ Montagegerüst n. ‖ échafaudage m. de montage. / ~ of poles and pullocks ‖ Stangengerüst n. mit Netzriegeln ‖ échafaudage m. d'échasses et de boulins.

scaffolding and tools ‖ Gerüst n. und Gerätschaften fpl.; Gerüstbauzubehör m. ‖ équipage m. pour échafaudage.

scaffolding enterprise ‖ Gerüstbauunternehmung f.; Gerüstbaugeschäft n. ‖ entreprise f. d'échafaudage. / ~ hole ‖ Rüstloch n. ‖ trou m. de boulin *ou* de traverse. / ~ imp *see* ~ pole. / ~ pole ‖ Rüststamm m.; Rüstbaum m.; Gerüststange f.; Richtstange f. ‖ baliveau m.; échasse f. d'échafaud; boulin m.; perche f. / ~ trestle ‖ Rüstbock m. ‖ tréteau m. d'échafaud; chevalet m.

scalar ‖ Skalar m. ‖ scalaire m.

scald, to (with boiling water) ‖ ausbrühen; brühen; abbrühen ‖ échauder.

scalding (Bleach) ‖ Nachbeuchen n. mit Zusatz von Seife ‖ débouillissage m. des toiles après l'immersion dans le chlore. / ~ tub ‖ Abbrühkessel m. ‖ échaudoir m.

scale, to ‖ abblättern; abbröckeln ‖ s'effeuiller; s'écailler; s'écaler. / ~ the boiler ‖ den Kessel m. ausklopfen; den Kesselstein m. klopfen ‖ piquer *ou* détartrer la chaudière. / ~ off ‖ (sich) schuppen; abschuppen; abblättern ‖ (s')écailler; (s')effeuiller. / ~ off the boiler ‖ den Kesselstein abkratzen *oder* abklopfen ‖ écailler *ou* détartrer la chaudière.

scale (Of fish etc.) ‖ Schuppe f. ‖ écaille f. / ~ of alburn ‖ Ukeleischuppe f. ‖ écaille f. d'ablette. / fish ~ ‖ Fischschuppe f. ‖ écaille f. de poisson. / in thin ~s pl. ‖ dünnschuppig ‖ en fines écailles.

scale (Boiler) *see* scales.

scale (Met) *see* scales.

scale (Balance) *see* scales. / ~ (Pan of a balance) ‖ Wageschale f.; Wagschale f. ‖ bassin m. de la balance. / ~ for assaying ‖ Probiernapf m.; Probierschale f. ‖ écaille f. de coupelation.

scale (Of degrees) ‖ Skale f.; Gradteilung f. ‖ échelle f.; ligne f. graduée; graduation f. / ~ (Mus) ‖ Tonleiter f. ‖ gamme f. / ~ (Page gauge) ‖ Kolumnenmaß n. ‖ longueur f; mesure f. d'une réglette. / ~ (Draw) ‖ Maßstab m. ‖ échelle f.; règle m. divisée. / to move the ~ up and down ‖ die Skale f. verschieben ‖ faire glisser l'échelle f. / to provide with a ~ ‖ mit einer Skale f. versehen ‖ pourvoir d'une échelle graduée. / to slide the ~ ‖ die Skale f. verschieben ‖ faire glisser l'échelle f. / ~ of charges ‖ Tarif m. ‖ tarif m. / according to ~ of charges ‖ nach *oder* laut Tarif m. ‖ selon le tarif; suivant le tarif. / circular ~ ‖ Kreisskale f.; Teilkreis m. ‖ cercle m. gradué. / concentric ~ ‖ konzentrische Skale f. ‖ échelle f. concentrique. / ~ of degrees ‖ Skale f.; Gradteilung f.; Teilung f. ‖ échelle f.; ligne f. graduée; graduation f. en degrés. / divided ~ ‖ Maßstab m. ‖ règle f. divisée. / ~ of x divisions ‖ Skale f. mit x Teilstrichen ‖ cadran m. de x divisions. / on an enlarged ~ ‖ in vergrößertem Maßstabe m. ‖ (sur une) échelle agrandie. / full ~ ‖ natürliche Größe f. ‖ vraie grandeur f. / ~ of hardness (Miner) ‖ Härteskale f. ‖ échelle f. de dureté. / ~ of height ‖ Höhenmaßstab m. ‖ échelle f. des hauteurs. / illuminated ~ ‖ beleuchtete Skale f. ‖ échelle f. éclairée. / on a large ~ ‖ in großem Maßstab m.; in großem Stile m.; in großem Umfang m.; im großen n. ‖ sur une grande échelle; sur un grand pied; en gros. / ~ of length ‖ Längenmaßstab m. ‖ échelle f. des longueurs. / ~ of the map ‖ Kartenmaßstab m. ‖ échelle f. de la carte. / millimetre ~ ‖ Millimeterskale f. ‖ échelle f. millimétrique. / plain ~ ‖ natürlicher Maßstab m. ‖ grandeur f. naturelle. / plotting ~ ‖ verkleinerter Maßstab m. ‖ échelle f. de réduction *ou* à rapporter. / projected ~ ‖ projizierte Skale f. ‖ échelle f. de projection. / reduced ~ ‖ verkleinerter *oder* verjüngter Maßstab m. ‖ échelle f. fuyante *ou* de réduction *ou* à rapporter. / on a reduced ~ ‖ in verjüngtem *oder* verkleinertem Maßstabe m. ‖ à échelle f. de réduction. / ~ of reduction *see* reduced ~. / scenographical ~ ‖ perspektivischer Maßstab m. ‖ échelle f. perspective. / on a small ~ ‖ in kleinem Maßstab m. ‖ en petit; sur une petite échelle. / ~ of sparks (Electr) ‖ Funkenskale f. ‖ échelle f. des étincelles. / thermometric ~ ‖ Thermometerskale f. ‖ échelle f. d'un thermomètre. / tonnage ~ ‖ Tonnengehaltskale f. ‖ échelle f. de tonnage. / triangular ~ ‖ dreikantiger Maßstab m. ‖ règle f. graduée triangulaire. / ~ for varying the distance between the eyes ‖ Skale f. zur Einstellung des Augenabstandes ‖ échelle f. pour le réglage de l'écartement des yeux. / ~ with vernier ‖ Skale f. mit Nonius ‖ échelle f. avec vernier. / ~ x to y ‖ Maßstab m. x : y ‖ échelle f. de x : y.

scale aræometer ‖ Skalenaräometer n. ‖ aréomètre m. à volume constant. / ~ beam ‖ Wagebalken m. ‖ fléau m. d'une balance. / ~ button (Tel) ‖ Skalenknopf m. ‖ bouton m. de cadran.

scaled tube (Chem) ‖ Einschmelzrohr n. ‖ tube m. à sceller.

scale disk (Tel) ‖ Skalenscheibe f. ‖ cadran m. / ~ division ‖ Skalenteilung f. ‖ graduation f. de l'échelle. / linear ~ division ‖ lineare Skalenteilung f. ‖ graduation f. d'échelle linéaire. / logarithmic ~ division ‖ logarithmische Skalenteilung f. ‖ graduation f. d'échelle logarithmique. / drawing to ~ ‖ Maßzeichnung f. ‖ dessin m. coté. / feeler ~ ‖ Tasterskale f. ‖ échelle f. du calibre.

scale-like ‖ schuppig ‖ écailleux.

scale line ‖ Teilstrich m. ‖ trait m. de graduation. / ~ maker ‖ Wagenmacher m. ‖ balancier m. / ~ microscope ‖ Skalenmikroskop n. ‖ microscope m. à micromètre.

scalene *see* scalenous.

scalenous (Geom) ‖ ungleichseitig ‖ scalène. / ~ triangle ‖ ungleichseitiges Dreieck n. ‖ triangle m. scalène.

scale pipette ‖ Meßpipette f. ‖ pipette f. jaugée.

sealer of boards (Sawm) ‖ Brettnachmesser m. ‖ mesureur m. de planches.

scale reading ‖ Skalenablesung f. ‖ lecture f. de l'échelle.

scale removing machine ‖ Entzunderungsmaschine f. ‖ machine f. à enlever les battitures.

scales pl. (Boiler) ‖ Kesselstein m. ‖ incrustations fpl.; sédiments mpl.; dépôts mpl.; vidanges fpl.; tartre m.; calcin m. / ~ (Met) ‖ Hammerschlag m.; Glühspan m.; Zunder m. ‖ pailles fpl. de fer; écailles fpl. *ou* battitures fpl. de fer; mâchefer m. / ~ of the liquated masses ‖ Pickschiefer m. ‖ écailles fpl. des masses ressuées.

scales pl. (Weighing device) ‖ Wage f. ‖ balance f. / automatic ~ ‖ selbsttätige Wage f. ‖ balance f. automatique. / beam ~ ‖ Balkenwage f. ‖ balance f. à fléau. / beam ~ (Steelyard) ‖ Schnellwage f.; Laufgewichtswage f. ‖ balance f. romaine; romaine f. / chemical ~ ‖ chemische Wage f. ‖ balance f. de chimiste. / express ~ ‖ Schnellwage f. ‖ balance f. pour pesées rapides. / ~ for the automatic weighing and sacking of flour ‖ Wage f. zum automatischen Verwiegen und Absacken von Mehl ‖ balance f. à peser et à ensacher la farine automatiquement. / gold ~ ‖ Goldwage f. ‖ balance f. d'orfèvre. / household ~ *see* kitchen ~. / kitchen ~ ‖ Haushaltwage f.; Küchenwage f.; Wirtschaftswage f. ‖ balance f. de cuisine *ou* de maison *ou* de ménage. / letter ~ ‖ Briefwage f. ‖ pèse-lettres m. / paper ~ ‖ Papierwage f. ‖ balance f. à papier. / ~ for persons ‖ Personenwage f. ‖ balance f. *ou* bascule f. pour personnes. / precision ~ ‖ Genauigkeitswage f.; Präzisionswage f. ‖ balance f. de précision. / Roberval ~ ‖ zweischalige Tafelwage f. ‖ balance f. de Roberval. / shop ~ ‖ Geschäftswage f. ‖ balance f. de comptoir. / special ~ for slaughter houses ‖ Schlachthausspezialwage f. ‖ bascule f. spéciale d'abattoirs. / sliding poise ~ ‖ Laufgewichtswage f. ‖ bascule f. à poids mobile. / spring ~ ‖ Federwage f. ‖ balance f. à ressort. / sensibility of the ~ ‖ Empfindlichkeit f. der Wage ‖ sensibilité f. de la balance. / turn of ~ ‖ Ausschlag m. der Wage ‖ trait m. *ou* don m. *ou* jeu m. de balance

scale solvent ‖ Kesselsteinlösemittel n. ‖ désincrustant m.

scaling ‖ Abblätterung f.; Abschuppen n.; Abschälen n. ‖ écaillage m.; écorçage m. / ~ of the rail ‖ Abblätterung f. der Schiene ‖ écaillage m. du rail.

scaling apparatus (Met) ‖ Entzunderungsvorrichtung f. ‖ appareil m. à enlever les battitures. / ~ hammer (Boiler) ‖ Abklopfer m.; Abklopfhammer m. ‖ marteau m. détartreur. / ~ tool for boilers ‖ Kesselsteinabklopfer m. ‖ écail-

leur m. à chaudière; outil m. de détartrage des chaudières.

scallop, to ‖ zacken; zähnen; auskerben; zackenförmig ausschneiden ‖ denteler; déchiqueter; festonner.

scalloping machine ‖ Lagnettiermaschine f. ‖ festonneuse f.; machine f. à festonner.

scalp, to (Mill) ‖ schälen ‖ décortiquer.

scalpel (Med) ‖ Skalpell n. ‖ scalpel m.

scalper (Mill) ‖ Schälmaschine f.; Abstreifmaschine f. ‖ décortiqueuse f.; machine f. à décortiquer.

scalping (Mill) ‖ Schälen n. ‖ décorticage m.

scalping machine *see* scalper.

scaly ‖ schuppig ‖ écailleux.

scammony resin ‖ Skammoniumharz n.; Skammonium n. ‖ scammonée f.

scamp, to ‖ schlecht arbeiten; pfuschen; (sich) übereilen ‖ bousiller; précipiter; gâcher.

scanty ‖ notdürftig ‖ à peine suffisant; indigent.

scapement (Watchm) *see also* escapement ‖ Hemmung f. ‖ échappement m.

scapolite ‖ Skapolit m.; Glaukolit m. ‖ scapolite f.; Wernérite f.

scar (In the paper) ‖ Narbe f. ‖ grain m.

scarbroite ‖ samische Erde f.; Kollyrit m. ‖ collyrite f.

scarce ‖ selten ‖ rare.

scarcity (Rareness) ‖ Seltenheit f. ‖ rareté f. / ~ (Curiosity) ‖ Rarität f. ‖ rareté f.; curiosité f. / ~ (Of funds) ‖ Knappheit f.; Mangel m. ‖ étroitesse f.; modicité f.; pénurie f. / ~ of provisions ‖ Teuerung f. ‖ cherté f.; disette f.; renchérissement m.

scarf, to (yoin; Carp) ‖ mit den Enden npl. zusammenblatten; zusammenlaschen ‖ assembler à mi-bois; empâter; faire une empâture. / ~ a beam with indents ‖ einen Balken m. einzahnen ‖ adenter *ou* entailler une poutre. / ~ two timbers pl. ‖ zwei Hölzer npl. bündig überschneiden ‖ assembler deux pièces fpl. de bois à mi-bois.

scarf (Cloth) ‖ Schärpe f.; Umschlagtuch n.; Schal m. ‖ écharpe f.; châle f.

scarf (Carp) ‖ Blattung f.; Laschung f. ‖ écart m. / ~ (Shipb) ‖ Scherbe f.; Lasch n.; Laschung f. ‖ écart m. / ~ (Groove) ‖ Kerbe f.; Stich m.; Marke f. ‖ encoche f.; coche f.; entaille f. / flat ~ ‖ horizontale Laschung f. ‖ écart m. pratiqué tout autour. / hook and butt ~ ‖ Hakenlaschung f. ‖ écart m. à croc *ou* à dent. / horizontal ~ *see* flat ~. / long ~ ‖ halber Spund m. ‖ refeuillure f. / oblique ~ ‖ vertikale Laschung f. ‖ écart m. pratiqué sur le droit. / tabled ~ *see* scarf and key. / vertical ~ *see* side ~.

scarf and key ‖ Hakenblatt n. ‖ assemblage m. à trait de Jupiter.

scarfed built beam ‖ verschränkter Balken m. ‖ poutre f. jointe par des encoches alternatives.

scarfing (Carp) ‖ Verblattung f. ‖ assemblage m. à mi-bois. / ~ angle ‖ Abschrägungswinkel m. ‖ angle m. d'écarvement. / ~ machine ‖ Anschärfmaschine f. ‖ machine f. à écarver.

scarf joint ‖ Verblattung f. ‖ assemblage m. à mi-bois.

scarf milling machine ‖ Ausschärffräsmaschine f. ‖ fraiseuse f. à amincir (les coins de tôles).

scarf planing machine ‖ Ausschärfhobelmaschine f. ‖ raboteuse f. à amincir (les coins de tôles).

scarf-pin ‖ Krawattennadel f.; Halstuchnadel f. ‖ épingle f. de cravate.

scarifier (Agr) ‖ Messeregge f.; Reißpflug m. ‖ scarificateur m. / ~ on wheels ‖ fahrbarer Straßenaufreißer m. ‖ piocheuse f. sur roues.

scarifier tool (For road building) ‖ Aufreißstahl m. ‖ pioche f.

scarlet (Dyer) ‖ Scharlach m.; Scharlachfarbe f. ‖ écarlate f. / ~ dyed in grain ‖ Kermesscharlach m.; Dunkelscharlach m. ‖ écarlate f. de France; écarlate f. de graine. / French ~ *see* ~ dyed in grain.

scarlet paper-hanging ‖ Scharlachtapete f. ‖ papier m. peint en écarlate.

scarph (Carp) *see* scarf (Carp).

scatter, to ‖ umherstreuen; zerstreuen ‖ disperser. / ~ (Met; Assay) ‖ spratzen ‖ rocher.

scattered ‖ zerstreut; verstreut ‖ dispersé. / houses pl. ~ about ‖ zerstreut liegende Häuser npl. ‖ maisons fpl. dispersées. / ~ radiation (Roentgen) ‖ Streustrahlung f. ‖ rayonnement m. dispersé. / ~ rays pl. ‖ Streustrahlen mpl. ‖ rayons mpl. dispersés.

scavenge, to ‖ a road ‖ eine Straße reinigen *oder* kehren ‖ débourber *ou* nettoyer une rue.

scavengering (of roads) ‖ Straßenreinigung f. ‖ nettoyage m. des rues; service m. de la voirie.

scavenging (Mot) ‖ Spülung f.; Spülen n.; Reinigung f. ‖ balayage m. / ~ air ‖ Spülluft f. ‖ air m. de balayage. / ~ air-pump ‖ Spülluftpumpe f. ‖ pompe f. de balayage d'air. / ~ duct ‖ Spülleitung f. ‖ tubulure f. de balayage. / ~ port ‖ Spülschlitz m. ‖ lumière f. de balayage. / ~ pump ‖ Spülpumpe f. ‖ pompe f. de balayage. / ~ system for bad cooling-water conditions ‖ Spülvorrichtung f. für schlechte Kühlwasserverhältnisse ‖ dispositif m. de balayage à recommander dans les conditions défavorables de l'eau de refroidissement. / ~ valve ‖ Spülventil n. ‖ soupape f. de balayage. / to examine the ~ valve ‖ das Spülventil prüfen ‖ vérifier la soupape de balayage.

scene (In the theatre) ‖ Kulisse f.; Theaterwand f. ‖ coulisse f. / ~ painter ‖ Theatermaler m. ‖ peintre m. de décorations; décorateur m. / ~ painting *see* scenery and ~ painting. / scenery and ~ painting ‖ Bühnenausstattung f.; Theaterdekoration f. ‖ décor m. et peinture f. de scène.

scenic painter ‖ Bühnenmaler m. ‖ peintre-décorateur m. de théâtre.

scenography (Draw) ‖ Perspektive f.; perspektivische Zeichnung f. ‖ dessin m. perspectif; perspective f.; scénographie f.

scent (Perfume) ‖ Riechstoff m. ‖ Riechmittel n. ‖ parfum m.; matière f. aromatique. / artificial ~ ‖ künstlicher Riechstoff m. ‖ matière f. aromatique artificielle. / fancy ~ ‖ Blütenöl n. ‖ essence f. de fleur. / synthetic ~ ‖ synthetischer Riechstoff m. ‖ parfum m. synthétique.

scent-bag ‖ Riechkissen n. ‖ sachet m. d'odeurs.

scent-bottle ‖ Parfümflasche f.; Riechfläschchen n. ‖ flaçon m. à parfum; fiole f. d'odeurs.

scented oil ‖ parfümiertes Öl n. ‖ huile f. parfumée. / ~ paste ‖ parfümierte Paste f. ‖ pâte f. parfumée. / ~ pomade ‖ parfümierte Pomade f. ‖ pommade f. par-

fumée. / ~ soap ‖ parfümierte Seife f. ‖ savon m. parfumé. / ~ water ‖ Riechwasser n. ‖ eau f. de senteur. / ~ wood ‖ wohlriechendes Holz n. ‖ bois m. odorant.

scent extract ‖ Riechstoffauszug m.; Duftextrakt m. ‖ extrait m. parfumé.

scent spray ‖ Parfümverdunster m.; Riechstoffverdunster m. ‖ évaporateur m. de parfum.

schappe (Silk) see also schappe silk ‖ Schappe f.; Florettseide f. ‖ schappe f.; bourre f. de soie. / artificial ~ ‖ künstliche Schappe f. ‖ schappe f. artificielle.

schappe silk see also schappe ‖ Florettseide f. ‖ bourre f. de soie. / ~ yarn ‖ Florettseidengarn n. ‖ fil m. de bourre de soie.

schedule ‖ Liste f.; Verzeichnis n. ‖ liste f.; relevé m.; spécification f. / ~ (of prices) ‖ Preisverzeichnis n. ‖ série f. ou bordereau m. de prix. / ~ of plant ‖ Verzeichnis n. der Betriebseinrichtung ‖ inventaire m. de l'installation. / ~ time (Railw) ‖ Fahrplan m. ‖ indicateur m. des chemins de fer; horaire m. (des trains).

scheelite ‖ Scheelit m.; Scheelerz n.; Schwerstein m.; Tungstein m. ‖ scheelin m. calcaire; scheelite f.

scheelitine ‖ Scheelbleierz n.; Scheelbleispat m.; Stolzit m.; Wolframbleierz n.; Scheelitin m. ‖ plomb m. tungstaté; scheelitine f.

scheererite ‖ Scheererit m. ‖ scheererite f.

scheme (Sketch) ‖ Entwurf m. ‖ esquisse f.; croquis m. / normal ~ of erection (Of a motor) ‖ Normalaufstellungszeichnung f. ‖ dessin m. normal de montage d'un moteur. / ~ of modulation (Tel) ‖ Modulationsschaltung f. ‖ modulation f. par absorption. / ~ of work (Build) ‖ Bauentwurf m. ‖ projet m. de construction.

schiefer spar ‖ schaliger Kalkspat m.; Schieferspat m. ‖ chaux f. carbonatée nacrée.

schiller spar ‖ Bastit m.; Schillerspat m.; Schillerstein m. ‖ bastite f.; diallage m. métalloïde.

schist ‖ Schiefer m. ‖ schiste m.; ardoise f. / argillaceous ~ ‖ Tonschiefer m. ‖ schiste m. argileux. / micaceous ~ ‖ Glimmerschiefer m. ‖ mica m. schiste ou schistoïde; schiste m. micacé. / siliceous ~ ‖ Kieselschiefer m. ‖ lydite f.; schiste m. siliceux; phthonite f.

schist oil ‖ Schieferöl n. ‖ huile f. ou essence f. de schiste.

schistosity ‖ Schieferung f. ‖ schistosité f.

schistous ‖ schieferhaltig; schieferig ‖ schisteux; ardoiseux.

schlich (Mine; Met) ‖ Schlamm m. ‖ schlich m.

schnaps ‖ Schnaps m.; Branntwein m. ‖ eau-de-vie f.

scholar's drawing set ‖ Schulreißzeug n. ‖ boîte f. de compas pour écoliers.

scholarships pl. for the advancement of technical education of apprentices ‖ Stipendienstiftung f. für die Weiterbildung von Lehrlingen ‖ bourses fpl. accordées aux apprentis afin de parfaire ou continuer leurs études professionnelles.

school ‖ Lehranstalt f.; Schule f. ‖ établissement m. d'instruction; école f.; instruction f. / ~ of arts and crafts ‖ Kunstgewerbeschule f. ‖ école f. des arts industriels. / ~ of commerce ‖ Handelsschule f. ‖ école f. de commerce. / cooking and housekeeping ~ ‖ Koch- und Haushaltungsschule f. ‖ école f. ménagère. / housekeeping ~ ‖ Haushaltungsschule f ‖ école f. ménagère. / industrial ~ ‖ Industrieschule f.; Gewerbeschule f. ‖ école f. d'industrie. / infant's ~ ‖ Kleinkinderschule f. ‖ école f. enfantine. / professional ~ ‖ Fachschule f. ‖ école f. professionnelle.

school aeroplane ‖ Schulflugzeug n. ‖ avion m. école. / ~ bag ‖ Schultasche f.; Schulmappe f. ‖ sac m. d'école; porte-cahiers m. / ~ bench ‖ Schulbank f. ‖ banc m. d'école. / ~ furniture ‖ Schulmöbel npl. ‖ mobilier m. scolaire; meubles mpl. pour écoles. / ~ microscope ‖ Schulmikroskop n. ‖ microscope m. pour écoles. / ~ pen ‖ Schulschreibfeder f. ‖ plume f. d'écolier. / ~ picture ‖ Schulbild n. ‖ image f. pour écoles. / ~ requisites pl. ‖ Schulbedarf m. ‖ fournitures fpl. scolaires. / ~ ship ‖ Schulschiff n. ‖ bâtiment-école m.; vaisseau-école m.

schooner (Shipb) ‖ Schoner m. ‖ goélette f. / fore and aft ~ ‖ Gaffelschoner m. ‖ goélette f. franche. / main topsail ~ ‖ Schonerbark f. ‖ trois-mâts goélette f. carrée. / three-masted ~ ‖ Dreimastschoner m. ‖ trois mâts-goélette f.

schooner bark ‖ Barkschoner m. ‖ barque f. voilée en goélette. / ~ sail ‖ Schonersegel n. ‖ voile f. de goélette. / ~ yacht ‖ Schonerjacht f. ‖ schooner m.

schorl (Miner) ‖ Schörl m.; schwarzer Turmalin m. ‖ schorl m.

schorlaceous ‖ schörlähnlich; schörlartig ‖ schorliforme.

schreibersite (Miner) ‖ Phosphornickeleisen n.; Schreibersit m. ‖ schreibersite f.

schuyt (Shipb) ‖ Schute f. ‖ gabare f.; chaland m.; barge f.; péniche f.

sciatic (Med) ‖ Hüftweh n. ‖ sciatique f.

science ‖ Wissenschaft f. ‖ science f. / natural ~ ‖ Naturwissenschaft f.; Naturwissenschaften fpl. ‖ sciences fpl. naturelles. / physical ~ see natural ~. / ~ of political economy ‖ Volkswirtschaftslehre f ‖ économie f. politique. / ~ of strength of material ‖ Festigkeitslehre f. ‖ science f. de la résistance des matériaux.

scientific ‖ wissenschaftlich ‖ scientifique. / ~ apparatus ‖ wissenschaftlicher Apparat m.; wissenschaftliches Gerät n. ‖ appareil m. scientifique. / ~ collection ‖ wissenschaftliche Sammlung f. ‖ collection f. scientifique. / ~ glassware ‖ wissenschaftliche Glaswaren fpl. ‖ verreries fpl. scientifiques. / ~ institute ‖ wissenschaftliche Anstalt f. ‖ institut m. scientifique. / ~ instrument ‖ wissenschaftliches Instrument n. oder Gerät n. ‖ instrument m. scientifique. / ~ optic glass ‖ wissenschaftliches optisches Glas n. ‖ verre m. d'optique scientifique. / ~ research ‖ wissenschaftliche Untersuchung f. oder Forschung(sarbeit) f. ‖ recherche f. scientifique. / ~ work ‖ wissenschaftliche Arbeiten fpl. ‖ travaux mpl. scientifiques.

scientifically ‖ in wissenschaftlicher Weise; wissenschaftlich ‖ scientifiquement.

scintillate, to ‖ funkeln ‖ scintiller; étinceler; briller.

scintillating ‖ schillernd; funkelnd ‖ à reflets changeants; scintillant.

scintillation (Met) ‖ Funkenwerfen n.; Funkensprühen n. ‖ scintillation f.

scintillometer ‖ Flimmermesser m. ‖ scintillomètre m.

scion (Gard) ‖ Ableger m.; Steckling m. ‖ jet m.; bouture f.; rejeton m.

scissor-like ‖ scherenförmig; scherenartig ‖ en forme de ciseaux.

scissors pl. see also shears ‖ (kleinere) Schere f. ‖ ciseaux mpl. / buttonhole ~ ‖ Knopflochschere f. ‖ ciseaux mpl. à boutonnière. / ~ for cardboard ‖ Kartonschere f. ‖ ciseaux mpl. à carton. / ~ curved on flat ‖ aufwärts gebogene Schere f. ‖ ciseaux mpl. recourbés. / hair ~ ‖ Haarschere f. ‖ ciseaux mpl. de coiffeur. / manicure ~ ‖ Nagelschere f. ‖ ciseaux mpl. à ongles. / pin-maker's ~ ‖ Knopfschere f. ‖ ciseaux mpl. camards. / straight ~ ‖ gerade Schere f. ‖ ciseaux mpl. droits. / tailor's ~ ‖ Schneiderschere f. ‖ ciseaux mpl. de tailleur. / umbilical ~ ‖ Nabelschnurschere f. ‖ ciseaux mpl. à cordon ombilical.

scissors blade ‖ Scherenklinge f. ‖ lame f. de ciseaux. / ~ grinder ‖ Scherenschleifer m. ‖ affûteur m. ou repasseur m. de ciseaux. / ~ hardener ‖ Scherenhärter m. ‖ trempeur m. de ciseaux. / ~ maker ‖ Scherenschmied m. ‖ coutelier m. en ciseaux. / ~ nailer ‖ Schraubeneinsetzer m. (in der Scherenfertigung) ‖ monteur m. de vis (en coutellerie). / ~ polisher ‖ Scherenpolierer m. ‖ polisseur m. de ciseaux. / ~ polishing ‖ Scherenpolieren n. ‖ polissage m. de ciseaux. / ~ regrinding ‖ Scherenschleifen n. ‖ repassage m. de ciseaux.

scissors-shape collector (Electr) ‖ Scherenstromabnehmer m. ‖ prise f. de courant à parallélogramme articulé.

scissors vice ‖ Scherenkluppe f. ‖ serreciseaux m.

sclerotic(a) ‖ Lederhaut f.; Augenhaut f.; Sklera f. ‖ sclérotique f. / ~ lamp ‖ Skleralampe f. ‖ lampe f. skléra.

sclerotium (Botany) ‖ Fruchtkörper m. ‖ sclérote m.

scobs pl. ‖ Sägespäne mpl. ‖ sciure f.

scolezite ‖ Skolezit m.; Faserzeolit m.; Mesolit m. ‖ scolésite f.; mésolite f.

scollop, to ~ see to scallop.

sconce ‖ Wandleuchter m. ‖ bras m.; bras m. de chandelier ou d'applique mural.

scoop, to ~ (To draw water etc.) ‖ ausschöpfen; schöpfen ‖ puiser; épuiser; vider. / ~ (To shovel) ‖ ausschaufeln; schaufeln ‖ enlever à la pelle.

scoop (Dipper) ‖ Schöpfkelle f.; Schöpfgefäß n.; Schöpfer m. ‖ écope f.; bidon m. à puiser. / ~ (Shovel) ‖ Schaufel f.; Schippe f.; Wurfschaufel f. ‖ pelle f.; écope f. / ~ (Med) ‖ Spatel m. ‖ spatule f. / boat's ~ ‖ Schöpfkette f.; Wasserschaufel f.; Ösfaß n. ‖ écope f.; escope f.; sasse f. / coal ~ ‖ Kohlenschaufel f. ‖ pelle f. à charbon ou à feu. / round coal ~ ‖ gewölbte Kohlenschaufel f. ‖ pelle f. à feu ronde. / square coal ~ ‖ Vierkantkohlenschaufel f. ‖ pelle f. à feu carrée. / dredging ~ ‖ Baggerschaufel f. ‖ cuillère f. de drague.

scooper (Engrav) ‖ Grabstichel m.; Stichel m.; Zeiger m. ‖ burin m. du graveur; matoire f.

scoop-type wagon tipping all round ‖ Schnabelrundkipper m. ‖ wagonnet m. avec caisse en pelle basculant dans

toutes les directions. / ~ tipping at end ‖ Schnabelvorderkipper m. ‖ wagonnet m. avec bec basculant en bout.

scoop wheel ‖ Schöpfrad n. mit Schaufeln; Heberad n. ‖ roue f. élévatoire à aubes. / frame for ~s ‖ Heberadgerüst n. ‖ charpente f. métallique pour roue élévatrice.

scooter (For children) ‖ Roller m. ‖ trottinette f. / ~ boat ‖ Gleitboot n. ‖ hydroglisseur m.

scope of business ‖ Geschäftskreis m. ‖ sphère f. d'activité.

score, to ‖ einschneiden; einkerben ‖ entailler.

score ‖ Kerbe f.; Einschnitt m.; Ritze f.; Rille f.; Rinne f. ‖ encoche f.; entaillecoche f.; gorge f.

scoria (Geol) ‖ vulkanische Schlacke f. ‖ scorie f. volcanique. / ~ (Met) see also slag and cinder ‖ Schlacke f.; crasse f.; laitier m.; scorie f. / to run off the ~ ‖ die Schlacken fpl. abwerfen ‖ haler le laitier ou les scories fpl. / bed of ~ ‖ Schlackenbett n. ‖ lit m. de scorie.

scoria lead ‖ Schlackenblei n. ‖ plomb m. de scories.

scorification ‖ Verschlackung f. ‖ scorification f. / ~ roasting ‖ Schlackenrösten n. (von oxydierten Bleierzen) ‖ calcinage m. des scories.

scorify, to ~ a metal ‖ ein Metall verschlacken ‖ scorifier un métal.

scoring ‖ Einkerben n.; Einritzen n. ‖ entaillage m.

scorious (Met) ‖ verschlackt ‖ scorifié.

scorza (Miner) ‖ Skorza f. ‖ scorza m.

scotch, to ‖ einkerben; einschneiden ‖ entailler. / ~ wheels pl. by wedges ‖ Räder verkeilen oder keilhemmen ‖ caler les roues.

scotch block ‖ Hemmschuh m.; Bremskeil m. ‖ taquet m. d'arrêt; cale f.; sabot m. d'enrayement ou de freinage. / ~ cross (Brew) ‖ Anschwänzer m.; Anschwänzkreuz n. ‖ croix f. écossaise.

Scotch elm ‖ Bergulme f. ‖ orme m. de montagne. / ~ twist ‖ Flor m. ‖ fil m. d'Écosse.

scour, to ‖ (blank) putzen; scheuern; blank reiben; polieren ‖ écurer; polir; décaper; blanchir. / ~ (Hydr arch) ‖ schlämmen; spülen; schwemmen; wegwaschen ‖ débourber; curer. / ~ (Spinn) ‖ entfetten ‖ dessuinter; dégraisser. / ~ metals pl. Metalle npl. blank machen ‖ décaper des métaux mpl. / ~ the silk ‖ die Seide entbasten oder entschälen oder degummieren ‖ décreuser ou dégommer la soie. / ~ the vessel (Mar) ‖ das Schiff schrubben ‖ balayer le bâtiment. / ~ the wire ‖ den Draht scheuern ‖ écurer le fil.

scoured silk ‖ entbastete oder entschälte oder linde Seide f. ‖ soie f. décreusée ou dégommée.

scourer (Dyer) ‖ Entfetter m. ‖ lessiveur m. / ~ (Hydr arch) ‖ Spüler m. ‖ appareil m. de curage. / ~ (Needl) ‖ Beizer m. ‖ décapeur m. / ~ (Wash) ‖ Fleckenreiniger m. ‖ dégraisseur m.

scouring ‖ Scheuern n.; Polieren n. ‖ polissage m.; décapage m. / chemical ~ ‖ chemische Reinigung f. ‖ nettoyage m. chimique. / ~ of metals ‖ Scheuern n. oder Blankputzen n. der Metalle ‖ décapage m. des métaux. / ~ of sheet iron before tinning ‖ Scheuern n. des Eisenblechs vor dem Verzinnen ‖ écurage m. ou frottement m. du fer en feuilles avant

l'étamage. / ~ of silk ‖ Entschälen n. der Seide ‖ décreusage m. de la soie. / ~ of wool ‖ Wollentschweißung f. ‖ dessuintage m. de la laine. / ~ of woollen rags ‖ Waschen n. der Abfallwolle ‖ nettoyage m. de chiffons.

scouring basin (Hydr arch) ‖ Spülbecken n. ‖ bassin m. de chasse. / ~ boiler (Bleach) ‖ Beuchkessel m. ‖ chaudière f. à coulage. / ~ cloth ‖ Scheuertuch n.; Putzlappen m. ‖ torchon m. / ~ cloth weaving ‖ Putzlumpenweberei f. ‖ tissage m. de torchons. / ~ cradle ‖ Spültisch m. ‖ évier m. / ~ drops pl. ‖ Fleckwasser n. ‖ eau f. à dégraisser ou de Javelle. / ~ drum ‖ Scheuerglocke f. ‖ cloche f. à récurer. /~ machine for wool ‖ Wollwaschmaschine f. ‖ dégraisseuse f. de laine; laveuse f. mécanique de laine. / ~ machine feeder ‖ Waschmaschinenarbeiter m. ‖ laveur m. à la machine. / ~ powder ‖ Scheuerpulver n. ‖ poudre f. à récurer. / ~ room (Met) ‖ Scheuerkammer f. ‖ récurage m.; écurage f. / ~ sluice (Hydr arch) ‖ Ablaßschütze f.; Grundablaß m. ‖ bonde f. ou décharge f. de fond. / ~ stick ‖ Fleckstift m. ‖ crayon m. dégraissant. / ~ vat ‖ Scheuerfaß n.; Scheuertonne f. ‖ tambour m. de décapage ou de décrassage. / ~ water see ~ drops.

scout aeroplane ‖ Aufklärungsflugzeug n. ‖ avion m. d'éclairage ou de reconnaissance. / ~ cruiser (Mar) ‖ Aufklärungskreuzer m. ‖ croiseur-éclaireur m.

scovel (Bak) ‖ Ofenwisch m.; Lochkehrer m. ‖ écouvillon m.

scow (Shipb) ‖ Prahm m.; Fährboot n. ‖ prame f.; bateau m. de passage.

scrap ‖ Abfall m. ‖ déchet. / ~s pl. of pottery ‖ Bruchstücke npl. von Tonwaren ‖ débris m. d'ouvrages en poteries. / steel ~ ‖ Stahlschrott m. ‖ mitraille f. d'acier.

scrap baling press ‖ Schrottpaketierpresse f. ‖ presse f. à paqueter la mitraille. / ~ bundling machine ‖ Schrotbündelmaschine f. ‖ machine f. à agglomérer la mitraille. / ~ coiling machine ‖ Schrotwickelmaschine f. ‖ machine f. à enrouler la mitraille. / ~ coke ‖ Abfallkoks m. ‖ déchets mpl. de coke.

scrape, to ‖ schaben; abschaben ‖ gratter; racler. / ~ off the flesh particles (Curr) ‖ ausfleischen ‖ écoller les peaux fpl. / ~ off the hair (Curr) ‖ abhaaren ‖ ébourrer; épiler; surtondre. / ~ off the pieces (Found) ‖ Werkstücke npl. abkratzen ‖ enlever les défectuosités fpl. des pièces. / ~ up the roll (Print) ‖ die Walze abstreichen ‖ nettoyer le cylindre.

scraper ‖ Kratzer m.; Schaber m. ‖ racloir m. / ~ (Bookb) ‖ Schaber m. ‖ gratteur m. / ~ (Drying kiln) ‖ Esel m. ‖ baudet m. / ~ (Join) ‖ Ziehklinge f. ‖ racloir m. / ~ (Mach) ‖ Abstreichvorrichtung f. ‖ dispositif m. de racloir. / ~ (Mar) ‖ Schraper m. ‖ gratte f.; racle m. / ~ (Mine: loading) ‖ Kratze f. ‖ râble m.; rasette f. / ~ (Mine: shooting) ‖ Krätzer m.; Löffelberäumer m.; Räumlöffel m. ‖ curette f. de mineur; tire-sable m. / ~ (Print) ‖ Schabklinge f. ‖ lame f. à racler. / ~ (Salt) ‖ Salzkrücke f. ‖ râble m. du saunier. / ~ chimney sweeper's ~ ‖ Kaminfegerscharre f. ‖ grappin m. de ramoneur. / flat ~ (Engr) ‖ Radiernadel f. ‖ échoppe f.; burin m. pour effacer. / fluted ~ ‖ Hohlschaber m. ‖ grattoir m.

cannelé. / round ~ see flat ~. / four-square ~ ‖ vierschneidiger Schaber m. ‖ ébarboir m. ou grattoir m. carré. / three-square ~ ‖ dreischneidiger Schaber m.; Dreikantschaber m. ‖ grattoir m. triangulaire. / toothed ~ ‖ gezahntes Kratzeisen n. ‖ ripe f. / two-edged ~ ‖ doppelter Schraper m. ‖ racle m. double.

scraper band (Loading mach) ‖ Kastenförderband n. ‖ bande f. à raclettes. / ~ brushes pl. ‖ Bürstenabstreicher m. ‖ racleur m. à brosses. /~ conveyor ‖ Kratzbandförderer m. ‖ ruban-transporteur m. à racloirs. / ~ conveyor for salt ‖ Salzkratzer m. ‖ gratteur m. de sel. / ~ disc ‖ Kratzerscheibe f. ‖ table f. à racloir. / ~ lithographic stone ~-off ‖ Steinschleifer m. ‖ préparateur m. de ou effaceur m. sur pierres lithographiques. / ~ ring (Mach) ‖ Abstreifung m, ‖ segment m. / ~ wire for ~s ‖ Kratzendraht m. ‖ fil m. pour grattoirs.

scrap heap ‖ Haufen m. alten Eisens ‖ ferrailles fpl. amoncelées ou entanées.

scraping (Parchment) ‖ Schaben n. ‖ rature f. / ~ band ‖ Kratzband n. ‖ bande f. à racloirs. / ~ device ‖ Schabevorrichtung f. ‖ dispositif m. de grattoir.

scraping iron ‖ Schabeisen n.; Kratzeisen n.; Kratze f. ‖ grattoir m.; ébarboir m.; ébardoir m. / square ~ ‖ vierkantiges Kratzeisen n. ‖ ébardoir m.; grattoir m. carré.

scraping knife (Curr) ‖ Schabmesser n.; Streichmesser n. ‖ lame f. à raturer. / ~ out tool ‖ Räumwerkzeug n. ‖ outil m. dégorgeoir.

scrapings pl. (Curr) ‖ Aas n.; Abschabsel n. ‖ écharnure f. / ~ (Print) ‖ Abfälle mpl. ‖ petits bouts mpl. / ~ of liquation (Metal) ‖ Seigerkrätz f. ‖ pailles fpl. de liquation.

scrap iron ‖ Alteisen n.; Schrot m. ‖ vieux fer m.; ferraille f.; fer m. de ferraille; mitraille f. / ~ crushing ‖ Verschrotung f. ‖ démolissage m. de machines ou de constructions métalliques.

scrap metals pl. ‖ Metallabfälle mpl. ‖ déchets mpl. métalliques.

scrap piling machine ‖ Schrotpaketiermaschine f. ‖ machine f. à paqueter la mitraille. / ~ shearing machine ‖ Schrotschere f. ‖ cisaille f. à mitraille ou pour riblons. / ~ shears pl. see ~ shearing machine. / ~ spattle ‖ Abstreichspatel m. ‖ spatule f. racloir. / ~ spool ‖ Schrothaspel m; f. ‖ dévidoir m. à riblons. / ~ steel and iron market ‖ Schrottmarkt m. ‖ marché m. aux ferrailles. / ~ value ‖ Altwert m. ‖ valeur f. de vieux fer ou de mitraille. / ~ wholesale ~ store ‖ Schrottgroßhandlung f. ‖ marschand m. de ferraille en gros. / ~ yard crane ‖ Schrottlagerkran m. ‖ grue f. de parc à riblons.

scratch, to (Gild) ‖ rauhmachen; ritzen ‖ hacher; rayer. / ~ the nap of cloth ‖ den Stoff aufkratzen oder aufrauhen oder kardätschen ‖ égratigner ou lainer ou garnir l'étoffe f.

scratch ‖ Abschürfung f. ‖ écorchure f. / superficial ~ (Material) ‖ Anriß m.; Hautriß m.; Oberflächenriß m. ‖ fissure f. superficielle.

scratch brush ‖ Drahtbürste f. ‖ grattebosse f.; saie f.; gratte-brosse f.; grattebrosse. / ~ hardness ‖ Ritzhärte f. ‖ ré-

sistance f. au striage; dureté f. sclérométrique.

scratching-brush motor ‖ Kratzmotor m. ‖ moteur m. à gratte-boësser. / ~ knife ‖ Ritzklinge f.; Ritzer m. ‖ couteau m. à gratter. / ~ machine (Bookb) ‖ Ritzmaschine f. ‖ machine f. à rayer. / ~ machine (Galv) ‖ Kratzmaschine ‖ machine f. à gratte-boësser. / ~ stone ‖ Abritzstein m. ‖ pierre f. à gratter.

scratch test ‖ Ritzversuch f. ‖ essai m. sclérométrique.

scrawl, to ‖ kritzeln ‖ griffonner.

scray for fabric (Weav) ‖ Warengleitmulde f. ‖ bac m. pour le tissu.

screen, to ‖ sieben; absieben; durchsieben ‖ tamiser; cribler; passer au crible *ou* au tamis. / ~ (Ore dress) ‖ klassieren ‖ classer; trier; séparer.

screen (Agr) ‖ Räder m.; Rätter m. ‖ crible. / ~ (Build) ‖ Gitter n.; Schranke f. ‖ écran m. / ~ (Cinema) ‖ Projektionsleinwand f. ‖ écran m. / ~ (Opt; Radio) ‖ Schirm m. ‖ écran m. / ~ (Phot) ‖ Blende f. ‖ diaphragme m.; écran m. / ~ (Ore dress) ‖ Sieb n. ‖ tamis m.; crible m. / ~ (Print) ‖ Raster m. ‖ trame f. / ~ (Sand) ‖ Sandsieb n.; Durchwurf m. ‖ tamis m de passage; crible m. à pied. / antidazzling ~ (Auto) ‖ Blendschutzscheibe f. ‖ écran m. antiéblouissant. / ~ of barium-platinacyanide ‖ Bariumplatincyanürschirm m. ‖ écran m. en platine-cyanure de barium. / coloured ~ (Opt) ‖ Lichtfilter m. ‖ écran m. coloré. / electrical ~ ‖ elektrischer Schirm m. ‖ écran m. électrique. / focussing ~ (Phot) ‖ Mattscheibe f. ‖ verre m. dépoli. / frontal ~ (Ore and coal dress) ‖ Vordersieb n. ‖ tamis m. de devant. / intensifying ~ (Röntgen) ‖ Verstärkungsschirm m. ‖ écran m. renforçateur. / metal ~ ‖ Metallschirm m. ‖ écran m. en métal. / ~ with narrow openings ‖ Feinsieb n. ‖ crible m. fin. / primary classifying ~ ‖ Vorklassiersieb n. ‖ crible m. avant-classeur. / reflecting ~ ‖ reflektierender Schirm m. ‖ écran m. réflecteur. / rinsing ~ ‖ Abbraussieb n. ‖ tamis m. d'arrosage. / shaking ~ ‖ Rüttelsieb n. ‖ tamis m. à secousse. / ~ for a stove ‖ Ofenschirm m. ‖ écran m. de cheminée. / yellow ~ ‖ Gelbfilter n. ‖ verre m. jaune.

screened (Mill) ‖ gesiebt ‖ criblé. / ~ (Opt) ‖ abgeschirmt ‖ cuirassé; muni d'un écran.

screen embroiderer ‖ Schirmstickerin f. ‖ brodeuse f. sur écrans.

screener, coke ~ ‖ Kokssieber m. ‖ cribleur m. de coke.

screening (Brew) ‖ Entkeimung f. ‖ dégermination f.; stérilisation f. / ~ (Ore dress) ‖ Klassierung f. ‖ classement f.; triage m.; séparation f. / dry ~ ‖ Trockensiebung f. ‖ criblage m. par voie sèche.

screening box ‖ Siebkasten m. ‖ boîte f. de criblage. / ~ circuit ‖ Siebkreis m. ‖ circuit m. filtre. / ~ device ‖ Absiebvorrichtung f. ‖ dispositif m. de criblage. / magnetic ~ device ‖ Magnetscheideanlage f. ‖ installation f. de triage magnétique. / ~ drum ‖ Siebtrommel f. ‖ tambour m. de tamisage.

screening effect ‖ Abschirmwirkung f. ‖ effet m. de parachute. / ~ of rail currents and cable-sheath currents ‖ Induktionsminderung f. durch Schienenstrom und Kabelmantelstrom ‖ effet m. compensateur du courant de rail et

d'enveloppe de câbles. / ~ of a range of trees ‖ Influenzschutzwirkung f. einer Baumreihe ‖ effet m. d'écran d'une rangée d'arbres.

screening frame ‖ Siebkranz m. ‖ bord m. de tamis.

screening machine (Brew) ‖ Entkeimungsmaschine f. ‖ dégermeuse f. / ~ (Mill) ‖ Siebmaschine f. ‖ machine f. à tamiser.

screening and scouring machine (Brew) ‖ Entkeimungs- und Poliermaschine f. ‖ machine f. à dégermer et à polir. / ~ plant (Mill) ‖ Sieberei f. ‖ criblage m. / ~ plant (Ore dress) ‖ Separationsanlage f. ‖ installation f. de triage. / building of the ~ plant ‖ Sieberereigebäude n. ‖ bâtiment m. de criblage. / ~ plate ‖ Siebblech n. ‖ tôle f. de triage. / ~ refuse ‖ Siebrückstände mpl. ‖ refus m. de crible *ou* de tamisage. / ~ wall ‖ Schirmwand f. ‖ chasse f.; paroi f. de l'écran. / ~ work ‖ Scheidearbeit f. ‖ opération f. de séparer.

screen opening ‖ Sieböffnung f. ‖ screen opening. / ~ wall *see* screening wall. / ~ wiper (Auto) ‖ Scheibenwischer m. ‖ essuie-glace m.

screw, to ‖ einschrauben; schrauben ‖ visser; boulonner. / ~ off ‖ losschrauben; abschrauben ‖ dévisser; desserrer *ou* défaire *ou* ôter les vis; déboulonner. / ~ on ‖ anschrauben; aufschrauben; festschrauben ‖ visser. / the bearings pl. are screwed on the frame ‖ die Lager npl. sind auf dem Rahmen aufgeschraubt ‖ les paliers mpl. sont vissés sur le châssis. / ~ together ‖ zusammenschrauben ‖ visser; boulonner ensemble; réunir par des boulons.

screw (Mach) ‖ Schraube f. ‖ vis f.; boulon m. / ~ (Shipb) ‖ Propellerschraube f.; Schiffsschraube f. ‖ hélice f.; hélice f. propulsive. / adjusting ~ ‖ Stellschraube f. ‖ vis f. de réglage. / archimedean ~ ‖ archimedische Schraube f. ‖ vis f. d'Archimède. / binding ~ *see* clamping ~. / ~ for carriages ‖ Wagenschraube f. ‖ vis f. de carrosserie. / clamping ~ ‖ Klemmschraube f. ‖ serre-fil m. / ~ for constructions ‖ Bauschraube f. ‖ vis f. pour constructions. / contact ~ ‖ Kontaktschraube f. ‖ vis f. de contact. / cork ~ ‖ Pfropfenzieher m.; Korkenzieher m. ‖ tire-bouchon m. / countersunk ~ ‖ Senkschraube f. ‖ vis f. à tête fraisée; vis f. noyée. / ~ for cover plate ‖ Deckplattenschraube f. ‖ boulon m. pour plaque de recouvrement. / ~ with curved blades ‖ Schraube f. mit gekrümmten Flügeln ‖ hélice f. à ailes cintrées. / to cut ~s pl. ‖ Schrauben fpl. schneiden ‖ fileter; tarauder. / ~ with cylindrical head ‖ Schraube f. mit walzenförmigem Kopf ‖ vis f. avec tête cylindrique. / differential ~ ‖ Differentialschraube f. ‖ vis f. à double pas de Prony; vis différentielle. / disconnecting ~ ‖ Schraube f. zum Entkuppeln ‖ hélice f. à débrayer. / double-thread ~ ‖ zweigängige Schraube f. ‖ vis f. à double pas. / eyeglass ~ ‖ Kneiferschraube f. ‖ vis f. pour pince-nez. / female ~ ‖ Schraubenmutter f. ‖ écrou m. / ~ with fixed blades ‖ Schraube f. mit festen Flügeln ‖ hélice f. à ailes fixes. / fixing ~ ‖ Befestigungsschraube f. ‖ boulon m. d'attache. / forcing ~ ‖ Abdrückschraube f. ‖ boulon m. de détente. / four-bladed ~ ‖ vierflügelige Schraube f. ‖ hélice f. à

quatre ailes. / hand-forged ~ ‖ handgeschmiedete Schraube f. ‖ vis f. forgée à la main. / helicopter ~ ‖ Hubschraube f. ‖ hélice f. sustentatrice. / knurled ~ ‖ Rändelschraube f. ‖ vis f. molletée. / lefthand ~ ‖ linksgängige Schraube f. ‖ vis f. filetée à gauche. / levelling ~ *see* adjusting ~. / lifting ~ ‖ Hebeschraube ‖ vérin m. / locking ~ ‖ Verschlußschraube f. ‖ vis f. de fermeture. / ~ for locomotives ‖ Lokomotivschraube f. ‖ vis f. pour locomotives. / machine ~ ‖ Maschinenschraube f. ‖ vis f. de machines. / measuring ~ ‖ Meßschraube f. ‖ vis f. de mesure. / metal ~ with turns ‖ Metallgewindeschraube f. ‖ vis f. de métal avec pas. / micrometer ~ ‖ Mikrometerschraube f. ‖ vis f. micrométrique *ou* à micromètre. / milled head ~ ‖ Kordelschraube f. ‖ vis f. molletée. / multiplex thread ~ ‖ mehrgängige Schraube f. ‖ vis f. à plusieurs filets. / ~ for oil feed ‖ Ölzuführungsschraube f. ‖ vis f. d'amenée d'huile. / oil hole ~ (Railw) ‖ Schmierschraube f. ‖ bouchon m. fileté de graisseur. / overhung ~ (Shipb) ‖ freitragende Schraube f. ‖ hélice f. en porte à faux. / perpetual ~ ‖ Schraube f. ohne Ende ‖ vis f. sans fin. / piston pin lock ~ ‖ Kolbenbolzensicherungsschraube f. ‖ vis f. de fixation du boulon de piston. / ~ for piston rod ‖ Kolbenstangenschraube f. ‖ vis f. de tige de piston. / platinum-tipped ~ ‖ Platinschraube f. ‖ vis f. platinée. / polished ~ ‖ blanke Schraube f. ‖ vis f. polie. / precision ~ ‖ Präzisionsschraube f. ‖ vis f. de précision. / pressed ~ ‖ gepreßte Schraube f. ‖ vis f. estampée. / pressing ~ ‖ Druckschraube f. ‖ vis f. de pression. / ~ for rails ‖ Schienenschraube f. ‖ vis f. pour rails. / raised ~ ‖ hochköpfige Schraube f. ‖ vis f. saillante. / right-hand ~ ‖ rechtsgängige Schraube f. ‖ vis f. filetée à droite. / semi-machined ~ ‖ preßblanke Schraube f. ‖ vis f. forgée. / set ~ *see* adjusting ~. / single-thread ~ ‖ eingängige Schraube f. ‖ vis f. à pas simple. / slotted ~ ‖ Schlitzschraube f. ‖ boulon m. fendu. / spectacle ~ ‖ Brillenschraube f. ‖ vis f. de lunettes. / square thread ~ ‖ flachgängige Schraube f. ‖ vis f. à filet rectangulaire. / steering ~ ‖ Lenkschraube f. ‖ vis f. de direction. / stretching ~ ‖ Zugschraube f. ‖ vis f. de traction. / to strip the thread of a ~ ‖ eine Schraube f. überdrehen ‖ déformer une vis. / sunk ~ ‖ versenkte Schraube f. ‖ vis f. à tête noyée; vis f. perdue. / tension ~ ‖ Zugschraube f. ‖ vis f. de tension. / ~ with x threads per inch ‖ Schraube f. mit x Gängen auf einem Zoll ‖ vis f. à x pas au pouce. / three-bladed ~ ‖ dreiflügelige Schraube f. ‖ hélice f. à trois ailes. / thumb ~ ‖ Flügelschraube f. ‖ vis f. à oreilles. / transport ~ ‖ Transportschraube f. ‖ vis f. de transport. / ~ with a triangular thread ‖ scharfgängige Schraube f. ‖ vis f. à filet triangulaire. / triple-thread ~ ‖ dreigängige Schraube f. ‖ vis f. à triple pas. / turbine ~ ‖ Turbinenschraube f. ‖ propulseur m. à turbine. / two-bladed ~ ‖ zweiflügelige Schraube f. ‖ hélice f. à deux ailes. / ~ for wagons ‖ Waggonschraube f. ‖ vis f. pour wagons. / winged ~ *see* thumb ~. / withdrawing ~ ‖ Abziehschraube f. ‖ vis f. à desserrer. / wood ~ ‖ Holzschraube f. ‖ vis f. à bois.

/ making wood ~s pl. ‖ Holzschrauben-
herstellung f. ‖ fabrication f. de vis à
bois. / wooden ~ ‖ Schraube f. aus Holz;
hölzerne Schraube f. ‖ vis f. en bois. / ~
and wheel ‖ Schneckenradgetriebe n. ‖
engrenage m. à vis sans fin.
screw assuring see ~ locking device. / attachment by ~s ‖ Schraubenbefestigung
f. ‖ assemblage m. par vis.
screw auger ‖ Spiralbohrer m. ‖ tarière f.
torse; foret m. hélicoïdal. / double-lipped
~ ‖ doppelt gewundener Spiralbohrer
m. ‖ tarière f. à vis double. / single-
lipped ~ ‖ einfach gewundener Spiralbohrer m. ‖ tarière f. à vis simple.
screw blade ‖ Schraubenflügel m. ‖ aile f.
d'hélice. / ~ attached to the boss ‖ an-
gesetzter Schraubenflügel m. ‖ aile f.
d'hélice rapportée au moyeu.
screw blowing machine ‖ Schraubenge-
bläse n. ‖ Cagniardelle f.; soufflerie f.
à vis sans fin.
screw bolt ‖ Schraubenbolzen m. ‖ boulon
m. fileté (ou à vis) ou taraudé. / hinged ~
‖ Klappschraubenbolzen m. ‖ boulon m.
articulé. / automatic ~ press ‖ selbsttätige Schraubenbolzenpresse f. ‖ presse
f. automatique à faire les boulons.
screw brake ‖ Spindelbremse f. ‖ frein m.
à vis. / truck with eight-shoe ~ ‖ Drehgestell n. mit achtklotziger Spindelbremse ‖ bogie f. avec frein à vis à huit
sabots. / horizontal ~ ‖ horizontale Spindelbremse f. ‖ frein m. à vis horizontale.
screw buffer ‖ Schraubenpuffer m. ‖ tampon m. taraudé ou à vis. / ~ bushing ‖
Schraubhülse f. ‖ vis-grain m. ou écrou
m. d'accouplement.
screw cap ‖ Überwurfmutter f. ‖ écrou m.
à chapeau. / ~ (Mot) ‖ Verschlußstück
n.; Gewindestopfen m. ‖ pièce f. de
fermeture; bouchon m. à vis. / ~ for
filling hole ‖ Füllochschraube f. ‖ bouchon m. fileté du trou de remplissage.
screw cartridge (Electr) ‖ Schraubpatrone
f. ‖ cartouche f. à pas de vis. / ~ chase
(Print) ‖ Schraubenrahmen m. ‖ châssis
m. à vis. / ~ chuck (Turn; Lathe) ‖
Schraubfutter n. ‖ mandrin m. à vis. /
~ clamp (Join) ‖ Leimzwinge f.; Schraubzwinge f. ‖ sergent m.; presse f. à main
ou à serrer; serre-joint m. / ~ compasses
pl. ‖ Schraubenzirkel m. ‖ compas m.
à vis ou à ressort ou élastique. / ~ conveyor ‖ Förderschnecke f. ‖ vis f. transporteuse.
screw coupling ‖ Schraubenkupplung f. ‖
attelage m. ou accouplement m. à vis. /
~ box ‖ Schraubenkupplungsmuffe m. ‖
manchon m. à vis. / machine for repairing
~s ‖ Maschine f. zur Ausbesserung von
Schraubenkupplungen ‖ machine f. pour
réparer les accouplements à vis.
screw cutter see screw cutting machine.
screw cutting ‖ Schraubenschneiden n.;
Gewindeschneiden n. ‖ taraudage m.;
filetage m.; opération f. de fileter ou
de tarauder. / ~ lathe ‖ Schraubendreh-
bank f.; Schraubenschneidebank f. ‖ tour
m. à fileter. / ~ machine ‖ Gewinde-
schneidemaschine f. ‖ machine f. à fileter.
/ ~ tool ‖ Gewindeschneidwerkzeug n. ‖
outil m. à fileter ou tarauder.
screw dies pl. ‖ Schneidbacken fpl. ‖ coins
mpl. à vis; coussinets mpl. de filière. /
~ of a screw stock ‖ Schneidbacken mpl.
einer Schraubenkluppe ‖ coussinets mpl.
(à fileter) d'une filière (brisée).

screw divider see ~ compasses pl. / ~ dividing machine ‖ Schraubenteilmaschine
f. ‖ machine f. à diviser à vis.
screw driver ‖ Schraubenzieher m. ‖ tournevis m. / ~ with handle ‖ Schraubenzieher m. mit Heft ‖ tournevis m. à
manche. / ~ with reversible bits ‖
Schraubenzieher m. zum Umstecken ‖
tournevis m. pouvant servir des deux
côtés ou aux deux bouts.
screwed dowel ‖ Schraubendübel m. ‖
cheville f. en bois vissée. / ~-on flange ‖
aufgeschraubter Flansch m. ‖ bride f.
vissée. / ~ pipe joint ‖ Rohrverschraubung f. ‖ assemblage m. de tuyaux par vis.
/ ~ plug ‖ verschraubbarer Verschlußstopfen m. ‖ bouchon m. fileté. / ~ rod ‖
Stange f. mit Gewinde ‖ tige f. filetée. /
~ spike (Railw) ‖ Schwellenschraube f. ‖
tirefond m.
screw eyelet ‖ Ösenschraube f. ‖ piton m.
fileté. / ~ factory ‖ Schraubenfabrik f. ‖
boulonnerie f.; fabrique f. de vis. /
~ gauge ‖ Schraubenlehre f. ‖ calibre m.
à vis.
screw head ‖ Schraubenkopf m. ‖ tête f.
de vis. / circular ~ ‖ Halbrundkopf m.
der Schraube ‖ tête f. de vis en goutte
de suif. / to countersink a ~ ‖ einen
Schraubenkopf m. versenken ‖ noyer la
tête d'une vis. / ~ slotting attachment ‖
Schraubenkopfschlitzeinrichtung f. ‖ dispositif m. à fendre les têtes de vis.
screw hook ‖ Schraubhaken m. ‖ crochet
m. à vis.
screwing machine, screw spike ~
‖ Schwellenschrauben-Eindrehmaschine
und -Ausdrehmaschine f. ‖ machine f.
à visser et dévisser les tirefonds.
screwing, part for ~ in ‖ Einschraubteil m.
‖ pièce f. à visser.
screw-in retaining ring for interchanging
lenses ‖ einschraubbarer Haltering m.
zum Auswechseln der Linsen ‖ bague f.
filetée permettant d'interchanger les lentilles. / ~ thermometer ‖ Einschraubthermometer n. ‖ thermomètre m. à
visser.
screw jack ‖ Schraubenwinde f. ‖ vérin m.;
verrin m.; cric m. à vis. / ~ key ‖ Schraubenschlüssel m. ‖ clef f. à vis ou à écrous
ou de service. / ~ knob ‖ Schraubenkopf
m. ‖ tête f. de vis. / ~ locking device ‖
Schraubensicherung f. ‖ dispositif m. de
sûreté pour vis; frein m. à vis. /
machine for making ~s ‖ Schraubenher-
stellungsmaschine f. ‖ machine f. à
fabriquer des vis. / ~ mill ‖ Schrauben-
mühle f. ‖ moulin m. à vis. / ~ milling
cutter ‖ Gewindefräser m. ‖ fraise f. à
fileter. / ~ nail ‖ Holzschraube f. ‖ vis f.
à bois. / ~ neck ‖ Schraubenhals m. ‖
collet m. de vis.
screw nut ‖ Schraubenmutter f. ‖ écrou m.
/ ~ bevelling machine ‖ Mutternabkant-
maschine f. ‖ machine f. à ébarber les
écrous. / ~ burr removing machine ‖
Mutternabgratmaschine f. ‖ machine f.
à ébarber les écrous. / ~ press ‖ Mutternpresse f. ‖ machine f. à estamper des écrous.
screw pile ‖ Schraubenpfahl m. ‖ pieu m.
à vis ou à extremité filetée.
screw pitch ‖ Gewindesteigung f. ‖ pas m.
de vis. / ~ gauge ‖ Gewindeschablone f.
‖ calibre m. de taraudage.
screw plate see ~ stock. / ~ plough plane ‖
Nuthobel m. mit Stellung ‖ bouvet m.
brisé ou de deux pièces.

screw plug ‖ Schraubverschluß m. ‖ fermeture f. à vis. / ~ box (Electr) ‖ Schraub-
steckdose f. ‖ bouchon m. de contact
à vis. / ~ washer ‖ Verschlußschrauben-
dichtung f. ‖ joint m. du bouchon graisseur fileté.
screw press ‖ Spindelpresse f. ‖ presse f.
à vis. / rapid-action ~ ‖ Schnellspindelpresse f. ‖ presse f. à vis rapide.
screw propeller (Shipb) ‖ Propellerschraube
f.; Schiffsschraube f. ‖ hélice f.; hélice
propulsive. / ~ rail anchor ‖ Schraubenklemme f. ‖ dispositif m. d'ancrage à vis.
/ ~ reversing gear ‖ Schraubensteuerung
f. ‖ changement m. de marche à vis. /
sections pl. for ~s ‖ Schraubeneisen n. ‖
profilés mpl. pour la fabrication de boulons et vis. / ~ shaft ‖ Schraubenwelle
f. ‖ arbre m. porte-hélice. / ~ slotting
cutter ‖ Schraubenschlitzfräser m. ‖ fraise
f. pour entailler les têtes de vis. / ~
spanner see screw wrench. / ~ spindle
steering gear ‖ Schraubenspindelsteuerung f. ‖ commande f. par tirant fileté. /
~ spring ‖ Schraubenfeder f. ‖ ressort m.
hélicoïdal ou à boudin. / ~ steamer ‖
Schraubendampfer m. ‖ vapeur m. à
hélice. / inverted ~ steamer marine engine ‖ stehende Schraubendampfer-
schiffsmaschine f. ‖ machine f. marine
verticale à hélice. / ~ stock ‖ Schneidkluppe f. ‖ filière f. à coussinets ou
simple. / ~ tap ‖ Gewindebohrer m. ‖
taraud m. / ~ terminal ‖ Befestigungsklemme f. ‖ borne f. de fixation.
screw thread ‖ Schraubengewinde n. ‖ filet
m. / to chase a ~ ‖ ein Gewinde n. nachschneiden ‖ aviver ou repasser un filet. /
left-handed ~ ‖ linksgängiges Gewinde n.;
Linksgewinde n. ‖ filet m. renversée. /
right-handed ~ ‖ rechtsgängiges Gewinde n.; Rechtsgewinde n. ‖ filet m.
à droite.
screw thread rolling machine ‖ Gewinde-
kaltwalzmaschine f. ‖ machine f. à laminer à froid les filets de vis.
screw tool see ~ cutting tool. / ~ vice ‖
Schraubstock m. ‖ étau m.; cadran m. /
~ well ‖ Schraubenbrunnen m. ‖ puits m.
de l'hélice.
screw wheel ‖ Schraubenrad n. ‖ roue f.
hélice. / half of ~ ‖ Schraubenradhälfte f.
‖ moitié f. d'une roue hélice. / ~ threading
machine ‖ Schraubenräderfräsmaschine
f. ‖ machine f. à fraiser les roues cylindriques hélicoïdales. / ~ wire ‖ Schrauben-
draht m. ‖ fil m. pour vis.
screw wrench ‖ Schraubenschlüssel m. ‖
clef f. à vis ou à écrous ou de service. /
universal ~ ‖ Universalschlüssel m. ‖ clef
f. anglaise ou universelle. / opening of
the ~ ‖ Maulweite f. des Schrauben-
schlüssels ‖ ouverture f. de clef à écrou.
scribble, to ‖ kritzeln ‖ griffonner. / ~
(Spinn) ‖ schrubbeln ‖ drosser; drous-
ser; scribler.
scribbler (Spinn) ‖ Reißkrämpel f.; Schrub-
belmaschine f. ‖ briseuse f.; carde f. à
nappes; drousse f.; droussette f.
scribbling (Spinn) ‖ Schrubbeln n. ‖
cardage m.; droussage m.; scriblage m.
/ ~ machine see scribbler. / ~ paper ‖
Konzeptpapier n. ‖ papier m. écolier
ou à minutes.
scribe awl ‖ Reißspitze f. ‖ pointe f. à
tracer.
scribing block ‖ Winkelstreichmaß n. ‖
trusquin m. à équerre.

scrip ‖ Interimsschein m. ‖ certificat m. provisoire.

script face, reflected ~ ‖ Spiegelschrift f. ‖ écriture f. reflétée.

scroll ‖ Schnörkel m. ‖ ornement m. en spirale; spirale f. / ~ spring ‖ Sprungfeder f. ‖ ressort m. à boudin.

scrub, to ~ the vessel ‖ das Schiff schrubben ‖ balayer le bâtiment.

scrubber (Gas) ‖ Gaswascher m. ‖ laveur à gaz. / ~ (Househ) ‖ Schrubber m. ‖ brosse f. à manche. / barrel ~ ‖ Faßwaschapparat m. ‖ appareil m. à laver les fûts.

scrubbing brush ‖ Scheuerbürste f. ‖ brosse f. à frotter. / ~ cloth ‖ Scheuertuch n. ‖ torchon m. / ~ machine for casks ‖ Faßbürstmaschine f. ‖ machine f. à brosser les fûts; brosseuse f. pour fûts. / cask ~ and cleaning machine ‖ Faßputz- und Schwenkmaschine f. ‖ machine f. à brosser à à rincer les fûts.

scrutinize, to ‖ eingehend prüfen ‖ scruter.

scrutinized ‖ genau geprüft ‖ scruté; examiné.

scud, to ‖ vor dem Winde m. segeln ‖ courir vent arrière ou en poupe.

scuffle ‖ Schürze f. ‖ tablier m.

sculper ‖ Grabstichel m.; Stichel m.; Zeiger m. ‖ burin m. du graveur; matoire f. / flat ~ ‖ Flachstichel m. ‖ échoppe f. plate.

sculpting machine ‖ Bildhaumaschine f. ‖ machine f. à sculpter.

sculptor ‖ Bildhauer m. ‖ sculpteur m.

sculptor's implements pl. ‖ Bildhauerwerkzeug n. ‖ outil m. de sculpteur. / ~ plaster ‖ Bildhauergips m. ‖ plâtre m. de moulage.

sculptural work ‖ Bildhauerarbeit f. ‖ travail m. de sculpture.

sculpture ‖ Skulptur f. ‖ sculpture f.

scum, to ~ off ‖ abschlacken; abschäumen ‖ écumer.

scum ‖ Schaum m. ‖ écume f. / ~ of yellow brass ‖ Messingabzug m.; Abstich m. ‖ écume f. de laiton.

scum basket ‖ Schaumkorb m. ‖ panier m. à écume.

scummer ‖ Schaumlöffel m. ‖ écumoir m.

scurvy grass ‖ Löffelkraut n. ‖ herbe f. de cochléaria.

scutcher (Spinn) ‖ Schwingmaschine f. ‖ teilleuse f.; moulin m. flamand.

scuttle (Shipb) ‖ Niedergang m.; Niedergangsluk n. ‖ écoutille f. / coaling ~ (Shipb) ‖ Kohlenluk n. ‖ sabord m. pour l'embarquement du charbon.

scythe ‖ Sense f. ‖ faux f. / English ~ ‖ Schleifsense f. ‖ faux f. façon anglaise. / German ~ ‖ Klopfsense f. ‖ faux f. façon d'Allemagne.

scythe anvil ‖ Dengelamboß m. ‖ enclumette f.; chaploir m. / ~ cane see ~ stick. / ~ grinder ‖ Sensenschleifer m. ‖ aiguiseur m. de faux. / ~ hammer ‖ Dengelhammer m. ‖ marteau m. à chapler les faux. / ~ ring ‖ Sensenring m. ‖ anneau m. de faux. / ~ rubber ‖ Sensenstein m. ‖ dalle f.; pierre f. à aiguiser la faux. / ~ smith ‖ Sensenmacher m. ‖ martineur m. de faux. / ~ steel ‖ Sensenstahl m. ‖ acier m. à faux. / ~ stick ‖ Sensenbaum m. ‖ manche m. de faux.

sea ‖ See f.; Meer n. ‖ mer f. / ~ (Motion of the sea) ‖ Seegang m.; See f. ‖ mouvement m. de la mer. / at ~ ‖ auf See; in

See; zur See ‖ à la mer; en mer. / now at ~ ‖ gegenwärtig auf See ‖ maintenant en mer. / the ~ breaks ‖ die See brandet oder bricht ‖ la mer brise ou falaise. / the ~ gets up ‖ die See kommt auf ‖ la mer se fait. / heavy or high ~ ‖ hohe See f.; hochgehende See f. ‖ mer f. grosse. / open ~ ‖ offene See f.; hohe See f. ‖ large m.; mer f. ouverte; haute ou pleine mer f. / rough ~ ‖ grobe See f. ‖ mer f. dure ou houleuse. / short ~ ‖ kurze See f. ‖ mer f. courte. / smooth ~ ‖ ruhige See f. ‖ mer f. calme.

sea air ‖ Seeluft f. ‖ vent m. de mer. / ~ amber ‖ Bernstein m. ‖ succin m. / ~ anchor ‖ Seeanker m. ‖ ancre f. du large. / ~ biscuit ‖ Schiffszwieback m.; Hartbrot n. ‖ biscuit m. (de mer). / ~ box ‖ Seekiste f. ‖ caisse f. de mâtelot; coffre m. de bord.

seabread ‖ Schiffszwieback m. ‖ biscuit m. de mer.

sea breeze (Mar) ‖ Seebrise f.; Seewind m. ‖ brise f. du large; vent m. de mer. / ~ chart ‖ Seekarte f.; Paßkarte f. ‖ carte f. marine ou nautique ou hydrographique. / ~ chest see ~ box. / ~ coast ‖ Meeresküste f.; Seeküste f. ‖ côte f.; bord m. de la mer. / ~ compass ‖ Seekompaß m. ‖ boussole f.; compas m. de mer ou de route. / ~ damage ‖ Seebeschädigung f.; Havarie f. ‖ avarie f. / ~ defence (Hydr arch) ‖ Schirmwehr n. ‖ brise-lames m. ; revêtement m. des côtes. / ~ defence work ‖ Hafenbau m. ‖ travail m. de port maritime. / ~ dike ‖ Seedeich m. ‖ digue f. à la mer. / ~ earth (Cable) ‖ See-Erde f. (im Kabelbetrieb) ‖ terre f. au large.

seafaring man ‖ Seemann m.; Seefahrer m. ‖ homme f. de mer; marin m.

sea-fast ‖ seefest ‖ exempt du mal de mer; pratiqué à la mer.

sea fish ‖ Seefisch m.; Meeresfisch m. ‖ poisson m. de mer. / ~ fisherman ‖ Seefischer m. ‖ pêcheur m. en mer; marin-pêcheur m.

seafishing ‖ Seefischerei f. ‖ pêche f. maritime ou en mer.

sea-foam (Miner) see also meerschaum. ‖ Meerschaum m. ‖ écume f. de mer.

sea gate of a sluice ‖ Sicherheitstor n. oder Fluttor n. einer Schleuse ‖ écluse f. de garde.

sea gauge ‖ Tiefenmesser m. ‖ bathomètre m.

sea-going craft ‖ Fahrzeug n. der Seeschiffahrt ‖ navire m. pour la navigation maritime. / ~ plane ‖ Überseeflugzeug n. ‖ avion m. ou hydravion m. de haute mer. / ~ ship ‖ Seeschiff n. ‖ vaisseau m. ou bâtiment m. de mer. / ~ vessel see ~ ship.

sea-grass ‖ Seegras n. ‖ varech m.

sea gravel ‖ Seekies m. ‖ gravier m. de mer.

sea-green ‖ Meergrün n. ‖ vert m. de mer; glauque m.

seal, to ‖ siegeln ‖ cacheter; sceller. / ~ legally ‖ gerichtlich versiegeln ‖ apposer des scellés. / ~ off the end of the glass tube by fusing it ‖ die Glasröhre f. zuschmelzen ‖ souder le tube de verre.

seal (Zoology) ‖ Seehund m.; Robbe f. ‖ chien m. marin; loup m. de mer; phoque m.

seal (Signet) ‖ Siegel n.; Petschaft n. ‖ sceau m.; cachet m.; scellé m. / ~ (Of

lead) ‖ Plombe f. ‖ plomb m. / adhesive ~ ‖ Siegelmarke f. ‖ étiquette f. pour lettres; pain m. à cacheter. / medicamental ~ ‖ Oblate f. ‖ pain m. azyme; cachet m. médicamenteux. / ~ of office ‖ Amtssiegel n. ‖ sceau m. officiel. / oil ~ (of crankcase) ‖ Ölfangring m. ‖ anneau m. (du carter) de retenue d'huile.

seal bread of paste ‖ Siegelbrot n.; Oblate f. aus Teig ‖ pain m. à cacheter en pâte.

sealed ‖ versiegelt ‖ scellé; cacheté. / ~ tube ‖ Druckrohr n. ‖ tube m. scellé.

sealer (Seal fisher) ‖ Robbenschläger m.; Seehundjäger m. ‖ chasseur m. ou pêcheur m. de phoques.

sea-level ‖ Meereshöhe f. ‖ niveau m. de la mer.

seal fisher see sealer.

sea light ‖ Kursfeuer n.; Seeleuchte f. ‖ feu m. de grand atterrage; feu m. de reconnaissance; phare m. de première ordre.

sealing (Mar) ‖ Robbenfang m. ‖ pêche f. de phoques.

sealing (With wax etc.) ‖ Siegeln n. ‖ apposition f. du cachet ou du sceau. / ~ flange (Mot) ‖ Verschlußflansch m. ‖ bride f. de fermeture. / ~ tongs pl. ‖ Plombierzange f. ‖ pince f. pour plombs. / ~ wafer ‖ Siegeloblate f.; Mundlack m. ‖ pain m. à cacheter.

sealing wax‖Siegellack m.‖cire f. à cacheter. / ~ heater ‖ Siegellackschmelzer m. ‖ chauffe-cire m. à cacheter. / ~ lamp ‖ Siegellampe f. ‖ lampe f. pour fondre la cire à cacheter.

sea lock ‖ Seeschleuse f. ‖ écluse f. maritime.

seal mark ‖ Siegelmarke f. ‖ pain m. à cacheter.

seal ring ‖ Siegelring m. ‖ bague f. à cachet.

sealskin ‖ Robbenfell n.; Sealskin n. ‖ peau f. de chien marin ou de veau marin; sealskin m. / ~ weaving mill ‖ Sealskinweberei f. ‖ tissage m. de sealskin.

seal wafer see sealing wafer.

seal wax see sealing wax.

seam, to ‖ säumen; mit Naht f. versehen ‖ border; ourler.

seam ‖ Naht f. ‖ couture f. / ~ (Hem of a dress etc) ‖ Nähkante f.; Saum m. ‖ ourlet m.; repli m.; lisière f. / ~ (Found) ‖ Gußnaht f. ‖ bavure f. / ~ (Tinm) ‖ Naht f.; Saum m. ‖ rapport m.; soudure f. / ~ (Welding) ‖ Schweißnaht f. ‖ soudure f. / to knock off the ~s from the castings ‖ die Gußnaht f. abschlagen ‖ enlever la bavure. / ~ of boiler ‖ Kesselnaht f. ‖ rivure f. de chaudière. / circumferential ~ ‖ Rundnaht f. ‖ rivure f. transversale. / flat ~ ‖ doppelte oder platte Naht f. ‖ couture f. cousue et rabattue. / herring-bone ~ ‖ Kreuznaht f. ‖ couture f. à arête de poisson. / longitudinal ~ ‖ Längsnaht f.; Längsfalz m. ‖ couture f. longitudinale. / ~ of a metal-plate ‖ Saum m. oder Rundkante f. oder Rundfalz m. an einem Blech ‖ membron m. ou ourlet m. d'une tôle. / mock ~ ‖ falsche Naht f. ‖ fausse couture f. / open ~ ‖ offene Naht f. ‖ couture f. ouverte. / reinforced ~ ‖ Wulstnaht f. ‖ joint m. renforcé ou surépaissé. / ridgeless ~ ‖ wulstlose Naht f. ‖ joint m. sans surépaisseur. / round ~ ‖ Kappnaht f.; runde Naht f. ‖ bigourelle f.; couture f.

ronde. / sharp ~ (Found) ‖ scharfer Grat m. ‖ bavure f.; arête f. vive. / soldered ~ ‖ Lötnaht f. ‖ soudure f. / staggered ~s pl. ‖ versetzte Nähte fpl. ‖ coutures fpl. contrariées *ou* croisées. / tube ~ ‖ Rohrnaht f. ‖ soudure f. de tube. / welding ~ ‖ Schweißnaht f. ‖ joint m. soudé; ligne f. de soudure. / breadth of the ~s ‖ Nahtbreite f. ‖ largeur f. de la couture. / width of ~ ‖ Nahtbreite f. ‖ largeur m. de la couture.

seam (Geol; Mine) ‖ Flöz n.; Lagerstätte f.; Lager n. ‖ veine f.; couche f.; filon m.; gîte m.; gisement m.; lit m. / to drain a ~ ‖ ein Flöz n. lösen ‖ saigner *ou* démerger une veine. / coal ~ ‖ Kohlenflöz n.; Flöz n.; Steinkohlenflöz n. ‖ veine f. *ou* couche f. de charbon. / dilated ~ ‖ schwebendes Flöz n. ‖ mine f. dilatée. / flat ~ ‖ flach fallendes Flöz n.; Flaches n. ‖ plat m.; plateur m.; plateure f. / gently dipping ~ ‖ flach einfallendes Flöz n. ‖ couche f. à faible pendage. / abundance of ~s ‖ Flözreichtum m. ‖ richesse f. en couches. / roof of a ~ ‖ Hangendes n. ‖ toit m. d'une couche; couche f. située au toit. / map showing ~s ‖ Flözkarte f. ‖ carte f. stratigraphique. / thickness of the ~s ‖ Mächtigkeit f. der Flöze ‖ puissance f. des couches.

seaman ‖ Seemann m.; Matrose m. ‖ marin m.; matelot m. / able ~ ‖ Vollmatrose m. ‖ matelot m. breveté. / ordinary ~ ‖ Leichtmatrose m. ‖ matelot m. de pont.

seam angle ‖ Saumwinkel m. ‖ cornière f. de bordure.

sea mark ‖ Seezeichen n. ‖ amer m.; reconnaissance f.; marque f.

seamen pl. *see also* seaman ‖ Seeleute mpl. ‖ gens mpl. de mer.

seam hammer ‖ Sickenhammer m. ‖ marteau m. de suage *ou* à soyer. / little ~ ‖ Abbindhammer m. ‖ petit marteau m. à soyer.

sea mile ‖ Seemeile f.; Knoten m. ‖ mille m. marin; nœud m.

seaming lace ‖ Abschlußborte f. ‖ galon m. couture. / ~ machine ‖ Verschließmaschine f.; Falzmaschine f. ‖ machine f. à sertir.

seamless ‖ nahtlos ‖ sans soudure f.; sans couture f. / ~ gas pipe ‖ nahtloses Gasrohr n. ‖ tuyau m. à gaz sans soudure. / ~ hosiery ‖ nahtlose Strumpfwaren fpl. ‖ bonneterie f. sans couture. / ~ pipe *see* ~ tube. / ~ rubber ‖ nahtloses Gummi n. ‖ caoutchouc m. sans couture. / ~ steel pipe ‖ nahtloses Stahlrohr n. ‖ tuyau m. en acier sans soudure. / ~ tube ‖ nahtloses Rohr n. ‖ tube m. sans soudure.

seamless-drawn ‖ nahtlos gezogen ‖ étiré sans soudure. / ~ hollow part ‖ nahtloser Hohlkörper m. ‖ pièce f. creuse emboutie sans soudure. / ~ tube ‖ nahtlos gezogenes Rohr n.; nahtlos gezogene Röhre f. ‖ tube m. étiré sans soudure.

seamless-pressed wheel ‖ nahtlos gepreßtes Rad n. ‖ roue f. matricée sans soudure.

seamless-rolled tube ‖ nahtlos gewalztes Rohr n. ‖ tuyau m. laminé sans soudure.

seamstress ‖ Näherin f.; Lohnnäherin f. ‖ couturière f. (à façon); couseuse f. / linen ~ ‖ Weißnäherin f. ‖ couturière f. en lingerie.

seam welding ‖ Nahtschweißung f. ‖ soudage m. des joints (pour tôles). / lapped

~ ‖ überlappte Nahtschweißung f. ‖ soudage m. de joints à recouvrement. / ridgeless ~ ‖ wulstlose Nahtschweißung f. ‖ soudage m. de joints sans surépaisseur.

seam welding machine ‖ Nahtschweißmaschine f. ‖ machine f. à souder les joints (pour tôles).

sea navigation ‖ Seeschiffahrt f. ‖ navigation f. maritime. / ~ company ‖ Seeschiffahrtsgesellschaft f. ‖ compagnie de navigation f. maritime.

seaplane ‖ Wasserflugzeug n.; Seeflugzeug n. ‖ hydravion m. / boat ~ ‖ Flugboot n. ‖ hydravion m. à coque. / commercial ~ ‖ Wasserverkehrsflugzeug n. ‖ hydravion m. commercial.

seaplane base ‖ Seeflughafen m. ‖ base f. aéro-navale. / ~ beaching trolley ‖ Wasserflugzeugschleppwagen m. ‖ remorque f. pour hydroplanes.

sea power ‖ Seemacht f. ‖ puissance f. maritime; force f. *ou* puissance f. navale.

sea pilot ‖ Seelotse m. ‖ hauturier m.; pilot m. hauturier.

sea protest ‖ Verklarung f. ‖ procès-verbal m. des avaries d'un bateau.

sea provisions pl. ‖ Salzproviant m.; Seeproviant m. ‖ vivres mpl. de campagne; salaison m.

sea quake ‖ Seebeben n. ‖ tremblement m. de mer.

sear, to ‖ abbrennen; sengen; versengen ‖ flamber.

search, to ‖ untersuchen; forschen; erforschen ‖ rechercher. / ~ into ‖ sichten ‖ mettre à part; examiner; trier. / ~ for minerals ‖ schürfen ‖ faire des recherches; fouiller; creuser.

search ‖ Untersuchung f.; Forschung f. ‖ recherche f.; examen n.

searcher ‖ Forscher m.; Sucher m.; Schürfer m. ‖ chercheur m. / ~ (Of a telescope) ‖ Sucher m.; Hilfsfernrohr n. (zum Einstellen eines Teleskopes) ‖ chercheur m.

searching ‖ Forschen n.; Forschung f.; Untersuchung f. ‖ recherche f. / ~ (Mine) ‖ Schurf m.; Schürfung f. ‖ fouille f. de recherche. / ~ works pl. (Mine) ‖ Schürfarbeiten fpl. ‖ travaux mpl. de recherche.

search-light (Auto etc.) ‖ Scheinwerfer m.; projecteur m.; phare m. / acetylene ~ ‖ Azetylenscheinwerfer m. ‖ phare m. à acétylène.

search-light car ‖ Scheinwerferwagen m. ‖ voiture f. à projecteur. / ~ carbon ‖ Scheinwerferkohle f. ‖ charbon m. pour projecteurs.

sea return (Tel) ‖ Seerückleitung f. ‖ conducteur m. de retour en mer.

sea risk (Insur) ‖ Seegefahr f. ‖ fortune f. de mer; péril m. de la mer.

sea room (Mar) ‖ Räumte f.; offene See f.; Seeraum m. ‖ belle dérive f.; eau f. à courir; haute mer f.; large m.

sea salt ‖ Seesalz n.; Siedesalz n. ‖ sel m. marin *ou* de mer; salmare f. / ~ work ‖ Salzteich m.; Salzgarten m.; Seesaline f. ‖ marais m. salant. / ~ workman ‖ Seesalinenarbeiter m. ‖ saunier m.

sea sand ‖ Seesand m.; Meeressand m. ‖ arène f. marine; sable m. de mer.

sea shipping *see* sea navigation.

sea-sickness ‖ Seekrankheit f. ‖ mal m. de mer.

sea signal ‖ Seesignal n. ‖ signal m. à mer.

sea snail ‖ Meeresschnecke f. ‖ escargot m. de mer.

season, to (Wine) ‖ lagern ‖ mettre sur chantier m.; enchanteler. / ~ in summer ‖ durchsommern ‖ conserver pendant l'été. / ~ in winter ‖ durchwintern ‖ conserver pendant l'hiver. / ~ wood ‖ Holz n. austrocknen ‖ dessécher le bois. / ~ wood in the open air ‖ Holz an der Luft austrocknen ‖ dessécher le bois à l'air.

season ‖ Jahreszeit f. ‖ saison f. / ~ for the manufacturing of beet sugar ‖ Kampagne f.; Zuckerrübenkampagne f. ‖ campagne f. (de sucrerie). / dry ~ ‖ trockene Jahreszeit f. ‖ saison f. sèche. / rainy ~ ‖ Regenzeit f. ‖ saison f. pluvieuse.

seasonal worker ‖ Saisonarbeiter m. ‖ ouvrier m. saisonnier.

season cracking (Met) ‖ Altersriß m. ‖ craquelure f. saisonnière.

seasoned (Wood) ‖ ausgetrocknet; trocken ‖ desséché; sec. / well ~ (Wine) ‖ abgelagert ‖ mis en cave *ou* sur chantier. / ~ wood ‖ trockenes Holz n. ‖ bois m. sec.

seasoning of sleepers in the open air ‖ Austrocknung f. der Schwellen an der Luft ‖ dessiccation f. des traverses à l'air libre. / ~ of wood in the open air ‖ Lufttrocknung f. des Holzes ‖ séchage m. du bois à l'air.

season ticket ‖ Zeitkarte f. ‖ carte f. d'abonnement.

seat (Furniture) ‖ Sessel m.; Sitzmöbel n.; Sitz m. ‖ siège m. / ~ (Sitting surface) ‖ Sitzfläche f. ‖ surface f. du siège. / ~ (Sitting place) ‖ Sitzplatz m. ‖ siège m. / ~ (of rails) ‖ Schienenlager n. ‖ assise f. *ou* chambre f. des rails. / ~ of a closet ‖ Abtrittbrille f. ‖ lunette f. (de privé). / ~s pl. crosswise ‖ Quersitze mpl. ‖ places fpl. en travers. / divided ~ (Carross) ‖ geteilter *oder* getrennter Sitz m. ‖ siège m. séparé. / driver's ~ (Auto) ‖ Führersitz m. ‖ siège m. de conducteur. / folding ~ (Auto) ‖ Klappsitz m.; Notsitz m. ‖ strapontin m. / hair ~ ‖ Roßhaarsitz m. ‖ siège m. en crin. / leather-bottomed ~ ‖ Ledersessel m. ‖ siège m. en cuir. / ~s pl. lengthwise ‖ Längssitze mpl. ‖ places fpl. disposées en long. / needle valve ~ ‖ Nadelventilsitz m. ‖ siège m. du pointeau. / plated ~ ‖ Furniersitz m. ‖ siège m. en plaqué. / ~ for the rail base ‖ Auflagefläche f. für den Schienenfuß ‖ appui m. du patin de rail. / revolving ~ of a telescope adjustable in height ‖ drehbarer und in der Höhe verstellbarer Sitz m. eines Fernrohres ‖ siège m. d'une lunette tournant et réglable en hauteur. / ~ for rowers ‖ Ruderbank f. ‖ banc m. de nage. / sliding ~ (Auto) ‖ verstellbarer Sitz m. ‖ siège m. réglable. / spring ~ (Furniture) ‖ Sprungfedersitz m. ‖ siège m. à ressorts. / spring ~ (Mach) ‖ Federsitz m. ‖ patin m. de ressort. / spring pivot ~ ‖ Federsattel m. ‖ patin m. de ressort tournant. / throw-back ~ ‖ Klappsitz m. ‖ siège m. à rabattement. / tilting ~ ‖ drehbarer Sitz m. ‖ siège m. pivotant. / ~ of a valve ‖ Ventilsitz m. ‖ siège m. d'une soupape. / well ~ (Auto) ‖ tiefer Sitz m. ‖ siège m. profond. / ~ of the wheel on the axis (Watchm) ‖ Sitz m. des Rades auf der Welle ‖ embase f. *ou* assiette f. sur l'arbre d'une roue. / ~ for writing table ‖ Schreibtischsessel m. ‖ siège m. de bureau.

seat back ‖ Rückenlehne f. ‖ dossier m. / ~ board ‖ Sitzbrett n. ‖ planche f. du siège. / ~ cane plaiting ‖ Rohrgeflecht

n. für Stuhlsitze; Sitzrohrgeflecht n. ‖ cannage m. de sièges. / ~ caner ‖ Stuhlflechter m. ‖ canneur m. de chaises. / ~ cushion ‖ Sitzkissen n. ‖ coussin m. de siège.

sea term ‖ Seemannsausdruck m. ‖ terme m. nautique.

seating capacity (Auto) ‖ Sitzzahl f.; Zahl f. der Sitze ‖ nombre m. de places.

seat pack parachute (Aero) ‖ Kissenfallschirm m. ‖ parachute m. sac-coussin. / ~ pillar ‖ Sattelstange f.; tige f. de selle. / width of ~ for the rail base ‖ Breite f. der Auflagefläche für den Schienenfuß ‖ largeur f. d'appui du patin de rail.

seats pl. see also seat (Furniture) ‖ Sitzmöbel pl. ‖ sièges mpl.

seat-type parachute (Aero) ‖ Sitzkissenfallschirm m. ‖ parachute m. sac-siège.

sea turn ‖ Seebrise f. ‖ brise f. du large.

sea valve ‖ Seeventil n. ‖ soupape f. de mer.

sea wall ‖ Strandmauer f. ‖ mur m. de quai.

seaward ‖ seewärts ‖ du côté de la mer; vers la mer; vers le large.

sea water ‖ Meerwasser n.; Seewasser n. ‖ eau f. de mer; eau-mer f.

seaweed ‖ Seegras n. ‖ zostère f.; crin m. marin; goémon m.; varech m. / ~ gatherer ‖ Seegrasfischer m. ‖ ramasseur m. de goémon ou de varech.

sea wind ‖ Seebrise f.; Seewind m. ‖ vent m. de mer ou du large.

sea worm ‖ Pfahlwurm m. ‖ taret m.; ver m. de digue ou de vaisseaux.

seaworthiness ‖ Seetüchtigkeit f.; Seefähigkeit f. ‖ qualité f. marine; navigabilité f.

seaworthy ‖ seetüchtig; seefähig ‖ en état de navigabilité; capable de tenir la mer. / the steamer is ~ ‖ der Dampfer ist seetüchtig ‖ le vapeur est en état de navigabilité. / ~ packed ‖ seemäßig verpackt ‖ en emballage m. maritime.

sea wrack see seaweed.

sebacic acid ‖ Talgsäure f.; Sebazinsäure f. ‖ acide m. sébacique.

secant ‖ Sekante f. ‖ sécante f. / ~ of an arc ‖ goniometrische Sekante f. ‖ sécante f. d'un arc.

second ‖ Sekunde f. ‖ seconde f. / ~ (Print) see second-working. / ~ of arc ‖ Bogensekunde f. ‖ seconde f. d'arc.

secondary ‖ sekundär ‖ secondaire. / ~ (Electr) ‖ sekundär; induziert ‖ secondaire; induit. / ~ action ‖ Nebenwirkung f. ‖ action f. secondaire. / ~ alternating current ‖ Sekundärwechselstrom m. ‖ courant m. alternatif secondaire. / ~ battery ‖ Sekundärbatterie f.; Sammlerbatterie f. ‖ batterie f. secondaire ou des accumulateurs. / ~ cell ‖ Sekundärelement n. ‖ élément m. secondaire. / ~ circuit ‖ Sekundärkreis m. ‖ circuit m. secondaire. / ~ circuit condenser (Radio) ‖ Kondensator m. im Sekundärkreis ‖ condensateur m. du circuit intermédiaire. / ~ coil ‖ Sekundärspule f. ‖ bobine f. secondaire. / ~ jigger (Radio) ‖ Sekundärjigger m. ‖ transformateur m. d'oscillations du circuit secondaire. / ~ line (Railw) ‖ Nebenbahn f. ‖ ligne f. secondaire. / ~ pressure ‖ Sekundärspannung f. ‖ tension f. secondaire. / ~ switching (Railw) ‖ Sekundärschaltung f. ‖ circuit m. secondaire. / ~ winding ‖ Sekundärwicklung f. ‖ enroulement m. secondaire.

second-cut file ‖ Halbschlichtfeile f. ‖ lime f. demi-douce.

second control for the driving clock ‖ Sekundenregelung f. des Uhrwerkantriebes ‖ contrôle m. des secondes du mouvement d'horlogerie. / electric ~ control ‖ elektrische Sekundenkontrolle f. ‖ contrôle m. électrique des secondes. / ten ~s division ‖ Zehnersekunde f. ‖ les dix secondes f. pl.; la dizaine f. de secondes.

second-hand ‖ aus zweiter Hand; gebraucht ‖ indirectement. / to buy ~ ‖ antiquarisch kaufen ‖ acheter d'occasion f. / ~ bookseller ‖ Altbuchhändler m.; Antiquar m. ‖ marchand m. de livres d'occasion; bouquiniste m. / ~ bookshop ‖ Antiquariat n.; Altbuchhandlung f. ‖ librairie f. d'occasion. / ~ car (Auto) ‖ (zum Verkauf angebotener) gebrauchter Wagen m. ‖ voiture f. d'occasion.

second hand (Of a watch) ‖ Sekundenzeiger m. ‖ aiguille f. des secondes.

second light ‖ indirektes Fenster n. ‖ fauxjour m. / ~ mate (Mar) ‖ Schiffsoffizier m. ‖ lieutenant m. / ~ pendulum ‖ Sekundenpendel n. ‖ pendule m. à secondes. / ~ pendulum clock ‖ Sekundenpenduluhr f. ‖ pendule f. battant la seconde. / ~ working (Print) ‖ Widerdruck m. ‖ retiration f.; réimpression f.; impression f. au verso.

secrecy of telegraph-service ‖ Telegrafengeheimnis n. ‖ secret m. des correspondances télégraphiques.

secret language ‖ geheime Sprache f.; Geheimsprache f. ‖ langage m. secret. / ~ service (Tel) ‖ Geheimverkehr m. ‖ service m. secret.

secretary ‖ Schriftführer m.; Sekretär m. ‖ secrétaire m. / ~ (Of state) ‖ Minister m.; Staatssekretär m. ‖ ministre m. / ~ (Furniture) ‖ Sekretär m.; Schreibschrank m. ‖ secrétaire m.

secretion ‖ Absonderung f.; Ausscheidung f.; Sekretion f. ‖ sécrétion f.

section (Geom) ‖ Schnitt m. oder Schnittlinie f. zweier Flächen ‖ intersection f.; section f. / ~ (Build) ‖ Profil n.; Schnitt m.; Querschnitt m. ‖ profil m.; coupe f. / ~ (Railw signalling) ‖ Gleisabschnitt m. ‖ section f. de bloc. / ~ (Tramway) ‖ Teilstrecke f. / made in ~s pl. ‖ zerlegbar ‖ démontable. / ~ showing areas (Railw) ‖ Flächenprofil n. ‖ profil m. des aires. / circular ~ ‖ kreisrunder Querschnitt m. ‖ section f. circulaire. / conic ~ ‖ Kegelschnitt m. ‖ section f. conique. / cross ~ ‖ Querschnitt m. ‖ coupe f. ou section f. transversale. / curve ~ (Railw) ‖ Kurvenjoch n. ‖ tronçon m. de voie courbe. / dangerous ~ ‖ gefährlicher Querschnitt m. ‖ section f. dangereuse. / ~ of earthworks (Railw) ‖ Massenprofil n. ‖ profil m. des terrassements. / equatorial ~ ‖ Äquatorschnitt m. ‖ section f. équatoriale. / fully protected ~ (El line) ‖ Induktionsschutzstrecke f. ‖ section f. de transposition complète. / geological ~ ‖ geologisches Profil n. ‖ profil m. géologique. / horizontal ~ (Build) ‖ Grundplan m.; Grundriß m. ‖ ichnographie f.; plan m. horizontal ou objectif; projection f. ichnographique. / the largest ~ of the shaft is x millimeters square ‖ der stärkste Querschnitt des Schaftes beträgt x mm im Quadrat ‖ la section la plus forte de la mèche est de x millimètres au carré. / ~ sufficiently

light to be easily laid down by hand (Railw) ‖ genügend leichtes Joch n., um von Hand verlegt zu werden ‖ tronçon m. de voie assez léger pour qu'un homme puisse le placer facilement à la main. / the ~s pl. may be chosen much lighter ‖ die Querschnitt(s)abmessungen fpl. können erheblich schwächer ausgeführt werden ‖ les mesures fpl. de la section se prennent plus faibles. / ~ of a line see ~ (Tramway). / longitudinal ~ ‖ Längsschnitt m. ‖ coupe f. longitudinale ou en long; section f. longitudinale. / meridian ~ ‖ Meridianschnitt m. ‖ section f. méridienne. / passage ~ (Railw) ‖ Durchgangsprofil n. ‖ profil m. de passage. / principal ~ ‖ Hauptschnitt m. ‖ section f. principale. / ~ of rails ‖ Schienenform f.; Schienenprofil n. ‖ section f. du rail. / ~ of railway ‖ Bahnprofil n. ‖ section f. de la voie du chemin de fer. / longitudinal ~ of a railway ‖ Längenprofil n. einer Eisenbahnstrecke ‖ profil m. longitudinal de la voie d'un chemin de fer. / rectangular ~ ‖ Querschnitt m. ‖ section f. droite. / ~ of track (Railw) ‖ Gleisrahmen m. ‖ tronçon m. de voie. / portable ~ of track ‖ tragbares Gleisjoch n. ‖ châssis m. de voie. / straight ~ of track ‖ gerades Gleisjoch n. ‖ châssis m. droit. / transverse ~ ‖ Querschnitt m. ‖ coupe f. ou section f. transversale. / ~ cut out of wing (Airpl) ‖ Flügelausschnitt m. ‖ secteur m. d'aile.

sectional boiler ‖ Sektionalkessel m.; Gliederkessel m. ‖ chaudière f. sectionnée. / ~ chamber water tube boiler ‖ Teilkammerwasserrohrkessel m. ‖ chaudière f. semitubulaire. / ~ drawing ‖ Querschnitt(s)zeichnung f. ‖ dessin m. de section. / ~ iron see also section iron ‖ Formeisen n.; Profileisen n. ‖ fer m. profilé. / ~ water tube boiler ‖ Sektionalwasserrohrkessel m. ‖ chaudière f. sectionale à tubes d'eau.

sectionals pl. with equal sides ‖ gleichschenkliges Winkeleisen n. ‖ cornière f. à ailes égales.

section block (ing) (Railw) ‖ Streckenblock m.; Durchgangsblockwerk n. ‖ bloc m. de pleine voie ou de section; blocage m. de section.

section cutter for barley examination ‖ Gerstenprüfer m.; Farinatom n. ‖ farinatome m.; coupe-grains m. pour l'orge.

section ingot ‖ vorgewalzter Block m. ‖ lingot m. profilé.

section iron ‖ Formeisen n.; Fassoneisen n.; Profileisen n. ‖ profilé m.; fer m. profilé; fer m. en barres façonnés. / ~ for screws ‖ Schraubeneisen n. ‖ profilés mpl. pour vis.

section iron bending machine ‖ Biegemaschine f. für Formeisen ‖ machine f. à cintrer les fers profilés. / ~ cutter ‖ Formeisenschere f. ‖ cisailles fpl. pour fers profilés. / ~ ring ‖ Profileisenring m. ‖ anneau m. en fer profilé. / ~ rolling mill ‖ Formeisenwalzwerk n. ‖ laminoir m. pour profilés. / ~ shears pl. ‖ Formeisenschere f. ‖ cisailles fpl. à fers profilés. / ~ shearing machine see ~ shears. / ~ shearing and mitre-cutting machine combined with universal punching and notching machine constructed of steel castings ‖ Profileisen- und Gehrungsschere f. mit Universalloch- und Ausklinkmaschine in Stahlgußausfüh-

rung || cisaille f. à profilés et à onglets combinée avec poinçonneuse universelle et grugeoir exécutée en acier coulé. / ~ straightening machine || Richtmaschine f. für Formeisen || machine f. à dresser les fers profilés.

section lifter || Schnittfänger m. || lève-coupe m.

sections pl. *see also* section iron || Profileisen n.; Formeisen n. || profilé m.; fer m. profilé.

section switch || Streckenschalter m. || interrupteur m. de section.

section wire || Formdraht m.; Fassondraht m. || fil m. profilé.

sector || Kreisausschnitt m.; Sektor m. || secteur m. / ~ (Instrument) || Proportionalzirkel m. || compas m. de proportion. / spherical ~ || Kugelausschnitt m.; Kugelsektor m. || secteur m. sphérique. / steering worm ~ (Auto) || Lenksegment n. || secteur m. de direction.

sectoral wire || Sektoraldraht m. || fil m. sectoral.

sector dam || Sektorwehr n.; Segmentwehr n. || barrage m. à secteurs.

sector diaphragm || Sektorenblende f. || diaphragme m. à secteurs. / ~ for focussing || Sektorenblende f. zum Fokussieren || diaphragme m. à secteurs pour la mise au point.

sector instantaneous shutter || Sektormomentsverchluß m. || obturateur m. instantané à secteurs.

sector weir || Segmentwehr n.; Sektorwehr n. || barrage m. à secteur *ou* à segment.

secure, to ~ (Mar) || seefest zurren; festmachen || amarrer à la mer.

secure || sicher || sûr; ferme. / . . . holds ~ly || . . . hält sicher || . . . tient ferme.

secured for sea || seefest (gemacht) || amarré.

secure screw || Befestigungsschraube f. || vis f. de fixation.

securing of the railway tire || Radreifensicherung f. || fixation f. du bandage.

securities pl. (Bank) *see also* security (Paper) || Effekten fpl.; Wertpapiere npl. || effets mpl. publics; valeurs fpl. (mobilières); titres mpl.

security || Zuverlässigkeit f.; Sicherheit f. || sécurité f.; sûreté f. / ~ (Trade: covering) || Sicherheit f.; Gewähr f.; Sicherheitsleistung f.; Deckung f. || couverture f.; garantie f.; sûreté f. / able to put up ~ || kautionsfähig || capable de fournir un cautionnement. / to give ~ || sicherstellen || assurer; garantir. / to hold ~ || gesichert sein || être à couvert; être nanti. / on the ~ of || gegen Sicherheit f. von || sur la garantie de / to stand ~ || avalieren || aval(is)er. / ~ of fire (Arm) || Treffsicherheit f. || sûreté f. du tir. / ~ of stability || Standsicherheit f. || sécurité f. de stabilité. / ~ against stresses due to sudden and repeated shocks ~ || Bruchsicherheit f. gegenüber stoßweiser Belastung || résistance f. à la rupture, les aciers étant soumis à des secousses répétées. / two-fold ~ || zweifache Sicherheit f. || sécurité f. double.

security (Paper) || Wertpapier n. || valeur m.; titre m. / Government ~ || Staatspapier n. || fonds m. public *ou* d'Etat. / ~ at regular interest || festverzinsliches Wertpapier n. || valeur f. à intérêts fixes. / transferable ~ || übertragbares Wertpapier n. || valeur f. mobilière.

security giving || Sicherheitsleistung f. || constitution f. de garantie.

security mortgage || Sicherungshypothek f. || hypothèque f. de garantie.

sedan (Auto) || Limusine f. mit Innenlenkung; Sedan m. || limousine f. à conduite intérieure; Sedan m.

sediment (Chem) || Bodensatz m.; Ablagerung f.; Sediment n. || dépôt m.; sédiment m. / ~ (Boiler) || Kesselstein m. || incrustations fpl.; sédiments mpl.; dépôts mpl.; tartres mpl. / ~ (Brew) || Geläger n. || dépôt m.; lie f.; fonds mpl. de foudres.

sedimentary || sedimentär || sédimentaire.

sedimentary rock || Sedimentgestein n. || roche f. sédimentaire. / chemical ~ || chemisches Sedimentgestein n. || roche f. sédimentaire chimique. / mechanical ~ || mechanisches Sedimentgestein n. || roche f. sédimentaire mécanique. / organogeneous ~ || organogenes Sedimentgestein n. || roche f. sédimentaire organogénique.

sedimentation (Chem) || Absetzen n.; Ablagern n. || sédimentation f. / to separate by ~ || durch Absetzen ausscheiden || séparer par sédimentation.

sedimentation process || Schlämmprozeß m.; Schlämmverfahren n. || procédé m. de sédimentation.

sediment filter press (Brew) || Gelägerfilterpresse f. || filtre-presse m. pour fonds de foudres.

see, to || sehen || voir.

seed, to || aussäen || semer; ensemencer.

seed (Botany) || Same m.; Saat f. || grains mpl. / ~ (Silk) || Grains pl.; Eier npl. des Seidenwurms || grains mpl.; graines fpl. / ~ (Fault in the glass) || Glasblase f.; bosse f. / ~ of forest trees || Waldbäumesamen mpl. || graines fpl. forestières. / garden ~ || Gärtnereisamen mpl. || graines fpl. horticoles. / vegetable ~ || Gemüsesamen mpl. || graines fpl. maraîchères.

seed bag || Samentüte f. || sachet m. pour graines. / ~ barley || Saatgerste f. || orge f. de semence. / ~ bed || Saatbeet n.; Samenbeet n.; Treibbeet n. || couche f. de semis; semis m. / ~ cleansing machine || Samenreinigungsmaschine f. || machine f. à nettoyer les grains de semences. / ~ corn || Aussaat f.; Saat f.; Saatkorn n.; semailles fpl. / ~ crops pl. || Saatgetreide n. || grains mpl. de semence. / ~ cultivating plant || Saatzüchterei f.; Saatgutzüchterei f.; Saatveredelungsanlage f. || culture f. spéciale de semences; installation f. d'amélioration des semences. / ~ envelope || Samentüte f. || enveloppe f. de graines. / ~ fibre || Samenfaser f.; Samenhaar n. || fibre f. extraite de la graine; duvet m. de la graine. / ~ grain cleansing plant || Saatgutreinigungsanlage f. || machine f. à nettoyer les semences. / ~ grower || Samenzüchter m. || jardinier-grainetier m. / ~ hair see ~ fibre. / ~ improving plant see ~ cultivating plant.

seeding machine || Sämaschine f. || semeuse f.; semoir m. / ~ stage (Forest) || Schlag m. für Besamung || coupe f. d'ensemencement.

seed lac || Körnerlack m.; Saatlack m. || laque f. en grains.

seedling (Gard) || Sämling m.; Samen-

pflanze f.; Pflänzling m. || jeune plante f. semée.

seed microscope || Getreidelupe f. || loupe f. pour l'examen des grains.

seed plot || Samenbeet n.; Samenschule f.; Treibbeet n. || semis m.

seed protective || Saatenschutzmittel n. || protectif m. de semences.

seeds pl. *see also* seed || Sämereien fpl. || semences fpl.

seed sorting || Kornsortierung f. || triage m. de grains *ou* de graines.

seed sorting machine || Samensortiermaschine f. || machine f. à trier les grains de semences.

seed vessel (Botany) || Fruchtknoten m. || ovaire m.

seed winnower || Getreidetriör m.; Kornsortierer m.; Getreideschwinge f. || trieur m. *ou* cribleur m. *ou* remueur m. de grains.

seeing || Sehen n. || vision f.; vue f. / ~ in relief || räumliches *oder* plastisches Sehen n. || vision f. stéréoscopique *ou* du relief. / total ~ capacity || volle Sehschärfe f. || acuité f. visuelle intégrale. / power of ~ || Sehschärfe f.; Sehvermögen f. || acuité f. visuelle.

seek, to || suchen; ermitteln; ausfindig machen || rechercher.

seeker earth (Tel) || Sucherde f. || terre f. de recherche.

seel, to ~ (Mar) || rollen; schlingern und stampfen || rouler.

seeler (Mar) || Schiff n., welches stark rollt || rouleur m.

seep, to || durchlaufen; durchsickern || suinter à travers; filtrer.

seepage || Versickerung f. || infiltration f.

seesaw || schaukelnde Bewegung f. || mouvement m. oscillatoire *ou* de bascule. / ~ pan (Sugar) || Kipppfanne f.; Schwungpfanne f. || chaudière f. à bascule.

seethe, to ~ with boiling water (Clothm) || abbrühen || échauder.

Seger cone || Brennkegel m.; Schmelzkegel m.; Segerkegel m. || cône m. pyrométrique *ou* de Seger.

segger (Porcel) || Kassette f.; Kapsel f. || cassette f.; gazette f. / ~ clay || Kapselton m. || terre f. à cassettes.

segment || Abschnitt m.; Segment n. || segment m. / ~ of a circle || Kreisabschnitt m. || segment m. d'un cercle. / ~ for combing machines || Segment n. für Krempelmaschinen || segment m. pour peigneuses. / commutator ~ || Kollektorlamelle f. || segment m. *ou* secteur m. d'un collecteur. / ~ of a sphere || Kugelabschnitt m.; Kugelsegment n. || segment m. de la sphère. / toothed ~ || Zahnradsegment n. || segment m. de roue dentée.

segmental || segmentartig || en forme de segment. / ~ grinding wheel || Segmentschleifscheibe f. || meule-segment m.

segment base || Segmentbock m. || support m. de secteur.

segment dam see segment weir.

segment weir || Segmentwehr n. || barrage m. à secteur *ou* à segment.

segregate, to ~ (Met) || seigern || se séparer; ségréger.

segregation (Met) || Seigerung f.; Absonderung f.; ségrégation f.; séparation f. / ~ of graphite || Ausscheiden n. des Graphits || ségrégation f. de graphite. / ~ of naphthalene || Naphthalinausschei-

dung f. ∥ séparation f. de la naphtaline. / ~ by oxydation ∥ oxydische Ausscheidung f. ∥ séparation f. par oxydation.

Seignette salt ∥ weinsaures Kalinatron n. ∥ sel m. de Seignette.

seine (Fish) ∥ Zugnetz n.; Schleppnetz n.; Kratzgarn n. ∥ seine f.; tressun m.; drège f.; traîneau m.

seining ∥ Schleppnetzfischerei f. ∥ pêche f. à la seine ou au traîneau.

seismic district ∥ Erdbebengebiet n. ∥ domaine m. séismique.

seismograph ∥ Erdbebenmesser m.; Seismograf m. ∥ sismographe m.

seismology ∥ Erdbebenkunde f. ∥ séismologie f.

seismometer ∥ Erdbebenmesser m. ∥ séismomètre m.

seizable ∥ greifbar ∥ saisissable; palpable.

seize, to ∥ ergreifen; fassen; packen ∥ saisir. / ~ a line (Tel) ∥ eine Leitung belegen ∥ saisir une ligne. / the piston seizes ∥ der Kolben frißt ∥ le piston se grippe.

seized-up (Mach) ∥ festgefressen ∥ grippé.

seizing line ∥ Bändselleine f.; Bindselleine f. ∥ petite corde f.; corde f. de saisine.

seizure ∥ Auspfändung f. ∥ saisie f.; saisie-exécution f.

selbite (Miner) ∥ Grausilber n.; Silberkarbonat n.; Selbit m. ∥ selbite f.; argent m. carbonaté.

select, to ∥ (aus)lesen; (aus)wählen; aussuchen ∥ choisir; assortir.

selecting and sifting machinery ∥ Auslese- und Sortiermaschine f. ∥ machine f. à trier et à assortir.

selecting needle (Acc) ∥ Abfühlnadel f. ∥ goujon m.; aiguille f.

selection ∥ Auswahl f. ∥ choix m. / ~ (Radio) ∥ Selektion f.; Selektivität f. ∥ sélectivité f. / careful ~ of material ∥ sorgfältige Materialauswahl f. ∥ choix m. judicieux des matériaux.

selective call in telephone lines ∥ wahlweiser Anruf m. in Fernsprechleitungen ∥ appel m. sélectif dans les lignes téléphoniques. / ~ ringing (Tel) ∥ Wahlanruf m. ∥ appel m. sélectif.

selectivity (Radio) ∥ Selektivität f.; Störungsfreiheit f. ∥ sélectivité f.

selector (Aut tel) ∥ Gruppenwähler m.; Wähler m. ∥ sélecteur m. / Ericsson ~ ∥ Kulissenwähler m. ∥ sélecteur m. Ericsson. / Gill ~ ∥ Zeitrelais n. von Gill ∥ relais m. à action différée de Gill. / ~ in manual telephone systems ∥ Wähler m. im Handbetrieb ∥ sélecteur m. dans les systèmes manuels. / ~ of periods ∥ Periodenwähler m. ∥ sélecteur m. de périodes. / ~ with spring device ∥ Wähler m. mit Federaufzug ∥ sélecteur m. avec ressort de mouvement.

selector stepping magnet ∥ Wählerschaltmagnet m. ∥ aimant m. pas-à-pas du sélecteur.

selenic acid ∥ Selensäure f. ∥ acide m. sélénique.

selenide of cobalt and lead ∥ Selenkobaltblei n.; Filkerodit m. ∥ séléniure m. de plomb et de cobalt. / ~ of copper ∥ Selenkupfer n.; Berzelin n.; Berzelianit m. ∥ cuivre m. sélénié. / ~ of copper and lead ∥ Selenkupferblei n.; Zorgit m. ∥ séléniure m. de plomb et de cuivre. / ~ of lead ∥ Selenblei n.; Clausthalit m. ∥ plomb m. sélénié; claustalite f. / ~ of mercury ∥ Selenquecksilber n.; Tiemannit m. ∥ séléniure m. de mercure. / ~ of mercury and lead ∥ Selenquecksilberblei n.; Lerbachit m. ∥ séléniure m. de plomb et de mercure. / ~ of silver ∥ Selensilber n.; Naumannit m.; Selensilberblei n. ∥ argent m. séléniuré; séléniure m. d'argent.

selenite ∥ Selenit n. ∥ sélénite f.

selenium ∥ Selen n. ∥ sélénium m. / ~ apparatus (Tel) ∥ Selenapparat m. ∥ appareil m. au sélénium. / ~ cell ∥ Selenzelle f. ∥ pile f. en sélénium; cellule f. en sélénium. / to illuminate and to darken the ~ cell alternatively ∥ die Selenzelle f. abwechselnd belichten und verdunkeln ∥ éclairer et obscurir tour à tour la cellule f. en sélénium. / ~ dioxide ∥ Selendioxyd n. ∥ anhydride m. sélénieux. / ~ monochloride ∥ Selenchlorür n. ∥ protochlorure f. de sélénium. / ~ ore ∥ Selenerz n. ∥ minerai m. de sélénium. / ~ tetrachloride ∥ Selenchlorid n. ∥ tétrachlorure f. de sélénium. / ~ water ∥ Selenitwasser n. ∥ eau f. séléniteuse.

seleno silver ∥ Selensilber n.; Naumannit m.; Selensilberblei n. ∥ argent m. séléniuré; séléniure m. d'argent.

selen sulphur ∥ Selenschwefel m. ∥ soufre m. sélénifère.

self-adjusting brake gear of a winding machine ∥ Bremsdruckregler m. einer Fördermaschine ∥ régulateur m. de pression de freinage pour machines de transport. / ~ line finder (Print; Typewr) ∥ selbsttätiger Zeilenanzeiger m. ∥ indicateur m. de ligne automatique.

sel-facting ∥ selbsttätig; automatisch ∥ automatique. / ~ brake ∥ selbsttätige Bremse f. ∥ frein m. automatique. / ~ disengaging of the feed (Mach tool) ∥ selbsttätige Auslösung f. des Vorschubs ∥ débrayage m. automatique de l'avancement. / ~ inclined plan (Mine) ∥ Bremsberg m. ∥ plan m. incliné de trainage. / ~ motion ∥ selbsttätige Bewegung f. ∥ mouvement m. automatique. / ~ movement see ~ motion. / ~ mule (Spinn) ∥ Selfaktor m. ∥ selfacteur m. / ~ press ∥ selbsttätige Handpresse f. ∥ Handpresse f. mit selbsttätigem Farbwerk ∥ presse f. d'imprimerie à encrage automatique / ~ spinner ∥ Selfaktorspinner m. ∥ fileur au self-acting m. / ~ valve ∥ Selbstschlußventil n. ∥ soupape f. à fermeture automatique.

selfactor (Spinn) ∥ Selfaktor m.; selbstspinnende Mulenmaschine f. ∥ mull-jenny m. renvideur; renvideur m. mull-jenny selfacting; mull-jenny m. automate.

self-aligning ring (Of a ball bearing) ∥ Einstellring m. ∥ bague f. de roulement à rotule.

self-balancing suspension tackle ∥ Ausgleichgehänge n. ∥ suspension f. à compensation.

self-bearing (Radio) ∥ Eigenpeilung f. ∥ relèvement m. par la propre station.

self-capacity of coils ∥ Eigenkapazität f. von Spulen ∥ capacité f. propre de bobines.

self-centering jaws pl. ∥ selbstzentrierende Klemmbacken fpl. ∥ mâchoires fpl. à centrage automatique.

self-clipping device (El line) ∥ Selbstklemme f. ∥ griffe f. de câble automatique.

self-closing inkstand ∥ Tintenfaß n. mit Selbstverschluß ∥ encrier m. à fermeture automatique.

self-combustion ∥ Selbstentzündung f. ∥ inflammation f. spontanée.

self-connecting telephone exchange ∥ Selbstanschlußamt n. ∥ bureau m. central téléphonique à service automatique. / ~ working (Tel) ∥ Selbstanschlußbetrieb m. ∥ exploitation f. automatique.

self-contained construction (Mach) ∥ geschlossener Aufbau m. ∥ construction f. serrée. / ~ house ∥ Einfamilienhaus n. ∥ maison f. pour une seule famille.

selfcost ∥ Herstellungswert m. ∥ prix m. de revient.

self-discharge (Acc) ∥ Selbstentladung f. ∥ auto-décharge f.; décharge f. spontanée.

self-discharging truck (Railw) ∥ Selbstentladewagen m.; Selbstentlader m. ∥ wagon m. à déchargement automatique; auto-déchargeur m. / ~ for standard gauge ∥ normalspuriger Selbstentlader m. ∥ auto-déchargeur m. ou déchargeur m. automatique à voie normale.

self-discharging wagon see self-discharging truck.

self-distributing piston ∥ selbststeuernder Kolben m. ∥ piston m. à distribution automatique.

self-dumping horse rake ∥ Pferderechen m. mit selbsttätiger Entleerung ∥ râteau m. à cheval à décharge automatique.

self-emptying lorry see also self-discharging truck ∥ Selbstentladewagen m. ∥ déchargeur m. automatique; auto-déchargeur m.

self-excitation ∥ Selbsterregung f.; Eigenerregung f. ∥ auto-excitation f.

self-excited dynamo ∥ selbsterregte Dynamomaschine f. ∥ dynamo f. auto-excitatrice.

self-exciter ∥ Selbsterreger m. ∥ auto-excitateur m.

self-feeding furnace with hopper above grate ∥ Schüttfeuerung f. ∥ foyer m. à alimentation supérieure continue.

self-filling fountain pen ∥ Selbstfüller m.; selbstfüllender Füllfederhalter m. ∥ porte-plume m. réservoir à remplissage automatique; stylographe m. à remplissage automatique.

self-help ∥ Selbsthilfe f. ∥ défense f. personnelle ou légitime.

self-heterodyne (Radio) ∥ Schwingaudion n. ∥ autodyne m.

self-hooped (Gun) ∥ kaltgereckt; autofrettiert ∥ autofretté.

self-hooping (Of guns) ∥ Kaltreckung f.; radiale Expansion f.; Selbstberingung f.; Selbstschrumpfung f. ∥ autofrettage m.

self-induction (Electr) ∥ Selbstinduktion f. ∥ self-induction f. / coefficient of ~ ∥ Selbstinduktionskoeffizient m. ∥ coefficient m. de self. / ~ coil ∥ Selbstinduktionsspule f. ∥ bobine f. de self-induction. / ~ standard ∥ Selbstinduktionsnormale f. ∥ étalon m. de self-induction.

self-inductivity ∥ Selbstinduktivität f. ∥ self m.

self-igniting ∥ selbstzündend ∥ pyrophorique; à allumage m. automatique.

self-ignition ∥ Selbstzündung f. ∥ allumage m. spontané.

self-lighting flame ∥ selbstleuchtende Flamme f. ∥ flamme f. autolumineuse.

self-locking ∥ Selbstsperrung f.; Selbstblockung f. ∥ encliquetage m. automatique. / ~ gear ∥ selbstsperrendes Getriebe n. ∥ mécanisme m. d'encliquetage auto-

matique. / ~ worm gear ‖ selbsthemmendes Schneckengetriebe n. ‖ transmission f. à vis sans fin à blocage automatique.

self-lubricating ‖ selbstschmierend ‖ (se) graissant automatiquement. / ~ bearing ‖ Ringschmierlager n. ‖ graisseur m. à bague.

self-management ‖ Selbstverwaltung f. ‖ gestion f. directe; autonomie f.

self-oiler ‖ Selbstöler m. ‖ graisseur m. automatique.

self-oiling bearing ‖ Schmierlager n. ‖ palier-graisseur m.

self-opening die head ‖ selbstöffnender Gewindeschneidkopf m. ‖ filière f. automatique.

self-propelled caterpillar gun mount ‖ Raupenkraftlafette f. ‖ affût m. à autopropulsion à chenilles. / ~ mount ‖ Kraftlafette f.; Motorlafette f. ‖ affût m. automoteur.

self-purification ‖ Selbstreinigung f. ‖ auto-purification f.

self-reading staff ‖ Nivellierlatte f. zum Selbstablesen ‖ mire f. parlante.

self-recording see self-registering.

self-registering ‖ selbstschreibend; selbstregistrierend ‖ auto-enregistreur; enregistreur; enregistrant automatiquement.

self-regulating dynamo ‖ selbstregelnde Dynamo f. ‖ dynamo f. auto-régulatrice.

self-starter (Electr) ‖ Selbstanlasser m. ‖ rhéostat m. autodémarreur. / ~ (Auto) ‖ Selbstanlasser m. ‖ (auto-)démarreur m. / ~ (Typewr) ‖ Kolonnensteller m. ‖ sélecteur m. de colonnes. / converter with ~ ‖ Umformer m. mit Selbstanlauf ‖ convertisseur m. avec auto-démarrage.

self-starting ‖ Selbstanlaufen n. ‖ auto-démarrage m. / to be ~ ‖ von selbst anlaufen ‖ démarrer seul.

self-supporting overhead cable ‖ freitragendes Luftkabel n. ‖ câble m. aérien à auto-support.

self-tipper ‖ Selbstentlader m. ‖ déchargeur m. spontané. / pneumatic operation of flaps for ~s ‖ Druckluftklappenbetätigung f. für Selbstentlader ‖ clapet m. pour déchargeurs automatiques commandé à l'air comprimé.

sell, to ‖ verkaufen; absetzen ‖ vendre; placer. / ~ by auction ‖ versteigern ‖ vendre aux enchères ou à l'encan. / ~ badly ‖ sich schwer verkaufen; schwer abgehen ‖ n'être pas de débit; être d'une vente difficile. / ~ below cost price ‖ schleudern; verschleudern ‖ vendre à vil prix; gâcher la marchandise. / ~ at a loss ‖ mit Schaden m. oder Verlust m. verkaufen; verschleudern ‖ vendre à perte f. / hard to sell ‖ schwer verkäuflich ‖ d'un écoulement difficile.

seller ‖ Lieferer m.; Verkäufer m. ‖ fournisseur m. / automatic(al) ~ ‖ Verkaufsautomat m.; Selbstverkäufer m. ‖ distributeur m. automatique.

selling ‖ Absatz m.; Verkauf m. ‖ débit m. / ~ by retail ‖ Kleinverkauf m.; Einzelverkauf m. ‖ vente f. au détail. / ~ at ruinous prices ‖ (Preis-)Schleuderei f. ‖ vente f. à vil prix; gâchage m.

selling commission ‖ Verkaufsprovision f.; Verkaufsvergütung f. ‖ commission f. de vente. / automatic ~ machine see seller, automatic.

selling-off ‖ Ausverkauf m. ‖ vente f. totale.

selling organization ‖ Verkaufsorganisation f. ‖ organisme m. de vente. / joint ~ organization ‖ gemeinsame Verkaufsorganisation f. ‖ organisme m. de vente commune. / ~ price ‖ Verkaufspreis m. ‖ prix m. de vente. / ~ rate (Of exchange bills) ‖ Briefkurs m. ‖ taux m. de vente; lettre f. / ~ terms pl. ‖ Verkaufsbedingungen fpl. ‖ conditions fpl. de vente.

selters see seltzer water.

seltzer water ‖ Selterwasser n. ‖ eau f. de Seltz.

selvage see selvedge.

selvedge (Clothm) ‖ Salband n.; Borte f.; Kante f.; Schrote f. ‖ lisière f.; cordon m.; cordeline f.

semaphore ‖ Semaphor m. ‖ sémaphore m. / ~ station ‖ Semaphorstelle f. ‖ poste m. sémaphorique. / ~ telegram ‖ Semaphortelegramm n. ‖ télégramme m. sémaphorique.

semi-automatic ‖ halbautomatisch ‖ semi-automatique. / ~ bar lathe ‖ Stangenhalbautomat m. ‖ machine f. semi-automatique pour le travail de la barre. / ~ interlocking station (Railw) ‖ halbselbsttätiges Stellwerk n. ‖ poste f. d'aiguillage semi-automatique.

semi-automatic machine ‖ Halbautomat m. ‖ machine f. semi-automatique. / ~ for chuck work ‖ Halbautomat m. für Futterarbeit ‖ machine f. semi-automatique pour le travail en mandrin. / multiple-spindle ~ ‖ Mehrspindelhalbautomat m. ‖ machine f. semi-automatique à broches multiples.

semi-automatic rifle ‖ Halbselbstladegewehr n. ‖ fusil m. mitrailleur.

semi-cantilever ‖ halbfreitragend ‖ demi-cantilever.

semicircle ‖ Halbkreis m. ‖ demi-cercle m.

semicircular ‖ halbkreisförmig ‖ demi-circulaire. / ~ cutter ‖ Pilzfräser m. ‖ fraise f. à champignon. / ~ lens ‖ Halbkugellinse f. ‖ lentille f. demi-boule.

semi-coke ‖ Halbkoks m. ‖ demi-coke m.; semi-coke m. / ~ with a calorific value approaching that of bituminous coal ‖ Halbkoks m. mit einem Heizwert, welcher dem der Fettkohle nahesteht ‖ semi-coke m. dont le pouvoir calorifique est voisin de celui de la houille grasse. / fine-grained ~ ‖ feinkörniger Halbkoks m. ‖ semi-coke m. sous forme de grains très fins. / ~ produced from lignite ‖ Halbkoks m. aus Braunkohle ‖ semi-coke m. provenant de lignite. / lumpy ~ ‖ stückiger Halbkoks m. ‖ semi-coke m. en morceaux. / ~ from xylit ‖ Halbkoks m. aus Xylit ‖ semi-coke m. provenant de xylite.

semicolon ‖ Semikolon n. (;) ‖ point m. et virgule; point-virgule m.

semi-conductor ‖ Halbleiter m. ‖ semi-conducteur m.

semi-crystal ware ‖ Halbkristallware f. ‖ (objets mpl. en) demi-cristal m.

semi-detached house for one family ‖ Einfamiliendoppelhaus n. ‖ maison f. de famille double.

semi-elliptic spring ‖ Halbelliptikfeder f. ‖ ressort m. demi-pincette.

semi-finished products pl. see semi-product.

semi-incandescent lamp ‖ Glimmlampe f. ‖ lampe f. semi-incandescente.

semi-jewel cutter ‖ Halbedelsteinschleifer m. ‖ tailleur m. de pierres semi-précieuses.

semi-machined screw ‖ preßblanke Schraube f. ‖ vis f. forgée. / ~ washer ‖ halbblanke Scheibe f. ‖ rondelle f. semi-usinée.

semi-manufacture ‖ Halbfabrikat n. ‖ produit n. à demi façonné; demi-produit m. / ~ of metal ‖ Halbfabrikat n. aus Metall ‖ demi-produit m. en métal.

semi-manufactured products pl. see semi-product.

semi-opal ‖ Halbopal m. ‖ demi-opale f.

semi-opaque ‖ halbdurchsichtig ‖ semi-opaque.

semipermeable ‖ halbdurchlässig ‖ semi-perméable.

semi-portable engine ‖ halbfeste Antriebsmaschine f. ‖ machine f. demifixe. / ~ track ‖ halbbewegliches Gleis n. ‖ voie f. demi-mobile.

semi-portal crane ‖ Halbportalkran m. ‖ grue f. à demi-portique. / ~ revolving crane ‖ Halbportaldrehkran m. ‖ grue f. pivotante à semi-portique ou tournante à demi-portique.

semi-precious stone ‖ Halbedelstein m. ‖ pierre f. demi-précieuse. / cut ~ ‖ geschliffener Halbedelstein m. ‖ pierre f. demi-précieuse taillée. / facetted ~ ‖ facettierter Halbedelstein m. ‖ pierre f. demi-précieuse à facettes. / polished ~ ‖ polierter Halbedelstein m. ‖ pierre f. demi-précieuse polie. / rounded ~ ‖ abgerundeter Halbedelstein m. ‖ pierre f. demi-précieuse arrondie. / unset ~ ‖ ungefaßter Halbedelstein m. ‖ pierre f. demi-précieuse non montée. / worked ~ ‖ bearbeiteter Halbedelstein m. ‖ pierre f. demi-précieuse travaillée.

semi-product ‖ Halbzeug n. ‖ demi-produit m. / ~ of electric steel ‖ Elektrostahlhalbzeug n. ‖ demi-produit m. en acier au four électrique. / ~ of mild steel ‖ Halbzeug n. aus Flußstahl ‖ demi-produit m. d'acier doux. / ~ of superrefined steel ‖ Edelstahlhalbzeug n. ‖ demi-produit m. en acier spécial.

semi-rigid dirigible ‖ halbstarres Luftschiff n. ‖ dirigeable m. semi-rigide. / ~ system (Aero) ‖ halbstarres System n. ‖ système m. semi-rigide.

semi-rotary pump ‖ Würgelpumpe f.; Flügelpumpe f. ‖ pompe f. à ailettes.

semi-solid ‖ halbfest ‖ semisolide.

semisteel ‖ Halbstahl m. ‖ demi-acier m.

semi-transparent see semi-opaque.

semi-universal suspension ‖ halbuniversale Aufhängung f. ‖ suspension f. semi-universelle.

semolina ‖ Grieß m. ‖ semoule f. / coarse ~ ‖ grober Grieß m. ‖ grosse semoule f. / fine ~ ‖ feiner Grieß m. ‖ fine semoule f. / purified ~ ‖ geputzter Grieß m. ‖ semoule f. sassée. / gravity purifier of ~ ‖ Grießsortiermaschine f. nach spezifischem Gewicht ‖ purifieur m. à gravité de la semoule.

semolina mill ‖ Grießmühle f. ‖ semoulerie f.

senarmontite ‖ Senarmontit m.; Antimonoxyd n. ‖ senarmontite f.

send, to ‖ schicken ‖ envoyer. / ~ after ‖ nachschicken ‖ faire suivre. / ~ back ‖ remittieren; zurückschicken ‖ faire des remises fpl.; renvoyer. / ~ current ‖ Strom m. senden ‖ envoyer du courant. / ~ to makers pl. ‖ an die Fabrik einsenden ‖ renvoyer à l'usine f. / ~ by rail ‖ mit der Eisenbahn f. senden ‖

expédier par chemin m. de fer. / ~ a thing to a person ‖ jemandem etwas zustellen ‖ remettre *ou* faire tenir quelque chose à quelqu'un.

sender (Aut tel) ‖ Zahlengeber m. ‖ émetteur m. d'impulsions. / ~ (Radio) ‖ Sender m. ‖ poste m. transmetteur; transmetteur m.; émetteur m. / to cut out the local ~ ‖ den Ortssender m. ausschalten ‖ éliminer le poste local. / ~ of a letter ‖ Briefschreiber m.; Absender m. auteur m. d'une lettre. / ~ driven by motor (Aut tel) ‖ Maschinenzahlengeber m. ‖ émetteur m. d'impulsations à moteur. / ~ with rotary switches (Aut tel) ‖ Drehwählerzahlengeber m. ‖ émetteur m. d'impulsations à sélecteurs rotatifs.

sender diagonal ‖ Sendediagonale f. ‖ diagonale f. du poste transmetteur. / ~ selector ‖ Dienstwähler m. ‖ sélecteur m. des tables auxiliaires. / ~ testing device (Aut tel) ‖ Zahlengeberprüfeinrichtung f. ‖ dispositif m. d'essai pour les claviers à numéros.

sending of telegram ‖ Aufgabe f. des Telegramms ‖ transmission f. du télégramme.

sending device (Radio) ‖ Sendevorrichtung f. ‖ dispositif m. de transmission; appareil m. transmetteur. / ~ tube ‖ Senderöhre f. ‖ lampe f. pour émetteurs.

Senegal gum ‖ Senegalgummi m. ‖ gomme f. du Sénégal.

Senega root ‖ Senegawurzel f. ‖ racine f. de Sénéga.

senna leaf ‖ Sennesblatt n. ‖ feuille f. de séné.

sennits pl. of wood ‖ Flechtholz n.; geflochtene Holzmatte f. ‖ nattes fpl. en bois.

sensation ‖ Empfindung f. ‖ sensation f. / chromatic ~ ‖ Farbenempfindung f. ‖ sensation f. chromatique. / ~ of light ‖ Lichtreiz m. ‖ impression f. lumineuse.

sense of duty ‖ Pflichtgefühl n. ‖ sentiment m. du devoir. / ~ of feeling ‖ Tastgefühl n. ‖ sens m. spécial au toucher.

sensibility ‖ Empfindlichkeit f. ‖ sensibilité f. / ~ of an instrument ‖ Empfindlichkeit f. eines Instrumentes ‖ sensibilité f. d'un instrument. / ~ to light ‖ Lichtempfindlichkeit f. ‖ sensibilité f. à la lumière. / ~ of plates ‖ Plattenempfindlichkeit f. ‖ sensibilité f. des plaques. / ~ to position ‖ Lageempfindlichkeit f. ‖ sensibilité f. de position. / ~ of a scale ‖ Empfindlichkeit f. einer Wage ‖ sensibilité f. d'une balance. / limit of ~ ‖ Empfindlichkeitsgrenze f. ‖ limite m. de sensibilité.

sensible to see sensitive.

sensitive ‖ empfindlich ‖ sensible. / extremely ~ ‖ höchstempfindlich ‖ extrêmement sensible. / ~ to the action of heat ‖ hitzempfindlich ‖ sensible à la chaleur. / ~ to light ‖ lichtempfindlich ‖ sensible à la lumière.

sensitive operation ‖ feinfühlige Führung f. ‖ commande f. précise. / ~ tube (Radio) ‖ empfindliche Röhre f. ‖ tube m. sensible.

sensitiveness see sensibility.

sensitize, to ‖ lichtempfindlich machen ‖ sensibiliser.

sensitized film ‖ lichtempfindlich gemachter Film m. ‖ pellicule f. sensibilisée. / ~ plate ‖ lichtempfindlich gemachte Platte f. ‖ plaque f. sensibilisée.

sensitizer ‖ Sensibilisator m. ‖ sensibilisateur m.

sentence ‖ Richterspruch m.; Urteilsspruch m. ‖ jugement m.; sentence f. / ~ of the court ‖ Gerichtsbeschluß m. ‖ décision f. du tribunal.

sentinel ‖ Schildwache f. ‖ sentinelle f.

separate, to ‖ trennen ‖ séparer. / ~ (Chem) ‖ ausfällen ‖ se déposer; être déposé. / ~ (Ore dress) ‖ klassieren ‖ classer; trier; séparer. / ~ the dross ‖ entschlakken ‖ séparer la crasse; écumer les scories fpl. / ~ by eliquation ‖ abseigern ‖ liquater; ressuer; achever de ressuer. / ~ as a floculent precipitate ‖ ausflocken ‖ se séparer en flocons mpl. / ~ the grain ‖ Pulver n. sortieren ‖ égaliser la poudre grenée. / ~ ores ‖ Erze npl. scheiden ‖ séparer les minerais mpl. / ~ ores (Screening) ‖ Erze npl. klassieren ‖ cribler les minerais mpl. / ~ out the excess water vapour ‖ den überschüssigen Wasserdampf m. ausscheiden ‖ séparer *ou* éliminer la vapeur d'eau en excès.

separate account ‖ Separatkonto n. ‖ compte m. spécial. / ~ motor with cylinders ‖ Motor m. mit einzeln stehenden Zylindern ‖ moteur m. à cylindres séparés. / ~ drive ‖ Einzelantrieb m. ‖ commande f. séparée *ou* individuelle. / ~ excitation (Tel) ‖ Fremderregung f. ‖ excitation f. indépendante. / ~ excitation of valve transmitters ‖ Fremderregung f. von Röhrensendern ‖ excitation f. séparée des émetteurs à lampes. / ~ focusing of the eyepieces of a field glass ‖ Einzeltriebeinstellung f. der Okulare eines Feldstechers ‖ mise f. au point indépendante par les oculaires d'une jumelle. / erection of ~ homes ‖ Bau m. von Eigenheimen ‖ construction f. de maisons de famille. / ~ oiling ‖ Einzelschmierung f. ‖ graissage m. séparé. / ~ operation (Chem) ‖ Einzeluntersuchung f.; Einzelbestimmung f. ‖ opération f. séparée.

separated (Ore and coal dress) ‖ klassiert ‖ classé; trié.

separately-excited dynamo ‖ fremderregte Dynamo f. ‖ génératrice f. à excitation indépendante.

separating apparatus ‖ Trennungsapparat m. ‖ appareillage m. séparateur. / ~ cylinder ‖ Sortierzylinder m. ‖ cylindre m. de tamisage. / ~ dam (Hydr arch) ‖ Separationswerk n.; Trennbuhne f. ‖ épi m. de séparation. / ~ layer (Cable) ‖ Separator m.; Trennschicht f. ‖ séparateur m.; couche f. de séparation. / ~ machine ‖ Triör m. ‖ trieur m.; machine f. à nettoyer le blé. / ~ plant ‖ Sichtungsanlage f. ‖ installation f. de triage; triage m. / sheet for ~ purposes ‖ Separationsblech n. ‖ tôle f. de séparation. / ~ switch ‖ Trennschalter m. ‖ interrupteur-séparateur m. / syrup ~ valve ‖ Siruptrennventil n. ‖ soupape f. de séparation pour sirop.

separation ‖ Trennung f. ‖ séparation f. / ~ (Ore dress) ‖ Scheidung f. ‖ séparation f.; triage m.; calibrage m. / dry ~ ‖ trockene Scheidung f. ‖ séparation f. à sec. / ~ by eliquation (Met) ‖ Saigerprozeß m.; Saigerung f. ‖ liquation f.; ressuage m. / ~ of gas mixtures ‖ Trennung f. von Gasgemischen ‖ séparation f. de mélanges de gaz. / ~ of gold with a large percentage of silver ‖ Scheidung f.

silberreichen Goldes ‖ séparation f. d'or à haute teneur en argent. / ~ of the humidity contained in the air ‖ Abscheiden n. des in der atmosphärischen Luft enthaltenen Wasserdampfes ‖ séparation f. des vapeurs d'eau contenue dans l'air. / ~ of the spectrum ‖ Spektrumteilung f. ‖ séparation f. du spectre. / wet ~ ‖ nasse Scheidung f. ‖ séparation f. par voie humide.

separation plant, coal ~ ‖ Kohlenseparation f. ‖ installation f. de séparation de charbon. / magnetic ~ ‖ magnetische Aufbereitungsanlage f. ‖ installation f. de séparation magnétique; séparateur m. magnétique.

separation, process of ~ ‖ Trennungsvorgang m. ‖ marche f. de la séparation. / ~ wall ‖ Scheidewand f. ‖ paroi f. mitoyenne.

separator ‖ Abscheider m.; Separator m. ‖ séparateur m. / ~ (Milk) ‖ Schleuder f.; Zentrifuge f. ‖ essoreuse f. / ~ (Wheat) ‖ Triör m. ‖ trieur m. / acid ~ ‖ Säureabscheider m. ‖ éliminateur m. d'acide. / centrifugal cream ~ ‖ Milchzentrifuge f. ‖ écrémeuse f. centrifuge. / cream ~ ‖ Milchzentrifuge f. ‖ écrémeuse f. centrifuge. / electromagnetic dry ~ ‖ elektromagnetischer Trockenscheider m. ‖ séparateur m. à sec électromagnétique. / electromagnetic wet ~ ‖ elektromagnetischer Naßscheider m. ‖ trieur m. électromagnétique par voie humide. / exhaust steam oil ~ ‖ Abdampfentöler m. ‖ déshuileur m. de vapeur d'échappement. / high-capacity ~ ‖ Hochleistungstriör m. ‖ trieur m. à grand débit. / iron ~ ‖ Eisenseparator m. ‖ séparateur m. électrique de ferraille. / magnetic ~ ‖ Magnetscheider m. ‖ séparateur m. magnétique. / magnetic drum ~ ‖ Trommelmagnetscheider m. ‖ tambour m. séparateur magnétique. / non-condensable gas ~ (Refrigerating plant) ‖ Entlüftungseinrichtung f. ‖ appareil m. désaérateur. / oil ~ ‖ Ölabscheider m. ‖ séparateur m. d'huile. / exhaust steam oil ~ ‖ Abdampfentöler m. ‖ déshuileur m. de vapeur d'échappement. / tar ~ ‖ Teerabscheider m. ‖ séparateur m. de goudron. / water ~ ‖ Wasserabscheider m. ‖ séparateur m. d'eau.

separator indent (Wheat) ‖ Triörzelle f. ‖ alvéole m. de trieur.

separatory funnel ‖ Scheidetrichter m. ‖ entonnoir m. séparateur.

sepia ‖ Sepia f. ‖ sepia f.

sepiolite ‖ Meerschaum m. ‖ écume f. de mer.

septic matter ‖ fäulniserregender Stoff m. ‖ matière f. septique.

septum ‖ Scheidewand f. ‖ cloison f.; mur m. de séparation.

sepulchral art ‖ Grabmalskunst f. ‖ art m. sépulcral. / ~ cross ‖ Grabkreuz n. ‖ croix f. de tombe. / ~ wreath of metal ‖ Metallgrabkranz m. ‖ couronne f. métall'que pour tombes.

sequence ‖ Folge f.; Reihenfolge f. ‖ suite f. / ~ of phantom crossings (Tel) ‖ Viererfolge f. ‖ distribution f. des transpositions de circuits combinés. / ~ switch (Aut tel) ‖ Steuerschalter m.; Folgeschalter m. ‖ combineur m. séquentiel.

serge ‖ Serge f. ‖ serge f. / cotton ~ ‖ Baumwollkörper m. ‖ croisé m. *ou* sergé m. coton. / Orleans ~ ‖ Orleansserge f. ‖ filin m.; serge f. d'Orléans; tourangette f.

sericite ‖ Sericit m. ‖ séricite f. / ~ slate ‖ Serizitschiefer m. ‖ schiste m. séricitique.

series ‖ Reihe f. ‖ série f. / converging ~ ‖ konvergierende Reihe f. ‖ serie f. convergente. / ~ of the cosine ‖ Kosinusreihe f. ‖ série f. du cosinus. / diverging ~ ‖ divergierende Reihe f. ‖ série f. divergente. / harmonical ~ ‖ harmonische Reihe f. ‖ série f. harmonique. / in ~ (Electr) ‖ hintereinander; in Reihenschaltung f. ‖ en série f. / in ~ (Trade) ‖ serienmäßig ‖ en série f. / infinite ~ ‖ unendliche Reihe f. ‖ série f. infinie. / logarithmic ~ ‖ logarithmische Reihe f. ‖ série f. logarithmique. / ~ of meshes ‖ Maschenreihe f. ‖ rangée f. de mailles. / ~ of observations ‖ Beobachtungsreihe f. ‖ série f. d'observations. / produced by ~ ‖ in Serien fpl. hergestellt ‖ fabriqué en série f. / recurring ~ ‖ rekurrierende Reihe f. ‖ série f. récurrente. / ~ of tests ‖ Versuchsreihe f. ‖ série f. d'essais.

series condenser ‖ Serienkondensator m. ‖ condensateur m. en série. / ~ connection (Electr) ‖ Reihenschaltung f. ‖ mise f. ou connexion f. ou couplage m. en série. / ~ dynamo ‖ Reihenschlußdynamo f. ‖ dynamo f. en série. / ~ fabrication *see* ~ manufacture. / ~ manufacture ‖ Reihenfertigung f. ‖ fabrication f. en série. / ~ motor ‖ Serienmotor m. ‖ moteur m. de série. / ~ parallel connection ‖ Gruppenschaltung f. ‖ montage m. en série-parallèle. / ~ parallel switch ‖ Serienparallelschalter m. ‖ interrupteur m. série-parallèle. / production in ~ ‖ Reihenfertigung f. ‖ fabrication f. en série. / ~ rheostat ‖ Serienwiderstand m. ‖ rhéostat m. en série.

serious ‖ schwerwiegend ‖ grave; sérieux. / ~ accident ‖ ernstlicher Unfall m. ‖ accident m. grave.

seriously sick person ‖ schwer Erkrankter m. ‖ homme m. gravement malade.

serpentaria root ‖ Schlangenwurzel f. ‖ racine f. de serpentaire.

serpentine ‖ Serpentin m. ‖ serpentine f. / ~ apparatus (Met) ‖ Schlangenrohrapparat m. ‖ appareil m. à serpentin. / ~ pipe ‖ Schlangenrohr n. ‖ serpentin m. / ~ pipe (Refrigerating plant) ‖ Kühlschlange f. ‖ serpentin m. de réfrigération. / ~ quarry ‖ Serpentinsteinbruch m. ‖ carrière f. de serpentine. / ~ tube *see* ~ pipe.

serpentinization ‖ Serpentinisierung f. ‖ serpentinisation f.

serpent mill ‖ Schlangenwalzwerk n. ‖ train m. à serpent.

serrated ‖ gezackt ‖ brettelé. / ~ edge of the piston ‖ Schlagkrone f. des Kolbens ‖ denture f. du piston perforateur.

serum ‖ Serum n. ‖ sérum m. / ~ glass *see* ~ tube. / ~ tube ‖ Serumglas n. ‖ tube m. pour sérum.

servant (Agr) ‖ Magd f. ‖ servante f. / civil ~ ‖ Beamte m. ‖ fonctionnaire m.; employé m.

serve, to ~ the furnace ‖ den Ofen m. beschicken ‖ charger le fourneau.

served quires pl. (Bookb) ‖ fadengehefteter Bogensatz m. ‖ ensemble m. de feuilles cousu au fil.

service ‖ Dienstleistung f.; Dienst m. ‖ service m. / ~ (Law) ‖ Zustellung f. ‖ remise f.; notification f. / ~ for the centralization of meteorological information ‖ Sammelwetterdienst m. ‖ service m. de centralisation des informations météorologiques. / ~ by draught animals ‖ Zugtierbetrieb m. ‖ service m. par bêtes de trait. / ~ with grabs ‖ Greiferbetrieb m. ‖ service m. par bennes. / high-speed ~ ‖ Schnellbetrieb m. ‖ fonctionnement m. à grande vitesse. / to leave the ~ ‖ aus dem Dienstverhältnis n. ausscheiden ‖ quitter le service. / ~ of post office vans ‖ Postkraftwagenverkehr m. ‖ service m. des automobiles postaux. / railway ~ ‖ Eisenbahndienst m. ‖ service m. des lignes de chemin de fer. / regular ~ ‖ regelmäßiger Verkehr m. ‖ service m. régulier. / to retain in ~ ‖ im Dienst m. behalten ‖ garder au service m. / ~ rendered in return ‖ Gegendienst m. ‖ service m. réciproque; revanche f. / ~ for the safety of aircraft ‖ Flugsicherungsdienst m. ‖ service m. de sécurité aérienne. / for ~ at table ‖ für den Tischgebrauch m. ‖ pour le service à table. / telegraphic ~ ‖ Telegrafendienst m. ‖ service m. télégraphique. / ~ of watchmen ‖ Wachtdienst m. ‖ service m. de garde.

serviceable ‖ zweckdienlich; zweckentsprechend; tauglich ‖ convenable; efficace; utile.

service advice ‖ Dienstnotiz f. ‖ avis m. de service. / ~ call (Tel) ‖ Dienstgespräch n. ‖ conversation f. de service. / ~ call key ‖ Dienstanrufschalter m. ‖ clef f. d'appel de service. / ~ call lamp ‖ Dienstanruflampe f. ‖ lampe f. d'appel de service. / ~ call relay ‖ Dienstanrufrelais n. ‖ relais m. d'appel de service. / ~ connection (Tel) ‖ Betriebsschaltung f. ‖ connexion f. de régime. / contract for ~ ‖ Dienstvertrag m. ‖ contrat m. de louage ou de service. / dismissal of ~ ‖ Dienstentlassung f. ‖ congé m. / disturbance of ~ ‖ Betriebsstörung f. ‖ dérangement m. de service. / ~ hour ‖ Dienststunde f. ‖ heure f. de service. / ~ indication ‖ Dienstvermerk m. ‖ indication f. de service. / ~ instructions pl. (Mach) ‖ Behandlungsvorschrift f. ‖ instructions fpl. de service ou d'entretien. / kind of ~ ‖ Betriebsart f. ‖ genre m. ou méthode m. de service. / length of ~ ‖ Dienstalter m. ‖ ancienneté f. / ~ lift (Kitchen) ‖ Speiseplattenaufzug m. ‖ monte-plat m. / ~ master clock ‖ Betriebshauptuhr f. ‖ horloge f. principale de service. / period of ~ on trial ‖ Probedienstzeit f. ‖ temps m. de service d'essai. / ~ platform (Mach) ‖ Bedienungsbühne f. ‖ plate-forme f. de service ou de commande. / putting into ~ ‖ Inbetriebsetzung f. ‖ mise f. en marche. / ~ quality ‖ Betriebsgüte f. ‖ qualité f. de service. / to commit a breach of contract by refusal of ~ ‖ vertragswidrig den Dienst m. verweigern ‖ refuser le service contrairement au contrat. / to continue ~ relations pl. ‖ das Arbeitsverhältnis fortsetzen ‖ prolonger le contrat de travail. / ~ reliability ‖ Betriebssicherheit f. ‖ sécurité f. de service. / ~ stress ‖ Betriebsbeanspruchung f. ‖ fatigue f. ou charge f. d'emploi. / ~ tank (Auto) ‖ Betriebsbehälter m. ‖ nourrice f. / ~ telegram ‖ Diensttelegramm n. ‖ télégramme m. de service. / ~ telephone plant ‖ Betriebsfernsprechanlage f. ‖ installation f. téléphonique de service. / to request termination of ~ before the expiry of the contractual period ‖ die Aufhebung des Dienstverhältnisses vor Ablauf der vertragsmäßigen Frist verlangen ‖ exiger la résiliation du contrat de service avant l'expiration de la période contractuelle. / ~ voltage ‖ Betriebsspannung f. ‖ tension f. de service. / ~ voltage of a current transformer ‖ Betriebsspannung f. eines Stromwandlers ‖ tension f. de service d'un transformateur d'intensité. / years pl. of ~ *see* length of ~.

serving (Cable) ‖ Umhüllung f. ‖ revêtement m. / ~ tray with painted glass ‖ Serviertablett n. mit bemalter Glaseinlage ‖ plateau m. de service à fond de verre peint.

servitude ‖ Dienstbarkeit f. ‖ servitude f. / real ~ ‖ Grunddienstbarkeit f. ‖ servitude f. réelle.

servo-motor ‖ Hilfsmotor m. ‖ servomoteur m.

sesame oil ‖ Sesamöl n. ‖ huile f. de sésame. / ~ seed ‖ Sesamsame m. ‖ graine f. de sésame.

sesquicarbonate ‖ anderthalb kohlensaures Salz n. ‖ sesquicarbonate m. / ~ of soda ‖ anderthalbkohlensaures Natrium n. ‖ sesquicarbonate m. de soude.

sesquioxide of cobalt ‖ Kobaltoxyd n. ‖ sesquioxyde m. de cobalt. / ~ of manganese ‖ Manganoxyd n. ‖ sesquioxyde m. de manganèse.

sesquiplane ‖ Eineinhalbdecker m. ‖ sesquiplan m.

session ‖ Sitzung f. ‖ séance f.; session f.

set, to ‖ setzen; stellen; aufstellen; legen ‖ mettre. / ~ (Mortar) ‖ abbinden ‖ prendre. / ~ (Razor) ‖ abziehen; schleifen ‖ repasser; affiler. / ~ in air (Chem) ‖ an der Luft f. erhärten ‖ durcir à l'air. / ~ the bricks in the kiln ‖ die Ziegel mpl. in den Ofen einsetzen; den Brand m. einfahren ‖ enfourner les briques fpl. / ~ the centers (Build) ‖ die Lehrbogen mpl. aufstellen ‖ poser les cintres mpl. / ~ a copy ‖ ein Manuskript n. absetzen ‖ composer une copie. / to give the glue the time to set (Bookb) ‖ dem Leim m. Zeit zum Anziehen geben ‖ donner à la colle le temps de prendre. / ~ going ‖ in Schwung m. bringen; in Gang m. setzen; anlaufen lassen ‖ mettre en train ou en marche; donner l'élan. / ~-in the plane iron ‖ das Hobeleisen n. einsetzen ‖ mettre en fût le rabot. / ~ to infinity ‖ auf Unendlich einstellen ‖ mettre au point sur l'infini. / ~ the iron (Forg) ‖ das Eisen richten ‖ parer le fer au marteau. / ~ by itself ‖ selbständig erhärten ‖ durcir de soi-même. / ~ itself in the direction of the field ‖ sich in die Feldrichtung f. einstellen ‖ se placer dans la direction du champ. / ~ the keel ‖ den Kiel m. strecken ‖ poser la quille. / ~ on the last ‖ auf den Leisten m. schlagen ‖ mettre ou monter sur forme f. / ~ more sails ‖ Segel npl. beisetzen oder setzen ‖ augmenter de voiles fpl. / ~-out on a journey ‖ eine Reise f. antreten ‖ partir en voyage. / ~-out straight lines on the ground ‖ gerade Linien fpl. im Gelände abstecken ‖ jalonner des lignes fpl. droites sur le terrain. / ~ the pots ‖ das Geschirr einsetzen ‖ enfourner les pots mpl. / ~ precious stones pl. ‖ Edelsteine mpl. fassen ‖ sertir ou monter des pierres fpl. précieuses. / ~ sails pl. ‖ Segel npl. setzen ‖ déployer les voiles fpl.; mettre les voiles fpl. au vent. / ~ taps

pl. (Brew) ‖ anzapfen ‖ mettre en perce f.; purger les robinets mpl. / ~ the teeth of a saw ‖ die Zähne mpl. einer Säge schränken *oder* verschränken ‖ contourner les dents d'une scie; donner de la voie à la scie. / ~ up (Build) ‖ aufrichten; errichten ‖ ériger; relever. / ~ up (Mach) ‖ zusammenbauen; montieren ‖ monter. / ~ up the roof of a house ‖ ein Haus n. richten ‖ poser le bouquet sur une maison. / ~ a varnish on a picture ‖ ein Gemälde n. mit Firnis überziehen ‖ passer le vernis sur un tableau. / ~ in water (Concrete) ‖ unter Wasser n. erhärten ‖ durcir sous l'eau f. / ~ to work ‖ in Betrieb m. setzen ‖ mettre en activité f. *ou* en marche f. / ~ to work the furnace ‖ den Hochofen m. anblasen ‖ mettre le fourneau à feu.

set ‖ Satz m. ‖ jeu m.; assortiment m. / ~ (Found) ‖ Gießloch n.; Gußloch n. ‖ godet m.; jet m. / ~ of artificial teeth ‖ künstliches Gebiß n. ‖ dentier m. *ou* râtelier m. artificiel. / bacteriological ~ ‖ bakteriologisches Besteck n. ‖ trousse f. bactériologique. / ~ of butts for the manufacture of cardboard ‖ Büttengarnitur f. für die Pappenfertigung ‖ garniture f. de tonneaux à cartonneries. / ~ of coke ovens ‖ Koksofenbatterie f. ‖ batterie f. de fours à coke. / cold ~ ‖ Schrotmeißel m. ‖ tranche f. (à froid). / ~ of drills ‖ Satz m. Bohrer ‖ jeu m. de mèches. / ~ of driving wheels ‖ Treibradsatz m. ‖ essieu m. moteur monté. / ~ of drums ‖ Trommelsatz m. ‖ tambours mpl. tamiseurs. / ~ of glass plates ‖ Glasplattensatz m. ‖ pile f. de glaces. / ~ of knives, forks and spoons ‖ Tafelbesteck n. ‖ couvert m. de table. / ~ for testing lightning conductors ‖ Blitzableiteruntersuchungsapparat m. ‖ appareil m. pour l'examen des paratonnerres. / ~ of millstones ‖ Mahlgang m. ‖ tournant m. / ~ of patterns ‖ Modellsatz m. ‖ jeu m. de modèles. / permanent ~ ‖ bleibende Formänderung f. ‖ allongement m. permanent. / deformation test by loads without a permanent ~ ‖ Belastungsprobe f. ohne bleibende Durchbiegung ‖ essai m. de flexion par charge morte sans déformation permanente. / ~ of piles (Build) ‖ Pfahlrost m.; Pfahlwerk n.; Pfählung m.; pilotage m.; piloté m.; ouvrage m. de pilotis. / ~ of points (Railw) ‖ Weichenstraße f. ‖ groupe m. de changements de voie. / ~ of prisms ‖ Prismensatz m. ‖ jeu m. de prismes. / ~ of pulleys ‖ Flaschenzug m. ‖ palan m. à moufles. / ~ of pumps (Mine) ‖ Pumpensatz m. ‖ jeu m. de pompes; pompe f. installée dans une colonne élévatoire. / ~ of riveters ‖ Nietkolonne f. ‖ équipe f. de riveurs. / riveting ~ (Boil) ‖ Döpper m. ‖ bouterolle f. / ~ of simple prisms of heavy flint glass ‖ Satz m. einfacher Prismen aus Schwerflintglas ‖ jeu m. de prismes simples en flint lourd. / ~ of soaked hides ‖ Satz m. von gebeizten Fellen ‖ encuvage m. de peaux. / permanent ~ of spring ‖ bleibende Durchbiegung f. der Feder ‖ flèche f. *ou* déformation f. permanente de ressort. / ~ of teeth ‖ Gebiß n. ‖ denture f. / ~ of timber (Mine) ‖ Geviere n.; Holzgeviere n. ‖ cadre m. de boisage. / ~ of tools ‖ Werkzeugsatz m. ‖ jeu m. d'outils. / ~ of tracks ‖

Bahnlinie f.; Fahrbahn f.; Gleis n.; Schienenstrang m. ‖ voie f.; voie ferreé; ligne f. d'un chemin de fer.

set of wheels ‖ Radsatz m. ‖ train m. de roues monté. / ~ for inside bearings ‖ Innenlagerradsatz m. ‖ essieu m. monté pour boîtes à graisse intérieures. / ~ for outside bearings ‖ Außenlagerradsatz m. / essieu m. monté pour boîtes à graisse extérieures.

set of workings (Mine) ‖ Abbaufeld n. ‖ lot m.

set back ‖ Rückschlag m. ‖ contre-coup m.; revirement m. / ~ bolt ‖ Kopfbolzen m. ‖ boulon m. à tête. / ~ collar ‖ Stellring m. ‖ bague f. d'arrêt. / ~ form ‖ Formular m.; Vordruck m. ‖ formulaire f. / ~ frame ‖ Wattenmaschine f. ‖ nappeuse f.

set hammer ‖ Setzhammer m.; Flachhammer m. ‖ chasse f. à parer; paroir m. / chamfered ~ ‖ schräger Setzhammer m. ‖ chasse f. en biseau. / half-round ~ ‖ runder Setzhammer m. ‖ chasse f. ronde; dégorgeoir m. / plain ~ ‖ gerader Setzhammer n. ‖ chasse f. carrée. / square ~ ‖ gerader Setzhammer m. ‖ chasse f. carrée.

set-off (Law) ‖ Gegenforderung f. ‖ créance f. en compensation; demande f. reconventionnelle. / ~ device (Print) ‖ Abschmutzvorrichtung f. ‖ bobine f. de décharge. / ~ sheet (Print) ‖ Schmutzbogen m. ‖ maculature f.

set pin ‖ Stellstift m.; Anschlagstift m.‖tige f. de fixation. / ~ screw ‖ Setzschraube f.; Hemmschraube f. ‖ vis f. de bloquage *ou* d'arrêt.

setter (Chisel) ‖ Setzmeißel m. ‖ chasse f. / ~ (Print) ‖ Setzer m. ‖ compositeur m. / ~ (Rivets) ‖ Nietstempel m. ‖ chasse-rivet m. / machine ~ ‖ Einrichter m. ‖ régleur m. de machines.

setting a boiler ‖ Kesseleinmauerung f. ‖ maçonnage m. de chaudière. / ~ the bubble ‖ Einstellung f. der Libelle ‖ mise f. au point de la nivelle. / ~ the eyepiece to the observer's sight ‖ das Okular auf die Sehschärfe des Beobachters einstellen ‖ adapter la mise au point de l'oculaire à la vue de l'observateur. / horizontal ~ ‖ Wagerechteinstellung f. ‖ mise f. de niveau. / initial ~ ‖ Bereitschaftsstellung f. ‖ position f. d'attente. / ~ of the mortar ‖ Binden n. des Mörtels ‖ prise f. du mortier. / ~ of pumps ‖ Einbau m. von Pumpen ‖ montage m. de pompes. / ~ the rails ‖ Geraderichten n. *oder* Richten n. der Schienen ‖ dressage n. *ou* dressement m. des rails. / ~ by spanners ‖ Schlüsseleinstellung f. ‖ réglage m. par clef. / ~ of a spring ‖ Spannen n. einer Feder ‖ serrage m. d'un ressort. / ~ taps pl. (Brew) ‖ Anzapfen n. ‖ mise f. en perce; purge f. des robinets. / valve ~ ‖ Ventileinstellung f. ‖ réglage m. de soupapes.

setting blow ‖ Setzschlag m. ‖ frappe f. appuyée. / ~ commencement of ~ (Concrete) ‖ Erhärtungsbeginn m. ‖ commencement m. de la prise. / ~ device ‖ Einstellvorrichtung f. ‖ dispositif m. de réglage *ou* d'ajustage. / ~ device for automatic spur gear cutting machines ‖ Aufspannvorrichtung f. für selbsttätige Stirnradfräsmaschinen ‖ dispositif m. de montage pour machines alternatives automatiques à tailler les

engrenages droits. / finish of ~ (Concrete) ‖ Erhärtungsende n. ‖ fin f. de la prise. / ~ gauge ‖ Einstellehre f. ‖ appareil m. de réglage.

setting-in (Pott) ‖ Einsetzen n. ‖ enfournage m.; enfournement m.

setting machine (Saw) ‖ Schränkmaschine f. ‖ machine f. à donner de la voie aux dents de scie. / type ~ ‖ Setzmaschine f. ‖ machine f. à composer.

setting-off of the train ‖ Abfahrt f. eines Zuges ‖ départ m. d'un train.

setting-out curves (Railw) ‖ Kurvenabstecken n. ‖ tracé m. de courbes. / instrument for ~ roads ‖ Instrument n. zum Abstecken von Straßen ‖ instrument m. destiné aux jalonnements de routes.

setting pole ‖ Bootshaken m. ‖ croc m. de batelier; gaffe f. / process of ~ (Concrete) ‖ Abbindeverlauf m.; Erhärtungsverlauf m. ‖ processus m. de la prise. / ~ strength (Concrete) ‖ Bindekraft f. ‖ puissance f. de prise. / ~ time (Concrete) ‖ Abbindezeit f. ‖ durée f. de la prise. / to shorten the time of ~ (Concrete) ‖ die Abbindezeit f. verkürzen ‖ raccourcir le temps de prise.

setting-up apparatus ‖ Aufsatzform f. ‖ forme f. d'assemblage. / ~ time ‖ Einstellzeit f. ‖ temps m. de réglage.

settle, to ‖ ansiedeln ‖ établir. / ~ (Build) ‖ sich setzen; sich senken ‖ se tasser. / ~ (Chem) ‖ sich abscheiden; sich absetzen; ausfallen ‖ se déposer; se précipiter; être déposé. / ~ (Dispute) ‖ schlichten ‖ arranger; aplanir. / ~ (Mar) ‖ sinken; untergehen ‖ couler; couler bas; sancir; sombrer. / ~ (Trade) ‖ liquidieren ‖ liquider. / to allow the liquid to settle ‖ die Flüssigkeit abstehen lassen ‖ laisser reposer le liquide. / the metal settles hard on the outside with a gradual transition from the tread into the middle ‖ das Eisen strahlt von der Lauffläche aus nach innen hart ein ‖ la couche blanche et dure à la surface de roulement se nuance graduellement vers l'intérieur.

settlement ‖ Niederlassung f. ‖ établissement m.; colonie f. / ~ (Trade) ‖ Aufrechnung f.; Abrechnung f. ‖ compensation f.; règlement m. / final ~ ‖ Schlußabrechnung f. ‖ règlement m. final. / to make a private ~ with one's creditors ‖ akkordieren; sich mit seinen Gläubigern mpl. abfinden ‖ s'arranger; passer un arrangement avec ses créanciers. / yearly ~ ‖ Jahresabrechnung f. ‖ règlement m. de la fin d'année.

settlement range (Stove) ‖ Siedelungsherd m. ‖ cuisine f. de colonies.

settler ‖ Ansiedler m.; Kolonist m. ‖ colon m. / ~ *see* settling cask.

settling (Building) ‖ Einsinken n.; Sacken n. ‖ affaissement m.; enfoncement m. / ~ (Earthw) ‖ Senkung f.; Sackung f. ‖ tassement m. / ~ (Liquid) ‖ Klärung f. ‖ clarification f. / ~ of dust ‖ Niederschlagen n. des Staubes ‖ captation f. de la poussière.

settling basin ‖ Klärbecken n. ‖ bac m. de décantation. / ~ bottom ‖ Setzbrett n. ‖ lit m. de lavage. / ~ cask ‖ Klärfaß n. ‖ cuve f. clarificatoire. / ~ day ‖ Stichtag m. ‖ jour m. de l'échéance. / ~ pool *see* ~ reservoir. / ~ reservoir ‖ Klärbassin n. ‖ réservoir m. à clarifier. / ~ sump ‖ Klärbassin n.; Klärbecken n.;

Klärsumpf m. ∥ bassin m. de décantation. / ~ tank *see* ~ reservoir. / ~ tub (Brew) ∥ Absetzbottich m. ∥ cuve f. guilloire. / ~ vat *see* ~ tub.

sew, to ∥ nähen ∥ coudre. / ~ (Bookb) ∥ broschieren ∥ brocher. / ~ a book ∥ ein Buch n. heften ∥ coudre un livre. / ~ by machine ∥ mit der Maschine f. nähen ∥ coudre à la machine.

sewage ∥ Abwasser n. ∥ eau f. résiduelle *ou* d'égout. / ~ clarifying *see* ~ purification. / ~ farm ∥ Rieselfeld n. ∥ champ m. d'épandage. / ~ filter ∥ Klärteich m. ∥ bassin m. de décantation. / ~ powder ∥ Düngpulver n. ∥ poudrette f.

sewage purification ∥ Abwasserreinigung f. ∥ clarification f. des eaux d'égouts. / ~ by filtration ∥ Schwemmfilterung f. ∥ filtration f. des eaux d'égouts. / ~ plant ∥ Kläranlage f. ∥ installation f. de clarification.

sewage removal enterprise ∥ Abfuhranstalt f. ∥ entreprise f. de vidanges. / ~ water *see* sewage.

sewed (Book) ∥ broschiert; geheftet ∥ broché. / ~ article ∥ genähter Gegenstand m. ∥ article m. confectionné.

sewer (Bookb) ∥ Hefter m. ∥ couseur m. / ~ (Build) ∥ Rinnstein m. ∥ culière f.; caniveau m.; rigole f. / ~ (Drain) ∥ Straßenkanal m.; Abzugschleuse f. ∥ égout m. / machine ~ (Bookb) ∥ Maschinenhefter m. ∥ couseur m. à la machine. / main ~ ∥ Hauptkanal m.; Sammelkanal m. ∥ collecteur m.; égout m. collecteur.

sewerage ∥ Entwässerung f.; Kanalisation f. ∥ canalisation f. / ~ plant *see* sewage purification plant.

sewer building ∥ Kanalisationsbau m. ∥ construction f. d'égouts. / ~ man ∥ Senkgrubenreiniger m. ∥ vidangeur m.

sewer-on of buttons (Shoem) ∥ Knopfannäher m. ∥ poseur m. de boutons.

sewing ∥ Näharbeit f. ∥ ouvrage m. de couture. / ~ (Bookb) ∥ Heften n. ∥ couture f. d'un livre. / ~ cotton ∥ Baumwollnähfaden m. ∥ coton m. à coudre.

sewing machine (Bookb) ∥ Heftmaschine f. ∥ coudreuse f. / ~ (Tail) ∥ Nähmaschine f. ∥ machine f. à coudre. / double-chain-stitch ~ ∥ Doppelkettenstichnähmaschine f. ∥ machine f. à coudre à point de chaînette double. / embroidering ~ ∥ Kurbelstickmaschine f. ∥ brodeuse f. à Bonnaz. / hand ~ ∥ Handnähmaschine f. ∥ machine f. à coudre à manivelle. / hosiery ~ ∥ Strumpfnähmaschine f. ∥ machine f. à coudre pour bonneterie. / straw hat ~ ∥ Strohhutnähmaschine f. ∥ machine f. à coudre les chapeaux de paille. / straw husks ~ ∥ Strohhülsennähmaschine f. ∥ machine f. à coudre les manchons de paille. / swift ~ ∥ Schnellnähmaschine f. ∥ machine f. à coudre rapide. / toy ~ ∥ Kindernähmaschine f. ∥ machine f. à coudre pour enfants.

sewing machine lighting ∥ Nähmaschinenbeleuchtung f. ∥ lampes fpl. pour machines à coudre. / ~ motor ∥ Nähmaschinenmotor m. ∥ moteur m. pour machines à coudre. / ~ needle ∥ Nähmaschinennadel f. ∥ aiguille f. pour machines à coudre. / ~ oil ∥ Nähmaschinenöl n. ∥ huile f. pour machine à coudre. / ~ part ∥ Nähmaschinenteil n. ∥ pièce f. détachée de machine à coudre. / ~ repairing workshop ∥ Nähmaschinenausbesserungswerk-

statt f. ∥ atelier m. de réparation de machines à coudre.

sewing needle ∥ Nähnadel f. ∥ aiguille f. à coudre. / surgical ~ ∥ chirurgische Nähnadel f. ∥ épingle f. à coudre pour usages chirurgicaux.

sewing press (Bookb) ∥ Heftlade f. ∥ cousoir m. / ~ silk ∥ Nähseide f. ∥ soie f. à coudre. / ~ table (Bookb) ∥ Heftlade f. ∥ cousoir m. / ~ thread ∥ Nähgarn n. ∥ fil m. à coudre. / ~ twine (Mar) ∥ Nähgarn n.; Segelgarn n. ∥ fil m. à coudre. / ~ wax ∥ Nähwachs n. ∥ cire f. de couture.

sextant ∥ Sextant m. ∥ sextant m.

sextuple system (Tel) ∥ Sechsfachbetrieb m. ∥ installation f. sextuple.

shabby ∥ schäbig ∥ usé; râpé; mesquin.

shabrack ∥ Schabracke f. ∥ shabraque f.

shackle ∥ Schäkel m.; Schake f.; Kettenglied n. ∥ manille f. / ~ (Hydr arch) ∥ Ankerbügel m.; Haltebügel m. ∥ bouche f. d'amarrage. / ~ (Locksm) ∥ Federbügel m.; Bügel m.; Schäkel m. ∥ anse f. de cadenas *ou* de serrure; manille f.; meuille f.; maille f. / ~ (Railw) ∥ Kupplungsbügel m. ∥ maille f. de tendeur. / ~ of a chain ∥ Kettenschäkel m. ∥ manille f. d'une chaîne. / ~ spring ~ ∥ Federlasche f. ∥ menotte f. du ressort.

shackle bar ∥ Kupplungsstange f. ∥ barre f. d'attelage. / ~ link (Railw) ∥ Kupplungslasche f. ∥ bielle f. d'attelage.

shackles pl. (Mach tool) ∥ Einspannvorrichtung f. ∥ dispositif m. de fixation *ou* de serrage.

shade, to ∥ beschatten; Schatten m. geben ∥ ombrayer; donner l'ombre. / ~ (Drawing) ∥ schattieren ∥ ombrer; nuancer; laver.

shade ∥ Schatten m. ∥ ombre f. / ~ (Lamp) ∥ Lampenschirm m. ∥ réflecteur m. / ~ (Paint) ∥ Schattierung f. ∥ dégradation f. des ombres; nuance f. / enamelled ~ (Lamp) ∥ emaillierter Schirm m. ∥ abat-jour m. émaillé. / lamp ~ *see* shade (Lamp).

shade effect ∥ Schattenwirkung f. ∥ effet m. d'ombre. / temperature in the ~ ∥ Schattentemperatur f. ∥ température f. à l'ombre.

shading, yarn ~ (Dyer) ∥ Flammgarnfärberei f. ∥ teinture f. de fils flammés.

shadow, to *see* to shade.

shadow ∥ Schatten m. ∥ ombre f. / to cast a ~ ∥ einen Schatten m. werfen ∥ faire de l'ombre f.; projeter une ombre. / complete ~ (Phys) ∥ Kernschatten m. ∥ ombre f. pure. / cone-shaped ~ ∥ kegelförmiger Schatten m. ∥ ombre f. en forme de cône. / to throw a ~ *see* to cast a ~.

shadow cone ∥ Schattenkegel m. ∥ cône m. d'ombre.

shaft (Coachm) ∥ Deichsel f. ∥ timon m. / ~ (Mach) ∥ Welle f. ∥ arbre m.; essieu m.; axe m. / ~ (Mine) ∥ Schacht m. ∥ puits m. / ~ (Pick) ∥ Stiel m. ∥ manche m. / air ~ (Mine) ∥ Wetterschacht m. ∥ puits m. d'aérage. / ~ of an axe ∥ axe m. / manche m. de hache. / axle ~ (Auto) ∥ Hinterachswelle f. ∥ arbre m. de roue motrice. / axle ~ (Railw) ∥ Achsschaft m.; Mittelachse f. ∥ corps m. de l'essieu. / the ~ is bent ∥ die Welle f. ist verbogen ∥ l'arbre m. est faussé. / bored ~ ∥ gebohrte Welle f. ∥ arbre m. foré. / ~ with bronze sleeve ∥ Welle f. mit Bronzebezug ∥ arbre m. à chemise en bronze. / ~ of

a calebasse-furnace ∥ Schacht m. eines Kesselofens ∥ puits m. d'un four calebasse. / cardan ~ ∥ Kardanwelle f. ∥ arbre m. cardan. / climbing ~ (Mine) ∥ Fahrschacht m. ∥ puits m. aux échelles. / ~ of a column ∥ Säulenschaft m. ∥ fût m. *ou* vif m. *ou* tronc m. d'une colonne. / commutator ~ ∥ Kommutatorwelle f. ∥ arbre m. de commutateur. / the ~ consists of three hole-bored cranks coupled together ∥ die Welle besteht aus drei miteinander gekuppelten durchbohrten Kurbeln ∥ l'arbre m. est assemblé de trois tronçons percés. / the ~ consists of two parts ∥ die Welle f. besteht aus zwei Teilen ∥ l'arbre m. se compose de deux tronçons. / counter ~ ∥ Vorgelegewelle f. ∥ arbre m. intermédiaire. / cranked ~ ∥ gekröpfte Welle f. ∥ arbre m. coudé. / to descend into a ~ (Mine) ∥ einen Schacht m. befahren ∥ descendre dans un puits. / double-cranked ~ ∥ doppelt gekröpfte Welle f. ∥ arbre m. doublement coudé. / ~ of dressing machine (Windmill) ∥ Klobenwelle f. ∥ arbre m. du babillard. / drive ~ (Auto) ∥ Antriebswelle f. ∥ arbre m. primaire *ou* de commande. / eccentric ~ ∥ Exzenterwelle f. ∥ arbre m. de dédoublement *ou* d'excentrique. / ~ with eccentrics forged on ∥ Welle f. mit angeschmiedeten Hubscheiben ∥ arbre m. avec plateaux d'excentriques venus de forge. / flexible ~ ∥ biegsame Welle f. ∥ arbre m. flexible. / ~ with forged flanges ∥ Welle f. mit angeschmiedeten Flanschen ∥ arbre m. à brides forgées. / forked steering arm ~ (Auto) ∥ Gabelhebelwelle f. ∥ axe m. du levier fourché. / ~ of a forked thill (Coachm) ∥ Gabelschaft m.; Gabelbaum m. ∥ bras m. de limonière. / freezing ~ sinking in quicksand ∥ Niederbringen n. von Gefrierschächten im Schwimmsand ∥ fonçage m. des puits de congélation à travers de couches de sables mouvants. / ~ of a furnace ∥ Ofenschacht m. ∥ cuve f. *ou* puits m. *ou* vide m. d'un fourneau. / gear shift lever ~ (Auto) ∥ Schaltwelle f. ∥ arbre m. de levier des vitesses. / hammer ~ ∥ Hammerstiel m. ∥ manche m. du marteau. / hoisting ~ (Mine) ∥ Förderschacht m. ∥ puits m. d'extraction. / ~ hole-bored from end to end ∥ auf die ganze Länge durchbohrte Welle f. ∥ arbre m. percé sur toute sa longueur. / horizontal ~ ∥ liegende Welle f. ∥ arbre m. horizontal. / inclined ~ (Mine) ∥ tonnlägiger Schacht m. ∥ puits m. incliné *ou* oblique. / inlet cam ~ ∥ Einlaßsteuerwelle f. ∥ arbre m. à cames d'admission. / intermediate ~ ∥ Zwischenwelle f.; Hilfswelle f. ∥ arbre m. intermédiaire *ou* de renvoi. / ~ actuating the knife (Agr; Mach) ∥ Messerwelle f. ∥ arbre-manivelle m. commandant les lames. / ~ of the knife-disk ∥ Messerscheibenwelle f. ∥ arbre m. porte-couteau. / middle ~ ∥ Mittelstück n. einer Welle ∥ arbre m. de couche intermédiaire. / motor ~ ∥ Motorwelle f. ∥ arbre m. de moteur. / mounted on a single ~ ∥ auf einer gemeinsamen Achse f. gelagert ∥ monté sur un arbre commun. / ~ of the oar ∥ Ruderstange f. ∥ perche f. d'aviron *ou* de rame. / oscillating ~ ∥ pendelnde Welle f. ∥ arbre m. pendulaire. / ~ passing from end to end ∥ durchgehende Welle f. ∥ arbre m. allant de bout en bout. / pedal ~

‖ Fußhebelwelle f. ‖ axe m. de pédale. / permanent ~ (Tunnel) ‖ Luftschacht m. ‖ puits m. d'aérage *ou* d'airage. / pinion ~ ‖ Ritzelwelle f. ‖ arbre m. à pignon. / pinion drive ~ ‖ Ritzelantriebswelle f. ‖ arbre m. de commande à pignon. / propeller ~ (Auto) ‖ Kardanwelle f. mit zwei Gelenken ‖ arbre m. à deux cardans. / pulley ~ *see* hoisting ~. / ~ put down by boring ‖ Bohrschacht m. ‖ puits m. de sondage. / ~ of a quarry ‖ Schacht m. einer Steingrube ‖ chemin m. de carrière. / reserve idler gear ~ (Auto) ‖ Rücklaufwelle f. ‖ arbre m. de marche arrière. / reversing ~ ‖ Umsteuerwelle f. ‖ arbre m. de changement de marche. / ~ of a rivet ‖ Nietschaft m. ‖ tige f. *ou* corps m. de rivet. / single-jointed drive ~ (Auto) ‖ Kardanwelle f. mit einem Gelenk ‖ arbre m. de cardan à un seul joint. / to sink a ~ ‖ einen Schacht m. abteufen ‖ foncer un puits. / sliding ~ (Auto) ‖ Nutenwelle f. ‖ arbre m. profilé. / ~ subjected to slight stress ‖ schwach beanspruchte Welle f. ‖ arbre m. fournissant un petit effort. / solid ~ ‖ volle Welle f. ‖ arbre m. plein. / spigot ~ ‖ Zentrierzapfen m. ‖ queue f. de centrage. / squared ~ ‖ Vierkantwelle f. ‖ arbre m. carré. / ~ of the stamp (Ore dress) ‖ Pochstempelschaft m.; Stempelstange f. ‖ tige f. de pilon. / steering ~ (Auto) ‖ Lenkstockspindel f. ‖ arbre m. de colonne de direction. / steering wheel ~ (Auto) ‖ Lenkspindel f. ‖ arbre m. de colonne de direction. / straight ~ ‖ gerade Welle f. ‖ arbre m. droit. / timer ~ (Auto) ‖ Verteilerwelle f. ‖ arbre m. de distributeur. / tip ~ ‖ Kippwelle f. ‖ arbre m. de versement. / transversal ~ ‖ durchgehende Welle f. ‖ arbre m. transversal. / trepanned marine ~ ‖ hohlgebohrte Schiffswelle f. ‖ arbre m. de couche foré creux. / tubular ~ ‖ hohle Welle f. ‖ arbre m. creux. / upcast ventilating ~ (Mine) ‖ ausziehender Schacht m. ‖ puits m. d'appel. / vertical ~ ‖ stehende Welle f. ‖ arbre m. vertical. / vertical ~ (Mine) ‖ seigerer Schacht m. ‖ puits m. perpendiculaire. / weigh ~ ‖ Steuerungswelle f.; Umsteuerwelle f. ‖ arbre m. de distribution *ou* de relevage.

shaft alley (Shipb) ‖ Schraubenwellentunnel m. ‖ tunnel m. *ou* coffre m. de l'arbre d'hélice. / ~ bender (Coachm) ‖ Gabelbieger m.; Deichselbieger m. ‖ cintreur m. de brancards. / ~ boring machine ‖ Wellenbohrbank f. ‖ tour m. à forer les arbres. / ~ bucket conveyance ‖ Schachtgefäßförderung f. ‖ transport m. de cages de montée. / ~ building (Mine) ‖ Schachtgebäude n. ‖ bâtiment m. de puits de mine. / ~ butt ‖ Wellenstumpf m. ‖ bout m. d'arbre. / ~ canting of the ~ ‖ Ecken n. der Welle ‖ coincement m. de l'arbre. / ~ closing (Mine) ‖ Schachtverschluß m. ‖ fermeture f. pour puits de mine. / ~ coupling ‖ Wellenkupplung f. ‖ accouplement m. des arbres. / ~ coupling flange ‖ Wellenkupplungsflansch m. ‖ plateau m. d'accouplement de l'arbre. / ~ covering (Mine) ‖ Schachtabdeckung f. ‖ couvercle m. de puits. / to measure the depth of a ~ with a plumb-line ‖ einen Schacht m. abseigern ‖ aplomber un puits. / ~ door (Mine) ‖ Schachttür f. ‖ porte f. de puits. / cardan ~ drive ‖ Kardanantrieb m. ‖ transmission f. à

cardan. / end of the ~ ‖ Wellenstumpf m. ‖ extrémité m. *ou* bout m. de l'arbre. / ~ fire ‖ Schachtbrand m. ‖ incendie m. d'un puits. / forged-on ~ flange ‖ angeschmiedeter Wellenflansch m. ‖ bride f. d'arbre venue de forge. / ~ frame (Mine) ‖ Schachtgeviere n. ‖ cadre m. pour le boisage des puits de mines. / ~ framework (Mine) ‖ Fördergerüst n. ‖ chevalement m. de mine. / ~ furnace ‖ Schachtofen m. ‖ fourneau m. à cuve. / ~ hoisting ‖ Schachtförderung f. ‖ transport m. dans le puits. / ~ hole of a hammer ‖ Hammerauge n. ‖ œil m. de marteau. / ~ horse power ‖ Wellenpferdestärke f. ‖ force f. mesurée sur l'arbre de la machine.

shafting ‖ Transmission f. ‖ transmission f. / complete ~ ‖ vollständige Wellenleitung f. ‖ ligne f. d'arbres complète.

shaft, key way in the ~ ‖ Wellennut f. ‖ rainure f. de l'arbre. / ~ kiln ‖ Schachtofen m. ‖ fourneau m. à cuve. / ~ leather (Curr) ‖ Schaftleder n. ‖ cuir m. mou pour tiges de bottes. / complete ~ line *see* shafting, complete. / ~ lining material (Mine) ‖ Schachtausbaumaterial n. ‖ matériel m. de cuvelage de puits de mine. / ~ mine ‖ Schachtanlage f. ‖ mine f. exploitée par puits. / pair of ~s (Coachm) ‖ Gabeldeichsel f.; Wagenschere f. ‖ limonière f.; limons mpl.; enrayoir m. à fourchette; brancard m. / part of ~ with forged-on eccentrics ‖ Wellenteil m. mit angeschmiedeten Exzentern ‖ tronçon m. d'arbre avec disques d'excentriques forgés. / ~ passage *see* shaft alley. / ~ pillar (Mine) ‖ Schachtpfeiler m.; Schachtsicherheitspfeiler m. ‖ massif m. d'un puits. / plumbing in a ~ ‖ Schachtlotung f. ‖ prise f. de l'aplomb dans un puits. / ~ pump ‖ Schachtpumpe f. ‖ pompe f. élévatoire *ou* d'épuisement d'un puits. / ~ reparer ‖ Schachtausbesserer m. ‖ repasseur de puits m. / ~ shed (Mine) ‖ Schachthalle f. ‖ hangar m. de puits. / ~ signalling plant ‖ Schachtsignalanlage f. ‖ installation f. de signaux pour puits de mine. / ~ sinker ‖ Schachtteufer m. ‖ puisatier m. / ~ sinking ‖ Schachtabteufung f. ‖ fonçage m. de puits. / ~ straightening machine ‖ Wellenrichtmaschine f. ‖ machine f. à dresser les arbres. / ~ straightening press ‖ Wellenrichtpresse f. ‖ presse f. à dresser les arbres. / ~ tower (Mine) ‖ Schachtgerüst n. ‖ chevalement m. d'extraction. / ~ turning lathe ‖ Wellendrehbank f. ‖ tour m. à tourner les arbres. / ~ weaving ‖ Schaftweberei f. ‖ tissage m. à lames.

shag ‖ Plüsch m. ‖ peluche f. / worsted ~ ‖ Wollplüsch m. ‖ peluche f. de laine.

shagreen, to ~ the leather ‖ das Leder mit Narben versehen ‖ chagriner le cuir.

shag tobacco ‖ Kraustabak m.; Shag m. ‖ tabac m. frisé.

shake, to ‖ schütteln; rütteln ‖ secouer; agiter. / ~ with the hand ‖ mit der Hand f. rühren ‖ agiter à la main. / ~ out with ether (Chem) ‖ ausäthern ‖ épuiser à l'éther. / ~ thoroughly ‖ durchschütteln ‖ battre; agiter fortement.

shake (Wood) ‖ Kernriß m. ‖ cadran m.; cadranure f. / external ~s (Wood) ‖ Spiegelklüfte fpl. ‖ fissures fpl. dans le sens du fil du bois fendu. / frost ~ (Wood) ‖ Frostriß m. ‖ gélivure f.

shaken (Wood) ‖ kernrissig ‖ cadrané; cadranuré.

shaker (Distill) ‖ Rüttelmeister m. ‖ remueur m. / tobacco ~ out ‖ Tabakaufhänger m. ‖ sécheur m. de tabacs.

shake trestle (Pap) ‖ Schüttelbock m. ‖ tréteau m. à secouer.

shakiness (Wood) ‖ Kernschäle f.; Ringkluft f. ‖ roulure f.

shaking ‖ Umschütteln n.; Schütteln n.; Rühren n. ‖ agitation f. / with constant ~ (Chem) ‖ unter beständigem Schütteln n. ‖ en agitant continuellement. / thorough ~ (Chem) ‖ Durchschütteln n. ‖ battage m.; forte agitation f.

shaking arm (Mill) ‖ Sichtearm m.; Beutelarm m. ‖ tige f. *ou* ailette f. de blutoir. / ~ apparatus (Mill) ‖ Sichtezeug n.; Hebezeug n. ‖ babillard m. / ~ channel ‖ Schüttelrinne f. ‖ gouttière f. à secousses. / ~ device ‖ Schüttelvorrichtung f. ‖ dispositif m. de secouage. / ~ feeder ‖ Schüttelaufgabevorrichtung f. ‖ feeder m. à secousses. / ~ grate ‖ Schüttelrost m. ‖ grille f. à secousses. / ~ machine ‖ Schüttelmaschine f. ‖ machine f. pour secouer; secoueuse f. / ~ motion ‖ Rüttelbewegung f.; Schüttelbewegung f. ‖ mouvement m. de secousse. / ~ motion of the bolting apparatus (Mill) ‖ rüttelnde Bewegung f. des Beutelzeuges ‖ mouvement m. de secousse du blutoir d'un moulin. / ~ movement *see* ~ motion. / sack ~ plant ‖ Sackausschüttelanlage f. ‖ installation f. d'époussetage des sacs. / ~ screen ‖ Rüttelsieb n.; Stoßsieb n. ‖ crible m. *ou* tamis m. à secousses. / ~ shoot *see* ~ trough. / ~ sieve *see* ~ screen. / ~ table ‖ Planrätter m. ‖ crible m. oscillant. / duplex ~ table ‖ Doppelplanrätter m. ‖ crible m. double. / ~ trough ‖ Schüttelrinne f.; Schüttelrutsche f. ‖ gouttière f. à secousses.

shale ‖ Schiefer m.; Schieferton m. ‖ argile f. (schisteuse). / ~ (Mine) ‖ Berge mpl. ‖ schistes mpl. / argillaceous ~ ‖ Tonschiefer m. ‖ schiste m. argileux. / bituminous ~ ‖ Brandschiefer m. ‖ schiste m. bitumineux. / combined pieces of coal and ~ ‖ verwachsene Kohle f. ‖ charbon m. barré. / fetid ~ ‖ Stinkschiefer m. ‖ schiste m. suant. / ground ~ ‖ Schiefermehl n. ‖ schiste m. en poudre.

shale burner ‖ Schieferdestillatör m. ‖ distillateur m. de schiste. / ~ oil ‖ Schieferöl n. ‖ huile f. de schiste.

shallop ‖ Pinasse f. ‖ péniche f.; pinasse f.

shallow ‖ flach; untief; seicht ‖ peu profond. / ~ eyepiece cup for spectacle wearers ‖ flache Okularmuschel f. für Brillenträger ‖ bonnette f. plate pour les porteurs de lunettes. / ~ plate ‖ flacher Teller m. ‖ assiette f. plate. / ~ sea ‖ Flachsee f. ‖ basse mer f.

shallowness ‖ Untiefe f.; Sandbank f.; Riff n. ‖ bas-fond m.; banc m. de sable; récif m.

sham ‖ falsch ‖ faux; imité; feint; orbe; simulé. / ~ dividend ‖ fiktive Dividende f. ‖ dividende m. fictif.

shamble ‖ Fleischbank f. ‖ étal m.

shammy *see* shamoy leather. / ~ leather *see* shamoy leather.

shamoy, to ‖ sämischgerben ‖ chamoiser.

shamoy leather ‖ Sämischleder n. ‖ cuir m. chamoisé.

shank (Button) ‖ Öhr n. ‖ attache f.; queue f. / ~ (Found) ‖ Gießpfanne f. ‖ poche f. de coulée. / ~ of an anchor ‖

Ankerschaft m. ‖ verge f. d'une ancre. / ~ of the bolt ‖ Schaft m. am Bolzen ‖ corps m. du boulon. / ~ of a hammer frame ‖ Hammersäule f. ‖ jambe f. d'un marteau pilon. / ~ of a key ‖ Schlüsselschaft m. ‖ bout m. *ou* tige f. d'une clef. / ~ of a letter ‖ Schriftkegel m. ‖ corps m. de lettre; force f. de corps d'un caractère. / ~ of a lever drill ‖ Schaft m. des Ratschbohrers; Bohrstange f. ‖ fût m. à levier du racagnac. / metric taper ~ ‖ metrischer Konus m. ‖ cône m. métrique. / ~ of a nail ‖ Nagelschaft m. ‖ tige f. d'un clou. / ~ of the rivet ‖ Schaft m. am Niet ‖ tige f. de rivet.

shank end-mill ‖ Schaftfräser m. ‖ fraise f. à queue.

shape, to ‖ formen; gestalten ‖ façonner; former; modeler; bosseler. / ~ the course for ‖ Kurs m. setzen auf ‖ tracer la route pour.

shape ‖ Form f.; Gestalt f. ‖ forme f.; façon f. / ~ of bar ‖ Stabform f. ‖ forme f. de barre. / to change the ~ ‖ die Gestalt ändern ‖ modifier la figure *ou* la forme. / conical ~ ‖ Kegelform f. ‖ conicité f.; forme f. conique. / ~ of the nose ‖ Nasenform f. ‖ forme f. du nez. / ~ of rails ‖ Schienenform f.; Schienenprofil n. ‖ section f. du rail.

shaped grinding wheel ‖ Formschleifscheibe f. ‖ meule f. de forme. / ~ piece ‖ Fassonteil m. ‖ pièce f. profilée.

shaper (Forg) ‖ Gesenk n. ‖ estampe f.; étampe f. / ~ (Mach tool) *see* shaping machine. / ~ (Wood) ‖ Holzfräsmaschine f. ‖ toupie f.

shape shears pl. ‖ Profileisenschere f. ‖ cisailles fpl. à profilés.

shaping ‖ Formgebung f. ‖ façonnage m. / ~ of machinery parts ‖ Formgebung f. der Maschinenteile ‖ façonnage m. donné aux éléments des machines.

shaping machine ‖ Shapingmaschine f.; Schnellhobelmaschine f. ‖ étau-limeur m. / ~ operation of parts ‖ Bestoßen n. von Werkstücken ‖ rabotage m. de pièces. / ~ press ‖ Biegepresse f. ‖ presse f. à cintrer.

shard cobalt ‖ Fliegenstein m.; Näpfchenkobalt m. ‖ arsenic m. natif.

share ‖ Anteil m. ‖ part f.; portion f. / ~ (Bank) ‖ Aktie f. ‖ action f. / to allot ~s pl. ‖ Aktien fpl. zuteilen ‖ attribuer des actions fpl. / bank ~ ‖ Bankaktie f. ‖ action f. de banque. / ~ in a business ‖ Geschäftsanteil m. ‖ part m. social. / ~ of capital ‖ Kapitalanteil m. ‖ part m. de capital. / deferred ~ ‖ Genußschein m. ‖ action f. de jouissance. / in equal ~s ‖ zu gleichen Teilen mpl. ‖ à parts mpl. égaux. / founder's ~ (Stocks) ‖ Gründeraktie f. ‖ action f. de fondateur. / founder's ~s pl. ‖ Gründeranteile pl. ‖ parts mpl. de fondateur. / ~ in a mine ‖ Kux m. ‖ action f. de mine. / oil ~ ‖ Erdölaktie f. ‖ pétrolifère f. / old ~ ‖ alte Aktie f. ‖ action f. ancienne. / exchange of old ~s for new ones ‖ Umtausch m. alter Aktien gegen neue ‖ échange m. de vieilles actions contre des nouvelles. / ordinary ~ ‖ Stammaktie f. ‖ action f. ordinaire. / to own ~s in a company ‖ Anteile mpl. einer Gesellschaft besitzen ‖ posséder des parts mpl. dans une société. / ~ paid-up in full ‖ voll eingezahlte Aktie f. ‖ action f. libérée *ou* complètement payée. / ~s pl. are at par ‖ die Aktien fpl.

stehen auf pari ‖ les actions fpl. sont au pair. / participating preference ~ ‖ gewinnbeteiligte Vorzugsaktie ‖ action f. privilégiée de participation. / preference ~ ‖ Vorzugsaktie f. ‖ action f. privilégiée. / ~ of the profits ‖ Gewinnanteil m. ‖ part m. dans les bénéfices. / ~s pl. are quoted at ‖ die Aktien fpl. sind notiert mit ‖ les actions fpl. sont cotées à. / registered ~ ‖ auf den Namen lautende Aktie f. ‖ titre m. nominatif. / shipping ~ ‖ Schiffahrtsaktie f. ‖ action f. de navigation. / ~s pl. are fully subscribed ‖ die Aktien fpl. sind voll gezeichnet ‖ les actions fpl. sont complètement souscrites. / ~s pl. are partly subscribed ‖ die Aktien fpl. sind teilweise gezeichnet ‖ les actions fpl. sont partiellement souscrites. / ~s pl. are over-subscribed ‖ die Aktien fpl. sind überzeichnet ‖ les actions fpl. sont souscrites au delà. / to take up ~s pl. ‖ Aktien fpl. zeichnen ‖ souscrire à des actions fpl.

share, block of ~s ‖ Aktienpaket n. ‖ tranche f. d'actions. / ~ capital ‖ Aktienkapital n. ‖ capital-actions m.; fond m. social. / ~ certificate ‖ Aktienschein m. ‖ bordereau m. d'actions.

shareholder ‖ Aktionär m. ‖ actionnaire m. / ~ in a limited liability company ‖ Kommanditär m. ‖ commanditaire m. / ~s' meeting ‖ Generalversammlung f. ‖ assemblée f. générale.

share, issue of ~s ‖ Aktienausgabe f. ‖ émission f. d'actions. / restricted issue of ~s ‖ beschränkte Ausgabe f. von Aktien ‖ émission f. limitée d'actions. / parcel of ~s ‖ Aktienpaket n. ‖ paquet m. d'actions. / redemption of ~s ‖ Aktieneinziehung f. ‖ rachat m. *ou* amortissement m. d'actions. / subscription in ~s ‖ Aktienzeichnung f. ‖ souscription f. d'actions. / ~ tenant ‖ Teilpächter m. ‖ colon m. partiaire.

shark hook ‖ Haifischhaken m. ‖ émerillon m. *ou* hameçon m. pour les requins.

sharp ‖ scharf ‖ tranchant; aigu; affilé. / ~ (Opt) ‖ deutlich; scharf ‖ net. / ~ angle of step ‖ scharf abgesetzte Stelle f.; scharfer Ansatz m. ‖ épaulement m. à angle aigu; entaille f. à angle droit. / ~-bottomed (Shipb) ‖ scharf gebaut ‖ pincé. / ~ curve ‖ scharfe Kurve f. ‖ courbe f. prononcée *ou* accentuée. / ~ edge ‖ scharfe Kante f. ‖ bord m. à vive arête. / ~-edged ‖ scharfkantig ‖ à bord m. aigu; à vive arête f. / ~-edged T-iron ‖ scharfkantiges T-Eisen n. ‖ fer m. T à angles vifs. / ~ file ‖ Stoßfeile ‖ lime f. à bouter. / ~ fire (Porcel) Scharffeuer n.; Glattfeuer n.; Glattbrand m. ‖ grand feu m. / ~ iron (Shipb) ‖ Schereisen n. ‖ fer m. taillant. / ~-pointed knife (Med) ‖ spitzes Messer n. ‖ bistouri m. à lame pointue. / ~ reading (Apparatus) ‖ Scharfeinstellung f. ‖ mise f. au point. / ~ tuning (Radio) ‖ scharfe Abstimmung f. ‖ syntonisation f. aiguë.

sharpen, to ‖ schärfen ‖ aiguiser; affûter. / ~ (Knife) ‖ abziehen; wetzen ‖ aiguiser. / ~ the teeth of a saw ‖ die Zähne mpl. einer Säge schärfen ‖ affûter *ou* affiler *ou* limer les dents fpl. d'une scie.

sharpener, pencil ~ ‖ Bleistiftspitzer m. ‖ taille-crayon m. / ~ for shaving blades ‖ Rasierklingenabziehapparat m. ‖ appareil m. à repasser les lames de rasoir.

sharpening ‖ Schärfen n.; Schleifen n. ‖ affûtage m. / ~ of cutting tools ‖ Schärfen n. von Schneidwerkzeugen ‖ affûtage m. d'outils coupants. / saw ~ ‖ Sägenschärfen n. ‖ affûtage m. de scies. / rolling mill for ~ wire ‖ Anspitzwalzwerk n. für Draht ‖ laminoir m. à appointer les fils de fer.

sharpening device ‖ Schleifvorrichtung f. ‖ appareil m. à affûter.

sharpening machine (Mach) ‖ Schleifmaschine f. ‖ machine f. à affûter *ou* à aiguiser. / ~ (Pencil) ‖ Anspitzmaschine f. ‖ machine f. à tailler (les crayons). / ~ for bars ‖ Stangenanspitzmaschine f. ‖ machine f. à appointer à froid les barres. / circular saw ~ ‖ Schärfmaschine f. für Metallkreissägeblätter ‖ affûteuse f. pour lames de scie circulaire à métaux. / drill ~ ‖ Bohrerschleifmaschine f. ‖ machine f. à aiguiser les forets. / ~ for knives ‖ Messerschleifmaschine f. ‖ machine f. à affiler les couteaux. / rock drill ~ ‖ Gesteinsbohrerschleifmaschine f. ‖ machine f. à aiguiser les forets à roches. / automatic ~ for cold and hot saws ‖ Kalt- und Warmsägenschärfautomat m. ‖ machine f. automatique à affûter les scies à chaud et à froid.

sharply focussing of the microscope ‖ Scharfeinstellung f. des Mikroskopes ‖ mise f. au point du microscope. / ~ of the spectre ‖ Scharfeinstellung f. des Spektrums ‖ rendre le spectre net.

sharply tuned transmitter ‖ scharf abgestimmter Sender m. ‖ transmetteur m. à syntonisation aiguë.

sharpness (Opt) ‖ Schärfe f. ‖ netteté f.

shave, to ‖ rasieren ‖ raser. / ~ (Techn) ‖ abhobeln; abschlichten ‖ raboter; planer; replanir.

shave ‖ Ziehmesser n. ‖ plane f.

shaving *see* shavings.

shaving apparatus ‖ Rasierapparat m. ‖ rasoir m. mécanique. / ~ blade ‖ Rasierklinge f. ‖ lame f. de rasoir. / sharpener for ~ blades ‖ Rasierklingenabziehapparat m. ‖ appareil m. à repasser les lames de rasoir. / ~ brush ‖ Rasierpinsel m. ‖ blaireau m. *ou* pinceau m. ou brosse f. à faire la barbe. / ~ cream ‖ Rasierkrem f. ‖ savon m. en pâte à barbe. / ~ knife ‖ Rasiermesser n. ‖ rasoir m. / ~ knife (Curr) ‖ Gerbeisen n.; Schabeisen n.; Streicheisen n.; Schabmesser n.; Ausfleischeisen n. ‖ couteau m. rond *ou* à écharner; écharnoir m.; drayoir m. / ~ mirror ‖ Rasierspiegel m. ‖ miroir m. à raser. / ~ set ‖ Rasiergarnitur f. ‖ garniture f. pour la barbe. / ~ soap ‖ Rasierseife f. ‖ savon m. à barbe.

shavings pl. (Join) ‖ Hobelspäne mpl. copeaux mpl.; planure f.; raboture f. / ~ (Turn) ‖ Drehspäne mpl. ‖ tournure f.; copeaux mpl. / all ~ are sucked off directly from the machine ‖ alle Späne mpl. werden unmittelbar aus der Maschine abgesaugt ‖ tous les copeaux mpl. sont enlevés directement de la machine par aspiration.

shavings basket ‖ Spankorb m. ‖ corbeille f. en copeaux de bois. / ~ catcher (Sugar) ‖ Schnitzelfänger m. ‖ corbeille f. à rognures. / ~ exhaust installation ‖ Späneabsauganlage f. ‖ installation f. de transport de copeaux et sciures. / ~ press ‖ Schnitzelpresse f. ‖ presse f. à

agglomérer les rognures. / ~ separator ‖ Spänefänger m. ‖ aspirateur m. *ou* séparateur m. de copeaux. / ~ suction plant ‖ Späneabsauganlage f. ‖ installation f. d'aspirateurs de copeaux. / apparatus for tearing the ~ ‖ Spänezerreißer m. ‖ déchireur m. de copeaux.

shawl ‖ Schal m. ‖ châle m. / damask ~ ‖ Damastschal m. ‖ châle m. damassé. / ~ maker ‖ Schalweber m. ‖ châlier m.

sheaf, to (Agr) ‖ in Garben fpl. binden ‖ engerber.

sheaf ‖ Garbe f. ‖ gerbe f. / ~ binding machine ‖ Garbenselbstbinder m. ‖ appareil m. à engerber.

shear, to ‖ abscheren ‖ couper. / ~ (Sheep) ‖ scheren ‖ tondre. / ~ the cloth ‖ das Tuch scheren ‖ tondre le drap. / ~ fine ‖ feinscheren ‖ affiner. / ~ roughly ‖ überscheren ‖ surtondre.

shear blade ‖ Scherblatt n. ‖ lame f.; tranchant m.; mâchoire f. ‖ cisaille f. à cutter ‖ Scherenmesser n. ‖ couteau m. à cisailles.

shearer ‖ Blechschneider m. ‖ cisailleur m.

shearing ‖ Abscheren n.; Scheren n. ‖ cisaillement m. / cloth ~ ‖ Scheren n. des Tuches ‖ tondage m. du drap. / detachable ~ attachment ‖ abnehmbare Schervorrichtung f. ‖ dispositif m. de cisaillage démontable. / ~ bolt ‖ Abscherbolzen m. ‖ goujon m. de cisaillement. / ~ breaker ‖ Scherenbrecher m. ‖ concasseur m. à cisailles. / ~ force ‖ Scherkraft f.; Schubkraft f. ‖ effort m. de cisaillement.

shearing machine (Clothm) ‖ Schermaschine f. ‖ ourdisseuse f.; tondeuse f. / ~ (Mach tool) ‖ Maschinenschere f.; Schere f. ‖ machine f. à cisailler *ou* à découper; cisaille f. / bar ~ ‖ Stabeisenschere f. ‖ cisaille f. à fers en barres. / billet ~ ‖ Knüppelschere f. ‖ cisaille f. à billettes. / bolt ~ ‖ Bolzenschere f. ‖ cisaille f. à boulons. / ~ with concave tables (Cloth) ‖ Schermaschine f. mit Hohltischen ‖ tondeuse f. avec tables à cuvette. / crank guillotine ~ ‖ Parallelkurbelblechtafelschere f. ‖ cisaille f. à guillotine pour tôles. / the ~ can cut up to x mm thick plates while cold ‖ die Schere f. vermag bis zu x mm dicke Bleche kalt zu schneiden ‖ la cisaille coupe à froid des tôles jusqu'à x mm d'épaisseur. / immoveable blade of a cylindrical ~ ‖ Lieger m. *oder* Untermesser n. einer Schermaschine ‖ contrecouteau m. *ou* femelle f. d'une tondeuse hélicoïde. / flat iron ~ ‖ Flacheisenschere f. ‖ cisaille f. à fers plats. / ~ with fly-wheel ‖ Schwungradschere f. ‖ cisaille f. à volant. / ~ for hoops ‖ Faßreifenabschermaschine f. ‖ machine f. à découper les cercles de fûts. / hoop shearing and punching machine ‖ Bandeisenschere- und Lochstanze f. ‖ machine f. à découper et à perforer les cercles. / ~ for horses (Cloth) ‖ Pferdeschermaschine f. ‖ tondeuse f. mécanique pour chevaux. / hydraulic ~ ‖ hydraulische Blockschere f. ‖ cisaille f. hydraulique pour lingots. / ~ with movable prismatic tables ‖ Schermaschine f. mit verschiebbaren Prismentischen ‖ tondeuse f. avec tables prismatiques réglables. / ~ with one cutter (Cloth) ‖ Schermaschine f. mit einem Schneidzeug ‖ tondeuse f. simple. / opengap plate ~ combined with section shearing and mitre cutting machine ‖

vereinigte Ausladungsblech-, Profileisen- und Gehrungsschere f. ‖ cisaille f. à tôles à col de cygne combinée avec cisaille à profiles et à onglets. / open-gap plate and scrap-metal ‖ Ausladungsblech- und Schrotschere f. ‖ cisaille f. col de cygne à mitraille / simple open gap plate ~ ‖ einfache Ausladungsblechschere f. ‖ cisaille f. à tôles simple à col de cygne. / oscillating ~ ‖ Schermaschine f. mit oszillierendem Zylinder ‖ tondeuse f. oscillatoire *ou* oscillante. / ~ for cutting the bars for reinforced concrete work ‖ Betoneisenschere f. ‖ coupe-fers m. pour béton armé. / rivet ~ ‖ Nietenschere f. ‖ cisaille f. à rivets. / rotary ~ ‖ umlaufende Schere f. ‖ cisaille f. circulaire. / round iron ~ ‖ Rundeisenschere f. ‖ cisaille f. à fers ronds. / scrap ~ ‖ Schrotschere f. ‖ cisaille f. à riblons. / section ~ ‖ Profileisenschere f. ‖ cisaille f. à profilés. / ~ for sheep ‖ Schafeschermaschine f. ‖ tondeuse f. mécanique pour moutons. / sheet billet ~ ‖ Platinenschere f. ‖ cisaille f. à largets. / square iron ~ ‖ Vierkanteisenschere f. ‖ cisaille f. à fers carrés. / transverse ~ ‖ Querschermaschine f. ‖ tondeuse f. transversale. / ~ with two cutters (Cloth) ‖ Schermaschine f. mit zwei Schneidzeugen ‖ tondeuse f. double longitudinale. / ~ for the close shearing of woollen fabrics ‖ Schermaschine f. zum Kahlscheren wollener Gewebe ‖ tondeuse f. pour tissus de laine. / ~ for woollen and half woollen fabrics ‖ Schermaschine f. für Wollwaren und Halbwollwaren ‖ tondeuse f. pour tissus de laine et mi-laine. / ~ for woollens and worsteds ‖ Schermaschine f. für Tuchschur und Feintuchschur ‖ tondeuse f. pour les tissus et les draps fins.

shearings pl. ‖ Scherwolle f.; Scherflocken fpl. ‖ tontisse f.; tonture f.

shearing strain ‖ Scherbeanspruchung f. ‖ effort m. tranchant *ou* de cisaillement. / ~ strength ‖ Scherfestigkeit f.; Abscherungsfestigkeit f. ‖ résistance f. au cisaillement. / ~ stress ‖ Scherspannung f. ‖ tension f. de cisaillement.

shear-jointed telescope ‖ Scherenfernrohr n. ‖ jumelle f. périscopique.

shears pl. ‖ Schere f. ‖ ciseaux mpl.; cisailles fpl. / ~ (Mach) *see also* shearing machine ‖ große Schere f. ‖ forces fpl.; cisaille f. / ~ (Clothm) ‖ Tuchschere f. ‖ forces fpl. / ~ (Gard) ‖ Baumschere f. ‖ sécateur m. / ~ (Sheet) ‖ Blechschere f. ‖ cisaille f. (à tôles) / billet ~ ‖ Knüppelschere f. ‖ cisaille f. à billettes. / bloom ~ *see* ingot ~. / cattle ~ ‖ Viehschere f. ‖ tondeuse f. pour animaux. / circular ~ ‖ Kreisschere f. ‖ cisaille f. circulaire. / ~ for clipping hedges ‖ Heckenschere f. ‖ ciseaux mpl. de jardinier. / cloth ~ ‖ Tuchschere f. ‖ forces fpl. à tondre les draps. / crank ~ ‖ Kurbelschere f. ‖ cisaille f. à guillotine. / crocodile ~ ‖ Hebelschere f. ‖ cisaille f. à mâchoire. / to cut off with the ~ ‖ beschneiden ‖ cisailler. / flying ~ ‖ fliegende Schere f. ‖ cisaille f. à porte-à-faux. / guillotine ~ *see* crank ~. / hand ~ ‖ Schere f. für Handbetrieb ‖ cisaille f. à main. / hydraulic ~ ‖ hydraulische Blockschere f. ‖ cisaille f. hydraulique. / ingot ~ ‖ Blockschere f. ‖ cisailles fpl. à lingots. / lever ~ ‖ Hebelschere f.; Tafelschere f. ‖ cisailles fpl. à

levier. / overhung ~ *see* flying ~. / plate ~ ‖ Platinenschere f. ‖ cisaille f. à largets. / pruning ~ ‖ Baumschere f. ‖ sécateur m. / reversible ~ ‖ Umkehrschere f. ‖ cisaille f. reversible. / ~ for rolling mills ‖ Walzwerkschere f. ‖ cisaille f. pour laminoirs. / rotating ~ ‖ rotierende Schere f. ‖ cisaille f. rotative. / scrap ~ ‖ Schrottschere f. ‖ cisaille f. à mitraille. / section iron ~ ‖ Formeisenschere f.; Profileisenschere f. ‖ cisaille f. à fers profilés. / sheep ~ ‖ Schafschere f. ‖ forces fpl. à tondre les moutons. / sheet billet ~ *see* plate ~. / slab bloom ~ ‖ Brammenschere f. ‖ cisaille f. à brammes. / steam plate ~ ‖ Dampfblechschere f. ‖ cisaille f. à tôles mues par la vapeur. / strip ~ ‖ Streifenschere f. ‖ cisaille f. pour couper en bandes. / strip ~ for dividing strips and plates ‖ Streifenschere f. zum Teilen von Streifen und Platten ‖ cisaille f. à découper les bandes et les plaques. / strip ~ for trimming strips and plates ‖ Streifenschere f. zum Besäumen von Streifen und Platten ‖ cisaille f. à bandes pour replier les bandes et les plaques.

shear steel ‖ Gerbstahl m. ‖ acier m. corroyé. / ~ strain *see* shearing strain. / ~ stress *see* shearing stress. ‖ ~ vice ‖ Scherenkluppe f. ‖ serreciseaux m.

sheath ‖ Scheide f. ‖ gaine f. / ~ of a knife ‖ Scheide f. eines Messers ‖ gaine f. *ou* fourreau m. d'un couteau. / ~ of a pile ‖ Pfahlschuh m. ‖ armature f. *ou* chaussure f. d'un pieu. / steel ~ ‖ Stahlscheide f. ‖ fourreau m. d'acier.

sheathing (Shipb) ‖ Außenhaut f.; Haut f.; Bekleidung f.; Beschlag m. ‖ revêtement m. / ~ of an observatory dome ‖ Außenverkleidung f. einer Sternwartenkuppel ‖ couverture f. extérieure d'une coupole d'observatoire. / ~ of yellow metal (Mar) ‖ Metallbeschlag m.; Kompositionsbeschlag m. ‖ doublage m. de composition.

sheathing loss (Electr) ‖ Bleimantelverlust m. ‖ perte f. dans l'enveloppe de plomb. / ~ plate for wagons ‖ Wagenbekleidungsblech n. ‖ tôle f. de revêtement pour wagons.

sheave ‖ Scheibe f.; Rolle f. ‖ poulie f.; disque m.; rouet m. / compensating ~ ‖ Ausgleichrolle f. ‖ galet m. de compensation. / eccentric ~ ‖ Hubscheibe f. ‖ disque m. d'excentrique. / jib-head ~ ‖ Schnabelrolle f. ‖ poulie f. de bec. / ~ of the key hole ‖ Schlüssellochdeckel m. ‖ cacheentrée m. / reversing ~ ‖ Umlenkscheibe f. ‖ poulie f. de renvoi. / rope balancing ~ ‖ Seilausgleichrolle f. ‖ galet m. compensateur de tension du câble.

sheave pulley ‖ Umführrolle f. ‖ galet m. de renvoi.

shed ‖ Schuppen m. ‖ échoppe f.; hangar m.; hutte f.; loge f. / aircraft ~ ‖ Flugzeughalle f. ‖ hangar m. d'avions. / airship ~ ‖ Luftschiffhalle f. ‖ hangar m. de ballons. / locomotive ~ ‖ Lokomotivschuppen m. ‖ remise f. de locomotives.

shed door ‖ Schuppentor n. ‖ porte f. de hangar. / ~ insulator ‖ Glockenisolator m. ‖ isolateur m. à simple cloche. / ~ roof ‖ Halbdach n. ‖ toit m. en appentis *ou* à un seul égout.

sheep ‖ Schaf n. ‖ mouton m. / electoral ~ ‖ sächsisches Merinoschaf n.; Elektoral-

schaf n. ‖ mérinos m. électoral. / Saxon
~ *see* electoral ~.

sheep backs pl. (Geol) ‖ Rundhöcker mpl.
‖ roches fpl. moutonnées. / ~ breeder ‖
Schafzüchter m. ‖éleveur n. de moutons.
/ ~ breeding ‖ Schafzucht f.; Schaf-
haltung f. ‖ élevage m. de moutons. /
~ leather ‖ Schafleder n. ‖ cuir m. de
mouton. / ~ rearing *see* ~ breeding. / ~
shearer ‖ Schafscherer m. ‖ tondeur m.
de moutons. / ~ shearing machine ‖
Schafschermaschine f. ‖ machine f. à
tondre les moutons. / ~ shears pl. ‖ Schafschere f. ‖ forces fpl. à tondre les moutons. / ~ skin ‖ Schafleder n. ‖ basane f.
/ ~'s trotter dresser ‖ Hammelfußbereiter
m. ‖ préparateur m. de pieds de mouton.
/ ~'s trotter grease ‖ Hammelklauenfett
n. ‖ huile f. de pied de mouton. / ~
wagon ‖ Schaftransportwagen m. ‖
bergerie f. / ~ wool ‖ Schafwolle f. ‖
laine f. de mouton.

sheer of a deck ‖ Sprung m. des Decks ‖
tonture f. de pont. / ~ of a ship ‖ Scheren
n. *oder* Gieren n. eines Schiffes ‖ embardée f. *ou* dérivation f. d'un navire.

sheer batten (Mar) ‖ Spreizlatte f. ‖ barra-
quette f. / ~ plan (Shipb) ‖ Längsriß m. ‖
plan m. vertical longitudinal. / ~ rail of
a boat ‖ Scheuerleiste f. eines Bootes ‖
cordon m. *ou* boudin m. *ou* bourrelet m.
d'un navire.

sheers pl. ‖ Scherenkran m. ‖ grue f.
mâture. / floating ~ ‖ schwimmender
Kran m. ‖ ponton-bigue m. / ~ for
masting ‖ Mastkran m. ‖ machine f. à
mâter; mâture f. / three-legged ~ ‖ Drei-
beinkran m. ‖ grue f. à trois os.

sheet (Bed) ‖ Bettuch n. ‖ drap m. de lit.
/ ~ (Mar) ‖ Schot f. ‖ écoute f. / ~ (Metal)
see also sheet iron *and* sheet metal ‖ Blech
n. ‖ tôle f. / automobile body ~ ‖ Karos-
serieblech n. ‖ tôle f. pour carrosseries. /
corrugated ~ ‖ Wellblech n. ‖ tôle f. ondulée. / to ease off the ~s pl. (Mar) ‖ die
Schoten fpl. abfieren ‖ filer les écoutes
fpl. / to let the ~s fly (Mar) ‖ die Schoten
fpl. fliegen lassen ‖ larguer les écoutes
fpl. en bande. / with flying ~s pl. ‖ mit
fliegenden Schoten fpl. ‖ à écoutes fpl.
larguées. / ~ folded in page size ‖ auf Sei-
tengröße gefalzter Bogen m. ‖ feuille f.
pliée en format d'une page. / gusset ~ ‖
Eckblech n. ‖ gousset m. / ~ of ice ‖
Eisscholle f. ‖ glaçon m. / lead-coated
~ ‖ ausgebleites Blech n. ‖ tôle f. plombée. / ~ of paper ‖ Papierbogen m. ‖
feuille f. de papier. / ~ of paper in quarto
‖ Quartbogen m. ‖ feuille f. in-quarto.
/ picture ~ ‖ Bilderbogen m. ‖ feuille f.
d'images. / ~ for separating purposes ‖
Separationsblech n. ‖ tôle f. de séparation. / ~ of snow ‖ Schneedecke f. ‖ couverture f. *ou* couche f. de neige. / spoilt
~ *see* waste ~. / stiffening ~ ‖ Verstärkungsblech n. ‖ tôle f. *ou* plaque f.
de renfort. / thick ~ *see* sheet iron,
thick. / thin ~ *see* sheet iron, thin. /
tightening ‖ Abdichtungsblech n. ‖ tôle
f. d'étanchéité. / waste ~ ‖ Makulatur-
bogen m. ‖ feuille f. de rebut.

sheet asbestos ‖ Asbestplatte f.; Asbest-
tafel f. ‖ carton m. d'amiante. / ~
double ended spanner ‖ Blechdoppelschraubenschlüssel m. ‖ clé f. double
en tôle.

sheet bar mill *see* sheet billet mill. / ~
shears pl.*see* sheet billet shearing machine.

sheet billet ‖ Platine f. ‖ larget m. / ~ mill ‖
Platinenwalzwerk n. ‖ laminoir m. à
largets. / ~ shearing machine ‖ Platinenschere f. ‖ cisaille f. à largets. / ~ shears
pl. *see* ~ shearing machine.

sheet bordering machine ‖ Blechbördel-
maschine f. ‖ machine f. à border la
tôle.

sheet brass ‖ Messingblech n. ‖ tôle f. *ou*
feuille f. de laiton. / ~ in rolls ‖ Messingblech n. in Rollen ‖ laiton m. en rouleaux.

sheet calendar ‖ Abreißkalender m. ‖
calendrier m. à effeuiller. / ~ clamping
device ‖ Blechspannvorrichtung f. ‖
mécanisme m. de serrage des tôles.

sheet copper ‖ Kupferblech n. ‖ plaque f.
ou feuille f. de cuivre; cuivre m. en
plaque *ou* en lame. / plated ~ ‖ plattiertes Kupferblech n. ‖ cuivre m. plaqué
ou doublé. / ~ in rolls ‖ Kupferblech n.
in Rollen ‖ cuivre m. en rouleaux.

sheet core coil ‖ Blechkernspule f. ‖ bobine f.
à noyau en tôle. / ~ counter ‖ Bogen-
zähler m. ‖ compteur m. de feuilles.

sheet cutter (Pap) ‖ Bogenschneider n . ‖
appareil m. coupe-feuilles. / longitudinal
~ (Bookb) ‖ Längsschneider m. ‖ coupeur m. longitudinal de feuilles.

sheet deliverer, flyer ~ (Print) ‖ Bogen-
ausleger m. ‖ receveur m. de feuilles. /
sheet doubler ‖ Blechdoppler m. ‖ machine
f. à doubler les tôles. / ~ elevator of a
calender ‖ Bogenhochführung f. eines
Kalanders ‖ ascenseur m. de feuilles
d'un calandre.

sheet fed offset machine ‖ Bogenoffsetpresse f. ‖ machine-offset f. pour la
marge de feuilles. / ~ photogravure
rotary machine ‖ Tiefdruckrotations-
maschine f. für Bogenanlage ‖ machine
f. rotative en creux pour la marge de
feuilles.

sheet feeder for printing presses ‖ Anlegeapparat m. für Druckpressen ‖ appareil
m. metteur de presses d'imprimerie.
/ ~ feeding apparatus ‖ Bogenanlegeapparat m. ‖ appareil m. margeur. / ~
galvanizing plant ‖ Blechgalvanisier-
anlage f. ‖ installation f. de galvanisation
de tôles. / ~ gauge ‖ Blechlehre f. ‖ jauge
f. des tôles. / ~ gelatine ‖ Gelatinefolie
f. ‖ feuille f. de gélatine. / ~ glass ‖
Tafelglas m. ‖ verre m. en tables. / ~
goods pl. for house-keeping and kitchen
‖ Blechwaren fpl. für Küche und Haus
‖ articles mpl. en tôle de ménage et de
cuisine.

sheeting ‖ Schalwand f. ‖ banche f. / ~
(Auto) ‖ Blechverkleidung f. ‖ blindage
m. en tôle. / ~ board ‖ Schaldiele f.;
Schalbrett n. ‖ planche f. de boisage. / ~
plate of copper ‖ Kupferbeschlagblech
n. ‖ cuivre m. en feuilles pour doublage.

sheet iron *see also* sheet *and* sheet metal ‖
Eisenblech n. ‖ tôle f. de fer. / annealed ~
‖ ausgeglühtes Eisenblech n. ‖ tôle f.
recuite. / bent ~ ‖ gebogenes Eisenblech
n. ‖ tôle f. courbée. / black ~ ‖ Schwarz-
blech n.; Sturzblech n. ‖ tôle f. de fer m.
noir; fer m. en feuilles noir; fer m. en
lames noir. / corrugated ~ ‖ Wellblech n.
‖ tôle f. ondulée. / corrugated ~ for roof
covering ‖ Wellblech n. für Dachdeckung
‖ tôle f. ondulée pour toitures. / decora-
tive ~ ‖ Zierblech n. ‖ tôle f. d'ornementation. / ~ for deep drawing ‖ Tiefzieh-
blech n.; Tiefstanzblech n. ‖ tôle f. à
estamper et à emboutir. / galvanized ~

‖ verzinktes Blech n. ‖ tôle f. galvanisée.
/ perforated ~ ‖ gelochtes Eisenblech n.
‖ tôle f. de fer perforée. / pressed and
punched pieces of ~ and steel plate ‖
Preß- und Stanzteile mpl. aus Eisen-
und Stahlblech ‖ pièces fpl. embouties
et estampées de tôle en fer et enacier.
/ riffled ~ ‖ geriffeltes Blech n.; Riffelblech n. ‖ tôle f. striée. / rolled ~ ‖
Walzblech n. ‖ tôle f. laminée. / ~ in rolls
‖ Rollenblech n. ‖ tôle f. en rouleaux. /
thick ~ ‖ Grobblech n. ‖ grosse tôle f. /
thin ~ ‖ dünnes Eisenblech n.; Feinblech
n. ‖ tôle f. fine. / tilted ~ ‖ gehämmertes
Eisenblech m. ‖ tôle f. martelée. / undulated ~ *see* corrugated ~.

sheet iron articles pl., japanned ~ ‖ lackierte
Blechwaren fpl. ‖ objets mpl. en tôle
vernie.

sheet iron basket ‖ Eisenblechkorb m. ‖
panier m. en tôle de fer. / ~ box ‖ Blech-
schachtel f. ‖ boîte f. en tôle. / ~ can ‖
Blechkanne f. ‖ bidon m. en tôle. / ~ case
‖ Blechkasten m. ‖ boîte f. en tôle. /
~ core ‖ Blechkern m. ‖ noyau m. en
tôle. / ~ cover ‖ Blechmantel m. ‖ capot
m. en tôle. / ~ goods pl. ‖ Blechwaren
fpl. ‖ tôlerie f.; articles mpl. en tôle. /
~ jacket ‖ Eisenblechmantel m. ‖ chemise
f. en tôle de fer. / ~ plate ‖ Blechtafel
f. ‖ feuille f. de tôle. / ~ puncher ‖ Blech-
locher m. ‖ poinçonneur m. de tôle. /
~ socle ‖ Blechsockel m. ‖ socle m. en
tôle. / ~ tube ‖ Blechrohr n. ‖ tuyau m.
en tôle. / ~ worker ‖ Blechschmied m. ‖
tôlier m. / ~ wrapping ‖ Blechpackung f.
‖ emballage f. en tôle.

sheet lead ‖ Bleiblech n.; Bleipapier n.;
Walzblei n. ‖ plomb m. en feuilles; feuille
f. de plomb. / ~ in rolls ‖ Rollenblei n. ‖
plomb m. en rouleaux.

sheet lead disk ‖ Bleiblechscheibe f. ‖ rondelle f. de plomb.

sheet lightning ‖ Flächenblitz m. ‖ éclair m.
diffus.

sheet metal *see also* sheet *and* sheet iron ‖
ˋBlech n. ‖ tôle f. / ~ for wall facing ‖
Wandkleidungsblech n. ‖ tôle f. pour
le revêtement des parois.

sheet metal card ‖ Blechkarte f. ‖ carton m.
métallique. / ~ corner piece ‖ Eckblech
n. ‖ cornière f. de tôle. / stiffened with ~
corner pieces ‖ mit Eckblechen npl. versteift ‖ renforcé par de cornières fpl. de
tôle. / ~ cover ‖ Blechmantel m. ‖ revêtement m. en tôle. / ~ covering ‖ Blech-
beplankung f. ‖ revêtement m. de tôle. /
~ cylinder ‖ Blechzylinder m. ‖ cylindre
m. de tôle. / power driven machine for ~
industries ‖ Blechbearbeitungsmaschine
f. für Kraftbetrieb ‖ machine f. à force
motrice à travailler les métaux en feuilles.
/ ~ packing ‖ Blechverpackung f. ‖ emballage m. en fer-blanc. / ~ polishing
machine ‖ Blechpoliermaschine f. ‖ machine f. à polir les tôles. / ~ printing ‖
Blechdruck m. ‖ impression f. sur fer
blanc. / ~ stamper ‖ Blechpresser m. ‖
emboutisseur m. de tôles. / ~ stamping ‖
Blechpressen n. ‖ emboutissage m. de
tôles. / ~ testing machine ‖ Blechprüf-
maschine f. ‖ machine f. à essayer les
tôles. / ~ work ‖ Blecharbeit f.; Klempnerarbeit f. ‖ tôlerie f. / ~ working machine ‖ Blechbearbeitungsmaschine f. ‖
machine f. à travailler les tôles.

sheet mica ‖ Plattenglimmer m. ‖ mica m.
en feuilles. / ~ mill ‖ Blechwalzwerk n. ‖

laminoir m. à tôle. / ~ pile ‖ Spundbohle f. ‖ palplanche f.

sheet piling ‖ Spundwand f. ‖ cloison f. avec palée; file f. de palplanches. / reinforced concrete ~ ‖ Eisenbetonspundwand f. ‖ cloison f. en béton armé. / wooden ~ ‖ Holzspundwand f. ‖ cloison f. de planches.

sheet roll ‖ Blechwalze f.; Glattwalze f. ‖ cylindre m. (de laminoir) à tôle. / ~ rolling mill see ~ mill. / ~ rubber ‖ Plattengummi m. ‖ caoutchouc m. en plaques.

sheets pl., loose ~ ‖ lose Blätter npl. ‖ feuilles fpl. détachées. / mackled ~ ‖ Ausschußpapier n.; Makulatur f.; Schmutzpapier n. ‖ maculature f.; papier m. de rebut. / ~ sewed in ‖ geheftete Blätter npl. ‖ feuillets mpl. brochés.

sheet steel ‖ Stahlblech n. ‖ tôle f. d'acier. / ~ tin ‖ Zinnblech n. ‖ feuille f. d'étain. / ~ transferring device of a calender ‖ Bogenüberführung f. eines Kalanders ‖ transporteur m. de feuilles d'un calandre.

sheet work ‖ Blecharbeit f. ‖ tôlerie f. / flanged ~ ‖ Bördelarbeit f. ‖ objet m. obtenu par bridage.

sheet working machine ‖ Blechbearbeitungsmaschine f. ‖ machine f. à travailler les tôles. / ~ zinc ‖ Zinkblech n. ‖ tôle f. de zinc; zinc m. laminé.

shelf ‖ Regal n.; Gestell n. ‖ tablette f.; rayon m. / ~ (Geol) ‖ Riff n.; Untiefe f.; Sandbank f. ‖ récif m.; bas-fond m.; banc m. de sable. / armour ~ ‖ Panzerträger m. ‖ appui m. ou chaise f. de la cuirasse. / ~ for drawings ‖ Zeichenregal n. ‖ tablette f. à dessins. / iron ~ ‖ eisernes Regal n. ‖ rayon m. ou étagère f. en fer. metal ~ for shop-windows ‖ eisernes Schaufenstergestell n. ‖ étagère f. en métal pour vitrines. / ~ for metal types ‖ Schriftregal n. ‖ rayon m. pour caractères métalliques. / ~ for plans ‖ Planregal n. ‖ étagère f. à plans. / steel ~ ‖ Stahlregal n. ‖ rayon m. en acier.

shell, to ‖ abschälen; ausschälen; enthülsen ‖ écaler; écosser; décortiquer.

shell (Arm) ‖ Granate f. ‖ obus m. / ~ (Corn) ‖ Schale f.; Hülse f. ‖ gousse f.; cosse f.; écale f. / ~ (Mach) ‖ Mantel m. ‖ enveloppe f. / ~ (Mussel) ‖ Muschel f. ‖ coquille f. / ~ (Musselshell) ‖ Muschelschale f. ‖ coquillage m. / ~ of blast furnace ‖ Rauhgemäuer n. des Hochofens ‖ enveloppe f. ou massif m. ou muraillement m. d'un haut-fourneau; manteau m. du haut-fourneau. / ~ of boiler ‖ Kesselmantel m. ‖ chemise f. de chaudière. / ~ of the breaker ‖ Brechmantel m. ‖ boiseau m. de concassage. / ~ of a building ‖ Gerippe n. eines Gebäudes ‖ carcasse f. ou squelette m. d'un bâtiment. / ~ of a button ‖ Oberboden m. oder Oberplatte f. eines Knopfes ‖ coquille f. d'un bouton. / ~ of a furnace ‖ Rauhschacht m.; Rauhschacht m. ‖ contre-paroi f. ou fausses-parois fpl. d'un haut-fourneau. / ~ of gun cartridge ‖ Kartuschenhülse f. ‖ douille f. de gargousse. / light ~ (Opt) ‖ blondes Schildpatt n. ‖ écaille f. blonde. / ~ of mould (Found) ‖ Kapsel f. oder Schale f. oder Mantel m. der Form; Formkappe f. ‖ enveloppe f. du moule. / ~ for percussion caps ‖ Sprengkapselhülse f. ‖ douille f. de cartouches explosibles. / ~ of a pulley ‖ Hülse f. einer Rolle ‖ chape f. d'une

moufle ou d'une poulie. / ~ of roll ‖ Walzenmantel m. ‖ enveloppe f. de cylindre. / ~ of a screening drum ‖ Mantel m. eines Trommelsiebes ‖ enveloppe f. du trommel-classeur. / shrapnel ~ ‖ Schrapnellgeschoß n. ‖ obus m. à balles ou à shrapnells. / ~ with tapered end (Arm) ‖ Geschoß n. mit verjüngtem Bodenteil; Stromliniengeschoß n. ‖ balle f. fuselée; obus m. à culot retreint. / wooden ~ ‖ Holzhülse f. ‖ douille f. en bois.

shellac ‖ Schellack m.; Scheibenlack m. ‖ gomme-laque f.; shellac m. / artificial ~ ‖ künstlicher Schellack m. ‖ gomme f. laque artificielle.

shellac varnish ‖ Schellackfirnis m. ‖ vernis m. à la gomme-laque. / ~ wax ‖ Schellackwachs n. ‖ cire f. de gomme laque. / ~ work installation ‖ Schellackfabrikeinrichtung f. ‖ installation f. de fabrique de gomme-laque.

shell almond ‖ Knackmandel f. ‖ coquemolle f. / ~ articles pl. see ~ goods pl. / ~ auger ‖ Löffelbohrer m.; Hohlbohrer m. ‖ tarière f. à cuiller; foret-cuiller m. / ~ barrow ‖ Geschoßtrage f. ‖ civière f. ou lanterne f. de chargement. / ~ bit see ~ auger. / ~ comb ‖ Schildpattkamm m. ‖ peigne m. en écaille. / ~ davit ‖ Geschoßdavit m. ‖ grue f. monte-projectiles.

shelled almond ‖ abgeschälte Mandel f. ‖ amande f. cassée.

shell extractor ‖ Patronenzieher m. ‖ tire-cartouches m. / ~ gatherer ‖ Muschelfischer m. ‖ pêcheur m. de coquilles ou de coquillages ou de moules. / ~ goods pl. ‖ Muschelwaren pl. ‖ articles mpl. en coquillages. / ~ hook ‖ Geschoßheber m. ‖ tire-obus m.; crochet m. à poignée. / ~ lime ‖ Muschelkalk m. ‖ chaux f. de coquilles. / ~ limestone (Geol) ‖ Muschelkalk m. ‖ calcaire m. conchylien. / ~ marl see ~ lime. / ~ plate (Loc) ‖ Mantelblech n. ‖ tôle f. à enveloppe. / boiler ~ plate ‖ Kesselwand f. ‖ paroi f. de la chaudière. / ~ ring (Boiler) ‖ Kesselschuß m. ‖ virole f. de chaudière. / taper ~ ring (Boiler) ‖ kegelförmiger Kesselschuß m. ‖ virole f. conique de chaudière. / ~ sand ‖ Muschelsand m. ‖ sable m. coquilleux. / ~ transformer ‖ Manteltransformator m. ‖ transformateur m. cuirassé. / ~ valve ‖ Muschelschieber m. ‖ soupape f. à coquilles. / ~ work ‖ Muschelwerk n. ‖ coquillage m. / ~ worker ‖ Muschelarbeiter m. ‖ ouvrier m. en coquillages.

shelly (Miner) ‖ muschelig ‖ conchoïde; coquillier. / ~ sandstone ‖ Muschelsandstein m. ‖ grès m. coquillier.

shelter ‖ Schutz m. ‖ abri m. / ~ (Build) ‖ Schirmdach n. ‖ hangar m.

sheltered zone (Aero) ‖ Windschatten m. ‖ ombre f. du vent; terrain m. abrité du vent.

shelter wood ‖ Schutzwald m. ‖ forêt f. de protection.

shelved wagon ‖ Etagenwagen m. ‖ wagonnet m. étagé.

shelving of a wall ‖ Böschung f. einer Mauer ‖ adossement m. ou pente f. ou talus m. d'un mur.

shelving coast ‖ Steilküste f. ‖ côte f. raide ou escarpée ou accore ou à pic.

shepherd ‖ Hirt m. ‖ berger m.

shepherd's pouch ‖ Hirtentäschelkraut n. ‖ herbe f. de panetière.

sherry ‖ Xereswein m. ‖ xérès m.

shield (Arm) ‖ Schutzwand f.; Schild m.; Panzer m. ‖ blindage m.; masque m.; carapace f. / ~ (Mach) ‖ Schild n.; Platte f. ‖ plaque f. / water gauge glass ~ ‖ Wasserstandsglasschutzgitter n. ‖ manchon m. ou douille f. de protection du tube de niveau d'eau.

shield marking machine ‖ Schutzringstempelmaschine f. ‖ machine f. à marquer les frettes. / ~ paper ‖ Schildpapier n. ‖ écu m.; papier m. écu.

shift, to ‖ schieben; verschieben; sich bewegen ‖ (se) déplacer. / ~ (Mach) ‖ einrücken; schalten; ausrücken ‖ embrayer; débrayer. / ~ (Wind) ‖ umschlagen; umschießen ‖ se changer; avoyer. / ~ the belt over from fast to loose pulley ‖ den Riemen m. von der festen auf die lose Scheibe verschieben ‖ passer la courroie de la poulie fixe à la poulie folle. / ~ the helm ‖ das Ruder überlegen ‖ changer la barre. / ~ the lading ‖ Güter fpl. umladen ‖ transborder les marchandises fpl.

shift ‖ Tagewerk n.; Schicht f. ‖ journée f. / to begin the ~ ‖ die Schicht antreten ‖ commencer le travail. / day ~ ‖ Tagesschicht f. ‖ équipe f. de jour. / night ~ ‖ Nachtschicht f. ‖ équipe f. de nuit. / to work out a ~ ‖ eine Schicht f. machen ‖ achever une journée.

shifter (Mach) ‖ Ausrücker ‖ arrêt m.; dispositif m. ou levier m. de débrayage. / belt ~ ‖ Riemenausrückvorrichtung f.; Riemenschalter m. ‖ embrayeur m. de courroie.

shifting ‖ Fortrückung f.; Platzänderung f. ‖ déplacement m. / ~ (Auto) ‖ Schaltung f. ‖ changement m. de vitesses. / ball and socket gear ~ (Auto) ‖ Kugelschaltung f. ‖ changement m. de vitesses par levier oscillant. / ~ of beach sand ‖ Küstenversetzung f. ‖ déplacement m. des côtes. / ~ clockwise (Aero) ‖ Rechtsablenkung f. ‖ déviation f. vers la droite. / ~ counter clockwise (Aero) ‖ Linksablenkung f. ‖ déviation f. vers la gauche. / ~ of phase ‖ Phasenverschiebung f. ‖ décalage m. de phases.

shifting ballast (Mar) ‖ fliegender Ballast m. ‖ lest m. volant. / ~ boards pl. (Shipb) ‖ fliegendes Schott n. ‖ cloison f. temporaire ou volante. / locomotive ~ crane ‖ Lokomotivversatzkran m. ‖ grue f. à déplacer les locomotives. / ~ device ‖ Verschiebevorrichtung f. ‖ dispositif m. de déplacement. / ~ gauge ‖ Reißmaß n. ‖ tracequin m.; trusquin m. / ~ platform ‖ Schiebebühne f. ‖ chariot m. transbordeur. / ~ preventers pl. ‖ Schlingerpardunen fpl. ‖ galhaubans mpl. volant. / ~ sand (Geol) ‖ bewegliche Sandbank f. ‖ sable m. mouvant. / ~ spanner ‖ Engländer m.; Franzose m.; englischer Schraubenschlüssel m.; verstellbarer Schraubenschlüssel m. ‖ clef f. anglaise.

shift key for capitals (Typewr) ‖ Umschalttaste f. für große Schrift ‖ touche f. des majuscules. / release of the ~ (Typewr) ‖ Freigabe f. der Umschaltetaste ‖ relâchement m. de la touche des majuscules.

shift lock key (Typewr) ‖ Feststelltaste f. ‖ touche f. de bloquage.

shim (Mach) ‖ Blechzwischenlage f.; Unterlegblättchen n. ‖ cale f. en tôle.

shin (Railw) ‖ Lasche f.; Stoßlasche f. ‖ éclisse f.

shin bone ‖ Schienbein n. ‖ tibia m.

shine, to ‖ leuchten ‖ luire.

shingle, to (Metal) ‖ zängeln ‖ cingler; ébaucher.

shingle ‖ Dachschindel f.; Schindel f. ‖ bardeau m.; échandole f. / ~ nail ‖ Schindelnagel m. ‖ clou m. à bardeaux. / ~ planer ‖ Schindelhobler m. ‖ raboteur m. de bardeaux.

shingler (Forg) ‖ Luppenschmied m. ‖ cingleur m.

shingle sawyer ‖ Schindelsäger m. ‖ scieur m. de bardeaux.

shingling ‖ Schindelbedachung f. ‖ couverture f. en bardeaux. / ~ hammer (Metal) ‖ Zänghammer m. ‖ marteau m. cingleur; cinglard m. / ~ layer ‖ Schindeldecker m.; Dachdecker m. ‖ poseur m. de voliges. / ~ roll ‖ Zängwalze f. ‖ laminoir m. à cingler. / ~ tongs pl. (Metal) ‖ Scherenzange f.; Rampfzange f. ‖ louperesse f.; écrevisse f.; cingleresse f.

shining ‖ glänzend ‖ brillant. / ~ blacking ‖ Glanzwichse f. ‖ cirage m. brillant. / ~ brush ‖ Polierbürste f. ‖ polissoire f. / ~ surface ‖ glänzende Oberfläche f. ‖ surface f. polie ou brillante.

ship, to (Passenger) ‖ einschiffen ‖ embarquer. / ~ (Sailor) ‖ heuern ‖ engager. / ~ (To send abroad) ‖ verschiffen ‖ expédier; envoyer. / ~ the rudder ‖ das Ruder einhaken oder einhängen oder einsetzen ‖ monter le gouvernail. / ~ a sea ‖ eine See f. übernehmen ‖ embarquer un coup de mer.

ship see also vessel ‖ Schiff n. ‖ navire m.; bâtiment m.; vaisseau m.; bateau m. / battle ~ ‖ Linienschiff n. ‖ cuirassé m. d'escadre. / bluff-headed ~ ‖ Schiff n. mit vollem Bug ‖ navire m. à avant renflé. / boarded ~ ‖ angesegeltes Schiff n. ‖ abordé m. / boarding ~ ‖ ansegelndes Schiff n. ‖ abordeur m. / to bring up a ~ see to capture a ~. / cable laying ~ ‖ Kabelleger m.; Kabelschiff n. ‖ navire m. câbler ou pose-câbles. / to capture a ~ ‖ ein Schiff n. aufbringen ‖ capturer un bateau. / coal ~ ‖ Kohlenschiff n. ‖ charbonnier m. / ~ under command see steering ~. / composite ~ ‖ Kompositschiff n. ‖ navire m. en bois et en fer. / condemned ~ ‖ kondemniertes Schiff n. ‖ navire m. condamné. / converted ~ ‖ umgebautes Schiff n. ‖ navire m. refondu. / Diesel ~ ‖ Dieselschiff n. ‖ navire m. à moteur Diesel. / fast going ~ ‖ gut segelndes Schiff n. ‖ navire m. bon-marcheur m. / ~ of the first reserve ‖ Schiff n. in erster Reserve ‖ navire m. en première catégorie. / ~ fitting-out ‖ Schiff n. in Ausrüstung ‖ navire m. en achèvement à flot. / flat-bottomed ~ ‖ flachgebautes oder flachbodiges Schiff n. ‖ navire m. à fond plat. / flush-deck ~ ‖ Glattdeckschiff n. ‖ navire m. à pont ras. / foreign trade ~ ‖ Schiff n. für lange Fahrt ‖ navire m. de long cours. / ~ in frames ‖ in Spanten stehendes Schiff n. ‖ navire m. boisé ou avec membrures montées. / home trade ~ ‖ Schiff n. für inländische Fahrt ‖ navire m. de cabotage. / ~ incapable of repair ‖ reparaturunfähiges Schiff n. ‖ navire m. irréparable. / iron ~ ‖ Eisenschiff n. ‖ navire m. en fer. / laboursome ~ ‖ Schiff n. mit schweren Bewegungen ‖ navire m. à mouvements durs. / leaky ~ ‖ leckes Schiff n. ‖ navire m. qui fait de

l'eau. / light ~ ‖ Leuchtschiff n. ‖ bateau-phare m. / ~ of the line see battle ~. / manned ~ ‖ bemanntes Schiff n. ‖ navire m. équipé. / merchant ~ ‖ Handelsschiff n. ‖ navire m. marchand ou de commerce; cargo m. / ~ in ordinary ‖ Schiff n. in zweiter Reserve ‖ navire m. en disponibilité. / paddle-wheel ~ ‖ Radschiff n. ‖ navire m. à roues. / ~ with raised quarterdeck ‖ Schiff n. mit erhöhtem Quarterdeck ‖ navire m. à pont surélevé. / ~ ready for sea ‖ segelfertiges oder seeklares Schiff n. ‖ navire m. prêt à sortir. / salvage crane ~ ‖ Bergungskranschiff n. ‖ bateau m. à grue de sauvetage. / sea-going ~ ‖ Seeschiff n. ‖ bâtiment m. ou navire m. de long cours. / seaworthy ~ ‖ seetüchtiges oder seefähiges Schiff n. ‖ navire m. en bon état de navigabilité; navire m. capable de tenir la mer. / sister ~ ‖ Schwesterschiff n. ‖ frèrejumeau m. / steam ~ ‖ Dampfschiff n.; Dampfer m. ‖ vapeur m.; navire m. à vapeur. / steering ~ ‖ manövrierfähiges Schiff ‖ navire qui gouverne ou qui obéit à son gouvernail. / to put a ~ on the stocks ‖ ein Schiff n. zum Bau auflegen ‖ mettre un vaisseau sur le chantier. / straight-sheered ~ ‖ Schiff n. ohne Sprung ‖ navire m. ras sur l'eau. / submarine depot ~ ‖ Hilfsschiff n. für Unterseeboote ‖ bâtiment m. de sauvetage de sous-marins. / tank ~ ‖ Tankschiff n. ‖ vapeur-citerne m.; navire m. à citernes. / towed ~ ‖ Schlepp n.; geschlepptes Schiff n. ‖ bâtiment m. remorqué; remorqué m. / training ~ ‖ Schulschiff n. ‖ navire-école m. / twin-screw ~ ‖ Doppelschraubenschiff n. ‖ navire m. à hélices jumelles. / two-decked ~ ‖ zweideckiges Schiff n. ‖ navire m. à deux ponts. / unseaworthy ~ ‖ seeuntüchtiges Schiff n. ‖ navire m. impropre à la navigation. / unsinkable ~ ‖ unsinkbares Schiff n. ‖ vaisseau m. insubmersible. / water logged ~ ‖ auf der Ladung treibendes Schiff n. ‖ navire m. engagé dans l'eau. / wooden ~ ‖ Holzschiff n. ‖ navire m. en bois.

ship anchor ‖ Schiffsanker m. ‖ ancre f. de bateau. / auxiliary engine of a ~ ‖ Schiffshilfsmaschine f. ‖ machine f. auxiliaire de bord. / bevelling machine for ~ beams ‖ Schmiegemaschine f. für Schiffsspanten ‖ machine f. à équerrer les couples. / ~ bedstead ‖ Schiffsbettstelle f. ‖ lit m. ou couche f. pour bateaux. / ~ bell ‖ Schiffsglocke f. ‖ cloche f. de bateaux. / ~ biscuit ‖ Schiffszwieback m.; Hartbrot n. ‖ biscuit m. / ~ block ‖ Schiffsblock m. ‖ bloc m. de bateaux. / ~ boiler ‖ Schiffskessel m. ‖ chaudière f. de bateau. / ~ breaker ‖ Schiffsausschlachter m. ‖ démolisseur m. de bateaux. / ~ breaking-up ‖ Schiffsabbruch m. ‖ démolition f. de bateau. / ~ broker ‖ Schiffsmakler m. ‖ courtier m. de navire.

shipbuilder ‖ Schiffbauer m. ‖ ingénieur m. ou constructeur de navires.

shipbuilding ‖ Schiffbau m. ‖ construction f. navale. / ~ crane ‖ Hellingkran m. ‖ grue f. de cale sèche ou pour cales de construction. / ~ hall ‖ Schiffbauhalle f. ‖ hall m. de construction navale. / ~ material ‖ Schiffbaumaterial n. ‖ matériaux mpl. de construction de navires.

~ plant ‖ Hellinganlage f. ‖ cale f. de construction. / special machine tool for ~ ‖ Schiffbausonderwerkzeugmaschine f. ‖ machine-outil f. spéciale pour la construction des bateaux. / ~ yard see shipyard.

ship cabin equipment ‖ Schiffskabineneinrichtung f. ‖ aménagement m. des cabines des bateaux. / ~ cable ‖ Schiffstauwerk n. ‖ cordage m. pour la marine. / ~ canal ‖ Schiffskanal n. ‖ canal m. de navigation. / ~ carpenter ‖ Schiffszimmermann m. ‖ charpentier m. de bateaux; fustier m. / ~ chandler ‖ Schiffslieferant m. ‖ fournisseur m. de navires. / ~ clock ‖ Schiffsuhr f. ‖ horloge f. pour navire. / ~ cooling plant ‖ Schiffskühlanlage f. ‖ installation f. frigorifique pour bateaux. / ~ davit ‖ Schiffsdavit m. ‖ davier m. de bateau. / ~ elevator ‖ Schiffsaufzug m. ‖ engin m. de levage de bateaux. / ~ engine ‖ Schiffsmaschine f. ‖ machine f. marine. / ~ engineer ‖ Schiffsmaschinist m. ‖ mécanicien m. de bord. / ~ equipment ‖ Schiffseinrichtung f. ‖ aménagement m. de bateau. / ~ fender ‖ Schiffsfender m. ‖ défense f. de navires. / ~ fittings pl. ‖ Schiffsarmatur f. ‖ armature f. pour bateaux. / flag of a ~ ‖ Schiffsflagge f. ‖ pavillon m. / machine for squaring ~ frames ‖ Schiffsspantenschmiegemaschine f. ‖ machine f. à équerrer les couples de navires. / ~ freighting ‖ Schiffsbefrachtung f. ‖ affrètement m. de navires. / ~ heating plant ‖ Schiffsheizung f. ‖ chauffage m. pour bateaux. / ~ heaver ‖ Schiffshebebock m. ‖ chèvre f. à bateau. / ~ hoisting tackle ‖ Schiffshebewerk n. ‖ élévateur m. de bateaux. / ~ hold see ship's hold. / ~ hull ‖ Schiffsrumpf m. ‖ coque f. de bateau. / ~ joiner ‖ Schiffsschreiner m. ‖ menuisier m. de bateaux. / ~ keeper ‖ Schiffswächter m. ‖ gardien m. de navire. / ~ ladder ‖ Schiffstreppe f. ‖ échelle f. d'un navire. / ~ language ‖ Schiffssprache f. ‖ argot m. maritime. / ~ lift ‖ Schiffshebewerk n. ‖ montecharge m. ou élévateur m. pour bateaux. / ~ lighting ‖ Schiffsbeleuchtung f. ‖ éclairage m. de navire.

shipload see ship's cargo.

ship loader ‖ Schiffsverlader m. ‖ chargeur m. de bateaux. / ~ loading ‖ Schiffsverladung f. ‖ chargement m. des bateaux.

shipmaster ‖ Schiffsführer m.; Seeschiffer m. ‖ capitaine m. de navire; patron m.

shipment, collective ~ (Railw) ‖ Sammelladung f. ‖ envoi m. en groupage.

ship mill ‖ Schiffsmühle f. ‖ moulin m. à nef ou sur bateau. / ~ mill's wheel ‖ Schiffsmühlenrad n. ‖ roue f. pendante de bateau. / ~ money ‖ Schiffsabgaben fpl. ‖ impôt m. de navire. / ~ mountings pl. ‖ Schiffsbeschläge mpl. ‖ armatures fpl. de navires. / ~ nail ‖ Schiffsnagel m. ‖ clou m. de navire. / ~ owner ‖ Reeder m. ‖ armateur m.

shipped ‖ versandt ‖ expédié.

shipping ‖ Verschiffung f. ‖ embarquement m. / ~ agency ‖ Spedition f. ‖ expédition f. / ~ agency (Mar) ‖ Reederei f. ‖ agence f. d'armement. / ~ board ‖ Seeamt n. ‖ bureau m. de la marine. / ~ cask ‖ Versandfaß n. ‖ fût m. d'exportation. / ~ dimensions pl. ‖ Umfangmaße

npl. bei seemäßiger Verpackung ‖ volume m. sous emballage maritime. / ~ instructions pl. ‖ Versandvorschriften fpl. ‖ instructions fpl. pour expédition. / ~ master ‖ Heuerbaas m. ‖ courtier m. de matelots. / ~ mole ‖ Landungsplatz m.; Schiffslände f.; Ladedamm m.; Anfurt f. ‖ embarcadère m.; embarcadeur m. / ~ note ‖ Anlieferungsschein m. ‖ permis m. d'embarquement. / ~ office ‖ Heueramt n.; Musterungsamt n. ‖ bureau m. de l'inscription maritime. / ~ share ‖ Schiffahrtsaktie f. ‖ valeur f. de navigation. / ~ weight ‖ Versandgewicht n. ‖ poids m. pour expédition.

ship plate ‖ Schiffsblech n. ‖ tôle f. de navire. / bending machine for ~ plates ‖ Schiffsplattenbiegemaschine f. ‖ machine f. à cintrer les plaques de navires. / true position of the ~ ‖ wahrer Schiffsort m. ‖ vraie position f. du navire. / ~ propeller ‖ Schiffsschraube f. ‖ hélice f. de bateau. / ~ pump ‖ Schiffspumpe f. ‖ pompe f. de bateau. / ~ provisions pl. ‖ Schiffsproviant m. ‖ provision f. de mer. / ~ radio station ‖ Bordfunkstelle f. ‖ station f. radiotélégraphique de bord. / ~ range ‖ Schiffsherd m. ‖ cuisinière f. de bord. / ~ repairing ‖ Schiffsausbesserung f. ‖ radoubage m. de navires. / ~ rigging ‖ Schiffstakelung f. ‖ montage m. d'agrès *ou* de manœuvres de navire.

ship's accounts pl. ‖ Schiffsverwaltung f. ‖ comptabilité f. de bord. / ~ articles pl. ‖ Musterrolle f. ‖ rôle f. d'équipage. / ~ ballast ‖ Schiffsballast m. ‖ lest m. d'un navire. / ~ books pl. ‖ Schiffsbücher npl. ‖ livres mpl. de bord. / ~ bottom ‖ Schiffsboden m. ‖ fond m. d'un navire. / ~ captain ‖ Schiffskapitän m. ‖ capitaine m. de vaisseau; patron m. / ~ cargo ‖ Schiffsladung f. ‖ chargement m.; cargaison f. / ~ compass ‖ Schiffskompaß m. ‖ boussole f. *ou* compas m. d'un navire. / ~ cook room ‖ Schiffsküche f. ‖ cuisine f. de bord.

ship's course, to determine the ~ according to direction ‖ den Schiffsweg m. nach der Richtung bestimmen ‖ déterminer la course du navire d'après la direction. / to find the ~ according to distance covered by dead reckoning ‖ den Schiffsweg m. nach der zurückgelegten Strecke bestimmen ‖ déterminer la course du navire d'après la distance parcourue.

ship screw ‖ Schiffsschraube f. ‖ hélice f. de navire.

ship's crew ‖ Bemannung f.; Mannschaft f. ‖ équipage m.; hommes mpl. / ~ doctor ‖ Schiffsarzt m. ‖ médecin m. du bord *ou* de vaisseau. / ~ dues pl. ‖ Schiffsabgaben fpl. ‖ impôt m. de navire. / ~ freight ‖ Schiffsfracht f. ‖ affrètement m.; fret m. / ~ freighter ‖ Schiffsbefrachter m. ‖ armateur m. / ~ gun ‖ Schiffsgeschütz n. ‖ canon m. de bord.

shipshape ‖ schiffsmäßig ‖ marin.

ship's hatchway ‖ Schiffsluke f. ‖ écoutille f.

ship sheathing ‖ Schiffsbeschlag m. ‖ feuille f. de doublage.

ship's hold ‖ Schiffsraum m. ‖ cale f.

ship, size of ~s pl. ‖ Schiffsgröße f. ‖ dimension f. des navires. / ~ slang ‖ Schiffssprache f. ‖ argot m. maritime.

ship's lantern ‖ Schiffslaterne f. ‖ fanal m. / ~ magnetism ‖ Schiffsmagnetismus m. ‖ magnétisme m. du navire. / ~ medical

officer ‖ Schiffsarzt m. ‖ médecin m. du bord. / ~ papers pl. ‖ Schiffspapiere npl. ‖ papiers mpl. de bord.

ship's place ‖ Schiffsort m.; Besteck n. ‖ point m. du navire. / ~ by bearings ‖ gepeiltes Besteck n. ‖ position f. du navire obtenue à l'aide de relèvement. / ~ by dead reckoning ‖ gegißtes Besteck n. ‖ point m. estimé. / ~ by observation ‖ astronomisches Besteck n. ‖ point m. observé.

ship's position, to check the ‖ den Schiffsort m. nachprüfen ‖ vérifier *ou* contrôler la position du navire. / ~ protest ‖ Verklarung f. ‖ procès-verbal m. des avaries d'un vaisseau. / ~ register ‖ Schiffsbrief m.; Schiffsregister n.; Schiffszertifikat n. ‖ registre m. du bord; certificat m. de construction d'un navire.

ship station (Radio) ‖ Schiffsstation f. ‖ station f. de bord. / ~ steering gear ‖ Schiffssteuer n. ‖ gouvernail m. d'un navire. / ~ telegraph ‖ Schiffstelegraf m. ‖ télégraphe m. de bateau. / ~ transfer lift ‖ Schiffsaufschleppe f. ‖ chariot m. de remorçage de bateaux. / ~ unloader ‖ Schiffslöscher m. ‖ déchargeur de bateaux m. / ~ unloading ‖ Schiffsausladung f. ‖ déchargement m. de bateaux. / ~ worm ‖ Bohrwurm m. ‖ taret m. / ~ winch ‖ Schiffswinde f. ‖ treuil m. de bateau.

shipwreck ‖ Schiffbruch m. ‖ naufrage m.

shipwrecked ‖ Schiffbrüchiger m. ‖ naufragé m. / to be ~ ‖ Schiffbruch m. leiden ‖ naufrager; faire naufrage m.

shipwright ‖ Schiffbauer m. ‖ ingénieur m *ou* constructeur de navires. / ~ yard *see* shipyard.

shipyard ‖ Schiffswerft f.; Werft f. ‖ chantier m. de construction navale. / ~ crane ‖ Werftkran m. ‖ grue f. de chantier. / ~ winch ‖ Hellingwinde f. ‖ treuil m. de cale de construction.

shirt ‖ Hemd n. ‖ chemise f. / day ~ ‖ Oberhemd n. ‖ chemise f. de jour *ou* de dessus. / flannel ~ ‖ Flanellhemd n. ‖ chemise f. de flanelle. / knitted ~ ‖ Trikothemd n. ‖ chemise f. en tricot. / ladies' ~ ‖ Damenhemd n. ‖ chemise f. de femme *ou* de dames. / linen-fronted ~ ‖ Hemd n. mit leinenem Brusteinsatz ‖ chemise f. à plastron toile. / worker's ~ ‖ Arbeiterhemd n. ‖ chemise f. pour ouvriers.

shirt cloth ‖ Hemdentuch n. ‖ drap m. pour chemises. / ~ collar ‖ Hemdkragen m. ‖ col m. de chemise; faux-col m. / ~ flannel ‖ Hemdenflanell n. ‖ flanelle f. pour chemises. / ~ front ‖ Hemdeneinsatz m. ‖ devant m. de chemise.

shirting ‖ Nessel m.; Schirting m.; Futterkattun m.; Hemdenkattun m. ‖ shirting m.; cretonne f.

shirt maker ‖ Hemdenmacher m. ‖ chemisier m. / ~ seamstress ‖ Hemdennäherin f. ‖ chemisière f.

shoal ‖ Sandbank f.; Kiesbank f. ‖ banc m. de sable; ensablement m.; barre f.; haut-fond m.

shock ‖ Stoß m.; Schlag m. ‖ choc m. / ~s pl. to building ‖ Gebäudeerschütterung f. ‖ vibration f. *ou* trépidation f. au bâtiment. / electric ~ ‖ elektrischer Schlag m. ‖ commotion f. électrique. / return ~ ‖ Rückschlag m. ‖ choc m. en retour. /

~ of the waves ‖ Wellenschlag m. ‖ coup m. de mer.

shock absorber ‖ Stoßdämpfer m. ‖ amortisseur m. de chocs. / adjustable ~ ‖ nachstellbarer Stoßfänger m. ‖ amortisseur m. réglable. / oleo-pneumatic ~ ‖ Ölstoßdämpfer m. ‖ amortisseur m. oléopneumatique.

shock absorption ‖ Stoßdämpfung f. ‖ amortissement m. de chocs. / ~ bending test ‖ Schlagbiegeprobe f. ‖ essai m. de flexion au choc. / ~ deflection test of axles ‖ Schlagbiegeprobe f. an Achsen ‖ essai m. de flexion par choc sur essieux.

shock effect ‖ Stoßwirkung f. ‖ effet m. du choc. / ~ of wheel against the head of the rail ‖ Stoßwirkung f. des Rades gegen den Schienenkopf ‖ effet m. du choc des roues contre le champignon.

shock equalizer with compressed air cushion ‖ Stoßausgleicher m. mit Druckluftbelastung ‖ compensateur m. de chocs à air comprimé. / ~ excitation (Radio) ‖ Stoßerregung f. ‖ excitation f. par choc. / resistance to ~ ‖ Schlagfestigkeit f. ‖ résistance f. au choc.

shock spring ‖ Stoßfeder f. ‖ ressort m. de choc. / ~ test of a wheel in vertical position ‖ senkrechte Schlagprobe f. an einem Rad ‖ essai m. au mouton d'une roue en position verticale.

shoddy ‖ Kunstwolle f.; Lumpenwolle f.; Schoddy f. ‖ laine f. artificielle *ou* renaissance. / ~ devil ‖ Lumpenwolf m. ‖ machine f. à défiler *ou* à rompre les chiffons. / machine for ~ ‖ Kunstwollmaschine f. ‖ effilocheuse f. / ~ wool *see* shoddy.

shoe, to ~ a horse ‖ ein Pferd n. beschlagen ‖ ferrer un cheval. / ~ a pile ‖ einen Pfahl beschuhen ‖ ferrer le bout d'un pieu; saboter le pieu.

shoe ‖ Schuh m. ‖ soulier m. / ~ (Bridge) ‖ Auflagerschuh m. ‖ coussinet m. à plat. / ~ (Mach) ‖ Gleitschuh m. ‖ coulisseau m. de crosse. / ~ (Plough) ‖ Stelze f. ‖ sabot m. / ~ (Horse) ‖ Hufeisen n. ‖ fer m. à cheval *ou* de cheval. / ~s pl. of asbestos ‖ Schuhwaren fpl. aus Asbest ‖ chaussures fpl. en amiante. / bathing ~ ‖ Badeschuh m. ‖ espadrille f. / brake ~ ‖ Bremsklotz m. ‖ sabot m. de frein. / cast iron ~ (Bridge) ‖ gußeiserner Schuh m. ‖ bout m. en fonte. / crosshead ~ ‖ Kreuzkopfschuh m. ‖ patin m. de crosse. / felt ~ ‖ Filzschuh m. ‖ chaussure f. de feutre. / gymnasium ~ ‖ Turnschuh m. ‖ chaussure f. pour gymnastes. / ~ with a half-wide cover ‖ halbbreites Hufeisen n. ‖ fer m. à cheval à glace. / hollow ~ ‖ gewölbtes Hufeisen n. ‖ fer m. à cheval voûté. / india rubber ~ ‖ Gummigalosche f. ‖ galoche f. *ou* chaussure f. en caoutchouc. / iron ~ ‖ eiserner Schuh m. ‖ sabot m. en fer. / low ~ ‖ Halbschuh m. ‖ soulier m. découvert. / machine-sewed ~ ‖ maschinengenähter Schuh m. ‖ chaussure f. cousue-machine. / nailed ~ ‖ genagelter Schuh m. ‖ chaussure f. clouée. / pegged ~ ‖ holzgenagelter Schuh m. ‖ chaussure f. chevillée-bois. / plaited ~ ‖ Strohpantoffel m. ‖ chausson m. de nattes. / seaside and sporting ~ ‖ Schuh m. für Strand und Sport ‖ chaussure f. de plage et de sport. / ~ of a sinking shaft walling ‖ Senkschuh m. einer Senkschachtmauerung ‖ sabot m. tranchant *ou* trousse f. coupante d'un puits descendant en maçonnerie. / ~s pl. of straw ‖ Schuhwaren fpl. aus Stroh ‖ chaussures

fpl. en paille. / travelling ~ || Reiseschuh m. || soulier m. de voyage. / vaulted ~ || gewölbtes Hufeisen n. || fer m. vouté. / welt stitched ~ || Rahmenschuh m. || chaussure f. cousue-trépointe. / wooden ~ || Holzschuh m. || sabot m.

shoe blacking || Schuhwichse f. || cirage m. / ~ brake || Backenbremse f.; Klotzbremse f. || frein m. à sabot. / double-~ brake || Doppelbackenbremse f. || frein m. à deux sabots. / ~ brush || Schuhbürste f. || brosse f. pour la chaussure. / ~ buckle || Schuhschnalle f. || boucle f. à chaussures. / ~ button || Schuhknopf m. || bouton m. de soulier. / ~ buttoner || Schuhknöpfer m. || crochet m. à chaussures. / ~ cast see ~ last. / ~ clasp see ~ buckle.

shoe cream || Schuhkrem m. || crème f. à soulier; pâte f. pour chaussure. / ~ emptying machine || Schuhkremabfüllmaschine f. || machine f. à soutirer la pâte pour chaussures.

shoe cutter || Schuhzuschneider m. || coupeur m. en chaussures. / ~ dressing see shoe cream. / ~ fabric || Schuhstoff m. || drap m. de chaussures. / ~ file rasp || Schusterraspel f. || râpe-lime f. pour bottiers. / ~ fittings pl. || Schuhbeschläge mpl. || ferrures fpl. de chaussures. / ~ horn || Schuhanzieher m. || chaussepied m. / ~ industry || Schuhindustrie f. || industrie f. de la chaussure.

shoeing of horses || Hufbeschlag m. || ferrure f. de chevaux.

shoeing forge || Hufschmiede f. || maréchalerie f.; forge f. maréchale. / ~ hammer || Hufhammer m. || brochoir m. / ~ pincers pl. || Hufzange f. || tricoises fpl. / ~ smith || Hufschmied m. || maréchal m. ferrant.

shoe lace || Schuhband n.; Schnürsenkel m. || cordon m. de chaussure. / ~ last || Schuhleisten m. || forme f. ou embouchoir m. pour chaussures. / ~ leather || Schuhleder n. || cuir m. pour chaussures. / ~ leg factory || Schäftefabrik f. || fabrique f. de tiges de chaussures.

shoemaker || Schuster m.; Schuhmacher m. || cordonnier m. / master ~ || Schuhmachermeister m. || maître-cordonnier m.

shoemaker's iron || Schuhmacherbügeleisen n. || fer m. à repasser pour cordonniers. / ~ knife || Schustermesser n. || tranchet m. / ~ pitch || Schusterpech n. || poix f. de cordonnier. / ~ thread || Pechdraht m. || fil m. poissé. / ~ tool || Schuhmacherwerkzeug n. || outil m. pour cordonniers.

shoe making machine || Schuhmaschine f. || machine f. pour la fabrication des chaussures. / fittings pl. for ~s || Schuhmaschinenarmatur f. || armature f. pour machines à chaussures. / ~ operator || Schuhmaschinenarbeiter m. || ouvrier m. en chaussures.

shoe making shop || Schuhmacherei f. || cordonnerie f. / ~ manufacture || Schuhfertigung f. || fabrication f. des chaussures. / ~ nail || Schuhnagel m. || clou m. pour chaussures. / ~ nailer || Schuhnagler m. || cloueur m. en chaussures.

shoe ornament buckle || Schuhbesatzschnalle f. || boucle f. pour l'ornementation des chaussures. / ~ clasp see shoe ornament buckle.

shoe pin see ~ nail. / ~ polish see shoe cream. / ~ repairer || Flickschuster m. || savetier m. / ~-shaped angular fish plate || Schuh-

winkellasche f. || éclisse f. cornière à sabot. / article for ~ smithies || Hufbeschlagartikel m. || article m. pour maréchalerie. / ~ sole || Schuhsohle f. || semelle f. / ~ tack || Schuhzwecke f.; Schuhstift m. || clou m. de cordonnier ou à souliers; cheville f. pour chaussures. / ~ thread || Pechdraht m. || fil m. poissé. / ~ trimmings pl. || Schuhposamenten mpl. || passementerie f. pour souliers.

shoot, to (Join) || fügen || dresser; corroyer. / ~ (Mine) || schießen || sprengen || faire sauter les rocs mpl. / ~ the edge of a board || ein Brett n. fügen oder säumen; die Kante eines Brettes abhobeln || dresser une planche sur la tranche; corroyer la tranche d'une planche.

shoot (Agr) || Förderrinne f. || rigole f. d'alimentation. / ~ (Mine) || Rutsche f. || plan m. incliné. / ~ (Weav) || Schußfaden m.; Eintragfaden m. || duite f. / ~ of an arch || Seitenschub m. eines Bogens || poussée f. horizontale d'un arc. / shaking ~ (Mine) || Schüttelrutsche f. || transporteur m. ou couloir m. à secousse.

shooting || Pirschjagd f. || chasse f. à tir. / ~ and blasting (Mine) || Schießarbeit f.; Sprengarbeit f. || tirage m. à la poudre.

shooting articles pl., leather ~ || Jagdrequisiten pl. aus Leder || articles mpl. de chasse en cuir.

shooting lead || Jagdblei n. || plomb m. de chasse. / ~ pocket || Jagdtasche f.; Weidtasche f. || gibecière f. / ~ powder || Jagdpulver n. || poudre f. de chasse. / ~ stick || Keiltreiber m. || chasse-coins m. / ~ suit || Jagdanzug m. || vêtements mpl. de chasse.

shop || Werkstatt f. || atelier m. / ~ (Store) || Geschäft n.; Laden m. || magasin m. / converter ~ || Konverterhalle f. || hall m. des convertisseurs. / the ~ consists of a cross bay and three long bays || die Werkstatt besteht aus einem Querschiff und dre Längsschiffen || l'atelier m. se compose d'une travée transversale à laquelle sont adossées trois travées longitudinales. / the ~ covers a useful area of about x square metres || die Werkstatt f. bedeckt eine nutzbare Fläche von x Quadratmetern || la surface f. utile de l'atelier est de x mètres carrés. / erecting ~ of a dynamo work || Montagehalle f. eines Dynamowerkes || hall m. de montage d'une usine de dynamos. / forming ~ || Formationsraum m. || salle f. de formation. / ~ for iron structures || Eisenkonstruktionswerkstatt f. || atelier m. pour charpentes métalliques. / ~ for the manufacture of puddled iron || Puddelwerk n. || atelier m. de puddlage. / ~ for ready made clothes || Konfektionsgeschäft n. || maison f. de confection. / ~ || Reparaturwerkstatt f. || atelier m. de réparations. / super-turbine ~ || Großturbinenhalle f. || atelier m. de grosses turbines.

shop assistant || Ladenangestellter m.; Ladendiener m. || commis m. ou employé m. de magasin. / ~ bay see ~ nave. / ~ furniture || Ladenmöbel npl. || meubles mpl. de magasin. / ~ girl || Ladenmädchen n. || fille f. de boutique. / ~ girl (Tail) || Nähmädchen n. || petite main f. / installation for ~ || Ladeneinrichtung f. || installation f. de magasin. / ~ nave || Werkstattschiff n. || travée f. d'un hall.

shopping bag || Einkaufstasche f.; Einkaufsbeutel m. || sac m. pour emplettes. /

~ basket || Einkaufskorb m. || panier m. pour emplettes. / ~ net || Einkaufsnetz n. || filet m. pour emplettes.

shop room equipment || Geschäftseinrichtung f.; Ladeneinrichtung f. || installation f. de magasin. / ~ scales pl. || Tafelwage f. || balance f. de comptoir. / ~ sign of glass || Glasfirmenschild n.; Firmenschild n. aus Glas || plaque f. de réclame en verre; affiche f. en verre. / ~ test || Werkstattprüfung f. || épreuve f. ou essai m. à l'atelier.

shop window || Schaufenster n.; Ladenfenster n. || fenêtre f. d'étalage; devanture f. / ~ advertising || Schaufensterreklame f. || article m. de réclame pour étalages. / ~ cleaning || Schaufensterputzen n. || nettoyage m. de vitrines fpl. ou de devantures de boutiques. / ~ decorating material || Schaufensterdekorationsstoff m. || étoffe f. décorative pour étalages. / ~ decoration || Schaufensterdekoration f. || décoration f. d'étalages. / ~ decoration stand || Schaufensterdekorationsständer m. || support m. pour la décoration d'étalages. / ~ figure || Schaufensterfigur f. || figurine f. pour étalages. / ~ heater || Schaufensterwärmer m. || chauffevitre m. / ~ lighting || Schaufensterbeleuchtung f. || éclairage m. d'étalage ou de vitrine. / fittings pl. for ~ stands || Schaufenstergestellbeschläge mpl. || garnitures fpl. de supports pour étalages.

shore || Land n.; Küste f. || terre f. / along ~ || längs der Küste f. || le long de la côte. / low ~ || flache oder niedrige Küste f. || côte f. basse.

shore bearing || Landpeilung f. || relèvement m. d'une terre. / ~ lighting || Küstenbeleuchtung f. || éclairage m. des côtes. / ~ line || Brandungslinie f. || ligne f. de démarcation des eaux sur les côtes.

short || kurz || court. / ~ (Metal) || spröde || cassant. / cold ~ || kaltbrüchig || cassant à froid. / ~ bill || Wechsel m. auf kurze Sicht || papier m. court. / ~ carriage knitting machine || Kurzschlittenstrickmaschine f. || machine f. à tricoter à chariot court. / ~-nap hat || kurzhaariger Hut m. || chapeau m. à poil ras. / at ~ notice || mit kurzer Lieferfrist f. || à délai m. court de livraison. / ~ sale || Blankogeschäft n. || marché m. à découvert. / ~ stocking || Socke f. || chaussette f. / ~ stocking knitting || Sockenstricken n. || tricotage m. de chaussettes. / ~ table || Kurztisch m. || table f. courte. / ~ wave || kurze Welle f. || onde f. courte. / ~-wave condenser || Verkürzungskondensator m. || condensateur m. pour onde raccourcie. / ~-wave ray || kurzwelliger Strahl m. || rayon m. de faible longueur d'onde. / ~-wave receiver || Kurzwellenempfänger m. || récepteur m. à ondes courtes. / ~-wave transmitter || Kurzwellensender m. || émetteur m. à ondes courtes. / ~-wave wireless telegraphy || Kurzwellentelegrafie f. || télégraphie f. sans fil à ondes courtes. / ~ weight || Untergewicht n. || manque m. de poids.

shortage || Ausfall m.; Mindereinnahme f. || perte f.; manque m. / ~ (Supplies) || Knappheit f. || étroitesse f.; modicité f.; pénurie f. / coal ~ || Kohlenknappheit f. || disette f. de charbons. / ~ of work || Arbeitsmangel m. || manque m. de travail. / ~ of workers || Arbeitermangel m. || manque m. d'ouvriers.

short-circuit, to ‖ kurzschließen ‖ court-circuiter. / ~ the resistance ‖ den Widerstand m. kurzschließen ‖ mettre la résistance en court-circuit.

short-circuit ‖ Kurzschluß m. ‖ court-circuit m. / ~ between plates ‖ Plattenkurzschluß m. ‖ court-circuit m. entre plaques.

short-circuit armature ‖ Kurzschlußanker m. ‖ induit m. à court-circuit. / ~ brake ‖ Kurzschlußbremse f. ‖ frein m. à court-circuit. / ~ brush ‖ Kurzschlußbürste f. ‖ balai m. à court-circuit. / ~ characteristic ‖ Kurzschlußcharakteristik f. ‖ caractéristique f. de court-circuit. / ~ current ‖ Kurzschlußstrom m. ‖ courant m. de court-circuit. / ~ device ‖ Kurzschlußvorrichtung f. ‖ dispositif m. de mise en court-circuit.

short circuiting device ‖ Kurzschließer m. ‖ dispositif m. de mise en court circuit. / ~ for the rotor ‖ Ankerkurzschlußvorrichtung f. ‖ court-circuiteur m. d'induit.

short circuit key ‖ Kurzschlußtaste f. ‖ clé f. de mise en court-circuit. / ~ plug ‖ Kurzschlußstöpsel m. ‖ plot m. de mise en court-circuit. / ~ rotor ‖ Kurzschlußanker m. ‖ induit m. en court-circuit. / ~ rotor motor ‖ Kurzschlußläufermotor m. ‖ moteur m. asynchrone. / ~ slip-ring rotor ‖ Kurzschlußschleifringläufer m. ‖ induit m. à cage d'écureuil. / ~ voltage ‖ Kurzschlußspannung f. ‖ tension f. de court-circuit. / ~ winding ‖ Kurzschlußwicklung f. ‖ enroulement m. à court-circuit.

short-dated ‖ mit kurzem Ziel n. ‖ à courte échéance f.

short-distance receiving (Radio) ‖ Nahempfang m. ‖ réception f. régionale. / ~ sender (Radio) ‖ Nahsender m. ‖ émetteur m. rapproché ou régional. / ~ traffic ‖ Nahverkehr m. ‖ trafic m. à petite distance. / ~ transporter ‖ Nahfördermittel n. ‖ transporteur m. à petite distance.

shorten, to ‖ verkürzen ‖ raccourcir; abréger. / ~ the belt ‖ den Riemen m. verkürzen ‖ raccourcir la courroie. / ~ the life (Mach) ‖ die Lebensdauer verkürzen ‖ raccourcir la durée d'usage. / ~ sails ‖ Segel npl. kürzen ‖ diminuer de voiles fpl. / ~ the time of construction ‖ die Bauzeit verkürzen ‖ raccourir le délai m. de construction.

shortening of type bars ‖ Abkürzen n. der Typenhebel ‖ raccourcissement m. des tiges à caractères.

shortening condenser ‖ Verkürzungskondensator m. ‖ condensateur m. de raccourcissement.

shorthand ‖ Stenografie f.; Kurzschrift f. ‖ stenographie f. / to write ~ ‖ stenografieren ‖ sténographier.

shorthand typewriter ‖ Stenografiermaschine f. ‖ machine f. à sténographier. / ~ typist ‖ Stenotypist m. ‖ sténodactylographe m. / ~ writer ‖ Stenograf m. ‖ sténographe m.

short-necked (Bottle) ‖ kurzhalsig ‖ à col m. court.

short-period service ‖ kurzzeitiger Betrieb m. ‖ service m. de courte durée.

short-sighted ‖ kurzsichtig ‖ myope.

shortsightedness ‖ Kurzsichtigkeit f. ‖ myopie f.

short-wave ... see under short.

shot (Textile) ‖ schillernd ‖ changeant; glacé.

shot ‖ Schuß m. ‖ coup m.; tir m. / ~ (Arm) ‖ Geschoß n. ‖ projectile m. / ~ (Weav) ‖ Schützenschlag m.; Schuß m.; Schlag m. ‖ passée f. / blown-out ~ ‖ Versager m. ‖ coup m. raté. / chilled ~ (Arm) ‖ Hartgußgeschoß n.; Panzergeschoß n. ‖ projectile m. en fonte dure. / ~ of distress (Mar) ‖ Notschuß m.; Notsignal n. ‖ coup m. de détresse. / to fire a ~ ‖ einen Schuß m. abgeben ‖ tirer un coup. / flat-headed ~ (Arm) ‖ flachköpfiges Geschoß n. ‖ projectile m. cylindrique ou de rupture. / hollow ~ (Arm) ‖ Hohlgeschoß n. ‖ projectile m. creux. / the ~ is lost (Mine) ‖ der Schuß hat nicht gewirkt ‖ le trou n'a pas travaillé. / missed ~ ‖ Fehlschuß m. ‖ coup m. bleu ou manqué. / oblong ~ (Arm) ‖ Langgeschoß n. ‖ projectile m. allongé. / to set ~s pl. (Mine) ‖ Sprengschüsse mpl. wegtun oder abfeuern ‖ tirer les coups mpl. de mine. / small ~ ‖ Flintenschrot n. ‖ plomb m. granulé ou de chasse; dragée f.; grenaille f.

shot bearer (Arm) ‖ Geschoßtrage f. ‖ civière f. ou lanterne f. de chargement. / ~ chamber ‖ Geschoßkammer f.; Geschoßraum m. ‖ chambre f. du projectile. / ~-coloured see shot. / ~ counter (Weav) ‖ Schußzähler m. ‖ compte-passées m. / ~ firer (Mine) ‖ Schießmeister m. ‖ canonnier m.; boutefeu m. / ~ gauge ‖ Kaliberring m.; Kaliberlehre f. ‖ lunette f. à calibrer ou de réception. / ~ grinder ‖ Kugelmühle f. ‖ broyeur m. à boulets.

shotgun ‖ Jagdgewehr n. ‖ fusil m. de chasse. / ~ cartridge ‖ Jagdpatrone f. ‖ cartouche f. de chasse.

shot mould ‖ Gußschale f. ‖ coquille f. / ~ taffety ‖ Schillertaft m. ‖ taffetas m. changeant.

shotted (Arm) ‖ scharf geladen ‖ chargé à boulet. / ~ (Metal) ‖ körnig ‖ granulaire; grenu. / ~ tower (Metal) ‖ Schrottturm m. ‖ tour f. à fondre la dragée. / ~ velvet ‖ Schillersamt m. ‖ velours m. glacé.

shoulder ‖ Schulter f. ‖ épaule f. / ~ (Mach) ‖ Ansatz m. ‖ saillie f.; épaulement m. / ~ of the axle collar ‖ Anlauf m. am Achsbund ‖ congé m. du collet de l'essieu. / ~ on crankshaft ‖ Kurbelwellenbund m. ‖ collet m. du vilebrequin; collet m. d'arbre à manivelle. / ~ of the letter ‖ Achsel f. des Buchstabens ‖ épaulement m. de la lettre. / ~ of rim ‖ Felgenschulter f. ‖ épaulement m. de la jante. / tapered ~ ‖ konischer Ansatz m. ‖ collet m. conique. / ~ of the tie plate (Railw) ‖ Randleiste f. der Unterlagsplatte ‖ rebord m. de la selle d'appui. / ~ of a tooth ‖ Zahnfuß m. ‖ pied m. d'une dent.

shoulder-blade ‖ Schulterblatt n. ‖ omoplate f.

shouldered T-head on the bolt ‖ Schaftansatz m. an der Hakenschraube ‖ mentonnet m. de crampon à vis.

shouldering wall ‖ Schultermauer f. ‖ épaulement m.

shoulder piece (Hoist) ‖ Frosch m. ‖ échantignole f. / ~ strap ‖ Epaulette f. ‖ épaulette f.

shovel, to ‖ ausschaufeln ‖ enlever avec la pelle.

shovel ‖ Schaufel f. ‖ pelle f. / ~ (Of a dredger) ‖ Löffel m. ‖ cuiller f.; cuillère f. / bent ~ (Mine) ‖ Kratze f. ‖ drague f. / blast furnace ~ ‖ Hochofenschaufel f. ‖ pelle f. de haut fourneau. / coal ~ ‖ Kohlenschaufel f. ‖ pelle f. à charbon. / corn ~ ‖ Kornschaufel f.; Getreideschaufel f. ‖ pelle f. à grains ou à blé. / crooked ~ see bent ~. / ~ with Diesel engine drive ‖ Diesellöffelbagger m. ‖ pelle f. Diesel. / ~ provided with drag line equipment ‖ mit Schleppschaufeleinrichtung f. versehener Löffelbagger ‖ pelle f. possédante un équipement de dragline. / grape ~ ‖ Traubenschaufel f. ‖ pelle f. à raisins. / hollow furnace ~ ‖ runde Kohlenschaufel f. ‖ pelle f. à foyer ronde. / liquid manure ~ ‖ Jaucheschöpfer m. ‖ pelle f. à purin. / malt ~ ‖ Malzschaufel f. ‖ pelle f. à malte. / mud ~ ‖ Schlammschaufel f. ‖ pelle f. à vase. / pointed nose ~ with long socket ‖ Schaufel f. mit langer Spitztülle ‖ pelle f. à longue douille pointue. / ~ fitted with rigidly guided crawlers ‖ Löffelbagger m. mit starr geführten Raupenketten ‖ pelle f. à chenilles guidées d'une façon rigide. / ~ fitted with spring crawlers ‖ Löffelbagger m. mit abgefederten Raupenketten ‖ pelle f. pourvue des chenilles suspendues sur ressorts. / square foundry ~ ‖ Formerschaufel f. ‖ pelle f. carrée de fonderie. / square furnace ~ ‖ flache Kohlenschaufel f. ‖ pelle f. à foyer carrée. / steam ~ ‖ Dampflöffelbagger m.; Löffelbagger m.; Trockenbagger m. ‖ pelle f. ou excavateur m. ou cuiller f. ou terrassier m. à vapeur; piocheuse f. / to turn with a ~ ‖ umschaufeln ‖ pelleter. / wooden ~ ‖ Holzschaufel f. ‖ pelle f. en bois.

shovel crane ‖ Kranschaufler m. ‖ pelle f. mécanique. / ~ dredger ‖ Schaufelbagger m. ‖ drague f. à pelles.

shovelling apparatus, portable ~ ‖ fahrbarer Kratzer m. ‖ excavateur m. mobile à câble de traction.

shovelling machine ‖ Schaufelwurfmaschine f. ‖ machine f. pour travaux de terrassement.

shovel maker ‖ Schaufelmacher m. ‖ emboutisseur m. de pelles. / machine for making ~s ‖ Schaufelherstellungsmaschine f. ‖ machine f. à fabriquer les pelles. / ~ stick ‖ Schaufelstiel m. ‖ manche m. de pelle.

show, to (Instrument) ‖ anzeigen ‖ indiquer.

show ‖ Schau f.; Ausstellung f. ‖ vue f.; inspection f.; revue f.; exposition f.; étalage m. / ~ (Ware) ‖ Aushang m.; Auslage f. ‖ devanture f.; étalage m. / ~ card ‖ Plakat n. ‖ tableau-réclame. / ~ case ‖ Schaukasten m. ‖ vitrine f.

shower, to ‖ abbrausen ‖ arroser.

shower (Bath) ‖ Brause f. ‖ douche f. / ~-action condensing ‖ Berieselungskondensation f. ‖ condensation f. par arrosage extérieur en pluie d'eau. / ~ bath ‖ Brausebad n. ‖ bain-douche m.; douche f. / ~ cooling ‖ Rieselkühlung f. ‖ rafraîchissement m. à pluie d'eau.

showering sieve ‖ Brausesieb n.; Bewässerungssieb n. ‖ tamis m. d'arrosage ou de rinçage.

showroom ‖ Ausstellungsraum m.; Ausstellungshalle f. ‖ salle f. de démonstration ou d'exposition. / ~ pattern ‖ Aus-

stellungsmodell n. ‖ modèle m. d'exposition.

show window ‖ Schaufenster n. ‖ fenêtre f. d'étalage; devanture f.; vitrine f. / ~ equipment ‖ Schaufenstereinrichtung f. ‖ installation f. de devanture. / ~ lamp ‖ Schaufensterlampe f. ‖ lampe-vitrine f.

shrapnel ‖ Schrapnell n. ‖ shrapnell m. / ~ bomb ‖ Splitterbombe f. ‖ bombe f. à shrapnell.

shred ‖ Lappen m. ‖ chiffon m.; loque f.; guenille f.; lobe m.

shredder (Pap) ‖ Lumpenzurichterin f. ‖ délisseuse f.

shredding machine, turnip ~ ‖ Rübenschnitzelmaschine f. ‖ machine f. à couper les betteraves en tranches.

shrimp ‖ Garnele f. ‖ crevette f. / ~ fisherman ‖ Garnelenfischer m. ‖ pêcheur m. de crevettes.

shrine ‖ Schrank m. ‖ armoire f.

shrink, to (Metal) ‖ schrumpfen ‖ (se) contracter. / ~ (Pott) ‖ schwinden ‖ s'amaigrir; se rétrécir. / ~ (Textile) ‖ einlaufen ‖ rétrécir; se gripper. / ~ the cloth ‖ Stoff m. dekatieren ‖ décatir le drap.

shrink, to ~ on (Forg) ‖ warm aufziehen ‖ emmancher à chaud. / ~ the crank ‖ die Kurbel warm aufziehen ‖ emmancher la manivelle à chaud sur l'arbre. / ~ a ring ‖ einen Ring m. aufschrumpfen ‖ emmancher ou fretter un anneau à chaud.

shrinkage (Metal) ‖ Schwund m. ‖ retrait m. / ~ (Textile) ‖ Einlaufen n. ‖ rentrée f. / ~ of clay ‖ Schwinden n. des Tons ‖ retraite f. de l'argile. / ~ of an embankment ‖ Setzen n. eines Dammes ‖ rechargement m. d'un remblai. / ~ of metal ‖ Schwinden n. des Metalles ‖ retraite f. ou retrait m. du métal. / the tyres were heated too high for ~ ‖ die Reifen mpl. wurden beim Aufziehen zu stark erwärmt ‖ on chauffait les bandages mpl. trop fort pour le serrage à chaud. / ~ of wood ‖ Schwinden n. des Holzes ‖ retraite f. du bois.

shrinkage allowance ‖ Schrumpfmaß n. beim Aufziehen ‖ retrait m. pour le serrage à chaud.

shrinker, cloth ~ ‖ Tuchdekatör m. ‖ décatisseur m. de drap.

shrinking see also shrinkage ‖ Schwinden n. ‖ contraction f. / ~ machine (Clothm) ‖ Dekatiermaschine f. ‖ machine f. à décatir. / ~-on ‖ Warmaufziehen n. ‖ serrage m. à chaud. / ~ stress ‖ Schrumpfspannung f. ‖ effort m. de contraction.

shrink ring ‖ Schrumpfring m. ‖ bague f. de serrage; frette f. posé à chaud ou calée à chaud. / ~ water ‖ Schwindwasser n. ‖ eau f. de retrait.

shroud (Shipb) ‖ Want n. ‖ hauban n. / ~ (Build) ‖ Lenkseil n.; Schwenkseil n.; Leitseil n.; Schwungseil n. ‖ hauban m.; écharpe f. / main ~ (Mar) ‖ Großwant n. ‖ hauban m. du grand mât.

shrouding of a water wheel ‖ Radboden m. eines Wasserrades ‖ plancher m. d'une roue hydraulique.

shrouds and burial linen ‖ Toten- und Sterbewäsche f. ‖ linge m. de morts.

shrub ‖ Strauch m. ‖ arbrisseau m.

shrunk-on collar ‖ aufgeschrumpfter Bundring m. ‖ collet m. rapporté. / ~ hoop ‖ Schrumpfband n.; Schrumpfring m. ‖ frette f. posée à chaud. / ~ tyre ‖ aufgeschrumpfter Reifen m. ‖ bandage m. posé à chaud.

shumac ‖ Farbholz n. ‖ bois m. de sumac.

shunt, to (Electr) ‖ nebenschließen ‖ shunter; dériver. / ~ (Railw) ‖ rangieren; verschieben ‖ manœuvrer les wagons.

shunt (Electr) ‖ Nebenschluß m. ‖ shunt m.; dérivation f. / electromagnetic ~ (Tel) ‖ Gegenstromrolle f. ‖ dérivation f. électromagnétique. / highly inductive ~ ‖ Nebenschluß m. mit hoher Selbstinduktion ‖ shunt m. à pouvoir inductif élevé. / non-inductive ~ ‖ induktionsfreier Nebenschluß m. ‖ shunt m. non inductif.

shunt admittance type equalization ‖ Querentzerrung f. ‖ compensation f. de la distorsion en parallèle. / ~ brake ‖ Nebenschlußbremse f. ‖ frein m. shunt ou de dérivation. / ~ circuit ‖ Nebenschlußstromkreis m. ‖ circuit m. dérivé. / ~ coil ‖ Nebenschlußspule f. ‖ bobine f. shunt. / ~ connection ‖ Parallelschaltung f. ‖ connexion f. en parallèle. / ~ current ‖ Zweigstrom m.; Nebenschlußstrom m. ‖ courant m. shunt. / ~ dynamo ‖ Nebenschlußdynamo f. ‖ dynamo-shunt f.

shunted buzzer ‖ Nebenschlußsummer m. ‖ vibrateur m. à dérivation. / ~ condenser (Tel) ‖ Maxwellerde f. ‖ terre f. de Maxwell.

shunt field coil ‖ Nebenschlußfeldspule f. ‖ bobine f. shunt de champ.

shunting station ‖ Rangier- oder Verschiebebahnhof m. ‖ gare f. de manœuvre. / ~ track ‖ Rangiergleis n. ‖ voie f. de service. / ~ trolley ‖ Rangierkarren m. ‖ wagonnet m. de manœuvre. / ~ winch ‖ Rangierwinde f. ‖ treuil m. de manœuvre de wagon. / ~ yard (Railw) ‖ Rangieranlage f. ‖ installation f. de manœuvre.

shunt method ‖ Nebenschlußmethode f. ‖ méthode f. de dérivation. / alternating current ~ motor ‖ Wechselstromnebenschlußmotor m. ‖ moteur m. à collecteur en dérivation. / ~ regulating resistance ‖ Nebenschlußregulierwiderstand m. ‖ résistance f. de réglage shunt. / ~ regulation ‖ Nebenschlußregelung f. ‖ réglage m. shunt. / ~ regulator ‖ Nebenschlußregler m. ‖ régulateur m. shunt. / ~ resistance ‖ Abzweigwiderstand m. ‖ résistance f. en dérivation; résistance f. du circuit dérivé. / ~ switch ‖ Umgehungsschalter m. ‖ interrupteur m. shunt. / ~-type arc lamp ‖ Nebenschlußbogenlampe f. ‖ lampe f. à arc en dérivation.

shunt-wound ‖ als Nebenschluß m. gewickelt ‖ enroulé ou bobiné en dérivation f. / ~ regulator ‖ Nebenschlußregler m. ‖ rhéostat-régulateur m. en dérivation.

shut, to ~ the diaphragm (Cash register) ‖ die Blende f. schließen ‖ fermer le diaphragm. / ~ down (Mach) ‖ zum Stillstand m. bringen; abstellen ‖ mettre au repos.

shut, to ~ off (Electr) ‖ unterbrechen ‖ interrompre; couper. / ~ a cock ‖ einen Hahn m. schließen ‖ fermer un robinet. / ~ the current ‖ den Strom m. unterbrechen ‖ couper le courant. / ~ the draft ‖ den Wind m. abstellen ‖ arrêter le vent. / ~ gas ‖ Gas n. abdrosseln ‖ étrangler le gaz. / ~ the pipe line ‖ die Leitung absperren ‖ fermer ou couper la conduite. / ~ the steam ‖ den Dampf m. absperren ‖ couper la vapeur.

shut (Welding) ‖ Schweißstelle f. ‖ soudure f.

shut-off device ‖ Absperrvorrichtung f. ‖ appareil m. de fermeture. / ~ valve ‖ Absperrventil n. ‖ soupape f. d'arrêt.

shutter (Camera) ‖ Verschluß m. ‖ obturateur m. / ~ (Found) ‖ Spund m. ‖ écluse f. / ~ (Railw) ‖ Schalter m.; Zahlschalter m. ‖ guichet m. / ~ (Window) ‖ Jalousie f. ‖ persienne f.; jalousie f. / automatic door ~ ‖ selbsttätiger Türschließer m. ‖ ferme-porte m. automatique. / instantaneous ~ (Phot) ‖ Momentverschluß m. ‖ obturateur m. instantané. / iron ~ ‖ eiserner Schutzladen m. ‖ devanture f. en fer. / ~ of the key hole ‖ Schlüssellochdeckel m. ‖ cache-entrée m. de la clé. / spring for revolving ~s (Window) ‖ Rolladenfeder f. ‖ ressort m. de jalousie f. / roll-up ~ ‖ Rolljalousie f. ‖ jalousie f. / time ~ (Phot) ‖ Zeitverschluß m. ‖ obturateur m. pour la pose. / Venetian ~ ‖ (Holz-)Jalousie f. ‖ jalousie f. (en lames de bois).

shutter drop (Tel) ‖ Fallklappe f. ‖ annonciateur m. avec clapet. / ~ release (Phot) ‖ Verschlußauslösung f. ‖ déclenchement m. de l'obturateur. / ~ sheet ‖ Jalousieblech n. ‖ lame f. de persienne.

shutting-off ‖ Abschluß m. ‖ arrêt m.

shuttle (Weav) ‖ Webschützen m.; Weberschiffchen n. ‖ navette f. / pirn ~ (Weav) ‖ Schützen m. mit Laufspule ‖ navette f. à dérouler. / ~ with spring rolling back the thread ‖ Schützen m. mit Federspannung ‖ navette f. rétrograde ou à retrait ou à renvidage. / wooden ~ ‖ Holzschützen m.; Holzweberschiffchen n. ‖ navette f. en bois.

shuttle armature (Electr) ‖ doppel-T-Anker m. ‖ armature f. en double T. / ~ box (Weav) ‖ Schützenkasten m. ‖ boîte f. à navette. / ~ conveyor ‖ Schüttelrinne f. ‖ gouttière f. à secousses. / ~ driver (Weav) ‖ Schützentreiber m. ‖ lanceur m. en tissus. / ~ guard (Weav) ‖ Schützenfänger m. ‖ garde-navette m. / ~ embroiderer or ~ loom ‖ Schiffchensticker m. ‖ brodeur m. au métier à navette. / ~ maker (Weav) ‖ Schützenmacher m. ‖ navetier m. / ~ net ‖ Schützennetz n. ‖ parenavette m. / ~ spool ‖ Weberschiffchenspule f. ‖ épeule f. de navette.

siccative (Paint; Print) ‖ Trockenmittel n.; Sikkativ n. ‖ siccatif m. / ~ extract ‖ Trockenmittelauszug m.; Sikkativextrakt m. ‖ extrait m. siccatif. / ~ oil ‖ Trockenöl n. ‖ huile f. siccative. / ~ powder ‖ Trockenpulver n.; Sikkativpulver n. ‖ siccatif m. en poudre. / ~ varnish ‖ Trockenfirnis m. ‖ vernis m. siccatif.

sick ‖ krank ‖ malade. / ~ (Ship) ‖ kopfschwer; rank ‖ faible du côté; volage. / seriously ~ ‖ schwerkrank; schwer erkrankt ‖ gravement malade. / slightly ~ ‖ leicht erkrankt ‖ légèrement malade.

sick fund ‖ Krankenkasse f. ‖ caisse f. d'assurance pour malades.

sickle see also scythe ‖ Sichel f.; Hippe f. ‖ faucille f. / ~ bill ‖ kleine Hippe f. ‖ serpillon f. / ~ smith ‖ Sichelschmied m. ‖ forgeur m. de faucilles.

sickness ‖ Krankheit f. ‖ maladie f.

sick room ‖ Krankenzimmer n.; Krankenraum m. ‖ chambre f. des malades.

side ‖ Seite f. ‖ côté m. / ~ of an angle ‖ Schenkel m. eines Winkels ‖ côté m.

d'un angle. / ~ of a barrage (Hydr arch) ‖ Hang m. einer Talsperre ‖ coteau m. d'un barrage. / ~ of belt (Mach) ‖ Trum n. ‖ brin m. / on both ~s pl. ‖ beiderseitig ‖ des deux côtés fpl. / ~ of a cone ‖ Seite f. eines Kegels ‖ génératrice f. d'un cône. / ~ of cross-head (Mach) ‖ Gleitbacken mpl. oder Gleitklotz m. bei der Geradführung ‖ coulisseau m. des glissoires; patin m. / ~ of a double reverberatory furnace ‖ hintere Seite f. eines Doppelflammofens ‖ face f. d'un fourneau à réverbère double. / ~ of the drive ‖ Antriebseite f. ‖ côté m. de la commande. / on each ~ ‖ an jeder Seite f. ‖ de ou à chaque côté m. / sectional iron with equal ~s ‖ gleichschenkliges Winkeleisen n. ‖ cornière f. à ailes égales. / ~ of an equation (Math) ‖ Seite f. einer Gleichung ‖ membre m. d'une équation. / ~ of a figure (Geom) ‖ Seite f. einer ebenen Figur ‖ côté m. d'une figure. / flat ~ of a hammer ‖ Hammerbahn f. ‖ panne f. ou plat m. ou table f. d'un marteau. / frontal ~ of a stone (Mas) ‖ Hauptseite f. oder Kopfseite f. eines Steines ‖ panneau m. de tête ou parement m. d'une pierre. / good ~ of a stuff ‖ rechte Seite f. oder Schönseite f. des Stoffes ‖ beau côté m. ou endroit m. d'une étoffe. / joint ~ (Shipb) ‖ Mallebene f. ‖ plan m. de gabariage ou du couple. / left ~ of an equation ‖ linke Seite f. einer Gleichung ‖ premier membre m. d'un équation. / moulding ~ (Shipb) see joint ~. / narrow ~ ‖ hohe Kante f.; schmale Seite f. ‖ carne f.; champ m.; tranche f. / on the narrow ~ ‖ hochkant; auf der schmalen Seite f. ‖ de champ m. / right ~ of an equation ‖ rechte Seite f. einer Gleichung ‖ second membre m. d'une équation. / right ~ of a stuff ‖ rechte Seite f. oder Schönseite f. eines Stoffes ‖ beau côté m. ou endroit m. d'une étoffe. / screwed-on at the ~ ‖ seitlich angeschraubt ‖ boulonné latéralement. / ~ of a shaft (Mine) ‖ Schachtstoß m. ‖ face f. ou paroi f. d'un puits. / ~ of a ship ‖ Seite f. eines Schiffes ‖ côte m. ou flanc f. d'un navire. / ~ of a stone (Mas) ‖ Fläche f. oder Fuge f. oder Seite f. eines Steines ‖ panneau m. d'une pierre. / ~ of the wagon (Railw) ‖ Wagenwand f. ‖ paroi f. d'un wagon. / ~ of a wedge ‖ Seitenfläche f. eines Keils ‖ face f. latérale ou côté m. d'un coin.

side aisle (Arch) ‖ Nebenschiff n.; Seitenschiff n. ‖ nef f. latérale ou basse; petite nef f.; collatéral m.; bas-côté m.; contre-allée f. / ~ arm ‖ blanke Waffe f. ‖ arme f. blanche. / ~ awnings (Railw) ‖ Wagenvorhang m. ‖ rideau m. / ~ axle compressor ‖ Seitenpresser m. ‖ presseur m. latéral. / ~ bearing ‖ Längslager n. ‖ palier m. longitudinal. / ~ board ‖ Schenktisch m.; Anrichte f.; Geschirrschrank m.; Büfett n. ‖ buffet m.; dressoir m. / ~ boat ‖ Seitenboot n. ‖ embarcation f. de côté. / ~ buffer (Railw) ‖ Seitenpuffer m. ‖ tampon m. latéral. / ~ car ‖ Seitenwagen m.; Beiwagen m. ‖ side-car m.; voiturette f. de remorque latérale. / ~ carriage see ~ car. / ~ chain (Railw) ‖ Notkette f.; Sicherheitskette f. ‖ chaîne f. de sûreté. / ~ channel (Road) ‖ Seitengraben m.; Seitenrinne f. ‖ canal m. latéral. / ~ circuit (Tel) ‖

Stammleitung f. ‖ ligne f. de base. / ~ circuit coil ‖ Stammspule f. ‖ bobine f. réelle. / ~ counter-timber (Shipb) ‖ Heckstütze f.; Windvierungsstütze f. ‖ cornière f.; allonge f. de cornière. / ~ cutting (Road; Earthw) ‖ Seitenentnahme f.; Füllgrube f. ‖ emprunt m. (de la terre). / ~ cutting nippers pl. ‖ Seitenschneider m. ‖ pince f. coupante de côté. / ~ delivery rake ‖ Schwadenrechen m. ‖ râteau m. à andain. / ~ door ‖ Seitentür f. ‖ entrée f. ou portière f. latérale. / ~ dump tip box wagon ‖ einseitiger Kastenkipper m. ‖ wagonnet m. à caisse basculante d'un seul côté. / ~ elevation see ~ face. / ~ face (Of a machine etc.) ‖ Seitenansicht f. ‖ profil m.; vue f. de côté. / ~ face (Of a crystal) ‖ Seitenfläche f. ‖ face f. latérale. / ~ face (Build) ‖ Seitenfront f. ‖ façade f. de côté. / ~ fin (Airpl) ‖ Seitenflosse f. ‖ quille f. de dérive. / ~ gable (Build) ‖ Seitengiebel m. ‖ traversier m. / ~ girder ‖ Längsträger m. ‖ longeron m. / suspended on one of the ~ girders ‖ an einem der Längsträger aufgehängt ‖ suspendu à un des longerons. / ~ gutter (Road) ‖ Seitengraben m.; Seitenrinne f. ‖ canal m. latéral. / ~ handle ‖ Seitengriff m. ‖ poignée f. latérale. / ~ hole of a pump (Shipb) ‖ Pumpengatt n. ‖ lumière f. de pompe. / ~ hook ‖ Spannkluppe f. ‖ mordache f.; crampon m. / ~ jaw ‖ Maulkeil m. ‖ joue f. latérale. / ~ joint ‖ senkrechte Fuge f.; Stoßfuge f. ‖ joint m. montant ou vertical. / ~ lamp ‖ Seitenlampe f.; Seitenlaterne f. ‖ lampe f. de côté; lanterne f. latérale ou de côté. / ~ lantern see ~ lamp. / ~ light (Nav) see also side lamp ‖ Positionslicht n.; Seitenlicht n. ‖ feu m. de route ou de côté. / ~ light (Shipb: window) ‖ Ochsenauge n.; Seitenlicht n.; Seitenfenster ‖ hublot m.; ventouse f.; jour m. latéral. / high ~ light (Build) ‖ Seitenoberlicht n. ‖ jour m. d'en haut. / ~ nail of a gun-lock ‖ Schloßschraube f. ‖ vis f. de platine. / ~ note (Print) ‖ Marginale n.; Randbemerkung f.; Randglosse f. ‖ glosse f. ou note f. marginale; manchette f. / ~ part ‖ Seitenteil n. ‖ partie f. latérale. / ~ piece ‖ Seitenstück n. ‖ pièce f. latérale. / ~ plate ‖ Seitenplatte f. ‖ plaque f. latérale. / ~ pond (Hydr arch) ‖ Sparteich m. ‖ bassin m. d'épargne. / ~ projection (Draw; Build) ‖ Profil n.; Schnitt m.; Querschnitt m.; Seitenriß m. ‖ profil m.; coupe f.; projection f. latérale. / ~ pull ‖ seitlich auftretender Zug m. ‖ force f. latérale. / ~ rail (Railw) ‖ Schutzschiene f.; Leitschiene f.; Sicherheitsschiene f. ‖ contre-rail m.

sidereal day ‖ Sterntag m. ‖ jour m. sidéral. / ~ time ‖ Sternzeit f. ‖ temps m. sidéral.

side rest (On a kitchen range) ‖ Abstellplatte f. ‖ plateau m. de côté.

side ring (Mill) ‖ Kippring m. ‖ couronne f. latérale.

siderite ‖ Spateisenstein m.; Eisenspat m.; Siderit m. ‖ sidérose f.; sidérite f.; fer m. spathique.

side rod (Loc) ‖ Kuppelstange f. ‖ bielle f. d'accouplement, / rear ~ ‖ hintere Kuppelstange f. ‖ bielle f. d'accouplement arrière.

side roll ‖ Seitenwalze f. ‖ cylindre m. latéral.

sideroscope ‖ Sideroskop n. ‖ sidéroscope m.

side rudder cable ‖ Seitensteuerkabel n. ‖ câble m. de commande du gouvernail de direction. / ~ screw ‖ Seitenschraube f. ‖ vis f. latérale. / ~ seam soldering machine ‖ Längsnahtlötmaschine f. ‖ machine f. à souder en long. / ~ shed ‖ Seitenbau m.; Seitenhalle f. ‖ hall m. latéral. / ~ sheet of a firebox ‖ Seitenwand f. einer Feuerbuchse ‖ flanc m. d'une boîte à feu.

side-slip, to ‖ schleudern ‖ déraper.

side slip (Of a vehicle) ‖ Schleudern n. ‖ dérapage m. / ~ street ‖ Seitenstraße f. ‖ chemin m. latéral ou retiré; voie f. détournée. / ~ switch (Tel) ‖ Seitenschalter m. ‖ commutateur m. latéral. / ~ tip truck ‖ Seitenkipper m. ‖ camion m. basculant de côté. / ~ tone (Tel) ‖ Nebenton m. ‖ bruit m. / ~ tones pl. (Tel) ‖ Nebengeräusche npl. ‖ bruits mpl. / ~ track (Railw) ‖ Nebengleis n.; Seitengleis n. ‖ voie f. secondaire ou accessoire ou de garage ou de service. / ~ tube ‖ Ansatzrohr n. ‖ tube m. latéral. / ~ valley ‖ Quertal n. ‖ vallée f. transversale. / ~ view ‖ Seitenansicht f. ‖ élévation f. longitudinale; vue f. de côté.

sidewalk ‖ Bürgersteig m.; Gehbahn f. ‖ trottoir m.

side wall ‖ Seitenwand f.; Seitenmauer f. ‖ paroi f. latérale ou de côté. / ~s pl. of a furnace ‖ Seitenmauern fpl. eines Ofens ‖ doublure f. d'un fourneau. / ~ of a lock ‖ Kammerwand f. einer Schleuse; Schleusenmauer f.; Schleusenwand f. ‖ bajoyer m. ou bajoyère f. d'une écluse. / removable ~ ‖ abnehmbare Seitenwand f. ‖ paroi f. de côté démontable ou amovible.

side way (Railw) ‖ Weiche f.; Abzweigung f. ‖ aiguille f.; raccordement m.

side wedge ‖ Seitenkeil m. ‖ coin m. latéral.

side wing (Build) ‖ Seitenflügel m. ‖ aile f. latérale. / ~ (Auto) ‖ Seitenwindschutzscheibe f. ‖ pare-brise m. de côté.

siding (Railw) ‖ Nebengleis n.; Anschlußgleis n. ‖ voie f. de garage; embranchement m. de voie. / standard-gauge ~ ‖ Vollspurgleisanschluß m. ‖ embranchement m. de chemin de fer à voie normale.

siding line ‖ Nebenlinie f.; Zweigbahn f. ‖ ligne f. latérale ou secondaire. / ~ track ‖ Gleisanschluß m. ‖ raccordement m. de vioe. / ~ way (Railw) ‖ Ausweichgleis n. ‖ voie f. de garage m.

siegenite (Miner) ‖ Siegenit m.; Kobaltnickelkies m. ‖ siégénite f.

Siemens-Martin steel ‖ Siemens-Martin-Stahl m. ‖ acier m. Siemens-Martin. / ~ works pl. ‖ Siemens-Martin-Stahlwerk n. ‖ aciérie f. Siemens-Martin.

Siemens high-speed telegraph printer ‖ Siemens-Schnelltelegraf m. ‖ appareil m. télégraphique automatique Siemens.

sienna ‖ Sienaerde f.; Siena f. ‖ (terre f. de) Sienne f.

sieve, to ~ see also to sift ‖ sieben; durchsieben; absieben ‖ tamiser; cribler; passer au crible; sasser. / ~ an ore ‖ ein Erz durch das Sieb setzen ‖ cribler un minerai.

sieve ‖ Sieb n. ‖ tamis m.; crible m. / double ~ (Ore dress; Coal dress) ‖ Doppelplanrätter m. ‖ double tamis m; sasseur m. double. / ~ with elongated holes ‖ Sieb n. mit länglichen Löchern ‖ crible m. muni de trous longitudinaux. / first ~

Schrotsieb n.; grobes Sieb n. ‖ guillaume m. / flat ~ ‖ Flachsieb n. ‖ tamis m. plat. / grain ~ ‖ Körnersieb n. ‖ tamis m. à graines. / hand ~ ‖ Handsieb n. ‖ crible m.; tamis m. à main. / horsehair ~ ‖ Haarsieb n. ‖ tamis m. de crin. / mechanical ~ ‖ mechanisches Sieb n. ‖ tamis m. mécanique. / metal ~ ‖ Metallsieb n.; Drahtsieb n. ‖ crible m. en toile métallique. / oat ~ ‖ Hafersieb n. ‖ crible m. à avoine. / ~ for corning powder ‖ Pulverkornsieb n. ‖ grenoir m.; grainoir m.; égalisoir m. / sand ~ ‖ Sandsieb n.; Durchwurf m. ‖ tamiseur m.; tamis m. à sable ou de passage; crible m. à pied. / shaking ~ ‖ Schüttelsieb n. ‖ sasseur m.; cribleur m.; tamis m. à secousses. / showering ~ ‖ Bewässerungssieb n. ‖ tamis m. d'arrosage. / standard ~ ‖ Einheitssieb n. ‖ tamis m. normal. / ~ of a stuff engine (Pap) ‖ Waschscheibe f. ‖ châssis laveur m. du chapiteau du moulin à cylindre. / wide-meshed ~ ‖ Grobsieb n. ‖ tamis m. grossier. / wire-gauze ~ ‖ Drahtsieb n. ‖ tamis m. de toile métallique.

sieve bottom of wire-gauze ‖ Siebboden m. aus Drahtgeflecht ‖ fond m. de tamis en toile métallique. / ~ brush ‖ Siebbürste f. ‖ brosse f. à tamis. / ~ drum ‖ Siebtrommel f. ‖ tambour m. tamiseur. / ~ frame ‖ Siebrahmen m. ‖ cadre m. à tamis. / ~ grate ‖ Siebrost m. ‖ grille f. tamiseur. / ~ hoop ‖ Siebzarge f.; Siebrand m. ‖ cercle m. ou rebord m. d'un tamis. / ~ maker ‖ Siebmacher m. ‖ tamisier m. / ~ maker's goods pl. ‖ Siebmacherwaren fpl. ‖ articles mpl. de tamiserie. / ~ netting ‖ Siebgewebe n. ‖ tissu m. à tamiser. / ~ residue ‖ Siebrückstand m. ‖ refus m. de tamisage. / set of ~s ‖ Siebsatz m. ‖ série f. de tamis. / ~ sheet ‖ Siebblech n. ‖ tôle f. à tamisage. / ~ shovel ‖ Siebschaufel f. ‖ pelle f. à grille. / size of ~ ‖ Siebgröße f. ‖ numéro m. de tamis. / ~ wire ‖ Siebdraht m. ‖ fil m. de fer pour tamis. / thickness of ~ wire ‖ Drahtstärke f. des Siebes ‖ grosseur f. du fil de fer du tamis. / wood for ~s ‖ Siebholz n. ‖ bois m. de tamis.

sift, to ~ see also to sieve ‖ sieben; durchsieben ‖ tamiser; passer au tamis; cribler. / ~ (Mill) ‖ sichten; beuteln ‖ bluter. / ~ sand ‖ den Sand sieben ‖ passer le sable au panier.

sifted (Ore and Coal dress) ‖ gesiebt; gerättert ‖ classé.

sifter (Mill: apparatus) ‖ Schüttelsieb n.; Rätter m. ‖ sasseur m.; cribleur m. / ~ (Mill: person) ‖ Sichtmaschinenwärter m. ‖ sasseur m. / ~ loft ‖ Sichterboden m. ‖ salle f. de blutage.

sifting (Mill) ‖ Sichterei f. ‖ sassage m.

sifting cylinder ‖ Siebzylinder m. ‖ cylindre m. tamiseur. / ~ drum ‖ Rundsieb n. ‖ tamis m. circulaire. / ~ machine ‖ Beutelmaschine f. ‖ bluterie f.; machine f. à tamiser. / selecting and ~ machinery ‖ Auslese- und Sichtmaschine f. ‖ machine f. à trier et à assortir.

sifting plant ‖ Sieberei f. ‖ installation f. de tamisage. / coal ~ ‖ Kohlensieberei f. ‖ installation f. à cribler en charbon. / flour ~ ‖ Mehlsiebanlage f. ‖ installation f. à tamiser la farine. / ~ for the pit coal dressing ‖ Siebanlage f. oder Sieberei f. für die Steinkohlenaufbereitung ‖ installation f. de criblage ou cribleur m. pour la préparation de la houille.

siftings pl. (Mill) ‖ Siebsel n.; Siebmehl n. ‖ criblure f.; tamisure f.

sifting wagon for coke sifting ‖ Kokssiebwagen m. ‖ wagon m. de criblage de coke.

sight, to ~ the land (Nav) ‖ Land n. in Sicht bekommen ‖ découvrir la terre; attérer; atterrer. / ~ out (Surv) ‖ abvisieren; abfluchten ‖ aligner; jalonner.

sight (Aspect) ‖ Ansicht f.; Anblick m. ‖ vue f.; aspect m. / ~ (Visual power) ‖ Sehvermögen n.; Sehkraft f.; Gesicht n. ‖ facultés fpl. visuelles; vue f. / to draw by ~ ‖ nach dem Augenmaß n. zeichnen ‖ dessiner à vue f. / panoramic ~ (Arm) ‖ Rundblickvisiervorrichtung f.; Rundblickaufsatz m. ‖ hausse f. panoramique. / weak ~ (Opt) ‖ Augenschwäche f. ‖ vue f. faible.

sight, angle of ~ ‖ Gesichtswinkel m. ‖ angle m. de visée. / ~ bill (Trade) ‖ Sichtwechsel m. ‖ billet m. à vue. / ~ correcting lens for eyes ametropia ‖ Korrektionsglas n. für Fehlsichtige ‖ verre m. correcteur pour amétropes. / ~ correcting lens for spectacle wearers ‖ Korrektionsglas n. für Brillenträger ‖ verre m. correcteur pour porteurs de lunettes. / defect of ~ ‖ Fehlsichtigkeit f.; amétropie f. / deficiency of ~ see defect of ~. / ~ draft (Trade) ‖ Sichtwechsel m. ‖ billet m. à vue.

sight feed (Mot) ‖ Ölschauglas n. ‖ viseur m. de graissage. / ~ lubricator ‖ Schautropföler m.; Tropföler m. ‖ graisseur m. compte-gouttes; graisseur m. à débit visible. / ~ nozzle ‖ Schautropfdüse f. ‖ gicleur m. de distribution visible. / ~ oiler see ~ feed lubricator.

sight hole ‖ Sehloch n.; Schauöffnung f. ‖ trou m. ou regard m. de visite.

sighting (Arm; Surv) ‖ Zielen n. ‖ visée f. / accuracy of ~ ‖ Zielgenauigkeit f. ‖ précision f. de la visée.

sighting line ‖ Ziellinie f. ‖ ligne f. de visée. / optical device for the parallel displacement of the ~ ‖ Einrichtung f. zur optischen Parallelverschiebung der Ziellinie ‖ dispositif m. pour le déplacement optique de la ligne de visée.

sighting slot ‖ Schauritze f. ‖ fente f. d'observation.

sighting telescope ‖ Zielfernrohr n.; Visierfernrohr n. ‖ lunette f. de visée.

sight, line of ~ ‖ Blickrichtung f. ‖ direction f. du regard. / principal line of ~ ‖ Hauptblickrichtung f. ‖ direction f. principale du regard. / loss of ~ ‖ Erblindung f.; Verlust m. des Augenlichts ‖ perte f. de la vue. / measuring by ~ ‖ Augenmaß n. ‖ estimation f. à vue d'œil; coup m. d'œil. / range of ~ ‖ Sehweite f. ‖ portée f. de vue.

sightseeing car ‖ Aussichtswagen m.; Rundfahrtwagen m. ‖ voiture f. d'excursion.

sight testing ‖ Sehprüfung f. ‖ examen m. de la vue. / ~ disk ‖ Sehprüfscheibe f. ‖ disque m. pour l'examen de la vue. / instrument for ~ and prescription of spectacles ‖ Instrument n. für Sehprüfung und Brillenverordnung ‖ instrument m. pour l'examen de la vue et l'ordonnance de lunettes.

sight vane (Surv) ‖ Kompaßdiopter n.; Diopter n. ‖ pinnule f.; dioptre f.

sign, to ~ ‖ zeichnen; signieren ‖ marquer. / ~ the agreement (Mar) ‖ anmustern ‖ signer l'engagement; enrôler. / ~ a bill ‖ einen Wechsel m. unterschreiben ‖ souscrire une lettre de change. / ~ per procuration

‖ per Prokura f. zeichnen ‖ signer ou souscrire par procuration f.

sign (Math) ‖ Vorzeichen n. ‖ signe m. / ~ (Distinctive mark) ‖ Kennzeichen n.; Merkzeichen n. ‖ marque f. distinctive; signe m. caractéristique. / ~ (Surv) ‖ Merkzeichen n.; Signal n. ‖ repère m.; repaire m.; signal m. / ~ (Signboard) ‖ Hausschild n.; Schild n.; Firmenschild n. ‖ enseigne f.; affiche f. peinte. / distinctive ~ ‖ Wahrzeichen n. ‖ marque f. distinctive. / ~ of distress see signal of distress. / enamel ~ ‖ Emailschild n. ‖ enseigne f. émaillée. / ~ of glass ‖ Glasfirmenschild n. ‖ plaque f. de réclame en verre; affiche f. en verre. / ~ of interrogation (Print) ‖ Fragezeichen n. ‖ point m. d'interrogation.

signal, to ‖ signalisieren; Zeichen npl. geben ‖ signaler.

signal ‖ Signal n.; Zeichen n. ‖ signal m. / to give ~ ‖ Signal n. geben ‖ faire signal m.; sonner. / acoustic ~ ‖ Schallsignal n. ‖ signal m. acoustique ou phonique. / answering ~ (Railw) see repeating ~. / back ~ (Tel) ‖ Rücksignal n. ‖ signal m. de retour. / balancing ~ (Radio) ‖ Balanziersignal n. ‖ signal m. équilibré / busy ~ ‖ Besetztzeichen n. ‖ signal m. d'occupation; signal m. de ligne occupée. / ~ for busy trunk lines ‖ Fernbesetztzeichen n. ‖ signal m. des lignes interurbaines occupées. / clear ~ (Railw) ‖ Freigabesignal n. ‖ signal m. de voie libre. / clearing ~ (Tel) ‖ Schlußzeichen n. ‖ signal m. de clôture. / detonating ~ (Railw) ‖ Knallsignal n. ‖ pétard m. / distant ~ (Railw) ‖ Vorsignal n. ‖ signal m. annonciateur. / ~ of distress (Mar) ‖ Notschuß m.; Notsignal n. ‖ coup m. de détresse. / ~ of distress (Railw) ‖ Notsignal n. ‖ signal m. de détresse. / ~ for mines ‖ Bergwerksignal n. ‖ signalisation f. de mines. / optical ~ ‖ optisches Signal n. ‖ signal m. optique. / repeating ~ (Railw) ‖ Rückmeldesignal n. ‖ signal m. de réplique. / return ~ (Tel) ‖ Rücksignal n. ‖ signal m. de retour. / sound ~ ‖ Schallsignal n.; akustisches Signal n. ‖ signal m. acoustique ou phonique. / starting ~ ‖ Abfahrtzeichen n.; Abfahrtsignal n. ‖ signal m. de départ.

signal alarm bell ‖ Läutwerk n. ‖ sonnerie d'annonce ou d'alarme. / ~ apparatus ‖ Signalvorrichtung f. ‖ appareil m. de signalisation. / ~ bell ‖ Signalglocke f. ‖ cloche f. de signal. / ~ clock ‖ Signaluhr f. ‖ horloge f. à signaux. / ~ colours pl. ‖ Signalfarben fpl. ‖ couleurs fpl. à signaux. / ~ corps (Military formation) ‖ Nachrichtentrupp m.; Telegrafentrupp m. ‖ section f. de transmission; formation f. de télégraphie militaire. / ~ disk (Railw) ‖ Signalscheibe f. ‖ voyant m. à disque; disque-signal m. / ~ feeder line ‖ Blockspeiseleitung f. ‖ ligne f. d'alimentation des postes. / ~ fire (Mar) ‖ Signalfeuer n. ‖ faux-feu m.; feu m. de signaux. / ~ flag (Railw) ‖ Signalflagge f. ‖ drapeausignal m. / ~ flag (Mar) ‖ Signalflagge f. ‖ pavillon m. à signaux. / ~ frame (Aut tel) ‖ Signalrahmen m. ‖ ensemble m. des signaux. / ~ gun (Mar) ‖ Signalschuß m. ‖ coup m. de canon de signal. / ~ instrument ‖ Signalinstrument n. ‖ instrument m. à avertir. / international code of ~s (Mar) ‖ internationales Signalbuch n. ‖ code m. international de signaux.

signalize, to ~ *see* to signal.
signal lamp ‖ Signallampe f. ‖ lampe f. de signaux. / ~ lantern (Mar) ‖ Signallaterne f.; Seitenlaterne f. ‖ fanal m.; lanterne f. de signal. / ~ light ‖ Signallicht n. ‖ feu m. de signal. / ~ light (Mar) *see also* fire ‖ Signalfeuer n. ‖ faux-feu m.; feu m. de signaux.
signalling of peak-load of traffic (Aut tel) ‖ Spitzensignalisierung f. ‖ tableaux mpl. lumineux des appels débordants. / submarine acoustic ~ ‖ Unterwasserschallsignal n. ‖ signal m. sous-marin à onde sonore.
signalling apparatus ‖ Signalvorrichtung f. ‖ appareil m. de signalisation. / electric ~ apparatus ‖ elektrisches Signal n. ‖ signal m. électrique. / ~ bells pl. ‖ Läutewerk n. ‖ sonnerie f. (électrique). / ~ device ‖ Signaleinrichtung f. ‖ dispositif m. de signalisation. / acoustic ~ device ‖ Schallsignalvorrichtung f. ‖ signaux-avertisseur m. à voix. / ~ gear for mines ‖ Grubensignalvorrichtung f. ‖ appareil m. de signaux de mine. / ~ installation ‖ Zeichengebungsanlage f. ‖ installation f. de signalisation. / remote ~ plant ‖ Fernsignalanlage f.; Fernmeldeanlage f. ‖ installation f. de télétransmission des signaux *ou* de signalisation sur grandes distances. / shaft ~ plant (Mine) ‖ Schachtsignalanlage f. ‖ installation f. de signaux pour puits de mine. / ~ shelf (Aut tel) ‖ Signalrahmen m. ‖ baie f. de signaux. / ~ system ‖ Zeichenübermittlungssystem n. ‖ système m. à signaler. / ~ testing apparatus (Tel) ‖ Rufprüfeinrichtung f. ‖ dispositif m. de vérification des appels. / ~ unit (Tel) ‖ Rufmaschine f. ‖ machine f. d'appel.
signalman (Mar) ‖ Signalgast m. ‖ timonier m. / ~ (Mine) ‖ Anschläger m. (Fahrtzeichengeber) ‖ sonneur m. / ~ (Railw) ‖ Signalwärter m. ‖ gardesignaux m.; signaleur m.
signal operating mechanism ‖ Signalantrieb m. ‖ commande f. de signal. / operation of ~ ‖ Bedienung f. eines Signals ‖ manœuvre f. de signal. / mechanical operation of ~s and switches ‖ mechanische Signal- und Weichenstellung f. ‖ commande f. mécanique des aiguilles et signaux. / ~ point indicator (lamp) ‖ Weichenlaterne f. ‖ lanterne f. du changement de voie. / normal position of ~s and switches ‖ Grundstellung f. von Signalen und Weichen ‖ position f. normale des signaux et aiguilles. / ~ post ‖ Signalmast m. ‖ poteau m. de sémaphore; mât-signal m. / ~ receiver (Tel) ‖ Signalempfänger m. ‖ poste m. récepteur. / ~ shelf *see* signalling shelf. / slot of ~ (Railw) ‖ Signalflügelkupplung f. ‖ désengageur m. / ~ transmitter (Tel) ‖ Signalgeber m.; Zeichengeber m. ‖ poste m. transmetteur. / ~ wire (Mine) ‖ Klopfgestränge n. ‖ fil m. à signaux.
signature (Mark) ‖ Signatur f. ‖ marque f.; étiquette f. / ~ (Sign manual) ‖ Unterschrift f.; Namenszug m. ‖ signature f.; souscription f.; chiffre m.; parafe m.
signboard ‖ Firmenschild n. ‖ enseigne f.; plaque f. de réclame. / advertising ~ ‖ Reklameschild n.; Werbeschild n. ‖ enseigne f. de réclame. / ~ for goods wagons ‖ Schild n. zum Zettelaufkleben an Güterwagen ‖ tableau m. au wagon à marchandises pour apposer les éti-

quettes. / ~ for sluice valves and hydrants ‖ Straßenschild n. für Wasserleitungsschieber und Hydranten ‖ plaque f. d'avertissement pour robinets-vannes et prises d'eau.
signet ‖ Petschaft n. ‖ cachet m. / ~ ring ‖ Siegelring m. ‖ bague-cachet f.
sign lighting ‖ Reklamebeleuchtung f.; Werbebeleuchtung f. ‖ éclairage m. de réclame *ou* de publicité.
sign plate for machines ‖ Maschinenschild n. ‖ plaque f. indicatrice de machines.
signpost ‖ Wegweiser m. ‖ poteau-guide m.; poteau m. indicateur.
sign type bar (Typewr) ‖ Zeichentypenhebel m. ‖ barre f. à signes.
silence ‖ Stille f.; Ruhe f. ‖ silence m.; tranquillité f.; repos m. / ~ (Radio) ‖ Funkstille f. ‖ silence m.
silencer (Auto) ‖ Schalldämpfer m.; Auspufftopf m. ‖ silencieux m.; pot m. d'échappement; sourdine f.
silent discharge (Electr) ‖ dunkle Entladung f. ‖ décharge f. obscure; effluve m. / ~ pictures pl. ‖ stummer Film m. ‖ film m. muet. / ~ running (Mach) ‖ ruhiger Gang m. ‖ marche f. silencieuse.
silex (Miner) ‖ Feuerstein m. ‖ silex m. (à feu). / ~ quarry ‖ Feuersteingrube f. ‖ carrière f. de silex. / ~ white ‖ Kieselweiß n. ‖ blanc m. de silex.
silhouette ‖ Schattenriß m.; Schattenbild n. ‖ silhouette f.
silica ‖ Kieselerde f.; Siliziumdioxyd n. ‖ silice f. / fused ~ ‖ Quarzglas n. ‖ verre m. quartzeux.
silica brick ‖ Silikastein m. ‖ brique f. acide silicique; brique f. silicate.
silicate ‖ kieselsaures Salz m.; Silikat n. ‖ silicate m. / ~ of alumina ‖ kieselsaure Tonerde f. ‖ silicate m. d'alumine. / fluoric ~ ‖ Fluorsilikat n. ‖ fluorsilicate m. / hydrous ~ of zinc ‖ Kieselzinkerz n.; Kieselzinkspat m.; Kieselgalmei m. ‖ hydro-silicate m. de zinc; zinc m. oxydé silicifère. / ~ of magnesia ‖ kieselsaure Magnesia f. ‖ silicate m. de magnésie. / ~ of manganese ‖ Kieselmangan n.; Manganaugit m.; Mangankiesel m. ‖ manganèse m. oxydé silicifère. / potassium ~ / kieselsaures Kalium n.; (Kalium-)Wasserglas n. ‖ (ortho)silicate m. de potasse. / sodium ~ ‖ kieselsaures Natrium n.; (Natrium-)Wasserglas n. ‖ (ortho)-silicate m. de sodium; verre m. soluble.
silicate bond ‖ Silikatbindemittel n. ‖ agglomérant m. au silicate. / ~ cotton ‖ Schlackenwolle f. ‖ laine f. de laitier *ou* de scories.
silicated soap ‖ Wasserglasseife f. ‖ savon m. au silicate alcalin.
silicate glass ‖ Silikatglas n. ‖ verre m. au silicate.
silicate plant ‖ Wasserglasanlage f. ‖ installation f. de silicates alcalins.
silicatine spangle ‖ Silikatinflitter m. ‖ paillette f. de silicatine.
siliceous *see* silicious.
silicic acid ‖ Kieselsäure f. ‖ acide m. silicique.
silicide ‖ Siliziumverbindung f. ‖ siliciure m. / ~ of carbon ‖ Siliziumkarbid n. ‖ carbure m. de silicium.
siliciferous *see also* silicious ‖ kieselhaltig ‖ silicifère. / ~ hydrate of alumina ‖ Kollyrit m.; samische Erde f. ‖ collyrite f.; alumine f. hydratée silicifère.

silicification ‖ Verkieselung f. ‖ silicatisation f.
silicified ‖ verkieselt ‖ silicaté.
silicify, to ‖ verkieseln ‖ silicater.
silicious ‖ kieselerdehaltig; kieselhaltig; kieselartig ‖ siliceux. / ~ calamine ‖ Kieselgalmei m.; Kieselzinkerz n. ‖ zinc m. oxydé silicifère. / ~ flux ‖ Kieselfluß m.; kieseliger Zuschlag m. ‖ fondant m. siliceux. / ~ limestone ‖ Kieselkalk m. ‖ calcaire m. siliceux. / ~ oxide of zinc ‖ Kieselgalmei f.; Zinkglas n. ‖ zinc m. oxydé silicifère; calamine f.; calamine f. électrique.
silicium *see also* silicon ‖ Silizium n. ‖ silicium m.
silicium carbide ‖ Siliziumkarbid n. ‖ carbure m. de silicium. / ~ brick ‖ Siliziumkarbidstein m. ‖ brique f. de siliciumcarbure.
silicium iron ‖ Siliziumeisen n. ‖ ferrosilicium m. / ~ anode ‖ Siliziumeisenanode f. ‖ anode f. de ferro-silicium.
silicium steel ‖ Siliziumstahl m. ‖ acier m. au silicium.
silicium tetrachloride ‖ Siliziumtetrachlorid n. ‖ tetrachlorure m. de silicium.
silico-fluoric acid ‖ Fluorwasserkieselsäure f.; Kieselflußsäure f. ‖ acide m. hydrofluosilicique; acide m. silicofluorique.
silico-fluorides pl. ‖ Kieselfluorsalze npl. ‖ silicofluorures mpl.
silico-manganese steel ‖ Mangansiliziumstahl m. ‖ acier m. mangano-siliceux.
silicon *see also* silicium ‖ Silizium n. ‖ silicium m.
silicon bronze ‖ Siliziumbronze f. ‖ bronze m. siliceux. / ~ wire ‖ Siliziumbronzedraht m. ‖ fil m. de bronze au silicium.
silicon carbide *see* silicium carbide.
silicon dressing plant ‖ Silikaaufbereitungsanlage f. ‖ installation f. de préparation de la silice.
silicon varnish ‖ Siliziumimprägnierungsmasse f.; Siliziumlack m. ‖ vernis m. au silicium.
silite ‖ Silit n. ‖ silite m. / ~ heating rod ‖ Silitheizstab m. ‖ crayon m. de chauffage en silite.
silk ‖ Seide f. ‖ soie f. / ~s pl. ‖ Seidenware f.; Seidenzeug n. ‖ soierie f. / artificial ~ ‖ Kunstseide f.; künstliche Seide f. ‖ soie f. artificielle. / artificial ~ factory ‖ Kunstseidefabrik f. ‖ fabrique f. de soie artificielle. / artificial ~ making machine ‖ Kunstseideherstellungsmaschine f. ‖ machine f. de fabrication de la soie artificielle. / boiled ~ ‖ entschälte *oder* gekochte *oder* linde *oder* sachte Seide f. ‖ soie f. cuite *ou* décreusée. / ~ broché ‖ broschierter Seidenstoff m. ‖ broché m. de soie. / cellulose acetate ~ ‖ Zelluloseazetatseide f. ‖ soie f. à l'acétate de cellulose. / ~ in cocoon ‖ Seidengehäuse n. ‖ soie f. en cocon. / collodion ~ ‖ Kollodiumseide f. ‖ soie f. de collodion. / crude ~ ‖ ungekochte *oder* unentschälte Seide f.; Ekrüseide f.; Bastseide f. ‖ soie f. écrue *ou* crue. / cuprammonium ~ ‖ Kupferseide f. ‖ soie f. au cuivre. / ~ in the gum *see* crude ~. / half-boiled ~ ‖ halbgekochte Seide f. ‖ soie f. mi-cuite *ou* souple. / natural ~ ‖ echte Seide f. ‖ soie f. pure. / raw ~ ‖ Rohseide f. ‖ soie f. grège; grèze f. / scoured ~ ‖ entschälte *oder* gekochte *oder* linde *oder* sachte Seide f. ‖ soie f. cuite *ou* décreusée. / sewing ~ ‖

Nähseide f. ‖ soie f. à coudre. / slack ~ ‖
Flockseide f. ‖ soie f. plate ou floche;
bourre f. de soie. / spun ~ ‖ Seidengarn
n. ‖ soie f. filée. / thrown ~ ‖ gezwirnte
oder filierte oder mulinierte Seide f.;
Seidenzwirn m. ‖ soie f. moulinée ou
ouvrée ou torse. / twisted ~ ‖ kordo-
nierte Seide f. ‖ cordonnet m. / unboiled
or unscoured ~ see crude ~. / unprepared
~ ‖ Naturelseide f. ‖ soie f. en mèches.
unthrown ~ ‖ Flockseide f.; ungezwirnte
Seide f. ‖ effiloches fpl.; effiloques fpl. /
viscose ~ ‖ Viskoseseide f. ‖ soie f.
viscose. / waste ~ ‖ Florettseide f. ‖
bourre f. de soie. / watered ~ ‖ Seiden-
mohr m. ‖ moirée f. de soie. / weft ~ ‖
Einschlagseide f. ‖ fil-trame m.; soie f.
trame.

silk boiling hand (Bleach) ‖ Seidenkocher
m. ‖ cuiseur m. de soie. / ~ bleaching ‖
Seidenbleiche f.; Seidenbleichung f. ‖
blanchiment m. de la soie. / ~ button ‖ mit
Seide überzogener Knopf m. ‖ bouton m.
recouvert de soie. / ~ canvas (For needle-
work) ‖ Seidengaze f.; Seidenstramin m.;
seidene Stickgaze f.; Stramin m. ‖ canevas m.; étamine f. / ~ cloth ‖ Seidentuch
n. ‖ drap m. de soie. / ~ clothing of a top
hat ‖ Zylinderüberzug m. von Seide ‖ garniture f. en soie d'un chapeau-claque. /
~ cotton ‖ Seidenbaumwolle f. ‖ soie-
coton f. / ~ damask ‖ Seidendamast m. ‖
damas m. de soie. / ~ dyer ‖ Seiden-
färber m. ‖ teinturier m. sur soie. / ~
dyeing plant ‖ Seidenfärberei f. ‖ teinturerie f. de soie. / ~ embroidery ‖ Sei-
denstickerei f. ‖ broderie f. en soie. /
~ farm ‖ Seidenzuchtanlage f. ‖ magnanerie f. / ~ farmer ‖ Seidenraupenzüchter
m. ‖ sériciculteur m.; éleveur m. de vers
à soie. / ~ feather shag ‖ echter oder
seidener Felbel m. ‖ panne f. de soie. /
~ filatory ‖ Filatorium n.; Seidenmühle
f. ‖ moulin m. à soie. / ~ finisher ‖ Seiden-
appretör m. ‖ apprêteur m. de soie. / ~
gauze ‖ Seidengaze f. ‖ gaze f. en soie. /
~ glove ‖ Seidenhandschuh m. ‖ gant m.
de soie. / ~ goods pl. ‖ Seidenwaren fpl. ‖
soieries fpl. / ~ green ‖ Seidengrün n. ‖
vert m. de soie. / ~ hat ‖ Seidenhut m. ‖
chapeau m. de soie ou de peluche de soie.
/ head of ~ ‖ Docke f. ‖ matteau m.;
bouin m. / ~ hose ‖ Seidenstrumpf m. ‖
bas m. de soie. / ~ hosiery ‖ Strickwaren
fpl. aus Seide; Seidenstrickerei f. ‖
bonneterie f. de soie. / ~ industry ‖
Seidenindustrie f. ‖ industrie f. de la
soie. / ~ laces pl. ‖ Seidenspitze f. ‖ dentelles fpl. de soie. / ~ loose threads pl. of ~
‖ Zupfseide f. ‖ soie f. effilée. / ~ manufactory see ~ mill. / ~ mill see also
~ spinning mill ‖ Seidenzwirnerei f.
‖ moulin m. à soie. / ~ noil ‖ Seiden-
kämmling m. ‖ blousse f. de soie. / ~
paper ‖ Seidenpapier n. ‖ papier m. ser-
pente ou de soie. / ~ piqué ‖ Seidenpikee m. ‖ tissu m. piqué soie.

silk plush ‖ Seidenplüsch m. ‖ peluche f.
soie. / ~ hat ‖ Seidenhut m. ‖ chapeau m.
de soie.

silk printing ‖ Seidendruck m. ‖ impression f. sur soie; impression f. des étoffes
de soie. / ~ produce ‖ Seidenerzeugnis n.
‖ produit m. en ou de soie. / ~ reel ‖ Spul-
rädchen m. ‖ escaladon m. / ~ reeler ‖
Seidenhaspeler m. ‖ tireur m. de soie. /
~ ribbon ‖ Seidenband n. ‖ ruban m. de
soie. / ~ shoddy spinning ‖ Seiden-

schoddyspinnerei f. ‖ filature f. du
shoddy de soie. / ~ sieve ‖ Seidensieb n. ‖
tamis m. en gaze de soie. / ~ spinner ‖ Seidenspinner m. ‖ fileur m. de soie; tireur m.
/ ~ spinning ‖ Filieren n.; Vorzwirnen n.
/ filage m. / ~ spinning mill ‖ Seiden-
spinnerei f.; Seidenwinderei f.; Seiden-
fabrik f. ‖ filature f. de soie; soierie f.;
fabrique f. de soie. / ~ spooler ‖ Seiden-
spuler m. ‖ bobineur m. ou canneteur m.
de soie. / ~ spool winder see ~ spooler. /
~ stretcher ‖ Seidenlüstrierer m. ‖ lustreur m. de soie. / ~ string ‖ Seidensaite f. ‖ corde f. harmonique en soie.

silk stuff ‖ Seidenstoff m. ‖ étoffe f. de soie.
/ ribbed ~ ‖ gerippptes Seidenzeug n. ‖
cannelé m.; tissu m. rayé.

silk and cotton stuff ‖ Halbseide f.; Seidenware f. mit baumwollenem Einschlag ‖
mignonette f.

silk thread ‖ Seidenfaden m. ‖ fil m. de
soie. / ~ thrower see also ~ spinner ‖
Seidenspinner m. ‖ fileur m. de soie;
tireur m. / ~ throwing ‖ Zwirnen n. der
Seide ‖ organsinage m. de soie. / ~
throwster ‖ Seidenspinnerin f.; Seiden-
hasplerin f. ‖ fileuse f. ou tireuse f. des
cocons. / ~ tissue ‖ Seidengewebe n. ‖
tissu m. de soie. / ~ twill ‖ Seidenserge f.
‖ sergé m. soie. / ~ twist ‖ Seidenzwirn m.
‖ fil m. de soie retors.

silk velvet ‖ Seidensamt m. ‖ velours m.
de soie. / light ~ ‖ leichter Seidensamt m.
‖ velours m. crevelle.

silk velvet maker ‖ Seidensamtweber m. ‖
veloutier m. en soie. / ~ shearer ‖ Seiden-
samtscherer m. ‖ tondeur m. de velours
de soie.

silk wadding ‖ Seidenwatte f. ‖ ouate f. de
soie.

silk waste ‖ Seidenabfall m.; Schappe f. ‖
bourre f. ou déchet m. de soie; chape f. /
~ carding ‖ Kardätschen n. des Seidenabfalls ‖ peignerie f. de bourre de soie. /
~ dyeing ‖ Färben n. des Seidenabfalls ‖
teinture f. de bourre de soie. / ~ spinning
‖ Spinnen n. des Seidenabfalls ‖ filature
f. de bourre de soie. / ~ yarn ‖ Flockseide-
garn n. ‖ fil m. de bourre.

silk waterer ‖ Seidenmoirör m. ‖ moireur m.
de soie. / ~ weaver ‖ Seidenweber m. ‖
tisserand m. en soie. / ~ weaving ‖
Seidenweberei f. ‖ tissage m. de soie. /
~ weighting ‖ Seidengarnbeschwerung f.
‖ charge f. des fils de soie. / ~ winder ‖
Seidenwickler m. ‖ dévideur m. de soie. /
~ winder-on of warps ‖ Seidenzettler m.
‖ plieur m. de soie. / ~ winding ‖ Wickeln
n. von Seide ‖ dévidage m. de la soie. /
~ winding machine ‖ Seidenspulmaschine
f.; Seidenhaspelmaschine f. ‖ bobinoir
m. mécanique pour soie; machine f. à
dévider la soie. / ~ wool ‖ Seidenwolle f.
‖ soie-laine f. / ~ worker ‖ Seidenarbeiter
m. ‖ ouvrier m. en soie.

silkworm ‖ Seidenraupe f. ‖ ver m. à soie. /
~ breeder ‖ Seidenraupenzüchter m. ‖
sériciculteur m.; éleveur m. de vers à
soie. / ~ breeding ‖ Seidenraupenzucht f. ‖
sériciculture f. / ~ house ‖ Seidenraupe-
rei f.; Seidenraupenhaus n. ‖ coconnière
f.; magnanerie f.; vererie f.

silk wringing ‖ Ausringen n. oder Schevillieren n. der Seide ‖ chevillage m. de la
soie.

silk yarn ‖ Seidengarn n. ‖ fil m. de soie;
soie f. filée.

silky lustre (Miner) ‖ Seidenglanz m. / éclat
m. soyeux; lustre m. brillant.

sill (Build) ‖ Schwelle f.; Schwellholz n. ‖
sablière f.; seuil m.; dormant m.; solive
f.; soliveau m. / ~ (of a window) ‖ Fenstersohle f.; Fensterschwelle f.; Fensterbank f. ‖ seuil m. ou banquette f. de fenêtre. / ~ (Mine) ‖ Liegendes n. eines Flözes; liegende Schicht f. ‖ mur m. d'une
couche; lit m.; couche f. inférieure. /
door ~ ‖ Türschwelle f. ‖ seuil m. de
porte. / longitudinal ~ ‖ Langschwelle f. ‖
longrine f.; longuerine f. / ~ of a pile-
driving engine ‖ Rammenschwelle f.;
Schwelle f. einer Ramme ‖ semelle f. de
sonnette. / ~ of a rising timber ‖ Knie-
hals m.; Sohlschenkel m. ‖ montant m.
de semelle. / roof ~ ‖ Dachunterzug m. ‖
sablière f. du toit. / ~ of a seam (Mine) ‖
Liegendes n. ‖ base f. d'une couche ou
d'un gisement.

sillimanite ‖ Sillimanit n. ‖ sillimanite f.

silo ‖ Silo m.; Getreidespeicher m. ‖ silo m.
/ automatic ~ ‖ automatischer Silo m. ‖
silo m. automatique.

silo bunker ‖ Silobunker m. ‖ trémie f. de
silo. / ~ compartment ‖ Silozelle f. ‖ compartiment m. de silo. / ~ process ‖ Silo-
verfahren n. ‖ procédé m. des silos.

silt, to ~ up ‖ verlanden; verschlammen;
versanden ‖ déposer le limon; envaser;
s'ensabler.

silt ‖ Verlandung f.; abgesetzter Schlamm
m. oder Moder m. ‖ dépôt m. de limon.

silting up (Hydr arch) ‖ Aufschlickung f. ‖
limonage m.

silty soil ‖ schlammiger oder mooriger Boden m. ‖ terrain m. limoneux ou marécageux.

silumin ‖ Silumin n. ‖ silumin m.

silundum ‖ Silundum n. ‖ silundum m.

Silurian (Geol) ‖ silurisch ‖ silurien.

silver, to ‖ versilbern; mit Silber m. überziehen ‖ argenter. / ~ (Glass) ‖ mit Folie
belegen; foliieren ‖ argenter.

silver ‖ Silber n. ‖ argent m. / antimonial ~
‖ Antimonsilber n.; Spießglanzsilber n. ‖
argent m. antimonial. / arsenical ~ ‖
Arseniksilber n .‖ argent m. arsenical. /
bar ~ ‖ Stangensilber n.; Barrensilber n.
‖ argent m. en barres. / beaten ~ ‖ Blattsilber n. ‖ feuilles fpl. d'argent; argent
m. battu. / burnished ~ ‖ Glanzsilber n. ‖
argent m. imité. / capillary ~ ‖ Haarsilber
n. ‖ argent m. vierge capillaire. / chased
~ ‖ getriebenes Silber n. ‖ argent m. embouti. / coined ~ ‖ gemünztes Silber n. ‖
argent m. monnayé. / colloidal ~ ‖ kolloidales Silber n. ‖ argent m. colloïdal. /
crude ~ ‖ rohes Silber n. ‖ argent m. brut.
/ dark red ~ ‖ Antimonsilberblende f.;
Silberblende f. ‖ argent m. antimonié ou
rouge; pyrargyrite f. / dentritic ~ ‖ den-
tritisches Silber n. ‖ argent m. dentriti-
que. / of due alloy ‖ lötiges Silber n. ‖
argent m. fin. / Dutch ~ ‖ Rauschsilber n.
‖ feuilles fpl. d'argent clinquant. / fine ~
‖ Kapellensilber n.; Feinsilber n. ‖ argent
m. de coupelle; argent du fin. / fulminating ~ ‖ Knallsilber n.; knallsaures Silberoxyd n. ‖ fulminate m. d'argent. / German
~ ‖ Neusilber n. ‖ maillechort m.; métal m. blanc. / gilt ~ ‖ vergoldetes Silber
n. ‖ argent m. doré. / grey ~ ‖ Grausilber
n.; Silberkarbonat n.; Selbit m. ‖ carbonate m. d'argent; selbite f. / in ingots ‖ Silber n. in Barren oder Stangen ‖
argent m. en barres ou en lingots.

/ light-red ~ ‖ Arsensilber n. ‖ argent m. arsénié; proustite f. / ~ in lumps ‖ Silber n. in Klumpen ‖ argent m. en masse. / metallic ~ ‖ metallisches Silber n. ‖ argent m. métallique. / native ~ ‖ gediegenes Silber n.; Jungfernsilber n. ‖ argent m. natif ou vierge. / oxydized ~ ‖ oxydiertes Silber n. ‖ argent m. oxydé. / powdered ~ ‖ Silberpulver n.; Silberstaub m. ‖ poudre f. d'argent; argent m. en poudre. / ~ quoted at ... ‖ Silber n. notiert mit ... ‖ argent m. coté à ... / red ~ ‖ Pyrargyrit m.; dunkles Rotgültigerz n.; Antimonsilberblende f. ‖ argent m. antimonié sulfuré. / refined ~ ‖ Raffinatsilber n.; Brandsilber n. ‖ argent m. raffiné. / rolled ~ ‖ Silber n. in Blechen; Silberblech n. ‖ argent m. laminé. / spun ~ ‖ Silber n. in Drähten; Silberdraht m. ‖ argent m. filé. / vitreous ~ (Miner) ‖ Argentit m.; Glanzerz n.; Silberglanz m. ‖ argentite m.; argent m. sulfuré. / worked ~ ‖ Werksilber n. ‖ argent m. d'œuvre.

silver alloy ‖ Silberlegierung f. ‖ alliage m. d'argent. / ~ rich in copper ‖ kupferreiche Silberlegierung ‖ alliage m. d'argent à haute teneur en cuivre.

silver amalgam ‖ Silberamalgam n. ‖ mercure m. argental; amalgame m. d'argent. / ~ amide ‖ Knallsilber n. ‖ amidure f. d'argent. / ~ apparatus ‖ Silberapparat m. ‖ appareil m. en argent. / ~ articles pl. for table ‖ Tafelartikel mpl. aus Silber ‖ argenterie f. de table. / ~ assay ‖ Silberprobe f. ‖ essai m. d'argent. / ~ bag ‖ Silbertäschchen n. ‖ porte-argent m. / ~ bath ‖ Silberbad n. ‖ bain m. (de nitrate) d'argent ou d'argenture.

silver-bearing ‖ silberführend ‖ argentifère.

silver beating ‖ Blattsilberschlägerei f. ‖ battage m. d'argent. / brittle sulphide of ~ ‖ Schwarzgültigerz n.; Melanglanz m.; Sprödglaserz n. ‖ argent m. antimonié sulfuré noir. / ~ brocade ‖ Silberbrokat m.; Silberstoff m. ‖ brocart m. en broché argent; drap m. d'argent. / ~ bromide paper ‖ Bromsilberpapier n. ‖ papier m. au bromure d'argent. / ~ bronze (Powder) ‖ Silberbronze f. ‖ bronze m. d'argent. / ~ bullion ‖ Silberbarren m. ‖ lingot m. d'argent. / ~ burnisher ‖ Silberglätter m. ‖ brunisseur m. en argent. / ~ button ‖ silberner Knopf m. ‖ bouton m. en argent.

silver case ‖ Silbergehäuse n. ‖ boîte f. d'argent. / ~ (For knife and fork) ‖ Besteckkasten m.; Silberkasten m. ‖ boîte f. ou coffret m. à couverts.

silver chain ‖ Silberkette f.; silberne Kette f. ‖ chaîne f. en argent. / ~ chlorate ‖ Silberchlorat n. ‖ chlorate m. d'argent. / ~ chloride ‖ Chlorsilber n.; Silberchlorid n.; Silberhornerz n. ‖ chlorure m. d'argent; argent m. corné ou muriaté. / ~ coin ‖ Silbermünze f. ‖ pièce f. ou monnaie f. d'argent; monnaie f. blanche.

silver-containing ‖ silberhaltig ‖ argentifère.

silver crucible ‖ Silbertiegel m. ‖ creuset m. en argent. / ~ currency ‖ Silberwährung f. ‖ étalon m. d'argent. / ~ cyanide ‖ Zyansilber n.; Silberzyanid n. ‖ cyanure m. d'argent. / ~ double salt ‖ Silberdoppelsalz n. ‖ sel m. double d'argent.

silvered ‖ versilbert ‖ argenté. / ~ glass reflector ‖ Glassilberspiegel m. ‖ miroir m. en verre argenté. / ~ mirror ‖ Silber-spiegel m. ‖ miroir m. argenté. / ~ parabolic glass reflector ‖ parabolischer Glassilberspiegel m. ‖ miroir m. parabolique en verre argenté.

silver electrode ‖ Silberelektrode f. ‖ électrode f. d'argent. / ~ embroidery ‖ Silberstickerei f. ‖ broderie f. d'argent. / ~ engraver ‖ Silbergravör m. ‖ graveur m. sur argent.

silverer ‖ Versilberer m. ‖ argenteur m. / ~ on wood ‖ Holzversilberer m. ‖ argenteur m. sur bois.

silver extraction ‖ Silbergewinnung f. ‖ extraction f. de l'argent. / ~ filigree ‖ Silbergespinst n. ‖ filigrane m. en argent / ~ fir ‖ Edeltanne f.; Weißtanne f.; Silbertanne f. ‖ sapin m. argenté. / ~ foil ‖ Silberbelag m.; Silberfolie f.; Blattsilber n. ‖ feuille f. d'argent. / ~ fulminate ‖ Knallsilber n.; knallsaures Silberoxyd n. ‖ fulminate m. d'argent. / ~ galloon ‖ Silberborte f. ‖ galon m. d'argent.

silver glance ‖ Argentit m.; Glanzerz n.; Silberglanz m. ‖ argentite m.; argent m. sulfuré. / antimonial ~ ‖ Melanglanz m.; Stephanit m.; Schwarzgültigerz n. ‖ argent m. noir; argent m. sulfuré fragile; argent m. antimonié sulfuré noir.

silver goods pl. ‖ Silberwaren fpl. ‖ argenterie f.

silver-graphite ‖ Silbergraphit m. ‖ plombagine f. argentifère.

silver horn ore ‖ Silberhornerz n. ‖ argent m. corné ou muriaté; chlorure m. d'argent.

silvering ‖ Versilberung f. ‖ argenture f. / electro ~ see galvanic ~. / galvanic ~ ‖ galvanische Versilberung f. ‖ argenture f. galvanique. / ~ of mirrors ‖ Versilbern n. oder Silberbelegen n. von Spiegeln ‖ argenture f. de glaces ou de miroirs. / ~ of the speculum see ~ of mirrors. / ~ of wood ‖ Holzversilberung f. ‖ argenture f. sur bois.

silvering shop ‖ Versilberungsraum m. ‖ atelier m. d'argenture. / ~ table (Mirror making) ‖ Belegtisch m. ‖ table f. pour l'étamage des glaces.

silver layer ‖ Silberschicht f. ‖ couche f. d'argent. / brazed-in ~ ‖ eingebrannte Silberschicht f. ‖ couche f. d'argent rapportée au feu.

silver leaves pl. ‖ Blattsilber n. ‖ feuilles fpl. d'argent; argent m. battu. / lightning of ~ (Met) ‖ Silberblick m. ‖ éclair m. de l'argent. / ~ litharge ‖ Silberglätte f. ‖ litharge f. d'argent. / ~ mine ‖ Silbergrube f.; Silberbergwerk n. ‖ mine f. d'argent. / ~ mirror ‖ Silberspiegel m. ‖ miroir m. argenté. ‖ ~ money ‖ Silbergeld n. ‖ monnaie f. d'argent ou blanche. / ~ muriate see silver horn ore. / ~ nitrate ‖ salpetersaures Silber n.; Höllenstein m. ‖ nitrate m. ou azotate d'argent; pierre f. infernale. / ~ nucléinate m. d'argent.

silver ore ‖ Silbererz n. ‖ minerai m. d'argent. / dark-red ~ ‖ dunkles Rotgültigerz n.; Antimonsilberblende f.; Pyrargyrit m. ‖ argent m. antimonié sulfuré / light-red ~ ‖ lichtes Rotgültigerz n.; Arseniksilberblende f.; Proustit m. ‖ argent m. arséniaté sulfuré / red ~ ‖ Rotgültigerz n.; Rotgulden n.; Silberblende f. ‖ argent m. rouge. / ruby ~ see red ~. / amalgamating device for ~s ‖ Amalgamiereinrichtung f. für Silber-erze ‖ installation f. à amalgamer les minerais d'argent.

silver oxide ‖ Silberoxyd n. ‖ oxyde m. ou protoxyde m. d'argent. / oxysalt of ~ ‖ Silbersalz n. ‖ sel m. d'argent. / ~ paper ‖ Silberpapier n. ‖ papier m. argenté. / ~ peroxide ‖ Silbersuperoxyd n. ‖ peroxyde m. d'argent.

silver plating ‖ Silberplattierung f.; Versilberung f. ‖ argenture f. / ~ of metals ‖ Metallversilberung f. ‖ argenture f. sur métaux. / ~ plant for ~ ‖ Versilberungsanlage f. ‖ installation f. d'argenture.

silver powder ‖ Silberpulver n. ‖ poudre f. d'argent. / ~ proteinate ‖ proteinsaures Silber n.; Protargol n. ‖ protéinate m. d'argent. / ~ purse ‖ silberne Börse f.: silberner Geldbeutel m. ‖ bourse f. en argent. / ~ refiner ‖ Silberscheider m. ‖ affineur m. d'argent. / ~ refining ‖ Silberscheidung f.; Silberraffination f. ‖ affinage m. ou raffinage m. d'argent. / ~ roller ‖ Silberwalzer m. ‖ lamineur m. d'argent. / ~ rolling ‖ Walzen n. von Silber ‖ laminage m. d'argent. / ~ rolling mill ‖ Silberwalzwerk n. ‖ laminoir m. d'argent.

silver salt (Chem) ‖ Silbersalz n. ‖ sel m. argentique ou d'argent. / ~ maker ‖ Silbersalzhersteller m. ‖ préparateur m. de sels d'argent.

silver setting ‖ Silberfassung f. ‖ sertissage m. en argent.

silver smelter ‖ Silbergießer m. ‖ fondeur m. d'argent.

silversmith ‖ Silberschmied m.; Silberarbeiter m. ‖ argentier m. / ~'s work ‖ Silberwaren fpl.; Silberschmiedewaren fpl. ‖ argenterie f.

silver soap ‖ Silberseife f. ‖ mélange m. de savon et de craie; pâte f. à nettoyer l'argent. / ~ solder ‖ Silberlot n.; Silberschlaglot n. ‖ soudure f. à l'argent. / ~ spangle ‖ Silberflitter m. ‖ paillette f. en argent. / ~ stamping works pl. ‖ Silberprägeanstalt f. ‖ établissement m. d'estampage d'argent. / ~ standard ‖ Silberwährung f. ‖ étalon m. d'argent. / ~ steel ‖ Silberstahl m. ‖ acier-argent m. / ~ suboxide ‖ Silbersuboxyd n. ‖ sous-oxide m. d'argent. / ~ sulphide ‖ Schwefelsilber n.; Silbersulfid n. ‖ sulfure m. d'argent. / ~ test ‖ Silberprobe f. ‖ essai m. d'argent. / ~ tissue ‖ Silbergewebe n. ‖ tissu m. d'argent. / ~ trimmings pl. ‖ Silberposamenten pl. ‖ passementerie f. argent. / ~ turner ‖ Silberdreher m. ‖ tourneur m. en argent. / ~ vessel ‖ Silberschale f. ‖ vase m. d'argent.

silverware ‖ Silberzeug n.; Silberwaren fpl. ‖ articles mpl. en argent; argenterie f.

silver weight-meter ‖ Silbergewichtszähler m. ‖ compteur m. du poids d'argent.

silver white ‖ Kremserweiß n. ‖ blanc m. d'argent.

silver wire ‖ Silberdraht m. ‖ fil m. d'argent. / alloyed ~ ‖ legierter Silberdraht m. ‖ fil m. allié d'argent. / genuine ~ ‖ echter Silberdraht m. ‖ fil m. d'argent fin. / gilt ~ ‖ Golddraht m.; vergoldeter Silberdraht m. ‖ fil m. ou trait m. d'argent doré; or m. trait. / imitation ~ ‖ unechter Silberdraht m. ‖ fil m. d'argent imitation.

silver wire drawing works pl. ‖ Silberdrahtzieherei f. ‖ tréfilerie f. d'argent.

silver wire fuse (Electr) ‖ Silberschmelz-sicherung f. ‖ fusible m. en argent.

silver works pl. ‖ Silberhütte f. ‖ fonderie f. d'argent.

silver wreath ‖ Silberkranz m. ‖ couronne f. d'argent.

silvery ‖ silberartig; silberig; silberfarben; silberhell ‖ argenté; argentin; clair argenté.

silviculture ‖ Forstkultur f.; Forstwirt-schaft f. ‖ économie f. forestière; sylvi-culture f.

similar (Electr) ‖ gleichnamig ‖ des mêmes signes. / ~ (Geom) ‖ ähnlich ‖ semblable. / ~ triangles pl. ‖ ähnliche Dreiecke npl. ‖ triangles mpl. semblables.

similarity (Geom etc.) ‖ Ähnlichkeit f. ‖ similitude f.

simili stone ‖ Similistein m. ‖ pierre f. de strass.

similitude see similarity.

simmer, to ‖ wallen; gelinde kochen ‖ bouillonner.

Simon-interrupter (Electr) ‖ Simon-Unter-brecher m. ‖ interrupteur m. Simon.

simple ‖ einfach ‖ simple. / ~ antenna ‖ einfacher Luftleiter m. ‖ antenne f. simple. / ~ battery switch ‖ Einfach-zellenschalter m. ‖ réducteur-adjoncteur m. simple. / ~ body (Chem) ‖ Grund-stoff m.; Element n. ‖ corps m. simple; élement m. / ~ lever cash register ‖ Ein-zählerhebelkontrollkasse f. ‖ totalisa-trice f. simple à leviers. / ~ operation (Mach) ‖ einfache Bedienung f. ‖ service m. simple. / ~ thrilling of the wires (El line) ‖ einfache Verdrehung f. der Drähte ‖ rotation f. simple des fils de fer.

simple (Weav) ‖ Sempel m.; Zampelzug m.; Zempel m. ‖ xemple m.; semple m.

simple-refracting (Opt) ‖ einfachbrechend ‖ monoréfringent.

simplex needle indicator (Tel) ‖ Einfach-nadelanzeiger m. ‖ simple aiguille f. / ~ operation (Tel) ‖ Einfachbetrieb m. ‖ exploitation f. sans multiplage.

simplification ‖ Vereinfachung f. ‖ simpli-fication f.

simplified roof support (El line) ‖ verein-fachter Dachstützpunkt m. ‖ appui m. de toit simplifié.

simplify, to ‖ vereinfachen ‖ simplifier. / ~ the calculation ‖ die Rechnung f. ver-einfachen ‖ simplifier le calcul.

simulate, to ‖ vortäuschen ‖ simuler.

simultaneity ‖ Gleichzeitigkeit f. ‖ simul-tanéité f.

simultaneous ‖ gleichzeitig ‖ simultané. / ~ existence of a number of disintegrated supply networks (Electr) ‖ Nebenein-anderbestehen n. kleiner vielfach zer-splitterter Versorgungsnetze ‖ un grand nombre m. de petits réseaux existants les uns à côté des autres. / ~ telegraph and telephone working ‖ Simultan-telegrafie f. ‖ télégraphie f. simultanée.

sinapism (Med) ‖ Senfpflaster n. ‖ sina-pisme m.

sine (Math) ‖ Sinus m. ‖ sinus m.

sine bar ‖ Sinuslineal n. ‖ règle f. inclinée.

sine curve ‖ Sinuslinie f.; Sinuskurve f. ‖ sinusoïde f. / maximum of a ~ ‖ Wellen-berg m. einer Sinuslinie ‖ maximum m. d'une sinusoïde. / minimum of a ~ ‖ Wellental n. einer Sinuslinie ‖ minimum m. d'une sinusoïde.

sine galvanometer ‖ Sinusbussole f. ‖ bous-sole f. des sinus.

sine line see sine curve.

sinew (Anatomy) ‖ Sehne f. ‖ tendon m.

sine wave current (Electr) ‖ Sinusstrom m. ‖ courant m. sinusoïdal.

sing, to ‖ singen; pfeifen ‖ chanter; siffler.

singe, to ‖ sengen; abbrennen; flambieren ‖ griller; flamber; passer à la flamme. / ~ the thread (Spinn) ‖ den Faden m. sengen ‖ flamber ou gazer le fil.

singeing (Weav) ‖ Sengen n.; Brennen n.; Flambieren n. ‖ grillage m.; flambage m.; gazage m. / ~ of railway sleepers ‖ Verkohlung f. der Schwellen ‖ flambage m. des traverses. / ~ of yarn ‖ Sengen n. des Garnes ‖ gazage m. de fils.

singeing machine (Weav) ‖ Sengmaschine f. ‖ machine f. à griller; grilleuse f. / ~ for yarns ‖ Garnsengmaschine f. ‖ machine f. à griller les fils.

singeing plate ‖ Sengeplatte f. ‖ plaque f. à griller.

singer (Weav) ‖ Senger m. ‖ flambeur m.

singing (Radio) ‖ Pfeifen n. ‖ chantage m.; sifflement m. / ~ of repeaters ‖ Eigen-tönen n. bei Verstärkern ‖ amorçage m. d'oscillations des répéteurs.

singing arc (Radio) ‖ musikalischer Licht-bogen m. ‖ arc m. musical.

single ‖ vereinzelt; einzeln; einzig ‖ seul; séparé; détaché.

single (Silk) ‖ Pelseide f.; Pelo f. ‖ poil m.

single-acting ‖ einfachwirkend ‖ de simple effet.

single apparatus ‖ Einzelgerät n. ‖ appa-reil m. séparé ou individuel.

single-axle vehicle ‖ einachsiges Fahrzeug n.; Einachsfahrzeug n. ‖ tracteur m. à essieu unique.

single battery-switch ‖ Einfachzellenschal-ter m. ‖ réducteur m. simple.

single-bay aeroplane ‖ Einstieler m. ‖ avion m. à une paire de mâts.

single-blow ‖ Einzelschlag m. ‖ frappe f. coup-par-coup. / ~ cold upsetting ma-chine ‖ Eindruckkaltpresse f. ‖ presse f. à froid à simple effet.

single-box plansifter ‖ Einkastenplansich-ter m. ‖ plansichter m. à caisse unique.

single-circuit transposition (El line) ‖ Schleifenkreuzung f. ‖ croisement m. des fils.

single-collar thrust bearing ‖ Einscheiben-drucklager n. ‖ palier m. de butée à disque unique.

single-cord operation (Tel) ‖ Einschnur-betrieb m. ‖ exploitation f. par mono-corde. / ~ switchboard ‖ Einschnur-schrank m. ‖ tableau m. monocorde. / ~ system ‖ Einschnursystem n. ‖ système m. monocorde.

single-crank compound steam engine ‖ Einkurbelverbunddampfmaschine f. ‖ machine f. à vapeur compound à simple manivelle.

single-cup insulator ‖ Einfachglockenisola-tor m. ‖ isolateur m. à simple cloche.

single-cylinder steam engine ‖ Einzylin-derdampfmaschine f. ‖ machine f. à vapeur monocylindrique ou à cylindre unique.

single-deck aeroplane ‖ Eindecker m. ‖ monoplan m.

single drive ‖ Einzelantrieb m. ‖ commande f. séparée ou individuelle.

single-ended spanner ‖ einfacher Schrau-benschlüssel m. ‖ clef f. simple (à vis).

single-engined ‖ einmotorig ‖ monomo-

teur. / ~ aeroplane ‖ einmotoriges Flug-zeug n. ‖ avion m. monomoteur.

single eyeglass ‖ Lorgnon n. ‖ lorgnon m.

single-flange wheel ‖ Rad n. mit einem Spurkranz ‖ roue f. à boudin unique.

single-flight wood stairs pl. ‖ einläufige Holztreppe f. ‖ escalier m. en bois à une volée.

single-frame hammer ‖ Einständerham-mer m. ‖ marteau m. pilon à single jambage. / ~ steam hammer ‖ Ein-ständerdampfhammer m. ‖ marteau m. pilon à vapeur à jambage unique.

single-gyro compass ‖ Einkreiselkompaß m. ‖ compas m. monogyroscopique; com-pas m. à un seul gyroscope.

single hood (Auto) ‖ Halbverdeck n. ‖ capote f.; demi-toit m.; demi-pavillon m.

single-housing grinding machine ‖ Ein-ständerschleifmaschine f. ‖ machine f. à rectifier à un montant.

single-key typewriter ‖ Eintasterschreib-maschine f. ‖ machine f. à écrire à une seule touche.

single lubricator ‖ Einzelöler m. ‖ graisseur m. séparé.

single magnifier ‖ Einzellupe f. ‖ loupe f. simple.

single measurement ‖ Einzelmessung f. ‖ mesure f. séparée ou individuelle.

single number (Of a journal etc.) ‖ Einzel-nummer f. ‖ numéro m. seul.

single observation ‖ Einzelbeobachtung f. ‖ observation f. isolée.

single parts pl. ‖ Einzelteile mpl. ‖ pièces fpl. détachées.

single-phase alternating current ‖ Ein-phasenwechselstrom m. ‖ courant m. alternatif monophasé. / ~ motor ‖ Ein-phasenwechselstrommotor m. ‖ moteur m. à courant alternatif monophasé. / ~ motor with auxiliary phase ‖ Ein-phasenwechselstrommotor m. mit Hilfs-phase ‖ moteur m. à courant alternatif monophasé avec phase auxiliaire.

single-phase current ‖ Einphasenstrom m. ‖ courant m. monophasé. / ~ small transformer ‖ Einphasenstromkleintrans-formator m. ‖ transformateur m. mono-phasé de faible puissance.

single-phase starter ‖ Einphasenanlasser m. ‖ démarreur m. monophasé.

single pickaxe ‖ Flachhacke f. ‖ pioche f. simple.

single-piece furnace ‖ Blasofen m. ‖ four-neau m. à loupe.

single plate (Tiler) ‖ Futterblech n. ‖ en-nusure f.; annusure f.; basque f.

single-prism spectrograph ‖ Einprisma-spektrograf m. ‖ spectrographe m. à un prisme.

single-pulley drive ‖ Einscheibenantrieb m.; Einzelscheibenantrieb m. ‖ com-mande f. par monopoulie.

single-purpose automatic (Turning lathe) ‖ Sonderzweckautomat m. ‖ tour m. auto-matique pour usages spéciaux. / ~ drilling machine ‖ Bohrmaschine f. für Sonderzwecke ‖ machine f. à percer spéciale. / ~ machine tool ‖ Eintypwerk-zeugmaschine f. ‖ machine outil f. pour modèle unique. / ~ planer ‖ Hobel-maschine f. für Sonderzwecke ‖ raboteuse f. spéciale. / ~ riveting machine ‖ Niet-maschine f. für Sonderzwecke ‖ riveuse f. spéciale. / ~ turret lathe ‖ Revolver-

drehbank f. für Sonderzwecke ‖ tour m. revolver spécial.

single-railed ‖ einspurig; Einschienen-... ‖ à une seule voie.

single-revolution machine ‖ Eintourenmaschine f. ‖ machine f. à un tour de cylindre.

single-roller drier ‖ Einwalzentrockner m. ‖ sécheur m. à cylindre unique.

single-row inner race with taper bore (Ball bearing)‖Innenring m. mit kegeliger Bohrung ‖ bague f. intérieure à alésage conique. / ~ rigid ball journal bearing ‖ einreihiges Querkugellager n. ‖ roulement m. à simple rangée de billes. / ~ rigid ball journal bearing with taper clamping sleeve ‖ einreihiges Spannhülsenkugellager n. ‖ roulement m. à simple rangée de billes et à manchon. / ~ rigid deep groove ball journal bearing ‖ Hochschulterkugellager n. ‖ roulement m. à simple rangée de billes à gorge profonde. / ~ self-aligning ball journal bearing ‖ einreihiges Querkugellager n. mit Einstellring ‖ roulement m. à simple rangée de billes et à rotule.

single-sashed window ‖ einflügeliges Fenster n. ‖ fenêtre f. à un battant.

single-seater fighter (Airpl) ‖ Kampfeinsitzer m. ‖ monoplace m. de combat.

single-shear rivet joint ‖ einschnittige Nietung f. ‖ rivure f. à une section de cisaillement.

single-slide bar guides pl. ‖ einschienige Kreuzkopfführung f. ‖ glissière f. à barre simple.

single spark gap ‖ Einzelfunkenstrecke f. ‖ éclateur m. simple.

single-stage ‖ einstufig ‖ à un seul étage. / ~ air compressor ‖ einstufige Luftpumpe f. ‖ pompe f. à air à une seule phase.

single-stroke bell system for fire alarm systems ‖ Einschlagglockensystem n. in Feuermeldeanlagen ‖ sonnerie f. à coups isolés dans les installations d'avertisseurs d'incendie.

single-track line ‖ eingleisige Eisenbahn f. ‖ chemin m. de fer à une seule voie.

single-way switch ‖ Einwegumschalter m. ‖ commutateur m. à une seule direction.

single window ‖ Einfachfenster n. ‖ fenêtre f. simple.

single-wire line (Electr) ‖ Eindrahtleitung f. ‖ ligne f. à fil unique.

singlings pl. (Distill) ‖ Lutter m.; Läuter m.; Nachlauf m. ‖ blanquette f.

singular ‖ einzeln; einfach ‖ simple; singulier.

sink, to ‖ sinken; einsinken; sich senken; rutschen ‖ s'affaisser; se tasser; se seller; glisser. / ~ (Mar) ‖ sinken; untergehen ‖ couler; couler bas; sancir; sombrer. / ~ a debt ‖ eine Schuld tilgen *oder* amortisieren ‖ amortir une dette. / ~ a hole in stone ‖ ein Loch n. in Stein einhauen ‖ refouiller un trou. / ~ in (Build) ‖ sich setzen *oder* senken ‖ s'affaisser; farder; prendre coup. / ~ in (Carp; Join etc.) ‖ einlassen ‖ enchâsser. / ~ into the bottom ‖ in den Boden versickern ‖ s'infiltrer dans le terrain. / ~ the loops pl. (Weav) ‖ kulieren ‖ cueillir. / ~ a shaft (Mine) ‖ einen Schacht abteufen *oder* niederbringen ‖ foncer *ou* avaler un puits.

sink (Hydr arch) ‖ Kloake f.; Abzugschleuse f. ‖ égout m.; cloaque m. / ~ (Road) ‖ Gossenstein m.; Rinnstein m. ‖ caniveau m.; dalle f. cuñière; lavoir m.

d'immondices. / ~ (For kitchen) ‖ Ausguß m.; Rinnstein m. ‖ évier m.; issue f.

sink basin ‖ Ausgußschale f. ‖ bassin m. d'évier.

sinker (Mine) ‖ Schachthauer m.; Hauer m. beim Schachtabteufen ‖ avaleur m. / ~'s pick ‖ Senkhaue f. ‖ pic m. d'avaleur *ou* en niveau.

sinker (Weav) ‖ Platine f. ‖ platine f. / ~ bar of a stocking loom ‖ Platinenbarre f. eines Strumpfwirkerstuhles ‖ barre f. à platines d'un métier à bas. / ~ ring ‖ Platinenkranz m. ‖ couronne f. à platines.

sink hole (Build) ‖ Senkgrube f.; Abzugsgrube f. ‖ puisard m.; trappe f. de nettoyage.

sinking (Build) ‖ Senkung f. ‖ enfoncement m.; fonture f.; affaissement m. / ~ (Met) ‖ Rohrfrischen n. ‖ travail m. de la loupe; affinage m. premier. / ~ (Mine) ‖ Abteufen n.; Abteufung f. ‖ fonçage m.; avaleresse f.; defonçement m. / ~ irregular ~ of the charge in the blast furnace ‖ Rutschen n. *oder* Kippen n. der Gichten im Hochofen ‖ éboulement m. des charges dans le haut-fourneau. / ~ through a layer (Mine) ‖ Durchteufung f. einer Schicht ‖ creusement m. d'une couche. / ~ of piles in miner's method ‖ bergmännische Absenkung f. von Pfeilern ‖ fonçage m. de piliers selon l'usage des mineurs. / ~ of a pit ‖ Schachtabteufen n.; Abteufen n. eines Schachtes ‖ avaleresse f. *ou* défoncement *ou* fonçage m. d'un puits. / ~ of a pit to the refrigerating process ‖ Schachtabteufung f. nach dem Gefrierverfahren ‖ creusement m. de puits de mine d'après le procédé de congélation. / ~ of a pit to the low temperature process ‖ Schachtabteufung f. nach dem Tiefkälteverfahren ‖ creusement m. de puits au procédé de congélation à basse température. / ~ of a pit to the sunk shaft process ‖ Schachtabteufung f. nach dem Senkschachtverfahren ‖ creusement m. de puits de mine à cuvelage. / ~ of a shaft *see* ~ of a pit. / ~ through a stratum ‖ Durchteufung f. einer Schicht ‖ creusement m. d'une couche. / ~ of the water level ‖ Sinken n. des Wasserstandes ‖ abaissement m. du niveau d'eau.

sinking bucket ‖ Abteufkübel m. ‖ benne f. de creusement. / cost of ~ ‖ Abteufkosten pl. ‖ coût m. du fonçage. / dynamic ~ force (Aero) ‖ dynamische Sinkkraft f. ‖ force f. descensionnelle dynamique. / ~ fund ‖ Tilgungsfonds m.; Amortisationsfonds m. ‖ fonds m. d'amortissement. / ~ hammer (Mine) ‖ Abteufhammer m. ‖ marteau m. à puits.

sinking-in (Build) ‖ Einsinken n.; Sacken n.; Senkung f. ‖ affaissement m.; enfoncement m.

sinking machine (Mine) ‖ Abteufmaschine f. ‖ machine f. de foncement. / ~ paper ‖ Löschpapier n. ‖ papier m. brouillard. / ~ pump (Mine) ‖ Abteufpumpe f.; Senkpumpe f. ‖ pompe f. de creusement; pompe f. suspendue *ou* volante. / ~ trestle (Mine) ‖ Abteufgerüst n. ‖ échafaudage m. pour creusements.

sink stone (Road) ‖ Gossenstein m. ‖ ruisseau m.

sinople (Miner) ‖ Sinopel m.; sinopische Erde f.; dunkelroter Eisenkiesel m. ‖ sinople m.

sinter, to ~ (Met: to clinker) ‖ sintern ‖ se concrétionner; se fritter; s'agglutiner. / ~ (To filter)‖sickern; sintern; aussickern ‖ suinter; filtrer.

sinter ‖ Sinter m. ‖ sinter m. / calcareous ~ ‖ Kalksinter m.; Stalaktit m.; Tropfstein m. ‖ stalactite f. / siliceous ~ ‖ Kieselsinter m.; Quarzsinter m. ‖ opale f. incrustante; quartz m. agate thermogène.

sinter brick ‖ Sinterstein m. ‖ brique f. frittée.

sintering coal ‖ Sinterkohle f. ‖ houille f. maigre à longue flamme; houille f. demigrasse. / ~ limit ‖ Sintergrenze f. ‖ limite f. de concrétion. / ~ plant ‖ Sinteranlage f. ‖ installation f. de concrétion.

sinuosity of a river ‖ Flußkrümmung f.; Rack m. ‖ coude m. *ou* sinuosité f. (du lit) d'une rivière.

siphon, to ‖ hebern; mit einem Siphon heben ‖ siphonner.

siphon ‖ Heber m.; Saugheber m.; Wassersackrohr n.; Siphon m.; Stechheber m. / ~ (Hydr arch) ‖ Düker m.; Kanaldüker m. ‖ passage m. d'eau; aquéduc-siphon m. / liquid ~ ‖ Flüssigkeitsheber m. ‖ siphon m. à liquides.

siphon acidometer ‖ Hebersäuremesser m. ‖ acidimètre m. à siphon. / ~ barometer ‖ Heberbarometer n.; Heberluftdruckmesser m. ‖ baromètre m. à siphon. / ~ bottle ‖ Siphonflasche f. ‖ siphon m. / ~ cup for oil ‖ Ölbüchse f. mit Heberdocht ‖ boîte f. à huile avec mèche à siphon. / ~ pipe ‖ Heberrohr n. ‖ tuyau m. à siphon. / ~ pipe of an hot-blast apparatus (Met) ‖ Hosenrohr n. eines Winderhitzungsapparates ‖ siphon m. *ou* tuyau m. à siphon d'un appareil à air chaud. / ~ recorder (Tel) ‖ Heberschreiber m. ‖ récepteur m. à siphon; siphon-recorder m. / ~ tap (Met) ‖ Selbststich m. ‖ siphon m. / ~ wick ‖ Heberdocht m. ‖ mèche f. à siphon.

siren ‖ Sirene f. ‖ sirène f. / electric ~ ‖ elektrische Sirene f. *oder* Hupe f. ‖ trompe f. *ou* corne f. électrique. / steam ~ ‖ Dampfsirene f. ‖ sirène f. à vapeur.

sirup ‖ Sirup m. ‖ sirop m. / caramelized ~ ‖ karamelisierter Sirup m. ‖ sirop m. caramélisé. / maltose ~ ‖ Maltosesirup m. ‖ sirop m. de maltose. / ~ of molasses ‖ Melassesirup m. ‖ syrop m. de mélasse. / sulphosot ~ ‖ Sulfosotsirup m. ‖ sirop m. sulfosot.

sirup discharging valve ‖ Sirupablaufventil n. ‖ soupape f. de vidange pour sirop. / ~ factory ‖ Sirupfabrik f. ‖ siroperie f. / ~ groove ‖ Siruprinne f. ‖ rigole f. à sirop. / ~ separating valve ‖ Siruptrennventil n. ‖ soupape f. de séparation pour sirop.

Sisal fibre ‖ Sisalfaser f. ‖ fibre f. de Sisal. / ~ hemp ‖ Sisal m.; Sisalhanf m.; Yukatansisal m. ‖ sisal m.; chanvre m. de Sisal.

sister ship ‖ Schwesterschiff n. ‖ frère-jumeau m.; navire-frère m. / ~ ships pl. ‖ Schwesterschiffe npl. ‖ navires mpl. jumeaux. / ~ vessel *see* sister ship.

site (Build) ‖ Bauplatz m.; Baugrundstück n. ‖ emplacement m. / ~ of the discovery ‖ Fundort m. ‖ lieu m. d'une découverte. / ~ of ore (Met) ‖ Rosthaufen m. ‖ tas m. de minerai grillé.

sitting ‖ Sitzung f. ‖ séance f.; session f.

sitting machine for the dressing industry ‖ Setzmaschine f. für die Aufbereitungsindustrie ‖ séparatrice f. des dépôts pour les préparations industrielles.

sitting room ‖ Wohnzimmer n. ‖ chambre f. de famille.

situated on ... ‖ liegend *oder* gelegen an ... ‖ situé à ... / ~ at an infinite distance ‖ in unendlicher Entfernung f. liegend ‖ situé à une distance f. infiniment grande.

situation (Appointment) ‖ Anstellung f.; Stellung f.; Posten m.; Stelle f. ‖ emploi m.; engagement m.; placement m. / ~ (State) ‖ Lage f.; Zustand m.; Situation f. ‖ situation f. / financial ~ ‖ finanzielle Lage f. ‖ situation f. financière. / ~ of works ‖ Lage f. *oder* Standort m. eines Werkes ‖ situation f. *ou* emplacement m. d'une usine. / making use of the ~ of works ‖ die Standortsvorteile mpl. eines Werkes ausnutzen ‖ mettre en valeur les avantages offerts par l'emplacement d'une usine.

six-angled ‖ sechseckig ‖ hexagonal.

six-chambered (Fire arm) ‖ sechsläufig; sechsschüssig ‖ à six coups.

six-foot way (Railw) ‖ Schienenzwischenraum m.; Mittelweg m. ‖ entre-voie f.

six-sided ‖ sechsseitig ‖ à six pans.

six-thread line (Mar) ‖ Sechsgarnleine f. ‖ ligne f. à six fils.

six-wheel brake ‖ Sechsradbremse f. ‖ frein m. à six roues.

six-wheeler (Auto) ‖ Sechsradwagen m.; Dreiachser m. ‖ automobile f. à six roues.

size, to ~ (To calibrate) ‖ kalibrieren ‖ calibrer.

size, to ~ (To glue) ‖ leimen ‖ encoller. / ~ the paper ‖ das Papier leimen *oder* planieren ‖ encoller *ou* coller le papier. / ~ the warp (Weav) ‖ die Kette stärken ‖ encoller la chaîne avec de l'empois.

size (Glue) ‖ Leimwasser n.; dünner Leim m.; Planierwasser n.; Kleister m. ‖ eau f. de) colle f. / ~ (Weav) ‖ Schlichte f.; Schlichtleim m. ‖ parement m.; breuvet m. / leather ~ ‖ Lederleim m. ‖ colle f. de cuir. / vegetable ~ ‖ Harzleim m. ‖ colle f. résineuse.

size (Dimensions)‖Abmessung f.; Größe f.; Format n. ‖ dimension f.; grandeur f. / ~ (Print) ‖Format n.; Satzgröße f. ‖ format m. / ~ of the bolter ‖ Siebgröße f. ‖ numéro m. du tamis. / ~ of a booklet ‖ Format n. einer Broschüre ‖ format m. d'une brochure. / ~ of coal ‖ Kohlengröße f.; Korngröße f. der Kohle ‖ grandeur f. de charbon; grosseur f. du grain. / full ~ ‖ natürlicher Maßstab m.; natürliche Größe f. ‖ grandeur f. naturelle. / ~ of grain ‖ Korngröße f. ‖ grosseur f. du grain. / ~ of the image ‖ Bildgröße f. ‖ grandeur f. de l'image. / ~ of an ingot ‖ Größe f. eines Blockes ‖ dimensions fpl. d'un lingot. / pieces up to largest ~s ‖ Stücke npl. bis zu den größten Abmessungen ‖ pièces fpl. jusqu'aux plus grandes dimensions. / medium ~ ‖ Mittelgröße f. ‖ dimension f. moyenne. / natural ~ ‖ natürliche *oder* wirkliche Größe f.‖ grandeur f. naturelle. / of natural ~ ‖ in natürlicher Größe f. ‖ de grandeur f. réelle; en grandeur naturelle. / ~ of paper ‖ Papierformat n. ‖ format m. du papier. / real ~ see natural ~. / ~ of sheet ‖ Papiergröße f.; Papierformat n. ‖ format m. de papier. / ~ of slot ‖ Spaltgröße f. ‖ largeur f. de la fente. / ~ at the

top end (Wood) ‖ Zopfstärke f. ‖ dimensions fpl. au sommet. / ~ of type (Print) ‖ Schriftgrad m. ‖ force f. de corps. / upright ~ (Of paper) ‖ Hochformat n. ‖ format m. normal. / ~ of yarn ‖ Feinheit f. *oder* Feinheitsgrad m. des Garnes ‖ titre m. du fil.

size box ‖ Kleisterkasten m. ‖ bac m. à colle.

size colour ‖ Leimfarbe f. ‖ peinture f. à la colle; détrempe f.

sizel (Coin) ‖ Münzgekrätz n. ‖ cisailles fpl.

size maker (Pap) ‖ Leimarbeiter m. ‖ encolleur m. / ~ (Weav) ‖ Schlichtekocher m. ‖ préparateur m. de colle.

size paint ‖ Leimfarbe f. ‖ peinture f. à la colle.

sizer (Bookb) ‖ Leimer m. ‖ encolleur m.; colleur m. / ~ (Weav) ‖ Schlichter m. ‖ encolleur m. / hand ~ (Weav) ‖ Handschlichter m. ‖ encolleur m. à la main. / paper ~ ‖ Papierleimer m. ‖ colleur m. de papier.

sizing (Pap) ‖ Leimung f.; Leimen n.; Planieren n. ‖ encollage m. / ~ (Weav) ‖ Schlichten n.; Leimen n. ‖ collage m.; encollage m.; gommage m. / double ~ of fabrics ‖ zweiseitiges Appretieren n. von Geweben ‖ apprêt m. de deux côtés des tissus. / ~ with glue water (Gild) ‖ Leimtränken n. ‖ encollage m. / ~ in the vat (Pap) ‖ Leimen n. in der Bütte ‖ collage m. à la cuve *ou* en pâte. / ~ of yarn ‖ Schlichten n. des Garnes ‖ encollage m. des fils.

sizing (Classification ‖ Sichtung ‖ f. *oder* Ordnung f. *oder* Sortierung f. nach der Größe ‖ classification f.; calibrage m. / ~ (Coal dress; Ore dress) ‖ Klassierung f.; Siebung f. ‖ classement m.; triage m.; séparation f.; criblage m. / ~ of the carded yarn ‖ Feinheitsbezeichnung f. des Streichwollgarns ‖ numérotage m. du fil de laine cardée.

sizing drum ‖ Siebtrommel f.; Klassiertrommel f.; Sortiertrommel f. ‖ tambour m. cribleur; trommel-classeur m.

sizing glue ‖ Klebelack m. ‖ collage m.

sizing machine (Coin) ‖ Justiermaschine f. ‖ colifichet m. / ~ (Textile) ‖ Schlichtmaschine f.; Gummierkalander m. ‖ encolleuse f.; pareuse f.; machine f. à gommer. / ~ for single and double-sizing of fabrics ‖Gummierkalander m. zum ein- und zweiseitigen Appretieren von Geweben ‖ machine f. à gommer pour l'apprêt d'un seul côté ou sur les deux côtés des tissus. / ~ for sizing the warp ‖ Stärkemaschine f. zum Stärken der Kette ‖ machine f. à encoller la chaîne avec de l'empois.

sizing material ‖ Klebstoff m. ‖ matière f. collante; colle f.; gomme f. / ~ (Textile) ‖ Schlichtmittel n. ‖ parement m.

sizing and crushing plant for coke ‖ Kokszerkleinerungs- und -sortieranlage f. ‖ installation f. de concassage et de triage du coke.

sizing preparation (Textile) ‖ Appreturmittel n. ‖ encollage m. pour tissus; apprêt m.

sizing (rolling) mill ‖ Maßwalzwerk n. ‖ laminoir m. de précision.

sizing rubber ‖ Klebgummi m. ‖ caoutchouc m. collant. / layer of ~ ‖ Klebgummischicht f. ‖ couche f. de caoutchouc collant.

sizing tooth of a broach ‖ Kalibrierzahn m. einer Räumnadel ‖ dent m. de calibrage d'une broche.

sizing trough (Pap) see sizing vat.

sizing vat (Pap) ‖ Leimtrog m. ‖ mouilloir m.

sizzle (Radio) ‖ Zischen n.; Knattern n.; Knistern n. ‖ friture f.

skate ‖ Schlittschuh m. ‖ patin m.

skating rink ‖ Eislaufbahn f. ‖ piste f. de patinage. / artificial ~ ‖ künstliche Eisbahn f. ‖ piste f. de patinage artificielle.

skein ‖ Docke f.; Strähne f. ‖ écheveau m.; flotte f.; bouin m. / bundle of ~s ‖ Docke f. ‖ matteau m.; bouin m. / ~ dyeing ‖ Strangfärberei f. ‖ teinture f. de fils en écheveaux. / ~ examiner ‖ Fitzenarbeiterin f. ‖ repasseuse f. d'écheveaux.

skeleton ‖ Skelett n.; Gerippe n. ‖ squelette m. / ~ of a building ‖ Gerippe n. eines Gebäudes ‖ carcasse f. *ou* squelette m. d'un bâtiment. / ~ of reinforced concrete ‖ Betongerippe n. ‖ ossature f. en béton. / ~ of a survey ‖ Netz n. einer Vermessung ‖ reseau m. *ou* canevas m. d'une ligne.

skeleton key ‖ Dietrich m.; Nachschlüssel m. ‖ fausse clef f.; rossignol m.

skelp ‖ Platine f.; Rohrschiene f. ‖ lame f. à canon; maquette f.

sketch, to (Drawing) ‖ skizzieren; aufzeichnen ‖ esquisser; croquer. / ~ (Surv) ‖ aufnehmen ‖ lever.

sketch (Drawing) ‖ Entwurf m.; Skizze f. ‖ esquisse f.; croquis m. / ~ (Surv) ‖ Aufnahme f. ‖ lever m. / constructional ~ ‖ Konstruktionsskizze f. ‖ croquis m. de construction. / dimensioned ~ ‖ Maßskizze f. ‖ esquisse f. cotée; croquis m. coté. / as shown in the ~ ‖ wie Skizze f. zeigt ‖ comme indiqué dans le croquis. / ~ to scale ‖ Maßskizze f. ‖ croquis m. coté. / ~ for a telegraph line plan ‖ Wegezeichnung f. zum Telegrafenwegeplan ‖ croquis m. d'un projet d'une ligne télégraphique.

sketch block ‖ Zeichenblock m. ‖ blocnotes m. à dessin.

sketching pad ‖ Skizzierblock m. ‖ carnet m. à croquis. / ~ paper ‖ Skizzierpapier n. ‖ papier m. à croquis.

skewed ‖ abgeschrägt ‖ biseauté; chanfreiné.

skewer (Mine) ‖ Schießnadel f.; Räumnadel f. ‖ épinglette f. / ~ turning machine ‖ Drehbank f. für Aufsteckspindeln ‖ machine f. à tourner les brochettes.

skew surface ‖ windschiefe Fläche f. ‖ surface f. gauche. / ~ wheel ‖ Hyperboloidenrad n. ‖ roue f. hyperbolique.

ski ‖ Ski m.; Schneeschuh m. ‖ ski m. / ~ (Aero) ‖ Schneekufe f. ‖ ski m. / ~ aeroplane ‖ Kufenflugzeug n. ‖ avion m. skieur *ou* à skis.

skid, to (Auto) ‖ schleudern ‖ déraper.

skid (Airplane) ‖ Kufe f. ‖ patin m. / ~ (Carr) ‖ Hemmschuh m. ‖ enrayure f.; chien m.; sabot m. d'enrayage. / auxiliary ~ (Aero) ‖ Hilfssporn m. ‖ béquille f. auxiliaire.

skid chain (Carr) ‖ Hemmkette f. ‖ chaîne f. d'enrayure; chaîne à enrayer *ou* d'enrayage; enrayoir m. / non-~ device ‖ Gleitschutzvorrichtung f. ‖ antidérapant m. / ~ runner (Aero) ‖ Schlittenkufe f. ‖ patin m.

skids pl. (Shipb) ‖ Reibholz n.; Freihalter m.; Fender m. ‖ défense f. en bois.

skiff ‖ Kahn m. ‖ canot m.; barque f.; nacelle f.; esquif m.

skill ‖ Geschicklichkeit f. ‖ aptitude f. / requiring great ~ ‖ große Übung f. erfordernd ‖ exigeant beaucoup de pratique f. ou d'habilité.

skilled labourer ‖ geübter oder gelernter Arbeiter m. ‖ artisan m.; homme m. de métier.

skillet (Met) ‖ (Gußstahl-) Tiegel m. ‖ creuset m.

skim, to ‖ abschöpfen ‖ écumer. / ~ (Brew) ‖ abschäumen ‖ écumer. / ~ (Milk) ‖ rahmen; entrahmen ‖ écrémer.

skim ‖ Schaum m. ‖ écume f.

skimmed off ‖ abgerahmt ‖ écremé. / ~ lead ‖ Abstrichblei n. ‖ plomb m. d'écumage.

skimmer ‖ Abhebelöffel m. ‖ écumoire f. / ~ (Person) ‖ Abschäumer m. ‖ écumeur m.

skim milk ‖ Magermilch f. ‖ lait m. écrémé.

skimming ‖ Abschöpfen n. ‖ écumage m. / ~ ‖ Abschäumen n. ‖ écumage m. / ~ ladle see skimmer.

skimmings pl. ‖ Schaum m. ‖ écume f.

skimming spoon see skimmer. / ~ system (Brew) ‖ Abschäummethode f. ‖ système m. d'écumage.

skin ‖ Fell n.; Haut f. ‖ peau f. / badger's ~ ‖ Dachsfell n. ‖ peau f. de blaireau. / ~ of casting ‖ Gußrinde f. ‖ croûte f. de la fonte. / pieces pl. without hard ~ of casting ‖ Gußstücke npl. ohne harte Kruste ‖ pièces fpl. sans croûte dure de coulée. / ~ of lamb ‖ Lammleder n. ‖ canepin m. / raw ~ ‖ rohes Fell n. ‖ peau f. brute. / ~ of a ship ‖ Außenhäute fpl. eines Schiffes ‖ bordé m. extérieur d'un navire. / smoothed ~ ‖ Blöße f. ‖ cuiret m.; peau f. planée. / tanned ~ ‖ gegerbtes Leder n. ‖ cuir m. corroyé. / untanned ~ ‖ ungegerbtes Fell n. ‖ peau f. non-tannée.

skin beater ‖ Fellklopfer m. ‖ batteur m. de peaux. / ~ cleaner ‖ Fellreiniger m. ‖ échardonneur m. de peaux. / ~ colourer ‖ Fellfärber m. ‖ coloriste m. en peau. / ~ comber ‖ Fellkämmer m. ‖ dégaleur m. de peaux. / ~ curing ‖ Häutekonservierung f. ‖ conservation f. des peaux brutes. / ~ cutter ‖ Fellhaarschneider m. ‖ coupeur m. de poils. / ~ drying ‖ Häutetrocknung f. ‖ séchage m. de peaux. / ~ effect (Electr) ‖ Hauteffekt m. ‖ effet m. Kelvin. / ~ friction ‖ Oberflächenreibung f. ‖ frottement m. superficiel. / ~ glue ‖ Lederleim m. ‖ colle f. de peau. / ~ microscope ‖ Hautmikroskop n. ‖ microscope m. dermatologique.

skinner ‖ Fellarbeiter m. ‖ peaussier m. / ~ (Knacker) ‖ Abdecker m. ‖ écorcheur m.; exoriateur m.

skinplate (Shipb) ‖ Außenhautplatte f. ‖ plaque f. d'enveloppe.

skin rug ‖ Fellteppich m. ‖ tapis m. en fourrure ou en peau. / ~ salter ‖ Häutesalzer m. ‖ saleur m. de peaux. / ~ scourer ‖ Fellreiniger m. ‖ sabreur m. / ~ sewer ‖ Fellnäherin f. ‖ bourseuse f. / ~ trampler ‖ Felltrampler m. ‖ fouleur m. de peaux. / ~ trimmer ‖ Fellbeschneider m. ‖ façonneur m. de peaux. / ~ unhairing the ~ ‖ Enthaarung f. der Felle ‖ délainage m. des peaux.

skip, to ‖ überspringen ‖ sauter. / ~ (Aut tel) ‖ übergreifen; verschränken ‖ changer.

skip (Crane) ‖ Kraneimer m. ‖ benne f.; baquet m. / ~ (Mine) ‖ Förderkübel m. ‖ baquet m.; tonne f.

skipper (Typewr) ‖ Überspringvorrichtung f. ‖ dispositif m. de sautage.

skipping of the type bars (Typewr) ‖ Tanzen n. der Typenhebel ‖ sautillement m. des tiges à caractères. / ~ motion ‖ hüpfende Bewegung f. ‖ mouvement m. sautillant.

skirt ‖ Rock m.; Frauenrock m. ‖ robe f.; jupe f. / ~ (Border) ‖ Einfassung f.; Rand m.; Saum m. ‖ bordure f.; bord m.; ourlet m. / ~ of a millstone ‖ Mahlbahn f. ‖ feuillure f. d'une meule. / ~ of a saddle ‖ kleine Satteltasche f. ‖ petit quartier m.; double quartier m. / ~ of a sail ‖ Saum m. oder Saumstreifen m. des Segels ‖ gaine f. de voile.

skirting board ‖ Scheuerleiste f.; Fußleiste f. ‖ socle m. de lambris; lambris m. d'appui bas.

skirt maker ‖ Rockarbeiterin f. ‖ jupière f.

skittle ‖ Spielkegel m.; Kegel m. ‖ billon m.

skiver (Curr) ‖ Abschaber m. ‖ racleur m. de peaux.

skiving cuts pl. (Curr) ‖ Schärfarbeit f. ‖ travail m. de parage.

skiving machine, stiffener ~ (Shoem) ‖ Kappenschärfmaschine f. ‖ machine f. à parer les contreforts.

skiving and splitting machine (Curr) ‖ Schärf- und Spaltmaschine f. ‖ machine f. à parer et à refendre.

skiving machine operator ‖ Maschinenspalter m. ‖ raseur m. à la machine.

sklerometer ‖ Sklerometer n.; Härtemesser m. ‖ scléromètre m.

skull ‖ Schädel m. ‖ crâne m. / ~ cap ‖ Käppchen n. ‖ calotte f. / ~ fracture ‖ Schädelbruch m. ‖ fracture f. du crâne.

skute ‖ Schute f. ‖ bateau-foncet m.; chaland m.

sky ‖ Himmel m. ‖ ciel m. / bright ~ ‖ heller Himmel m. ‖ ciel m. brillant. / clear ~ ‖ klarer Himmel m. ‖ ciel m. clair. / cloudy ~ ‖ bewölkter Himmel m. ‖ ciel m. nuageux. / lamb's wool ~ ‖ Himmel m. mit Schäfchenwolken ‖ ciel m. pommelé. / overcast ~ ‖ bedeckter Himmel m. ‖ ciel m. couvert.

sky blue ‖ himmelblau ‖ bleu ciel.

skylight ‖ Deckenlicht n.; Oberlicht n. ‖ jour m. d'en haut; hypèthre m.; jour m. à plomb; claire-voie f. / ~ above a door ‖ Oberlichtfensterchen n. über einer Tür ‖ fenêtrelle f.; imposte f. vitrée. / half ~ ‖ Seitenoberlicht n. ‖ jour m. d'en haut.

skylight fittings pl. ‖ Oberlichtbeschläge mpl. ‖ ferrures fpl. de claire-voie. / ~ grating (Shipb) ‖ Schutzgitter n. der Deckfenster ‖ grille f. de protection de claire-voie. / ~ turret ‖ Dachaufsatz m. ‖ lanterne f.; tourelle f. ouverte.

skyscraper ‖ Wolkenkratzer m.; Hochhaus n.; Turmhaus n. ‖ gratte-ciel m. / ~ of reinforced concrete ‖ Hochhaus n. aus Eisenbeton ‖ gratte-ciel m. en béton armé.

sky vault ‖ Himmelswölbung f. ‖ voûte f. céleste.

slab (Build) ‖ Platte f.; Tafel f. ‖ dalle f.; lame f.; planche f. / ~ (Metal) see slab bloom. / ~ (Spinn) ‖ Lunte f.; Vordergespinst n. ‖ boudin m.; mèche f. / ~ of cement ‖ Zementplatte f. ‖ car-reau m. ou dalle f. en ciment. / landing ~ ‖ Treppenabsatz m. ‖ palier m. / majolica ~ ‖ Majolikafliese f. ‖ carreau m. de majolique. / ~ of marble ‖ Marmorplatte f. ‖ dalle f. de marbe. / sawed ~ of marble ‖ geschnittene oder gesägte Marmorplatte f. ‖ tranche f. de marbre. / ~ for platforms ‖ Laufbühnenbelag m. ‖ planche f. de plate-forme. / ~ of slate ‖ Schieferplatte f.; Schiefertafel f. ‖ table f. d'ardoise. / ~ of timber ‖ Schalbrett n.; Schwartenbrett n.; Schwarte f. ‖ dosse f.; dosse-flache f.; flache f.

slabbing (Spinn) ‖ Luntenspinnen n. ‖ filage m. en gros. / ~ mill ‖ Luppenwalzwerk n. ‖ laminoir m. à blooms.

slab bloom ‖ Bramme f. ‖ brame f. / ~ shears pl. ‖ Brammenschere f. ‖ cisaille f. à brames.

slab, dressing with ~s ‖ Betäfelung f. mit Platten ‖ tablement m. / ~ line (Mar) ‖ Schlappgording n. ‖ cargue-à-vue f. / putting-on device for ~s ‖ Auflegevorrichtung f. für Brammen ‖ dispositif m. de chargement pour lingots. / ~ shears pl. see ~ bloom shears pl.

slack, to (Lime) see to slake. / ~ off see to slacken.

slack ‖ schlaff; entspannt; locker; lose ‖ flasque; lâche; mou; détendu. / to be ~ ‖ lose sein; schlaff sein ‖ avoir du mou. / to be ~ (Mach) ‖ Spiel n. oder Spielraum m. haben ‖ jouer; avoir du jeu.

slack hours (Tel) ‖ verkehrsschwache Zeit f. ‖ heures fpl. de faible trafic. / ~ rope release (Mach) ‖ Schlaffseilausrückung f. ‖ mécanisme m. de débrayage par câble mou. / ~ rope test ‖ Schlappseilprüfung f. ‖ essai m. du mou de câble. / ~ wind ‖ schwache Brise f. ‖ brise f. molle.

slack (Coal) ‖ Feinkohle f.; Kohlenklein n.; Grus m. ‖ menu charbon m.; charbon m. fin. / ~ of a cable see slackening of the rope. / inferior ~ (Coal) ‖ minderwertige Kohle f. ‖ charbon m. de rebut.

slack adjuster ‖ Gestängenachstellvorrichtung f. ‖ ajusteur m. de timonerie. / ~ coal see slack.

slacken, to ‖ lockern ‖ relâcher. / ~ the fire ‖ das Feuer mäßigen ‖ modérer le feu. / ~ the screws ‖ losschrauben ‖ dévisser; desserrer ou défaire ou ôter les vis fpl. / ~ speed ‖ die Fahrt vermindern ‖ diminuer la vitesse.

slackening of the rope ‖ Schlappwerden n. des Seiles ‖ mou m. du câble. / ~ of the speed ‖ Abnahme f. der Geschwindigkeit ‖ diminution f. de la vitesse.

slackening motion ‖ abnehmende Bewegung f. ‖ mouvement m. de relâche.

slack silk ‖ Stickseide f.; Flachseide f.; Plattseide f. ‖ soie f. à broder.

slag, to ‖ sintern; zusammensintern ‖ se fritter; se concréter. / ~ out ‖ abschlacken ‖ laisser couler le laitier.

slag ‖ Schlacke f.; crasse f.; laitier m.; scorie f. / ~ (Forg) ‖ Herdschlacke f.; Schmiedeschlacke f. ‖ mâchefer m. / ~ accumulated in dump hills ‖ Schlacke f., zu Halden aufgehäuft ‖ laitier m. entassé sur des terrils. / acid ~ ‖ saure Schlacke f. ‖ laitier m. acide. / basic ~ ‖ basische Schlacke f. ‖ scorie f. ou laitier m. basique. / blast furnace ~ ‖ Hochofenschlacke f. ‖ laitier m. de haut-fourneau. / blast furnace ~ in pieces ‖ Hochofenstückschlacke f. ‖ laitier m. du

haut-fourneau en morceaux. / crystalline ~ || kristallinische Schlacke f. || scorie f. crystalline. / devitrified ~ || entglaste Schlacke f. || scorie f. dévitrifiée. / to do away the ~s || die Schlacke abwerfen || retirer les scories fpl. avec le fourgon. / to draw the ~s || die Schlacke ziehen || faire sortir les crasses. / earthy ~ || erdige Schlacke f. || scorie f. terreuse. / finery ~ || Frischschlacke f.; Lacht m. || scorie f. d'affinerie. / to free from ~s || entschlacken || débarasser des scories fpl. / granulated blast furnace ~ || granulierte Hochofenschlacke f. || laitier m. granulé de haut-fourneau. / honey-comb ~ || bimssteinartige Schlacke || scorie f. poreuse. / ~ of liquation || Krätzschlacke f.; Seigerschlacke f. || scorie f. de liquation. / neutral ~ || neutrale Schlacke f. || scorie f. neutre. / phosphatic ~ || phosphorhaltige Schlacke f. || scorie f. phosphatée. / grinding the ~ into phosphate meal || Verarbeitung f. von Schlacke zu Phosphatmehl || broyage m. des laitiers en farine de phosphate. / raw ~ || Rohschlacke f. || scorie f. brute ou crue; laitier m. pauvre. / rich ~ (Metal) || Garschlacke f. || scorie f. douce ou riche; laitier m. riche. / to run off the ~s pl. see to do away the ~s. / stone line ~ || steinige Schlacke f. || scorie f. pierreuse. / strongly oxidizing ~ || stark oxydierende Schlacke f. || scorie f. riche en oxygène. / to take off the ~s || abschlacken; die Schlacke abziehen || faire couler les scories fpl. / Thomas ~ || Thomasschlacke f. || scorie f. Thomas. / ~ of viscous fluidity || zähflüssige Schlacke f. || scorie f. consistante. / vitrous ~ || glasige Schlacke f. || scorie f. vitreuse.

slag analysis || Schlackenuntersuchung f. || analyse f. du laitier. / ~ bank || Schlackenhalde f. || crassier m. / ~ bed see ~ bottom. / ~ bottom || Schwahlboden m. || fond m. ou sole f. de sorne. / ~ bottom process || Schwahlarbeit f. || affinage m. au bain des scories. / ~ breaker || Schlackenbrecher m. || concasseur m. de scorie.
slag brick || Schlackenziegelstein m. || brique f. de laitier. / ~ manufacturing plant || Schlackensteinherstellungsanlage f. || installation f. pour la fabrication des pierres en scorie.
slag bucket || Schlackenkübel m. || baquet m. à scories. / ~ car || Schlackenwagen m. || chariot m. à laitier. / ~ cart see ~ car. / ~ cement || Schlackenzement m. || ciment m. de laitier ou de scorie. / ~ channel || Schlackenlauf m. || voie f. de scorie; rigole f. à laitier. / ~ concrete || Schlackenbeton m. || beton m. de laitier. / ~ duct see ~ channel. / ~ dump || Schlackenhalde f. || terril m.; crassier m.; halde f. / ~ fibre || Schlackenfaser f. || fibre f. de scorie. / ~ fining process see ~ bottom process.
slagged out || abgeschlackt || décrassé.
slagger || Schlackenarbeiter m. || décrasseur m.
slagging || Verschlackung f. || scorification f.
slag granulation || Schlackengranulation f. || granulation f. du laitier.
slaggy iron || schlackenreiches Eisen n. || fer m. gras.
slag hair see slag wool. / ~ hearth || Schlackenherd m. || fourneau m. à scories. / ~ hole || Schlackenloch n. || ouverture f. à

la scorie; porte f. de scorie. / ~ ladle || Schlackenpfanne f.; Schlackenkümpel m. || poche f. à laitier; pot m. à scorie. / ~ lead || Krätzblei n. || plomb m. de crasse. / ~ lifting device || Schlackenhebewerk n. || appareil m. de levage de laitier. / ~ man || Schlackenläufer m. || brouetteur m. de crasses. / ~ mill to grind down Thomas slag into phosphate meal || Schlackenmühle f. für die Verarbeitung der Thomasschlacke zu Phosphatmehl || moulin m. à laitiers pour le broyage des laitiers Thomas en farine de phosphate. / ~ mould || Schlackenform f. || moule f. à laitier. / ~ Portland cement || Eisenportlandzement m. || ciment m. de laitier Portland. / ~ process || Sinterprozeß m. || procédé m. de scorification. / realization of ~ || Schlackenverwertung f. || utilisation f. de scorie. / ~ removing plant || Entschlackungsanlage f. || installation f. d'enlèvement de scories. / ~ sand || Schlackensand m. || sable m. de laitier. / ~ stone || Schlackenstein m.; Schlackenziegel m. || brique f. de scories ou de laitier. / ~ tapping side || Schlackenabstichseite f. || côté f. coulée des scories. / ~ tongs pl. || Schlackenzange f. || pince f. à scorie. / ~ treatment || Schlackenverarbeitung f. || traitement m. du laitier. / ~ truck || Schlackenwagen m. || chariot m. à laitier. / ~ wagon see ~ truck. / ~ washer || Schlackenwäscher m. || laveur m. de laitier. / ~ washery || Schlackenwäsche f. || laverie f. de scories. / ~ wheeler || Schlackenfahrer m. || rouleur m. de cendres.
slag wool / Schlackenwolle f. || laine f. de laitier ou de scorie. / ~ covering || Schlackenisolierung f. || enveloppe f. en laine de scorie.
slake, to || löschen || éteindre. / ~ lime || Kalk m. löschen || éteindre la chaux vive. / ~ the lime with too much water || den Kalk m. ersaufen lassen || noyer la chaux.
slaked || gelöscht || éteint.
slake pan || Kalklöschpfanne f. || caisse f. à chaux.
slaking of lime || Löschen n. des Kalkes || extinction f. de la chaux. / ~ basket (Mas) || Löschkorb m. || panier m. de maçon ou à claire-voie. / ~ basket (Sugar) || Löschkorb m. || panier m. à clairée.
slaking drum || Löschtrommel f. || tambour m. extincteur. / ~ battery || Löschtrommelbatterie f. || batterie f. de tambours extincteurs. / ~ process || Löschtrommelverfahren n. || procédé m. par tambour extincteur.
slant || schief; schräg; abschüssig || oblique; incliné; en pente f. / to cut on the ~ || keilförmig zuschneiden || biaiser; couper en biais.
slanting see slant.
slant rolling mill || Schrägwalzenwerk n. || laminoir m. oblique.
slasher (Weav) || Maschinenschlichter m. || encolleur m. à la machine.
slashing (Weav) || Schlichten n. || parage m.
slashings pl. (Forest) || Schlagabraum m. || déchets mpl. de coupes.
slash saw || Schlitzsäge f.; Schießsäge f. || scie f. à tenon.
slate || Schiefer || ardoise f.; schiste m. / ~ (Writing) || Schiefertafel f. || table f.

en ardoise. / adhesive ~ || Klebschiefer m. || argile f. feuilletée. / argillaceous ~ || Tonschiefer m. || schiste m. argileux. / artificial ~ || künstlicher Schiefer m. || ardoise f. artificielle. / asbestos ~ manufacturing plant || Asbestschieferherstellungsanlage f. || installation f. pour la fabrication des ardoises d'asbeste. / ~ for billiard table || Billardschiefer m. || ardoise f. pour billards. / bituminous ~ || Brandschiefer m. || pyroschiste m.; schiste m. bitumineux. / ~ in blocks || Schiefer m. in Blöcken || ardoise f. en blocs. / ~ for buildings || Bauschiefer m. || ardoise f. pour la construction. / cross-grained ~ || Griffelschiefer m. || ardoise f. à écrire. / cupriferous ~ || Kupferschiefer m. || schiste m. cuivreux. / cut ~ || geschnittener Schiefer m. || ardoise f. taillée. / drawing ~ || schwarze Kreide f.; Zeichenschiefer m. || schiste m. graphique; crayon m. noir; ampélite f. graphique. / flinty ~ || Probierstein m.; edler Kieselschiefer m. || pierre f. de touche; silex m. corné; lydienne f. / metal ~ || Metallschiefer m. || ardoise f. métallique. / patched ~ || Fleckschiefer m. || schiste m. tacheté ou noyeux. / ~ in plates || Schiefer m. in Platten || ardoise f. en dalles. / polished ~ || polierter Schiefer m. || ardoise f. polie. / sawn ~ || gesägter Schiefer m. || ardoise f. sciée. / ~ in sheets || Schiefer m. in Tafeln || ardoise m. en tables. / thin leaved and veiny ~ || dünnblättriger und adriger Schiefer m. || feuilletis m.
slate black || Schieferschwarz n. || noir m. de schiste. / ~ board see slate (Writing). / ~ bucket elevator || Schieferbecherwerk n. || noria f. à schistes. / ~ clay || Schieferton m. || argile f. schisteuse. / ~ clay quarry || Tonschieferbruch m. || carrière f. de schistes argileux. / ~ coal || Schieferkohle f. || houille f. schisteuse. / contents pl. of ~ || Schiefergehalt m. || teneur f. en schiste. / ~ cutter || Schieferschneider m. || tailleur m. d'ardoises. / ~ finisher || Schieferabrunder m. || arrondisseur m. d'ardoises. / ~ gray (Chem) || Schiefergrau n. || gris m. de schiste. / ~ like see slaty. / ~ note block || Schiefernotizblock m. || bloc-notes m. en ardoise. / ~ oil || Schieferöl n. || huile f. de schiste. / ~ paper || Schieferpapier n. || papierardoise m. / ~ peg || Schiefernagel m. || clou m. ou pointe f. à ardoise. / ~ pencil || Griffel m. || crayon m. d'ardoise. / ~ pit see slate quarry. / ~ quarrier || Schiefergrubenarbeiter m.; Schieferbrecher m. || ardoisier m.; piqueur m. d'ardoise.
slate quarry || Schieferbruch m. || ardoisière f.; carrière f. d'ardoise. / ~ foreman || Schiefergrubenaufseher m. || écaillou m.
slater (Build) || Schieferdecker m. || couvreur m. en ardoise.
slate roof || Schieferdach n. || couverture f. en ardoise. / ~ roofing || Eindecken n. mit Schiefer || couverture f. en ardoise.
slater's anvil || Schieferdeckeramboß m. || enclume f. du couvreur. / ~ hammer || Schieferdeckerhammer m. || martelet m. de couvreur. / ~ nail see slate peg.
slate slab for roofs || Dachschiefer m. || table f. en ardoise pour toits; schiste m. tégulaire. / ~ spar || schaliger Kalkspat m.; Schieferspat m. || chaux f. carbonatée nacrée. / ~ splitter || Schiefer-

spalter m. || répartonneur m. / ~ stone || Schieferstein m. || pierre f. d'ardoise.

slating see slate roof and slate roofing.

slaty || schieferig || ardoiseux; schisteux. / ~ coal see slate coal. / ~ marl || Mergelschiefer m. || schiste m. marneux.

slaughter, to || schlachten || abattre; égorger; tuer.

slaughter animal || Schlachttier n. || animal m. de boucherie.

slaughterer || Schlächter m. || boucher m.; tueur m. de bestiaux. / home ~ || Hausmetzger m.; Hausschlächter m. || abatteur m. ou boucher m. à domicile.

slaughterhouse || Schlachthaus n. || abattoir m. / ~ equipment || Schlachthof einrichtung f. || installation f. d'abattoir. / ~ keeper || Schlachthausaufseher m. || surveillant m. d'abattoir. / ~ manager || Schlachthausverwalter m. || directeur m. d'abattoir.

slaughtering || Schlachtung f. || abatage m. / ~ hall see slaughterhouse. / ~ mask || Schlachtmaske f. || masque m. d'abattoir.

slavering chain || Schaumkette f. || chaînette f. d'un mors de bride.

sleave silk || Flockseide f. || bourre f. de soie.

sled || Schlitten m. || traîneau m. / motor ~ || Motorschlitten m. || traîneau m. automobile.

sledge || Schlitten m.; Lastschlitten m. || traîneau m. / ~ (Ropem) || Schlitten m.; Toppschlitten m. || carrosse m.; chariot m. / ~ (Tool) see sledge hammer. / ~ brake (Railw) || Schlittenbremse f.; Schuhbremse f. || frein m. de traîneau. / ~ chime || Schlittengeläut n. || clochette f. pour traîneaux.

sledge hammer (Forg) || Zuschlaghammer m.; Vorschlaghammer m. || masse f.; marteau m. à devant. / ~ (Mine) || schweres Treibfäustel n. || masse f. de fer. / ~ (Railw) || Schienenhammer m.; Schwellenhammer m. || marteau m. à rails. / ~ driven by compressed air || Preßluftschmiedehammer m. || marteau m. de forge à air comprimé. / ~ sharp-faced ~ (Forg) || Senkhammer m. || fonçoir m.

sled plane || Kufenflugzeug n. || avion m. skieur ou à skis.

sleek, to || glätten; Glanz geben || lisser; lustrer.

sleek || glatt; glänzend || lisse; brillant; lustré.

sleeking machine (Pap) || Glättmaschine f. || lissoir m.; machine f. à lisser. / ~ stick (Shoem) || Glättholz n.; Glättschiene m. || buis m. / ~ tool see ~ stick.

sleeper (Bridge) || Brückenbaum m. || poutrelle f. / ~ (Build) || Schwelle f.; Grundschwelle f.; Bodenschwelle f.; Grundbalken m. || racinal m.; dormant m.; semelle f. / ~ (Railw) || Eisenbahnschwelle f.; Schwelle f. || traverse f.; bille f. / bevelled rectangular ~ (Railw) || trapezförmige Schwelle f. || traverse f. trapéziforme. / bottom ~ (Railw) || Bodenschwelle f. || traverse f. de plancher. / channelled ~ (Railw) || Rillenschwelle f. || traverse f. à gorge. / the ~s are closed at the end by pressing || an den Enden sind die Schwellen fpl. durch Zusammenpressen geschlossen || les extrémités fpl. des traverses ont été fermées sous la presse. / fir wood ~ (Railw) || Tannenschwelle f. || traverse f. en sapin. / full-squared ~ || vollkantige Schwelle f. || traverse f. bien

équarrie. / ~ of a ground floor || Lagerschwelle f. oder Unterzug m. eines Fußbodens || racinal m. ou sole f. de plancher. / half-round ~ || halbkreisförmige Schwelle f. || traverse f. semi-circulaire ou demironde. / impregnated ~ || getränkte oder imprägnierte Schwelle f. || traverse f. imprégnée. / iron ~ || eiserne Schwelle f. || traverse f. en fer. / joint ~ || Stoßschwelle f. || traverse f. de joint. / longitudinal ~ || Langschnelle f. || longrine f. / lopped ~ || gekappte Schwelle f. || traverse f. avec fermeture de tête. / to pack the ~s pl. || Schwellen fpl. unterstopfen || bourrer les traverses fpl. / rectangular ~ || rechteckige Schwelle f. || traverse f. rectangulaire. / reinforced concrete ~ || Eisenbetonschwelle f. || traverse f. en béton armé. / square ~ || behauene Schwelle f. || traverse f. en bois équarri. / ~ of stairs || Treppensohle f. || patin m. d'escalier. / steel ~ with welded-on rib plates || eiserne Schwelle f. mit aufgeschweißten Rippenplatten || traverse f. en fer avec des selles à nervures soudées dessus. / sunk ~ || gesunkene Schwelle f. || traverse f. déversée. / timber ~ || Holzschwelle f. || traverse f. en bois. / wooden ~ see timber ~.

sleeper, distance of ~s between centres || Schwellenentfernung f. von Mitte zu Mitte || écartement . m. des traverses d'axe en axe. / ~ drilling machine || Schwellenbohrmaschine f. || perceuse f. de traverse. / ~ fastening with gauge rods || Schwellenbefestigung f. mit Spurstangen || fixation f. des traverses par tringles d'écartement. / ~ fish plate || Schwellenlasche f. || éclisse f. de traverse. / impregnation of ~s || Eisenbahnschwellenimprägnierung f. || imprégnation f. des traverses. / ~ press || Schwellenpresse f. || presse f. à traverses. / ~ screw screwing machine || Schwellenschrauben-Eindrehmaschine und -Ausdrehmaschine f. || machine f. à visser et dévisser les tirefonds. / ~ straightening machine || Schwellenrichtmaschine f. || machine f. à dresser les traverses. / top of ~ || Schwellendecke f. || table f. supérieure de la traverse. / width of ~ || Schwellenbreite f. || largeur f. de la traverse. / width of ~ at the foot || untere Schwellenbreite f. || largeur f. de la traverse en bas. / width of ~ at the top || obere Schwellenbreite f. || largeur f. de la traverse en haut. / ~ working machine || Schwellenbearbeitungsmaschine f. || machine f. à travailler les traverses.

sleeping blanket || Schlafdecke f. || couverture f. de nuit. / ~ car || Schlafwagen m. || wagon-lit m. / ~ compartment || Schlafwagenabteil n. || coupé m. lit. / ~ partner || stiller Teilhaber m. || assoicé m. commanditaire. / ~ room || Schlafraum m. || dortoir m.

sleet || Glatteis n. || verglas m.

sleeve || Muffe f.; Hülse f.; Buchse f. || manchon m.; douille f. / braided ~ (Light) || geklöppelter Strumpf m. || manchon m. tricoté. / ~ shaft with bronze ~ || Welle f. mit Bronzebezug || arbre m. à chemise en bronze. / clamping ~ || Klemmuffe f. || manchon m. de (raccordement à) serrage. / deflating ~ (Aero) || Entleerungsschlauch m. || manche f. de dégonflement. / fixed ~ for welding and fusing burners || feste Schlauchtülle f.

für Schweiß- und Schneidbrenner || douille f. fixe pour brûleurs à souder et à découper. / governor ~ || Reglermuffe f. || manchon m. de régulateur. / ~ for grip cocks || Schlauchtülle f. für Griffhähne || douille f. pour robinets à poignée. / light-proof connecting ~ || Lichtmanschette f. || manchon m. de raccordement étanche à la lumière. / loose ~ || lösbare Schlauchtülle f. || douille f. mobile. / packing ~ || Rillenbuchse f. || fourrure f. à rainures. / removable ~ || lösbare Schlauchtülle f. || douille f. mobile. / sliding ~ || Schiebehülse f. || douille f. coulissante ou de mise au point.

sleeve brick || Hohlstein n. || brique f. creuse. / filling up of the ~ joints || Vergießen n. der Muffenkupplungen || coulage m. des joints à manchon. / ~ link || Manschettenknopf m. || bouton m. de manchette. / ~ maker (Tail) || Ärmelarbeiter m. || confectionneur m. de manches. / cradle assuming the shape of a ~ || muffenartige Wiege f. || berceau m. à moufles. / ~ silk || Galletseide f.; Seidenabfall m. || bourre f. de soie. / ~ socket || Reduziereinsatz m.; Paßeinsatz m. || coussinet m. de réduction. / ~ valve || Muffenventil n. || soupape f. à manchon.

sleigh see also sled and sledge. / ~ for children || Kinderschlitten m. || traîneau m. pour enfants.

slew, to || schwenken || tourner; converser.

slewing bracket || Gelenkarm m. || applique f. articulée.

slewing crane || Drehkran m. || grue f. pivotante ou tournante. / portable ~ || fahrbarer Drehkran m. || grue f. pivotante transportable. / wall ~ || Wandkran m. || grue f. murale pivotante. / radius of ~ || Drehkreis m. eines Drehkranes || rayon m. balayé par une grue tournante fixe.

slewing gear, pinion of ~ || Drehwerkritzel n. || pignon m. du dispositif de rotation.

slice (Dyer; Print) || Schabeisen n.; Farbeisen n. || grattoir m. / ~ (Glazier) || Spatel m.; Spachtel m. || truelle f.; spatule f. / regulating ~ (Fly press) || regulierendes Schabeisen n. || grattoir m. régulateur.

slicer, vegetable ~ || Gemüseschneider m. || coupe-légume m.; ustensile m. servant à triturer les légumes.

slicing lathe || Abstechbank f. || tour m. à tronçonner ou à décolleter. / pipe ~ || Rohrabstechbank f. || banc m. à tronçonner les tubes.

slicing machine || Abstechmaschine f. || machine f. à tronçonner. / meat, ham and sausage ~ || Aufschnittschneidemaschine f. || machine f. à couper la charcuterie.

slick || Schlamm m. || schlich m. / ~ of waste metal || Krätzschlich m. || schlich m. du déchet des métaux.

slide, to || gleiten || glisser. / ~ on a graduation || auf einer Teilung f. gleiten || glisser sur une graduation.

slide (Action) || Gleiten n. || glissement m. / ~ (Device) || Schieber m.; Schiebevorrichtung f. || curseur m.; coulisseau m.; coulisse f. / ~ (Belt) || Schnalle f. || boucle f. / ~ (Geol) || Verwerfung f. || faille f.; rejet m. / ~ (Mach) || Führung f.; guide f.; coulisse f. / ~ (Tel) || Läufer m.; Schlitten m. || chariot m. / ~ (Typewr; Balance) || Reiter m. || curseur m. / axle ~ || Achslagerführung f. || guide m. de la boîte d'essieu. / ~ of bottom discharging

wagon ‖ Schieber m. des Bodenentleerers ‖ trappe f. de déchargement par le fond. / glass ~ ‖ Glasschieber m. ‖ pièce f. coulissante en verre. / ~ of a lathe ‖ Schlitten m. *oder* Support m. einer Drehbank ‖ chariot m. *ou* support m. à chariot d'un tour. / reversing ~ ‖ Wendesupport m. ‖ support m. à renversement. / ~ of a slide rule ‖ Schieber m. eines Rechenschiebers ‖ réglette f. d'une règle à calcul. / ~ of a sliding door ‖ Türlaufschiene f. ‖ glissière f. de porte roulante.

slide, adjustment of ~ ‖ Stößelverstellung f. ‖ réglage m. du coulisseau. / ~ armature brake ‖ Verschiebeankerbremse f. ‖ frein m. à induit balladeur.

slide bar ‖ Geradführung f.; Gleitbahn f.; Gleitschiene f. ‖ glissière f.; guide m. / ~ bracket ‖ Gleitbahnträger m. ‖ porteglissière f. / single ~ guide ‖ einschienige Kreuzkopfführung f. ‖ glissière f. à barre simple.

slide bearing *see* sliding bearing. / ~ block ‖ Gleitblock m. ‖ coulisseau m. / ~ bolt (Arm) ‖ Sicherung f. ‖ verrou m. de sûreté. / ~ brake ‖ Schlittenbremse f.; Schuhbremse f. ‖ frein m. de traîneau. / ~ chair ‖ Gleitstuhl m. ‖ plaque f. de glissement. / ~ channel (Lathe) ‖ Schlittenfalz m. ‖ coulisse f. du support. / clamping surface on ~ ‖ Stößelspannfläche f. ‖ surface f. de serrage de la semelle du coulisseau. / guide of ~ (Lathe) ‖ Stößelführung f. ‖ guidage m. du coulisseau. / ~ induction apparatus ‖ Schlittenapparat m. ‖ appareil m. d'induction à curseur. / ~ lathe ‖ Supportdrehbank f. ‖ tour m. à support. / ~ part ‖ Verschiebekörper m. ‖ pièce f. déplaçable.

slide plate ‖ Gleitfläche f. ‖ plan m. de glissement. / ~ for crossings of lines ‖ Zungenplatte f. für Gleiskreuzungen ‖ plaque f. de pointe pour croisements de voies.

slide preserver for motor cars ‖ Gleitschutz m. für Kraftfahrzeuge ‖ antidérapant m. pour automobiles. / leather ~ preventing device ‖ Ledergleitschutz m. ‖ antidérapant m. en cuir; cuir m. pour empêcher le glissage.

slider *see also* slide ‖ Schieber m. ‖ curseur m. / ~ (Buckle) ‖ Schiebering m. ‖ coulant m. / ~ (Radio) ‖ Reiter m.; Schleifer m. ‖ glisseur m.

slide rail (Mach) ‖ Spannschiene f.; Gleitschiene f.; Stellschiene f. ‖ glissière f. / ~ rest ‖ Support m.; Werkzeugschlitten m. ‖ support m. / ~ rest tool ‖ Drehstahl m. ‖ couteau m.; outil m. de tour. / ~ ring ‖ Gleitring m. ‖ anneau-guide m.; curseur m. / ~ rule ‖ Rechenschieber m. ‖ règle f. à calcul. / ~ spring ‖ Gleitfeder f. ‖ ressort m. coulissant *ou* de glissement.

slide valve ‖ Schieber m.; Flachschieber m. ‖ tiroir m. / auxiliary ~ ‖ Hilfsschieber m. ‖ tiroir m. auxiliaire. / balanced ~ ‖ Entlastungsschieber m. ‖ tiroir m. équilibré. / ~ for chimneys ‖ Kaminschieber m. ‖ registre m. de cheminée. / circular ~ ‖ Rundschieber m. ‖ tiroir m. rond. / superheated ~ for locomotives ‖ Heißdampfschieber m. für Lokomotiven ‖ tiroir m. à locomotives pour vapeur surchauffée. / tubular ~ ‖ Rohrschieber m. ‖ tiroir m. tubulaire.

slide valve box ‖ Schieberkasten m. ‖ boîte f. à tiroir *ou* à vapeur; boîte f. de distri-

bution. / distribution by ~ ‖ Flachschiebersteuerung f. ‖ distribution f. par tiroir plan. / distribution by tubular ~ ‖ Rohrschiebersteuerung f. ‖ distribution f. par valve tubulaire. / ~ face ‖ Schieberfläche f. ‖ bande f. de frottement; plane m. de tiroir; glissière f. / ~ main face ‖ Schieberspiegel m. ‖ table f. de tiroir. / ~ position ‖ Schieberstellung f. ‖ position f. du tiroir.

slide window ‖ Schiebefenster n.; Schubfenster n.; Aufziehfenster n. ‖ fenêtre f. à coulisse *ou* à guillotine.

slide wire bridge (Electr) ‖ Schleifdrahtbrücke f. ‖ pont m. de mesure à fil. / ~ resistance ‖ Gleitdrahtwiderstand m. ‖ rhéostat m. à curseur.

sliding axle ‖ Lenkachse f. ‖ essieu m. mobile. / ~ bar *see* sliding bolt. / ~ bearing ‖ Gleitlager n. ‖ palier m. lisse *ou* à glissement.

sliding block ‖ Kulissenstein m. ‖ coulisseau m. / expansion ~ ‖ Expansionsgleitbacke f. ‖ coulisseau m. de détente.

sliding bolt ‖ Schubriegel m. ‖ barre f. *ou* targette f. *ou* verrou m. glissant. / ~ bottom (Slide valve) ‖ Schieberboden m. ‖ fond m. à coulisse. / ~ bow ‖ Bügelschleifkontakt m. ‖ cadre m. de glissement. / ~ caliper ‖ Schieblehre f. ‖ pied m. à coulisse. / ~ cam ‖ verschiebbarer Nocken m. ‖ came f. de déplacement. / ~ change gear ‖ Schieberädergetriebe n. ‖ engrenage m. baladeur. / ~ clutch ‖ Rutschkupplung f. ‖ accouplement m. à glissement. / ~ contact ‖ Schleifkontakt m.; Gleitkontakt m. ‖ curseur m. / ~ damper ‖ Fallschieber m. ‖ registre m. à guillotine. / ~ diaphragm ‖ Schiebeblende f.; Blendschieber m. ‖ diaphragme m. coulissant.

sliding door ‖ Schiebetür f. ‖ porte f. à coulisses *ou* coulissante. / ~ fittings pl. ‖ Schiebetürbeschläge mpl. ‖ garnitures fpl. pour portes à coulisse.

sliding file ‖ Scharnierfeile f. ‖ lime f. plate à coulisse *ou* à charnière. / ~ fitting shop work ‖ Bandmontage f. ‖ montage m. à ruban sans fin. / ~ frame ‖ Spannplatte f. ‖ châssis-tendeur m. / ~ friction ‖ gleitende Reibung f. ‖ frottement m. de glissement. / ~ gate ‖ Schiebetor n. ‖ porte f. coulissante *ou* à glissière *ou* à coulisse. / ~ gear box ‖ Schalträderkasten m. ‖ boîte f. d'engrenages. / ~ key ‖ Gleitfeder f. ‖ languette f. de guidage. / ~ ladder ‖ Rollleiter f. ‖ échelle f. roulante. / ~ and surfacing lathe ‖ Zugspindeldrehbank f. ‖ tour m. à charioter. / ~, surfacing and screw-cutting lathe ‖ Leit- und Zugspindeldrehbank f. ‖ tour m. à charioter, surfacer et fileter. / ~ lever bar (Tel) ‖ Wanderhebelmaschine f. ‖ barre f. pour clé glissante. / ~ objective changer ‖ Schlittenobjektivwechsler m. ‖ changeur m. d'objectifs à coulisse. / ~ plane ‖ Gleitbahn f. ‖ glissière f.; voie f. de glissement. / ~ platform ‖ Schiebebühne f. ‖ chariot m. transporteur. / ~ puppet (Lathe) ‖ Reitstock m.; Spitzdocke f. ‖ poupée f. mobile *ou* à pointe; contrepoupée f. / ~ sash window *see* ~ window. / ~ saw ‖ Schlittensäge f. ‖ scie f. à chariot. / ~ scale ‖ Gleitskala f. ‖ échelle f. mobile. / ~ seat ‖ verstellbarer Sitz m. ‖ siège m. réglable. / ~ shaft ‖ Nutenwelle f. ‖ arbre m. profilé. / ~ sleeve ‖ Schiebhülse f. ‖ manchon m. à frottement. / ~

spur wheel ‖ ein- und ausrückbares Zahnrad n. ‖ roue f. dentée baladeuse. / ~ staff (Surv) ‖ Nivellierlatte f. zum Verschieben ‖ mire f. à coulisse. / ~ switch (Aut tel) ‖ Gleitwerk n. ‖ sélecteur m. glissant. / ~ table ‖ Ausziehtisch m. ‖ table f. à coulisses. / ~ tripod ‖ zusammenschiebbares Dreibeinstativ n. ‖ trépied m. à branches coulissantes. / ~ tube ‖ Schiebrohr n. ‖ tube m. coulissant. / ~ valve (Sluice) ‖ Schütz n.; Ziehschütz n.; Stauschütz n.; Schott n.; Falle f. ‖ vanne f. / ~ vane (Surv) ‖ Nivellierscheibe f.; Tafel f. einer Nivellierlatte ‖ voyant m.; plaque f.

sliding weight ‖ Laufgewicht n. ‖ poids m. curseur *ou* mobile. / ~ balance ‖ Laufgewichtswage f. ‖ balance f. à poids curseur.

sliding wheel ‖ Schlepprad n. ‖ roue f. d'entraînement. / ~ window ‖ Schiebefenster n. ‖ châssis m. glissant; fenêtre f. à coulisse.

slight accident ‖ leichter Unfall m. ‖ accident m. peu important.

slightly sick person ‖ leicht Erkrankter m. ‖ homme m. légèrement malade.

slime ‖ Schlamm m. ‖ limon m. / indurated ~ peat ‖ Lebertorf m. ‖ tourbe f. de saprocolle. / ~ pit (Ore dress) ‖ Schlammsumpf m. ‖ bassin m. de dépôt.

slimes pl. (Ore dress) ‖ Schlamm m.; Schlämme f. ‖ matières fpl. *ou* mouces fpl. fines; schlamm m. / coal ~ ‖ Kohlenschlamm m. ‖ boue f. de charbon; schlamms mpl.

slime separator ‖ Schlammscheider m. ‖ trieur m. de schlamms.

slimy ‖ schlammig ‖ limoneux. / ~ bacterium ‖ schleimbildendes Bakterium n. ‖ bactérie f. visqueuse. / ~ fermentation ‖ Schleimgärung f.; schleimige Gärung f. ‖ fermentation f. visqueuse. / ~ ground ‖ Schlammboden m. ‖ vase f. / ~ sand ‖ Schlammsand m.; Moddersand m. ‖ sable m. vasard.

sling ‖ Schlinge f.; Stropp m. ‖ élingue f. / ~ (Transportation) ‖ Tragriemen m. ‖ sangle f.; bretelle f. / arm ~ ‖ Armgurt m. ‖ brassière f. / boat's ~ (Shipb) ‖ Heißstropp m. ‖ patte f. d'embarcation.

slip, to ‖ rutschen ‖ glisser. / ~ a cable ‖ ein Tau n. schießen lassen ‖ mouliner un câble. / ~ the chain ‖ die Kette auslaufen lassen ‖ filer le câble.

slip (Gard) ‖ Steckling m.; Ableger m. ‖ bouture f.; jet m. / ~ (Grinding) ‖ Schliff m.; Schleifsel n.; Schlips m. ‖ moulée f. / ~ (Mach) ‖ Schlupf m. ‖ recul m.; glissement m. / ~ (Porcel) ‖ Schlicker m. ‖ barbotine f. / ~ (Print) ‖ Fahne f. ‖ placard m. / ~ (Sawn) ‖ Holzspan m. ‖ copeau m. de bois. / ~ (Shipb) ‖ Helling f. ‖ cale f. / ~ (Whetstone) ‖ Abziehstein m.; Schleifstein m.; Wetzstein m. ‖ pierre f. à adoucir *ou* à aiguiser *ou* à repasser; repassoir m. / ~ armature ‖ Ankerschlüpfung f.; Ankerschlupf m. ‖ glissement m. de l'induit. / ~ of the propeller ‖ Slip m. *oder* Rücklauf der Schraube ‖ recul m. du propulseur. / to pull in ~s (Print) ‖ in Fahnen fpl. abziehen ‖ placarder; tirer des épreuves. / printing receipt on sale ~s ‖ Quittungsdruck m. ‖ impression f. sur fiches de vente *ou* de caisse. / side ~ ‖ Schleudern n. ‖ dérapage m. / ~ for vessels of special shape ‖ Helling f. für Bauten von besonderer Form ‖ cale f. spé-

ciale pour la construction de bâtiments de formes peu courantes.

slip bolt ‖ Schubriegel m. ‖ verrou m. monté sur platine. / ~ **breech** (Arm) ‖ Scheibe f. *oder* Basküle f. am Gewehrschaft ‖ bascule f.; fausse-culasse f. / ~ **curve** ‖ Schlüpfungskurve f. ‖ courbe f. de glissement. / ~ **fuel tank** (Aero) ‖ Abwurfbehälter m. ‖ réservoir m. décrochable. / ~**gauge** ‖ Endmaß n. ‖ jauge f. / ~ **kiln** (Pott) ‖ Abdampfofen m. ‖ caisse f. *ou* cuve f. pour raffermir la barbotine. / ~ **meter** ‖ Schlüpfungsmesser m. ‖ indicateur m. de glissement. / ~**-on mount of a spectacle lens** ‖ Vorhängerfassung f. eines Brillenglases ‖ face f. supplémentaire d'un verre correcteur.

slipper ‖ Hausschuh m.; Pantoffel m. ‖ pantoufle f.; chausson m. / ~ (Car) ‖ Gleitschuh m. ‖ patin m. / **felt** ~ ‖ Filzpantoffel m. ‖ pantoufle f. en feutre.

slipper, cramp for ~**s** ‖ Pantoffelschnalle f. ‖ agrafe f. à pantoufles. / ~ **handle** (Shipb) ‖ Ankerfallhebel m. ‖ levier m. de mouilleur.

slipping ‖ Gleiten n. ‖ glissement m. / ~ **of clutch** ‖ Gleiten n. der Kupplung ‖ patinage m. de l'embrayage. / ~ **of earthwork** ‖ Einstürzen n. von Erdmassen ‖ mouvement m. des terres. / ~ **of the embankment** ‖ Dammrutsch m. ‖ éboulement m. du remblai. / ~ **of the rails** ‖ Wandern n. der Schienen ‖ chasse f. *ou* glissement m. *ou* marche f. des rails. / ~ **of a slope** ‖ Abbröckelung f. einer Böschung ‖ éboulement m. d'un talus.

slipping area ‖ Rutschfläche f. ‖ surface f. de glissement. / ~ **coupling** ‖ Rutschkup(p)elung f. ‖ accouplement m. à glissement. / ~**-off of the lens attachment** ‖ Abnehmen n. des Vorhängers ‖ enlèvement m. de la face supplémentaire. / ~**-on of the lens attachment** ‖ Aufsetzen n. des Vorhängers ‖ mise f. en place de la face supplémentaire. / ~ **resistance** ‖ Schlupfwiderstand m. ‖ résistance f. de glissement.

slip points pl. (Railw) ‖ Kreuzungsweiche f. ‖ traversée f. à aiguille. / **double** ~ (Railw) ‖ beiderseitige Kreuzungsweiche f. ‖ traversée-jonction f. double. / **single** ~ (Railw) ‖ einfache Kreuzungsweiche f. ‖ traversée-jonction f. simple.

slip-ring (Electr) ‖ Schleifring m. ‖ bague f. collectrice. / ~ **armature** ‖ Schleifringanker m. ‖ induit m. à bagues collectrices. / ~ **motor** ‖ Motor m. mit Schleifringanker m. ‖ moteur m. à bagues collectrices.

slip shackle ‖ Schlippschäkel m. ‖ manille f. à échappement.

slipway (Shipb) ‖ Helling f.; Ablaufgerüst n. ‖ cale f.; cale f. sèche *ou* de construction *ou* de lancement. / **breast of** ~ ‖ Vorhelling f. ‖ avant-cale f. / ~ **crane** ‖ Helling(dreh)kran m. ‖ grue f. (pivotante) pour cales sèches.

slit, to ‖ spalten; schlitzen ‖ fendre; refendre. / ~ **the iron** ‖ das Eisen schneiden ‖ fendre *ou* refendre fe fer.

slit ‖ Schlitz m.; Spalt m. ‖ fente f. / **eye** ~ **of tank** ‖ Sehschlitz m. am Kampfwagen ‖ épiscope m. du tank. / **movable in a** ~ ‖ in einem Schlitz m. beweglich ‖ coulissant dans une fente. / ~ **of the type bar** ‖ Auszahnung f. am Typenhebel ‖ évidement m. de la tige à caractères.

slit and tongue joint (Carp) ‖ Verzapfung f.; Verzahnung f. ‖ joint m. à rainure et languette. / ~ **arc lamp** ‖ Spaltbogenlampe f. ‖ lampe f. à fente à arc. / ~**-type burner** ‖ Schlitzbrenner m.; Schnittbrenner m. ‖ bec m. à fente. / ~ **cutter** ‖ gespaltener Keil m. ‖ clavette f. fendue. / ~ **diaphragm** ‖ Spaltblende f.; Schlitzblende f. ‖ diaphragme m. à fente. / ~ **guide** ‖ Schlitzführung f. ‖ conduite f. à fente. / ~ **image** ‖ Spaltbild n. ‖ image f. de fente.

slit lamp ‖ Spaltlampe f. ‖ lampe f. à fente. / ~ **lens** ‖ Spaltlampenlinse f. ‖ lentille f. de la lampe à fente. / ~ **light** ‖ Spaltlampenlicht n. ‖ lumière f. de la lampe à fente. / ~ **microscopy** ‖ Spaltlampenmikroskopie f. ‖ étude f. microscopique au moyen de la lampe à fente.

slit mechanism ‖ Spalteinrichtung f. ‖ dispositif m. de fente. / ~ **Nitra lamp** ‖ Spaltnitralampe f. ‖ lampe f. Nitra à fente. / ~ **nose bit** (Carp) ‖ Löffelbohrer m.; Hohlbohrer m. mit Zahn ‖ mèche-cuiller. / ~ **stop** *see* ~ **diaphragm**.

slitting apparatus ‖ Schlitzapparat m. ‖ raineuse f. / ~ **cutter** ‖ Schlitzfräser m. ‖ fraise f. à rainures.

slitting machine ‖ Schlitzmaschine f. ‖ machine f. à rainurer. / **gang** ~ ‖ Rollenschere ‖ cisaille f. circulaire multiple.

slitting rollers pl. (Metal) ‖ Schneidwerk n. ‖ trousse f. de fenderie. / **automatic sharpening machine for** ~ **saws** ‖ Scheibenfräserselbstschärfer m. ‖ machine f. automatique à affûter les fraises-disques.

slit tube lens ‖ Spaltröhrlinse f. ‖ lentille f. du tube à fente. / ~ **ultra microscope** ‖ Spaltultramikroskop n. ‖ ultramicroscope m. à fente.

sliver box (Spinn) ‖ erste Spulmaschine f. ‖ bobinoir m. réunisseur.

slobbering chain ‖ Schaumkette f. ‖ chaînette f. d'un mors de bride.

sloe berry ‖ Schlehenfrucht f. ‖ baie f. de prune sauvage.

sloop ‖ Jacht f. ‖ chaloupe f. / ~ **of war** ‖ Korvette f. ‖ corvette f.

slope, to (Road) ‖ böschen ‖ adosser. / ~ **steeply** ‖ steil abböschen ‖ escarper.

slope ‖ Gefälle n.; Neigung f.; Steigung f. ‖ inclinaison m.; pente f.; chute f.; égout m. / ~ (Road) ‖ Böschung f. ‖ talus m. / ~ **of the boshes** ‖ Rastwinkel m. ‖ angle m. des étalages. / ~ **of the boshes of a blast furnace** ‖ Neigung f. der Rast eines Hochofens ‖ inclinaison f. des étalages d'un haut-fourneau. / ~ **of the curve** ‖ Steilheit f. der Kurve ‖ pente f. de la courbe. / **down stream** ~ (Hydr arch) ‖ Landabdachung f.; Binnenböschung f. ‖ talus m. intérieur. / ~ **of the embankment** ‖ Dammböschung f. ‖ talus m. du remblai. / **natural** ~ (Road) ‖ natürliche Böschung f. ‖ talus m. naturel. / ~ **of roof** ‖ Dachneigung f. ‖ pente f. du toit. / ~ **of the side** (Railw) ‖ Neigung f. der Kastenwand ‖ pente f. de la paroi. / ~ **of valve characteristics** ‖ Steilheit f. von Röhrenkennlinien ‖ roideur f. des caractéristiques de valve. / ~ **of a wall** ‖ Böschung f. *oder* Schmiege f. *oder* Schräge f. einer Mauer ‖ adossement m. *ou* pente f. *ou* talus m. d'un mur.

slope, gradient of ~ ‖ Böschungswinkel m. ‖ inclinaison f. de talus. / ~ **stone** ‖ Bordstein m. ‖ pierre f. de talus.

sloping (Road) ‖ abfallend; abdachig; geneigt; schräg ‖ incliné; en pente; en talus. / ~ **beam** (Carp) ‖ Sparren m. ‖ arbalétrier m. / ~ **clamp** ‖ Feilkluppe f. ‖ étau m. à main. / ~ **end** (Railw) ‖ konische Stirnwand f. ‖ face f. frontale incliné. / ~ **position** ‖ Schräglage f. ‖ position f. inclinée. / ~ **terrace** ‖ Rampe f.; Steigung f.; Auffahrt f. ‖ rampe f.; chemin m. taluté. / ~ **wall** ‖ geböschte Mauer f. ‖ mur rampant.

sloping (Subst) *see* slope.

slot, to (Metal) ‖ bestoßen; stoßen ‖ mortaiser.

slot ‖ Spalt m.; Schlitz m. ‖ fente f. / ~ (Mach) ‖ Keilnut f.; Nut f. ‖ mortaise f.; rainure f. / **closed** ~ ‖ geschlossene Nut f. ‖ rainure f. fermée. / **dovetail** ~ ‖ schwalbenschwanzförmige Nut f. ‖ entaille f. en queue d'arronde. / **fixing** ~ ‖ Aufspannschlitz m. ‖ rainure f. de fixation. / **half-open** ~ ‖ halbgeschlossene Nut f. ‖ rainure f. sémi-ouverte. / ~ **of a letter box** ‖ Briefeinwurf m. ‖ ouverture f. pour les lettres. / **longitudinal** ~ ‖ Längsnut f. ‖ rainure f. longitudinale. / **open** ~ ‖ offene Nut f. ‖ rainure f. ouverte. / **sighting** ~ ‖ Schauritze f. ‖ fente f. d'observation.

slot coin ‖ Automatenmünze f. ‖ jeton m. de distributeur automatique. / ~ **cutter** *see* ~ **milling cutter**. / ~ **drilling machine** ‖ Langlochbohrmaschine f. ‖ machine f. à percer les trous longs. / ~ **machine** ‖ Münzenautomat m. ‖ distributeur m. automatique. / ~ **milling cutter** ‖ Nutenfräser m. ‖ fraise f. pour cannelures *ou* rainures. / ~ **milling machine** ‖ Nutenfräsmaschine f.; Nutstoßmaschine f. ‖ machine f. à fraiser les rainures. / ~ **mortising machine** (Wood working) ‖ Schlitzlochmaschine f. ‖ mortaiseuse f. / ~ **paying mechanism** ‖ Geldautomateneinrichtung f. ‖ déclenchement m. monétaire automatique; mécanisme m. de distributeur automatique. / **telescope with** ~ **paying mechanism** ‖ mit Geldautomateneinrichtung ausgerüstetes *oder* ausgestattetes Fernrohr n. ‖ lunette f. munie d'un déclenchement monétaire automatique. / ~ **plate** ‖ Schlitzplatte f. ‖ plaque f. à fente. / ~ **and window sights** pl. ‖ Diopter n. ‖ dioptre f.; viseur m.; pinnule f. / ~ **size of** ~ ‖ Spaltgröße f. ‖ dimensions fpl. de la fente.

slotted crosshead ‖ Langlochkreuzkopf m. ‖ crosse f. à coulisse. / ~ **cylinder** ‖ geschlitzter Zylinder m. ‖ cylindre m. avec fente. / ~ **jaw for brakes** ‖ Bremsgabel f. ‖ étrier m. de frein. / ~ **screw** ‖ Schlitzschraube f. ‖ vis f. fendue; boulon m. *ou* écrou m. fendu. / ~ **wing** ‖ Spaltflügel m. ‖ aile f. à fentes.

slotter *see* **slot milling machine**.

slotting (Forg) ‖ Aufschroten n.; Aufspalten n. ‖ fendage m. / ~ **forged bars** ‖ Aufspalten n. geschmiedeter Stäbe m. des barres forgées. / ~ **hammer** ‖ Keillochhammer m. ‖ marteau m. pour trous à clavette.

slotting machine ‖ Stoßmaschine f. ‖ mortaiseuse f.; machine f. à mortaiser. / ~ **for choppers** ‖ Bestoßmaschine f. für Schnitzelmesser ‖ machine f. à écorner les couteaux à découper. / **vertical** ~ ‖ Senkrechtstoßmaschine f. ‖ mortaiseuse f. verticale.

slotting tool || Stoßmeißel m.; Stoßstahl m. || outil m. à mortaiser.

slot welding || Schlitzschweißung f. || soudage m. des fentes.

slough || Schlauch m.; Balg m. || boyau m. / ~ of a snake || abgeworfene Haut f. einer Schlange || dépouille f. de serpent.

slovenly built wall || fluchtlose Mauer f. || mur m. bâti par épaulées.

slow, to ~ down the engine || den Gang m. der Maschine verlangsamen || ralentir la marche de la machine.

slow || langsam; träge; säumig; schwerfällig || lent; languissant; lourd; retardataire / ~ burning stove || Dauerbrandofen m. || poêle f. à feu continu. / ~ growing || langsam wachsend || à croissance f. lente. / ~ match || Zündschnur f. || mèche f.; étoupille f.

slow motion || Feinbewegung f.; Feinverstellung f. || mouvement m. lent. / ~ in altitude (Telescope) || Höhenfeinbewegung f. || mouvement m. lent vertical. / ~ in azimuth (Telescope) || Horizontalfeinbewegung f. || mouvement m. lent horizontal. / ~ gear || Zahnradfeinbewegung f. || mouvement m. lent à roues dentées. / ~ screw || Feinbewegungsschraube f. || vis f. à déplacement lent.

slowness || Langsamkeit f. || lenteur f.

slow releasing relay || langsam abfallendes Relais n. || relais m. à relâchement retardé. / ~ running nozzle (Auto) || Nebendüse f. für Langsamgang || gicleur m. de ralenti. / ~-speed motor || langsam laufender Motor m. || moteur m. à faible vitesse. / ~ train || Personenzug m. || train m. omnibus.

slub (Spinn) || Lunte f.; Vordergespinst n. || boudin m.; mèche f.

slubbing (Spinn) || Luntespinnen n. || filage m. en gros; béliage m. (de laine). / ~ frame (Spinn) || Grobfleier m. || banc m. à broches en gros. / ~ frame tenter (Spinn) || Vorfleier m. || bancbrocheur m. en gros.

sludge || Schlamm m. || boue f. / salt grained ~ with admixture of clay || breiartige Salzmasse f. mit Tonbeimengung || boue f. saline avec addition d'argile.

sludge draining || Schlammentwässerung f. || drainage m. du schlamm.

sludger || Schlammlöffel m. || tarière f. à clapet.

sluggish || strengflüssig || réfractaire; difficilement fusible. / ~ fermentation || träge Gärung f. || fermentation f. lente ou paresseuse.

sluice || Schleuse f. || écluse f. / ~ (Gate) || Schütz n. || vanne f. / ~ with check gates || Drempelschleuse f.; Schlagschleuse f.; Schleuse f. mit Stemmtoren || écluse f. busquée ou en éperon. / ~ with circular chamber || Kesselschleuse f. || Trommelschleuse f. || écluse f. à tambour. / coupled ~ || Doppelschleuse f. || écluse f. double. / ~ for damming plants || Schütz n. für Stauanlagen || pale f. de barrage. / ~ with screws || Schraubenschleuse f. || écluse f. à vis. / square ~ || Kastenschleuse f. || écluse f. carrée. / ~ with turning doors || Drehtorschleuse f. || écluse f. à portes tournantes.

sluice board || Ablaßschütz n.; Grundablaß m. || bonde f. ou décharge f. de fond. / ~ chamber || Schleusenkammer f.; Schleusenkessel m. || chambre f. d'écluse; neptune m. / ~ door || Drehtor n. einer Schleuse || porte f. d'écluse tournante.

sluice gate || Schleusentor n. || porte f. d'écluse. / roller ~ which can be raised and lowered || versenkbares Rollschütz n. || vannette f. à rouleaux submersible.

sluice keeper see ~ master. / ~ master || Schleusenmeister m. || éclusier m. / ~ stay || Ablaß m. || empellement m.; bonde f. / ~ valve || Abzugschieber m. || tiroir m. ou vanne f. d'évacuation. / ~ weir || Schützenwehr n. || barrage m. à vanne. / double-~ weir || Doppelschützenwehr n. || barrage m. à vannes doubles. / roller ~ weir || Walzenschützenwehr n. || barrage m. à cylindres. / ~ work || Schleusenwerk n. || vannage m.

slump (Stocks) || Sturz m. der Kurse || débâcle f. / ~ (Trade) || Preissturz m. || baisse f. (de prix) soudaine.

slur, to (Print) || schmieren || barbouiller.

slur page || Schmutzseite f. || page f. de décharge.

slush (Mach) || Schmierfett n. || graisse f. de lubrification.

small || klein; schmal; verengt; petit; resserré. / ~ beer || Einfachbier n. || bière f. de table; petite bière f. / ~ bell || Schelle f. || grelot m. / ~ board || Brettchen n. || planchette f. / ~ bran || feine Kleie f. || petit son m. / ~ capacity meter (Tel) || Kleinkapazitätsmesser m. || appareil m. pour la mesure de faibles capacités. / ~ change || Kleingeld n. || monnaie f. / ~ content || schwacher Gehalt m. || faible teneur f. / ~ current || schwacher Strom m. || courant m. faible. / ~ furniture || Kleinmöbel npl. || petits meubles mpl. / ~ hammer || Fäustel m.; Handfäustel m. || massette f. / ~ hand fire engine || Handfeuerspritze f. || pompe f. à incendie à main. / ~ holder || landwirtschaftlicher Kleinbetrieb m. || petit cultivateur m. / ~ house || Kleinhaus n. || maisonnette f.; chalet m. / ~ matter || Kleinigkeit f. || bagatelle f. / ~ package || Päckchen n. || petit paquet m. / ~ parcel || Päckchen n. || petit paquet m. / ~ piece || Stückchen n.; Schnipsel n.; || petit morceau m. / ~ piston compressor || Kleinkolbenkompressor m. || compresseur m. à piston de faible puissance. / ~-power electromotor || Kleinelektromotor m. || petit électromoteur m. / ~ power station (Radio) || Kleinstation f. || station f. à faible puissance. / ~ sack || Beutel m. || sachet m.; bourse f.; petit sac m. / ~ shovel and table brush || Schippchen m. und Tafelbürste f. || pelette f. et brosse f. de table; ramassemiettes m.

small arm || Handfeuerwaffe f. || arme f. (à feu) portative.

small arms committee || Handwaffenkomitee n. || comité m. des armes portatives. / ~ factory || Waffenfabrik f. || manufacture f. d'armes.

small art furniture || Kleinkunstmöbel npl. || meubles mpl. mignons.

small-berried coffee || kleinbohniger Kaffee m. || café m. à petits grains.

small bronze articles pl. || Bronzegalanteriewaren fpl. || petits bronzes mpl.

small charcoal || Kohlenlösche f. || poussière f. de charbon.

small coal || Feinkohle f.; Grießkohle f.; Kohlenklein n.; Grus m. || charbon m. menu ou fin; menu m.; fins mpl.

small coke || Koksklein n. || coke m. menu.

smaller amount || Minderzahl f. || nombre m. inférieur; minorité f.

small-fry breeding || Fischbrutzucht f. || élevage f. d'alevins ou de nourrains.

small-gauge locomotive || Schmalspurlokomotive f. || locomotive f. pour voie étroite.

small industry || Kleinindustrie f. || petite industrie f.

small iron || Kleineisen n. || petit-fer m. / ~ bar rolling mill || Feineisenwalzwerk n. || laminoir m. à petits fers. / ~ fittings pl. of the permanent way (Railw) || Kleineisenzeug n. für die Strecke || accessoires mpl. ou petit matériel m. de la voie. / ~ mill || Feineisenwalzwerk n. || laminoir m. à petits fers.

small iron trade || Kleineisenindustrie f. || industrie f. de quincaillerie. / ~ ware || Kleineisenindustrieerzeugnis n. || produit m. de l'industrie de quincaillerie.

small iron-ware || Kleineisenwaren fpl. || quincaillerie f. / ~ industry || Kleineisenindustrie f. || industrie f. des objets de quincaillerie.

small-meshed || engmaschig || à petites mailles fpl.

smallness || Kleinheit f. || petitesse f.

small ore || Schlich m. || schlich m.

small pica || Garmond f. || corps m. dix.

smalls pl. || Grießkohle f. || charbon m. menu. / coking ~ || Koksfeinkohle f. || fines fpl. ou charbon m. à coke. / sorted ~ (Min) || sortierte Förderkohle f. || toutvenant m. assorti.

small section rolling mill || Feineisenstraße f.; Feinstrecke f. || train m. à petits fers; petit train m.

small trade || Kleinhandel m. || commerce m. de détail.

small ware weaver || Bandweber m. || tissutier-passementier m.

small wire goods pl. || Drahtkurzwaren fpl. || quincailleries fpl. en fil métallique.

smalt || Schmalte f.; Schmaltblau n. || smalt m.; bleu m. de cobalt.

smaltine || Smaltin m.; Speisekobalt m. || smaltine f.; cobalt m. arsenical.

smaragdite || Smaragdit m. || smaragdite f.; diallage f. verte.

smart || geschäftskundig || versé dans les affaires fpl.

smash, to || zertrümmern || briser.

smear || Schmiere f.; Schmiermittel n. || enduit m.; graisse f.; cambouis m.

smell || Geruch m. || odeur f. / of disagreable ~ || übelriechend || d'odeur f. désagréable.

smelling, offensively || übelriechend || fétide.

smelling bottle || Riechfläschchen n. || flacon m. à odeurs. / ~ salt || Riechsalz n. || sel m. volatil.

smell lock || Geruchverschluß m. || fermeture f. d'odeurs. / ~ salt || Riechsalz n.; Hirschhornsalz n. || sel m. volatile.

smelt, to see also to melt || schmelzen || fondre; liquéfier. / ~ ore || Erz n. schmelzen || fondre le minerai.

smelted || geschmolzen || fondu.

smelter || Schmelzer m. || fondeur m.

smelting || Schmelzen n. || fonte f.; fusion f. / concentration ~ || Konzentrationsschmelzen n. || fusion f. de concentration. / electric ~ || elektrisches Schmelzen n. || fonte f. électrique. / oxidizing blast ~ || oxydierendes Schmelzen n. || fusion f. oxydante. / purifying ~ || solvierendes Schmelzen n. || fusion f. de dépuration. / raw ~ || Rohschmelzen n. || travail m.

cru de fonte. / reducing ~ ‖ reduzierendes Schmelzen n. ‖ fusion f. de réduction. / ~ of small ore ‖ Schlichschmelzen n. ‖ fonte f. des schlichs.

smelting apparatus ‖ Schmelzapparat m. ‖ petit four m. / ~ bath ‖ Schmelzbad n. ‖ bain m. de fusion. / ~ boiler ‖ Schmelzkessel m. ‖ marmite f. à fusion. / ~ coke ‖ Schmelzkoks m. ‖ coke m. d'usine. / ~ crucible ‖ Schmelztiegel m. ‖ creuset m. / ~ flux electrolysis ‖ Schmelzflußelektrolyse f. ‖ électrolyse f. pour la fusion des métaux.

smelting furnace ‖ Schmelzofen m. ‖ four m. à fusion. / three-phase arc ~ ‖ Drehstromlichtbogenofen m. ‖ four m. à arc électrique à courant triphasé.

smelting house ‖ Schmelzhütte f.; Schmelzerei f. ‖ fonderie f.; usine f. sidérurgique. / ~ crane ‖ Hüttenwerkskran m. ‖ grue f. pour usines métallurgiques. / ~ requirements pl. ‖ Hüttenbedarf m. ‖ articles mpl. d'usine métallurgique.

smelting pan ‖ Schmelzpfanne f. ‖ bassin m. de fusion. / ~ plant ‖ Schmelzanlage f. ‖ installation f. de fonderie. / ~ pot ‖ Schmelztiegel n. ‖ creuset m.; pot m. de fusion. / science of ~ ‖ Hüttenkunde f. ‖ métallurgie f. / ~ works pl. see smelting house. / ~ zone ‖ Schmelzzone f. ‖ zone f. ou région f. de fusion.

smith ‖ Schmied m. ‖ forgeron m.; forgeur m. / country ~ ‖ Dorfschmied m. ‖ forgeron m. de village. / gold and silver ~ ‖ Goldschmied m. ‖ bijoutier m. / hammer ~ ‖ Hammerschmied ‖ marteleur m. / tool ~ ‖ Werkzeugschmied m. ‖ forgeron m. en outils.

smithery ‖ Schmiede f. ‖ forge f.

smith's anvil ‖ Schmiedeamboß m. ‖ enclume f. de forge. / ~ tools ‖ Schmiedeamboßwerkzeuge npl. ‖ outillage m. d'enclume de forge.

smith's assistant ‖ Zuschläger m. ‖ frappeur m. / ~ bellows pl. ‖ Schmiedefeuergebläse n.; Schmiedeblasebalg m. ‖ soufflet m. de forge. / ~ blowing machine see ~ bellows. / ~ coal ‖ Schmiedekohle f. ‖ charbon m. de forge.

smith's forge ‖ Schmiedefeuer n. ‖ forge f.; feu m. de forge; chaufferie f. / portable ~ ‖ Feldschmiede f. ‖ forge f. portative ou de campagne.

smith's hearth ‖ Schmiedeesse f. ‖ forge f.; chaufferie f. / ~ fan ‖ Schmiedefeuerventilator m. ‖ ventilateur m. de forge.

smithsonite ‖ edler Galmei m.; Zinkspat m.; Zinkkarbonat m. ‖ oxyde m. de zinc carbonaté hydraté.

smith's tongs pl. ‖ Schmiedezange f. ‖ pince f. ou tenailles fpl. de forgeron. / ~ tools pl. ‖ Schmiedewerkzeug n. ‖ outil m. de forge. / ~ vice ‖ Feuerschraubstock m. ‖ étau m. à chaud. / ~ work ‖ Schmiedearbeit f. ‖ travail m. de forgeron.

smithy ‖ Schmiede f. ‖ forge f. / article for shoe smithies ‖ Hufbeschlagartikel m. ‖ article m. pour maréchalerie. / ~ coal ‖ Schmiedekohle f. ‖ fine f. forge. / ~ doubles pl. ‖ Schmiedenußkohle f. ‖ braisettes fpl. lavées pour forges. / ~ peas see ~ doubles. / ~ smalls pl. ‖ Schmiedefördergrus m. ‖ menu m. grenu.

smock-frock ‖ Kittel m. ‖ sarreau m.

smoke, to ‖ rauchen ‖ fumer. / ~ (Meat) ‖ räuchern ‖ fumer. / ~ the mould ‖ die Gußform f. anblaken oder anrauchen ‖ flamber ou noircir le moule à la fumée.

/ ~ a vessel ‖ ein Schiff ausräuchern ‖ fumiger ou parfumer un navire.

smoke ‖ Rauch m. ‖ fumée. / ~ (Met) ‖ Flugstaub m.; Dunst m.; Hüttenrauch m. ‖ fumée f. d'usine. / ~ (Meteor) ‖ Nebel m. ‖ brouillard m.; brume f.; nonvue f. / to emit thick ~ ‖ qualmen ‖ dégager d'épaisses fumées fpl.

smoke black ‖ Kienrußfarbe f.; Ruß m. ‖ noir m. de fumée; suie f. / ~ bomb ‖ Rauchbombe f. ‖ bombe f. fumigène.

smoke box ‖ Rauchkammer f. ‖ boîte f. à fumée. / ~ door ‖ Rauchkammertür f. ‖ porte f. de la boîte à fumée. / ~ drilling machine ‖ Rauchkammerbohrmaschine f. ‖ perceuse f. pour boîtes à fumée. / ~ front ‖ Rauchkammertürwand f. ‖ fond m. ou paroi f. avant de boîte à fumée. / ~ plate ‖ Rauchkammerwand f. ‖ plaque f. de boîte à fumée.

smoke burning ‖ Rauchverbrennung f.; Rauchverzehrung ‖ combustion f. de la fumée; fumivorité f. / ~ chamber (Metal) ‖ Flugstaubkammer f. ‖ chambre f. de condensation. / projectile with coloured cloud of ~ ‖ Geschoß n. mit farbiger Rauchwolke ‖ projectile m. à nuage d'éclatement coloré. / ~ combustion ‖ Rauchverbrennung f. ‖ combustion f. de la fumée. / ~ consumer ‖ Rauchverzehrer m. ‖ appareil m. fumivore; fumivore.

smoke consuming ‖ rauchverzehrend ‖ fumivore. / ~ apparatus ‖ Rauchverbrennungsapparat m. ‖ appareil m. fumivore. / ~ furnace ‖ rauchverzehrende Feuerung f. ‖ foyer m. fumivore.

smoke, consumption of ~ ‖ Rauchverzehrung f. ‖ fumivorité f. / ~ curtain ‖ Rauchschleier m.; Rauchvorhang m. ‖ écran m. de fumée.

smoked glass ‖ Rauchglas n. ‖ verre m. enfumé. / ~ meat ‖ Rauchfleisch n. ‖ viande f. fumée.

smoke development ‖ Rauchentwicklung f. ‖ dégagement m. ou émission f. de fumée. / ~ drum ‖ Dunsttrommel f. ‖ caisse f. à fumée.

smoke-dry, to ‖ räuchern ‖ fumer.

smoke, exhausting plant for ~ ‖ Rauchabsaugeanlage f. ‖ installation f. d'aspiration des fumées. / ~ flue ‖ Rauchrohr n.; Schornstein m. ‖ cheminée f.; tuyau m. de cheminée. / formation of ~ ‖ Rauchentwicklung f. ‖ dégagement m. de fumée. / ~ gas testing apparatus ‖ Rauchgasprüfungsapparat m. ‖ appareil m. à analyser les gaz de fumée. / ~ gauge ‖ Rauchgasmesser m. ‖ capnoscope m. / ~ house (Curr) ‖ Schwitzkammer f. ‖ étuve f. de fermentation.

smokeless ‖ rauchfrei ‖ sans fumée f. / ~ combustion ‖ rauchfreie Verbrennung f. ‖ combustion f. sans fumée. / ~ fuel rich in gas ‖ gasreicher, rauchloser Brennstoff m. ‖ combustible m. riche en gaz et ne produisant pas de fumée. / ~ furnace ‖ rauchfreie Feuerung f. ‖ foyer m. fumivore. / ~ powder ‖ rauchschwaches Pulver n. ‖ poudre f. brûlant sans fumée.

smoke nucleus ‖ Rauchkern m. ‖ noyau m. de fumée.

smoke pipe ‖ Kaminrohr n.; Schornstein m. ‖ tuyau m. de cheminée; cheminée f. / ~ of sheet-iron ‖ Blechschornstein m. ‖ chéminée f. en tôle.

smoke producing device ‖ Raucherzeuger m. ‖ appareil m. fumigène.

smoker, articles pl. for ~s ‖ Gegenstände mpl. für Raucher ‖ articles mpl. pour fumeurs.

smoke sail (Shipb) ‖ Rauchsegel n. ‖ masque m. pour la fumée. / ~ slide valve ‖ Rauchschieber m. ‖ tiroir m. de cheminée.

smoke stack ‖ Schornstein m. ‖ cheminée f. / the ~ is fitted with a spark catcher ‖ der Schornstein m. ist mit einem Funkenfänger versehen ‖ la cheminée est coiffée d'un pare-étincelles. / telescopic ~ ‖ Teleskopschornstein m. ‖ cheminée f. à coulisse.

smoke stack top ‖ Schornsteinaufsatz m. ‖ chapiteau m. ou capote f. de cheminée.

smoke tube boiler ‖ Rauchröhrenkessel m. ‖ chaudière f. à tubes de fumée. / combination of flue and ~ boiler ‖ kombinierter Flammrohrrauchrohrkessel m. ‖ chaudière f. combinée de tube-foyer et de tubes de fumée. / boiler tube and ~ working machine ‖ Siede- und Rauchrohrbearbeitungsmaschine f. ‖ machine f. pour le travail des tubes bouilleurs et des tubes à fumée.

smoke writing plane ‖ Rauchschreibflugzeug n. ‖ avion m. émetteur de fumées; avion m. de réclame céleste.

smoking ‖ Schmauchfeuer n. ‖ étuvage m. / ~ of the moulds ‖ Schwärzen n. der Lehmformen ‖ noircissement m. des moules. / ~ prohibited ‖ Rauchen n. verboten ‖ défense de fumer.

smoking apparatus ‖ Räucherapparat m. ‖ appareil m. à fumer. / ~ car ‖ Wagen m. für Raucher ‖ voiture f. pour fumeurs. / ~ compartment ‖ Raucherabteil n. ‖ compartiment m. de fumeurs; wagon-tabagie m. / ~ grate (Meat) ‖ Räucherrost m. ‖ boucan m. / ~ pipe ‖ Tabakspfeife f. ‖ pipe f. / ~ requisites pl. ‖ Raucherwaren fpl.; Rauchutensilien fpl. ‖ articles mpl. pour fumeurs. / ~ room ‖ Rauchzimmer n. ‖ fumoir m. / ~ tobacco ‖ Rauchtabak m. ‖ tabac m. à fumer.

smoky (Chem) ‖ rußend; rauchend ‖ fuligineux. / ~ quartz ‖ Morion m.; (schwarzbrauner) Rauchquarz m. ‖ quartz m. enfumé.

smooth, to (Join) ‖ ebnen; glätten ‖ dresser; planer. / ~ (Mach) ‖ Glanz m. geben; schleifen; polieren ‖ brunir; lustrer; polir; lisser. / ~ the cloth ‖ das Tuch ausrecken ‖ détirer le drap. / ~ by emery and tripoli ‖ polieren; schleifen ‖ polir à l'émeri; adoucir; doucir; frotter. / ~ the glass plates pl. ‖ die Spiegelplatten fpl. feinschleifen ‖ savonner les glaces fpl. / ~ on a grinding mill ‖ blankschleifen ‖ polir à la meule; écacher. / ~ the joints pl. of planks ‖ die Bretter npl. fügen ‖ dresser les planches fpl. sur la tranche; dresser les carnes fpl. des planches. / ~ linen ‖ die Wäsche bügeln oder plätten ‖ repasser le linge. / ~ the rabbets pl. of planks see ~ the joints of planks.

smooth ‖ glatt ‖ lisse. / ~ as a mirror ‖ spiegelglatt ‖ parfaitement lisse.

smooth change of driving direction without shocks ‖ stoßfreier Fahrtrichtungswechsel m. ‖ changement m. de direction (de passage) doux et exempt de chocs. / specially ~ cut ‖ besonders glatter Schnitt m. ‖ coupe f. spécialement nette. / ~ fracture ‖ feinkörniges Gefüge n. ‖ fracture f. à grain fin. / ~ leather ‖ abgenarbtes Leder n. ‖ cuir m. lissé.

/ ~ line (Tel) ‖ homogene Leitung f. ‖ ligne f. homogène. / ~ roll ‖ glatte Walze f. ‖ cylindre m. lisse. / ~ running ‖ ruhiger Lauf m. ‖ roulement m. doux. / ~ sail (Wind mill) ‖ glatter Flügel m. ‖ palette f. plane.

smooth-cutter ‖ Scherenschnitt m. ‖ coupeuse f. à coupe droite.

smoothed ‖ geglättet ‖ lissé; poli; glacé.

smoother ‖ Spachtel m.; f.; Spatel m. ‖ spatule f.

smooth file ‖ Schlichtfeile f.; Abziehfeile f. ‖ lime f. à planer; lime f. douce.

smooth-grinding ‖ Blankschleifen n. ‖ polissage m.

smoothing ‖ Schlichten n.; Glatthobeln n. ‖ planage m.; replanage m. / ~ machine ~ with pumice-stone ‖ Bimsmaschine f. ‖ ponceuse f. / rolling mill for ~ tubes ‖ Röhrenglättwalzwerk n. ‖ laminoir m. à polir les tubes.

smoothing broach ‖ Glättahle f. ‖ alésoir m. à polir. / ~ condenser ‖ Abflachungskondensator m. ‖ condensateur m. d'aplatissement. / ~ drift ‖ Schlichtdorn m. ‖ mandrin m. à planer. / ~ iron ‖ Bügeleisen n.; Plätteisen n. ‖ fer m. à repasser; carreau m.

smoothing machine (Pap) ‖ Glättmaschine f. ‖ machine f. à lisser; satineuse f. / ~ (Tail) ‖ Bügelmaschine f. ‖ machine f. à repasser. / stone ~ for glazed board and pressing board ‖ Steinplättmaschine f. für Glanzpappe und Preßspan ‖ calandre f. à pierre pour carton glacé et presspan.

smoothing plane ‖ Schlichthobel m. ‖ rabot m. plat ou à repasser. / ~ iron ‖ Schlichthobeleisen n. ‖ fer m. à planer ou à replaner ou à recaler.

smoothing planer ‖ Abrichtmaschine f. ‖ machine f. à corroyer.

smoothness ‖ Schliff m. ‖ poli m. / ~ of running·(Mach) ‖ Ruhe f. des Ganges ‖ douceur f. de roulement.

smooth-planing machine see smoothing planer. / ~-roller ‖ Ackerwalze f. ‖ rouleau m. de labourage. / ~-roller mill ‖ Glattwalzenstuhl m. ‖ convertisseur m.

smother, to ‖ ersticken ‖ étouffer. / ~ with boiling water ‖ abbrühen ‖ échauder. / ~ the chrysalids in the cocoons ‖ die Kokons mpl. töten ‖ étouffer les cocons mpl.

smoulder, to ‖ schwelen; qualmen ‖ couver; brûler sans flamme; distiller à basse température.

smouldering plant ‖ Schwelanlage f. ‖ installation f. de distillation à basse température.

smuggler ‖ Schmuggler m. ‖ interlope m.; contrebandier m. / ~ ship ‖ Schmugglerschiff n. ‖ interlope m.; aventurier m.

smuggling (Comm; Mar) ‖ Schleichhandel m.; Schmuggelhandel m. ‖ commerce m. interlope, contrebande f.

smut, to (Print) ‖ unsauber abziehen ‖ mâchurer.

smuth ‖ erdige Kohle f. ‖ terre f. houille.

smut mill ‖ Kornreinigungsmaschine f. ‖ nettoyeur m. de blé; vanneuse f.; tarare m.

smutter ‖ Aspirationsreinigungsmaschine f. ‖ nettoyeur-aspirateur m.

smutterman ‖ Getreideputzer m. ‖ nettoyeur m. de blé.

snacket ‖ Fensterreiber m.; Vorreiber m. ‖ happe f.

snaffle ‖ Trense f. ‖ filet m. / ~ bit ‖ Trensengebiß n. ‖ mors m. de bridon.

snag in wood ‖ Holzknorren m. ‖ malandre f. ou nœud m. dans le bois.

snaggy ‖ knorrig; knotig; knästig ‖ malandreux.

snail ‖ Schnecke f. ‖ escargot m.

snake, to (Cable) ‖ schlangenförmig verlegen ‖ faire serpenter. / ~ (Mar) ‖ schwichten ‖ brider; serpenter.

snake ‖ Schlange f.; Natter f. ‖ serpent m. / paper ~ ‖ Papierluftschlange f. ‖ serpentin m. de carnaval.

snake root oil ‖ Schlangenwurzelöl n. ‖ essence f. d'aristoloche. / ~ track of tank ‖ Raupenkette f. des Kampfwagens ‖ chenille f. (serpentaire) du char de combat.

snap, to ~ in ‖ einschnappen ‖ se fermer à ressort ou au loquet. / ~ out to release ‖ ausschnappen ‖ déclencher.

snap ‖ Schnapper m. ‖ loqueteau m. / ~ (Rivet) ‖ Nietstempel m.; Schelleisen n.; Döpper m. ‖ chasse-rivet m.; bouterolle f. à river. / ~ die ‖ Nietentreiber m. ‖ bouterolle f.; chasse-rivet m. / ~ frequency meter (Electr) ‖ Zungenfrequenzmesser m. ‖ fréquencemètre m. à lames vibrantes. / ~ gauge ‖ Rachenlehre f. ‖ calibre m. mâchoire ou en fer à cheval. / ~ hammer ‖ Schellhammer m. ‖ bouterolle f. à œil. / ~ head ‖ Schellkopf m. ‖ tête f. de bouterolle. / ~ hook ‖ Karabinerhaken m. ‖ crochet m. porte-mousqueton. / ~ piston ring ‖ federnder Kolbenring m. ‖ segment-ressort m. de piston. / ~ riveting ‖ Schellkopfnietung f. ‖ rivure f. bombée ou bouterollée.

snapshot, to ‖ eine Momentaufnahme f. machen ‖ prendre un instantané.

snapshot (Phot) ‖ Momentaufnahme f. ‖ instantané m.

snapshotter ‖ Schnellfotograf m. ‖ opérateur m. pour l'instantané.

snap spectacle case ‖ Klappdeckelbrillenfutteral n. ‖ étui m. de lunettes à bascule. / ~ switch ‖ Schnappschalter m. ‖ commutateur m. à ressort. / ~ tool ‖ Nietdöpper m. ‖ bouterolle f.; chasse-rivet m.

snare ‖ Schleife f.; Schlinge f. ‖ lacet m. / ~ head (Drum) ‖ Saitenfell n. der Trommel; unteres Trommelfell n. ‖ peau f. inférieure de tambour.

snatch brake ‖ Schnappbremse f. ‖ frein m. cliquet.

sneak, to ‖ schleichen; kriechen ‖ ramper; glisser.

sneak current ‖ Irrstrom m.; Kriechstrom m. ‖ courant m. vagabond ou rampant.

sniffle valve see snifting valve.

snifting valve ‖ Schnarchventil n.; Schnüffelventil n. ‖ soupape f. reniflante ou d'évent.

snip ‖ Schnippel m.; Schnipsel n. ‖ petit morceau m.; petite tranche f. / tinman's ~s pl. ‖ Handblechschere f. ‖ cisailles fpl. de ferblantier.

snore hole of a bucket-lift ‖ Saugloch n. einer Schachtpumpe zum Abteufen ‖ narine f. du tuyau aspirateur d'une pompe d'avaleur.

snore piece ‖ Saugrohr n. ‖ tuyau m. aspirateur. / ~ of a sinking-pump ‖ Senkkorb m. einer Pumpe ‖ reniflard m. ou grenouillère ou crépine f. du tuyau d'une pompe.

snorer ‖ Schnarcher m.; Wasserläufer m. ‖ courrier m.

snout (Found) ‖ Auslaufrinne f. ‖ bec m.; tuyère f.

snow, to ‖ schneien ‖ neiger.

snow ‖ Schnee m. ‖ neige f. / free from ~ ‖ schneefrei ‖ sans neige f.; libre de neige f. / fresh ~ ‖ Neuschnee m. ‖ neige f. tombée récemment. / ~ of glass ‖ Schnee m. aus Glas ‖ neige f. de verre. / perpetual ~ ‖ ewiger Schnee m. ‖ neige f. éternelle.

snow chain ‖ Schneekette f. ‖ chaîne f. à neige. / covering of ~ see layer of ~. / ~ drift ‖ Schneetrift f. ‖ accumulation f. ou amas m. de neige; enneigement m. / ~ drift of cutting ‖ Schneeverwehung f. der Einschnitte ‖ enneigement m. des tranchées. / drifting of ~ ‖ Schneetreiben n. ‖ chasse-neige m.

snowed up ‖ eingeschneit ‖ pris par les neiges fpl.

snow fall ‖ Schneefall m. ‖ chute f. de neige.

snow fence (Railw) ‖ Schneezaun m. ‖ écran m. pare-neige. / ~ for roofs ‖ Schneefanggitter n. für Dächer ‖ grille f. pare-neige de toiture.

snow field ‖ Schneefeld n. ‖ champ m. de neige. / ~ flake ‖ Schneeflocke f. ‖ flocon m. de neige. / ~ hoop ‖ Schneereifen m. ‖ raquette f. de neige. / layer of ~ ‖ Schneedecke f. ‖ couverture f. ou couche f. de neige.

snowless ‖ schneefrei ‖ sans neige f.; libre de neige f.

snow load ‖ Schneelast f. ‖ charge f. de neige. / melting of ~ ‖ Schneeschmelze f. ‖ fonte f. de la neige.

snowplough ‖ Schneepflug m. ‖ chasse-neige m. / rotary ~ ‖ Schneeschleuder f.; Schneeräumer m. ‖ chasse-neige m. rotatif.

snowplow see snowplough.

snow protection ‖ Schneeschutzanlagen fpl. ‖ mesures fpl. préventives contre les neiges. / ~ bank ‖ Schneeschutzdamm m. ‖ écran m. pare-neige.

snow rake ‖ Schneerechen m.; Kratzbrett n. ‖ râteau m. à neige. / apparatus for the removal of ~ ‖ Schneeräumvorrichtung f. ‖ chasse-neige m. / ~ retting ‖ Schneerotte f. ‖ rouissage m. à la neige. / sheet of ~ see layer of ~. / ~ shelter see snow fence. / ~ shoe ‖ Schneeschuh m. ‖ raquette f.; ski m. / ~ shovel ‖ Schneeschaufel f. ‖ pelle f. à neige.

snow-slide ‖ Schneelawine f. ‖ avalanche f. de neige.

snow-slip ‖ Schneeverschüttung f. ‖ avalanche f.

snowstorm ‖ Schneesturm m. ‖ tempête f. de neige.

snow sweeper ‖ Schneepflug m. ‖ chasse-neige m. / ~ water ‖ Schneeschmelzwasser n. ‖ eau f. de fonte de la neige. / ~ white ‖ schneeweiß ‖ blanc comme neige f. / wreath of ~ ‖ Schneeschanze f. ‖ amas m. ou banc m. de neige.

snowy ‖ schneeig ‖ neigeux.

snuff ‖ Schnupftabak m. ‖ tabac m. à priser. / ~ box ‖ Schnupftabakdose f. ‖ tabatière f.

snuffers pl. ‖ Putzschere f. ‖ mouchettes fpl.

snuff grinder ‖ Schnupftabakmahler m. ‖ râpeur m. de tabac.

snug (Mar) ‖ handlich ‖ commode.

snug (Mach) ‖ Stift m.; Anschlag m. ‖ ergot m.; oreille f. / ~ bolt ‖ Nasenschraube f. ‖ boulon m. à ergot.

soak, to (Curr) ‖ einweichen ‖ tremper; ramollir en trempant. / ~ casks ‖ Fässer npl. quellen ‖ combuger les futailles fpl. / ~ the hides ‖ die Häute fpl. in die Beizkufe legen ‖ encuver les peaux fpl. / ~ malt ‖ das Malz einmaischen ‖ encuver le malt. / ~ a wall ‖ eine Mauer f. netzen ‖ abreuver un mur.

soaker ‖ Einweicher m. ‖ trempeur m.

soaking (Curr) ‖ Einweichen ‖ trempe f.; lavage m. / ~ apparatus ‖ Tränkapparat m. ‖ appareil m. à imbiber. / tobacco ~ machine ‖ Röstmaschine f. für Tabak ‖ machine f. à rouir le tabac.

soaking pit ‖ Durchweichungsgrube f. ‖ pit m. *ou* puits m. chauffé. / ~ crane ‖ Tiefofenkran m. ‖ pont m. roulant pour four pit. / ~ furnace ‖ Tiefofen m. ‖ four m. à sous-sol.

soaking tank ‖ Einweichtrog m.; Einweichkasten m. ‖ bac m. à tremper. / trough for ~ barley ‖ Gerstenweiche f. ‖ auge f. de trempage pour l'orge. / ~ tub ‖ Quellbottich m.; Weichkufe f. ‖ cuve f. mouilloire. / ~ wheel ‖ Einweichrad n. ‖ roue f. à tremper.

soap ‖ Seife f. ‖ savon m. / automaton ~ ‖ Automatenseife f. ‖ savon m. pour automates. / ~ in balls ‖ Seife f. in Kugeln ‖ savon m. en boules. / ~ in bars ‖ Riegelseife f. ‖ savon m. en briques. / benzine ~ ‖ Benzinseife f. ‖ savon m. à la benzine. / ~ in blades ‖ Seife f. in Blättern ‖ savon m. en feuilles. / brown ~ ‖ Schmierseife f. ‖ savon m. mou. / ~ in cakes ‖ Seife f. in Handstücken ‖ savon m. en pains. / cold-stirred ~ ‖ kaltgerührte Seife f. ‖ savon m. agité à froid. / fancy ~ ‖ Toiletteseife f. ‖ savon m. de toilette. / floating bath ~ ‖ Schwimmbadeseife f. ‖ savon m. léger. / fuller's ~ ‖ Walkseife f. ‖ savon m. à fouler. / glass ~ ‖ Glasmacherseife f. ‖ savon m. de verrerie. / green ~ ‖ grüne Seife f. ‖ savon m. vert *ou* de potasse. / hard ~ ‖ harte Seife f. ‖ savon m. dur. / light ~ ‖ schaumige Seife f. ‖ savon m. léger. / liquid ~ ‖ flüssige Seife f. ‖ savon m. liquide. / marbled ~ ‖ marmorierte Seife f. ‖ savon m. marbré. / medical ~ ‖ Arzeneiseife f. ‖ savon m. médicamental. / medicinal ~ ‖ medizinische Seife f. ‖ savon m. médicinal. / mottled ~ ‖ marmorierte Seife f. ‖ savon m. marbré. / oil ~ ‖ Ölseife f. ‖ savon m. d'huile. / palm ~ ‖ Palmseife f. ‖ savon m. d'huile de palme. / pasty ~ ‖ Seife f. in Teigform ‖ savon m. en pâte. / planed ~ ‖ gehobelte Seife f.; Seifenspäne mpl. ‖ savon m. rapé. / potash ~ *see* green ~. / ~ in powder ‖ Seife f. in Pulverform ‖ savon m. en poudre. / resin ~ ‖ Harzseife f. ‖ savon m. de résine. / scented ~ ‖ parfümierte Seife f. ‖ savon m. parfumé. / shaving ~ ‖ Rasierseife f. ‖ savon m. pour la barbe. / silicated ~ ‖ Wasserglasseife f. ‖ savon m. au silicates alcalins. / soft ~ ‖ Schmierseife f. ‖ savon m. mou. / transparent ~ ‖ Transparentseife f. ‖ savon m. transparent. / white ~ ‖ weiße Seite f. ‖ savon m. blanc. / yellow ~ ‖ gelbe Harztalgseife f. ‖ savon m. jaune de résine.

soap bark ‖ Seifenrinde f.; Panamarinde f. ‖ écorce f. de Panama *ou* de quillai. / ~ beating press ‖ Seifen(schlag)presse f. ‖ presse f. à comprimer les pains de savon. / ~ boiler (Apparatus) ‖ Seifenkocher m. ‖ chaudière f. à savon. / ~ boiler (Person)

‖ Seifensieder m. ‖ savonnier m. / ~ boiling man *see* ~ boiler. / ~ box ‖ Seifendose f. ‖ boîte f. à savon. / cake of ~ ‖ Stück n. Seife ‖ pain m. de savon. / ~ chip container ‖ Seifenspänekasten m. ‖ caisse f. à copeaux de savon. / ~ colours pl. ‖ Farben fpl. für Seifen ‖ couleurs fpl. pour savons. / ~ consumption of ~ ‖ Seifenverbrauch m. ‖ consommation f. de savon. / ~ cooling press ‖ Seifenkühlpresse f. ‖ presse f. à refroidir les savons. / ~ cream ‖ Seifenkrem m. ‖ crème f. de savon. / ~ cutting machine ‖ Seifenschneidmaschine f. ‖ machine f. à couper le savon. / ~ distributor ‖ Seifenspender m. ‖ distributeur m. de savon. / ~ drying apparatus ‖ Seifentrockenapparat m. ‖ séchoir m. à savon; appareil m. à sécher les savons. / ~ extract ‖ Seifenextrakt m. ‖ extrait m. de savon. / ~ factory plant ‖ Seifenfabrikeinrichtung f. ‖ installation f. de fabrication du savon. / ~ flakes pl. ‖ Seifenflocken fpl. ‖ flocons mpl. de savon. / ~ frame ‖ Seifenform f. ‖ mise f. *ou* moule m. à savon. / ~ froth ‖ Seifenschaum m. ‖ mousse f. de savon.

soaping ‖ Einseifen n. ‖ savonnage m.

soaping machine ‖ Seifmaschine f. ‖ savonneuse f. / ~ for cloth ‖ Tuchseifmaschine f. ‖ machine f. à savonner les tissus.

soap maker ‖ Seifensieder m. ‖ savonnier m.

soap maker's boiler ‖ Seifensiederkessel m. ‖ campane f. / ~ machinery ‖ Seifenherstellungsmaschine f. ‖ machine f. pour savonneries.

soap mill ‖ Piliermaschine f. ‖ broyeuse f. à savon. / ~ nut ‖ Seifennuß f. ‖ noix f. de savon. / ~ paper ‖ Seifenpapier n. ‖ papier m. savonné. / ~ paste ‖ Seifenleim m. ‖ colle f. de savon.

soap powder ‖ Seifenpulver n. ‖ poudre f. de savon; savon m. en poudre. / ~ manufacturing machinery ‖ Seifenpulverherstellungsmaschine f. ‖ machine f. pour la fabrication de savon en poudre. / ~ packing engine ‖ Seifenpulverpackmaschine f. ‖ machine f. d'emballage de savon en poudre.

soap refining outfit ‖ Pilieranlage f. ‖ installation f. pour le broyage de savon. / ~ root ‖ Seifenwurzel f. ‖ racine f. saponnaire. / ~ shaving planing machine ‖ Seifenspänehobelmaschine f. ‖ raboteuse f. de copeaux de savon.

soap slab cooling plant ‖ Seifenplattenkühlanlage f. ‖ refroidisseur m. de plaques de savon. / ~ cutting machine ‖ Seifenplattenschneidmaschine f. ‖ coupeuse f. de plaques de savon.

soap solution ‖ Seifenlösung f. ‖ solution f. de savon. / ~ maker ‖ Laugenbereiter m. ‖ préparateur m. de lessive.

soap, spirit of ~ ‖ Seifenspiritus m. ‖ esprit m. de savon; liniment m. savonneux alcoolique; opoldeldoch m.

soapstone ‖ Speckstein m.; Steatit m. ‖ stéatite f.; pierre f. de lard. / pulverized ~ ‖ Talkum n. ‖ talc m. / ~ insulator ‖ Specksteinisolator m. ‖ isolateur m. en stéatite. / ~ machine ‖ Talkumiermaschine f. ‖ machine f. à talquer. / ~ quarry ‖ Specksteingrube f. ‖ carrière f. de saponite.

soap substitute ‖ Seifenersatzmittel n. ‖ succédané m. de savon. / ~ suds pl. ‖ Seifenwasser n. ‖ eau f. de savon. / ~ superfattening ‖ Seifenüberfettung f. ‖ surgraissage m. de savons. / ~ test ‖

Seifenprobe f. ‖ essai m. au savon. / ~ vat ‖ Seifenfaß n. ‖ bugadière f. / ~ works pl. ‖ Seifensiederei f. ‖ savonnerie f.

soapwort ‖ Seifenkraut n. ‖ saponnaire f.

soapy ‖ seifig; seifenartig ‖ savonneux. / ~ water ‖ Seifenwasser n. ‖ eau f. de savon.

social game ‖ Gesellschaftsspiel n. ‖ jeu m. de société. / ~ status ‖ Lebensstellung f. ‖ position f. (sociale). / ~ task ‖ soziale Aufgabe f. ‖ tâche f. sociale.

society of miners ‖ Knappschaft f. ‖ corps m. des mineurs. / mutual ~ ‖ Gesellschaft f. auf Gegenseitigkeit ‖ société f. mutuelle. / polytechnical ~ ‖ Gewerbeverein m. ‖ société f. industrielle. / ~ game ‖ Gesellschaftsspiel n. ‖ jeu m. de société.

sock (Agr) ‖ Pflugsech n. ‖ coutre m. de charrue; sep m. / ~ (Shoem) ‖ Einlegesohle f. ‖ chausson m. / ~ (Textile) ‖ Socke f. ‖ bas m.; chaussette f. / ~ of cork ‖ Korkeinlegesohle f. ‖ semelle f. intérieure en liège. / fancy ~ ‖ Fantasiesocke f. ‖ chaussette f. fantaisie.

sock blade *see* sock (Agr).

socket (Electr) ‖ Steckhülse f. ‖ douille f. / ~ (Incandescent electric lamp) ‖ Fassung f. ‖ douille f. / ~ (Mach) ‖ Muffe f. ‖ manchon m. / ~ (Radio) ‖ Sockel m.; Röhrensockel m. ‖ culot m. / ~ for brake circuit ‖ Bremsanschlußdose f. ‖ boîte f. de jonction pour circuit de freinage. / conical ~ ‖ Kegelhülse f. ‖ douille f. conique. / ~ metal ~ of a candlestick ‖ Metalleuchtertülle f. ‖ bobèche f. en métal. / ~ of the pivot ‖ Türangelpfanne f. ‖ crapaudine f.; piton m.

socket chisel ‖ Rohrstechbeitel m. ‖ ciseau m. à douille. / ~ key ‖ Aufsteckschlüssel m. ‖ clef f. à douille. / ~ pipe ‖ Muffenrohr n. ‖ tuyau m. à manchon *ou* à emboîtement. / ~ turning and screwing machine ‖ Muffendreh- und Gewindeschneidbank f. ‖ tour m. à façonner et fileter les manchons.

socket wrench (Mach) ‖ Steckschlüssel m.; Aufsteckschlüssel m. ‖ clef f. à canon. / ~ (Mot) ‖ Rohrschlüssel m. ‖ clef f. à douille. / ~ (Railw) ‖ Schienenschraubenschlüssel m. ‖ clef f. à tire-fonds. / ~ with hexagon head for a width of jaw of x mm ‖ Steckschlüssel m. mit oberem Sechskant für x mm Maulweite ‖ clé f. à canon avec hexagone supérieur pour ouverture de x mm. / square ~ ‖ Vierkantsteckschlüssel m. ‖ clé f. à canon carrée. / square ~ for spindles and screws ‖ Vierkantsteckschlüssel m. für Spindeln und Schrauben ‖ clé f. à canon carrée pour goupilles et vis. / ~ of tube ‖ Steckschlüssel m. aus Rohr ‖ clé f. à canon en tuyau. / universal ~ ‖ Universalsteckschlüssel m. ‖ clé f. à douille universelle. / ~ for a width of jaw of x mm ‖ Steckschlüssel m. für x mm Maulweite ‖ clé f. à canon pour ouverture de x mm.

sock holder ‖ Sockenhalter m. ‖ jarretelle f. / ~ maker ‖ Sockenmacher m. ‖ chaussetier m. / ~ machine for making ~s ‖ Sockenmaschine f. ‖ machine à chaussettes. / ~ stitches pl. ‖ versetzte Nähte fpl. ‖ coutures fpl. contrariées *ou* croisées.

socle of a column ‖ Fuß m. *oder* Basis f. einer Säule ‖ base f. d'une colonne. / ~ for pianos ‖ Pianosockel m. ‖ socle m. pour pianos. / ~ for poles ‖ Stangenfuß m. ‖ socle m. pour poteaux en bois.

socle diameter ‖ Sockeldurchmesser m. ‖ diamètre m. du socle. / ~ wainscotting in a room ‖ Fußsockel m. *oder* Brüstungsverkleidung f. rings um ein Zimmer ‖ lambris m. d'appui d'une chambre.

sod, to ‖ mit Rasen m. belegen; berasen ‖ gazonner.

sod ‖ Rasen m.; Sode f.; Rasenstück m. ‖ gazon m.; tranche f. de gazon.

soda ‖ Soda f.; Natriumkarbonat n. ‖ carbonate m. de soude; soude f. / artificial ~ ‖ künstliche Soda f. ‖ soude f. artificielle. / calcinated ~ ‖ kalzinierte Soda f. ‖ sel m. de soude calcinée; carbonate m. de sodium calciné. / calcined ~ ‖ entwässerte Soda f. ‖ soude f. calcinée. / caustic ~ ‖ kaustische Soda f. ‖ soude f. caustique. / crude ~ (Chem) ‖ rohe Soda f. ‖ soude f. brute. / crystallized ~ ‖ kristallisierte Soda f. / soude f. cristallisée. / native ~ ‖ natürliche Soda f. ‖ soude f. native. / refined ~ ‖ raffinierte *oder* gereinigte Soda f. ‖ soude f. raffinée. / tartarated ~ ‖ Natronweinstein n.; Rochellesalz n. ‖ sel m. de la Rochelle; tartrate m. sodicopotassique.

soda alum ‖ Natronalaun n. ‖ alun m. de soude. / ~ aluminate ‖ Tonerdenatron n. ‖ aluminate m. de soude. / ~ arsenite ‖ arsenigsaures Natron n.; Natriumarsenik m. ‖ arsénite m. de soude. / ~ ash ‖ kalzinierte Soda f. ‖ sel m. de soude calcinée; carbonate m. de sodium calciné. / ~ boiler ‖ Sodasieder m. ‖ soudier m. / ~ crusher ‖ Sodabrecher m. ‖ broyeur m. de soude. / ~ crystals pl. ‖ Kristallsoda f. ‖ soude f. cristallisée. / ~ drum ‖ Sodatrommel f. ‖ tambour m. pour soude. / ~ furnace ‖ Sodaofen m. ‖ four m. à soude. / ~ glass ‖ Natronglas n. ‖ verre m. de soude. / hydrate of ~ ‖ Ätznatron n.; kaustisches Natron n. ‖ hydrate m. de soude; soude f. caustique. / ~ lake ‖ Natronsee m. ‖ lac m. natron. / ~ lime ‖ Natronkalk m. ‖ chaux f. sodée. / ~ lye ‖ Natronlauge f. ‖ lessive f. de soude.

sodammonium ‖ Natriumammonium n. ‖ sodammonium n.

soda pan ‖ Sodakessel m. ‖ chaudière f. à soude. / ~ powder ‖ Sodapulver n. ‖ soda-powder m.; sel m. de soude en poudre. / ~ producing plant ‖ Sodaherstellungsanlage f. ‖ installation f. de fabrication de la soude. / ~ selenite ‖ selenigsaures Natron n. ‖ sélénite m. de soude. / ~ soap ‖ Sodaseife f.; Natronseife f. ‖ savon m. dur *ou* de soude. / ~ vat ‖ Sodaküpe f. ‖ cuve f. allemande. / ~ water ‖ Sodawasser n. ‖ eau f. de Seltz *ou* de soude. / ~ works pl. ‖ Sodawerk n.; Sodafabrik f. ‖ soudière f.

sodding ‖ Rasenarbeit f. ‖ gazonnage m.

sodium ‖ Natrium n. ‖ sodium m. / ~ acetate ‖ essigsaures Natrium n.; Natriumazetat n. ‖ acétate m. de sodium. / ~ aluminate ‖ Natriumaluminat n. ‖ aluminate m. de sodium. / ~ antimoniate ‖ antimonsaures Natrium n. ‖ antimoniate m. de sodium. / ~ arseniate ‖ arsensaures Natron n. *oder* Natrium n.; Natriumarseniat n. ‖ arséniate m. sodique *ou* de soude. / ~ benzoate ‖ benzoesaures Natrium n.; Natriumbenzoat n. ‖ benzoate m. de sodium. / ~ bicarbonate ‖ doppeltkohlensaures Natron n.; Natriumbikarbonat n. ‖ bicarbonate m. de sodium. / ~ bichromate ‖ Natriumbichromat n.; doppeltchromsaures Natrium n. ‖ bi-

chromate m. de sodium. / ~ bisulphite ‖ Natriumbisulfit n.; Mononatriumsulfit n. ‖ bisulfite m. de sodium. / ~ borate ‖ borsaures Natrium n.; Natriumborat n. ‖ borate m. de sodium. / ~ burner ‖ Natriumbrenner m. ‖ brûleur m. sodique. / ~ carbonate ‖ kohlensaures Natrium n.; Soda f. ‖ carbonate m. de sodium; (sel m. de) soude f. / pot for adding carbonate of ~ ‖ Sodazusatztopf m. ‖ pot m. à soude. / ~ cellulose ‖ Natronzellstoff m. ‖ cellulose f. sodique. / ~ chlorate ‖ chlorsaures Natrium n.; Natriumchlorat n. ‖ chlorate m. de sodium. / ~ chloride ‖ Kochsalz n.; Natriumchlorid n.; Chlornatrium n. ‖ chlorure m. de sodium; sel m. commun *ou* de cuisine. / chloroaurate of ~ ‖ Natriumgoldchlorid n.; Goldsalz n. ‖ sel m. d'or; oxysel m. aureux. / ~ citrate ‖ zitronensaures Natrium n.; Natriumzitrat n. ‖ citrate m. de sodium. / ~ flame ‖ Natriumflamme f. ‖ flamme f. de sodium. / ~ fluoride ‖ Fluornatrium n.; Natriumfluorid n. ‖ fluorure m. de sodium. / ~ fluosilicate ‖ Kieselfluornatrium m. ‖ fluosilicate m. de sodium. / ~ formiate ‖ ameisensaures Natrium n.; Natriumformiat n. ‖ formiate m. de sodium. / ~ hydrate ‖ Ätznatron n.; Natronlauge f. ‖ lessive f. de soude *ou* sodique; soude f. caustique; hydrate m. de sodium. / ~ hydrosulphite ‖ Natriumhydrosulfit n.; unterschwefligsaures Natrium n. ‖ hydrosulfite m. de sodium. / ~ hydroxide ‖ Ätznatron n.; Natriumhydroxyd n. ‖ hydroxyde m. de sodium; soude f. caustique. / ~ hypochlorite ‖ unterchlorigsaures Natrium n.; Natriumhypochlorit n. ‖ hypochlorite m. de sodium. / ~ hypophosphite ‖ unterphosphorigsaures Natrium n.; Natriumhypophosphit n. ‖ hypophosphite m. de sodium. / ~ hyposulphate ‖ unterschwefelsaures Natrium n.; Natriumhyposulfat n. ‖ hyposulfate m. de sodium. / ~ hyposulphite ‖ unterschwefligsaures Natrium n.; Natriumhyposulfit n. ‖ hyposulfite m. de sodium. / ~ iodide ‖ Jodnatron n.; Natriumjodid n. ‖ iodure m. de sodium. / ~ manganate ‖ mangansaures Natrium n. ‖ manganate m. de sodium. / ~ metal ‖ Natriummetall n. ‖ sodium m. métallique. / ~ nitrate ‖ salpetersaures Natrium n.; Natronsalpeter m.; Chilesalpeter m. ‖ nitrate m. *ou* azotate m. de sodium; salpètre m. de soude *ou* du Chili. / ~ nitrite ‖ salpetrigsaures Natrium n.; Natriumnitrit n. ‖ nitrite m. de sodium. / ~ oxide ‖ Natron n.; Natriumoxyd n. ‖ oxyde m. de sodium. / ~ perborate ‖ Natriumperborat n. ‖ perborate m. de sodium. / ~ peroxide (Chem) ‖ Natriumsuperoxyd n. ‖ peroxyde m. de sodium. / ~ phosphate ‖ phosphorsaures Natron n.; Natriumphosphat n. ‖ phosphate m. de sodium. / ~ press ‖ Natriumpresse f. ‖ presse f. à sodium. / ~ pyrophosphate ‖ pyrophosphorsaures Natron n.; Natriumpyrophosphat n. ‖ pyrophosphate m. de sodium. / ~ pyrosulphite ‖ Natriumpyrosulfit n. ‖ pyrosulfite m. de sodium. / ~ salicylate ‖ salizylsaures Natrium n. ‖ salicylate m. de sodium. / ~ seleniate ‖ selensaures Natrium n. ‖ séléniate m. de sodium. / ~ silicate ‖ Natronwasserglas n.; Natriumsilicat n.; kieselsaures Natrium n. ‖ silicate m. de soude. / ~ silicofluoride ‖

Kieselfluornatrium n. ‖ silicofluorure m. de sodium. / ~ stannate ‖ zinnsaures Natrium n.; Natriumstannat n. ‖ stannate m. de sodium. / ~ sulphate ‖ schwefelsaures Natron n.; Natriumsulfat n.; Glaubersalz n. ‖ sulfate m. de sodium; sel m. de Glauber. / ~ sulphide ‖ Schwefelnatrium n. ‖ sulfure m. de sodium. / ~ sulphite ‖ schwefligsaures Natrium n.; Natriumsulfit n. ‖ sulfite m. de sodium. / ~ superoxide ‖ Natriumdioxyd n.; Natriumsuperoxyd n. ‖ peroxyde m. de sodium. / ~ thiosulphate ‖ Natriumthiosulfat n. ‖ thiosulfate m. de sodium. / ~ tungstate ‖ wolframsaures Natrium n.; Natriumwolframat n. ‖ tungstate m. de sodium. / ~ uranate ‖ unransaures Natrium n. ‖ uranate m. de sodium. / ~ wood-pulp ‖ Natronzellulose f. ‖ cellulose f. de soude.

sod knife ‖ Rasenstecher m.; Sodenpflug m. ‖ coupe-gazon m.; tranche-gazon m. / ~ lifter ‖ Rasenhacke f.; Rasenheber m. ‖ lève-gazon m.; pioche f. aux gazons. / ~ plough *see* sod knife. / ~ revetment ‖ Rasenbekleidung f. ‖ gazonnement m.; revêtement m. en gazons.

sod work ‖ Sodendecke f.; Rasenbekleidung f. ‖ gazonnement m. / builder of ~ ‖ Rasenleger m. ‖ gazonneur m.

sofa ‖ Sofa n. ‖ sofa m.; canapé m.; divan m. / ~ ‖ Ruhebett n. ‖ lit m. de repos. / ~ beadstead ‖ Schlafsofa n.; Patentsofa n. ‖ canapé-lit m.; divan m. / ~ cushion ‖ Sofakissen n. ‖ coussin m. de canapé.

soffit (Build) ‖ Laibung f. eines Gewölbes; innere Gewölbfläche f. ‖ intrados m.; douelle f. intérieure; dessous m. de voûte. / ~ (Theatre) ‖ Soffitte f. ‖ soffite m.

soft (Chem) ‖ weich; locker ‖ mou; doux. / ~ (Iron) ‖ ungehärtet ‖ non trempé. / ~ (Metal) ‖ schmiedbar ‖ malléable; ductile. / ~ (Miner) ‖ gebrech; bröcklig ‖ cassant, fragile. / ~ coal ‖ Weichkohle f. ‖ charbon m. gras; houille f. grasse. / ~ grain ‖ weiche Frucht f. ‖ blé m. tendre.

soft iron ‖ Weicheisen n. ‖ fer m. doux. / ~ instrument ‖ Weicheiseninstrument n. ‖ instrument m. de mesure à fer doux.

soft leather ‖ Weichleder n. ‖ cuir m. mou. / ~ metal ‖ Weichmetall n. ‖ métal m. doux. / ~ mud ‖ Schlick m. ‖ vase f. molle. / ~ packing ‖ Weichpackung f. ‖ bourrage m. mou. / ~ porcelain ‖ Weichporzellan n. ‖ porcelaine f. tendre. / ~ rubber ‖ Weichgummi n. ‖ caoutchouc m. mou. / ~ soap ‖ Schmierseife f.; Kaliseife f. ‖ savon m. mou *ou* de potasse. / ~ solder ‖ Zinnlot n.; Weichlot n.; Weißlot n. ‖ soudure f. à l'étain *ou* tendre. / ~ soldering ‖ Weichlöten n. ‖ soudure f. tendre. / ~ spring suspension ‖ weiche Federung f. ‖ suspension f. douce sur ressorts. / ~ water ‖ weiches Wasser n. ‖ eau f. douce. / ~ wood ‖ Weichholz n. ‖ bois m. tendre.

soften, to (Chem) ‖ enthärten; erweichen ‖ adoucir; ramollir. / ~ (Phot) ‖ abtönen ‖ rendre une nuance. / ~ iron ‖ Eisen weich machen ‖ adoucir le fer m. / ~ the steel ‖ Stahl ausglühen *oder* nachlassen *oder* weichmachen ‖ recuire l'acier m.

softener (Curr) ‖ Strecker m. ‖ recasseur m. / ~ (Glassm) ‖ Einschmelzer m. ‖ rebrûleur m.

softening ‖ Enthärtung f. ‖ adoucissement m.; amollissement m. / ~ of colours ‖

Mildern n. der Farben ‖ adoucissage m. des couleurs. / ~ of the steel ‖ Entkohlen n. des Stahls ‖ décarbonisation f. de l'acier. / water ~ ‖ Wasserenthärtung f. ‖ adoucissement m. de l'eau.

softening point ‖ Erweichungspunkt m. ‖ point m. de ramollissement.

soft-sized paper ‖ halbgeleimtes Papier n. ‖ papier m. demi-collé.

soggy envelope (Aero) ‖ schlappe Hülle f. ‖ enveloppe f. flasque *ou* détendue.

soil, to ‖ beschmutzen ‖ maculer.

soil ‖ Baugrund m.; Erdboden m. ‖ terrain m.; sol m. / clayey ~ ‖ Lehmboden m. ‖ terre f. argileuse. / stratum of firm ~ ‖ guter Baugrund m. ‖ sol m. résistant. / ~ for forest growth ‖ Waldboden m. ‖ terre f. à bois. / friable ~ ‖ krümeliger Boden m. ‖ sol m. friable. / frozen ~ ‖ gefrorener Boden m. ‖ sol m. congelé. / ~ formed by glacial action ‖ Gletscherboden m. ‖ sol m. glaciaire. / ~ of a pavement ‖ Planum n. einer zu pflasternden Straße ‖ aire f. d'un pavé; plate-forme f. / poor ~ ‖ ärmlicher Boden m. ‖ sol m. pauvre. / rich ~ ‖ reicher Boden m. ‖ sol m. riche. / sandy ~ ‖ sandiger Boden m. ‖ terre f. sablonneuse. / solid ~ ‖ festes Gebirge n. ‖ terrain m. ferme. / top ~ ‖ Oberboden m. ‖ couche f. superficielle du sol. / vegetable ~ ‖ Ackerkrume f.; Humusboden m. ‖ terre f. végétale.

soil, covering of the ~ ‖ Bodenbedeckung f. ‖ recouvrement m. du sol. / ~ draining plant ‖ Bodenentwässerungsanlage f. ‖ installation f. de drainage du terrain. / ~ forming process ‖ bodenbildender Vorgang m. ‖ processus m. de formation de terre végétale. / ~ pipe ‖ Kanalisationsröhre f.; Abflußrohr n. ‖ tuyau m. de chute; chausse f. d'aisance. / pressure of the ~ ‖ Gebirgsdruck m. ‖ pression f. du terrain. / sample of the ~ ‖ Bodenprobe f. ‖ échantillon m. du terrain. / surface constitution of the ~ ‖ Oberflächenbeschaffenheit f. des Bodens ‖ constitution f. de la surface du sol.

soil-tilling, draught machines pl. for ‖ Bodenbearbeitungszuggerät n. ‖ machines fpl. à attelage pour cultiver le sol. / tools pl. for ~ ‖ Bodenbearbeitungsgerät n. ‖ outils mpl. pour cultiver le sol.

soil, usufruct of ‖ Nutzung f. von Grund und Boden ‖ utilisation f. du fonds.

soil ventilating tools pl. ‖ Bodenlüftungsgerät n. ‖ outils mpl. à ventiler le sol.

soja bean oil (Chem) ‖ Sojabohnenöl n. ‖ huile f. de soja.

sol ‖ Bundschwelle f. ‖ semelle f. d'assemblage; sablière f.

solar camera ‖ Sonnenkamera f. ‖ chambre f. noire pour le soleil. / ~ day ‖ Sonnentag m. ‖ jour m. solaire. / ~ heat ‖ Sonnenwärme f. ‖ chaleur f. solaire. / ~ nucleus ‖ Sonnenkern m. ‖ noyau m. solaire. / ~ oil ‖ Solaröl n. ‖ huile f. solaire. / polarizing ~ prism ‖ Polarisationssonnenprisma n. ‖ hélioscope m. de polarisation. / ~ projection screen ‖ Sonnenprojektionsschirm m. ‖ écran m. pour la projection du soleil; écran m. de projection pour le soleil. / ~ spectrum ‖ Sonnenspektrum n. ‖ spectre m. solaire.

solarization ‖ Einstrahlung f. auf die Erde ‖ insolation f.; rayonnement m. du soleil sur la terre.

sold ‖ vergriffen ‖ vendu. / ~ for account of … ‖ verkauft für Rechnung f. von …

vendu pour compte m. de … / ~ by auction ‖ meistbietend verkauft ‖ vendu aux enchères fpl. / to be ~ by auction ‖ zur Auktion f. kommen ‖ passer aux enchères fpl. / goods pl. ~ for cash ‖ gegen bar verkaufte Ware f. ‖ marchandise f. vendue au comptant. / ~ free on rail ‖ verkauft franko Waggon m. ‖ vendu franco sur wagon m. / ~ for immediate delivery ‖ verkauft für sofortige Lieferung f. ‖ vendu pour livraison f. immédiate. / ~ subject to inspection ‖ verkauft vorbehaltlich der Besichtigung f. ‖ vendu subordonné à inspection f. / ~ for joint account ‖ verkauft für gemeinschaftliche Rechnung ‖ vendu en compte m. à demi. / ~ at a loss of … ‖ verkauft mit einem Verlust m. von … ‖ vendu avec une perte de … / ~ out ‖ ausverkauft ‖ tout vendu. / ~ privately ‖ freihändig verkauft ‖ vendu à l'amiable. / ~ at a profit of … ‖ verkauft mit einem Gewinn m. von … ‖ vendu avec un bénéfice de … / ~ standing ‖ auf dem Halm m. verkauft ‖ vendu sur pied m. / ~ ex warehouse ‖ verkauft ab Lager ‖ vendu pris en magasin.

solder, to ‖ löten ‖ souder. / ~ with resin ‖ mit Kolofonium n. löten ‖ souder à la résine. / ~ on top of ‖ auflöten ‖ dessouder; appliquer à la soudure. / ~ into a turned groove ‖ in eine Ausdrehung f. einlöten ‖ souder dans une gorge tournée.

solder ‖ Lot n.; Lötmittel n. ‖ brasure f.; soudure f. / hard ~ ‖ Hartlot n.; Messinglot n.; Schlaglot n. ‖ soudure f. dure; brasure f. / poor in tin ‖ zinnarmes Lot n. ‖ soudure f. maigre. / rich ~ ‖ zinnreiches Lot n. ‖ soudure f. grasse. / soft ~ ‖ Schnellot n.; Weichlot n.; Weißlot n.; Zinnlot n. ‖ soudure f. tendre *ou* à l'étain. / tin – *see* soft ~.

solderable ‖ lötbar ‖ soudable.

soldered seam ‖ Lötnaht f. ‖ soudure f. / ~ up ‖ verlötet ‖ soudé.

solderer ‖ Löter m. ‖ soudeur m. / blow pipe ~ ‖ Rohrlöter m. ‖ soudeur m. au chalumeau. / blow pipe lead ~ ‖ Bleilöter m. ‖ soudeur m. de plomb au chalumeau. / tin box ~ ‖ Blechdosenlöter m. ‖ soudeur-boîtier m.

soldering ‖ lötbar ‖ soudant; soudable. / non-~ ‖ unlötbar ‖ insoudable.

soldering ‖ Lötung f.; Löten n. ‖ soudure f.; soudage m.; brasure f. / autogenous ~ ‖ autogenes Schweißen n. ‖ soudure f. autogène. / brass ~ ‖ Messinglötung f. ‖ soudure f. au laiton. / electric ~ ‖ elektrisches Löten n. ‖ soudure f. électrique. / hard ~ ‖ Hartlöten n. ‖ soudure f. forte; brasure f. / to join by ~ ‖ anlöten ‖ souder; braser; assembler par la soudure. / ~ of longitudinal seams ‖ Längsnahtlötung f. ‖ soudure f. longitudinale. / soft ~ ‖ Weichlöten n. ‖ soudure f. tendre.

soldering acid ‖ Lötwasser n. ‖ acide m. à souder. / ~ apparatus ‖ Lötapparat m. ‖ appareil m. à souder. / ~ bit *see* ~ iron. / ~ blowers pl. ‖ Lötgebläse n. ‖ souffleries fpl. pour soudure. / ~ board ‖ Lötbrett n. ‖ étamoir m. / ~ box ‖ Lötmuffe f. ‖ manchon m. de soudage. / ~ copper *see* iron.

soldering device ‖ Löteinrichtung f. ‖ installation f. à souder. / electric band saw ~ ‖ elektrische Bandsägenlötvorrichtung f. ‖ dispositif m. électrique à braser les scies à ruban.

soldering furnace ‖ Lötofen m. ‖ four m. à souder. / ~ iron ‖ Lötkolben m. ‖ soudoir m.; fer m. de soudure *ou* à souder. / ~ lamp ‖ Lötlampe f. ‖ lampe f. à souder. / side seam ~ machine ‖ Längsnahtlötmaschine f. ‖ machine f. à souder en long. / ~ material ‖ Lötmaterial n. ‖ matériel m. à souder. / ~ means pl. ‖ Lötmittel n. ‖ (matériel m. de) soudure f. / ~ metal ‖ Lötmetall n. ‖ métal m. de soudure. / ~ pan ‖ Lötpfanne f. ‖ polastre m. / ~ pewter ‖ Lötzinn n. ‖ étain m. marcassite. / ~ plate ‖ Lötblech n. ‖ plaque f. à souder. / ~ powder ‖ Lötpulver n. ‖ poudre f. à braser. / ~ preparation ‖ Lötmittel n. ‖ produit m. à souder. / ~ seam ‖ Lötstelle f. ‖ brasure f.; soudure f. / ~ stone ‖ Lötstein n. ‖ roche f. à souder. / ~ stove see ~ furnace. / ~ terminal ‖ Lötklemme f. ‖ borne f. à souder. / ~ tin ‖ Lötzinn n. ‖ étain m. de soudure. / ~ tongs pl. ‖ Lötzange f. ‖ pince f. à souder.

soldering tools pl. ‖ Lötwerkzeug n. ‖ outils mpl. de soudage. / electric ~ ‖ elektrisches Lötwerkzeug n. ‖ fers mpl. à souder électriques.

soldering tweezers pl. ‖ Lötzange f. ‖ pince f. à souder. / ~ utensil ‖ Lötwerkzeug n. ‖ appareil m. à souder. / ~ water ‖ Lötwasser n. ‖ eau f. à souder; esprit m. de sel décomposé.

soldier, lead ~ ‖ Bleisoldat m. ‖ soldat m. de plomb.

sole, to ‖ besohlen ‖ resemeller; mettre des semelles fpl.

sole ‖ allein; einzig ‖ seul; unique; exclusif. / ~ agent ‖ Alleinvertreter m. ‖ représentant m. exclusif. / ~ bill of exchange ‖ Solawechsel m. ‖ seule f. de change. / ~ distributor ‖ Alleinvertreter m. ‖ représentant m. exclusif.

sole (Fish) ‖ Scholle f.; Seezunge f. ‖ sole f. / ~ (Furnace) ‖ Herd m. ‖ sole f.; aire f.; âtre m. / ~ (Mine) ‖ Sohle f. ‖ sol m.; fond m.; base f. / ~ (Shoem) ‖ Sohle f.; Schuhsohle f. ‖ semelle f. / ~ (Windmill) ‖ Schwelle f. ‖ semelle f. / ~ cardboard ~ ‖ Pappsohle f. ‖ semelle f. en carton. / ~ of caoutchouc ‖ Kautschuksohle f. ‖ semelle f. en caoutchouc. / ~ of a crane ‖ Grundbalken m. eines Kranes ‖ racinal m. d'une grue. / felt ~ ‖ Filzsohle f. ‖ semelle f. en feutre. / ~ of the foot ‖ Fußsohle f. ‖ plante f. du pied. / horse hair ~ ‖ Haarsohle f. ‖ semelle f. en crin. / inner ~ (Shoem) ‖ Brandsohle f. ‖ semelle f. intérieure; première f. / paper ~ ‖ Papiersohle f. ‖ semelle f. en papier. / ~ of the pivot ‖ Türangelpfanne f. ‖ crapaudine f.; piton m. / ~ of a plane ‖ Sohle f. *oder* Bahn f. eines Hobels ‖ plan m. *ou* semelle f. d'un rabot. / ~ of the reverberatory furnace ‖ Herd m. des Flammofens ‖ foyer m. du four à réverbère. / straw ~ ‖ Strohsohle f. ‖ semelle f. en paille. / wooden ~ ‖ Holzsohle f. ‖ semelle f. en bois.

sole blocker ‖ Sohlenstanzer m. ‖ brocheur m. de semelles. / ~ cutter ‖ Sohlenzuschneider m. ‖ coupeur m. de semelles. / ~ cutting machine ‖ Sohlenausschneidmaschine f. ‖ machine f. à découper les semelles. / ~ filler ‖ Sohlenbeleger m. ‖ remplisseur m. de semelles. / ~ layer (Shoem) ‖ Sohlenaufleger m. ‖ afficheur m. de semelles. / ~ leather ‖ Sohlleder n. ‖ cuir m. à semelles.

solenoid ‖ Solenoid n. ‖ solénoïde m. / ~ brake ‖ Solenoidbremse f. ‖ frein m. à solénoïde. / ~ core ‖ Solenoidkern m. ‖ noyau m. de solénoïde.

sole plate ‖ Fußplatte f. ‖ plaque f. d'assise / ~ of wind wheel ‖ Windradsockel m. ‖ socle m. de fondation de la roue éolienne.

sole repairing ‖ Besohlanstalt f. ‖ ressemelage m. de chaussures. / ~ rounder ‖ Sohlenbeschneider m. ‖ fraiseur m. de semelles. / ~ saving ‖ Sohlenschoner m. ‖ couverture f. protectrice de semelle. / ~ splicing ‖ Sohlenverstärkung f. ‖ renforcement m. de semelle. / ~ splitter ‖ Sohlenspalter m. ‖ refendeur m. de semelles.

solfatara (Geol) ‖ Solfatara f. ‖ solfatare f.; soufrière f.

solicitor ‖ Anwalt m.; Rechtsanwalt m. ‖ avoué m.; avocat m.

solid (Bank) ‖ kapitalkräftig ‖ disposant d'importants capitaux mpl. / ~ (Build) ‖ massiv; stark; massig ‖ solide; massif. / ~ (Chem) ‖ fest ‖ ferme; concret; solide; consistant.

solid angle, lateral ~ (Crystal) ‖ Randecke f. oder Seitenecke f. eines Kristalls ‖ angle m. ou sommet m. latéral. / vertical ~ (Crystal) ‖Polecke f. eines Kristalls ‖ angle m. terminal; angle m. du sommet.

solid axle ‖ Vollachse f. ‖ essieu m. massif. / ~ bearing ‖ einteiliges Lager n. ‖ coussinet m. d'une seule pièce.

solid brick ‖ Vollziegel m. ‖ brique f. pleine. / porous ~ ‖ poriger Vollziegel m. ‖ brique f. pleine poreuse.

solid crank disc ‖ volles Kurbelblatt n. ‖ flasque m. de manivelle plein. / ~-drawn tube ‖ nahtlos gezogenes Rohr n. ‖ tube m. étiré sans soudure. / ~ ground (Mine) ‖ festes Gebirge n. ‖ terrain m. ferme. / ~ injection vertical two-cycle engine with pistons working in opposite directions ‖ kompressorloser Zweitaktmotor m. stehender Bauart mit gegenläufigen Kolben ‖ moteur m. vertical sans compresseur à pistons convergents. / ~ jacket ‖ Vollmantel m. ‖ corps m. à paroi pleine. / ~ journal bearing ‖ Augenlager n. ‖ palier m. à œillet. / ~ material obtained ‖ gewonnenes Schleudergut n. ‖ résidu m. sec obtenu par essorage. / ~ matter (Print) ‖ kompresser Satz m. ‖ composition f. pleine ou non-interlignée. / ~ measure ‖ Festmaß n. ‖ mesure f. de solidité. / ~ residue of evaporation ‖ Abdampfrückstand m. ‖ résidu m. solide d'évaporation. / ~ rock ‖ anstehendes Gestein n. ‖ roche f. vive; minerai m. en vue. / press for mounting ~ rubber tires ‖ Presse f. zum Aufziehen von Vollgummireifen ‖ presse f. à monter les bandages pleins. / ~ section ‖ Vollquerschnitt m. ‖ profil m. plein / ~ soil see ~ ground. / cut from ~ steel blanks ‖ aus dem Vollen n. herausgearbeitet ‖ façonné ou travaillé dans la masse. / ~ tire see solid tyre. / ~ tool ‖ Vollstahl m. ‖ outil m. en acier plein. / ~ tubbing ‖ wasserdichter Schachtausbau m. durch Ringe ‖ cuvelage m. ou cuvellement m. en fer circulaire.

solid tyre ‖ Vollgummireifen m.; Vollreifen m. ‖ bandage m. plein; plein m. / felloe for ~s ‖ Felge f. für Vollreifen ‖ jante f. à bandages pleins.

solid wheel ‖ vollwandiges Rad n. ‖ roue f. pleine.

solid ‖ Körper m. ‖ solide m.; corps m. / ~ of a machine ‖ Körper m. einer Maschine ‖ corps m. ou bâti m. d'une machine. / machined from the ~ ‖ aus dem Vollen n. herausgearbeitet ‖ façonné ou taillé dans la masse. / ~ of rotation ‖ Rotationskörper m. ‖ solide m. de révolution. / ~ generated by rotation see ~ of rotation.

solidification ‖ Erstarrung f. ‖ solidification f. / partial ~ ‖ Teilerstarrung f. ‖ solidification f. partielle. / quick ~ of the rubber and insulating mass ‖ rasche Erstarrung f. der Gummi- und Isoliermasse ‖ solidification f. rapide du caoutchouc et de la masse isolante. / heat of ~ ‖ Erstarrungswärme f. ‖ chaleur f. de solidification. / point of ~ ‖ Erstarrungspunkt m. ‖ point m. de solidification. / product of ~ ‖ Erstarrungsgebilde n. ‖ produit m. de solidification.

solidify, to ‖ erstarren ‖ solidifier. / ~ the glue jelly ‖ die Gallerte f. erstarren lassen ‖ solidifier la colle.

solidifying point of the oil ‖ Stockpunkt m. des Öles ‖ point m. de figeaison de l'huile.

solidity (Build) ‖ Festigkeit f. ‖ solidité f. / ~ (Geom) ‖ Volumen n.; Rauminhalt m. ‖ volume m.

solitary ‖ vereinzelt ‖ isolément.

solo flight ‖ Einzelflug m. ‖ vol m. seul.

solstice ‖ Sonnenwende f. ‖ solstice m. / summer ~ ‖ Sommersonnenwende f. ‖ solstice m. d'été.

solubility ‖ Löslichkeit f. ‖ solubilité f. / low ~ ‖ Schwerlöslichkeit f. ‖ faible solubilité f.

soluble ‖ löslich ‖ soluble. / difficultly ~ ‖ schwerlöslich ‖ difficilement soluble. / easily ~ ‖ leichtlöslich ‖ facilement soluble. / to make ~ ‖ löslich machen ‖ solubiliser.

soluble glass colour ‖ Wasserglasfarbe f. ‖ couleur f. au silicate d'alcali. / ~ oil ‖ wasserlösliches Öl n. ‖ huile f. soluble.

solution ‖ Lösung f.; Auflösung f. ‖ solution f.; dissolution f. / ~ of alum and salt ‖ Alaunbrühe f. ‖ solution f. d'alun et de sel. / ~ of aniline colour ‖ Anilinfarbstofflösung f. ‖ solution f. de couleur d'aniline. / ~ of barium chloride ‖ Bariumchloridlösung f. ‖ solution f. de chlorure de barium. / ~ by calculation ‖ rechnerische Auswertung f. ‖ évaluation f. numérique. / ~ of caustic potash ‖ Kalilauge f.; Ätzkalilösung f. ‖ lessive f. ou solution f. de potasse caustique. / chemical ~ used in the manufacturing process ‖ im Fabrikationsprozeß benötigte chemische Lösung f. ‖ solution f. chimique nécessaire à la fabrication. / colloidal ~ ‖ gallertartige Lösung f. ‖ solution f. colloïdale. / graphical ~ ‖ grafische Auswertung f. ‖ évaluation f. ou solution f. graphique. / india rubber ~ ‖ Gummilösung f. ‖ solution f. de caoutchouc. / normal ~ ‖ Normallösung f. ‖ solution f. normale. / to pass into ~ ‖ in Lösung f. geben ‖ passer en solution f.; dissoudre. / ~ of tartar ‖ Weinsteinlösung f. ‖ solution f. tartrique.

solution, heat of ~ ‖ Lösungswärme f. ‖ chaleur f. de solution. / ~ pressure ‖ Lösungsdruck m. ‖ tension f. de la solution. / strength of ~ ‖ Lösungsstärke f. ‖ concentration f. de la solution. / ~ tension ‖

Lösungstension f. ‖ tension f. de dissolution.

solvency ‖ Solvenz f.; Zahlungsfähigkeit f. ‖ solvabilité f.

solvent ‖ solvent; zahlungsfähig ‖ solvable. / ~ (Chem) ‖ auflösend; lösend ‖ dissolvant. / ~ electrode ‖ Lösungselektrode f. ‖ électrode f. dissolvante. / ~ power ‖ Lösungsfähigkeit f.; Lösungsvermögen n. ‖ pouvoir m. dissolvant.

solvent (Chem) ‖ Lösungsmittel n. ‖ (dis)solvant m. / celluloid ~ ‖ Zelluloidlösungsmittel n. ‖ dissolvant m. de celluloïd. / the ~ escapes ‖ das Lösungsmittel entweicht ‖ le dissolvant fuit ou s'échappe. / volatile ~ ‖ flüchtiges Lösungsmittel n. ‖ dissolvant m. ou solvant m. volatil; solvante f. volatile.

sonometer ‖ Schallmesser m.; Tonmesser m. ‖ sonomètre m.; échomètre m.

sonorous ‖ wohlklingend ‖ harmonieux; sonore; euphonique.

soot, to ‖ verrußen ‖ encrasser; se couvrir de suie.

soot ‖ Ruß m. ‖ suie f. / ~ arrester see ~ catcher. / ~ black ‖ Kienruß m. ‖ noir m. de fumée. / ~ blower ‖ Rußbläser m. ‖ soufflet m. à suie. / ~ catcher ‖ Rußfänger m. ‖ pare-suie m. / ~ coal ‖ Rußkohle f. ‖ houille f. fuligineuse.

soporific ‖ Schlafmittel n. ‖ soporatif m.

sorbite (Met) ‖ Sorbit m.; Temperit m. ‖ sorbite f.; tempérite f.

sordawalite ‖ Sordawalit m. ‖ sordawalite f.

sordid ‖ schmutzig ‖ malhonnête.

sore ‖ Wunde f. ‖ blessure f.; plaie f.

sorghum ‖ Sorghum n. ‖ sorgho m.

sorrel-salt ‖ Kleesalz n.; Sauerkleesalz n.; oxalsaures Kalium n. ‖ sel m. d'oseille; bioxalate m. de potasse.

sort, to ‖ sichten; sortieren ‖ assortir; mettre à part; examiner; trier. / ~ the rags pl. ‖ die Lumpen mpl. auslesen oder sortieren ‖ délisser ou séparer ou trier les chiffons mpl. / ~ tobacco ‖ Tabak m. sortieren ‖ époularder les feuilles fpl. de tabac. / ~ wool ‖ Wolle f. sortieren ‖ assortir ou détricher la laine.

sort ‖ Sorte f.; Marke f. ‖ sorte f.; marque f. / ~s pl. of ingot steel ‖ Flußstahlsorten fpl. ‖ genres mpl. d'acier.

sorter ‖ Sortierer m. ‖ classeur m.; sorteur m. / letter ~ ‖ Briefsortierer m. ‖ classeur de lettres m.

sorting ‖ Sortierung f. ‖ classification f.; calibrage m. / ~ (Mill) ‖ Sonderung f. ‖ triage m. / hand ~ (Mine) ‖ Handscheidung f. ‖ triage m. à la main. / ~ of wool ‖ Lesen n. der Wolle; Wollewerei f. ‖ assortissage m. de laines.

sorting apparatus ‖ Sortierer m. ‖ trieur m. / ~ belt (Ore dress) ‖ Leseband n. ‖ bande f. de triage. / ~ board ‖ Klaubbühne f. ‖ épluchoir m. / ~ cylinder ‖ Sortierzylinder m. ‖ tambour m. à assortir.

sorting machine ‖ Auslesemaschine f.; Sortiermaschine f. ‖ machine f. de triage; trieuse f. / ~ for corn ‖ Sortiermaschine f. für Getreide ‖ machine f. à trier les grains de céréales. / ~ for seeds ‖ Sortiermaschine f. für Samen ‖ machine f. à trier les grains de semences.

sorting and cleaning machine for coffee and coffee surrogates ‖ Sortier- und Reinigungsmaschine f. für Kaffee und Kaffee-Ersatz ‖ machine f. pour le triage et le nettoyage du café et des succédanés de café.

sorting plant, waste paper ~ ‖ Altpapiersortieranlage f. ‖ installation f. de triage des vieux papiers.

sorting table ‖ Verlesetisch m. ‖ table f. collectrice. / ~ wool ‖ Wollsortieren n. ‖ détrichage m.

sought, to be much ~ after ‖ großen Zulauf m. haben ‖ être en vogue f.

sound, to (Mus) ‖ auf den Klang m. prüfen ‖ examiner par le son. / ~ (Nav) ‖ loten ‖ sonder. / ~ (Tel) ‖ tönen ‖ résonner. / ~ the pump ‖ die Pumpe peilen ‖ sonder la pompe.

sound (Capital) ‖ kapitalkräftig ‖ disposant d'importants capitaux. / ~ (Trust) ‖ vertrauenswürdig ‖ digne de toute confiance f. / ~ firm ‖ reelles Geschäft n. ‖ affaire f. sérieuse; maison f. respectable. / ~ wood ‖ gesundes Holz n. ‖ bois m. sain.

sound (Fish) ‖ Luftblase f. ‖ vessie f. natatoire. / ~ (Mus) ‖ Ton m.; Klang m. ‖ son m. / ~ (Med) ‖ Sonde f. ‖ sonde f. / ~ (Phys) ‖ Schall m. ‖ son m. / concave ~ ‖ Hohlsonde f. ‖ sonde f. cannelée. / to pick up the ~ ‖ den Ton m. aufnehmen ‖ recueillir le son. / to take by ~ (Tel) ‖ nach dem Gehör n. aufnehmen ‖ lire au son.

sound, perfect ~ absorber ‖ vollkommener Schalldämpfer m. ‖ absorbant m. parfait du bruit. / ~ amplifier ‖ Lautverstärker m. ‖ amplificateur m. acoustique. / ~ board (Mus) ‖ Resonanzboden m. ‖ résonnance f.; table f. d'harmonie. / ~ boarding (Carp) ‖ Einschubbretter npl. ‖ lucet m. / ~ box ‖ Schalldose f. ‖ diaphragme m. / ~ conductivity of the water ‖ Schalleitfähigkeit f. des Wassers ‖ conductivité f. de l'eau pour le son. / ~ damper ‖ Schalldämpfer m. ‖ sourdine f.; modérateur m. de son; amortisseur m. ou absorbant m. de bruit. / ~ deadening ‖ Schalldämpfung f. ‖ amortissement m. du bruit. / ~ detector ‖ Horchgerät n. ‖ appareil m. de repérage par le son.

sounder (Electr) ‖ Läutewerk n. ‖ sonneur m. / ~ (Tel) ‖ Klopferapparat m. ‖ parleur m.

sound hole of the violin ‖ Schalloch n. der Violine ‖ ouïe f. du violon.

sound film ‖ Tonfilm m.; Klangfilm m. ‖ film m. parlant ou sonore. / ~ projector ‖ Tonfilmapparatur f. ‖ appareil m. de projection pour films sonores. / ~ reproducing apparatus ‖ Tonfilmwiedergabevorrichtung f. ‖ appareillage m. pour la projection de films sonores. / ~ technics pl. ‖ Tonfilmtechnik f. ‖ technique f. du film sonore.

sound floor (Build) ‖ Fehlboden m.; Einschub m.; Schragboden m. ‖ plancher m. à remplissage. / ~ board (Carp) ‖ Schalbrett n. ‖ entrevous m.; ais m. d'entrevous.

sounding ‖ tönend ‖ sonore.

sounding ‖ Lot n.; Senkel m. ‖ plomb m.; fil m. à plomb; sonde f. / to be in ~s (Nav) ‖ auf lotbarem Grunde m. sein ‖ être sur la sonde; avoir ou trouver fond m. / deep-sea ~ ‖ Tiefseelotung f. ‖ sondage m. des grandes profondeurs. / to get ~s ‖ Grund m. werfen ‖ avoir ou trouver fond m.

sounding apparatus (Ship) ‖ Tiefenlotapparat m. ‖ appareil m. de sondage à plomb. / ~ balloon ‖ Registrierballon m. ‖ ballon m. enregistreur ou explorateur. / ~ board wood ‖ Klangholz n.; Resonanz-

holz n. ‖ bois m. de résonnance. / ~ lead ‖ Lotblei n. ‖ plomb m. de sonde; plomb m. / ~ machine ‖ Lotmaschine f. ‖ machine f. de sondage. / ~ plummet see ~ lead. / ~ rod ‖ Peilstock m. ‖ sonde f. de pompe.

sound intensity ‖ Lautstärke f. ‖ intensité f. du son. / ~ listening device see sound locator.

sound locator ‖ Horchgerät n. ‖ appareil m. de repérage par le son. / synchronized ~ ‖ Richtungshörer m. ‖ artophone m.

sound oscillation ‖ Schallschwingung f. ‖ vibration f. sonore. / ~ picture projector ‖ Tonbildprojektor m. ‖ projecteur m. pour la reproduction sonore. / ~ projector ‖ Tonprojektor m. ‖ projecteur m. pour l'émission sonore.

sound-proof box (Tel) ‖ schalldichte Zelle f. ‖ cabine f. insonore ou sourde ou isolante du son.

sound proofness ‖ Schalldichtigkeit f. ‖ insonorité f.; aphonicité f.

sound ranging ‖ Schallmessen n. ‖ repérage m. par le son. / ~ troop ‖ Schallmeßtrupp m. ‖ section f. de repérage par le son.

sound ray ‖ Schallstrahl m. ‖ rayon m. sonore. / ~ screen (Tel) ‖ Schallkammer f. ‖ abat-son m. / ~ signal ‖ akustisches Signal n. ‖ signal m. phonique. / ~ velocity ‖ Schallgeschwindigkeit f. ‖ vélocité f. du son. / ~ vibration ‖ Tonschwingung f. ‖ vibration f. sonore. / ~ wave ‖ Schallwelle f. ‖ onde f. sonore ou acoustique.

soup, preserved ~ ‖ Suppenkonserve f. ‖ soupe f. conservée.

soup extract ‖ Suppenextrakt m. ‖ extrait m. pour la soupe. / ~ ladle ‖ Suppenlöffel m. ‖ louche f.; cuiller f. à potage. / ~ plate ‖ Suppenteller m. ‖ assiette f. à soupe.

sour ‖ sauer ‖ aigre; sur.

source ‖ Quelle f. ‖ source f. / ~ of current ‖ Stromquelle f. ‖ source f. de courant. / ~ of electricity ‖ Elektrizitätsquelle f. ‖ source f. d'électricité. / ~ of energy ‖ Kraftquelle f. ‖ source f. d'énergie. / ~ of errors ‖ Fehlerquelle f. ‖ source f. d'erreurs. / ~ of heat ‖ Wärmequelle f. ‖ source f. de chaleur. / ~ of income ‖ Einnahmequelle f. ‖ source f. de revenus.

source of light ‖ Lichtquelle f. ‖ source f. lumineuse. / ~ of great intensity ‖ Lichtquelle f. von großer Intensität ‖ source f. lumineuse très intense, / ~ exhibiting structural markings ‖ Lichtquelle f. mit Struktur ‖ source f. lumineuse présentant une structure. / structureless ~ ‖ Lichtquelle f. ohne Struktur ‖ source f. lumineuse unie. / ~ for viewing the fundus of eye by entirely red free light ‖ Lichtquelle f. für die völlig rotfreie Untersuchung des Augenhintergrundes ‖ source f. lumineuse pour l'examen du fond de l'œil avec une lumière absolument exempte de rouge.

source of natural gas ‖ Erdgasquelle f. ‖ source f. de gaz naturel.

source of power ‖ Kraftquelle f.; Energiequelle f. ‖ source f. d'énergie; source f. énergétique. / to combine sources pl. of power to form one closed system ‖ Kraftquellen fpl. zu einem System zusammenfassen ‖ réunir des sources fpl. énergétiques dans un seul système. / current which is generated close to the ~ ‖ unmittelbar auf der Energie-

grundlage erzeugter Strom m. ‖ courant m. obtenu sur les lieux-mêmes des sources d'énergie. / to tap a ~ in an economical manner ‖ eine Kraftquelle f. in wirtschaftlichster Weise erschließen ‖ exploiter une source f. d'énergie à la façon la plus rationnelle.

source, quiescent ~ of sound ‖ ruhende Schallquelle f. ‖ source f. sonore immobile. / these mines pl. constitute one of the chief ~s of supply of ore ‖ diese Gruben fpl. bilden eine der wichtigsten Quellen für den Erzbezug ‖ ces mines fpl. constituent une des sources principales de l'alimentation en minerai.

sourdine (Mus) ‖ Dämpfer m.; Schalldämpfer m. ‖ sourdine f.; amortisseur m. de bruit.

sours pl. ‖ Sauerbad n. ‖ eau f. sure.

sour water (Curr) ‖ Sauerbeize f.; Sauerwasser n. ‖ confit m. / fat ~ ‖ fettes Sauerwasser n. ‖ eau f. sure grasse. / ~ of starch-makers ‖ Sauerwasser n. für die Stärkeherstellung ‖ eau f. sure des amidonniers.

Southern Cross ‖ Kreuz n. des Südens ‖ Croix f. Australe. / southern fruit ‖ Südfrucht f. ‖ fruit m. du midi.

south-magnetic ‖ südmagnetisch ‖ magnétique sud.

souvenir for travellers ‖ Reiseandenken n. ‖ souvenir m. de voyage.

sow, to ‖ säen ‖ semer.

sow (Metal) ‖ Sau f.; Eisenklumpen m. ‖ loup m.; bloc m. / ~ channel (Metal) ‖ Leistengraben m.; Masselgraben m. ‖ lit m. de gueuse; rigole f. de coulée.

sower (Machine) ‖ Sämaschine f. ‖ semeuse f. / ~ (Person) ‖ Sämann m.; Säer m. ‖ semeur m.

sowing ‖ Saat f.; Aussaat f. ‖ ensemencement m. / ~ in furrows ‖ Furchensaat f. ‖ semis m. en sillons. / ~ in lines ‖ gedrillte Saat f. ‖ semis m. en lignes. / ~ in rills ‖ Rillensaat f. ‖ semis m. en rigoles.

sowing barley ‖ Saatgerste f. ‖ orge f. de semence. / ~ machine ‖ Sämaschine f. ‖ semeuse f.; semoir m.; machine f. à semer. / ~ and weeding machine ‖ Sä- und Jätmaschine f. ‖ semoir m. et sarcloir m. mécanique.

sowing machine driver ‖ Sämaschinenführer m. ‖ semeur m. à la machine; conducteur m. du semoir ou de la machine à semer.

sow iron ‖ Schaleneisen n. ‖ fonte f. de rigole; maître-calle m.

soya see soy (bean).

soy (bean) ‖ Sojabohne f. ‖ (fève f. de) soja m.

soy bean oil ‖ Sojaöl n. ‖ huile f. de soja.

sozoiodol salt ‖ Sozojodolsalz n. ‖ sozoiodolate m.

space, to ~ the letters ‖ die Schrift sperren ‖ espacer les lettres fpl.

space ‖ Raum m. ‖ espace m. / ~ (Mech) ‖ Bahn f. ‖ espace m. parcouru. / ~ in the clear ‖ lichter Raum m. ‖ aire f. / ~ between the grate bars ‖ Rostspalte f. ‖ intervalle f. entre des barreaux de grille. / ~ between the holes (Perforated Sheets) ‖ Steg m. ‖ espace f. entre les trous. / ~ between the lines (Railw) ‖ Schienenzwischenraum m. ‖ entre-voie f. / ~ between the links ‖ Laschenteilung f. ‖ division f. des maillons. / noxious ~ see waste ~. / open ~ ‖ Freifläche f. ‖

plein air m. / required ~ ‖ Platzbedarf m. ‖ encombrement m. / ~ of the sea ‖ Seestrich m. ‖ parage m. / ~ of time ‖ Zeitabschnitt m. ‖ période f. / ~ between two lines (Print) ‖ Linienbreite f. ‖ réglure f. / ~ of a vaultïng ‖ Gewölbfach n.; Fach zwischen Gewölbgurten ‖ surface f. d'une lunette ou d'un triangle de voûte. / waste ~ ‖ schädlicher Raum m. ‖ espace m. mort. / ~ between words (Print) ‖ Wortzwischenraum m. ‖ espace m. entre les mots.

space bar see space key.

space charge (Radio) ‖ Raumladung f. ‖ charge f. spatiale. / ~ effect of the repeater valves ‖ Raumladungswirkung f. der Verstärkerröhren ‖ effet m. de la charge d'espace des lampes amplificatrices. / ~ grid (Radio) ‖ Raumladungsgitter n. ‖ grille f. de contrôle .

space current source ‖ Raumstromquelle f. ‖ source f. de courant spatial. / ~ discharge current ‖ Raumentladestrom m. ‖ courant m. de décharge de l'espace. / ~ key (Typewr) ‖ Zwischenraumtaste f. ‖ touche f. d'espacement. / ~ line (Print) ‖ Durchschuß m. ‖ interligne f.

spacer (Auto) ‖ Abstandhülse f. ‖ douille f. d'écartement. / back ~ (Typewr) ‖ Rücktaste f. ‖ touche f. de marche arrière.

space ship ‖ Raumschiff n. ‖ navire m. d'espace. / rocket-propelled ~ ‖ Raketenflugzeug n. ‖ avion m. à réaction; avionfusée m.

space vector ‖ Raumvektor m. ‖ vecteur m. spacial.

spacial radiation ‖ Raumwelle f. ‖ onde f. spatiale.

spacing ‖ Intervall n. ‖ intervalle m. / close ~ (Print) ‖ gedrängte Schrift f. ‖ composition f. serrée ou nourrie. / ~ of the frames (Shipb) ‖ Spantdistanz f. ‖ espacement m. des couples. / mechanism for adjusting the ~ between lines ‖ Zeileneinstellvorrichtung f. ‖ mécanisme m. réglant l'espacement entre les lignes.

spacing piece ‖ Zwischenstück n. ‖ pièce f. d'espacement. / temporary ~ ring ‖ provisorischer Distanzring m. ‖ bague f. de distance provisoire.

spacious ‖ geräumig ‖ spacieux.

spade ‖ Spaten m. ‖ bêche f. / ~ (Curr) ‖ gerader Rührhaken m. ‖ gâche f. droite. / back-treaded ~ ‖ Spaten m. mit rückgebogenem Rand ‖ bêche f. à rebord en arrière. / crooked ~ (Curr) ‖ gebogener Rührhaken m. ‖ gâche f. coudée. / ~ with cross handle ‖ Spaten m. mit Knaufstiel ‖ bêche f. à manche à crosse. / curved ~ (Mine) ‖ Minenkratze f. ‖ drague f. / English treaded small ~ ‖ englischer Schmalspaten m. mit Tritt ‖ bêche f. anglaise étroite et appui pied. / flat American ~ ‖ amerikanischer Flachspaten m. ‖ bêche f. américaine plate. / front-treaded ~ ‖ Spaten m. mit vorgebogenem Tritt ‖ bêche f. à rebord appui-pied en avant. / point ~ ‖ Rundspaten m. ‖ bêche f. ronde.

spade-type chisel ‖ Schaufelmeißel m. ‖ bêche f. / handle for ~s ‖ Spatenstiel m. ‖ manche m. pour bêches. / machine for making ~s ‖ Spatenherstellungsmaschine f. ‖ machine f. à fabriquer les bêches.

spagnolet ‖ Drehriegel m. ‖ espagnolette f.

spalling of ores ‖ Scheidung f. der Erze ‖ scheidage m. des minerais.

span ‖ Spannweite f. ‖ empan m.; ouverture f.; vide m.; portée f.; travée f. / ~ (El line) ‖ Abspannung f. ‖ haubanage m. / ~ (Hand) ‖ Spanne f. ‖ empan m. / ~ of poles ‖ Standweite f. der Telegrafenstangen ‖ portée f. des poteaux de télégraphe. / ~ of the wing ‖ Flügelspannweite f. ‖ envergure f. de l'aile.

span bracket (Tel) ‖ Abspannmauerbügel m. ‖ potelet m. d'arrêt de ligne en façade. / ~ ceiling ‖ Balkendecke f. ‖ plafond m. enfoncé; lambris m.

spandrel (Arch) ‖ Gewölbzwickel m. ‖ reins mpl. de voûte. / ~ wall ‖ schwebende oder fliegende oder auf Bogen ruhende Mauer f. ‖ mur m. en l'air ou portant à faux ou monté sur des voûtes.

spangle ‖ Flitter m. ‖ paillette f. / gelatine ~ ‖ Gelatineflitter m. ‖ paillette f. en gélatine. / mica ~ ‖ Glimmerflitter m. ‖ paillette f. en mica. / silicatine ~ ‖ Silikatinflitter m. ‖ paillette f. de silicatine. / viscose ~ ‖ Viskoseflitter m. ‖ paillette f. en viscose.

spangle maker ‖ Flitterarbeiter m. ‖ pailleteur m. / machine for making ~s ‖ Flitterherstellungsmaschine f. ‖ machine f. à fabriquer des paillettes. / setter ‖ Flitternäherin f. ‖ pailleteuse f.

Spanish cane ‖ spanisches Rohr n. ‖ jonc m. des Indes. / ~ white ‖ Schlämmkreide f. ‖ blanc m. de Meudon ou de Troyes ou d'Espagne.

spanker (Shipb) ‖ Briggsegel n. ‖ brigantine f.

spanner ‖ Schraubenschlüssel m. ‖ clef f. (à vis ou à écrous). / adjustable ~ ‖ verstellbarer Schraubenschlüssel m. ‖ clef f. réglable. / ~ for caps ‖ Kappenschlüssel m. ‖ clef f. à canon. / crocodile ~ ‖ Löwenmaul n. ‖ clef f. à mâchoires dentées. / double-ended ~ ‖ Doppelschraubenschlüssel m. ‖ clef f. double. / hexagon ~ ‖ Sechskantschraubenschlüssel m. ‖ clef f. hexagonale. / ~ for magneto ‖ Magnetschlüssel m. ‖ clef f. de la magnéto. / medium heavy ~ ‖ mittelschwerer Schraubenschlüssel m. ‖ clef f. semilourde. / octagonal ~ ‖ Achtkantschraubenschlüssel m. ‖ clef f. à huit pans. / setting by ~s ‖ Schlüsseleinstellung f. ‖ réglage m. par clef. / sheet arm double-ended ~ ‖ Blechdoppelschraubenschlüssel m. ‖ clef f. double en tôle. / single-ended ~ ‖ einfacher Schraubenschlüssel m. ‖ clef f. simple. / ~ for square head screws ‖ Vierkantschraubenschlüssel m. ‖ clef f. carrée. / standard ~ ‖ Normschlüssel m. ‖ clef f. normale. / ~ of the vice ‖ Schlüssel m. des Schraubstocks ‖ manivelle f. de l'étau.

spanner board ‖ Schraubenschlüsselbrett n. ‖ planche f. ou ratelier m. à clefs. / ~ wrench ‖ Hakenschlüssel m. ‖ clef f. à crochet.

spanning ‖ Verspannung f. ‖ montage m. / ~ power of the tank ‖ Überschreitungsvermögen n. des Kampfwagens ‖ franchissement m. du char de combat.

span press plant, hydraulic ~ ‖ hydraulische Spanpreßanlage f. ‖ presse f. hydraulique pour carton comprimé.

span saw ‖ Gestellsäge f.; Spannsäge f. ‖ scie f. à chassis ou montée ou à queue. / ~ wire ‖ Abspanndraht m. ‖ fil m. d'arrêt.

spar (Airplane) ‖ Holm m. ‖ longeron m.

/ ~ (Build) ‖ Sparren m. ‖ chevron m. / ~ (Mar) ‖ Spiere f. ‖ espar m.; mâtereau m. / ~ (Miner) ‖ Spat m. ‖ spath m. / blue ~ ‖ Blauspat m.; Lazulith m. ‖ feldspath m. bleu; klaprothine f.; lazulite m. / box ~ (Mar) ‖ Kastenspiere f. ‖ longeron-caisson m. / calcareous ~ ‖ Kalkspat m.; Calcit m.; Doppelspat m. ‖ chaux f. carbonatée ou cristallisée; calcaire m.; spath m. calcaire ou d'Islande. / extruded ~ (Aero) ‖ gezogener Holm m. ‖ longeron m. profilé à chaud. / heavy ~ ‖ Schwerspat m. ‖ spath m. pesant. / Iceland ~ ‖ isländischer Doppelspat m. ‖ spath m. d'Islande. / iron ~ ‖ Spateisenstein m.; Eisenspat m. ‖ fer m. spathique. / main ~ (Aero) ‖ Hauptholm m. ‖ longeron m. principal. / manganese ~ ‖ Manganspat m. ‖ carbonate m. de manganèse. / rear ~ (Aero) ‖ Hinterholm m. ‖ longeron m. arrière. / ~ of symmetrical cross section (Aero) ‖ Holm m. mit gleichmäßig verteiltem Querschnitt ‖ longeron m. à section droite symétrique ou symétriquement répartie. / wooden ~ (Aero) ‖ Holzholm m. ‖ longeron m. en bois.

spar box, wing ‖ Holmschuh m. ‖ embout m. du longeron.

spare ‖ Spitzkeil m. ‖ picot m.

spare ‖ sparsam ‖ ménager; économe. / ~ anchor ‖ Raumanker m. ‖ ancre f. de la cale. / ~ bar ‖ Ersatzhebel m. ‖ tige f. de rechange. / ~ blade ‖ Ersatzklinge f. ‖ lame f. de rechange. / ~ capital ‖ flüssiges Kapital n. ‖ fonds mpl. disponibles. / ~ cell (Electr) ‖ Abschaltzelle f. ‖ élément m. pouvant être mis hors du circuit. / ~ commutator ‖ Ersatzkommutator m. ‖ collecteur m. de rechange. / ~ conductor in the cable ‖ Ersatzader f. im Kabel ‖ conducteur m. de rechange dans le câble. / ~ current source ‖ Ersatzstromquelle f. ‖ installation f. d'énergie de secours. / ~ dyke ‖ Schlafdeich m. ‖ digue f. de réserve. / ~ heater ‖ Ersatzheizkörper m. ‖ élément m. de chauffage de rechange. / ~ isolator ‖ Ersatzisolator m. ‖ isolateur m. de rechange. / ~ lead (Pencil) ‖ Ersatzmine f. ‖ mine f. de réserve. / ~ link (Mar) ‖ Notschake f.; Notschäkel m. ‖ esse f. / ~ mast ‖ Reservemast m. ‖ mât m. de rechange. / ~ master clock ‖ Reservehauptuhr f. ‖ horloge f. principale de réserve. / ~ part ‖ Ersatzstück n.; Ersatzteil m.; Reserveteil m. ‖ pièce f. de réserve; élément m. de rechange. / ~ parts pl. are always kept in stock ‖ Ersatzteile mpl. sind stets auf Lager ‖ pièces fpl. de rechange sont toujours en magasin. / ~ piece for spinning machines ‖ Spinnereimaschinenersatzteil m. ‖ pièce f. de réserve de métier à filer. / ~ rotor ‖ Ersatzanker m. ‖ induit m. de réserve. / ~ sail ‖ Reservesegel n.; Ersatzsegel n. ‖ voile f. de rechange. / ~ spar (Mar) ‖ Reservespier n. ‖ espar m. de rechange. / ~ tyre ‖ Ersatzreifen m. ‖ pneu m. de réserve. / ~ tyre equipment ‖ Ersatzbereifung f. ‖ équipement m. de pneu de rechange. / ~ wheel carrier (Auto) ‖ Ersatzradhalter m. ‖ porte-roue m. de secours. / ~ yard (Mar) ‖ Reserverahe f. ‖ vergue f. de rechange.

spar flange (Aero) ‖ Holmgurt m. ‖ bride f. de longeron. / ~ gate ‖ Gattertor n.; Gittertor n. ‖ porte f. à claire-voie.

sparge, to ~ the draff ‖ die Treber pl. anschwänzen ‖ arroser *ou* laver les drêches fpl.

sparger (Draff) ‖ Anschwänzer m.; Anschwänzkreuz n. ‖ croix f. écossaise.

sparging (Draff) ‖ Anschwänzen n. ‖ arrosage m.; lavage m. / ~ temperature (Draff) ‖ Anschwänztemperatur f. ‖ température f. de lavage. / ~ water (Draff) ‖ Anschwänzwasser n. ‖ eau f. *ou* trempe de lavage.

spark, to ‖ feuern; funken ‖ produire des étincelles; cracher.

spark ‖ Funke m. ‖ étincelle f. / ~ at breaking contact ‖ Öffnungsfunke m. ‖ étincelle f. d'ouverture. / ~ before contact ‖ Schließungsfunken m. ‖ étincelle f. de fermeture. / early ~ ‖ Vorzündung f. ‖ avance f. à l'allumage. / electric ~ ‖ elektrischer Funken m. ‖ étincelle f. électrique. / ignition ~ ‖ Zündfunke f. ‖ étincelle f. d'allumage. / jump ~ ‖ überspringender Funke m. ‖ étincelle f. disruptive. / musical ~ ‖ musikalischer Funke m. ‖ étincelle f. musicale. / quenched ~ ‖ Löschfunke m. ‖ étincelle f. étouffée. / slow ~ (Radio) ‖ Knallfunke m. ‖ étincelle f. explosive. / timed ~ (Radio) ‖ gesteuerte Schwingung f. ‖ oscillation f. guidée. / touch ~ ‖ Abreißfunken m. ‖ étincelle f. à rupture.

spark arrester (Electr) ‖ Funkenlöscher m. ‖ éclateur m. pare-étincelles. / ~ arrester (Loc) ‖ Funkenfänger m. ‖ pare-étincelles m. / ~ blow-out *see* ~ arrester (Electr). / ~ catcher *see* ~ arrester (Loc). / ~ chamber ‖ Funkenkammer f. ‖ chambre f. à étincelles.

spark coil ‖ Funkeninduktor m. ‖ bobine f. d'induction. / ~ with hammer break ‖ Funkeninduktor m. mit Hammerunterbrecher ‖ bobine f. d'induction à interrupteur à marteau.

spark control ‖ Zündregelung f. ‖ commande f. d'allumage. / ~ decrement ‖ Funkendekrement n. ‖ décrément m. d'étincelles. / ~ discharge ‖ Funkenentladung f. ‖ décharge f. disruptive *ou* à étincelle. / ~ drawer ‖ Funkenzieher m. ‖ excitateur m. / ~ extinguishing ‖ Funkenlöschung f. ‖ extinction f. de l'arc. / ~ frequency ‖ Funkenfrequenz f. ‖ fréquence f. d'étincelles.

spark gap ‖ Funkenstrecke f. ‖ éclateur m. déchargeur *ou* à étincelle. / auxiliary ~ ‖ Hilfszündung f. bei Löschfunkensendern ‖ aide-éclateur m. à étincelle. / disk ~ ‖ Plattenfunkenstrecke f. ‖ éclateur m. à plaques. / hydrogen ~ ‖ Wasserstofffunkenstrecke f. ‖ éclateur m. à atmosphère d'hydrogène. / multiple ~ ‖ unterteilte Funkenstrecke f.; Serienfunkenstrecke f. ‖ éclateur m. en série. / nonsynchronous ~ ‖ asynchrone Funkenstrecke f. ‖ éclateur m. asynchrone. / quenched ~ ‖ Löschfunkenstrecke f. ‖ éclateur m. pour étincelle étouffée. / rotating ~ ‖ rotierende Funkenstrecke f. ‖ déchargeur m. rotatif. / safety ~ ‖ Sicherheitsfunkenstrecke f. ‖ parcours m. d'étincelle de sûreté. / single ~ ‖ Einzelfunkenstrecke f. ‖ éclateur m. simple. / synchronous ~ ‖ synchrone Funkenstrecke f. ‖ éclateur m. synchrone.

spark hand lever (Auto) ‖ Zündhebel m. ‖ manette f. d'allumage. / ~ tube (Auto) ‖ Zündspindel f. ‖ tige f. de la manette d'allumage.

spark ignition ‖ Funkenzündung f. ‖ allumage m. par étincelles. / image of the ~ ‖ Funkenbild n. ‖ image m. de l'étincelle.

sparking ‖ Funken n. ‖ crachement m. / non ~ ‖ Funkenlosigkeit f. ‖ absence f. d'étincelles.

sparking advance ‖ Vorzündung f.; Frühzündung f. ‖ avance f. à l'allumage. / ~ lever ‖ Frühzündungshebel m. ‖ manette f. d'avance à l'allumage.

sparking distance see spark gap. / ~ paper (Motor) ‖ Zündlunte f. ‖ mèche f. d'allumage. / ~ printer ‖ Lichtbildtypendrucker m. ‖ appareil m. imprimeur à étincelle. / ~ retard ‖ Nachzündung f.; Spätzündung f. ‖ retard m. à l'allumage; allumage m. retardé.

sparkle, to ‖ glänzen ‖ étinceler.

spark length (Radio) ‖ Schlagweite f.; Funkenlänge f. ‖ distance f. explosive *ou* de l'étincelle.

sparkler ‖ Wunderkerze f. ‖ cierge f. miraculeuse.

sparkless (Electr) ‖ funkenlos ‖ sans étincelles fpl. / ~ commutation ‖ funkenfreies Arbeiten n. des Kollektors ‖ commutation f. sans crachement. / ~ controller for mining railway ‖ funkenfreier Grubenbahnfahrschalter m. ‖ combinateur m. de travail sans étincelles pour chemins de fer de mine.

sparkling (Miner) ‖ schillernd ‖ chatoyant.

sparkling wine ‖ Schaumwein m. ‖ vin m. mousseux; champagne m. / ~ apparatus ‖ Schaumweinapparat m. ‖ appareil m. pour vins mousseux. / ~ fabrication ‖ Schaumweinbereitung f. ‖ fabrication f. de vins mousseux. / ~ manufacture ‖ Schaumweinfabrikation f.; Sektkellerei f. ‖ préparation f. de vins mousseux.

spark micrometer ‖ Funkenmikrometer n. ‖ micromètre m. à étincelle. / number of ~s ‖ Funkenzahl f. ‖ nombre m. d'étincelles.

spark plug ‖ Zündkerze f. ‖ bougie f. d'allumage. / mica ~ ‖ Glimmerkerze f. ‖ bougie f. en mica.

spark plug gasket ‖ Zündkerzendichtung f. ‖ joint m. de bougie. / ~ ignition ‖ Kerzenzündung f. ‖ allumage m. par bougie. / ~ tester ‖ Zündkerzenprüfer m. ‖ appareil m. de contrôle pour l'allumage des bougies. / ~ thread ‖ Kerzengewinde n. ‖ filet(age) m. de la bougie.

spark potential ‖ Funkenpotential n. ‖ potentiel m. des étincelles. / ~ quencher *see* ~ arrester. / ~ reducer ‖ Funkenschwächer m. ‖ affaiblisseur m. des étincelles. / ~ resistance ‖ Funkenwiderstand m. ‖ résistance f. des étincelles. / scale of ~s ‖ Funkenskale f. ‖ échelle f. des étincelles. / spectrum of ~s ‖ Funkenspektrum n. ‖ spectre m. des étincelles. / ~ transmitter ‖ Funkensender m. ‖ émetteur m. à étincelles.

sparry fluor ‖ Flußspat m. ‖ spath m. fluor. / ~ iron stone see spathic iron.

spar section (Aero) ‖ Holmquerschnitt m. ‖ section f. transversale du longeron.

spartaite ‖ Spartait m. ‖ spartaïte f.

spar web (Aero) ‖ Holmsteg m. ‖ âme f. du longeron.

spasm ‖ Krampf m. ‖ spasme m.

spathic iron ‖ Spateisenstein m. ‖ fer m. carbonaté; sidérose f.

spathiopyrite ‖ Spathiopyrit m. ‖ spathiopyrite f.

spatter, to ‖ verspritzen ‖ éclabousser.

spattering leather (Coachm) ‖ Spritzleder n. ‖ mantelet m.; garde-boue m.

spattle (Print; Paint) ‖ Spatel m.; Spachtel f. ‖ spatule f. / scrap ~ ‖ Abstreichspatel m. ‖ spatule f. racloir. / stirring ~ ‖ Rührspatel m. ‖ spatule f. agitateur.

spatula (Med) ‖ Spatel m. ‖ spatule f.

spatular strut ‖ Löffelstrebe f. ‖ jambe f. de force *ou* contrefiche f. en forme de cuiller.

spawn ‖ Fischlaich m. ‖ frai m. de poisson. / ~ breeding ‖ Fischbrutzucht f. ‖ élevage f. d'alevins *ou* de nourrains.

spawning ‖ Laichen n. ‖ frai m. / ~ time ‖ Laichzeit f. ‖ temps m. *ou* saison m. du frai.

speaking circuit ‖ Sprechkreis m. ‖ circuit m. de conversation.

speaking current supply ‖ Mikrofonspeisung f. ‖ alimentation f. du microphone. / ~ supply tube ‖ Mikrofonspeiseröhre f. ‖ tube m. à gaz spécial pour l'alimentation de microphones.

speaking key (Tel) ‖ Abfrageschalter m. ‖ clef f. d'écoute. / method of ~ ‖ Sprechtechnik f. ‖ art f. de parler. / ~ oscillation ‖ Sprachschwingung f. ‖ oscillation f. vocale. / ~ relation ‖ Sprechbeziehung f. ‖ relation f. téléphonique. / ~ staff ‖ Nivellierlatte f. zum Selbstablesen ‖ mir f. parlante. / motor car ~ trumpet ‖ Kraftwagensprachrohr n. ‖ tuyau m. acoustique pour automobiles. / ~ tube ‖ Sprachrohr n. ‖ tuyau m. acoustique; porte-voix m.

spearmint ‖ Krauseminze f. ‖ menthe f. crépue. / ~ oil ‖ Krauseminzöl n. ‖ essence f. de menthe crépue.

special arrangement ‖ Sondervorrichtung f. ‖ dispositif m. spécial. / ~ design (Mach) ‖ Sonderausführung f. ‖ construction f. particulière. / ~ drive ‖ Sonderantrieb m. ‖ commande f. spéciale. / ~ edition ‖ Extraausgabe f.; Extrablatt n. ‖ édition f. spéciale. / ~ eyepiece ‖ Sonderokular n. ‖ oculaire m. spéciale. / ~ installation ‖ Spezialeinrichtung f.; Sondereinrichtung f. ‖ installation f. spéciale. / ~ iron ‖ Façoneisen n. ‖ fer m. spécial. / ~ line ‖ Spezialität f. ‖ spécialité f.

special machine ‖ Sondermaschine f. ‖ machine f. spéciale. / ~ for the construction of automobiles ‖ Automobilbausondermaschine f. ‖ machine f. spéciale pour la construction des automobiles. / ~ for the construction of locomotives ‖ Lokomotivbausondermaschine f. ‖ machine f. spéciale pour la construction des locomotives. / ~ for the construction of wagons ‖ Waggonbausondermaschine f. ‖ machine f. spéciale pour la construction des wagons. / ~ for the wood industry ‖ Holzindustriesondermaschine f. ‖ machine f. spéciale pour l'industrie du bois.

special machine tool ‖ Sonderwerkzeugmaschine f. ‖ machine-outil f. spéciale. / ~ for shipbuilding ‖ Schiffbausonderwerkzeugmaschine f. ‖ machine-outil f. speciale pour la construction des bateaux.

special mount ‖ Sonderfassung f. ‖ monture f. spéciale. / ~ notching and mitre cutting machine ‖ Spezialausklink- und Gehrungsstanzmaschine f. ‖ machine f. spéciale à gruger et découper les onglets. / ~ oil ‖ Sonderöl n. ‖ huile f. spéciale. / ~ periodical ‖ Fachzeitschrift f. ‖ revue f. spéciale.

special press ‖ Spezialpresse f.; Sonderpresse f. ‖ presse f. spéciale. / **electro hydraulic ~** ‖ elektro-hydraulische Sonderpresse f. ‖ presse f. spéciale électro-hydraulique.

special price ‖ Ausnahmepreis m. ‖ prix m. d'exception. / **~ profile** ‖ Spezialprofil n.; Sonderprofil n. ‖ profil m. spécial. / **~ purpose machine for producing threads** ‖ Sondermaschine f. für Gewindeherstellung ‖ machine f. spéciale à produire les filets (de vis). / **~ purpose welding machine** ‖ Schweißmaschine f. für Sonderzwecke ‖ machine f. à souder pour ouvrages spéciaux. / **~ quotation** ‖ Sonderangebot n. ‖ offre f. spéciale ou de faveur. / **~ reserve** ‖ Sonderrücklage f. ‖ réserves fpl. spéciales. / **~ spring steel** ‖ Sonderfederstahl m. ‖ acier m. spécial à ressort. / **~ standard** (Mach) ‖ Sonderständer m. ‖ montant m. spécial.

special steel ‖ Spezialstahl m.; Edelstahl m.; Sonderstahl m. ‖ acier m. spécial. / **~ of high tenacy** ‖ besonders zäher Sonderstahl m. ‖ acier m. spécial à haute résistance. / **tungsten ~** ‖ Wolframsonderstahl m. ‖ acier m. spécial au tungstène.

special tire steel ‖ Radreifensonderstahl m. ‖ acier m. spécial pour les bandages. / **~ tissue** ‖ Spezialgewebe n.; Sondergewebe n. ‖ tissu m. spécial. / **~ trimming machine** ‖ Spezialbeschneidmaschine f. ‖ machine f. spéciale à rogner. / **~ wagon** ‖ Sonderwagen m.; Spezialwagen m. ‖ wagon m. spécial. / **~ wire** ‖ Fassondraht ‖ fil m. spézial ou façonné.

specialist ‖ Fachmann m.; Spezialist m. ‖ spécialiste m.; expert m.

speciality ‖ Spezialität f. ‖ spécialité f. / **pharmaceutical ~** ‖ pharmazeutische Spezialität f. ‖ spécialité f. pharmaceutique.

specialize, to ‖ spezialisieren ‖ spécialiser.

specialized ‖ spezialisiert ‖ spécialisé.

species pl., **leading ~** ‖ gewöhnliche oder gangbarste Sorten fpl. ‖ marques fpl. les plus usitées. / **~ of stone** (Geol) ‖ Gebirgsart f. ‖ espèce f. de terrain ou de minérai.

specific ‖ spezifisch ‖ spécifique. / **~ electric resistance** ‖ spezifischer elektrischer Widerstand m. ‖ résistance f. électrique spécifique. / **~ heat** ‖ spezifische Wärme f. ‖ chaleur f. spécifique. / **~ inductive capacity** ‖ Dielektrizitätskonstante f. ‖ pouvoir m. inducteur spécifique; constante f. diélectrique.

specification ‖ genaue Aufstellung f.; Spezifizierung f. ‖ spécification f. / **~ (List)** ‖ Stückliste f. ‖ liste f. de pièces. / **according to ~** ‖ laut Spezifizierung f. ‖ selon spécification f. / **exact ~ (List)** ‖ genaues Verzeichnis n. ‖ spécification f. précise. / **~ of prices** ‖ Preisverzeichnis n. ‖ série f. ou bordereau m. de prix.

specifications pl. (Contract) ‖ Lastenheft n.; Verdingungsunterlagen fpl. ‖ cahier m. des charges. / **~ (Mach)** ‖ Vorschrift f. ‖ prescriptions fpl.; conditions fpl. prescrites. / **~ for acceptance** ‖ Abnahmevorschrift f. ‖ prescription f. de réception. / **~ for quality** ‖ Gütevorschrift f. ‖ prescription f. de qualité. / **standard ~** ‖ Normalbedingungen fpl. ‖ conditions fpl. normales.

specification test ‖ Werkstoffuntersuchung f. ‖ essai m. de matériaux.

specimen ‖ Muster n.; Probe f. ‖ spécimen m.; échantillon m. / **~ (Print)** ‖ Druck-

probe f. ‖ épreuve f. d'impression. / **~ of the bottom** ‖ Grundprobe f. ‖ **specimen** m. du fond. / **to cut-out test ~s** pl. (Met) ‖ Probestäbe mpl. entnehmen ‖ prélever des éprouvettes fpl. / **~ of type** / Schriftprobe f. ‖ échantillon m. de caractères.

specimen book ‖ Musterbuch n. ‖ collection f. de dessins. / **~ number** (Print) ‖ Probenummer f. ‖ spécimen m. / **~ page** ‖ Probeseite f. ‖ page f. d'essai.

speck ‖ Fleck m. ‖ paille f.; tache f.

speckled ‖ gefleckt, gesprenkelt ‖ marbré, marqué, marqueté, tacheté, tigré; moucheté. / **~ wood** ‖ Maserholz n. ‖ bois m. madré ou tapiré.

spectacle see spectacles. / **~ axis** ‖ Brillenachse f. ‖ axe m. de la lunette. / **~ case** ‖ Brillenfutteral n. ‖ étui m. pour lunettes.

spectacle frame ‖ Brillengestell n.; Brillenfassung f. ‖ châsse f. ou monture f. de lunettes. / **~ of pantoscopic shape** ‖ Brillenfassung f. in pantoskopischer Scheibenform ‖ monture f. de lunette à drageoir pantoscopique. / **~ with protecting sides** ‖ Brillenfassung f. mit Seitenschutz ‖ monture f. de lunette à œillères. / **~ gauge** ‖ Brillengestellmaß n. ‖ calibre m. pour montures de lunettes. / **~ welder** ‖ Brillengestellöter m. ‖ soudeur m. en montures de lunettes.

spectacle glass ‖ Brillenglas n. ‖ verre de lunettes. / **~ cutter** ‖ Brillenglasschneider m. ‖ tailleur de verres de lunettes f. / **~ fitter** ‖ Brilleneinglaser m. ‖ monteur m. de verres de lunettes.

spectacle grinder ‖ Brillenschleifer m. ‖ lunettier m.

spectacle lens ‖ Brillenglas n. ‖ verre m. de lunettes. / **~ of crescent shape** ‖ halbmondförmiges Brillenglas n. ‖ verre-correcteur m. en demi-lune. / **~ which is free from strains** ‖ spannungsfreies Brillenglas n. ‖ verre m. correcteur de lunette exempt de tensions. / **ground ~** ‖ geschliffenes Brillenglas n. ‖ verre m. de lunettes taillé. / **~ of the meniscus form which has no refractive power** ‖ durchgebogenes Brillenglas n. ohne lichtbrechende Wirkung ‖ verre m. correcteur neutre en forme coquille. / **point focal ~** ‖ punktuell abbildendes Brillenglas n. ‖ verre m. correcteur à image ponctuelle. / **polished ~** ‖ poliertes Brillenglas n. ‖ verre m. de lunettes poli. / **strained ~** ‖ gespanntes Brillenglas n. ‖ verre m. correcteur serré. / **refraction of ~** ‖ Scheitelbrechwert m. von Brillengläsern ‖ puissance f. frontale de verres correcteurs.

spectacle magnifier ‖ Brillenlupe f. ‖ loupe-lunettes f. / **adjustable ~** ‖ verstellbare Brillenlupe f. ‖ loupe-lunettes f. réglable.

spectacle maker ‖ Brillenmacher m. ‖ lunettier m. / **~ mount** see spectacle frame.

spectacle rim ‖ Brillenrand m. ‖ cercle m. des lunettes. / **~ mount with side protectors** ‖ mit Seitenschutz versehene Randfassung f. einer Brille ‖ monture f. de lunette à œillères.

spectacles pl. ‖ Brille f. ‖ lunettes fpl. / **aluminium-bronze ~** ‖ Aluminiumbronzebrille f. ‖ lunettes fpl. en bronze d'aluminium. / **~ with bifocal lenses** ‖ Brille f. mit Bifokalgläsern ‖ lunette f. à double foyer. / **celluloid ~** ‖ Zelluloidbrille f. ‖ lunettes fpl. en celluloïd. / **distance ~** ‖ Fernbrille f. ‖ lunettes fpl. pour les lointains ou destinée aux lointains. / **framed ~** ‖ gefaßte Brille f. ‖

lunettes fpl. à verres cerclés ou montés. / **frameless ~** ‖ ungefaßte Brille f. ‖ lunettes fpl. à verres non cerclés. / **gold-rimmed ~** ‖ Goldbrille f. ‖ lunettes fpl. d'or. / **horn ~** ‖ Hornbrille f. ‖ lunettes fpl. en corne. / **leather-rimmed ~** ‖ Lederbrille f. ‖ lunettes fpl. en cuir. / **nickel-rimmed ~** ‖ Nickelbrille f. ‖ lunettes fpl. nickel. / **nickeled steel ~** ‖ vernickelte Stahlbrille f. ‖ lunettes fpl. en acier nickelé. / **presbyopic ~** ‖ Altersbrille f. ‖ verres mpl. aux presbytes. / **prescription of ~** ‖ Brillenverordnung f. ‖ ordonnance f. de lunettes. / **prismatic ~** ‖ prismatische Brille f. ‖ lunettes fpl. à verres prismatiques. / **protective ~** ‖ Schutzbrille f. ‖ lunettes fpl. protectrices. / **reading ~** ‖ Lesebrille f. ‖ lunettes fpl. pour lecture. / **rimmed ~** ‖ gefaßte Brille f. ‖ lunettes fpl. à verres montés. / **rolled gold ~** ‖ Golddoublébrille f. ‖ lunettes fpl. en doublé or. / **safety ~** ‖ Schutzbrille f. ‖ lunettes fpl. de protection. / **side of the ~** ‖ Brillenbügel m. ‖ branche f. des lunettes. / **silver-rimmed ~** ‖ Silberbrille f. ‖ lunettes fpl. à monture d'argent. / **~ for sports** ‖ Sportbrille f. ‖ lunettes fpl. pour sport. / **steel-rimmed ~** ‖ Stahlbrille f. ‖ lunettes fpl. à monture en acier. / **telescopic ~** ‖ Fernrohrbrille f. ‖ télélunettes fpl.; lunettes fpl. grossissantes. / **telescopic ~ for extremely weak sights** ‖ Fernrohrbrille f. für hochgradig Schwachsichtige ‖ lunettes fpl. grossissantes pour amblyopes. / **working ~** ‖ Arbeitsbrille f. ‖ lunettes fpl. de travail.

spectacle screw ‖ Brillenschraube f. ‖ vis f. de lunettes. / **~ side arm** ‖ Brillenbügel m. ‖ branche f. des lunettes. / **~ trial frame** ‖ Brillenprobgestell n. ‖ monture f. de verres d'essai. / **~ vertex** ‖ Brillenscheitel m. ‖ sommet m. des lunettes. / **~ wearer** ‖ Brillenträger m. ‖ porteur m. de lunettes.

spectro comparator ‖ Spektrokomparator m. ‖ spectrocomparateur m.

spectrogram, measuring ~s pl. ‖ Ausmessung f. von Spektrogrammen ‖ mesurage m. des photographies spectrales.

spectrograph ‖ Spektrograf m. ‖ spectrographe m. / **astronomical ~** ‖ astronomischer Spektrograf m. ‖ spectrographe m. astronomique. / **prismatic objective ~** ‖ Objektivprismaspektrograf m. ‖ spectrographe m. à prisme-objectif. / **quadruple-prism quartz ~** ‖ Vierprismenquarzspektrograf m. ‖ spectrographe m. à quatre-prismes en quartz. / **quartz ~** ‖ Quarzspektrograf m. ‖ spectrographe m. à optique en quartz. / **~ with test tube condenser** ‖ Spektrograf m. mit Reagensglaskondensor ‖ spectrographe m. muni d'un condensateur pour tubes à essai.

spectrometer ‖ Spektrometer n. ‖ spectromètre m. / **concave mirror ~** ‖ Hohlspiegelspektrometer n. ‖ spectromètre m. à miroirs concaves. / **reflecting ~ for heat rays** ‖ Spiegelspektrometer n. für Wärmestrahlen ‖ spectromètre m. à miroirs pour les radiations calorifiques.

spectrometric ‖ spektrometrisch ‖ spectrométrique. / **~ measurement** ‖ spektrometrische Messung f. ‖ mesure f. spectrométrique.

spectrophotometer ‖ Spektralfotometer n. ‖ spectrophotomètre m.

spectrophotometric ‖ spektrofotometrisch ‖ spectrophotométrique. / **~ measurement** ‖

spektrofotometrische Messung f. ‖ mesure f. spectrophotométrique.

spectrophotometry ‖ Spektralfotometrie f. ‖ photométrie f. spectrale.

spectroscope ‖ Spektroskop n. ‖ spectroscope m. / astro ~ ‖ Astrospektroskop n. ‖ spectroscope m. astronomique. / autocollimating ~ ‖ Autokollimationsspektroskop n. ‖ spectroscope m. autocollimateur. / comparison ~ ‖ Vergleichsspektroskop n. ‖ spectroscope m. de comparaison. / ~ with concave mirrors ‖ Hohlspiegelspektroskop n. ‖ spectroscope m. à miroirs concaves. / directvision ~ ‖ gradsichtiges Spektroskop n. ‖ spectroscope m. à vision directe. / eyepiece ~ ‖ Okularspektroskop n. ‖ spectroscope m. oculaire. / fixed arm ~ ‖ festarmiges Spektroskop n. ‖ spectroscope m. à bras fixes. / hand ~ ‖ Handspektroskop n. ‖ spectroscope m. à main. / hand ~ with test tube condenser ‖ Handspektroskop n. mit Reagensglaskondensor ‖ spectroscope m. à main muni du condensateur pour tubes à essai. / ~ set to infinity ‖ auf Unendlich eingestelltes Spektroskop n. ‖ spectroscope m. mis au point sur l'infini. / prism ~ ‖ Prismenspektroskop n. ‖ spectroscope m. à prismes. / prominence ~ ‖ Protuberanzenspektroskop n. ‖ spectroscope m. à protubérances. / reflecting ~ ‖ Spiegelspektroskop n. ‖ spectroscope m. à miroirs. / tele-~ ‖ Fernspektroskop n. ‖ téléspectroscope m. / ~ without telescope ‖ Spektroskop n. ohne Fernrohr ‖ spectroscope m. sans lunette. / ~ tube ‖ Spektroskoprohr n. ‖ tube m. du spectroscope.

spectroscope, calibration of a ~ in terms of wave lengths ‖ Eichung f. eines Spektroskops nach Wellenlängen ‖ étalonnage m. d'un spectroscope d'après les échelles de longueurs d'ondes.

spectroscopic ‖ spektroskopisch ‖ spectroscopique.

spectroscopic analysis ‖ Spektralanalyse f. ‖ analyse f. spectrale. / ~ of dye stuffs ‖ Spektralanalyse f. von Farbstoffen ‖ analyse f. spectrale des colorants. / quantitative ~ ‖ quantitative Spektralanalyse f. ‖ analyse f. spectrale quantitative.

spectroscopic apparatus ‖ spektroskopischer Apparat m. ‖ appareil m. de spectroscopie. / ~ camera ‖ spektrografische Kamera f. ‖ chambre f. spectroscopique. / ~ eyepiece ‖ Spektralokular n.; Spektroskopokular n. ‖ oculaire m. spectroscopique. / ~ illuminator ‖ Spektralbeleuchtungsapparat m. ‖ appareil m. pour l'éclairage monochromatique. / ~ investigation ‖ spektroskopische Untersuchung f. ‖ recherche f. spectroscopique. / ~ reaction ‖ spektroskopische Reaktion f. ‖ réaction f. spectroscopique.

spectroscopist ‖ Spektroskopiker m. ‖ spectroscopiste m.

spectroscopy ‖ Spektroskopie f. ‖ spectroscopie f. / ~ of blood ‖ Blutspektroskopie f. ‖ spectroscopie f. du sang.

spectrum ‖ Spektrum n. ‖ spectre m. / absorption ~ ‖ Absorptionsspektrum n. ‖ spectre m. d'absorption. / continuous ~ ‖ kontinuierliches Spektrum n. ‖ spectre m. continu. / diffraction ~ ‖ Beugungsspektrum n. ‖ spectre m. de

diffraction. / electrical ~ ‖ elektrisches Spektrum n. ‖ spectre m. électrique. / emission ~ ‖ Emissionsspektrum n. ‖ spectre m. d'émission. / fluorescence ~ ‖ Fluoreszenzspektrum n. ‖ spectre m. fluorescent *ou* de fluorescence. / general ~ ‖ Übersichtsspektrum n. ‖ spectre m. synoptique. / grating ~ *see* diffraction ~. / line ~ (Phys) ‖ Linienspektrum n. ‖ spectre m. linéaire. / ~ exhibiting a great number of lines ‖ besonders linienreiches Spektrum n. ‖ spectre m. particulièrement riche en raies. / prismatic ~ ‖ Brechungsspektrum n. ‖ spectre m. prismatique. / secondary ~ ‖ sekundäres Spektrum n. ‖ spectre m. secondaire. / ~ of sparks (Electr) ‖ Funkenspektrum n. ‖ spectre m. des étincelles. / visible ~ ‖ sichtbares Spektrum n. ‖ spectre m. visible.

spectrum analysis *see* spectroscopic analysis.

spectrum apparatus ‖ Spektralapparat m.; Spektroskop n. ‖ appareil m. spectral; spectroscope m. / ~ of great light transmitting power ‖ lichtstarker Spektralapparat m. ‖ appareil m. spectral lumineux.

spectrum camera ‖ Spektralkamera f. ‖ chambre f. spectroscopique. / device for comparison of ~s consisting of a comparison wave length scale and a comparison spectrum ‖ Einrichtung f. zur Vergleichung des Spektrums mit einer Wellenlängenskale und einem Vergleichsspektrum ‖ dispositif m. pour comparer le spectre avec une échelle de longueurs d'onde et un autre spectre.

spectrum line ‖ Spektrallinie f. ‖ raie f. du spectre ou spectrale.

spectrum lines pl. of an incandescent vapour ‖ Spektrallinien fpl. eines leuchtenden Dampfes ‖ raies fpl. spectrales d'une vapeur incandescente. / group of ~ ‖ Spektralliniengruppe f. ‖ groupe m. de raies de tous les spectres. / measurement of displacements of ~ ‖ Messung f. von Spektrallinienverschiebungen ‖ mesurement m. des déplacements des raies spectroscopiques. / apparatus for measuring displacements of ~ ‖ Apparat m. zur Messung von Spektrallinienverschiebungen ‖ appareil m. pour mesurer les déplacements des raies spectrales.

spectrum, passing in review the ~ ‖ Durchmusterung f. des Spektrums ‖ exploration f. du spectre. / ~ plate ‖ Spektrumplatte f. ‖ plaque f. de spectre. / region of the ~ ‖ Spektralbereich m. ‖ région f. du spectre. / separation of the ~ ‖ Spektrumteilung f. ‖ séparation f. du spectre. / ~ test ‖ Spektralprobe f. ‖ essai m. spectrométrique.

specular ‖ spiegelnd ‖ miroitant. / ~ iron ‖ Spiegeleisen n. ‖ fonte f. miroitante; fonte-spiegel n. ‖ metal ‖ Spiegelmetall n. ‖ métal m. spéculaire.

speculate, to ‖ spekulieren ‖ spéculer.

speculation ‖ Spekulation f. ‖ spéculation f. / ~ in stocks ‖ Aktienspekulation f. ‖ agiotage m.

speculatius factory ‖ Spekulatiusfabrik f. ‖ fabrique f. de speculatius.

speculator, buying ~ ‖ Aufkäufer m. ‖ acheteur m.; accapareur m.

speculum metal *see* specular metal.

speechlessness ‖ Stummheit f. ‖ mutisme m.

speed, to ~ up ‖ beschleunigen ‖ accélérer. /

~ the exciter (Electr) ‖ die Drehzahl des Erregers erhöhen ‖ augmenter le nombre m. de tours de l'excitateur.

speed ‖ Geschwindigkeit f. ‖ vitesse f. / ~ (Auto) ‖ Gang m. ‖ vitesse f. / angular ~ variation ‖ Veränderung f. der Winkelgeschwindigkeit ‖ variation f. de la vitesse angulaire. / ~ of answering (Tel) ‖ Abfragegeschwindigkeit f. ‖ vitesse f. de réponse. / average ~ ‖ Durchschnittsgeschwindigkeit f.; mittlere Geschwindigkeit f. ‖ vitesse f. moyenne. / the ~ was between x and y kilometers per hour ‖ die Geschwindigkeit f. bewegte sich zwischen x bis y km je Stunde ‖ la vitesse se tenait entre x et y km par heure. / to change over from first to second ~ (Auto) ‖ vom ersten auf den zweiten Gang übergehen ‖ passer de la première à la deuxième vitesse. / commercial ~ (Aero) ‖ Verkehrsgeschwindigkeit f. ‖ vitesse f. commerciale *ou* d'utilisation. / to keep the ~ approximately constant (Mach) ‖ die Drehzahl annähernd konstant halten ‖ maintenir la vitesse à peu près constante. / critical ~ ‖ Grenzgeschwindigkeit f. ‖ vitesse f. critique. / cruising ~ (Aero) ‖ Reisegeschwindigkeit f. ‖ vitesse f. commerciale. / (variable) cutting ~ ‖ (veränderliche) Schnittgeschwindigkeit f. ‖ vitesse f. (variable) de coupe. / ~ of diffusion ‖ Diffusionsgeschwindigkeit f. ‖ vitesse f. de diffusion. / the ~ drops (Mach) ‖ die Drehzahl fällt ab ‖ la vitesse diminue. / economical ~ (Mach) ‖ wirtschaftliche Drehzahl f. ‖ vitesse f. économique. / flight ~ ‖ Fahrtgeschwindigkeit f. ‖ vitesse f. d'avancement. / forward ~ (Auto) ‖ Vorwärtsgang m. ‖ vitesse f. d'avance. / fourth ~ (Auto) ‖ vierter Gang ‖ quatrième vitesse f. / full ~ *see* maximum ~. / in full ~ (Mach) ‖ in vollem Gang m. ‖ en pleine marche f. / high ~ ‖ hohe Geschwindigkeit f. ‖ grande vitesse f. / high ~ movement ‖ Eilbewegung f. ‖ mouvement m. rapide. / low ~ ‖ geringe Geschwindigkeit f. ‖ faible vitesse f. / low ~ (Auto) ‖ erster Gang m. ‖ première vitesse f. / lowest ~ ‖ kleinste Geschwindigkeit f. ‖ vitesse f. minimum. / maximum ~ ‖ Höchstgeschwindigkeit m. ‖ limite f. de vitesse; vitesse maximum. / ~ of motion ‖ Fahrgeschwindigkeit f. ‖ vitesse f. de translation. / normal ~ ‖ Normalgeschwindigkeit f. ‖ vitesse f. normale. / ~ of paper ‖ Papiergeschwindigkeit f. ‖ vitesse f. du papier. / ~ of the piston ‖ Kolbengeschwindigkeit f. ‖ vitesse f. du piston. / ~ of pressure transmission ‖ Druckübertragungsgeschwindigkeit f. ‖ vitesse f. de transmission de la poussée. / to prevent the ~ from exceeding the maximum permissible (Mach) ‖ die Überschreitung f. der höchstzulässigen Drehzahl verhindern ‖ empêcher de dépasser le nombre de tours maximum admis. / ~ of propagation ‖ Fortpflanzungsgeschwindigkeit f. ‖ vitesse f. de propagation *ou* de translation. / proper ~ ‖ Normalgeschwindigkeit f. ‖ vitesse f. normale. / ~ of rotation ‖ Umdrehungsgeschwindigkeit f. ‖ vitesse f. de rotation *ou* rotatoire. / ~ of telegraphing ‖ Telegrafiergeschwindigkeit f. ‖ rapidité f. de transmission; vitesse f. de transfert. / ~ of translation *see* ~ of propagation.

speed adjustment ‖ Geschwindigkeitsregelung f. ‖ réglage m. de la vitesse. / ~ change ‖ Geschwindigkeitswechsel m. ‖ changement m. de vitesse. / ~ control *see* speed adjustment. / ~ counter *see* speedometer. / dropping of ~ ‖ Abfallen n. der Geschwindigkeit ‖ diminution f. de la vitesse. / ~ governor ‖ Geschwindigkeitsregler m. ‖ régulateur m. de vitesse. / ~ governor for automobiles ‖ Fahrtregler m. für Automobile ‖ régulateur m. de vitesse pour automobiles.

speed indicator *see also* speedometer ‖ Geschwindigkeitsmesser m. ‖ indicateur m. de vitesse. / cup anemometer type ~ (Aero) ‖ Schalenkreuzfahrtmesser m. ‖ indicateur m. de vitesse à coquilles. / Pitot static air ~ ‖ Staudruckfahrtmesser m. ‖ indicateur m. de vitesse à pression dynamique. / pressure ~ ‖ Staudruckfahrtmesser m. ‖ indicateur m. de vitesse à pression dynamique. / registering ~ ‖ schreibender Geschwindigkeitsmesser m. ‖ enregistreur m. de vitesse.

speedometer *see also* speed indicator ‖ Geschwindigkeitsmesser m. ‖ compteur m. *ou* indicateur m. de vitesse. / centrifugal type ~ ‖ Fliehkraftgeschwindigkeitsmesser m. ‖ indicateur m. de vitesse type centrifuge. / ~ for trains ‖ Geschwindigkeitsmesser m. für Eisenbahnzüge ‖ indicateur m. de vitesse pour trains.

speed reducer ‖ Geschwindigkeitsminderapparat m. ‖ réducteur m. de vitesse. / ~ variation ‖ Geschwindigkeitsänderung f. ‖ changement m. de vitesse.

speiss ‖ Speiß m. ‖ speiss m.

spell, to ‖ buchstabieren ‖ épeler.

spelling table ‖ Buchstabiertafel f. ‖ tableau m. d'épellation.

spelt ‖ Spelz m. ‖ épeautre m. / ~ determination ‖ Spelzbestimmung f. ‖ détermination f. de la quantité d'épeautres.

spelter ‖ Zink n. ‖ spiauter m.; zinc m. / ~ solder ‖ Messingschlaglot n. ‖ soudure f. de laiton.

spelt flour ‖ Spelzmehl n. ‖ farine f. d'épeautre.

spend, to ‖ ausgeben ‖ débourser; dépenser.

spent grains pl. (Brew) ‖ Treber pl. ‖ drêches fpl.; drague m. ‖ drêche f. / ~ to remove ~ ‖ austrebern ‖ sortir les drêches fpl. *ou* les dragues. / ~ conveying screw ‖ Austreberschnecke f. ‖ vis f. sans fin pour l'évacuation des drêches. / ~ remover ‖ Austrebermaschine f. ‖ machine f. à sortir les drêches. / ~ reraking and removing machine ‖ Aufhack- und Austrebermaschine f. ‖ piocheur-dédrêcheur m. / vent for ~ ‖ Treberausstoßloch n. ‖ trappe f. de décharge des drêches.

spermaceti ‖ Walrat n. ‖ blanc m. de balaine / ~ oil ‖ Walfischtran m. ‖ huile f. de baleine.

sperm whale ‖ Pottfisch m. ‖ cachalot m.

spessartite ‖ Mangangranat m. ‖ spessartite f.

sphene ‖ Titanit m. ‖ titanite f.

sphere (Geom) ‖ Kugel f. ‖ sphère. / ~ (Glassm) ‖ konvexe Schleifschale f. ‖ boule f.; sphère f. / ~ of action ‖ Wirkungskreis m.; Arbeitsfeld n. ‖ rayon m. (d'action). / ~ of explosion ‖ Sprengsphäre f. ‖ sphère f. d'explosion. / hollow ~ ‖ Hohlkugel f. ‖ sphère f. creuse. / ~ of interest ‖ Interessengebiet n. ‖ périmètre m. *ou* rayon m. d'influence. /

whole ~ ‖ gesamtes Gebiet n. ‖ domaine m. entier.

spherical ‖ sphärisch; kugelförmig ‖ sphérique. / non-~ ‖ asphärisch ‖ asphérique.

spherical aberration ‖ sphärische Abweichung f. ‖ aberration f. de sphéricité. / ~ of the object ‖ sphärische Abweichung f. des Objektivs ‖ aberration f. de sphéricité de l'objectif. / chromatic difference of the ~s ‖ chromatische Differenz f. der sphärischen Abweichungen ‖ différence f. chromatique des aberrations de sphéricité.

spherical angle ‖ sphärischer Winkel m. ‖ angle m. sphérique. / ~ bearing ‖ Kugellager n. ‖ coussinet m. sphérique; roulement m. à billes. / ~ calotte ‖ Kugelkalotte f.; Kugelschale f. ‖ calotte f. sphérique. / ~ correction ‖ sphärische Korrektion f. ‖ correction f. sphérique *ou* de l'aberration de sphéricité. / ~ crank with handle ‖ Kugelhandkurbel f. ‖ manivelle f. à main boule. / ~ double bottom ‖ Doppelkugelboden m. ‖ doublefond m. sphérique. / ~ flask ‖ Glaskugel f. ‖ ballon m. *ou* boule f. de verre. / ~ lens ‖ sphärische Linse f.; Kugellinse f. ‖ lentille f. sphérique. / ~ mirror ‖ sphärischer Hohlspiegel m. ‖ miroir m. concave sphérique. / ~ pot ‖ Rundkessel m. ‖ chaudière f. ronde. / ~ shape ‖ Kugelgestalt f. ‖ forme f. sphérique. / ~ spectacle lens ‖ sphärisches Dioptrienglas n. ‖ verre m. de lunette à surfaces sphériques. / ~ surface ‖ Kugelfläche f. ‖ surface f. sphérique. / ~ thrust bearing ‖ Kugelspurlager n. ‖ crapaudine f. à billes. / ~ turning (Lathe) ‖ Balligdrehen n.; Kugeldrehen n. ‖ tournage m. de sphères. / ~ vault ‖ Kugelgewölbe n. ‖ dôme m.; cul-de-four m.; voûte f. sphérique.

spherically curved glass cover ‖ kugelförmig gekrümmter Glasdeckel m. ‖ couvercle m. de verre sphérique.

spherocylindrical ‖ sphärisch zylindrisch ‖ sphéro-cylindrique.

spheroid ‖ Rotationsellipsoid n.; Sphäroid n. ‖ ellipsoïde m. de révolution *ou* de rotation. / oblate ~ ‖ flaches Rotationsellipsoid n. ‖ ellipsoïde m. de rotation engendré par la rotation d'une ellipse autour de son petit axe. / oblong ~ ‖ verlängertes Rotationsellipsoid n. ‖ ellipsoïde m. de révolution prolongé. / prolate ~ *see* oblong ~.

spheroidal concretion (Geol) ‖ Niere f. ‖ rognon m.; globule m. oblong. / ~ of marl ‖ Niere n. von Kalkmergelstein ‖ marne f. sphéroïdale cloisonnée.

spherometer ‖ Sphärometer n. ‖ sphéromètre m.

spherosiderite ‖ Sphärosiderit m. ‖ sphérosidérite f.

sphygmograph ‖ elektrischer Pulsmesser m. ‖ sphygmographe m.

spice ‖ Gewürz n. ‖ épice f.

spiced wine ‖ Bowle f. ‖ bol m.

spice extract ‖ Gewürzextrakt m. ‖ extrait m. d'épices. / ~ grinding mill ‖ Gewürzmühle f. ‖ broyeur m. d'épices.

spicery *see* spice.

spider (Mach) ‖ Armkreuz n. ‖ croisillon m. / ~ differential ‖ Ausgleichstern m. ‖ axe m. des satellites du différentiel. / ~ wheel ‖ Radstern m. ‖ centre m. de roue.

spiegeleisen *see* spiegel iron.

spiegel iron ‖ Spiegeleisen n. ‖ fonte f. spiegel.

spigot (Coop) ‖ Faßhahn m. ‖ robinet m.; chantepleure f.

spike, to ‖ nageln ‖ clouer. / ~ (Shipb) ‖ spikern ‖ clouer.

spike ‖ langer Nagel m. ‖ broche f. / ~ (Locksm) ‖ Dorn m. ‖ chevillette f. / ~ (Mar) ‖ Spiker m. ‖ clou m. / ~ (Railw) ‖ Hakennagel m.; Schienennagel m. ‖ crampon m.; tirefond m. / dog ~ *see* spike (Railw). / dog-headed ~ ‖ Hakenstift m. ‖ crampon m. / fluted ~ ‖ eingekerbter Nagel m. ‖ crampon m. à encoche. / screwed ~ ‖ Schwellenschraube f. ‖ tirefond m.

spike box (Mar) ‖ Spikerback f.; Gerätekasten m. ‖ équipet m. / ~ drawer (Railw) ‖ Schienennagelklaue f.; Geißfuß m. ‖ pince f. à pied-de-biche. / ~ drawing winch (Railw) ‖ Schienennagelwinde f. ‖ vis f. à tirer les crampons de rail. / ~ hammer (Railw) ‖ Schienennagelhammer m. ‖ chassecrampon m. / chain with ~ hooks (Railw) ‖ Kette f. mit Einschlagkeilen ‖ chaîne f. avec crampons. / ~ iron (Shipb) ‖ Spikereisen n. ‖ calfat m. à clous. / ~ oil ‖ Spiköl n. ‖ essence f. de spic. / ~ tongs pl. *see* ~ drawer.

spilepin ‖ Faßzwickel m. ‖ fausset m.

spillage (Mar) ‖ Spillage f. ‖ balayures.

spiller (Fish) ‖ Angelschnur f. ‖ ligne f. (à pêcher); palancre f.; palangre f.

spill valve ‖ Überströmventil n. ‖ soupape f. de décharge.

spillway ‖ Überfallwehr n. ‖ déversoir m.

spin, to ‖ spinnen ‖ filer. / to begin to spin ‖ anspinnen ‖ commencer à filer. / ~ cotton ‖ Baumwolle f. spinnen ‖ filer le coton. / ~ over ‖ überspinnen ‖ guiper. / ~ proper ‖ feinspinnen ‖ filer en fin. / ~ a wick ‖ einen Docht m. spinnen ‖ filer une mèche.

spinach ‖ Spinat m. ‖ épinard m.

spindle ‖ Spindel f. ‖ broche f.; axe m.; pivot m.; arbre m. / ~ (Textile) ‖ Spindel f. ‖ broche f. / carburettor float ~ ‖ Schwimmernadel f. ‖ pointeau m. de carburateur. / ~ clutch ‖ Kupplungswelle f. ‖ arbre m. d'embrayage. / complete ~ ‖ vollständige Spindel f. ‖ broche f. complète. / ~ of the core ‖ Kernspindel f. ‖ gabarit m. trousseau; fuseau m. gabarit. / ~ cutter ‖ Frässpindel f. ‖ arbre m. porte-fraise. / disengagement lever ~ ‖ Ausrückhebelwelle f. ‖ arbre m. du levier de débrayage. / dividing ~ ‖ Teilspindel f. ‖ broche f. de la poupée à diviser. / driving ~ ‖ Antriebsspindel f. ‖ broche f. de commande. / float ~ (Auto) ‖ Schwimmerstange f. ‖ tige f. de flotteur. / ~ for fly frames ‖ Fleierspindel f. ‖ broche f. pour continus à ailettes. / insulator ~ (Electr) ‖ Isolatorstütze f. ‖ porte-isolateur m.; console f. *ou* support m. d'isolateur. / bore of the lathe ~ ‖ Drehbankspindelbohrung f. ‖ alésage m. de la broche de tour. / ~ of the model ‖ Spindel f. des Kerns ‖ fuseau-gabarit m.; gabarit-trousseau m. / ~ of the mould ‖ Formspindel f. ‖ trousseau m. / multiple-~d ‖ mehrspindlig ‖ à plusieurs broches. / ~ with one ~ ‖ einspindlig ‖ à une seule tige; à broche unique. / ~ of a printing press ‖ Druckpressenspindel f. ‖ vis f. de la presse à imprimer. / pump ~ ‖ Pumpenwelle

f. ‖ axe m. de pompe. / ~ for self-acting mules ‖ Selfaktorspindel f. ‖ broche f. pour métiers selfacting. / ~ for spinning machines ‖ Spindel f. für Spinnereimaschinen ‖ broche f. pour machines de filature. / stirrer ~ ‖ Rührwerkswelle f. ‖ arbre m. de l'agitateur. / table ~ ‖ Tischspindel f. ‖ arbre m. de table. / work ~ ‖ Werkstückspindel f. ‖ broche f. porte-pièce.

spindle, hand operated ~ adjustment (Roll) ‖ Spindelanstellung f. von Hand ‖ réglage m. à main de la broche.. / ~ axle ‖ Spindelachse f. ‖ essieu m. de broche. / ~ bearing ‖ Zapfenlager n. ‖ palier m. du tourillon. / ~ bolster ‖ Spindeluntersatz m. ‖ support m. de broche. / ~ brake ‖ Spindelbremse f. ‖ frein m. à vis. / ~ guide ‖ Spindelführung f. ‖ guide m. de broche. / ~ loom ‖ Klöppelmaschine f. ‖ métier m. à fuseaux. / ~ pitch (Spinn) ‖ Spindelteilung f. ‖ espace f. de broche. / ~ press ‖ Spindelpresse f. ‖ presse f. à vis. / ~ sleeve ‖ Spindelhülse f. ‖ douille f. de broche. / ~ speed ‖ Spindeldrehzahl f. ‖ vitesse f. de broche. / ~ step ‖ Spindelbuchse f. ‖ gaine f. de broche. / ~ switchboard ‖ Drehschalterschrank m. ‖ tableau m. à levier tournant. / ~ transmission ‖ Spindelübertragung f. ‖ transmission f. à fuseau. / ~ wharve ‖ Spindelwirtel m. ‖ noix f. de broche. / ~ and hand wheel ‖ Spindel f. mit Handrad ‖ vis f. commandée par roue à main.

spine ‖ Rückgrat n. ‖ épine f. dorsale.

spinel ‖ Spinell m. ‖ spinelle m.; alumine f. magnésiée. / black ~ ‖ schwarzer Spinell m.; Pleonast m.; Ceylanit m. ‖ ceylanite f.; pléonaste m. / blue ~ ‖ Saphirspinell m. ‖ spinelle m. (d'un) gris bleuâtre. / chromic ~ ‖ Chromspinell m.; Picotit m. ‖ picotite f. / ruby ~ ‖ Rubinspinell m. ‖ rubis spinelle m. ou balais m. / zinc ~ ‖ Zinkspinell m.; Automolit m. ‖ spinelle m. zincifère; gahnite f.

spine wool ‖ Kernwolle f.; Oberwolle f.; Rückenwolle f. ‖ laine f. mère ou prime.

spinner ‖ Spinner m. ‖ fileur m.

spinneret ‖ Spinndüse f. ‖ filière f.

spinner helper ‖ Spinnergehilfe m. ‖ aide-fileur m.

spinning ‖ Spinnerei f.; Spinnen n. ‖ filature f.; filage m. / dry ~ ‖ Trockenspinnen n. ‖ filage m. au sec. / ~ with hot water ‖ Warmnaßspinnerei f. ‖ filature f. à décomposition. / long wool ~ ‖ Kammgarnspinnerei f. ‖ filature f. de laine à peigner ou de laine longue. / vigogne ~ ‖ Halbwollspinnerei f. ‖ filature f. de la laine vigogne ou de la laine mixte. / worsted ~ see long wool ~.

spinning, accessories for ~ ‖ Spinnereigeräte npl. ‖ accessoires mpl. pour la filature. / ~ bobbin turning machine ‖ Spinnspulendrehbank f. ‖ tour m. à bobine de filature. / ~ can ‖ Spinntopf m. ‖ pot m. de filature. / ~ frame see spinning machine. / ~ lathe ‖ Planierbank f. ‖ tour m. à lisser ou à dresser.

spinning machine ‖ Spinnmaschine f. ‖ machine f. à filer. / ~ (Cable) ‖ Umspinnungsmaschine f. ‖ machine f. à guiper. / carded yarn ~ ‖ Streichgarnspinnmaschine f. ‖ machine f. à filer la laine cardée. / cotton ~ ‖ Baumwollspinnmaschine f. ‖ machine f. à filer le coton. / ~ for covering electric wires ‖ Umspinnmaschine f. für elektrische Drähte ‖ ma-

chine f. à guiper les fils électrotechniques. / hemp hards ~ ‖ Wergspinnmaschine f. ‖ machine f. à filer l'étoupe. / horse hair ~ ‖ Roßhaarspinnmaschine f. ‖ machine f. à filer le crin de cheval. / jute ~ ‖ Jutespinnmaschine f. ‖ machine f. à filer la jute. / paper yarn plate ~ ‖ Papiergarntellerspinnmaschine f. ‖ machine f. à plateau pour le filetage du fil en papier. / straw rope ~ ‖ Strohseilspinnmaschine f. ‖ machine f. à filer les liens de paille. / wool ~ ‖ Wollspinnmaschine f. ‖ machine f. à filer la laine. / worsted ~ ‖ Kammgarnspinnmaschine f. ‖ machine f. à filer la laine peignée.

spinning machine fitter ‖ Spinnereimechaniker m. ‖ mécanicien m. pour filatures. / spare piece for ~s ‖ Spinnereimaschineneinzelteil m. ‖ pièce f. détachée de métier à filer. / ~ tender ‖ Spinnstuhlführer m. ‖ fileur m. au métier.

spinning master ‖ Spinnmeister m. ‖ contremaître m. de filature.

spinning material ‖ Spinnstoff m. ‖ matière f. textile. / vegetable ~ ‖ pflanzliche Spinnstoffe mpl. ‖ matière f. textile végétales.

spinning mill ‖ Spinnerei f. ‖ filature f. / ~ for paper yarns ‖ Papiergarnspinnerei f. ‖ filature f. de ficelles en papier. / ~ for twisting silk ‖ Seidenmühle f. ‖ moulin m. à soie. / owner of a ~ ‖ Spinnereibesitzer m. ‖ filateur m.; propriétaire m. de filature.

spinning mule ‖ Mulemaschine f.; Mulespinnmaschine f. ‖ mull-jenny m. en fin. / ~ nozzle see spinneret. / ~ oil ‖ Spinnöl n. ‖ huile f. de filage.

spinning pot drive for artificial silk ‖ Spinnzentrifuge f. für Kunstseide ‖ centrifugeuse f. à soie artificielle. / ~ process ‖ Topfspinnverfahren n. ‖ méthode f. de filature avec pot tournant.

spinning pump ‖ Spinnpumpe f. ‖ pompe f. de filature. / ~ regulator ‖ Spinnregler m. ‖ régulateur m. de vitesse pour machine à filer. / ~ ring ‖ Spinnring m. ‖ anneau m. de filature. / ~ solution ‖ Spinnlösung f. ‖ solution f. à filer. / ~ wheel ‖ Spinnrad n. ‖ rouet m. à filer.

spiral antenna for suspension in rooms ‖ Spiralantenne f. für Zimmeraufhängung ‖ antenne f. en boudin à suspendre dans la chambre. / ~ chute ‖ Spiralrutsche f.; Wenderutsche f. ‖ glissière f. en spirale. / ~ conveyor ‖ Förderschnecke f. ‖ transporteur m. à vis sans fin. / ~ diagram of sine waves on lines ‖ Spiralendiagramm n. von Sinuswellen auf Leitungen ‖ diagramme m. en spirale d'ondes sinusoïdales sur lignes. / ~ dowel ‖ Spiraldübel m. ‖ goujon m. à spirale. / ~ drill ‖ Spiralbohrer m. ‖ foret m. hélicoïdal. / ~ drilling machine ‖ Drehbohrmaschine f. ‖ perforatrice f. rotative. / ~ groove ‖ Spiralnut f. ‖ rainure f. hélicoïdale. / ~ heating cartridge ‖ Spiralheizpatrone f. ‖ cartouche f. de chauffage en spirale. / ~ line see spiral. / ~ moulding machine (Wood) ‖ Spiralformmaschine f. ‖ machine f. à moulurer en spirale. / ~ pump ‖ Wasserschraube f.; Wasserschnecke f.; archimedische Schraube ‖ limace f.; escargot m.; vis f. d'Archimède ou hydraulique. / ~ quad (Cable) ‖ Sternvierer m. ‖ quadrette f. ou quadruple m. en étoile. / ~ separator (Agr) ‖ Schneckentriör m. ‖ trieur-tobogan m.

spiral spring ‖ Spiralfeder f. ‖ ressort m. en spirale ou à boudin ou hélicoïdal. / ~ of brasswire ‖ Messingspiralfeder f. ‖ ressort m. à boudin en fil de laiton. / conical ~ ‖ kegelförmige Schraubenfeder f. oder Spiralfeder f. ‖ ressort m. hélicoïdal conique.

spiral stairs pl. ‖ Wendeltreppe f. ‖ escalier m. en limaçon ou à vis. / ~ tooth ‖ Spiralzahn m. ‖ dent f. hélicoïdale. / ~ tube ‖ Spiralrohr n. ‖ serpentine f. / ~ weave cable ‖ Litzendraht m.; Hochfrequenzlitze f. ‖ brin m. pour haute fréquence; litzendraht m.

spiral ‖ Spirale f.; Spirallinie f. ‖ spirale f. / ~ of Archimedes ‖ archimedische Spirale f. ‖ spirale f. linéaire ou d'Archimède. / heating ~ ‖ Heizspirale f. ‖ spirale f. de chauffe. / hyperbolic ~ ‖ hyperbolische Spirale f. ‖ spirale f. hyperbolique. / iron wire ~ ‖ Eisendrahtspirale f. ‖ spirale f. en fil de fer. / logarithmic ~ ‖ logarithmische Spirale f. ‖ spirale f. logarithmique.

spirally wound ‖ schraubenförmig gewunden ‖ enroulé en spirale f.

spire (Build) ‖ Helm m.; Spitze f. ‖ flèche f.; aiguille f. / ~ (Screw) ‖ Schraubenwindung f. ‖ spire f.

spirit ‖ Spiritus m. ‖ esprit m. / ~ of ammonia ‖ Salmiakgeist m. ‖ hydrate m. d'ammonium. / methylated ~ ‖ vergällter oder denaturierter Spiritus m. ‖ alcool m. dénaturé. / pyroligneous ~ see wood ~. / ~ of sal-ammonia see ~ of ammonia. / ~ of salt ‖ Salzsäure f. ‖ acide m. chlorhydrique. / ~ of wine ‖ Alkohol m.; Weingeist m. ‖ esprit m. de vin; alcool m. / wood ~ ‖ Holzgeist m. ‖ esprit m. de bois ou pyroxylique.

spirit, adulteration of ~ ‖ Spiritusdenaturierung f. ‖ dénaturation f. de l'alcool. / ~ colour ‖ Zinnfarbe f. ‖ couleur f. à base d'étain. / ~ distiller ‖ Spiritusbrenner m. ‖ distillateur m. d'alcool. / ~ distillery ‖ Spiritusbrennerei f. ‖ distillerie f. d'alcool. / ~ gas stove ‖ Spiritusgasherd m. ‖ fourneau m. au gaz d'alcool. / ~ gauge ‖ Weingeistmesser m. ‖ pèse-esprit m. / ~ heating apparatus ‖ Spiritusofen m. ‖ foyer m. à alcool. / incandescent light for ~ ‖ Spiritusglühlicht n. ‖ lampe f. incandescente à l'alcool. / ~ lamp ‖ Spirituslampe f. ‖ lampe f. à alcool.

spirit level ‖ Setzwage f.; Wasserwage f.; Libelle f. ‖ niveau m. à bulle d'air. / to place the ~ ‖ die Wasserwage aufsetzen ‖ poser le niveau à bulle d'air.

spirit manufacture ‖ Spiritusherstellung f. ‖ fabrication f. d'alcool. / ~ mordant ‖ Spiritusbeize f. ‖ mordant m. à l'alcool. / ~ motor ‖ Spiritusmotor m. ‖ moteur m. à alcool. / ~ orange ‖ Orangezinnfarbe f. ‖ couleur f. d'orange à sel d'étain. / ~ poise ‖ Alkoholometer n. ‖ alcoolomètre m. / ~ rectification ‖ Spiritusrektifikation f. ‖ rectification f. de l'alcool. / ~ rectifier ‖ Spiritusrektifikatör m. ‖ raffineur m. d'alcool. / ~ refinery ‖ Spiritusraffinerie f. ‖ raffinerie f. d'alcool.

spirits pl. ‖ Spirituosen pl. ‖ spiritueux mpl. / manufacture of ~ ‖ Spirituosenfabrikation f. ‖ fabrication f. de spiritueux.

spiritus see spirit.

spirit varnish || Weingeistfirnis m. || vernis m. à l'alcool ou spiritueux.

spit, to || spucken || cracher.

spiteful || übelwollend || malveillant.

spitting (Chem) || Spratzen n. || rochage m.

spittoon || Spucknapf m. || crachoir m.

spitzkasten || Spitzkasten m. || caisse f. pointue.

splash board || Schutzblech n. || garde-crotte f.

splashing || Tauchbad n. || barbotage m.

splash lubrication || Tauchbadschmierung f. || graissage m. par barbotage. / ~ ring || Spritzring m. || bague f. centrifuge.

splay (Build) || Laibungsschräge f.; Fensterschmiege || embrasement m.; ébrasement.

splayed arch (Build) || ausgeschrägter Bogen m. || arc m. ébrasé.

spleen || Milz f. || rate f.

splice, to || spleißen || épisser. / ~ the cable || das Ankertau spleißen || épisser le câble. / ~ cords pl. || Schnüre fpl. flechten || épisser des lacets mpl. / ~ straight through (Cable) || glatt durchschalten || coupler directement.

splice see also splicing || Spleißung f.; Spleißstelle f. || épissure f. / ~ of a cable || Kabelspleißung f. || épissure f. du câble. / long ~ || Langspleißung f. || épissure f. en long.

splice bar to prevent creeping of the rails || Stemmlasche f. gegen Schienenwandern || éclisse f. épaulée pour empêcher le cheminement des rails.

splicer || Spleißer m. || épisseur m.

splicing see also splice (Cable) || Spleißung f.; Spleißstelle f. || épissure f. / ~ (Textile) || Verstärkung f. || renforcement m. / ~ of belt || Spleißstelle f. des Riemens; Riemenverbindung f. || couture f. ou joint m. de courroie. / ~ clamp || Verbindungsklammer f. || agrafe f. de joint.

spline || Nutung f. || cannelure f. / ~ shaper || Keilnuthobler m. || machine f. à raboter les rainures des arbres à clavettes.

splint see also split pin || Splint m. || clavette f. / wooden ~ || Holzspan m. || copeau m. de bois.

splint bolt || versplinteter Bolzen m. || boulon m. goupillé. / ~ coal || Schieferkohle f. || houille f. schisteuse.

splinter, to || zersplittern || fractionner.

splinter || Splitter m. || éclat m. / ~ (Carp) || Span m.; Hobelspan m. || copeau m. / ~ of wood || Holzsplitter m. || éclat m. ou copeau m. de bois.

splinter catcher || Splitterfänger m. || arrête-éclats m. / ~ forceps pl. || Splitterpinzette f. || pince f. à écharde.

splinter-proof || splittersicher || résistant à l'éclatement; incassable. / ~ glass || splittersicheres Glas n. || verre m. incassable.

splinter screen (Arm) || Schild m. || masque f.; pare-éclats m. / ~ wood || Spanholz n. || bois m. en éclisses; copeaux mpl. de bois.

split, to || spalten || refendre; fendre. / ~ (Wood) || aufreißen; springen || se fendre. / ~ a cable || ein Kabel n. spleißen || épisser un câble. / ~ hides pl. || Leder n. spalten || dédoubler ou refendre les peaux fpl. / ~ into two halves || in zwei Hälften fpl. zerfallen || se diviser en deux parties fpl. / ~ the iron || das Eisen schneiden || fendre ou refendre le fer.

split || gesprungen; gespalten; crevassé; éclié; fendu. / ~ cotter || gespaltener Keil m. || clavette f. fendue. / ~ lath (Build) || Reißlatte f.; Spaltlatte f.; Waldlatte f. || latte f. fendue ou de fente. / ~ nut || aufgeschnittene Mutter f. || écrou m. fendu. / ~ pattern || geteiltes Modell n. || modèle m. en deux parties. / ~ pulley || geteilte Riem(en)scheibe f. || poulie f. divisée ou en deux moitiés. / ~ stave || gespaltene Daube f. || douve f. fendue.

split || Spalt m. || fêlure f. / ~ burner || Schlitzbrenner m. || bec m. fendu. / ~ firewood || Scheitholz n. || bois m. de quartier. / ~ order wire (Tel) || Sammeldienstleitung f. || ligne f. auxiliaire de concentration.

split pin see also splint || Splint m.; Schlitzstift m. || goupille f.; broche f. fendue. / to fix a nut with a ~ || eine Mutter f. versplinten || goupiller un écrou. / ~ extractor || Splintauszieher m. || chasse-goupille m. / machine for making ~s || Splintherstellungsmaschine f. || machine f. à fabriquer les goupilles. / ~ pliers pl. || Splintzieher m. || arrache-goupille m. / ~ wire || Splintdraht m. || fil m. pour goupilles.

split spring (Watchm) || Druckfeder f.; Springfeder f. || pas m. d'âne.

splitter || Spalter m. || fendeur m. / hand ~ (Weav) || Handaufwinder m. || ensoupleur m. à la main.

split timber || Kluftholz n.; Spaltholz n. || bois m. fendu ou de fente.

splitting, fat ~ || fettspaltend || dédoublant les graisses. / ~ of fire wood || Spalten n. von Brennholz || fendage m. de bois de chauffage. / ~ of hides || Spalten n. der Häute || dédoublage m. des peaux. / ~ of plates pl. || Schneiden n. von Blechtafeln || coupe f. de feuilles de tôles.

splitting cuts pl. (Curr) || Spaltarbeit f. || travail m. à refendre.

splitting machine for hides || Lederspaltmaschine f. || machine f. à refendre les peaux. / plate ~ || Blechschere f. || cisaille f. à tôles. / wood ~ || Holzspaltmaschine f. || machine f. à fendre le bois.

splitting-off (Chem) || Abspaltung f. || dissociation f.

splitting property of wood || Spaltbarkeit f. des Holzes || fissilité f. du bois. / ~ test || Spaltprobe f. || essai m. de fendage.

splitting-up of current || Stromteilung f. || division f. de courant.

split trail carriage (Arm) || Spreizlafette f. || affût m. biflèche ou à flèches ouvrantes. / ~ wood || Spaltholz n. || bois m. de fente.

spluttering of arc || Sprühen n. des Lichtbogens || crachement m. de l'arc.

spoil, to || verderben || gâter. / ~ by careless work || verpfuschen || gâter; gâcher.

spoil, height from which the ~ is dumped || Fallhöhe f. des Baggergutes || hauteur f. de chute du matériel dragué. / ~ bank || Aufstürzung f.; Seitenablagerung f. || dépôt m. à la terre.

spoke, to ~ a wheel || ein Rad n. verspeichen || enchausser une roue.

spoke || Speiche f. || rais m.; rayon m. / ~ (Ladder) || Leitersprosse f. || échelon m. / cast-in ~ || eingegossene Speiche f. || rayon m. enrobé dans la fonte. / double ~ || Doppelspeiche f. || rayon m. double. / hollow ~ || hohle Speiche f. || rayon m. creux. / ~ of oval section || Speiche f. von ovalem Querschnitt || rayon m. de section ovale. / round ~ || Rundspeiche f. || rayon m. rond. / tubular ~ || rohrförmige Speiche f. || rayon m. tubulaire. / wire ~ || Drahtspeiche f. || rayon m. d'acier ou métallique.

spoke fitter (Coachm) || Speicheneinsetzer m. || enrayeur m. / ~ flange || Speichenträger m. || flasque f. de rayonnage. / ~ hammer || Speichenhammer m. || masse f. à enrayer. / ~ iron (Join) || Speicheisen n. || fer m. à rais. / ~ maker || Speichenmacher m. || ouvrier m. en rayons. / ~ nipple || Speichennippel m. || raccord m. de rais. / ~ shave || Speichenhobel m.; Schabhobel m. || bastringue m.; wastringle m.; racloir m. / ~ tightener || Speichenspanner m. || serre-rayon m. / ~ and handle turning machine || Speichen- und Griffdrehbank f. || tour m. pour rayons de roues et manches d'outils.

spoke wheel, cast steel ~ || Stahlgußspeichenrad n. || roue f. à rayons en acier moulé. / double ~ || Doppelspeichenrad n. || roue f. à rayons doubles. / welded ~ || geschweißtes Speichenrad n. || roue f. à rayons soudée. / welded iron ~ || Speichenrad n. aus Schweißeisen || roue f. à rayons en fer soudé.

spoke wheel centre || Radstern m. || centre m. de roue à rayons.

sponge, to ~ the cloth || den Stoff m. dekatieren || décatir le drap.

sponge || Schwamm m. || éponge f. / ~ (Mach) || Saugkopf m. || crépine f.; pomme f. d'arrosoir. / prepared ~ || zubereiteter Schwamm m. || éponge f. préparée. / raw ~ || roher Schwamm m. || éponge f. brute. / rubber ~ || Gummischwamm m. || éponge f. de caoutchouc.

sponge cup || Schwammschale f. || coupe f. à éponge. / ~ fisherman || Schwammfischer m. || pêcheur m. d'éponges. / ~ pouch || Schwammbeutel m. || sac m. à éponge. / ~ reef || Schwammriff n. || récif m. d'éponges.

sponger (Clothm) || Dekatierer m. || décatisseur m.

sponging (Cloth) || Dekatieren n. || décatissage m. / ~ establishment || Dekatieranstalt f. || atelier m. à décatir.

spongious || schwammig || spongieux; poreux.

spongy iron || schwammiges Eisen n. || fer m. spongieux. / ~ lead || Bleischwamm m. || plomb m. spongieux. / ~ nickel || Nickelschwamm m. || nickel m. spongieux ou poreux.

sponson (Shipb) || Schwalbennest n. || encorbellement m.

spontaneous || freiwillig; spontan || spontané. / ~ fermentation || spontane Gährung f. || fermentation f. spontanée. / ~ ignition || Selbstzündung f. || allumage m. spontané. / ~ slaking of lime in the air || Selbstlöschung f. des Kalks || extinction f. spontanée de la chaux.

spool || Haspel f. || enrouleur m. / ~ for copying ribbon || Farbbandspule f.; bobine f. de rubans-encreurs. / hoop iron ~ || Bandeisenhaspel m.; f. || dévidoir m. de feuillard. / ~ of paper || Papierspule f. || bobine f. de papier. / scrap ~ || Schrothaspel m.; f. || dévidoir m. à riblons. / shuttle ~ (Weav) || Weberschiffchenspule f. || épeule f. de navette. / wire ~ || Drahthaspel f. || dévidoir m. de fil. / wooden ~ || Holzspule f. || bobine f. ou bobinot m. en bois. / wound on ~s

‖ gespult ‖ enroulé sur bobines fpl.; bobiné.

spool cover ‖ Spulendeckel m. ‖ couvercle m. de la bobine. / ~ **dyeing** ‖ Kopsfärberei f. ‖ teinture f. de fils en cannettes.

spooler ‖ Abspulerin f. ‖ dévideuse f. / ~ (Weav) ‖ Spuler m. ‖ bobineur m. / **hand** ~ ‖ Handspuler m. ‖ bobineur m. à la main.

spool feeder ‖ Spulenaufstecker m. ‖ rechargeur m. de broches.

spooling ‖ Spulen n. ‖ bobinage m. / ~ **frame** see spooling machine.

spooling machine ‖ Spulmaschine f. ‖ bobineuse f.; machine à bobiner. / **multiple** ~ ‖ Fachtmaschine f. ‖ machine f. à bobinage multiple. / ~ **for textiles** ‖ Spulmaschine f. für Textilien ‖ bobineuse f. pour textiles.

spooling wheel ‖ Spulrad n. ‖ rouet m. à bobiner.

spoon ‖ Löffel m. ‖ cuiller f.; cuillère f. / **alumin(i)um** ~ ‖ Aluminiumlöffel m. ‖ cuiller f. en aluminium. / **gold** ~ ‖ Goldlöffel m. ‖ cuiller f. en or. / **salt** ~ ‖ Salzlöffel m. ‖ cuiller f. à sel. / **silver** ~ ‖ Silberlöffel m. ‖ cuiller f. en argent. / **table** ~ ‖ Tischlöffel m.; ‖ cuiller m. de table. / **tea** ~ ‖ Teelöffel m.; Kaffeelöffel m. ‖ petite cuiller f.; cuiller f. à café. / **tin** ~ ‖ Zinnlöffel m. ‖ cuiller f. d'étain. / **wooden** ~ ‖ Holzlöffel m.; Kelle f. ‖ cuiller f. en bois.

spoon bit ‖ Löffelbohrer m. ‖ mèche f. à cuiller. / ~ **brake** ‖ Schuhbremse f. ‖ frein m. à sabot. / **wooden** ~ **maker** ‖ Holzlöffelschnitzer m.; Kellenmacher m. ‖ fabricant m. de cuillers en bois.

spoon rolling mill ‖ Löffelwalzwerk n. ‖ laminoir m. à cuillers. / **hunting** ~ ‖ pendelnde Löffelwalzmaschine f. ‖ laminoir m. à pendule à faire des cuillers ou cuillères. / **rotating** ~ ‖ rotierende Löffelwalzmaschine f. ‖ laminoir m. tournant à faire des cuillers ou cuillères.

spoon-shaped strut (Aero) ‖ Löffelstrebe f. ‖ jambe f. de force ou contrefiche f. en forme de cuiller.

sporangium carrier ‖ Fruchthyphe f.; Fruchtträger m. ‖ tube m. sporangifère; porte-sporange m. / ~ **wall** ‖ Fruchtwand f. ‖ membrane f. de sporange.

spore ‖ Sporen f. ‖ spore f. / **bacteria** ~ ‖ Bakterienspore f. ‖ spore f. de bactérie.

sport ‖ Sport m.; Spiel n. ‖ sport m. / **aquatic** ~ ‖ Wassersport m. ‖ sport m. nautique; yachting m.

sporting aeroplane ‖ Sportflugzeug n. ‖ avion m. de sport. / ~ **articles** pl. see ~ **goods** pl. / ~ **belt** ‖ Sportgürtel m. ‖ ceinture f. de sport.

sporting boot ‖ Sportstiefel m. ‖ chaussure f. de sport. / **stout** ~ ‖ derber Sportstiefel m. ‖ soulier m. robuste de sport.

sporting cap ‖ Sportmütze f. ‖ casquette f. de sport. / ~ **goods** pl. ‖ Sportartikel mpl. ‖ articles mpl. de sport. / ~ **ground** ‖ Sportplatz m. ‖ place f. destinée au sport.

sporting gun ‖ Jagdflinte f.; Jagdgewehr n. ‖ fusil m. de chasse. / ~ **with three barrels** ‖ Drilling m. ‖ fusil m. de chasse à trois coups.

sporting powder ‖ Jagdpulver n. ‖ poudre f. de chasse. / ~ **shirt** ‖ Sporthemd n. ‖ chemise f. de sport. / ~ **shoe** see sporting boot. / ~ **suit** ‖ Sportanzug m. ‖ costume

m. de sport. / ~ **tenant** ‖ Jagdpächter m. ‖ fermier m. de chasse. / ~ **tricot** ‖ Sporttrikot n. ‖ tricot m. de sport.

sports ball ‖ Sportball m. ‖ balle f. de sport. / ~ **bandage** ‖ Sportbandage f. ‖ bandage m. de sport. / **special boots and shoes for** ~ ‖ Spezialschuhwerk n. für Sportzwecke ‖ chaussure f. spéciale pour les sports.

sportsman ‖ Jäger m. ‖ chasseur m.

sports model (Auto) ‖ Sportwagen m. ‖ torpédo m. sport. / ~ **outfit** ‖ Sportausrüstung f. ‖ équipement m. de sport. / ~ **vehicle for children** ‖ Sportfahrzeug n. für Kinder ‖ véhicule m. de sport pour enfants. / ~ **wear** ‖ Sportkleidung f. ‖ costumes mpl. de sport.

spot, to ‖ sprenkeln ‖ moucheter.

spot ‖ Fleck m. ‖ tache f.; paille f. / **blind** ~ (Eye) ‖ blinder Fleck m. ‖ papille f. / ~ **of the sun** ‖ Sonnenfleck m. ‖ tache f. du soleil.

spot electrode ‖ Punkt(schweiß)elektrode f. ‖ électrode f. punctiforme. / ~ **goods** pl. ‖ greifbare Ware f. ‖ marchandise f. disponible. / ~ **light** ‖ Sucherlampe f. ‖ projecteur m. auxiliaire orientable. / ~ **price** ‖ Preis m. für greifbare Mengen ‖ cours m. en disponibles.

spotted ‖ fleckig; gefleckt; gesprenkelt ‖ moucheté. / ~ **hemlock** ‖ Schierlingskraut n. ‖ ciguë f. / ~ **mahogany** ‖ geflecktes Mahagoniholz n. ‖ acajou m. moucheté.

spotter see spotting device.

spotting device ‖ Fleckgerät n.; Flecker m. ‖ instrument m. d'observation du tir; appareil m. pour la détermination du but. / ~ **and plotting system** (Military) ‖ Feuerleitverfahren n. ‖ système m. ou méthode f. de conduite de tir.

spot transactions pl. ‖ Platzgeschäft n. ‖ commerce m. local. / ~ **welder** ‖ Punktschweißmaschine f. ‖ machine f. à souder par points. / ~ **welding** ‖ Punktschweißung f. ‖ soudage m. par point. / **staggered** ~ **welding** ‖ Zickzackpunktschweißung f. ‖ soudage m. par points en quinconce. / ~ **welding machine** see ~ **welder.**

spout (Glass) ‖ Ausgußschnauze f. ‖ bec m. / ~ **of a gutter** ‖ Schnauze f. einer Dachrinne ‖ canon m. de gouttière. / ~ **closing apparatus for varnish cans** ‖ Tüllenverschlußapparat m. für Lackkannen ‖ appareil m. de fermeture à douille pour bidons à laque. / ~ **tube** (Brew) ‖ Flaschenfüllrohr n. ‖ canule f.

sprag (Auto) ‖ Bergstütze f. ‖ béquille f. / ~ **brake** ‖ Knüppelbremse f. ‖ freinage m. à gourdin.

sprain ‖ Verrenkung f. ‖ dislocation f.

spray (Found) ‖ Gußröhre f. ‖ coulée f. du jet de fonte. / **compressed air** ~ **apparatus** ‖ Druckluftstrahlapparat m. ‖ appareil m. à jet d'air comprimé. / ~ **carburettor** ‖ Einspritzvergaser m.; Zerstäubungsvergaser m. ‖ carburateur m. à pulvérisation. / ~ **cooling** ‖ Kühlung f. durch Zerstäubung ‖ rafraîchissement m. par pulvérisation. / ~ **graphiting device** ‖ Graphitierapparat m. ‖ appareil m. pour graphiter.

sprayer ‖ Zerstäuber m. ‖ pulvérisateur m.; vaporisateur m. / ~ **for dyes** ‖ Spritzapparat m. für Farben ‖ pistolet m. ou appareil m. pulvérisateur pour couleurs. / **revolving** ~ (Water) ‖ Drehsprenger m. ‖ arrosoir m. rotatif.

spray and hopper granary ‖ Rieselspeicher m. ‖ magasin m. à ruissellement.

spraying ‖ Abspritzen n. ‖ arrosage m. / **magnetic** ~ ‖ magnetische Streuung f. ‖ dispersion f. magnétique. / ~ **nozzle** ‖ Spritzdüse f. ‖ gicleur m.

spraying pistol ‖ Spritzpistole f. ‖ pistolet m. de vernissage. / ~ **for painting** ‖ Spritzpistole f. zum Anstreichen ‖ pistolet m. à peinturer ou à peindre.

spraying system, paint ~ (Auto) ‖ Lackierspritzverfahren n. ‖ peinture f. pneumatique.

spray nozzle ‖ Streudüse f. ‖ tuyère f. / ~ **painting** ‖ Spritzmalerei f. ‖ peinture f. au pistolet. / ~ **water** ‖ Spritzwasser n. ‖ eau f. projetée.

spread, to ‖ ausstreuen ‖ disséminer. / ~ **the glass plates** ‖ das Tafelglas strecken ‖ étendre le verre à vitres. / ~ **the gold varnish** ‖ den Goldfirnis m. aufstreichen ‖ emplâtrer le vernis d'or. / ~ **the hides** ‖ die Häute fpl. versetzen ‖ coucher les peaux fpl. gonflées.

spread of the wing ‖ Spannweite f. des Flügels ‖ envergure f. de l'aile.

spreader, helicoidal ~ (Spinn) ‖ Spiralwattenmaschine f. ‖ batteur m. hélicoïde. / **manure** ~ ‖ Düngerstreuer m. ‖ distributeur m. d'engrais.

spreader tube ‖ Quetschtube f. ‖ tube m. à presser.

spreading brush (Pap) ‖ Streichbürste f. ‖ brosse f. à tirer. / ~ **hammer** ‖ Treibhammer m. ‖ marteau m. à chasser.

spreading machine (Spinn) ‖ Wattenmaschine f. ‖ batteur-étaleur m. / ~ (Rubber) ‖ Streichmaschine f. ‖ cardeuse f. / ~ **for flax spinning** ‖ Anlegemaschine f. für Flachsspinnereien ‖ machine f. étaleur pour le filage du chanvre.

spring, to ‖ abfedern; federn; mit Federn fpl. versehen; federnd aufhängen ‖ suspendre sur ressorts mpl. / ~ (River) ‖ entspringen ‖ prendre sa source. / ~ **an arch** ‖ einen Bogen m. wölben ‖ arquer. / ~ **a well** ‖ einen Brunnen m. graben ‖ foncer un puits.

spring (Geol) ‖ Quelle f. ‖ source f. / ~ **due to damming of underground water** ‖ Stauquelle f. ‖ source f. d'eau arrêtée. / **hot** ~ ‖ heiße Quelle f. ‖ source f. chaude.

spring (Mach) ‖ Feder f. ‖ ressort m. / **adjusting** ~ ‖ Paßfeder f. ‖ ressort m. d'ajustage; languette f. ajustée. / **arched** ~ ‖ gewölbte Feder f. ‖ ressort m. demi-jonc. / ~ **of the balance wheel** (Watch) ‖ Unruhefeder f. ‖ ressort m. du balancier. / **bearing** ~ see suspension ~. / **body** ~ see car ~. / **buffer** ~ ‖ Pufferfeder f. ‖ ressort m. de tampon ou supplémentaire. / **car** ~ ‖ Wagenfeder f. ‖ ressort m. de carrosserie ou de voiture. / **catch** ~ ‖ Einschnappfeder f. ‖ ressort m d'encliquetage / **chair** ~ ‖ Stuhlfeder f. ‖ ressort m. de chaise. / **chimney** ~ ‖ Kaminschnäpper m. ‖ ressort m. de cheminée. / **clutch** ~ ‖ Kupplungsfeder f. ‖ ressort m. d'embrayage. / **compression** ~ ‖ Druckfeder f. ‖ ressort m. de pression. / **conical spiral** ~ ‖ kegelförmige Schraubenfeder f. ‖ ressort m. hélicoïdal conique. / **contact** ~ ‖ Kontaktfeder f. ‖ ressort m. de contact. / **controlling** ~ **for locomotive trucks** ‖ Rückstellfeder f. für Lokomotivdrehgestelle ‖ ressort m. de rappel pour bogies de locomotives. / **cylindrical** ~ ‖ zylindrische Schraubenfeder f. ‖ ressort

m. à boudin *ou* en spirale à boudin *ou* hélicoïdal. / double cone ~ || Doppelkegelfeder f. || ressort m. biconique. / elliptic ~ || elliptische Feder f. || ressort m. elliptique. / elliptic ~ for tenders || Doppelfeder f. für Tender || ressort m. à pincette pour tenders. / equalizer ~ || Einstellfeder f.; Balancierfeder f. || ressort m. de rappel *ou* de balancier. / ~ for eyeglasses || Kneiferfeder f. || ressort m. de pince-nez. / ~ of flat rectangular section || Feder f. von flachrechteckigem Querschnitt || ressort m. de section de lame rectangulaire. / front ~ || Vorderfeder f. || ressort m. avant. / ~ for furniture || Sprungfeder f. für Polstermöbel || ressort m. pour meubles. / governor ~ || Reglerfeder f. || ressort m. de régulateur. / half-elliptic ~ || Halbfeder f. || ressort m. à demi-pincette. / helical ~ || Schraubenfeder f. || ressort m. hélicoïdal *ou* à boudin. / laminated ~ *see* leaf ~. / leaf ~ || Blattfeder f. || ressort m. feuilleté *ou* à lames. / ~ with left-handed helice || links gewundene Feder f. || ressort m. enroulé à gauche. / the ~ allows of extraordinary loads || die Feder f. läßt eine außergewöhnlich hohe Belastung zu || le ressort m. peut supporter une très grande charge. / longitudinal ~ || Längsfeder f. || ressort m. longitudinal. / lower ~ || Unterfeder f. || ressort m. inférieur. / main drive ~ || Hauptantriebsfeder f. || ressort m. principal de commande. / ~ for motor cars || Automobilfeder f. || ressort m. d'automobile. / multi-coupled ~s pl. || mehrfach gekuppelte Federn fpl. || accouplement m. de plusieurs ressorts. / plate ~ *see* leaf ~. / quarter elliptic ~ || Viertelelliptikfeder f. || ressort m. quart-elliptique. / rear ~ || Hinterfeder f. || ressort m. arrière. / rectangular ~ || Rechteckfeder f. || ressort m. rectangulaire. / release ~ || Rückzugfeder f. || ressort m. de rappel. / ~ for relieving || Entlastungsfeder f. || ressort m. de décharge. / removable ~ || bewegliche Feder f. || ressort m. mobile. / ~ for revolving shutters || Rolladenfeder f. || ressort m. de jalousies. / ribbed ~ || Rippenfeder f. || ressort m. à nervure. / ~ with rolled end || Feder n. mit gerolltem Ende || ressort m. à bout roulé. / safety ~ for locomotive valves || Lokomotivsicherheitsventilfeder f. || ressort m. de sûreté pour soupapes de locomotives. / shock ~ || Stoßfeder f. || ressort m. de choc. / the ~ grows slack || die Feder f. entspannt sich || le ressort se détende. / slide ~ || Gleitfeder f. || ressort m. de glissement. / spiral ~ || Spiralfeder f.; Schraubenfeder f. || ressort m. à boudin *ou* hélicoïdal. / ~ of square section || Feder f. von quadratischem Querschnitt || ressort m. de section carrée. / starting ~ (Aero) || Abschnellfeder f. || ressort m. de lancement. / supplementary ~ (Auto) || Dämpfungsfeder f.; Hilfsfeder f. || ressort m. compensateur. / suspension ~ || Tragfeder f. || ressort m. de suspension. / ~ under tension || Feder f. unter Spannung || ressort m. sous pression. / three-armed ~ || dreiflügelige Feder f. || ressort m. à trois branches. / three-coupled elliptical ~ || dreifach gekuppelte Doppelfeder f. || ressort m. elliptique couplé par trois. / transverse ~ || Querfeder f. || ressort m. transversal. / under-

slung ~ || unterbaute Feder f. || ressort m. sous l'essieu. / ~s pl. united in one set || zu einem Satz vereinigte Federn || ressorts mpl. formant un jeu. / upper ~ || Oberfeder f. || ressort m. supérieur. / ~ with upset end || Feder f. mit gestauchtem Ende || ressort m. à bout refoulé. / valve ~ || Ventilfeder f. || ressort m. de soupape. / volute ~ || Schneckenfeder f. || ressort m. à volute. / watch ~ || Uhrfeder f. || ressort m. d'horlogerie *ou* de montre. / without ~s || federlos || sans ressort m.

spring balance || Dynamometer n.; Federwage f. || dynamomètre m.; balance f. à ressort. / ~ for the household || Wirtschaftsfederwage f. || balance f. à ressort de ménage.

spring band || Federbund m. || bride f. de ressort. / ~ dismantling hammer || Federbundabziehhammer m. || marteau m. à démonter les brides de ressorts. / ~ dismantling press || Federbundabziehpresse f. || presse f. à démonter les brides de ressort. / ~ dismounting hammer *see* ~ dismantling hammer. / ~ mounting press || Federbundaufziehpresse f. || presse f. à monter les brides de ressorts. / ~ steel || Federbundstahl m. || acier m. pour brides de ressort.

spring bearing || federndes Lager n. || palier m. à ressort.

spring bolt || Federbolzen m. || boulon m. de ressort. / ~ (Lock) || Schnäpper m.; Schnappriegel m. || pêne m. coulant.

spring bow compasses || Nullenzirkel m. || compas m. à pompe. / ~ divider || Teilzirkel m. || compas m. à pression.

spring bracket || Federbock m. || support m. de ressort. / rear ~ || Hinterfederbock m. || support m. de ressort arrière.

spring bridge || Federbrücke f. || ressort m. formant pont. / ~ buffer || Federpuffer m. || tampon m. à ressort. / ~ callipers pl. || Federzirkel m. || compas m. à ressort. / ~ cap || Federteller m. || cuvette f. de ressort. / ~ carriage *see* sprung carriage. / denture ~ carrier || Gebißfederträger m. || porte-ressort m. de dentier. / ~ case || Federgehäuse n. || boîte f. du ressort; barillet m. à ressort. / ~ catch (Mach) || Hebelsteuerung f. || encliquetage m. / ~ chuck || Spannpatrone f. || pince f. de serrage. / ~ clamp || federnde Klemme f. || pince f. à ressort. / ~ clip || Federbügel m. || bride f. de ressort. / ~ clock || Federuhr f. || horloge f. à ressort. / ~ coiling machine || Federwindmaschine f. || machine f. à enrouler les ressorts. / ~ contact || Federkontakt m. || contact m. à ressort. / ~ coupling || Federkupplung f. || accouplement m. à ressort. / ~ cover || Federverkleidung f. || carénage m. du ressort. / ~ divider *see* ~ callipers pl.

spring-driven clock || Uhrwerk n. mit Federantrieb || mouvement m. d'horlogerie à ressort. / winding key for ~ || Schlüssel m. zum Aufziehen des Federuhrwerks || clef f. pour remonter le mouvement d'horlogerie à ressort.

spring drum || Federtrommel f. || tambour m. à ressort. / ~ dynamometer || Federdynamometer m. || dynamomètre m. à ressort. / end of the ~ || Federende n. || bout m. du ressort.

springer (Arch) || Bogenkämpfer m. || imposte f.; coussinet m.

spring, exchange of the ~ in case of a fracture || Auswechslung f. der Feder bei Federbruch || remplacement m. du ressort en cas de rupture. / ~ eye || Federauge n. || œil m. de ressort. / hollow flyer with ~ finger || hohler Flügel m. mit Preßfinger || ailette f. creuse avec doigt comprimeur. / ~ fish plate || Federlasche f. || éclisse f. à ressort. / ~ fork || Federgabel f. || fourche f. à ressort. / ~ frame || Federrahmen m. || cadre m. à ressorts. / ~ gaiter || Federschutz m. || gaine f. de ressort. / ~ governor || Federregler m. || régulateur m. à ressort. / ~ hammer || Federhammer m. || marteau m. à ressort.

spring hard || federhart || écroui. / to become ~ || federhart werden || s'écrouir.

spring hook || Karabinerhaken m. || porte-mousqueton m.

springing, undercarriage ~ || Fahrgestellfederung f. || amortisseur m. de choc du train d'atterrissage. / ~ of a vault (Build) || Kämpferlinie f. || naissance f. de voute.

springing course (Build) || Kämpferschicht f. || assise f. des sommiers. / ~ stone (Build) || Kämpferstein m. || imposte f.; sommier m.

spring leaf || Federblatt n. || lame f. de ressort. / ribbed ~ || Rippenfeder f. || ressort m. à nervures.

spring leaf retainer || Federklammer f. || étrier m. de ressort.

spring lever || federnder Hebel m. || levier m. à ressort.

spring-loaded (Valve) || federbelastet || commandé par un ressort. / ~ valve || federbelastetes Ventil n. || soupape f. commandée par un ressort.

spring lobster || Languste f. || langouste f.

spring lock || Schnappschloß n. || serrure f. à houssette. / German ~ || deutsches Schloß; offenes Schloß; Schnappschloß; Halbtourschloß || serrure f. à ressort; bec m. de cane ; demi-tour m. / half turning ~ *see* German ~.

spring lubrication || Federschmierung f. || graissage m. des ressorts. / machine for making ~s || Federherstellungsmaschine f. || machine f. à fabriquer les ressorts. / ~ manometer || Federmanometer m. || manomètre m. à ressort. / ~ manufacturing machine || Federherstellungsmaschine f. || machine f. pour ressorts. / ~ mattress || Sprungfedermatratze f. || sommier m. élastique; matelas m. à ressorts. / ~ motor of talking machine || Laufwerk n. einer Sprechmaschine || mécanisme m. de machines parlantes. / ~ mounting || Abfederung f. || suspension f. à ressorts. / ~ needle (Lace) || Hakennadel f.; Spitzennadel f. || aiguille f. à bec. / ~ needle machine || Spitzennadelmaschine f. || machine f. à faire les aiguilles à bec. / ~ number of plates in a ~ || Lagenzahl f. einer Blattfeder || nombre m. de lames d'un ressort. / ~ oiler || Federschmierbüchse f. || graisseur m. de ressort. / ~ pin || federnder Stift m. || broche f. à ressort. / ~ pivot seat || Federsattel m. || patin m. de ressort tournant.

spring plate || Federblatt n. || lame f. de ressort. / ~ (Railw) || Federplatte f. || rondelle-ressort f. / ~ bending and cutting press || Federblattbiege- und Schneidepresse f. || presse f. à cintrer et couper les lames de ressorts. / ~ bending roll ||

Federblattbiegewalze f. ‖ laminoir m. à cintrer les lames de ressorts.

spring-pressed ‖ durch Federdruck m. gehalten ‖ tenu par pression f. de ressort. / ~ pressure gauge ‖ Federmanometer m. ‖ manomètre m. à ressort.

spring ring ‖ Sprengring m. ‖ agrafe f. circulaire; bague f. de retenue. / hook ~ ‖ Hakensprengring m. ‖ bague f. de sûreté à crochet.

spring ring section ‖ Sprengringeisen n. ‖ profilé m. pour agrafes-circulaires.

spring safety brake ‖ Federsicherheitsbremse f. ‖ frein m. de sûreté à ressort. / ~ salt ‖ Quellsalz n. ‖ sel m. des sources salées. / ~ scales pl. ‖ Federwage f. ‖ peson m. ou bascule f. à ressort. / ~ screw ‖ Federschraube f. ‖ vis f. à tête fendue. / ~ seat ‖ Federsitz m. ‖ patin m. de ressort; siège m. à ressorts. / ~ set ‖ Federsatz m. ‖ jeu m. de ressorts. / ~ shackle ‖ Federlasche f. ‖ menotte f. de ressort. / ~ shackle bolt ‖ Federbolzen m. ‖ boulon m. de ressort. / ~ shaft (Weav) ‖ Wage f. ‖ tire-lisse f. / ~ shock absorber ‖ Federstoßfänger m. ‖ ressort m. compensateur./ ~ shop ‖ Federwerkstatt f. ‖ atelier m. aux ressorts. / ~ slider ‖ federnder Schieber m. ‖ pièce f. coulissante à ressort. / ~ spoke ‖ federnde Speiche f. ‖ rayon m. élastique.

spring steel ‖ Federstahl m. ‖ acier m. à ressorts. / ribbed ~ ‖ gerippter Federstahl m. ‖ acier m. à ressorts cannelé.

spring stone (Met) ‖ Prellstein m. ‖ chappe f. / ~ stop ‖ Federanschlag m. ‖ butée f. à ressort; arrêt m. de ressort. / ~ suspension (Auto) ‖ Federung f. ‖ suspension f. à ressort. / ~ suspension of front axle ‖ Vorderachsfederung f. ‖ suspension f. à ressort de l'essieu avant. / tapping a ~ ‖ Erschließung f. einer Quelle ‖ ouverture f. ou fonçage m. d'une source. / ~ tension ‖ Federspannung f. ‖ tension f. du ressort. / ~-tensioned driving plate (Mach tool) ‖ Dorn m. mit federnder Mitnehmerscheibe ‖ dispositif m. entraîneur à ressort. / ~ testing machine ‖ Federprüfmaschine f. ‖ machine f. à essayer les ressorts. / ~ tide ‖ Springflut f. ‖ marée f. de syzygie ou des vives eaux. / ~ timber ‖ Frühjahrsholz n. ‖ bois m. de printemps. / ~ valve ‖ Federventil n. ‖ soupape f. à ressort. / ~ van ‖ Federwagen m. ‖ suspendue f.; voiture f. suspendue (sur ressorts). / ~ washer ‖ federnde Unterlagscheibe f.; Federteller m. ‖ rondelle f. à ressort ou élastique. / ~ washer lock nut ‖ Federtellersicherung f. ‖ arrêt m. de sûreté de godet. / ~ water ‖ Quellwasser n. ‖ eau f. de source. / ~ weight ‖ Federbelastung f. ‖ charge f. sur le ressort./ ~ wheel ‖ federndes Rad n. ‖ roue f. à ressort ou élastique. / ~ wire ‖ Federdraht m. ‖ fil m. à ressort. / polishing device for ~ wires ‖ Federdrahtpoliereinrichtung f. ‖ installation f. à polir le fil pour ressorts.

sprinkle, to ‖ begießen ‖ arroser; asperger.

sprinkle (Forg) ‖ Löschwedel m.; Sprengwedel m. ‖ goupillon m.

sprinkled ‖ gesprenkelt ‖ jaspé.

sprinkler ‖ Regner m.; Brause f. ‖ appareil m. de pluie artificielle; crépine f. / ~ (Bookb) ‖ Sprenkler m. ‖ jaspeur m. sur tranches. / cask ~ ‖ Faßausspritzer m. ‖ injecteur m. pour fûts. / motor ~ with brush and washing roll ‖ Motorsprengmaschine f. mit Kehr- und Waschwalze ‖ automobile f. d'arrosage avec cylindre à brosse et cylindre-laveur.

sprinkling apparatus ‖ Spritzapparat m. ‖ appareil m. injecteur. / ~ (Brew) ‖ Aufgußapparat m. ‖ croix f. écossaise.

sprinkling brush (Mas) ‖ Sprengpinsel m. ‖ goupillon m. / ~ can ‖ Gießkanne f. ‖ arrosoir m. / ~ machine (Clothm) ‖ Einsprengmaschine f. ‖ machine f. à arroser. / motor ~ machine ‖ Motorsprengwagen m. ‖ arroseuse f. automobile. / ~ pipe (Draff) ‖ Anschwänzer m.; Anschwänzkreuz n. ‖ croix f. écossaise.

spritsail ‖ Sprietsegel n. ‖ voile f. à baleston ou à livarde.

sprocket chain ‖ Laschenkette f.; Gelenkkette f. ‖ chaîne f. articulée. / ~ wheel ‖ Kettenrad n.; Kettennuß f. ‖ pignon m. ou roue f. à chaîne. / ~ wheel bearing ‖ Kettenradlager n. ‖ palier m. du pignon à chaîne.

sproud, to ‖ keimen ‖ germer.

sprouding ‖ Keimung f. ‖ germination f.

spruce ‖ Fichte f.; Rottanne f. ‖ épicéa m,; sapin rouge. / ~ wood ‖ Fichtenholz n.; Rottannenholz n. ‖ bois m. de sapin rouge.

sprue (Found) ‖ Eingußtrichter m. ‖ entonnoir m. de coulée.

sprung (Wood) ‖ gerissen; gespalten; rissig ‖ lézardé. / ~ carriage ‖ abgefederter Wagen m. ‖ voiture f. suspendu sur ressorts.

spun cotton ‖ baumwollenes Garn n. ‖ fil m. de coton. / ~ glass ‖ Glasfaden m.; Glasgespinst n. ‖ fil m. de verre. / ~ gold ‖ Goldgespinst n.; Goldfaden m. ‖ filet m. d'or; or m. filé. / ~ hair ‖ Krollhaar n. ‖ crin m. frisé. / ~ hemp ‖ Hanfgarn n. ‖ fil m. de chanvre. / ~ yarn ‖ Gespinst n. ‖ filé m.; filure f. / ~ yarn (Mar) ‖ Schiemannsgarn n. ‖ bitord m.

spunge, to ‖ see to sponge.

spunger (Cloth) ‖ see sponger.

spur ‖ Sporn m. ‖ éperon m. / ~ gear see also spur wheel ‖ Stirnrad n. ‖ engrenage m. droit.

spur gear cutting machine, automatic ~ ‖ selbsttätige Stirnradfräsmaschine f. ‖ machine f. automatique à tailler les engrenages droits. / automatic ~ with vertical work arbor ‖ selbsttätige Stirnradfräsmaschine f. mit vertikalem Aufspannbolzen ‖ machine f. automatique à tailler les engrenages droits avec mandrin porte-pièce vertical.

spur gear grinding machine, automatic ~ ‖ selbsttätige Stirnradschleifmaschine f. ‖ machine f. automatique à rectifier les engrenages droits.

spur gear hobbing machine ‖ Stirnräderabwälzfräsmaschine f. ‖ machine f. à tailler les engrenages cylindriques par fraise-mère.

spur gearing ‖ Zahnradgetriebe n. ‖ train m. d'engrenages.

spur gear planing machine, automatic ~ ‖ selbsttätige Stirnradhobelmaschine f. ‖ machine f. automatique à raboter les engrenages cylindriques à denture droite.

spur maker ‖ Sporenmacher m. ‖ fabricant m. d'éperons. / ~ pinion ‖ Stirngetriebe n. ‖ pignon m. droit. / ~ post (Road) ‖ Radstößer m. ‖ bouteroue f.; bouterue f. / ~ stone ‖ Prellstein m. ‖ borne f.; bouteroue f.

spurt cork ‖ Spritzkorken m. ‖ bouchonverseur m.

spur wheel see also spur gear ‖ Stirnrad n. ‖ roue f. dentée cylindrique; engrenage m. droite. / ~ differential ‖ Stirnraddifferential n. ‖ différentiel m. à engrenages droits.

sputter, to ~ the solution ‖ Masse f. spritzen ‖ filer la pâte.

sputum flask ‖ Spuckflasche f. ‖ crachoir m. de poche.

spy-glass ‖ Fernglas n. ‖ lunette f.

squabble, to (Print) ‖ die Lettern fpl. verrücken ‖ déranger les lettres fpl.

squadron (Mar) ‖ Geschwader n. ‖ escadre f.

squall, black ~ ‖ Gewitterbö f. ‖ grain m. noir. / heavy ~ ‖ schwere Bö f. ‖ rafale f. violente; grain m. violent. / sudden ~ ‖ Bö f. ‖ rafale f.; grain m. de vent.

squall cloud ‖ Böenwolke f. ‖ nuage m. de grain.

squamous ‖ schuppig ‖ écailleux.

squander, to ‖ verzetteln ‖ égarer; éparpiller.

square, to (Carp) ‖ abvieren; behauen ‖ équarrir; carrer. / ~ (Forg) ‖ viereckig schmieden ‖ équarrir. ~ (Trade) ‖ saldieren ‖ balancer; solder; arrêter un compte. / ~ an ashlar ‖ einen Stein m. winkeln ‖ équarrir une pierre. / ~ a number ‖ eine Zahl f. quadrieren ‖ former le carré d'un nombre; élever un nombre au carré. / ~ the timber ‖ das Rundholz vierkantig zuschneiden ‖ carrer le bois. / ~ wood ‖ Holz n. abvieren oder beschlagen ‖ carrer ou équarrir le bois.

square ‖ quadratisch ‖ carré. / ~ bar (Met) ‖ Vierkantbarren m. ‖ barreau m. carré. / ~ bar iron ‖ Quadrateisen n.; Vierkanteisen n. ‖ fer m. carré ou quarré ou en barres carrées. / ~ bit ‖ Kronenbohrer m. ‖ perçoir m. à couronne. / ~ block ‖ Vierkantblock m. ‖ bloc m. rectangulaire. / ~ brick ‖ Backsteinplatte f.; Ziegelplatte f. ‖ carreau m. de brique. / ~ centimeter ‖ Quadratzentimeter n. ‖ centimètre m. carré. / ~ coal scoop ‖ Vierkantkohlenschaufel f. ‖ pelle f. à feu carrée. / ~ decimeter ‖ Quadratdezimeter n. ‖ décimètre m. carré. / ~ I-washer for I-irons ‖ Vierkant-I-Scheibe f. für I-Träger ‖ éclisse f. carrée en I sur fer I. / ~ file ‖ Vierkantfeile f. ‖ lime-carrée f. / ~ foot ‖ Quadratfuß m. ‖ pied m. carré. / ~ foundry shovel ‖ Formerschaufel f. ‖ pelle f. de fonderie carrée. / ~ frame of a joiner's press ‖ viereckiger Rahmen m. einer Tischlerpresse ‖ châssis m. carré d'une presse de menuisier. / ~ furnace shovel ‖ flache Kohlenschaufel f. ‖ pelle f. à foyer carrée. / ~ groove ‖ Quadratfurche f. ‖ cannelure f. carrée. / ~ head ‖ Vierkant n. ‖ carré m. / ~ head bolt ‖ Vierkantschraube f.; Vierkantbolzen m. ‖ vis f. carrée; boulon m. à tête carrée. / ~ hole ‖ Vierkantloch n. ‖ trou m. carré. / ~ iron ‖ Quadrateisen n.; Vierkanteisen n. ‖ fer m. carré. / ~ iron shearing machine ‖ Vierkanteisenschere f. ‖ cisaille f. à fers carrés. / ~ joint ‖ rechtwinklige Fuge f. ‖ joint m. carré. / ~ kilometer ‖ Quadratkilometer n. ‖ myriare m. / ~ lath ‖ Dachlatte f. ‖ latte f. carrée. / ~-legged compasses pl. ‖ Zirkel m. mit Vierkantschenkeln ‖ compas m. droit à branches

carrées. / ~ measure ‖ Flächenmaß n. ‖ mesure f. de superficie *ou* carrée. / ~ meter ‖ Quadratmeter n. ‖ mètre m. carré. / ~ mile ‖ Quadratmeile f. ‖ mille m. carré; lieue f. carrée. / ~ number / ~ Quadratzahl f. ‖ nombre m. carré. / ~ nut ‖ Vierkantmutter f. ‖ écrou m. carré. / ~ perforation ‖ Quadratlochung f.; Vierkantlochung f. ‖ perforation f. carrée. / ~ root ‖ Quadratwurzel f. ‖ racine f. carrée. / ~ sail ‖ Rahesegel n. ‖ voile f. carrée. / ~ set-hammer ‖ Setzhammer m. ‖ chasse-carrée m. / ~ shackle padlock ‖ Vorhängeschloß n. mit Vierkantbügel ‖ cadenas m. avec anse carrée. / ~ socket wrench for spindles and screws ‖ Vierkantsteckschlüssel m. für Spindeln und Schrauben ‖ clef f. à canon carrée pour goupilles et vis. / ~ steel ‖ Vierkantstahl m. ‖ acier m. carré. / ~ stone (Build) ‖ Quader m.; Quaderstein m. ‖ pierre f. carrée; carreau m. / ~ stove tile ‖ quadratische Kachel f. ‖ carreau m. glacé carré. / ~ thread ‖ Flachgewinde n. ‖ filet m. plat *ou* carré. / ~ timber ‖ Kantholz n. ‖ bois m. équarri. / ~ upper beam ‖ Vierkantoberwange f. ‖ sommier m. supérieur rectangulaire. / ~ U-washer for U-irons ‖ Vierkant-U-Scheibe f. für U-Träger ‖ éclisse f. carrée en U pour fer U. / ~ washer for wood joints ‖ Vierkantscheibe f. für Holzverbindungen ‖ éclisse f. carrée pour joints en bois. / ~ yard ‖ Quadratyard n. ‖ yard m. carré.

square ‖ Quadrat n. ‖ carré m. / ~ (Drawing) ‖ Winkel m. ‖ équerre f. / ~ (Gold-b) ‖ Quartier n. ‖ quartier m. / ~ of glass ‖ Glasscheibe f. ‖ panneau m. *ou* plateau m. *ou* table f. de verre. / in the ~ ‖ im Quadrat n. ‖ au carré m.; d'équarrissage m. / method of least ~s ‖ Verfahren n. der kleinsten Quadrate ‖ méthode f. des plus petits carrés. / ~ of a number ‖ zweite Potenz f. *oder* Quadrat n. einer Zahl ‖ deuxième puissance *ou* carré m. d'un nombre. / optical ~ (Surv) ‖ Winkelspiegel m. ‖ goniomètre m. à réflecteur. / out of ~ (Angle) ‖ schiefwinklig ‖ à fausse équerre f. / ~ for regulating the sag (El line) ‖ Winkelhaken m. für Durchhangsprüfung ‖ équerre f. pour le réglage des fils. / tapered ~ ‖ kegeliges Vierkant n. ‖ carré m. conique.

squared beam with shots ‖ vollkantiger Balken m. ‖ poutre f. à vive arête. / ~ shaft ‖ viereckige Welle f. ‖ arbre m. carré. / ~ shank ‖ quadratische Stange f. ‖ tige f. forgée carrée.

square-forging ‖ Viereckigschmieden n. ‖ équarrissage m.; forgeage m. carré.

squareness (Carp) ‖ Ausvierung f. ‖ équerrissage m.

square-shaped magnetic needle ‖ rautenförmige Magnetnadel f. ‖ aiguille f. aimantée en forme de losange. / ~-sterned vessel ‖ Schiff mit flachem Heck *oder* Spiegel ‖ navire m. à poupe carrée.

squaring (Carp) ‖ Abvierung f. ‖ équarrissement m.

squaring axe ‖ Zurichtaxt f. ‖ hache f. à équarrir. / ~ machine (Carp) ‖ Abkantmaschine f. ‖ machine f. à équarrir. / machine for ~ ship frames ‖ Schiffsspantenschmiegemaschine f. ‖ machine f. à équerrer les couples de navires. / ~ roll ‖ Schmiegerolle f. ‖ rouleau m. à équerrer.

squatter ‖ Holzfäller m. ‖ bûcheron m.; bocquillon m.; bocanier m.

squaze, to ‖ einpressen ‖ enformer.

squeak ‖ Knirschen n. ‖ grincement m.

squeeze, to (Techn) ‖ quetschen; pressen ‖ broyer; presser. / ~ out ‖ ausquetschen; abpressen ‖ faire sortir par pression; pressurer; exprimer.

squeezer (Iron) ‖ Preßwerk n. ‖ cingleur m.; machine f. à cingler. / ~ (Mach) ‖ Auspreßmaschine f. ‖ machine f. à pressurer. / crocodile ~ (Met) ‖ Luppenmühle f. ‖ presse m. à cingler. / rotary ~ *see* crocodile ~.

squeezing apparatus, rope ~ (Weav) ‖ Strangausquetschapparat m. ‖ appareil m. d'exprimage en boyaux.

squeezing cock ‖ Quetschhahn m. ‖ pince f. à ressort. / ~ machine *see also* squeezer ‖ Quetschwalzwerk n. ‖ squeezer m.; machine f. à cingler; cingleur m. / ~ machine minder ‖ Quetschwalzenführer m. ‖ conducteur m. de squeezer. / ~-out ‖ Ausspressen n. ‖ pressurage m.; expression f. / ~ rollers pl. *see* ~ machine.

squill ‖ Meerzwiebel f. ‖ scille f.

squint, angle of ~ for one eye ‖ Schielablenkung f. für ein Auge ‖ strabisme m. monoculaire. / ~-eyed ‖ schielend ‖ strabique.

squirrel, artificial tails pl. of grey ~s ‖ künstliche Schwänze mpl. von grauen Eichhörnchen ‖ queues fpl. artificielles de petit-gris.

squirrel cage (Spinn) ‖ Staubtrommel f. ‖ roteur m. / ~ armature (Electr) ‖ Käfiganker m.; Kurzschlußanker m. ‖ induit m. à cage d'écureuil.

squirt, to ~ the solution ‖ Masse f. spritzen ‖ filer la pâte.

squirting machine (Cable) ‖ Spritzmaschine f. ‖ pompe f. foulante à profiler le caoutchouc.

St. Andrew's cross brace ‖ Diagonalstrebe f. ‖ bras m. de croix St. André.

St. John's bread ‖ Johannisbrot n.; Karobe f.; Karube f. ‖ caroube f.; carouge f.

stabile ‖ stabil; haltbar; dauerhaft ‖ stable.

stabilite ‖ Stabilit n. ‖ stabilite f.

stability ‖ Dauerhaftigkeit f.; Beständigkeit f.; Haltbarkeit f. ‖ stabilité f.; solidité f. / ~ (Aero) ‖ dynamisches Gleichgewicht n. ‖ équilibre m. dynamique. / ~ (Mach) ‖ Standfestigkeit f. ‖ stabilité f. / ~ (Mech) ‖ Stabilität f. ‖ stabilité f. / directional ~ (Aero) ‖ Seitenstabilität f. ‖ stabilité f. de direction. / ~ of a firm ‖ Solidität f. einer Firma ‖ solidité f. d'une maison. / inherent ~ ‖ Eigenstabilität f. ‖ stabilité f. propre *ou* de la forme. / lateral ~ ‖ Querstabilität f. ‖ stabilité f. latérale. / longitudinal ~ ‖ Längsstabilität f. ‖ stabilité f. longitudinale.

stability calculation ‖ Stabilitätsberechnung f. ‖ calcul m. de stabilité. / curve of ~ ‖ Stabilitätskurve f. ‖ courbe f. de la stabilité. / limit of ~ ‖ Stabilitätsgrenze f. ‖ limite f. de stabilité. / security of ~ ‖ Standsicherheit f. ‖ securité f. de stabilité.

stabilization ‖ Beständigmachen n.; Stabilizieren n. ‖ stabilisation f.

stabilizer (Aero) ‖ Höhenflosse f. ‖ plan m. fixe stabilisateur.

stabilizing flap ‖ Stabilisierungsfläche f. ‖ surface f. stabilisatrice. / fellows ~ ma-

chine ‖ Felgenstabilisiermaschine f. ‖ machine f. à équilibrer les jantes.

stable ‖ standfest ‖ stable. / ~ in the air (Chem) ‖ luftbeständig ‖ stable à air m. / ~ when boiling ‖ kochbeständig ‖ stable à l'ébullition. / ~ under heat ‖ hitzbeständig ‖ stable à chaud m. / ~ to light ‖ lichtbeständig ‖ stable à la lumière. / ~ at red heat (Chem) ‖ glühbeständig ‖ stable au rouge. / ~ equilibrium ‖ stabiles Gleichgewicht n. ‖ équilibre m. stable.

stable ‖ Stall m. ‖ écurie f.; étable m. / ~ for horses ‖ Pferdestall m. ‖ écurie f. pour chevaux.

stable boy ‖ Stallknecht m. ‖ palefrenier m. / ~ contrivance ‖ Stalleinrichtung f. ‖ installation f. d'écurie; équipment m. d'étable. / ~ fittings pl. see ~ contrivance. / ~ manure ‖ Stalldünger m. ‖ fumier m. d'étable. / ~ watchman ‖ Stallwächter m. ‖ gardien m. d'étables.

stack, to ‖ stapeln ‖ empiler; entasser. / ~ bricks ‖ die Ziegel mpl. aufschichten ‖ enhayer les briques fpl. / ~-up wood ‖ Holz n. aufstapeln ‖ entasser le bois.

stack (Forg) ‖ Schmiedeesse f. ‖ forge f.; chaufferie f.

stacked ‖ aufgestapelt ‖ mis en meule f. / ~ wood ‖ Schichtholz n. ‖ bois m. empilé.

stacker, wood ~ ‖ Holzstapler m. ‖ empileur m. de bois.

stacking-up the rails ‖ Aufsetzen n. der Schienen ‖ empilage m. des rails. / ~ the sleepers ‖ Aufstapeln n. der Schwellen ‖ empilage m. des traverses.

stadia line ‖ Entfernungsmeßfaden m. ‖ fil m. stadiométrique; stadia.

stadiometric straightedge ‖ Entfernungslineal n. ‖ règle f. des distances.

staff ‖ Personal n. ‖ personnel m. / ~ (Clockm) ‖ Unruhewelle f. ‖ axe m. du balancier. / ~ (Mus) ‖ Notenlinien fpl.; Notensystem n. ‖ portée f. / ~ of a cart ‖ Bergstütze f. ‖ Schleppstock m.; Hemmstütze f. ‖ servante f. d'un chariot. / ~ of railway board ‖ Zugpersonal n. ‖ hommes mpl. d'équipe; personnel m. du train. / round ~ ‖ Rundstab m. ‖ bâton m. rond. / trained ~ ‖ geschultes Personal n. ‖ personnel m. habile *ou* expérimenté.

staff demand ‖ Personalbedarf m. ‖ besoin m. en personel. / ~ man (Surv) ‖ Meßgehilfe m. ‖ aide m.; jalonneur m. / ~ member of the factory ‖ Fabrikbeamter m. ‖ employé m. d'usine. / ~ wood (Coop) ‖ Daubenholz n. ‖ merrain m.; douvain m.; bourdillon m.; longailles fpl.

stage, to (Build) ‖ rüsten; berüsten ‖ échafauder.

stage (Aut tel) ‖ Stufe f. ‖ étage m. / ~ (Metal) ‖ Gichtbühne f. ‖ plate-forme f. / ~ (Print) ‖ Farbläufer m.; Reibstein m. ‖ marbre m.; molette f.; broyon m. / ~ (Theater) ‖ Bühne f. ‖ scène m. / by ~s ‖ absatzweise ‖ par étapes fpl. / hanging ~ (Build) ‖ fliegendes Gerüst n.; Hängegerüst n. ‖ échafaud m. *ou* pont m. volant. / lowering ~ ‖ Senkbühne f. ‖ chargeur m. descendant. / ~ of manufacture ‖ Verarbeitungsstufe f. ‖ état m. de fabrication. / mechanical ~ ‖ Kreuztisch m. ‖ platine f. à chariot. / metal ~ ‖ Metalltisch m. ‖ platine f. en métal. / ~ of a microscope ‖ Tisch m. eines

Mikroskopes ‖ platine f. d'un microscope. / object ~ ‖ Objekttisch m. ‖ platine f. / revolving ~ ‖ drehbarer Tisch m. ‖ platine f. *ou* table f. tournante.

stage builder (Carp) ‖ Gerüstbauer m. ‖ échafaudeur m. / ~ cable ‖ Bühnenkabel n. ‖ câble m. de scène. / ~ coach ‖ Postkutsche f. ‖ diligence f. / ~ dimmer ‖ Bühnenlichtregler m. ‖ régulateur m. de l'éclairage de la scène. / ~ drying ‖ Stufentrocknung f. ‖ séchage m. par gradins. / ~ heating ‖ Stufenerwärmung f. ‖ chauffage m. par gradins. / ~ lighting apparatus ‖ Bühnenbeleuchtungsapparat m. ‖ appareil m. pour l'éclairage de la scène. / ~ lighting plant ‖ Bühnenbeleuchtungsanlage f. ‖ installation f. d'éclairage de la scène. / ~ light regulator ‖ Bühnenlichtregler m. ‖ régulateur m. de l'éclairage de la scène; jeu m. d'orgue. / ~ machine ‖ Bühnenmaschine f. ‖ machine f. pour scènes de théâtre. / ~ micromter ‖ Objektmikrometer n. ‖ micromètre-objectif m. / hydraulic plant for ~s ‖ hydraulische Bühnenanlage f. ‖ installation f. hydraulique pour scènes.

stages pl. **of appeal** ‖ Instanzenweg m. ‖ voie f. hiérarchique.

stag fat (Chem) ‖ Hirschtalg m. ‖ graisse f. de cerf.

stagger, to ‖ staffeln ‖ échelonner; graduer. / ~ (Rivet) ‖ versetzt anordnen ‖ placer alternativement *ou* en quinconce.

staggered (Aero) ‖ versetzt angeordnet ‖ installé alternativement *ou* en quinconce. / ~ driving pinions pl. ‖ gegeneinander versetzte Zahnscheiben fpl. ‖ disques mpl. à dents placés alternativement. / ~ row of rivets ‖ versetzte *oder* versetzt angeordnete Nietreihe f. ‖ rang m. de rivets alternés. / ~ spot welding ‖ Zickzackpunktschweißung f. ‖ soudage m. par points en quinconce.

stag horn ‖ Hirschhorn n. ‖ corne f. de cerf.

stagnant ‖ leblos ‖ inanimé. / ~ (Market) ‖ lustlos; geschäftslos; matt ‖ calme; stagnant; lourd.

stagnation ‖ Stockung f. ‖ stagnation f. / ~ of business ‖ Geschäftsstille f.; Geschäftsstockung f. ‖ morte-saison f.; calme m.

stagnation point (Mech) ‖ Staupunkt m. ‖ point m. mort.

stain, to (Glassm) ‖ bemalen; malen ‖ peindre. / ~ (Wood) ‖ färben; beizen ‖ teindre.

stain (Mordant) ‖ Beize f. ‖ caustique m. / ~ (Spot) ‖ Fleck m. ‖ tache f.

stained ‖ fleckig; gefleckt; gesprenkelt ‖ moucheté. / ~ glass ‖ Kathedralglas n. ‖ verre m. cathédrale. / ~ paper ‖ Marmorpapier n. ‖ papier m. peigne *ou* marbré.

staining (Brew) ‖ Färben n. ‖ coloration f. / ~ of glass ‖ Glasmalerei f. ‖ peinture f. sur verre. / ~ machine for paper and cardboard ‖ Färbmaschine f. für Papier und Karton ‖ machine f. à teindre le papier et le carton. / ~ method (Brew) ‖ Färbungsmethode f. ‖ méthode f. de coloration.

stainless ‖ rostfrei ‖ non-corrosive. / ~ steel ‖ rostsicherer *oder* nichtrostender Stahl m. ‖ acier m. inoxydable.

stair see stairs pl.

stair carpet ‖ Treppenläufer m. ‖ tapis m. d'escalier; tapis m. chemin. / ~ rod ‖ Treppenläuferstange f. ‖ barre f. de fixation pour les tapis d'escaliers.

staircase *see also* stairs pl. ‖ Treppe f. ‖ escalier m. / cast iron ~ ‖ gußeiserne Treppe f. ‖ escalier m. en fonte. / iron ~ ‖ eiserne Treppe f. ‖ escalier m. en fer. / wrought iron ~ ‖ Treppe f. aus Schmiedeeisen ‖ escalier m. en fer forgé.

staircase fittings pl. ‖ Treppenbeschläge mpl. ‖ garnitures fpl. pour escaliers. / ~ lamp ‖ Treppenlampe f. ‖ lampe f. d'escalier. / ~ lighting ‖ Treppenbeleuchtung f. ‖ éclairage m. d'escalier. / ~ well ‖ Treppenhaus n. ‖ cage f. d'escalier. / ~ window ‖ Treppenfenster n. ‖ fenêtre f. d'escalier.

stair cover rod see stair carpet rod. / ~ horse ‖ Treppenseitenstück n. ‖ limon m. de l'escalier.

stair railing ‖ Treppengeländer n. ‖ rampe f. *ou* balustrade f. d'escalier. / wooden ~ ‖ Holzgeländer n. einer Treppe ‖ rampe f. d'escalier en bois.

stairs pl. *see also* staircase ‖ Treppe f. ‖ escalier m. / back ~ ‖ Hintertreppe f. ‖ escalier m. dérobé *ou* dégagé. / cellar ~ ‖ Kellertreppe f. ‖ escalier m. de cave. / centered ~ ‖ halbringförmige Treppe f. ‖ escalier m. cintré. / ~ with fixed railing ‖ Treppe f. mit festem Geländer ‖ escalier m. à rampe fixe. / flat ~ ‖ flache Treppe f. ‖ escalier m. à girons rampants. / inner ~ ‖ eingebaute Treppe f. ‖ escalier m. dans l'œuvre. / ~ with landing places ‖ gebrochene Treppe f. ‖ escalier m. à repos. / main ~ ‖ Haupttreppe f. ‖ escalier m. principal. / overhanging ~ ‖ freitragende Treppe f. ‖ escalier m. suspendu. / ~ with removable railing ‖ Treppe f. mit abnehmbarem Geländer ‖ escalier m. à rampe détachable. / ~ with rises of great or little height ‖ Treppe f. mit hoher oder niedriger Steigung ‖ escalier m. à montée de grande ou de faible hauteur. / saddled ~ ‖ aufgesattelte Treppe f. ‖ escalier m. à cheval. / single-flight ~ ‖ einläufige Treppe f. ‖ escalier m. à une volée. / stone ~ ‖ Steintreppe f. ‖ escalier m. en pierre. / ~ with two parallel flights ‖ Treppe f. mit zwei parallelen Läufen ‖ escalier m. à deux rampes parallèles. / wooden ~ ‖ Holztreppe f. ‖ escalier m. en bois.

stair step ‖ Treppentritt m. ‖ marche f. de l'escalier.

stake, to ~ out ‖ abpfählen ‖ piqueter. / ~ a railway-line ‖ eine Bahnlinie f. abpflöcken ‖ jalonner le tracé d'un chemin de fer.

stake (Railw) ‖ Runge f. ‖ rancher m. / ~ (Surv) ‖ Pflock m. ‖ piquet m. / ~ folding down to horizontal position (Railw) ‖ horizontal umlegbare Runge f. ‖ rancher m. à rabattre horizontalement. / ~ marked with a number ‖ Nummerpfahl m. ‖ pieu m. numéroté.

stake socket (Railw) ‖ Rungenhalter m. ‖ bride f. de rancher. / ~ to make a box wagon (Raiw) ‖ Rungenhalter m. zur Herstellung eines Kastenwagens ‖ mortaise f. de rancher pour construire un wagon à caisse.

staking-out a telegraph line ‖ Abpfählung f. einer Telegraphenlinie ‖ jalonnage m. d'une ligne télégraphique.

stalactite ‖ Tropfstein m.; Stalaktit m. ‖ stalactite f. / ~ quarry ‖ Tropfsteinbruch m. ‖ carrière f. de stalactites.

stalagmite ‖ (stehender) Tropfstein m.; Stalagmit m. ‖ stalagmite f.

stale cheque ‖ verjährter Scheck m. ‖ chèque m. prescrit.

stalk ‖ Stengel m.; Halm m.; Stiel m. ‖ tige f. / flexible ~ of plant ‖ biegsamer Pflanzenstengel m. ‖ tige f. végétale flexible. / herbaceous ~ ‖ krautartiger Stengel m. ‖ tige f. herbacée. / vegetable ~ ‖ Pflanzenstengel m. ‖ tige f. végétale.

stalk fibre ‖ Stielfaser f. ‖ fil m. raide; fibre f. de la tige.

stall ‖ Stand m.; Verkaufsstand m. ‖ étal m.; stand m. / butcher's ~ ‖ Fleischerstand m. ‖ étal m. de boucher. / ~ at a fair ‖ Messestand m. ‖ stand m. d'exposition.

stalling speed (Aero) ‖ Mindestauftriebsgeschwindigkeit f. ‖ vitesse f. minimum de sustension.

stallion owner ‖ Hengsthalter m. ‖ étalonnier m.

stamp, to ‖ prägen ‖ estamper; étamper; matricer; frapper. / ~ (Documents) ‖ stempeln; abstempeln ‖ marquer; timbrer; estampiller. / ~ (Post) ‖ stempeln ‖ estampiller; oblitérer. / ~ (Print) ‖ prägen ‖ empreindre; gaufrer. / ~ (Punch) ‖ stanzen ‖ estamper; étamper; faire l'estampage. / each blade is stamped with one dovetailed end (Turbine) ‖ jede Schaufel f. ist mit einem schwalbenschwanzförmigen Ende ausgestanzt ‖ le pied de chaque aube f. est matricée en queue d'aronde. / ~ the earth ‖ die Erde feststampfen ‖ battre *ou* damer la terre. / ~ money ‖ münzen ‖ frapper des monnaies; monnayer. / ~ the ore ‖ das Erz pochen *oder* stampfen ‖ bocarder un minerai. / ~ out ‖ ausstanzen ‖ découper à l'emporte-pièce. / parts pl. stamped out in the cold state ‖ kalt gestanzte Teile mpl. ‖ pièces fpl. estampées à froid. / ~ railway tickets pl. ‖ Fahrkarten fpl. abstempeln ‖ timbrer des billets.

stamp (Coin) ‖ Fallwerk n. ‖ mouton m. / ~ (Engr) ‖ radierte Platte f. ‖ estampe f. gravée à l'eau-forte. / ~ (Punch) ‖ Stanze f.; Stempel m. ‖ matrice f.; estampe f.; poinçon m. / ~ (Punching machine) ‖ Stanze f. ‖ poinçonneuse f. / ~ (Office) ‖ Stempel m. ‖ timbre m.; cachet m. / ~ (Ore dress) ‖ Pochstempel m. ‖ pilon m. / ~ (Post) ‖ Briefmarke f. ‖ timbre-poste m. / ~ (Print) ‖ Stempel m. ‖ coin m. / burning ~ for wood printing ‖ Brennstempel m. für Holzdruck ‖ fer m. à marquer le bois au feu. / ~ of caoutchouc see rubber ~. / ~ for making crucibles ‖ Kern m. zur Tiegelfertigung ‖ noyau m. pour la fabrication des creusets. / discount ~ ‖ Rabattmarke f. ‖ marque f. de rabais. / glue ~ ‖ Leimstempel m. ‖ tampon m. à la colle. / ~ for parcel closing ‖ Paketverschlußmarke f. ‖ marque f. à fermer les paquets. / postage ~ see stamp (Post). / rubber ~ ‖ Kautschukstempel m. ‖ timbre m. *ou* cachet m. *ou* marquoir m. en caoutchouc. / self-inking ~ ‖ selbstfärbender Stempel m. ‖ timbre m. à encrage automatique.

stamp album ‖ Briefmarkenalbum n. ‖ album m. pour timbres-poste. / ~ apparatus ‖ Stempelapparat m. ‖ appareil m. à estampiller. / ~ battery (Ore dress) ‖ Pochwerk n. ‖ bocard m. / ~ bearer ‖ Stempelhalter m. ‖ porte-tampon m. / ~ cutter (Coin) ‖ Stempelschneider m. ‖

graveur m. de timbre. / ~ die (Ore dress) ‖ Pochsohle f. ‖ dé m. de bocard.

stamp duty ‖ Stempelgebühr f. ‖ droit m. de timbre. / free from ~ ‖ stempelfrei ‖ exempt de timbre. / subject to ~ ‖ stempelpflichtig ‖ soumis au timbre.

stamped sheet of paper ‖ Stempelbogen m. ‖ feuille f. de papier timbré.

stamp engraver ‖ Petschaftstecher m.; Stempelgravör m. ‖ graveur m. de cachets.

stamper ‖ Präger m. ‖ estampeur m. / ~ (Brew) ‖ Stempler m. ‖ estampilleur m. / ~ (Hatt) ‖ Treibeisen n. ‖ avaloire f. du chapelier. / ~ (Pap) ‖ Stampfe f. ‖ maillet m.; pilon m. / ~ press (Oil) ‖ Ölpresse f.; Öllade f. ‖ harnard m.

stamp factory ‖ Stempelfabrik f. ‖ fabrique f. de cachets. / ~ hammer ‖ Fallhammer m. ‖ marteau m. à mouton; pilon m. / ~ hole (Pap) ‖ Stampfloch n. ‖ pile f. aux chiffons.

stamping ‖ Ausstanzen n. ‖ estampage m.; découpage m. / ~ (Piece) ‖ Gesenkschmiedestück n. ‖ pièce f. estampée. / ~ (Print) ‖ Prägung f. ‖ empreinte f. / ~ of the hands (Watch) ‖ Ausstanzen n. der Zeiger ‖ découpage m. d'aiguilles. / ~ in lead ‖ Bleiprägung f. ‖ empreinte f. au plomb. / metal (cold) ~ ‖ Metallprägerei f. ‖ estampage m. (à froid) de métaux. / ~ of an ore ‖ Stampfen n. eines Erzes ‖ bocardage m. d'un minerai.

stamping appliances pl. ‖ Prägegerät n. ‖ outils mpl. à estamper. / ~ box ‖ Stampfkasten m. ‖ caisse f. à piloner. / ~ cardboard ‖ Stanzpappe f. ‖ carton m. à poinçonner. / ~ ink ‖ Stempelfarbe f. ‖ encre f. à estampiller. / ~ ink cushion ‖ Stempelkissen n. ‖ encreur m. pour timbres; tampon m. / ~ knife ‖ Stanzmesser n. ‖ poinçon m.; couteau m. d'estampage; lame f. de découpage.

stamping machine (Office) ‖ Stempelmaschine f. ‖ machine f. à timbrer ou à estampiller. / ~ (Post) ‖ Frankiermaschine f. ‖ machine f. à affranchir. / ~ (Punch) ‖ Stanzmaschine f. ‖ machine f. à estamper; estampeuse f. / ~ for converter bottoms ‖ Konverterbodenstampfmaschine f. ‖ machine f. à damer les fonds de convertisseurs. / ~ for paper and pasteboard ‖ Prägemaschine f. für Papier und Pappe ‖ machine f. à empreindre le papier et le carton. / ~ plant ‖ Stampfanlage f. ‖ installation f. de pilonnage.

stamping matrice ‖ Gesenk n. ‖ matrice f. à estamper. / ~ mill (Ore dress) ‖ Pochwerk n. ‖ bocard m. / ~ mould ‖ Prägeform f.; Matrize f. ‖ forme f. à estamper; matrice f. / ~ pad ‖ Stempelkissen n. ‖ tampon m. à timbres; coussin m. à encrer.

stamping press ‖ Stanzpresse f. ‖ presse f. à étamper; étampeuse f. / ~ (Print) ‖ Prägepresse f. ‖ presse f. à matricer. / hydraulic ~ ‖ hydraulische Gesenkpresse f. oder Schmiedepresse f. ‖ presse f. hydraulique à estamper. / ~ for metal ‖ Metallprägepresse f. ‖ presse f. à estamper le métal.

stamping sheet ‖ Stanzblech n. ‖ tôle f. à estamper. / ~ tool ‖ Prägewerkzeug n. ‖ outil m. à estamper. / ~ trough (Ore dress) ‖ Pochladen m.; Pochtrog m. ‖ huche f.

stamping works pl., sheet metal ~ ‖ Blechpresserei f. ‖ atelier m. d'emboutissage de tôles.

stampman (Ore dress) ‖ Erzpocher m. ‖ bocardeur m. or broyeur m. or pileur m. de minerai.

stamp mill see stamping mill. / ~ printing ‖ Stempeldruck m. ‖ timbrage m. / ~ shoe (Ore dress) ‖ Pochschuh m. ‖ sabot m. de bocard. / ~ stand ‖ Stempelständer m. ‖ porte-cachet m.; porte-sceau m. / ~ tax ‖ Stempelsteuer f. ‖ impôt m. du timbre.

stanchion (Shipb) ‖ Relingsstütze f. ‖ allonge f. du pavois; batayole f.; chandelier m. ou montant m. de bastingage.

stanchion (Railw) ‖ Runge f. ‖ rancher m. / end ~ ‖ Eckrunge f. ‖ rancher m. cornier ou de bout. / loose ~ ‖ lose Runge f. ‖ rancher m. démontable. / disengaging bracket under the ~s ‖ Auslöseknagge f. unter den Rungen ‖ support m. de déclenchement sous les ranchers.

stanchion strap ‖ Rungenhalter m. ‖ bride f. de rancher.

stand, to ~ to sea (Nav) ‖ seewärts anliegen ‖ porter au large. / ~ security ‖ Sicherheit f. leisten ‖ avalier; avaliser. / ~ upon the course (Nav) ‖ Kurs m. halten ‖ aller ou porter à route f.

stand ‖ Gestell n. ‖ chantier m.; étagère f.; support m. / ~ (Tripod) ‖ Stativ n. ‖ support m.; trépied m. / ~ (Of an exhibition) ‖ (Ausstellungs-)Stand m. ‖ stand m. / ~ (Forest) ‖ Bestand m. ‖ peuplement m. / bottle ~ ‖ Flaschenständer m. ‖ porte bouteilles m. / ~ with casters (Office) ‖ Gestell n. auf Laufrollen ‖ support m. sur roulettes. / ~ for centering cylindrical pieces on the machine tool ‖ Stativ n. zum Ausrichten von zylindrischen Körpern auf der Werkzeugmaschine ‖ support m. à centrer des pièces cylindriques sur la machine-outil. / ~ for checking the true running of cylindrical pieces ‖ Stativ n. zum Prüfen des Rundlaufens zylindrischer Körper ‖ support m. à vérifier l'exactitude de la cylindricité des pièces cylindriques. / ~ for flowers ‖ Blumenständer m.; Blumenkrippe f. ‖ jardinière f. pour fleurs. / horse shoe ~ ‖ Hufeisenfuß m. ‖ pied m. en fer à cheval. / pyramid ~ ‖ Pyramidenstativ n. ‖ pied-pyramide f. / ~ for a rolling mill ‖ Walzenständer m. ‖ montant m. de laminoir. / ~ for smoothing iron ‖ Untersatz m. für Bügeleisen ‖ porte-fer m. pour fers à repasser. / testing ~ ‖ Prüfstand m. ‖ banc m. d'essai. / travelling ~ ‖ Reisestativ n. ‖ statif m. de voyage. / wooden ~ ‖ Holzuntersatz m. ‖ socle m. ou support m. en bois.

standard ‖ Norm f. ‖ norme f. / ~ (Flag) ‖ Fahne f. ‖ drapeau m. / ~ (Stand) ‖ Gestell n. ‖ piedestal m.; montant m. / ~ (Fineness of gold etc.) ‖ Feingehalt m. ‖ titre m. / ~ (Measure) ‖ Normalmaß n.; Eichmaß n. ‖ étalon m. / above the ~ ‖ über der Norm f. ‖ audessus du type. / below the ~ ‖ unter der Norm f. ‖ au-dessous du type. / below the ~ (Coin) ‖ zu geringhaltig ‖ échars. / ~ of a couple of millstones ‖ Gestell n. eines Mahlganges ‖ bâti m. d'une paire de meules. / ~ of cutters ‖ Scherenständer m. ‖ montant m. de cisaille. / ~ for a forging press ‖ Schmie-

depressenständer m. ‖ bâti m. d'une presse à forger. / single ~ of a hammer ‖ einseitiger Hammerständer m. ‖ montant m. simple d'un marteau-pilon. / ~ of measurement see stand (Measure). / ~ of mills ‖ Mühlenständer m. ‖ montant m. de moulin. / ~ of silver ‖ Feingehalt m. des Silbers ‖ titre m. de l'argent. / ~ of a solution ‖ Gehalt m. einer Lösung ‖ titre m. d'une solution.

standard atmosphere ‖ Normalatmosphäre f. ‖ atmosphère f. normale ou standard. / ~ barometer ‖ Vergleichsbarometer n. ‖ baromètre m. d'essai ou de contrôle. / ~ boiler ‖ Normalkessel m. ‖ chaudièretype. / ~ brick size ‖ Normalziegelformat n. ‖ dimensions fpl. normales des briques. / ~ candle power ‖ Normalkerze f. ‖ bougie f. normale. / ~ cell (Electr) ‖ Standardelement n. ‖ élément m. normal. / ~ construction ‖ Normalausführung f. ‖ construction f. normale. / ~ copper ‖ Normalkupfer n. ‖ cuivre m. étalon. / ~ cube ‖ Normalwürfel m. ‖ cube m. normal. / ~ current ‖ Normalstrom m. ‖ courant m. normal. / ~ cylinder ‖ Kaliberzylinder m. ‖ cylindre m. vérificateur ou à calibrer. / ~ design ‖ gewöhnliche Ausführung f.; Normalausführung f. ‖ construction f. normale. / ~ drieer ‖ Normaltrockner m. ‖ sécheur m. normal. / ~ electrode ‖ Normalelektrode f. ‖ électrode f. normale. / ~ element see ~ cell. / ~ file ‖ Normalfeile f. ‖ lime f. normale.

standard gauge ‖ Normallehre f. ‖ calibre m. étalon. / ~ (Railw) ‖ Normalspur f. ‖ voie f. normale; écartement m. normal.

standard gauged length of test bar ‖ Normalmeßlänge f. des Probestabes ‖ longueur f. normale de l'éprouvette.

standard gauge locomotive ‖ Normalspurlokomotive f. ‖ locomotive f. à voie normale. / ~ railway ‖ Normalspurbahn f. ‖ chemin m. de fer à voie normale. / ~ sack ‖ Normalsack m. ‖ sac m. normal. / ~ siding ‖ Vollspurgleisanschluß m. ‖ embranchement m. de voie normale de chemin de fer.

standard gold ‖ Münzgold n. ‖ or m. de vaisselle ou de monnaie. / ~ heating element ‖ Normalheizelement n. ‖ élément m. de chauffage normal.

standardization ‖ Vereinheitlichung f.; Normung f. ‖ standardisation f.; normalisation f. / ~ (Gauging) ‖ Eichung f. ‖ jaugeage m.; étalonnage m. / ~ committee ‖ Normenausschuß m. ‖ comité m. de normalisation.

standardize, to ‖ vereinheitlichen; normen ‖ standariser; normaliser. / ~ (To gauge) ‖ eichen ‖ étalonner; jauger.

standardizing sheet ‖ Normenblatt n. ‖ tableau m. de normalisation.

standard length ‖ Normallänge f. ‖ longueur f. type ou normale. / ~ letter ‖ Normalbuchstabe m. ‖ caractère m. ou lettre f. étalon. / ~ measure ‖ Normalmaßstab m. ‖ mesure f. normale. / ~ officer ‖ Eichinspektor m. ‖ inspecteur m. des poids et mesures. / ~ paper ‖ Normalpapier n. ‖ papier m. normal. / ~-pitch cutter bar ‖ Schneidbalken m. für Normalschnitt ‖ barre f. coupeuse pour coupe normale. / ~ price ‖ Normalpreis m. ‖ prix m. normal. / ~ railway ‖ Normalbahn f. ‖ chemin m. de fer normal. / ~ reel (Silk) ‖ Probehaspel f. ‖ éprouvette f.

/ ~ resistance ‖ Vergleichswiderstand m. ‖ résistance f. étalon. / ~ rule ‖ Normallineal n. ‖ règle f. normale. / ~ screw ‖ Normalschraube f. ‖ vis f. de filet normal. / ~ section of inner span (Railw) ‖ Normalprofil n. des lichten Raumes ‖ gabarit m. / ~ section of tyre ‖ normales Radreifenprofil n. ‖ profil m. normal du bandage. / ~ silver ‖ lötiges Silber n. ‖ argent m. au titre. / ~ size ‖ Normalgröße f. ‖ grandeur f. type. / ~ size of paper ‖ Papier n. im Normalformat ‖ format m. courant de papier. / ~ solution ‖ Normallösung f. ‖ solution f. normale. / ~ spanner ‖ Normschlüssel m. ‖ clef f. normale. / ~ specifactions pl. (Mach) ‖ Normen fpl. ‖ règles fpl. / ~ test bar ‖ Normalprobestab m. ‖ éprouvette f. normale. / ~ thread ‖ Normalgewinde n. ‖ filet m. normal. / ~ tin ‖ lötiges Zinn n. ‖ étain m. au titre. / ~ truss (Carp) ‖ erstes Gebinde n.; Lehrgebinde n. ‖ ferme f. d'échantillon. / ~ type ‖ Normalbauart f.; Einheitsbauart f. ‖ type m. standardisé ou normal. / ~ tyre section ‖ normales Radreifenprofil n. ‖ profil m. normal du bandage. / ~ value of focal length of object glasses ‖ Normalbrennweite f. der Objektive ‖ focale f. normale d'objectifs. / ~ weight ‖ Normalgewicht n. ‖ poids m. type ou normal. / ~ wheel set ‖ Radsatz m. normaler Bauart ‖ essieu m. monté normal.

stand-by, instantaneous ~ ‖ Momentanreserve f. ‖ réserve f. momentanée.

standing ‖ stehend ‖ debout. / every thing ~ (Mar) ‖ unter vollen Segeln fpl. ‖ à pleines voiles fpl.; toutes voiles fpl. dehors. / sold ~ (Crop) ‖ auf dem Halm m. verkauft ‖ vendu sur pied m.

standing balance of millstone ‖ ruhendes Gleichgewicht n. des Mühlsteins ‖ équilibre m. de la meule en repos. / ~ clock ‖ Standuhr f. ‖ horloge f. de cheminée; pendule f. (de table). / ~ crop ‖ Getreide n. auf dem Halm ‖ grains mpl. sur pied. / ~ matter (Print) ‖ stehender Satz m.; Stehsatz m. ‖ composition f. conservée ou permanente. / ~ pendulum clock ‖ Pendelstanduhr f. ‖ pendule f. (statuaire). / ~ room ‖ Stehplatz m. ‖ place f. debout.

stand ophthalmoscope ‖ Standophthalmoskop n. ‖ ophtalmoscope m. sur pied.

stand pipe ‖ Standrohr n. ‖ tuyau m. de montée.

standpoint of national economy ‖ volkswirtschaftlicher Gesichtspunkt m. ‖ point m. de vue d'économie national.

standstill ‖ Stillstand m. ‖ arrêt m.; cessation f. des mouvements. / the affair is at a ~ ‖ die Angelegenheit stockt ‖ l'affaire f. languit. / ~ of business ‖ Stillstand m. der Geschäfte ‖ stagnation f. des affaires.

stannate ‖ zinnsaures Salz n. ‖ stannate m. / ~ of soda ‖ zinnsaures Natron n. ‖ stannate m. de soude.

stannic chloride ‖ Chlorzinn n. ‖ chlorure m. d'étain. / ~ oxide ‖ Zinnoxyd n.; Zinnasche f. ‖ bioxyde m. d'étain.

stannite ‖ Stannit m. ‖ stannite f.

stannous chloride ‖ Zinnchlorür n. ‖ protochlorure m. d'étain; chlorure m. stanneux. / ~ hydroxide ‖ Zinnoxydulhydrat n. ‖ hydrate m. stannique.

staple (Bookb) ‖ Öse f.; Drahtöse f.; Kramme f. ‖ cavalier m. / ~ (El line) ‖ Krampe f.; Kramme f. ‖ cavalier m.; crochet m. / ~ (Market) ‖ Markt m. ‖ marché m. / ~ (Locksm) ‖ Schließhaken m. ‖ mentonnet m.; fermoir m.; nappe f. / ~ for a bolt ‖ Haken m. eines Querriegels ‖ crampon m. d'une targette. **staple** machine (Office) ‖ Heftmaschine f. ‖ brocheuse f. mécanique. / ~ plate (Carp) ‖ Riegelblech n.; Hakenblatt n. ‖ auberonnière f. / ~ products pl. ‖ Gebrauchsgüter npl. ‖ objets mpl. d'utilité. / ~ trade ‖ Stapelhandel m.; Haupthandel m. ‖ commerce m. d'étape ou de marchandises en entrepôt.

stapling machine (Bookb) ‖ Drahtheftmaschine f. ‖ brocheuse f. à fil métallique.

star ‖ Stern m. ‖ étoile f. / ~ of the xth magnitude ‖ Stern m. xter Größenklasse ‖ étoile f. de la xième grandeur. / ~ of spokes (of a wheel) ‖ Radstern m. ‖ étoile f. de la roue ou du moyeu.

star aerial ‖ sternförmiger Luftleiter m.; sternförmige Antenne f. ‖ antenne f. en étoile.

star anise ‖ Sternanis m. ‖ badiane m. / ~ oil ‖ Sternanisöl n. ‖ essence f. de badiane.

starboard ‖ Steuerbord n. ‖ tribord m. / ~ engine ‖ Steuerbordmotor m. ‖ moteur m. de tribord.

starch, to ‖ stärken ‖ amidonner.

starch ‖ Stärke f. ‖ amidon m. / maize ~ ‖ Maisstärke f. ‖ amidon m. de maïs. / potato ~ ‖ Kartoffelstärke f. ‖ amidon m. ou fécule f. de pommes de terre. / rice ~ ‖ Reisstärke f. ‖ amidon m. de riz. / roasted ~ ‖ geröstete Stärke f. ‖ amidon m. torréfié. / tapioca ~ ‖ Tapiokastärke f. ‖ amidon m. Tapioca. / wheat ~ ‖ Weizenstärke f. ‖ amidon m. de froment.

starch blue ‖ Neublau n.; Waschblau n. ‖ bleu m. de toilette. / contents pl. of ~ ‖ Stärkegehalt m. ‖ teneur f. en amidon.

starcher (Cloth) ‖ Schlichter m. ‖ colleur m.

starch factory ‖ Stärkefabrik f. ‖ amidonnerie f.

starch flour ‖ Stärkemehl n.; Kartoffelmehl n. ‖ fécule f. / roasted ~ ‖ geröstetes Stärkemehl n. ‖ fécule f. torréfiée.

starch gum ‖ Stärkegummi n.; Dextrin n. ‖ gomme f. artificielle; dextrine f.

starching (Of cloth) ‖ Schlichten n. ‖ collage m.; encollage m.; gommage m. / ~ clay (Weav) ‖ Stärkeglanz m. ‖ lustre m. d'amidon. / ~ machine (Cloth) ‖ Schlichtmaschine f. ‖ machine f. à apprêter. / ~ machine operator (Cloth) ‖ Maschinenschlichter m. ‖ colleur m. à la machine.

starch manufacturing plant see starch factory. / ~ paste ‖ Stärkekleister m. ‖ colle f. d'amidon. / ~ sirup ‖ Stärkesirup m. ‖ sirop m. de fécule. / ~ sugar ‖ Stärkezucker m. ‖ sucre m. de la fécule. / ~ works pl. see starch factory.

star connection (Electr) ‖ Sternschaltung f. ‖ connexion f. en étoile.

star cluster ‖ Sternhaufen m. ‖ amas m. d'étoiles.

star-delta switch ‖ Sterndreieckschalter m. ‖ commutateur m. ou interrupteur m. étoile-triangle.

star indicator (Tel) ‖ Sternschauzeichen n. ‖ voyant m. en étoile.

starling (Bridge) ‖ Pfeilerhaupt n.; Pfeilerkopf m. ‖ bec m. de pile. / back ~ ‖

Pfeilerhinterhaupt n.; Pfeilersterz m.; Talpfeilerkopf m. ‖ arrière-bec m.; bec m. d'aval. / fore ~ ‖ Kronpfeilerkopf m.; Pfeilervorhaupt n.; Pfeilervorspitze f. ‖ avant-bec m.; bec m. d'amont.

star map ‖ Sternkarte f. ‖ carte f. des étoiles.

star-mesh switch see star-delta switch.

starry heaven ‖ Sternhimmel m. ‖ ciel m. étoilé.

star scale, comparison ‖ Vergleichssternskale f. ‖ échelle f. de comparaison des astres.

star shake (Of wood) ‖ Strahlenriß m. ‖ fente f. étoilée.

star-shaped roller bars pl. ‖ Rollenstern m. ‖ étoile f. de galets. / ~ wheel ‖ Sternrad n. ‖ roue f. d'engrenage centrale.

start, to ~ (Mot) ‖ anlassen ‖ démarrer; mettre en marche f. / ~ (Railw) ‖ abfahren ‖ partir. / ~ a factory ‖ ein Werk n. eröffnen ‖ mettre en activité une usine. / ~ without jerk ‖ stoßfrei anlaufen ‖ démarrer sans àcoups. / ~ under load ‖ belastet anlaufen ‖ démarrer en charge m. / ~ without load ‖ leer anlaufen ‖ démarrer sans charge m. / ~ a train ‖ einen Zug m. abfertigen oder ablassen ‖ expédier ou lancer un train. / the engine does not start ‖ die Maschine springt nicht an ‖ le moteur ne s'amorce pas.

start (Aero) ‖ Abflug m. ‖ envol m. / ~ (Auto; Railw) ‖ Abfahrt f. ‖ départ m. / ~ (Mach) ‖ Anlaufen n.; Anlassen n. ‖ démarrage m.; mise f. en marche.

starter ‖ Anlasser m.; Starter m. ‖ démarreur m.; starter m. / automatic ~ ‖ Selbstanlasser m. ‖ démarreur m. automatique. / compressed air ~ ‖ Preßluftanlasser m. ‖ démarreur m. à air comprimé. / drum ~ ‖ Schaltwalzenanlasser m. ‖ démarreur m. tambour tournant. / emergency ~ ‖ Aushilfsanlasser m. ‖ démarreur m. auxiliaire. / foot ~ (Auto) ‖ Fußanlasser m. ‖ pédale f. de mise en marche. / kick-~ ‖ Kickstarter m. ‖ démarreur m. à pied. / primary ~ (Electr) ‖ Gehäuseanlaßwiderstand m. ‖ démarreur m. primaire. / regulating ~ ‖ Regelanlasser m. ‖ démarreur m. de réglage. / self ~ ‖ Selbstanlasser m. ‖ démarreur m. automatique. / ~ combined with shunt regulator ‖ Anlasser m. mit Nebenschluß ‖ démarreur m. avec rhéostat de champ. / simple ~ ‖ einfacher Anlasser m. ‖ démarreur m. simple. / single-phase ~ ‖ Einphasenanlasser m. ‖ démarreur m. monophasé. / ~ for extra slow starting ‖ Anlasser m. für sehr langsames Anlassen ‖ démarreur m. pour démarrage très lent. / ~ with switching in jerks ‖ Anlasser m. mit ruckweiser Schaltung ‖ démarreur m. à degrés. / three-phase ~ ‖ Dreiphasenanlasser m. ‖ démarreur m. triphasé.

starter cable ‖ Anlaßkabel n. ‖ câble m. de démarrage. / ~ check valve ‖ Anlaßrückschlagventil n. ‖ soupape f. de retenue de mise en marche. / ~ clutch ‖ Andrehklaue f. ‖ griffe f. de mise en marche. / ~ gear ‖ Anlaßsteuerung f. ‖ distribution f. de la mise en marche. / ~ step ‖ Anlaßstufe f. ‖ plot m. de démarrage.

starting ‖ Anlassen n. ‖ démarrage m.; mise f. en marche. / asynchronous ~ ‖ asyn

chrones Anlassen n. || démarrage m. asynchrone. / ~ by means of compressed air || Anlassen n. mittels Druckluft || démarrage m. pneumatique *ou* à air comprimé. / ~ without jerk || stoßfreier Anlauf m. || démarrage m. sans à-coup. / ~ under load || Anlassen n. unter Last || démarrage m. sous charge. / ~ with full load || Anlauf m. unter Vollast || démarrage m. en pleine charge. / ~ with half load || Anlauf m. unter Halblast || démarrage m. à demi-charge. / ~ with low load || Anlauf m. mit geringer Belastung || démarrage m. à charge réduite. / ~ without load || Anlauf m. ohne Belastung; leerer Anlauf m. || démarrage m. sans charge *ou* à vide. / ~ the motor || Anlassen n. des Motors || démarrage m. du moteur. / self ~ || Selbstanlassen n. || auto-démarrage m. / ~ without shock || stoßfreies Anspringen n. || mise f. en marche sans choc. / spark ~ || Anspringen n. auf Zündfunken || départ m. au contact. / ~ of a train || Abfahrt f. eines Zuges || départ m. d'un train. / ~ of work || Betriebseröffnung f. || ouverture f. du service. / ~ of the wort (Brew) || Ankommen n. der Würze || entrée f. en fermentation du moût.

starting air || Anlaßluft f. || air m. de mise en marche. / distributing gear controlling the ~ || Anlaßluftsteuerung f. || distribution f. d'air de mise en marche. / distributing valve controlling the ~ || Anlaßluftsteuerventil n. || soupape f. de distribution pour l'air de mise en marche. / ~ pump || Anlaßluftpumpe f. || pompe f. pour l'air de mise en marche.

starting apparatus (Mot) || Anlaßvorrichtung f. || appareil m. démarreur. / ~ arrangement (Aero) || Anlaufvorrichtung f. || dispositif m. de lancement *ou* de départ. / ~ bar of a locomotive engine || Handhebel m. *oder* Anlaßhebel m. *oder* Steuerungshebel m. einer Lokomotive || levier m. de mise en marche d'une locomotive. / ~ button (Auto) || Starterknopf m. || contact m. de départ. / ~ cam || Anlaßnocken m. || came f. de mise en marche. / ~ carriage (Aero) || Anlaufgestell n. || chariot m. de lancement. / ~ circuit || Anlaßstromkreis m. || circuit m. de démarrage. / ~ cock || Anlaßhahn m. || robinet m. de mise en marche. || ~ connection (Electr) || Anlaßschaltung f. || couplage m. de démarrage. / ~ cradle (Aero) || Anlaufgestell n. || chariot m. de lancement.

starting crank (Auto) || Andrehkurbel f. || manivelle f. de mise en marche. / ~ handle || Andrehkurbelgriff m. || poignée f. de la manivelle de mise en marche. / ~ jaw || Klaue f. der Andrehkurbel || griffe f. de la manivelle de mise en marche. / ~ shaft || Andrehkurbelwelle f. || arbre m. de la manivelle de mise en marche.

starting current || Anlaufstrom m. || courant m. de démarrage. / ~ deck (Aero) || Startdeck n. || plateforme f. de départ. / ~ device || Anlaßvorrichtung f. || dispositif m. de démarrage. / ~ handle see starting crank. / ~ instructions pl. || Inbetriebsetzungsvorschriften fpl. || instructions fpl. de mise en marche.

starting lever || Anlaßhebel m.; Bedienungshebel m. || levier m. de mise en marche *ou* de manœuvre. / ~ of the driving clock || Auslösung f. des Uhrwerkganges || déclenchement m. du mouvement d'horlogerie.

starting load || Anlauflast f. || charge f. de démarrage. / ~ loss || Anlaufverlust m. || perte f. au démarrage. / ~ magneto || Anlaßmagnet m. || magnéto m. de mise en marche *ou* de démarrage f. / ~ material || Ausgangsmaterial n. || matière f. première. / ~ motor (Auto) || Anlaßmotor m. || démarreur m. électrique. / ~ output || Anlaufleistung f. || puissance f. de démarrage. / ~ period || Anlaufperiode f. || période f. de démarrage. / ~ plunger || Anlaßkolben m. || piston m. de la mise en marche. / ~ point || Ausgangspunkt m. || point m. de départ.

starting position || Anlaufstellung f. || position f. de mise en marche. / to turn the crank to the ~ || die Kurbel in die Anlaßstellung einstellen || placer la manivelle dans la position de mise en marche.

starting power || Anlaufkraft f. || force f. de démarrage. / ~ relay || Anlaßrelais n. || relais m. de démarrage. / ~ resistance *see* ~ rheostat.

starting rheostat || Anlaßwiderstand m. || résistance f. de démarrage *ou* de mise en marche. / liquid ~ || Flüssigkeitsanlaßwiderstand m. || rhéostat m. de démarrage à liquide. / metallic ~ || Metallanlaßwiderstand m. || rhéostat-démarreur m. métallique.

starting and reversing rheostat || Umkehranlasser m. || rhéostat démarreur-inverseur m.

starting signal (Railw) || Abfahrtsignal n. || signal m. de départ. / ~ spring (Aero) || Abschnellfeder f. || ressort m. de lancement. / ~ station (Railw) || Abgangsbahnhof m. || station f. de départ. / ~ switch || Anlaßschalter m. || interrupteur m. *ou* conjoncteur m. de démarrage. / ~ time || Anlaufzeit f. || durée f. de démarrage. / ~ torsional moment || Anlaufdrehmoment n. || couple m. de démarrage. / ~ track || Abfahrtgleis n. || voie f. de départ. / ~ transformer || Anlaßtransformator m. || transformateur m. de démarrage. / ~ tub (Brew) || Anstellbottich m. || cuve f. guilloire. / ~ valve || Anlaßventil n. || soupape f. de mise en marche. / ~ valve fork || Anlaßventilgabel f. || chape f. de soupape de mise en marche. / ~ vessel *see* ~ tub. / ~ voltage || Anlaßspannung f. || tension f. de démarrage. / ~ wheel || Anlaufrad n. || roue f. de mise en marche. / ~ winding || Anlaßwicklung f. || enroulement m. *ou* bobinage m. de démarrage.

start stop apparatus (Tel) || Springschreiber m. || appareil m. start-stop.

star twisting (Cable) || Sternverseilung f. || câblage m. en étoile.

star wheel || Arretierungsscheibe f.; Rastenscheibe f. || disque m. d'arrêt *ou* à crans.

state || Zustand m.; Beschaffenheit f. || état m. / ~ (Government) || Staat m.; État m. / ~ of affairs || Sachlage f. || circonstances fpl. / ~ of aggregation || Aggregatzustand m. || état m. (de matière). / ~ of equilibrium || Gleichgewichtszustand m. || état m. d'équilibre. / gaseous ~ || gasförmiger Zustand m. || état m. gazeux. / liquid ~ || flüssiger Zustand m. || état m. liquide. / ~ of the market || Geschäftslage f. || situation f. des affaires. / ~ of saturation || Sättigungszustand m. || état m. de saturation. / solid ~ || fester *oder* starrer Zustand m. || état m. solide. / steady ~ || stationärer Zustand m. || état m. de régime. / ~ of the weather || Wetterlage f. || état m. *ou* situation f. atmosphérique. / ~ of the work || Stand m. der Arbeit || état m. des travaux.

state forest || staatliche Forst f.; Staatsforst f. || forêt f. de l'État.

statement || Angabe f. || allégation f.; déclaration f. / ~ of account || Rechnungsauszug m. || relevé m. *ou* extrait m. de compte. / ~ of the aerial lines and wires of a local telephone plant || Nachweisung f. der oberirdischen Linien und Leitungen eines Ortsfernsprechnetzes || carnet m. des lignes aériennes d'un réseau téléphonique. / annual ~ of accounts || Jahresabschluß m.; Jahresausweis m. || fin f. de l'année; balance f. annuelle; bilan m. / ~ of contents (In a book) || Inhaltsangabe f. || résumé m.; index m. / monthly ~ || Monatsauszug m. || relevé m. mensuel.

state property || Staatsbesitz m. || possession f. de l'État. / to turn into ~ || verstaatlichen || racheter pour l'État; laïciser.

state room (Shipb) || Kabine f. || cabine f.

state service || Staatsbetrieb m. || exploitation f. par l'État.

static(al) || statisch || statique. / ~ calculation || statische Berechnung f. || calcul m. statique. / ~ equilibrium || statisches Gleichgewicht n. || équilibre m. statique. / ~ experiments pl. || statische Untersuchung f. || vérifications fpl. statiques. / ~ lift || statischer Auftrieb m. || poussée f. statique. / ~ pressure || statischer Druck m. || pression f. statique. / ~ strength || statische Festigkeit f. || résistance f. statique. / ~ stress || statische Beanspruchung f. || effort m. statique. / ~ transformer || ruhender Transformator m. || transformateur m. statique.

statically determinate || statisch bestimmt || statiquement déterminable. / ~ indeterminate || statisch unbestimmt || statiquement indéterminable. / ~ stable || statisch stabil || statiquement stable.

statics pl. || Statik f. || statique f. / ~ of rigid bodies || Statik f. fester Körper || statique f. des corps solides.

station || Station f.; Stelle f. || station f. / ~ (Railw) || Bahnhof m.; Haltestelle f.; Haltepunkt m. || gare f.; station f.; halte m. / ~ (Surv) || Netzpunkt m.; Richtpunkt m.; Fixpunkt m. || point m. fixé; point m. de repère. / branching-off ~ (Tel) || Knotenstation f. || station f. de bifurcation. / ~ for changing carriages || Umsteigestation f. || gare f. de transbordement des voyageurs. / discharging ~ || Entladestation f. || station f. de déchargement. / electric power ~ || Elektrizitätswerk n. || usine f. électrique. / ~ for experiments || Versuchsanstalt f. || station f. d'essai. / ~ on the ground floor || Station f. zu ebener Erde || station f. à fleur du sol. / intermediate ~ (Tel) || Zwischenanstalt f. || poste m. intermédiaire. / loading ~ (Railw) || Verladebahnhof m. || gare f. de transbordement. / ~ of miners (Radio) || Grubengerät n. || poste m. de mine. / portable military ~ (Radio) || tragbare Militärstation f. || poste f. militaire transportable. / radiotelegraphic long distance ~ || radiotelegraphische Großstation f. || poste f. radiotélégrafique de grande distance. / railway ~ || Bahnhof

m. ‖ gare f. / receiving ~ (Radio) ‖ Empfangsstation f. ‖ poste m. récepteur. / small power ~ (Electr) ‖ Kleinstation f. ‖ usine f. à faible puissance. / transmitting ~ (Radio) ‖ Sendestation f.; Sender m. ‖ poste m. émetteur.

stationary ‖ stationär; ortsfest; feststehend ‖ stationnaire. / ~ accumulator ‖ stationärer *oder* ortsfester Sammler m. ‖ accumulateur m. stationnaire. / ~ battery ‖ stationäre Batterie f. ‖ batterie f. fixe. / ~ boiler ‖ feststehender Dampfkessel m. ‖ chaudière f. fixe. / ~ engine ‖ ortsfeste *oder* stationäre Maschine f. ‖ machine f. fixe. / ~ point of a curve ‖ Umkehrpunkt m. *oder* Rückkehrpunkt m. einer Kurve ‖ point m. de rebroussement d'une courbe. / ~ riveting machine ‖ ortsfeste Nietmaschine f. ‖ riveuse f. fixe. / ~ source of sound ‖ ruhende Schallquelle f. ‖ source f. sonore immobile. / ~ steam-driven revolving crane ‖ feststehender Dampfdrehkran m. ‖ grue f. à vapeur pivotante et fixe. / ~ wave (Phys) ‖ stehende Welle f. ‖ onde f. stationnaire. / ~ wharf revolving crane ‖ feststehender Hafendrehkran m. ‖ grue f. fixe pivotante de port.

station block (Railw) ‖ Befehlsstellwerk n. ‖ bloc m. de gare. / ~ building ‖ Bahnhofsgebäude n. ‖ bâtiment m. de gare. / ~ clock ‖ Bahnsteiguhr f. ‖ horloge f. de quai.

stationery ‖ Schreibwaren fpl.; Papierwaren fpl. ‖ articles mpl. de papeterie *ou* à écrire *ou* en papier. / ~ for offices ‖ Büroschreibwaren fpl. ‖ articles mpl. à écrire pour bureaux. / ~ for schools ‖ Schulschreibwaren fpl. ‖ articles mpl. à écrire pour écoles.

stationery shop ‖ Papierwarenhandlung f. ‖ papeterie f.

station-house *see* station building.

station-master *see* station superintendent.

station superintendent ‖ Bahnhofsvorstand m. ‖ chef m. de gare *ou* de station.

statistical balance ‖ statistische Bilanz f. ‖ balance f. statistique. / ~ data ‖ statistische Angaben fpl. ‖ données fpl. statistiques. / ~ work ‖ statistische Arbeiten fpl. ‖ travaux mpl. de statistique.

statistician ‖ Statistiker m. ‖ statisticien m.

statistics pl. ‖ Statistik f. ‖ statistique f. / commercial ~ ‖ Handelsstatistik f. ‖ statistique f. commerciale.

statue ‖ Statue f. ‖ statue f. / ~ for churches ‖ Heiligenbild n. ‖ statue f. réligieuse.

statuette ‖ Statuette f. ‖ statuette f.

statute ‖ Statut n.; Satzung f. ‖ statut m.; règlement m. / ~s pl. of a company ‖ Gesellschaftsstatuten npl. ‖ statuts mpl. d'une société.

Stauffer grease ‖ Staufferfett n. ‖ graisse f. Stauffer. / ~ lubricator ‖ Staufferbüchse f. ‖ graisseur m. Stauffer.

stave ‖ Faßdaube f.; Daube f. ‖ douve f.; douelle f. / to joint ~s pl. ‖ Dauben fpl. fügen ‖ dresser *ou* jointer les douves fpl. / to straighten ~s pl. ‖ Dauben fpl. abrichten ‖ dresser des douves fpl. / bent ~ for repairs ‖ Flickdaube f. ‖ douve f. brute; merrain m. / beer barrel ~ ‖ Bierfaßdaube f. ‖ douve f. pour tonneaux à bière. / cask ~ ‖ Faßdaube f. ‖ douve f. de tonneau. / cleft ~ ‖ gespaltene Faßdaube f. ‖ douve f. fendue. / riven ~ *see* cleft ~. / sawn ~ ‖ gesägte Faßdaube f.

oder Daube f. ‖ douve f. sciée. / split ~ *see* cleft ~.

stave and head jointing machine ‖ Dauben- und Bodenfügemaschine f. ‖ machine f. à jointer les douves et les pièces de fond. / ~ backing machine ‖ Daubenhobelmaschine f. ‖ machine f. à doler *ou* à planer les douves. / ~ backing and hollowing machine ‖ Daubenhobel- und Aussparrmaschine ‖ machine f. à doler et à évider les douves. / ~ band saw ‖ Daubenbandsäge f. ‖ scie f. à ruban pour douves. / ~ bending machine ‖ Daubenbiegemaschine f. ‖ machine f. à cintrer les douves; plieuse f. de douves. / ~ cylinder saw ‖ Daubentrommelsäge f. ‖ scie f. à tambour pour douves.

staved end ‖ angestauchter Rand m. ‖ bord m. refoulé.

stave dressing machine ‖ Daubenabrichtmaschine f. ‖ machine f. à dresser les douves. / ~ end ‖ Daubenkopf m. ‖ tête f. de douve. / ~ hole ‖ Daubenloch n. ‖ trou m. de douve. / ~ jointing ‖ Daubenfügen n. ‖ action f. de dresser *ou* de jointer les douves. / ~ jointing machine ‖ Daubenfügemaschine f. ‖ machine f. à jointer les douves. / ~ jointing saw ‖ Daubenfügesäge f. ‖ scie f. à jointer les douves. / ~ maker ‖ Daubenmacher m. ‖ douvellier m. / ~ planing machine ‖ Daubenhobelmaschine f. ‖ machine f. à doler *ou* à planer les douves. / ~ press ‖ Daubenpresse f.; Faßdaubenpresse f. ‖ presse f. à douves. / ~ shortening and grooving machine ‖ Daubenabkürz- und Krösemaschine f. ‖ machine f. à raccourcir et à jâbler les douves. / ~ shortening saw ‖ Daubenabkürzsäge f. ‖ scie f. à raccourcir les douves. / ~ wood ‖ Daubenholz n. ‖ bois m. douvain m.

stay ‖ Strebe f. ‖ contrefiche f.; jambe f. de force. / ~ (Shipb) ‖ Stag n. ‖ étai m. / diagonal ~ ‖ Diagonalstrebe f. ‖ entretoise f. diagonale. / ~ of the link of a chain cable (Mar) ‖ Steg m. des Gliedes einer Ankerkette ‖ étançon m. de la maille d'une chaîne. / steel-cast frame plate ~ ‖ Zylinderversteifung f. in Stahlguß ‖ caissonnement m. en acier moulé. / ~ of a tripod ‖ Spreize f. eines Stativs ‖ traverse f. *ou* tirant m. d'un pied.

staybolt ‖ Stehbolzen m. ‖ entretoise f. / to bore out the ~s pl. ‖ die Stehbolzen mpl. abbohren ‖ enlever les entretoises fpl. / perforated ~ ‖ durchbohrter Stehbolzen m. ‖ entretoise f. percée.

stay crutch (Electr line) ‖ Ankerstütze f. ‖ console f. de hauban.

stay head (Mach) ‖ Zugstangenkopf m. ‖ tête f. de tige de traction.

stay hook (Electr line) ‖ Ankerhaken m. ‖ crochet m. de hauban.

staying of a beam ‖ Absteifung f. eines Balkens ‖ étayement m. d'une poutre.

stay lace ‖ Korsettschnur f. ‖ lacet m. de corsets.

stay pin ‖ Kettensteg m. ‖ étai m. du maillon.

stays pl. ‖ Korsett n. ‖ corset m.

stays pl. and guys pl. ‖ Abspannmaterial n. ‖ matériel m. d'étayage.

stay tackle (Mar) ‖ Fußtau n.; Stagtakel n. ‖ palan m. d'étai.

stay tightener (Electr line) ‖ Ankerspannschraube f. ‖ tendeur m.

stay wire (Electr line) ‖ Ankerseil n. ‖ corde f. d'acier pour l'hauban.

steadily falling ‖ stetig fallend ‖ baissant de façon continue. / ~ rising ‖ stetig steigend ‖ haussant de façon continue.

steadiness ‖ Stetigkeit f. ‖ continuité f.; constance f.

steady ‖ stetig ‖ continu; constant. ‖ ~ current ‖ gleichmäßiger Strom m. ‖ courant m. constant. / ~ deflection ‖ gleichmäßiger Ausschlag m. ‖ déviation f. soutenue. / ~ flow (Aero) ‖ stationäre Strömung f. ‖ écoulement m. *ou* courant m. stationnaire.

steady pin (Clockm) ‖ Quadraturstift m. ‖ tenon m.

steady position, to oscillate about the ‖ um die Ruhelage schwingen ‖ osciller autour de la position de repos. / ~ state ‖ stationärer Zustand m. ‖ état m. de régime. / ~ voltage ‖ gleichmäßige Spannung f. ‖ tension f. constante.

steadying resistance ‖ Beruhigungswiderstand m. ‖ résistance f. de stabilisation. / ~ wheel (Aero) ‖ Stoßrad n. ‖ roue f. de choc.

steam, to ‖ dämpfen ‖ préparer à l'aide de vapeur. / ~ the cloth ‖ Stoff m. dekatieren ‖ décatir le drap.

steam ‖ Dampf m.; Wasserdampf m. ‖ vapeur f. / apparatus for counter ~ ‖ Dampfbremsgerät n. ‖ appareil m. pour la contre-vapeur. / to cut off ~ ‖ den Dampf m. absperren ‖ couper la vapeur. / dead ~ ‖ Abdampf m. ‖ vapeur f. épuisée *ou* passive. / to distil with ~ ‖ mit Wasserdampf m. abblasen ‖ distiller à la vapeur. / dry ~ ‖ trockener Dampf m. ‖ vapeur f. sèche. / at full ~ ‖ mit voller Kraft f. ‖ à toute volée f.; à toute vapeur f. / to work with full ~ ‖ mit vollem Dampf m. arbeiten ‖ être en pleine vapeur f. / high-pressure ~ ‖ Hochdruckdampf m. ‖ vapeur m. à haute pression. / live ~ ‖ Frischdampf m. ‖ vapeur f. fraîche *ou* vive *ou* d'admission. / low-pressure ~ ‖ Niederdruckdampf m. ‖ vapeur f. à basse pression. / overheated ~ *see* superheated ~. / superheated ~ ‖ überhitzter Dampf m. ‖ vapeur m. surchauffée. / superheated ~ plant ‖ Heißdampfanlage f. ‖ installation f. à chauffage à vapeur surchauffée. / throttled ~ ‖ gedrosselter Dampf m. ‖ vapeur f. étranglée. / under ~ ‖ unter Dampf m. ‖ sous vapeur f. / wet ~ ‖ nasser Dampf m. ‖ vapeur f. humide.

steam accumulator ‖ Dampfspeicher m. ‖ accumulateur m. de vapeur. / ~ ausbrenner ‖ Dampfentpichmaschine f. ‖ dégoudronneur m. à vapeur.

steam baking oven ‖ Dampfbackofen m. ‖ four m. de boulangerie à vapeur. / draw plate ~ ‖ Auszugdampfbackofen m. ‖ four m. de boulangerie à vapeur amovible *ou* avec chariot sortant.

steam bath ‖ Schwitzbad n.; Dampfbad n. ‖ étuve f.; bain m. russe. / ~ bell ‖ Dampfpfeife f. ‖ sifflet m. à vapeur. / ~ blast *see also* steam jet ‖ Dampfstrahl m. ‖ jet m. de vapeur. / ~ blower ‖ Dampfgebläse n. ‖ tirage m. à jet de vapeur; soufflerie f. à vapeur. / ~ blow-off device ‖ Dampfabblasevorrichtung f. ‖ appareil m. pour l'évacuation de vapeur. / ~ boat *see* steamship.

steam-boiler *see also* boiler ‖ Dampfkessel m. ‖ chaudière f. à vapeur; chaudière f. / high-pressure ~ ‖ Hochdruckdampfkessel m. ‖ chaudière f. à

vapeur à haute pression. / ~ with mechanical stokers ‖ Dampfkessel m. mit mechanischer Feuerung ‖ chaudière f. à vapeur avec grilles mécaniques. / vertical ~ ‖ Steilrohrkessel m. ‖ chaudière f. à tubes verticaux.

steam-boiler accessories pl. ‖ Dampfkesselzubehör n. ‖ accessoires mpl. pour chaudières. / ~ cleaner ‖ Dampfkesselreiniger m. ‖ nettoyeur m. de chaudières à vapeur. / ~ cleaning ‖ Dampfkesselreinigung f. ‖ nettoyage m. de chaudières à vapeur. / ~ cleaning tool ‖ Dampfkesselreinigungswerkzeug n. ‖ outil m. pour le nettoyage de chaudières à vapeur. / ~ fittings pl. ‖ Dampfkesselarmaturen fpl. ‖ accessoires mpl. de chaudières à vapeur. / ~ furnace ‖ Dampfkesselfeuerung f. ‖ foyer m. de chaudière. / ~ inspection association ‖ Dampfkesselrevisionsverein m.; Dampfkesselüberwachungsverein m. ‖ association f. pour la surveillance des générateurs de vapeur.

steam boiling ‖ Dampfkochen n. ‖ cuisson f. à la vapeur. / ~ boiling apparatus ‖ Dampfkochgerät n. ‖ appareil m. de cuisson à la vapeur. / ~ brake ‖ Dampfbremse f. ‖ frein m. à vapeur. / ~ brewing pan ‖ Dampfbraupfanne f. ‖ poile f. à vapeur pour brasserie. / ~ brushing machine ‖ Dampfbürstmaschine f. ‖ brosseuse f. à vapeur. / ~ bubble ‖ Dampfblase f. ‖ bulle f. de vapeur. / ~ chamber ‖ Dampfsammler m. ‖ réservoir m. collecteur de vapeur.

steam chest ‖ Schieberkasten m. ‖ boîte f. à vapeur *ou* de tiroir. / ~ cover ‖ Schieberkastendeckel m. ‖ couvercle m. de la boîte à vapeur.

steam coal ‖ Dampfkohle f. ‖ charbon m. à vapeur.

steam cock ‖ Dampfhahn m. ‖ robinet m. à vapeur. / ~ grease ‖ Dampfhahnschmiere f. ‖ graisse f. pour robinets à vapeur.

steam coil ‖ Dampfschlange f. ‖ serpent m. *ou* serpentin m. à vapeur. / ~ compression ‖ Dampfkompression f. ‖ compression f. de la vapeur. / ~ condenser ‖ Dampfkondensator m. ‖ condensateur m. à vapeur.

steam conduit ‖ Dampfleitung f. ‖ conduit m. de vapeur. / ~ accessories pl. ‖ Dampfleitungszubehör n. ‖ accessoirs mpl. pour conduits de vapeur.

steam cone of the injector ‖ Ansaugeraum m. der Dampfstrahlpumpe ‖ cheminée f. de l'injecteur.

steam consumption ‖ Dampfverbrauch m. ‖ consommation f. de vapeur. / guaranteed ~ ‖ gewährleisteter Dampfverbrauch m. ‖ consommation f. garantie de vapeur.

steam consumption meter ‖ Dampfmesser m. ‖ instrument m. pour mesurer la consommation de vapeur. / ~ cooking installation ‖ Dampfkochanlage f. ‖ installation f. à cuire à la vapeur. / ~ cooling device ‖ Dampfkühler m. ‖ réfrigérant m. à vapeur. / ~ crane ‖ Dampfkran m. ‖ grue f. à vapeur. / cutting off of the ~ ‖ Dampfabsperrung f. ‖ interruption f. de la vapeur.

steam cylinder ‖ Dampfzylinder m. ‖ cylindre m. à vapeur. / ~ body ‖ Dampfzylinderkörper m. ‖ corps m. du cylindre à vapeur. / ~ jacket ‖ Dampfzylinderverkleidungsblech n.; Dampfzylinder-

mantel m. ‖ enveloppe f. *ou* chemise f. du cylindre à vapeur.

steam dairy ‖ Dampfmolkerei f. ‖ laiterie f. à vapeur. / ~ diagram ‖ Dampfdiagramm n. ‖ diagramme m. de vapeur. / ~ distribution ‖ Dampfverteilung f. ‖ distribution f. de la vapeur. / ~ douche ‖ Dampfsprudler m. ‖ fontaine f. de vapeur. / ~ dome (Boil) ‖ Dampfstutzen m.; Dampfdom m. ‖ dôme m. de vapeur *ou* de chaudière. / ~ dredger ‖ Dampfbagger m. ‖ drague f. à vapeur. / ~ drier ‖ Dampftrockner m. ‖ sécheur m. à vapeur. / ~ drive ‖ Dampfantrieb m. ‖ commande f. à vapeur.

steam-driven floating crane ‖ Dampfschwimmkran m. ‖ grue f. flottante à vapeur. / ~ plant ‖ Anlage f. mit Dampfbetrieb ‖ installation f. à (service par) vapeur. / ~ revolving crane ‖ Dampfdrehkran m. ‖ grue f. à vapeur pivotante.

steam drum ‖ Oberkessel m. ‖ chaudière f. supérieure. / ~ dryer ‖ Kondenstopf m. ‖ séparateur m. d'eau de condensation. / ~ dynamo ‖ Dampfdynamo f. ‖ dynamo f. à vapeur. / eduction of the ~ ‖ Dampfausströmung f. ‖ échappement m. de la vapeur.

steam engine ‖ Dampfmaschine f. ‖ machine f. à vapeur. / atmospheric ~ ‖ atmosphärische Dampfmaschine f. ‖ machine f. à vapeur atmosphérique. / beam ~ ‖ Balanzierdampfmaschine f. ‖ machine f. à vapeur à balancier. / compound ~ ‖ Verbunddampfmaschine f. ‖ machine f. à vapeur compound. / condensing ~ ‖ Kondensationsdampfmaschine f. ‖ machine f. à vapeur à condensation. / double-acting ~ ‖ doppeltwirkende Dampfmaschine f. ‖ machine f. à vapeur à double effet. / double-cylinder ~ ‖ Zweizylinderdampfmaschine f. ‖ machine f. à vapeur à deux cylindres. / expansion ~ ‖ Expansionsdampfmaschine f. ‖ machine f. à vapeur à détente. / horizontal ~ ‖ liegende Dampfmaschine f. ‖ machine f. à vapeur horizontale. / middle-pressure ~ ‖ Mitteldruckdampfmaschine f. ‖ machine f. à vapeur à moyenne pression. / non-condensing ~ ‖ Auspuffdampfmaschine f.; Dampfmaschine f. ohne Kondensation ‖ machine f. à vapeur sans condensation; machine f. à vapeur à échappement libre. / oscillating ~ ‖ oszillierende Dampfmaschine f. ‖ machine f. à vapeur oscillante *ou* à cylindre oscillant. / portable ~ ‖ Lokomobile f. ‖ locomobile f. à vapeur. / steam-heated portable ~ ‖ Dampflokomobile f. mit Überhitzer ‖ locomobile f. à vapeur surchauffée. / single-acting ~ ‖ einfachwirkende Dampfmaschine f. ‖ machine f. à vapeur à simple effet. / single-crank ~ ‖ Einkurbeldampfmaschine f. ‖ machine f. à vapeur à une seule manivelle. / single-crank compound ~ ‖ Einkurbelverbunddampfmaschine f. ‖ machine f. à vapeur compound à manivelle unique. / single-cylinder ~ ‖ Einzylinderdampfmaschine f. ‖ machine f. à vapeur monocylindrique *ou* à cylindre unique. / stationary ~ ‖ feststehende Dampfmaschine f. ‖ machine f. à vapeur fixe. / superheated ~ ‖ Heißdampfmaschine f. ‖ machine f. à vapeur surchauffée. / twin ~ ‖ Zwillingsdampfmaschine f. ‖ machine f. à vapeur jumelle.

steam engine driver ‖ Maschinist m. ‖ conducteur m. d'une machine à vapeur. / ~ plant ‖ Dampfmaschinenanlage f. ‖ installation f. *ou* établissement m. d'une machine à vapeur.

steamer ‖ Dampfer m. ‖ bateau m. à vapeur; vapeur m. / cargo ~ ‖ Frachtdampfer m. ‖ vapeur m. de transport. / cattle ~ ‖ Viehdampfer m. ‖ vapeur m. à bétail. / express ~ ‖ Schnelldampfer m. ‖ vapeur m. rapide. / mail ~ ‖ Postdampfer m. ‖ bateau m. poste; paquebot m. à vapeur. / merchant ~ ‖ Handelsdampfer m. ‖ vapeur m. marchand. / paddlewheel ~ ‖ Raddampfer m. ‖ vapeur m. à roues. / passenger ~ ‖ Fahrgastdampfer m. ‖ vapeur m. à passagers. / screw ~ ‖ Schraubendampfer m. ‖ vapeur m. à hélice. / sea-going ~ ‖ Seedampfer m. ‖ vapeur m. de long cours. / sternwheel ~ ‖ Heckraddampfer m. ‖ vapeur m. avec roue à l'arrière; vapeur m. monorue. / tank ~ ‖ Tankdampfer m. ‖ vapeur m. à réservoirs. / tramp ~ ‖ Trampdampfer m. ‖ tramp m. / transatlantic ~ ‖ Überseedampfer m. ‖ vapeur m. transatlantique. / the ~ has special facilities for transport ‖ der Dampfer weist für den Transport besondere Einrichtungen auf ‖ le vapeur est aménagé pour le transport. / turret-deck ~ ‖ Turmdeckdampfer m. ‖ vapeur m. à turret-deck.

steam escape see ~ exhaust. / ~ exhaust ‖ Dampfentweichung f. ‖ fuite f. *ou* échappement m. de vapeur. / exhausting plant for ~ ‖ Absaugeanlage f. für Dampf ‖ installation f. d'aspiration de vapeur. / ~ exhaust pipe ‖ Dampfauslaßrohr n. ‖ tuyau m. d'échappement à vapeur. / ~ extraction ‖ Dampfentnahme f. ‖ prise f. de vapeur. / ~ feed-pump ‖ Dampfspeisepumpe f. ‖ pompe f. d'alimentation à vapeur. / ~ ferry boat ‖ Fährdampfer m. ‖ vapeur m. de passage.

steam fire engine ‖ Dampffeuerspritze f. ‖ pompe f. à incendie à vapeur. / ~ with implements wagon ‖ Dampffeuerspritze f. mit Gerätewagen ‖ pompe f. à incendie à vapeur avec chariot d'agrès.

steam gate valve ‖ Dampfabsperrschieber m. ‖ vanne f. à vapeur *ou* d'arrêt de vapeur.

steam gauge ‖ Manometer n.; Dampfdruckmesser m. ‖ manomètre m. / ~ case ‖ Manometergehäuse n. ‖ boîte f. du manomètre. / ~ hand ‖ Manometerzeiger m. ‖ aiguille f. du manomètre.

steam-generating heat ‖ Dampfbildungswärme f. ‖ chaleur f. de vaporisation.

steam generation ‖ Dampfbildung f. ‖ vaporisation f. / ~ gas producer ‖ Gaserzeuger m. mit Dampferzeugung ‖ gazogène m. à production de vapeur.

steam generator ‖ Dampfkessel m. ‖ chaudière f. à vapeur. / ~ grainer ‖ Dampfgrainer m. ‖ steam grainer.

steam hammer ‖ Dampfhammer m. ‖ marteau m. à vapeur. / single-frame ~ ‖ Einständerdampfhammer m. ‖ marteau m. pilon à vapeur à montant unique.

steam-heated oven ‖ Dampfofen m. ‖ étuve f. à vapeur.

steam heater ‖ Dampfheizung f.; Heizkörper m. ‖ calorifère f. à vapeur. / ~ heating ‖ Dampfheizung f. ‖ chauffage m. à vapeur. / ~ heating apparatus see steam heater.

steam-hydraulic ‖ dampfhydraulisch ‖ vapo-hydraulique. / ~ forging press ‖ dampf-hydraulische Schmiedepresse f. ‖ presse f. hydraulique à vapeur à forger. / ~ machine ‖ dampfhydraulisch betriebene Maschine f. ‖ machine f. vapo-hydraulique. / ~ press ‖ dampfhydraulische Presse f. ‖ presse f. hydraulique à vapeur. / ~ system ‖ dampfhydraulisches System n. ‖ système m. combiné à la vapeur et à l'eau.

steaming (Cloth) ‖ Dekatieren n. ‖ décatissage m.; délustrage m. / dry ~ (Cloth) ‖ Trockendekatieren n. ‖ décatissage m. par vapeur sèche.

steaming apparatus ‖ Dämpfapparat m. ‖ appareil m. à vaporiser ou à étuver; étuve f. / ~ (Cloth) ‖ Dekatiergerät n. ‖ appareil m. à décatir. / bone coal ~ ‖ Knochenkohledämpfer m. ‖ évaporateur m. à charbon d'os.

steaming capacity ‖ Verdampfungsfähigkeit f. ‖ pouvoir m. évaporatif. / ~ machine (Cloth) ‖ Dekatiermaschine f. ‖ machine f. à décatir. / conditioning and tentering ~ machine ‖ Dämpf-, Egalisier- und Spannmaschine f. ‖ machine f. à vaporiser, égaliser et élargir. / ~ period ‖ Dämpfzeit f. ‖ temps m. de daube. / ~ plant ‖ Dämpfanlage f. ‖ installation f. d'étuvage. / ~ table for woollen fabrics ‖ Dämpf-tisch m. zum Abdämpfen wollener Stoffe ‖ table f. à vaporiser les tissus de laine.

steam inlet pipe ‖ Dampfeinströmungsrohr n. ‖ tuyau m. d'admission de la vapeur. / ~ intensifier ‖ Dampftreibgerät n. ‖ multiplicateur m. à vapeur. / ~ jacket ‖ Dampfmantel m.; Dampfhemd n. ‖ enveloppe f. de (cylindre à) vapeur; chemise f. du cylindre.

steam-jacketed cylinder ‖ Zylinder m. mit Dampfmantel ‖ cylindre m. chemisé de vapeur.

steam jet ‖ Dampfstrahl m. ‖ jet m. de vapeur. / ~ bilge pump ‖ Dampfstrahl-bilgepumpe f. ‖ pompe f. de cale à jet de vapeur. / ~ blower ‖ Dampfstrahl-gebläse n. ‖ soufflerie f. à vapeur; machine f. soufflante à jet de vapeur. / ~ ejector ‖ Dampfstrahlejektor m. ‖ éjecteur m. à vapeur. / ~ flue cleaning apparatus ‖ Dampfstrahlrauchrohrreiniger m. ‖ nettoyeur m. de tubes à fumée à jet de vapeur. / ~ sand blast ‖ Dampfsand-strahlgebläse n. ‖ sableuse f. à vapeur.

steam kitchen ‖ Dampfküche f. ‖ cuisine f. à vapeur. / ~ launch (Mar) ‖ Dampf-barkasse f. ‖ chaloupe f. à vapeur. / ~ laundry ‖ Dampfwäscherei f. ‖ blanchisserie f. à vapeur. / ~ locomotive ‖ Dampflokomotive ‖ locomotive f. à vapeur. / ~ mill ‖ Dampfmühle f. ‖ moulin m. à vapeur. / ~ mixing jet ‖ Dampf-strahlrührgebläse n. ‖ soufflerie f. d'agitateur à jet de vapeur. / ~ motor ‖ Dampfmotor m. ‖ moteur m. à vapeur. / ~ motor lorry ‖ Dampflastwagen m. ‖ camion m. à vapeur. / ~ navigation ‖ Dampfschiffahrt f. ‖ navigation f. à vapeur. / ~ navigation company ‖ Dampf-schiffahrtsgesellschaft f. ‖ compagnie de navigation f. à vapeur.

steam-operated plant see steam-driven plant.

steam pile driver ‖ Dampframme f. ‖ sonnette f. ou mouton m. à vapeur.

steam pipe ‖ Dampfleitungsrohr n.; Dampf-rohr n. ‖ conduit m. de vapeur; tuyau m.

à vapeur. / waste ~ ‖ Dampfableitungs-rohr n. ‖ tuyau m. d'échappement de la vapeur. / welded ~ ‖ geschweißtes Dampfrohr n. ‖ tube m. à vapeur soudé.

steam pipe closing valve, self-acting ~ ‖ Selbstschlußdampfabsperrventil n. ‖ soupape f. d'arrêt de vapeur à fermeture automatique.

steam piping ‖ Dampfrohrleitung f. ‖ tuyauterie f. de vapeur; conduite f. ou canalisation f. de vapeur. / ~ piston ‖ Dampfkolben m. ‖ piston m. à vapeur. / ~ pitching machine ‖ Dampfentpich-maschine f. ‖ dégoudronneur m. à vapeur. / ~ plant which is situated right on top of the lignite beds ‖ unmittelbar auf der Braunkohle errichtetes Dampf-kraftwerk n. ‖ usine f. thermique (qui est) établie sur le carreau des mines de lignites. / ~ plate shears pl. ‖ Dampfblech-schere f. ‖ cisaille f. à tôles commandée par la vapeur.

steam plough ‖ Dampfpflug m. ‖ charrue f. à vapeur. / ~ machinist ‖ Dampfpflug-maschinist m. ‖ conducteur m. de la charrue à vapeur.

steam port ‖ Dampfkanal m. ‖ passage m. de vapeur. / ~s pl. without angles and sharp curves ‖ glatte Dampfkanäle mpl. ohne Winkel und scharfe Krümmungen ‖ passages mpl. de vapeur lisses sans angles et courbes abruptes.

steam pot ‖ Autoklav m.; Dampfbehälter m. ‖ réservoir m. à vapeur; autoclave m. / ~ with agitator ‖ Autoklav m. mit Rührwerk ‖ réservoir m. à vapeur avec agitateur.

steam power ‖ Dampfkraft f. ‖ force f. de vapeur. / ~ electrical plant ‖ Elektrizitätswerk n. mit Dampfkraft; Dampf-elektrizitätswerk n. ‖ usine f. électrique à vapeur. / ~ plant ‖ Dampfkraftzentrale f. ‖ centrale f. d'énergie à vapeur.

steam press (Print) ‖ Schnellpresse f. ‖ presse f. à la mécanique ou mécanique. / ~ and air hydraulic universal press ‖ dampf- und lufthydraulische Universal-presse f. ‖ presse f. universelle vapo-hydraulique et hydro-pneumatique.

steam pressure ‖ Dampfdruck m.; Dampf-spannung f. ‖ pression f. ou tension f. de la vapeur. / ~ above atmospheric ‖ Dampfüberdruck m. ‖ surpression f. de vapeur. / ~ injurious ‖ schädlicher Dampfdruck m. ‖ pression f. de vapeur dangereuse.

steam pressure boiler ‖ Dampfdruckkessel m. ‖ chaudière f. à vapeur comprimée. / ~ intensifier ‖ Dampfdruckübersetzer m. ‖ multiplicateur m. de pression à vapeur. / ~ reducing valve ‖ Druckminderventil n. ‖ soupape f. réductrice de vapeur. / ~ reservoir see steam pot.

steam pump ‖ Dampfpumpe f. ‖ pompe f. à vapeur. / automatically working ~ ‖ selbsttätig wirkende Dampfpumpe f. ‖ pompe f. à vapeur (à action) automatique.

steam purifier ‖ Dampfreiniger m. ‖ épurateur m. de vapeur. / ~ ram ‖ Dampf-ramme f. ‖ mouton m. à vapeur. / ~ recorder ‖ Schreibgerät n. des Dampf-messers ‖ appareil m. enregistreur de la consommation de vapeur. / ~ rivetter ‖ Dampfnietmaschine f. ‖ riveuse f. à vapeur. / ~ road roller see steam roller.

steam roller ‖ Dampf(straßen)walze f.; Straßenwalze f. ‖ rouleau m. compres-

seur à vapeur; écraseur m. à vapeur. / three-wheeled ~ ‖ Dreiraddampfwalze f. ‖ rouleau m. compresseur à vapeur à un rouleau et deux roues.

steam room ‖ Dampfkammer f. ‖ chambre f. ou boîte f. à vapeur. / ~ rose ‖ Dampf-brause f. ‖ crépine f. à vapeur. / ~ saw mill ‖ Dampfsägemühle f. ‖ scierie f. à vapeur.

steam separator ‖ Kondenswasserabscheider m.; Kondenstopf m. ‖ séparateur m. ou purgeur m. d'eau de condensation. / ~ controlling device ‖ Kon-denstopfkontrollapparat m. ‖ appareil m. contrôleur de séparateur d'eau de condensation. / ~ recorder ‖ Kondens-topfkontrollapparat m. ‖ appareil m. de contrôle pour pots de condensation.

steamship see also steamer ‖ Dampfer m. ‖ bateau m. à vapeur; vapeur m. / ~ line ‖ Dampferlinie f. ‖ ligne f. de navigation à vapeur.

steam shovel ‖ Dampflöffelbagger m.; Löffelbagger m.; Trockenbagger m.; Exkavator m. ‖ pelle f. ou cuiller f. ou excavateur m. ou terrassier m. à vapeur; piocheuse f. / ~ siren ‖ Dampfsirene f. ‖ sirène f. à vapeur. / ~ slide valve ‖ Dampfschieber m. ‖ tiroir m. de vapeur. / ~ steering apparatus (Shipb) ‖ Dampf-steuergerät n. ‖ appareil m. pour gouvernail à vapeur. / ~ steering wheel (Shipb) ‖ Dampfsteuerrad n. ‖ gouvernail m. à vapeur. / ~ stop valve ‖ Dampfabsperrventil n. ‖ soupape f. d'arrêt de vapeur. / turbo set for Ruths ~ storage operation ‖ Turbo-satz m. für Ruthsspeicherbetrieb ‖ groupe m. turbo-générateur à accumulateurs de chaleur Ruths. / Ruths ~ storage plant with vertical Ruths accumulators ‖ Ruthsdampfspeicheranlage f. mit stehenden Ruthsspeichern ‖ installation f. d'accumulateurs de chaleur Ruths comportant des accumulateurs Ruths verticaux. / ~ superheater ‖ Dampfüberhitzer m. ‖ surchauffeur m. de vapeur. / ~ test ‖ Dampfprobe f. ‖ essai m. à la vapeur. / ~ thrasher ‖ Dampfdreschmaschine f. ‖ batteuse f. à vapeur. / ~ thrashing machine see ~ thrasher.

steamtight ‖ dampfdicht ‖ imperméable à la vapeur; étanche de vapeur.

steam tilery ‖ Dampfziegelei f. ‖ briqueterie f. à vapeur. / ~ traction (Hydr arch) ‖ Treidelung f. durch Dampfkraft ‖ halage m. à vapeur. / ~ tramway ‖ Dampf-straßenbahn f. ‖ tramway m. à vapeur. / ~ trap see steam separator. / ~ trough drier ‖ Dampfmuldentrockner m. ‖ sécheur m. à auge à vapeur. / ~ tug ‖ Schleppdampfer m. ‖ remorqueur m. à vapeur.

steam turbine ‖ Dampfturbine f. ‖ turbine f. à vapeur. / high-pressure ~ ‖ Hoch-druckdampfturbine f. ‖ turbine f. à vapeur à haute pression.

steam turbine blade ‖ Dampfturbinen-schaufel f. ‖ aube f. de turbine à vapeur. / ~ material ‖ Dampfturbinenschaufel-material n. ‖ matériel m. pour aubes de turbines à vapeur.

steam turbine dynamo ‖ Dampfturbinen-dynamo f. ‖ turbine f. à vapeur dynamo-électrique. / ~ locomotive ‖ Dampf-turbinenlokomotive f. ‖ locomotive f. à turbine à vapeur. / ~ power station ‖ Dampfturbinenkraftwerk n. ‖ centrale f. turbo-électrique. / blading of a ~ rotor ‖

Beschaufelung f. eines Dampfturbinenläufers ‖ aubage m. d'une roue motrice de turbine à vapeur. / ~ test bed ‖ Dampfturbinenprüfstand m. ‖ plate-forme f. d'essais pour turbines à vapeur.

steam utilization for heating and cooking purposes ‖ Dampfverwendung f. für Heiz- und Kochzwecke ‖ utilisation f. de la vapeur pour le chauffage et la cuisson. / intermediate ~ ‖ Zwischendampfentnahme f. ‖ prise f. intermédiaire de vapeur; soutirage m. de vapeur.

steam valve ‖ Dampfventil n. ‖ soupape f. à vapeur. / ~ vessel see steamship. / ~ vulcanizer ‖ Dampfvulkanisator m. ‖ vulcanisateur m. à vapeur. / waste of ~ ‖ Dampfverlust m. ‖ perte f. de vapeur. / ~ whistle ‖ Dampfpfeife f. ‖ sifflet m. à vapeur. / ~ winch ‖ Dampfwinde f. ‖ treuil m. à vapeur.

steam-working see steam-driven.

steam yacht ‖ Dampfjacht f. ‖ yacht m. à vapeur.

stearate ‖ stearinsaures Salz n. ‖ stéarate m.

stearic acid ‖ Stearinsäure f. ‖ acide m. stéarique.

stearin ‖ Stearin n. ‖ stéarine f.

stearin candle ‖ Stearinkerze f. ‖ chandelle f. en stéarine. / ~ maker ‖ Stearinkerzenfabrikant m. ‖ mouleur m. de bougies en stéarine.

stearin factory ‖ Stearinfabrik f. ‖ stéarinerie. f. / ~ match ‖ Stearinzündholz n. ‖ allumette f. en stéarine. / ~ oil ‖ Stearinöl n. ‖ huile f. de stéarine. / ~ pitch ‖ Stearinpech n. ‖ brai m. stéarique. / ~ plant see ~ factory. / ~ works pl. see ~ factory.

statite ‖ Speckstein m. ‖ stéatite f. / ~ burner ‖ Specksteinbrenner m. ‖ bec m. en stéatite. / ~ fuse ‖ Steatitsicherung f. ‖ fusible m. en stéatite. / ~ quarry ‖ Specksteingrube f. ‖ carrière f. de talc ou de stéatite.

steel, to ‖ verstählen; stählen ‖ aciérer; acérer.

steel ‖ Stahl m. ‖ acier m. / the ~ will not corrode under the action of sea water ‖ der Stahl wird durch Seewasser nicht angegriffen ‖ l'acier m. est résistant contre l'influence de l'eau de mer. / the ~ is uniformly hard throughout ‖ der Stahl m. ist durchgehend gleichmäßig hart ‖ l'acier m. a une dureté uniforme. / the grades of ~ are classed according to their hardness and toughness ‖ die Stähle mpl. sind nach Härte und Zähigkeit abgestuft ‖ les aciers mpl. sont gradués d'après leur dureté et leur ténacité. / the ~ can be supplied with minimum characteristics ‖ der Stahl m. kann mit Mindestfestigkeit geliefert werden ‖ nous pouvons livrer l'acier m. avec un minimum de résistance. / to purify the ~ of its residues of slag by remelting it ‖ den Stahl m. durch Umschmelzen von den Schlackenresten reinigen ‖ purifier par la fusion l'acier des restes de scorie. / to quench the ~ in water ‖ Stahl m. härten oder ablöschen ‖ tremper l'acier m. dans l'eau. / the ~ is particularly distinguished by its great toughness ‖ der Stahl zeichnet sich ganz besonders durch große Zähigkeit aus ‖ l'acier m. est surtout remarquable par sa grande ténacité. / acid-proof ~ ‖ säurebeständiger Stahl m. ‖ acier m. inattaquable aux acides. / alloy ~ ‖ legierter

Stahl m. ‖ acier m. allié. / alloyed superrefined ~ ‖ legierter Edelstahl m. ‖ acier m. spécial allié. / ~ almost unaffectable by dilute acid ‖ gegen den Angriff von verdünnter Säure widerstandsfähiger Stahl ‖ acier m. résistant aux effets de l'acide étendu. / baby Bessemer ~ ‖ Kleinbessemerstahl m. ‖ acier m. Bessemer de petit convertisseur. / ball bearing ~ ‖ Kugellagerstahl m. ‖ acier m. pour roulements à billes. / Bessemer ~ ‖ Bessemerstahl m. ‖ acier m. Bessemer. / blistered ~ ‖ Blasenstahl m. ‖ acier m. boursoufflé. / not blistered ~ ‖ blasenloser Stahl m. ‖ acier m. sans ampoules. / bright drawn ~ ‖ blank gezogener Stahl m. ‖ acier m. étiré à brillant. / burnt ~ ‖ verbrannter Stahl m. ‖ acier m. brûlé. / carbon ~ ‖ Kohlenstoffstahl m. ‖ acier m. au carbone. / ~ for case-hardening ‖ Stahl m. für Einsatzhärtung ‖ acier m. de cémentation (en paquet). / cast ~ ‖ Gußstahl m. ‖ acier m. fondu ou au creuset. / cement ~ ‖ Zementstahl m. ‖ acier m. cémenté. / charcoal ~ ‖ Holzkohlenstahl m. ‖ acier m. au charbon de bois. / chromium ~ ‖ Chromstahl m. ‖ acier m. au chrome. / containing ~ ‖ stahlhaltig ‖ aciéreux. / cold-rolled ~ ‖ kalt gewalzter Stahl m. ‖ acier m. laminé à froid. / ~ for corset springs ‖ Korsettfederstahl m. ‖ acier m. pour baleines de corsage. / crucible ~ ‖ Tiegelstahl m. ‖ acier m. au creuset. / damask ~ ‖ Damaszenerstahl m. ‖ lame f. damasquinée; damas m. / drill ~ ‖ Bohrmeißelstahl m. ‖ acier m. pour forets. / dynamo ~ ‖ Dynamostahl m. ‖ acier m. pour dynamo. / electric ~ ‖ Elektrostahl m. ‖ acier m. au four électrique. / electric ~ in ingots. ‖ Elektrostahl m. in Blöcken ‖ acier m. électrique en lingots. / fibrous ~ ‖ sehniger Stahl m. ‖ acier m. nerveux. / forged ~ ‖ geschmiedeter Stahl m. ‖ acier m. forgé. / hard ~ ‖ Hartstahl m. ‖ acier m. dur. / hardened ~ ‖ gehärteter Stahl m. ‖ acier m. durci ou trempé. / ~ capable of being hardened ‖ härtbarer Stahl m. ‖ acier m. pouvant être trempé ou prenant la trempe. / heat-resisting ~ ‖ hitzebeständiger Stahl m. ‖ acier m. allant au feu. / highest quality ~ see high grade ~. / high-grade ~ ‖ hochwertiger Stahl m. ‖ acier m. de haute qualité. / high-speed ~ ‖ Schnelldrehstahl m. ‖ acier m. (à coupe) rapide. / ~ for high-speed lathes see high-speed ~. / industrial ~ (Militär-A) ‖ handelsüblicher Stahl m. ‖ acier m. de type commercial. / knife ~ ‖ Messerstahl m. ‖ acier m. pour couteaux. / magnet ~ ‖ Magnetstahl m. ‖ acier m. pour aimants. / manganese ~ ‖ Manganstahl m. ‖ acier m. au manganèse. / mild ~ ‖ Flußeisen n. ‖ acier m. doux. / natural ~ see raw ~. / naturally hard ~ ‖ naturharter Stahl m. ‖ acier m. de dureté naturelle. / nickel ~ ‖ Nickelstahl m. ‖ acier m. au nickel; ferro-nickel m. / non-magnetic ~ ‖ unmagnetischer Stahl m. ‖ acier m. nonmagnétique. / non-rusting ~ see stainless ~. / open-hearth ~ ‖ Siemens-Martinstahl m. ‖ acier m. Siemens-Martin. / ~ of particular physical characteristics ‖ Stahl m. mit besonderen physikalischen Eigenschaften ‖ acier m. de propriétés physiques spéciales. / plain superrefined ~ ‖ unlegierter Edelstahl m. ‖ acier

m. spécial non allié. / puddled ~ ‖ Puddelstahl m. ‖ acier m. puddlé. / raw ~ ‖ Rohstahl m. ‖ acier m. naturel ou brut. / ~ ready for casting ‖ unmittelbar vergießbarer Stahl m. ‖ acier m. liquide instantanément prêt à la coulée. / refined ~ ‖ Edelstahl m. ‖ acier m. raffiné ou fin. / remelted ~ ‖ umgeschmolzener Stahl m. ‖ acier m. refondu. / ~ in the shape of ribbon ‖ Stahl m. in Bandform ‖ acier m. sous forme de ruban. / rolled ~ ‖ gewalzter Stahl m. ‖ acier m. laminé. / round ~ ‖ Rundstahl m. ‖ acier m. rond. / rustless ~ see stainless ~. / rustproof ~ see stainless ~. / semifinished ~ ‖ Halbzeug n. ‖ demi-produits mpl. / shear ~ see refined ~. / sheet ~ ‖ Stahlblech n. ‖ tôle f. d'acier. / Siemens Martin ~ ‖ Siemens-Martin-Stahl m. ‖ acier m. Siemens-Martin. / silicium ~ ‖ Siliziumstahl m. ‖ acier m. au silicium. / silico-manganese ~ ‖ Mangansiliziumstahl m. ‖ acier m. mangano-silicieux. / special ~ ‖ Sonderstahl m. ‖ acier m. spécial. / special ~ for motor cars ‖ Spezialstahl m. für den Kraftwagenbau ‖ aciers mpl. spéciaux pour la construction d'automobiles. / stainless ~ ‖ nichtrostender oder rostfreier Stahl m. ‖ acier m. inoxydable ou non-corrossive. / structural ~ ‖ Konstruktionsstahl m.; Baustahl m. ‖ acier m. de construction. / superrefined ~ ‖ Edelstahl m. ‖ acier m. spécial ou fin. / superrefined ~ in ingots ‖ Edelstahl m. in Blöcken ‖ acier m. spécial en lingots. / tempered ~ ‖ gehärteter oder angelassener Stahl m. ‖ acier m. trempé. / tool ~ ‖ Werkzeugstahl m. ‖ acier m. (fin) à outil ou pour outils. / tungsten ~ ‖ Wolframstahl m. ‖ acier m. au tungstène. / unhardened ~ ‖ ungehärteter Stahl m. ‖ acier m. non trempé. / unweldable ~ ‖ nicht schweißbarer Stahl m. ‖ acier m. non soudable. / vanadium ~ ‖ Vanadiumstahl m. ‖ acier m. au vanadium. / weldable ~ ‖ Schweißstahl m. ‖ acier m. soudable. / well tempered ~ ‖ gut gehärteter Stahl m. ‖ acier m. bien trempé. / wolfram ~ see tungsten ~. / wrought ~ ‖ Schmiedestahl m. ‖ acier m. forgé.

steel alloy ‖ Stahllegierung f. ‖ alliage m. d'acier. / ~ aluminium cable ‖ Stahlaluminiumseil n. ‖ câble m. en acier et en aluminium. / ~ analysis ‖ Stahluntersuchung f.; Stahlanalyse f. ‖ analyse f. de l'acier.

steel-armoured conduit ‖ Stahlpanzerrohr n. ‖ tube m. protecteur en acier.

steel articles pl. ‖ Stahlwaren fpl. ‖ objets mpl. ou articles mpl. en acier; coutelleries fpl. / polished ~ ‖ polierte Stahlwaren fpl. ‖ objets mpl. en acier poli.

steel ball ‖ Stahlkugel f. ‖ bille f. d'acier. / hardened ~ ‖ gehärtete Stahlkugel f. ‖ bille f. en acier trempé.

steel band ‖ Stahlband n. ‖ ruban m. ou bande f. d'acier.

steel bar ‖ Stahlstange f. ‖ tige f. ou barre f. d'acier. / flat ~ ‖ flache Stahlstange f. ‖ acier m. en barre plate. / square ~ ‖ viereckige Stahlstange f. ‖ acier m. en barre carrée.

steel barrel ‖ Stahlfaß n. ‖ baril m. en tôle d'acier. / ~ bead ‖ Stahlperle f. ‖ perle f. d'acier. / ~ belt see steel band.

steel blade ‖ Stahlklinge f. ‖ lame f. en acier. / nickel-plated ~ ‖ vernickelte

Stahlklinge f. ‖ lame f. en acier nickelé.

steel bolt coupling ‖ Stahlbolzenkupplung f. ‖ accouplement m. à clavettes en acier.

steel bottle ‖ Stahlflasche f. ‖ bouteille f. en acier. / ~ for compressed gases ‖ Stahlflasche f. für komprimierte Gase ‖ bouteille f. en acier pour gaz comprimés. / cart for ~s ‖ Transportkarren m. für Stahlflaschen ‖ chariot m. de transport pour bouteilles en acier.

steel bronze ‖ Stahlbronze f. ‖ bronze m. acier. / ~ buckle ‖ Stahlschnalle f. ‖ boucle f. en acier. / ~ cable ‖ Stahldrahtseil n. ‖ corde f. ou câble m. en acier. / ~ carriage ‖ Stahlwagen m. ‖ chariot m. en acier.

steel-cast frog (Railw) ‖ Stahlgußherzstück n. ‖ cœur m. en acier moulé.

steel castings pl. ‖ Stahl(form)guß m. ‖ acier m. moulé; moulage m. d'acier. / to produce a denser texture in ~ ‖ den Stahlgüssen ein dichteres Gefüge geben ‖ donner aux lingots une texture plus compacte. / dynamo ~ ‖ Dynamostahlguß m. ‖ moulage m. d'acier pour dynamos. / ~ of softest quality to replace forgings ‖ Stahlguß m. weichster Beschaffenheit zum Ersatz von Schmiedestücken ‖ moulages mpl. d'acier très doux pour remplacer des pièces forgées. / construction of ~ ‖ Stahlgußausführung f. ‖ exécution f. en acier coulé.

steel cheek ‖ Stahlbacke f. ‖ mâchoire f. en acier. / ~ chips pl. ‖ Stahlspäne pl. ‖ riblons mpl. ou éclats mpl. ou tournures fpl. d'acier. / ~ colour ‖ Stahlfarbe f. ‖ couleur f. d'acier.

steel-coloured ‖ stahlfarbig ‖ bleu m. d'acier.

steel concrete ‖ Stahlbeton m. ‖ béton m. en acier. / ~ construction ‖ Eisenbauwerk n.; Eisenkonstruktion f. ‖ construction f. ou charpente f. en fer. / ~ converter ‖ Stahlkonverter m. ‖ convertisseur m. d'aciéries.

steel cylinder ‖ Stahlflasche f. ‖ cylindre m. en acier. / ~ valve ‖ Stahlflaschenventil n. ‖ soupape f. pour des bouteillles en acier.

steel dealer ‖ Stahlhändler m. ‖ marchand m. d'acier. / ~ disk ‖ Stahllamelle f. ‖ lamelle f. en tôle d'acier. / ~ disk wheel ‖ Stahlvollrad n. ‖ roue f. en acier d'une seule pièce. / ~ dowel ‖ Stahldübel m. ‖ goujon m. d'acier. / ~ engraver's ink ‖ Stahlstichschwärze f. ‖ encre f. à gravure sur acier. / ~ engraving ‖ Stahlstich m. ‖ gravure f. sur ou en acier; estampe f. sur acier. / ~ fining process ‖ Stahlfrischverfahren n. ‖ procédé m. d'affinage de l'acier. / ~ fork ‖ Stahlgabel f. ‖ fourche f. en acier. / ~ foundry ‖ Stahlgießerei f. ‖ fonderie f. d'acier. / tubular ~ frame ‖ Stahlrohrgestell n. ‖ châssis m. en tubes d'acier.

steel-framed hall construction ‖ Eisenhallenbau m. ‖ construction f. de halles en charpente métallique.

steel frying pan ‖ Stahlbratpfanne f. ‖ poêle f. à frire en tôle d'acier. / ~ furniture ‖ Stahlmöbel pl. ‖ meubles mpl. en acier. / ~ goods whetting device ‖ Stahlwarenschleiferei f. ‖ installation f. à repasser la coutellerie. / ~ grey ‖ stahlgrau ‖ gris d'acier; chalybé. / ~ hollow ware ‖ Stahlküchengeräte npl. ‖ usten-

siles mpl. de ménage en acier. / ~ house ‖ Stahlhaus n. ‖ maison f. en acier.

steeling ‖ Verstählung f. ‖ aciération f. / ~ of dies ‖ Verstählen n. von Gesenken ‖ aciérage m. d'étampes.

steel ingot ‖ Stahlblock m. ‖ lingot m. d'acier. / crude ~ ‖ Rohstahlblock m. ‖ lingot m. d'acier brut. / ~ rolling ‖ Auswalzen n. von Stahlblöcken ‖ laminage m. de lingots d'acier.

steel jewels pl. ‖ Stahlschmuck m. ‖ bijoux mpl. d'acier. / way of judging ~ by the appearance of the fracture ‖ Stahlbeurteilung f. nach dem Bruchaussehen ‖ appréciations fpl. de l'acier basées sur l'aspect de la cassure. / ~ magnet ‖ Stahlmagnet m. ‖ aimant m. en acier. / ~ mortar ‖ Stahlmörser m. ‖ mortier m. en acier. / ~ moulding ‖ Stahlgußformerei f. ‖ moulage m. d'acier. / ~ pen ‖ Stahlschreibfeder f. ‖ plume f. d'acier. / ~ pig ‖ Stahlroheisen n.; Rohstahleisen n. ‖ fonte f. aciéreuse. / ~ pin ‖ Stahlstecknadel f. ‖ épingle f. en acier. / ~ pipe ‖ Stahlröhre f. ‖ tube m. en acier. / electric ~ plant ‖ Elektrostahlwerk n. ‖ aciérie f. à four électrique.

steel plate ‖ Stahlblech n. ‖ tôle f. d'acier. / galvanized ~ ‖ verzinktes Stahlblech n. ‖ tôle f. en acier galvanisée. / lead-coated ~ ‖ verbleites Stahlblech n. ‖ tôle f. en acier plombée. / perforated ~ ‖ gelochtes Stahlblech n. ‖ tôle f. d'acier perforée.

steel plate engraver ‖ Stahlstecher m. ‖ graveur m. sur acier. / ~ printer ‖ Stahldrucker m. ‖ imprimeur m. sur acier. / ~ printing shop ‖ Stahldruckerei f. ‖ imprimerie f. sur acier. / ~ work ‖ Stahlblechkonstruktion f. ‖ construction f. en tôle d'acier.

steel plating plant ‖ Verstählungsanlage f. ‖ installation f. d'aciérage. / ~ pourer ‖ Stahlgießer m. ‖ verseur m. d'acier ou de poches. / half-finished ~ product ‖ Halbzeug n. ‖ demi-produit m. en acier.

steel production ‖ Stahlerzeugung f.; Stahlherstellung f. ‖ fabrication f. de l'acier. / ~ by the Bessemer process ‖ Herstellung f. von Stahl im Bessemerverfahren ‖ fabrication f. de l'acier par le procédé Bessemer. / ~ by the acid Bessemer process ‖ Stahlerzeugung f. nach dem sauren Bessemerverfahren ‖ fabrication f. de l'acier d'après le procédé Bessemer acide. / ~ in the electric furnace ‖ Stahlgewinnung f. im Elektrofen ‖ production f. de l'acier dans un four électrique. / ~ by the Martin process ‖ Herstellung f. von Stahl im Martinverfahren ‖ fabrication f. de l'acier par le procédé Martin.

steel puddling ‖ Stahlpuddeln n. ‖ puddlage m. d'acier. / ~ pulley ‖ Stahlriemenscheibe f. ‖ poulie f. en acier. / ~ quality ‖ of ‖ Stahlgüte f.; Stahlqualität f. ‖ qualité f. d'acier. / ~ rail ‖ Stahlschiene f. ‖ rail m. en acier. / ~ rim ‖ Stahlfelge f. ‖ jante f. en acier. / ~ roll ‖ Stahlwalze f. ‖ rouleau m. en acier. / ~ rolling mill ‖ Stahlwalzwerk n. ‖ usine f. de laminage d'acier.

steel rope ‖ Stahlseil n. ‖ câble m. en acier. / ~ with moderate twist ‖ drallarmes Stahlseil n. ‖ câble m. souple en acier à faible torsion.

steel scrap ‖ Stahlschrott m. ‖ mitraille f. d'acier. / ~ sectional mast ‖ Stahlmast m. in Teilen ‖ mât m. d'acier en sections.

/ ~ semi-product of electric steel ‖ Elektrostahlhalbzeug n. ‖ demi-produit m. en acier au four électrique. / ~ shavings pl. see steel chips pl. / ~ sheath ‖ Stahlscheide f. ‖ fourreau m. d'acier. / ~ shelf ‖ Stahlregal n. ‖ rayon m. en acier. / ~ ship ‖ Stahlschiff n. ‖ navire m. en acier. / ~ spar (Aero) ‖ Stahlholm m. ‖ longeron m. en acier. / ~ spectacles pl. ‖ Stahlbrille f. ‖ lunettes fpl. en acier. / ~ spring ‖ Stahlfeder f. ‖ ressort m. en acier. / ~ stamp ‖ Stahlstempel m. ‖ étampe f. en acier. / ~ string ‖ Stahlsaite f. ‖ corde f. en acier. / ~ structural work ‖ Eisenbauwerk n.; Eisenkonstruktion f. ‖ charpente f. métallique. / ~ structure for architectural buildings ‖ Eisenkonstruktion f. für architektonische Bauwerke ‖ charpente f. métalliques pour bâtiments architectoniques. / ~ tape ‖ Stahlbandmaß n. ‖ mesure f. à ruban en acier. / ~ tapping side ‖ Stahlabstichseite f. ‖ côté f. coulée de l'acier. / ~ tool ‖ Stahlwerkzeug n. ‖ outil m. en acier. / ~ tube see ~ pipe. / ~ turning ‖ Stahldrehspan m. ‖ tournure f. d'acier. / ~ type ‖ Stahltype f. ‖ caractère m. en acier. / ~ vessel ‖ Stahlgefäß n. ‖ vase m. en acier. / ~ ware ‖ Stahlware f. ‖ quincaillerie f. en acier. / ~ wheel ‖ Stahlrad n. ‖ roue f. en acier.

steel wire ‖ Stahldraht m. ‖ fil m. d'acier. / drawn ~ ‖ gezogener Stahldraht m. ‖ fil m. d'acier tréfilé. / galvanized ~ ‖ verzinkter Stahldraht m. ‖ fil m. d'acier étamé. / ~ for needles ‖ Nadelstahldraht m. ‖ fil m. d'acier pour aiguilles. / nickel-plated ~ ‖ vernickelter Stahldraht m. ‖ fil m. d'acier nickelé.

steel wire armouring ‖ Stahldrahtarmierung f. ‖ armature f. en fil d'acier. / ~ band ‖ Stahldrahtband n. ‖ ruban m. en fil d'acier. / ~ brush ‖ Stahldrahtbürste f. ‖ brosse f. à fils d'acier. / ~ rope ‖ Stahldrahtseil n. ‖ câble m. métallique. / ~ strand ‖ Stahldrahtlitze f. ‖ toron m. en fils d'acier.

steel works pl. ‖ Stahlwerk n. ‖ fonderie f. d'acier; aciérie f. / Bessemer ~ ‖ Bessemerstahlwerk n. ‖ aciérie f. Bessemer. / Siemens-Martin ~ ‖ Siemens-Martin-Stahlwerk n. ‖ aciérie f. Siemens-Martin. / Thomas ~ ‖ Thomasstahlwerk n. ‖ aciérie f. Thomas.

steel works blower ‖ Stahlwerkgebläse n. ‖ soufflante f. d'aciérie. / ~ crane ‖ Stahlwerkkran m. ‖ grue f. d'aciérie. / ~ equipment ‖ Stahlwerkseinrichtung f. ‖ équipement m. d'aciérie. / ~ laboratory ‖ Stahlwerkslaboratorium n. ‖ laboratoire m. d'aciérie.

steely malt ‖ Glasmalz n. ‖ malt m. vitreux.

steelyard ‖ Laufgewichtswage f.; Schnellwage f. ‖ balance f. romaine.

steep, to ‖ eintauchen ‖ immerger. / ~ (Brew) ‖ einweichen; einquellen ‖ tremper; mettre en trempe f.; mouiller. / ~ in alum (Dyer) ‖ in Alaunwasser n. sieden ‖ aluner une étoffe. / ~ in lye ‖ laugen ‖ lessiver.

steep ‖ steil ‖ raide; escarpé. / ~ coast ‖ Steilküste f. ‖ falaise f. / ~ coast covered with dunes ‖ Dünensteilküste f. ‖ côte f. escarpée couverte de dunes. / ~ dipping seam ‖ steiles Flöz n. ‖ couche f. en dressant.

steep (Brew) ‖ Getreideweiche f. ‖ cuve f. mouilloire ou à tremper; auge f. de

trempage. / barley ~ ‖ Gerstenweiche f. ‖ cuve f. à tremper l'orge; auge f. de trempage pour orge.

steeped (Brew) ‖ geweicht; eingeweicht ‖ mouillé; trempé.

steeper (Dyer) ‖ Eintaucher m. ‖ baigneur m.

steeping ‖ Einquellen n.; Einweichen n. ‖ trempage m.; mouillage m. / ~ of wood ‖ Holzimprägnierung f. ‖ imprégnation f. *ou* pénétration f. *ou* injection f. des bois.

steeping bowl, acid ‖ Säureeinweichbottich m. ‖ cuve f. de trempage acide. / ~ cistern (Brew) ‖ Quellbottich m. ‖ cuve f. mouilloire. / ~ trough (Brew) ‖ Quellbottich m.; Weichkufe f. ‖ cuve f. mouilloire. / ~ vat (Dyer) ‖ Gärungsküpe f. ‖ pourriture f.; trempoire f.

steeple ‖ Kirchturm m. ‖ clocher m.

steeply ascending ‖ steil ansteigend ‖ à front raide.

steer, to ‖ lenken ‖ diriger. / ~ (Mar) ‖ steuern ‖ gouverner. / ~ a relay ‖ ein Relais n. steuern ‖ contrôler un relais. / ~ towards... (Mar) ‖ ansteuern ‖ faire aborder.

steerage ‖ Zwischendeck n. ‖ entrepont m. / ~ passenger ‖ Zwischendeckpassagier m. ‖ passager m. de l'avant.

steer breeder ‖ Stierzüchter m. ‖ propriétaire m. d'un taureau étalon.

steering ‖ Steuerung f.; Lenkung f. ‖ direction f. / ~ apparatus (Shipb) ‖ Dampfsteuerapparat m. ‖ machine f. pour gouvernail à vapeur. / forked ~ arm ‖ Gabelsteuerhebel m. ‖ levier m. fourché de direction. / ~ axle ‖ Steuerachse f.; Steuerwelle f. ‖ essieu m. directeur. / ~ column ‖ Steuersäule f.; Lenksäule f. ‖ colonne f. de direction. / ~ connecting rod ‖ Lenkschubstange f. ‖ bielle f. de direction. / ~ device (Mar) ‖ Steuervorrichtung f. ‖ appareil m. gouvernail.

steering gear ‖ Lenkvorrichtung f.; Steuerungsmechanismus m. ‖ appareil m. *ou* mécanisme m. de direction. / screw spindle ~ ‖ Schraubenspindelsteuerung f. ‖ commande f. par arbre fileté.

steering gear arm ‖ Lenkstockhebel m. ‖ levier m. de direction. / ~ case ‖ Steuerungsgehäuse n. ‖ boîte f. de direction. / ~ connecting rod (Auto) ‖ Lenkstange f. ‖ barre f. de direction.

steering guide ‖ Steuerführung f. ‖ guide m. de direction.

steering knuckle ‖ Achsschenkel m.; Achszapfen m.; Achshals m. ‖ fusée f. d'essieu. / ~ arm ‖ Lenkschenkel m. ‖ levier m. de commande de la fusée. / ~ pivot ‖ Vorderachszapfen m. ‖ pivot m. de l'essieu avant. / ~ spindle ‖ Achsschenkel m. ‖ fusée f. de l'essieu avant. / ~ tie rod ‖ Spurstange f. ‖ barre f. d'accouplement.

steering lever ‖ Steuerhebel m. ‖ levier m. de commande. / ~ light (Mar) ‖ Positionslicht n. ‖ feu m. de route. / ~ lock (Auto) ‖ Ausschlag m. der Räder; Steuerungsausschlag m. ‖ braquage m. / ~ nut ‖ Lenkmutter f. ‖ écrou f. de direction. / ~ pivot pin ‖ Achsschenkelbolzen m. ‖ tourillon m. de direction. / ~ screw ‖ Lenkschraube f. ‖ vis f. de direction.

steering shaft ‖ Schaltwelle f. ‖ arbre m. d'avance. / ~ (Auto) ‖ Steuerspindel f. ‖ arbre m. de colonne de direction.

steering stop ‖ Lenkungsanschlag m. ‖ arrêt m. de braquage. / ~ tackle (Shipb)

‖ Rudertalje f. ‖ palan m. de la barre *ou* du gouvernail.

steering wheel ‖ Lenkrad n.; Steuerrad n. ‖ volant m. de direction. / ~ (Shipb) ‖ Steuerrad n. ‖ roue f. du gouvernail. / ~ shaft ‖ Lenkspindel f. ‖ arbre m. de colonne de direction.

steering worm ‖ Lenkschnecke f. ‖ vis f. sans fin de direction. / ~ sector ‖ Lenksegment n. ‖ secteur m. de direction.

steeve of the bow sprit ‖ Steigung f. des Bugspriets ‖ apiquage m. du beaupré.

stellar time ‖ Sternzeit f. ‖ temps m. sidéral.

stem, to ~ the water ‖ das Wasser anstauen ‖ hausser *ou* élever les eaux.

stem ‖ Stamm m. ‖ tige f.; tronc m. / ~ (Shipb) ‖ Steven m. ‖ étrave m.; étambot m. / ~ of a girder ‖ Trägersteg m. ‖ âme f. de la poutre. / ~ of a rail ‖ Steg m. einer Eisenbahnschiene ‖ tige f. d'un rail. / rough-hewn ~ (Forest) ‖ waldrecht behauener Holzstamm m. ‖ tronc m. dégrossi à la forêt. / screw-on ~ ‖ abschraubbarer Stiel m. ‖ manche m. démontable. / ~ of tree ‖ Baumstamm m. ‖ corps m. *ou* tronc m. d'arbre.

stem pipe (Shipb) ‖ Vorstevenklüse f. ‖ écubier m. d'étrave. / ~ shoe (Shipb) ‖ Vorstevenschuh m. ‖ sabot m. d'étrave. / ~ winding watch ‖ Remontoiruhr f. ‖ montre f. à remontoir.

stench ‖ Gestank m. ‖ mauvaise odeur f.

stencil ‖ Schablone f. ‖ patron m.; stencil m. / ~ for spray painting ‖ Schablone f. für Spritzmalerei ‖ patron m. pour la peinture au pistolet.

stencil cutting machine ‖ Schablonenschneidemaschine f. ‖ machine f. à couper les stencils *ou* les patrons. / ~ paper ‖ Schablonenpapier n. ‖ papier m. à stencils *ou* à patrons. / ~ plate (Weav) ‖ Schablone f. ‖ gabari m.

stenographer ‖ Stenograph m. ‖ sténographe m.

stenographer's note book ‖ Stenogrammblock m. ‖ bloc-notes m. de sténographe. / loose-leaf ~ ‖ Stenogrammblock m. mit losen Blättern ‖ bloc-notes m. de sténographe à feuillets mobiles.

stenopaic slit ‖ stenopäischer Spalt m. ‖ fente f. stenopéique.

step, to ~ down (Electr) ‖ herabtransformieren ‖ réduire le voltage. / ~ up (Electr) ‖ hinauftransformieren ‖ survolter.

step ‖ Maßnahme f.; Maßregel f. ‖ mesure f. / ~ (Ladder) ‖ Sprosse f. ‖ échelon m. / ~ (Railw) ‖ Trittbrett n.; Auftritt m.; Wagentritt m. ‖ marchepied m. / ~ (Stairs) ‖ Stufe f. ‖ marche f. / ~ of axle bearing ‖ Achslagerschale f. ‖ coussinet m. de la boîte d'essieu. / ~ of the capstan (Shipb) ‖ Spillbett n.; Spillspur f. ‖ carlingue f. de cabestan. / to come into ~ ‖ in Tritt m. kommen ‖ atteindre le synchronisme. / the motor comes out of ~ ‖ der Motor fällt außer Tritt m. ‖ le moteur ne tourne plus synchroniquement. / curved ~ (Stair) ‖ geschweifte Stufe f. ‖ marche f. courbe. / to fall out of ~ ‖ außer Tritt m. fallen ‖ se décrocher. / ~ of spindle ‖ Spindelbuchse f. ‖ gaine f. de broche.

step-back welding ‖ Pilgerschrittschweißung f. ‖ soudage m. à pas de pèlerin.

step bearing ‖ Spurlager n. ‖ crapaudine f. / ~ board see step (Railw). / ~ bracket ‖

Auftrittstütze f.; Tritthalter m. ‖ support m. de marchepied.

step-by-step seam welding ‖ Schweißen n. im Rollenschrittverfahren ‖ soudage m. pas à pas par électrode roulante. / ~ system (Aut tel) ‖ Schrittschaltsystem n. ‖ système m. à pas à pas.

step cone ‖ Stufenkonus m. ‖ cône m. à gradins.

step-down transformer ‖ Abwärtstransformator m. ‖ dévolteur m.; transformateur-réducteur m.

step grate (Fuel) ‖ Treppenrost m. ‖ grille f. à gradins *ou* en escalin *ou* à étages. / ~ groove (Carp) ‖ Stufennut f. ‖ emmarchement m. / ~ hanger ‖ Trittbretthalter m. ‖ patte f. de marchepied.

stephanite ‖ Antimonsilberglanz m. ‖ stéphanite f.

step ladder ‖ Stufenleiter f. ‖ échelle f. de meunier. / ~ mat ‖ Trittbrettmatte f. ‖ tapis-décrottoir m.; matte f. de marchepied.

stepped ‖ abgestuft ‖ gradué. / ~ bore hole ‖ abgesetzte Bohrung f. ‖ alésage m. étagé. / ~ resistance ‖ abgestufter Widerstand m. ‖ résistance f. à degrés.

stepping line (Aut tel) ‖ Impulsleitung f. ‖ ligne f. d'impulsion. / ~ magnet (Electr) ‖ Fortschaltemagnet m. ‖ électro-aimant m. d'avancement. / ~ place of a shaft ‖ Schachtbühne f. ‖ repos m. d'un puits. / ~ relay ‖ Stufenrelais m. ‖ relais m. à action graduée *ou* à double action. / ~ switch (Aut tel) ‖ Schrittschaltwerk n. ‖ commutateur m. pas-à-pas.

step pulley ‖ Stufenscheibe f. ‖ cône m. étagé *ou* à gradins. / ~ rail ‖ Trittbrettschiene f. ‖ rail m. de marchepied. / ~ roll ‖ Stufenwalze f. ‖ cylindre m. à cones. / ~ spring ‖ Stufenfeder f. ‖ ressort m. à feuilles étagées.

steps teller ‖ Schrittzähler m. ‖ compte-pas m.; pédomètre m.

step-up transformer ‖ Aufwärtstransformator m. ‖ survolteur m.; transformateur-élévateur m.

steradian ‖ räumlicher Winkel m. ‖ angle m. solide; stéradian m.

stercorite ‖ Phosphorsalz n. ‖ stercorite f.

stereo camera ‖ Stereokamera f. ‖ chambre f. stéréoscopique.

stereo comparator with flicker microscope ‖ Stereokomparator m. mit Blinkmikroskop ‖ stéréocomparateur m. avec microscope à éclipses.

stereometry ‖ Geometrie f. des Raumes; Stereometrie f. ‖ géométrie f. à trois dimensions; stéréométrie f.

stereo-microscopy ‖ Stereomikroskopie f. ‖ stéréo-microscopie f.

stereophotograph, instantaneous ~ ‖ Augenblicksstereoaufnahme f. ‖ photographie f. stéréoscopique instantanée.

stereophotographic surveying ‖ Raumbildmessung f. ‖ mesure f. stéréophotographique.

stereophotographing, apparatus for ~ the anterior segment of the eye ‖ Apparat m. zur Stereofotografie des vorderen Augenabschnittes ‖ chambre f. stéréophotographique pour la partie avant de l'œil.

stereoscope ‖ Stereoskop n. ‖ stéréoscope m. / ~ picture see stereoscopic picture.

stereoscopic ‖ stereoskopisch ‖ stéréoscopique. / ~ dissecting microscope ‖ stereoskopisches Präpariermikroskop n. ‖ microscope m. à dissection stéréoscopique.

/ ~ effect of the view ‖ erhöhte Plastik f. des Bildes ‖ relief m. rehaussé de l'image. / ~ image ‖ Raumbild n. ‖ image f. en relief. / ~ magnifier ‖ stereoskopische Lupe f. ‖ loupe f. stéréoscopique. / ~ picture ‖ Stereoaufnahme f.; Stereoskopbild n. ‖ image f. ou photographie f. stéréoscopique. / ~ range finder ‖ Stereotelemeter n. ‖ télémètre m. stéréoscopique. / ~ telemeter ‖ Raumbildentfernungsmesser m. ‖ télémètre m. stéréoscopique; stéréo-télémètre m.

stereoscopically viewing of eye ‖ stereoskopische Lupenbeobachtung f. des Auges ‖ examen m. stéréoscopique de l'œil.

stereostop ‖ Stereoblende f. ‖ stéréo-diaphragme m.

stereotype, to ‖ stereotypieren; abklatschen ‖ clicher.

stereotype ‖ Plattenschrift f.; Stereotypplatte f. ‖ stéréotype f.; cliché m. /~ block ‖ Klischee n. ‖ cliché m. / ~ founder ‖ Stereotypengießer m. ‖ fondeur-stéréotypeur m. / ~ foundry ‖ Stereotypengießerei f. ‖ fonderie f. de stéréotypes. / ~ plate holder ‖ Stereotypieplattenhalter m. ‖ porte-plaques m. de stéréotypie. / ~ printing ‖ Plattendruck m.; Stereotypendruck m. ‖ stéréotypage m.; stéréotypie f.

stereotyper ‖ Stereotyparbeiter m. ‖ stéréotypeur m.; clicheur m.

stereotype tempering plant ‖ Stereotypieverhärtungsanlage f. ‖ installation f. de durcissage de stéréotypies.

stereotyping ‖ Stereotypieren n.; Klischieren n. ‖ clichage m.; stéréotypage m. / ~ ‖ Stereotypie f. ‖ stéréotypie f. / ~ apparatus ‖ Stereotypieapparat m. ‖ appareil m. de stéréotypie. / auxiliary printing machine for ~ ‖ Druckhilfsmaschine f. für Stereotypie ‖ machine f. auxiliaire d'impression pour stéréotypie. / ~ workshop ‖ Stereotypieranstalt f. ‖ atelier m. de stéréotypie.

stereotypography ‖ Plattendruck m.; Stereotypendruck m. ‖ stéréotypage m.; stéréotypie f.

stereotypy ‖ Stereotypie f. ‖ stéréotypie f. / ~ installation ‖ Stereotypieeinrichtung f. ‖ installation f. de stéréotypie.

sterilization ‖ Sterilisation f.; Sterilisierung f.; Entkeimung f. ‖ stérilisation f.; dégermination f. / ~ of water ‖ Wassersterilisation f. ‖ stérilisation f. de l'eau.

sterilization apparatus for house keeping ‖ Sterilisierapparat m. für den Haushalt ‖ appareil m. de stérilisation pour le ménage. / chloride gas apparatus for ~ ‖ Chlorgassterilisierapparat m. ‖ appareil m. de stérilisation au gaz de chlorure.

sterilize, to (Chem) ‖ sterilisieren ‖ stériliser.

sterilized cotton ‖ sterilisierende Watte f. ‖ coton m. stérilisé. / ~ water manufacture ‖ Wasserfiltrierunternehmung f. ‖ entreprise f. d'eaux filtrées.

sterilizer ‖ Sterilisator m. ‖ stérilisateur m.

sterilizing apparatus ‖ Sterilisierapparat m. ‖ appareil m. à stériliser; stérilisateur m. / ~ and washing apparatus for filterstuff ‖ Filtermassensterilisier- und Waschapparat m. ‖ appareil m. à stériliser et à laver la masse filtrante.

stern (Shipb) ‖ Heck n.; Achterschiff n.; Hinterschiff n. ‖ arrière m.; poupe f. / heavy by the ~ ‖ achterlastig; steuerlastig ‖ lourd sur l'arrière m.; lourd sur cul m.

stern chaser ‖ Heckgeschütz n. ‖ canon m. ou pièce f. de retraite. / ~ cleat ‖ Heckklampe f. ‖ taquet m. de poupe. / ~ davit ‖ Heckdavit m. ‖ bossoir m. du portemanteau. / ~ flag ‖ Heckflagge f. ‖ pavillon m. ou enseigne f. de poupe; grande enseigne f. / ~-heavy ‖ hecklastig ‖ chargé à la poupe; lourd de la queue; à queue f. lourde. / ~ light ‖ Hecklampe f. ‖ feu m. arrière. / ~ mouldings pl. ‖ Heckverzierungen fpl. ‖ sculptures fpl. de l'arrière. / ~ post ‖ Hintersteven m.; Achtersteven m. ‖ étambot m. / ~ tube ‖ Wellenaustrittsrohr n. ‖ tube m. d'arbre d'hélice.

stern wheel ‖ Heckrad n. ‖ roue f. arrière. / ~ steamer ‖ Heckraddampfer m. ‖ vapeur m. avec roue arrière. / ~ tug ‖ Heckradschlepper m. ‖ remorqueur m. à roue arrière.

stevedore ‖ Stauer m.; Staumeister m. ‖ arrimeur m.

steward ‖ Proviantmeister m.; Bottelier m. ‖ distributeur m.; sommelier m.

stewer, fodder ~ ‖ Futterdämpfer m. ‖ étuve f. pour fourrage.

stew pan ‖ Schmortopf m. ‖ marmite f. à ragout.

stibium ‖ Antimon n. ‖ antimoine m.

stick, to ‖ auf Grund m. sitzen ‖ être échoué. / the punches pl. stick (Cash register) ‖ die Stifte mpl. klemmen ‖ les poinçons mpl. coincent. / the valve sticks ‖ das Ventil hängt ‖ la soupape reste accrochée.

stick ‖ Stock m.; Spazierstock m. ‖ canne f. / ~ (Coachm) ‖ Spriegel m. ‖ cerceau m. / ~ (Print) ‖ Steg m. ‖ bois m. de garniture. / ~ (Pyrot) ‖ Raketenstab m.; Rakenrute f. ‖ baguette f. de direction; panaceau m. / ~ (Tool) ‖ Stiel m. ‖ bâton m.; manche m. / resting ~ ‖ Malerstock m. ‖ appui-main m. / short ~ ‖ Knebel m. ‖ garot m.; bâillon m. / ~ of solder ‖ Stangenlötzinn n. ‖ tige f. de soudure. / walking ~ ‖ Spazierstock m. ‖ canne f.

stick bending machine ‖ Stockbiegemaschine f. ‖ machine f. à plier les cannes. / ~ candy spinner ‖ Zuckerspinner m. ‖ fileur m. de sucre. / ~ control ‖ Knüppelsteuerung f. ‖ commande f. par manche à balai. / ~ force (Aero) ‖ Steuerkraft f. ‖ force f. sur manche à balai. / ~ handle ‖ Stockgriff m. ‖ poignée f. de canne.

sticking (Bookb) ‖ Aufziehen n. (des Buchrückens) ‖ couvrure f. / ~ patch see ~ plaster. / ~ plaster ‖ Heftpflaster n. ‖ emplâtre m. adhésif; sparadrap m. / ~ wax ‖ Klebwachs n.; Wachskitt m. ‖ cire f. à luter.

stick lac ‖ Stocklack m. ‖ laque f. en bâtons. / ~ maker (Tool) ‖ Stockmacher; Stielschneider m. ‖ cannier m.; bâtonnier m. / ~ potash ‖ Stangenkali n. ‖ potasse f. en crayons.

sticks pl. ‖ Reiser npl. ‖ ramilles fpl. / round ~ ‖ Knüppelholz n. ‖ rondins mpl.; rondinage m.

stick space (Print) ‖ Quadrat n.; breites Spatium n. ‖ cadrat m. / ~ sulphur ‖ Stangenschwefel m. ‖ soufre m. en canon. / ~ umbrella ‖ Stockschirm m. ‖ canneparapluie m.

sticky ‖ pappig ‖ pâteux.

stiffen, to ‖ versteifen ‖ renforcer.

stiffened ‖ versteift ‖ renforcé. / ~ inwards and outwards ‖ von innen und außen verstärkt ‖ avec plaques fpl. de renfort en dedans et au dehors. / ~ with sheetmetal corner pieces ‖ mit Eckblechen npl. versteift ‖ renforcé par des goussets mpl. de tôle.

stiffener skiving machine (Shoem) ‖ Kappenschärfmaschine f. ‖ machine f. à parer les contreforts.

stiffening (Shoem) ‖ Hinterkappe f. ‖ contrefort m. / ~ of the frame ‖ Rahmenversteifung f. ‖ renforcement m. du châssis. / longitudinal ~ ‖ Längsversteifung f. ‖ renforcement m. ou raidissement m. longitudinal. / ~ of the web ‖ Stegversteifung f. ‖ raidisseur m. de l'âme.

stiffening angle ‖ Versteifungswinkel m. ‖ gousset m. ou équerre f. de renforcement. / ~ bar for one-sided traverses ‖ Versteifungsschiene f. für einseitige Querträger ‖ jambe f. de force spéciale pour traverses coupées. / ~ brace ‖ Verstärkungsstrebe f. ‖ entretoise f. de renforcement. / ~ cutter (Shoem) ‖ Hinterkappenzuschneider m. ‖ coupeur m. de contreforts. / ~ iron ‖ Versteifungseisen n. ‖ pièce f. de renforcement. / cylinder ~ ‖ Zylinderverstrebung f. ‖ renforcement m. de cylindre. / ~ plate ‖ Verstärkungsblech n.; Stehblech n. ‖ tôle f. ou plaque f. de renfort; âme f. / ~ rib ‖ Verstärkungsrippe f. ‖ nervure f. de renforcement. / ~ ring ‖ Verstärkungsring m. ‖ bague f. de renforcement. / ~ sheet see ~ plate. / ~ strap ‖ Versteifungsbügel m. ‖ bride f. ou étrier m. de renfort.

stiffness ‖ Steifheit f. ‖ raideur f.

stifle, to ~ frequencies ‖ Frequenzen fpl. unterdrücken ‖ étouffer des fréquences fpl.

stilbuite ‖ Schwefelantimon n.; Grauspießglanz m.; Antimonglanz m. ‖ antimoine m. sulfuré.

stile of a frame (Join) ‖ stehendes Rahmstück n.; Höhschenkel m. eines Rahmens ‖ montant m. d'un châssis.

still, in ~ air ‖ bei Windstille f. ‖ par vent m. nul. / ~ champagne ‖ nicht moussierender Champagner m. ‖ champagne m. non-mousseux.

still (Chem) ‖ Destillierapparat m.; Kolben m. ‖ appareil m. à distiller; alambic m. / benzine ~ ‖ Benzindestillierapparat m. ‖ appareil m. de distillation de benzine. / secondary ~ ‖ Nachdestiller m. ‖ appareil m. de redistillation; appareil m. redistillateur.

stillage ‖ Faßlager n. ‖ chantier m.

stillion ‖ Gärfaß n. ‖ fût m. ou tonne f. de fermentation.

stillman ‖ Brenner m. ‖ distillateur m.

stilt (Build) ‖ Gerüststange f. ‖ boulin m.; perche f.

stimulate, to ‖ antreiben; anregen ‖ pousser; stimuler; animer.

stimulus ‖ Reizmittel n. ‖ stimulant m.

stipple graver ‖ Punktierstichel m. ‖ burin m. à pointiller.

stippling ‖ Punktierung f. ‖ pointillage m.

stipulate, to ‖ vereinbaren; ausbedingen ‖ convenir; stipuler. / ~ in writing ‖ schriftlich vereinbaren ‖ stipuler par écrit m.

stipulated load ‖ vorgeschriebene Belastung f. ‖ charge f. prescrite.

stir, to (Chem) ‖ umrühren ‖ agiter; remuer; brasser. / ~ briskly ‖ stark umrühren ‖ remuer énergiquement. / ~ the dough ‖

den Teig m. rühren ‖ malaxer la pâte. / ~ the fire ‖ das Feuer anschüren ‖ tisonner *ou* râbler le feu.

stirrer ‖ Rührwerk n. ‖ agitateur m. mécanique; remueur m. / wooden ~ ‖ Rührholz n. ‖ spatule f. en bois.

stirrer blade ‖ Rührflügel m. ‖ palette f. de l'agitateur. / ~ spindle ‖ Rührwerkswelle f. ‖ arbre m. de l'agitateur.

stirring (Brew) ‖ Aufrühren n. ‖ vaguage m. / ~ apparatus ‖ Rührwerk n. ‖ agitateur m. mécanique. / ~ arm ‖ Rührarm m. ‖ bras m. de mélangeur *ou* d'agitateur. / ~ blades pl. working in opposite direction ‖ gegeneinander arbeitendes Rührwerk n. ‖ agitateur m. à mouvement contraire. / ~ boiler ‖ Rührkessel m. ‖ chaudière f. à agitateur.

stirring device ‖ Rührwerk n. ‖ agitateur m. mécanique; remueur m. / brewery ~ ‖ Brauereirührwerk n. ‖ agitateur m. de brasserie.

stirring machine, dough ~ ‖ Teigrührmaschine f. ‖ machine f. à remuer la pâte.

stirring mechanism, planetary ~ ‖ Planetenrührwerk n. ‖ agitateur m. planétaire.

stirring pan ‖ Rührpfanne f. ‖ marmite f. à agitateur. / ~ rabble ‖ Krücke f. ‖ râble m. / ~ spattle ‖ Rührspatel m. ‖ spatule f. / ~ stand ‖ Rührstativ n. ‖ support m. de mélangeur. / ~ tub ‖ Rührbottich m. ‖ cuve f. à agitateur.

stirrup (Build) ‖ Steigeisen n. ‖ échelon m. / ~ (Carp) ‖ Bügel m. ‖ cerceau m.; étrier m. / ~ (Saddl) ‖ Steigbügel m. ‖ étrier m. / ~ (Shoem) ‖ Knieriemen m. ‖ tire-pied m. / U-shaped ~ ‖ U-förmiger Bügel m. ‖ étrier m. en forme d'U.

stitch, to (Bookb) ‖ heften; brochieren ‖ piquer; coudre; brocher. / ~ (Weav) ‖ ketteln ‖ remmailler; entrelacer. / ~ the cap peaks pl. ‖ die Mützenschirme mpl. steppen ‖ piquer les visières fpl. de casquettes. / ~ leather (Saddl) ‖ Leder n. mit Lederriemen nähen ‖ brédir. / ~ on ‖ anketteln ‖ remmailler.

stitch ‖ Stich m. (beim Nähen) ‖ point m. d'aiguille. / double ~ ‖ Doppelmasche f. ‖ maille f. gardée.

stitched (Bookb) ‖ broschiert; geheftet ‖ broché. / ~ book ‖ broschiertes Buch n.; Heft n. ‖ livre m. broché; fascicule m. / ~ pack (Bookb) ‖ gehefteter Block m. ‖ bloc m. cousu *ou* piqué. / ~ stuff and knitting machine ‖ Wirk- und Strickmaschine f. ‖ machine f. à mailler et à tricoter.

stitcher (Bookb) ‖ Hefter m. ‖ brocheur m.; couseur m. / ~ (Tail) ‖ Stepper m. ‖ piqueur m. / machine-~ ‖ Maschinenhefter m. ‖ couseur m. à la machine. / wire ~ ‖ Drahthefter m. ‖ couseur m. au fil de fer.

stitch glass ‖ Maschenzähler m. ‖ comptemaille m.

stitching (Bookb) ‖ Broschüre f.; Heften n. brochure f.; brochage m. / ~ (Tail) ‖ Steppung f. ‖ piqûre f. / book ~ ‖ Buchheften n. ‖ brochage m. d'un livre. / machine-~ ‖ Maschinensteppen n. ‖ piquage m. à la machine.

stitching awl ‖ Sattlerahle f. ‖ alêne f. de sellier. / ~ hook ‖ Heftklammer f. ‖ clavette f. de brochage. / ~ machine (Bookb) ‖ Heftmaschine f. ‖ machine f. à brocher. / ~ machine (Weav) ‖ Steppmaschine f. ‖ machine f. à piquer. / ~ needle ‖ Heftnadel f. ‖ aiguille f. à relier. / ~

thread ‖ Heftfaden m. ‖ fil m. à brocher. / ~ wire ‖ Heftdraht m. ‖ fil m. métallique à brocher.

stoak, to ‖ tunken ‖ saucer.

stock (Bank) ‖ Betriebskapital n.; Fonds m. ‖ fonds mpl.; capital m. / ~ (Forg) ‖ Hammerstock m. ‖ billot m.; tronchet m.; chabotte f. de l'enclume. / ~ (Inventory) ‖ Lagerbestand m. ‖ stock m.; inventaire m. / ~ (Mach) ‖ Schaft m. ‖ fût m.; tige f. / ~ (Storage place) ‖ Speicher m. ‖ magasin m.; entrepôt m. / ~ (Trade) ‖ Warenvorräte mpl. ‖ approvisionnements mpl.; magasin m. / the capital ~ is almost entirely in the hands of . . . ‖ das Aktienkapital n. ist fast völlig im Besitz von . . . ‖ la presque totalité des actions appartient à . . . / circulation ~ ‖ Betriebsmaterial n. ‖ matériel m. d'exploitation. / to detach the ~ ‖ das Mahlgut auflösen ‖ désagréger le produit. / government ~ ‖ Anleihepapier n. ‖ titre m. d'emprunt. / gun ~ ‖ Gewehrschaft m. ‖ fût m. de fusil. / ~ in hand ‖ Warenbestand m. ‖ stock m. / in ~ ‖ vorrätig ‖ en magasin m. / not in ~ ‖ nicht auf Lager n. ‖ pas en magasin m. / to own the majority of the ~s of a joint-stock company ‖ die Majorität f. einer Aktiengesellschaft besitzen ‖ être titulaire m. de la majorité des parts d'une société anonyme. / to keep in ~ ‖ auf Lager n. haben ‖ avoir en magasin m. / to lay a ship on the ~s pl. ‖ ein Schiff n. auf Stapel legen ‖ mettre un navire sur le chantier. / to loosen the ~ ‖ das Mahlgut lockern ‖ ramollir la marchandise. / ~ of a plane ‖ Hobelgehäuse n. ‖ fût m. de rabot. / preferred ~ ‖ Vorzugsaktie f. ‖ action f. privilégiée. / rolling ~ ‖ rollendes Material n. ‖ matériel m. roulant. / screw ~ ‖ Gewindekluppe f. ‖ filière f. brisée *ou* à coussinets. / to take ~ ‖ Inventur f. aufnehmen; Bestandaufnahme f. machen; inventarisieren ‖ inventorier. / to take delivery of ~s ‖ Aktien fpl. beziehen ‖ prendre livraison f. des titres. / to take-over part of the ~s ‖ einen Aktienanteil m. übernehmen ‖ signer un certain nombre d'actions. / ~ of wood ‖ Holzvorrat m. ‖ provision f. de bois.

stock account ‖ Kapitalkonto n. ‖ compte m. capital.

stockade ‖ Einpfählung f.; Einzäunung f.; Spalier n.; Staket n. ‖ clôture f. de palis; espalier m.; palissade f.

stock and die maker ‖ Kluppenmacher m. ‖ filiériste m.; fabricant m. de filières. / ~ book ‖ Bestandbuch n. ‖ livre m. de stock; état m. / ~ brick ‖ hartgebrannter Ziegel m. ‖ brique f. fortement cuite. / ~ broker ‖ Börsenmakler m.; Effektenmakler m.; Aktienhändler m. ‖ agent m. de change; courtier m. de fonds publics; agioteur m. / ~ capital ‖ Stammkapital n. ‖ fonds mpl. / ~ car ‖ Serienwagen m. ‖ voiture f. de série. / dealer in ~s ‖ Aktienhändler m. ‖ agioteur m.; courtier m. / ~ department ‖ Effektenabteilung f. ‖ département m. pour valeurs mobilières.

stock exchange ‖ Effektenbörse f.; Fondsbörse f. ‖ bourse f. des fonds publics *ou* de valeurs mobilières. / ~ call (Tel) ‖ Börsengespräch n. ‖ conversation f. boursière. / ~ list ‖ Kurszettel m. ‖ cote de la Bourse. / ~ official list ‖ Kursbericht

m. ‖ bulletin m. de la Bourse. / ~ rules pl. ‖ Börsenordnung f. ‖ règlement m. de la Bourse. / ~ rumour ‖ Börsengerücht n. ‖ bruit m. de Bourse. / ~ securities pl. ‖ Börseneffekten pl. ‖ valeurs fpl. en Bourse. / ~ station ‖ Börsensprechstelle f. ‖ poste m. de Bourse. / ~ tax ‖ Börsensteuer f. ‖ impôt m. sur les opérations de la Bourse.

stock fish dryer ‖ Stockfischtrockner m. ‖ sécheur m. de morue.

stockholder ‖ Aktieninhaber m. ‖ porteur m. *ou* détenteur d'actions.

stocking (In stock) ‖ auf Lager ‖ en magasin m.

stocking ‖ Strumpf m. ‖ bas m. / elastic ~ ‖ Gummistrumpf m. ‖ bas m. élastique. / ribbed ~ ‖ gerippter Strumpf m. ‖ bas m. à côtes. / short ~ ‖ Socke f. ‖ chaussette f. / worsted ~ ‖ wollener Strumpf m. ‖ bas m. de laine.

stocking automaton ‖ Strumpfautomat m. ‖ métier automatique m. pour bas. / ~ cutter ‖ Vorfräser m. ‖ fraise f. ébaucheuse. / ~ embroiderer ‖ Strumpfstickerei f. ‖ brodeuse f. de coins de bas. / ~ factory ‖ Strumpffabrik f. ‖ fabrique f. de bas. / ~ finishing ‖ Strumpfappretur f. ‖ apprêt m. de bas. / ~ footing ‖ Strumpfanstricken n. ‖ entage m. de bas. / ~ former ‖ Strumpfformer m. ‖ metteur m. de bas en forme.

stocking frame ‖ Strumpfmaschine f. ‖ machine f. à bas. / ~ needle ‖ Strickmaschinennadel f. ‖ aiguille f. de tricoteuse; aiguille f. pour métiers à bas.

stocking holder ‖ Strumpfhalter m. ‖ jarretelle f.; jarretière f. / ~ knitter ‖ Strumpfwirker m. ‖ tricoteur m. de bas. / ~ knitting ‖ Strumpfstrickerei f. ‖ tricotage m. de bas. / ~ loom ‖ Strumpfwirkerstuhl m. ‖ métier m. à tricoter. / ~ mender ‖ Strumpfstopfer m. ‖ repriseur m. de bas / ~ suspender ‖ Strumpfhalter m. ‖ jarretelle f. / ~ turner ‖ Strumpfwender m. ‖ retourneur m. de bas.

stock jobber ‖ Aktienhändler m. ‖ agioteur m.; courtier m. / ~ lifter (Pap) ‖ Stoffpacker m. ‖ chargeur m. de pâte.

stock-list (Exchange) ‖ Kurszettel m. ‖ bulletin m. de la Bourse. / ~ (Trade) ‖ Lagerliste f. ‖ liste f. de stocks.

stock printing works pl. ‖ Effektendruckerei f. ‖ imprimerie f. de titres. / ~ protection (El line) ‖ Stockschutz m. ‖ préservation f. du pied des poteaux. / ~ rail ‖ Anschlagschiene f. ‖ rail m. fixe. / ~ raiser ‖ Tierzüchter m.; Viehzüchter m. ‖ éleveur m. (d'animaux). / ~ rooms pl. ‖ Lagerhaus n.; Warenlager n. ‖ entrepôt m.; magasin m.; dépôt m.

stocks pl. ‖ Effekten pl. ‖ effets mpl. publics; valeurs fpl. (mobilières). / ~ (Mar) ‖ Helling f.; Stapel m. ‖ cale f.; chantier m. / ~ and dies pl. ‖ Schneidkluppe f. ‖ Windeisen n. ‖ porte-filière m.; filière f. / ~ and shares pl. ‖ Börsenpapiere npl. ‖ valeurs fpl. de bourse.

stock smith ‖ Kluppenschmied m. ‖ forgeur m. de filières. / speculation in ~s ‖ Aktienspekulation f. ‖ agiotage m.

stock taking ‖ Inventur f.; Inventuraufnahme f.; Bestandaufnahme f.; Aufnahme f. eines Warenlagers ‖ inventaire f.; levée f. d'inventaire. / ~ sale ‖ Inventurausverkauf m. ‖ liquidation f. après inventaire.

stock ticker ∥ Börsendrucker m. ∥ télégraphe m. commercial. / ~ wood ∥ Stockholz n. ∥ bois m. du tronc. / ~ yard ∥ Viehhof m. ∥ cour f. aux bestiaux.

stoke, to ∥ das Feuer schüren ∥ attiser ou pousser le feu.

stokehold (Shipb) ∥ Heizraum m. ∥ chambre f. de chauffe.

stoke hole ∥ Feuerloch n.; Heizloch n.; Schürloch n. ∥ chauffe f.; ouverture f. du foyer.

stoker (Boil) ∥ Schüreisen n. ∥ lance f. à feu; ringard m. / ~ (Metal) ∥ Ofenkrücke f. ∥ croc m. à feu; râble m. / ~ (Person) ∥ Heizer m. ∥ chauffeur m. / mechanical ~ ∥ Schürvorrichtung f. ∥ chargeur m. mécanique.

stoker's poker bar ∥ Schürhaken m. ∥ pique-feu f. / ~ shovel ∥ Feuerschaufel f ∥ pelle f. de chauffeur.

stoking ∥ Heizung f. ∥ chauffage m.

stole ∥ Stola f. ∥ étole f.

stomach ∥ Magen m. ∥ estomac m. / animal ~ ∥ Tiermagen m. ∥ estomac m. d'animal.

stone, to (Fruits) ∥ auskernen ∥ cerner; égrener; ôter les noyaux. / ~ (Pav) ∥ besteinen ∥ empierrer.

stone ∥ Stein m. ∥ pierre f. / artificial ~ (Mas) ∥ Kunststein m. ∥ pierre f. artificielle ou factice. / machine for making artificial ~s ∥ Kunststeinherstellungsmaschine f. ∥ machine f. à fabriquer les pierres artificielles. / artificial ~ plant ∥ Fabrik f. für künstliche Steine ∥ installation f. pour la fabrication de pierres artificielles. / boundary ~ ∥ Grenzstein m. ∥ borne f.; borne f. limitrophe. / broken ~ ∥ Bruchstein m. ∥ moellon m. / broken ~s pl. ∥ Steinschlag m.; Schotter m.; Kleinschlag m. ∥ pierres mpl. concassées; pierraille f. / ~ for building ∥ Mauerstein m. ∥ moellon m.; pierre f. à bâtir ou de construction. / to cut a ~ ∥ einen Stein m. behauen ∥ tailler une pierre. / cut ~ ∥ Haustein m. ∥ pierre f. taillée ou de taille. / ~ drinking trough ∥ Schwemmstein m. ∥ abreuvoir m. en pierre. / expletive ~ (Mas) ∥ Füllstein m. ∥ blocaille f. / flagging ~ ∥ Fliese f.; Platte f. ∥ carreau m. en pierre naturelle; dalle f. / hard ~ (Min) ∥ Knauer m.; Hartklamm m. ∥ roche f. très dure. / joining ~ ∥ Satzstein m. ∥ pierre f. d'ajoute. / lithographic ~ ∥ Lithografiestein m. ∥ pierre f. lithographique. / Lydian ~ ∥ Probierstein m.; Kieselschiefer m.; lydischer Stein m.; pierre f. de touche; silex m. corné; lydienne f. / mere ~ ∥ Grenzstein m. ∥ borne f.; borne limitrophe. / metallic ~ ∥ erzhaltiges Gestein n. ∥ minerai m. / ~s pl. for metalling roads ∥ Steine mpl. zur Beschotterung von Straßen ∥ pierres fpl. pour l'empierrement des routes. / natural ~ ∥ Naturstein m. ∥ pierre f. naturelle. / ~ pared on every side ∥ allseitig bearbeiteter Stein m. ∥ pierre f. retournée. / pebble ~ ∥ Gerölle n.; Geschiebe n.; grober Kies m.; Rollstein m.; cailloux mpl. roulés; galet m. / precious ~ ∥ Edelstein m. ∥ pierre f. précieuse. / radial ~ ∥ Schachtstein m. ∥ pierre f. radiale. / ~ for razors ∥ Stein m. für Rasiermesser ∥ pierre f. à rasoir. / rubble ~ see pebble ~. / semi-precious ~ ∥ Halbedelstein m. ∥ pierre f. demi-précieuse. / ~ for sharpening by hand ∥ Handschärfstein m. ∥ pierre f. à affûter à la main. / soap ~ ∥ Speckstein m. ∥ stéatite f. /

synthetic ~ ∥ synthetischer Edelstein m. ∥ pierre f. fine synthétique. / unhewn ~ ∥ unbehauener Stein m. ∥ pierre f. brute.

stone articles pl. ∥ Grobsteinwaren fpl. ∥ gros objets mpl. en pierre. / ~ band (Mine) ∥ Bergemittel n. ∥ laie f.; banc m. de schiste. / ~ batter ∥ Steinböschung f. ∥ pierrée f. ∥ ~ bit ∥ Steinbohrer m. ∥ mèche f. à pierre. / ~ block ∥ behauener Stein m. ∥ pierre f. de taille. / handling ~ blocks ∥ Steinblocktransport m. ∥ transport m. de blocs de pierre. / ~ board ∥ Steinpappe f. ∥ carton-pierre m. / ~ bond ∥ Steinverband m. ∥ appareil m. ou liaison f. de pierres. / ~ boring machine ∥ Steinbohrmaschine f. ∥ perceuse f. à pierres. / ~ bottle ∥ steinerner Krug m.; Kruke f. ∥ bouteille f. de grès; cruchon m. / ~ box ∥ Mühlbuchse f. ∥ boîtard f.

stone breaker ∥ Steinbrecher m. ∥ concasseur m. de pierres; casse-pierres m. / combined ~ and roller mill ∥ Steinbrecherwalzenmühle f. ∥ concasseur m. à pierres et broyeur m. à cylindres combinés. / ~ that can be taken to pieces ∥ zerlegbarer Steinbrecher m. ∥ casse-pierres m. démontable.

stone breaking plant ∥ Schotterwerk n. ∥ usine f. de ballast; usine f. de cailloutis.

stone bridge ∥ Steinbrücke f. ∥ pont m. de pierre. / ~ butter ∥ Bergbutter f. ∥ beurre m. de roche; halotrichite f. / ~ carrier ∥ Steinträger m. ∥ bardeur m. / ~ channel (Road) ∥ Steinrinne f. ∥ rigole f. pavée. / ~ chisel ∥ Gesteinsmeißel m. ∥ ciseau m. minéralogique. / ~ crane ∥ Steinkran m. ∥ grue f. à meules. / ~ crusher ∥ Steinbrecher m. ∥ concasseur m. de pierres; casse-pierres m. / ~ crushing machine ∥ Steinspaltmaschine f. ∥ machine f. à fendre les pierres. / ~ cutter ∥ Steinmetz m.; Steinhauer m. ∥ tailleur m. de pierres.

stone-cutter's saw ∥ Steinhauersäge f. ∥ scie f. à pierre. / ~ shop ∥ Steinhauerwerkstatt f. ∥ atelier m. de tailleur de pierres.

stone cutting ∥ Steinschneiden n.; Steinmetzerei f. ∥ taille f. de pierres. / ~ dike ∥ gepflasterter Deich m. ∥ digue f. maçonnée. / ~ drain (Hydr arch) ∥ Steinrinne f. ∥ pierrée f. souterraine. / ~ dresser see ~ cutter. / ~ dressing machine ∥ Steinbearbeitungsmaschine f. ∥ machine f. à travailler les pierres. / ~ dust ∥ Steingrus m. ∥ blocaille f. / ~ engraver ∥ Steingravör m. ∥ graveur m. sur pierre. / ~ faucet ∥ Hahn m. aus Steinzeug ∥ robinet m. en grès. / ~ feller ∥ Steinhauer m. ∥ rocteur m. / ~ fence ∥ Steineinfriedigung f. ∥ clôture f. en pierres. / ~ filler ∥ Bergeverlader m.; Bergabzieher m. ∥ chargeur m. de pierres. / ~ fruit ∥ Steinobst n. ∥ fruits mpl. à noyau. / ~ grainer ∥ Steinkörner m. ∥ graineur m. en lithographie. / ~ grinding machine ∥ Steinschleifmaschine f. ∥ machine f. à planer les pierres. / ~ grinding mill ∥ Steinmühle f. ∥ moulin m. à pierres. / ~ header ∥ Querschlaghauer m. ∥ bacneur m.; bouveleur m. / ~ lintel ∥ steinerner Sturz m. ∥ linteau en pierre.

stoneman ∥ Gesteinshauer m. ∥ ouvrier m. aux étreintes.

stone mason see also stone cutter ∥ Steinmetz m.; Steinhauer m. ∥ tailleur m. de

pierres. / ~ masonry ∥ Bruchsteinmauerwerk n. ∥ maçonnerie f. en moellons.

stone mason's tool ∥ Steinmetzgerät n. ∥ outil m. du tailleur de pierres. / ~ work ∥ Steinmetzarbeit f. ∥ travail m. du tailleur de pierres.

stone mill ∥ Mahlgang m. ∥ moulin m. à meules; meules fpl. / ~ milling machine ∥ Steinfräsmaschine f. ∥ machine f. à fraiser les pierres. / ~ miner ∥ Gesteinshauer m. ∥ ouvrier m. à la pierre; rocheur m.; coupeur m. de voie. / ~ mug ∥ Steinkrug m. ∥ cruche f.; cruchon m. / ~ pavement ∥ Steinpflaster n. ∥ pavé m. de pierre. / artificial ~ pavement ∥ Kunststeinpflaster n. ∥ pavé m. en pierres artificielles. / ~ paving ∥ Steinpflasterung f. ∥ pavage m. en pierre. / ~ picker ∥ Steinsucher m.; Bergeausklauber m. ∥ ramasseur m. de pierres. / ~ picking ∥ Steinbehauen n. ∥ smillage m. de moellons. / ~ pincers pl. ∥ Steinzange f. ∥ louvre f. à tenailles. / ~ polishing ∥ Steinschleiferei f. ∥ polissage m. de pierres. / ~ polishing machine ∥ Steinpoliermaschine f. ∥ machine f. à polir les pierres. / ~ powder ∥ Steinmehl n. ∥ pierre f. pulvérisée; poussier m. de pierres. / pneumatic conveying plant for ~ powder ∥ Druckluftgesteinsstaubförderanlage f. ∥ installation f. pour le transport pneumatique de poussier de pierres. / ~ pulverizing plant ∥ Gesteinstaubmahlanlage f. ∥ installation f. pour moudre la poussière de pierres. / ~ putter ∥ Bergeschlepper m. ∥ remeneur m. de terres.

stoner ∥ Schleifer m. ∥ doucisseur m. / ~ with rotating drum ∥ Steinausleser m. mit rotierender Trommel ∥ épierreur m. avec tambour rotatif.

stoner-out ∥ Abbimser m. ∥ ponceur m. de peaux.

stone roller ∥ steinerne Walze f. ∥ cylindre m. de pierre. / ~ for facette-grinding shops ∥ Steinwalze f. für Facettschleifereien ∥ rouleau m. de pierre pour facetter.

stone runner ∥ Steinläufer m. ∥ meule f. en pierre.

stone saw ∥ Steinsäge f. ∥ scie f. à pierre. / circular ~ ∥ Steinkreissäge f. ∥ scie f. circulaire pour pierres. / ~ with distanced teeth ∥ Steinsäge f. mit weiten Zähnen ∥ scie f. à pierre avec dents écartées. / ~ with inserted teeth ∥ Steinsäge f. mit eingesetzten Zähnen ∥ scie f. à pierre avec dents rapportées. / ~ with narrow teeth ∥ Steinsäge f. mit engen Zähnen ∥ scie f. à pierre avec dents serrées.

stone sawing ∥ Steinsägerei f. ∥ sciage m. de pierres. / ~ mill ∥ Steinsägerei f. ∥ scierie f. de pierre.

stone scaling machine ∥ Steinschälmaschine f. ∥ machine f. à planer les pierres. / ~ screw ∥ Steinschraube f. ∥ boulon m. de scellement. / ~ separator ∥ Steinausleser m. ∥ épierreur m. / ~ setter ∥ Steinsetzer m. ∥ pinceur m. / ~ smoothing machine for glazed board and pressing board ∥ Steinglättmaschine f. für Glanzpappe und Preßspan ∥ calandre f. à pierre pour carton glacé et presspan. / ~ spindle ∥ Mühleisen n. ∥ arbre m. de meule. / ~ splitting hammer ∥ Keillochhammer m. ∥ marteau m. pour trous coniques. / ~ stopper for carboys ∥ Steinstopfen m. für Säureballons ∥ bouchon m. en grès pour bonbonnes. / ~ stud ∥ Eckpfeiler m. ∥ pilastre m. cornier; cornière f. / ~ toys pl. ∥

Steinspielwaren fpl. ‖ jouets mpl. en pierre. / ~ tubbing ‖ wasserdichte Schachtausmauerung f. oder Schachtmauerung f. ‖ cuvelage m. en maçonnerie; muraillement m. d'un puits. / ~ turner ‖ Steindreher m. ‖ tourneur m. en pierres. / ~ turning ‖ Steindreherei f. ‖ tournage m. de pierres. / ~ turning lathe ‖ Steindrehbank f. ‖ tour m. à pierres.

stoneware ‖ Steingut n.; Steinzeug n. ‖ faïence f.; grès m. / ~ for electric purposes ‖ Steinzeugwaren fpl. für die Elektrotechnik ‖ articles mpl. en grès pour installations électriques. / enamelled ~ ‖ emailliertes Steinzeug n. ‖ grès m. émaillé.

stoneware jug ‖ Steinkrug m. ‖ cruche f. en grès. / ~ maker ‖ Steinzeugformer m. ‖ potier m. en grès-cérame. / ~ pipe ‖ Steinzeugrohr n. ‖ tuyau m. en grès. / ~ pot ‖ Steinzeugtopf m. ‖ pot m. en grès.

stone working ‖ Steinbearbeitung f. ‖ travail m. de la roche ou de la pierre. / ~ machine ‖ Steinbearbeitungsmaschine f. ‖ machine f. à travailler la pierre. / milling machine for ~ ‖ Fräsmaschine f. für die Steinbearbeitung ‖ fraiseuse f. pour travailler la pierre. / ~ tool ‖ Steinbearbeitungswerkzeug n. ‖ outil m. à travailler la pierre.

stoning machine, cherry ~ ‖ Kirschenentkernungsmaschine f. ‖ dénoyauteuse f. de cerises.

stony ground ‖ Steingrund m. ‖ enrochement m.

stool ‖ Stuhl m. (ohne Lehne); Schemel m. ‖ escabeau m. / ~ (Med) ‖ Stuhlgang m. ‖ selle f. / ~ (Shipb) ‖ Rüst f. ‖ portehaubans m. / camp ~ ‖ Feldstuhl m. ‖ siège m. de campagne. / folding ~ ‖ Klappstuhl m. ‖ siège m. pliant. / turning tip ~ ‖ Drehkippstuhl m. ‖ chaise f. basculante et à vois.

stooper (Mine) ‖ Schrämmhauer m. ‖ dépileur m.

stop, to (Bottle) ‖ zukorken ‖ boucher. / ~ (Hydr arch) ‖ dichten ‖ étancher; rendre étanche.

stop, to ‖ anhalten; aufhalten ‖ stopper; arrêter. / ~ the admission of gas ‖ die Gaszufuhr drosseln ‖ étrangler l'admission de gaz. / ~ the blast furnace ‖ den Hochofen m. dämpfen ‖ arrêter le hautfourneau. / ~ an engine ‖ eine Maschine f. abstellen ‖ arrêter une machine. / ~ the furnace ‖ einen Ofen m. kaltlegen; niederblasen ‖ mettre le fourneau hors feu. / ~ a leak ‖ ein Leck n. stopfen ‖ aveugler ou boucher une voie d'eau. / ~ the time ‖ die Zeit f. abstoppen ‖ arrêter le temps. / ~ the way of a ship ‖ die Fahrt eines Schiffes stoppen ‖ arrêter le cours d'un navire. / ~ the work ‖ den Betrieb m. einstellen ‖ faire cesser les travaux mpl.; arrêter le travail ou l'usine.

stop ‖ Halt m. ‖ arrêt m. / ~ (Chem) ‖ Scheidewand f.; Membran f.; Diaphragma n. ‖ diaphragme m. / ~ (Factory) ‖ Betriebspause f. ‖ récréation f. / ~ (Organ) ‖ Register n.; Orgelzug m. ‖ registre m.; jeu m. / ~ (Tool mach) ‖ Anschlag m. ‖ arrêt m.; taquet m.; butoir m.; butée f.; ergot m.; équerre f. / additional ~ (Organ) ‖ Nebenzug m.; Nebenregister n. ‖ jeu m. accessoire d'orgues. / battery ~ ‖ Batteriekontakt m. ‖ contact m. d'une pile. / to come to a ~ ‖ zum Stillstand m. kommen ‖ s'arrêter. / full ~ (Print) ‖

Punkt m. ‖ point m. / graded ~s pl. ‖ stufenförmig angeordnete Blenden fpl. ‖ diaphragmes mpl. échelonnés. / ~ of ice ‖ Eisstopfung f. ‖ barrage m. de glace. / limit ~ (Mach) ‖ Anschlag m. ‖ équerre f.; butée f.; arrêt m. / movable ~ ‖ verschiebbarer Anschlag m. ‖ équerre f. mobile. / organ ~ ‖ Orgelregister n. ‖ jeu m. d'orgues. / steering ~ ‖ Lenkungsanschlag m. ‖ arrêt m. de braquage. / to turn the ring until it comes to the ~ ‖ den Ring m. bis zum Anschlag drehen ‖ tourner la bague jusqu'à la butée.

stop bar ‖ Arretierstange f. ‖ tige f. d'arrêt. / ~ block ‖ Prellklotz m.; Hemmschuh m. ‖ taquet m. d'arrêt; buttoir m.; sabot m. de frein. / ~ block cast in steel ‖ Hemmschuh m. aus Stahlguß ‖ sabot m. d'enrayage en acier moulé.

stop cock ‖ Absperrhahn m. ‖ robinet m. d'arrêt. / ~ of pressure gauge with air discharge ‖ Manometerabsperrhahn m. mit Entlüftung ‖ robinet m. d'arrêt de manomètre avec évacuation d'air. / ~ of a wind pipe ‖ Lufthahn m. einer Röhrenleitung ‖ robinet m. d'une ventouse.

stop face ‖ Anschlagfläche f. ‖ surface f. d'arrêt. / ~ gap ‖ Notbehelf m. ‖ pisaller m.; expédient m. / ~ gear ‖ Sperrvorrichtung f. ‖ encliquetage m. / ~ link ‖ Ausrückglied n. ‖ chaînon m. à grain.

stop motion ‖ Abstellvorrichtung f. ‖ dispositif m. de soulèvement. / ~ for drawing frames ‖ Streckenausrückvorrichtung f. für Ziehbänke ‖ dispositif m. d'arrêt pour bancs d'étirage.

stoppage (Hydr arch) ‖ Verstopfung f. ‖ engorgement m. / ~ (Mach) ‖ Außerbetriebsetzen n. ‖ arrêt m. / ~ (Trade) ‖ Sperre f. ‖ suspension f.; blocus m. / quick ~ ‖ schnelles Anhalten n. ‖ arrêt m. rapide. / ~ of work ‖ Betriebseinstellung f. ‖ arrêt m. du travail.

stopped ‖ abgestoppt ‖ arrêté.

stopper (Bottle) ‖ Stöpsel m.; Stopfen m.; Korken m. ‖ bouchon m. / ~ (Found) ‖ Stopfen m. ‖ quenouille f. / ~ (Mar) ‖ Stopper m. ‖ bosse f. / ~ (Mach tool) ‖ Ausrücker m. ‖ arrêt m.; dispositif m. de débrayage; passe-courroie m. / aluminium ~ ‖ Aluminiumstöpsel m. ‖ bouchon m. en aluminium. / pinch ~ ‖ Stöpsel m. mit Griff ‖ bouchon m. à poignée.

stopper borer ‖ Korkbohrer m. ‖ foreur m. de bouchons. / ~ bottle ‖ Stöpselflasche f. ‖ bouteille f. à bouchon. / ~ grinding machine ‖ Stopfenschleifmaschine f. ‖ machine f. à roder les bouchons. / ~ maker (Met) ‖ Stopfenmacher m. ‖ mouleur m. de tampons.

stop pin ‖ Anschlagstift m. ‖ goujon m. d'arrêt. / ~ for pianos ‖ Klavierstift m. ‖ goujon m. pour pianos.

stopping ‖ Halt m. ‖ arrêt m. / automatic ~ ‖ Selbstabstellung f. ‖ arrêt m. automatique. / ~ of the motor ‖ Stillstand m. des Motors ‖ arrêt m. du moteur. / ~ of payment ‖ Zahlungssperre f. ‖ arrêt m. sur les versements.

stopping cam ‖ Auflaufnocken m. ‖ came f. d'arrêt.

stopping device (Mach) ‖ Anhaltevorrichtung f. ‖ dispositif m. d'arrêt. / ~ (Typewr) ‖ Bremsvorrichtung f. ‖ dispositif m. de freinage. / ~ for fluids and gases ‖ Absperrvorrichtung f. für Flüssigkeiten und Gase ‖ appareil m. de fermeture pour liquides et gaz.

stopping distance ‖ Bremsweg m. ‖ longueur f. d'arrêt. / ~-down the image of the crystal ‖ Abblendung f. des Kristallbildes ‖ diaphragme m. agissant sur l'image du cristal.

stopping machine, bottle ~ ‖ Flaschenverschließmaschine f. ‖ machine f. à boucher les bouteilles.

stopping motion for thread breaking (Knitting) ‖ Abschlagabsteller m. ‖ cassefil m. d'abattage. / ~ piece ‖ Stemmstück n. ‖ pièce f. de détention. / ~ point of the carriage (Typewr) ‖ Anschlagpunkt m. des Wagens ‖ point m. d'arrêt de la course du chariot. / ~ power (Projectile) ‖ Aufhaltekraft f.; Schmetterkraft f. ‖ force f. d'arrêt. / ~ signal ‖ Haltesignal n. ‖ signal m. d'arrêt. / ~ signal disk ‖ Haltescheibe f. ‖ disque m. d'arrêt. / ~ valve ‖ Absperrventil n. ‖ soupape f. d'arrêt.

stopple see also stopper ‖ Stöpsel m. ‖ bouchon m. / caoutchouc ~ ‖ Kautschukstöpsel m. ‖ bouchon m. de caoutchouc. / cork ~ ‖ Korkstöpsel m. ‖ bouchon m. de liège.

stop screw ‖ Anschlagschraube f. ‖ vis f. d'arrêt ou de butée. / ~ spring ‖ Rastfeder f. ‖ ressort m. à cran. / ~ stoppage ‖ Stillstand m. ‖ arrêt m.

stop valve ‖ Absperrventil n. ‖ soupape d'arrêt. / ~ of pressure gauge with round test flange ‖ Manometerabsperrventil n. mit rundem Prüfflansch ‖ soupape f. d'arrêt du manomètre à bride de contrôle ronde.

stop watch ‖ Stoppuhr f.; Sekundenuhr f. ‖ montre f. ou chronomètre m. à déclic; compte-secondes m. / ~ wire ‖ Anschlagdraht m. ‖ fil m. d'arrêt. / ~ work (Watchm) ‖ Feststellvorrichtung f. ‖ arrêtage m.

storage ‖ Lagerung f.; Aufspeicherung f. ‖ emmagasinage m.; accumulation f. / ~ of compressed air ‖ Aufspeichern n. von Druckluft ‖ emmagasinage m. d'air comprimé. / ~ of energy ‖ Energieaufspeicherung f. ‖ accumulation f. d'énergie.

storage basin ‖ Staubecken n. ‖ réservoir m. de barrage. / elevated ~ ‖ Hochspeicherbecken n. ‖ réservoir m. surélevé; château m. d'eau.

storage battery ‖ Sammlerbatterie f. ‖ batterie f. d'accumulateurs. / ~ for lighting ‖ Beleuchtungsbatterie f. ‖ batterie f. d'éclairage. / ~ traction ‖ Akkumulatorenfahrbetrieb m. ‖ traction f. par accumulateurs.

storage bin (Mill) ‖ Vorratsbehälter m. ‖ trémie f. / ~ (Mine) ‖ Erztasche f. ‖ poche f. à minerais.

storage cask ‖ Lagerfaß n. ‖ foudre m. de garde. / ~ cellar ‖ Lagerkeller m. ‖ cave f. de garde.

storage hopper ‖ Füllrumpf m. ‖ chambre f. de remplissage; tremie f. / closing trap of ~ (Mach) ‖ Füllrumpfverschluß f. ‖ fermeture f. de trémie.

storage pocket ‖ Vorratstasche f. ‖ poche f. d'approvisionnement. / ~ power station ‖ Speicherkraftwerk n. ‖ usine f. d'accumulation. / ~ vessel ‖ Standgefäß n. ‖ bidon m. de magasin ou d'exposition.

storax ‖ Storax m. ‖ styrax m.; baume m. storax. / ~ oil ‖ Storaxöl n. ‖ huile f. de styrax.

storching reservoir ‖ Speisebecken n. ‖ bassin m. d'alimentation.

store, to ‖ lagern ‖ mettre en magasin m.; entreposer. / ~ **the gas** ‖ das Gas aufspeichern ‖ stocker *ou* emmagasiner le gaz. / ~ **the heat** ‖ Wärme f. aufspeichern ‖ conserver *ou* emmagasiner la chaleur. / ~ **up** ‖ aufspeichern; aufstapeln ‖ emmagasiner. / **do not store in a damp place!** ‖ trocken aufspeichern! ‖ pas emmagasiner en lieu m. humide!

store (Emporium) ‖ Kaufladen m.; Verkaufslokal n. ‖ boutique f. (de marchand); magasin m. / ~ (Sluice) ‖ Rolltafel f. ‖ rideau m. articulé. / ~ (Storage place) ‖ Speicher m.; Lagerhaus n.; Vorratsplatz m. ‖ dépôt m.; parc m.; magasin m.; entrepôt m. / ~ (Window) ‖ Store m.; Vorhang m. ‖ store m.; rideau m. / ~ **for bottled beer** ‖ Flaschenbierniederlage f. ‖ dépôt m. de bière en bouteilles. / **cold** ~ ‖ Kühlhalle f. ‖ entrepôt m. frigorifique. / **general** ~ ‖ Kramladen m. ‖ boutique f. (de mercier); échoppe f. / ~ **for keg beer** ‖ Faßbierlagerraum m.; Faßbierniederlage f. ‖ entrepôt m. *ou* dépôt m. de bière en fûts.

store book ‖ Bestandbuch n. ‖ état m.; livre m. de stock. / ~ **cask** ‖ Lagerfaß n. ‖ foudre m.; tonneau m. de chantier. / **to wash** ~ **casks** pl. ‖ faßschlupfen ‖ Fässer npl. spülen ‖ nettoyer les foudres mpl. intérieurement. / **washing of** ~ **casks** ‖ Faßschlupfen n. ‖ nettoyage m. intérieur des foudres. / ~ **cellar** ‖ Lagerkeller m. ‖ cave f.; dépôt m. / ~ **clerk** ‖ Lagerhalter m.; Lagerist m. ‖ magasinier m. / ~ **equipment** ‖ Ladeneinrichtung f. ‖ installation f. de magasin.

storehouse ‖ Niederlage f.; Warenlager n.; Lagerhaus n. ‖ dépôt m.; entrepôt m.; magasin m.

storekeeper ‖ Lageraufseher m. ‖ gardemagasin m.; magasinier m.

store keeping accounts pl. ‖ Lagerbuchführung f. ‖ comptabilité f. des stocks de magasin.

storeman *see* storekeeper.

store pond ‖ Einsatzteich m. ‖ vivier m.

store reservoir ‖ Speicherbecken n. ‖ bassin m. de barrage; barrage-réservoir m. / ~ **of rain-water** ‖ Regenwasseraufspeicherungswerk n. ‖ réservoir m. pour recueillir les eaux pluviales.

store room (Mar) ‖ Hellegat n.; Vorratskammer f. ‖ soute f. / ~ **rooms** pl. *see* storehouse. / ~ **room windlass** ‖ Speicherwinde f. ‖ treuil m. de magasin.

stores pl. (Mar) ‖ Proviant m. ‖ vivres mpl.; provisions fpl.

store ship ‖ Proviantschiff n. ‖ vaisseau m. ravitailleur. / ~ **tank** ‖ Vorratsbehälter m. ‖ réservoir m. / ~ **timber** ‖ Nutzholz n.; Stapelholz n. ‖ bois m. de chantier *ou* de charpente *ou* de construction.

storing ‖ Aufstapelung f.; Aufbewahrung f.; Speicherung f. ‖ emmagasinage m.; conservation f.; garde f. / **the plant is capable of** ~ **over x tons of material** ‖ bei dem Werk n. können über x Tonnen eingelagert werden ‖ le dépôt a une capacité de plus de x tonnes. / ~ **of inflammable liquids** ‖ Lagerung f. für feuergefährliche Flüssigkeiten ‖ magasin m. de liquides inflammables. / ~ **room** ‖ Lagerraum m. ‖ dépôt m. / ~ **up** ‖ Aufspeicherung f. ‖ emmagasinage m.; amas m.; entassement m.

storm ‖ Sturm m. ‖ tempête f. / **magnetic**

~ ‖ magnetisches Gewitter n. ‖ orage m. magnétique.

storm bank ‖ Gewitterbank f. ‖ banc m. d'orage. / ~ **damage** ‖ Sturmschaden m. ‖ dommage m. causé par la tempête. / ~ **flap** (Wind mill) ‖ Sturmtür f. ‖ pale f. de tempête. / ~ **flood** ‖ Sturmflut f. ‖ grande marée f. de tempête. / ~ **lantern** ‖ Sturmlaterne f. ‖ lanterne-tempête f.

stormsail ‖ Sturmsegel n. ‖ voile f. de cape.

storm warning ‖ Sturmwarnung f. ‖ avertissement m. de tempêtes. / ~ **service** ‖ Sturmwarnungsdienst m. ‖ service m. d'avertissement de tempêtes. / ~ **signal** ‖ Sturmzeichen n. ‖ signal m. d'avertissement de tempête.

storm water overflow (hydr arch) ‖ Notauslaß m. einer Kanalisation ‖ bonde f. de pluie.

stormy ‖ stürmisch ‖ orageux.

story (Build) ‖ Wohngeschoß n. ‖ étage m.

stove ‖ Ofen m.; Stubenofen m. ‖ poêle m. / ~ (Techn) ‖ Ofen m. ‖ four m.; fourneau m. / ~ **baker's** ~ ‖ Backofen m. ‖ four m. pour pâtissiers *ou* de boulangerie. / **cast iron** ~ ‖ gußeiserner Ofen m. ‖ poêle m. de fonte. / **cast iron kitchen** ~ ‖ Kochherd m. aus Gußeisen ‖ fourneau m. de cuisine en fonte. / ~ **made of clay** ‖ tönerner Ofen m. ‖ poêle m. en terre. / ~ **for drying** ‖ Darrofen m.; Trockenofen m. ‖ four m. à sécher. / **electric** ~ ‖ elektrischer Ofen m. ‖ four m. électrique. / **gas** ~ ‖ Gasofen m. ‖ fourneau m. à gaz. / ~ **imitating a fire-place** ‖ Kaminofen m. ‖ cheminée f. à la prussienne. / **iron** ~ ‖ eiserner Ofen m. ‖ four m. en métal. / ~ **for large rooms** ‖ großer Zimmerofen m. ‖ poêle m. à grande surface de chauffe. / **portable electric** ~ **with heating chamber** ‖ elektrische Kleinküche f. mit Wärmehaube ‖ four m. électrique à rôtir et à cuivre avec hotte de chaleur. / **purifying** ~ ‖ Desinfektionskammer f. ‖ étuve f. de désinfection. / **slow burning** ~ ‖ Dauerbrandofen m. ‖ poêle m. à feu continu; calorifère m. à chauffage constant.

stove door ‖ Ofentür f. ‖ porte f. de poêle. / ~ **maker** ‖ Ofenmacher m. ‖ poêlier m. constructeur m. de fours.

stoveman ‖ Heizer m. ‖ chauffeur m.; homme m. de four.

stove manufacture ‖ Herdfertigung f. ‖ fabrication f. de fourneaux. / ~ **pipe** ‖ Ofenrohr n. ‖ tuyau m. de poêle. / ~ **polish** ‖ Ofenputzmittel n. ‖ pâte f. à fourneaux. / ~ **tile for clay stoves** ‖ Kachel f. für Tonöfen ‖ carreau m. glacé pour poêles en argile. / ~ **varnish** ‖ Ofenlack m. ‖ vernis m. pour poêles.

stowage ‖ Stauung f.; Packen n. ‖ arrimage m.

stower (Mine) ‖ Versatzarbeiter m. ‖ remblayeur m.; restapleur m.; releveur m. de terres.

stowing (Mine) ‖ Bergeversatz m. ‖ stappes fpl.; remblai m. / **hydraulic** ~ ‖ Spülversatz m. ‖ remblayage m. hydraulique.

stowing arrangement ‖ Stauvorrichtung f. ‖ dispositif m. de retenue.

strabism ‖ Schielen n. ‖ strabisme m. / **angle of** ~ ‖ Schielwinkel m. ‖ angle m. strabique.

strabismus hook (Med) ‖ Schielhaken m. ‖ crochet m. à strabisme. / ~ **scissors** pl. (Med) ‖ Schielschere f. ‖ ciseaux mpl. à strabisme.

straddle gauge ‖ Reiterlehre f. ‖ calibre m. cavalier.

straight ‖ gerade ‖ droit. / ~ **axle** ‖ gerade Achse ‖ essieu m. droit. / ~ **bar linking machine** ‖ Flachkettelmaschine f. ‖ remailleuse f. rectiligne. / **internal grinding of** ~ **bores** ‖ Innenschliff m. zylindrischer Bohrungen ‖ rectification f. intérieure d'alésages cylindriques. / ~ **cut** (Saw teeth) ‖ Geradschliff m. ‖ affûtage m. droit. / ~ **fishplate** ‖ gerade Lasche f. ‖ éclisse f. plate. / ~ **forceps** pl. ‖ gerade Pinzette f. ‖ pince f. droite. / ~ **hand knitting machine** ‖ Flachstrickmaschine f. für Handbetrieb ‖ machine f. à tricoter rectiligne à la main. / ~ **length** ‖ Flucht f.; Baulinie f.; Baufluchtlinie f. ‖ alignement m. / ~ **line** ‖ Gerade f. ‖ droite f.; ligne f. droite. / **to set out** ~ **lines** pl. **on the ground** ‖ gerade Linien fpl. im Gelände abstecken ‖ jalonner des lignes fpl. droites sur le terrain. / ~ **link** ‖ gerades Glied n. ‖ membre m. droit. / ~ **scissors** pl. ‖ gerade Schere f. ‖ ciseaux mpl. droits. / ~ **side rim** ‖ Felge f. für Drahtpneumatik ‖ jante f. pour enveloppe à fil.

straightedge (Join) ‖ Richtscheit n. ‖ réglet m. / **revolving** ~ **table** (Textile) ‖ drehbarer Linealtisch m. ‖ table f. à règle tournante. / ~ **tooth sharpening machine** (Saw) ‖ Geradschliffmaschine f. ‖ machine f. pour l'affûtage droit.

straighten, to ‖ ausrichten; richten ‖ dresser; redresser. / ~ **the iron** ‖ das Eisen richten ‖ parer *ou* dresser le fer au marteau. / ~ **staves** pl. ‖ Dauben fpl. abrichten ‖ dresser des douves fpl.

straightened border ‖ Randversteifung f. ‖ bord m. renforcé.

straightener (Roll mill) ‖ Strecker m. ‖ dresseur m. / **plate** ~ ‖ Blechrichter m. ‖ dresseur m. de tôles. / **rail** ~ ‖ Richtmaschine f. für Schienen ‖ appareil m. à dresser des rails.

straightening ‖ Geradeschlagen n.; Ausrichten n.; Richten n. ‖ redressage m. / ~ **of rails** ‖ Geraderichten n. der Schienen ‖ dressage m. *ou* redressage m. des rails. / ~ **of the staves** ‖ Abrichten n. der Dauben ‖ dressage m. des douves. / ~ **of wire** (Needl) ‖ Richten n. des Drahts ‖ redressage m. du fil de fer.

straightening board for pins (Needl) ‖ Richteisen n. für Stecknadeln ‖ dressoir m. *ou* planche f. à épingles. / **device for** ~ **the cut-off portion** ‖ Vorrichtung f. zum Geradebiegen des abgetrennten Schnittteils ‖ dispositif m. pour redresser la partie coupée. / ~ **hammer** ‖ Richthammer m.; Schlagwerkzeug n. ‖ marteau m. à dresser. / ~ **iron** (Locksm) ‖ Richteisen n. ‖ grattoir m. / ~ **iron-rod** (Glassm) ‖ Richteisen n.; Ausstreichlinieal n. ‖ fer m. à dresser.

straightening machine ‖ Richtmaschine f. ‖ machine f. à dresser. / ~ (Needl) ‖ Richtmaschine f. ‖ règle f. à bascule. / ~ (Silk) ‖ Seidenstreckmaschine f. ‖ machine f. à cheviller. / ~ **angle iron** ‖ Winkeleisenrichtmaschine f. ‖ machine f. à dresser les cornières. / ~ **for corrugated iron** ‖ à cheviller. / ~ **for corrugated iron** ‖ Wellblechrichtmaschine f. ‖ machine f. à dresser les tôles ondulées. / ~ **head** ~ **and jointing machine** ‖ Abrichthobel- und Fügemaschine f. ‖ machine f. à dresser et à jointer. / **high-capacity plate** ~ ‖ Hochleistungsblechrichtmaschine f. ‖ machine

f. à planer les tôles à grand rendement. / **plate ~** || Richtmaschine f. für Bleche || machine f. à planer les tôles. / **sectional iron ~** | Richtmaschine f. für Formeisen || machine f. à dresser les fers profilés. / **shaft ~** || Richtmaschine f. für Wellen || machine f. à dresser les arbres. / **sleeper ~** || Schwellenrichtmaschine f. || machine f. à dresser les traverses. / **tram ~** | Förderwagenrichtmaschine f. || machine f. à diriger les wagonnets de transport. / **tube ~** || Richtmaschine f. für Rohre || machine f. à dresser les tubes. / **wire ~** | Richtmaschine f. für Drähte || machine f. à dresser les fils métalliques.

straightening plate || Richtplatte f. || plaque f. à dresser.

straightening press || Richtpresse f. || presse f. à (re)dresser. / **pipe ~** || Rohrrichtpresse f. || presse f. à dresser les tubes. / **shaft ~** | Wellenrichtpresse f. || presse f. à dresser les arbres.

straightening roll || Streckwalze f. || cylindre m. de dressage / **~ shed** || Richtbude f. || atelier m. de dressage. / **~ tongs** pl. || Richtzange f. || pince f. à dresser. / **~ trough** || Richtrinne f. || gouttière f. de dressage.

straightforward working || glatte Fabrikation f. || bonne marche f. de la fabrication.

straight-glued joint || Leimfuge f.; stumpfe Fuge f. || joint m. à plat-point.

straightline(d) || gerade; geradlinig || rectiligne. / **~ spot welding** || Reihenpunktschweißung f. || soudage m. par points en ligne droite.

straight-line (Subst) *see under* straight.

straightway cock || Durchgangshahn m. || robinet m. droit. / **~ valve** || Durchgangsventil n. || soupape f. droite.

strain, to || überanstrengen; spannen || surmener; tendre. / **~ (Brew)** || abläutern || filtrer. / **~ every nerve** || alle Kräfte fpl. anspannen || travailler à toute force / **~ off (Brew)** || abseihen || tirer les fonds mpl. des foudres. / **cock with long pipe to ~ off store casks** || Abseihhahn m.; Abseihwechsel m. || robinet m. pour soutirer les foudres à fond. / **~ to the utmost limit of the capacity** || bis zur Grenze f. der Leistungsfähigkeit ausnützen || employer jusqu'à la limite f. de la puissance. / **~ wires** pl. || Drähte mpl. spannen || tendre les fils mpl.

strain || Spannung f.; Anstrengung f.; Kraft f. || tension f.; effort m. / **casting ~s** pl. || Gußspannung f. || tension f. de coulée. / **compressing ~** || Druckkraft f. || force f. de compression. / **evergrowing ~** | wachsende Belastung f. || exigence f. ou charge toujours croissante. / **to subject to a great ~** || großer Beanspruchung f. aussetzen || soumettre à un grand effort. / **internal ~** || innere Beanspruchung f. || effort m. de tensions moléculaires. / **lateral ~** || Schubkraft f. || force f. de cisaillement. / **longitudinal ~** *see* tensile **~**. / **parts** pl. **subjected to moderate ~s** || mäßig beanspruchte Teile mpl. || pièces fpl. exposées à des efforts modérés. / **~ of pole** || Gestängebelastung f. || charge f. d'appui. / **pulling ~** *see* tensile **~**. / **shearing ~** || Scherkraft f. || force f. ou effort m. de cisaillemnet. / **~ of springs** || Spannung f. der Federn || tension f. de ressorts / **stretching ~** *see* tensile. / **tensile ~** || Zugkraft f. || force f. de traction; effort m. de tension. /

torsional **~** || Beanspruchung f. auf Torsion | force f. de torsion. / **transverse ~** *see* tensile **~**. / **twisting ~** *see* torsional **~**. / **unusually high ~** || außergewöhnlich hohe Beanspruchung f. || effort m. extraordinaire. / **without ~** || spannungsfrei || sans tension f.

strained spectacle lens || gespanntes Brillenglas n. || verre m. correcteur serré.

strainer || Sieb n.; Seiher m. || tamis m.; sas m.; crible m. / **air ~** || Luftfilter n.; Luftreinigungssieb n. || filtre m. d'air. / **gasoline ~** || Benzinfilter n. || filtre m. à essence. / **oil ~** || Ölfilter n. || filtre m. d'huile. / **~ for oil filter** || Ölfiltersieb n. || tamis m. à filtre l'huile. / **tea ~** || Teesieb n. || passoir m. à thé.

strainer texture (Pap) || Siebgewebe n. || tissu m. à tamis.

straining || Filtern n. || filtration f. / **end ~** || Endverankerung f. || ancrage m. dans le sol. / **intermediate ~** || Zwischenverankerung f. || ancrage m. intermédiaire. / **~ of a self-supporting overhead cable** || Abspannung f. eines Luftkabels || ancrage m. d'un câble aérien à autosupport.

straining bag of trellis || Seihesack m. von Zwillich || chausse f. en toile de treillis pour filtrer. / **~ basket** || Seihekorb m. || panier m. à passer. / **~ beam** || Spannbalken m.; Spannriegel m.; Zange f. || poutre f. traversière; traversière f.; entrait m. / **~ chamber** || Siebkammer f. || chambre f. de tamisage. / **~ device** || Abseihvorrichtung f. || appareil m. à tirer les fonds de foudres. / **~ drum** || Siebtrommel f. || tambour m. tamiseur. / **~ hose** || Abseihschlauch m. || (tube m. en) caoutchouc m. à tirer les fonds de foudres. / **~ machine** || Passiermaschine f. || machine f. à passer. / **~ pipe** || Abseihrohr n. || tube m. à tirer les fonds de foudres. / **~ press** || Seiherpresse f. || presse f. à passoir ou à filtrer. / **~ strap** || Ziehband n. || bride f. de fixation.

strain tester for spectacle lenses || Spannungsprüfer m. für Brillengläser || tensiscope m. pour des verres correcteurs.

straits pl. || Meerenge f. || détroit m.

strake (Shipb) || Plattengang m. || virure f.; file f. / **inside ~** || innerer *oder* anliegender Plattengang m. || tôle f. formant clin intérieur, virure f. d'en dedans. / **outside ~** || abliegender *oder* äußerer Plattengang m. || tôle f. formant clin extérieur; virure f. d'en dehors. / **~ of plates** || Plattengang m. || virure f. de tôles.

stramonium leaves pl. || Stechapfelkraut n. || herbe f. de stramoine.

strand || Strand m. || plage f. / **~ (Textile)** || Litze f. || cordon m.; toron m.; brin m. / **bronze ~** || Bronzelitze f. || toron m. de cuivre ou en bronze. / **~ of a cable-laid rope** || Kardeel n. eines Kabels || cordon m. d'une aussière. / **rocky ~** || felsiger Strand m. || plage f. rocheuse.

stranded (Cable) || verseilt || câblé. / **~ aluminium wire** || Aluminiumseil n. || câble m. en aluminium. / **to be ~ (Mar)** || auf Grund sitzen || être échoué.

stranding machine || Verlitzmaschine f. || toronneuse f.

strangler, air ~ || Lufteinlaßklappe f. || clapet m. d'entrée d'air.

strangling of gas || Gasdrosselung f. || étranglement m. de gaz.

strap || Riemen m.; Gurt m. || courroie f.; sangle f. / **~ for blinds** || Rolladengurt m. || sangle f. de persiennes. / **cotton ~** || Baumwollgurt m. || ceinture f. de coton. / **coupling ~** || Schelle f. || agrafe f. à serrage. / **eccentric ~** || Exzenterring m. || collier m. *ou* bride f. d'excentrique. / **elastic ~** || Gummigurt m. || cordon m. élastique. / **endless ~** || Riemen m. ohne Ende; Transmissionsriemen m.; Treibriemen m. || courroie f. sans fin. / **leather ~** || Ledergürtel m. || ceinture f. en cuir. / **stiffening ~** || Versteifungsbügel m. || bride f. de renfort. / **~ for wristlet watch** || Uhrarmband n. || bracelet m. à montre.

strap arbor || Riemenwelle f. || tige f. à courroie. / **~ belt** || Gurt m. || sangle f. / **~ driving pulley** || Antriebriemenscheibe f. || poulie f. de commande à courroie. / **~ end of main rod** || Treibstangenbügel m. || tête f. de bielle en étrier. / **~ guide** || Riemenleiter m. || guide-courroie m. / **~ rod** || Riemenwelle f. || tige f. ou arbre m. à courroie. / **~ saw** || Bandsäge f. || scie f. sans fin; scie f. à lame continue. / **~ weaver** || Gurtweber m. || tisseur m. de sangles. / **~ wheel** || feste Riemenscheibe f. || poulie f. fixe.

strass || Straß m. || strass m.

stratification || Schichtung f. || stratification f.

stratified || geschichtet || stratifié. / **~ cloud** || Schichtenwolke f. || nuage m. stratifié. / **~ rocks** pl. || Schichtgestein n. || roches fpl. stratifiées. / **~ stones** pl. || Schichtgestein n. || pierres fpl. stratifiées.

stratum (Mine) || Schicht f.; Flöz n. || couche f.; veine f. / **~ of air** || Luftschicht f. || couche f. d'air; région f. d'atmosphère. / **cold air ~** || kalte Luftschicht f. || couche f. d'air froid. / **~ of constant temperature** || wärmebeständige Schicht f. || couche f. à température constante. / **~ of firm soil** || guter Baugrund m. || sol m. résistant. / **lower ~ of the concrete road surface** || Unterschicht f. der Betonstraßendecke || couche f. inférieure du revêtement de route en béton. / **~ of metallic vapour** || Metalldampfschicht f. || couche f. de vapeur métallique. / **to sink through a ~ (Mine)** || eine Schicht f. durchteufen || creuser ou traverser une couche. / **~ of warm air** || warme Luftschicht f. || couche f. d'air chaud.

straw, to || mit Stroh m. beflechten || empailler.

straw || Stroh n. || paille f. / **barley ~** || Gerstenstroh n. || paille f. d'orge. / **chopped ~** || Häcksel m. || paille f. hachée. / **~ of corn** || Stroh n. von Getreide || paille f. de céréales. / **drinking ~** || Trinkhalm m. || chalumeau m. / **maize ~** || Maisstroh n. || paille f. de maïs. / **oat ~** || Haferstroh n. || paille f. d'avoine. / **rice ~** || Reisstroh n. || paille f. de riz. / **rye ~** || Roggenstroh n. || paille f. de seigle. / **wheat ~** || Weizenstroh n. || paille f. de blé.

straw, articles pl. **of ~** || Strohwaren fpl. || ouvrages mpl. en paille. / **~ band** || Strohseil n. || corde f. ou lien m. de paille.

strawberry || Erdbeere f. || fraise f. / **~ grower** || Erdbeerbauer m. || cultivateur m. de fraises. / **~ leaves** pl. || Erdbeerblätter npl. || feuilles fpl. de fraisier.

straw, blade of ~ || Strohhalm m. || brin m. de paille. / **~ blade for drinking** || Strohröhrchen n. || chalumeau m. en paille. / **~ board** || Strohpappe f. || carton m. (de)

paille. / corrugated ~ board || Wellpappe f. || carton-paille m. ondulé. / ~ boiler (Pap) || Strohkocher m. || lessiveur m. de paille. / ~ bonnet Strohhut m. (für Damen) || chapeau m. de paille. / ~ bottle-envelope || Strohhülse f. für Flaschen || paillon m. à bouteilles. / ~ boxer || Polsterholz n.; Stopfholz n. || rembourroir m. / ~ cardboard || Strohpappe f. || carton m. de paille. / ~ chopping factory || Häckselfabrik f. || fabrique f. de paille hachée. / ~ colour || Strohgelb n. || jaune m. de paille. / ~ cutter || Häckselbank f. || coupe-paille m. / ~ cutting machine || Futterschneidemaschine f. || hache-pailles f.; machine f. à hacher le fourrage. / ~ dressing (Basket) || Strohzurichtung f. || préparation f. de la paille. / ~ dyeing || Strohfärben n. || teinture f. de paille. / ~ electrometer || Strohhalmelektrometer n. || électromètre m. à brins de paille. / ~ embroiderer || Strohstickerin f. || brodeuse f. sur paille. / ~ envelope || Strohhülse f. || paillon m. / ~ file || Strohfeile f. || lime f. au paquet. / ~ flattener (Basket) || Strohplätter m. || lamineur m. de paille. / ~ goods pl. || Strohwaren fpl. || articles mpl. en paille.

straw hat || Strohhut m. || chapeau m. de paille. / knotted ~ || geknüpfter Strohhut m. || chapeau m. de paille noué. / plaited ~ || geflochtener Strohhut m. || chapeau m. de paille tressé. / sewn ~ || genähter Strohhut m. || chapeau m. de paille cousu.

straw hat bleacher || Strohhutbleicher m. || blanchisseur m. de chapeaux de paille. / ~ cleaning agent || Strohhutreinigungsmittel n. || produit m. à nettoyer les chapeaux de paille. / ~ dyes pl. || Farben fpl. für Strohhüte || couleurs fpl. pour chapeaux de paille. / ~ lake see ~ varnish. / ~ plaiter || Strohhutflechter m. || tresseur m. de chapeaux de paille. / ~ presser || Strohhutpresser m. || presseur m. de chapeaux de paille. / ~ sewer || Strohhutnäher m. || couseur m. de chapeaux de paille. / ~ sewing machine || Strohhutnähmaschine f. || machine f. à coudre les chapeaux de paille. / ~ varnish || Strohhutlack m. || laque f. pour chapeaux de paille. / ~ varnisher || Strohhutlackierer m. || vernisseur m. de chapeaux de paille.

straw husk || Strohhülse f. || enveloppe f. de paille. / ~ husks sewing machine || Strohhülsennähmaschine f. || machine f. à coudre les manchons de paille. / ~ insulator for heat || Wärmeschutz m. aus Stroh || isolation f. calorifuge en paille. / ~ knife || Häckselmesser n.; Häckselklinge f.; lame f. ou couteau m. du hache-paille. / ~ loam || Strohlehm m.; torchis m.; bauge f. / ~ mat || Strohmatte f. || natte f. de paille; paillasson m. / ~ material || Strohstoff m. || matière f. de paille. / ~ paper || Strohpapier n. || papier m. de paille. / ~ plait || Strohgeflecht n. || tresse f. de paille. / ~ press || Strohpresse f. || presse f. à paille. / ~ pulp || Strohstoff m. || pâte f. de paille. / ~ pulp factory || Strohstoffabrik f. || fabrique f. de pâte de paille.

straw rope || Strohseil n. || corde f. de paille; éclisse f. en paille. / ~ spinning machine || Strohseilspinnmaschine f. || machine f. à filer les liens de paille.

straw screen || Strohschüttler m. || secoueur m. de paille. / ~ shaker || Strohschütt-ler m. || secoueur m. de paille. / ~ sheaf || Strohschaube f. || javelle f. / ~ shed || Strohschuppen m. || paillier m. / ~ shoe plaiter || Strohschuhflechter m. || tresseur m. d'espadrilles. / ~ sole || Strohsohle f. || semelle f. en paille. / ~ splitter || Strohspalter m. || fendeur m. de paille. / ~ splitting || Strohspalten n. || fendage m. de paille. / ~ stack || Strohschober m. || tas m. de paille. / ~ stuff || Strohstoff m. || tissu m. de paille. / ~ tress dyer || Strohgeflechtfärber m. || teinturier m. de tresses de paille. / ~ worker || Stroharbeiterin f. || confectionneuse f. en paille. / ~ wrapper for bottles || Strohhülse f. für Flaschen || enveloppe f. de bouteilles en paille.

stray (Electr) || Streuung f. || dispersion f. / ~ (Tel) || Nebengeräusch n. || bruit m. parasite. / clear of ~s || störungsfrei || débarrassé d'ondes fpl. intruses.

stray current || vagabundierender Strom m. || courant m. vagabond.

streak || Streifen m. || bande f.; raie f.; strie f. / ~ (Miner) || Strich m. || tachure f. / ~ of wood || Holzfaser f. || fil m. du bois.

streaked || gestreift || rayé; strié; zébré.

streaks pl. of mist || Nebelschwaden m. || brouillard m. flottant.

stream (Current) || Strömung f. || courant m. / ~ (River) || Fluß m. || rivière f.; fleuve m. / ingoing ~ || Flutstrom m. || courant m. d'entrée ou de flot. / casting in rising ~ || Gießen n. in steigendem Strom || coulée f. en source. / thin ~ || feiner Strahl m. || filet m. mince. / tributary ~ || Nebenfluß m. || confluent m.

streamer || Bergarbeiter m. || ouvrier m. à veine.

streamline || Stromlinie f. || ligne f. de courant. / ~ body || Stromlinienkarosserie f. || carrosserie f. à lignes fuyantes.

stream-lined, the engine is mounted in a ~ cowling || der Motor ist windschnittig verkleidet eingebaut || le moteur revêtu d'un capot est carrossé à lignes fuyantes.

streamline section || Stromlinienquerschnitt m. || section f. fuselée ou à lignes fuyantes.

stream tin || Seifenzinn n. || étain m. d'alluvion. / ~ washer || Stromsetzmaschine f. || machine-laveuse f. à courant d'eau; crible m. hydraulique à courant. / ~ work || Seifenwerk n. || laverie f. ou lavoir m. de minerais d'alluvion; mine f. d'alluvion.

street || Straße f. || rue f. / one-way ~ || Einbahnstraße f. || rue f. à sens unique. / main ~ || Hauptstraße f. || avenue f. / tarred ~ || geteerte Straße f. || roue f. goudronnée.

street building machine || Straßenbaumaschine f. || machine f. pour la construction des routes.

street car || Straßenbahnwagen m. || wagon m. de tramway.

street cleaner, motor ~ || Motorstraßenkehrmaschine f. || autobalayeuse f.

street cleaning machine || Straßenreinigungsmaschine f. || machine f. à nettoyer les rues. / ~ current || Netzstrom m. || courant m. du secteur.

street door || Haustür f. || porte f. d'entrée ou de rue ou de la maison. / ~ contact || Haustürkontakt m. || contact m. de la porte d'entrée de maison.

street lamp || Straßenlaterne f. || réverbère m. / ~ lantern for gas lighting || Straßen-laterne f. für Gasbeleuchtung || réverbère m. pour l'éclairage au gaz. / ~ macadam || Straßenschotter m. || macadam m. / ~ organ || Drehorgel f.; Leierkasten m. || orgue f. de Barbarie ou à manivelle. / ~ railway || Straßenbahn f. || tramway m. / ~ railway motor || Straßenbahnmotor m. || moteur m. de tramway.

street roller || Straßenwalze f. || rouleau m. compresseur; cylindre m. de route. / ~ shell || Straßenwalzenmantel m. || enveloppe f. de rouleau compresseur.

street steam tractor || Dampfstraßenzugmaschine f. || locomotive f. routière à vapeur. / ~ sweeper (Car) see ~ sweeping vehicle. / ~ sweeper (Person) || Straßenkehrer m. || balayeur de rues f. / ~ sweeping vehicle || Straßenreinigungsfahrzeug n. || voiture f. pour le nettoyage des rues. / ~ tarring machine || Straßenteermaschine f. || machine f. à goudronner les rues. / ~ transformer || Straßentransformator m. || transformateur m. de rue. / ~ watering car || Straßensprengwagen m. || arroseuse f. de route.

strength || Festigkeit f.; Widerstand m.; Widerstandsfähigkeit f.; Kraft f. || solidité f.; résistance f. / bending ~ || Biegungsfestigkeit f. || résistance f. à la flexion. / breaking ~ || Bruchfestigkeit f. || résistance f. à la rupture. / compression ~ || Druckfestigkeit f. || résistance f. à la compression. / ~ of the current || Stromstärke f. || intensité f. du courant. / dielectric ~ || dielektrische Festigkeit f. || rigidité f. diélectrique. / disruptive ~ || Durchschlagfestigkeit f. || rigidité f. diélectrique. / dynamic ~ || Schwingungsfestigkeit f.; dynamische Festigkeit f. || résistance f. à l'oscillation. / ~ of extension || Zugfestigkeit f. || résistance f. à la traction. / ~ of flexure || Biegungsfestigkeit f. || résistance f. à la flexion. / the ~ guarantees a safe working || die Festigkeit reicht aus || la résistance est jugée suffisante. / ~ at high temperature || Festigkeit f. bei hoher Temperatur || résistance f. à haute température. / ~ of material || Festigkeitslehre f, || science f. de la résistance des matériaux. / physical ~ || körperliche Anstrengung f. || fatigue f. corporelle. / ~ of pole || Polstärke f. || intensité f. de pôle. / to reduce the ~ (Chem) || verarmen || appauvrir. / static ~ || statische Festigkeit f. || résistance f. statique. / sufficient ~ || ausreichende Festigkeit f. || force f. de résistance suffisante. / tensile ~ || Zugfestigkeit f. || résistance f. à la traction. / torsional ~ || Drehungsfestigkeit f. || résistance f. à la torsion. / transverse ~ || Biegungsfestigkeit f. || résistance f. à la flexion.

strength, calculation of || Festigkeitsberechnung f. || calcul m. de la résistance. / doctrine of ~ || Festigkeitslehre f. || théorie f. de la résistance des matériaux.

strengthen, to || verstärken || renforcer.

strengthened by ribs || mit Rippen fpl. versteift || renforcé par des nervures fpl.

strengthener (Radio) || Verstärker m. || amplificateur m.; répéteur m. / counter contact ~ || Gegentaktverstärker m. || amplificateur m. en va-et-vient. / end ~ || Endverstärker m. || amplificateur m. final.

strengthening of the field || Feldverstärkung f. || renforcement m. de champ.

strengthening curve ‖ Verstärkungskurve f. ‖ courbe f. d'amplification. / ~ preparation ‖ Kräftigungsmittel n. ‖ produit m. de régime. / ~ ring ‖ Verstärkungsring m. ‖ anneau m. de renforcement. / ~ screen ‖ Verstärkungsschirm m. ‖ diaphragme m. de renforcement. / ~ tube ‖ Verstärkerröhre f. ‖ lampe f. amplificatrice. / ~ value ‖ Verstärkungsziffer f. ‖ facteur m. d'amplification.

strength, reduction of ~ (Chem) ‖ Verarmung f. ‖ appauvrissement m. / ~ test ‖ Zerreißprobe f. ‖ essai m. de rupture. / thread ~ tester ‖ Fadenfestigkeitsprüfer m. ‖ dynamomètre m. pour fils.

stress, to ‖ pressen; drücken; beanspruchen ‖ charger; fatiguer; faire travailler. / ~ the fabric excessively ‖ den Stoff m. übermäßig beanspruchen ‖ faire travailler le tissu d'une manière excessive.

stress ‖ Beanspruchung f.; Spannung f.; Kraft f. ‖ effort m.; tension f. / additional ~ ‖ Zusatzspannung f. ‖ tension f. additionnelle. / ~ in bending ‖ Biegebeanspruchung f. ‖ effort m. de flexion. / casting ~es pl. ‖ Gußspannung f. ‖ tension f. de coulée. / the pieces pl. show absence of casting ~es ‖ die Gußstücke npl. sind vollkommen spannungsfrei ‖ les pièces fpl. ne présentent pas des tensions de coulée. / dangerous ~es pl. in the neck of the pin ‖ gefährliche Beanspruchung f. in der Hohlkehle des Zapfens ‖ effort m. dangereux au congé du tourillon. / dynamic ~ ‖ dynamische Beanspruchung f.; Schwingungsbeanspruchung f. ‖ effort m. d'oscillation. / electric ~ ‖ elektrische Beanspruchung f. ‖ effort m. ou sollicitation f. électrique. / ever-growing ~ ‖ wachsende Belastung f. ‖ exigence f. toujours croissante. / excessive ~ ‖ zu hohe Beanspruchung f. ‖ surcharge f. exagérée. / engine part which has to undergo high ~ ‖ schwer beanspruchter Maschinenteil m. ‖ élément m. de machines subissant des efforts augmentés. / highest ~ ‖ Höchstbeanspruchung f. ‖ effort m. maximum. / intermittent ~ ‖ stoßweise Beanspruchung f. ‖ effort m. intermittent ou pulsatoire. / internal ~ ‖ innere Spannung f. ‖ tension f. interne. / material ~ ‖ Materialbeanspruchung f. ‖ effort m. des matériaux. / pulsating ~ see intermittent ~. / generation of ~es at rest ‖ Erzeugung f. von Ruhespannungen ‖ génération f. de tension au repos. / service ~ ‖ Betriebsbeanspruchung f. ‖ fatigue f. de service; charge f. d'emploi. / machine parts subjected to ~ of a simple nature ‖ einfach beanspruchte Maschinenteile mpl. ‖ organes mpl. de machines soumis à des efforts simples et uniformes. / static ~ ‖ statische Beanspruchung f. ‖ effort m. statique. / ~ of wire ‖ Drahtspannung f. ‖ tension f. du fil.

stress analysis ‖ Kräftebestimmung f. ‖ détermination f. des efforts. / ~ diagram ‖ Kräfteplan m. ‖ polygone m. des forces. / distribution of ~es ‖ Verteilung f. der Spannungen ‖ distribution f. des tensions. / ~ relieving system to telescope ‖ Entlastungssystem n. des Fernrohres ‖ dispositif m. pour décharger la lunette. / security against ~es due to sudden and repeated shocks ‖ Bruchsicherheit f. gegenüber stoßweiser Belastung ‖ sûreté

f. contre la rupture, les aciers étant soumis à des secousses répétées.

stretch, to ‖ strecken; dehnen; spannen ‖ (s')étendre; (se) dilater; (s')allonger; étirer. / ~ (Curr) ‖ strecken ‖ craminer. / ~ (Shoes) ‖ weiten ‖ élargir. / ~ cold ‖ kaltstrecken ‖ tréfiler ou étirer à froid.

stretched ‖ gespannt ‖ tendu.

stretcher (Boat) ‖ Fußlatte f. ‖ marchepied m. de nage. / ~ (Build) ‖ Läufer m.; Strecker m. ‖ panneresse f.; pierre f. placée en parement. / ~ (Litter) ‖ Tragbahre f. ‖ brancard m.; cifière f. / ~ for microphone cases ‖ Spannvorrichtung f. für Kapselmikrophone ‖ tendeur m. pour les capsules microphoniques. / ~ for paper mill cylinder felt sleeves ‖ Walzenbezugspanner m. für Papiermaschinen ‖ tendeur m. à manchon pour la fabrication des cylindres à papier. / self-acting ~ ‖ Selfaktor m.; selbstspinnende Mulemaschine f. ‖ mull-jenny m. renvideur; renvideur m. mull-jenny self-acting.

stretcher rod ‖ Zungenverbindungsstange f.; Zungenangriffstange f. ‖ tringle f. de connexion ou d'attaque d'aiguille. / ~ support ‖ Tragbahrenstütze f. ‖ support m. de civière.

stretching ‖ Ziehen n.; Zug m. ‖ extension f. / ~ of wire ‖ Recken n. des Drahtes ‖ rectification f. du fil; tréfilage m.

stretching arrangement ‖ Spannvorrichtung f. ‖ disposition f. de tension. / ~ course ‖ Läuferschicht f.; Streckerschicht f. ‖ assise f. en panneresse ou en parement ou par carreaux. / ~ device ‖ Spannvorrichtung f. ‖ installation f. de tension. / ~ hammer ‖ Spannhammer m. ‖ marteau m. à étirer ou à dresser.

stretching machine (Rubber) ‖ Spannmaschine f. ‖ machine f. tendeur. / ~ (Weav) ‖ Spannmaschine f. ‖ rame f. continue. / metal gauze ~ ‖ Metalltuchstreckmaschine f. ‖ machine f. à étirer les toiles métalliques. / spiral ~ (Clothm) ‖ Spiraltrockenmaschine f.; Spiralspannmaschine f. ‖ ténoxère f.

stretching, conditioning and winding-on machine ‖ Breitstreck-, Egalisier- und Aufwickelmaschine f. ‖ machine f. à élargir, égaliser et enrouler. / ~ mule ‖ Vorspinnmaschine f. ‖ machine f. à filer en doux. / ~ plate for mirrors and table glass ‖ Streckplatte f. für Spiegel- und Tafelglas ‖ pierre f. à étendre le verre à vitres et à glaces. / ~ pulley ‖ Spannrolle f. ‖ poulie f. de tension; tendeur m. de courroie. / ~ ring (Rubber) ‖ Spannring m. ‖ anneau m. tendeur. / ~ rod ‖ Verbindungsstange f. der Weiche ‖ tringle m. d'écartement. / ~ rolls pl. ‖ Reckwalzwerk n.; Stabwalzwerk n. ‖ cylindres fpl. étireurs ou finisseurs. / ~ screw (Aero) ‖ Spannschloß n. ‖ manchon m. de serrage; écrou-tendeur m.; tendeur m. / ~ screw (Mach) ‖ Zugwinde f. ‖ cric m. de traction.

strewing meal ‖ Streumehl n. ‖ farine f. en poudre. / ~ powder ‖ Streupulver n. ‖ poudre f. vulnéraire.

striae measuring apparatus ‖ Schlierenapparat m. ‖ appareil m. pour l'étude des stries.

striated ‖ gestreift ‖ rayé; strié.

strickle (Carp; Mas) ‖ Lehre f.; Streichmodel m.; Stichmaß n. ‖ échandillon m. / ~ (Found) ‖ Abstreichlatte f. ‖ trousse f.; gabarit m.

strickling plant ‖ Rieselanlage f. ‖ bâtiment m. de graduation.

stricture coil ‖ Striktionsspule f. ‖ bobine f. de stricture.

strigs pl. ‖ Reiser fpl. ‖ ramilles fpl.

strike, to ‖ streiken ‖ faire grève f.; chômer. / ~ (Bell) ‖ anschlagen ‖ frapper. / ~ a balance ‖ eine Bilanz f. aufstellen ‖ dresser un bilann. / ~ bottom (Mar) ‖ Grund m. werfen ‖ avoir ou trouver fond m. / ~ the gin ‖ das Hebezeug n. zerlegen ‖ démonter la chèvre. / ~ ground see ~ bottom. / ~ a key ‖ eine Taste f. anschlagen ‖ toucher une clef. / ~ a line (Carp) ‖ das Holz abschnüren oder schnüren ‖ aligner le bois. / ~-off a proof with the beating brush ‖ einen Bürstenabzug m. machen ‖ tirer une épreuve à la brosse. / ~-off a proof-sheet ‖ einen Korrekturbogen m. abziehen ‖ tirer une épreuve. / ~ a sail ‖ ein Segel n. streichen ‖ amener une voile. / ~ the tents ‖ ein Zeltlager n. abbrechen ‖ plier les tentes fpl.

strike ‖ Streik m. ‖ grève f. / ~ (Brick) ‖ Streichholz n.; Streichbrett n. ‖ plane f. / to call a ~ ‖ in den Streik m. treten ‖ se mettre en grève f. / ~ of seam ‖ Streichen n. des Flözes ‖ direction f. de la couche. / to settle the ~ ‖ den Streik m. beilegen ‖ terminer la grève. / sympathetic ~ ‖ Sympathiestreik m. ‖ grève f. de sympathie.

strike breaker ‖ Streikbrecher m. ‖ antigréviste m.; briseur m. de grève. / general direction of ~ (Mine) ‖ Hauptstreichrichtung f. ‖ ligne f. de direction principale.

strike funds pl. ‖ Streikfonds m. ‖ fond m. de grève. / to collect ~ ‖ Streikgelder npl. sammeln ‖ recueillir de l'argent pour la grève.

strike, length of ~ of the mine (Mine) ‖ streichende Länge f. des Grubenfeldes ‖ longueur f. du champ d'exploitation. / the line of ~ and direction of dip vary (Geol) ‖ Streichrichtung f. und Fallrichtung f. wechseln ‖ la direction f. et l'inclinaison f. changent. / to set out ~ pickets pl. ‖ Streikposten mpl. ausstellen ‖ placer des postes de grève. / ~ purse ‖ Streikkasse f. ‖ caisse f. de grève.

striker ‖ Streikender m.; Ausständiger m. ‖ gréviste m.; chômeur m. / ~ (Electr) ‖ Zünder m. ‖ allumeur m.; appareil m. d'allumage. / ~ (Forg) ‖ Zuschläger m. ‖ frappeur m. / ~ fork ‖ Verschiebegabel f. ‖ fourche f. déplaçable.

striking of the type bars (Typewr) ‖ Aufeinanderschlagen n. der Typenhebel ‖ choc m. des tiges à caractères.

striking clock ‖ Schlaguhr f. ‖ montre f. à sonnerie. / ~ distance (Electr) ‖ Schlagweite f.; Funkenlänge f. ‖ longueur f. d'étincelle. / ~ effect (Typewr) ‖ Anschlagswirkung f. ‖ effet m. de la frappe. / ~ mechanism for tower clocks ‖ Schlagwerk n. für Turmuhren ‖ (mouvement m. de) sonnerie f. pour horloges de tours. / ~ moment ‖ Schlagmoment n. ‖ moment m. de choc. / ~ movement ‖ schlagende Bewegung f. ‖ mouvement m. percutant. / ~ surface of the hammer ‖ Schlagfläche f. des Hammers ‖ surface f. de frappe du marteau. / ~ voltage ‖ Zündspannung f. ‖ tension f. d'allumage.

string, to ‖ bespannen; besaiten ‖ garnir de cordes fpl. / ~ (Pearls etc.) ‖ anreihen;

aufreihen ‖ joindre; enfiler. / ~ the pin ‖ einen Stecknadelkopf m. aufspießen ‖ enfiler les têtes fpl. d'épingle.

string ‖ Schnur f.; Band n.; Bindfaden m.; Kordel f. ‖ cordon m.; ficelle f. / ~ (Bot) ‖ Blattrippe f. ‖ côte f. d'une feuille. / ~ (Bow) ‖ Sehne f. ‖ corde f. / ~ (Fibre) ‖ Faser f.; Fiber f. ‖ fibre f. / ~ (Mus) ‖ Saite f. ‖ corde f. / gut ~ ‖ Darmsaite f. ‖ corde f. en boyau. / gut ~ making machine ‖ Darmsaitenherstellungsmaschine f. ‖ machine f. à fabriquer des cordes en boyaux. / bow ~ for hatter's ‖ Bogenschnur f. für Hutmacher ‖ corde f. d'arçon pour chapeliers. / ~ for musical instruments ‖ Instrumentensaite f. ‖ corde f. pour instruments de musique; corde f. harmonique. / ~ for pianos ‖ Klaviersaite f. ‖ corde f. à piano. / ~ for rackets ‖ Tennissaite f.; Darmsaite f. für Tennisschläger ‖ corde f. à raquettes. / silk ~ (Mus) ‖ Seidensaite f. ‖ corde f. en soie. / spun ~ ‖ besponnene Saite f. ‖ corde f. filée ou guipée. / wire ~ (Mus) ‖ Drahtsaite f. ‖ corde f. métallique.

stringed, long-~ brick moulding machine ‖ Strangziegelpresse f. ‖ machine f. de façonnage m. mécanique de briques à la filière. / ~ instrument see string instrument.

string factory ‖ Schnurfabrik f. ‖ ficellerie f. / ~ fitter (Mus) ‖ Saitenaufspanner m. ‖ monteur m. de cordes. / ~ galvanometer ‖ Saitengalvanometer n.; Fadengalvanometer n. ‖ galvanomètre m. à fil.

string instrument ‖ Saiteninstrument n.; Streichinstrument n. ‖ instrument m. à cordes. / parts pl. for ~s ‖ Bestandteile mpl. für Saiteninstrumente ‖ accessoires mpl. pour instruments à cordes.

string maker (Rope) ‖ Kordelmacher m. ‖ ficellier m. / ~ making machine ‖ Bindfadenherstellungsmaschine f. ‖ machine f. à fabriquer les ficelles. / ~ piece (Build) ‖ Langschwelle f. ‖ longrine f.; longuerine f. / ~ tag ‖ Anhängezettel m. ‖ étiquette f. à ficelle; attache f. / ~ test (Sug) ‖ Fadenprobe f. ‖ preuve f. ou essai m. au filet.

string wire (Mus) ‖ Saitendraht m. ‖ corde f. à instrument. / polishing device for ~s ‖ Saitendrahtpoliereinrichtung f. ‖ installation f. à polir le fil pour cordes à musique.

stringy (Sugar) ‖ fadenziehend ‖ filant.

strip, to ‖ abstreifen; abnehmen ‖ arracher. / ~ (To file) ‖ schlichtfeilen ‖ finir à la lime. / ~ the bark from the wood ‖ abborken ‖ écorcer. / ~ the leaves pl. of tobacco ‖ die Tabakblätter npl. entrippen ‖ écôter les feuilles fpl. de tabac. / ~-off the burrs pl. ‖ putzen ‖ ébarber. / ~ the quills pl. ‖ die Federn fpl. schleißen ‖ ébarber les plumes fpl. / ~ the thread of a screw ‖ eine Schraube f. überdrehen ‖ déformer une vis. / ~ a wire ‖ die Isolation f. vom Draht entfernen; einen Draht m. blank machen ‖ mettre le fil à nu.

strip ‖ Streifen m. ‖ bande f. / ~ of carpet ‖ Läufer m. ‖ chemin m. de velours. / ~ of chips ‖ Spanband n. ‖ bande f. de copeaux. / ~ of drops (Tel) ‖ Klappenstreifen m. ‖ réglette f. d'annonciateurs. / endless ~ ‖ endloser Streifen m. ‖ bande m. sans fin. / ~ of fabric ‖ Stoffstreifen m. ‖ bande f. d'étoffe. / thin ~ of glass ‖ Glasstab m. ‖ baguette f. de verre. / ~ of lead ‖ Bleistreifen m. ‖ bande f. de plomb.

/ ~ of paper ‖ Papierstreifen m. ‖ bande f. de papier. / wearing ~ ‖ Abnutzungsschiene f. ‖ rail m. d'usure.

strip cutting machine ‖ Streifenschere f. ‖ machine f. à découper en bandes.

stripe, to ‖ streifen ‖ rayer.

striped ‖ streifig; gestreift ‖ rayé. / vertically ~ ‖ langgestreift ‖ rayé en long. / ~ goods pl. (Textile) ‖ gestreifte Ware f. ‖ tricot m. rayé. / ~ heel ‖ Ringelferse f. ‖ talon m. rayé en travers.

stripe machine (Textile) ‖ Ringelmaschine f. ‖ machine f. à raies horizontales.

strip fuse ‖ Streifensicherung f. ‖ fusible m. à lame. / disconnectable ~ ‖ abschaltbare Streifensicherung ‖ lame f. fusible pouvant être mise hors circuit.

striping chain (Textile) ‖ Ringelkette f. ‖ chaîne f. pour rayures horizontales.

strip iron see also hoop iron ‖ Bandeisen n. ‖ fer m. feuillard; fer m. en rubans; feuillard m. / cold-rolled ~ ‖ kaltgewalztes Bandeisen n. ‖ fer m. feuillard laminé à froid.

strip mill ‖ Bandwalzwerk n. ‖ laminoir m. à bandes. / reversing ~ ‖ Umkehrbandwalze f. ‖ laminoir m. à feuillards réversible.

strip mine ‖ Tagebau m. ‖ exploitation f. à ciel ouvert.

stripped leaf ‖ ausgeripptes Blatt n. ‖ feuille f. écôtée. / ~-off ‖ abgeschält ‖ pelé; écorcé.

stripper (Spinn) ‖ Putzgerät n. ‖ débourreur m. / ~ (Tabacco) ‖ Abripper m.; Tabakripper m. ‖ écôteur m. / ~ adjustable in the height (Agr) ‖ hochkippbarer Abstreifer m. ‖ dévêtisseur m. basculant vers le haut.

stripping a paved road ‖ Aufbrechen n. einer gepflasterten Straße ‖ dépavage m. d'une route pavée.

stripping crane ‖ Stripperkran m.; Abstreifkran m. ‖ grue-démouleuse f. / ~ panel ‖ Reißbahn f. ‖ panneau m. de déchirure. / ~ tongs pl. (Electr) ‖ Isolationsabziehzange f.; Isolationsabziehpinzette f. ‖ pincette f. écorchante.

strip radiator (Heater) ‖ Streifenheizkörper m. ‖ radiateur m. lamellaire.

strips pl. ‖ Streifenmaterial n. ‖ matériel m. découpé en bandes.

strip shears pl. ‖ Streifenschere f. ‖ cisailles fpl. pour couper en bandes. / ~ for dividing strips and plates ‖ Streifenschere f. zum Teilen von Streifen und Platten ‖ cisailles fpl. à découper les bandes et les plaques. / ~ for trimming strips and plates ‖ Streifenschere f. zum Besäumen von Streifen und Platten ‖ cisaille f. à border ou à rogner les bandes et les plaques.

strip steel ‖ Bandstahl m. ‖ feuillard m. en acier. / ~ winding machine ‖ Bandwickelmaschine f. ‖ enrouleuse f. de bandes.

strobile (Bot) ‖ Dolde f. ‖ cône m.; ombelle f.

stroboscope ‖ rotierende Schartenblende f.; Stroboskop n. ‖ appareil m. stroboscopique.

stroke ‖ Schlag m.; Stoß m. ‖ coup m.; choc m. / ~ (Mach) ‖ Hub m.; Kolbenhub m. / ~ (Mar) ‖ Ruderschlag m.; Riemenschlag m. ‖ coup m. de rame. / ~ (Lightning) ‖ Blitz m.; Blitzstrahl m. ‖ foudre f. / back ~ (Mach) ‖ Rückgang m.; Rückwärtsbewegung f. ‖ marche f. (en) arrière. / ~ of broaching

machine ‖ Ziehlänge f. der Räummaschine ‖ longueur m. de traction d'une machine à brocher. / ~ of carriage ‖ Schlittenhub m. ‖ course f. du chariot. / cold ~ (Lightning) ‖ kalter Blitzschlag m. ‖ coup m. de foudre froid. / compression ~ ‖ Kompressionshub m. ‖ course f. de compression. / downward ~ ‖ Abwärtshub m. ‖ course f. descendante. / at the end of each ~ ‖ bei jedem Hubwechsel m. ‖ à chaque changement m. de course. / expansion ~ ‖ Explosionshub m. ‖ course f. d'explosion. / four-~ ‖ Viertakt m. ‖ à quatre temps mpl. / hard ~ (Typewr) ‖ starker Anschlag m. ‖ frappe f. violente. / induction ~ ‖ Einlaßhub m. ‖ course f. d'aspiration. / piston ~ ‖ Kolbenhub m. ‖ course f. de piston. / the ~ of the press can be adjusted to any required measure ‖ der Hub m. der Presse ist in den weitesten Grenzen verstellbar ‖ la course du piston comprimeur est réglable dans des limites très larges. / return ~ (Lightening) ‖ Rückschlag m. ‖ choc m. en return. / ~ of a scale ‖ Gradstrich m. einer Skale oder einer Teilung ‖ barre f. d'une graduation. / two-~ ‖ Zweitakt m. ‖ à deux temps mpl. / upward ~ ‖ Aufwärtshub m. ‖ course f. ascendante. / working ~ ‖ Arbeitshub m. ‖ course f. d'explosion.

stroke compensator ‖ Stoßausgleicher m. ‖ compensateur m. de choc. / ~ counter ‖ Hubzähler m. ‖ compteur m. de courses. / ~ counting apparatus see ~ counter. / ~ engraving ‖ Stichelschneiden n. ‖ gravure f. au burin. / height of ~ (Forg) ‖ Hub m. ‖ hauteur f. de chute. / length of ~ ‖ Hublänge f. ‖ longueur f. de course. / indicator showing the length of ~ ‖ Hublängenanzeiger m. ‖ indicateur m. de la longueur de course. / ~ volume ‖ Hubvolumen n. ‖ cylindrée f.

strong ‖ fest ‖ fixe; ferme; fort; robuste; solide. / ~ acid ‖ starke Säure f. ‖ acide m. fort. / ~ beer ‖ Lagerbier n.; Starkbier n. ‖ bière f. forte ou double. / ~ current see also heavy current ‖ Starkstrom m. ‖ courant m. fort. / generation of ~ current ‖ Erzeugung f. von Starkstrom ‖ production f. de courant fort. / ~ design ‖ widerstandsfähige Ausführung f. ‖ exécution f. résistante. / ~ leather ‖ Starkleder n. ‖ cuir m. dur. / ~ satin ‖ schwerer Atlas m. ‖ satin m. fort. / ~ ticking (Weav) ‖ Drell m. ‖ treillis m.; linge m. ouvré. / ~ wall ‖ Brandmauer f. ‖ mur m. massif ou mitoyen.

strontia see strontium oxide and strontian.

strontian ‖ Strontian m.; Strontianerde f. ‖ strontiane f.

strontianite (Miner) ‖ Strontianit m. ‖ strontianite f.

strontium ‖ Strontium ‖ strontium m. / ~ carbonate ‖ Strontiumkarbonat n. ‖ carbonate m. de strontium. / ~ chloride ‖ Strontiumchlorid n. ‖ chlorure m. de strontium. / ~ hydrate ‖ Strontiumhydrat n. ‖ hydrate m. de strontium m. / ~ hydroxide ‖ Strontiumhydroxyd n. ‖ hydroxide m. de strontium. / ~ nitrate ‖ Strontiumnitrat n.; Strontiumsalpeter m. ‖ azotate m. ou nitrate de strontium. / ~ oxide ‖ Strontiumoxyd n. ‖ oxyde m. de strontium; strontiane f. caustique. / ~ salt ‖ Strontiumsalz n. ‖ sel m. de strontium. / ~ sulphate ‖ Strontiumsulfat n. ‖ sulfate m. de strontium.

/ ~ sulphide ‖ Schwefelstrontium n. ‖ sulfure m. de strontium. / ~ yellow ‖ Strontiumchromat n. ‖ chromate m. de strontium.

strop ‖ Streichriemen m.; Abziehriemen m. ‖ cuir m. à rasoir. / ~ (Aero) ‖ Stropp m. ‖ élingue f.

strophantus seed ‖ Strophantussamen m. ‖ semence f. de strophantus.

structural ‖ structurell ‖ structural. / ~ casting ‖ Bauguß m. ‖ fonte f. pour constructions. / machine for making ~ fittings pl. ‖ Baubeschlagherstellungsmaschine f. ‖ machine f. à fabriquer des garnitures de constructions. / ~ iron work contractor ‖ Eisenbauunternehmer m. ‖ entrepreneur m. de constructions en fer. / ~ material ‖ Baustoff m. ‖ matériaux mpl. de construction. / ~ part subjected to highest stresses ‖ stark beanspruchter Konstruktionsteil m. ‖ pièce f. de construction exposée à de grands efforts. / ~ steel ‖ Konstruktionsstahl m.; Baustahl m. ‖ acier m. de construction. / ~ steelwork ‖ Eisenbauwerk n.; Eisenkonstruktion f. ‖ charpente f. métallique. / ~ steelwork for conveying plants ‖ eiserne Gerüstbauten mpl. für Transportanlagen ‖ charpentes fpl. métalliques pour installations de transport. / ~ strength ‖ Baufestigkeit f. ‖ résistance f. de construction.

structure ‖ Struktur f.; Gefüge n.; Aufbau m. ‖ structure f. / ~ (Build) ‖ Gebäude n.; Bauwerk n. ‖ bâtiment m.; édifice m.; construction f. / anatomical ~ of the eye ‖ anatomischer Bau m. des Auges ‖ anatomie f. de l'œil. / columnar ~ ‖ stenglige Struktur f. ‖ structure f. aciculaire ou bacillaire. / fibrous ~ ‖ faserige Struktur f. ‖ structure f. fibreuse. / fluidal ~ ‖ Fluidalstruktur f. ‖ structure f. fluidale. / globular ~ ‖ kugelige Struktur f. ‖ structure f. globulaire. / granular ~ ‖ körniges Gefüge n.; körnige Struktur f. ‖ texture f. granulée; structure f. grenue. / iron ~ ‖ Eisenkonstruktion f. ‖ charpente f. métallique. / shop for iron ~s pl. ‖ Eisenkonstruktionswerkstatt f. ‖ atelier m. pour charpentes métalliques. / lamellar ~ ‖ blättrige Struktur f. ‖ structure f. lamellaire. / ~ of the material ‖ Werkstoffstruktur f. ‖ structure f. de la matière. / optical ~ of the eye ‖ optischer Bau m. des Auges ‖ système m. optique de l'œil. / ~ of the steel ‖ Gefüge n. des Stahls ‖ structure f. de l'acier. / ~ castings pl. for ~s pl. ‖ Bauguß m. ‖ fonte f. moulée pour la construction de bâtiments.

strut, to ‖ ausstreben ‖ entretoiser. / ~ a telegraph pole ‖ eine Telegraphenstange f. verstreben ‖ mettre une contrefiche à un poteau télégraphique.

strut ‖ Spreizholz n.; Spreize f.; Verstrebung f.; Strebe f. ‖ contre-fiche f. / rear undercarriage ~ (Aero) ‖ hintere Fahrgestellstrebe f. ‖ mât m. arrière du train d'atterrissage. / tubular ~ ‖ Rohrstrebe f. ‖ montant m. ou entretoise f. ou contrefiche f. tubulaire.

strut attachment ‖ Strebenknotenstück n. ‖ ferrure f. nodale du mât de suspension. / ~ beam ‖ Stützbalken m. ‖ arbalétrier m. / ~ brace ‖ Sturmband n. ‖ attache f. en contre-fiche.

strut-braced wing ‖ verstrebte Zelle f. ‖ aile f. maintenue par des contrefiches ou par des montants.

strut distribution ‖ Strebenteilung f. ‖ écartement m. des jambes de force ou des contrefiches. / fineness ratio of the ~ ‖ Schlankheitsverhältnis n. der Strebe ‖ allongement m. du mât.

strut-frame (Carp) ‖ Sprengwerk n. ‖ assemblage m. à contre-fiches. / ~ bridge ‖ Sprengbrücke f. ‖ pont m. à contre-fiches.

strut repartition see strut distribution. / ~ section (Aero) ‖ Strebenquerschnitt m. ‖ section f. transversale des mâts.

strutted wing ‖ verstrebter Flügel m. ‖ aile f. entretoisée ou à mât ou à montant.

strutting ‖ Verstreben n.; Verstrebung f. ‖ mâture f. / engine ~ ‖ Motorverstrebung f. ‖ mâture f. du bâti-moteur.

strychnine ‖ Strychnin n. ‖ strychnine f. / ~ corn (Chem) ‖ Strychningetreide n. ‖ blé m. à la strychnine.

stubborn ‖ starr ‖ raide; engourdi. / ~ (Met) ‖ schwer schmelzbar ‖ réfractaire.

stub mast ‖ Stumpfmast m. ‖ mât m. tronqué. / ~ switch (Railw) ‖ Schleppwechsel m. ‖ changement m. sans contre-rails.

stucco ‖ Gipsstuck m.; Stuck m. ‖ enduit m. en plâtre; plâtre m.; stuc m. / ligneous ~ ‖ Holzstuck m. ‖ stuc m. en pâte de bois; bois m. coulé.

stucco ceiling ‖ Gipsdecke f. ‖ plafond m. en plâtre. / ~ work ‖ Stukkaturarbeit f. ‖ ouvrage m. de stuc. / workshop for ~ ‖ Stuckwerkstatt f. ‖ atelier m. de plafonnier ou pour ouvrages en stuc.

stud (Bolt) ‖ Stehbolzen m. ‖ goujon m. / ~ (Chain) ‖ Kettensteg m. ‖ étai m. du maillon. / ~ (Clockm) ‖ Spiralklötzchen n. ‖ piton m. / ~ (Mach) ‖ Stiftschraube f. ‖ goujon m.; prisonnier m. / ~ (Shirt) ‖ Manschettenknopf m. ‖ bouton m. de manchettes. / ~ (of horses) ‖ Gestüt n.; haras m. / ~ bearing ‖ Anschlagnocken m. ‖ butée f. d'aiguille; came f. de butée. / machined ~ ‖ blanke Stiftschraube f. ‖ prisonnier m. décolleté. / metal ~ ‖ Metallsteg m. ‖ bande f. métallique. / root ~ ‖ Wurzelstift m. ‖ cheville f. à racine. / threaded end of the ~ ‖ Einschraubende n. der Stiftschraube ‖ extrémité f. filetée du prisonnier.

stud bolt see stud (Bolt) and (Mach).

stud chain ‖ Kette f. mit länglichen Gliedern und Steg ‖ chaîne f. à mailles étançonnées.

studded disc discharger (Radio) ‖ rotierende Scheibenfunkenstrecke f. mit Zähnen ‖ éclateur m. à disque muni de prisonniers latéraux.

studding sail ‖ Leesegel n. ‖ bonnette f. / main ~ ‖ Großleesegel n.; Großunterleesegel n. ‖ grande bonnette f.

studdle ‖ Webgeschirr n. ‖ accessoires mpl. pour tissage.

student ‖ Student m. ‖ étudiant m. / ~'s note book ‖ Kollegheft n. ‖ calepin m. pour étudiants. / ~'s ribbon ‖ Studentenband n. ‖ ruban m. pour étudiants.

stud-farm ‖ Gestüt n.; Beschälstation f. ‖ haras m.

studio ‖ Atelier n.; Werkstatt f.; Studio n. ‖ atelier m. / broadcasting ~ ‖ Rundfunkaufnahmeraum m. ‖ studio m. pour radiodiffusion.

studio performance (Radio) ‖ Sendespiel n. ‖ représentation f. dans le studio. / ~ photo camera ‖ Atelierkamera f. ‖ caméra f. d'atelier.

stud keeper ‖ Beschälwärter m. ‖ surveillant m. de haras. / ~ lathe ‖ Stehbolzendrehbank f. ‖ tour m. pour goujons. / ~ link ‖ Kettenglied n. mit Steg ‖ maillon m. à étai. / ~ link chain ‖ Stegkette f. ‖ chaîne f. à étançons. / ~ manager ‖ Gestütsverwalter m. ‖ administrateur m. de haras. / ~ master ‖ Gestütsmeister m. ‖ maître m. de haras. / ~ owner ‖ Gestütsbesitzer m. ‖ propriétaire m. de haras. / ~ staves pl. (Coachm) ‖ Wagenrungen fpl. ‖ cornes fpl. d'un chariot à ridelles. / ~ work (Carp) ‖ Riegelwerk n. ‖ clayonnage m. ou colombage m. de charpente; cloisonnage m. de bois.

study, to ~ thoroughly ‖ eingehend studieren ‖ étudier à fond.

studying with the polarizing microscope ‖ polarisationsmikroskopische Untersuchung f. ‖ examen m. microscopique en lumière polarisée.

stuff, to ‖ polstern; ausstopfen ‖ rembourrer.

stuff ‖ Stoff m.; Material n.; Werkstoff m.; Masse f. ‖ matière f.; matériel m.; matériau m. / ~ (Carp) ‖ Hölzer npl.; Holz n.; Zimmerhölzer npl. ‖ charpente f.; bois m. de charpente. / ~ (Curr) ‖ Gerberfett n.; Weißbrühe f. ‖ dégras m. / ~ (Pap) ‖ Feinzeug n. ‖ raffiné m.; pâte f. raffinée. / ~ (Weav) ‖ Gewebe n.; Stoff m.; Zeug n. ‖ étoffe f.; tissu m. / checked ~ (Weav) ‖ gewürfeltes Zeug n. ‖ étoffe f. à carreaux. / fulled ~ (Weav) ‖ Walkware f. ‖ tissus mpl. foulés. / ~ for gentlemen's clothing ‖ Herrenanzugstoff m. ‖ étoffe f. pour habillement d'homme. / ~ for shirts ‖ Hemdenstoff m. ‖ étoffe f. pour chemises. / striped ~ (Weav) ‖ gestreiftes Zeug n. ‖ étoffe f. rayée. / tweeled ~ ‖ Keper m.; Köper m. ‖ croisé m.

stuff chest (Pap) ‖ Ganzzeugkasten m. ‖ caisse f. de dépôt. / ~ cutting machine ‖ Stoffschneidemaschine f. ‖ machine f. à découper l'étoffe. / ~ grinder (Pap) ‖ Holzschleifmühle f. ‖ défibreur m. / ~ grinding (Pap) ‖ Holzschleiferei f. ‖ déchiquetage m. du bois.

stuffing ‖ Polstern n. ‖ rembourrage m. / ~ (Mach) ‖ Dichtung f. ‖ étoupage m.; bourrage m.; garniture f. / ~ for motor cars ‖ Automobilpolsterwerkstoff m. ‖ matériaux mpl. de rembourrage pour automobiles.

stuffing box ‖ Stopfbuchse f. ‖ boîte f. à bourrage; presse-étoupe m. / expansion ~ ‖ Ausgleichstopfbuchse f. ‖ boîte f. à bourrage compensatrice; presse-étoupe m. compensateur.

stuffing box bearing ‖ Stopfbuchsenlager n. ‖ palier m. de presse-étoupes. / ~ cover ‖ Stopfbuchsendeckel m. ‖ couvercle m. de presse-étoupe. / ~ packing ‖ Stopfbüchsendichtung f. ‖ garniture f. de boîte à bourrage.

stuffing coat (Upholstery) ‖ Polsterleinwand f. ‖ rembourrure f. toile f. à rembourrure. / ~ hair (Upholstery) ‖ Füllhaar n. ‖ bourre f. / ~ material ‖ Polstermaterial n. ‖ matière f. de rembourrage.

stuff printer ‖ Zeugdrucker m. ‖ imprimeur m. sur tissus. / ~ printing ‖ Zeugdruck m. ‖ impression f. sur étoffes.

stuff shoe ‖ Stoffschuh m. ‖ soulier m. en étoffe. / ~ factory ‖ Stoffschuhfabrik f. ‖ fabrique f. de chaussures en étoffe.

stump, to ‖ abstumpfen; stutzen ‖ raccourcir.

stump ‖ Stumpf m.; Stummel m. ‖ tronçon m.; bout m. / ~ (Draw) ‖ Wischer m. ‖ estompe f. / ~ (Tree) ‖ Baumstumpf m. ‖ estoc m.; souche f.; chicot m.

stunted growth ‖ Strauchwuchs m. ‖ arbre m. rabougri.

sturdy ‖ fest; stark; derb; hart; starr; unbiegsam ‖ ferme; solide; compact; fort; raide. / ~ design of the instrument ‖ unverwüstliche Bauart f. des Instrumentes ‖ solidité f. à toute épreuve de l'instrument.

sturdy (Sheep) ‖ Drehkrankheit f. ‖ tournis m.

sturgeon ‖ Stör m. ‖ esturgeon m. / ~ fisherman ‖ Störfischer m. ‖ pêcheur m. d'esturgeons.

sty ‖ Schweinestall m. ‖ étable f. à porcs; porcherie f.

style, to ‖ benennen; betiteln ‖ dénommer; intituler.

style ‖ Stil m.; Baustil m. ‖ style m. / ~ (Carp) ‖ Ständer m.; Pfeiler m.; Säule f.; Pfosten m. ‖ montant m.; pilier m.; poteau m. / ~ (Engr) ‖ Grabstichel m.; Stichel m. ‖ burin m.; ciselet m.; poinçon m. / ~ (Etching needle) ‖ Radiernadel f.; Nadel f. ‖ échoppe f.; pointe f. (de graveur). / ~ (Mach) ‖ Ausführung f.; Bauart f. ‖ construction f. / ~ (Med) ‖ Sonde f. ‖ sonde f.; stylet m. / ~ (Sun-dial) ‖ Zeiger m. der Sonnenuhr; Sonnenzeiger m. ‖ style m. / ~ (For writing) ‖ Schreibgriffel m.; Griffel m. / ~ crayon m. en ardoise; style m. / epistolary ‖ Briefstil m. ‖ style m. épistolaire. / ~ for gramophone ‖ Stift m.; Grammofonstift m.; Sprechmaschinennadel f.; Nadel f. ‖ aiguille f. de gramophone.

stylet (Med) ‖ kleine Sonde f.; Senknadel f. ‖ stylet m.

stylograph ‖ Füllfederhalter m.; Stylograf m. ‖ stylographe m.

stylographic ink ‖ Füllfederhaltertinte f. ‖ encre f. pour porteplumes-réservoir.

styptic cotton ‖ blutstillende Watte f. ‖ coton m. styptique ou astringent.

subacetate (Chem) ‖ basisch essigsaures Salz n. ‖ sousacétate m.

subarch (Arch) ‖ Tragbogen m. ‖ archivolte f.

sub-audio telegraphy ‖ Unterlagerungstelegrafie f. ‖ télégraphic f. infraacoustique.

sub-base (Mach) ‖ Grundplatte f. ‖ plaque f. de fondation.

subcarbonate (Chem) ‖ basisch kohlensaures Salz n. ‖ souscarbonate m.

sub-compartment ‖ Unterabteilung f. ‖ subdivision f.; branche f.

subcontractor (Trade) ‖ Unterlieferant m. ‖ soustraitant m.

subdivision ‖ Unterteilung f.; Einteilung f. ‖ subdivision f.; classification f. / ~ of aeroplane types ‖ Einteilung f. der Flugzeuge (nach Typen) ‖ classification f. des avions. / ~ of plant (Electr) ‖ Netzunterteilung f. ‖ division f. des réseaux. / ~ of work ‖ Arbeitsteilung f.; Aufteilung f. der Arbeit ‖ division f. ou répartition f. du travail. / the ~ of work is so organized that the pieces go systematically forward until they are finished and fitted ‖ die Arbeitsteilung ist so geordnet, daß die Teile in einem geregelten Arbeitsgang vollständig bearbeitet und montiert werden ‖ la division du travail est organisée de telle façon que ces éléments sont

usinés et montés suivant une marche parfaitement bien réglée.

suberic acid ‖ Korksäure f.; Suberylsäure f. ‖ acide m. subérique.

sub-frame (Auto) ‖ Hilfsrahmen m.; Zwischenrahmen m. ‖ faux-châssis m.

subhead(ing) (Print) ‖ Nebentitel m.; Untertitel m. ‖ sous-titre m.

subject to atmospheric agencies ‖ den Einflüssen mpl. der Atmosphäre ausgesetzt ‖ exposé aux influences des agents atmosphériques. / ~ to being sold ‖ freibleibend ‖ sans engagement m.

sublimable (Chem) ‖ sublimierbar ‖ sublimable.

sublimate, to ‖ sublimieren ‖ sublimer.

sublimate ‖ Sublimat n. ‖ sublimé m. / corrosive ~ ‖ Quecksilberchlorid n.; Quecksilbersublimat n. ‖ sublimé m. corrosif; protochlorure m. de mercure.

sublimate pastils pl. ‖ Sublimatpastillen fpl. ‖ pastilles fpl. au sublimé

sublimating apparatus ‖ Sublimiervorrichtung f. ‖ appareil m. de sublimation.

sublimation ‖ Sublimation f. ‖ sublimation f. / ~ furnace (Met) ‖ Kapellenofen m. zur Sublimation ‖ fourneau m. de sublimation. / ~ point ‖ Sublimationspunkt m. ‖ point m. de sublimation.

submarine ‖ unterseeisch ‖ sous-marin.

submarine (boat) ‖ Unterseeboot n.; Tauchboot n.; U-Boot n. ‖ sous-marin m.; submersible m.

submarine bell signal ‖ Unterwasserglockenzeichen n. ‖ signal m. à cloche sous-marine.

submarine cable ‖ Seekabel n.; Unterseekabel n. ‖ câble m. sous-marin. / ~ box ‖ Seekabelmuffe f. ‖ manchon m. de câble sous-marin. / ~ communications pl. ‖ Seekabelnetz n. ‖ réseau m. de câbles sous-marins. / ~ company ‖ Seekabelgesellschaft f. ‖ compagnie f. de câbles sous-marins. / ~ factory ‖ Seekabelfabrik f. ‖ usine f. de câbles sous-marins. / ~ laying and repairing ‖ Seekabellegung f. und -instandsetzung f. ‖ pose f. et réparation f. de câbles sous-marins.

submarine cruiser ‖ Unterseekreuzer m. ‖ croiseur m. submersible. / ~ depot ship ‖ Hilfsschiff n. für Unterseeboote ‖ bâtiment m. de sauvetage de sous-marins. / ~ signal ‖ Unterwasserzeichen n. ‖ signal m. sous-marin. / ~ sound receiver ‖ Unterwasserschallempfänger m. ‖ récepteur m. sous-marin du son. / ~ sound signal ‖ Unterwasserschallzeichen n. ‖ signal m. acoustique sous-marin. / ~ sound signalling apparatus ‖ Unterwasserschallsignalvorrichtung f. ‖ appareil m. de signaux acoustiques sous-marins. / ~ telephone cable ‖ Seefernsprechkabel n. ‖ câble m. téléphonique sous-marin. / ~ vessel see submarine boat. / ~ warfare ‖ Unterseekrieg m.; U-Boot-Krieg m. ‖ Unterseebootkrieg m. ‖ guerre f. sous-marine.

submerge, to ~ a village ‖ ein Dorf n. überstauen ‖ submerger un village.

submerged ‖ untergetaucht ‖ plongé m.; submergé. / ~ (Mine) ‖ ersoffen ‖ noyé; submergé. / in the ~ state the stability is good (Shipb) ‖ die Stabilität f. im untergetauchten Zustande ist gut ‖ la stabilité en plongé est bonne sous tous les rapports.

submergence (Geol) ‖ geologische Senkung f.; Bodensenkung f. ‖ affaissement m.

submersible (Mar) see also submarine (boat) ‖ Tauchboot n.; Unterseeboot n.; U-Boot n. ‖ submersible m.; sous-marin m. / ~ cruiser ‖ Unterseekreuzer m. ‖ croiseur m. submersible. / ~ inspection lamp ‖ Untersäurelampe f. ‖ lampe f. sous-acide.

submersion roller (Hydr arch) ‖ Versenkwalze f. ‖ rouleau m. submersible.

submission ‖ Verdingung f.; Ausschreibung f. ‖ (mise f. en) adjudication f. / official ~ ‖ öffentliche Ausschreibung f. ‖ adjucation f. publique.

subnormal (Geom) ‖ Subnormale f. ‖ sous-normale f.

sub-office (Tel) ‖ Unteramt n. ‖ bureau m. satellite ou secondaire; sous-central m.

subphosphate ‖ basisch phosphorsaures Salz n. ‖ sousphosphate m.

subsalt ‖ basisches Salz n. ‖ sel m. basique.

subscribe, to ‖ unterschreiben; unterzeichnen ‖ signer; souscrire. / ~ (For a periodical) ‖ beziehen; abonnieren ‖ s'abonner.

subscriber (Of a periodical) ‖ Bezieher m.; Abonnent m. ‖ abonné m. / ~ (Tel) ‖ Fernsprechteilnehmer m. ‖ abonné m. / called ~ (Tel) ‖ angerufener Teilnehmer m. ‖ abonné m. demandé. / calling ~ (Tel) ‖ anrufender Teilnehmer m. ‖ abonné m. demandeur.

subscriber's cable ‖ Fernsprechanschlußkabel n.; Anschlußkabel n. ‖ câble m. d'abonné. / ~ line (Tel) ‖ Anschlußleitung f. ‖ ligne f. d'abonné. / ~ number (Tel) ‖ Rufnummer f. ‖ numéro m. de l'abonné. / ~ station (Tel) ‖ Teilnehmerstelle f. ‖ poste f. d'abonné.

subscription ‖ Bezug m.; Abonnement n. ‖ abonnement m. / annual ~ ‖ Jahresbeitrag m. ‖ contribution f. annuelle. / ~ in shares ‖ Aktienzeichnung f. ‖ souscription f. d'actions.

subscription form (Bank) ‖ Zeichnungsformblatt n. ‖ bulletin m. de souscription. / ~ right (Of shares) ‖ Bezugsrecht n. ‖ droit m. d'acheter ou de prendre ou d'achat.

subsequent ‖ nachträglich ‖ supplémentaire; ultérieur. / ~ grant ‖ Nachbewilligung f. ‖ crédit m. supplémentaire. / ~ order ‖ Nachbestellung f. ‖ commande f. ou ordre m. supplémentaire.

subside, to ~ (Mar) ‖ sacken; sinken ‖ caler; fondre; descendre. / ~ (Mine) ‖ niedergehen; zusammenbrechen ‖ crouler; s'enfoncer.

subsidence (Geol) ‖ geologische Senkung f. ‖ affaissement m. / ~ of the earth see ~ of the ground. / ~ of the ground ‖ Bodensenkung f.; Erdsenkung f. ‖ affaissement m. ou tassement m. du sol.

subsiding (Build) ‖ Senkung f.; Einsinken n.; Sacken n. ‖ enfoncement m.; affaissement m.

subsoil ‖ Untergrund m.; Unterboden m. ‖ sous-sol m. / ~ plough ‖ Untergrundpflug m. ‖ charrue f. à sous-sol. / ~ water ‖ Grundwasser n. ‖ eau f. souterraine; nappe f. d'eau.

substance ‖ Stoff m.; Masse f.; Substanz f. ‖ substance f.; matière f. / adhesive ~ ‖ Klebstoff m. ‖ matière f. collante; colle f. / admixed ~ (Chem) ‖ Beimengung f. ‖ corps m. entraîné; addition f. / ~ which is difficult to volatilize ‖ schwer verdampfbarer Stoff m. ‖ substance f. difficile à vaporiser. / foreign ~ ‖ Fremdkörper m. ‖ corps m. étranger. / mother ~ ‖

Stammsubstanz f. || substance f. génératrice. / plastic ~ || plastische Masse f. || substance f. plastique.

substantial || körperlich; substantiell; materiell || substantiel. / ~ (Trade) || greifbar || saisissable; palpable.

sub-station (Electr) || Unterwerk n. || sous-station f. / electric ~ || elektrisches Unterwerk n. || sous-station f. d'électricité. / ~ of the telephone system || Anschluß m. an das Fernsprechnetz || poste m. de réseau.

sub-station switchboard (Tel) || Nebenstellenumschalter m. || commutateur m. de centrale d'abonné.

substitute, to || ersetzen || remplacer. / ~ (Chem) || ersetzen || substituer.

substitute || Ersatzstoff m.; Surrogat n. || succédané m. / ~s pl. || Ersatzwaren fpl. || succédanés mpl.

substitute antenna || Ersatzluftleiter m.; Ersatzantenne f. || antenne f. provisoire.

substitution (Chem) || Ersetzung f. || substitution f. / ~ (Math) || Einsetzung f.; Substituierung f. || substitution f.

substructure (Build) || Unterbau m. || substruction f.; soubassement m. / ~ (Mach) || Untergestell n. || partie f. inférieure du bâti. / ~ (Mas) || Grundmauer f.; Stützmauer f. || mur m. de fondation; chaîne f. ou jambage m. de pierres. / ~ (Railw) || Unterbau m.; Eisenbahnunterbau m. || infrastructure f.; terrassement m.

subtangent || Subtangente f. || soustangente f.

subtense (Geom) || Sehne f. || corde f.; soustendante f. / telescope provided with a ~ attachment || mit einem Entfernungsmesser ausgerüstetes Fernrohr n. || lunette f. munie de traits stadimétriques.

subterranean || unterirdisch || souterrain. / ~ cable || unterirdisches Kabel n.; Erdkabel n. || câble m. souterrain. / ~ water || Tiefenwasser n. || eau f. souterraine ou profonde.

subterraneous see subterranean.

subtract, to ~ (Math) || abziehen; subtrahieren || soustraire.

subtract cam (Of a calculating machine) || Subtraktionszahnrad n. || came m. de soustraction.

subtracting device || Subtraktionsvorrichtung f. || dispositif m. de soustraction.

subtraction (Math) || Abziehen n.; Subtraktion f. || soustraction f.

suburb || Vorstadt f.; Vorort m. || banlieue f.; faubourg m.

suburb(an) railway || Vorortbahn f. || ligne f. suburbaine; chemin m. de fer de banlieue.

suburban service || Vorortverkehr m. || trafic m. suburbain.

subvention || Geldbeitrag m.; Subvention f.; geldliche Unterstützung f. oder Beihilfe f. || contribution f.; cotisation f.

subway || Tunnelgang m.; Unterführung f.; unterirdischer Gang m. || souterrain m. / ~ (Underground railway) || Untergrundbahn f.; U-Bahn f. || chemin m. de fer souterrain; métro m. / ~ crossing || Wegunterführung f. || passage m. souterrain; subterraneau m.

successful || erfolgreich || couronné de succès m.; avec succès m. / ~ in practice || in der Praxis f. bewährt || éprouvé dans la pratique.

succession || Reihenfolge f. || suite f. / ~ of beds (Geol) || Schichtenfolge f. || suite

f. des couches. / ~ of cuttings (Forest) || Hiebfolge f. || succession f. des coupes. / ~ of the winds (Meteor) || Aufeinanderfolge f. der Winde || suite f. des vents.

successive delivery || allmähliche Lieferung f.; Sukzessivlieferung f. || livraison f. successive.

successor || Nachfolger m. || successeur m. / ~ in law || Rechtsnachfolger m. || ayant-cause m.

succinate || bernsteinsaures Salz n. || succinate m.

succinic acid || Bernsteinsäure f. || acide m. succinique.

suck, to || saugen; einsaugen || aspirer. / ~ (Found) || saugen || se tasser. / ~-in || einsaugen || sucer; aspirer. / ~-off air || Luft f. absaugen || aspirer de l'air m. / all shavings pl. are sucked-off direct from the machine || alle Späne mpl. werden unmittelbar aus der Maschine abgezogen || tous les copeaux mpl. sont enlevés directement de la machine par aspiration. / ~ through || durchsaugen || aspirer à travers. / ~-up || ansaugen || aspirer.

sucker, to ~ (The tobacco) || ausgeizen || rejetonner.

sucker of a pump || Pumpenkolben m.; Pumpenschuh m. || piston m. de pompe.

sucker apparatus || Ansaugvorrichtung f. || dispositif m. à succion ou d'aspiration.

sucking || saugend || aspirant.

sucking see also suction || Saugen n. || aspiration f.; succion f. / ~ (Found) || Saugen n. || tassement m.

sucking bottle for infants || Kindersaugflasche f. || biberon m. pour enfants.

sucking-in || Einsaugen n. || aspiration f.

sucking-jet pump || Saugstrahlpumpe f. || pompe f. à jet aspirant.

sucking pipe || Ansaugrohr n. || tuyau m. d'aspiration. / ~ port (Of a pump) || Saugöffnung f. || orifice m. d'aspiration. / ~ pump see also suction pump || Saugpumpe f. || pompe f. aspirante. / ~ and forcing pump || Saug- und Druckpumpe f. || pompe f. aspirante et foulante. / ~ table || Saugkasten m.; Saugtisch m. || suçon m. / ~ wick || Saugdocht m. || mèche f. aspirante.

sucrate || Saccharat n. || saccharate m.

suction || Saugung f.; Ansaugung f.; Ansaugen n. || succion f.; aspiration f. / ~ (Mill) || Saugwind m. || aspiration f.; succion f.

suction air || Saugluft f. || air m. aspiré. / conveying plant operating with ~ || durch Saugluft betriebene Förderanlage f. || installation f. (d'engin) de transport fontionnant au moyen d'air aspiré. / ~ conveyor || Saugluftförderer m. || transporteur m. à aspiration d'air.

suction apparatus || Saugvorrichtung f.; Saugapparat m.; Nutschvorrichtung f. || appareil m. d'aspiration ou de succion. / ~ basket || Saugkorb m. || crépine f. d'aspiration. / ~ box (Pap) || Saugvorrichtung f. || aspirateur m. / ~ cell filter || Saugzellenfilter n. || filtre m. à cellules d'aspiration. / ~ chamber || Saugekammer f. || chambre f. à succion. / ~ circuit (Tel) || Saugkreis m. || circuit m. piège d'onde. / ~ conveyor || Saugluftförderer m. || transporteur m. aspirateur. / ~ drain || Saugleitung f.; Saugdrain m. || drain m. d'aspiration. / ~ draught plant for firings || Saugzuganlage f. für

Feuerungen || installation f. de tirage par aspiration pour foyers. / ~ dredger for emptying barges || Schutensauger m. || aspirateur m. pour gabares. / ~ drum || Saugtrommel f. || tambour m. d'aspiration. / ~ dryer || Saugtrockner m. || sécheur m. aspirateur ou par succion. / ~ duct || Saugkanal m. || conduite f. d'aspiration. / ~ dust collector || Saugfilter n. || filtre m. à air aspiré. / effect of ~ || Saugwirkung f. || effet m. d'aspiration. / ~ filter (Sugar) || Nutsche f.; Nutschenfilter m. || aspirateur m. de sucrerie; filtre m. de succion. / ~ flask (Chem) || Absaugekolben m. || ballon m. à faire le vide. / ~ frame || Saugrahmen m. || cadre m. à suçoirs. / crane with ~ frames for the transport of plate glass || Saugrahmenkran m. für Spiegelglastransporte || grue f. à cadre à suçoirs pour le transport des glaces.

suction gas || Sauggas n. || gaz m. aspiré; gaz m. pauvre. / ~ engine see ~ motor. / ~ motor || Sauggasmotor m. || moteur m. à gaz aspiré. / ~ plant || Sauggasanlage f. || installation f. de gazogène par aspiration ou de gazogène aspiré.

suction height || Saughöhe f. || hauteur f. d'aspiration. / ~ hood || Saughutze f. || cheminée f. ou hotte f. d'aspiration. / ~ machine || Absaugmaschine f. || essoreuse f. (à succion). / ~ pipe of a pump || Saugrohr n. einer Pumpe || tuyau m. aspirateur d'une pompe; aspirateur m. d'une pompe. / ~ pipe || Saugrohr n. || tuyau m. d'aspiration. / ~ and delivery pipes pl. || Saug- und Druckröhren fpl. || tuyaux mpl. d'aspiration et de refoulement. / ~ pipe socket (Mot) || Saugstutzen m. || tubulure f. d'aspiration. / ~ piping || Saugleitung f. || tuyauterie f. d'aspiration. / ~ plant || Absaug(ungs)-anlage f. || installation f. d'aspiration. / ~ plate (Of a pump) || Saugplatte f.; Wechselplatte f. || plaque f. d'aspiration. / ~ pressure recorder || Saugdruckschreiber m. || enregistreur m. de la pression d'aspiration.

suction pump see also sucking pump || Saugpumpe f. || pompe f. aspirante. / automatic ~ for the removal of feces || Selbstsauger m. für die Fäkalienabfuhr || aspirateur m. automatique pour l'enlèvement des matières fécales.

suction strainer || Saugsieb n. || crépine f. d'aspiration.

suction valve || Saugventil n.; Einlaßventil n. || soupape f. d'aspiration; clapet m. d'aspiration. / ~ of a pump || Saugventil n. oder Saugklappe f. einer Pumpe || soupape f. d'aspiration d'une pompe.

suction ventilator || Luftabsauger m. || aspirateur m.

sudatory || Schwitzbad n. || étuve f.; bain m. russe.

sudden || plötzlich; unvermittelt || subit; soudain; imprévu; brusque. / ~ change of weather || Witterungsumschlag m. || changement m. subit ou brusque du temps. / ~ fall in prices || Preissturz m. || baisse f. soudaine. / ~ fall of rates || Kurssturz m.; Sturz m. der Kurse || débâcle f. / ~ squall || Bö f.; kurzer Windstoß m. || rafale f.; grain m. de vent.

sue, to || verklagen || intenter une action contre; porter plainte f. contre. / ~ for the permission of working a mine ||

muten ‖ demander la concession d'une mine.

suet ‖ Talg m. ‖ suif m. / ~ of deer ‖ Hirschtalg m. ‖ suif m. *ou* axonge f. de cerf.

suffer, to ~ from . . . ‖ leiden an . . . ‖ souffrir de . . .

sufficient ‖ reichlich ‖ ample; suffisant. / ~ articulation (Tel) ‖ hinreichende Sprachgüte f. ‖ bonne qualité f. de transmission.

sufficiently fine ‖ ausreichend fein ‖ de finesse f. suffisante.

suffocating ‖ erstickend ‖ suffocant.

suffocation ‖ Erstickung f. ‖ suffocation f.

sugar, to ~ over ‖ überzuckern ‖ saupoudrer de sucre m.; candir.

sugar ‖ Zucker m. ‖ sucre m. / amorphous ~ ‖ amorpher Zucker m. ‖ sucre m. d'orge. / beet ~ ‖ Rübenzucker m. ‖ sucre m. de betterave. / brown ~ ‖ brauner Zucker m.; Rohzucker m.; gedeckter Zucker m. ‖ sucre m. noir; cassonade f. / cane ~ ‖ Rohrzucker m. ‖ sucre m. de canne. / clayed ~ *see* brown ~. / cube ~ ‖ Würfelzucker m. ‖ sucre m. concassé; sucre m. en morceaux. / ~ of gelatine ‖ Leimzucker m. ‖ sucre m. de gélatine; glycocolle f. / incrystallizable ~ ‖ unkristallisierbarer Zucker m. ‖ sucre m. incristallisable. / ~ of lead ‖ Bleizucker m. ‖ acétate m. de plomb. / lump ~ ‖ Kochzucker m. ‖ lumps m. / mucilaginous ~ ‖ Schleimzucker m. ‖ glucose f. mucilagineuse. / pounded ~ ‖ gestoßener Zucker m. ‖ sucre m. râpé. / raw ~ ‖ Rohzucker m. ‖ moscouade f.; sucre m. brut. / refined ~ ‖ raffinierter *oder* harter Zucker m.; Raffinade f. ‖ sucre m. raffiné; raffiné m. / unrefined ~ *see* raw ~. / white ~ ‖ weißer Zucker m. ‖ sucre m. blanc.

sugar beet ‖ Zuckerrübe f.; Runkelrübe f. ‖ betterave m. / ~ cellar ‖ Zuckerrübenkeller m. ‖ cave f. de betteraves. / ~ grower ‖ Zuckerrübenbauer m. ‖ cultivateur m. de betteraves. / ~ seed ‖ Zuckerrübensamen m. ‖ graine f. de betterave (sucrière).

sugar blue paper ‖ Zuckerpapier n. ‖ papier m. à pains de sucre. / ~ boiler ‖ Zuckerkessel m. ‖ chaudière f. à sucre. / ~ breaker ‖ Zuckerschneider m. ‖ casseur m. de sucre. / ~ breaking works pl. ‖ Zuckerschneiderei f. ‖ casserie f. de sucre. / ~ candy ‖ Kandiszucker m. ‖ sucre m. candi; candi m.

sugar cane ‖ Zuckerrohr n. ‖ canne f. à sucre. / ~ car ‖ Zuckerrohrwagen m. ‖ wagonnet m. pour canne à sucre. / ~ car with wood-covered platform and cross-barred ends ‖ Zuckerrohrwagen m. mit Holzbelag und Gitterkopfwänden ‖ wagonnet m. pour canne à sucre recouvert de bois te avec parois de tête en treillis. / ~ crusher ‖ Zuckerrohrquetsche f. ‖ broyeur m. à canne à sucre. / ~ mill ‖ Zuckerrohrwalzwerk n. ‖ moulin m. à canne à sucre.

sugar coal ‖ Zuckerkohle f. ‖ charbon m. de sucre. / ~ colour ‖ Zuckercouleur f. ‖ teinture f. de caramel; couleurs fpl. de sucre. / ~ compound (Chem) ‖ Zuckerverbindung f.; Saccharat n. ‖ saccharate m. / ~ determination ‖ Zuckerbestimmung f. ‖ dosage m. du sucre. / extraction of ~ ‖ Entziehung f. von Zucker; Entzuckerung f. ‖ désucration f. / rapid and thorough extraction of the ~ from

the molasses ‖ schnelle und gründliche Entzuckerung f. der Melasse ‖ désucration f. rapide et complète de la mélasse.

sugar factory ‖ Zuckerfabrik f. ‖ sucrerie f. / ~ labourer ‖ Zuckerfabrikarbeiter m. ‖ ouvrier m. de sucrerie. / ~ plant ‖ Zuckerfabrikanlage f. ‖ installation f. de sucrerie. / ~ turbine ‖ Turbine f. für Zuckerfabriken ‖ turbine f. pour sucreries.

sugar goods pl. ‖ Zuckerwaren fpl. ‖ articles mpl. en sucre; sucreries fpl. / ~ grinding plant ‖ Zuckervermahlungsanlage f. ‖ moulin m. à sucre. / ~ industry ‖ Zuckerindustrie f. ‖ industrie f. du sucre.

sugar loaf ‖ Zuckerhut m. ‖ pain m. de sucre. / ~ mould ‖ Zuckerhutform f. ‖ forme f. à pain de sucre. / ~ press ‖ Zuckerhutpresse f. ‖ presse f. à pain de sucre.

sugar machine ‖ Zuckerfabrikmaschine f. ‖ machine f. pour sucrerie. / manufacture of ~ ‖ Zuckerherstellung f.; Zuckerfabrikation f. ‖ fabrication f. du sucre; sucrerie f. / ~ manufacturer ‖ Zuckerfabrikant m. ‖ fabricant m. de sucre; sucrier m. / ~ manufactory *see* sugar factory. / ~ maple ‖ Zuckerahorn m. ‖ érable m. à sucre. / melting of ~ ‖ Zuckerschmelzen n. ‖ fonte f. du sucre. / ~ mill ‖ Zuckermühle f. ‖ moulin m. à sucre. / ~ rasp ‖ Reibemaschine f. ‖ râpe f. à sucre. / ~ refinement ‖ Zuckerraffination f. ‖ raffinage m. du sucre. / ~ refiner ‖ Zuckersieder m. ‖ raffineur m. de sucre. / ~ refinery ‖ Zuckerraffinerie f. ‖ raffinerie f. de sucre. / ~ refractometer ‖ Zuckerrefraktometer n. ‖ réfractomètre m. à sucre. / ~ sirup ‖ Rübensirup m. ‖ caramel m. liquide; sirop m. de sucre. / ~ sorting plant ‖ Zuckersortieranlage f. ‖ tamiseur m. à sucre. / ~ tongs pl. ‖ Zuckerzange f. ‖ pince f. à sucre. / ~ touch ‖ Zuckerprobe f. ‖ épreuve f. du sucre. / ~ works pl. *see also* sugar factory ‖ Zuckerfabrik f. ‖ sucrerie f.

sugary ‖ zuckerig ‖ sucré.

suint (Of sheep wool) ‖ Schafschweiß m.; Wollschweiß m. ‖ suint m.

suit ‖ Reihe f. ‖ suite f.; série f.; succession f. / ~ of appartments (Build) ‖ Zimmerflucht f. ‖ enfilade f.

suit (Cloth) ‖ Anzug m. ‖ vêtement m. / protection ~ for miners ‖ Grubenanzug m.; Schachtanzug m.; Schutzanzug m. ‖ vêtement m. de protection de mineurs.

suitable ‖ zweckmäßig; empfehlenswert ‖ convenable; conforme au but m.; recommandable. / ~ for industrial use ‖ zur gewerblichen Verwendung geeignet ‖ susceptible d'une utilisation industrielle.

sullage piece (Found) ‖ Anguß m. *oder* Ansatz m. am Gußstück ‖ jet m.; masselotte f.; saumon m.

sulphanilic acid ‖ Sulfanilsäure f. ‖ acide m. sulfanilique.

sulphate, to ‖ sulfatieren ‖ se sulfater.

sulphate ‖ Sulfat n.; schwefelsaures Salz n. ‖ sulfate m. / ~ of alumina ‖ schwefelsaure Tonerde f. ‖ sulfate m. d'alumine. / ~ of aluminium ‖ Aluminiumsulfat n. ‖ sulfate m. d'aluminium. / ~ of ammonium ‖ Ammoniumsulfat n. ‖ sulfate m. d'ammoniaque. / ~ of barium ‖ schwefelsaurer Baryt m. ‖ sulfate m. de baryte. / ~ of copper ‖ Kupfervitriol n. ‖ sulfate m. de cuivre; cuivre m. sulfaté *ou*

vitriolé. / ~ of ferrous ammonium ‖ schwefelsaures Eisenoxydulammon n.; Ammoniumferrosulfat n. ‖ sulfate m. ferreux ammoniacal; ferro-sulfate m. d'ammonium; sulfate m. de fer et d'ammonium. / hydric ~ ‖ Schwefelsäure f.; Vitriol(öl) n. ‖ acide m. sulfurique; sulfacide m.; huile f. de vitriol. / ~ of indigo ‖ indigschwefelsaures Salz n. ‖ céruléosulfate m.; sulfindigotate m. / ~ of lead ‖ Bleisulfat n. ‖ sulfate m. de plomb. / hydrated ~ of lime ‖ Gips m. ‖ chaux f. sulfatée hydratée; gypse m.; plâtre m. / ~ of magnesia ‖ Bittersalz n.; schwefelsaure Magnesia f.; Magnesiumsulfat m. ‖ sulfate m. de magnésie. / ~ of potassium ‖ schwefelsaures Kali n. ‖ sulfate m. de potasse. / ~ of soda ‖ schwefelsaures Natron n.; Natriumsulfat n.; Glaubersalz n. ‖ sulfate m. de soude; sel m. de Glauber. / ~ of zinc ‖ Zinksulfat n.; schwefelsaures Zink n.; Zinkvitriol n. ‖ sulfate m. de zinc; zinc m. sulfaté.

sulphate-containing ‖ sulfathaltig ‖ sulfaté.

sulphate pan ‖ Sulfatpfanne f. ‖ bassin m. *ou* cuve f. pour sulfate.

sulphate plant ‖ Sulfatanlage f. ‖ installation f. de sulfate.

sulphate plate ‖ Sulfatplatte f. ‖ plaque f. de sulfate.

sulphating (Acc) ‖ Sulfatation f.; Sulfatieren n. ‖ sulfatation f.

sulphide ‖ Schwefelmetall n.; Sulfid n. ‖ sulfure m. / ~ of ammonium ‖ Schwefelammonium n.; Ammoniumsulfuret n. ‖ hydrosulfate m. d'ammoniaque. / argentic ~ ‖ Schwefelsilber n. ‖ sulphure m. d'argent. / yellow ~ of arsenic ‖ Rauschgelb n. ‖ deutosulfure m. d'arsenic. / ~ of calcium ‖ Schwefelkalzium n. ‖ sulfure m. de calcium. / ~ of copper ‖ Korellin m. ‖ corelline f. / cuprous ~ ‖ Schwefelkupfer n.; Kupfersulfür n. ‖ sulfure m. de cuivre. / ~ of iron ‖ Schwefeleisen n. ‖ sulfure m. de fer. / mercuric ~ ‖ Schwefelquecksilber n.; Zinnober m. ‖ mercure m. sulfuré; cinabre m. / ~ of potassium ‖ Schwefelkalium n. ‖ sulfure m. de potassium. / ~ of silver ‖ Schwefelsilber n. ‖ sulfure m. d'argent. / ~ of sodium ‖ Schwefelnatrium n. ‖ sulfure m. de sodium. / ~ of zinc ‖ Schwefelzink n.; Zinkblende f.; Blende f.; Sphalerit m. ‖ zinc m. sulfuré; blende f.

sulphindigotate of potassium ‖ indigschwefelsaures Kalium n.; Indigkarmin m.; blauer Karmin m. ‖ indigo m. soluble; céruléosulfate m. *ou* sulf-indigotate m. de potasse.

sulphite ‖ schwefligsaures Salz n.; Sulfit n. ‖ sulfite m. / ~ of soda ‖ schwefligsaures Natron n. ‖ sulfite m. de soude.

sulphite alcohol ‖ Sulfitsprit m. ‖ alcool m. au sulfite. / ~ lye ‖ Sulfitlauge f. ‖ lessive f. de sulfite. / ~ pulp ‖ Sulfitzellstoff m. ‖ pâte f. de bois au sulfite.

sulphocarbide ‖ Schwefelkohlenstoff m. ‖ sulfure m. de carbone; sulfocarbure m.

sulphocarbonate of soda ‖ Natriumthiokarbonat n. ‖ sulfocarbonate m. de soude.

sulphocyanite ‖ Rhodanmetall n.; Rhodanverbindung f.; Schwefelzyanverbindung f.; Rhodansalz n. ‖ sulfocyanide m.; sulfocyanure m.

sulphosot sirup ‖ Sulfosotsirup m. ‖ sirop m. sulfosot.

sulpho-urea ‖ Sulfoharnstoff m. ‖ sulf- urée f.

sulphur, to ‖ schwefeln; einschwefeln ‖ soufrer; ensoufrer.

sulphur ‖ Schwefel m. ‖ soufre m. / to dip into ~ ‖ schwefeln; in Schwefel tauchen ‖ soufrer. / amorphous ~ ‖ amorpher Schwefel m. ‖ soufre m. amorphe. / earthy ~ ‖ Mehlschwefel m. ‖ soufre m. pulvérulent. / native ~ ‖ Jungfern- schwefel m. ‖ soufre m. gris ou vierge. / precipitated ~ ‖ Schwefelmilch f. ‖ lait m. de soufre. / sublimated ~ ‖ Schwefel- blumen fpl.; Schwefelblüten fpl. ‖ fleurs fpl. de soufre; soufre m. en fleurs. / virgin ~ see native ~.

sulphurated hydrogen see also sulphuretted hydrogen ‖ Schwefelwasserstoff m. ‖ hydrogène m. sulfuré.

sulphurating agent ‖ Schwefelungsmittel n. ‖ agent m. de sulfuration ou de soufrage.

sulphuration ‖ Schwefeln n.; Einschwefeln n. ‖ soufrage m.; ensoufrage m. / basket for ~ see room for ~. / room for ~ ‖ Schwefelkorb m.; Schwefelkammer f. ‖ ensoufroir m.

sulphur balsam ‖ Schwefelbalsam m. ‖ baume m. de soufre. / ~ black ‖ Schwefel- schwarz n. ‖ noir m. au soufre. / ~ burner ‖ Schwefelzieher m. ‖ raffineur m. de soufre. / ~ calciner (Furnace) ‖ Schwefel- brennofen m. ‖ cubilot m. ou four m. à soufre. / ~ carbon ‖ Schwefelkohlenstoff m. ‖ sulfure m. de carbone. / ~ chamber ‖ Schwefelkammer f. ‖ soufroir m. / ~ chloride ‖ Chlorschwefel m. ‖ chlorure m. de soufre.

sulphur-containing ‖ schwefelhaltig; schwe- felig ‖ sulfureux; sulfurifère.

sulphur dichloride ‖ Schwefeldichlorid n. ‖ dichlorure m. de soufre.

sulphur dioxide ‖ schweflige Säure f. ‖ acide m. sulfureux (anhydre).

sulphur dyes pl. ‖ Schwefelfarbstoffe mpl. ‖ colorants mpl. au soufre.

sulphured dyes pl. see sulphur dyes. / ~ wick ‖ Schwefelfaden m. ‖ mèche f. soufrée.

sulphuret of cobalt ‖ Schwefelkobalt n.; Kobaltkies m. ‖ cobalt m. sulfuré. / ~ of iron ‖ Schwefeleisen n. ‖ sulfure m. de fer. / ~ of nickel ‖ Schwefelnickel m.; Haarkies m. ‖ nickel m. sulfuré; pyrite f. capillaire.

sulphuret antimony ‖ Goldschwefel m.; Schwefelantimon n. ‖ soufre m. doré d'antimoine.

sulphuretted hydrogen ‖ Schwefelwasser- stoff m.; Wasserstoffsulfid n. ‖ hydro- gène m. sulfuré; acide m. hydrosulfuri- que. / ~ water ‖ Schwefelwasser n.; Schwefelleberwasser n. ‖ eau f. sulfurée ou hépatique. / ~ wood ‖ Schwefelholz n. ‖ bois m. soufré.

sulphuretting agent ‖ Schwefelungsmittel n. ‖ agent m. de soufrage.

sulphur, flowers pl. of ~ ‖ Schwefelblumen fpl.; Schwefelblüte f. ‖ fleur f. de soufre; soufre m. en fleur.

sulphur furnace ‖ Schwefelofen m. ‖ four m. à soufre.

sulphuric acid ‖ Schwefelsäure f.; Vi- triol(öl) n. / acide m. sulfurique; sulf- acide m.; huile f. de vitriol. / dilute(d) ~ / verdünnte Schwefelsäure f. ‖ acide m. sulfurique dilué. / English ~ ‖ englische Schwefelsäure f. ‖ acide m. sulfurique

anglais. / fuming ~ ‖ rauchende Schwe- felsäure f. ‖ acide m. sulfurique fumant.

sulphuric acid concentration plant ‖ Schwe- felsäurekonzentrationsanlage f. ‖ installa- tion f. de concentration d'acide sulfuri- que. / ~ determination ‖ Schwefelsäure- bestimmung f. ‖ dosage m. de l'acide sulfurique. / ~ producing plant ‖ Schwefel- säureherstellungsanlage f. ‖ installation f. à fabriquer l'acide sulfurique.

sulphuric ether ‖ Schwefeläther m. ‖ éther m. sulfurique. / ~ apparatus ‖ Schwefel- ätherapparat m. ‖ appareil m. à éther sulfurique.

sulphuring ‖ Einschwefeln n.; Schwefeln n. ‖ ensoufrage m.; soufrage m.; ensoufre- ment m. / ~ chamber ‖ Schwefelkammer f. ‖ soufroir m. / ~ room see ~ chamber. / ~ stove see ~ chamber.

sulphur kiln ‖ Schwefelofen m. ‖ four m. à soufre.

sulphur liver ‖ Schwefelleber f. ‖ foie f. de soufre; sulfure m. de potassium.

sulphur match ‖ Schwefelhölzchen n. ‖ allumette f. soufrée.

sulphurous ‖ schweflig ‖ sulfureux. / ~ acid ‖ schweflige Säure f. ‖ acide m. sulfureux. / ~ acid ice machine ‖ Schwef- ligsäureeismaschine f. ‖ machine f. à glace à acide sulfureux. / ~ anhydrid ‖ Schwef- ligsäureanhydrid n. ‖ anhydride m. sul- fureux. / ~ carbon oils pl. ‖ Sulfuröle npl. ‖ huiles fpl. de grignons. / ~ chlorid ‖ Schwefelchlorür n. ‖ chlorure m. de soufre. / ~ ore ‖ schwefelhaltiges Erz n.; Schwefelerz n. ‖ minerai m. sulfureux ou sulfurifère. / ~ pyrites ‖ Schwefelkies m. ‖ sulfure m. de fer jaune; pyrite m. de fer. / ~ water ‖ Schwefelwasser n.; Schwefel- leberwasser n. ‖ eau f. sulfurée ou hé- patique.

sulphur pit ‖ Schwefelgrube f. ‖ mine f. de soufre. / ~ præcipitatum ‖ Schwefel- milch f. ‖ lait m. de soufre. / ~ purifying ‖ Schwefelraffinierung f. ‖ raffinage m. de soufre. / ~ remover ‖ Entschwefe- lungsmittel n. ‖ désulfürant m. / ~ soap ‖ Schwefelseife f. ‖ savon m. au soufre. / ~ sticks pl. ‖ Stangenschwefel m. ‖ soufre m. en canons.

sulphuryl chloride ‖ Sulfurylchlorid n. ‖ chlorure m. de sulfuryle.

sultry (Meteor) ‖ schwül ‖ étouffant; lourd.

sum, to ~ **up** (To add up) ‖ zusammenzäh- len; summieren; addieren ‖ additionner. / ~ (To calculate) ‖ berechnen ‖ calculer; compter.

sum ‖ Summe f. ‖ somme f.; montant m. / ~ of the contract ‖ Verdingungssumme f. ‖ montant m. du forfait. / lump ~ ‖ ein- malige Abfindung f. ‖ somme f. globale. / ~ of money ‖ Geldsumme f.; Geldbe- trag m.; Geldposten m. ‖ somme f. d'ar- gent; article m.

suma(ch) ‖ Sumach m.; Schmack m. ‖ sumac m. / ~ bark ‖ Sumachrinde f. ‖ écorce f. de sumac. / ~ bud ‖ Sumach- knospe f. ‖ feuille f. de brindille de su- mac. / ~ extract ‖ Sumachauszug m. ‖ extrait m. de sumac. / ~ wood ‖ Sumach- holz n. ‖ bois m. de sumac.

sumbuly root ‖ Sumbulwurzel f. ‖ racine f. de sumbul.

summary ‖ summarisch ‖ sommaire.

summary (Print) ‖ Angabe f. des Haupt- inhaltes; Hauptinhalt m. ‖ sommaire(s) mpl.

summer (Carp) ‖ Oberschwelle f.; Saum- schwelle f.; Tragbalken m. ‖ sommier m.

summer (Season) ‖ Sommer m. ‖ été m. / to season in ~ ‖ durchsommern ‖ conserver pendant l'été.

summer dike (Hydr arch) ‖ Sommerdeich m. ‖ digue f. de bordage; digue f. sub- mersible. / ~ road ‖ Sommerweg m. ‖ chemin m. retiré ou de terre. / ~ solstice ‖ Sommersonnenwende f. ‖ solstice m. d'été.

summit of an angle ‖ Scheitel m. eines Winkels ‖ sommet m. d'un angle. / ~ of a dam ‖ Kappe f. eines Dammes; Deich- krone f. ‖ crête f. ou couronnement m. ou sommet m. d'une digue. / dome shaped ~ (Of a mountain) ‖ Bergkuppe f. ‖ som- met m. en forme de dôme. / ~ of the mountain ‖ Berggipfel m. ‖ sommet m. ou cime f. de montagne. / ~ of a railway line ‖ höchster Punkt m. einer Eisen- bahnstrecke ‖ point m. culminant ou point m. de portage d'un chemin de fer. / ~ of a road ‖ Krone f. einer Straße ‖ couronne f. ou couronnement m. d'une route.

summons ‖ gerichtliche Vorladung f. ‖ as- signation f.; citation f.

sump (Mine) ‖ Sumpf m. ‖ puisard m.; pa- hage m. / ~ (Mach) ‖ Sumpf m. ‖ puisard m. / ~ in a furnace ‖ Sumpf m. eines Ofens ‖ fond m. d'un fourneau. / ~ of a pit ‖ Schachtsumpf m.; Sumpf m. ‖ puis- ard m. / settling ~ ‖ Klärbecken n.; Klärsumpf m. ‖ bassin m. de décantation. / shaft ~ ‖ Schachtsumpf m. ‖ puisard m.

sump shaft ‖ Pumpenschacht m.; Wasser- haltungsschacht m. ‖ puits m. d'épuise- ment ou d'exhaure; bure f. aux pompes.

sumpter (horse) ‖ Saumtier n. ‖ bête f. de somme.

sum total ‖ Gesamtbetrag m. ‖ somme f. totale.

sun ‖ Sonne f. ‖ soleil m. / to protect against the ~ ‖ vor Sonnenschein m. schützen ‖ abriter contre le soleil. / artificial moun- tain ~ ‖ künstliche Höhensonne f. ‖ so- leil m. d'altitude artificiel.

sun's altitude ‖ Sonnenhöhe f. ‖ hauteur f. du soleil.

sun and moon camera ‖ Sonnemondka- mera f. ‖ chambre f. noire pour le soleil et la lune.

sun and planets gear ‖ Planetengetriebe n. ‖ engrenage m. planétaire ou épicyc- loïdal.

sun's corona photograph ‖ Sonnenkorona- aufnahme f. ‖ photographie f. de la cou- ronne solaire.

sunday current (Electr) ‖ Sonntagsstrom m. ‖ courant m. de dimanche.

sundew herb ‖ Sonnentau m. ‖ herbe f. à la rosée.

sun-dial ‖ Sonnenuhr f. ‖ cadran m. so- laire.

sun-dried ‖ an der Sonne getrocknet ‖ séché au soleil.

sundries pl. in lots ‖ Massenartikel mpl. ‖ articles mpl. faits en masse ou de grande consommation.

sunflower oil ‖ Sonnenblumenöl n. ‖ huile f. de tournesol ou de hélianthe.

sun glass (Opt) ‖ Sonnenglas n. ‖ verre m. noir pour le soleil.

sunk in a shaft (Mine) ‖ Senkzimmerung f. ‖ cuvelage m. dans des terrains ébou- leux.

sunk hole (For screw heads etc.) ‖ Versenk n. ‖ noyure f.; noyon m. / ~ line (Railw) ‖ versenktes Gleis n. ‖ voie f. noyée. / ~ screw ‖ versenkte Schraube f. ‖ vis f. noyée *ou* perdue *ou* à tête noyée.

sunk shaft process (Mine) ‖ Senkschachtverfahren n. ‖ procédé m. de cuvelage. / sinking to the ~ ‖ Schachtabteufung f. nach dem Senkschachtverfahren ‖ creusement m. de puits de mine à cuvelage.

sunk well ‖ Senkbrunnen m. ‖ puits m. foncé.

sunless day ‖ sonnenloser Tag m. ‖ jour m. sans soleil.

sunlight ‖ Sonnenlicht n. ‖ lumière f. solaire. / distribution of energy in ~ ‖ Energieverteilung f. im Sonnenlicht ‖ répartition f. des radiations solaires.

Sunn hemp ‖ Sun m.; Sunhanf m.; ostindischer Hanf m. ‖ sunn m.; chanvre m. de Bengale.

sun prism ‖ Sonnenprisma n. ‖ hélioscope m.; prisme m. pour le soleil.

sun protection screen ‖ Sonnenprojektionsschirm m. ‖ écran m. pour la projection du soleil; écran m. de projection pour le soleil.

sunrise ‖ Sonnenaufgang m. ‖ lever m. du soleil.

sunset ‖ Sonnenuntergang m. ‖ coucher m. du soleil.

sunshade ‖ Sonnenschirm m. ‖ parasol m. / small ~ ‖ kleiner Sonnenschirm m. ‖ ombrelle f.

sunshine ‖ Sonnenschein m. ‖ soleil m. / duration of ~ ‖ Sonnenscheindauer f. ‖ durée f. de l'insolation.

sunshine motor car body ‖ Limusine f. mit vorn zurückschiebbarem Dach ‖ carrosserie f. à avant transformable.

sun-stone ‖ Sonnenstein m.; Oligoklas m. ‖ pierre f. du soleil; héliolithe m.

sunstroke ‖ Sonnenstich m.; Hitzschlag m. ‖ coup m. de soleil.

superannuated (Law term) ‖ verjährt ‖ prescrit; suranné.

superannuation allowance ‖ Alterszulage f. ‖ haute-paye f. d'ancienneté. / ~ funds pl. ‖ Pensionskasse f. ‖ caisse f. des retraites.

supercargo ‖ Frachtaufseher m. ‖ subrécargue m.

supercharged engine ‖ überverdichtender Motor m. ‖ moteur m. suralimenté *ou* surcomprimé.

supercharger ‖ Überverdichter m. ‖ surcompresseur m. / ~ engine ‖ Gebläsemotor m. ‖ moteur m. à alimentation sous pression.

supercharging ‖ Vorverdichtung f.; Überverdichtung f. ‖ suralimentation f.

supercooled ‖ überkühlt; unterkühlt ‖ surréfrigéré. / (Surfused) ‖ überschmolzen ‖ surfondu; coulé à température trop élevé.

supercooling ‖ Überkühlung f. ‖ surfusion f.

superelevation ‖ Überhöhung f. ‖ surhaussement m.; surélévation f. / ~ of the outer rail ‖ Überhöhung f. der äußeren Schiene ‖ surélévation f. du rail extérieur.

superficial ‖ oberflächlich ‖ superficiel. / ~ area ‖ Flächenraum m.; Oberfläche f. ‖ surface f.; aire f.; superficie f. / ~ area of shops ‖ Flächenraum m. der Werkstätten ‖ superficie f. des ateliers. / ~ content *see* ~ area. / ~ measure ‖

Flächenmaß n. ‖ mesure f. de superficie; mesure f. carrée. / ~ measuring instrument ‖ Flächenmeßinstrument n. ‖ instrument m. de mesure en plan.

superficies *see also* surface ‖ Fläche f.; Oberfläche f. ‖ surface f.; superficie f. / curved ~ of a cone ‖ Mantel m. eines Kegels ‖ nappe f. du cône; surface f. convexe du cône. / ~ of a hyperboloid ‖ Mantelfläche f. eines Hyperboloids ‖ nappe f. d'un hyperboloïde.

superfine (File) ‖ doppelschlicht; extraschlicht ‖ extra doux.

superfine hardening ‖ Weißerde f. ‖ terre f. blanche.

superheated ‖ überhitzt ‖ surchauffé. / ~ express locomotive ‖ Heißdampfschnellzuglokomotive f. ‖ locomotive f. d'express à vapeur surchauffée. / ~ locomotive ‖ Heißdampflokomotive f. ‖ locomotive f. à vapeur surchauffée. / ~ slide for locomotives ‖ Heißdampfschieber m. für Lokomotiven ‖ tiroir m. de locomotives à vapeur surchauffée.

superheated steam ‖ überhitzter Dampf m.; Heißdampf m. ‖ vapeur f. surchauffée. / ~ engine ‖ Heißdampfmaschine f. ‖ machine f. à vapeur surchauffée. / ~ goods locomotive with separate tender ‖ Heißdampfgüterzuglokomotive f. mit getrenntem Tender ‖ locomotive f. à vapeur surchauffée pour trains de marchandises avec tender séparé. / ~ goods tank locomotive with radial gear-wheel coupled axles ‖ Heißdampfgüterzugtenderlokomotive f. mit kurvenbeweglichen, zahnradgekuppelten Achsen ‖ locomotive-tender f. à vapeur surchauffée pour trains de marchandises à essieux à orientation radiale, accouplés par roues dentées. / ~ locomotive ‖ Heißdampflokomotive f. ‖ locomotive f. à vapeur surchauffée. / ~ plant ‖ Heißdampfanlage f. ‖ chauffage m. à vapeur surchauffée. / ~ tandem roller ‖ Heißdampftandemwalze f. ‖ rouleau m. compresseur tandem à vapeur surchauffée. / ~ tank locomotive for passenger service ‖ Heißdampfpersonenzugtenderlokomotive f. ‖ locomotive-tender f. à vapeur surchauffée pour trains de voyageurs. / ~ thermometer ‖ Heißdampfthermometer n. ‖ thermomètre m. pour vapeur surchauffée. / ~ threewheel roller ‖ Heißdampfdreiradwalze f. ‖ rouleau m. compresseur à trois rouleaux à vapeur surchauffée.

superheater ‖ Überhitzer m. ‖ surchauffeur m. / steam ~ ‖ Dampfüberhitzer m. ‖ surchauffeur m. de vapeur.

superheater chamber ‖ Überhitzerkammer f. ‖ chambre f. de surchauffe.

superheating (Chem) ‖ Überheizen n.; Siedeverzug m. ‖ surchauffe f. / ~ of the steam ‖ Dampfüberhitzung f. ‖ surchauffe f. de la vapeur.

superheating plant ‖ Überhitzeranlage f. ‖ installation f. de surchauffe.

super-heterodyne (Radio) ‖ Zwischenfrequenz f. beim Funkempfang ‖ superhétérodyne m. / ~ receiver ‖ Superheterodynempfänger m. ‖ récepteur m. superhétérodyne. / ~ reception ‖ Superheterodynempfang m. ‖ réception f. par super-hétérodyne.

superimpose, to ‖ überlagern ‖ superposer.

superimposing ‖ Überlagerung f. ‖ superposition f.

superintend, to ‖ überwachen ‖ surveiller.

superintendence ‖ Beaufsichtigung f. ‖ surveillance f.

superintendent ‖ Geschäftsführer m. ‖ gérant m.

superior ‖ ober(er) ‖ supérieur. / ~ to other systems in respect of efficiency and safety of working ‖ anderen Systemen npl. an Leistungsfähigkeit f. und Betriebssicherheit f. überlegen ‖ supérieur aux autres systèmes mpl. par sa puissance et sûreté de service. / ~ to all other types ‖ allen anderen Arten überlegen ‖ supérieur à toutes les autres sortes. / ~ product ‖ Edelerzeugnis n. ‖ produit m. supérieur.

superior (Print) ‖ Notenbuchstabe m.; Spaltenbuchstabe m.; Verweisungsbuchstabe m. ‖ lettrine f. supérieure; renvoi m. de notes; supérieure f.

superiority ‖ Überlegenheit f. ‖ superiorité f.

superior product ‖ Edelerzeugnis n. ‖ produit m. supérieur.

supernumerary ‖ überzählig ‖ surnuméraire; excédant; en trop. / ~ call (Tel) ‖ überzähliger Anruf m. ‖ appel m. en surnombre.

superoxide *see also* peroxide ‖ Superoxyd n. ‖ peroxyde m.

superphosphate ‖ Superphosphat n. ‖ superphosphate m. / ~ dissolving machine ‖ Aufschließmaschine f. für Superphosphate ‖ machine f. à préparer des superphosphates. / ~ plant ‖ Superphosphatanlage f. ‖ installation f. de fabrication de superphosphates.

superposed ‖ übereinanderliegend ‖ superposé. / ~ circuit (Tel) ‖ Vierersprechkreis m. ‖ circuit m. de conversation quadruple *ou* combiné.

superposition ‖ Überlagerung f. ‖ superposition f.

super power distribution ‖ Großstromversorgung f. ‖ distribution f. de l'énergie électrique dans un grand réseau. / undertaking engaged in ~ ‖ Unternehmung f. für Stromgroßverteilung ‖ entreprise f. de distribution de courant électrique dans un grand réseau.

super power station ‖ Großkraftwerk n. ‖ supercentrale f.

superrefined steel ‖ Edelstahl m. ‖ acier m. spécial; acier m. fin (pour outils). / ~ in ingots ‖ Edelstahl m. in Blöcken ‖ acier m. spécial en lingots. / ~ plate ‖ Edelstahlblech n. ‖ tôle f. d'acier spécial. / semi-product of ~ ‖ Edelstahlhalbzeug n. ‖ demi-produit m. en acier spécial.

superregenerative receiver (Radio) ‖ Superregenerativempfänger m. ‖ **superrégénérateur** m.

supersalt ‖ saures Salz n. ‖ sel m. acide.

supersaturate, to ~ (Chem) ‖ übersättigen ‖ sursaturer.

supersaturated ‖ übersättigt ‖ sursaturé. / ~ air ‖ übersättigte Luft f. ‖ air m. sursaturé.

supersaturation ‖ Übersättigung f. ‖ sursaturation f.

supersteel *see also* superrefined steel ‖ Edelstahl m. ‖ acier m. fin.

superstructure (Railw) ‖ Oberbau m. ‖ voie f. permanente; superstructure f. / iron ~ ‖ eiserner Oberbau m. ‖ voie f. perma-

nente en traverses en fer. / iron ~ of a bridge (Railw) ‖ eiserner Oberbau m. einer Brücke ‖ tablier m. métallique d'un pont. / ~ with longitudinal sleepers ‖ Langschwellenoberbau m. ‖ voie f. permanente à longrines. / ~ for railways ‖ Eisenbahnoberbau m. ‖ superstructure f. de chemins de fer. / steel-frame ~ ‖ Eisenhochbau m. ‖ charpente f. métallique; construction f. en fer au-dessus du sol.

superstructure work (Build) ‖ Hochbau m. ‖ travail m. de superstructure.

super-turbine ‖ Großturbine f. ‖ turbine f. à grande puissance. / ~ shop ‖ Großturbinenhalle f. ‖ atelier m. de turbines à grande puissance.

supervise, to ~ ‖ überwachen ‖ surveiller.

supervising see also supervision ‖ Überwachung f. ‖ surveillance f. / thermal control board for ~ the circulation of water ‖ Wärmewarte f. zur Überwachung des Wasserkreislaufes ‖ poste m. de contrôle thermique destiné à surveiller la circulation d'eau.

supervision see also supervising ‖ Kontrolle f.; Aufsichtsdienst m. ‖ contrôle m.; surveillance f. / ~ of answering (Tel) ‖ Rufüberwachung f. ‖ surveillance f. de l'appel. / ~ of calls (Tel) ‖ Gesprächsüberwachung f. ‖ contrôle m. des communications. / ~ of railways ‖ Eisenbahnoberaufsicht f.; Eisenbahnüberwachungsdienst m. ‖ surveillance f. des chemins de fer.

supervision table (Tel) ‖ Aufsichtstisch m. ‖ table f. de surveillance.

supervisor ‖ Aufsichtsbeamter m.; Aufsichtsperson f. ‖ surveillant m.; agent m. de surveillance.

supervisory apparatus ‖ Überwachungseinrichtung f. ‖ installation f. de surveillance. / ~ board with recorders for registering active and reactive voltamperes ‖ Überwachungstafel f. mit Schreibgeräten zur Aufzeichnung von Wirkleistung und Blindleistung ‖ tableau m. de contrôle avec enregistreurs de puissances active et réactive. / ~ enquiry (Tel) ‖ Überwachungsfrage f. ‖ demande f. de contrôle. / ~ lamp (Tel) ‖ Schlußlampe f. ‖ lampe f. de fin de conversation. / ~ relay ‖ Kontrollschütz n.; Kontrollrelais n. ‖ relais m. pilote.

supplement (Of an angle) ‖ Supplement n.; Ergänzungswinkel m. ‖ supplément m.; angle m. supplémentaire. / ~ (Of a book) ‖ Nachtrag m. ‖ supplément m. / ~ (Of a newspaper) ‖ Beiblatt n.; Beilage f. ‖ supplément m.

supplementary allowance of a credit ‖ Nachbewilligung f. eines Kredites ‖ crèdit m. supplémentaire. / ~ apparatus ‖ Zusatzgerät n. ‖ appareillage m. de complément. / ~ car for motor cycles ‖ Beiwagen m. für Motorräder ‖ voiture f. latérale ou sidecar m. pour motocyclettes. / ~ glass (For spectacles) ‖ Aufsteckglas n. ‖ verre m. additionnel. / ~ heating ‖ Aushilfsheizung f.; Zusatzheizung f. ‖ chauffage m. auxiliaire. / ~ lens ‖ Zusatzlinse f. ‖ lentille f. additionnelle. / to attach to the field glass a ~ lens ‖ auf den Feldstecher eine Zusatzlinse f. aufstecken ‖ emboîter sur l'objectif une lentille additionnelle. / ~ order ‖ Nachbestellung f. ‖ commande f. supplémentaire. / ~ patent specification ‖ Nachtrag m. zu einer Patentschrift ‖

supplément m. à une lettre patente. / ~ printing unit of three rollers ‖ dreiwalziges Eindruckwerk n. ‖ groupe m. d'impression supplémentaire à trois rouleaux. / ~ set (Electr) ‖ Zusatzaggregat n. ‖ groupe m. survolteur.

supplied see under to supply.

supplier ‖ Lieferer m.; Lieferant m. ‖ fournisseur m.

supply, to ~ (Trade) ‖ liefern ‖ livrer; fournir; alimenter. / ~ current ‖ Strom m. liefern; mit Strom m. speisen ‖ fournir du courant. / ~ with provisions ‖ mit Proviant m. versehen oder versorgen ‖ fournir, en vivres mpl.; avitailler. / being supplied in bulk ‖ lieferbar in großen Mengen ‖ livrable par grandes quantités. / supplied by continuous current ‖ mit Gleichstrom m. gespeist ‖ alimenté de courant continu.

supply (Delivery) ‖ Lieferung f. ‖ livraison f. / ~ (Of provisions) ‖ Vorrat m.; Proviant m. ‖ provision f.; approvisionnement m. / to tender and contract for a ~ ‖ einen Lieferungsvertrag m. abschließen ‖ entreprendre une fourniture. / ~ of coal ‖ Kohlenvorrat m. ‖ provision f. de charbon.

supply of current (Electr) ‖ Stromlieferung f.; Stromversorgung f. ‖ fourniture f. de courant. / to take a ~ ‖ Strom m. abnehmen ‖ acheter du courant. / ~ from other sources assured by contract ‖ vertraglich festgelegte Fremdlieferung f. von Strom ‖ courant m. étranger livré par d'autres sociétés à base de contracts passés. / agreement already in force for the ~ ‖ der laufende Stromlieferungsvertrag m. ‖ contrat m. de livraison de courant en cours.

supply of drinking water ‖ Trinkwasserversorgung f. ‖ distribution f. d'eau à boire. / ~ of money ‖ Geldvorrat m. ‖ encaisse f. / ~ of provisions ‖ Proviantlieferung f. ‖ fourniture f. de vivres. / ~ of steam ‖ Dampfentnahme f. ‖ prise f. de vapeur. / ~ of water ‖ Wasservorrat m. ‖ approvisionnement m. d'eau.

supply area of current (Electr) ‖ Verbrauchsgebiet n. des Stromes ‖ centre m. de consommation du courant électrique. / to deliver the current economically and reliably to the ~ ‖ den Strom wirtschaftlich und betriebssicher den Verbrauchsgebieten zuführen ‖ conduire le courant jusqu'aux centres de consommation avec la plus grande économie et sûreté possibles. / fresh grouping of the ~ ‖ Neugliederung f. des Stromversorgungsgebietes ‖ réorganisation f. du centre de consommation.

supply circuit ‖ Speiseleitung f.; Zuleitungsstromkreis m. ‖ circuit m. d'alimentation. / ~ gallery (Mine) ‖ Förderstrecke f. ‖ galerie f. de roulage. / ~ meter ‖ Verbrauchsmesser m. ‖ compteur m. de consommation.

supply network (Electr) ‖ Versorgungsnetz n. ‖ réseau m. d'alimentation. / simultaneous existence of a number of disintegrated ~s ‖ Nebeneinanderbestehen m. kleiner vielfach zersplitterter Versorgungsnetze ‖ un grand nombre m. de petits réseaux existants les uns à côté des autres.

supply passage (Steam) ‖ Einströmungskanal m. ‖ canal m. d'admission ou d'introduction (de la vapeur).

supply pipe ‖ Einlaufrohr n. ‖ tuyau m. d'arrivée. / ~ for steam ‖ Dampfeinströmungsrohr n. ‖ tuyau m. d'admission de la vapeur.

supply plant for long distance (Electr) ‖ Überlandwerk n. ‖ centrale f. interurbaine de distribution d'énergie. / ~ pressure ‖ Verbrauchsspannung f. ‖ tension f. de consommation. / ~ source of ~ ‖ Bezugsquelle f. ‖ source f. d'alimentation. / ~ stores pl. ‖ Konsumanstalt f. ‖ magasins mpl. économiques. / ~ tank (War impl) ‖ Nachschubkampfwagen m. ‖ char m. supplémentaire. / ~ way see supply passage.

support, to ‖ abstützen; stützen ‖ étançonner; étayer; entretoiser.

support ‖ Stütze f.; Träger m.; Gestell n.; Auflage f.; Untersatz m. ‖ support m.; appui m. / ~ for aerial lines (Electr) ‖ Stützpunkt m. für Freileitungen ‖ appui m. ou support m. pour lignes aériennes. / ~ arm ‖ Armstütze f. ‖ appui-bras m. / ~ for brake hangers ‖ Bremsgehängeträger m. ‖ support m. de la suspension de frein. / ~ of a bridge ‖ Auflager n. einer Brücke ‖ appui m. de ponts. / ~ for compressed air ‖ Preßluftgegenhalter m. ‖ bouterolle f. à air comprimé. / counter ~ ‖ Gegenständer m. ‖ montant m. à supports. / insulating ~ (Acc) ‖ Isolierfuß m. ‖ pied m. isolant. / license ~ (Auto) ‖ Tragstütze f. für das Nummernschild ‖ support m. pour la plaque de license. / ~ of mines ‖ Grubenausbau m. ‖ fortification f. de mines. / ~ of the muffle (Porcel) ‖ Muffelblatt n. ‖ support m. de moufle. / outer ~ ‖ Gegenlager n. ‖ contre-support m. / ~ of travelling crane ‖ Laufkranständer m. ‖ support m. de pont roulant. / ~ of a windmill ‖ Tragebank f. einer Windmühle ‖ support m. d'un moulin à vent. / wooden ~ ‖ Holzstütze f. ‖ support m. en bois.

supported ‖ aufruhend; aufliegend ‖ supporté. / to be ~ ‖ aufliegen; sich stützen (auf ...) ‖ aufruhen ‖ être couché (sur ...). / ~ by ... ‖ gelagert auf ... oder in ... ‖ supporté par ...; appuyé sur ...; logé dans ...; monté sur ... / not ~ ‖ freitragend ‖ non-soutenu; saillant.

supported joint (Railw) ‖ aufliegender Schienenstoß m. ‖ joint m. des rails soutenu.

supporting angle piece ‖ Winkelstütze f. ‖ équerre f. de support. / ~ band ‖ Tragband n. ‖ tirant m.; ruban m. porteur. / ~ bars pl. (Acc) ‖ Halteschiene f. ‖ ferrure f. de support. / ~ board (Typewr etc.) ‖ Untersatzbrett n. ‖ planche-support f. / ~ bracket (Of a windmill) ‖ Auflageknagge f. ‖ tasseau m. de support. / ~ column ‖ Stützträger m. ‖ poutre f. de soutènement. / ~ construction ‖ Unterstützungskonstruktion f. ‖ construction f. de support. / ~ disk (Acc) ‖ Stützscheibe f. ‖ disque m. de support. / ~ frame (Build) ‖ Stützgerüst n.; Verlagerungsgerüst n. ‖ charpente f. de support. / ~ frame-work ‖ Traggerüst n. ‖ charpente f. de support; ossature-support f. / ~ gas ‖ Traggas n. ‖ gaz m. de gonflement ou de sustentation. / ~ girder ‖ Träger m. ‖ poutre f. de support. / set of ~ girders ‖ Trägerlager n. ‖ ensemble m. des poutres de support. / ~ nave ‖ Tragnabe f. ‖

moyeu m. de support. / ~ plate || Auflagerplatte f. || plaque f. d'appui *ou* d'assise. / ~ pole (El line) || Stützpfahl m. || soutien m. / ~ post (Of a cableway) || Stütze f. || pylône m. / ~ roller || Stützwalze f. || rouleau m. d'appui. / ~ roller pedestal in box-form || kastenartiger Stützrollenträger m. || support m. des rouleaux d'appui en forme de caisse. / ~ strand for overhead cables || Tragseil n. für Luftkabel || corde f. de suspension pour câbles aériens. / ~ structure of iron || eisernes Tragwerk n. || appareil m. porteur en fer.

support stock (Join) || Knecht m.; Stehknecht m.; Bankknecht m. || servante f.; valet m. de pied.

support truss (Aeropl) || Abstützblock m. || support m. d'entretoise.

suppository (Med) || Suppositorium n.; Stuhlzäpfchen n.; Seifenzäpfchen n. || suppositoire m. / ~ mould || Suppositorienpresse f. || presse f. pour suppositoires.

suppression of the noise || Abdämpfung f. der Geräusche || amortissement m. du bruit.

suppressor of harmonics (Radio) || Wellensauger m. || dispositif m. pour supprimer les harmoniques.

supra-conductivity (Electr) || Überleitfähigkeit f. || superconductibilité f.

Supreme Court || Oberster Gerichtshof m. || Cour m. Suprême.

surbased (Arch) || gedrückt; flach gebaut || à corniche; surbaissé. / ~ vault || Stichbogengewölbe n. || voûte f. basse.

surcharge, to || überladen; überlasten || surcharger. / ~ steam || Dampf m. überhitzen || surchauffer la vapeur.

surcharge (Mail) || Strafporto n. || surtaxe f. / ~ (Steam) || Überhitzung f. || surchauffe f.

surety || Bürgschaft f.; Sicherheit f. || caution f.; garantie f. / to act as ~ || Bürgschaft f. übernehmen || se porter garant m.

surety bond || Verpflichtungsschein m. || cautionnement m.

surf (of the sea) || Brandung f. der See || brisement m. des flots.

surface, to ~ (Turn) || plandrehen; flachdrehen || dresser une surface au tour; surfacer; façonner une surface plane au tour.

surface || Oberfläche f.; Fläche f. || surface f.; superficie f.; aire f. / active ~ of the carbon || aktive Oberfläche f. der Kohle || surface f. active du charbon. / ~ of the backwater behind a weir || Stauspiegel m. || niveau m. de l'eau en amont d'un barrage. / bright ~ || glänzende Oberfläche f. || surface f. polie *ou* brillante. / ~ of a circular cylinder || Kreiszylinderfläche f. || surface f. cylindrique circulaire. / clean and flawless ~ ready for machining || reine porenfreie Bearbeitungsfläche f. || surface f. de façonnement nette et sans pores. / ~ of cloth || Filzdecke f.; Tuchdecke f. || couverture f. de drap. / ~ of the concrete road cover || Oberschicht f. der Betonstraßendecke || couche f. supérieure du revêtement de la route en béton. / ~ of a cone || Mantel m. eines Kegels || nappe f. du cône; surface f. convexe du cône. / conical ~ || Kegelfläche f. || surface f. conique. / curved ~ || krumme *oder* gekrümmte Fläche f. || sur-

face f. courbe. / cylindrical ~ || Zylinderfläche f. || surface f. cylindrique. / ~ of the earth || Erdoberfläche f. || surface f. de la terre. / edged ~ || kantige Oberfläche f. || surface f. anguleuse. / even ~ || glatte Fläche f. || surface f. lisse *ou* unie. / ~ of the formation (Railw) || Planum n.; Bahnoberfläche f. unter dem Bettungsmaterial || plate-forme f. de l'infrastructure. / ~ of friction of the guide-bars (Loc) || Reibfläche f.; Reibungsfläche f. der Gleitbahnen || surface f. de frottement des glissières. / hard ~ || harte Oberfläche f. || surface f. dure. / ~ as hard as glass || glasharte Oberfläche f. || surface f. dure comme le verre. / heating ~ || Heizfläche f. || surface f. de réchauffement. / interior ~ || Innenfläche f. || paroi f. intérieure. / plane ~ || ebene Fläche f. || surface f. plane. / convex ~ of a pyramid || Mantel m. einer Pyramide || surface f. convexe *ou* latérale d'une pyramide. / ~ of revolution || Rotationsfläche f. || surface f. de révolution *ou* de rotation. / ~ of separation || Trennungsfläche f. || surface f. de séparation. / shining ~ || glänzende Oberfläche f. || surface f. polie *ou* brillante. / skew ~ || windschiefe Fläche f. || surface f. gauche *ou* dévers *ou* oblique. / superior ~ of a dam || Kappe f. *oder* Kamm m. *oder* Krone f. eines Deiches || crête f. *ou* couronnement m. *ou* sommet m. d'une digue. / ~ of the underground water || Grundwasserspiegel m. || surface f. de la nappe souterraine. / uneven ~ || unebene Oberfläche f. || surface f. inégale *ou* âpre. / ~ of water || Wasserspiegel m. || surface f. *ou* niveau m. de l'eau. / ~ of a water line (Shipb) || Oberfläche f. einer Wasserlinie || surface f. d'une ligne d'eau.

surface blow-off cock || Ablaßhahn m.; Abschäumhahn m. || robinet m. de vidange à hauteur de niveau. / ~ capacity || Oberflächenkapazität f. || capacité f. surfacique. / ~ carburettor || Oberflächenvergaser m. || carburateur m. par surface. / cleanliness of ~ || Reinheit f. der Oberfläche || netteté f. de la surface. / ~ combustion || Oberflächenverbrennung f. || combustion f. sans flamme. / ~ condensation (Steam) || Oberflächenkondensation f.; trockene Kondensation f. || condensation f. à *ou* par surface. / ~ condenser || Flächenberieselungskondensator m.; Berieselungskondensator m.; Oberflächenkondensator m. || condenseur m. à ruissellement (à de la surface); condensateur m. à surface. / ~ constitution of the soil || Oberflächenbeschaffenheit f. des Bodens || constitution f. de la surface du sol. / ~ cooler || Berieselungskühler m. || réfrigérant m. à ruissellement de la surface. / ~ cooling || Oberflächenkühlung f. || refroidissement m. superficiel. / ~ crack (Met) || Hautriß m. || crevasse f. superficielle. / ~ density || Flächendichte f. || densité f. superficielle. / ~ dose (Med) || Oberflächendosis f. || dose f. superficielle. / element of ~ || Flächenelement n. || élément m. de surface. / ~ evaporative cooler || Berieselungskondensator m.; condensateur m. à ruissellement. / ~ fermentation (Brew) || Obergärung f. || fermentation f. ordinaire. / ~ grinding machine || Flächenschleifmaschine f. || ma-

chine f. à rectifier les surfaces planes. / ~ grinding and polishing machine || Flächenschleif- und -poliermaschine f. || machine f. à mouler et polir à disques. / ~ hardness || Oberflächenhärte f. || dureté f. à la surface. / ~ hardness on parts of the ~ || teilweise Oberflächenhärte f. || dureté f. à certaines parties de la surface. / ~ integral || Oberflächenintegral n. || intégrale f. de surface. / ~ lathe || Plan(scheiben)drehbank f.; Scheibendrehbank f. || tour m. à plateau *ou* à surfacer *ou* en l'air. / ~ milling attachment || Planfräsvorrichtung f. || dispositif m. à surfacer à la fraise. / ~ milling cutter for the wood industry || Oberfräser m. für Holz || défonceuse f. pour bois. / ~ mining || Tagebau m. || ouvrage m. *ou* exploitation f. à ciel ouvert.

surface plate (Forg) || Richtplatte f. || plaque f. à dresser; marbre m. de dressage. / ~ (Turn) || Planscheibe f. || plateau m. (de tour en l'air). / ~ for rail straightening || Schienenrichtplatte f. || table f. en fonte pour le dressage des rails.

surface pressure || Flächendruck m. || pression f. de surface.

surface printing (Text print) || Hautwalzendruck m.; Flächendruck m. || impression f. par le métier à surface; impression f. en relief. / ~ machine || Reliefwalzendruckmaschine f. || métier m. à surface; hernetine f.; plombine f.

surface rib (Arch) || Zierrippe f. || nervure f. décorative. / ~ temperature || Oberflächenwärme f. || température f. superficielle. / ~ tension || Oberflächenspannung f. || tension f. superficielle. / ~ traverser || unversenkte Schiebebühne f. || transbordeur m. à niveau. / ~ unit || Oberflächeneinheit f. || unité f. de surface. / ~ water (Mine) || Tagwasser n. || eau f. du jour. / ~ wave (Radio) || Oberflächenwelle f. || onde f. de surface. / ~ and underground workers pl. || Belegschaft f. über und unter Tage || ensemble m. *ou* équipe f. du fond et du jour.

surge chamber (Hydr arch) || Wasserschloß n. || château m. d'eau.

surgeon || Chirurg m.; Wundarzt m. || chirurgien m. / veterinary ~ || Tierarzt m. || vétérinaire m. / ~ india-rubber goods pl. for ~s || chirurgische Gummiwaren fpl. || articles mpl. de chirurgie en caoutchouc.

surgeon's saw || Wundarztsäge f. || scie f. de chirurgien.

surgeous agaric || Wundschwamm m. || agaric m.

surge pressure (Electr) || Stoßspannung f. || tension f. à choc. / high-tension equipment for generating ~s || Hochspannungsprüfeinrichtung f. zur Erzeugung von Stoßspannungen || dispositif m. d'essais à haute tension pour la production d'à-coups de tension.

surgery || Chirurgie f.; Wundarzneikunst f. || chirurgie f. / apparatus pl. and instruments pl. for ~ || Apparate mpl. und Instrumente npl. für die Chirurgie || appareils mpl. et instruments mpl. pour la chirurgie.

surgical bandage (Chem) || Verbandstoff m. || matière f. à pansement. / ~ case || Verbandkasten m.; boîtier m.; coffret m. à pansement. / ~ forceps pl. || chirurgische Pinzette f. || pince f. de chirurgie. / ~ glass article || chirurgischer Glasartikel m. || article m. de

chirurgie en verre. / ~ instrument ‖ chirurgisches *oder* ärztliches Instrument n. ‖ instrument m. chirurgical *ou* de chirurgie. / ~ instrument of rustless steel ‖ chirurgisches Instrument n. aus nichtrostendem Stahl ‖ instrument m. de chirurgie en acier inoxydable. / ~ knife ‖ chirurgisches *oder* ärztliches Messer n. ‖ bistouri m. / ~ sewing needle ‖ chirurgische Nähnadel f. ‖ épingle f. à coudre pour usages chirurgicaux. / ~ support ‖ Krankenstütze f. ‖ pelviphore m. / ~ wool ‖ Verbandwatte f. ‖ ouate f. à pansement.

surmounted (Build) ‖ überhöht ‖ surhaussé.

surmounting (Arch) ‖ Überhöhung f. ‖ surhaussement m.

surpass, to ‖ übertreffen ‖ surpasser.

surplus ‖ überzählig ‖ excédant; en trop.

surplus ‖ Zugabe f. ‖ supplément m. / ~ of orders ‖ Überfluß m. an Aufträgen ‖ ordres mpl. en abondance. / ~ of power ‖ Kraftüberschuß m. ‖ excès m. de force.

surplus dividend ‖ außerordentliche Dividende f. ‖ superdividende m. / ~ fund ‖ Reservefonds m.; Rücklagefonds m. ‖ fonds m. de réserve.

surprise box ‖ Attrappe f. ‖ attrape f.

surrender value ‖ Rückkaufswert m. ‖ valeur f. de rachat. / ~ of an insurance policy ‖ Rückkaufswert m. einer Versicherungspolice ‖ valeur f. de rachat d'une police d'assurance.

surrogate (Substitute) ‖ Ersatzmittel n.; Surrogat n. ‖ substitut m.; succédané m.

surround, to ‖ umschließen; umgeben; einschließen ‖ entourer. / ~ (To wash round) ‖ umspülen ‖ lécher; entourer d'eau. / ~ with a mantle ‖ ummanteln ‖ envelopper; chemiser. / ~ with walls ‖ ummauern ‖ emmurer.

surrounding air ‖ Außenluft f.; umgebende Luft f. ‖ air m. extérieur *ou* ambiant; atmosphère f.

survey, to ‖ vermessen ‖ arpenter. / ~ an area ‖ ein Gebiet n. aufnehmen *oder* vermessen ‖ lever un terrain. / ~ the concession area (Mine) ‖ das Grubenfeld vermessen ‖ mesurer le gîte minier. / ~ underground ‖ markscheiden ‖ lever *ou* tracer des plans de mine.

survey ‖ Überblick m. ‖ résumé m.; coup m. d'œil. / ~ (Surv) ‖ Vermessung f. ‖ arpentage m. / cadastral ~ ‖ Katastervermessung f. ‖ levé m. du cadastre. / check ~ ‖ Kontrollvermessung f. ‖ arpentage m. de contrôle. / geological ~ ‖ geologische Landesaufnahme f. ‖ relèvement m. géologique. / ~ of land ‖ Geländeaufnahme f. ‖ levé m. du terrain. / ~ of parcels of land ‖ Grundstücksvermessung f.; Parzellenvermessung f.; Stückvermessung f. ‖ arpentage m. parcellaire. / ~ of small areas *see* ~ of parcels of land. / surface ~ (Mine) ‖ Vermessung f. über Tage ‖ arpentage m. au jour. / tachometric ~ ‖ Tachometrie f. ‖ levé m. tachymétrique; tachéométrie. / trigonometrical ~ ‖ trigonometrisches Netz n. ‖ canevas m. trigonométrique. / underground ~ ‖ Vermessung f. unter Tage ‖ arpentage m. souterrain.

survey book (El line) ‖ Trassierbuch n.; Abpfählbuch n. ‖ état m. descriptif. / ~ data ‖ Vermessungsangabe f. ‖ indication f. topographique.

surveying ‖ Vermessungswesen n.; Feldmessen n. ‖ géodésie f. pratique; arpentage m. / ~ (El line) ‖ Auskundung f. ‖ étude f. détaillée de la ligne. / ~ the land ‖ Landesaufnahme f.; Landmessung f.; Landvermessung f. ‖ levé m. du plan d'un terrain; arpentage m.; mesurage m. du terrain. / photographic ~ ‖ Meßbildverfahren n.; Photogrammetrie f. ‖ photogrammétrie f. / stereo-photographic ~ ‖ Raumbildmessung f. ‖ mesure f. stéréophotographique.

surveying compass ‖ Patentbussole f. ‖ bussole f. à réflexion. / ~ instrument ‖ Feldmeßgerät n.; Vermessungsinstrument n. ‖ instrument m. géodésique *ou* de géodésie *ou* d'arpentage. / ~ instrument tripod ‖ geodätisches Dreibeinstativ n. ‖ trépied m. géodésique. / ~ office ‖ Vermessungsinstitut n. ‖ institut m. de levés photogrammétriques. / ~ operations pl. ‖ Feldmeßarbeiten fpl. ‖ travaux mpl. géodésiques. / ~ technology ‖ Vermessungstechnik f. ‖ technique f. de mesurage. / ~ vessel (Mar) ‖ Vermessungsschiff n. ‖ navire m. hydrographe.

survey line on surface ‖ Zug m. über Tage ‖ levé m. de plan au jour.

survey photograph for aerial surveying ‖ Luftmeßbild n. ‖ photographie f. de mesure aérienne; aérophotogramme m.

surveyor ‖ Feldmesser m.; Geometer m.; Landmesser m. ‖ arpenteur m.; géomètre m. / ~ of a dike ‖ Deichmeister m. ‖ surintendant m. des digues. / ~ of mines ‖ Markscheider m. ‖ géomètre-souterrain m. / assistant of the ~ of mines ‖ Markscheidergehilfe m. ‖ jambot-niveleur m.; posteur m. de chaîne. / ~ of taxes ‖ Steuerinspektor m. ‖ inspecteur m. des contributions directes.

surveyor's compass *see* ~ dial. / ~ dial ‖ Feldkompaß m. ‖ boussole f. d'arpenteur. / ~ lamp (Mine) ‖ Markscheiderlampe f. ‖ lampe f. d'arpenteur de mine. / ~ level ‖ Fernglaslibelle f. ‖ niveau m. à lunette. / ~ rod (Mine) ‖ Lachterstab m. ‖ mesure f. d'arpenteur. / ~ table ‖ Meßtisch m. ‖ plateau m. pour tracer des plans.

survive, to ‖ überdauern; überleben ‖ survivre (à . . .).

susceptance (Electr) ‖ Blindleitwert m. ‖ susceptance f.

susceptibility ‖ Empfänglichkeit f.; Erregbarkeit f.; Suszeptibilität f. ‖ susceptibilité f. / ~ magnetic ~ ‖ Suszeptibilität f.; magnetische Aufnahmefähigkeit f.; Magnetisierungsfähigkeit f. ‖ susceptibilité f. (magnétique).

suspend, to ‖ aufhängen ‖ suspendre. / ~ (Chem) ‖ suspendieren; schwebend halten ‖ tenir en suspension. / ~ freely ‖ frei beweglich aufhängen ‖ suspendre librement. / ~ on gimbals ‖ kardanisch aufhängen ‖ suspendre à la cardan.

suspended ‖ aufgeschoben; unterbrochen; eingestellt ‖ suspendu. / ~ temporarily ~ ‖ zeitweilig eingestellt *oder* unterbrochen ‖ temporairement suspendu.

suspended (Hanging) ‖ aufgehängt; hängend ‖ suspendu. ‖ ~ contact box ‖ Hängeanschlußdose f. ‖ prise f. de courant suspendu. / ~ electric railway ‖ Elektrohängebahn f. ‖ transporteur m. aérien électrique; chemin m. de fer électrique à suspension; électro-chemin m. suspendu. / ~ gas lamp ‖ Hängelichtgaslampe f. ‖

lampe f. à gaz à éclairage renversé. / ~ instrument ‖ Hängeinstrument n. ‖ instrument m. à suspension. / ~ joint (Railw) ‖ schwebender Schienenstoß m. ‖ joint m. en porte-à-faux. / ~ matter ‖ Sinkstoff m. ‖ matière f. en suspension. / ~ railway ‖ Hängebahn f. ‖ chemin m. de fer à suspension; voie f. suspendue; transporteur m. aérien à câble. / ~ roof (Build) ‖ Hängedecke f. ‖ ciel m. suspendu.

suspenders pl. (For stockings) ‖ Strumpfhalter mpl. ‖ jarretelles fpl.; jarretières fpl. / ~ (For trousers) ‖ Hosenträger mpl. ‖ bretelles fpl.

suspension (Chem) ‖ Aufschlemmung f. ‖ suspension f. / ~ (Mach) ‖ Aufhängung f.; Abfederung f. ‖ suspension f. (en ressorts). / bifilar ~ ‖ Doppelfadenaufhängung f. ‖ suspension f. bifilaire. / ~ in double bows ‖ Doppelbügelaufhängung f. ‖ suspension f. à étriers doubles. / elastic ~ ‖ federnde Aufhängung f. ‖ suspension f. élastique. / front-axle ~ ‖ Vorderachsaufhängung f. ‖ suspension f. de l'essieu avant. / ~ of payment ‖ Zahlungseinstellung f. ‖ suspension f. de payement. / revolving-shackle ~ ‖ Drehbügelaufhängung f. ‖ suspension f. à étrier tournant. / semi-universal ~ ‖ halbuniversale Aufhängung f. ‖ suspension f. semi-universelle. / ~ by spring ‖ Abfederung f.; Federaufhängung f. ‖ suspension f. à ressort *ou* en ressorts. / three-point ~ ‖ Dreipunktaufhängung f. ‖ suspension f. triangulaire *ou* par trois points. / ~ by a toothed bow ‖ Zahnbügelaufhängung f. ‖ suspension f. à étrier denté. / circular ~ of the type bars (Typewr) ‖ kreisförmige Aufhängung f. der Typenhebel ‖ suspension f. circulaire des tiges à caractères. / single ~ of the type bars (Typewr) ‖ Einzelaufhängung f. der Typenhebel ‖ suspension f. séparée pour chaque tige à caractère.

suspension arrangement ‖ Aufhängevorrichtung f. ‖ dispositif m. de suspension.

suspension bridge ‖ Hängebrücke f. ‖ pont m. suspendu.

suspension chain ‖ Hängekette f. ‖ chaîne f. de suspension. / ~ of the cage (Mine) ‖ Quenzelkette f. ‖ chaîne f. de suspension de la cage. / ~ for clothes ‖ Rockaufhänger m. ‖ chaînette f. de suspension d'habit.

suspension clamp (El line) ‖ Seilschelle f. ‖ mâchoire m.; plaque f. de suspension. / ~ eye hoist ‖ Ösenzug m. ‖ suspension f. par œillets. / ~ ferry ‖ Schwebefähre f. ‖ bac m. suspendu. / ~ gear ‖ Gehänge n. ‖ appareil m. de suspension. / ~ iron ‖ Aufhängeeisen n. ‖ fer m. de suspension. / ~ loop ‖ Aufhängebügel m. ‖ étrier m. de suspension. / ~ point ‖ Aufhängepunkt m.; Unterstützungspunkt m. ‖ point m. de suspension *ou* d'appui. / ~ railway ‖ Hängebahn f. ‖ chemin m. de fer suspendu; chemin m. de fer aérien. / ~ rods pl. ‖ Aufhängegestänge n. ‖ tiges fpl. de suspension. / ~ spring ‖ Tragfeder f. ‖ ressort m. de suspension. / self-balancing ~ tackle ‖ Ausgleichgehänge f. ‖ suspension f. à compensation. / ~ wire winding apparatus ‖ Seilwinde f. ‖ treuil m. à câble.

suspensor (Med) *see* suspensory bandage.

suspensory bandage ‖ Suspensorium n.; Tragebinde f. ‖ suspensoir m.

sustain, to ~ a loss ‖ Verlust m. erleiden ‖ subir des pertes fpl. / ~ the roof (Mine) ‖

die Firste abfangen ‖ arc-bouter le gradin renversé.

sustained wave (Radio) ‖ ungedämpfte Welle f. ‖ onde f. inamortie *ou* entretenue.

sustaining wall ‖ Stützmauer f. ‖ mur m. de soutènement.

suttle-weight ‖ Nettogewicht n. ‖ poids m. net.

swab, to ~ (Mar) ‖ schwabbern; dweilen; abdweilen ‖ fauberter.

swab ‖ Haarpinsel m. ‖ blaireau m. / ~ (Mar) ‖ Schwabber m. ‖ faubert m.; écoupe f.

swage, to ‖ im Gesenk n. schmieden ‖ estamper; étamper; matricer.

swage ‖ Gesenk n. ‖ estampe f.; étampe f.; matrice f. / ~ **anvil** ‖ Gesenkamboß m. ‖ enclume f. à enfonçures *ou* à estamper. / ~ **block** ‖ Gesenkplatte f.; Lochplatte f. ‖ tas-étampe f.; étampe f.

swaged ‖ gesenkgeschmiedet ‖ forgé en matrice; matricée; étampé. / ~ **piece** ‖ Gesenkschmiedestück n. ‖ estampe f. de forge; pièce f. matricée *ou* estampée.

swager ‖ Gesenkschmied m. ‖ étampeur m.

swaging ‖ Gesenkarbeit f. ‖ travail m. d'étampage. / ~ **of base plates** (Dent) ‖ Pressen n. von Gebißplatten ‖ repoussage m. de plaques base pour dentiers. / ~ **of iron and steel** ‖ Gesenkschmieden n. von Eisen und Stahl ‖ étampage m. à chaud du fer et de l'acier.

swaging machine (Forg) ‖ Anspitzmaschine f. ‖ machine f. à appointer. / ~ (Tinm) ‖ Siekenmaschine f. ‖ machine f. à suager *ou* à border.

swaging steam hammer ‖ Gesenkdampfhammer m. ‖ marteau-pilon m. à vapeur d'estampage.

swallow of a pair of stones (Mill) ‖ Schluck m. der Mühlsteine ‖ entrée f. des meules.

swallow tail (Carp) ‖ Schwalbenschwanz m.; Zinke f. ‖ queue f. d'aronde *ou* d'hironde; tenon m. à queue.

swamp, to ~ (Mar) ‖ vollschlagen ‖ se remplir.

swamp ‖ Morast m.; Moor n.; Sumpf m. ‖ terrain m. vaseux; vase f.; marais m.

swamp ore ‖ Raseneisenstein m.; Sumpferz n. ‖ limonite f.; fer m. limoneux.

swampy ‖ schlammig; moorig; morastig; sumpfig ‖ limoneux; marécageux; vaseux; bourbeux. / ~ **ground** ‖ mooriger Boden m. ‖ terrain m. marécageux. / ~ (building) **plot** ‖ versumpftes Grundstück n. ‖ terrain m. inondé.

swan's-down (Weav) ‖ rauher Barchent m. ‖ futaine f. à poil.

swan neck (Techn) ‖ Schwanenhals m. ‖ col m. de cygne. / ~ **bearer** (Railw) ‖ Schwanenhalsträger m. ‖ traverse f. en col de cygne.

swanskin (Weav) ‖ Molton m.; Multon m. ‖ molleton m.

sward ‖ Rasen m. ‖ gazon m. / ~ **of a meadow** ‖ Grasnarbe m. *oder* Rasendecke f. einer Wiese ‖ gazon m. d'une prairie.

sward cutter (Agr) ‖ Rasenpflug m. ‖ charrue f. à peler; dégazonnoir m.

swath (Agr) ‖ Schwad m.; Schwaden m.; andain m.; javelle f. / **to lay in ~s** pl. ‖ in Schwaden mpl. ablegen ‖ javeler. / ~ **of mist** ‖ Nebelschwaden m. ‖ brouillard m. flottant.

swath board (Of a mowing machine) ‖ Schwadenblech n. ‖ planche f. à andain. / ~ **height** ‖ Schwadenhöhe f. ‖ hauteur f. d'andain.

sway, to ~ (Mar) ‖ hissen; aufhissen; heißen ‖ hisser; haler; palanquer. / ~ **away** (Mar) ‖ laufend holen ‖ haler à courir *ou* en courant.

sway bar (Coachm) ‖ Lenkscheit n.; Reibscheit n. ‖ sassoire f.

sweat, to ‖ schwitzen ‖ ressuer.

sweat ‖ Schweiß m. ‖ sueur f.; transpiration f.

sweater (Cloth) ‖ Sportjacke f.; Wolljacke f. ‖ sweater m.

sweating ‖ Schwitzen n. ‖ ressuage m.; transpiration f. / ~ **of the charcoal pile** ‖ Schwitzen n. des Meilers ‖ exsudation f. de la meule de carbonisation. / ~ **of indigo** ‖ Schwitzen n. des Indigos ‖ ressuage m. de l'indigo.

sweating apparatus ‖ Schwitzapparat m. ‖ étuve f. / ~ **bath** ‖ Schwitzbad n. ‖ étuve f.; bain m. russe. / ~ **gutter** (Met) ‖ Seigergasse f. ‖ puits m. perpendiculaire. / ~ **house** (Tobacco) ‖ Schwitzhaus n.; Schwitzstapel m. ‖ suerie f. / ~ **kiln** (Brew) ‖ Gerstendarre f.; Darre f. zum Trocknen der Gerste ‖ touraille f. à sécher l'orge. / ~ **pile** *see* ~ **house**.

Swedish balance ‖ Schnellwage f. ‖ balance f. romaine. / ~ **pig** ‖ schwedisches Eisen n. ‖ fonte f. suédoise.

sweep, to ~ (Join) ‖ schweifen ‖ échancrer; évider. / ~ **the funnel** ‖ den Schornstein m. fegen ‖ ramoner la cheminée.

sweep (Mar) ‖ langes Ruder n. ‖ aviron m. / ~ (Met) *see also* **sweepings** (Goldsm) ‖ Gekrätz n.; Krätze f.; Metallabfälle mpl. ‖ déchet m. de métal; crasse f. / ~ (Mine) ‖ Gestänge n. ‖ tige f.; tirant m. / **metal** ~ (Mowing machine) ‖ Anhaublech n. ‖ tôle f. à moissonner en andain.

sweep-bar (Coachm) ‖ Lenkscheit n.; Reibscheit n. ‖ sassoire f.

sweeper (Railw) ‖ Abräumer m.; Schienenräumer m.; Aufräumer m. ‖ chasse-pierres m. / ~ (Road) ‖ Straßenkehrer m.; Straßenreiniger m.; Straßenfeger m. ‖ balayeur m. (des rues); boueur m. / **motor** ~ ‖ Straßenkehrmaschine f. ‖ balayeuse f. automobile.

sweep gauge ‖ Übergreiflehre f.; Rachenlehre f. ‖ calibre m.

sweeping (Join) ‖ Schweifung f. ‖ échancrure f.; cambrure f.; chantournement m.

sweeping force of the water ‖ Schleppkraft f. des Wassers ‖ force f. d'entraînement de l'eau. / ~ **machine for gold leaf** ‖ Abkehrmaschine f. für Blattgold ‖ machine f. à brosser l'or en feuilles. / ~ **mill** (Agr) ‖ Kornschwinge f.; Staubmühle f.; Schwingwanne f.; Wurfmaschine f. ‖ van m. émotteur. / ~ **motor car for streets** ‖ Straßenreinigungsmotorfahrzeug n. ‖ voiture f. automobile pour le nettoyage des rues.

sweepings pl. ‖ Kehricht m. ‖ balayures fpl. / ~ (Goldsm) ‖ Goldkrätze f. ‖ cendres fpl. *ou* lavure f. d'or. / ~ (Mar) ‖ Spillage f.; Verlust m. bei trockenen Gütern durch undichte Verpackung ‖ balayures fpl.

sweep net (Fish) ‖ Wurfnetz n. ‖ épervier m.; ressaut m.

sweep rod (For cables) ‖ Einführungsgestänge n. zum Kabeleinziehen ‖ aiguille f. de tirage pour l'introduction des câbles.

sweep saw ‖ Schweifsäge f.; Stellsäge f. ‖ scie f. à tourner *ou* à tourne-fond *ou* à chantourner *ou* à échancrer.

sweet ‖ süß ‖ doux; sucré.

sweet ‖ Süßigkeit f.; Leckerei f.; Zuckerbonbon m. ‖ bonbon m. de sucre.

sweeten, to ‖ versüßen ‖ adoucir; sucrer.

sweetening chemical for brewers ‖ Versüßungsmittel n. für Brauer ‖ édulcorant m. pour brasseurs.

sweeting (Sugar) ‖ Gewinnung f. des Saftes; Saftgewinnung f. ‖ extraction f. du jus *ou* des jus de diffusions.

sweet malt ‖ Melassetreber pl. ‖ drèches fpl. mélassées.

sweetmeat *see* **sweetmeats** pl. / ~ **glass** ‖ Süßwarenglas n.; Bonbonglas n. ‖ verre m. à bonbons. / ~ **machine** ‖ Bonbonmaschine f. ‖ machine f. à bonbons.

sweetmeats pl. *see also* **sweets** ‖ Zuckerwaren fpl.; Zuckerwerk n.; Süßwaren fpl.; Konfekt n. ‖ sucreries fpl.; dragées fpl.; confitures fpl.

sweet orange ‖ Apfelsine f. ‖ orange m. (douce). / ~ **oil** ‖ Apfelsinenschalenöl n. ‖ essence f. de Portugal.

sweets pl. ‖ Zuckerwerk n.; Süßigkeiten fpl.; Konfekt n.; Konfitüre f. ‖ confitures; sucreries fpl.; dragées fpl. / **glazer of** ~ ‖ Zuckerwerkglasierer m. ‖ glaceur m. de dragées. / **manufacture of** ~ ‖ Süßigkeitsgewerbe n. ‖ fabrication f. de sucreries. / ~ **wrapper** ‖ Bonboneinwickler m. ‖ habilleur m. de bonbons.

sweet-sounding ‖ wohlklingend ‖ harmonieux; sonore; euphonique.

sweet water ‖ Süßwasser n. ‖ eau f. douce. / ~ **tank** ‖ Süßwasserbehälter m. ‖ réservoir m. d'eau douce.

sweet wine ‖ Ausbruch m. ‖ vin m. doux.

swell, to ‖ anschwellen; blähen; aufquellen; aufblähen ‖ (s')enfler; gonfler; foisonner. / ~ (Curr) ‖ schwellen; treiben ‖ gonfler; travailler à l'orge. / ~ (Said of wood) ‖ quellen ‖ se gonfler. / ~ **up** ‖ quellen ‖ gonfler.

swell *see also* **swelling** ‖ Anschwellung f.; Geschwulst f.; Beule f. ‖ gonflement m.; tumeur m. / **having a slight** ~ ‖ leicht gekrümmt ‖ bombé.

swell (Of the water) ‖ Anschwellen n.; Schwellen n.; Steigen n. ‖ crue f. (des eaux). / ~ **of the ocean** ‖ Dünung f. ‖ agitation f. *ou* houle f. de l'océan. / **slight** ~ ‖ leichte Dünung f. ‖ faible *ou* petite houle f.

swell (Hydr arch) ‖ Stauung f. ‖ remous m. / ~ **of a river** ‖ Stauung f. eines Flusses ‖ remous m. d'une rivière. / **amplitude of** ~ ‖ Stauweite f. ‖ amplitude f. du remous. / **height of** ~ ‖ Stauhöhe f. ‖ hauteur f. de remous.

swelling *see also* **swell** ‖ Schweifung f.; Ausbauchung f.; Anschwellung f.; Quellung f. ‖ bombement m.; renflement m.; enflement m.; enflure f.; gonflement m. / ~ (Of the lime) ‖ Aufgehen n.; Gedeihen n.; Wachsen n. ‖ foisonnement m. / ~ **of wood** ‖ Quellen n. des Holzes ‖ gonflement m. du bois.

swelling heat ‖ Quellungswärme f. ‖ chaleur m. de gonflement.

swells pl. (Weav) ‖ Schützenhalter m. ‖ serre-navette m.

swift, to ~ (Mar) ‖ schwichten ‖ brider; serpenter. / ~ **the shrouds** pl. ‖ die Wanten fpl. schwichten ‖ brider les haubans mpl.

swifter (Mar) ‖ Schwichtleine f. ‖ bridure f.

swift sewing machine ‖ Schnellnähmaschine f. ‖ machine f. à coudre rapide.

swill box ‖ Müllkasten m.; Mülleimer m. ‖ boîte f. à ordures; caisse f. ou panier m. au balayures; poubelle f.

swim, to ‖ schwimmen ‖ nager.

swimmer (Techn) *see also* float ‖ Schwimmer m. ‖ flotteur m. / ~ for fermenting tuns ‖ Gärbottichschwimmer m. ‖ flotteur m. réfrigérant pour cuves de fermentation.

swimmer switch ‖ Schwimmerschalter m. ‖ interrupteur m. à flotteur.

swimming (Said of a ship) ‖ flott ‖ à flot.

swimming basin ‖ Schwimmbecken n.; Badebecken n. ‖ bassin m. de natation. / ~ bath ‖ Schwimmbad n.; Schwimmbadeanstalt f. ‖ grand bain m.; établissement m. de bains./~ bladder ‖ Schwimmblase f. ‖ vessie f. natatoire.

swindling (Trade) ‖ schwindelhaft ‖ vertigineux; trompeur.

swine ‖ Schwein n. ‖ porc m.; cochon m.

swine-stone (Miner) ‖ bituminöser Kalkstein m.; Stinkstein m; Stinkkalkstein m. ‖ chaux f. carbonatée fétide; calcaire m. fétide.

swing, to ‖ schwingen; schleudern ‖ osciller; vibrer. / ~ (Mach) ‖ schleudern ‖ tourner (irrégulièrement). / ~ (Ship) ‖ schwaien; schwoien; schwingen ‖ éviter; se tourner. / ~ clear ‖ frei schwingen ‖ osciller librement.

swing ‖ Schaukel f. ‖ escarpolette f.; balançoire f. / ~ (Oscillation) ‖ Schwingung f.; Schwingen n. ‖ oscillation f. / ~ of the pendulum ‖ Pendelschwingung f. ‖ oscillation f. du pendule.

swing bar (Coachm) ‖ Ortscheit n.; Zugscheit n.; Schwengel m.; Wagenschwengel m. ‖ palonnier m. / ~ bolster ‖ Pendelwiege f. ‖ balancier m. transversal. / ~ bridge ‖ Drehbrücke f. ‖ pont m. tournant. / ~ bucket elevator ‖ Schaukelbecherwerk n. ‖ noria f. à godets oscillants. / ~ crane ‖ Drehkran m. ‖ grue f. pivotante. / ~ grinding machine ‖ Hängeschleifmaschine f.; Pendelschleifmaschine f. ‖ machine f. à meuler oscillante.

swinging ‖ schwingend ‖ oscillant.

swinging ‖ Schwingung f.; Oszillation f. ‖ oscillation f.; excursion f.

swinging air column ‖ pendelnde Luftsäule f. ‖ colonne f. d'air alternative. / ~ arm ‖ Schwenkarm m. ‖ bras m. oscillant. / ~ berth (Shipb) ‖ Schwingkoje f. ‖ couchette f. à suspension. / ~ crane ‖ Schwenkkran m. ‖ grue f. à flèche pivotante. / ~ isolator with suspending hook ‖ Pendelisolator m. ‖ isolateur m. à suspension. / ~ lever (Electr) ‖ Pendelanker m.; Schwinganker m. ‖ ancre f. oscillante. / ~ leverwork ‖ Schwingankerwerk n. ‖ mouvement m. à ancre oscillante. / ~ platform ‖ Schwenkbühne f. ‖ pont-levis m. / ~ roller ‖ schwenkbare Rolle f. ‖ galet m. ou poulie f. à pivot. / ~ room (Mar) ‖ Schwairaum m. ‖ évitage m.

swing lamp ‖ Hängelampe f. ‖ suspension f.; lampe f. de suspension.

swingle ‖ Hanfschwinge f.; Flachsschwinge f. ‖ échanvroir m. / ~ braces pl. (Coachm) ‖ Ortscheitriemen mpl. ‖ courroies fpl. de palonnier.

swingler (Agr) ‖ Schwinger m. ‖ espadeur m.

swingle tree (Coachm) ‖ Ortscheit n.; Zugscheit n.; Schwengel m.; Wagen-

schwengel m. ‖ palonnier m. / ~ clasp ‖ Ortscheitkappe f.; Mittelkappe f.; Ortscheitblech n. ‖ lamette f. de palonnier.

swingling machine (Spinn) ‖ Schwingmaschine f. ‖ machine f. à teiller.

swing-out‖ausklappbar‖pivotant; articulé; pouvant s'écarter; s'écartant. / ~ condenser ‖ ausklappbarer Kondenser m. ‖ condensateur m. à bascule. / ~ mandrel ‖ ausschwenkbarer Einlagedorn m. ‖ mandrin m. supplémentaire articulé.'/~ punch holder ‖ ausschwenkbarer Stempelhalter m. ‖ porte-poinçon m. oscillant. / ~ table ‖ Drehtisch m. ‖ table f. pivotante; plateau m. tournant. / ~ top clamping bar ‖ ausschwenkbare Oberwange f. ‖ coulisse f. supérieure articulée.

swing pan ‖ Kipppfanne f.; Schwungpfanne f. ‖ chaudière f. à bascule. / ~ pipe ‖ Schwenkrohr n. ‖ tube m. articulé. / ~ plough ‖ Schwingpflug m. ‖ sochet m. / ~ screen ‖ Schwingsieb n. ‖ crible m. oscillant. / ~ sieve (Ore dress; Coal dress) ‖ Rätter m.; Schüttelsieb n. ‖ crible m. à bascule ou à manivelle. / ~ support (Bridge) ‖ Kipplager n. ‖ appui m. à bascule. / ~ tow (Spinn) ‖ Schwinghede f. ‖ repérants mpl.; ~ tree (Coachm) *see* swingle tree. / ~ wheel (Watchm) ‖ Hemmungsrad n.; Steigrad n. ‖ roue f. de rencontre.

swipe of a draw-bridge ‖ Schwengel m. oder Wippe f. einer Aufzugbrücke ‖ fléau m. ou flèche f. d'un pontlevis.

switch, to ~ (Railw) ‖ eine Weiche stellen ‖ aiguiller. / ~ (Electr) ‖ schalten ‖ commuter. / ~-in ‖ einschalten ‖ mettre en circuit. /~-in resistance ‖ Widerstand m. einschalten ‖ mettre de la résistance en circuit. / ~-off ‖ den Strom unterbrechen; ausschalten ‖ couper le circuit ou le contact. / ~-on ‖ einschalten ‖ mettre le contact.

switch (Railw) ‖ Weiche f. ‖ aiguille f.; changement m. (de voie). / automatic ~ ‖ selbsttätige Weiche f. ‖ changement m. de voie automatique. / blunt-ended ~ ‖ Schleppwechsel m. ‖ changement m. sans contre-rails. / contractor's ~ ‖ Schleppweiche f. ‖ changement m. à rails mobiles. / ~ with counter poise ‖ Weiche f. mit Gegengewicht ‖ aiguille f. à contre-poids. / ~ curve ‖ Kurvenweiche f. ‖ changement m. en courbe. / double ~ ‖ Doppelweiche f. ‖ changement m. à double aiguille. / inclined-plane ~ (Railw) ‖ Kletterweiche f.; Kletterkreuzung f. ‖ changement m. de voie à plans inclinés ou sans discontinuité de la voie principale. / left-hand ~ ‖ Linksweiche f. ‖ changement m. à gauche. / ~ for overhead railways ‖ Weiche f. für Hängebahnen ‖ aiguille f. pour voies suspendues. / right-hand ~ ‖ Rechtsweiche f. ‖ branchement m. à droite. / ~ with self-acting slide rail ‖ Weiche f. mit selbsttätiger Einstellung der Zunge ‖ changement m. avec fonctionnement automatique de l'aiguille. / ~ with sliding plates ‖ Plattenweiche f. ‖ changement m. à plaque d'assise. / ~ with spring tongs ‖ Federweiche f. ‖ changement m. de voie à aiguilles flexibles. / symmetrical double ~ ‖ symmetrische Doppelweiche f. ‖ changement m. double à voies symétriques. / three-throw ~ ‖ Dreiwegeweiche f. ‖ changement m. à

trois voies. / trail ~ ‖ Schleppweiche f. ‖ rail m. mobile.

switch (Aut tel) ‖ Wähler m. ‖ sélecteur m. / final ~ ‖ Linienwähler m.; Leitungswähler m. ‖ sélecteur m. final; connecteur m. / individual line ~ ‖ Vorwähler m. ‖ présélecteur m. / rotary ~ ‖ Drehwähler m. ‖ sélecteur m. rotatif.

switch (Electr) ‖ Schalter m.; Einschalter m.; Ausschalter m. ‖ conjoncteur m.; interrupteur m.; disjoncteur m.; commutateur m. / aerial wire (change-over) ~ ‖ Luftdrahtschalter m. ‖ commutateur m. d'antenne. / automatic ~ ‖ Selbstschalter m. ‖ interrupteur m. automatique. / automatic field-break ~ ‖ selbsttätiger Magnetausschalter m. ‖ interrupteur m. automatique de l'excitation. / auxiliary ~ ‖ Nebenschalter m. ‖ commutateur m. supplémentaire. / battery ~ ‖ Zellenschalter m. ‖ réducteur-adjoncteur. / double battery ~ ‖ Doppelzellenschalter m. ‖ réducteur-adjoncteur m. double. / hand-operated battery ~ ‖ Handzellenschalter m. ‖ réducteur-adjoncteur à main. / simple battery ~ ‖ Einfachzellenschalter m. ‖ réducteur-adjoncteur m. simple. / change-over ~ ‖ Umschalter m. ‖ commutateur m.; inverseur m. / change-tune ~ ‖ Wellenumschalter m. ‖ commutateur m. de longueur d'ondes. / charging ~ ‖ Ladeschalter m. ‖ interrupteur m. de charge. / combined ~ and fuse ‖ Schalter m. mit Sicherung ‖ interrupteur m. avec coupe-circuit. / commutator ~ ‖ Fahrtrichtungsschalter m. ‖ commutateur m. de sens de marche. / compressed-air ~ ‖ Druckluftschalter m. ‖ interrupteur m. à air comprimé. / controlling ~ ‖ Steuerschalter m. ‖ appareil m. de commande. / ~ operated by crank ‖ Kurbel(aus)-schalter m. ‖ interrupteur m. à manivelle. / disconnecting ~ ‖ Trennschalter m. ‖ sectionneur m. / disconnecting double-throw ~ ‖ Trennumschalter m. ‖ sectionneur-inverseur m. / distant control ~ ‖ Schalter m. mit Fernsteuerung ‖ téléinterrupteur m. / double-bladed knife ~ ‖ Doppelmesserschalter m.; Doppelschalter m. ‖ interrupteur m. bipolaire à lames. / double-pole ~ ‖ zweipoliger Schalter m. ‖ interrupteur m. bipolaire. / double-pole double-throw ~ ‖ zweipoliger Umschalter m. ‖ commutateur m. bipolaire à deux directions. / double-throw ~ ‖ Hebelumschalter m.; Umschalter m. ‖ commutateur m. à levier. / field-break ~ ‖ Magnetausschalter m. ‖ interrupteur m. de l'excitation. / float ~ for pumps ‖ Schwimmerschalter m. für Pumpen ‖ interrupteur m. à flotteur pour pompes. / foot ~ ‖ Fußschalter m. ‖ commutateur m. à pédale. / pneumatic foot ~ ‖ pneumatischer Fußschalter m. ‖ commutateur m. pneumatique à pédale. / heating current testing ~ ‖ Heizstrommeßschalter m. ‖ interrupteur m. de mesure du courant de chauffage. / high-tension ~ ‖ Hochspannungsschalter m. ‖ interrupteur m. pour haute tension. / high-tension remote control ~ ‖ Hochspannungsfernschalter m. ‖ téléinterrupteur m. pour haute tension. / ignition ~ ‖ Magnetschalter m. ‖ commutateur m. de la magnéto. / inter-through ~ (Tel) ‖ Zwischenstellenumschalter m. ‖ commutateur m. intermédiaire. / isolating ~

|| Trennschalter m. || sectionneur m. / knife ~ || Messerschalter m. || interrupteur m. (unipolaire) à lames. / lateral ~ || Seitenschalter m. || commutateur m. latéral. / lever ~ || Hebelschalter m. || interrupteur m. à levier. / limit ~ || Endausschalter m. || interrupteur m. de fin de course. / listening ~ || Mithörschalter m. || clef f. d'écoute. / main ~ see master ~. / master ~ || Hauptschalter m. || interrupteur m. principal. / multi-circuit ~ || Serienschalter m. || commutateur m. multiple ou à combinaison. / multiple-throw ~ || Vielfachumschalter m. || commutateur m. multiple. / no-load ~ || Nullausschalter m. || interrupteur m. à zéro. / oil break ~ || Ölschalter m. || interrupteur m. à bain d'huile. / pear ~ || Birnenausschalter m. || interrupteur m. à poire. / pedal ~ || Tretschalter m. || commutateur m. à pédale. / plug ~ || Stöpselschalter m. || commutateur m. à fiche. / plug seat ~ || Stöpselsitzschalter m. || commutateur m. à fiche reposeuse. / pole ~ || Mastschalter m. || interrupteur m. de poteau. / principal ~ || Hauptschalter m. || interrupteur m. principal. / pull ~ || Zugschalter m. || interrupteur m. à tirette. / pushbutton ~ || Druckknopfschalter m. || interrupteur m. à poussoir. / quick-break ~ || Schnappschalter m.; Momentschalter m. || interrupteur m. à rupture brusque. / remote ~ || Fernschaltapparat m.; Fernschalter m. || interrupteur m. à commande à distance. / reverse-current ~ || Rückstromausschalter m. || interrupteur m. à retour de courant. / reversible ~ || Kippschalter m. || interrupteur m. à bascule. / reversing ~ || Stromwender m. || inverseur m. de courant. / reversing ~ (Tel) || Richtungswechsler m. || clef f. (ou clé) d'inversion. / rotary ~ || Drehschalter m. || interrupteur m. rotatif. / section ~ || Streckenschalter m. || interrupteur m. de section. / semi-automatic ~ || Gruppenumschalter m. || commutateur m. semi-automatique. / sequence ~ || Stufenschalter m.; Folgeschalter m. || commutateur m. à cascade; combineur m. / shunt ~ || Umgehungsschalter m. || interrupteur m. tournant. / star-delta ~ || Sterndreieckschalter m. || commutateur m. étoile-triangle. / testing ~ || Prüfschalter m. || clef f. d'essais. / three-phase ~ || Drehstromschalter m. || interrupteur m. pour courant triphasé. / three-position ~ || Schalter m. für drei Stellungen || commutateur m. à trois positions. / three-way ~ || Dreiwegeumschalter m. || commutateur m. à trois directions. / timing ~ || Zeitschaltwerk n. || commutateur m. horaire ou à temps. / timing sequence ~ || Zeiteinstellfolgeschalter m. || conjoncteur m. séquentiel à temps. / transfer sequence ~ || Übertragungsfolgeschalter m. || commutateur m. de la suite des transmissions. / tumbler ~ || Kippschalter m. || interrupteur m. tumbler ou à bascule. / two-way ~ || Umschalter m.; Kommutator m. || commutateur m. / universal ~ || Generalumschalter m. || commutateur m. universel. / ~ for weak current || Schwachstromschalter m. || interrupteur m. à faible intensité.

switch and crossing (Railw) || Ausweich-vorrichtung f.; Weiche f. und Kreuzung f. || changement m. et croisement m. de voie.

switch blade || Weichenzunge f. || aiguille f. de changement de voie. / pair of ~s pl. || Zungenpaar n. || paire f. d'aiguilles. / ~ planing machine || Weichenzungenhobelmaschine f. || raboteuse f. à rail d'aiguilles de changement de voie.

switchboard (Electr) || Schalttafel f.; Schaltbrett n. || tableau m. de distribution. / direct current and alternating current ~ || Schalttafel f. für Gleichstrom und Wechselstrom || tableau m. de distribution pour courant continu et alternatif. / flat ~ || Flachschalttafel f. || tableau m. de distribution plat. / main ~ with distribution bus bars || Hauptschalttafel f. mit Kreuzschienenverteiler || tableau m. de couplage principal avec répartiteur à barres croisées. / main experimental ~ || Hauptexperimentierschalttafel f. || tableau m. principal d'expérimentation. / ~ in marble || Schalttafel f. aus Marmor || tableau m. de distribution en marbre. / ~ in oak || Schalttafel f. aus Eichenholz || tableau m. de distribution en bois de chêne. / thermal and remote control ~ || Wärmewarte f. || poste m. de contrôle thermique.

switchboard (Tel) || Klappenschrank m. || tableau m. (commutateur) à volets. / cordless ~ || schnurloser Klappenschrank m. || tableau m. commutateur à volets sans cordons. / drop ~ see switchboard (Tel). / ~ for intercommunication sets || Klappenschrank m. für Reihenanlagen || tableau m. à volets pour les postes en série. / little ~ || Vermittlungskästchen n. || petite boîte f. pour établir la communication. / ~ with plug restored indicators || Rückstellklappenschrank m. || tableau m. à volets à relèvement automatique.

switchboard attendant || Schalttafelwärter m. || électricien m. de service au tableau de distribution. / ~ drop (Tel) || Klappe f.; Anrufklappe f. || annonciateur m. à clapet ou d'appel. / ~ fastening screw || Schalttafelbefestigungsschraube f. || boulon m. de fixation pour tableau de distribution. / ~ frame || Schalttafelgerüst n. || chassis m. du tableau de distribution. / ~ gallery || Schalttafelbühne f. || plate-forme f. du tableau de distribution. / ~ instrument || Schaltbrettinstrument n. || instrument m. du tableau de distribution. / ~ measuring instrument || Schalttafelmeßgerät n. || instrument m. de mesure du tableau de distribution. / ~ terminal || Schalttafelklemme f. || borne f. de jonction du tableau de distribution.

switch box (Electr) || Schaltkasten m. || tableau m. ou coffret m. de distribution; boîte f. de manœuvre; combinateur m. / ~ (Railw) || Weichenbock m. || boîte f. de manœuvre d'aiguille.

switchboy (Mine) || Weichensteller m. || aiguilleur m.

switch case || Schaltkasten m. || coffret m. de distribution. / ~ clock || Schaltuhr f. || minuterie f. de contact; commutateur m. à temps. / ~ column || Schaltsäule f. || colonne f. de distribution. / ~ construction works pl. (Railw) || Weichenbauanstalt f. || atelier m. de construction d'aiguillages. / ~ cover || Schalterdeckel m. || couvercle m. d'interrupteur. / ~ cupboard || Schaltschrank m. || cabine f. ou armoire f. de distribution. / ~ curve of ~ (Railw) || Weichenbogen m. || courbe f. du changement. / ~ desk || Schaltpult n. || pupitre m. de distribution. / ~ diagram || Schaltbild n.; Schaltplan m. || plan m. de montage ou de câblage.

switch frame (Electr) || Schaltrahmen m. || cadre m. du tableau de distribution. / ~ (Aut tel) || Wählergestell n. || bâti m. des sélecteurs.

switchgear || Schaltwerk n.; Schaltanlagengerät n. || appareillage m. de distribution; mécanisme m. de commande (d'un interrupteur). / enclosed ~ || eisengekapselte Schaltanlage f. || distribution f. sous enveloppe en tôle. / manufacture of electrical ~ || Herstellung f. von elektrischem Schaltzeug || fabrication f. de tout l'équipement de couplage électrique.

switch hut || Schalthäuschen n. || kiosque m. de distribution.

switching (Electr) || Schalten n.; Schaltung f. || distribution f.; commutation f. / electric brake ~ || elektrische Bremsschaltung f. || freinage m. rhéostatique. / parallel ~ || Parallelschalten n. || montage m. en parallèle. / primary ~ (Radio) || Primärschaltung f. || circuit m. primaire. / secondary ~ (Radio) || Sekundärschaltung f. || circuit m. secondaire.

switching device || Schaltwerk n. || dispositif m. de commutation ou de distribution. / ~ device for connection with Morse lines || Anschaltgerät n. für Morseleitungen || appareil m. pour intercepter une ligne Morse. / ~ engine (Railw) || Verschiebemaschine f.; Rangierlokomotive f. || coucou m.; machine f. de manœuvre. / ~ equipment see ~ device. / ~ handle || Arbeitskurbel f. || manivelle f. de manœuvre. / ~ key || Kippschalter m. || clef f. / ~ locomotive || Verschiebelokomotive f.; Rangierlokomotive f. || locomotive f. à faire les manœuvres ou à ranger; coucou m. / ~ member || Schaltorgan n. || membre de connexion. / ~ relay || Durchschalterelais n.; Durchschalteschutz n. || relais m. à relier. / ~ repeater selector (Aut tel) || Mitlaufwerk n. || sélecteur m. à permutation ou de commutation. / ~ selector repeater || Überbrückungsverstärker m. || répétiteur m. de commutation.

switching wheel || Schaltrad n. || roue f. de distribution ou de réglage. / ~ shaft || Schaltradwelle f. || axe m. de la roue de réglage.

switching work (Railw) || Verschiebedienst m. || service m. de manœuvre.

switch lever (Railw) || Stellvorrichtung f.; Stellbock m.; Stellhebel m.; Umleghebel m. || levier m. ou appareil m. de manœuvre. / ~ outside the rails || Umstellhebel m. außerhalb des Gleises || levier m. de manœuvre du côté extérieur de la voie.

switchman (Mine) || Weichensteller m. || aiguilleur m.

switch-off position (Electr) || Ausschaltstellung f. || position f. de disjonction.

switch-on position (Electr) || Einschaltstellung f. || position f. de fermeture.

switch operating mechanism (Railw) || Weichenstellvorrichtung f. || appareil m. de manœuvre d'aiguille.

switch, operation of a ~ ‖ Bedienung f. einer Weiche ‖ manœuvre m. d'aiguille.

switch plate (Electr) ‖ Schaltplatte f. ‖ plaque f. d'interrupteur. / ~ (Railw) ‖ Grundplatte f. der Weiche ‖ plaque f. d'assise d'aiguille.

switch plug ‖ Steckkontakt m. ‖ bouchon m. de contact; contact m. à fiches.

switch-rails pl. ‖ Versetzschienen fpl. ‖ rails mpl. mobiles.

switch resistance ‖ Vorschaltwiderstand m. ‖ résistance f. à intercaler.

switch rod (Electr) ‖ Schaltstange f. ‖ bielle f. d'attaque ou de commutation. / ~ (Railw) ‖ Verbindungsweiche f. ‖ aiguille f. de raccordement.

switch room ‖ Schalttafelraum m.; Schaltraum m. ‖ salle f. du tableau de distribution.

switch shaft ‖ Schaltarmstange f. ‖ barre f. du bras porte-balais.

switch shelf (Aut tel) ‖ Wählerrahmen m. ‖ rangée f. des sélecteurs.

switch signal (Railw) ‖ Weichensignal n. ‖ signal m. de branchement.

switch sleeper (Railw) ‖ Weichenschwelle f. ‖ traverse f. d'aiguille ou de changement.

switch sleepers pl. (Railw) ‖ Weichenrost m. ‖ châssis m. d'aiguille.

switch stand handle (Railw) ‖ Stellvorrichtung f. einer Weiche ‖ mécanisme m. de manœuvre des aiguilles.

switch station ‖ Schaltwarte f. ‖ poste m. de manœuvre. / main ~ ‖ Hauptschaltwarte f. ‖ poste m. principal de manœuvre.

switch step ‖ Schaltstufe f. ‖ degré m. d'interruption ou de commutation.

switch system ‖ Schaltsystem n. ‖ système m. d'interruption. / step-by-step ~ ‖ Schrittschaltesystem n. ‖ système m. pas-à-pas.

switch tongue ‖ Weichenzunge f.; Zunge f.; Weichenschiene f. ‖ aiguille f. (de changement de voie).

switch vessel ‖ Schaltertopf m. ‖ bâche f.

swivel ‖ Drehring m. ‖ rotule f. / ~ of a chain (Mar) ‖ Kettenwirbel m. ‖ émerillon m.

swivel arm chair ‖ Armstuhl m. mit verstellbarer Lehne ‖ fauteuil m. basculant. / ~ bridge ‖ Drehbrücke f. ‖ pont m. tournant. / ~ chair ‖ Drehstuhl m.; Drehsessel m. ‖ chaise f. à vis. / ~ doll ‖ Gelenkpuppe f. ‖ poupée f. articulée. / ~ frame (Railw) ‖ bewegliches Radgestell n. ‖ train m. de roues mobile autour d'une cheville. / ~ hook ‖ Wirbelhaken m. ‖ croc m. à émerillon.

swivelling bolster (Railw) ‖ Drehschemel m. / traverse f. mobile. / ~ constant speed pulley drive ‖ schwenkbarer Einscheibenantrieb m. ‖ commande f. par monopoulie à position variable. / ~ feeler ‖ schwingender Taster m. ‖ touche f. pivotante. / ~ V-block ‖ Schwenkprisma n. ‖ vé m. à bascule.

swivel link ‖ Kettenglied n. mit Wirbel ‖ maillon m. à émerillon. / ~ loom ‖ Broschierwebstuhl m. ‖ métier m. à brocher. / ~ ring ‖ Ringwirbel m. ‖ anneau m. à émerillon. / ~ roller ‖ schwenkbare Rolle f. ‖ galet m. à rotule; poulie f. à pivot. / ~ table ‖ Schwenktisch m. ‖ table f. pivotante.

sword ‖ Degen m.; Säbel m. ‖ épée f.; sabre m. / ~ blade ‖ Säbelklinge f.; Degenklinge f.; Fechtklinge f. ‖ lame f.

de sabre ou d'épée. / ~ cane ‖ Stockdegen m. ‖ canne f. à épée. / ~ furbisher ‖ Schwertfeger m. ‖ polisseur m. d'armes blanches.

sword-like ‖ schwertförmig ‖ ensiforme.

sword stick ‖ Degenstock m. ‖ canne f. épée.

sworn appraiser ‖ gerichtlich vereidigter Gutachter m. ‖ expert m. juré.

syenite (Miner) ‖ Syenit m. ‖ syénite f.

syllable ‖ Silbe f. ‖ syllabe f. / ~ typewriter ‖ Silbenschreibmaschine f. ‖ machine f. à écrire les syllabes.

sylvanite (Miner) ‖ Schrifterz n.; Schrifttellur n.; Sylvanit m.; Weißtellur n. ‖ or m. graphique; sylvane m.; tellure m. natif auro-argentifère.

sylvinite ‖ Sylvinit m. ‖ sylvinite f.

symbiotic fermentation ‖ symbiotische Gährung f. ‖ fermentation f. symbiotique.

symbol (Chem etc.) ‖ Zeichen n.; Symbol n.; Kurzzeichen n.; Sinnbild n. ‖ symbole m.; abréviation f. / ~ (Electr) ‖ Bildzeichen n. ‖ symbole m. graphique.

symbolt (Weav) ‖ Zampel m.; Zampelzug m. ‖ xemple m.; semple m.

symmetrical ‖ symmetrisch; spiegelgleich ‖ symétrique. / axial ~ ‖ achsensymmetrisch ‖ symétrique par rapport à l'axe.

symmetrical circuit (Electr) ‖ symmetrischer Stromkreis m. ‖ circuit m. symétrique. / ~ line (Electr) ‖ symmetrische Leitung f. ‖ ligne f. symétrique. / ~ profile ‖ spiegelgleiches Profil n. ‖ profil m. symétrique. / ~ section see ~ profile.

symmetry ‖ Symmetrie f.; Ebenmaß n.; Spiegelgleichheit f. ‖ symétrie f. / mirror ~ ‖ Spiegelgleichheit f. ‖ symétrie f. par rapport à un miroir.

symmetry axis ‖ Symmetrieachse f. ‖ axe m. de symétrie.

symmetry test ‖ Symmetrieprüfung f. ‖ mesure f. de symétrie.

sympathetic ink ‖ sympathetische Tinte f. ‖ encre f. sympathique. / ~ strike ‖ Sympathiestreik m. ‖ grève f. de sympathie.

symphytum root ‖ Schwarzwurzel f. ‖ racine f. de consoude; salsifis m. noir; scorsonère f.

synchronical see synchronous.

synchronism (Electr) ‖ Gleichzeitigkeit f.; Gleichlauf m.; Synchronismus m. ‖ synchronisme m. / to keep in ~ ‖ in Synchronismus m. bleiben ‖ être au synchronisme.

synchronization ‖ Synchronisierung f. ‖ synchronisation. / ~ of relay master clocks ‖ Synchronisierung f. von Relaishauptuhren ‖ synchronisation f. d'horloges principales à relais.

synchronize, to ‖ synchronisieren ‖ correspondre; synchroniser.

synchronized sound locator ‖ Richtungshörer m. ‖ artophone m.

synchronizing apparatus ‖ Synchronisierungseinrichtung f. ‖ installation f. de synchronisation. / ~ current ‖ synchronisierender Strom m. ‖ courant m. synchronisant. / ~ device ‖ Synchronisiereinrichtung f. ‖ appareil m. synchroniseur. / ~ power ‖ synchronisierende Kraft f. ‖ puissance f. synchronisante.

synchronous ‖ synchron ‖ synchrone. / to run ~ly ‖ synchron laufen ‖ marcher synchrone.

synchronous converter ‖ Einankerumformer m. ‖ commutatrice f.; convertisseur m. / ~ gap ‖ synchrone Funkenstrecke f. ‖ éclateur m. synchrone. / ~ measurement ‖ Messung f. zu gleichen Zeiten ‖ mesure f. à des intervalles de temps égaux. / ~ motor ‖ Synchronmotor m. ‖ moteur m. synchrone. / ~ position indicator ‖ Synchronoskop n. ‖ synchronoscope m. / ~ record (Talking picture) ‖ zeitgleich ablaufende Schallplatte f. ‖ disque m. tournant en synchronisme.

synclinal flexure (Of strata) ‖ Muldenfalte f. ‖ synclinal m. / ~ formation ‖ Muldenbildung f. ‖ formation f. synclinale.

syncopal ‖ synkopisch ‖ syncopique.

syndic ‖ Syndikus m. ‖ syndic m.

syndicate ‖ Konsortium n.; Syndikat n. ‖ consortium m.; syndicat m.; groupe m. / workmen's ~ ‖ Betriebsvertretung f. ‖ syndicat m. des ouvriers; représentation f. des employés.

synopsis ‖ Übersicht f. ‖ aperçu m.

synthesis (Chem) ‖ Synthese f.; Aufbau m. ‖ synthèse f.

synthetic(al) ‖ synthetisch; aufbauend ‖ synthétique. / ~ ammonia ‖ synthetisches Ammoniak m. ‖ ammoniaque f. synthétique. / ~ perfume ‖ synthetischer Riechstoff m. ‖ parfum m. synthétique. / ~ Peruvian balsam ‖ künstlicher Perubalsam m. ‖ baume f. du Pérou synthétique. / ~ precious stone ‖ synthetischer Edelstein m. ‖ pierre f. précieuse scientifique. / ~ product ‖ synthetisches Erzeugnis n. ‖ produit m. synthétique. / ~ stone ‖ synthetischer Stein m. ‖ pierre f. de synthèse.

synthetize, to ‖ synthetisieren ‖ faire la synthèse.

syntonic (Radio) ‖ abgestimmt ‖ syntonisé; accordé.

syntonization (Radio) ‖ Abstimmung f. ‖ syntonisation f.

syntonize, to ‖ abstimmen ‖ syntoniser.

syntonized wireless telegraphy ‖ abgestimmte drahtlose Telegraphie f. ‖ télégraphie f. sans fil syntonisée.

syntonizing coil ‖ Abstimmspule f. ‖ bobine f. de syntonisation; bobine f. syntonisatrice. / ~ inductance ‖ Abstimmindukتanz f. ‖ inductance f. de syntonisation.

syphilization (Med) ‖ Impfung f. mit Syphilisgift ‖ syphilisation f.

syphon see siphon.

syringe ‖ Spritze f.; Handspritze f. ‖ seringue f. / ~ (for enema) see rectal ~. / glass ~ ‖ Glasspritze f. ‖ seringue f. en verre. / lubricating ~ ‖ Ölspritze f. ‖ seringue f. à graissage. / rectal ~ ‖ Klistierspritze f. ‖ seringue f. à lavement. / rubber ~ ‖ Gummispritze f. ‖ seringue f. en caoutchouc. / ~ for vineyards ‖ Weinbergspritze f. ‖ seringue f. pour vignes.

syrup see also sirup ‖ Sirup m. ‖ sirop m.

system ‖ System n. ‖ système m. / ~ of accounts ‖ Rechnungswesen n. ‖ comptabilité f. / ~ of beds (Geol) ‖ Formation f.; Schichtenreihe f. ‖ formation f.; terrain m. / ~ of boring-rods ‖ Bohrgestänge n.; Gestänge n. für Tiefbohrungen ‖ tiges fpl. pour les sondages. / by-path ~ ‖ Kreislaufsystem n. ‖ système m. circulaire. / central nervous ~ ‖ Zentralnervensystem n. ‖ système m. nerveux central. / ~ of coordinates ‖ Achsenkreuz n.; Koordinatensystem n. ‖ système m.

d'axes *ou* de coordonnées. / to refer points to a ~ of coordinates ‖ die Punkte mpl. auf ein Koordinatensystem beziehen ‖ rapporter des points mpl. à un système de coordonnées. / ~ of crystallization ‖ Kristallsystem n. ‖ système m. cristallin. / ~ of horizontal and inclined sections (Mine) ‖ Querbau m. mit horizontalen geneigten Abbaustrecken ‖ ouvrage m. par tranches horizontales et inclinées. / ~ of lenses ‖ Linsensystem n. ‖ système m. de lentilles. / movable erecting ~ of lenses ‖ verstellbares bildumkehrendes Linsensystem n. ‖ système

m. de lentilles redresseur réglable *ou* mobile. / metric ~ ‖ metrisches System n. ‖ système m. métrique. / mixed ~ ‖ gemischtes System n.; gemischter Betrieb m. ‖ système m. mixte. / monometric ~ of crystals ‖ reguläres Kristallsystem n. ‖ système m. régulier *ou* cubique. / nervous ~ ‖ Nervensystem n. ‖ système m. nerveux. / ~ of pipes ‖ Rohrnetz n. ‖ réseau m. de tuyaux. / a ~ of pipes distributes the gas ‖ zur Verteilung des Gases dient ein Rohrnetz ‖ le service de distribution du gaz est fait par un réseau de tuyaux. / ~ of railroads ‖ Eisen-

bahnnetz n. ‖ réseau m. de chemins de fer. / semi-rigid ~ (Airship) ‖ halbstarres System n. ‖ système m. semi-rigide.

systematic ‖ planmäßig; systematisch ‖ systématique; méthodique.

systematically ‖ planmäßig ‖ méthodiquement.

systemless ‖ planlos; systemlos; unsystematisch ‖ sans systéme.

system stopping ‖ Systemarretierung f. ‖ arrêt m. du système.

syzygy (Astron) ‖ Konjunktion f. ‖ conjonction f.

T

T-antenna ‖ T-Antenne f. ‖ antenne f.

T-beam *see* T-iron.

T-head bolt (Build) ‖ Hammerschraube f. ‖ boulon m. de fondation. / ~ (Mach) ‖ Hakenschraube f. ‖ crampon m. à vis; boulon m. à crochet. / ~ with nose ‖ Hammerschraube f. mit Nase ‖ boulon m. de fondation avec nez.

T-headed bolt *see* T-head bolt.

T-iron ‖ T-Eisen n. ‖ fer m. (en) T. / double ~ ‖ Doppel-T-Eisen n.; T-Eisen n. ‖ fer m. en T double. / sharp-edged ~ ‖ scharfkantiges T-Eisen n. ‖ fer m. T à angles vifs.

T-joint (El line) ‖ Knotenverbindung f. ‖ raccordement m. en y.

T-slot for securing of . . . ‖ Aufspannschlitz m. zum Befestigen von . . . ‖ rainure f. destinée à la fixation de . . .

T-square (Draw) ‖ Reißschiene f. ‖ té m. (à dessiner); équerre f. / wooden ~ ‖ hölzerne Reißschiene f. ‖ té m. en bois.

tabby, to (Weav) ‖ moirieren; wässern ‖ moirer; tabiser.

table ‖ Tisch m. ‖ table f. / ~ (Mach tool) ‖ Auflagetisch m. ‖ plateau m. *ou* table f. porte-pièce. / ~ (Index) ‖ Tabelle f.; (übersichtliches) Verzeichnis n.; Register n. ‖ index m.; tableau m. / ~ (Plank) ‖ Planke f.; Diele f.; Brett n. ‖ planche f.; ais m. / alphabetical ~ ‖ abeceliches *oder* alphabetisches Verzeichnis n.; alphabetisches Register n. ‖ registre m. *ou* table alphabétique. / amalgamating ~ ‖ Amalgamiertisch m. ‖ table f. d'amalgamation. / ~ of contents ‖ Sachverzeichnis n. ‖ sommaire m. / distributing ~ ‖ Verteilertisch m. ‖ table f. doseuse. / fixed ~ (Mach tool) ‖ feststehender Aufspanntisch m. ‖ table f. de serrage fixe. / iron ~ ‖ eiserner Tisch m. ‖ table f. en fer. / lifting ~ (Roll mill) ‖ Hebetisch m. ‖ tablier m. releveur. / ~ of magnification ‖ Vergrößerungstabelle f. ‖ table f. des grossissements. / ~ of pictures ‖ Bildertafel f. ‖ tableau m. des illustrations. / picking ~ (Ore dress) ‖ Lesetisch m. ‖ table f. de triage. / plate supporting ~ ‖ Blechauflagetisch m. ‖ table f. pour placer la tôle. / printing ~ ‖ Drucktisch m. ‖ table f. d'impression. / ~ of rates ‖ Tarif m.; Preistabelle f. ‖ tarif m. / ~ of reduction ‖ Reduktionstabelle f. ‖ table f. de réduction. / roller ~ (Glassm) ‖ Walztisch m. ‖ table f. de laminoir. / roller gear ~ ‖ Roll-

tisch m. ‖ table f. à roulette. / simple adjustable ~ for supporting material ‖ verstellbarer einfacher Auflagetisch m. ‖ table f. de travail réglable et simple. / sleeping ~ *see* washing ~. / sorting ~ ‖ Verlesetisch m. ‖ table f. collectrice. / sweep ~ *see* washing ~. / swinging ~ (Shipb) ‖ Hängetisch m. ‖ table f. à roulis. / travelling ~ ‖ Lauftisch m. ‖ banc m. à glissières. / travelling ~ for sliding fitting shop work ‖ Wandertisch m. für Gleitmontagen ‖ table f. mobile pour montages au ruban. / turnover top ~ ‖ Wendetisch m. ‖ table f. renversable. / washing ~ (Dress ore) ‖ Glauchherd m.; Kehrherd m. ‖ table f. allemande *ou* à balais; table f. dormante.

table anvil ‖ Tischamboß m. ‖ enclume f. de serrurier. / ~ area (Mach toll) ‖ Tischfläche f. ‖ surface f. de table. / ~ beer ‖ Einfachbier n.; Dünnbier n. ‖ bière f. de table; petite bière f. / ~ board ‖ Tischplatte f. ‖ tablette f. de table. / ~ boiling apparatus ‖ Tischkocher m. ‖ réchaud m. de table. / ~ card ‖ Tischkarte f.; menu m. / ~ centre piece ‖ Tafelaufsatz m. ‖ surtout m. de table. / ~ clock ‖ Standuhr f. ‖ pendule f. de cheminée.

table cloth ‖ Tischtuch n. ‖ nappe f. / paper ~ ‖ Papiertischtuch n. ‖ nappe f. en papier.

table cloth clamp ‖ Tischtuchklammer f. ‖ pince-nappe f.; crampon m. pour nappes. / ~ clip *see* ~ clamp.

table cover ‖ Tischdecke f. ‖ tapis m. de table. / ~ (Knife and fork) ‖ Tischbesteck n. ‖ couvert m. de table. / coloured ~ ‖ bunte Tischdecke f. ‖ nappe f. de table en couleur.

table crockery ‖ Tischgeschirr n. ‖ vaisselle f. / ~ drill press ‖ Tischbohrmaschine f. ‖ machine f. à percer d'établi. / ~ fan ‖ Tischfächer m.; Tischventilator m.; Tischwindflügel m. ‖ ventilateur m. de table. / ~ flag ‖ Tischflagge m. ‖ pavillon m. pour table. / ~ floor (Build) ‖ Plattenbelag m. ‖ carrelage m.; dallage m. / ~ furniture ‖ Eßbesteck n.; Tischbesteck n. ‖ couvert m. (de table). / ~ glass ‖ Tafelglas n. ‖ verre m. à vitres; plaque f. de verre. / ~ hand vice ‖ Bankkloben m. ‖ étau m. à main avec agrafe. / ~ knife ‖ Tischmesser n.; Tafelmesser n. ‖ couteau m. de table. / ~ lamp ‖ Tischlampe f. ‖

lampe f. de table. / ~ leg ‖ Tischbein n. ‖ pied m. de table.

table linen ‖ Tafelwäsche f.; Tafelleinen n.; Tischwäsche f.; Tischzeug n. ‖ linge m. de table; lingerie f. de table. / damask ~ ‖ damastenes Tafelzeug n. ‖ linge m. damassé de table. / embroidered ~ ‖ gestickte Tischwäsche f. ‖ lingerie f. brodée de table.

table man (Roll mill) ‖ Walztischumsteller m. ‖ basculeur m. de table. / ~ mat ‖ Untersetzer m. ‖ dessous m. de plats. / ~ match box ‖ Tischzündholzbehälter m. ‖ porte-allumettes m. de table. / ~ napkin ‖ Mundtuch n.; Serviette f. ‖ serviette f. / ~ planing machine ‖ Tischhobelmaschine f. ‖ raboteuse f. à table mobile. / ~ requisites pl. *see* ~ service. / ~ saddle ‖ Tischträger m. ‖ support m. de la table. / ~ service ‖ Tafelgeschirr n.; Tischgeschirr n. ‖ service m. de table; vaisselle f. / ~ spindle ‖ Tischspindel f. ‖ arbre m. fileté de table.

table stand ‖ Tischstativ n. ‖ pied m. de table. / disjointing ~ ‖ zerlegbares Tischstativ n. ‖ pied m. de table démontable. / ~ for telescopes ‖ Tischstativ n. für Fernrohre ‖ pied m. de table pour lunettes.

tablet (Med) ‖ Tablette f. ‖ tablette f. / ~ (For writing) ‖ Schreibtafel f. ‖ tablette f. / ~ of slate ‖ Schieferplatte f.; Schiefertafel f. ‖ table f. d'ardoise; ardoise f.

tablet compressing machine ‖ Tablettenmaschine f. ‖ machine f. à faire les tablettes *ou* à comprimer les tablettes.

table telephone station ‖ Tischfernsprecher m. ‖ poste m. téléphonique mobile. / ~ top ‖ Tischplatte f. ‖ planche f. de la table.

tablet press ‖ Tablettenpresse f. ‖ presse f. à comprimer les tablettes.

table utensil ‖ Tafelgerät n. ‖ service m. de table. / ~ vice (Locksm) ‖ Tischkloben m.; Bankschraubstock m. ‖ étau m. d'établi. / ~ wine ‖ Tafelwein m. ‖ vin m. de table.

tabular ‖ tabellarisch; listenförmig ‖ tabulaire; en forme de tableau. / in ~ form *see* tabular. / ~ work (Print) ‖ Tabellensatz m. ‖ composition f. en forme de table.

tabulate, to ‖ tabellarisch zusammenstellen; übersichtlich einordnen ‖ cataloguer.

tabulated *see* tabular.

tabulating machine ‖ Tabulatormaschine f. ‖ machine f. à tabulateur. / ~ work ‖ tabellarische Arbeit f. ‖ exécution f. de tableaux.

tabulator ‖ Tabulator m. ‖ tabulateur m. / ~ button ‖ Tabulatorknopf m. ‖ bouton m. du tabulateur. / ~ stop rack ‖ Tabulatorzahnstange f. ‖ crémaillère f. d'arrêt du tabulateur.

tacheometer see tachometer.

tachograph ‖ Tachograf m. ‖ tachygraphe m.

tachometer ‖ Umlaufzähler m.; Geschwindigkeitsmesser m.; Tachometer n.; Drehzahlmesser m. ‖ compteur m. de tours ou de vitesse; tachymètre m.; comptetours m.; tachéomètre m. / distant-reading ~ ‖ Drehzahlmesser m. mit Fernablesung ‖ tachymètre m. à distance. / electric remote ~ ‖ elektrisches Ferntachometer n. ‖ télétachymètre m. électrique. / recording ~ ‖ Drehzahlschreiber m. ‖ tachymètre m. enregistreur. / reduction ~ ‖ Reduktionstachometer n. ‖ tachéomètre m. auto-réducteur.

tachometer connection ‖ Tachometeranschluß m. ‖ prise f. de tachymètre. / ~ drive ‖ Tachometerantrieb m. ‖ commande f. de tachymètre. / ~ level ‖ Nivelliertachometer n. ‖ niveau-tachymètre m.

tachometric survey ‖ Tachometrie f. ‖ levé m. tachymétrique; tachéométrie f.

tachygraphy ‖ Schnellschreiben n.; Geschwindschrift f. ‖ tachygraphie f.

tack ‖ Zwecke f.; Stift m.; kleiner Nagel m. ‖ pointe f.; cheville f. / ~ (Mar) ‖ Hals m.; Halse f. ‖ amure f.; lof m. / opposite ~ (Mar) ‖ Gegenbord m. ‖ contre-bord m. / ~ for paper hangers ‖ Tapetennagel m.; Schloßnagel m. ‖ clou m. de tapissier. / ~ for pipes ‖ Rohrnagel m. ‖ clou m. à roseaux. / slate ~ (Til) ‖ Schiefernagel m. ‖ clou m. ou pointe f. à ardoise.

tacker (Tail) ‖ Säumer m. ‖ ourleur m. à la main.

tackle (Mar; Hoisting gear) ‖ Takel n.; Talje f.; Flaschenzug m.; Kloben m. ‖ palan m. (à moufles); moufle f. / ~ (Mar; Rigging) ‖ Tauwerk n.; Takelwerk n.; Takelage f. ‖ cordage m. / ~ (Mine) ‖ Haspel m.; Förderhaspel m. ‖ treuil m.; cabestan m. / to underrun a ~ ‖ ein Takel n. klaren ‖ détordre un palan; défaire les tours mpl. d'un palan. / auxiliary (hoisting) ~ ‖ Hilfshubwerk n. ‖ treuil m. auxiliaire. / ground ~ (Mar) ‖ Grundgeschirr n. ‖ apparaux mpl. de mouillage. / main (hoisting) ~ ‖ Haupthubwerk n. ‖ treuil m. principal. / with only one fixed pulley ‖ Potenzflaschenzug m. ‖ moufle f. à une seule poulie fixe.

tackle block ‖ Flaschenzugkloben m. ‖ poulie f. à moufle. / ~ hook (Mar) ‖ Blockhaken m. ‖ crochet m. de palan. / ~ rope ‖ Hißtau n. ‖ corde f. de poulie.

tack-tackle (Mar) ‖ Halstalje f. ‖ palan m. d'amure.

taffeta ‖ Taft m.; Taffet m.; glatter Seidenstoff m. ‖ taffetas m.; foulard m. / flowered ~ ‖ geblümter Schwertaffet m. ‖ prussienne f. / plain ~ ‖ glatter Taft m. ‖ taffetas m. uni. / printed (warp) ~ ‖ bedruckter Taft m. ‖ taffetas m. imprimé. / shot ~ ‖ Glanztaft m. ‖ taffetas m. glacé. / watered shot ~ ‖ Schillertaft m. ‖ taffetas m. changeant.

taffeta ribbon ‖ Taftband n. ‖ ruban m. taffetas.

taffrail (Shipb) ‖ Heckreling f.; Heckbord m. ‖ bandinet m.

tag (Label) ‖ Anhängeschildchen n.; Anhängeetikett n. ‖ étiquette f. à attacher. / ~ (Needl) ‖ Schnürnadel f. ‖ passelacet m. / ~ (Shoe lace) ‖ Schnürsenkel m. ‖ cordelière f.; lacet m. / ~ (Tel) ‖ Lötöse f. ‖ lame f. de raccord. / ~ strip (Tel) ‖ Lötösenstreifen m. ‖ réglette f. de connexion ou de broches.

tail ‖ Schwanz m. ‖ queue f. / ~ (Hand vice) ‖ Stielklöbchen n. ‖ étau m. à queue. / ~ of a dressing machine (Mill) ‖ Überschlag m. der Sichtmaschine ‖ issue f. ou queue f. d'une bluterie. / ~ of a forge hammer ‖ Hammerschwanz m. ‖ queue f. d'un marteau ou d'un martinet. / ~ bay (Build) ‖ Ortfach n.; Balkenfach n. ‖ travée f. contiguë au mur. / ~ bay of a sluice (Hydr arch) ‖ Unterhaupt n. einer Schleuse ‖ queue f. d'écluse. / ~ block (Shipb) ‖ Schwanzblock m.; Steertblock m. ‖ poulie f. à fouet. / ~ edger (Sawm) ‖ Holzbesäumer m. ‖ coupeur m. de bouts. / ~ fin (Airpl) ‖ Kielflosse f.; Schwanzflosse f. ‖ plan m. de dérive. / ~-first aeroplane ‖ Entenflugzeug n. ‖ avion-canard m. / ~ flow of a barrage ‖ Abfluß m. einer Talsperre ‖ écoulement m. d'un barrage. / ~ gate of a sluice ‖ unteres Schleusentor n.; Untertor n. einer Schleuse ‖ porte f. d'aval ou de mouille (d'une écluse). / ~ hammer ‖ Schwanzhammer m. ‖ marteau m. à queue. / ~ heaviness ‖ Schwanzlastigkeit f. ‖ lourdeur f. de queue ou d'empennage.

tail-heavy (Nav; Aero) ‖ hecklastig; schwanzlastig ‖ chargé à la poupe; lourd de la queue; à queue lourde.

tailing motion (Loc) ‖ Hin- und Herschwanken n.; Schlängeln n. ‖ mouvement m. en lacet.

tailings pl. (Ore dress) ‖ Schlamm m.; Erzabfälle mpl.; Abgänge mpl. ‖ schlamm m.; matières fpl.; résidu m. de minerai; pertes fpl.

tail lamp ‖ Schlußlampe f.; Schlußlaterne f. ‖ lanterne f. arrière ou de queue. / ~ light ‖ Schlußlaterne f.; Schlußlicht n. ‖ lampe f. arrière.

tailor ‖ Schneider m. ‖ tailleur m. / ladies' ~ ‖ Damenschneider m. ‖ tailleur m. pour dames. / ~'s chalk ‖ Schneiderkreide f. ‖ craie f. pour tailleurs. / ~'s goose ‖ Schneiderplätteisen n.; Schneiderbügeleisen n. ‖ carreau m.; fer m. à repasser pour tailleurs. / ~'s presser ‖ Kleiderbügler m. ‖ presseur m. / ~'s scissors pl. ‖ Schneiderschere f. ‖ ciseaux mpl. de tailleur. / ~'s smoothing iron ~ ‖ goose. / ~'s trimming ‖ Kleiderposamenten pl. ‖ passementerie f. pour vêtements.

tail piece (Curr) ‖ Schwanzstück n. ‖ émouchet m. / ~ piece (Print) ‖ Zierleiste f.; Schlußleiste f.; Schlußvignette f.; Finalstock m. ‖ bordure f.; cul-de-lampe m. / ~ pin (Of gun) ‖ Schwanzschraube f. ‖ culasse f. de fusil. / ~ plane (Airpl) ‖ Schwanzfläche f.; Leitwerk n. ‖ surface f. de queue; empennage m. / ~ plane area ‖ Schwanzflächeninhalt m. ‖ surface f. de dérive. / ~ plane fin ‖ Leitwerksflosse f. ‖ dérive f. d'empennage. / ~ shaft bracket (Shipb) ‖ Schraubenwellenbock m.; Wellenbock m. ‖ sup-

port m. d'arbre porte-hélice. / ~ ski (Airpl) ‖ Schneesporn m. ‖ béquille f. à neige ou en forme de ski.

tailskid (Airpl) ‖ Sporn m.; Schwanzsporn m.; Schwanzkufe f. ‖ béquille f. / steerable ~ ‖ steuerbarer Sporn m. ‖ béquille f. réglable.

tailskid plate ‖ Spornteller m. ‖ sabot m. évasé de la béquille. / ~ shoe ‖ Spornschuh m. ‖ sabot m. de la béquille. / ~ spring ‖ Spornfeder f. ‖ ressort m. amortisseur de la béquille. / ~ springing ‖ Spornfederung f. ‖ amortisseur m. de la béquille.

tail spin (Aero) ‖ Abtrudeln n. ‖ vrille f.

tail starling (Bridge) ‖ Pfeilerhinterhaupt n.; Pfeilersterz m.; Talpfeilerkopf m. ‖ arrière-bec m.; bec m. d'aval.

tailstock (Mach tool) ‖ Reitstock m. ‖ contre-poupée f.; poupée f. mobile.

tail strut ‖ Leitwerkstrebe f. ‖ mât m. d'empennage. / ~ surface ‖ Leitwerk n. ‖ empennage m. / ~ type (Print) ‖ geschwänzte Schrift f. ‖ lettre f. à queue. / ~ unit see ~ surface. / ~ water port ‖ Unterhafen m. ‖ port m. inférieur. / ~ wheel (Airpl) ‖ Spornrad n. ‖ roue f. de béquille.

take, to ~ an angle ‖ einen Winkel m. messen oder aufnehmen ‖ mesurer ou relever ou observer un angle. / ~ apart ‖ zerlegen ‖ démonter; désassembler. / ~ bearings pl. ‖ peilen; visieren ‖ faire des relèvements mpl.; relever. / ~ a clock to pieces ‖ eine Uhr f. auseinandernehmen ‖ démonter une pendule. / ~ down a building ‖ einen Bau m. abbrechen ‖ démolir un bâtiment. / ~ the finishing cut ‖ nachdrehen ‖ repasser au tour. / ~ a heave (Mine) ‖ verworfen sein; verschoben sein ‖ être dérangé. / ~-in loading (Mar) ‖ laden; Ladung f. einnehmen ‖ charger; prendre chargement. / ~-in a sail ‖ ein Segel bergen oder einnehmen ‖ serrer une voile. / ~ legal steps against . . . ‖ verklagen; den Klageweg m. beschreiten gegen . . . ‖ intenter une action contre . . .; porter plainte contre . . . / ~ off (Aero) ‖ starten; aufsteigen; loskommen ‖ monter; s'élever. / ~ off (Prices) ‖ ablassen ‖ rabattre. / ~ off the edges ‖ die Kanten fpl. brechen ‖ chanfreiner. / ~ off the furniture (Print) ‖ das Format n. abschlagen ‖ abattre ou enlever la garniture. / ~ off a sheet (Print) ‖ einen Korrekturbogen m. abziehen ‖ tirer une épreuve. / ~ off the slags ‖ die Schlacken abstechen ‖ retirer les scories (avec le fourgon). / ~ off the receiver (Tel) ‖ den Hörer m. abnehmen oder abheben ‖ enlever ou décrocher le récepteur. / ~ off the tenter (Clothm) ‖ das Tuch vom Rahmen abnehmen ‖ dérainer le drap. / ~ off on water (Aero) ‖ abwassern ‖ demerrir. / ~ on rapidly (Trade) ‖ reißend abgehen ‖ être d'un bon débit. / ~ out (Found) ‖ herausnehmen ‖ démouler. / ~ out of the press (Print) ‖ aus der Presse nehmen ‖ dépresser. / ~ out the wedge ‖ den Keil m. herausziehen ‖ décoincer. / ~ over a business ‖ ein Geschäft n. übernehmen ‖ se charger d'une affaire. / ~ over part of the stock ‖ einen Aktienanteil m. übernehmen ‖ signer un certain nombre d'actions. / ~ over a power station. ‖ ein Kraftwerk n. übernehmen ‖ prendre à sa charge une usine. / ~ a plan ‖ einen

Grundriß m. aufnehmen ‖ lever un plan m. / ~ a purchase with the lever ‖ mit dem Hebebaum kanten *oder* umlegen ‖ faire un abatage. / ~ samples pl. ‖ Probe nehmen ‖ échantillonner. / ~ sheers (Mar) ‖ (vor Anker liegend) scheren ‖ rôder. / ~ the size ‖ kalibrieren ‖ calibrer. / ~ up (bill) ‖ honorieren ‖ honorer; rémunérer.

Take care! ‖ Achtung!; Obacht!; Vorsicht! ‖ attention!; garde!

take-in grips pl. ‖ Einspannvorrichtung f. mit Klauen ‖ pinces fpl. de serrage.

take-off distance (Aero) ‖ Startlänge f. ‖ distance f. parcourue en vol.

taker of a bill ‖ Wechselnehmer m. ‖ preneur m. d'une lettre de change.

take-up roller (Spinn) ‖ Wickelwalze f.; Aufwickelvorrichtung f. ‖ rouloir m. / ~ (Weav) ‖ Abzugswalze f.; Zugbaum m. ‖ rouleau m. de tirage *ou* d'appel.

taking of an inventory *see* stock ~. / stock ~ ‖ Inventur f.; Inventuraufnahme f.; Bestandaufnahme f.; Aufnahme f. eines Warenlagers ‖ inventaire f.; levée f. d'inventaire.

taking-in (Weav) ‖ Einzug m. ‖ rentrée f. du chariot.

taking-off of the gases of blast-furnaces ‖ Gichtgasentziehung f. ‖ prise f. des gaz du haut-fourneau.

taking-over of material ‖ Werkstoffabnahme f. ‖ réception f. de matériaux.

taking-over test ‖ Abnahmeprüfung f. ‖ épreuve f. de réception.

taking-up much time ‖ zeitraubend ‖ qui prend du temps.

taking-up of the load by the individual power stations ‖ Einsatz m. der einzelnen Kraftwerke ‖ interconnection f. des différentes usines. / ~ of the wire over the poles ‖ Auflegen n. des Drahtes auf die Telegraphenstangen ‖ montage m. du fil sur poteaux.

talc ‖ Talk m.; Talkum n.; spanische Kreide f. ‖ talc m.; craie f. d'Espagne. / earthy ~ ‖ Erdtalk m. ‖ talc m. terreux.

talc alum ‖ Talkerdealaun m.; Magnesiaalaun m. ‖ alun m. de magnésie. / ~ quarry ‖ Talksteingrube f. ‖ carrière f. de talc *ou* de stéatite. / ~ schist ‖ Talkschiefer m. ‖ schiste m. talqueux. / ~ stone ‖ Talkstein m. ‖ pierre f. talqueuse.

talcum powder ‖ Talkum n. ‖ talc m.

talk, to ‖ sprechen ‖ parler. / to cross-talk ‖ übersprechen ‖ diaphoner.

talkie *see also* talking film ‖ Tonfilm m.; Klangfilm m. ‖ film m. parlant *ou* sonore. / ~ cinema projector ‖ Tonfilmvorführungsvorrichtung f.; Tonfilmapparatur f. ‖ appareil m. de projection pour films sonores.

talking connection (Tel) ‖ Sprechverbindung f. ‖ communication f. de conversation. / ~ current ‖ Sprechstrom m. ‖ courant m. de conversation. / ~ film *see also* talkie ‖ Sprechfilm m.; Klangfilm m. ‖ film m. parlant *ou* sonore.

talking machine ‖ Sprechmaschine f.; Sprechapparat m.; Grammofon n. ‖ machine f. parlante; phonographe m. / ~ case ‖ Sprechmaschinengehäuse n. ‖ armoire f. *ou* coffret m. de machine parlante. / ~ needle ‖ Nadel f. für Sprechmaschinen ‖ aiguille f. de machine parlante. / ~ parts pl. ‖ Sprechmaschinenbestandteile mpl.; Bestandteile mpl. von Sprechmaschinen ‖ pièces fpl. détachées

de phonographes. / ~ spring ‖ Grammofonfeder f. ‖ ressort m. de machine parlante *ou* de grammophone. / ~ spring motor ‖ Sprechmaschinenlaufwerk n. ‖ mécanisme m. de machine parlante.

talking set (Tel) ‖ Sprechvorrichtung f.; Sprechapparat m. ‖ appareil m. téléphonique.

talking station ‖ Sprechstelle f. ‖ poste m. téléphonique.

tallow, to ‖ eintalgen ‖ ensuifer. / ~ leather (Over a charcoal fire) ‖ Leder n. abflammen ‖ donner le suif au cuir.

tallow ‖ Talg m. ‖ suif m. / ~ (Mar) ‖ Schmiere f. ‖ graisse f.; gras m. / beef ~ ‖ Rindstalg m.; Rindertalg m. ‖ suif f. de bœuf. / rendered ~ ‖ ausgelassener Talg m. ‖ suif m. fondu. / smelted ~ ‖ Speisefett n. ‖ graisse f. alimentaire. / unmelted ~ ‖ Fleischertalg m. ‖ suif m. de boucher.

tallow candle ‖ Talglicht n.; Talgkerze f.; Unschlittkerze f. ‖ chandelle f. en suif. / ~ moulder ‖ Talglichtgießer m. ‖ mouleur m. de chandelles en suif.

tallow cup ‖ Talgnapf m. ‖ godet m. à suif.

tallowed leather ‖ Sattlerleder n. ‖ cuir m. en suif.

tallow grieves pl. ‖ Talggrieben fpl. ‖ cretons mpl. de suif.

tallowing ‖ Eintalgen n. ‖ action f. d'ensuifer.

tallow melter ‖ Talgsieder m. ‖ fondeur m. de suif. / ~ melting house ‖ Talgschmelzerei f. ‖ fondoir m. de suif. / ~ melting plant ‖ Talgschmelzanlage f. ‖ installation f. de fonderie de suif. / ~ pencil ‖ Fettstift m. ‖ crayon m. de suif. / ~ preparation ‖ Talgpräparat n. ‖ préparation f. de suif. / ~ refinery ‖ Talgraffinerie f. ‖ raffinerie f. de suif. / ~ smelter ‖ Talgsieder m. ‖ fondeur m. de suif. / ~ soap ‖ Talgseife f. ‖ savon m. au suif.

talmi gold ‖ Talmigold n. ‖ talmi m.

talon ‖ Erneuerungsschein m. ‖ talon m.

talus ‖ Böschung f.; Abdachung f. ‖ talus m.; pente f. / angle of the ~ ‖ Abdachungswinkel m. ‖ angle m. de la pente.

tamarind ‖ Tamarinde f.; Tamarindenfrucht f. ‖ tamarin m. / ~ jam ‖ Tamarindenmus n. ‖ marmelade f. de tamarins.

tambour ‖ Trommel f. ‖ tambour m.

tambourine ‖ Tamburin n. ‖ tambourin m.

tambour needle (Textile) ‖ Häkelnadel f.; Tamburiernadel f. ‖ aiguille f. à crochet.

tambour weir (Hydr arch) ‖ Trommelwehr n. ‖ barrage m. à tambour.

taminy ‖ Etamin m. ‖ étamine f.

tam o'shanter ‖ Matrosenmütze f.; Tellermütze f. ‖ béret m.

tamp, to ‖ stampfen; festtreten ‖ damer; fouler. / ~ the bore hole ‖ das Bohrloch versetzen ‖ bourrer le trou du pétard. / ~ a mine ‖ eine Mine verdämmen ‖ bourrer une mine.

tamp ‖ Stampfer m. ‖ dame f. / iron ~ ‖ eiserner Stampfer m. ‖ dame f. en fonte. / wooden ~ ‖ Holzstampfer m. ‖ dame f. en bois.

tamped concrete ‖ Stampfbeton m. ‖ béton m. damé. / ~ sleeper ‖ unterstopfte Schwelle f. ‖ traverse f. bourrée.

tamper, to ‖ verfälschen ‖ falsifier; altérer.

tamper (Railw) ‖ Stampfer m.; Stopfer m. ‖ bourroir m. / cylinder of the ~ ‖ Stopferkolben m. ‖ cylindre m. de bourroir.

tampering pick ‖ Stopfhacke f. ‖ pic m. à bourrer.

tamping of the bore hole ‖ Versetzen n. des Bohrloches ‖ bourrage m. du trou du pétard. / ~ of a mine ‖ Verdämmung f. einer Mine ‖ bourrage m. d'une mine. / ~ of railway sleepers ‖ Stopfen n. von Eisenbahnschwellen ‖ bourrage m. des traverses.

tamping clay ‖ Stampfmasse f. ‖ pisé m. réfractaire damé. / (ordinary) ~ pick ‖ Handstopfhacke f. ‖ pioche f. ordinaire. / ~ speed ‖ Stopferleistung f. ‖ vitesse f. de bourrage. / ~ tool ‖ Stopfwerkzeug n. ‖ bourroir m.

tan, to ‖ rotgerben; gerben; lohen; lohgerben ‖ tanner; passer en tan.

tan ‖ Eichenlohe f.; Gerberlohe f.; Lohe f. ‖ tan m. / lixiviated ~ ‖ ausgelaugte Gerberlohe f. ‖ tan m. lessivé.

tan ball ‖ Lohkuchen m.; Lohballen m. ‖ motte f. à brûler *ou* de tan; briquette f. de tan. / ~ bark *see* tan. / ~ brick *see* ~ ball. / ~ cake *see* ~ ball. / ~ colour / Lohfarbe f. ‖ couleur f. de tan.

tandem ‖ hintereinander; einer hinter dem anderen ‖ en tandem. / placed in ~ ‖ hintereinander angeordnet ‖ installé en tandem.

tandem cylinders pl. ‖ Zylinder mpl. in Tandemanordnung ‖ cylindres mpl. en tandem. / ~ motor roller ‖ Motortandemwalze f. ‖ rouleau m. tandem à moteur. / ~ office (Tel) ‖ Durchgangsamt n.; Knotenamt n. ‖ bureau m. tandem; bureau m. de concentration. / ~ operation (Tel) ‖ Tandembetrieb m. ‖ exploitation f. en tandem. / ~ reversing rolling mill engine ‖ Tandemreversierwalzenzugmaschine f. ‖ machine f. tandem réversible de laminoir.

tandem roller ‖ Tandemwalze f. ‖ rouleau m. tandem. / superheated steam ~ ‖ Heißdampftandemwalze f. ‖ rouleau m. compresseur à tandem à vapeur surchauffée.

tandem seat ‖ Doppelsitz m. ‖ siège f. tandem. / ~ steam engine ‖ Tandemdampfmaschine f. ‖ machine f. à vapeur tandem. / ~ steam road roller ‖ Dampftandemwalze f. ‖ rouleau m. tandem à vapeur. / ~ system (Steam eng) ‖ Tandemsystem n. ‖ système m. en tandem.

tang (Letter-f) ‖ Gußbart m. ‖ talus m. / ~ (File) ‖ Angel f. ‖ soie f. / the ~ is that part of a file, which is inserted into the handle ‖ die Angel f. ist der Teil einer Feile, der in den Griff eingesetzt ist ‖ la soie f. est la partie d'une lime sur laquelle se monte la poignée *ou* le manche.

tangent ‖ Berührende f.; Berührungslinie f.; Tangente f. ‖ tangente f. / ~ compass (Electr) ‖ Tangentenbussole f. ‖ boussole f. des tangentes. / ~ galvanometer *see* ~ compass.

tangential ‖ tangential ‖ tangentiel. / ~ admission ‖ tangentiale Beaufschlagung f. ‖ adduction f. tangentielle. / ~ admission of scavenging air ‖ tangentiale Zuführung f. der Spülluft ‖ amenée f. tangentielle de l'air de balayage. / ~ stress ‖ Schubspannung f. ‖ cisaillement m.; tension f. tangentielle.

tangent key(ing) ‖ Tangentkeil m. ‖ clavette f. tangentielle. / ~ keyway for shock-like alternating thrust ‖ Tangentkeilnut f. für stoßartigen Wechseldruck ‖ rainure f. de clavette tangentielle pour pression alternative par à-coups.

/ ~ point of a curve ‖ Tangentialpunkt m. einer Kurve ‖ point m. de tangente d'une courbe. / ~ screw (Of an instrument) ‖ Tangentenschraube f. ‖ vis f. tangente.

tangible (Goods) ‖ greifbar ‖ saisissable; palpable.

tangle ‖ Verwicklung f.; Gewirr n.; Verknotung f. ‖ entortillement m.; embrouillement m. / to form ~s pl. ‖ Schlingen fpl. bilden ‖ former des serpentins mpl.

tan house (Curr) ‖ Lohhaus n. ‖ écorcier m.

tank, to ~ (Beer) ‖ fassen ‖ entonner; traverser.

tank (Vessel) ‖ Tank m.; Behälter m.; Gefäß n. ‖ réservoir m.; cuve f.; tank m. / ~ (Cistern) ‖ Zisterne f.; Wasserbchälter m. ‖ citerne f. / ~ (Coop) ‖ Äscher m. ‖ bac m. / ~ (Mach) ‖ Wasserbehälter m. ‖ bâche f. / ~ (War mat) ‖ Kampfwagen m.; Raupenkampfwagen m.; Tank m. ‖ char m. d'assaut ou de combat; tank m. / benzine ~ ‖ Benzintank m. ‖ réservoir m. à essence. / ~ for chemical industries ‖ Gefäß n. für Chemikalien ‖ récipient m. pour produits chimiques. / clarifying ~ ‖ Läuterungsbecken n. ‖ bassin m. de décantation. / compound ~ for cable manufacturing ‖ Massebehälter m. für die Kabelherstellung ‖ récipient m. de masse pour la fabrication de câbles. / compressed-air ~ ‖ Druckluftkessel m. ‖ réservoir m. d'air comprimé. / decanting ~ ‖ Schlammbehälter m. ‖ cuve f. de décantation. / emergency ~ ‖ Nottank m. ‖ réservoir m. de secours. / feed (water) ~ ‖ Speisewasserbehälter m. ‖ réservoir m. (d'eau) d'alimentation. / female ~ (Arm) ‖ weiblicher Kampfwagen m. oder Tank m.; Maschinengewehrtank m. ‖ char m. mitrailleuse; tank m. femelle. / forming ~ (Acc) ‖ Formiergefäß n. ‖ récipient m. de formation. / fuel ~ ‖ Brennstoffbehälter m. ‖ réservoir m. à combustible. / fuel reserve ~ (Aut) ‖ Kraftstoffhilfsbehälter m. ‖ réservoir m. à combustible de réserve. / gravity ~ (Airpl) ‖ Hochbehälter m. ‖ réservoir m. en charge. / half-track ~ (War mat) ‖ Räderraupenkampfwagen m. ‖ char m. d'assaut à roues et chenilles; char m. de combat composite; tank m. composite. / heavy ~ (War mat) ‖ schwerer Kampfwagen m.; Durchbruchtank m. ‖ char m. de combat lourd ou de rupture; tank m. écraseur. / hermaphrodite ~ see half-track ~. / high-level service ~ (for water) (Hydr arch) ‖ Hochbehälter m. ‖ réservoir m. à distribution d'eau. / light ~ (War mat) ‖ leichter Kampfwagen m. ‖ char m. (de combat) leger. / ~ for liquids ‖ Flüssigkeitsbehälter m. ‖ réservoir m. à liquides. / machine-gun ~ (War mat) ‖ weiblicher Kampfwagen m. oder Tank m.; Maschinengewehrtank m. ‖ char m. mitrailleuse; tank m. femelle. / main ~ ‖ Hauptbehälter m.; Hauptbecken n. ‖ réservoir m. principal. / male ~ (War mat) ‖ männlicher Tank m. oder Kampfwagen m.; mit Kanonen bestückter Kampfwagen m. ‖ char m. canon. / medium ~ (War mat) ‖ mittelschwerer Kampfwagen m. ‖ char m. médium; médium tank m. / oil ~ ‖ Ölbehälter m. ‖ réservoir m. à huile. / radiator ~ ‖ Wasserkasten m. des Kühlers ‖ réservoir

m. du radiateur. / ~ equipped with radio ‖ Rundfunktank m.; Radiokampfwagen m. ‖ char m. équipé de télégrapie sans fil. / reconnaissance ~ ‖ Erkundungskampfwagen m. ‖ char m. de reconnaissance. / ~ of reinforced concrete ‖ Behälter m. aus Eisenbeton ‖ réservoir m. en béton armé. / reserve ~ ‖ Hilfsbehälter m.; Vorratsbehälter m. ‖ réservoir m. de secours ou de réserve. / riveted iron ~ ‖ eiserner genieteter Behälter m. ‖ réservoir m. en fer riveté. / settling ~ ‖ Klärbecken n.; Klärbehälter m.; Klärteich m.; Klärbassin n. ‖ bassin m. de décantation ou de curage. / water ~ ‖ Wasserbehälter m.; Wasserbecken n.; Wasserkasten m. ‖ réservoir m. à eau. / ~ mounted on wheel trains ‖ auf Radsätze montierter Behälter m. ‖ réservoir m. monté sur trains de roues.

tank armour ‖ Kampfwagenpanzer m. ‖ cuirasse f. de char de combat. / ~ capacity ‖ Behälterinhalt m. ‖ contenance f. du réservoir. / ~ car (Railw) ‖ Kesselwagen m.; Zisternenwagen m. ‖ wagon-citerne m. / ~ destroyer (War mat) ‖ Kampfwagenjäger m. ‖ Tankzerstörer m.; contre-tank m. / ~ engine ‖ Tenderlokomotive f. ‖ locomotive f. à tender.

tanker ‖ Tankschiff n. für Petroleum; Petroleumtankdampfer m. ‖ pétrolier m.; bateau-citerne m.; bateau m. pétrolier; navire-tank m. à pétrole.

tankette (War mat) ‖ Kleinkampfwagen m. ‖ tankette f.

tank furnace (Glassm) ‖ Wanneofen m. ‖ four m. à bassin.

tanking (Of beer) ‖ Fassen n. ‖ entonnement m.; traversage m.

tank locomotive ‖ Tenderlokomotive f. ‖ locomotive f. tender. / articulated ~ ‖ Drehgestelltenderlokomotive f. ‖ locomotive-tender f. à essieux couplés articulés.

tank maker ‖ Tankmacher m. ‖ constructeur m. de réservoirs. / ~ motor lorry ‖ Kesselauto n.; Kesselkraftwagen m. ‖ camion m. citerne automobile. / ~ plant ‖ Tankanlage f. ‖ installation f. de réservoirs. / ~ sheet iron ‖ Behälterblech n. ‖ tôle f. à réservoir. / ~ steamer see tanker. / ~ system (Loc) ‖ Tanksystem n. ‖ système m. à tender. / ~ tracks pl. ‖ Tankraupen fpl.; Kampfwagenraupen fpl. ‖ chenilles fpl. de char de combat. / elasticity of ~ tracks ‖ elastische Nachgiebigkeit f. der Tankraupen ‖ souplesse f. des chenilles de char de combat. / ~ truck ‖ Kesselwagen m.; Behälterwagen m. ‖ voiture f. réservoir ou citerne. / ~ wagon (Railw) see also ~ car ‖ Kesselwagen m.; Tankwagen m.; Faßwagen m. ‖ wagon m. citerne; wagon m. foudre.

tan mill ‖ Lohmühle f. ‖ moulin m. à tan.

tanned ‖ gegerbt ‖ tanné.

tannentite (Miner) ‖ Kupferwismuterz n.; Kupferwismutglanz m. ‖ bismuth m. sulfuré cuprifère; emplectite f.

tanner ‖ Lohgerber m.; Gerber m.; Rotgerber m. ‖ tanneur m. / ~'s mordant ‖ Gerbereibeize f. ‖ mordant m. de tannerie. / ~'s tools pl. ‖ Lohgerbergeräte npl.; Lohgerberwerkzeuge npl. ‖ outils mpl. du tanneur.

tannery ‖ Gerberei f.; Lohgerberei f. ‖ tannerie f. / electric ~ ‖ elektrische Gerberei f. ‖ tannage m. électrique.

tannery machine ‖ Gerbereimaschine f. ‖ machine f. pour tanneries.

tannic acid ‖ Galläpfelsäure f.; Gerbsäure f.; Tannin n.; Gerbstoff m. ‖ acide m. gallotannique ou gallique ou tannique; tannin m.

tannin see also tannic acid ‖ Gerbstoff m.; Tannin n. ‖ tannin m. / vegetable ~ ‖ Pflanzentannin n. ‖ tannin m. végétal.

tannin acid see tannic acid. / ~ extract ‖ Gerbstoffauszug m. ‖ extrait m. tannant. / ~ producing device ‖ Gerbstofferzeugungseinrichtung f. ‖ installation f. pour la production des matières tannantes.

tanning ‖ Gerben n.; Lohgerben n.; Lohgerberei f. (Verfahren) ‖ Rotgerben n.; tannage m. / taken off by ~ ‖ abgegerbt ‖ tanné. / bark ~ ‖ Lohgerben n. ‖ tannage m. aux écorces. / chrome ~ ‖ Chromgerbung f. ‖ tannage m. au chrome. / Danish ~ ‖ dänische Gerberei f. ‖ sippage m.; apprêt m. à la danoise. / electric ~ ‖ elektrisches Gerben n. ‖ tannage m. à l'électricité. / rapid ~ ‖ Schnellgerben n. ‖ tannage m. rapide.

tanning bark ‖ Gerbrinde f. ‖ écorce f. à tan.

tanning extract ‖ Gerbstoff m.; Gerbstoffauszug m. ‖ extrait m. tannique. / chromium ~ ‖ Chromgerbauszug m. ‖ extrait m. de chrome tannant.

tanning machine ‖ Gerbereimaschine f. ‖ machine f. de tannerie. / ~ material ‖ Gerbstoff m. ‖ matériel m. tannant. / materials pl. for ~ ‖ Rohstoffe mpl. zum Gerben ‖ matières fpl. premières pour le tannage. / ~ vat ‖ Gerbfaß n. ‖ tonneau m. de tannage.

tan pit ‖ Versetzgrube f. ‖ fosse f. de tanneur.

tansy herb ‖ Rainfarnkraut n. ‖ herbe f. de tanaisie. / ~ oil ‖ Rainfarnöl n. ‖ essence f. de tanaisie.

tantalum ‖ Tantal n. ‖ tantale m. / ~ lamp ‖ Tantallampe f. ‖ lampe f. au tantale.

tan turf see tan ball ‖ Lohkuchen m. ‖ motte f. de tan. / ~ vat ‖ Lohkufe f. ‖ emprimerie f. / ~ yards pl. ‖ Lohefabrik f. ‖ fabrique f. de tan.

tap, to ~ (Brew) ‖ anstecken; anzapfen; anbohren ‖ mettre en perce f.; ~ (Found) ‖ abstechen ‖ couler; percer. / ~ (To cut a thread) ‖ Gewinde n. schneiden ‖ tarauder. / ~ (Typewr) ‖ tippen; Schreibmaschine schreiben ‖ taper. / ~ the blastfurnace ‖ den Hochofen m. stechen oder abstechen ‖ percer le haut-fourneau; faire la percée du haut-fourneau; faire couler la fonte. / ~ the cinder ‖ die Schlacke abstechen ‖ faire écouler le laitier. / ~ the iron ‖ Eisen abfangen ‖ recevoir la fonte. / ~ a source of power in an economical manner ‖ eine Kraftquelle f. in wirtschaftlichster Weise erschließen ‖ soumettre une source f. énergétique à l'exploitation la plus rationnelle.

tap (Cock) ‖ Anstichhahn m.; Ansteckhahn m.; Anzapfhahn m.; Faßhahn m. ‖ robinet m. de mise en perce. / ~ (For cutting a thread) ‖ Gewindebohrer m. ‖ taraud m. / ~ (Found) ‖ Abstich m. ‖ coulée. / ~ (Pin) ‖ Zapfen m. ‖ cheville f. / to set ~s pl. see to tap (Brew). / to start ~s pl. see to tap (Brew). / setting ~s pl. (Brew) ‖ Anzapfen n. ‖ mise f. en perce; purge f. des robinets. / starting ~s pl. see setting ~s. / adjustable ~ ‖ nachstellbarer Gewinde-

bohrer m. ‖ taraud m. ajustable. / master ~ ‖ Normalbohrer m.; Backenbohrer m.; Originalbohrer m. ‖ taraud-mère m. / original ~ see master ~. / plug ~ see master ~. / screw ~ ‖ Gewindebohrer m. ‖ taraud m. / taper ~ see screw ~. / three-fluted ~ ‖ dreimal genuteter Gewindebohrer m. ‖ taraud m. à trois rainures.

tap bar (Found) ‖ Stange f. zum Verschließen des Stichloches ‖ serrière f. / ~ binding (Aerial lines) ‖ Abzweigbund m. ‖ ligature f. de raccordement. / ~ borer ‖ Zapfenbohrer m. ‖ vrille f. / ~ cinder ‖ Rohschlacke f.; Puddelschlacke f. ‖ scorie f. pauvre ou crue; laitier m. pauvre scorie f. des fours à puddler.

tape ‖ Band n. ‖ ruban m.; bande f. / insulating ~ ‖ Isolierband n. ‖ ruban m. isolant ou isolateur. / adhesive insulating ~ ‖ klebriges Isolierband n. ‖ ruban m. isolant adhésif; bande f. isolante collante. / ~ of paper ‖ Papierstreifen m. ‖ bande f. de papier. / tracing ~ (Build) ‖ Absteckleine f. ‖ cordeau m. à tracer.

tape antenna ‖ Bandantenne f. ‖ antenne f. à ruban. / ~ band stitcher ‖ Riemennäherin f. ‖ couseuse f. de courroies. / ~ galvanizing plant ‖ Bandgalvanisieranlage f. ‖ installation f. de dépôts galvaniques sur feuillards.

tapeless rotary machine ‖ bänderlose Rotationsmaschine f. ‖ machine f. rotative sans cordon.

tape-line (Surv) ‖ Meßband n. ‖ mesure f. en ruban; ruban-mesure m.; cordeau m. d'arpenteur.

tape-measure ‖ Bandmaß n.; Rollmaß n. ‖ mesure f. en ruban; mètre m. à ruban. / spring ~ ‖ Bandmaß n. oder Rollmaß n. mit Feder ‖ mesure f. en ruban avec ressort.

taper, to ‖ (sich) verjüngen; (kegelförmig oder konisch) zuspitzen ‖ (s')effiler; (se) rétrécir; ajuster ou faire cône; donner du cône. / the shaft tapers toward the top and the bottom ends ‖ der Schaft verjüngt sich nach oben und unten ‖ l'arbre va en se diminuant vers le haut et le bas.

taper ‖ spitz; spitz zulaufend; angespitzt ‖ pointu; appointé.

taper ‖ Verjüngung f. ‖ rétrécissement m. / ~ (Of a socket) ‖ Muffenübergang m. ‖ raccord m. conique. / ~ (Wax candle) ‖ Wachskerze f. ‖ bougie f. de cire. / drawn ~ ‖ Wachsstock m. ‖ bougie f. filée. / ~ of key ‖ Keilanzug m. ‖ serrage m. de la clavette.

taper bore ‖ konische Bohrung f. ‖ alésage m. conique. / internal grinding of ~s ‖ Innenschliff m. konischer Rundflächen ‖ rectification f. intérieure de surfaces circulaires coniques.

tapered see also tapering ‖ verjüngt; zugespitzt ‖ effilé; conique; diminué. / ~ end ‖ verjüngtes Ende n. ‖ rétrécissement m. de l'extrémité. / ~ machinehandle ‖ Kegelgriff m. ‖ poignée f. conique. / ~ pin with threaded end ‖ Kegelstift m. mit Gewindezapfen ‖ goupille f. conique avec tourillon fileté. / ~ shoulder ‖ konischer Ansatz m. ‖ collet m. conique. / ~ square ‖ kegeliges Vierkant n. ‖ carré m. conique. / ~ top ‖ Kegelkuppe f. ‖ coupe f. conique.

taper file ‖ Spitzfeile f. ‖ spitze Feile f. lime f. pointue. / ~ flat-file ‖ spitzflache Feile f. ‖ lime f. plate pointue. / ~ gauge

‖ Konuslehre f. ‖ calibre m. de conicité. / ~ handfile see ~ flat-file.

taper hole ‖ konische Bohrung f. ‖ forage m. conique. / attachment for boring ~s ‖ Konischbohrvorrichtung f. ‖ appareil m. à aléser cône.

tapering see also tapered ‖ verjüngt ‖ fuyant; rétréci. / mandril ~-off x mm to each y mm ‖ konischer Dorn m., der sich auf je x mm Länge um y mm verjüngt ‖ mandrin m. d'une conicité de x/y. / section ~ at the top and the bottom ends ‖ Querschnittsverjüngung f. nach oben und unten ‖ section f. allant en se diminuant vers le haut et le bas.

tapering-in section ‖ Profilverjüngung f. ‖ amincissement m. du profil.

tapering scale (Draw) ‖ Verjüngungsmaßstab m. ‖ échelle f. fuyante ou de réduction. / ~ shape ‖ sich verjüngende Gestalt f. ‖ forme f. s'effilant.

taper pin ‖ kegeliger Stift m.; Keilstift m. ‖ goupille f. conique. / ~ pin making machine ‖ Keilstiftherstellungsmaschine f. ‖ machine f. à fabriquer les clavettes. / ~ roller bearing ‖ Kegelrollenlager n. ‖ roulement m. à rouleaux coniques. / web of ~ section ‖ sich verjüngender Steg m. ‖ âme f. de section décroissante. / ~ shank ‖ Konusdorn m. ‖ mandrin m. conique. / ~ shell ring ‖ kegelförmiger Kesselschuß m. ‖ virole f. conique de chaudière. / ~ sleeve ‖ Kegelhülse f. ‖ douille f. conique.

taper-sunk key ‖ Treibkeil m. ‖ clavette f. chassée ou conique.

taper tap see also tap (For cutting thread) ‖ Gewindebohrer m.; Schneidbohrer m. ‖ taraud m. / ~ turning attachment ‖ Einrichtung f. zum Konischdrehen ‖ dispositif m. à tourner conique.

tapestry ‖ gewebte Tapete f.; Wandteppich m.; Wirktapete f. ‖ tapisserie f. ou tenture f. (tissée). / high-warp ~ ‖ hochschäftige Tapete f. ‖ tapisserie f. de haute-lisse. / low-warp ~ ‖ tiefschäftige Tapete f. ‖ tapisserie f. de basse-lisse. / ~ from wool ‖ Wandbehang m. aus Wolle ‖ tapisserie f. de laine.

tapestry beating ‖ Teppichklopfen n. ‖ battage m. de tapis. / ~ designer ‖ Tapetenvorzeichner m. ‖ dessinateur m. en maquettes. / ~ embroiderer ‖ Teppichsticker m. ‖ brodeur m. en tapisserie. / ~ loom ‖ Gobelinwebstuhl m.; Wirktapetenwebstuhl m. ‖ métier m. pour tapisserie. / ~ manufactory ‖ Wirktapetenfabrik f. ‖ fabrique f. de tapisseries. / ~ mending ‖ Wirktapetenausbesserung f.; Teppichausbesserung f. ‖ rentraiture f. de tapisseries. / ~ work ‖ Tapisserien fpl. ‖ tapisseries fpl.

tape telephonograph ‖ Bandtelefonograf m. ‖ téléphonographe m. à ruban.

tap handle ‖ Absperrgriff m. ‖ manette f. de fermeture.

tap-hole (Met) ‖ Stichloch n. oder Abstichloch m. eines Ofens ‖ trou m. de coulée. / ~ (Brew) ‖ Zapfloch n. ‖ trou m. de bonde; trou m. de mise en perce. / ~ plug ‖ Lehmpfropf m.; Stichpfropf m. ‖ tampon m. de coulé. / ~ stopping machine ‖ Stichlochstopfmaschine f. ‖ machine f. à boucher le trou de coulée.

tapica (Build) ‖ Lehmstampfbau m.; Piseebau m.; Hastenwerk n. ‖ manière f. à bâtir en pisé; œuvre f. pisée; coffre m.; construction f. en pisé.

taping ‖ Hanftrensen fpl. ‖ guipage m. / ~ machine (Weav) ‖ Bandwickler m. ‖ rubaneuse f.; machine f. à rubaner. / ~ machine (For cables) ‖ Umwicklungsmaschine f. ‖ machine f. à envelopper.

tapioca ‖ Tapioka f. ‖ tapioca m. / ~ flocks pl. ‖ Tapiokaflocken fpl. ‖ tapioca m. en flocons. / ~ starch ‖ Tapiokastärke f. ‖ amidon m. tapioca.

tapped coil (Tel) ‖ Abzweigspule f. ‖ bobine f. de bifurcation. / ~ housings pl. (Roll mill) ‖ Walzenständer m. mit Querhaupt ‖ cage f. à chapeau rapporté. / ~ turbine ‖ Entnahmeturbine f. ‖ turbine f. à prise de vapeur.

tapper (Radio; Tel) ‖ Klopfer m.; Entfritter m. ‖ frappeur m.; décohéreur m. / ~ (For cutting threads) ‖ Gewindebohrer m. ‖ taraud m.

tappet (Mach) ‖ Knagge f.; Daumen m.; Nase f.; Mitnehmer m. ‖ taquet m.; came f. / ~ of an arbor ‖ Hebedaumen m.; Welldaumen m. ‖ mentonnet m. de l'arbre. / coupling ~ ‖ Auslösungsknagge f. ‖ taquet m. d'embrayage et de débrayage.

tapping (Of a cask) ‖ Anstecken n. ‖ mise f. en perce. / ~ (Of beer) ‖ Ausschank m. ‖ débit m. / ~ (El line) ‖ Abzweigbund m. ‖ ligature f. de raccordement. / ~ (Met) ‖ Abstechen n. ‖ percée f. / ~ of inflammable liquids ‖ Zapfstelle f. für feuergefährliche Flüssigkeiten ‖ poste m. de débit des liquides inflammables. / ~ of resin ‖ Harzgewinnung f. ‖ résinage m. / ~ of a spring ‖ Erschließung f. einer Quelle ‖ ouverture f. d'une source.

tapping bar (Met) ‖ Räumeisen n.; Stecheisen n.; Loseisen n. ‖ ringard m.; percefournaise f. / ~ and straining device ‖ Zapf- und Abseihvorrichtung f. ‖ installation f. pour soutirer et filtrer. / ~ hole (Found) ‖ Gußloch n. ‖ trou m. de coulée. / ~ machine (For cutting threads) ‖ Gewindeschneidmaschine f.; Gewindebohrmaschine f. ‖ machine f. à fileter les vis ou à tarauder; taraudeuse f.

tapping-off the blast-furnace ‖ Abstich m. des Hochofens ‖ coulée f. du hautfourneau.

tapping point (Electr) ‖ Anzapfung f. fraction f.

tapping side ‖ Abstichseite f. ‖ côté f. coulée. / slag ~ ‖ Schlackenabstichseite f. ‖ côté f. coulée des scories. / steel ~ ‖ Stahlabstichseite f. ‖ côté f. coulée de l'acier.

tap water ‖ Leitungswasser n.; Brunnenwasser n. ‖ eau f. de fontaine.

tar, to ‖ teeren ‖ goudronner. / ~ down (Mar) ‖ labsalben ‖ brayer ou enduire de goudron; goudronner. / ~ over (Tiler) ‖ teeren ‖ goudronner.

tar ‖ Teer m. ‖ goudron m. / to convert ~ into motor spirit and Diesel oil by means of the cracking process ‖ Teer m. durch das Krackverfahren in Treiböle überführen ‖ traiter le goudron par cracking de manière à le transformer en essences pour moteurs. / ~ which carries a large amount of dust ‖ stark staubhaltiger Teer m. ‖ goudron m. contenant beaucoup de poussières. / mineral ~ ‖ Mineralteer m.; Bergteer m. ‖ goudron m. minéral. / vegetable ~ ‖ Holzteer m. ‖ pflanzlicher Teer m. ‖ goudron m. végétal. / plant for eliminating water from ~ ‖ Teerent-

wässerungsanlage f. ‖ installation f. d'élimination de l'eau du goudron.

tar barrel ‖ Teerfaß n. ‖ gonne f. *ou* tonneau m. à goudron. / ~ board ‖ Teerpappe f. ‖ papier m. bitumé. / ~ boiler ‖ Teerkochkessel m. ‖ chaudière f. à goudron. / ~ boiling man ‖ Teerkocher m. ‖ goudronneur m. / ~ brush ‖ Teerquast m.; Teerquaste f. ‖ guipon m.; brosse f. à goudron. / ~ bucket (Mar) ‖ Teerbütte f. ‖ auge f. à goudron. / ~ cistern ‖ Teerzisterne f. ‖ réservoir m. à goudron. / ~ coke ‖ Pechkoks m. ‖ coke m. de goudron. / ~ colours pl. ‖ Teerfarben fpl. ‖ matières fpl. colorantes de goudron. / ~ and asphalt cooking boiler ‖ Teer- und Asphaltkochkessel m. ‖ chaudière f. à faire bouillir le goudron et l'asphalte. / ~ distilling ‖ Teerdestillation f. ‖ distillation f. du goudron. / ~ distilling plant ‖ Teerdestillationsanlage f. ‖ installation f. de distillation de goudron. / dregs pl. of ~ ‖ Teersatz m. ‖ rache f. de goudron.

tare, to ‖ tarieren ‖ tarer.

tare ‖ Tara f. ‖ tare f. / to deduct ~ *see* to tare. / additional ~ ‖ Supertara f. ‖ surtare f.

tare beam ‖ Wägebalken m. ‖ fléau m. / tariff-rate of ~ (Duty) ‖ Tarasatz m. ‖ taxe f. de tare. / ~ weight ‖ Eigengewicht n. ‖ poids m. mort.

tar extraction plant ‖ Teergewinnungsanlage f. ‖ installation f. pour l'extraction de goudron.

tar extractor ‖ Teerextraktionsvorrichtung f. ‖ extracteur m. de goudron.

tar fat oil ‖ Teerfettöl n. ‖ huile f. grasse de goudron.

tar-fired furnace ‖ Teerfeuerung f. ‖ foyer m. à goudron.

target ‖ Schießscheibe f. ‖ cible f.; disque m. pour stands de tir. / ~ (Electr) ‖ Antikathode f. ‖ anticathode f. / ~ (Railw) ‖ Scheibensignal n. ‖ disque m. / advancing ~ *see* approaching ~. / air ~ ‖ Luftziel n. ‖ objectif m. aérien. / approaching ~ ‖ kommendes Ziel n. ‖ but m. venant vers l'observateur. / flying ~ ‖ Flugscheibe f. ‖ cible f. planeuse *ou* volante. / moving ~ ‖ gehendes Ziel n. ‖ but m. (allant en) s'éloignant de l'observateur. / receding ~ *see* moving ~.

target distance ‖ Zielentfernung f. ‖ distance f. du but. / ~ plate distance ‖ Fokusplattenabstand m. ‖ distance f. foyer-plaque. / ~ practice ‖ Scheibenschießen n.; Schießübung f. ‖ tir m. à la cible. / ~ practice ground ‖ Scheibenstand m.; Schießstand m. ‖ tir m.; stand m. / ~ raft ‖ Scheibenfloß n. ‖ radeau m. pour cible.

tariff, to ‖ tarifieren; einen Tarif aufstellen ‖ taxer; tarifier.

tariff ‖ Tarif m. ‖ tarif m. / adjustable ~ ‖ Staffeltarif m. ‖ tarif m. mobile. / ad valorem ~ ‖ Wertzolltarif m. ‖ tarif m. ad valorem. / custom's ~ ‖ Zolltarif m. ‖ tarif m. douanier. / decreasing ~ ‖ nach unten abgestufter Tarif m. ‖ tarif m. dégressif. / ~ of duties ‖ Zolltarif m. ‖ tarif m. douanier. / measured rate ~ (Tél) ‖ Gesprächsgebührentarif m. ‖ tarif m. à conversation taxée. / specific ~ ‖ spezifischer Tarif m. ‖ tarif m. spécifique.

tariff agreement ‖ Tarifvertrag m. ‖ convention f. au tarif; contrat m. d'emploi.

/ application of the ~ ‖ Anwendung f. des Zolltarifs ‖ application f. du tarif de douane. / formation of ~ ‖ Tarifgestaltung f. ‖ établissement m. des tarifs; tarification f. / improvement of ~ ‖ Tarifverbesserung f. ‖ amélioration f. du tarif. / double ~ meter ‖ Doppeltarifzähler m. ‖ compteur m. à tarifs alternatifs *ou* à double tarif. / ~ nomenclature ‖ Tarifnomenklatur f. ‖ nomenclature f. tarifaire. / ~ rate ‖ Zollsatz m. ‖ classe f. de tarif douanier. / ~ rate of tare (Duty) ‖ Tarasatz m. ‖ tare f. / ~ system ‖ Zollsystem n. ‖ système m. douanier. / ~ treaty ‖ Tarifvertrag m. ‖ traité m. sur les tarifs.

taring ‖ Tarieren n. ‖ tarage m. / ~ of wagons (Railw) ‖ Tarieren n. der Eisenbahnwagen ‖ tarage m. des wagons.

tar macadam ‖ Teermakadam m. ‖ goudronmacadam m. / ~ plant ‖ Beteerungsanlage f. ‖ installation f. à macadamiser.

tar mist ‖ Teernebel m. ‖ brouillard m. de goudron.

tarnish, to ~ (Met) ‖ abblicken; den Glanz verlieren ‖ cesser de faire l'éclair; perdre son éclat; se ternir.

tarnisher (Engr) ‖ Mattpunze f. ‖ matoir m.

tar oil ‖ Teeröl n. ‖ huile f. de goudron. / mineral ~ ‖ Steinkohlenteeröl n. ‖ huile f. de goudron d'houille.

tar paper ‖ Teerpapier n. ‖ papier m. goudronné.

tarpaulin ‖ Teerleinwand f.; Teertuch n. ‖ toile f. goudronnée. / ~ (Coachm) ‖ Wagenplane f.; Wagendecke f. ‖ bâche f.; prélart m.; tarpaulin m. / ~ (Shipb) ‖ Persenning f. ‖ prélart m.

tar producing plant ‖ Teergewinnungsanlage f. ‖ installation f. d'extraction du goudron.

tar product ‖ Teererzeugnis n.; Teerprodukt n. ‖ produit m. de goudron.

tar pump ‖ Teerpumpe f. ‖ pompe f. pour goudron.

tarragon herb ‖ Estragonkraut n. ‖ herbe f. d'estragon.

tarred board ‖ Teerpappe f. ‖ carton m. bitumé.

tar residue ‖ Teerrückstand m. ‖ résidu m. de goudron.

tarring of roads ‖ Teeren n. von Straßen ‖ goudronnage m. des routes.

tarring device ‖ Teervorrichtung f.; Vorrichtung f. zum Teeren ‖ dispositif m. à goudronner.

tar roofing ‖ Asphaltdachpappe f.; Teerdachpappe f. ‖ carton m. bitumé.

tarry ‖ teerig; teerartig ‖ goudronneux.

tar separator ‖ Teerabscheider m.; Teerausscheider m. ‖ séparateur m. de goudron.

tar soap ‖ Teerseife f. ‖ savon m. au goudron.

tar and sulphur soap ‖ Teerschwefelseife f. ‖ savon m. au goudron et au soufre.

tar spraying ‖ Teerung f. ‖ goudronnage m.

tart ‖ Torte f.; Pastete f. ‖ tarte f.

tartar ‖ Weinstein m.; doppeltweinsteinsaures Kalium n. ‖ tartre m.; bitartrate m. de potasse; crème f. de tartre. / crude ~ ‖ roher Weinstein m. ‖ tartre m. brut. / emetic ~ ‖ Brechweinstein m. ‖ émétique m. / cream of ~ *see* tartar.

tartaric acid ‖ Weinsteinsäure f.; Weinsäure f. ‖ acide m. tartrique. / ~ producing plant ‖ Weinsäureherstellungsanlage f. ‖

installation f. pour la fabrication de l'acide tartrique.

tartar maker ‖ Weinsteinfabrikant m. ‖ tartrier m.

tartrate ‖ Tartrat n.; weinsaures *oder* weinsteinsaures Salz n. ‖ tartrate m.

tart shovel ‖ Tortenschaufel f. ‖ pelle f. à tarte.

tart tray ‖ Tortenplatte f. ‖ plateau m. à tartes.

tar water ‖ Teerwasser n. ‖ eau f. de goudron.

tar working plant ‖ Teerverarbeitungsanlage f. ‖ installation f. de préparation du goudron.

tasimeter ‖ Mikrotasimeter n. ‖ microtasimètre m.

task (Working time) ‖ Schicht f. ‖ journée f.; tâche f. / to work out the ~ (Mine) ‖ die Schicht verfahren ‖ faire sa tâche. / extra ~ ‖ Überschicht f. ‖ heures fpl. supplémentaires; tâche f. extraordinaire.

taskwork ‖ Akkordarbeit f. ‖ travail m. à la tâche *ou* à forfait.

tassel ‖ Quaste f.; Troddel f. ‖ gland m.; houppe f.

taste ‖ Geschmack m. ‖ goût m.

tasteful ‖ schmackhaft ‖ savoureux.

tasteless ‖ geschmacklos ‖ sans goût m.

tauro-colla ‖ Leim m. aus tierischen Abfällen ‖ taurocolle f.

tautening block ‖ Spannbock m. ‖ support m. de tendeur.

tautochronous curve ‖ Tautochrone f.; Isochrone f. ‖ ligne f. *ou* courbe f. isochrone; ligne f. *ou* courbe f. tautochrone.

tavern ‖ Schenke f. ‖ cabaret m.

taw, to ‖ weißgerben ‖ mégisser; passer en mégie.

tawed ‖ weißgar ‖ mégissé.

tawer ‖ Weißgerber m. ‖ mégissier m.

tawery ‖ Weißgerberei f. ‖ mégisserie f.

tawing ‖ Weißgerben n. ‖ mégisserie f.

tax, to ‖ abschätzen; schätzen; veranschlagen ‖ estimer. / ~ (To impose taxes) ‖ besteuern; mit Steuern belegen ‖ imposer; taxer.

tax ‖ Abgabe f.; Gebühr f.; (unmittelbare) Steuer f. ‖ impôt m.; droit m.; taxe f. / ~ on buildings ‖ Gebäudesteuer f. ‖ impôt m. sur la propriété bâtie. / exchange stamp ~ ‖ Wechselstempelsteuer f. ‖ impôt m. du timbre des effets de commerce. / gross ~ ‖ Rohbesteuerung f. ‖ imposition f. brute. / income ~ ‖ Einkommensteuer f. ‖ impôt m. sur le revenu. / ~ on increment values ‖ Wertzuwachssteuer f. ‖ taxe f. sur les plusvalues. / land ~ ‖ Grundsteuer f. ‖ impôt m. foncier. / ~ on payments *see* turnover ~. / sales ~ *see* turnover ~. / ~ of shares ‖ Aktiensteuer f. ‖ impôt m. sur les actions. / trade ~ ‖ Gewerbesteuer f. ‖ impôt m. sur le revenu professionnel; impôt m. des patentes. / turnover ~ ‖ Umsatzsteuer m. ‖ impôt m. sur les transactions; impôt m. sur le chiffre des affaires *ou* des opérations. / ~ on wages ‖ Lohnsteuer f. ‖ impôt m. sur les salaires *ou* sur les gages.

taxable ‖ steuerpflichtig ‖ imposable; contribuable.

taxation ‖ Abschätzung f. ‖ estimation f.; taxation f.

taxed costs pl. ‖ (ab)geschätzte *oder* taxierte Kosten pl. ‖ frais mpl. *ou* dépens mpl. taxés. / ~ goods pl. ‖ besteuerte Waren fpl. ‖ marchandises fpl. imposées. / heavily ~ goods pl. ‖ hoch besteuerte

Waren fpl. ‖ marchandises fpl. lourdement imposées.

taxer ‖ Taxator m.; Abschätzer m. ‖ taxateur m.; commissaire-priseur m.

tax-free ‖ steuerfrei ‖ exempt d'impôt m.

taxicab ‖ Kraftdroschke f.; Autodroschke f. ‖ taxi m.; auto-taxi m. / ~ call ‖ Autoruf m.; Kraftdroschkenanruf m. ‖ téléphone m. pour l'appel de taxis *ou* de voitures de place. / ~ call system ‖ Droschkenfernsprechsystem n. ‖ système m. téléphonique pour l'appel de voitures de place.

taxi-car *see* taxicab.

taximeter ‖ Taxameter n. ‖ taximètre m.

taxing authority ‖ Steuerbehörde f. ‖ administration f. des contributions.

tax law ‖ Steuergesetz n. ‖ loi f. fiscale.

taxpayer ‖ Steuerzahler m. ‖ contribuable m.

tea ‖ Tee m. ‖ thé m. / ~ in leaves ‖ Tee m. in Blättern ‖ thé m. en feuilles.

teacher ‖ Lehrer m. ‖ professeur m.; maître m.

teaching ‖ Anlernen n. ‖ éducation f. / article for ~ ‖ Lehrmittel n. ‖ article m. pour l'enseignement. / ~ laboratory ‖ Unterrichtslaboratorium n. ‖ laboratoire m. d'enseignement. / ~ spectrograph ‖ Lehrspektrograf m. ‖ spectrographe m. d'enseignement.

tea cosy doll ‖ Teepuppe f. ‖ couvre-théière m. / ~ dressing plant ‖ Teeaufbereitungsanlage f. ‖ installation f. pour la préparation du thé. / ~ glass ‖ Teeglas n. ‖ verre m. à thé. / ~ glass cup of metal ‖ Teeglashalter m. aus Metall ‖ support m. métallique pour verres à thé. / ~ grounds pl. ‖ Teesatz m. ‖ marc m. de thé.

teak wood ‖ Teakholz n. ‖ bois m. de teck.

tea merchant ‖ Teehändler m. ‖ négociant m. en thé.

teamsman ‖ Koppelknecht m. ‖ charretier m. agricole.

teapot ‖ Teekanne f.; Teekessel m. ‖ théière f.

tear, to ‖ zerreißen; aufreißen; einreißen; reißen ‖ déchirer.

tear ‖ Riß m. ‖ déchirure f.; rupture f. / ~ (Glassm) ‖ Rampe f.; Träne f.; Tropfen m. ‖ larme f. / wear and ~ ‖ Abnutzung f. ‖ usure f.; détérioration f.

tear and wear *see* wear and tear.

tear-exciting ‖ tränenerregend ‖ lacrymogène.

tear gas ‖ Tränengas n. ‖ lacrymogène m.

tearing ‖ Zerreißen n. ‖ déchirement m.; effilochage m.

tearing-off apparatus for shavings ‖ Spänezerreißer m. ‖ déchireur m. à copeaux.

tearing strength ‖ Zerreißfestigkeit f. ‖ résistance f. à la déchirure.

tear-off calendar ‖ Abreißkalender m. ‖ calendrier m. à feuilles détachables.

tease, to ~ cloth ‖ das Tuch rauhen ‖ garnir *ou* lainer *ou* tirer la perche. / ~ the fire (Met) ‖ das Feuer schüren ‖ attiser *ou* pousser *ou* tisonner le feu.

teasel (Botany) ‖ Kardendistel f. ‖ chardon m. (cardère). / ~ (Clothm) ‖ Rauhkarde f.; Tuchkarde f.; Karde f. ‖ chardon m. (à foulon). / steel ~ ‖ Stahlkratze f. ‖ chardon m. en acier.

teasel cleaning brush ‖ Kardenreinigungsbürste f. ‖ brosse f. à nettoyer les chardons.

teaseler (Clothm) ‖ Rauher m. ‖ laineur m.

teasel raising machine ‖ Rollkardenrauhmaschine f. ‖ laineuse f. à chardons roulants. / ~ rod gig ‖ Rauhmaschine f. für Strichrauherei ‖ laineuse f. pour poil long et couché. / ~ spindle ‖ Kardenspindel f. ‖ broche f. de chardon.

tea set ‖ Teegeschirr n. ‖ service m. à thé.

teasing *see also* teasling (Spinn) ‖ Kämmen n. ‖ peignage m.

teasling (Clothm) ‖ Rauhen n. ‖ lainage m.; grattage m. / cloth ~ ‖ Rauhen n. des Tuches ‖ lainage m. *ou* garnissage m. de drap.

teasling machine ‖ Rauhmaschine f. ‖ laineuse f. / ~ workshop ‖ Rauhanstalt f. ‖ atelier m. de grattage.

tea spoon ‖ Teelöffel m.; Kaffeelöffel m. ‖ petite cuiller f.; cuiller f. à thé *ou* à café.

tea strainer ‖ Teesieb n. ‖ passe-thé m.; passoir m. à thé.

teat (Of a child's bottle) ‖ Sauger m. ‖ tétine f.

tea table ‖ Teetisch m. ‖ table f. à thé.

teaze, to ~ *see* to tease.

teazle *see* teasel.

technic *see* technics.

technical ‖ technisch ‖ technique. / ~ article ‖ technischer Gegenstand m. ‖ article m. technique. / ~ college ‖ technische Hochschule f. ‖ école f. polytechnique. / ~ college ‖ Gewerbeschule f. ‖ école f. industrielle *ou* professionnelle; école f. des arts et métiers. / ~ dictionary ‖ technisches Wörterbuch n. ‖ dictionnaire m. technique. / ~ library ‖ technische *oder* fachwissenschaftliche Bücherei f. ‖ bibliothèque f. technique. / ~ office ‖ technisches Büro n. ‖ bureau m. technique. / ~ photography ‖ fotografische Technik f. ‖ technique f. photographique. / ~ press ‖ Fachpresse f. ‖ presse f. technique. / for ~ purposes ‖ für technische Zwecke mpl. ‖ pour buts mpl. techniques. / ~ school *see* ~ college. / ~ term ‖ technischer Ausdruck m. ‖ terme m. technique. / ~ world ‖ Fachwelt f. ‖ monde m. technique.

technicalities pl. ‖ technische Einzelheiten fpl. ‖ details mpl. techniques.

technically pure ‖ technisch rein ‖ techniquement pur.

technician ‖ Techniker m. ‖ ingénieur m.

technics pl. ‖ Technik f. ‖ technique f. / ~ of dressing ‖ Aufbereitungstechnik f. ‖ technique f. du traitement. / electrical communication ~ ‖ Fernmeldetechnik f. ‖ technique f. de la télécommunication. / ~ of the fibre materials ‖ Faserstofftechnik f. ‖ technique f. des matières de fibre. / ~ of flying ‖ Flugtechnik f. ‖ aviation f. dynamique. / ~ of measurement ‖ Meßtechnik f. ‖ technique f. de mesure.

technique ‖ Technik f.; Fingerfertigkeit f.; Kunstfertigkeit f. ‖ technique f.

techno-chemical product ‖ chemisch-technisches Erzeugnis n. ‖ produit m. technochimique.

technological institution *see* technical college.

technology ‖ Technologie f. ‖ technologie f. / chemical ~ ‖ chemische Technologie f. ‖ technologie f. chimique.

tedder ‖ Heuwender m.; Heuwendemaschine f. ‖ râteau-faneur m.; faneuse f. / fork-type ~ ‖ Gabelheuwender m. ‖ faneuse f. à fourches. / hay ~ *see* tedder. /

reel ~ ‖ Trommelheuwender m. ‖ râteau-faneur m. à tambour.

tedding ‖ Heuwenden n. ‖ fanage m. / ~ fork ‖ Gabel f. zum Heuwenden; Wendegabel f. ‖ fourche f. faneuse. / ~ machine ‖ Heuwendemaschine f.; mechanischer Heuwender m. ‖ tourne-foin m.; faneuse f. mécanique.

teddy bear ‖ Teddybär m. ‖ ours m. Teddy. / ~ wool cloth ‖ Teddybärenfellstoff m. ‖ tissu m. peau d'ours Teddy.

teeth pl. (of gear) *see also* tooth ‖ Radzähne mpl. ‖ dents fpl. de roue. / ~ in steps ‖ Stufenzähne mpl. ‖ dents fpl. en étages; dents mpl. étagées d'une roue à étages.

teething ring (For children) ‖ Beißring m. ‖ anneau m. de dentition.

teeth-of-gear-cutter ‖ Zahnradschneider m. ‖ tailleur m. d'engrenages.

telautograph ‖ Fernschreiber m.; Telautograf m. ‖ télautographe m.

telautography ‖ Schriftfernübertragung f. ‖ télautographie f.

telecounter for watermeters ‖ Fernzählwerk n. für Wassermesser ‖ totalisateur m. à distance pour compteurs d'eau.

telegram ‖ Telegramm n.; Drahtnachricht f.; Drahtung f.; Drahtbericht m. ‖ télégramme m. / to repeat a ~ ‖ ein Telegramm n. wiederholen ‖ collationner un télégramme. / collated ~ ‖ Telegramm n. mit Vergleichung ‖ télégramme m. avec collationnement. / deferred ~ ‖ zurückgestelltes Telegramm n. ‖ télégramme m. différé. / maritime ~ ‖ Überseetelegramm n. ‖ télégramme m. maritime. / mutilated ~ ‖ verstümmeltes Telegramm n. ‖ télégramme m. mutilé. / press ~ ‖ Pressetelegramm n. ‖ télégramme m. de la presse. / reproduction of press ~s ‖ Vervielfältigung f. von Pressetelegrammen ‖ reproduction f. des télégrammes de la presse. / handing-in of ~s ‖ Telegrammaufgabe f. ‖ dépôt m. des télégrammes. / mutilation of a ~ ‖ Verstümmelung f. eines Telegramms ‖ mutilation f. d'une dépêche. / non-delivery of ~s ‖ Unzustellbarkeit f. von Telegrammen ‖ la non-remise de télégrammes. / preamble of the ~ ‖ Kopf m. eines Telegramms ‖ préambule m. du télégramme. / preparation of ~s (Tel) ‖ Abfassung f. von Telegrammen ‖ rédaction f. de télégrammes. / reading of a ~ by sound ‖ Aufnahme f. eines Telegrammes nach dem Gehör ‖ lecture f. d'une dépêche au son. / stoppage of ~ ‖ Anhalten n. des Telegramms ‖ arrêt m. du télégramme.

telegraph, to ‖ drahten; telegrafieren ‖ télégraphier.

telegraph ‖ Telegraf m.; Fernschreiber m. ‖ télégraphe m. / acoustic ~ ‖ akustischer Telegraf m. ‖ télégraphe m. acoustique. / alarm ~ ‖ Glockentelegraf m. ‖ télégraphe m. à sonnette. / bell ~ *see* alarm ~. / copying ~ ‖ Kopiertelegraf m. ‖ télégraphe m. copieur *ou* autographique. / dial ~ *see* needle ~. / electric ~ ‖ elektrischer Telegraf m. ‖ télégraphe m. électrique. / electro-chemical ~ ‖ elektrochemischer Bildschreiber m.; Chemograf m.; chemischer Telegraf m. ‖ chémographe m.; télégraphe m. électrochimique. / multiple ~ ‖ mehrfacher Telegraf m. ‖ télégraphe m. multiple. / needle ~ ‖ Zeigertelegraf m. ‖ télégraphe m. à cadran. / optical ~ ‖ optischer Telegraf m. ‖ télégraphe m. optique. / picture ~ ‖ Bildübertragungsgerät

n.; Bildtelegraf m. ‖ appareil m. téléphotographique. / printing ~ ‖ Drucktelegraf m. ‖ télégraphe m. imprimeur. / submarine ~ ‖ unterseeischer Telegraf m. ‖ télégraphe m. sous-marin. / type-printing ~ ‖ Typendrucktelegraf m. ‖ télégraphe m. imprimeur. / underground ~ unterirdischer Telegraf m. ‖ télégraphe m. souterrain.

telegraph accessories pl. ‖ Telegrafenbaumaterial n.; Zubehör n. für den Telegrafenbau ‖ matériel m. de construction des télégraphes. / ~ act ‖ Telegrafengesetz n. ‖ loi f. des télégraphes. / ~ alphabet ‖ Telegrafenalphabet n. ‖ alphabet m. télégraphique. / ~ apparatus ‖ Telegrafenapparat m.; Telegrafengerät n. ‖ appareil m. télégraphique. / ~ cable ‖ Telegrafenkabel n. ‖ câble m. télégraphique. / ~ clerk ‖ Telegrafenbeamter m. ‖ employé m. des télégraphes. / ~ construction department see ~ construction office. / ~ construction office ‖ Telegrafenbauamt n. ‖ bureau m. de construction des télégraphes. / ~ construction tool ‖ Telegrafenbaugerät n. ‖ outillage m. de construction des télégraphes. / ~ craftsman ‖ Telegrafenbauhandwerker m. ‖ poseur m. des télégraphes. / ~ crossfire ‖ Induktionsstörung f. zwischen benachbarten Telegrafeneinzelleitungen ‖ trouble m. inductif des lignes télégraphiques. / ~ equation ‖ Telegrafengleichung f. ‖ équation f. des télégraphistes.

telegraphic ‖ drahtlich; telegrafisch ‖ télégraphique. / ~ address ‖ Drahtanschrift f.; Telegrammadresse f. ‖ adresse f. télégraphique. / ~ charges pl. ‖ Telegrammkosten pl. ‖ frais mpl. de télégramme. / ~ code ‖ Telegrammschlüssel m.; Kodeschlüssel m.; Kode m. ‖ code m. télégraphique.

telegraph instrument see telegraph apparatus.

telegraphist ‖ Telegrafist m. ‖ télégraphiste m.

telegraph line ‖ Telegrafenleitung f.; Telegrafenlinie f. ‖ ligne f. télégraphique. / ~ line plan ‖ Telegrafenwegeplan m. ‖ projet m. ou plan m. de ligne télégraphique. / ~ material ‖ Telegraphenbauzeug n. ‖ matériel m. de construction des télégraphes. / ~ material store ‖ Telegrafenzeugamt n. ‖ dépôt m. de matériel de construction des télégraphes. / ~ messenger ‖ Telegrammausträger m. ‖ porteur m. de télégrammes.

telegraph office ‖ Telegrafenanstalt f.; Telegrafenamt n. ‖ bureau m. télégraphique. / general ~ ‖ Haupttelegrafenamt n. ‖ bureau m. principal des télégraphes. / ~ of origin ‖ Aufgabetelegrafenanstalt f. ‖ bureau m. d'origine télégraphique.

telegraphone ‖ Telegrafon n. ‖ télégraphone m.

telegraph order wire working ‖ Summermeldebetrieb m. ‖ préparation f. télégraphique. / ~ plant ‖ Telegrafenanlage f. ‖ installation f. télégraphique. / ~ plant (Network) ‖ Telegrafennetz n. ‖ réseau m. télégraphique. / ~ pole ‖ Telegrafenstange f. ‖ poteau m. télégraphique ou de télégraphe. / ~ pole impregnation ‖ Tränken n. oder Imprägnieren n. von Telegrafenstangen ‖ injection f. de poteaux télégraphiques. / ~ printing-ink ‖ Öltinte f.; Apparatfarbe f. ‖ encre f. oléique. / ~ rates pl. ‖ Telegrafentarif m. ‖

tarif m. télégraphique. / ~ regulation ‖ Telegrafenordnung f. ‖ règlement m. télégraphique. / ~ repeater ‖ Übertragung f. in Telegrafenleitungen ‖ translation f. télégraphique. / ~ system ‖ Telegrafennetz n. ‖ réseau m. télégraphique. / ~ technics pl. ‖ Telegrafentechnik f. ‖ technique f. télégraphique. / development of the ~ technics pl. ‖ Entwicklung f. der Telegrafentechnik ‖ développement m. de la technique télégraphique. / ~ terminal cable ‖ Telegrafenabschlußkabel n. ‖ câble m. télégraphique de fermeture. / ~ wire ‖ Telegrafendraht m. ‖ fil m. télégraphique. / ~ workman see ~ craftsman.

telegraphy ‖ Telegrafie f.; Fernschreibung f. ‖ télégraphie f. / acoustic ~ ‖ Gehörtelegrafie f.; akustische Telegrafie f. ‖ télégraphie f. acoustique. / duplex ~ ‖ Gegensprechtelegrafie f. ‖ télégraphie f. duplex. / ~ not to be listened to ‖ unabhörbare Telegrafie f. ‖ télégraphie f. non interceptable. / metallic polar duplex ~ ‖ (Gleichstrom-) Unterlagerungstelegrafie f. ‖ télégraphie f. infraacoustique (par courant continu). / military ~ ‖ Militärtelegrafie f. ‖ télégraphie f. militaire. / multiplex ~ ‖ Mehrfachtelegrafie f. ‖ télégraphie f. multiplex. / sub-audio ~ ‖ Unterlagerungstelegraphie f. ‖ télégraphie f. infraacoustique. / voice frequency ~ ‖ Wechselstromtelegrafie f. ‖ télégraphie f. à fréquences vocales. / wireless ~ ‖ drahtlose Telegrafie f.; Funkentelegrafie f. ‖ télégraphie f. sans fil; T. S. F. / syntonized wireless ~ ‖ abstimmbare drahtlose Telegrafie f. ‖ télégraphie f. sans fil syntonisée.

telemeter ‖ Entfernungsmesser m. ‖ télémètre m. / coincidence ~ ‖ Schnittbildentfernungsmesser m.; Koinzidenzentfernungsmesser m. ‖ télémètre m. à coïncidence. / depression ~ ‖ Entfernungsmesser m. mit senkrechter Basis; Depressionsentfernungsmesser m. ‖ télémètre m. de dépression. / inversion ~ ‖ Invertentfernungsmesser m. ‖ télémètre m. à renversement. / stereoscopic ~ ‖ Raumbildentfernungsmesser m. ‖ télémètre m. stéréoscopique; stéréo-télémètre m.

tele-metering ‖ Fernmessung f. ‖ télémesure f. / impulse frequency ~ ‖ Impulsfrequenzfernmessung f. ‖ télémesure f. à fréquence d'impulsions.

tele-metering network ‖ Fernmeßnetz n. ‖ système m. ou réseau m. de télémesure.

telemetry ‖ Entfernungsmessen n.; Entfernungsmessung f. ‖ télémétrie f. / apparatus for ~ ‖ Fernmeßgerät n.; Gerät n. zum Fernmessen; Fernmeßapparat m. ‖ appareil m. télémétrique.

telephone, to ‖ fernsprechen; telefonieren ‖ téléphoner.

telephone ‖ Fernsprecher m.; Telefon n.; Fernsprechgerät n. ‖ téléphone m. / electromagnetic ~ ‖ elektromagnetischer Fernsprecher m. ‖ téléphone m. électromagnétique. / electrostatic ~ ‖ elektrostatischer Fernsprecher m. ‖ téléphone m. électrostatique. / ~ for fire alarm systems ‖ Fernsprecher m. in Feuermeldeanlagen oder für Feuermeldeanlagen ‖ téléphone m. dans les installations d'avertisseurs d'incendie. / ~ with high-tension protection ‖ Fernsprecher m. mit Schutz gegen Hochspannung

téléphone m. avec protection contre les hautes tensions. / intercommunication ~ for telephone lines with protection against high tension ‖ Wahlfernsprecher m. für hochspannungsgeschützte Fernsprechleitungen ‖ poste m. téléphonique à appel sélectif, pour lignes téléphoniques protégées contre (l'action de) la haute tension. / long-distance ~ ‖ Fernverkehrfernsprecher m. ‖ téléphone m. à longue distance. / loudspeaking ~ ‖ Lautfernsprecher m. ‖ téléphone m. haut-parleur. / mast ~ fitted with high-tension protection device ‖ Mastfernsprecher m. mit Schutz gegen Hochspannung ‖ poste m. téléphonique sur poteau avec protection spéciale contre la haute tension.

telephone apparatus ‖ Fernsprechgerät n.; Fernsprecher m.; Fernsprechapparat m.; Telefongerät n. ‖ appareil m. téléphonique ou à téléphone. / ~ area ‖ telefonischer Anschlußbereich m.; Sprechbereich m. ‖ circonscription f. ou district m. téléphonique. / ~ booth ‖ Fernsprechzelle f. ‖ cabine f. téléphonique. / ~ box see ~ booth. / ~ bridge ‖ Telefonbrücke f. ‖ pont m. de téléphone. / ~ cabin ‖ Fernsprechzelle f. ‖ cabine f. téléphonique.

telephone cable ‖ Fernsprechkabel n.; Telefonkabel n. ‖ câble m. téléphonique. / long-distance ~ ‖ Fernsprechkabel n. für den Fernverkehr. ‖ câble m. téléphonique à grande distance. / loaded ~ ‖ pupinisiertes Fernsprechkabel n. ‖ câble m. téléphonique pupinisé.

telephone car ‖ Fernsprechbauwagen m. ‖ caisson m. pour le matériel téléphonique. / ~ cell see ~ booth. / ~ circuit (Tel) ‖ Fernsprechleitung f.; Fernsprechkreis m. ‖ circuit m. téléphonique. / ~ clock ‖ Fernsprecheruhr f.; Telefonuhr f. ‖ chronomètre m. pour téléphones. / ~ condenser ‖ Fernsprecherkondensator m. ‖ condensateur m. pour téléphone. / ~ connection ‖ Fernsprechverbindung f. ‖ communication f. téléphonique. / urgent ~ connection ‖ dringendes Telefongespräch n. ‖ communication f. téléphonique urgente. / ~ construction car ‖ Fernsprechbauwagen m. ‖ voiture f. pour constructions téléphoniques. / ~ directory ‖ Fernsprechbuch n. ‖ annuaire m. officiel des abonnés au téléphone. / ~ disturbing effect ‖ Fernsprechstörwirkung f. ‖ effet m. perturbateur téléphonique. / ~ earth cable ‖ Fernsprecherdkabel n. ‖ câble m. téléphonique enterré. / ~ end cable ‖ Fernsprechabschlußkabel n. ‖ câble m. téléphonique de fermeture. / ~ engineering ‖ Fernsprechtechnik f. ‖ technique f. téléphonique.

telephone exchange ‖ Fernsprechamt n.; Fernsprechvermittlungsstelle f. ‖ central m. téléphonique. / automatic ~ ‖ Selbstanschlußamt n. ‖ central m. téléphonique automatique. / public ~ ‖ öffentliche Vermittlungsstelle f.; öffentliches Fernsprechamt n. ‖ central m. public.

telephone installation ‖ Fernsprechanlage f. ‖ installation f. téléphonique. / private automatic ~ ‖ selbsttätige Hausfernsprechanlage f. ‖ installation f. téléphonique privée automatique.

telephone lamp ‖ Fernsprechglühlampe f. ‖ lampe f. téléphonique.

telephone line ‖ Fernsprechleitung f.; Telefonleitung f. ‖ ligne f. téléphonique. / asymmetrical ~ ‖ unsymmetrische Fernsprechleitung f. ‖ ligne f. téléphonique asymétrique. / ~ with protection against high tension ‖ hochspannungsgeschützte Fernsprechleitung f. ‖ ligne f. téléphonique protégée contre l'action de la haute tension.

telephone network ‖ Fernsprechnetz n. ‖ réseau m. téléphonique. / ~ office ‖ Fernsprechanstalt f. ‖ bureau m. téléphonique. / ~ olive for the ear-way ‖ Gehörgangfernhörer m. ‖ récepteur m. à insérer dans le pavillon de l'oreille.

telephone plant ‖ Fernsprechanlage f.; Telefonanlage f. ‖ installation f. téléphonique ou de téléphone. / ~ (Network) ‖ Fernsprechnetz n. ‖ réseau m. téléphonique. / project of a ~ ‖ Entwurf m. oder Plan m. oder Projekt n. einer Fernsprechanlage ‖ projet m. d'une installation téléphonique.

telephone rates pl. ‖ Fernsprechtarif m.; Fernsprechgebühren fpl. ‖ tarif m. téléphonique. / ~ receiver ‖ Fernhörer m.; Telefonhörer m. ‖ récepteur m. téléphonique. / ~ repeater ‖ Fernsprechverstärker m. ‖ relais m. amplificateur téléphonique. / ~ section (Military) ‖ Fernsprechabteilung f. ‖ section f. des téléphonistes de campagne. / ~ service ‖ Fernsprechbetrieb m.; Vermittlungsdienst m. ‖ service m. ou exploitation f. téléphonique. / withdrawal of the ~ service ‖ Fernsprechsperre f. ‖ suspension f. (du service téléphonique). / ~ short-circuiting contact ‖ Fernsprecherkurzschlußkontakt m. ‖ contact m. de mise en court-circuit du téléphone. / ~ signal ‖ Telefonsignal n.; Fernsprecherzeichen n. ‖ signal m. téléphonique. / ~ station ‖ Sprechstelle f. ‖ poste m. téléphonique (d'abonnés). / automatic ~ station ‖ Fernsprechautomat m. ‖ poste m. téléphonique automatique. / ~ switchboard ‖ Fernsprechumschalter m. ‖ commutateur m. téléphonique. / ~ switchboard for private branch exchanges ‖ Klappenschrank m. für Nebenstellen ‖ tableau m. à volets pour postes supplémentaires. / ~ system ‖ Fernsprechsystem n. ‖ système m. téléphonique ou de téléphone. / ~ table ‖ Fernsprechertisch m. ‖ table f. pour téléphone. / ~ technics pl. ‖ Fernsprechtechnik f. ‖ technique f. téléphonique. / ~ traffic ‖ Fernsprechverkehr m. ‖ trafic m. téléphonique. / ~ transmission reference system ‖ Fernsprecheichkreis m. ‖ système m. de référence pour la transmission téléphonique. / ~ wire ‖ Fernsprecherdraht m.; Telefondraht m. ‖ fil m. téléphonique. / ~ wire system see also ~ network ‖ Telefonnetz n. ‖ réseau m. téléphonique.

telephonic ‖ telefonisch; fernmündlich ‖ téléphonique; par téléphone m. / ~ communication ‖ Telefongespräch n.; Fernspruch m. ‖ conversation f. ou communication f. téléphonique. / ~ conversation see ~ communication. / ~ frequency ‖ Sprechfrequenz f. ‖ fréquence f. du courant de transmission.

telephonist ‖ Telefonbeamtin f.; Platzbeamtin f. ‖ téléphoniste f.; opératrice f.

telephonograph ‖ Fernsprechschreiber m.; Telefonograf m. ‖ téléphonographe m.

telephony ‖ Fernsprechen n.; Telefonie f. ‖ téléphonie f. / duplex ~ ‖ Duplexfernsprechen n. ‖ téléphonie f. duplex. / interurban ~ ‖ Ferntelefonie f.; Fernsprechen n. im Fernverkehr ‖ téléphonie f. interurbaine. / line ~ ‖ Drahtfernsprechen n. ‖ téléphonie f. par fil. / multiplex ~ ‖ Mehrfachfernsprechen n. ‖ téléphonie f. multiplex. / phantom ~ ‖ Doppelsprechen n. ‖ double conversation f. / radio ~ see wireless ~. / train ~ ‖ Zugtelefonie f. ‖ téléphonie f. avec un train. / wire ~ see line ~. / wireless ~ ‖ drahtloses Fernsprechen n.; drahtlose Telefonie f.; Funkfernsprechen n. ‖ téléphonie f. sans fil.

telephoto attachment ‖ Teleansatz m. ‖ téléaccord m.

telephotographic transmission ‖ bildtelegrafische Übertragung f. ‖ transmission f. phototélégraphique.

telephotography ‖ Fernfotografie f. ‖ téléphotographie f.

teleprinter ‖ Ferndrucker m. ‖ téléimprimeur m.

telepsychrometer ‖ Fernfeuchtigkeitsmesser m.; Fernpsychrometer n. ‖ télépsychromètre m.

telescope to, the parts pl. have been telescoped ‖ die Stücke npl. sind durch Ineinanderstecken vereinigt ‖ les pièces fpl. sont assemblées en télescope.

telescope ‖ Fernrohr n.; Teleskop n. ‖ télescope m.; longue-vue f.; lunette f. d'approche. / to blacken the inside of the ~ ‖ das Innere des Fernrohres matt schwärzen ‖ noircir en mat l'intérieur de la lunette. / to level a ~ ‖ ein Fernrohr n. richten ‖ braquer un télescope. / to point a ~ see to level a ~. / to sight an object with the ~ ‖ einen Gegenstand m. mit dem Fernrohr anzielen ‖ viser un objet au moyen de la lunette. / aiming ~ ‖ Zielfernrohr n.; Visierfernrohr n. ‖ lunette f. viseur; lunette f. de visée. / altazimuth ~ ‖ azimutales Fernrohr n. ‖ lunette f. azimutale. / ~ in altazimuth mount ‖ azimutal montiertes Fernrohr n. ‖ lunette f. à monture azimutale. / astronomical ~ ‖ astronomisches Fernrohr n. ‖ lunette f. ou télescope m. astronomique. / automatic ~ ‖ Automatenfernrohr n. ‖ lunette f. d'approche munie d'un déclenchement automatique. / ~ balanced for motion in altitude ‖ Balanzierung f. eines Fernrohres in Höhenbewegung ‖ lunette f. équilibrée pour le mouvement vertical. / biaxial ~ with reversible spirit levels ‖ biaxiales Fernrohr n. mit Wendelibelle ‖ lunette f. biaxiale munie d'une libelle réversible. / binocular prism ~ ‖ Doppelprismenfernrohr n. ‖ jumelles fpl. à prismes. / ~ for binocular use ‖ Fernrohr n. für zweiäugigen Gebrauch ‖ lunette f. binoculaire. / celestial ~ see astronomical ~. / collimator ~ ‖ Kollimatorfernrohr n. ‖ lunette f. collimatrice. / combination ~ for astronomical and terrestrial observation ‖ Fernrohr n. für Himmels- und Erdbeobachtung ‖ lunette f. pour les observations célestes et terrestres. / double ~ with the object glasses farther apart than the eyes ‖ Doppelfernrohr n. mit erweitertem Achsenabstand ‖ lunette f. binoculaire avec amplification de l'écart des axes. / equatorial ~ ‖

parallaktisches Fernrohr n. ‖ lunette f. parallactique. / equatorially mounted ~ ‖ parallaktisch montiertes Fernrohr n. ‖ lunette f. à monture parallactique. / ~ with erecting prisms ‖ Fernrohr n. mit Prismenumkehrsatz ‖ lunette f. munie du système de prismes redresseurs. / ~ for hand use ‖ Fernrohr n. für den Handgebrauch ‖ lunette f. (disposée pour servir) à la main. / highly magnifying ~ ‖ stark vergrößerndes Fernrohr n. ‖ lunette f. à fort grossissement. / ~ with internal focusing ‖ Fernrohr n. mit Innenfokussierung ‖ lunette f. à mise au point interne. / ~ of great light-transmitting power ‖ sehr lichtstarkes Fernrohr n. ‖ lunette f. très lumineuse. / long-focus ~ ‖ langbrennweitiges Fernrohr n. ‖ lunette f. à long foyer. / look-out ~ ‖ Aussichtsfernrohr n. ‖ lunette f. d'approche. / look-out ~ on stand for terrestrial observations ‖ Standaussichtsfernrohr n. für terrestrische Beobachtungen ‖ longue-vue f. d'approche pour des observations terrestres. / binocular look-out ~ ‖ binokulares Aussichtsfernrohr n. ‖ lunette f. d'approche binoculaire. / monocular look-out ~ ‖ monokulares Aussichtsfernrohr n. ‖ lunette f. d'approche monoculaire. / plumbing ~ ‖ Fernrohrlot n. ‖ instrument m. de prise d'aplomb à lunette. / portable ~ ‖ Handfernrohr n. ‖ lunette f. à main. / reading ~ ‖ Ablesefernrohr n. ‖ lunette f. de lecture. / reflecting ~ ‖ Spiegelteleskop n. ‖ télescope m. (réflecteur). / reversible ~ ‖ umlegbares Fernrohr n. ‖ lunette f. réversible ou pouvant être retournée bout pour bout. / ~ of short focal length ‖ kurzbrennweitiges Fernrohr n. ‖ lunette f. à foyer court. / sighting ~ ‖ Zielfernrohr n.; Visierfernrohr n. ‖ lunette f. de visée; lunette-viseur f. / single-tube ~ ‖ einfaches Fernrohr n. ‖ lunette f. simple. / ~ for viewing the slit ‖ Fernrohr n. zur Spaltbeobachtung ‖ lunette f. pour observer la fente. / ~ with slot-paying mechanism ‖ mit Geldautomateneinrichtung ausgerüstetes oder ausgestattetes Fernrohr n. ‖ lunette f. munie d'un déclenchement monétaire automatique. / ~ provided with a subtense attachment ‖ mit einem Entfernungsmesser ausgerüstetes Fernrohr n. ‖ lunette f. munie de traits stadimétriques. / terrestrial ~ ‖ Erdfernrohr n.; terrestrisches Fernrohr n.; Fernrohr n. für Erdbeobachtung ‖ lunette f. terrestre; longue-vue f. / transit ~ ‖ durchschlagbares Fernrohr n. ‖ lunette f. révolutionnant complètement; lunette f. pouvant être retournée sur elle-même. / travelling ~ ‖ Reisefernrohr n. ‖ lunette f. de voyage. / ~ mounted on trunnions in fork bearing ‖ in Gabelaufhängung gelagertes Fernrohr n. ‖ lunette f. suspendue dans une fourche. / ~ for use in the unsupported hand ‖ Fernrohr n. für freihändigen Gebrauch ‖ lunette f. s'employant à la main.

telescope automaton ‖ Automateneinrichtung f. am Fernrohr ‖ déclenchement m. automatique de la lunette d'approche. / ~ base ‖ Fernrohrfuß m. ‖ pied m. de la lunette. / coarse motion of a ~ in altitude ‖ Grobverstellung f. eines Fernrohrs in Höhe ‖ mise f. au point vertical à la main de la lunette. / ~ cradle ‖ Fernrohrwiege f. ‖ berceau m. de lunette.

telescope fork ‖ Fernrohrgabel f. ‖ fourche f. pour recevoir la lunette. / ~ **gas-holder** ‖ Teleskopgasbehälter m. ‖ gazomètre m. à lunette. / ~ **head** ‖ Fernrohrobjektivkopf m. ‖ portée f. de lunette. / **illuminating arrangement on** ~ ‖ Beleuchtungseinrichtung f. am Fernrohr ‖ appareil m. d'éclairage de la lunette. / ~ **lens** ‖ Fernrohrlinse f. ‖ lentille f. de télescope. / ~ **lens combination made of special optical glasses** ‖ aus optischen Spezialgläsern hergestelltes Fernrohrobjektiv n. ‖ objectif m. de lunette à lentilles taillées dans des verres spéciaux. / ~ **level** ‖ Nivellierinstrument n. mit Fernrohr ‖ niveau m. à lunette. / **motion of a** ~ **by wormwheel with disengaging worm** ‖ Bewegung f. eines Fernrohres durch Schneckenrad mit auslösbarer Schnecke ‖ mouvement m. d'une lunette par vis tangente qu'on peut débrayer. / ~ **mounting** ‖ Fernrohrmontierung f. ‖ monture f. de lunette. / **equatorial** ~ **mounting** ‖ parallaktische Fernrohrmontierung f. ‖ monture f. parallactique de lunette.
telescope objective ‖ Fernrohrobjektiv n. ‖ objectif m. de lunette. / **apochromatic** ~ ‖ apochromatisches Fernrohrobjektiv n. ‖ objectif m. de lunette apochromatique. / **astronomical** ~ ‖ astronomisches Fernrohrobjektiv n. ‖ objectif m. de lunette astronomique. / **two-lens apochromatic** ~ **of glasses without secondary spectrum** ‖ zweiteiliges apochromatisches Fernrohrobjektiv n. ‖ objectif m. de lunette apochromatique à deux lentilles en verre exempts de spectre secondaire.
telescope parts pl. ‖ Fernrohrteile mpl. ‖ pièces fpl. détachées de la lunette. / **to accomodate all parts of a** ~ **in one case** ‖ die sämtlichen Teile npl. eines Fernrohres in einem Aufbewahrungskasten unterbringen ‖ caler dans une boîte toutes les pièces d'une lunette. / **keeping case for all** ~ **parts** ‖ Aufbewahrungskasten m. für sämtliche Fernrohrteile ‖ coffret m. pour loger toutes les pièces de la lunette. / **position of the** ~ ‖ Fernrohrlage f. ‖ position f. de la lunette. / **slot-paying mechanism of** ~ ‖ Geldautomateneinrichtung f. am Fernrohr ‖ déclenchement m. monétaire automatique de la lunette. / **slow motion of the** ~ **in altitude** ‖ Feinbewegung f. des Fernrohres in Höhe ‖ mouvement m. lent vertical de la lunette. / **slow motion of the** ~ **in azimuth** ‖ Feinbewegung f. des Fernrohres in Azimut ‖ mouvement m. lent horizontal de la lunette. / ~ **standard with equatorial head** ‖ Fernrohrsäule f. mit parallaktischem Achsensystem ‖ colonne f. de lunette portant la monture parallactique. / **stress relieving system of** ~ ‖ Entlastungssystem n. des Fernrohres ‖ dispositif m. de mise au repos de la lunette. / ~ **tripod** ‖ Fernrohrständer m.; Fernrohrstativ n. ‖ pied m. de lunette. / ~ **tripod elevating gear** ‖ Hochstellvorrichtung f. eines Fernrohrstativs ‖ dispositif m. élévateur d'un pied de lunette. / ~ **trunnion sleeve** ‖ Fernrohrschelle f. ‖ collier m. de la lunette.
telescopic ‖ teleskopisch ‖ télescopique. / ~ **alidade with vertical circle for topographic mapping** ‖ Kippregel f. mit Höhenkreis für topografische Arbeiten ‖ alidade f. munie d'un cercle vertical pour les levés topographiques. / ~ **alidade and cross-sectioning apparatus** ‖ Profilzeichner m. und Kippregel f. ‖ alidade f. tachygraphe pour le levé des profils. / ~ **leg** (Airpl) ‖ Federstrebe f. ‖ guide f. d'amortisseur.
telescopic magnifier ‖ Fernrohrlupe f. ‖ téléloupe f. / **binocular** ~ ‖ binokulare Fernrohrlupe f. ‖ téléloupe f. binoculaire. / ~ **magnifying up to x times** ‖ xfach vergrößernde Fernrohrlupe f. ‖ téléloupe f. d'un grossissement atteignant x fois le diamètre.
telescopic mast ‖ Teleskopmast m. ‖ mât m. télescopique. / ~ **microscope** ‖ Fernrohrmikroskop n. ‖ télémicroscope m. / ~ **sight** ‖ Zielfernrohr n. ‖ lunette f. viseur. / ~ **sight for machine-guns** ‖ Zielfernrohr n. für Maschinengewehre ‖ lunette f. viseur pour mitrailleuses. / ~ **span wing** ‖ Ausziehflügel m. ‖ aile f. à allongement ou à étirement; aile f. étirable.
telescopic spectacles pl. ‖ Fernrohrbrille f. ‖ télélunettes fpl.; lunettes fpl. grossissantes. / ~ **for extremely weak sights** ‖ Fernrohrbrille f. für hochgradig Schwachsichtige ‖ lunettes fpl. grossissante pour amblyopes. / ~ **trial combination** ‖ Fernrohrbrillenprobsystem n. ‖ système m. d'essais de lunettes grossissantes.
telespectroscope ‖ Fernspektroskop n. ‖ téléspectroscope m.
tele-transmitter for differential pressure meters ‖ Ferngeber m. für Druckdifferenzmesser ‖ télé-transmetteur m. pour compteurs de différence de la pression.
teletyper ‖ Ferndrucker m. ‖ téléimprimeur m.; téléscripteur m.
television ‖ Fernsehen n. ‖ télévision f.; radiovision f.
telewriter ‖ Fernschreiber m.; Ferndrucker m. ‖ télautographe m.
telltale ‖ Zähler m.; Messer m.; selbsttätige Anzeigevorrichtung f; Registriervorrichtung f. ‖ compteur m.; indicateur m.; enregistreur m. / **watchman's** ~ see ~ **watch**.
telltale board (Electr) ‖ Anzeigebrett n. ‖ tableau m. d'indication ou de contrôle. / ~ **compass** ‖ Hängekompaß m.; Kajütskompaß m. ‖ compas m. de chambre; compas m. suspendu. / ~ **watch** ‖ Wächterkontrolluhr f.; Wächteruhr f.; Kontrolluhr f. ‖ contrôleur m. des rondes.
telluric acid (Chem) ‖ Tellursäure f. ‖ acide m. tellurique.
telluric ochre see **tellurite**.
telluride of lead ‖ Tellurblei n. ‖ plomb m. telluré; altaïte f.
tellurite ‖ tellurige Säure f. ‖ acide m. tellureux.
tellurium ‖ Tellur n. ‖ tellure m. / **black** ~ ‖ Blättertellur n.; Nagyagererz n.; Nagyagit m. ‖ tellure m. natif auro-plombifère. / **graphic** ~ see **yellow** ~. / **native** ~ ‖ gediegenes Tellur n. ‖ tellure m. natif. / **yellow** ~ ‖ Schrifterz n.; Schrifttellur n.; Sylvanit m.; Weißtellur n. ‖ or m. graphique ou sylvane m.; tellure m. natif auro-argentifère.
tellurium-bearing ‖ tellurführend ‖ tellurifère.
tellurium ore ‖ Tellurerz n. ‖ minerai m. de tellure.
tellurous acid ‖ tellurige Säure f. ‖ acide m. tellureux.
telpherage ‖ (elektrische) Lastenbeförderung f. ‖ telphérage m.

telpher line ‖ elektrische Seilbahn f.; Elektrohängebahn f. ‖ ligne f. de telphérage (électrique); électro-chemin m. suspendu.
telpher way see **telpher line**.
temper, to (Met) ‖ härten; abschrecken; tempern; vergüten; veredeln ‖ tremper; recuire. / ~ **the steel** ‖ Stahl m. anlassen oder härten ‖ recuire ou faire revenir ou tremper l'acier.
temper ‖ Härtegrad m. ‖ trempe f.; degré m. de dureté. / ~ **of steel** ‖ Anlaßhärte f. des Stahles ‖ revenu m. de l'acier.
tempera colour ‖ Temperafarbe f. ‖ couleur f. à détrempe.
temperament ‖ Temperament n. ‖ tempérament m.
temperate climate ‖ gemäßigtes Klima n. ‖ climat m. tempéré.
temperature ‖ Temperatur f.; Wärmegrad m. ‖ température f. / **at a** ~ **of** ‖ bei einer Temperatur von ‖ à une température de. / **absolute** ~ ‖ absolute Temperatur f.; absolute Wärme f. ‖ température f. absolue. / **ambiant** ~ ‖ Umgebungstemperatur f.; Umgebungswärmegrad m. ‖ température f. ambiante. / **breaking-down** ~ (Chem) ‖ Abbautemperatur f. ‖ température f. de dégradation ou de peptonisation. / **constant** ~ ‖ gleichbleibende Temperatur f.; gleichbleibender Wärmegrad m. ‖ température f. constante. / **final** ~ ‖ Endtemperatur f. ‖ température f. finale. / **finishing** ~ (Malt kiln) ‖ Abdarrtemperatur f. ‖ température f. finale de touraillage. / **hardening** ~ ‖ Härtetemperatur f. ‖ température f. de trempe. / **high** ~ ‖ hohe Temperatur f.; hoher Wärmegrad m. ‖ température f. élevée. / **high** ~ **of test piece** ‖ hohe Temperatur f. des Probestabes ‖ température f. élevée de l'éprouvette. / **initial** ~ ‖ Anfangstemperatur f.; Anfangswärmegrad m. ‖ température f. initiale. / **inlet** ~ ‖ Eintrittstemperatur f.; Eintrittswärmegrad m. ‖ température f. d'admission. / **internal** ~ ‖ Innentemperatur f.; Innenwärmegrad m. ‖ température f. régnant à l'intérieur. / **kilning** ~ ‖ Darrtemperatur f. ‖ température f. de touraillage. / **low** ~ ‖ niedrige oder tiefe Temperatur f.; niedriger Wärmegrad m. ‖ température f. basse. / **technics** pl. **of the low** ~**s** ‖ Kältetechnik f. ‖ technique f. des basses températures. / **lowest** ~ ‖ tiefste Temperatur f.; niedrigster Wärmegrad m. ‖ température f. la plus basse. / **mean** ~ **of the water to be frozen** ‖ mittlere Gefrierwassertemperatur f. ‖ température f. moyenne de l'eau à congeler. / **mean annual** ~ ‖ mittlere Jahrestemperatur f. oder Jahreswärme f. ‖ moyenne f. annuelle de température. / **outdoor** ~ ‖ Außentemperatur f. ‖ température f. extérieure. / **outside** ~ see **outdoor** ~. / **peptonization** ~ ‖ Abbautemperatur f. ‖ température f. de dégradation ou de peptonisation. / **pitching** ~ (Brew) ‖ Anstelltemperatur f.; Anstellwärmegrad m. ‖ température f. de mise en levain. / **room** ~ ‖ Zimmertemperatur f. ‖ température f. intérieure ou du laboratoire. / **at room** ~ ‖ bei Zimmertemperatur ‖ à la température du laboratoire. / ~ **in the shade** ‖ Wärmegrad m. im Schatten; Schattentemperatur f. ‖ température f. à l'ombre. / **sparging** ~ (Brew) ‖ Anschwänztemperatur f. ‖

température f. de lavage. / tempering ~ ‖ Anlaßtemperatur f.; Anlaßwärmegrad m. ‖ température f. de recuit.
temperature bath ‖ Temperierbad n. ‖ four m. tient-chaud. / ~ coefficient ‖ Wärmezahl f.; Temperaturkoeffizient m. ‖ coefficient m. de température. / ~ difference ‖ Wärmegradunterschied m.; Temperaturgefälle n. ‖ différence f. de température; écart m. de température. / distribution of ~ ‖ Temperaturverteilung f. ‖ distribution f. de la température. / ~ fuse (Electr) ‖ Temperatursicherung f. ‖ protecteur m. thermique. / ~ gradient ‖ Wärmegefälle n. ‖ chute f. de température; gradient m. de la température. / increase in ~ ‖ Temperaturzunahme f.; Wärmegradzunahme f. ‖ augmentation f. de température. / ~ inversion ‖ Temperaturumkehr f. ‖ inversion f. de température. / ~ meter ‖ Temperaturmesser m.; Wärmegradmesser m. ‖ appareil m. à mesurer la température. / ~ regulator ‖ Temperaturregler m.; Wärmegradregler m.; Wärmeregler m. ‖ régulateur m. de température. / rise in ~ ‖ Wärmezunahme f. ‖ hausse f. ou augmentation f. de température. / variation of ~ ‖ Temperaturschwankung f.; Schwankung f. des Wärmegrades ‖ fluctuation f. de température.
temper carbon ‖ Temperkohle f. ‖ carbone m. de recuit; graphite m.
tempered ‖ gemäßigt; abgemildert; temperiert ‖ tempéré. / ~ (Met) ‖ getempert; gehärtet ‖ trempé; recuit.
tempering (Met) ‖ Härtung f.; Härten n.; Tempern n.; Vergüten n. ‖ trempe f.; refroidissement m. lent; traitement m. thermique. / ~ of steel ‖ Stahlhärtung f. ‖ trempe f. de l'acier.
tempering bath ‖ Härtebad n. ‖ bain m. de trempe. / ~ box ‖ Glühtopf m. ‖ pot m. à recuit. / ~ chemical ‖ Chemikalie f. für die Härtetechnik ‖ produit m. chimique à tremper. / ~ colour (Paint) ‖ Temperafarbe f. ‖ couleur f. à détrempe. / ~ colour (Met) ‖ Anlauffarbe f. ‖ couleur f. de recuit. / crack from ~ ‖ Härteriß m. ‖ fissure f. de trempe. / ~ flame furnace ‖ Härteflammofen m. ‖ four m. à réverbère à tremper. / flaw from ~ see crack from ~. / ~ furnace ‖ Härteofen m.; Temperofen m.; Anlaßofen m. ‖ four m. de cémentation ou à tremper ou à recuire ou à faire revenir. / electrically heated ~ furnace ‖ elektrischer Anlaßofen m. ‖ four m. électrique à recuire. / ~ powder ‖ Härtepulver n.; Zementierpulver n. ‖ poudre f. à tremper (les métaux). / ~ stove see ~ furnace. / ~ temperature ‖ Anlaßtemperatur f.; Anlaßwärmegrad m. ‖ température f. de recuit. / ~ water ‖ Härtewasser n. ‖ eau f. ou liquide m. de trempe. / ~ workman ‖ Härter m. ‖ trempeur m.
tempest ‖ Unwetter n.; Gewitter n. ‖ gros temps m.; orage m.; tempête f.
template ‖ Stichmaß n. ‖ gabarit m.; pige f.
temple (Anatomy) ‖ Schläfe f. ‖ tempe f. / ~ (Arch) ‖ Tempel m. ‖ temple m. / ~ (Weav) ‖ Zeugspanner m. ‖ temple m. / self-adjusting ~ (Weav) ‖ selbstwirkender Zeugspanner m. ‖ temple m. continu.
templet (Gauge) ‖ Lehre f. ‖ jauge f.; lunette f.; calibre m. / ~ (Pattern) ‖ Schablone f. ‖ gabarit m.; sabot m.; étalon m. / ~ (Carp) ‖ Dachpfette f.; Pfette f. ‖ filière f.

de comble; panne f. / ~ (Found) ‖ Lehre f.; Schablone f. ‖ calibre m.; échantillon m.; gabarit m.; panneau m. / ~ (Pott) ‖ Schablone f.; Lehre f.; Drehbrett n. ‖ estèque f.; calibre m.; échantillon m. / ~ for the industry of artificial stones ‖ Schablone f. für die Kunststeinindustrie ‖ étalon m. pour l'industrie de pierres artificielles. / ~ for boring ‖ Bohrlehre f. ‖ calibre m. de perçage. / ~ for scoring sleepers (Railw) ‖ Lehre f. zum Einschneiden der Schwellen ‖ gabarit m. pour l'entaillage des traverses.
templet, arm support for ~s ‖ Schablonenhalter m. ‖ support m. d'étalon. / ~ moulding ‖ Schablonenformerei f. ‖ moulage m. à la trousse.
temporary ‖ zeitweilig; vorübergehend; einstweilig; aushilfsweise ‖ provisoire. / ~ bridge ‖ Notbrücke f. ‖ pont m. de circonstances; pont m. provisoire. / ~ end sleeve (Cable) ‖ Transportverschluß m. ‖ capote f. d'extrémité de transport. / ~ railway ‖ provisorische Eisenbahn f.; Interimsbahn f. ‖ chemin m. de fer provisoire. / ~ rudder (Mar) ‖ Notruder n. ‖ gouvernail m. de fortune.
tenacity ‖ Zähigkeit f. ‖ ténacité f.
tenancy ‖ Pacht f. ‖ ferme f.; bail m.; fermage m.
tenant ‖ Mieter m.; Pächter m. ‖ locataire m.
tend, to ~ to . . . ‖ neigen zu . . . ‖ tendre à . . .
tendency ‖ Neigung f.; Bestreben n.; Tendenz f. ‖ tendance f. / downward ~ ‖ fallende Tendenz f.; Bestreben n. zu fallen ‖ tendance f. à baisser. / weaker ~ of prices ‖ Abschwächung f. der Preise; Preisabschwächung f. ‖ affaiblissement m. ou baisse f. des prix. / ~ to thunderstorm ‖ Gewitterneigung f. ‖ tendance f. à l'orage. / upward ~ ‖ steigende Tendenz f. ‖ tendance f. à la hausse.
tender (Loc) ‖ Tender m. ‖ tender m. / ~ (Shipb) ‖ Begleitschiff n.; Beischiff n. ‖ navire m. matelot; bâtiment m. de servitude; navire m. annexe ou de servitude. / ~ (Trade) ‖ Offerte f.; Lieferungsangebot n. ‖ soumission f. / to give out in ~ ‖ in Submission f. vergeben ‖ mettre à la soumission. / to make a ~ ‖ ein Submissionsangebot n. einreichen ‖ concourrir à une soumission. / to fix the minimum admissible ~ ‖ das niedrigst zulässige Gebot feststellen ‖ fixer l'offre la plus basse. / to send in a ~ ‖ bei der Submission f. konkurrieren ‖ soumissioner. / ~ for coral fishers ‖ Hilfsschiff n. für die Korallenfischerei ‖ navire m. auxiliaire pour la pêche du corail. / highest ~ ‖ Meistgebot n.; Höchstgebot n. ‖ offre f. la plus élevée ou la plus haute. / legal ~ ‖ gesetzliches Zahlungsmittel n. ‖ valeur f. légale. / locomotive ~ ‖ Lokomotivtender m. ‖ tender m. de locomotive. / separate ~ (Loc) ‖ getrennter Tender m. ‖ tender m. séparé.
tender axle (Loc) ‖ Tenderachse f. ‖ essieu m. de tender. / ~ brake ‖ Tenderbremse f. ‖ frein m. du tender. / ~ coupling bar ‖ Tenderkupplungsstange f. ‖ barre f. d'attelage du tender. / ~ date of receipt of ~ ‖ Versteigerungstermin m. ‖ terme m. ou délai m. pour les enchères. / ~ engine ‖ Lokomotive f. mit Schlepptender; Tenderlokomotive f. ‖ locomotive f. et tendeur m.

tenderer ‖ Bewerber m.; Submittent m. ‖ concurrent m.; soumissionnaire m.
tender locomotive ‖ Tenderlokomotive f.; locomotive f. (à) tender. / ~ water tank plate ‖ Tenderwasserkastenblech n. ‖ tôle f. pour caisses à eau du tender. / ~ wheel ‖ Tenderrad n. ‖ roue f. de tender.
tendon ‖ Sehne f. ‖ tendon m. / ~ glue ‖ Sehnenleim m. ‖ colle f. de nerf. / ~ powder ‖ Sehnenpulver n. ‖ poudre f. de tendons.
tennantite ‖ Kupferblende f.; zinkhaltiger Tennantit m.; Arsenfahlerz n. ‖ tennantite f.; cuivre m. gris arsenifère.
tennis ball ‖ Tennisball m. ‖ balle f. tennis. / ~ boot ‖ Tennisschuh m. ‖ chaussure f. tennis. / ~ court ‖ Tennisplatz m. ‖ court m. (de tennis). / ~ equipment ‖ Tennisgerät n. ‖ ustensile m. pour lawn-tennis. / ~ net ‖ Tennisnetz n. ‖ filet m. de tennis.
tennis racket ‖ Tennisschläger m. ‖ raquette f. de tennis. / ~ gut ‖ Tennisschlägersaite f. ‖ corde f. de raquette pour jeu de tennis. / ~ press ‖ Tennisschlägerpresse f. ‖ presse f. pour raquettes de tennis.
tenon (Join; Carp) ‖ Zinke f.; Zapfen m. ‖ tenon m. / to mortise a ~ ‖ einen Zapfen m. einlochen faire la mortaise d'un tenon; trouer un tenon. / bolted ~ ‖ vernagelter Zapfen m. ‖ tenon m. enlacé. / bored ~ ‖ verbohrter Zapfen m. ‖ tenon m. à clef. / ~ to be driven in ‖ Jagdzapfen m. ‖ tenon m. à chasse. / passing ~ ‖ durchgehender Zapfen m. ‖ tenon m. traversant. / square ~ ‖ viereckiger Zapfen m. ‖ tenon m. carré ou à oulices.
tenon cutting machine see tenoning machine.
tenoning machine ‖ Zapfenschneidmaschine f. ‖ machine f. à (faire les) tenons. / ~ machinist ‖ Maschinenzapfenschneider m. ‖ tenoneur à la machine. / ~ tenter ‖ Zapfenschneider m. ‖ tenoneur m.
tenon saw ‖ Zapfensäge f.; Furniersäge f. ‖ scie f. à arraser ou à placage ou à refendre.
tenor violin ‖ Viola f.; Bratsche f. ‖ alto m.
tenotome (Med) ‖ Tenotom n. ‖ ténotome m.
tens pl. ‖ Zehner mpl. ‖ dizaines fpl.
ten-seconds division ‖ Zehnersekunde f. ‖ les dix secondes fpl.; dizaine f. de secondes.
tensile ‖ dehnbar; streckbar ‖ ductile; extensible.
tensile boom (Bridge) ‖ Zuggurtung f. ‖ semelle f. à tension. / ~ strain ‖ Zugbeanspruchung f. ‖ effort m. de tension; travail m. à la traction.
tensile strength ‖ Zugfestigkeit f. ‖ résistance f. à la traction. / ultimate ~ ‖ Zerreißfestigkeit f.; Bruchfestigkeit f. ‖ résistance f. à la rupture. / the ~ is x kg per square millimeter ‖ die Zugfestigkeit f. beträgt x kg je Quadratmillimeter ‖ la résistance à la traction est de x kg par millimètre carré.
tensile strength test ‖ Zerreißprobe f. ‖ essai m. à la rupture. / ~ testing machine ‖ Zerreißmaschine f. ‖ machine f. pour l'essai à la rupture.
tensile stress ‖ Zugspannung f. ‖ tension f. de traction.
tensile test ‖ Zugversuch m.; Zerreißprobe f. ‖ épreuve f. de traction; essai m. à la rupture. / machine for ~s ‖ Zerreißmaschine f. ‖ machine f. à essais à la traction.

tensile testing machine || Festigkeitsprüf-maschine f.||machine f. à vérifier la résistance mécanique.

tension (Electr) || Spannung f. || tension f. / ~ (Mech) || Zug m.; Spannung f.; An-spannung f. || traction f.; extension f.; tension f. / ~ (Met) || Spannung f. || ser-rage m. / ~ (Steam eng) || Spannung f.; Druck m. || pression f.; tension f. / to be under ~ || unter Spannung f. stehen || être sous pression f. / to increase the ~ of the spring || die Feder f. nachspannen || retendre *ou* resserrer le ressort. / ~ of the belt || Riemenspannung f. || tension f. de la courroie. / high ~ || Hochspannung f. || haute tension f. / low ~ || Niederspan-nung f. || basse tension f.

tension arrangement || Nachstellvorrich-tung f. || dispositif m. de tension. / ~ brace (Bridge) || Zugdiagonale f. || tige f. inclinée; tirant m. incliné diagonale. / ~ cable || Spanndraht m. || câble m. de tension. / ~ difference (Electr) || Span-nungsunterschied m.; Spannungsdiffe-renz f. || différence f. de tension. / ~ disk (Of a sewing machine) || Bremsscheibe f. || disque m. de freinage. / ~ dynamometer || Zugmesser m. || dynamomètre m. de traction. / ~ flange || Zuggurtung f. || membrure f. tendue. / ~ lever || Spann-hebel m. || levier m. de tension. / ~ pul-ley || Spannrolle f. || galet m. *ou* poulie f. *ou* rouleau m. de tension. / ~ regulation table (El line) || Spannungstafel f. || table f. de tension des fils. / ~ rod || Spann-stange f. || tirant m. / ~ roller (For belt tension) || Spannrolle f. || poulie f. de ten-sion; tendeur m. de courroie. / ~ screw || Zugschraube f. || vis f. de tension. / ~ shackle || Spannschloß n. || tendeur m. / ~ spring || Spannfeder f. || ressort m. ten-deur. / ~ test || Zugversuch m. || essai m. à la traction. / ~ testing machine || Zer-reißmaschine f. || machine f. à essayer (les matériaux) à la traction. / ~ and compression testing machine || Prüfma-schine f. für Zug- und Druckversuche || machine f. à essayer (les matériaux) à la traction et à la compression.

tent || Zelt n. || tente f. / ~ for aeroplane || Flugzeugzelt n. || tente f. pour aéro-plane.

tent canvas || Zeltleinwand f. || toile f. à voile. / ~ cloth see tent canvas.

tenter, to (Clothm) || aufrahmen || arramer; ramer.

tenter (Frame for stretching cloth) || Spann-rahmen m.; Trockenrahmen m. || rame f. / ~ (One who takes care of machines) || Maschinenwärter m. || machiniste m.; mécanicien m. / cupola ~ || Kupolofen-schmelzer m. || fondeur m. au cubilot.

tenter hook (Nailsm) || Hakennagel m. || clou m. à crochet. / ~ (Weav) || Kluppe f.; Spannhaken m. || tenaille f.

tentering (Weav) || Aufrahmen n. || ramage m. / ~ frame || Rahmenmaschine f.; Spannrahmen m. || machine f. à ramer; étendoir m. / ~ limit || Spannfeld n. || champ m. d'élargissement. / ~ machine see ~ frame. / ~ and drying machine || Rahm- und Trockenmaschine f. || ra-meuse-sécheuse f.

tent fittings pl. || Zeltbeschläge mpl. || garnitures fpl. de tentes.

tent pole || Zeltstange f.; Zeltpfahl m. || arbre m. *ou* mât m. de tente.

tent roof || Zeltdach n. || toit m. en pavillon.

tenure || Pachtvertrag m. || contrat m. de fermage.

tepid || lau; lauwarm || tiède.

terchloride of gold || Goldchlorid n. || or m. potable; sesqui-chlorure m. d'or.

term (Time) || Frist f.; Termin m. || delai m.; terme m. / ~ of delivery || Liefer-frist f.; Lieferzeit f. || délai m. *ou* terme m. de livraison. / ~ of payment || Zah-lungstermin m. || terme m. de payement; échéance f.

term (Math) || Glied n. || terme m. || mean ~s pl. || mittlere Glieder npl. || termes mpl. moyens. / extreme ~s pl. (Math) || äußere Glieder npl. || termes mpl. extrêmes.

term (Word) || Ausdruck m.; Wort n. || terme m. / technical ~ || technischer Ausdruck m. || terme m. technique.

terms pl. (Bedingungen) || conditions fpl.; termes mpl. / to come to ~ || über-einkommen || tomber d'accord. / to make ~ || Bedingungen fpl. aufstellen || établir des conditions fpl. / the ~ are unreason-able || die Bedingungen fpl. sind un-billig || les conditions fpl. ne sont pas raisonnables. / advantageous ~ || vor-teilhafte Bedingungen fpl. || conditions fpl. avantageuses. / ~ of agreement || Vertragsbedingungen fpl. || conditions fpl. de contrat. / best ~ || beste Be-dingungen fpl. || les meilleures conditions fpl. / on best possible ~ || unter best-möglichen Bedingungen || aux condi-tions les meilleures possibles. / ~ of contract || Submissionsbedingungen fpl. || cahier m. des charges. / ~ of delivery || Lieferbedingungen fpl.; Lieferungsbe-dingungen fpl.; Bezugsbedingungen fpl. || conditions fpl. de livraison. / favourable ~ || günstige Bedingungen fpl. || condi-tions fpl. favorables. / impossible ~ || unmögliche Bedingungen fpl. || condi-tions fpl. impossibles. / ~ of renewal || Erneuerungsbedingungen fpl. || termes mpl. de renouvellement. / special ~ || besondere Bedingungen fpl. || conditions fpl. spéciales. / unreasonable ~ || un-billige Bedingungen fpl. || conditions fpl. pas raisonnables. / ~ as usual || Be-dingungen fpl. wie üblich || conditions fpl. d'usage.

terminable || kündbar || limité.

terminal || begrenzend; die Grenze bildend; End... || terminal; final.

terminal (Electr) || Klemme f.; Polklemme f. || borne f.; serre-fils m. / accumulator ~ || Akkumulatorklemme f. || borne f. d'accumulateur. / concentric ~ || kon-zentrische Klemme f. || serre-fils m. con-centrique. / electrode ~ || Polklemme f. || borne f. d'élément. / flat ~ || Flach-klemme f. || borne f. plate. / intermediate connecting ~ || Verbindungsklemme f. || serre-fils m. / ~ with wing nut || Flügel-klemme f. || borne f. à oreilles.

terminal bar (Acc) || Bleileiste f. || bande f. de plomb. / ~ board (Electr) || Klemm-brett n. || planche f. *ou* tablette f. à bornes de jonction. / ~ box || Kabelend-verschluß m.; Abschlußmuffe f.; Klemm-schutzkasten m. || tête f. de câble; man-chon m. de fermeture; boîte f. de pro-tection de bornes de jonction. / ~ bracket (El line) || Eisenkonsole f. || console f. d'arrêt. / ~ cable || Abschlußkabel n. || câble m. de fermeture. / ~ cell || Endzelle f. || cellule f. extrême. / ~ charge (Tel) ||

Endgebühr f. || taxe f. terminale. / covered-in ~ connection || verdeckter Klemmenanschluß m. || bornes fpl. re-couvertes. / ~ country (Tel) || Endland n. || pays m. terminal. / ~ double-pin || Abspanndoppelstütze f.; W-förmige Dop-pelstütze f. || console f. double d'arrêt. / ~ edge of a crystal || Polkante f. eines Kristalls || arête f. culminante d'un cristal. / ~ exchange (Tel) || Endanstalt f. || bureau m. terminal. / ~ face (Of a crystal) || Endfläche f. || face f. terminale; base f. / ~ office of an international cir-cuit || Grenzausgangsanstalt f. || bureau m. tête de ligne internationale. / ~ pole || Ab-spanngestänge n. || appui m. d'arrêt. / ~ potential difference || Klemmen-spannung f. || tension f. aux bornes. / ~ relay || Endrelais n. || relais m. terminal. / ~ repeater || Endverstärker m. || répé-teur m. terminal; amplificateur m. d'extrémité. / ~ repeater with sending and receiving amplification || Endver-stärker m. mit Sende- und Empfangs-verstärkung || amplificateur m. d'extré-mité amplifiant les courants trans-metteur et récepteur. / ~ station || End-station f.; Kopfbahnhof m. || (gare f.) terminus m. / ~ traffic (Tel) || Endverkehr m. || trafic m. de départ *ou* d'arrivée. / ~ velocity || Endgeschwindigkeit f. || vitesse f. finale. / ~ voltage || Klemmen-spannung f. || tension f. aux bornes.

terminate, to || begrenzen; bestimmen || borner; limiter. / ~ the employment unlawfully || das Arbeitsverhältnis un-berechtigt lösen || rompre sans justifica-tion le contrat de travail. / ~ the engage-ment || das Arbeitsverhältnis beendigen; (Bergb:) abkehren || s'écarter de la con-vention de travail; mettre fin f. à la convention.

terminated || abgeschlossen || terminé.

terminating (Aerial lines) || Abspannung f. || ligature f. à l'isolateur de tension.

termination || Ende n.; Beendigung f. || fin f. / to give notice of the ~ of an engagement || einen Arbeitsvertrag m. kündigen || résilier un contrat de travail.

terminology || Terminologie f.; Fach-sprache f.; Fachausdrücke mpl. || termi-nologie f. / ~ of customs || Zollterminolo-gie f. || terminologie f. douanière.

terminus (Railw) || Kopfbahnhof m.; End-bahnhof m.; Endstation f.; Ausgangs-bahnhof m. || station f. terminus; ter-minus m. / central ~ || Sammelbahnhof m. || gare f. centrale.

terms pl. *see under* term.

ternary (Chem) || ternär; dreistoffig; aus drei Elementen bestehend || ternaire. / ~ compound || ternäre *oder* dreistoffige Verbindung f. || combinaison f. ternaire.

terpene of essential oils || Terpen n. äthe-rischer Öle || terpène m. des huiles essentielles.

terpeneless || terpenfrei || déterpéné.

terrace, to ~ (Railw) || Abhänge mpl. terrassieren *oder* stufenweise erhöhen || ménager des banquettes fpl.

terrace || Terrasse f. || terrasse f. / ~ on a roof || Plattform f. auf einem Dach || terrasse f. sur un toit.

terrace walk || gemauerte Terrasse f. || terrasse f. maçonnée.

terracotta || Terrakotta f.; Kunsttonware f.; gebrannte Tonware || terre f. cuite. / ~ floor || Terrakottafußboden m. ||

dallage m. en terre cuite. / ~ goods pl. ‖ Terrakottawaren fpl. ‖ articles mpl. en terre cuite. / ~ statue ‖ Terrakottafigur f. ‖ statuette f. en terre cuite.

terra japonica ‖ Katechu n. ‖ cachou m.

terralith ‖ Terralith m. ‖ sidérolithe m.

terrestrial anti-aircraft defence ‖ Flugabwehr f. von der Erde aus ‖ défense f. terrestre antiaérienne. / ~ globe ‖ Erdglobus m. ‖ globe m. terrestre. / ~ observation ‖ Erdbeobachtung f.; Geländebeobachtung f.; terrestrische Beobachtung f. ‖ observation f. terrestre. / ~ telescope ‖ Erdfernrohr n.; Fernrohr n. für terrestrische Beobachtung ‖ lunette f. terrestre.

terrine ‖ Schüssel f. ‖ terrine f.

territorial ‖ territorial ‖ territorial. / ~ sea ‖ Küstenmeer n. ‖ mer f. territoriale.

territory ‖ Gebiet n.; Territorium n. ‖ territoire m. / ~ (Sphere of interest) ‖ Interessengebiet n. ‖ zône f. d'intérêt. / ~ proper of a company ‖ das engere Arbeitsgebiet n. einer Gesellschaft ‖ rayon m. de travail d'une société. / ~ of the undertaking ‖ Arbeitsgebiet n. der Unternehmung ‖ rayon m. de l'entreprise.

terry velvet ‖ Halbsamt m.; Ritzer m.; gezogener oder ungerissener oder ungeschnittener Samt m. ‖ velours m. frisé ou épinglé.

tertiary ‖ tertiär ‖ tertiaire.

test, to ‖ (aus)proben; prüfen; versuchen; untersuchen; erproben ‖ essayer; éprouver; vérifier; rechercher. / ~ a line (Tel) ‖ eine Leitung f. prüfen ‖ examiner ou contrôler une ligne; faire le test d'une ligne. / ~ silk ‖ konditionieren ‖ conditionner la soie. / tested for x atmospheres pl. ‖ auf x Atmosphären fpl. geprüft ‖ éprouvé sur ẋ atmosphères fpl.

test ‖ Probe f.; Versuch m.; Prüfung f.; Untersuchung f. ‖ essai m.; épreuve f.; contrôle m. recherche f. / ~ (Chem) ‖ Reagens n. ‖ réactif m. / ~ (Goldsm) ‖ Probierscherben m.; Text m. ‖ casse f.; coupelle f. d'essai. / the ~ was executed in accordance with the specifications. ‖ der Versuch m. wurde nach den Bedingungen durchgeführt ‖ l'essai m. fut exécuté d'après les conditions du cahier des charges. / to stand the ~ ‖ sich bewähren ‖ donner de bons résultats mpl. / acceptance ~ ‖ Abnahmeprüfung f. ‖ essai m. de réception. / ~ of accuracy ‖ Genauigkeitsprüfung f. ‖ épreuve f. de précision. / bending ~ by shock ‖ Schlagbiegeversuch m. ‖ essai m. de flexion par choc. / boiler ~ ‖ Kesselprobe f. ‖ épreuve f. des chaudières. / brake ~ ‖ Bremsprüfung f. ‖ essai m. au frein. / ~ with a cast-on piece ‖ Probe f. mit einem angegossenen Stück ‖ essai m. sur une barre venue de fonte avec la pièce. / deflection ~ ‖ Durchbiegeversuch m. ‖ essai m. de flexion. / endurance ~ ‖ Dauerversuch m. ‖ essai m. de longue durée. / hardness ~ ‖ Härteprüfung f. ‖ examen m. de la dureté. / hot bending ~ ‖ Warmbiegeprobe f. ‖ essai m. de ployage à chaud. / impact ~ ‖ Kerbschlagprobe f. ‖ essai m. de choc sur barre entaillée. / ~ of iron ‖ Eisenprobe f. ‖ essai m. du fer. / ~ of a line (Tel) ‖ Prüfung f. einer Leitung ‖ essai m. d'une ligne. / longduration ~ ‖ Dauerversuch m.; Dauerprüfung f. ‖ essai m. d'endurance. / mandril ~ ‖ Dornprobe f. ‖ essai m. de

mandrinage. / mechanical ~ ‖ mechanische Festigkeitsprobe f. ‖ essai m. de résistance mécanique. / slack-rope ~ ‖ Schlappseilprüfung f. ‖ essai m. de mou du câble. / ~ of steel ‖ Stahlprobe f. ‖ essai m. de l'acier. / taking-over ~ ‖ Abnahmeprüfung f. ‖ épreuve f. de réception. / thorough ~ ‖ eingehender Versuch m. ‖ essai m. à fond. / ~ by torsion ‖ Verdrehungprobe f.; Torsionsversuch m.; Verwindeprobe f. ‖ essai m. de torsion. / upsetting ~ ‖ Stauchversuch m. ‖ essai m. au refoulement. / X-ray ~ ‖ Röntgenuntersuchung f. ‖ radiodiagnostic m.; diagnostic m. par les rayons X.

test balance ‖ Prüfwage f. ‖ trébuchet m.

test bar see also test piece ‖ Zerreißstab m.; Probestab m. ‖ éprouvette f. / ~ cast solid with the piece ‖ angegossener Probestab m. ‖ éprouvette f. venue de fonte avec la pièce moulée; éprouvette f. attenant à la pièce. / V-notched ~ ‖ Probestab m. mit scharfer Einkerbung ‖ éprouvette f. à entaille en V. / ~ diameter and gauged length of ~s ‖ Probestababmessungen fpl. ‖ dimensions fpl. de l'éprouvette. / shop for preparing ~s ‖ Probestabwerkstatt f. ‖ atelier m. pour tourner les éprouvettes.

test bar milling machine ‖ Probestabfräsmaschine f. ‖ fraiseuse f. pour éprouvettes.

test barometer ‖ Prüfungsluftdruckmesser m.; Prüfungsbarometer m. ‖ baromètre m. d'essai ou de contrôle. / ~ bed ‖ Prüfstand m. ‖ plate-forme f. d'essais; banc m. d'épreuve.

test board (Tel) ‖ Klinkenumschalter m. ‖ commutateur m. à jacks. / ~ for lost calls ‖ Abwerfeinrichtung f. ‖ montage m. d'épreuve pour les appels défectueux. / ~ of repeater stations ‖ Prüfschrank m. für Verstärkerämter ‖ table f. de mesures dans les stations de répéteurs.

test box (Tel) ‖ Prüfkasten m.; Untersuchungskasten m. ‖ boîte f. d'essai ou de coupure. / ~ for private branch exchanges ‖ Postprüfeinrichtung f. für Privatnebenstellenanlagen ‖ appareil m. de contrôle pour installations privées.

test cable stub ‖ Prüfstumpf m. ‖ bout m. d'essai de câble. / ~ certificate ‖ Werkprüfzeugnis n. ‖ tableau m. d'essai. / ~ circuit ‖ Prüfstromkreis m. ‖ circuit m. d'épreuve. / ~ circuit (Tel) ‖ Prüfleitung f. ‖ ligne f. d'essai. / ~ clerk's desk (Tel) ‖ Prüfschrank m. ‖ table f. d'essai. / most favourable condition of ~ ‖ günstigste Versuchsbedingung f. ‖ condition f. très favorable de mesure. / ~ cube ‖ Probewürfel m. ‖ cube m. d'essai. / ~ desk (Tel) ‖ Prüftisch m. ‖ table f. d'essai.

tester for calibration circuit ‖ Eichleitungsprüfer m. ‖ dispositif m. d'essai de ligne étalon. / flue gas ~ ‖ Rauchgasprüfer m. ‖ analyseur m. des gaz de fumée. / grain ~ ‖ Farinatom n. ‖ coupe-grains m.; farinatome m. / ~ for incandescent lamps ‖ Glühlampenprüfer m. ‖ instrument m. à essayer des lampes à incandescence. / machine ~ ‖ Maschinenprüfer m. ‖ ouvrier m. aux essais de machines.

test flight ‖ Probeflug m.; Probefahrt f. ‖ vol m. d'essai. / ~ frame (Tel) ‖ Prüfgestell n. ‖ table f. ou support m. d'essai.

test-glass see also test-tube ‖ Probierglas n.; Reagensglas n. ‖ éprouvette f. / ~

holder ‖ Reagensglashalter m. ‖ porte-éprouvette m.

test hut (Tel) ‖ Untersuchungshäuschen n. ‖ guérite f. de coupure.

testify, to ‖ bezeugen ‖ attester; témoigner.

testimonial ‖ Zeugnis n. ‖ certification f.; attestation f. / to be entitled to a ~ ‖ Anspruch m. auf ein Zeugnis haben ‖ avoir droit m. à un certificat.

testimony ‖ Zeugenaussage f. ‖ déposition f.

testing (Techn) ‖ Prüfung f. ‖ épreuve f. / ~ of great quantities ‖ Massenprüfung f. ‖ essai m. de grandes quantités. / ~ of load (Electr) ‖ Belastungsprüfung f. ‖ contrôle f. de la charge. / ~ of materials ‖ Werkstoffprüfung f. ‖ essai m. ou épreuve f. des matériaux. / machine for mechanical ~ of materials ‖ Maschine f. für die mechanische Prüfung von Werkstoffen ‖ machine f. pour l'essai mécanique des matériaux. / ~ of the nozzle ‖ Düsenprüfung f. ‖ essai m. du gicleur. / ~ of steel balls for resistance of rupture ‖ Prüfung f. der Stahlkugeln auf Bruchfestigkeit ‖ essai m. de la résistance à la rupture des billes en acier. / (dynamometric) ~ of textiles ‖ (dynamometrische) Prüfung f. der Gewebe ‖ essai m. (dynamométrique) des textiles.

testing apparatus ‖ Prüfgerät n. ‖ appareil m. à essayer les matériaux. / flue gas ~ ‖ Rauchgasprüfer m. ‖ appareil m. à analyser les gaz de fumée. / ~ for glow lamps ‖ Glühlampenprüfgerät n. ‖ appareil m. d'essai pour lampes à incandescence. / physical ~ ‖ physikalisches Prüfgerät n. ‖ appareil m. d'essai physique. / ~ for testing tapers ‖ Kegelmeßgerät n. ‖ appareil m. pour la vérification de cônes. / ~ for water meters ‖ Wassermesserprüfgerät n. ‖ appareil m. de contrôle des compteurs d'eau.

testing battery ‖ Meßbatterie f. ‖ batterie f. ou pile f. de mesure. / ~ body (Auto) ‖ Probefahrtkarosserie f. ‖ carrosserie f. d'essai. / ~ circuit (Electr) ‖ Meßstromkreis m.; Prüfstromkreis m. ‖ circuit m. de mesure ou d'essai. / ~ department ‖ Prüffeld n. ‖ plate-forme f. d'essais. / ~ directions pl. ‖ Versuchsvorschriften fpl.; règlement m. des essais. / drilling machine table for ~ ‖ Versuchsbohrtisch m. ‖ table f. pour essais de perçage sur les foreuses. / ~ floor ‖ Meßzimmer n. ‖ salle f. d'épreuves. / ~ glass see test glass. / ~ height ‖ Prüfhöhe f. ‖ hauteur f. d'essai. / ~ house to control the current manufacture ‖ Probeanstalt f. zur Überprüfung der laufenden Fertigung ‖ station f. d'essais pour contrôler la fabrication courante. / ~ instrument see testing apparatus. / ~ jack for main distributing boards ‖ Prüfklinke f. für Hauptverteiler ‖ jack m. d'essai au répartiteur d'entrée. / ~ key (Tel) ‖ Meßtaste f. ‖ touche f. d'essai. / ~ lamp ‖ Prüflampe f. ‖ lampe f. d'essai. / ~ load ‖ Probebelastung f. ‖ charge f. d'épreuve.

testing machine ‖ Prüfmaschine f.; Werkstoffprüfmaschine f. ‖ machine f. à essayer les matériaux; machine f. d'essai. / bending vibration ~ ‖ Biegeschwingungsprüfmaschine f. ‖ machine f. à mesurer les vibrations dues à la flexion. / ~ for the brick making industry ‖ Prüfmaschine f. für die Ziegelindustrie ‖ machine f. d'essai pour l'industrie des bri-

ques. / buckling ~ || Knickprüfmaschine f. || machine f. à essayer les matériaux au flambage. / chain ~ || Kettenprüfmaschine f. || machine f. à essayer les chaînes. / chain and rope ~ || Ketten- und Seilprüfmaschine f. || machine f. à essayer les chaînes et les câbles. / file ~ || Feilenprüfmaschine f. || machine f. à essayer les limes. / gear ~ || Zahnräderprüfmaschine f. || machine f. à vérifier les engrenages. / hydraulic ~ || hydraulisch betriebene Prüfmaschine f. || machine f. hydraulique d'essai. / ~ for insulators || Isolatorenprüfmaschine || machine f. à essayer les isolateurs. / material ~ || Werkstoffprüfmaschine f. || machine f. à essayer la résistance des matériaux. / ~ for metals || Prüfmaschine f. für Metalle || machine f. à essayer les métaux. / ~ for non-metals || Prüfmaschine f. für Nichtmetalle || machine f. à essayer les matières non-métalliques. / pipe ~ || Rohrprüfmaschine f. || machine f. à essayer les tubes. / spring ~ || Federprüfmaschine f. || machine f. à essayer les ressorts. / ~ for testing tensile strength || Festigkeitsprüfmaschine f. || machine f. pour essais de traction. / tension and compression ~ || Prüfmaschine f. für Zug- und Druckversuche || machine f. à essayer les matériaux à la traction et à la compression.

testing magnet (Electr) || Prüfelektromagnet m. || électro m. de test ou d'essai.

testing method || Prüfverfahren n. || méthode f. d'essai. / ~ for fire-proof building material || Prüfverfahren n. für feuerfeste Baustoffe || méthode f. d'essai pour matériaux réfractaires de construction.

testing point (El line) || Untersuchungsstelle f. || point m. de coupure. / ~ with remote control || Untersuchungsstelle f. mit elektrischer Fernsteuerung || relais m. de coupure à distance.

testing power || Prüfkraft f. || puissance f. d'épreuve. / ~ press for cement || Zementprüfpresse f. || presse f. à essayer le ciment. / ~ record sheet || Prüfungsprotokoll n. || procès-verbal m. de l'épreuve. / ~ report || Prüfungsbericht m. || rapport m. de l'épreuve. / ~ result || Prüf(ungs)ergebnis n. || résultat m. d'essai. / ~ stand || Prüfstand m. || banc m. d'essai. / ~ stand for locomotives || Lokomotivprüfstand m. || plateforme f. d'essai pour locomotives. / ~ station of an electricity work || Prüfamt n. eines Elektrizitätswerkes || station f. d'essais d'une usine électrique. / ~ switch (Electr) || Prüfschalter m. || interrupteur m. d'essai. / ~ time || Probezeit f. || stage m. / ~ van for testing power current cables || Kabelprüfwagen m. zum Prüfen von Starkstromkabeln || fourgon m. d'essai de câbles pour courant fort. / ~ wire || Prüfdraht m. || fil m. de contrôle.

test lamp || Prüflampe f. || lampe f. témoin ou d'essai. / ~ for ringing current || Rufstromprüflampe f. || lampe f. d'essai de courant d'appel.

test load || Probebelastung f. || charge f. d'essai ou d'épreuve. / ~ mark || Prüfzeichen n. || marque f. d'épreuve. / ~ paper || Reaktionspapier n.; Reagenspapier n.; Indikatorpapier n. || papier m. réactif ou indicateur.

test piece see also test bar || Probestab m. || éprouvette f. / broken ~ || durchgeschlagener Probestab m. || éprouvette f. cassée

en deux. / round-notched ~ || Probestab m. mit Rundkerb || barreau m. à entaille en ∪.

test plug (Tel) || Prüfstöpsel m. || bouchon m. d'essai. / ~ portion || Probenahme f.; Probe f. || prise f. d'essai; prélèvement m. / ~ pressure || Probedruck m. || épreuve f. de pression. / ~ relay || Prüfrelais m. || relais m. de test. / ~ result || Prüfungsergebnis n. || résultat m. d'essai. / ~ run (Mach) || Probelauf m. || marche f. d'essai. / ~ sample || Probekörper m. || éprouvette f. / ~ selector (Aut Tel) || Prüfwähler m. || sélecteur m. d'essai. / ~ series of ~s || Versuchsreihe f. || série f. d'essais. / (automatic) ~ set (Aut Tel) || Prüfsatz m. || dispositif m. automatique d'essai. / ~ splicing (Cable) || Kreuzungsverfahren n. || procédé m. de croisement. / ~ station || Versuchsstation f. || station f. d'études. / ~ tree (Forest) || Versuchsstamm m. || arbre m. d'expérience.

test-tube see also test glass || Prüfröhre f.; Probierröhre f.; Probierglas n.; Reagensglas n.; Reagensröhre f. || éprouvette f.; tube m. d'essai ou à essayer. / ~ condenser || Reagensglaskondensator m.; Prüfröhrenkondensator m. || condensateur m. pour tubes à essai.

test-type chart || Sehprobentafel f. || tableau m. d'essai de vue. / ~ for weak-sighted eyes || Sehprobentafel f. für Schwachsichtige || tableau m. d'essai pour amblyopes.

test-type holder || Sehprobenhalter m. || porte-optotype m. / ~ voltage of a current transformer || Prüfspannung f. eines Stromwandlers || tension f. d'essai d'un transformateur d'intensité. / ~ wiper (Aut Tel) || Prüfarm m. || frotteur m. privé ou de test. / ~ wire (Tel) || Prüfader f.; Prüfdraht m. || fil m. d'essai ou de contrôle.

tetrachloræthan || Tetrachloräthan n. || tetrachloréthane m.

tetrachloride of platinum || Platinchlorid n.; Platinichlorid n.; Platintetrachlorid n. || tetrachlorure m. de platine.

tetrahedrite || Antimonfahlerz n. || panabase f.; cuivre m. gris.

tetrahedron || Tetraeder n. || tétraèdre m.

tetra-hexahedron || Pyramidenwürfel m. || cube m. pyramidé; hexatétraèdre m.

tetraline (Tel) || Tetralin n. || tetraline m.

text || Text m. || texte m. / large ~ (Print) || große Schrift f. || écriture f. grosse; gros caractère m. / round ~ || mittlere Schrift f. || écriture f. en moyenne. / condensing of a ~ (Print) || Zusammenziehung f. eines Textes || condensation f. d'un texte.

textbook || Leitfaden m. || guide m.; manuel m.

textile || Gewebe n. || tissu m. / ~ calender || Textilkalander m. || calandre m. textile. / ~ fabric || Gewebe n.; Stoff m.; Zeug n. || étoffe f.; tissu m. / ~ factory || Textilfabrik f.; Spinnerei f. || fabrique f. textile; filature f. / ~ goods pl. || Textilwaren fpl.; Webwaren fpl. || tissus mpl. textiles mpl. / ~ industry || Textilindustrie f.; Webereiindustrie f. || industrie f. textile. / ~ machine || Textilmaschine f.; Webereimaschine f. || machine f. textile. / ~ soap || Textilseife f. || savon m. textile.

textilose || Textilose f. || textilose f. / ~ yarn || Textilosegarn n. || fil m. en textilose.

texture (Structure) || Gefüge n.; Struktur f.; Textur f. || texture f. / ~ (Textile) || Gewebe n.; Stoff m.; Zeug n. || tissu m.; étoffe f. / amygdaloidal ~ (Geol) || Mandelsteintextur f. || texture f. amygdaloïdale. / close ~ uniform throughout || gleichmäßig dichtes Gefüge n. || texture f. serrée et uniforme. / fibrous ~ || faseriges Gefüge n. || structure f. fibreuse. / fine(-grained) ~ || Kleingefüge n.; Feingefüge n.; feinkörniges Gefüge n.; feine Struktur f. || texture f. fine; grain m. fin; structure f. à grain fins. / granular ~ || körnige Struktur f. || structure f. granulée. / homogeneous ~ || gleichmäßiges Gefüge n. || texture f. parfaitement homogène. / strainer ~ (Pap) || Siebgewebe n. || tissu m. à tamis. / tough fibrous ~ distinctly fibrous in a longitudinal sense || zähes sehniges Gefüge n. mit ausgesprochener Längsfaser || texture f. très tenace et nerveuse nettement fibreuse en sens longitudinal.

thallium salt || Thalliumsalz n. || sel m. de thallium.

thank (Letter of thanks) || Dankschreiben n. || (lettre f. de) remerciement m.

thatch || Dachstroh n. || chaume m.; paille f. de toiture. / ~ (roof) || Rohrdach n.; Strohdach n.; Strohbedachung f. || couverture f. en roseau ou en paille.

thatcher || Rohrdachdecker m. || couvreur m. en roseau.

thatching || Eindecken n. mit Stroh || couverture f. en paille ou en chaume.

thaw, to || (auf)tauen || dégeler. / ~ off || abtauen || dégeler.

thaw (Thawing weather) || Tauwetter n. || dégel m.

thawing || Wiederauftauen n. || dégel m. / ~ tank || Auftaugefäß n. || cuve f. à dégeler. / ~ weather || Tauwetter n. || dégel m.

thaw-off tank || Auftaugefäß n. || bac m. à démouler; cuve f. à dégeler.

theatre || Schaubühne f.; Theater n.; Schauspielhaus n. || scène f.; théâtre m. / ~ building || Theaterbau m. || construction f. de théâtre. / ~ glass || Theaterglas n. || jumelle f. de théâtre. / ~ glass of the Galilean type || Theaterglas n. galileischer Bauart || jumelle f. de théâtre du type Galilée.

theatrical costumer || Theaterschneider m. || costumier m. de théâtre. / ~ joiner || Theaterschreiner m. || menuisier m. en décors de théâtre.

theine (Chem) see also teine || Tein n. || théine f.

Thenard's blue || Kobaltblau n.; Kobaltultramarin n. || bleu m. de cobalt ou de Thenard.

theobromine || Theobromin n. || théobromine f.

theodolite || Theodolit m. || théodolite m. / ~ with optical micrometer || Theodolit m. mit optischem Mikrometer || théodolite-tachéomètre m. à micromètre optique. / ~ with reading microscopes || Mikroskoptheodolit m. || théodolite m. microscope. / repeating ~ || Repetitionstheodolit m. || théodolite m. répétiteur. / ~ with stadia lines || Theodolit m. mit Abstandmeßeinrichtung || théodolite m. avec dispositif de mesure des distances. / vernier ~ || Nonientheodolit m. || théodolite m. à vernier.

theorem (Math) ‖ Lehrsatz m.; Satz m. ‖ théorème m.; proposition f.

theoretical ‖ theoretisch ‖ théorique. / ~ law for receiving energy in acoustics ‖ Tiefempfangsgesetz n. der Akustik und Elektrodynamik ‖ loi f. théorique d'énergie acoustique arrivante. / ~ yield ‖ theoretische Ausbeute f. ‖ rendement m. théorique.

theory ‖ Theorie f.; Lehre f. ‖ théorie f. / ~ of continuous action ‖ Nahewirkungstheorie f. ‖ théorie f. d'action continue. / electromagnetic ~ of light ‖ elektromagnetische Lichttheorie f. ‖ théorie f. électromagnétique de la lumière. / electromagnetic ~ of light and electric waves ‖ elektromagnetische Theorie f. des Lichtes und der elektrischen Wellen ‖ théorie f. électromagnétique de la lumière et des ondes électriques. / ~ of heat ‖ Wärmelehre f.; Wärmetheorie f. ‖ théorie f. de la chaleur. / ~ of ions ‖ Ionenlehre f.; Ionentheorie f. ‖ théorie f. des ions. / ~ of management ‖ Betriebslehre f. ‖ théorie f. d'aménagement. / ~ of probability ‖ Wahrscheinlichkeitslehre f.; Wahrscheinlichkeitsrechnung f. ‖ calcul m. des probabilités. / ~ of relativity ‖ Relativitätstheorie f. ‖ théorie f. de la rélativité. / ~ of solution ‖ Lösungstheorie f. ‖ théorie f. de dissolution. / ~ of stability ‖ Stabilitätslehre f. ‖ théorie f. de la stabilité.

therapeutics pl. *see* therapy.

therapy ‖ Therapie f.; (praktische) Heilkunde f.; Heilverfahren n. ‖ thérapie f.; thérapeutique f. / penetrating ~ ‖ Tiefentherapie f. ‖ thérapie f. pénétrante. / semi-penetrating ~ ‖ leichte Tiefentherapie f. ‖ thérapie f. semi-pénétrante. / skin ~ ‖ Oberflächentherapie f. ‖ thérapeutique f. superficielle. / surface ~ *see* skin ~.

thermal analysis (Chem) ‖ thermische Analyse f. ‖ analyse f. thermique. / ~ central ‖ Wärmezentrale f. ‖ centrale f. thermique. / ~ conductivity ‖ Wärmeleitfähigkeit f. ‖ conductibilité f. calorifique. / ~ control board for supervising the circulation of water ‖ Wärmewarte f. zur Überwachung des Wasserkreislaufes ‖ poste m. de contrôle thermique destiné à surveiller la circulation d'eau. / ~ cycle ‖ thermischer Kreisprozeß m. ‖ évolution f. *ou* cycle m. thermique. / ~ efficiency ‖ Wärmewirkungsgrad m. ‖ rendement m. thermique *ou* calorifique. / ~ equator ‖ Wärmeäquator m. ‖ équateur m. thermique. / ~ insulator *see* ~ non-conductor. / ~ non-conductor ‖ Wärmeisoliermittel n. ‖ calorifuge m.; matière f. isolante. / ~ spring(s pl.) ‖ warme *oder* heiße Quelle f.; Therme f.; Thermalquelle f. ‖ source f. chaude *ou* thermale. / ~ unit ‖ Einheit f. der Wärme; Wärmeeinheit f.; Kalorie f. ‖ unité f. thermique *ou* de chaleur; calorie f. / ~ water ‖ Thermalwasser n. ‖ eau f. thermale.

thermic protection ‖ Wärmeschutz m. ‖ isolation f. calorifuge; isolement m. thermique.

thermionic current ‖ Thermionenstrom m. ‖ courant m. thermoionique. / ~ generator ‖ thermionischer Generator m. ‖ générateur m. thermoionique d'oscillations. / ~ tube (Radio) ‖ Verstärkerröhre f. ‖ tube m. amplificateur. / ~

valve ‖ Elektronenröhre f. ‖ lampe f. à trois électrodes; triode f.

thermit(e) ‖ Thermit n. ‖ thermite f. / ~ bond ‖ Thermitverbindung f. ‖ joint m. à la thermite. / ~ iron ‖ Thermiteisen n. ‖ fer m. à la thermite. / ~ welding ‖ Thermitschweißung f. ‖ soudage m. à la thermite *ou* à l'aluminium. / ~ welding process ‖ Thermitschweißverfahren n. ‖ procédé m. de soudage à la thermite.

thermo-cautery ‖ Thermokauter m. ‖ thermocautère m.

thermo-chemistry ‖ Thermochemie f. ‖ thermochimie f.

thermo-couple (Electr) ‖ Thermoelement n. ‖ élément m. *ou* couple m. *ou* pile f. thermo-électrique; thermo-élément m. / single ~ with gilt concave mirror ‖ einfaches Thermoelement n. mit vergoldetem Hohlspiegel ‖ élément m. thermoélectrique simple muni d'un réflecteur concave doré. / ~ for spectro-photometry ‖ Thermoelement n. für Spektrallichtstärkemessung ‖ élément m. thermoélectrique pour la photométrie spectrale. / ~ carrier ‖ Thermoelementträger m. ‖ porte-élément m. thermo-électrique. / closing window of a ~ ‖ Abschlußfenster n. eines Thermoelementes ‖ fenêtre f. de fermeture pour un élément thermoélectrique.

thermodynamics pl. ‖ Thermodynamik f.; Wärmemechanik f. ‖ thermodynamique f.

thermoelectric ‖ thermoelektrisch ‖ thermoélectrique. / ~ cell *see* thermo-couple. / ~ current ‖ Thermostrom m. ‖ courant m. thermoélectrique. / ~ detector ‖ thermoelektrischer Detektor m. ‖ détecteur m. thermoélectrique. / ~ force ‖ Thermokraft f. ‖ force f. thermoélectrique.

thermo-electrical *see* thermoelectric.

thermo-electricity ‖ Thermoelektrizität f. ‖ thermoélectricité f.

thermoelement *see* thermo-couple.

thermogalvanometer ‖ Thermogalvanometer n. ‖ galvanomètre m. à thermoélément.

thermometer ‖ Thermometer n.; Wärmegradmesser m. ‖ thermomètre m. / alarm ~ ‖ Alarmthermometer n. ‖ thermomètre m. avertisseur. / alcohol ~ ‖ Alkoholthermometer n. ‖ thermomètre m. à alcool. / atmospherical ~ ‖ Luftthermometer n. ‖ thermomètre m. à air. / baro-~ ‖ Barothermometer n. ‖ barothermomètre m. / bent ~ ‖ Winkelthermometer n. ‖ thermomètre m. coudé. / bimetallic ~ ‖ Doppelmetallthermometer n. ‖ thermomètre m. bimétallique. / black bulb ~ ‖ Schwarzkugelthermometer n. ‖ thermomètre m. à boule noire. / distance ~ ‖ Fernthermometer n. ‖ téléthermomètre m. / dry bulb ~ ‖ trockenes Thermometer n. ‖ thermomètre m. sec. / electrical ~ ‖ Fernthermometer n.; elektrischer Wärmegradmesser m. ‖ thermomètre m. électrique. / ~ with enclosed scale ‖ Einschlußthermometer n. ‖ thermomètre m. gradué sur verre opale *ou* divisé sur plaque en verre opale. / ~ graduated on the stem ‖ Stabthermometer n. ‖ thermomètre m. gradué sur tige. / gyrostatic ~ ‖ Schleuderthermometer n. ‖ thermomètre-fronde m. / maximum ~ ‖ Höchstwärmegradmesser m.; Maxi-

mumthermometer n. ‖ thermomètre m. à maximum. / maximum and minimum ~ ‖ Maximum- und Minimumthermometer n.; Höchst- und Niedrigstwärmegradmesser m. ‖ thermomètre m. à maxima et à minima. / mercurial ~ ‖ Quecksilberthermometer n. ‖ thermomètre m. à mercure. / metallic ~ ‖ Metallthermometer n. ‖ thermomètre m. métallique. / minimum ~ ‖ Niedrigstwärmegradmesser m.; Minimumthermometer n. ‖ thermomètre m. à minimum. / platinum resistance ~ ‖ Platinwiderstandsthermometer n. ‖ thermomètre m. à résistance de platine. / radiator ~ (Auto) ‖ Kühlwasserthermometer n.; Kühlwasserwärmegradmesser m. ‖ thermomètre m. de refroidissement. / remote ~ *see* distance ~. / screwed-in ~ ‖ Einschraubthermometer n. ‖ thermomètre m. à raccord fileté. / ~ for superheated steam ‖ Heißdampfthermometer n.; Heißdampfwärmegradmesser m. ‖ thermomètre m. pour vapeur surchauffée. / tubular ~ ‖ Rohrthermometer n. ‖ thermomètre m. à tuyau. / wet bulb ~ ‖ feuchtes Thermometer n. ‖ thermomètre m. mouillé.

thermometer bulb ‖ Thermometerkugel f. ‖ boule f. du thermomètre.

thermometric ‖ thermometrisch ‖ thermométrique. / ~ column ‖ Thermometerfaden m. ‖ colonne f. thermométrique. / ~ scale ‖ Thermometerskale f.; Thermometergradteilung f. ‖ échelle f. d'un thermomètre.

thermo-pile ‖ Thermosäule f. ‖ pile f. thermoélectrique. / ~ with ten couples ‖ Thermosäule f. mit zehn Thermoelementen ‖ pile f. thermo-électrique de dix éléments.

thermo-regulator ‖ Wärmeregler m.; Thermoregler m. ‖ thermorégulateur m.

thermoscope ‖ Thermoskop n. ‖ thermoscope m.

thermos flask ‖ Thermosflasche f. ‖ bouteille f. isolante.

thermosiphon ‖ Thermosiphon m. ‖ thermosyphon m.

thermostat (Phys) ‖ Thermostat n.; (selbsttätiger) Wärmeregler m. ‖ thermostat m. / ~ (Auto) ‖ Kühlwasserregler m.; Thermostat m. ‖ thermostate m.

thick ‖ dick ‖ épais; large. / ~ (Coal seam) ‖ mächtig ‖ épais; large.

thick board *see* ~ plank. / ~ fuel oil ‖ dickflüssiger Brennstoff m.; dickflüssiges Treiböl m. ‖ combustible m. épais. / ~ mash ‖ Dickmaische f. ‖ dickmaische f.; trempe f. épaisse. / ~ paper ‖ kartenstarkes Papier n. ‖ papier m. fort pour cartes. / ~ plank ‖ Bohle f.; Diele f. ‖ madrier m.; ais m.; planche f. épaisse. / ~ printing paper ‖ Dickdruckpapier n.; dickes Druckpapier n. ‖ papier m. fort pour impression. / ~ sheet-iron ‖ Grobblech n. ‖ grosse tôle f. / ~ slurry process ‖ Dickschlammverfahren n. ‖ procédé m. à pâte épaisse.

thicken, to (To make thick) ‖ verdicken; eindicken ‖ épaissir; concentrer. / ~ (To become thick) ‖ dickflüssig werden; sich verdicken; dicker werden ‖ (s')épaissir. / ~ (Forg) ‖ stauchen ‖ refouler.

thickening ‖ Eindicken n.; Eindickung f. ‖ épaississement m.; concentration f. / ~ drum for wood pulp ‖ Eindicktrommel f. für Zellulose ‖ tambour m. d'épaississe-

ment pour la cellulose. / ~ substance ‖ Eindickungsmittel n.; Verdickungsmittel n. ‖ agent m. d'épaississement; épaississant m.

thickly liquid (Met) ‖ dickflüssig ‖ consistant. / ~ (Met) ‖ matt ‖ pâteux; consistant.

thickness ‖ Dicke f. ‖ épaisseur f. / ~ (Found) ‖ Modell n.; Eisenstärke f. ‖ chemise f. d'un moule en terre; fausse-cloche f. / if the ~ goes down beyond a certain limit ‖ wenn die Stärke f. ein gewisses Maß unterschreitet ‖ si l'épaisseur f. passe en-dessous d'une certaine dimension. / ~ of·x mm ‖ Stärke f. von x mm ‖ épaisseur f. de x mm. / ~ of key ‖ Keilhöhe f. ‖ épaisseur f. de clavette. / leather ~ ‖ Lederdicke f. ‖ épaisseur f. de cuir. / ~ of metal (Found) ‖ Fleisch n. eines Gußstückes ‖ épaisseur f. en métal. / ~ of metal of a tube ‖ Wandstärke f. eines Rohrs ‖ épaisseur m. de paroi d'un tuyau. / ~ of plates (Shipb) ‖ Plattendicke f. ‖ épaisseur f. des tôles. / prescribed ~ ‖ vorgeschriebene Dicke f. ‖ épaisseur f. prescrite. / of a prescribed ~ ‖ von vorgeschriebener Dicke f. ‖ d'une épaisseur prescrite. / ~ of the seams ‖ Mächtigkeit f. der Flöze ‖ puissance f. des couches. / ~ of a stratum ‖ Mächtigkeit f. einer Lagerstätte ‖ épaisseur f. ou puissance f. d'un gisement. / ~ of a tooth ‖ Stärke f. eines Radzahnes ‖ épaisseur f. d'une dent de roue. / ~ of the tube plate ‖ Stärke f. der Rohrwand ‖ épaisseur f. de la plaque tubulaire. / ~ of web ‖ Stegstärke f. ‖ épaisseur f. de l'âme.

thickness board (Found) ‖ Hemdbrett n.; Schablone f. zum Gußhemd ‖ échantillon m. de chemise.

thickness micrometer ‖ Dickenmesser m. ‖ appareil m. à mesurer les épaisseurs.

thicknessing machine for planks ‖ Dicktenhobelmaschine f.; Bretterhobelmaschine f. ‖ machine f. à tirer d'épaisseur; raboteuse f. pour planches.

thickset ‖ Kord m.; gestreifter Manchester m. ‖ velours m. de coton rayé.

thick-skinned ‖ dickhülsig; dickschalig ‖ cossu; à écorce f. épaisse.

thick-walled ‖ dickwandig; starkwandig ‖ à paroi f. épaisse.

thigh (Anatomy) ‖ Schenkel m. ‖ cuisse f.

thill (Coachm) see forked ~. / ~ (Mine) ‖ Liegendes n. eines Flözes ‖ mur m. d'une couche. / forked ~ ‖ Gabeldeichsel f. ‖ limonière f.; limons mpl.; enrayoir m. à fourchette; brancard m.

thill rail ‖ Riegel m. der Deichselarme ‖ épars m.; épart m. / ~ wagon ‖ Gabelwagen m. ‖ fourgon m.

thimble ‖ Zwinge f.; Kausche f. ‖ cosse f. / ~ (Finger pad) ‖ Fingerhut m. ‖ dé m. à coudre. / ~ (For line poles) ‖ Ankerkausche f. ‖ cosse f. de hauban. / cable ~ ‖ Kabelschuh m. ‖ cosse f. de câble. / grummet ~ ‖ Seilkausche f. ‖ cosse f. pour cordages. / heart-shaped ~ ‖ Herzkausche f. ‖ cosse f. en forme de cœur. / wooden ~ ‖ hölzerne Kausche f. ‖ cosse f. de bois.

thimble-hook (Mar) ‖ Haken m. mit Kausche f. ‖ croc m. à cosse.

thin, to ‖ dünn machen; verdünnen ‖ amincir; délayer. / ~ (A forest) ‖ durchforsten ‖ faire une éclaircie.

thin ‖ dünn ‖ mince. / ~ flame ‖ Stichflamme f. ‖ jet m. de flamme; flamme f. en coup de feu. / ~ plate ‖ Fein-

blech n. ‖ tôle f. mince ou fine. / ~ printing paper ‖ Dünndruckpapier n.; dünnes Druckpapier n. ‖ papier m. mince pour impression. / ~ printing paper making machine ‖ Dünndruckpapierherstellungsmaschine f. ‖ machine f. à fabriquer du papier pour impression délicate. / ~ rock section (Miner) ‖ Dünnschliff m. ‖ lame f. mince. / ~ sheet of iron ‖ dünne Blechscheibe f. ‖ disque m. en tôle fine. / ~ sheet-iron ‖ dünnes Eisenblech n.; Feinblech n. ‖ tôle f. fine ou mince. / ~ sheet rolling mill ‖ Feinblechwalzwerk n. ‖ laminoir m. à tôles minces.

thin-ground stone plate ‖ Gesteinsdünnschliff m. ‖ lame f. de pierre polie.

things pl. in sea ‖ Seetriften fpl. ‖ débris mpl. ou épaves fpl. de mer.

thin-liquid ‖ dünnflüssig ‖ très liquide.

thinning (Of a forest) ‖ Durchforstung f. ‖ éclaircie f. / ~ (Met) ‖ Ausstrecken n.; Nachstrecken n. ‖ allongement m.; étirage m. sous le marteau. / ~ agent ‖ Verdünnungsmittel n. ‖ délayant m.

thin-skinned ‖ dünnschalig ‖ à écorce f. fine.

thin-walled ‖ dünnwandig ‖ à paroi f. mince. / ~ casting see ~ piece. / ~ piece (Found) ‖ dünnwandiges Stück n. oder Gußstück n. ‖ pièce f. à parois de faibles épaisseurs.

thioantimonic acid ‖ Schwefelantimonsäure f. ‖ acide m. thioantimonique.

thioarsenic acid ‖ Schwefelarsensäure f. ‖ acide m. thioarsénique.

thiosulphate ‖ Hyposulfit n. ‖ hyposulfite m.; thiosulfate m.

thiosulphuric acid ‖ unterschweflige Säure f.; Thioschwefelsäure f. ‖ acide m. hyposulfureux.

third of a sheet (Print) ‖ Drittelsbogen m. ‖ tiers m. de la feuille.

third-rail system ‖ System n. der dritten Schiene ‖ système m. à rail de contact.

third wire (Electr) ‖ Mittelleiter m. ‖ conducteur m. médian.

This side up! (Packing) ‖ oben! ‖ haut!

thole board (Shipb) ‖ Rojebord m.; Rojebrett n.; Rojeplanke f. ‖ planche f. des tolets. / ~ strings pl. ‖ Rojeklampen fpl. ‖ dames fpl.; porte-ame m.; taquets mpl. de nage; toletières fpl.

Thomas iron ‖ Thomaseisen n. ‖ fer m. Thomas. / ~ meal ‖ Thomasmehl n. ‖ farine f. de scorie Thomas. / ~ pig ‖ Thomasroheisen n. ‖ fonte f. crue Thomas. / ~ process ‖ Thomasverfahren n. ‖ procédé m. Thomas. / production of steel by the basic ~ process ‖ Stahlerzeugung f. nach dem basischen Thomasverfahren ‖ production f. d'acier d'après le procédé Thomas (basique). / ~ slag ‖ Thomasschlacke f. ‖ scorie f. Thomas. / ~ slag grinding ‖ Mahlen n. der Thomasschlacke ‖ broyage m. de scories de déphosphoration ou de scories Thomas. / ~ slag mill ‖ Thomasschlackenmühle f. ‖ moulin m. à scories Thomas. / ~ slag works plant ‖ Thomasschlackenwerksanlage f. ‖ installation f. pour l'utilisation du laitier Thomas. / ~ steel ‖ Thomasstahl m. ‖ acier m. Thomas. / ~ steel works pl. ‖ Thomasstahlwerk n. ‖ aciérie f. Thomas.

Thomson cable ‖ Thomsonkabel n. ‖ câble m. sans inductivité.

thong ‖ Binderiemen m.; Riemen m. ‖ courroie f.; lanière f.

thorium ‖ Thorium n. ‖ thorium m.

thorium-coated filament ‖ Thordraht m. ‖ filament m. au thorium ou thorié.

thorium salt ‖ Thoriumsalz n. ‖ sel m. de thorium.

thorn ‖ Dorn m.; Stachel m. ‖ épine f.

thorn-house (Salt) ‖ Gradierwerk n. ‖ bâtiment m. de graduation.

thorough ‖ gründlich; durch und durch; vollkommen ‖ radical; complet; à fond. / ~ repair ‖ gründliche Ausbesserung f. ‖ raccommodage m. radical.

thoroughfare (Passage) ‖ Durchfahrt f.; Durchgang m. ‖ passage m.; traversée f. / ~ (Chief artery of a town) ‖ Hauptstraße f.; Verkehrsader f.; große Verkehrsstraße f. ‖ avenue f.; chemin m. de grande communication; artère f.

thrash, to ~ see to thresh.

thrasher see thresher.

thrashing see threshing.

thread, to ~ (A needle) ‖ einfädeln ‖ enfiler. / ~ (A screw) ‖ mit Gewinde n. versehen ‖ fileter.

thread (For sewing) ‖ Nähgarn n.; Faden m.; Garn n.; Zwirn m. ‖ fil m. (à coudre ou à voile); retors m.; filament m. / ~ (For packing) ‖ Bindfaden m. ‖ ficelle f. / to carry ~s (Spinn) ‖ einziehen ‖ enfiler. / coarse ~ ‖ grober Bindfaden m. ‖ ficelle f. / glazed ~ ‖ Glanzzwirn m. ‖ coton-cordonnet m.; fil m. glacé. / linen ~ ‖ Leinenzwirn m. ‖ fil m. retors de lin. / ~ of oakum (Mar) ‖ Wergzopf m. ‖ quenillon m. d'étoupe. / packing ~ ‖ Bindfaden m. ‖ ficelle f. (d'emballage). / main ~ ‖ Hauptfaden m. ‖ fil m. principal. / ~ for stitching ‖ Heftfaden m. ‖ fil m. à brocher. / vegetative ~ ‖ vegetativer Faden m. ‖ filament m. végétatif. / ~ of the weft ‖ Schußfaden m.; Eintragfaden m. ‖ duite f.

thread (of a screw) ‖ Gewinde n.; Schraubengewinde n. ‖ filet m. ou pas m. (d'une vis). / angular ~ ‖ Spitzgewinde n.; scharfes oder dreieckiges Gewinde n. ‖ filet m. triangulaire. / double ~ ‖ doppelgängiges oder doppeltes Gewinde n. ‖ double filet m.; double pas m. / female ~ ‖ Innengewinde n. ‖ filet m. femelle ou intérieur. / flat ~ ‖ flaches oder flachgängiges Gewinde n.; Flachgewinde n. ‖ filet m. carré. / gaspipe ~ ‖ Gasgewinde n. ‖ filet m. des tuyaux à gaz. / inside ~ see female ~. / left-hand ~ ‖ Linksgewinde n. ‖ filet m. à gauche. / male ~ ‖ männliches Gewinde n.; Außengewinde n. ‖ pas m. de vis mâle. / metric ~ ‖ metrisches Gewinde n. ‖ filetage m. métrique. / metric fine ~ ‖ metrisches Feingewinde n. ‖ filetage m. métrique fin. / multiplex ~ ‖ mehrfaches Gewinde n. ‖ vis f. à plusieurs filets. / right- and left-handed ~ ‖ Rechts- und Linksgewinde n. ‖ vis f. à filets contraires. / right-hand ~ ‖ Rechtsgewinde n.; rechtsgängiges Gewinde n. ‖ filet m. à droite. / round(ed) ~ ‖ rundes Schraubengewinde n. ‖ filet m. arrondi. / single ~ ‖ einfaches Gewinde n. ‖ pas m. simple. / square ~ ‖ Flachgewinde n.; flaches oder viereckiges Gewinde n. ‖ filet m. carré. / standard ~ ‖ Normalgewinde n. ‖ filet m. normal. / triangular ~ see angular ~.

thread bacterium ‖ Fadenbakterium n. ‖ bactérie f. filamenteuse. / ~ **brake** ‖ Fadenbremse f. ‖ frein m. de fil. / **breakage of** ~ ‖ Fadenbruch m. ‖ rupture f. de fil. / **stopping motion for** ~ **breaking** ‖ Abschlagabsteller m. ‖ casse-fil m. d'abattage. / ~ **bulging machine** ‖ Gewindedrückmaschine f. ‖ machine f. à presser les filets de vis. / ~ **cal(l)ipers** pl. ‖ Gewindetaster m. ‖ compas m. d'épaisseur pour pas de vis.

thread carrier (Spinn) see also thread guide ‖ Fadenführer m. ‖ guide-fil m. / ~ **brake** ‖ Fadenführerbremse f. ‖ frein m. de guide-fil. / ~ **lever** ‖ Fadenführerhebel m. ‖ levier m. de guide-fil. / ~ **screw** ‖ Fadenführerschraube f. ‖ vis f. de guide-fil.

thread cleaner ‖ Fadenreiniger m. ‖ épurateur m. de fil. / ~ **counter** ‖ Fadenzähler m. ‖ compte-fils m. / ~ **cutting machine** ‖ Gewindeherstellungsmaschine f. ‖ machine f. à fileter ou à faire les filets. / ~ **cutting tool** ‖ Gewindeschneidwerkzeug n.; Schneidkluppe f. ‖ outil m. à faire les filets (de vis); filière f.

threaded bolt with head not-machined ‖ Gewindebolzen m. mit rohem Kopf ‖ boulon m. fileté avec tête brute. / ~ **bush** ‖ Gewindebüchse f. ‖ tampon m. fileté. / ~ **end of the stud** ‖ Einschraubende n. der Stiftschraube ‖ extrémité f. filetée du prisonnier. / ~ **nipple** ‖ Gewindenippel f. ‖ chapeau m. fileté. / ~ **pin** ‖ Gewindestift m. ‖ goupille f. filetée.

thread end ‖ Fadenende n. ‖ bout m. de fil. / **end of** ~ (of a screw) ‖ Gewindeauslauf m. ‖ fin m. de filetage. / ~ **feeder** ‖ Fadeneinleger m. ‖ accrocheur m. / ~ **fillet** ‖ Gewindegang m. ‖ pas m. de filet. / ~ **gauge** ‖ Gewindelehre f.; Gewindekaliber n. ‖ calibre m. pour filets ou de filetage. / ~ **glazer** ‖ Garnpolierer m. ‖ lustreur ou glaceur de fil. / ~ **glove** ‖ Zwirnhandschuh m. ‖ gant m. de fil. / ~ **grinding machine** ‖ Gewindeschleifmaschine f. ‖ machine f. à rectifier les pas de vis. / **groove of** ~ ‖ Gewinderille f. ‖ rainure f. de filetage.

thread guide see also thread carrier ‖ Fadenleiter m.; Fadenführer m. ‖ conducteur m. de fil; guide-fil m. / **porcelain** ~ ‖ Porzellanfadenführer m.; Fadenführer m. aus Porzellan ‖ guide-fil m. en porcelaine. / ~ **coupling** ‖ Fadenführerkupplung f. ‖ embrayage m. de guide-fil.

thread guider see thread guide. / ~ **hose** ‖ Florstrumpf m. ‖ bas m. en crêpe ou en fil d'Ecosse.

threading of the needle ‖ Einfädeln n. der Nadel ‖ enfilage m. de la ficelle. / ~ **attachment** ‖ Gewindeschneideinrichtung f. ‖ dispositif m. de filetage. / ~ **die** ‖ Gewindeschneidbacke f. ‖ peigne m. à fileter. / **grinding attachment for** ~ **dies** ‖ Gewindeschneidbackenschleifvorrichtung f. ‖ dispositif m. à affûter les peignes à fileter. / ~ **lathe** ‖ Gewindedrehbank f. ‖ tour m. à fileter. / ~ **machine** ‖ Gewindeschneidmaschine f. ‖ machine f. à fileter ou à tarauder. / ~ **machine for nipples** ‖ Nippelgewindeschneidmaschine f. ‖ machine f. à fileter les raccords ou nipples. / ~, **beading and trimming machine** ‖ Gewindedrück-, Sicken- und Beschneidemaschine f. ‖ machine f. à

rogner et à moulurer et à imprimer les filets de vis.

thread, kind of ‖ Gewindeart f. ‖ genre m. de filet. / **length of** ~ ‖ Gewindelänge f. ‖ longueur f. du filet.

thread-like ‖ fadenförmig ‖ filiforme; en forme f. de fil.

thread milling cutter ‖ Gewindefräser m. ‖ fraise f. à fileter. / ~ **milling machine** ‖ Gewindefräsmaschine f. ‖ machine f. à fileter à la fraise; machine f. à fraiser les vis ou à fraiser les filets. / **passing of** ~**s into combs** ‖ Blattstechen n.; Kammeinzug m.; Riedstechen n. ‖ empeignage m.; piquage m. / ~ **picker** ‖ Fadenklauber m. ‖ éplucheuse f. à filaments.

thread pitch ‖ Gewindesteigung f. ‖ pas m. du filetage ou d'une vis. / **difference of the** ~ **from the real size** ‖ Abweichung f. der Gewindesteigung vom Sollmaß ‖ écart m. entre le pas du filetage et la dimension théorique. / **measuring machine for** ~ ‖ Meßmaschine f. für Gewindesteigungen ‖ machine f. à mesurer les pas de filetages. / **minimetre instrument to check the** ~ ‖ Minimetergerät n. zum Prüfen der Steigung an Gewinden ‖ appareil m. à minimètre pour la vérification du pas de filetages.

thread polisher ‖ Garnpolierer m. ‖ lustreur ou glaceur de fil. / ~ **rolling machine** ‖ Gewindewalzmaschine f. ‖ machine f. à laminer les filets de vis. / **cold** ~ **rolling machine** ‖ Kaltgewindewalze f. ‖ machine f. à laminer à froid les filets de vis. / ~ **stitching machine** ‖ Fadenheftmaschine f. ‖ machine f. à piquer au fil. / ~ **strength tester** ‖ Fadenfestigkeitsprüfer m. ‖ dynamomètre m. pour fils. / ~ **take-up lever** see also thread guide ‖ Fadenleiter m. ‖ conducteur m. de fil; guide-fil m. / ~ **tension device** ‖ Fadenspanner m. ‖ tendeur m. de fil. / ~ **tube** ‖ Fadenführeröse f. ‖ anneau m. de guide-fil. / ~ **worm** see thread (of a screw).

three ‖ drei ‖ trois. / **of** ~ **threads** (Yarn) ‖ dreidrähtig ‖ à trois fils; à triple fil; à trois bouts.

three-bayed (Arch) ‖ dreischiffig ‖ à trois baies.

three-bladed screw (Shipb) ‖ dreiflügelige Schraube f. ‖ hélice f. à trois ailes.

three-coat work (Mas) ‖ Putz m. aus drei Lagen ‖ enduit m. en trois couches.

three-coil transformer (Tel) ‖ Symmetrieübertrager m. ‖ transformateur m. équilibré.

three-colour chromotypogravure ‖ Dreifarbenätzung f. ‖ chromotypogravure f. à trois couleurs. / ~ **photography** ‖ Dreifarbenfotografie f. ‖ photographie f. trichrome. / ~ **printing** ‖ Dreifarbendruck m. ‖ trichromie f.; impression f. trichrome. / ~ **printing process** ‖ Dreifarbendruckverfahren n. ‖ procédé m. d'impression trichrome. / ~ **projection** ‖ Dreifarbenprojektionsverfahren n. ‖ procédé m. trichrome à la projection.

three-conductor cable see three-core cable.
three-core cable ‖ Dreileiterkabel n.; Dreifachkabel n.; dreiaderiges Kabel n. ‖ câble m. à trois conducteurs.

three-cornered scraper ‖ Dreikantschaber m. ‖ râcloir m. triangulaire.

three-decade resistance ‖ Dreidekadenstufenwiderstand m. ‖ résistance f. de trois décades.

three-electrode tube (Radio) ‖ Dreielektrodenröhre f. ‖ lampe f. triode; triode f.

three-engined (Airpl) ‖ dreimotorig ‖ trimoteur.

three-engine set (Electr) ‖ Dreimaschinensatz m. ‖ groupe m. de trois machines.

three-figure exchange (Aut tel) ‖ Tausenderamt n. ‖ bureau m. à trois figures.

three-filament lamp ‖ Dreifadenlampe f. ‖ lampe f. à trois filaments.

three-finger rule ‖ Dreifingerregel f. ‖ règle f. des trois doigts.

three-floored kiln (Brew) ‖ dreihordige Darre f.; Dreihordendarre f. ‖ touraille f. à trois plateaux.

threefold ‖ dreifältig; dreifach ‖ triple. / ~ **purchase** (Mar) ‖ Schwerttakel n. ‖ caliorne f.

three-grid valve (Radio) ‖ Dreigitterröhre f. ‖ lampe f. trigrille.

three-hay (Airpl) ‖ Dreistieler m. ‖ avion m. à trois pairs de mâts de chaque côté du fuselage.

three-high mill ‖ Dreiwalzenstraße f. ‖ train m. trio.

three-leg cal(l)ipers pl. ‖ Dreispitzzirkel m. ‖ compas m. à trois branches.

three-legged tongs pl. ‖ Dreiarmklemme f. ‖ pince f. à trois branches.

three-mash process (Brew) ‖ Dreimaischverfahren n. ‖ procédé m. de brassage à trois trempes.

three-masted bark ‖ Dreimastbark f. ‖ trois-mâts barque f.

three-membered (Chem) ‖ dreigliedrig ‖ ternaire.

three-necked flask (Chem) ‖ Dreihalskolben m. ‖ ballon m. à trois tubulures.

three-needle frame (Weav) ‖ Dreinadelstuhl m. ‖ métier m. à trois aiguilles.

three-nozzle atomizer ‖ Dreidüsenzerstäuber m. ‖ pulvérisateur m. à trois gicleurs.

three-parted ‖ dreiteilig ‖ partagé ou divisé en trois; tiercé.

three-phase (Electr) ‖ dreiphasig ‖ triphasé. / ~ **alternating current** ‖ Drehstrom m. ‖ courant m. triphasé. / ~ **alterno-motor** ‖ Dreiphasenwechselstrommotor m. ‖ moteur m. à courant alternatif triphasé. / ~ **commutator motor** ‖ Drehstromkollektormotor m. ‖ moteur m. triphasé à collecteur.

three-phase current ‖ Drehstrom m.; Dreiphasenstrom m. ‖ courant m. (alternatif) triphasé. / ~ **continuous current converter** ‖ Drehstromgleichstromumformer m. ‖ groupe m. convertisseur de courant triphasé en continu; convertisseur m. triphasé-continu. / ~ **motor** ‖ Drehstrommotor m. ‖ moteur m. électrique triphasé. / **compensated** ~ **motor** ‖ kompensierter Drehstrommotor m. ‖ moteur m. triphasé compensé. / ~ **oil transformer** ‖ Drehstromölumformer m. ‖ transformateur m. à courant triphasé refroidi par l'huile. / ~ **output meter** ‖ Drehstromleistungsmesser m. ‖ wattmètre m. pour courant triphasé. / ~ **plant** ‖ Drehstromanlage f. ‖ installation f. de courant triphasé. / ~ **side** ‖ Drehstromseite f. ‖ côté m. alternatif ou triphasé. / ~ **synchronous generator** ‖ Drehstrom-Synchrongenerator m. ‖ génératrice f. synchrone triphasée.

three-phase generator ‖ Drehstromdynamo f.; Dreiphasendynamo f. ‖ alternateur m. triphasé; dynamo f. ou génératrice f. triphasée. / ~ **low-voltage plant** ‖ Dreh-

stromniederspannungsanlage f. ‖ installation f. à courant triphasé à basse tension. / ~ mains pl. ‖ Drehstromnetz n. ‖ réseau m. triphasé. / ~ motor ‖ Dreiphasenmotor m. ‖ moteur m. triphasé. / direct coupled ~ motors pl. ‖ unmittelbar gekuppelte Drehstrommotoren mpl. ‖ moteurs mpl. triphasés accouplés directement. / ~ network with four wires ‖ Dreiphasenvierleiternetz n. ‖ réseau m. triphasé à quatre fils. / ~ repulsion motor ‖ Drehstromrepulsionsmotor m. ‖ moteur m. triphasé à répulsion. / ~ serieswound motor ‖ Drehstromreihenschlußmotor m. ‖ moteur m. série triphasé. / ~ series-wound short-circuit motor ‖ Drehstromreihenschlußkurzschlußmotor m. ‖ moteur m. série triphasé en court-circuit. / ~ squirrel cage motor ‖ Drehstrommotor m. mit Kurzschlußläufer ‖ moteur m. triphasé à induit en court-circuit. / ~ starter ‖ Dreiphasenanlasser m. ‖ démarreur m. triphasé. / ~ switch ‖ Drehstromschalter m. ‖ interrupteur m. pour courant triphasé. / ~ transformer ‖ Drehstromtransformator m. ‖ transformateur m. triphasé.

three-point contact on the ground (Gun carriage) ‖ Dreipunktauflage f. ‖ portage m. de trois points d'appui sur le sol. / ~ suspension ‖ Dreipunktaufhängung f. ‖ suspension f. triangulaire. / ~ system ‖ Dreipunktsystem n. ‖ système m. à triple articulation.

three-quarters pl. (Build) ‖ Dreiviertelstein m. ‖ trois quartiers mpl.

three-rail track ‖ Dreischienengleis n. ‖ voie f. à trois rails.

three-roller plate bending machine ‖ Dreiwalzenblechbiegemaschine f.; Dreiwalzenblechrundmaschine f. ‖ machine f. à cintrer ou à rouler les tôles à trois cylindres. / ~ refiner for colours ‖ Dreiwalzwerkfarbenreibmaschine f. ‖ broyeuse f. à trois rouleaux à broyer les couleurs.

three-skin work see three-coat work.

three-square ‖ dreikantig; dreiseitig ‖ à trois bords; triquètre. / ~ file ‖ Dreikantfeile f. ‖ lime f. triangulaire. / ~ scraper ‖ dreischneidiger Schaber m.; Dreikantschaber m. ‖ grattoir m. triangulaire.

three-stage amplifier (Radio) ‖ Dreifachverstärker m. ‖ amplificateur m. à trois étages. / ~ compressor ‖ dreistufiger Kompressor m. ‖ compresseur m. à trois étages.

three-storied ‖ dreistöckig ‖ de trois étages; à triple étage.

three-threaded yarn ‖ Dreifachgarn n. ‖ fil m. à trois bouts.

three-throw hydraulic pump ‖ hydraulische Drillingspreßpumpe f. ‖ pompe f. de compression à trois corps. / ~ switch ‖ Dreiwegeweiche f. ‖ changement m. à trois voies.

three-way cock ‖ Dreiwegehahn m. ‖ robinet m. à trois voies. / ~ flange cock ‖ Dreiwegeflanschenhahn m. ‖ robinet m. à bride à trois voies. / ~ flange cock (without and) with stuffing box ‖ Dreiwegeflanschenhahn m. (ohne und) mit Stopfbuchse ‖ robinet m. à bride à trois voies (sans et) avec boîte à bourrage. / ~ switch ‖ Dreiwegeumschalter m. ‖ commutateur m. à trois directions.

three-wheeled steam road roller ‖ DreiradDampfstraßenwalze f. ‖ rouleau m. com-

presseur à vapeur à un rouleau et deux roues.

three-wheel vehicle ‖ dreiräderiges Fahrzeug n.; Dreirad n. ‖ véhicule m. à trois roues; tricycle m.

three-wire (Tel) ‖ Dreileiter m. ‖ trois conducteurs mpl. / ~ network ‖ Dreileiternetz n. ‖ réseau m. à trois fils. / ~ system (Tel) ‖ Dreidrahtsystem n.; Dreileitersystem n.; Erdsystem n. ‖ système m. à trois fils.

thresh, to ‖ dreschen ‖ battre (le blé).

treshed corn ‖ gedroschenes Getreide n. ‖ grains mpl. battus.

thresher see also threshing machine ‖ Dreschmaschine f. ‖ machine f. à battre; batteuse f. / horse gear ~ ‖ Dreschmaschine f. mit Göpelantrieb; Göpeldreschmaschine f. ‖ batteuse f. à manège. / motor ~ ‖ Dreschmaschine f. mit Motorantrieb; Motordreschmaschine f. ‖ batteuse f. à moteur; moto-batteuse f. / steam ~ ‖ Dampfdreschmaschine f. ‖ batteuse f. à vapeur.

threshing ‖ Dreschen n.; Ausdreschen n.; Drusch m. ‖ battage m. / enterprise of ~ ‖ Lohndreschen n. ‖ entreprise f. de battage. / ~ flail ‖ Dreschflegel m. ‖ fléau m. (à battre). / ~ floor ‖ Tenne f.; Dreschtenne f. ‖ aire f. (d'une grange).

threshing machine see also thresher ‖ Dreschmaschine f. ‖ batteuse f.; machine f. à battre. / ~ broad ‖ Breitdreschmaschine f. ‖ batteuse f. large. / ~ for motor and steam use ‖ Dreschmaschine f. für Motor- und Dampfbetrieb ‖ batteuse f. à moteur et à vapeur.

threshing machine attendant see ~ driver. / ~ driver ‖ Dreschmaschinenführer m. ‖ conducteur m. de batteuse. / ~ labourer ‖ Dreschmaschinenarbeiter m. ‖ ouvrier m. à la batteuse. / ~ manager ‖ Dreschunternehmer m. ‖ entrepreneur m. de battage. / ~ owner ‖ Dreschmaschinenbesitzer m. ‖ propriétaire m. de batteuse.

thrift box ‖ Sparbüchse f. ‖ tirelire f.

thrifty ‖ sparsam; wirtschaftlich ‖ économe; économique.

thrilling machine ‖ Dessinwalzwerk n.; Rändelmaschine f. ‖ moletteuse f. / ~ tool ‖ Rändelgabel f. ‖ porte-molette m.

throat (Anatomy) ‖ Kehle f.; Speiseröhre f.; Gurgel f. ‖ gorge f. / ~ (Carp) ‖ Kehle f.; Dünnung f.; Hals m. ‖ gorge f. (de démaigrissement). / ~ (Blast-furnace) ‖ Hochofengicht f.; Hochofenkranz m. ‖ gueulard m.; gueule f. / ~ of seam (Welding) ‖ Stärke f. der Schweißnaht ‖ épaisseur f. de la soudure. / ~ of a shaft ‖ Hals m. einer Welle ‖ gorge f. d'un arbre.

throat band (Saddl) ‖ Kehlriemen m. ‖ sous-gorge f. / ~ bolt (Shipb) ‖ Halsbolzen m. eines Knies ‖ cheville f. de gorge d'un coude. / ~ brush ‖ Halspinsel m. ‖ pinceau m. pour la gorge. / depth of ~ (Mach) ‖ Maultiefe f. ‖ portée f. de gorge. / attachment for grinding the ~ on threading dies ‖ Vorrichtung f. zum Schleifen des Anschnittes an Gewindeschneidbacken ‖ dispositif m. à rectifier les entrées des peignes à fileter. / ~ lash see throat band. / ~ ring of the head collar (Saddl) ‖ Halfterring m. ‖ anneau m. du licou; porte-barres mpl. / ~ seizing (Mar) ‖ Hartbindsel n.; Herzbindsel n. ‖ amarrage m. en étrive. / ~ stopper for blast furnaces ‖ Hochofengichtverschluß

m. ‖ fermeture f. de gueulard pour hautsfourneaux. / ~ stopper winch ‖ Gichtglockenwinde f. ‖ treuil m. de manœuvre pour cloches à gueulard.

throstle frame (Spinn) ‖ Drosselstuhl m.; Drosselmaschine f. ‖ continu m. à filer. / ~ frame tenter ‖ Drosselspinner m. ‖ fileur m. au continu. / ~ spindle ‖ Drosselspindel f. ‖ broche f. de continu.

throttle, to ~ (Mot) ‖ drosseln; abdrosseln ‖ étrangler.

throttle (Tel) ‖ Drossel f.; Drosselspule f. ‖ bobine f. de réactance ou de self; réactance f.; self m. / to open out the ~ ‖ Gas n. geben ‖ mettre les gaz. / foot ~ (Auto) ‖ Akzelerator m.; Fußgashebel m. ‖ accélérateur m.

throttle chain (Tel) ‖ Drosselkette f. ‖ chaîne f. de réactances. / ~ coil ‖ Drosselspule f. ‖ self m. de choc. / ~ coupling (Tel) ‖ Drosselkopplung f. ‖ accouplement m. de réactance.

throttled ‖ abgedrosselt ‖ étranglé. / ~-down see throttled. / ~ steam ‖ gedrosselter Dampf m. ‖ vapeur f. étranglée.

throttle hand lever ‖ Gashebel m.; Drosselhebel m.; Drosselhandhebel m. ‖ manette f. d'étranglement. / ~ hand lever tube ‖ Gasspindel f. des Gashebels ‖ tige f. de manette d'étranglement. / ~ lever ‖ Drosselhebel m. ‖ levier m. d'étranglement. / ~ pedal ‖ Gaspedal n.; Fußgashebel m. ‖ pédale f. d'accélérateur. / ~ shaft ‖ Drosselspindel f. ‖ tige f. de manette d'étranglement. / ~ slide ‖ Drosselschieber m. ‖ tiroir m. d'étranglement.

throttle valve ‖ Drosselklappe f.; Drosselventil n.; Absperrschieber m. ‖ robinet m. modérateur; soupape f. d'admission; clapet m. de réglage d'étranglement; soupape f. d'arrêt. / gas ~ ‖ Gasdrosselklappe f. ‖ papillon m. de commande des gaz.

throttle valve adjustment ‖ Drosselklappenregelung f. ‖ réglage m. du papillon. / ~ spindle ‖ Drosselklappenachse f. ‖ axe m. du papillon.

throttling ‖ Abdrosselung f.; Drosselung f. ‖ étranglement m.

through see also thorough ‖ durch; hindurch ‖ à travers. / ~ bolt ‖ durchgehender Bolzen m. ‖ boulon m. traversant. / ~ call (Tel) ‖ Durchgangsgespräch n. ‖ communication f. de transit.

through-carved ‖ durchbrochen gearbeitet ‖ travaillé à jour.

through communication of a carriage (Railw) ‖ Übergangsbrücke f. eines Wagens ‖ passerelle f. d'intercommunication d'une voiture.

through connection (Tel) ‖ Durchgangsverbindung f.; Dauerverbindung f. ‖ communication f. de transit; liaison f. directe.

throughfreight ‖ Durchfracht f. ‖ fret m. de transit.

through-going shaft ‖ durchgehende Welle f. ‖ arbre m. traversant.

through line (Railw) ‖ durchgehendes Gleis n.; Hauptgleis n. ‖ voie f. principale. / ~ line repeater (Tel) ‖ fester Zwischenverstärker m. ‖ amplificateur m. embroché. / ~ message (Tel) ‖ Durchgangstelegramm n.; Durchgangsdrahtung f. ‖ télégramme m. de transit. / ~ position (Tel) ‖ Durchsprechstelle f. ‖ position f. de communication directe.

/ ~ rate (of freight) ‖ Durchfrachtsatz m. ‖ tarif m. de transit. / ~ switching board (Tel) ‖ Durchgangsschrank m. ‖ table f. de transit interurbain. / ~ switching exchange ‖ Durchgangsanstalt f. ‖ bureau m. de transit. / ~ switching position ‖ Ferndurchgangsplatz m. ‖ position f. de transit interurbain. / ~ traffic ‖ Durchgangsverkehr m. ‖ commerce m. de transit; transit m. / ~ train (Railw) ‖ durchgehender Zug m.; Durchgangszug m. ‖ train m. parcourant toute la ligne.
through-way valve ‖ Durchgangsventil n. ‖ soupape f. droite.
throw, to ‖ werfen; schleudern ‖ jeter; lancer. / ~ (Pott) ‖ formen; drehen ‖ tourner. / ~ (Silk) ‖ zwirnen; mulinieren; drehen ‖ mouliner. / ~ a bridge ‖ eine Brücke f. bauen *oder* schlagen ‖ construire *ou* établir *ou* jeter un pont m. / ~ a harpoon ‖ mit der Harpune f. schießen ‖ darder le harpon m. / ~ gas into . . . ‖ Gase npl. einblasen in . . . ‖ souffler des gaz dans . . . / ~ into gear (Mot) ‖ anwerfen; in Gang setzen ‖ mettre en train *ou* en mouvement. / ~ off (To tip) ‖ kippen; umkippen ‖ renverser; culbuter. / ~ off the brakes pl. ‖ die Bremsen fpl. lockern ‖ défreiner. / ~ the coupling out ‖ entkuppeln; loskuppeln ‖ débrayer. / ~ out of gear see ~ the coupling out. / ~ out the piece ‖ das Werkstück ausstoßen ‖ refouler la pièce; expulser la matière. / ~ out of work ‖ den Betrieb m. einstellen; stillegen ‖ faire chômer les travaux mpl. / ~ water (with a pump) ‖ spritzen; sprengen ‖ arroser.
throw (Of an instrument) ‖ Ausschlag m. ‖ déviation f. / ~ (Geol) ‖ Verwerfung f.; Sprung m. ‖ faille f.; rejet m. / ~ of crankshaft ‖ Kröpfung f. der Kurbelwelle ‖ coude m. du vilebrequin. / ~ of the piston ‖ Hub m. *oder* Hubhöhe f. *oder* Hublänge f. eines Kolbens ‖ coup m. *ou* course f. *ou* levée f. *ou* volée f. du piston.
thrower (Pott) ‖ Former m.; Dreher m. ‖ tourneur m. / ~ (Silk) ‖ Seidenzwirner m.; Mulinör m. ‖ moulineur m. / ~ of crockery ‖ Steingutdreher m. ‖ tourneur m. en faïence.
throwing of silk ‖ Zwirnen n. der Seide ‖ organsisage m. *ou* moulinage m. de soie. / ~ out-of-gear ‖ Ausrückung f. ‖ désaccouplement m.; débrayage m.
throwing-in gear ‖ Einrückvorrichtung f. ‖ appareil m. d'embrayage.
throwing-out of sparks (Loc) ‖ Funkenwurf m.; Funkenflug m. ‖ jet m. de flammèches.
throwing wheel (Pott) ‖ Drehscheibe f.; Töpferscheibe f. ‖ tour m. *ou* roue f. de potier.
throw-off truck ‖ Abwurfwagen m. ‖ déchargeur m. mobile; chariot m. de déchargement.
throw-over switch with break ‖ Umschalter m. mit Unterbrechung ‖ commutateur m. avec interruption. / ~ without break ‖ Umschalter m. ohne Unterbrechung ‖ commutateur m. sans interruption.
throw rod (Railw) ‖ Weichenstellstange f. ‖ tige f. de manœuvre.
throwster (Silk) ‖ Seidenzwirner m. ‖ moulinier m.; ovaliste m.
thrum (Mach) ‖ Riementrum n.; Trum n. ‖ brin m. de courroie. / ~ (Spinn) ‖ Faden-

trum n.; Fadenstück n. ‖ portion f. *ou* partie f. de fil.
thrust (Mech) ‖ Druck m. ‖ pression f.; butée f.; poussée f. / ~ of earth ‖ Erddruck m. ‖ poussée f. des terres.
thrust ball bearing ‖ Kugeldrucklager n. ‖ roulement m. de butée.
thrust bearing ‖ Drucklager n. ‖ palier m. de butée. / single-collar ~ ‖ Einscheibendrucklager n. ‖ palier m. de butée à disque unique. / spherical ~ ‖ Kugelspurlager n. ‖ crapaudine f. à billes.
thrust block (Shipb) ‖ Lagerstuhl m.; Drucklagerstuhl m. ‖ palier m. de butée. / ~ bridge ‖ Drucklagerbügel m. ‖ collier m. de palier de butée. / ~ chest ‖ Drucklagergehäuse n. ‖ cage f. de palier de butée.
thrust pin ‖ Federstütze f. ‖ support m. du ressort.
thrust plate (Railw) ‖ Stoßpufferplatte f. ‖ plaque f. de choc.
thrust shaft ‖ Druckwelle f. ‖ arbre m. de butée.
thumb (Anatomy) ‖ Daumen m. ‖ pouce m. / ~ (Locksm) ‖ Türdrücker m. ‖ loquet m.
thumbknot ‖ einfacher Knoten m. ‖ nœud m. simple.
thumb lock ‖ Drückerschloß n. ‖ serrure f. à ressort. / ~ nut ‖ Flügelmutter f. ‖ écrou m. à oreilles. / ~ screw ‖ Flügelschraube f.; Ohrenschraube f.; Knebelschraube f. ‖ vis f. à oreilles *ou* à clef; vis f. ailée. / ~ stall ‖ Fingerling m. ‖ doigtier m. / ~ tack ‖ Reißnagel m. ‖ punaise f. à dessin. / ~ tack with brass head ‖ Reißnagel m. mit Messingkopf ‖ punaise f. à dessin à tête en laiton.
thunder cloud ‖ Gewitterwolke f. ‖ nuage m. orageux. / bank of ~s ‖ Gewitterbank f. ‖ banc m. d'orage.
thunder squall ‖ Gewitterbö f. ‖ grain m. orageux *ou* noir.
thunderstorm ‖ Gewitter n. ‖ orage m.; tempête f. / day ~ ‖ Tagesgewitter n. ‖ orage m. de jour.
thunderstorm station ‖ Gewitterstation f. ‖ station f. d'orage. / tendency to ~ ‖ Gewitterneigung f. ‖ tendance m. à l'orage. / ~ warning ‖ Gewitterwarnung f. ‖ avertissement m. d'orages. / zone of frequent ~s ‖ gewitterreiche Zone f. ‖ zone f. riche en orages. / zone of rare ~s ‖ gewitterarme Zone f. ‖ zone f. pauvre en orages.
thwart ‖ Ruderbank f.; Ducht f. ‖ banc m. de nage; traversier m. (de chaloupe). / middle ~ ‖ Segelducht f. ‖ banc m. du milieu.
thwart-ships ‖ querschiffs; dwarsschiffs ‖ par le travers.
thyme (Botan) ‖ Quendel m.; Thymian m. ‖ thym m.; serpolet m. / ~ extract ‖ Thymianauszug m. ‖ essence f. de thym. / ~ leaf ‖ Thymianblatt n. ‖ feuille f. de thym. / ~ oil ‖ Thymianöl n.; Quendelöl n. ‖ essence f. de thym *ou* de serpolet.
thymol ‖ Thymol n. ‖ thymole m.
tibia ‖ Schienbein n. ‖ tibia m.
tick, to (Watch) ‖ ticken ‖ faire tic-tac.
tick (Weav) see ticking.
ticken see ticking.
ticker (Electr) ‖ Ticker m. ‖ trembleur m.; contact m. à trembleur.
ticket, to ‖ auszeichnen; etikettieren ‖ étiqueter.
ticket (Of admission) ‖ Eintrittskarte f.; Einlaßkarte f. ‖ billet m. d'entrée;

ticket m. / ~ (Railw etc.) ‖ Fahrkarte f.; Fahrschein m. ‖ billet m.; coupon m.; ticket m. / ~ (Label) ‖ Auszeichnung f.; Etikett n. ‖ marque f.; étiquette f. / ~ (Cash register) ‖ Scheck m. ‖ ticket m. / ~ (Tel) ‖ Gesprächsblatt n. ‖ fiche f. / railway ~ ‖ Eisenbahnfahrkarte f. ‖ billet m. *ou* ticket m. de chemin de fer. / ~ for a reserved seat ‖ Platzkarte f. ‖ billet m. spécial *ou* réservé.
ticket counting machine ‖ Fahrkartenzählmaschine f. ‖ machine f. compteuse pour billets. / ~ factory ‖ Fahrkartenfabrik f.; Eintrittskartenfabrik f. ‖ fabrique f. de billets. / ~ label and ~ fastener ‖ Kollianhänger m. ‖ étiquette f. pour colis. / ~ nippers pl. (Railw) ‖ Kartenlochzange f.; Kartenlocher m.; Knipszange f. ‖ pince f. de contrôle. / ~ printer ‖ Fahrkartendrucker m. ‖ imprimeur m. de billets. / ~ printing establishment ‖ Fahrkarten- und Eintrittskartendruckerei f.; Fahrscheindruckerei f. ‖ imprimerie f. de billets. / ~ printing machine ‖ Fahrscheindruckmaschine f.; Eintrittskartendruckmaschine f. ‖ machine f. à imprimer les billets. / ~ printing rotation machine ‖ Rotationsmaschine f. für den Druck von Fahrscheinen *oder* Einlaßkarten ‖ machine f. rotative à imprimer des billets. / ~ sewer (Clothing) ‖ Etikettennäher m. ‖ couseur m. d'étiquettes. / ~ stamp ‖ Fahrkartenstempel m.; Eintrittskartenstempel m. ‖ timbre m. du billet. / ~ time (Tel) ‖ Gesprächsminuten fpl. ‖ minutes fpl. de durée de la conversation.
ticking ‖ Drell m.; Drillich m. ‖ treillis m. / ~ for mattresses ‖ Matratzendrell n. ‖ treillis m. pour matelas.
ticking maker ‖ Drillichweber m. ‖ coutier m. / ~ weaving ‖ Zwillichweberei f. ‖ tissage m. de coutil *ou* de treillis.
tickler (Radio) ‖ Rückkopplungsspule f. ‖ bobine f. de réaction. / ~ (Weav) ‖ Decker m. ‖ porte-poinçon m. / ~ machine ‖ Deckmaschine f. ‖ diminueuse f.
tidal power plant ‖ Gezeitenkraftwerk n. ‖ usine f. marémotrice. / ~ signal ‖ Wasserstandzeichen n. ‖ signal m. de marée. / ~ stream ‖ Gezeitenstrom m. ‖ courant m. de marée. / ~ wave ‖ Gezeitenwelle f.; Flutwelle f. ‖ ondemarée f.; onde f. de marée.
tidal and coast works pl. (Build) ‖ Seebauten pl. ‖ travaux mpl. à la mer.
tide ‖ Ebbe f. und Flut f.; Gezeit f.; Tide f. ‖ marée f.; flux m. et reflux m. / the ~ rises and falls ‖ die Gezeit f. steigt und fällt ‖ la marée monte et descend.
tide gate (Sluice) ‖ Fluttor n.; Obertor n.; oberes Schleusentor n. ‖ porte f. d'amont *ou* de tête. / ~ gauge ‖ Flutmesser m. ‖ marémètre m.; marégraphe m. / self-registering ~ gauge ‖ Limnigraf ‖ marégraphe m. enrégistreur. / height of the ~ ‖ Fluthöhe f. ‖ grandeur f. de la marée. / ~ hour ‖ Hafenzeit f. ‖ établissement m. du port. / ~ light ‖ Gezeitenfeuer n. ‖ feu m. de marée. / ~ lock (Sluice) ‖ Flutschleuse f. ‖ écluse f. de marée. / ~ tables pl. ‖ Gezeitentafeln fpl. ‖ annuaire m. des marées.
tideway ‖ Gezeitenstrom m. ‖ courant m. de marée.
tie, to ‖ anknüpfen; anbinden; binden; knüpfen ‖ attacher; nouer; lier. / ~ up

zuschnüren; festbinden ‖ ficeler; lacer. / ~ up (Tel) ‖ unnütz belegen ‖ occuper inutilement. / ~ up the page (Print) ‖ die Kolumne f. ausbinden ‖ lier la page.

tie ‖ Band n.; ficelle f.; cordon m. / ~ (Build) ‖ Anker(bolzen) m.; Verbindungsstück n. ‖ tirant m. / ~ (Railw) ‖ Schwelle f. ‖ traverse f. / diagonal ~s pl. (Carp) ‖ Kreuzzangen fpl.; Kreuzgurtung f. ‖ moises fpl. inclinées ou en écharpe. / hanging ~ (Build) ‖ Hängeband n.; Hängeschiene f.; Zange f. ‖ moise f. pendante. / iron ~ (Build) ‖ Anker m.; Zugband n. ‖ ancre f.; tirant m. en fer; entretoise f.

tie anchor (Build) ‖ Gewölbeanker m. ‖ ancre f. de voûte. / ~ band ‖ Hängeeisen n. ‖ étrier m.

tie bar (Railw) ‖ Spurstange f. ‖ entretoise f.; tringle f. d'écartement des rails. / rear-axle ~ ‖ Hinterachsstrebe f. ‖ contre-fiche f. du pont arrière.

tie beam (Build) ‖ Gebindsparren m. ‖ chevron m. / short ~ (Build) ‖ Stichbalken m. ‖ entrait m. / ~ of a truss frame ‖ Bindebalken m. eines Hängewerkes ‖ maître-entrait m.

tie block (Mar) ‖ Plattblock m. ‖ poulie f. plate. / ~ bolt ‖ Spannriegel m.; Anker m.; Ankerbolzen m. ‖ poutre f. traversière; tirant m. / ~ brace (Bridge) ‖ Zugdiagonale f. ‖ tige f. inclinée; lien m. incliné diagonale. / ~ envelope ‖ Schnurbandumschlag m. ‖ enveloppe f. à ruban. / ~ line (Tel) ‖ Querverbindung f. zwischen Nebenstellen ‖ ligne f. d'intercommunication ou transversale.

tie plate (Railw) ‖ Unterlagsplatte f.; Stoßplatte f.; Stuhlplatte f. ‖ selle f. d'appui; platine f. pour rails; plaque f. d'assise. / hooked ~ ‖ Hakenplatte f. ‖ selle f. à crochet. / hooked ~ with tenon ‖ Zapfenplatte f. ‖ selle f. à crochet et à crampon. / ~ with inclined surface ‖ Unterlagsplatte f. mit geneigter Oberfläche ‖ selle f. d'appui avec surface inclinée. / ordinary plain ~ with level surface ‖ offene Unterlagsplatte f. mit gerader Oberfläche ‖ selle f. d'appui simple avec surface horizontale. / ~ with shoulder (or rib) on one side ‖ einköpfige Unterlagsplatte f. mit Ansatz (oder Randleiste) an einer Seite ‖ selle f. d'appui avec rebord d'un côté. / ~ with shoulders on either side ‖ zweiköpfige Unterlagsplatte f. mit Ansatz an beiden Seiten ‖ selle f. d'appui avec rebord des deux côtés.

tier ‖ Reihe f.; Lage f. ‖ assise f.; couche f. / ~ of boxes (Theatre) ‖ Logenreihe f. ‖ rang m. de loges.

tier frame ‖ Etagengestell n. ‖ étagère f.

tie ring ‖ Bindering m. ‖ anneau m. d'attache.

tie rod ‖ Zugstange f.; Spannstange f.; tirant m. / ~ (Railw) ‖ Spurstange f.; Spurhalter m. ‖ tringle f. d'écartement de rails. / rear-axle ~ ‖ Hinterachsunterzug m. ‖ tendeur m. du pont arrière.

tier stand ‖ Etagengestell n. ‖ étagère f. à plusieurs planches.

tier-up (Weav) ‖ Ausschnürer m. ‖ empouteur m.

tie-up for weaving names ‖ Schnürung f. für Namenweberei f.; Monogrammschnürung f. ‖ empoutage m. pour noms ou monogrammes.

tie-up jacquard ‖ Schaftmaschine f. ‖ métier m. à pédale genre Jacquard.

tiffany (Weav) ‖ Flor m.; Gaze f. ‖ crêpe m.; gaze f. / silk ~ ‖ Seidengaze f.; Seidenflor m. ‖ gaze f. de soie; canevas m. en soie.

tight ‖ dicht; dichtschließend ‖ étanche. / ~ (Stretched) ‖ straff; gespannt ‖ tendu; raide. / to make ~ ‖ dichten; abdichten ‖ étancher; rendre étanche; étanchéifier. / air-~ ‖ luftdicht ‖ étanche à l'air. / oil-~ ‖ öldicht ‖ étanche à huile. / pressure-~ ‖ druckdicht ‖ tenant la pression.

tighten, to ‖ abdichten; dichten; dicht machen ‖ étanchéifier; étancher; rendre étanche; imperméabiliser. / ~ (To stretch) ‖ straffen; spannen ‖ tendre; raidir. / ~ the brake-shoes ‖ die Bremsbacken fpl. anziehen ‖ serrer les sabots mpl. du frein. / ~ a screw ‖ eine Schraube f. anziehen ‖ serrer une vis. / ~ a spring ‖ eine Feder f. spannen ‖ tendre un ressort m.

tightened ‖ abgedichtet ‖ étanchéifié. / ~ (Stretched) ‖ festgezogen ‖ serré.

tightener (El line) ‖ Spannschloß n. ‖ tendeur m.

tightening key ‖ Gegenkeil m. ‖ clavette f. de calage ou de dressage. / ~ sheet ‖ Abdichtungsblech n. ‖ tôle f. d'étanchéité.

tightening-up device ‖ Spannvorrichtung f. ‖ tendeur m.

tightening wedge ‖ Kreuzkopfkeil m.; Stellkeil m. ‖ clavette f. de réglage; coin m. d'ajustage.

tight-fitting screw ‖ Paßschraube f. ‖ boulon m. ajusté.

tightness ‖ Dichtigkeit f. ‖ étanchéité f. / ~ (Of clothes) ‖ Knappheit f. ‖ étroitesse f. / defective ~ ‖ schlechte Abdichtung f.; Undichtigkeit f. ‖ étanchéité f. défectueuse. / ~ of money ‖ Geldnot f. ‖ pénurie f. d'argent ou de numéraire.

tightness testing apparatus for gas pipes ‖ Dichtigkeitsprüfer m. für Gasleitungen ‖ appareil m. à contrôler l'étanchéité des conduites de gaz.

tilbury ‖ Tilbury n. ‖ tilbury m.

tile, to ‖ (das Dach) mit Ziegeln eindecken ‖ couvrir en tuile.

tile (Roof) ‖ Dachziegel m. ‖ tuile f. / ~ for the borders ‖ Ortziegel m. ‖ tuile f. gironnée. / double ~ ‖ doppelter Dachziegel m. ‖ tuile f. double. / Dutch ~ ‖ Kachel f.; Fliese f. ‖ carreau m. (de terre cuite). / encaustic ~ ‖ farbig glasierter Ziegel m. ‖ tuile f. vernie. / end ~ ‖ Schlußziegel m. ‖ tuile f. recourbée. / flat ~ ‖ Plattziegel m.; Zungenstein m.; Flachziegel m.; Ochsenzunge f. ‖ tuile f. plate; tuile f. à crochet. / ~ for floorings ‖ Fußbodenplatte f. ‖ dalle f. / glazed ~ see Dutch ~. / gutter ~ ‖ Hohlziegel m.; Kehlziegel m. ‖ tuile f. creuse ou gouttière. / hard-baked ~ ‖ Klinker m. ‖ brique f. hollandaise. / hollow ~ see gutter ~. / metal ~ ‖ Metallziegel m. ‖ tuile f. métallique. / plain ~ ‖ Biberschwanz m. ‖ tuile f. à crochet. / ridge ~ ‖ Firstziegel m. ‖ tuile f. faîtière. / stove ~ ‖ Kachel f.; Ofenkachel f.; Ofenziegel m. ‖ carreau m. de poêle. / rectangular stove ~ ‖ rechteckige Kachel f. ‖ carreau m. de poêle rectangulaire. / square stove ~ ‖ quadratische Kachel f. ‖ carreau m. de poêle carré. / tower ~ ‖ Turmziegel m. ‖ tuile f. pour tours.

tile burner ‖ Dachziegelbrenner m. ‖ tuilier m. / ~ cramp ‖ Ziegelklammer f. ‖

crochet m. à tuiles. / ~ drain ‖ Dränierung f. mit Ziegeln ‖ sous-doublis m. / ~ factory see also tilery ‖ Ziegelei f. (für Dachziegel) ‖ tuilerie f. / ~ factory machine ‖ Ziegeleimaschine f. ‖ machine f. pour tuileries. / ~ hearth ‖ Kachelherd m. ‖ poêle m. de cuisine en carreaux glacés ou de poterie ou de faïence. / ~ kiln see tilery. / ~ machine ‖ Dachziegelmaschine f. ‖ machine f. à tuiles. / ~ maker ‖ Dachziegelbrenner m. ‖ tuilier m. / nose of the ~ ‖ Nase f. des Dachziegels ‖ crochet m. de la tuile. / ~ ore ‖ Kupferbraun n.; Ziegelerz n.; Rotkupfererz n. ‖ cuivre m. oxydulé ferrifère ou oxydulé terreux. / ~ pin ‖ Mauernagel m. ‖ clou m. à tuile. / ~ press ‖ Fliesenpresse f. ‖ presse f. à carrelages.

tiler ‖ Ziegeldecker m.; Dachdecker m. ‖ tuileur m.; couvreur m. en tuiles.

tile roof ‖ Ziegeldach n. ‖ toit m. couvert en tuiles.

tilery ‖ Dachziegelei f.; Ziegelei f.; Ziegelofen m. ‖ tuilerie f.; fabrique f. de tuiles. / mechanical ~ ‖ Maschinenziegelei f. ‖ tuilerie f. mécanique.

tile stove ‖ Kachelofen m. ‖ poêle m. faïence. / ~ works pl. see tilery.

tiling ‖ Ziegelbedachung f. ‖ toiture f. en tuile.

till, to ~ (Agr) ‖ ackern; pflügen ‖ labourer; remuer avec la charrue.

tillage ‖ Ackerbestellung f. ‖ labourage m.

tillage cutter ‖ Ackerfräser m. ‖ fraiseuse f. de labour. / motor ~ ‖ Motorackerfräser m. ‖ fraiseuse f. de labour à moteur.

tiller (Agricultural labourer) ‖ Landarbeiter m.; Pflüger m.; Ackersmann m. ‖ laboureur m. / ~ (Tillage cutter) ‖ Bodenfräse f.; Ackerfräse f. ‖ fraiseuse f. de labour. / rotary ~ for gardens ‖ Gartenfräse f. ‖ motoculteur m. ou motofraise f. de jardin.

tiller (Handle of a tool) ‖ Griff m.; Handhabe f. ‖ manche m.; manette f. / double-pronged ~ ‖ Gabelheft n. ‖ queue f. fourchure. / socket ~ ‖ Augenheft n. ‖ queue f. à douille.

tiller (Shipb) ‖ Ruderpinne f. ‖ timon m. ou barre f. du gouvernail. / crooked ~ ‖ gebogene Ruderpinne f. ‖ barre f. courbée du gouvernail.

tilt, to ‖ kippen ‖ culbuter; faire la bascule. / ~ (To incline) ‖ schrägstellen; neigen; schief legen ‖ incliner. / ~ about its axis ‖ um seine Achse f. kippen ‖ basculer autour de son axe. / ~ the blooms ‖ die Schirbel mpl. recken oder strecken ‖ étirer les maquettes de fer au marteau.

tilt ‖ Schirmtuch n.; Zeltdach n.; Plan m.; Plane f.; Wagenplane f. ‖ banne f.; bâche f.

tilted iron ‖ gehämmertes Stabeisen n. ‖ fer m. forgé.

tilter (Roll mill) ‖ Kantvorrichtung f. ‖ culbuteur m. / ~ for ingots ‖ Blockkipper m.; Blockwendevorrichtung f. ‖ culbuteur m. de lingots; appareil m. de renversement pour lingots. / ~ for plates ‖ Blechwendevorrichtung f. ‖ appareil m. de renversement pour tôles.

tilt-frame (Met) ‖ Hammergebälk n.; Hammergerüst n. ‖ soucherie f.; bâti m. du pilon.

tilt hammer ‖ Schwarzhammer m.; Reckhammer m. ‖ marteau m. à queue ou à bascule.

tilting (Inclining) ‖ Kantung f. ‖ inclinaison f. latérale. / ~ of blooms ‖ Recken n. *oder* Schmieden n. der Schirbel ‖ étirage m. des maquettes de fer au marteau. / ~ of table up to x degrees (Mach tool) ‖ Schrägstellbarkeit f. des Tischschlittens um x Grad ‖ inclinaison f. possible de la table de x degrés.

tilting back ‖ verstellbare Lehne f. ‖ dossier m. réversible. / ~ bearing ‖ Kipplager n.; Klapplager n. ‖ palier m. basculant *ou* articulé. / ~ cart ‖ Kippkarren m. ‖ tombereau m.

tilting device ‖ Wippvorrichtung f.; Kippvorrichtung f. ‖ dispositif m. basculant. / ~ for rolling mills ‖ Kantvorrichtung f. für Walzwerke ‖ culbuteur m. pour laminoirs.

tilting and moving device for rolling mills ‖ Kant- und Verschiebevorrichtung f. für Walzwerke ‖ culbuteur m. et dispositif m. de déplacement pour laminoirs.

tilting furnace ‖ Kippofen m. ‖ four m. basculant *ou* culbutant. / ~ moment ‖ Kippmoment n. ‖ moment m. de renversement. / ~ movement ‖ Kippbewegung f. ‖ mouvement m. basculant. / ~ seat ‖ drehbarer Sitz m. ‖ siège m. pivotant. / mechanically operated ~ support ‖ maschinell verstellbare Kippvorrichtung f. ‖ inclinaison f. mécanique du bâti. / ~ table ‖ Kipptisch m. ‖ table f. inclinable. / ~ table with live rollers (Roll mill) ‖ Schwenktisch m. mit angetriebenen Rollen ‖ releveur m. basculant avec rouleaux d'amenée. / ~ wagon ‖ Kippwagen m. ‖ wagon m. basculant.

tilt linen ‖ Planleinwand f. ‖ toile f. à bâches.

tilt mill ‖ Hammerwerk n.; Hammerschmiede f. ‖ forge f. de martineur.

tilt van *see* tilt wagon.

tilt wagon ‖ Zeltwagen m.; Planwagen m. ‖ chariot m. couvert d'une bâche; voiture f. à bâche.

timber, to ‖ verzimmern; zimmern; bauen ‖ revêtir de charpente. / ~ (Mine) ‖ Stempel mpl. setzen; verzimmern ‖ étançonner; boiser. / ~ a shaft ‖ einen Schacht verzimmern *oder* ausbauen ‖ cuveler *ou* tuber un puits. / ~ wood ‖ Holz n. abvieren *oder* beschlagen ‖ carrer *ou* équarrir le bois.

timber ‖ Bauholz n.; Nutzholz n.; Zimmerholz n.; Stapelholz n. ‖ bois m. d'œuvre *ou* de construction *ou* de charpente. / to measure the ~ ‖ das Holz vermessen ‖ cuber le bois. / to work ~ ‖ zimmern ‖ charpenter. / arched ~ ‖ Krummholz n.; Knieholz n. ‖ bois m. courbé *ou* tortu; courbe f. / autumn ~ ‖ Herbstholz n. ‖ bois m. d'automne. / back-sided ~ ‖ windschiefes Holz n. ‖ bois m. gauchi. / cant ~ ‖ abgefaßtes *oder* abgekantetes Holz n.; abgekanteter Balken m. ‖ bois m. écorné. / clean ~ ‖ astfreies Holz n. ‖ bois m. net. / compass ~ *see* arched ~. / constructional ~ ‖ Bauholz n. ‖ bois m. de construction. / crooked ~ *see* arched ~. / cross-grained ~ ‖ Hirnholz n. ‖ bois m. de bout. / curve-~ *see* arched ~. / cut (or cleft) with the grain ‖ Langholz n.; Aderholz n. ‖ bois m. de fil. / ~ beginning to decay ‖ rückgängiges *oder* überständiges Holz n. ‖ bois m. surâgé *ou* suranné. / decayed ~ (Shipb) ‖ Wrackholz n. ‖ bois m. de démolition *ou* de tins. / dull-edged ~ ‖ Schalholz n.; waldkantiges Holz n. ‖ bois m. flacheux *ou* dévers. / endwise ~ ‖ Hirnholz n. ‖ bois m. de bout. / exotic ~ ‖ Überseeholz n. ‖ bois m. d'outremer. / filling ~ (Shipb) ‖ Füllspant n. ‖ couple m. de remplissage. / full-edged ~ ‖ vollkantiges Holz n. ‖ bois m. à vives arêtes; bois m. vif. / heavy ~ ‖ Starkholz n. ‖ gros bois m.; hewn ~ ‖ glattbehauenes Bauholz n. ‖ bois m. dégrossi. / knee ~ *see* arched ~. / lofty ~ ‖ hochstämmiges Holz n. ‖ bois m. de haute futaie. / long ~ ‖ Langholz n. ‖ bois m. en longues poutres; bois m. de brin *ou* de charpente. / long-tailed ~ *see* long ~. / ~ for mines ‖ Grubenholz n. ‖ bois m. des mines. / offal ~ ‖ Abfallholz n. ‖ bois m. en déchets. / old ~ ‖ Altholz n. ‖ vieille futaie f. / old ~ (Shipb) ‖ Wrackholz n. ‖ bois m. de démolition ou de tins. / planed ~ ‖ gehobeltes Bauholz n. ‖ bois m. de construction raboté. / rectangular ~ *see* squared ~. / rolled ~ ‖ Windbruch m.; windbrüchiges Holz n. ‖ bois m. gras *ou* chablé *ou* chablis; chablis m. / rough-edged ~ ‖ Schalholz n.; waldkantiges Holz n. ‖ bois m. flacheux *ou* dévers. / rough-hewn ~ ‖ behauenes Holz n. ‖ bois m. dressé à la hache. / round ~ ‖ Rundholz n. ‖ bois m. rond. / round ~ (Unhewn) ‖ Ganzholz n.; unbehauenes Holz n.; Rundholz n. ‖ bois m. de brin *ou* en grume. / sawed ~ ‖ Schnittholz n.; Sägeholz n. ‖ bois m. de sciage. / sawn ~ *see* sawed ~. / seasoned ~ ‖ trockenes Holz n. ‖ bois m. desséché. / ~ of a ship ‖ Schiffspant f. ‖ couple f. d'un bateau. / ~ for shipbuilding ‖ Schiffbauholz n. ‖ bois m. de construction navale. / ~s pl. of a ship ‖ Inhölzer npl. eines Schiffes ‖ membrure f. d'un navire. / split ~ ‖ Spaltholz n.; Kluftholz n. ‖ bois m. de fente; bois m. refendu *ou* de refend. / spring ~ ‖ Frühjahrsholz n. ‖ bois m. de printemps. / square ~ (Shipb) ‖ Winkelspant n. ‖ couple m. carré. / squared ~ ‖ Kantholz n.; (vierkantig) beschlagenes Holz n. ‖ bois m. d'équarrissage; bois m. carré. / stern ~ (Shipb) ‖ Heckspant n. ‖ montant m. de poupe. / straight ~ ‖ geradstämmiges Holz n. ‖ bois m. d'un beau brin. / tall ~ ‖ hochstämmiges Holz n. ‖ bois m. de haute futaie. / ~ for trade ‖ Handelsholz n. ‖ bois m. de commerce. / unbarked ~ ‖ Holz n. ohne Rinde; entrindetes Holz n. ‖ bois m. décortiqué. / unhewn ~ ‖ Ganzholz n.; Rundholz n.; unbehauenes Holz n.; Rohholz n.; Holz n. mit Rinde ‖ bois m. en grume *ou* de brin; bois m. en état brut. / white ~ ‖ Weißholz n.; Pappelholz n. ‖ bois m. blanc; bois m. de peuplier.

timber auction ‖ Holzversteigerung f.; Holzauktion f. ‖ vente f. de bois aux enchères. / ~ bond ‖ Zimmerverband m. ‖ assemblage m. de bois. / cargo of ~ ‖ Holzladung f. ‖ charge f. de bois. / ~ cart ‖ Brückenlangholzwagen m.; Langholzwagen m. ‖ wagon m. pour bois de construction. / ~ and iron-cased concrete foundations pl. ‖ Mantelgründung f. ‖ fondation f. par encaissement. / ~ crest (Arch) ‖ Helmzierat m. ‖ crête f. ornée. / ~ cross sleeper ‖ Holz(quer)schwelle f. ‖ traverse f. en bois. / ~ cutter ‖ Holzsäger m. ‖ scieur de bois m. / ~ destroyer ‖ Holzzerstörer m. ‖ destructeur m. de bois. / ~ distributor (Mine) ‖ Holzverteiler m. ‖ distributeur de bois m. / ~ facing ‖ Holzverkleidung f. ‖ revêtement m. en bois. / ~ framing ‖ Holzfachwerk n.; Riegelwerk n. ‖ cloisonnage de bois; pan m. en charpente. / piled ~ grating ‖ Pfahlrost m.; Pfahlwerk n.; Pfählung f. ‖ pilotage m.; piloté m.; ouvrage m. de pilotis. / ~ harbour ‖ Holzhafen m. ‖ port m. des bois. / ~ hewer ‖ Holzfäller m. ‖ bûcheron m.; bocquillon m.; bocanier m.

timbering ‖ Verzimmerung f.; Zimmerung f.; Holzverkleidung f.; Verschalung f.; boisage m. / ~ of the galleries of a mine ‖ Verkleidung f. der Minengänge ‖ coffrage m. des galeries de mines. / ~ of a mine ‖ Grubenzimmerung f.; Zimmerung f. der Grube ‖ charpente f. de mine. / ~ and walling of the mines ‖ Ausbau m. der Gruben ‖ construction f. en bois et en maçonnerie des mines. / ~ of a shaft ‖ Schachtzimmerung f.; Küvelage f.; Austonnung f. ‖ cuvelage m.; cuvellement m. / ~ and walling of a shaft ‖ Schachtausbau m. ‖ cuvelage m. et muraillement m. d'un puits.

timber lining (Mine) ‖ Pfändung f. ‖ charpente f. à coins. / ~ loader (Mine) ‖ Holzanschlepper m. ‖ meneur de bois m. / ~ loading bridge ‖ Verladebrücke f. für Holzbeförderung ‖ pont m. de transbordement pour la manutention de bois. / ~ man (Mine) ‖ Stempelsetzer m. ‖ étançonneur m. / ~ partition ‖ Holzverschlag m.; Holzwand f. ‖ cloison f. en bois *ou* en charpente. / ~ pile ‖ Holzpfahl m. ‖ pieu m. en bois. / ~ planking ‖ Bohlenbelag m.; Holzbelag m. ‖ couverture f. en madriers. / covered by ~ planking ‖ durch Holzbelag m. abgedeckt ‖ garni d'une couverture en madriers. / ~ platform (Build) ‖ Schwellrost m.; Schwellwerk n. ‖ grillage m.; platin m. de charpente. / ~ preserving ‖ Holzkonservierung f. ‖ conservation f. des bois. / ~ seasoning ‖ Ablagern n. des Holzes ‖ séchage m. du bois. / ~ setter (Mine) ‖ Zimmerhauer m. ‖ boiseur m. / ~ sleeper ‖ Holzschwelle f. ‖ traverse f. en bois. / ~ taker (Mine) ‖ Holzabnehmer m. ‖ recueilleur de bois m. / ~ trade ‖ Holzhandel m. ‖ commerce m. du bois. / ~ wagon (Railw) ‖ Langholzwagen m. ‖ wagon m. pour (le transport des) grandes pieces de bois. / ~ wholesale merchant ‖ Holzgroßhändler m. ‖ marchand m. de bois en gros. / ~ wholesale trade ‖ Holzgroßhandel m. ‖ commerce m. de bois en gros. / ~ wood (Forest) ‖ Zimmerholz n. ‖ pile f. / ~ work (Build) ‖ Gebälk n.; Gezimmer n.; Zimmerwerk n.; Holzwerk n. ‖ charpente f. / ~ work (Bridge) ‖ Pfahlwerk n. ‖ estacade f. en charpente. / construction of a ~ work ‖ Holzverband m.; Zimmerverband m. ‖ assemblage m. des bois. / ~ yard ‖ Zimmerplatz m.; Bauhof m. ‖ chantier m. (de charpentier).

timbre (Acoust) ‖ Tonfarbe f.; Klangfarbe f. ‖ timbre m. (acoustique).

time, to ‖ die Zeit f. abstoppen *oder* abmessen ‖ mesurer le temps.

time ‖ Zeit f. ‖ temps m. / to be behind ~ ‖ sich verspäten ‖ s'attarder. / to check the ~ ‖ die Zeit f. vergleichen ‖ collationner le temps. / to determine the ~ ‖ die Uhrzeit f. feststellen ‖ déterminer le

temps de la montre. / to give a certain ~ ‖ eine Frist setzen ‖ fixer un terme. / to measure the ~ ‖ die Zeit f. ausmessen ‖ mesurer le temps. / in order to save ~ ‖ um Zeit zu sparen ‖ afin de gagner du temps. / apparent ~ ‖ wahre Zeit f. ‖ temps m. vrai. / ~ of arrival (Railw) ‖ Ankunftszeit f. ‖ arrivée f.; heure f. d'arrivée. / astronomical ~ ‖ astronomische Zeit f.; Sternwartezeit f. ‖ temps m. astronomique. / ~ for centrifuging ‖ Schleuderzeit f. ‖ durée f. d'essorage. / ~ to get connection (Aut Tel) ‖ Einstellzeit f. ‖ temps m. de mouvement. / ~ of construction ‖ Bauzeit f. ‖ délai m. de construction. / ~ of delivery ‖ Ablieferungszeit f.; Lieferzeit f. ‖ terme m. ou délai m. de livraison. / to observe the ~ of delivery ‖ die Lieferzeit f. innehalten ‖ respecter le délai de livraison. / estimated ~ ‖ veranschlagte Zeit f. ‖ temps m. estimé. / ~ of exposure ‖ Aufnahmezeit f. ‖ temps m. de pose. / ~ of flight ‖ Flugzeit f. ‖ durée f. de vol. / in good ~ ‖ rechtzeitig ‖ bien à l'heure. / ~ of high-water ‖ Hochwasserzeit f. ‖ heure f. de la pleine mer. / ~ of learning ‖ Lehrzeit m. ‖ temps m. d'apprentissage. / ~ for loading ‖ Ladezeit f.; Ladefrist f. ‖ jours mpl. de planche pour le chargement; temps m. de charge; délai m. de chargement. / local sideral ~ ‖ Ortssternzeit f. ‖ temps m. sidéral du lieu. / for a long ~ ‖ langfristig ‖ à longs jours. / taking much ~ ‖ zeitraubend ‖ avec dissipation f. du temps. / ~ of oscillation ‖ Schwingungsdauer f. ‖ durée f. d'une oscillation. / paid ~ (Tel) ‖ bezahlte Sprechzeit f. ‖ temps m. payé. / ~ of payment (Trade) ‖ Verfallzeit f. ‖ échéance f. / to allow ~ for payment ‖ stunden ‖ accorder un délai. / ~ allowed for payment ‖ Stundungsfrist f. ‖ délai m. accordé pour le payement. / ~ of performance ‖ Erfüllungszeit f. ‖ époque f. de l'exécution; jour m. de l'échéance. / ~ of revolution (Mach) ‖ Umlaufzeit f. ‖ durée f. d'une révolution. / ~ of rotation in hours ‖ Stundenumlaufzeit f. ‖ durée f. horaire de révolution. / at the same ~ ‖ zugleich ‖ en même temps. / ~ of setting (Cement) ‖ Abbindezeit f. ‖ durée f. de la prise. / for a short ~ ‖ kurzfristig ‖ à courts jours. / ~ of stay (Railw) ‖ Aufenthalt m. ‖ arrêt m. / within the ~ ‖ innerhalb der Frist ‖ dans le temps; au délai.

time adjustment ‖ Zeiteinstellung f. ‖ réglage m. horaire. / arc of ~ (Astron) ‖ Zeitbogen m. ‖ arc m. du temps. / ~ ball (Mar) ‖ Zeitball m. ‖ ballon m. de temps. / ~ ball and luminous time signal station ‖ Zeitball- und Zeitlichtsignalstation f. ‖ station f. d'émission de signaux lumineux et à boules. / ~ bargain (Exchange) ‖ Termingeschäft n.; Terminhandel m. ‖ affaire f. à terme. / ~ card ‖ Kontrollkarte f. ‖ carte f. de contrôle. / ~ check (Tel) ‖ Kurzzeitmesser m.; Gesprächsuhr f.; Telefonometer n. ‖ compteur m. horaire; compteur m. de la durée des conversations. / ~ constant ‖ Zeitkonstante f. ‖ constante f. de temps. / ~ control apparatus for secondary clock systems ‖ Rückkontrolleinrichtung in sympathischen Uhrenanlagen ‖ dispositif m. de contrôle inverse dans les installations à horloges

secondaires. / ~ correction ‖ Zeitverbesserung f. ‖ correction f. de temps. **timed sparks** pl. ‖ gesteuerte Funken mpl. ‖ oscillations fpl. guidées.

time delay relay ‖ Zeitrelais n.; Zeitschütz n. ‖ relais m. rétardeur ou à temps. / equation of ~ (Astron) ‖ Zeitgleichung f. ‖ équation f. du temps. / ~ exposition (Phot) ‖ Zeitaufnahme f. ‖ exposition f. ou pose f. à temps. / ~ flux of light ‖ Lichtmenge f. ‖ éclairage m. / ~ fuse (Electr) ‖ Zeitsicherung f. ‖ coupe-circuit m. à retardation. / ~ fuse (Arm) ‖ Zeitzünder m. ‖ fusée f. à temps. / ~ impulse (Tel) ‖ Zeitimpuls m. ‖ impulsion f. horaire. / ~ insurance ‖ Versicherung f. auf Zeit ‖ assurance f. à temps. / ~ integral ‖ Zeitintegral n. ‖ intégrale f. de temps. / ~ keeper (Instrument) ‖ Chronometer n.; Zeitmesser m. ‖ chronomètre m. / ~ keeper (Person) ‖ Zeitnehmer m. ‖ chronomètreur m. / loss of ~ ‖ Zeitverlust m. ‖ perte f. de temps.

timely günstig; passend ‖ opportun; de circonstance; favorable.

time meter ‖ Zeitmesser m. ‖ chronomètre m. / ~ meter in telephony ‖ Zeitmesser m. im Fernsprechbetrieb ‖ compteur m. de durée d'une conversation pour la téléphonie. / ~ and zone metering ‖ Zeitzonenzähler m. ‖ compteur m. de temps et zone. / observation for ~ (Mar) ‖ Zeitbestimmung f. (durch Stundenwinkel) ‖ observation f. de longitude.

timepiece ‖ Standuhr f. ‖ pendule f. / (Mar) Chronometer n.; Seeuhr f. ‖ chronomètre m.; montre f. marine.

time policy ‖ Zeitversicherungspolice f. ‖ police f. d'assurance à terme.

timer see also time keeper ‖ Sekundenuhr f. ‖ montre f. à secondes. / ~ for eggs ‖ Eieruhr f. ‖ sablier m.

time recording device ‖ Zeitregistriervorrichtung f.; Zeitzähleinrichtung f. ‖ dispositif m. de chronométrage; enregistreur m. de temps.

time-relay ‖ Zeitrelais n.; Zeitschütz n. ‖ relais m. temporisé ou à temps.

timer shaft ‖ Verteilerwelle f. ‖ arbre m. de distributeur.

time shutter (Phot) ‖ Zeitverschluß m. ‖ obturateur m. pour la pose.

time signal ‖ Zeitzeichen n.; Zeitsignal n. ‖ signal m. horaire. / ~ distributor ‖ Zeitzeichenübertrager m. ‖ distributeur m. des signaux horaires. / ~ service ‖ Zeitzeichendienst m. ‖ service m. de signaux horaires. / ~ transmitter ‖ Zeitzeichengeber m. ‖ émetteur m. de signaux horaires.

time signalling connection ‖ Zeitzeichenschaltung f. ‖ schéma m. des signaux horaires. / ~ system ‖ Zeitsignalanlage f. ‖ installation f. d'émission de signaux horaires.

time stamp ‖ Zeitstempel m. ‖ timbre m. horaire; timbre m. marquant la date ou le temps; autodateur m. / ~ (Control clock) ‖ Arbeitszeitkontrolluhr f.; Stechuhr f. ‖ horloge f. de contrôle du temps de travail.

time switch (Electr) ‖ Zeitschalter m. ‖ interrupteur m. à temps; interrupteur m. horaire.

time table (Railw) ‖ Fahrplan m. ‖ horaire m. (des trains); indicateur m. / railway ~ ‖ Eisenbahnfahrplan m. ‖ indicateur m. de chemin de fer.

time ticket ‖ Arbeitszettel m. ‖ fiche f. de travail. / unit (or unity) of ~ ‖ Zeiteinheit f. ‖ unité f. de temps. / ~ vector ‖ Zeitvektor m.; Zeitlinie f. ‖ vecteur m. temporel ou de temps. / ~ wages pl. ‖ Zeitlohn m. ‖ salaire m. au temps.

timing of the ignition ‖ Zündeinstellung f. ‖ réglage m. de l'allumage. / ~ device for watches and clocks ‖ Regelvorrichtung f. für Uhren ‖ dispositif m. de réglage des horloges. / ~ gear (Mot) ‖ Steuerung f. ‖ distribution f. / ~ gear case ‖ Steuerrädergehäuse n. ‖ carter m. de distribution. / ~ machine ‖ Zeitregelmaschine f. ‖ machine f. à régler le temps. / ~ switch (Electr) ‖ Zeitschaltwerk n. ‖ commutateur m. horaire ou à temps.

tin, to (Chem) ‖ verzinnen ‖ étamer.

tin ‖ Zinn n. ‖ étain m. / ~ (Box) ‖ Dose f.; Blechdose f.; Blechbüchse f. ‖ boîte f. en fer-blanc. / to take off ~ ‖ entzinnen ‖ ôter l'étamure f. / made of ~ ‖ zinnern; aus Zinn ‖ d'étain. / best ~ ‖ Hutzinn m.; Malakkazinn m. ‖ étain m. de Malacca ou en chapeau. / block ~ ‖ Körnerzinn n.; Kornzinn n. ‖ étain m. en lârmes ou en grains. / cap ~ see best ~. / common stamped ~ with an alloy of lead ‖ Pfundzinn n. ‖ étain m. commun. / fine ~ ‖ Klangzinn n.; Feinzinn n. ‖ étain m. fin ou sonnant. / finest smelted ~ see block ~. / grained ~ see block ~. / laminated ~ ‖ Walzzinn n. ‖ étain m. laminé. / Malacca ~ ‖ Hutzinn m.; Malakkazinn m. ‖ étain m. de Malacca ou en chapeau. / old ~ ‖ Altzinn n. ‖ étain m. à refondre. / ordinary ~ ‖ Blockzinn n. ‖ étain m. en saumons ou en bloc. / ringing ~ ‖ Klangzinn n.; Feinzinn n. ‖ étain m. fin ou sonnant. / sonorous ~ see ringing ~. / ~ of standard alloy ‖ Probezinn n. ‖ étain m. au titre. / worked ~ ‖ Werkzinn n. ‖ étain m. d'œuvre ou travaillé.

tin ashes pl. ‖ Zinnkrätze f.; Zinnasche f. ‖ crasse f. d'étain.

tin-bearing ‖ zinnführend ‖ stannifère.

tin box ‖ Blechdose f.; Blechbüchse f.; Konservendose f.; Zinndose f.; Blechemballage f.; Blechpackung f. ‖ boîte f. en tôle ou en fer-blanc ou en étain. / ~ maker ‖ Blechbüchsenmacher m. ‖ ferblantier-boîtier m.; boîtier m. / ~ making machine ‖ Blechdosenherstellungsmaschine f.; Maschine f. zur Herstellung von Blechdosen ‖ machine f. pour la fabrication des boîtes en fer-blanc. / ~ opener ‖ Büchsenöffner m. ‖ couteau m. à ouvrir les boîtes à conserves. / ~ solderer ‖ Blechdosenlöter m. ‖ soudeur-boîtier m.

tin butter ‖ Zinnchlorid n.; Zinnbutter f. ‖ tétrachlorure m. d'étain.

tincal ‖ borsaures Natrium n.; Natron n.; Borax m.; Tinkal m. ‖ soude f. boratée; borate m. de soude; borax m.; tincal m.

tin can see tin box. / ~ cash box ‖ Metallkassette f. ‖ caissette f. métallique. / ~ chloride ‖ Chlorzinn n. ‖ chlorure m. d'étain.

tin-containing ‖ zinnhaltig ‖ stannifère.

tin control plate ‖ Blechetikett n.; Blechmarke f. ‖ cachet m. en tôle; étiquette f. en fer-blanc. / crackling of ~ ‖ Zinnschrei m. ‖ cri m. de l'étain. / ~ crystals pl. ‖ Zinnpräparat n. ‖ sel m. d'étain.

tinctorial power ‖ Färbekraft f. ‖ pouvoir m. colorant.

tincture (Chem) ‖ Tinktur f. ‖ teinture f. / ~ **press** ‖ Tinkturenpresse f. ‖ presse f. pour teintures (pharmaceutiques).

tin cutting and edging machine ‖ Dosenabschneide- und -bördelmaschine f. ‖ machine f. à couper et à border les boîtes de conserves.

tinder ‖ Feuerschwamm m.; Zunder m. ‖ amadou m. / ~ **box** ‖ **Feuerzeug** n. ‖ briquet m. / ~ **paper** ‖ Zunderpapier n. ‖ papier-amadou m.

tin dichloride ‖ Zinnsalz n. ‖ protochlorure m. d'étain. / ~ **drum** ‖ Blechtrommel f. ‖ tambour m. en fer-blanc. / ~ **dust** ‖ Zinnstaub m. ‖ poussière f. d'étain. / ~ **file** ‖ Zinnfeile f. ‖ lime f. à étain.

tin-foil ‖ Stanniol n.; Blattzinn n.; Silberpapier n.; Zinnfolie f. ‖ étain m. battu *ou* laminé; papier m. *ou* feuilles fpl. d'étain; papier m. argenté. / ~s pl. in rolls ‖ Rollzinn n. ‖ étain m. en rouleaux.

tin-foil cable ‖ Stanniolkabel n. ‖ câble m. sous feuille d'étain. / ~ **condenser** ‖ Stanniolkondensator m. ‖ condensateur m. à feuilles d'étain. / ~ **hammer** ‖ Stanniolhammer m. ‖ marteau m. pour battre l'étain. / ~ **making machine** ‖ Stanniolherstellungsmaschine f. ‖ machine f. à fabriquer les feuilles d'étain. / ~ remover see ~ removing machine. / ~ **removing machine** (or bottles) ‖ Entkapselmaschine f. ‖ machine f. à enlever les capsules. / ~ **tape** ‖ Stanniolband n. ‖ ruban m. de feuille d'étain. / ~ **working machine** ‖ Stanniolbearbeitungsmaschine f. ‖ machine f. à travailler les feuilles d'étain. / ~ **works** pl. ‖ Stanniolschlägerei f. ‖ usine f. à battre l'étain.

tin founder ‖ Zinngießer m. ‖ fondeur m. *ou* potier m. d'étain. / ~ **foundry** ‖ Zinngießerei f. ‖ fonderie f. d'étain.

tinge, to ~ **wood** ‖ das Holz färben ‖ teindre le bois.

tin glazing ‖ Zinnglasur f. ‖ glaçure f. stannifère. / ~ **goods** pl. ‖ Zinnwaren fpl. ‖ articles mpl. *ou* marchandises fpl. en étain.

tinker ‖ Kesselflicker m. ‖ chaudronnier m. ambulant.

tin making machine see tin box making machine.

tinman ‖ Klempner m.; Blechschmied m. ‖ ferblantier m. / ~'s **maschine** ‖ Klempnereimaschine f. ‖ machine f. de ferblantier. / ~'s **snips** pl. ‖ Handblechschere f.; Blechschere ‖ cisailles fpl. de ferblantier. / ~'s **work** ‖ Klempnerarbeit f. ‖ ouvrage m. du ferblantier.

tin mine ‖ Zinnbergwerk n.; Zinngrube f. ‖ mine f. d'étain m.

tinned ‖ verzinnt ‖ étamé. / **internally** ~ ‖ innen verzinnt ‖ étamé intérieurement.

tinned fish ‖ Fischkonserve f. ‖ conserve f. de poisson. / ~ **fruit** ‖ Obstkonserven fpl. ‖ conserves fpl. de fruits. / ~ **iron plate** ‖ Weißblech n. ‖ fer-blanc m.; feuille f. de fer-blanc; tôle f. de fer étamée. / ~ **meat** ‖ Büchsenfleisch n. ‖ conserve f. de viande. / ~ **vegetables** pl. ‖ Büchsengemüse n. ‖ légumes mpl. conservés.

tinner ‖ Verzinner m. ‖ étameur m.

tinning ‖ Verzinnung f. ‖ étamage m. / ~ **of copper conductors** (El line) ‖ Verzinnung f. der Kupferleiter ‖ étamage m. des conducteurs en cuivre.

tinning furnace ‖ Verzinnungsofen m. ‖

fourneau m. d'étamage. / ~ **tank** ‖ Zinnkessel m. ‖ chaudière f. d'étamage.

tinol ‖ Tinol n. ‖ tinol m.

tin opener ‖ Dosenöffner m.; Büchsenöffner m. ‖ couteau m. à ouvrir les boîtes de conserves. / ~ **ore** ‖ Zinnerz n. ‖ minerai m. d'étain. / **alluvial** ~ **ore** ‖ Seifenzinn n.; Seifenzinnerz n.; Stromzinn n. ‖ étain m. d'alluvion. / ~ **packing** ‖ Blechpackung f.; Blechemballage f.; Blechverpackung f. ‖ emballage m. en tôle. / ~ **pipe** ‖ Zinnröhre f. ‖ tuyau m. d'étain.

tin plate ‖ Zinnplatte f.; Zinnblech n. ‖ plaque f. d'étain; étain m. en plaques. / ~ **(Tinned sheet iron)** ‖ Weißblech n. ‖ tôle f. de fer étamée. / **crystallized** ~ ‖ Metallmoor m. ‖ moiré m. métallique.

tin plate bucket ‖ Blecheimer m. ‖ seau m. en fer-blanc. / ~ **goods** pl. ‖ Weißblechwaren fpl. ‖ ferblanterie f.; articles mpl. en fer-blanc. / ~ **ink** ‖ Blechdruckfarbe f. ‖ couleur f. pour impression sur fer-blanc. / ~ **machine** ‖ Blechbearbeitungsmaschine f. ‖ machine f. pour la ferblanterie. / ~ **mould** ‖ Blechform f. ‖ moule m. en fer-blanc. / ~ **placard** ‖ Blechplakat n. ‖ affiche f. en tôle. / ~ **printing machine** ‖ Blechdruckmaschine f. ‖ machine f. pour l'impression sur tôle. / ~ **puncher** ‖ Weißblechstanzer m. ‖ estampeur m. de ferblanc. / ~ **scraps** pl. ‖ Weißblechabfälle mpl. ‖ déchets mpl. de fer-blanc. / ~ **toys** pl. ‖ Blechspielwaren fpl. ‖ jouets mpl. en fer-blanc. / ~ **varnish** ‖ Blechlack m. ‖ vernis m. pour fer-blanc. / ~ **vessel** ‖ Blechgefäß n. ‖ bidon m. en tôle *ou* en fer-blanc. / ~ **ware** see ~ goods pl. / ~ **warming pan** ‖ Wärmepfanne f. *oder* Wärmekessel m. ‖ aus Weißblech ‖ chaudron m. en fer-blanc.

tin plating plant ‖ Verzinnungsanlage f. ‖ installation f. d'étamage.

tin preparation ‖ Zinnpräparat n. ‖ préparation f. d'étain. / ~ **printing** ‖ Blechdruck m. ‖ impression f. sur fer-blanc. / ~ **printing machine** ‖ Blechdruckmaschine f. ‖ machine f. pour impression sur fer blanc. / ~ **putty** ‖ Zinnasche f.; unreines Zinnoxyd n. ‖ potée f. d'étain. / ~ **salt** ‖ Zinnsalz n. ‖ sel m. d'étain; chlorure m. d'étain. / ~ **scraps** pl. see tin plate scraps.

tinsel ‖ Rauschgold n.; Flitter m. ‖ clinquant m.; paillette f. / ~ **(Lacem)** ‖ Lahn m. ‖ lame f. / ~ **cord** (El line) ‖ Litzenschnur ‖ tresse f. corde. / ~ **dust** (For postcards) ‖ Brillantstaub m. ‖ poudre f. brillante. / ~ **rollers** pl. ‖ Lahnwalzen fpl. ‖ rouleaux mpl. guimpiers.

tinsmith see tinman.

tin solder ‖ Zinnlot n.; Weichlot n.; Weißlot n. ‖ soudure f. à l'étain; soudure f. tendre. / ~ **soldier** ‖ Zinnsoldat m. ‖ soldat m. d'étain. / ~ **spoon** ‖ Zinnlöffel m. ‖ cuiller f. d'étain.

tint, to ‖ abtönen; (eintönig) färben; einen Anstrich m. geben ‖ teindre.

tint ‖ Farbton m.; Tönung f. ‖ teinte f.

tin tack ‖ Tapeziernagel m. ‖ broquette f.

tinted glass ‖ Rauchglas n. ‖ verre m. fumé. / ~ **paper** ‖ Buntpapier n.; Tonpapier n. ‖ papier m. de couleur; papier m. peint.

tin tetrachloride ‖ Zinnchlorid n.; Zinnbutter f. ‖ tetrachlorure m. d'étain.

tinting brush ‖ Tuschpinsel m. ‖ pinceau m. / **tin toys** pl. ‖ Zinnspielwaren fpl. ‖ jouets mpl. en étain. / ~ **tube** ‖ Zinntube f. ‖

tube f. en étain. / **making of** ~ **tubes** ‖ Zinntubenherstellung f. ‖ fabrication f. des tubes en étain. / ~ **vessels** pl. ‖ Zinngeschirr n. ‖ poterie f. d'étain. / ~ **ware** see tinplate goods. / ~ **wire** ‖ Zinndraht m. ‖ fil mpl. d'étain. / ~ **work** ‖ Blecharbeit f. ‖ ferblanterie f.; ouvrage m. en fer-blanc. / ~ **works** pl. ‖ Zinnhütte f. ‖ fonderie f. de minerai d'étain.

tip, to ‖ ausschütten; abstürzen; stürzen; kippen ‖ déverser; culbuter; renverser. / ~ (To lose one's balance) ‖ kippen; sich (seitwärts) neigen ‖ basculer; faire la bascule. / ~ **up** ‖ umkippen; kippen ‖ culbuter; faire la bascule. / **Not to be tipped!** ‖ nicht kanten! ‖ ne pas culbuter! / the box of the wagon is arranged for being tipped ‖ der Wagenkasten m. ist aufkippbar ‖ la caisse du wagon est à rabattement. / **tipped position of the wagon box** ‖ gekippte Stellung f. des Wagenkastens ‖ caisse f. de wagonnet basculée; position f. de déversement de la caisse de wagonnet.

tip (Extremity) ‖ Spitze f.; äußerstes Ende f.; Griff m. ‖ pointe f.; bout m.; extrémité f.; poignée f. / ~ (Ferrule) ‖ Zwinge f. ‖ virole f.; ferrure f. annulaire. / ~ (Light blow) ‖ leichter Schlag m. ‖ tape f.; coup m. léger. / ~ (Goods yard) ‖ Abladeplatz m. ‖ debarcadère f. / ~ (Waste heap) ‖ Halde f. ‖ halde f.; crassier m. / ~ (Tilting device) ‖ Kippvorrichtung f. ‖ dispositif m. de basculement. / **black steel** ~ ‖ Griff m. aus schwarzem Stahl ‖ bout m. en acier noir. / ~ **of blade** ‖ Schaufelspitze f. ‖ tête f. d'aube. / **corrugated** ~ ‖ gewellter Griff m. ‖ bout m. ondulé. / ~ **of the finger** ‖ Fingerspitze f. ‖ bout m. du doigt. / ~ **of a shoe** ‖ Schuhkappe f. ‖ bout m. d'un soulier.

tip beam ‖ Kippwelle f. ‖ arbre m. de versement. / ~ **box wagon** ‖ Kastenkipper m. ‖ wagonnet m. à caisse basculante. / **side** ~ **box wagon** ‖ einseitiger Kastenkipper m. ‖ wagonnet m. à caisse basculante d'un seul côté. / ~ **car** see ~ lorry *and* tipping car *and* tip wagon. / ~ **chute** ‖ Sturzrinne f.; Sturzrutsche f. ‖ couloir m. *ou* glissière f. de déversement. / ~ **electrode** ‖ Punkt(schweiß)-elektrode f. ‖ électrode f. punctiforme. / ~ **lorry** ‖ Kippwagen m. ‖ voiture f. basculante; tombereau m. automobile. / **end** ~ **lorry** ‖ Stirnkipper m. ‖ benne f. basculante longitudinalement. / ~ **marker** (Shoem) ‖ Kappenmarkierer m. ‖ pointeur m. de bouts. / ~ **pan** ‖ kippbarer Trog m.; Kipptrog m. ‖ auge f. basculante. / ~ **paster** (Shoem) ‖ Kappenleimer m. ‖ colleur m. de bouts.

tipper (Person) ‖ Kipper m. ‖ culbuteur m.; verseur m. / ~ (Wagon) see also tippler ‖ Kippwagen m.; Kipper m. ‖ wagon m. basculeur; culbuteur m. / **automatic** ~ ‖ Selbstentlader m. ‖ wagonnet m. à déchargement automatique. / **end** ~ ‖ Hinterkipper m. ‖ wagonnet m. *ou* camion m. basculant en arrière. / **front** ~ ‖ Vorderkipper m. ‖ wagonnet m. basculant en bout. / **rotary** ~ (Wagon) ‖ Kreiselkipper m.; Kreiselwipper m. ‖ basculeur m. circulaire. / **rotary** ~ **for trams** (Mine) ‖ Kreiselwipper m. für Grubenwagen ‖ culbuteur m. rotatif pour wagonnets de mine. / **rotary** ~ **on wheels** ‖ fahrbarer Kreiselwipper m. ‖ culbuteur m.

rotatif sur roues. / wagon ~ ‖ Wagenkipper m. ‖ culbuteur m. de wagons.

tippet (Build) ‖ Schaube f.; Dachschaube f.; Strohschaube f. ‖ javelle f.

tipping ‖ kippend ‖ basculant. / dump car ~ to either side ‖ zweiseitiger Kippwagen m. ‖ wagonnet m. à double bascule.

tipping ‖ Kippen n. ‖ mouvement m. de bascule. / wagon for ~ towards either side ‖ nach beiden Seiten kippender Wagen m. ‖ wagonnet m. basculant des deux côtés.

tipping angle ‖ Kippwinkel m. ‖ angle m. de bascule. / ~ automobile ‖ Kippkraftwagen m. ‖ automobile m. à benne basculante. / ~ bucket ‖ Kippbecher m.; Klappkübel m. ‖ auget m. basculant; benne f. basculante. / ~ car ‖ Kippwagen m. ‖ wagonnet m. basculeur; camion m. à benne basculante. / allround ~ car ‖ Rundkipper m. ‖ wagonnet m. à benne basculante dans toutes les directions. / side ~ car ‖ Seitenkipper m. ‖ camion m. basculant de côté. / ~ cart ‖ Kippkarren m. ‖ tombereau m. / ~ chair ‖ Kippstuhl m. ‖ chaise f. à bascule. / ~ chute ‖ Fallrohr n. ‖ tuyau m. de descente. / ~ cradle ‖ Kippwiege f. ‖ berceau m. de déversement.

tipping device see also tipper and tippler ‖ Kippvorrichtung f. ‖ dispositif m. de basculement; appareil m. à bascule; basculeur m.; basculateur m. / automatic ~ ‖ selbsttätige Kippvorrichtung f. ‖ déversement m. automatique. / ingot ~ ‖ Blockkipper m. ‖ culbuteur m. de lingots. / mechanical ~ ‖ mechanische Kippvorrichtung f. ‖ dispositif m. mécanique de bascule. / wagon with mechanical ~ ‖ Wagen m. mit mechanischer Kippvorrichtung ‖ wagon m. basculeur à dispositif mécanique de bascule. / ~ for mine rubbish ‖ Bergekipper m. ‖ tombereau m. à éboulis. / ~ for mixers ‖ Kippvorrichtung f. für Mischer ‖ culbuteur m. pour mélangeurs. / ~ for motor cars ‖ Kippvorrichtung f. für Kraftwagen ‖ dispositif m. de basculage pour automobiles. / ~ for railway trucks ‖ Kippvorrichtung f. für Eisenbahnwagen ‖ appareil m. à bascule pour wagons de chemin de fer. / ~ wagon ~ ‖ Wagenkippvorrichtung f.; Wagenkipper m. ‖ basculeur m. pour wagons.

tipping hopper ‖ Kippmulde f. ‖ benne f. basculante; benne-griffe f. / ~ jetty ‖ Sturzgerüst n.; Stürzgerüst n. ‖ charpente f. ou estacade f. de déversement. / ~ plant for harbours ‖ Kippanlage f. für Häfen ‖ installation f. d'appareils basculants pour ports. / ~ plate (Shoem) ‖ Stoßplatte f. ‖ plaque f. de ferrure. / ~ platform ‖ Plattformkipper m. ‖ plateforme f. à bascule. / ~ position of the box ‖ Kippstellung f. des Wagenkastens ‖ position f. en bascule de la caisse de wagonnet; position f. de déversement de la caisse de wagonnet. / ~ stage ‖ Sturzbühne f.; Sturzgerüst n.; Kippbühne f. ‖ pont m. de décharge; déchargeur m. à bascule; plateforme f. de déversement. / ~ table ‖ Wipptisch m. ‖ releveur m. à rouleaux; table f. basculante. / ~ table installation ‖ Wipptischanlage f. ‖ installation f. de table basculante.

tipping trough ‖ Mulde f.; Kippwagenmulde f.; Kippmulde f. ‖ benne f. ou

auge f. basculante; benne-griffe f. / ~ with sloping ends ‖ konische Kippmulde f. ‖ auge f. (basculante) conique. / automatic fastening device for ~ ‖ selbsttätige Muldenfeststellung f. ‖ dispositif m. de retenue automatique de la benne basculante. / ~ mixer ‖ Kipptrogmischmaschine f. ‖ malaxeur m. ou mélangeur m. à auge basculante. / ~ mixing machine see ~ mixer.

tipping trunnion ‖ Kippzapfen m. ‖ tourillon m. du wagonnet basculant. / ~ tub ‖ Kippkübel m. ‖ bâche f. basculante. / ~ wagon see tip wagon.

tippler see also tipper and tipping device ‖ Kipper m.; Wipper m.; Kreiselwipper m. ‖ culbuteur m.; basculeur m. / ~ floor ‖ Wipperbühne f. ‖ plancher m. du culbuteur. / ~ frame ‖ Kipperrahmen m. ‖ cage m. du culbuteur.

tip shaft ‖ Kippwelle f. ‖ arbre m. de déversement. / ~ stitcher (Shoem) ‖ Kappennäher m. ‖ piqueur m. de bouts. / ~ trough see tipping trough. / ~ truck see tipping car and tip wagon.

tip-up tread of a ladder ‖ aufklappbare Stufe f. einer Leiter ‖ marche f. relevable d'une échelle.

tip vortex ‖ Randwirbel m. ‖ tourbillon m. de bord.

tip wagon see also tip lorry and tipping car ‖ Wagen m. mit Kippvorrichtung; Kippwagen m. ‖ camion m. à benne basculante; benne f. basculante; wagon m. basculeur. / all-round box ~ ‖ Rundkipper m.; Kastenrundkipper m. ‖ wagonnet m. à caisse basculante dans toutes les directions. / box ~ ‖ Kastenkipper m. ‖ wagonnet m. à caisse basculante. / cradle ~ ‖ Wiegenkipper m. ‖ wagonnet m. à berceaux de versement. / end ~ ‖ Vorkipper m. ‖ wagon m. basculant d'extrémité. / ~ moved by locomotives ‖ Kippwagen m. für Lokomotivbetrieb ‖ wagon m. à bascule pour traction par locomotive. / side ~ ‖ Seitenkipper m. ‖ wagon m. à bascule de côté; benne f. basculante latéralement. / trough ~ ‖ Muldenkipper m. ‖ wagonnet m. basculeur à auge. / trough ~ for mines ‖ Grubenmuldenkipper m. ‖ wagonnet m. basculeur pour mines. / trunnion ~ ‖ Zapfenkipper m. ‖ wagonnet m. culbuteur à tourillons.

tire, to ‖ ermüden; ermatten; müde machen ‖ fatiguer.

tire (Railw; Auto) see tyre.

tiredness ‖ Mattigkeit f. ‖ lassitude f.

tissue ‖ Gewebe n.; Weberei f.; feines Zeug n. ‖ tissu m. / ~ of basketry ‖ Gewebe n. aus Flechtstoffen ‖ tissu m. de vannerie. / ~ from camel hair ‖ Gewebe n. aus Kamelhaar ‖ tissu m. en poils de chameau. / cellular ~ ‖ Zellengewebe n. ‖ tissu m. cellulaire. / crape-like ~ ‖ krepppartiges Gewebe n. ‖ crépon m. / elastic ~ ‖ elastisches Gewebe n. ‖ tissu m. élastique. / ~ of floss silk ‖ Flockseidegewebe n. ‖ tissu m. de bourre. / ~ from horsehair ‖ Gewebe n. aus Roßhaar ‖ tissu m. en crins. / impregnated ~ for roofing or clothing of walls ‖ getränktes Gewebe n. zur Bedachung oder Wandbekleidung ‖ tissu m. imprégné pour toitures ou pour revêtement de murs. / light ~ ‖ leichtes Gewebe n. ‖ tissu m. léger. / special ~ ‖ Spezialgewebe n. ‖ tissu m. spécial. / tight ~ ‖ dichtes Gewebe n. ‖ tissu m.

serré. / transparent ~ ‖ durchsichtiges oder undichtes Gewebe n. ‖ tissu m. clair. / water-proof ~ ‖ wasserdichtes Zeug n. ‖ tissu m. imperméable.

tissue paper ‖ Seidenpapier n. ‖ papier m. de soie.

titanate ‖ titansaures Salz n.; Titanat n. ‖ titanate m.

titanic acid ‖ Titansäure f. ‖ acide m. titanique. / ~ ore see titaniferous ore.

titaniferous ‖ titanführend; titanhaltig ‖ titanifère. / ~ iron ‖ Titaneisen n. ‖ fer m. oxydulé titanifère. / ~ ore ‖ Titanerz n. titanhaltiges Erz n. ‖ minerai m. titanifère.

titanite ‖ Titanit m.; Sphen m. ‖ titanite f.; sphène f.

titanium ‖ Titan n.; Titanium n. ‖ titane m. / ~ chloride ‖ Chlortitanium n. ‖ chlorure m. de titanium. / ~ steel ‖ Titanstahl m. ‖ acier m. au titane. / ~ white ‖ Titanweiß n. ‖ blanc m. de titane.

titer (Chem) see titre.

title (Print) ‖ Titel m.; Buchtitel m.; Aufschrift f. ‖ titre m. / bastard ~ ‖ Schmutztitel m. ‖ faux-titre m. / capital ~ ‖ Haupttitel m. ‖ grand-titre m.; frontispice m.

title (Legal) ‖ Anrecht n. ‖ droit m.; titre m.; légitimité f. / ~ to acquire ‖ Erwerbstitel m. ‖ titre f. d'acquisition. / ~ to allow water to flow off through land belonging to others ‖ Anspruch m. auf Durchleitung von Abwässern durch fremde Grundstücke ‖ droit m. au passage des eaux d'écoulement à travers des terrains étrangers. / ~ indisputable ‖ nicht bestreitbarer Anspruch m. ‖ droit m. incontestable. / ~ to realty ‖ Realrecht n. ‖ droit m. réel. / ~ to work a mine ‖ Berechtigung f. zum Bergwerksbetriebe ‖ autorisation f. d'exploitation d'une mine.

title leaf see ~ page. / ~ page ‖ Titelblatt n. ‖ feuille f. ou page f. de titre; frontispice m. / ~ paper ‖ Titelpapier n. ‖ papier m. verni.

titrate, to ‖ titrieren; den Titer oder den Feingehalt bestimmen ‖ titrer.

titration ‖ Titrierung f.; Feingehaltbestimmung f. ‖ titrage m.

titre (Chem) ‖ Titer m.; Feingehalt m. ‖ titre m. / ~ of a solution ‖ Gehalt m. einer Flüssigkeit ‖ titre m. (d'une solution).

to-and-fro ‖ hin und her; vorwärts und rückwärts ‖ va-et-vient.

to-and-fro (Motion) ‖ Hin- und Herbewegung f. ‖ mouvement m. de va-et-vient.

tobacco ‖ Tabak m. ‖ tabac m. / ~ (Plant) ‖ Tabakpflanze f.; Tabak m. ‖ tabac m.; nicotiane f. / chewing ~ ‖ Kautabak m. ‖ tabac m. à chiquer. / ~ for cigarettes ‖ Zigarettentabak m. ‖ tabac m. à cigarettes. / common ~ (Bot) ‖ virginischer Tabak m. ‖ virginie m. / cut ~ see ~ for smoking. / ~ in dried leaves ‖ Tabak m. in getrockneten Blättern ‖ tabac m. en feuilles séchées. / ~ in fresh leaves ‖ Tabak m. in frischen Blättern ‖ tabac m. en feuilles fraîches. / ~ in powder ‖ Tabak m. in Pulverform ‖ tabac m. en poudre. / raw ~ ‖ Rohtabak m. ‖ tabac m. brut. / ~ for smoking ‖ Rauchtabak m. ‖ tabac m. à fumer. / ~ for snuffing ‖ Schnupftabak m. ‖ tabac m. à priser. / unribbed ~ ‖ entrippter Tabak m. ‖ tabac m. écôté.

tobacco box ‖ Tabakdose f. ‖ boîte f. à tabac; tabatière f. / bundle of ~ ‖ Tabakballen m. ‖ ballot m. de tabac. / ~ buyer ‖ Tabakeinkäufer m. ‖ acheteur m. de tabacs. / ~ cask ‖ Tabakfaß n. ‖ boucaut m. à tabac. / cultivation of ~ ‖ Tabakbau m. ‖ culture f. du tabac. / ~ cutter (Person) ‖ Tabakschneider m. ‖ hacheur m. de tabac (à la machine). / ~ cutter (Machine) see ~ cutting machine. / ~ cutting machine ‖ Tabakschneidemaschine f. ‖ machine f. à couper le tabac. / ~ dipper ‖ Tabakeinweicher m. ‖ mouilleur m. de tabac. / ~ drier ‖ Tabaktrockner m. ‖ sécheur m. de tabac. / ~ extract ‖ Tabakextract m.; Tabakauszug m. ‖ extrait m. de tabac. / ~ grower ‖ Tabakpflanzer m.; Tabakbauer m. ‖ planteur m. de tabac. / ~ hanger ‖ Tabakaufhänger m. ‖ sécheur m. de tabacs. / ~ industry ‖ Tabakindustrie f. ‖ industrie f. du tabac. / ~ juice ‖ Tabaksaft m. ‖ jus m. de tabac. / ~ leaf ‖ Tabakblatt n. ‖ feuille f. de tabac. / ~ leaf classer ‖ Tabakblättersortierer m. ‖ trieur m. de feuilles de tabac. / ~ leaf opener ‖ Tabakblattzurichter m. ‖ ouvreur m. des feuilles de tabac. / ~ liquorer ‖ Tabakanfeuchter m. ‖ mouilleur m. de tabac. / ~ machine ‖ Tabakindustriemaschine f. ‖ machine f. pour l'industrie du tabac. / ~ manipulator ‖ Tabakzurichter m. ‖ manipulateur m. de tabacs. / ~ manufactory ‖ Tabakfabrik f. ‖ manufacture f. de tabac. / ~ manufacture ‖ Tabakbereitung f.; Tabakzubereitung f.; Tabakfabrikation f. ‖ fabrication f. du tabac; manufacture f. de tabacs. / ~ mixing plant ‖ Tabakmischanlage f. ‖ installation f. de mélangeurs pour tabac. / ~ monopoly ‖ Tabakmonopol n. ‖ monopole m. du tabac. / ~ packing machine ‖ Tabakpackmaschine f. ‖ machine f. à empaqueter les tabacs. / ~ pipe ‖ Tabakpfeife f. ‖ pipe f. (à tabac). / ~ pipe mountings pl. ‖ Tabakpfeifenbeschläge mpl. ‖ ferrures fpl. pour pipes à tabac. / ~ potter see ~ wringer / ~ pouch ‖ Tabakbeutel m. ‖ blague f. (à tabac). / ~ press ‖ Tabakpresse f. ‖ presse f. à tabac. / ~ presser see ~ wringer. / ~ roasting machine ‖ Tabakröstmaschine f. ‖ torréfacteur m. à tabacs. / roll of ~ ‖ Tabakrolle f. ‖ carotte f. (de tabac). / ~ sauce ‖ Tabakbrühe f. ‖ sauce f. de tabac. / ~ shaker-out ‖ Tabakaufhänger m. ‖ sécheur m. de tabac. / ~ soaking machine ‖ Beizmaschine f. oder Röstmaschine f. für Tabak ‖ machine f. à rouir ou à faire fermenter le tabac. / ~ sorter ‖ Tabaksortierer m. ‖ trieur m. de feuilles de tabac. / ~ spinner ‖ Tabakspinner m. ‖ fileur m. de tabac. / ~ stalk ‖ Tabakstengel m. ‖ côte f. de tabac. / ~ stover see ~ drier. / ~ substitute ‖ Tabakersatzstoff m. ‖ succédané m. du tabac. / ~ tearering machine ‖ Tabakreißmaschine f. ‖ machine f. à déchirer le tabac. / ~ twister see ~ spinner. / ~ weigher ‖ Tabakwäger m. ‖ peseur m. de tabac. / ~ worm ‖ Tabakwurm m. ‖ ver m. de tabac. / ~ wringer ‖ Tabakpresser m. ‖ presseur m. de tabac.

toe (Of stocking) ‖ Fußspitze f. ‖ bout m. de pied. / ~ of a horse-shoe ‖ Griff m. am Hufeisen ‖ griffe f. ou pince f. d'un fer à cheval.

toecap (of a shoe) ‖ Kappe f.; Schuhkappe f. ‖ bout m. (d'un soulier). / ~ cutter ‖ Kappenzuschneider m. ‖ coupeur m. de bouts.

toeing knife ‖ Hufmesser n.; Wirkmesser n. ‖ rogne-pied m.; rainette f. simple; couteau m. anglais du maréchal.

toffee maker ‖ Karamellenfabrikant m. ‖ fabricant m. de caramels.

toggle ‖ Knebel m. ‖ garot m.; bâillon m.; gabillot m. / to fix with a ~ ‖ knebeln ‖ trésillonner.

toggle collapsible tube press ‖ Kniehebeltubenpresse f. ‖ presse f. à genouillère pour tubes. / ~ drawing press ‖ Kniehebelziehpresse f. ‖ presse f. à emboutir à genouillère.

toggle joint ‖ Knebelgelenk n.; Kniegelenk n. ‖ genouillère f. / ~ brake ‖ Kniehebelbremse f. ‖ frein m. à genouillère. / ~ riveting machine ‖ Kniehebelnietmaschine f. ‖ riveuse f. à col de cygne.

toggle lever ‖ Kniehebel m. ‖ levier m. coudé. / ~ mechanism ‖ Kniehebelbewegung f. ‖ commande f. par levier coudé. / ~ press ‖ Kniehebelpresse f. ‖ presse f. à genouillère.

toggle plate ‖ Druckplatte f. ‖ plaque f. de compression.

toggle switch ‖ Druckschalter m. ‖ interrupteur m. à pression.

toilet articles pl. ‖ Toilettengegenstände mpl. ‖ articles mpl. de toilette. / ~ fittings pl. ‖ Abortbeschläge mpl. ‖ ferrures fpl. de cabinets d'aisance. / ~ paper ‖ Klosettpapier n. ‖ papier m. hygiénique ou de toilette. / ~ scented powder ‖ parfümierter Toilettepuder m. ‖ poudre f. de toilette parfumée. / ~ seat ‖ Abortsitz m.; Klosettsitz m. ‖ siège m. de cabinet. / ~ soap ‖ Toiletteseife f. ‖ savon m. de toilette; savonnette f. / ground ~ soap ‖ pilierte Toiletteseife f. ‖ savon m. de toilette pilé. / ~ vinegar ‖ Toiletteessig m. ‖ vinaigre m. de toilette. / ~ water ‖ Toilettewasser n. ‖ eau f. de toilette.

token ‖ Kennzeichen n. ‖ marque f. distinctive; signe m. caractéristique. / ~ (Print) ‖ Papierzeichen n.; Zeichen n. ‖ marque f. ou corne f. de papier.

tolerance (Techn) ‖ Maßtoleranz f.; Toleranz f.; Spielraum m. ‖ tolérance f. / ~ of screw bolts ‖ Schraubentoleranz f. ‖ tolérance f. des vis.

toll, to ~ a bell ‖ eine Glocke läuten ‖ sonner une cloche.

toll ‖ Zollgeld n.; Zollgebühr f.; Wegezoll m.; Brückengeld n. ‖ péage m.

toll answering jack (Tel) ‖ Fernabfrageklinke f. ‖ jack m. interurbain. / ~ bridge ‖ Zollbrücke f. ‖ pont m. à péage. / ~ cable (Tel) ‖ (gewöhnliches) Fernsprech-Fernkabel n. ‖ câble m. interurbain. / ~ collector ‖ Zolleinnehmer m. ‖ receveur m. de péage. / ~ line (Tel) ‖ Fernleitung f. ‖ circuit m. interurbain. / ~ line conversation ‖ Ferngespräch n. ‖ communication f. interurbaine. / ~ line dialing (Aut tel) ‖ Wählerfernsteuerung f. ‖ installation f. d'une connection interurbaine par cadran d'appel. / ~ recording operator (Tel) ‖ Fernamtsbeamtin f.; Fernamtbeamtin f. ‖ opératrice f. interurbaine. / ~ switching trunk (Tel) ‖ Fernvermittlungsleitung f. ‖ ligne f. intermédiaire. / ~ test board (Tel) ‖ Klinkenumschalter m. ‖ commutateur m. de jacks. / ~ traffic (Tel) ‖ Fernverkehr m. ‖ trafic m. interurbain ou à longue distance.

tolly stick (Mine) ‖ Kerbholz n. ‖ taille f. des mines; bois m. entaillé.

tolu (balsam) ‖ Tolubalsam m. ‖ baume m. de Tolu.

toluene ‖ Toluol n.; Toluen n. ‖ toluène m.

toluidine ‖ Toluidin n. ‖ toluidine f.

toluol see toluene.

tomato ‖ Tomate f. ‖ tomate f. / ~ sauce ‖ Tomatentunke f. ‖ sauce f. à la tomate.

tomb (Stone) see tomb stone.

tombac (Met) ‖ Rotguß m.; Tombak m.; Rotmessing n. ‖ tombac m.; laiton m. rouge. / ~ bronze ‖ Tombakbronze f. ‖ bronze m. tombac. / ~ hose ‖ Tombakschlauch m. ‖ tuyau m. en tombac. / ~ rod ‖ Tombakstange f. ‖ barre f. de tombac. / ~ sheet ‖ Tombakblech n. ‖ tombac m. en feuilles. / ~ wire ‖ Tombakdraht m. ‖ tombac m. en fils; fil m. de tombac.

tomb stone ‖ Grabstein m.; Grabmal n. ‖ pierre f. tombale; monument m. funèbre ou funéraire.

tommy bar ‖ Brechstange f.; Hebel n.; Brecheisen n. ‖ levier m. de fer.

tommy screw ‖ Knebelschraube f. ‖ vis m. à clé.

ton (Measure) ‖ Tonne f. ‖ tonne f. / ~ burden ‖ Tonnentragfähigkeit f. ‖ poids m. par tonne.

tone, to (Phot) ‖ tonen ‖ virer. / ~ down (Paint) ‖ aufhellen; abtönen ‖ éclaircir; dégrader.

tone (Acoust) ‖ Ton m. ‖ ton m. / low ~ ‖ tiefer Ton m. ‖ ton m. grave ou sonore. / sharp ~ ‖ hoher Ton m. ‖ ton m. haut-aigu.

toneband (Sound film) ‖ Tonstreifen m. ‖ bande f. sonore.

tone filter ‖ Tonfilter n. ‖ filtre m. acoustique. / ~ fixing salt ‖ Tonfixiersalz n. ‖ sel m. de fixage ou de virage-fixage. / ~ frequency ‖ Tonfrequenz f. ‖ fréquence f. musicale. / ~ selection ‖ Tonselektion f. ‖ sélection f. du ton. / ~ signal (Tel) ‖ Summerzeichen n. ‖ signal m. de ronfleur. / ~ wheel (Radio) ‖ Einstimmungsrad n. ‖ roulette f. de mise au diapason.

tongs pl. ‖ (größere) Zange f. ‖ tenailles fpl.; tenaille f.; pince f. / ~ (Forg) ‖ Schmiedezange f. ‖ pince f. ou tenailles fpl. de forgeron. / assayer's ~ ‖ Probierzange f. ‖ pince f. d'essayeur ou à essai. / fire ~ ‖ Feuerzange f. ‖ pince f. à feu. / flat-bit ~ (Forg) ‖ flache Schmiedezange f. ‖ pince f. plate (à forger). / large ~ (Forg) ‖ große Zange f.; Schrötlingszange f.; Scherenzange f. ‖ Luppenzange f. ‖ étangue f.; louperesse f.; grande tenaille f. / leading ~ ‖ Plombierzange f. ‖ pince f. à plomber. / ~ of a pile-engine ‖ Schere f. oder Scherenhaken m. einer Ramme ‖ pince f. de déclic. / small ~ ‖ kleine Zange f. ‖ tenettes fpl. / three-legged ~ ‖ Dreiarmklemme f. ‖ pince f. à trois branches.

tongsman (Roll mill) ‖ Zangenmann m. ‖ tenailleur m.

tongs riveting machine ‖ Zangennietmaschine f. ‖ riveuse f. à pince.

tongue, to (Carp) ‖ verzapfen; durch einen Scherzapfen m. verbinden ‖ languetter.

tongue (Join; Carp) ‖ Scherzapfen m.; Schlitzzapfen m. ‖ languette f. / ~ (Railw) ‖ Weichenzunge f. ‖ aiguille f. de changement de voie. / ~ (For stretching saw-blades) ‖ Knebel m. ‖ garrot m.; languette f. / ~ of a buckle ‖ Schnallendorn m.; Schnallenzunge f. ‖ ardillon m. d'une boucle. / ~ of a crossing (Railw) ‖ Herzstück n. einer Kreuzung ‖ pointe f. de cœur d'un croisement. / ~ of a sliding door ‖ Führungsfeder f. einer Schiebetür ‖ coulisseau m. d'une porte à coulisse.

tongued wood ‖ gezapftes Holz n. ‖ bois m. languetté.

tongue attachment (Railw) ‖ Zungenkloben m. ‖ patte f. d'attache d'aiguille. / ~ depressor (Med) ‖ Zungenhalter m. ‖ abaisse-langue m. / ~ file ‖ Gabelfeile f.; Zungenfeile f. ‖ langue f. de carpe. / heel of ~ (Railw) ‖ Zungenwurzel f. ‖ racine f. de l'aiguille. / ~ instrument (Mus) ‖ Zungeninstrument n. ‖ instrument m. à anches. / ~ plane ‖ Federhobel m.; Spundhobel m. ‖ bouvet m. mâle; bouvet ou rabot m. à languette. / ~ rail ‖ Zungenschiene f. ‖ aiguille f. mobile. / switch with spring ~s ‖ Federweiche f. ‖ changement m. de voie à aiguilles flexibles.

tonguing-and-grooving machine ‖ Nutmaschine f.; Spundmaschine f. ‖ machine f. à rainer et à languetter; machine f. à faire les rainures et languettes.

toning bath (Phot) ‖ Tonbad n. ‖ bain m. de virage.

toning-down (Paint) ‖ Abtönung f. ‖ dégradation f.

tonka bean ‖ Tonkabohne f. ‖ fève f. Tonka.

ton measurement ‖ Tonnenmaß n. ‖ mesure f. à la tonne.

tonnage (Mar) ‖ Tonnengehalt m.; Tonnage f. ‖ tonnage m. / gross ~ ‖ Bruttotonnengehalt m. ‖ tonnage m. brut. / net ~ ‖ Nettotonnengehalt m. ‖ tonnage m. net. / ~ of output (Mine) ‖ Fördermenge f. ‖ tonnage m. / ~ of a wagon ‖ Ladegewicht n. oder Belastungsfähigkeit f. eines Wagens ‖ poids m. de charge ou tonnage m. d'un wagon.

tonnage deck (Mar) ‖ Vermessungsdeck n. ‖ pont m. de tonnage. / depth for ~ ‖ Vermessungstiefe f. ‖ creux m. de la cale pour le jaugeage. / ~ dues pl. ‖ Tonnengeld n. ‖ (droit m. de) tonnage m. / ~ law ‖ Vermessungsgesetz n. ‖ loi m. sur le jaugeage des navires.

tonometer ‖ Tonmesser m.; Stimmgabel f. ‖ tonomètre m.

tonquin bean see tonka bean.

ton register ‖ Registertonne f. ‖ tonneau m. de jauge.

tonsil probe (Med) ‖ Mandelquetscher m. ‖ sonde f. à amygdales.

tool see also tools ‖ Werkzeug n. ‖ outil m. / ~ (Turn) ‖ Drehstahl m.; Drehstichel m.; Drehmeißel m.; Drehwerkzeug n. ‖ outil m. à charioter ou de tour; acier m.; outil m. / to clamp the ~ ‖ den Drehstahl m. einspannen ‖ serrer l'acier m. / ~ with a curved face of lip ‖ Drehstahl m. mit gekrümmter Brustfläche ‖ outil m. de tour à tranchant arrondi. / cutting ~ ‖ Schneidewerkzeug n. ‖ outil m. tranchant. / cutting-off ~ (Turn) ‖ Einstechstahl m. ‖ outil m. à saigner. / depthing ~ ‖ Eingriffzirkel m. ‖ compas m. aux engrenages. / dressing ~ ‖ Richtwerk-

zeug n.; Abrichtwerkzeug n. ‖ outil m. à dresser. / finishing ~ (Turn) ‖ Schlichtmeißel m. ‖ outil m. à planer. / hand ~s pl. ‖ Handwerkszeug n. ‖ outillage m. / iron ~ ‖ eisernes Werkzeug n.; Eisenwerkzeug n. ‖ outil m. en fer. / marking ~ ‖ Anreißwerkzeug n. ‖ outil m. à marquer. / measuring ~ ‖ Meßwerkzeug n. ‖ outil m. de mesure. / ~ for working metal ‖ Werkzeug n. für die Metallbearbeitung ‖ outil m. à travailler le métal. / planing ~ (Turn) ‖ Hobelmeißel m. ‖ outil m. à raboter. / pneumatic ~ ‖ Preßluftwerkzeug n. ‖ outil m. pneumatique ou à air comprimé. / rotating ~ ‖ umlaufendes Werkzeug n. ‖ outil m. tournant./ ~ for rough turning ‖ Schruppmeißel m. ‖ outil m. dégrossisseur; outil m. à dégrossir. / saddler's ~ ‖ Sattlerwerkzeug n. ‖ outil m. pour selliers. / steel ~ ‖ Stahlwerkzeug n. ‖ outil m. en acier. / ~ of wood ‖ Werkzeug n. aus Holz ‖ outil m. en bois. / ~ for wood working ‖ Werkzeug n. für die Holzbearbeitung ‖ outil m. à travailler le bois; outil m. pour le travail du bois. / adjusting of ~s ‖ Einstellen n. der Werkzeuge ‖ placement m. ou ajustage m. des outils.

tool-and-cutter grinder see tool grinder.

tool-and-gear wagon (Of a fire brigade) ‖ Gerätewagen m. ‖ voiture f. d'agrès.

tool bag ‖ Werkzeugtasche f. ‖ sacoche f.; trousse f. à outils. / ~ box ‖ Werkzeugkasten m.; Werkzeugkiste f. ‖ boîte f. ou caisse f. ou coffre m. à outils. / ~ box (Turn) ‖ Stichelhaus n. ‖ porte-outil m. / ~ cabinet ‖ Werkzeugschrank m. ‖ armoire f. pour outils. / ~ carrier ‖ Stahlträger m.; Stichelträger m.; Meißelträger m.; Werkzeugaufnahme f.; Support m. ‖ porte-outil m. / ~ case see ~ box. / ~ chest (Mach) see ~ box. / ~ chest (Mine) ‖ Gezähekasten m. ‖ boîte f. pour outils. / ~ feed ‖ Werkzeugvorschub m. ‖ avance f. des outils. / ~ fitter ‖ Werkzeugmontör m. ‖ monteur m. d'outils.

tool grinder (Person) ‖ Werkzeugschleifer m. ‖ affûteur m. ou meuleur m. d'outils. / ~ (Machine) ‖ Werkzeugschleifmaschine f. ‖ machine f. à affûter les outils. / plain wet ~ ‖ Werkzeugnaßschleifmaschine f. ‖ machine f. à affûter les outils à l'eau.

tool grinding machine (For turning tools) ‖ Werkzeugschleifmaschine f. ‖ affûteuse-profileuse f. pour outils.

tool handle ‖ Werkzeuggriff m. ‖ manche m. d'outil.

tool holder (Turn) ‖ Stahlhalter m.; Stichelhalter m.; Werkzeughalter m.; Stichelhaus n. ‖ porte-outil m. / circular form ~ ‖ Rundstahlhalter m. ‖ porte-outil m. pour outil rond. / double ~ ‖ doppelter Werkzeughalter m. oder Support m. ‖ support m. porte-outil double. / hinged ~ ‖ klappbarer Werkzeughalter m. oder Support m. ‖ support m. porte-outil à charnière; boîte f. porte-outil à charnière. / plain ~ ‖ einfacher Werkzeughalter m. oder Support m. ‖ support m. porte-outil simple. / special ~ for cutting threads without clearance for the tools ‖ Rückzugstahlhalter m. für Gewinde mit Auslauf ‖ porte-outil m. à récul pour filets sans dégagement.

tool maker ‖ Werkzeugmacher m .‖ outilleur m.; fabricant m. d'outils. / ~ manu-

factory ‖ Werkzeugfabrik f. ‖ atelier m. de fabrication d'outils. / ~ post see ~ holder.

tools pl. see also tool ‖ Handwerkszeug n.; Gerät n. ‖ outillage m.; outils mpl. / ~ (Mine) ‖ Gezähe n. ‖ outils mpl. / hand ~ see tools. / ~ for soil-tilling ‖ Bodenbearbeitungsgerät n. ‖ outils mpl. pour cultiver le sol.

tool slide for facing ‖ Plandrehschlitten m. ‖ chariot m. transversal de tour. / ~ for turning ‖ Langdrehschlitten m. ‖ chariot m. longitudinal à charioter.

tool smith ‖ Werkzeugschmied m.; Zeugschmied m.; Kleinschmied m. ‖ forgeron m. en outils; taillandier m. / ~ steel ‖ Werkzeugstahl m. ‖ acier m. pour outils ou à outils. / high-speed ~ steel ‖ Schnelldrehstahl m. ‖ acier m. pour tours à grande vitesse; acier m. rapide.

tooth, to ~ (Mach) ‖ bezahnen ‖ endenter; fraiser les dents. / ~ (Forg) ‖ krausschmieden ‖ créneler ou crêper le fer.

tooth ‖ Zahn m. ‖ dent f. / ~ (Of a cog wheel) see cog ~. / to strip the teeth pl. ‖ die Radzähne mpl. abbrechen ‖ casser les dents fpl. / animal ~ ‖ Tierzahn m. ‖ dent f. d'animal. / artificial ~ ‖ künstlicher Zahn m.; Kunstzahn m. ‖ dent f. artificielle. / artificial ~ of enamel ‖ künstlicher Zahn m. aus Schmelz ‖ dent f. artificielle en émail. / artificial ~ of porcelain ‖ künstlicher Zahn m. aus Porzellan ‖ dent f. artificielle en porcelaine. / set of artificial teeth pl. ‖ künstliches Gebiß n. ‖ dentier m. ou râtelier m. artificiel. / cast ~ ‖ gegossener Zahn m. ‖ dent f. brute de fonte. / cog ~ ‖ Radzahn m. ‖ dent f. de roue ou d'engrenage. / ~ of a fork ‖ Zinke f. einer Gabel ‖ dent f. ou fourchon m. d'une fourche. / gold pin ~ (Dent) ‖ Goldknopfzahn m. ‖ dent f. à pointe en or. / inclined ~ ‖ schiefer Zahn m. ‖ dent f. couchée. / machine-made ~ ‖ maschinengearbeiteter Zahn m. ‖ dent f. formée à la machine. / ~ with pin of not genuine metal (Dent) ‖ Zahn m. mit unechtem Stift ‖ dent f. à pointe en métal non précieux. / pivot ~ ‖ Stiftzahn m. ‖ dent f. à pivot. / platinum pin ~ (Dent) ‖ Platinstiftzahn m. ‖ dent f. à pointe en platine. / porcelain ~ ‖ Porzellanzahn m. ‖ dent f. en porcelaine. / ratched ~ ‖ Sperrzahn m. ‖ dent f. d'arrêt. / ~ imbedded in rubber ‖ in Kautschuk eingesetzter Zahn m. ‖ dent f. posée en caoutchouc. / ~ of a saw ‖ Sägezahn m.; Zahn m. einer Säge ‖ dent f. d'une scie. / straight ~ ‖ gerader Zahn m. ‖ dent f. droite. / ~ of trepan ‖ Messer n. des Bohrmeißels ‖ tranchant m. de fleuret. / V-shaped ~ ‖ Winkelzahn m. ‖ dent f. en V.

toothache ‖ Zahnschmerz m. ‖ mal m. de dents. / ~ cure ‖ Zahnschmerzmittel n. ‖ médicament m. odontique.

tooth brush ‖ Zahnbürste f. ‖ brosse f. à dents. / ~ brush glass ‖ Zahnbürstenglas n. ‖ verre m. pour brosses à dents. / ~ chamfering machine with two milling spindles ‖ Zahnabrundfräsmaschine f. mit zwei Frässpindeln ‖ machine f. à arrondir les entrées de dents à deux broches. / ~ crown ‖ Zahnkrone f. ‖ couronne f. dentaire ou à dents.

toothed ‖ gezahnt ‖ denté; endenté. / ~ arc ‖ Zahnbogen m. ‖ arc m. denté.

/ ~ disk ∥ gezahnte Scheibe f. ∥ disque m. denté.

toothed gear ∥ Zahnradgetriebe n.; Zahnrädergetriebe n. ∥ engrenage m. (à roues dentées). / ~ cutting ∥ Zahnräderschneiden n. ∥ taille f. d'engrenages. / ~ drive ∥ Zahnradantrieb m. ∥ commande f. par engrenage.

toothed gearing ∥ Zahnradgetriebe n. ∥ train m. d'engrenages.

toothed gear transmission ∥ Zahnräderübertragung f. ∥ transmission f. à engrenage.

toothed quadrant lever ∥ Zahnsegmenthebel m. ∥ levier m. pour secteur denté. / ~ rack-and-pinion drive ∥ Zahnstangenantrieb m. ∥ commande f. par crémaillère. / ~ rim ∥ Zahnkranz m. ∥ couronne f. dentée. / hollow-forged ~ rim ∥ hohlgeschmiedeter Zahnkranz m. ∥ couronne f. dentée forgée à creux. / ~ roll ∥ Stachelwalze f. ∥ cylindre m. denté. / ~ segment ∥ Zahnsegment n.; Zahnsektor m. ∥ secteur m. denté. / ~ transmission gear ∥ Zahnradvorgelege n. ∥ transmission f. par engrenage; roues fpl. d'engrenage.

toothed wheel see also cog wheel ∥ Zahnrad n. ∥ roue f. dentée ou d'engrenage. / ~ box ∥ Zahnradkasten m. ∥ carter m. des engrenages. / ~ drive ∥ Zahnradantrieb m. ∥ commande f. par engrenage ou à roues dentées. / ~ gear ∥ Zahnradgetriebe n. ∥ engrenage m. (à roues dentées). / ~ gearing ∥ Zahnradvorgelege n. ∥ transmission f. par engrenage; roues f. dentées.

toothed wheel-work (Mach) ∥ Zahnräderwerk n.; Zahnrädergetriebe n. ∥ engrenage m.

tooth flank ∥ Zahnflanke f. ∥ flanc m. de la dent. / ~ form ∥ Zahnform f. ∥ profil m. de la dent(ure). / ~ gear see toothed gear. / ~ gum ∥ Zahnkitt m. ∥ mastic m. pour les dents.

toothing ∥ Verzahnung f. ∥ endentement m.; crénelage m.; denture f. / bevel wheel ~ ∥ Kegelradverzahnung f. ∥ denture f. à roues coniques. / external ~ ∥ Außenverzahnung f. ∥ denture f. extérieure. / horizontal ~ ∥ liegende Verzahnung f. ∥ denture f. en retraite. / involute ~ ∥ Evolventenverzahnung f. ∥ denture f. à développante. / upright ~ (Build) ∥ stehende Verzahnung f. ∥ chaîne f. de pierres d'attente. / upright ~ (Mach) ∥ stehende Verzahnung f. ∥ attente f.

toothing iron ∥ Zackeisen n. ∥ fer m. à déchiqueter. / ~ plane ∥ Zahnhobel m. ∥ rabot m. à dents.

toothless saw ∥ zahnlose Säge f. ∥ scie f. sans dents.

tooth line ∥ Zahnreihe f.; Zackenreihe f. ∥ rangée f. de dents. / ~ lotion ∥ Zahnwasser n. ∥ eau f. dentifrice. / (standard) ~ outline ∥ (normale) Zahnform f. ∥ forme f. (normale) des dents. / ~ paste ∥ Zahnpaste f. ∥ pâte f. dentifrice. / ~ paste kneading machine ∥ Zahnpastenknetmaschine f. ∥ pétrin m. mécanique pour pâtes dentifrices.

toothpick ∥ Zahnstocher m. ∥ cure-dent m. / ~ making machine ∥ Zahnstocherfertigungsmaschine f. ∥ machine f. pour la fabrication des cure-dents.

tooth powder ∥ Zahnpulver n. ∥ poudre f. dentifrice. / scented ~ powder ∥ parfümierter Zahnpuder m. ∥ poudre f. dentifrice parfumée. / width of ~ space of spur gears ∥ Zahnlückenbreite f. an Stirnrädern ∥ largeur m. du creux des roues dentées droites. / ~ wash ∥ Zahnwasser n.; Mundwasser n.; Zahnreinigungswasser n. ∥ eau f. dentifrice. / ~ wax ∥ Zahnwachs n. ∥ cire f. pour dents. / ~ work (Build) ∥ Verzahnung f. ∥ denture f. / ~ wheel see toothed wheel.

top, to ~ a tree ∥ einen Baum m. kappen ∥ étêter ou épointer ou écimer un arbre.

top ∥ Spitze f. ∥ pointe f. / ~ (Upper side) ∥ Oberseite f. ∥ dessus m. / ~ (Auto; Coachm) ∥ Verdeck n. ∥ capote f. / ~ (Build) ∥ Dachfirst m.; First m. ∥ faîte m. / ~ (Road) ∥ Krone f. ∥ couronne f.; couronnement m. / ~ (Shipb) ∥ Mars m. ∥ hune f. / ~ (Spinn) ∥ Kammzug m. ∥ peigné m.; trait m. / ~ (Of wood) ∥ Zopf m. ∥ sommet m. / ~! (Packing) ∥ oben! ∥ dessus! / ~ boot ~ ∥ Stiefelschaft m. ∥ tige f. de botte. / ~ of a dike ∥ Deichkrone f. ∥ crête f. ou couronne f. ou sommet m. d'une digue. / folding ~ (Auto) ∥ zusammenklappbares Verdeck n. ∥ capote f. ployante. / ~ of the mountain ∥ Berggipfel m. ∥ sommet m. ou cime f. de la montagne. / ~ of rail ∥ Schienenoberkante f. ∥ dessus m. de rail. / measure from the ~ of rail to the centre of buffer ∥ Maß n. von Schienenoberkante bis Mitte Puffer ∥ distance f. entre le dessus du rail et le centre du tampon. / removable ~ (Roll mill) ∥ abnehmbare Kappe f. ∥ chapeau m. amovible. / ~ of sleepe ∥ Schwellendecke f. ∥ traverse f.; table f. supérieure de la traverse. / ~ of a wave ∥ Wellenkamm m. ∥ crête f. d'une lame. / ~ of wire netting ∥ Drahtaufsatz m. ∥ parois fpl. d'exhaussement en tissue métallique.

topaz ∥ Topas m. ∥ topaze f. / false (or smoky) ~ ∥ Rauchquarz m.; Rauchtopas m. ∥ fausse-topaze f.; quartz m. hyalin enfumé; topaze f. enfumée ou occidentale.

topaz pebbles pl. ∥ Topasgeschiebe npl. ∥ topazes fpl. roulées.

top ballasting (Road) ∥ obere Schüttung f. ∥ balastage m. seconde couche. / ~ beam (Carp) ∥ Hahnebalken m. ∥ fauxentrait m. / ~ beam (Bridge) ∥ Holm m.; Kappe f.; Kopfbalken m. (eines Brückenjochs) ∥ chapeau m.; chapiteau m.; chape f. / ~ binding (El line) ∥ Kopfbindung f. ∥ ligature f. supérieure. / ~ boot ∥ Stulpenstiefel m. ∥ botte f. à retroussis. / ~ box (Found) ∥ Oberform f.; Oberkasten m. ∥ contre-châssis m. / ~ camber (Airpl) ∥ Saugseite f. ∥ surface f. dorsale. / ~ casing (Of a bucket lever) ∥ Kopfhaube f. ∥ chapeau m. d'élévateur. / swing-out ~ clamping bar ∥ ausschwenkbare Oberwange f. ∥ coulisse f. supérieure articulée. / ~ course (Mas) ∥ Mauerabdeckung f.; Kappe f.; Mauerdeckplatte f. ∥ chaperon m.; tablettes fpl. / ~ cover of a cylinder ∥ Zylinderdeckel m. ∥ couvercle m. de cylindre. / ~ edge ∥ Außenkante f. ∥ face f. extérieure. / ~ end (Of wood) ∥ Zopfende n.; Zopf m. ∥ petit bout m.; sommet m. / size at the ~ end ∥ Zopfstärke f. ∥ dimensions fpl. au sommet. / ~ fencing ∥ Gitteraufsatz m. ∥ treillage m. de surhaussement. / ~ flame (Met) ∥ Gichtflamme f. ∥ flamme f. du gueulard. / ~ flange (Mach) ∥ Oberflansch m. ∥ bride f. supérieure. / ~ flange (Bridge) ∥ Obergurt m. ∥ membrure f. supérieure. / ~ flask (Found) see ~ box. / ~ frame (Mine) ∥ Flügelrahmen m.; Ohrjoch n.; Ohrrahmen m. ∥ cadre m. à oreilles. / ~ framing (Shipb) ∥ Mastkorb m.; Marskorb m. ∥ cage f. de mât de hune. / ~ fuller (Forg) ∥ runder Setzhammer m. ∥ chasse f. ronde; dégorgeoir m.

topgallant sail ∥ Bramsegel n. ∥ perroquet m.

top gallery (Met) ∥ Gichtbühne f.; Gichtebene f.; Plattform f. ∥ plate-forme f. du gueulard. / ~ gases pl. (Met) ∥ Gichtgas n. ∥ gaz m. des hauts-fourneaux; gaz m. perdu. / ~ gas valve ∥ Manövrierventil n. ∥ soupape f. de manœuvre. / ~ gear ∥ hohe Übersetzung f. ∥ prise f. directe. / ~ gig (Weav) ∥ Barchentrauhmaschine f.; Rauhmaschine f. ∥ laineuse f. à futaine à poil.

top-heavy (Shipb) ∥ oberlastig; überstürzig; kopflastig ∥ jaloux; lourd de la proue.

top lantern (Mar) ∥ Topplaterne f.; Marslaterne f. ∥ lanterne f. de hune. / ~ limit of lift ∥ Hubbegrenzung f. ∥ limite f. supérieure de levage.

topmast ∥ Marsstänge f.; Toppmast m. ∥ mât m. de hune.

topographical map ∥ topografische Karte f. ∥ carte f. topographique. / ~ particular ∥ Situationsgegenstand m. ∥ objet m. de situation.

topography ∥ Topografie f.; Ortsbeschreibung f. ∥ topographie f.

top part (Found) ∥ obere Kastenhälfte f.; Oberkasten m. ∥ contre-châssis m. d'un châssis de moulage.

topping machine for turnips ∥ Rübenköpfer m. ∥ écimeur m. de betteraves.

topple, to ~ (down) ∥ fallen; stürzen; umfallen; umstürzen ∥ tomber. / ~ backwards ∥ rückwärts umfallen ∥ tomber à la renverse.

top rail of a door-frame ∥ Oberfries m. einer eingestemmten Tür ∥ traverse f. supérieure d'une porte encadrée. / ~ of window valve ∥ Flügelweite f. eines Fensterflügels ∥ traverse f. supérieure d'un battant de fenêtre.

top ripper (Mine) ∥ Reparaturhauer m. ∥ recarreur m.

top roll ∥ obere Walze f.; Oberwalze f. ∥ cylindre m. supérieur. / exchangeable ~ ∥ auswechselbare Oberwalze f. ∥ cylindre m. supérieur échangeable. / in height tilting ~ ∥ hochkippbare Oberwalze f. ∥ cylindre m. supérieur basculant en hauteur.

top roll bearing body ∥ Oberwalzenlagerkörper m. ∥ corps m. de palier du cylindre supérieur.

topsail ∥ Marssegel n.; Toppsegel n. ∥ hunier m. / upper ~ ∥ Obermarssegel n. ∥ volant m.

topsail yard ∥ Marsrahe f. ∥ vergue f. du hunier.

top screw ∥ Schraubdeckel m. ∥ couvercle m. ou bouchon m. fileté. / ~ shaft frame (Mine) ∥ Flügelrahmen m.; Ohrjoch n.; Ohrrahmen m. ∥ cadre m. à oreilles. / ~ smoke (Blast furnace) ∥ Gichtrauch m. ∥ fumée f. du haut-fourneau. / ~ soil ∥ Oberboden m. ∥ couche f. superficielle du sol. / ~ stairs pl. (Blast furnace) ∥ Gichttreppe f. ∥ escalier m. d'un haut-fourneau. / ~ swage (Forg) ∥ Obergesenk n.; Oberteil

m. des Gesenks ‖ dessus m. d'une étampe.

topsy-turvy ‖ in Unordnung ‖ en gavauche.

top tumbler (Of a dredger) ‖ obere Eimertrommel f. ‖ tambour m. à godets supérieur. / ~ valve (Mach) ‖ Oberventil n. ‖ soupape f. ou clapet m. supérieur. / ~ view ‖ Draufsicht f. ‖ vue f. d'en haut. / ~ weight (Shipb) ‖ Obergewicht n. ‖ poids m. des hauts.

toque ‖ Zipfelmütze f.; Barett n. ‖ toque f.

torberite (Miner) ‖ Kupferuranglimmer m. ‖ chalcolite f.

torch ‖ Fackel f.; Leuchtfackel f. ‖ flambeau m. / cutting ~ ‖ Schneidbrenner m. ‖ brûleur m. à découper. / pitch ~ ‖ Pechfackel f. ‖ torche f. de poix; torchère f. / wax ~ ‖ Wachsfackel f. ‖ torche f. en cire.

torch lamp ‖ Lötlampe f. ‖ lampe f. à souder.

toric ‖ torisch ‖ torique.

torn ‖ zerrissen ‖ déchiré. / ~ ribbon (Typewr) ‖ abgenutztes Farbband n. ‖ ruban m. épuisé.

tornado ‖ Zyklon m. ‖ tempête f. tournante; cyclone m.

torpedo ‖ Torpedo m. ‖ torpille f. / aerial ~ ‖ Lufttorpedo m. ‖ torpille f. aérienne. / automatical ~ ‖ Kontaktmine f. ‖ torpille f. automatique. / marine ~ ‖ Seetorpedo n. ‖ torpille f. marine.

torpedo aeroplane ‖ Torpedoflugzeug n. ‖ avion m. torpilleur. / ~ boat ‖ Torpedoboot n. ‖ torpilleur m. / ~ boat destroyer ‖ Torpedobootjäger m.; Torpedobootzerstörer m.; Zerstörer m. ‖ contre-torpilleur m. / compressed air chamber for ~es ‖ Torpedoluftkessel m. ‖ réservoir m. à air comprimé des torpilles. / ~ gyro ‖ Kreiselgeradlaufgerät n. für Torpedos ‖ instrument m. gyroscopique pour torpilles.

torque (Mech) ‖ Zugkraft f. ‖ force f. de traction. / ~ rod ‖ Kardanwelle f. ‖ arbre m. à cardan m. / ~ stand ‖ Prüfstand m. ‖ banc m. d'épreuve.

torrent ‖ Wildwasser n.; Wildbach m.; Sturzbach m. ‖ torrent m. / damming of ~s pl. ‖ Wildbacheindämmung f. ‖ correction f. des torrents; défenses fpl. contre les torrents. / ~ works pl. see damming of ~s.

torsade ‖ schraubenförmig gewundene Franse f.; Torsade f. ‖ torsade f.

torsion ‖ Torsion f.; Windung f.; Drehung f. ‖ torsion f.

torsional moment ‖ Drehmoment n.; Verdrehungsmoment n. ‖ moment m. de torsion. / ~ stress ‖ Drehungsbeanspruchung f. ‖ effort m. de torsion. / ~ tension ‖ Drehspannung f. ‖ tension f. de torsion. / ~ tester ‖ Torsionsprüfmaschine f. ‖ machine f. à (essayer la) torsion.

torsion balance ‖ Drehwage f.; Torsionswage f. ‖ balance f. de torsion. / ~ galvanometer ‖ Torsionsgalvanometer n. ‖ galvanomètre m. de torsion. / ~ head ‖ Torsionskopf m. ‖ bouton m. de torsion. / ~ strength ‖ Dreh(ungs)festigkeit f. ‖ résistance f. à la torsion. / ~ suspension ‖ Torsionsaufhängung f. ‖ suspension f. à torsion. / ~ test ‖ Torsionsversuch m.; Verwindeprobe f.; Verdrehungsprobe f. ‖ essai m. de torsion. / ~ testing machine ‖ Torsionsprüfmaschine f. ‖ machine f. à vérifier la torsion. / ~ and tensile testing

machine ‖ Torsionszugmaschine f. ‖ machine f. pour essais de torsion et de traction. / ~ vibration testing machine ‖ Drehschwingungsprüfmaschine f. ‖ machine f. à mesurer les vibrations dues à la torsion. / ~ wire ‖ Torsionsfaden m. ‖ fil m. de torsion.

tortoise shell ‖ Schildpatt n.; Schildkrötenschale f. ‖ écaille f. (de tortue); carapace f. / to cut ~ ‖ Schildpatt n. schneiden ‖ fendre l'écaille (de tortue).

tortoise shell comb ‖ Schildpattkamm m. ‖ peigne m. en écaille. / ~ cutter ‖ Schildpattschneider m. ‖ fendeur m. d'écaille. / ~ goods pl. ‖ Schildpattwaren fpl. ‖ articles mpl. en écaille (de tortue); tabletterie f. d'écaille. / ~ paper ‖ Schildpattpapier n. ‖ papier-écaille m. / ~ turner ‖ Schildpattdrechsler m. ‖ tourneur m. sur écaille. / ~ worker ‖ Schildpattarbeiter m. ‖ tabletier m. en écaille.

toss, to ~ the oars pl. (Mar) ‖ die Riemen mpl. hochnehmen ‖ mâter les avirons mpl.

toss ‖ Wurf m.; Werfen n. ‖ jet m.; coup m.

tossing tub (Ore dress) ‖ Rührfaß n.; Schlämmfaß m. ‖ cuve f. à rincer.

total amount ‖ Gesamtbetrag m. ‖ montant m. total. / ~ amplification (Tel) ‖ Gesamtverstärkung f. ‖ amplification f. totale. / ~ area ‖ Gesamtoberfläche f. ‖ surface f. totale. / ~ attenuation (Tel) ‖ Gesamtdämpfung f. ‖ affaiblissement m. total. / ~ coal consumption ‖ Gesamtkohlenverbrauch m. ‖ dépense f. totale en charbon. / ~ delivery of current ‖ Gesamtstromabgabe f. ‖ vente f. de courant totale. / ~ efficiency ‖ Gesamtwirkungsgrad m. ‖ rendement m. total. / ~ hardness ‖ Gesamthärte f. ‖ dûreté f. totale. / ~ heat ‖ Gesamtwärme f. ‖ chaleur f. totale. / ~ heating surface ‖ Gesamtheizfläche f. ‖ surface f. de chauffe totale. / ~ lift (Aero) ‖ Gesamtauftrieb m. ‖ poussée f. ascensionnelle totale. / ~ number ‖ Gesamtzahl f. ‖ nombre m. total. / ~ output ‖ Gesamtleistung f. ‖ débit m. total. / ~ reflection ‖ totale Reflexion f. oder Rückstrahlung f. ‖ réflexion f. totale. / ~ residue ‖ Gesamtrückstand m. ‖ résidu m. total. / ~ resistance ‖ Gesamtwiderstand m. ‖ résistance f. totale. / ~ weight ‖ Gesamtgewicht n. ‖ poids m. total. / ~ weight of a dutiable ware ‖ Gesamtgewicht n. einer zollpflichtigen Ware ‖ poids m. total d'une marchandise soumise aux droits de douane. / ~ width ‖ Gesamtbreite f. ‖ largeur f. totale.

total ‖ Gesamtbetrag m.; Endsumme f. ‖ somme f. totale; total m. / ~ of employees ‖ Gesamtbelegschaft f. ‖ ensemble m. du personel ouvrier.

totalizer (Mach) ‖ Zählwerk n. ‖ totalisateur m. / ~ (Racing) ‖ Totalisator m.; Toto m. ‖ totalisateur m.

totally reflected ray ‖ total reflektierter Strahl m. ‖ rayon m. totalement réfléchi.

tottering contact (Tel) ‖ Wackelkontakt m. ‖ contact m. intermittent.

touch, to ~ ‖ antasten; betasten; anfühlen ‖ toucher; tâter; palper. / ~ up ‖ nacharbeiten; nachbessern ‖ retoucher; réparer.

touch (Build; Mach) ‖ Riß m.; Vorzeichnung f. auf dem Werkstück ‖ trait m. de repère. / ~ (Goldsm) ‖ Strichprobe f. ‖

touche f. / ~ (Sugar) ‖ Fingerprobe f. ‖ preuve f. du filet. / ~ of the type bar ‖ Anschlag m. des Typenträgers ‖ frappe f. de la barre à caractères.

touching needle see touch needle.

touch needle ‖ Probiernadel f.; Probierstift m.; Streichnadel f. ‖ aiguille f. d'essai; touchau m. ou toucheau m. d'essayeur. / ~ spark (Electr) ‖ Abreißfunken m. ‖ étincelle f. à rupture.

touchstone (Assay) ‖ Probierstein m.; Strichstein m. ‖ pierre f. d'essai ou de touche. / ~ (Miner) ‖ Probierstein m.; Kieselschiefer m.; lydischer Stein m. ‖ pierre f. de Lydie; lydite f. lydienne f.

tough ‖ zähe; strengflüssig ‖ tenace. / charge making the metal-bath not too ~ ‖ Beschickung f., die das Metall nicht zu strengflüssig macht ‖ charge f. qui ne rend pas le métal trop difficilement fusible. / ~ and fibrous fracture (Met) ‖ sehniger und zäher Anbruch m. ‖ cassure f. nerveuse et tenace. / ~ hardness ‖ zähe Härte f. ‖ dureté f. tenace.

toughen, to ~ copper ‖ das Kupfer hammergar machen; das Kupfer zähe polen ‖ raffiner ou affiner le cuivre.

toughening of copper ‖ Garmachen n. des Rohkupfers ‖ raffinage m. de cuivre.

toughness ‖ Zähigkeit f.; Festigkeit f. ‖ ténacité f.; dureté f. / the steel is particularly distinguished by its great ~ ‖ der Stahl zeichnet sich ganz besonders durch große Zähigkeit aus ‖ l'acier m. est surtout remarquable par sa grande ténacité. / the ~ is greater than before ‖ die Zähigkeit ist größer als zuvor ‖ la ténacité est devenue plus grande. / ~ as shown by a notched bar under the action of impact ‖ die an eingekerbten Stäben durch Schlag festgestellte Zähigkeit f. ‖ la ténacité du métal définie par l'essai au choc sur barreaux entaillés. / insufficient ~ ‖ unzureichende Zähigkeit f. ‖ ténacité f. insuffisante.

tough-pitch, to be at ~ ‖ hammergar sein ‖ être raffiné.

tough-pitch copper ‖ Raffinatkupfer n.; hammergares Kupfer n. ‖ cuivre m. raffiné ou fin.

touring ‖ Wanderfahren n.; Reisen n. ‖ tourisme m. / ~ car ‖ Reisewagen m.; Tourenwagen m. ‖ voiture f. de tourisme.

tourist article of aluminium ‖ Touristengerät m. aus Aluminium ‖ article m. de tourisme en aluminium. / ~ ticket ‖ Rundreisekarte f. ‖ billet m. circulaire.

tourmaline (Miner) ‖ Turmalin m.; ‖ tourmaline f. / red ~ ‖ roter Turmalin m.; Rubellit m. ‖ tourmaline f. apyre.

tow, to ~ (Mar) ‖ schleppen; bugsieren; verholen ‖ remorquer; haler. / ~ abreast ‖ längsseit schleppen ‖ remorquer à couple. / ~ astern ‖ achteraus schleppen ‖ remorquer en arbalète.

tow (Rope) ‖ Schlepptau n.; Schleppseil n. ‖ (câble m. de) remorque f.; touline f.; traîne f. / to take a vessel in ~ ‖ ein Schiff n. ins Schlepptau nehmen ‖ prendre un navire à la remorque; donner la remorque à un navire.

tow (Vessel) ‖ Bugsierboot n.; Schlepper m. ‖ (bateau m.) remorqueur m.

tow (Spinn) ‖ Werg n.; Hede f. ‖ étoupe f.; rebut m. de filasse. / caulking ~ ‖ Schiffswerg n. ‖ étoupe f. à calfater. / hackle ~ ‖ Hechelhede f.; Hechelwerg n. ‖ peignon m.; étoupe f. de peignage. / scutching ~

‖ Schwinghede f.; Schwingwerg n. ‖ étoupe f. d'espadage; repérants mpl. / stuffing ~ ‖ Polsterwerg n. ‖ étoupe f. à rembourrer.

towage ‖ Schleppschiffahrt f.; Schleppen n.; Bugsieren n. ‖ remorquage m.; touage m. / ~ (Fee) ‖ Schlepplohn m.; Bugsierlohn m. ‖ remorquage m.

towboat ‖ Schleppschiff n.; Schlepper m.; Schleppboot n.; Bugsierboot n.; Seilschleppschiff n. ‖ remorqueur m.; bateau m. de remorque; bateau m. toueur.

tow breaking ‖ Wergreiben n. ‖ battage m. d'étoupes.

towed boat ‖ Schleppkahn m.; Schleppschiff n. ‖ bateau m. remorqué.

towel ‖ Handtuch n. ‖ essuie-main m. / bath ~ ‖ Badehandtuch n.; Badelaken n.; Badetuch n. ‖ peignoir m. / rubber ~ ‖ Frottier(hand)tuch n. ‖ serviette f. floconneuse à frotter. / ~ clip ‖ Tuchklemme f. ‖ pince f. de nappe. / ~ horse ‖ Handtuchständer m.; Handtuchgestell n.; Handtuchhalter m. ‖ porte-essuie-main m.; séchoir m.

tower ‖ Turm m. ‖ tour f. / ~ for aerial lines ‖ Leitungsmast m. ‖ pylône m. métallique pour conduites. / ~ for boring purposes ‖ Bohrturm m. ‖ tour f. à forer ou de sondage. / coal ~ ‖ Kohlenturm m. ‖ trémie f. à charbon. / cooling ~ ‖ Kühlturm m. ‖ tour f. de réfrigération. / water ~ ‖ Wasserturm m. ‖ chateau m. (à reservoir) d'eau. / water ~ of x m³ capacity ‖ Hochbehälter m. von x cbm Fassung für Wasser ‖ château m. d'eau de x m³ de capacité.

tower car (Auto) ‖ Turmwagen m. ‖ wagon m. à échafaudage automoteur. / ~ clock ‖ Turmuhr f. ‖ horloge m. (pour tours ou de clocher). / ~ crane ‖ Turmkran m. ‖ grue f. à pylône. / ~ mill ‖ holländische Windmühle f.; Turmwindmühle f. ‖ moulin m. à vent hollandais. / ~ revolving crane ‖ Turmdrehkran m. ‖ grue f. tournante sur pylône. / stationary ~ revolving crane ‖ feststehender Turmdrehkran m. ‖ grue f. pivotante géante et fixe. / ~ slater ‖ Turmdecker m. ‖ couvreur m. de clochers. / ~ tile ‖ Turmziegel m. ‖ tuile f. pour tours. / ~ wagon ‖ Turmwagen m. ‖ camion m. à échafaudage.

tow flannel for cleaning ‖ Putzlappen m.; Putzwerg n.; Putzwolle f. ‖ étoupe f. ou chiffons mpl. ou coton m. à nettoyer.

towing ‖ Schleppen n.; Bugsieren n.; Schleppschiffahrt f.; Treideln n. ‖ remorquage m.; halage m. / ~ on canals and rivers ‖ Schleppschiffahrt f. auf Binnengewässern ‖ remorquage m. de bateaux sur canaux et rivières. / ~ by horses ‖ Pferdetreidelung f. ‖ halage m. par chevaux. / ~ by men ‖ Treidelung f. mit Menschen ‖ halage m. à bras d'hommes.

towing boat see towboat. / ~ drum (Mar) ‖ Zugtrommel f. ‖ tambour m. de halage. / ~ gear ‖ Schleppgeschirr n. ‖ dispositif m. de remorquage. / ~ line see tow line. / ~ path see tow path.

tow line (Mar) ‖ Schleppleine f.; Treidelleine f. ‖ câble m. ou cordage m. ou aussière f. de remorque; ligne f. de halage. / ~ linen ‖ Wergleinwand f.; Hedeleinen n. ‖ toile f. d'étoupe. / ~ linen weaving ‖ Hedeleinenweberei f. ‖

tissage m. d'étoupes. / ~ line winch (Mar; Aero) ‖ Verholwinde f. ‖ treuil m. de rappel. / ~ maker ‖ Wergmacher m. ‖ étoupier m.

town ‖ Stadt f. ‖ ville f. / ~ car (Auto) ‖ Stadtwagen m. ‖ voiture f. de ville. / ~ dues pl. ‖ Gemeindeabgabe f. ‖ contribution f. municipale. / ~ fog ‖ Stadtnebel m. ‖ brouillard m. de ville. / ~ line (Tel) ‖ Stadtleitung f. ‖ ligne f. urbaine. / ~ plan ‖ Stadtplan m. ‖ plan m. de ville. / ~ reservoir ‖ Hauptbehälter m. ‖ réservoir m. principal. / ~ service (Traffic) ‖ Stadtverkehr m. ‖ communication f. de banlieue. / ~ survey ‖ Stadtmessung f.; Stadtvermessung f. ‖ levé m. dans la ville. / ~ traveller ‖ Platzreisender m. ‖ voyageur m. de place.

tow packing ‖ Wergliderung f.; Liderung f. mit Werg ‖ garniture f. d'étoupe. / ~ path ‖ Leinpfad m.; Treidelpfad m.; Schleppweg m. ‖ chemin m. de halage; tirage m. / ~ picker ‖ Wergzupfer m. ‖ éplucheur m. d'étoupes. / ~ rope see also tow line ‖ Schlepptau n.; Trosse f.; Schlepptrosse f.; Treidelleine f.; Bugsiertau n. ‖ câble m. de remorque; grélin m.; remorque f. / ~ sorter ‖ Wergsortierer m. ‖ trieur m. d'étoupes. / ~ spinning ‖ Wergspinnerei f.; Hedespinnerei f. ‖ filature f. d'étoupes. / ~ weaver ‖ Wergweber m. ‖ tisseur m. d'étoupe.

toxic ‖ giftig ‖ toxique.

toxicity ‖ Giftigkeit f. ‖ toxicité f.

toy see also toys pl. ‖ Spielzeug n. ‖ jouet m. / ~ animal ‖ Spieltier n. ‖ animal m. jouet. / ~ animal of plush ‖ Plüschspieltier n. ‖ animal m. jouet en peluche. / ~ cart ‖ Spielwagen m. ‖ voiture f. jouet. / ~ colours pl. ‖ Farben fpl. für Spielwaren ‖ couleurs fpl. pour jouets. / ~ cooking range ‖ Kinderkochherd m. ‖ fourneau m. de cuisine pour enfants. / ~ factory ‖ Spielwarenfabrik f. ‖ fabrique f. de jouets. / ~ fire-arm ‖ Kinderschußwaffe f. ‖ arme f. à feu pour enfants. / ~ flag ‖ Kinderfahne n. ‖ drapeau m. pour enfants. / ~ furniture ‖ Kindermöbel pl. ‖ meubles mpl. d'enfants. / ~ gun ‖ Kindergewehr n. ‖ fusil m. d'enfants. / ~ maker ‖ Spielwarenmacher m. ‖ monteur m. de jouets. / ~ maker (Carver) ‖ Kunstschnitzer m. ‖ débiteur m. pour tabletterie; tabletier m. / ~ making machine ‖ Spielwarenherstellungsmaschine f. ‖ machine f. à fabriquer des jouets. / ~ money ‖ Spielgeld n. ‖ argent m. du jeu. / ~ musical instrument ‖ Kindermusikinstrument n. ‖ instrument m. de musique pour enfants. / ~ optical lantern ‖ Spielprojektionsapparat m. ‖ appareil m. jouet de projection. / ~ painter ‖ Spielwarenmaler m. ‖ coloriste m. de jouets. / ~ piano ‖ Kinderklavier n. ‖ piano m. pour enfants. / ~ pistol ‖ Kinderpistole f. ‖ pistolet m. d'enfants.

toys pl. see also toy ‖ Spielwaren fpl.; Spielzeug n. ‖ jouets mpl. (d'enfants); articles mpl. de bimbeloterie. / cardboard ~ ‖ Spielwaren fpl. aus Pappe ‖ jouets mpl. en carton. / celluloid ~ ‖ Zelluloidspielwaren fpl. ‖ jouets mpl. en celluloïd. / chenille ~ ‖ Spielwaren fpl. aus Raupe oder Schenille ‖ jouets mpl. en chenille. / children's ~ ‖ Kinderspielzeug n. ‖ jouet m. pour enfants. / China ~ ‖ Spielwaren fpl. aus Porzellan ‖ jouets mpl. en porcelaine. / electrical ~ ‖ elek-

trische Spielwaren fpl. ‖ jouets mpl. électriques. / felt ~ ‖ Spielwaren fpl. aus Filz ‖ jouets mpl. en feutre. / lead ~ ‖ Bleispielwaren fpl. ‖ jouets mpl. en plomb. / mechanical ~ ‖ mechanisches Spielzeug n. ‖ jouets mpl. mécaniques. / ~ of metal ‖ Spielwaren fpl. aus Metall ‖ jouets mpl. en métal. / moulded cardboard ~ ‖ Spielwaren fpl. aus Papierstoff oder aus Papiermaché ‖ jouets mpl. en carton moulé. / musical ~ ‖ Spielwaren fpl. mit Musik ‖ jouets mpl. à musique. / optical ~ ‖ optische Spielwaren fpl. ‖ jouets mpl. optiques. / paper ~ ‖ Spielwaren fpl. aus Papier ‖ jouets mpl. en papier. / pasteboard ~ ‖ Pappspielwaren fpl. ‖ jouets mpl. en carton. / pasteboard and paper ~ ‖ Pappe- und Papierspielwaren fpl. ‖ jouets mpl. de carton et de papier. / plush ~ ‖ Spielwaren fpl. aus Plüsch ‖ jouets mpl. en peluche. / rubber ~ ‖ Gummispielwaren fpl. ‖ jouets mpl. en caoutchouc. / tin ~ ‖ Spielwaren fpl. aus Zinn ‖ jouets mpl. en étain. / tin-plate ~ ‖ Blechspielwaren fpl. ‖ jouets mpl. en fer-blanc. / wooden ~ ‖ Spielwaren fpl. aus Holz ‖ jouets mpl. en bois.

toy sabre ‖ Kindersäbel m. ‖ sabre m. d'enfants. / ~ sewing machine ‖ Kindernähmaschine f. ‖ machine f. à coudre pour enfants. / ~ ship ‖ Spielschiff n. ‖ bateau m. jouet. / ~ shop ‖ Kinderkaufladen m. ‖ magasin m. pour enfants. / ~ store ‖ Spielwarengeschäft n. ‖ magasin m. de jouets. / ~ tea set ‖ Kinderteegeschirr n. ‖ service m. de thé pour enfants. / ~ torpedo ‖ Knallerbse f. ‖ bombe f. pétard; pois m. fulminant. / ~ weapon ‖ Kinderwaffe f. ‖ arme f. pour enfants.

trace, to ~ (Draw) ‖ pausen; durchzeichnen; durchpausen ‖ calquer; poncer. / ~ (Join; Carp) ‖ anreißen ‖ marquer; tracer. / ~ (Surv) ‖ (eine Linie) abstecken; trassieren ‖ tracer. / ~ in full size (Carp) ‖ aufschnüren ‖ épurer. / ~ the ground ‖ den Boden trassieren ‖ tracer un ouvrage sur le sol.

trace ‖ Räderspur f.; Geleise n.; Wagenspur f. ‖ ornière f.; trace f. / ~ on paper (Of a mineral) ‖ Strich m. auf Papier ‖ tachure f. ou trace f. sur le papier. / ~ of the rays ‖ Strahlengang m. ‖ marche f. des rayons.

traced design ‖ Pause f.; durchgepauste Zeichnung f. ‖ calque m.

trace press ‖ Strangpresse f. ‖ boudineuse-peloteuse f.; boudineuse f.

tracer ‖ Vorzeichner m.; Anreißer m. ‖ traceur m. / ~ bullet ‖ Rauchspurgeschoß n. ‖ projectile m. traceur.

tracery (Arch) ‖ Maßwerk n. ‖ broderie f.; découpure f.; réseau m.; tracé m.

tracheitis ‖ Luftröhrenentzündung f. ‖ trachéite f.

tracheotomy tube ‖ Tracheotubus m. ‖ canule f. trachéale.

trachyte ‖ Trachyt m. ‖ trachyte m. / ~ tuff ‖ Trachyttuff m. ‖ trachyte-tuf m.

tracing (Copy) ‖ Pause f.; Kopie f.; Durchzeichnung f. ‖ calque m. / ~ (Plan) ‖ Grundriß m.; Plan m.; Riß m. ‖ délinéation f.; plan m.; tracé m. / to make a ~ ‖ durchpausen ‖ calquer. / ~ of a map ‖ Kartierung f. ‖ confection f. d'une carte.

tracing cloth ‖ Pausleinwand f.; Pausleinen n.; Kopierleinwand f. ‖ toile f. à

calquer; papier-toile m. / ~ paper ‖ Pauspapier n. ‖ papier m. à calquer ou à poncer. / oiled ~ paper ‖ Ölpapier n. zum
Durchzeichnen; Transparentpapier n. ‖
papier m. huile à calquer; papier m.
transparent. / ~ picket ‖ Absteckpfahl m.
‖ piquet m. / ~ punch ‖ Ziehpunze f. ‖
traçoir m. / ~ tape (Build) ‖ Absteckleine f. ‖ cordeau m. à tracer.

track, to ~ (A ship) ‖ treideln ‖ haler a la
corde.

track (Of ball bearing) ‖ Laufrille f. ‖ gorge
f. de roulement. / ~ (Of a crane) ‖ Fahr-
bahnstütze f.; Laufbahn f. ‖ voie f. de
roulement. / ~ (Maritime route) ‖ See-
weg m.; Route f. ‖ route f. maritime. /
~ (Mine) ‖ Gefährt n.; Rollbahngeleise
n. im Bergwerk ‖ voie f. de roulement. / ~
(Railw) ‖ Gleis n.; Eisenbahngleis n.;
Schienenstrang m. ‖ voie f. (ferrée); rails
mpl./~(Rail gauge) see also ~ gauge ‖ Spurweite f. ‖ écartement m. de rails; entre-
rails m. / ~ (Of a wheel) ‖ Radspur f.;
Spur f. ‖ voie f. des voitures; ornière f. /to
lay the ~ ‖ das Gleis verlegen ‖ poser la
voie. / arranging ~ (Railw) ‖ Verschiebe-
gleis n.; Rangiergleis n. ‖ voie f. de formation des trains; voie f. de service. /
branch ~ ‖ Nebengleis n. ‖ voie f. de branchement / double ~ ‖ Doppelgleis n. ‖ double voie f. / elevated discharging ~ ‖ Sturz-
bahn f. ‖ voie f. surélevée de déchargement. / fixed ~ ‖ festes Gleis n. ‖ voie
f. fixe. / grinding ~ (Mill) ‖ Mahlbahn
m. ‖ surface f. de mouture. / hook-
plate ~ (Railw) ‖ Hakenplattenoberbau
m. ‖ superstructure f. à selles à crochet.
/ horizontal ~ (Railw) ‖ horizontale
Strecke f. ‖ palier m. / main ~ ‖ Hauptgleis n. ‖ voie f. principale. / monorail
~ ‖ Einschienenlaufbahn f. ‖ voie f. de
roulement monorail. / overhead ~ (Crane)
‖ Hochbahn f. ‖ voie f. surélevée. /
portable ~ (for a frequent and quick
displacement) ‖ leicht bewegliches Gleis
n. (für häufiges und schnelles Verlegen)
‖ voie f. mobile et facilement transportable (pour poser et changer souvent
et rapidement). / railway ~ ‖ Gleisstrang m. ‖ (file f. de) rails mpl.; voie
f.; tronçon m. / rib-plate ~ ‖ Rippenplattenoberbau m. ‖ superstructure f. à
selles à nervures. / ~ of a swing bridge
(On which the rollers travel) ‖ Laufkranz
m. einer Drehbrücke ‖ couronne f. d'un
pont tournant. / ~ for rolling-stock in
repair ‖ Reparaturgleis n. ‖ voie f. pour
la réparation du matériel. / ~ of the
sea ‖ Seestrich m. ‖ parage m. / semi-
portable ~ (is to be displaced occasionally
only) ‖ halbbewegliches Gleis n. (wird
nur hin und wieder verlegt) ‖ voie f.
demi-mobile (n'est posée et changée que
rarement). / side ~ ‖ Nebengleis n.;
Seitengleis n. ‖ voie f. secondaire ou
latérale ou supplémentaire; voie f. de
garage. / ~ of a station ‖ Bahnhofsgleis
n. ‖ voie f. de la gare. / ~ having steel
sleepers ‖ Eisenbahnschwellenoberbau m.
‖ superstructure f. sur traverses en fer.
/ straight ~ ‖ gerader Strang m. ‖ voie f.
droite. / (endless) ~ of the tank (Arm) ‖
Raupe f. oder Kette f. des Kampfwagens ‖ chenille f. du char de combat.
/ elastic ~ of the tank ‖ federnde Raupe
f. des Kampfwagens ‖ chenille f. à
ressort du char de combat. / three-rail
~ ‖ Dreischienengleis n. ‖ voie f. à trois

rails. / transfer ~ ‖ Übergangsgleis n. ‖ voie
f. de raccordement. / ~ of the types
(Typewr) ‖ Typenbahn f. ‖ trajectoire f.
des caractères. / ~ having wooden sleepers ‖ Holzschwellenoberbau m. ‖ superstructure f. sur traverses en bois.

track boat see tow boat. / ~ bolt for
fastening fish plates ‖ Laschenschraube
f.; Laschenbolzen m. ‖ boulon m. à vis
à fixer les éclisses. / ~ bolt hole ‖
Laschenloch n. ‖ trou m. pour boulon
d'éclisse. / ~ chart (Mar) ‖ Kurskarte
f.; Kursskizze f. ‖ routier m.; trace f.
des routes parcourues. / ~ circuit ‖ Gleis-
stromkreis m. ‖ circuit m. de la voie. / ~
circuiting for interlocking ‖ Gleis-
besetzungsanlage f.; Gleisfreimeldean-
lage f. ‖ blocage m. par des circuits de
voie. / ~ curve ‖ Gleiskrümmung f. ‖
courbe f. de la voie. / illuminated ~
diagram ‖ Gleistafel f. ‖ diagramme m.
lumineux. / fastening of ~ ‖ Gleisbefestigung f. ‖ fixation f. des rails. /
~ fighting machine ‖ Raupenkampfma-
schine f. ‖ engin m. de combat muni de
chenilles. / ~ gauge ‖ Spurweite f. ‖ écartement m. de rails; entre-rails m.; largeur f. de voie.

tracking rope see tow line and tow rope.
track layer ‖ Schienenleger m.; Oberbauarbeiter m. ‖ poseur m. (de la voie). /
~ laying machine ‖ Gleisbaumaschine f.
‖ machine f. à poser les voies ferrées. / lifting of the ~ ‖ Freilegen n.
der Gleise ‖ enlèvement m. des rails. /
~ link of the tank ‖ Kettenglied n. oder
Raupenglied n. des Kampfwagens ‖ patin
m. de la chenille du char de combat. /
~ lock ‖ Gleissperre f. ‖ taquet m. d'arrêt.
/ ~ plant ‖ Gleisanlage f. ‖ installation
f. de voie. / ~ relay (Railway signalling) ‖
Blockrelais n. ‖ relais m. de voie. / ~ scales pl. ‖ Gleiswage f. ‖ bascule f. à wagons. / ~ section of ~ ‖ Gleisrahmen m.;
Gleisjoch n.; Gleisabschnitt m. ‖ châssis
m. de voie. / bevelled section of ~ ‖ Tra-
pezjoch n. ‖ châssis m. trapéziforme. /
~ section for contractor's switches ‖
Schlepprahmen m. für Schleppweichen ‖
tronçon m. mobile pour le changement
de voie. / portable section of ~ ‖ Gleisjoch n. ‖ châssis m. de voie. / straight
section of ~ ‖ gerades Gleisjoch n. ‖
châssis m. droit. / ~ shoe of the tank ‖
Kettenglied n. oder Raupenglied n. des
Kampfwagens ‖ patin m. de la chenille
du char de combat. / ~ tamping ‖ Unterstopfen n. der Schwellen ‖ bourrage m.
des traverses. / ~ tamping machine ‖
Gleisstopfmaschine f. ‖ machine f. pour
le bourrage de la voie ou à bourrer les
rails.

trackway (Tow path) see also tow path ‖
Treidelweg m.; Treidelpfad m.; Leinpfad m. ‖ chemin m. de halage.
trackway (Railw; Mine) ‖ Schienenbahn
f.; Schienenweg m. ‖ chemin m. à rails;
voie f. ferrée. / ~ with rails of timber ‖
Holzbahn f. ‖ route f. à ornières en bois.
traction (Mech) ‖ Zug m. ‖ traction f. /
electric ~ ‖ elektrischer Fahrbetrieb m.
‖ traction f. électrique.
traction bar ‖ Zugstange f. ‖ tige f. de
traction. / ~ bar adjustable in its length
‖ in ihrer Länge verstellbare Zugstange
f. ‖ tige f. de traction réglable sur sa
longueur. / ~ chain ‖ Zugkette f. ‖ chaîne
f. de traction. / ~ coefficient ‖ Trak-

tionskoeffizient m. ‖ coefficient m. de
traction. / ~ current (Electr) ‖ Fahr-
strom m. ‖ courant m. de traction. / ~
dynamometer ‖ Zugkraftmesser m. ‖
dynamomètre m. de traction. / ~ engine
‖ Traktor m.; Zugmaschine f. ‖ tracteur
m.; machine f. de traction. / ~ eye ‖ Zugöse f. ‖ anneau m. d'attelage. / cross-
bar ~ eye ‖ Kreuzlappenzugöse f. ‖
anneau m. d'attelage à joint en croix.
/ ~ lever ‖ Zughebel m. ‖ levier m. de
traction. / ~ load ‖ Zugbelastung f. ‖
charge f. de traction. / ~ network ‖
Straßenbahnnetz n. ‖ réseau m. de tramway. / ~ rod see also traction bar ‖ Zugstange f. ‖ tige f. de traction. / elastic
~ rod going right through ‖ durchgehende
federnde Zugstange f. ‖ tige f. de traction
à ressort allant de part en part. / ~
rope ‖ Zugseil n. ‖ câble m. tracteur. /
~ station (Tramway) ‖ Bahnzentrale f. ‖
centrale f. de tramway.
tractive effort ‖ Zugkraft f. ‖ effort m.
ou force f. de traction. / ~ force see
~ effort. / ~ output ‖ Zugleistung f. ‖
puissance f. de traction.
tractor (Screw) ‖ Zugschraube f. ‖ hélice
f. tractive. / ~ (Auto) ‖ Trecker m.;
Zugmaschine f.; Kraftschlepper m.;
Schlepper m.; Traktor m. ‖ tracteur m.
/ ~ (Tramway) ‖ Triebwagen m.; Zug-
wagen m. ‖ automotrice f.; voiture f.
motrice; tracteur m. / armoured ~ ‖
gepanzerter Schlepper m. ‖ tracteur m.
blindé. / caterpillar ~ ‖ Raupenschlepper
m. ‖ tracteur m. à (commande) chenille. /
chain ~ ‖ Kettenschlepper m. ‖ tracteur
m. à chaîne. / creeper-type ~ see caterpillar ~. / electric ~ ‖ Elektroschlepper
m. ‖ remorqueur m. électrique. / ~
driven by electricity ‖ Triebwagen m.
mit elektrischem Antrieb ‖ automotrice
f. à électricité. / farm ~ ‖ Zugmaschine
f. für die Landwirtschaft ‖ tracteur m.
agricole. / motor ~ ‖ Motorzugwagen
m.; Kraftschlepper m.; Motortrecker m.
‖ tracteur m. à moteur ou à pétrole. /
~ driven by motor (Tramway) ‖ Triebwagen m. mit Motorantrieb ‖ automotrice f. à essence. / single-axle ~ ‖
Einachsschlepper m. ‖ remorque f. à
essieu unique. / small ~ ‖ Kleinkraft-
schlepper m. ‖ tracteur m. léger. / ~
driven by steam (Railw) ‖ Dampfzug-
wagen m.; Triebwagen m. mit Dampfantrieb ‖ automotrice f. ou tracteur m.
à vapeur. / street ~ ‖ Straßenzug-
maschine f. ‖ locomotive f. routière.
tractor airscrew ‖ Zugschraube f. ‖ hélice
f. tractive. / ~ caisson ‖ Schleppermunitionswagen m. ‖ caisson-tracteur m.
/ ~ dragbar ‖ Schlepperdeichsel f. ‖ barre
f. d'attelage du tracteur. / ~ drawbar
see ~ dragbar.
tractor-drawn artillery ‖ kraftgeschleppte
Artillerie f. ‖ artillerie f. à tracteur.
tractor driver ‖ Schlepperführer m.;
Treckerführer m. ‖ conducteur m. du
tracteur. / ~ haulage ‖ Schlepperzug m. ‖
remorquage m. par tracteur. / ~ housing
‖ Traktorengehäuse n. ‖ carcasse f. de
tracteur. / ~ propeller see ~ airscrew. / ~
screw see tractor (Screw).
trade ‖ Handel m. ‖ commerce m. / ~ (Commercial intercourse) ‖ Handelsverkehr
m. ‖ relations fpl. commerciales; marché
m. des affaires. / ~ (Profession) ‖ Gewerbe n.; Gewerbszweig m.; Handwerk

n. ‖ métier m.; profession f. / ~ (Mine) ‖ taubes Gestein n.; Gangmasse f. ‖ gangue f.; matière f. *ou* roche f. stérile. / to follow a ~ ‖ ein Gewerbe n. treiben ‖ exercer un métier. / ~ across the frontier ‖ Grenzverkehr m. ‖ relations fpl. de frontière; trafic m. à la frontière. / foreign ~ ‖ Außenhandel m.; Auslandhandel m. ‖ commerce m. extérieur. / import ~ ‖ Einfuhr f.; Einfuhrhandel m. ‖ importation f. / international ~ ‖ Welthandel m.; Weltmarkt m. ‖ commerce m. mondial; marché m. mondial. / ~ on a monetary basis ‖ Geldwirtschaft f. ‖ finances fpl. / transatlantic ~ ‖ überseeischer Handel m.; Überseehandel m. ‖ commerce m. transatlantique. / transit ~ ‖ Transithandel m.; Durchgangshandel m. ‖ commerce m. de transit.

trade balance ‖ Handelsbilanz f. ‖ balance f. commerciale. / to burden the ~ balance ‖ die Handelsbilanz f. belasten ‖ grever la balance commerciale. / ~ bill ‖ Warenwechsel m.; Kundenwechsel m. ‖ effet m. de commerce; papier m. commercial. / Board of Trade ‖ Handelsministerium n. ‖ ministère m. de commerce. / ~ cask ‖ Versandfaß n.; Transportfaß n. ‖ fût m. d'expédition *ou* d'exportation; tonneau m. de transport; barrique f. / ~ clouds pl. ‖ Passatwolken fpl. ‖ nuages mpl. qui courent en direction contraire aux alizés. / ~ directory ‖ Firmenbuch n.; Firmenregister n. ‖ répertoire m. d'adresses; annuaire m. / ~ expenses pl. ‖ Geschäftsunkosten pl. ‖ frais mpl. de commerce. / new facilities pl. for ~ ‖ Verkehrserleichterung f. ‖ facilité f. de communication *ou* de transport. / ~ license ‖ Gewerbesteuer f. ‖ (impôt m. de la) patente f.; impôt m. sur le revenu professionnel. / ~ mark ‖ Fabrikmarke f.; Schutzmarke f.; Handelsmarke f.; Warenzeichen n. ‖ marque f. de fabrique; marque f. déposée. / registered ~ mark ‖ eingetragenes Warenzeichen n. ‖ marque f. de fabrique enregistrée; marque f. déposée. / ~ outlook ‖ Konjunktur f. ‖ conjoncture f. / ~ report ‖ Geschäftsbericht m. ‖ compte m. rendu. / ~ returns pl. *see* ~ report.

tradesman ‖ Gewerbetreibender m.; Handwerker m. ‖ personne f. exerçant une profession; industriel m.; homme m. de métier; artisan m.

trade society ‖ Gewerksgenossenschaft f. ‖ association f. ouvrière. / ~ statistics pl. ‖ Handelsstatistik f. ‖ statistique f. du commerce. / ~ terms pl. ‖ Bezugsbedingungen fpl. ‖ conditions fpl. de livraison. / ~ union ‖ Gewerkschaft f.; Arbeiterverband m. ‖ union f. des ouvriers; corps m. de métier; syndicat m. ouvrier. / miners' ~ union ‖ Bergarbeiterverband m. ‖ union f. des mineurs. / war upon ~ ‖ Handelskrieg m. ‖ guerre f. de destruction du commerce. / ~ ware ‖ Industrieerzeugnis n. ‖ produit m. de l'industrie. / ~ wind ‖ Passat m.; Passatwind m. ‖ vent m. alizé; alizé m. / region of the ~ winds ‖ Passatgebiet n. ‖ région f. des alizés; région f. alizée.

trading ‖ Handelsverkehr m. ‖ commerce m. / ~ fleet ‖ Handelsflotte f. ‖ flotte f. marchande. / ~ house ‖ Kaufhaus n. ‖ maison f. de commerce; magasin m.; bazar m. / ~ settlement ‖ Handelsniederlassung f. ‖ comptoir m.; factorerie

f. / ~ surplus ‖ Betriebsüberschuß m. ‖ bénéfices mpl. d'exploitation. / ~ vessel ‖ Handelsschiff n.; Frachtschiff n. ‖ navire m. de commerce; navire m. marchand; bâtiment m. de transport.

traffic ‖ Verkehr m.; Verkehrswesen n.; Verkehrsdienst m.; Betrieb m. ‖ trafic m.; service m.; circulation f. / ~ (Trade) ‖ Handelsverkehr m. ‖ trafic m.; relations fpl. commerciales; marché m. des affaires; commerce m. / ~ to open to ~ ‖ dem Verkehr übergeben ‖ livrer à la circulation. / ~ on canals ‖ Kanalschiffahrt f. ‖ navigation f. sur canaux. / ~ across the frontier ‖ Grenzverkehr m. ‖ relations fpl. de la frontière; trafic m. à la frontière. / incoming ~ ‖ ankommender Verkehr m. ‖ trafic m. d'arrivée. / international ~ ‖ Weltverkehr m. ‖ trafic m. international; relations fpl. internationales. / outgoing ~ ‖ abgehender Verkehr m. ‖ trafic m. de départ. / two-way ~ ‖ doppelseitiger Verkehr m.; wechselseitiger Verkehr m. ‖ trafic m. des deux côtés.

traffic block ‖ Verkehrsstockung f.; Verkehrsstörung f. ‖ interruption f. du service. / ~ chart ‖ Verkehrskurve f. ‖ courbe f. du trafic. / ~ dispatch ‖ Verkehrsabwicklung f. ‖ écoulement m. de trafic. / interruption of ~ ‖ Verkehrsstockung f.; Verkehrsstörung f. ‖ interruption f. du service. / ~ manager (Railw) ‖ Betriebsingeniör m. ‖ ingénieur m. de service. / ~ peak ‖ Verkehrsspitze f. ‖ pointe f. du trafic. / ~ plan ‖ Verkehrsplan m. ‖ plan m. de communication. / ~ sign ‖ Verkehrstafel f. ‖ poteau m. de signalisation.

tragacanth (Gum) ‖ Tragantgummi m. ‖ gomme f. adragante.

trail (Hydr arch) ‖ Fährseil n.; Giertau n. ‖ traille f. / ~ of a swing bridge ‖ Laufkranz m. einer Drehbrücke ‖ couronne f. de roulement d'un pont tournant. / ~ car *see* trailer.

trailer ‖ Anhängewagen m.; Anhänger m. ‖ voiture f. remorquée; remorque f. / ~ of the freight motor car ‖ Lastkraftwagenanhänger m. ‖ remorque f. de camion automobile. / ~ for motor cars *or* motor cycles ‖ Anhänger m. für Kraftwagen *oder* Krafträder ‖ remorque f. pour automobiles *ou* pour motocycles. / small ~ ‖ kleiner Schleppwagen m. ‖ voiturette f. remorque.

trailer caisson (War mat) ‖ Anhängermunitionswagen m. ‖ caisson-remorque m. / ~ car *see* trailer. / ~ coupling ‖ Anhängerkupplung f. ‖ dispositif m. d'attelage de remorque. / lead to the ~ ‖ Anhängerleitung f. ‖ conduit m. pour la remorque.

trail flying bridge ‖ Gierbrücke f.; Gierfähre f. ‖ gleitende Fähre f. am Spanntau ‖ bac m. à traille; traille f. (sur pontons).

trailing aerial ‖ Hängeantenne f.; Hängeluftleiter m. ‖ antenne f. suspendue. / ~ antenna *see* ~ aerial. / ~ axle (Loc) ‖ Hinterachse f. ‖ axe f. arrière. / ~ edge strip ‖ Abschlußleiste f. ‖ arêtier m. du bord de fuite. / two-wheel ~ truck ‖ hinterer Laufradsatz m. ‖ essieu m. porteur arrière. / ~ wheel ‖ Hinterrad n. ‖ roue f. arrière; arrière-roue f.

trail rope ‖ Schleppseil n. ‖ guide-rope m.; amarre f. de touage.

trail switch (Railw) ‖ Schleppweiche f. ‖ aiguille f. mobile.

train, to ‖ üben; trainieren ‖ entraîner.

train (Nav) ‖ Schleppzug m. ‖ convoi m.; train m. de remorquage *ou* de touage. / ~ (Railw) ‖ Zug m.; Eisenbahnzug m. ‖ train m. / ~ (Roll mill) ‖ Walzenstraße f. ‖ train m. de laminoir. / to start a ~ (Railw) ‖ einen Zug m. ablassen ‖ expédier un train. / ambulance ~ *see* Red-Cross ~. / arriving ~ (Railw) ‖ ankommender Zug m. ‖ train m. arrivant. / ~ of a carriage ‖ Wagengestell n. ‖ train m. de voiture. / corridor ~ (Railw) ‖ D-Zug m.; Durchgangszug m. ‖ train m. accordéon; train m. à wagons couloir. / delayed ~ (Railw) ‖ verspäteter Zug m. ‖ train m. en retard. / down-and-up ~ (Railw) ‖ Pendelzug m. ‖ convoi m. d'aller et retour. / express ~ ‖ Schnellzug m.; Eilzug m. ‖ train m. à grande vitesse; express m. / extra ~ (Railw) ‖ Sonderzug m. ‖ train m. spécial. / fast ~ *see* express ~. / goods ~ ‖ Güterzug m. ‖ train m. de marchandises. / ~ of impulses (Aut tel) ‖ Stromstoßserie f. ‖ train m. d'impulsions. / load ~ ‖ Lastenzug m. ‖ train m. routier. / medium iron ~ (Roll mill) ‖ Mitteleisenstraße f. ‖ laminoir m. *ou* train m. à fers moyens. / mixed ~ (Railw) ‖ gemischter Zug m. ‖ train m. mixte. / ~ of railway workmen ‖ Bauzug m. ‖ train m. de la route. / Red-Cross ~ ‖ Sanitätszug m. ‖ train m. de Croix Rouge *ou* d'ambulance. / ~ with refrigerator cars ‖ Kühlwagenzug m. ‖ train m. à voitures frigorifiques. / regular ~ ‖ fahrplanmäßiger Zug m. ‖ train m. régulier. / ~ of rolls ‖ Walzenstraße f. ‖ train m. de laminoir. / ~ adapted to the reception of the sick ‖ Sanitätszug m. ‖ train m. d'ambulance. / slow ~ ‖ Bummelzug m.; Personenzug m. ‖ train m. omnibus *ou* de petite vitesse. / starting ~ ‖ abfahrender Zug m. ‖ train m. partant. / through ~ ‖ durchgehender Zug m. ‖ train m. direct. / ~ of waves (Phys) ‖ Wellenzug m. ‖ train m. d'ondes.

train accident ‖ Zugunfall m. ‖ accident m. de train. / ~ call (Tel) ‖ Zuggespräch n. ‖ communication f. avec un train. / ~ conductor ‖ Zugführer m. ‖ conducteur m. de train. / ~ control ‖ Signalübertragung f. auf den Zug ‖ transmission f. des signaux sur les trains en marche. / ~ crew ‖ Zugpersonal n. ‖ personel m. du train. / ~ dispatching ‖ Zugmeldedienst m. ‖ service m. de signalisation (des trains). / ~ disaster ‖ Eisenbahnunglück n. ‖ accident m. de chemin de fer.

trained staff ‖ geschultes Personal n. ‖ personnel m. habile. / ~ workman ‖ Facharbeiter m. ‖ ouvrier m. spécial *ou* entraîné.

train end (Railw) ‖ Zugschluß m. ‖ queue f. du train. / ~ exchange (Tel) ‖ Zugvermittlungsstelle f. ‖ bureau m. intermédiaire aux trains. / ~ ferry ‖ Eisenbahnfähre f. ‖ bac m. porte-train; ferryboat m. / electric heating of ~s ‖ elektrische Zugheizung f. ‖ chauffage m. électrique des wagons.

training aeroplane ‖ Schulflugzeug n.; Übungsflugzeug n. ‖ avion m. école *ou* d'entraînement. / ~ boiler ‖ Lehrkessel m. ‖ chaudière f. d'instruction. / control for ~ purposes (Aero) ‖ Schulsteuerung f. ‖ commande f. d'instruction. / ~ shop for apprentices ‖ Lehrlingswerkstatt f. ‖

taelier m. d'apprentissage. / ~ ship ‖ Schulschiff n. ‖ bâtiment-école m.; vaisseau-école m.; navire-école m.

train lighting (Railw) ‖ Zugbeleuchtung f. ‖ éclairage m. des trains. / electric ~ ‖ elektrische Zugbeleuchtung f. ‖ éclairage m. électrique des trains.

train lighting battery ‖ Zugbeleuchtungsbatterie f. ‖ batterie f. pour éclairage des trains. / ~ dynamo ‖ Zugbeleuchtungsdynamo f. ‖ dynamo f. pour l'éclairage des trains.

train oil ‖ Fischtran m. ‖ huile f. de poisson. / ~ radio telephony ‖ Zugfunk m. ‖ téléphonie f. radiotrain. / ~ service ‖ Zugdienst m. ‖ service m. des trains. / ~ staff ‖ Zugpersonal n. ‖ personnel m. du train. / ~ stop (Railw; Tel) ‖ Fahrsperre f. ‖ arrêt m. (automatique) du train. / ~ telephony ‖ Zugfernsprechen n.; Zugtelefonie f. ‖ téléphonie f. avec un train. / ~ traffic line (Tel) ‖ Zugverkehrsleitung f. ‖ circuit m. servant à la téléphonie avec un train.

traject (Shipb) ‖ Fähre f.; Fährschiff n.; Trajekt n. ‖ trajet m.

trajectory f. ‖ Flugbahn f.; Bahn f. ‖ trajectoire f. / ascendent branch of the ~ ‖ aufsteigender Ast m. der Flugbahn ‖ branche f. ascendante de la trajectoire. / descendent branch of the ~ ‖ absteigender Ast m. der Flugbahn ‖ branche f. descendante de la trajectoire.

tram (Tramway car) ‖ Straßenbahnwagen m. ‖ wagon m. de tramway. / ~ (Mine) ‖ Förderwagen m.; Hund m. ‖ berline f.; wagonnet m. de transport. / ~ (Silk) ‖ Tramseide f.; Einschlagseide f. ‖ soie f. de trame; trame f. / wire ~ ‖ Drahtseilriese f.; Riese f. ‖ transporteur m. aérien.

tram-board ‖ Laufbrett n.; Laufbohle f. ‖ planche f. de marchepied. / ~ bracket ‖ Laufbrettstütze f. ‖ porte-planchette f.; porte-planche f.

tramcar ‖ Straßenbahnwagen m. ‖ voiture f. de tramway. / ~ traffic ‖ Straßenbahnverkehr m. ‖ trafic m. des tramways.

tram cleaning machine (Mine) ‖ Förderwagenreinigungsmaschine f. ‖ machine f. à nettoyer les wagonnets de transport.

trammel (Compasses) ‖ Ellipsenzirkel m.; Ovalzirkel m.; Stangenzirkel m. ‖ ellipsographe m.; compas m. à ovale ou à ellipse; compas m. à verge. / ~ (Net) ‖ dreimaschiges Standnetz n.; Schleppnetz n. ‖ tramail m.; trémail f.; drège f.

trammer (Mine) ‖ Schlepper m.; Wagenstößer m. ‖ traîneur m.; rouleur m.; esclauneur m.

tram pushing device (Mine) ‖ Förderwagenaufschieber m. ‖ appareil m. à remonter les wagonnets de transport.

tram rail ‖ Straßenbahnschiene f.; Rillenschiene f. ‖ rail m. de tramway; rail m. à ornière.

tramroad see also tramway ‖ Straßenbahn f. ‖ tramway m. / ~ (Mine) ‖ Schienenbahn f.; Schienenweg m. ‖ voie f. (de rails); chemin m. à rails. / ~ with rails of timber ‖ Holzbahn f. ‖ voie f. à ornières en bois.

tram silk see tram (Silk).

tram straightening machine ‖ Förderwagenlenkvorrichtung f. ‖ machine f. à diriger les wagonnets de transport.

tramway ‖ Straßenbahn f. ‖ tramway m. /

aerial ~ ‖ Schwebebahn f.; Drahtseilbahn f. ‖ tramway m. aérien; chemin m. de fer suspendu. / electric ~ ‖ elektrische Straßenbahn f.; Elektrische f. ‖ tramway m. électrique. / steam ~ ‖ Dampfstraßenbahn f. ‖ tramway m. à vapeur.

tramway car ‖ Straßenbahnwagen m. ‖ wagon m. ou voiture f. de tramway. / ~ clearer (Mine) ‖ Schienenwärter m. ‖ raccomodeur m. / ~ depot ‖ Straßenbahnbahnhof m. ‖ dépôt m. à voitures de tramway. / ~ material ‖ Straßenbahnbetriebsmittel npl. ‖ matériel m. de tramways. / ~ motor ‖ Straßenbahnmotor m. ‖ moteur m. de tramways. / ~ plant ‖ Straßenbahnanlage f. ‖ installation f. (d'une ligne) de tramway. / ~ point setting device ‖ Weichenstellvorrichtung f. für Straßenbahnen ‖ dispositif m. d'aiguillage pour tramways. / ~ rail ‖ Straßenbahnschiene f. ‖ rail m. de tramway ou à ornière. / ~ repair workshop ‖ Straßenbahnreparaturwerkstatt f. ‖ atelier m. de réparation de tramways. / ~ switch ‖ Straßenbahnweiche f.; Pflasterweiche f. der Straßenbahn ‖ changement m. de voie de tramway.

tramway-type reversing controller ‖ Schaltwalzenumkehranlasser m. ‖ démarreur m. de renversement à cylindres.

tramway warning bell ‖ Straßenbahnklingel f.; Straßenbahnglocke f.; Straßenbahnläutewerk n. ‖ timbre m. avertisseur pour tramways.

transaction ‖ Geschäft n.; Geschäftsabschluß m.; Verkauf m. ‖ affaire f.; opération f. commerciale; vente f.

transatlantic ‖ transatlantisch ‖ transatlantique. / ~ cable ‖ Überseekabel m.; transatlantisches Kabel n. ‖ câble m. transatlantique. / ~ steamer ‖ Überseedampfer m. ‖ vapeur m. transatlantique. / ~ telegraphy installation ‖ überseeische Telegrafenanlage f. ‖ installation f. de télégraphie transatlantique. / ~ trade ‖ überseeischer Handel m.; Überseehandel m. ‖ commerce m. transatlantique.

transfer, to ‖ (Bank) ‖ überweisen; transferieren; anweisen ‖ transférer; assigner; céder. / ~ (Weav) ‖ aufstoßen ‖ rebrousser. / ~ (Found) ‖ abdrücken ‖ empreindre; reporter. / ~ one account to another ‖ stornieren ‖ contre-passer; résilier; ristorner.

transfer ‖ Übertragung f. ‖ transmission f.; transposition f. ‖ (Trade) ‖ Transfer m. ‖ transfert m. / ~ (Document) ‖ Abtretungsurkunde f.; Zessionsurkunde f. ‖ acte m. d'abandon ou de délaissement. / ~ (Picture) see also transfer picture ‖ Abziehbild n. ‖ décalcomanie f. / ~ of balance ‖ Saldoübertrag m. ‖ solde m. à nouveau. / ~ of a business ‖ Geschäftsübertragung f. ‖ cession f. d'un commerce. / ~ of a right subject to compensation ‖ Übertragung f. eines Rechtes gegen Entgelt ‖ transfert m. d'un droit contre redevance. / ~ of a right for a prescribed period ‖ Übertragung f. eines Rechtes auf Zeit ‖ transfert m. d'un droit à temps. / ~ of shares ‖ Übertragung f. von Aktien ‖ transfert m. de titres.

transferable security ‖ übertragbares Wertpapier n. ‖ valeur f. mobilière.

transfer copying book ‖ Durchschreibebuch n. ‖ livre m. à calquer. / ~ deed ‖

Übertragungsurkunde f. ‖ feuille f. de transfert.

transferee (Of a bill) ‖ Indossat m.; Indossatar m. ‖ endossé m.

transference ‖ Indossament n. ‖ endossement m.

transfer exchange (Tel) ‖ Überleitungsamt n. ‖ bureau m. de transfert. / ~ key (Tel) ‖ Überspringtaste f. ‖ touche f. de transfert. / ~ line (Tel) ‖ Überweisungslinie f. ‖ ligne f. de transfert. / ~ needle (Weav) ‖ Aufstoßnadel f. ‖ aiguille f. de rebrousseuse. / ~ paper ‖ Umdruckpapier n. ‖ papier m. pour reproductions.

transfer picture ‖ Abziehbild n. ‖ décalcomanie f. / ceramic ~ ‖ einbrennbares Abziehbild n. ‖ décalcomanie f. pour porcelaine. / ~ for children ‖ Kinderabziehbild n. ‖ décalcomanie f. pour enfants.

transfer port (Mot) ‖ Überströmkanal m. ‖ canal m. d'admission des gaz.

transferrer (Of a bill) ‖ Indossant m. ‖ endosseur m.

transferring device (Weav) ‖ Aufstoßeinrichtung f. ‖ mécanisme m. à rebrousser. / ~ machine ‖ Aufstoßmaschine f. ‖ machine f. à rebrousser.

transfer sequence switch (Tel) ‖ Übertragungsfolgeschalter m. ‖ commutateur m. de la suite des transmissions. / ~ system (Tel) ‖ Transfersystem n. ‖ système m. transfer. / ~ table (Railw) ‖ Schiebebühne f. ‖ pont m. roulant. / ~ ticket ‖ Überweisungsscheck m. ‖ mandat m. de virement. / ~ track ‖ Übergangsgleis n. ‖ voie f. de raccordement. / ~ trunk exchange (Tel) ‖ Überweisungsfernamt n. ‖ bureau m. de transfert.

transform, to ‖ umgestalten; umwandeln ‖ transformer. / ~ (Electr) ‖ umspannen; umformen; transformieren ‖ transformer. / ~ the current ‖ den Strom m. umspannen ‖ transformer le courant. / ~ by reckoning ‖ umrechnen ‖ ramener par le calcul.

transformation ‖ Umwandlung f. ‖ transformation f. / ~ (Electr) ‖ Umspannung f.; Transformation f. ‖ transformation f. / chemical ~ ‖ chemische Umsetzung f. ‖ transformation f. chimique. / ~ into coal ‖ Inkohlung f. ‖ carburation f. / ~ of coordinates ‖ Umrechnung f. von Koordinaten ‖ transformation f. des coordonnées. / ~ of energy ‖ Energieumwandlung f. ‖ transformation f. d'énergie. / molecular ~ ‖ Molekülumlagerung f. ‖ déplacement m. moléculaire. / ratio of ~ (Electr) ‖ Transformierungsverhältnis n.; Umwandlungsverhältnis n.; Übersetzungsverhältnis n. ‖ rapport m. de transformation.

transformator see transformer.

transformer ‖ Transformator m.; Umwandler m.; Umspanner m. ‖ transformateur m. / adjustable ~ ‖ Drehumformer m. ‖ transformateur m. réglable. / air-cooled ~ ‖ luftgekühlter Transformator m. ‖ transformateur m. à air. / balanced differential ~ ‖ Ausgleichsübertrager m. ‖ transformateur m. d'équilibre. / balanced three-winding ~ see balanced differential ~. / balancing ~ ‖ Ausgleichtransformator m.; Spartransformator m. ‖ autotransformateur-égalisateur m. / booster ~ ‖ Spannungserhöher m. ‖ autotransformateur-élévateur m. / ~ with compressed-air cooling ‖ Transformator

m. mit Preßluftkühlanlage ‖ transformateur m. avec refroidissement par l'air
comprimé. / continuous-current ~ ‖
Gleichstromtransformator m. ‖ transformateur m. à courant continu. / core ~ ‖
Kerntransformator m. ‖ transformateur
m. à noyau. / current ~ ‖ Stromtransformator m.; Stromwandler m. ‖ transformateur m. de courant. / electric ~ ‖ elektrischer Transformator m. *oder* Umwandler m. ‖ transformateur m. électrique. / frequency ~ ‖ Frequenztransformator m. ‖ transformateur m. de fréquence. / hedgehog ~ ‖ Igeltransformator
m. ‖ transformateur m. hérisson. / intervening ~ ‖ Zwischentransformator m. ‖
transformateur m. intermédiaire. / lighting ~ ‖ Lichttransformator m. ‖ transformateur m. pour l'éclairage. / measuring ~ ‖ Meßtransformator m. ‖ transformateur m. de mesure. / oil ~ ‖ Öltransformator m. ‖ transformateur m. à huile.
/ phase ~ ‖ Phasentransformator m. ‖
transformateur m. de phase. / pole ~ ‖
Masttransformator m. ‖ transformateur
m. à mât. / portable ~ ‖ fahrbarer Transformator m. ‖ transformateur m. transportable. / potential ~ ‖ Spannungswandler m. ‖ transformateur m. de
tension. / power ~ ‖ Leistungstransformator m. ‖ transformateur m. à grande
puissance. / pressure ~ ‖ Spannungstransformator m. ‖ transformateur m. de
tension. / regulating ~ ‖ Regeltransformator m. ‖ transformateur m. de régulation. / ring ~ ‖ Ringtransformator m. ‖
transformateur m. annulaire. / rotary ~
see rotating ~. / rotary-current ~ ‖ Drehstromtransformator m. ‖ transformateur
m. à champ tournant. / rotating ~ ‖
rotierender Transformator m.; Motorgenerator m. ‖ transformateur m. rotatif.
/ shell ~ ‖ Manteltransformator m. ‖
transformateur m. cuirassé. / small ~ ‖
Kleintransformator m. ‖ petit transformateur m.; transformateur m. de faible
puissance. / spreader current ~ in an
open air transformer station ‖ Stützerstromwandler m. in einer Freiluftanlage
‖ transformateur m. d'intensité support
installé dans un poste d'extérieur. /
starting ~ ‖ Anlaßtransformator m. ‖
transformateur m. de démarrage. / static
~ ‖ ruhender Transformator m. ‖ transformateur m. statique. / step-down ~ ‖
spannungserniedrigender Transformator
m. ‖ transformateur m. abaisseur. / step-
up ~ ‖ spannungserhöhender Transformator m. ‖ transformateur m. élévateur.
/ three-phase ~ ‖ Drehstromtransformator m. ‖ transformateur m. triphasé.
transformer, burning-out of the ~ ‖ Durchschlag m. des Transformators ‖ claquage
m. du transformateur. / ~ capacity ‖
Umspannerleistung f. ‖ puissance f. de
transformation. / ~ carriage ‖ Transfor-
mator(en)wagen m. ‖ chariot m. de transformateur. / ~ coil ‖ Transformatorspule
f. ‖ bobine f. de transformateur. / ~ core
‖ Transformatorkern m. ‖ noyau m. de
transformateur. / ~ coupling (Radio) ‖
Transformatorenkopplung f. ‖ accouplement m. par transformateurs. / ~ oil ‖
Transformatorenöl n. ‖ huile f. de transformateur *ou* pour transformateurs. /
~ sheet ‖ Transformatorenblech n. ‖ tôle
f. à transformateurs. / annealing of ~
sheets ‖ Ausglühen n. von Transforma-

torenblechen ‖ recuit m. des tôles à
transformateurs. / ~ stamping *see* ~ sheet.
/ ~ station ‖ Transformatorenstation f.;
Abspannwerk n.; Umspannwerk n. ‖
station f. de transformation; poste m.
abaisseur de tension. / open-air ~ station
‖ Freilufttransformatorenstation f. ‖
poste m. extérieur de transformation.
/ ~ testing device ‖ Transformatoren-
prüfvorrichtung f. ‖ installation f. de
vérification de transformateurs. / ~
winding ‖ Transformatorwicklung f. ‖
enroulement m. de transformateur.
transforming plant ‖ Umspannungsanlage
f. ‖ station f. transformatrice *ou* de transformation.
transhipping ‖ Umladung f.; Umschlag m.
‖ transbordement m. / ~ of goods at the
waterside ‖ Uferumschlag m. ‖ déchargement m. de bateaux dans d'autres
(bateaux).
transhipping device ‖ Umladeeinrichtung
f. ‖ appareil m. de transbordement. /
~ plant for corn ‖ Getreideumschlag-
anlage f. ‖ installation f. de déchargement de grain *ou* de manutention de
grains.
transient (Tel) ‖ Einschwingvorgang m.;
Ausgleichsvorgang m. ‖ phénomène m.
transitoire. / ~ effect *see* transient. / ~
effects pl. in coil-loaded toll-cable circuits ‖ Einschwingvorgänge mpl. in pu-
pinisierten Fernkabelleitungen ‖ phénomènes mpl. transitoires en circuits
de câbles pupinisés. / ~ phenomenon *see*
transient.
transit (Trade) ‖ Durchgangsverkehr m. ‖
transit m.; trafic m. *ou* commerce m.
de transit. / ~ administration (Tel) ‖
Durchgangsverwaltung f. ‖ administration f. de transit. / ~ book (Tel) ‖ Durch-
gansbuch n. ‖ liste f. de transit. / ~ charge
(Tel) ‖ Durchgangsgebühr f. ‖ taxe f. de
transit. / ~ circle (Astro) ‖ Richtkreis m.
‖ cercle m. d'alignement. / ~ country
(Tel) ‖ Durchgangsland n. ‖ pays m. de
transit. / ~ duty ‖ Durchgangszoll m.;
Transitzoll m. ‖ droits mpl. de transit. /
~ goods pl. ‖ Durchgangsgüter npl.;
Transitgüter npl. ‖ marchandises fpl.
en transit. / ~ instrument (Astro) ‖
Passageinstrument n.; Meridianfernrohr
n. ‖ instrument m. de passages; lunette f.
méridienne.
transition ‖ Übergang m. ‖ transition f.;
passage m. / with gradual ~ ‖ mit all-
mählichem Übergang m. ‖ se nuanceant
graduellement.
transitional stage *see* transition period.
transition coupling (Railw) ‖ Übergangskupplung f. ‖ attelage m. temporaire. /
~ period ‖ Übergangszeit f. ‖ période f.
de transition.
transit telescope ‖ durchschlagbares Fernrohr n. ‖ lunette f. révolutionnant
complètement; lunette f. pouvant être
retournée sur elle-même. / ~ ticket (Tel)
‖ Durchgangsblatt n. ‖ fiche f. de transit.
/ ~ trade ‖ Transithandel m.; Durchgangshandel m. ‖ commerce m. de
transit. / ~ traffic ‖ Durchgangsverkehr
m. ‖ trafic m. de transit.
translate, to ~ (A language) ‖ übersetzen
‖ traduire.
translated ‖ übersetzt ‖ traduit.
translation ‖ Übersetzung f. ‖ traduction f.
/ right of ~ ‖ Übersetzungsrecht n. ‖
droit m. de traduction.

translocation diastase (Brew) ‖ Translokationsdiastase f. ‖ diastase f. de translocation.
translucent ‖ durchscheinend ‖ translucide;
transparent; diaphane.
translumination lamp ‖ Durchleuchtungslampe f. ‖ lampe f. pour l'éclairage
interne pénétrant.
transmarine ‖ überseeisch ‖ d'outre-mer;
transatlantique. / ~ wood ‖ Überseeholz n. ‖ bois m. d'outre-mer.
transmission ‖ Übersetzung f.; Transmission f.; Übertragung f. ‖ transmission f. / chain ~ ‖ Kettenübertragung
f. ‖ transmission f. par chaîne. / ~ by
compressed air ‖ Druckluftübertragung
f. ‖ transmission f. par air comprimé. /
directive ~ (Tel) ‖ gerichtete Übertragung f. ‖ transmission f. dirigée. /
double ~ ‖ zweifache Übersetzung f. ‖
double transmission f. / ~ of energy ‖
Energieübertragung f. ‖ transmission f.
d'énergie. / ~ of force ‖ Kraftübertragung f. ‖ transmission f. de force.
/ ~ of heat by conduction ‖ Wärmeleitung f. ‖ transmission f. de chaleur
par conduction. / intermediate ~ ‖
Zwischentransmission f. ‖ transmission
f. intermédiaire. / main ~ ‖ Haupttransmission f. ‖ transmission f. principale. / ~ of motion ‖ Bewegungsübertragung f. ‖ transmission f. de mouvement. / multiple ~ ‖ Vielfachübertragung
f. ‖ transmission f. multiple. / ~ of opera
music ‖ Opernübertragung f. ‖ transmission f. de la musique de l'opéra. /
~ of power ‖ Kraftübertragung f. ‖
transmission f. de force. / ~ of telegram
‖ Abgabe f. des Telegramms ‖ transmission f. du télégramme. / ~ telephotographic
~ ‖ bildtelegrafische Übertragung f. ‖
transmission f. phototélégraphique.
transmission case ‖ Getriebegehäuse n. ‖
carter m. de la boîte à vitesse. / ~ case
cover ‖ Getriebegehäusedeckel m. ‖
couvercle m. de la boîte à vitesse. / ~
chain ‖ Transmissionsgliederkette f. ‖
chaîne f. de transmission. / ~ coefficient
(Tel) ‖ Durchlässigkeitsfaktor m. ‖
coefficient m. de passage. / ~ drive
(Mach) ‖ Transmissionsantrieb m. ‖ commande f. par courroie. / ~ efficiency
(Tel) ‖ Übertragungswirksamkeit f. ‖
efficacité f. de transmission. / ~ equivalent (Tel) ‖ Übertragungsäquivalent n.;
Dämpfungsmaß n. ‖ équivalent m. de
transmission; rendement m. de référence.
/ ~ gear (Mach) ‖ Vorgelege n. ‖ engrenage
m. ou transmission f. intermédiaire; renvoi m. (de mouvement); contre-arbre m.
/ ~ hammer ‖ Transmissionshammer m.;
marteau m. pilon de transmission. / ~ level indicator (Tel) ‖ Pegelzeiger m. ‖ indicateur m. de niveau de transmission. / ~
level meter ‖ Pegelmesser m. ‖ appareil
m. à mesurer le niveau de transmission.
/ ~ parameter of lines ‖ Übertragungs-
parameter n. für Leitungen ‖ paramètre
m. de transmission. / ~ plant ‖ Trans-
missionsanlage f. ‖ installation f. de
transmission. / quality of ~ (Tel) ‖ Güte
f. der Übertragung ‖ qualité f. de transmission. / ~ rods pl. ‖ Übertragungs-
gestänge n. ‖ tiges fpl. de transmission.
/ ~ rope ‖ Transmissionsseil n. ‖ câble m.
/ ~ shaft ‖ Übertragungswelle f. ‖ arbre m. de transmission. / ~ shaft from clock to helical

drive ‖ Übertragungswelle f. vom Uhrwerk zum Schneckenantrieb ‖ tige f. transmettant le mouvement d'horlogerie à la vis tangente. / brake on the ~ shaft ‖ Vorgelegebremse f. ‖ frein m. sur l'arbre de transmission. / ~ theory (Tel) ‖ Leitungstheorie f. ‖ théorie f. de transmission. / ~ unit (Tel) ‖ Übertragungseinheit f.; Übertragungsmaß n. ‖ unité f. de transmission. / ~ velocity (Of waves) ‖ Fortpflanzungsgeschwindigkeit f. ‖ vitesse f de propagation.

transmit, to ‖ übertragen ‖ transmettre. / ~ (Bank) ‖ überweisen ‖ assigner; céder. / ~ (Electr) ‖ fortleiten; leiten ‖ conduire. / ~ (Opt) ‖ durchlassen ‖ transmettre. / ~ heat ‖ Wärme f. durchlassen ‖ laisser passer la chaleur. / ~ a telegram ‖ ein Telegramm abgeben ‖ transmettre ou expédier un télégramme.

transmitted light ‖ durchfallendes Licht n. ‖ lumière f. transmise ou transparente.

transmitter (Radio) ‖ (drahtloser) Sender m.; Mikrofon n. ‖ transmetteur m.; microphone m. / automatic ~ ‖ Maschinensender m.; selbsttätige Sendevorrichtung f. ‖ transmetteur m. automatique. / ~ of a combined fire alarm and watchman control installation ‖ Melder m. einer vereinigten Feuermeldeanlage und Wächterkontrollanlage ‖ avertisseur m. d'une installation combinée d'avertissement d'incendie et de contrôle de rondes. / ~ for continuous oscillations ‖ Sender m. für kontinuierliche Schwingungen ‖ transmetteur m. pour oscillations continues. / granular ~ ‖ Körnermikrofon n. ‖ microphone m. à granules sphériques. / inductive ~ ‖ (induktiv) gekoppelter Sender m. ‖ transmetteur m. à couplage inductif. / intermediate-circuit ~ ‖ Zwischenkreissender m. ‖ émetteur m. à circuit intermédiaire. / long-wave ~ ‖ Langwellensender m. ‖ émetteur m. à ondes longues. / quenched-spark ~ ‖ Löschfunkensender m. ‖ émetteur m. à étincelles chantantes étouffées. / sharply-tuned ~ ‖ scharf abgestimmter Sender m. ‖ transmetteur m. à syntonisation aiguë. / short-wave ~ ‖ Kurzwellensender m. ‖ émetteur m. à ondes courtes. / simple ~ ‖ einfacher Sender m. ‖ dispositif m. d'émission directe. / spark ~ ‖ Funkensender m. ‖ émetteur m. à étincelles. / ~ for water-level tele-indicators ‖ Geber m. für Wasserstandsfernmelder ‖ transmetteur m. pour téléindicateurs de niveau d'eau.

transmitter diaphragm ‖ Mikrofonmembran f. ‖ plaque f. microphonique. / ~ station for a picture telegraph installation ‖ Gebeeinrichtung f. einer Bildtelegrafenanlage ‖ équipement m. d'émission d'une installation de téléphotographie.

transmitting aerial ‖ Sendeantenne f.; Sendeluftleiter m. ‖ antenne f. d'émission. / ~ amplifier ‖ Sendeverstärker m. ‖ amplificateur m. émetteur. / ~ apparatus (Radio) ‖ Sender m. ‖ appareil m. de transmission. / ~ arrangement ‖ Sendeanordnung f. ‖ dispositif m. d'émission. / ~ basis of earth telegraphy ‖ Sendebasis f. der Erdtelegrafie ‖ base f. d'émission de télégraphie par le sol. / ~ change of connection for ~ (Radio) ‖ Umschaltung f. für Sendung ‖ commutation f.

pour la transmission. / ~ circuit ‖ Sendekreis m. ‖ circuit m. d'envoi. / ~ current ‖ Telegrafierstrom m.; Sendestrom m. ‖ courant m. de transmission. / ~ gearing (Mach) ‖ Transmissionsgetriebe n. ‖ engrenage m. de transmission. / ~ insulator ‖ Isolator m. für die Sendeantenne ‖ isolateur m. de transmission. / ~ jigger ‖ Sendejigger m. ‖ jigger m. d'émission. / ~ room ‖ Senderaum m. ‖ chambre f. des appareils de transmission. / ~ station ‖ Sendestation f. ‖ poste m. émetteur. / ~ system ‖ Sendersystem n. ‖ système m. d'émission. / auxiliary ~ system ‖ Hilfsübertragersystem n. ‖ système m. de transformateur auxiliaires. / ~ tap (Tel) ‖ Senderstreifen m. ‖ bande f. de transmission.

transom (Build) ‖ Latteiholz n.; Querbalken m.; Querholz n. ‖ dormant m.; traverse f. / ~ of a gin ‖ Riegel m. eines Hebebocks ‖ épart m. ou traverse f. de chèvre. / ~ of a window ‖ Fensterkämpfer m. ‖ dormant m. de fenêtre.

transom bed (Coachm) ‖ Lenkschemel m.; Wendeschemel m. ‖ lisoir m. / ~ window ‖ Querfenster n. ‖ fenêtre f. gisante.

transparency ‖ Durchsichtigkeit f.; Lichtdurchlässigkeit f. ‖ transparence f.; diaphanéité f. / ~ of the air ‖ Durchsichtigkeit f. der Luft ‖ transparence f. de l'air.

transparent ‖ durchsichtig; lichtdurchlässig; durchscheinend ‖ transparent; diaphane; translucide. / ~ air ‖ klares oder sichtiges Wetter n. ‖ temps m. clair. / ~ cloth ‖ durchscheinende Leinwand f. ‖ toile f. transparente. / ~ colour ‖ Saftfarbe f. ‖ couleur f. de sève. / ~ envelope ‖ Fensterbriefumschlag m. ‖ enveloppe f. à guichet ou à fenêtre. / ~ ice ‖ Klareis n. ‖ glace f. transparente. / ~ membrane ‖ durchsichtige Membran f. ‖ membrane f. transparente. / ~ object (Opt) ‖ durchscheinendes Objekt n. ‖ objet m. translucide. / ~ paper ‖ Ölpapier n.; durchsichtiges Papier n.; Transparentpapier n. ‖ papier m. huilé ou transparent. / ~ positive ‖ Diapositiv n. ‖ diapositif m.; positif m. transparent. / ~ poster ‖ durchscheinendes Werbeschild n.; Transparent n.; Transparentplakat n. ‖ affiche f. transparente. / ~ rubber ‖ durchsichtiges oder transparentes Gummi n. ‖ caoutchouc m. transparent. / ~ soap ‖ Transparentseife f. ‖ savon m. transparent. / ~ varnish ‖ Lasurlack m. ‖ vernis m. transparent.

transport see also transportation ‖ Beförderung f.; Förderung f.; Transport m. ‖ transport m. / ~ by ice ‖ Eistransport m. ‖ transport m. par la glace. / ~ of plate-glass ‖ Spiegelglasbeförderung f. ‖ transport m. des glaces. / ~ by water ‖ Wassertransport m. ‖ transport m. par eau. / ~ by wind ‖ Windtransport m. ‖ transport m. éolien ou par le vent.

transportable ‖ transportfähig; beförderbar ‖ transportable. / ~ battery (Acc) ‖ transportable Batterie f. ‖ batterie f. transportable. / ~ house ‖ transportables Gebäude n.; transportables Haus n. ‖ maison f. transportable.

transport airplane ‖ Transportflugzeug n. ‖ avion m. de transport.

transportation see also transport ‖ Beförderung f.; Versand m.; Transport m. ‖ transport m. / level ~ ‖ Wagerechtförderung f. ‖ transport m. horizontal. / ~ by

motor lorry ‖ Beförderung f. mit Lastkraftwagen ‖ transport m. par camion automobile.

transportation plant ‖ Transportanlage f. ‖ installation f. de transport.

transport band ‖ Förderband n. ‖ courroie f. de transport; bande f. transporteuse. / ~ barrel ‖ Versandfaß n.; Transportfaß n. ‖ fût m. d'expédition; tonneau m. de transport; barrique f. / ~ barrel of iron ‖ eisernes Transportfaß n. ‖ tonneau m. de transport en fer. / ~ bridge (Min) ‖ Förderbrücke f.; Transportbrücke f. ‖ pont m. de transport. / ~ bucket ‖ Förderbecher m.; Transportbecher m. ‖ godet m. transporteur. / ~ car ‖ Beförderungswagen m.; Lastwagen m.; Transportwagen m. ‖ chariot m. de transport. / ~ car for trunks ‖ Langholzwagen m.; Langholztransportwagen m. ‖ chariot m. de transport pour troncs. / ~ cart ‖ Transportkarren m. ‖ charrette f. de transport. / ~ cycle ‖ Gepäckfahrrad n.; Transportrad n. ‖ bicyclette f. de transport.

transporter ‖ Befördernder m.; Förderer m.; Transportör m. ‖ transporteur m. / ~ (Measuring instrument) ‖ Winkelmesser m.; Transportör m. ‖ transporteur m. / short-distance ~ ‖ Nahfördermittel n. ‖ engin m. transporteur à petite distance.

transporting device for factories ‖ Fördervorrichtung f. für Fabriken ‖ engin m. de transport pour usines. / ~ worm ‖ Förderschnecke f.; Transportschnecke f. ‖ vis f. sans fin de transport.

transport kettle ‖ Transportkessel m. ‖ réservoir m. de transport. / ~ means pl. ‖ Fördervorrichtung f.; Fördermittel n.; Transportvorrichtung f. ‖ dispositif m. ou moyen m. de transport. / ~ plant ‖ Beförderungsanlage f.; Transportanlage f. ‖ installation f. de transport. / ~ rails pl. ‖ Fördergleis n.; Transportgleis n.; voie f. de transport. / right of ~ ‖ Durchfahrtsrecht n. ‖ droit m. de passage. / ~ screw ‖ Förderschraube f.; Transportschraube f. ‖ vis f. de transport. / ~ service of meals by heated vans ‖ Speisentransport m. oder Speisenbeförderung f. in geheizten Wagen ‖ (service m. de) transport m. de repas en fourgons chauffés. / ~ ship ‖ Frachtschiff n.; Transportschiff n. ‖ bâtiment m. de transport. / ~ trough ‖ Transportmulde f. ‖ auge f. de transport. / ~ utensils pl. ‖ Fördergeräte npl.; Transportgeräte npl. ‖ ustensiles mpl. de transport. / ~ vessel ‖ Beförderungsgefäß n.; Transportgefäß n. ‖ bidon m. de transport. / ~ wagon see ~ car.

transpose, to ‖ umstellen; verstellen ‖ transposer; déplacer. / ~ (El line) ‖ kreuzen ‖ croiser; transposer.

transposition (El line) ‖ Kreuzung f. (am Gestänge) ‖ transposition f.; changement m. de position. / ~ insulator ‖ Doppelisolator m. ‖ isolateur m. à deux poupées. / ~ point ‖ Abschnittspunkt m.; Kreuzungspunkt m. ‖ point m. de transposition ou de croisement. / ~ pole ‖ Kreuzungsstange f.; Abschnittsgestänge n. ‖ appui m. ou poteau m. de transposition ou de croisement. / scheme of ~s ‖ Kreuzungsfolge f.; Kreuzungsschema n. ‖ schéma m. de croisements ou de transpositions. / ~ section ‖ Kreuzungsabschnitt m. ‖ élément m. de transposition. / complete ~ section (Tel) ‖ vollständiger

Kreuzungsabschnitt m. ‖ section f. d'anti-induction complète. / ~ step ‖ Kreuzungsabstand m. ‖ pas m. de transposition. / ~ system ‖ Kreuzungssystem n. ‖ armement m. de transposition.

transshipping *see* transhipping.

transversal *see* transverse.

transverse ‖ quer; transversal; diagonal ‖ transversal. / ~ axis ‖ Querachse f. ‖ axe m. transversal. / ~ ball bearing ‖ Querkugellager m. ‖ palier m. transversal à billes. / ~ bench ‖ Querbank f. ‖ banquette f. transversale. / ~ brushes pl. (Weav) ‖ Querbürsteinrichtung f. ‖ brosse f. transversale. / ~ bulkhead (Shipb) ‖ Querschott n. ‖ cloison f. transversale. / ~ crack ‖ Querriß m. ‖ fissure f. transversale. / ~ equalizer (Loc) ‖ Querausgleichhebel m. ‖ balancier m. transversal. / ~ girder ‖ Binderquerträger m. ‖ transversale f. dans les fermes. / ~ motion ‖ Querbewegung f. ‖ mouvement m. transversal. / ~ planing machine ‖ Querhobelmaschine f. ‖ raboteuse f. transversale. / ~ rail ‖ Querschiene f. ‖ rail m. transversal. / ~ rib ‖ Querrippe f. ‖ nervure f. transversale. / ~ seat (In a vehicle) ‖ Quersitz m. ‖ place f. assise transversale. / ~ shearing machine ‖ Querschermaschine f. ‖ tondeuse f. transversale. / ~ sleeper (Railw) ‖ Querschwelle f. ‖ traverse f. ‖ stay (El line) ‖ Windanker m. ‖ hauban m. de consolidation perpendiculaire à la ligne. / ~ strength ‖ Bruchfestigkeit f. ‖ résistance f. à la rupture. / ~ subdivision ‖ Querteilung f. ‖ subdivision f. transversale. / ~ wall ‖ Quermauer f.; Zwerchmauer f. ‖ mur m. en traverse *ou* de batardeau.

transversely ‖ quer ‖ transversalement; en travers. / ~ ribbed ‖ quer gerippt ‖ à nervures transversales.

trap ‖ Falle f.; Fangeisen n. ‖ piège m.; trappe f.; engin m. / ~ (Hydr arch) ‖ Klappe f.; Wasserabschluß m. ‖ clapet m.; clôture f. / live bird ~ for pigeon shooting ‖ Ablaßkäfig m. für Taubenschießen ‖ boîte f. à pigeons. / mouse ~ ‖ Mausefalle f. ‖ souricière f. / rat ~ ‖ Rattenfalle f. ‖ (nasse f.) ratière f. / steam ~ ‖ Kondenstopf m. ‖ séparateur m. d'eau de condensation.

trap board ‖ Fallklappe f. ‖ trappe f.

trap door ‖ Falltor n.; Falltür f. ‖ porte f. à coulisse; trappe f. / ~ of a pit (Mine) ‖ Schachtdeckel m. ‖ trappe f. d'un puits.

trapeziform ‖ trapezförmig ‖ trapéziforme; trapézoïde.

trapezium *see* trapezoid.

trapezoid ‖ Trapez n. ‖ trapèze m.

trapezoidal ‖ trapezähnlich; trapezförmig ‖ trapézoïdale; trapézoïde. / ~ load ‖ Trapezbelastung f. ‖ charge f. trapézoïdale.

trap hole ‖ Senkgrube f.; Abzugsgrube f. ‖ puisard m.; trappe f. à nettoyage.

trap maker ‖ Fallenschmied m. ‖ forgeur m. de pièges.

trapper (Mine) ‖ Wettertürschließer m. ‖ fermeteur m. de porte.

trap ring with bolt ‖ Fallklappenring m. mit Bolzen ‖ anneau m. de trappe à boulon.

trap-tuff (Miner) ‖ Basalttuff m. ‖ tuf m. basaltique.

trash (Mine) ‖ Abraum m. ‖ couche f.; lit m. de terre; lit m. de décombres. / ~ (Trade) ‖ Ramsch m.; Ramschware f. ‖ marchandises fpl. de pacotilles; camelote f.

trass ‖ Tuffstein m.; Traß m. ‖ trass m.; tuf m. calcaire. / ~ concrete ‖ Traßbeton m. ‖ béton m. au trass.

trassee (Of a bill) ‖ Trassat m.; Bezogener m. ‖ tiré m.

trass mill ‖ Traßmühle f. ‖ moulin m. à trass.

trass mortar ‖ Traßmörtel m. ‖ mortier m. au trass.

travel of the centre of pressure ‖ Druckpunktwanderung f. ‖ déplacement m. du centre de pression. / ~ of drill spindle ‖ Bohrspindelhub m. ‖ course f. de la broche de perçage. / workman free to ~ ‖ freizügiger Arbeiter m. ‖ ouvrier m. ayant la liberté de résidence.

traveler; traveling *see* traveller; travelling.

traveller (Mach) *see* travelling crab. / ~ (Spinn) ‖ Fliege f.; Läufer m.; Reiter m. ‖ curseur m.; traveller m.; voyageur m. / ~ (Trade) ‖ Reisender m.; Handelsreisender m. ‖ commis m. voyageur.

traveller's expense book ‖ Reisespesenbuch n. ‖ livre m. de dépenses pour voyageurs de commerce. / ~ flat iron ‖ Reisebügeleisen n. ‖ fer m. à repasser de voyage. / ~ guide *see* ~ handbook. / ~ handbook ‖ Reisehandbuch n. ‖ guide m. (du voyageur). / ~ sample ‖ Gebrauchsmuster n. ‖ brevet m. pour modèles d'utilité; modèle m. déposé.

travelling allowances pl. ‖ Reisekosten pl. ‖ frais mpl. de voyage. / ~ article ‖ Reiseartikel m. ‖ article m. de voyage. / ~ bag ‖ Reisekoffer m.; Reisetasche f. ‖ valise f.; sac m. (de) voyage. / ~ band conveyor ‖ fahrbarer Bandförderer m. ‖ transporteur m. à bande sur roues. / ~ boiling device ‖ Reisekocher m. ‖ marmite f. de voyage. / ~ bridge ‖ Schiebebrücke f. ‖ pont m. transbordeur. / ~ camera ‖ Reisekamera f. ‖ caméra f. de voyage. / ~ cap ‖ Reisemütze f. ‖ casquette f. de voyage. / ~ carriage (Typewriter) ‖ beweglicher Schlitten m. ‖ chariot m. mobile. / ~ case ‖ Reisebesteck n. ‖ nécessaire m. de voyage. / ~ cloak ‖ Reisemantel m. ‖ cache-poussière m. / ~ clock ‖ Reiseuhr f. ‖ pendule f. de voyage. / ~ compass ‖ Reisekompaß m. ‖ compas m. de route.

travelling crab ‖ Laufkatze f.; Katze f. ‖ chariot m. (roulant); chariot m. portecrochet. / monorail ~ ‖ Einschienenhängekatze f. ‖ chariot m. monorail suspendu. / casing for ~ ‖ Laufkatzengehäuse n. ‖ bâti m. de chariot roulant porte-crochet.

travelling crane ‖ Laufkran m.; fahrbarer Kran m. ‖ pont m. roulant; grue f. à chariot; grue f. roulante. / bracket ~ ‖ Wandlaufkran m. ‖ grue-console f. latérale. / ~ operated by hand ‖ Handlaufkran m. ‖ pont m. roulant actionné à bras. / ~ arranged under the roof of the shop ‖ unter dem Hallendach angebrachter Laufkran m. ‖ pont m. roulant disposé sous la toiture des halls. / ~ on the top of the slipway superstructure ‖ Hellinggerüstlaufkran m. ‖ pont m. roulant de cales sèches. / ~ for transporting tubes ‖ Rohrtransportkran m. ‖ pont m. roulant pour le transport des tuyaux. / ~ for transporting tubes fitted with self-acting regulating magnets and safety gripper ‖ Rohrtransportkran m. mit selbsttätig einstellbaren Magneten und Sicherheitsbügel ‖ pont m. roulant pour le transport des tuyaux avec aimants réglés automatiquement et avec étrier de sûreté. / ~ support of ~ ‖ Laufkranständer m. ‖ support m. de pont roulant.

travelling equipage ‖ Reisebedarf m. ‖ articles mpl. *ou* effets mpl. de voyage. / ~ expenses pl. ‖ Reisekosten pl.; Reisespesen pl. ‖ frais mpl. de voyage. / allowance for ~ expenses ‖ Vergütung f. für Reisekosten ‖ allocation f. pour frais de voyage. / ~ forge ‖ Feldschmiede f. ‖ forge f. portative. / ~ frame ‖ fahrbares Gerüst n. ‖ charpente f. sur roues. / ~ gear ‖ Fahrvorrichtung f.; Fahrmechanismus m. ‖ appareillage m. de roulement. / ~ grate ‖ Wanderrost m. ‖ barreau m. de grille mobile; grille f. mobile. / firing with ~ grate ‖ Wanderrostfeuerung f. ‖ foyer m. à grille mobile. / ~ hoist ‖ Laufwinde f. ‖ treuil m. roulant. / ~ keepsake ‖ Reiseandenken n. ‖ souvenir m. de voyage. / ~ microscope ‖ Reisemikroskop n. ‖ microscope m. de voyage. / ~ motion ‖ Fahrbewegung f. ‖ mouvement m. de translation. / ~ motor hoist ‖ Motorlaufwinde f. ‖ treuil m. roulant à moteur. / ~ nécessaire ‖ Reisenecessaire n. ‖ nécessaire m. de voyage. / ~ outfit ‖ Reisegarnitur f.; Reiseausrüstung f. ‖ garniture f. de voyage. / ~ platform ‖ Schiebebühne f. ‖ plateforme f. roulante. / ~ platform (Railw) ‖ Gleisrückmaschine f. ‖ transbordeur m. / ~ pouch ‖ Reisetasche f. ‖ sac m. de voyage. / ~ ranging pole ‖ Reisefluchtstab m. ‖ jalon m. portatif. / ~ receiver (Radio) ‖ Reiseempfänger m. ‖ récepteur m. de voyage. / ~ revolving crane ‖ fahrbarer Drehkran m. ‖ grue f. pivotante roulante. / ~ rug ‖ Reisedecke f. ‖ couverture f. de voyage. / hot-iron ~ saw ‖ Heißeisenschlittensäge f. ‖ scie f. à chaud à chariot. / ~ shoe ‖ Reiseschuh m. ‖ soulier m. de voyage. / ~ speed ‖ Fahrgeschwindigkeit f. ‖ vitesse f. (de translation *ou* de déplacement). / ~ stand ‖ Reisestativ n. ‖ statif m. de voyage. / ~ table ‖ Lauftisch m. ‖ banc m. à glissières. / ~ table for sliding fitting shop work ‖ Wandertisch m. für Gleitmontagen ‖ table f. mobile pour montages sur ruban. / ~ telescope ‖ Reisefernrohr n. ‖ lunette f. de voyage. / ~ trunk ‖ Reisekoffer m. ‖ malle f. *ou* valise f. de voyage. / ~ typewriter ‖ Reiseschreibmaschine f. ‖ machine f. à écrire de voyage. / ~ winch ‖ Laufwinde f. ‖ treuil m. roulant.

traverse, to (Electr) ‖ durchfließen ‖ parcourir. / ~ (Crane) ‖ verfahren ‖ déplacer.

traverse ‖ Traverse f. ‖ traverse f. / ~ of saddle on arm ‖ Bohrschlittenweg m. ‖ course f. du chariot porte-broche. / tubular ~ of square iron ‖ röhrenförmiger Querträger m. aus Vierkanteisen ‖ traverse f. tubulaire en fer carré.

traverse beam (Build) ‖ Querbalken m.; Querschwelle f.; Sattelholz n. ‖ poutre f. transversale; traverse f.; potence f. / ~ joist *see* traverse beam. / ~ problem ‖ polygoniometrische Aufgabe f. ‖ problème m. polygonométrique.

traverser ‖ Fahrbühne f.; Schiebebühne f. ‖ chariot m. transbordeur; transbordeur m. / half-sunk ~ ‖ halbversenkte Schiebebühne m. ‖ transbordeur m. à canivaux.

traverse sleeper ‖ Querschwelle f. ‖ traverse f. / to make a ~ survey ‖ polygonisieren ‖ mesurer des polygones mpl. / ~ trolley ‖ Rollwagen m. ‖ truck m. roulant. / ~ winding frame ‖ Kreuzspulmaschine f. ‖ bobinoir m. à fil croisé.

traversing, precise ~ ‖ Polygonierung f. ‖ tachéométrie.

traversing beam (Bridge) ‖ Laufbalken m.; Portalbalken m. ‖ traverse f. de la potence. / ~ gear (Of a crane) ‖ Fahrwerksantrieb m.; Laufwerk n. ‖ commande f. du déplacement du chariot; mécanisme m. d'avancement. / ~ speed of the crab ‖ Fahrgeschwindigkeit f. der Laufkatze ‖ vitesse f. de déplacement du chariot. / ~ table (Railw) ‖ Übergangsscheibe f. ‖ tableau m. de traversé. / ~ wheel ‖ Laufrad n. ‖ galet m.; roue f. portante.

trawl, to ‖ mit dem Schleppnetz n. fischen ‖ draguer; chaluter.

trawl (Fish) ‖ Schleppnetz n.; Grundnetz n.; Grundschleppnetz n. ‖ chalute f.; drague f.; rets m. de fond.

trawler (Person) ‖ Schleppnetzfischer m.; Zesenfischer m. ‖ pêcheur m. au chalut; chalutier m. / ~ (Boat) ‖ Schleppnetzfischerboot n.; Fahrzeug n. für die Schleppnetzfischerei ‖ bateau m. de pêche; chalutier m. / steam ~ ‖ Dampfschleppnetzfischerboot n.; Schleppnetzfischdampfer m. ‖ chalutier m. à vapeur.

trawling ‖ Schleppnetzfischerei f.; Grundnetzfischerei f. ‖ pêche f. au chalut ou à la drague. / ~ fisherman ‖ Schleppnetzfischer m.; Zesenfischer m. ‖ pêcheur m. au chalut; chalutier m.

trawl rope ‖ Treilfischleine f.; Fischleine f. mit Schwimmern, an welcher Angelhaken befestigt sind ‖ palancre f.; palangre f.

tray ‖ Untersatz m.; Untersetzer m.; Kaffeebrett n.; Teebrett n.; Präsentierteller m.; Servierbrett n. ‖ plateau m. (à servir); cabaret m. / ~ (for mortar) ‖ Mörteltrog m.; Mörtelkübel m. ‖ auge f. ou bac m. à mortier. / ~ of acid-proof sandstone ‖ Trog m. aus säurefestem Sandstein; säurefester Sandsteintrog m. ‖ bac m. en grès résistant aux acides.

tread, to ‖ clay ‖ den Ton m. treten ‖ piétiner l'argile.

tread (Of a pneumatic tyre) ‖ Laufband n.; Rolloberfläche f.; Lauffläche f. ‖ bande f. ou surface f. de roulement. / ~ (Of a railway tyre) ‖ Lauffläche f.; Laufkranz m. ‖ surface f. ou cercle m. de roulement. / the metal settles hard on the outside with a gradual transition from the ~ into the middle ‖ das Eisen strahlt von der Lauffläche aus nach innen hart ein ‖ la couche blanche et dure à la surface de roulement se nuance graduellement vers l'intérieur. / the ~ may be worn down to a greater extent ‖ der Laufkranz kann weiter abgenutzt werden ‖ la surface de roulement permet de pousser l'usure plus loin. / ~ wheel with hard ~ (Railw) ‖ Rad n. mit harter Lauffläche ‖ roue f. avec surface de roulement dure. / ~ of a ladder ‖ Leiterstufe f.; Trittstufe f. ‖ marche f. d'une échelle; giron m. / tip-up ~ of a ladder ‖ aufklappbare Stufe f. einer Leiter ‖ marche f. relevable d'une échelle. / ~ of rails ‖ Schienenlauffläche f. ‖ surface f. de roulement des rails. / ~ of a step ‖ Trittstufe f. einer Treppe; Trittfläche f. einer Treppenstufe ‖ giron m.; marche f. / ~ of a tripod ‖ Fußwinkel m.

eines Stativs ‖ fer m. saillant ou équerre f. d'un trépied.

tread board ‖ Stufe f.; Trittstufe f. ‖ marche f.

treading contact (Electr) ‖ Tretkontakt m. ‖ contact m. à pédale. / ~ wheel (Spinn) ‖ Trittrad n.; Tretrad n. ‖ rouet m. à filer à pédale.

treadle ‖ Fußtritt m.; Pedal n. ‖ pédale f. / ~ of a lathe ‖ Fußtritt m. einer Drehbank ‖ marche f. ou pédale f. ou pas m. d'un tour.

treadle arrangement ‖ Tretvorrichtung f. ‖ dispositif m. à pédale. / ~ drive ‖ Fußantrieb m. ‖ commande f. à pédale. / ~ operation ‖ Fußbetrieb m. ‖ marche f. à pédale. / ~ switch (Electr) ‖ Fußtrittschalter m. ‖ interrupteur m. à pédale.

tread mill ‖ Tretmühle f. ‖ treuil m. à tambour. / ~ wheel ‖ Tretrad n. ‖ roue f. à échelons.

treasury ‖ Fiskus m. ‖ fisc m. / ~ bill see ~ bond. / ~ bond ‖ Staatsschuldschein m.; Schatzanweisung f. ‖ bon m. du trésor. / ~ warrant see ~ bond.

treat, to ‖ verarbeiten; behandeln ‖ traiter.

treating see also treatment ‖ Verarbeitung f. ‖ traitement m. / the most advantageous methods of ~ ‖ die vorteilhafteste Verarbeitung f. ‖ le procédé le plus avantageux de traitement.

treatment ‖ Verarbeitung f.; Bearbeitung f.; Behandlung f. ‖ traitement m. / careful ~ ‖ schonende Behandlung f. ‖ traitement m. avec ménagement. / ~ of ceilings ‖ Deckenbehandlung f. ‖ traitement m. de plafonds. / chemical ~ ‖ chemische Behandlung f. ‖ traitement m. chimique. / dry ~ ‖ Trockenverfahren n.; trocknes Verfahren n. ‖ voie f. sèche (de traitement). / half-wet ~ ‖ Halbnaßverfahren n. ‖ voie f. demi-sèche. / ~ of the motor ‖ Wartung f. des Motors ‖ soins mpl. à donner au moteur. / ~ of ore ‖ Aufbereitung f. oder Verhüttung f. der Erze ‖ traitement m. des minerais. / experimental station for the ~ of ores ‖ Erzaufbereitungsversuchsanstalt f. ‖ atelier m. d'essai du traitement des minerais. / thermal ~ (Met) ‖ thermische Behandlung f. ‖ traitement m. thermique. / before machining the pieces undergo a thermal ~ ‖ vor der Bearbeitung f. werden die Stücke npl. einer thermischen Behandlung unterworfen ‖ avant d'être finies les pièces sont soumises à un traitement thermique. / ~ of vegetable oils and fats ‖ Veredelung f. vegetabilischer Öle und Fette ‖ raffinage m. des huiles et graisses végétales. / method of ~ ‖ Behandlungsweise f.; Verarbeitungsweise f. ‖ mode m. de traitement.

treaty ‖ Vertrag m. ‖ traité f. / commercial ~ ‖ Handelsvertrag m. ‖ traité m. de commerce. / ~ of navigation ‖ Schiffsvertrag m.; Schiffahrtsvertrag m. ‖ traité m. de navigation. / ~ of reciprocity ‖ Gegenseitigkeitsvertrag m. ‖ traité m. de réciprocité.

treble ‖ dreifach ‖ triple. / ~ draught (Of a vehicle) ‖ mehrfache Gabelbespannung f. ‖ attelage m. à trois files; attelage m. à trois chevaux de front.

trebled ‖ verdreifacht ‖ triplé.

trebles pl. (Coal) ‖ Nußkohle f. ‖ gailletteries fpl.; noisettes fpl.

treblet see triblet.

tree ‖ Baum m. ‖ arbre m. / ~ (Saddl) ‖ Sattelbaum m. ‖ arçon m. / ~ of a pump ‖ Pumpenrohr n. ‖ tuyau m. de pompe. / standing ~ (Forest) ‖ Holz n. auf dem Stamme ‖ bois m. en état ou en étant; bois m. sur pied ou à tige.

tree bark for medicinal usage ‖ Baumrinde f. für Heilzwecke ‖ écorce f. d'arbre à usage médicinal. / ~ felling contractor ‖ Holzschlagunternehmer m. ‖ entrepreneur m. d'abatage de bois sur pied; entrepreneur m. de coupes de bois. / ~ felling machine ‖ Baumfällmaschine f. ‖ machine f. à abattre les arbres. / ~ fence ‖ Baumrost m. ‖ grille f. à racines d'arbres. / ~ forest ‖ Baumholz n. ‖ arbre m. de futaie. / ~ gauge ‖ Baumkluppe f. ‖ compas m. forestier.

treenail see also dowel ‖ Dübel m.; Holznagel m.; Pflock m. ‖ cheville f. en bois; goujon m. / ~ driver ‖ Dübelnagler m. ‖ chevilleur m.

tree nursery ‖ Baumschule f. ‖ pépinière f. / ~ prop ‖ Baumpfahl m.; Baumstütze f. ‖ tuteur m. (d'arbre); étançon m. / ~ protector of expanded metal ‖ Baumschützer m. aus Streckmetall ‖ grillage m. de protection en métal étiré pour arbres.

treer (Shoem) ‖ Schäfteglätter m. ‖ astiqueur m. de tiges.

tree resin ‖ Baumharz n. ‖ poix f. résine. / ~ stump ‖ Baumstumpf m. ‖ tronc m. d'arbre. / ~ stump grubber ‖ Baumstumpfrodemaschine f. ‖ machine f. à déroder des troncs d'arbres. / ~ wart ‖ Holzkropf m. ‖ excroissance f. du tronc.

trefoil ‖ Klee m. ‖ trèfle m.

trellis ‖ Gitter n.; Gatter n.; Gitterwerk n. ‖ grillage m.; grille f.; treillis m. / ~ (Weav) ‖ Glanzleinwand f.; Sackzwillich m. ‖ treillis m. / ~ for flowers ‖ Blumengitter n. ‖ treillis m. pour fleurs. / ~ for glass roofs ‖ Glasdachsprosse f. ‖ treillage m. pour toits en verre. / ~ iron ‖ eisernes Gitter n. ‖ treillis m. en fer.

trellis fence ‖ Gitterzaun m. ‖ clôture f. à claire-voie ou de treillage.

trellis girder ‖ Gitterträger m. ‖ poutre f. à claire-voie ou à treillis. / boltless ~ ‖ nietloser Gitterträger m. ‖ poutre f. à treillis sans rivets.

trellis-work see also trellis ‖ Gitterwerk n. ‖ treillis m.

tremble, to ‖ zittern ‖ trembler.

trembler (Electr) ‖ Selbstunterbrecher m. ‖ interrupteur m. automatique. / ~ (Radio) ‖ Schwingungshammer m. ‖ trembleur m. / ~ bell ‖ Gleichstromwecker m.; Wecker m. mit Selbstunterbrechung ‖ sonnerie f. d'appel à courant continu; sonnerie f. à trembleur.

trembling poplar ‖ Zitterpappel f.; Espe f. ‖ (peuplier m.) tremble m.

tremolite (Miner) ‖ edle oder weiße Hornblende f.; Tremolit m. ‖ trémolite f.

trenail see treenail.

trench, to ‖ (tief) graben; mit Gräben mpl. durchziehen ‖ fouiller.

trench ‖ Graben m.; Rinne f. ‖ fossé m.; tranchée f.; rigole f. / ~ (El line) ‖ Kabelgraben m. ‖ tranchée f. / ~ (Mine) ‖ Schramhieb m.; Schramme f. ‖ couche f. / ~ (War) ‖ Laufgraben m.; Schützengraben m. ‖ tranchée f. / to cut ~es ‖ Gräben mpl. ziehen ‖ faire des tranchées. / ~ made for discovering mineral beds ‖

Schürfgraben m. ‖ tranchée f. à la recherche du minerai.

trench artillery ‖ Grabenartillerie f. ‖ artillerie f. de tranchée.

trench-coat ‖ Trenchcoat m.; Wettermantel m.; Regenmantel m. ‖ imperméable m.; waterproof· m.

trenched (Coal) ‖ unterschrämt ‖ entaillé.

trenching (El line) ‖ Kabelgraben m. ‖ tranchée f. / ~ plough *see* trench plough. / ~ shovel ‖ Planierschaufel f.; Skarpierschaufel f. ‖ louchet m.; pelle f. tranchante.

trench mortar (War mat) ‖ Grabenmörser m. ‖ mortier m. de tranchée. / ~ shell ‖ Grabenmine f. ‖ mine f. de tranchée.

trench-plough, to ‖ rajolen; rigolen ‖ effondrer; ouvrir la terre; défoncer.

trench plough ‖ Rajolpflug m.; Rigolpflug m. ‖ charrue f. à effondrer.

trench warfare ‖ Schützengrabenkrieg m. ‖ guerre f. de tranchée. / ~ work ‖ Grabarbeit f. ‖ travail m. dans les tranchées. / working of ~es (Mine) ‖ Schrämarbeit f. ‖ travail m. d'entailles.

trend, to ~ (Geol) ‖ streichen ‖ s'étendre en direction.

trepan, to ‖ hohlbohren ‖ aléser; trépaner.

trepan (Mine: tool) ‖ Bohrmeißel m.; Erdbohrer m.; Steinbohrer m.; Bergbohrer m. ‖ trépan m.; sonde f.; tarière f.; fleuret m.; pointeau m. / ~ (Mine: machine) ‖ Bohrmaschine f. ‖ perforatrice f. / ~ (Med) ‖ Schädelbohrer m.; Trepan m. ‖ trépan m. / common design of ~ ‖ gewöhnlicher Bohrmeißel m. ‖ trépan m. ordinaire. / ~ with central watre clearing ‖ Bohrmeißel m. mit zentraler Wasserspülung ‖ fleuret m. à rinçage à l'eau central. / ~ with lateral water clearing ‖ Bohrmeißel m. mit seitlicher Wasserspülung ‖ fleuret m. à circulation d'eau latéral.

trepanned marine shaft ‖ hohlgebohrte Schiffswelle f. ‖ arbre m. de couche foré creux.

trepanning machine ‖ Hohlbohrmaschine f. ‖ machine f. à forer.

trepan tooth ‖ Schneide f. des Bohrmeißels ‖ tranchant m. du trépan.

treshold (Hydr arch) ‖ Schleusendrempel m.; Stemmgeschwell n. ‖ busc m. d'une écluse. / ~ value (Radio) ‖ Schwellenwert m. ‖ valeur f. seuil.

tress ‖ Zopf m.; Haarflechte f.; Locke f. ‖ natte f. *ou* tresse f. de cheveux. / plaited ~ *see* tress. / ~ of wool (Spinn) ‖ Zopf m. ‖ tortillon m.

trestle ‖ Gestell n.; Gerüst n.; Bock m.; Stützbock m.; Schragen m. ‖ tréteau m.; chevalet m. / sinking ~ (Mine) ‖ Abteufgerüst n. ‖ échafaudage m. de creusement.

trestle work ‖ Balkengerüst n. ‖ échafaudage m. de poutres.

tret ‖ Gewichtsvergütung f.; Refaktie f. ‖ réfaction f.

trevet *see* trivet.

triakis-octahedron ‖ Pyramidenoktaëder n.; Triakisoktaëder n. ‖ octaèdre m. pyramidé, octatrièdre m.

trial ‖ Probe f.; Versuch m. ‖ essai m.; épreuve f.; recherche f. / the ~s pl. were continued ‖ die Versuche mpl. wurden fortgesetzt ‖ les essais mpl. furent continués. / ~ boiler ‖ Kesselprobe f. ‖ épreuve f. de chaudière. / preliminary ~ ‖ Vorversuch m. ‖ essai m. préliminaire.

/ ~ upon a small scale ‖ Probe f. im Kleinen ‖ essai m. en petit.

trial-and-error (Math) ‖ Regula f. falsi ‖ règle f. de fausse position.

trial balance ‖ Rohbilanz f.; Bruttobilanz f. ‖ bilan m. brut *ou* de vérification. / ~ connection (Tel) ‖ Probeverbindung f. ‖ connexion f. d'essai. / ~ cruise (Sailship) ‖ Probekreuzen n.; Versuchsfahrt f. ‖ croisière f. d'essai. / ~ frame for spectacles ‖ Probebrille f.; Probierbrille f. ‖ lunettes fpl. d'essai. / ~ order (Trade) ‖ Probeauftrag m. ‖ commande f. d'essai. / ~ piece (Porcel) ‖ Probescherben m. ‖ montre f. / ~ pit ‖ Schurf m.; Schürfung f. ‖ fouille f. de recherche. / ~ ride (Auto) ‖ Probefahrt f.; Versuchsfahrt f. ‖ parcours m. de garantie. / ~ rod (Met) ‖ Probestab m.; Probestange f.; Gareisen n. ‖ témoin m.; éprouvette f.; verge f. d'essai. / ~ run *see* ~ trip. / case for ~ spectacles pl. ‖ Probebrillengläserkasten m.; Probegläserkasten m. ‖ boîte f. de verres d'essai. / ~ speed ‖ Probefahrtgeschwindigkeit f. ‖ vitesse f. réalisée à l'essai. / ~ spring ‖ Probefeder f. ‖ ressort-échantillon m. / ~ station (Mach) ‖ Prüfstand m. ‖ banc m. d'épreuve; station f. d'essais. / Diesel engine at the ~ station ‖ Dieselmotor m. auf dem Prüfstand ‖ moteur m. Diesel à la station d'essais. / ~ track (Railw) ‖ Versuchsstrecke f. ‖ ligne f. d'épreuves *ou* d'essai. / ~ trip ‖ Probefahrt f.; Versuchsfahrt f. ‖ voyage m. *ou* course f. d'essai; parcours m. de garantie.

triangle ‖ Dreieck n. ‖ triangle m. / ~ (Draw) ‖ Winkel m.; Dreieck n. ‖ équerre f.; triangle m. / ~ (Mach) ‖ Gestängekreuz n. ‖ levier m. en croix. / acute-angled ~ ‖ spitzwinkliges Dreieck n. ‖ triangle m. acutangle. / equilateral ~ ‖ gleichseitiges Dreieck n. ‖ triangle m. équilatéral. / ~ of forces ‖ Kräftedreieck n. ‖ triangle m. de forces. / obtuse-angled ~ ‖ stumpfwinkliges Dreieck n. ‖ triangle m. à angle obtus; triangle m. obtusangle *ou* amblygone. / isosceles ~ ‖ gleichschenkliges Dreieck n. ‖ triangle m. isoscèle *ou* isocèle. / right-angled ~ ‖ rechtwinkliges Dreieck n. ‖ triangle m. rectangle *ou* rectangulaire. / right-angled isosceles ~ ‖ gleichschenklig-rechtwinkliges Dreieck n. ‖ triangle m. rectangle isoscèle. / scalene ~ ‖ ungleichseitiges Dreieck n. ‖ triangle m. scalène. / similar ~s pl. ‖ ähnliche Dreiecke npl. ‖ triangles mpl. semblables.

triangle gin ‖ dreischenkliges Hebezeug n.; Dreibein n. ‖ chèvre f. à trois branches; support m. à trois pieds.

triangular ‖ dreieckig ‖ triangulaire. / ~ frame ‖ Dreieckbügel m. ‖ bride m. *ou* étrier m. triangulaire. / ~ perforation ‖ Dreikantlochung f. ‖ perforation f. triangulaire. / ~ section ‖ Dreieckquerschnitt m. ‖ section f. triangulaire. / ~ stirrup *see* ~ frame.

triangulated structure ‖ Dreieckverband m. ‖ section f. triangulaire.

triangulation (Network) ‖ trigonometrisches Netz n.; Triangulation f.; Landesvermessungsnetz n. ‖ triangulation f.; réseau m. de triangles géodésiques; canevas m. trigonométrique.

tribasic (Chem) ‖ dreibasisch ‖ tribasique.

triblet ‖ Reibahle f.; Dorn m.; Steckdorn m. ‖ alésoir m.; estampe f.; étampe f.; poinçon m.; mandrin m. / flat ~ ‖ flacher Dorn m. ‖ mandrin m. méplat. / square ~ ‖ viereckiger Dorn m. ‖ mandrin m. carré. / ~ of a tube press ‖ Kern m. *oder* Dorn m. einer Röhrenpresse ‖ mandrin m. *ou* âme f. d'une presse à tuyaux.

tribolet *see* triblet.

tribromacetic acid ‖ Tribromessigsäure f. ‖ acide m. tribromacétique.

tribromphenylate of bismuth ‖ Xeroform n. ‖ xéroforme m.; tribromphénylate m. de bismuth.

tribune ‖ Tribüne f.; Rednerbühne f. ‖ tribune f.

tribute work (Mine) ‖ Akkordarbeit f. ‖ travail m. à forfait *ou* à la tâche.

tricar ‖ dreiräderiger Wagen m. ‖ tricar m.

trichinoscope ‖ Trichinoskop n. ‖ trichinoscope m.

trichlorethylene ‖ Trichloräthylen n. ‖ trichloréthylène m.

trichloride of gold ‖ Goldchlorid n. ‖ or m. potable; sesqui-chlorure m. d'or.

trichlor-methane ‖ Chloroform n. ‖ chloroforme m.

trick ‖ Handwerkskniff m. ‖ coup m. de main; secret m. de polichinelle; truc m. / ~ flying ‖ Kunstflug m. ‖ vol m. acrobatique.

trickle, to ‖ rieseln; tropfen; tröpfeln; durchsickern ‖ tomber goutte à goutte; ruisseler; suinter. / ~ (Brew) ‖ berieseln ‖ ruisseler. / ~ down drop by drop ‖ aufträufeln; langsam tropfen lassen ‖ verser goutte à goutte. / ~ into ‖ einsickern ‖ s'imbiber; se perdre dans...; s'infiltrer.

trickling (Brew) ‖ Berieselung f.; Berieseln n. ‖ ruissellement m. / ~ apparatus ‖ Berieselungsapparat m. ‖ appareil m. à ruissellement. / ~ water ‖ Rieselwasser n. ‖ eau f. de ruissellement.

trick-track board ‖ Triktrakspiel n. ‖ jeu m. de trictrac.

tricot ‖ Trikot n.; Stickstoff m.; Wirkstoff m.; tricot m. / sporting ~ ‖ Sporttrikot n. ‖ tricot m. de sport.

tricot tissue ‖ Trikotstoff m. ‖ étoffe f. en tricot. / ~ weaving mill ‖ Trikotwirkerei f. ‖ tissage m. de tricots.

tricycle ‖ Dreirad n. ‖ tricycle m. / front-steerer ~ ‖ Dreirad n. mit vorderem Steuerrad ‖ tricycle f. à roue directrice avant.

tried for . . . *see also* to try ‖ geprüft auf . . . ‖ essayé pour . . .

trift, to ~ wood ‖ Holz n. triften ‖ faire flotter le bois.

trigger (Of a gun) ‖ Abzug m. ‖ gâchette f. / ~ (On a carriage) ‖ Hemmschuh m.; Hemmvorrichtung f. ‖ enrayure f.; chien m.; sabot m. d'enrayage. / ~ (Phot) ‖ Auslöser m.; Auslösungshebel m. ‖ déclencheur m. / ~ chain ‖ Hemmkette f. ‖ chaîne f. d'enrayure. / ~ pin ‖ Sperrklinke f. ‖ cliquet m. / ~ release (Of a gun) ‖ Gewehrabzug m. ‖ transmission f. de la gâchette (de la mitrailleuse).

trigonometrical ‖ trigonometrisch ‖ trigonométrique. / ~ problem ‖ trigonometrische Aufgabe f. ‖ problème m. trigonométrique. / ~ signal ‖ trigonometrisches Zeichen n. ‖ signal m. trigonométrique.

45*

trigonometry ‖ Trigonometrie f. ‖ trigonométrie f. / plane ~ ‖ ebene Trigonometrie f. ‖ trigonométrie f. rectiligne. / spherical ~ ‖ sphärische Trigonometrie f. ‖ trigonométrie f. sphérique.

trilateral ‖ dreiseitig ‖ trilatéral.

trill machine ‖ Fältelmaschine f.; Plissiermaschine f. ‖ machine f. à plisser.

trim, to ~ (Lacem) ‖ besetzen ‖ orner. / ~ (Wood working) ‖ schneiteln; ausästen ‖ émonder; élaguer. / ~ with the axe ‖ mit der Axt f. behauen ‖ équarrir à la hache. / ~ the electrotype plate ‖ das Galvano bestoßen ‖ couper le galvano. / ~ in ‖ in einpassen ‖ adapter; assembler. / ~ in (Join; Carp) ‖ einlassen ‖ enchâsser; assembler. / ~ the sails pl. ‖ die Segel npl. richten *oder* stellen ‖ orienter les voiles fpl. / ~ a ship ‖ die Ladung f. eines Schiffes stauen ‖ arrimer un vaisseau.

trim (Mar) ‖ in richtiger Lage f. ‖ assis. / ~ by the bow (Shipb; Airpl) ‖ buglastig ‖ chargé à la proue; lourd du nez; à nez lourd. / ~ by the stern ‖ hecklastig ‖ chargé à la poupe; lourd de la queue; à queue lourde.

trim (Mar; Aero) ‖ Gleichgewichtslage f.; Trimm m.; richtige Lage f. (im Wasser *oder* in der Luft) ‖ assiette f.; centrage m. / in very good ~ ‖ in sehr gutem Trimm m. ‖ très-bien assis. / out of ~ ‖ vertrimmt ‖ déséquilibré. / lateral ~ ‖ Seitenlastigkeit f. ‖ centrage m. latérale. / longitudinal ~ ‖ Längslastigkeit f. ‖ centrage m. longitudinal. / change of ~ ‖ Trimmänderung f. ‖ changement m. d'assiette.

trimetal ‖ Trimetall n. ‖ tri-métal m.

trimmed (Found) ‖ bestoßen ‖ taillé. / ~ in bunkers ‖ in den Bunkern mpl. gestaut ‖ arrimé dans les soutes fpl. / ~ edge (Found) ‖ bestoßene Kante f. ‖ arête f. taillée. / ~ hat ‖ Hut m. mit Garnitur ‖ chapeau m. garni.

trimmer (Carp) ‖ Schlüssel m.; Trumm n.; Wechselbalken m. ‖ linçoir m.; linsoir m.; chevêtre m. / ~ (Lacem) ‖ Besatznäherin f. ‖ agrémaniste en passementerie; enjoliveuse f. / ~ (Mar) ‖ Trimmer m. ‖ soutier m. / ~ (Pap) ‖ Papierbeschneider m.; Trimmer m. ‖ rogneur m.; massicoteur m. / ~ (Tail) ‖ Nopper m. ‖ pareur m.

trimming (Found) ‖ Abgratarbeit f. ‖ ébarbage m. / ~ of buffers ‖ Abgraten n. von Puffern ‖ ébarbage m. de tampons. / ~ of draw-hooks ‖ Abgraten n. von Zughaken ‖ ébarbage m. de crochets de traction. / ~ of sheets ‖ Besäumen n. von Blechtafeln ‖ affranchissement m. des tôles. / strip shear for ~ strips and plates ‖ Streifenschere f. zum Besäumen von Streifen und Platten ‖ cisailles fpl. à bandes à replier les bandes et les plaques.

trimming (Lacem) *see also* trimmings ‖ Besatz m.; Borte f.; Einfassung f. ‖ garniture f.; galon m. / gold ~ ‖ Goldtresse f. ‖ galon m. d'or. / ~ for linen ‖ Wäschebesatz m. ‖ garniture f. de linge. / silver ~ ‖ Silbertresse f. ‖ galon m. d'argent.

trimming button ‖ Besatzknopf m. ‖ bouton m. de garniture. / ~ cord ‖ Tresse f. ‖ tresse f.; soutache f. / ~ embroiderer ‖ Posamentensticker m. ‖ passementierbrodeur m.

trimming machine (Bookb) ‖ Beschneidmaschine f. ‖ rogneuse f.; machine f. à

rogner; massicoteuse f. / ~ (Found) ‖ Abgratmaschine f. ‖ ébarbeuse f. / ~ (Textile) ‖ Steppstichnähmaschine f. ‖ machine f. à piquer *ou* à garnir.

trimming maker ‖ Posamentier m.; Bortenwirker m.; Nestler m. ‖ passementier m.

trimming manufacture ‖ Posamentenfertigung f. ‖ fabrication f. de passement.

trimming-off machine (Wood working) ‖ Holzkantenstoßmaschine f. ‖ scie f. circulaire dégauchisseuse; scie f. à dresser les fonds *ou* à rafraîchir les champs.

trimming press (Found) ‖ Abgratpresse f. ‖ presse f. à ébarber; ébarbeuse f. / ~ (Wood working) ‖ Abkantpresse f. ‖ presse f. à écorner.

trimming ribbon ‖ Bordürenband n. ‖ galon-ruban m.

trimmings pl. ‖ Posamentierwaren fpl.; Besatzartikel mpl.; Posamenten pl. ‖ (garnitures fpl. de) passementeries fpl.

trim saw ‖ Kerbsäge f.; Schrotsäge f.; Trummsäge f. ‖ scie f. à débiter *ou* de travers; scie f. à deux mains.

tringle (Covering a joint) ‖ Deckleiste f.; Fugenleiste f. ‖ lisse f. de recouvrement; couvre-joint m.

trinitrophenol ‖ Pikrinsäure f.; Bittersäure f.; Trinitrophenol n. ‖ acide m. picrique.

trinitrotoluene ‖ Trinitrotoluol n. ‖ trinitrotoluol m.

trinkets pl. ‖ Nippes pl.; Kram m.; Bijouteriewaren fpl.; Flitterwerk n. ‖ bibelots mpl.; (articles mpl. de) bijouterie f.

trio mill ‖ Dreiwalzwerk n.; Triowalzwerk n.; Drillingswalzwerk n. ‖ laminoir m. trio; train m. trio; trio m. / ~ rollers pl. *see* trio mill.

trioxyde of chromium ‖ Chromsäure f. ‖ acide m. chromique.

trioxymethylene ‖ Trioxymethylen n. ‖ trioxyméthylène m.

trip ‖ (kleine) Reise f.; Ausflug m.; Vergnügungsfahrt f. ‖ excursion f.; voyage m. (d'agrément). / ~ (Mar) ‖ Seereise f.; Seefahrt f. ‖ voyage m. (par mer).

tripe dressing ‖ Kaldaunenbereitung f. ‖ préparation f. des tripes.

tripe-stone (Miner) ‖ Gekrösestein m. ‖ pierre f. de tripes; anhydrite f. compacte.

trip gearing ‖ Ausklinkmechanismus m. ‖ mécanisme m. de déclenchement.

trip hammer ‖ Schmiedehammer m. ‖ marteau m. de forge.

triphase current *see also* three-phase current ‖ Drehstrom m.; Dreiphasenstrom m. ‖ courant m. triphasé.

triphaser *see also* threephaser ‖ Drehstromdynamo f.; Dreiphasendynamo f. ‖ alternateur m. triphasé; dynamo f. *ou* génératrice f. triphasée.

triphenylmethan dye ‖ Triphenylmethanfarbstoff m. ‖ colorant m. du triphénylméthane.

triplane (Airpl) ‖ Dreidecker m. ‖ triplan m.

triple ‖ dreifach ‖ triple. / ~ crane ‖ Dreifachkran m. ‖ grue f. triple.

triple-cylinder engine ‖ Dreizylindermaschine f. ‖ machine f. à trois cylindres.

triple-gyro compass ‖ Dreikreiselkompaß m. ‖ compas m. trigyroscopique *ou* à trois gyroscopes.

triple nozzle for atomization of ‖ water Dreifachdüse f. für Wasserzerstäubung ‖ gicleur m. triple pour la pulvérisation de l'eau.

triple prism ‖ dreiteiliges Prisma n. ‖ prisme m. à trois verres.

triple-pully drive ‖ Dreiriemenscheibenantrieb m. ‖ commande f. à trois poulies.

triple-rail line ‖ Dreischienengleis n. ‖ voie f. à trois rails.

triple valve ‖ Steuerventil n. ‖ triple valve f. / quick-acting ~ ‖ schnellwirkendes Steuerventil n. ‖ triple valve f. à action rapide.

triplicate ‖ dritte Ausfertigung f.; Triplikat n. ‖ troisième copie f.; triplicate m. / in ~ ‖ in dreifacher Ausfertigung ‖ en triplicate.

triplite (Miner) ‖ Eisenapatit m. ‖ triplite f.

tripod ‖ Dreifuß m.; Dreibein n.; Dreibeinstativ n. ‖ trépied m.; support m. à trois pieds; pied m. / field ~ ‖ Feldstativ n. ‖ trépied m. de campagne. / ~ with foot screws and casters ‖ Dreibeinstativ n. mit Fußschrauben und Fußrollen ‖ trépied m. avec vis calantes et galets. / sliding ~ ‖ zusammenschiebbares Dreibeinstativ n. ‖ trépied m. à branches coulissantes. / wooden sliding ~ ‖ zusammenschiebbares Holzstativ n. ‖ pied m. en bois à branches coulissantes. / steel (tube) ~ ‖ Stahlrohrstativ n. ‖ pied m. en tubes d'acier. / ~ of steel tubes with fixing stays ‖ Stativ n. aus Stahlrohren mit verstellbaren Spreizen ‖ pied m. muni de branches en métal et de traverses réglables. / wooden ~ ‖ Holzstativ n. ‖ pied m. en bois. / light wooden ~ ‖ leichtes Holzstativ n. ‖ trépied m. léger en bois. / strong wooden ~ ‖ kräftiges Holzstativ n. ‖ pied m. robuste en bois.

tripod base *see* tripod. / ~ head with polar adjustment ‖ Stativkopf m. mit Polhöhenverstellung ‖ tête f. de pied à latitude variable. / ~ leg ‖ Gestellbein n.; Bein n. des Dreibeinstativs ‖ pied m. de support à trois pieds. / ~ magnifier ‖ Dreifußlupe f. ‖ loupe f. sur trépied. / ~ mast ‖ Dreifußmast m. ‖ mât m. tripode. / ~ ring ‖ Dreifußring m. ‖ bague f. du trépied. / ~ tread of a ~ ‖ Fußwinkel m. eines Stativs ‖ fer m. saillant *ou* équerre f. d'un pied.

tripoli (Miner) ‖ Tripel m.; Tripelerde f. ‖ tripoli m.

tripping relay ‖ Ausklinkerelais n. ‖ relais m. de détente.

trip rider (Mine) ‖ Schienenleger m. ‖ poseur m. de voie.

trip spindle ‖ Bürstenwelle f.; Schaltspindel f. ‖ arbre m. porte-balais *ou* de connexion.

trisoctahedron ‖ Ikositetraeder n.; Trapezoeder n. ‖ icositetraèdre m.; trapezoèdre m.

trisulphide of antimony ‖ Antimonglanz m.; Antimonit m.; Grauspießglanzerz n. ‖ antimoine m. gris *ou* sulfuré; stibine f. / ~ of arsenic ‖ Rauschgelb n. ‖ deutosulfure m. d'arsenic; orpiment m.

triturate, to ‖ mahlen; zerreiben ‖ triturer; broyer.

trituration ‖ Zerreibung f. ‖ trituration f.

trivalent ‖ dreiwertig ‖ trivalent.

trivet *see also* tripod ‖ Dreifuß m. ‖ trépied m. / ~ (Weav) ‖ Samthaken m.; Samt-

messer n. ‖ rabot m.; rasoir m.; taille-rolle f.

trolley (Mine) ‖ Hund m.; Förderwagen m. ‖ berline f.; wagonnet m. / ~ (Railw) ‖ Dräsine f. ‖ draisienne f.; lorry m.; wagonnet m. de tournée. / ~ (Overhead apparatus of electric tramway) ‖ Stromabnehmerrolle f. ‖ trolley m.; trôlet m. / ~ (Runner of an overhead travelling crane) ‖ Laufkatze f. ‖ chariot m. roulant. / ~ (Truck for hand traction) ‖ Rollbock m.; Förderkarren m.; Karren m. ‖ charrette f.; diable m.; truc-transbordeur m. / electric ~ ‖ Elektrokarren m. ‖ charrette f. électrique. / inspection ~ (Railw) ‖ Bahnmeisterwagen m. ‖ wagonnet m. d'inspection. / shunting ~ ‖ Rangierkarren m. ‖ wagonnet m. de manœuvre.

trolley arm ‖ Stromabnehmer m. ‖ (perche f. de) trolley. / ~ bus ‖ Oberleitungsomnibus m. ‖ trolley-autobus m. / ~ pole ‖ Rollenkontaktstange f.; Stromabnehmerstange f. ‖ perche f. du trôlet ou du trolley. / ~ wire ‖ Fahrdraht m.; Fahrleitung f.; Oberleitung f. ‖ fil m. de trôlet ou de trolley ou de contact.

trolly see trolley.

trombone ‖ Posaune f. ‖ trombone f.

troostite (Met) ‖ Troostit m.; Hartperlit m. ‖ troostite f.; perlite f. dure.

tropic (Geogr) ‖ Wendekreis m. ‖ tropique m. / the ~s pl. ‖ die Tropen pl. ‖ les tropiques mpl.

tropical ‖ tropisch ‖ tropical. / ~ climate ‖ Tropenklima n.; tropisches Klima n. ‖ climat m. des tropiques. / equipment for ~ climate ‖ Tropenausrüstung f. ‖ vêtement m. pour les tropiques. / ~ helmet ‖ Tropenhelm m. ‖ casque m. pour pays chauds; casque m. colonial ou de liège.

troposphere (Meteor) ‖ Wolkenzone f.; Tropensphäre f. ‖ zone f. de nuages; troposphère f.

trouble (Geol) ‖ Verwerfung f. ‖ faille f.; rejet m. / ~ (Techn) ‖ Störung f. ‖ dérangement m. / ~ desk (Tel) ‖ Störungsmeldetisch m. ‖ table f. d'avis de dérangement. / ~ junction (Tel) ‖ Störungsmeldeleitung f. ‖ ligne f. d'avis de dérangement. / ~ man (Tel) ‖ Störungsbeamter m. ‖ agent m. du service des dérangements. / ~ prevention (Tel) ‖ Störungsvermeidungsdienst m. ‖ mesures fpl. en vue d'éviter les dérangements. / ~ time (Tel) ‖ Störungszeit f. ‖ temps m. de dérangement.

trough (Channel) ‖ Abflußrinne f.; Gerinne n. ‖ couloir m. ou chenal m. d'écoulement. / ~ (Vat) ‖ Trog m.; Mulde f.; Bottich m. ‖ auge f.; cuve f. / the ~ is locked in charging position ‖ die Mulde f. wird in Ladestellung festgehalten ‖ la caisse est retenue en position de chargement. / barley soaking ~ ‖ Gerstenweiche f. ‖ auge f. de trempage pour orge. / cold-water ~ ‖ Kaltwassertrog m. ‖ cuve f. à eau froide. / collecting ~ ‖ Sammelmulde f. ‖ benne f. collectrice. / conical ~ ‖ konische Mulde f. ‖ benne f. conique. / conveying ~ see conveyer ~. / conveyer ~ ‖ Förderrinne f.; Rinnentrog m. ‖ rigole f. ou gouttière f. de transport. / grinding (wheel) ~ ‖ Schleifsteintrog m. ‖ auge f. de la meule; auge f. de meule à aiguiser. / large ~ (Coop) ‖ Beute f. ‖ auge f. / pneumatic ~ ‖ pneumatische Wanne f. ‖ cuve f. pneu-

matique. / shaking ~ ‖ Schüttelrutsche f.; Schüttelrinne f. ‖ plan m. de transport incliné à secousses; gouttière f. à secousses. / smith's ~ ‖ Kühltrog m.; Löschtrog m. ‖ auge f. de rafraîchissement; étang m. d'enclume. / straightening ~ ‖ Richtrinne f. ‖ gouttière f. de dressage. / tipping ~ ‖ Kippmulde f. ‖ benne f. basculante; benne-griffe f.

trough charging crane ‖ Muldenbeschickkran m. ‖ grue f. d'enfournement. / ~ dryer ‖ Muldentrockner m. ‖ sécheur m. à auge. / ~ mixer ‖ Trogmischer m.; mélangeur m. à auge. / ~ roller ‖ Muldenrolle f. ‖ rouleau m. à plier la bande. / ~ terminal (Cable) ‖ Trogendverschluß m. ‖ tête f. de câble à auge. / ~ tip wagon ‖ Muldenkipper m. ‖ basculeur m. à auge; culbuteur m. à cuiller. / ~ tip wagon for mines ‖ Grubenmuldenkipper m. ‖ basculeur m. à auge de mine.

trouser buckle ‖ Hosenschnalle f. ‖ boucle f. de pantalons. / ~ button ‖ Hosenknopf m. ‖ bouton m. de pantalons.

trousering ‖ Hosenzeug n.; Hosenstoff m.; Beinkleiderzeug n. ‖ étoffe f. de pantalon(s). / twilled linen ~ ‖ Hosendrell m. ‖ coutil m. de pantalon.

trousers pl. ‖ Hose f.; Beinkleid n.; Hosen fpl. ‖ pantalon m.

trousseau ‖ Brautausstattung f.; Aussteuer f. ‖ trousseau m. / doll's ~ ‖ Puppenausstattung f. ‖ trousseau m. de poupées.

trout ‖ Forelle f. ‖ truite f.

trowel (Mas) ‖ Kelle f.; Maurerkelle f. ‖ truelle f. / a ~ full ‖ eine Kelle voll ‖ une truellée. / assay ~ ‖ Probekelle f. ‖ truelle f. d'essai. / notched ~ ‖ Schabkelle f.; Kratzkelle f. ‖ truelle f. brettée. / plastering ~ ‖ Gipskelle f. ‖ platronoir m.

truck, to (To carry with trucks) ‖ in Güterwagen mpl. versenden ‖ rouler.

truck (Small strong wheel) ‖ Blockrad n.; Rollrad n.; Laufrad n.; Rolle f. ‖ poulie f.; galet m. de roulement.

truck (Hand cart) ‖ Handkarren m.; Handrollwagen m. ‖ charrette f. à bras; haquet m. à main; galopin m.; diable m.; diable-brouette m.; cabrouet m. / ~ (Small car for moving luggage) ‖ Gepäckkarren m. ‖ cabrouet m. à bagages. / ~ (Strong car for heavy hauling) ‖ Lastwagen m.; Rollwagen m. ‖ camion m. / caterpillar ~ ‖ Raupenwagen m. ‖ chariot m. à chenilles. / ~ for charging the coke kiln ‖ Koksofenfüllwagen m. ‖ benne f. roulante à remplir pour fours à coke. / ~ for converters ‖ Bodeneinsatzwagen m. für Konverter ‖ wagonnet m. de transport pour fonds de convertisseurs. / electric ~ ‖ Elektrokarren m. ‖ (elektrischer) Kraftkarren m. ‖ chariot m. électrique. / electric crane ~ ‖ elektrischer Krankarren m. ‖ truck m. électrique à grue. / electric runabout ~ see electric ~. / founder's ~ ‖ Rollwagen m.; Schleppe f. ‖ chariot m. de transport dans les fonderies. / garbage removal ~ ‖ Müll(kraft)wagen m. ‖ camion m. pour ordures ménagères. / industrial ~ see electric ~. / ingot charging ~ (Roll mill) ‖ Blockwagen m. ‖ chariot m. à lingots. / light ~ ‖ leichter oder kleiner Lastwagen m. ‖ camionette f. / miner's ~ ‖ Hund m.; Förderhund m.; Förderwagen m.; Grubenwagen m. ‖ berline f.; wagonnet m.; chien m. / motor ~ ‖ Lastkraftwagen m.;

Lastauto n. ‖ camion m. automobile. / (small) motor ~ ‖ Kraftkarren m. ‖ charrette f. automobile. / motor tank ~ ‖ Motortankwagen m. ‖ automobile f. à réservoir; tonneau m. d'eau automobile. / repair ~ ‖ Werkstattwagen m. ‖ camion m. atelier. / side tip ~ ‖ Seitenkipper m. ‖ camion m. basculant de côté. / small ~ see light ~. / stove ~ see founder's ~. / wooden ~ ‖ hölzerne Sackkarre f. oder Speicherkarre f. ‖ diable m. en bois.

truck (Open railway freight car) ‖ offener Güterwagen m.; Lore f. ‖ wagon m. (découvert) à marchandises; wagon m. en plate-forme; truc m. / covered ~ ‖ bedeckter Wagen m. ‖ wagon m. couvert. / ~ with swivelling bolster ‖ Drehschemelwagen m. ‖ wagon m. avec traverses mobiles.

truck (Bogie for railway rolling stock) ‖ Drehgestell n.; Untergestell n. ‖ bogie m.; truck m. (pivotant); avant-train m. / ~ of any wheel base ‖ Drehgestell n. für beliebigen Radstand ‖ bogie m. à écartement de roues quelconque. / ~ with eight-shoe screw brake ‖ Drehgestell n. mit achtklotziger Spindelbremse ‖ bogie m. avec frein à vis à huit sabots. / ~ with rolling ring ‖ Drehgestell n. mit Laufring ‖ bogie m. avec cercle de roulement. / ~ for tramcars ‖ Untergestell n. für Straßenbahnwagen ‖ bogie m. pour voitures de tramways.

truck (Runner of an overhead travelling crane) see trolley.

truck (Wooden cap at the summit of a flagstaff etc.) ‖ Flaggenknopf m. ‖ pomme f. de girouette ou de mât ou de pavillon.

truck (Vegetables raised for the market) ‖ Gemüse n. ‖ légume m.

truck (Bartering trade) ‖ Tausch m.; Tauschhandel m. ‖ troc m.; commerce m. d'échange.

truck arrangement (articulated) ‖ (bewegliche) Drehgestellanordnung f. ‖ système m. (articulé) des bogies. / ~ balance (Railw) ‖ Gleiswage f. ‖ bascule f. à wagons. / ~ body ‖ Wagenkasten m. ‖ caisse f. de wagon. / ~ bolster ‖ Drehpfannenträger m. ‖ traverse f. de la crapaudine. / ~ cart (Found) ‖ Rollwagen m.; Schleppe f. ‖ chariot m. de transport (dans les fonderies). / ~ centre pin ‖ Drehzapfen m. ‖ pivot m. du bogie. / ~ chassis (Auto) ‖ Kraftwagenrahmen m. ‖ châssis m. d'automobile. / ~ counter (Mine) ‖ Wagenzähler m.; Karrenzähler m. ‖ compteur m. de berlines. / ~ driver (Mine) ‖ Wagenführer m. ‖ conducteur m. de berlines. / ~ filler ‖ Wagenlader m. ‖ chargeur m. de wagonnets. / ~ gun (Arm) ‖ Geschütz n. auf Radlafette ‖ pièce f. sur affût marin. / ~ load ‖ Wagenladung f. ‖ charge f. de wagon.

truckman ‖ Rollkutscher m. ‖ roulier m.; camionneur m.

trucks pl. ‖ Hose f.; Beinkleider npl. ‖ pantalon m.

truck setter (Mine) ‖ Rangierer m. ‖ avanceur m. de berlnes. / ~ tippler ‖ Wagenkipper m. ‖ basculeur m. de wagon. / ~ train ‖ Schleppzug m. ‖ train m. de voitures. / ~ wheel ‖ Blockrad n.; Rollrad n.; Laufrad n.; Rolle f. ‖ poulie f.; galet m.

true, to ~ a wheel ‖ ein Rad n. zentrieren oder mitten ‖ centrer une roue. / ~ a

bearing ‖ ein Lager n. ausrichten ‖ aligner un palier. / not properly trued ‖ schlecht ausgerichtet ‖ mal dressé.

true ‖ echt; richtig ‖ vrai. / ~ (Compass needle) ‖ rechtweisend ‖ corrigé. / ~ to profile (Road) ‖ profilgerecht ‖ d'un profil bien dressé. / ~ position of the ship ‖ wahrer Schiffsort m. ‖ vraie position f. du navire.

trueing see truing.

truffle ‖ Trüffel f. ‖ truffe f.

truing device for grinding wheels ‖ Schleifsteinabrichtvorrichtung f. ‖ dispositif m. à dégrossir les pierres à affûter.

trumpery (ware) ‖ Ramschware f.; Plunder m. ‖ marchandises fpl. de pacotille; camelote f.

trumpet ‖ Trompete f. ‖ trompette f.

truncate, to ~ (Met) ‖ abstechen ‖ tronçonner. / ~ an angle (Miner) ‖ eine Kante brechen ‖ couper l'angle ou le coin.

truncheon see trunk (of a tree).

trunk (of a tree) ‖ Baumstamm m.; Stamm m.; Stumpf m.; Stock m.; Stubben m.; Klotz m. ‖ tige f.; tronc m.; tronche f.; souche f. / baulked ~ ‖ waldrecht behauener Holzstamm m. ‖ brin m. de bois; tronc m. dégrossi. / ~ of a column ‖ Säulenschaft m. ‖ fût m. ou vif m. ou tronc m. d'une colonne. / rough-hewn ~ see baulked ~.

trunk (Telephonic junction line) ‖ Verbindungsleitung f. ‖ ligne f. de jonction. / defective ~ ‖ gestörte Verbindungsleitung f. ‖ ligne f. de jonction dérangée.

trunk (Chest) ‖ Lade f.; Truhe f. ‖ caisse f.; coffre m.; tiroir m. / ~ (Travelling box) ‖ Koffer m. ‖ malle f.; valise f.; coffre m. / leather ~ ‖ Lederkoffer m. ‖ malle f. en cuir. / motor-car ~ ‖ Automobilkoffer m. ‖ coffre m. d'automobile. / motor touring ~ see motor car ~. / ~ of pasteboard covered with leather ‖ Koffer m. aus Pappe, mit Leder überzogen ‖ malle f. en carton recouverte de cuir. / pointed ~ (Dress ore) ‖ Spitzkasten m. ‖ caisse f. pointue. / rush plate ~ ‖ Rohrplattenkoffer m. ‖ malle f. à parois en jonc. / travelling ~ ‖ Reisekoffer m. ‖ malle f. de voyage. / vulcan fibre ~ ‖ Vulkanfiberkoffer m. ‖ malle f. en fibre vulcanique. / wardrobe ~ ‖ Schrankkoffer m. ‖ malle-armoire f. / wooden ~ ‖ Holzkoffer m. ‖ malle f. en bois.

trunk answering jack (Tel) ‖ Fernabfrageklinke f. ‖ jack m. interurbain. / ~ barrow ‖ Kastenkarren m.‖ charrette f. à caisse. / ~ bow ‖ Kofferbügel m. ‖ monture f. de valise.

trunk-busy (Tel) ‖ fernbesetzt ‖ occupé par une conversation interurbaine.

trunk cable ‖ Fernleitungskabel n. ‖ câble m. interurbain. / ~ call ‖ Ferngespräch n. ‖ communication f. interurbaine. / ~ calling lamp ‖ Fernanruflampe f. ‖ lampe f. d'appel interurbain. / ~ calling signal ‖ Fernanrufzeichen n. ‖ signal m. d'appel interurbain. / ~ case ‖ Schrankkoffer m.; valise-armoire f. / ~ circuit (Tel) ‖ Fernamtsschaltung f. ‖ circuit m. interurbain. / ~ connection (Tel) ‖ Fernverbindung f. ‖ communication f. interurbaine. / ~ drop (Tel) ‖ Fernklappe f. ‖ annonciateur m. interurbain. / ~ exchange (Tel) ‖ Fernamt n.; Fernstelle f. ‖ bureau m. central interurbain. / ~ fittings pl. ‖ Kof-

ferbeschläge mpl. ‖ ferrures fpl. de malles. / ~ frame ‖ Koffergestell n. ‖ châssis m. de malle.

trunking, direct ~ (Tel) ‖ Abfragebetrieb m.; Anrufbetrieb m. ‖ exploitation f. par ligne de conversation.

trunking traffic (Tel) ‖ Verbindungsleitungsverkehr m. ‖ trafic m. sur les lignes auxiliaires.

trunk jack key (Tel) ‖ Fernklinkentaste f. ‖ bouton m. pour les lignes de jack multiple interurbaines. / ~ jack lamp ‖ Fernklinkenlampe f. ‖ lampe f. de fin de conversation interurbaine. / ~ jack multiple ‖ Fernklinkenleitung f. ‖ ligne f. de jack multiple interurbaine. / ~ jack panel ‖ Fernklinkenfeld n. ‖ champ m. de jacks interurbains. / ~ junction (circuit) (Tel) ‖ Fernvermittlungsleitung f. ‖ ligne f. intermédiaire; ligne f. (auxiliaire) interurbaine. / ~ junction jack ‖ Fernvermittlungsklinke f. ‖ jack m. additionnel. / ~ junction position ‖ Fernvermittlungsplatz m. ‖ position f. intermédiaire. / ~ junction section ‖ Fernvermittlungsschrank m. ‖ table f. des lignes auxiliaires interurbaines. / ~ line (Tel) ‖ Fernleitung f. ‖ circuit m. interurbain. / ~ line dialling (Aut tel) ‖ Wählerfernsteuerung f. ‖ installation f. d'une connection interurbaine par cadran d'appel. / ~ line relay ‖ Fernanrufrelais n. ‖ relais m. d'appel interurbain. / ~ line system ‖ Fernleitungsnetz n. ‖ réseau m. de lignes interurbaines. / ~ lock ‖ Kofferschloß n.; Fallschloß n. ‖ serrure f. pour malles; serrure f. à pêne dormant et loquet. / ~ making machine ‖ Kofferherstellungsmaschine f. ‖ machine f. à fabriquer des coffres. / ~ mill ‖ Kochermühle f. ‖ moulin m. de support ou à colonne creuse. / ~ mountings pl. ‖ Kofferbeschläge mpl. ‖ ferrures fpl. pour malles. / ~ operating observation (Tel) ‖ Fernbetriebsüberwachung f. ‖ surveillance f. de l'exploitation interurbaine. / ~ operator (Tel) ‖ Vermittlungsbeamtin f. ‖ operatrice f. / ~ order wire (Tel) ‖ Ferndienstleitung f. ‖ ligne f. de service interurbain. / ~ order wire jack ‖ Ferndienstklinke f. ‖ jack m. de ligne de service interurbain. / ~ plug (Tel) ‖ Fernstöpsel m. ‖ fiche f. interurbaine. / ~ position (Tel) ‖ Fernplatz m. ‖ position f. interurbaine. / ~ rack ‖ Gepäckhalter m.; Gepäckträger m. ‖ porte-bagages m.; portepaquet m. / ~ rot (Of wood) ‖ Stockfäule f. ‖ pourriture f. complète. / ~ section (Tel) ‖ Fernschrank m. ‖ table f. interurbaine. / ~ system for magneto boards ‖ Fernleitungssystem n. für Klappenschränke ‖ système m. interurbain pour commutateurs à clapets. / ~ table (Tel) ‖ Ferntisch m. ‖ table f. interurbaine. / ~ telephone station ‖ Fernamtsanschluß m. ‖ poste m. pour le trafic interurbain. / ~ test board (Tel) ‖ Klinkenumschalter m. ‖ commutateur m. de jacks. / ~ timber ‖ Stockholz n. ‖ bois m. du tronc. / ~ traffic (Tel) ‖ Fernverkehr m. ‖ trafic m. interurbain. / ~ wood ‖ Stammholz n. ‖ bois m. de tronc.

trunnel see treenail.

trunnion ‖ Zapfen m.; Drehzapfen m.; Schildzapfen m. ‖ tourillon m. / ~ of an oscillating cylinder ‖ Schwungzapfen m. eines oszillierenden Zylinders ‖ tourillon m. d'un cylindre oscillant. / tipping ~

(Of a tip wagon) ‖ Kippzapfen m. ‖ tourillon m. (de wagonnet basculant).

trunnion belt of the converter (Met) ‖ Schildzapfenring m. der Bessemerbirne ‖ ceinture f. du convertisseur. / ~ cap ‖ Scharnierdeckel m. ‖ chapeau m. à charnière. / ~ screw ‖ Zapfenschraube f. ‖ vis f. à pivot. / ~ sleeve of telescope ‖ Fernrohrschelle f. ‖ collier m. de la lunette. / ~ tip wagon ‖ Zapfenkipper m. ‖ wagonnet m. culbuteur à tourillons. / width over the ~s ‖ Länge f. über Zapfen ‖ longueur f. sur les tourillons.

truss, to ‖ ausstreben; armieren ‖ entretoiser; armer. / ~-up the hoops pl. of casks ‖ Fässer npl. antreiben ‖ serrer les cercles mpl. de fûts. / ~-up staves pl. ‖ Reifen mpl. auftreiben ‖ serrer des cercles mpl. de fûts.

truss (Bridg) ‖ Hängewerk n. ‖ armature f. (établie sur les poutres). / ~ (Carp) ‖ Bock m.; Gerüst n.; Gestell n.; Hängewerk n. ‖ chevalet m.; tréteau m.; armature f. / ~ (Forg) ‖ Paket n.; Gespann n.; Zange f. ‖ paquet m.; trousse f. ~ (Med) ‖ Bruchband n. ‖ bandage m. (herniaire). / ~ (Shipb) ‖ Rack n. ‖ drosse f.; racage m.; raque f. / cantilever ~ ‖ Konsolträger m. ‖ poutre f. en saillie. / end ~ ‖ Endträger m. ‖ poutre f. de rive; poutre f. de tête. / half-~ ‖ Halbbinder m.; halbes Gebinde n. ‖ demi-ferme f. de croupe. / intermediate ~ ‖ Freigebinde n.; Leergebinde n.; Zwischengesperre n. ‖ ferme f. de remplage. / king-post ~ ‖ Hängewerk n. mit einer Hängesäule ‖ armature f. à un seul poinçon. / ~ exposed to longitudinal strains ‖ Längsträger m. ‖ longrine f.; longeron m. / principal ~ ‖ Dachbinder m. ‖ maîtresse f. ferme. / queen-post ~ ‖ Hängewerk n. mit zwei Hängesäulen ‖ armature f. à clefs pendantes. / ~ of roofing ‖ Hängewerk n. im Dach ‖ armature f. de ferme. / simple roof ~ ‖ Sparrendach n. ‖ armature f. simple. / standard ~ ‖ erstes Gebinde n.; Lehrgebinde n. ‖ ferme f. d'échantillon.

truss bar of brake ‖ Bremsdreieck n.; Bremsstange f. ‖ levier m. en triangle de frein; tirant m. de frein.

trussed pole (El line) ‖ verspannte Stange f. ‖ appui m. haubanné sur lui-même.

truss frame (Carp) see also truss ‖ Hängewerk n.; Hängebock m. ‖ ferme f. à clefs pendantes; armature f. de ferme. / ~ hoop (Coop) ‖ Arbeitsreif m. ‖ cercle m. de travail. / ~ hoop driving machine ‖ Arbeitsreifenanziehmaschine f. ‖ machine f. à serrer les cercles de travail.

trussing (Carp) ‖ Gebinde n.; Armierung f. ‖ armature f. / ~ of frame (Auto) ‖ Rahmenunterzug m. ‖ renforcement m. du châssis.

trussing frame see truss frame.

trussing machine for hoops of casks ‖ Reifenanziehmaschine f.; Reifenantreibmaschine f. ‖ machine f. à serrer les cercles de fûts; serreuse f. pour cercles de fûts.

trussing-up (Of hoops) ‖ Antreiben n. ‖ serrage m.

truss maker ‖ Bandagist m.; Bruchbandmacher m. ‖ bandagiste m.

truss rod (Build) ‖ Anker(bolzen) m. ‖ tirant m.

trust (Trade) ‖ Trust m. ‖ trust m. / ~ company ‖ Treuhandgesellschaft f. ‖ société f. financière.

trustee ‖ Treuhänder m.; Bevollmächtigter m.; Syndikus m. ‖ curateur m.; mandataire m.; plénipotentiaire m.; syndic m. / ~ of a bankrupt's estate ‖ Konkursverwalter m. ‖ syndic m. de faillite.

trust money ‖ Mündelgeld n. ‖ deniers mpl. pupillaires. / sufficiently safe for investment of ~ ‖ mündelsicher ‖ d'une sûreté absolue; pupillaire.

trustworthiness (Trade) ‖ Zuverlässigkeit f.; Kreditfähigkeit f.; Vertrauenswürdigkeit f. ‖ certitude f.; authenticité f.; sûreté f.; solidité f.; réputation f.

trustworthy ‖ zuverlässig; vertrauenswürdig; kreditfähig ‖ éprouvé; certain; solide.

truth ‖ Richtigkeit f. ‖ exactitude f.; régularité f. / ~ of running (Turn) ‖ Rundlaufen n. ‖ concentricité f.

try, to ‖ prüfen; feststellen; erproben; revidieren ‖ vérifier; examiner; éprouver; essayer. / ~ on ‖ anproben ‖ essayer.

trying ‖ Probieren n. ‖ essayage m. / ~ (Met) ‖ Erzprobe f. ‖ essai m. du minerai. / ~ plane (Join) ‖ Nachfügehobel m.; Rauhbank f.; (langer) Schlichthobel m. ‖ varlope f. (à repasser).

trysail (spencer) ‖ Gaffelsegel n. ‖ voilegoélette f.

try square ‖ Anschlagwinkel m. ‖ équerre f. épaulée.

tschermigite ‖ Ammonalaun m. ‖ tschermigite f.

T-square see under T-.

tub ‖ Bottich m.; Kübel m.; Wanne f.; Zuber m.; Bütte f. ‖ cuve f.; seau m.; baquet m. / ~ (Mine: tram) ‖ Hund m.; Förderwagen m. ‖ berline f.; chien m.; wagonnet m. / ~ (Mine: kibble) ‖ Förderkübel m. ‖ bâche f.; seau m.; tonne f.; tine f. / ~ (Mar) ‖ Balje f.; baille f. / bath(ing) ~ ‖ Badewanne f. ‖ baignoire f. / ~ of butter ‖ Butterfaß n. ‖ baratte f.; tine f. de beurre. / ~ for flowers ‖ Blumenkübel m. ‖ cache-pot m. pour fleurs. / ~ for lixiviating ‖ Laugenfaß n. ‖ cuvier m. / stirring ~ ‖ Rührbottich m. ‖ cuve f. à agitateur.

tubbing (Mine) ‖ Schachtausbau m. mit Tübbings; Tübbingausbau m. ‖ tubage m.; cuvelage m.

tube see also pipe ‖ Rohr n.; Röhre f. ‖ tuyau m.; tube m.; conduit m. / ~ (Envelope for colours etc.) ‖ Tube f. ‖ tube m. / ~ (Of a telescope etc.) ‖ Tubus m.; Rohr n. ‖ tube m. / ~ (Hose) ‖ Schlauch m. ‖ tuyau m.; tuyau m. flexible. / ~ (Radio) ‖ Röhre f. ‖ lampe f.; valve f. / ~ (Siphon) ‖ Heber m.; Siphon m.; Saugheber m. ‖ siphon m. / the ~s pl. were to be furnished bored in the rough ‖ die Rohre npl. sollten vorgebohrt geliefert werden ‖ les tubes devaient être livrés avec un forage d'ébauchage. / the ~s pl. terminate partly below and partly above the water level ‖ die Wasserrohre npl. münden zum Teil unter, zum Teil über dem Wasserspiegel ‖ les tubes mpl. aboutirent en partie au-dessous et en partie au-dessus du niveau de l'eau.

tube, acid-resisting ~ ‖ säurebeständiges Rohr n. ‖ tuyau m. résistant aux acides. / air-supply ~ ‖ Luftansaugrohr n. ‖ tube m. d'aspiration d'air. / binocular ~ (Opt) ‖ binokularer Tubus m.; Doppeltubus m. ‖ tube m. binoculaire; doubletube m. brass ~ ‖ Messingröhre f. ‖

tube m. en laiton. / butt-welded ~ ‖ stumpfgeschweißtes Rohr n. ‖ tuyau m. soudé à rapprochement; tube m. soudé par contact. / caoutchouc ~ ‖ Kautschukschlauch m.; Gummischlauch m. ‖ tube m. de caoutchouc. / capillary ~ ‖ Haarröhrchen n. ‖ tube m. capillaire. / clogged ~ ‖ verstopftes Rohr n. ‖ tube m. bouché. / closed ~ (Chem) ‖ Bajonettrohr n. ‖ tube m. fermé. / cold-drawn ~ ‖ kaltgezogenes Rohr n. ‖ tube m. étiré à froid. / collapsible ~ ‖ Tube f. ‖ tube m. / conical ~ ‖ konisches Rohr n. ‖ tube m. conique. / conveyor ~ ‖ Förderrohr n. ‖ tuyau m. de transport. / copper ~ ‖ Kupferrohr n. ‖ tube m. en cuivre. / corrugated ~ ‖ Wellenrohr n. ‖ tuyau m. ondulé. / of current redresser ‖ Gleichrichterröhre f. ‖ ampoule f. ou tube m. (de) redresseur de courant. / cylindrical ~ ‖ zylindrisches Rohr n. ‖ tube m. cylindrique. / dipping ~ ‖ Tauchrohr n. ‖ tuyau-plongeur m. / drawn ~ ‖ gezogenes Rohr n. ‖ tube m. étiré. / flattened(-out) at the ends ‖ an den Enden flachgedrücktes Rohr n. ‖ tube m. aplati aux extrémités. / flexible ~ ‖ Schlauch m. ‖ tuyau m. (flexible). / flexible metal ~ ‖ Metallschlauch m.; biegsame Metallröhre f. ‖ tuyau m. métallique flexible. / flared ~ of rear axle ‖ Hinterachstrichter m. ‖ cône m. tubulaire du pont arrière. / flue ~ ‖ Feuerrohr n. ‖ tube-foyer m. / corrugated flue ~ ‖ gewelltes Feuerrohr n. ‖ tubefoyer m. ondulé. / straight flue ~ ‖ glattes Feuerrohr n. ‖ tube-foyer m. lisse. / galvanized ~ ‖ verzinktes Rohr n. ‖ tuyau m. galvanisé. / glass ~ ‖ Glasrohr n.; Glasröhre f. ‖ tube m. en verre. / graduated ~ ‖ Meßrohr n. ‖ tube m. gradué. / heat-resisting ~ ‖ hitzebeständiges Rohr n. ‖ tuyau m. allant au feu. / high-pressure ~ ‖ Hochdruckrohr n. ‖ tuyau m. à haute pression. / Hittorf ~ ‖ Hittorfsche Röhre f. ‖ tube m. ou tuyau m. d'Hittorf. / image erecting ~ (Opt) ‖ bildaufrichtender Tubus m. ‖ tube m. redresseur. / india-rubber ~ ‖ Gummirohr n.; Gummischlauch m. ‖ tube m. de caoutchouc. / inner ~ (Cycle) ‖ Luftschlauch m. ‖ chambre f. à air; pneu m. / insulating ~ ‖ Isolierrohr n. ‖ tuyau m. isolant. / interchangeable ~ (Opt) ‖ auswechselbarer Tubus m. ‖ tube m. interchangeable. / lap-welded ~ ‖ überlappt geschweißtes Rohr n. ‖ tube m. soudé à recouvrement. / lead ~ ‖ Bleirohr n. ‖ tube m. en plomb. / leather ~ ‖ Lederschlauch m. ‖ tuyau m. ou boyau m. de cuir. / medullary ~ (In the wood) ‖ Markröhre f. ‖ tube m. médullaire. / of the microscope ‖ Mikroskoptubus m. ‖ tube m. du microscope. / monocular ~ (Opt) ‖ monokularer Tubus m. ‖ tube m. monoculaire. / paper ~ ‖ Papierhülse f. ‖ tube m. en papier. / pasteboard ~ ‖ Papprohr n. ‖ tube m. cylindrique en carton. / Pitot ~ ‖ Stauröhre f.; Pitotsche Röhre f. ‖ tube m. de Pitot. / pneumatic ~ ‖ Rohrpost f. ‖ poste f. pneumatique. / ~ with push ‖ Steckkapselglas n. ‖ tube m. à mise en capsule. / punch pliers ~ ‖ Lochhülse f. für Lochzange ‖ tube m. pour pince à emporte-pièce. / ~ of rectifier ‖ Gleichrichterröhre f. ‖ ampoule f. ou tube m. (de) redresseur de courant. / of reinforced concrete ‖ Eisenbeton-

rohr n. ‖ tuyau m. en béton armé. / rolled ~ ‖ gewalztes Rohr n. ‖ tube m. laminé ou cylindré. / rubber ~ see indiarubber ~. / rust-resisting ~ ‖ nichtrostendes Rohr n. ‖ tuyau m. inoxydable. / ~ with screw cap (Chem) ‖ Glas n. mit Gewindeverschluß ‖ tube m. à bouchon fileté. / seamless ~ ‖ nahtloses Rohr n. ‖ tube m. sans soudure. / seamless drawn ~ ‖ nahtlos gezogenes Rohr n. ‖ tube m. étiré sans soudure. / seamless rolled ~ ‖ nahtlos gewalztes Rohr n. ‖ tuyau m. laminé sans soudure. / self-oscillating ~ (Radio) ‖ selbsterregende Röhre f. ‖ tubegénérateur m. / sensitive ~ (Radio) ‖ empfindliche Röhre f. ‖ tube m. sensible. / sheet-iron ~ ‖ Blechrohr n. ‖ tuyau m. en tôle. / slotted ~ ‖ aufgeschlitztes Rohr n. ‖ tube m. avec fente. / spiral-welded ~ ‖ spiralgeschweißtes Rohr n. ‖ tube m. soudé en spirale. / tin ~ ‖ Zinntube f.; Tube f. ‖ tube m. (én etain). / tyre ~ ‖ Luftschlauch m. ‖ chambre f. à air. / water ~ ‖ Siederohr n. ‖ tube m. à eau. / weld(ed) ~ ‖ Schweißrohr n.; geschweißtes Rohr n. ‖ tuyau m. soudé.

tube amplifier (Radio) ‖ Vakuumröhrenverstärker m. ‖ amplificateur m. à tube à vide. / ~ attachment (Opt) ‖ Tubusaufsatz m. ‖ rallonge f. du tube. / ~ beader ‖ Rohreinwalzapparat m. ‖ appareil m. à dudgeonner. / ~ bend ‖ Rohrbogen m. ‖ tube m. compensateur. / ~ bender (Person) ‖ Rohrbieger m. ‖ cintreur m. de tubes. / ~ bender (Machine) ‖ Rohrbiegemaschine f. ‖ machine f. à cintrer les tubes. / ~ bit ‖ Kanonenbohrer m. ‖ mèche f. à forer les canons / ~ blank (Met) ‖ Hohlkörper m. ‖ corps m. creux; ébauche f. / ~ boiler ‖ Röhrenkessel m. ‖ chaudière f. tubulaire. / vertical ~ boiler ‖ stehender Röhrenkessel m. ‖ chaudière f. tubulaire verticale. / ~ boiler system ‖ Röhrenkesselsystem n. ‖ système m. de chaudière tubulaire. / ~ boring machine ‖ Röhrenbohrmaschine f. ‖ machine f. à percer les tubes ou à aléser les tubes. / ~ broaching press ‖ Rohraufweitepresse f. ‖ presse f. à élargir les tubes. / ~ brush (Loc) ‖ Rohrwischer m. ‖ balai m. ou raclette f. ou tringle f. à nettoyer les tubes. / ~ buffer press ‖ Röhrenpufferpresse f. ‖ presse f. à faire des tampons tubulaires. / ~ capacity (Radio) ‖ Röhrenkapazität f. ‖ capacité f. de la lampe. / ~ clasp ‖ Schlauchklemme f. ‖ pince f. pour tuyau. / ~ cleaner (Apparatus) ‖ Rohrreiniger m. ‖ appareil m. à nettoyer les tubes. / ~ closing ‖ Rohrverschluß m. ‖ fermeture f. de tuyau. / ~ compensator ‖ Rohrausgleicher m. ‖ compensateur m. à tuyaux. / ~ curvature of the ~s ‖ Biegen n. der Röhren ‖ cintrage m. des tubes. / the curvature of the ~s is kept as slight as possible ‖ das Biegen der Röhren geschieht unter Einhaltung möglichst schlanker Bogen ‖ les tubes sont cintrés au plus faible rayon. / ~ cutter ‖ Rohrabschneider m. ‖ coupe-tuyau m.; découpeur m. de tubes. / ~ drawing bench ‖ Rohrziehbank f. ‖ banc m. à étirer les tubes; dragon m. / hydraulic ~ drawing press ‖ hydraulische Röhrenziehpresse f. ‖ presse f. hydraulique à étirer les tuyaux. / ~ end plug ‖ Rohrverschlußstück n. ‖ bouchon m. de tuyau.

/ ~ expander ‖ Rohraufweiter m. ‖ cylindre m. à mandriner les tubes. / ~ ferrule ‖ Rohrring m. ‖ bague f.; virole f. / ~ filling machine ‖ Tubenfüllmaschine f. ‖ machine f. à remplir les tubes. / ~ filter of a pump ‖ Saugkorb m.; Pumpenkorb m.; Seiher m. *oder* Saugkopf m. einer Pumpe ‖ couloir m. du tuyau d'aspiration d'une pompe. / ~ founding ‖ Röhrengießerei f. ‖ fonderie f. des tuyaux. / ~ fuse (Electr) ‖ Röhrensicherung f. ‖ coupe-circuit m. à cartouche. / ~ heating member ‖ Röhrenheizkörper m. ‖ radiateur m. (de chauffe) tubulaire. / ~ holder (Radio) ‖ Röhrensockel m. ‖ support m. de lampes. / ~ hole ‖ Rohrloch n. ‖ trou m. de tube. / when the plates glow, the ~ hole becomes oval ‖ beim Ausglühen n. des Bleches wird das Rohrloch oval ‖ au recuit de la tôle le trou du tube s'ovalise. / ~ ignition ‖ Glührohrzündung f. ‖ allumage m. par tube incandescent. / ~ joint ‖ Rohrverbindung f. ‖ jonction f. des tuyaux. / ~ lathe ‖ Rohrdrehbank f. ‖ tour m. à tourner les tubes. / ~ length (Opt) ‖ Tubuslänge f. ‖ longueur f. du tube. / mechanical ~ length ‖ mechanische Tubuslänge f. ‖ longueur f. mécanique du tube. / optical ~ length ‖ optische Tubuslänge f. ‖ longueur f. optique du tube. / ~ lightning arrester ‖ Luftleerblitzableiter m. ‖ paratonnerre m. à vide. / ~ making machine ‖ Tubenherstellungsmaschine f. ‖ machine f. à fabriquer des tubes (à pâtes). / ~ mill ‖ Rohrmühle f. ‖ moulin m. à tube; broyeur m. à tambour; tube-broyeur m.; moulin m. tubulaire. / ~ mill (Roll mill) *see* ~ rolling mill. / compound ~ mill ‖ Verbundrohrmühle f. ‖ moulin m. tubulaire combiné. / ~ mounting of a telescope without optical equipment ‖ Rohrmontierung f. eines Fernrohres ohne Optik ‖ tube m. de lunette sans optique. / ~ notching machine ‖ Röhreneinkerbmaschine f. ‖ machine f. à entailler les tubes. / ~ nut ‖ Rohrmutter f. ‖ écrou m. de tuyau. / ~ packing ‖ Rohrdichtung f. ‖ garniture f. pour tuyaux. / ~ plate (Boiler) ‖ Rohrwand f.; Feuerrohrwand f.; Heizrohrwand f. ‖ plaque f. tubulaire *ou* à tubes. / the ~ plate is lap-welded to the furnace ‖ das Anschweißen der Rohrwand erfolgt mittels Überlappung ‖ le raccordement de la plaque tubulaire est fait par soudage à recouvrement. / ~ plug (Steam eng) ‖ Rohrstopfen m.; Feuerrohrpfropfen m. ‖ tampon m. de tube. / ~ plug ram ‖ Rohrstopfstange f. ‖ tringle f. à tamponner les tubes. / ~ pole (El line) ‖ Rohrständer m. ‖ montant m. en fer; potelet m. / ~ pole cap ‖ Verschlußkappe f. des Rohrständers ‖ chapeau m. de montant en fer. / ~ pole shoe ‖ Schuh m. für Rohrständer ‖ étrier m. *ou* socle m. de potelet. / ~ press ‖ Rohrpresse f. ‖ presse f. pour tubes.

tuber (Botany) ‖ Knolle f. ‖ tubercule m. / ~ edible ‖ eßbare Knolle f. ‖ tubercule m. alimentaire.

tubercular ‖ tuberkulös ‖ tuberculeux. / ~ tissues pl. ‖ tuberkulöse Gewebe npl. ‖ tissus mpl. atteints de tuberculose.

tube reaming (rolling) mill ‖ Rohraufweitewalzwerk n. ‖ laminoir m. à élargir les tubes / ~ receiver (Radio) ‖ Röhrenempfänger m. ‖ récepteur m. à lampe(s). / ~ re-

ducing (rolling) mill ‖ Reduzierwalzwerk n. für Rohre ‖ laminoir m. à réduire les tubes *ou* pour la réduction des tuyaux. / ~ rincing machine ‖ Röhrennachbohrmaschine f. ‖ machine f. à repercer les tubes. / ~ rolling machine ‖ Rohrwalzmaschine f. ‖ laminoir m. à tubes. / ~ rolling mill ‖ Rohrwalzwerk n. ‖ laminoir m. à tubes *ou* à tuyaux. / equipment for ~ rolling mills ‖ Rohrwalzwerkeinrichtung f. ‖ installation f. de laminoirs à tubes. / machine for ~ rolling mills ‖ Rohrwalzwerkmaschine f. ‖ machine f. pour laminoirs à tubes. / ~ rolling works pl. *see* ~ rolling mill. / ~ roughing machine ‖ Röhrenschruppmaschine f. ‖ machine f. à dégrossir les tubes. / ~ rounding ‖ Rundbiegen n. von Röhren ‖ cintrage m. de tubes. / row of ~s (Boiler) ‖ Rohrreihe f. ‖ rangée f. de tubes. / ~ sandpapering machine ‖ Röhrenabschleifmaschine f. ‖ machine f. à polir les tubes. / ~ scouping machine ‖ Röhrenaushöhlmaschine f. ‖ machine f. à évider les tubes. / ~ scraper ‖ Rohrauskratzer m. ‖ grate-tubes m.; racle m. pour tubes. / ~ sealing machine ‖ Tubenschließmaschine f. ‖ machine f. à fermer les tubes. / ~ seam ‖ Rohrnaht f. ‖ couture f. de tube. / ~ sender (Radio) ‖ Röhrensender m. ‖ émetteur m. à lampes. / ~ sharpening machine ‖ Rohranspitzmaschine f. ‖ machine f. à appointer à froid les tubes. / ~ sharpening (rolling) mill ‖ Anspitzwalzwerk n. für Rohre ‖ laminoir m. à appointer des tubes. / ~ sheet (Boiler) *see* ~ plate. / ~ shortening ‖ Rohrverkürzung f. ‖ raccourcissement m. du tube. / ~ slot ‖ Rohrnute f. ‖ rainure f. à tubes. / ~ smoothing (rolling) mill ‖ Röhrenglättwalzwerk n. ‖ laminoir m. à polir les tubes. / ~ solder ‖ Röhrenlötzinn n. ‖ soudure f. en tubes / ~ stand (Of a Rœntgen apparatus) ‖ Röhrenhalter m.; Röhrenständer m. ‖ porte-ampoule m.; support m. d'ampoule. / ~ straightening machine ‖ Richtmaschine f. für Rohre ‖ machine f. à dresser les tubes. / ~ support (Chem) ‖ Rinne f.; Schiene f. ‖ gouttière f. / ~ switch (Electr) ‖ Röhrenschalter m. ‖ interrupteur m. à tubes. / ~ system ‖ Rohrsystem n. ‖ faisceau m. tubulaire. / ~ transmitter (Radio) ‖ Röhrensender m. ‖ émetteur m. à lampe(s). / ~ turning machine ‖ Röhrendrehbank f. ‖ tour m. à tourner les tubes. / upper edge of the ~ (Opt) ‖ oberer Tubusrand m. ‖ bord m. supérieur du tube. / ~ upsetting press ‖ Rohrstauchpresse f. ‖ presse f. à refouler les tubes. / ~ vice ‖ Rohrschraubstock m. ‖ étau m. pour tubes. / ~ voltmeter ‖ Röhrenvoltmeter m. ‖ voltmètre m. à tube. / ~ wall (Boiler) ‖ Rohrwand f. ‖ paroi f. de tubes. / ~ welding ‖ Rohrschweißung f. ‖ soudage m. de tubes. / ~ welding machine ‖ Rohrschweißmaschine f. ‖ machine f. à souder les tubes. / ~ working machine ‖ Rohrbearbeitungsmaschine f. ‖ machine f. à travailler des tuyaux. / ~ wrapping machine (Pap) ‖ Hülsenwickelmaschine f. ‖ machine f. à rouler les douilles.

tub greaser (Mine) ‖ Wagenöler m. ‖ graisseur m. de berlines.

tubiform *see also* tubular ‖ röhrenförmig ‖ en forme de tuyau; tubulaire; en tube.

tubing *see also* tube ‖ Rohr n.; Röhre f.; Schlauch m. ‖ tube m.; tuyau m. / ~ (Tube conduit) ‖ Rohrleitung f. ‖ conduite f.; tuyauterie f. / borehole ~ ‖ Bohrlochverrohrung f. ‖ tuyau m. de sondage. / rubber ~ ‖ Gummischlauch m. ‖ tuyau m. en caoutchouc. / ~ machine (For cables) ‖ Schlauchmaschine f. ‖ machine f. à tubes.

tubular ‖ röhrenartig; röhrenförmig ‖ tubulaire; en forme de tube; en tube. / ~ axle ‖ Rohrachse f. ‖ essieu m. tubulaire. / ~ boiler ‖ Röhrenkessel m. ‖ chaudière f. tubulaire. / ~ box (For cables) ‖ Röhrenmuffe f. ‖ manchon m. à tube. / ~ cooler ‖ Batteriekühler m. ‖ réfrigérant m. tubulaire. / ~ extension of the Rœntgen tube ‖ Hals m. der Röntgenröhre ‖ col m. du tube Rœntgen. / ~ fabric ‖ Schlauchware f. ‖ tricot m. tubulaire. / plaited ~ fabric ‖ plattierte Schlauchware f. ‖ tricot m. tubulaire en maille vanisée. / ~ felt ‖ Rundfilz m.; geschlossener Filz m.; Manchon m. ‖ manchon m. de feutre; feutre m. tubulaire. / ~ felt from entangled woollen goods ‖ Manchon m. aus verfilztem Wollgewebe ‖ manchon m. en tissu de laine feutré. / ~ felt stretcher ‖ Rundfilzspanner m. ‖ tendeur m. à manchon de feutre. / ~ frame ‖ Rohrrahmen m. ‖ châssis m. tubulaire. / ~ girder ‖ Rohrträger m. ‖ poutre f. en tube; poutre f. tubulaire. / ~ goods pl. (Knitting) *see* ~ fabric. / ~ guide ‖ Führungsrohr n. ‖ tube m. de guidage. / ~ heating-surface of a boiler ‖ Rohrheizfläche f. eines Kessels ‖ surface f. de chauffe tubulaire d'une chaudière. / ~ hosiery mill ‖ Rundwirkmaschine f. ‖ métier m. à tricoter circulaire; métier m. circulaire à tricot. / ~ lock machine (Textile) ‖ Schlauchschloßmaschine f. ‖ machine f. à serrure tubulaire. / ~ mount ‖ röhrenförmige Fassung f. ‖ monture f. en forme de tube. / ~ mount with triple prism ‖ Prismenrohr m. mit dreiteiligem Prisma ‖ tube m. contenant un prisme à trois verres. / ~ pole *see also* tube pole ‖ Rohrmast m. ‖ poteau m. tubulaire. / ~ shaft ‖ hohle Welle f. ‖ arbre m. creux. / ~ slide valve ‖ Rohrschieber m. ‖ valve f. tubulaire. / ~ stand (Mach) ‖ Röhrengestell n. ‖ bâti m. tubulaire. / ~ steel frame ‖ Stahlrohrgestell n. ‖ châssis m. en tubes d'acier. / ~ steel pole ‖ Stahlrohrmast m. ‖ poteau m. tubulaire en acier. / ~ strut ‖ Rohrstrebe f. ‖ montant m. tubulaire. / ~ tower *see* ~ pole.

tubulated flask (Chem) ‖ Kolben m. mit Ansatzrohr ‖ ballon m. à tubulure.

tubulure (Chem) *see* tubulus.

tubulus (Chem) ‖ Tubus m.; Tubulus m. ‖ tubulure f.

tub washer (Dyer) ‖ Küpenwäscher m. ‖ nettoyeur m. de cuves.

tuck pattern (Textile) ‖ Preßmuster n. ‖ dessin m. de presse. / ~ presser ‖ Musterpresse f.; Preßmaschine f. ‖ presse f. à guillocher.

tufa *see* tuff.

tufaceous limestone ‖ Tuffkalk m. ‖ tuf m. calcaire.

tuff ‖ Tuff m.; Tuffstein m. ‖ tuf m. / basaltic ~ ‖ Basalttuff m. ‖ tuf m. basaltique. / calcareous ~ ‖ Kalktuff m.; Kalksinter m. ‖ tuf m. calcaire. / palagonitic ~ ‖

Palagonittuff m. ‖ tuf m. palagonitique. / ~ quarry ‖ Tuffsteinbruch m. ‖ carrière f. de tuff. / ~ stone see tuff.

tuft ‖ Büschel m.; Quaste f. ‖ houppe f. / ~ of hair ‖ Haarbusch m. ‖ crinière f.

tug, to ~ (Mar) ‖ schleppen; ruckweise holen ‖ remorquer; hauspiller.

tug (Mar) ‖ Schlepper m.; Schleppschiff n.; Schleppboot n.; Bugsierboot n. ‖ remorqueur m. / chain steam ~ ‖ Kettenschleppdampfer m. ‖ toueur m. à chaîne à vapeur. / stern wheel ~ ‖ Heckradschlepper m. ‖ remorqueur m. à roue arrière. / ~ boat see tug.

tugging device for ships ‖ Schiffsschleppeinrichtung f. ‖ installation f. de touage pour bateaux.

tula metal ‖ Tulametall n.; Tulasilber n. ‖ (argent m. de) Toula m.

tulle ‖ Tüll m. ‖ tulle m. / ~ embroiderer ‖ Tüllsticker m. ‖ brodeur m. sur tulle. / ~-like net tissue ‖ tüllartiges Netzgewebe n. ‖ tissu m. à mailles façon tulle. / ~ machine ‖ Tüllmaschine f. ‖ métier m. à tulle.

tumble, to ~ (Build) ‖ einstürzen; einfallen ‖ s'écrouler; ébouler.

tumbler (Locksm) ‖ Zuhaltung f. ‖ arrêt m. / ~ (Ore dress) ‖ Wipper m. ‖ culbuteur m. ~ (Shipb) ‖ Schlipphaken m. ‖ loquet m. du mouilleur. / dredging ~ ‖ Baggertrommel f.; Kettentreibscheibe f.; Turas m. ‖ tambour m. de drague. / ~ file ‖ Ovalfeile f.; Vogelzunge f. ‖ lime f. ovale. / ~ spring (Locksm) ‖ Zuhaltungsfeder f. ‖ ressort m. d'arrêt. / ~ switch (Electr) ‖ Kippschalter m.; Wippe f. ‖ interrupteur m. tumbler; commutateur m. à bascule. / ~ toe (Locksm) ‖ Zuhaltungslappen m. ‖ levée f.

tumbling barrel ‖ Scheuerfaß n. ‖ tonneau m. de polissage. / ~ device for hoop iron ‖ Schleudervorrichtung f. für Bandeisen ‖ appareil m. centrifuge pour fabriquer des bandes de fer.

tun, to ‖ eintonnen; in Fässer npl. füllen oder einpacken ‖ entonner; embariller; enfutailler.

tun ‖ Faß n.; Tonne f.; Kübel m. ‖ baril m.; fût m.; tonneau m.; futaille f.; seau m. / ~ for a buoy ‖ Fahrwassertonne f. ‖ tonneau m. de bouée.

tune, to ~ (Radio etc.) ‖ abstimmen; abtönen ‖ syntoniser; accorder.

tune ‖ Ton m.; Laut m.; Klang m. ‖ ton m. / in ~ with ‖ in Rhythmus m. mit ‖ en rythme m. avec. / to be out of ~ ‖ verstimmt sein ‖ être désaccordé.

tuned ‖ abgestimmt ‖ syntonisé, accordé. / ~ to resonance ‖ auf Resonanz abgestimmt ‖ accordé à résonnance.

tuned buzzer (Radio) ‖ abgestimmter Summer m. ‖ vibrateur m. syntonisé. / ~ circuit ‖ abgestimmter Kreis m. ‖ circuit m. accordé. / ~ ringing (Tel) ‖ abgestimmter Anruf m. ‖ appel m. harmonique.

tuner (Person) ‖ Stimmer m. ‖ accordeur m. / instrument ~ ‖ Instrumentenstimmer m. ‖ accordeur m. d'instruments de musique. / piano ~ ‖ Klavierstimmer m. ‖ accordeur m. de piano.

tuner (Apparatus) ‖ Abstimmvorrichtung f. ‖ syntonisateur m. / multiple ~ ‖ Vielfachabstimmvorrichtung f. ‖ syntonisateur m. multiple.

tungstate of lead ‖ Scheelbleierz n.; Scheelbleispat m.; Stolzit m.; Wolfram-

bleierz n.; Scheelitin m. ‖ plomb m. tungstaté; scheelitine f. / ~ of lime ‖ Scheelit m.; Scheelerz n.; Schwerstein m.; Tungstein m. ‖ scheelin m. calcaire; scheelite f.

tungsten ‖ Wolfram n.; Tungsten n. ‖ tungstène m. / ~ arc lamp ‖ Punktlichtlampe f. ‖ lampe f. ponctuelle. / ~ electrode ‖ Wolframelektrode f. ‖ électrode f. en tungstène. / ~ filament ‖ Wolframfaden m. ‖ filament m. (au) tungstène. / ~ filament valve ‖ Wolframröhre f. ‖ lampe f. filament de tungstène. / ~ lamp ‖ Wolframlampe f. ‖ lampe f. au tungstène. / ~ mine ‖ Wolframbergwerk n. ‖ mine f. de tungstène. / ~ ore ‖ Scheelsäure f.; Wolframsäure f. ‖ acide m. tungstique ou scheelique ou wolframique; ocre f. ou peroxyde m. au tungstène. / ~ ore ‖ Wolframerz n. ‖ minerai m. de tungstène. / ~ rectifier ‖ Wolframgleichrichter m. ‖ redresseur m. au tungstène. / ~ salt ‖ Wolframsalz n. ‖ tungstate m. / ~ special steel ‖ Wolframsonderstahl m. ‖ acier m. spécial au tungstène. / ~ steel ‖ Wolframstahl m. ‖ acier m. au tungstène. / ~ steel for permanent magnets ‖ Wolframstahl m. für Dauermagnete ‖ acier m. au tungstène pour aimants permanents.

tungstic acid ‖ Scheelsäure f.; Wolframsäure f. ‖ acide m. tungstique ou scheelique ou wolframique; ocre f. ou peroxyde m. de tungstène. / ~ monohydrate see ~ acid. / ~ ochre see ~ acid. / ~ steel see tungsten steel.

tuning (Radio) ‖ Abstimmung f.; Abstimmen n. ‖ syntonisation f. / flat ~ ‖ unscharfes Abstimmen n. ‖ syntonisation f. non-aiguë. / note and wave ~ ‖ Abstimmung f. von Tonhöhe und Welle ‖ syntonisation f. de la note et de l'onde. / sharp ~ ‖ scharfe Abstimmung f. ‖ syntonisation f. aiguë.

tuning coil ‖ Abstimmspule f. ‖ bobine f. syntonisatrice. / ~ condenser ‖ Abstimmkondensator m. ‖ condensateur m. d'accord ou de syntonisation. / ~ fork ‖ Stimmgabel f. ‖ diapason m.; fourchette f. tonique. / ~ fork oscillator (Electr) ‖ Stimmgabelsummer m. ‖ oscillateur m. à diapason. / ~ lamp ‖ Abstimmlampe f. ‖ lampe f. de syntonisation. / ~ lamp and choke ‖ Syntonisierlampe f. mit Impedanz ‖ lampe f. de syntonisation avec bobine de réactance. / ~ means pl. for ~ ‖ Abstimmungsmittel n. ‖ moyen m. de syntonisation. / ~ method ‖ Abstimmungsverfahren n. ‖ procédé m. de syntonisation. / ~ precision ‖ Abstimmschärfe f.; Schärfe f. der Abstimmung f. ‖ sélectivité f.; finesse f. de syntonisation. / sharpness of ~ see ~ precision. / ~ wave ‖ Abstimmungswelle f. ‖ onde f. de syntonisation. / ~ wire (In an organ pipe) ‖ Stimmkrücke f. ‖ rasette f.

tunnel, to ‖ durchbohren; durchtunneln; einen Tunnel bauen ‖ percer un tunnel.

tunnel ‖ Tunnel m.; Durchstich m. ‖ tunnel m. / ~ (Mine) ‖ unterirdischer Kanal m.; Rösche f. ‖ canal m. souterrain. / ~ (High-furnace) ‖ Schacht m.; Seele f. ‖ cuve f. ou cheminée f. intérieure.

tunnel baking oven for bread factory ‖ Tunnelbackofen m. für Brotfertigung ‖ four m. à tunnel pour la panification. / ~ cellar ‖ Felsenkeller m. ‖ cave f.

creusée dans le roc. / ~ construction ‖ Tunnelbau m. ‖ construction f. de tunnel(s). / ~ dryer ‖ Kanaltrockner m. ‖ canal-sécheur m. / ~ entrance ‖ Tunneleingang m. ‖ tête f. de tunnel. / ~ head (Met) ‖ Gichtturm m. ‖ tour f. de chargement.

tunneller ‖ Querschlaghauer m. ‖ bacneur m.; bouveleur m.

tunnel line ‖ Tunnellinie f. ‖ ligne f. à travers un tunnel.

tunnelling ‖ Tunnelbau m. ‖ construction f. ou percement m. de tunnels. / earthshake due to ~ ‖ Tunnelbeben n. ‖ tremblement m. de terre causé par le percement d'un tunnel.

tunnel lining ‖ Tunneleinfassung f. ‖ paroi f. de tunnel. / ~ miner ‖ Tunnelbohrer m. ‖ terrassier-mineur m. / ~ mining (Mine) ‖ Stollenbergbau m. ‖ travaux mpl. de percement. / ~ pit ‖ Tunnelschacht m. ‖ puits m. de tunnel. / ~ ring for underground railways ‖ Tunnelring m. für Untergrundbahnen ‖ virole f. de tunnel pour voies souterraines. / ~ timbering ‖ Tunnelzimmerung f. ‖ boisage m. pour tunnels. / ~ tube dryer ‖ Schachttrockner m. ‖ sécheur m. à couloir.

tun room ‖ Gärraum m. ‖ cave f. de fermentation. / ~ man ‖ Gärkellerbursche m. ‖ garçon m. de cave de fermentation.

tunny ‖ Thunfisch m. ‖ thon m.

tup (Ram) ‖ Rammbär m.; Schlagwerk n. ‖ mouton m. / bent under ~ ‖ unter dem Fallwerk geschlagen ‖ essayé au mouton. / hammer ~ (Forg) ‖ Bär m.; Hammerklotz m.; Hammerbär m. ‖ mouton m. / free fall of the ~ ‖ freier Fall m. des Bären ‖ chute f. libre du mouton. / weight of ~ ‖ Bärgewicht n. ‖ poids m. du mouton. / steam hammer with a weight of ~ of x kg ‖ Dampfhammer m. mit x kg Bärgewicht ‖ marteau-pilon m. à vapeur (poids du mouton x kg).

turbid ‖ trübe ‖ trouble.

turbidity of the air ‖ Lufttrübung f. ‖ obscurcissement m. ou assombrissement m. atmosphérique; obscurcissement m. de l'air. / artificial ~ of the air ‖ künstliche Trübung f. der Luft ‖ obscurcissement m. artificiel de l'air; assombrissement m. mécanique de l'air. / albumen ~ ‖ Eiweißtrübung f. ‖ trouble m. d'albumine ou des matières azotées. / bacterial ~ ‖ Bakterientrübung f. ‖ trouble m. de bactéries. / proteid ~ see albumen ~.

turbine ‖ Turbine f. ‖ turbine f. / action ~ ‖ Gleichdruckturbine f.; Aktionsturbine f. ‖ turbine f. à action. / ahead ~ ‖ Vorwärtsturbine f. ‖ turbine f. à (ou de) marche avant. / astern ~ ‖ Rückwärtsturbine f. ‖ turbine f. à (ou de) marche arrière. / axial (flow) ~ ‖ Axialturbine f. ‖ turbine f. axiale. / compressed-air ~ ‖ Preßluftturbine f. ‖ turbine f. à air comprimé. / condensing ~ ‖ Kondensationsturbine f. ‖ turbine f. à condensation. / ~ running forward see ahead ~. / fractional supply ~ ‖ Turbine f. mit teilweiser Beaufschlagung ‖ turbine f. (à injection) partielle. / free-deviation ~ ‖ Freistrahlturbine f. ‖ turbine f. à libre déviation. / full-supply ~ ‖ Turbine f. mit voller Beaufschlagung ‖ turbine f. à injection pleine. / high-pressure ~ ‖ Hochdruckturbine f. ‖ turbine f. à haute pression. / hot-air ~ ‖ Heißluftturbine f. ‖ turbine f. à air chaud. / hydraulic ~ ‖

Wasserturbine f. || turbine f. hydraulique. / impulse ~ see action ~. / inward-flow ~ || Turbine f. mit äußerer Beaufschlagung || turbine f. à injection. / live-steam ~ || Frischdampfturbine f. || turbine f. à vapeur fraîche ou vive. / low-pressure ~ || Niederdruckturbine f. || turbine f. à basse pression. / outward-flow ~ || Turbine f. mit innerer Beaufschlagung || turbine f. Fourneyron. / pressure ~ || Druckturbine f. || turbine f. à pression. / ~ of propeller type with vertical shaft || Propellerturbine f. mit senkrechter Welle || turbine f. à hélice à arbre vertical. / radial (flow) ~ || Radialturbine f. || turbine f. radiale. / steam ~ || Dampfturbine f. || turbine f. à vapeur. / tangential-flow ~ || Tangentialturbine f. || turbine f. tangentielle. / tapped ~ || Entnahmeturbine f. || turbine f. à prise de vapeur. / two-cylinder ~ || zweigehäusige Turbine f. || turbine f. à deux corps. / water ~ || Wasserturbine f. || turbine f. hydraulique.

turbine blade || Turbinenschaufel f. || aube f. à turbine. / ~ blade rolling mill || Turbinenschaufelwalzwerk n. || laminoir m. à aubes de turbine. / ~ boat || Turbinenschiff n. || bateau m. à turbines. / ~ casing see also ~ cylinder || Turbinengehäuse n. || enveloppe f. de turbine. / ~ casing in halves || Turbinengehäuse n. in Hälften || enveloppe f. de turbine en moitiés. / ~ compass || Kreiselkompaß m. || boussole f. gyroscopique. / ~ cylinder see also ~ casing || Turbinengehäuse n. || cylindre m. de turbine. / ~ disk || Turbinenscheibe f. || disque m. de turbine. / ~ drum || Turbinentrommel f. || tambour m. de turbine. / forged conical ~ drum || konische geschmiedete Turbinentrommel f. || tambour m. de turbine forgé conique. / ~ exhaust steam || Turbinenabdampf m. || vapeur f. d'échappement de la turbine. / ~ generator see turbo generator. / ~ governor || Turbinenregler m. || régulateur m. de turbine. / ~ interrupter (Tel) || Turbinenunterbrecher m. || interrupteur m. à turbine. / ~ locomotive || Turbinenlokomotive f. || locomotive f. à turbine. / ~ locomotive boiler || Turbinenlokomotivkessel m. || chaudière f. de locomotive à turbine. / ~ pipe || Turbinenrohr n. || tuyau m. de turbine. / ~ pipe line || Turbinenrohrleitung f. || conduite f. de turbines. / ~ plant || Turbinenanlage f. || usine f. ou installation f. de turbine. / control board for ~ plant || Turbinenüberwachungstafel f. || tableau m. de contrôle d'une installation de turbines. / ~ pump || Turbinenpumpe f. || pompe-turbine f. / ~ regulator || Turbinenregler m. || régulateur m. de turbines. / ~ screw || Turbinenschraube f. || propulseur m. à turbine. / ~ shaft || Turbinenwelle f. || arbre m. de turbine. / ~ steamer || Turbinendampfer m. || vapeur m. à turbine. / ~ wheel || Turbinenrad n. || roue f. de turbine.

turbith mineral || Mineralturpeth n. || turbith m. minéral.

turbo blower || Turbogebläse n. || soufflerie f. centrifuge ou rotative; turbo-soufflante f. / ~ cannon see ~ gun. / ~ compressor || Turbokompressor m. || turbo-compresseur m. / ~ dynamo || Turbodynamo f. || turbo-dynamo f. / ~ generator || Turbogenerator m. || turbo-génératrice

f. / ~ gun || Turbokanone f.; Turbinengeschütz n. || turbo-canon m. / ~ pump || Turbopumpe f. || turbo-pompe f. / high-pressure ~ pump || Hochdruckturbopumpe f. || turbo-pompe f. à haute pression. / ~ set for Ruths steam storage operation || Turbosatz m. für Ruthsspeicherbetrieb || groupe m. turbo-générateurs à accumulateurs de chaleur Ruths.

turbulence || Wirbelbewegung f.; Durchwirbelung f.; Turbulenz f. || turbulence f.; mouvement m. tourbillonnaire. / ~ of the air || Durchwirbelung f. der Luft || turbulence f. de l'air; mouvement m. turbulent de l'air. / energy of ~ || Turbulenzenergie f. || énergie f. de turbulence.

turbulent motion see turbulence. / ~ movement see turbulence.

turf || Rasen m. || gazon m. / ~ (Peat) || Torf m. || tourbe f. / ~ earthy ~ || Erdtorf m. || tourbe f. terreuse. / grass ~ || Rasen m. || gazon m. / ~ for litter || Torfstreu f. || litière f. de tourbe.

turf cake || Backtorf m. || motte f. de tourbe. / ~ charcoal || Torfkohle f. || tourbe f. carbonisée. / ~ cutter (Plough) || Plaggenpflug m.; Rasenstecher m.; Sodenpflug m. || coupe-gazon m.; tranchegazon m. / ~ digging || Torfgewinnung f.; Torfstecherei f. || extraction f. de tourbe. / ~ drying || Torftrocknerei f. || séchage m. de tourbe. / ~ dust || Torfstreu f.; Torfmull m. || poussier m. de mottes; litière f. de tourbe. / ~ dust machine || Torfstreumaschine f. || machine f. pour la litière de tourbe. / ~ heating with ~ || Torffeuerung f. || combustion f. de la tourbe. / ~ machine || Torfmaschine f. || machine f. pour tourbe. / ~ oil || Torföl n. || huile f. de tourbe. / ~ pit || Torfstich m. || tourbière f. / ~ plough see ~ cutter. / ~ tar || Torfteer n. || goudron m. de tourbe.

Turkey carpet || türkischer Teppich m. || tapis m. de Turquie ou à nœuds.

Turkey coffee || türkischer Kaffee m.; Mokka m. || café m. de Moka.

Turkey-red || Türkischrot n.; Merinorot n.; Oxydrot n. || rouge m. des Indes; rouge m. turc; rouge m. oxyde. / ~ dyeing || Türkischrotfärberei f. || teinture f. en rouge turc. / dyeing of cotton-yarn with ~ || Krappen n. der Baumwolle || garançage m. du coton. / ~ oil || Türkischrotöl n. || huile f. pour rouge turc; huile f. rouge d'Andrinople.

turmeric (Chem) || Kurkuma f. || curcuma m. / ~ paper || gelbes Reagenspapier n.; Kurkumapapier n. || papier m. de curcuma. / ~ plant || Gelbwurzel f.; Kurkuma f. || safran m. ou souchet m. des Indes; curcuma m.

turn, to (Glassm) || marbeln; rollen || marbrer. / ~ (Pott) || abdrehen || tournasser. / ~ (Turn) || drehen; drechseln || tourner. / ~ (To rotate) || sich drehen; rotieren || pivoter; tourner. / ~ (Balance: to incline) || ausschlagen || pencher; pousser. / ~ back || hinterdrehen || détalonner; façonner à profil invariable. / ~ backwards (Auto) || zurückkurbeln || tourner en arrière. / ~ a carriage || ein Fuhrwerk n. lenken || tourner ou braquer une voiture. / ~ on circulation || in Umlauf m. bringen || mettre en circulation. / ~ the crank to the starting position || die Kurbel f. in die Anlaßstellung einstellen ||

placer la manivelle dans la position de mise en marche. / ~ a curve (Auto) || eine Kurve f. fahren || virer. / ~ hollow (Turn) || ausdrehen || aléser au tour. / ~ letters (Print) || blockieren || bloquer. / ~ off (Electr) || abdrehen || déclencher. / ~ off (Mine) || abkehren || se désister. / ~ off (Railw) || eine Weiche f. stellen || aiguiller. / ~ off railway tyres || Radkränze mpl. abdrehen || rafraîchir les bandages mpl. de roues. / ~ on (Electr) || aufdrehen; andrehen || ouvrir; tourner. / ~ on the blast (Mach) || das Gebläse anlassen || donner le vent. / ~ out (Brew) || ausstoßen || débiter; vendre. / ~ out (Electr) || ausschalten; ausdrehen || tourner le bouton. / ~ out (Turn) || ausdrechseln || creuser au tour. / ~ the outside of a gun || ein Geschützrohr außen abdrehen || tourner la surface extérieure d'une bouche à feu. / ~ over || umwenden || retourner. / ~ upon a pivot || sich auf einem Zapfen m. drehen || pivoter. / ~ a rim at the back surface || auf der Rückseite einen Rand m. andrehen || tourner un bord sur la face arrière. / ~ with a shovel || umschaufeln || pelleter. / ~ with the template || mit der Schablone abdrehen || trousser. / ~ through x⁰ || um x⁰ drehen || tourner de x⁰. / ~ up (Carp) || kanten; umkanten || rouler sur la carne; cabaner; renverser. / ~ up a hat || einen Hut aufkrämpen || retrousser ou retaper un chapeau. / ~ up the paper (Print) || das Papier umschlagen || remanier le papier. / ~ (rails) upside down || (Schienen) umkehren oder umlegen || renverser ou retourner (les rails). / ~ wood || Holz n. drechseln || tourner le bois.

turn (Revolution) || Umdrehung f.; Umlauf m. || tour m.; rotation f.; revolution f. / ~ (Winding) || Windung f. || tour m.; spire f.; tour m. de spire. / ~ (Twist of a cable etc.) || Drall m. || tors m. / ~ (In the road) || Kehre f.; Wendung f.; Kurve f. || virage m. / best ~ || wendigste Kurve f. || virage m. dans les meilleures conditions. / ~ of a carriage || Lenkbarkeit f. eines Fuhrwerks || tournant m. ou rayonnement m. des voitures. / ~ of the market || Konjunktur f. || conjoncture f. / ~ of the scale || Ausschlag m. der Wage || trait m. ou don m. de balance. / ~ of the tide || Gezeitenwechsel m. || changement m. des marées.

turnable || drehbar || mobile autour d'un axe; tournant sur un tourillon.

turn bench see also (turning) lathe || Drehbank f. || tour m. / ~ with faceplate (Watchm) || Scheibendrehstuhl m. || tour m. à plateau.

turn bridge (Bridge) || Drehbrücke f. || pont m. tournant. / ~ (Railw) || Drehscheibe f. || pont m. tournant.

turnbuckle || Spannschloß n.; Spannmutter f.; Spannschraube f.; Schraubenschloß n. || manchon m. de serrage; écroutendeur m.; tendeur m.; écrou m. de réglage; vis f. de tension. / ~ of closed pattern || Spannschloß n. von geschlossener Form || tendeur m. fermé. / ~ of open pattern || Spannschloß n. von offener Form || tendeur m. à lanterne. / ~ of a window || Fensterwirbel m. || tourniquet m.; targette f. tournante; birloir m.

turn button || Drehverschluß m.; (Fenster-)Wirbel m. || tourniquet m.; targette f. tournante; birloir m.

turned ‖ gedrechselt; gedreht ‖ tourné. / profile-~ ‖ formgedreht; fassongedreht ‖ façonné ou décolleté au tour. / profile-~ piece ‖ Formdrehteil m. ‖ pièce f. façonnée au tour.

turned goods pl. of wood ‖ Drechslerwaren fpl.; Drechslerarbeiten fpl. ‖ articles mpl. en bois tourné. / ~ wood ‖ gedrechseltes Holz n. ‖ bois m. tourné. / artistic ~ work ‖ Kunstdrechslerarbeit f. ‖ travail m. d'art tourné.

turned-off ‖ abgedreht ‖ renvoyé; fermé.

turned-on ‖ angedreht ‖ tourné; dirigé.

turned-out ‖ ausgeräumt ‖ mis dehors; déblayé.

turner (Metal) ‖ Dreher m. ‖ tourneur m. / ~ (Wood) ‖ Drechsler m. ‖ tourneur m. / amber ~ ‖ Bernsteindrechsler m. ‖ tourneur m. sur ambre. / bone ~ ‖ Knochendrechsler m. ‖ tourneur m. sur os. / button ~ ‖ Knopfdreher m. ‖ tourneur m. de boutons. / fancy goods ~ ‖ Kunstdrechsler m. ‖ tourneur m. en tabletterie. / horn ~ ‖ Horndrechsler m. ‖ tourneur m. sur corne. / iron ~ ‖ Eisendreher m. ‖ tourneur m. en fer. / ivory ~ ‖ Elfenbeindrechsler m. ‖ tourneur m. sur ivoire. / job ~ ‖ Fassondreher m. ‖ tourneur m. à façon; décolleteur m. / metal ~ ‖ Metalldreher m. ‖ tourneur m. sur métaux. / pattern ~ ‖ Modelldrechsler m. ‖ modeleur-tourneur m. / tortoise shell ~ ‖ Schildpattdrechsler m. ‖ tourneur m. sur écaille. / wood ~ ‖ Holzdrechsler m. ‖ tourneur m. sur bois. / wooden pipe ~ ‖ Pfeifendrechsler m. ‖ tourneur m. de pipes.

turner-over (Glue fabrication) ‖ Leimumwender m. ‖ retourneur m. de colle forte.

turner's bit-pointed nose ‖ Drechslerspitzbohrer m. ‖ mèche f. pour tourneur langue-de-vipère. / ~ chisel ‖ Drehstahl m.; Abdrehstahl m.; Schrotstahl m. ‖ couteau m. ou ciseau m. du tourneur; gouge f. / ~ shop ‖ Dreherei f. ‖ atelier m. de tours. / ~ tool see ~ chisel.

turnery (Shop) ‖ Drechslerei f. ‖ tournerie f. / ~ (Goods) ‖ Drechslerwaren fpl. ‖ articles mpl. de tournerie. / ~ of hard caoutchouc ‖ Kunstdrechslerwaren fpl. aus Hartkautschuk ‖ tabletterie f. de caoutchouc durci.

turnery machine ‖ Drechslereimaschine f. ‖ machine f. de tournerie.

turning ‖ drehend; drehbar ‖ tournant; mobile. / ~ on a pivot ‖ um einen Zapfen m. drehbar ‖ pivotant ou mobile autour d'un tourillon.

turning ‖ Drechseln n.; Drehen n. ‖ tournage m. / ~ (Pott) ‖ Abdrehen n. ‖ tournassage m. / ~ (Of a road) see turn (In the road). / ~ (Turning chip) see also turnings pl. ‖ Drehspan m. ‖ copeau m. / long ~ ‖ Langdrehen n. ‖ chariotage m. / metal ~ ‖ Metalldreherei f. ‖ tournage m. sur métaux. / spherical ~ ‖ Balligdrehen n.; Kugeldrehen n. ‖ tournage m. de sphères.

turning arbor ‖ Drehstift m. ‖ arbre m. tournant. / ~ with disk ‖ Scheibendrehstift m.; Drehstift m. mit Kittscheibe ‖ arbre m. tournant à cire. / ~ of a hammer shaft ‖ Hammerwelle f. ‖ arbre m. moteur d'un marteau de forge.

turning bar ‖ Wendestange f. ‖ tringle f. de retournement. / ~ barrier with counterpoise ‖ Drehschraube f. mit Gegengewicht ‖ barrière f. à lisse pivotante. /

~ capacity ‖ Drehbereich m. ‖ capacité f. de tournage. / ~ change-over switch (Electr) ‖ Drehumschalter m. ‖ commutateur-permutateur m. tournant. / ~ chisel ‖ Plattmeißel m.; Schlichtmeißel m.; Drehmeißel m. ‖ ciseau m. à planer; plane f. / ~ device (Steam eng etc.) ‖ Andrehvorrichtung f. ‖ appareil m. de démarrage. / ~ device for the commutator ‖ Abdrehvorrichtung f. für den Kommutator ‖ appareil m. à rafraîchir le collecteur. / ~ door ‖ Drehtür f. ‖ porte f. tournante. / ~ drum ‖ Wendetrommel f. ‖ tambour m. de retournement. / ~ furnace ‖ Drehofen m. ‖ fourneau m. ou four m. tournant. / ~ gouge ‖ Drechslerrohr n.; Hohlmeißel m.; Rohrmeißel m.; Schrotmeißel m. der Drechsler ‖ gouge f. du tourneur; gouge f. à ébaucher. / ~ graver ‖ Drehstichel m.; Grabstichel m. ‖ burin m. du tourneur. / ~ handle ‖ Drehbaum m.; Drehgriff m. ‖ levier m. de manœuvre. / ~ hook ‖ Kettelnadel f. ‖ aiguille f. à manche. / ~ joint ‖ Gelenk n.; Scharnier n. ‖ articulation f.; jointure f.; charnière f.

turning lathe see also lathe ‖ Drehbank f. ‖ tour m. / ~ for chuck jobs ‖ Drehbank f. für Futterarbeiten ‖ tour m. à manchon ou à decolleter. / ~ of x mm height of centre ‖ Drehbank f. von x mm Spitzenhöhe ‖ tour m. de x mm de hauteur de poupée. / oval ~ ‖ Ovaldrehwerk n. ‖ mécanisme m. à tourner en ovale. / ~ for rod jobs ‖ Drehbank f. für Stangenarbeiten ‖ tour m. pour travaux dans la barre.

turning lever with collar (Screw coupling) ‖ Schwengel m. mit Schwengelbund ‖ tige f. de la vis de tendeur avec collet rapporté. / ~ lock door ‖ Drehtor n. der Schleuse ‖ porte f. d'écluse tournante. / ~ machine see also turnery machine ‖ Drehereimaschine f. ‖ machine f. de tournerie. / ~ and boring machine with horizontal face plate ‖ Drehwerk n. und Bohrwerk n. mit wagerechter Planscheibe ‖ tour m. et aléseuse f. à plateau horizontal. / vertical boring and ~ machine ‖ Karuselldrehbank f. ‖ tour m. vertical; tour m. à plateau horizontal. / ~, boring and cutting-off machine ‖ Dreh-, Bohr- und Abstechbank f. ‖ tour m. à charioter, percer et tronçonner. / ~ moment of the crank ‖ Kurbeldrehmoment n. ‖ moment m. de rotation de la manivelle.

turning-out per day ‖ Tagesleistung f. ‖ capacité f. de production journalière; débit m. journalier.

turning platform (Railw) see also turntable ‖ Drehscheibe f. ‖ plaque f. tournante; plateforme f. tournante.

turning point ‖ Wendepunkt m. ‖ point m. critique. / ~ of colour ‖ Farbenumschlag m. ‖ virage m. de couleur. / ~s pl. (Railw) ‖ Drehweiche f. ‖ aiguille f. pivotante.

turnings pl. ‖ Drehspäne mpl. ‖ copeaux mpl.; tournure f.

turning saw ‖ Schweifsäge f.; Stellsäge f.; Zuschneidesäge f.; Örtersäge f. ‖ scie f. à débiter ou à tourner. / ~ shackle padlock ‖ Vorhängeschloß n. mit Drehbügel ‖ cadenas m. à anse tournante. / ~ shop ‖ Drehereiwerkstatt f. ‖ atelier m. de tours. / ~ socket ‖ Drehhülse f. ‖ douille f. du levier. / ~ square ‖ Tiefenmaß n.; Lochwinkel m.; Schubwinkel m. ‖ équerre f. coulante ou à coulisse. / ~ suspension

~ shoe ‖ Hängedrehschuh m. ‖ sabot m. de suspension pivotant. / ~ thread (Weav) ‖ Schlingfaden m.; Polfaden m.; Dreherfaden m. ‖ fil m. de tour. / ~ tip stool (Railw) ‖ Drehkippschemel m. ‖ banquette f. giratoire à bascule. / ~ tool see also tool (Turn) ‖ Drehmeißel m.; Drehstahl m. ‖ outil m. à tourner ou de tour; acier m. de tour. / ~ tool of short overhang ‖ kurz gespannter Drehstahl m. ‖ outil m. de tour serré court. / ~ tools pl. ‖ Drehzeug n.; Drechslerwerkzeug n. ‖ outils mpl. du tourneur. / ~ tub for plash-mortar ‖ Mörtelmühle f. ‖ tonneau m. mélangeur à mortier. / ~ work ‖ Dreharbeit f.; Drechslerarbeit f. ‖ travail m. de tournage. / ~ work with rotating tools ‖ Dreharbeit f. mit umlaufenden oder rotierenden Werkzeugen ‖ travail m. de tournage au moyen d'outils rotatifs.

turnip ‖ Rübe f.; weiße Rübe f.; Runkelrübe f.; Steckrübe f. ‖ rave f.; navet m.; betterave f.; turnep m. / ~ cutter ‖ Rübenschneider m. ‖ coupoir m. à betteraves; coupe-navets m. / ~ digging machine see ~ gathering machine. / ~ gathering machine ‖ Rübenerntemaschine f. ‖ machine f. à récolter les betteraves; arracheur m. de turneps. / ~ lifter ‖ Rübenheber m. ‖ arracheur m. de betteraves. / ~ mill ‖ Rübenmühle f. ‖ moulin m. à betteraves. / ~ planting machine ‖ Rübenpflanzmaschine f. ‖ planteur m. mécanique de raves. / ~ rooter ‖ Rübenroder m. ‖ essarteuse f. de betteraves. / ~ shredding machine ‖ Rübenschnitzelmaschine f. ‖ machine f. à couper les betteraves en tranches. / ~ topping machine ‖ Rübenköpfer m. ‖ écimeur m. de betteraves. / ~ tops plucking machine ‖ Rübenblätterzerreißmaschine f. ‖ machine f. à déchiqueter les feuilles de betteraves. / ~ top washery ‖ Wäsche f. für Rübenblätter ‖ laveur m. pour feuilles de betterave. / ~ washer ‖ Rübenwäsche f. ‖ laveur m. (mécanique) pour betteraves.

turn mould (Found) ‖ Stürzform f. ‖ moule m. à renverser.

turnoff (Railw) see also turnout ‖ Abzweigung f. ‖ branchement m.

turnout (Railw) see also turnoff ‖ Ausweichgleisanlage f. ‖ voie f. de branchement m. / ~ in a canal ‖ Ausweiche f. in einem Kanal ‖ élargissement m. d'un canal. / left-hand ~ ‖ Linksweiche f. ‖ branchement m. à gauche. / right-hand ~ ‖ Rechtsweiche f. ‖ branchement m. à droite.

turnout track ‖ Ausweichegleis n. ‖ voie f. d'évitement.

turnover (Of money) ‖ Geldumsatz m.; Geldverkehr m. ‖ mouvement m. ou maniement m. de fonds; roulement m. ou mouvement m. ou circulation f. monétaire. / ~ top table (Mach tool) ‖ Wendetisch m. ‖ table f. renversable.

turnpike, to ~ (A road) ‖ beschottern ‖ empierrer; macadamiser.

turnpike (Barrier) ‖ Zollschranke f.; Wegeschranke f. ‖ barrière f. de péage. / ~ engineering enterprise ‖ Straßenbauunternehmung f. ‖ entreprise f. de construction de routes. / ~ money ‖ Wegesteuer f.; péage m. / ~ road ‖ Kunststraße f.; (große) Landstraße f.; Chaussee f. ‖ chaussée f.; grande route f.; route f. nationale.

turn-plate (Railw) *see also* turntable ‖ Wendeplatte f.; Drehplatte f. ‖ plaque f. tournante *ou* de manœuvre.

turn-screw ‖ Schraubenzieher m. ‖ tournevis m. / ~ gas pliers pl. ‖ Gasrohrzange f. mit Schraubenzieher ‖ pince f. à gaz à tournevis.

turnsick (Of sheep) ‖ Drehkrankheit f. ‖ tournis m.

turnspit ‖ Bratenwender m.; Bratspießdrehvorrichtung f. ‖ tourne-broche m.

turntable (Railw) ‖ Drehscheibe f. ‖ plaque f. tournante. / (Mach tool) ‖ Schwenktisch m. ‖ table f. pivotante. / the ~ is covered by a double timber planking ‖ die Drehscheibe ist durch einen doppelten Bohlenbelag abgedeckt ‖ la plaque tournante est recouverte d'une couche double en madriers. / articulated ~ ‖ Gelenkdrehscheibe f. ‖ plaque f. tournante à articulation. / ball-bearing ~ ‖ Kugeldrehscheibe f. ‖ plaque f. tournante à billes. / climbing ~ ‖ Kletterdrehscheibe f. ‖ plaque f. tournante en saillie. / ~ with cross guide ledges ‖ Drehscheibe f. mit Kreuzspurleisten ‖ plaque f. tournante avec rails-guides entretoisés. /~ with flush rails ‖ Drehscheibe f. mit eingegossener Kreuzspur ‖ plaque f. tournante avec ornières entrecroisées de fonte. / locomotive ~ ‖ Lokomotivdrehscheibe f. ‖ plaque f. tournante pour locomotives. / ~ with plain disk ‖ Drehscheibe f. mit glattem Teller ‖ plaque f. tournante à disque lisse. / ~ on rollers ‖ Rollendrehscheibe f. ‖ plaque f. tournante sur galets. / two-way ~ ‖ Kreuzdrehscheibe f. ‖ plaque f. tournante pour croisement.

turntable base plate ‖ Grundplatte f. der Drehscheibe ‖ plaque f. de base de la plaque tournante. / central pivot of the ~ ‖ Drehzapfen m. der Drehscheibe ‖ pivot m. central de la plaque tournante. / ~ press ‖ Revolverpresse f. ‖ presse f. révolver.

turn window ‖ Drehfenster n. ‖ fenêtre f. tournante.

turpentine ‖ Terpentin n. ‖ térébenthine f. / crude ~ ‖ Jungfernharz n. ‖ gème f.; gomme f. molle; résine-vierge f. / essence of ~ *see* ~ oil. / ~ oil ‖ Terpentinöl n.; Terpentingeist m. ‖ essence f. de térébenthine *ou* de térébinthine. / adulterated ~ oil ‖ Terpentinölersatz m. ‖ essence f. de térébenthine artificielle. / ~ oil varnish ‖ Terpentinölfirnis m. ‖ vernis m. (à l'essence) de térébenthine. / ~ spirits pl. *see* ~ oil.

turpeth (mineral) ‖ Mineralturpeth n. ‖ turbith m. minéral.

turps *see* turpentine oil.

turret (Arch) ‖ Türmchen n.; kleiner Turm m. ‖ tourelle f. / ~ (Mach tool) ‖ Revolverkopf m. ‖ tourelle f. revolver; revolver m.; touret m. / gun ~ ‖ Geschützturm m.; (drehbarer) Panzerturm m. ‖ tourelle f. de canons; tourelle f. cuirassée (tournante).

turret clock ‖ Turmuhr f. ‖ horloge f. de clocher *ou* de tour. / ~-decker *see* ~-deck steamer. / ~-deck steamer ‖ Turmdeckdampfer m. ‖ vapeur m. à turret-deck. / length of ~ feed ‖ Revolverkopfdrehlänge f. / course f. de chariotage de la tourelle. / ~ head *see* turret (Mach tool). / ~ lathe ‖ Revolverdrehbank f. ‖ tour m. à revolver *ou* à tourelle. / single purpose ~ lathe ‖ Revolverdrehbank f. für Son-

derzwecke ‖ tour m. revolver spécial. / ~ slide (Mach tool) ‖ Revolverkopfschlitten m. ‖ chariot m. porte-tourelle. / turning engine (Shipb) ‖ Turmdrehmaschine f. ‖ machine f. pour tourner une tourelle.

turtle ‖ Schildkröte f.; Seeschildkröte f. ‖ tortue f. / ~ meat ‖ Schildkrötenfleisch n. ‖ viande f. de tortue. / ~ shell *see also* tortoise shell ‖ Schildkrötenschale f.; Schildpatt n. ‖ écaille f. de tortue.

tut bargain *see* tutwork.

tutia *see* tutty.

tutty (Met) ‖ Ofenbruch m.; Gichtschwamm m.; Tutia f. ‖ cadmie f.; calamine f. de fourneau; tutie f.

tutwork (Mining) ‖ Gedinge n.; Akkordarbeit f. ‖ travail m. à la tâche *ou* à forfait.

tuyère (Forg) (Met) *see also* twyer ‖ Windform f. ‖ tuyère f. / closed ~ ‖ geschlossene Windform f. ‖ tuyère f. fermée. / open ~ ‖ offene Windform f. ‖ tuyère f. ouverte. / eye of the ~ *see* tuyère hole.

tuyère hole ‖ Öffnung f. der Windform; Formauge n. ‖ bouche f. *ou* œil m. de la tuyère.

tweel, to ~ *see* to twill.

tweel (Weav) *see* twill.

'tween deck (Shipb) ‖ Zwischendeck n. ‖ entrepont m.

tweezers pl. ‖ Pinzette f.; Federzange f.; Haarzange f. ‖ pince f.; pincette f. / ~ (Watchm) ‖ Spiralzange f. ‖ brucelles fpl. / weaver's ~ ‖ Noppzange f.; Weberzange f.; Klüppchen n. ‖ pincette f. du tisserand.

twibill ‖ Stichaxt f.; Queraxt f.; Zwergaxt f. ‖ bisaiguë f.; tire-boucher m.

twig ‖ Zweig m. ‖ rameau m. / ~s pl. ‖ Reiser npl.; Reisig n. ‖ ramilles fpl.

twig gatherer ‖ Reisigsammler m. ‖ ramasseur m. de ramilles.

twilight ‖ Zwielicht n.; Dämmerung f. ‖ demi-jour m.; crépuscule m. / ~ before sun-rise ‖ Morgendämmerung f. ‖ crépuscule m. du matin. / ~ after sunset ‖ Abenddämmerung f. ‖ crépuscule m. du soir.

twill, to ~ (Weav) ‖ köpern; kepern ‖ croiser (une étoffe).

twill ‖ Köper m. (von Baumwolle) ‖ croisé m. / cotton ~ ‖ Baumwollköper m. ‖ croisé m. coton; sergé m. coton. / silk ~ ‖ Seidenserge f. ‖ sergé m. soie.

twilled ‖ geköpert ‖ croisé (à grains d'orge).

twin-armature rolling mill motor ‖ Doppelankerwalzmotor m. ‖ moteur m. de laminoir à double induit. / ~ calender (for glazing paper) ‖ Doppelkalander m.; calandre-jumelle m. (pour papier en feuilles). / ~ compressor ‖ Zwillingsverdichter m.; Zwillingskompressor m. ‖ compresseur m. duplex. / ~ cylinder (Mot) ‖ Doppelzylinder m. ‖ cylindre m. double. / ~-cylinder drying machine ‖ Zweiwalzentrockner m. ‖ séchoir m. à cylindres conjugués. / ~-cylinder engine with triple bearing ‖ Dreilagerzwillingsmaschine f. ‖ moteur m. jumelé à trois paliers. / ~ drive ‖ Zwillingsantrieb m. ‖ commande f. symétrique double.

twine, to ~ (Spinn) ‖ zwirnen ‖ retordre. / ~ a rope ‖ ein Tau schlagen ‖ commettre un cordage.

twine (Pack thread) ‖ Bindfaden m.; Kordel f.; Schnur f. ‖ ficelle f. / ~ (Twisted yarn) ‖ gezwirntes Garn n.; Zwirn m. ‖

fil m. retors; retors m. / Dutch ~ (Mar) ‖ Nähgarn n.; Segelgarn n. ‖ fil m. à coudre; ‖ Papierbindfaden m.; Papiergarn n. ‖ ficelle f. en papier. / roping ~ (Mar) ‖ Liekgarn n. ‖ fil m. à ralingue. / sail ~ (Mar) ‖ Segelgarn n. ‖ fil m. à voile. / seaming ~ *see* sail ~ ‖ (Mar) ‖ Takelgarn n. ‖ fil m. goudronné.

twined ‖ gedreht; gezwirnt ‖ câblé (en trois, en quatre etc.).

twine holder ‖ Bindfadenkorb m.; Schnurhalter m. ‖ dévidoir m. *ou* reteneur m. pour ficelle. / ~ knife (Of a harvester and binding machine) ‖ Fadenmesser n. ‖ couteau m. à ficelle. / ~ machine ‖ Zwirnmaschine f. ‖ retordoir m. / ~ manufacture ‖ Schnurfabrik f. ‖ ficellerie f.

twin engine ‖ Zwillingsmaschine f. ‖ machine f. jumelle.

twin-engined ‖ zweimotorig ‖ bimoteur m.

twin grinder ‖ Zwillingsschleifer m. ‖ défibreur m. jumeau.

twining (Spinn) ‖ Zwirnen n. ‖ retordage m. / ~ mule ‖ Mulezwirnmaschine f. ‖ mull-jenny f. à retordre; (métier m.) renvideur m. à retordre.

twinkle, to ‖ funkeln ‖ scintiller; étinceler; briller.

twin pump ‖ Zwillingspumpe f. ‖ pompe f. jumelle.

twin-ram impact testing machine for continuous work ‖ Zwillingsdauerschlagwerk n. ‖ machine f. jumelle pour essais à chocs répétés.

twin screws pl. ‖ Zwillingsschraube f. ‖ hélices fpl. jumelles.

twin-screw steamship ‖ Doppelschraubendampfer m. ‖ vapeur m. à double hélice.

twin steam engine ‖ Zwillingsdampfmaschine f. ‖ machine f. à vapeur jumelle.

twin tyre ‖ Zwillingsluftreifen m. ‖ pneumatique m. jumelé.

twin wire (Tel) ‖ Doppelader f. ‖ fil m. double.

twirl ‖ Wirbel m. ‖ tournoiement m.; tourbillon m.; tourbillonnement m.

twist, to ‖ zusammendrehen; verwinden; winden; flechten; zwirnen ‖ tordre (ensemble); natter. / ~ (Cable) ‖ verdrillen ‖ torsader. / ~ a rope ‖ ein Tau schlagen ‖ commettre un cordage. / twisted while cold ‖ kalt verwunden ‖ tordu à froid.

twist (Torsion of yarn) ‖ Drall m.; Draht m. ‖ torsion f.; tors m. / ~ (Spiral of the grooves in a gun) ‖ Drall m. ‖ pas m. des rayures. / ~ of a cable ‖ Drall m. eines Kabels ‖ tors m. d'un câble. / steel rope with moderate ~ ‖ drallarmes Stahlseil n. ‖ câble m. souple en acier à faible torsion. / right-handed ~ (Gun) ‖ Rechtsdrall m. ‖ pas m. allant à droite.

twist (Cotton yarn) ‖ Twist m.; Baumwollgarn n. ‖ coton m. filé; fil m. de coton. / Scotch ~ ‖ Flor m. ‖ fil m. d'Écosse.

twist belt ‖ geflochtener Riemen m. ‖ courroie f. torse. / ~ button ‖ Stoffknopf m.; bouton m. d'étoffe *ou* de lingerie. / ~ clamp ‖ Würgeklemme f.; Würgezange f. ‖

twist drill ‖ Spiralbohrer m. ‖ foret m. hélicoïdale *ou* en spirale; mèche f. hélicoïdale. / ~ grinder (Machine) ‖ Spiralbohrerschleifmaschine f. ‖ ma-

chine f. à affûter les forets hélicoïdaux. / automatic ~ grinder ‖ Spiralbohrerschärfautomat m. ‖ machine f. automatique à affûter les forets hélicoïdaux. / ~ milling machine ‖ Spiralbohrerfräsmaschine f. ‖ machine f. à fraiser les forets hélicoïdaux. / machine for the production of ~s ‖ Maschine f. zur Herstellung von Spiralbohrern ‖ machine f. pour la fabrication des forets hélicoïdaux. / ~ relief grinding machine ‖ Spiralbohrerhinterschleifmaschine f. ‖ machine f. à dépouiller les forets hélicoïdaux à la meule. / ~ relief milling machine ‖ Spiralbohrerhinterfräsmaschine f. ‖ machine f. à dépouiller à la fraise les forets hélicoïdaux. / ~ square shank ‖ Schneckenbohrer m. mit Vierkantkopf ‖ mèche f. spirale à tête carrée.

twisted filament (Of artificial silk) ‖ gezwirnter Faden m. ‖ fil m. organsiné. / to put the ~ filament into hank form ‖ den gezwirnten Faden m. in Strangform überführen ‖ former des écheveaux mpl. du fil organsiné. / ~ fringe ‖ gedrehte Franse f. ‖ frange f. tordue. / ~ growth (Of a tree) ‖ Drehwuchs m. ‖ croissance f. *ou* contournement m. en spirale. / ~ rope ‖ geschlagene Leine f. ‖ corde f. en fils tordus; corde f. à torons. / ~ silk ‖ kordonnierte Seide f. ‖ cordonnet m. / ~ wire ‖ verseilter Draht m. ‖ fil m. câblé.

twister (Ropem) ‖ Seildreher m. ‖ commetteur m. / ~ (Spinn) ‖ Garnzwirner m. ‖ retordeur m. de fil. / warp-yarn ~ ‖ Kettengarnzwirner m. ‖ tordeur m. de chaînes.

twister-in (Weav) ‖ Kettenabdreher m. ‖ noueur m. de chaînes.

twisting (Spinn) ‖ Zwirnen n. ‖ retordage m. / ~ (Of a wire) ‖ Verseilung f. ‖ câblage m. / ~ (Of a cable) ‖ Verdrillung f.; Verdrallung f. ‖ toronnage m. / to fix the ~ ‖ die Zwirnung f. fixieren ‖ fixer le tors. / ~ of chains (Weav) ‖ Andrehen n. *oder* Anknoten n. der Ketten ‖ nouage m. de chaînes. / ~ of the grab (Dredger) ‖ Drehen n. des Greifers ‖ contournage du grappin.

twisting frame (Spinn) ‖ Zwirnmaschine f. ‖ tordoir m.; métier m. à retordre.

twisting machine (Cable) ‖ Verseilmaschine f. ‖ machine f. à câbler; toronneuse f. / ~ (Spinn) ‖ Zopfdrehmaschine f. ‖ machine f. à tortiller. / bar iron ~ ‖ Stabeisenverwindemaschine f. ‖ machine f. à torsader les fers en barres.

twisting mill *see* twisting frame.

twisting moment ‖ Drehmoment n. ‖ moment m. de torsion.

twist joint ‖ Wickellötstelle f. ‖ ligaturesoudure f. / length of ~ (Gun) ‖ Drallänge f. ‖ longueur f. de pas. / ~ machine ‖ Twistmaschine f. ‖ machine f. à doubler le fil. / ~ mill (Spinn) ‖ Zwirnerei f. ‖ retorderie f. / number of ~s ‖ Zahl f. der Verwindungen ‖ nombre m. de torsions. / ~ test ‖ Verwindungsversuch m.; Torsionsversuch m. ‖ essai m. de torsion. / ~ tobacco ‖ Kautabak m. ‖ tabac m. à chiquer.

two-beaked anvil (Locksm) ‖ Bankhorn n. ‖ bigorneau m.

two-bladed mincing knife ‖ Doppelschneidewiegemesser n.; Zweischneidewiegemesser n. ‖ hachoir m. à deux lames.

two-coat work (Mas) ‖ Putz m. aus zwei Lagen ‖ enduit m. en deux couches.

two-coloured ribbon (Typewr) ‖ Zweifarbenband n. ‖ ruban m. encreur bicolore. / ~ writing device ‖ Zweifarbenschreibeinrichtung f. ‖ dispositif m. à ruban encreur bicolore.

two-colour printing ‖ Zweifarbendruck m. ‖ impression f. bicolore.

two-column drilling machine ‖ Doppelsäulenbohrmaschine f. ‖ perceuse f. à deux montants.

two-cycle engine *see* ~ motor. / ~ motor ‖ Zweitaktmotor m. ‖ moteur m. à deux temps. / solid injection ~ motor ‖ kompressorloser Zweitaktmotor m. ‖ moteur m. à deux temps sans compresseur. / ~ reversible motor ‖ umsteuerbarer Zweitaktmotor m. ‖ moteur m. réversible à deux temps.

two-decker (Shipb) ‖ Zweidecker m. ‖ vaisseau m. à deux ponts.

two-floored (malt) kiln ‖ Doppel(horden)darre f.; doppelhordige Darre f. ‖ touraille f. à deux plateaux.

two-high mill (Roll mill) ‖ Zweiwalzenstraße f. ‖ train m. duo; duo m. / ~ reversing mill ‖ Reversierduostraße f. ‖ duo m. réversible.

two-hole range ‖ Zweilochherd m. ‖ cuisinière f. à deux trous.

two-horse draught ‖ zweispänniger Pferdezug m. ‖ attelage m. à deux chevaux.

twohundredweight ‖ Doppelzentner m. ‖ quintal m. métrique.

two-lens chromatic (Opt) ‖ zweiteiliger Chromat m. ‖ chromate m. à deux lentilles. / ~ condenser ‖ zweilinsiger Kondensator m. ‖ condensateur m. à deux lentilles. / ~ objective ‖ zweiteiliges Objektiv n. ‖ objectif m. à deux lentilles.

two-line service (Tel) ‖ Zweileitungsbetrieb m. ‖ transmission f. à deux lignes.

two-part ring ‖ zweiteiliger Ring m. ‖ bague f. en deux pièces.

two-phase alternator ‖ Zweiphasenwechselstromgenerator m. ‖ alternateur m. diphasé. / ~ current motor ‖ Zweiphasenstrommotor m. ‖ moteur m. biphasé. / ~ generator *see* ~ alternator.

two-point measuring system ‖ Zweipunktmeßverfahren n. ‖ système m. de mesure avec deux points de contact. / to apply the ~ ‖ nach dem Zweipunktmeßverfahren n. arbeiten ‖ travailler avec deux points de contact.

two-prism spectrograph ‖ Zweiprismenspektrograf m. ‖ spectrographe m. à deux prismes.

two-rate meter (Electr) ‖ Doppeltarifzähler m. ‖ compteur m. à double tarif.

two-reel rotary printing machine ‖ Zweirollenrotationsdruckmaschine f. ‖ machine f. à imprimer rotative à deux bobines.

two-revolution press (Print) ‖ Zweitourenschnellpresse f. ‖ machine f. à imprimer à deux tours.

two-rowed barley ‖ zweizeilige Gerste f. ‖ orge f. à deux rangs.

two-sashed window ‖ zweiflügeliges Fenster n. ‖ fenêtre f. à deux battants.

two-skin work (Mas) ‖ Putz m. aus zwei Lagen ‖ enduit m. en deux couches.

two-span (Arch) ‖ zweischiffig ‖ à double travée.

two-spindle centering machine ‖ zwei-

spindlige Zentriermaschine f. ‖ machine f. à centrer à deux broches.

two-stage air pump ‖ zweistufige Luftpumpe f. ‖ pompe f. à air à deux phases. / ~ amplifier (Tel) ‖ zweistufiger Verstärker m.; Zweifachverstärker m. ‖ amplificateur m. à deux étages. / ~ compressor ‖ zweistufiger Verdichter m. *oder* Kompressor m. ‖ compresseur m. à deux étages. / ~ mixed pressure steam turbine ‖ Zweidruckdampfturbine f. ‖ turbine f. mixte à deux pressions.

two-stroke cycle ‖ Zweitaktvorgang m.; Zweitaktprozeß m. ‖ cycle m. à deux temps. / ~ motor *see* two-cycle motor.

two-stroke motor *see* two-cycle motor.

two-strutter (Airpl) ‖ Zweistieler m. ‖ avion m. à deux paires de montants de chaque côté du fuselage.

two-threads pl. (Spinn) ‖ zweidrähtiger Zwirn m. ‖ fil m. double.

two-valve intermediate amplifier ‖ Doppelrohrzwischenverstärker m. ‖ amplificateur m. intermédiaire à deux tubes.

two-voltage change-over switch ‖ Zweispannungsstecker m. ‖ fiche f. commutatrice pour deux tensions.

two-way cock ‖ Zweiwegehahn m.; Durchgangshahn m. ‖ robinet m. droit *ou* à deux orifices *ou* canaux. / ~ cock without stuffing box ‖ Durchgangshahn m. ohne Stopfbuchse ‖ robinet m. droit sans boîte à bourrage. / ~ switch ‖ Zweiwegumschalter m. ‖ commutateur m. à deux directions. / ~ switch for hotels ‖ Hotelschalter m. ‖ montage m. va-et-vient pour hôtel; interrupteur m. d'escalier. / ~ turntable ‖ Kreuzdrehscheibe f. ‖ plaque f. tournante pour croisement. / ~ valve ‖ Durchgangsventil n. ‖ soupape f. droite.

two-wheel trailing truck ‖ hinterer Laufradsatz m. ‖ essieu m. porteur arrière.

two-wire network (Electr) ‖ Zweileiternetz n. ‖ réseau m. à deux fils. / ~ system (Tel) ‖ Zweidrahtsystem n.; Schleifensystem n. ‖ système m. à deux fils. / ~ working (Aut tel) ‖ Schleifenleitungsbetrieb m. ‖ service m. téléphonique automatique à deux conducteurs.

twybil(l) *see* twibill.

twyer (Forg; Met) *see also* tuyère ‖ Windform f.; Ofenform f. ‖ tuyère f. / the ~ becomes dark ‖ die Form nast stark *oder* setzt viel Schlacken an ‖ lo scorie obscurcit la tuyère. / the ~ becomes unclean ‖ die Form nast ‖ la scorie adhère *ou* s'attache à la tuyère. / open ~ ‖ offene Form f. ‖ tuyère f. ouverte.

twyer arch (Of blast furnace) ‖ Formgemäuer n.; Windgewölbe n. ‖ voûte f. *ou* encorbellement m. *ou* arche f. des soufflets; mureau m. / ~ hole ‖ Formauge n.; Formöffnung f. ‖ œil m. *ou* bouche f. de la tuyère. / ~ lip ‖ Formlippe f. ‖ lèvre f. de la tuyère. / ~ nose ‖ Formnase f. der Gebläseform ‖ nez m. de la tuyère. / ~ wall *see* twyer arch.

ty(e), to ~ ores pl. ‖ die Erze im Schlämmgraben waschen ‖ laver les minerais mpl.

tye (Ore dress) ‖ Schlämmgraben m.; geneigter Herd m. ‖ caisse f. *ou* table allemande; caisson m.; table f. servante au lavage des sables.

tying (Washing the ores in a tye) ‖ Schlämmung f. der Erze in einem Schlämmgraben ‖ lavage m. des minerais dans

une caisse à débourber *ou* sur une table inclinée.

tying-up ‖ Ausschnürung f.; Schnürung f. ‖ empoutage m.; armure f.; encordage m.

tympan (Anatomy) ‖ Trommelfell n.; tympan m. / ~ (Print) ‖ Preßdeckel m.; Einlegedeckel m. ‖ tympan m.

tympanites (Of sheep) ‖ Trommelsucht f. ‖ tympanité f.

tympan sheet (Print) ‖ Einstechbogen m. ‖ feuille f. d'imposition.

tympanum *see* tympan.

type, to ~ (To write with the typewriter) ‖ tippen; mit der Schreibmaschine schreiben ‖ écrire à la machine; taper.

type (Letter) ‖ Druckbuchstabe m.; Buchstabe m.; Letter f.; Type f. ‖ lettre f.; caractère m. d'imprimerie; type m. / ~s pl. (Print) ‖ Schrift f.; Druck m.; Typen fpl. ‖ caractères mpl.; types mpl. / being in ~ ‖ abgesetzt ‖ composé, en composition. / to cleanse the ~ ‖ die Schrift waschen ‖ brosser la lettre. / ~ for addressing machines ‖ Adressiermaschinentype f. ‖ type m. pour machines à imprimer des adresses. / broken ~ ‖ abgefallene Schrift f. ‖ lettre f. tombée *ou* marchée. / copper ~ ‖ Kupferbuchstabe m. ‖ lettre f. en cuivre. / fat ~s pl. ‖ Fettdruck m. ‖ caractères mpl. gras. / Italic ~ ‖ Kursivschrift f.; Kursiv f. ‖ cursive f.; italique m. / large ~ ‖ große Schrift f. ‖ écriture f. grosse; gros caractère m.; grosse f. anglaise. / lean ~s pl. ‖ gemeine Schrift f. ‖ petites lettres fpl. / ~s pl. of metal for printing machines ‖ Schrift f. in Metall für Druckmaschinen ‖ lettres fpl. en métal pour machines à imprimer. / Roman ~ ‖ gerade Schrift f. ‖ caractère m. romain. / wooden ~ ‖ Holztype f. ‖ caractère m. (mobile) en bois. / worn-out ~ ‖ abgenutzte Schrift f. ‖ caractère m. d'impression vieux.

type (Model) ‖ Typ m.; Modell n.; Typus m.; Bauart f.; Ausführung f. ‖ modèle m.; type m. / elongated ~ ‖ verlängerte Bauart ‖ modèle m. allongé. / enclosed ~ ‖ geschlossene Bauart f. ‖ construction f. fermée. / heaviest ~ ‖ schwerstes Modell n. ‖ modèle m. extra robuste. / primitive ~ ‖ einfache Ausführung f. ‖ construction f. primitive. / standard ~ ‖ Einheitsbauart f. ‖ type m. standardisé. / in x ~s pl. ‖ in x-facher Ausführung f. ‖ en x modèles mpl.

type adjusting pliers pl. (Typewr) ‖ Typenrichtzange f. ‖ pince f. pour le redressage des caractères. / precise adjustment of the ~ ‖ genaue Einstellung f. der Type ‖ amenée f. du caractère au point précis.

type bar (Typewr) ‖ Typenhebel m.; Typenträger m. ‖ tige f. à caractères; porte-caractères m. / to bend a ~ ‖ einen Typenhebel m. verbiegen ‖ fausser une tige à caractères. / direct acting ~ ‖ unmittelbar wirkender Typenhebel m. ‖ tige f. à caractères à action directe. / divided ~ ‖ geteilter Typenhebel m. ‖ tige f. à caractères divisée. / pivoted ~ ‖ Zapfentypenhebel m. ‖ tige f. à caractères montée sur pivot.

type bar, arrangement of the ~s ‖ Typenhebelanordnung f. ‖ disposition f. des tiges à caractères. / ~ bit ‖ Bart m. des Typenhebels ‖ barbe f. du levier à caractères. / circular suspension of the ~s ‖ kreisförmige Aufhängung f. der Typenhebel ‖ suspension f. circulaire des tiges à caractères. / ~ guide segment ‖ Typenstangenführungsbogen m. ‖ segment m. à sections servant au guidage des barres porte-caractères. / lifting of the ~s ‖ Lüften n. der Typenhebel ‖ soulèvement m. des tiges à caractères. / play of the ~s ‖ Spiel n. der Typenhebel ‖ jeu m. des tiges à caractères. / radial arrangement of the ~s ‖ sternartig gelagerte Typenhebel mpl ‖ tiges fpl. à caractères disposées en étoile. / return of the ~s ‖ Zurückfallen n. der Typenhebel ‖ retour m. des tiges à caractères. / shortening of the ~s ‖ Abkürzen n. der Typenhebel ‖ raccourcissement m. des tiges à caractères. / single suspension of the ~s ‖ Einzelaufhängung f. der Typenhebel ‖ suspension f. séparée pour chaque tige à caractères. / skipping of the ~s ‖ Tanzen n. der Typenhebel ‖ sautillement m. des tiges à caractères. / slit of the ~ ‖ Auszahnung f. am Typenhebel ‖ évidement m. de tige à caractères. / ~ straightening pliers pl. ‖ Richtzange f. für Typenhebel ‖ pince à redresser les barres à caractères. / striking of the ~s ‖ Aufeinanderschlagen n. der Typenhebel ‖ choc m. des tiges à caractères. / ~ suspension ‖ Typenhebelaufhängung f. ‖ suspension f. des tiges à caractères. / ~ typewriter ‖ Typenhebelschreibmaschine f. ‖ machine f. à écrire avec tiges à caractères.

type body (Print) ‖ Schriftkegel m.; corps m. / ~ brush ‖ Typenbürste f. ‖ brosse f. à caractères. / ~-carrying cylinder of the rotary machine ‖ Formzylinder m. der Rotationsmaschine ‖ cylindre m. porte-clichés de la presse rotative. / ~ case (Print) ‖ Setzkasten m. ‖ casse f. (de compositeur). / ~ case for keeping the types ‖ Setzkasten m. zur Aufbewahrung der Typen ‖ casse f. pour contenir les caractères non-utilisés. / ~ casting ‖ Schriftgießerei f.; Schriftgießen n. ‖ fonderie f. de caractères; fondage m. de lettres d'imprimerie. / ~ casting machine ‖ Schriftgießmaschine f. ‖ machine f. à fondre les caractères; fondeuse f. de caractères. / ~ cleaning ‖ Typenreinigung f. ‖ nettoyage m. des caractères. / ~ cleaning brush ‖ Typenreinigungsbürste f. ‖ brosse f. pour le nettoyage des caractères. / ~ cleaning device ‖ Typenreinigungsvorrichtung f. ‖ dispositif m. de nettoyage des caractères.

type cylinder ‖ Typenwalze f. ‖ cylindre m. à caractères. / to remove the ~ ‖ die Typenwalze herausnehmen ‖ retirer le cylindre à caractères. / bearing of the ~ ‖ Typenzylinderlager n. ‖ palier m. du cylindre à caractères.

typed (Typewr) ‖ mit der Maschine geschrieben; getippt ‖ écrit à la machine.

type disc (Typewr) ‖ Typenscheibe f. ‖ disque m. à caractères. / ~ distributor (Print) ‖ Schriftableger m. ‖ distributeur m. de caractères. / ~ founder ‖ Schriftgießer m. ‖ fondeur m. de caractères; fondeur-typographe m. / ~ foundry ‖ Schriftgießerei f. ‖ fonderie f. de caractères; fonderie f. typographique. / ~ gauge (Print) ‖ Zeilenmesser m.; lignomètre m. / ~ high gauge ‖ Schrifthöhenmesser m. ‖ typomètre m. / ~ inking brush ‖ Bürste f. zum Färben der Typen

brosse f. pour l'encrage des caractères. / ~ inking recording attachment ‖ Typendruckschreibeinrichtung f. ‖ dispositif m. enregistreur par caractères d'impression. / ~ justifier ‖ Schriftberichtiger m.; Schriftjustierer m. ‖ globuleur m. / ~ lever *see* type bar. / ~ magazine (Print) ‖ Typenmagazin n.; Schriftmagazin n. ‖ magasin m. à caractères. / ~ metal ‖ Schriftmetall n.; Letternmetall n. ‖ matière f.; métal m. à lettres. / ~ plate ‖ Typenplatte f. ‖ plaque f. à caractères. / ~ printer *see* ~ printing telegraph. / ~ printing receiver (Tel) ‖ Typendruckempfangsgerät n. ‖ appareil m. imprimeur-récepteur. / ~ printing telegraph ‖ Typendrucker m.; Typendrucktelegraf m.; Drucktelegraf m. ‖ télégraphe-imprimeur m.; typotélégraphe m. / ~ rod (Typewr) *see also* type bar ‖ Typenstab m. ‖ baguette f. à caractères.

type setting machine ‖ Setzmaschine f. ‖ machine f. à composer. / ~ for setting single types ‖ Buchstabensetzmaschine f. ‖ machine f. à composer des types mobiles. / motor for ~ ‖ Setzmaschinenmotor m. ‖ moteur m. de machine à composer.

type-setting and distributing machine ‖ Setz- und Ablegemaschine f. ‖ machine f. à composer et à distribuer.

type shank (Print) ‖ Schriftkegel m. ‖ corps m. (de la lettre). / ~ (Typewr) *see also* type bar ‖ Typenschaft m.; Typenhebel m. ‖ tige f. à caractères.

type shuttle ‖ Typenschiffchen n. ‖ navette f. à caractères. / size of ~ ‖ Schriftgrad m. ‖ force f. de corps. / ~ washing lye ‖ Typenwaschlauge f. ‖ lessive f. à laver les types.

type wheel ‖ Typenrad n. ‖ roue f. à caractères *ou* à types. / complementary strip for the foreign ~ ‖ Ergänzungsblatt n. für das fremdsprachliche Typenrad ‖ bande f. servant à completer le jeu de caractères de la roue à caractères en langue étrangère. / rotation plane of ~ ‖ Bewegungsebene f. des Typenrades ‖ plan m. de rotation de la roue à caractères. / ~ typewriter ‖ Typenradschreibmaschine f. ‖ machine f. à écrire à roue à caractères.

typewrite, to ‖ mit der Schreibmaschine f. schreiben; Maschine schreiben; tippen ‖ écrire à la machine *ou* avec la dactylotype; taper.

typewriter ‖ Schreibmaschine f. ‖ machine f. à écrire. / ~ for the blind ‖ Blindenschreibmaschine f. ‖ machine f. à écrire pour aveugles. / calculating ~ ‖ rechnende Schreibmaschine f. ‖ machine f. à écrire combinée avec machine à calculer. / cheque ~ ‖ Scheckschreibmaschine f. ‖ machine f. pour écrire les chèques. / child's ~ ‖ Kinderschreibmaschine f. ‖ machine f. à écrire pour enfants. / cipher code ~ ‖ Geheimschreibmaschine f. ‖ machine f. à écrire pour écriture chiffrée. / invoice ~ ‖ Maschine f. zum Schreiben von Rechnungen ‖ machine f. à écrire les factures. / key board ~ ‖ Tastenbrettschreibmaschine f.; Klaviaturschreibmaschine f. ‖ machine f. à écrire à clavier. / multiple-type ~ ‖ Vieltypenschreibmaschine f. ‖ machine f. à écrire à caractères multiples. / music ~ ‖ Notenschreibmaschine f. ‖ machine f. à écrire les notes à musique. / office ~ ‖ Büroschreibmaschine f. ‖ machine f. à écrire

de bureau. / ~ for one-armed persons ‖ Schreibmaschine f. für Einarmige ‖ machine f. à écrire pour manchots. / pocket ~ ‖ Kofferschreibmaschine f. ‖ machine f. à écrire de voyage. / pointer ~ ‖ Zeigerschreibmaschine f. ‖ machine f. à écrire à tige indicatrice. / rapid ~ ‖ Schnellschreibmaschine f. ‖ machine f. à écrire rapide. / ~ for several languages ‖ Vielsprachschreibmaschine f. ‖ machine f. à écrire plusieurs langues. / short-hand ~ ‖ Stenografiermaschine f. ‖ machine f. à sténographier. / single-key ~ ‖ Eintasterschreibmaschine f. ‖ machine f. à écrire à une seule touche. / syllable ~ ‖ Silbenschreibmaschine f. ‖ machine f. à écrire les syllabes / tele-~ ‖ mechanische Fernschreibmaschine f. ‖ téléimprimeur m. mécanique (dit: machine à écrire à distance). / travelling ~ ‖ Reiseschreibmaschine f. ‖ machine f. à écrire de voyage. / type bar ~ ‖ Typenhebelschreibmaschine f. ‖ machine f. à écrire à tiges à caractères. / type wheel ~ ‖ Typenradschreibmaschine f. ‖ machine f. à écrire à roue à caractères. / visible ~ ‖ sichtbar schreibende Schreibmaschine f. ‖ machine f. à écriture visible. / ~ with a wide carriage ‖ Schreibmaschine f. mit breitem Wagen ‖ machine f. à écrire à long chariot.

typewriter accessories pl. ‖ Schreibmaschinenzubehör n. ‖ accessoires mpl. de machine à écrire. / ~ chair ‖ Schreibmaschinenstuhl m. ‖ chaise f. de dactylographe. / ~ desk ‖ Schreibmaschinenpult n. ‖ pupitre n. de machine à écrire. / ~ factory ‖ Schreibmaschinenfabrik f. ‖ manufacture f. de machines à écrire. / ~ frame ‖ Schreibmaschinengestell n. ‖ bâti m. de machine à écrire. / ~ industry ‖ Schreibmaschinenindustrie f. ‖ industrie f. des machines à écrire. / ~ inking tape ‖ Schreibmaschinenfarbband n. ‖ ruban m. encreur pour machines à écrire. / ~ mechanic ‖ Schreibmaschinenmechaniker m. ‖ mécanicien m. spécialisé dans la construction de la machine à écrire. / ~ paper ‖ Schreibmaschinenpapier n. ‖ papier m. pour machine à écrire. / ~ tools pl. for ~ repair ‖ Werkzeug n. für Schreibmaschinenausbesserung ‖ outils npl. pour la réparation des machines à écrire. / ~ ribbon ‖ Farbband n. ‖ ruban m. encreur. / ~ spare part ‖ Schreibmaschinenersatzteil m. ‖ pièce f. de rechange pour machines à écrire. / ~ system ‖ Schreibmaschinensystem n. ‖ système m. de machine à écrire. / ~ table ‖ Schreibmaschinentisch m. ‖ table f. pour machine à écrire. / ~ tool box ‖ Werkzeugtasche f. für Schreibmaschinen ‖ trousse f. d'outils pour machines à écrire. / ~ type factory ‖ Schreibmaschinentypenfabrik f. ‖ fabrique f. de caractères pour machines à écrire.

typewriting ‖ Maschinenschrift f.; Maschinenschreiben n.; Tippen n. ‖ dactylographie f.; écriture f. à la machine. / ~ instruction ‖ Schreibmaschinenunterricht m.; Unterricht m. im Maschinenschreiben ‖ enseignement m. de la machine à écrire. / ~ ribbon ‖ Farbband n. ‖ ruban m. encreur. / ~ speed ‖ Schreibmaschinengeschwindigkeit f. ‖ vitesse f. de machine à écrire. / ~ telegraph see type printing telegraph.

typewritten ‖ mit der Maschine geschrieben; getippt ‖ dactylographié; écrit à la ma-

chine. / ~ impression ‖ Schreibmaschinendruck m. ‖ impression f. faite à la machine à écrire.

typhlo-typography ‖ Blindenschrift f. ‖ lettres fpl. en relief; caractères mpl. pour aveugles.

typing (Typewr) see typewriting.

typist (Male) ‖ Maschinenschreiber m. ‖ dactylographe m. / ~ (Female) ‖ Maschinenschreiberin f.; Tippfräulein n. ‖ (dame f.) dactylographe f.

typographer ‖ Typograf m.; Buchdrucker m.; Setzer m. ‖ typographe m. / ~'s requirements pl. ‖ Buchdruckereibedarfsartikel mpl. ‖ fournitures fpl. pour la typographie.

typographical ‖ typografisch ‖ typographique. / ~ machine ‖ Buchdruckereimaschine f. ‖ machine f. typographique. / ~ printing ‖ Buchdruck m. ‖ imprimerie f. typographique.

typography ‖ Typografie f.; Buchdruckerkunst f. ‖ typographie f.

typometer ‖ Letternmesser m. ‖ typomètre m.

tyre also **tire** (Railw; Coachm) ‖ Radreifen m.; Bandage f.; Radkranz m. ‖ bandage m. ou bande f. ou cercle m. de roue. / ~ (Pneumatic) ‖ Radreifen m.; Reifen m.; Luftreifen m.; Pneumatik m. ‖ pneu m.; pneumatique m. / the ~ crept x millimeters and sounded loose ‖ der Radreifen m. ist um x mm gewandert und klingt lose ‖ le bandage m. s'est déplacé de x millimètres et sonne détaché. / to fasten the ~ by a steel ring and hammering it down over the inner edge ‖ den Reifen m. durch einen über den inneren Rand der Bandage gestemmten Ring festhalten ‖ fixer le bandage par un anneau maté sur le bord intérieur du bandage. / to fit-on the ~ ‖ den Luftreifen m. aufziehen ‖ monter le pneu. / the ~ is secured to the wheel centre by a retaining ring ‖ der Radreifen m. ist mittels Sprengring befestigt ‖ le bandage est fixé au corps de roue au moyen d'une bague de retenue. / airplane ~ ‖ Flugzeugluftreifen m. ‖ pneu m. d'avion. / balloon ~ ‖ Ballonreifen m. ‖ pneu m. ballon. / cast-on ~ (Railw) ‖ angegossener Radreifen m. ‖ bandage m. venu de fonte avec la roue. / clincher ~ ‖ Wulstreifen m. ‖ pneu m. à talons. / cord ~ ‖ Kordreifen m. ‖ pneu m. à corde. / cushion ~ ‖ hochelastischer Reifen m. ‖ bandage m. creux. / dual ~ ‖ Doppelreifen m. ‖ pneu m. jumelé. / fabric tread ~ ‖ Kreuzgewebereifen m. ‖ pneu m. à entoilage normal. / giant (air) ~ ‖ Riesenluftreifen m.; Ballonreifen m. ‖ pneu m. gros ou ballon. / high-pressure ~ ‖ Hochdruckreifen m. ‖ pneu m. à haute pression. / ~ exposed to the highest strains (Railw) ‖ besonders stark beanspruchter Reifen m. ‖ bandage m. exposé aux plus grands efforts. / low-pressure ~ ‖ Niederdruckreifen m. ‖ pneu m. à basse pression. / motor-car ~s pl. ‖ Kraftwagenbereifung f.; Automobilbereifung f. ‖ pneumatiques mpl. d'automobiles. / non-skid ~ ‖ Gleitschutzreifen m. ‖ pneu m. antidérapant. / oversize ~ ‖ Luftreifen m. in Übergröße ‖ pneu m. surprofilé. / pneumatic ~ ‖ Luftreifen m.; Pneumatik m. ‖ pneumatique m.; pneu m. / rubber ~ ‖ Gummireifen m. ‖ bandage m. en caoutchouc. / rubber

studded ~ ‖ Gummigleitschutzreifen m. ‖ antidérapant m. tout en caoutchouc. / shrunk-on ~ ‖ aufgeschrumpfter Reifen m. ‖ bandage m. posé à chaud; bandage m. embattu. / smooth ~ ‖ glatter Luftreifen m. ‖ pneu m. lisse. / solid ~ ‖ Vollgummireifen m.; Vollreifen m. ‖ bandage m. plein ou solide. / double solid ~ ‖ Doppelvollgummireifen m. ‖ bandage m. plein double. / spare ~ ‖ Ersatzreifen m. ‖ pneu m. de réserve. / steel-studded ~ ‖ Gleitschutzreifen m. ‖ enveloppe f. antidérapante. / straightside ~ ‖ Geradseitreifen m. ‖ pneu m. à tringles. / twin ~ ‖ Zwillingsluftreifen m. ‖ pneumatique m. jumelé. / weldless ~ (Railw) ‖ nahtloser Eisenbahnradreifen m. ‖ bandage m. sans soudure.

tyre bending machine ‖ Radreifenbiegemaschine f. ‖ machine f. à cintrer les bandages. / ~ case (Auto) ‖ Reifenkoffer m. ‖ coffre m. à pneumatique. / ~ casing ‖ Laufdecke f. des Reifens; Mantel m. des Luftreifens ‖ enveloppe f. de roulement; bandage m. pneumatique. / ~ centering machine ‖ Maschine f. zum Zentrieren der Radreifen ‖ machine f. à centrer les bandages. / ~ chain ‖ Schneekette f. ‖ chaîne f. à neige. / ~ dog ‖ Reifhaken m. zum Aufziehen der Radreifen ‖ diable m. / ~ equipment (Auto) ‖ Bereifung f. ‖ équipement m. de bandages. / spare ~ equipment ‖ Ersatzbereifung f. ‖ équipement m. de pneu de rechange.

tyre fastening ‖ Radreifensicherung f. ‖ fixation f. de bandage. / ~ by double lip retaining ring ‖ Radreifensicherung f. mittels zweier Klammerringe ‖ fixation f. de bandage au moyen d'un cercle à double agrafage. / ~ by double lip retaining ring, single flange and rivets ‖ Radreifensicherung f. mittels zweier Klammerringe und Untergriff und Nieten ‖ attache f. de bandage au moyen d'un cercle à double agrafage, talon et rivets. / ~ by retaining rings on both sides and rivets ‖ Radreifensicherung f. mittels zweier Klammerringe und Nieten ‖ fixation f. de bandage par cercle à double agrafage et rivets. / ~ with ring ‖ Radreifensicherung f. mittels Schließringes ‖ fixation f. par cercle à simple agrafage. / ~ by set screws ‖ Radreifensicherung f. mittels Kopfschrauben ‖ fixation f. de bandage par broches filetées. / ~ by single flange ‖ Radreifensicherung f. mittels Untergriffs ‖ fixation f. par talon du bandage. / ~ by single flange and one retaining ring ‖ Radreifensicherung f. mittels Untergriffs und eines Klammerringes ‖ fixation f. par cercle de retenue d'un seul côté et talon du bandage. / ~ by single flange and set screws ‖ Radreifensicherung f. mittels Untergriffs und Kopfschrauben ‖ fixation f. par talon du bandage et broches filetées. / ~ by spring ring ‖ Radreifensicherung f. mittels Sprengrings ‖ fixation f. de bandage par agrafe annulaire. / ~ by means of through bolts ‖ Radreifensicherung f. mittels durchgehender Schrauben ‖ fixation f. au moyen des boulons traversant le bandage.

tyre flap ‖ Felgenband n. ‖ toile f. cerclant la jante. / ~ gauge ‖ Reifendruckprüfer m. ‖ vérificateur m. de pression des

pneumatiques. / ~ heating furnace (For railway tyres) || Bandagenglühofen m. || four m. à chauffer les bandages. / ~ holder / Reifenhalter m. || ferrure f. pour pneumatiques. / ~ inner tube || Luftschlauch m. || chambre f. à air. / ~ iron || Radreifeisen n. || fer m. de bandage. / ~ lathe || Radreifendrehbank f.; Bandagendrehbank f. || tour m. à tourner les bandages. / lip of ~ || Radreifenansatz m. || talon m. du bandage. / ~ mill see ~ rolling mill. / ~ press || Radreifenpresse f.; Bandagenpresse f. || presse f. à (caler les) bandages. / ~ protector || Reifenschutz m.; Reifenschützer m. || protecteur m. de bandage.

/ ~ pump || Reifenpumpe f. || pompe f. à pneumatique. / ~ remover || Reifenheber m. || démonte-bandage m. / ~ rim || Felge f.; Reifenfelge f. || jante f. / steel ~ rim || Stahlfelge f. || jante f. d'acier. / ~ rolling mill || Radreifenwalzwerk n.; Radkranzwalzwerk n.; Bandagenwalzwerk n. || laminoir m. à bandages. / ~ shoe (Of a pneumatic tyre) || Mantel m.; Decke f.; Reifendecke f. || chape f.; enveloppe f. / ~ smith || Radreifenschmied m. || forgeur m. de cercles de roues. / standard section of ~ || normales Radreifenprofil n. || profil m. normal du bandage. / ~ tread || Lauffläche oder Lauf-

band n. des Reifens || surface f. ou bande f. de roulement. / ~ valve || Schlauchventil n. || valve f. de gonflement ou de chambre à air. / ~ welder || Radreifenschweißer m. || soudeur m. de bandages. / ~ welding || Reifenschweißung f. || soudage m. de bandages.

tyrolite || Kupferschaum m.; Tyrolit m. || tyrolite f.; kupaphrite f.

tyrosine || Tyrosin n. || tyrosine f.

tyrotoxicone || Tyrotoxikon n. || tyrotoxicone f.

tysonite || Tysonit m. || tysonite f.

U

U-bolt || U-Stütze f. || support m. ou console f. en U. / ~ double pin || U-förmige Doppelstütze f. || console f. double en U. / ~ iron || U-Eisen n. || fer m. à côtes ou à U. / ~-shaped plate || Kanalblech n. || tôle f. en U. / ~-shaped pressed girder || U-förmiger Preßträger m. || longeron m. embouti en U. / ~ stirrup || U-förmiger Bügel m. || étrier m. en forme d'U.

udometer || Regenmesser m. || pluviomètre m.; udomètre m.

ulexite || Bornatronkalzit m. || ulexite f.; natroborocalcite f.

ullmannite || Nickelantimonglanz m.; Nickelspießglanz m.; Ullmannit m. || nickel m. antimonié sulfuré; ullmannite f.

ulmic acid || Humussäure f. || acide m. humique.

ulterior forging of special steels already rough-forged || Weiterschmieden n. bereits vorgeschmiedeter Sonderstähle || second procédé m. de forgeage d'aciers spéciaux préalablement forgés. / ~ manufacture || Weiterverarbeitung f. || fabrication f. ultérieure.

ultimate || endgültig || définitif. / ~ load || Bruchbelastung f.; Bruchbeanspruchung f. || charge f. de rupture. / ~ straining see ~ load. / ~ strength || Bruchfestigkeit f. || résistance f. à la rupture.

ultra-audion || Ultraaudion n. || ultra-audion m.

ultramarine || Ultramarin n. || outremer m. / artificial ~ || künstliches Ultramarin n. || outremer m. artificiel. / ~ blue || Ultramarinblau n. || bleu m. d'outremer. / ~ black || Ultramarinschwarz n. || noir m. d'outremer. / ~ green || Ultramaringrün n. || vert m. d'outremer. / ~ red || Ultramarinrot n. || rouge m. d'outremer. / ~ substitute || Ultramarinersatz m. || succédané m. d'outremer. / ~ violet || Ultramarinviolett n. || violet m. d'outremer.

ultra-microscope || Ultramikroskop n. || ultramicroscope m. / slit ~ || Spaltultramikroskop n. || ultramicroscope m. à fente.

ultra-red || Ultrarot n.; Infrarot n. || infrarouge m. / ~ radiation || ultrarote Strahlung f. || radiation f. infrarouge.

ultra violet || Ultraviolett n. || ultraviolet m.

ultra-violet || ultraviolett || ultraviolet. / glass of unusual transparency for ~ light || besonders gut ultraviolett-durchlässiges Glas n. || verre m. particulièrement transparent aux radiations ultraviolettes. / ~ prism spectrum || ultraviolettes Prismenspektrum n. || spectre m. ultraviolet à prisme. / ~ radiation || ultraviolette Strahlung f. || radiation f. ultraviolette. / ~ rays pl. || ultraviolette Strahlen mpl. || rayons mpl. ultraviolets. / enhanced permeability of the glass to ~ rays || gesteigerte Durchlässigkeit f. des Glasmaterials gegen ultraviolette Strahlen || transparence f. plus grande du verre pour les rayons ultraviolets. / ~ spectrum || ultraviolettes Spektrum n. || spectre m. ultraviolet.

ultra vires || statutenwidrig || antistatuaire.

umbel || Dolde f. || cône m.; ombelle f.

umber (Miner) || Umbra f.; Umbererde f. || ombre f.; terre f. d'ombre.

umbilical scissors pl. (Med) || Nabelschnurschere f. || ciseaux mpl. à cordon ombilical.

umbra || Schatten m.; Kernschatten m.; Schattenkegel m. || ombre f. pure; cône m. d'ombre.

umbrella || Regenschirm m. || parapluie m. / carriage ~ || Wagenschirm m. || ombrelle f. pour voitures.

umbrella aerial see ~ antenna. / ~ antenna || Schirmantenne f. || antenne f. en parapluie. / ~ case || Schirmhülle f. || fourreau m. de parapluie. / ~ cloth || Baumwolltaft m.; Regenschirmstoff m. || silésienne f. / ~ fittings pl. || Schirmbeschlag m. || garniture f. pour parapluies. / ~ frame || Gestell n. eines Regenschirmes; Schirmgestell n. || carcasse f. de parapluie. / ~ frame maker || Schirmgestellmacher m. || carcassier m. de parapluies. / ~ handle || Schirmgriff m. || poignée f. pour parapluies. / ~ making machine || Schirmfabrikationsmaschine f. || machine f. pour fabriquer des parapluies. / ~ material || Schirmstoff m. || étoffe f. pour parapluies. / ~ mender || Schirmflicker m. || réparateur m. de parapluies. / ~ mending || Schirmausbesserung f. || réparation f. de parapluies. / ~ sewer || Schirmnäher m. || cou-

seur m. de parapluies. / ~-shaped roof || schirmförmiges Dach n. || toit m. ou toiture f. conique. / ~ stick || Schirmstock m. || canne-parapluie f. / ~ and parasole trimmings pl. || Schirmposamenten pl. || passementerie f. pour parapluies.

umpire (Sport) || Schiedsrichter m. || arbitre m.; dispacheur m.

unable || unfähig || incapable.

unaided eye || unbewaffnetes Auge n. || œil m. nu.

unalloyed || unlegiert || non-allié.

unaltered || unverändert || inaltéré.

unassuming || anspruchslos || sans prétentions fpl.; modeste.

unbalance, to || aus dem Gleichgewicht n. bringen || déséquilibrer.

unbalance || Abgleichfehler m. || défaut m. d'équilibrage; déséquilibre m. duplex. / ~ (Radio) || kapazitive Kopplung f. || accouplement m. électrostatique. / ~ of the circuits || Unsymmetrie f. der Leitungen || déséquilibre m. des circuits.

unbalanced || nicht ausbalanziert || déséquilibré. / ~ section (El line) || unkompensierter Abschnitt m. || section f. non-compensée.

unballast, to || den Ballast m. ausladen || délester.

unbarked timber || Holz n. mit Rinde || bois m. en grume.

unbend, to ~ the cable of the anchor || das Ankertau losmachen || détalinguer ou démarrer le câble de l'ancre. / ~ a rope (Mar) || ein Tau n. losstecken oder abstecken || démarrer un cordage.

unbiassed || sachlich || objectif.

unbleached || ungebleicht || écru. / ~ linen || ungebleichte Leinwand f. || toile f. écrue.

unbraced (Airpl) || verspannungslos; unverspannt || sans haubanage m.

unbuckle, to || losschnallen || déboucler.

unbur, to (Wool) || auszupfen || égrateronner.

unburnt || ungebrannt || cru. / ~ brick || ungebrannter Stein m. || brique f. séchée à l'air. / ~ gas || unverbranntes Gas n. || gaz m. non-brûlé.

uncalcined || roh; ungeröstet || cru; non calciné.

uncalled capital ‖ noch nicht eingezahltes Kapital n. ‖ capital m. non-appelé.

uncemented ‖ unverkittet ‖ non-accollé. / ~ window ‖ kittfrei eingesetztes Fenster n. ‖ fenêtre f. adaptée sans mastic.

unchanged see unaltered.

unclaimed dividend ‖ nicht erhobene Dividende f. ‖ dividende m. non-réclamé.

unclean coffee ‖ unreiner Kaffee m. ‖ café m. gras.

uncoagulable albumine ‖ ungerinnbares Eiweiß n. ‖ albumine m. incoagulable.

uncoil, to ‖ abwickeln ‖ dérouler. / ~ wire ‖ Draht m. abrollen ‖ développer le fil de ligne.

uncombined ‖ ungebunden ‖ non combiné.

uncompleted ‖ unvollendet ‖ incomplet.

unconfirmed credit ‖ widerruflicher Kredit m. ‖ crédit m. non-confirmé.

uncorking device, bottle ~ ‖ Flaschenentkorker m. ‖ débouchoir m. de bouteilles; tire-bouchon m.

uncouple, to ‖ auskuppeln ‖ déclencher; débrayer. / ~ wheels pl. ‖ die Räder npl. ausrücken ‖ désengrener les roues fpl.

uncoupling ‖ Auskupplung f. ‖ déclenchement m.; débrayage m.

uncover, to (Mine) ‖ erschürfen ‖ découvrir en creusant. / ~ a house ‖ ein Haus n. abdecken ‖ découvrir une maison.

uncovered (Bank) ‖ blanko ‖ en blanc. / ~ (Electr) ‖ blank ‖ dénudé.

uncrystallizable ‖ unkristallisierbar ‖ incrystallisable.

unctuous ‖ fettig; ölig ‖ onctueux; graisseux; adipeux.

uncultivated ‖ unangebaut ‖ incultivé.

undamaged ‖ unbeschädigt ‖ intact.

undamped gyro oscillation ‖ ungedämpfte Kreiselschwingung f. ‖ oscillation f. de gyroscope non amortie. / ~ waves pl. ‖ ungedämpfte Wellen fpl. ‖ ondes fpl. non amorties.

under leads pl. ‖ unter Bleiverschluß m. ‖ plombé.

under steam ‖ unter Dampf m. ‖ sous vapeur f.

underback (Brew) ‖ Ausschlagbottich m. ‖ panier m. à houblon.

underbid, to ‖ unterbieten ‖ offrir moins que. . . .

underbridge (Railw) ‖ Unterführung f. ‖ passage m. inférieur ou souterrain.

undercarriage (Aero) ‖ Fahrgestell n. ‖ châssis m. ou train m. d'attérrissage. / axleless ~ ‖ achsenloses Fahrgestell n. ‖ châssis m. sans essieu. / ~ which can be raised ‖ hochziehbares Fahrgestell n. ‖ châssis m. escamotable.

undercarriage bracing ‖ Fahrgestellverspannung f. ‖ haubanage m. du train d'atterrissage. / ~ socket ‖ Fahrgestellschuh m. ‖ embout m. inférieur de jonction des mâts du train d'atterrissage. / ~ springing ‖ Fahrgestellfederung f. ‖ amortisseur m. de choc du train d'atterrissage. / rear ~ strut ‖ hintere Fahrgestellstrebe f. ‖ mât m. postérieur de train d'atterrissage.

underclothes pl. ‖ Unterwäsche f. ‖ habits mpl. de dessous. / hygienic ~ ‖ Reformwäsche f. ‖ lingerie f. hygiénique; sous-vêtements mpl. hygiéniques.

undercut, to ~ types ‖ Lettern fpl. unterschneiden ‖ créner les caractères mpl.

undercutter see underseller.

undercutting of a bank in a quarry ‖ Unterhöhlung f. einer Bank im Steinbruch ‖ évidement m. sous une couche de la carrière.

under-drain pipe ‖ Drainröhre f. ‖ tuyau m. de drainage ou de dessèchement.

underexposure ‖ Unterbelichtung f. ‖ sous-exposition f.

underfarmer ‖ Unterpächter m. ‖ sous-fermier m.

underfloor channel ‖ Unterflurkanal m. ‖ canal m. souterrain. / ~ way (Roll mill) ‖ Tieflauf m. ‖ voie f. de profondeur.

underframe ‖ Untergestell n. ‖ châssis m. / ~ of cars ‖ Wagenuntergestell n. ‖ châssis m. des voitures. / ~ for motor cars with Z shaped pressed girders ‖ Untergestell n. für Motorwagen mit Z-förmigen Preßträgern ‖ châssis m. pour voitures motrices à longerons emboutis en Z. / part of ~ ‖ Untergestellteil m. ‖ partie f. de châssis.

undergrate blast ‖ Unterwind m. ‖ soufflage m. sous grille. / ~ blower ‖ Unterwindgebläse n. ‖ ventilateur m. sous grille de foyer.

underground ‖ unterirdisch ‖ souterrain. / ~ (Mine) ‖ unter Tage ‖ sous terre f. / to go ~ (Mine) ‖ einfahren ‖ descendre; entrer.

underground cable ‖ unterirdisches Kabel n.; Erdkabel n. ‖ câble m. souterrain ou sous-terre. / ~ combustion see ~ fire. / ~ coupling (Cable) ‖ Schutzmuffe f. ‖ manchon m. de (fer pour la) protection de raccords. / ~ fire (Mine) ‖ Grubenbrand m. ‖ incendie m. dans une mine. / ~ haulage ‖ Streckenförderung f.; Förderung f. unter Tage ‖ roulage m. intérieur; traction f. dans les galeries. / ~ hauling see ~ haulage. / ~ hydrant ‖ Unterflurhydrant m. ‖ bouche f. à eau sous sol. / ~ line ‖ unterirdische Leitung f. ‖ ligne f. souterraine. / ~ mine ‖ Tiefbaugrube f.; Grube f.; Zeche f. ‖ mine f. souterraine. / ~ piping ‖ Erdleitung f. ‖ conduite f. souterraine. / ~ pole reinforcement (El line) ‖ Bodenverstärkung f. von Masten ‖ renforcement m. des poteaux souterrains. / ~ railway ‖ Untergrundbahn f. ‖ chemin m. de fer souterrain. / ~ survey ‖ Grubenvermessung f. ‖ arpentage m. souterrain. / ~ tank plant for liquid fuels ‖ unterirdische Tankanlage f. für flüssige Brennstoffe ‖ installation f. de réservoirs souterrains pour combustible liquides. / ~ ventilation (Mine) ‖ Bewetterung f. ‖ aérage m. des mines.

underground water ‖ Grundwasser n. ‖ eaux fpl. souterraines. / to draw off ~ ‖ das Grundwasser abführen ‖ évacuer l'eau f. du fond. / ~ level ‖ Grundwasserspiegel m. ‖ niveau m. des eaux souterraines. / ~ packing ‖ Grundwasserabdichtung f. ‖ calfeutrage m. de nappe souterraine.

underground winning see ~ working. / ~ work see ~ working. / ~-worked allotment ‖ Tiefbaubetrieb m. ‖ exploitation f. souterraine; concession f. avec excavation souterraine. / ~ workers pl. ‖ Belegschaft f. unter Tage ‖ ensemble m. du fond. / ~ working ‖ Grubenbetrieb m. unter Tage; Untertagebau m. ‖ exploitation f. souterraine. / ~ working enterprise ‖ Tiefbauunternehmen n. ‖

entreprise f. de construction au-dessous du sol.

undergrowth ‖ mangelhafter Wuchs m. ‖ végétation f. inférieure.

underlay, to (Geol) ‖ einfallen ‖ plonger; s'incliner. / ~ (Print) ‖ füttern; unterlegen ‖ taquonner; rehausser.

underlay (Geol) ‖ Einfallen n. ‖ pente f.; inclinaison f.; pendage m.

underlayer (Mine) ‖ Richtschacht m. ‖ puits m. perpendiculaire.

underlaying (Mine) ‖ Liegendes n. ‖ couche f. inférieure; mur m.

underlie, to (Geol) ‖ einfallen ‖ plonger; s'incliner.

underlinen ‖ Unterwäsche f. ‖ linge m. de dessous. / ~ marking ink ‖ Wäschezeichentinte f. ‖ encre f. à marquer le linge.

underlining, arrangement for ~ (Typewr) ‖ Unterstreichvorrichtung f. ‖ dispositif m. pour souligner.

underload (Electr) ‖ Unterbelastung f. ‖ charge f. incomplète ou partielle.

undermined (Build) ‖ unterwaschen ‖ déchaussé.

underneath ‖ unterhalb ‖ dessous. / to place ~ ‖ unterlegen ‖ placer dessous.

underpin, to (Build) ‖ unterbauen; unterfangen ‖ reprendre.

underpinning of a building ‖ Unterfangung f. eines Gebäudes ‖ reprise f. des fondements d'un édifice.

underrate, to ‖ unterschätzen ‖ sous-évaluer.

underrun, to ~ a tackle ‖ ein Takel n. klaren ‖ détordre un palan; défaire les tours mpl. d'un palan.

underseller ‖ Preisverderber m.; Schleuderer m. ‖ gâte-métier m.; gâcheur m.

underselling ‖ Schleudern n. ‖ vente f. à vil prix; gâchage m.

undershot water wheel ‖ unterschlächtiges Wasserrad n. ‖ roue f. hydraulique en dessous.

underslung radiator ‖ Hängekühler m. ‖ radiateur m. à éclipse.

understanding, verbal ~ ‖ mündlich getroffene Vereinbarung f. ‖ arrangement m. verbal.

understructure (Blast furnace) ‖ Sockel m. ‖ piliers mpl. de cœur.

undertaking ‖ Unternehmung f. ‖ entreprise f. de production. / ~ engaged in current generation ‖ Unternehmung f. für Stromerzeugung ‖ entreprise f. de production de courant électrique. / ~ engaged in power distribution ‖ Unternehmung f. für Stromverteilung ‖ entreprise f. de distribution de courant électrique.

undervalue, to see to underrate.

undervest ‖ Unterjacke f. ‖ camisole f.

underwagon for inclined planes (Mine) ‖ Unterwagen m. für Bremsberge ‖ truck m. pour le transport des wagonnets dans les plans inclinés.

underwater cutting burner ‖ Unterwasserschneidbrenner m. ‖ brûleur m. à découper au-dessous de l'eau. / ~ foundation ‖ Gründung f. unter Wasser ‖ fondation f. sous l'eau. / ~ speed ‖ Geschwindigkeit f. unter Wasser ‖ vitesse f. sous l'eau. / ~ structure see ~ foundation. / ~ travelling ‖ Unterwasserfahrt f. ‖ marche f. en plongée.

underwear ‖ Unterkleidung f.; Unterzeug n. ‖ sous-vêtements mpl.; dessous m. / ladies' ~ ‖ Damenwäsche f. ‖ linge

m. pour dames. / manufacture of ~ ‖ Trikotagenfabrikation f.; Unterwäschefabrikation f. ‖ fabrication f. de tricotages.

underweight ‖ Mindergewicht n. ‖ manque m. de poids.

underwood ‖ Unterholz n. ‖ bois m. de taille.

undestroyable ‖ unzerstörbar ‖ indestructible.

undeterminable, statically ~ ‖ statisch unbestimmt ‖ undéterminable par la statique.

undeveloped coal deposits pl. ‖ unaufgeschlossene Kohlenfelder npl. ‖ gisement m. houiller non mis en exploitation. / ore deposits ~ pl. which are at present ~ ‖ zur Zeit unaufgeschlossene Erzfelder npl. ‖ filon m. métallifère pas encore mis en exploitation.

undisturbed by reflexes pl. ‖ reflexfrei ‖ sans réflexe m.

undo, to ‖ zerlegen ‖ décomposer. / ~ the plaits pl. (Cloth) ‖ Stoffe mpl. ausstreichen oder glätten ‖ défroncer; déplisser; écrancher les faux plis d'un drap. / ~ a screw ‖ eine Schraube f. lösen ‖ défaire un boulon.

undock, to ~ a ship ‖ ein Schiff n. ausdocken ‖ faire sortir un navire du bassin.

undue wear ‖ übermäßiger Verschleiß m. ‖ usure f. exagérée.

undulated ‖ gewellt ‖ ondulé.

undulating motion ‖ Wellenbewegung f. ‖ mouvement m. ondulatoire; ondulation f.

undulation (Phys) ‖ Welle f. ‖ ondulation f. / ~ (Ship) ‖ Schwanken n. ‖ tangage m.; balancement m. / ~ of ground ‖ Bodenerhebung f. ‖ rideau m.

undulator ‖ Undulator m. ‖ ondulateur m.

undulatory ‖ wellenförmig ‖ ondulatoire. / ~ theory ‖ Undulationstheorie f.; Wellentheorie f. ‖ théorie f. des ondulations. / ~ theory of light ‖ Wellentheorie f. des Lichtes ‖ théorie f. ondulatoire de la lumière.

unemployed ‖ arbeitslos ‖ sans-travail. / ~ stock ‖ totes Kapital n. ‖ argent m. mort; fonds m. inoccupé.

unemployed ‖ Arbeitsloser m.; Erwerbsloser m. ‖ chômeur m.; sans-travail m.

unemployment ‖ Arbeitslosigkeit m.; Erwerbslosigkeit m. ‖ chômage m. / ~ pay ‖ Arbeitslosenunterstützung f. ‖ dédommagement m. de chômage.

unenclosed ‖ offen; frei ‖ ouvert; découvert.

unencumbered ‖ unbelastet; schuldenfrei ‖ non encombré; sans dettes fpl.

unequal angle ‖ ungleichschenkliges Winkeleisen n. ‖ cornière f. à ailes inégales.

unequally loaded ‖ ungleich belastet ‖ à charge f. inégalement repartie.

uneven ‖ ungleich; uneben ‖ inégal. / of ~ ages ‖ von ungleichem Alter n. ‖ d'âges mpl. différents. / ~ surface ‖ Unebenheit f. des Bodens ‖ aspérité f. du terrain.

unexamined ‖ ungeprüft ‖ non examiné.

unexecuted ‖ unausgeführt ‖ inexécuté.

unexpected ‖ plötzlich; unerwartet ‖ subit; soudain.

unexplored ‖ unerforscht ‖ inexploré.

unexposed ‖ unbelichtet ‖ non sensibilisé.

unfair competition ‖ unlauterer Wettbewerb m. ‖ concurrence f. déloyale.

unfavorable ‖ ungünstig ‖ défavorable; désavantageux.

unfeasible ‖ unausführbar ‖ inexécutable; impossible d'exécuter.

unfermented ‖ ungegoren ‖ muet; non fermenté.

unfinished ‖ unvollendet; unfertig ‖ inachevé; imparfait.

unfit for work ‖ arbeitsunfähig ‖ impropre au travail m.

unfixed ‖ unbefestigt ‖ mobile. / ~ (Mine) ‖ rollig ‖ mouvant. / ~ masses pl. (Mine) ‖ rollige Massen fpl. ‖ terrains mpl. mouvants.

unfounded ‖ grundlos ‖ dénué de fondement.

unfrozen, to reach down into ~ ground ‖ in frostfreie Tiefe f. reichen ‖ atteindre la profondeur à l'abri de la gelée.

unfurl, to ~ a sail ‖ ein Segel n. losmachen ‖ déferler ou larguer une voile.

ungasifiable ‖ nicht vergasbar; unvergasbar ‖ ingazéifiable.

ungear, to (Mach) ‖ auslösen ‖ dégager; déclencher.

ungild, to ‖ entgolden ‖ dédorer.

ungrease, to ‖ entfetten ‖ dégraisser.

ungreasing agents pl. ‖ Entfettungsmittel npl. ‖ produits mpl. à dégraisser.

unground ‖ ungeschliffen ‖ non-rectifié.

unguent (Mach) ‖ Schmiermittel n. ‖ enduit m.; graisse f.

ungula (Geom) ‖ schräger Zylinderabschnitt m. oder Kegelabschnitt m. ‖ onglet m.

ungum, to (Dyer) ‖ ansieden ‖ décruser.

unhair, to ‖ abhaaren ‖ ébourrer; épiler; surtondre. / ~ the hide ‖ die Haut enthaaren oder abpälen ‖ débourrer ou dépiler les peaux fpl.

unhairing machine ‖ Enthaarungsmaschine f. ‖ machine f. de dépilage.

unhang, to ~ the rudder ‖ das Ruder aushaken ‖ démonter le gouvernail.

unhardened (Steel) ‖ ungehärtet ‖ non trempé.

unhealthy ‖ ungesund ‖ maladif; insalubre. / ~ climate ‖ ungesundes Klima n. ‖ climat m. insalubre.

unheard ‖ ungehört ‖ non entendu.

unhook, to ‖ abhaken ‖ décrocher.

uniaxial (Crystal) ‖ einachsig ‖ à un axe; uniaxe. / ~ crystal ‖ einachsiger Kristall m. ‖ cristal m. uniaxe.

uniaxiality (Crystal) ‖ Einachsigkeit f. ‖ uniaxie f.

unicellular ‖ einzellig ‖ unicellulaire.

unidirectional ‖ in einer Richtung f. wirkend ‖ agissant à sens m. unique.

unifilar suspension ‖ Einfadenaufhängung f. ‖ suspension f. unifilaire.

uniflow steam engine ‖ Gleichstromdampfmaschine f. ‖ machine f. à vapeur équicourant.

uniform ‖ gleichförmig; gleichmäßig ‖ uniforme. / ~ (Material) ‖ gleichförmig; homogen ‖ homogène; uniforme. / ~ decreasing motion ‖ gleichförmig verzögerte Bewegung f. ‖ mouvement m. uniformément retardé. / of ~ density ‖ von gleichförmiger Dichte f. ‖ uniformément compacte. / ~ duty ‖ Einheitszoll m. ‖ droit m. unitaire. / ~ mixture ‖ gleichmäßige Mischung f. ‖ mélange m. uniforme. / ~ quality of material ‖ gleichmäßige Werkstoffgüte f. ‖ qualité f. uniforme d'un matériel. / ~ running ‖ regelmäßiger Gang m. ‖ marche f. normale.

uniform ‖ Uniform f. ‖ uniforme m.; tenue f. (militaire). / ~ button ‖ Uniformknopf m. ‖ bouton m. pour uniformes. / ~ cloth ‖ Uniformtuch n. ‖ drap m. pour uniformes.

uniformity ‖ Gleichförmigkeit f.; Gleichheit f. ‖ égalité f. / to adjust to ~ ‖ auf Gleichheit f. einstellen ‖ mettre au point d'égalité. / ~ of all teeth ‖ Gleichheit f. aller Zähne ‖ égalité f. de toutes les dents. / ~ of the zinc coating ‖ Gleichförmigkeit f. des Zinküberzuges ‖ uniformité f. de la couche de zinc.

uniformly accelerated motion ‖ gleichförmig beschleunigte Bewegung f. ‖ mouvement m. uniformément accéléré. / ~ distributed load ‖ gleichmäßig verteilte Belastung f. ‖ charge f. uniformément répartie. / ~ variable motion ‖ gleichförmig veränderliche Bewegung f. ‖ mouvement m. uniformément variable.

uniform outfit ‖ Uniformausrüstungsgegenstand m. ‖ garniture f. d'uniformes. / ~ tailor ‖ Uniformschneider m. ‖ tailleur-uniformier m. / ~ trimmings pl. ‖ Uniformbeschläge mpl. ‖ garnitures fpl. d'uniformes.

unilateral conductivity ‖ einseitige Leitfähigkeit f. ‖ conductance f. unilatérale.

uninhabited ‖ unbewohnt ‖ inhabité.

unintelligible telegram ‖ unverständliches Telegramm n. ‖ télégramme m. inintelligible.

uninterrupted, the rails of the line are ~ throughout ‖ die Schienen fpl. sind ohne Unterbrechung durchgeführt ‖ les rails mpl. sont posés sans interruption de la ligne.

union ‖ Verbindung f. ‖ union f.; jonction f. / ~ (Trade) ‖ Gewerkschaft f.; Verband m. ‖ syndicat m.; union f. / ~ for fighting purposes ‖ Kampfverband m. ‖ union f. de combat ou de lutte. / ~ of interests ‖ Interessengemeinschaft f. ‖ convention f.; syndicat m.; communauté f. d'intérêts. / trade ~ see workmen's ~. / workmen's ~ ‖ Arbeiterverein m. ‖ union f. d'ouvriers; syndicat m. ouvrier.

union hose ‖ Schlauchverschraubung f. ‖ joint m. pour manches (de pompe).

unionist ‖ Gewerkschafter m.; Gewerkschaftsmitglied n. ‖ membre m. d'un syndicat.

union joint hose ‖ Schlauchverbindung f. ‖ tuyau m. de raccordement. / ~ nut ‖ Schlauchverschraubung f. ‖ joint m. ou raccord m. pour manches de pompe. / ~ screw ‖ Schraubverbindung f. ‖ raccord m. à vis.

unipolar ‖ einpolig ‖ unipolaire. / ~ dynamo ‖ Unipolardynamo f.; Einpolgerdynamo m. ‖ dynamo f. unipolaire.

unison ‖ Einklang m.; Gleichklang m. ‖ unisson m.

unit ‖ Einheit f. ‖ unité f. / absolute ~ (Electr) ‖ absolute Einheit f. ‖ unité f. absolue. / caloric ~ see thermal ~. / ~ of coinage ‖ Münzeinheit f. ‖ unité f. monétaire. / derived ~ ‖ abgeleitete Einheit f. ‖ unitée f. dérivée. / ~ of force ‖ Krafteinheit f. ‖ unité f. de force. / fundamental ~ ‖ Grundeinheit f. ‖ unité f. fondamentale. / ~ of heat see thermal ~. / ~ of length ‖ Längeneinheit f. ‖ unité f. de longueur. / ~ of light ‖ Lichteinheit f. ‖ unité f. de lumière. / ~ of measure ‖ Maßeinheit f. ‖ unité f. de

mesure. / ~ of output || Leistungseinheit f. || unité f. de débit *ou* de puissance. / practical ~ || praktische Einheit f. || unité f. pratique. / ~ of a steam plant || Maschinensatz m. eines Dampfkraftwerkes || groupe m. de machines génératrices d'une usine thermique. / ~ of surface (Surv) || Flächeneinheit f. || unité f. de surface *ou* de superficie. / ~ of tension || Spannungseinheit f. || unité f. de tension. / thermal ~ || Wärmeeinheit f.; Kalorie f. || calorie f.; unité f. thermique *ou* de chaleur. / ~ of time || Zeiteinheit f. || unité f. de temps. / ~ of weight || Gewichtseinheit f.; Wichte f. || unité f. de poids. / ~ of work || Arbeitseinheit f. || unité f. de travail.

unit construction || Blockkonstruktion f. || construction f. en une pièce.

unite, to || zusammenfügen || joindre; réunir; assembler.

unit load || Flächeneinheitslast f. || charge f. par mètre carré. / ~ measure || einheitliches Maß n. || mesure f. unitaire. / ~ price || Einheitspreis m. || prix m. unitaire.

unity *see* unit.

universal blade with adjustable and removable cutters for shearing U-and double T-iron || Universalmesser n. mit einsetzbaren und verschiebbaren Schneidzungen für U- und Doppel-T-Eisen || lame f. universelle avec languettes rapportables et déplaçables pour fer à U et en double T. / ~ bridge || Universalbrücke f. || pont m. universel. / ~ chuck (Mach) || Universalfutter n. || mandrin m. universel. / ~ clamping chuck || Klemmfutter n. || mandrin m. de fixation universel. / ~ compasses pl. || Universalzirkel m. || compas m. universel. / ~ field glass || Universalfeldstecher m. || jumelle f. universelle. / ~ folding, rounding and box-forming machine || Universalabkant-, Rund- und Kastenbiegemaschine f. || machine f. universelle à plier, rouler et former des boîtes. / ~ forging machine || Universalschmiedemaschine f. || machine f. à forger universelle. / ~ grinding machine || Universalschleifmaschine f. || machine f. à rectifier universelle. / gun carriage || Rundumlafette f. || affût m. universel. / ~ hoisting appliance || Universalhebezeug n. || appareil m. universel de levage. / ~ joint || Kreuzgelenk n. || joint m. de cardan. / ~ lathe || Universaldrehbank f. || tour m. universel. / ~ mill || Universalwalzwerk n. || laminoir m. universel. / ~ milling machine || Universalfräsmaschine f. || fraiseuse f. universelle. / ~ motor || Universalmotor m. || moteur m. électrique universel. / ~ notching machine || Universalausklinkmaschine f. || grugeoir m. universel. / ~ pliers pl. || Universalzange f. || pince f. universelle. / steam and air hydraulic ~ press || dampf- und lufthydraulische Universalpresse f. || presse f. universelle vapo-hydraulique et hydropneumatique. / ~ punching machine || Universallochmaschine f. || poinçonneuse f. universelle. / horizontal ~ punching machine || horizontale Universallochmaschine f. || poinçonneuse f. universelle **ho**rizontale. / horizontal ~ punching machine combined with bending and straightening press || horizontale Universallochmaschine f.

vereinigt mit Biege- und Richtpresse || poinçonneuse f. universelle horizontale combinée avec presse à cintrer et à dresser. / ~ double-ended punching and notching machine || doppelte Universalloch- und Ausklinkmaschine f. || poinçonneuse f. double universelle-grugeoir. / ~ punching and notching machine combined with open-gap plate shearing machine || vereinigte Universalloch- und Ausklinkmaschine f. mit Ausladungsblechschere || poinçonneuse f. universelle-grugeoir combinée avec cisaille à tôles à col de cygne. / ~ radial drilling machine || Universalradialbohrmaschine f. || perceuse f. radiale universelle. / ~ rail anchor (Railw) || Einheitsklemme f. || dispositif m. d'ancrage universel. / ~ revolving stage || Universaldrehtisch m. || plateau rotatif universel. / ~ rolling mill *see* ~ mill. / ~ time clock || Weltzeituhr f. || horloge f. indiquant les différentes heures du globe. / ~ tool milling machine || Universalwerkzeugfräsmaschine f. || fraiseuse f. d'outils universelle. / ~ varnish || Universallack m. || vernis m. universel.

universe || Weltraum m. || univers m.

unknown || Unbekannte f. || inconnue f.

unlash, to ~ a rope || ein Tau n. abstecken *oder* losstecken || démarrer un cordage.

unlay, to ~ a rope || ein Tau n. in Kardeele zerlegen || décommettre un cordage.

unleavened bread || ungesäuertes Brot n. || pain m. azyme.

unlevelled || nicht wagerecht || hors de niveau. / ~ (Mine) || nicht söhlig || hors -d'eau.

unlike (Electr) || ungleichnamig || de signes mpl. contraires.

unliming || Entkalkung f. || lavage m.; déchaudage.

unlimited credit || Blankokredit m. || crédit m. à découvert. / ~ power of attorney || Blankovollmacht f. || blanc-seing m. / ~ quantity || beliebig große Menge f. || quantité f. illimitée.

unload, to || abladen; entladen || décharger; débarquer. / ~ a kiln (Malt) || eine Darre f. abräumen || décharger une touraille. / ~ a ship || ein Schiff n. löschen || décharger un navire.

unloaded || unbelastet || sans charge f.

unloader || Auslader m.; Entlader m. || déchargeur m.

unloading || Entladen n. || déchargement m. / ~ (Mar) || Löschung f. || déchargement m. / ~ of the cargo || Löschen n. der Ladung || débarquement m. de la cargaison. / ~ of a kiln (Malt) || Abräumen n. einer Darre || déchargement m. d'une touraille.

unloading apparatus || Löschapparat m.; Entladeapparat m. || appareil m. de déchargement. / ~ crane || Entladekran m.; Umladekran m.; Überladekran m. || grue f. de déchargement *ou* de transbordement. / the ~ efficiency has been raised by new discharging appliances || die Entlademöglichkeit f. wurde durch Anlage weiterer Ausladevorrichtungen verstärkt || le transbordement fut facilité par des installations de déchargement à capacité plus forte. / ~ plant for bulk goods || Umladeanlage f. für Massengüter || installation f. de transbordement pour matières en amas. / ~

time || Entladezeit f. || temps m. nécessaire au déchargement.

unlock, to || aufschließen; öffnen || ouvrir. / ~ (Print) || aufschließen || desserrer.

unmagnetizing apparatus || Entmagnetisierungsapparat m. || appareil m. à désaimanter.

unmalleable || undehnbar || immalléable.

unmalted grain || ungemälztes Getreide n. || grain m. non malté.

unmarried workman || unverheirateter Arbeiter m. || ouvrier m. célibataire.

unmoor, to (Mar) || das Tauwerk losmachen || démarrer; désamarrer. / ~ a ship || den Anker m. lichten || lever *ou* déplanter *ou* déraper l'ancre f.; démarrer un vaisseau.

unmounted mirror || ungefaßter Spiegel m. || miroir m. non monté.

unnail, to (Railw) || die Schienennägel mpl. ausziehen || arracher les tirefonds mpl.

unnavigable || nicht schiffbar || innavigable.

unpack, to || auspacken || déballer. / ~ **unpacking** the machine || die Maschine auspacken || déballer la machine.

unpaid item || unbeglichener Posten m. || solde m. restant à régler.

unpleasant odour || unangenehmer Geruch m. || odeur f. désagréable.

unpolarizable || unpolarisierbar || susceptible de ne pas se polariser; impolarisable.

unpolished stoneware || einfaches Steinzeug n. || grès m. mat.

unprecedented || beispiellos || sans exemple m.

unpretentious || anspruchslos || sans prétentions fpl.; modeste.

unproductive || unproduktiv || improductif.

unpunctual || unpünktlich || inexact.

unquiet running || unruhiger Gang m. || marche f. par à-coups.

unrecognizable || unkenntlich || méconnaissable.

unreliability of result || Unsicherheit f. des Ergebnisses || insécurité f. du résultat.

unrib, to || entrippen || écôter. / ~ **leaves** pl. of tobacco || Tabakblätter npl. ausrippen || éjamber les feuilles fpl. de tabac.

unrigged || abgetakelt || dégréé.

unrigging || Abtakelung f. || dégréement m.; dégréage m.

unrivet, to || losnieten || dériver.

unroasted (Ore) || roh; ungeröstet || cru; non grillé.

unroof, to ~ a house || ein Haus n. abdecken || découvrir une maison.

unrust, to || entrosten; den Rost m. entfernen || dérouiller.

unsaleable article || Ladenhüter m. || fond m. de magasin; garde-boutique m.

unsaponifiable || unverseifbar || insaponifiable.

unsatisfactory || unbefriedigend || peu satisfaisant.

unsaturated || ungesättigt || non saturé. / ~ air || ungesättigte Luft f. || air m. non saturé.

unscrew, to || losschrauben; abschrauben || dévisser.

unseaworthiness || Seeuntüchtigkeit f. || incapabilité f. de pendre la mer.

unseaworthy || nicht seetüchtig || innavigable.

unsecured loan || offener Kredit m. || avance f. à découvert.

unserviceable ‖ unbrauchbar ‖ impropre au service; inutilisable.

unshackle, to ‖ losschäkeln ‖ démaniller.

unship, to ‖ von Bord gehen ‖ débarquer.

unsightly ‖ unansehnlich ‖ peu apparent; insignifiant.

unsilvered clearance at the centre of the parabolic reflector ‖ unversilberte Stelle f. in der Mitte des Parabolspiegels ‖ place f. non argentée dans le centre du réflecteur parabolique.

unsized paper ‖ ungeleimtes Papier n. ‖ papier m. non-collé.

unskilled labourer ‖ ungelernter Arbeiter m. ‖ simple ouvrier m. / ~ worker see ~ labourer.

unsolder, to ‖ loslöten ‖ dessouder.

unspaced (Print) ‖ undurchschossen ‖ non interligné.

unspool, to ‖ abspulen ‖ dévider; débobiner.

unsprung ‖ ungefedert ‖ non-suspendu à ressort.

unstable ‖ unbeständig ‖ instable.

unstick time (Airpl) ‖ Startzeit f. ‖ longueur f. de roulement au décollage.

unstitch, to ‖ eine Naht f. auftrennen ‖ découdre une couture.

unsuccessful ‖ erfolglos ‖ sans succès m.

unsuitable ‖ verfehlt; nicht passend ‖ manqué.

unsymmetrical ‖ unsymmetrisch ‖ dissymétrique.

untangler, yarn ~ (Spinn) ‖ Entwirrer m. ‖ démêleur m. de fils.

untanned ‖ ungegerbt ‖ non tanné; cru.

untearable ‖ unzerreißbar ‖ indéchirable.

untempered steel ‖ ungehärteter Stahl m. ‖ acier m. non trempé.

unthrifty ‖ unwirtschaftlich ‖ dépensier.

untie, to (Print) ‖ abbinden ‖ détacher; délier. / ~ the form (Print) ‖ die Form auflösen ‖ délier la forme. / ~ the page cord ‖ die Kolumnenschnur abbinden ‖ délier les pages mpl. ou la ficelle.

untight ‖ undicht ‖ inétanche.

untiled ‖ abgedeckt ‖ découvert.

untin, to ‖ entzinnen ‖ ôter l'étamure f.

untouched (Min) ‖ unverritzt ‖ intact.

untreated water, evaporator for ‖ Rohwasserverdampfer m. ‖ vaporisateur m. d'eau brute.

untrue, the flywheel runs ~ ‖ das Schwungrad schlägt ‖ le volant tourne à faux.

unvitrified ‖ unverglast ‖ invitrifié.

unwashed produce (Mine) ‖ ungewaschenes Produkt n. ‖ produit m. non lavé.

unweldable ‖ unschweißbar ‖ non soudable.

unwieldly part (Railw) ‖ sperriges Stück n. ‖ pièce f. encombrante ou volumineuse ou de dimensions démesurées.

unwinding of the wire ‖ Abwickeln n. des Drahtes ‖ déroulement m. du fil. / ~ mechanism ‖ Abwickelvorrichtung f. ‖ mécanisme m. dérouleur.

unwrought ‖ unbearbeitet ‖ à l'état m. brut.

up ‖ hinauf ‖ vers le haut.

up and down line (Electr) ‖ Hin- und Rückleitung f. ‖ ligne f. d'aller et de retour. / ~ motion ‖ Auf- und Abbewegung f. ‖ mouvement m. montant et descendant. / ~ stroke ‖ Doppelhub m. ‖ course f. double.

upcast (Mine) ‖ Sprung m. ins Hangende ‖ relèvement m. / ~ ventilating shaft ‖ ausziehender Schacht m. ‖ puits m. d'appel.

up-current of air (Aero) ‖ Aufwind m. ‖ courant m. d'air ascendant.

up-grate ‖ Steigung f. ‖ rampe f.

uphill ‖ bergaufwärts ‖ en amont m.

upholstered ‖ bezogen; gepolstert ‖ rembourré. / ~ furniture ‖ Polstermöbel npl. ‖ meubles mpl. rembourrés. / ~ goods pl. ‖ Polsterwaren fpl. ‖ rembourrages mpl. / ~ seat ‖ Polstersitz m. ‖ siège m. rembourré.

upholsterer ‖ Tapezierer m. ‖ tapissier m.

upholsterer's claw-handled hammer ‖ Tapeziererhammer m. mit Nagelklaue ‖ marteau m. de tapissier avec tireclou. / ~ hammer ‖ Tapeziererhammer m. ‖ marteau m. de tapissier. / ~ trimming ‖ Möbelposamenten pl. ‖ passementerie f. d'ameublement.

upholstering material ‖ Polsterwerkstoff m. ‖ matériel m. à rembourrer. / ~ nail ‖ Polsternagel m. ‖ clou m. de tapisserie. / stuff for ~ ‖ Tapezierstoff m. ‖ étoffe f. pour tapisserie.

upholstery ‖ Polsterung f. ‖ capitonnage m. / ~ (Shop) ‖ Tapeziergeschäft n. ‖ tapisserie f.; tenture f. / ~ (Ware) see upholstered goods. / ~ weaver ‖ Möbelstoffweber m. ‖ tisseur m. d'étoffes d'ameublement.

upkeep ‖ Instandhaltung f.; Unterhaltung f. ‖ entretien m. / ~ of the rolling stock ‖ Unterhaltung f. der Betriebsmittel ‖ l'entretien m. du matériel roulant. / to be capable of operating at a minimum of ~ expenses pl. ‖ mit einem Minimum an Unterhaltungskosten pl. arbeiten können ‖ pouvoir fonctionner avec le minimum de frais d'entretien.

uplift of strata ‖ Aufrichtung f. der Schichten ‖ redressement m. des couches.

upper ‖ ober(er) ‖ supérieur. / ~ case (Print) ‖ Kapitalkasten m. ‖ haut m. de casse. / ~ casing ‖ Gehäuseoberteil m. ‖ partie f. supérieure du carter. / ~ chamber (Sluice) ‖ Oberhaupt n. ‖ chambre f. d'amont; tête f. d'amont. / ~ chord (Beam) ‖ Obergurt m. ‖ bride f. supérieure.

upper-deck ‖ Oberdeck n. ‖ pont m. supérieur. / ~ beam ‖ Oberdecksbalken m. ‖ barrot m. du pont supérieur.

upper edge ‖ Oberkante f. ‖ bord m. supérieur. / ~ of a rail ‖ Oberkante f. einer Schiene ‖ dessus m. du rail.

upper end of the heel post of the lock gate of a sluice ‖ Hals m. einer Schleusentorwendesäule ‖ tourillon m. du poteau d'une porte d'écluse. / ~ fermentation ‖ Obergärung f. ‖ fermentation f. ordinaire. / ~ frog of the hinge (Door) ‖ Oberpfanne f. für die Türangel ‖ femelle f. de penture de porte. / ~ guard (Railw) ‖ Zugführer m. ‖ chef m. de train. / ~ layer ‖ Deckenschicht f. ‖ couche f. de couverture.

upper leather (Curr) ‖ Schmalleder n. ‖ cuir m. à œuvre; cuir m. mou pour tiges de bottes. / ~ (Shoem) ‖ Oberleder n. ‖ (cuir m. de) dessus m.; empeigne f. / ~ cutter (Shoem) ‖ Oberlederzuschneider m. ‖ coupeur m. de dessus. / ~ manufacturing (Shoem) ‖ Schäftefertigung f. ‖ fabrication f. de tiges.

upper lip ‖ Oberlippe f. ‖ lèvre f. supérieure. / ~ mast ‖ Stänge f. ‖ mât m. supérieur.

upper part ‖ Oberteil m. ‖ partie f. supérieure. / ~ of the boshes (Metal) ‖ Kohlensack m. eines Schachtofens ‖ ventre m. d'un four à cuve. / ~ of a blast furnace ‖ Obergestell n. eines Hochofens ‖ ouvrage m. d'un haut-fourneau. / ~ of a precious stone ‖ Pavillon m. eines Edelsteins ‖ dessus m. d'une pierre précieuse.

upper pond (Sluice) ‖ Oberwasser n. ‖ bief m. supérieur ou d'amont.

upper roller system ‖ Oberwalzensystem n. ‖ système m. des cylindres supérieurs.

upper rope traction ‖ Oberseilführung f. ‖ passage m. du câble par dessus.

uppers pl. ‖ Gamaschen fpl. ‖ guêtres fpl. / leather ~ ‖ Ledergamaschen fpl. ‖ guêtres fpl. en cuir.

upper shaft of a blast furnace ‖ Oberschacht m. eines Hochofens ‖ cuve f. ou vide m. d'un haut-fourneau. / ~ shell of a blast furnace ‖ Hochofenkegel m. ‖ tour m. d'un haut-fourneau. / ~ spring ‖ Oberfeder f. ‖ ressort m. supérieur. / ~ stitcher (Shoem) ‖ Schäftestepper m. ‖ piqueur m. de tiges. / ~ story ‖ Obergeschoß n. ‖ étage m. supérieur.

upper surface (Mine) ‖ Hangendes n. ‖ couche f. supérieure; toit m. / ~ of a rail ‖ Bahn f. oder Lauffläche f. einer Schiene ‖ surface f. ou table f. de roulement d'un rail.

upper tile ‖ Oberziegel m. ‖ tuile f. de dessus. / ~ work (Shipb) ‖ Oberschiff n.; totes Werk n. ‖ accastillage m.; encastillage m.; œuvre f. morte.

upright ‖ aufrecht; gerade; senkrecht ‖ debout; en ligne f. droite. / Keep ~ ! ‖ nicht stürzen! ‖ ne pas renverser!; ne pas laisser tomber!

upright projection ‖ Aufriß m. ‖ projection f. verticale. / ~ shell (Build) ‖ Wölbring m. ‖ assise f. arquée debout. / ~ size (Bookb) ‖ Hochformat n. ‖ format m. normal.

upright of a gantry (Crane) ‖ Pendelstütze f. ‖ montant m. articulé.

uprooter ‖ Rübenheber m. ‖ arrache-betteraves m.

uprooting machine ‖ Rodemaschine f. ‖ déracineuse f.

upset (Wood) ‖ wimmerig ‖ à fibres fpl. tordues.

upset, to (Forg) ‖ stauchen ‖ refouler. / the ingots pl. will be upset ‖ die Knüppel mpl. werden gestaucht ‖ les lingots mpl. sont refoulés.

upsetting ‖ Stauchung f.; Stauchen n. ‖ refoulement m. / cold press for forming and ~ rivets ‖ Kaltpresse f. zum Formen und Stauchen von Nieten ‖ presse f. à froid pour former et fouler les rivets.

upsetting device ‖ Stauchvorrichtung f. ‖ dispositif m. refouleur. / ~ die ‖ Stauchkaliber n. ‖ calibre m. à refouler.

upsetting machine ‖ Stauchmaschine f. ‖ machine f. à refouler. / ~ for working cold ‖ Kaltvorstauchmaschine f. ‖ machine f. à refouler à froid. / drill ~ ‖ Bohrerstauchmaschine f. ‖ machine f. à refouler les forets. / rock drill ~ ‖ Gesteinsbohrerstauchmaschine f. ‖ machine f. à refouler les forets à roches. / ~ for tyres ‖ Stauchmaschine f. für Radreifen ‖ machine f. à serrer les bandages sur les roues.

upsetting operation ‖ Stauchprozeß m.; Haucharbeit f. ‖ travail m. de refoulement.

upsetting press, mandrel ~ ‖ Dornstauchpresse f. ‖ presse f. à fouler les mandrins. / tube ~ ‖ Rohrstauchpresse f. ‖ presse f. à fouler les tubes.

upsetting test ‖ Stauchversuch m. ‖ essai m. au refoulement.

upside ‖ Oberseite f. ‖ dessus m.

upsignal (Railw) ‖ Einfahrtsignal n. ‖ signal m. avancé *ou* à l'arrivée.

upstream ‖ stromaufwärts ‖ en amont m. de.

upstroke (Print) ‖ Haarstrich m. ‖ délié m. / ~ (Mech) *see* upward movement.

up-to-date ‖ zeitgemäß ‖ opportun; de circonstance; de notre époque. / ~ (Fashion) ‖ modern ‖ moderne; à la mode; dernier genre m. / to bring ~ ‖ modernisieren ‖ moderniser.

upward, amount of ~ current of air ‖ Aufwindwert m. ‖ coëfficient m. d'ascendance du vent. / ~ folding wing ‖ hochklappbarer Flügel m.; Hochklappflügel m. ‖ aile f. repliable en hauteur. / dynamic ~ force ‖ dynamische Steigkraft f. ‖ force f. ascensionnelle dynamique. / ~ movement ‖ Aufwärtsbewegung f. ‖ mouvement m. ascendante.

upwards ‖ aufwärts ‖ en haut; vers le haut.

upward tendency ‖ Haussetendenz f. ‖ tendance f. à la hausse. / ~ of prices ‖ Steigen n. der Preise ‖ renchérissement m.; hausse f. de prix.

up wind (Aero) ‖ Aufwind m. ‖ courant m. d'air ascendant.

uraninite ‖ Uranpecherz n. ‖ uraninite f.

uranium ‖ Uran n. ‖ uranium m.; urane m. / ~ containing glass ‖ Uranglas n. ‖ verre m. d'urane. / ~ glass strip ‖ Uranglasstreifen m. ‖ bande f. en verre d'urane. / ~ oxide ‖ Uranoxyd n. ‖ oxyde m. d'urane.

urate ‖ harnsaures Salz n. ‖ urate m.

urea ‖ Harnstoff m. ‖ urée f.

urge, to ‖ auffordern; antreiben ‖ inviter; engager; pousser; stimuler. / ~ the fire ‖ das Feuer anschüren ‖ activer *ou* aviver *ou* tisonner le feu.

urgent call (Tel) ‖ dringendes Gespräch n. ‖ conversation f. urgente.

uric acid ‖ Harnsäure f. ‖ acide m. urique. / ~ derivative ‖ Harnsäureabkömmling m.; Harnsäurederivativ n. ‖ dérivé m. de l'acide urique.

urine ‖ Harn m.; Urin m. ‖ urine f. / ~ vat (Dyer) ‖ Urinküpe f. ‖ cuve f. à l'urine.

usage ‖ Verwendungszweck m. ‖ usage m.

usance ‖ Gebrauch m. ‖ usance f. / local ~ ‖ Platzgebrauch m. ‖ usance f. de la place.

use, to ‖ anwenden; gebrauchen ‖ utiliser; employer; faire usage m. de. / ~ up ‖ verbrauchen; aufbrauchen ‖ consommer; épuiser. / ~ up (Mach) ‖ verschleißen ‖ s'user.

use ‖ Anwendung f. ‖ emploi m.; utilité f.; application f. / ~ (Custom) ‖ Gebrauch m. ‖ usage m. / ~ (Industrial works) ‖ Nutzung f. ‖ mise f. à profit; exploitation f.; jouissance f.; rapport m. / annual ~ ‖ jährlicher Bedarf m. ‖ usage m. annual. / article of daily ~ ‖ Gebrauchsgegenstand m. ‖ objet m. d'usage. / extensive ~ ‖ ausgedehnte Verwendung f. ‖ emploi m. vaste. / for external ~ (Med) ‖ zum äußerlichen Gebrauch m. ‖ pour usage m. externe. / for immediate ~ ‖ zum sofortigen Gebrauch m. ‖ pour usage m. immédiat. / ready for immediate ~ ‖ sofort gebrauchsfertig ‖ prêt immédiatement à l'usage. / in ~ ‖ im Gebrauch m. ‖ en usage m. / to make ~ of ‖ gebrauchen; benutzen ‖ faire usage m. de; employer; utiliser. / to make ~ of the situation of a power station ‖ die Standortsvorteile mpl. eines Kraftwerkes ausnutzen ‖ mettre en valeur les avantages qu'offre l'emplacement d'une usine.

useful ‖ nützlich; brauchbar; zweckdienlich; zweckentsprechend ‖ utile; utilisable; convenable; effiace. / to be ~ ‖ taugen; wert sein ‖ être bon *ou* utile; valoir. / the shop covers a ~ area of about x square metres ‖ die Werkstatt f. bedeckt eine nutzbare Fläche von x qm ‖ la surface f. utile de l'atelier est de x-mètres carrés. / ~ height ‖ Nutzhöhe f. ‖ hauteur f. utile. / ~ output ‖ Nutzleistung f. ‖ effet m. utile. / ~ power *see* ~ output. / ~ resistance (Radio) ‖ Nutzdämpfung f. ‖ amortissement m. utile.

useless ‖ unbrauchbar; zwecklos ‖ inutile; impropre au service; inutilisable. / making ~ ‖ Unbrauchbarmachen n. ‖ mise f. hors usage.

user ‖ Konsument m.; Verbraucher m. ‖ consommateur m. / ~ (Electr) ‖ Abnehmer m. ‖ consommateur m.

using, right of ~ ‖ Nutzungsrecht n. ‖ droit m. d'usage.

usual ‖ üblich ‖ usuel. / ~ commercial quality ‖ handelsübliche Güte f. ‖ qualité f. courante.

usufruct of surface and soil ‖ Nutzung f. von Grund und Boden ‖ utilisation f. du fonds. / to withdraw the ~ ‖ die Nutzung f. entziehen ‖ retirer la faculté de jouissance. / possessor of the ~ of a property ‖ Nutzungsberechtigter m. eines Grundstückes ‖ usufruitier m. d'un terrain.

utensils pl. ‖ Gerät n.; Handwerkzeug n.; Zeug n.; Gezähe n. ‖ ustensiles mpl.; outils mpl. / ~ for manufacturing cigarettes ‖ Utensilien pl. für Zigarettenfertigung ‖ ustensiles mpl. pour la fabrication de cigarettes. / ~ for courses of instruction ‖ Lehrwerkzeug n. ‖ outils mpl. pour l'enseignement. / ~ for the domestic usage ‖ Gerät n. für den häuslichen Gebrauch ‖ ustensiles mpl. servant à l'usage domestique.

utility ‖ Nützlichkeit f. ‖ utilité f. / ~ model *see* ~ patent. / ~ patent ‖ Gebrauchsmuster n. ‖ modèle m. d'ubilité.

utilization ‖ Verwertung f.; Ausnutzung f. ‖ réalisation f.; mise f. en valeur; utilisation f. / clever ~ ‖ geschickte Ausnutzung f. ‖ utilisation f. habile. / economical ~ ‖ wirtschaftliche Ausnutzung f. ‖ utilisation f. économique. / ~ of exhaust gases *see* ~ of waste gases. / ~ of fuel ‖ Brennstoffausnützung f. ‖ utilisation f. du combustible. / ~ of heat ‖ Wärmeausnutzung f. ‖ utilisation f. de la chaleur. / ~ of heat of exhaust gases ‖ Abgaswärmeverwertung f. ‖ utilisation f. de la chaleur des gaz perdus. / industrial ~ ‖ gewerblicher Gebrauch m. ‖ utilisation f. industrielle. / plant for the ~ of old rubber ‖ Einrichtung f. zur Verarbeitung von Altgummi ‖ installation f. pour l'utilisation de vieux cautchouc. / ~ per cent ‖ Nutzprozente npl. ‖ taux m. du rendement. / ~ of waste gases ‖ Abgasverwertung f. ‖ utilisation f. des gaz brûlés *ou* d'échappement. / ~ of waste heat ‖ Abwärmeverwertung f. ‖ utilisation f. de la chaleur perdue. / ~ of the wind ‖ Windausnutzung f. ‖ utilisation f. du vent.

utilize, to ‖ ausnutzen ‖ utiliser; tirer profit m.; exploiter. / ~ the motor fully ‖ den Motor m. voll ausnutzen ‖ utiliser la pleine puissance du moteur.

utilizing plant ‖ Verwertungsanlage f. ‖ installation f. d'utilisation. / carcass ~ ‖ Kadaververwertungsanlage f. ‖ installation f. de préparation industrielle des cadavres d'animaux. / waste ~ ‖ Abfallverwertungsanlage f. ‖ installation f. pour l'utilisation des déchets.

utmost ‖ Äußerste n. ‖ extrême m.

utterance ‖ Äußerung f. ‖ expression f.; propos m.; déclaration f.

V

V-aerial ‖ V-Antenne f. ‖ antenne f. en V. / ~ dump car with automatic locking device ‖ Muldenkipper m. mit selbsttätiger Muldenfeststellung ‖ basculeur m. à auge avec dispositif de fixation automatique de l'auge. / ~ notch *see* V-shaped notch. / ~-shaped bearing ‖ wiegenartiges Zapfenlager n. ‖ palier m. angulaire *ou* en berceau. / ~-shaped notch ‖ Scharfkerb m. ‖ entaille f. en V. / ~-toothed ‖ mit Pfeilzähnen mpl. versehen ‖ taillé à chevrons mpl.

vacancy ‖ offene Stelle f. ‖ place f. vacante.

vacant post *see* vacancy. / ~ situation *see* vacancy.

vacate, to ‖ leeren ‖ vider; évacuer.

vacation ‖ Ferien pl. ‖ vacations fpl.

vaccinating lancet ‖ Impflanzette f. ‖ lancette f. à vacciner.

vaccination, instrument for ~ ‖ Impfinstrument n. ‖ instrument m. pour la vaccination.

vaccine lymph ‖ Lymphe f. ‖ vaccin m. / ~ matter *see* vaccine lymph.

vachette leather ‖ Vachette f. ‖ vachette f.

vacuum ‖ Vakuum m.; luftleerer Raum m. ‖ vide m. / in absolute ~ ‖ im absoluten Vakuum m. ‖ dans le vide absolu. / Torricellian ~ ‖ Torricellische Leere f. ‖ vide m. de Torricelli.

vacuum air pump ‖ Vakuumluftpumpe f. ‖ pompe f. à faire le vide. / ~ apparatus ‖ Vakuumapparat m. ‖ appareil m. à vide. / ~ bell jar ‖ Vakuumglocke f. ‖ cloche f. à vide. / ~ brake ‖ Vakuumbremse f. ‖ frein m. à vide. / ~ bulb ‖ Vakuumbirne f. ‖ ampoule f. à vide. / ~ cleaner ‖ Staubsauger m. ‖ aspirateur m. de poussiére. / ~ cleaning ‖ Staubabsaugen n. ‖ nettoyage m. par le vide. / ~ distillation ‖ Vakuumdestillation f. ‖ distillation f. dans le vide. / ~ drier ‖ Vakuumtrockner m. ‖ sécheur m. *ou* étuve f. à vide. / ~ drying apparatus *see* ~ drier. / ~ drying chamber ‖ Vakuumtrockenschrank m. ‖ armoire f. de séchage par le vide. / ~ drying drum ‖ Vakuumtrockentrommel f. ‖ tambour m. de séchage par le vide. / ~ drying plant ‖ Vakuumtrockenanlage f. ‖ installation f. de séchage par le vide. / ~ evaporation plant ‖ Vakuumdampfungsanlage f. ‖ installation f. de vaporisation dans le vide. / ~ evaporator ‖ Vakuumverdampfer m. ‖ vaporisateur m. par le vide. / ~ fermentation ‖ Vakuumgärung f. ‖ fermentation f. dans le vide. / ~ fermenting process ‖ Vakuumgärverfahren n. ‖ procédé m. de fermentation dans le vide. / ~ flask ‖ Isolierflasche f. ‖ bouteille f. isolante. / ~ fuel feed device ‖ Unterdruckbrennstofförderer m. ‖ élévateur m. d'essence à vide. / ~ fuse ‖ Vakuumsicherung f. ‖ parafoudre m. à vide. / ~ gauge ‖ Vakuummesser m. ‖ vacuomètre m. / ~ ice machine ‖ Vakuumeismaschine f. ‖ machine f. à froid à vide. / ~ impregnation ‖ Imprägnierung f. im Vakuum ‖ imprégnation f. par le vide. / ~ pan (Sugar)

vacuumpfanne f. ‖ appareil m. à cuire (le sucre) dans le vide. / ~ panman (Sugar) ‖ Zuckerkocher m. ‖ cuiseur m. / ~ pump ‖ Vakuumpumpe f. ‖ pompe f. à vide. / ~ receiver ‖ Saugluftempfänger m. ‖ poste m. récepteur à air aspiré. / ~ sender ‖ Saugluftsender m. ‖ poste m. émetteur à air aspiré. / ~ tube *see* vacuum valve.

vacuum valve ‖ Vakuumröhre f. ‖ valve f. à vide. / ~ arrester ‖ Vakuumblitzableiter m. ‖ parasurtension m. à tube à vide. / ~ receiver ‖ Vakuumröhrenempfänger m. ‖ récepteur m. à valve d'oscillations. / ~ rectifier ‖ Richtverstärker m. ‖ amplificateur m. redresseur.

vaginal capsule ‖ Vaginalkapsel f. ‖ capsule f. vaginale. / ~ retractor (Med) ‖ Scheidenhalter m. ‖ valve f. vaginale. / ~ speculum (Med) ‖ Scheidenspiegel m. ‖ spéculum m. vaginal.

vagrant current ‖ vagabundierender Strom m. ‖ courant m. vagabond.

valence ‖ Wertigkeit f.; Valenz f. ‖ valence f. / secondary ~ ‖ Nebenvalenz f. ‖ valence f. secondaire.

valency *see* valence.

valentinite ‖ Antimonblüte f. ‖ valentinite f.

valerianate ‖ Baldriansalz n. ‖ valérianate m.

valerian acid ‖ Baldriansäure f. ‖ acide m. valérianique. / ~ oil ‖ Baldrianöl n. ‖ essence f. de valériane. / ~ root ‖ Baldrianwurzel f. ‖ racine f. de valériane.

valid ‖ gültig; vollgültig ‖ valable. / ~ in law ‖ rechtsgültig ‖ valide; légal. / to render ~ ‖ legalisieren ‖ légaliser.

validation ‖ Gültigkeitserklärung f. ‖ validation f.

validity ‖ Rechtsgültigkeit f. ‖ authenticité f.; validité f. / ~ of a claim ‖ Rechtsgültigkeit f. einer Forderung ‖ validité f. d'une créance.

valise ‖ Mantelsack m. ‖ valise f.

valonia ‖ Knopper f. ‖ gallon m.

valley ‖ Tal n. ‖ vallée f. / longitudinal ~ ‖ Längstal n. ‖ vallée f. longitudinale. / rift ~ (Geol) ‖ Graben m. ‖ renforçage m. / round deep ~ ‖ Kesseltal n. ‖ vallée f. encaissée. / transversal ~ ‖ Quertal n. ‖ vallée f. transversale. / bottom of ~ ‖ Talsohle f. ‖ étage m. de vallée.

valley breeze ‖ Talwind m. ‖ brise f. *ou* vent m. de vallée. / ~ channel *see* ~ gutter. / ~ gutter (Tiler) ‖ Dachziegel m. mit Kehlrinne ‖ noue f. cornière; noulet m. / ~ rafter ‖ Kehlsparren m.; Kehlschifter m. ‖ noulet-chevron m.; chevron m. à noulet. / ~ wind *see* valley breeze.

valuable ‖ wertvoll ‖ précieux.

valuation ‖ Schätzung f.; Wertbestimmung f.; Bewertung f. ‖ valuation f.; évaluation f. / ~ (Estimate) ‖ Kostenanschlag m. ‖ devis m. estimatif; estimation f. / expert's ~ ‖ Sachverständigengutachten n. ‖ évaluation f. par expert.

valuator ‖ Schätzer m. ‖ estimateur m.

value, to ‖ veranschlagen; abschätzen ‖ estimer; évaluer; taxer.

value ‖ Wert m. ‖ valeur f. / ~ (Bank) ‖ Valuta f. ‖ valeur f. / admissible ~ ‖ zulässiger Wert m. ‖ valeur f. admissible; tolérance f. / approximate ~ ‖ Näherungswert m. ‖ valeur f. rapprochée. / to declare the ~ ‖ den Wert m. angeben ‖ déclarer la valeur. / declared ~ ‖ Wertangabe f. ‖ déclaration f. de valeur. / to determine a ~ ‖ einen Wert m. ermitteln ‖ fixer une valeur. / estimated ~ ‖ Schätzungswert m. ‖ valeur f. estimative. / final ~ ‖ Endwert m. ‖ valeur f. finale. / determination of the ~ of fuels ‖ Brennstoffwertbestimmung f. ‖ détermination f. de la puissance calorifique des combustibles. / full ~ ‖ vollgültig ‖ qui a la valeur requise. / gross ~ ‖ Bruttowert m. ‖ valeur f. brute. / ~ of ground ‖ Bodenwert m. ‖ valeur f. du terrain. / mean ~ ‖ mittlerer Wert m.; mittlere Größe f.; Mittelwert m. ‖ moyenne f.; valeur f. moyenne. / minimum ~ ‖ Mindestwert m. ‖ valeur f. minimum. / nominal ~ ‖ Nennwert m. ‖ valeur f. nominale. / normal ~ ‖ Normalwert m. ‖ valeur f. normale. / surrender ~ ‖ Ablösungswert m. ‖ valeur f. de cession. / article of ~ ‖ Wertgegenstand m. ‖ objet m. précieux. / difference of ~ ‖ Wertunterschied m. ‖ différence f. de valeur.

valuer ‖ Abschätzer m.; Taxator m. ‖ taxateur m.; commissaire-priseur m.

valve, to (Gas) ‖ abblasen ‖ dégonfler.

valve ‖ Ventil n. ‖ soupape f. / ~ (Radio) ‖ Röhre f. ‖ tube m.; valve f. / to adjust the ~s pl. ‖ die Ventile npl. einstellen ‖ ajuster la distribution. / admission ~ ‖ Einlaßventil n. ‖ soupape f. d'admission. / air pump ~ ‖ Luftpumpenventil n. ‖ soupape f. de pompe à air. / air suction ~ ‖ Luftsaugeventil n. ‖ soupape f. d'aspiration d'air. / angle ~ ‖ Eckventil n. ‖ soupape f. d'équerre *ou* à coude. / automatic ~ *see* self-acting ~. / automatically operated inlet ~ ‖ selbsttätiges Einlaßventil n. ‖ soupape f. d'admission automatique. / ball ~ ‖ Kugelventil n. ‖ soupape f. à boulet. / ball retaining ~ ‖ Kugelrückschlagventil n. ‖ soupape f. de retenue à boulet. / bell-shaped ~ ‖ Kronenventil n.; Glockenventil n. ‖ soupape f. à cloche; clapet m. à couronne. / blast ~ ‖ Windschieber m. ‖ vanne f. à air. / blow-off ~ ‖ Ausblasventil n. ‖ soupape f. de purge. / the ~ blows off with reports ‖ das Ventil n. bläst unter Knallen ab ‖ la soupape f. laisse échapper l'air bruyamment. / bullet ~ *see* ball ~. / carburettor float ~ *see* float ~. / charging ~ ‖ Ladeventil n. ‖ soupape f. de charge. / check ~ ‖ Rückschlagventil n. ‖ soupape f. d'arrêt *ou* de retenue. / clack ~ *see* ball ~. / conical ~ ‖ Kegelventil n. ‖ soupape f. conique. / cup ~ *see* bell shaped ~. / cut-off ~ ‖ Absperrventil n. ‖ soupape f. de détente. / double-seat ~

|| Doppelsitzventil n. || soupape f. à double siège. / economical pressure water ~ || Druckwassersparventil n. || soupape f. à économie d'eau de pression. / excess pressure ~ || Überdruckventil n. || soupape f. de pression excessive *ou* de surpression. / exhaust ~ || Auslaßventil n.; Auspuffventil n. || soupape f. d'échappement. / expansion ~ || Regelventil n. || vanne f. de réglage. / expansion slide ~ || Expansionsschieber m. || tiroir **m.** à expansion *ou* à détente. / float ~ || Schwimmerventil n. || soupape f. à flotteur. / fuel ~ || Brennstoffventil n. || soupape f. à combustible. / gas ~ || Gasventil n. || soupape f. à gaz. / ~ of the gas container (Aero) || Gasflaschenventil n. || soupape f. de la bouteille à gaz. / globe ~ *see* ball ~. / graduating ~ || Abstufungsventil n. || valve f. de graduation. / inlet ~ || Einlaßventil n. || soupape f. d'admission. / interchangeable ~ || auswechselbares Ventil n. || soupape f. interchangeable. / inverted vertical ~ || hängendes Ventil n. || soupape f. renversée. / main ~ || Hauptventil n. || soupape f. principale. / manœuvring ~ || Manövrierventil n. || soupape f. de manœuvre. / mechanically operated ~ || gesteuertes Ventil n. || soupape f. commandée. / mechanically operated inlet ~ || gesteuertes Einlaßventil n. || soupape f. d'admission mécanique. / needle ~ || Nadelventil n. || pointeau m. / oil pump safety ~ || Ölpumpensicherheitsventil n. || soupape f. de sûreté de la pompe à huile. / one-grid ~ (Radio) || Eingitterröhre f. || lampe f. à une seule grille. / outlet ~ || Auslaßventil n. || soupape f. de purge *ou* d'échappement. / overhead inlet ~ || hängendes Einlaßventil n. || soupape f. d'admission commandée par le haut. / piston (slide) ~ || Kolbenschieber m. || piston-tiroir m.; vanne f. de distribution à tiroir-piston. / pitted ~ || angefressenes Ventil n. || soupape f. piquée. / pneumatic control ~ || Druckluftkontrollventil n. || soupape f. de contrôle pour air comprimé. / preliminary feed ~ for hydraulic presses || Vorfüllventil n. für hydraulische Pressen || soupape f. d'admission préalable pour presses hydrauliques. / pressure balancing ~ || Druckausgleichventil n. || soupape f. de compensation de pression. / pressure reducing ~ || Druckminderventil n. || soupape f. réductrice *ou* à réduction de pression. / pump ~ || Pumpenventil n. || soupape f. de pompe. / reducing ~ *see* pressure reducing ~. / regulating ~ || Regulierventil n. || soupape f. régulatrice. / release ~ || Auslösventil n. || valve f. de purge. / relief ~ || Überdruckventil n. || soupape f. de soulagement. / reversing ~ || Umsteuerventil n. || soupape f. de renversement de marche. / rotary slide ~ || Drehschieber m. || tiroir m. rotatif. / safety ~ || Sicherheitsventil n. || soupape f. de sûreté. / self-acting ~ || Selbstschlußventil n. || soupape f. à fermeture automatique. / self-acting pipe isolating ~ || Rohrbruchventil n. || soupape f. à rupture de conduite. / shutoff ~ || Absperrventil n. || soupape f. d'arrêt. / sleeve ~ || Muffenventil n. || soupape f. à manchon. / slide ~ || Schieber m. || tiroir m. / slide ~ chest || Schieberkasten m.; Schiebergehäuse n. || boîte f. à tiroir *ou* de distribution. / spherical ~

see ball ~. / spring loaded ~ || federbelastetes Ventil n. || soupape f. commandée par un ressort. / starter check ~ || Anlaßrückschlagventil n. || soupape f. de retenue de mise en marche. / starting ~ || Anlaßventil n. || soupape f. de mise en marche. / steam ~ || Dampfventil n. || soupape f. à vapeur. / steam pressure reducing ~ || Dampfdruckminderventil n. || soupape f. réductrice à vapeur. / steel cylinder ~ || Stahlflaschenventil n. || soupape f. pour des bouteilles en acier. / the ~ sticks || das Ventil hängt || la soupape reste accrochée. / stop ~ || Absperrventil n. || soupape f. d'arrêt. / stop ~ of pressure-gauge with round test flange || Manometerabsperrventil n. mit rundem Prüfflansch || soupape f. d'arrêt de manomètre à bride de contrôle ronde. / ~ with straight passage || Ventil n. mit geradem Durchgang || robinet m. (de prise) à passage droit. / tyre ~ || Schlauchventil n. || valve f. de gonflement; soupape f. de chambre à air. valve adjustment || Ventileinstellung f. || réglage m. de soupape. / ~ arrangement || Ventilanordnung f. || position f. des soupapes. / ~ bonnet || Ventilhaube f. || chapeau m. de soupape. / ~ box || Ventilgehäuse n. || boîte f. de soupape. / ~ bush (Loc) || Schieberbuchse f. || fourreau m. de tiroir. / ~ cage || Ventilkorb m. || cloche f. de soupape. / ~ cap gasket || Ventilverschraubungsdichtung f. || joint m. de bouchon de soupape. / ~ case *see* ~ box. / ~ chain (Cycle) || Ventilkette n. || chaîne f. de valve. / ~ chamber || Ventilkammer f. || chambre f. de soupape. / ~ characteristics || Anodenstromkennlinie f. || caractéristique f. de plaque. / ~ chest || Ventilkasten m. || boîte f. *ou* chapelle f. de soupape. / slide ~ chest || Schieberkasten m. || boîte f. à tiroir. / ~ cock || Ventilhahn m. || robinet-valve m. / ~ cone || Ventilkegel m. || clapet m. / ~ cover || Ventildeckel m. || couvercle m. de soupape. / ~ disk || Ventilteller m. || plateau m. de soupape. / ~ face || Schieberfläche f. || bande f. de frottement *ou* plane des tiroirs; glissière f. / ~ gear || Ventilantrieb m.; Ventilsteuerung f. || commande f. de soupapes. / accurate and silent ~ gear || genau und ruhig arbeitende Steuerung f. || distribution f. marchant avec précision et sans bruit. / ~ gear housing || Ventilverkleidung f. || cache-soupape m. / ~ generator (Radio) || Senderöhre f. || générateur m. à lampes. / ~ globe || Ventilkugel f. || boulet m. de soupape. / ~ grinding || Ventileinschleifen n. || rodage m. de soupapes. / ~ guard || Hubbegrenzung f. eines Ventils || buttoir m. de soupape. / ~ handle || Ventilhebel m. || levier m. de soupape. / ~ head || Ventilteller m.; Ventilkopf m. || plateau m. *ou* tête f. de soupape. / ~ holder (Radio) *see* ~ socket. / ~-in-head || hängende Ventilanordnung f. || soupapes fpl. à culbuteurs. valveless || ventillos || sans soupapes fpl. / ~ air-cooled internal combustion engine operating on the two-stroke cycle || ventilloser luftgekühlter Zweitaktverbrennungsmotor m. || moteur m. à explosion fonctionnant (suivant le cycle) à deux temps et sans soupapes. / pump || ventillose Pumpe f. || pompe

f. sans clapet. / adoption of the ~ type of two-stroke engine || ventillose Zweitaktbauart f. || adoption f. d'un moteur à deux temps (du système) sans soupapes. **valve lift** || Ventilhub m. || levée f. de soupape. / ~ lifter || Ventilheber m. || lève-soupape f. / ~ exhaust ~ lifter || Auspuffventilheber m. || lève-soupape f. d'échappement. / ~ needle || Ventilnadel f. || pointeau m. de soupape. / ~ nut || Ventilkegelmutter f. || écrou m. de clapet. / ~ oscillator for measuring purposes (Radio) || Röhrenoszillator m. für Meßzwecke || oscillateur m. de tubes-à-vide pour des mesures. / ~ packing ring || Ventildichtungsscheibe f. || rondelle f. de joint de soupape. / ~ petticoat (Balloon) || Stoffschürze f. || jupon m. / ~ piston || Ventilkolben m. || piston m. à clapets. / ~ plug || Ventilgehäuseverschluß m. || bouchon m. de soupape. / ~ push rod *see* ~ rod. / ~ rocker || Schwinghebel m. || balancier m. **valve rod** || Ventilspindel f. || tige f. de soupape. / ~ (Slide valve) || Schieberstange f. || bielle f. du tiroir. / ~ guide || Schieberstangenführung f. || guide m. de la tige du tiroir. **valve seat** || Ventilsitz m. || siège m. de soupape. / needle ~ || Nadelventilsitz m. || siège m. du pointeau. / ~ grinding machine || Ventilsitzschleifmaschine f. || machine f. à rectifier les sièges de soupapes. / ~ ring || Ventilsitzring m. || anneau m. de siège de clapet; bague f. de siège de soupape. **valve setting** || Ventileinstellung f. || réglage m. de soupapes. / ~ socket (Radio) || Röhrensockel m. || socle m. d'une lampe amplificatrice. **valve spring** || Ventilfeder f. || ressort m. de soupape. / ~ cotter || Ventilfederkeil m. || clavette f. de ressort de soupape. / ~ retainer || Ventilfederteller m. || cuvette f. pour le bas du ressort de soupape. / ~ retainer lock || Ventilkeil m. || clavette f. de soupape. **valve tappet** || Ventilstößel m. || tige-poussoir m. de soupape. / ~ thumb screw || Ventilflügelschraube f. || boulon-valve m. à ailettes. / ~ transmitter || Röhrensender m. || émetteur m. à lampes. / ~ voltmeter || Röhrenvoltmeter n. || voltmètre m. à lampe. / ~ well || Ventilbrunnen m. || borne-fontaine f. à soupape. **vamp** (Shoem) || Oberleder n. || avant-pied m.; empeigne f. / ~ cutter (Shoem) || Oberlederzuschneider m. || coupeur m. d'empeignes. **van** || Frachtwagen m.; Rollwagen m. || fourgon m.; camion m.; fardier m.; chariot m.; voiture f. de roulage. / ~ (Agr) || Getreideschwinge f. || van m. / ~ (Furniture) *see* removing ~. / ~ (Mine) || Schwingschaufel f. || van m.; pelle f. à vanner. / ~ (Railw) || Güterwagen m. || wagon m. à marchandises. / delivery ~ || Lieferwagen m. || voiture f. de livraison. / furniture ~ *see* removing ~. / luggage ~ || Gepäckwagen m. || fourgon m.; wagon m. à bagages. / postal ~ || Postwagen m. || voiture f. postale. / removing ~ || Möbelwagen m. || voiture f. de déménagement; tapissière f. **vanadiate** || vanadinsaures Salz n. || vanadiate m. **vanadic acid** || Vanadinsäure f. || acide m. vanadique.

vanadium ‖ Vanadium n. ‖ vanadium m. / ~ chloride ‖ Chlorvanadium n. ‖ chlorure m. de vanadium. / ~ steel ‖ Vanadiumstahl m. ‖ acier m. au vanadium.

Vandyke-brown ‖ Vandyckbraun n. ‖ brun m. Vandyck.

vane (Arch) ‖ Wetterfahne f. ‖ girouette f. / ~ (Surv) ‖ Diopter n.; Visiertafel f. ‖ voyant m. / ~ (Windmill) ‖ Flügel m. ‖ aile f.; palette f.; bras m. / air ~ see ~ of a ventilating fan. / rigid ~ (Windmill) ‖ unbeweglicher *oder* fester Flügel m. ‖ aube f. *ou* palette f. fixe. / ~ of a turbine ‖ Turbinenschaufel f. ‖ aube f. *ou* palette f. d'une turbine. / ~ of a ventilating fan ‖ Ventilatorflügel m. ‖ aile *ou* ailette f. d'un ventilateur. / ~ of a windmill ‖ Windmühlenflügel f. ‖ aile f. *ou* bras m. d'un moulin à vent. / wooden ~ (Windmill) ‖ Holzflügel m. ‖ palette f. en bois.

vane beam (Windmill) ‖ Flügelbalken m. ‖ volée f. de l'aile. / group of ~s (Windmill) ‖ Flügelgruppe f. ‖ groupe m. de palettes. / pitch of the ~s (Windmill) ‖ Flügelabstand m. ‖ écartement m. des palettes. / ~ setting (Windmill) ‖ Flügelstellung f. ‖ orientation f. des palettes. / ~ spindle (Windmill) ‖ Flügelwelle f. ‖ arbre m. des palettes. / total ~ surface (Windmill) ‖ Gesamtflügelfläche f. ‖ surface f. totale des palettes.

vanillin ‖ Vanillin n. ‖ vanilline f.

vanilla ‖ Vanille f. ‖ vanille f.

van loader ‖ Wagenauflader m. ‖ chargeur m. de voitures.

vapor *see* vapour.

vaporization ‖ Verdampfung f. ‖ évaporation f.; vaporisation f. / activity of ~ ‖ Verdampfungsschnelligkeit f. ‖ activité f. de la vaporisation. / ~ heat (Chem) ‖ Verdampfungswärme f. ‖ chaleur f. de vaporisation. / ~ point ‖ Verdampfungspunkt m. ‖ point m. de vaporisation.

vaporize, to ‖ verdampfen ‖ vaporiser; évaporer.

vaporizer ‖ Zerstäuber m.; Verdampfungsapparat m. ‖ pulvérisateur m.; vaporisateur m.

vaporizing carburettor ‖ Verdampfungsvergaser m. ‖ carburateur m. à évaporation.

vapour ‖ Dampf m. ‖ vapeur f. / dry ~ ‖ trockener Dampf m. ‖ vapeur f. sèche. / wet ~ ‖ nasser Dampf m. ‖ vapeur f. humide.

vapour, atmosphere of ~ ‖ Dampfatmosphäre f. ‖ atmosphère f. de vapeur. / ~ bath ‖ Dampfbad n. ‖ bain m. de vapeur. / ~ channel of a charcoal pile ‖ Luftkanal m. eines Kohlenmeilers ‖ évent m. en maçonnerie d'un fourneau de carbonisation. / ~ cloud ‖ Nebelwolke f. ‖ couche f. nuageuse. / ~ density ‖ Dampfdichte f. ‖ densité f. de vapeur. / ~ escape ‖ Dunstabzug m. ‖ cheminée f. / exhausting plant for ~s ‖ Absaugeanlage f. für Dünste ‖ installation f. d'aspiration des exhalaisons. / ~ pressure ‖ Dampfdruck m. ‖ tension f. de vapeur.

variable ‖ veränderlich, wechselnd ‖ variable. / ~ in a very wide range ‖ in sehr weiten Grenzen fpl. veränderlich variable *ou* mobile dans de très larges limites fpl.

variable admission ‖ veränderliche Füllung f. ‖ admission f. variable. / ~ condenser ‖ variabler Kondensator m. ‖ condensateur m. réglable. / ~ increasing motion ‖ veränderlich beschleunigte Bewegung f. ‖ mouvement m. variable accéléré. / ~ motion ‖ veränderliche Bewegung f. ‖ mouvement m. variable. / ~ span wing ‖ Ausziehflügel m. ‖ aile f. à allongement *ou* à étirement *ou* étirable. / ~ speed gear for cutting ‖ veränderliche Schnittgeschwindigkeit f. ‖ vitesse f. variable de coupe.

variable speed motor ‖ Regelmotor m. ‖ moteur m. à vitesse réglable. / pressbutton-controlled ~ ‖ druckknopfgesteuerter Regelmotor m. ‖ moteur m. réglable manœuvré par bouton-poussoir.

variable ‖ Veränderliche f. ‖ variable f. / dependent ~ ‖ abhängige Veränderliche f. ‖ variable f. dépendante. / independent ~ ‖ unabhängige Veränderliche f. ‖ variable f. indépendante.

variation ‖ Abweichung f.; Schwankung f. ‖ variation f.; différence f. / ~ (Agr) ‖ Spielart f. ‖ variété f. / ~ (Math) ‖ Variation f. ‖ variation f. / ~ of the barometer ‖ Schwankung f. des Barometers ‖ variation f. du baromètre. / ~ of the intensity of a current ‖ Schwankung f. in der Stromstärke ‖ variation f. de la force du courant. / ~ in the intensity of light ‖ Helligkeitsunterschied m. ‖ différence f. de clarté. / ~ of the light intensity ‖ Lichtstärkeschwankung f. ‖ variation f. de l'intensité lumineuse. / ~ of the magnetic needle ‖ Mißweisung f. der Magnetnadel ‖ déviation f. *ou* déclinaison f. de l'aiguille aimantée. / negative ~ ‖ negative Schwankung f. ‖ variation f. négative. / ~ of temperature ‖ Temperaturschwankung f. ‖ fluctuation f. de température. / ~ of velocity ‖ Geschwindigkeitsschwankung f. ‖ fluctuation f. de vitesse. / ~ of voltage ‖ Spannungsschwankung f. ‖ fluctuation f. de tension.

variation compass (Mar) ‖ Peilkompaß m.; Variationskompaß m. ‖ compas m. de variation.

vari-coloured coat of war material ‖ Buntfarbenanstrich m. des Heeresgeräts ‖ peinture f. du matériel de guerre en plusieurs couleurs.

varied ‖ mannigfach ‖ varié; divers.

variegated carpet ‖ geflammter Teppich m. ‖ tapis m. jaspé.

variety lathe ‖ Drehbank f. für verschiedene Zwecke ‖ tour m. à différents usages.

variometer ‖ Variometer n. ‖ variomètre m.

various ‖ verschieden ‖ différent. / ~ purpose manufacturing milling machine ‖ Mehrzweckfräsmaschine f. ‖ fraiseuse f. servant à différentes fabrications.

varnish, to ‖ lackieren; firnissen ‖ laquer; vernir.

varnish ‖ Firnis m.; Lack m. ‖ vernis m.; laque m. / acid-proof ~ ‖ säurefester Lack m. ‖ laque m. *ou* vernis m. anti-acide *ou* inattaquable aux acides. / air-proof ~ ‖ Luftlack m. ‖ vernis m. résistant à l'air. / alkali-proof ~ ‖ alkalifester Lack m. ‖ vernis m. résistant aux alcalis. / carriage ~ ‖ Wagenlack m. ‖ vernis m. pour voitures. / cellulose acetate ~ ‖ Zelluloseazetatlack m. ‖ laque m. à l'acétate de cellulose. / cycle ~ ‖ Fahrradlack m. ‖ vernis m. pour bicyclettes. / dipping ~ ‖ Tauchlack m. ‖ vernis m. à plonger. / drying ~ see varnish siccative. / ~ for fermenting tubs ‖ Gärbottichglasur f. ‖ vernis m. pour cuves de fermentation. / flatting ~ ‖ Politurlack m. ‖ vernis m. à polir. / ~ for heating bodies ‖ Heizkörperlack m. ‖ vernis m. pour radiateurs. / ~ for incandescent lamps ‖ Glühlampenlack m. ‖ laque f. pour lampes à incandescence. / insulating ~ ‖ Isolierlack m. ‖ vernis m. isolant. / iron ~ ‖ Eisenlack m. ‖ vernis m. à fer. / maple ~ ‖ Ahornlack m. ‖ vernis m. acéracé. / ~ for motor cars ‖ Kraftwagenlack m. ‖ vernis m. pour automobiles. / protecting ~ ‖ Deckfirnis m. ‖ vernis m. de finissage; crème f. vernis. / ~ for sewing machines ‖ Nähmaschinenlack m. ‖ vernis m. pour machines à coudre. / siccative ~ ‖ Trockenfirnis m. ‖ vernis m. siccatif. / sounding board ~ ‖ Resonanzbodenlack m. ‖ vernis m. pour tables d'harmonie. / tinplate ~ ‖ Blechlack m. ‖ vernis m. pour fer-blanc. / violin ~ ‖ Geigenlack m. ‖ vernis m. pour violons. / ~ for window envelopes ‖ Fensterbriefumschlaglack m. ‖ vernis m. pour enveloppes à guichet.

varnish boiling plant ‖ Firnisküche f. ‖ installation f. à fabriquer le vernis. / ~ can ‖ Lackkanne f. ‖ bidon m. à laque. / spout closing apparatus for ~ cans ‖ Tüllenverschlußapparat m. für Lackkannen ‖ appareil m. de fermeture à douille pour bidons à laque. / ~ colours pl. ‖ Farben fpl. für Lacke ‖ couleurs fpl. pour vernis. / ~ drier ‖ Sikkativ n. ‖ siccatif m.

varnished ‖ lackiert ‖ verni(ssé). / ~ paper goods pl. ‖ Papierlackwaren fpl. ‖ articles mpl. en carton verni *ou* en papier-mâché. / ~ sail canvas ‖ gefirnißtes Segeltuch n. ‖ toile f. enduite de vernis. / ~ wooden art work ‖ mit Lack überzogene Holzkunstarbeit f. ‖ travail m. d'art en bois verni.

varnisher ‖ Lackierer m. ‖ vernisseur m. / wood ~ ‖ Holzwarenlackierer m. ‖ vernisseur m. sur bois.

varnish extract ‖ Lackextrakt m. ‖ extrait m. de vernis. / ~ gum crusher ‖ Harzmühle f. ‖ moulin m. à résine.

varnishing ‖ Lackieren n.; Firnissen n. ‖ vernissage m. / furniture ~ ‖ Lackieren n. von Möbeln ‖ vernissage m. de meubles. / ~ of wood articles ‖ Lackieren n. von Holzwaren ‖ vernissage m. d'articles en bois.

varnishing brush ‖ Lackierpinsel m. ‖ pinceau m. à vernir. / ~ drum ‖ Lackiertrommel f. ‖ tambour m. à vernir. / ~ machine ‖ Lackiermaschine f. ‖ machine f. à laquer *ou* à vernir *ou* de vernissage.

varnish manufacturing plant ‖ Lackfabrikationsanlage f. ‖ installation f. à fabriquer le vernis. / ~ preparation ‖ Firnispräparat n. ‖ préparation f. pour vernis. / ~ red ‖ Lackrot n. ‖ laque m. rouge. / ~ remover ‖ Lackabbeizmittel n. ‖ décapant m. de vernis. / ~ standards pl. ‖ Lacknormen fpl. ‖ règles fpl. *ou* normes fpl. pour le vernis. / ~ substitute ‖ Firnisersatz m. ‖ substitut m. de vernis.

vary, to ~ colours pl. ‖ schillern ‖ chatoyer; miroiter; jeter des reflets mpl. variés. / ~ continuously ‖ sich stetig ändern ‖ varier d'une façon continue.

vase ‖ Vase f. ‖ vase m.

vaseline ‖ Vaseline f.; Vaselin n. ‖ vaseline f. / ~ oil ‖ Vaselinöl n. ‖ huile f. de vaseline. / ~ plant ‖ Vaselineherstellungs-

anlage f. ‖ installation f. à fabriquer la vaseline.

vat ‖ Bottich m.; Wanne f.; Küpe f.; Trog m. ‖ cuve f.; bac m.; bidon m. / ~ (Pap) ‖ Back m. des Holländers; Holländerkasten m. ‖ pile f.; bac m.; cuve f. / cold ~ *see* copperas ~. / copperas ~ ‖ Vitriolküpe f. ‖ cuve f. à la couperose. / decomposing ~ ‖ durchgehende Küpe f. ‖ cuve f. qui soufre. / German ~ ‖ deutsche Küpe f.; Sodaküpe f. ‖ cuve f. allemande. / made ~ *see* prepared ~. / to prepare the ~ ‖ die Küpe zurichten *oder* ansetzen *oder* anstellen ‖ monter *ou* poser la cuve. / prepared ~ ‖ fertige Küpe f. ‖ assiette f. *ou* cuve f. garnie. / preparing ~ (Pot) ‖ Anmachebottich m. ‖ gâchoir m. / ~ under pressure ‖ Autoklav m. ‖ autoclave m. / sharp ~ ‖ scharfe Küpe f. ‖ cuve f. forte *ou* raide *ou* usée. / soft ~ ‖ leise Küpe f. ‖ cuve f. douce. / thrown back ~ ‖ schwarze Küpe f. ‖ cuve f. rebutée. / warm ~ ‖ Gärungsküpe f.; warme Küpe f. ‖ cuve f. à chaud.

vat brick ‖ Wannenstein m. ‖ brique f. pour Ruves. / ~ dye ‖ Küpenfarbstoff m. ‖ colorant m. pour cuve.

vatman (Pap) ‖ Büttgeselle m. ‖ plongeur m.

vault, to ‖ wölben; einwölben; ein Gewölbe n. aufführen ‖ voûter. / ~ in ‖ überwölben ‖ envoûter.

vault ‖ Gewölbe n.; Wölbung f.; Gewölbebogen m. ‖ voûte f. / ~ (Glassm) ‖ Kappe f. ‖ voûte f.; couronne f. / acoustic ~ ‖ Schallgewölbe n. ‖ voûte f. acoustique. / annular ~ ‖ Ringgewölbe n. ‖ voûte f. annulaire *ou* sur le noyau; berceau m. tournant. / domical ~ ‖ Kesselgewölbe n.; Kuppelgewölbe n. ‖ dôme m.; cul-de-four m.; voûte f. sphérique. / family ~ ‖ Erbbegräbnis n. ‖ tombeau m. de famille. / funeral ~ ‖ Grabgewölbe n. ‖ caveau m.; sépulcre m.; sépulture f. / groined ~ ‖ Kreuzgewölbe n. ‖ voûte f. croisée. / helical ~ *see* spiral ~. / ~ of the sky ‖ Himmelswölbung f. ‖ voûte f. céleste. / snail formed ~ ‖ Schneckengewölbe n. ‖ voûte f. en limaçon. / spherical ~ *see* domical ~. / spiral ~ ‖ Schneckengewölbe n. ‖ voûte f. en limaçon. / surbased ~ ‖ Stichbogengewölbe n. ‖ voûte f. basse. / ~ of a tunnel ‖ Tunnelgewölbe n. ‖ voûte f. d'un tunnel. / upright conical ~ ‖ Trichtergewölbe n. ‖ voûte f. conique verticale.

vaulted cell ‖ Gewölbfach n.; Fach zwischen Gewölbgurten ‖ surface f. d'une lunette *ou* d'une triangle de voûte.

vaulting *see* vault. / ~ ruler ‖ Wölbrichtscheit n. ‖ échasse f.

vault support ‖ Gewölbeträger m. ‖ support m. de voûte.

vector ‖ Vektor m. ‖ vecteur m. / radius ~ ‖ Radiusvektor m. ‖ rayon m. vecteur. / time ~ ‖ Zeitvektor m. ‖ vecteur m. de temps.

vector analysis ‖ Vektoranalysis f. ‖ analyse f. vectorielle. / ~ calculus ‖ Vektorrechnung f. ‖ calcul m. vectoriel. / ~ diagram ‖ Vektordiagramm n. ‖ diagramme m. vectoriel. / ~ direction of ~ *see* sense of ~. / ~ equation ‖ Vektorgleichung f. ‖ équation f. vectorielle. / ~ field ‖ Vektorfeld n. ‖ champ m. vectoriel. / ~ function ‖ Vektorfunktion f. ‖ fonction f. vectorielle. / ~ potential ‖ Vektorpotential n. ‖ potentiel m. vecteur. / ~ product ‖

Vektorprodukt n. ‖ produit m. vectoriel. / ~ quantity ‖ Vektorgröße f. ‖ grandeur f. *ou* quantité f. vectorielle. / sense of ~ ‖ Vektorsinn m.; Vektorrichtung f. ‖ sens m. du vecteur. / ~ tube ‖ Vektorröhre f. ‖ tube m. vectoriel.

veer, to ‖ abvieren; behauen ‖ équarrir; carrer. / ~ aloft ‖ hissen; heißen; aufziehen ‖ hisser; palanquer; haler. / ~ a cable (Mar) ‖ ein Tau n. nachlassen *oder* schießen lassen ‖ filer *ou* larguer *ou* mollir un câble. / ~ and haul ‖ stoßweise holen ‖ haler à secousses fpl.

veering clockwise ‖ Rechtsablenkung f. ‖ déviation f. vers la droite.

vegetable ‖ pflanzlich ‖ végétal. / ~ acid ‖ Pflanzensäure f. ‖ acide m. végétal. / ~ black ‖ Pflanzenschwarz n. ‖ noir m. végétal. / goods pl. of ~ braiding material ‖ Waren fpl. aus pflanzlichem Flechtstoff ‖ objets mpl. en matière végétale à tresser. / ~ butter ‖ Pflanzenbutter f. ‖ beurre m. végétal; végétaline f. / ~ colouring matter ‖ Pflanzenfarbstoff m. ‖ matière f. colorante (naturelle d'origine) végétale. / ~ extract ‖ Pflanzenauszug m. ‖ extrait m. végétal. / ~ fat ‖ Pflanzenfett n. ‖ graisse f. végétale. / curling machine for ~ fibres pl. ‖ Pflanzenfaserkräuselmaschine f. ‖ ratineuse f. pour filaments végétaux. / ~ glue ‖ Pflanzenleim m. ‖ colle f. végétale. / ~ grease *see* ~ fat. / ~ humus deposit ‖ vegetabilische Humuserde f. ‖ humus m. végétal. / ~ kingdom ‖ Pflanzenreich n. ‖ règne m. végétal. / ~ matter ‖ Pflanzenstoff m. ‖ substance f. végétale. / ~ mould *see* ~ soil. / ~ oil ‖ Pflanzenöl n. ‖ huile f. végétale. / fixed ~ oil ‖ festes Pflanzenöl n. ‖ huile f. fixe végétale. / ~ oil refining ‖ Pflanzenölreinigung f. ‖ épuration f. d'huiles végétales. / ~ origin ‖ pflanzlicher Ursprung m. ‖ origine f. végétale. / ~ plaiting material ‖ pflanzlicher Flechtstoff m. ‖ matière f. végétale à tresser. / ~ produce *see* ~ product. / ~ product ‖ Pflanzenerzeugnis n. ‖ produit m. végétal. / ~ soil ‖ Humus m. ‖ humus m.; terreau m. / ~ stalk ‖ Pflanzenstengel m. ‖ tige f. végétale. / ~ tar ‖ Holzteer m. ‖ goudron m. végétal. / ~ wax ‖ Pflanzenwachs n. ‖ cire f. végétale.

vegetable ‖ Pflanze f.; Gemüsepflanze f.; Küchenpflanze f. ‖ plante f.; plant m.; plante f. oléracée. / ~ cutter ‖ Gemüsehobel m. ‖ coupe-légumes m. / ~ cutting machine ‖ Gemüseschneidemaschine f. ‖ machine f. à couper les légumes. / ~ dryer ‖ Gemüsetrockner m. ‖ sécheur m. pour légumes. / ~ grinding mill ‖ Gemüseschroterei f. ‖ décortication f. de légumes. / ~ hot-house ‖ Gemüsetreibhaus n. ‖ serre f. à légumes. / ~ peeler ‖ Gemüseklauberin f.; Gemüseverleserin f. ‖ éplucheuse f. de légumes. / ~ picking ‖ Gemüseklauben n.; Gemüselesen n. ‖ épluchage m. de légumes. / ~ plane *see* vegetable cutter. / ~ seeds pl. ‖ Gemüsesamen m. ‖ graines fpl. maraîchères.

vegetables pl. ‖ Gemüse n. ‖ légume m. / canned ~ ‖ Büchsengemüse n.; Gemüsekonserven fpl. ‖ légume m. conservé; conserves fpl. de légumes. / canned ~ in natural state ‖ naturreine Gemüsekonserven fpl. ‖ conserves fpl. de légume au naturel. / dried ~ ‖ Dörrgemüse n. ‖ légume m. séché. / tinned ~ *see* canned ~.

vegetables pl., culture of ~ ‖ Gemüsebau

m. ‖ horticulture f. maraîchère. / washery for ~ ‖ Wäsche f. für Gemüse ‖ laveur m. pour légumes.

vegetation ‖ Vegetation f. ‖ végétation f.

vegetative growth ‖ vegetatives Wachstum n. ‖ génération f. végétative. / ~ thread ‖ Pflanzenfaden m. ‖ filament m. végétal.

vehicle ‖ Fahrzeug n. ‖ véhicule m.; voiture f. / commercial ~ ‖ Nutzfahrzeug n. ‖ véhicule m. industriel. / horse-drawn ~ ‖ mit Pferden bespanntes Fahrzeug n. ‖ voiture f. attelée. / motor ~ ‖ Kraftfahrzeug n. ‖ véhicule m. à moteur. / ~ for rails ‖ Schienenfahrzeug n. ‖ véhicule m. pour voies ferrées.

vehicle battery ‖ Fahrzeugbatterie f. ‖ batterie f. de voiture. / ~ wheel ‖ Fahrzeugrad n. ‖ roue f. de véhicule.

veil ‖ Schleier m. ‖ voile m.

veiling ‖ Schleierstoff m. ‖ voile f. / ~ spotter ‖ Schleierpunktiererin f. ‖ moucheteuse f. sur tulle.

vein, to ‖ adern ‖ veiner.

vein (Anatomy) ‖ Vene f. ‖ veine f. / ~ (Geol) ‖ Ader f.; Flöz n.; Gang m. ‖ filon m.; couche f.; gîte m. / ~ following bedding planes of ‖ Lagergang m. ‖ filoncouche f. / ~ of coal ‖ Kohlenflöz n. ‖ couche f. de charbon *ou* de houille. / the ~ crops ‖ das Flöz streicht zu Tage aus ‖ la couche f. se montre *ou* affleure. / cross ~ ‖ Kreuzgang m.; Quergang m. ‖ filon m. croiseur. / exhausted ~ (Mine) ‖ abgebauter Gang m. ‖ filon m. épuisé *ou* exploité. / in ~s pl. (Mine) ‖ gangartig; gangförmig ‖ en filons mpl. / ~ in the marble ‖ Marmorader f. ‖ fil m. dans le marbre. / ~ of quartz ‖ Quarzgang m. ‖ veine f. de quartz. / ~ of rock ‖ Gesteinsgang m. ‖ filon m. rocheux.

veined ‖ geadert; geädert ‖ veiné; marbré. / ~ growth ‖ Maserwuchs m. ‖ croissance f. d'un arbre veiné. / ~ mahogany ‖ geadertes Mahagoniholz n. ‖ acajou m. veiné. / ~ paper ‖ geädertes Papier n. ‖ papier m. veiné. / ~ wood ‖ geadertes Holz n. ‖ bois m. marbré.

vein, infilling of ~s ‖ Gangfüllung f. ‖ remplissage m. du filon.

veining gouge ‖ Kanneliergutsche f. ‖ gouge f. à bretter.

vein stone (Mine) ‖ Gangmasse f. ‖ gangue f.; matière f. stérile d'un gîte.

veiny *see* veined.

vellum ‖ Schreibpergament n.; Kalbspergament n. ‖ vélin m.; parchemin m. à écrire. / ~ cloth ‖ Pausleinen n. ‖ toile f. à calquer; papier-toile f. / ~ paper ‖ Pergamentpapier n. ‖ papier m. parchemin *ou* parcheminé.

velocimeter ‖ Geschwindigkeitsmesser m. ‖ indicateur m. de vitesse; tachymètre m.

velocipede car (Railw) ‖ Draisine f. ‖ wagonnet m. de tournée. / ~ crane ‖ Einschienenkran m. ‖ grue-vélocipède f.

velocity ‖ Geschwindigkeit f. ‖ vitesse f. / angular ~ ‖ Winkelgeschwindigkeit f. ‖ vitesse f. angulaire. / angular ~ (Electr) ‖ Kreisfrequenz f. ‖ vitesse f. angulaire. / final ~ ‖ Endgeschwindigkeit f. ‖ vitesse f. finale. / ~ of flow ‖ Strömungsgeschwindigkeit f. ‖ vitesse f. du flux. / ~ of flow of a liquid ‖ Ausflußgeschwindigkeit f. einer Flüssigkeit ‖ vitesse f. d'écoulement d'un liquide. / ~ of formation of ions ‖ Bildungsgeschwindigkeit f. der Ionen ‖ vitesse f. de formation des ions. / hoisting ~ ‖ Hubgeschwindigkeit f. ‖ vitesse f.

de levage. / ~ of impact (Arm) ‖ Auftreffgeschwindigkeit f. ‖ vitesse f. au choc. / initial ~ ‖ Anfangsgeschwindigkeit f. ‖ vitesse f. initiale. / little ~ ‖ geringe Geschwindigkeit f. / faible vitesse f. / mean ~ ‖ mittlere Geschwindigkeit f. ‖ vitesse f. moyenne. / retarded ~ ‖ verzögerte Geschwindigkeit f. ‖ vitesse f. retardée. / ~ of rotation ‖ Drehgeschwindigkeit f. ‖ vitesse f. de rotation. / terminal ~ see final ~. / uniform ~ ‖ gleichförmige Geschwindigkeit f. ‖ vitesse f. uniforme. / virtual ~ ‖ virtuelle Geschwindigkeit f. ‖ vitesse f. virtuelle. / working ~ ‖ Arbeitsgeschwindigkeit f. ‖ vitesse f. de travail.

velocity, height of ~ ‖ Geschwindigkeitshöhe f. ‖ hauteur f. de la vitesse. / ~ potential ‖ Geschwindigkeitspotential n. ‖ potentiel m. de vitesse. / variation of ~ ‖ Geschwindigkeitsschwankung f. ‖ fluctuation f. de vitesse.

velour paper ‖ Velourpapier n. ‖ papier m. imitation velours.

velvet ‖ Samt m. ‖ velours m. / cotton ~ see velveteen. / cut ~ ‖ gerissener oder geschnittener Samt m. ‖ velours m. coupé. / fancy ~ ‖ gemusterter Samt m. ‖ velours m. façonné ou figuré. / figured ~ see fancy ~. / ~ for furniture ‖ Möbelsamt m. ‖ velours m. d'ameublement. / Genoa back ~ ‖ Köpersamt m. ‖ velours m. de Gênes; velours m. croisé. / jean back ~ see Genoa back. / Manchester ~ ‖ Manchester m. ‖ velours m. de Manchester. / plain ~ ‖ glatter Samt m. ‖ velours m. plein. / plain back ~ ‖ glatter Samt m. mit Leinwandgrund ‖ velours m. à toile ou uni. / ribbed ~ ‖ gerippter Samt m.; Kord m. ‖ velours m. à côtes; velours m. cordelet; velours m. de chasse. / silk ~ ‖ Seidensamt m. ‖ velours m. de soie. / tabby ~ see plain ~. / terry ~ see uncut ~. / tweeled ~ see Genoa back ~. / uncut ~ ‖ Halbsamt m.; gezogener oder ungerissener Samt m. ‖ velours m. épinglé ou frisé. / Utrecht ~ ‖ Möbelplüsch m. ‖ velours m. d'Utrecht. / warp ~ ‖ Kettsamt m. ‖ velours m. par chaîne. / weft ~ ‖ Schußsamt m. ‖ velours m. par trame. / worsted ~ ‖ Wollsamt m. ‖ velours m. de laine.

velveted ‖ samtartig ‖ velouté.

velveteen ‖ Samtmanchester m.; Baumwollsamt m. ‖ velours m. lisse ou de coton.

velvet, filling of empty spaces in ~ ‖ Ausbesserung f. fehlerhafter Stellen im Samt ‖ resarcissage m. de velours. / ~ finishing machine ‖ Velvetausrüstungsmaschine f. ‖ machine f. à velvet. / ~ hat ‖ Velourhut m. ‖ chapeau m. en velours. / ~-lined ‖ mit Samt m. gefüttert ‖ revêti de velours m. / ~ maker ‖ Samtweber m. ‖ veloutier m. / ~ mender ‖ Samtausbesserin f. ‖ replanteuse f. de velours. / ~ paper ‖ Samtpapier n. ‖ papier m. velouté. / ~ printing ‖ Samtdruck m. ‖ impression f. sur velours. / ~ ribbon ‖ Samtband n. ‖ ruban m. de velours. / ~ shearer ‖ Samtscherer m.; raseur m. de velours. / ~ shearing ‖ Scheren n. des Samtes ‖ tondage m. de velours. / ~ weaver's loom ‖ Samtstuhl m.; Samtwebstuhl m. ‖ métier m. à velours. / ~ weaving ‖ Samtweberei f. ‖ tissage m. de velours.

veneer, to ‖ furnieren ‖ plaquer. / ~ on both sides ‖ auf beiden Seiten fpl. furnieren; gegenfurnieren ‖ contre-plaquer.

veneer ‖ Furnier n.; Furnierblatt n. ‖ plaque f. en bois; feuillet m.; feuille f. de placage. / sawed ~ ‖ gesägtes Furnierblatt n. ‖ feuille f. de placage scié. / superposed and glued ~s pl. ‖ übereinandergelegte und geleimte Furnierblätter npl. ‖ feuilles fpl. de placage superposées et collées. / unrolled ~ ‖ ausgerolltes Furnierblatt n. ‖ feuille f. de placage déroulé.

veneer cutter ‖ Furnierschneider m. ‖ trancheur m. de feuilles de placage. / cutter of ~s for brushes ‖ Bürstenfurniermacher m. ‖ débiteur m. de placages pour brosses. / ~ cutting machine ‖ Furnierschneidmaschine f. ‖ scie f. de placage; machine f. à découper les feuilles de placage.

veneerer ‖ Furnierschreiner m. ‖ plaqueur-ébéniste m.

veneer frame saw ‖ Furnierrahmensäge f. ‖ scie f. de placage à chassis. / ~ gluer ‖ Furnierleimer m. ‖ colleur m. de placage.

veneering ‖ Furnieren n. ‖ placage m. en bois. / ~ on both sides ‖ Gegenfurnierung f. ‖ contre-placage m. / ~ machine ‖ Furniermaschine f. ‖ machine f. à plaquer. / ~ press ‖ Furnierpresse f.; Sperrholzpresse f. ‖ presse f. à plaquer. / ~ wood ‖ Furnierholz n. ‖ bois m. de placage.

veneer pattern mahogany ‖ Mahagonifurnierholz n. ‖ feuille f. d'acajou. / ~ peeling machine ‖ Furnierschälmaschine f. ‖ machine f. à dérouler les bois ronds. / ~ planing machine ‖ Furnierhobelmaschine f. ‖ raboteuse f. à bois de placage. / ~ press see veneering press. / ~ saw ‖ Furniersäge f. ‖ scie f. à bois de placage ou à refendre. / ~ wood see veneering wood.

venetian see Venetian blind.

Venetian blind ‖ Jalousie f.; Rolladen m. ‖ jalousie f.; persienne f. / ~ boat ‖ Gondel f. ‖ gondole f. / ~ carpet ‖ Treppenläufer m. ‖ tapis m. d'escalier. / ~ red ‖ Venezianischrot n. ‖ rouge m. de Venise. / ~ shutter see ~ blind.

vent ‖ Luftloch n. ‖ évent m.; ventouse f. / ~ for spent grains ‖ Treberausstoßloch n. ‖ trappe f. de décharge des drèches. / ~ hole see vent.

ventilate, to ‖ lüften; auslüften ‖ ventiler; aérer. / ~ a mine ‖ die Grubenbaue mpl. mit Wettern versorgen ‖ ventiler une mine.

ventilated, porous ~ brick ‖ poriger Lochstein m. ‖ brique f. creuse poreuse. / enclosed ~ ‖ ventiliert gekapselt ‖ blindé ventilé.

ventilating air (Mine) ‖ Bewetterungsluft f. ‖ air m. de ventilation. / cooling the ~ (Mine) ‖ Kühlung f. der Bewetterungsluft ‖ réfrigération f. de l'air de ventilation.

ventilating aperture ‖ Ventilationsklappe f. ‖ registre m. de ventilation. / ~ blade ‖ Ventilationsflügel m. ‖ ailette f. à ventilation. / ~ chimney ‖ Dunstkamin m. ‖ cheminée f. d'appel. / ~ duct ‖ Luftkanal m. ‖ évent m. de ventilation. / ~ fan ‖ Ventilator m.; Windflügel m. ‖ ventilateur m. ~ installation see ventilation plant. / ~ machine ‖ Ventilations-

maschine f. ‖ machine f. à ventiler. / ~ motor see ~ machine. / ~ pipe ‖ Lüftungsrohr n. ‖ tuyau m. d'aération. / main ~ pipe ‖ Hauptlüftungsrohr n. ‖ tuyau m. de ventilation principal; tuyau m. principal d'arrivée d'air. / ~ plant see ventilation plant. / ~ shaft ‖ Luftschacht m. ‖ puits m. d'aérage. / upcast ~ shaft ‖ ausziehender Schacht m. ‖ puits m. d'appel.

ventilation ‖ Ventilation f.; Lüftung f.; Ventilierung f. ‖ aération f.; ventilation f. / ~ (Mine) ‖ Wetterhaltung f. ‖ ventilation f. / ~ of carriages ‖ Wagenlüftung f. ‖ ventilation f. des voitures. / ~ of the core ‖ Kernlüftung f. ‖ aérage m. du noyau. / natural ~ ‖ natürlicher Luftwechsel m. ‖ ventilation f. naturel.

ventilation bin (Mine) ‖ Wetterlutte f. ‖ tuyau m. ou buse f. d'aérage. / ~ dam (Mine) ‖ Wetterdamm m. ‖ barrage m. contre gaz. / ~ door (Mine) ‖ Wettertür f. ‖ porte f. de ventilation. / ~ flap ‖ Entlüftungsklappe f. ‖ régistre m. d'aérage. / ~ lock (Mine) ‖ Wetterschleuse f. ‖ écluse f. d'aération. / ~ nozzle (Mine) ‖ Bewetterungsdüse f. ‖ tuyère f. d'aération. / ~ plant ‖ Lüftungsanlage f.; Ventilationsanlage f. ‖ installation f. d'aération ou de ventilation. / ~ plate ‖ Lüftungsblech n. ‖ tôle f. de ventilation. / ~ top ‖ Ventilationsaufsatz m. ‖ chapeau m. du tuyau de ventilation.

ventilator ‖ Ventilator m. ‖ ventilateur m. / ~ for the air conduit (Mine) ‖ Luttenventilator m. ‖ ventilateur m. de galerie. / suction ~ ‖ Saugventilator m. ‖ ventilateur m. aspirant. / ~ for table and ceiling ‖ Ventilator m. für Tisch und Decke ‖ ventilateur m. de table et de plafond.

ventilator blade ‖ Ventilatorflügel m. ‖ aile f. ou ailette de ventilateur. / ~ ring ‖ Ventilatorring m. ‖ anneau m. pour ventilateurs. / ~ vane see ventilator blade.

venting wire (Sand moulding) ‖ Luftspieß m.; Spieß m. ‖ aiguille f.; dégorgeoir. m.

vent peg (Wine cask) ‖ Zwicker m. ‖ fausset m.

venture, to ‖ spekulieren; riskieren ‖ spéculer; risquer.

Venturi meter for water, steam, gas and air ‖ Venturimesser m. für Wasser, Dampf, Gas und Luft ‖ compteur m. Venturi pour l'eau, la vapeur, le gaz et l'air. / ~ tube ‖ Venturirohr n. ‖ trompe f. de Venturi.

venue ‖ Gerichtsort m. ‖ voisinage m.

verbal ‖ wörtlich; buchstäblich ‖ textuel; littéral; mot m. à mot. / ~ agreement ‖ mündliche Vereinbarung f. ‖ convention f. ou accord m. verbale. / ~ process of approbation ‖ Abnahmeverhandlung f. ‖ procès m. verbal de réception.

verbatim ‖ wortgetreu ‖ littéral; verbal.

verbena herb ‖ Eisenkraut n. ‖ herbe f. de vervéine.

verderer ‖ Förster m. ‖ garde forestier m.

verdict ‖ Rechtsspruch m. ‖ verdict m.

verdigris ‖ Grünspan m. ‖ vert-de-gris m.; verdet m. / crystallized ~ ‖ kristallisierter Grünspan m. ‖ verdet m. cristallisé ou vert m. distillé. / spirit of ~ ‖ Grünspanessig m. ‖ acide m. acétique cristallisable; esprit m. de vert de gris; vinaigre m. radical.

verdigrized ‖ mit Grünspan m. beschlagen ‖ érugineux.

verge of the balance (Watchm) ‖ Schwungstift m. ‖ renversement m. / ~ **riveting tool** (Clockm) ‖ Spindelnieter m. ‖ noisette f. à river les verges.

verification ‖ Prüfung f. ‖ examen m.; épreuve f.; essai m.; vérification f.

verifier (Bookb) ‖ Kollationnierer m. ‖ collationneur m.

verify, to ‖ prüfen; feststellen; revidieren ‖ vérifier; examiner.

verifying *see* verification.

vermeil ‖ vergoldete Silberware f. ‖ vermeil n .

vermicelli ‖ Fadennudeln fpl. ‖ vermicelle m. / ~ **maker** ‖ Fadennudelmacher m. ‖ vermicellier m.

vermifuge ‖ Bandwurmmittel n. ‖ remède m. contre les vers solitaires.

vermilion ‖ Zinnober m. ‖ vermillon m. / **substitute for** ~ ‖ Zinnoberersatz· m. ‖ vermillion m. factice.

vermin ‖ Ungeziefer n. ‖ vermine f. / ~ **destroyer** ‖ Ungeziefervertilgungsmittel n. ‖ produit m. à exterminer les vermines. / ~ **killing product** *see* ~ destroyer.

vermouth ‖ Wermut(wein) m. ‖ vermout m.

vernier ‖ Nonius m. ‖ vernier m. / ~ **with magnifying reader** ‖ Nonius m. mit Lupe ‖ vernier m. muni d'une loupe. / ~ **reading to one minute** ‖ auf eine Minute ablesbarer Nonius m. ‖ vernier m. donnant la minute. / ~ **sliding on the graduation** ‖ auf einer Teilung gleitender Nonius m. ‖ vernier m. coulissant sur une graduation.

vernier division ‖ Noniusteilstrich m. ‖ trait m. de division de vernier. / ~ **graduation** *see* vernier division. / ~ **theodolite** ‖ Nonientheodolit m. ‖ théodolite m. à vernier. / ~ **zero** ‖ Noniusnullpunkt m. ‖ zéro m. du vernier. / **zero line of** ~ ‖ Noniusnullstrich m. ‖ zéro m. du vernier.

Verona green ‖ Veronesererde f. ‖ vert m. de Vérone.

veronica herb ‖ Ehrenpreis m. ‖ herbe f. de véronique.

vertebra ‖ Wirbel m. ‖ vertèbre f.

vertex ‖ Scheitel m. ‖ sommet m. / ~ **of spectacles** ‖ Brillenscheitel m. ‖ sommet m. des lunettes. / ~ **refractometer** ‖ Scheitelbrechwertmesser m. ‖ frontofocomètre m.

vertical ‖ senkrecht; lotrecht ‖ à plomb; vertical; perpendiculaire. / ~ (Mine) ‖ seiger; senkrecht ‖ d'aplomb; perpendiculaire. / ~ **adjustment** ‖ Senkrechtverstellung f. ‖ déplacement m. vertical. / ~ **brace** ‖ Vertikalverband m. ‖ entretoisement m. vertical. / ~ **drilling machine** ‖ Vertikalbohrmaschine f. ‖ foreuse f. verticale. / ~ **line** ‖ Lotriß m.; lotrechte *oder* vertikale Linie f. ‖ ligne f. à plomb. / ~ **machine** (Typewr) ‖ Senkrechtschreibmaschine f. ‖ machine f. à écrire verticalement. / ~ **milling machine** ‖ Senkrechtfräsmaschine f. ‖ fraiseuse f. verticale. / ~ **motor** ‖ Vertikalmotor m. ‖ moteur m. vertical. / ~ **motor for direct coupling** ‖ Vertikalmotor m. für unmittelbare Kupplung ‖ moteur m. vertical pour accouplement direct. / ~ **movement of milling spindle** ‖ Senkrechtverstellung f. der Frässpindel ‖ déplacement m. vertical de la broche portefraise. / ~ **planetary spindle grinding machine** ‖ Vertikalschleifmaschine f. mit Planetenspindel ‖ machine f. à rectifier à arbre vertical planétaire. / ~ **range** (Gun) ‖ Steighöhe f.; senkrechte Reichweite f. ‖ portée f. verticale. / ~ **shaft** ‖ Königswelle f. ‖ maître-arbre m. / ~ **slotting machine** ‖ Senkrechtstoßmaschine f. ‖ mortaiseuse f. verticale. / ~ **steam boiler** ‖ Steilrohrkessel m. ‖ chaudiere f. à tubes verticaux. / ~ **stripe machine** ‖ Langstreifenmaschine f. ‖ machine f. à rayures en long. / ~ **stripe pattern** ‖ Langstreifenmuster n. ‖ dessin m. à rayures verticales. / ~ **tail surfaces** pl. (Aero) ‖ Seitenleitwerk n. ‖ empennage m. vertical. / ~ **tube boiler** ‖ stehender Röhrenkessel m. ‖ chaudière f. tubulaire verticale. / ~ **turning and boring mill** ‖ Karusselldrehbank f. ‖ tour m. vertical *ou* à plateau horizontal.

vertically striped ‖ langgestreift ‖ rayé en long *ou* verticalement.

vesicatory ‖ Senfpflaster n.; Senfpapier n. ‖ sinapisme m.; vésicant m.

vesicle ‖ Bläschen n. ‖ petite vessie f.

vessel ‖ Kessel m.; Gefäß n. ‖ chaudron m.; vase m.; vaisseau; récipient m. / ~**s** pl. (Household) ‖ Geschirr n.; Küchengeschirr n. ‖ vaisselle f.; batterie f. de cuisine. / ~ (Mar) ‖ Schiff n. ‖ navire m.; bâtiment m.; vaisseau m.; bateau m. / **large** ~ **of aluminium** ‖ Aluminiumgroßgefäß n. ‖ vaisseau m. de grande capacité en aluminium. / ~ **at anchor** ‖ Schiff n. vor Anker ‖ navire m. sur le fer. / ~ **employed in the arctic fisheries** (Mar) ‖ Grönlandsfahrer m. ‖ baleinier m. arctique. / **badly built** ~ (Mar) ‖ schlechtgebautes Schiff n. ‖ navire m. manqué. / **ball-shaped** ~ ‖ kugelförmiges Gefäß n. ‖ vase m. sphérique. / **blood** ~ ‖ Blutgefäß n. ‖ vaisseau m. sanguin. / **broad-built** ~ (Mar) ‖ vollgebautes Schiff n. ‖ navire m. à formes pleines. / **captured** ~ ‖ aufgebrachtes Schiff n. ‖ navire m. capturé. / ~ **with a clean run aft** ‖ Schiff n. mit scharfem Achterende ‖ navire m. bien évidé *ou* taillé de l'arrière. / **coppered** ~ (Mar) ‖ gekupfertes Schiff n. ‖ navire m. doublé en cuivre. / **copper-fastened** ~ (Mar) ‖ kupferfestes Schiff n. ‖ navire m. chevillé en cuivre. / **crank** ~ (Mar) ‖ oberlastiges *oder* rankes Schiff n. ‖ navire m. jaloux *ou* de faible côté. / **crazy** ~ (Mar) ‖ gebrechliches Schiff n. ‖ navire m. en mauvais état. / **crystallizing** ~ ‖ Kristallisationsgefäß n. ‖ cristallisoir m. / **decked** ~ (Mar) ‖ Fahrzeug n. mit einem Deck ‖ bâtiment m. ponté. / **deep** ~ (Mar) ‖ tiefgehendes Schiff n. ‖ navire m. profond. / **deep-laden** ~ *see* deep-loaded ~. / **deep-loaded** ~ (Mar) ‖ tiefgeladenes Schiff n. ‖ navire m. très-chargé. / **Diesel-engined sailing** ~ ‖ Dieselmotorsegelschiff n. ‖ voilier m. à moteur auxiliaire Diesel. / **dismantled** ~ ‖ abgetakeltes Schiff n. ‖ navire m. dégrée. / **dismasted** ~ ‖ entmastetes Schiff n. ‖ navire m. démâté. / ~ **with double-acting two-stroke Diesel engines** ‖ Schiff n. mit doppeltwirkenden Zweitaktdieselmotoren ‖ navire m. équipé de moteurs Diesel à deux temps et à double effet. / ~ **down by the stern** ‖ achterlastiges *oder* steuerlastiges Schiff n. ‖ navire m. sur l'arrière; bâtiment m. sur cul. / **the** ~ **drives** ‖ das Schiff treibt ‖ le navire va en dérive. / **enamelled** ~

emailliertes Gefäß n. ‖ vase m. émaillé. / **evaporating** ~ ‖ Abdampfkessel m. ‖ chaudière f. *ou* bassine f. d'évaporation. / ~ **on an even keel** ‖ gleichlastiges Schiff n. ‖ navire m. sans différence de tirant d'eau. / **expanding** ~ ‖ Expansionsgefäß n. ‖ pot m. à détente. / **fast sailing** ~ ‖ Schnellsegler m. ‖ fin voilier m.; navire m. bon marcheur à voiles. / ~ **with a flat bottom** *see* flat-floored ~. / **flat-floored** ~ ‖ flachgebautes *oder* flachbodiges Schiff n. ‖ navire m. à fond plat. / **flush-decked** ~ ‖ Schiff n. mit glattem Deck ‖ navire m. ras. / ~ **with forecastle and poop** ‖ Schiff n. mit Back und Schanze ‖ navire m. accastillé. / **foreign** ~ ‖ fremdes Schiff n. ‖ bâtiment m. de propriété étrangère. / ~ **with a foul bottom** ‖ bewachsenes Schiff n. ‖ navire m. sale *ou* de fond sale. / **full** ~ ‖ vollgeladenes Schiff n. ‖ navire m. barroté. / ~ **which is full aft** ‖ Schiff n. mit vollem Achterende ‖ navire m. rond *ou* plein à l'arrière. / ~ **with a full bow** ‖ Schiff n. mit vollem Bug ‖ navire m. à avant renflé. / **full built** ~ ‖ vollgebautes Schiff n. ‖ navire m. à formes pleines. / ~ **of full forms** *see* full built ~. / **full loaded** ~ *see* full ~. / ~ **with a good bilge** ‖ Schiff n. mit scharfer Kimm ‖ navire m. de belles fleurs. / **graduated** ~ ‖ Meßgefäß n. ‖ vase m. jaugé. / ~ **of great draught** ‖ tiefgehendes Schiff n. ‖ navire m. de grand tirant d'eau. / **to hail a** ~ ‖ ein Schiff n. preien ‖ hêler un bâtiment. / ~ **by the head** ‖ vorlastiges Schiff n. ‖ navire m. sur le nez *ou* trop sur l'avant. / **high-built** ~ *see* deep ~. / ~ **fastened with galvanized iron bolts** ‖ eisenfestes Schiff n. ‖ navire m. chevillé en fer zingué. / **the** ~ **labours much** ‖ das Schiff arbeitet heftig ‖ le navire se fatigue. / **lapsided** ~ ‖ Schiff n. mit Schlagseite ‖ bordier m.; bâtiment m. bordier. / **leeward** ~ ‖ Schiff n. mit viel Abtrift ‖ mauvais boulinier m.; navire m. qui dérive beaucoup. / **lengthened** ~ ‖ verlängertes Schiff n. ‖ navire m. rallongé. / ~ **with a light cargo** ‖ leicht beladenes Schiff n. ‖ navire m. léger à l'eau. / **loaded** ~ ‖ beladenes Schiff n. ‖ navire m. chargé. / **the** ~ **is lost** ‖ das Schiff ist verloren ‖ le navire a péri *ou* est perdu. / **merchant** ~ ‖ Handelsschiff n. ‖ navire m. à marchand. / **missing** ~ ‖ verschollenes Schiff n. ‖ navire m. qui manque. / **motor** ~ ‖ Motorschiff n. ‖ vaisseau m. *ou* bâtiment m. à moteurs. / **native** ~ ‖ einheimisches Schiff n. ‖ bâtiment m. indigène. / **one-deck** ~ ‖ Eindeckschiff n. ‖ navire m. à un pont. / **overloaded** ~ ‖ überladenes Schiff n. ‖ navire m. surchargé *ou* chargé à couler bas. / **pink-sterned** ~ ‖ Schiff n. mit hohem schmalen Heck ‖ navire m. à arrière pointu. / **the** ~ **plies well to windward** ‖ das Schiff kreuzt gut ‖ le navire est bon boulinier. / ~ **in repairs** ‖ Schiff n. in Ausbesserung ‖ navire m. mis en cran. / **rigged** ~ ‖ aufgetakeltes Schiff n. ‖ navire m. en furin. / ~ **with a round bilge** ‖ Schiff n. mit runder Kimm ‖ navire m. à fleurs rondes. / **round-sheered** ~ ‖ Schiff n. mit viel Sprung ‖ navire m. gondolé. / **to run-down a** ~ ‖ ein Schiff n. übersegeln ‖ passer dessus un bâtiment. / **sailing** ~ ‖ Segelschiff n. ‖ bâtiment m. à voiles. / **sea-going** ~ ‖ Seeschiff n. ‖ bâti-

ment m. de mer. / hull of a sea-going ~ ‖ Seeschiffsrumpf m. ‖ coque f. de bâtiment de mer. / ~ in a seaworthy condition ‖ seetüchtiges *oder* seefähiges Schiff n. ‖ navire m. en bon état de navigabilité; navire m. capable de tenir la mer. / ~ of shallow draught ‖ flach gehendes Schiff n. ‖ navire m. à faible tirant d'eau. / ~ with a sharp bow ‖ Schiff n. mit scharfem Bug ‖ navire m. bien taillé de l'avant. / sharp-built ~ ‖ scharfgebautes Schiff n. ‖ navire m. fin. / ~ sheathed with copper ‖ gekupfertes Schiff n. ‖ navire m. doublé en cuivre. / ~ with a sheer ‖ Schiff n. mit Sprung ‖ navire m. tonturé. / ~ with single-acting four-stroke Diesel engines ‖ Schiff n. mit einfachwirkenden Viertaktdieselmotoren ‖ navire m. équipé de moteurs Diesel à quatre temps et à simple effet. / sister ~ ‖ Schwesterschiff n. ‖ frère-jumeau m. / spar-decked ~ ‖ Spardeckschiff n. ‖ navire m. à spardeck. / stiff ~ ‖ steifes Schiff n. ‖ navire m. fort de côté *ou* dur à abattre. / straight-sheered ~ ‖ Schiff n. ohne Sprung ‖ navire m. sans tonture *ou* ras sur l'eau. / strong-built ~ ‖ starkes Schiff n. ‖ navire m. qui a de bonnes liaisons. / submarine ~ ‖ Unterseeboot n. ‖ navire m. sous-marin. / tight ~ ‖ dichtes Schiff n. ‖ navire m. étanche. / top-heavy ~ *see* crank ~. / unmoored ~ ‖ Schiff n., das vom Anker los ist ‖ navire m. démarré. / unrigged ~ *see* dismantled ~. / unsinkable ~ ‖ unsinkbares Schiff n. ‖ bâtiment m. insubmersible. / ~s pl. for use ‖ Gebrauchsgeschirr n. ‖ vaisselle f. de service. / ~ used all up *see* worn out ~. / wall-sided ~ ‖ Schiff n. mit geraden Seiten ‖ navire m. sans rentrée *ou* aux côtés droits. / ~ of war ‖ Kriegsschiff n. ‖ bâtiment m. de guerre. / the ~ makes water ‖ das Schiff macht Wasser ‖ le navire fait de l'eau. / weak-built ~ ‖ schwachgebautes Schiff n. ‖ navire m. qui manque de liaison. / well-built ~ ‖ gutgebautes Schiff n. ‖ navire m. bien construit. / to work a ~ ‖ ein Schiff n. manövrieren ‖ évoluer. / worn-out ~ ‖ verbrauchtes Schiff ‖ navire consommé; bâtiment mal usé. / ~ with zinc sheathing ‖ Schiff n. mit Zinkboden ‖ navire m. doublé en zink.

vessel, breadth of ~ ‖ Breite f. eines Schiffes ‖ largeur f. d'un navire. / capacity of a ~ ‖ Fassungsvermögen n. eines Schiffes ‖ capacité f. d'un navire. / hull of a ~ ‖ Schiffsrumpf m. ‖ coque f. d'un navire.

vessel-shaped ground-out ‖ tonnenförmig ausgeschliffen ‖ évasé en forme f. de tonneau.

vest ‖ Weste f. ‖ gilet m.

vesta ‖ Wachszündholz n. ‖ allumette f. en cire.

vest backer (Clothm) ‖ Aufschlagnäher m. ‖ poseur m. de revers.

vestibule ‖ Vorhalle f. ‖ vestibule m.

vest maker ‖ Jackenschneider m. ‖ vestonnier m. / ~ pocket ‖ Westentasche f. ‖ poche f. de gilet.

vetch ‖ Wicke f. ‖ vesce f.

veterinary academy ‖ tierärztliche Hochschule f. ‖ école f. vétérinaire supérieure. / ~ instrument ‖ tierärztliches Instrument n. ‖ instrument m. pour vétérinaires. / ~ remedy ‖ Tierheilmittel n. ‖ médicament m. vétérinaire. / ~ surgeon ‖ Tierarzt m. ‖ vétérinaire m.

vetiver oil ‖ Vetiveröl n. ‖ essence f. de vétiver. / ~ root ‖ Vetiverwurzel f. ‖ racine f. de vétiver.

viable ‖ lebensfähig ‖ viable.

viaduct ‖ Straßenbrücke f.; Viadukt m. ‖ viaduc m.

vial ‖ Phiole f. ‖ fiole f.

vibrate, to ‖ vibrieren; schwingen; zittern ‖ trembler; vibrer; osciller.

vibrating diaphragm (Tel) ‖ Membran f. ‖ membrane f. phonique. / ~ trough ‖ Schüttelrinne f.; Schüttelrutsche f. ‖ gouttière f. à secousse. / ~ tuning fork ‖ Stimmgabel f. ‖ diapason m.

vibration ‖ Schwingung f.; oscillation f.; vibration f. / ~ of aerial (Radio) ‖ Antennenschwingung f. ‖ oscillation f. de l'antenne. / constrained ~ ‖ erzwungene Schwingung f. ‖ vibration f. contrainte. / longitudinal ~ ‖ Längsschwingung f. ‖ vibration f. longitudinale. / natural ~ ‖ Eigenschwingung f. ‖ oscillation f. propre. / safe from ~s ‖ erschütterungsfrei ‖ exempte de vibrations fpl. / transversal ~ ‖ Querschwingung f. ‖ vibration f. transversale.

vibration damper ‖ Schwingungsdämpfer m. ‖ amortisseur m. de vibration *ou* de trépidation. / ~ damper (Tel) ‖ Tondämpfer m. ‖ sourdine f. / ~ galvanometer ‖ Vibrationsgalvanometer n. ‖ galvanomètre m. de système vibrateur.

vibrationless operation ‖ erschütterungsfreier Gang m. ‖ marche f. exempte de vibrations.

vibration measuring apparatus *see* vibrograph. / ~ tachometer ‖ Vibrationstachometer n. ‖ tachymètre m. à oscillations.

vibrator (Tel) ‖ Summer m. ‖ vibrateur m.

vibrograph ‖ Vibrograf m. ‖ vibrographe m.; appareil m. à mesurer les vibrations.

vibro massage ‖ Vibrationsmassage f. ‖ massage m. vibratoire.

viburnum ‖ Schneeball m. ‖ viorne f.

vice ‖ Schraubstock m. ‖ étau m. / automatic clamping ~ ‖ selbsttätiger Schraubstock m. ‖ étau m. parallèle automatique. / ~ with ball and socket joint ‖ Schraubstock m. mit Kugelgelenk ‖ étau m. à genou. / bench ~ ‖ Bankschraubstock ‖ étau m. d'établi *ou* à agrafe. / bench ~ with block ‖ Bankschraubstock m. mit Amboß ‖ étau m. à agrafe avec tas. / cross-chap hand ~ ‖ Feilkloben m. mit breitem Maul ‖ étau m. à main à grandes mâchoires. / filing ~ ‖ Feilkloben m. ‖ étau m. limeur *ou* à main. / hand ~ ‖ Feilkloben m. ‖ étau m. à main. / hand ~ with handle ‖ Stielfeilkloben m. ‖ étau m. à main avec manche. / parallel ~ ‖ Parallelschraubstock m. ‖ étau m. (à mouvement) parallèle. / parallel ~ with straight and angular tightening ‖ Parallelschraubstock m. für gerade und Winkelspannung ‖ étau m. parallèle à serrage droit et angulaire. / pin ~ ‖ Stiftklöbchen n. ‖ étau m. à goupilles. / rapid tightening ~ ‖ Schnellspannschraubstock m. ‖ étau m. à serrage instantané. / smith's ~ ‖ Feuerschraubstock m. ‖ étau m. à chaud. / table hand ~ ‖ Bankkloben m. ‖ étau m. à main avec agrafe.

vice bench ‖ Arbeitstisch m.; Werkbank f. ‖ établi m. / ~ clamps pl. ‖ Spannkluppe f. ‖ crampon m.; mordache f. / ~ jaw ‖ Schraubstockspannbacke f. ‖

machoire f. d'étau. / ~ man ‖ Schraubstockarbeiter m. ‖ ouvrier m. à l'étau.

victualling motor car ‖ Verpflegungskraftwagen m. ‖ camion m. à vivres. / ~ ship ‖ Verproviantierungsschiff n. ‖ navire-vivrier m. / ~ yard ‖ Verproviantierungsmagazin n. ‖ dépôt m. de vivres.

victuals pl. ‖ Proviant m.; Lebensmittel npl. ‖ provisions fpl.; vivres mpl.

vicuña down weaver ‖ Vigogneweber m. ‖ tisseur m. de vigogne. / ~ hair ‖ Vigognehaar n. ‖ poil m. de vigogne. / ~ spinner ‖ Vigognespinner m. ‖ fileur m. de vigogne. / ~ spinning ‖ Vigognewollspinnerei f. ‖ filature f. de la laine vigogne. / ~ wool ‖ Vigognewolle f. ‖ laine f. de vigogne. / ~ yarn ‖ Vigognewollgarn n. ‖ fil m. de vigogne.

Vienna chalk ‖ Wiener Putzkalk m. ‖ chaux f. de Vienne.

view ‖ Ansicht f. ‖ vue f.; perspective f. / ~ (Opinion) ‖ Anschauung f. ‖ vue f.; opinion f. / ~ from above (Drawing) ‖ Aufsicht f.; Grundriß m. ‖ vue f. d'en haut. / back ~ ‖ Hinteransicht f. ‖ élévation f. de derrière. / bird's eye ~ ‖ Ansicht f. aus der Vogelschau ‖ vue f. à vol d'oiseau. / ~ of cross section ‖ Ansicht f. im Querschnitt ‖ vue f. de coupe en travers. / general ~ ‖ Gesamtansicht f. ‖ vue f. d'ensemble. / interior ~ ‖ Innenansicht f. ‖ intérieur m. / limited ~ ‖ beschränkte Fernsicht f. ‖ vue f. limitée. / side ~ ‖ Seitenansicht f. ‖ vue f. de côté. / top ~ ‖ Ansicht f. von oben ‖ vue f. d'en haut.

view, angle of ~ ‖ Blickwinkel m. ‖ angle m. de vue *ou* de regard. / direction of ~ ‖ Blickrichtung f. ‖ direction f. du regard. / field of ~ ‖ Gesichtsfeld n. ‖ champ m. de vision. / ~ finder ‖ Visiervorrichtung f. ‖ viseur m.

viewing apparatus for negatives ‖ Betrachtungsapparat m. für Negative ‖ appareil m. pour examiner les négatifs. / ~ tube ‖ Beobachtungsrohr n. ‖ tube m. *ou* lunette f. d'observation.

view post-card ‖ Ansichtspostkarte f.; Ansichtskarte f. ‖ carte f. postale illustrée; carte f. illustrée.

vignette ‖ Letter f.; Vignette f. ‖ vignette f.; caractère m.; fleuron m. / ~ in the title ‖ Titelvignette f. ‖ vignette f. du frontispice.

Vignoles rail ‖ Vignoleschiene f.; Breitfußschiene f. ‖ rail m. Vignole *ou* à patin.

vigorous ‖ kraftvoll ‖ vigoureux.

village ‖ Dorf n. ‖ village m.

villager ‖ Dorfbewohner m. ‖ villageois m.

vine ‖ Wein m.; Weinrebe f.; Weinstock m. ‖ vigne f. / ~ bill ‖ Winzerhippe f. ‖ serpe f. de vigneron. / ~ black ‖ Rebenschwarz n. ‖ noir m. de vigne. / ~ cultivation ‖ Weinbau m. ‖ culture f. de la vigne; viticulture f.; viniculture f. / ~ culture utensil ‖ Rebkulturgerät n. ‖ ustensile m. de viticulture. / ~ dresser ‖ Winzer m.; Weinbergbesitzer m. ‖ vendangeur m.

vinegar ‖ Essig m. ‖ vinaigre m. / aromatic ~ ‖ wohlriechender Essig m. ‖ vinaigre m. parfumé. / concentrated ~ *see* essence. / radical ~ ‖ Eisessig m. ‖ acide m. acétique cristallisable; vinaigre m. radical. / ~ of wood ‖ Holzessig m. ‖ vinaigre m. de bois.

vinegar brewer || Essigbereiter m. || vinaigrier m. / ~ brewery || Essigfabrik f. || vinaigrerie f. / ~ essence || Essigessenz f. || essence f. de vinaigre. / ~ merchant || Essighändler m. || vinaigrier m. / quick ~ process || Schnellessigfabrikation f. || vinaigrerie f. rapide. / ~ tank || Essigkessel m. || appareil m. mycodermique.

vine knife || Winzermesser n. || serpette f.; couteau m. à vendanges. / ~ pole || Rebstecken m.; Rebpfahl m. || échalas m. / ~ prop see ~ post. / ~ sprayer || Weinbergspritze f.; Rebenspritze f. || seringue f. pour vignes. / ~ stake see ~ post. / ~ sulphuring || Weinrebenspritzen n. || sulfatage m. des vignes phylloxérées.

vineyard || Weinberg m. || vigne f.; vignoble m. / ~ owner || Weinbergbesitzer m. || propriétaire m. de vignes; viticulteur m. / syringe for ~s || Weinbergspritze f. || seringue f. pour vignes.

vinosity || Weingehalt m. || vinosité f.

vinous fermentation || geistige Gärung f. || fermentation f. vineuse ou spiritueuse.

vintage || Weinernte f. || vendange f.

vintager || Winzer m. || viticulteur m.; vendangeur m. / ~'s machine || Winzermaschine f. || machine f. pour viticulteurs.

violation (Law) || Gesetzübertretung f. || contravention f.

violent proof || Gewaltprobe f. || épreuve f. à outrance.

violet || violett || violet. / ~ black || violettschwarz || noir violet. / ~ region of the spectrum || violetter Spektralbereich m. || région f. du violet du spectre.

violin || Geige f.; Violine f. || violon m. / ~ for children || Kindergeige f. || violon m. pour enfants. / ~ bow || Geigenbogen m.; Violinbogen m. || archet m. de violon. / ~ case || Geigenkasten m. || boîte f. à violon m. / ~ maker || Geigenbauer m. || luthier m.

violoncello || Violoncell n.; Cello n. || violoncelle m. / ~ bow || Violoncellobogen m.; Cellobogen m. || archet m. de violoncelle.

viper catcher || Vipernfänger m. || chasseur m. de vipères.

virgin (Met) || gediegen || natif; vierge; pur. / ~ gold || Jungferngold n. || or m. vierge. / ~ honey || Jungfernhonig m. || miel m. vierge. / ~ land || ungepflügtes Land n.; Neuland n. || novale f. / ~ wax || Jungfernwachs n. || cire f. vierge.

Virginia (tobacco) || Virginiatabak m.; virginischer Tabak m. || virginie m.

virtu, article of ~ || Kunstgegenstand m. || objet m. d'art.

virtual || virtuell || virtuel. / principle of ~ moments || Prinzig n. der virtuellen Momente || principe m. des moments virtuels. / ~ value || Effektivwert m. || valeur f. efficace ou virtuelle. / ~ velocity || virtuelle Geschwindigkeit f. || vitesse f. virtuelle.

viscid see viscous.

viscose || Viskose f. || viscose f. / ~ silk || Viskoseseide f. || soie f. en viscose. / ~ spangle || Viskoseflitter m. || paillette f. en viscose.

viscosity || Zähflüssigkeit f.; Flüssigkeitsgrad m.; Viskosität f. || viscosité f. / determination of ~ || Viskositätsbestimmung f. || détermination f. de la viscosité.

viscous || zähflüssig; viskos || visqueux. / very ~ || hochviskos || très visqueux.

viscous fermentation || Schleimgärung f. || fermentation f. visqueuse.

visibility || Sicht f.; Sichtbarkeit f. || visibilité f.

visible || wahrnehmbar; sichtbar || visible; perceptible. / ~ to the naked eye || für das bloße Auge sichtbar || visible à l'œil m. nu. / ~ radiation || sichtbare Strahlung f. || radiation f. visible. / ~ writing typewriter || sichtbar schreibende Schreibmaschine f. || machine f. à écriture visible.

vis inertiæ || Beharrungsvermögen n. || force f. d'inertie.

vision || Sehen n.; Sehvermögen n. || vue f.; vision f. / binocular ~ || beidäugiges Sehen n. || vision f. binoculaire. / axis of ~ || Sehachse f. || axe m. visuel. / distance of ~ || Sehweite f. || distance f. ou portée de la vue. / distance of distinct ~ || deutliche Sehweite f. || distance f. de la vision distincte. / field of ~ || Blickfeld n.; Gesichtsfeld n. || champ m. visuel ou de vue. / range of ~ (Eye) || Augenweite f. || écartement m. des yeux.

visiting card || Besuchskarte f. || carte f. de visite. / ~ time || Besuchzeit f. || heure f. de visite.

visual acuity, ascertaining the ~ || Sehschärfenbestimmung f. || détermination f. de l'acuité visuelle.

visual capacity || Sehvermögen n. || acuité f. visuelle. / determining the ~ || Feststellung f. des Sehvermögens || détermination f. de l'acuité visuelle.

visual observation || visuelle Beobachtung f. || observation f. visuelle. / ~ power || Sehkraft f. || vue f.

vis viva || lebendige Kraft f. || force f. vive.

vital matter || Lebensfrage f. || question f. vitale. / ~ part || lebenswichtiger Teil m. || partie f. vitale.

vitamine preparation || Vitaminpräparat n. || préparation f. de vitamine.

viticulture see vine cultivation.

vitreous || glasig; glasartig || vitreux. / ~ arsenic trioxide || Arsenikglas n. || anhydride m. arsénieux vitreux. / ~ body || Glaskörper m. || corps m. vitré. / ~ sand pit || Glassandgrube f. || extraction f. de sable pour verrerie.

vitrifiable pigment || Schmelzfarbe f. || couleur f. vitrifiable.

vitrification || Verglasung f. || vitrification f.

vitrifications pl. || Glasschmelzwaren fpl. || vitrifications fpl.

vitrified bond || keramische Binde f. || aggloméran m. céramique. / ~ brick || Glasurstein m. || brique f. gobetée ou glacée. / ~ malt || Glasmalz n. || malt m. vitreux.

vitriol || Vitriol n.; Sulfat n. || vitriol m.; sulfate m. / blue ~ see ~ of copper. / ~ of copper || Kupfervitriol n. || vitriol m. bleu ou de cuivre; couperose f. bleue. / cupreous ~ || kupferhaltiges Vitriol n. || vitriol m. cuivreux. / green ~ || Eisenvitriol n. || couperose f. verte; sulfate m. de fer. / white ~ || Zinkvitriol n. || sulfate m. de zinc.

vitriolic acid || Schwefelsäure f. || acide m. sulfurique.

vitriol, oil-of ~ see vitriolic acid.

vivianite || Vivianit m. || vivianite f.

vocational education || Berufsausbildung f. || instruction f. professionelle. / ~ training see vocational education.

voice || Stimme f.; Sprache f.; Laut m.; Ton m. || voix f.; parole f.; son m.

voice frequency generator || Tonfrequenzmaschine f. || alternateur m. à fréquences vocales. / ~ ringing in four-wire circuits || Tonfrequenzanruf m. in Vierdrahtleitungen || appel m. utilisant des fréquences musicales en circuits à quatre fils.

voice, lowering the ~ || Senken n. der Stimme || abaissement m. de la voix. / ~ pipe || Sprachrohr n. || porte-voix m. / raising the ~ || Heben n. der Stimme || élévation f. de la voix.

void || nichtig || nul; vain.

volatile || ätherisch || essentiel; éthéré. / ~ fuel || leicht entzündlicher Brennstoff m. || combustible m. très inflammable. / ~ matter || flüchtige Bestandteile mpl. || matières fpl. volatiles. / ~ oil || ätherisches Öl n. || huile f. essentielle ou volatile. / ~ solvent || flüchtiges Lösungsmittel n. || dissolvant m. volatil.

volatility || (Leicht-)Flüchtigkeit f. || volatilité f.

volatilizable || flüchtig || volatil(isabl)e.

volatilization || Verdunsten n.; Verflüchtigen n. || volatilisation f.

volatilize, to || sich verflüchtigen; verdunsten || se volatiliser; s'évaporer. / substance which is difficult ~ || schwer verdampfbarer Stoff m. || substance f. difficile à vaporiser.

volatilizing tube, collar of ~ || Riechrohrschelle f. || collier m. pour tuyaux à substances volatiles.

volcanic || vulkanisch || volcanique. / ~ rock || vulkanisches Gestein n. || roche f. volcanique. / ~ sedimentary rock || vulkanisches Sedimentgestein n. || roche f. sédimentaire volcanique.

volcano || Vulkan m. || volcan m. / burning ~ || tätiger Vulkan m. || volcan m. en activité. / extinct ~ || erloschener Vulkan m. || volcan m. éteint.

Volta cell || Voltasches Element n. || élément m. Volta.

voltage || Spannung f. || tension f.; voltage m. / additional ~ || Zusatzspannung f. || tension f. additionnelle. / ~ of the circuit || Betriebsspannung f. || tension f. de régime. / to keep the ~ constant || die Spannung konstant halten || tenir constante la tension. / excess ~ || Überspannung f. || tension f. supplémentaire. / ~ of exciter || Erregerspannung f. || voltage m. d'excitation. / external ~ || Fremdspannung f. || tension f. d'origine étrangère. / final ~ on charge (Acc) || Endspannung f. der Ladung || tension f. à la fin de la charge. / final ~ on discharge (Acc) || Endspannung f. der Entladung || tension f. à la fin de la décharge. / to impress a ~ || eine Spannung f. aufdrücken || imprimer une tension. / to increase the ~ || die Spannung erhöhen || élever ou augmenter la tension. / initial ~ on charge (Acc) || Anfangsspannung f. der Ladung || tension f. initiale de la charge. / initial ~ on discharge (Acc) || Anfangsspannung f. der Entladung || tension f. initiale de la décharge. / interlinked ~ || verkettete Spannung f. || tension f. composée. / invariable ~ || unveränderliche Spannung f. || tension f. invariable. / nominal ~ ||

Nennspannung f. ‖ tension f. nominale. / non-interlinked phase ~ ‖ unverkettete Phasenspannung f. ‖ tension f. du courant à phases non reliées. / phase ~ ‖ Phasenspannung f. ‖ tension f. de phase. / steady ~ ‖ gleichmäßige Spannung f. ‖ tension f. constante. / terminal ~ ‖ Klemmenspannung f. ‖ tension f. aux bornes.

voltage amplification factor ‖ Spannungsverstärkungsfaktor m. ‖ coefficient m. d'amplification de voltage. / ~ curve ‖ Spannungskurve f. ‖ courbe f. de tension. / ~ degree ‖ Spannungsstufe f. ‖ degré m. de voltage; prise f. de voltage. / ~ divider ‖ Spannungsteiler m. ‖ potentiomètre m.; limiteur m. de tension. / drop in ~ ‖ Spannungsabfall m. ‖ chute f. de tension. / ~ fuse ‖ Spannungssicherung f. ‖ fusible m. pour tension. / ~ level ‖ Spannungspegel m. ‖ niveau m. de transmission de la tension. / ~ recorder ‖ Spannungsschreiber m. ‖ voltmètre m. enregistreur. / ~ regulator ‖ Spannungsregler m. ‖ graduateur m. de tension. / ~ step up ‖ Aufwärtstransformator m. ‖ transformateur m. élévateur.

voltaic current ‖ galvanischer Strom m. ‖ courant m. galvanique.

voltameter, recording ~ ‖ schreibendes Voltameter n. ‖ voltamètre m. enregistreur; voltamétrographe m.

volt-hour meter ‖ Voltstundenzähler m. ‖ volt-heuremètre m.

voltmeter ‖ Spannungsmesser m.; Voltmeter n. ‖ voltmètre m. / alternating current ~ ‖ Voltmeter n. für Wechselstrom ‖ voltmètre m. pour courant alternatif. / direct current ~ ‖ Voltmeter n. für Gleichstrom ‖ voltmètre m. pour courant continu. / moving iron ~ ‖ Dreheisenspannungsmesser m. ‖ voltmètre m. à fer doux.

voltmeter switch ‖ Voltmesserumschalter m. ‖ commutateur m. du voltmètre.

volume ‖ Volumen n.; Rauminhalt m. ‖ volume m. / ~ (Print) ‖ Band m. ‖ volume m.; tome m. / ~ of the reservoir of a barrage ‖ Stauinhalt m. einer Talsperre ‖

capacité f. de retenu d'un barrage de vallée. / ~ of sound ‖ Lautstärke f. ‖ sonorité f.; audition f. / to work for ~ ‖ auf Menge f. arbeiten ‖ travailler en débit m.

volume, contraction of ~ ‖ Volumenkontraktion f.; Raumverminderung f. ‖ contraction f. de volume. / ~ indicator (Balloon) ‖ Prallanzeiger m. ‖ indicateur m. de plénitude ou de remplissage. / ~ integral ‖ Raumintegral n. ‖ intégrale f. de volume. / ~ part ‖ Raumteil m. ‖ partie f. en volume. / per cent by ~ ‖ volumprozentig ‖ pour cent en volume.

volumetric ‖ volumetrisch ‖ volumétrique. / ~ analysis ‖ Maßanalyse f. ‖ analyse f. volumétrique. / ~ apparatus ‖ Titrierapparat m. ‖ appareil m. de titration.

voluminous ‖ umfangreich; voluminös ‖ volumineux. / ~ part (Railw) ‖ sperriges Stück n. ‖ pièce f. encombrante ou volumineuse ou de dimensions démesurées.

volute buffer spring ‖ Pufferfeder f.; Wickelfeder f.; Schneckenfeder f. ‖ ressort m. conique pour tampons. / ~ compasses pl. ‖ Spiralenzirkel m. ‖ compas m. à volute. / ~ spring ‖ Schneckenfeder ‖ ressort m. à volute.

vomiting ‖ Erbrechen n.; Kotzen n.; Sichübergeben n. ‖ vomissement m.

vomit nut ‖ Brechnuß f. ‖ noix f. vomique.

vortex ‖ Wirbel m. ‖ tourbillon m. / air ~ ‖ Luftwirbel m. ‖ tourbillon m. ou remous m. d'air. / partial ~ ‖ Teilwirbel m. ‖ tourbillon m. composant. / water ~ ‖ Wasserwirbel m. ‖ tournant m. ou remous m. d'eau.

vortex centre see ~ core. / ~ core ‖ Wirbelkern m. ‖ noyau m. du tourbillon; partie f. centrale du tourbillon. / ~ motion ‖ Wirbelbewegung f. ‖ mouvement m. tourbillonnaire.

vorticity ‖ Turbulenz f. ‖ turbulence f.

vote, to ‖ abstimmen ‖ voter. / entitled ~ ‖ stimmberechtigt ‖ qui a le droit de vote.

voting, to carry by ~ ‖ beschließen ‖ résoudre; décider.

voucher ‖ Rechnungsbeleg m. ‖ pièce f. justificative.

voussoir (Build) ‖ Wölbkeil m. ‖ brique f. de voûte; claveau m. / indented ~ (Build) ‖ Kropfstein m.; verzahnter Wölbstein m. ‖ voussoir m. engrenant.

vox ‖ Stimme f. ‖ voix f.

voyage ‖ Seereise f. ‖ voyage m. / ~ out and in ‖ Hin- und Rückreise f. ‖ voyage m. d'aller et de retour. / outward ~ ‖ Ausreise f. ‖ voyage m. d'aller.

vulcan fibre trunk ‖ Vulkanfiberkoffer m. ‖ malle f. en fibre vulcanisée.

vulcanite ‖ Hartgummi n.; Ebonit n. ‖ caoutchouc m. durci; ébonite f. / ~ mount ‖ Hartgummifassung f. ‖ monture f. caoutchouc durci. / ~ stage ‖ Hartgummitisch m. ‖ platine f. en ébonite.

vulcanization ‖ Vulkanisierung f.; Vulkanisieren n. ‖ vulcanisation f. / ~ works see vulcanizing factory.

vulcanizator for motor car tyres ‖ Vulkanisator m. für Autoreifen ‖ vulcanisateur m. pour bandages d'automobiles.

vulcanize, to ‖ vulkanisieren ‖ vulcaniser.

vulcanized ‖ vulkanisiert ‖ vulcanisé. / ~ fibre ‖ Vulkanfiber f. ‖ fibre f. vulcanisée. / ~ rubber ‖ vulkanisiertes Gummi n. ‖ gomme f. vulcanisée; caoutchouc m. vulcanisé.

vulcanizer ‖ Vulkanisierapparat m. ‖ appareil m. de vulcanisation; vulcanisateur m. / ~ (Person) ‖ Vulkanisör m. ‖ vulcaniseur m. / steam ~ ‖ Dampfvulkanisierapparat m. ‖ vulcanisateur m. à vapeur.

vulcanizing apparatus for dentists ‖ Vulkanisierapparat m. für Zahnärzte ‖ appareil m. à vulcaniser pour dentistes. / ~ boiler ‖ Vulkanisierkessel m. ‖ chaudière f. à vulcanisation. / ~ factory ‖ Vulkanisieranstalt f. ‖ atelier m. à vulcaniser. / ~ machine see vulcanizer. / ~ plant see ~ factory. / ~ press ‖ Vulkanisierpresse f. ‖ presse f. à vulcaniser. / ~ stove ‖ Vulkanisierofen m. ‖ four m. de vulcanisation.

vulnerary powder ‖ Wundpulver n. ‖ poudre f. vulnéraire.

W

W-type engine ‖ Fächermotor m. ‖ moteur m. en flèche.

wabble, to (Wheel) ‖ flattern ‖ flageoler.

wacke, basaltic ~ ‖ Basaltwacke f. ‖ wackite f. basaltique. / gray ~ (Geol) ‖ Grauwacke f. ‖ grauwacke f.

wad, to ‖ wattieren ‖ ouater.

wad see wadding.

wadding ‖ Watte f. ‖ ouate f. / antiseptic ~ ‖ antiseptische Watte f. ‖ ouate f. antiseptique. / ~ of cellulose ‖ Zellulosewatte f.; Zellstoffwatte f. ‖ ouate f. de cellulose. / ~ of cotton ‖ Baumwollwatte f. ‖ ouate f. de coton. / ~ of glass ‖ Glaswolle f. ‖ ouate f. de verre. / ~ imbued with glue or starch ‖ mit Leim oder Stärke getränkte Watte f. ‖ ouate f. enduite ou imbibée de colle ou d'amidon. / ~ for packing ‖ Packwatte f. ‖ ouate f. d'emballage. / ~ for the ready-made clothes trade ‖ Konfektionswatte f. ‖

ouate f. de confection. / ~ of sheep's wool ‖ Schafwollwatte f. ‖ ouate f. de laine de mouton. / silk ~ ‖ Seidenwatte f. ‖ ouate f. de soie. / water-absorbing ~ ‖ wasseraufsaugende Watte f. ‖ ouate f. hydrophile. / ~ for upholstering ‖ Polsterwatte f. ‖ ouate f. de rembourrage; rembourrure f.

wadding linen ‖ Wattierleinen n. ‖ toile f. d'ouatage. / ~ manufacturing machine ‖ Watteherstellungsmaschine f. ‖ machine f. à fabriquer l'ouate. / roll of ~ ‖ Rolle f. aus Watte ‖ bourrelet m. d'ouate.

wafer (Bak) see waffle. / ~ (Church) ‖ Hostie f.; Oblate f. ‖ pain m. à chanter; hostie f. / ~ (Med) ‖ Oblate f. ‖ pain m. azyme; cachet m. médicamenteux. / flour ~ ‖ Mehloblate f. ‖ hostie f. de farine. / seal ~ ‖ Siegeloblate f. ‖ cachet m. en oublie / sweet ~ ‖ süße Eßoblate f. ‖ oublie f. douce.

waffle ‖ Waffel f. ‖ gaufre f. / ~ baker ‖ Waffelbäcker m. ‖ fabricant m. de gaufres. / ~ mould ‖ Waffelform f. ‖ moule m. à gaufres. / ~ wafer ‖ Waffeloblate f. ‖ oublie f. à cacheter.

wage earner ‖ Lohnempfänger m. ‖ salarié m.

wages pl. ‖ Lohn m. ‖ paye f.; salaire m.; gages mpl. / ~ (Mar) ‖ Heuer f. ‖ loyer m.; salaire m.; paye f.; solde f. / additional ~ ‖ Lohnzulage f. ‖ supplément m. de solde. / daily ~ ‖ Tagelohn m. ‖ journée f. / daily ~ (Mine) ‖ Schichtlohn m. ‖ salaire m. fixe. / labourer's ~ ‖ Arbeitslohn m. ‖ salaire m.; main f. d'œuvre; façon f. / minimum ~ ‖ Mindestlohn m. ‖ salaire minimum. / ~ for piecework ‖ Stücklohn m. ‖ forfait m. / weekly ~ ‖ Wochenlohn m. ‖ semaine f.; paye f. de semaine.

wages pl., to combine in order to obtain better conditions pl. of ~ ‖ sich zur Erlan-

gung besserer Lohnbedingungen fpl. zusammenschließen ‖ se coaliser pour obtenir de meilleures conditions fpl. de salaire. / ~ conference ‖ Lohnverhandlung f. ‖ négociation f. sur les salaires. / decrease of ~ ‖ Lohnherabsetzung f.; Lohnabbau m. ‖ réduction f. de salaire. / increase of ~ ‖ Lohnerhöhung f. ‖ augmentation f. de salaire. / register of ~ ‖ Lohnzettel m. ‖ fiche f. de paye. / tax on ~ ‖ Lohnsteuer f. ‖ impôt m. sur les salaires ou sur les gages.

wagon ‖ Wagen m. ‖ wagon m.; chariot m.; voiture f. / all-round box tip ~ ‖ Rundkipper m.; Kastenrundkipper m. ‖ wagonnet m. à caisse basculante dans toutes les directions. / ~ with an awning ‖ Zeltwagen m.; Planwagen m. ‖ chariot m. couvert d'une bâche; voiture f. à bâche. / bin ~ ‖ Kübelwagen m. ‖ wagon m. à benne. / bottom discharging ~ ‖ Bodenentleerer m. ‖ wagon m. déchargeant par le fond. / ~ for bulky goods ‖ Wagen m. für sperrige Stücke ‖ wagon m. pour pièces encombrantes. / cable ~ ‖ Kabelwagen m. ‖ chariot m. à câble. / cask and barrel ~ ‖ Faßtransportwagen m. ‖ wagonnet m. à transporter des barils. / closed ~ ‖ geschlossener Wagen m. ‖ wagon m. fermé. / covered ~ (Railw) ‖ bedeckter Güterwagen m. ‖ wagon m. couvert. / cradle tip ~ ‖ Wiegenkipper m. ‖ wagonnet m. à berceaux de versement. / eight-wheeled ~ ‖ vierachsiger Wagen m. ‖ wagon m. à huit roues. / 50 tons goods ~ ‖ Großraumgüterwagen m. ‖ wagon m. à marchandises de 50 tonnes. / flat ~ ‖ Plateauwagen m. ‖ wagon m. à plateforme ou à plateau. / ~ with gable bottom ‖ Sattelwagen m.; Wagen m. mit Sattelboden ‖ wagon m. à fond en dos d'âne. / goods ~ ‖ Güterwagen m. ‖ wagon m. à marchandises. / goods ~ hanging down between the cross girders ‖ Tiefladewagen m. für durchhängende Ladungen ‖ wagon m. pour pièces pendantes entre les traverses. / general tool and gear ~ ‖ Gerätewagen m. ‖ voiture f. d'agrès. / high-capacity ~ ‖ Großgüterwagen m. ‖ wagon m. à marchandises de grande capacité. / ~ with hopper bottom ‖ Wagen m. mit Bodentrichter ‖ wagon m. à fond en trémie. / ~ for iron and steel works ‖ Hüttenwagen m. ‖ wagonnet m. pour usines sidérurgiques. / covered lime ~ ‖ Kalkdeckelwagen m. ‖ wagon m. à couvercle pour chaux. / luggage ~ ‖ Gepäckwagen m. ‖ fourgon m.; wagon m. à bagages. / ~ with mechanical tipping device ‖ Wagen m. mit mechanischer Kippvorrichtung ‖ wagonnet m. basculeur à dispositif mécanique de bascule. / open ~ (Railw) ‖ offener Wagen m.; Kastenwagen m. ‖ wagon m. ouvert ou à caisse. / open goods ~ ‖ offener Güterwagen m. ‖ wagon m. découvert. / open-sided ~ see open spar ~. / open spar ~ ‖ Leiterwagen m. ‖ chariot m. à ridelles. / railway ~ ‖ Eisenbahnwagen m. ‖ wagon m. de chemin de fer. / to replace a ~ upon the track ‖ Aufgleisung f. ‖ remise f. sur rails. / roofed ~ ‖ bedeckter Wagen m. ‖ wagon m. couvert. / ~ running on rails ‖ Gleisfahrzeug n. ‖ véhicule m. sur rails. / ~ running up ‖ anlaufender Wagen m. ‖ wagon m. en approche. / ~ running without load ‖ leergehender Wagen m. ‖ wagon m. mar-

chant à vide. / scoop-type tip ~ ‖ Schnabelvorderkipper m. ‖ wagonnet m. avec bec basculant en bout. / self-discharging ~ ‖ Selbstentlader m. ‖ wagon m. à déchargement automatique. / ~, self-discharging from the bottom ‖ Bodenselbstentlader m. ‖ wagon m. à déchargement automatique par le fond. / ~ with shafts ‖ Gabelwagen m. ‖ fourgon m. / side dump tip box ~ ‖ einseitiger Kastenkipper m. ‖ wagonnet m. à caisse basculante d'un seul côté. / ~ for spraying salt ‖ Salzstreuwagen m. ‖ wagon m. pour le répandage du sel. / ~ for suspension railway ‖ Hängebahnwagen m. ‖ wagonnet m. suspendu. / tank ~ ‖ Kesselwagen m. ‖ wagon m. citerne. / tip ~ ‖ Kippwagen m. ‖ wagon m. basculeur. / ~ for tipping towards either sides ‖ nach beiden Seiten kippender Wagen m. ‖ wagonnet m. basculant des deux côtés. / trough tip ~ for mines ‖ Grubenmuldenkipper m. ‖ basculeur m. à auge de mine. / trunnion tip ~ ‖ Zapfenkipper m. ‖ wagonnet m. culbuteur à tourillons. / ~ of two four-wheeled bogies ‖ Wagen m. mit zwei zweiachsigen Drehgestellen ‖ wagon m. à deux bogies à deux essieux. / ~ on two trucks each of two axles see ~ of two four-wheeled bogies.

wagon balance ‖ Waggonwage f. ‖ bascule f. à wagons. / ~ box ‖ Wagenkasten m. ‖ caisse f. de wagon. / ~ checker ‖ Wagenkontrollör m. ‖ contrôleur m. de wagons. / automatic ~ circulation ‖ selbsttätiger Wagenumlauf m. ‖ dispositif m. automatique de circulation pour wagonnets. / special machine for the construction of ~s ‖ Waggonbausondermaschine f. ‖ machine f. spéciale pour la construction de wagons. / ~ counting device for railways ‖ Wagenzähleinrichtung f. für Bahnanlagen ‖ compteur m. de wagons pour chemins de fer. / ~ door ‖ Wagentür f. ‖ porte f. de wagon.

wagonet ‖ Förderwagen m. ‖ wagonnet m.; berline f.

wagon fat ‖ Wagenfett n. ‖ graisse f. pour voitures. / ~ fittings pl. ‖ Wagenbeschlagteile mpl. ‖ ferrures fpl. de wagons. / ~ hoist ‖ Wagenaufzug m. ‖ grue f. pour soulever les wagons; élévateur m. de wagons. / ~ lifting appliance see ~ hoist. / ~ load ‖ Fuhre f. ‖ charroi m.; charge f. / ~ load (Railw) ‖ Frachtgut n. in Wagenladung; Wagenladung f. ‖ marchandise f. par charge complète. / ~ loading ‖ Wagenbeladen m. ‖ chargement de wagons m. / ~-maker's machine ‖ Stellmachermaschine f. ‖ machine f. pour travaux de carrosserie. / ~ pushing device ‖ Wagenstoßvorrichtung f. ‖ dispositif m. de choc pour wagons. / ~ road (Mine) ‖ Förderstrecke f. ‖ galerie f. de roulage. / ~ rolling stock ‖ Wagenpark m. ‖ parc m. à voitures ou de wagons. / ~ roof ‖ Wagendach n. ‖ toiture f. de wagon. / ~ set (Radio) ‖ Karrenstation f.; fahrbare Station f. ‖ station f. du type sur voiture. / ~ shed ‖ Wagenschuppen m. ‖ hangar m. à voitures. / ~ spring ‖ Wagenfeder f. ‖ ressort m. de wagons. / ~ tipper ‖ Wagenkipper m. ‖ basculeur m. ou culbuteur m. de wagons. / rotary ~ tipper ‖ Kreiselkipper m. ‖ basculeur m. circulaire. / ~ tipper for a useful load of x tons ‖ Wagenkipper m. für x t Nutzlast ‖ culbuteur m.

de wagons pour x t de charge utile. / ~ tipping device see ~ tipper. / ~ unloading ‖ Wagenausladung f. ‖ déchargement de wagons m. / ~ wheel ‖ Wagenrad n. ‖ roue f. de wagon. / ~ wheel set ‖ Wagenradsatz m. ‖ essieu m. monté pour wagons.

wainscot, to ‖ vertäfeln ‖ lambrisser.

wainscot (Build) ‖ Holzverschalung f.; Täfelung f. ‖ lambris m.; revêtement m. en bois. / ~ (Wood) ‖ Täfelholz n. ‖ bois m. de lambrissage.

wainscotting see wainscot.

waistcoat ‖ Weste f. ‖ gilet m. / cotton ~ ‖ Baumwollweste f. ‖ gilet m. de coton. / flannel ~ ‖ Flanellunterjacke f. ‖ gilet m. de flanelle. / woollen ~ ‖ Wollweste f. ‖ gilet m. de laine.

waistcoat clasp ‖ Westenschnalle f. ‖ boucle f. de gilets. / ~ cloth ‖ Westenstoff m. ‖ étoffe f. pour gilets. / ~ maker ‖ Westenschneider m. ‖ giletier m.

waist sheet of locomotive boiler ‖ Sattelplatte f.; Stehkesselvorderwand f. ‖ selle f. avant d'une chaudière de locomotive.

waiter key (Cash register) ‖ Kellnertaste f. ‖ touche f. du garçon.

waiting room (Railw) ‖ Wartesaal m. ‖ salle f. d'attente. / ~ tank ‖ Sammelkasten m. ‖ bassin m. intermédiaire. / ~ time (Tel) ‖ Wartezeit f. ‖ délai m. d'attente.

waker ‖ Wecker m. ‖ réveil-matin m.

wale (Shipb) ‖ Gurtholz n. ‖ ceinture f. d'un navire. / ~ of a boat ‖ Randgeer n. oder Wallschiene f. eines Bootes ‖ préceinte f. d'une chaloupe.

walk, side ~ ‖ Bürgersteig m. ‖ trottoir m.

walkable deck ‖ begehbares Deck n. ‖ pont m. sur lequel on peut marcher.

walker bridge ‖ Brücke f. für Fußgänger; Fußgängerbrücke f. ‖ passerelle f.

walking cane see walking stick.

walking stick ‖ Spazierstock m. ‖ canne f. / ~ for children ‖ Kinderspazierstock m. ‖ canne f. pour enfants. / ~ fittings pl. ‖ Stockbeschlag m. ‖ garniture f. pour cannes. / ~ handle ‖ Stockgriff m. ‖ poignée f. pour cannes. / ~ maker ‖ Stockmacher m. ‖ fabricant m. de cannes.

walkway (Airship) ‖ Laufgang m. ‖ passerelle f.

wall, to ‖ mauern ‖ maçonner. / ~ in bond ‖ verbandsmäßig oder in gutem Verband m. mauern ‖ murer en liaison f.; poser en bonne liaison f.; liaisonner. / ~ the bricks for drying ‖ Ziegel mpl. zum Trocknen aufsetzen ‖ mettre en haie les briques fpl. pour les sécher. / ~ in the form of stairs ‖ eine Mauer f. abtreppen ‖ maçonner par retraites ou en degrés. / ~ roughly ‖ roh mauern ‖ limosiner; limosiner; hourder. / ~ up ‖ vermauern ‖ murer.

wall ‖ Mauer f.; Wand f. ‖ mur m.; paroi f. / ~ (Mine) ‖ Abbaustoß m. ‖ paroi m.; front m. ou fond m. de taille. / baked ~ ‖ Füllmauer f. ‖ mur m. de remplage ou de blocage ou rempli de hourdage. / bare-based ~ ‖ Mauer f. mit entblößtem Grund ‖ mur m. déchaussé. / battering ~ ‖ ausbauchende oder bauchige Mauer f. ‖ mur m. bouché ou gauchissant. / the ~ batters ‖ die Mauer baucht aus ‖ le mur gauchit. / blind ~ ‖ blinde Mauer f. ‖ mur m. orbe ou aveugle. / ~ of bricks ‖ Mauer f. aus Ziegelsteinen ‖ mur m. de

briques. / common ~ ‖ Grenzmauer f.; gemeinschaftliche Mauer f. ‖ mur m. commun *ou* mitoyen. / ~ of the cornea ‖ Hornhautwand f. ‖ paroi f. de la cornée. / the ~ cracks *see* the ~ gapes. / ~ upon discharging arches ‖ auf einem Schildbogen ruhende Mauer f. ‖ mur m. en décharge. / dry ~ ‖ trockene *oder* kalte Mauer f.; Steinpackung f. ‖ mur m. en pierres sèches; perré m.; pierré m. / fire-proof ~ ‖ Brandmauer f. ‖ mur m. massif *ou* réfractaire. / ~ of freestones ‖ Mauer f. aus Hausteinen ‖ mur m. de pierres franches. / ~ founded upon grating ‖ auf einen Rost gegründete Mauer f. ‖ mur m. planté. / the ~ gapes ‖ die Mauer bekommt Risse ‖ le mur se fend. / ~, half a brick thick ‖ ½ Stein starke Mauer f. ‖ mur m. de l'épaisseur de ½ brique. / ~ of a high-furnace ‖ Schacht m. eines Hochofens ‖ chemise f. d'un haut-fourneau. / ~ leaning against an earth body ‖ an das Erdreich angelehnte Mauer f. ‖ mur m. adossé à un terre-plein. / ~ of a lode ‖ Salband n. eines Ganges ‖ éponte f. *ou* paroi f. *ou* salbande f. d'un filon. / mean ~ *see* common ~. / overhanging ~ (Mine) ‖ überhängender Stoß m. ‖ paroi m. surplombant. / partition ~ *see* wall, common. / principal ~ ‖ Hauptmauer f. ‖ maîtresse-muraille f. / ~ of rough stone ‖ Mauer f. aus Bruchsteinen ‖ mur m. de moëllon. / the ~ settles ‖ die Mauer senkt sich *oder* setzt sich *oder* sackt ‖ le mur s'affaise *ou* prend coup. / ~ of a shaft ‖ Schachtstoß m. ‖ face f. *ou* paroi f. d'un puits. / shrinking ~ *see* battering ~. / the ~ sinks in *see* the ~ settles. / to soak a ~ ‖ eine Mauer f. netzen ‖ abreuver un mur. / the ~ splits *see* the ~ gapes. / ~ with timber holes ‖ Mauer f. mit Balkenlöchern ‖ mur m. coupé. / the middle rows are protected against the flue gases by a compact of tubes ‖ die mittleren Röhren fpl. werden durch eine dichte Rohrwand vor den Feuergasen geschützt ‖ on protège les tubes mpl. médians contre l'action des gaz chauds au moyen d'une paroi de tubes serrés. / water stop ~ ‖ Wehrmauer f. ‖ mur m. bâtardeau.

wall anchor plate ‖ Wandankerplatte f. ‖ plaque f. d'ancrage pour parois. / ~ arch ‖ Mauerbogen m. ‖ arc m. *ou* arceau m. d'un mur. / back of a ~ ‖ innere Mauerfront f. *oder* Mauerflucht f. ‖ rez-mur m.; parement m. intérieur d'un mur. / ~ bracket ‖ Wandkonsole f.; Wandarm m.; Wandstütze f.; Mauerbügel m.; potelet m.; console m. *ou* bras mural. / ~ bracket bearing ‖ Mauerlager n. ‖ palier m. mural. / ~ calender ‖ Wandkalender m. ‖ calendrier m. bloc. / ~ cap of a ~ ‖ Mauerkappe f.; Mauerhut m.; Mauerkrone f. ‖ chape f. d'un mur. / ~ channel ‖ Mauerdurchführung f. ‖ percement m. de mur. / ~ chest ‖ Wandschrank m. ‖ placard m. / ~ chisel ‖ Steinbohrer m. ‖ perce-meule m.; bonnet m. de prêtre. / ~ clamp ‖ Maueranker m.; Stichanker m. ‖ lien m. tirant; tirant m. / ~ coffee mill ‖ Wandkaffeemühle f. ‖ moulin m. à café mural. / ~ column ‖ Mauersäule f. ‖ colonne f. murale. / ~ crane ‖ Wandkran m. ‖ grue f. murale. / decay of a ~ by efflorescence ‖ Mauerfraß m. ‖ carie f. des murailles. / ~ diagram ‖ Wandtafel

f. ‖ tableau m. mural. / ~ dowel ‖ Mauerdübel m. ‖ goujon m. mural. / ~ dresser ‖ Fassadenputzer m. ‖ ravaleur m. / ~ drill ‖ Wandbohrmaschine f. ‖ perceuse f. murale. / ~ duct ‖ Durchführungstülle f. ‖ douille f. de traversée. / ~ duct (Cable) ‖ Hochführungsschacht m. ‖ cheminée f. d'ascension.

walled, double-~ ‖ doppelwandig ‖ à double paroi f.

wallet ‖ Reisetasche f. ‖ sac m. de voyage.

wall facing ‖ Wandbekleidung f. ‖ revêtement m. de parois *ou* de mur; boiserie f. / ~ flag (Build) ‖ Wandfliese f. ‖ carreau m. pour murs. / ~ fruit ‖ Spalierobst n. ‖ fruits mpl. d'espalier. / ~ hook ‖ Wandhaken m. ‖ crochet m. à applique murale.

walling ‖ Mauerwerk n.; Gemäuer n.; Mauerung f. ‖ maçonnerie f.; ouvrage m. de maçonnerie; maçonnage m.; murage m.; muraillement m. / baked ~ ‖ Füllmauerwerk n. ‖ maçonnerie f. en blocage *ou* de remplage *ou* remplie de hourdage; murage m. bloqué. / decayed ~ ‖ verfallenes Mauerwerk n. *oder* Gemäuer ‖ masure f. / ~ of a mine ‖ Grubenmauerung f.; Stollenmauerung f. ‖ muraillement m. des galeries et d'un puits. / ~ of a shaft ‖ Schachtmauerung f. ‖ muraillement m. d'un puits; cuvelage m. de maçonnerie.

wall knot (Mar) ‖ Schauermannsknoten m. ‖ cul-de-porc m. / ~ lamp ‖ Wandlampe f. ‖ lampe f. murale *ou* applique. / ~ match box ‖ Wandzündholzbehälter m. ‖ porte-allumettes m. mural. / ~ mirror ‖ Wandspiegel m. ‖ miroir m. mural. / ~ motto ‖ Wandspruch m. ‖ pancarte f. murale.

wall-paper ‖ Tapete f. ‖ papier m. peint *ou* de tenture. / hand-printed ~ ‖ Handdrucktapete f. ‖ papier m. peint à la main. / machine-printed ~ ‖ Maschinendrucktapete f. ‖ papier m. peint mécanique.

wall-paper brusher ‖ Tapetenglätter m. ‖ glaceur m. de papiers peints. / ~ colours pl. ‖ Farben fpl. für Tapeten ‖ couleurs fpl. pour papiers peints. / ~ dealer ‖ Tapetenhändler m. ‖ marchand m. de papiers peints. / ~ embosser ‖ Tapetenpräger m. ‖ frappeur m. de papiers peints. / ~ engraver ‖ Tapeten(muster)stecher m. ‖ graveur m. en papiers peints. / ~ gilder ‖ Tapetenvergolder m. ‖ doreur m. sur papiers peints. / ~ glazer ‖ Tapetensatinierer m. ‖ satineur m. de papiers peints. / ~ gofferer ‖ Tapetengaufrierer m. ‖ gaufreur m. de papiers peints. / ~ machine-printer ‖ Tapetenmaschinendrucker m. ‖ imprimeur m. à la machine sur papiers peints. / ~ making machine ‖ Tapetenherstellungsmaschine f. ‖ machine f. pour la fabrication de papiers peints. / ~ printer ‖ Tapetendrucker m. ‖ imprimeur m. sur papiers peints. / ~ printer's hand ‖ Tapetendruckergehilfe m. ‖ tireur m. de papiers peints. / ~ roller-up ‖ Tapetenroller m. ‖ rouleur m. en papiers peints. / machine for ~ stamping ‖ Tapetenprägemaschine f. ‖ machine f. à empreindre des papiers peints.

wall plate ‖ Wandplatte f.; Wandteller m. ‖ plaque f. *ou* assiette f. pour murs.

wall plug (Electr) ‖ Wandstecker m. ‖ contact m. mural. / ~ with movable contact pins ‖ Wandstecker m. mit be

weglichen Kontaktstiften ‖ contact m. mural à contacts élastiques. / two-pin ~ ‖ Steckkontakt m. ‖ contact m. à fiches.

wall rock (Mine) ‖ Nebengestein n. des Ganges ‖ roche f. des parois. / ~ saltpetre ‖ Mauersalpeter m. ‖ aphronitre m. / ~ screw ‖ Steinschraube f. ‖ boulon m. de scellement. / ceramic ~ slab ‖ keramische Wandplatte f. ‖ carreau m. céramique mural. / ~ slewing crane ‖ Wandkran m. ‖ grue f. murale pivotante. / ~ socket (Electr) ‖ Steckdose f. ‖ boîte f. à fiche. / ~ stove ‖ Wandofen m. ‖ cheminée f. / ~ switch board ‖ Wandschalttafel f. ‖ tableau m. de distribution mural. / ~ thickness ‖ Wandstärke f. ‖ épaisseur m. de paroi. / ~ tie ‖ Maueranker m. ‖ tirant m. de mur.

wall ventilator ‖ Wandlüfter m. ‖ ventilateur m. type mural. / ~ with flap shutter ‖ Wandlüfter m. mit Klappenverschluß ‖ ventilateur type mural avec obturateur. / ~ with protection against moisture ‖ Wandlüfter m. mit Feuchtschutz ‖ ventilateur m. type mural avec obturateur étanche.

wall winch ‖ Wandwinde f. ‖ treuil m. d'applique.

walnut (Fruit) ‖ Walnuß f. ‖ noix f. / ~ (Wood) ‖ Walnußholz n. ‖ noyer m. / ~ case ‖ Nußbaumkasten m. ‖ boîte f. en noyer. / husk of ~ ‖ grüne Walnußschale f. ‖ brou m. de noix. / ~ oil ‖ Walnußöl n. ‖ huile f. de noix. / ~ shell ‖ Walnußschale f. ‖ brou m. de noix. / ~ sheller ‖ Walnußauskernerin f. ‖ dénoiseuse f. / ~ wood *see* ~ (Wood).

wandering of the arc ‖ Wandern n. des Lichtbogens ‖ migration f. de l'arc.

want of agreement ‖ Unstimmigkeit f. ‖ discordance f. / for ~ of ‖ mangels ‖ à défaut m. de; faute f. de. / ~ of money ‖ Geldmangel m. ‖ disette f. d'argent. / ~ of physical uniformity ‖ ungleichmäßig in der Herstellung f. ‖ manque f. d'uniformité. / ~ of power ‖ Kraftbedarf m. ‖ besoin m. de force. / to supply a ~ ‖ eine Lücke f. ausfüllen ‖ combler une lacune.

wanted ‖ gesucht ‖ demandé; recherché.

war, industrial ~ ‖ Industriekrieg m. ‖ guerre f. industrielle. / ~ upon trade ‖ Handelskrieg m. ‖ guerre f. de destruction du commerce. / world ~ ‖ Weltkrieg m. ‖ guerre f. mondiale.

war arms pl. ‖ Kriegswaffen fpl. ‖ armes fpl. de guerre.

wardrobe ‖ Kleiderschrank m. ‖ garderobe f. / ~ for factories ‖ Arbeiterkleiderschrank m. ‖ armoire f. pour vestiaires ouvriers. / ~ holder ‖ Garderobenhalter m. ‖ portemanteau m. / ~ trunk ‖ Schrankkoffer m. ‖ coffre-armoire m.; malle-armoire f.

wardroom (Mar) ‖ Offiziersmesse f. ‖ carré m. des officiers. / warrant officers' ~ (Mar) ‖ Deckoffiziersmesse f. ‖ poste m. des maîtres.

ware ‖ Ware f. ‖ marchandise f. / cast-iron ~ ‖ Gußeisenwaren fpl. ‖ articles mpl. en fonte. / plated ~ (Textile) ‖ plattierte Ware f. ‖ doublé m.

warehouse, to ‖ einlagern ‖ emmagasiner.

warehouse ‖ Warenspeicher m.; Lagerhaus n. ‖ magasin m.; entrepôt m. / bonded ~ ‖ Zollagerhaus n. ‖ magasin m. d'entrepôt public. / principal ~ ‖ Hauptniederlage f. ‖ dépôt m. central.

warehouseman ‖ Lagerist m. ‖ magasinier m.; manutentionnaire m.; garde-magasin m.

warfare, chemical ~ ‖ chemische Kriegführung f. ‖ guerre f. chimique. / chemical ~ committee ‖ Ausschuß m. für chemische Kriegführung ‖ comité m. de la guerre chimique. / corsair ~ ‖ Piratenkrieg m. ‖ guerre f. de course. / mobile ~ ‖ Bewegungskrieg m. ‖ guerre f. de mouvement. / stabilized ~ ‖ Stellungskrieg m.; Schützengrabenkrieg m. ‖ guerre f. de position *ou* de tranché. / submarine ~ ‖ Unterseebootkrieg m. ‖ guerre f. sous-marine. / trench ~ *see* stabilized ~.

war gas ‖ Kampfgas n. ‖ gaz m. de combat. / industrial preparedness for ~ ‖ industrielle Kriegsbereitschaft f. ‖ préparation f. industrielle en vue de la guerre.

war industry ‖ Kriegsindustrie f.; Rüstungsindustrie f. ‖ industrie f. de guerre. / ~ board ‖ Kriegsindustrieamt n. ‖ commission f. des industries de guerre. / ~ plants pl. ‖ Rüstungsindustriewerke fpl. ‖ usines fpl. du matériel de guerre.

warlike service, incapable of further ~ ‖ unbrauchbar für weitere Kriegszwecke mpl. ‖ incapable de servir dorénavant à la guerre.

warm, to ‖ erhitzen ‖ réchauffer.

warm ‖ warm ‖ chaud. / to keep ~ ‖ warmhalten ‖ tenir chaud.

war material ‖ Kriegsgerät n. ‖ matériel m. de guerre. / ~ industry *see* war industry. / vari-coloured coat of ~ ‖ Buntfarbenanstrich m. des Heeresgeräts ‖ peinture f. en plusieurs couleurs du matériel de guerre.

warm bed ‖ Warmbett n. ‖ lit m. à tenir chaud. / tractor for ~s ‖ Schleppapparat m. für Warmbetten ‖ appareil m. de remorque pour lits à tenir chaud.

war measuring box (Tel) ‖ Feldmeßkästchen n. ‖ petite boîte f. de mesures de campagne.

warming ‖ Erwärmung f. ‖ échauffement m.; caléfaction f. / ~ of a dyeing-vat ‖ Aufwärmen n. einer Küpe ‖ réchauffement m. d'une cuve de teinture.

warming cupboard ‖ Wärmeschrank m. ‖ compartiment m. de tient-chaud. / ~ cushion ‖ Heizkissen n. ‖ chauffe-corps m. / ~ furnace ‖ Anwärmeherd m. ‖ four m. à rechauffer. / ~ table ‖ Wärmetisch m. ‖ étuve f. forme table.

warm and moist temperature ‖ feuchtwarm ‖ humide et chaud. / ~ pressing ‖ Warmpreßarbeit f. ‖ travail m. d'estampage à chaud. / ~ saw ‖ Warmsäge f. ‖ scie f. à chaud. / ~ water heating ‖ Warmwasserheizung f. ‖ chauffage m. à eau chaude. / ~ water supply installation ‖ Warmwasserversorgungsanlage f. ‖ installation f. d'approvisionnement en eau chaude.

warning ‖ Warnung ‖ avertissement m. / ~ of danger ‖ Gefahrmeldung f. ‖ avis m. *ou* avertissement m. de danger.

warning apparatus for crossings ‖ Warnvorrichtung f. für Kreuzungen ‖ appareil m. avertisseur pour passages. / ~ bell ‖ Läutwerk n. ‖ sonnerie f. d'avertissement. / ~ board ‖ Warnungstafel f. ‖ écriteau m. *ou* plaque f. d'avertissement. / ~ ring (Cable) ‖ Warnungsring m. ‖ anneau m. de couleur avertisseur. / air ~ service ‖ Luftwarnungsdienst m. ‖ ser-

vice m. d'avertissement météorologique pour la navigation aérienne.

warp, to ‖ sich werfen; sich verziehen ‖ se déjeter; se courber; gauchir. / ~ (Weav) ‖ scheren; die Kette scheren ‖ ourdir. / the rail warps ‖ die Schiene f. wirft sich ‖ le rail se déjette.

warp (Weav) ‖ Schweif m.; Kette f.; Zettel m. ‖ chaîne f. / ground ~ ‖ Grundkette f.; Bodenkette f. ‖ chaîne f. de fond. / ~ wound into rings ‖ Ringkette f. ‖ chaînette f.; chaîne f. pliée en boucles.

warp and filling ‖ Kette f. und Schuß m. ‖ chaîne f. et trame f. / adjusting of the ~ on the loom ‖ Kettenaufziehen n. ‖ montage m. de la chaîne sur le métier. / ~ beam ‖ Kettenbaum m.; Garnbaum m. ‖ ensouple f. de derrière. / ~ beam adjusting apparatus ‖ Kettenbaumregelapparat m. ‖ appareil m. de réglage pour ensouples. / ~ cop ‖ Kettengarnklötzer m.; Zettelklötzer m. ‖ cannette f. *ou* fusée f. à chaîne. / ~ drawer ‖ Ketteneinzieher m. ‖ remetteur m. de chaînes. / ~ drawing ‖ Ketteneinziehen n. ‖ remettage m. de chaînes. / ~ dressing ‖ Kettenschlichten n. ‖ dressage m. de chaînes. / ~ drier ‖ Kettentrockner m. ‖ sécheur m. de chaînes. / ~ dyeing ‖ Kettengarnfärben n. ‖ teinture f. de chaînes pour tissus.

warped ‖ windschief ‖ gauchi; déjeté.

warper ‖ Kettenscherer m.; Zettler m. ‖ ourdisseur m. / ~ (Bookb) ‖ Rundpresser m.; Rundschläger m. ‖ arrondisseur m. / hand-~ ‖ Handkettenscherer m. ‖ ourdisseur m. à la main. / ~'s bobbin end press ‖ Endpresse f. für Kettgarnspulen ‖ presse f. à extrémités pour bobines d'ourdissage.

warp, double-rib ~ goods pl. ‖ Fangkettenware f. ‖ tricot m. chaîne à double fond.

war phone ‖ Feldfernsprecher m. ‖ téléphone m. de campagne.

warping ‖ Kettenscheren n.; Zetteln n. ‖ ourdissage m.; warpage m. / ~ block (Shipb) ‖ Scherblock m. ‖ poulie f. de halage. / ~ drum (Weav) ‖ Schertrommel f. ‖ tambour m. à ourdir. / ~ frame *see* ~ machine. / ~ machine ‖ Schermaschine f.; Kettenmaschine f.; Zettelmaschine f. ‖ machine f. à ourdir. / ~ machine minder ‖ Maschinenscherer m. ‖ ourdisseur m. à la machine. / ~ mill *see* ~ machine. / ~ pipe (Shipb) ‖ Verholklüse f. ‖ écubier m. de halage. / ~ wing (Aero) ‖ verwindbarer Flügel m. ‖ aile f. gauchissable.

warp knit goods pl. ‖ Kettenwirkware f. ‖ tricot m. chaîne. / ~ loom ‖ Kettenstuhl m. ‖ métier m. à chaîne. / ~ machine knitter ‖ Trikotweber m. ‖ tricoteur m. au métier à chaîne. / ~ plush ‖ Kettplüsch m. ‖ peluche f. par chaîne. / ~ printing ‖ Warpdruck m. ‖ impression f. de chaînes pour tissus. / ~ protector ‖ Schußwächter m. ‖ garde-trame f. / ~ sizer ‖ Kettenleimer m. ‖ encolleur m. de chaînes. / ~ thread ‖ Kettenfaden m.; Kettengarn n. ‖ fil m. de chaîne. / spare ~ thread ‖ Ersatzkettengarn n. ‖ reneuil m. / ~ thread spinner ‖ Kettengarnspinner m. ‖ fileur m. de chaîne. / ~ velvet ‖ Kettensamt m. ‖ velours m. par chaîne. / ~ yarn twister ‖ Kettengarnzwirner m. ‖ tordeur m. de chaînes.

warrant, to ‖ garantieren ‖ garantir; cautionner.

warrant, dividend ~ ‖ Kupon m. ‖ coupon m.

warranty ‖ Garantie f.; Gewährleistung f. ‖ garantie f.; sûreté f. / certificate of ~ ‖ Garantieschein m. ‖ cautionnement m.; promesse f. de garantie.

war repeater unit (Tel) ‖ Feldverstärkersatz m. ‖ jeu m. d'amplificateurs de campagne.

war-ship, obsolete ~ ‖ veraltetes Kriegsschiff n. ‖ bâtiment m. de guerre vieilli. / ~ to be scrapped ‖ zu verschrottendes Kriegsschiff n. ‖ navire f. de guerre à démolir.

war signalling gear ‖ Feldsignalgerät n. ‖ appareil m. de signalisation de campagne. / ~ strength ‖ Kriegsstärke f. ‖ effectif m. de guerre. / ~ switchboard ‖ Feldklappenschrank m. ‖ tableau m. de distribution de campagne. / ~ telegraph instrument ‖ Feldtelegrafenapparat m. ‖ appareil m. télégraphique de campagne. / ~ telegraph pole ‖ Feldtelegrafenstange f. ‖ poteau m. télégraphique de campagne. / ~ telegraphy ‖ Feldtelegrafie f. ‖ télégraphie f. de campagne. / ~ telephone hand set ‖ (Hand-)Feldhörer m. ‖ appareil m. à main de campagne. / ~ test case (Tel) ‖ Feldprüfschrank m. ‖ armoire f. de contrôle de campagne. / ~ testing box (Tel) ‖ Feldprüfkasten m. ‖ boîte f. de contrôle de campagne.

wash, to ‖ waschen; spülen ‖ laver. / ~ by decantation ‖ durch Dekantieren n. auswaschen ‖ laver par décantation f. / ~ off ‖ abspülen ‖ rincer. / ~ ores pl. ‖ die Erze npl. waschen ‖ laver les minerais mpl. / ~ out ‖ auswaschen ‖ nettoyer par épuisement m. / ~ store casks pl. ‖ faßschlupfen ‖ nettoyer les foudres mpl. intérieurement. / ~ after a treatment ‖ nachwaschen ‖ laver après un traitement.

wash (Geol) ‖ Ablagerung f. ‖ alluvion f.; dépôt m. / ~ (Nav) ‖ Sog m. ‖ remous m. / furnace for distiller's ~ ‖ Schlempeofen m. ‖ four m. à vinasse. / ~ of an oar ‖ Ruderblatt n. *oder* Schaufel f. eines Riemens ‖ pale f. *ou* pelle f. d'aviron. / strong ~ ‖ starke Lauge f. ‖ eaux fpl. fortes. / weak ~ ‖ schwache Lauge f. ‖ eaux fpl. de lessivage *ou* faibles *ou* petites.

washable ‖ waschecht ‖ bon teint m. / ~ leather glove ‖ Waschlederhandschuh m. ‖ gant m. de peau lavable.

wash board (Build) ‖ Fußleiste f.; Scheuerleiste f. ‖ lambris m. d'appui *ou* de socle. / ~ cylinder ‖ Waschtrommel f. ‖ tambour-laveur m.

washed cane for seats ‖ gewaschenes Stuhlrohr n. ‖ jonc m. lavé pour sièges. / ~ drawing ‖ getuschte Zeichnung f. ‖ dessin m. au lavis. / ~ out ‖ abgeschlämmt ‖ débourbé *ou* lavé *ou* décrassé. / ~ produce (Ore dress) ‖ Waschprodukt n. ‖ produit m. lavé.

washer ‖ Waschapparat m.; Wäsche f. ‖ laveur m. / ~ (Jigger) ‖ Setzmaschine f. ‖ crible m. hydraulique. / ~ (Mach) ‖ Unterlagscheibe f. ‖ rondelle f. / ~ (Pap) ‖ Halbholländer m. ‖ pile f. défileuse; cylindre m. effilocheur *ou* dégrossisseur. / barrel ~ ‖ Faßwaschapparat m. ‖ appareil m. à laver les fûts. / bone coal ~ ‖ Knochenkohlewäsche f. ‖ laverie f. à charbon d'os. / carbonic acid ~ ‖ Kohlensäurewäsche f. ‖ laveur m. d'acide carbonique. / cask ~ *see* barrel ~. / elastic ~ ‖ Sprungring m. ‖ rondelle f. élastique.

/ electric automatic ~ || elektrischer Sprudelwascher m. || lessiveuse f. électrique à jet d'eau chaude. / gas ~ || Gaswascher m. || laveur m. de gaz. / hurdle ~ || Hordenwascher m. || laveur m. à claie. / keg ~ see washer, barrel. / leather ~ || Lederunterlagscheibe f. || rondelle f. de cuir. / machined ~ || blanke Scheibe f. || rondelle f. décolletée. / raw ~ with hole || rohe Scheibe f. mit Loch || rondelle f. brute perforée. / semi-machined ~ || halbblanke Scheibe f. || rondelle f. semiusinée. / spring ~ (Mach) || Federring m.; federnde Unterlegscheibe f. || rondelle f. Belleville; anneau-ressort m. / ~ with square hole || Scheibe f. mit Vierkantloch || rondelle f. avec trou carré. / ~ with submerged piston || Kolbensetzmaschine f. || bac m. à piston. / unpolished ~ || rohe Scheibe f. || rondelle f. brute. / turbo ~ || Turbowascher m. || lessiveuse f. turbo.

washer-woman || Waschfrau f.; Wäscherin f. || blanchisseuse f.; laveuse f. de linge.

washery || Wäsche f. || lavoir m.; installation f. de lavage. / ~ for broken stone plant || Wäsche f. für Schotteranlagen || laveur m. pour installations de cailloutis. / coal ~ || Kohlenwäsche f. || installation f. de lavage du charbon. / ~ for liquids || Waschanlage f. für Flüssigkeiten || installation f. de lavage de liquides. / locomotive ~ || Lokomotivauswaschanlage f. || installation f. à nettoyer les locomotives. / ~ for pebble works || Wäsche f. für Kieswerke || laveur m. pour usines de gravier. / ~ of sludge || Schlammwäsche f. || lavage m. de schlamm. / ~ for turnips || Rübenwäsche f. || laveur m. pour betteraves. / ~ for turnip tops || Wäsche f. für Rübenblätter || laveur m. pour feuilles de betterave. / ~ for vegetables || Wäsche f. für Gemüse || laveur m. pour légumes.

washery building || Wäschereigebäude n. || bâtiment m. de lavoir. / ~ foreman || Wäschereimeister m. || chef-laveur m. / ~ slag || Waschberge mpl. || schiste m. provenant du lavage.

wash gilding || Vergoldung f. auf Bronze || dorure f. sur bronze. / ~ hand stand see wash-stand.

wash-house || Waschhaus n. || lavoir m. / ~ (Brew) || Banzenbrücke f.; Schwankhalle f. || rince-fûts m. / ~ (Househ) || Waschküche f. || buanderie f.

washing || Waschen n.; Auswaschen n. || lavage m. / dry ~ || trockenes Waschen n. || lavage m. par voie sèche. / ~ by (means of) jets || Strahlwäsche f. || lavage m. à jet. / ~ of nettings || Netzwaschen n. || lavage m. de filets. / preliminary ~ || Vorwäsche f. || lavage m. préparatoire. / ~ of rags || Waschen n. von Lumpen; Lumpenwaschen n. || blanchissage m. de chiffons. / removable by ~ || auswaschbar || lavable. / ~ of store casks || Faßschlupfen n. || nettoyage m. intérieur des foudres. / ~ of waste || Waschen n. von Abfällen; Abfallwaschen n. || blanchissage m. de déchets. / wet ~ || nasses Waschen n. || lavage m. par voie humide.

washing accomodation || Wascheinrichtung f. || installation f. de lavoir.

washing apparatus, automatic ~ || Waschautomat m. || lessiveuse f. automatique. / ~ for automobile components || Auto-

mobilteilewaschapparat m. || appareil m. à laver des pièces d'automobiles. / filtering mass ~ || Filtermassewaschapparat m. || appareil m. de lavage de la masse filtrante.

washing away of a foundation || Unterwaschung f. oder Unterspülung f. eines Fundaments || déchaussement m. d'une fondation. / ~ by a torrent || Abspülung f. durch Wildwasser || ravinement m. ou corrosion f. par un torrent.

washing basin || Waschbecken n. || cuvette f. à laver. / ~ benches pl. for workmen || Arbeiterreihenwaschanlage f. || lavoirs mpl. en ligne pour ouvriers. / ~ board || Waschbrett n. || planche f. à laver ou à lessives. / ~ boiler || Waschkessel m. || chaudron m. pour bouillir le linge. / ~ bottle || Waschflasche f.; Spritzflasche f. || flaconlaveur m.; fiole f. à jet. / ~ brush (Bleach) || Waschbürste f. || chien m. / ~ drum || Waschtrommel f. || tambour m. de lavage. / ~ engine (Pap) || Waschholländer m. || barbotteuse f. / ~ establishment || Waschanstalt f. || établissement m. de lavage; lavoir m. / ~ flask see ~ bottle. / ~ liquor || Auslaugflüssigkeit f. || liquide m. laveur.

washing machine || Waschmaschine f. || machine f. à laver; laveuse mécanique; lessiveuse f. / ~ bottle || Flaschenspülmaschine f. || machine f. à rincer les bouteilles. / high-production ~ for cloth and worsted yarn || Hochleistungswaschmaschine f. für Tuche und Kammgarnstoffe || machine f. à laver de grand rendement pour draps et tissus peignés. / electric ~ || elektrische Waschmaschine f. || lessiveuse f. électrique. / ~ in full width || Breitwaschmaschine f. || machine f. à laver au large. / ~ in full width with rotation in the liquor tanks and self-acting apparatus regulating circulation || Breitwaschmaschine f. mit Flottenzirkulation und selbsttätigem Warenlaufregelapparat || machine f. à laver au large avec circulation d'eau et introducteur automatique. / ~ in full width with self-acting regulating devise for the fabric || Breitwaschmaschine f. mit selbsttätigem Warenlaufregelapparat || laveuse f. au large avec dispositif automatique pour le réglage de la marche du tissu. / ~ in full width with triple rolling system || Breitwaschmaschine f. mit Dreiwalzensystem || machine f. à laver au large à trois rouleaux ou cylindres. / ~ for gravel || Kieswaschmaschine f. || laveur m. à gravier. / rope ~ || Strangwaschmaschine || machine f. à laver en boyaux. / white ~ || Anstreichmaschine f. || pulvérisateur m. à blanchir. / ~ with winch || Waschmaschine f. mit Abzughaspel || machine f. à laver avec tournette de sortie. / wool ~ || Wollwaschmaschine f. || machine f. à laver les laines.

washing material || Waschmittel n. || lessive f.; matière f. à laver. / ~ pan || Waschpfanne f. || cuve f. de lavage. / ~ plant see washery. / ~ powder || Waschpulver n. || poudre f. à lessiver. / ~ process || Waschverfahren n. || procédé m. de lavage. / ~ refuse || Wäscheabgänge mpl. || refus m. de lavage.

washings pl., coal ~ || Kohlenschlamm m. || boue f. de charbon; schlamms mpl. / **washing** system || Waschsystem n. ||

système f. de lavage. / ~ table (Galv) || Waschtisch m. || table f. de lavage. / dish ~ table || Geschirraufwaschtisch || table f. à laver la vaisselle. / ~ tank || Waschgefäß n. || récipient m. à laver. / ~ tank (Jigger) || Setzkasten m. || bac m. de lavage. / ~ tower || Waschturm m. || tour f. de lavage. / ~ trough see ~ tub. / ~ tub || Waschbütte f.; Waschfaß n. || cuvier m.; baquet m. / ~-up machine || Aufwaschmaschine f. || machine f. à laver la vaisselle. / ~ vat see ~ tub. / ~ water || Waschwasser n. || eau f. de lavage.

wash leather || Waschleder n. || cuir m. rosette ou chamoisé. / ~ liquor || Laugeflüssigkeit f. || lessive f.; liquide m. / ~-off linen || Dauerwäsche f. || linge m. durable ou permanent. / ~ ore || Wascherz n. || minerai m. de lavage. / ~-out plug || Waschbolzen m. || bouchon m. de lavage. / ~ plate (Shipb) || Schlingerplatte f.; Schlingerblech n. || tôle f. de roulis.

wash-stand || Waschtisch m. || lavabo m.; table f. de toilette.

wastage see waste (Metal).

waste, to verzetteln || égarer; éparpiller.

waste || Abfall m. || déchets mpl.; résidus mpl. / ~ (Carp) || Verschnitt m. || déchet m. de coupe. / ~ (Metal) || Abbrand m. || perte f. au feu. / ~ (Mine) || alter Mann m. || vieux travaux mpl. / ~ from combing (Wool) || Abfall m. vom Kämmen || déchet m. de peignage. / to eliminate sources of ~ || Verlustquellen fpl. beseitigen || éliminer les sources de perte. / ~ of energy || Energieverschwendung f. || gaspillage m. d'énergie. / ~ of hides || Fellschnitzel npl. || effleurures fpl. / ~ of labour || Arbeitsverlust m. || perte f. de travail. / ~ of leather manufacture || Lederabfall m. || déchet m. de cuir. / ~ in mining || Abbauverlust m. || déchet m. de mine. / ~ of sheets || Blechabfall m. || déchet m. de tôle. / ~ of tannery || Gerbereiabfall m. || déchet m. de tannerie. / ~ of time || Zeitverlust m.; Zeitvergeudung f. || perte f. de temps.

waste basket || Papierkorb m. || panier m. à papier. / ~ fibre ~ || Fiberpapierkorb m. || panier m. à papier en fibre. / steel ~ || Papierkorb m. aus Stahl || panier m. à papier en acier.

waste book || Kladde f. || brouillon m.; brouillard m. / ~ cotton || Putzbaumwolle f. || coton m. à polir. / ~ fat || Abfallfett n. || graisse f. de déchet. / furnace for ~ || Feuerung f. für Abfälle || foyer m. à déchets.

waste gas || Abgas n. || gaz m. brûlé ou d'échappement. / ~ (Blast furnace) || Gichtgas n. || gaz m. de hauts-fourneaux. / industrial ~es pl. || industrielle Abgase npl. || gaz mpl. industriels perdus. / to lead off the ~es pl. over the roof || die Abgase npl. über Dach abführen || faire échapper les gaz mpl. brûlés au-dessus du toit. / ~ of a rotary kiln || Drehofenabgas n. || gaz m. d'échappement d'un four tournant.

waste gas duct || Abgaskanal m. || canal m. d'échappement des gas brûlés. / ~ heating || Abgasheizung f. || chauffage m. par les gaz perdus. / ~ purifying plant || Abgasreinigungsanlage f. || installation f. d'épuration des gaz d'échappement. / utilization of the ~es ||

Abgasverwertung f. ‖ utilisation f. des gaz brûlés *ou* d'échappement. / ~ utilizing plant ‖ Abgasverwertungsanlage f. ‖ installation f. pour l'utilisation des gaz d'échappement.

waste gate (Sluice) ‖ Freiarche f. ‖ auge f.; conduit m. / ~ hand (Spinn) ‖ Abfallarbeiter m. ‖ ouvrier m. aux déchets. / ~ heap ‖ Halde f. ‖ halde f.; crassier m. / ~ heap ore ‖ Haldenerz n. ‖ minerai m. de halde.

waste heat ‖ Abwärme f.; Abhitze f. ‖ chaleur f. perdue. / to utilize the ~ from the furnaces for the generation of steam ‖ die Abwärme der Öfen für die Dampferzeugung benutzen ‖ employer la chaleur s'échappant des fours pour la production de vapeur.

waste heat boiler ‖ Abhitzekessel m. ‖ chaudière f. à chaleur d'échappement. / ~ economy ‖ Abhitzeverwertung f.; Abwärmeverwertung f. ‖ économie f. de la chaleur d'échappement. / ~ recovering ‖ Abhitzerückgewinnung f. ‖ récupération f. de la chaleur perdue. / ~ recuperator ‖ Abwärmeverwerter m. ‖ récupérateur m. de la chaleur d'échappement. / ~ utilizing plant ‖ Abwärmeverwertungsanlage f. ‖ installation f. de recouvrement de pertes de chaleur.

waste matter ‖ Geschur n.; Hüttenafter n. ‖ scorie f. / ~ metal ‖ Krätze f.; Gekrätz n.; Metallabfälle mpl. ‖ déchet m. de métal. / ~ oil ‖ Abfallöl n. ‖ résidu m. d'huile. / ~ oil screw ‖ Ölablaßschraube f. ‖ vis f. de purge pour l'huile de graissage.

waste paper ‖ Makulatur f.; Papierabgang m. ‖ maculature f. / ~ cleaning plant ‖ Altpapierreinigungsanlage f. ‖ installation f. de nettoyage des vieux papiers. / ~ cutting plant ‖ Altpapierschneideanlage f. ‖ installation f. de découpage des vieux papiers. / ~ sorting plant ‖ Altpapiersortieranlage f. ‖ installation f. de triage des vieux papiers.

waste pickling water ‖ Beizabwässer pl. ‖ eaux fpl. évacuées. / ~ pipe ‖ Abflußrohr n. ‖ tuyau m. de décharge *ou* de dégorgement. / ~ pipe (Build) ‖ Fallrohr n.; Dachröhre f. ‖ tuyau m. de descente. / ~ place ‖ schädlicher Raum m. ‖ espace m. nuisible. / ~ products pl. *see* waste.

waster ‖ Fehlguß m. ‖ rebut m.; pièce f. manquée. / ~s pl. ‖ Berge mpl.; Versatzberge mpl. ‖ remblai m.

waste silk ‖ Seidenabfall m. ‖ bourre f. de soie. / ~ sorter (Spinn) ‖ Abfallsortierer m. ‖ trieur m. de déchets *ou* bouts.

waste steam heat utilization ‖ Abdampfverwertung f. ‖ utilisation f. de la vapeur d'échappement. / ~ pipe ‖ Dampfableitungsrohr n. ‖ tuyau m. d'échappement de la vapeur.

waste utilizing plant ‖ Abfallverwertungsanlage f. ‖ installation f. pour l'utilisation des déchets.

waste water ‖ Abwasser n. ‖ eaux fpl. résiduaires *ou* de décharge. / ~ (Steam) ‖ Kondensationswasser n. ‖ eau f. de condensation. / ~ cleaning plant for slaughter houses ‖ Abwasserreinigungsanlage f. für Schlachthöfe ‖ installation f. d'épuration des eaux résiduaires d'abattoirs. / ~ funnel ‖ Abwassertrichter m. ‖ entonnoir m. d'eau d'écoulement. / ~ purifyer ‖ Abwasserklär-

apparat m. ‖ appareil m. d'épuration des eaux d'égouts. / ~ purifying plant ‖ Abwasserreinigungsanlage f. ‖ installation f. d'épuration des eaux ménagères *ou* résiduaires.

waste weir ‖ Überfallwehr n. ‖ déversoir m. / ~ well (Build) ‖ Senkgrube f.; Senkloch n.; Schwindgrube f. ‖ puisard m.; puits m. absorbant; égougeoir m. / ~ wood ‖ Holzabfall m. ‖ bois m. de rebut. / ~ wool ‖ Putzwolle f. ‖ déchet m. de coton.

wasting ‖ Abnutzung f. ‖ usure f.

watch ‖ Taschenuhr f.; Uhr f. ‖ montre f. / ~ (Mar) ‖ Quartier n. ‖ quart m. / alarm ~ ‖ Taschenuhr f. mit Weckvorrichtung; Wecker m. ‖ montre-réveil f. / automobile ~ ‖ Kraftwagenuhr f. ‖ horloge f. d'automobiles. / calendar ~ ‖ Uhr f. mit Taganzeiger ‖ montre f. à quantième. / chronograph ~ ‖ Uhr f. mit Beobachtungssekunden ‖ montre f. chronographe. / date-indicating ~ ‖ Datumanzeigeuhr f. ‖ montre-quantième f. / ~ with independent second-hand ‖ Uhr f. mit unabhängigem Sekundenzeiger ‖ montre f. à seconde indépendante. / keyless ~ ‖ Remontoiruhr f. ‖ montre f. à remontoir. / repeating ~ ‖ Repetieruhr f. ‖ montre f. à répétition. / in the shape of a ~ ‖ in Gestalt f. einer Taschenuhr ‖ en forme f. de montre. / striking ~ ‖ Schlaguhr f. ‖ montre f. à sonnerie. / wrist ~ ‖ Armbanduhr f. ‖ montre-bracelet f.

watch board (Mar) ‖ Logbrett n.; Logtafel f. ‖ table f. de loch. / ~ box *see* watchman's house.

watch case ‖ Taschenuhrgehäuse n.; Uhrgehäuse n. ‖ boîte f. de montre. / enamel foil of the ~ bottom ‖ Emailbelag m. des Uhrgehäusedeckels ‖ couche f. d'émail sur le fond de la boîte d'une montre. / ~ factory ‖ Uhrgehäusefabrik f. ‖ fabrique f. pour boîtes de montres. / ~ repairing ‖ Uhrgehäuseausbesserung f. ‖ rhabillage m. de boîtes de montres.

watch chain ‖ Uhrkette f. ‖ chaîne f. de montre. ‖ ~ cover ‖ Uhrdeckel m. ‖ cuvette f. de montre. / ~ dish ‖ Uhrschälchen n. ‖ petite écuelle f. de montre. / ~ factory ‖ Taschenuhrenfabrik f. ‖ fabrique f. de montres. / ~ file ‖ Uhrmacherfeile f. ‖ lime f. d'horloger.

watchfulness ‖ Wachsamkeit f.; Aufmerksamkeit f. ‖ attention f.

watch glass ‖ Uhrglas n. ‖ verre m. de montre. / ~ (Mar) ‖ Glas n.; Sandglas n.; Sanduhr f.; Stundenglas n. ‖ ampoulette f.; horloge f. de sable; sablier m. / fitting and drilling of the ~es ‖ Einsetzen n. und Durchbohren n. der Uhrgläser ‖ pose f. et perçage m. des glaces de montres.

watch hand ‖ Uhrzeiger m. ‖ aiguille f. de montre. / ~ key ‖ Uhrschlüssel m. ‖ clé f. à remonter.

watchmaker ‖ Uhrmacher m. ‖ horloger m.

watchmaker's eyeglass ‖ Uhrmacherlupe f. ‖ loupe f. d'horloger. / ~ lathe ‖ Uhrmacherdrehstuhl m. ‖ tour m. d'horloger. / ~ tools pl. ‖ Uhrmacherwerkzeug n. ‖ outil m. d'horlogerie. / ~ utensils pl. ‖ Uhrmacherbedarf m. ‖ ustensiles mpl. d'horlogers.

watchmaking ‖ Uhrmacherkunst f. ‖ horlogerie f. / ~ school ‖ Uhrmacherschule f. ‖ école f. d'horlogerie. / ~ shop ‖ Uhrmacherei f. ‖ horlogerie f.

watchman ‖ Wächter m. ‖ gardien m. / ~ of the line ‖ Bahnwärter m. ‖ gardeligne m. / ~ of the mine ‖ Grubenwächter m. ‖ garde m. de mine. / ~ of oyster beds ‖ Austernbankwächter m. ‖ garde-parc m. à huîtres.

watchman's control clock ‖ Wächterkontrolluhr f. ‖ contrôleur m. de rondes. / ~ portable control clock ‖ tragbare Wächterkontrolluhr f. ‖ contrôleur m. de rondes portable.

watchman's control installation, transmitter of a combined fire alarm and ‖ Melder m. einer vereinigten Feuermeldeanlage und Wächterkontrollanlage ‖ avertisseur m. d'une installation combinée d'avertissement d'incendie et de contrôle de rondes. / ~ controlling advertiser ‖ Wächterkontrollmelder m. ‖ avertisseur m. de contrôle pour gardiens. / ~ control system (Tel) ‖ Wächterkontrollanlage f. ‖ contrôleur m. de rondes. / ~ dwelling ‖ Wärterwohnung f. ‖ logement m. de garde. / ~ house (Railw) ‖ Bahnwärterhaus n. ‖ maison f. de gardes; guérite f.

watch manufacturing ‖ Uhrenherstellung f. ‖ fabrication f. d'horloges. / ~ materials pl. ‖ Uhrenbestandteile mpl. ‖ fournitures fpl. d'horlogerie. / ~ oil ‖ Uhrenöl n. ‖ huile f. pour horloges. / ~ overcase ‖ Uhrkapsel f. ‖ boîtier m. protecteur pour montres. / ~ pivot ‖ Uhrzapfen m. ‖ pivot m. d'horloge. / ~ repairer ‖ Uhrmacher m. ‖ rhabilleur m. de montres. / ~ shield ‖ Uhrschild n. ‖ écusson m. de montres.

watch spring ‖ Uhrfeder f. ‖ ressort m. de montre *ou* d'horlogerie. / ~ steel ‖ Uhrfederstahl m. ‖ acier m. pour ressorts de montre. / ~ testing machine ‖ Dauerprüfmaschine f. für Uhrfedern ‖ machine f. à essayer les ressorts de montre.

watch station of the fire brigade ‖ Wache f. der Feuerwehr ‖ poste f. du corps de pompiers. / ~ stone ‖ Taschenuhrstein m. ‖ pierre f. de montres. / ~ vessel ‖ Hafenwachtschiff n. ‖ bâtiment m. garde-port; navire m. de surveillance. / ~ work ‖ Uhrwerk n. ‖ mouvement m. d'une montre. / ~ wheel ‖ Uhrrad n. ‖ roue f. d'horlogerie.

water, to ‖ begießen ‖ arroser. / ~ (Eye) ‖ tränen ‖ verser des larmes. / ~ (Weav) ‖ moirieren; wässern ‖ moirer; tabiser. / ~ the flax ‖ den Flachs m. rösten *oder* roten ‖ naiser *ou* rouir le lin. / ~ a stuff ‖ einen Stoff m. flammen ‖ chiner une étoffe. / ~ tobacco with sauce ‖ den Tabak m. beizen ‖ mouiller le tabac.

water ‖ Wasser n. ‖ eau f. / ~ above ‖ über Wasser n. ‖ au-dessus de l'eau. / acidulated ~ *see* acidulous ~. / acidulous ~ ‖ kohlensaures Wasser n.; kohlensäurehaltiges Wasser n. ‖ eau f. acidulée *ou* gazeuse *ou* carbonique. / aerated ~ *see* water, acidulous. / artificial mineral ~ ‖ künstliches Mineralwasser n. ‖ eau f. minérale artificielle. / brackish ~ ‖ brakkiges Wasser n. ‖ eau f. saumâtre. / broken ~ ‖ Kreuzsee f.; kreuzweis laufende See f. ‖ mer f. contraire *ou* creuse. / by land and by ~ ‖ zu Wasser n. und zu Lande n. ‖ par mer f. et par terre f. / carbonated ~ *see* acidulated ~. / chalybeate ~ ‖ eisenhaltiges Wasser n. ‖ eau f. ferrugineuse *ou* martiale. / chlorinated ~ *see* muriated ~. / clean ~ free

from scale ‖ reines, kesselsteinfreies Wasser n. ‖ eau f. pure et exempte de chaux. / ~ of condensation ‖ Kondenswasser n. ‖ eau f. condensée. / apparatus pl. for returning ~ of condensing ‖ Kondenswasserrückspeiseanlage f. ‖ installation f. alimentaire à recouvrement de l'eau de condensation. / cooling ~ ‖ Kühlwasser n. ‖ eau f. de refroidissement. / ~ of crystallization ‖ Kristallwasser n. ‖ eau f. de cristallisation. / to cut-off the ~ ‖ das Wasser abstellen ‖ couper l'eau f. / dead ~ (Mar) ‖ Kielwasser n. ‖ remous m. / distilled ~ ‖ destilliertes Wasser n. ‖ eau f. distillée. / ~ for domestic purposes ‖ Nutzwasser n. ‖ eau f. pour les besoins domestiques. / drinkable ~ ‖ Trinkwasser n. ‖ eau f. potable. / to fix ~ with avidity ‖ begierig Wasser n. anziehen; hygroskopisch sein ‖ être avide d'eau f. / flushing ~ ‖ Schwemmwasser n. ‖ eau f. d'arrosage. / fresh ~ ‖ Süßwasser n. ‖ eau f. de fraiche ou source. / gaseous ~ ‖ gashaltiges Wasser n. ‖ eau f. gazeuse. / ~ in gaseous state ‖ gasförmiges Wasser n. ‖ eau f. sous forme de gaz. / hard ~ ‖ hartes Wasser n. ‖ eau f. dure ou crue. / to keep ~ free from impurities ‖ die Gewässer npl. rein halten ‖ tenir les eaux fpl. propres. / inland ~ ‖ Binnengewässer n. ‖ eaux fpl. continentales. / to lead ~ on to a property ‖ einem Grundstück n. Wasser zuführen ‖ amener l'eau f. à un terrain. / by loss of ~ ‖ unter Abgabe f. von Wasser ‖ par départ m. d'eau. / low ~ ‖ seichtes Wasser n. ‖ eau f. maigre. / mean ~ ‖ mittlerer Wasserstand m. ‖ eau f. moyenne. / muddy ~ ‖ schlammiges Wasser n. ‖ eau f. mouceuse ou boueuse. / muriated ~ ‖ Salzsole f.; Salzwasser n. ‖ eau f. salée. / oxygenated ~ ‖ Wasserstoffsuperoxyd n. ‖ eau f. oxygénée. / ~ under pressure ‖ Druckwasser n. ‖ eau f. sous pression. / public ~s pl. ‖ öffentliches Gewässer n. ‖ eaux fpl. du domaine public. / rinsing ~ ‖ Spülwasser n. ‖ eau f. d'arrosage. / running ~ ‖ fließendes Gewässer n. ‖ eau f. courante. / scented ~ ‖ Riechwasser n. ‖ eau f. de senteur. / smooth ~ ‖ stilles Wasser n. ‖ eau f. calme. / soft ~ ‖ weiches Wasser n. ‖ eau f. douce. / stagnant ~ ‖ stehendes Gewässer n. ‖ eau f. stagnante. / subterranean ~ ‖ Tiefenwasser n. ‖ eau f. souterraine ou profonde. / surface ~ ‖ Tagwasser n. ‖ eau f. du jour. / thick ~ ‖ trübes Wasser n. ‖ eau f. troublée. / under ~ ‖ unter Wasser n. ‖ inondé. / underground ~ ‖ Grundwasser n. ‖ eau f. souterraine. / useful ~ ‖ Nutzwasser n. ‖ eaux fpl. utiles. / waste ~ ‖ Abwässer npl. ‖ eaux fpl. résiduaires. / to withdraw ~ from a property ‖ einem Grundstück n. Wasser entziehen ‖ enlever de l'eau f. à un terrain.

water absorbing capacity ‖ Wasseraufnahmefähigkeit f. ‖ capacité f. d'absorption d'eau. / absorption of ~ ‖ Wasseraufnahme f. ‖ absorption f. d'eau. / amount of ~ required ‖ Wasserbedarf m. ‖ quantité f. d'eau nécessaire. / ~ bailiff ‖ Fischereiaufseher m. ‖ garde-pêche m. / ~ ballast ‖ Wasserballast m. ‖ lest m. en eau; lest-eau m. / ~ bath ‖ Wasserbad n. ‖ bain-marie m. / boiling ~ bath ‖ kochendes Wasserbad n. ‖ bain-marie m.

bouillant. / ~ bath boiler ‖ Wasserbadkessel m. ‖ chaudière f. pour bains-marie. / ~ board ‖ Wasserpolizeibehörde f. ‖ autorité f. exerçant la police des eaux. / ~ boiler with overflow ‖ Wasserkessel m. mit Sieb ‖ bouilloire f. à tamis. / ~ bottle ‖ Wasserflasche f. ‖ carafe f. à eau / ~ bridge ‖ Feuerbrücke f. mit Wasserkühlung ‖ autel m. d'eau. / ~ bucket ‖ Wassereimer m. ‖ seau m. à eau. / ~ cable ‖ Flußkabel n. ‖ câble m. sous-fluvial. / ~ calender ‖ Naßkalander m.; Wasserkalander m. ‖ calandre f. à essorer. / ~ can ‖ Wasserkanne f. ‖ bidon m. à eau. / ~ carrier ‖ Wasserträger m. ‖ porteur m. d'eau. / ~ cart ‖ Wasserkarren m. ‖ charrette f. à eau. / ~ cart (Sprinkling) ‖ Sprengwagen m. ‖ voiture f. d'arrosage. / ~ cask ‖ Wasserfaß n. ‖ tonneau m. à eau. / ~ cement ‖ Roman-Zement m. ‖ ciment m. romain. / ~ chamber ‖ Wasserkammer f. ‖ chambre f. à eau. / ~ channel (Mach) ‖ Wasserrinne f. ‖ rigole f. à eau. / ~ circulation ‖ Wasserumlauf m. ‖ circulation f. d'eau. / the boilers pl. are not provided with any downcomers for the ~ circulation with the lower middle drum ‖ die Kessel mpl. haben für den Wasserumlauf mit dem mittleren Unterkessel keine besonderen Fallröhren ‖ les chaudières fpl. n'ont pas de tubes de retour spéciaux pour assurer la circulation d'eau avec le corps inférieur médian. / ~ clarifying plant ‖ Wasserkläranlage f. ‖ installation f. de décantation d'eau. / trepan with central ~ clearing ‖ Bohrmeißel m. mit zentraler Wasserspülung ‖ trépan m. à nettoyage à l'eau central. / trepan with lateral ~ clearing ‖ Bohrmeißel m. mit seitlicher Wasserspülung ‖ fleuret m. à nettoyage à l'eau latéral.

water closet ‖ Wasserklosett n.; Abort m. mit Wasserspülung ‖ water-closet m. / ~ for pits ‖ Grubenabort m. oder Grubenklosett n. mit Wasserspülung ‖ watercloset m. pour mines. / ~ flushing apparatus ‖ Klosettspülapparat m. ‖ appareil m. à chasse d'eau pour water-closets. / ~ installation ‖ Wasserklosettanlage f. ‖ installation f. de water-closet.

water closing ‖ Wasserverschluß m. ‖ clôture f. à l'eau. / ~ with bell attachment ‖ Glockenwasserverschluß m. ‖ clôture f. à l'eau à cloche.

water cock ‖ Wasserhahn m. ‖ robinet m. d'eau. / ~ colour ‖ Wasserfarbe f.; Aquarellfarbe f. ‖ couleur f. à l'eau ou aquarelle. / ~ colour printing ‖ Aquarelldruck m. ‖ impression f. à l'aquarelle. / ~ column ‖ Wassersäule f. ‖ colonne f. d'eau. / ~ column pressure ‖ Wassersäulendruck m. ‖ pression f. en colonne d'eau. / ~ company ‖ Wassergenossenschaft f. ‖ association f. pour l'usage des eaux. / ~ condenser ‖ Wasserkondensator m. ‖ condensateur m. d'eau.

water conduit ‖ Wasserleitung f. ‖ conduite f. d'eau. / ~ fittings pl. ‖ Wasserleitungsarmaturen fpl. ‖ armatures fpl. de conduites d'eau. / welded ~ pipe ‖ geschweißtes Wasserleitungsrohr n. ‖ tuyau m. de conduite d'eau soudé.

water conservancy ‖ Wasserhaushalt m. ‖ économie f. des eaux. / ~ containing sludge see muddy water. / ~ conveying by wind power ‖ Wasserförderung f. mit

Windbetrieb ‖ élévation f. d'eau par moulin à vent.

water-cooled ‖ wassergekühlt ‖ à refroidissement m. à l'eau. / ~ bearing ‖ wassergekühltes Lager n. ‖ coussinet m. à refroidissement par eau. / ~ cylinder ‖ wassergekühlter Zylinder m. ‖ cylindre m. à chemise d'eau. / ~ machine-gun ‖ wassergekühltes Maschinengewehr n. ‖ mitrailleuse f. à refroidissement à l'eau. / ~ motor ‖ wassergekühlter Motor m. ‖ moteur m. à refroidissement par eau.

water cooling ‖ Wasserkühlung f. ‖ refroidissement m. par eau.

water course ‖ Wasserlauf m. ‖ lit m. de rivière. / to dam a ~ ‖ einen Wasserlauf m. stauen ‖ barrer un cours d'eau. / to direct a ~ ‖ einen Wasserlauf m. verändern ‖ modifier un cours d'eau.

water craft ‖ Wasserfahrzeug n. ‖ bâtiment. / ~ crane ‖ Wasserkran m. ‖ grue f. hydraulique.

watercress cultivator ‖ Kressenzüchter m. ‖ cressiculteur m. / ~ cutter ‖ Kressenschneider m. ‖ ramasseur m. de cressons.

water cushion ‖ Wasserkissen n. ‖ matelas m. d'eau. / ~ decoction apparatus ‖ Abkochgerät n. ‖ appareil m. à faire bouillir l'eau. / ~ de-ironing ‖ Wasserenteisenung f. ‖ déferrisation f. de l'eau. / ~ demanganesing ‖ Wasserentmanganung f. ‖ séparation f. de manganèse de l'eau. / discharge of ~ ‖ Wasserauslauf m. ‖ sortie f. de l'eau. / ~ disoxidation ‖ Wasserentsäuerung f. ‖ désoxydation f. de l'eau. / ~ distillation apparatus ‖ Wasserdestillierapparat m. ‖ appareil m. de distillation d'eau. / ~ distribution ‖ Wasserwirtschaft f. ‖ distribution f. des eaux. / ~ diviner see ~ finder. / ~ drip (Build) ‖ Unterschneidung f. ‖ gouttière f. du larmier; mouchette f. / ~ droplet ‖ Wassertröpfchen n. ‖ gouttelette f. d'eau. / ~ drum (Mach) ‖ Unterkessel m. ‖ chaudière f. inférieure.

watered ‖ gewässert ‖ arrosé. / ~ (Weav) moiriert; geflammt; gewässert ‖ moiré. / ~ paper ‖ Metallpapier n. ‖ papier m. moiré métallique. / ~ stuff (Weav) ‖ Moiré m. ‖ moiré m.; moirée f.; étoffe f. moirée. / ~ yarn (Spinn) ‖ Moirégarn n. ‖ fil m. moiré.

water eliminating plant for tar ‖ Teerentwässerungsanlage f. ‖ installation f. d'élimination de l'eau du goudron.

waterfall ‖ Wasserfall m. ‖ cascade f.

water fascine ‖ Senkfaschine f. ‖ saucisson m. / ~ finder ‖ Rutengänger m. ‖ sourcier m.; rhabdomant m. / ~ fittings pl. ‖ Wasserarmatur f. ‖ armature f. à eau. / ~ foam ‖ Wasserschaum m. ‖ écume f. d'eau. / ~-forming fermentation ‖ Wassergärung f. ‖ fermentation f. produisant de l'eau. / ~ gas ‖ Wassergas n. ‖ gaz m. à l'eau. / lighting by ~ gas ‖ Wassergasbeleuchtung f. ‖ éclairage m. au gaz à l'eau. / ~ gas plant ‖ Wassergasanlage f. ‖ installation f. de gaz à l'eau. / ~ gate (Sluice) ‖ Fluttor n.; Obertor n.; oberes Schleusentor n. ‖ porte f. d'amont ou de tête.

water gauge ‖ Wasserstandsanzeiger m. ‖ indicateur m. du niveau d'eau. / ~ (Hydr arch) ‖ Pegel n.; Peil m. ‖ échelle f. d'eau ou fluviale; marque m. d'eau. / self-registering ~ (Hydr arch) ‖ selbstregistrierendes Pegel n.; Maregraf m.; Flut-

zeiger m. ‖ échelle f. enregistrante; marégraphe m.

water gauge cock see ~ test cock. / ~ glass ‖ Wasserstandsglas n. ‖ niveau m. d'eau à tube de verre. / ~ pocket (Boil) ‖ Wasserstandsstutzen m. ‖ tubulure f. de niveau d'eau. / ~ test cock ‖ Wasserstandshahn m. ‖ robinet m. d'épreuve de niveau d'eau.

water glass ‖ Wasserglas n. ‖ verre m. soluble. / ~ handling plant ‖ Wasserreinigungsanlage f. ‖ installation f. d'épuration d'eau. / ~ hardening (Met) ‖ Wasserhärtung f. ‖ trempe f. à l'eau. / ~ hold (Shipb) ‖ Wasserlast f. ‖ cale f. de l'eau. / ~ hose ‖ Wasserschlauch m. ‖ tuyau m. de caoutchouc à eau.

watering (Hydr arch) ‖ Bewässerung f. ‖ arrosage m. / ~ (Weav) ‖ Wässern n. ‖ moirage m. / ~ the lawn ‖ Rasensprengung f. ‖ arrosage m. de la pelouse.

watering apparatus ‖ Wasserspritze f. ‖ appareil m. d'arrosage. / ~ can ‖ Gießkanne f. ‖ arrosoir m. / ~ car ‖ Sprengwagen m. ‖ voiture f. arroseuse. / ~ crane see water crane. / ~ device (Stable) ‖ Tränkanlage f. ‖ abreuvoir m. / ~ ditch ‖ Berieselungsgraben m. ‖ ru m.; fossé m. d'irrigation. / ~ place ‖ Tränke f. für Tiere ‖ abreuvoir m. / ~ plant ‖ Berieselungsanlage f. ‖ installation f. d'arrosage. / ~ pot (Build) ‖ Wasserluke f. ‖ chantepleure f. / ~ pot (Gard) see ~ can.

water inlet ‖ Wassereinlaß m. ‖ entrée f. de l'eau. / ~ of radiator ‖ Kühlereinlaßstutzen m. ‖ entrée f. de l'eau du radiateur.

water inspector ‖ Wassermeister m. ‖ surveillant m. des eaux. / ~ interferometer ‖ Wasserinterferometer n. ‖ interféromètre m. à eau.

water jacket ‖ Wassermantel m. ‖ chemise f. d'eau. / ~ cooling ‖ Mantelkühlung f. ‖ rafraîchissement m. de la chemise. / ~ furnace ‖ Wassermantelofen m. ‖ four m. à chemise d'eau.

water jet ‖ Wasserstrahl m. ‖ jet m. d'eau. / ~ air pump ‖ Wasserstrahlluftpumpe f. ‖ pompe f. à air par jet d'eau. / ~ injector ‖ Wasserstrahlpumpe f. ‖ pompe f. à jet d'eau; trompe f. à eau. / ~ (type) lightning arrester ‖ Wasserstrahlerder m. ‖ déchargeur m. continu à jet d'eau. / ~ pump see ~ injector.

water jug ‖ Wasserkanne f. ‖ aiguière f. / owing to lack of ~ ‖ wegen Wassermangel m. ‖ en raison du manque d'eau. / ~ landing ‖ Wasserlandung f. ‖ atterrissage m. sur l'eau; amerrissage m.

water level ‖ Wasserstand m.; Wasserspiegel m. ‖ niveau m. d'eau. / ~ (Mar) ‖ Fluthöhe f. ‖ grandeur f. de la marée. / ~ (Surv) ‖ Wasserwage f.; Libelle f. ‖ niveau d'eau f. ou à bulle d'air. / to lower the ~ ‖ den Wasserspiegel m. absenken ‖ abaisser le niveau de l'eau.

water level drift ‖ Grundstrecke f. ‖ chasse f. ou voie f. de fond. / ~ indicator ‖ Wasserstandsanzeiger m. ‖ indicateur m. de niveau d'eau. / remote controlled ~ indicator see ~ tele-indicator. / ~ tele-indicator ‖ Wasserstandsfernmelder m. ‖ télé-indicateur m. électrique du niveau d'eau. / ~ transmitter ‖ Wasserstandsmelder m. ‖ transmetteur m. de niveau d'eau. / remote operating ~ transmitter see ~ tele-indicator.

water lime ‖ Wasserkalk m. ‖ chaux f. hydraulique. / ~ burner ‖ Wasserkalkbrenner m. ‖ cuiseur m. de chaux hydraulique.

water line ‖ Wasserstandslinie f.; Wasserlinie f. ‖ niveau m. d'eau. / ~ (Shipb) ‖ Wasserlinie f. ‖ ligne f. de flottaison; ligne d'eau. / length at the ~ (Shipb) ‖ Länge f. in der Wasserlinie ‖ longueur f. à la ligne de flottaison.

water lodge (Mine) ‖ Sumpf m. ‖ pahage; puisard n.

water-main pressure ‖ Wasserleitungsdruck m. ‖ pression f. dans la canalisation d'eau.

watermark (Mar) ‖ Flutzeichen n. ‖ marque f. de la hauteur de marée. / ~ (Pap) ‖ Wasserzeichen n. ‖ filigrane m. / mean ~ ‖ mittlerer Wasserstand m. ‖ niveau m. d'eau moyen. / ~ post see water gauge.

watermelon oil ‖ Wassermelonenöl n. ‖ huile f. de gros béraff.

water meter ‖ Wassermesser m.; Wasseruhr f. ‖ compteur m. d'eau. / ~ with tele-indicating apparatus ‖ Wassermesser m. mit Fernmeldeeinrichtung ‖ compteur m. d'eau télé-indicateur. / recording apparatus for ~s ‖ Registriergerät n. für Wassermesser ‖ enregistreur m. pour compteurs d'eau. / ~ testing apparatus ‖ Wassermesserprüfapparat m. ‖ appareil m. de contrôle des compteurs d'eau.

watermill ‖ Wassermühle f. ‖ moulin m. à eau.

water motor ‖ hydraulischer Motor m.; Wassermotor m. ‖ moteur m. hydraulique. / ~ for cleaning bottles ‖ Wassermotor m. zur Flaschenreinigung ‖ moteur m. hydraulique pour le rinçage des bouteilles.

water outlet ‖ Wasserauslaß m. ‖ sortie f. d'eau. / ~ of a dam ‖ Wasserdurchlaßeinrichtung f. einer Sperrmauer ‖ disposition f. pour le passage de l'eau par le mur d'un barrage. / ~ of radiator ‖ Kühlerausflußstutzen m. ‖ sortie f. d'eau du radiateur.

water ozonizing ‖ Wasserozonisierung f. ‖ ozonisage m. d'eau. / cold ~ paints pl. ‖ Kaltwasserfarben fpl. ‖ couleurs fpl. à eau froide. / ~ particle ‖ Wasserteilchen n. ‖ particule f. d'eau. / percentage of ~ ‖ Wassergehalt m. ‖ teneur f. en eau.

waterplant ‖ Wasserpflanze f. ‖ plante f. aquatique.

water pipe ‖ Wasserrohr n. ‖ tuyau m. à eau. / wooden ~ (Hydr arch) ‖ Holzröhre f. ‖ tuyau m. de bois.

water piping see water conduit.

water power ‖ Wasserkraft f. ‖ force f. hydraulique. / available ~ ‖ verfügbare Wasserkraft f. ‖ force f. hydraulique disponible. / ~ plant ‖ Wasserkraftanlage f. ‖ installation f. de force motrice hydraulique.

water press ‖ hydraulische Presse f. ‖ presse f. hydraulique ou hydrostatique. / ~ pressure ‖ Wasserdruck m. ‖ pression f. hydraulique.

waterproof ‖ wasserdicht ‖ imperméable ou étanche à l'eau. / ~ boot ‖ Wasserstiefel m. ‖ botte f. imperméable. / ~ cement ‖ wasserfester Kitt m. ‖ mastic m. résistant à l'action de l'eau. / ~ cloth ‖ wasserdichter Stoff m. ‖ tissu m. imperméable. / ~ clothing ‖ wasserdichtes Kleidungsstück n. ‖ vêtement m. imperméable. / ~ cover ‖ wasserdichter

Deckel m. ‖ couvercle m. étanche. / ~ leather ‖ wasserdichtes Leder n. ‖ cuir m. imperméable. / ~ motor ‖ wasserdichter Motor m. ‖ moteur m. étanche. / ~ paint ‖ wasserdichter Anstrich m. ‖ enduit m. ou peinture f. hydrofuge. / ~ tissue ‖ wasserdichter Webstoff m. ‖ tissu m. imperméable.

waterproof ‖ Gummimantel m.; Regenmantel m. ‖ imperméable m. / stuff for ~s ‖ Regenmantelstoff m. ‖ étoffe f. pour imperméables.

water propeller ‖ Rotationspumpe f. ‖ pompe f. à rotation ou rotative. / ~ pump ‖ Wasserpumpe f. ‖ pompe f. à eau. / ~ pump wind mill ‖ Wasserschöpfwindmühle f. ‖ moulin m. à vent pour pomper l'eau. / ~ pumping by wind power ‖ Wasserförderung f. mit Windbetrieb ‖ élévation f. d'eau par moulin à vent. / ~ purifier ‖ Speisewasserreiniger m. ‖ épurateur m. d'eau alimentaire. / ~ purifying plant ‖ Wasserreinigungsanlage f. ‖ installation f. pour l'épuration des eaux. / quantity of ~ ‖ Wassermenge f. ‖ volume m. d'eau. / ~ quantity registering apparatus ‖ Wassermengenregistrierapparat m. ‖ appareil m. enregistreur de la quantité d'eau. / ~ raising apparatus ‖ Wasserhebungsmaschine f. ‖ appareil m. élévatoire. / ~ ram ‖ hydraulischer Widder m. ‖ bélier m. hydraulique. / ~ resistance ‖ Wasserwiderstand m. ‖ résistance f. hydraulique. / ~ rights pl. ‖ Wasserrechte pl. ‖ droits mpl. à l'usage de l'eau; droits mpl. sur l'eau. / ~ room ‖ Wasserraum m. ‖ chambre f. à eau. / ~ scooping machine ‖ Wasserschöpfmaschine f. ‖ machine f. à puiser l'eau ou d'épuisement d'eau. / ~ screw ‖ Wasserschraube f.; Wasserschnecke f.; archimedische Schraube f. ‖ limace f.; escargot m.; vis f. d'Archimède ou hydraulique. / ~ separator ‖ Wasserabscheider m. ‖ séparateur m. d'eau. / ~ shed ‖ Wasserscheide f. ‖ ligne f. de partage des eaux. / ~ shock ‖ Wasserschlag m. ‖ coup m. d'eau. / ~ sluice valve ‖ Wasserschieber m. ‖ robinet-vanne m. d'eau. / ~ softening ‖ Wasserenthärtung f. ‖ adoucissement m. d'eau. / ~ softening apparatus ‖ Wasserenthärtungsapparat m. ‖ appareil m. pour l'adoucissement de l'eau.

water-soluble ‖ wasserlöslich ‖ soluble dans l'eau. / ~ fat see ~ grease. / ~ grease ‖ wasserlösliches Fett n. ‖ graisse f. soluble dans l'eau. / ~ oil ‖ wasserlösliches Öl n. ‖ huile f. soluble à l'eau.

water, sound conductivity of the ~ ‖ Schalleitfähigkeit f. des Wassers ‖ conductivité f. de l'eau pour le son. / ~ spray diffuser ‖ Wasserzerstäubungsdüse f. ‖ vaporisateur m. d'eau. / ~ stain (Pap) ‖ Wasserfleck m. ‖ goutte f. / ~ stop wall ‖ Wehrmauer f. ‖ mur m. bâtardeau. / hot ~ storage tank ‖ Heißwasserspeicher m. ‖ accumulateur m. d'eau chaude.

water supply ‖ Wasserversorgung f. ‖ alimentation f. et distribution d'eau. / domestic ~ ‖ Hauswasserversorgung f. ‖ adduction f. des eaux de maison.

water supply enterprise ‖ Wasserversorgungsunternehmen n. ‖ entreprise f. d'adduction d'eau. / ~ plant ‖ Wasserversorgungsanlage f. ‖ installation f. ou établissement m. d'adduction d'eau. / ~ plant for single house service ‖ Wasserversorgungsanlage f. für Einzelgebäude ‖

installation f. de distributions d'eau pour bâtiments séparés. / ~ pumping station ‖ Wasserhebewerk n. ‖ usine f. élévatoire pour les eaux. / ~ station (Railw) ‖ Wasserstation f. ‖ station f. de prise d'eau. / decanting plant for ~ works pl. ‖ Wasserwerkkläranlage f. ‖ installation f. de clarification pour installations d'eau.

water surface ‖ Wasserspiegel m. ‖ surface f. d'eau. / useful area of ~ ‖ nutzbare Wasserfläche f. ‖ surface f. d'eau utile.

water tank ‖ Wasserbehälter m. ‖ réservoir m. *ou* récipient m. à eau. / ~ with cover ‖ Wasserkasten m. mit Deckel ‖ caisse f. à eau à couvercle. / reserve ~ ‖ Aushilfswasserbehälter m. ‖ récipient m. de réserve d'eau. / ~ of the tender ‖ Wasserbehälter m. des Tenders ‖ caisse f. à eau du tender.

water tank truck (Railw) ‖ Wasserwagen m. ‖ wagon m. à eau.

water temperature ‖ Wassertemperatur f. ‖ température f. de l'eau. / ~ test ‖ Wasserprobe f. ‖ preuve f. à l'eau.

watertight *see* waterproof.

water tower ‖ Wasserturm m. ‖ château m. d'eau. / ~ of x m³ capacity ‖ Hochbehälter m. von x cbm Fassung für Wasser ‖ château m. à réservoir de x m³ de capacité.

water trough ‖ Wasserrinne f. ‖ rigole f. d'eau. / ~ tube ‖ Wasserrohr n. ‖ tube m. d'eau.

water-tube boiler ‖ Wasserrohrkessel m. ‖ chaudière f. à tubes d'eau *ou* aquatubulaire. / bent tube type of ~ ‖ Wasserrohrkessel m. mit gebogenen Rohren ‖ chaudière f. aquatubulaire à tubes cintrés. / sectional chamber ~ ‖ Teilkammerwasserrohrkessel m. ‖ chaudière f. semitubulaire. / ~ with small tubes ‖ engrohriger Wasserrohrkessel m. ‖ chaudière f. aquatubulaire à tubes étroits. / ~ with straight tubes ‖ geradrohriger Wasserrohrkessel m. ‖ chaudière f. aquatubulaire à tubes droits. / ~ with two headers ‖ Zweikammerwasserrohrkessel m. ‖ chaudière f. aquatubulaire à deux collecteurs.

water turbine ‖ Wasserturbine f. ‖ turbine f. hydraulique.

water vapour, the ~ condenses ‖ der Wasserdampf schlägt sich nieder *oder* kondensiert ‖ la vapeur d'eau se condense.

water vat ‖ Wassergefäß n. ‖ cuve f. à eau.

water way ‖ Wasserstraße f.; Schiffahrtsweg m. ‖ voie f. navigable. / ~ of a bridge ‖ Flutraum m. *oder* Durchflußprofil m. einer Brücke ‖ débouché m. d'un pont. / ~ structure ‖ Wasserstraßenbau m. ‖ construction f. de voies navigables.

water wheel ‖ Wasserrad n. ‖ roue f. hydraulique. / ~ working without channels ‖ ohne Kropf arbeitendes Wasserrad n. ‖ roue f. hydraulique travaillant sans coursier. / breast ~ *see* middleshot ~. / high-breast ~ ‖ rückschlächtiges Wasserrad n. ‖ roue f. hydraulique par derrière. / middleshot ~ ‖ mittelschlächtiges Wasserrad n. ‖ roue f. hydraulique de côté. / overshot ~ ‖ oberschlächtiges Wasserrad n. ‖ roue f. hydraulique en dessus. / screw ~ ‖ Schneckenrad n. ‖ roue f. hélice. / undershot ~ ‖ unterschlächtiges Wasserrad n. ‖ roue f. hydraulique en dessous.

water wheel paddle ‖ Wasserradschaufel f. ‖ aube f. de la roue hydraulique.

water willow ‖ Bachweide f. ‖ osier m. bleu.

waterworks pl. ‖ Wasserwerk n. ‖ entreprise f. de distribution d'eau. / pumping station of regional ~ ‖ Förderstation f. einer Landeswasserversorgung ‖ poste m. de mise en charge d'une distribution d'eau d'une région.

waterworks labourer ‖ Wasserwerksarbeiter m. ‖ ouvrier m. à la distribution d'eau. / ~ machine ‖ Wasserwerksmaschine f. ‖ machine f. de la distribution d'eau.

watery ‖ wässerig ‖ aqueux; humide.

watt ‖ Watt n. ‖ watt m. / ~s pl. per candle ‖ Watt npl. je Kerze ‖ watts mpl. par bougie. / one-~ lamp ‖ Einwattlampe f. ‖ lampe f. d'un watt.

wattage of cooking and heating apparatus ‖ Wattverbrauch m. *oder* Energieverbrauch m. für Koch- und Heizapparate ‖ consommation f. d'énergie des appareils de cuisine et de chauffage.

watt component ‖ Wattkomponente f. ‖ composante f. wattée. / ~ consumption ‖ Wattverbrauch m. ‖ consommation f. en watts.

watt-hour meter ‖ Wattstundenzähler m. ‖ watt-heuremètre m. / integrating ~ ‖ integrierender Wattstundenzähler m. ‖ watt-heuremètre m. intégrateur.

wattless component of electric values ‖ Blindwert m. elektrischer Größen ‖ composante f. déwattée *ou* réactive des valeurs électriques. / ~ current ‖ wattloser Strom m. ‖ courant m. déwatté. / ~ current consumption ‖ Blindstromverbrauch m.; Verbrauch m. an wattlosem Strom ‖ consommation f. déwattée. / ~ power ‖ Blindleistung f.; wattlose Leistung f. ‖ puissance f. réactive *ou* déwattée.

wattling ‖ Geflecht n. ‖ clayonnage m.

wattmeter ‖ Wattmesser m.; Leistungsmesser m. ‖ wattmètre m. / precision ~ ‖ Genauigkeitswattmesser m. ‖ wattmètre m. de précision. / recording ~ ‖ Wattschreiber m.; registrierender Wattmesser m. ‖ wattmètre m. enregistreur.

wave ‖ Welle f. ‖ onde f. / abruptly rising ~ ‖ steil ansteigende Welle f. ‖ onde f. à front raide. / cold ~ ‖ Kältewelle f. ‖ onde f. *ou* vague f. de froid. / continuous ~ ‖ kontinuierliche Welle f. ‖ onde f. continue. / damped ~ ‖ gedämpfte Welle f. ‖ onde f. amortie. / electromagnetic ~ ‖ elektromagnetische Welle f.; Hertz'sche Welle f. ‖ onde f. électromagnétique *ou* Hertzienne. / flat-front ~ ‖ flach ansteigende Welle f. ‖ onde f. à front aplati. / fundamental ~ ‖ Grundwelle f. ‖ onde f. fondamentale. / heat ~ ‖ Hitzewelle f. ‖ onde f. *ou* vague f. chaude *ou* de chaleur. / Hertzian ~ *see* electromagnetic ~. / non-damped ~ ‖ ungedämpfte Welle f. ‖ onde f. inamortie *ou* entretenue. / non-sustained ~ *see* damped ~. / oscillating ~ ‖ oszillierende Welle f. ‖ onde f. oscillante. / shock ~ ‖ Stoßwelle f. ‖ onde f. de chock. / standing ~ *see* stationary ~. / stationary ~ ‖ stehende Welle f. ‖ onde f. stationnaire. / steep-front ~ *see* abruptly rising ~. / sustained ~ *see* non-damped ~. / tidal ~ ‖ Flutwelle f.; Gezeitenwelle f. ‖ onde f. de marée; ondemarée f. / transient ~ ‖ Wanderwelle f. ‖ onde f. transitoire. / travelling ~ ‖ fortschreitende Welle f.

‖ onde f. courante. / tuning ~ (Radio) ‖ Abstimmungswelle f. ‖ onde f. de syntonisation. / undamped ~ *see* non-damped ~.

wave changing switch (Radio) ‖ Wellenumschalter m. ‖ commutateur m. de changement d'onde. / classification of ~s ‖ Welleneinteilung f. ‖ classification f. des ondes. / ~ crest ‖ Wellenberg m. ‖ crête f. d'une onde; point m. haut d'une onde.

waved (Weav) ‖ moiriert; geflammt; gewässert ‖ moiré.

wave, designation of ~s ‖ Wellenbezeichnung f. ‖ désignation f. des ondes. / ~ detector ‖ Wellendetektor m.; Detektor m. (d'ondes). ‖ détecteur m. (d'ondes). / ~ front ‖ Wellenfront f.; Wellenkopf m. ‖ front m. de l'onde. / ~ indicator *see* ~ detector.

wave length ‖ Wellenlänge f. ‖ longueur f. d'onde. / of long ~ ‖ langwellig ‖ de grande longueur d'onde. / of short ~ ‖ kurzwellig ‖ de faible longueur f. d'onde.

wave length prolongation ‖ Wellenverlängerung f. ‖ augmentation f. de la longueur d'onde. / ~ scale ‖ Wellenlängenskale f. ‖ échelle f. des longueurs d'onde. / ~ shortening ‖ Wellenverkürzung f. ‖ diminution f. de la longueur d'onde.

wave-like ‖ wellenförmig ‖ ondulatoire. / ~ line ‖ Wellenlinie f. ‖ ligne f. ondulée. / ~ meter ‖ Wellenmesser m. ‖ ondemètre m. / ~ motion ‖ Wellenbewegung f. ‖ mouvement m. ondulatoire. / ~ parameter ‖ Wellenparameter n. ‖ paramètre m. d'ondes. / ~ passage button (Radio) ‖ Welleneinstellknopf m. ‖ bouton m. de réglage d'ondes.

wave propagation along the earth's surface ‖ Ausbreitung f. drahtloser Wellen längs der Erdoberfläche ‖ propagation f. des ondes le long de la terre. / ~ on lines ‖ Wellenausbreitung f. auf Leitungen ‖ propagation f. des ondes le long de conducteurs.

waver, to ‖ zaudern ‖ hésiter.

wave, radiation of ~s ‖ Ausstrahlung f. der Wellen ‖ radiation f. des ondes.

wave range ‖ Wellenbereich m. ‖ gamme f. / to switch over from one ~ to another ‖ von einem Wellenbereich m. auf den andern übergehen ‖ aller d'une gamme f. à l'autre.

wave reflection ‖ Reflexion f. von Wellen ‖ réflexion f. des ondes.

waves pl. dashing against cliffs ‖ Klippenbrandung f. ‖ ressac m. contre les roches.

wave screen ‖ Wellensieb n. ‖ filtre m. d'ondes. / ~ shape ‖ Wellenform f. ‖ forme f. d'ondes. / ~ train ‖ Wellenzug m. ‖ train m. d'ondes. / ~ trough ‖ Wellental n. ‖ point m. bas d'une onde. / ~ zone ‖ Wellenzone f. ‖ zone f. de radiation.

wavy ‖ wellig ‖ ondulé.

wavy-fibred growth ‖ Wellenfaserigkeit f.; wimmeriger Wuchs m. ‖ croissance f. ondulée; ondulation f. des fibres. / ~ wood ‖ verwundenes Holz n. ‖ bois m. à fibre contournée.

wax, to ‖ mit Wachs n. überziehen ‖ encirer.

wax ‖ Wachs n. ‖ cire f. / animal ~ ‖ tierisches Wachs n. ‖ cire f. animale. / bees' ~ ‖ Bienenwachs n. ‖ cire f. d'abeilles. / bleached ~ ‖ gebleichtes Wachs n. ‖ cire f. blanchie. / Chinese ~ ‖ japanisches Wachs n. ‖ cire f. du

Japon. / cobbler's ~ ‖ Schuhmacherwachs n.; Schusterpech n. ‖ cire f. de
cordonnerie. / ~ in combs ‖ Wachs n.
in Waben ‖ cire f. en rayons. / to impress
on ~ ‖ in Wachs n. abdrucken ‖ faire
une empreinte dans la cire. / insect ~
‖ Insektenwachs n. ‖ cire f. d'insectes. /
mineral ~ ‖ Mineralwachs n. ‖ cire f.
minérale. / modelling ~ ‖ Modellierwachs n. ‖ cire f. à modeler. / ~ in plates
‖ Wachs n. in Platten ‖ cire f. en plaques.
/ sealing ~ ‖ Siegellack m. ‖ cire f. à cacheter. / sewing ~ ‖ Nähwachs n. ‖ cire f.
de couture. / tooth ~ ‖ Zahnwachs n. ‖
cire f. pour dents. / unbleached ~ ‖ un-
gebleichtes Wachs n. ‖ cire f. jaune. /
vegetable ~ ‖ Pflanzenwachs n. ‖ cire
f. végétale. / virgin ~ ‖ Jungfernwachs
n. ‖ cire f. vierge. / white ~ ‖ weißes
Wachs n. ‖ cire f. blanche. / yellow ~ ‖
gelbes Wachs n. ‖ cire f. jaune.

wax aroma ‖ Wachsaroma n. ‖ arome m.
de cire. / ~ articles pl. ‖ Wachsware f.
‖ articles mpl. en cire. / ~ bead ‖ Wachs-
perle f. ‖ perle f. en cire. / ~ bleaching
plant ‖ Wachsbleichanlage f. ‖ installation f. de blanchiment de cire. / ~ bust
‖ Wachsbüste f. ‖ buste m. en cire. /
~ cake ‖ Wachskuchen m. ‖ pain m.
de cire. / ~ candle ‖ Wachslicht n. ‖
chandelle f. en cire. / ~ chandler ‖ Wachs-
zieher m. ‖ cirier m.

wax cloth ‖ Wachstuch n. ‖ toile f. cirée.
/ ~ for flooring ‖ Wachstuch n. für
Fußbodenbelag ‖ toile f. cirée pour
parquets. / ~ for wall hangings ‖ Wachstuch n. für Wandbehang ‖ toile f. cirée
pour tentures. / ~ for wrappings ‖ Wachstuch n. für Verpackungszwecke ‖ toile
f. cirée pour emballage.

wax cloth finishing machine ‖ Wachstuchfertigungsmaschine f. ‖ machine f. pour
le finissage de toile cirée. / ~ hood ‖
Wachstuchhaube f. ‖ dessus m. en toile
cirée.

wax colours pl. ‖ Wachsfarben fpl. ‖
couleurs fpl. pour cire. / ~ crayon ‖
Wachsstift m. ‖ crayon m. en cire. / ~
cylinder ‖ Wachswalze f. ‖ cylindre m.
en cire.

waxed silk fabric ‖ gewachstes Seidengewebe n. ‖ tissu m. de soie ciré. / ~
wire ‖ Wachsdraht m. ‖ fil m. ciré.

wax figure ‖ Wachsfigur f. ‖ buste f. en cire.
/ ~ floor polish ‖ Bohnerwachs n.; Par-
kettwachs n. ‖ cire f. pour parquets. / ~
flower ‖ Wachsblume f. ‖ fleur f. en
cire. / ~ fruit ‖ Wachsfrucht f. ‖ fruit
m. en cire. / ~ goods pl. see ~ articles pl.
/ ~ head ‖ Wachskopf m. ‖ tête f. en
cire. / ~ kettle ‖ Wachskessel m. ‖ perrau
m.; perreau m. / ~ leather ‖ Wichsleder
n. ‖ cuir m. à cirer. / ~ mask ‖ Wachs-
maske f. ‖ masque m. en cire. / ~ match
‖ Wachszündholz n. ‖ allumette f. en
cire. / ~ melting plant ‖ Wachsschmelz-
anlage f. ‖ installation f. à fondre la
cire. / ~ melting table ‖ Wachsgieß-
tisch m. ‖ table f. à couler la cire. / ~
model ‖ Wachsmodell n. ‖ modèle m.
en cire. / ~ moulding ‖ Wachsformen
n. ‖ moulage m. en cire. / ~ paper ‖
Wachspapier n. ‖ papier m. ciré. / ~
paper manufacturing machine ‖ Wachs-
papierherstellungsmaschine f. ‖ machine
f. à fabriquer le papier ciré. / ~ pearl
see ~ bead. / plastic article of ~ ‖
Formerarbeit f. aus Wachs ‖ article m.

plastique en cire. / ~ polishing ‖ Wachs-
politur f. ‖ poli m. à la cire. / ~ presser
‖ Wachspresser m. ‖ presseur m. de cire.
/ to make ribbons of ~ ‖ Wachs n. bän-
dern ‖ grêler la cire. / ~ seal ‖ Wachs-
siegel n. ‖ cachet m. de cire. / ~ sealing
machine ‖ Siegelmaschine f. ‖ machine
f. à cacheter à la cire. / ~ shavings pl.
‖ Wachsspäne mpl. ‖ copeaux mpl. de
cire. / ~ smelting ‖ Wachsschmelzen n. ‖
fonderie f. de cire. / ~ smelting boiler ‖
Wachsausschmelzkessel m. ‖ chaudière
f. à fondre la cire. / ~ soap ‖ Wachsseife
f. ‖ savon m. à la cire. / ~ sprayer and
atomizer ‖ Bohnerwachszerstäuber m. ‖
pulvérisateur m. de cire à parquet. / ~
stick ‖ Siegellackstange f. ‖ bâton m. de
cire à cacheter. / ~ taper ‖ Wachsstock
m. ‖ pelote-bougie f. / ~ torch ‖ Wachsfackel f. ‖ torche f. en cire. / mineral
~ works pl. ‖ Montanwachsfabrik f. ‖
fabrique f. de cire minérale.

waxy ‖ wachsartig ‖ cireux.

way ‖ Weg m.; Bahn f. ‖ voie f.; route f.;
chemin m. / ~ (Mar) ‖ Lauf m. ‖ erre f.
/ ~ (Railw) ‖ Gleis n.; Geleise n.; Schienenstrang m. ‖ voie f.; voie ferrée. / in
the accustomed ~ ‖ in gewohnter Weise
f. ‖ comme d'usage m. / ~ of craftsmen
‖ Handwerksbrauch m. ‖ usage m. des
gens du métier. / double ~ (Railw) ‖
doppelspuriges Geleise n. ‖ chemin m.
(de fer) à deux voies. / to find one's ~
about ‖ sich zurechtfinden ‖ s'orienter. /
to give ~ (Build) ‖ einbrechen; einfallen;
einstürzen ‖ s'enfoncer; s'écrouler. / ~
of manufacturing ‖ Herstellungsweise f.
‖ mode f. de fabrication. / permanent ~
‖ Bahnoberbau m. ‖ superstructure f. /
single ~ (Railw) ‖ eingleisige oder ein-
spurige Bahn f. ‖ chemin m. de fer à
simple voie.

way bill ‖ Frachtbrief m. ‖ lettre f. de
voiture. / ~-out ‖ Ausgang m. ‖ sortie f.

weak ‖ schwach ‖ faible. / ~ acid ‖ schwache
Säure f. ‖ acide m. faible.

weak current ‖ Schwachstrom m. ‖ courant m. faible. / ~ cable ‖ Schwachstromkabel n. ‖ câble m. à faible intensité. /
generation of ~ ‖ Schwachstromerzeu-
gung f. ‖ production f. de courant faible.

weak solution ‖ verdünnte Lösung f. ‖ solution f. diluée. / ~ springing ‖ weiche
Federung f. ‖ suspension f. douce sur
ressorts.

weaken, to ‖ abschwächen ‖ affaiblir.

weakening the colour ‖ Verwaschen n. der
Farbe ‖ délavage m. / ~ of the field
(Electr) ‖ Feldschwächung f. ‖ affaiblis-
sement m. de champ.

weaker tendency of prices ‖ Preisabschwächung f. ‖ affaiblissement m. des prix.

weakness ‖ Schwäche f. ‖ faiblesse f.

wealth, national ‖ Volkswohlstand m.;
Volksvermögen n. ‖ fortune f. publique. / to open the subterranean ~
‖ Bodenschätze mpl. erschließen ‖ découvrir des richesses souterraines fpl.
/ underground ~ ‖ Bodenschatz m. ‖ richesse f. souterraine.

weapon ‖ Waffe f. ‖ armure f.; arme f.

wear, to ‖ sich abnutzen; verschleißen ‖
s'user. / ~ (Mar) ‖ halsen ‖ virer vent m.
arrière. / the tread may be worn to a
greater extent ‖ der Laufkranz kann weiter abgenutzt werden ‖ le cercle de roulement permet de pousser l'usure plus
loin. / only x inches were worn away from

the chisel ‖ der Meißel zeigte nur eine
Abnutzung von x mm ‖ le burin ne montrait qu'une usure de x millimètres. / ~
down to the utmost ‖ bis auf das äußerste
Maß ausnutzen ‖ user à la dernière limite.
/ ~ out see to wear. / ~ up ‖ aufbrauchen
‖ consommer; épuiser.

wear see also wear and tear ‖ Abnutzung f.;
Verschleiß m. ‖ usure f. / admissible ~ ‖
zulässige Abnützung f. ‖ usure f. admissible. / ~ of belt ‖ Riemenverschleiß m. ‖
usure f. de la courroie. / ~ on the brushes
(Electr) ‖ Bürstenverschleiß m. ‖ usure
f. des balais. / ~ of cylinder ‖ Unrundwerden n. des Zylinders ‖ ovalisation f.
du cylindre. / even ~ ‖ gleichmäßige Abnutzung f. ‖ usure f. uniforme. / exceedingly uniform ~ ‖ außerordentlich gleichmäßiger Verschleiß m. ‖ usure f. excessivement uniforme. / an attempt was
made to remedy the rapid ~ ‖ man
suchte der schnellen Abnutzung f. ent-
gegen zu arbeiten ‖ on essaya d'éviter
la rapide usure. / ~ on the rope ‖
Seilverschleiß m. ‖ usure f. de câble. /
to suffer a ~ ‖ dem Verschleiß m. unterworfen sein ‖ être sujet m. à usure. / ~ on
swage ‖ Gesenkverschleiß m. ‖ usure f.
des matrices.

wear and tear see also wear ‖ Verschleiß m.;
Abnutzung f. ‖ usure f. / excessive ~ ‖
übermäßig starker Verschleiß m. ‖ usure
f. trop rapide. / parts exposed to great ~
‖ starkem Verschleiß unterworfene Teile
mpl. ‖ pièces fpl. exposées à une usure
rapide. / normal ~ of rails ‖ normaler
Schienenverschleiß m. ‖ usure f. normale
des rails. / small ~ ‖ geringe Abnutzung
f. ‖ usure f. faible. / the ~ is very small ‖
der Verschleiß m. ist sehr gering ‖ l'usure
f. est minime. / allowance for ~ ‖ Abzug
m. für Verschleiß ‖ réduction f. pour
usure. / resistance against ~ ‖ Verschleißhärte f.; Verschleißfestigkeit f. ‖
résistance f. à l'usure.

wearing see also wear and wear and tear ‖
Abnutzung f. ‖ usure f.; déchet m. / ~
piece ‖ Schleißstück n. ‖ pièce d'usure. /
~ plate ‖ Schleißblech n. ‖ tôle f. d'usure.
/ interchangeable ~ plate ‖ auswechselbare Verschleißplatte f. ‖ plaque f.
d'usure amovible. / ~ rail ‖ Abnutzungsschiene f. ‖ rail m. d'usure. / ~ strip see
~ rail.

wearisome experiment ‖ mühevoller Versuch m. ‖ expérience f. laborieuse.

weary ‖ überdrüssig ‖ dégoûté.

weather ‖ Witterung f.; Wetter n. ‖ temps
m. / clear ~ (Mar) ‖ sichtiges Wetter n. ‖
temps m. clair. / to influence the ~ ‖ das
Wetter beeinflussen ‖ influencer le temps.

weather board (Shipb) ‖ Setzbord n. ‖
falque f.; farque f. / ~ box ‖ Wetterhäus-
chen n. ‖ maisonnette f. à hygromètre.
/ ~ bureau ‖ Wetterwarte f. ‖ bureau
m. météorologique. / ~ cock ‖ Wetter-
hahn m. ‖ girouette f. / ~ conditions
pl. ‖ Witterungszustand m. ‖ état m.
du temps. / ~ flag ‖ Windfahne f. ‖ girouette f.

weather forecast ‖ Wettervorhersage f. ‖
prévision f. ou pronostic m. du temps. /
local ~ ‖ örtliche Wettervorhersage f. ‖
prévision f. régionale du temps.

weather forecasting see weather forecast. /
~ station ‖ Wetterdienststelle f. ‖ station
f. de prévision du temps.

weather gage, to keep the ~ ‖ luv halten ‖ tenir le lof *ou* le vent.

weather glass ‖ Wetterglas n. ‖ baromètre m. / ~ groove (Build) ‖ Unterschneidung f. ‖ gouttière f. du larmier; mouchette f. / ~ guard ‖ Regenschutzscheibe f. ‖ plateau m. de protection contre la pluie. / ~ helm (Mar) ‖ Luvruder n. ‖ gouvernail m. arrivé. / ~ information ‖ Wetterberatung f. ‖ information f. météorologique.

weathering ‖ Lüftung f.; Lüften n. ‖ évent m. / ~ (Geol) ‖ Verwitterung f. ‖ décomposition f.; effrittement m.; désagrégation f.

weather message, radio synoptic ~ ‖ Wettersammelfunkspruch m. ‖ radiogramme m. météorologique collectif.

weather moulding (Build) ‖ Kranzleiste f.; Rinnleiste f. ‖ larmier m.

weatherproof ‖ wetterfest ‖ résistant m. aux intempéries; à l'épreuve f. des intempéries. / ~ paint ‖ wetterfeste Farbe f. ‖ peinture f. à l'épreuve des intempéries. / ~ wire ‖ wetterbeständiger Draht m. ‖ fil m. résistant aux intempéries.

weather protection ‖ Wetterschutz m. ‖ protection f. contre les intempéries.

weather radiogram, collective ~ *see* weather message, radio synoptic.

weather report ‖ Wetterbericht m. ‖ bulletin m. météorologique; avis m. du temps. / airway ~ ‖ Streckenwettermeldung f. ‖ avis m. *ou* rapport m. météorologique de la ligne aérienne.

weather resisting *see* weatherproof.

weather service ‖ Wetterdienst m. ‖ service m. de prévision du temps. / aviation ~ ‖ Luftwarnungsdienst m. ‖ service m. d'avertissement météorologique pour la navigation aérienne.

weather sheet (Mar) ‖ Luvschot f. ‖ écoute f. de revers. / ~ shore ‖ Luvküste f.; Luvwall m. ‖ terre f. ou côte f. au vent. / ~ side (Mar) ‖ Luvseite f. ‖ côté m. du vent. / ~ signal ‖ Wetterzeichen n. ‖ signal m. de temps.

weave, to ‖ weben ‖ tisser; tresser. / ~ bone lace ‖ klöppeln ‖ travailler au fuseau./ ~ in ‖ einweben ‖ tisser dans; damasser.

weaver ‖ Weber m. ‖ tisseur m.; tisserand m. / spare-~ ‖ Ersatzweber m. ‖ tisseur m. supplémentaire.

weaver's beam ‖ Kettbaum m. ‖ ensouple f. / ~ glass ‖ Fadenzähler m.; Leinwandprober m.; Leinwandmikroskop n. ‖ compte-fil m.; loupe f. du tisserand. / ~ knot ‖ Weberknoten m. ‖ nœud m. plat *ou* de tisserand. / ~ loom *see* weaving loom. / ~ trade ‖ Weberei f. ‖ tisseranderie f. / ~ tweezers pl. ‖ Weberzange f. ‖ pince f. de tisserand.

weaving ‖ Weberei f.; Weben n. ‖ tissage m. / fancy ~ ‖ Gebildweberei f. ‖ tissage m. au Jacquard *ou* d'étoffes grand façonnées. / figured ~ *see* fancy ~. / hand loom ~ ‖ Handweben n.; Handweberei f. ‖ tissage m. à main. / Jacquard ~ *see* fancy ~. / plain linen ~ ‖ Flachsleinwandweberei f. ‖ tissage m. de toiles de lin. / power loom ~ ‖ mechanische Weberei f. ‖ tissage m. mécanique.

weaving accessories pl. ‖ Webereigeschirr n. ‖ accessoires mpl. de tissage. / ~ factory ‖ Webwarenfabrik f. ‖ atelier m. ou manufacture f. de tissage. / ~ factory for paper stuff ‖ Papierstoffweberei f. ‖ atelier m. de tissage d'étoffes en papier.

weaving loom ‖ Webstuhl m. ‖ métier m.

à tisser. / wire ~ ‖ Drahtwebstuhl m. ‖ métier m. à tisser le fil métallique.

weaving loom fitter ‖ Webstuhlsetzer m. ‖ monteur m. de métiers à tisser. / ~ harness ‖ Harnisch m. für Webstühle ‖ harnais m. pour métiers à tisser.

weaving machine ‖ Webereimaschine f. ‖ machine f. de tissage. / ~ industry ‖ Wirkmaschinenindustrie f. ‖ industrie f. des machines à tisser.

weaving mill ‖ Weberei f. ‖ atelier de tissage m. / tricot ~ ‖ Trikotwirkerei f. ‖ tissage m. de tricots. / wire ~ ‖ Drahtweberei f. ‖ atelier m. de tissage de fil métallique. / ~ for worsted yarn ‖ Kammgarnweberei f. ‖ tissage m. de laine longue.

weaving, preparatory machine for ~ ‖ Webereivorbereitungsmaschine f. ‖ machine f. préparatoire aux tissages. / ~ product ‖ Webereierzeugnis n. ‖ produit m. de tissage. / ~ requirements pl. for ~ ‖ Webereibedarfsartikel m. ‖ objet m. nécessaire aux tissages. / ~ yarn ‖ Webgarn n. ‖ fil m. à tisser.

web (Mach) ‖ Scheibe f.; Rippe f. ‖ âme f.; disque m.; coude m. / ~ (Textile) ‖ Gewebe n. ‖ tissu m. / ~ (Print) ‖ Papierbahn f. ‖ bande f. de papier. / crank ~ ‖ Kurbelarm m. ‖ bras m. de manivelle. / ~ of girder ‖ Steg m. am Träger ‖ âme f. d'une poutrelle. / hollow ~ ‖ Hohlgewebe n. ‖ tissu m. double-étoffe. / provided with x ~s pl. on its inside ‖ an der Innenseite mit x Rippen fpl. versehen ‖ muni de x nervures fpl. du côté intérieur.

web of rail ‖ Schienensteg m. ‖ âme f. *ou* tige f. du rail. / thickness of ~ ‖ Schienenstegstärke f. ‖ épaisseur f. de l'âme du rail.

web of rib ‖ Rippensteg m. ‖ âme f. de nervure. / ~ of a saw ‖ Sägeblatt n. ‖ lame f. *ou* feuille f. de scie. / ~ of taper section (Railw) ‖ sich verjüngender Schienensteg m. ‖ âme f. de rail de section conique.

webbed, deep-~ ∪-iron ‖ hochstegiges ∪-Eisen n. ‖ fer m. en ∪ avec âme de grande hauteur.

webbing ‖ Gurtgewebe n. ‖ sangle m.

web breaking roller (Print) ‖ Abreißwalze f. ‖ cylindre m. de rupture. / ~ glass *see* weaver's glass. / ~ holder ‖ Einschließkamm m. ‖ barre f. de décolleteuse. / thickness of ~ ‖ Stegstärke f. ‖ épaisseur f. de l'âme.

wedding ring ‖ Trauring m. ‖ alliance f.

wedge, to ‖ festkeilen ‖ claveter. / ~ (Print) ‖ füttern; unterlegen ‖ taquonner; rehausser. / ~ a mast in the partners ‖ einen Mast m. verkeilen ‖ coincer un mât aux étambrais. / ~ the rails pl. ‖ die Schienen fpl. verkeilen ‖ coincer les rails mpl. / ~ up *see* wedge, to.

wedge ‖ Keil m. ‖ coin m. / ~ (Print) ‖ Keil m.; Unterlage f.; Fütterung f. ‖ tacon m.; taquon m. / absorbing ~ ‖ Absorptionskeil m. ‖ coin m. absorbant. / adjusting ~ ‖ Stellkeil m. ‖ clavette f. de réglage; coin m. d'ajustage. / axle box ~ ‖ Achslagerstellkeil m. ‖ coin m. de serrage des boîtes d'essieu. / ~ for cleaving trees ‖ Keil m. zum Holzspalten ‖ ébuard m. / ~ of the curb piece (Hydr arch) ‖ Reitelkeil m.; Rödelkeil m.; Rötelkeil m. ‖ coin m. de guindage. / to fasten with a ~ *see* to wedge. / ~ for fastening the helve to the hammer ‖

Helmkeil m. ‖ angrois m. d'un marteau. / iron ~ ‖ Eisenkeil m. ‖ coin m. en fer. / to loosen the ~ ‖ loskeilen ‖ déclaveter. / tightening ~ *see* adjusting ~. / wooden ~ ‖ Holzkeil m. ‖ coin m. en bois.

wedge axe ‖ Spaltaxt f. ‖ hache f. à fendre. / ~ characters pl. ‖ Keilschrift f. ‖ caractères mpl. cunéiformes. / ~ cleaving method ‖ Keilspaltmethode f. ‖ méthode f. à fendre au moyen de coins. / ~ closing contrivance ‖ Keilverschluß m. ‖ fermeture f. à coin. / ~-formed part ‖ Keilstück n. ‖ pièce f. en forme de coin. / ~ friction gear ‖ Keilrädergetriebe n. ‖ transmission f. à friction par poulies à gorge. / ~ friction wheel ‖ Keilrad n. ‖ roue f. à gorges *ou* à coin. / ~ man (Print) ‖ Keiltreiber m. ‖ homme m. de bois. / ~ pass-gap setting device ‖ Keilanstellung f. ‖ réglage m. par coin. / ~ rail anchor ‖ Keilklemme f. ‖ dispositif m. d'ancrage à coin. / ~ secured by ~ ‖ aufgekeilt ‖ fixé par cheville. / ~-shaped ‖ keilförmig ‖ cunéiforme; en forme de coin. / ~-shaped plate ‖ keilförmige Platte f. ‖ plaque f. à forme conique. / ~-sized *see* ~-shaped. / ~ writing *see* ~ character.

wedging (Mach) ‖ Verkeilen n. ‖ calage m. / ~ (Pott) ‖ Kneten n. ‖ pétrissage m. / ~ of rails ‖ Verkeilen n. der Schienen ‖ coinçage m. des rails.

weed, to ‖ jäten ‖ sarcler.

weed ‖ Unkraut n. ‖ mauvaises herbes fpl. / free from ~s ‖ unkrautfrei ‖ débarrassé des mauvaises herbes fpl.

weeder ‖ Jäter m. ‖ sarcleur m.

weeding hook ‖ Jäthacke f. ‖ serfouette f.; sarclet m.; sarcloir m.

weed killer ‖ Unkrautvertilgungsmittel n. ‖ produit m. à exterminer les mauvaises herbes.

week day ‖ Werktag m.; Wochentag m. ‖ jour m. ouvrable.

weekend ‖ Wochende n. ‖ semaine f. anglaise; fin f. de semaine. / ~ letter telegram ‖ Wochenendtelegramm n. ‖ télégramme m. de fin de semaine.

weekly ‖ wöchentlich ‖ hebdomadaire. / ~ wages pl. ‖ Wochenlohn m. ‖ salaire m. de la semaine; semaine f.

weel ‖ Fischreuse f.; Reuse f. ‖ bire f.; nasse f.; panier m. / ~ without neck ‖ Reuse f. ohne Einkehle ‖ panier m. de bonde.

weft (Stuff) ‖ Gewebe n.; Stoff m.; Zeug n. ‖ étoffe f.; tissu m. / ~ (Weav) ‖ Einschuß m.; Schuß m.; Einschlag m. ‖ trame f. / ~ counter ‖ Schußzähler m. ‖ compteur m. de duites. / ~ distributor ‖ Spulenverteiler m. ‖ distributeur m. d'époules. / ~ plush ‖ Schußplüsch m. ‖ peluche f. par trame. / ~ preparation of the ~ ‖ Vorbereitung f. des Einschlags ‖ tramage m.; préparation f. du fil de trame. / ~ ring frame ‖ Ringspinnmaschine f. für Schußgarn ‖ continue f. à anneaux à filer la trame. / ~ room hand ‖ Eintrager m. ‖ trameur m. / ~ silk ‖ Einschlagseide f. ‖ fil-trame m.; soie f. trame. / ~ store hand ‖ Webegarnausgeber m. ‖ distributeur m. de trames. / ~ velvet ‖ Schußsamt m. ‖ velours m. par trame. / ~ winding ‖ Einschlaggarnspulen n. ‖ cannetage m.; époulage m. / ~ winding machine ‖ Schußspulmaschine f. ‖ machine f. à cannettes; cannetière f.;

trameuse f. / ~ yarn ‖ Einschußgarn n. ‖ trame f.; fil m. de trame.

weigh, to ‖ wiegen ‖ peser. / ~ again ‖ nachwiegen ‖ repeser; vérifier le poids. / ~ the anchor ‖ den Anker m. lichten ‖ lever *ou* déplanter *ou* déraper l'ancre f.; démarrer un vaisseau.

weigh-boy tenter (Spinn) ‖ Kannenwieger m. ‖ échantillonneur m.

weighbridge ‖ Brückenwage f. ‖ pont m. *ou* balance f. à bascule.

weighed-up ‖ ausgewogen ‖ équilibré.

weigher ‖ Wieger m. ‖ peseur m.; basculeur m.

weigh house ‖ Stadtwage f. ‖ balance f. publique.

weighing ‖ Wägung f.; Wiegen n. ‖ pesée f.; pesage m. / ~ the anchor ‖ Lichten n. des Ankers ‖ désancrage m. / double ~ ‖ Doppelwägung f. ‖ double pesée f.

weighing apparatus ‖ Wiegevorrichtung f. ‖ appareil m. de pesage.

weighing machine ‖ Wage f. ‖ bascule f. / ~ for cars ‖ Fuhrwerkswage f. ‖ balance f. à voitures. / cattle ~ ‖ Viehwage f. ‖ balance f. à bestiaux. / crane ~ ‖ Kranwage f. ‖ balance f. pour grues. / decimal ~ ‖ Dezimalwage f. ‖ bascule f. *ou* balance décimale. / gas ~ ‖ Gaswage f. ‖ balance f. pour gaz. / ~ for persons ‖ Personenwage f. ‖ bascule f. pour personnes. / pneumatic balancing for ~s ‖ Luftdruckentlastung f. für Wagen ‖ soulagement m. à air comprimé pour bascules.

weighing room ‖ Wägezimmer n. ‖ salle f. des balances. / ~ scale ‖ Wagschale f. ‖ bassin m. *ou* plateau m. de balance.

weigh shaft ‖ Steuerungswelle f.; Umsteuerwelle f. ‖ arbre m. de relevage *ou* de distribution.

weight, to ‖ belasten ‖ charger avec des poids mpl.

weight ‖ Gewicht n. ‖ poids m / ~ of the air ‖ Gewicht n. der Luft ‖ poids m. de l'air. / ~ of the air displaced ‖ Gewicht n. der verdrängten Luft ‖ poids m. de l'air déplacé. / approximate ~ ‖ ungefähres Gewicht n. ‖ poids m. approximatif. / ~ of armature copper ‖ Ankerkupfergewicht n. ‖ poids m. du cuivre de l'induit. / ~ of an atom ‖ Atomgewicht n. ‖ poids m. atomique. / small ~ of an atom ‖ niedriges Atomgewicht n. ‖ poids m. atomique faible. / average ~ ‖ durchschnittliches Gewicht n. ‖ poids m. moyen. / ~ on the axle ‖ Achsbelastung f. ‖ charge f. par essieu. / ~ of a body ‖ Schwere f. *oder* Gewicht n. eines Körpers ‖ poids m. d'un corps. / ~ of a bridge ‖ Eigengewicht n. einer Brücke ‖ poids m. propre d'un pont. / ~ of the car ‖ Wagengewicht n. ‖ poids m. de la voiture. / counter-balance ~ ‖ Ausgleichgewicht n. ‖ contrepoids m. / dead ~ *see* net ~. / driving ~ for clock ‖ Antriebsgewicht n. für das Uhrwerk ‖ poids m. moteur *ou* entraîneur du mouvement d'horlogerie. / effective ~ ‖ Nutzlast f. ‖ poids m. absolu *ou* effective; charge f. utile. / ~ when empty ‖ Leergewicht n. ‖ poids m. à vide. / extra ~ ‖ Übergewicht n. ‖ surpoids m. / ~ of fish plates per metre run ‖ Laschengewicht n. je laufenden Meter ‖ poids m. des éclisses par mètre courant. / ~ of fuel ‖ Brennstoffgewicht n. ‖ poids m. du combustible. / gross ~ ‖ Rohgewicht; Bruttogewicht n. ‖ poids m. brut. / half-gross ~ ‖ Halbrohgewicht

poids m. demi-brut. / ~ per HP ‖ Gewicht n. je PS ‖ poids m. par CV. / ~ of the hull (Shipb) ‖ Eigengewicht n. des Schiffskörpers ‖ poids m. de la coque. / individual ~ ‖ Stückgewicht n. ‖ poids m. de la pièce. / invoiced ~ ‖ fakturiertes Gewicht n. ‖ poids m. facturé. / total lifted ~ ‖ gesamtes gehobenes Gewicht n. ‖ charge f. globale élevé. / loaded ~ (Aero) ‖ geladenes Gewicht n.; Zuladung f. ‖ poids m. chargé. / ~ of a locomotive when empty ‖ Leergewicht n. einer Lokomotive ‖ poids m. d'une locomotive non compris l'eau dans la chaudière. / to lose ~ ‖ an Gewicht n. abnehmen ‖ perdre du poids m. / ~ of machine with accessories ‖ Gewicht n. der Maschine mit Zubehör ‖ poids m. de la machine avec accessoires. / ~ per metre run of rail ‖ Schienengewicht n. je laufenden Meter ‖ poids m. au mètre courant de rail. / theoretic ~ per metre ‖ rechnungsmäßiges Metergewicht n. ‖ poids m. théorique par mètre courant. / net ~ ‖ Eigengewicht n. ‖ poids m. net. / ~ of packing ‖ Gewicht n. der Verpackung ‖ poids m. de l'emballage. / paying ~ *see* real ~. / ~ of plate ‖ Blechgewicht n. ‖ poids m. de la tôle. / ~ in pounds ‖ Gewicht n. in Pfunden ‖ poids m. en livres. / precision ~ ‖ Präzisionsgewicht n. ‖ poids m. de précision. / real ~ ‖ Nutzlast f. ‖ poids m. absolu *ou* effective; charge f. utile. / rough ~ ‖ ungefähres *oder* abgeschätztes Gewicht n. ‖ poids m. approximatif. / set of ~s ‖ ineinander passende Gewichte npl.; Einsatzgewichte npl. ‖ jeu m. de poids à cuvette. / short ~ ‖ Mindergewicht n.; Gewichtsmanko n. ‖ poids m. insuffisant. / sliding ~ ‖ verschiebbares Gewicht n. ‖ poids m. curseur *ou* coulissant. / specific ~ ‖ spezifisches Gewicht n.; Wichte f. ‖ poids m. spécifique. / specific ~ of air ‖ Wichte f. der Luft ‖ poids m. spécifique de l'air. / stamped ~ ‖ geeichtes *oder* gestempeltes Gewicht n. ‖ poids m. poinçonné. / standard ~ ‖ Normalgewicht n. ‖ poids m. étalon. / tare ~ *see* real ~. / total ~ ‖ Gesamtgewicht n. ‖ poids m. total. / total ~ of a dutiable ware ‖ Gesamtgewicht n. einer zollpflichtigen Ware ‖ poids m. total d'une marchandise soumise aux droits. / ~ of tup ‖ Fallgewicht n.; Bärgewicht n. ‖ poids m. de mouton. / ~ per unit of volume ‖ Gewicht n. je Raumeinheit ‖ poids m. par unité de volume. / ~ by volume ‖ Volumgewicht n. ‖ poids m. par volume. / ~ on the wheel ‖ Radbelastung f. ‖ poids m. sur la roue. / whole ~ *see* total ~. / ~ in working order ‖ Dienstgewicht n. ‖ poids m. en ordre de service.

weight areometer ‖ Gewichtsaräometer n. ‖ aréomètre m. à poids constant. / bill of ~ ‖ Gewichtsnota f. ‖ note f. du poids. / ~ brake ‖ Gewichtsbremse f. ‖ frein m. à contrepoids. / ~ capacity ‖ Gewichtskapazität f. ‖ capacité f. massique. / ~-driven clock ‖ Uhrwerk n. mit Gewichtsantrieb ‖ mouvement m. d'horlogerie à poids.

weighted safety valve ‖ gewichtbelastetes Sicherheitsventil n. ‖ soupape f. de sûreté à contre-poids.

weight equivalent ‖ Gewichtsäquivalent

n. ‖ poids m. équivalent. / ~ holder ‖ Gewichtsträger m. ‖ porte-poids m. / ~ indication ‖ Gewichtsangabe f. ‖ indication f. du poids. / loss in ~ ‖ Gewichtsverlust m. ‖ perte f. de poids. / part of ~ ‖ Gewichtsteil n. ‖ partie f. en poids. / per cent by ~ ‖ gewichtsprozentig ‖ pour cent en poids m. / saving of ~ ‖ Gewichtsersparnis f. ‖ économie f. de poids. / set of ~s ‖ Gewichtssatz m. ‖ série f. de poids. / unity of ~ ‖ Gewichtseinheit f. ‖ unité f. de poids. / ~ winding gear of the clock ‖ Gewichtsaufzug m. der Uhr ‖ remontage m. du poids d'horloge.

weighty ‖ schwer ‖ pesant; lourd.

weir ‖ Wehr n.; Überfall m. ‖ déversoir m.; barrage m. / carved ~ ‖ durchbrochenes Wehr n. ‖ écluse f. à pertuis. / double-sluice ~ ‖ Doppelschützenwehr n. ‖ barrage m. à vannes doubles. / ~ with a lock ‖ Schleusenwehr n. ‖ barrage m. à écluse. / needle ~ ‖ Nadelwehr n. ‖ barrage m. à aiguilles *ou* à fermettes mobiles. / overflowing ~ ‖ Überfallwehr n. ‖ déversoir m. à trop-plein. / roller sluice ~ ‖ Walzenschützenwehr n. ‖ barrage m à cylindres. / sector ~ ‖ Segmentwehr n. ‖ barrage m. à secteur *ou* à segment. / sluice ~ ‖ Schützenwehr n. ‖ barrage m. à vanne. / stop plank ~ *see* needle ~. / valve ~ ‖ Schleusenwehr n.; Schützenwehr n. ‖ barrage m. en éperon. / ~ with waste-sluices ‖ Schützenwehr n. ‖ barrage m. à vannes. / ~ building ‖ Wehrbau m. ‖ construction f. de barrages. / ~ fisherman ‖ Reusenfischer m. ‖ pêcheur m. à la masse. / ~ plant ‖ Wehranlage f. ‖ installation f. de barrage.

weld, to ‖ schweißen ‖ souder. / ~ autogeniously ‖ autogen schweißen ‖ souder par le procédé autogène. / ~ electrically ‖ elektrisch *oder* elektrolytisch schweißen ‖ souder électriquement. / ~ tyres directly on wheels ‖ Bandagen fpl. mit dem Rad unmittelbar verschweißen ‖ souder le bandage directement avec la roue.

weld ‖ Schweißstelle f. ‖ soudure f.

weldability ‖ Schweißbarkeit f. ‖ soudabilité f.

weldable ‖ schweißbar ‖ soudable. / ~ steel ‖ schweißbarer Stahl m.; Schweißstahl m. ‖ acier m. soudable.

welded ‖ geschweißt ‖ soudé. / double ~ ‖ doppelt geschweißt ‖ à soudure f. double. / ~ ice can ‖ geschweißte Eiszelle f. ‖ mouleau m. à glace soudé. / ring that has to be ~-in ‖ einzuschweißender Ring m. ‖ anneau m. qui doit être soudé dans les pièces. / ~ iron spoke wheel ‖ Speichenrad n. aus Schweißeisen ‖ roue f. à rayons en fer soudabl·. / ~ joint ‖ Schweißverbindung f. ‖ joint m. soudé. / ~ wrought iron plate with a layer of chrome steel ~-on ‖ schmiedeeiserne Platte f. mit aufgeschweißter Chromstahlschicht ‖ plaque f. en fer forgé avec une couche d'acier chromé soudée dessus. / electrically ~ rail joints pl. ‖ elektrisch geschweißte Schienenstöße mpl. ‖ joints mpl. de rails soudés électriquement. / ~ seam *see* welding seam. / ~ steel pipe *see* ~ steel tube. / ~ steel tube ‖ geschweißtes Stahlrohr n. ‖ tube m. *ou* tuyau m. d'acier soudé. / ~ tube ‖ geschweißtes Rohr n. ‖ tuyau m. soudé.

welder ‖ Schweißer m. ‖ soudeur m. / ~'s mask see welding helmet. / ~'s outfit ‖ Schweißerausrüstung f. ‖ équipement m. de soudeur.

welding ‖ Schweißen n.; Schweißarbeit f. ‖ soudure f.; soudage m. / alternating current ~ ‖ Wechselstromschweißung f. ‖ soudage m. à courant alternatif. / arc ~ ‖ Lichtbogenschweißung f. ‖ soudure f. à l'arc. / autogenous ~ ‖ autogenes Schweißen n. ‖ soudure f. autogène. / automatic ~ ‖ Maschinenschweißen n. ‖ soudage m. automatique ou mécanique. / butt ~ ‖ Stumpfschweißen n. ‖ soudure f. par rapprochement. / butt seam ~ see welding, butt. / caulk ~ ‖ Dichtungsschweißung f. ‖ soudure f. étanche. / ~ of chains ‖ Kettenschweißung f. ‖ soudage m. de chaînes. / cold ~ ‖ Kaltschweißung f. ‖ soudage m. sans échauffement préalable des pièces. / cross bar ~ ‖ Kreuzschweißung f. ‖ soudage m. de pièces en croix. / ~ with current from the public electric supply ‖ Netzstromschweißung f. ‖ soudage m. par le courant du réseau. / direct current ~ ‖ Gleichstromschweißung f. ‖ soudage m. à courant continu. / ~ of edges ‖ Kantenschweißung f. ‖ soudage m. sur arêtes. / joints which are to be effected by ~ ‖ Verbindungen f. die durch Schweißen hergestellt werden sollen ‖ jonctions fpl. qui doivent s'effectuer par soudage. / electric ~ ‖ elektrisches Schweißen n. ‖ soudure f. électrique. / forge ~ ‖ Hammerschweißung f. ‖ soudage m. au marteau. / fusion ~ ‖ Schmelzschweißung f. ‖ soudage m. par fusion. / gas ~ ‖ Gasschweißung f. ‖ soudage m. au gaz. / hot ~ ‖ Warmschweißung f. ‖ soudage m. à chaud. / lap ~ ‖ Überlapptschweißen n. ‖ soudage m. par recouvrement. / pipe ~ ‖ Rohrschweißung f. ‖ soudage m. de tubes. / point ~ ‖ Punktschweißung f. ‖ soudage m. par points. / ~ without preheating see cold ~. / pressure ~ ‖ Preßschweißung f. ‖ soudage m. sous pression. / rail ~ ‖ Schienenschweißung f. ‖ soudure f. de rails. / resistance ~ ‖ Widerstandsschweißung f. ‖ soudage m. par résistance. / ridgeless seam ~ ‖ wulstlose Nahtschweißung f. ‖ soudage m. de joints sans surépaisseur. / ring ~ ‖ Ringschweißung f. ‖ soudage m. d'anneaux. / seam ~ ‖ Nahtschweißung f. ‖ soudage m. des joints. / staggered spot ~ ‖ Zickzackpunktschweißung f. ‖ soudage m. par points en quinconce. / step-back ~ ‖ Pilgerschrittschweißung f. ‖ soudage m. à pas de pèlerin. / shielded arc ~ ‖ Schutzgasschweißung f. ‖ soudage m. à arc protégé. / slot ~ ‖ Schlitzschweißung f. ‖ soudage m. des fentes. / spot ~ see point ~. / step by step seam ~ ‖ Schweißung f. nach dem Rollenschrittverfahren ‖ soudage m. pas à pas par électrode roulante. / straight line spot ~ ‖ Reihenpunktschweißung f. ‖ soudage m. par points en ligne droite. / strength ~ see welding, strong. / strong ~ ‖ Festigkeitsschweißung f. ‖ soudure f. de force ou solide. / thermit(e) ~ ‖ Thermitschweißung f. ‖ soudage m. à la thermite ou à l'aluminium. / tire ~ ‖ Reifenschweißung f. ‖ soudage m. de bandages. / tube ~ ‖ Rohrschweißung f. ‖ soudage m. de tubes. / vertical ~ ‖ senkrechte Schweißung f. ‖ soudage m. vertical.

welding apparatus ‖ Schweißapparat m. ‖ appareil m. de soudage. / ~ bar ‖ Schweißstab m. ‖ barre f. à souder. / ~ bead ‖ Schweißperle f. ‖ perle f. de soudure. / blooming mill for ~ see welding mill. / ~ burner ‖ Schweißbrenner m. ‖ chalumeau m. à souder. / connexion for ~ and fusion burner ‖ Anschluß m. für Schweiß- und Schneidebrenner ‖ raccord m. pour chalumeau à souder et à découper. / ~ characteristic ‖ Schweißcharakteristik f. ‖ caractéristique f. de soudure. / ~ current ‖ Schweißstrom m. ‖ courant m. de soudure. / defect in ~ ‖ Schweißfehler m. / cran m. ou défaut m. de soudure. / ~ dynamo ‖ Schweißgenerator m. ‖ génératrice f. ou dynamo f. de soudure. / ~ edge ‖ Schweißkante f. ‖ bord m. à souder.

welding electrode ‖ Schweißelektrode f.; Schweißstab m. ‖ électrode f. ou barre f. à souder. / ~ holder ‖ Schweißzange f. ‖ poignée f. pour électrodes de soudage.

welding engine see ~ machine. / ~ equipment ‖ Schweißeinrichtung f. ‖ équipement m. ou installation f. de soudage. / ~ furnace ‖ Schweißofen m. ‖ four m. à souder. / ~ globule see ~ bead. / ~ handle see ~ pistol. / ~ heat ‖ Schweißhitze f.; Schweißwärme f. ‖ chaude f. soudante. / ~ helmet ‖ Schweißkappe f. ‖ capot m. protecteur de soudeur. / ~ installation ‖ Schweißeinrichtung f. ‖ équipement m. ou installation f. de soudage. / ~ job ‖ Schweißarbeit f. ‖ travail m. de soudure.

welding machine ‖ Schweißmaschine f. ‖ machine f. à souder. / arc ~ ‖ Lichtbogenschweißmaschine f. ‖ appareil m. à souder à l'arc électrique. / autogenous ~ ‖ autogene Schweißmaschine f. ‖ machine f. pour la soudure autogène. / automatic ~ ‖ Schweißautomat m. ‖ machine f. automatique à souder. / butt ~ ‖ Stumpfschweißmaschine f. ‖ machine f. à souder par rapprochement. / chain ~ ‖ Kettenschweißmaschine f. ‖ machine f. à souder les chaînes. / circumferential seam ~ ‖ Rundnahtschweißmaschine f. ‖ machine f. à souder les joints circonférentiels ou ronds. / combined ~ see universal ~. / electrical ~ ‖ elektrische Schweißmaschine f. ‖ machine f. à souder électrique. / ~ for hollow bodies ‖ Hohlkörperschweißmaschine f. ‖ machine f. à souder les pièces creuses. / longitudinal seam ~ ‖ Längsnahtschweißmaschine f. ‖ machine f. à souder les joints longitudinaux. / multiple-operator ~ ‖ Mehrfachschweißmaschine f. ‖ machine f. à souder multiple. / resistance ~ ‖ Widerstandsschweißmaschine f. ‖ machine f. à souder par résistance. / ~ for rolled sections ‖ Profileisenschweißmaschine f. ‖ machine f. à souder le fer profilé. / seam ~ ‖ Nahtschweißmaschine f. ‖ machine f. à souder les bords. / special purpose ~ ‖ Schweißmaschine f. für Sonderzwecke ‖ machine f. à souder pour buts speciaux. / tube ~ ‖ Rohrschweißmaschine f. ‖ machine f. à souder les tubes. / universal ~ ‖ Universalschweißmaschine f. ‖ machine f. à souder universelle.

welding means pl. ‖ Schweißmittel n. ‖ matériel m. de soudure. / metal repair part for ~ ‖ Anschweißende n. ‖ pièce f. à souder sur une autre. / ~ mill ‖ Schweiß-walzwerk n. ‖ laminoir-soudeur m.; train-soudeur m. / ~ operator see welder. / ~ pistol ‖ Schweißpistole f. ‖ pistolet m. à souder. / ~ piston see welding rod. / ~ place ‖ Anschweißstelle f. ‖ soudure f.

welding plant ‖ Schweißanlage f. ‖ installation f. de soudage ou à souder. / autogeneous ~ ‖ autogene Schweißanlage f. ‖ installation f. de soudure autogène. / autogenous ~ and cutting plant ‖ Autogenschweiß- und -schneidanlage f. ‖ installation f. de soudage et de coupage autogène.

welding pole ‖ Schweißpol m. ‖ pôle m. de soudure. / ~ powder ‖ Schweißpulver n. ‖ poudre f. à souder. / ~ press ‖ Aufschweißpresse f. ‖ presse f. à souder. / ~ pressure ‖ Schweißdruck m. ‖ pression f. de soudage.

welding process ‖ Schweißverfahren n. ‖ procédé m. de soudage ou de soudure. / electric ~ ‖ elektrisches Schweißverfahren n. ‖ procédé m. de soudure électrique. / electrolytic ~ ‖ elektrolytisches Schweißverfahren n. ‖ procédé m. de soudure électrolytique. / gas ~ ‖ Gasschweißverfahren n. ‖ procédé m. de soudage au gaz. / thermit(e) ~ ‖ Thermitschweißverfahren n. ‖ procédé m. de soudage à la thermite.

welding rod ‖ Schweißstab m. ‖ barre f. à souder. / ~ rolling mill see ~ mill. / ~ sand ‖ Schweißsand m. ‖ sable m. à souder. / to sprinkle with ~ sand ‖ mit Schweißsand bestreuen ‖ sablonner. / ~ seam ‖ Schweißnaht f. ‖ joint m. soudé; ligne f. de soudure. / ~ set ‖ Schweißaggregat n. ‖ groupe m. d'appareils à souder.

welding shop ‖ Schweißwerkstatt f.; Schweißerei f. ‖ atelier m. de soudage. / autogenous ~ ‖ Werk n. für autogene Schweißung ‖ atelier m. de soudure autogène. / electric ~ ‖ Elektroschweißwerk n. ‖ atelier m. de soudage électrique.

welding transformer ‖ Schweißumformer m. ‖ convertisseur m. de soudure. / ~ voltage ‖ Schweißspannung f. ‖ tension f. de soudure. / ~ wire ‖ Schweißdraht m. ‖ fil m. à souder. / ~ workshop see ~ shop.

weld iron ‖ Schweißeisen n. ‖ fer m. soudable.

weldless tyre ‖ nahtloser Radreifen m. ‖ bandage m. de roue sans soudure.

weld stalk ‖ Waustengel m. ‖ tige f. de gaude. / ~ steel ‖ Schweißstahl m. ‖ acier m. soudable ou à souder. / ~ tube rolling mill ‖ Schweißrohrwalzwerk n. ‖ laminoir m. pour tubes soudés.

welfare ‖ Wohlfahrt f. ‖ bien-être m. / ~ institution ‖ Wohlfahrteinrichtung f. ‖ institution f. patronale.

well ‖ Brunnen m. ‖ puits m. / ~ (Mine) ‖ Bohrloch n. ‖ trou m. de sondage. / ~ (Build) ‖ Treppenhaus n. ‖ cage f. d'escalier. / ~ (Spring) ‖ Quelle f. ‖ source f.; nappe f. / artesian ~ ‖ artesischer Brunnen m. ‖ puits m. artésien. / ~ of the centre-board ‖ Schwertkasten m. eines Schwertbootes ‖ puits m. de dérive.

well boring ‖ Brunnenbohren n. ‖ fonçage m. de puits. / ~ borer ‖ Bohrmeister m. ‖ foreur m. de sondage. / ~ boring ‖ Brunnenbau m. ‖ construction f. de puits. / boring tool for ~s ‖ Brunnenbohrgerät n. ‖ outil m. de creusage pour puits. / ~ building ‖ Brunnenbau m. ‖ construction f. de puits. / ~ cylinder ‖ Brunnenzylin-

der m. ‖ cylindre m. de puits. / artesian ~ engineer ‖ Fachingeniör m. für artesische Brunnen ‖ ingénieur m. spécialiste de puits artésiens. / ~ foundation ‖ Brunnenfundierung f. ‖ fondation f. de puits.

well-founded (Trade) ‖ kapitalkräftig ‖ disposant d'importants capitaux mpl.

well house ‖ Brunnenhaus n.; Brunnenstube f. ‖ chambre f. de puits. / ~ pump ‖ Brunnenpumpe f. ‖ pompe f. à puits. / ~ room (Shipb) ‖ Sodraum m.; Pumpsod m. ‖ sentine f.; ousseau m.

well-seasoned cigar ‖ gut abgelagerte Zigarre f. ‖ cigare m. bien sec.

well sinker ‖ Brunnenmacher m. ‖ fontenier m.; fontainier m.

well sinking ‖ Brunnenbau m. ‖ fontainerie f. / ~ enterprise ‖ Tiefbohrgeschäft n. ‖ entreprise f. de sondages. / ~ plant ‖ Bohranlage f. ‖ installation f. de sondage. / ~ tool (Min) ‖ Bohrgerät n. ‖ appareil m. de sondage.

well wagon (Railw) ‖ Tiefladewagen m. ‖ wagon m. pour pièces pendantes entre les traverses. / ~ water ‖ Brunnenwasser n. ‖ eau f. de puits. / ~ water (Spring) ‖ Quellwasser n. ‖ eau f. de source.

welt ‖ Rand m.; Saum m.; Rahmen m. ‖ ourlet m.; bordure f.; renforcement m.

welted edge ‖ eingefaßter Saum m. ‖ bord m. replié *ou* doublé.

welt leather ‖ Rahmenleder n. ‖ contrefort m. / ~-stitched shoe ‖ Rahmenschuh m. ‖ chaussure f. cousue-trépointe.

west ‖ westlich ‖ ouest; de l'ouest m.; occidental.

west ‖ Westen m. ‖ ouest m.; occident m.; couchant m.

westing ‖ westlicher Kurs m. ‖ chemin m. ouest.

wet, to ‖ befeuchten; anfeuchten; begießen ‖ humecter; mouiller; arroser. / ~ the casing ‖ die Schalung begießen ‖ arroser le coffrage. / ~ thoroughly ‖ durchfeuchten ‖ pénétrer d'humidité; mouiller; tremper.

wet ‖ naß ‖ mouillé. / ~ (Meteor) ‖ niederschlagsreich ‖ riche de condensations *ou* en pluies; à fortes condensations. / ~ brushing machine ‖ Naßbürstmaschine f. ‖ machine f. à brosser au mouillé. / ~ bulb thermometer ‖ feuchtes Thermometer n. ‖ thermomètre m. mouillé. / ~ calender ‖ Naßkalander m. ‖ calandre m. au mouillé. / ~ chamber ‖ Feuchtkammer f. ‖ chambre f. humide. / ~ decatizing machine ‖ Naßdekatiermaschine f. ‖ décatisseuse f. au mouillé. / ~ universal decatizing machine with one closable trough ‖ Universalnaßdekatiermaschine f. mit einem verschließbaren Trog ‖ décatisseuse f. au mouillé universelle à cuve fermante. / ~ dressing ‖ nasse Aufbereitung f. ‖ traitement m. de séparation par voie humide. / ~ finishing machine ‖ Naßappreturmaschine f. ‖ machine f. pour l'apprêt au mouillé.

wet grinding ‖ Naßschleifen n. ‖ affûtage m. à l'eau; meulage m. en mouillant. / ball mill for ~ ‖ Naßkugelmühle f. ‖ moulin m. à boulets pour le broyage par voie humide. / machine for grinding machine-knives when wet ‖ Maschinenmessernaßschleifmaschine f. ‖ machine f. à affûter à l'eau les couteaux des machines. / ~ mill ‖ Naßmühle f. ‖ moulin m. par voie humide.

wet iron ‖ schlackenreiches Eisen n. ‖ fer m. gras. / ~ method ‖ Halbnaßverfahren n. ‖ voie f. demi-sèche.

wetness of the steam ‖ Nässe f. des Dampfes ‖ humidité f. de la vapeur.

wet parting *see* ~ separation. / ~ press ‖ Naßpresse f. ‖ presse f. humide. / ~ process ‖ nasses Verfahren n. ‖ procédé m. par voie humide. / ~ rot ‖ Naßfäule f. ‖ putréfaction f. humide. / ~ separation ‖ nasse Scheidung f. ‖ triage m. par voie humide. / ~ shrinking (Clothm) ‖ Wasserkrumpe f. ‖ décatissage m. à l'eau. / ~ spinning ‖ Naßspinnen n. ‖ filage m. au mouillé. / ~ spinning frame ‖ Naßspinnmaschine f. ‖ métier m. à filer au mouillé. / ~ stamps pl. (Ore dress) ‖ Naßpochwerk n. ‖ bocard m. mouillé.

wetted surface ‖ benetzte Oberfläche f. ‖ surface f. mouillée.

wetter (Print) ‖ Papierfeuchter m. ‖ trempeur m.

wetting (Print) ‖ Anfeuchten n. ‖ trempage m.; mouillage m.; trempe f. / ~ the clay ‖ Einsumpfen n. des Tones ‖ détrempage m. de l'argile. / ~ the coal ‖ Nässen n. der Kohle ‖ arrosage m. du charbon.

wetting machine ‖ Feuchtmaschine f. ‖ humecteur m.; machine f. à humecter. / ~ for paper ‖ Feuchtmaschine f. für Papier ‖ machine f. à humecter le papier. / ~ for yarns ‖ Feuchtmaschine f. für Garne ‖ machine f. à humecter les fils.

wetting, air ~ plant ‖ Luftbefeuchtungsanlage f. ‖ installation f. pour humecter l'air. / ~ room (Print) ‖ Feuchtkammer f. ‖ temperie f.

wet, half ~ treatment *see* half ~ method. / ~ vapour ‖ nasser Dampf m. ‖ vapeur f. humide. / ~ washing ‖ nasses Waschen n. ‖ lavage m. par voie humide. / in the ~ way ‖ auf nassem Wege m. ‖ par voie f. humide.

whale ‖ Wal m.; Walfisch m. ‖ baleine f.

whaleboat ‖ Walfischboot n. ‖ baleinière f.

whalebone ‖ Fischbein n. ‖ baleine f. / imitation ~ ‖ Hornfischbein n. ‖ baleine f. de corne. / ~ dresser ‖ Fischbeinbearbeiter m. ‖ apprêteur m. de baleines. / ~ dressing ‖ Fischbeinbearbeitung f. ‖ apprêt m. de baleines. / ~ sawyer ‖ Fischbeinsäger m. ‖ scieur m. de baleines. / ~ splitter ‖ Fischbeinreißer m. ‖ refendeur m. de baleines.

whale fin ‖ Walfischbarte f. ‖ fanon m. de baleine. / ~ fisher *see* whaler. / ~ fishery ‖ Walfischfang m. ‖ pêche f. de la baleine.

whale oil ‖ Walfischtran m. ‖ huile f. de baleine. / ~ boiler ‖ Transieder m. ‖ fondeur m. d'huile de baleine.

whaler ‖ Walfischfänger m. ‖ pêcheur m. de baleines.

wharf ‖ Flußdamm m.; Kai m.; Pier m.; Hafenkai m. ‖ quai m. / customs ~ ‖ Zollmole f. ‖ môle m. de la douane. / ex ~ ‖ ab Kai ‖ ex quai.

wharfage ‖ Kaigeld n. ‖ droits mpl. de quai.

wharfinger ‖ Kaimeister m. ‖ maître m. de quai.

wharf, stationary revolving crane for ~ use ‖ feststehender Hafendrehkran m. ‖ grue f. fixe pivotante pour port.

whatnot, object of ~ ‖ Etagerengegenstand m. ‖ objet m. d'étagère.

wheat ‖ Weizen m. ‖ froment m. / ~ brush ‖ Getreidebürste f. ‖ brosse f. à blé. / ~ brushing of ~ *see* ~ cleaning. / ~ cleaner

‖ Getreidereiniger m. ‖ nettoyeur m. de blé. / ~ cleaning ‖ Getreidereinigung f. ‖ nettoyage m. du blé. / ~ conditioner ‖ Weizenvorbereiter m. ‖ conditionneur m. à blé. / ~ crop ‖ Weizenernte f. ‖ récolte f. de blé. / ~ damper ‖ Getreidenetzapparat m. ‖ mouilleur m. du blé.

wheaten flour ‖ Weizenmehl n. ‖ farine f. de froment. / ~ mill ‖ Weizenmühle f. ‖ moulin m. à froment.

wheaten groats pl., coarse ~ ‖ grobes Weizenschrot n. ‖ froment m. égrugé.

wheat grader ‖ Weizensortiermaschine f. ‖ trieur m. à grain. / ~ mill ‖ Weizenmühle f. ‖ moulin m. à froment. / ~ polishing machine ‖ Getreidepoliermaschine f. ‖ polisseur m. de blé. / ~ sack ‖ Getreidsack m. ‖ sac m. à blé. / ~ scourer ‖ Getreidespitz- und Schälmaschine f. ‖ épointeuse-décortiqueuse f. de blé. / ~ sorter *see* wheat grader. / ~ starch ‖ Weizenstärke f. ‖ amidon m. de blé.

Wheatstone bridge ‖ Wheatstonesche Brücke f. ‖ pont m. de Wheatstone. / ~ receiver ‖ Wheatstoneempfänger m. ‖ appareil récepteur m. Wheatstone. / ~ transmitter ‖ Wheatstonesender m. ‖ transmetteur m. automatique. / ~ working (Tel) ‖ Wheatstonebetrieb m. ‖ exploitation f. avec appareil de Wheatstone.

wheat straw ‖ Weizenstroh n. ‖ paille f. de blé. / ~ washing machine ‖ Getreidewaschmaschine f. ‖ laveuse f. de blé.

wheel, to ‖ rollen; karren ‖ rouler. / ~ the ground ‖ die Erde abkarren ‖ brouetter la terre.

wheel ‖ Rad n. ‖ roue f. / angular ~ *see* bevel ~. / annular ~ ‖ Rad n. mit innerer Verzahnung ‖ roue f. dentée intérieure. / axle drive bevel ~ ‖ Tellerrad n. ‖ couronne f. d'angle du différentiel. / bevel ~ ‖ Kegelrad n.; konisches Rad n. ‖ roue f. conique. / blade ~ (Turbine) ‖ Laufrad n. ‖ roue f. mobile; rotor m. / bogie ~ (Loc) ‖ Laufrad n. ‖ roue f. porteuse. / bucket ~ ‖ Schöpfrad n. ‖ roue f. élévatoire. / built-up ~ ‖ zusammengesetztes Rad n. ‖ roue f. avec le bandage embattu. / cast steel ~ ‖ Stahlgußrad n. ‖ roue f. en acier moulé. / cast steel one-piece ~ ‖ gegossenes Vollrad n. ‖ roue f. d'une seule pièce en acier moulé. / cellular ~ (Hydr arch) ‖ Kastenrad n.; Schöpfrad n. ‖ roue f. à godets *ou* à seaux. / central ~ ‖ Mittelrad n. ‖ roue f. centrale. / chain ~ ‖ Kettenrad n. ‖ roue f. à chaîne. / change ~ ‖ Ersatzrad n. ‖ roue f. de rechange. / chilled iron ~ ‖ Schalengußrad n. ‖ roue f. de fonte en coquille. / coiled disc ~ ‖ Wickelrad n. ‖ roue f. enroulée. / conical ~ *see* bevel ~. / of cork (Glassm) ‖ Korkscheibe f. ‖ meule f. de liège. / coupled ~ ‖ Kuppelrad n. ‖ roue f. accouplée. / coupled ~ with lateral play ‖ verschiebbares Kuppelrad n. ‖ roue f. couplée à jeu latéral. / ~ on cover ‖ Deckelrad n. ‖ roue f. portée par le couvercle. / crown ~ (Auto) ‖ Kettenkranz m. ‖ couronne f. de chaîne. / ~ with curved float-boards ‖ Schaufelrad n. mit gekrümmten Schaufeln ‖ roue f. à aubes courbes. / detachable ~ ‖ abnehmbares Rad n. ‖ roue f. amovible. / dividing worm ~ ‖ Teilschneckenrad n. ‖ roue f. hélicoïdale diviseur. / double-

disk ~ || Doppelscheibenrad n. || roue f. à double disque. / double-flange ~ || Rad n. mit zwei Spurkränzen; zweiflanschiges Rad n. || roue f. à deux boudins. / double-plate ~ see double-disk ~. / double-spoke ~ || Doppelspeichenrad n. || roue f. à rayons doubles. / driving ~ || Antriebsrad n. || roue f. moteur. / set of driving ~s || Treibradsatz m. || essieu m. moteur monté. / elastic ~ || elastisches Rad n. || roue f. élastique. / flangeless ~ || Rad n. ohne Spurkranz m. || roue f. à surface cylindrique ou sans boudin. / float-board ~ || Schaufelrad n. || roue f. à aubes. / forged and rolled steel disc ~ || gewalztes Stahlvollrad n. || roue f. forgée et laminée d'une seule pièce. / friction ~ || Reibrad n. || galet m. de friction. / front ~ || Vorderrad n. || roue f. avant. / gear ~ || Getrieberad n. || roue f. d'engrenage. / great ~ (Watchm) || Schloßscheibe f. || limaçon m. / grinding ~ || Schleifscheibe f. || meule f. à affûter. / half-double plate ~ || Halbdoppelscheibenrad n. || roue f. à disque semi-double. / hand ~ || Handrad n. || roue f. à main. / ~ with hard tread || Rad n. mit harter Lauffläche || roue f. avec surface de roulement dure. / hoisting ~ || Heberad n. || roue f. élévatrice. / hyperbolical ~ || Hyperboloidenrad n. || roue f. dentée hyperbolique. / ~ with integral tread and flanges || Einstückrad n. mit angewalztem Lauf- und Spurkranz || roue f. laminée dont le bandage fait corps avec le disque. / intermediate ~ || Zwischenrad n. || roue f. intermédiaire. / internal spur ~ || Rad n. mit Innenverzahnung || roue f. à engrenage intérieur. / iron ~ || eisernes Rad n. || roue f. en fer. / middle ~ see central ~. / ~ with middle flange || Rad n. mit Mittelflansch || roue f. avec bride centrale. / milling ~ || Rändelrad n. || molette f. / minute ~ see central ~. / mounted ~ || montiertes Rad n. || roue f. montée. / turning shop for mounted ~s || Radsatzwerkstatt f. || atelier m. à tourner les essieux montés. / ~ with one flange || Rad n. mit einem Spurkranz || roue f. avec un boudin. / one-piece ~ || Vollrad n. || roue f. d'une seule pièce. / the ~ is out of truth || das Rad schlägt || la roue flotte. / the ~ is overhung || das Rad ist fliegend auf der Achse angeordnet || la roue est montée en porte-à-faux. / the ~s pl. are arranged to be pressed on the axle || die Räder npl. werden zum Aufpressen eingerichtet || les roues fpl. sont arrangées pour être calées à la presse. / ratched ~ || Sperrad n. || roue f. à rochet ou à cliquet. / rear ~ || Hinterrad n. || roue f. arrière. / road ~ || Wagenrad n. || roue f. de voiture. / rotor ~ see blade ~. / rough-pressed ~ || vorgepreßtes Rad n. || roue f. ébauchée à la presse. / ~ running loose on its axle || Rad n., das lose auf der Achse läuft || roue f. folle sur l'essieu. / scoop ~ see hoisting ~. / ~ with separate tyre || Reifenrad n. || roue f. à bandage. / ~ with a shrunk-on tyre || Rad n. mit aufgeschrumpftem Reifen || roue f. à bandage calé. / single-flange ~ see ~ with one flange. / solid ~ see one-piece ~. / spare ~ || Ersatzrad n.; Reserverad n. || roue f. de secours ou de rechange. / ~ with spokes || Speichenrad n. || roue f. avec rais ou à rayons. / spring ~ ||

federndes Rad n. || roue f. à ressort. / sprocket ~ (Auto) || Kettennuß f. || pignon m. à chaîne. / starting ~ || Anlaufrad n. || roue f. de mise en marche. / steering ~ || Steuerrad n. || volant m.; volant m. de direction; roue f. directrice. / ~ with stepped teeth || Zahnrad n. mit Stufenzähnen || roue f. à denture croisée. / ~ in steps || Stufenrad n.; Stufenscheibe f. || roue f. en étages. / sun and planets ~s pl. || Planetengetriebe n. || mouche f.; engrenage m. planétaire. / ~ of a swing-bridge || Laufrad n. einer Drehbrücke || roue f. sur l'arbre d'un pont tournant. / ~ with tyre cast-on || Rad n. mit angegossenem Reifen || roue f. avec bandage venu de fonte. / ~ with tyre shrunk-on || Rad n. mit aufgeschrumpftem Reifen || roue f. à bandage serré au chaud. / toothed ~ || Zahnrad n. || roue f. dentée. / traversing ~ || Laufrad n. || galet m.; roue f. portante. / ~ with tread and flange rolled all in one piece || Einstückrad n. mit angewalztem Lauf- und Spurkranz || roue f. laminée dont le cercle de roulement fait corps avec le disque. / ~ with tubular spokes || Rohrspeichenrad n. || roue f. à rayons tubulaires / ~ with two flanges || Rad n. mit zwei Spurkränzen || roue f. avec deux boudins. / ~ of vehicle || Fahrzeugrad n. || roue f. de véhicule. / ~ of vulcanized fibre || Vulkanfiberrad n. || roue f. en fibre vulcanisée. / water ~ || Wasserrad n. || roue f. hydraulique. / welded iron spoke ~ || Speichenrad n. aus Schweißeisen || roue f. à rayons en fer soudable. / wire ~ see ~ with wire spokes. / ~ with wire spokes || Drahtspeichenrad n. || roue f. à rayons en fil métallique. / wooden ~ || Holz(speichen)rad n. || roue f. en bois. / worm ~ || Schneckenrad n. || roue f. hélice ou à vis sans fin.

wheel and axle set grinder || Radsatzschleifmaschine f. || machine f. à rectifier les trains de roues. / ~ and caterpillar tank || Räderraupenkampfwagen m. || char m. d'assaut à roues et chenilles. / ~ axle || Radachse f. || axe m. de la roue. / ~ barometer || Zeigerbarometer n. || baromètre m. à cadran.

wheelbarrow || Schubkarren m.; Laufkarre f. || brouette f. / iron ~ || eiserner Schubkarren m. || bouette f. en fer. / ~ maker || Schubkarrenmacher m. || brouettier m.

wheel base || Radstand m.; Achsstand || écartement m. des roues; empattement m. / truck of any ~ || Drehgestell n. für beliebigen Radstand || bogie m. à écartement quelconque des roues.

wheel, beehive on ~s || Bienenwanderwagen m. || ruche f. / ~ blank || Radrohling m. || ébauche f. de roue. / ~ body || Radkörper m. || corps m. de roue. / ~ boss || Radnabe f. || moyeu m. de roue. / ~ boss boring mill || Radnabenbohrbank f. || tour m. vertical à aléser les moyeux. / ~ box see ~ case. / ~ brake || Radbremse f. || frein m. de roue. / ~ cap || Nabenkappe f. || chaperon m. de moyeu. / spare ~ carrier || Ersatzradhalter m. || porte-roue m. de secours. / ~ case || Radkasten m. || couvre-engrenage m. / ~ casing see ~ case. / ~ centre || Radkörper m.; Radstern m. || centre m. de roue. / cast steel ~ centre for sets of locomotive and tender wheels || Radstern

m. aus Stahlformguß für Lokomotiv- und Tenderradsätze || centre m. de roues en acier moulé pour des essieux montés de locomotives et de tenders. / ~ chair || Rollstuhl m. || fauteuil m. roulant. / ~ choke || Bremsklotz m. || cale f. de roue. / ~ cleaning shop || Räderputzerei f. || atelier m. d'ébarbage des roues. / ~ cog || Radzahn m. || dent f. de roue. / ~ cutter || Zahnradfräser m. || fraise f. pour engrenages. / ~ diameter || Raddurchmesser m. || diamètre m. de roue. / ~ disk || Radverblendscheibe f. || couvre-rayons m. de roue. / ~ drag || Hemmzeug n.; Hemmvorrichtung f. || enrayement m.; enrayage m.; enrayure f. / front ~ drive || Vorderradantrieb m. || commande f. à ou sur roue avant. / rear ~ drive || Hinterradantrieb m. || transmission f. à roue arrière.

wheeler (Horse) || Stangenpferd || cheval m. de derrière ou de timon; timonier m. / slag ~ || Schlackenfahrer m. || rouleur m. de cendres.

wheel, face width of ~ || Radbreite f. || largeur f. de roue. / ~ felloe || Radfelge f. || jante f. de roue. / ~ fit || Radsitz m.; Nabensitz m. || portée f. de calage du moyeu.

wheel flange || Radflansch m.; Radkranz m.; Spurkranz m. || boudin m. de roue. / to prevent the ~ from bearing too heavily against the outer rail || zum Schutze der äußeren Schiene f. gegen die seitlichen Angriffe der Spurkränze || pour protéger le rail extérieur contre le frottement latéral des boudins.

wheel forge || Radschmiede f. || forge f. de roues. / ~ fork bearing || Radgabellager n. || articulation f. de la fourche de roue. / ~ foundry || Radgießerei f.; Rädergießerei f. || fonderie f. de roues. / ~ movable ~ frame || bewegliches Radgestell n. || train m. de quatre roues mobile. / ~ gear || Getriebe n.; Rädergetriebe n. || engrenage m. / ~ grease || Wagenfett n. || graisse f. pour roues. / ~ guard || Prellpfahl m. || chasse-roue m. / ~ horse see wheeler. / ~ hub || Radnabe f. || moyeu m. de roue. / ~ hub cap || Radkappe f. || chapeau m. de moyeu de roue. / ~ key (Electr) || Radtaster m. || pedale f. / admissible ~ load || zulässiger Raddruck m. || charge f. de roue admissible. / ~ making machine || Radfertigungsmaschine f. || machine f. pour la fabrication de roues.

wheel milling machine || Radfräsmaschine f. || machine f. à fraiser les roues. / ~ shop || Räderfräserei f. || atelier m. de fraisage de roues.

wheel mounting and dismounting press || Räderauf- und -abziehpresse f. || presse f. à caler et décaler les roues. / ~ plough || Räderpflug m. || charrue f. à roues. / ~ press || Räderaufziehpresse f. || presse f. à caler les roues ou pour le calage des roues. / ~ pressure || Raddruck m. || pression f. exercée par la roue. / ~ press for putting-off ~s || Räderabziehpresse f. || presse f. à décaler les roues. / ~ rim || Radfelge f. || jante f. de roue. / ~ rimmer || Felgenmacher m. || ouvrier m. en jantes. / ~ rolling mill || Räderwalzwerk n. || laminoir m. à roues. / ~ rut see ~ track. / ~ seat see ~ fit.

wheel set (Railw) || Radsatz m. || train m. de roues; essieu m. monté. / coupled ~ ||

Kuppelradsatz m. ‖ essieu m. accouplé monté. / ~ for inside bearings ‖ Innenlagerradsatz m. ‖ essieu m. monté pour boîtes à graisse intérieures. / locomotive ~ ‖ Lokomotivradsatz m. ‖ essieu m. monté pour locomotives. / movable ~ ‖ Einstellradsatz m. ‖ essieu m. monté mobile. / ~ for outside bearings ‖ Außenlagerradsatz m. ‖ essieu m. monté pour boîtes à graisse extérieures. / radial ~ ‖ radial einstellbarer Lenkradsatz m. ‖ essieu m. radial monté. / rear ~ ‖ Hinterradsatz m. ‖ train m. de roues *ou* essieu m. monté arrière. / standard ~ ‖ Radsatz m. normaler Bauart ‖ essieu m. monté normal.

wheel set lathe ‖ Radsatzdrehbank f. ‖ tour m. pour trains de roues. / ~ working machine ‖ Radsatzbearbeitungsmaschine f. ‖ machine f. à travailler les essieux montés.

wheel spider *see* ~ centre. / ~ spoke ‖ Radspeiche f. ‖ rayon m. *ou* rais m. de la roue.

wheel tire ‖ Radreifen m. ‖ bandage m. de roue. / ~ lathe ‖ Radreifendrehbank f. ‖ tour m. pour bandages de roues. / ~ press ‖ Radreifenpresse f. ‖ presse f. pour bandages de roues.

wheel tooth ‖ Radzahn m. ‖ dent f. de roue *ou* d'engrenage. / wooden ~ tooth ‖ Radzahn m. aus Holz ‖ dent f. d'engrenage en bois. / ~ track ‖ Radspur f. ‖ ornière f. / ~ tyre *see* wheel tire. / ~ vane ‖ Radflügel m. ‖ ailette f. de roue. / ~ window ‖ Radfenster n. ‖ fenêtre f. de roue.

wheel work ‖ Räderwerk n.; Getriebe n. ‖ rouage m.; engrenage m. / ~ with bevel gearing ‖ konisches Getriebe n.; Kegelradgetriebe n. ‖ engrenage m. à roues coniques. / ~ with cylindrical gearing ‖ Stirnradgetriebe n. ‖ engrenage m. cylindrique.

wheelwright ‖ Stellmacher m. ‖ charron m.

wheelwright's machine ‖ Stellmachermaschine f.; Radmaschine f. für Stellmacher ‖ machine f. de charronnage. / ~ tools pl. ‖ Stellmacherwerkzeug n. ‖ outils mpl. de charron. / ~ wood ‖ Stellmachereiholz n. ‖ bois m. de charronnage. / ~ work ‖ Stellmacherarbeit f. ‖ pièce f. de charronnage.

whelp of the capstan ‖ Spillklampe f. ‖ taquet m. du cabestan.

wherry boat ‖ Ewer m. ‖ éver m.

whet, to ‖ wetzen ‖ aiguiser.

whetiron ‖ Wetzstahl m. ‖ aiguisoir m.

whetstone ‖ Wetzstein m. ‖ pierre f. à aiguiser. / ~ quarry ‖ Wetzsteinbruch m. ‖ carrière f. de pierres à aiguiser.

whetted ‖ scharf ‖ tranchant; aigu; affilé.

whetting, steel goods ~ installation ‖ Stahlwarenschleiferei f. ‖ installation f. à repasser les articles d'acier. / ~ material ‖ Schleifmittel n. ‖ produit m. à aiguiser.

whey ‖ Molke f. ‖ petit lait m.

whim ‖ liegende Winde f.; Göpel m. ‖ treuil m. / steam ~ ‖ Dampfwinde f. ‖ manège m. à vapeur.

whim drive ‖ Göpelantrieb m. ‖ commande f. à manège. / ~ gin ‖ Göpel m.; Pferdegöpel m. ‖ manège m.; vargue m.

whip, to (Mar) ‖ heißen; hissen ‖ hisser. / ~ (Tail) ‖ überwendlich nähen ‖ surjeter.

whip ‖ Peitsche f. ‖ cravache f.; fouet m. / ~ (Mar) ‖ Jollentau n. ‖ agui m.; cartahu m. / ~ (Windmill) ‖ Mast m. *oder* Rute f. der Flügel ‖ bras m. des ailes. / ~ for children ‖ Kinderpeitsche f. ‖ fouet m. d'enfants.

whip cord ‖ Peitschenschnur f. ‖ corde f. *ou* mèche f. de fouet. / ~ handle ‖ Peitschenstiel m. ‖ manche m. de fouet. / ~ lash *see* ~ cord. / ~ maker ‖ Peitschenmacher m. ‖ fouettier m.

whipper (Mine) ‖ Wipper m. ‖ culbuteur m.

whipping (Mar) ‖ Takelung f. ‖ surliure f. / ~ of shaft ‖ Schlagen n. der Welle ‖ rotation f. folle de l'arbre.

whipping crane ‖ Wippkran m. ‖ bossoir m.; potence f. de transbordement. / ~ gear ‖ Hilfskran m. ‖ crochet m. *ou* moufle f. de petite levée.

whip saw ‖ Schrotsäge f. ‖ scie de long; passe-partout m. / ~ without frame ‖ Fuchsschwanz m.; Blattsäge f.; Biberschwanz m. ‖ scie f. à manche d'égohine; sciotte f.

whirl, to ‖ wirbeln ‖ tourbillonner.

whirl ‖ Wirbel m. ‖ tourbillon m. / air ~ ‖ Luftwirbel m. ‖ tourbillon m. *ou* remous m. d'air.

whirlpool ‖ Wasserwirbel m.; Strudel m. ‖ tournant m. *ou* remous m. d'eau.

whirlwind ‖ Wirbelwind m.; Windhose f. ‖ tourbillon m.

whisk ‖ Federwisch m. ‖ plumeau m.

whisky ‖ Branntwein m.; Whisky m. ‖ eau-de-vie f.; whisky m.

whistle, to ‖ pfeifen ‖ siffler.

whistle ‖ Pfeife f. ‖ sifflet m. / alarm ~ ‖ Alarmpfeife f. ‖ sifflet m. d'alarme. / ~ with double tone ‖ Zweiklangpfeife f. ‖ sifflet m. à deux notes. / steam ~ ‖ Dampfpfeife f. ‖ sifflet m. à vapeur.

whistle buoy ‖ Heultonne f. ‖ bouée f. à sifflet. / ~ call (Mar) ‖ Bootsmannspfeife f. ‖ sifflet m.; rossignol m.

white ‖ weiß ‖ blanc. / ~ arsenic ‖ arsenige Säure f. ‖ arsenic m. blanc. / ~ bread ‖ Weißbrot n. ‖ pain m. blanc. / ~ cooper ‖ Schäffler m.; Böttcher m. ‖ boisselier m. / ~ copper ore ‖ Weißkupfererz n. ‖ cubane m. / ~ embroidery ‖ Weißstickerei f. ‖ broderie f. blanche. / ~ hawthorn ‖ Mehlbeerbaum m. ‖ alizier m.

white heat ‖ Weißglut f. ‖ chaude f. *ou* chaleur f. blanche. / to give a ~ ‖ weißglühend machen ‖ chauffer au blanc.

white hoarhound leaves ‖ Andornkraut n. ‖ feuilles fpl. de marrube. / ~-hot ‖ weißglühend ‖ de chaleur f. blanche. / ~ ice ‖ Matteis n. ‖ glace f. opaque. / ~ iron plate ‖ Weißblech n. ‖ fer-blanc m.; feuille f. de fer-blanc.

white lead ‖ Bleiweiß n. ‖ blanc m. *ou* carbonate m. de plomb; céruse f.; blanc m. de céruse. / ~ maker ‖ Bleiweißmacher m. ‖ cérusier m. / ~ substitute ‖ Bleiweißersatz m. ‖ céruse f. imitée. / ~ trade ‖ Bleiweißhandel m. ‖ commerce m. de céruserie.

white line (Print) ‖ Durchschuß m. ‖ interligne f.

white metal ‖ Weißmetall n.; Lagermetall n. ‖ métal m. blanc; régule m. antifriction. / extracting the tin from ~ ‖ Entzinnung f. von Weißblech ‖ désétamage m. du fer blanc.

white metal lining ‖ Weißmetallausguß m. ‖ garniture f. de métal blanc.

withe pine ‖ Edeltanne f.; Silbertanne f. ‖ sapin m. blanc. / ~ wood ‖ Weißtannenholz n. ‖ bois m. de sapin blanc.

white print ‖ Weißpause f. ‖ blanc m. / ~ rot ‖ Weißfäule f. ‖ putréfaction f.; pourriture f. sèche *ou* sous l'écorce. / ~-rotted ‖ weißfaul ‖ pourri sous l'écorce. / ~ sugar ‖ Weißzucker m. ‖ sucre m. blanc. / ~ timber ‖ Weißholz n. ‖ bois m. blanc. / ~ vitriol ‖ Zinkvitriol n. ‖ sulfate m. de zinc; vitriol m. blanc. / ~ vitriol plant ‖ Zinkvitriolanlage f. ‖ installation f. de sulfate de zinc. / ~ wine ‖ Weißwein m. ‖ vin m. blanc. / ~ zinc ‖ Zinkoxyd n. ‖ blanc m. de zinc.

white ‖ Weiß n.; Weiße n. ‖ blanc m. / brillant~ ‖ Glanzweiß n. ‖ blanc m. brillant. / ~ of an egg ‖ Eiweiß n. ‖ blanc m. d'œuf. / permanent ~ ‖ Barytweiß n.; Permanentweiß n. ‖ blanc m. fixe.

whiteness ‖ weiße Farbe f.; Weiße f. ‖ blancheur f.

whitening ‖ Schlämmkreide f. ‖ blanc m. minéral *ou* chaux lavée.

whites pl. (Print) ‖ Füllmaterial n. ‖ blancs mpl.

white sizer (Pap) ‖ Papierweißer m. ‖ blanchisseur m. sur papier.

whitesmith ‖ Klempner m. ‖ ferblantier m.

whitewash, to ‖ weißen; tünchen ‖ blanchir; échauder; peindre à la chaux.

whitewash ‖ Tünche f.; Kalkmilch f. ‖ enduit m. de chaux f.; échaudage m.; badigeon m.

whitewashing ‖ Tünchen n. ‖ badigeonnage m. / ~ by compressed air ‖ Tünchen n. durch Preßluft ‖ badigeonnage m. *ou* blanchiment m. à l'air comprimé. / ~ machine ‖ Anstreichmaschine f. ‖ pulvérisateur m. à blanchir.

whiting *see* whitening.

whitish ‖ weißlich ‖ blanchâtre. / ~ smoke ‖ weißlicher Qualm m. ‖ fumée f. blanchâtre.

Whitworth fine thread ‖ Whitworthfeingewinde n. ‖ filetage m. Whitworth-fin. / ~ pipe thread ‖ Whitworthrohrgewinde n. ‖ filetage m. Whitworth-de-tuyau. / ~ thread ‖ Whitworthgewinde n. ‖ filetage m. Whitworth.

whizz, to ‖ schleudern ‖ centrifuger.

whizzer ‖ Zentrifugaltrockenmaschine f. ‖ colonne f. centrifuge *ou* sécheuse *ou* laveuse.

whole ‖ Ganzes n.; Ganzheit f. ‖ ensemble m. / rigid ~ ‖ starres Ganzes n. ‖ ensemble m. rigide.

whole milk ‖ Vollmilch f. ‖ lait m. complet.

wholesale ‖ en gros ‖ en gros.

wholesale ‖ Großverkauf m. ‖ vente f. en gros. / ~ articles pl. ‖ Massenartikel mpl. ‖ articles mpl. en grandes quantités. / ~ business ‖ Großhandelsgeschäft n. ‖ commerce m. de gros. / ~ coal merchant ‖ Kohlengroßhändler m. ‖ négociant m. en gros de charbons. / ~ dealer ‖ Großhändler m. ‖ commerçant m. *ou* négociant m. en gros; grossiste m. / conveying plant for ~ goods pl. ‖ Förderanlage f. für Massengüter ‖ installation f. de transport pour machandises en masses. / ~ gut dealers pl. ‖ Darmgroßhandlung f. ‖ commerce m. de boyaux en gros. / ~ industry ‖ Großgewerbe n. ‖ grande industrie f. / ~ manufacture ‖ Massenfertigung f. ‖ fabrication f. en grandes quantités. / ~ merchant *see* ~ dealer. / ~ price ‖

Engrospreis m.; Großhandelspreis m. || prix m. de gros.

wholesaler *see* wholesale dealer.

wholesale scrap store || Schrottgroßhandlung f. || magasin m. de ferraille en gros. / ~ **trade** || Großhandel m. || commerce m. en gros.

wholesome coffee || Gesundheitskaffee m. || café m. de santé.

whortleberry, red ~ || Preiselbeere f. || airelle f. rouge *ou* ponctuée.

whorl || Wirtel m.; Nuß f. am Spinnrade || portée f. de la broche.

whorler || Töpferscheibe f. || tour m. de potier.

wick || Docht m. || mèche f. / **candle ~** || Kerzendocht m. || mèche de bougie. / **circular ~** || Hohldocht m. || mèche f. ronde. / **cotton ~** || Baumwolldocht m. || mèche f. de coton. / **lamp ~** || Lampendocht m. || mèche f. de lampe. / **lubricating ~** || Schmierdocht m. || mèche f. de graissage. / **plaited ~** || geflochtener Docht m. || mèche f. tressée. / **to spin a ~** || einen Docht m. spinnen || filer une mèche. / **stearin-coated ~** || mit Stearin überzogener Docht m. || mèche f. enduite de stéarine. / **sulphured ~** || Schwefelfaden m. || mèche f. soufrée. / **weaved ~** || gewebter Docht m. || mèche f. tissée.

wick carburettor || Dochtvergaser m. || carburateur m. à mèche.

wicker || Korbweide f. || osier m. / **~ basket** || Weidenkorb m. || nacelle f. en osier. / **~ bottle** || Korbflasche f. || bonbonne f. clissée. / **~ bottle plaiter** || Korbflaschenflechter m. || clissier m. / **~ framework** || rohrgeflochtenes Fachwerk n. || treillis m. d'osier. / **~ furniture** || Korbmöbel npl. || meubles mpl. en osier. / **~ goods** pl. || Flechtwaren fpl. || articles mpl. en matières à tresser. / **~ hedge** || geflochtene Hecke f. || haie f. de branches, haie sèche *ou* morte. / **~ lamp** || Korblampe f. || lampe f. de parquet en moelle de rotin.

wickerwork || Korbflechterwaren fpl. || articles mpl. de vannerie. / **~ toys** pl. || Korbspielwaren fpl. || jouets mpl. d'osier.

wicket || Schalter m.; Schiebfenster n. || gueule f. de croisée; guichet m. / **~ of a lock gate** || Klinket n. eines Schleusentores || guichet m. d'une porte d'écluse.

wick lamp || Dochtlampe f. || lampe f. à mèche. / **~ lubrication** || Dochtschmierung f. || graissage m. à mèche. / **~ lubricator** || Dochtschmierbüchse f.; Dochtöler m. || graisseur m. à mèche. / **~ preparer** || Dochtmacher m. || préparateur m. de mèches. / **~ spinner** || Dochtspinner m. || tresseur m. de mèches. / **~ spinning** || Dochtspinnen n. || filature f. de mèches. / **~ weaving factory** || Dochtweberei f. || atelier m. de tissage de mèches. / **~ wire** || Dochtspieß m. || baguette f. / **~ yarn** || Lichtgarn n.; Dochtgarn || fil m. à mèche.

wide || weit || large.

wide-angle field glass || Weitwinkelfeldstecher m. || jumelle f. grand-angulaire. / **~ of great light transmitting capacity** || lichtstarker Weitwinkelfeldstecher m. || jumelle f. lumineuse grand-angulaire.

wide-meshed || weitmaschig || à alvéoles mpl. larges. / **~-mouthed bottle** || weithalsige Flasche f. || bouteille f. à col large.

widen, to || weiten || élargir. / **~ a vessel by hammering** || ein Gefäß n. aushämmern || écolleter un vase m.

widening || Ausbauchung f. || bombement m.; renflement m.

widening test of the hub || Nabenaufweiteprobe f. || essai m. de mandrinage du moyeu. / **~ by mandril** of forged rings || Aufweiteprobe f. von geschmiedeten Ringen || essai m. de mandrinage sur anneaux forgés.

width || Weite f. || largeur f. / **~ (Bridge)** || Spannweite f. || travée f. / **~ of an arc** || Bogenbreite f. || vide m. *ou* joue d'une voûte. / **~ of the carriage (Typewr)** || Wagenbreite f. || largeur f. du chariot. / **~ in the clear** *see* inner **~.** / **~ to investigate the ~ of a deposit** || die Lagerstätte auf ihre Mächtigkeit untersuchen || explorer le gisement sous le rapport de sa puissance. / **grinding ~** || Schleifbreite f. || largeur f. à rectifier. / **inner ~** || lichte Weite f. || largeur f. intérieure. / **~ of jaws** (Vice) || Backenbreite f. || largeur f. des mors. / **~ of key** || Keilbreite f. || largeur f. de la clavette. / **~ of key-way** || Nutbreite f. || largeur f. de la rainure. / **overall ~** || Baubreite f. || largeur f. de l'installation. / **~ between the rails** || Spurweite f. || largeur f. de voie. / **~ of scoop** || Löffelbreite f. || largeur f. de la cuiller. / **~ of the thread** || Gangbreite f. der Schraube || largeur f. du filet. / **~ of tooth** || Zahnbreite f. || largeur f. de dent. / **~ of tooth space of spur gears** || Zahnlückenbreite f. an Stirnrädern || largeur m. du creux des roues dentées droites. / **~ over the trunnions** || Länge f. über Zapfen || longueur f. par dessus les tourillons. / **~ of the vault** || Gewölbespannweite f. || échappée f. de voûte. / **~ of wheel rim** || Radkranzbreite f. || largeur m. de la jante de roue. / **working ~** || Nutzbreite f. || largeur f. utile.

wig || Perücke f. || perruque f. / **doll's ~** || Puppenperücke f. || perruque f. de poupées. / **theatrical ~** || Theaterperücke f. || perruque f. de théâtre.

wig maker || Perückenmacher m. || perruquier m.

wild boar's bristle || Wildschweinsborste f. || soie f. de sanglier. / **~ oats** pl. || Taubhafer m. || folle avoine f.

wileyer *see* willower.

willow (Bot) || Weidenholz n.; Weide f. || saule m.; osier m. / **~ (Spinn)** || (Reiß-)Wolf m. || diable m.; loup m. / **~ bark removing machine** || Weidenschälmaschine f. || machine f. à écorcer les saules.

willower (Spinn) || Wolfer m.; Teufler m.; Lumpenreißer m. || diableur m.; ouvrier m. de willow.

willowing drum || Reißtrommel f. || tambour m. à crocs. / **~ machine** *see* willow.

willow twig || Weidenrute f. || verge f. d'osier *ou* de saule. / **~ wood** || Weidenholz n. || bois m. de saule.

Wilton carpet || sammetartiger Teppich n.; Tournaivelourssteppich m. || tapis m. velouté *ou* de Tournai; moquette f. veloutée.

wimble || Drillbohrer m. || porte-foret m.; foret m. à vis. / **~ (Carp)** || Holzbohrer m.; Zimmermannsbohrer m. || foret m. à bois *ou* de charpentier. / **~ scoop** || Löffelbohrer m.; Schappe f. || cuiller f.; tarière f.; tarrière f.; tarière à glaise.

winch, to || mittels einer Winde f. in die Höhe ziehen || guinder; hisser au treuil m.; élever.

winch || Winde f.; Haspel m. || treuil m. / **~ (Ropem)** || Knüppel m. || gatton m.; gaton m. / **builder's ~** || Bauwinde f. || treuil m. pour constructions en bâtiment. / **crab ~** || Katzenwinde f. || treuil m. roulant. / **~ of the dragline for pulling-in the ropes** || Einziehwinde f. des Schleppschaufelbaggers || treuil m. de halage de la pelle-dragline. / **~ for drawing-in rope** || Seilwinde f. || treuil m. à câble de tirage. / **field ~** || Ackerwinde f. || treuil m. de labourage. / **~ for the grab** || Greiferwinde f. || treuil m. du grappin. / **hand ~** || Handwinde f. || cric m. simple. / **~ for hand and motor force** || Winde f. für Hand- und Kraftbetrieb || treuil m. à main et à force motrice. / **lifting ~** || Hebewinde f. || treuil m. de levage. / **lifting ~ for wagons** || Wagenwinde f. || engin m. de levage pour wagons. / **pneumatic ~** || Lufthaspel m. || treuil m. à air comprimé. / **shunting ~** || Rangierwinde f. || treuil m. de manœuvre. / **throat stopper ~** || Gichtglockenwinde f. || treuil m. de manœuvre pour cloches de gueulard. / **tow-line ~** || Verholwinde f. || treuil m. de rappel. / **travelling ~** || Laufwinde f. || treuil m. roulant. / **~ with worm gear** || Schneckenwinde f. || treuil m. à vis sans fin.

winch handle || Windenkurbel f. || manivelle f. *ou* coude m. du treuil. / **~ house** || Windenhaus n. || cabane f. *ou* cabine f. du treuil. / **~ hut** *see* ~ house. / **~ rope** || Knüppelband n. || livarde f. de gatton. / **~ wagon** || Windenwagen m. || chariot m. à treuil.

wind, to (Electr) || wickeln || enrouler. / **~ (Spinn)** || aufspulen || bobiner. / **~ in balls** || in Knäuel npl. wickeln || empeloter. / **~-off** || abspulen || dévider; débobiner.

wind, to ~ up || aufwickeln || enrouler. / **~ (Crane)** || mittels einer Winde f. in die Höhe ziehen || hisser au treuil m. / **~ (Watchm)** || aufziehen || remonter. / **~ into balls** || zu Knäueln npl. aufwickeln || empeloter. / **~ into banks** || zu Docken fpl. *oder* zu Strähnen fpl. aufwickeln || enrouler; mettre en écheveau.

wind || Wind m. || vent m. / **by the ~** || am Winde m. || près du vent m. / **contrary ~** || Gegenwind m. || vent m. contraire. / **cross ~** || Querwind m. || vent m. de travers. / **down the ~** || mit dem Winde m. || avec le vent. / **dry ~** || trockener Wind m. || vent m. sec. / **~ on the surface of the earth** || Bodenwind m. || vent m. du sol. / **foul ~** *see* contrary ~. / **head ~** || Gegenwind m. || vent m. debout *ou* contraire. / **leading ~** || Rückwind m. || vent m. arrière. / **mountain ~** || Bergwind m. || brise f. *ou* vent m. de montagne. / **principal ~** || Hauptwind m. || vent m. dominant. / **~ right astern** || Rückenwind m. || vent m. arrière. / **side ~** || Seitenwind m. || vent m. latéral. / **strong ~** || steife Brise f. || grand frais m. / **the ~ turns** || der Wind dreht || le vent m. tourne.

wind area || dem Wind ausgesetzte Fläche f. || surface f. exposée à la pression du vent. / **backing of the ~** || Zurückspringen n. des Windes || rebroussement m. du vent. / **~ beam** || Sturmlatte f. || contre-latte f. / **~ board (Windmill)** || Bordbrett n.; Windbrett n. || planche f. du bord.

/ ~ brace ‖ Windstrebe f. ‖ hauban m. /
~ bracing (Aero) ‖ Querverspannung f. ‖
contreventement m. / ~ bracing (Build)
‖ Windverband m'.; Windversteifung f. ‖
contreventement m. / ~ break ‖ Windbruch m. ‖ volis m.; bois m. chablis;
chablis m. / ~ cap ‖ Windschutzkappe f.
‖ chapeau m. paravent. / ~ chart ‖ Wind-
karte f. ‖ carte f. des vents. / ~ chest
(Organ) ‖ Windkasten m. ‖ secret m.;
sommier m. / ~ cloud ‖ Windwolke f. ‖
nuage m. venteux. / ~ course of the ~ ‖
Gang m. des Windes ‖ marche f. du vent.
/ ~ cutter (Organ) ‖ Oberlabium n. ‖
biseau m. du tuyau à bouche. / deflection
of the ~ by mountains ‖ Windablenkung
f. durch das Gebirge ‖ déviation f. du
vent par la montagne. / density of ~ ‖
Winddichte f. ‖ densité f. du vent.

wind direction, free ~ ‖ freie Windrichtung
f. ‖ direction f. libre du vent. / prevailing ~ ‖ herrschende Windrichtung f.
‖ direction f. dominante du vent.

wind-driven electric power station ‖ Wind-
elektrizitätswerk n. ‖ centrale f. aéro-
électrique. / ~ saw mill ‖ Schneidemühle
f. mit Windantrieb ‖ moulin m. à vent
pour scierie; scierie f. mécanique à vent.

wind, driving force of the ~ ‖ treibende
Wirkung f. des Windes ‖ poussée f. du
vent. / ~ engine ‖ Windmotor m. ‖ moteur
m. à vent; aéromoteur m.; épuise f.
volante; éolienne f.

winder (Build) ‖ Wendelstufe f. ‖ marche f.
dansante. / ~ (Mach) ‖ Hakeneisen n. ‖
fer m. à crochet. / ~ (Pap) ‖ Aufwinder
m. ‖ bobineur m. / ~ (Spinn) ‖ Haspeler
m. ‖ dévideur m. / cop ~ ‖ Kötzerspulmaschine f. ‖ bobinoir m. pour canette;
cannetière f. / doubler ~ ‖ Trommel-
wickler m. ‖ bobinoir-doubleur m.;
doubleuse f. / pirn ~ see cop ~. / weft
~ ‖ Schußspulmaschine f. ‖ bobinoir
m. pour trame.

wind fall (Forest) ‖ Windbruch m. ‖ bois
m. chablé ou chablis; chablis m. / force
of ~ ‖ Windstärke f. ‖ force f. de vent. /
~ gate ‖ Wettertür f. ‖ porte f. d'aérage.
/ ~ gauge ‖ Anemometer n.; Windmesser
m. ‖ anémomètre m. / ~ governor ‖
Windrose f. ‖ régulateur m. à vent. /
~ governor controlling gear ‖ Wind-
rosensteuerung f. ‖ dispositif m. de
réglage par le régulateur à vent. / ~ guard
‖ Windschutzscheibe f. ‖ paravent m.;
pare-brise m. / ~ gun ‖ Windbüchse f. ‖
fusil m. à air. / illuminated ~ indicator
‖ beleuchteter Windanzeiger m. ‖ indicateur m. de vent éclairé.

winding (Electr) ‖ Wicklung f. ‖ enroulement m.; bobinage m. / ~ (Spinn) ‖
Spulen n.; Haspeln n. ‖ bobinage m.;
dévidage m. / counter-clockwise ~ (Electr)
‖ Linkswicklung f. ‖ enroulement m. à
gauche. / creeping ~ ‖ fortschreitende
Wicklung f. ‖ bobinage m. progressif. /
differential compound ~ (Electr) ‖ Ge-
genwicklung f. ‖ contre-enroulement m. /
end ~ ‖ Endwicklung f. ‖ enroulement m.
frontal. / full-power ~ ‖ Wicklung f. für
volle Kraft ‖ enroulement m. pour pleine
force. / high-speed ~ ‖ Wicklung f. für
schnellen Gang ‖ enroulement m. pour
marche rapide. / lapped ~ ‖ überlappte
Wicklung f. ‖ enroulement m. imbriqué ou
en boucles. / low-speed ~ ‖ Wicklung f.
für langsamen Gang ‖ enroulement m.
pour marche lente. / low-tension ~ ‖

Niederspannungswicklung f. ‖ enroulement m. à basse tension. / primary
~ ‖ Primärwicklung f. ‖ enroulement
m. primaire. / ring ~ ‖ Ringwicklung
f. ‖ bobinage m. ou enroulement m.
en anneau. / short-circuited ~ ‖ Kurzschlußwicklung f. ‖ bobinage m. en
court-circuit. / slot ~ ‖ Nutenwicklung f.
‖ bobinage m. en rainures. / spiral ~ ‖
Spiralwicklung f. ‖ enroulement m. en
spirale. / wool ~ ‖ Haspeln n. der Wolle ‖
dévidage m. de la laine.

winding cable (Mine) ‖ Förderseil n. ‖
câble m. d'extraction ou de trainage. /
~ drum ‖ Fördertrommel f.; Seiltrommel
f. ‖ tambour m. d'enroulement; cylindre
m. enrouleur. / ~ drum (Print) ‖ Wickelwalze f. ‖ cylindre m. d'enroulement;
rouleau m. bobineur. / ~ engine ‖ Fördermaschine f. ‖ machine f. d'extraction.
/ ~ engine man ‖ Fördermaschinist m.
‖ machiniste m. d'extraction. / ~ form
(Electr) ‖ Wicklungsschablone f. ‖ gabarit
m. d'enroulement ou de bobinage. / traverse ~ frame ‖ Kreuzspulmaschine f. ‖
bobinoir m. à fil croisé. / layer of ~ ‖
Lage f. der Wicklung ‖ couche f. d'enroulement.

winding machine ‖ Spulmaschine f.; Garn-
winde f.; Haspel m. ‖ bobineuse f.; machine f. à bobiner; dévidoir m. / ball ~ ‖
Knäuelwickelmaschine f. ‖ peloteuse f. /
~ for bobbins ‖ Bobinenspulmaschine f.
‖ machine f. à bobiner. / double-faced ~
‖ doppelseitige Spulmaschine f. ‖ bobinoir m. à deux faces. / ~ for spinning
mills ‖ Windemaschine f. für Spinnereien
‖ peloteuse f. pour filatures. / strip ~
(Cable) ‖ Bandwickelmaschine f. ‖ enrouleuse f. de bandes. / wire ~ ‖ Drahtwickelmaschine f. ‖ enrouleuse f. à fil. /
~ for wire galvanizing ‖ Wickelmaschine
f. für Drahtverzinkerei ‖ enrouleuse f.
pour le zinguage du fil. / ~ for wire
weaving mills ‖ Spulmaschine f. für
Drahtwebereien ‖ machine f. à bobiner
pour tissages de fil métallique.

winding mechanism of the shovel dredger
‖ Löffelbaggerwindwerk n. ‖ mécanisme
m. de treuil de la drague à cuillers. / ~
motor ‖ Fördermotor m. ‖ moteur m.
d'extraction. / number of ~s ‖ Win-
dungszahl f. ‖ nombre m. de spires. /
~-on beam ‖ Kettbaum m. ‖ ensouple f. /
permanent load of ~ ‖ Dauerbelastung f.
der Wicklung ‖ charge f. permanente
du bobinage. / ~ pipe ‖ Schlangenrohr n.
‖ tube m. à serpentin. / ~ pit ‖ Förderschacht m. ‖ puits m. d'extraction. / ~
pitch ‖ Wicklungsschritt m.; Steigung f.
der Wicklung ‖ pas m. de l'enroulement.
/ ~ process ‖ Spulvorgang m. ‖ procédé m.
de bobinage. / ~ pump ‖ Haspelpumpe f.
‖ pompe f. à tourniquet. / ~ roller see ~
drum. / ~ room of a stocking factory ‖
Spulereisaal m. einer Strumpffabrik ‖
salle f. des bobinoirs d'une fabrique de
bas. / ~ rope ‖ Windetau n. ‖ câble m. du
treu l. / ~ shaft ‖ Windewelle f. ‖ arbre
m. du treuil. / ~ shop in the meter
works ‖ Wickelei f. in der Zählerfabrik ‖
bobinage m. à la fabrique de compteurs.
/ ~ speed ‖ Fördergeschwindigkeit f. ‖
vitesse f. d'extraction. / ~ stairs pl. ‖
Wendeltreppe f. ‖ escalier m. tournant
ou en limaçon. / ~ support (Electr) ‖
Wicklungsträger m. ‖ support m. d'enroulement. / ~ table (Electr) ‖ Wicklungs-

tabelle f. ‖ tableau m. de bobinage. /
~ turnspit ‖ Bratspieß m. ‖ tourne-
broche m.

winding-up of the estate ‖ Liquidations-
verfahren n. ‖ vérification f. des créances.
/ electric ~ device for clocks ‖ elektrischer Uhrenaufzug m. ‖ dispositif m.
de remontage m. d'horloges électrique.
/ ~ mechanism (Watchm) ‖ Aufziehvorrichtung f. ‖ mécanisme m. de remontage; remontoir m.

winding wing (Aero) ‖ verwindbarer Flügel
m. ‖ aile f. gauchissable.

wind instrument ‖ Blasinstrument n. ‖ instrument m. à vent. / ~ of brass ‖ Blasinstrument n. aus Messing ‖ instrument
m. à vent en laiton. / ~ of wood ‖ Blasinstrument n. aus Holz ‖ instrument m. à
vent en bois.

wind jet ‖ Windstrahl m. ‖ jet m. de vent.

windlass ‖ Winde f. ‖ treuil m. / ~ (Mar) ‖
Schiffswinde f.; Gangspill n. ‖ cabestan
m. / ~ (Mine) ‖ Haspel m. ‖ treuil m. / ~
(Spinn) ‖ Kreuzhaspel m. ‖ treuil m. à
leviers. / builder's ~ ‖ Bauwinde f. ‖
treuil m. pour constructions en bâtiment.
/ cask ~ ‖ Faßwinde f. ‖ monte-fûts m.;
treuil m. à fûts. / chain ~ (Shipb) ‖ Kettenspill m. ‖ vireveau m. / hydraulic ~ ‖
hydraulische Winde f. ‖ treuil m. hydraulique. / ~ for motor cars ‖ Kraftfahr-
zeugwinde f. ‖ cric m. pour automobiles.
/ store room ~ ‖ Speicherwinde f. ‖ treuil
m. de magasin.

windlass driver (Mine) ‖ Bremser m. ‖ conducteur m. de treuil. / ~ housing ‖ Win-
dengehäuse n. ‖ enveloppe f. de crics. / ~
shed ‖ Windenhaus n. ‖ cabane f. ou cabine f. de treuil.

windmill ‖ Windmühle f. ‖ moulin m. à
vent. / ~ for drainage ‖ Paltrockmühle f.;
Poldermühle f. ‖ moulin m. à assécher ou
pour assèchements. / Dutch ~ ‖ holländische Windmühle f. ‖ moulin m. à vent
hollandais. / ~ with four sails ‖ vierflügelige Windmühle f. ‖ moulin m. à vent à
quatre ailes. / German ~ ‖ deutsche
Mühle f.; Bockmühle f.; Ständermühle
f. ‖ moulin m. à vent allemand ou à pile
ou à pylone. / post ~ see German ~. / ~
standing clear ‖ freistehende Windmühle
f. ‖ moulin m. à vent isolé.

windmill aeroplane ‖ Windmühlenflugzeug
n. ‖ autogyre m. / ~ carpenter ‖ Wind-
mühlenmacher m. ‖ charpentier m. en
moulins à vent.

wind motor ‖ Windmotor m. ‖ moteur m.
à vent. / ~ movement ‖ Windbewegung
f. ‖ mouvement m. ou déplacement m.
du vent.

window, to ‖ mit Fenstern npl. versehen ‖
fenêtrer.

window ‖ Fenster n. ‖ fenêtre f. / arched ~ ‖
Bogenfenster n. ‖ fenêtre f. cintrée ou
voutée. / ~ blank ‖ blindes Fenster n. ‖
fenêtre f. feinte ou borne ou aveugle. /
burglar-proof ~ ‖ einbruchsicheres Fenster n. ‖ croisée f. à l'abri de l'effraction. /
cast iron ~ ‖ gußeisernes Fenster n. ‖
fenêtre f. en fonte. / circular ~ ‖ Rund-
fenster n. ‖ fenêtre f. circulaire. / dead ~
see blank ~. / drop ~ ‖ versenkbares Fenster n. ‖ fenêtre f. mobile. / Flemish ~ ‖
flämisches Fenster n.; Halbgeschoßfenster n. ‖ fenêtre f. mezzanine. / gable-
headed ~ ‖ Fenster n. mit Giebel oder mit
Spitzverdachung ‖ fenêtre f. mitrée
ou pignonnée; fenêtre couronnée d'un

fronton. / iron ~ ‖ eisernes Fenster n. ‖ fenêtre f. en fer. / leaning ~ ‖ liegendes Fenster n.; Querfenster n. ‖ fenêtre f. / oblique ~ ‖ Fenster n. mit schräg eingehender Laibung ‖ fenêtre f. biaise. / opening outwards ‖ nach außen aufgehendes Fenster n. ‖ fenêtre f. s'ouvrant à l'extérieur. / rope ~ ‖ Seilfenster n. ‖ fenêtre f. du passage du câble. / round-headed ~ ‖ Rundbogenfenster n. ‖ fenêtre f. cintrée à demi-cercle; fenêtre voûtée en plein cintre. / secular ~ ‖ Profanfenster n. ‖ vitrail m. profane. / semicircular ~ ‖ halbkreisförmiges Fenster n. ‖ fenêtre f. à lunette. / shop ~ ‖ Ladenfenster n.; Schaufenster n. ‖ vitrine f. ou devanture f. de boutique. / skylight ~ ‖ schräges Fenster n. ‖ abat-jour m.; tabatière f. / sliding ~ ‖ Schiebefenster n. ‖ fenêtre f. à coulisse. / three-pane ~ ‖ dreischeibiges Fenster n. ‖ fenêtre f. à trois vitres. / turnable ~ ‖ Drehfenster n. ‖ fenêtre f. tournante. / two-sashed ~ ‖ zweiflügeliges Fenster n. ‖ fenêtre f. à deux battants. / uncemented ~ ‖ kittfrei eingesetztes Fenster n. ‖ fenêtre f. adaptée sans mastic. / wooden ~ ‖ Holzfenster n. ‖ fenêtre f. en bois. / wrought iron ~ ‖ schmiedeeisernes Fenster n. ‖ fenêtre f. en fer forgé.

window arch ‖ Fensterbogen m. ‖ remenée f. de fenêtre; arrière-voûte f. / ~ bar ‖ Fensterstange f. ‖ vitrière f. ou barlotière f. de fenêtre. / wooden ~ bar ‖ Fenstersprosse f. ‖ barlotière f. en bois; éparselle f. de fenêtre. / ~ bench ‖ Fensterbank f. ‖ seuil m. ou banquette f. de fenêtre.

window blind ‖ Fenstervorhang m. ‖ rideau m. de fenêtre; jalousie f. / rolling ~ ‖ Rollladen m.; Rollvorhang m. ‖ jalousie f.; fermeture f. à rouleau. / wood ~ ‖ Holzrolladen m.; Holzjalousie f. ‖ jalousie f. en bois.

window blind lath ‖ Jalousiebrett n. ‖ lame f. de jalousies.

window case ‖ Fenstereinfassung f. ‖ croisée f. ou huisserie f. de fenêtre; châssis m. dormant. / stone ~ ‖ Fenstereinfassung f. von Stein ‖ jambage m. de fenêtre.

window catch ‖ Fensterriegel m. ‖ espagnolette f.; tourniquet m. / ~ cleaning leather ‖ Fensterleder n. ‖ peau f. de chamois pour nettoyage des vitres; époussetoir m. en cuir pour fenêtres. / ~ contact ‖ Fensterkontakt m. ‖ contact m. de fenêtre. / ~ cramp iron ‖ Fensterband n. ‖ penture f. de fenêtre. / ~ curtain ‖ Fenstervorhang m. ‖ marquise f.; rideau m. de fenêtre. / ~ edge ‖ Fensterkante f. ‖ arête f. de la fenêtre. / ~ envelope ‖ Fensterbriefumschlag m. ‖ enveloppe f. à guichet. / ~ fittings pl. ‖ Fensterbeschläge mpl. ‖ garnitures fpl. de fenêtre.

window frame ‖ Fensterrahmen m. ‖ cadre m. ou châssis m. de fenêtre. / concrete ~ ‖ Fensterrahmen m. aus Beton ‖ châssis m. de fenêtre en béton. / iron ~ ‖ eiserner Fensterrahmen m. ‖ châssis m. de fenêtre en fer.

window glass ‖ Tafelglas n.; Fensterglas n. ‖ verre m. à vitres; plaque f. de verre. / curved ~ ‖ gewölbtes Fensterglas n. ‖ vitre f. bombée. / ribbed ~ ‖ geschupptes oder gerieftes oder kannelieltes Fensterglas n. ‖ vitre f. cannelée.

window glass cutter ‖ Glasschneider m. ‖ coupeur m. de vitres.

window grate ‖ Fenstergitter n. ‖ cage f. ou treillis m. de fenêtre. / ~ head ‖ Fensterschluß m.; Überdeckung f. eines Fensters ‖ fermeture f. d'une croisée. / iron fittings ‖ eiserne Fensterbeschläge mpl. ‖ ferrures fpl. pour fenêtres. / ~ knob ‖ Fenstergriff m. ‖ bouton m. de fenêtre. / ~ lead ‖ Fensterblei n. ‖ plomb m. à vitres. / cutting out ~ openings pl. ‖ Ausschneiden n. der Fensteröffnungen ‖ découpage m. des ouvertures de fenêtre. / ~ pane ‖ Fensterscheibe f. ‖ vitre f. de fenêtre. / ~ pier ‖ Fensterpfeiler m. ‖ pied-droit m. de fenêtre. / ~ post ‖ Fensterpfosten m. ‖ poteau m. d'huisserie; montant m. de croisée. / rabbet of a ~ ‖ Fensteranschlag m.; Fensterfalz m. ‖ feuillure f. / ~ sash see window frame. / ~ seat see ~ bench. / ~ shutter ‖ Fensterladen m.; Laden m. ‖ contrevent m.; volet m. / ~ sill see ~ bench. / ~ square ‖ Fensterfach n.; Fensterfeld n. ‖ panneau m. à verre ou à vitre ou de fenêtre. / ~ stay ‖ Fensterhaken m. ‖ crochet m. de fenêtre. / ~ trellis see window grate.

wind pipe (Anatomy) ‖ Luftröhre f. ‖ trachée-artère f. / ~ power ‖ Windkraft f. ‖ force f. du vent. / ~ power engine ‖ Windkraftmaschine f. ‖ moteur m. éolien.

wind pressure ‖ Winddruck m. ‖ pression f. du vent. / normal ~ ‖ Normaldruck m. des Windes ‖ pression f. normale du vent. / ~ on plane surfaces ‖ Winddruck m. auf ebene Flächen ‖ pression f. du vent sur surfaces planes.

wind resistance ‖ Windwiderstand m. ‖ résistance f. au vent. / ~ resisting power ‖ Sturmsicherheit f. ‖ stabilité f. au vent.

wind screen ‖ Windschirm m. ‖ paravent m. / ~ (Auto) ‖ Windschutzscheibe f. ‖ paravent m.; pare-brise m. / ~ (Metal) ‖ Gichtschirm m.; Windschirm m. ‖ brisevent m. / ~ cleaner ‖ Sch(utzsch)eibenwischer m. ‖ essuie-pare-brise m.

wind shield see wind screen. / shift of ~ ‖ Windwechsel m. ‖ saut m. du vent. / ~ sifter ‖ Windsichter m. ‖ séparateur m. à vent. / ~ sifting machine ‖ Windsichtmaschine f. ‖ blutoir m. / ~ stream ‖ Windstrom m. ‖ courant m. de vent. / succession of the ~s ‖ Aufeinanderfolge f. der Winde ‖ suite f. des vents. / ~ surface ‖ Windangriffsfläche f. ‖ surface f. frappée par le vent. / ~ transport ‖ Windtransport m. ‖ transport m. éolien ou par le vent. / ~ tube (Organ) ‖ Windkanal m. ‖ porte-vent m. / ~ tunnel ‖ Windkanal m. ‖ soufflerie f. aérodynamique. / ~ turbine ‖ Windturbine f. ‖ turbine f. aérienne. / automatic regulation of the ~ turbine ‖ Selbstregelung f. der Windturbine ‖ régulation f. automatique de la turbine aérienne.

wind-up and wind-off reel ‖ Auf- und Abwickelhaspel m. ‖ enrouleur m. et dévidoir m.

wind, utilization of the ~ ‖ Windausnutzung f. ‖ utilisation f. du vent. / ~ valve ‖ Windklappe f. ‖ bascule f. de cheminée; éolipyle m. / ~ vane ‖ Windfahne f. ‖ girouette f. / ~ velocity in metres per second ‖ Windgeschwindigkeit f. in Metern je Sekunde ‖ vitesse f. du vent en mètres par seconde.

windward ‖ windwärts; luvwärts ‖ au vent m. / ~ shore ‖ Luvküste f. ‖ terre f. ou côte f. au vent.

wind way (Mine) ‖ Wetterstrecke f. ‖ galerie f. d'aérage.

wind wheel ‖ Windrad n. ‖ roue f. éolienne. / ~ with horizontal wheel disc ‖ Windrad n. mit wagerechter Radscheibe ‖ roue f. éolienne horizontale. / ~ with movable sails ‖ Windrad n. mit beweglichen Flügeln ‖ roue f. éolienne à palettes mobiles. / ~ with radial sails ‖ Windrad n. mit radialen Flügeln ‖ roue f. éolienne à palettes radiales. / ~ with rigid sails ‖ Windrad n. mit festen Flügeln ‖ roue f. éolienne à palettes fixes. / ~ with vertical wheel disc ‖ Windrad n. mit senkrechter Radscheibe ‖ moteur m. éolien avec roue verticale.

wind wheel drum ‖ Windradzylinder m. ‖ embase f. cylindrique de la roue éolienne.

wind wing ‖ Windflügel m. ‖ aile f. à vent.

wine ‖ Wein m. ‖ vin m. / ~ produced with the aid of aromatic plants ‖ aus aromatischen Pflanzen hergestellter Wein m. ‖ vin m. préparé à l'aide de plantes aromatiques. / artificial ~ ‖ Kunstwein m. ‖ vin m. artificiel. / Cyprus ~ ‖ Zyperwein m. ‖ vin m. de Chypre. / to draw off ~ ‖ Wein m. abziehen ‖ tirer du vin m. / dry ~ ‖ abgelagerter Wein m. ‖ vin m. mûr ou reposé. / fruit ~ ‖ Obstwein m. ‖ vin m. de fruits. / harsh ~ ‖ herber Wein m. ‖ vin m. vert. / medicated ~ ‖ Würzwein m. ‖ vin m. aromatisé. / Moselle ~ ‖ Moselwein m.; Mosel m. ‖ vin m. de la Moselle. / mulled ~ ‖ Würzwein m. ‖ vin m. aromatisé. / red ~ ‖ Rotwein m. ‖ vin m. rouge. / Rhenish ~ see Rhine ~. / Rhine ~ ‖ Rheinwein m. ‖ vin m. du Rhin. / sparkling ~ ‖ Schaumwein m.; Sekt m. ‖ vin m. mousseux. / spiced ~ Bowle f. ‖ bol m. / strong ~ ‖ starker Wein m. ‖ vin m. fort. / sweet ~ ‖ Süßwein m. ‖ vin m. doux. / table ~ ‖ Tafelwein m. ‖ vin m. de table. / well seasoned ~ ‖ abgelagerter Wein m. ‖ vieux vin m. / white ~ ‖ Weißwein m. ‖ vin m. blanc. / wormwood ~ ‖ Wermutwein m. ‖ vermout m.

wine barrel ‖ Weinfaß n. ‖ baril m. ou tonneau m. à vin. / ~ bottle ‖ Weinflasche f. ‖ bouteille f. à vin. / ~ brandy ‖ Weinbrand m. ‖ eau-de-vie f. naturelle de vin; cognac m. / ~ cask see wine barrel. / ~ cellar ‖ Weinkeller m. ‖ cave f. à vin. / ~ cellarage machine ‖ Weinkellereimaschine f. ‖ machine f. pour caves de vins. / ~ dosser ‖ Kelterbutte f. ‖ hotte f. battue ou poissée. / ~ dresser ‖ Winzer m. ‖ vigneron m. / ~ dressing ‖ Weinpflege f. ‖ conservation f. du vin. / ~ filter ‖ Weinfilter m. ‖ filtre m. à vin. / ~ fittings pl. ‖ Weinarmatur f. ‖ armature f. à vin. / ~ glass ‖ Weinglas n. ‖ verre m. à vin. / ~ growing ‖ Weinbau m. ‖ culture f. de la vigne; viticulture f.; viniculture f. / ~ industry ‖ Weinindustrie f. ‖ industrie f. vinicole. / good judge of ~ ‖ Weinkenner m. ‖ connaisseur m. en vin. / ~ lees pl. ‖ Weinhefe f. ‖ lie f. de vin. / ~ lees ashes pl. ‖ Weinhefenasche f. ‖ cendres fpl. gravelées. / ~ making ‖ Weinbereitung f. ‖ vinification f. / ~ mash ‖ Weinmaische f. ‖ vidange f. / ~ measurer ‖ Weinmesser m. ‖ mesureur m. de vins. / ~ packer ‖ Weinflascheneinpacker m. ‖ emballeur m. de vins. / ~ press ‖ Weinpresse f. ‖ presse f. à vin. / ~ prop ‖ Weinpfahl m. ‖

échalas m. / ~ pump ‖ Weinpumpe f. ‖ pompe f. à vin. / ~ spirit ‖ Weingeist m. ‖ esprit m. *ou* eau-de-vie f. de vin. / ~ stone ‖ Weinstein m. ‖ tartre m.; bitartrate m. de potasse; crème f. de tartre. / ~ tunner ‖ Faßzieher m.; Weinabzieher m. ‖ entonneur m. de vins. / ~ vaults pl. ‖ Weinkeller m. ‖ cave f. à vin. / ~ vinegar ‖ Weinessig m. ‖ vinaigre m. de vin. / ~ yeast ‖ Weinhefe f. ‖ lie f. de vin.

wing ‖ Flügel m. ‖ aile f. / ~ (Aero) ‖ Tragdeck n.; Tragflügel m.; Tragfläche f. ‖ aile f.; surface f. portante. / ~ (Auto) ‖ Kotflügel m. ‖ garde-boue m. / ~ auxiliary ~ ‖ Hilfsflügel m. ‖ aile f. auxiliaire. / cambered ~ ‖ gewölbter Flügel m. ‖ aile f. courbe *ou* bombée *ou* cambrée. / cantilever ~ ‖ freitragender Flügel m. ‖ aile f. en porte à faux. / ~ folding upwards ‖ Hochklappflügel m.; hochklappbarer Flügel m. ‖ aile f. repliable en hauteur. / lower ~ ‖ unterer Flügel m. ‖ aile f. inférieure. / metal ~ ‖ Metallflügel m. ‖ aile f. métallique. / one-spar ~ ‖ einholmiger Flügel m. ‖ aile f. à un seul longeron. / receding ~ ‖ rückwärts geschweifter Flügel m. ‖ aile f. fuyante *ou* en retraite. / sheet ~ (Auto) ‖ Blechkotflügel m. ‖ aile f. en tôle. / single-sparred ~ *see* one-spar ~. / slotted ~ ‖ Spaltflügel m. ‖ aile f. à fentes. / strut-braced ~ *see* strutted ~. / strutted ~ ‖ verstrebter Flügel m. ‖ aile f. entretoisée *ou* à mât *ou* à montant. / two-spar ~ ‖ zweiholmiger Flügel m. ‖ aile f. à deux longerons. / upper ~ ‖ oberer Flügel m. ‖ plan m. supérieur. / warping ~ ‖ verwindbarer Flügel m. ‖ aile f. gauchissable. / winding ~ *see* warping ~. / ~ of window ‖ Fensterflügel m. ‖ battant m. *ou* vantail m. de fenêtre. / wire-braced ~ ‖ verspannter Flügel m. ‖ aile f. haubanée *ou* maintenue par des haubans. / wooden ~ (Auto) ‖ Holzkotflügel m. ‖ aile f. en bois.

wing area ‖ Flügelfläche f. ‖ surface f. d'aile. / bottom side of ~ ‖ Unterseite f. des Flügels ‖ face f. ventrale de l'aile. / ~ burner ‖ Flachbrenner m.; Schlitzbrenner m. ‖ bec m. à papillon. / ~ cam ‖ Nadelsenker m. ‖ descente f. d'aiguilles. / ~ car (Aero) ‖ Seitengondel f. ‖ nacelle f. latérale. / ~ chord ‖ Flügeltiefe f. ‖ profondeur f. de l'aile. / contour of ~ ‖ Flügelumriß m. ‖ contour m. de l'aile.

winged tap ‖ Flügelhahn m. ‖ robinet m. à papillon.

wing float (Aero) ‖ Gleitflosse f. ‖ nageoire f. latérale. / ~ stump (Aero) ‖ Flossenstummel m. ‖ nageoire f. latérale.

wing framework ‖ Flügelgerippe n. ‖ ossature f. *ou* carcasse f. de l'aile. / ~ gap (Aero) ‖ Flächenabstand m. ‖ écartement m. des plans. / ~ hand hold *see* ~ tip hand grip. / ~-heavy ‖ querlastig ‖ lourd sur l'aile f. / ~ loading ‖ Flächenbelastung f. ‖ charge f. à la surface. / ~ nut ‖ Flügelmutter f. ‖ écrou m. à oreilles *ou* à ailettes. / ~ pivot ‖ Flügelgelenk n. ‖ articulation f. de la pale. / ~ pump ‖ Flügelpumpe f. ‖ pompe f. à ailette *ou* mi-rotative. / ~ radiator ‖ Tragdeckkühler m. ‖ radiateur m. de plan sustentateur. / ~ rib ‖ Flügelrippe f. ‖ nervure f. de l'aile. / main ~ rib ‖ Haupttrippe f. des Flügels ‖ nervure f. principale de l'aile; nervure f. de compression de l'aile.

/ ~ root ‖ Flügelwurzel f. ‖ emplanture f. de l'aile. / ~ section ‖ Flügelquerschnitt m. ‖ section f. transversale des ailes. / section cut-out of ~ ‖ Flügelausschnitt m. ‖ secteur m. d'aile. / ~ shape ‖ Flügelform f.; Flügelprofil n. ‖ profil m. de l'aile. / ~ socket ‖ Flügelbefestigungsschuh m. ‖ emplanture f. d'une aile. / ~ skid ‖ Flügelkufe f. ‖ protège-aile m. / span of the ~ ‖ Flügelspannweite f. ‖ envergure f. de l'aile. / ~ spar ‖ Tragflächenholm m.; Flügelholm m. ‖ longeron m. *ou* bras m. de l'aile. / ~ spar box ‖ Holmschuh m. ‖ embout m. du longeron. / ~ structure *see* ~ framework. / ~ strut ‖ Flügelstiel m. ‖ mât m. de l'aile. / ~ stump ‖ Bootsstummel m. ‖ nageoire f. / ~ surface ‖ Flügelfläche f. ‖ surface f. alaire *ou* d'aile.

wing tip ‖ Flügelspitze f. ‖ bout m. d'aile. / curved ~ ‖ Abschlußbogen m. des Flügels ‖ cintre m. d'extrémité de l'aile.

wing tip flare ‖ Flügelendfackel f. ‖ torche f. de bout de l'aile. / ~ float ‖ Hilfsschwimmer m.; Seitenschwimmer m. ‖ ballonnet m. de bout d'aile; nageoire f. latérale. / ~ hand grip ‖ Flügelhandgriff m. ‖ poignée f. d'aile.

wing, top-side of ~ ‖ Oberseite f. des Flügels ‖ dos m. de l'aile; face f. dorsale de l'aile. / ~ transom (Shipb) ‖ Heckbalken m. ‖ grande barre f. d'arcasse; lisse f. de hourdi. / ~ trussing ‖ Flügelfachwerk n. ‖ charpente f. des ailes. / type of ~ ‖ Flügelart f. ‖ type m. d'aile. / ~ wall of a lock crown ‖ Flügelmauer f. eines Schleusenhauptes ‖ musoir m. d'une tête d'écluse.

winning of a seam ‖ Abbau m. eines Flözes ‖ pourchasse f. *ou* exploitation f. d'une veine. / underground ~ ‖ Gewinnung f. unter Tage ‖ exploitation f. souterraine.

winning costs pl. (Mine) ‖ Förderkosten pl. ‖ frais mpl. d'exploitation.

winnow, to ‖ Korn n. schwingen; worfeln ‖ vanner; éventer le blé.

winnowing ‖ Kornschwingen n.; Worfeln n. ‖ vannage m. / ~ fan ‖ Kornreinigungsmaschine f.; Worfler m. ‖ vanneuse f. / ~ machine *see* ~ fan.

winter, to ‖ überwintern ‖ hiverner.

winter ‖ winterlich ‖ hivernal.

winter ‖ Winter m. ‖ hiver m. / to season in ~ (Agr) ‖ durchwintern; überwintern ‖ conserver pendant l'hiver m.

winter barley ‖ Wintergerste f. ‖ escourgeon m. / ~ cherry ‖ Judenkirschenbeere f. ‖ baie f. d'alkékenge. / ~ quarters pl. ‖ Winterhafen m. ‖ port m. d'hivernage. / ~ solstice ‖ Wintersonnenwende f. ‖ solstice m. d'hiver. / ~ sports article ‖ Wintersportartikel m. ‖ article m. pour les sports d'hiver.

wipe, to ~ off ‖ auswischen ‖ essuyer; effacer.

wiper (Aut tel) ‖ Kontaktarm m. ‖ frotteur m. / ~ (Mach) ‖ Hebedaumen m.; Nocken m. ‖ came f.; mentonnet m. / ~ ring ‖ Ölabstreifring m. ‖ collier m. racloir *ou* lécheur d'huile.

wipper ‖ Kipper m. ‖ culbuteur m.; verseur m.

wire, to (Electr) ‖ Spulen fpl. wickeln ‖ bobiner. / ~ (El line) ‖ Leitungen fpl. verlegen ‖ canaliser; poser les conducteurs mpl. / ~ (Mail) ‖ telegrafieren;

drahten ‖ télégraphier. / ~ a house ‖ in ein Haus n. einen Draht ziehen *oder* eine Leitung legen ‖ poser des fils mpl. dans une maison.

wire ‖ Draht m. ‖ fil m. (métallique). / ~ (Electr) ‖ Leitungsdraht m.; Leiter m. ‖ fil m. conducteur. / acid-proof ~ ‖ säurebeständiger Draht m. ‖ fil m. à isolement inattaquable aux acides. / aerial ~ ‖ Freileitung f. ‖ fil m. aérien. / ~ for aeroplanes ‖ Flugzeugdraht m. ‖ fil m. pour aéroplanes. / aluminium ~ ‖ Aluminiumdraht m. ‖ fil m. d'aluminium. / annealed ~ ‖ geglühter Draht m. ‖ fil m. recuit. / barbed ~ ‖ Stacheldraht m. ‖ fil m. barbelé *ou* à ronces. / barbed ~ making machine ‖ Stacheldrahtmaschine f. ‖ machine f. à fabriquer le fil de fer barbelé. / bare ~ ‖ nackter *oder* blanker Draht m. ‖ fil m. (métallique) nu. / bell ~ ‖ Klingeldraht m. ‖ fil m à sonnerie. / bimetallic ~ ‖ Doppelmetalldraht m. ‖ fil m. bimétallique. / binding ~ ‖ Bindedraht m. ‖ fil m. à lier *ou* d'amarrage. ‖ bookbinder's ~ ‖ Heftdraht m. ‖ fil m. à relier. / bracing ~ ‖ Spanndraht m. ‖ fil m. tendeur. / braided ~ ‖ Litzendraht m.; Klöppeldraht m. ‖ litzendraht m.; fil m. sous tresse. / branch ~ ‖ Drahtabzweigung f. ‖ fil m. de déviation. / brass ~ ‖ Messingdraht m. ‖ fil m. de laiton *ou* d'archal. / bright ~ ‖ blanker Draht m. ‖ fil m. clair *ou* poli. / bright iron ~ ‖ blanker Eisendraht m. ‖ fil m. de fer clair. / bronze ~ ‖ Bronzedraht m. ‖ fil m. de bronze. / ~ of the cable ‖ Kabelseele f. ‖ âme m. du câble. / ~ carrying current ‖ stromdurchflossene Leitung f. ‖ ligne f. *ou* canalisation f. traversée par un courant. / coarse ~ ‖ Grobdraht m. ‖ écotage m.; fil m. d'écotage. / coated ~ ‖ umsponnener Leitungsdraht m. ‖ fil m. garni; conducteur m. guipé. / conducting ~ ‖ Leitungsdraht m. ‖ fil m. conducteur. / copper ~ ‖ Kupferdraht m. ‖ fil m. de cuivre. / copper-covered steel ~ ‖ verkupferter Stahldraht m. ‖ fil m. en acier cuivré. / coppered ~ ‖ verkupferter Draht m. ‖ fil m. cuivré. / cotton-covered ~ ‖ mit Baumwolle besponnener Draht m. ‖ fil m. guipé de coton. / double cotton-covered ~ ‖ Draht m. mit doppelter Baumwollumspinnung ‖ fil m. à guipage double en coton. / to cover ~ ‖ den Draht m. umspinnen ‖ guiper le fil. / to cover ~ with braid ‖ den Draht m. umklöppeln ‖ tresser le fil. / covered ~ ‖ besponnener *oder* umsponnener Draht m. ‖ fil m. recouvert *ou* guipé. / double-covered ~ ‖ doppelt umsponnener Draht m. ‖ fil m. à guipage double. / drag ~ (Aero) ‖ Schleppdraht m. ‖ câble m. de trainée *ou* de recul. / to draw ~ *see* to wiredraw. / drawn ~ ‖ gezogener Draht m. ‖ fil m. étiré *ou* tréfilé. / electric ~ ‖ Draht m. für elektrische Leitungen ‖ fil m. électrique. / enamelled ~ ‖ Emaildraht n. ‖ fil m. émaillé. / figured ~ *see* shaped ~. / fire-tinned ~ ‖ feuerverzinnter Draht m. ‖ fil m. étamé au feu. / flat ~ ‖ Flachdraht m. ‖ fil m. plat. / flower ~ ‖ Blumendraht m. ‖ fil m. pour fleurs. / galvanized ~ ‖ verzinkter Draht m. ‖ fil m. galvanisé *ou* zingué. / gold-covered ~ ‖ goldplattierter Draht m. ‖ fil m. plaqué d'or. / ground ~ ‖ Erd-

leitung f. ‖ fil m. de masse. / guttapercha-insulated ~ ‖ Guttaperchaader f. ‖ fil m. isolé à la gutta-percha. / hard-drawn ~ ‖ hartgezogener Draht m. ‖ fil m. écroui *ou* étiré à froid. / hard-drawn copper ~ ‖ Hartkupferdraht m. ‖ fil m. de cuivre écroui. / heating ~ ‖ Heizdraht m. ‖ fil m. de chauffage. / heel tip ~ ‖ Stiefeleisendraht m. ‖ fil m. pour fers de bottes. / ~ for indoor installations ‖ Zimmerleitungsdraht m. ‖ fil m. d'appartements. / insulated ~ ‖ isolierter Leitungsdraht m. ‖ fil m. isolé. / interrupted feed ~ ‖ unterbrochene Zuleitung f. ‖ conducteur m. d'amenée interrompu. / iron ~ ‖ Eisendraht m. ‖ fil m. de fer. / isolated ~ *see* insulated ~. / lead ~ ‖ Bleidraht m. ‖ fil m. de plomb. / live ~ *see* ~ carrying current. / ~ for mandibular jaw ‖ Unterkieferbügeldraht m. ‖ fil m. pour arc de prothèse dentaire partielle inférieure. / music ~ *see* ~ string. / needle ~ ‖ Nadeldraht m. ‖ fil m. pour aiguilles. / neutral ~ ‖ Nulleiter m. ‖ fil m. neutre. / ~ for nuts ‖ Mutterndraht m. ‖ fil m. pour écrous. / phosphor-bronze ~ ‖ Phosphorbronzedraht m. ‖ fil m. de bronze phosphoreux. / piano ~ ‖ Klaviersaitendraht m. ‖ fil m. pour corde de piano. / platinum ~ ‖ Platindraht m. ‖ fil m. de platine. / resistance ~ ‖ Widerstandsdraht m. ‖ fil m. de résistance. / rivet ~ ‖ Nietendraht m. ‖ fil m. pour rivets. / rolled ~ ‖ Walzdraht m. ‖ fil m. laminé. / rope ~ ‖ Seildraht m. ‖ fil m. pour câbles. / round ~ ‖ Runddraht m. ‖ fil m. à section circulaire. / rubber-insulated ~ ‖ Gummiader f. ‖ fil m. caoutchouté. / ~ for scrapers ‖ Kratzendraht m. ‖ fil m. pour cardes. / second ~ (Cable) ‖ Meßader f. ‖ fil m. sain. / shaped ~ ‖ Formdraht m.; Profildraht m.; Fassondraht m. ‖ fil m. profilé *ou* façonné. / soft ~ ‖ weicher Draht m. ‖ fil m. doux. / span ~ ‖ Spanndraht m. ‖ fil m. raidisseur *ou* tendeur. / special ~ *see* shaped ~. / split pin ~ ‖ Splintdraht m. ‖ fil m. pour goupilles. / spring ~ ‖ Federdraht m. ‖ fil m. à ressort. / standard ~ ‖ Normaldraht m. ‖ fil m. normal. / steel ~ ‖ Stahldraht m. ‖ fil m. d'acier. / ~ for stitching *see* bookbinder's ~. / to put ~s pl. straight ‖ Telegraphendrähte mpl. richten ‖ redresser les fils mpl. télégraphiques. / stranded ~ ‖ Litze f.; Litzendraht m. ‖ toron m. / stranded aluminium ~ ‖ Aluminiumseil n. ‖ câble m. en aluminium. / telegraph ~ ‖ Telegrafendraht m. ‖ fil m. télégraphique. / telephone ~ ‖ Fernsprecherdraht m.; Telefondraht m. ‖ fil m. téléphonique. / thick ~ *see* coarse ~. / third ~ ‖ Mittelleiter m. ‖ conducteur m. médium. / three ~ (system) ‖ Dreileiter m. ‖ trois conducteurs mpl. / tinned ~ ‖ verzinnter Draht m. ‖ fil m. étamé. / tube ~ ‖ Rohrdraht m. ‖ fil m. à tuyau. / twisted ~ ‖ verseilter Draht m. ‖ fil m. câblé. / varnished ~ ‖ Lackdraht m.; Leitungsdraht m. mit Lackisolierung *oder* Lacküberzug ‖ fil m. verni *ou* gommelaqué. / wrapped ~ ‖ Manteldraht m. ‖ fil m. sous enveloppe.

wire aerial ‖ Drahtantenne f. ‖ aérien m. filiforme. / ~ articles pl. *see* ~ goods. / ~ bar ‖ Drahthebel m. ‖ tige f. en fil métallique. / ~ bar (Electr) ‖ Leitungskupfer n. ‖ cuivre m. pour fil. / ~ basket ‖ Drahtkorb m. ‖ panier m. en fil métallique. / ~ belt ‖ Drahtgurt m. ‖ sangle f. en fil de fer. / ~ bending apparatus ‖ Drahtbiegeapparat m. ‖ appareil m. à plier les fils. / ~ bending machine ‖ Drahtbiegemaschine f. ‖ machine f. à plier le fil; plieuse f. pour fils. / ~ binding pliers pl. ‖ Drahtbindezange f. ‖ pince f. à lier les fils. / ~-braced wing (Aero) ‖ verspannter Flügel m. ‖ aile f. haubanée *ou* maintenue par des haubans. / ~ braiding machine ‖ Drahtflechtmaschine f. ‖ machine f. à tresser le fil de fer. / ~ brazer ‖ Drahtlöter m. ‖ soudeur m. de fils. / ~ brush ‖ Drahtbürste f. ‖ brosse f. métallique; gratte-brosse f. / ~ cable *see also* wire rope ‖ Drahtseil n. ‖ câble m. métallique. / ~ cable of a telegraph ‖ Telegrafenkabel n. ‖ câble m. télégraphique. / ~ cable suspension bridge ‖ Drahtseilbrücke f. ‖ pont m. suspendu en fil de fer. / ~ cage ‖ Drahtkäfig m. ‖ cage f. en fil de fer. / ~ carrier (Electr) ‖ Leitungsträger m. ‖ support m. de conducteur. / ~ chain ‖ Schuppenkette f. ‖ chaîne f. colonne. / ~ clamp ‖ Drahtklemme f. ‖ borne f. *ou* serre-lame m. en fil de fer. / ~ clamp making machine ‖ Drahtklammerherstellungsmaschine f. ‖ machine f. à fabriquer des agrafes en fil. / ~ cloth ‖ Drahtgaze f.; Drahtgewebe n. ‖ toile f. *ou* tissu m. métallique. / ~ cloth woven in brass ‖ Messingdrahtgewebe n. ‖ tissu m. en fil de laiton. / ~ cloth loom ‖ Drahtwebestuhl m. ‖ métier m. à tissu métallique. / ~ coating apparatus ‖ Drahtumspulgerät n. ‖ appareil m. à guiper de fil métallique. / ~ coiling machine ‖ Drahtaufrollmaschine f. ‖ machine f. à enrouler le fil. / contact of ~s pl. ‖ Drahtberührungsstelle f. ‖ contact m. de fils. / copper for ~ ‖ Leitungskupfer n. ‖ cuivre m. pour fil. / ~ core ‖ Drahteinlage f. ‖ armature f. intérieure en fil métallique. / ~ core coil ‖ Drahtkernspule f. ‖ bobine f. à noyau en fil. / silk ~ coverer ‖ Seidendrahtwickler m. ‖ guipeur m. de fil à soie. / ~ covering machine ‖ Drahtumspinnmaschine f. ‖ machine f. à guiper les fils; guipeuse f. à fil. / ~ cramp *see* ~ clamp. / ~ cross section ‖ Leitungsquerschnitt m. ‖ section f. du fil conducteur. / ~ cutter ‖ Drahtzange f. ‖ coupe-fil m. / ~ cutting machine ‖ Drahtabschneidemaschine f. ‖ machine f. à couper le fil de fer. / ~ cylinder bridge ‖ Drahtwalzenbrücke f. ‖ pont m. de cylindre en fil métallique.

wired rolled glass ‖ Drahtglas n. ‖ verre m. ‖ fil. / ~ tyre ‖ mit Stahldraht verstärkter Reifenmantel m. ‖ pneu m. renforcé en fil d'acier.

wire diameter ‖ Drahtstärke f. ‖ épaisseur m. du fil.

wiredraw, to ‖ Draht m. ziehen ‖ tréfiler; étirer le fil.

wiredrawer ‖ Drahtzieher m. ‖ tréfileur m.

wiredrawing ‖ Drahtziehen n. ‖ tréfilerie f. / metal ~ ‖ Metalldrahtziehen n. ‖ tréfilage m. de fils métalliques.

wiredrawing bench ‖ Drahtzug m.; Drahtziehbank f. ‖ banc m. de tréfilerie. / ~ machine ‖ Drahtziehmaschine f. ‖ machine f. de tréfilerie. / ~ plate ‖ Drahtzieheisen n. ‖ filière f. à tréfiler. / ~ roller ‖ Drahtziehwalze f. ‖ cylindre m. du laminoir à fil. / ~ stone ‖ Drahtziehstein m. ‖ pierre f. à tréfiler.

wire dresser ‖ Drahtzurichter m. ‖ dresseur m. de fils de fer. / ~ fence ‖ Drahtzaun m. ‖ clôture f. métallique. / ~ finder ‖ Leitungssonde f.; Kabelsonde f. ‖ explorateur m. de fil. / ~ foot mat ‖ Drahtfußmatte f. ‖ décrottoir m. en fil métallique. / ~ frame (Opt) ‖ Drahtfassung f. ‖ monture f. à fil. / ~ frame for lamp shades ‖ Drahtgestell n. für Lampenschirme ‖ monture f. en fil de fer pour abat-jours de lampes. / ~ galvanizing plant ‖ Drahtverzinkerei f. ‖ installation f. de zinguage des fils. / ~ gauge ‖ Drahtlehre f. ‖ jauge f. à fils métalliques. / ~ gauze ‖ Drahtgaze f.; Drahtgewebe n. ‖ gaze f. *ou* toile f. métallique. / ~ gauze brush ‖ Drahtgewebebürste f. ‖ balai m. en toile métallique. / ~ gauze loom ‖ Drahtwebestuhl m. ‖ métier m. à tissu métallique. / ~ glass ‖ Drahtglas n. ‖ verre m. armé. / ~ glass building stone ‖ bewehrter Glasbaustein m. ‖ brique f. *ou* carreau m. en verre armé. / ~ goods pl. ‖ Drahtwaren fpl. ‖ articles mpl. en fil métallique. / ~ goods pl. for housekeeping and kitchen ‖ Drahtwaren fpl. für Küche und Haus ‖ articles mpl. de ménage et de cuisine en fil métallique. / ~ grate *see* ~ lattice. / ~ guard ‖ Drahtschutzkorb m. ‖ panier m. protecteur. / ~ hardening plant ‖ Drahthärterei f. ‖ installation f. de trempe des fils. / ~ iron train ‖ Drahteisenstraße f. ‖ laminoir m. *ou* train m. à fils de fer. / ~ joint ‖ Drahtverbindungsstelle f. ‖ joint m. *ou* liaison f. des fils. / ~ jointing sleeve ‖ Drahtverbindungshülse f. ‖ manchon m. de fil. / ~ kiln floor ‖ Drahthorde f. ‖ plateau m. de touraille en toile métallique. / ~ kiln hurdle *see* ~ kiln floor. / ~ lacquering machine ‖ Drahtlackiermaschine f. ‖ machine f. à laquer le fil. / ~ lattice ‖ Drahtgitter n. ‖ grillage m. métallique. / iron ~ lattice ‖ eisernes Drahtgitter n. ‖ grillage m. en fil de fer. / ~ lattice maker ‖ Drahtgitterflechter m. ‖ grillageur m. / ~ lead fused into the glass ‖ in Glas eingeschmolzener Zuführungsdraht m. ‖ fil m. d'amenée fondu dans le verre.

wireless ‖ drahtlos ‖ sans fil m. / ~ aircraft service ‖ Flugfunkdienst m. ‖ service m. radioélectrique aérien. / ~ apparatus ‖ Funkgerät n. ‖ appareil m. radiotélégraphique. / giant loudspeaker of a ~ broadcasting station for music ‖ Großlautsprecher m. einer Musikübertragungsanlage ‖ haut-parleur m. géant d'une installation de transmission de musique sans fil. / ~ compass ‖ Radiokompaß m.; Richtungsfinder m. ‖ radioboussole f.; boussole f. sans fil. / ~ direction finder ‖ Richtungsanzeiger m. für drahtlose Telegrafie ‖ indicateur m. de direction de sans fil. / ~ direction finding ‖ drahtlose Richtungsbestimmung f. ‖ radiogoniométrie f. / ~ equipment ‖ Funkentelegrafieausrüstung f. ‖ équipement m. de télégraphie sans fil. / ~ high-speed Morse reception ‖ drahtloser Schnellmorseempfang m. ‖ réception f. rapide de signes Morse radiodiffusés. / ~ installation ‖ Funkentelegrafieanlage f. ‖ installation f. radiotélégraphique. / ~ message ‖ Radiogramm n. ‖ radio m.; sans-fil m. / ~ meteorological service ‖ Funkwetterdienst m. ‖ service m. météorologique

de télégraphie sans fil. / ~ officer ‖ Funkeroffizier m. ‖ officier m. d'antenne. / ~ operator ‖ Funker m. ‖ opérateur m. de télégraphie sans fil. / ~ picture telegraph ‖ Bildrundfunkempfänger m. ‖ téléphotographie f. sans fil. / ~ picture telegraphy ‖ Bildfunk m. ‖ téléphotographie f. sans fil. / ~ position finding ‖ drahtlose Ortsbestimmung f. ‖ détermination f. de la position sans fil. / ~ reception ‖ drahtloser Empfang m. ‖ réception f. sans fil. / directional ~ reception ‖ gerichteter drahtloser Empfang m. ‖ réception f. dirigée radiotélégraphique. / ~ service with ship stations ‖ Seefunkdienst m. ‖ service m. radioélectrique maritime. / decree of protection of ~ service ‖ Funkverordnung f. ‖ décret m. sur la radiotélégraphie. / ~ station ‖ Funkstation f. ‖ poste f. de télégraphie sans fil. / ~ telegram ‖ drahtloses Telegramm n. ‖ télégramme m. sans fil. / ~ telegraphy ‖ Funkentelegrafie f.; Radiotelegrafie f.; drahtlose Telegrafie ‖ télégraphie f. sans fil; radiotélégraphie f. / directed ~ telegraphy ‖ gerichtete drahtlose Telegrafie f. ‖ radiotélégraphie f. dirigée. / syntonized ~ telegraphy ‖ abgestimmte drahtlose Telegrafie f. ‖ télégraphie f. sans fil syntonisée. / apparatus for ~ telegraphy ‖ Apparat m. für drahtlose Telegrafie; Radioapparat m. ‖ appareil m. de télégraphie sans fil. / lamp for ~ telegraphy ‖ Lampe f. für drahtlose Telegrafie ‖ lampe f. pour télégraphie sans fil. / monopoly of the State relating to the ~ telegraphy ‖ Funkhoheitsrecht n. ‖ monopole m. de l'État en matière de télégraphie sans fil. / portable station for ~ telegraphy ‖ fahrbare Station f. für drahtlose Telegrafie ‖ poste m. radiotélégraphique roulant. / ~ telephone traffic ‖ drahtloser Fernsprechverkehr m. ‖ trafic m. téléphonique sans fil. / ~ telephony ‖ drahtlose Telefonie f. ‖ téléphonie f. sans fil. / ~ tower ‖ Funkturm m. ‖ tour m. de radio-diffusion. / ~ weather service ‖ funkentelegrafischer Wetterdienst m. ‖ service m. radiotélégraphique de prévision du temps.

wireless see wireless telegraphy.

wire, single ~ line ‖ Eindrahtleitung f. ‖ ligne f. à fil unique. / ~ loop ‖ Drahtschlinge f. ‖ agrafe f. en fil de fer; boucle f. de fil métallique.

wireman ‖ Elektriker m.; Installatör m. ‖ installateur m.; monteur m. électricien.

wireman's tent ‖ Kabellöterzelt n. ‖ tente f. de soudeur.

wire marking ‖ Leitungsbezeichnung f. ‖ dénomination f. du fil. / ~ mat ‖ Fußmatte f. aus Draht ‖ natte f. de toile métallique. / ~ mattress ‖ Drahtmatratze f. ‖ matelas m. en fils métalliques. / ~ mill ‖ Drahtwalzwerk n. ‖ laminoir m. à fils métalliques.

wire nail ‖ Drahtstift m. ‖ clou m.; pointe f.; pointe f. de Paris. / edged ~ ‖ kantiger Drahtstift m. ‖ pointe f. carrée; clou m. carré. / ~ with deep countersunk head ‖ Drahtstift m. mit tief versenktem Kopf ‖ pointe f. ou clou m. à tête noyée et profonde. / ~ with flat countersunk head ‖ Drahtstift m. mit flach versenktem Kopf ‖ pointe f. ou clou m. à tête plate noyée. / ~ with semi-circular head ‖ Drahtstift m. mit Halbrundkopf ‖ pointe f. ou clou

m. à tête demi-ronde. / ~ with upset head ‖ Drahtstift m. mit gestauchtem Kopf ‖ pointe f. ou clou m. à tête aplatie. / ~ without head ‖ Drahtstift m. ohne Kopf ‖ pointe f. ou clou m. sans tête.

wire nail machine ‖ Drahtstiftmaschine f. ‖ machine f. à fabriquer les pointes.

wire net fencing ‖ Drahteinfriedigung f. ‖ clôture f. en fil métallique.

wire netting ‖ Drahtgeflecht n. ‖ treillis m. ou toile f. métallique; tissu m. à mailles métalliques. / ~ for arc-lamp globes ‖ Drahtgeflecht n. für Bogenlampenglocken ‖ filet m. protecteur de globes de lampe à arc. / ~ for trees ‖ Schutzgitter n. für Bäume ‖ armure f. pour arbres.

wire netting automatic machine ‖ Drahtgeflechtautomat m. ‖ automate m. à tresser le fil métallique. / crimping machine for ~ ‖ Krippmaschine f. für die Drahtweberei ‖ machine à plier le fil gros de tissage. / ~ dyeing machine ‖ Drahtgewebefärbmaschine f. ‖ machine f. à peindre les toiles métalliques. / ~ machine ‖ Drahtflechtmaschine f. ‖ machine f. à faire les grillages.

wire noose for corked bottles ‖ Drahtschleife f. für verkorkte Flaschen ‖ boucle f. de fil pour bouteilles à bouchon. / pair of ~s (Tel) ‖ Doppelleitung f. ‖ ligne f. bifilaire. / ~ pickler ‖ Drahtbeizer m. ‖ décapeur m. de fils. / ~ pin ‖ Splint m. ‖ goupille f.

wire polishing device for spring wires ‖ Federdrahtpoliereinrichtung f. ‖ installation f. à polir le fil à ressorts. / ~ for wire strings ‖ Musikdrahtpoliereinrichtung f. ‖ installation f. à polir le fil pour cordes à musique.

wire press ‖ Drahtpresse f. ‖ presse f. à tréfiler les fils. / ~ puller see ~ pulling tank. / ~ pulling tank ‖ Hindernisräumer m. ‖ char m. chasse-fil. / ~ rail bond ‖ Schienenverbinder m. aus Draht ‖ joint m. à fils pour rails. / ~ reel ‖ Drahthaspel f. ‖ dérouleuse f. / regulation of the ~s ‖ Einstellung f. der Drähte ‖ réglage m. des fils. / ~ riddle (Sand moulding) ‖ Luftspieß m.; Spieß m. ‖ aiguille f.; dégorgeoir m. / ~ roll ‖ Drahtwalze f. ‖ cylindre m. de laminoir à fil. / ~ rolling mill see ~ mill.

wire rope ‖ Drahtseil n. ‖ câble m. métallique. / ~ block ‖ Drahtseilkloben m. ‖ moufle f. à câble. / ~ grease ‖ Drahtseilschmiere f. ‖ graisse f. pour câbles métalliques. / ~ making ‖ Drahtseilerei f. ‖ fabrication f. des câbles métalliques. / ~ making machine ‖ Drahtseilereimaschine f. ‖ machine f. à câbler le fil. / ~ railway ‖ Drahtseilbahn f. ‖ funiculaire f.; chemin m. de fer funiculaire.

wire sharpening machine ‖ Drahtanspitzmaschine f. ‖ machine f. à appointer (à froid) le fil. / rolling mill for sharpening ~ ‖ Anspitzwalzwerk n. für Draht ‖ laminoir m. à appointer des fils. / ~ shears pl. ‖ Drahtschere f. ‖ cisaille f. à fil métallique. / ~ sieve ‖ Drahtsieb n. ‖ tamis m. en toile métallique. / ~ spark catcher ‖ Funkenfänger m. aus Draht ‖ pare-étincelles m. en toile métallique. / ~ spiral ‖ Drahtspirale f. ‖ cannelure f. / ~ spoke ‖ Drahtspeiche f. ‖ rayon m. en fil d'acier. / ~ spoke wheel ‖ Drahtspeichenrad n. ‖ roue f. à rayons en fil métallique. / ~

spool ‖ Drahthaspel f. ‖ dévidoir m. à fil métallique. / ~ spooling machine ‖ Drahtspulmaschine f. ‖ machine f. à bobiner le fil métallique. / ~ stay ‖ Drahtanker m. ‖ hauban m. en fil. / ~ stitch ‖ Heftklammer f. ‖ agrafe f. métallique. / ~-stitched quires pl. ‖ drahtgehefteter Bogensatz m. ‖ ensemble m. de feuilles piqué au fil de fer. / ~ stitching machine ‖ Drahtheftmaschine f. ‖ brocheuse f. à fil métallique. / ~ straightening machine ‖ Drahtrichtmaschine f. ‖ machine f. à dresser le fil métallique. / ~ strainer ‖ Drahtspanner m. ‖ tendeur m. pour fil de fer. / ~ stranding machine ‖ Drahtlitzenmaschine f. ‖ toronneuse f. à fils métalliques. / ~ stretcher see ~ strainer. / ~ string ‖ Drahtsaite f. ‖ corde f. métallique. / ~ tack see wire nail. / ~ telephonograph ‖ Drahttelefonograf m. ‖ téléphonographe m. à fil. / ~ tension testing ‖ Drahtspannungsprüfung f. ‖ vérification f. de la tension des fils métalliques. / ~ thread guide (Textile) ‖ Sauschwanz m. ‖ guide-fil m. métallique. / ~ tinning plant ‖ Drahtverzinnerei f. ‖ installation f. d'étamage des fils. / ~ tissue belt ‖ Drahtgurtband n. ‖ bande f. de tissu métallique. / ~ train ‖ Drahtstraße f. ‖ laminoir m. ou train m. à fil. / ~ trellis see ~ lattice. / ~ twisting apparatus ‖ Drahttorsionsapparat m.; Drahtverwindungsapparat m. ‖ appareil m. à tordre les fils. / ~ washery ‖ Drahtwäsche f. ‖ laveur m. à fil. / ~ weaving loom ‖ Drahtwebstuhl m. ‖ métier m. à tisser le fil de fer. / ~ weaving mill ‖ Drahtweberei f. ‖ atelier m. de tissage de fil métallique. / ~ wholesale dealers pl. ‖ Drahtgroßhandlung f. ‖ commerce m. de fils métalliques en gros / ~ winch ‖ Drahtwinde f. ‖ cric m. tendeur. / ~ winding machine ‖ Drahtwickelmaschine f. ‖ enrouleuse f. à fil de fer.

wirework see also wire netting ‖ Drahtgeflecht n. ‖ treillis m. en fil de fer. / plaited ~ for enclosures ‖ Drahtgeflecht n. zur Einfriedigung ‖ treillis m. en fil métallique pour clôtures.

wire working machine ‖ Drahtbearbeitungsmaschine f. ‖ machine f. à travailler les fils métalliques.

wiring (Aero) ‖ Verspannung f. ‖ haubanage m. / ~ (El line) ‖ Herstellen n. der Drahtleitung; Leitungsverlegung f. ‖ pose f. de fils; canalisation f. / concealed ~ ‖ Verlegung f. von Leitungen unter Putz ‖ canalisation f. dérobée ou cachée ou masquée. / ~ in conduit ‖ Verlegung f. des Leitungsdrahtes in Rohr ‖ canalisation f. sous tubes isolateurs. / exposed ~ ‖ sichtbare Verlegung f. von Leitungen ‖ pose f. des fils à découvert. / home ~ ‖ Hausinstallation f. ‖ canalisation f. des immeubles. / temporary ~ ‖ provisorische Verlegung f. von Leitungen ‖ canalisation f. provisoire.

wiring diagram ‖ Schaltschema n. ‖ schéma m. des connexions. / ~ machine (Electr) ‖ Drahteinlegemaschine f. ‖ machine f. à insérer le fil de fer dans les bords.

wish ‖ Anliegen n. ‖ demande f.

wisp of straw for tying sheaves ‖ Garbenband n. ‖ lien m. à gerbes.

witch meal ‖ Bärlappsamen m. ‖ soufre m. végétal.

withdraw, to ‖ herausziehen; zurückziehen ‖ retirer. / ~ a bridge ‖ eine Brücke f. ab-

brechen ‖ enlever *ou* replier un pont. / ~ horizontally ‖ wagerecht herausziehen ‖ enlever horizontalement. / ~ the usufruct ‖ die Nutzung f. entziehen ‖ retirer la faculté de jouissance f.

withdrawal (Money) ‖ Entnahme f. ‖ remboursement m. / ~ (Retirement) ‖ Austritt m. ‖ retraite f. / ~ roller ‖ Ausgangswalze f. ‖ rouleau m. d'appui.

withdrawing screw ‖ Abziehschraube f. ‖ vis f. d'extraction *ou* de décalage.

wither, to (Brew) ‖ abschwelken ‖ faner; sécher.

withered (Brew) ‖ geschwelkt ‖ fané.

withering (Brew) ‖ Abschwelken ‖ fanage m.; séchage m. / ~ of the trunk (Wood) ‖ Absterben n. des Stammes ‖ dépérissement m. du tronc.

witherite ‖ Witherit m. ‖ withérithe f.

withstand, to ~ a rain of machine-gun bullets ‖ widerstandsfähig gegen Maschinengewehrgarben fpl. sein ‖ résister aux gerbes fpl. des balles de mitrailleuse.

witness ‖ Zeuge m. ‖ témoin m. / to bear ~ ‖ bezeugen ‖ attester; témoigner. / ~ for the defence ‖ Entlastungszeuge m. ‖ témoin m. à décharge. / eye ~ ‖ Augenzeuge m. ‖ témoin m. oculaire.

wolfram ‖ Wolfram n. ‖ tungstène m.; wolfram m.

wolframic acid ‖ Wolframsäure f. ‖ acide m. tungstique *ou* wolframique.

wolfram mine ‖ Wolframgrube f. ‖ mine f. de wolfram. / ~ steel ‖ Wolframstahl m. ‖ acier m. au tungstène.

woman worker ‖ Arbeiterin f. ‖ ouvrière f. / ~'s hair ‖ Frauenhaar n. ‖ cheveu m. de femme. / ~'s hose ‖ Frauenstrumpf m. ‖ bas m. de femme.

wonder candle ‖ Wunderkerze f. ‖ bougie f. merveilleuse.

wood *see also* timber ‖ Holz n. ‖ bois m. / air-dried ~ ‖ lufttrockenes Holz n. ‖ bois m. séché à l'air. / artificial ~ ‖ Kunstholz n. ‖ bois m. artificiel. / barked ~ see disbarked ~. / ~ bent by steam ‖ in Dampf gebogenes Holz n. ‖ bois m. cintré à la vapeur. / birch ~ ‖ Birkenholz n. ‖ bois m. de bouleau. / ~ for brushes ‖ Bürstenholz n. ‖ bois m. à brosses. / ~ for buildings ‖ Bauholz n. ‖ bois m. de constructions. / burnt ~ ‖ Brandholz n. ‖ bois m. arsin. / carved ~ ‖ Holzschnitzerei f.; Holzschnitzwerk n. ‖ bois m. sculpté. / the ~ is casting ‖ das Holz n. wirft sich ‖ le bois se déjette *ou* se gauchit. / ~ in chips for clarifying liquids ‖ Holz n. in Spänen zum Klären von Flüssigkeiten ‖ bois m. en copeaux pour la clarification de liquides. / cleavable ~ ‖ spaltbares Holz n. ‖ bois m. fendable. / close-grained ~ ‖ engfaseriges Holz n. ‖ bois m. compact *ou* à fibre serrée. / colty ~ ‖ kernschäliges *oder* splintrissiges Holz n. ‖ bois m. roulé *ou* crevassé circulairement. / coniferous ~ ‖ Nadelholz n. ‖ bois m. conifère. / cooper's ~ ‖ Faßholz n. ‖ bois m. à tonneaux. / the ~ cracks ‖ das Holz n. reißt auf ‖ le bois crevasse *ou* se fend. / ~ with crooked fibres ‖ widerwüchsiges Holz n. ‖ bois m. tranché *ou* versé. / cross-cut ~ ‖ Stirnholz n.; Querholz n.; Hirnholz n. ‖ bois m. de bout. / cross-grained ~ ‖ verwachsenes Holz n. ‖ bois m. rebours. / curled ~ ‖ Maserholz n. ‖ bois m. madré *ou* tapiré; madrure f. / ~ for the manufacturing of curtains ‖

Holz n. für die Herstellung von Rollladen ‖ bois m. pour la fabrication des stores. / damp ~ ‖ feuchtes Holz n. ‖ bois m. humide. / dead ~ ‖ abgestorbenes Holz n. ‖ bois m. mort. / decayed ~ ‖ angefaultes Holz n. ‖ bois m. piqué. / ~ free from defects ‖ Holz n. ohne Fehler ‖ bois m. sans défauts. / defective ~ ‖ fehlerhaftes Holz n. ‖ bois m. défectueux. / disbarked ~ ‖ geschältes Holz n. ‖ bois m. écorcé. / drifted ~ ‖ Flößholz n. ‖ bois m. flotté. / dry-rotten ~ ‖ brandiges Holz n. ‖ bois m. échauffé *ou* pouilleux *ou* pourri. / dyer's ~ ‖ Farbholz n.; Färberholz n. ‖ bois m. colorant *ou* de teinture. / end-grained ~ see cross-cut ~. / fine ~ ‖ feines Holz n. ‖ bois m. fin. / fir ~ ‖ Tannenholz n.; Fichtenholz n. ‖ bois m. de sapin. / fire ~ ‖ Brennholz n. ‖ bois m. à brûler. / fire-proofed ~ ‖ feuerbeständig imprägniertes Holz n. ‖ bois m. ignifuge *ou* à l'épreuve du feu. / ~ having fissures ‖ eisklüftiges Holz n. ‖ bois m. gélif *ou* gélivé. / ~ free from flaws ‖ Holz n. ohne Risse ‖ bois m. sans roulures. / ~ with flaws on one side ‖ einwüchsiges Holz n. ‖ bois m. cautiban *ou* cantiban. / floated ~ see drifted ~. / ~ for fuel see fire ~. / fully stocked ~ ‖ geschlossener Waldbestand m. ‖ partie f. de bois en massif. / ~ glued together ‖ zusammengeleimtes Holz n. ‖ bois m. collé. / ~ granted free of charge ‖ Deputatholz n. ‖ bois m. délivré gratuitement. / green ~ ‖ Grünholz n. ‖ bois m. vert *ou* vif. / grooved ~ ‖ genutetes Holz n. ‖ bois m. rainé. / ground ~ ‖ gemahlenes Holz n. ‖ bois m. moulu. / half-round ~ ‖ Halbholz n. ‖ demi-bois m. / hard ~ ‖ Hartholz n. ‖ bois m. dur. / heart ~ ‖ Kernholz n. ‖ bois m. pris du cœur. / ~ impregnated with creosote ‖ mit Kreosot imprägniertes Holz n. ‖ bois m. imbibé de créosote. / ~ for heating see fire ~. / ~ for hoops ‖ Reifenholz n. ‖ bois m. feuillard *ou* de cerclage. / ~ for inlaid work ‖ Holz n. für Einlegearbeiten ‖ bois m. de marqueterie. / joinery ~ ‖ Tischlerholz n. ‖ bois m. de menuiserie. / kiln-dried ~ ‖ Darrholz n. ‖ bois m. séché au four. / knaggy ~ ‖ knorriges *oder* knotiges Holz n. ‖ bois malandreux *ou* rebours. / knotty ~ ‖ ästiges Holz n. ‖ bois m. noueux. / ~ without knots ‖ astfreies Holz n. ‖ bois m. sans nœuds *ou* non noueux. / live ~ ‖ lebendes Holz n. ‖ bois m. vivant. / long-cut ~ ‖ Langholz n. ‖ bois m. de long. / made entirely of ~ ‖ ganz aus Holz n. hergestellt ‖ fait entièrement en bois m. / ~ for manufacturing matches ‖ Holz n. für die Herstellung von Zündhölzern ‖ bois m. pour la fabrication des allumettes. / medicinal ~ ‖ Arzneiholz n. ‖ bois m. médicinal. / ~ for mines ‖ Grubenholz n. ‖ bois m. de mines. / ~ for musical instruments ‖ Klangholz n.; Resonanzholz n. ‖ bois m. de résonnance. / natural ~ ‖ Naturholz n. ‖ bois m. naturel. / oak ~ ‖ Eichenholz n. ‖ bois m. de chêne; chêne m. / of ~ ‖ hölzern ‖ de bois. / overseasoned ~ ‖ überständiges Holz n. ‖ bois m. sur le retour. / ~ paid for according to the cubic measurement ‖ Holz n. nach Kubikmaß bezahlt ‖ bois m. payé au cube. / palisander ~ ‖ Palisander m.; Palisanderholz n. ‖ palissandre m.; jacaranda m.

/ ~ for paper manufacture ‖ Papierholz n. ‖ bois m. pour la fabrication de papier. / planed ~ for cabinet-making ‖ gehobeltes Kunsttischlereiholz n. ‖ bois m. d'ébénisterie raboté. / the ~ gets pricked by worms ‖ das Holz n. wird wurmstichig ‖ le bois se pique. / to raft ~ ‖ Holz n. flößen ‖ flotter du bois. / raw ~ ‖ rohes Holz n. ‖ bois m. brut. / refuse ~ ‖ Holzabfälle mpl. ‖ déchets mpl. de bois. / resinous ~ ‖ harziges Holz n. ‖ bois m. résineux. / the ~ rifts ‖ das Holz n. reißt auf ‖ le bois crevasse *ou* se fend. / root-rot ~ ‖ wurzelfaules Holz n. ‖ bois m. charmé. / rotten ~ ‖ krankes *oder* faules Holz n. ‖ bois m. détérioré. / ~ in round ‖ unbehauenes Holz n. ‖ bois m. en grume. / ~ in the sap ‖ Holz n. im Saft ‖ bois m. en sève. / sappy ~ ‖ saftreiches Holz n. ‖ bois m. riche en sève. / scented ~ ‖ wohlriechendes Holz n. ‖ bois m. odorant. / to season ~ ‖ Holz n. austrocknen ‖ dessécher le bois. / seasoned ~ ‖ trockenes Holz n. ‖ bois m. sec. / ~ with short fibre ‖ kurzfaseriges Holz n. ‖ bois m. à fibre courte. / silicified ~ ‖ Holzstein m. ‖ silex m. corné. / ~ with small annual rings ‖ engringiges Holz n. ‖ bois m. aux cernes étroits. / soft ~ ‖ Weichholz n. ‖ bois m. tendre. / sound ~ ‖ gesundes Holz n. ‖ bois m. sain. / speckled ~ see curled ~. / stacked ~ ‖ Schichtholz n. ‖ bois m. empilé. / standing ~ ‖ Holz n. auf dem Stamm geschnitten ‖ bois m. sur pied. / teak ~ ‖ Teakholz n. ‖ bois m. de teck. / tender ~ see soft ~. / tough ~ ‖ zähes Holz n. ‖ bois m. tenace. / upshot ~ ‖ hochgeschoßtes Holz n. ‖ bois m. de haut revenu. / veined ~ ‖ geadertes Holz n.; Aderholz n. ‖ bois m. veiné *ou* marbré. / the ~ is warping see the ~ is casting. / waste ~ see refuse ~. / wavy-fibred ~ ‖ verwundenes Holz n. ‖ bois m. à fibre contournée. / ~ for wheelwrights ‖ Holz n. für Stellmacher ‖ bois m. de charronnage. / the ~ works ‖ das Holz arbeitet ‖ le bois joue. / worm-eaten ~ ‖ wurmstichiges Holz n. ‖ bois m. mouliné *ou* vermoulu. / yellow ~ ‖ Gelbholz n. ‖ bois m. jaune.

wood... *see also* wooden...

wood acid ‖ Holzessig m. ‖ acide m. pyroligneux; vinaigre m. de bois.

wood alcohol ‖ Holzgeist m. ‖ alcool m. méthylique. / ~ acetone ‖ Holzgeistazeton n. ‖ acétone m. d'alcool méthylique. / ~ oil ‖ Holzgeistöl n. ‖ huile f. d'alcool méthylique.

wood ashes pl. ‖ Holzasche f. ‖ cendre f. de bois. / ~ back ‖ Holzlehne f. ‖ dossier m. en bois. / ~ bailiff ‖ Holzschlagaufseher m. ‖ garde-vente m. des bois / ~ barn ‖ Holzschuppen m. ‖ hangar m. à bois. / ~ base ‖ Holzfuß m. ‖ pied m. de bois. / ~ beetle ‖ Bockkäfer m. ‖ capricorne m. / ~ bender ‖ Holzbieger m. ‖ courbeur m. de bois. / ~ bending machine ‖ Holzbiegemaschine f. ‖ machine f. à courber le bois; cintreuse f. pour bois. / ~ bending plant ‖ Holzbiegerei f. ‖ fabrique f. de bois courbés. / ~ board ‖ Holzpappe f. ‖ carton m. en pâte de bois. / ~ bobbin ‖ Holzspule f. ‖ bobine f. de bois. / ~ border ‖ Holzleiste f. ‖ bordure f. en bois. / ~ bow ‖ Holzbügel m. ‖ archet m. de bois. / ~ branding apparatus ‖ Holzbrandgerät n. ‖ appareil m. de pyrogravure.

wood building ‖ Holzbau m. ‖ construction f. en bois. / portable ~ ‖ zerlegbares Holzhaus n. ‖ maison f. de bois démontable.

wood building blocks pl. ‖ Holzbaukasten m. ‖ boîte f. de construction en bois.

wood-built aeroplane ‖ Holzflugzeug n. ‖ avion m. en bois.

wood bundle ‖ Reisigbündel n.; Holzbündel n. ‖ bourrée f.; fagot m.

woodbury print ‖ Woodburydruck m. ‖ photoglyptie f.; photoplastographie f.

wood button ‖ Holzknopf m. ‖ bouton m. en bois. / ~ **card** ‖ Holzkarte f. ‖ carton m. en bois. / ~ **cart** ‖ Holzwagen m. ‖ fardier m. / ~ **carver** ‖ Holzschnitzer m. ‖ sculpteur m. sur bois; xylographe m. / ~ **carving** ‖ Holzbildhauerei f.; Holzschnitzkunst f. ‖ sculpture f. en bois. / ~ **cellulose** see wood pulp. / ~ **cement** ‖ Holzzement m. ‖ ciment m. en pâte de bois. / ~ **cement roof** ‖ Holzzementdach n. ‖ toit m. en ciment à pâte de bois. / ~ **charring plant** ‖ Holzverkohlungsanlage f. ‖ installation f. à carboniser le bois. / ~ **chopper** ‖ Holzhacker m. ‖ fendeur m. de bois / ~ **cleaver** ‖ Keil m. zum Holzspalten ‖ ébuard m. / ~ **cleaver** (Person) see ~ **chopper.** / horse collar of ~ ‖ Holzkumt m. ‖ collier m. de cheval en bois. / ~ **colour** ‖ Holzfarbe f. ‖ couleur f. pour le bois. / ~-**coloured** ‖ holzfarbig ‖ de couleur f. de bois. / ~ **colouring** ‖ Holzfärben n. ‖ coloration f. du bois. / ~ **comb** ‖ Holzkamm m. ‖ peigne m. en bois. / ~ **conduit** ‖ Holzkanal m. ‖ canal m. en bois. / ~ **conveying plant** ‖ Holzförderanlage f. ‖ installation f. à transporter les bois. / core of ~ ‖ Kern m. aus Holz ‖ âme f. en bois. / ~ **cover of the mould** ‖ Mantelholz n. ‖ bois m. du moule. / ~ **covering** ‖ Holzbelag m. ‖ revêtement m. en bois. / ~ **crusher** ‖ Holzmehlmaschine f. ‖ machine f. pour la fabrication de la farine de bois. / cultivation of ~ ‖ Holzpflanzung f. ‖ culture f. de bois. / ~ **cut** see ~ **engraving.** / ~ **cutter** ‖ Holzfäller m. ‖ bûcheron m. / ~ **cutter** (Engr) see ~ **engraver.**

woodcutter's axe ‖ Holzfälleraxt f. ‖ hache f. de bûcheron. / ~ **bill** ‖ Holzhauerhippe f. ‖ serpe f. de bûcheron.

wood cutting (Engr) see wood engraving. / ~ (Forest) ‖ Holzfällen n. ‖ abattage f. du bois. / ~ (Sawmill) ‖ Holzschneiden n. ‖ sciage m. de bois.

wood dealer ‖ Holzhändler m. ‖ marchand m. de bois. / ~ **dipper** ‖ Holzfärber m. ‖ trempeur m. de bois. / ~ **distillation** ‖ Holzdestillation f. ‖ distillation f. du bois. / ~ **distiller** ‖ Holzdestillatör m. ‖ distillateur m. de bois. / ~ **distilling plant** ‖ Holzdestillationsanlage f. ‖ installation f. de distillation du bois. / ~ **drilling machine** ‖ Holzbohrmaschine f. ‖ perceuse f. pour bois.

wood drying ‖ Holztrocknung f. ‖ séchage m. ou dessiccation f. des bois. / ~ **plant** ‖ Holztrocknungsanlage f. ‖ installation f. à sécher le bois.

wood dust ‖ Holzstaub m.; Holzmehl n. ‖ poudre f. de bois. / ~ **machine for** ~ **factories** ‖ Maschine f. für Holzmehlfabriken ‖ machine f. pour fabriques de sciure de bois. / ~ **making machine** see wood crusher.

wood dyeing ‖ Holzbeizen n. ‖ teinture f. du bois.

wooded ‖ bewaldet ‖ boisé.

wooden ‖ hölzern ‖ de bois m. / ~ ... see also wood ...

wooden articles pl. ‖ Holzwaren fpl. ‖ articles mpl. en bois. / ~ **for house keeping and kitchen** ‖ Holzwaren fpl. für Küche und Haus ‖ articles mpl. en bois de ménage et de cuisine.

wooden backed brushes pl. ‖ in Holz gebundene Bürstenbinderwaren fpl. ‖ brosserie f. montée sur bois. / ~ **bead** ‖ Holzperle f. ‖ perle f. en bois. / ~ **beam** ‖ Holzbalken m. ‖ poutre f. en bois. / ~ **board** ‖ Holzbrett n. ‖ planche f. de bois. / ~ **bobbin** ‖ Holzspule f. ‖ bobine f. ou bobinot m. en bois. / ~ **box** ‖ Holzschachtel f.; Holzkiste f. ‖ boîte f. ou caisse f. en bois. / ~ **broom** ‖ Strauchbesen m. ‖ balai m. de brindilles de bois. / ~ **buffer** ‖ Holzpuffer m. ‖ tampon m. en bois. / ~ **building** ‖ Holzbau m. ‖ construction f. en bois. / ~ **bung** ‖ Holzspund m. ‖ bonde f. en bois. / ~ **case** ‖ Holzkasten m. ‖ coffre m. en bois. / ~ **dome of an observatory** ‖ Holzkuppel f. einer Sternwarte ‖ coupole f. en bois d'un observatoire. / ~ **dowel** ‖ Holzdübel m. ‖ cheville f. en bois. / ~ **fork** ‖ Holzgabel f. ‖ fourche f. en bois. - ~ **furniture for children** ‖ Kindermöbel npl. aus Holz ‖ meubles mpl. en bois pour enfants. / ~ **greenhouse** ‖ hölzernes Gewächshaus n. ‖ serre f. en bois. / ~ **hammer** ‖ Holzhammer m. ‖ maillet m. / ~ **house** ‖ Holzhaus n. ‖ maison f. en bois. / ~ **lattice work making machine** ‖ Holzgittermaschine f. ‖ machine f. à fabriquer des grillages en bois. / ~ **lock** ‖ Holzverschluß m. ‖ cage f. en bois. / ~ **model** ‖ Holzmodell n. ‖ modèle m. en bois. / ~ **mould** ‖ Holzform f. ‖ moule m. en bois. / ~ **pail** ‖ Holzeimer m. ‖ seau m. en bois. / ~ **partition** ‖ Verschlag m. ‖ cloison f. de charpente. / ~ **pedestal lamp** ‖ Ständerlampe f. aus Holz ‖ lampe f. de parquet en bois. / ~ **pegs making machine** ‖ Holzstiftmaschine f. ‖ machine f. à faire des goupilles en bois. / ~ **pillar** ‖ Holzträger m. ‖ pilier m. en bois. / ~ **pin** see ~ **dowel.** / ~ **pipe** ‖ Holzpfeife f. ‖ pipe f. en bois. / ~ **pole** ‖ Holzstange f. ‖ poteau m. en bois. / ~ **revolving shutter** see wood window blind. / ~ **roof** ‖ Holzdach n. ‖ toit m. en bois. / ~ **screw** ‖ Holzschraube f.; hölzerne Schraube f. ‖ vis f. en bois. / ~ **ship** ‖ Holzschiff n. ‖ bateau m. en bois. / **building of** ~ **ships** ‖ Holzschiffbau m.; Bootsbauerei f. ‖ construction des bateaux en bois. / ~ **shoe** ‖ Holzschuh m. ‖ sabot m. / **maker of** ~ **shoes** ‖ Holzschuhmacher m. ‖ sabotier m. / **machine for making** ~ **shoes** ‖ Holzschuhmaschine f. ‖ machine f. à faire les sabots. / ~ **shovel** ‖ Holzschaufel f. ‖ pelle f. en bois. / ~ **sleeper track** ‖ Holzschwellenoberbau m. ‖ superstructure f. sur traverses en bois. / ~ **sole** ‖ Holzsohle f. ‖ semelle f. de bois. / ~ **spool** see ~ **bobbin.**

wooden spoon ‖ Holzlöffel m.; Kelle f. ‖ cuiller f. en bois. / ~ **maker** ‖ Holzlöffelschnitzer m.; Kellenmacher m. ‖ fabricant m. de cuillers en bois.

wooden stand ‖ Holzgestell n. ‖ support m. ou carcasse f. en bois. / ~ **tamp** ‖ Holzstampfer m. ‖ dame f. en bois. / ~ **toys** pl. ‖ Holzspielwaren fpl. ‖ jouets mpl. en bois. / ~ **trunk** ‖ Holzkoffer

m. ‖ malle f. en bois. / ~ **ware** see wooden articles.

wood engraver ‖ Holzschneider m. ‖ graveur m. sur bois; xylographe m. / ~ **engraving** ‖ Holzschnitt m. ‖ gravure f. sur bois. / ~ **envelope** ‖ Holzhülle f. ‖ enveloppe f. en bois. / ~ **exportation** ‖ Holzausfuhr f. ‖ exportation f. de bois. / ~ **fibre** ‖ Holzfaser f. ‖ fibre f. de bois. / ~ **firing** ‖ Holzfeuerung f. ‖ chauffage m. au bois. / ~ **floating** ‖ Flößen n. ‖ flottage m. de bois. / ~ **frame** ‖ Holzrahmen m. ‖ châssis m. en bois. / ~ **fretter** ‖ Holzwurm m. ‖ artison m. / ~ **fungus** ‖ Hausschwamm m.; Holzschwamm m. ‖ champignon m. du bois; bolet m. destructeur. / **bent** ~ **furniture** ‖ Möbel npl. aus gebogenem Holz ‖ meubles mpl. en bois courbé. / ~ **gilder** ‖ Holzvergolder m. ‖ doreur m. sur bois. / ~ **gilding** ‖ Holzvergoldung f. ‖ dorure f. sur bois. / ~ **grinder** ‖ Holzschleifer m. ‖ défibreur m. ou râpeur m. de bois. / ~ **grinding plant** ‖ Holzschleifereianlage f. ‖ installation f. à râper le bois.

wood handle ‖ Holzgriff m. ‖ manche m. de bois. / ~ **of a tool** ‖ Holzstiel m. eines Werkzeuges ‖ manche m. d'outil en bois.

wood, portable house of ~ ‖ zerlegbares Holzhaus n. ‖ maison f. de bois portative. / ~ **impregnation** ‖ Holzimprägnierung f. ‖ imprégnation f. ou injection f. des bois.

wood industry ‖ Holzindustrie f. ‖ industrie f. du bois. / **special machine for the** ~ ‖ Holzindustriesondermaschine f. ‖ machine f. spéciale pour l'industrie du bois.

wood, kind of ~ see species of ~. / ~ **land** ‖ Holzboden m. ‖ terre f. à bois. / ~ **market** ‖ Holzmarkt m. ‖ marché m. au bois. / ~ **meal** see ~ **powder.** / ~ **mordant** ‖ Holzbeize f. ‖ mordant m. pour bois. / ~ **moulder** ‖ Holzleistenmacher m. ‖ moulurier m.

wood moulding ‖ Holzleiste f. ‖ moulure f. en bois. / ~ **for furniture** ‖ Holzleiste f. für Möbel ‖ moulure f. en bois pour meubles. / ~ **machine** ‖ Kehlhobelmaschine f. ‖ machine f. à faire les moulures en bois.

wood oil ‖ Holzöl n. ‖ huile f. de bois. / ~ **varnish** ‖ Holzölfirnis m. ‖ vernis m. à l'huile de bois.

wood-opal ‖ Holzopal m. ‖ bois m. opalisé. / ~ **painter** ‖ Holzmaler m. ‖ peintre m. sur bois. / ~ **pattern maker** ‖ Modellschreiner m. ‖ menuisier-modeleur m.; menuisier-mécanicien m. / ~ **pavement** ‖ Holzpflaster n. ‖ pavé m. en bois. / ~ **paving** ‖ Holzpflasterung f. ‖ pavage m. en bois. / ~ **peeling machine** ‖ Holzschälmaschine f. ‖ machine f. à décortiquer le bois. / ~ **pile** ‖ Holzstoß m. ‖ bûcher m.; chale f. / ~ **pin for shoemaking** ‖ Holzstift m. für die Schuhmacherei ‖ cheville f. en bois de cordonnerie. / ~ **pitch** ‖ Holzpech n. ‖ poix f. végétale. / ~ **plait** ‖ Holzgeflecht n. ‖ tresse f. en fil de bois. / ~ **planing machine** ‖ Holzhobelmaschine f. ‖ machine f. à raboter le bois. / ~ **plantation labourer** ‖ Waldarbeiter m. ‖ ouvrier forestier m. / ~ **powder** ‖ Holzmehl n. ‖ poudre f. ou farine f. de bois. / ~ **possessor** ‖ Waldbesitzer m. ‖ propriétaire m. de forêts ou de bois. / ~ **preservative** ‖ Holzkonservierungsmittel n. ‖ produit m. à conserver le bois.

wood printing, burning stamp for ~ || Brennstempel m. für Holzdruck || fer m. à marquer au feu les bois. / printing stamp for ~ || Druckstempel m. für Holzdruck || timbre m. à imprimer sur bois.

wood pulley || Holzriemenscheibe f. || poulie f. en bois.

wood pulp || Holzzellulose f.; Holzstoff m.; Zellstoff m.; Zellulose f. || cellulose f.; pâte f. de bois. / mechanical ~ || Holzschliff m. || pâte f. mécanique de bois. / sodium ~ || Natronzellulose f. || cellulose f. à la soude. / sulphured ~ || Sulfitzellulose f. || cellulose f. au bisulfite.

wood pulp bleacher || Holzstoffbleicher m. || blanchisseur m. de pâte de bois. / ~ desiccating plant || Zelluloseentwässerungsanlage f. || installation f. de déshydratation de la cellulose. / machine for the manufacture of ~ || Zellstoffherstellungsmaschine f. || machine f. à fabriquer la cellulose. / ~ paper || Zellulosepapier n. || papier m. de cellulose. / thickening drum for ~ || Eindicktrommel f. für Zellulose || tambour m. d'épaississement de cellulose. / ~ wadding making machine || Zellulosewatteherstellungsmaschine f. || machine f. à fabriquer l'ouate de cellulose. / ~ works pl. || Holzstoffabrik f.; Zellulosefabrik f. || fabrique f. de cellulose ou de pâte de bois.

wood rasp || Holzraspel f. || râpe f. à bois. / ~ rim || Holzfelge f. || jante f. en bois. / ~ road || Holzweg m. || chemin m. de vidange.

Woodruff keying || Scheibenfeder f. || clavette f. Woodruff.

wood saw || Holzsäge f. || scie f. à bois. / ~ saw dust drying drum || Sägespänetrockentrommel f. || tambour m. à sécher les sciures.

wood sawing, mechanical ~ || mechanisches Holzsägen n. || sciage m. mécanique du bois. / ~ mill || Sägewerk n. || scierie f.

wood sawyer || Holzsäger m. || scieur m. de bois.

wood screw || Holzschraube f. || vis f. à bois. / round-head ~ || Halbrundholzschraube f. || vis f. à bois à tête goutte de suif. / manufacture of ~s || Holzschraubenherstellung f. || fabrication f. des vis à bois.

wood sculpture || Holzbildhauerarbeit f. || sculpture f. en bois. / ~ seat || hölzerner Sitz m. || siège m. en bois. / ~ shaving || Hobelspan m. || copeau m. de bois. / basket made of ~ shavings || Spankorb m. || corbeille f. en bois tressé. / box made of ~ shavings || Holzspanschachtel f. || boîte f. en copeaux de bois. / ~ ship see wooden ship. / ~ silverer || Holzversilberer m. || argenteur m. sur bois. / ~ silvering || Holzversilberung f. || argenture f. sur bois. / ~ slipper || Holzpantoffel m. || pantoufle f. en bois. / ~ species of ~ || Holzart f. || espèce f. de bois.

wood spirit || Holzgeist m. || esprit m. de bois; alcool m. méthylique. / raw ~ || Rohholzgeist m. || alcool m. méthylique brut.

wood splitting machine || Holzspaltmaschine f. || machine f. à fendre le bois. / ~ stack see ~ pile. / ~ stain see ~ mordant. / ~ stairs pl. || Holztreppe f. || escalier m. en bois. / ~ stair railing || Holzgeländer n. || rampe f. d'escalier en

bois. / steaming of ~ || Dämpfen n. des Holzes || étuvage m. du bois. / ~ stereotype || Inkunabel f. || édition f. incunable; ouvrage m. stéréotype sur bois. / ~ store || Holzlager n. || entrepôt m. de bois. / ~ straw || Holzstroh n. || paille f. de bois. / ~ sugar || Holzzucker m. || sucre m. de bois. / ~ tar || Holzteer n. || goudron m. de bois. / ~ tar oil || Holzteeröl n. || huile f. de goudron végétal. / ~ tin || Holzzinn n. || étain m. de bois. / ~ tissue || Holzgewebe n. || tissu m. de bois. / ~ tower || Holzturm m. || tour f. en bois. / ~ transport || Holzabfuhr f. || vidange f. de bois. / ~ trimmer || Holzkantenbestoßmaschine f. || machine f. à tronçonner les bois. / ~ tubbing || Senkzimmerung f. || cuvelage m. dans des terrains ébouleux. / ~ turner || Holzdrechsler m. || tourneur m. sur bois. / ~ turning lathe || Holzdrehbank f. || tour m. à bois. / ~ varnisher || Holzwarenlackierer m. || vernisseur m. sur bois. / ~ varnishing || Lackieren n. von Holzwaren || vernissage m. sur bois. / ~ vinegar || Holzessig m. || acide m. pyroligneux; vinaigre m. de bois.

woodward || Förster m. || garde-forestier m.

wood wasp || Holzwespe f. || sirèce f. / ~ waste || Holzabfälle mpl. || déchets mpl. de bois. / ~ wind instrument || Holzblasinstrument n. || instrument m. à vent en bois. / ~ window || Holzrahmenfenster n.; Holzfenster n. || fenêtre f. en bois. / ~ window blind || Holzrolladen m. || store m. (de boutique) en bois.

wood wool || Holzwolle f. || laine f. de bois. / ~ in cords || Holzwolle f. in Schnüren || laine f. de bois en cordes.

wood wool machine || Holzwollemaschine f. || machine f. pour la fabrication de la laine de bois. / ~ rope || Holzwollseil n. || tresse f. en laine de bois. / ~ working machine || Holzwolleverarbeitungsmaschine f. || machine f. à préparer la laine de bois.

woodwork || Täfelholz n.; Täfelwerk n. || boiserie f.; lambrissage m.; parquetage m. / ~ construction (Build) || Zimmerverband m. || boisage m.

woodworker || Holzarbeiter m. || boiseron m.

woodworking || Holzbearbeitung f. || travail m. du bois. / ~ copying lathe || Kopiermaschine f. für Holzbearbeitung || tour m. à bois à copier. / ~ industry || Holzindustrie f. || industrie f. du bois. / ~ machine || Holzbearbeitungsmaschine f. || machine f. à travailler le bois.

woodworm exterminator || Holzwurmmittel n. || exterminateur m. de gâtebois.

woody || bewaldet || boisé.

woof (Weav) || Schuß m.; Einschuß, m.; Einschlag m.; Eintrag m. || trame f. / ~ yarn (Weav) || Einschußgarn n. || trame f.; fil m. de trame.

wool || Wolle f. || laine f. / ~ of animal origin || Wolle f. tierischer Herkunft || laine f. de provenance animale. / artificial ~ || Kunstwolle f.; Lumpenwolle f. || laine f. artificielle. / artificial ~ making machine || Kunstwolleherstellungsmaschine f. || machine f. à fabriquer la laine artificielle. / asbestos ~ || Asbestwolle f. || laine f. d'amiante. / bleached ~ || gebleichte Wolle f. || laine f. blanchie. / carded ~ || gekrempelte

Wolle f. || laine f. cardée. / ~ for cleaning see waste ~. / crude ~ || rohe Wolle f. || laine f. brute. / dyed ~ || gefärbte Wolle f. || laine f. teinte. / embroidery ~ || Stickwolle f. || laine f. à broder. / finest ~ || Kernwolle f.; Oberwolle f.; Rückenwolle f. || laine f. mère. / forest ~ || Waldwolle f. || laine f. de pins ou de bois. / good long ~ || gute lange Wolle f. || cœur m. de laine. / greasy ~ || Fettwolle f. || laine f. surge ou en suint; surge f. / knitting ~ || Strickwolle f. || laine f. à tricoter. / lamb's ~ || Lammwolle f. || laine f. d'agneau. / longstapled ~ || langschürige Wolle f. || laine f. de longue soie. / to loosen ~ by arsenic || Wolle f. abgiften || délainer à l'arsenic. / native ~ || Landwolle f. || laine f. indigène. / natural coloured ~ || naturfarbene Wolle f. || laine f. beige. / pine needle ~ see forest ~. / reel ~ || Kammwolle f. || laine f. de roseaux. / sheared ~ || Schurwolle f. || laine f. de toison. / sheep ~ || Schafwolle f. || laine f. de mouton. / short-stapled ~ || kurzschürige Wolle f. || laine f. de courte soie. / skinner ~ || Gerberwolle f.; Raufwolle f. || avalies fpl.; écouailles fpl. / torn-up ~ || Schoddywolle f. || effiloché m. de laine. / unspun ~ || ungesponnene Wolle f. || laine f. non filée. / unwashed ~ || ungewaschene oder fette Wolle f. || laine f. en suint ou grasse ou surge. / washed ~ || gewaschene Wolle f. || laine f. lavée. / waste ~ || Flockwolle f. || bourre f. / waste ~ (Mach) || Putzwolle f. || déchet m. de laine. / wood ~ || Holzwolle f. || laine f. de bois.

wool blanket || Wolldecke f. || couverture f. en laine. / ~ bleacher || Wollbleicher m. || blanchisseur m. de laine. / ~ bleaching || Bleichen n. der Wolle || blanchiment m. de la laine. / ~ blender || Wollmischer m. || mélangeur m. de laine. / ~ breaking || Schlagen n. der Wolle || battage ou louvetage de laines. / machine for carbonizing ~ || Wollkarbonisiermaschine f. || machine f. à carboniser la laine. / ~ carder || Wollkrempler m. || cardeur m. de laine. / ~ carding || Krempeln n. der Wolle || cardage m. ou droussage m. de laine. / ~ carpet || Wollteppich m. || tapis m. de laine. / ~ classer || Wollsortierer m. || détricheur m. de laine. / ~ cleaning machine || Wollreinigungsmaschine f. || machine f. à laver la laine. / ~ comb || Wollkamm m. || peigne m. à laine. / ~ comber || Wollkämmer m. || cardeur m. ou peigneur m. de laine. / ~ combing || Kämmen n. der Wolle || peignage m. de la laine. / combing machine for ~ || Kämmaschine f. für Wolle || machine f. à peigner la laine. / ~ disk || Wollscheibe f. || disque m. en laine. / ~ dressing machine || Wollaufbereitungsmaschine f. || machine f. à préparer la laine. / ~ drier || Wolltrockner m. || sécheur m. de laine. / ~ drying || Wolltrocknung f. || séchage m. de la laine. / ~ drying plant || Wolltrocknungsanlage f. || installation f. à sécher la laine. / ~ dyeing plant || Wollfärberei f. || teinturerie f. de laine.

wool fat || Wollfett n. || graisse f. de suint ou de laine. / ~ extracting plant || Wollfettgewinnungsanlage f. || installation f. d'extraction de graisse de laine.

/ ~ pitch ‖ Wollfettpech n. ‖ poix f. de graisse de laine. / ~ stearin ‖ Wollfettstearin n. ‖ stéarine f. de graisse de laine.

wool felt ‖ Wollfilz m. ‖ feutre m. de laine. / ~ cloth for paper machines ‖ Filztuch n. für Papiermaschinen ‖ feutre m. de laine pour papeteries.

wool frame tenter ‖ Wollstrecker m. ‖ défeutreur m. de laine. / ~ grease *see* wool fat. / ~ hair ‖ Wollhaar n. ‖ brin m. de laine. / ~ hosiery ‖ wollene Strumpfwaren fpl. ‖ bonneterie f. de laine. / ~ industry ‖ Wollindustrie f. ‖ industrie f. lainière.

woollen blanket ‖ Wolldecke f. ‖ couverture f. de laine. / ~ goods pl. ‖ Wollwaren fpl. ‖ articles mpl. en laine; lainage m. / ~ goods printing ‖ Wollwarendruck m. ‖ impression f. sur laine. / ~ rug *see* woolen blanket. / ~ satin ‖ Wollatlas m. ‖ satin m. de laine *ou* zéphyr. / ~ yarn ‖ Wollgarn n. ‖ fil m. en laine.

woolly ‖ wollig ‖ laineux.

wool oiling ‖ Wolleinfettung f. ‖ ensimage m. de laine. / ~ picker ‖ Wollputzer m. ‖ débourgeonneur m. de laine. / ~ picking ‖ Entkletten n. der Wolle ‖ épaillage m. *ou* échardonnage m. *ou* égratteronnage m. de laine. / ~ picking machine ‖ Wollzupfmaschine f. ‖ machine f. échardonneuse. / ~ rags pl. ‖ Putzlumpen mpl.; Putzwolle f.; Abfallwolle f. ‖ déchets mpl. *ou* chiffons mpl. de laine. / ~ rag scouring ‖ Waschen n. der Putzwolle ‖ nettoyage m. des déchets de laine. / ~ rag sorter ‖ Wollumpensortierer m.; Wollhadernsortierer m. ‖ repasseur m. de chiffons de laine. / ~ scouring ‖ Wollentschweißung f. ‖ dessuintage m. de laine. / ~ sorter ‖ Wollsortierer m. ‖ affineur m. de laine. / ~ sorting ‖ Lesen n. der Wolle ‖ assortissage m. de laines. / ~ spinning machine ‖ Wollspinnereimaschine f. ‖ machine f. à filer la laine. / ~ spinning mill ‖ Wollspinnerei f. ‖ filature f. de laine. / ~ supplier ‖ Wollaufleger m.; alimenteur m. de laine. / ~ tissue ‖ Wollgewebe n. ‖ tissu m. de laine. / ~ twister ‖ Wollgarnzwirner m. ‖ retordeur m. de laine. / ~ twisting ‖ Wollzwirnen n. ‖ retorderie f. de laine. / ~ ware *see* woollen goods. / ~ washer ‖ Wollwäscher m. ‖ nettoyeur m. de laine.

wool washing ‖ Wollwäsche f. ‖ lavage m. des laines. / ~ machine ‖ Wollwäschereimaschine f. ‖ machine f. pour le lavage de laines. / ~ plant ‖ Wollwäscherei f. ‖ installation f. de lavage de laine.

wool waste ‖ Wollabfälle mpl. ‖ déchets mpl. de laine. / ~ weaver ‖ Wollweber m. ‖ tisseur m. de laine. / ~ weaving plant ‖ Wollweberei f. ‖ atelier m. de tissage de laine. / ~ winder ‖ Wollwickler m.; Wollhaspeler m. ‖ dévideur m. *ou* haspeleur m. de laine. / ~ winding ‖ Haspeln n. der Wolle ‖ dévidage m. de la laine.

wool yarn ‖ Wollgarn n. ‖ fil m. de laine. / carded ~ ‖ Streichgarn n. ‖ fil m. de laine cardée. / combed ~ ‖ Kammgarn n. ‖ fil m. de laine peignée; laine f. filée; peigné m. / ~ made-up for retail ‖ Wollgarn n. in Aufmachung für den Kleinverkauf ‖ fil m. de laine conditionné pour la vente au détail.

word counter ‖ Wortzähler m. ‖ compteur

m. de mots. / ~ counting ‖ Wortzählung f. ‖ compte m. des mots.

work, to ‖ arbeiten ‖ travailler. / ~ (Mach) ‖ bearbeiten ‖ façonner; usiner; dresser. / ~ (Chem) ‖ gären; aufgehen ‖ fermenter; entrer en fermentation. / ~ (Engine) ‖ laufen; im Gange sein ‖ fonctionner; marcher. / ~ concrete ‖ betonieren ‖ bétonner. / ~ by contract *see* ~ by the piece. / ~ correctly ‖ einwandfrei arbeiten ‖ fonctionner d'une façon irréprochable. / the engine ceases working without obvious reason ‖ der Motor m. bleibt ohne erkennbaren Grund stehen ‖ le moteur s'arrête sans cause apparente. / cylinders pl. working independently from each other ‖ unabhängig voneinander steuerbare Zylinder ‖ cylindres mpl. à manœuvrer indépendamment l'un de l'autre. / ~ in the drift (Mine) ‖ vor Ort arbeiten ‖ travailler au front m. de taille. / ~ into each other ‖ ineinander greifen ‖ engrener. / ~ hard ‖ angestrengt arbeiten ‖ bûcher; piocher. / ~ hard (Mach) ‖ schwer gehen ‖ marcher durement. / hard worked element of machinery ‖ schwer beanspruchter Maschinenteil m. ‖ élément m. de machines subissant l'effet des travaux augmentés. / ~ the ink on the table ‖ die Farbe auf die Form auftragen ‖ toucher la forme. / ~ from the level upwards (Mine) ‖ aufhauen ‖ travailler au-dessous de la tête. / ~ loose (Mach) ‖ Spiel n. *oder* Spielraum m. haben ‖ jouer; avoir du jeu m. / ~ a mine ‖ eine Grube f. bauen ‖ exploiter une mine. / ~ off a debt ‖ eine Schuld f. abarbeiten ‖ acquitter une dette par son travail. / ~ off the form (Print) ‖ die Form ausdrucken ‖ achever l'impression f. / ~ off a proof sheet (Print) ‖ einen Korrekturbogen m. abziehen ‖ tirer une épreuve. / ~ open (Mine) ‖ Tagebau f. treiben ‖ exploiter à ciel ouvert. / ~ out (Beer) ‖ ausstoßen ‖ débiter; vendre. / ~ out (Metal) ‖ ausarbeiten ‖ développer. / ~ out of the cask bunghole ‖ im Lagerfaß n. ausstoßen ‖ cracher. / ~ overtime ‖ Überstunden fpl. machen ‖ faire des heures fpl. supplémentaires; travailler au delà du temps prescrit. / all parts pl. are working as they should ‖ alle Teile mpl. arbeiten richtig ‖ toutes les pièces fpl. travaillent convenablement. / ~ by the piece ‖ in Akkord m. arbeiten ‖ travailler à la pièce *ou* à la tâche. / ~ the process on a large scale ‖ das Verfahren im Großbetrieb durchführen ‖ mettre en œuvre le procédé sur une vaste échelle. / ~ at red heat ‖ rotwarm bearbeiten ‖ travailler les pièces fpl. à la chaleur rouge. / ~ a seam ‖ ein Flöz abbauen ‖ exploiter une veine. / it pays ~ it even in small units ‖ die Verarbeitung f. lohnt sich selbst in kleinen Einheiten ‖ l'industrie f. rend même si elle est entreprise sur une petite échelle. / ~ timber ‖ zimmern ‖ charpenter. / ~ the traverses pl. (Mar) ‖ die Kurse mpl. koppeln ‖ réduire les routes fpl. / ~ unsteadily (Mot) ‖ unruhig arbeiten ‖ faire du bruit m. / ~ for volume ‖ auf Menge f. arbeiten ‖ travailler en débit m. / ~ worthy of being worked (Seam) ‖ baufähig ‖ labourable.

work ‖ Arbeit f. ‖ travail m. / ~ (Task) ‖ Aufgabe f. ‖ tâche f. / ~ agreed upon (Mine) ‖ Gedingarbeit f. ‖ travail m. à

la tâche. / at ~ (Mach) ‖ in Gang m. ‖ en marche f. / to be at ~ (Mine) ‖ in Betrieb m. stehen ‖ être en exploitation. / ~ of a building ‖ Rumpf m. eines Gebäudes; Baukörper m. ‖ corps m. de bâtiment. / carved ~ ‖ Schnitzarbeit f. ‖ sculpture f. en bois. / to cease ~ ‖ die Arbeit niederlegen ‖ abandonner le travail. / chased ~ (Met) ‖ getriebene Arbeit f. ‖ articles mpl. repoussés. / ~ by contract ‖ Akkordarbeit f. ‖ travail m. à la tâche. / cramped ~ (Print) ‖ gedrängte Schrift f. ‖ composition f. serrée *ou* nourrie. / dead ~ (Shipb) ‖ totes Werk n. ‖ œuvres fpl. mortes. / dinged ~ *see* embossed ~. / earth ~ ‖ Erdbewegung f. ‖ transport m. des terres. / effective ~ ‖ travail m. utile. / embossed ~ *see* chased ~. / ~ is expedited and cheapened ‖ die Arbeit wird beschleunigt und verbilligt ‖ le travail est accéléré et rendu à meilleur marché. / ~ of forged copper ‖ Kupferschmiedearbeit f. ‖ travail m. de chaudronnerie. / hammered ~ ‖ getriebene Arbeit f. ‖ ouvrage m. martelé. / to leave the ~ before the expiration of the contractual period of service ‖ die Arbeit vor Ablauf der vertragsmäßigen Arbeitszeit verlassen ‖ quitter le travail avant la fin du temps de travail stipulé au contrat. / to leave ~ without authority ‖ die Arbeit unbefugt verlassen ‖ quitter le travail sans permission. / to leave the ~ without notice ‖ die Arbeit ohne Aufkündigung verlassen ‖ quitter le travail sans préavis. / out of ~ ‖ arbeitslos ‖ sans travail. / ~ of resistance ‖ Widerstandsarbeit f. ‖ travail m. résistant. / ~ under a roof ‖ Arbeit f. unter Dach ‖ travail m. sous toit. / scientific ~ ‖ wissenschaftliche Arbeit f. ‖ travail m. scientifique. / to set to ~ ‖ in Betrieb m. setzen ‖ mettre en activité f. / ~ has been simplified by the adoption of machines ‖ die Arbeit ist durch die Anwendung von Maschinen erleichtert worden ‖ le travail a été facilité par l'emploi de machines. / steel structural ~ ‖ Eisenbauwerk n.; Eisenkonstruktion f. ‖ charpente f. métallique. / useful ~ *see* effective ~.

workable (Metal) ‖ schmelzwürdig ‖ exploitable. / ~ (Mine) ‖ abbauwürdig; bauwürdig ‖ exploitable. / ~ ore field ‖ abbauwürdiges Erzvorkommen n. ‖ gisement m. exploitable de minerai. / to occur in ~ quantity ‖ in bauwürdiger Menge f. vorkommen ‖ apparaître en quantité valant l'exploitation.

work arbor ‖ Aufspanndorn m. ‖ mandrin m. porte-pièce. / ~ bar (Textile) ‖ Scheuerblech n. ‖ plaque f. à abattre. / ~ basket ‖ Nähkorb m. ‖ corbeille f. à ouvrage. / ~ bench ‖ Arbeitsbank f.; Werkbank f. ‖ établi m. / ~ dress ‖ Arbeitsanzug m. ‖ vêtement m. de travail.

worked, not ~ ‖ unbearbeitet ‖ non travaillé; brut.

worker *see also* workman ‖ Arbeiter m. ‖ ouvrier m. / class of ~ ‖ Arbeiterkategorie f. ‖ catégorie f. d'ouvrier.

workers pl. (Mine) ‖ Belegschaft f. ‖ ensemble m.; personnel m.

work holder ‖ Werkstückauflage f. ‖ support m. de la pièce.

working ‖ in Betrieb m. ‖ en marche f. / ~ without load ‖ leerlaufend ‖ marchant à vide. / ~ quickly ‖ schnellarbeitend ‖ travaillant vite *ou* rapide. / ~ rapidly *see*

~ quickly. / ~ satisfactorily || zufriedenstellend arbeitend || travaillant de façon f. satisfaisante.

working (Mach) || Gang m.; Arbeiten n. || activité f.; fonction f.; marche f.; jeu m. / cold ~ (Metal) || kalter Gang m. || allure f. *ou* marche f. froide. / defective ~ || fehlerhaftes Arbeiten n. || fonctionnement m. irrégulier. / ~ due (Patent) || Ausübung f. fällig || mise f. en exploitation due. / ~ of a furnace || Ofengang m.; Gang m. || marche f. *ou* allure f. d'un fourneau. / hot ~ (Met) || heißer *oder* hitziger Gang m. || allure f. *ou* marche f. chaude. / ~ on a large scale || Großbetrieb m. || fabrication f. sur une grande échelle. / ~ of a machine || Gang m. einer Maschine || marche f. d'une machine. / manual ~ (Tel) || Handbetrieb m. || travail m. *ou* service m. manuel. / metallurgical ~ of ores || Verhütten n. der Erze || fusion f. des minerais.

working of a mine || Grubenbetrieb m.; Bergbau m. || exploitation f. d'une mine. / ~ careless of the future || Raubbau m. || exploitation f. par grappillage.

working, long-wall ~ (Mine) || Strebbau m. mit breitem Blick || tailles fpl. chassantes à front continu. / ~ a patent || Ausbeutung f. *oder* Ausführung f. eines Patents || exploitation f. d'un brevet. / quiet ~ (Cutting || ruhiger Schnitt m. || coupe f. sans broutage. / regular ~ (Metal) || guter *oder* regelmäßiger *oder* garer Gang m.; Gargang m. || bonne allure f.; allure f. *ou* marche f. régulière. / reliable ~ || zuverlässiges Arbeiten n. || fonctionnement m. sûr. / ~ of a seam || Abbau m. eines Flözes || exploitation f. d'une veine. / ~ by steam || Dampfbetrieb m. || exploitation f. à la vapeur. / straight forward ~ || glatte Fabrikation f. || bonne marche f. de la fabrication. / ~ of switches || Umlegen n. der Weichen || déplacement m. des aiguilles. / ~-up materials || Rohstoffverarbeitung f. || emploi m. des matières premières.

working agreement || Interessengemeinschaft f. || convention f. / ~ beam (Mach) || Balanzier m. || balancier m. / ~ bench *see* work bench. / Board of Workings || Bauamt n. || intendance f. des bâtiments. / ~ capital || Betriebskapital n. || fonds mpl. *ou* capital m. de roulement *ou* d'exploitation. / ~ cloth || Arbeitsanzug m. || vêtement m. de travail. / ~ clothes pl. || Arbeitskleidung f. || habit m. d'atelier; vêtements mpl. de travail. / ~ conditions pl. || Arbeitsbedingungen fpl. || conditions fpl. de travail. / in ~ condition || in betriebsfähigem Zustand m. || en état m. d'exploitation; en condition f. de service. / ~ control (Clockm) || Gangkontrolle f. || contrôle m. de la marche.

working costs pl. || Betriebskosten pl.; Gestehungskosten pl. || frais mpl. d'exploitation; prix m. coûtant *ou* de revient. / live ~ || laufende Betriebsausgaben fpl. || frais mpl. réguliers d'exploitation. / to operate at a minimum of ~ || mit einem Minimum an Betriebskosten arbeiten || fonctionner avec le minimum de frais de main d'œuvre. / to reduce ~ || die Gestehungskosten pl. verringern || diminuer le prix de revient.

working current || Betriebsstrom m. || courant m. de régime. / ~ intensity || Be-

triebsstromstärke f. || intensité f. du courant de transmission.

working cylinder || Arbeitszylinder m. || cylindre m. de travail. / ~ day || Arbeitstag m. || jour m. ouvrable. / ~ distance (Opt) || Arbeitsabstand m. || distance f. frontale. / ~ distance of automatic systems || Reichweite f. bei Wählersystemen || distance f. obtenue par des systèmes automatiques. / ~ drawing || Werkzeichnung f.; Werkstattzeichnung f. || dessin m. d'exécution. / ~ dress || Arbeitsanzug m. || vêtement m. *ou* habit m. de travail. / ~ element || arbeitender Bestandteil m. || organe m. travaillant *ou* actif. / ~ expenses pl. *see* ~ costs. / ~ face *see* ~ place. / ~ floor || Arbeitsboden m. || plancher m. de travail. / ~ furnace || Schmelzofen m. || fourneau m. de fusion. / frame of convenient ~ height || gebrauchshoher Ständer m. || montant m. en hauteur convenable. / daily ~ hours pl. || tägliche Arbeitszeit f. || journée f. du travail. / weekly ~ hours pl. || wöchentliche Arbeitszeit f. || temps m. de travail hebdomadaire. / ~ hypothesis || Arbeitshypothese f. || hypothèse f. de travail. / ~ instructions pl. || Bedienungsvorschrift f. || instruction f. de service *ou* de maniement. / ~ load || zulässige Belastung f. || charge f. admissible. / ~ machine || Arbeitsmaschine f. || machine f. opératrice *ou* à travailler. / ~ magnet || Arbeitselektromagnet m. || électro-aimant m. de marche. / ~ manner *see* method. / ~ material || Betriebsmaterial n. || matériel m. d'exploitation. / ~ method || Arbeitsweise f. || mode m. de travail *ou* de fonctionnement. / ~ model || Arbeitsmodell n. || modèle m. mécanique.

working moment, absolute ~ || absolutes Arbeitsmoment n. || moment m. absolu de travail. / effective ~ || nutzbares Arbeitsmoment n. || moment m. utile de travail. / ~ of the sails (Windmill) || Arbeitsmoment n. der Flügel || moment m. moteur des ailes.

working motion || Arbeitsbewegung f. || course f. de travail. / ~ objective || Arbeitsobjektiv n. || objectif m. de travail. / in ~ order || betriebsfertig || prêt à fonctionner. / the machine is in good ~ order || die Maschine ist in betriebsfähigem Zustande || la machine est en bon ordre de service. / ~ organisation || Betriebsorganisation f. || organisation f. de l'exploitation. / ~ part || arbeitender (Bestand-)Teil m. || partie f. ouvrière; organe m. actif. / ~ period *see* ~ season. / ~ pit (Mine) || Förderschacht m. || puits m. d'extraction. / ~ pit (Railw) || Arbeitsgrube f. || fosse f. de travail ou de visite. / ~ place (Mine) || vor Ort m. || front m. de taille; lieu m. de travail d'une galerie. / ~ plan || Arbeitsplan m.; Betriebsplan m. || plan m. des travaux; règlement m. d'exploitation. / ~ point || Angriffspunkt m. || point m. d'application d'une force. / ~ position || Arbeitsstellung f. || position f. de travail. / ~ practice *see* ~ method. / ~ pressure (Electr) || Betriebsspannung f. || tension f. du régime. / ~ pressure (Mach) || Arbeitsdruck m.; Betriebsdruck m. || pression f. de travail *ou* effective *ou* motrice *ou* de service. / ~ process || Arbeitsvorgang m. || procédé m. du travail.

/ high ~ rate || starke Beanspruchung f. || charge f. forte; engagement m. élevé. / low ~ rate || leichte Beanspruchung f. || charge f. légère; engagement m. léger. / ready for ~ || betriebsfähig || en ordre de service; prêt à marcher. / ~ regulations pl. || Arbeitsordnung f. || règlement m. de travail. / determining the ~ resistance || Bestimmung f. des Arbeitswiderstandes || détermination f. de la résistance des matériaux. / ~ room || Arbeitsraum m. || atelier m.

workings pl. (Mine) || Grubenbaue mpl. || ouvrages mpl. / old ~ (Mine) || alter Mann m.; alte Baue mpl. || anciens mpl.; vieux hommes mpl. *ou* ouvrages mpl. *ou* travaux mpl.

working, safety of ~ || Betriebssicherheit f. || sécurité f. de service. / ~ scope || Arbeitsbereich m. || rayon m. d'action. / ~ scope on slopes || Arbeitsbereich m. bei Böschungen || rayon m. d'action avec talus. / ~ season || Betriebszeit f.; Kampagne f. || roulement m.; campagne f. / ~ section (Forest) || Betriebsklasse f. || série f. d'exploitation. / ~ spectacles pl. || Arbeitsbrille f. || lunettes fpl. de travail. / ~ speed || Arbeitsgeschwindigkeit f. || vitesse f. de travail *ou* de marche. / ~ state of a furnace || Ofengang m.; Gang m. || marche f. *ou* allure f. d'un fourneau. / good ~ state (Met) || Gargang m. || bonne allure f.; allure f. *ou* marche f. régulière. / ~ stroke || Arbeitshub m.; Krafthub m. || course f. motrice. / ~ suit || Arbeitsanzug m. || vêtement m. de travail. / ~ surface || Angriffsfläche f. || surface f. d'attaque. / ~ surface of cylinder || Zylinderlauffläche f. || portée f. du cylindre. / system of ~ (Mine) || Abbausystem n. || méthode f. d'exploitation. / ~ time || Arbeitszeit f. || temps m. de travail. / ~ tun (Brew) || Gärbottich m.; Döse f. || cuve f. guilloire; guilloire f. / ~ value || Betriebswert m. || valeur f. effective. / ~ velocity || Arbeitsgeschwindigkeit f. || vitesse f. de travail. / ~ voltage || Betriebsspannung f. || voltage m. de régime. / ~ width (Weav) || Arbeitsbreite f. || largeur f. de travail. / ~ year || Betriebsjahr n. || année f. d'exercice; exercice m.

workman || Arbeiter m.; Handarbeiter m.; Handwerker m. || ouvrier m.; travailleur m.; manœuvre m. / ~ above ground (Mine) || Arbeiter m. über Tage || ouvrier m. du jour. / to accommodate the workmen || die Belegschaft unterbringen || loger la population ouvrière. / building ~ || Bauarbeiter m. || ouvrier m. en bâtiments. / disabled ~ || invalider Arbeiter m. || ouvrier m. invalide. / to dismiss a ~ without notice || einen Arbeiter m. ohne Kündigung entlassen || congédier un ouvrier sans préavis. / dismissed ~ (Mine) || abgekehrter Arbeiter m. || ouvrier m. congédié. / ~ free to travel || freizügiger Arbeiter m. || ouvrier m. ayant la liberté de résidence. / ~ of full age || volljähriger Arbeiter m. || ouvrier m. adulte. / local ~ || ansässiger Arbeiter m. || ouvrier m. domicilié. / to lock out workmen || Arbeiter mpl. aussperren || mettre dehors des ouvriers. / to come to a decision to lock out workmen || eine Entschließung f. über Aussperrung von Arbeitern treffen || concerter un accord pour renvoyer des ouvriers. / married ~ || ver-

heirateter Arbeiter m. ∥ ouvrier m. marié.
/ organized ~ ∥ organisierter Arbeiter m.
∥ ouvrier m. syndiqué. / seasonal ~ ∥
Saisonarbeiter m. ∥ ouvrier m. saisonnier. / unmarried ~ ∥ lediger Arbeiter m.
∥ ouvrier m. célibataire. / unskilled ~ ∥ ungelernter Arbeiter m. ∥ simple ouvrier m.
workman's family ∥ Arbeiterfamilie f. ∥
famille f. ouvrière. / ~ dwelling in the
colonies ∥ Arbeiterwohnhaus n. in den
Siedlungen ∥ maison f. ouvrière des
colonies. / ~ pass ∥ Arbeitsbuch n. ∥
livret m. de travail.
workmanship ∥ Werkstattarbeit f. ∥ travail
m. d'atelier. / defective ~ ∥ mangelhafte
Arbeit f. ∥ travail m. défectueux; construction f. défectueuse *ou* mal faite. /
exact ~ ∥ genaue Werkstattarbeit f. ∥
usinage m. précis.
workmen pl. employed in a mine ∥ Belegschaft f. einer Grube ∥ ensemble m. des
mineurs d'un puits.
workmen's clothes pl. ∥ Arbeiterkleidung
f.; Arbeitskleidung f. ∥ vêtement m.
d'ouvriers. / ~ clothing factory ∥ Arbeiter-
bekleidungsfabrik f. ∥ fabrique f. d'habits
d'ouvriers. / ~ club ∥ Arbeiterverein m.
∥ union f. d'ouvriers; syndicat m. ouvrier.
/ ~ colony *see* ~ dwellings pl. / ~ congress
∥ Arbeiterkongreß m. ∥ congrès m.
ouvrier. / ~ domestic conditions ∥ häus-
liche Verhältnisse npl. der Arbeiter ∥
économie f. *ou* condition f. domestique des ouvriers. / ~ dwellings pl. ∥
Werksiedlung f. ∥ colonie f. ouvrière.
/ to put up ~ dwellings pl. ∥ Werk-
wohnungen fpl. beschaffen ∥ construire
des logements mpl. pour la population ouvrière. / ~ insurance ∥ Arbeiterver-
sicherung f. ∥ assurance f. des ouvriers. /
sums pl. paid under the ~ insurance act
∥ Ausgaben fpl. für die gesetzliche Ar-
beiterversicherung ∥ sommes fpl. payées
pour l'assurance ouvrières prescrite par
la loi. / ~ linen ∥ Arbeitswäsche f. ∥ linge
m. pour ouvriers. / ~ shirt ∥ Arbeits-
hemd m. ∥ chemise f. pour ouvriers. /
~ shortage ∥ Arbeitermangel m. ∥ manque
m. d'ouvriers. / ~ syndicate ∥ Betriebsvertretung f. ∥ syndicat m. des ouvriers;
représentation f. des employés. / ~ union
∥ Arbeiterverein m.; Gewerkschaft f. ∥
union f. d'ouvriers; syndicat m. ouvrier.
/ ~ wardrobe ∥ Arbeiterkleiderschrank
m. ∥ armoire f. pour vestiaires d'ouvriers.
work piece ∥ Arbeitsstück n.; Werkstück
n. ∥ pièce f. d'ouvrage. / ~ room ∥ Ar-
beitssaal m. ∥ salle f. de travail.
works pl. ∥ Fabrik f.; Betrieb m.; Anlage
f. ∥ fabrique f.; usine f.; établissement m.
/ carriage ~ (Railw) ∥ Waggonfabrik f. ∥
atelier m. de fabrication de wagons. /
drainage ~ ∥ Kanalisationswerk n. ∥ installation f. de drainage. / ~ for electric
hoists ∥ Elektrozugfabrik f. ∥ fabrique f.
de palans électriques. / gas ~ ∥ Gaswerk
n.; Gasanstalt f. ∥ usine f. *ou* centrale f. de
gaz. / ~ for hammered silver ∥ Silberham-
merwerk n. ∥ forge f. d'argent. / industrial
~ ∥ industrielle Anlage f. ∥ usine f.; établissement m. industriel; installation f. industrielle. / running ~ control ∥ laufende
Betriebskontrolle f. ∥ contrôle m. courant
de l'exploitation.
works engineer ∥ Betriebsingeniör m. ∥
ingénieur m. chef de service.
workshop ∥ Werkstätte f.; Werkstatt f.;
Werkstelle f. ∥ atelier m. / ~ for works

of art ∥ Kunstwerkstätte f.; kunstgewerbliche Werkstätte f. ∥ atelier m.
pour ouvrages artistiques. / ~ for building cars ∥ Wagenbauwerkstätte f. ∥
atelier m. de construction de voitures.
/ ~ of a chamoiser ∥ Gerberei f.
∥ chamoiserie f.; mégisserie f.; tannerie f. / farmer's ~ ∥ Farmerwerkstatt f. ∥ atelier m. de fermier. / ~ for
industrial art ∥ kunstgewerbliche Werkstätte f. ∥ atelier m. d'art industriel. /
instructional ~ ∥ Lehrlingswerkstatt f. ∥
atelier m. d'apprentissage. / joiner's ~
∥ Tischlerwerkstatt f. ∥ atelier m. de
menuiserie; menuiserie f.; ébénisterie f.
/ locksmith's ~ ∥ Schlosserwerkstatt f. ∥
atelier m. de serrurier. / mechanical
~ ∥ mechanische Werkstatt f. ∥ atelier m.
mécanique *ou* de constructions mécaniques. / modern ~ ∥ neuzeitlich eingerichtete Werkstätte f. ∥ atelier m.
moderne. / principal ~ ∥ Zentralwerkstatt f.; Hauptwerkstatt f. ∥ atelier m.
principal. / ~ for repairing works of art
∥ Werkstatt f. zur Wiederherstellung
von Kunstwerken ∥ atelier m. de restauration d'objets d'art. / tanner's ~ ∥ Gerberei f. ∥ tannerie f.; mégisserie f.; chamoiserie f.; atelier m. du tanneur.
workshop car ∥ Werkstattwagen m. ∥
voiture-atelier f. / ~ costs pl. ∥ Werk-
stattkosten pl. ∥ frais m. d'atelier. /
~ hall ∥ Werkstatthalle f. ∥ hall m.
d'atelier. / ~ measuring instrument ∥
Werkstattmeßinstrument n. ∥ instrument m. de mesure d'atelier. / ~ result ∥
Betriebsergebnis n. ∥ résultat m. de travail de l'atelier.
works laboratory ∥ Betriebslaboratorium
n. ∥ laboratoire m. d'usine. / ~ management ∥ Geschäftsleitung f. ∥ direction f. /
~ manager (Build) ∥ Bauleiter m. ∥ directeur m. des travaux. / ~ manager (Mach)
∥ Betriebsleiter m. ∥ directeur m. technique; chef m. de fabrication. / ~ organization ∥ Betriebsgliederung f.; Betriebsorga-
nisation f. ∥ organisation f. des ateliers. /
~ outfit ∥ Fabrikeinrichtung f. ∥ équipement m. *ou* installation f. de l'usine.
work spindle ∥ Werkstückspindel f. ∥
broche f. porte-pièce.
works railway ∥ Werkbahn f. ∥ chemin m.
de fer de l'usine. / ~ standards pl. ∥
Fabriknormen fpl. ∥ standards mpl. d'é-
tablissement.
work table ∥ Werktisch m. ∥ table f.
de travail. / vertical adjustable ~ ∥
vertikal verschiebbarer Arbeitstisch m.
∥ table f. de travail déplaçable verticalement.
world ∥ Welt f. ∥ monde m. / ~-famous ∥
weltberühmt ∥ universellement connu.
world's commerce ∥ Welthandel m. ∥
commerce m. mondial. / ~ fair ∥ Welt-
ausstellung f. ∥ exposition f. universelle.
/ ~ market ∥ Weltmarkt m. ∥ marché m.
mondial. / ~ trade and industry ∥ Welt-
wirtschaft f. ∥ économie f. mondiale. /
~ traffic ∥ Weltverkehr m. ∥ relations
fpl. internationales.
world war ∥ Weltkrieg m. ∥ guerre f.
mondiale.
worm (Mach) ∥ Schnecke f.; Schraube f.
ohne Ende ∥ vis f. sans fin. / condensing
~ ∥ Kühlschlange f. ∥ serpentin m.
refroidisseur. / cooling ~ *see* condensing
~. / grinding ~ ∥ Brechschnecke f. ∥ vis

f. sans fin broyeuse. / steering ~ ∥ Lenkschnecke f. ∥ vis f. sans fin de direction.
worm and pinion *see* worm gear.
worm casing ∥ Schneckengehäuse n. ∥
boîte f. de la vis sans fin. / ~ conveyer
∥ Transportschnecke f. ∥ vis f. transporteuse. / ~ drive ∥ Schneckenantrieb m.
∥ transmission f. par vis sans fin. / ~
-eaten ∥ wurmstichig ∥ piqué des vers
mpl. / ~-eaten wood ∥ wurmstichiges
Holz n. ∥ bois m. vermoulu. / ~ eateness
∥ Wurmfraß m. ∥ vermination f. / ~
feeder ∥ Schneckenaufgabevorrichtung
f. ∥ appareil m. chargeur à vis sans fin.
worm gear ∥ Schneckengetriebe n. ∥ engrenage m. *ou* transmission f. à vis
sans fin. / self-locking ~ ∥ selbsthemmendes Schneckengetriebe n. ∥ transmission
f. à vis sans fin à blocage automatique.
worm groove (Wood) ∥ Wurmgang m. ∥
gale f. du bois. / ~ hob ∥ Schnecken-
fräser m. ∥ fraise f. hélicoïdale ou à
vis-mère. / ~ milling hob arbor carriage
∥ Schneckenräderfrässupport m. ∥ chariot
m. à tailler les roues à vis sans fin. / ~
milling machine ∥ Schneckenfräsmaschi-
ne f. ∥ machine f. à fraiser les vis sans
fin. / ~ seed ∥ Wurmsamen m. ∥ absinthe
f. de Judée. / ~ tablet ∥ Wurmtablette
f. ∥ tablette f. de santonine. / ~ trough
(Mach) ∥ Schneckentrog m. ∥ auge f.
de la vis sans fin.
worm wheel ∥ Schneckenrad n. ∥ roue f.
hélicoïdale *ou* à vis sans fin. / dividing
~ ∥ Teilschneckenrad n. ∥ roue f. hélicoïdale diviseur. / ~ threading machine
∥ Schneckenräderfräsmaschine f. ∥ machine f. à fraiser les roues hélicoïdales.
wormwood ∥ Wermut m. ∥ absinthe f. /
~ oil ∥ Wermutöl n. ∥ essence f. d'absinthe. / ~ wine ∥ Wermutwein m. ∥
vermout m.
worn-out ∥ abgenutzt ∥ usé. / to get ~ ∥
verschleißen; sich abnutzen ∥ s'user. / ~
type ∥ abgenutzte Schrift f. ∥ caractère
m. d'impression vieux.
worsted ∥ Kammgarn n. ∥ (fil m.) peigné
m. / ~ dyeing ∥ Färben n. von Kammgarn
∥ teinture f. de laine peignée. / ~ fabric
∥ Kammgarngewebe n. ∥ étoffe f. de
laine peignée. / washing machine for ~
fabrics ∥ Waschmaschine f. für Kamm-
garnstoffe ∥ machine f. à laver les
tissus peignés. / ~ shag ∥ Wollplüsch
m. ∥ peluche f. de laine. / ~ spinner ∥
Kammgarnspinner m. ∥ fileur m. de
laine peignée.
worsted spinning machine ∥ Kammgarn-
spinnereimaschine f. ∥ machine f. à filer
la laine peignée. / ~ mill ∥ Kammgarnspinnerei f. ∥ filature f. de laine peignée.
worsted stocking ∥ wollener Strumpf m. ∥
bas m. de laine. / ~ velvet ∥ Wollsamt
m. ∥ velours m. de laine. / ~ weaver ∥
Kammgarnweber m. ∥ tisseur m. de laine
peignée. / ~ yarn ∥ Kammgarn n. ∥ (fil
m. de) laine f. peignée; peigné m.
wort ∥ Würze f. ∥ moût m. / hot ~ ∥ warme
Würze f. ∥ moût m. chaud. / new ~ ∥
frische Würze f. ∥ surmoût m.
wort cooling ∥ Würzekühlung f. ∥ refroidissement m. du moût. / ~ copper ∥
Würzepfanne f. ∥ chaudière f. à moût.
/ ~ gelatine ∥ Würzegelatine f. ∥ gélatine
f. de moût; moût m. gélatinisé.
worth mentioning ∥ nennenswert ∥ notable;
appréciable. / ~ seeing ∥ sehenswert;
sehenswürdig ∥ digne d'être vu; curieux.

worthiness of being worked (Mine) ‖ Bauwürdigkeit f. ‖ exploitabilité f.

worthless ‖ wertlos ‖ sans valeur f.

worthlessness ‖ Wertlosigkeit f. ‖ peu m. de valeur; futilité f.

wort sink *see* ~ trough. / ~ trough ‖ Würzegrand m. ‖ reverdoir m.

wound (Electr) ‖ gewickelt ‖ bobiné; enroulé. / ~ condenser ‖ Wickelkondensator m. ‖ condensateur m. enroulé.

wound ‖ Wunde f. ‖ blessure f.; plaie f.

wound hook ‖ Wundhaken m. ‖ écarteur m. / ~ with blunt teeth ‖ stumpfzinkiger Wundhaken m. ‖ écarteur m. à griffes obtuses. / ~ with sharp teeth ‖ scharfzinkiger Wundhaken m. ‖ écarteur m. à griffes tranchantes.

woven bandage ‖ Leibbinde f. ‖ bande f. à maillot. / ~ cotton fabric ‖ Baumwollgewebe n. ‖ tissu m. de coton. / ~ goods pl. ‖ Wirkware f. ‖ tissu m. à maille; bonneterie f. / ~ label ‖ gewebtes Etikett n. ‖ étiquette f. tissée. / ~ material ‖ Gewebe n. ‖ tissu m.

wrap, to ‖ einwickeln ‖ emballer. / ~ the cover to a booklet (Bookb) ‖ den Umschlag m. an eine Broschüre einhängen ‖ appliquer la couverture contre la brochure.

wrapped wire ‖ Manteldraht m. ‖ fil m. sous enveloppe.

wrapper ‖ Packleinwand f.; Packleinen n. ‖ toile f. d'emballage. / ~ (Person) ‖ Packer m. ‖ emballeur m. / ~ (Weav) ‖ Mitläufer m. ‖ doublier m. / leather ~ ‖ Ledermappe f. ‖ portefeuille m. en cuir.

wrapper arm ‖ Schnurführung f. ‖ guideficelle m. / ~ renovating device (Weav) ‖ Mitläuferpflegeeinrichtung f. ‖ dispositif m. à soigner le doublier. / ~ selector (Tobacco) ‖ Deckblattsortierer m. ‖ trieur m. d'enveloppes. / ~ stripper (Tobacco) ‖ Deckblattripper m. ‖ écôteur m. d'enveloppes.

wrapping ‖ Umwicklung f.; Verpackung f. ‖ enveloppe f. / wire with braided ~ ‖ umklöppelter Draht m. ‖ fil m. sous tresse.

wrapping machine ‖ Wickelmaschine f. ‖ machine f. d'enroulement. / ~ (Packing) ‖ Einwickelmaschine f. ‖ machine f. d'emballage. / ~ for cold rolling mills ‖ Wickelmaschine f. für Kaltwalzmaschinen ‖ machine f. d'enroulement pour laminoirs à froid. / ~ for the rubber trade ‖ Wickelmaschine f. für die Gummiindustrie ‖ machine-réunisseuse f. pour l'industrie du caoutchouc. / ~ for the textile industry ‖ Wickelmaschine f. für die Textilindustrie ‖ réunisseuse f pour l'industrie textile. / tube ~ ‖ Hülsenwickelmaschine f. ‖ machine f. à rouler les douilles.

wrapping paper ‖ Einschlagpapier n.; Einwickelpapier n. ‖ papier m. d'emballage *ou* à emballer. / coloured ~ ‖ farbiges Packpapier n. ‖ papier m. d'emballage coloré.

wreath ‖ Kranz m. ‖ couronne f. / ~ of smoke ‖ Rauchwirbel m. ‖ tourbillon m. de fumée. / ~ of snow ‖ Schneeschanze f. ‖ amas m. *ou* banc m. de neige.

wreck, to (Ship) ‖ abwracken ‖ démolir. / ~ (Shipwreck) ‖ Schiffbruch m. leiden ‖ naufrager; faire naufrage m.

wreck ‖ Wrack n. ‖ épave f. / ~ (Wrecking) ‖ Schiffbruch m. ‖ naufrage m.

wrench, to ‖ verdrehen; verbiegen ‖ fausser.

wrench ‖ Schraubenschlüssel m. ‖ clef f. à écrou. / adjustable ~ ‖ verstellbarer Schraubenschlüssel m.; Engländer m. ‖ clef f. anglaise *ou* à molettes *ou* à ouverture réglable. / flange ~ ‖ Flanschenschlüssel m. ‖ clef f. pour brides. / fork ~ ‖ Gabelschlüssel m. ‖ clef f. à fourche. / hexagonal ~ ‖ Sechskantschlüssel m. ‖ clef f. hexagonale. / monkey ~ ‖ Universalschlüssel m. ‖ clef f. anglaise. / piston ~ ‖ Kolbenschlüssel m. ‖ clef f. du piston. / socket ~ ‖ Steckschlüssel m.; Aufsteckschlüssel m. ‖ clef f. à douilles *ou* à canon. / spanner ~ ‖ Hakenschlüssel m. ‖ clef f. à crochet.

wring, to ~ **out** ‖ (aus)wringen ‖ tordre.

wringer ‖ Wringmaschine f. ‖ essoreuse f. à rouleaux.

wringing, silk ‖ Schevillieren n. der Seide; Ausringen n. des Seidengarnes ‖ chevillage m. de la soie.

wringing bar ‖ Ringholz n. ‖ chevilleau m. / ~ machine *see* wringer. / ~ pole (Dyer) ‖ Wringeisen n. ‖ épart m.

wrinkle, free from ~s ‖ faltenfrei ‖ sans plis mpl.

wrinkled ‖ gefaltet ‖ plissé.

wrist ‖ Zapfen m.; Wellenzapfen m. ‖ tourillon m.; axe m.; fusée f. / ~ joint ‖ Handgelenk n. ‖ poignet m. / ~ pin ‖ Kolbenbolzen m. ‖ axe m. de piston. / ~ watch ‖ Armbanduhr f. ‖ montre-bracelet m.

write, to ~ the dimensions into a design ‖ die Maße npl. in eine Zeichnung eintragen ‖ coter un dessin. / ~-down ‖ anschreiben ‖ noter. / ~-off ‖ abschreiben ‖ amortir.

writer ‖ Autor m.; Urheber m.; Verfasser m. ‖ auteur m. / short-hand ~ ‖ Stenograf m. ‖ sténographe m.

writing ‖ Schrift f. ‖ écriture f. / blind ~ ‖ unsichtbare Schrift f. ‖ écriture f. invisible. / ciphered ~ ‖ chiffrierte Schrift f. ‖ écriture f. chiffrée. / ~ of ciphered documents ‖ Abfassung f. chiffrierter Schriftstücke ‖ rédaction f. de documents chif-

frés. / in ~ ‖ brieflich ‖ par lettre f. / invisible ~ *see* blind ~. / multi-coloured ~ ‖ mehrfarbige Schrift f. ‖ écriture f. polychrome. / readable ~ ‖ lesbare Schrift f. ‖ écriture f. lisible. / ~ with reversed hair strokes (Print) ‖ Schrift f. mit vertauschten Grund- und Haarstrichen ‖ lettres mpl. en chemise *ou* à la duchesse. / susceptible ~ *see* blind ~.

writing book ‖ Heft n. ‖ cahier m. / ~ case ‖ Schreibzeug n. ‖ écritoire f. / ~ chalk ‖ Schreibkreide f. ‖ craie f. à écrire. / ~ cloth ‖ Pauseleinwand f. ‖ papier m. toile; toile f. à calquer. / ~ diamond ‖ Schreibdiamant m.; Glaserdiamant m. ‖ diamant m. pour écrire *ou* dessiner sur le verre; diamant m. de vitrier. / ~-down ‖ Niederschrift f. ‖ procès-verbal m. / ~ ink ‖ Schreibtinte f. ‖ encre f. à écrire. / autographic ~ ink ‖ lithografische Tusche f. ‖ encre f. autographique. / ~ lever ‖ Schreibhebel m. ‖ levier m. écrivant. / ~ magnet ‖ Schreibmagnet m. ‖ aimant m. écrivant. / ~ materials pl. ‖ Schreibwaren fpl. ‖ articles mpl. de papeterie. / ~ mechanism ‖ Schreibapparat m. ‖ appareil m. à écrire. / ~-off ‖ Abschreibung f. ‖ déduction f.; amortissement m. / ~ pad ‖ Schreibunterlage f. ‖ sous-main m. / ~ paper ‖ Schreibpapier n. ‖ papier m. à écrire. / ~ pen ‖ Schreibfeder f. ‖ plume f. à écrire. / ~ slate ‖ Schiefertafel f. ‖ ardoise f. à écrire. / ~ spring (Tel) ‖ Schreibfeder f. ‖ ressort m. écrivant. / ~ table ‖ Schreibtisch m. ‖ bureau m.; secrétaire m. / ~ table calendar ‖ Schreibtischkalender m. ‖ calendrier m. de table bureau. / ~ telegraph ‖ Schreibtelegraf m. ‖ appareil m. *ou* télégraphe m. écrivant. / ~ work ‖ Schreibarbeit f. ‖ travail m. d'écriture.

written-off ‖ abgeschrieben ‖ déduit.

wrong manipulation ‖ falsche Handhabung f. ‖ fausse manipulation f. / ~ side (Textile) ‖ Kehrseite f.; Rückseite f.; linke Seite f. ‖ revers m.; envers m. / ~ use ‖ Mißbrauch m. ‖ abus m.

wrought ‖ bearbeitet ‖ façonné.

wrought iron ‖ Schmiedeeisen n. ‖ fer m. de forge *ou* forgé. / made of ~ ‖ schmiedeeisern ‖ de fer m. forgé.

wrought iron bottle ‖ schmiedeeiserne Flasche f. ‖ bouteille f. en fer forgé. / ~ plate ‖ schmiedeeiserne Platte f. ‖ plaque f. en fer forgé.

wrought silver ‖ verarbeitetes Silber n. ‖ argent m. travaillé. / ~ steel ‖ Schmiedestahl m. ‖ acier m. corroyé.

wrung heads pl. ‖ Kimm f.; Kimmung f. ‖ fleurs fpl. (du vaisseau).

wulfenite ‖ Gelbbleierz n.; Molybdänblei n.; Wulfenit m. ‖ molybdate m. de plomb; plomb m. jaune *ou* molybdaté.

X

x meshes pl. per unit of area ‖ x-maschig ‖ x mailles fpl. par unité de surface.

X-ray, to ‖ mit Röntgenstrahlen durchleuchten; röntgen ‖ examiner au moyen des rayons X.

X-ray apparatus ‖ Röntgenapparat m. ‖ appareil m. à rayons X. / ~ diagnosis ‖ Röntgenuntersuchung f. ‖ radiodiagnostic m.; diagnostic m. par les rayons X. / ~ equipment for research laboratories ‖ Röntgenanlage f. für Werkstoffuntersuchung ‖ installation f. de métallographie aux rayons X.

X-rays pl. ‖ Röntgenstrahlen mpl.; X-Strahlen mpl.‖rayons mpl. X ou Rœntgen.

X-ray test ‖ Röntgenuntersuchung f. ‖ radiodiagnostic m.; diagnostic m. par les rayons X. / ~ tube ‖ Röntgenröhre f. ‖ tube m. de Rœntgen.

x-stopper (Radio) ‖ Nebengeräuschunterdrücker m. ‖ suppresseur m. de bruits parasites.

xanthate ‖ xanthogensaures Salz n. ‖ xanthate m.; xanthogénate m. / potassium ~ ‖ xanthogensaures Kalium n. ‖ xanthogénate m. de potassium. / ~ of sodium ‖ xanthogensaures Natrium n. ‖ xanthate m. de soude.

xantheine ‖ Xanthein n. ‖ xanthéine f.

xanthene ‖ Xanthen n. ‖ xanthène m.

xanthic ‖ xanthogensauer ‖ xanthique; xanthogénique. / ~ acid ‖ Xanthogensäure f. ‖ acide m. xanthique.

xanthine ‖ Krappgelb n. ‖ xanthine f.

xanthite ‖ Xanthit m. ‖ xanthite f.

xanthocarthanic acid ‖ Xanthokarthaminsäure f. ‖ acide m. xanthocarthamique.

xanthogallolic acid ‖ Xanthogallolsäure f. ‖ acide m. xanthogallolique.

xanthogenate see xanthate.

xanthogenic see xanthic.

xanthone colouring matter ‖ Xanthonfarbstoff m. ‖ colorant m. de xanthone.

xanthophyllite ‖ Xanthophyllit m. ‖ xanthophyllite f.; waluéwite f.

xanthoprotein ‖ Xanthoprotein n. ‖ xanthoprotéine f.

xanthosiderite ‖ Gelbeisenstein m. ‖ xanthosidérite f.

xanthosterine ‖ Xanthosterin n. ‖ xanthostérine f.

xanthotoxic acid ‖ Xanthotoxinsäure f. ‖ acide m. xanthotoxique.

xanthylic acid ‖ Xanthylsäure f. ‖ acide m. xanthylique.

xebec ‖ Schebecke f. ‖ chebec m.; chébec m.; chabec m.

xenolite ‖ Xenolith m. ‖ xénolithe f.

xenon ‖ Xenon n. ‖ xénon m.

xeroform ‖ Xeroform n. ‖ xéroforme m.; tribromphénylate m. de bismuth.

X-rays pl. see under X.

xylane ‖ Xylan n. ‖ xylane m.

xylene ‖ Xylol n. ‖ xylol m.; xylène m. / ~ red ‖ Xylenrot n. ‖ rouge m. (de) xylène.

xylidine red ‖ Xylidinrot n. ‖ rouge m. (de) xylidine.

xylit, semi-coke from ~ ‖ Halbkoks m. aus Xylit ‖ semi-coke m. provenant de xylite.

xylochlore ‖ Xylochlor m. ‖ xylochlore m.

xylogen ‖ Steinholz n.; Xylolith m. ‖ bois m. factice; xylolit(h)e f.

xylogram ‖ Holzschnitt m. (als Abdruck) ‖ gravure f. en bois; xylogramme m.

xylographer ‖ Xylograf m.; Holzschneider m. ‖ graveur m. sur bois; xylographe m.

xylography ‖ Xylografie f.; Holzschneidekunst f. ‖ gravure f. sur bois; xylographie f.

xylohydroquinone ‖ Xylohydrochinon n. ‖ xylohydroquinone f.

xylol ‖ Xylol n. ‖ xylol m.; xylène m.

xylolin yarn ‖ Xylolin n. ‖ fil m. (de) xyloline.

xylolit(e) ‖ Xylolith m.; Steinholz n. ‖ xylolit(h)e f. / ~ colours pl. ‖ Farben fpl. für Steinholz ‖ couleurs fpl. pour xylolithe. / equipment for ~ factory ‖ Steinholzfabrikeinrichtung f. ‖ installation f. pour fabriques de xylolithe. / partition wall of ~ ‖ Xylolithwand f. ‖ cloison f. en xylolithe.

xylonic acid ‖ Xylonsäure f. ‖ acide m. xylonique.

xylophone m. (Mus) ‖ Xylophon n. ‖ xylophone m.

xyloquinone ‖ Xylochinon n. ‖ xyloquinone f.

xylyl ‖ Xylyl n. ‖ xylyle m.

xylylene ‖ Xylylen n. ‖ xylylène m. / ~ urea ‖ Xylylenharnstoff m. ‖ xylylène-urée f.

xylylic acid ‖ Xylylsäure f. ‖ acide m. xylylique.

xyster ‖ Schab(e)messer n. ‖ rugine f.

xystus ‖ Laubengang m. ‖ xyste m.

Y

Y-connected three-phase system with neutral ‖ Sternschaltung f.; Y-Schaltung f. ‖ montage m. en étoile ou en Y; connexion f. ou couplage m. en étoile.

Y-connection ‖ Sternschaltung f.; Y-Schaltung f. ‖ montage m. en étoile ou en Y; connexion f. ou couplage m. en étoile.

Y-drain ‖ Y-förmiges Entwässerungsrohr n. ‖ drain m. en Y.

Y-pipe ‖ Y-Rohr n. ‖ tube m. en Y.

Y-tube see Y-pipe.

yacht ‖ Jacht f. ‖ yacht m. / ~ with auxiliary motor ‖ Segeljacht f. mit Hilfsmotor ‖ yacht m. mixte. / fancy ~ ‖ Luxusjacht f. ‖ yacht m. de luxe. / motor ~ ‖ Motorjacht f. ‖ yacht m. à moteur. / motor ~ with auxiliary sails ‖ Motorjacht f. mit Hilfssegeln ‖ yacht m. automobile avec voilure auxiliaire. / racing ~ ‖ Rennjacht f.; Rennboot n. ‖ racer m.; bateau m. de régates ou de course. / sailing ~ ‖ Segeljacht f. ‖ yacht m. à voile. / schooner ~ ‖ Schonerjacht f. ‖ schooner m. / steam ~ ‖ Dampfjacht f. ‖ yacht m. à vapeur.

yachting ‖ Lustschiffahrt f.; Wettsegeln n. ‖ navigation f. de plaisance.

yak ‖ Yak m. ‖ yak m.; yack m.

yam ‖ Yamswurzel f. ‖ igname f.

yamamaï silk ‖ Yamamayseide f. ‖ soie f. du yama-maï.

yangona acid ‖ Yangonasäure f. ‖ acide m. de yangona.

yaquilla fibre ‖ Yaquillafaser f. ‖ fibre f. du yaquilla.

yard (Build) ‖ Hof m. ‖ cour f. / ~ (Factory) ‖ Vorratsplatz m.; Fabrikhof m. ‖ dépôt m.; parc m. / ~ (Measure) ‖ Yard n. ‖ yard m. / ~ (Sailship) ‖ Rahe f. ‖ vergue f. / ~ (Shipb) ‖ Werft f. ‖ chantier m. / cattle ~ ‖ Viehhof m. ‖ cour f. ou parc m. aux bestiaux. / coal ~ ‖ Kohlenbahnhof m. ‖ gare f. aux charbons. / customs ~ ‖ Zollhof m. ‖ entrepôt m. de douane. / erecting ~ ‖ Werkplatz m. für den Zusammenbau ‖ cour f. de montage. / finished stuff ~ ‖ Lager n. für fertige Waren ‖ magasin m. des produits finis. / floated timber ~ ‖ Floßholzplatz m. ‖ chantier m. de bois flotté. / freight ~ ‖ Güterbahnhof m. ‖ gare f. à marchandises. / goods ~ see freight ~. / lateen ~ ‖ lateinische Rahe f. ‖ antenne f. / main ~ ‖ große Rahe f. ‖ grande vergue f. / naval ~ ‖ Werft f. ‖ chantier m. (de constructions navales). / ore ~ ‖ Erzlagerplatz m. ‖ dépôt m. de minerais. / railway ~ see freight ~. / ~ run of track ‖ laufendes Meter n. Gleis ‖ mètre m. courant de voie ferrée. / ship ~ ‖ Werft f. ‖ chantier m. (naval). / ship ~ for motor boats ‖ Motorbootswerft f. ‖ chantier m. de construction de bateaux automobiles. / small court ~ ‖ Lichthof m. ‖ cour f. vitrée. / timber ~ ‖ Zimmerplatz m.; Zulage f. ‖ chantier m. de charpente. / top-sail ~ ‖ Marsrahe f. ‖ vergue f. du hunier. / wheel ~ ‖ Räderhof m. ‖ parc m. aux roues. / work ~ ‖ Arbeitsplatz m. ‖ place f. de travail.

yard arm ‖ Rahennock n. ‖ bout m. de vergue. / ~ crutches pl. ‖ Raheträger mpl. ‖ porte-lofs mpl.; porte-vergues mpl. / ~ foot-rope ‖ Paard n.; Rahepferd n. ‖ marchepied m. de vergue. / ~ labourer ‖ Platzarbeiter m. ‖ homme m. de cour. / ~ line valve ‖ Erdleitungsventil n. ‖ soupape f. de la conduite souterraine. / ~ locomotive ‖ Rangierlokomotive f. ‖ locomotive f. de gare ou de manœuvre.

yardman (Factory) ‖ Hofarbeiter m. ‖ ouvrier m. de cour. / ~ (Mar) ‖ Rahegast m. ‖ homme m. de vergue.

yard pump ‖ Hauspumpe f.; Hofpumpe f. ‖ pompe f. à colonne.

yards pl. **and booms** pl. (Shipb) ‖ Rundholz n. ‖ espars mpl.

yarn ‖ Garn n. ‖ fil m. / artificial-silk ~ ‖ Kunstseidengarn n. ‖ fil m. de soie artificielle. / asbestos ~ with core ‖ Asbestgarn n. mit Seele ‖ fil m. d'amiante avec âme. / barchant ~ ‖ Barchentgarn n.; Baumwollstreichgarn n.; Zweiwalzengarn n. ‖ fil m. à futaine ou de condenseur; fil m. coton carde. / bleached ~ ‖ gebleichtes Garn n. ‖ fil m. blanchi. / blend ~ ‖ Mischgarn n. ‖ fil m. mixte. / ~ on the bobbin ‖ Spulengarn n. ‖ fil m. en bobine. / bourrette ~ ‖ Bourrettegarn n. ‖ fil m. de bourrette de soie. / braiding ~ ‖ Flechtgarn n. ‖ fil m. pour tresser. / to bundle the ~ ‖ das Garn bündeln ‖ empaqueter les fils mpl. / cable ~ see rope ~. / cable filling ~ ‖ Kabelfüllgarn n. ‖ matière f. de fil pour remplir les câbles. / carded ~ ‖ Streichgarn n.; Halbkammgarn n.; Krempelgarn n. ‖ fil m. de laine cardée; peignécardé m. / carded cotton ~ ‖ Baumwollstreichgarn n. ‖ fil m. de coton cardé; fil m. coton carde. / carded wool ~ see carded ~. / carded worsted see carded ~. / carpet ~ ‖ Teppichgarn n. ‖ fil m. de tapisserie. / Cashmere ~ ‖ Kaschmirgarn n. ‖ fil m. cachemire. / cellulose ~ ‖ Zellstoffgarn n.; Zellulosegarn n. ‖ fil m. en cellulose. / cheviot ~ ‖ Cheviotgarn n. ‖ fil m. cheviot. / clean ~ ‖ reines Garn n. ‖ fil m. propre. / ~ of coarse animal hairs ‖ Garn n. aus groben Tierhaaren ‖ fil m. en poils grossiers d'animaux. / coloured ~ ‖ gefärbtes Garn n. ‖ fil m. teint. / spinning of coloured ~s ‖ Baumwollbuntspinnerei f. ‖ filature f. de fils de couleur. / coloured twist ~ ‖ jaspiertes Garn n. ‖ fil m. jaspé. / combed ~ see worsted ~. / combed cotton ~ ‖ Baumwollkammgarn n. ‖ fil m. de coton peigné. / condensed ~ see barchant ~. / condenser ~ see barchant ~. / cop ~ see cop-spun ~. / cop-spun ~ ‖ Kötzergarn n. ‖ fil m. en cannette. / cordonnet ~ ‖ Litzengarn n. ‖ fil m. cordonnet. / cotton ~ ‖ Baumwollgarn n. ‖ fil m. de coton. / crochet ~ ‖ Häkelgarn n. ‖ fil m. à crocheter. / ~ for crochet work see crochet ~. / doubled ~ ‖ gezwirntes Garn n.; Zwirn m. ‖ fil m. retors; retors m. / dyed ~ see coloured ~. / Egyptian ~ ‖ Makogarn n. ‖ fil m. Jumel; fil m. de coton d'Egypte. / endless ~ ‖ endloser Faden m. ‖ fil m. continu. / fancy ~ ‖ Zierfaden m.; Ziergarn n. ‖ fil m. (de) fantaisie. / felted ~ ‖ Filzgarn n. ‖ fil m. feutré. / fine ~ ‖ feines Garn n. ‖ fil m. fin. / flake ~ ‖ geflammtes oder flammiertes Garn n.; Flammgarn n. ‖ fil m. flammé ou chiné. / flax ~ ‖ Flachsgarn n.; Leinengarn n. ‖ fil m. de lin. / flaxline ~ see linen ~. / gassed cotton ~ ‖ Florgarn n.; gasiertes Baumwollgarn n. ‖ fil m. de coton gazé. / glazed ~ ‖ Glanzgarn n. ‖ fil m. glacé. / goat's hair ~ ‖ Ziegenhaargarn n. ‖ fil m. de poil de chèvre. / greasy ~ ‖ fetthaltiges Garn n. ‖ fil m. gras ou graisseux. / hair ~ ‖ Haargarn n. ‖ fil m. de poil. / half-wool(len) ~

‖ Halbwollgarn n.; Halbwollgespinst n. ‖ fil m. mi-laine. / hand-spun ~ ‖ Handgarn n. ‖ fil m. à fuseau ou à rouet ou de main. / hemp ~ ‖ Hanfgarn n. ‖ fil m. de chanvre. / high-quality ~ ‖ Primagarn n. ‖ fil m. de première qualité; fil m. extra; fil m. de choix. / home-spun ~ ‖ Homespungarn n. ‖ fil m. home spun. / hosiery ~ see knitting ~. / imitation ~ ‖ Imitatgarn n. ‖ fil m. imitation. / jute ~ ‖ Jutegarn n. ‖ fil m. de jute. / knitting ~ ‖ Strickgarn n.; Strumpfgarn n. ‖ fil m. à tricoter; retors m. pour bonneterie. / knot-work ~ ‖ Knüpfgarn n. ‖ fil m. pour filet. / lama ~ ‖ Lamagarn n. ‖ laine f. de lama. / linen ~ ‖ Langflachsgarn n. ‖ fil m. de lin. / loosely-spun ~ ‖ loses Gespinst n. ‖ fil m. à torsion floche. / machine-spun ~ ‖ Maschinengarn n. ‖ fil m. mécanique. / medio ~ ‖ Halbkettgarn n.; Mediogarn n. ‖ fil m. mi-chaîne ou petitechaîne. / mélange ~ ‖ meliertes oder plattiertes Garn n. ‖ fil m. mélangé. / mercerized ~ ‖ merzerisiertes Garn n. ‖ fil m. mercerisé. / merino ~ ‖ Merinogarn n. ‖ fil m. mérinos. / mill-spun ~ see machine-spun ~. / mixed ~ see blend ~. / mule-spun ~ ‖ Selbstspinnergarn n.; Mulegarn n.; Selfaktorgarn n. ‖ fil m. de métier renvideur. / nitrated ~ ‖ Nitragarn n. ‖ fil m. nitré. / oily ~ ‖ ölhaltiges Garn n. ‖ fil m. huileux. / paper ~ ‖ Papiergarn n. ‖ fil m. en papier. / partly coloured ~ see mélange ~. / polished ~ ‖ Glanzgarn n.; Eisengarn. n. ‖ fil m. glacé. / printed ~ ‖ bedrucktes Garn n. ‖ fil m. imprimé. / ramie ~ ‖ Ramiegarn n. ‖ fil m. de ramie. / raw ~ ‖ Rohgarn n. ‖ fil m. brut. / reeled ~ ‖ gehaspeltes oder geweiftes Garn n. ‖ fil m. dévidé. / re-manufactured ~ ‖ Kunstwollgarn n. ‖ fil m. de laine renaissance. / ~ made up for the retail trade ‖ Garn n. in Aufmachung für den Kleinverkauf ‖ fil m. conditionné pour la vente au détail. / ring-spun ~ ‖ Ringspinnergarn n.; Drosselgarn n. ‖ fil m. de chaîne du continu. / rope ~ ‖ Kabelgarn n.; Taugarn n.; Seilergarn n. ‖ fil m. de caret. / schappe silk ~ ‖ Florettseidengarn n. ‖ fil m. de bourre de soie. / second-quality ~ ‖ Sekundagarn n. ‖ fil m. de seconde qualité; fil m. moyen. / semi-worsted ~ see carded ~. / shoddy ~ ‖ Shoddygarn n. ‖ fil m. de shoddy. / silk ~ ‖ Seidengarn n. ‖ fil m. de soie. / spun ~ ‖ Gespinst n. ‖ filé m.; filure f. / spun-silk ~ ‖ Abfallseidengarn n. ‖ fil m. de déchets de soie. / stocking ~ see knitting ~. / strong ~ ‖ starkes Garn n. ‖ fil m. fort. / tambour-work ~ ‖ Rahmengarn n. ‖ fil m. de broderie au tambour. / textilose ~ ‖ Textilosegarn n. ‖ fil m. en textilose. / third-quality ~ ‖ Tertiagarn n. ‖ fil m. de troisième qualité; fil m. de qualité inférieure. / three-cord ~ ‖ dreifädiges oder dreidrähtiges Garn n. ‖ fil m. triple. / threefold ~ see three-cord ~. / three-threads ~ see three-cord ~. / tow ~ ‖ Werggarn n.; Hedegarn n. ‖ fil m. d'étoupe. / ~ for trimmings ‖ Posamentiergarn n. ‖ fil m. de passementerie. / ~ on tubes ‖ Pfeifengarn n. ‖ fil m. sur tube. / twisted ~ ‖ Zwirn m. ‖ fil m. retors. / unwashed ~ ‖ ungewaschenes Garn n. ‖ fil m. non lavé. / vicuña ~ ‖ Vicuñagarn n. ‖ fil m. de vigogne. /

vigogne ~ ‖ Vigognegarn n.; Vigognegespinst n.; Halbwollgespinst n. ‖ fil m. (de) vigogne; fil m. mixte. / vigogne imitation ~ ‖ Vigogne-Imitatgarn n. ‖ fil m. imitation de vigogne. / warp ~ ‖ Kettgarn n.; Zettelgespinst n.; Watergarn n. ‖ fil m. de chaîne. / washed ~ ‖ gewaschenes Garn n. ‖ fil m. lavé. / ~ for weaving purposes ‖ Webgarn n. ‖ fil m. à tisser. / wool(len) ~ ‖ Schafwollgarn n.; Wollgarn n. ‖ fil m. de laine; laine f. filé. / worsted ~ ‖ Kammgarn n. ‖ fil m. peigné. / zephyr ~ ‖ Zephirgarn n. ‖ fil m. zéphir.

yarn balance ‖ Garnnummerbestimmungswage f.; Garnsortierwage f. ‖ balance f. de numérotage de fils. / ~ beam ‖ Kettenbaum m.; Garnbaum m. ‖ ensouple f. de derrière. / ~ bleacher ‖ Garnbleicher m. ‖ blanchisseur m. de fils. / ~ brake ‖ Fadenbremse f. ‖ frein m. de fil. / ~ brushing machine ‖ Garnbürstmaschine f. ‖ machine f. à brosser les fils. / ~ bundling press ‖ Garnpackpresse f.; Garnbündelpresse f.; Garnpresse f. ‖ presse f. à empaqueter les fils; presse f. à fil(s). / ~ carrier (Weav) ‖ Garnträger m. ‖ porteur m. d'époules. / ~ case ‖ Garnkiste f. ‖ caisse f. de fil. / ~ changing attachment ‖ Garnwechseleinrichtung f. ‖ dispositif m. pour le changement du fil. / ~ conductor ‖ Kordelader f. ‖ conducteur m. sous fil. / ~ consumption ‖ Garnverbrauch m. ‖ consommation f. de fil. / ~ counter ‖ Fadenprüfer m. ‖ échantillonneur m. / ~ counting ‖ Garnnumerierung f. ‖ numérotage m. ou titrage m. des fils. / ~ covering ‖ Zwirnumspinnung f. ‖ guipage m. en fils retors. / ~ dyeing ‖ Garnfärberei f. ‖ teinture f. de fils. / ~ dyer ‖ Garnfärber m. ‖ teinturier m. de fils. / ~ finishing ‖ Garnappretur f. ‖ apprêt m. de fils. / ~ glazing ‖ Glänzen n. des Zwirnes ‖ lustrage m. de fils.

yarn guide see also thread guide ‖ Fadenführung f. ‖ guidage m. du fil. / ~ (Device) ‖ Fadenführer m. ‖ guide-fil m.

yarn guiding arm ‖ Fadenführerhebel m. ‖ bras m. du guide-fil. / ~ maker ‖ Garner m. ‖ filetier m. / ~ manufacture ‖ Garnfabrik f.; Zwirnfabrik f.; Leinenzwirnfabrik f. ‖ fileterie f.; fabrique f. de fil de lin. / ~ mender ‖ Garnanknüpferin f. ‖ partisseuse f. / ~ numbering see ~ counting. / ~ press see ~ bundling press. / accurate ~ quadrant ‖ Genauigkeitsgarnwage f. ‖ romaine f. de précision. / ~ reel ‖ Garnrolle f. ‖ bobine f. pour fils à coudre. / ~ reeler ‖ Garnhaspeler m. ‖ dévideur m. de fil. / ~ scale ‖ Garnsortierwage f. ‖ balance f. de numérotage des fils. / ~ scale with reel ‖ Garnsortierwage f. mit Weife ‖ balance f. de numérotage avec dévidoir. / ~ shading (Dyer) ‖ Flammgarnfärberei f. ‖ teinture f. de fils flammés. / ~ singeing ‖ Sengen n. des Garnes ‖ gazage m. de fils. / ~ sizing ‖ Schlichten n. des Garnes ‖ encollage m. de fils. / ~ spinning ‖ Spinnen n. des einfachen Fadens ‖ filage m. du fil simple ou du fil de caret. / spinning the single ~ see ~ spinning. / ~ spinning machine ‖ Garnspinnereimaschine f. ‖ machine f. pour filanderies. / ~ spinning mill ‖ Garnspinnerei f. ‖ filanderie f. / ~ spooler ‖ Garnspuler m. ‖ bobineur m. de fil. / ~ steamer ‖ Garn-

dämpfer m. ‖ vaporiseur m. d'époules. / ~ strength tester ‖ Garnfestigkeitsprüfer m. ‖ dynamomètre m. pour fils. / ~ taker-off ‖ Garnaufhänger m. ‖ déchargeur m. de fil. / ~ tentering machine ‖ Garnstreckmaschine f. ‖ machine f. à tendre les fils. / ~ untangler ‖ Entwirrer m. ‖ démêleur m. de fils. / ~ waste ‖ Spinnenden npl. ‖ bouts mpl. durs; déchets mpl. de fils. / wetting machine for ~s ‖ Befeuchtmaschine f. für Garne ‖ machine f. à humecter les fils. / ~ winding machine ‖ Garnhaspelmaschine f. ‖ machine f. à dévider le fil. / ~ wringer ‖ Einweicher m. ‖ assouplisseur m.

yarrow oil ‖ Schafgarbenöl n. ‖ essence f. de millefeuille. / ~ tree ‖ Jarrah m. ‖ eucalyptus m. marginata.

yaw, to (Aero) ‖ wenden; scheren ‖ voler en lacet m. / ~ (Mar) ‖ scheren ‖ embarder.

yaw (Mar) ‖ Scheren n. ‖ embardée f. / angle of ~ ‖ Gierungswinkel m. ‖ angle m. de lacet.

yawing ‖ Gieren n. ‖ lacet m.; dérivation f.; embardée f.

yawl ‖ Jolle f. ‖ yole f.

year ‖ Jahr n. ‖ an m.; année f. / ~ (Print) ‖ Jahrgang m. ‖ année f. / astronomical ~ ‖ astronomisches Jahr n.; Sonnenjahr n. ‖ année f. astronomique ou solaire. / bissextile ~ ‖ Schaltjahr n. ‖ année f. intercalaire ou bissextile. / business ~ ‖ Betriebsjahr n. ‖ exercice m. / civil ~ ‖ bürgerliches Jahr n. ‖ année f. civile. / financial ~ see business ~. / of last ~ ‖ vorjährig ‖ de l'année f. passée. / solar ~ see astronomical ~.

yearly balance of accounts ‖ Jahresabschluß m. ‖ fin f. de l'année; balance f.; bilan m. / ~ consumption ‖ Jahreskonsum m.; Jahresverbrauch m. ‖ consommation f. annuelle. / ~ contract ‖ Jahreskontrakt m. ‖ contrat m. annuel. / ~ income ‖ Jahreseinkommen n. ‖ revenu m. annuel.

years pl. **of service** ‖ Dienstalter n. ‖ ancienneté f.

yeast ‖ Hefe f. ‖ levure f. / alcoholic ~ ‖ Alkoholhefe f. ‖ levure f. alcoolique. / brewery ~ ‖ Brauereihefe f. ‖ levure f. de brasserie. / dry ~ ‖ Trockenhefe f. ‖ levure f. sèche. / pitching ~ ‖ Anstellhefe f. ‖ levain m. / pressed ~ ‖ Preßhefe f. ‖ levure f. pressée ou comprimée. / wine ~ ‖ Weinhefe f. ‖ lie f. de vin.

yeast bitter ‖ Hefenbitter n. ‖ amertume f. de la bière. / ~ cultivating apparatus ‖ Hefeneinzuchtapparat m. ‖ appareil m. pour la culture de la levure. / degeneration of the ~ ‖ Ausarten n. der Hefe ‖ dégénération f. de la levure. / ~ dividing machine ‖ Hefenteilmaschine f. ‖ machine f. à diviser la levure. / ~ factory ‖ Hefenfabrik f. ‖ fabrique f. de levure. / ~ forming machine ‖ Hefenformmaschine f. ‖ machine f. à mouler la levure. / ~ forming and dividing machine ‖ Hefenform- und -teilmaschine f. ‖ machine f. à mouler et à diviser la levure. / ~ kiln ‖ Hefentrockner m. ‖ séchoir m. de levure. / ~ maker ‖ Hefenfabrikant m. ‖ fabricant m. de levure. / ~ powder ‖ Hefenmehl n. ‖ levure f. en poudre. / preserving the ~ ‖ Konservieren n. der Hefe ‖ conservation f. de la levure. / ~ producing plant ‖ Hefenherstellungsanlage f. ‖ installation f. à fabriquer la levure. / ~ stopper ‖

Hefepfropfen m. ‖ bouchon m. de levure.

yeast works pl. ‖ Hefefabrik f. ‖ fabrique f. de levure.

yelk see also yolk (Wool) ‖ Schweiß m. (in der Wolle) ‖ suint m.

yellow ‖ gelb ‖ jaune. / to become ~ ‖ vergilben ‖ jaunir. / ~ amber ‖ Bernstein m. ‖ ambre m. jaune; succin m.

yellow brass ‖ Gelbguß m.; Messing n. ‖ cuivre m. jaune; laiton m. / impure ~ ‖ unreines Messing n. ‖ cuivre-potin m.

yellow-cross shell ‖ Gelbkreuzgeschoß n. ‖ obus m. à croix jaune.

yellow earth ‖ Ocker m. ‖ ocre f. / ~ quarry ‖ Ockergrube f. ‖ carrière f. d'ocre.

yellow electric light ‖ gelbes elektrisches Licht n. ‖ lumière f. électrique jaune-rouge.

yellow glass, attachable ~ ‖ aufsteckbares Gelbglas n. ‖ verre m. jaune adaptable.

yellow-green region of radiation ‖ gelbgrüner Strahlenbereich m. ‖ région f. vert-jaune des radiations.

yellow lead ‖ Bleigelb n. ‖ oxyde m. jaune de plomb; wulfénite f. / ~ ore ‖ Gelbbleierz n.; Molybdänblei n.; Wulfenit m. ‖ molybdate m. de plomb; plomb m. jaune ou molybdaté.

yellow light ‖ gelbes Licht n. ‖ feu m. jaune. / ~ mason's colour ‖ Mauergelb n. ‖ badigeon m.; badigeon m. jaune.

yellow metal ‖ Komposition f.; Metall n. ‖ composition f. / ~ founder ‖ Messinggießer m.; Gelbgießer m. ‖ fondeur m. en cuivre jaune.

yellow mineral ‖ Bernstein m. ‖ ambre f. jaune; succin m. / ~ oak bark ‖ Quercitronrinde f. ‖ quercitron m. / ~ ochre ‖ Gelberde f.; Ocker m. ‖ gelberde m.; ocre f.; argile f. ocreuse. / ~ pine ‖ Yellowpine f. ‖ pin m. jaune. / ~ prussiate of potash ‖ gelbes Blutlaugensalz n. ‖ prussiate m. jaune de potasse; ferrocyanure m. de potassium; cyanoferrure m. de potassium jaune; lessive f. du sang. / ~ tellurium ‖ Schrifterz n.; Schrifttellur n.; Sylvanit m.; Weißtellur n. ‖ or m. graphique ou sylvane; tellure m. natif auro-argentifère.

yellow ‖ Gelb n.; gelbe Farbe f. ‖ jaune m. / mineral ~ ‖ Kaisergelb n. ‖ jaune m. minéral. / ~ as straw ‖ Strohgelb n. ‖ jaune m. de paille.

yellowish ‖ gelblich ‖ jaunâtre. / ~ green ‖ gelbgrün ‖ vert jaunâtre.

yellow-wood ‖ Gelbholz n. ‖ bois m. jaune. / Cuba ~ ‖ kubanisches Gelbholz n. ‖ bois m. jaune de Cuba. / ~ extract ‖ Gelbholzextrakt m. ‖ extrait m. de bois jaune.

yenite ‖ Yenit m. ‖ yénite f.

yeoman ‖ kleiner Grundbesitzer m.; Büdner m. ‖ petit propriétaire m. foncier ou terrien.

yew ‖ Eibe f.; Eibenbaum m.; Eibenholz n. ‖ if m. / ~ tree see yew. / ~ wood ‖ Eibenholz n.; Taxholz n. ‖ bois m. d'if.

yield, to (Metal) ‖ enthalten ‖ contenir; tenir. / ~ (Soil) ‖ sich geben; nachgeben; weichen ‖ céder; plier; se relâcher; se détendre. / ~ a cable ‖ ein Tau n. nachlassen oder schießen lassen ‖ larguer ou mollir un câble. / ~ an income ‖ sich rentieren ‖ rapporter; rendre. / ~ metal ‖ an Metall n. ausbringen ‖ produire métal m. / ~ a revenue see ~ an income.

yield ‖ Ertrag m.; Nutzung f.; Ausbeute f. ‖ profit m.; exploitation f.; jouissance f.; rapport m.; rendement m. / ~ (Metal) ‖ Gehalt m.; Metallgehalt m. ‖ teneur f.; richesse f. / brew-house ~ ‖ Sudhausausbeute f. ‖ rendement m. en chaudière. / current ~ ‖ Stromausbeute f. ‖ rendement m. en courant. / fibre ~ of the cocoons ‖ Ausbeute f. der Kokons an Faserstoff ‖ rendement m. en soie des cocons. / laboratory ~ ‖ Laboratoriumsausbeute f. ‖ rendement m. de laboratoire. / ~ of the mines ‖ Ausbeute f. der Gruben ‖ bénéfice m. en minerais ou des mines. / practical ~ ‖ praktische Ausbeute f. ‖ rendement m. pratique. / theoretical ~ ‖ theoretische Ausbeute f. ‖ rendement m. théorique.

yield, calculation of the ~ ‖ Ausbeuteberechnung f. ‖ calcul m. du rendement. / determination of ~ ‖ Ertragfeststellung f. ‖ calcul m. de rendement.

yielding (Metal) ‖ enthaltend ‖ contenant; tenant.

yielding (Soil) ‖ Nachgeben n. ‖ cédement m.; fléchissement m.

yieldingness ‖ Ergiebigkeit f. ‖ fertilité f.

yield limit ‖ Streckgrenze f. ‖ limite f. apparente d'élasticité. / ~ point ‖ Streckgrenze f. ‖ limite f. d'allongement ou d'élasticité. / ~ ratio see ~ point. / ~ table ‖ Ausbeutetabelle f. ‖ table f. de rendement.

ylang-ylang ‖ Ylang-Ylang n. ‖ ylangylang m. / ~ oil ‖ Ylang-Ylang-Öl n. ‖ essence f. d'ylang-ylang.

yog(ho)urt ‖ Joghurt n. ‖ yoghourt m.

yohimbine ‖ Yohimbin n. ‖ yohimbine m.

yohimboa acid ‖ Yohimboasäure f. ‖ acide m. de yohimboa.

yoke ‖ Joch n. ‖ joug m. / ~ (Aut) ‖ Bügel m. ‖ étrier m. / ~ (Shipb) ‖ Ruderjoch n. ‖ joug m. du gouvernail; barre f. brisée. / adjustable ~ ‖ beweglicher Tragbügel m. ‖ étrier m. d'armature à jeu de dilatation. / axle box ~ ‖ Achsbuchsjoch n. ‖ bride f. oscillante de la boîte à graisse. / carrying ~ ‖ Tragbügel m. ‖ étrier m. de suspension. / drill ~ ‖ Bohrbügel m. ‖ agrafe f. de perceuse. / guide ~ ‖ Gleitbahnträger m. ‖ porteglissière m. / ~ of magnet ‖ Magnetjoch n. ‖ culasse f. de l'aimant.

yoked bottle ‖ Doppelflasche f. ‖ bouteille f. jumelle.

yoke ring (Mach tool) ‖ Halsring m. ‖ bague f. à gorge. / ~ suspension ‖ Jochaufhängung f. ‖ suspension f. du joug ou à deux supports.

yolk (Egg) ‖ Eigelb n.; Eidotter m. ‖ jaune m. d'œuf. / ~ (Wool) ‖ Wollfett n.; Wollschweiß m.; Schafschweiß m. ‖ suint m. / waxy ~ ‖ wachsartiger Fettschweiß m. ‖ suint m. cireux.

yolk wax ‖ Schweißwachs n. ‖ cire f. du suint.

Yperit ‖ Senfgas n.; Yperit n.; Ypérite m.; gaz m. moutarde; sulfure m. d'éthyle dichloré.

ytterbia see yterbium oxide.

ytterbium (Chem) ‖ Ytterbium n. ‖ ytterbium m. / ~ carbonate ‖ Ytterbiumkarbonat n. ‖ carbonate m. d'ytterbium. / ~ nitrate ‖ Ytterbiumnitrat n. ‖ nitrate m. d'ytterbium. / ~ oxide ‖ Ytterbin n.; Ytterbinerde f. ‖ ytterbine f.; oxyde m. d'ytterbium. / ~ selenate ‖ Ytterbiumselenat n. ‖ sélénate m. d'ytterbium.

yttria ‖ Yttererde f. ‖ yttria m.; yttrine f.; oxyde m. d'yttrium.

yttrialite ‖ Yttrialith m. ‖ yttrialithe f.

yttric ‖ ytterhaltig ‖ yttrifère; yttreux.

yttriferous *see* yttrious.

yttrious ‖ ytterhaltig ‖ yttrifère; yttreux. / ~ salt ‖ Yttersalz n. ‖ sel m. yttreux.

yttrium ‖ Yttrium n. ‖ yttrium m. / ~ chloride ‖ Yttriumchlorid n. ‖ chlorure m. d'yttrium. / ~ oxide ‖ Ytterrde f. ‖

yttria m.; yttrine f.; oxyde m. d'yttrium. / ~ pyrophosphate ‖ Yttriumpyrophosphat n. ‖ pyrophosphate m. d'yttrium. / ~ sulphate ‖ Yttriumsulfat n. ‖ sulfate m. d'yttrium.

yttrocerite ‖ Ytterflußspat m. ‖ yttrocérite f.

yttrofluorite ‖ Yttrofluorit m. ‖ yttrofluorite f.

yttroilmenite ‖ Yttroilmenit m. ‖ yttroilménite f.; samarskite f.

yttrotantalite ‖ Yttrotantalit m. ‖ yttrotantalite f.

yttrotitanite ‖ Yttrotitanit m. ‖ yttrotitanite f.

yucca ‖ Yukka f.; Palmlilie f. ‖ yucca m. / ~ fibre ‖ Yukkafaser f. ‖ fibre f. de yucca.

yuccasaponine ‖ Yukkasaponin n. ‖ yuccasaponine f.

yufts pl. ‖ Juchten m.; Juchtenleder n. ‖ cuir m. de Russie; youftes mpl.

Z

Z-axis ‖ Z-Achse f. ‖ axe m. de Z.

Z-bar ‖ Z-Eisen n. ‖ fer m. en Z.

Z-fish plate ‖ Z-Lasche f. ‖ éclisse f. Z.

Z-iron ‖ Z-Eisen n. ‖ fer m. en Z.

Z-mouthed chisel ‖ Z-Bohrer m. ‖ fleuret m. à tranchant en Z.

zaffer ‖ Zaffer m.; Eschel m. ‖ zafre m.

zanella weaving mill ‖ Zanellaweberei f. ‖ tissage m. de zanella.

zantewood ‖ Fisettholz n. ‖ bois m. de fustet.

zaratite ‖ Nickelsmaragd m.; Zaratit m. ‖ émeraude f. de nickel; zaratite f.

zedoary root ‖ Zitwerwurzel f. ‖ racine f. de zédoaire.

zellophan ‖ Zellophan n. ‖ zellophan m.

zenith ‖ Zenit m.; Scheitelpunkt m. ‖ zénith m.; point m. vertical. / ~ prism ‖ Zenithprisma n. ‖ prisme m. pour observer au zénith *ou* pour le zénith; prisme m. zénithal.

zeolite ‖ Kupfonspat m.; Zeolit m. ‖ zéolite f. / foliated ~ ‖ Heulandit m.; Stilbit m.; Blätterzeolit m. ‖ heulandite f.; stilbite f.

zero (Math) ‖ Null f. ‖ zéro m. / ~ (Thermometer) ‖ Gefrierpunkt m.; Nullpunkt m. ‖ zéro m. / above ~ ‖ über Null f. ‖ audessus de zéro m. / absolute ~ ‖ absoluter Nullpunkt m. ‖ zéro m. absolu. / to adjust to ~ ‖ auf Null f. einstellen ‖ mettre au zéro m. / below ~ ‖ unter Null f.; unter dem Nullpunkt m. ‖ audessous de zéro m. / resetting to ~ (Cash register) ‖ Stellen n. auf Null ‖ mise f. à zéro. / free for resetting to ~ (Cash register) ‖ Nullstellen n. frei ‖ libre pour la remise à zéro. / set to ~ (Cash register) *see* resetting to ~.

zero conductor (Electr) ‖ Nulleiter m. ‖ conducteur m. mis à terre. / ~ cut-out ‖ Nullausschalter m. ‖ disjoncteur m. *ou* interrupteur m. à zéro. / ~ error ‖ Nullpunktabweichung f. ‖ déviation f. du zéro. / ~ line of the vernier ‖ Noniusnullstrich m. ‖ zéro m. du vernier. / ~ mark ‖ Nullstrich m. ‖ marque f. de zéro. / ~ method ‖ Nullmethode f. ‖ méthode f. de réduction à zéro.

zero point ‖ Nullpunkt m. ‖ zéro m. / ~ in the centre ‖ Nullpunkt m. in der Mitte ‖ zéro m. central. / lateral ~ ‖ seitlicher Nullpunkt m. ‖ zéro m. unilatéral. / without ~ ‖ unterdrückter Nullpunkt m. ‖ zéro m. supprimé.

zero position ‖ Nullage f. ‖ position f. zéro. / ~ resistance ‖ Nullpunktswiderstand m. ‖ resistance f. de point neutre. / ~ terminal ‖ Nullklemme f. ‖ borne f. de raccordement de fil neutre.

zero voltage cut-out switch ‖ Nullspannungsausschalter m. ‖ interrupteur m. à voltage nul. / ~ release ‖ Nullspannungsauslösung f. ‖ déclenchement m. à voltage nul.

zigzag ‖ Zickzack m. ‖ zigzag m. / row of tubes arranged in ~ ‖ Rohrreihen fpl. im Zickzack ‖ rangée f. tubulaire disposée en zigzag.

zigzag connection ‖ Zickzackschaltung f. ‖ connexion f. zig-zag. / ~ course ‖ Zickzackweg m. ‖ chemin m. en zig-zag. / ~ lap joint ‖ Zickzacküberlappung(snietung) f. ‖ (rivure f. à) recouvrement m. en zigzag. / ~ line ‖ Zickzacklinie f. ‖ ligne f. en zig-zag.

zigzag riveting ‖ Zickzacknietung f. ‖ rivure f. en zig-zag *ou* en quinconce. / double-row double-butt-strap ~ ‖ zweireihige Zickzackdoppellaschennietung f. ‖ rivure f. en zig-zag à double couvrejoint et à deux-rangs.

zinc, to ‖ verzinken ‖ zinguer; galvaniser (au zinc).

zinc ‖ Zink n. ‖ zinc m. / commercial ~ ‖ Handelszink n. ‖ zinc m. commercial. / rough ~ ‖ Rohzink n. ‖ zinc m. brut *ou* d'œuvre. / white ~ ‖ Zinkweiß n. ‖ blanc m. de zinc.

zinc acetate ‖ essigsaures Zink n. ‖ acétate m. de zinc. / ~ amalgam ‖ Zinkamalgam n. ‖ amalgame m. de zinc. / ~ bath ‖ Zinkwanne f. ‖ baignoire f. en zinc. / ~ bearing ‖ zinkführend ‖ zincifère.

zinc-blende ‖ Zinkblende f. ‖ zinc-blende f.; zinc m. sulfuré.

zinc-bloom ‖ Zinkblüte f. ‖ fleur f. de zinc.

zinc borate ‖ borsaures Zink n. ‖ borate m. de zinc. / ~ carbonate ‖ kohlensaures Zink n. ‖ carbonate m. de zinc.

zinc chloride ‖ Chlorzink n. ‖ chlorure m. de zinc. / ~ ammonia (Chem) ‖ Chlorzinkammoniak n. ‖ chlorure m. de zinc ammoniacal. / ~ plant ‖ Chlorzinkanlage f. ‖ installation f. de chlorure de zinc. / ~ solution ‖ Chlorzinklauge f. ‖ solution f. de chlorure de zinc.

zinc chromate ‖ chromsaures Zink n.; Zinkchromat n. ‖ chromate m. de zinc. / ~ coat ‖ Zinkniederschlag m. ‖ dépôt m. de zinc.

zinc coating, the ~ is applied by galvanic means ‖ die Verzinkung f. erfolgt auf galvanischem Wege ‖ le (procédé de) zingage se fait par galvanisage. / uniformity of the ~ ‖ Gleichförmigkeit f. des Zinküberzuges ‖ égalité f. parfaite de la couche de zinc.

zinc coffin ‖ Zinksarg m. ‖ cercueil m. en zinc. / ~ colour ‖ Zinkfarbe f. ‖ couleur f. de zinc. / ~-containing ‖ zinkhaltig ‖ zincique. / ~-covered roof ‖ Zinkdach n. ‖ toit m. couvert en zinc. / ~ covering ‖ Zinkbedachung f. ‖ toiture f. en zinc. / ~ cyanide ‖ Zyanzink n. ‖ cyanure m. de zinc. / ~ cylinder ‖ Zinkzylinder m. ‖ cylindre m. de zinc. / ~ deposit ‖ zinkischer Anbruch m.; Gichtschwamm m. ‖ cadmies fpl.; dépôt m. de zinc. / ~ dust ‖ Zinkstaub m. ‖ poussière f. *ou* poudre f. de zinc; cadmies fpl. fines.

zinced iron ‖ verzinktes Eisen n. ‖ fer m. galvanisé.

zinc electrode ‖ Zinkelektrode f. ‖ électrode f. de zinc.

zincenite ‖ Bleiantimonglanz m. ‖ zinkénite f.

zinc fluoride ‖ Fluorzink n. ‖ fluorure m. de zinc. / ~ founder ‖ Zinkgießer m. ‖ zinquier m. / ~ foundry ‖ Zinkgießerei f. ‖ fonderie f. de zinc. / ~ goods pl. ‖ Zinkwaren fpl. ‖ articles mpl. en zinc. / ~ gray ‖ Zinkgrau n. ‖ gris m. de zinc. / ~ green ‖ Zinkgrün n. ‖ vert m. de zinc. / ~ hammer ‖ Zinkhammer m. ‖ marteau m. en zinc. / ~ hydrosulphite ‖ Zinkhydrosulfit n. ‖ hydrosulfite m. de zinc.

zinciferous ‖ zinkhaltig ‖ zincifère.

zincification ‖ Verzinken n. ‖ zingage m.; galvanisation f. / ~ wire ~ plant ‖ Drahtverzinkungsanlage f. ‖ installation f. pour le zingage de fil de fer.

zincify, to ‖ verzinken ‖ zinguer; galvaniser (au zinc).

zincing *see* zincking.

zinc ingot, crude ~ ‖ Rohzinkbarren m. ‖ saumon m. de zinc brut *ou* de zinc de première fusion.

zincite ‖ Rotzinkerz n.; Zinkit m. ‖ zinc m. oxydé *ou* rouge; zinkite f.

zincking of iron ‖ (galvanische) Verzinkung f. des Eisens ‖ zingage m. (galvanique) du fer; galvanisation f. du fer. / ~ fittings pl. ‖ Verzinkungseinrichtung f. ‖ installation f. de zingage. / ~ plant ‖ Verzinkungsanlage f. ‖ atelier m. à zingage.

zincky ‖ zinkartig ‖ de la nature du zinc.

zinc-lined ‖ mit Zinkeinlage f. versehen ‖ doublé en zinc. / ~ case ‖ Kiste f. mit Zinkeinlage ‖ caisse f. doublée de zinc.

zinc linoleate ‖ leinölsaures Zink n. ‖ linoléate m. de zinc. / ~ manufacture ‖ Zinkgewinnung f. ‖ métallurgie f. du zinc. / ~ mine ‖ Zinkerzgrube f. ‖ mine f. de zinc.

zincographic printer ‖ Zinkdrucker m. ‖ zincographe m.

zincography ‖ Zinkdruckerei f.; Zinkätzung f.; Zinkografie f. ‖ zincographie f.; gravure f. sur zinc.

zincopaper ‖ Zinkopapier n. ‖ papier m. zinc.

zinc ore ‖ Zinkerz n. ‖ minerai m. de zinc. / manganiferous ~ ‖ manganhaltiges Zinkerz n. ‖ minerai m. de zinc manganésé. / red ~ see zincite. / siliceous ~ ‖ Kieselzinkerz n. ‖ hydrosilicate m. de zinc.

zinc ore foundry ‖ Zinkschmelze f. ‖ fonderie f. de minerai du zinc.

zincous see zincky.

zinc oxide ‖ Zinkoxyd n. ‖ oxyde m. de zinc.

zinc paint ‖ Zinkfarbe f. ‖ couleur f. de zinc. / ~ permanganate ‖ Zinkpermanganat n. ‖ permanganate m. de zinc. / ~ peroxide ‖ Zinksuperoxyd n. ‖ peroxyde m. de zinc. / ~ phosphate ‖ Zinkphosphat n. ‖ phosphate m. de zinc. / ~ pipe ‖ Zinkröhre f. ‖ tuyau m. en zinc. / ~ plate ‖ Zinkblech n. ‖ feuille f. de zinc. / ~ plate for cells ‖ Zinkplatte f. für Elemente ‖ plaque f. d'élément en zinc. / ~ plating plant ‖ Verzinkungsanlage f. ‖ installation f. de zingage. / ~ powder ‖ Zinkstaub m. ‖ poudre f. de zinc.

zinc printing ‖ Zinkdruckverfahren n. ‖ impression f. sur zinc; zincographie f. / ~ high-speed machine ‖ Zinkdruckschnellpresse f. ‖ presse f. rapide pour zincographie. / ~ plate ‖ Zinkdruckplatte f. ‖ plaque f. de zinc pour l'imprimerie.

zinc, electrothermic ~ recovery ‖ elektrothermische Zinkgewinnung f. ‖ extraction f. électrothermique du zinc. / ~ resinate ‖ harzsaures Zink n. ‖ résinate m. de zinc. / ~ rod ‖ Zinkstab m. ‖ baguette f. ou tige f. de zinc.

zinc rolling mill ‖ Zinkwalzwerk n. ‖ laminoir m. à zinc. / equipment for ~s ‖ Zinkwalzwerkeinrichtung f. ‖ installation f. de laminoir à zinc.

zinc sheathing ‖ Zinkbeschlag m. ‖ doublage m. en zinc. / ~ silicate ‖ Zinksilikat n. ‖ silicate m. de zinc. / ~ spar ‖ Zinkspat m ‖ smithsonide m.; zinc m. carbonaté.

/ ~ stamping works pl. ‖ Zinkprägeanstalt f. ‖ atelier m. d'estampage sur zinc. / ~ stearate ‖ stearinsaures Zink n. ‖ stéarate m. de zinc. / ~ sulphate ‖ Zinksulfat n. ‖ sulfate m. de zinc. / ~ trough ‖ Zinkblecheinsatz m. ‖ gaine f. en tôle de zinc. / ~ vitriol ‖ Zinkvitriol n. ‖ sulfate m. de zinc; vitriol m. blanc.

zinc white ‖ Zinkweiß n.; Deckweiß n. ‖ blanc m. de zinc ou couvrant. / ~ plant ‖ Zinkweißanlage f. ‖ installation f. de blanc de zinc.

zinc worker ‖ Zinkarbeiter m. ‖ zingueur m. / ~ working plant ‖ Zinkhüttenanlage f. ‖ installation f. d'usines à zinc. / ~ works pl. ‖ Zinkhütte f. ‖ usine f. à zinc.

zinc yellow ‖ Zinkgelb n. ‖ jaune m. de zinc.

zinkiferous see zinciferous.

zinkification see zincification.

zinkify, to see to zincify.

zinkite see zincite.

zinky see zincky.

zip fastener ‖ Reißverschluß m. / fermoiréclair m.; curseurs-bloqueurs mpl.; fermeture f. à glissière ou à tirer.

zircite (Miner) ‖ natürliches Zirkoniumoxyd n. ‖ oxyde m. natif de zirconium.

zircon ‖ Zirkon m. ‖ zircon m.

zirconium ‖ Zirkonium n. ‖ zirconium m. / ~ glass ‖ Zirkonglas n. ‖ verre m. au zirconium. / ~ preparation ‖ Zirkoniumpräparat n. ‖ préparation f. au zirconium.

zither ‖ Zither f. ‖ cithare f.

zodiac ‖ Tierkreis m. ‖ zodiaque m.

zone, to ‖ in Zonen fpl. einteilen ‖ zoner.

zone ‖ Zone f. ‖ zone f. / ~ of combustion ‖ Verbrennungszone f. ‖ zone f. de combustion. / producer with two ~s of combustion ‖ Generator m. mit doppelter Brennzone ‖ gazogène m. à double zone de combustion. / ~ of eddies ‖ Wirbelzone f. ‖ zone f. de remous ou de tourbillons. / ~ of equilibrium ‖ Gleichgewichtszone f. ‖ zone f. d'équilibre. / free from ~s pl. ‖ zonenfrei exempt de zones fpl. / ~ of fusion ‖ Schmelzzone f. ‖ zone f. de fusion. / ~ of incandescence ‖ Glühzone f. ‖ zone f.

d'incandescence. / ~ of incomplete combustion ‖ Schwelzone f. ‖ zone f. de combustion incomplète. / neutral ~ ‖ Indifferenzzone f.; neutrale Zone f. ‖ région f. ou zone f. neutre. / ~ of oxidation ‖ Oxydationszone f. ‖ zone f. d'oxydation. / ~ of preheating ‖ Vorwärmezone f. ‖ zone f. de réchauffage. / ~ of pressure ‖ Druckzone f. ‖ zone f. de compression. / ~ within which prospecting is free ‖ Freischürfgebiet n. ‖ domaine m. de liberté des sondages. / refined ~ ‖ Übergangszone f. ‖ zone f. raffinée. / signal cabin ~ ‖ Stellwerkbezirk m. ‖ rayon m. d'action du poste de manœuvre. / smouldering ~ see ~ of incomplete combustion. / spherical ~ ‖ Kugelzone f. ‖ zone f. sphérique. / ~ of frequent thunderstorms ‖ gewitterreiche Zone f. ‖ zone f. riche en orages. / ~ of rare thunderstorms ‖ gewitterarme Zone f. ‖ zone f. pauvre en orages.

zone tariff ‖ Zonentarif m. ‖ tarif m. par zones.

zoological ‖ zoologisch ‖ zoologique. / ~ collection ‖ zoologische Sammlung f. ‖ collection f. de zoologie. / ~ preparation ‖ zoologisches Präparat n. ‖ préparation f. zoologique.

zoologist ‖ Zoologe m. ‖ zoologiste m.

zoology ‖ Zoologie f.; Tierkunde f. ‖ zoologie f.

zores iron ‖ Belageisen n.; Quadranteisen n. ‖ fer m. zorès.

zymase ‖ Zymase f. ‖ zymase f.

zymogenic ‖ gärungserregend ‖ zymogène.

zymology ‖ Gärungslehre f.; Zymotechnik f. ‖ zymologie f.

zymoma ‖ Gärungsstoff m.; Ferment n. ‖ ferment m.; zumine f.

zymometer ‖ Gärungsmesser m. ‖ zymosimètre m.

zymotechnical ‖ gärungsphysiologisch ‖ zymotechnique. / ~ analysis ‖ gärungsphysiologische oder zymotechnische Analyse f. ‖ analyse f. zymotechnique.

zymotechnology ‖ Gärungstechnik f.; Gärungsphysiologie f. ‖ zymotechnie f.; technique f. de la fermentation.

Druck von Oscar Brandstetter in Leipzig.